12th International Conference on Nucleus-Nucleus Collisions 2015

EPJ Web of Conferences Volume 117 (2016)

Catania, Italy
21 - 26 June 2015

Part 1 of 2

Editors:

V. Greco
M. La Cognata
S. Pirrone
F. Rizzo
C. Spitaleri

ISBN: 978-1-5108-2405-8

Printed from e-media with permission by:

Curran Associates, Inc.
57 Morehouse Lane
Red Hook, NY 12571

Some format issues inherent in the e-media version may also appear in this print version.

Printed by Curran Associates, Inc. (2016)

For additional information, please contact EDP Sciences – Web of Conferences at the address below.

EDP Sciences – Web of Conferences
17, Avenue du Hoggar
Parc d'Activité de Courtabœuf
BP 112
F-91944 Les Ulis Cedex A
France

Phone: +33 (0) 1 69 18 75 75
Fax: +33 (0) 1 69 28 84 91

contact-edps@webofconferences.org

Additional copies of this publication are available from:

Curran Associates, Inc.
57 Morehouse Lane
Red Hook, NY 12571 USA
Phone: 845-758-0400
Fax: 845-758-2633
Email: curran@proceedings.com
Web: www.proceedings.com

TABLE OF CONTENTS

PART 1

PART 2

EPJ Web of Conferences **117**, 00001 (2016) DOI: 10.1051/epjconf/201611700001
NN2015

Preface

It is the 12th Conference of this triennial series. Previous editions were held in San Antonio (NN2012), Beijing (NN2009), Rio de Janeiro (NN2006) and Moscow (NN2003). NN2015 maintained the general spirit and intentions of this Conference series. This is the second occasion to have this series of the conference in Italy (the first one was held on 1994 in Taormina).

The NN2015 Conference was organized by the Department of Physics and Astronomy of the University of Catania, and by INFN - Laboratori Nazionali del Sud and Sezione di Catania, under the auspices of the International Union of Pure and Applied Physics (IUPAP).

Catania represented an exceptionally suited venue. From a scientific point of view, Catania hosts one of the oldest Universities in Europe and one of the four National Laboratories of INFN. From a historic, artistic and natural point of view, downtown Catania is registered in the UNESCO World Heritage List owing to its beautiful Baroque buildings, while the nearby Mount Etna is included in same List as one of the world's most active and iconic volcano.

The Conference maintained its general spirit and the intentions of the previous ones. The scientific subjects covered a rather broad range of topics relevant to nucleus-nucleus collisions with special emphasis on applied aspects. The Scientific Program included the following sessions:

- Heavy and Superheavy Elements
- QCD and Hadron Physics
- Relativistic Heavy-Ion Collisions
- Nuclear Structure
- Nuclear Energy and Applications of Nuclear Science and Technologies
- Reactions and Structure - Unstable Nuclei
- Equation of State of Neutron-Rich Nuclear Matter, Clusters in Nuclei and Nuclear Reactions

- Fusion and Fission

- Nuclear Astrophysics

- New Facilities and Detectors

A total of 360 registered participants from 37 countries attended the conference. Among the participants were a large number of graduate students and post-doctoral fellows for whom special financial support was provided. As in the previous editions, the presented papers were of very high level, often representing breakthrough in the respective research fields.

The keynote talk: "Some selected highlights since NN2012" was given by A. Bracco, NuPECC Chair, and the summary talk by T. Motobayashi, RIKEN.

There were 38 invited talks in the plenary sessions and 58 invited talks in the parallel sessions and 113 oral presentations in the parallel sessions and 34 poster presentations. All the contributions sent by the authors are included in the conference proceedings.

We would like to take this opportunity to extend our sincere thanks to all participants for their attending and contributions to the success of the Conference and to gratefully acknowledge the International Advisory Committee, Program Committee and Organizing Committee for their support and excellent advices.

Special thanks are due to the Conference Secretariat for the efficient work. Further, we would like to thank the staff and students for their contributions to the preparation and organization of the Conference.

Claudio Spitaleri
Chairman NN2015
December 2015

edp sciences

EPJ Web of Conferences

Statement of Peer review

In submitting conference proceedings to EPJ Web of Conferences, I certify to the Publisher that I adhere to the **Policy on Publishing Integrity** of the journal in order to safeguard good scientific practice in publishing.

1. All articles have been subjected to peer review administered by the proceedings editors.
2. Reviews have been conducted by expert referees, who have been requested to provide unbiased and constructive comments aimed, whenever possible, at improving the work.
3. Proceedings editors have taken all reasonable steps to ensure the quality of the materials they publish and their decision to accept or reject a paper for publication has been based only on the merits of the work and the relevance to the journal.

Title, date and place of the conference

12 th INTERNATIONAL CONFERENCE ON NUCLEUS-NUCLEUS COLLISIONS 2015
21-26 JUNE 2015, CATANIA, ITALY

Proceedings editor(s):

V. GRECO, M. LA COGNATA, S. PIRRONE, F. RIZZO, C. SPITALERI

on behalf of the editors

1/03/2016

Date and editor's signature

17, avenue du Hoggar - PA de Courtabœuf – BP 112 - 91944 Les Ulis Cedex A (France)
Tél. : 33 (0)1 69 18 75 75 - Fax : 33(0)1 69 07 45 17 – www.edpsciences.org

EPJ Web of Conferences **117**, 01001 (2016) DOI: 10.1051/epjconf/201611701001
NN2015

Superheavy-element spectroscopy: Correlations along element 115 decay chains

D. Rudolph, U. Forsberg, L.G. Sarmiento, P. Golubev, and C. Fahlander

Department of Physics, Lund University, 22100 Lund, Sweden

Abstract

Following a brief summary of the region of the heaviest atomic nuclei yet created in the laboratory, data on more than hundred α-decay chains associated with the production of element 115 are combined to investigate time and energy correlations along the observed decay chains. Several of these are analysed using a new method for statistical assessments of lifetimes in sets of decay chains.

1 Introduction

"The quest for superheavy elements is driven by the desire to find and explore one of the extreme limits of existence of matter" [1]. It relates to fundamental questions concerning the properties of the strong nuclear force capable of forming the heaviest atomic nuclei yet known. The question of whether superheavy elements can be formed in stellar processes is also of fundamental nature. In chemistry superheavy elements pose questions such as: What are their properties? How do they fit into the existing scheme of the periodic table? In fact, the international superheavy-element community will discuss their past, presence and future in the framework of an approved Nobel Symposium on *Chemistry and Physics of Heavy and Superheavy Elements* near Lund in Spring 2016.

"More than 40 years ago, an island of stability of superheavy nuclei was predicted by nuclear structure theorists (see, e.g., Ref. [2]). This island rests on increased nuclear stability due to anticipated magic proton and neutron numbers at spherical nuclear shape (see, e.g., Ref. [3]), while surrounding archipelagos may also arise from increased nuclear binding of deformed atomic nuclei [4,5]. Even today theoreticians face the challenge – or have the freedom – to extrapolate over several tens of mass units towards the island, based on quite reliable spectroscopic information in the region around nobelium and rutherfordium [6].

Along their way, experimentalists have not only been producing heavier and heavier atomic nuclei, but also added more and more chemical elements to the Periodic Table. Present-day research builds upon one-atom-at-a-time creation and observation. In this context, 'time' refers to typically a day, a week, or as much as several months – and it is at this tiny rate that detailed spectroscopic information needs to be obtained.

During the past 15 years, correlated α-decay chains have been observed in a number of experiments performed at the Flerov Laboratory of Nuclear Reactions (FLNR) in Dubna, Russia. Beams of ^{48}Ca ($Z = 20$) impinged on actinide targets ($Z = 92$-98). The decay chains were interpreted to start from isotopes of elements $Z = 112$-118, produced via the mechanism of fusion-evaporation reactions [7,8]. The α-decays of nuclei with even proton number, Z, and even mass number, A, typically proceed from ground-state to ground-state. The decay energies along such decay chains are thus characteristic for a given decay step and can be used as a fingerprint for a given isotope of a certain element [9].

Decay patterns are usually much more complex for α decays of odd-A or especially odd-odd nuclei. Their α decays usually proceed to excited states in the daughter nucleus [10]. On the one hand, decay into excited states hampers *indirect* methods of determining Z, not least because decay chains along the same isotopes may follow different decay paths depending on the starting point of a certain isotopic decay sequence. On the other hand, the possibility arises to observe X-ray photons in the course of electromagnetic internal conversion decays of the excited states in α-decay daughter nuclei [11]. X-ray energies are predicted with high precision even for the heaviest elements since long (see, e.g., Ref. [12]), and they are characteristic for a given proton, hence atomic, number [13]. In addition, high-resolution α-photon coincidence spectroscopy – such as pioneered on element $Z = 115$ decay chains – closes in towards nuclear structure studies near the island of stability [14]" [15].

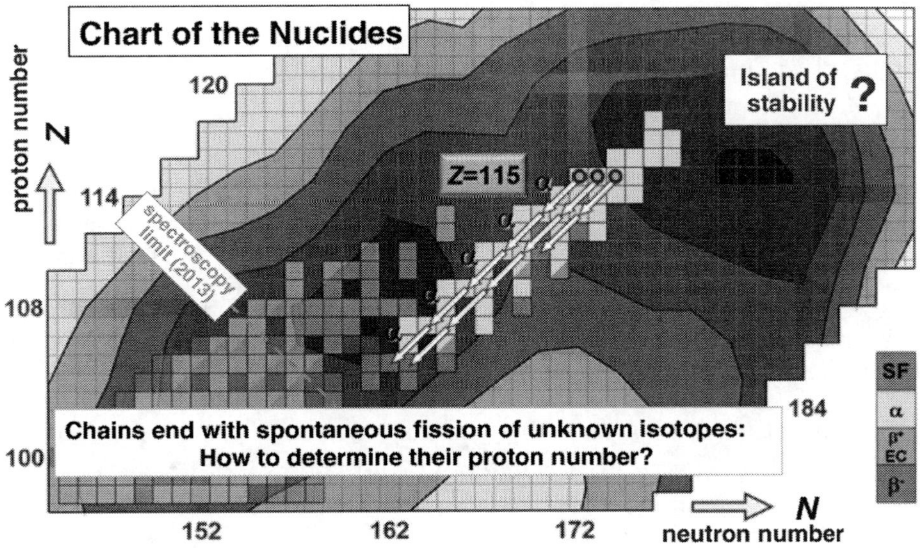

Figure 1: Upper end of the chart of atomic nuclei. Each square in this two-dimensional chart represents a certain isotope comprising Z protons and N neutrons. Red lines denote possible 'magic numbers'. The combination of Z=114 and N=184 is one of several long-standing predictions for the centre of the island of stability [2]. Bright shading indicates observed isotopes, with α-decay chains of element 115 indicated by diagonal arrows. The darker the blue background, the more stable the isotopes are predicted to be. A coarse '2013 α-photon spectroscopy limit' is indicated by a dashed orange line.

The status of the field is coarsely illustrated in fig. 1, and detailed from the point of view of the TASCA collaboration[1] in, for instance, Refs. [16,17]. Independent confirmation of the production of elements up to $Z = 117$ has been achieved [18], months-long search experiments for elements 119 and 120 conducted, and gas-phase chemistry of flerovium, $Z = 114$, being tackled [19]. The subject of this article is an extended analysis of correlations of decay times and decay energies along element 115 decay chains.

2 Correlation analysis

Until now, experimental data on in total 113 α-decay chains associated with the direct production of isotopes of element 115 (placeholder name ununpentium, Uup) have been published [14,20-23]. Following the first observation of four decay chains of element 115 at the Dubna gas-filled

[1] "TransActinide Separator and Chemistry Apparatus" at GSI Helmholtzzentrum für Schwerionenforschung, Darmstadt, Germany, see also https://www-windows.gsi.de/tasca/

recoil separator (DGFRS) [20], decay energies and correlation times from 37 element 115 "Dubna chains" are summarized in tables II, III, and IV of ref. [21]. More recently, high-resolution α-photon coincidence data taken at the TASCA facility at GSI Darmstadt resulted in 30 element 115 "TASCA chains" (cf. supplemental material of ref. [14] and table I of ref. [22]), while a similar experiment conducted at the Berkeley gas-filled separator gave rise to additional 46 element 115 "Berkeley chains" (cf. supplemental material of ref. [23]) — and $37 + 30 + 46 = 113$. The most relevant observables are decay energies, E_i, for a given decay step i along an α-decay chain, and correlation times between

- the implantation of a recoiling element 115 nucleus into a position sensitive silicon-detector arrangement and the first α decay, Δt_1,
- between n subsequent α decays, Δt_i, $i = 2, ..., n$,
- and between the last correlated α decay and a spontanous fission (SF) event terminating the chain after n α-decay steps, Δt_{n+1}.

Note that decay energies, E_i, can involve summing related to α-particle emission and subsequent prompt emission of conversion electrons in case α decay feeds excited states in the daughter nucleus, which can quickly decay into lower-lying states by electromagnetic decay processes [24].

Out of these in total 113 decay chains of directly produced evaporation residues (ER) of element 115, $4 + 7 + 3 = 14$ chains are of type ER-α(-α)-SF ($n = 1,2$), while the remaining $33 + 23 + 43 = 99$ chains comprise five subsequent α decays ($n = 5$), before SF events conclude the chains after typically several hours [14, 21, 23]. Note that it until now has not been possible to experimentally decide whether or not these SF events are preceded by an electron capture (EC) decay.

Inspecting the correlation times and decay energies observed for each individual decay step and eventually applying statistical measures such as the "Schmidt test" [25] yields that these 99 $n = 5$ chains are consistent with two different decay sequences. See, for instance, ref. [15] for the latest numerical compilation. Folding the mean excitation energy of the compound nucleus $^{291}115$ into the picture, $2+1+0 = 3$ decay chains are associated with the production of $^{287}115$, i.e. the 4-neutron ($4n$) evaporation channel, while the vast majority of $31 + 22 + 43 = 96$ chains are considered to represent the production of $^{288}115$ in the $3n$-evaporation channel.

Given that $^{288}115$ is an odd-odd nucleus, it is noteworthy that the description of such a comprehensive ensemble of 96 long α-decay chains is consistent with the observation of just one decay sequence. In turn, more detailed inspections of partly broad distributions of decay energies and, for

DOI: 10.1051/epjconf/201611701001

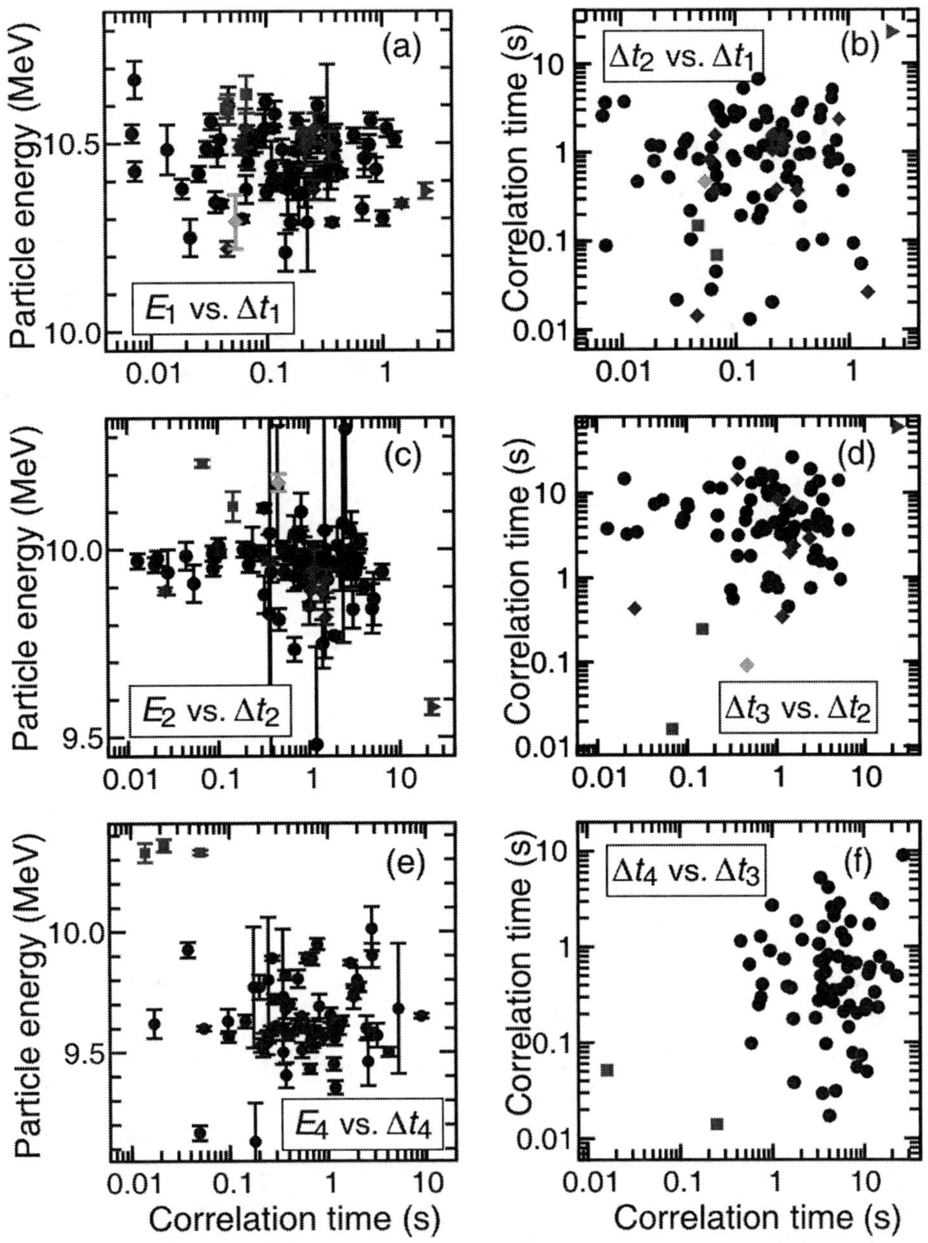

Figure 2: Two-dimensional histograms of measured decay energies and correlation times of decay step 1 [panel (a)], decay step 2 [panel (c)], and decay step 4 [panel (e)] of in total 113 α-decay chains associated with isotopes of element 115 [14, 21-23]. Filled black circles correspond to data from 96 five-α-decay long chains of $^{288}115$ and filled green squares to data from 3 five-α-decay long chains of $^{287}115$. Filled blue diamonds represent data from 12 out of 14 recoil-α(-α)-fission chains. The chains denoted D3 and D4 (cf. table III in ref. [21]) are emphasized as red triangles and turquoise diamonds, respectively. Panels (b),(d),(f) illustrate relations of correlation times of subsequent decay steps along element 115 chains.

the first time in this superheavy regime, α-photon coincidence spectroscopy, give rise to first glimpses of solid experimental information for detailed nuclear structure assessments of the heaviest atomic nuclei. This implies decay schemes including excited states, γ-ray decays, and α-decay hindrance factors, to name but a few [14, 15, 23, 26].

Prior to more comprehensive statistical assessments of decay chains as a whole (cf. next section and ref. [22]), relations between correlation times and decay energies of a given decay step or two correlation times of, for instance, subsequent decay steps can be used to identify candidates for different ensembles of decay chains. The six panels shown in fig. 2 are examples of such a graphical analysis.

The data points of the 96 chains associated with ^{288}115, displayed as filled black circles in fig. 2, can be considered to form a reference for any of the other element 115 decay chains. For instance, due to the broad range of decay energies and similar correlation times – and, thus, half-lives, $T_{1/2}$ [14, 15, 21, 23] – data for decay step 1 [panels (a) and (b)] does not allow to distuingish the three chains associated with ^{287}115, displayed as filled green squares, from the ^{288}115 reference. However, a clear trend becomes visible in decay step 2 [panels (c) and (d)] due to somewhat higher decay energies as well as shorter correlation times. Finally, data associated with decay step 4 [panels (e) and (f)] allow for an apparent visual distinction, which can also be turned into a significant statistical distinction due to significantly higher decay energies *and* shorter correlation times. The statement is backed up by the fact that the three ^{287}115 chains are observed at higher excitation energies of the compound nucleus ^{291}115, as mentioned earlier [14, 21].

While this situation for the long $n = 5$ element 115 decay chains is rather straightforward and undisputed, the interpretation of the remaining 14 "short chains" ($0+2+2 = 4$ ER-α-SF, $n = 1$, and $4+5+1 = 10$ ER-α-α-SF, $n = 2$) is less clear. Twelve of these chains are shown in fig. 2 as blue diamonds, while the chains denoted D3 and D4 (cf. table III in ref. [21]) are emphasized as red triangles and turquoise diamonds, respectively.

Looking at the relevant panels (a)-(d) of fig. 2 and comparing the distributions of the black circles and blue diamonds, there is no apparent difference. In fact, this correspondence also holds numerically [15, 22]. Thus, for any of these twelve short chains it is impossible to decide whether it represents a possible EC or SF branch in either ^{284}113 or ^{280}Rg [17], or whether it belongs to an independent sequence starting from ^{289}115, i.e. the isotope representing the $2n$ evaporation channel of the reaction ^{48}Ca+^{243}Am. Of course, such an α-decay sequence would have to have very similar decay characteristics as the first three decay steps of the long ^{288}115 reference.

In fact, arguments based on reaction cross sections and compound nucleus excitation energies *suggest* that these short chains are shared about 50:50 between $^{288}115$ and $^{289}115$ [22].

Two short chains deserve special attention. Chain D4 [21], represented by turquoise diamonds in fig. 2, is indistinguishable from the black circles and blue diamonds as far as decay step 1 is concerned [panels (a) and (b)]. However, the energy of decay step 2 [panel (c)] as well as the decay-time correlation between steps 2 and 3 [panel (d)] places this chain in the vicinity of the three long decay chains associated with $^{287}115$ (green squares). A definite assignment to any of the isotopes of element 115 is not possible though. Finally, the decay characteristics reported for chain D3 [21], represented by red triangles in fig. 2(a)-(d), suggest that it has a different origin than any other of the 113 chains associated with element 115. The red triangles show up in the upper right corner of panels (b) and (d) and in the lower right corner of panel (c), This is due to three exceptionally long correlation times and one exceptionally low decay energy, i.e. due to four unusual out of five possible parameters, clearly indicating that this chain is distinct. It may form a second, independent decay sequence of $^{289}115$ [17,22].

3 Statistical assessments

An alternative approach to assess the chains is the relatively "new test" suggested by Schmidt [25]. Here, the standard deviation of the logarithms of measured lifetimes $\theta_i = \ln(t_i)$ in a set of m decays of one decay step in a given decay chain is compared with expected values thereof. Too large standard deviations indicate that the set of decays might not originate from one single radioactive species, and too small values indicate a lack of sensitivity in the experiment, or that the data set is unjustly pruned. The fundamental principle behind this test is the fixed shape of a probability density function (PDF) on a logarithmic scale for an exponential decay, ensuring that the expected standard deviation is independent from the half-life. When the number m of decays is large, the expectation value of the standard deviation is 1.28, and is found in the interval $1.28 \pm 2.15/\sqrt{m}$ with 90% probability. For small values of m, the upper and lower limits are tabulated in ref. [25].

Applied to the short chains, the standard deviations are 1.20, 1.75, and, 1.84 for the first, second and third decay step, respectively. For steps one and two the number of chains is $m = 14$, and for the third step $m = 10$. Comparing with the 90% confidence intervals, [0.73,1.77], [0.73,1.77] and [0.65,1.82], respectively, the assignment to only one radioactive decay species is seemingly justified.

However, this measure only applies to one decay step at the time. Hence, not all of the available information is considered. For example, a data set containing one exceptionally long correlation time in each step can be correctly judged sound using the test, while there might still be reasons to question the assignment if the long correlation times all come from the very same chain. Therefore, we propose a generalisation of the test.

Assume that m chains have been observed and that each chain i contains n decay steps. Let θ_{i_j} be the logarithm of the j:th lifetime in chain i. Then, the measure

$$\xi_{m,n} = \sqrt[2]{\frac{\sum_{i=1}^{m} \sqrt[n]{\prod_{j=1}^{n}(\theta_{i_j} - \bar{\theta_j})^2}}{m}}, \quad \bar{\theta_j} = \frac{\sum_{i=1}^{m} \theta_{i_j}}{m} \tag{1}$$

will take into account correlation times observed along decay *chains* rather than in single decay steps. By multiplying the deviations $(\theta_{i_j} - \bar{\theta_j})^2$ with each other, all data on correlation times from one chain are considered jointly. Introducing the roots of order n allows for a further generalisation to be introduced later.

Figure 3 shows simulated distributions of $\xi_{m,n}$ for $m = 14$ and $n = 1$,2 and 3. Two million sets of correlation times have been randomly drawn from exponential PDFs. This measure, just like the one proposed by Schmidt, is independent from the half-life. When more decay steps n are considered, the distributions of $\xi_{m,n}$ get more narrow. Varying the number of chains m modifies the shape as well, with larger sets of chains leading to more narrow distributions. For each m and n the expectation value and the 90% confidence intervals can be tabulated. Some values are given in the left part of table 1.

Applying this generalised test to the short $n = 1$,2 chains associated with element 115 is unfortunately not straight forward due to the fact that the chains are of different length. This motivates a further generalisation of eq. 1, where the number of decays in chain i is n_i:

$$\xi_{m,n_i} = \sqrt[2]{\frac{\sum_{i=1}^{m} \sqrt[n_i]{\prod_{j=1}^{n_i}(\theta_{i_j} - \bar{\theta_j})^2}}{m}} \tag{2}$$

The averages θ_{i_j} are calculated using the respective numbers of available correlation times in that decay step. To construct relevant confidence intervals, the structure of each chain, i.e., its length and the number and position of potentially missing members need to be taken into account in the simulation. For the purpose of evaluating the short chains, it is sufficient to

Table 1: Left: Expectation values $\langle\xi\rangle$, 90% confidence intervals (5% of the probability density lies below "5% ↓" and 5% lies above "↑") of ξ for various values of m, n, and l. The upper left part of the table relates to $\xi_{m,n}$, i.e. sets where all chains have equally many decay steps. The lower left part refers to $\xi_{m,n,l}$, where l chains end earlier. The right part presents ξ_{m,n_i} values when the respective chain is *excluded* from the data set. Those marked with * should be compared with $\xi_{m=13,n=3,l=3}$, and the others with $\xi_{m=13,n=3,l=4}$. See text for details.

m	n	l	$<\xi>$	5% ↓	5% ↑	Chain	ξ_{m,n_i}	Chain	ξ_{m,n_i}
14	2	-	0.93	0.65	1.27	D1	1.34	C4	1.22
13	2	-	0.93	0.64	1.28	D2	1.34	C5	1.35
10	3	-	0.82	0.57	1.12	D3	1.03	C6	1.35
9	3	-	0.81	0.55	1.13	D4	1.33	C7	1.34
14	3	4	0.86	0.62	1.15	C1	1.36*	B1	1.35
13	3	4	0.86	0.61	1.16	C2	1.34*	B2	1.34*
13	3	3	0.85	0.61	1.15	C3	1.34	B3	1.18*
12	3	3	0.85	0.60	1.16				
11	3	3	0.85	0.58	1.17				

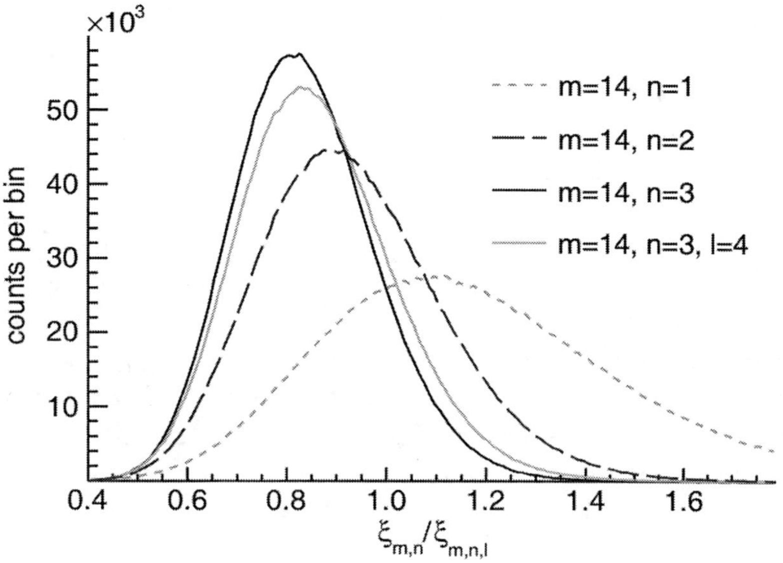

Figure 3: Simulated distributions of $\xi_{m,n,l}$ for various values of m, n and l.

consider the case when l out of m chains have only two decays, while the rest are "complete" ($n_i = n = 3$) with three decays. In fig. 3 the distribution of $\xi_{m,n,l}$ is shown for $m = 14$, $n = 3$, and $l = 4$. Increasing l from zero to m, the distribution gradually changes from the respective distributions from $\xi_{m,n=3}$ to $\xi_{m,n=2}$. Some relevant 90% confidence intervals are presented in the left part of table 1.

For the complete set of $m = 14$ short chains the value of ξ_{m,n_i} is 1.30 – well above the corresponding 90% confidence interval [0.62,1.15] obtained for $\xi_{14,3,4}$. Therefore, it is very unlikely that it is a mistake to conclude that not all chains originate from the same radioactive species. To investigate the data set more carefully, one chain at the time was excluded from the full set. The resulting values of ξ_{m,n_i} are given in the first column in table 1 behind the respective *excluded* chain. When interpreting the values, attention must be paid to the fact that l changes when a chain with only two decays is excluded. The values marked with * should thus be compared with the interval [0.61,1.15] ($\xi_{13,3,3}$) and the others with [0.61,1.16] ($\xi_{13,3,4}$). The *only* way to reach the 90% confidence interval by excluding *only one* chain is to exclude chain D3.

Excluding chains B3 or C4 also diminishes ξ_{m,n_i} somewhat, which draws the attention to those chains. If chains C4 and B3 are *both* excluded, the twelve remaining chains lead to $\xi_{m,n_i} = 1.06$. This is within the 90% confidence interval [0.60,1.16] for $\xi_{12,3,3}$. Hence the exclusion of these *two* chains pose an alternative to excluding D3.

When excluding all three chains D3, C4 and B3 the value becomes $\xi_{m,n_i} = 0.71$. This is below the expectation value 0.85. Since it is within the corresponding interval [0.58,1.17], there is no strong argument *not* to exclude all of them. However, since smaller changes to the data set are sufficient to account for the 90% confidence interval, such trimming of the data cannot be motivated by only an analysis of the spread of correlation times.

The conclusion from the statistical assessment is that the short chains most likely do not come from the same radioactive species. There are several different subsets of the data that fulfill the requirement of a ξ_{m,n_i} within the 90% confidence interval. This test cannot unambigously tell *how* the data set should be divided, but it shows that separation into different species is necessary. A proper inclusion of decay energies in the statistical analysis should be considered as well.

4 Summary

Modern spectroscopy provide first glimpses of solid experimental information for detailed nuclear structure assessments of the heaviest atomic nuclei available today. Furthermore, the increasing number of decay chains associated with element 115 allows extended assessments of the data, which is relevant for assigning a given decay chain to a certain isotope (or isomer of an isotope). Such studies lead to possible EC or SF branches of otherwise long element 115 decay chains and/or parallel decay sequences starting from the same isotope. Ultimately, all these features have to find their explanation in the underlying nuclear structure of the heaviest elements.

Acknowledgements

NN2015 attendance and these Proceedings are supported by the Swedish Research Council and the Royal Physiographic Society in Lund. The assembly of TASISpec would have been impossible without generous grants from the Royal Physiographic Society in Lund and support from the Euroball Owners Committee. The GSI experiment on element 115 was supported by the European Community FP7 – Capacities ENSAR No. 262010.

References

[1] M. Schädel, Phil. Trans. R. Soc. A **373**, 20140191 (2015).
[2] S.G. Nilsson *et al.*, Nucl. Phys. **A131**, 1 (1969).
[3] M. Bender, W. Nazarewicz, and P.-G. Reinhard, Phys. Lett. **B515**, 42 (2001).
[4] S. Ćwiok, P.H. Heenen, and W. Nazarewicz, Nature **433**, 705 (2005).
[5] J. Dvorak *et al.*, Phys. Rev. Lett. **97**, 242501 (2006).
[6] R.-D. Herzberg and P.T. Greenlees, Prog. Part. Nucl. Phys. **61**, 674 (2008).
[7] Y.T. Oganessian, J. Phys. G **34**, R165 (2007); Radiochim. Acta **99**, 429 (2011).
[8] Y.T. Oganessian *et al.*, Phys. Rev. C **87**, 054621 (2013).
[9] R.C. Barber *et al.*, Pure Appl. Chem. **83**(7), 1485 (2011).
[10] G.T. Seaborg and W.D. Loveland, *The Elements Beyond Uranium* (Wiley-Interscience, New York, 1990).
[11] C.E. Bemis, Jr. *et al.*, Phys. Rev. Lett. **31**, 647 (1973).
[12] T.A. Carlsson *et al.*, Nucl. Phys. **A135**, 57 (1969); At. Data Nucl. Data Tables **19**, 153 (1977).
[13] H.G.J. Moseley, Phil. Mag. **26**, 1024 (1913).
[14] D. Rudolph *et al.*, Phys. Rev. Lett. **111**, 112502 (2013).
[15] D. Rudolph, L.G. Sarmiento and U. Forsberg, submitted to AIP Conf. Proc.

[16] Ch.E. Düllmann, Proc. 5th *International Conference on Fission and Properties of Neutron-Rich Nuclei*, 2012, Sanibel Island, Florida, Eds. J.H. Hamilton and A.V. Ramayya, World Scientific, Singapore (2013), p. 271.
[17] D. Rudolph *et al.*, JPS Conf. Proc. **6**, 010026 (2015).
[18] J. Khuyagbaatar *et al.*, Phys. Rev. Lett. **112**, 172501 (2014).
[19] A. Yakushev *et al.*, Inorg. Chem. **53**, 1624 (2014).
[20] Yu. Ts. Oganessian *et al.*, Phys. Rev. C **69**, 021601(R) (2004).
[21] Yu. Ts. Oganessian *et al.*, Phys. Rev. C **87**, 014302 (2013).
[22] U. Forsberg *et al.*, submitted to Nucl. Phys. A; http://arxiv.org/abs/1502.03030.
[23] J.M. Gates *et al.*, Phys. Rev. C **92**, 021301(R) (2015).
[24] F.P. Heßberger *et al.*, Nucl. Instrum. Meth. A **274**, 522 (1989).
[25] K.-H. Schmidt, Eur. Phys. J. A **8**, 141 (2000).
[26] D. Rudolph *et al.*, Acta Phys. Pol. **B45**, 263 (2014).

EPJ Web of Conferences **117**, 01002 (2016) DOI: 10.1051/epjconf/201611701002
NN2015

Quasifission dynamics in TDHF

A.S. UMAR[1], V.E. OBERACKER[1] and C. SIMENEL[2]

[1]Department of Physics and Astronomy, Vanderbilt University, Nashville, TN 37235, USA
[2]Department of Nuclear Physics, RSPE, Australian National University, Canberra, ACT 0200, Australia

Abstract

For light and medium mass systems the capture cross-section may be considered to be the same as that for complete fusion, whereas for heavy systems leading to superheavy formations the evaporation residue cross-section is dramatically reduced due to the quasifission (QF) and fusion-fission processes thus making the capture cross-section to be essentially the sum of these two cross-sections, with QF occurring at a much shorter time-scale. Consequently, quasifission is the primary reaction mechanism that limits the formation of superheavy nuclei. Within the last few years the time-dependent Hartree-Fock (TDHF) approach has been utilized for studying the dynamics of quasifission. The study of quasifission is showing a great promise to provide insight based on very favorable comparisons with experimental data. In this article we will focus on the TDHF calculations of quasifission observables for the ^{48}Ca+^{249}Bk system.

1 Introduction

One of the most fascinating research areas involving low-energy nuclear reactions is the search for superheavy elements. Experimentally, two approaches have been used for the synthesis of these elements, one utilizing targets in the lead region (cold-fusion) [1,2], the other utilizing deformed actinide targets with ^{48}Ca projectiles (hot-fusion) [3–5]. While both methods have been successful in synthesizing new elements the evaporation residue cross-sections of the hot-fusion reactions were found to be as much as three times larger than those of the cold fusion ones. To pinpoint the root of this difference it is important to understand the details of the entrance channel dynamics of these systems since the properties of the dinuclear system at the capture point will strongly influence the outcome of the reaction. For light and medium mass systems the capture cross section may be considered to be the same as that for complete fusion, whereas for heavy systems leading to superheavy formations the evaporation residue cross-section is dramatically reduced due to the quasi-fission [6] and fusion-fission processes, thus making the capture cross-section to be essentially the sum of these two cross-sections. What is also difficult to ascertain is the configuration of the composite system, namely, whether the system has a single-center compound-like configuration or a dinuclear configuration accompanied by particle exchange. Most dynamical models [7–11] argue that for heavy systems a dinuclear complex is formed initially and the barrier structure and the excitation energy of this precompound system will determine its survival to breaking up via quasi-fission. Furthermore, if the nucleus survives this initial state and evolves to a compound system it can still fission due to its excitation.

Within the last few years the time-dependent Hartree-Fock (TDHF) approach [12, 13] has been utilized for studying the dynamics of quasifission [14–18] and scission dynamics [19–21]. Particularly, the study of quasifission is showing a great promise to provide insight based on very favorable comparisons with experimental data. Similarly, an extension of TDHF called the density-constrained TDHF [22] (DC-TDHF) has been used to obtain microscopic potential barriers and capture cross-sections for superheavy [23] and lighter systems [24–26]. In this article we will focus on the TDHF studies of quasifission for the ^{48}Ca+^{249}Bk system.

2 Results

During the past several years it has become feasible to perform TDHF calculations on a three-dimensional (3D) Cartesian grid with no symmetry restrictions and with much more accurate numerical methods [27, 28]. In the present TDHF calculations we use the Skyrme SLy4d energy density functional (EDF) [29] including all of the relevant time-odd terms in the mean-field Hamiltonian. First we generate very accurate static HF wave functions for the two nuclei on the 3D grid. The initial separation of the two nuclei is 30 fm. In the second step, we apply a boost operator to the single-particle wave functions. The time-propagation is carried out using a Taylor series expansion (up to orders $10-12$) of the unitary mean-field propagator, with a time step $\Delta t = 0.4$ fm/c. By virtue of long contact-times for quasifission and the energy and impact parameter dependence these calculations require extremely long CPU times. In Fig. 1a we plot the microscopic DC-TDHF

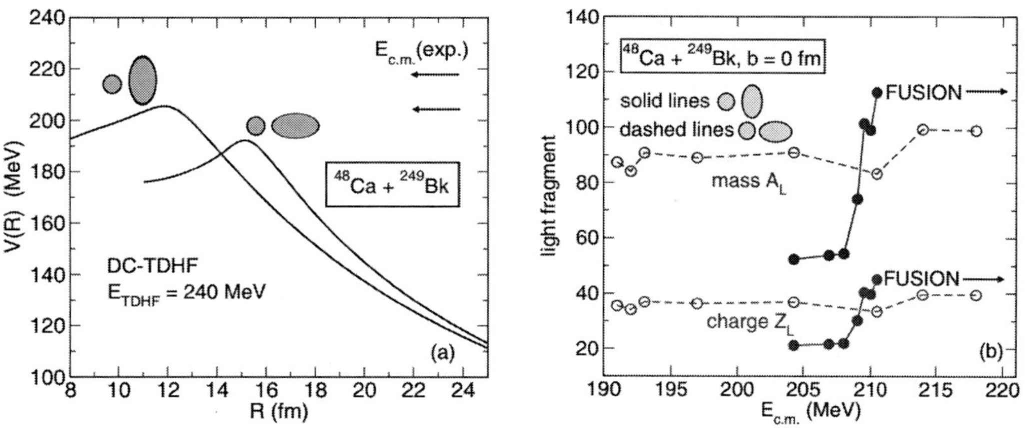

Figure 1: (a) Nucleus-nucleus potential, $V(R)$, for the $^{48}Ca+^{249}Bk$ system obtained from DC-TDHF calculation for selected orientation angles of the ^{249}Bk nucleus. Also shown are the experimental c.m. energies. (b) Mass and charge of the light fragment as a function of $E_{\text{c.m.}}$ for central collisions of ^{48}Ca with the side and tip orientations of ^{249}Bk.

potential barriers obtained for the $^{48}Ca+^{249}Bk$ system. The two barriers depict the two extreme orientations of the ^{249}Bk nucleus. Also, shown are the experimental energies at which this reaction has been studied [4, 30]. As expected the polar or tip orientation of ^{249}Bk results in a significantly lower barrier. The highest experimental energy is above both barriers but the lowest experimental energy is below the barrier for the equatorial or side orientation of ^{249}Bk. Figure 1b shows the mass and charge of the light fragment as a function of $E_{\text{c.m.}}$ for central collisions of ^{48}Ca with the side and tip orientation of ^{249}Bk. For the side orientation of ^{249}Bk for energies below $E_{\text{c.m.}} = 204$ MeV we get quasielastic collisions whereas for energies above 211 MeV we see fusion, which we define arbitrarily as reactions with contact times exceeding 35 zs. Naturally, non-central impact parameters can show quasifission in the range where we see fusion. We observe that for central collisions with the side orientation of ^{249}Bk quasifission is limited to a small range of energies $E_{\text{c.m.}} = 209 - 211$ MeV. The quasifission results are very different for the tip orientation of ^{249}Bk, ranging over a much wider energy domain, from $E_{\text{c.m.}} = 191$ MeV to $E_{\text{c.m.}} = 218$ MeV, which is the highest energy we have computed. The tip orientation also leads to smaller maximum mass and charge transfer.

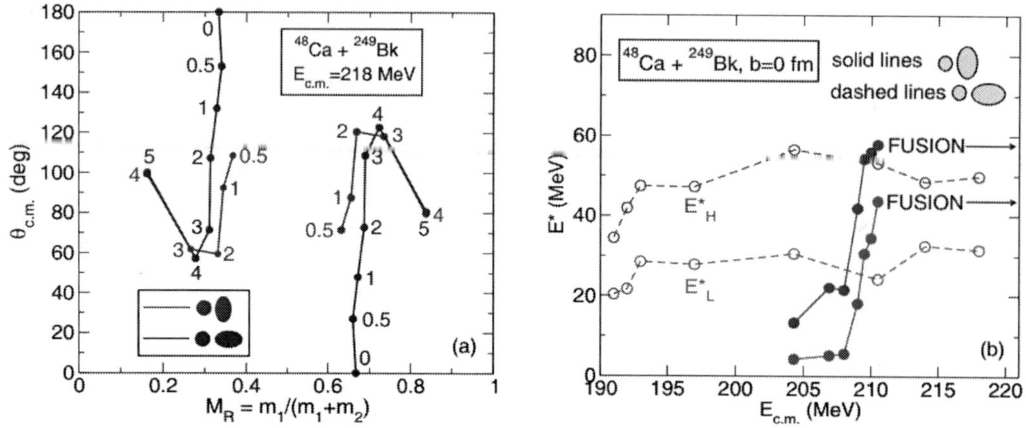

Figure 2: (a) Mass-angle distribution for the $^{48}Ca+^{249}Bk$ system for the tip and side orientations of the ^{249}Bk nucleus. (b) Excitation energy, E^*, of the heavy and light fragments as a function of $E_{\mathrm{c.m.}}$ for central collisions and for two orientations of the ^{249}Bk nucleus.

In Fig. 2(a) we show the TDHF calculations of quasifission MADs for $^{48}Ca+^{249}Bk$ at $E_{\mathrm{c.m.}} = 218$ MeV, corresponding to the two orientations of the ^{249}Bk nucleus. The regions of MAD's near $M_R = 0.2$ and $M_R = 0.8$ correspond to elastic and quasielastic reactions, followed by transition to deep-inelastic reactions and subsequently quasifission. The TDHF calculations predict where the transition from deep-inelastic to quasifission occurs, as well as the general behavior of the MADs. However, due to the fact that TDHF is a deterministic theory, it will only give us the most probable outcome or path for the MADs rather than a full distribution.

Recently, we have developed an extension to TDHF theory via the use of a density constraint to calculate the excitation energy of *each fragment* directly from the TDHF density evolution. This gives us new information on the repartition of the excitation energy between the heavy and light fragments which is not available in standard TDHF calculations. In Fig. 2b we show that the heavy and light fragments contain up to 60 MeV and 45 MeV (side orientation) and up to $\sim$ 55 MeV and $\sim$ 30 MeV (tip orientation) of excitation energy, respectively. It is interesting to note that the excitation energy follows qualitatively the mass and charge transfer behaviour observed in Fig. 1-b.

Figure 3a shows the contact time as a function of impact parameter at $E_{\mathrm{c.m.}} = 218$ MeV. We see that the contact times are $8-30$ zs for impact parameters resulting in quasifission and falls sharply for fragments produced in deep-inelastic collisions. These contact times are in agreement with typical quasi-fission times obtained from interpretations of experimental mass-angle distributions of the fragments [6, 31, 32].

Figure 3b show the mass and charge of the light fragment for the ^{48}Ca+^{249}Bk system as a function of impact parameter for the two orientations of the ^{249}Bk nucleus calculated at $E_{\mathrm{c.m.}} = 218$ MeV. As expected quasifission is identified with large mass and charge transfer, in this case corresponding to the doubling of the charge from 20 to 40 and mass from 48 to 100. It is also interesting to note the slightly atypical value of the contact time at impact parameter $b = 2$ fm in Fig. 3a in comparison to the neighboring impact parameters. Fig. 3b shows that in this region the light fragment is a neutron rich Zr isotope with $A \approx 102 - 106$. The microscopic evolution of the shell structure seems to have a tendency to form a composite with a longer lifetime when the light fragment is in this region. Similar observations were made in ^{40,48}Ca+^{238}U quasifission study [17]. A possible explanation is the presence of strongly bound deformed isotopes of Zr in this region [33, 34].

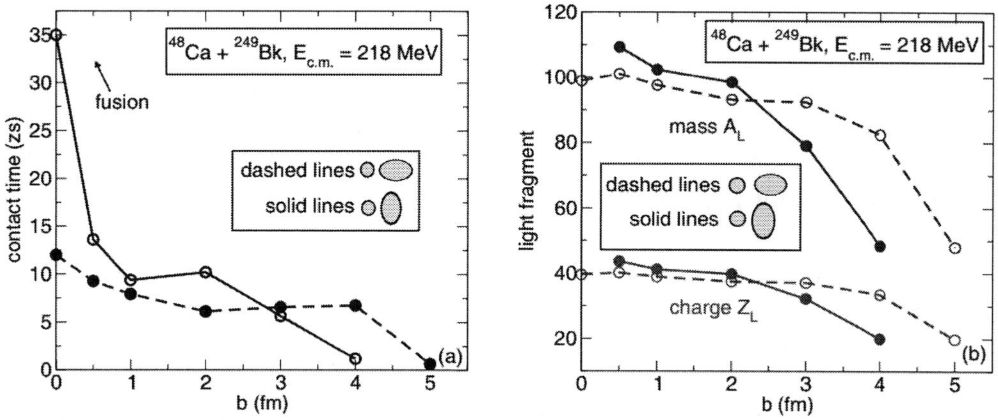

Figure 3: (a) Contact time and (b) mass and charge of the light fragment for the ^{48}Ca+^{249}Bk system as a function of impact parameter for the two orientations of the ^{249}Bk nucleus calculated at the c.m. energy of $E_{\mathrm{c.m.}} = 218$ MeV.

3 Summary

We have shown recent quasifission results for the ^{48}Ca+^{249}Bk system. Further calculations are underway to obtain a full range of observables including mass-angle distributions and fragment TKE's. Recent TDHF calculations of phenomena related to superheavy element searches show that TDHF can be a valuable tool for elucidating some of the underlying physics for these reactions. As a microscopic theory with no free parameters, where the effective nucleon-nucleon interaction is only fitted to the static properties of a few nuclei, these results are very promising.

Acknowledgments

This work has been supported by the U.S. Department of Energy under grant No. DE-SC0013847 with Vanderbilt University and by the Australian Research Council Grant No. FT120100760.

References

[1] S. Hofmann and G. Münzenberg, Rev. Mod. Phys., **72** (2000) 733.

[2] S. Hofmann, F. P. Heßberger, D. Ackermann, G. Münzenberg, S. Antalic, P. Cagarda, B. Kindler, J. Kojouharova, M. Leino, B. Lommel, R. Mann, A. G. Popeko, S. Reshitko, S. Śaro, J. Uusitalo, and A. V. Yeremin, Eur. Phys. J. A, **14** (2002) 147.

[3] Yuri Oganessian, J. Phys. G, **34** (2007) R165.

[4] Yu. Ts. Oganessian, F. Sh. Abdullin, C. Alexander *et al.*, Phys. Rev. C, **87** (2013) 054621.

[5] S. Hofmann, D. Ackermann, S. Antalic *et al.*, Eur. Phys. J. A, **32** (2007) 251.

[6] J. Töke and R. Bock and G.X. Dai *et al.*, Nucl. Phys. A **440** (1985) 327.

[7] G. Fazio, G. Giardina, G. Mandaglio, R. Ruggeri *et al.*, Phys. Rev. C, **72** (2005) 064614.

[8] G. G. Adamian, N. V. Antonenko, and W. Scheid, Phys. Rev. C, **68** (2003) 034601.

[9] G. G. Adamian, N. V. Antonenko, and W. Scheid, Eur. Phys. J. A, **41** (2009) 235.

[10] A. K. Nasirov, G. Giardina, G. Mandaglio *et al.*, Phys. Rev. C, **79** (2009) 024606.

[11] Zhao-Qing Feng, Gen-Ming Jin, Jun-Qing Li, and Werner Scheid, Nucl. Phys. A, **816** (2009) 33.

[12] J. W. Negele, Rev. Mod. Phys., **54** (1982) 9135.

[13] C. Simenel, Eur. Phys. J. A, **48** (2012) 152.

[14] C. Golabek and C. Simenel, Phys. Rev. Lett., **103** (2009) 042701.

[15] D. J. Kedziora and C. Simenel, Phys. Rev. C, **81** (2010) 044613.

[16] A. Wakhle, C. Simenel, D. J. Hinde *et al.*, Phys. Rev. Lett., **113** (2014) 182502.

[17] V. E. Oberacker, A. S. Umar, and C. Simenel, Phys. Rev. C, **90** (2014) 054605.

[18] K. Hammerton, Z. Kohley, D. J. Hinde *et al.*, Phys. Rev. C, **91** (2015) 041602(R).

[19] C. Simenel and A. S. Umar, Phys. Rev. C, **89** (2014) 031601.

[20] G. Scamps, C. Simenel, and D. Lacroix, Phys. Rev. C, **92** (2015) 011602.

[21] P. M. Goddard, P. D. Stevenson, and A. Rios, arXiv:1504.00919, (2015).

[22] A. S. Umar and V. E. Oberacker, Phys. Rev. C, **74** (2006) 021601.

[23] A. S. Umar, V. E. Oberacker, J. A. Maruhn, and P.-G. Reinhard, Phys. Rev. C, **81** (2010) 064607.

[24] V. E. Oberacker, A. S. Umar, J. A. Maruhn, and P.-G. Reinhard, Phys. Rev. C, **82** (2010) 034603.

[25] R. Keser, A. S. Umar, and V. E. Oberacker, Phys. Rev. C, **85** (2012) 044606.

[26] C. Simenel, R. Keser, A. S. Umar, and V. E. Oberacker, Phys. Rev. C, **88** (2013) 024617.

[27] C. Bottcher, M. R. Strayer, A. S. Umar, and P.-G. Reinhard, Phys. Rev. A, **40** (1989) 4182; Phys. Rev. A, **41** (1990) 1399.

[28] J. A. Maruhn, P.-G. Reinhard, P. D. Stevenson, and A. S. Umar, Comp. Phys. Comm., **185** (2014) 2195.

[29] Ka-Hae Kim, Takaharu Otsuka, and Paul Bonche, J. Phys. G, **23** (1997) 1267.

[30] J. Khuyagbaatar, A. Yakushev, Ch. E. Düllmann *et al.*, Phys. Rev. Lett., **112** (2014) 172501.

[31] R. du Rietz, D. J. Hinde, M. Dasgupta *et al.*, Phys. Rev. Lett. **106** (2011) 052701.

[32] R. du Rietz, E. Williams, D. J. Hinde *et al.*, Phys. Rev. C **88** (2013) 054618.

[33] V. E. Oberacker, A. S. Umar, E. Terán, and A. Blazkiewicz, Phys. Rev. C, **68** (2003) 064302.

[34] A. Blazkiewicz, V. E. Oberacker, A. S. Umar, and M. Stoitsov, Phys. Rev. C, **71** (2005) 054321.

EPJ Web of Conferences **117**, 01003 (2016) DOI: 10.1051/epjconf/201611701003
NN2015

A novel approach to the island of stability of super-heavy elements search

A. WIELOCH[1*], M. ADAMCZYK[1], M. BARBUI[2], N. BLANDO[2], G. GIULIANI[2], K. HAGEL[2], E-J. KIM[2], S. KOWALSKI[3], Z. MAJKA[1], J. NATOWITZ[2], K. PELCZAR[1], R. PŁANETA[1], K. SCHMIDT[3], Z. SOSIN[1†], S. WUENSCHEL[2], K. ZELGA[1] and H. ZHENG[2]

[1] M. Smoluchowski Institute of Physics, Jagiellonian University, Łojasiewicza 11, 30-348, Kraków, Poland
[2] Cyclotron Institute, Texas A&M University, TX, USA
[3] University of Silesia, Katowice, Poland

Abstract

It is expected that the cross section for super-heavy nuclei production of $Z > 118$ is dropping into the region of tens of femto barns. This creates a serious limitation for the complete fusion technique that is used so far. Moreover, the available combinations of the neutron to proton ratio of stable projectiles and targets are quite limited and it can be difficult to reach the island of stability of super heavy elements using complete fusion reactions with stable projectiles. In this context, a new experimental investigation of mechanisms other than complete fusion of heavy nuclei and a novel experimental technique are invented for our search of super- and hyper-nuclei. This contribution is focused on that technique.

1 Introduction

The history of super-heavy elements (SHEs) begins in the late 50-s. At that time the impressive progress in accelerator technology started and highly

*email: andrzej.wieloch@uj.edu.pl
†Deceased 24 September 2015.

efficient heavy ion sources were developed. New accelerators offered irradiation of target nuclei with high intensity beams of heavy ions which opened the door for the creation of SHEs [1]. In year 1957 G. Scharff-Goldhaber suggested a new region of existence of SHEs, the so called "island of stability", centered at nucleus $^{310}_{126}X$ [2] and theoreticians W.D. Myers and W.J. Świtecki [3] calculated in 1966 that the center of the island should be located at $Z = 126$ and neutron number $N = 184$ while at the same time A. Sobiczewski, F.A. Gareev and B.N. Kalinkin [4] located this center at $Z = 114$, $N = 184$. Such a region of enhanced stability of nuclei, was possible solely due to the quantum effects. In contrast,a widely used liquid drop model was predicting disintegration of a nucleus already at $Z = 104$.

Inspired by theory and new technological possibilities experimentalists initiated searches for SHEs and for the island of stability. To reach these goals a "cold" [5,6] and "hot" [7] complete fusion reactions in nuclear collisions were used where the excitation energies were $10-15$ MeV and $30-50$ MeV, respectively. Unfortunately, the cross section for SHEs production in fusion reactions is decreasing rapidly by a factor of 4 with each Z reaching about 1 pb for element $Z = 112$ [8]. This is a very serious limitation to the synthesis of the next elements. Moreover, half-lives of the heaviest nuclei were becoming very short, a few tens of μs. For even heavier nuclei one can expect that the cross section for their production is dropping into the region of tens of fb. Presently it is difficult to reach such low cross sections experimentally because of accessible intensities of ion beams. Radioactive ion beams (RIB) facilities are not a remedy as the intensities of neutron reach secondary beams delivered by RIB facilities are far too low.

So far we know fifteen SHEs starting from atomic number $Z = 104$ and ending on $Z = 118$. LBNL Berkely produced elements with $Z = 104-106$ [9,10], GSI Darmstadt created elements with $Z = 107-112$ [5,11,12], JINR Dubna where elements with $Z = 104-107$, $113-118$ [7,13–16] were produced and RIKEN Tokyo discovered element with Z=113 [6].

However, the question How heavy can an atomic nucleus be? is still under debate and the existence of the stability island of SHEs is not yet experimentally confirmed.

2 Alternative reaction mechanism for super- and hyper-heavy element production

In order to overcome the constraints on SHEs production described in the introduction our groups began in 2002 to investigate massive transfer (multi nucleon transfer) reactions between heavy projectile (e.g. Au) and heavy fissile target nuclei (e.g. Th) [17].

More detailed discussion on the reaction scenario for the SHEs formation can be found in [18] and references quoted therein. Here we recall only the main points. If a large portion of mass is transferred from one reaction partner to the other, conditions similar to those of the fusion reactions might be met e.g. small angular momenta of the combined masses and a very heavy system can be formed. The majority of products originating from low-energy nuclear fission are neutron-rich nuclei. If such a fission fragment is transferred to the projectile nucleus as in the proposed collision scenario the fusion probability can be strongly enhanced due to the neutron excess. For more neutron rich heavy nuclei with proton/neutron numbers near the closed shells, then the survival probability of the formed SHE will increase as the fission barrier of the resulting SHE will be high. Besides, if the transfer process of the fission (or massive) fragment takes place at the peripheral collision some of the resulting systems may be formed with small excitation energy. In that case its survival probability again increases. Both factors may play a major role in an expected production of the SHEs and the fission or multinucleon transfer can be a kind of "ion source" delivering a wide spectrum of different heavy ions. Therefore, nature itself can choose the most appropriate fission fragment that will fuse with other reaction partner in order to produce a SHE. Using for example the projectile with $Z = 79$ (Au) and the target nucleus with $Z = 90$ (Th) we may be able to explore production of SHEs and possibly hyper-heavy nuclear systems in the regions of atomic numbers $Z \approx 114$ and $Z \approx 126$ where the shell effects might play an important rule [19,20].Even higher regions of Z can be explored with such a technique.

However, one should remember that in the proposed reactions a relatively long interaction time will be involved. As a result, quite high excitation energy of the reaction partners can be generated and in those cases the survival probability of the produced SHE nucleus can decrease dramatically.The survival will depend strongly on the excitation energy distribution.

3 Experimental setup

One can expect that in the proposed reaction mechanism SHEs will be produced with a wide spectrum of atomic numbers and velocities and have relatively broad angular distributions. Thus the "classical" Wien filters applied for a complete fusion SHEs production where Z and velocity of synthesized SHEs are well defined can not be utilized. Instead, a completely new experimental setup was introduced in research presented here.

Our experiments started in 2002 at K500 accelarator of Cyclotron Institute of TAMU using an efficient magnetic velocity filter, the superconducting solenoid (BigSol) [17]. We investigated reaction ^{197}Au (7.5 $A.MeV$)+^{232}Th. Obtained results showed [21] that nuclei with Z above 100 can be produced via heavy mass transfer described in the former chapter.

In 2012, a new detection system was introduced into our SHE search [18]. The active catcher (AC) construction was based on an idea to identify production of SHEs via characteristic alpha decay. Indeed, many SHEs are expected to decay by alpha emission with unusually high energies [22] and this delivers a method to distinguish such decays from that of lighter elements. A prototype of the AC detection system was used for the first time in August 2013 in a test experiment. A final construction/geometry of the AC detection system (see Fig. 1) was tested in measurements in the summer of 2014.

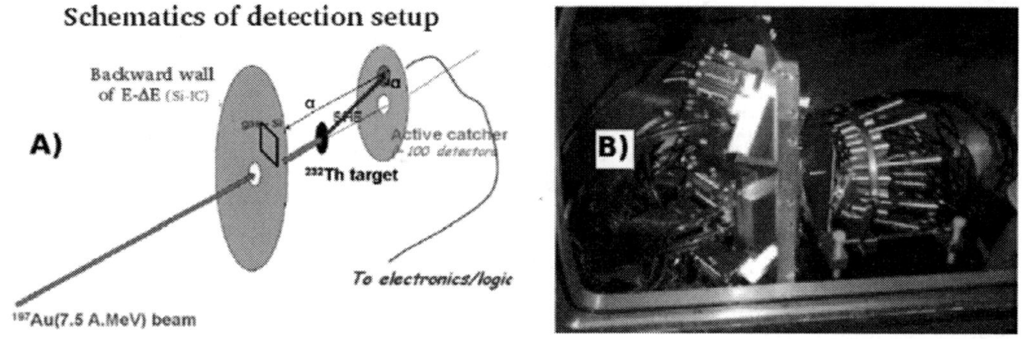

Figure 1: A) Schematics drawing of the AC detection system. B) Photo of the constructed AC system mounted inside the vacuum chamber at TAMU (on the left $\Delta E - E$ telescopes and on right the AC).

In our experimental setup the detection and identification of SHEs is a two stage process. In the first step the heavy reaction products are implanted and registered in the AC units which are located downstream from the target. In the second stage the alpha particles emitted are detected either by the AC detection unit in which the heavy reaction product was implanted and decayed and, for those emerging from the AC in the backward hemisphere, by an array of ΔE(gas)$-E$(Si) telescopes. and/or Spontaneous fission decays of implanted heavy nuclei can be recognized by the AC, also.

Each unit of the AC detector consists of a fast plastic scintillator (BC-418) where the reaction products are implanted, an aluminum cylinder which serves as a light guide and the photomultiplier tube (PMT R9880U-110) with an active window diameter of $\phi = 8\ mm$ (for a schematic view see [18]). To get optimal granularity of the AC detector system two sets of aluminum cylinders were applied (see Fig.1B). At the angles closest to the beam 18 units are mounted $\phi = 8\ mm$ diameter (long cylinders) and at angles larger than 15° 36 units are used with diameter equal 19 mm (short cylinders). For a better covering of the available space at the largest angles 9 units with long cylinders are added. The whole detection setup is quite compact, units of the AC system and $\Delta E - E$ telescopes of the backward wall are located less then 10 cm from the target. Such a configuration gives capability of measuring decays of very short lived SHEs of the order of 10 ns.

The other advantages of the AC detection system are a very good time resolution (the scintillation pulse width is about 5 ns with rise time 1 ns) and a dynamical range which allows to distinguish between deposited heavy reaction products and decayed alpha particles.

A high background is a natural consequence of the approach we are using for SHE search. In order to eliminate unwanted events such as scattered beam, projectile like fragments and a few nucleon transfer reaction products from data acquisition (ACQ) a dedicated trigger of an ACQ was built. The first block of electronics shown in Fig. 2A produces logical signals from analog pulses of the PMT modules of the AC detector. A final formation of the trigger is done by two FPGA boards (see Fig.3).

A very important role in the trigger formation is played by the electronics which converts the RF cyclotron signal to a logical pulse in the way that one logical level (in our case it is logical "0") has a duration of the beam burst from the cyclotron which is 5 ns at the target and level logical "1" of 50 ns width equal to a time period with no beam particles at the target (see Fig.2B).

The FPGA board shown in Fig.3 produces logical OR of all of the PMT pulses. Then the logical AND vetoes logical OR during the time period when the beam particles hit the target (5 ns). The logical OR (marked in the Fig. 3 as IC-Si trigger) allows generation of the ACQ trigger when the alpha particle is detected by a $\Delta E - E$ backward detector. This final trigger activates the ACQ system that is composed of two 32 Channel fast Flash ADCs V1742. The pulse shapes of the PMT modules are recorded by Flash ADCs while signals from $\Delta E - E$ detectors are picked up by another independent acquisition. To assure event agreement for both ACQs an event counter is followed and marked as "EvtCount" in Fig. 3. Module V1742 was working with sampling rate of 1 Gs/s while the width of the common time window for the detection systems was 1 μ. Location of the ACQ trigger in V1742 time window enables to record interesting event time structure backwards in time upto 1 μs.

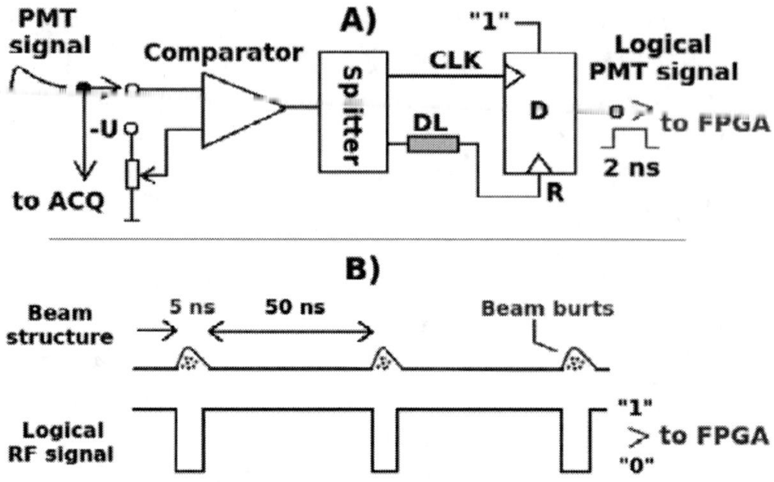

Figure 2: A) PMT spectroscopic signal from each module of the AC is converted by dedicated electronics to the fast logical signal. B) Logical RF pulses produced from the sinus shape of the original RF signal are converted in a such way that the logical "0" fits to the beam burst duration. Both types of signals are delivered to the trigger production logic build on the FPGA module.

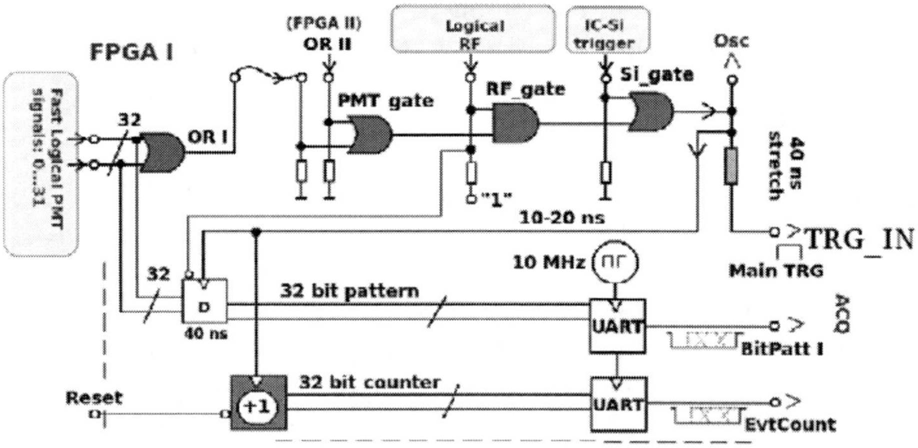

Figure 3: Logic constructed on FPGA boards served to produce main trigger. Schematics of only one FPGA is shown while the other has the same logic.

Fig.4 presents example of two interesting events extracted from measurement in summer 2014. In both cases acquisition was triggered by an α particle detected in one of $\Delta E - E$ detectors. With such a trigger flash ADC recorded pulse shapes generated by the PMTs of the AC units number 11 (panel A) and 24 (panel B). For the first event the distance between peaks of the pulses is $(113 - 22)$ ns while for the second event this distance is $(275 - 133)$ ns. If these time differences are divided by the characteristic time of beam bunches equal to 55 ns we obtain 1.65 and 2.58 values. These two numbers prove that in both cases at least one of the detected particles came between beam bursts. These examples clearly show that the AC detection system is able to register α decays of very short lived isotopes.

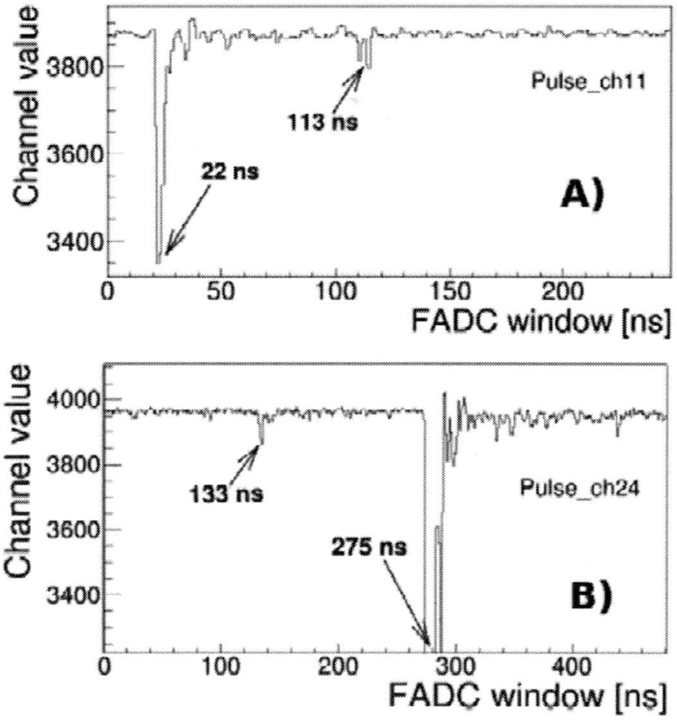

Figure 4: Flash ADC recorded pulses for two events A) and B) while ACQ system was triggered by signal of α particle detected in of $\Delta E - E$ detector. A) Energy of α particle was 9.90 MeV while in case of event B) α particle has energy 7.60 MeV.

Presently further improvements in the active catcher detection system are underway. Especially, to get better quality pulses from the scintillator of the AC units we installed lucite light guides with the shapes optimized for a short and long aluminum cylinders. Additionally computer driven thresholds of the first electronic block are being implemented. The next measurements are scheduled in early November of 2015.

Acknowledgments

This work is supported by the National Science Center in Poland, contract no. UMO-2012/04/A/ST2/00082, by the U.S. Department of Energy under Grant No. DE-FG03-93ER40773 and by the Robert A. Welch Foundation under Grant A0330.

References

[1] See e.g. Kumar K., in *Super Heavy Elements*, edited by Adam Hilger, (Bristol and New York) 1989.

[2] Scharff-Goldhaber G., *Nucleonics*, **15** (1957) 122.

[3] Myers W.D., Świtecki W.J., *Nucl. Phys.*, **81** (1966) 1.

[4] Sobiczewski A. et al., *Phys. Lett.*, **22** (1966) 500.

[5] Hofmann S. and Münzenberg G., *Rev. Mod. Phys.*, **75** (2000) 733.

[6] Morita K. et al., *J. Phys. Soc. Jpn.*, **73** (2004) 2593.

[7] Oganessian Yu., *Phys. Rev.*, **C74** (2006) 044602.

[8] Hofmann S. et al., *Eur. Phys. J.*, **A14** (2002) 147.

[9] Ghiorso A. et al., *Phys. Rev. Lett.*, **22** (1969) 1317.

[10] Ghiorso A. et al., *Phys. Rev. Lett.*, **33** (1974) 1490.

[11] Hofmann H. et al., *Nucl. Phys.*, **A496** (1989) 269.

[12] Hofmann S. et al., *Z. Phys.*, **A350** (1995) 281.

[13] Oganessian Yu. et al.,*Nucl. Phys.*, **A273** (1976) 505.

[14] Oganessian Yu., *Phys. Rev.*, **C63** (2001) 011301R.

[15] Oganessian Yu., *J. Nucl. Radiochem. Sci.*, **3** (2002) 5.

[16] Oganessian Yu., *LNL Report, Berkeley,* **UCRL-ID-151619** (2003).

[17] Materna T. et al., Progress in Research April 1, 2003-March 31, 2004, Cyclotron Institute, Texas A&M University, 2004 pp. II-17.

[18] Majka Z. et al., *Acta Phys. Pol.*, **B45** No 2 (2014) 279.

[19] Zagrebaev V.I. et al. *J. Phys.: Conference Series*, **420** (2013) 012001.

[20] Zagrebaev V.I. and Greiner W., *Phys. Rev.*, **C87** (2013) 034608.

[21] Barbui M. et al., *Int. J. Mod. Phys*, **E18** (2009) 1036.

[22] Bender M. and Heenen P.-H., *J. Phys.: Conference Series*, **420** (2013) 012002.

EPJ Web of Conferences **117**, 01004 (2016) DOI: 10.1051/epjconf/201611701004
NN2015

Recent experiments in inverse kinematics with the magnetic spectrometer PRISMA

E. Fioretto[1], L. Corradi[1], D. Montanari[2], S. Szilner[3], G. Pollarolo[4], F. Galtarossa[1,5], D. Ackermann[6], G. Montagnoli[7], F. Scarlassara[7], A.M. Stefanini[1], S. Courtin[2], A. Goasduff[1], F. Haas[2], D. Jelavić-Malenica[3], C. Michelagnoli[8], T. Mijatović[3], N. Soić[3], C. Ur[9] and J.J. Valiente-Dobon[1]

[1]INFN, Laboratori Nazionali di Legnaro, Legnaro (PD), Italy
[2]IPHC, CNRS-IN2P3 and Université de Strasbourg, Strasbourg, France
[3]Ruđer Bošković Institute, Zagreb, Croatia
[4]Dip. di Fisica Teorica, Università di Torino and INFN, Torino, Italy
[5]Dip. di Fisica e Scienze della Terra, Università di Ferrara, Ferrara, Italy
[6]GSI, Darmstadt, Germany
[7]Dip. di Fisica e Astronomia, Univ. di Padova and INFN, Padova, Italy
[8]GANIL, CEA/DSM-CNRS/IN2P3, Caen Cedex 05, France
[9]Horia Hulubei National Institute of Physics and Nuclear Engineering, Bucharest, Romania

Abstract

In the last period, two classes of experiments have been carried out with the large acceptance magnetic spectrometer PRISMA. In particular, the one- and two-neutron transfer processes at energies ranging from the Coulomb barrier to deep below it and the population of exotic neutron rich nuclei in the A$\sim$130 and A$\sim$200 mass regions have been studied. Both kinds of experiments have been performed in inverse kinematics identifying in A, Z and velocity the light target-like recoils with PRISMA placed at very forward angles in order to have, at the same time, high efficiency and good energy and mass resolutions.

1 Introduction

Recent studies on transfer processes benefited from the construction of new generation tracking spectrometers, based on the trajectory reconstruction, and the use of the state-of-art large area particle detectors. The large acceptances of these spectrometers and the high resolving powers of their detection systems allowed to investigate the transfer process around and well below the Coulomb barrier with high efficiency and good fragment identification.

In this work we report on studies on the interplay between single and pair transfer modes at near- and sub-barrier energies and on the population of neutron rich nuclei in the A$\sim$130 and A$\sim$200 mass regions around the Coulomb barrier. The measurements have been performed in a very efficient way making use of inverse kinematics reactions and the large solid angle magnetic spectrometer PRISMA [1–3] installed at Laboratori Nazionali di Legnaro.

2 Sub-barrier transfer measurements

In the sub-barrier region interacting nuclei are only slightly influenced by the nuclear potential and follow almost pure Coulomb trajectories. Therefore, only few transfer channels are open, excitation energies are restricted to few MeV and uncertainties in calculations associated with optical potentials can be minimized [4]. Available data on transfer reactions in this energy range are extremely scarce because of various kinds of significant experimental difficulties which can be overcome making use of inverse kinematics detecting the lighter target-like fragments with magnetic spectrometers at very forward angles [5, 6].

Transfer processes were investigated at LNL in closed shell nuclei (or nearly so) for both neutrons and protons with the ^{96}Zr+^{40}Ca reaction [7] and in superfluid nuclei with the ^{116}Sn+^{60}Ni system [8]. The ^{96}Zr and ^{116}Sn beams were delivered by the Tandem/PIAVE-ALPI accelerator complex at energies ranging from the Coulomb barrier down to 25% below with an average current of about 2 pnA. Light target-like ^{40}Ca and ^{60}Ni ions were detected at a very forward angle (θ_{lab}=20°) with the large acceptance magnetic spectrometer PRISMA.

Excitation functions of the main transfer channels were measured and yields were detected at the level of 10^{-4} with respect to the elastic channel. In fig. 1 we show an example of two-dimensional plot velocity (reconstructed via time-of-flight) vs in-plane scattering angle for the ^{96}Zr+^{40}Ca reaction at

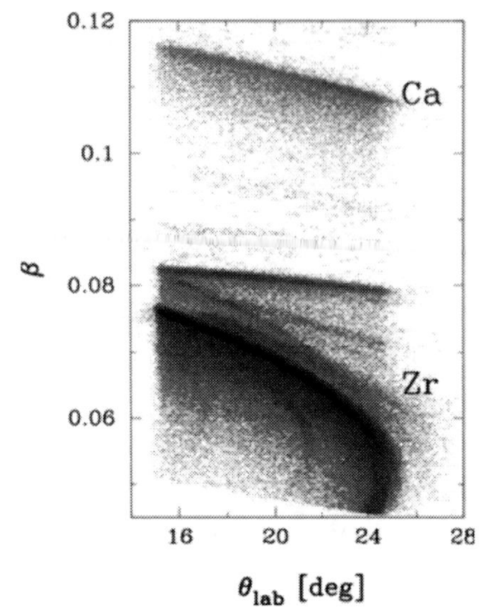

Figure 1: Velocity (β = v/c) vs in-plane scattering angle (θ_{lab}) for the reaction ^{96}Zr+^{40}Ca at E_{lab} = 330 MeV.

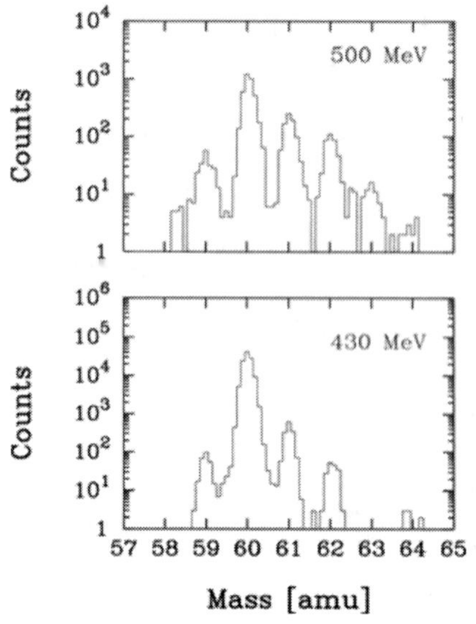

Figure 2: Mass spectra of Ni isotopes in the ^{116}Sn+^{60}Ni system at two representative energies, above and below the Coulomb barrier.

E_{lab}=330 MeV. One sees the events corresponding to Ca-like recoils as well as those corresponding to Zr-like ions entering into PRISMA. The ridges of these Zr-like ions are due to scattering on Ca (main ridge presenting a bending around the kinematic limiting angle (θ_{lab} = 24.6°)), on lighter elements (with smaller limiting angles) and on other heavier elements (target contaminants). The almost horizontal ridge, with high yield and with a $\beta \sim 0.08$, is due to Zr scattering in direct kinematics. The very good mass resolution $\Delta A/A \sim 1/240$ is guaranteed by the high kinetic energy of the recoils due to the inverse kinematics. As an example, mass spectra for Ni isotopes, measured for ^{116}Sn+^{60}Ni reaction at two bombarding energies, are displayed in fig. 2.

The experimental transfer probabilities as a function of the distance of closest approach for the (+1n) and (+2n) neutron transfer channels have been compared with results of microscopic calculations. The theoretical two-neutron transfer probabilities have been computed including nucleon-nucleon correlations (see Ref [7,8] for details). For the first time with heavy ions, experimental transfer probabilities have been well reproduced (in absolute values and in slope) by microscopic calculations which include nucleon-nucleon correlations.

3 Population of heavy neutron rich nuclei

Certain neutron rich regions of the nuclide chart, like those below ^{208}Pb or in the actinides and transactinides, can be hardly accessed by fragmentation or fission reactions, and multinucleon transfer represents a suitable mechanism to approach these areas. We remind that nuclear properties of neutron rich nuclei around N=126, relevant for r-process, play a critical role for theoretical predictions of the synthesis of the heaviest elements [9], for disentangling a variety of astrophysical scenarios [10, 11] and to study the competition between Gamow-Teller and First-Forbidden β transitions [11].

It is worthwhile to mention the experiment very recently carried out in GANIL using the large acceptance magnetic spectrometer VAMOS coupled to the γ-array EXOGAM. Its main goal was to measure the yield distribution of light and associated heavy transfer products populated in the ^{136}Xe+^{198}Pt reaction [12] at an energy of about 8 MeV/A. Another (integral) experiment was performed at Jyväskylä using a gas cell system to extract and study, off-line by means of β-γ coincidences, nuclei in the Os region populated through the reaction ^{136}Xe+^{208}Pb in deep inelastic processes. Both experiments were carried out during the preparatory phases of the facilities KISS (KEK Isotope Separator System) at RIKEN and GaLS (Gas cell and Laser ion Separation) at the Flerov Laboratory in Dubna [13] aiming to the study of β-decay spectroscopy of heavy neutron rich nuclei in the A$\sim$200 and N$\sim$126 region populated via deep inelastic reactions.

Within this context, an experiment was recently carried out at LNL in which the yields of light and heavy fragments were measured by using the kinematic coincidence technique. To this end the spectrometer (the main arm of the coincidence) was coupled to a second arm (for the direct detection of heavy partner) which consists of a position sensitive Parallel Plate Avalanche Counter (PPAC) followed by an axial field ionization chamber (IC). In order to overcome the experimental difficulties related to the direct detection of the heavy partner, multineutron and multiproton transfer channels were populated via the inverse kinematics reaction ^{197}Au+^{130}Te at E_{lab}=1070 MeV. The ^{197}Au beam, delivered by PIAVE-ALPI accelerator complex with an intensity of 2 pnA, represents the presently heaviest beam available at LNL. The PRISMA spectrometer was placed around the grazing angle (θ_{lab}=+37°) with respect to the beam axis in order to identify the lighter target-like fragments populated in the reaction. Projectile-like fragments were detected with the second arm placed at about 90 cm of distance from the target and at the kinematically correlated angle θ_{lab}=-37°.

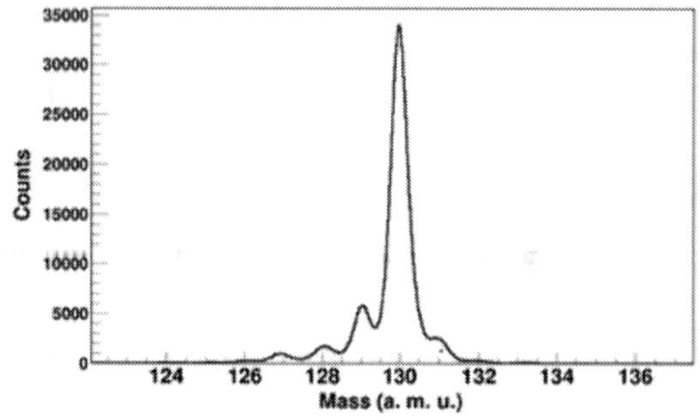

Figure 3: Mass distribution of tellurium isotopes at the focal plane of PRISMA measured in the ^{197}Au+^{130}Te reaction at E_{lab}= 1070 MeV and θ_{lab}=+37°.

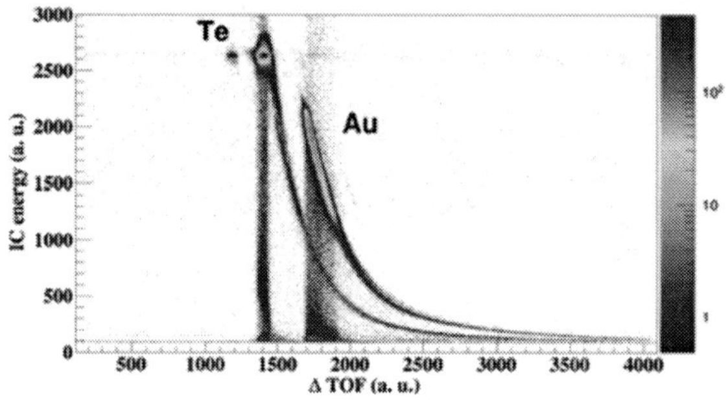

Figure 4: Energy vs ΔTOF spectrum for fragments produced in the ^{197}Au+^{130}Te reaction at E_{lab}= 1070 MeV (see text for details).

The main goal of the measurement was to study the population of neutron rich nuclei in the A∼130 and A∼200 mass regions. In particular, via the two-proton stripping ($-2p$) and four-neutron pick-up ($+4n$) channel one should be able to populate ^{132}Sn, which represents a benchmark neutron rich nucleus for different physics cases, reported in the proposals of forthcoming radioactive beam facilities. In addition, proton pick-up and neutron stripping channels from ^{130}Te would lead, from the heavy partner side, to the population of neutron rich nuclei in the Pt-Os region of astrophysics interest. Secondary processes should play a non negligible role for these heavy nuclei, thus it will be interesting to compare the final yields with those expected from theoretical models, already successfully applied for lower mass systems.

The PRISMA spectrometer was set for the selection and identification of the target-like fragments thus measuring A, Z and Q-value distributions for the light partner. As a preliminary result, the yields of tellurium isotopes measured with PRISMA are shown in fig. 3. Using PRISMA events as trigger of the DAQ, coincident fragments were detected in the second arm. As additional parameter for the fragment discrimination, we used the difference ΔTOF of the time-of-flight of the fragments measured between the start detector of PRISMA and the PPAC of the second arm, respectively. Typical Energy (measured with IC) vs ΔTOF scatter plot is shown in fig. 4 where one can observe a clear separation between projectile-like and target-like fragments.

The data analysis for the heavy partner is still in progress from which we expect to get important information on the production yield of heavy neutron rich nuclei and to identify any significant contribution from the onset of secondary processes such as the transfer induced fission.

Acknowledgement

This work was partly supported by European Union Seventh Framework Programme FP7/2007-2013 under Grant No. 262010 - ENSAR.

References

[1] A.M. Stefanini *et al.*, Nucl. Phys. A **701**, 217c (2002).

[2] S. Szilner *et al.*, Phys. Rev. C **76**, 024604 (2007).

[3] L. Corradi *et al.*, Nucl. Instr. and Meth. in Phys. Res. B **317**, 743 (2013).

[4] L. Corradi, G. Pollarolo, and S. Szilner, J. of Phys. G **36**, 113101 (2009).

[5] C.L. Jiang *et al.*, Phys. Lett. B **337**, 59 (1994).

[6] C.L. Jiang *et al.*, Phys. Rev. C **57**, 2393 (1998).

[7] L. Corradi *et al.*, Phys. Rev. C **84**, 034603 (2011).

[8] D. Montanari *et al.*, Phys. Rev. Lett. **113**, 052501 (2014).

[9] V. Zagrebaev and W. Greiner, Phys. Rev. Lett. **101**, 122701 (2008).

[10] K.-L. Kratz *et al.*, Astrophys. J. **403**, 216 (1994).

[11] H. Grawe et al., Rep. Prog. Phys. **70**, 1525 (2007).

[12] Y.X. Watanabe *et al.*, Phys. Rev. Lett., accepted (2015).

[13] K. Novikov *et al.*, J. of Phys.: C.S. **515**, 012016 (2014).

EPJ Web of Conferences **117**, 02001 (2016) DOI: 10.1051/epjconf/201611702001
NN2015

Chiral nucleon-nucleon forces in nuclear structure calculations

L. Coraggio[1], A. Gargano[1], J. W. Holt[2], N. Itaco[1], R. Machleidt[3], L. E. Marcucci[4], and F. Sammarruca[3]

[1] INFN, Sezione di Napoli, Via Cintia - I-80126 Napoli, Italy
[2] Physics Department, University of Washington, Seattle, WA 98195, USA
[3] Department of Physics, University of Idaho, Moscow, ID 83844, USA
[4] Dipartimento di Fisica "Enrico Fermi", Università di Pisa, and INFN, Sezione di Pisa, Largo B. Pontecorvo 3 - I-56127 Pisa, Italy

Abstract

Realistic nuclear potentials, derived within chiral perturbation theory, are a major breakthrough in modern nuclear structure theory, since they provide a direct link between nuclear physics and its underlying theory, namely the QCD. As a matter of fact, chiral potentials are tailored on the low-energy regime of nuclear structure physics, and chiral perturbation theory provides on the same footing two-nucleon forces as well as many-body ones. This feature fits well with modern advances in ab-initio methods and realistic shell-model. Here, we will review recent nuclear structure calculations, based on realistic chiral potentials, for both finite nuclei and infinite nuclear matter.

1 Introduction

The last decade has seen chiral EFT emerging as a powerful tool to describe the interaction among hadrons in a low-energy regime in a systematic and model-independent way (see Refs. [1, 2] for recent reviews).

Generally, in nuclear many-body systems the nuclear momenta are on the order of the pion mass, so the construction of chiral nuclear potentials is based on an expansion in powers of this soft scale ($Q \sim m_\pi$) over the hard scale, which is set by the typical hadron masses $\Lambda_\chi \sim m_\rho \sim 1$ GeV. The latter is also known as the chiral-symmetry breaking scale. The EFT must have a firm link with quantum chromodynamics (QCD) so to emerge above the level of phenomenology. This is obtained having the EFT undergo all relevant symmetries of the underlying theory, in particular, the broken chiral symmetry of low-energy QCD [3].

Nuclear potentials based on chiral perturbation theory (ChPT) [1,4,5] are at present largely employed so to bridge the QCD, that is the fundamental theory of strong interactions, to the physics of the atomic nucleus. A relevant apect of ChPT is that nuclear two-body forces, many-body forces, and currents [1,6,7] are generated on an equal footing. As a matter of fact, some low-energy constants (LECs) which appear in the two-nucleon-force (2NF), fitted to two-nucleon data, belong also to the expressions of the three-nucleon forces (3NF), four-nucleon forces (4NF), and electroweak currents.

In this contribution we will review about calculations for many-body systems, both finite nuclei and infinite nuclear matter, performed starting from nuclear forces derived within the ChPT. More precisely, we consider a low-momentum chiral potential constructed at next-to-next-to-next-to-leading order (N^3LO) using a sharp cutoff $\Lambda = 414$ MeV [8], dubbed as N^3LOW.

The paper is organized as follows. In Sec. 2, we briefly outline the theoretical framework of our many-body calculations. More precisely, we will present a few details of the chiral potential we have employed, of the theory underlying the derivation of an effective hamiltonian for a shell-model calculation, and of the perturbative expansion of the energy per particle in infinite neutron- and symmetric-nuclear matter.

Sec. 3 is dedicated to the presentation of results of realistic shell-model calculation for medium-mass nuclei outside ^{14}C and ^{16}O closed-shell nuclei, starting from N^3LOW chiral potential. The results of the calculation of the equation of state (EOS) for neutron- and symmetric-infinite matter are reported in Sec. 4, where we also focus on the comparison of the calculated symmetry energy with the empirical values.

Finally, in Sec. 5 our conclusions are reported.

2 Outline of calculations

First of all, we start this section with a brief description of the chiral potential N^3LOW. This potential has been introduced in Ref. [8], so to be compared directly with a $V_{\text{low-k}}$ derived from the chiral potential of Entem and Machleidt [4], renormalized within a cutoff momentum $\Lambda = 2.1$ fm^{-1} [9].

In the $V_{\rm low-k}$ approach, the original potential V_{NN} is smoothed by integrating out the high-momentum modes of V_{NN} down to a cutoff momentum Λ. This is achieved by a unitary transformation as the one, for example, suggested by Lee and Suzuki [9, 10].

Basically, the $V_{\rm low-k}$ approach to nuclear structure physics is successful because the dynamics which rules nuclear physics can be described in the framework of a low-energy EFT.

As already mentioned in Sec. 1, this nuclear EFT is characterized by the spontaneously broken chiral symmetry, and the degrees of freedom which are relevant for nuclear physics, nucleons and pions.

In order to reach a proper convergence rate, the ChPT expansion is valid only for momenta $Q < \Lambda_\chi \simeq 1$ GeV, where Λ_χ denotes the chiral symmetry breaking scale. Nucleon-nucleon (NN) potentials derived from ChPT are typically multiplied by a (non-local) regulator function $f(p',p) = \exp[-(p'/\Lambda)^{2n} - (p/\Lambda)^{2n}]$, where $\Lambda \simeq 0.5$ GeV is a typical choice for the cutoff scale. In terms of a relative-momentum cutoff, this means that present chiral NN potentials [4, 5] typically apply values for Λ around 2.5 fm^{-1}.

On these grounds, in Ref. [8] a chiral N^{3}LO NN potential has been considered, using the regulator function in the above expression with $n = 10$ and $\Lambda = 414$ MeV, so to be confronted directly with a $V_{\rm low-k}$ whose cutoff momentum is $\Lambda = 2.1$ fm^{-1}. We have dubbed this potential N^{3}LOW.

One advantage of this potential, respect to a $V_{\rm low-k}$ is that it is given in analytic form. The analytic expressions are the same as for the "hard" N^{3}LO potential by Entem and Machleidt of 2003 [4].

The N^{3}LOW potential reproduces accurately the empirical deuteron binding energy, the experimental low-energy scattering parameters, and the empirical phase-shifts of NN scattering up to at least 200 MeV laboratory energy (see fig. 1).

Another important issue are many-body forces. One great advantage of ChPT is that it generates nuclear two- and many-body forces on an equal footing. Most interaction vertices that appear in the 3NF and in the four-nucleon force (4NF) also occur in the two-nucleon force (2NF). The parameters carried by these vertices are fixed in the construction of the chiral 2NF. Consistency requires that for the same vertices the same parameter values are used in the 2NF, 3NF, 4NF, ...

If one considers the perturbative expansion of the 3NF at N^{2}LO, then only two extra-parameters need to be fixed, namely the low-energy constants (LECs) dealing with the one-pion exchange three-nucleon force $V_{3N}^{1\pi}$ and the contact three-nucleon force $V_{3N}^{\rm cont}$ at N^{2}LO. These 3NF LECs have been adjusted to $A = 3$ observables, more precisely so to reproduce the ^{3}H

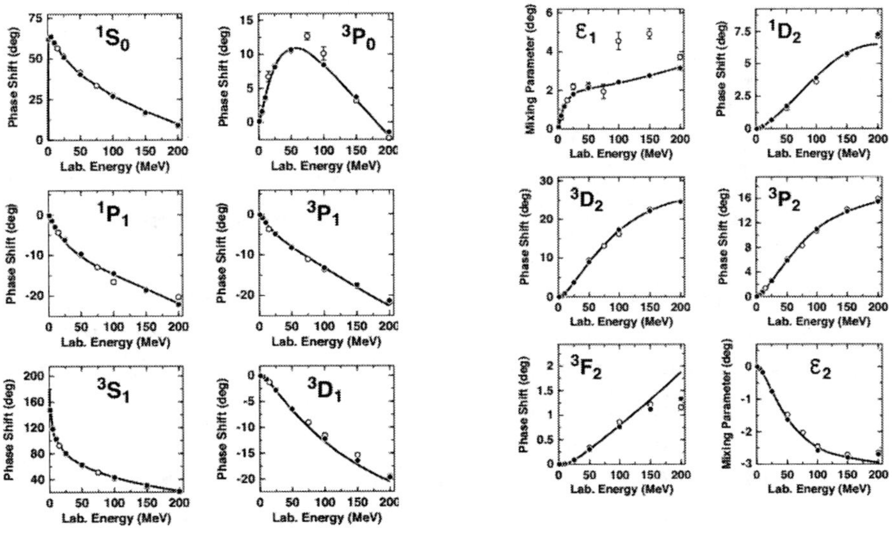

Figure 1: Phase parameters of neutron-proton scattering up to 200 MeV laboratory energy. The solid lines show the predictions by the N^{3}LOW potential. The solid dots and open circles represent the Nijmegen multienergy np phase shift analysis and the GWU/VPI single-energy np analysis SM99, respectively.

and ^{3}He binding energies together with the triton half-life (specifically the Gamow-Teller matrix element) [11, 12].

We have employed the N^{3}LOW potential in calculations for both finite nuclei [8, 13, 14] - within the framework of the realistic shell model - and for infinite nuclear matter [12, 15].

Let us start introducing the realistic shell-model.

We intend as a realistic shell-model, calculations for finite nuclei where the effective shell-model hamiltonian, namely the single-particle energies (s.p.e.) and two-body matrix elements (TBME) of the residual interaction, has been derived from a realistic NN force [16], such as the N^{3}LOW potential.

Starting from this potential, the shell-model effective hamiltonian is derived within the many-body perturbation theory, as developed by Kuo and coworkers through the 1970s [17]. More precisely, we have used the well-known $\hat{Q}$-box plus folded-diagram method [16, 17], where the $\hat{Q}$-box is a collection of one- and two-body irreducible valence-linked Goldstone diagrams. In the following calculations the $\hat{Q}$-box includes all diagrams up to third order [16], and the folded-diagram series is summed up to all orders using the Lee-Suzuki iteration method [10]. We sum over a number of intermediate states between successive vertices whose unperturbed excitation energy is less than a value $E_0 = E_{max}$, which is sufficiently large to ensure that both s.p.e. and TBME are almost independent from the choice of E_{max}.

As regards the calculation of EOS of infinite nuclear matter, we calculate the ground state energy (g.s.e.) per particle within the framework of many-body perturbation theory. More precisely, the g.s.e. is expressed as a sum of Goldstone diagrams up to third order.

The effects of the 3NF are taken into account via a density-dependent two-body potential $\overline{V}_{3N}$, that is added to the chiral N^{3}LO potential V_{2N}. This potential is obtained by integrating one nucleon up to the Fermi momentum k_F, thus leading to a density-dependent two-nucleon interaction $\overline{V}_{3N}(k_F)$. Presently, the analytic expressions for $\overline{V}_{3N}$ [18] has been derived only for the N^{2}LO 3NF, which is the one we take into account in the calculation of our EOS. In order to consider the correct combinatorial factors of the normal-ordering at the two-body level of the 3NF, the matrix elements of $\overline{V}_{3N}(k_F)$ have to be multiplied by a factor 1/3 in the first-order Hartree-Fock (HF) diagram, and by a factor 1/2 in the calculation of the HF single-particle energies [12, 15].

The analytic expressions of first-, second-, and third-order particle-particle (*pp*) and hole-hole (*hh*) contributions, together with the one of single-particle HF potential, are reported in Ref. [15]. The implicit expression of the third-order particle-hole *ph* diagram - also known as the ring diagram - can be found in Ref. [19].

3 Shell-model results

In Ref. [13], we have presented a shell-model description of heavy carbon isotopes, using a fully microscopic approach, starting from N^{3}LOW potential and using ^{14}C as a closed core. As mentioned in sec. 2, both the single-particle energies and residual two-body interaction have been derived within the many-body perturbation theory. Our calculations have led to a successful description of these isotopes when approaching the neutron dripline. More precisely, our results have reproduced the disappearance of the $N = 14$ subshell closure that is present in the oxygen chain, and predicted the $N = 16$ one.

Moreover, the shell-model calculations have predicted that ^{21}C is unstable against one-neutron decay, fitting the picture of ^{22}C as a Borromean nucleus. It is worth mentioning that the latter is one of the most exotic nuclei of the 3000 known isotopes, its N/Z ratio being 2.67, and can be

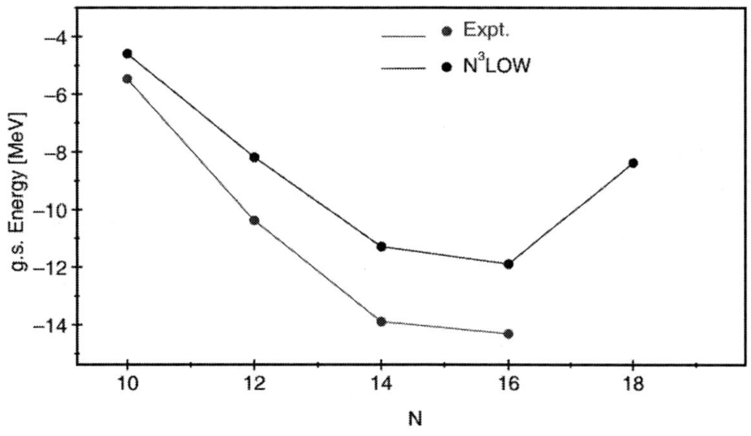

Figure 2: (Color online) Experimental and calculated ground-state energies for carbon isotopes from $A = 16$ to 24. N is the number of neutrons. See text for details.

considered the heaviest Borromean nucleus ever observed. We reproduce successfully the fact that ^{22}C is the last bound isotope (see fig. 2).

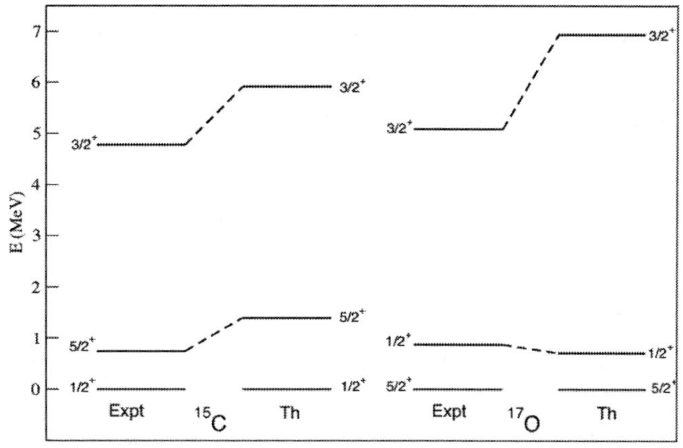

Figure 3: Experimental [20] and theoretical single-particle states of ^{15}C and ^{17}O.

In ref. [13] we investigated only the identical-particle channel of the effective hamiltonian. Here, we point out that also the proton-neutron channel features well, as testified in fig. 3 where the experimental and theoretical single-particle spectra of ^{15}C and ^{17}O are reported.

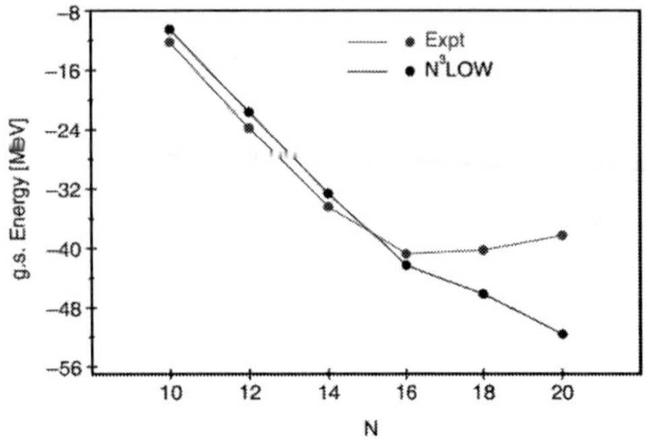

Figure 4: (Color online) Experimental and calculated ground-state energies for oxygen isotopes from $A = 18$ to 24. N is the number of neutrons.

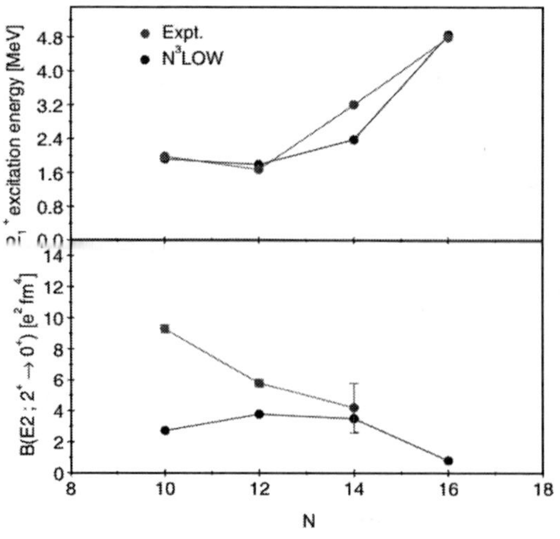

Figure 5: (Color online) Experimental [20] and calculated excitation energies of the yrast $J^{\pi} = 2^{+}$ states and $B(E2; 2_1^{+} \rightarrow 0_1^{+})$ transition rates for oxygen isotopes.

As can be seen the inversion of the $J^{\pi} = 1/2^{+}$ and $5/2^{+}$ states is correctly reproduced, thanks to the proton-neutron monopole component of the effective hamiltonian.

We have employed the N^{3}LOW potential also for calculations of nuclei belonging to the *sd*-shell, using ^{16}O as a closed core. In ref. [23] we reported only results for two-valence-particle nuclei ^{18}O and ^{18}F. Here we report in fig. 4 the calculated ground-state energies, with respect to ^{16}O, of even-mass oxygen isotopes and compare them with the experimental ones. As can be observed, the agreement is quite remarkable up to ^{24}O, then the theory predicts a bound ^{26}O, at variance with experiment. This probably traces back to the fact that we have not included the effect of three-body forces in the derivation of our effective hamiltonian [24]

In fig. 5 we have also reported the experimental and calculated excitation energies of the yrast $J^{\pi} = 2^{+}$ states and $B(E2; 2_1^{+} \rightarrow 0_1^{+})$ transition rates.

4 Infinite nuclear matter

As mentioned previously, we calculate the energy per particle of infinite nuclear matter in the framework of the Goldstone expansion taking into account the effects of the N^{2}LO 3NF via the introduction of a density-dependent two-body potential $\overline{V}_{NNN}(k_F)$.

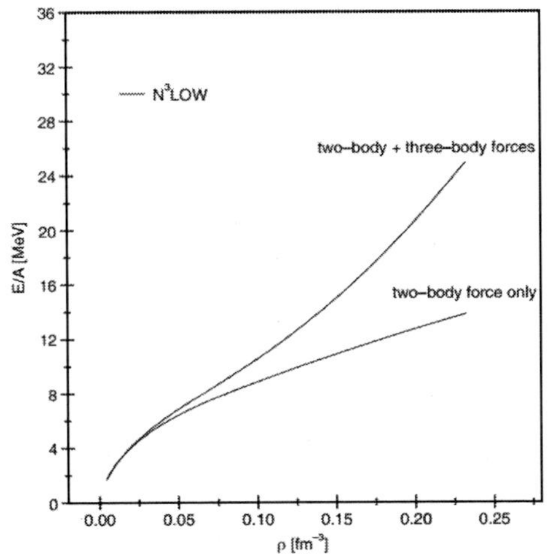

Figure 6: Results obtained for the PNM energy per nucleon, with and without 3NF effects.

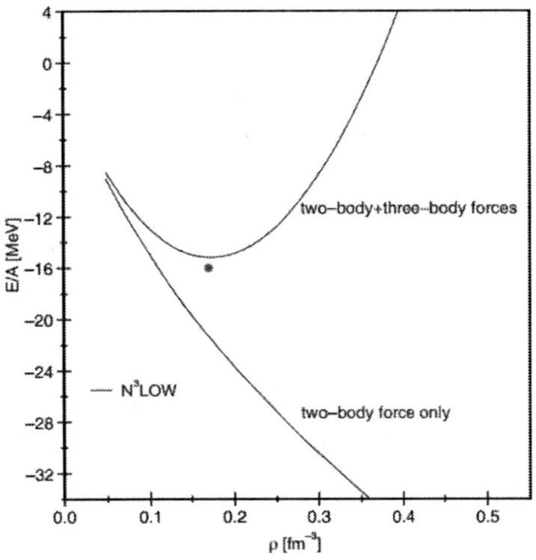

Figure 7: (Color online) Same as in Fig. 7, but for the PNM energy per nucleon.

In Fig. 6 and 7 we show, respectively, the pure-neutron matter (PNM) and symmetric-nuclear matter (SNM) EOS as a function of density, calculated with and without including the 3NF contributions.

From the inspection of the figures, it can be seen that the repulsive effect arising from 3NF is consistent in both cases. In particular it can be observed that the 3NF contribution to the energy per nucleon in SNM is larger than that in PNM, and that its inclusion proves to be crucial to reproduce the empirical saturation properties.

The symmetry energy E_{sym} is defined as the strength of the quadratic term in an expansion of the energy per particle ($\overline{E}$) in asymmetric matter with respect to the asymmetry parameter $\alpha = (\rho_n - \rho_p)/(\rho_n + \rho_p)$:

$$\overline{E}(\rho, \alpha) \simeq \overline{E}(\rho, \alpha = 0) + E_{sym}\alpha^2 + \mathcal{O}(\alpha^4) \ . \tag{1}$$

Many microscopic calculations provide a nearly linear behavior of $\overline{E}(\rho, \alpha)$ as a function of α^2, and this justifies the common approximation of defining the symmetry energy as the difference between the energy per particle in PNM and SNM, neglecting powers beyond α^2 in the $\overline{E}(\rho, \alpha)$ expansion. The density dependence of E_{sym} can be correlated with the neutron skin thickness of nuclei and the radius of neutron stars, and systematic efforts are ongoing to get better empirical constraints on its value, through both laboratory and astrophysical measurements.

In Fig. 8, our results for the symmetry energy as a function of density are reported. In this connection, of particular interest is the slope of the

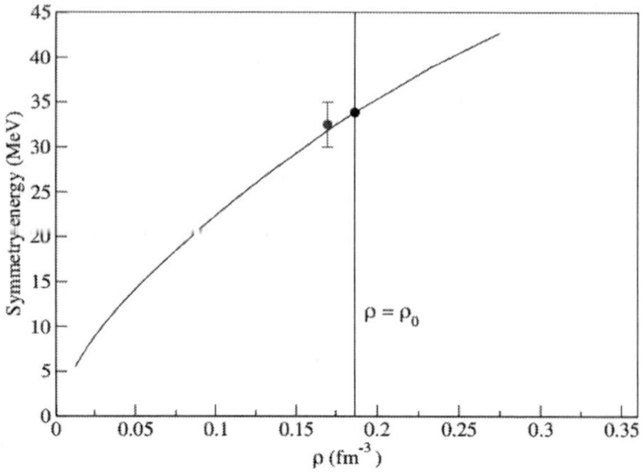

Figure 8: (Color online) Calculated symmetry energy as a function of density. The red error bar represents the empirical constraint on E_{sym} [21].

symmetry energy at the saturation density L defined as

$$L = 3\rho_0 \left(\frac{\partial E_{sym}}{\partial \rho}\right)_{\rho_0} \quad . \tag{2}$$

Our calculated value for L is 68 MeV, that has to be compared with the not yet firmly constrained empirical value of $L = 70 \pm 25$MeV [22].

5 Conclusions

In this paper, we have presented nuclear structure calculations, based on the realistic N^{3}LOW chiral potential, for both finite nuclei and infinite nuclear matter.

The quality of the results obtained for finite nuclei, and the ability to provide realistic nuclear matter predictions, makes us confident that the approach to nuclear structure, based on chiral potentials, may be a valuable tool to study properties of stable and unstable nuclear systems.

This excellent agreement paves the way towards an exciting future for the study of the physics of nuclei.

References

[1] R. Machleidt and D. R. Entem, *Phys. Rep.*, **503** (2011) 1.

[2] E. Epelbaum, *et al.*, *Rev. Mod. Phys.*, **81** (2009) 1773.

[3] S. Weinberg, *Physica*, **96A** (1979) 327.

[4] D. R. Entem and R. Machleidt, *Phys. Rev. C*, **68** (2003) 041001(R).

[5] E. Epelbaum, *et al.*, *Nucl. Phys. A*, **747** (2005) 362.

[6] S. Weinberg, *Phys. Lett. B*, **295** (1992) 114.

[7] U. van Kolck, *Phys. Rev. C*, **49** (1994) 2932.

[8] L. Coraggio, *et al.*, *Phys. Rev. C*, **75** (2007) 024311.

[9] S. Bogner, *et al.*, *Phys. Rev. C*, **65** (2002) 051301(R).

[10] S. Y. Lee and K. Suzuki, *Phys. Lett. B*, **91** (1980) 173.

[11] L. E. Marcucci, *et al.*, *Phys. Rev. Lett.*, **108** (2012) 052502.

[12] L. Coraggio, *et al.*, *Phys. Rev. C*, **89** (2014) 044321.

[13] L. Coraggio, *et al.*,*Phys. Rev. C*, **81** (2010) 064303.

[14] L. Coraggio, *et al.*, *Nucl. Phys. A*, **928** (2014) 43.

[15] L. Coraggio, *et al.*, *Phys. Rev. C*, **87** (2013) 014322.

[16] L. Coraggio, *et al.*, *Ann. Phys.*, **327** (2012) 2125.

[17] T. T. S. Kuo and E. Osnes, *Lecture Notes in Physics*, **364** (1990).

[18] J. W. Holt, *et al.*, *Phys. Rev. C*, **81** (2010) 024002.

[19] James J. MacKenzie, *Phys. Rev.*, **179** (1969) 1002.

[20] *Data extracted using the NNDC On-line Data Service from the ENSDF database*, file revised as of September 30, 2015.

[21] M. Dutra *et al.*, *Phys. Rev. C*, **85** (2012) 035201.

[22] M. B. Tsang *et al.*, *Phys. Rev. C*, **86** (2012) 015803.

[23] L. Coraggio, *et al.*, *Nucl. Phys. A*, **928** (2014) 43.

[24] T. Otsuka, *Phys. Rev. Lett.*, **105** (2010) 032501.

EPJ Web of Conferences **117**, 02002 (2016) DOI: 10.1051/epjconf/201611702002
NN2015

Nuclear reactions at intermediate energies

Radhey Shyam

Saha Institute of Nuclear Physics, 1/AF Bidhan Nagar, Kolkata 700064, India

Abstract

In the domain of Nuclear reactions at intermediate energies, the QCD coupling constant α_s is large enough ($\sim$ 0.3 - 0.5) to render the perturbative calculational techniques inapplicable. In this regime the quarks are confined into colorless hadrons and it is expected that effective field theories of hadron interactions via exchange of hadrons, provide useful tools to describe such reactions. In this contribution we discuss the application of one such theory, the effective Lagrangian model, in describing the hadronic reactions at intermediate energies whose measurements are the focus of a vast international experimental program.

In the quest for understanding the quantum chromodynamics (QCD), the hadron-hadron, hadron-photon and hadron-lepton reactions have played a major role. From these studies two quite distinct regimes of QCD have emerged. The first regime is of those processes that involve very high momentum transfers. Here the asymptotic freedom property of QCD has enabled the perturbative QCD (pQCD) methods to describe [1] a large set of high energy large momentum transfer reaction observables with great precision, providing the most precise test of QCD to date [2, 3].

The second regime that includes nuclear reactions at intermediate energies, is characterized by the confinement of quarks inside the hadrons. This remarkable feature, along with the running of the QCD coupling constant (α_s) towards larger values makes the pQCD methods inapplicable - one requires techniques that can solve the theory exactly. To date, the only fully predictive non-perturbative method for studying QCD at low energies is lattice QCD (LQCD). The progress in LQCD studies of the low-lying spectrum of QCD, hadronic structure and interactions has been significant in the recent years (see, eg., Ref. [4]).

Nevertheless, LQCD calculations are still far from being routinely amiable to solutions of intermediate energy nuclear reactions. It is therefore, necessary to use effective methods to describe the dynamics of such processes. The effective field theories where hadrons interact via exchange of hadrons, provide a viable alternative to LQCD for the description of the intermediate energy nuclear reactions (see, e.g., Refs. [5–11]). These theories implement the symmetries of the underlying theory, QCD, by using proper effective Lagrangians. However, many of these models invariably require *a priory* knowledge of unknown parameters related to the coupling constants associated with the effective Lagrangians and the masses of the hadrons. Because the effective theories provide flexible theoretical approaches that enable a rapid understanding of the complexities involved in the scattering problems at intermediate energies and have the capability to guide to newer experimental studies, steady progress has been made towards determining the realistic input parameters involved in these theories.

In this presentation we discuss the application of effective-Lagrangian based theoretical models, to describe the meson production reactions (both non-strange and strange) in photon and meson induced reactions on nucleons at intermediate energies. We also present some results on the charmed baryon and meson production in the antiproton-proton annihilation process.

1 Meson production in photon-nucleon reactions

The photon induced reactions on nucleons where mesons and nucleons are measured as final decay products, have revealed a rich excitation spectrum of nucleons that reflects their complicated multi-quark inner dynamics. The determination of the properties of the nucleon excited states (or resonances) from the data of such experiments is a major challenge of hadron physics [12].

The effective methods that are used to describe the dynamics of meson production reactions, include explicitly the baryon resonance states whose properties are extracted by comparing the predictions of the theory with the experimental data [9–11, 13–17]. However, for a reliable realization of this goal, a model is required that can analyze different reactions over the entire energy range using a single Lagrangian density, which generates all non-resonance contributions from Born, u- and t-channel contributions without introducing new parameters. At the same time, the Lagrangian should

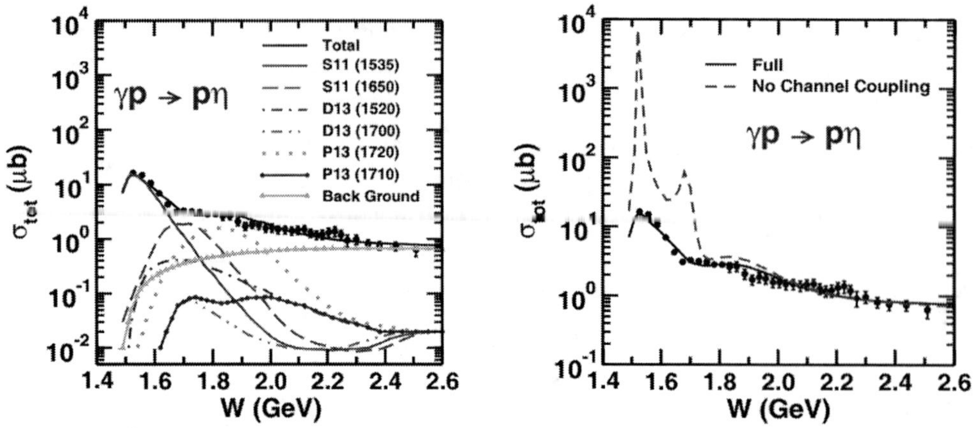

Figure 1: (Left panel) Total cross section for the $\gamma p \to \eta p$ reaction as a function of γp invariant mass W. Individual contributions of various resonances and of background terms are shown by different curves as indicated in the figure. (Right panel) Effect of channel coupling on the total cross section for the $\gamma p \to \eta p$ reaction as a function of W. The solid line represents the results of the full coupled-channel calculations while the dashed line shows the one where channel coupling is switched off.

also satisfy the symmetries of the fundamental theory (i.e. QCD) while retaining only mesons and baryons as effective degrees of freedom. The coupled-channels method within the K-matrix approximation is one model that respects these constraints while providing a way to analyze simultaneously all the reaction data for a multitude of observables in different reaction channels. It also leads to a convenient way of imposing the unitarity constraint because the S matrix in this approach is unitary provided the K-matrix is taken to be real and Hermitian (see, e.g., Ref. [11,16]).

The experimental data on the η meson and associated strangeness ($K^+\Lambda$ and $K^+\Sigma^0$) production in photon-nucleon reactions are available for photon energies (E_γ) in the range of threshold to 3 GeV. Thus they cover not only the entire resonance region but also the region where the background contributions (t-channel amplitudes mainly) are expected to be dominant. The data for the η channel are reviewed recently in Ref. [20]. The coupled-channels K-matrix method has been used extensively in Refs. [11, 16] to describe the data of both these channels in the entire range of E_γ with a single set of parameters. These reaction channels have also been studied within the Giessen coupled-channels K-matrix model in Refs. [18,19] for E_γ between threshold to 2.0 GeV.

As an example of the success of the coupled-channels K-matrix model of Refs. [11,16] in describing the meson production in photon-nucleon reactions, we show in the left panel of Fig. 1 the contribution of various resonances to the total cross section (σ_{tot}) of the $\gamma p \to \eta p$ reaction. The experimental data are taken from Ref. [21]. We see that the general features of the data are described reasonably well by the calculations. While contributions of the $S_{11}(1535)$ resonance dominate the σ_{tot} from near threshold to invariant mass (W) values of 1.7 GeV (corresponding to $E_\gamma \sim 1.1$ GeV), those of the $S_{11}(1650)$ and P_{13} (1720) resonance are important for E_γ between 0.950 GeV to 1.28 GeV and 1.1 GeV to 2.2 GeV, respectively. The contributions of other resonances are quite weak in the entire range of E_γ. The cross sections beyond 2 GeV are mostly governed by the contributions of the background terms.

The effect of channel coupling on the σ_{tot} is shown in the right panel of Fig. 1. In the no coupling case (NCC), the amplitudes of various processes are simply added together, ignoring the modifications to the widths of the resonances introduced by the channel couplings. We notice that for $W > 1.8$ GeV the differences between the full and the NCC results are very small because the contributions of the resonances are almost negligible in this region. However, at lower energies, the channel-coupling effects are large. In fact, at some energies, the resonance propagators can develop poles in the absence of channel couplings. Thus modifications introduced to the widths of the resonances due to channel couplings are indeed vital for reproducing the energy dependence of the experimental cross sections. Therefore, the use of a full coupled-channels approach is essential to describe the meson-production in photo-nucleon reactions.

2 Double strangeness ($S = -2$) production with K^- induced reactions

The investigations of the cascade (Ξ) hypernuclei (having strangeness $S = -2$) are of great importance. The binding energies and widths of the Ξ hypernuclear states determine the strengths of the ΞN and $\Xi N \to \Lambda\Lambda$ interactions, which are difficult to explore directly in a laboratory. This input is vital for understanding the multi-strange hadronic and quark matter. Since strange quarks are negatively charged they are preferred in charge neutral dense matter. Thus these studies are capable of probing the role of strangeness in the equation of state at high density, eg. in the cores of neutron stars [22].

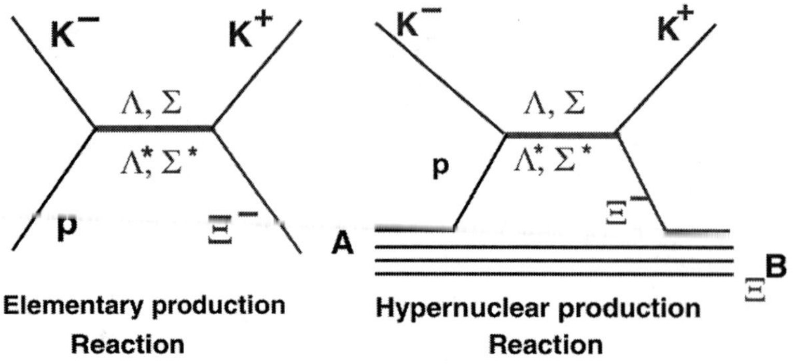

Figure 2: (Left panel) Graphical representation of the s-channel diagrams used to describe the $p(K^-, K^+)\Xi^-$ reaction. The u-channel diagrams are not explicitly shown here but their contributions are included in the calculations. (Right panel) Diagram used to calculate the amplitudes of the $A(K^-, K^+)_{\Xi^-}B$ reaction. A represents the target nucleus and $_{\Xi^-}B$ the final hypernucleus.

Because the (K^-, K^+) reaction leads to the transfer of two units of both charge and strangeness to the target nucleus, it provides one of the most promising ways of studying the $S = -2$ systems such as Ξ hypernuclei and a dibaryonic resonance (H), which is a near stable six-quark state with spin parity of 0^+ and isospin 0 [23–25]. The (K^-, K^+) reaction implants a Ξ hyperon in the nucleus through the elementary $p(K^-, K^+)\Xi^-$ process. The cross sections for this reaction were measured in the 1960s and early 1970s using hydrogen bubble chambers.

In Refs. [26,27], the (K^-, K^+) reaction on a proton target leading to the production of a free Ξ^- hyperon has been studies within a single-channel effective Lagrangian model (SCELM) (see, eg., Refs. [28, 29]). A coupled-channels study of this reaction is not feasible at this stage due to the scarcity of the data on the K^- interaction channels. In the SCELM the full field theoretic structure of the interaction vertices are retained and baryons are treated as Dirac particles. The initial state interaction of the incoming K^- with a free or bound target proton leads to the excitation of intermediate Λ and Σ resonant states, which propagate and subsequently decay into Ξ^- and K^+. In case of the reaction on nuclei, Ξ^- gets captured into one of the nuclear orbits, while the K^+ meson goes out. We have considered as intermediate states the Λ and Σ hyperons and eight of their resonances with masses up to 2.0 GeV, which are represented by Λ^* and Σ^* in Figs. 2. The effective Lagrangians and the coupling constants at various vertices are taken to be the same as those described in Ref. [26].

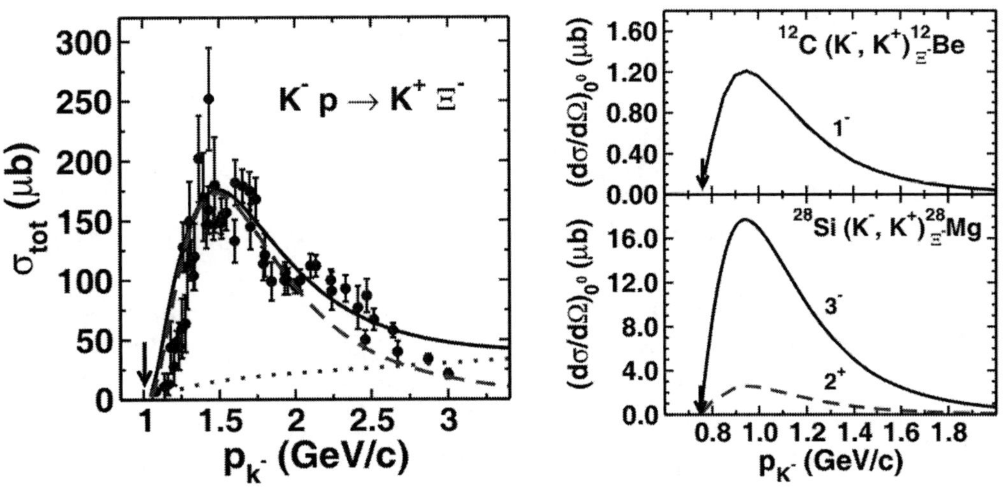

Figure 3: (Left panel) Total cross section for the $p(K^-, K^+)\Xi^-$ reaction as a function of incident K^- momentum. Individual contributions of of s- and u-channel diagrams to the total cross sections are also shown. (Right panel) p_{K^-} dependence of the zero degree differential cross section for the $^{12}C(K^-, K^+)^{12}_{\Xi^-}$Be and $^{28}Si(K^-, K^+)^{28}_{\Xi^-}$Mg reactions. The spin-parity of the final hypernuclear states are indicated on each curve. Arrows show the thresholds for the corresponding reactions.

In the left panel of Fig. 3, we show comparisons of the SCELM calculations with the data for the σ_{tot} of the $p(K^-, K^+)\Xi^-$ reaction as a function of K^- beam momenta (p_{K^-}). It is clear that the data are reproduced reasonably well by the model within the statistical errors. The zero degree differential cross sections for the hypernuclear production reactions $^{12}C(K^-, K^+)^{12}_{\Xi^-}$Be and $^{28}Si(K^-, K^+)^{28}_{\Xi^-}$Mg calculated by using the same vertex parameters are shown in the right panel of Fig. 3. In these calculations we have have employed pure single-particle-single-hole ($p^{-1}\Xi$) configurations to describe the nuclear structure part. The bound proton-hole and Ξ^--particle state spinors were generated in a phenomenological model, where they are obtained by solving the Dirac equation with scalar and vector fields having Woods-Saxon radial forms. With a set of radius and diffuseness parameters, the depths of these fields are searched to reproduce the binding energy (BE) of a given state. The BEs for the Ξ^- bound states were taken to be those predicted in the latest version of the quark-meson coupling (QMC) model [30, 31] as their experimental values are still unknown.. In this figure, we have shown results for populating the hypernuclear state with maximum spin of natural parity for each configuration. We see that the zero degree differential cross sections for the Ξ^- hypernuclear production reactions peak around the beam momentum of 1.0 GeV/c with peak cross section of more than 1 μb. They closely follow the trends of the elementary Ξ^- production cross sections. Our predictions will be useful for the future JPARC experiments.

3 Charmed hadron production in $\bar{p}p$ annihilation

The heavy flavored hadrons consisting of relatively heavier mass charm quarks provide an additional handle for the understanding of QCD. The large mass of the charm quark introduces a mass scale that is much larger than the confinement scale $\Lambda_Q \approx 300$ MeV. In contrast, the energy scale of the lighter quarks is $\ll \Lambda_Q$. The presence of two scales in such systems naturally leads to the construction of an effective theory where one can actually calculate a big portion of the relevant physics using perturbation theory and renormalization-group techniques.

In this context, the investigations of the production of heavy flavor hadrons are of great interest. Since the discovery of J/ψ in 1974 [32, 33], the production of charmonium ($c\bar{c}$) states has been extensively studied in electron-positron and proton-antiproton ($\bar{p}p$) annihilation experiments. Yet a substantial part of the charmonium spectrum is still to be precisely measured. The first charmed baryon states were detected in 1975 in neutrino interactions [34]. Although many new excited charmed baryon states have been discovered since then, the studies of the production and spectroscopy of the charmed baryons have not been carried out in the same detail as the charmonium states.

In the near future, charmed hadron production will be studied in the $p\bar{p}$ annihilation in the ($\bar{P}ANDA$) experiment at the FAIR in GSI, Darmstadt [35]. The advantage of using antiprotons in the study of the charmed baryon is that in $p\bar{p}$ collisions the production of no extra particle is needed for the charm conservation, which reduces the threshold energy. Moreover, all the the charmonium states can be reached directly in this method, which is in contrast to the electron-positron annihilation where direct formation of the only those final charmonium states is allowed that have the quantum numbers of the photon ($J^{PC} = 1^{--}$).

For the planning of these experiments at the $\bar{P}ANDA$ facility, reliable theoretical estimates of the cross section of $\bar{p}p \to \bar{\Lambda}_c^- \Lambda_c^+$, $\bar{p}p \to \bar{D}^0 D^0$ and $\bar{p}p \to D^- D^+$ reactions are of crucial importance [36, 37]. In Ref. [38], the SCELM has been used to study the $\bar{p}p \to \bar{\Lambda}_c^- \Lambda_c^+$ reaction as a sum of the t-channel D^0 and D^{*0} meson-exchange diagrams [Fig. 4(a)]. A similar model

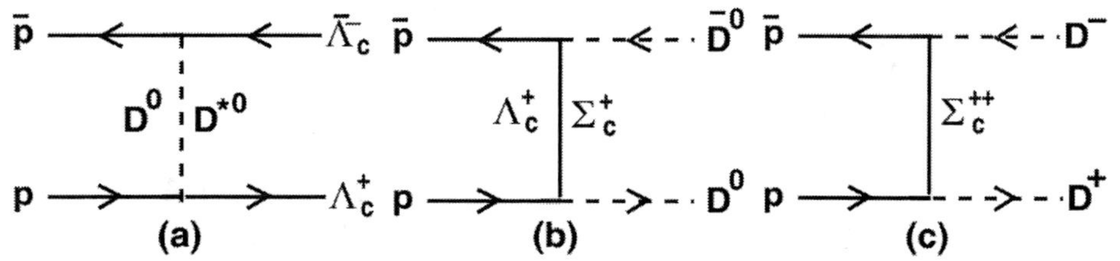

Figure 4: Graphical representations of the t-channel meson exchange diagrams used to describe the $\bar{p}+p \rightarrow \bar{\Lambda}_c^- + \Lambda_c^+$ reaction (a), and the t-channel baryon exchange diagrams to describe the $\bar{p}+p \rightarrow \bar{D}^0+D^0$ and $\bar{p}+p \rightarrow D^-+D^+$ reactions (b and c).

is employed in Ref. [39] to calculate the cross sections of $\bar{p}p \rightarrow \bar{D}^0 D^0$, and $\bar{p}p \rightarrow D^- D^+$ reactions as a sum of the t-channel Λ_c^+ and Σ_c^+, and Σ_c^{++} baryon-exchange diagrams, respectively [Figs. 4(b) and 4(c)]. Also included in these calculations are the s-channel excitation, propagation and decay of the $\Psi(3770)$ resonance into the $\bar{D}^0 D^0$ and $D^- D^+$ channels.

It is found that the σ_{tot} of the $\bar{p}p \rightarrow \bar{\Lambda}_c^- \Lambda_c^+$ and $\bar{p}+p \rightarrow \bar{D}^0 D^0$ reactions at the $\bar{p}$ beam momentum of 15 GeV/c (which the regime of the $\bar{P}ANDA$ experiment) are around 25 μb and 550 nb, respectively. These values are drastically larger than those predicted in previous calculations based on models that invoke the quark degrees of freedom. The future $\bar{P}ANDA$ experiment will clarify the adequacy of various models at these relatively lower beam momenta.

In summary, we can say that effective Lagrangian models are able to describe resonably well the intermediate energy nuclear reactions. The coupled-channels K-matrix method allows for the simultaneous calculation of the observables for a large multitude of reactions with considerably fewer parameters than would be necessary if each reaction channel were fitted separately. For the strangeness and charm production reactions, a single-channel effective Lagrangian method with suitably chosen parameters is able to make important predictions that will be tested in the future experiments planned at upcoming facilities.

Acknowledgements: This work has been supported by the Helmholtz International Center (HIC) for FAIR and the Council of Scientific and Industrial Research (CSIR), India. The author acknowledges useful discussions with H. Lenske, O. Scholten, A.W. Thomas and K. Tsushima.

References

[1] S. Bethke, Progr. Part. & Nucl. Phys. **58** (2007) 351.

[2] John Collins, *Foundations of Perturbative QCD*, Cambridge, 2013.

[3] A.H. Müller, Phys. Rep. **73** (1981) 237.

[4] R.A. Briceno, Z. Davoudi, and Thomas C. Luu, arXiv:1406.5673 [hep-lat].

[5] M. Nielsen, F.S. Navarra and S.H. Lee, 2010 Phys. Rept. **497** (2010) 41.

[6] I.C. Cloet and C.D. Roberts 2014 Prog. Part. Nucl. Phys. **77** (2014) 1.

[7] J. de Vries, U.-G. Meissner, E. Epelbaum, N. Kaiser, Eur. Phys. J. A **49** (2013) 149.

[8] D. Siemens, V. Bernard, E. Epelbaum, H. Krebs, U.-G. Meissner, Phys.Rev. C **89** (2014) 6, 065211.

[9] T. Feuster and U. Mosel, Phys. Rev. C **58** (1998) 457; *ibid.* **59** (1999) 460.

[10] G. Penner and U. Mosel, Phys. Rev. C **66** (2002) 055211; *ibid.* **66** (2002) 055212.

[11] R. Shyam and O. Scholten, Phys. Rev. C **78** (2008) 065201.

[12] S. Capstick and W. Roberts, Prog. Part. Nucl. Phys. **45** (2000) 241.

[13] B. Saghai and Z. Li, Eur. Phys. J. A **11** (2001) 217.

[14] B. Borasoy, E. Marco, and S. Wetzel, Phys. Rev. C **66** (2002) 055208.

[15] J. Durand, B. Julia-Diaz, T.-S.H. Lee, A. Matsuyama, T. Sato, Phys. Rev. C **78** (2008) 025204.

[16] R. Shyam, O. Scholten and H. Lenske, Phys. Rev. C **81** (2010) 015204.

[17] D. Rönchen, M. Döring, H. Haberzettl, J. Haidenbauer, U.-G. Meiner, K.Nakayama, Eur. Phys. J. A **51** (2015) 70.

[18] V. Shklyar, H. Lenske, and U. Mosel, Phys. Rev. C **87** (2013) 015201, and references of previous studies therein.

[19] Xu Cao, V. Shklyar and H. Lenske, Phys. Rev. C **88** (2013) 055204.

[20] V.L. Kashevarov, arXiv:1506.02546 [nucl-ex].

[21] V. Crede *et al.*, Phys. Rev. Lett. **94** (2005) 012004.

[22] J. Schaffner-Bielich, Nucl. Phys. A **804** (2008) 309; *ibid.* Nucl. Phys. A **835** (2010) 279c.

[23] R. L. Jaffe, Phys. Rev. Lett. **38** (1977) 195

[24] P. J. Mulders and A. W. Thomas, J. Phys. G **9** (1983) 1159.

[25] R. Shyam, O. Scholten and A.W. Thomas, Phys. Rev. C **88** (2013) 025209.

[26] R. Shyam, O. Scholten and A. W. Thomas, Phys. Rev. C **84** (2011) 042201(R).

[27] B.C. Jackson, Y. Oh, H.Haberzettl, and K. Nakayama, Phys. Rev. C **91** (2015) 065208.

[28] R. Shyam, Phys. Rev. C **60** (1999) 055213.

[29] R. Shyam and U. Mosel, Phys. Rev. C **67** (2003) 065202.

[30] K. Tsushima, R. Shyam and A. W. Thomas, Few-Body Syst. **54** (2013) 1271.

[31] R. Shyam, Nucl. Phys. A **914** (2013) 79.

[32] J. Aubert *et al.*, Phys. Rev. Lett. **33** (1974) 1404.

[33] J. Augustin *et al.*, Phys. Rev. Lett. **33** (1974) 1406.

[34] E. Cazzoli, A. Cnops, P. Connolly, R. Louttit, M. Murtagh, R. Palmer, N. Samios, T. Tso, and H. Williams, Phys. Rev. Lett. **34** (1975) 1125.

[35] U. Wiedner, Prog. Part. Nucl. Phys. **66**, 477 (2011).

[36] J. Haidenbauer, and G. Krein, Phys. Lett. B **678**, 314 (2010).

[37] J. Haidenbauer, and G. Krein, Phys. Rev. D **89**, 114003 (2014); *ibid.* **91**, 114022 (2015).

[38] R. Shyam and H. Lenske, Phys. Rev. D **90** (2014) 014017.

[39] R. Shyam and H. Lenske, arXiv:1509.04786 [hep-ph].

EPJ Web of Conferences **117**, 02003 (2016) DOI: 10.1051/epjconf/201611702003
NN2015

Emission of neutrino-antineutrino pairs by hadronic bremsstrahlung processes

Sonia Bacca[1,2], Rishi Sharma[3] and Achim Schwenk[4,5]

[1] TRIUMF, Vancouver, British Columbia, V6T 2A3, Canada
[2] Department of Physics and Astronomy, University of Manitoba, Winnipeg, MB, R3T 2N2, Canada
[3] TIFR, Homi Bhabha Road, Navy Nagar, Mumbai 400005, India
[4] Institut für Kernphysik, Technische Universität Darmstadt, 64289 Darmstadt, Germany
[5] ExtreMe Matter Institute EMMI, GSI Helmholtzzentrum für Schwerionenforschung GmbH, 64291 Darmstadt, Germany

Abstract

We review our recent calculations of neutrino-antineutrino pair production from bremsstrahlung processes in hadronic collisions at temperature and densities relevant for core-collapse supernovae. We focus on neutron-neutron and neutron-α collisions.

1 Introduction

Understanding core-collapse supernovae and neutron star formation requires input from nuclear physics regarding not only the equation of state, but also transport processes. In core-collapse supernovae, neutrinos of all flavors are liberated seconds after the collapse and take away 99% of the total gravitational energy. These are dominantly produced in the proto-neutron star created after the collapse and propagate in the dense material around the neutrinosphere while undergoing interactions with themselves and with protons, neutrons and nuclei in this material [1,2] before eventually decoupling.

Interactions of neutrinos with this material affect their spectrum and play a role in the dynamics of the shock wave passing through it.

Consequently, the identification and quantitative determination of the processes that lead to neutrino production, scattering and absorption for the relevant astrophysical conditions is crucial. Here, we concentrate on the calculation of the neutrino-antineutrino pair production and absorption by bremsstrahlung from hadronic collisions. This process is induced by the neutral current interaction and is the only hardonic process that produces muon and tau neutrinos, which unlike electron neutrinos are not produced in charged-current reactions like β-decays and electron captures.

Supernova explosions are most sensitive to neutrino processes near the surface of the proto-neutron star, where matter is very neutron-rich, and at densities of the order of one tenth of the nuclear saturation density $\rho_0 = 2.8 \times 10^{14}$ g cm^3. The bremsstrahlung process involving a neutron-neutron collision is the dominant mechanism for production of muon and tau neutrinos in this region and involves nuclear interactions. While this process has been relatively well studied (see Refs. [3, 4] and references therein), the question about the relevance of neutrino production from other hadronic collisions involving neutrons and protons [5] and neutrons and finite nuclei [6] has only been explored recently. Below we review our recent calculations [3, 4, 6] of the neutrino-antineutrino pair production from bremsstrahlung processes and compare results from neutron-neutron [3, 4] and neutron-α collisions [6].

2 Results

We consider the $\nu\bar{\nu}$ bremsstrahlung from neutron (n) collisions with other neutrons or α-particles (denoted as N). The Feynman diagram of this process is shown in Fig. 1. The weak vertex radiating the Z^0 boson can come either from the incoming neutron or the outgoing neutron. Moreover, in case that N in Fig. 1 is another neutron, two additional diagrams associated with the Z^0 boson emitted from the right in-going or outgoing neutron exist (see Ref. [3, 6] for more details).

In case of $n\alpha$ collisions, the scattering amplitude $\mathcal{M}$ for this process can be written as

$$i\mathcal{M} = -\frac{i\,G_F C_A}{\sqrt{2}}\frac{1}{\omega}\sum_{j=1,2,3} l^j \chi_1^\dagger\,[\boldsymbol{\sigma}^j, \mathcal{T}(\mathbf{k})]\,\chi_3\,, \tag{1}$$

where G_F is the Fermi coupling constant and $C_A = -g_A/2$ is the axial-

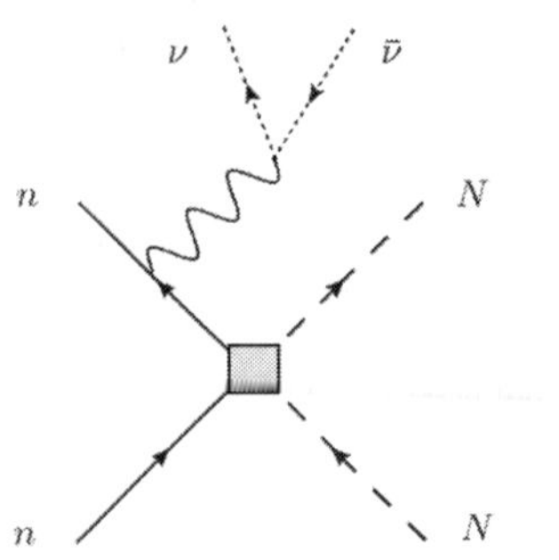

Figure 1: Feynman diagram for the neutrino-antineutrino (dotted lines) bremsstrahlung process from hadronic collisions: solid lines are neutrons, while dashed lines are either neutrons or α-particles and the wiggly line is the Z^0 boson. The square box represents the hadronic interactions taking place in the collision.

vector coupling for neutrons. Here, ω is the energy transferred from the neutrino-pair to the neutron, $\mathbf{k}$ is the momentum transfer, l^j is the leptonic current, $\boldsymbol{\sigma}^j$ are Pauli matrices associated with the neutron spin, $\chi_{1,3}$ are neutron spinors, and $\mathcal{T}(\mathbf{k})$ denotes the strong interaction scattering vertex. From the fact that the commutator of the neutron spin operator with the strong vertex appears in Eq. (1), one can readily see that only noncentral terms of the strong force will contribute to the amplitude. Similarly, also for nn collisions only noncentral terms of the strong force contribute.

For nn bremsstrahlung, this process was studied first by Friman and Maxwell [7] using the Born approximation to calculate the scattering vertex $\mathcal{T}$ with the nucleon-nucleon interaction described by the one-pion-exchange potential, which gives rise to the long-range tensor force. In Ref. [8], it was calculated for degenerate conditions using nucleon-nucleon phase shifts. We extended these studies in Ref. [3] by using chiral effective field theory (EFT) interactions at different orders [9, 10] and solving the Boltzmann equation with collisions in a relaxation time approximation. Our main result is shown in Fig. 2, where we give the spin relaxation rate, a quantity proportional to the bremsstrahlung rate to a good approximation (see Ref. [3] for more details), as a function of the Fermi momentum k_F of the nucleons in degenerate conditions. Note that $k_F = 1.2\ \text{fm}^{-1}$ corresponds to a density of about $1 \times 10^{14}\ \text{g cm}^{-3}$. It is worth remarking that, given that central contact terms do not contribute to this process, the leading order (LO) in EFT corresponds to the case of the one-pion-exchange potential, typically used in supernovae simulations [11]. Calculations at next-to-next-to-next-to-leading order (N^3LO) differ by about a factor of 2-3 from the LO results. The largest decrease of the strength is observed in going from the LO to

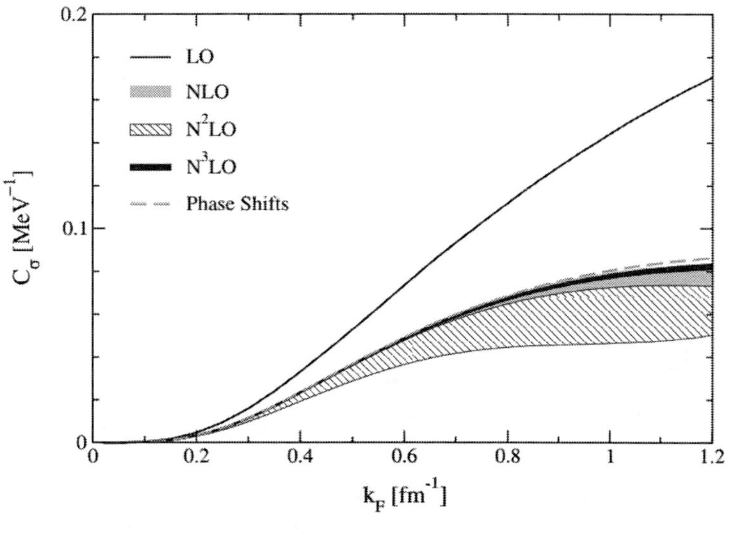

Figure 2: (Color online) Spin relaxation rate in nn collisions as a function of Fermi momentum k_F obtained from chiral EFT interactions at different orders. The leading order corresponds to the one-pion-exchange potential. The calculation obtained from nucleon-nucleon phase shifts [12] is also shown for comparison.

next-to-leading order (NLO), where two-pion-exchange contributions introduce further tensor forces.

In Fig. 2 we also show the comparison of our N^{3}LO calculation with the result obtained using the Nijmegen partial-wave analysis of experimental nucleon-nucleon scattering data [12]. The fact that the latter is very similar to the Born approximation of the N^{3}LO potentials indicate that the involved nn amplitudes are perturbative.

Motivated by nondegenerate conditions encountered in supernovae, we have also investigated this regime [4], essentially describing nucleons with Boltzmann distributions instead of Fermi distributions. We have observed also in this case that N^{3}LO calculations are lower than LO results by a factor of 2-3.

While for systems consisting of neutrons only, the noncentral part of nuclear forces give non-zero rates, in the case of proton admixture also the central part contributes, because the axial charge of the neutron and proton are unequal. Thus, also LO contact terms and S-wave scattering are important. This was recently investigated in Ref. [5], where it was observed that using the $\mathcal{T}$ matrix from phase shifts leads to different results than N^{3}LO at low densities, indicating that central interactions are nonperturbative. Moreover, relative to the one-pion-exchange potential, the $\mathcal{T}$ matrix leads

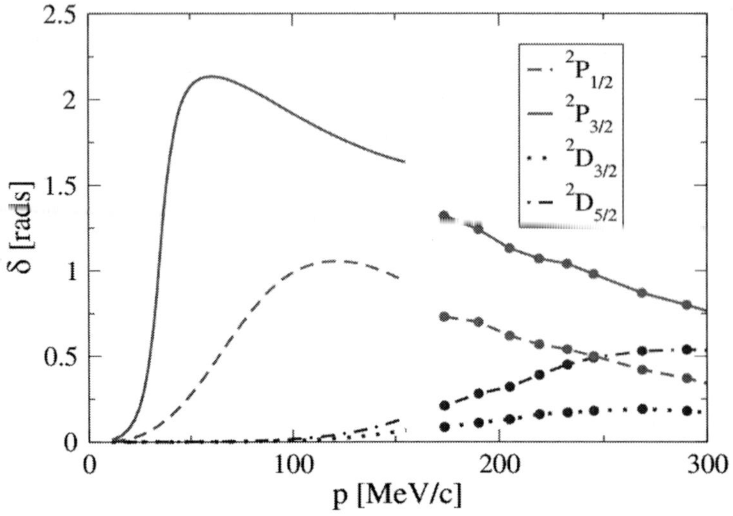

Figure 3: (Color online) Neutron-α scattering phase shifts δ as a function of the relative momentum p. The different partial waves are labeled by the standard notation $^{2S+1}\ell_j$. Phase shifts are shown following Ref. [15]: for low momenta from the fits to data of Ref. [16] and for higher momenta (lines with points) from optical model calculations [17]. S-waves are not shown because they do not contribute to the bremsstrahlung process.

to enhanced neutrino rates at low densities and reduced rates at higher densities, which could lead to interesting mechanisms for energy transport in the outer layers of the supernova, due to competing neutrino processes at different temperatures and densities.

After nucleon-nucleon collisions have been addressed, a natural follow-up question is: What is the effect of neutrino-antineutrino pair production from collisions of nucleons with finite nuclei? To tackle this question, we recently have addressed the case of neutron-α collisions [6]. Some α particles (^{4}He nuclei) are in fact present in the outer layers of the proto-neutron star [13, 14]. While the density of α particles is smaller than the density of neutrons, given that $n\alpha$ scattering features a resonance at low energies in the P-waves as shown in Fig. 3, one expects the bremsstrahlung rate per particle pair from this process to be enhanced due to a larger $\mathcal{T}(k)$, as is clear from Eq. (1). The P-wave resonance in the $n\alpha$ system can be seen as a single particle excitation on top of the filled s-shell of the α particle. The spin-orbit interaction splits the $^2P_{3/2}$ from the $^2P_{1/2}$ channel, hence the P-waves are expected to contribute to the bremsstrahlung process.

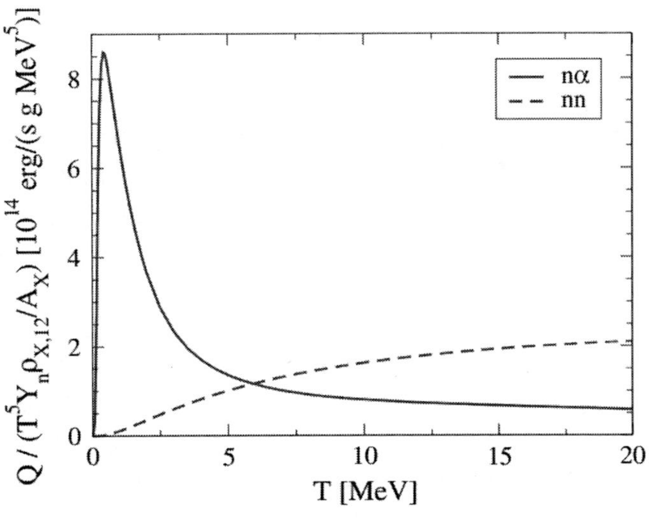

Figure 4: (Color online) Energy-loss rates as a function of temperature T. The solid (dashed) line is for n-α (n-n) bremsstrahlung.

As for the nn case, we estimate the rate of the $n\alpha$ process by evaluating the $\mathcal{T}$ matrix starting from the experimental phase shifts shown in Fig. 3 as

$$\langle p\ell Sj|\mathcal{T}|p\ell Sj\rangle = (2\pi)^3 \frac{p^2}{\pi\, m_{\rm red}} \frac{1 - e^{2i\delta(p,\ell,S,j)}}{2ip} \, . \qquad (2)$$

Here ℓ, S, and j are the relative orbital angular momentum, the neutron spin $S = 1/2$, and the total angular momentum, respectively, and $\delta(p, \ell, S, j)$ are the phase shifts, while $m_{\rm red}$ is the reduced mass for the $n\alpha$ system.

Indeed, we find that the resonance leads to an enhanced contribution in the neutron spin structure function at temperatures in the range of $0.1 - 4$ MeV. A comparison with the nn scattering case is shown in Fig. 4. Specifically, the energy loss Q is shown as a function of the temperature T. Q is divided by T^5 (in MeV), by the neutron mass fraction f_n, and by the mass density over the mass number $\rho_{X,12/A_X}$ of $X = n$ or α particles (in 10^{12} g cm^3), with $A_X = 1$ or 4, respectively. We observe that, for the same neutron mass fraction f_n and if the α-particle density $\rho_\alpha/4 = m_N\, n_\alpha$ is comparable to $\rho_n = m_N\, n_n$, $n\alpha$ bremsstrahlung dominates over nn for $T < 6$ MeV.

We have included the D-wave contribution in our calculations and find that for the temperatures of interest they can be neglected. Finally, even though for energies greater than 20 MeV $n\alpha$ scattering has inelastic channels, which can be parameterized by an imaginary part in the phase shifts, the Boltzmann factors are small at such large energy, making their effects negligible.

3 Conclusions

We have presented results for bremsstrahlung rates for $\nu\bar{\nu}$ production in nn scattering processes for degenerate neutrons [3], and nn [4] and $n\alpha$ [6] scattering processes when neutrons and α-particles are nondegenerate. For the nn case, the used chiral N^3LO interactions generally decreases the bremsstrahlung rates compared to the one-pion-exchange potential by a factor of about $2-3$.

The P-wave resonance for $n\alpha$ scattering near 1 MeV center of mass energy leads to an enhancement in the bremsstrahlung rate for $T < 6$ MeV. For $T \lesssim 2$ MeV, for α number densities comparable to neutron number densities, the $n\alpha$ constitution is over an order of magnitude larger than the contribution from nn bremsstrahlung. Since these temperatures are relevant for proto-neutron stars, it would be interesting to check whether this enhanced rate affects the final predicted spectrum of neutrinos and/or the dynamics of the dense matter outside the proto-neutron star. This can be accomplished by including the $n\alpha$ process in the simulations of core-collapse supernovae and proto-neutron start cooling.

Acknowledgments. This work was supported in parts by the Natural Sciences and Engineering Research Council (NSERC), the National Research Council of Canada, and the ERC Grant No. 307986.

References

[1] H.-T. Janka, Annu. Rev. Nucl. Part. Sci. **62**, 407 (2012).

[2] A. Burrows, Rev. Mod. Phys. **85**, 245 (2013).

[3] S. Bacca, K. Hally, C. J. Pethick, and A. Schwenk, Phys. Rev. C, **80** 032802 (2009).

[4] S. Bacca, K. Hally, M. Liebendörfer, A. Perego, C. J. Pethick, and A. Schwenk, Astrophys. J. **758**, 34 (2012).

[5] A. Bartl, C.J. Pethick, and A. Schwenk, Phys. Rev. Lett. **113**, 081101 (2014).

[6] R. Sharma, S. Bacca, and A. Schwenk, Phys. Rev. C **91** 042801(R) (2015).

[7] B. Friman and O. V. Maxwell, Astrophys. J. **232**, 541 (1979).

[8] C. Hanhart, D. R. Phillips, and S. Reddy, Phys. Lett. B **499**, 9 (2001).

[9] E. Epelbaum, Prog. Part. Nucl. Phys. **57**, 654 (2006).

[10] D.R. Entem and R. Machleidt, Phys. Rev. C **68**, 041001 (2003).

[11] S. Hannestad and G. Raffelt, Astrophys. J. **507**, 339 (1998).

[12] NN-Online, http://www.nn-online.org.

[13] H.-T. Janka *et al.*, Phys. Rept. **442** 38-74 (2007).

[14] A. Arcones, G. Martinez-Pinedo, E. O'Connor, A. Schwenk, H.-Th. Janka, C. J. Horowitz, and K. Langanke, Phys. Rev. C **78**, 015806 (2008).

[15] C. J. Horowitz and A. Schwenk, Nucl. Phys. A **776**, 55 (2006).

[16] R. A. Arndt and L. D. Roper, Phys. Rev. C **1**, 903 (1970).

[17] K. Amos, P. J. Dortmans, H. V. von Geramb, S. Karataglidis, and J. Raynal, Adv. Nucl. Phys. **25**, 275 (2000); K. Amos and S. Karataglidis (private communication).

Hermes results on 3D imaging of the nucleon

LUCIANO L. PAPPALARDO
(on behalf of the HERMES COLLABORATION)

Dipartimento di Fisica e Scienze della Terra, Università degli Studi di Ferrara and INFN, Sezione di Ferrara, Via Saragat 1, 44122 Ferrara, Italy

Abstract

The study of the quantum phase-space distribution of quarks and gluons inside nucleons in terms of TMDs and GPDs has become, in the last decade, a cutting-edge research field in hadron physics. These non-perturbative objects, respectively measurable in semi-inclusive deep-inelastic scattering and exclusive processes, allow to obtain 3-dimensional representations of the nucleon in the momentum and spatial coordinates as well as indirect insights into the still unknown parton orbital angular momentum. The HERMES experiment at HERA has been a precursor in this field. A selection of HERMES results sensitive to both TMDs and GPDs is presented.

Theoretical developments of the *parton model* in the framework of non-perturbative QCD followed by decades of deep-inelastic scattering (DIS) experiments have led to a relatively detailed description of the internal structure of nucleons in terms of quark and gluon degrees of freedom. In particular, very high-precision results on the longitudinal[1] momentum distribution of quarks, described by the spin-independent parton distribution function (PDF) $f_1(x)$, are provided by global fits of world data [1] that span several orders of magnitude in x and Q^2, the longitudinal momentum fraction of the nucleon carried by the partons and the squared four-momentum transfer of the hard probe to the struck quark, respectively. In the last decades an

[1]With respect to the direction of the exchanged virtual photon, the hard probe.

increasing focus was put on the spin degrees of freedom of quarks and gluons, described by the spin-dependent *helicity* PDF $g_1(x)$. While a relatively precise knowledge of g_1 has been achieved for the valence quarks, large uncertainties still affect the results for sea quarks and gluons [1]. More recently, the HERMES experiment provided first evidences of a non-zero transverse spin distribution of *up* and *down* quarks, described by the *transversity* PDF $h_1(x)$, through the measurement of specific azimuthal asymmetries in the production of π^+ and π^- (the so-called *Collins effect*) in semi-inclusive DIS (SIDIS) measurements off a transversely polarized hydrogen target [2]. The three fundamental PDFs $f_1(x)$, $g_1(x)$ and $h_1(x)$ are said to be *collinear*, in the sense that they neglect any parton transverse momentum degree of freedom[2], thus providing 1-dimensional descriptions of the nucleon structure in terms of the longitudinal momentum fraction x.

In recent years, huge efforts are being focused towards a more comprehensive multi-dimensional description of the nucleon. The final goal is to map the full quantum phase-space distribution of quarks and gluons inside the nucleon, as formally encoded by the Wigner function (depending on both spatial and momentum coordinates). Although, due to the uncertainty principle, the Wigner function is not directly measurable, two independent classes of partially integrated objects, the Transverse Momentum Dependent PDFs (TMDs) and the Generalized Parton Distributions (GPDs), are being intensively investigated both theoretically and experimentally. TMDs and GPDs allow for complementary descriptions of the nucleon in three dimensions (*nucleon tomography*), spanned by the quark's longitudinal momentum and, respectively, by its transverse momentum components and transverse spatial coordinates. Furthermore, they provide complementary insights into the yet unmeasured quark orbital angular momentum. Experimentally, TMDs and GPDs can be accessed through the analysis of specific azimuthal asymmetries measured, respectively, in SIDIS and hard exclusive processes, such as hard lepto-production of real photons or mesons.

At leading-twist, eight TMDs, each with a specific probabilistic interpretation in terms of quark number densities, enter the SIDIS cross-section in conjunction with a fragmentation function (FF) [3]: the poorly known chiral-odd Collins function $H_1^\perp$, describing left-right asymmetries in the fragmentation of transversely polarized quarks, or the relatively well known spin-independent chiral-even D_1 FF. Three of them are the transverse-

[2]More precisely, they are the only three leading-twist PDFs that survive integration over the parton momentum components that are transverse with respect to the direction of the hard probe.

momentum-dependent versions of f_1, g_1 and h_1. The other five are the Sivers function $f_{1T}^{\perp}$, the Boer-Mulders function $h_1^{\perp}$, the *pretzelosity* function $h_{1T}^{\perp}$ and the so-called *worm-gear* functions $h_{1L}^{\perp}$ and $g_{1T}^{\perp}$.

The HERMES experiment at HERA has played a pioneering role in these studies, by measuring observables (azimuthal asymmetries) sensitive to these (and also some higher twist) TMDs, in some cases providing the first evidence of non-zero effects. Most of the HERMES results are confirmed by other experiments (COMPASS, CLAS, etc), though intriguing differences were observed in some cases. High-precision results from the next generation high-luminosity and large-acceptance experiments (at present or upgraded facilities and, possibly, at the future Electron-Ion Collider) will greatly improve our understanding of the nucleon structure in terms of TMDs.

Among the TMDs, particularly interesting is the Sivers function. It describes the correlation between the quark transverse momentum and the transverse spin of the nucleon. The interest in this TMD, proposed more than 20 years ago [4] to explain the large single-spin asymmetries observed in pion production in the collision of unpolarized with transversely polarized nucleons, suddenly increased after it was demonstrated [5] to be linked to the quark and gluon orbital angular momentum, the main still unmeasured contribution to the nucleon spin. It is naive-T-odd, i.e. it requires final-state interactions (FSIs) of the ejected quark with the color field of the nucleon remnant. These FSIs are formally introduced in the definition itself of the Sivers function through a gauge link. In general, gauge links are process-dependent and this leads to the remarkable fact that naive-T-odd TMDs (such as the Sivers and the Boer-Mulders functions) are not universal. In particular, they are expected to have opposite sign when measured in Drell-Yan processes. The experimental verification of this direct QCD prediction is eagerly awaited. The first evidence for a non-zero Sivers function was reported by the HERMES Collaboration in 2005 [2], using an unpolarized 27.5 GeV electron beam and a transversely polarized hydrogen target. These amplitudes are sensitive to the convolution of the Sivers function $f_{1T}^{\perp}$ and the unpolarized fragmentation function D_1:

$$A_{UT}^{\sin(\phi-\phi_S)} \propto \sum_q e_q^2 \, f_{1T}^{\perp,q}(x, p_T^2) \otimes D_1^q(z, k_T^2) \ . \qquad (1)$$

Here, ϕ and ϕ_S are, respectively, the azimuthal angle of the detected hadron and of the target transverse polarization with respect to the lepton scattering plane and about the virtual-photon direction, z denotes the fraction of the energy of the exchanged virtual photon carried by the produced hadron (in this case a pion), and p_T and k_T represent the transverse momentum of the struck quark and that of the fragmenting quark, respectively.

In particular, HERMES observed large positive amplitudes for π^+ and amplitudes consistent with zero for π^-. The results were provided in 1-dimensional (1D) projections as a function of x, z or $P_{h\perp}$. The vanishing amplitudes for π^- require cancellation effects, e.g., from a d-quark Sivers function opposite in sign to the u-quark Sivers function. The final HERMES results [6], based on the full statistics available, confirm the main features of the early results for charged pions and include results for neutral pions and charged kaons (Fig. 1). The amplitudes for K^+ are large and positive while those for π^0 and K^- are slightly positive. Unexpectedly, the K^+ amplitudes are larger than the π^+ ones. This observation can be ascribed, for instance, to a significant role of the sea quarks or to possible higher-twist contributions in kaon production. Recently, the data analysis was refined and the relevant asymmetry amplitudes were extracted in a 3-dimensional (3D) binning in x, z and $P_{h\perp}$. As an example, Fig. 2 shows the Sivers amplitude for π^+ as a function of $P_{h\perp}$ in simultaneous bins in x and z. The 3D binning allows to map the kinematic dependence of the asymmetry amplitudes in a greatly detailed way. For instance, the rise of the Sivers amplitude for π^+ as a function of $P_{h\perp}$, as observed in the conventional 1D projection (Fig. 1), reveals non-trivial kinematic dependencies in x and z when observed in the 3D representation (Fig. 2), where most of the rise is clearly localized in the high-x and high-z regions. Recently, the HERMES experiment has also measured single and double-spin asymmetries in SIDIS for protons and antiprotons. Figure 3 shows the Sivers amplitudes for protons and antiprotons in 1D bins in x, z and $P_{h\perp}$. While the antiproton amplitude is consistent with zero within the experimental uncertainties, the proton data exhibit a small but significantly positive amplitude. This constitutes, at present, the first measurement of a non-zero Sivers effect for protons in SIDIS.

The experimental access to GPDs is so far obtained from the measurements of hard exclusive lepton-nucleon scattering processes. These processes are described by four leading-twist, quark chirality conserving GPDs (H, E, $\widetilde{H}$, and $\widetilde{E}$). The cleanest process to probe GPDs is Deeply Virtual Compton Scattering (DVCS), $\ell+N \to \ell+N'+\gamma$, where a real photon is emitted by the struck quark in the nucleon. During the last decade the DVCS process was extensively studied both theoretically and experimentally. Currently there exist several measurements by HERMES, H1, ZEUS, CLAS and Hall-A experiments. The Bethe-Heitler (BH) process, where a real photon is radiated

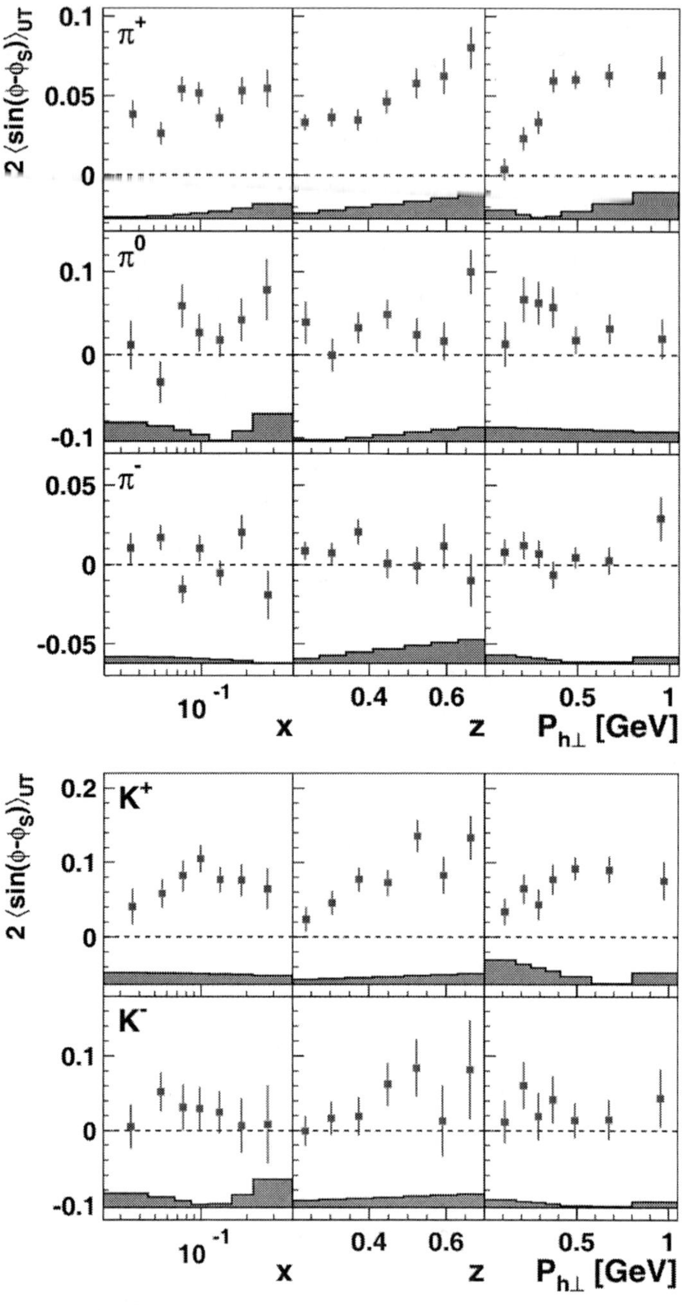

Figure 1: Sivers amplitudes for pions (top) and charged kaons (bottom) as a function of x, z, or $P_{h\perp}$. A 7.3% scale uncertainty, not shown, arises from the accuracy in the measurement of the (transverse) target polarization.

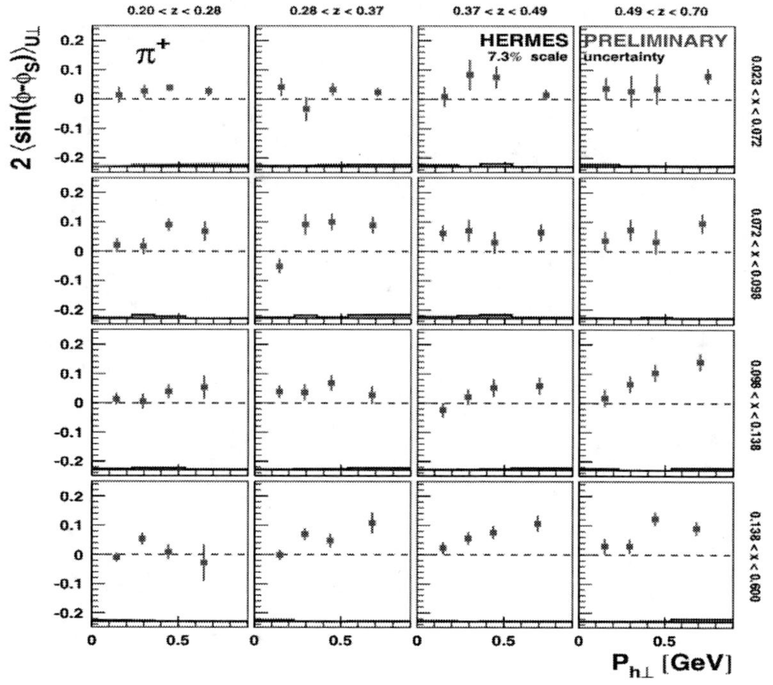

Figure 2: Sivers amplitudes for π^+ in 3-dimensional binning in x, z and $P_{h\perp}$.

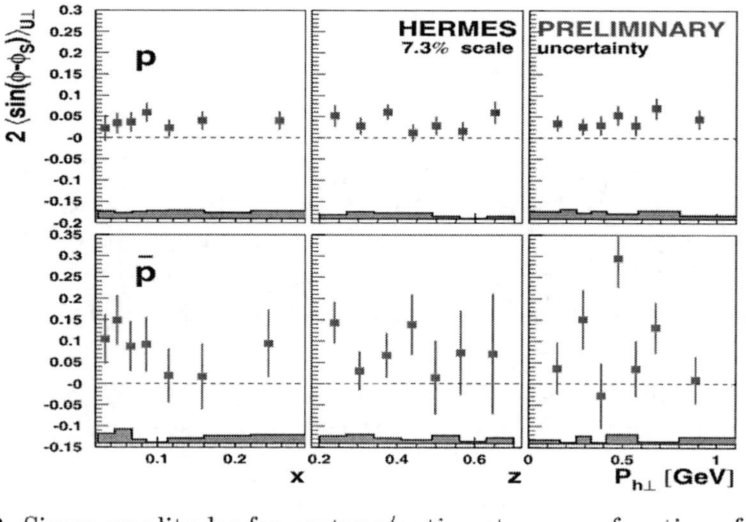

Figure 3: Sivers amplitudes for protons/anti-protons as a function of x, z and $P_{h\perp}$.

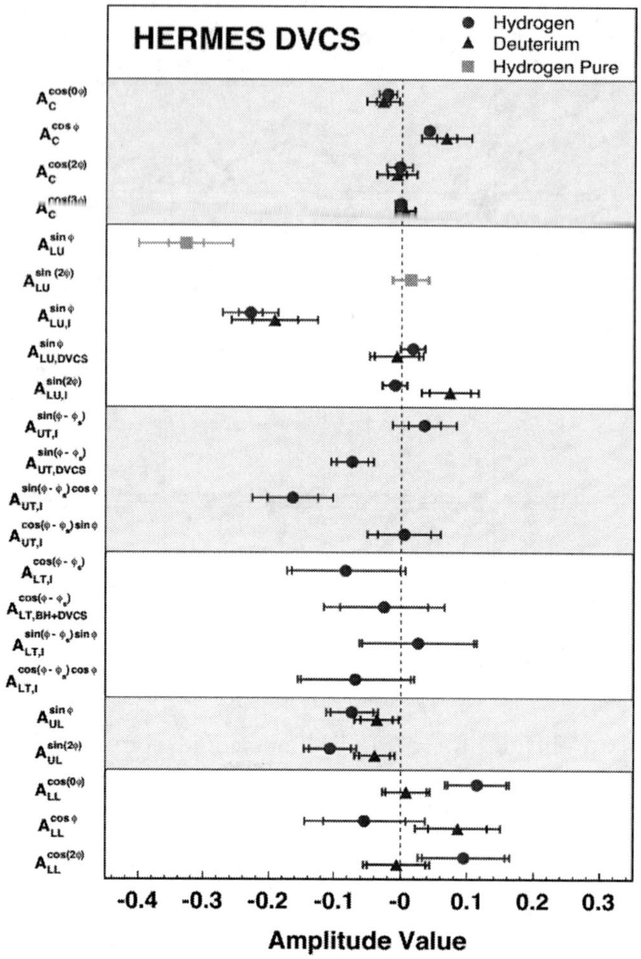

Figure 4: Overview of DVCS amplitudes over the entire HERMES kinematic range.

by the incoming or outgoing lepton, has the same initial and final state as the DVCS process. These two processes are thus experimentally indistinguishable, and therefore the total cross-section of hard lepto-production of real photons contains an interference term $\mathcal{I}$, which depends on the charge of the lepton beam: $\sigma \propto |\mathcal{T}_{BH}|^2 + |\mathcal{T}_{DVCS}|^2 + \mathcal{I}$. The individual terms of the cross-section can be decomposed into Fourier harmonics in the relevant azimuthal angles. Different Fourier components depend on different Compton Form Factors (CFFs), which in turn are convolutions of hard scattering amplitudes with the corresponding GPDs.

At HERMES, the DVCS process is accessed through measurements of cross-section asymmetries that appear in the azimuthal distributions of final state photons. Using data collected with longitudinally polarized electron and positron beams and unpolarized, longitudinally and transversely polarized hydrogen and deuterium targets, it was possible to measure asymmetries with respect to beam charge, beam polarization and target polarization alone and also with respect to their different combinations. An overview of all extracted azimuthal asymmetry amplitudes corresponding to the entire HERMES kinematics is presented in Fig. 4 for both hydrogen and deuterium targets. The amplitudes of the beam–helicity and beam–charge asymmetries $A_{LU}^{DVCS}(\phi)$, $A_{LU}^{I}(\phi)$ and $A_C(\phi)$ are presented in the top panels. Significant non-zero $\cos(\phi)$ and $\sin(\phi)$ amplitudes are observed respectively for the beam-charge $A_C(\phi)$ and beam–helicity $A_{LU}^{I}(\phi)$ asymmetries. The results for the $\sin(\phi)$ amplitude of the charge-averaged beam-helicity asymmetry $A_{LU}^{DVCS}(\phi)$ are consistent with zero. In the bottom panels, the results of the longitudinal single-target-spin $A_{UL}(\phi)$ and double–spin $A_{LL}(\phi)$ asymmetries are presented. Also shown are the leading amplitudes of transverse single–target–spin and double–spin asymmetries, measured on Hydrogen target. The amplitudes $A_{UT,I}^{\sin(\phi-\phi_S)\cos(\phi)}$ and $A_{LT,I}^{\sin(\phi-\phi_S)\sin(\phi)}$ are significantly negative and consistent with zero, respectively. Full references for the HERMES DVCS results can be found in [7–9].

References

[1] K. A. Olive et al. (PDG), Chin. Phys. C **38**, 090001 (2014).

[2] A. Airapetian *et al.* (HERMES), Phys. Rev. Lett. **94**, 012002 (2005).

[3] A. Bacchetta *et al.*, JHEP **02**, 093 (2007).

[4] D. W. Sivers, Phys. Rev. D **41**, 83 (1990).

[5] S. J. Brodsky *et al.* Phys. Lett. B **530**, 99 (2002).

[6] A. Airapetian *et al.* (HERMES), Phys. Rev. Lett. **103**, 152002 (2009).

[7] A. Airapetian *et al.* (HERMES), *JHEP* **07**, 032 (2012), *JHEP* **06**, 019 (2010), *JHEP* **06**, 066 (2008), *Phys. Lett.* **B 704**, 15 (2011).

[8] A. Airapetian *et al.* (HERMES) *Nucl. Phys.* **B 829**, 1 (2010), *Nucl. Phys.* **B** 842, 265 (2011).

[9] A. Airapetian *et al.* (HERMES), *JHEP* **10**, 042 (2012).

EPJ Web of Conferences **117**, 02005 (2016) DOI: 10.1051/epjconf/201611702005
NN2015

Search for the η-mesic Helium bound state with the WASA-at-COSY facility

MAGDALENA SKURZOK[1], WOJCIECH KRZEMIEŃ[2], OLEKSANDR RUNDEL[1] and PAWEL MOSKAL[1]
for the WASA-at-COSY COLLABORATION

[1] M. Smoluchowski Institute of Physics, Jagiellonian University, Cracow, Poland
[2] National Centre of Nuclear Research, Świerk, Poland

Abstract

We performed a search for ^{4}He-η bound state with high statistics and high acceptance with the WASA-at-COSY facility using a ramped beam technique. The signature of η-mesic nuclei is searched for in $dd \to {}^3\mathrm{He}n\pi^0$ and $dd \to {}^3\mathrm{He}p\pi^-$ reactions by the measurement of the excitation functions in the vicinity of the η production threshold. This paper presents the experimental method and the preliminary results of the data analysis for $dd \to {}^3\mathrm{He}n\pi^0$ process.

1 Introduction

In 1986 Haider and Liu [1], based on the coupled-channel analysis of the $\pi N \to \pi N$, $\pi N \to \pi\pi N$ and $\pi N \to \eta N$ reactions [2], postulated the existence of η-mesic nuclei in which the η meson is bound within a nucleus via the strong interaction. Since then, the search for η- and η'-mesic bound states was performed in many laboratories as e.g.: COSY [3–9], ELSA [10], GSI [11], JINR [12], JPARC [13], LPI [14], and MAMI [15, 46], however, till now none of the experiments confirmed its existence. The status of the search was recently described in the following reviews [17–19].

Recent theoretical studies e.g. [18, 20–26] support the search for η and η'-mesic nuclei. Phenomenological and theoretical investigations of η production in hadronic- and photo- induced reactions result in a wide range of possible scattering lengths from $a_{\eta N} = (0.18 + 0.16i)$ fm to $a_{\eta N} = (1.03 + 0.49i)$ fm [18], which do not exclude the formation of a bound states for the helium [27,28] and even the deuteron [29]. The observation of an steep rise in the total cross-section of the $dp \rightarrow {}^3\mathrm{He}\eta$ and $dd \rightarrow {}^4\mathrm{He}\eta$ reactions close to the kinematic threshold [8, 9, 30–35], gives a strong evidence for the existence of a bound state. Similar observations were carried out in case of the cross section for the photoproduction process $\gamma^3\mathrm{He} \rightarrow {}^3\mathrm{He}\eta$ [45,46]. It shows that the rise of the cross section above threshold is independent of the initial channel giving a strong argument for the existence of the pole in the scattering matrix which could be associated with a bound state. Therefore, we consider the ^{4}He-η and ^{3}He-η as a good candidates for the η-mesic bound states.

The discovery of postulated η-mesic nuclei would be interesting on its own as well as would be very important for (i) better understanding of the η and η' meson properties [36] and their interaction with nucleons inside nuclear matter, (ii) providing information about the $N^*(1535)$ resonance [26, 37] and (vi) about flavour singlet component of the quark-gluon wave function of the η and η' mesons [38, 39].

2 Search for ^{4}He-η bound state with WASA

We performed the search for the ^{4}He-η bound state with unique accuracy with the WASA facility, installed at the COSY synchrotron in Forschungszentrum Jülich (Germany). The advantage of this detection system is the possibility of a simultaneous measurement of all ejectiles with high acceptance while continuously changing the beam momentum around the η production threshold. The signature of the η-mesic nuclei is searched for by studying the excitation function for the chosen decay channels of the ^{4}He-η system, formed in deuteron-deuteron collision [5, 40]. Till now two experiments were carried out focusing on the bound state decay into ^{3}He and a nucleon-pion pair [4–6, 41, 42].

The first experiment was carried out in June 2008, by measuring the excitation function of the $dd \rightarrow {}^3\mathrm{He}p\pi^-$ reaction near the η production threshold covering the excess energy range from -51.4 MeV up to 22 MeV. In the excitation function no structure which could be interpreted as a resonance originating from decay of the η-mesic ^{4}He was observed. The upper limit of the total cross-section for the bound state formation and decay in the $dd \rightarrow ({}^4\mathrm{He}\text{-}\eta)_{bound} \rightarrow {}^3\mathrm{He}p\pi^-$ process was determined and varies from

20 nb to 27 nb at the confidence level 90% [6,41].

In the second experiment, in November 2010, about 10 times higher statistics were collected with respect to the previous measurement. The search for the ^{4}He-η bound state was performed for two reactions $dd \rightarrow {}^3\mathrm{He}n\pi^0$ and $dd \rightarrow {}^3\mathrm{He}p\pi^-$ via the measurement of the excitation function for each of them in the vicinity of the η production threshold. The deuteron beam momentum was varying continuously within each acceleration cycle from 2.127 GeV/c to 2.422 GeV/c, crossing the kinematic threshold for the η production in the $dd \rightarrow {}^4\mathrm{He}\eta$ reaction at 2.336 GeV/c, which corresponds to a range of excess energy Q∈(-70,30) MeV.

Here we show result for the channel $dd \rightarrow {}^3\mathrm{He}n\pi^0$ [42].

The ^{3}He was identified in the Forward Detector based on the ΔE-E method. The neutral pion π^0 was reconstructed in the Central Detector from the invariant mass of two gamma quanta originating from its decay while the neutron was identified via the missing mass technique.

Events corresponding to the bound states production were selected by applying cuts in the ^{3}He center of mass (CM) momentum, nucleon CM kinetic energy, pion CM kinetic energy and the opening angle between neutron-pion pair in the CM based on Monte Carlo simulations.

The excitation functions for the reaction were determined for a region rich in signal corresponding to momenta of the ^{3}He in the CM system in range $p^{cm}_{3He} \in (0.1, 0.2)$ GeV/c. The excitation function was obtained by normalizing the events selected in individual excess energy intervals by the corresponding integrated luminosities (the detailed description of the luminosity determination one can find in Ref. [42,44]) and corrected for acceptance and efficiency is presented in the left panel of Fig. 1.

No narrow resonance-like strucutre, which could be interpreted as the indication of the η-mesic nuclei, is observed [42]. Therefore, the upper limit of the total cross section for the $dd \rightarrow ({}^4\mathrm{He}\text{-}\eta)_{bound} \rightarrow {}^3\mathrm{He}n\pi^0$ process was determined on the 90% confidence level. For this purpose the excitation function was fitted with a quadratic function describing the background combined with the Breit-Wigner function which can describe the signal from the bound state. In performing the fit, the binding energy and the bound state width are fixed parameters while the polynomial coefficients and the amplitude of the Breit-Wigner distribution are treated as free parameters. The fitting procedure was performed for various values of binding energy and width. An example of the fit for $\Gamma = 40$ MeV and $B_s = 30$ MeV is presented in Fig. 1. The upper limit of the total cross section for considered process varies from 21 to 36 nb for the bound state width ranging from 5 to 50 MeV, while the systematic error varies from 12% to 27% (See Fig. 1).

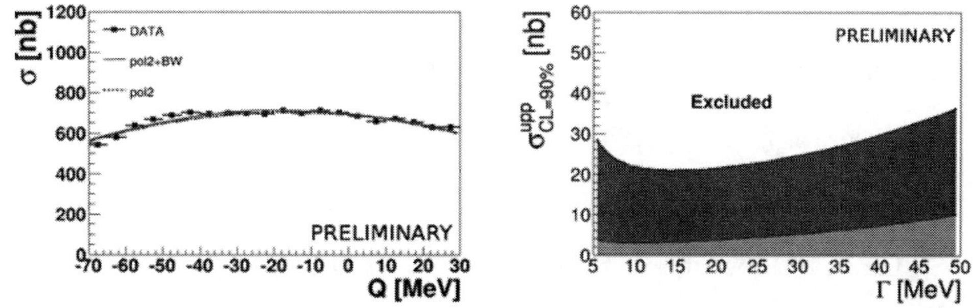

Figure 1: Left: Excitation function for the $dd \to {}^3\text{He}n\pi^0$ reaction. The red solid line represents a fit with second order polynomial combined with a Breit-Wigner function with fixed binding energy and width equal to 30 and 40 MeV, respectively. The blue dotted line shows the second order polynomial corresponding to the background. The figure is adopted from [42]. Right: Upper limit of the total cross-section for $dd \to ({}^4\text{He-}\eta)_{bound} \to {}^3\text{He}n\pi^0$ reaction as a function of the width of the bound state. The binding energy was set to 30 MeV. The green area denotes the systematic uncertainties. The figure is adopted from [42].

3 Conclusion

The performed analysis allows for the determination of the excitation function for $dd \to ({}^4\text{He-}\eta)_{bound} \to {}^3\text{He}n\pi^0$ process and the estimation of the upper limit of the cross section for the η-mesic ^{4}He formation and decay. The obtained excitation function does not show any narrow structure which could be a signature of the bound state with width less than 50 MeV. The upper limit is by a factor of five larger than theoretically predicted value [21] therefore, the current measurement does not exclude the existence of bound state in this process.

Acknowledgements

We acknowledge support by the Foundation for Polish Science - MPD program, co-financed by the European Union within the European Regional Development Fund, by the Polish National Science Center through grants No. 2013/11/N/ST2/04152, 2011/01/B/ST2/00431, 2011/03/B/ST2/01847 and by the FFE grants of the Forschungszentrum Jülich.

References

[1] Q. Haider, L. C. Liu, *Phys. Lett.* **B172**, 257 (1986).

[2] R. S. Bhalerao, L. C. Liu, *Phys. Rev. Lett.* **54**, 865 (1985).

[3] P. Moskal, J. Smyrski, *Acta Phys. Pol.* **B 41**, 2281 (2010)

[4] M. Skurzok, P. Moskal, W. Krzemien, *Prog. Part. Nucl. Phys.* **67**, 445 (2012).

[5] W. Krzemien, *PhD Thesis, Jagiellonian Univ., arXiv:1101.3103* (2011).

[6] P. Adlarson *et al., Phys. Rev.* **C87**, 035204 (2013).

[7] A. Budzanowski *et al., Phys. Rev.* **C79**, 012201 (2009).

[8] T. Mersmann *et al., Phys. Rev. Lett.* **98**, 242301 (2007).

[9] J. Smyrski *et al., Phys. Lett.* **B649**, 258 (2007).

[10] M. Nanova *et al., Phys. Lett.* **B727**, 417 (2013).

[11] Y. K. Tanaka *et al., Few Body Syst.* **54**, 1263 (2013).

[12] S. V. Afanasiev, *Phys. Part. Nucl. Lett.* **8**, 1073 (2011).

[13] H. Fujioka, *Acta Phys. Pol.* **B41**, 2261 (2010).

[14] V. A. Baskov *et al., PoS Baldin-ISHEPP-XXI 102* (2012).

[15] B. Krusche *et al., J. Phys. Conf. Ser.* **349**, 012003 (2012).

[16] F. Pheron *et al., Phys. Lett.* **B709**, 21 (2012).

[17] H. Machner, *J. Phys.*, **G42**, 043001 (2015).

[18] N. G. Kelkar *et al., Rept. Prog. Phys.* **76**, 066301 (2013).

[19] B. Krusche, C. Wilkin, *Prog. Part. Nucl. Phys.*, **80**, 43 (2014).

[20] S. D. Bass, A. W. Thomas, *Acta Phys. Pol.* **B45**, 627 (2014).

[21] S. Wycech, W. Krzemien, *Acta Phys. Pol.* **B45**, 745 (2014).

[22] S. Hirenzaki, H. Nagahiro, *Acta Phys. Pol.* **B45**, 619 (2014).

[23] E. Friedman, A. Gal, J. Mares, *Phys. Lett.* **B725**, 334 (2013).

[24] C. Wilkin, *Phys. Lett.* **B 654**, 92 (2007).

[25] H. Nagahiro *et al.*, *Phys. Rev.* **C87**, 045201 (2013).

[26] S. Hirenzaki, *et al.*, *Acta Phys. Pol.* **B41**, 2211 (2010).

[27] C. Wilkin, *Phys. Rev.* **C47**, 938 (1993).

[28] S. Wycech, A. M. Green, J. A. Niskanen, *Phys. Rev.* **C52**, 544 (1995).

[29] A. M. Green, J. A. Niskanen, S. Wycech, *Phys. Rev.* **C54**, 1970 (1996).

[30] J. Berger *et al.*, *Phys. Rev. Lett.* **61**, 919 (1988).

[31] B. Mayer *et al.*, *Phys. Rev.* **C53**, 2068 (1996).

[32] R. Frascaria *et al.*, *Phys. Rev.* **C50**, 537 (1994).

[33] N. Willis *et al.*, *Phys. Lett.* **B406**, 14 (1997).

[34] H. Machner, *Acta. Phys. Polon.* **B41**, 2221 (2010).

[35] A. Wrońska *et al.*, *Eur. Phys. J.* **A26**, 421 (2005).

[36] T. Inoue, E. Oset, *Nucl. Phys.* **A710**, 354 (2002).

[37] D. Jido, H. Nagahiro, S. Hirenzaki, *Phys. Rev.* **C66**, 045202 (2002).

[38] S. D. Bass, A. W. Thomas, *Acta Phys. Pol.* **B41**, 2239 (2010).

[39] S. D. Bass, A. W. Thomas, *Phys. Lett.* **B634**, 368 (2006).

[40] P. Moskal *et al.*, *Int. J. Mod. Phys.* **A 22**, 305 (2007).

[41] W. Krzemien, P. Moskal, M. Skurzok, *Acta Phys. Pol.* **B45**, 689 (2014).

[42] M. Skurzok, *PhD Thesis, Jagiellonian Univ., arXiv:1509.01385* (2015).

[43] M. Skurzok, *arXiv:1009.5503, Diploma Thesis, Jagiellonian Univ.*, Berichte des FZ-Jülich **Jül-4332** (2010).

[44] M. Skurzok, W. Krzemien, P. Moskal, *Acta Phys. Pol.* **B46**, 133 (2015).

[45] M. Pfeiffer *et al.*, *Phys. Rev. Lett.* **92**, 252001 (2004).

[46] F. Pheron *et al.*, *Phys. Lett.* **B709**, 21 (2012).

EPJ Web of Conferences **117**, 02006 (2016) DOI: 10.1051/epjconf/201611702006
NN2015

Elastic nucleon-deuteron scattering and breakup with chiral forces

HENRYK WITAŁA, JACEK GOLAK, ROMAN SKIBIŃSKI and
KACPER TOPOLNICKI

Institute of Physics, Jagiellonian University, PL-30348 Kraków

Abstract

Results on three-nucleon (3N) elastic scattering and breakup below the pion production threshold are discussed. The large discrepancies found between a theory based on numerical solutions of 3N Faddeev equations with standard nucleon-nucleon (NN) potentials only and data point to the need for three-nucleon forces (3NF's). This notion is supported by the fact that another possible reason for the discrepancies in elastic nucleon-deuteron (Nd) scattering, relativistic effects, turned out to be small. Results for a new generation of chiral NN forces (up to N^4LO) together with theoretical truncation errors are shown. They support conclusions obtained with standard NN potentials

1 Introduction

Traditionally in nuclear physics the Hamiltonian has been taken in a non-relativistic form in which pairwise interactions between nucleons are supplemented by 3NF's for systems with more than two nucleons. The 3N system is the first nontrivial case where NN potentials can be tested. In that system also for the first time 3NF's come into play making it a valuable source of information on 3NF properties and their significance in the nuclear Hamiltonian.

The need for 3NF's was established when three- and four-nucleon bound states have been solved exactly using high precision semi(phenomenological) NN potentials [1–3]. It was natural to look for an explanation of resulting underbinding by introducing 3NF's such as the Urbana IX (UIX) [4] or the Tucson-Melbourne (TM) model [5] into the nuclear Hamiltonian.

Effective field theoretical methods in a form of chiral perturbation theory (χPT) provided a solid basis for a construction of nuclear forces. Consistent chiral NN [6,7] and 3N [8,9] forces have been derived in the χPT framework. Recently improved chiral NN potentials have been constructed by using regularization in the coordinate space [10, 11]. This significantly reduced finite-cutoff artefacts of the nonlocal momentum-space regulator used in [6,7] allowing us to apply the improved forces to higher energy Nd scattering.

In the next section we briefly review comparison of 3N scattering data to various NN potential predictions alone or combined with different 3NF's. These examples are chosen to show the importance of the 3NF in the 3N system. The 3NF effects seem to grow with increasing energy of the 3N system. We also discuss the importance of relativistic effects in elastic Nd scattering and their significance in the study of 3NF effects. The small size of relativistic effects indicates that one should focus on 3NF's, specifically of short-range nature. In the following section we present examples of results for improved chiral NN potentials. We summarize in the last section.

2 3N reactions with standard forces

All observables for elastic Nd scattering can be obtained from solutions of 3N Faddeev equations. Presently they can be solved numerically exactly for any NN potential and 3NF [12]. Using the realistic NN forces: AV18 [1], CD Bonn [2], Nijm1, Nijm2, and Nijm93 [3] one gets, in general, predictions for 3N scattering observables which agree well with data at energies below $\approx$ 30 MeV. A fairly complete overview of those theoretical predictions in comparison to data is presented in [12, 13]. At higher energies discrepancies develop. They are exemplified for the neutron-deuteron (nd) elastic scattering cross section in Figure 1A. The large discrepancy in the minimum of the elastic scattering cross section obtained with NN forces only, seen for energies above $\approx$ 60 MeV, is removed for energies below $\approx$ 140 MeV when 3NF's, which reproduce the experimental triton binding energy, are included [13]. A similar behavior shows up for the high energy deuteron vector analyzing power $A_y(d)$ [13]. But there are many spin observables for which large 3NF effects are predicted and where the TM and the Urbana IX do not reproduce the data [13]. This is the case e.g. for the nucleon analyzing power A_y [13] and for the deuteron tensor analyzing powers [13]. In none of these cases the data can be reproduced by pure 2N force predictions. Also for the Nd

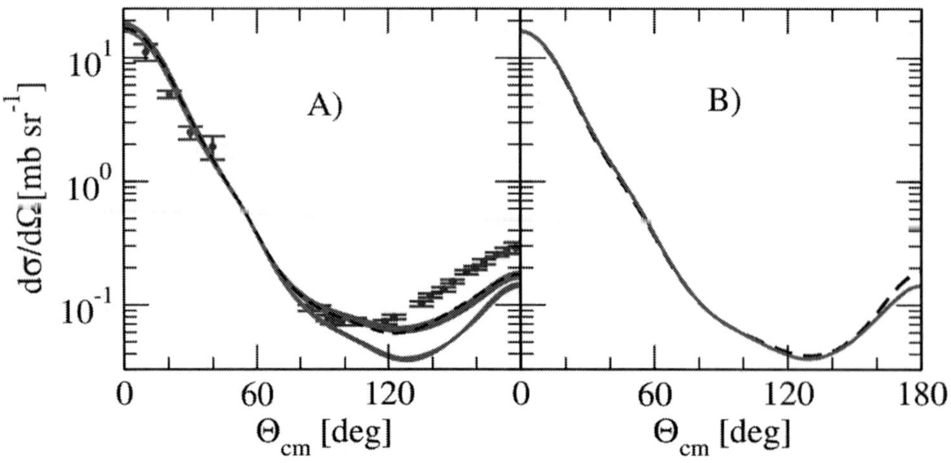

Figure 1: The angular distribution for nd elastic scattering at neutron energy $E_{lab} =$ 250 MeV. The nd experimental data (solid dots) are from [19]. In A) the lower and upper bands result from predictions of (semi)phenomenological potentials alone (AV18, CD Bonn, Nijm1, Nijm2) and combined with TM99 3NF [20], respectively. The dashed line is the prediction of the AV18+UIX 3NF. In B) the solid and dashed lines are nonrelativistic and relativistic predictions based on the CD Bonn potential.

breakup reaction there are spin observables which require 3NF's for their explanation [14, 15]. Therefore observables, which are sensitive to the 3NF structure, can be identified for elastic Nd scattering and deuteron breakup.

There are large discrepancies at energies above ≈ 140 MeV between data and theory in elastic Nd scattering which cannot be removed by adding standard 3NF's (see Figure 1A). They require to study the magnitude of relativistic effects. We used an instant form relativistic approach which encompasses relativistic kinematics, boost corrections and Wigner spin rotations [16, 17]. It turned out that relativistic effects lead to a rather small effects in the cross section, mostly restricted to the backward angles [16] (see Figure 1B). Higher energy elastic scattering spin observables are only slightly modified by relativity [16, 17].

3 Results with chiral NN forces

The small size of relativistic effects indicates that very likely the short range contributions to the 3NF are responsible for the higher energy elastic scattering discrepancies. The recently constructed new generation of chiral NN potentials up to N^4LO with an appropriate regularization in the coordinate space [10, 11] made it possible to reduce significantly finite-cutoff artefacts

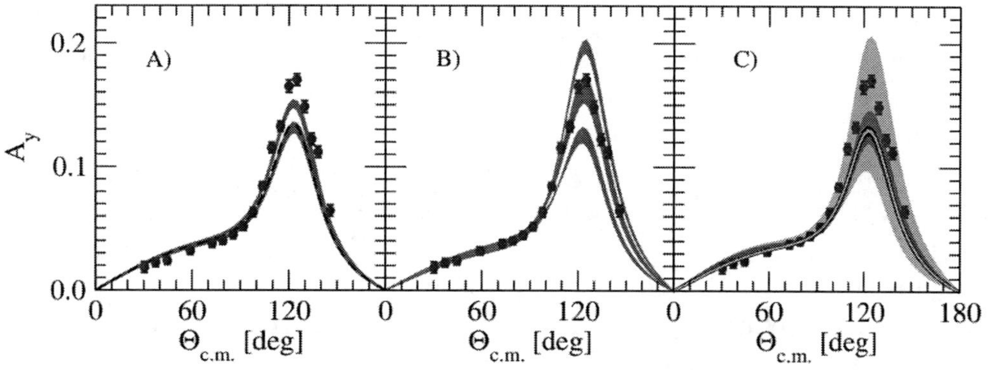

Figure 2: The nd elastic scattering analyzing power A_y at $E_{lab} = 10$ MeV. For description of bands and lines in A) see Figure 1A. In B) bands of predictions for five versions of the old Bochum chiral NN potentials of Ref. [6] at different orders of the chiral expansion are shown: NLO - the upper band, N^{2}LO - the middle band, and N^{3}LO - the bottom band. In C) predictions based on the improved chiral NN potentials of Refs. [10, 11] with local regulator and parameter $R = 1.0$ fm are shown. The bands of increasing width show estimated theoretical uncertainty: the narrowest at N^{4}LO , slightly wider at N^{3}LO, then N^{2}LO and widest at NLO. The full circles are nd data from Ref. [21].

present when using the nonlocal momentum-space regulator employed in the chiral NN potentials of Refs. [6, 7]. These artefacts are clearly seen in strong variations around $p = 2$ fm^{-1} of the $l = 0$ component of the deuteron wave function calculated with N^{3}LO NN potentials of Refs. [6] and [7]. They are absent when these new forces are used. Applications of these new NN potentials do not lead to distortions in the cross section minimum of the higher energy elastic Nd scattering that were found in Ref. [18]. Also convergence with respect to the order of chiral expansion is improved when the new chiral forces are applied, as exemplified in Figure 2C for the analyzing power A_y. At low energies of the incoming neutron, theoretical predictions of standard NN potentials fail to explain the experimental data for A_y (see Figure 2A). When instead of the standard forces the old chiral NN interactions with nonlocal momentum space regulator are used, the predictions for A_y vary with the order of chiral expansion. In particular, the NLO results overestimate the A_y data while N^{2}LO NN forces seem to provide quite a good description of them (see Figure 2B). However, when N^{3}LO NN chiral forces are used, a clear discrepancy between theory and data emerges in the region of A_y maximum, which is similar to the one for the standard forces. Using a new procedure for estimating the theoretical truncation errors introduced in [10, 11], we show in Figure 2C theoretical uncertainties at different orders of chiral expansion for A_y predictions with the improved potentials. With increasing order the uncertainties become very small and very precise predictions are provided at N^{3}LO and N^{4}LO. It is interesting to note that the width of the uncertainty band at N^{2}LO is of the same size as the discrepancy with respect to the data. At that order for the first time 3NF, which is neglected in the present calculations, appears. Therefore one can expect that including consistent N^{3}LO chiral 3NF's will probably provide an explanation for A_y puzzle.

That new procedure for estimating the theoretical truncation errors provides also evidence for missing 3NF effects at higher energy elastic Nd scattering, supporting thus the results found with the standard NN potentials. It remains to be seen whether consistent 3NF's at N^{3}LO will resolve higher energy discrepancies found in Nd elastic scattering.

4 Summary

Solving 3N scattering exactly up to energies below the pion production threshold allows one to test the 3N Hamiltonian based on modern NN potentials and 3NF's. At the higher energies for some observables large 3NF effects are predicted when using models such as TM and Urbana IX. Some Nd elastic scattering cross section and polarization data support these predictions. In some other cases, however, defects of the (semi)phenomenological 3NF's are demonstrated. Relativistic effects are found to be small for the elastic scattering cross section and negligible for higher energy spin-observables. The discrepancies at high energies, which remain even when Urbana IX or TM 3NF's are included, point to the importance of short-range contributions to the 3NF. Application of improved chiral NN interactions up to N^{4}LO order of chiral expansion and estimation of theoretical uncertainties due to truncation of higher orders supports conclusions obtained with the standard NN potentials. Higher order chiral 3NF's comprise a number of shorter-range terms. It can be expected that an application of consistent chiral NN and 3N forces will play an important role in understanding of elastic scattering and breakup reactions at higher energies.

Acknowledgments

This work was supported by the Polish National Science Center under Grant No. DEC-2013/10/M/ST2/00420. Some part of the work was performed within LENPIC collaboration. The numerical calculations have been performed on the supercomputer cluster of the JSC, Jülich, Germany.

References

[1] Wiringa R. B. et al., Phys. Rev. C, **51** (1995) 38.

[2] Machleidt R., Phys. Rev. C, **63** (2001) 024001.

[3] Stoks V. G. J. et al., Phys. Rev. C, **49** (1994) 2950.

[4] Pudliner B. S. et al., Phys. Rev. C, **56** (1997) 1720.

[5] Coon S. A. et al., Nucl. Phys. A, **317** (1979) 242.

[6] Epelbaum E., Prog. Part. Nucl. Phys., **57** (2006) 654.

[7] Machleidt R. and Entem D. R., Phys. Rep., **503** (2011) 1.

[8] Epelbaum E. et al., Phys. Rev. C, **66** (2002) 064001.

[9] Bernard V. *et al.*, Phys. Rev. C, **84** (2011) 054001.

[10] Epelbaum E. et al., Eur. Phys. J. A, **51** (2015) 53.

[11] Epelbaum E. et al., arXiv:1412.4623 [nucl-th].

[12] Glöckle W. et al., Phys. Rep., **274** (1996) 107.

[13] Witała H. et al., Phys. Rev. C, **63** (2001) 024007.

[14] Sekiguchi K. et al., Phys. Rev. C, **79** (2009) 054008.

[15] Stephan E. et al., Eur. Phys. Journal A, **42** (2009) 13.

[16] Witała H. et al., Phys. Rev. C, **71** (2005) 054001.

[17] Witała H. et al., Phys. Rev. C, **77** (2008) 034004.

[18] Witała H. et al., J. Phys. G, **41** (2014) 094011.

[19] Maeda Y. et al., Phys. Rev. C, **76** (2007) 014004.

[20] Coon S. A. and Han H. K., Few-Body Systems, **30** (2001) 131.

[21] Tornow W. et al., Phys. Rev. Lett., **49** (1982) 312.

EPJ Web of Conferences **117**, 03001 (2016) DOI: 10.1051/epjconf/201611703001
NN2015

Collective dynamics, fluctuations and instabilities in relativistic heavy-ion collisions

László P. Csernai

Department of Physics and Technology,
University of Bergen, Allegaten 55, 5007 Bergen, Norway

Abstract

Collective processes occurring in high energy heavy ion collisions are discussed. The observed fluctuations and asymmetries have different causes, which is shown clearly in the experimental data for peripheral collisions. We discuss how these can be separated and which are the most interesting collective instabilities, which do not arise from random statistical fluctuations.

The angular fluctuations of Cosmic Microwave Background are well studied, with the Planck satellite we observed these fluctuations with a very high resolution beyond multipole moment 1000, but the dominant peak is around 200, which means that a typical fluctuation peak is a bit larger than one degree.

The same type of study for central heavy ion collisions at the CERN LHC for the azimuth angle in the direction transverse to the beam showed a maximum for multipolarity three, or in other words the largest flow harmonics was v_3 and exceeded 0.30 [1]. It is important that these collisions were almost exactly central ones, with zero impact parameter, thus the initial condition could not lead to a collective azimuthal asymmetry. So the observed asymmetry could be originated from random fluctuations only. Studying fluctuation systematics for both central and different peripheral collisions, shows that the $v_2(p_t^{large})$ vs. $v_2(p_t^{low})$, v_2 vs. v_3, v_2 vs. v_4, v_2

vs. v_5, show distinctly different correlations for central and peripheral collisions [2]. Random fluctuations can have several reasons, e.g. the random initial nucleon configurations, or the critical fluctuations in a phase transition at the hadronization of the Quark Gluon Plasma (QGP). If we study the shape of density or energy density distribution in a finite system during hadronization, we can see that the Kurtosis of the distribution is becoming negative during the transition (critical opalescence), while the Skewness of the distribution changes from negative to positive as the matter hadronizes starting from QGP [3]. If we compare this with the observed negative Skewness of the multipoilarity distribution in central heavy ion collisions, which is negative, this could indicate that the fluctuation arise from flow fluctuations from the QGP phase. This does not contradict with the observed constituent quark number scaling of flow harmonics of different hadrons, which also indicates that the flow harmonics should have developed mainly in the QGP phase.

If we want to analyse peripheral reactions we have to separate flow from random fluctuations and from initial global symmetries. These latter ones are correlated with the center of mass and reaction plane of the event, which can and should be determined Event by Event (EbE). This is in part already done in experiments [4], while part of the complete procedure was proposed to be performed [5].

Thus, if we know the EbE reaction plane (RP) and the center of mass (CM) rapidity y_{cm}, then we can analyse the both the collective and fluctuating flow separately. This separation can be done if all data are provided in the EbE RP,CM system in the format usually presented recently as if only fluctuations would exist. This way of analysis works for exact central collisions but as mentioned it is inadequate for non-central ones [2].

The way how to make this separation is given in Ref. [6] Recently flow data are parametrized in the form

$$\frac{d^3N}{dydp_td\phi} = \frac{1}{2\pi}\frac{d^2N}{dydp_t}\Big[1 + 2v_1(y,p_t)\cos(\phi - \Psi_1^{EP}) + 2v_2(y,p_t)\cos(2(\phi - \Psi_2^{EP})) + \cdots\Big], \tag{1}$$

where v_n and Ψ_n^{EP} are measured averages for a set of events in a given impact parameter, b, transverse momentum, p_t, and rapidity, y, bin. The Ψ_n^{EP} values should be measured with respect to a given angle in a given event, most reasonably with respect to the EbE RP angle, Ψ_{RP}. Similarly the rapidity variable should also be given with respect to the EbE CM rapidity, y_{CM}. In most publications Ψ_{RP} and y_{CM} are not evaluated and thus the

Ψ_n^{EP} values are not published. This is possible with the "cumulative" methods, which can give the *absolute values* of the v_n parameters, but present a rather constrained information about the flow process.

Assuming that the complete experimental data set is given (published), $v_n(y - y_{CM}, p_t)$, Ψ_n^{EP}, with respect to Ψ_{RP}, then one can decompose the flow parametrization into a fluctuating and into a collective part [6]:

$$v_{n\,\frac{even}{odd}}^{Coll.} \cos[n(\phi - \Psi_n^{EP})] = \frac{1}{2}\left[{}^c v_n(\mathbf{y}, p_t) \pm {}^c v_n(-\mathbf{y}, p_t)\right] \cos(n\phi) \quad (2)$$

$$v_{n\,\frac{even}{odd}}^{Fluct.} \cos[n(\phi - \Psi_n^{EP})] = \frac{1}{2}\left[{}^c v_n(\mathbf{y}, p_t) \mp {}^c v_n(-\mathbf{y}, p_t)\right] \cos(n\phi) + {}^s v_n(\mathbf{y}, p_t)\ \sin(n\phi) \quad (3)$$

where $\mathbf{y} = y - y_{CM}$, and new coefficients

$${}^c v_n \equiv v_n \cos(n(\Psi_n^{EP})) \ \text{ and } \ {}^s v_n \equiv v_n \sin(n(\Psi_n^{EP}))\ ,$$

are constructed, so that ${}^s v_n$ can be due to fluctuations only and the Colelctive part can have only azimuthal angle dependence $\cos(n\phi)$ because of the mirror symmetry accross the RP.

We can this way see that the Collective flow components are such that these have cosine azimutal dependence only and the even harmonics are rapidity even functions, while the odd harmonics are rapidity odd functions. The fluctuating flow components have both sine and cosine azimuthal dependence. This separation of the flow enables us to discuss and analyze the two types of flow processes separately. On should also anticipate that for the Fluctuationg flow component the sine and cosine components in the flow should be of similar magnitude.

This separation is important for analysing both fluctuations and collective flow in peripheral collisions. For example viscosity usually damps flow structure and asymmetries, while critical fluctuations may build up with time. Thus, to evaluate viscosity with precision, requires the separation of collective and fluctuating flows. Initial random fluctuations are usually damped with time due to viscosity. On the other hand initial shear flow can feed the development of Kelvin Helmholtz Instability, thus increasing rotation and turbulence. For converting shear flow into rotation and KHI som, (but low) viscosity is required.

Recently there is an increasing theoretical effort to study the effects of increasing angular momentum with increasing beam energy in peripheral collisions [7–15]. Here the initial shear plays an important role. The development of strong fields (e.g. Color Glass Condensate, CGC) leads to a more compact and more energetic initial state, and streak by streak momentum conservation leads to strong shear. Still in simplified models the matter is assumed to propagate to infinity without hindrance, leading to a uniform Bjorken type flow, without shear. Thus, the realistic assumptions on the initial state are vital in getting a realistic view about the collision process. If initial shear is neglected, vorticity is not formed and no turbulence or KHI could develop.

The initial shear leads to a much larger (3 c/fm) vorticity [16] than random fluctuations (0.2 c/fm) [17]. This can lead to different observable effects: Two particle correlations or particle polarizatons. The large vorticity, i.e. local orbital motion equilibrates with the spins of the constituent particles of matter due to the mechanical equipartition principle [18]. The arising Λ polarization is evaluated [19] and it was obtained that for particles emitted in the reaction plane, transverse to the beam (x -direction) one can observe a transverse ($-y$-directed) polarization of 8-9 %. In other azimutal directions the polarization vanishes.

Although, this was measured in an experiment already in 2007 [20], all emission angles for all azimuth were averaged over, and the EbE CM and the RP were not precisely determined. One has to emphasize that for the polarization measurements the Projectile and Target side must be determined otherwise this effect obviously vanishes. So the usual cumulative methods are not adequate at all for these measurements.

Another important consequence of this mechanical polarization that being a paramagnetic type of effect, with high temperature the polarization decreases. Thus, the characteristic quantity is the thermal vorticity $\nabla \times \boldsymbol{\beta}$ where,

$$\beta^{\mu} = u^{\nu}(x)/T(x),$$

i.e. the flow velocity is divided by the temperature. This is important because although the angular momentum and vorticity increase with increasing beam energy the freeze-out temperature is also higher. Thus, it turned out that the polarization is somewhat larger at RHIC energies than at the LHC, furthermore still ar NICA and FAIR energies one can obtain significant thermal vorticity [21].

It was also shown that the Hanbury Brown and Twiss effect is aslo sensitive on rotation [22], and interestingly the azimuthal asymmetry of the source as well as the rotation of the source may interfere with each other, so one has to find the adequate azimuth angle where one can measure exclusively the rotation.

The interest in rotation in high energy heavy ion collisions led to the development of an Exact expanding and rotating fluid dynamical model [23]. The model parameters were compared to the fluid dynamical model of a peripheral collision, and the initial sizes and angular and expansion velocities were determind to obtain a best fit to the realistic fluid dynamical solutions [24]. In this model the vorticity was evaluated [25], the arising two particle correlations [26], as well as the arising polarization [27].

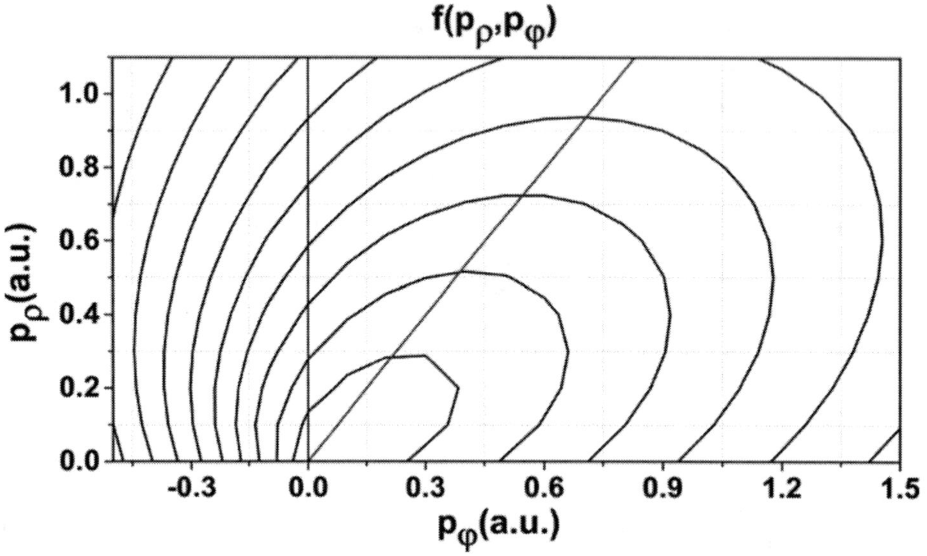

Figure 1: The schematic phase space distribution of the rotating and expanding source in the momentum space. The momentum of the expansion increases with the radius just as the momentum arising from the rotation. So, higher radial flow momenta correspond to higher rotation momenta, indicated by the red line. The distribution is thermally smeared as indicated by the contour lines. From Ref. [26].

It is interesting that the two particle correlations also indicate the rotation of the system. As pointed out in [26] the expanding and rotating sustem is azimuthally symmetric in the configuration space, but not in the momentum space as shown in Fig. 1. The momentum of the fluid motion is characterized by the flow velocity, this is indicated by the red full line in Fig. 1. This characteristic flow velocity distribution determines the final momentum distribution of the emitted particles.

The results of the latest work [27,28], are discussed in the contribution of Yilong Xie, which can be found in this proceedings. These results included the evatuation of the polarization arising from the time derivative of the thermal vorticty, which is turned out to be not negligible. These results were also compared to the fluid dynamical model result of the Itanian ECHO-QGP code [11].

Thus, both fluctuations and collective flow processes carry important information on the dynamics of the system and thus on the propoerties of QGP. For a precise analysis, forst of all the fluctuatinf and collective flow components should be separated and for this we need the accurate determination of the EbE collision symmetry axes, the Reaction plane and the Center of mass.

The most sensitive measurement of viscosity can be achieved by phenomena which are directly dependent on the shear viscosity like the Kelvin Helmholtz Instability. To study these processes high resolution relativistic fluid dynamical codes are required, which can handle shear and turbulence. Necessarily, the initial shear cannot be neglected if we intend to study the shear viscosity.

References

[1] K. Aamodt et al. (ALICE Collaboration), Phys. Rev. Lett. 107, 032301 (2011), and CERN Courier, October 2011, p. 6.

[2] Michael Weber (CERN) talk (ATLAS data) at the "X Workshop on Particle Correlations and Femtoscopy WPCF 2014", 25-29.08.2014, Kroly Rbert College, Gyngys, Hungary.

[3] L.P. Csernai, G. Mocanu, Z. Neda, Phys. Rev. C 85, 068201 (2012).

[4] R. Snellings, J. Phys. G 41, 24007 (2014).

[5] L.P. Csernai, G. Eyyubova, V.K. Magas, Phys. Rev. C 86, 024912 (2012).

[6] L.P. Csernai, H. Stöcker, J. Phys. G 41, 124001 (2014).

[7] L.P. Csernai, V.K. Magas, H. Stöcker, D.D. Strottman, Phys. Rev. C 84, 024914 (2011).

[8] L.P. Csernai, D.D. Strottman, Cs. Anderlik, Phys. Rev. C 84, 054901 (2012).

[9] D.J. Wang, Z. Néda, and L.P. Csernai, Phys. Rev. C **87**, 024908 (2013).

[10] G. Graef, M. Bleicher, M. Lisa, Phys. Rev. C **89**, 014903 (2014).

[11] F. Becattini, G. Inghirami, V. Rolando, A. Beraudo, L. Del Zanna, A. De Pace, M. Nardi, G. Pagliara, and V. Chandra, arXiv:1501.04468v2 [nucl-th] (2015).

[12] Guangyao Chen, Rainer J. Fries, Phys. Lett. B **723** 417 (2013).

[13] G. Chen, R.J. Fries, J.I. Kapusta, Y. Li, arXiv:1507.03524v1 [nucl-th]

[14] Brett McInnes, Yen Chin Ong, Nucl. Phys. B 898, 197 (2015), arXiv:1504.07344 [hep-th].

[15] Brett McInnes Nucl. Phys. B 887, 246 (2014), arXiv:1403.3258 [hep-th].

[16] L.P. Csernai, V.K. Magas, D.J. Wang, Phys. Rev. C 87, 034906 (2013).

[17] S. Floerchinger, U.A. Wiedemann, JHEP 100,1111 (2011); J. Phys. G 38, 124171 (2011).

[18] F. Becattini, V. Chandra, L. Del Zanna, and E. Grossi, Ann. Phys. **338**, 32 (2013); arXiv: 1303.3431 [nucl-th].

[19] F. Becattini, L.P. Csernai, D.J. Wang, Phys. Rev. C 88, 034905 (2013).

[20] B.I. Abelev et al., (STAR), Phys. Rev. C 76, 024915 (2007).

[21] L.P. Csernai, D.J. Wang, M. Bleicher, H. Stöcker, Phys. Rev. C 90, 021904(R) (2014).

[22] L.P. Csernai, S. Velle, D.J. Wang, Phys. Rev. C 89, 034916 (2014).

[23] T. Csörgő and M.I. Nagy, Phys. Rev. C **89**, 044901 (2014).

[24] L.P. Csernai, D.J. Wang, and T. Csörgő Phys. Rev. C **90**, 024901 (2014).

[25] L.P. Csernai, J.H. Inderhaug, Int. J. Modern Physics E **24**, 1550013 (2015), arXiv:1503.03247v1 [nucl-th].

[26] S. Velle, L.P. Csernai, Phys. Rev. C 92, 024905 (2015).

[27] Y.L. Xie, R.C. Glastad, L.P. Csernai, arXiv: 1505.07221 [nucl-th].

[28] Yilong Xie, Contribution to this volume (2015).

EPJ Web of Conferences **117**, 03002 (2016) DOI: 10.1051/epjconf/201611703002
NN2015

An overview of resonance measurements at the ALICE experiment

A. G. Knospe (for the ALICE Collaboration)

The University of Texas at Austin, Austin, TX, USA

Abstract

Resonances play a unique role in the study of ultra-relativistic heavy-ion collisions. Resonance yields, which may be modified by rescattering and regeneration after hadronization, can be used to study the properties of the hadronic phase of the collision. The transverse-momentum spectra of the proton and the ϕ(1020) can be used to study the mechanisms of particle production. In addition, resonance measurements in pp and p–Pb collisions help to distinguish initial-state effects from the effects of the hot and dense final state. The ALICE Collaboration has studied the $K^{*}(892)^{0}$ and ϕ(1020) mesons in pp, p–Pb, and Pb–Pb collisions. Measurements of many resonance properties, including p_{T} spectra, integrated yields, masses, widths, mean p_{T} values, and the nuclear modification factors R_{AA} and R_{pPb}, are presented and compared to measurements from other experiments, non-resonances, and the predictions of theoretical models.

In relativistic heavy-ion collisions, strongly interacting matter is studied under extreme conditions. Based on quantum chromodynamics (QCD), it is expected that when hadronic matter reaches a high enough energy density a transition will occur in which quarks become deconfined [1–3]. In this state of matter, the quark-gluon plasma (QGP), the relevant degrees of freedom are not hadrons, but rather quarks and gluons. For low baryochemical potentials (typical of heavy-ion collisions at the LHC), the transition between hadronic matter and the QGP is expected to occur around a temperature of $T_{c} \approx 160$ MeV [3–5]. Experimentally, this is achieved by colliding nuclei at very high energies (a center-of-mass energy per nucleon-nucleon pair, $\sqrt{s_{\rm NN}}$, greater than a few tens of GeV). An ultra-relativistic heavy-ion collision is expected to produce a QGP, which thermalizes, expands, and cools for a few fm/c before making the transition back to hadronic matter. The hadronic system continues to expand and cool. It first passes chemical freeze-out, after which inelastic interactions cease and the yield of (most) particle species are fixed. Elastic interactions continue until kinetic freeze-out, after which the system is too diffuse to interact. Typical estimates of the time between chemical and kinetic freeze-out are ~ 10 fm/c [6, 7].

These proceedings focus on the use of hadronic resonances as probes of heavy-ion collisions. In addition to deconfinement, strongly interacting matter is expected to undergo a transition in which chiral symmetry is restored and the quark masses approach 0. Lattice QCD calculations [8] suggest that the deconfinement and chiral transitions occur around the same temperature. Resonances that decay when chiral symmetry was (partially) restored may exhibit shifted masses or increased widths [9–12].

The yields of stable particles are expected to be fixed at chemical freeze-out, but the measured yields of resonances, which can decay during the hadronic phase, may be changed by scattering processes involving their decay products [13–15]. Resonance yields can be regenerated when hadrons undergo pseudo-elastic scattering through a resonance state. Experimentally, resonances are reconstructed in invariant-mass analyses; if one of the resonance decay products undergoes elastic scattering in the hadronic medium, the resonance invariant-mass peak may be smeared out, reducing the reconstructible resonance yield. The decay product of one resonance may also undergo pseudo-elastic scattering through a different resonance state, removing information about the original resonance and reducing its measured yield [6]. The strengths of these processes depend on the chemical freeze-out temperature, the time between chemical and kinetic freeze-out, the lifetime of the resonance, and the scattering cross sections of its decay products. Regeneration and re-scattering are expected to be most important for $p_{\rm T} \lesssim 2$ GeV/c [6, 13].

In addition, resonances can be used along with stable particles to study the properties of the collision system and their effects on hadrons with different flavors, masses, and baryon numbers. Specifically, resonances can be used to study partonic energy loss in the deconfined system, which is quantified using the nuclear modification factor (discussed below). They can also be used to study the mechanisms through which particles are produced and the effects that modify the shapes of particle $p_{\rm T}$ spectra (see, in particular, the discussion of the p/ϕ(1020) ratio below).

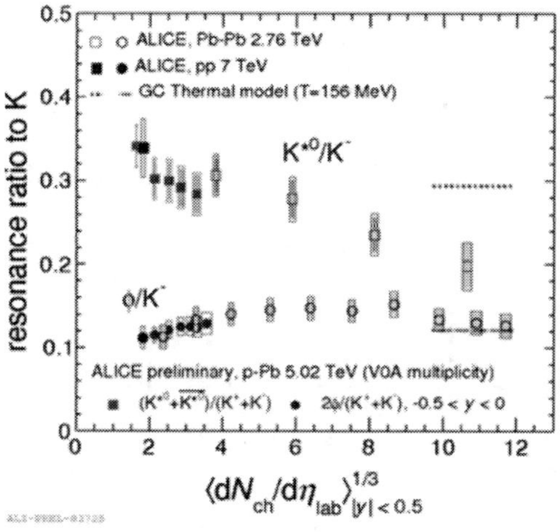

Figure 1: K^{*0}/K^- and ϕ/K^- ratios as a function of the mid-rapidity $\langle dN_{ch}/d\eta\rangle^{1/3}$ for pp, p–Pb, and Pb–Pb collisions [16–18]. Bars represent statistical uncertainties, open boxes represent total systematic uncertainties, and shaded boxes represent the parts of the systematic uncertainties that are uncorrelated between multiplicity or centrality intervals.

These proceedings focus on measurements by the ALICE Collaboration of the $K^*(892)^0$ and $\overline{K}^*(892)^0$ resonances (sum or average denoted as K^{*0}), as well as the $\phi(1020)$ meson (denoted by ϕ). These resonances are reconstructed via their decay to charged hadrons: $K^{*0} \to \pi^\pm K^\mp$ (branching ratio 0.667) and $\phi \to K^-K^+$ (branching ratio 0.489). For a description of the technique, see [16, 17].

The ALICE Collaboration has measured the masses and widths of the K^{*0} and ϕ mesons by reconstructing their decays in Pb–Pb collisions at $\sqrt{s_{NN}} = 2.76$ TeV [17]. The measured masses and widths of these particles are consistent with their vacuum values. The lack of the expected signatures of chiral symmetry restoration may be due to the loss of resonances produced early in the collision due to re-scattering and the regeneration of additional resonances late in the collision.

Figure 1 shows the p_T-integrated K^{*0}/K and ϕ/K ratios measured in pp, p–Pb, and Pb–Pb collisions at LHC energies [16–18]. These ratios are plotted as funcitons of $\langle dN_{ch}/d\eta\rangle^{1/3}$ (measured at mid-rapidity: $|\eta| < 0.5$), which is taken as a proxy for the system size. The K^{*0}/K ratio is significantly suppressed in central Pb–Pb collision with respect to smaller collision systems (pp collisions at $\sqrt{s} = 7$ TeV and p–Pb collisions at $\sqrt{s_{NN}} = 5.02$ TeV). (A "central" collision is one with a small impact paramter, in which the volume of the system is largest; a "peripheral" collision has a large impact parameter and smaller system size.) This suppression may be due to signal loss resulting from re-scattering of the K^{*0} decay products in the hadronic medium (which would be dominant over regeneration). In contrast, the ϕ/K ratio depends only weakly on system size and is not suppressed in central Pb–Pb collisions, suggesting that the longer-lived ϕ meson is affected less

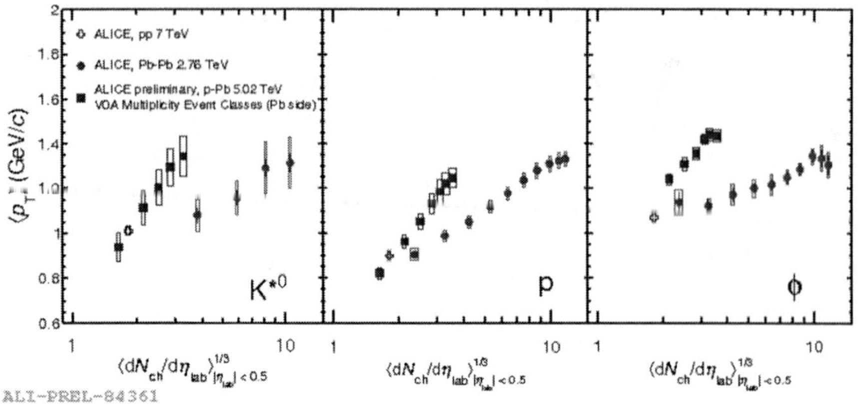

Figure 2: Measurements of $\langle p_{\mathrm{T}} \rangle$ of K^{*0}, p, and ϕ in pp, p–Pb, and Pb–Pb collisions [18]. Bars (boxes) represent statistical (systematic) uncertainties.

by re-scattering in the hadronic phase. For both ratios, the values measured in p–Pb collisions are similar to the values measured in pp and Pb–Pb collision with similar multiplicities. These ratios are also compared to the values given by a thermal-model fit to particle yields measured by ALICE in central Pb–Pb collisions with a chemical freeze-out temperature of 156 MeV (the fit includes the ϕ yield, but not the K^{*0}). The $\mathrm{K}^{*0}/\mathrm{K}$ ratio is suppressed with respect to the thermal-model value, while the ϕ is not supressed. As described in [17], the measured $\mathrm{K}^{*0}/\mathrm{K}$ ratio in central Pb–Pb collisions can be compared to an extended thermal model (modified to include re-scattering effects) [14, 19, 20] to extract a model-dependent estimate of the time between chemical and kinetic freeze-out. When this is done, a lower limit of 2 fm/c is found (only a lower limit can be extracted because the model does not include regeneration).

The K^{*0} and ϕ mesons can also be used along with other particles in studies of how flavor, mass, and baryon number affect the shapes of particle p_{T} spectra, which can be characterized using the mean transverse momentum $\langle p_{\mathrm{T}} \rangle$. Figure 2 shows the $\langle p_{\mathrm{T}} \rangle$ of three particles with similar masses, K^{*0}, p, and ϕ, measured in pp, p–Pb, and Pb–Pb collisions [18]. In central Pb–Pb collisions, the $\langle p_{\mathrm{T}} \rangle$ values of these three particles are consistent with each other. Hydrodynamic models have been found to provide good descriptions of the matter produced in heavy-ion collisions [21–23]. In such models, the particle mass tends to be the dominant factor in determining the shape of the p_{T} spectrum; the similarity of the $\langle p_{\mathrm{T}} \rangle$ values for K^{*0}, p, and ϕ therefore appears to be consistent with what would be expected from hydrodynamic models. In contrast, in smaller collision systems (pp, p–Pb, and peripheral

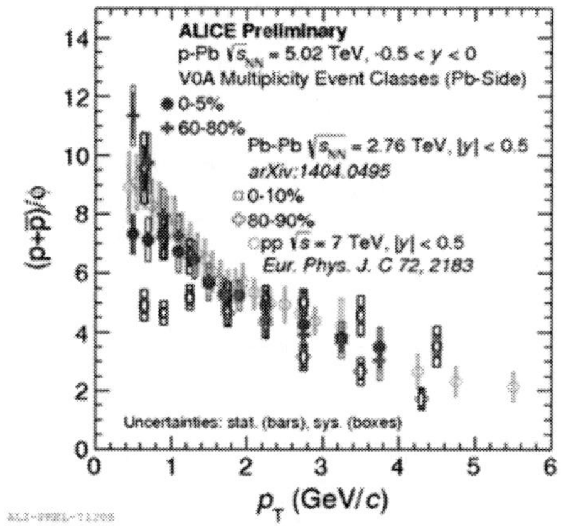

Figure 3: Ratio p/ϕ as a function of p_T for p, and ϕ in pp, p–Pb, and Pb–Pb collisions. Bars (boxes) represent statistical (systematic) uncertainties.

Pb–Pb collisions) a splitting is observed between the $\langle p_T \rangle$ values of p and ϕ, suggesting that factors other than the particle mass are playing important roles in determining the shapes of the p_T spectra. It should also be noted that (unlike the K^{*0}/K and ϕ/K ratios) the $\langle p_T \rangle$ values in p–Pb collisions follow different trends than observed in Pb–Pb collisions. The $\langle p_T \rangle$ values for K^{*0}, p, and ϕ rise much faster as functions of $\langle dN_{ch}/d\eta \rangle^{1/3}$ in Pb–Pb collisions and reach (or even exceed) the values observed in central Pb–Pb collisions, despite the very different system sizes.

When central Pb–Pb collisions are compared to pp collisions, an enhancement is observed in p_T-dependent baryon-to-meson ratios such as p/π and Λ/K^0_S [24, 25] for $1.5 \lesssim p_T \lesssim 6$ GeV/c at RHIC and LHC energies. Various explanations have been proposed to explain this enhancement. In a hydrodynamic picture the enhancement can be explained by the difference in the particle masses: all particles would have a common flow velocity, so more massive particles would be pushed out to higher p_T, resulting in an enhancement in their ratios to less massive particles (see, for example, [21–23]). In recombination models [26, 27], the quark content and the number of valence (anti)quarks may play a role in determining the shapes of particle p_T spectra. Such models might therefore yield different shapes for the p_T spectra of baryons and mesons (even those with similar masses). Since the (anti)proton and ϕ have similar masses, their ratio can be used to study this effect, as shown in fig. 3 [18]. The p/ϕ ratio is constant as a function of p_T for central Pb–Pb collisions at $\sqrt{s_{NN}} = 2.76$ TeV, while developing a slope for more peripheral collision (as well as for p–Pb and pp collisions; not shown). The constant ratio in central collisions is consistent with a hydrodynamic interpretation, but it should be noted that some recombination models can also reproduce the observed behavior [28].

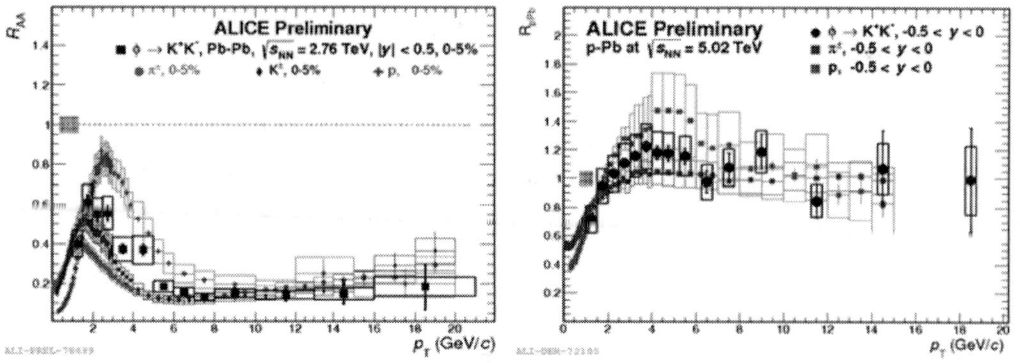

Figure 4: Nuclear modification factors R_{AA} (left) and R_{pPb} (right) of various hadrons. Bars (boxes) represent statistical (systematic) uncertainties.

It is also possible to use the ϕ meson along with other particles in studies of the flavor, mass, and baryon-number dependence of parton energy loss in the QGP. The effect of nuclear matter R_{AA} on particle yields is quantified using the nuclear modification factor: the ratio of the yield in A–A collisions to the same yield in pp collisions scaled by the number of binary nucleon-nucleon collisions in the A–A coillison. If the A–A collision is a simple superposition of NN collisions, R_{AA} would be unity. Figure 4 (left) shows R_{AA} as a function of p_{T} for the ϕ meson and for light hadrons; the suppression at intermediate and high p_{T} is attributed to parton energy loss in the QGP. The high-p_{T} suppression is the same for almost all hadron species (including those containing c and b quarks). For $2 < p_{\mathrm{T}} < 6$ GeV/c, there appears to be mass ordering in the R_{AA} values for mesons. At RHIC, different shapes for R_{AA} of (anti)protons and ϕ was taken as evidence in favor of recombination models, but in the present data the different R_{AA} values for these particles are due to differences in the reference (pp) spectra. Figure 4 (right) shows R_{pPb}, the nuclear modification factor for p–Pb collisions, which quantifies the effect of nuclear matter in the absence of a QGP (so-called "cold nuclear matter"). The R_{pPb} for $\pi^{\pm}$ is consistent with unity for $p_{\mathrm{T}} > 2.5$ GeV/c, while R_{pPb} for protons exhibits a moderate enhancement. The R_{pPb} values for ϕ are consistent with unity for intermediate and high p_{T}, though there are hints of a small Cronin enhancement. Taken together, these R_{AA} and R_{pPb} suggest the presence of mass dependence and/or baryon/meson differences for the processes that determine the shapes of particle spectra in pp, p–Pb, and Pb–Pb collisions.

Hadronic resonances are useful probes in the study of heavy-ion collisions. Their short lifetimes make them well suited to study the properties of the hadronic phase. They can also be compared to other particles as part of systematic studies of the flavor, mass, and baryon-number dependence of the process that determine the shapes of particle p_{T} spectra. The ALICE Collaboration has measured the K^{*0} and ϕ mesons in multiple collision systems [16–18] and has also measured the $\Sigma(1385)^{\pm}$ and $\Xi(1530)^{0}$ baryonic resonances in pp collisions [29]. There are also ongoing measurements of several particles, including $\rho(770)^{0}$, Σ^{0}, $\Delta(1232)^{++}$, $\Sigma(1385)^{\pm}$, $\Lambda(1520)$, and $\Xi(1530)^{0}$ in pp, p–Pb, and Pb–Pb collisions at LHC energies. The p_{T}-integrated $\mathrm{K}^{*0}/\mathrm{K}$ ratio is observed to be suppressed in central Pb–Pb collisions, consistent with signal loss due to re-scattering of the K^{*0} decay products, while the longer-lived ϕ meson is not suppressed. Measurements of $\langle p_{\mathrm{T}} \rangle$ and the p_{T}-dependent p/ϕ ratio indicate that hydrodynamics (in which particle masses are a dominant factor in determining the shapes of p_{T} spectra) is well suited to describe central Pb–Pb collisions, while effects depending on quark content may be important for smaller collision systems. As quantified by the nuclear modification factor R_{AA}, particle suppression at high p_{T} is the same for most particle species; at intermediate p_{T} there may be mass and/or baryon-number dependence.

References

[1] Petreczky P., *Proc. of Science (Confinement X)* (2012) 028.

[2] Borsányi S. *et al.*, *J. High Energy Phys.*, **11** (2010) 077.

[3] Borsányi S. *et al.*, *J. High Energy Phys.*, **09** (2010) 073.

[4] Aoki Y. *et al.*, *J. High Energy Phys.*, **06** (2009) 088.

[5] Bazavov A. *et al.*, *Phys. Rev.* D, **85** (2012) 054503.

[6] Bleicher M. and Aichelin J., *Phys. Lett.* B, **530** (2002) 81–87.

[7] Bass S. A. *et al.*, *Phys. Rev.* C, **60** (1999) 021902.

[8] Petreczky P., *Nucl. Phys.* A, **785** (2007) 10–17.

[9] Brown G. E. and Rho M., *Rhys. Rep.*, **363** (2002) 85–171.

[10] Rapp R. *et al.*, *Relativistic Heavy Ion Physics*, edited by Stock R., (Springer Berlin Heidelberg) 2010, pp. 134–175.

[11] Brodsky S. J. and de Teramond G. F., *Phys. Rev. Lett.*, **60** (1988) 1924.

[12] Eletsky V. L. *et al.*, *Phys. Rev.* C, **64** (2001) 035202.

[13] Bleicher M. and Stöcker H., *J. Phys.* G, **30** (2004) S111–S118.

[14] Markert C. *et al.*, *AIP Conf. Proc.*, **631** (2002) 533-552.

[15] Vogel S. and Bleicher M., *Ricerca Scientifica ed Educazione Permanente*, edited by I. Iori and A. Bortolotti, Supplemento N. 124 (Università degli Studi di Milano, Milan) 2005, pp. 116–119, nucl-th/0505027 Preprint.

[16] Abelev B. *et al.* (ALICE Collaboration), *Eur. Phys. J.* C, **72** (2012) 2183.

[17] Abelev B. *et al.* (ALICE Collaboration), *Phys. Rev.* C, **91** (2015) 024609.

[18] Bellini F. (for the ALICE Collaboration), *Nucl. Phys.* A, **931** (2014) 846–850.

[19] Torrieri G. and Rafelski J., *Phys. Lett.* B, **509** (2001) 239–245.

[20] Rafelski J. *et al.*, *Phys. Rev.* C, **64** (2001) 054907.

[21] Qui Z. *et al.*, *Phys. Lett.* B, **707** (2012) 151–155.

[22] Shen C. *et al.*, *Phys. Rev.* C, **84** (2011) 044903.

[23] Bożek P. and Wyskiel-Piekarska I., *Phys. Rev.* C, **85** (2012) 064915.

[24] Abelev B. *et al.* (ALICE Collaboration), *Phys. Rev.* C, **88** (2013) 044910.

[25] Abelev B. *et al.* (ALICE Collaboration), *Phys. Rev. Lett.*, **111** (2013) 222301.

[26] Fries R. *et al.*, *Phys. Rev. Lett.*, **90** (2003) 202303.

[27] Fries R. *et al.*, *Annu. Rev. Nucl. Part. Sci*, **58** (2008) 177–205.

[28] Minissale V. *et al.*, *Phys. Rev.* C, **92** (2015) 054904.

[29] Abelev B. *et al.* (ALICE Collaboration), *Eur. Phys. J.* C, **75** (2015) 1.

EPJ Web of Conferences **117**, 03003 (2016) DOI: 10.1051/epjconf/201611703003
NN2015

(Anti)hypertriton lifetime puzzle

YU-GANG MA

Shanghai Institute of Applied Physics, Chinese Academy of Sciences, Shanghai 201800, China

Abstract

Most calculations on the lifetime of (anti)hypertriton gave a similar lifetime which is close to the lifetime of free Λ decays. However, recent measurements on (anti)hypertriton lifetime demonstrate a much short lifetime. All results for (anti)hypertriton lifetime by two-body decay channel of ${}^3He + \pi$ for Au+Au collision at RHIC, Pb+Pb collision at LHC and Li + C collisions at GSI show a significant short lifetime in comparison with lifetime of free Λ decays. However, theoretical interpretation remains puzzle.

Hypernucleus provides an ideal environment to learn the hyperon-nucleon interaction, which is responsible partly for the binding of hypernuclei and lifetime. Therefore, a meaurement of the precise lifetime of hypernucleus could provide a tool to investigate of hyeron-nucleon interaction [1–7]. The most simple hypernucleus is hypertriton, which is a bound state of a proton, a neutron, and a hyperon (Λ or Ξ), it is bound with respect to the Λ-d threshold by 0.136±0.05 MeV. In 2010, ${}^3_\Lambda\overline{\mathrm{H}}$, the antimatter partner of ${}^3_\Lambda$H, has been discovered at RHIC-STAR as the first antimatter hypernucleus [8] . The hypertriton decays weakly into mesonic and nonmesonic channels. The nonmesonic decay channels are ${}^3_\Lambda\mathrm{H}\rightarrow$ d + n and ${}^3_\Lambda\mathrm{H}\rightarrow$ p + n +n. In the mesonic decay mode there are more channels: ${}^3_\Lambda\mathrm{H}\rightarrow \pi^-(\pi^0) + {}^3\mathrm{He}({}^3\mathrm{H})$, ${}^3_\Lambda\mathrm{H}\rightarrow \pi(\pi^0)$ + d + p(n), and ${}^3_\Lambda\mathrm{H}\rightarrow \pi(\pi^-)$ + p + n + p(n). In contrast to heavier hypernuclei, where mesonic decays are Pauli blocked, here in the hypertriton they are by far the dominant ones. Further there are experimental data on the branching ratio $R = \Gamma({}^3_\Lambda\mathrm{H}\rightarrow \pi^- + {}^3\mathrm{He})/\Gamma({}^3_\Lambda\mathrm{H}\rightarrow$ all π^- meson modes) ranging between 0.30 ± 0.07 [9] and 0.39 ± 0.07 [10]. From the technique viewpoint in heavy-ion collision, a simple identification method of ${}^3_\Lambda$H (${}^3_\Lambda\overline{\text{H}}$) can be achieved by reconstructing their secondary vertex via the decay channel of ${}^3_\Lambda\text{H} \rightarrow {}^3\text{He} + \pi^-$ (${}^3_\Lambda\overline{\text{H}} \rightarrow {}^3\overline{\text{He}} + \pi^+$), which occurs with a branching ratio of 25% (assuming that this branching fraction is the same as that for ${}^3_\Lambda$H [2]) [8,11,12]. In contrast, reconstruction of ${}^3_\Lambda$H (${}^3_\Lambda\overline{\text{H}}$) from three-body decay, i.e. ${}^3_\Lambda\text{H} \rightarrow \pi(\pi^0) + \text{d} + \text{p(n)}$, is more complicate, but it deserves to check with two-body decay results, and it is under progress well [13].

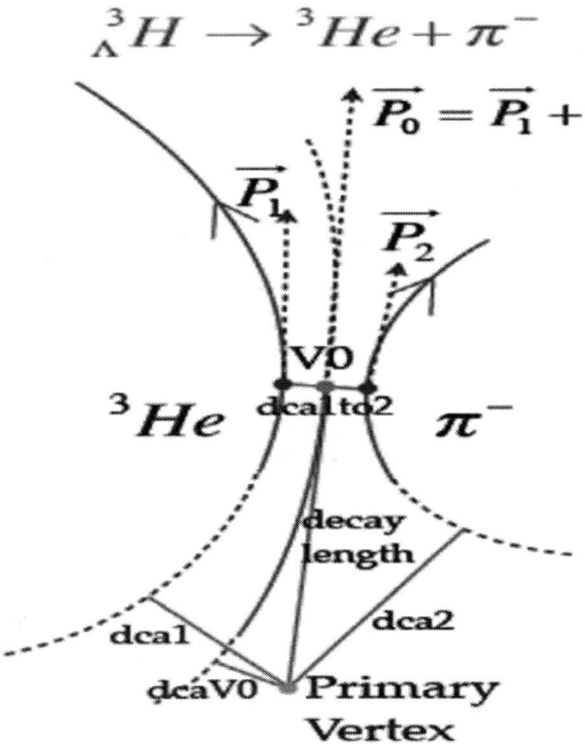

Figure 1: Topological decay map for 2-body channel of ${}^3_\Lambda$H.

Theoretically, most of calculations predicted that the hypertriton lifetime is a littler shorter than the free Λ lifetime since it could exhibit the modification of the Λ wavefunction in the nuclear medium. For an example, the π-mesonic decay of the hypertriton is calculated based on a hypertriton wave function and 3N scattering states, which are solutions of three-body Faddeev equations using realistic NN and hyperon-nucleon interactions [14]. They predicted that the total lifetime of ${}^3_\Lambda$H is 256 ps which is 3% smaller than for the free Λ particle [14].

In this work, the lifetime measurement data for ${}^3_\Lambda$H (${}^3_{\overline{\Lambda}}\overline{\text{H}}$) two-body decay from Au + Au collisions at RHIC is reported. The data was collected by the STAR experiment at RHIC, using the cylindrical TPC, which is 4 meters in diameter and 4.2 meters long in the beamline direction [15]. The identification of tracks can be achieved by correlating their ionization energy loss $\langle dE/dx \rangle$ in TPC with their magnetic rigidity.

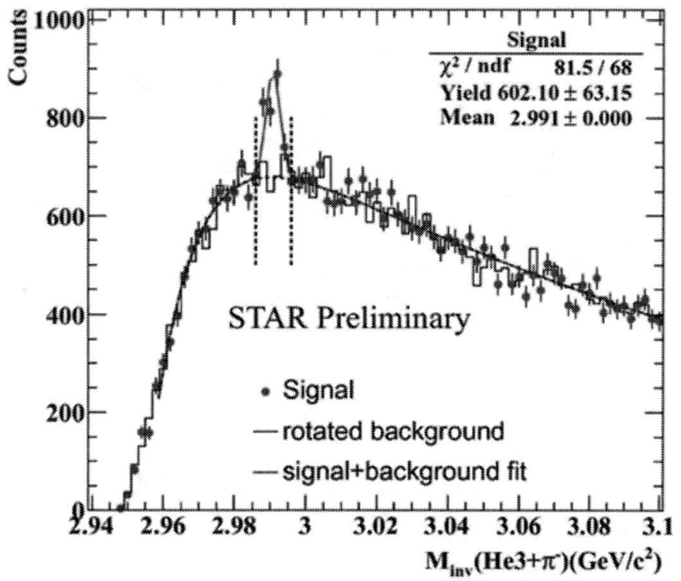

Figure 2: Reconstructed invariant mass distribution of ${}^3_\Lambda$H from two-body decay channel of ^{3}He. Solid circles stand for the signal distribution, while dashed lines are the rotated combinatorial background. Vertical dashed lines represent the mass range we use for bin counting of ${}^3_\Lambda$Hyield. Adapted from Ref. [12].

For two-body decay channel, topological cuts include the distance between two daughter tracks ^{3}He (${}^3\overline{\text{He}}$) and $\pi^-(\pi^+)$ (<1cm), distance of closest approach (DCA) between ${}^3_\Lambda$H (${}^3_{\overline{\Lambda}}\overline{\text{H}}$) and primary vertex (<1cm), decay length of ${}^3_\Lambda$H (${}^3_{\overline{\Lambda}}\overline{\text{H}}$) (>2.4cm), and the DCA of π track (>0.8cm), are employed to enhance the signal to background ratio [12]. Figure 1 shows the topological decay maps for both 2-body channel. The invariant mass of ${}^3_\Lambda$H (${}^3_{\overline{\Lambda}}\overline{\text{H}}$) was calculated based on the conservation of momentum and energy in the decay process. The result is shown in Figure 2. The successfully reproduced combinatorial background with a rotation strategy can be described by double exponential function: $f(x) \propto exp[-(x/p_1)]-exp[-(x/p_2)]$, where $x = m - m({}^3He) - m(\pi)$, and p_1, p_2 are the free parameters. Finally, the signals are counted by subtracting the double exponential background of ${}^{3}_{\Lambda}\mathrm{H}$ and ${}^{3}_{\Lambda}\overline{\mathrm{H}}$.

To demonstrate a visual scenario for ${}^{3}_{\Lambda}\mathrm{H}$ decay, a carton picture is depicted to help us understand the possible decay mechanism. Figure 3 shows a possible decay process of two-body decay of ${}^{3}_{\Lambda}\mathrm{H}$, which can be understood by a capture of decayed proton from Λ by the existed neutron and proton inside ${}^{3}_{\Lambda}\mathrm{H}$, resulting in a final decayed products ${}^{3}He$ and π .

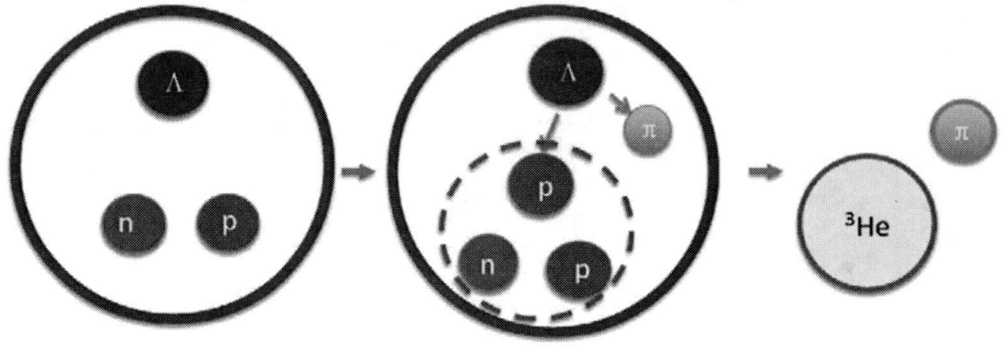

Figure 3: Cartons for ${}^{3}_{\Lambda}\mathrm{H}$ (${}^{3}_{\Lambda}\overline{\mathrm{H}}$) two-body decay channel of ${}^{3}\mathrm{He}+\pi$.

The above secondary vertex reconstruction of ${}^{3}_{\Lambda}\mathrm{H}$ (${}^{3}_{\Lambda}\overline{\mathrm{H}}$) allows ones to perform a calculation of its lifetime, via equation $N(t) = N(0)exp(-t/\tau)$, where $t = l/(\beta\gamma c)$, $\beta\gamma c = p/m$, l is the decay length of ${}^{3}_{\Lambda}\mathrm{H}$, p is their momentum, m is their mass value, while c is the speed of light. ${}^{3}_{\Lambda}\mathrm{H}$ and ${}^{3}_{\Lambda}\overline{\mathrm{H}}$ samples are combined together to get a better statistics, with the assumption of the same lifetime of ${}^{3}_{\Lambda}\mathrm{H}$ and ${}^{3}_{\Lambda}\overline{\mathrm{H}}$ based on the CPT symmetry theory. The measured yield is corrected for the tracking efficiency and acceptance of TPC, as well as the reconstruction efficiency of ${}^{3}_{\Lambda}\mathrm{H}$ and ${}^{3}_{\Lambda}\overline{\mathrm{H}}$. Then, the $l/(\beta\gamma)$ distribution can be fitted with an exponential function to extract the lifetime parameter $c\tau$.

The RHIC beam energy scan program in 2010-2011 allowed STAR to collect data for Au+Au collisions over a broad range of energies. To get an even better statistics, datasets are combined in the lifetime measurement. For two-body decay, the best fitting with χ^2 minimization method gives a value of $c\tau = 3.69\pm^{0.77}_{0.65}$ cm, which corresponds to a lifetime of $123 \pm ^{26}_{22}$ (stat) $\pm$ 10(sys) ps (Fig. 4) [12]. As a comparison, STAR 2010 ${}^{3}_{\Lambda}\mathrm{H}$ lifetime measurement [8] and the STAR 2010+2012 combined results are also provided. The current measurement is consistent with the STAR 2010 measurement within 1.5σ and is statistically improved. In this lifetime measurement, two kinds of sources for systematic study are considered: 1. choice of V0 topol-

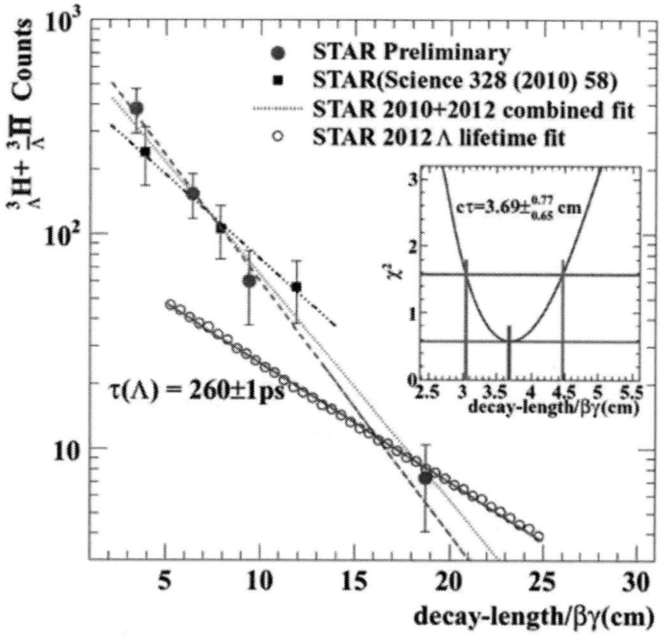

Figure 4: The yields of ${}^{3}_{\Lambda}$H+ ${}^{3}_{\Lambda}\overline{\mathrm{H}}$ and Λ (open circles) vs $c\tau$ distribution. The solid lines stand for the $c\tau$ fits, and the insert plot describes χ^2 distribution of the best fits. Comparison between the present measurement and the previous measurements is also demonstrated. Adapted from Ref. [12].

ogy cuts; 2. choice of bin width and invariant mass range. These effects contribute to the final systematic error. Additional sources of loss, like the interaction between ${}^{3}_{\Lambda}$H and material (air+detector) are also considered, which can be neglected due to its less than 1.5% effect [12]. As a further cross-check, Λ is reconstructed via the $\Lambda \rightarrow p + \pi^-$ decay channel. To this end, the same method is adopted to obtain the Λ lifetime and the result is 260 $\pm$ 1 ps which is consistent with the τ = 263 $\pm$ 2 ps compiled by the Particle Data Group [16]. In this way, we confirm that the method to extract ${}^{3}_{\Lambda}$H (${}^{3}_{\Lambda}\overline{\mathrm{H}}$) lifetime is reliable.

In summary, with the current highest statistics of (anti-)hypertriton samples, the results of STAR measurement definitely gives a short lifetime from two-body decay channel, $123^{+26}_{-22}ps$, which is much shorter than the free Λ decay lifetime. Also, recent measurements of ALICE collaboration for Pb + Pb collisions at $\sqrt{s_{NN}}$ = 2.76 TeV show τ = 181^{+54}_{-39} (stat.) $\pm$ 33 (syst.) ps for ${}^{3}_{\Lambda}$H lifetime [17] and HypHI Collaboration for ^{6}Li projectiles at 2A GeV on a carbon target also gave τ = $183^{+42}_{-32} \pm 37$ ps for ${}^{3}_{\Lambda}$H lifetime [18]. All these results are consistent within the errors and demonstrated that

${}^{3}_{\Lambda}$H (${}^{3}_{\overline{\Lambda}}\overline{\text{H}}$) decay has a significant short lifetime than theoretical expected values. Therefore an open question remains for theoretical interpretation, which still leave us a puzzle for understanding the structure of ${}^{3}_{\Lambda}$H (${}^{3}_{\overline{\Lambda}}\overline{\text{H}}$) and hyperon-nucleon interaction.

This work is partially supported by the 973 program under contract no. 2014CB845401 and NSFC under contract Nos. 11421505, 11035009 and 11220101005.

References

[1] R. H. Dalitz, G. Rajasekharan, Phys. Lett. **1**, 58 (1962)

[2] H. Kamada, J. Golak, K. Miyagawa, H. Witala, W. Glockle, Phys. Rev. C **57**, 1595 (1998)

[3] B. F. Gibson, E.V. Hungerford III, Phys. Rep. **257**, 349 (1995)

[4] E. Oset, P. Fernandez de Cordoba, L.L. Salcedo, R. Brockmann, Phys. Rep. **188**, 79 (1990)

[5] E. Hiyama, T. Motoba, T. A. Rijken and Y. Yamamoto, Progress of Theoretical Physics Supplement **185**, 1 (2010)

[6] A. S. Botvinaa, J. Steinheimer, E. Bratkovskaya, M. Bleicher, J. Pochodzalla, Phys. Lett. B **742**, 7 (2015)

[7] H. Shen, J. S. Liang, Nuclear Techniques (in Chinese) **37** , 100507 (2014)

[8] B. I. Abelev *et al.* (STAR Collaboration), Science **328** , 58 (2010)

[9] G. Keyes *et al.*, Phys. Rev. D **1**, 66 (1970); Phys. Rev. Lett. **20**, 819 (1968); Nucl. Phys. B **67**, 269 (1973)

[10] M. Block *et al.*, in Proceedings of the Sienna International Conference on Elementary Particles, edited by G. Bernandini and G. P. Puppi Societa de Fisica, Bologna, 1963, p. 62.

[11] J. H. Chen, Nucl. Phys. A **835**, 117 (2010); Nucl. Phys. A **830** , 761c (2009)

[12] Y. H. Zhu, Nucl. Phys. A **904-905**, 551c (2013); PhD Thesis (2013), Shanghai Institute of Applied Physics, CAS, 2013.

[13] Y. F. Xu for the STAR Collaboration, the talk at the 12th International Conference on Hypernuclear and Strange Particle Physics (HYP2015), Sept 2015, Sendai, Japan.

[14] H. Kamada, J. Golak, K. Miyagawa, H. Witala, and W. Glöckle, Phys. Rev. C **57**, 1595 (1998)

[15] M. Anderson *et al.*, Nucl. Instrum. Methods A **499**, 659 (2003)

[16] C. Amsler *et al.*, Phys. Lett. B **667**, 1 (2008)

[17] J. Adam *et al.* (ALICE Collaboration), arXiv:1506.08453.

[18] C. Rappold, E. Kim, D. Nakajima, T. Saito, O. Bertini *et al.* (HypHI Collaboration), Nucl. Phys. A **913**, 170 (2013)

EPJ Web of Conferences **117**, 03004 (2016) DOI: 10.1051/epjconf/201611703004
NN2015

Shear viscosity of QGP and the anisotropic flows within an event by event transport approach

S. Plumari[1,2], G. L. Guardo[2], A. Puglisi[1,2], F. Scardina[1,2] and V. Greco[1,2]

[1]Department of Physics and Astronomy, University of Catania, Via S. Sofia 64, I-95123 Catania, Italy
[2]INFN-Laboratori Nazionali del Sud, Via S. Sofia 62, I-95123 Catania, Italy

Abstract

We have employed a relativistic kinetic transport approach that incorporates initial state fluctuations to study the effect of a temperature dependent shear viscosity to entropy density ratio $\eta/s(T)$ on the build-up of the anisotropic flows $v_n(p_T)$. We find that at LHC energies and for ultra-central collisions $(0-0.2\%)$ the $v_n(p_T)$ have a stronger sensitivity to the T dependence of η/s in the QGP phase and this sensitivity increases with the order of the harmonic n. Moreover we have studied the correlation between the initial spatial anisotropies ϵ_n and the final flow coefficients $\langle v_n \rangle$ for different centralities and for the two beam energies. The study shows that at LHC energies there is more correlation than at RHIC energies. In particular at LHC energies and for ultra-central collisions the linear correlation coefficient $C(\epsilon_n, v_n) \approx 1$ for $n = 2, 3, 4$ and 5 suggesting that the $\langle v_n \rangle$ are strongly related to the initial value of ϵ_n.

1 Introduction

One of the most surprising results obtained in the experiments conducted at RHIC and more recently at the LHC is that the matter created in these heavy ion collisions behaves as an almost perfect fluid with a very low η/s ratio. The observable indicating such a low value for η/s is the so called elliptic flow, v_2. The elliptic flow is the second coefficient in the Fourier expansion of the azimuthal angle distribution of emitted particles and it a measure of the azimuthal asymmetry in momentum space. The determination of the η/s ratio of the QGP is one of the main challenges in the relativistic heavy ion physics. The comparison between experimental data and theoretical calculations within viscous hydrodynamics and transport approach [1–4] have shown that this large value of v_2 is consistent with a very low η/s. The average value extracted is close to $1/(4\pi)$ which is the lower bound conjectured for strongly interacting systems, $\eta/s = 1/4\pi$ [5].

The possibility to measure event-by-event the angular distribution of emitted particle has made possible to extend this analysis to high order Fourier coefficients v_n with $n > 2$. A shown by theoretical calculations the sources for these momentum anisotropies v_n are fluctuations in the initial geometry [6–9]. The comparison between experimental data and event-by-event viscous hydrodynamic calculations have shown that the expected range of viscosity explored from RHIC to LHC energies is about $4\pi \langle\eta/s\rangle \sim 1 - 3$ [8, 9]. However there are several theoretical indications that η/s should have a particular temperature dependence [10, 11, 13–15]. As an example in Fig.1 shows a collection of theoretical results about the T dependence of η/s. Therefore a phenomenological estimation of the T dependence of η/s could give more information information if the matter created in these collisions undergoes a phase transition [10, 11, 16]. More recently within viscous hydrodynamic calculation have been studied the correlations between the initial anisotropies, measured by the eccentricities ϵ_n, and the final anisotropies in momentum space v_n [17–19]. In general it has been shown that the degree of correlation between ϵ_n and v_n for a fixed centrality decrease with the order of the harmonic n.

In this proceeding we show the effects of the $\eta/s(T)$ on the $v_n(p_T)$ and we study the correlations between the initial ϵ_n and the final v_n by using a cascade approach with initial state fluctuations [20].

2 Kinetic approach at fixed $\eta/s(T)$

We have employed a relativistic transport code developed in these years to perform studies of the dynamics of heavy-ion collisions at both RHIC and LHC energies [3, 20, 25–28]. We solve numerically the RBT equation using the so called test particle method while for the collision integral we use the Monte Carlo methods, for details see [3, 25, 29].

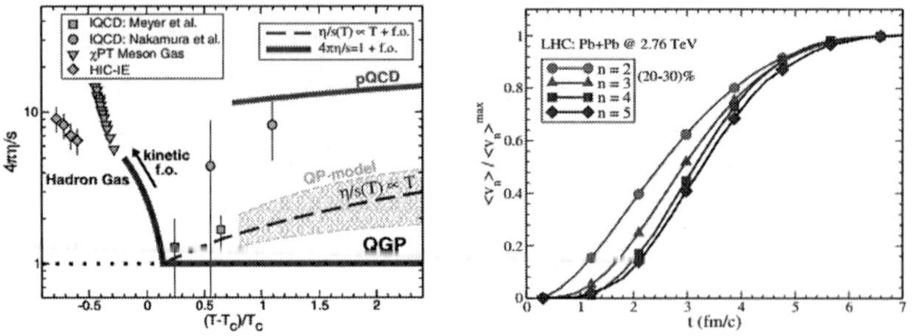

Figure 1: Left panel: $\eta/s(T)$ parametrizations. The light blue symbols are lQCD data [14, 21]. The orange symbols are for chiral perturbation theory [12, 13] and the phenomenological analysis for HIC at intermediate energy (HIC-IE) [22, 23]. The green line refers to pQCD calculations [24]. Right panel: $\langle v_n \rangle / \langle v_n \rangle^{max}$ as a function of time for $(20-30)\%$. Different symbols are for different harmonics n.

In the standard use of the transport theory one fixes the microscopical details of the scattering like cross sections of the processes to study the effect on the observables. In the following discussion we invert the description not in terms of fixed cross section σ_{tot} but in terms of fixed η/s. We simulate a system at fixed $\eta/s(T)$ by evaluating in each grid cell the total cross section σ_{tot} needed to have the wanted local viscosity. For a pQCD inspired cross section where $d\sigma/dt \sim \alpha_s^2/(t-m_D^2)^2$ with m_D being the screening mass regulating the angular dependence of the cross section, the η/s in the Chapmann-Enskog theory is given by the following expression:

$$\eta/s = \frac{1}{15} \langle p \rangle \, \tau_\eta = \frac{1}{15} \frac{\langle p \rangle}{g(a) \sigma_{tot} \rho}, \tag{1}$$

with $a = m_D/2T$ while $g(a)$ is a function that take into account for the pertinent relaxation time $\tau_\eta^{-1} = g(a)\sigma_{tot}\rho$ associated to the shear transport coefficient. The function $g(a)$ it is given by: $g(a) = \frac{1}{50} \int dy y^6 \left[(y^2 + \frac{1}{3}) K_3(2y) - y K_2(2y) \right] h\left(\frac{a^2}{y^2}\right)$ where K_n-s are the Bessel functions. The function h relate the transport cross section to the total cross section $\sigma_{tr}(s) = \sigma_{tot} \, h(m_D^2/s)$ with $h(\zeta) = 4\zeta(1+\zeta)[(2\zeta+1)ln(1+1/\zeta)-2]$.

In Fig.1 it shown shown the T dependence expected for η/s. The solid and dashed lines are the two parametrization considered in this work. As shown the η/s ratio is expected to increase at low temperature towards the estimated value for hadronic matter $4\pi\eta/s \approx 6$. Within this approach we have a self-consistent implementation of the kinetic freeze-out with a smooth switching-off of the scattering rates at low temperature, see [3, 20, 27].

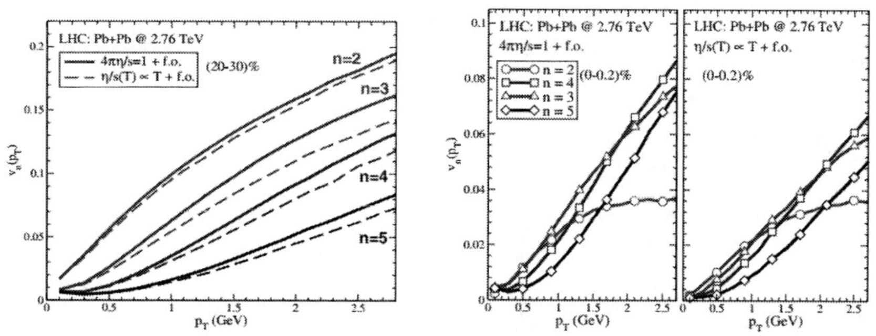

Figure 2: Left panel: $v_n(p_T)$ at mid rapidity for $(20-30)\%$. Solid lines correspond to $\eta/s = (4\pi)^{-1}$ in QGP phase while dashed lines to $\eta/s \propto T$ in QGP phase. Right panel: comparison between $v_n(p_T)$ at mid rapidity for $(0-0.2)\%$ and for different T dependence of η/s. Different symbols are for different harmonics n.

3 Anisotropic flows $v_n(p_T)$ and correlations

We will show the results for $Au+Au$ collisions at $\sqrt{s_{NN}} = 200\,GeV$ produced at RHIC and $Pb+Pb$ collisions at $\sqrt{s_{NN}} = 2.76\,TeV$ at LHC. The implementation of initial state fluctuations in a transport cascade approach is made by using a Monte-Carlo Glauber model. Within this approach we determine initial distribution of the participant nucleons. The discrete distribution for the nucleons is converted in a smooth one by assuming for each nucleon a gaussian distribution. The width it has been fixed to $\sigma_{xy} = 0.5\,fm$. The density in the transverse plane $\rho_T(x,y)$ is given by the following sum $\rho_T(x,y) \propto \sum_{i=1}^{N_{part}} \exp\big[-\frac{(x-x_i)^2+(y-y_i)^2}{2\sigma_{xy}^2}\big]$. In this way we generate an initial density profile that change event-by-event. For the initialization in momentum space at RHIC (LHC) energies we have considered a thermalized spectrum for partons with transverse momentum $p_T \leq p_0 = 2\,GeV$ $(3\,GeV)$. The initial local temperature in the transverse plane $T(x,y)$ is given by standard thermodynamical relation $\rho_T(x,y) = \gamma T^3/\pi^2$ with $\gamma = 40$. The central region of the fireball for mid peripheral collision can reach temperature $T \approx 300\,MeV$ at RHIC and $T \approx 400\,MeV$ at LHC. While for partons with $p_T > p_0$ we assume the spectrum of non-quenched minijets according to standard NLO-pQCD calculations [30]. The initial time of the simulation is $\tau_0 = 0.6\,fm/c$ for RHIC and $\tau_0 = 0.3\,fm/c$ for LHC. For a detailed discussion about the implementation of the initial condition see [20]. In the right panel of Fig.1 it is shown the time evolution of the v_n normalized to their maximum. As shown different v_n have a different formation time. In particular harmonics with larger n develop later [20]. This imply that differ-

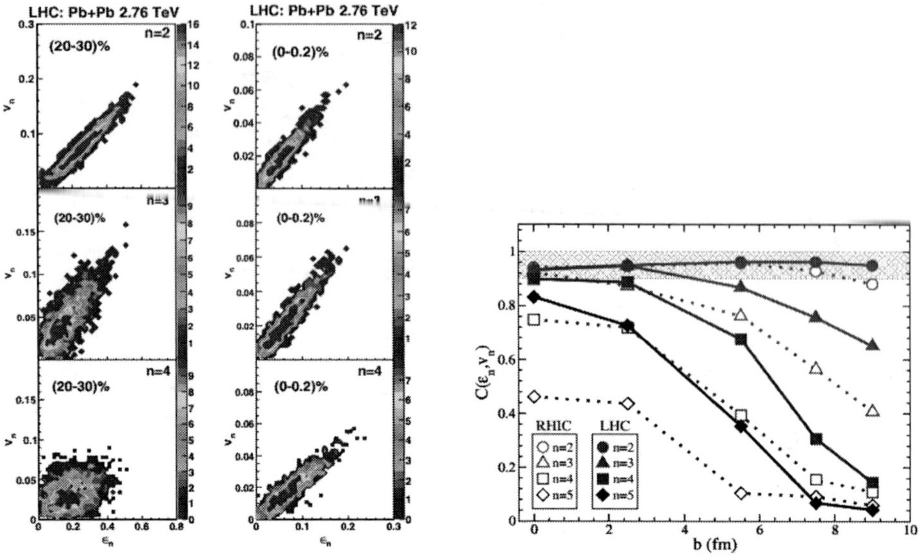

Figure 3: Left panel: Correlation plots between the initial eccentricities ϵ_n and final v_n at LHC energies and for two centralities $(20-30)\%$ and $(0-0.2)\%$. In these calculations we have considered $\eta/s = 1/(4\pi)$ for hight temperature and the kinetic f.o. at lower temperature (see red solid line in Fig.1). Right panel: $C(\epsilon_n, v_n)$ as a function of the impact parameter b. Different symbols refer to different harmonics n. The dotted (solid) lines are for RHIC (LHC).

ent v_n develop in different stages of the fireball evolution and therefore they can probe different values of the η/s of the QGP. This means that each v_n have a different sensitivity to the T dependence of $\eta/s(T)$. As shown in the left panel of Fig.2 for mid-peripheral collisions the effect of the increase of η/s at high temperature is to reduce the $v_n(p_T)$ and the reduction is about of $10\,\%$. In ultra-central collisions (see right panel of Fig.2) the sensitivity to $\eta/s(T)$ is strongly enhanced and we observe a reduction of $v_n(p_T)$ of about $30-35\%$ for $n > 3$. This sensitivity increase with the order of the harmonics n. Moreover the p_T dependence of $v_n(p_T)$ change with the centrality. In fact as shown in Fig.2 for mid-peripheral collisions we observe that larger is n smaller is the corresponding anisotropic flow. In ultra-central collisions the ordering occurs only at low $p_T < 1\,GeV$ where $v_n(p_T) \propto p_T^n$. At higher p_T the $v_n(p_T)$ are approximatively linear in p_T for $n \geq 3$ while we observe a saturation for $v_2(p_T)$ similar to that observed experimentally.

In the following discussion we show the results of correlations between the initial ϵ_n and the final v_n. The following results have been obtained with $N_{event} = 1000$ events for each centrality class with a total number of test particle per event of $N_{test} = 2 \cdot 10^6$. In the correlation plots shown in left panel of Fig.3 we compare the degree of correlation between ϵ_n and v_n for $(0 - 0.2)\%$ and $(20 - 30)\%$. The results are for LHC energies. In the upper panel we observe a strong linear correlation between ϵ_2 and v_2 for both centralities. In agreement with viscous hydrodynamics [19]. In the middle and lower panel of Fig.3 it has been shown the correlation between ϵ_3 and v_3 and between ϵ_4 and v_4 respectively. For both cases we observe a reduction of the degree of correlation with the centrality. The correlation between ϵ_3 and v_3 for all the centralities is weaker with respect to that obtained for ϵ_2 and v_2. Furthermore the v_4 shows a even a weaker degree of correlation with the initial ϵ_4 where for mid-peripheral collisions approximatively they are decorrelated. Similar behavior we observe at RHIC energies. To quantify the degree of correlation between ϵ_n and v_n we have studied the linear correlation coefficient $C(\epsilon_n, v_n)$ as a function of the impact parameter for both RHIC (dotted lines) and LHC energies (solid lines). Where $C(\epsilon_n, v_m)$ is given by:

$$C(\epsilon_n, v_m) = \frac{\sum_i (\epsilon_n^i - \langle \epsilon_n \rangle)(v_m^i - \langle v_m \rangle)}{\sqrt{\sum_i (\epsilon_n^i - \langle \epsilon_n \rangle)^2 \sum_i (v_m^i - \langle v_m \rangle)^2}} \tag{2}$$

the ϵ_n^i and v_m^i are the values of ϵ_n and v_m corresponding to the given event i. $C(\epsilon_n, v_n)$ is a decreasing function of the impact parameter for all the harmonics n, see Fig.3. We observe that for the second harmonic the correlation coefficient $C(\epsilon_2, v_2) \approx 1$ for all the range of impact parameter studied and for both RHIC and LHC. A lower correlation is observed for $n > 3$ where for example $C(\epsilon_5, v_5) < 0.5$ for all centralities. The comparison between dotted an solid lines show that at LHC there is more correlation then at RHIC. In particular at LHC and for ultra-central collisions $C(\epsilon_n, v_n) > 0.9$ for $n = 2, 3, 4$ as shown by the orange band. This behavior suggest that the value obtained for $\langle v_n \rangle$ are strongly related to the initial value of ϵ_n.

4 Conclusions

We have studied the effect of the η/s on the anisotropic flows $v_n(p_T)$ for $n = 2, 3, 4$ and 5 within an event-by-event transport approach. We found that at LHC energies and for mid-peripheral collisions there is a weak sensitivity of $v_n(p_T)$ to the T dependence in the QGP phase. An interesting result we observe for ultra-central collisions where found an enhancement of the sensitivity of the $v_n(p_T)$. This sensitivity increase with the order of the harmonics n and for $n = 5$ it reaches about a 30%.

Finally, the study of the correlations between the initial eccentricities ϵ_n, and the final anisotropic flows $\langle v_n \rangle$ shows that $C(\epsilon_n, v_n)$ is a decreasing function of the impact parameter. Moreover larger is the collision energy larger is the degree of correlation and at LHC there is significantly more correlation than at RHIC. In particular at LHC energies and for ultra-central collisions $(0-0.2\%)$ we found that $C(\epsilon_n, v_n) > 0.9$ up to the 4-th harmonic. This strong linear correlation imply that the $v_n \propto \epsilon_n$ and it suggests that the study of v_n can give information about the initial geometry of the fireball.

Acknowledgments

V.Greco, S. Plumari, F. Scardina and G.L. Guardo acknowledge the support of the ERC-StG Grant under the QGPDyn project.

References

[1] Romatschke P and Romatschke U 2007 *Phys.Rev.Lett.* **99** 172301.

[2] Song H and Heinz U W 2008 *Phys.Rev.* **C78** 024902.

[3] Ferini G, Colonna M, Di Toro M and Greco V 2009 *Phys.Lett.* **B670** 325.

[4] Xu Z and Greiner C 2009 *Phys.Rev.* **C79** 014904.

[5] Kovtun P, Son D and Starinets A 2005 *Phys.Rev.Lett.* **94** 111601.

[6] Qin G Y, Petersen H, Bass S A and Muller B 2010 *Phys.Rev.* **C82** 064903.

[7] Holopainen H, Niemi H and Eskola K J 2011 *Phys.Rev.* **C83** 034901.

[8] Schenke B, Jeon S and Gale C 2012 *Phys.Rev.* **C85** 024901.

[9] Gale C, Jeon S, Schenke B, Tribedy P and Venugopalan R 2013 *Phys.Rev.Lett.* **110** 012302.

[10] Csernai L P, Kapusta J and McLerran L D 2006 *Phys.Rev.Lett.* **97** 152303.

[11] Lacey R A, Ajitanand N, Alexander J, Chung P, Holzmann W *et al.* 2007 *Phys.Rev.Lett.* **98** 092301.

[12] Prakash M, Prakash M, Venugopalan R and Welke G 1993 *Phys.Rept.* **227** 321.

[13] Chen J W, Li Y H, Liu Y F and Nakano E 2007 *Phys.Rev.* **D76** 114011.

[14] Meyer H B 2007 *Phys.Rev.* **D76** 101701.

[15] Das S K and Alam J e 2011 *Phys.Rev.* **D83** 114011.

[16] Plumari S, Greco V and Csernai L 2013 arXiv:1304.6566.

[17] Gardim F G, Grassi F, Luzum M and Ollitrault J Y 2012 *Phys.Rev.* **C85** 024908.

[18] Chaudhuri A, Haque M R, Roy V and Mohanty B 2013 *Phys.Rev.* **C87** 034907.

[19] Niemi H, Denicol G, Holopainen H and Huovinen P 2013 *Phys.Rev.* **C87** 054901.

[20] Plumari S, Guardo G L, Scardina F and Greco V 2015 arXiv:1507.05540.

[21] Nakamura A and Sakai S 2005 *Phys.Rev.Lett.* **94** 072305.

[22] Schmidt W, Katscher U, Waldhauser B, Maruhn J, Stocker H *et al.* 1993 *Phys.Rev.* **C47** 2782.

[23] Danielewicz P, Barker B and Shi L 2009 *AIP Conf.Proc.* **1128** 104.

[24] Arnold P B, Moore G D and Yaffe L G 2003 *JHEP* **0305** 051.

[25] Plumari S, Puglisi A, Scardina F and Greco V 2012 *Phys.Rev.* **C86** 054902.

[26] Ruggieri M, Scardina F, Plumari S and Greco V 2013 *Phys.Lett.* **B727** 177.

[27] Ruggieri M, Scardina F, Plumari S and Greco V 2014 *Phys.Rev.* **C89** 054914.

[28] Plumari S, Guardo G L, Greco V and Ollitrault J Y 2015 *Nucl. Phys.* **A941** 87.

[29] Xu Z and Greiner C 2005 *Phys.Rev.* **C71** 064901.

[30] Greco V, Ko C and Levai P 2003 *Phys.Rev.* **C68** 034904.

EPJ Web of Conferences **117**, 03005 (2016) DOI: 10.1051/epjconf/201611703005
NN2015

Particle production in p–Pb collisions with ALICE at the LHC

ALBERICA TOIA, on behalf of the ALICE COLLABORATION

Goethe University, Frankfurt, Germany and GSI, Darmstadt, Germany

Abstract

Measurements of the transverse momentum spectra of light flavor particles at intermediate and high p_T are an important tool for QCD studies. In pp collisions they provide a baseline for perturbative QCD, while in Pb–Pb they are used to investigate the suppression caused by the surrounding medium. In p–Pb collisions, such measurements provide a reference to disentangle final from initial state effects and thus play an important role in the search for signatures of the formation of a deconfined hot medium. While the comparison of the p–Pb and Pb–Pb data indicates that initial state effects do not play a role in the suppression of hadron production observed at high p_T in heavy ion collisions, several measurements of particle production in the low and intermediate p_T region indicate the presence of collective effects.

1 Introduction

The p–Pb physics program has developed from crucial control-experiment to study cold nuclear effects and to establish a baseline for Pb–Pb to an area where to find groundbreaking discoveries but also new challenges.

Nuclear modification factors measured by ALICE in minimum bias (MB) p–Pb collisions for charged particles [1], heavy flavor and jets show no deviations from unity at high-p_T, demonstrating that the observed strong suppression in Pb–Pb collisions is due to final state effects. However several measurements [2–5] of particle production in the low and intermediate p_T region can not be explained by an incoherent superposition of pp collisions,

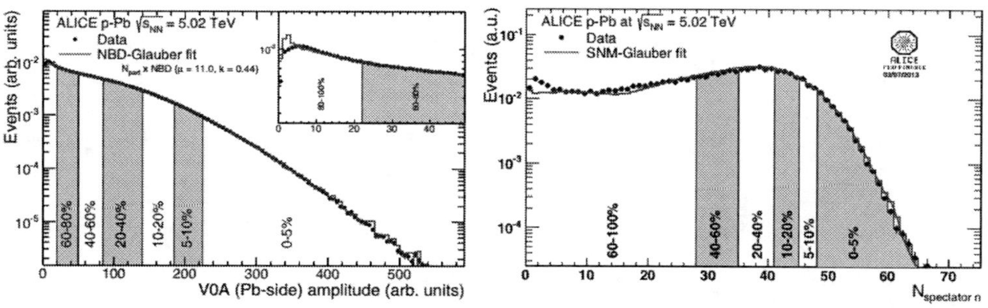

Figure 1: Distribution of the sum of amplitudes in the V0 hodoscopes (left) and of the neutron energy spectrum measured in the ZN calorimeter (right) in the Pb-remnant side (A). (Pb-going). Centrality classes are indicated by vertical lines. The distributions are fitted with a Glauber model coupled to a NBD (left) or with the SNM-model (right) and are shown as a line.

but rather call for coherent and collective effects. As their strength increases with multiplicity, a more detailed characterisation of the collision geometry is needed. Moreover, a knowledge of the geometry dependence is necessary to interpret the suppression pattern of charmonia (J/ψ ψ(2S)) and in Pb–Pb collisions relative to the effects observed in the nuclear medium produced in p–Pb collisions.

ALICE has carried out detailed studies of the centrality determination in p–Pb collisions, and the possible biases induced by the event selection on the scaling of hard processes in a selected event sample. The centrality determination consists in relating a Glauber model, which calculates the geometric properties of the event ($N_{\rm coll}$), to a measured observable related to the event activity in a specific rapidity region [6], via the conditional probability to observe a certain activity for a given $N_{\rm coll}$. More specifically, particle production measured by detectors at mid-rapidity can be modeled with a negative binomial distribution (NBD). The zero-degree energy is related to the number of the slow nucleons emitted in the nucleus fragmentation process, which we model with a Slow Nucleon Model (SNM) [7]. Fits of both models to our data are shown in Fig. 1.

However, the connection of the measurement to the collision geometry has to be validated, eg by correlating observables from kinematic regions causally disconnected after the collision, or by comparing a Glauber MC with data for a known process, as eg the deuteron dissociation probability at RHIC [8].

In addition the consistency of the approach must be demonstrated. As in general the selection in a system with large relative fluctuations can in-

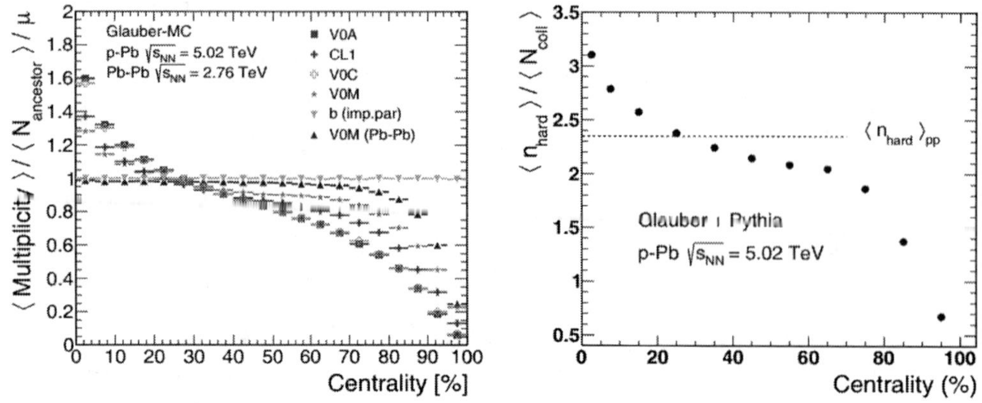

Figure 2: Left: Multiplicity fluctuation bias quantified as the mean multiplicity per $\langle N_{\mathrm{part}}\rangle/\mu$ from the NBD-Glauber MC in p–Pb and Pb–Pb calculations. Right: Number of hard scatterings (MSTI(31) in PYTHIA6) per N_{coll} as a function of the centrality calculated with a toy MC that couples a pp PYTHIA6 calculation to a p–Pb Glauber MC.

duce a bias, one needs to identify the physics origin of the bias in order to correct centrality dependent measurements. In p–Pb collisions, the relative large size of the multiplicity fluctuations has the consequence that a centrality selection based on multiplicity may select a biased sample of nucleon-nucleon collisions. In essence, by selecting high (low) multiplicity one chooses not only large (small) average N_{part} but also positive (negative) multiplicity fluctuations per nucleon-nucleon collision. This is shown on the left in Fig. 2. These fluctuations are partly related to qualitatively different types of collisions, described in all recent Monte Carlo generators by impact parameter dependence of the number of particle sources via multiparton interaction. Hence, the biases on the multiplicity discussed above correspond to a bias on the number of hard scatterings (n_{hard}). As a consequence, for peripheral (central) collisions we expect a lower (higher) than average number of hard scatterings per binary collision, corresponding to a nuclear modification factor less than one (greater than one), see Fig. 2, right. However also other types of biases have an influence on the nuclear modification factor: the jet-veto effect, due to the trivial correlation between the centrality estimator and the presence of a high-p_{T} particles originating from jets in the event; the geometric bias, resulting from the mean impact parameter between nucleons rising for the most peripheral events [9].

 DOI: 10.1051/epjconf/201611703005

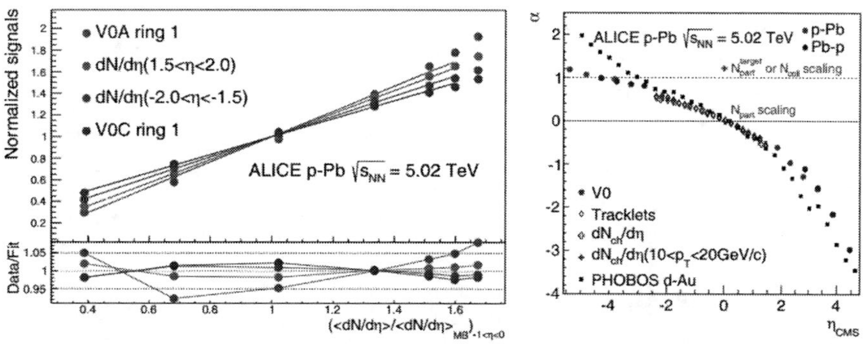

Figure 3: Left: Normalized signal from various observables versus the normalized charged-particle density, fit with a linear function of N_{part}. Right: Results from the fits as a function of the pseudorapidity covered by the various observables. The red horizontal lines indicate the ideal N_{part} and N_{coll} geometrical scalings. The most central point is excluded from the fit, to avoid pile-up effects. PHOBOS points taken from [10].

2 The ALICE approach

The ALICE approach aims to a centrality selection with minimal bias and, therefore, uses the signal in the Zero Degree Calorimeter (ZNA). In this case we cannot establish a direct connection using a well established model to the collision geometry but we can study the correlation of two or more observables that are causally disconnected after the collision, e.g because they are well separated in rapidity.

In centrality classes selected by ZNA, we study the dependence of various observables in different η and p_{T} regions on the charged particle density at mid-rapidity. Fig.3 indicate a monotonic change of the scaling with rapidity. The correlation between the ZDC energy and any variable in the central part shows unambiguously the connection of these observables to geometry.

Exploiting the findings from the correlation analysis described, we make use of observables that are expected to scale as a linear function of N_{coll} or N_{part}, to calculate N_{coll}:

- $N_{\text{coll}}^{\text{mult}}$: the multiplicity at mid-rapidity proportional to the N_{part};
- $N_{\text{coll}}^{\text{Pb-side}}$: the target-going multiplicity proportional to $N_{\text{part}}^{\text{target}}$;
- $N_{\text{coll}}^{\text{high-}p_{\text{T}}}$: the yield of high-$p_{\text{T}}$ particles at mid-rapidity is proportional to N_{coll}.

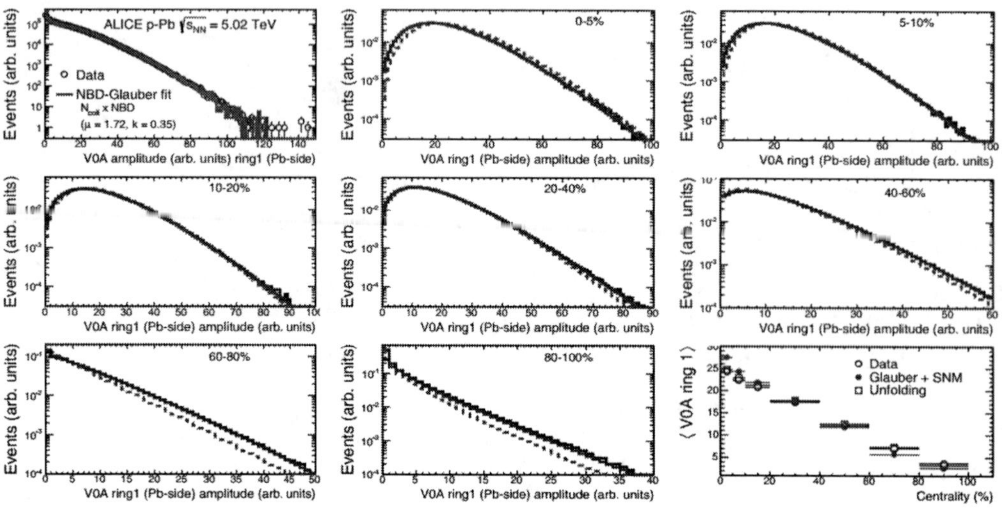

Figure 4: V0A ring1 signal distributions. The top left panel shows the distribution for MB events together with a NBD-Glauber fit. The remaining panels show the distributions and mean values for centrality classes selected with ZNA. These are compared to those obtained by the convolution of the $\mathcal{P}(N_{\text{coll}}|cent_{\text{ZNA}})$ distributions from the SNM with the NBD from the NBD-Glauber fit to V0A ring1. Data are also compared to the distributions obtained with an unfolding procedure, where the N_{coll} distributions have been fitted to the data using the parameters from the NBD-Glauber fit. The bottom right panel compares the mean values of these distributions as a function of the centrality.

These scalings are used as an ansatz, in the ALICE socalled *hybrid method* to calculate N_{coll}, rescaling the MB value $N_{\text{coll}}^{\text{MB}}$ by the ratio of the normalized signals to the MB one. We therefore obtain 3 sets of values of N_{coll}. The relative difference does not exceed 10%.

We have performed a consistency check, correlating ZNA and V0A centrality measurements, to establish their relation to the centrality. This is shown in Fig. 4. The N_{coll} distributions for centrality classes selected with ZNA, obtained from the SNM-Glauber fit, are convolved with the NBD from the fit to the V0A distribution. These are compared to the data and a good agreement is found. Moreover, a good agreement is also achieved with an unfolding procedure, where the N_{coll} distributions have been fitted to the data using the parameters from the NBD-Glauber fit. For each V0A distribution selected by ZNA, we find the N_{coll} distribution that, convolved with the NBD_{MB}, fits the data, i.e. the parameters of the fit are the relative contributions of each N_{coll} bin. The existence of N_{coll} distributions that folded with NBD agree with measured signal distributions is a necessary condition for ZNA to behave as an unbiased centrality selection. This

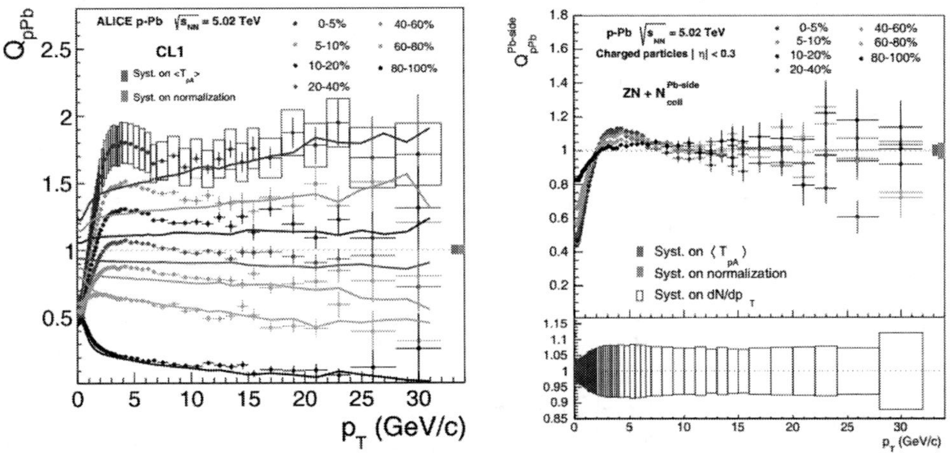

Figure 5: $Q_{\rm pPb}$ calculated with CL1 estimator (left), the lines are the G-PYTHIA calculations; with the hybrid method (right), spectra are measured in ZNA-classes and $N_{\rm coll}$ are obtained with the assumption that forward multiplicity is proportional to $N_{\rm part}^{\rm target}$.

procedure, which actually does not work for a biased centrality selection (eg CL1) shows that the energy measured by ZN is connected to the collision geometry.

3 Physics results

3.1 Nuclear modification factors

The nuclear modification factors $Q_{\rm pPb}$ calculated with multiplicity based estimator (shown in Fig.5 for CL1, where centrality is based on the tracklets measured in $|\eta| < 1.4$) widely spreads between centrality classes. They also exhibit a negative slope in $p_{\rm T}$, mostly in peripheral events, due to the jet veto bias, as jet contribution to particle production increases with $p_{\rm T}$. The $Q_{\rm pPb}$ compared to G-PYTHIA, a toy MC which couples Pythia to a p-Pb Glauber MC, show a good agreement, everywhere in 80-100%, and in general at high-$p_{\rm T}$, demonstrating that the proper scaling for high-$p_{\rm T}$ particle production is an incoherent superposition of pp collisions, provided that the bias introduced by the centrality selection is properly taken into account, as eg in G-PYTHIA.

With the hybrid method, using either the assumption on mid-rapidity multiplicity proportional to $N_{\rm part}$, or forward multiplicity proportional to $N_{\rm part}^{\rm target}$, the $Q_{\rm pPb}$ shown in Fig. 5, are consistent with each other, and also

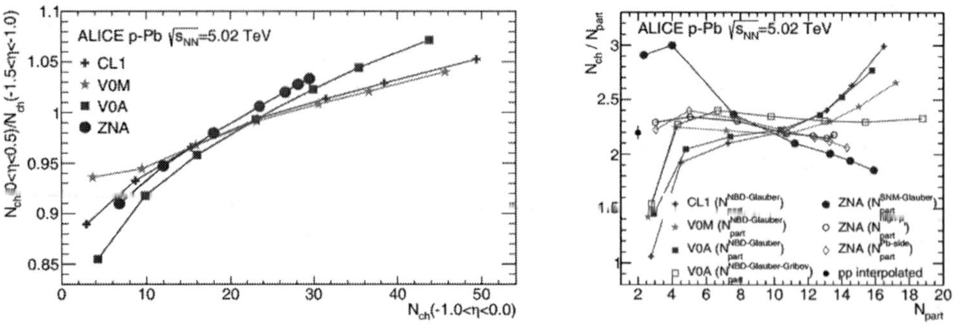

Figure 6: Left: Asymmetry of particle yield, as a function of the pseudorapidity density at mid-rapidity for various centrality classes and estimators. Right: Pseudorapidity density of charged particles at mid-rapidity per participant as a function of N_{part} for various centrality estimators.

consistent with unity for all centrality classes, as observed for MB collisions, indicating the absence of initial state effects. The observed Cronin enhancement is stronger in central collisions and nearly absent in peripheral collisions. The enhancement is also weaker at LHC energies compared to RHIC energies.

3.2 Charged particle density

Charged particle density is also studied as a function of η, for different centrality classes, with different estimators. In peripheral collisions the shape of the distribution is almost fully symmetric and resembles what is seen in proton-proton collisions, while in central collisions it becomes progressively more asymmetric, with an increasing excess of particles produced in the direction of the Pb beam. We have quantified the evolution plotting the asymmetry between the proton and lead peak regions, as a function of the yield around the midrapidity (see Fig. 6, left): the increase of the asymmetry is different for the different estimators. Fig. 6 right shows N_{ch} at mid-rapidity divided by N_{part} as a function of N_{part} for various centrality estimators. For Multiplicity-based estimators (CL1, V0M, V0A) the charged particle density at mid rapidity increases more than linearly, as a consequence of the strong multiplicity bias. This trend is absent when N_{part} is calculated with the Glauber-Gribov model, which shows a relatively constant behavior, with the exception of the most peripheral point. For ZNA, there is a clear sign of saturation above $N_{\text{part}} = 10$, due to the saturation of forward neutron emission. None of these curves points towards the pp data point. In contrast, the results obtained with the hybrid method, using either $N_{\text{part}}^{\text{target}}$-scaling at forward rapidity or N_{coll}-scaling for high-p_{T} particles give very similar trends, and show a nearly perfect scaling with N_{part}, which naturally reaches the pp point. This indicates the sensitivity of the N_{part}-scaling behavior to the Glauber modeling, and the importance of the fluctuations of the nucleon-nucleon collisions themselves.

4 Conclusions

Multiplicity Estimators induce a bias on the hardness of the pN collisions. When using them to calculate centrality-dependent Q_{pPb}, one must include the full dynamical bias. However, using the centrality from the ZNA estimator and our assumptions on particle scaling, an approximate independence of the multiplicity measured at mid-rapitity on the number of participating nucleons is observed. Furthermore, at high p_{T} the p–Pb spectra are found to be consistent with the pp spectra scaled by the number of binary nucleon–nucleon collisions for all centrality classes. Our findings put strong constraints on the description of particle production in high-energy nuclear collisions.

References

[1] B. Abelev et al. (ALICE Coll.), arXiv:1405.2737, 2014.

[2] B. Abelev et al. (ALICE Coll.), Phys.Lett. B719, 29-41, 2013.

[3] B. Abelev et al. (ALICE Coll.), Phys.Lett. B727, 371-380, 2013.

[4] B. Abelev et al. (ALICE Coll.), Phys.Lett. B728, 25-38, 2014.

[5] B. Abelev et al. (ALICE Coll.), Phys.Lett. B726, 164-177, 2013.

[6] B. Abelev et al. (ALICE Coll.), Phys.Rev. C88 (2013) 044909.

[7] J. Adam et al. (ALICE Coll.), Phys.Rev. C91 (2015) 064905

[8] A. Adare et al. (PHENIX Coll.), Phys.Rev. C90 (2014) 3, 034902

[9] J. Jia, Phys.Lett. B681, 320-325, 2009.

[10] B.B. Back et al. (PHOBOS Coll.), Phys.Rev. C72, 031901, 2005.

EPJ Web of Conferences **117**, 03006 (2016) DOI: 10.1051/epjconf/201611703006
NN2015

Light flavour hadron production in the ALICE experiment at LHC

Angela Badalà for the ALICE Collaboration

INFN, Sezione di Catania, Via S. Sofia 64, I95123, Catania, Italy

Abstract

Unique among the LHC experiments, ALICE has excellent particle identification capabilities for the measurement of light-flavour hadrons. A large number of hadron species from pions to multi-strange baryons and light nuclei have been measured over a large transverse momentum region. The measurement of the production of these particles is a valuable tool to study the properties of the medium formed in heavy-ion collisions. In particular they give information on the collective phenomena of the fireball, on the parton energy loss in the hot QCD medium and on the hadronization mechanisms such as recombination and statistical hadronization. The measurements in pp and in p-nucleus collisions provide the necessary baseline for heavy-ion data and help to investigate the effects of the ordinary nuclear matter. In this paper some of the main ALICE results on identified light-flavour hadron production in Pb–Pb collisions at $\sqrt{s_{\mathrm{NN}}} = 2.76$ TeV and p–Pb collisions at $\sqrt{s_{\mathrm{NN}}} = 5.02$ TeV will be presented.

1 Introduction

Heavy-ion collisions at ultra-relativistic energies allow one to study the physics of strongly interacting matter and to characterize the Quark-Gluon Plasma (QGP), a state of deconfined quarks and gluons. The hot and dense matter created in such collisions rapidly expands collectively and cools down, until the (pseudo)-critical temperature is reached and the phase transition (or a crossover at very low μ_B) to a hadron gas takes place. Hadron yields measured in central heavy-ion collisions from AGS up to LHC energies can be described very well within a hadro-chemical equilibrium model [1, 2]. Then in high-energy interactions hadrons are expected to be produced in approximate chemical and thermal equilibrium. Chemical freeze-out takes place when inelastic collisions stop, while kinetic freeze-out happens after the chemical freeze-out, once also elastic collisions stop.

The production of particles formed only by light quarks (u, d, s) is an important tool not only for the understanding of particle production mechanisms in high energy collisions but also to study the collective phenomena characterizing the dynamical evolution of the fireball. In particular, the study of particles with strangeness helps in understanding of the hadrochemistry of the matter created in the collisions since no net strangeness is present in the colliding nuclei. Moreover, high energy heavy-ion collisions offer the opportunity to study light (anti-)nuclei (such as d, ^{3}He, ^{4}He) and hypernuclei (such as $^3_\Lambda$H) and to test their production mechanisms. Furthermore, production of hadronic resonances (such as $K^*(892)^0$, $\Lambda(1520)$) can help in understanding the late hadronic phase.

In the next section, following a brief description of the ALICE detector and the analysis techniques used to identify particles, selected results on light-flavour hadrons and nuclei in Pb–Pb collisions at $\sqrt{s_{\rm NN}} = 2.76$ TeV and p–Pb collisions at $\sqrt{s_{\rm NN}} = 5.02$ TeV are presented and discussed.

2 Particle identification in ALICE and selected results

A detailed review of the ALICE detector and its particle identification capabilities can be found in [3,4]. Charged hadrons in the low to intermediate $p_{\rm T}$ region (below 5 GeV/c) are identified using the energy loss $\mathrm{d}E/\mathrm{d}x$ from the Inner Tracking System (ITS) and Time Projection Chamber (TPC) detectors, the time-of-flight measurement from the Time Of Flight (TOF) detector and the Cherenkov light from the High Momentum Particle Identification Detector (HMPID). At higher $p_{\rm T}$ particles are identified via the $\mathrm{d}E/\mathrm{d}x$ in the relativistic rise range of the Bethe-Bloch curve in the TPC. Also nuclei and anti-nuclei are identified over a wide range of momentum using the combined information of the specific energy loss in the TPC and the velocity measured by the TOF. More information can be found in [5], which reports the recent CPT invariance test in the systems bound by nuclear forces obtained by ALICE measuring the difference between the ratios of the mass and the charge of d and $\overline{\rm d}$, ^{3}He and $^3\overline{\rm He}$ nuclei. Strange hadrons are measured through the reconstruction of their weak decay into charged particles. K^0_S, Λ, $\overline{\Lambda}$, ${}^3_\Lambda$H, ${}^3_{\overline{\Lambda}}\overline{\text{H}}$ candidates are found by combining two charged tracks in a V-shaped decay pattern while for multi-strange baryons (charged Ξ and Ω) a selected Λ candidate is further combined with a third track. Candidates are required to satisfy topological and kinematical restrictions. In addition, each of the daughter tracks is checked for compatibility with the pion, kaon, proton or ^{3}He hypotheses using their energy loss in the TPC. A description of the identification of hadronic resonances (as K* and ϕ) can be found in [6].

The transverse momentum ($p_{\rm T}$) spectra of identified hadrons $\pi^\pm$, $K^\pm$ and p($\overline{\rm p}$) in Pb–Pb collisions show a clear evolution with the centrality, with an almost exponential behaviour at high $p_{\rm T}$. The spectrum hardens at high $p_{\rm T}$ and flattens at low $p_{\rm T}$ as a function of increasing centrality. The $\langle\beta_T\rangle$ parameter extracted from fits to the blast-wave parameterization indicates a radial flow of about 10% higher than at RHIC at $\sqrt{s_{\rm NN}} = 200$ GeV in central collisions [7]. The $p_{\rm T}$ spectra are succesfully described [7] by hydrodynamic models (VISH2+1 [8], HKM [9], Kraków [10], EPOS [11]). Comparison with prediction from hydrodynamic calculations, based on the same models used for π, K and p have also been carried out for the heavier multi-strange baryons [12]. For the most central events (0-10%), all the available models describe quite well the shape of the Ξ spectrum in the $p_{\rm T}$ range up to 3 GeV/c, although only the Kraków model correctly reproduces the yield. The description is less successful with the Ω. Considering that for the hydrodynamic description the shape of the transverse momentum distribution is mainly determined by particle mass, it is interesting to note that for $p_{\rm T} < 3$ GeV/c the ϕ-mesons and protons $p_{\rm T}$ spectra have a similar shape [6].

The $p_{\rm T}$ spectra of $\pi^\pm$, $K^\pm$, p($\overline{\rm p}$) and $\Lambda(\overline{\Lambda})$ have been measured in p–Pb collisions in different VZERO multiplicity classes [13]. For $p_{\rm T} < 2$ GeV/c the spectra become harder as the multiplicity increases and the effect is stronger for heavier particles. This pattern is similar to the one observed in Pb–Pb collisions, where it is attributed to the hydrodynamical evolution of the medium and indeed also $p_{\rm T}$ spectra measured in high-multiplicity p–Pb collisions are better described by hydrodynamical models (EPOS [11], Kraków [10]) than QCD inspired models (DPMJET [14]).

Ultrarelativistic heavy-ion collisions offer also the opportunity to study light anti-(hyper) nuclei. Even if the measurement is challenging as the production probability decreases with increasing mass, the data collected with the ALICE detector during the first LHC run phase allow for the identification and measurement of d, $\overline{\rm d}$, ^{3}He, $^3\overline{\rm He}$ [15], $^4\overline{\rm He}$ and (anti-)hypertriton ${}^3_\Lambda$H [16] in Pb–Pb collisions at $\sqrt{s_{\rm NN}}$ =2.76 TeV. As expected in a hydrodynamic description of the fireball as a radially expanding source, a hardening of the spectra with increasing centrality is observed [15]. The decrease with the increasing mass of the nuclei of the $p_{\rm T}$-integrated yield of p, d, ^{3}He and $^{4}\overline{\rm He}$ in 0-20% centrality nicely follows the exponential fall predicted by the thermal model, with a penalty factor of about 300 for each added baryon. The (anti-)hypertriton ($^{3}_{\overline{\Lambda}}\overline{\rm H}$) $^{3}_{\Lambda}$H is the lightest observed hypernucleus and is formed by a (anti-)proton, a (anti-)neutron and a (anti-)Λ. The statistics reached for the hypertriton (particle + anti-particle) has permitted to estimate the yield in four ct bins. From a fit of these measurements a lifetime of 181^{+54}_{-39}(stat.) $\pm$ 33 (syst.) ps for the $^{3}_{\Lambda}$H was determined which is compatible within 1σ to the world average value [16].

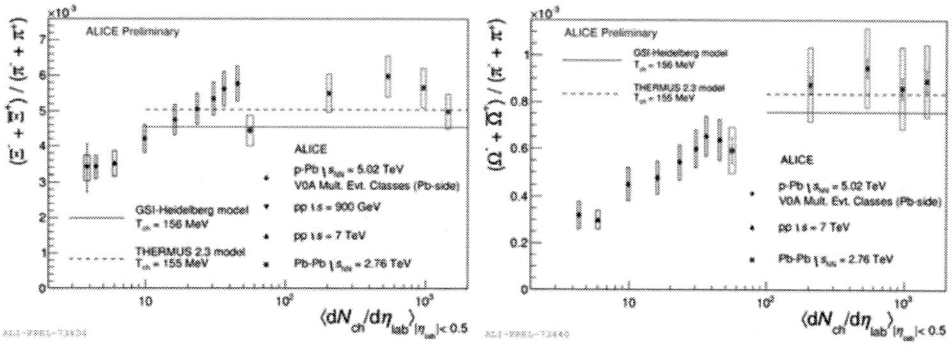

Figure 1: Multi-strange hyperon-to-pion ratios as a function of multiplicity (centrality) for pp (minimum-bias), p–Pb and Pb–Pb collisions shown together with predictions from statistical hadronizations models.

Since the production rate of charged particles does not scale linearly with $\langle N_{\rm part}\rangle$ [17] a way to factor out the enhancement due to strangeness content from the overall increase in charged-particle production is to compute the ratio of strange particle to pions. A clear increase of strangeness production is observed from minimum-bias pp to central Pb–Pb collisions [12]. In Fig. 1 the $\frac{\Xi^{-}+\overline{\Xi}^{+}}{\pi^{-}+\pi^{+}}$ and $\frac{\Omega^{-}+\overline{\Omega}^{+}}{\pi^{-}+\pi^{+}}$ ratios for the three collisions systems are reported as a function of the averaged charged multiplicity. The multi-strange baryon production relative to pions increases by up to a factor of about 3 when going from minimum-bias pp to central Pb–Pb. It saturates in the most central Pb–Pb collisions at values predicted by statistical hadronization models. In p–Pb collisions these ratios increase with multiplicity, with Ξ/π reaching the saturation level observed in central Pb–Pb collisions and Ω/π being compatible with the value for peripheral Pb–Pb collisions. A similar trend is observed for the d/p ratio (not shown). It increases with multiplicity in p–Pb collisions bridging the pp and Pb–Pb values. However, within errors, no significant centrality dependence is observed in Pb–Pb collisions.

In high-energy interactions hadrons are expected to be produced in approximate chemical and thermal equilibrium. A grand-canonical model is usually employed to describe the observed yields of each particle species in nucleus-nucleus collisions. The mid-rapidity p_T-integrated yield of light flavour hadrons (including nuclei) measured by ALICE in central Pb–Pb collisions at $\sqrt{s_{NN}} = 2.76$ TeV have been fitted to three statistical hadronization models (THERMUS 2.3 [18], GSI-Hidelberg [19] and SHARE [20]) with the baryochemical potential set to 1 MeV. The yields are described over 7 orders of magnitude with a common $T_{chemical} = 156 \pm 2$ MeV, while fits to RHIC data resukted in a slightlt higher temperature ($\sim$ 164 MeV). However the models largely overestimate the proton and anti-proton yields. Some of the suggested interpretations of this 'anomaly' are: baryon-antibaryon annhihilation after chemical freeze-out, non-equilibrium statistical hadronization and effects due to pre-hadronic flavour-dependent bound states above the QCD transition temperature (see ref. [21] for details). Probably due to the importance of rescattering effects in the late hadronic phase the $K^*(892)^0$ resonance yield, excluded from the fit, is also overestimated.

3 Summary and conclusions

The ALICE experiment has performed a comprehensive study of the light flavour hadrons and light nuclei, in a wide range of p_T for events collected in pp, p–Pb and Pb–Pb collisions during the first LHC run phase. Bulk particle production in p–Pb shows a clear evolution with multiplicity similar to the pattern observed in Pb–Pb collisions, where it is considered a signature of collectivity. However we should be aware that also non-collective effects can mimic flow-like patterns [13]. Particle production changes with increasing the colliding system size. In particular, the relative suppression of baryon and $K^*(892)^0$ and the relative enhancement of strangeness and deuteron production are observed. In central Pb–Pb collisions p_T-integrated yields of light flavour particles fit the predictions from statistical hadronization with a chemical freeze-out temperature around 156 MeV. Production of light (hyper) nuclei is found to be in agreement with thermal model expectation, despite their low binding energies. Their inclusion in chemical equilibrium fits, which might seem questionable, can be understood if one consideres that the hadronic interactions do not change the entropy per baryon and the statistical hadronization [1] and coalescence [22] approach yield compatible results [15].

References

[1] A. Andronic et al., *Nucl. Phys.* A **772** (2006) 167; J. Stachel et al, *J. Phys. Conf. Ser.* **509** (2014) 012019, arXiv 1311.4662.

[2] F. Becattini et al., *Phys. Rev.* C **90** (2014) 054907.

[3] K. Aamodt et al. (ALICE Coll.), *J. Instrum.* **3** (2008) S08002.

[4] B. Abelev et al. (ALICE Coll.), *Int. J. Mod. Phys.* A **29** (2014) 1430044.

[5] B. Abelev et al. (ALICE Coll.), arXiv:1506.08951; B. Abelev et al. (ALICE Coll.), *Nat. Phys. Lett.* **NPHYS34432** (2015).

[6] B. Abelev et al. (ALICE. Coll.), *Phys. Rev.* C **91** (2015) 024609.

[7] B. Abelev et al. (ALICE. Coll.), *Phys. Rev.* C **88** (2013) 044910.

[8] C. Shen et al., *Phys. Rev.* C **84** (2011) 044903.

[9] Y. Karpenko and Y. Senyukov, *J. Phys.* G **38** (2011) 124059.

[10] P. Bozek, *Phys. Rev.* C **85** (2012) 034901.

[11] K. Werner, *Phys. Rev.* C **85** (2012) 064907.

[12] B. Abelev et al. (ALICE. Coll.), *Phys. Lett.* B **728** (2014) 216.

[13] B. Abelev et al. (ALICE. Coll.), *Phys. Lett.* B **728** (2014) 25.

[14] S. Roesler et al., arXiv:hep-ph 0012252.

[15] J. Adam et al. (ALICE Coll.), arXiv: 1506.08951.

[16] J. Adam et al. (ALICE Coll.), arXiv: 1506.08453.

[17] K. Aamodt et al. (ALICE Coll.), *Phys. Rev. Lett.* **106** (2011) 032301.

[18] S. Wheaton et al., *Comp. Phys. Commun.* **180** (2009) 8.

[19] A. Andronic et al., *Phys. Lett.* B **673** (2009)14.

[20] M. Petran et al., *Comp. Phys. Comm.* **185** (2014) 2056.

[21] M. Floris, *Nucl. Phys.* A **931** (2014) 103.

[22] P. Siemens and J.I. Kapusta, *Phys. Rev. Lett.* **43** (1979) 1486.

Light fragments from (C + Be) interactions at 0.6 GeV/nucleon

B.M. Abramov[1,2], P.N. Alekseev[1], Yu.A. Borodin[1], S.A. Bulychjov[1], I.A. Dukhovskoy[1], A.I. Khanov[1], A.P. Krutenkova[1], V.V. Kulikov[1], M.A. Martemianov[1], S.G. Mashnik[3], M.A. Matsyuk[1], E.N. Turdakina[1] and P.I. Zarubin[4]

[1]Institute for Theoretical and Experimental Physics (ITEP) SRC "Kurchatov institute", Moscow, 117218, Russia,
[2]Moscow Institute of Physics and Technology (MIPT), Dolgoprudnyy, 117303, Russia,
[3]Los Alamos National Laboratory (LANL), Los Alamos, NM, 87545, USA
[4]Joint Institute for Nuclear Research, Dubna, 141980, Russia

Abstract

Nuclear fragments emitted at 3.5° in ^{12}C fragmentation at 0.6 GeV/nucleon have been measured. The spectra obtained are used for testing the predictions of four ion-ion interaction models: INCL++, BC, LAQGSM03.03 and QMD as well as for the comparison with the analytical parametrization in the framework of thermodynamical picture of fragmentation.

1 Introduction

The study of emission of light fragments is important to understand the nature of ion-ion interactions. Different reaction mechanisms contribute to this rather complicated process which can hardly be described in analytical way. For this reason we tested a few Monte-Carlo transport codes against the data of the FRAGM experiment [1–3].

2 The FRAGM experiment and the test of the models of ion-ion interactions

In the FRAGM experiment at ITEP TWA heavy ion accelerator, we have measured the fragment yields from the reaction

$$^{12}\mathrm{C} + ^{9}\mathrm{Be} \rightarrow \mathrm{f} + X \qquad (1)$$

with a beamline spectrometer set at 3.5° to carbon beam. Here f stands for all fragments from protons up to isotopes of projectile nucleus. The projectile kinetic energies were T_0= 0.2–3.2 GeV/nucleon. In this report we present data at T_0= 0.6 GeV/nucleon for fragments: hydrogen, helium and two lithium (^{6}Li, ^{7}Li) isotopes. The fragments were measured at a wide momentum region which include the midrapidity, the fragmentation peak and the cumulative regions. In the last one the fragment momenta per nucleon are much higher than momentum per nucleon of the projectile. This gives a good testing ground for a comparison with predictions of different ion-ion interaction models.
The fragment yields were measured by scanning the beamline spectrometer momentum with a step of 50–100 MeV/c and counting the number of events corresponding to different fragments and normalizing to the monitor. The fragments were well separated on time-of-flight *vs* dE/dx plots. The relative cross sections $d^2\sigma/(d\Omega dp)$, where p is the fragment momentum in a laboratory frame, were calculated. They are shown for hydrogen, helium and lithium isotopes in comparison with the calculations by four models: INCL++ (Fig. 1(a)), BC (Fig. 1(b)), LAQGSM (Fig. 1(c)) and QMD (Fig. 1(d)). We used the INCL++, BC and QMD models from a GEANT4-package.

Our measurements cover three-to-six orders in the cross section magnitude, depending on the fragment. Qualitatively, all the models reproduce well the energy dependence of the differential cross sections of the fragment yields. Model prediction for the cross section at fragmentation peak maxima differ by no more than 2–3 times[1]. The largest differences are observed for the QMD model, which predicts smaller width for the fragmentation peaks. LAQGSM reproduces the energy dependence of the cross sections at high energy part of the fragmentation peaks, but underestimates the cross section in the low-energy part. The predictions of BC and INCL++ models are very

[1]The data of the FRAGM experiment were normalized to BC calculations at the maximum of the proton peak. This normalization factor was used in all figures.

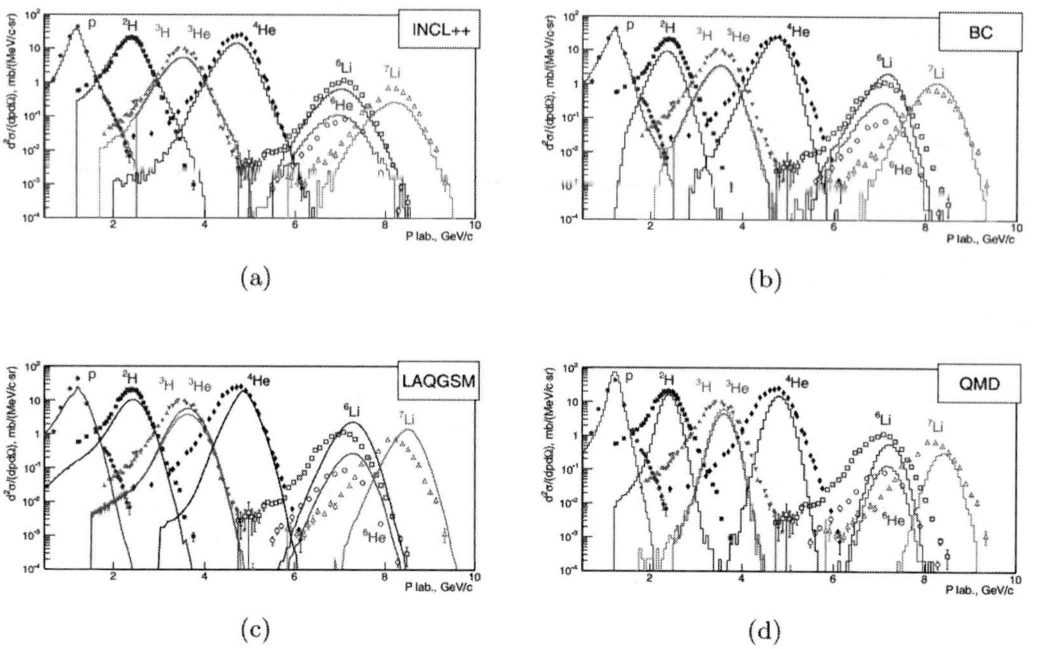

Figure 1: Relative yields of H, He and Li isotopes in ^{12}C + ^{9}Be interaction at 0.6 GeV/nucleon and at 3.5° as a functions of fragment laboratory momenta. Data (points), model calculations (histograms/lines): (a) INCL++ model [7], (b) BC model [4], (c) LAQGSM model [5], (d) QMD model [6].

close, but INCL++ gives better description of the experimental data, which is especially noticeable in the areas far from the fragmentation peak maxima.

3 Slope parameters from kinetic energy spectra

In the framework of thermodynamical picture of nuclear fragmentation, the fragment kinetic energy (T) distribution in the rest frame of the carbon nucleus should be of Maxwell-Boltzmann type and not depend on the fragment type. The distributions of invariant cross section for fragment yields $Ed^3\sigma/d^3p = (E/p^2)d^2\sigma/(d\Omega dp)$, where E is a total energy, are shown in Fig. 2 as function of T. Both experimental data and model calculations are presented. The INCL++ model gives a good description of the experimental data much better than the others. The spectra were parameterized by a sum of two Maxwell-Boltzmann distributions

$$Ed^3\sigma/d^3p = E(A_S exp(-T/T_S) + A_C exp(-T/T_C)), \quad (2)$$

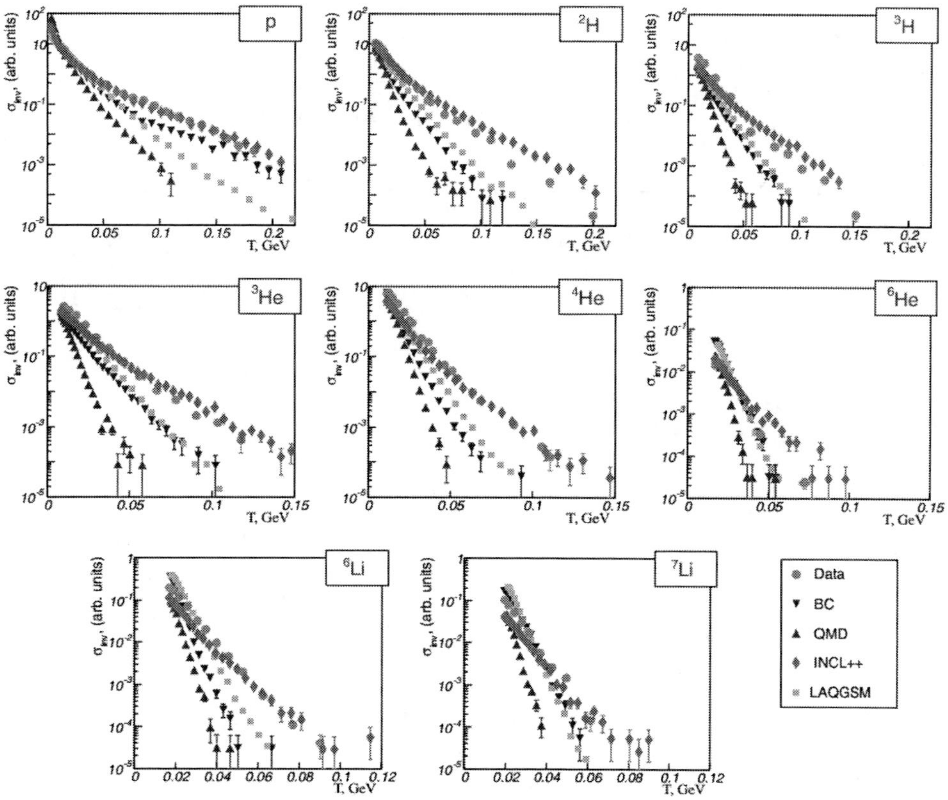

Figure 2: Invariant cross sections as functions of fragment kinetic energies in the ^{12}C rest frame: measured data *vs* model calculations.

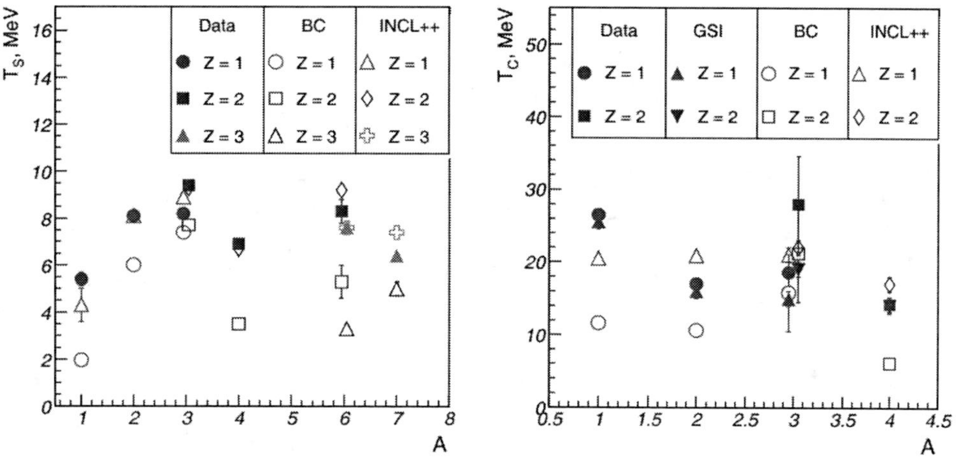

Figure 3: Temperature parameters T_S (left) and T_C (right) as a function of fragment atomic number. Closed points are ours, and GSI results are from [9], open points are calculations in INCL++ and BC models.

where A_S and A_C are normalization factors for low and high energy regions, and the slope parameters T_S and T_C are "temperatures" defined in these regions. The function (2) gives a good description of both the data and the calculations for all models and all fragments. The obtained values of T_S and T_C are shown in Fig. 3. The T_S values are in a reasonable agreement with those obtained in [8] for 1–2 GeV/nucleon carbon ions. The experimental results for T_C from [9] obtained at GSI at 1 GeV/nucleon for ^{197}Au + ^{197}Au collisions are in a reasonable agreement with our results. The INCL++ model describes the data on T_S well and better than the other models.

4 Conclusion

Fragment yields from the reaction ^{9}Be(^{12}C, f)X (f – fragments from p to ^{7}Li) at 0.6 GeV/nucleon were measured and compared to prediction of four models of ion-ion interactions. The INCL++ describes all momentum spectra rather well, both in the region of fragmentation peak and in the cumulative region while all other models underestimate the experimental results in the cumulative (high momentum) region. Kinetic energy spectra in the projectile rest frame can be parameterized as $E(A_S \exp(-T/T_S) + A_C \exp(-T/T_C))$, where both T_S and T_C values are in satisfactory agreement with the predictions of the INCL++ model. Other models strongly underestimates the data at high kinetic energies; T_C values are higher for protons than for other fragments.

Authors would like to thank I.I. Tsukerman for help. We are also indebted to the personnel of TWAC-ITEP and technical staff of the FRAGM experiment. The work has been supported in part by the RFBR (grant No. 15-02-06308). Part of the work performed at LANL by S.G.M. was carried out under the auspices of the National Nuclear Security Administration of the U.S. Department of Energy at Los Alamos National Laboratory under Contract No. DE-AC52-06NA25396.

References

[1] Abramov B.M. et al., JETP Letters, **97** (2013) 439. For English version see also arXiv:1304.6220v2.

[2] Abramov B.M. et al., EPJ Web of Conferences, **95** (2015) 04035.

[3] Abramov B.M. et al., Physics of Atomic Nuclei, **78** (2015) 373.

[4] Folger G. et al, Europ. Phys. J., **A21** (2004) 407.

[5] Mashnik S.G. et al., LANL Report LA-UR-08-2931, 2008; arXiv:0805.0751 and LANL Report LA-UR-07-6198, 2007; arXiv:0709.1736.

[6] Koi T. et al, AIP Conf. Proc., **896** (2007) 21.

[7] Dudouet J. et al., Phys. Rev., **C89** (2014) 054616; Dudouet J. et al., Phys. Rev., **C88** (2013) 024606.

[8] Greiner E.D. et al., Phys. Rev. Lett., **35** (1975) 152.

[9] Odeh T. et al., Phys. Rev. Lett., **84** (2000) 4557.

EPJ Web of Conferences **117**, 03008 (2016) DOI: 10.1051/epjconf/201611703008
NN2015

MCNP6 simulation of light and medium nuclei fragmentation at intermediate energies

Stepan G. Mashnik[1] and Leslie M. Kerby[1,2]

[1]Los Alamos National Laboratory, Los Alamos, NM 87545, USA
[2]University of Idhao, Moscow, ID 83844-4264, USA

Abstract

Fragmentation reactions induced on light and medium nuclei by protons and light nuclei of energies around 1 GeV/nucleon and below are studied with the Los Alamos transport code MCNP6 and with its CEM03.03 and LAQGSM03.03 event generators. CEM and LAQGSM assume that intermediate-energy fragmentation reactions on light nuclei occur generally in two stages. The first stage is the intranuclear cascade (INC), followed by the second, Fermi breakup disintegration of light excited residual nuclei produced after the INC. CEM and LAQGSM account also for coalescence of light fragments (complex particles) up to ^{4}He from energetic nucleons emitted during INC. We investigate the validity and performance of MCNP6, CEM, and LAQGSM in simulating fragmentation reactions at intermediate energies and discuss possible ways of further improving these codes.

1 Introduction and theoretical background

Fragmentation reactions induced by protons and light nuclei of energies around 1 GeV/nucleon and below on light target nuclei are involved in different applications, like cosmic-ray-induced single event upsets (SEU's), radiation protection, and cancer therapy with proton and ion beams, among others. It is impossible to measure all nuclear data needed for such applications; therefore, Monte Carlo transport codes are usually used to simulate impacts associated with fragmentation reactions. It is important that available transport codes simulate such reactions as well as possible.

The Los Alamos Monte Carlo transport code MCNP6 [1] uses by default the latest version of the cascade-exciton model (CEM) as incorporated in its event generator CEM03.03 to simulate fragmentation of light nuclei at intermediate energies for reactions induced by nucleons, pions, and photons, and the Los Alamos version of the quark-gluon string model (LAQGSM) as implemented in the code LAQGSM03.03 (see [2] and references therein) to simulate fragmentation reactions induced by nuclei and by particles at energies above ~ 3.5 GeV, up to about 1 TeV/nucleon.

Generally, both CEM and LAQGSM assume that nuclear reactions occur in three stages. The first stage is the IntraNuclear Cascade (INC), completely different in CEM and LAQGSM, in which primary particles can be re-scattered and produce secondary particles several times prior to absorption by, or escape from the nucleus. When the cascade stage of a reaction is complete, CEM and LAQGSM use the coalescence model to "create" high-energy d, t, ^{3}He, and ^{4}He via final-state interactions among emitted cascade nucleons, already outside of the target. The subsequent relaxation of the nuclear excitation is treated in terms of an improved version of the modified exciton model of preequilibrium decay followed by the equilibrium evaporation/fission stage of the reaction. But if the residual nuclei after the INC have atomic numbers with $A < 13$, CEM and LAQGSM use the Fermi breakup model to calculate their further disintegration instead of using the preequilibrium and evaporation/fission models. Thus, for targets with $A < 13$, reactions are assumed to occur only in two stages.

The "standard" version of CEM and LAQGSM account for possible multiple emission of only n, p, d, t, ^{3}He, and ^{4}He during the preequilibrium stage of reactions (see Ref. [2]). Their latest, "F", version (see Refs. [3–5]) considers a possibility of preequilibrium emission of light fragments (LF) heavier than ^{4}He, up to ^{28}Mg. It also simulates coalescence of LF heavier than ^{4}He, up to $A = 7$, in CEM03.03F (see [3–5]), and up to $A = 12$, in LAQGSM03.03F (see [6]).

In recent years, MCNP6, with its CEM and LAQGSM event generators, has been extensively validated and verified (V&V) against a large variety of nuclear reactions on both thin and thick targets (see, e.g. Refs. [3] - [8] and references therein). In Ref. [3], it was tested specifically on fragmentation of light nuclei at intermediate energies. Here, we present a few results from our recent work [3] and investigate further the performance of MCNP6, CEM, and LAQGSM in simulating fragmentation reactions at intermediate energies and discuss possible ways of further improving these codes.

2 Results and conclusion

Figs. 1, 2, 3, and 5 show examples of fragmentation reactions on light nuclei simulated by our codes. Figs. 6 and 4 shows examples for medium targets, ^{48}Ca and ^{nat}Ag. Many more similar results, their discussion, and useful details can be found in Refs. [3] - [9] and references therein.

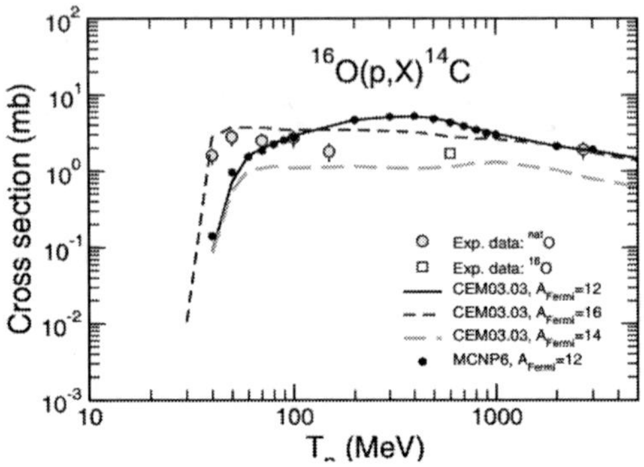

Figure 1: Excitation function for the production of ^{14}C from p + ^{16}O calculated with CEM03.03 using the "standard" version of the Fermi breakup model (A_{Fermi} = 12) and with cut-off values for A_{Fermi} of 16 and 14, as well as with MCNP6 using CEM03.03 (A_{Fermi} = 12) compared with experimental data, as indicated. Experimental data are from the T16 Lib compilation [10] (see details in [3]).

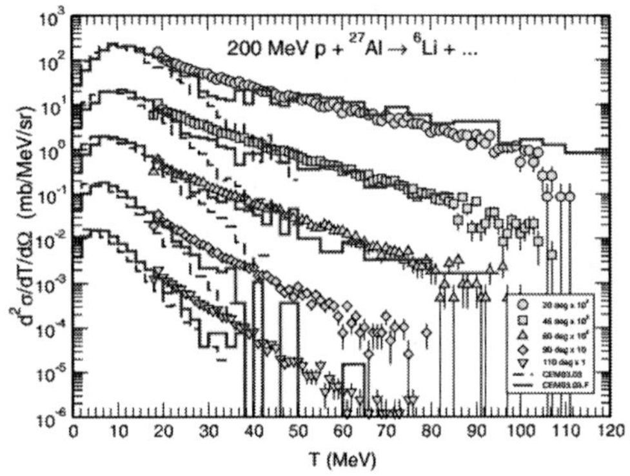

Figure 2: Comparison of experimental ^{6}Li spectra at 20, 45, 60, 90, and 110 degrees by Machner et al. [11] (symbols) with calculations by the unmodified CEM03.03 (dashed histograms) and results by CEM03.03.F (solid histograms), as indicated (see more details in [3]).

Our results indicate that MCNP6 using CEM03.03 and LAQGSM03.03 simulates fragmentation reactions on light and

medium-light nuclei at intermediate energies well, in a satisfactory agreement with experimental data. The recent "F" version of codes (see Refs. [3] – [6]) is even better, as it allows us to describe emission of energetic LF from practically arbitrary reactions.

However, MCNP6 is not yet ready to predict well heavy fragments from reactions with heavier nuclei, with mass numbers $A \sim 100$. Such nuclear targets are considered too light to fission in conventional codes. Similarly, the fragments are too light to be produced as spallation residues and too heavy to be produced via standard evaporation and/or preequilibrium models, or via coalescence.

DOI: 10.1051/epjconf/201611703008

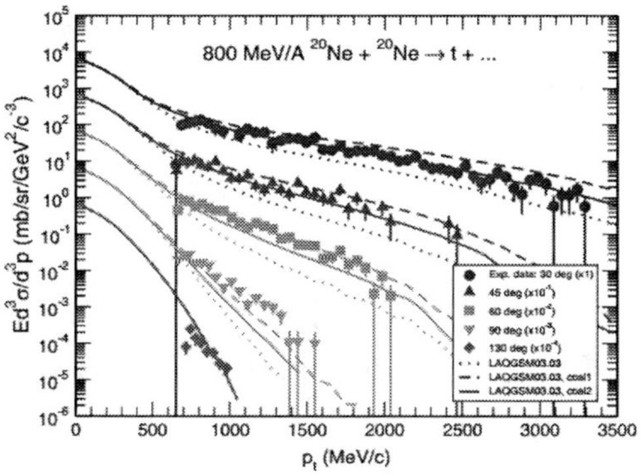

Figure 3: Comparison of measured [12] t spectra at 45, 60, 90, and 130 degrees from 800 MeV/nucleon ^{20}Ne + ^{20}Ne with calculations by LAQGS03.03 using its "standard" version of the coalescence model (p_0 = 0.108 GeV/c for t and ^{3}He; dotted lines) and with modified values of p_0 labeled in legend as "coal1" (dashed lines) and "coal2" (solid lines), as indicated in legend and discussed in detail in Ref. [3].

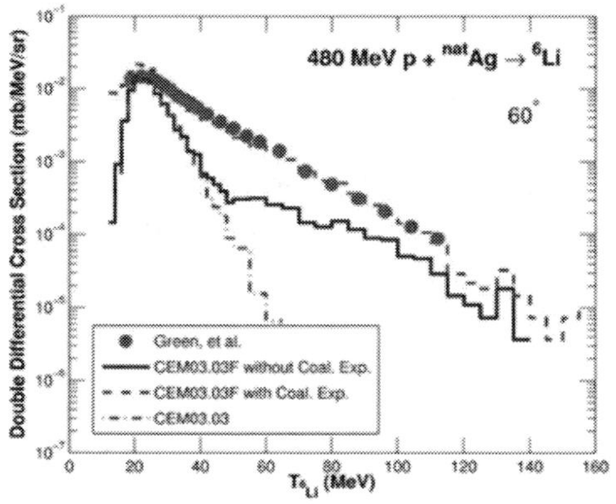

Figure 4: Comparison of experimental data by Green *et al.* [13] (circles) for the production of ^{6}Li at an angle of 60° from the reaction 480 MeV p + ^{nat}Ag, with results by standard CEM03.03 (dot-dashed line) from CEM03.03F without coalescence expansion (solid line) and CEM03.03F with coalescence expansion (dashed line) (see details in [5, 19]).

One way to approach this problem would be to employ after the INC stage of reactions a fission-like sequential-binary-decay model, like the code GEMINI by Charity *et al.* [16] to describe the compound nuclear decay. In our case, this means separately merging CEM and LAQGSM with GEMINI. Actually, we already have done so more than a decade ago, producing the "G" versions of CEM and LAQGSM we had at that time (see, *e.g.*, Ref. [18] and references therein).

Another way to address this problem is to implement in CEM and LAQGSM the Statistical Multifragmentation Model (SMM) by Botvina *et al.* [17]. Thus, we would consider multifragmentation as a mode competitive to evaporation of particles and light fragments, when the excitation energy E^* of a compound nucleus produced after the preequilibrium stage of a reaction is above a certain value, E^*_{tr}, e.g., $E^*_{tr} = 2 \times A$ MeV, as we did in the "S" versions of CEM03.01 and LAQGSM03.01 (see, e.g., Ref. [18] and references therein).

As of today, neither the "S" nor the "G" versions of CEM and LAQGSM have been implemented in MCNP6. We plan to incorporate them in our event generators used by MCNP6 after we tune several parameters in SMM and GEMINI, that are essential in chosing the excitation energy (or temper-

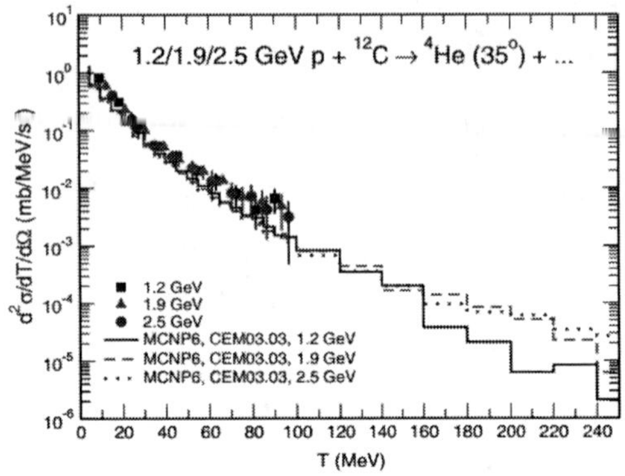

Figure 5: ^{4}He spectra (at 35°) from 1.2/1.9/2.5 GeV p + ^{12}C measured by M. Fidelus of the PISA collaboration [14] (symbols) with calculations by MCNP6 using CEM03.03 (see details in [5]).

ature) of nuclei when reaction mechanisms change from "usual evaporation", to binary decays described by GEMINI, and/or to multifragmentation simulated with SMM (see details in [19]).

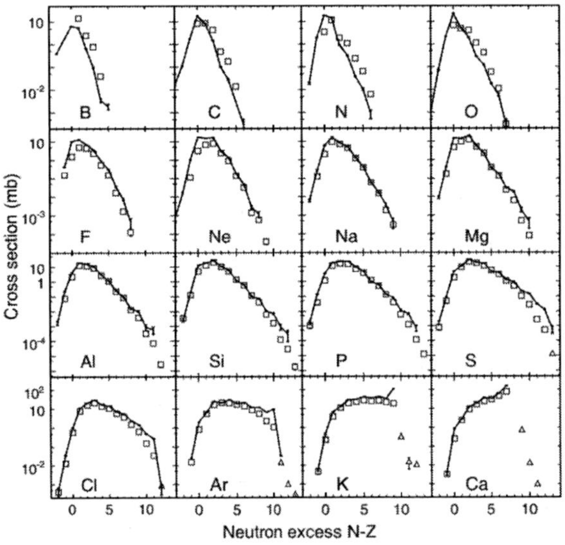

Figure 6: Measured cross sections for ^{48}Ca fragmentation on ^{9}Be at 140 MeV/nucleon [15] compared with LAQGSM03.03 predictions (see details in [9]).

This study was carried out under the auspices of the National Nuclear Security Administration of the U.S. Department of Energy at Los Alamos National Laboratory under Contract No. DE-AC52-06NA25396. This work is supported in part (for L.M.K) by the M. Hildred Blewett Fellowship of the American Physical Society, www.aps.org.

References

[1] T. Goorley *et al.* Nucl. Technol. **180** (2012) 298.

[2] S. G. Mashnik *et al. CEM03.03 and LAQGSM03.03 Event Generators for the MCNP6, MCNPX, and MARS15 Transport Codes*, LANL Report LA-UR-08-2931, Los Alamos (2008); arXiv:0805.0751.

[3] S. G. Mashnik and L. K. Kerby, NIM A **764** (2014) 59.

[4] L. K. Kerby and S. G. Mashnik, NIM B **356-357** (2015) 135.

[5] L. M. Kerby, *Precompound emission of energetic light fragments in spallation reactions*, PhD thesis, University of Idaho, 2015.

[6] K. K. Gudima *et al.*, *Fragmentation of Light Nuclei at Intermediate Energies Simulated with MCNP6*, LANL Report LA-UR-15-20953, invited talk at NUFRA2015, 4 – 11 October 2015, Kemer, Turkey.

[7] S. G. Mashnik, Eur. Phys. J. Plus **126** (2011) 49; arXiv:1011.4978

[8] S. G. Mashnik, *Validation and Verification of MCNP6 Against High-Energy Experimental Data and Calculations by Other Codes. Primers I to IV*, LANL Reports LA-UR-11-05129, LA-UR-11-05627, LA-UR-13-26944, LA-UR-14-27018 Los Alamos (2011, 2013, and 2014); https://mcnp.lanl.gov/.

[9] S. G. Mashnik *et al. LAQGSM03.03 Upgrade and Its Validation*, LANL Report LA-UR-07-6198 Los Alamos (2007); arXiv:0709.1736.

[10] S. G. Mashnik *et al.* in *Proc. SARE-4, Knoxville, TN, September 13-16, 1998* (ORNL, 1999, pp. 151-162); arXiv:nucl-th/9812071; our T-16 Library "T-16 Lib" is updated permanently when new experimental data became available to us.

[11] H. Machner *et al.* Phys. Rev. C **73** (2006) 044606.

[12] M.-C. Lemaire *et al.* Lawrence Berkeley Laboratory Report LBL-8463 (1978).

[13] R. Green *et al.* Phys. Rev. C, **29** (1984) 1806.

[14] M. Fidelus, *Model Description of Proton Induced Fragmentation of Atomic Nuclei*, Ph.D. Thesis, Jagiellonian University, Cracow, 2010.

[15] Michal Mocko, *Rare Isotope Production*, PhD thesis, Michigan State University, 2006.

[16] R. J. Charity *et al.*, Nucl. Phys. A **483** (1988) 371.

[17] J. P. Bondorf, A. S. Botvina *et al.*, Phys. Rep. **257** (1995) 133.

[18] S. G. Mashnik *et al.*, *CEM03.S1, CEM03.G1, LAQGSM03.S1, and LAQGSM03.G1 Versions of CEM03.01 and LAQGSM03.01 Event-Generators*, LANL Report LA-UR-06-1944, Los Alamos (2006); E-print: nucl/th-0603046.

[19] S. G. Mashnik, *Possible Improvements to MCNP6 and its CEM/LAQGSM Event-Generators*, LANL Report LA UR UR-15-26166, Los Alamos (2015).

EPJ Web of Conferences **117**, 03009 (2016) DOI: 10.1051/epjconf/201611703009
NN2015

Overview of jet physics with ALICE at the LHC

Mengliang Wang* for the ALICE Collaboration

Central China Normal University, Wuhan, China
SUBATECH, Ecole des Mines de Nantes, Université de Nantes, CNRS-IN2P3, Nantes, France

Abstract

In these proceedings, a selection of recent results on jet physics from ALICE experiment is presented, including jet spectra, nuclear modification factors, jet structure and fragmentation observables, obtained in different collision systems (pp, p–Pb and Pb–Pb).

1 Physics motivations

According to Lattice Quantum Chromodynamics, a phase transition from hadron gas to a deconfined phase of quarks and gluons, the Quark Gluon Plasma (QGP), is expected at high temperature or high baryon density. The QGP is studied experimentally using ultra-relativistic heavy-ion collisions.

ALICE (A Large Ion Collider Experiment) [1], one of the four large LHC experiments, is mainly dedicated to the study of the QGP properties. In the last six years, ALICE has collected a large amount of data. In this paper we will discuss results from proton-proton (pp) and lead–lead (Pb–Pb) collisions at $\sqrt{s_{\mathrm{NN}}} = 2.76$ TeV as well as from p–Pb collisions at $\sqrt{s_{\mathrm{NN}}} = 5.02$ TeV. Measurements in pp collisions can be used as a test of perturbative QCD (pQCD) calculations as well as a baseline to quantify effects that may occur in more complex systems, such as proton-nucleus (p–A) or nucleus-nucleus

*Email: mengliang.wang@subatech.in2p3.fr

(A–A). p–A collisions are used to study possible Cold Nuclear Matter (CNM) effects, while A–A collisions are used to explore the properties of the QGP.

As the lifetime of the QGP is very short, its properties cannot be measured directly. Hard probes, such as jets (collimated showers of particles), are created at the early stage of the collision, before the QGP formation. The potential modification of jets, while they traverse and interact with the medium, can be used to probe the properties of the QGP. Comparing experimental results to theoretical models, quantities such as the transport coefficients can be extracted, providing information on the medium density or about energy loss mechanisms inside the QGP [2].

2 How do we measure jets in ALICE?

"Charged jets" are reconstructed using the ALICE mid-rapidity tracking system. The Inner Tracking System (ITS) and Time Projection Chamber (TPC) are used to measure charged particles down to very low transverse momentum ($p_{\mathrm{T}} \geq 150$ MeV/c), in full azimuth and pseudo-rapidity $|\eta| < 0.9$. To measure the electromagnetic component of a "full jet", clusters are reconstructed using the Electromagnetic Calorimeter (EMCal), also used as a trigger [3]. It covers 108° in azimuth ($80° < \phi < 188°$) and ∼1.4 units of rapidity ($|\eta| < 0.67$) with an energy resolution of $11\%/\sqrt{E} + 1.7\%$ [1,3].

Jets have been reconstructed with different resolution parameters R using the sequential recombination algorithm, anti-k_{T} from the FASTJET package, which has the advantage of being infrared and collinear safe [4].

3 A selection of jet results from ALICE

3.1 Results from pp collisions

In pp collisions at 7 TeV, ALICE has measured the differential "charged jet" production cross-sections for $20 < p_{\mathrm{T}}^{\mathrm{jet,ch}} < 100$ GeV/c using several R ($R \in [0.2, 0.6]$) [5]. Our measurements, shown for both $R = 0.4$ and $R = 0.6$ in Fig. 1, are in good agreement with the measurements from ATLAS [6].

The full jet cross-section has also been measured from $p_{\mathrm{T}}^{\mathrm{jet}} = 20$ to 120 GeV/c at $\sqrt{s} = 2.76$ TeV [7]. The corresponding result is shown on the right panel of Fig. 1 and is in good agreement with NLO pQCD calculations.

As an indirect measurement of the jet structure, ratios of the charged jet spectra obtained with $R = 0.2, 0.4$ and 0.6, are shown Fig. 2 [5]. For $p_{\mathrm{T}}^{\mathrm{jet,ch}} >$ 30 GeV/c, those ratios are in good agreement with the ones obtained using

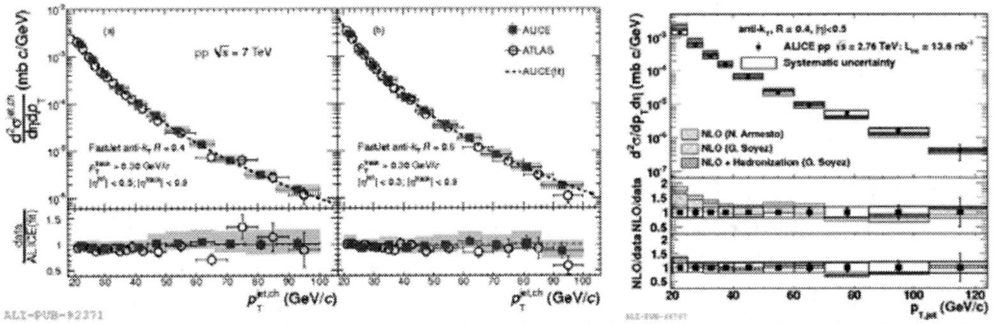

Figure 1: Left: Charged jet cross sections from ALICE (squares) and ATLAS (circles) in pp collisions at 7 TeV for $R = 0.4$ and 0.6. Right: Full jet p_T cross-section in pp collisions at 2.76 TeV for $R = 0.4$ compared to several pQCD predictions.

PYTHIA Perugia-2011 and HERWIG [8]. The distributions are increasing with jet p_T as the corresponding jets are more collimated. A similar study performed using full jets at $\sqrt{s} = 2.76$ TeV shows that the data are in good agreement with NLO pQCD calculations including hadronization [7].

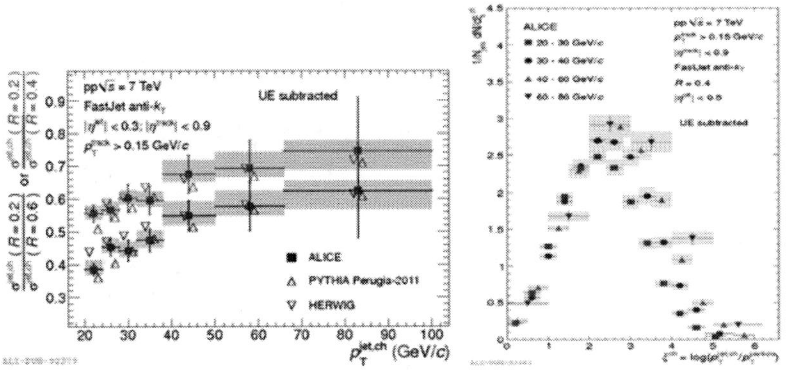

Figure 2: Left: Ratio of charged jet p_T cross-sections: $\sigma^{\mathrm{jet,ch}}(R = 0.2)/\sigma^{\mathrm{jet,ch}}(R = 0.4)$ and $\sigma^{\mathrm{jet,ch}}(R = 0.2)/\sigma^{\mathrm{jet,ch}}(R = 0.6)$ (squares), compared with PYTHIA and HERWIG (triangles). Right: Particle ξ distributions in leading charged jet.

In order to learn more about the internal structure of jets, ALICE has measured the distributions of the charged constituent transverse momentum in pp collisions [5]. The evolution of the measured leading jet fragmentation distributions with jet p_T (4 bins from $p_T^{\mathrm{jet,ch}} = 20$ to 80 GeV/c) can be seen in Fig. 2 (right) as a function of $\xi = log(1/z)$ ($z = p_T^{\mathrm{track}}/p_T^{\mathrm{jet,ch}}$) and is in agreement with jet angular ordering expected from QCD color coherence [9].

3.2 Results from p–A collisions

The p_T-differential charged jet production cross-sections in p–Pb collisions at $\sqrt{s} = 5.02$ TeV, for $R = 0.2$ and 0.4, and $20 < p_T^{\mathrm{jet,ch}} < 120$ GeV/c, are in good agreement with boosted NLO pQCD calculations (taking into account the rapidity shift of the nucleon-nucleon center-of-mass system in p–Pb with a boosted parton system) without and with nuclear modification of the Parton Distribution Function (nPDF) [10]. Within error bars the effect of nPDF can be neglected.

The ALICE charged jet R_{pPb} is presented in the left two panels of Fig. 3 [10]. Because of the lack of pp data at $\sqrt{s} = 5.02$ TeV, the R_{pPb} has been calculated using a pp reference which has been obtained by scaling the spectrum measured at $\sqrt{s} = 7$ TeV. The result is consistent with unity withing experimental uncertainties, which indicates the absence of strong Cold Nuclear Matter (CNM) effects on jet production.

The ratio of the charged jet cross-sections obtained with $R = 0.2$ and 0.4 in p–Pb collisions (Fig. 3 right panel), which does not show any strong centrality dependence [10], is compatible with the ratio obtained using 7 TeV pp data (Fig. 2, left). The same behavior with respect to the collimation of jets thus applies for p–Pb data. A good agreement with predictions from PYTHIA6 Perugia 2011 and POWHEG+PYTHIA8 simulations at 5.02 TeV is also observed. Consequently, no significant energy dependence or change with collision species are observed. We observe no redistribution of the jet energy while comparing $R = 0.2$ to $R = 0.4$ in p–Pb at $\sqrt{s_{\mathrm{NN}}} = 5.02$ TeV.

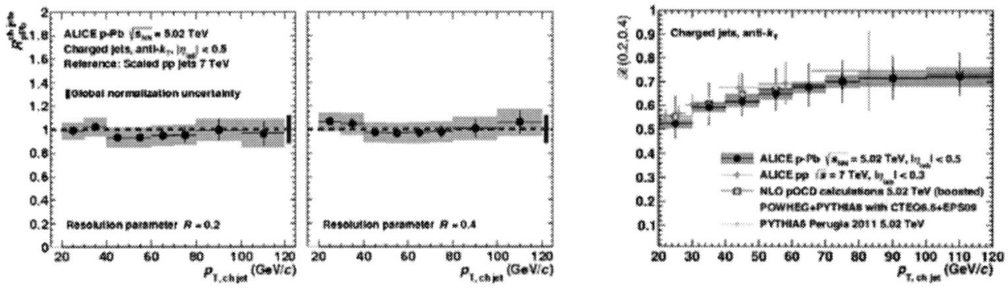

Figure 3: Left: Charged jet R_{pA} in p–Pb collisions at 5.02 TeV measured using $R = 0.2$ and 0.4 jets. Right: Ratio of the jet spectra $R = 0.2$ / $R = 0.4$ in p–Pb collisions compared with results obtained using different collision systems or energies, and with PYTHIA Perugia 2011 and POWHEG+PYTHIA8 simulations.

3.3 Results from A–A collisions

In Pb–Pb collisions at $\sqrt{s_{\mathrm{NN}}} = 2.76$ TeV, the full jet p_{T} spectra and the corresponding R_{AA} (Fig. 4 left panels) have been measured using $R = 0.2$, in the centrality bins 0-10% for $40 < p_{\mathrm{T}}^{\mathrm{jet}} < 120$ GeV/c and, 10-30% for $30 < p_{\mathrm{T}}^{\mathrm{jet}} < 100$ GeV/c [11]. A large suppression of the jet nuclear modification factor is observed in both cases. Since the jet R_{pPb} is consistent with unity (section 3.2), the observed suppression in Pb–Pb is most likely due to final state effects: energy loss in the hot and dense medium.

The right panel of Fig. 4 presents a comparison of the measurements of jet R_{AA} at LHC (using $R = 0.2$). For the results compared here, different jet reconstructions (full jets for ALICE, particle-flow jets for CMS and calorimeter jets† for ATLAS) and background subtraction methods are used, and the kinematics of the measurements are quite different [12]. However ALICE, CMS and ATLAS measurements are compatible within errors.

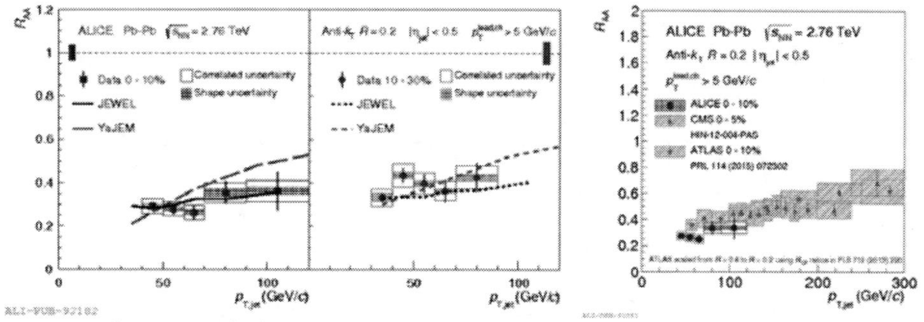

Figure 4: Left: R_{AA} of "full jets" measured in two centrality bins. Predictions from JEWEL and YaJEM are both in agreement with the data. Right: comparisons of the measured jet suppression at LHC (ALICE, CMS and ATLAS).

Doing such an inclusive jet measurement using jets with a larger resolution parameter R would certainly be very interesting. However, in heavy ion collisions at LHC, jet measurements are affected by the large and fluctuating background, increasing with R and centrality as discussed in [13].

In order to overcome those limitations, correlations between a high p_{T} trigger hadron and the corresponding recoil-jet can be used. Doing so, the parton producing the recoil-jet is biased toward higher in-medium path length, while the leading hadron is close to the surface of the fireball [14]. This analysis technique (at LHC, unique to ALICE) has several advantages: it provides us with a good handle on the combinatorial background which is statistically subtracted when computing the difference between two recoil-jet spectra, corresponding to different values of trigger hadron p_{T}. It allows us to use jets with a larger R, without any fragmentation bias on the recoil-jet side. Recoil jet spectra were measured for $20 < p_{\mathrm{T}}^{\mathrm{jet,ch}} < 100$ GeV/c using R = 0.2, 0.4 and 0.5 [15]. The recoil jet yield is found to be suppressed when compared to reference pp distributions at $\sqrt{s} = 2.76$ TeV using Monte-Carlo or NLO pQCD methods (validated using pp data at $\sqrt{s} = 7$ TeV). No signif-

†The ATLAS R_{AA} was scaled from $R = 0.4$ to $R = 0.2$ using their jet R_{CP} ratio [12].

icant evidence of intra-jet broadening nor of medium-induced acoplanarity are observed using jets with resolution parameters up to $R = 0.5$.

4 Outlook

During the first LHC long shutdown, an extension of EMCal called Di-jet Calorimeter (DCal) was installed in ALICE. It will allow jet-jet and photon-jet measurements while providing us with increased triggering capabilities. With the start of the LHC run 2, many new exciting results are expected.

References

[1] K. Aamodt *et al.* (ALICE Coll.), JINST **3** (2008) S08002; B. Alessandro *et al.* (ALICE Coll.), J. Phys. G **32** (2006) 1295

[2] M. Gyulassy and M. Plumer, Phys. Lett. B **243** (1990) 432

[3] U. Abeysekara *et al.* (ALICE EMCal Coll.), arXiv:1008.0413

[4] M. Cacciari, G. P. Salam and G. Soyez, Eur. Phys. J.C **72** (2012) 1896

[5] B. Abelev *et al.* (ALICE Coll.), Phys. Rev. D **91** (2015) 112012

[6] G. Aad *et al.* (ATLAS Coll.), Phys. Rev. D **84** (2011) 054001

[7] B. Abelev *et al.* (ALICE Coll.), Phys. Lett. B **722** (2013) 262

[8] PYTHIA: T. Sjostrand, S. Mrenna and P. Skands, JHEP **05** (2006) 026; HERWIG: G. Corcella *et al.*, JHEP **01** (2001) 010

[9] B. Ermolaev and V. S. Fadin, JETP Lett. **33** (1981) 269. A. H. Mueller, Phys. Lett. B **104**: (1981) 161

[10] J. Adam *et al.* (ALICE Coll.), Phys. Lett. B **749** (2015) 68

[11] J. Adam *et al.* (ALICE Coll.), Phys. Lett. B, **746** (2015) 1

[12] CMS HIN-12-004-PAS; G. Aad *et al.* (ATLAS Coll.), Phys. Rev. Lett. **114** (2015) 072302, Phys. Lett. B **719** (2013) 220

[13] B. Abelev *et al.* (ALICE Coll.), JHEP **03** (2012) 053

[14] G. De Barros *et al.*, Nucl. Phys. A **910-911** (2013) 314

[15] J. Adam *et al.* (ALICE Coll.), JHEP **09** (2015) 170

EPJ Web of Conferences **117**, 03010 (2016) DOI: 10.1051/epjconf/201611703010
NN2015

Hadronization via coalescence at RHIC and LHC

V. Minissale[1,2], F. Scardina[1,2] and V. Greco[1,2]

[1]Department of Physics and Astronomy, University of Catania, Catania, Italy
[2]INFN - Laboratori Nazionali del Sud, Catania, Italy

Abstract

An hadronization model that includes coalescence and fragmentation is used in this work to obtain predictions at both RHIC and LHC energy for light and strange hadrons transverse momentum spectra (π, p, k, Λ) and baryon to meson ratios (p/π, Λ/k) in a wide range of p_T. This is accomplished without changing coalescence parameters. The ratios p/π and Λ/K shows the right behaviour except for some lack of baryon yield in a limited p_T range around 6 GeV. This would indicate that the AKK fragmentation functions is too flat at $p_T < 8$ GeV.

1 Introduction

In the study of the Quark Gluon Plasma partonic dynamics is not directly related to the observables measured in heavy ion collision experiments. The choice of the model for the hadronization process is a crucial point in order to have a comparison with experimental data. A decade ago the observation of the enhancement of baryon to meson ratio at intermediate transverse momenta and the scaling of elliptic flow according to the constituent quark number found in Au+Au collisions at RHIC were a great surprise. The coalescence hadronization process can explain both this features that appears in the same intermediate p_T region [1]. The coalescence approach born from the idea that in presence of QGP matter comoving partons combine their momenta to produce a final-state meson or baryon, this can happen at a time scale in which the partons rescatter and generate a thermal medium expanding with a collective radial flow [2]. Initial hard processes between colliding nucleons producing minijet partons provide a dominant contribution to hadron spectra at high transverse momentum.

1.1 Coalescence model

The coalescence approach is based on a Wigner function formalism. The hadron spectrum can be written as:

$$\frac{dN_H}{d^2P_T} = g_H \int \prod_{i=1}^{n} \frac{d^3p_i}{(2\pi)^3 E_i} p_i \cdot d\sigma_i \; f_q(x_i, p_i) f_H(x_i..x_n, p_i..p_n) \delta^{(2)} \left(P_T - \sum_{i=1}^{n} p_{T,i} \right) \tag{1}$$

where $f_H(x_i..x_n, p_i..p_n)$ is the hadron Wigner distribution function giving the probability for n quark to form an hadron, $d\sigma$ denotes an element of a space-like hypersurface, g_H is the probability of forming a color neutral object with the spin of the hadron considered from n coloured quarks. The function $f_q(x,p)$ is the covariant distribution function of quarks (and antiquarks) in the phase space. For the light hadrons Wigner function we take a sphere in both space and momentum, with radii Δ_r and Δ_p, respectively, which in the Wigner formalism are related by $\Delta_r \cdot \Delta_p = 1$. For meson we have:

$$f_M(x_1, x_2; p_1, p_2) = \frac{9\pi}{2(\Delta_x \Delta_p)^3} \, \Theta(\Delta_x^2 - (x_1 - x_2)^2) \Theta(\Delta_p^2 - (p_1 - p_2)^2 + (m_1 - m_2)^2) \tag{2}$$

A good description of pion, kaon, proton, antiproton and Λ spectra can be obtained with a radius parameter $\Delta_p = 0.19$ GeV for mesons, 0.33 GeV for proton, 0.38 GeV for Lambdas [3].

2 Fireball and model parameters

In this study is considered a fireball of thermalized particles which includes light quarks, antiquarks and gluons at a temperature corresponding to $T_c = 160$ MeV, that is about the cross-over transition temperature. Parton are distributed uniformly in the transverse plane. Longitudinal positions are distributed according to $z = \tau \sinh y$. Assuming a constant acceleration, we connect the radial expansion $R_\perp = R_0 + 0.5\beta_{max}\tau$ with the radial flow β_{max}. In such manner the radial flow and the volume has constraints due to the

experimental value of total multiplicity dN/dy and total transverse energy dE_T/dy. The values obtained for this quantities are $\beta_{max} = 0.37$, $R_\perp = 8.7$ fm, $\tau = 4.5\,\mathrm{fm}/c$ at RHIC, and $\beta_{max} = 0.60$, $R_\perp = 10.2$ fm, $\tau = 7.8\,\mathrm{fm}/c$ at LHC. For their longitudinal momentum distribution we assume a uniform rapidity distribution in $y \in (-0.5, +0.5)$.
For partons in the quark-gluon plasma we take a thermal distribution for transverse momenta up to $p_0 = 2$ GeV, in particular the momentum spectra for quark and antiquarks is given by:

$$\frac{dNq, \bar{q}}{d^2r_T\, d^2p_T} = \frac{g_{q,\bar{q}}\tau m_T}{(2\pi)^3} \exp\left(-\frac{\gamma_T(m_T - p_T \cdot v_T \mp \mu_q)}{T}\right) \tag{3}$$

where $g_q = g_{\bar{q}} = 6$ are the spin-color degeneracy of light quarks and antiquarks, and the minus and plus signs are for quarks and antiquarks, respectively. The slope parameter T is taken to be $T = 165\ MeV$. Masses of light quarks and antiquarks are taken to be $m_{u,d,\bar{u},\bar{d}} = 300\ MeV$ and $m_{s,\bar{s}} = 475\ GeV$, similar to the masses of constituent quarks. For the quark chemical potential we use a value of light antiquark to quark ratio which lead to the antiproton to proton ratio observed at midrapidity in heavy ion collisions experiments that is $\mu_q = 10$MeV at RHIC and vanishing at LHC. Partons at high transverse momenta (greater than 2 GeV) are mainly from the minijets produced in initial hard collisions among nucleons and are described by a power-law parametrization.

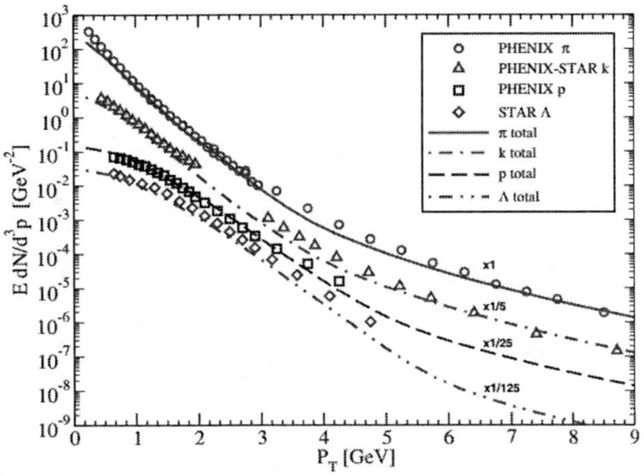

Figure 1: Particle transverse momentum spectrum at RHIC from Au+Au collisions at $\sqrt{s} = 200$ GeV. Pion (solid), kaon (dash-dotted), proton (dashed), Lambda (dashed and double dotted). The spectra are scaled for better visibility. [4,6]

3 Spectra

In Fig.1 are shown the transverse momentum spectra of pions, kaons, protons and Lambdas at RHIC. In all the spectra the contribution that comes from resonances decay are included with their decay channel and the pertinent branching ratios (e.g. for π spectra a dominant contribution comes, in low momentum region, from ρ decay). Higher resonances (K^*,Δ in π case) are less relevant and only contribute to some little better description at very low momenta [3]. We find a good description in a wide range of momentum. We can see in Fig.2 that the ratios $\bar{p}/\pi^-$, Λ/K are well predicted, from the rise at low p_T up to the peak region and then the falling-down behaviour. However in both cases it is clear that there is an underestimation in the predicted ratio in the region of $p_T \sim 5-7$ GeV. Looking at the particle spectra this lack can be imputed to an insufficient production of baryons. In Fig.3 are shown all the particle transverse momentum spectra at LHC. The results obtained in this case, occur without any change or addition of model parameters respect to the one at RHIC. Only the radial flow and volume of the hadronizing fireball are changed, and are constrained by comparison to the multiplicity and energy density experimentally measured.
There is a good agreement with the experimental data in all the transverse momentum region, except some lack of yield for pions at very low p_T due to absence of all the possible resonances. The model is able to correctly predict the evolution, from RHIC to LHC, of the absolute yield and especially its p_T shape. For baryons in the p_T region where the fragmentation dominates,

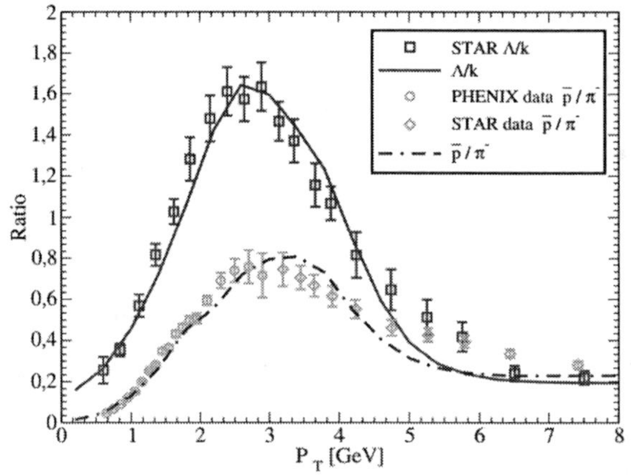

Figure 2: Antiproton to negative pion ratio (dash-dotted line) and Lambda to kaon ratio (solid line) at RHIC from Au+Au collisions at $\sqrt{s} = 200$ GeV. [4,5].

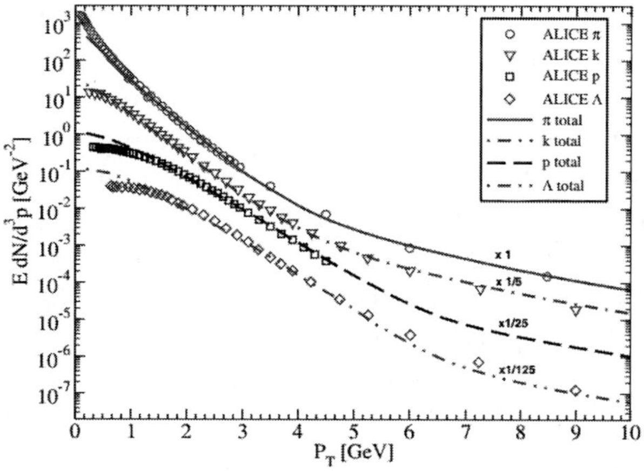

Figure 3: Particle transverse momentum spectrum at LHC in Pb+Pb collisions at $\sqrt{s} = 2.7$ TeV. Pion (solid), kaon (dash-dotted), proton (dashed), Lambda (dashed and double dotted). The spectra are scaled for better visibility. [6, 7]

i.e. $p_T \sim 6 - 7$ GeV, there is some lack of yield. The baryon over meson ratios at LHC are shown in Fig.4. The model overestimate the ratio in the low p_T region, due to the production excess of p and Λ, but there is a good description especially in the peak region. These ratios show also that the region where fragmentation becomes dominant is characterized by an underestimation of baryons.

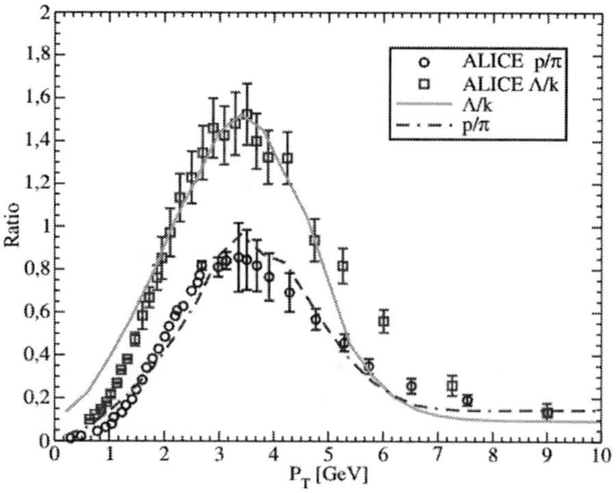

Figure 4: Proton to pion (dash-dotted line) and Lambda to kaon ratio (solid line) in Pb+Pb collisions at $\sqrt{s} = 2.7$ TeV. [7]

4 Conclusions

We have studied the evolution of the transverse momentum spectra from RHIC to LHC energy in a coalescence plus fragmentation approach. In order to predict the features at LHC no one adjustment of the coalescence paramotor, like the width of Wigner function, has been made. The results obtained are in good agreement with the experimental data from LHC in a wide range of p_T. It appear that the independent fragmentation approach (at least using the AKK) gives too hard spectra at least up to $p_T \simeq 8$ GeV or it could be that coalescence contribution should extend to large p_T with respect to the present modeling. There are studies that, recently, investigate a process that within the coalescence plus fragmentation approach could be quite important in solving this issue [8]. They essentially describe the in-medium fragmentation as a quark recombination of shower partons taking into account also the gluon splitting in quark pairs that recombine.

References

[1] STAR Collaboration, J. Adams *et al.*, Phys.Rev. **C72**, 014904 (2005), nucl-ex/0409033.

[2] R. Fries, V. Greco, and P. Sorensen, Ann. Rev. Nucl. Part. Sci. **58**, 177-205 (2008).

[3] V. Minissale, F. Scardina and V. Greco, arXiv:1502.06213, (2015), nucl-th.

[4] PHENIX Collaboration, S. Adler *et al.*, Phys.Rev. **C69**, 034909 (2004), nucl-ex/0307022.

[5] STAR Collaboration, G. Agakishiev *et al.*, Phys.Rev.Lett. **108**, 072301 (2012), 1107.2955.

[6] ALICE Collaboration, B. Abelev *et al.*, Phys.Rev.Lett. **109**, 252301 (2012), 1208.1974.

[7] ALICE Collaboration,Abelev, Betty Bezverkhny and others, Phys.Rev.Lett. **111**, 222301 (2013),1307.5530.

[8] K. C. Han, R. J. Fries, and C. M. Ko, J.Phys.Conf.Ser. **420**, 012044 (2013), 1209.1141.

EPJ Web of Conferences **117**, 03011 (2016) DOI: 10.1051/epjconf/201611703011
NN2015

Analysis of angular distribution of fragments in relativistic heavy-ion collisions by quantum molecular dynamics

TATSUHIKO OGAWA, TATSUHIKO SATO, SHINTARO HASHIMOTO, KOJI NIITA

Japan Atomic Energy Agency, Japan, Research Organization for Information Science and Technology, Japan

Abstract

To predict angular distribution of fragments produced in nucleus-nucleus collisions, JAERI quantum molecular dynamics model (JQMD) was improved. Because JQMD underestimated fragments in the forward angle, which were mainly produced by peripheral collisions, JQMD was revised so as to simulate peripheral collisions accurately. Density-dependent in-medium effect and relativistic effect on nucleon-nucleon interactions were incorporated for this purpose. The revised version of JQMD coupled with a statistical decay model was used to calculate differential fragment production cross sections measured in earlier studies. Comparison of the measured data and calculation by the revised and old JQMD showed that the revised JQMD can predict fragment angular distribution better than old JQMD. Particularly, agreement of fragment yield in the forward angle is substantially improved.

1 Introduction

The angular distribution of the fragments produced in nucleus-nucleus collisions is essential for cosmic ray physics, radiation safety of space exploration, and dosimetry in heavy-ion cancer therapy.

Among the microscopic reaction models developed to simulate fragment production in nucleus-nucleus reactions, quantum molecular dynamics (QMD) models are generally adopted in general-purpose radiation transport simulation codes [1, 2] and used in various applications. One of the QMD models, JAERI-QMD (JQMD) [3] was developed as an event generator for general purpose radiation transport code PHITS (Particle and Heavy Ion Transport code System) [4] and widely used to simulate nucleus-nucleus collision (e.g., [5,6]). However, it was suggested that QMD models overestimated the width of the fragment angular distributions [7, 8] and underestimated production at the forward angle. This trend was probably attributed to inaccurate treatment of peripheral collisions, in which projectiles are little deflected and therefore fragments are ejected in the forward angle.

In this study, we revise JQMD to simulate the fragment angular distribution by accurate treatment of peripheral collisions.

Fragment production in the forward angle was underestimated by the previous version of JQMD; however, revised JQMD can reproduce fragment yield angular distribution measured in earlier studies.

2 Methods

In nucleus-nucleus collisions, the trajectory of the projectile ion is curved owing to the Coulomb repulsion by the target nucleus. The larger the impact parameter, the less is the fragment deflection.

In the previous version of JQMD; however, nuclei were sometimes spuriously disintegrated even without interacting with other nuclei owing to the instability of the nuclei simulated by JQMD. Therefore, nucleus-nucleus collisions at a large impact parameter simulated by JQMD may result in spurious reactions. To cut-off spurious reactions, impact parameter sampling range was intentionally truncated at a threshold defined by Eq. 1,

$$b_{\mathrm{max}} = 1.15 \times (A_{\mathrm{t}}^{1/3} + A_{\mathrm{p}}^{1/3}) - 0.4(\mathrm{fm}), \tag{1}$$

where A_{t} is target mass number and A_{p} is projectile mass number. This truncation cuts off not only spurious disintegration but also true peripheral collisions. As a result, fragments in the forward direction was underestimated.

The spurious decay of nuclei was occurred because the description of interaction between nucleons was frame-dependent. Nucleus configured to be stable in the rest frame became unstable in the center-of-mass frame of projectile and target nucleus. Therefore, Lorentz-covariant description of nucleon-nucleon interaction proposed by Mancusi [9] was adopted in the revised version of JQMD (hereafter referred to as JQMD-2.0). The equation of motion is described as follows;

$$\dot{\mathbf{r}}_{\mathbf{i}} = \frac{\mathbf{p}_{\mathbf{i}}}{2p_i^0} + \sum_j^N \frac{m}{p_j^0}\frac{\partial\langle\hat{V}_j\rangle}{\partial\mathbf{p}_{\mathbf{i}}}, \tag{2}$$

$$\dot{\mathbf{p}}_{\mathbf{i}} = -\sum_j^N \frac{m}{p_j^0}\frac{\partial\langle\hat{V}_j\rangle}{\partial\mathbf{r}_{\mathbf{i}}}.$$

$$p_i^0 = \sqrt{\mathbf{p}_{\mathbf{i}}{}^2 + m^2 + 2m\langle\hat{V}_i\rangle},$$

where $\mathbf{r}_{\mathbf{i}}$ is the spatial coordinate of the centroid of the i-th nucleon, $\mathbf{p}_{\mathbf{i}}$ is the momentum of the i-th nucleon, m is the rest mass of nucleons, $\langle\hat{V}_j\rangle$ is the potential of j-th particle, and N is the number of particles in the system. In both JQMD and JQMD-2.0, the potential term of the Hamiltonian V is a sum of the Skyrme-type force term, Coulomb interaction term, and symmetry term. The potential V_i was inherited from previous JQMD without any modifications. Using the new equation of motion, nuclei at the ground state stayed stable over a typical reaction period of 150 fm/c .

The impact parameter was sampled from zero to a threshold calculated by Eq. 1 in JQMD; however, impact parameter cut-off was not adopted in JQMD-2.0. Projectile and target stayed without losing nucleons or being substantially excited if the impact parameter was large enough. In this case, reaction simulation was started afresh after newly sampling the impact parameter.

In addition to the above improvements, density dependence of in-medium effect on nucleon-nucleon scattering cross sections was incorporated. Previously, we used neutron-proton scattering suppressed by in-medium Pauli-blocking In JQMD; however, in-medium effect should get weaker near the nuclear surface, where nucleon-nucleon collisions are occurred in peripheral collisions. Therefore in JQMD-2.0, the neutron-proton scattering cross section in free space was adopted in peripheral collisions.

In actual simulation, JQMD and JQMD-2.0 were coupled with the statistical decay model GEM [10] to determine the final products. More details on JQMD-2.0 is provided elsewhere [11].

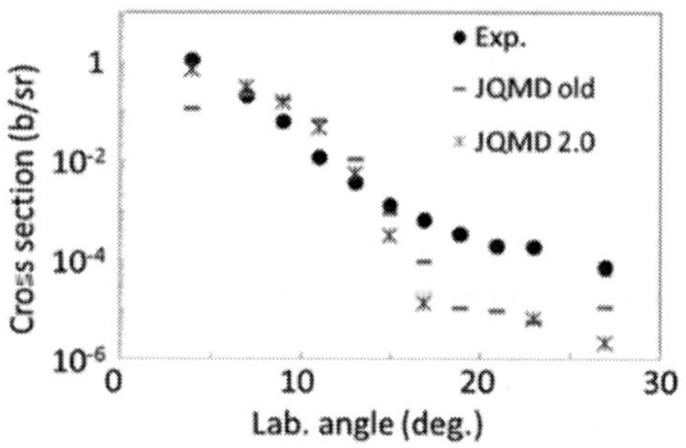

(a) ^{12}C(^{Nat}C,x)^{10}B

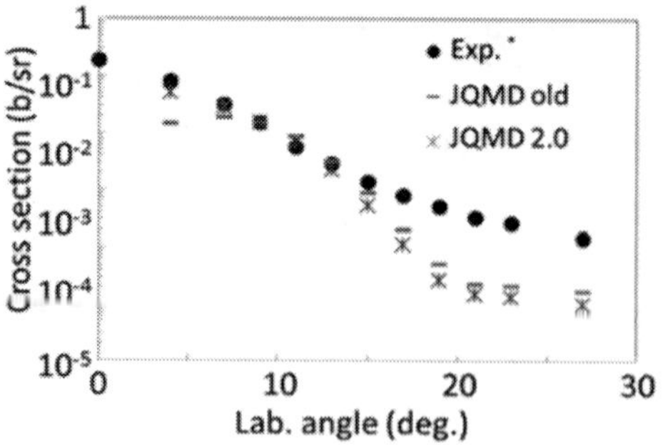

(b) ^{12}C(^{Nat}C,x)^{7}Be

Figure 1: Differential fragment production cross section of ^{12}C(^{Nat}C,x)X at 95 MeV/nucleon calculated by JQMD, JQMD-2.0 and that measured by Dudouet *et al.* [12, 13].

3 Results and discussion

To evaluate the effect of the revisions, fragment production cross sections calculated by JQMD-2.0 were compared with measured data. Fragment production cross sections in ^{12}C(^{Nat}C,x) reactions, which are important for cosmic ray studies and cancer therapy, were used for this benchmark. Fig. 1 shows comparison of isotopic differential fragment production cross sections at 95 MeV/nucleon calculated using JQMD, JQMD-2.0 and those measured by Dudouet *et al.* [12, 13].

The fragment yields calculated by JQMD have a peak at about 7 degrees and the yields are smaller at 4 degrees because peripheral collisions are disregarded by the impact parameter cut-off. In contrast, the fragment yields calculated by JQMD-2.0 are in better agreement with the measured data in this angular range. Particularly, ^{10}B is produced by peripheral collisions in which two nucleons are knocked out, therefore the increase in the yield in the forward angle is more pronounced.

Above 15 degrees, yield is underestimated by both JQMD and JQMD-2.0. Because the deflection of projectiles beyond 15 degrees cannot be explained by the dynamic reaction phase simulated by JQMD or the Coulomb repulsion, it is probably attributed to the statistical decay phase.

The increase in the yields at the forward angle and deficiency above 15 degrees were observed for the other isotopes.

Comparison of the differential fragmentation cross sections at a higher energy is shown in Figs. 2. Isotopic fragmentation cross sections of

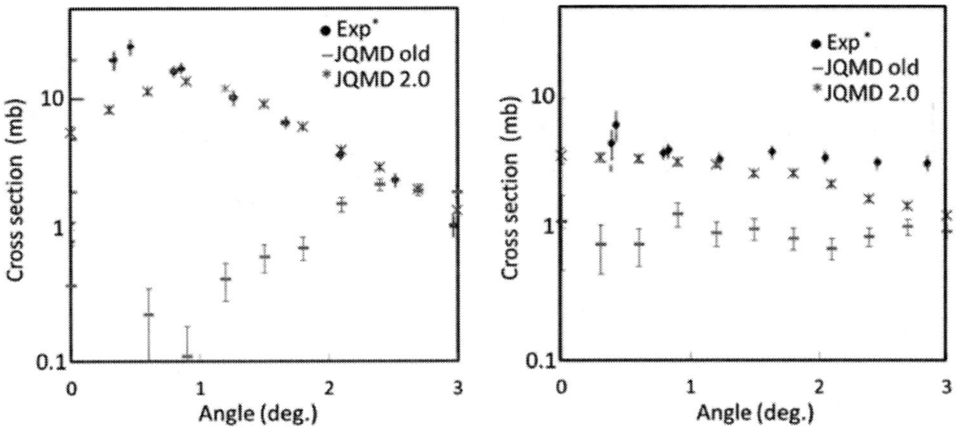

(a) ^{12}C(^{Nat}C,x)^{11}B (b) ^{12}C(^{Nat}C,x)^{6}Li

Figure 2: Differential fragment production cross section of ^{12}C(^{Nat}C,x)X at 250 MeV/nucleon calculated by JQMD, JQMD-2.0 and that measured by Kidd *et al.* [14].

^{12}C(^{Nat}C,x) reactions measured at 250 MeV/nucleon [14] were compared with those calculated using the old JQMD and JQMD-2.0.

At higher energies, fragment angular distribution is more forward peaked owing to the decrease in the impulse of Coulomb repulsion.

In Fig. 2 (a), JQMD-2.0 reproduce the angular dependence of the yield below 2 degrees whereas the yield in the forward angle is underestimated by JQMD as with the case of Fig. 1.

On the other hand, the yield of ^{6}Li in Fig. 2 (b) is more angular independent compared to that of ^{11}B. This trend is reproduced both by JQMD and JQMD-2.0; however, JQMD underestimates the absolute yield by factor of 4.

Above benchmark shows that JQMD-2.0 is more accurate than JQMD in reproducing the fragment yields at the forward direction.

4 Conclusion

Reaction event generator JQMD was revised for accurate simulation of fragment angular distribution in nucleus-nucleus collisions.

JQMD underestimated fragment yields at the forward angle owing to inaccurate treatment of peripheral collisions.

By introducing the Lorentz-covariant formulation of interaction, the ground state of target and projectile was stabilized. In addition, the density dependence of nuclear medium effect on proton-neutron scattering cross section was taken into account and the limitation of impact parameter sampling range was eliminated.

Comparison of the simulated results with earlier measurements showed that JQMD-2.0 reproduce fragment production in the forward angle more accurately. On the other hand, improvement of the statistical decay model is necessary for accurate simulation of fragment yield at large angles.

JQMD-2.0 has been incorporated to the general purpose radiation transport simulation code PHITS Ver. 2.76 and later. It can serve for a variety of applications such as cosmic ray physics and heavy-ion dosimetry.

Acknowledgments

We wish to thank Dr. Davide Mancusi for providing the source code of R-JQMD. We also thank the operation team of the Center for Computational Science and E-systems (CCSE) in the Japan Atomic Energy Agency. Monte-Carlo simulations reported in this paper were executed on a PC-cluster system of CCSE. This work was partly supported by JSPS KAKENHI Grant Numbers 24860072 (Grant-in-Aid for Research Activity Start-up), and 26790072 (Grant-in-Aid for Young Scientists B).

References

[1] F. Ballarini, G. Battistoni, F. Cerutti, A. Empl, A. Fasso, A. Ferrari, E. Gadioli, M.V. Garzelli, A. Ottolenghi, L.S. Pinsky, J. Ranft, S. Roesler, P. R. Sala, and G. Smirnov. *Nuclear Models in FLUKA: Present Capabilities, Open Problems and Future Improvements*, volume SLAC-PUB.

[2] T. Koi. New native QMD code in Geant4. *Proceedings of Joint International Conference on Supercomputing in Nuclear Applications and Monte Carlo 2010 (SNA + MC2010).*

[3] Koji Niita, Satoshi Chiba, Toshiki Maruyama, Tomoyuki Maruyama, Hiroshi Takada, Tokio Fukahori, Yasuaki Nakahara, and Akira Iwamoto. Analysis of the (N,xN') reactions by quantum molecular dynamics plus statistical decay model. *Phys. Rev. C*, 52:2620–2635, Nov 1995.

[4] Tatsuhiko Sato, Koji Niita, Norihiro Matsuda, Shintaro Hashimoto, Yosuke Iwamoto, Shusaku Noda, Tatsuhiko Ogawa, Hiroshi Iwase, Hi-

roshi Nakashima, Tokio Fukahori, Keisuke Okumura, Tetsuya Kai, Satoshi Chiba, Takuya Furuta, and Lembit Sihver. Particle and Heavy Ion Transport code System, PHITS, version 2.52. *Journal of Nuclear Science and Technology*, 50(9):913–923, 2013.

[5] L. Sihver, D. Mancusi, K. Niita, T. Sato, L. Townsend, C. Farmer, L. Pinsky, A. Ferrari, F. Cerutti, and I. Gomes. Benchmarking of calculated projectile fragmentation cross-sections using the 3-D, MC codes PHITS, FLUKA, HETC-HEDS, MCNPX-HI, and NUCFRG2. *Acta Astronautica*, 63(7-10):865 – 877, 2008.

[6] H. Yashima, Y. Uwamino, H. Iwase, H. Sugita, T. Nakamura, S. Ito, and A. Fukumura. Cross sections for the production of residual nuclides by high-energy heavy ions. *Nuclear Instruments and Methods in Physics Research Section B: Beam Interactions with Materials and Atoms*, 226(3):243 – 263, 2004.

[7] B. Alpat, E. Pilicer, S. Blasko, D. Caraffini, F. Di Capua, V. Postolache, G. Saltanocchi, M. Menichelli, L. Desorgher, M. Durante, R. Pleskac, and C. La Tessa. Total and partial fragmentation cross-section of 500 MeV/nucleon carbon ions on different target materials. *Nuclear Science, IEEE Transactions on*, 60(6):4673–4682, Dec 2013.

[8] J. Dudouet, D. Cussol, D. Durand, and M. Labalme. Benchmarking Geant4 nuclear models for hadron therapy with 95 MeV/nucleon carbon ions. *Phys. Rev. C*, 89:054616, May 2014.

[9] Davide Mancusi, Koji Niita, Tomoyuki Maruyama, and Lembit Sihver. Stability of nuclei in peripheral collisions in the JAERI quantum molecular dynamics model. *Phys. Rev. C*, 79:014614, Jan 2009.

[10] S. Furihata. Statistical analysis of light fragment production from medium energy proton-induced reactions. *Nuclear Instruments and Methods in Physics Research Section B: Beam Interactions with Materials and Atoms*, 171(3):251 – 258, 2000.

[11] T. Ogawa, T. Sato, S. Hashimoto, D. Satoh, S. Tsuda, and K. Niita. Energy-dependent fragmentation cross sections of relativistic ^{12}C. *Phys. Rev. C*, 92:024614, Aug 2015.

[12] J. Dudouet, D. Juliani, M. Labalme, D. Cussol, J. C. Angélique, B. Braunn, J. Colin, Ch. Finck, J. M. Fontbonne, H. Guérin, P. Henriquet, J. Krimmer, M. Rousseau, M. G. Saint-Laurent, and S. Salvador.

Double-differential fragmentation cross-section measurements of 95 MeV \ nucleon ^{12}C beams on thin targets for hadron therapy. *Phys. Rev. C*, 88:024606, Aug 2013.

[13] J. Dudouet, M. Labalme, D. Cussol, C. Finck, R. Rescigno, M. Rousseau, S. Salvador, and M. Vanstalle. Zero-degree measurements of ^{12}C fragmentation at 95 MeV \ nucleon on thin targets. *Phys. Rev. C*, 89:064615, Jun 2014.

[14] J. M. Kidd, P. J. Lindstrom, H. J. Crawford, and G. Woods. Fragmentation of carbon ions at 250 Mev/nucleon. *Phys. Rev. C*, 37:2613–2623, Jun 1988.

EPJ Web of Conferences **117**, 03012 (2016) DOI: 10.1051/epjconf/201611703012
NN2015

Evolution of heavy quark distribution function on quark-gluon plasma: Using the Iterative Laplace Transform Method

Sharareh Mehrabi Pari[1,2*], Kurosh Javidan[1**], Fatemeh Taghavi Shahri[1***]

[1] Department of Physics, Ferdowsi University of Mashhad, 91775-1436 Mashhad, Iran
[2] Department of Physics and Technology, University of Bergen, 5007 Bergen, Norway

Abstract

The "Iterative Laplace Transform Method" is used to solve the Fokker-Plank equation for finding the time evolution of the heavy quarks distribution functions such as charm and bottom in quark gluon plasma. These solutions will lead us to calculation of nuclear suppression factor R_{AA}. The results have good agreement with available experiment data from the PHENIX collaboration.

1 Introduction

Heavy partons, which are produced in the early stage of heavy ion collisions could induce a large amount of energy loss for hard patrons produced in the initial stage of the collision. So, the observed jet quenching in such collisions can indicate the formation of this strongly interacting dense matter [1, 2]. Charm and bottom quarks play crucial role in this endeavor because they are

*Email: sharareh.mehrabi.pari@gmail.com
**Email: Javidan@um.ac.ir
***Email: taghavishahri@um.ac.ir

the witness to the entire space-time evolution of the system and due to their large masses, a memory of their interaction history may be preserved [3].

The Fokker-Plank equation (FPE) provides an appropriate framework for studying the evolution of the heavy quark distribution functions [4].

Heavy quarks dissipate their energy during their propagation through QGP via two processes [5]: The first one is the collisional process and the second one is the radiative process [6].These processes can be described well by the Fokker-Plank equation. In this paper our main task is to study the evolution of heavy quarks in quark gluon plasma by solving the Fokker-Plank equation with a new method named "Iterative Laplace Transform Method" [7]. This method gives numerical solutions in the form of convergent series. To verify our solutions, we calculate the nuclear suppression factor, $R_{AA}(p_T)$, of heavy quarks using simulated data of heavy ion collisions [8].To do this we consider both the collisional and radiation parts of energy loss in FPE and finally we compare our results with experimental data from PHENIX [9].

2 The heavy quark momentum evolution in FPE dynamics

The Quark Gluon Plasma (QGP) is formed at the time τ_i after the impact. Time evolution of the momentum distribution of non-equilibrated heavy quarks due to the interaction with the equilibrated QGP during the time interval $\tau_i < \tau < \tau_{HQ}$ can be calculated by the Fokker-Plank equation as follow:

$$\frac{\partial f(p,t)}{\partial t} = A(p)\frac{\partial}{\partial p}(pf(p,t)) + D(p)\left[\frac{\partial}{\partial p}\right]^2 f(p,t) \quad . \tag{1}$$

Where A is drag and D is diffussion coefficients.The drag coefficient is an important quantity carrying information about the dynamics of elastic collisions. It is expected that the drag coefficient should be determined by the properties of the bath and not by the nature of the HQ. In this work we use the drag coefficient as follows:

$$A(p,t) = -\frac{1}{p}\frac{dE}{dL} \quad . \tag{2}$$

If coupling between HQs and background QGP is weak, then the diffusion coefficient can be calculated using the Einstein relation, $D(p) = ETA(p)$, where M is the mass of the heavy quark and T is the thermal bath temperature [10]. In our calculations, the collisional energy loss of HQs in QGP by considering "Hard and Soft Thermal Loops" are given as follows [11]. For $E \ll \frac{M^2}{T}$:

$$-\frac{dE}{dL} = \frac{8\pi\alpha_s^2 T^2}{3}\left(1+\frac{n_f}{6}\right)\left[\frac{1}{v} - \frac{1-v^2}{2v^2} ln(\frac{1+v}{1-v})\right] ln\left[2^{\frac{n_f}{6+n_f}} B(v)\frac{ET}{m_g M}\right] \quad , \tag{3}$$

for $E \gg \frac{M^2}{T}$:

$$-\frac{dE}{dL} = \frac{8\pi\alpha_s^2 T^2}{3}\left(1+\frac{n_f}{6}\right) ln\left[2^{\frac{1}{2}(\frac{n_f}{6+n_f})} 0.92\frac{\sqrt{ET}}{m_g}\right] \quad , \tag{4}$$

where v is the HQ speed, α_s is the strong coupling constant, n_f is the number of quark flavors in the medium, E and M are energy and mass of the propagated HQ respectively, $m_g = \sqrt{(1+n_f/6)g^2T^2/3}$ is the thermal gluon mass, $g = \sqrt{4\pi\alpha_s}$ is the gauge coupling parameter and $B(v)$ is a smooth velocity function, which can be taken approximately as 0.7 [11]. Finally the total energy loss due to both collision and radiation precesses is:

$$-\frac{dE_{total}}{dL} = \frac{10\alpha_s^2 T^2}{3} ln\left(\frac{3E}{8\pi\alpha_s T}\right)\left[ln\left(\xi+\sqrt{1+\xi^2}\right)+\xi ln\left(\frac{1}{\xi}+\sqrt{1+\frac{1}{\xi^2}}\right)+\frac{3}{8\pi}\right] \tag{5}$$

where $\xi = \frac{9E}{2\pi^3 T}$. Using equation (4), (5), we will calculate the drag and diffusion coefficients for HQ energy loss during the interaction with QGP medium. In the next section we will find the solution of the Fokker-plank equation, eq(1).

3 The Iterative Laplace Transform Method

In this section we want to introduce a new method for solving the FP equation, namely the "Iterative Laplace Transform Method" (ILTM) [7]. This new method was proposed by Daftardar-Gejji and Jafari to seek numerical solutions of nonlinear functional equations. This method is called the Iterative Laplace Transform Method, which is a combination of two powerful methods, namely, the Laplace transform method and the Iterative method. The method gives numerical solutions in the form of convergent series with easily computable components. The most outstanding feature of this method is that it provides an analytical solution using the initial conditions only, without any discretization or restrictive assumptions. In general, the Fokker-Plank equation can be written as [7]:

$$D_t^\alpha f = \left[D_p^\beta A(p,t,f) + D_p^{2\beta} B(p,t,f)\right] f(p,t) \quad . \tag{6}$$

Here $D_t^{\alpha}(.), D_p^{\beta}(.), D_p^{2\beta}(.)$ are the Caputo fractional derivative with respect to t and p.

In our calculations we use the following definitions:

I. The Caputo fractional derivative of function (p, t) is defined as

$$D_t^{\alpha} f(p,t) = \frac{1}{\Gamma(m-\alpha)} \int (t-\eta)^{m-\alpha-1} f^m(p,\eta) d\eta \tag{7}$$

for $m-1 < \alpha \leq m$,$m \in N$.

II. Laplace transform of $D_t^{\alpha} f(p,t)$ is given by

$$\mathscr{L}[D_t^{\alpha} f(p,t)] = x^{\alpha} \mathscr{L}[f(p,t)] - \sum_{k=0}^{m-1} f^{(k)}(p,0) x^{\alpha-1-k}; \quad m-1 < \alpha \leq m, \tag{8}$$

where $f^{(k)}(p,0)$ is the kth derivative of $f(p,t)$ at $t=0$.

To illustrate how the iterative Laplace transform method works, the general space-time fractional partial differential equation is considered

$$D_t^{\alpha} f = \mathscr{A}(f, D_p^{\beta} f, D_p^{2\beta} f, ...) \qquad m-1 < \alpha \leq m$$
$$n-1 < \beta \leq n \quad m,n \in N, \tag{9}$$

where $\mathscr{A}(f, D_p^{\beta} f, D_p^{2\beta} f, ...)$ is a linear or nonlinear operator of $f, D_p^{\beta} f, D_p^{2\beta} f, ...$

With the initial condition

$$f^{(k)}(p,0) = h_k(p); \quad k = 0,1,...,m-1, \tag{10}$$

and $f(p,t)$ will be determined later.

First of all we take the Laplace transform of both sides of (9)

$$x^{\alpha} \mathscr{L}[f(p,t)] - \sum_{k=0}^{m-1} x^{\alpha-1-k} f^{(k)}(p,0) = \mathscr{L}[\mathscr{A}(f, D_p^{\beta} f, D_p^{2\beta} f, ...)]. \tag{11}$$

After simplification, we have

$$\mathscr{L}[f(p,t)] = \sum_{k=0}^{m-1} x^{-1-k} f^{(k)}(p,0) + x^{-\alpha} \mathscr{L}[\mathscr{A}(f, D_p^{\beta} f, D_p^{2\beta} f, ...)]. \tag{12}$$

By inverse Laplace transforms of both sides of Eq. (12) we have:

$$f(p,t) = \mathscr{L}^{-1}[\sum_{k=0}^{m-1} x^{-1-k} f^{(k)}(p,0)] + \mathscr{L}^{-1}[x^{-\alpha}\mathcal{L}[\mathscr{A}(f, D_p^{\beta} f, D_p^{2\beta} f, ...)]] , \tag{13}$$

Then we set

$$f_0 = \mathscr{L}^{-1}\left[\sum_{k=0}^{m-1} x^{-1-k} f^{(k)}(p,0)\right],$$
$$f_1 = \mathscr{L}^{-1}[x^{-\alpha}\mathcal{L}[\mathscr{A}(f, D_p^{\beta} f, D_p^{2\beta} f, ...)]]$$

If we want to calculate the next term in the series we have to consider f_1 as initial condition and putting it in the equation (9) and repeat all the above steps. Therefore the solution of (9) with initial condition (10) is given by:

$$f(p,t) \cong f_0(p,t) + f_1(p,t) + ... + f_m(p,t), m = 1, 2, ... \tag{14}$$

If we set $\alpha = \beta = 1$ in Equation (9) we will reach the FPE. We solved the FPE using both finite difference method and ILTM with different types of initial conditions and compared their results. For initial conditions with non vanishing values at the borderer, the ILTM successfully solves the FPE while results of finite difference method diverges and therefore it failes to find the solution. We have taken transverse momentum distribution functions of charm and bottom quarks at $\sqrt{s} = 200$GeV in Au-Au and also P-P collisions from [12]. ILTM has been performed up to $m = 2$.

4 Calculating the nuclear suppression factor $R_{AA}(P_T)$

The dissipation coefficients of the thermal bath and initial momentum distribution of HQs, are basic inputs required to solve the FPE. In our calculation the time dependence of temperature is taken from [13]. Here we considered a fixed value for coupling constant as $\alpha_s = 0.3$.

Figure 1 demonstrates the drag coefficient for b and c quarks due to collision and radiation separately. This figure clearly shows that the energy loss for c quark is greater than that for b quark.

One of the most important experimental observables to quantify the nuclear medium is the depletion of high p_T particles (D and B mesons

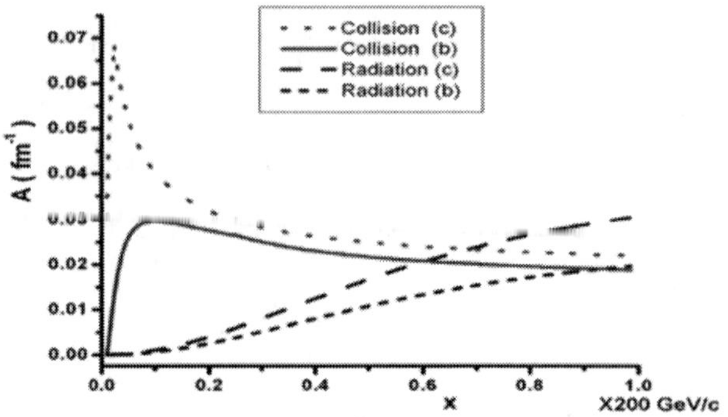

Figure 1: The drag coefficient due to collision and radiation as functions of momentum for Bottom and Charm quarks in QGP.

or non-photonic single e$^-$ spectra resulting from semi leptonic decays of hadrons containing charm and bottom quarks) produced in Nucleus-Nucleus collisions with respect to those produced in proton-proton collisions which is formulated in nuclear suppression factor R_{AA}.

Here the nuclear modification factor in relativistic heavy ion collisions is calculated for demonstrating the power of ILTM applied to solve the FP equation. The solution of the FP equation have been convoluted with the fragmentation functions of the heavy quarks to obtain the P_T distribution of the heavy mesons [13]. Similar simulations also have been performed to find

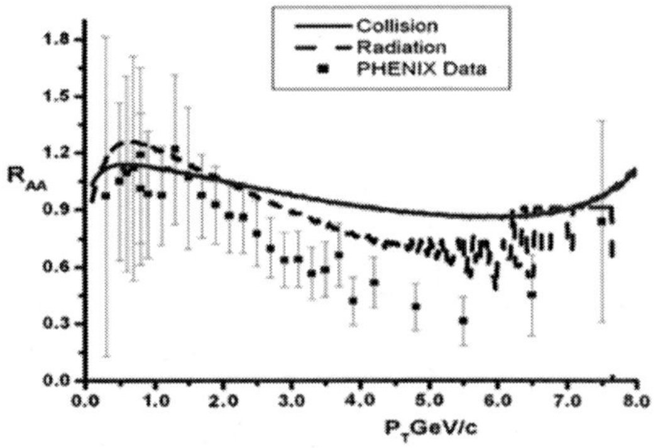

Figure 2: Nuclear suppression factor R_{AA} due to collisional and radiation energy loss separately as functions of P_T. Used parameters are: $\tau_0 = 0.33$ fm/c, $T_0 = 0.375$ GeV, $\alpha_s = 0.3, T_c = 0.175$ GeV [13].

the electron spectrum created in the P-P collisions. For this purpose HQ distribution functions of proton at $\sqrt{s} = 200\text{GeV}$ have been taken as initial conditions to solve FPE again. The ratio of these two quantities is proportional to the nuclear suppression factor, R_{AA} measured in experiments:

$$R_{AA}(P_T) \propto \frac{\frac{dN^e}{dP_T d\eta}^{Au-Au}}{\frac{dN^e}{dP_T d\eta}^{P-P}} \tag{15}$$

A small value of R_{AA} indicates a strong suppression and therefore, large energy loss of heavy quarks in the medium. It is clear that this ratio is equal to one in the absence of any medium.

Figure 2 presents contribution of collision and radiation energy loss in suppression factor R_{AA} separately. Displayed experimental data has been obtained from the PHENIX collaborations for $Au+Au$ collisions at $\sqrt{s_{NN}} = 200\text{GeV}$. This figure shows that suppression at higher values of P_T can not be explained by collisional energy loss. In other words, contribution of radiation in energy loss at higher P_T is more important than collisional energy loss only.

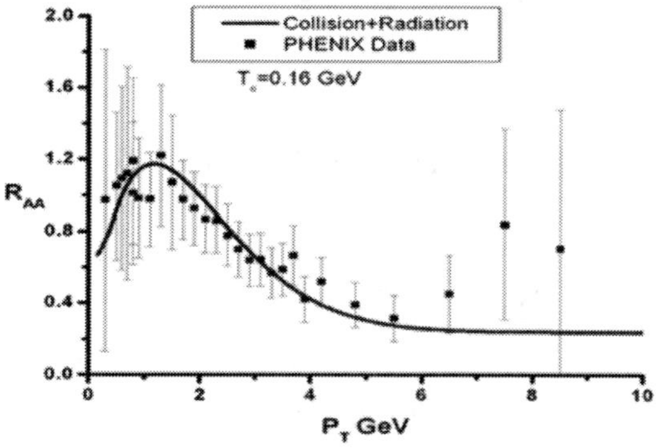

Figure 3: Nuclear suppression factor R_{AA} as function of P_T. Energy loss due to collision and radiation have been considered in calculating the drag coefficient.

5 Conclusions and remarks

The "Iterative Laplace Transform Method" has been introduced as an effective method to solve the Fokker-Plank equation for initial conditions with non-zero values at the boundaries. Our numerical algorithm shows that this method is able to solve time evolution of distribution function of HQs successfully. To verify the ability of ILTM and demonstrating its application, we calculated the nuclear suppression factor R_{AA}. It is shown that there is a good agreement between simulation results and reported experimental data (fig. 3).

References

[1] W. Scheid, M. Muller and W. Greiner, PRL **32**, 741 (1974). G. F. Chapline, M. H. Johnsou, E. Teller, M. S. Weiss PRD **8**, 135 (1973). L. P. Csernai, et al., PRC **25**, 2482 (1982).

[2] J.Y. Ollitrault, Eur. J. Phys. **29**, 275 (2008).

[3] B. Svetitsky Phys. Rev. D **37**, 2484 (1988).

[4] M. L. Mangano, P. Nason and G. Ridolfi, Nucl. Phys. B **538**, 282 (2002).

[5] S. Mazumder, T. Bhattacharyya, J. Alam and S.K. Das, Phys. Rev. C **84**, 044901 (2011).

[6] B. Z. Kopeliovich, I. K. Potashnikova, I. Schmidt, Phy. Rev. C **82**, 037901 (2010).

[7] L. Yan, Hindawi Publishing Corporation, Volume 2013, Article ID 465160.

[8] S.K. Das, F. Scardina, S. Plumari, and V. Greco, Phys. Rev. C **90**, 044901 (2014).

[9] S. S. Adler et al., (PHENIX Collaboration), Phys. Rev. C **84**, 044905 (2011).

[10] A.K. Chaudhuri, V.P. Goncalves, and L. F. Mackedanz, Brazilian Journal of Physics, 37,no 2B (2007).

[11] M. B. Gay Ducati, V. P. Goncalves and L. F. Mackedanz, Brazilian Journal of Physics, 37, 656-660 (2007).

[12] M. Cacciari, P. Nason, R. Vogt, PRL 95, 122001 (2005)

[13] S. K. Das, J. Alam, and P. Mohanty, Phys. Rev. C **82**, 014908 (2010).

EPJ Web of Conferences **117**, 03013 (2016) DOI: 10.1051/epjconf/201611703013
NN2015

Shear viscosity to electric conductivity ratio of the QGP

A. Puglisi[1,2], S. Plumari[1,2] and V. Greco[1,2]

[1] Department of Physics and Astronomy, University of Catania, Via S. Sofia 64, I-95125 Catania, Italy
[2] Laboratorio Nazionale del Sud, INFN-LNS, Via S. Sofia 63, I-95125 Catania, Italy

Abstract

The transport coefficients of strongly interacting matter are currently subject of intense theoretical and phenomenological studies due to their relevance for the characterization of the quark-gluon plasma produced in ultra-relativistic heavy-ion collisions (uRHIC). We predict that $(\eta/s)/(\sigma_{el}/T)$, independently on the running coupling $\alpha_s(T)$, should increase up to about ~ 20 for $T \to T_c$, while it goes down to a nearly flat behavior around $\simeq 4$ for $T \geq 4\,T_c$. Therefore we find a stronger T-dependence of σ_{el}/T with respect to η/s that in a quasiparticle approach is constrained by lQCD thermodynamics. A conformal theory, instead, predicts a similar T dependence of η/s and σ_{el}/T.

1 Introduction

Relativistic Heavy Ion Collider (RHIC) at BNL and Large Hadron Collider (LHC) at CERN have produced a very hot and dense system of strongly interacting particles as in the early universe with energy densities and temperatures largely above the transition temperature $T_c \simeq 160$MeV [1] expected for the transition from nuclear matter to the Quark-Gluon Plasma (QGP) [2]. The phenomenological studies by viscous hydrodynamics [3, 4] and parton transport [5–7] of the collective behavior of such a matter has shown that the QGP has a very small shear viscosity to entropy density ratio η/s, quite close to the conjectured lower-bound limit $\eta/s = 1/4\pi$ [8].

Another key transport coefficient of interest is the electric conductivity σ_{el} which represents the linear response of the system to an applied external electric field. Several processes occurring in uRHIC as well as in the Early Universe are regulated by the electric conductivity. Very high electric and magnetic fields ($eE \simeq eB \simeq m_\pi^2$, with m_π the pion mass) are expected to be produced in the very early stage of the collisions [9, 10]. First preliminary studies in lQCD has extracted only few estimates with large uncertainties [11,12] and only recently more safe extrapolation has been developed [13–15]. In this work we emphasize the connection with the η/s [16].

2 Shear viscosity and electric conductivity

In this Section we report the general formulas for shear viscosity and electric conductivity. For a system with different species, shear viscosity can be written as [17]:

$$\frac{\eta}{s} = \frac{1}{15Ts}\left\langle \frac{p^4}{E^2} \right\rangle \left(\tau_q \rho_q^{tot} + \tau_g \rho_g\right) \qquad (1)$$

where T is the temperature, $\rho_{q(g)}$ the quark (gluon) density, $\tau_{q(g)}$ relaxation time and $\langle\cdots\rangle$ the thermale average, being E the energy and p the momentum of particles.

Electric conductivity can be written as [18]:

$$\sigma_{el} = \frac{e^2}{3T}\left\langle \frac{\vec{p}^{\,2}}{E^2} \right\rangle \sum_{j=q,\bar{q}} f_j^2\, \tau_j \rho_j = \frac{e_\star^2}{3T}\left\langle \frac{\vec{p}^{\,2}}{E^2} \right\rangle \tau_q \rho_q \qquad (2)$$

where $e_\star^2 = e^2 \sum_{j=u,d,s}^{\bar{u},\bar{d},\bar{s}} f_j^2 = 4e^2/3$ with f_j the fractional quark charge. The thermal average $\langle p^4/E^2\rangle$ and $\langle p^2/E^2\rangle$ will be fixed employing a quasi-particle (QP) scheme tuned to reproduce the bulk thermodynamics evaluated by lQCD [19]. This means that thermodynamical terms in Eqs. (1)-(2) are determined by the Lattice QCD thermodynamics and do not rely on the detailed value of $m_{q,g}(T)$ in the QP model. The quark and gluon masses are given by $m_g^2 = 3/4\, g^2T^2$ and $m_q^2 = 1/3\, g^2T^2$ in terms of a running coupling $g(T)$ that is determined by a fit to the lattice energy density. In Ref. [19] we have obtained: $g^2_{QP}(T) = 48\pi^2/\,(11N_c - 2N_f)\ln[\lambda\left(\frac{T}{T_c} - \frac{T_s}{T_c}\right)]^2$ with $\lambda = 2.6$, $T_s/T_c = 0.57$, $T_c = 160\,\mathrm{MeV}$. We also notice that a self-consistent dynamical model has been developed in [6,20] and leads to nearly the same behavior of the strong coupling $g(T)$. For its general interest and asymptotic validity for $T \to \infty$, we also consider the behavior of the pQCD running coupling constant: $g_{pQCD}(T) = \frac{8\pi^2}{9} \ln^{-1}\left(\frac{2\pi T}{\Lambda_{QCD}}\right)$.

The last term to be defined and fixed in Eqs. (1)-(2) are the relaxation times for quarks and gluons:

$$\tau_q^{-1} = \sum_{i=q,\bar{q},g} \langle \rho_i v_{rel}^{iq} \sigma_{tr}^{iq} \rangle = \langle \sigma(s)_{tr} v_{rel} \rangle (\rho_q \sum_{i=u,d,s}^{\bar{u},\bar{d},\bar{s}} \beta^{qi} + \rho_g \beta^{qg})$$
$$\tau_g^{-1} = \sum_{i=q,\bar{q},g} \langle \rho_i v_{rel}^{ig} \sigma_{tr}^{ig} \rangle = \langle \sigma(s)_{tr} v_{rel} \rangle \left(\rho_q^{tot} \beta^{qg} + \rho_g \beta^{gg} \right) \quad (3)$$

where σ_{tr}^{ij} is the transport cross-section, v_{rel}^{ij} is the relative velocity of the two scattering particles. As done within the Hard-Thermal-Loop (HTL) approach, we will consider the total transport cross section regulated by a screening Debye mass $m_D = g(T)T$: $\sigma_{tr}^{ij}(s) = \int \frac{d\sigma}{dt} \sin^2\Theta \, dt = \beta^{ij} \frac{\pi \alpha_s^2}{m_D^2} \frac{s}{s+m_D^2} h(a)$ where $\alpha_s = g^2/4\pi$, $h(a)$ regulates the anisotropy of scatterings (for more details see Ref. [16,21,22]). The coefficient β^{ij} depends on the pair of interacting particles: $\beta^{qq} = 16/9$, $\beta^{qq'} = 8/9$, $\beta^{qg} = 2$, $\beta^{gg} = 9$ which are directly related to the quark and gluon Casimir factor, for example $\beta^{qq}/\beta^{gg} = (C_F/C_A)^2 = (4/9)^2$. It is not obvious that relaxation times are those evaluated with the same coupling $g(T)$ from the QP model. We fix $\tau_{q,g}$ in order to reproduce the minimum $\eta/s = 1/4\pi$. In Fig. 1 we show η/s as a function of T/T_c: red thick line is obtained using g_{QP} rescaled to the minimum $\eta/s = 1/4\pi$, blue dot-dashed line using g_{pQCD} and symbols several lQCD results (open and full circles [23], full squares [24], diamonds and triangles [25]).

We want to stress that the σ_{el}/T predicted, with the same τ_q that reproduces $\eta/s = 1/4\pi$, is in quite good agreement with most of the lQCD data, shown by symbols in Fig. 2 (grey squares [11], triangles [14], circle [15], yellow diamond [12], orange square [26] and red diamonds [13]). Therefore a low σ_{el}/T is obtained at variance with the early lQCD estimate, Ref. [11]. In Fig.2, we also plot the $\mathcal{N} = 4$ Super Yang Mills (green dotted line) electric conductivity [27] that predicts a constant behavior for $\sigma_{el}/T = e^2 N_c^2/(16\pi) \simeq 0.0164$. We note that in our framework one instead expects that the σ_{el} should still have a strong T-dependence: using simple considerations one can obtain $\sigma_{el}/T \simeq \eta/s \, T/m(T)$ [16] which shows that, even if η/s is constant, σ_{el}/T has an extra T-dependence that generates the steep decrease close to T_c.

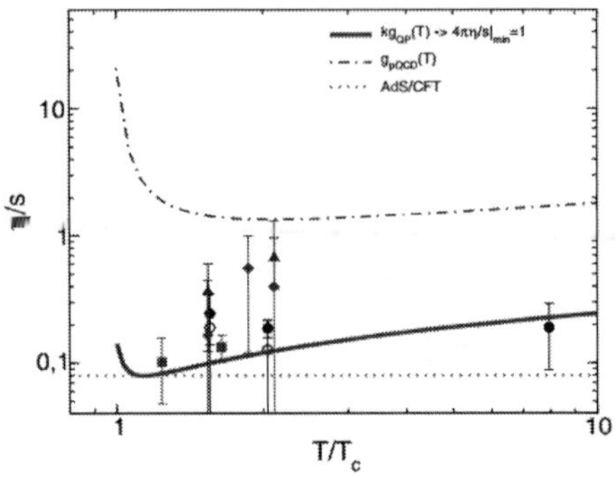

Figure 1: Shear Viscosity as a function of temperature T/T_c. See the text for details.

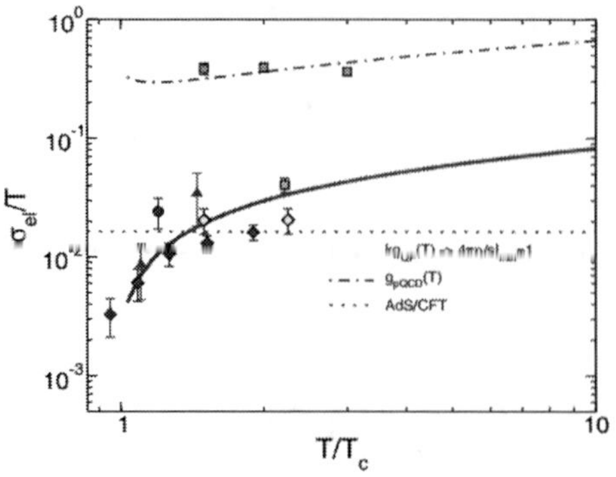

Figure 2: Electric conductivity as a function of temperature T/T_c. See the text for details.

An interesting quantity under our investigation is to consider the ratio between η/s and σ_{el}/T which can be written as:

$$\frac{\eta/s}{\sigma_{el}/T} = \frac{6}{5}\frac{1}{T s\, e_\star^2}\frac{\langle p^4/E^2\rangle}{\langle p^2/E^2\rangle}\left(1+\frac{\tau_g}{\tau_q}\frac{\rho_g}{\rho_q^{tot}}\right), \qquad \frac{\tau_g}{\tau_q} = \frac{C^q+\frac{\rho_g}{\rho_q}}{6+\frac{\rho_g}{\rho_q}C^g}. \tag{4}$$

The previous equation is written in terms of generic relaxation times and we note that the ratio τ_g/τ_q is proportional to the coefficient $C^q = (\beta^{qq}+\beta^{q\bar{q}}+2\beta^{q\bar{q}'}+2\beta^{qq'})/\beta^{qg}$ and $C^g = \beta^{gg}/\beta^{qg}$ which represent the relative magnitude between quark-(anti-)quark and gluon-gluon with respect to gluon-quark scatterings ($C^q|_{pQCD} = \frac{28}{9} \simeq 3.1$ and $C^g|_{pQCD} = \frac{9}{2}$).
In Fig. 3 we show $(\eta/s)/(\sigma_{el}/T)$ as a function of T/T_c: the red solid line is the the prediction for the ratio using $g_{QP}(T)$, but it is clear from the Eq. (4) that the ratio is independent on the running coupling itself (blue dashed line using g_{pQCD}). The ratio is sensitive only to the relative strength of the quark scatterings: as we can see comparing the orange curve (obtained increasing the quark scatterings $C^q = 10\, C^q_{pQCD}$) with the black curve (increasing the gluon scatterings $C^g = 10 C^g_{pQCD}$). As $T \rightarrow T_c$ a steep increase is predicted that is essentially regulated by $\langle p^2/E^2\rangle$.

In this work we point out the direct relation between the shear viscosity η and the electric conductivity σ_{el}. In particular, we have discussed why most recent lQCD data [13–15] predicting an electric conductivity $\sigma_{el} \simeq 10^{-2}T$ (for $T < 2\,T_c$) , appears to be consistent with a fluid at the minimal conjectured viscosity $4\pi\eta/s \simeq 1$. The increasing behavior of σ_{el}/T as a function of temperature supports AdS/QCD predictions [28] (violet dot-dashed line in Fig. 3).

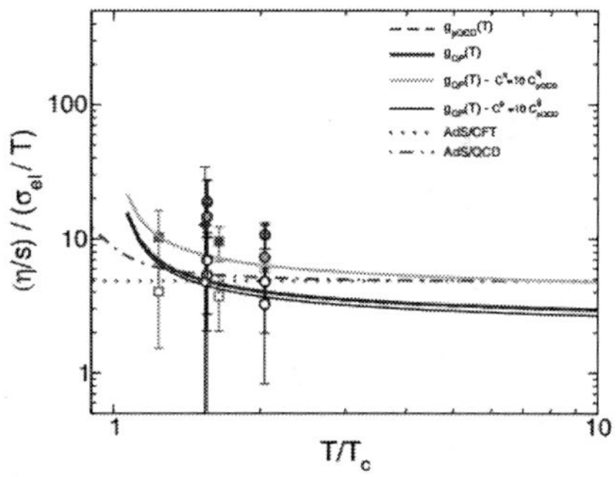

Figure 3: Ratio $(\eta/s)/(\sigma_{el}/T)$ as a function of temperature T/T_c. See the text for details.

We found the ratio $(\eta/s)/(\sigma_{el}/T)$ is independent of the uncertainties of the running coupling $g(T)$ and it is regulated by the relative strength and chemical composition of the QGP through the term $(1 + \tau_g\rho_g/\tau_q\rho_q^{tot})$. Our study provides a criterion to interpret the ratio and understand the relative role of quarks and gluons in the QGP thanks to the developments of lQCD techniques.

V.G. acknowledges the support of the ERC-StG Grant under the QGP-Dyn project.

References

[1] B. V. Jacak and B. Muller, *Science* **337**, 310 (2012).

[2] M. P. Lombardo, PoS LATTICE **2012** (2012) 016

[3] P. Romatschke and U. Romatschke, Phys.Rev.Lett. **99**, 172301 (2007).

[4] H. Niemi, G. S. Denicol, P. Huovinen, E. Molnar and D. H. Rischke, Phys. Rev. Lett. **106** (2011) 212302.

[5] Z. Xu, C. Greiner, and H. Stocker, Phys. Rev. Lett. **101**, 082302 (2008).

[6] W. Cassing and E. Bratkovskaya, Nucl. Phys. **A831**, 215 (2009).

[7] S. Plumari and V. Greco, AIP Conf. Proc. **1422**, 56 (2012).

[8] P. Kovtun, D. Son, and A. Starinets, Phys. Rev. Lett. **94**, 111601 (2005).

[9] K. Tuchin, *Adv.High Energy Phys.* 2013 (2013) 490495.

[10] Y. Hirono, M. Hongo, T. Hirano, *arXiv*:1211.1114.

[11] S. Gupta, *Phys. Lett.* B **597**, 57 (2007).

[12] G. Aarts, C. Allton, J. Foley, S. Hands, and S. Kim, *Phys. Rev. Lett.* **99**, 022002 (2007).

[13] A. Amato, G. Aarts, C. Allton, P. Giudice, S. Hands, and J. Skullerud, *Phys. Rev. Lett.* **111**, 172001 (2013).

[14] H.-T Ding, A. Francis, O. Kaczmarek, F. Karsch, E. Laermann et al., PoS 185 (2012).

[15] B. B. Brandt, A. Francis, H. B. Meyer, and H. Witting, PoS 186 (2012).

[16] A. Puglisi, S. Plumari and V. Greco, arXiv:1407.2559 [hep-ph].

[17] C. Sasaki and K. Redlich, *Phys. Rev. C* **79**, 055207 (2009).

[18] D. Fernandez-Fraile and A. Gomez Nicola, Phys. Rev. D **73** (2006) 045025.

[19] S. Plumari, W. M. Alberico, V. Greco and C. Ratti, Phys. Rev. **D84** (2011) 094004.

[20] R. Marty, E. Bratkovskaya, W. Cassing, J. Aichelin, and H. Berrehrah, Phys. Rev. C **88**, 045204 (2013).

[21] S. Plumari, A. Puglisi, F. Scardina and V. Greco, *Phys. Rev. C.* **86**, 054902 (2012).

[22] A. Puglisi, S. Plumari and V. Greco, Phys. Rev. D **90** (2014) 11, 114009

[23] S. Sakai and A. Nakamura, Lattice calculation of QGP viscosities: Present results and next project Proc. Sci. LAT2007(2007)221.

[24] H. B. Meyer, *Phys. Rev. D* **76** 101701 (2007).

[25] A. Nakamura and S. Sakai, *Phys. Rev. Lett.* **94** 072305 (2005).

[26] P. V. Buividovich, M. N. Chernodub, D. E. Kharzeev, T. Kalaydzhyan, E. V. Luschevskaya and M. I. Polikarpov, *Phys. Rev. Lett.* **105**, 132001 (2010)

[27] S. C. Huot, P. Kovtun, G. D. Moore, A. Starinets and L. G. Yaffe, *JHEP*12 (2006) 015.

[28] S. I. Finazzo and J. Noronha, *Phys.Rev. D* **89** 106008 (2014).

EPJ Web of Conferences **117**, 03014 (2016) DOI: 10.1051/epjconf/201611703014
NN2015

Modelling early stages of relativistic heavy-ion collisions

M. RUGGIERI[1], A. PUGLISI[1,2], L. OLIVA[1,2], S. PLUMARI[1,2], F. SCARDINA[1,2] and V. GRECO[1,2]

[1]Department of Physics and Astronomy, University of Catania, Via S. Sofia 64, I-95125 Catania
[2]INFN-Laboratori Nazionali del Sud, Via S. Sofia 62, I-95123 Catania, Italy

Abstract

In this study we model early time dynamics of relativistic heavy ion collisions by an initial color-electric field which then decays to a plasma by the Schwinger mechanism. The dynamics of the many particles system produced by the decay is described by relativistic kinetic theory, taking into account the backreaction on the color field by solving self-consistently the kinetic and the field equations. Our main results concern isotropization and thermalization for a $1+1$D expanding geometry. In case of small η/s ($\eta/s \lesssim 0.3$) we find $\tau_{isotropization} \approx 0.8$ fm/c and $\tau_{thermalization} \approx 1$ fm/c in agreement with the common lore of hydrodynamics.

1 Introduction

According to the general understanding of relativistic heavy ion collisions, in the overlap region of the two colliding nuclei a peculiar ensemble of (almost) longitudinal color fields named the Glasma is produced soon after the collision. One of the main problems is then how the Glasma, which corresponds to a configuration of classical fields, evolves to a quark-gluon plasma. In this talk we present one possible approach to this problem, presenting our recent results about modelling early time dynamics of relativistic heavy ion collisions by coupling the dynamics of the initial color field to that of the quark-gluon plasma. More specifically we assume that initially the system consists of a classical color field only, which then decays according to the Schwinger mechanism into a plasma of quarks and gluons; the dynamics of the system particles+field is studied self-consistently by coupling evolution equations for distribution function and field.

2 The flux tube model

Here we briefly summarize the flux tube model which we implement in our simulations of the initial stage of relativistic heavy ion collisions [1]. Due to lack of space we skip all the unnecessary details, referring to the original literature for a more complete treatment [1–3]. In the present study we limit ourselves to a simple geometrical initial condition with just one flux tube of a given transverse area. Although this is an oversimplified version of the appropriate initial condition for high energy nuclear collisions, this model permits to understand the main physical mechanisms which lead to particle production, thermalization and isotropization, as well as to compute the time scales for such processes. Hence our work is a necessary step for the study of a more realistic initial condition which will be the subject of forthcoming work.

The main assumptions of the model are that in the initial condition a color electric field is present, which decays into particle quanta by the Schwinger mechanism; the equations of motion for the classical fields are assumed to be the Maxwell equations, in which we introduce the backreaction to the field due to polarization and conduction currents. We assume the electric field, which we denote by E in the following, to be purely longitudinal, and limit ourselves to a one dimensional expansion calculation. Within this simplified geometry, given a pure longitudinal initial field, E remains longitudinal along all the time evolution since transverse currents are not produced.

3 Relativistic transport theory

Our calculation scheme is based on the Relativistic Transport Boltzmann equation which, in the presence of a gauge field $F^{\mu\nu}$, can be written as follows:

$$\left(p^\mu \partial_\mu + gQ_{jc}F^{\mu\nu}p_\nu \partial^p_\mu\right) f_{jc}(x,p) = \frac{dN_{jc}}{d\Gamma} + \mathcal{C}_{jc}[f] \ , \tag{1}$$

where $f_{jc}(x,p)$ is the distribution function for flavour j and color c, $F^{\mu\nu}$ is the field strength tensor. On the right hand side we have the source term $dN/d\Gamma$ which describes the creation of quarks, antiquarks and gluons due to the Schwinger effect [1] and $\mathcal{C}[f]$ which represents the collision integral, which is responsible of the change of distribution function due to stochastic collisions in the system and which is responsible for viscous corrections to ideal hydrodynamics. In our calculations we solve Eq. (1) by mapping the phase space of the fluid by means of test particles; the collision integral is computed using Monte Carlo methods based on the stochastic interpretation of transition amplitude [4–12].

At variance with the standard use of transport theory, in which one fixes a set of microscopic processes into the collision integral, we have developed an approach that fixes the total cross section cell by cell in configuration space in order to have the wished η/s. By means of this scheme we are able to use the Boltzmann equation to simulate the dynamical evolution of a fluid with specified shear viscosity, in analogy to what is done within hydrodynamical simulations. Once again, due to lack of space we refer to [7–12] for more details.

We assume the dynamics of the classical field is invariant for boosts along the longitudinal directions; in this case the relevant Maxwell equations can be combined to give a boost invariant equation

$$\frac{dE}{d\tau} = -j \ , \tag{2}$$

where τ corresponds to proper time and j is the total current computed in the local rest frame of the fluid which is given by the sum of two contributions, $j = j_D + j_M$: in fact the Schwinger effect can be described as a dielectric breakdown in which dipoles are produced by quantum tunneling hence changing the local dipole moment of the vacuum, with the corresponding current given by j_D; then the charges move in the medium due to the residual electric field giving rise to a conductive current, j_M. The color charge and current densities depend on the particle distribution function: hence they link the Maxwell equations (2) to the kinetic equation (1). We solve self-consistently the field and kinetic equations: in this way we take into account the back reaction of particle production and propagation on the color field.

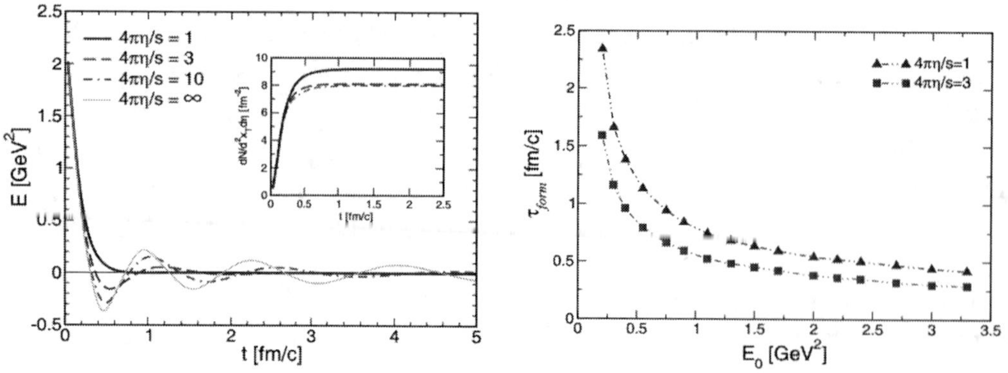

Figure 1: *Left panel.* Chromoelectric field strength (main panel) and particle number produced per unit of transverse area and rapidity (inset panel) as a function of time. *Right panel.* Proper formation time of quark-gluon plasma as a function of the initial field strength.

4 Results and discussion

In the right panel of Fig. 1 we plot the color electric field strength averaged in the central rapidity region $|\eta| < 0.5$ (main panel) and particle number produced per unit of transverse area and rapidity (inset panel) as a function of time. The calculations are shown for several values of η/s. The chromoelectric field experiences a rapid decay for small values of η/s, which in fact is a power law decay as it can be proved analytically [1]. On the other hand for intermediate and large values of η/s the electric field experience fluctuations during time evolution, due to the formation of plasma modes which affect the late time evolution of the system. In the inset of Fig. 1 we plot the number of produced gluons per unit of transverse area and rapidity versus time. We find that regardless of the value of η/s the particles are produced at very early time, approximately within 0.5 fm/c, with the only expection of very few particles produced at later times in the case $4\pi\eta/s = 10$.

On the right panel of Fig. 1 we plot the proper formation time of the quark-gluon plasma, for two values of η/s, as a function of the initial electric field strength, E_0. These data complete the ones shown in the left panel of Fig. 1 in which only one value of E_0 is considered. It is interesting that unless E_0 is too small, quark-gluon plasma is formed by the field decay within the first fm/c.

On the left panel of Fig. 2 we plot the ratio among longitudinal and transverse pressures, respectively P_L and P_T, as a function of time, for several values of η/s. These quantities are computed cell by cell in the local

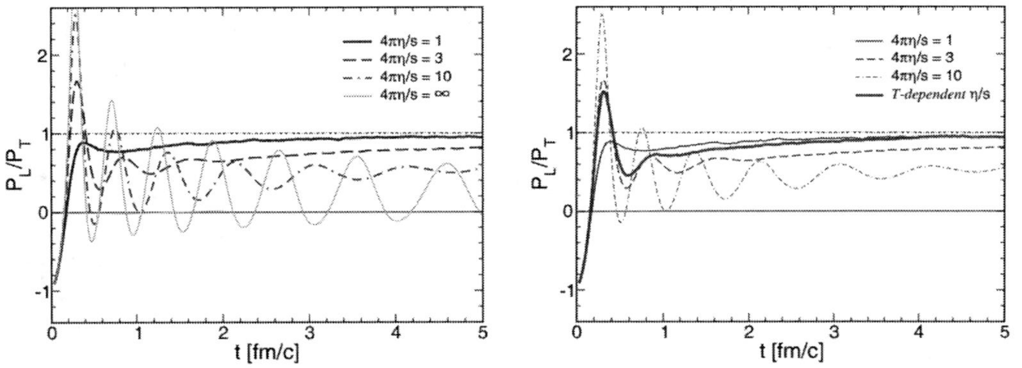

Figure 2: *Left panel.* Ratio P_L/P_T versus time for several values of η/s. *Right panel.* Same quantity for the case of a temperature dependent η/s (few results with fixed η/s are shown to ease comparison).

rest frame of the fluid, then averaged in the rapidity range $|\eta| < 0.5$; both particles and field contributions are taken into account in this calculation. A perfect isotropy requires $P_L = P_T$. The initial longitudinal pressure at initial time is negative and $P_L/P_T = -1$ because at initial time the system is made of pure longitudinal chromoelectric field. Hence the initial condition is highly anisotropic. However as soon as particles are produced, they give a positive contribution to the longitudinal pressure and the field magnitude decreases, eventually leading to a positive pressure. For all the value of η/s we consider in our simulations we find that the time needed to the total longitudinal pressure to be positive is about 0.2 fm/c. Moreover in the case $4\pi\eta/s = 1$ the strong interactions among the particles remove the initial pressure anisotropy quite efficiently and quickly: in this case $P_L/P_T \approx 0.7$ within 0.6 fm/c, then the ratio tends to increase towards 1 within the time evolution of the system. This would justify the use of viscous hydrodynamics with $\tau_0 \approx 0.6$ fm/c as commonly done. On the other hand for larger values of η/s plasma oscillations govern the evolution of P_L/P_T; moreover at large times P_L/P_T is quite smaller than 1, hence in these cases the system is quite far from the isotropic regime.

On the right panel of Fig. 2 we superimpose P_L/P_T obtained with a temperature dependent η/s. This temperature dependence is borrowed from Ref. [14], which is linearly increasing for $T \geq T_c$ while linearly decreasing for $T \leq T_c$; such a temperature dependence might be relevant for collisions at the LHC energy in which a large temperature range is expected to be experienced by the quark-gluon plasma; moreover it permits to implement a kinetic freezout at $T \approx T_c$ then mimicking some milder interaction in the hadron phase. We notice that besides a bump at initial times, the behaviour of P_L/P_T is comparable to the one we obtain for $4\pi\eta/s = 1$ after ≈ 1.5 fm/c, signaling that the larger initial η/s delays isotropization time, but the delay does not change drastically the picture obtained for $4\pi\eta/s = 1$.

5 Conclusions

In conclusions, in this work we have discussed a model for early stages of relativistic heavy ion collisions, coupling the dynamics of a color field to that of the quark-gluon plasma produced by the decay of the field itself. We have studied self-consistently the dynamics of the system field+particles by coupling field and relativistic Boltzmann equations, assuming a simplified initial condition consisting of a single color flux tube and a 1+1 dimensional expansion. Our main results concern about isotropization, thermalization and formation of the quark-gluon plasma: we have found that unless we consider values of η/s very large, thermalization, isotropization and particle creation occur within a time of ≈ 1 fm/c. Improvements of the present results include a realistic initial condition as well as 3+1 dimensional expansion, which will permit the computation of collective flows: these interesting improvements will be the subject of forthcoming publications.

References

[1] M. Ruggieri, A. Puglisi, L. Oliva, S. Plumari, F. Scardina and V. Greco, arXiv:1505.08081 [hep-ph].

[2] R. Ryblewski and W. Florkowski, Phys. Rev. D **88**, 034028 (2013).

[3] W. Florkowski, *Phenomenology of Ultra-Relativistic Heavy-Ion Collisions*, Singapore, Singapore: World Scientific (2010) 416 p.

[4] Z. Xu and C. Greiner, Phys. Rev. C **79** (2009) 014904

[5] Z. Xu, C. Greiner and H. Stocker, Phys. Rev. Lett. **101** (2008) 082302

[6] E. L. Bratkovskaya, W. Cassing, V. P. Konchakovski and O. Linnyk, Nucl. Phys. A **856** (2011) 162

[7] G. Ferini, M. Colonna, M. Di Toro and V. Greco, Phys. Lett. B **670**, 325 (2009)

[8] S. Plumari and V. Greco, AIP Conf. Proc. **1422** (2012) 56

[9] S. Plumari, A. Puglisi, M. Colonna, F. Scardina and V. Greco, J. Phys. Conf. Ser. **420**, 012029 (2013).

[10] S. Plumari, A. Puglisi, F. Scardina and V. Greco, Phys. Rev. C **86**, 054902 (2012).

[11] M. Ruggieri, F. Scardina, S. Plumari and V. Greco, Phys. Lett. B **727**, 177 (2013).

[12] M. Ruggieri, F. Scardina, S. Plumari and V. Greco, Phys. Rev. C **89**, no. 5, 054914 (2014).

[13] A. Puglisi, S. Plumari and V. Greco, Phys. Rev. D **90**, no. 11, 114009 (2014).

[14] S. Plumari, V. Greco and L. P. Csernai, arXiv:1304.6566 [nucl-th].

EPJ Web of Conferences **117**, 03015 (2016) DOI: 10.1051/epjconf/201611703015
NN2015

Heavy quark dynamics in the QGP: Towards a solution of the R_{AA} and v_2 puzzle

F. Scardina[1,2], S. K. Das[2], S. Plumari[1,2] and V. Greco[1,2]

[1]Department of Physics and Astronomy, University of Catania, Via S. Sofia 64, I-95123 Catania, Italy
[2]INFN-Laboratori Nazionali del Sud, Via S. Sofia 62, I-95123 Catania, Italy

Abstract

The two key observables related to heavy quarks that have been measured in experiments are the nuclear suppression factor R_{AA} and the elliptic flow v_2. The simultaneous reproduction of these two observables is a puzzle which have challenged all the existing models. We discuss two ingredients responsible for addressing a large part of such a puzzle: the temperature dependence of the energy loss and the full solution of the Boltzmann collision integral for the scattering between the heavy quarks and the particle of the bulk.

1 Introduction

One of the primary aims of the ongoing nuclear collisions at Relativistic Heavy Ion Collider (RHIC) and Large Hadron Collider (LHC) energies is to create a Quark Gluon Plasma (QGP). Heavy Quarks (HQ), charm and bottom, created in ultra-Relativistic Heavy Ion Collisions (uRHIC) represents ideal probes to study the QGP [1,2]. An essential feature in analyzing Heavy quarks motion in a QGP is that their mass is much larger than the typical momentum exchanged with the plasma particles entailing that many soft scatterings are necessary to change significantly the momentum and the

trajectory of the heavy quarks. Therefore the propagation of heavy quarks has been usually treated as a Brownian motion that is described by means of the Fokker-Planck (FP) equation.
The two key observables related to HQ that have been measured in experiments are the nuclear suppression factor R_{AA} and the elliptic flow v_2 [3–5]. Several theoretical efforts have been made to study the R_{AA} and the v_2 measured in experiments within the Fokker Planck (FP) approach [6–9, 11, 12] and the relativistic Boltzmann approach (BM) [13–16]. However all the approaches show some difficulties to describe both the nuclear modification factor and the elliptic flow simultaneously.
To study the time evolution of R_{AA} and v_2 we have solved the Fokker Planck (FP) equation stochastically in terms of the Langevin equation, for detail we refer to our previous work [11]. The evolution of the bulk is provided by a 3+1D relativistic transport code tuned at fixed η/s [18] which is able to reproduce the same results of hydrodynamical simulations. We have carried out simulations of $Au + Au$ collisions at $\sqrt{s} = 200$ AGeV for the minimum bias. The HQ distribution in momentum space is in accordance with the one in proton-proton rescaled by the number of binary collisions. We have used three different modelings to calculate the drag coefficient. The diffusion coefficient is instead calculated in accordance with the Einstein relation $D = \Gamma E T$.
In the first modeling we have evaluated the drag coefficient from (pQCD) and we have considered elastic interaction among HQ and the bulk (light quarks and gluons). The scattering matrix related to these processes $\mathcal{M}_{gHQ}$, $\mathcal{M}_{qHQ}$ and $\mathcal{M}_{\bar{q}HQ}$ in leading order are the well known Combridge matrix. The infrared singularity in the t-channel is regularized introducing a Debye screening mass $m_D = \sqrt{4\pi\alpha_s}T$ with a running coupling.
We have also considered another modeling in which the drag coefficient is evaluated considering a bulk consisting of particles with a T-dependent quasi-particle masses, $m_q = 1/3g^2T^2$, $m_g = 3/4g^2T^2$. This model is able to reproduce the thermodynamics of lattice QCD [20] (see also [21]) by fitting the coupling $g(T)$. Such a fit leads to the following coupling [20]:

$$g^2(T) = \frac{48\pi^2}{[(11N_c - 2N_f)ln[\lambda(\frac{T}{T_c} - \frac{T_s}{T_c})]^2} \tag{1}$$

where λ=2.6 and T/T_s=0.57.
Finally we have considered a model in which the light quarks and gluons are massless but the coupling is from the QPM as indicated in Eq. 1. This last case is indicated in the figures as $(\alpha_{QPM}(T), m_q = m_g = 0)$ and has

DOI: 10.1051/epjconf/201611703015

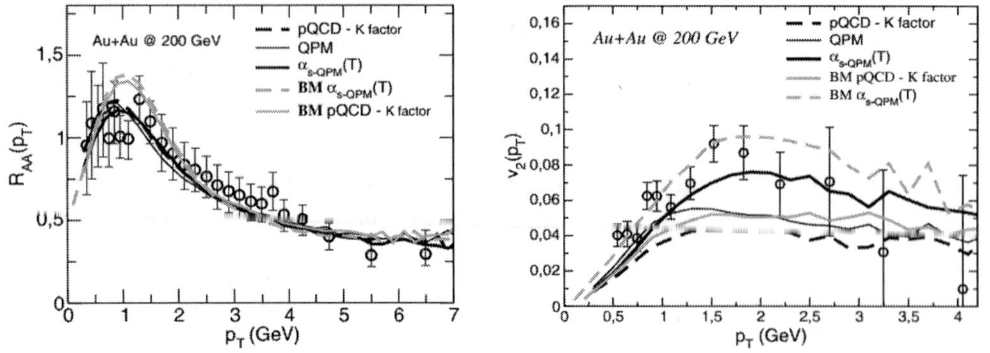

Figure 1: Comparison of the experimental data for the R_{AA} (left) and v_2 (right)at $(200AGeV)$ with the results we get using the different models. See text.

to be considered as an expedient to have a drag which increases when the temperature decreases.

For all the three cases considered the interaction has been rescaled to reproduce the R_{AA} observed in experiments. We have performed simulations of HQ propagation with the Langevin dynamics for the three different models presented above. The Langevin equation gives as output the momentum distributions of HQ at the quark-Hadron transition temperature T_c. The momenta distributions are convoluted with the Peterson fragmentation functions of the heavy quark to get the momentum distribution of D and B mesons.

In Fig. 1 (left) the nuclear modification factor R_{AA} of the D and B mesons is shown as a function of p_T for RHIC $(200AGeV)$. Instead in Fig. 1 (right) the elliptic flow (v_2) at the same energy as a function of p_T is depicted. We observe that the larger is the interaction in the region of low temperature the larger is the elliptic flow. The same conclusions has been discussed also in the light flavor sector as shown in Ref. [22]. The reason of such a strong dependence of the elliptic flow on the temperature dependence of the drag coefficient is due to the fact that the elliptic flow is generated in the final stage of the evolution of the fireball when the temperature is lower. This is shown in the Fig. 2 where the R_{AA} (left) and v_2 (right) are depicted at different times. We observe that the R_{AA} is generated in the early stage of the QGP when the temperature is larger and is not sensitive to the final stage of the evolution, while the elliptic flow is generated later. This results refer to the pQCD case, however the behavior is similar also for the other models.

We have also used the transport approach to study the propagation of Heavy quarks. The Boltzmann equation for the HQ distribution function is

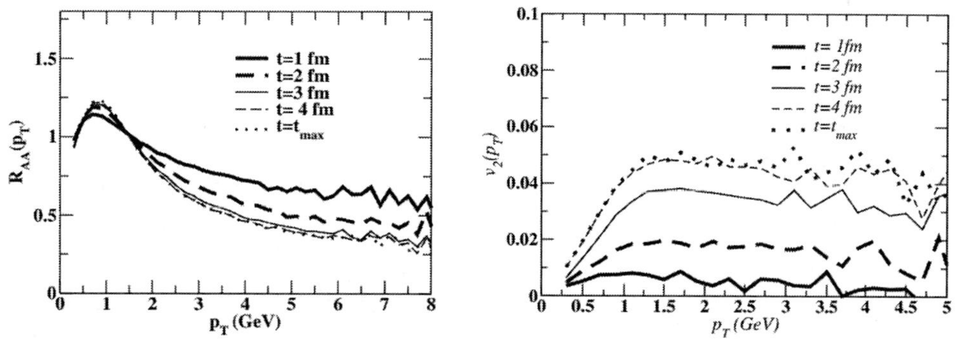

Figure 2: R_{AA} (left) and v_2 (right) evaluated at different time at RHIC energies ($200AGeV$).

indicated here

$$p^{\mu}\partial_{\mu}f_Q(x,p) = \mathcal{C}[f_Q](x,p) \tag{2}$$

where $\mathcal{C}[f_Q](x,p)$ is the relativistic Boltzmann-like collision integral which is solved by means of a stochastic algorithm. We use the Boltzmann equation to describe the propagation of the heavy quark as well as the evolution of the bulk.

The comparison between LV and BM approach has been thoroughly studied in these references [15,16] where it is shown that for charm quark the results that one gets using the Fokker-Planck approach deviate significantly from those obtained using the Boltzmann approach and such a deviation significantly depends on the the values of the Debye screening mass, whereas for bottom quarks the FP is a very good approximation. We considered in references [15, 16] three values of m_D: 0.4 GeV, 0.83 GeV and 1.6 GeV. Here we have not considered a fixed value of the Debye screening mass but a value which depends on the temperature according to $m_D = gT$. In Fig. 1 the comparison for the R_{AA} (left) and v_2 (right) at RHIC between the BM (light gray lines) and the FP (black lines) are shown. We found that using the BM for the same values of the R_{AA} we get larger values for v_2 with respect to those obtained using the FP.

Our results show a non-negligible impact of the approximation in the collision integral involved in the Fokker Plack equation on the relation between R_{AA} and v_2. We found that for the same R_{AA} we get a larger v_2 using the BM with respect to the v_2 we get with the FP.

To summarize our results, regarding the different values of elliptic flow that can be obtained using the different models of energy loss, we have introduced in Fig. 3 a new plot in which R_{AA} vs v_2 at a given momentum ($p_T = 1.3$ GeV) is shown.

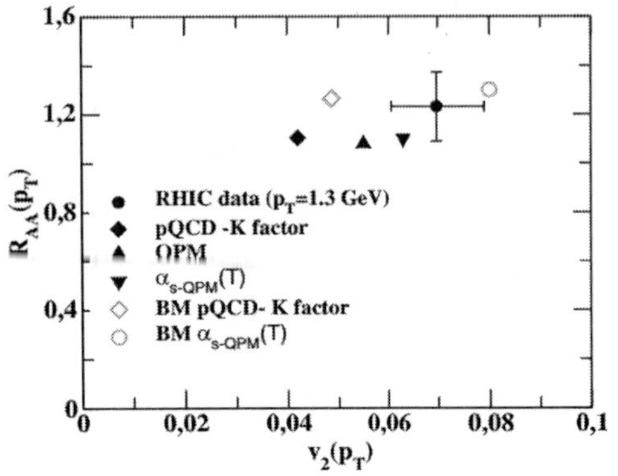

Figure 3: R_{AA} vs v_2 obtained with the FP for the three different T-dependences of the drag coefficient with the experimental data at RHIC energy at $p_T = 1.3GeV$. The open light gray simbols indicates the results obtained using the BM approach.

This plot clearly shows how the building up of the v_2 can differ up to a factor 3 for the same R_{AA} depending on the temperature dependence of the energy loss and on the approach used to describe the HQ propagation. Other two ingredients that can have an impact on the relation between R_{AA} and v_2 are the hadronization via coalescence that we are going to include in our description and the role of the hadronic phase that we have studied in [23]. We have found in such a study that the hadronic medium further enhance the v_2 by around 20% without affecting the nuclear modification factor.

Acknowledgments

The authors acknowledge the support by the ERC StG under the QGPDyn Grant n. 259684.

References

[1] Svetitsky B 1988 *Phys. Rev.* D **37**, 2484

[2] R.Rapp and H van Hees, R. C. Hwa, X. N. Wang (Ed.) Quark Gluon Plasma 4, 2010, *World Scientific*, **111**

[3] B. I. Abeleb (STAR Collaboration), Phys. Rev. Lett. **98**, 192301 (2007).

[4] A. Adare*et al.* (PHENIX Collaboration) A. Adare et al. (PHENIX Collaboration), Phys. Rev. C **84** , 044905 (2011)

[5] B. Abelev *et al.*,(ALICE Collaboration) JHEP **1209** (2012) 112

[6] Hees H van, Greco V and Rapp R 2006 *Phys. Rev.* C **73**, 034913; Hees H van, Mannearelli M, Greco V and Rapp R 2008 *Phys. Rev. Lett.* **100** 192301

[7] Das S K, Alam J, Mohanty 2009 *Phys. Rev.* C **80** 054916;

[8] Alberico W M *et al.* 2011 *Eur. Phys. J.* C **71** 1666;

[9] S. Cao, G. Y. Qin and S. A. Bass, Phys. Rev. C **88** (2013) 4, 044907

[10] M. He, R. J. Fries and R. Rapp, Phys. Rev. Lett. **110**, 112301 (2013)

[11] S. K. Das, F. Scardina, S. Plumari and V. Greco, Phys Lett. B747 (2015) 260-264

[12] F. Scardina,S. K. Das, S. Plumari,J. I. Bellone and V. Greco, arXiv:1509.01551

[13] Gossiaux P B, Aichelin J 2008 *Phys. Rev.* C **78** 014904

[14] Uphoff J, Fochler O, Xu Z and Greiner C 2011 *Phys. Rev.* C **84** 024908

[15] S. K. Das, F. Scardina, S. Plumari and V. Greco,Phys. Rev. C **90** 044901 (2014)

[16] F. Scardina,S. K. Das, S. Plumari and V. Greco, J.Phys.Conf.Ser. **535** (2014) 012019; F. Scardina,S. K. Das, S. Plumari and V. Greco, J.Phys.Conf.Ser. **636** (2015) 012017

[17] Younus M, Coleman-Smith C E,Bass S A and Srivastava D K, Phys.Rev. C **91** (2015) 2, 024912

[18] S. Plumari,A. Puglisi, F. Scardina and V. Greco, Phys. Rev. C **86** (2012) 054902

[19] V. Greco, C. M. Ko and P. Levai, Phys. Rev. Lett. **90** (2003) 202302; V. Greco, C. M. Ko and P. Levai, Phys. Rev. C **68** (2003) 034904

[20] S. Plumari, W. M. Alberico, V. Greco and C. Ratti, Phys. Rev. D, **84**, 094004 (2011)

[21] H. Berrehrah, E. Bratkovskaya, W. Cassing, P.B. Gossiaux, J. Aichelin, and M. Bleicher, Phys.Rev. C **89**, 054901 (2014),

[22] F. Scardina, M. Di Toro and V. Greco, Phys. Rev. C **82** (2010) 054901.

[23] S. K. Das, F. Scardina, S. Plumari and V.Greco, arXiv:1509:06307

Λ polarization in an exact fluid dynamical model for heavy-ion collisions

Yilong Xie*

Department of Physics and Technology,
University of Bergen, Allegaten 55, 5007 Bergen, Norway

Abstract

Λ polarization is calculated in an exact analytical, rotating model based on parameters from a high resolution (3+1)D Particle-in-Cell Relativistic hydrodynamics calculation. The polarization is attributed to effects from thermal vorticity and for the first time the effects of the radial and axial acceleration are also studied separately.

At finite impact parameters, the initial state (IS) has non-vanishing angular momentum. Early studies neglected effects arising from the non-vanishing angular momentum, but interest increased recently. With the development of hydrodynamic modeling, rotation and its consequences were studied as well.

Thermal vorticity arises from the inverse temperature field in heavy ion collisions, and due to the non-vanishing angular momentum and shear in the initial stages. We look at polarization in effects arising from thermal vorticity in an exact rotating model [1], corresponding to an appropriate time-period of the collision based on a (3+1)D fluid dynamical model

Conventionally, $[x, z]$-plane is the reaction plane, with y being the axis of rotation. Then the initial angular momentum points into the negative y-direction, with an absolute value of approximately $1.45 \cdot 10^4\hbar$.

*Email: yilong.xie@uib.no

Following [2], the polarization arises from the thermal velocity field, $\beta^\mu(x) = u^\mu(x)/T(x)$, due to equipartition between vorticity and spin as

$$\Pi_\mu(p) = \hbar\epsilon_{\mu\sigma\rho\tau}\frac{p^\tau}{8m}\frac{\int d\Sigma_\lambda p^\lambda n_F(x,p)(1-n_F(x,p))\partial^\rho\beta^\sigma}{\int d\Sigma_\lambda p^\lambda n_F(x,p)} \ . \tag{1}$$

where $\epsilon_{\mu\rho\sigma\tau}$ is the completely antisymmetric Levi-Civita symbol, n_F the Fermi-Jüttner distribution for spin-1/2 particles, $(1-n_F)$ is the Pauli blocking factor and p is the four-momentum of the Λ.

The Λ polarization is determined by measuring the angular distribution of the decay protons in the Λ's rest frame. By Lorentz boosting the polarization vector, $\mathbf{\Pi}(\boldsymbol{p})$, in the participant frame, one can obtain the polarization vector $\mathbf{\Pi}_0(\boldsymbol{p})$ in Λ's rest frame:

$$\mathbf{\Pi}_0(\boldsymbol{p}) = \mathbf{\Pi}(p) - \frac{\boldsymbol{p}}{p^0(p^0+m)}\mathbf{\Pi}(p)\cdot\boldsymbol{p} \ , \tag{2}$$

where $(p^0, \boldsymbol{p})$ is the Λ four-momentum and m its mass.

As the Λ is transversely polarized, $\Pi^\mu p_\mu = 0$, one can confine himself to the spatial part of Π^μ. The simplified spatial part of polarization vector is:

$$\begin{aligned}\mathbf{\Pi}(p) &= \mathbf{\Pi_1}(p) + \mathbf{\Pi_2}(p) \\ &= \frac{\hbar\epsilon}{8m}\frac{\int dV n_F(x,p)\,(\nabla\times\boldsymbol{\beta})}{\int dV n_F(x,p)} \\ &\quad + \frac{\hbar\boldsymbol{p}}{8m}\times\frac{\int dV n_F(x,p)\,(\partial_t\boldsymbol{\beta}+\nabla\beta^0)}{\int dV n_F(x,p)} \ .\end{aligned} \tag{3}$$

Using the vorticity evaluated in [3], for the non-relativistic Exact model, we deduced the analytical solution for the Λ polarization:

$$\begin{aligned}\mathbf{\Pi}(p) = \frac{\hbar}{8mT}\Bigg[&\frac{p_y c_9}{c_3\sqrt{c_4}}\frac{M_{-1,\frac{1}{2}}}{M_{-\frac{1}{2},0}}\boldsymbol{e}_x + \left(2\epsilon\omega - \frac{|p_x|c_9}{c_3\sqrt{c_4}}\times\frac{M_{-1,\frac{1}{2}}}{M_{-\frac{1}{2},0}}\right)\boldsymbol{e}_y \\ &+ \left(\frac{|p_x|c_7c_1}{2c_2} - \frac{p_y c_8}{c_3\sqrt{c_4}}\frac{M_{-1,\frac{1}{2}}}{M_{-\frac{1}{2},0}}\right)\boldsymbol{e}_z\Bigg] \ ,\end{aligned} \tag{4}$$

where c_i $(i = 1-9)$ are parameters in terms of scaling variables and $M_{\mu,\nu}(z)$, is the Whittaker function, the confluent hypergeometric function [1].

As seen in eq. (4), the polarization consists of two terms, $\mathbf{\Pi_1}(p)$ and $\mathbf{\Pi_2}(p)$, which arise from local vorticity $(\nabla\times\boldsymbol{\beta})$ and expansion $(\partial_t\boldsymbol{\beta})$. One can see from Fig. 1, that the second term in the polarization is of comparable magnitude to the term arising from local vorticity. In a previous calculation [4], the p dependence of n_F, was considered negligible in the integral, and the time derivative and gradient terms were also assumed to be smaller. The present calculation shows that in general these terms are not negligible.

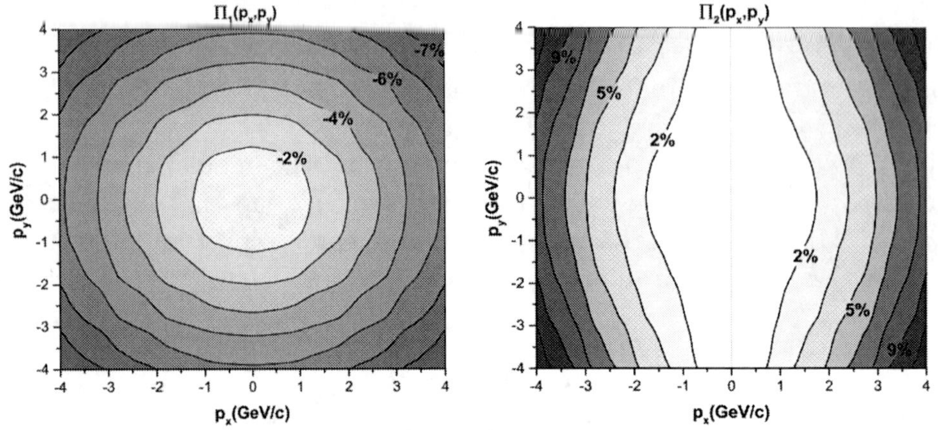

Figure 1: (color online) The two terms of Λ polarization $\mathbf{\Pi}_1(\boldsymbol{p})$ (left panel), $\mathbf{\Pi}_2(\boldsymbol{p})$ (right panel) in the participant frame at time $t = 0.5$ fm/c after the equilibration of the rotation, in the Exact model. Based on Ref. [1].

The Λ polarization is measured via the angular distribution of the decay protons in the Λ's rest frame, as shown in Eq. (2). The resulting distribution is shown in Fig. 2. The structure of $\Pi_{0y}(\boldsymbol{p})$ is similar to the one obtained in [4], but it reaches 12% at high p_x values, greater than 9% in Ref. [4], due to the contribution of the "second", $\partial_t \boldsymbol{\beta}$ term. These new studies indicate that the dynamics of the expansion may also lead to non-negligible contribution to the observable polarization.

Recently the vorticity and polarization were also studied in another fluid dynamical model [5], where the initial shear flow is neglected. This results in negligible thermal vorticity (Figs. 3 and 13 of Ref. [5]), and consequently a negligible polarization from the vorticity, i.e. from the "first term" discussed here. On the other hand, there is qualitative agreement between Fig. 12 of Ref. [5] and this work in the sense that only the y-directed (i.e. $[x, z]$ or $[x, \eta]$) component of the vorticity leads to an overall average net polarization. This arises in both models from the initial angular momentum and points into the $-y$-direction. In Ref. [5] this arises as a consequence of viscous evolution of the initial, vorticity-less flow, while in our Exact model it is present in the initial state.

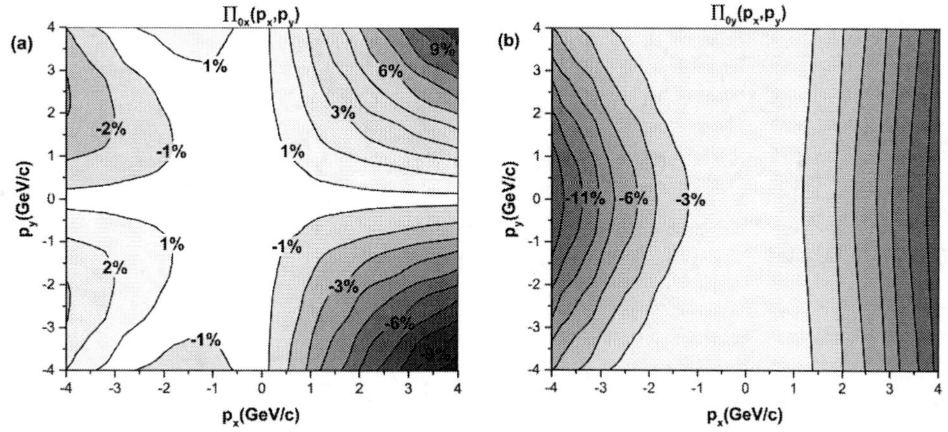

Figure 2: (color online) The radial, x, and axial, y, components of Λ-polarization, $\mathbf{\Pi}_0(\boldsymbol{p})$, in the Λ's rest frame. Both plots are asymmetric due to the Lorentz boost to the Λ rest frame. From [1].

In this work we analyzed and compared the two terms, and the Exact model. Including both rotation, expansion, and vorticity arising from both effects. This study indicates that the assumptions regarding the initial state are influencing the predictions on the observed vorticity.

Acknowledgements

Enlightening discussions with Francesco Becattini are gratefully acknowledged. This author is supported by the China Scholarship Council.

References

[1] Yilong Xie, Robert C. Glastad, Laszlo P. Csernai, arXiv: 1505.07221v2.

[2] F. Becattini, V. Chandra, L. Del Zanna, E. Grossi, Annals of Physics **338**, 32 (2013).

[3] L.P. Csernai, J.H. Inderhaug, Int. J. Mod. Phys. E **24**, 1550013 (2015).

[4] F. Becattini, L.P. Csernai, and D.J. Wang, Phys. Rev. C **88**, 034905 (2013).

[5] F. Becattini, G. Inghirami, V. Rolando, A. Beraudo, L. Del Zanna, A. De Pace, M. Nardi, G. Pagliara, and V. Chandra, arXiv: 1501.04468v2.

EPJ Web of Conferences **117**, 03017 (2016) DOI: 10.1051/epjconf/201611703017
NN2015

Overview of anisotropic flow measurements from ALICE

YOU ZHOU on behalf of the ALICE COLLABORATION

Niels Bohr Institute, University of Copenhagen, Denmark

Abstract

Anisotropic flow is an important observable to study the properties of the hot and dense matter, the Quark Gluon Plasma (QGP), created in heavy-ion collisions. Measurements of anisotropic flow for inclusive and identified charged hadrons are reported in Pb–Pb, p–Pb and pp collisions with the ALICE detector. The comparison of experimental measurements to various theoretical calculations are also presented in these proceedings.

1 Introduction

The study of anisotropic flow in relativistic heavy collisions at RHIC suggested that the produced system is described as a strongly coupled quark-gluon plasma (sQGP) [1]. Recent anisotropic flow measurements at the LHC indicate that the system created in Pb–Pb collisions at $\sqrt{s_{\rm NN}} = 2.76$ TeV also behaves as a strongly interacting liquid [2]. Meanwhile, additional constraints on the value of shear viscosity over entropy density ratio η/s are obtained by studying the anisotropic flow of identified particles. The most pronounced component of anisotropic flow is second order harmonic, called elliptic flow v_2. In these proceedings, we will discuss the elliptic flow of both charged particles and identified particles measured in ALICE.

2 Anisotropic flow measurements in Pb–Pb collisions

The centrality dependence of elliptic flow v_2, measured with different methods, is shown in Fig. 1 (left). The results fall into two groups: (a) results from two-particle correlations have high values; (2) results from multi-particle correlations have smaller v_2. The clear difference between the two groups was observed for all centrality classes. This could be due to the flow fluctuations, as well as due to the so-called non-flow effects, which is the azimuthal correlation not associate to the common symmetry planes. In addition, the v_2 measurements with the LYZ method at STAR are smaller than those at ALICE, at the level of 30%. This 30% difference is bigger than the ideal hydrodynamic predictions [3] but agrees with viscous hydrodynamics calculations [4].

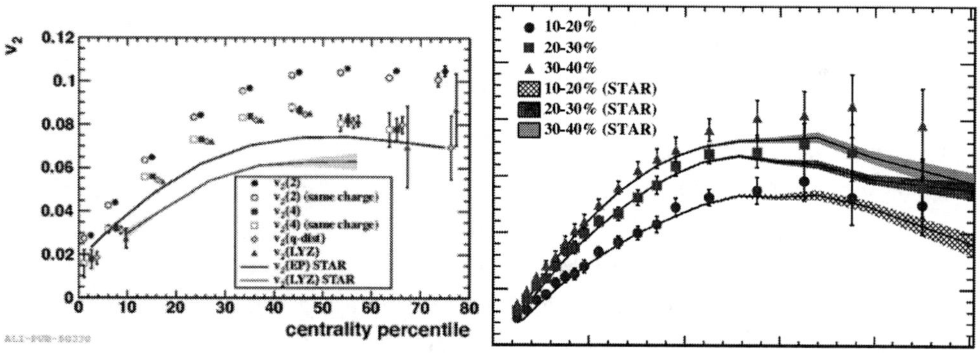

Figure 1: Centrality dependence of elliptic flow v_2 measured with various methods for RHIC and the LHC; p_{T}-differential $v_2\{4\}$ of charged particles for RHIC and the LHC (right).

In order to understand the 30% increasing of integrated v_2 from RHIC to the LHC, we study the v_2 in more details with the p_{T} differential flow $v_2(p_{\mathrm{T}})$ measurements. Figure 1 (right) shows the $v_2(p_{\mathrm{T}})$ for the centrality class 10-20%, 20-30% and 30-40%, using a 4-particle cumulant method, denoted as $v_2\{4\}$. The comparison for the same measurements in STAR are presented in the same centrality class. It is found that the differential $v_2\{4\}$ measurement does not change within uncertainties from top RHIC to LHC energies. It suggests that a large fraction of the 30% increasing integrated v_2 is due to an increase in the $\langle p_{\mathrm{T}} \rangle$.

In addition, the $v_2(p_{\mathrm{T}})$ is also measured for identified particles in Pb–Pb collisions at $\sqrt{s_{\mathrm{NN}}} = 2.76$ TeV [8]. It was reported that for the low p_{T} region (i.e. $p_{\mathrm{T}} \leq 3$ GeV/c), one can see a mass ordering feature ($v_2^{\pi} < v_2^{\mathrm{K}} < v_2^{\mathrm{p}} \leq$

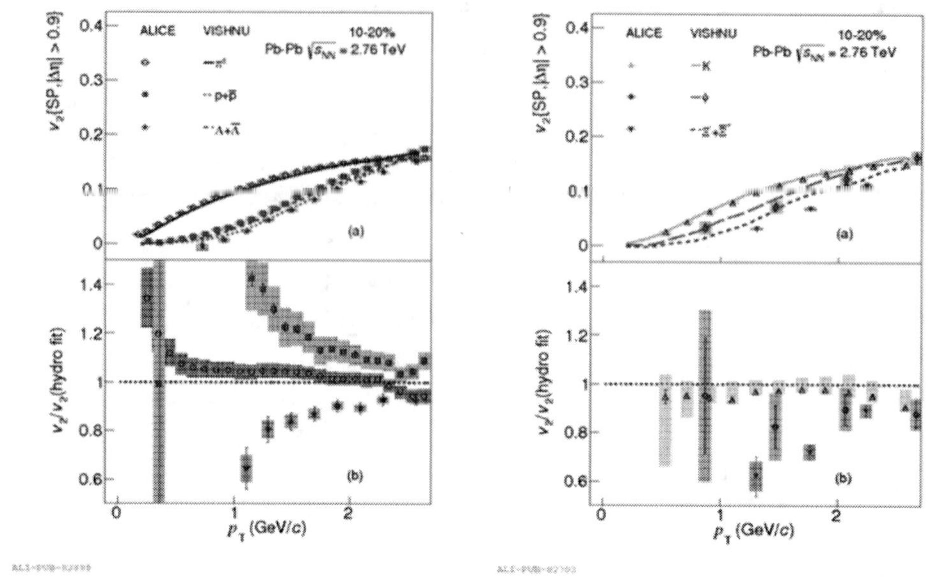

Figure 2: Measurements of dentified particle v_2 and the comparisons to the VISHNU calculations.

$v_2^{\phi} < v_2^{\Lambda} < v_2^{\Xi} < v_2^{\Omega}$). This is expected due to the interplay between elliptic and radial flow [5,6]. For higher values of p_{T} (i.e. $p_{\mathrm{T}} > 3$ GeV/c), particles tend to group according to their type, i.e. mesons and baryons.

Hydrodynamic calculations reveal the fact that the information of η/s of the QGP, the hadronic chemical compositions as well as off-equilibrium kinetics can be extracted from the comparison of identified particle v_2 measurements. Figure 2 shows a comparison between data and the calculations from VISHNU [9], which is based on initial conditions modeled by MC-KLN and a shear viscosity over entropy density ratio $\eta/s = 0.16$. It was found that the VISHNU calculations generate the main feature of the v_2 mass ordering like data at low p_{T}. However, a closer look reveals that VISHNU calculations of $\pi^{\pm}$ and p($\overline{\mathrm{p}}$) v_2 are systematically below the measured v_2 from ALICE, while the calculations of $\mathrm{K}^{\pm}$ v_2 are compatible with the data. Also v_2 of heavier particles, including ϕ meson, strange and multi-strange particles, are overestimated in this hybrid hydrodynamic calculations compared to data. Furthermore, it is noticed that in the VISHNU calculations, the results of ϕ and $\Lambda(\overline{\Lambda})$ v_2 do not follow the precise mass ordering at low p_{T} region. All these results suggest that ϕ meson, strange and multi-strange particles might either freeze-out later than what the current implementation of VISHNU expects, and/or the hadronic cross sections of these parti-

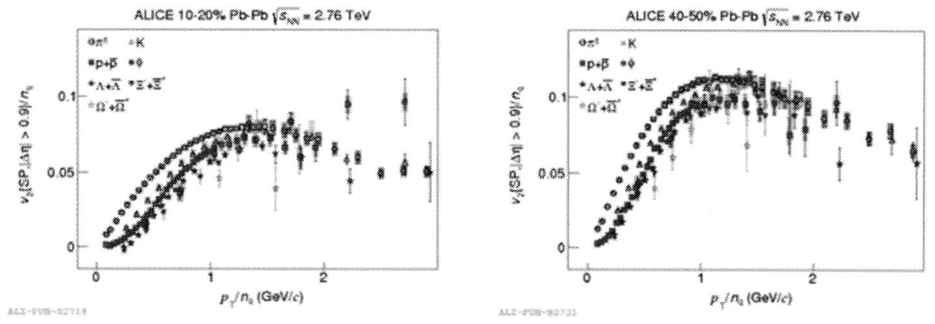

Figure 3: The p_{T}/n_q dependence of v_2/n_q of identified particles in the 10-20% and 40-50% centrality intervals of Pb-Pb collisions at $\sqrt{s_{\mathrm{NN}}} = 2.76$ TeV.

cles are underestimated in UrQMD model. Thus, further knowledge about freeze-out conditions, including the radial flow, freeze-out temperature, and better constraints of the hadronic cross section of the ϕ meson, strange and multi-strange particles are still necessary to improve the hydrodynamic calculations.

The number of constituent quarks scaling (NCQ scaling), was used to support the picture that the partonic collectivity already develops at top RHIC energy [7]. It serves as an examination of hadron production via the quark coalescence/recombination mechanism. The v_2 scaled by the number of constituent quarks (n_q) as a function of p_{T}/n_q for all particle species are shown in Fig. 3. It is clear that the NCQ scaling does not hold perfectly, the deviations of this scaling from unity are at the level of $\pm 20\%$ for all centrality classes.

3 Anisotropic flow measurements in p–Pb and pp collisions

The measurements of anisotropic flow in p–Pb collisions at the LHC were originally aimed to provide reference data for the Pb–Pb collisions, e.g. study the cold nuclear matter effects. However, unexpected collective behaviors have been discovered in experiments. Figure 4 show the two- and four-particle cumulants as a function of the charged particle multiplicity in $\sqrt{s_{\mathrm{NN}}} = 5.02$ TeV p–Pb collisions. It was observed that the two-particle cumulant decreases with increasing charged particle multiplicity. Various pseudorapidity gaps between the two particles have been examined. It was realized that using a pseudorapidity gap $|\Delta\eta| > 1.4$, the two-particle cumulant is

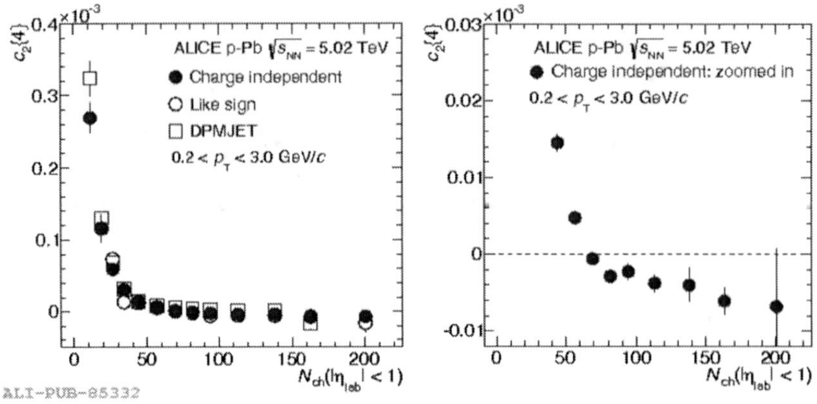

Figure 4: Two- and multi-particle cumulant as a function of the charged particle multiplicity in p–Pb collisions.

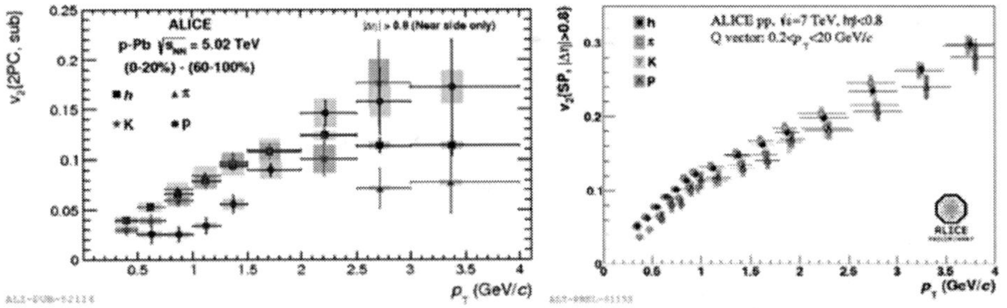

Figure 5: p_T-differential v_2 for all charged and identified particles in p–Pb collisions for the 0-20% multiplicity class (left) and in minimum biased pp collisions (right).

strongly suppressed, and an increase with increasing multiplicity is observed. In addition, the four-particle cumulant, which is less sensitive to the non-flow effects, was measured for the same multiplicity region. It was observed that for the events with a multiplicity above 70, the four-particle cumulant becomes negative, which is a prerequisite to extract $v_2\{4\}$, due to $v_2\{4\} = \sqrt[4]{-c_2\{4\}}$. Similar analysis was also performed in pp collisions at 7 TeV [11], no negative sign of $c_2\{4\}$ has been observed. The anisotropic flow signal in pp collisions, has not been observed for the presented multiplicity range.

The identified particle v_2 as a function of p_T was also investigated in p–Pb collisions. This result was obtained using a two-particle correlation technique. It was shown that after applying the subtraction procedure using 60-100 % multiplicity events to remove the non-flow contributions, a mass ordering is observed, which is in qualitatively agreement with hydrodynamic calculations. It was noticed that the mass ordering feature can be also reproduced in UrQMD simulations in p-Pb collisions. This is the consequence of hadronic interactions [12]. Figure 5 (right) shows the identified particle v_2 measurements in minimum bias pp collisions at $\sqrt{s} = 7$ TeV. It seems that the v_2 of pion is consistent with that of kaon for $p_T < 3$ GeV/c, while it is higher than v_2 of proton for the entire p_T range. This behavior is different from the mass ordering feature observed in both Pb–Pb and p–Pb collisions, presented in Fig. 2 and 5 (right).

References

[1] I. Arsene *et al.* [BRAHMS Collaboration], Nucl. Phys. A **757**, 1 (2005); K. Adcox *et al.* [PHENIX Collaboration], Nucl. Phys. A **757**, 184 (2005); J. Adams *et al.* [STAR Collaboration], Nucl. Phys. A **757** (2005) 102

[2] K. Aamodt *et al.* [ALICE Collaboration], Phys. Rev. Lett. **105**, 252302 (2010); Phys. Rev. Lett. **107**, 032301 (2011)

[3] G. Kestin and U. W. Heinz, Eur. Phys. J. C **61**, 545 (2009)

[4] H. Niemi, K. J. Eskola and P. V. Ruuskanen, Phys. Rev. C **79**, 024903 (2009)

[5] P. Huovinen, P. F. Kolb et. al., Phys. Lett. B **503**, 58 (2001)

[6] D. Teaney, J. Lauret and E. V. Shuryak, Phys. Rev. Lett. **86**, 4783 (2001)

[7] J. Adams *et al.* [STAR Collaboration], Phys. Rev. Lett. **92**, 052302 (2004) A. Adare *et al.* [PHENIX Collaboration], Phys. Rev. Lett. **98**, 162301 (2007)

[8] B. B. Abelev *et al.* [ALICE Collaboration], JHEP **1506**, 190 (2015)

[9] H. Song, S. Bass and U. W. Heinz, Phys. Rev. C **89**, 034919 (2014)

[10] B. B. Abelev *et al.* [ALICE Collaboration], Phys. Rev. C **90**, 054901 (2014)

[11] Y. Zhou [ALICE Collaboration], arXiv:1201.4671 [nucl-ex].

[12] Y. Zhou, X. Zhu, P. Li and H. Song, Phys. Rev. C **91**, 064908 (2015)

EPJ Web of Conferences **117**, 03018 (2016) DOI: 10.1051/epjconf/201611703018
NN2015

Is hadronic flow produced in p–Pb collisions at the Large Hadron Collider?

You Zhou[1*], Xiangrong Zhu[2], Pengfei Li[2], Huichao Song[2]

[1] Niels Bohr Institute, University of Copenhagen, Denmark
[2] Department of Physics and State Key Laboratory of Nuclear Physics and Technology, Peking University, China

Abstract

Using the Ultra-relativistic Quantum Molecular Dynamics (`UrQMD`) model, we investigate the azimuthal correlations in p–Pb collisions at $\sqrt{s_{\rm NN}} = 5.02$ TeV. It is shown that the simulated hadronic p–Pb system can not generate the collective flow signatures, but mainly behaves as a non-flow dominant system. However, the characteristic $v_2(p_{\rm T})$ mass-ordering of pions, kaons and protons is observed in `UrQMD` simulations, which is the consequence of hadronic interactions and not necessarily associated with strong fluid-like expansions.

The relativistic heavy ion collisions at the Large Hadron Collider (LHC) have provided strong evidences for the creation of the Quark–Gluon Plasma (QGP). One of the crucial observables is the azimuthal anisotropy of the transverse momentum distribution for produced hadrons [1], usually characterized by the Fourier flow-coefficients [2]. The second Fourier flow-coefficient v_2 is called elliptic flow, which has been systematically measured and studied at the LHC [3]. These results, together with the comparisons to theoretical model calculations, provides important information on the Equation of State (EoS) and the transport properties of the QGP. The azimuthal correlations in $\sqrt{s_{\rm NN}} = 5.02$ TeV p–Pb collisions were also measured at the LHC with the original purposes of providing reference data for the high

* you.zhou@cern.ch

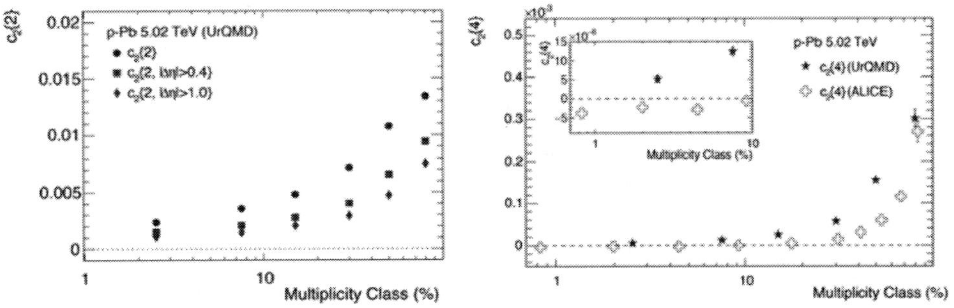

Figure 1: Multiplicity class dependence of $c_2\{2\}$ (left) and $c_2\{4\}$ (right).

energy Pb–Pb collisions. However, a large amount of unexpected collective behaviors have been discovered [9] and the results can be semi-quantitatively described by (3+1)-d hydrodynamic simulations [4,5]. It might suggest that the large collective flow has been developed in the small p–Pb systems. In this study [6], we utilize a hadron cascade model Ultra-relativistic Quantum Molecular Dynamics (`UrQMD` version 3.4) [7] to simulate the evolution of the hadronic matter and then study the azimuthal correlations of final produced hadrons.

Figure 1 (left) presents the centrality dependence of the 2-particle cumulant $c_2\{2\}$ from `UrQMD` model. With various pseudorapidity gaps, $c_2\{2\}$ exhibits decreasing trends from low multiplicity events to high multiplicity events. As the pseudorapidity gap increases, the magnitudes of $c_2\{2\}$ become weaker for `UrQMD`. These results agree with the suppression of the so-called non-flow effects, which is the azimuthal correlations not associated with the symmetry plane. When the pseudorapidity gap $|\Delta\eta|$ is larger than 1.0, $c_2\{2\}$ from `UrQMD` still presents strong centrality dependence, showing a typical non-flow behavior. It indicates that `UrQMD` hadronic expansion could not generate enough flow in a small p–Pb system, and non-flow effects still dominate the 2-particle correlations even with $|\Delta\eta| > 1.0$. To better understand the hadronic systems simulated by `UrQMD`, we investigate the 4-particle cumulant of the second Fourier flow-coefficient $c_2\{4\}$, which is less sensitive to non-flow effects. Figure 1 (right) shows that the $c_2\{4\}$ measurement exhibits a transition from positive to negative values for the most central collisions, indicating the creation of flow-dominated systems in the high multiplicity events. However, $c_2\{4\}$ calculations from `UrQMD` simulations keep positive for all available multiplicity classes, including central collisions. As a result, real values of $v_2\{4\}$ can not be extracted. This comparison further illustrates the difference between the p–Pb systems created

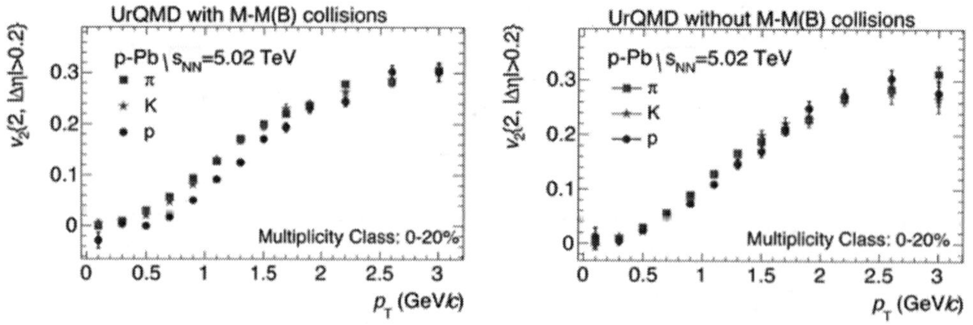

Figure 2: $v_2(p_T)$ of pions, kaons and protons in p–Pb collisions at $\sqrt{s_{NN}} = 5.02$ TeV, calculated from `UrQMD` with (left) and without (right) M-M and M-B collisions.

in experiment and simulated by `UrQMD`. Without the contributions from the initial stage and/or the QGP phase, the measured flow-like 4-particle correlations in high multiplicity events can not be reproduced by a microscopic transport model with only hadronic scatterings and decays.

In addition, the characteristic feature of $v_2(p_T)$ mass-ordering among pions, kaons and protons was observed in p–Pb collisions [9]. This mass ordering feature is similar with what observed in Pb–Pb collisions [10] and can be roughly reproduced by hydrodynamic calculations [4,5]. It was token as a strong evidence for the collective expansion of the p–Pb systems created in $\sqrt{s_{NN}} = 5.02$ TeV collisions. Figure 2 shows that a clear v_2 mass-ordering among pions, kaons and protons below 2 GeV is also observed in `UrQMD`. Such mass-ordering pattern, caused by only hadronic interactions, qualitatively agrees with the ALICE measurement [9].In `UrQMD`, the unknown hadronic cross sections are calculated by the additive quark model (`AQM`) through counting the number of constituent quarks within two colliding hadrons. As a result, the main meson-baryon (M-B) cross sections from `AQM` are about 50% larger than the meson-meson (M-M) cross sections, leading to the v_2 splitting between mesons and baryons after the evolution of hadronic matter. It is also observed in Fig. 2 (right) that when switch off the M-B and M-M interaction channels in `UrQMD`, the v_2 mass-ordering almost disappears. The comparison of Fig. 2 (left) and (right) illustrates that the hadronic interactions could lead to a v_2 mass-ordering feature, even for small p–Pb systems without enough flow generation.

In summary, using `UrQMD` hadron cascade model, we studied azimuthal correlations in p–Pb collisions at $\sqrt{s_{NN}} = 5.02$ TeV. It is found that hadronic interactions alone could not generate sufficient collective flow as observed in experiment. In order to fit the azimuthal correlations measurements, the contributions from the initial stage and/or the QGP phase can not be neglected. In addition, we extended our study of azimuthal correlations to identified hadrons. A v_2 mass-ordering was generated by `UrQMD`, which is similar to the ALICE measurements. This characterize feature is mainly caused by hadronic interactions. The experimentally observed v_2 mass-ordering alone is not necessarily explained as a flow signal associated with the strong fluid-like expansions.

References

[1] J. -Y. Ollitrault, Phys. Rev. D **46**, 229 (1992).

[2] S. Voloshin and Y. Zhang, Z. Phys. C **70**, 665 (1996)

[3] KAamodt *et al.* [ALICE Collaboration], Phys. Rev. Lett. **105**, 252302 (2010); Phys. Rev. Lett. **107**, 032301 (2011); S. Chatrchyan *et al.* [CMS Collaboration], Eur. Phys. J. C **72**, 2012 (2012); G. Aad *et al.* [ATLAS Collaboration], Phys. Lett. B **707**, 330 (2012).

[4] P. Bozek, W. Broniowski and G. Torrieri, Phys. Rev. Lett. **111**, 172303 (2013).

[5] K. Werner, M. Bleicher, B. Guiot *et al.*, Phys. Rev. Lett. **112**, 232301 (2014).

[6] Y. Zhou, X. Zhu, P. Li and H. Song, Phys. Rev. C **91**, 064908 (2015)

[7] S. A. Bass, M. Belkacem, M. Bleicher *et al.*, Prog. Part. Nucl. Phys. **41**, 255 (1998)

[8] B. B. Abelev *et al.* [ALICE Collaboration], Phys. Rev. C **90**, 054901 (2014)

[9] B. B. Abelev *et al.* [ALICE Collaboration], Phys. Lett. B **726**, 164 (2013)

[10] B. B. Abelev *et al.* [ALICE Collaboration], JHEP **1506**, 190 (2015)

EPJ Web of Conferences **117**, 04001 (2016) DOI: 10.1051/epjconf/201611704001
NN2015

Interplay between collective and single particle excitations around neutron-rich doubly-magic nuclei

S. Leoni

Università degli Studi di Milano and INFN, Sezione di Milano,
Milano, Italy

Abstract

The excitation spectra of nuclei with one or two particles outside a doubly-magic core are expected to be dominated, at low energy, by the couplings between phonon excitations of the core and valence particles. A survey of the experimental situation is given for some nuclei lying in close proximity of neutron-rich doubly-magic systems, such as ^{47,49}Ca, ^{133}Sb and ^{210}Bi. Data are obtained with various types of reactions (multinucleon transfer with heavy ions, cold neutron capture and neutron induced fission of ^{235}U and ^{241}Pu targets), with the employment of complex detection systems based on HPGe arrays. A comparison with theoretical calculations is also presented, in terms of large shell model calculations and of a phenomenological particle-phonon model. In the case of ^{133}Sb, a new microscopic "hybrid" model is introduced: it is based on the coupling between core excitations (both collective and non-collective) of the doubly-magic core and the valence nucleon, using the Skyrme effective interaction in a consistent way.

1 Introduction

One of the greatest challenges of contemporary nuclear physics is the achievement of a unified microscopic description of all nuclei through the entire periodic table, on basis of the common symmetry principle of the nuclear force. A major difficulty is related to the very large number of nuclear configurations one has to deal with, when distributing nucleons in the available orbitals of the proton and neutron shells. As an example, in an N=Z nucleus like ^{44}Ti such a number is of the order of 10^4, and it increases quickly to 10^{10} in ^{56}Ni and 10^{28} in ^{88}Ru: this makes microscopic calculations computationally challenging in medium mass nuclei and simply impossible in heavy systems. As a consequence, different theoretical approaches have been developed to describe the nuclear many body quantum system [1]. In particular, large scale Shell Model calculations can be performed for nuclei up to mass 100 and around closed shells, assuming a frozen core as a truncation scheme. Following this approach, one restricts excitations to valence nucleons within one/two oscillator shells, neglecting almost completely the excitations of the core. This become a relevant limitation particularly in nuclei with one or two particles outside a doubly-magic core, since in these systems the lowest structure should be dominated by the couplings between phonon excitations of the core (with a high degree of collectivity) and valence particles, giving rise to series of multiplets [2]. The identification of these multiplets can provide precise, quantitative information on the phonon-particle couplings, a phenomen of primary importance, being at the origin of the quenching of spectroscopic factors [3] and of the damping of giant resonances [4,5].

This paper will present the experimental situation in nuclei lying in close proximity of doubly-magic systems, such as ^{47,49}Ca [6,7], ^{133}Sb [8] and ^{210}Bi [8]. Additional examples around semi-magic systems can be found in Refs. [9,10]. Various types of reactions will be discussed: from multinucleon transfer with heavy ions, to cold neutron capture and neutron induced fission of ^{235}U and ^{241}Pu targets, using complex detection systems based on HPGe arrays coupled to magnetic spectrometers or fast $LaBr_3$ scintillator detector arrays for lifetime measurements. The experimental data will be then compared with theoretical calculations either based on a phenomenological particle-phonon coupling approach [2] or on a shell model employing realistic effective nucleon-nucleon interactions. A newly developed microscopic model which takes into account couplings between the valence particle and core excitations (both collective and non-collective), using the Skyrme effective interaction in a consistent way, will be also presented in connection with the interpretation of ^{133}Sb.

2 Around the stable doubly-magic ^{48}Ca: the cases of ^{47,49}Ca

The nucleus ^{48}Ca is a double shell closure at the valley of stability, exhibiting only one sizable collective excitation, the 3^- phonon at 4.507 MeV (B(E3)=6.8±1 W.u.), while the 2^+ state at 3.832 MeV is little collective (B(E2)<2 W.u.). The collective character of the 3^- suggests the existence of multiplets of positive parity states arising by coupling the 3^- excitation to the unpaired $p_{3/2}$ neutron of ^{49}Ca and the $f^{-1}_{7/2}$ neutron hole of ^{47}Ca. Experimentally, candidates for this type of states were identified, in the past, both in ^{49}Ca and ^{47}Ca, in low resolution particle spectroscopy studies [11] and in γ-spectroscopy analysis of heavy-ion transfer reactions with thick targets [12]. However, only recently firm spin and parity assignments and lifetimes of these states were obtained in experiments performed at Legnaro National Laboratory of INFN [6, 7]. The nuclei of interest were populated by a ^{48}Ca beam (≈6 MeV/A) impinging on a thin ^{64}Ni target, employing the detection setup consisting of the large acceptance magnetic spectrometer PRISMA combined with the high efficiency Ge array CLARA [13].

Figure 1 shows, in the bottom, the γ-spectrum measured in coincidence with ^{49}Ca ions detected in PRISMA. All strong transitions correspond to decays from states identified as single particle excitations, besides the 660- and 3357-keV lines which depopulate the levels at 4017 keV and 3357 keV, with $9/2^+$ and $7/2^-$ tentative spin assignment, respectively [6]. The 4017 keV state is expected to originate from the $p_{3/2}$ neutron coupled to the 3^- phonon of ^{48}Ca. The spin assignments were proven by measuring the angular distribution and polarization of both transitions, which provided values consistent with electric stretched dipole for the 660-keV line and electric stretched quadrupole for the 3357-keV γ ray (cf. Fig. 1 a) and b)). In addition, the nature of these states was established by a lifetime analysis performed with the differential plunger technique, using the reaction ^{48}Ca+^{208}Pb (see Fig. 1 c) and d)). In particular, the lifetime of the 4017 keV level (depopulated by the 660-keV γ-ray) was found to be 8.5±2.0 ps, which gives a reduced transition probability B(E3) = 7.9 ± 2.0 Wu, in good agreement with the measured strength of the 3^- phonon of ^{48}Ca. A similar analysis was performed on the ^{47}Ca reaction product and the state at 3999 keV was suggested to be the $11/2^+$ or $13/2^+$ member of the $3^- \otimes f^{-1}_{7/2}$ multiplet. This identification was supported by the lifetime analysis, which provided a reduced transition probability equal to B(E3)=7.4±1.9 W.u., again very similar to the 3^- phonon strength of ^{48}Ca.

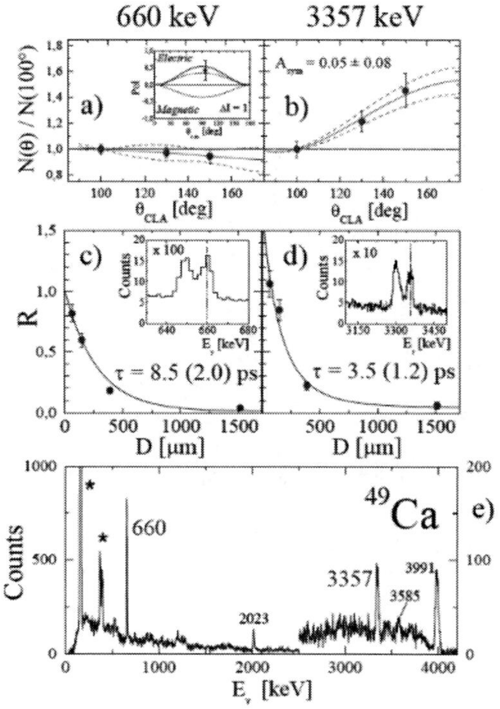

Figure 1: Panel e): γ spectrum measured in coincidence with ^{49}Ca ions. Panels a) and b): angular distribution and polarization analysis of the 660 and 3357 keV transitions, depopulating the $9/2^+$ and $7/2^-$ states. Panels c) and d): analysis of the 660 and 3357 keV lines by the differential recoil distance Doppler shift method, providing the lifetimes of the $9/2^+$ and $7/2^-$ levels of ^{49}Ca [6].

A weak coupling model [2] can be applied to predict the excitation energy spectrum arising from coupling a particle/hole to a phonon. This type of calculation was performed for both ^{47}Ca and ^{49}Ca, resulting in multiplets of states with energies shifted with respect to the 3^- phonon, with reasonable agreement with the experimental energies, especially in the case of the $9/2^+$ state of ^{49}Ca (see Fig. 2). At the same perturbative order, the calculations predict that the reduced transition probability B(EL) to the ground state should be associated to the same multipole of the vibrational state and should have the same value of the core nucleus, as observed experimentally in both cases. This gives strong evidence that these states have a particle-phonon coupled nature.

Large scale shell model calculations were also performed for ^{49}Ca using a truncated model space, since the full sd-pf (sd-pf-sdg) shell calculation is

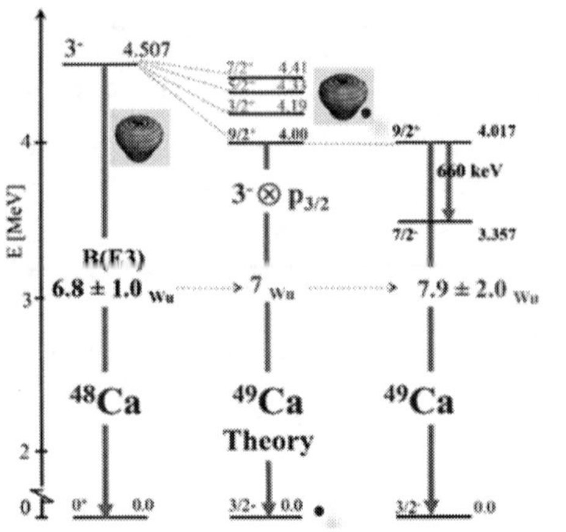

Figure 2: Right: experimental level scheme of ^{49}Ca, with the $9/2^+$ state as the lowest member of the $3^- \otimes p_{3/2}$ multiplet, arising by coupling one neutron $p_{3/2}$ to the collective 3^- phonon of ^{48}Ca (left). The theoretical predictions (center) are obtained by a weak coupling model of Bohr and Mottelson [2] using the SkX Skyrme interaction.

still beyond the present computational capabilities [14]. By employing an effective interaction based on the monopole-based universal interaction, the energy and strength of the 3^- phonon of the ^{48}Ca core are well reproduced, as well as the ones of the $9/2^+$ state of ^{49}Ca, while a too high spectroscopic factor (0.42, with 51% contribution from the $g_{9/2}$ orbital) was obtained in comparison with the experimental value (0.14) of Ref. [15]. All together, the shell model calculations can be considered rather successful in reproducing the ^{49}Ca data, demonstrating their capability in calculating complex core excitations up into the mass A≈50 region.

3 Around the exotic doubly-magic ^{132}Sn: the one-proton nucleus ^{133}Sb

In the case of the exotic doubly magic system ^{132}Sn, three core excitations coexist, around 4-4.5 MeV, i.e. the 2^+, 3^- and 4^+ phonons, with a similar strength (≈ 7 W.u.). As a consequence, in the ^{133}Sb nucleus – with one proton outside the doubly magic ^{132}Sn core – the lowest excited states are single-proton excitations, while around 4 MeV states arising from the coupling between core excitations and the valence proton are expected. With these characteristics, ^{133}Sb is an ideal playground for testing shell model predictions and/or different approaches taking into account collective excitations, in particular phonon excitations of the core. So far, the knowledge about the structure of ^{133}Sb was limited to the γ-decay of the $21/2^+$ long-lived isomeric state at 4545 keV (with $\tau_{1/2} = 16.6\mu s$), which prevented to easily investigate the feeding from higher-spin states [16, 17].

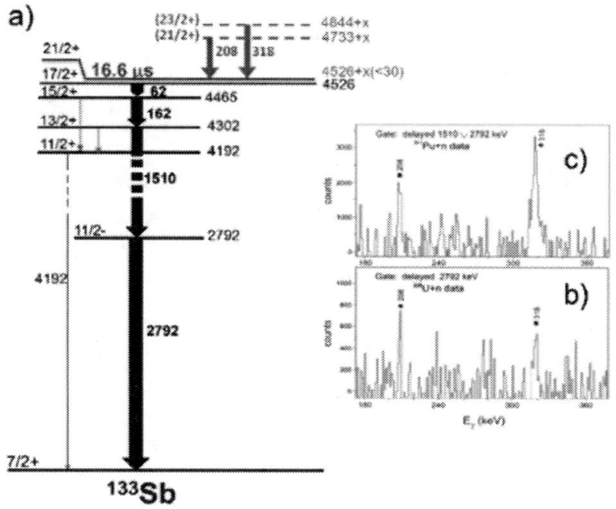

Figure 3: a) Partial level scheme of ^{133}Sb, showing (in black) the known level scheme below the long lived $21/2^+$ isomer [16, 17]. New states are given in pink, with tentative spin assignments. They were observed by studying coincidences between between delayed and prompt transitions, as shown in panel b) and c) for ^{235}U and ^{241}Pu targets, respectively.

Recent γ-spectroscopy data on ^{133}Sb were obtained in the EXILL campaign at the PF1B [18] cold-neutron facility at Istitut Laue Langevine (Grenoble, France), employing an highly collimated beam with a capture flux of 10^8 cold neutrons s^{-1} cm^{-2} on target. Two detectors setups were used, the first consisting of 8 EXOGAM clovers, 6 large coaxial detectors from GASP and 2 ILL-Clover detectors, with a total photopeak efficiency of about 6%. In the second setup, the GASP and ILL detectors were replaced by 16 $LaBr_3$ crystals for lifetime measurements by fast-timing techniques. A digital data acquisition, triggerless, allowed event rates up to 0.84 MHz to be handled and to study coincidences among γ transitions separated in time by several tens of μs [19]. ^{133}Sb was populated in two long runs of neutron induced fission of ^{235}U and ^{241}Pu targets. The data analysis was based on double and triple γ coincidence histograms constructed either requiring correlations among γ-rays prompt with fission events (within 200 ns) or requesting correlations between prompt and delayed transitions (within 20 μs). By setting gates on the strong transitions deexciting the isomer, candidates for γ-rays populating the isomer (the more intense with energies of 207 and 318 keV) were observed in the prompt coincidence spectrum, consistently for both fissioning targets (see Fig. 3 b) and c)). These new transitions are marked in pink in the level scheme of Fig. 3 a), while in black a partial level scheme of ^{133}Sb is shown, as follows from the known γ-decay of the 16.6 μs isomer.

In the EXILL campaign, fission data from ^{235}U and ^{241}Pu targets, taken with the setup including the LaBr$_3$ scintillators, were used to extract the lifetimes of the $13/2^+$ and $15/2^+$ states ^{133}Sb, by fast-timing techniques [20,21]. The analysis was based on triple coincidence events, within a time window of 200 ns, in which two γ-rays were detected in the LaBr$_3$ scintillators and the third one in the Ge array. The half lives values $T_{1/2}$ = 31(8) ps and $<$ 20 ps were obtained for the $13/2^+$ and $15/2^+$ states, after averaging over the results from the ^{235}U and ^{241}Pu targets. This gives B(M1) values, for the $15/2^+ \rightarrow 13/2^+$ and $13/2^+ \rightarrow 11/2^+$ transitions, >0.24 W.u. and equal to 0.0042(15) W.u., respectively. Such a large difference, almost two orders of magnitude, clearly indicates that the $13/2^+$ and $15/2^+$ states of ^{133}Sb do not arise from similar, simple configurations. Instead, they very likely involve the coupling of the valence proton to various complex excitations of the ^{132}Sn core.

In order to interpret the experimental findings, a new microscopic model has been developed: the traditional phenomenological particle-vibration model [2] has been transformed into a "hybrid" model that includes non-collective states as well. This is very much needed, since shell model calculations in the configuration space including shells below and above the ^{132}Sn core, which must be considered to describe one-valence particle nuclei around ^{132}Sn, are computationally not possible. The hybrid model calculations are self-consistent in the sense that both single-particle states and phonons come out of Hartree-Fock (HF) and Random Phase Approximation (RPA) calculations performed with the Skyrme SkX interaction introduced in Ref. [22]. In each of these subspaces one has both one-particle states and so-called one particle-one phonon states, and some of the so-called phonons turn out to be pure 1p-1h states, as it can be expected. In the calculations for ^{133}Sb, excitations of ^{132}Sn are considered up to 5.5 MeV together with the proton states of the 50-82 shell. In addition, the correction for the non-orthonormality of the basis is taken into account [23]. Without any free parameter, the model does reproduce the energy sequence of the high-spin states observed experimentally, within 200 keV. Also, a gradual transition from more complex towards non-collective excitations is seen, with increasing spin: the importance of the proton states admixed with the low-lying 2^+ decreases, while the relevance of the proton coupled with the single neutron $h^{-1}_{11/2}f_{7/2}$ increases. In particular, the states located, experimentally, above the $21/2^+$ isomer are found to arise from almost pure (>95%) configuration of $\pi g_{7/2}\ \nu f_{7/2}h^{-1}_{11/2}$ character, while a much stronger admixture is found for the $13/2^+$ and $15/2^+$ states, below the isomer, in agreement with the experimental results of the lifetime analysis.

At present, this hybrid model is the only tool which describes, in medium-heavy mass systems, the low-lying spectra of odd nuclei made of a magic core and an unpaired nucleon, where states having real particle-vibration character co-exist with states that have less collective or non-collective character.

4 Around the stable doubly-magic ^{208}Pb: the one-proton, one-neutron nucleus ^{210}Bi

The ^{208}Pb nucleus is considered one of the best known doubly-magic nuclei, therefore nuclei in its proximity are ideal candidates for testing interactions among valence particles. This is the case of ^{210}Bi, a one-valence-proton, one-valence-neutron nucleus, which is expected to provide information on proton-neutron excitations and couplings of particles with excitations of the ^{208}Pb core. The low-energy structure of ^{210}Bi, up to approx. ~2 MeV, should arise exclusively from couplings of a proton on the $1h_{9/2}$, $2f_{7/2}$, $1i_{13/2}$, $2f_{5/2}$, $3p_{3/2}$, $3p_{1/2}$ and a neutron on the $2g_{9/2}$, $1i_{11/2}$, $1j_{15/2}$, $3d_{5/2}$, $4s_{1/2}$, $2g_{7/2}$, $3d_{3/2}$ orbitals, while at higher energies states that originate from the couplings between valence particles and the very strong 3^- octupole vibration of ^{208}Pb (at 2615 keV with B(E3)≈ 34 W.u.) should appear.

High resolution, low spin data on ^{210}Bi were obtained during the recent EXILL campaign (discussed in Sec. 3), with the cold neutron capture reaction ^{209}Bi(n,γ)^{210}Bi [8]. From the capture state, at 4605 keV, the γ decay of ^{210}Bi was found to proceeds through many paths, feeding a large number of levels. This decay was investigated in great details: 64 primary γ rays were observed (40 new) and a total number of 70 discrete states (33 new) were located below the neutron binding energy. The analysis of the angular correlations of γ rays provided information about transitions multipolarities, which allowed to confirm most of the previously known spin-parity assignments as well as helped establishing new ones. Such a detailed experimental investigation, together with previous works [11], allows to approach a complete spectroscopy of ^{210}Bi at low spins (J $<$ 8). This provides a unique opportunity for a one-to-one comparison with theoretical calculations over the entire energy range, from the ground state to the neutron binding energy.

In a heavy system such as ^{210}Bi, a detailed comparison with shell model calculations can only be made assuming a frozen ^{208}Pb core. The calculations were performed by the computer code OXBASH [24], using the residual interaction of Kuo and Herling [25] with modifications introduced by Warburton and Brown (KHPE [26]). It is found that the experimental states, in large majority, correspond to excitations of one proton and one neutron valence particles. Below 2 MeV of excitation energy the agreement between experiment and theory is very good and almost all states are reproduced by the calculations within a few keV. This is also true above 2 MeV, where the states observed experimentally have configurations which arise from neutron and proton promotions to almost all orbitals available above the ^{208}Pb core, allowed by the Kuo-Herling model space. Only 12 levels, above 2 MeV, do not have their theoretical counterparts, and they are most naturally considered as arising from the couplings of one-proton one-neutron excitations to the octupole vibration of the ^{208}Pb core at 2615 keV, which is known to largely contribute to the structure of the states in neighbouring nuclei.

5 Conclusions

Experimental results around the doubly-magic neutron-rich cores of ^{48}Ca, ^{132}Sn and ^{208}Pb were discussed. In particular, data on ^{47,49}Ca, ^{133}Sb and ^{210}Bi were presented, focusing on the investigation of excited states arising from the coupling of valence particles/holes to core excitations: they are key ingredients for the understanding of basic phenomena, such as the quenching of spectroscopic factors and the damping mechanism in giant resonances. The present data provide an excellent testing ground for state of the art theoretical approaches, such as shell model calculations or newly developed extended microscopic particle-phonon coupling models.

References

[1] D. Lacroix, 2009 International Joliot-Curie School (EJC2009).

[2] A. Bohr, B.R. Mottelson, *Nuclear Structure, I and II, W.A. Benjamin, 1975.*

[3] M. B. Tsang et al., Phys. Rev. Lett. 102, 062501 (2009).

[4] P.F. Bortignon, A. Bracco, and R. A. Broglia, *Giant Resonances: Nuclear Structure at Finite Temperature (Harwood Academic, New York, 1998).*

[5] Y.F. Niu et al., Phys. Rev. Lett. 114, 142501 (2015).

[6] D. Montanari et al., Phys. Lett. B **697**, 288 (2011).

[7] D. Montanari et al., Phys. Rev. C **85**, 044301 (2012).

[8] S. Leoni et al., Acta Phys. Pol. B **46**, 637 (2015) and to be published.

[9] G. Bocchi et al., Phys. Rev. C **89**, 054302 (2014).

[10] C.R. Niţă et al., Phys. Rev. C **89**, 064314 (2014).

[11] Nuclear Data Base, http://www.nndc.bnl.gov/nudat2/

[12] R. Broda, J. Phys. G: Nucl. Part. Phys. 32 (2006) R151.

[13] D. Montanari, et al., Eur. Phys. J. A 47 (2011) 4 and references therein.

[14] Y. Utsuno et al., Prog. Theor. Phys. Supp. 196 (2012) 304.

[15] Y. Uozumi et al., Nucl. Phys. A 576, 123 (1994)

[16] W. Urban et al., Phys. Rev. C **62**, 027301 (2000).

[17] W. Urban et al., Phys. Rev. C **79**, 037304 (2009).

[18] H. Abele et al., Nucl. Instr. Meth. Phys. Res. A **562**, 407 (2006).

[19] G. De France et al., Eur. Ph. J. Web of Conferences **66**, 02010 (2014).

[20] J.-M. Regis et al., Nucl. Instr. Meth. Phys. Res. A **726**, 191 (2013).

[21] J.-M. Regis et al., Nucl. Instr. Meth. Phys. Res. A **763**, 210 (2014).

[22] B. A. Brown, Phys. Rev. C 58, 220 (1998).

[23] J. Rowe, J. Math. Phys. 10, 1774 (1969).

[24] B. A. Brown et al., MSU-NSCL Report No. 1289, 2004 (unpublished).

[25] G. Herling and T. T. S. Kuo, Nucl. Phys. A **181**, 113 (1972).

[26] E. K. Warburton, B. A. Brown, Phys. Rev. C **43**, 602 (1991).

EPJ Web of Conferences **117**, 04002 (2016) DOI: 10.1051/epjconf/201611704002
NN2015

Nuclear shape phase transitions

D. Vretenar

Physics Department, Faculty of Science, University of Zagreb,
Zagreb, Croatia

Abstract

The evolution of shapes and shape (phase) transitions, including regions of short-lived exotic nuclei that are becoming accessible in experiments at radioactive-beam facilities, are governed by the shell structure of single-nucleon orbitals. In most cases the transition between different shapes is gradual but in a number of examples, with the addition or subtraction of only few nucleons, signatures of abrupt changes in observables are noticed. A quantitative analysis necessitates accurate modelling of the underlying nucleonic dynamics. Important advances have been reported in theoretical studies of complex shapes, especially in the "beyond mean-field" framework based on density functionals.

1 Introduction and theory framework

The evolution of equilibrium shapes and the corresponding excitation dynamics present some of the most studied low-energy nuclear phenomena, both experimentally and theoretically [1–4]. Different shapes, shape coexistence, and shape transitions take place in light, medium-heavy, and heavy nuclei, and have their origin in the evolution of single-nucleon shell structure that reflects the underlying confining single-particle potential. In contrast to spherical shapes that occur for the particular case of single or doubly closed-shell nuclei, deformed equilibrium shapes arise due to strong proton-neutron correlations in open-shell nuclei. Far from the β-stability line, in particular, the energy spacings between single-particle levels can be very different from those found in stable nuclei. This can result in reduced spherical shell gaps, modifications of magic numbers, occurrence of islands of inversion, deformed equilibrium shapes and coexistence of shapes with different deformations, and shape (phase) transitions.

As the number of nucleons changes from nucleus to nucleus, in general one observes a gradual evolution of different shapes – spherical, axially quadrupole deformed, shapes that are soft with respect to triaxial deformations, octupole shapes [1,2]. In a number of cases the addition or subtraction of only a few nucleons leads to rapid changes in equilibrium shapes and shape phase transitions may occur [4]. Phase transitions in the equilibrium shapes of nuclei correspond to first- and second-order quantum phase transitions (QPTs) between competing ground-state phases induced by variation of a non-thermal control parameter at zero temperature. Particularly important are studies of observables that can be related to order parameters and the degree to which discontinuities at a phase transitional point are smoothed out in finite nuclei. In recent years nuclear QPTs have been investigated extensively, both in experimental studies and employing a variety of theoretical models. An integral microscopic description of these universal phenomena must be based on a theoretical framework that can be extended from light to heavy nuclei, and from the line of β-stability to the nucleon drip lines. Such a framework is the one based on nuclear density functional theory (DFT).

Nuclear energy density functionals (EDFs) enable a global and accurate microscopic description of ground-state properties and collective excitations over the entire nuclide chart. The basic implementation of the EDF framework is in terms of self-consistent mean-field (SCMF) models, in which an EDF is constructed as a functional of one-body nucleon densities that correspond to a single product state. This approach to nuclear structure is analogous to Kohn-Sham DFT. Nuclear SCMF models effectively map the many-body problem onto a one-body problem, and the exact EDF is approximated by functionals of powers and gradients of ground-state nucleon densities and currents [5,6]. When considering applications that go beyond the simple mean-field level, that is, to calculate excitation spectra and electromagnetic transition rates in individual nuclei, it is necessary to extend the Kohn-Sham EDF framework to include correlations that arise from symmetry restoration and fluctuations around the mean-field minimum [5–7]. The second level of implementation of nuclear EDFs takes into account collective correlations through the restoration of broken symmetries and configuration mixing of symmetry-breaking product states. In this work we consider two examples of applications of structure models based on relativistic EDFs in studies of complex shapes, shape transitions and the corresponding excitation spectra and decay patterns.

Over the last two decades a number of successful relativistic mean-field (RMF) models have been constructed based on the framework of quantum hadrodynamics, in which a nucleus is described as a system of Dirac nucleons coupled to exchange mesons through an effective Lagrangian [8]. However, at energies characteristic for nuclear structure phenomena, meson exchange (σ, ω, ρ, ...) is just a convenient representation of the effective inter-nucleon interaction. The exchange of heavy mesons is not resolved at low energies, and can be replaced by the corresponding local contact interactions between nucleons.

The basic building blocks of a relativistic nuclear energy density functional are the densities and currents bilinear in the Dirac spinor field ψ of the nucleon: $\bar{\psi}\mathcal{O}_\tau\Gamma\psi$, $\mathcal{O}_\tau \in \{1, \tau_i\}$, $\Gamma \in \{1, \gamma_\mu, \gamma_5, \gamma_5\gamma_\mu, \sigma_{\mu\nu}\}$. Here τ_i are the isospin Pauli matrices. The nuclear ground-state density and energy are determined by the self-consistent solution of relativistic linear single-nucleon Kohn-Sham equations. To derive those equations it is useful to construct an interaction Lagrangian with contact interaction terms in the various isospace-space channels: isoscalar-scalar $(\bar{\psi}\psi)^2$, isoscalar-vector $(\bar{\psi}\gamma_\mu\psi)(\bar{\psi}\gamma^\mu\psi)$, isovector-scalar $(\bar{\psi}\vec{\tau}\psi)\cdot(\bar{\psi}\vec{\tau}\psi)$, and isovector-vector $(\bar{\psi}\vec{\tau}\gamma_\mu\psi)\cdot(\bar{\psi}\vec{\tau}\gamma^\mu\psi)$. A general Lagrangian can be written as a power series in the currents $\bar{\psi}\mathcal{O}_\tau\Gamma\psi$ and their derivatives, with higher-order terms representing in-medium many-body correlations. Alternatively, one constructs a Lagrangian that comprise second-order interaction terms only, with many-body correlations encoded in density-dependent coupling functions. This type of RMF-based models has been successfully used in analyses of a variety of structure phenomena, not only in nuclei along the valley of β-stability, but also in exotic nuclei with extreme isospin values and close to the particle drip lines. In the present study illustrative calculations are performed with the EDF DD-PC1 [9]. Starting from microscopic nucleon self-energies in nuclear matter, the density-dependent coupling parameters of the DD-PC1 point-coupling Lagrangian were fine-tuned to the experimental masses of a set of 64 deformed nuclei in the mass regions $A \approx 150-180$ and $A \approx 230-250$. The functional has been further tested in a number of mean-field and beyond-mean-field calculations in different mass regions. For the examples considered here, pairing correlations are taken into account by employing an interaction that is separable in momentum space, and is completely determined by two parameters adjusted to reproduce the empirical bell-shaped pairing gap in symmetric nuclear matter [10].

The analysis of ground-state properties and low-energy collective excitations starts with a calculation of deformation energy surfaces with mass multipole moments as constrained quantities. The static nuclear mean-field is characterised by the breaking of symmetries of the underlying Hamiltonian – translational, rotational, particle number. To calculate spectroscopic properties it is necessary to extend the SCMF scheme to include collective correlations that arise from symmetry restoration and fluctuations around the mean-field minima. Many studies of the evolution of low-energy collective states have been carried out by employing the Generator Coordinate Method (GCM) in configuration mixing calculations of angular-momentum and particle-number projected states based on relativistic energy density functionals [11, 12]. The application of this method to medium-heavy and heavy nuclei presents a more involved and technically difficult problem and, therefore, only few studies of collective states in heavier systems have been reported.

In an alternative approach a collective Hamiltonian can be formulated, with deformation-dependent parameters determined by self-consistent mean-field calculations. The dynamics of the collective Hamiltonian is governed by the vibrational inertial functions and the moments of inertia, and these functions are determined by the energy density functional and the effective pairing interaction [6]. For instance, in the case of quadrupole degrees of freedom:

$$\hat{H}_{\mathrm{coll}} = \hat{T}_{\mathrm{vib}} + \hat{T}_{\mathrm{rot}} + V_{\mathrm{coll}} \,, \tag{1}$$

where the vibrational kinetic energy is parameterized by the the mass parameters $B_{\beta\beta}$, $B_{\beta\gamma}$, $B_{\gamma\gamma}$, the three moments of inertia $\mathcal{I}_k$ determine the rotational kinetic energy, and V_{coll} is the collective potential. The self-consistent mean-field solution for the single-quasiparticle energies and wave functions for the entire energy surface, as functions of the quadrupole deformations β and γ, provide the microscopic input for calculation of the mass parameters, moments of inertia and the collective potential. The Hamiltonian describes quadrupole vibrations, rotations, and the coupling of these collective modes. The diagonalization gives the excitation energies and collective wave functions that can be used to calculate various observables. The details of the particular implementation of the relativistic EDF-based collective Hamiltonian used in the present study can be found in Ref. [6].

2 Triaxial shapes in germanium isotopes

Most deformed nuclei display axially-symmetric quadrupole equilibrium shapes but in a number of cases axial symmetry is explicitly broken. Triaxial quadrupole shapes can be described in terms of two polar deformation parameters: β (proportional to the intrinsic quadrupole moment) and the

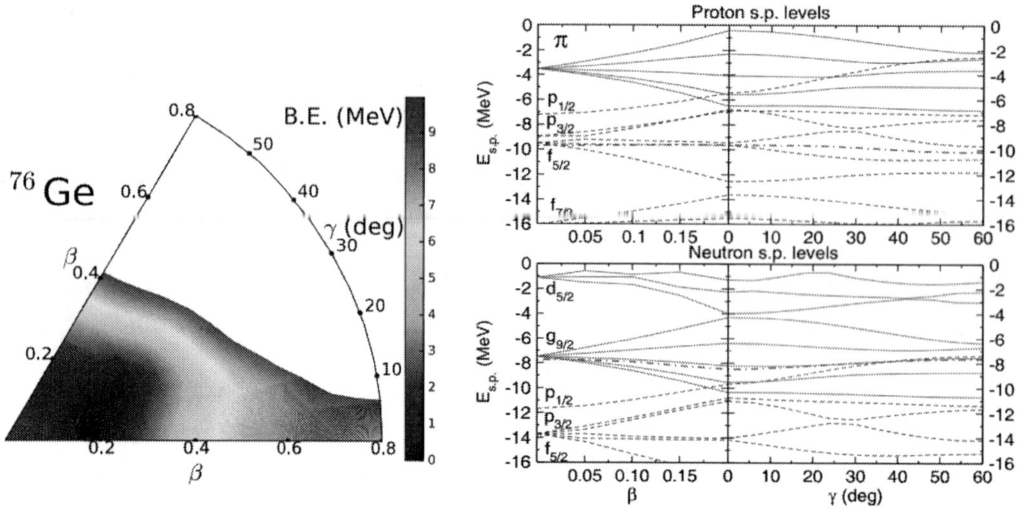

Figure 1: Left: Self-consistent RHB deformation energy surface of ^{76}Ge in the $\beta-\gamma$ plane ($0 \leq \gamma \leq 60^0$). Right: Proton and neutron energy levels of ^{76}Ge as functions of the deformation parameters.

angular variable γ. Two simple geometrical models describe the limiting cases of a triaxial shape: in the rigid-triaxial rotor model the collective potential has a stable minimum at a particular value of γ (γ-rigid potential), whereas in the γ-unstable rotor model the potential is independent of γ (γ-soft potential). A recent analysis [13] based on nuclear density functional theory, has shown that neither the limit of rigid-triaxial rotor nor the the γ-unstable rotor are actually realised in finite nuclei. The systematics of low-lying collective spectra and transition rates indicate that non-axial medium-heavy and heavy nuclei lie almost exactly in the middle between the two geometrical limits.

Empirical evidence for rigid triaxial deformation at low energy in ^{76}Ge was recently reported in [14], where it was shown that the phase of the odd-even staggering of the γ-band is consistent with the assumption of a γ-rigid structure, although the amplitude of the staggering is considerably smaller than that predicted by the rigid-triaxial rotor model. In the left panel of Fig. 1 we thus plot the triaxial quadrupole energy map of ^{76}Ge in the $\beta - \gamma$ plane ($0 \leq \gamma \leq 60^0$). The deformation energy surface displays an axially-symmetric minimum on the prolate axis, but one also notices a considerable softness in the γ direction. Mean-field shapes are governed by the evolution of the underlying shell structure of single-nucleon orbitals, and the formation of deformed minima is related to the occurrence of regions of

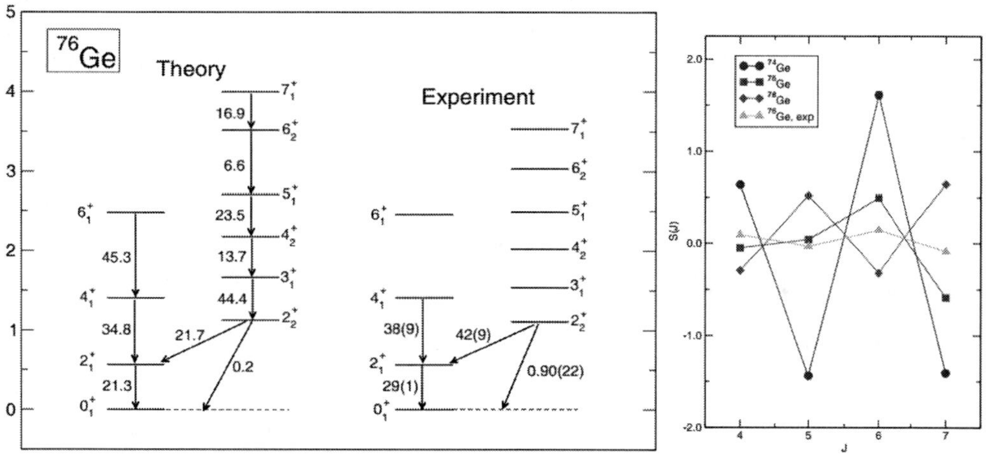

Figure 2: Left: The excitation spectrum of ^{76}Ge calculated with the collective Hamiltonian and compared to data. $B(E2)$ values are in Weisskopf units. Right: The staggering $S(J)$ in the γ-bands of ^{74,76,78}Ge isotopes.

low single-particle level density around the Fermi surface. The right panel of Fig. 1 displays the proton and neutron single-particle energy levels in ^{76}Ge. Solid (blue) and dashed (red) curves correspond to levels with positive and negative parity, respectively. The dot-dashed (green) curves denote the Fermi level. The single-particle levels are plotted as functions of the deformation parameters along a path in the $\beta - \gamma$ plane. Starting from the spherical configuration, we follow the single-nucleon levels on a path along the prolate axis up to the approximate position of the minimum and then, for this fixed value of β, we trace the levels along the path from $\gamma = 0^0$ to $\gamma = 60^0$. Proton levels show a pronounced gap between the last occupied and first unoccupied levels in the triaxial region at $\gamma \approx 30°$, whereas for the neutron levels the gap appears to be almost independent of γ. Because of small energy differences between different mean-field configurations, dynamical effects related to restoration of broken symmetries and fluctuations in collective coordinates can have a decisive role in the eventual formation of a triaxial ground state [15].

Figure 2 displays the spectrum of ^{76}Ge calculated with the collective Hamiltonian based on the functional DD-PC1. Levels that belong to the ground-state band and the γ-band are compared to the experimental spectrum [14], and assigned either to the ground-state band or the γ-band according to the distribution of the projection K of the angular momentum J on the z axis in the body-fixed frame. Yrast states have a predominant $K = 0$ component in the wave functions, whereas the γ-band comprises states above the yrast characterized by dominant $K = 2$ components. The level of K-mixing is reflected in the staggering in energy between odd- and even-spin states in the γ-band, and it can be quantified by considering the differential quantity:

$$S(J) = \frac{E[J_\gamma^+] - 2E[(J-1)_\gamma^+] + E[(J-2)_\gamma^+]}{E[2_1^+]}. \tag{2}$$

For a deformed γ-soft potential, $S(J)$ has negative values for even-spin states and positive values for odd-spin states, with the magnitude slowly increasing with spin. For a triaxial potential the level clustering in the γ-band is opposite, and $S(J)$ oscillates between positive values for even-spin states and negative values for odd-spin states. In the right panel of Fig. 2 we plot the calculated values for the staggering $S(J)$ in the γ-bands of ^{74,76,78}Ge, and the experimental values for ^{76}Ge [14]. The phase of $S(J)$ for ^{74}Ge is consistent with a γ-rigid triaxial shape, whereas ^{78}Ge displays the opposite pattern for $S(J)$, that is, the γ-band indicates a soft shapes. ^{76}Ge is at the (shape) transition between the triaxial ^{74}Ge and the γ-soft heavier isotopes. For this nucleus the amplitudes of $S(4)$ and $S(5)$ almost vanish, whereas $S(6)$ and $S(7)$ follow a pattern characteristic for triaxial shapes, but with considerably smaller amplitudes. The theoretical results are in very good agreement with available data, both for the ratio $E(2_2^+)/E(2_1^+) = 2$ and the pattern and amplitude of the staggering $S(J)$. The self-consistent mean-field calculation has shown that the energy surface of ^{76}Ge is γ-soft. The inclusion of collective correlations drives the nucleus toward triaxiality, but the correlations are not strong enough to stabilise a $\gamma \approx 30°$ triaxial shape.

3 Octupole shape-phase transitions in Thorium

Reflection-asymmetric shapes are characterized by the presence of negative-parity bands, and by pronounced electric dipole and octupole transitions. In the case of static octupole deformation the lowest $\pi = +1$ even-spin states and the $\pi = -1$ odd-spin states form an alternating-parity band. Octupole deformations develop through a coupling of orbitals in the vicinity of the Fermi surface with quantum numbers (l, j) and an intruder unique-parity orbital with $(l+3, j+3)$ [2]. For instance, in the case of the ($N \approx 134$ and $Z \approx 88$) nuclei in the region of light actinides, the coupling of the neutron orbitals $g_{9/2}$ and $j_{15/2}$, and that of the proton single-particle states $f_{7/2}$ and $i_{13/2}$, can result in octupole mean-field deformations.

In light actinides an interesting structure phenomenon is the simultaneous transition between quadrupole equilibrium shapes, and from stable

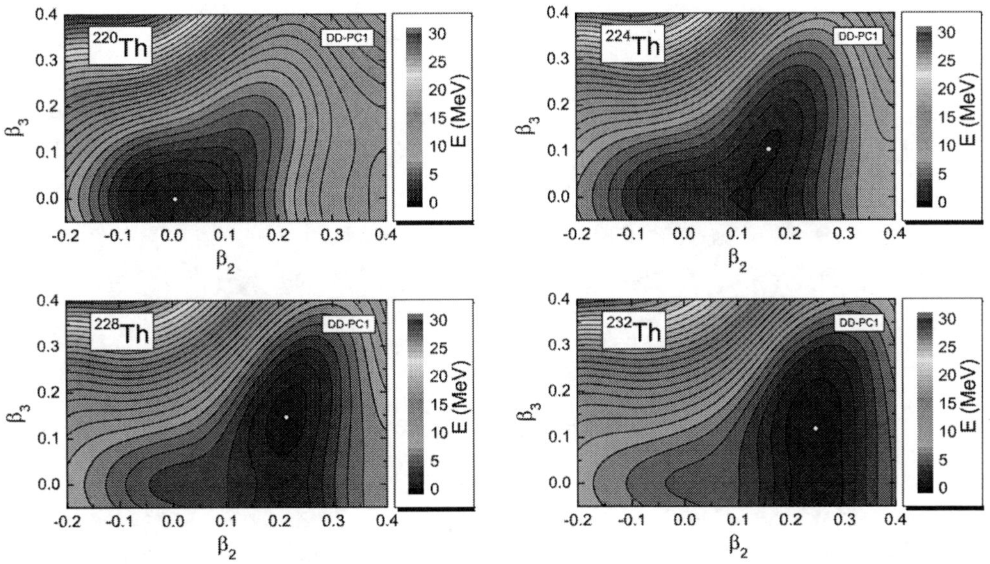

Figure 3: Microscopic DD-PC1 self-consistent axially symmetric energy surfaces of the nuclei $^{220-232}$Th in the (β_2, β_3) plane, symmetric with respect to $\beta_3 = 0$ axis. The contours join points on the surface with the same energy, and the separation between neighbouring contours is 1 MeV.

octupole deformations to octupole vibrations [16–18]. In Fig. 3 we plot the axially symmetric deformation energy surfaces in the (β_2,β_3) plane for the isotopes $^{220-232}$Th, calculated with the functional DD-PC1. A soft spherical energy surface is predicted for ^{220,222}Th. The quadrupole deformation becomes more pronounced in ^{224}Th and, in addition, octupole deformation develops. The energy minimum is found in the $\beta_3 \neq 0$ region, located at $(\beta_2, \beta_3) \approx (0.15, 0.1)$. From ^{224}Th to ^{228}Th a distinct octupole minimum is calculated whereas, starting from ^{228}Th, the minimum becomes softer in β_3 direction. Soft octupole surfaces are predicted for ^{230,232}Th.

A quadrupole-octupole collective Hamiltonian [17] can be used to analyse simultaneous quadrupole and octupole shape transitions. An alternative and more economic approach is the interacting-boson model (IBM), based on the EDF framework [16,18]. The *sdf* IBM Hamiltonian is constructed by mapping the microscopic quadrupole-octupole fermion energy surface onto the equivalent one in the corresponding boson system. Figure 4 displays the IBM excitation spectra of the ground-state band ($K^\pi = 0^+_1$) and the lowest negative-parity band ($K^\pi = 0^-_1$) in $^{220-232}$Th, obtained starting from the DD-PC1 constrained self-consistent mean-field solutions (cf. Fig. 3). The energies of the $\pi = +1$ levels systematically decrease in energy with mass

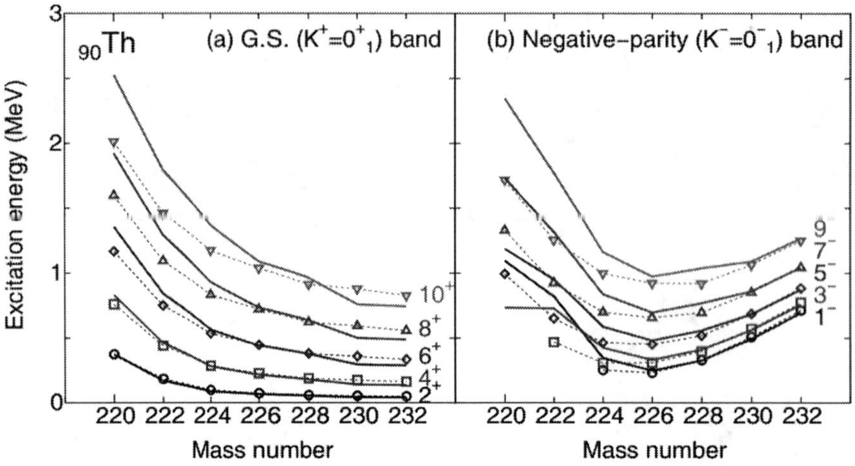

Figure 4: Isotopic dependence of the excitation energies of levels of the positive-parity ground-state band ($K^\pi = 0_1^+$) (a), and the lowest negative-parity band ($K^\pi = 0_1^-$) (b), for $^{220-232}$Th. In each panel solid lines and symbols denote the theoretical and experimental values, respectively.

number, reflecting the increase of quadrupole collectivity. ^{220,222}Th exhibit a quadrupole vibrational structure, while ground-state rotational bands are obtained in $^{226-232}$Th. The $\pi = -1$ bands form a parabolic structure centred at ^{226}Th, in which the octupole deformed minimum is most pronounced. Starting from ^{226}Th, the energies of negative-parity states systematically increase and the band becomes more compressed. A rotational-like collective band based on the octupole vibrational state 1_1^- develops.

The vertex of the parabola of the $\pi = -1$ states can be associated with a phase transition between stable octupole deformations and octupole vibrations characteristic for β_3-soft potentials, with the excitation energy of the 1_1^- band-head as the order parameter. Another signature of the phase transition is the odd-even staggering $E(J)/E(2_1^+)$. In lighter Th isotopes the staggering is negligible, and it only becomes more pronounced from ^{228}Th, indicating that negative-parity states form a separate rotational band built on the octupole vibration. Thorium isotopes therefore exhibit a transition from spherical shapes near ^{220}Th to stable octupole and quadrupole deformations around ^{226}Th, and the development of octupole vibrations characteristic for β_3-soft potentials in heavier Th nuclei, with the occurrence of a shape phase transition near ^{226}Th.

References

[1] Heyde K. and Wood J. L., *Rev. Mod. Phys.*, **83** (2011) 1467.

[2] Butler P. A. and Nazarewicz W., *Rev. Mod. Phys.*, **68** (1996) 349.

[3] Sorlin O. and Porquet M.-G., *Prog. Part. Nucl. Phys.*, **61** (2008) 602.

[4] Cejnar P., Jolie J., and Casten R. F., *Rev. Mod. Phys.*, **82** (2010) 2155.

[5] Bender M., Heenen P.-H., and Reinhard P.-G., *Rev. Mod. Phys.*, **75** (2003) 121.

[6] Nikšić T., Vretenar D., and Ring P., *Prog. Part. Nucl. Phys.*, **66** (2011) 519.

[7] Duguet T., in *Lecture Notes in Physics*, **879** (Springer, Heidelberg) 2014, pp. 293-350.

[8] Vretenar D., Afanasjev A.V., Lalazissis G.A., and Ring P., *Phys. Rep.*, **409** (2005) 101.

[9] Nikšić T., Vretenar D., and Ring P., *Phys. Rev. C*, **78** (2008) 034318.

[10] Tian Y., Ma Z. Y., and Ring P., *Phys. Lett. B*, **676** (2009) 44.

[11] Yao J. M., Meng J., Ring P., and Vretenar D., *Phys. Rev. C*, **81** (2010) 044311.

[12] Yao J. M., Mei H., Chen H., Meng J., Ring P., and Vretenar D., *Phys. Rev. C*, **83** (2011) 014308.

[13] Nomura K., Shimizu N., Vretenar D., Nikšić T., and Otsuka T., *Phys. Rev. Lett.*, **108** (2012) 132501.

[14] Toh Y. et al., *Phys. Rev. C*, **87** (2013) 041304(R).

[15] Nikšić T., Marević P., and Vretenar D., *Phys. Rev. C*, **89** (2014) 044325.

[16] Nomura K., Vretenar D., and Lu B.-N., *Phys. Rev. C*, **88** (2013) 021303(R).

[17] Li Z. P., Song B. Y., Yao J. M., Vretenar D., and Meng J., *Phys. Lett. B* **726**, (2013) 866.

[18] Nomura K., Vretenar D., Nikšić T., and Lu B.-N., *Phys. Rev. C*, **89**, (2014) 024312.

EPJ Web of Conferences **117**, 04003 (2016) DOI: 10.1051/epjconf/201611704003
NN2015

Spatial properties of pairing and quarteting correlations in nuclei

V. V. Baran[1,4] and D.S. Delion[1,2,3]

[1] "Horia Hulubei" National Institute of Physics and Nuclear Engineering, Bucharest-Măgurele, România
[2] Academy of Romanian Scientists, Bucharest, Romania
[3] Bioterra University, Bucharest, Romania
[4] Department of Physics, University of Bucharest, Romania

Abstract

Interest in the study of the spatial properties of pairing correlations in nuclei has recently been revived. The main interesting physical quantity is the coherence length (CL). We investigate the CL as obtained from schematic pairing potentials. We then generalize to quarteting correlations by considering the CL build out of proton and neutron pairs. We point out the main differences between the pairing and quarteting CLs. Finally we comment on the important role of proton-neutron correlations as they strongly modify the behaviour of the CL when taken into account.

The coherence properties of the spatial distribution of the two-particle density are known to give information about nuclear correlations [1, 2]. The main physical quantity that describes the coherence property is the CL, defined as the root mean square relative distance averaged over the pairing density, in the case of superfluid nuclei. It has been found [3, 4] that this quantity is of the order of the nuclear radius inside the nucleus and smaller around the nuclear surface. In contrast, the correlations between proton and neutron pairs by the analogous quarteting CL.

1 Theoretical background

The pairing interaction generates the most important two-body correlations beyond the mean field in even-even nuclei. We describe such systems within the standard HF+BCS approach, where the averaged particle number is conserved, separately for protons and neutrons. We consider in our basis bound single particle (sp) states with negative energy, as well as relatively narrow sp resonances with positive energy up to $e_{max} = 10$ MeV with a decay width $\Gamma \leq 1$ MeV (the background contribution is not relevant [6]), given by a Woods Saxon central field with universal parameterization [8].

The information about the spatial properties of pairing correlations is contained in the anomalous density $\kappa(\mathbf{r_1}, \mathbf{s_1}; \mathbf{r_2}, \mathbf{s_2}) = \langle BCS|\mathcal{A}\{\hat{\psi}(\mathbf{r_1}, \mathbf{s_1})\hat{\psi}(\mathbf{r_2}, \mathbf{s_2})\}|BCS\rangle$. After passing to the spherical basis of states and recoupling from the j-j to the L-S scheme, we retain only the largest singlet component, $\kappa_1(\mathbf{r_1}, \mathbf{r_2})$. By changing to the relative $\mathbf{r} = \mathbf{r_1} - \mathbf{r_2}$ and center of mass coordinate $\mathbf{R} = (\mathbf{r_1} + \mathbf{r_2})/2$ and averaging over their relative angle θ one obtains the averaged pairing tensor $\bar{\kappa}_1^2(r, R)$. The CL is then defined to be the relative distance averaged over this pairing tensor:

$$\xi(R) = \sqrt{\frac{\int \mathrm{d}r\ r^4\ \bar{\kappa}_1^2(r,R)}{\int \mathrm{d}r\ r^2\ \bar{\kappa}_1^2(r,R)}} \equiv \sqrt{\int \mathrm{d}r\ r^2\ w(r,R)}\ . \qquad (1)$$

Intuitively it gives the size of the Cooper pair at a certain point inside the nucleus.

In order to investigate medium to heavy α-decaying nuclei having the valence protons and neutron in different major shells it is customary to construct a quartet by two separate proton and neutron pairs [10, 11]. By considering the overlaps of the proton-proton and respectively neutron-neutron wave functions (given by the corresponding pairing tensors) with the two-proton and respectively two-neutron part of the α-particle wave-function, we construct the quartet tensor $\kappa_q(\mathbf{R}_\pi, \mathbf{R}_\nu) = \langle\kappa_\pi(\mathbf{r_1}, \mathbf{r_2})|\phi_{00}^{(\beta_\alpha/2)}(r_\pi)\rangle \cdot \langle\kappa_\nu(\mathbf{r_3}, \mathbf{r_4})|\phi_{00}^{(\beta_\alpha/2)}(r_\nu)\rangle$, where $\mathbf{r}_{\pi,\nu} = \mathbf{r}_{1,3} - \mathbf{r}_{2,4}$, $\mathbf{R}_{\pi,\nu} = (\mathbf{r}_{1,3} + \mathbf{r}_{2,4})/2$, $\phi_{nl}^{(\beta)}$ is the standard radial spherical Harmonic Oscillator function and $\beta_\alpha \approx 0.5 fm^{-2}$. This quantity plays the role of the pairing tensor in the quarteting case. The Quartet CL ξ_q is now defined by proceeding similarly with the pairing case (see eq. 1). It is a function of the center of mass of the two pairs $\mathbf{R}_\alpha = (\mathbf{R}_\pi + \mathbf{R}_\nu)/2$.

By taking into account proton-neutron correlations, described by the part in the α-particle wave function dependent on the relative distance $\mathbf{r}_\alpha = \mathbf{R}_\pi - \mathbf{R}_\nu$, we may also define the so-called alpha CL. The corresponding correlation tensor is $\kappa_\alpha(r_\alpha, R_\alpha) = \kappa_q(r_\alpha, R_\alpha)\phi_{00}^{(\beta_\alpha)}(r_\alpha)$, and the CL is $\xi_\alpha(R_\alpha)$. The alpha correlation tensor is directly related to the amplitude to find a quartet in the alpha state, namely the formation amplitude [10]: $\mathcal{F}(R_\alpha) = \int_0^\infty \mathrm{d}r_\alpha\, r_\alpha^2\, \kappa_\alpha(r_\alpha, R_\alpha)$.

2 Numerical application and results

We analyzed all even-even nuclei with $20 < Z < 100$ and known experimental pairing gaps, determined by the binding energies of neighboring nuclei [7]. We present here the results obtained by solving the BCS equations with the Gaussian interaction defined by $v(r_{12}) = -v_0 e^{-[r_{12}/r_0]^2}$ depending on the relative radius r_{12}. There is no a priori reason to consider that the parameters of the in-medium effective interaction are equal to those of the bare interaction. Thus we compared the case r_0=2 fm, corresponding to the singlet "bare" value in the free space, with the larger $r_0 = 4.5$ and even $r_0 = R_N = 1.2A^{1/3}$, corresponding to the nuclear radius. The strength v_0 is always chosen as for the Fermi level gap to reproduce the experimental gap. We note that the corresponding CLs are all similar in shape for all considered interactions, and in agreement with Ref. [3].

For the bare Gaussian, the ratio of the mean CL to the nuclear radius decreases from 1.4 for light nuclei down to around unity for heavy nuclei. The systematic study also revealed that the proton CL approximately scales as $2.4A^{1/5}$ fm, while for the neutrons the scaling goes as $2.0A^{1/4}$ fm. As a general trend, the CL is larger for neutrons, but shell effects are stronger for protons. We also investigated the temperature dependence of the CL by solving the finite temperature BCS equations. One would normally expect a strong variation induced by thermal pair breaking. By going up to the critical temperature, the CL saw almost no variation. It was only slightly affected in the internal region by the variable mixing between the odd and even parts of the pairing tensor [4].

Moving on to the quarteting case, we first checked that our treatment correctly evidences the surface nature of α condensation. In fact, $\mathcal{F}(R_\alpha)$ defined above has a pronounced maximum on the nuclear surface. Given the way we constructed the quartet by putting together proton and neutron pairs, we find that the formation amplitude is strongly dependent on the strength of the pairing interaction. We find that the quartet CL is larger in the internal region, similarly to the pairing CL. After taking into account proton-neutron correlations the shape is significantly altered: the α-CL oscillates about the geometrical dimension of the α-particle of 1.9 fm.

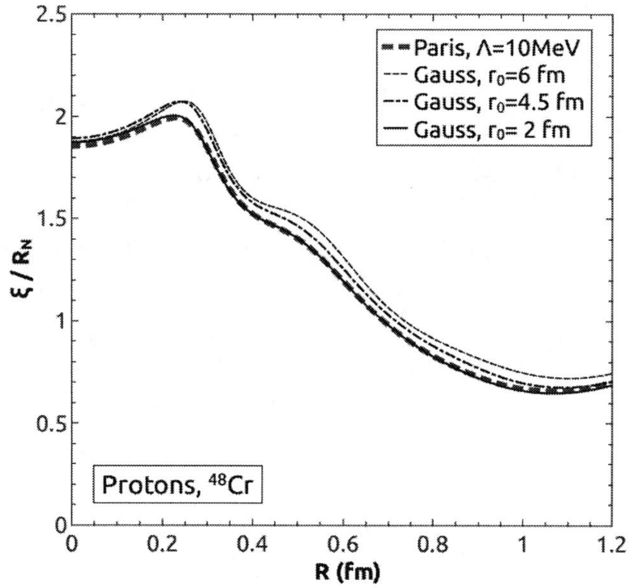

Figure 1: Proton CL in units of the nuclear radius in ^{48}Cr for Gaussian potentials with various width parameters and for the realistic Paris potential renormalized down to the cutoff energy $\Lambda = 10$ MeV (see [9]).

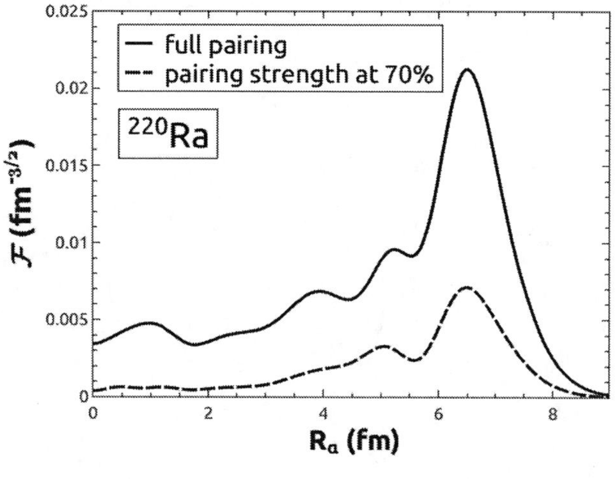

Figure 2: Formation amplitude $\mathcal{F}$ vs the quartet center of mass R_α for the nucleus ^{220}Ra.

3 Conclusions

The pairing CL has similar characteristics for all considered interactions. It is larger than the geometrical radius for light nuclei and approaches this value for heavy nuclei. Our analysis provides evidence of strong shell effects. The pairing CL also is insensitive to the gradual decrease in the intensity of the pairing correlation due to incresing temperature. The quarteting CL describes correlations between proton and neutron pairs and resembles in shape the pairing CL. We evidenced the important role played by proton-neutron correlations, as they change completely the behavior of the quarteting CL. Namely, the new alpha CL has oscillating values around the α-particle geometrical radius. Its mean value weakly depends on the nuclear mass.

References

[1] Ferreira L. et al. *Nucl. Phys.* A **426** 276 (1984)

[2] Hagino K. and Sagawa H., *Phys. Rev.* C **72** 044321 (2005)

[3] Pillet N., Sandulescu N. and Schuck P., *Phys. Rev.* C **76** 024310 (2007)

[4] Pillet N., Sandulescu N., Schuck P. and Berger J.-F., *Phys. Rev.* C **81** 034307 (2010)

[5] Delion D. S., *Theory of Particle and Cluster Emission*, Springer-Verlag, New York-Berlin (2010)

[6] Betan R. Id *Nucl. Phys.* A **879** 14 (2012)

[7] Möller P. and Nix J. R., *Nucl. Phys.* A **272** 502 (1995)

[8] Dudek J., Nazarewicz W. and Werner T., *Nucl. Phys.* A **341** 253 (1980)

[9] Delion D. S., Baldo M. and Lombardo U. *Nucl. Phys.* A **593** 151 (1995)

[10] Mang H. J., *Phys. Rev.* 119, **1069** (1960)

[11] Sandulescu A., *Nucl. Phys.* 37, **332** (1962).

[12] Delion D. S., Baran V.V., *Phys. Rev.* C **91**, 024312 (2015)

EPJ Web of Conferences **117**, 04004 (2016) DOI: 10.1051/epjconf/201611704004
NN2015

Extracting the cross section angular distributions for ^{15}C high-energy resonance excited via the (^{18}O,^{16}O) two-neutron transfer reaction

D. Carbone[1], C. Agodi[1], F. Cappuzzello[1,2], M. Cavallaro[1], A. Foti[2,3] and R. Linares[4]

[1]INFN - Laboratori Nazionali del Sud, Catania, Italy
[2]Dipartimento di Fisica e Astronomia, Università di Catania, Catania, Italy
[3] INFN, Sezione di Catania, Catania, Italy
[4]Instituto de Fisica, Universidade Federal Fluminense, Niteroi, Brazil

Abstract

The $^{13}C(^{18}O,^{16}O)^{15}C$ reaction has been studied at 84 MeV incident energy. The ejectiles have been momentum analized by the MAGNEX spectrometer and ^{15}C excitation energy spectra have been obtained up to about 20 MeV. In the region above the two-neutron separation energy, a bump has been observed at 13.7 MeV. The extracted cross section angular distribution for this structure, obtained by using different models for background, displays a clear oscillating pattern, typical of resonant state of the residual nucleus.

1 Introduction

In the last few years, a study of the structure of different nuclei was pursued at the Catania INFN-LNS laboratory by the (^{18}O,^{16}O) two-neutron transfer reaction at 84 MeV on different targets (^{12}C,^{13}C,^{9}Be,^{11}B,^{16}O), using the MAGNEX large acceptance magnetic spectrometer to detect the ejectiles. Thanks to its high resolution and large acceptance, high quality inclusive spectra were obtained, even in a largely unexplored region above the two-neutron emission threshold in the residual nucleus [1] [2] [3]. New phenomena appeared, such as the dominance of the direct one-step transfer of the two neutrons [4] and the presence of broad resonances at high excitation energy in the ^{14}C and ^{15}C spectra. These structures were recently identified as the first experimental signature of the Giant Pairing Vibration (GPV) [5] [6], predicted long time ago [7]. One of the key aspect in the investigation of the origin of such resonances was the multipolarity analysis on the cross section angular distributions, which suggested an $L = 0$ angular momentum transfer. In this paper the methods used to extract the cross section angular distributions for the ^{15}C GPV are described. In particular, since the GPV is found above the two-neutron separation energy ($S_{2n} = 9.394$ MeV) a careful evaluation of the background underneath it is mandatory.

2 The experiment

The ^{15}C nucleus was populated by the ^{13}C(^{18}O,^{16}O) two-neutron transfer reaction at 84 MeV incident energy. The $^{18}O^{6+}$ beam was produced and accelerated by the Tandem Van der Graaff facility of the INFN-Laboratori Nazionali del Sud in Catania. A 50 μg/cm^2 99 % enriched ^{13}C self-supporting target was used. Supplementary runs with ^{12}C were recorded for estimating the background coming from ^{12}C impurities in the ^{13}C target. The ^{16}O ejectiles were momentum analyzed by the MAGNEX large acceptance magnetic spectrometer [8] covering a solid angle $\Omega \sim 50$ msr and momentum range $\Delta p/p \sim 24\%$. The ejectiles were identified event by event in atomic number (Z), atomic mass (A) and charge (q) by combining two techniques [9]: the standard $\Delta E - E$ correlation plot for the Z identification and the correlation between the horizontal position at the focal plane (X_{foc}) and the ejectile residual energy (E_{resid}) for the the mass identification, which exploits the properties of the Lorentz force. The horizontal and vertical positions and angles of the ^{16}O ejectiles at the focal plane were used as input for a 10th order ray-reconstruction, based on the differential algebraic method implemented in MAGNEX [10]. The ray-reconstruction technique allows for an effective compensation of the high order aberrations connected with the large acceptance of the spectrometer. As a result, the initial phase space parameters at the target point are obtained, which are directly related to the momentum modulus and the scattering angle of the detected ejectiles. The corresponding Q-values, or equivalently the excitation energy $E_x = Q_0 - Q$ (where Q_0 is the ground to ground state Q-value), were obtained by a missing mass determination using relativistic energy and momentum conservation laws, assuming a binary reaction. Examples of the reconstructed energy spectra for the ^{15}C nucleus can be found in refs. [5] [6]. In the region above S_{2n}, a new resonance appears, recently identified as the GPV [5] [6].

In order to extract the spectroscopic features of the resonance observed in the ^{15}C spectra above S_{2n}, a best-fit procedure with Gaussian shapes on a locally adjusted linear model for the 3-body continuum background (as shown in Fig.1-a) was assumed. The resonance was thus identified at E_x = 13.7 ± 0.1 MeV with a full-width at half-maximum (FWHM) = 1.9 ± 0.3 MeV.

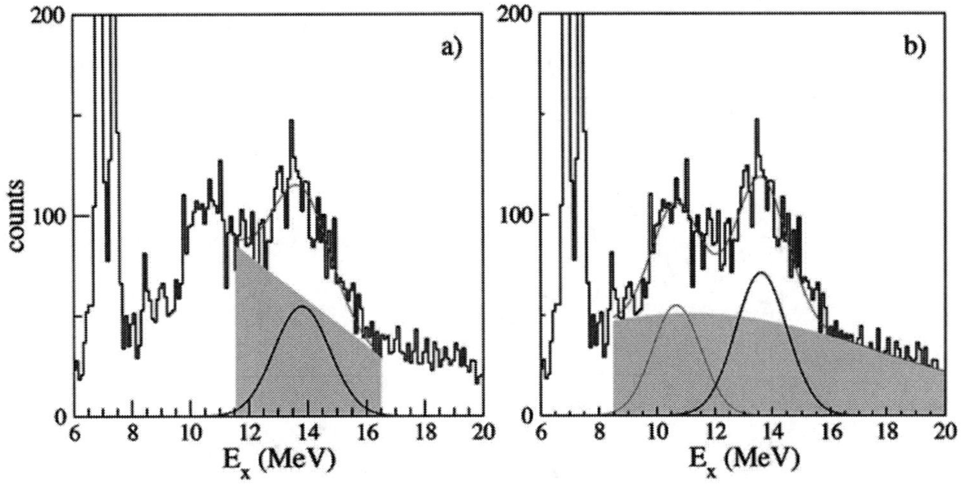

Figure 1: (Colour online) Models for the background underneath the ^{15}C GPV. a) linear background model (grey area), fitted bump (black gaussian) and the sum of them (red line); b) gaussian background model (grey area), fitted bumps (green and black gaussians) and the sum of them (red line).

3 GPV cross section angular distribution

The angular distributions of the absolute cross-section were deduced for the most intense transitions, including that populating the wide resonance observed above the two-neutron separation energy. The differential solid angle for the full spectrometer acceptance was carefully determined taking into account the overall transport efficiency, as described in ref. [11]. A dead-time coefficient of ~30% was measured. An angular bin of 1° in the laboratory reference frame was chosen for the GPV angular distributions in order to achieve a good compromise between the statistical uncertainties in the number of counts, the background subtraction and the angular resolution. The contribution to the angular distributions due to the continuous background in the spectra was estimated at each angle by a least-squared approach with a Gaussian model shape superimposed on a linear background as shown in Fig.1-a. The adopted background model is consistent with two-neutron break-up calculations, performed considering an independent removal of the two neutrons, as described in ref. [12]. In order to carefully look at the projectile break-up contribution to the angular distribution, a comparison between the GPV angular distribution obtained without any background subtraction, the subtracted background and the final GPV distributions is shown in Fig.2. The oscillating pattern of the GPV angular distribution is slightly visible in the total one, since it is smoothed by the flat background underneath. Such a background has the same shape of that at about 18 MeV excitation energy (green left triangles in Fig.2), where no resonances are present.

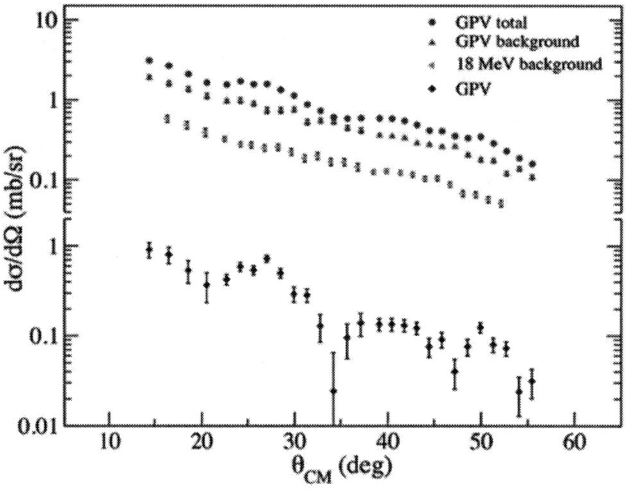

Figure 2: (Colour online) Comparison between the GPV distribution without any background subtraction (blue circles), the subtracted background underneath the GPV (red up triangles), the background present in the region at 18 MeV (green left triangles) and the final GPV (black diamonds) angular distributions for the ^{15}C residual nucleus.

We used different models for the background subtraction and the obtained results for the centroid and width of the resonances and also the shape of the angular distributions did not change within the quoted uncertainties. As an example, a second model for the background subtraction is show in Fig.1-b , which assumes a wide Gaussian (centroid = 11.2 MeV, FWHM = 16 MeV) for the background underneath the GPV (grey area) and a Gaussian model (green curve) for the structure at ~11 MeV (centroid = 10.7 MeV, FWHM = 2 MeV) . A comparison between the GPV angular distributions obtained using different background models is shown in Fig.3. The first model (black diamonds) assumes the background shown in Fig.1-a and the second model (red triangles) correspond to that shown in Fig.1-b. The average between the two models is also shown (green squares). This comparison demonstrates that both the shape and the absolute value of the GPV cross section angular distribution are stable within the error bars. The observation of such clear and stable oscillations in the resonance angular distribution indicates that the structure correspond to a resonance of the residual ^{15}C nucleus, since it is characterized by a well-defined angular momentum. Moreover the presence of oscillations represents a first indication of an $L = 0$ transfer, as discussed in refs. [5] [6], in relation to the phenomenon described in ref. [13].

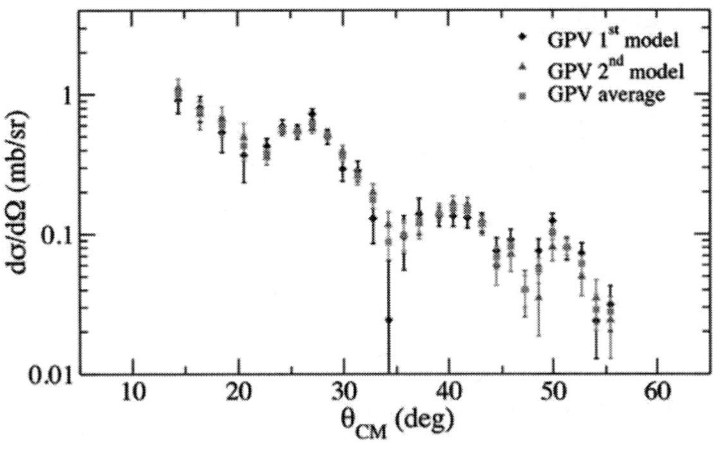

Figure 3: (Colour online) Comparison between the ^{15}C GPV angular distribution obtained assuming two different models (black diamonds and red triangles) for the background subtraction. The first model is shown in Fig.1-a and the second one is shown in Fig.1-b. The weighted average between the two models is shown as the green squares.

Aknowledgments

The work of D.C. was supported by the Italian Ministry of Education, Universities and Research (MIUR) under the grant LNS-Astrofisica Nucleare (fondi premiali).

References

[1] D. Carbone *et al.*, Phys. Rev. C 90 (2014) 064621.

[2] M. Cavallaro *et al.*, Phys. Rev. C 88 (2013) 054601.

[3] D. Nicolosi *et al.*, Acta Phys. Pol. B 44 (2013) 657.

[4] D. Carbone *et al.*, J. Phys.: Conf. Ser. 312, 082016 (2011).

[5] F. Cappuzzello *et al.*, Nat. Commun. 6, 6743 (2015) doi: 10.1038/ncomms7743.

[6] D. Carbone, Eur. Phys. J. Plus 130, 143 (2015).

[7] R. A. Broglia and D.R. Bes, Phys. Lett. B 69 (1977) 129.

[8] F. Cappuzzello, D. Carbone, M. Cavallaro and A. Cunsolo, Magnets: Types, Uses and Safety (Nova Publisher Inc., New York, 2011) pp. 1-63.

[9] F. Cappuzzello *et al.*, Nucl. Instr. Methods A 621 (2010) 419.

[10] F. Cappuzzello, D. Carbone and M. Cavallaro, Nucl. Instr. Methods A 638 (2011) 74.

[11] M. Cavallaro *et al.*, Nucl. Instr. Methods A 637 (2011) 77.

[12] F. Cappuzzello *et al.*, Phys. Lett. B 711 (2012) 347.

[13] S. Kahana and A. J. Baltz, Adv. Nucl. Phys. 9 (1977) 1.

EPJ Web of Conferences **117**, 04005 (2016) DOI: 10.1051/epjconf/201611704005
NN2015

Interplay between proton-neutron pairing and deformation in self-conjugated medium mass nuclei

Danilo Gambacurta [1] and Denis Lacroix[2]

[1]Extreme Light Infrastructure - Nuclear Physics (ELI-NP), "Horia Hulubei" National Institute for Physics and Nuclear Engineering, 30 Reactorului Street, RO-077125 Măgurele, Jud. Ilfov, Romania
[2]Institut de Physique Nucléaire, IN2P3-CNRS, Université Paris-Sud, F-91406 Orsay Cedex, France

Abstract

We employ a model combining self-consistent mean-field and shell-model techniques to study the competition between particle-like and proton-neutron pairing correlations in fp-shell even-even self-conjugate nuclei. Deformation effects are realistically and microscopically described. The resulting approach can give a precise description of pairing correlations and eventually treat the coexistence of different condensate formed of pairs with different total spin/isospin. The standard BCS calculations are systematically compared with approaches including correlation effects beyond the independent quasi-particle picture. The competition between proton-neutron correlations in the isoscalar and isovector channels is also analyzed, as well as their dependence on the deformation properties.

1 Introduction

The effect of correlation between nucleons of different spin and isospin, is expected to be more pronounced in self-conjugate nuclei, i.e., nuclei with the same number of protons (Z) and neutrons (N). High intensity radioactive beams will offer new possibilities to study the interplay of the isoscalar (T = 0) and isovector (T = 1) pairing interaction between protons and neutrons in these nuclei. By analyzing the relative energies of the T=0 and T=1 states in even-even and odd-odd nuclei [1] and the T = 0 band in ^{74}Rb [2], the role of isovector proton-neutron (p-n) pairing has been recently singled out in a quite clear way. The same kind of analysis does not support the existence of the isoscalar pairing. At the same time, recent experiments seem to manifest the possibility to observe exotic structures of aligned pairs [3] that could be explained in terms of isoscalar p-n pairing correlations. For a recent review, se for example [4].

From a theoretical point of view, several frameworks have been proposed to incorporate p-n pairing correlations [5–9]. Most often, these approaches, based on a mean field-type approximation, lead to a non-coexistence of particle-like and particle unlike pairing, as a consequence of the inherent particle number and isospin symmetries violations. Alternatively, shell-model calculations can go beyond the independent quasi-particle picture and provide a particle/isospin number conserving framework able to solve the pairing problem including all spin/isospin channels.

The aim of the present work is to combine mean-field and shell model techniques to study of the role of deformation on particle-like and p-n pairing. More precisely, a microscopic mean-field is used to obtain realistic single-particle energies and two-body residual pairing interactions. Then, pairing correlations are studied in spherical and deformed nuclei through direct diagonalization of the Hamiltonian in a restricted space. This framework allows for a realistic description of nuclear deformation necessary to analyze its influence on the role of pairing correlations in both channels.

2 Methodology and results

The first step of our approach consists in solving the the self-consistent Hartree-Fock + BCS equations in a three-dimensional mesh by using the *EV8* code [10, 11]. In Fig. 1 the mean-field energy for different even-even $N = Z$ nuclei is plotted (solid-red lines), as a function of the quadrupole deformation parameter β. Starting from the mean-field results, we construct then a two-body pairing Hamiltonian written as:

$$H = H_{\text{s.p.}} + H_{10} + H_{01} \tag{1}$$

where $H_{\text{s.p.}}$ is the independent particle contribution. H_{10} and H_{01} correspond to a pairing two-body Hamiltonian acting respectively in the

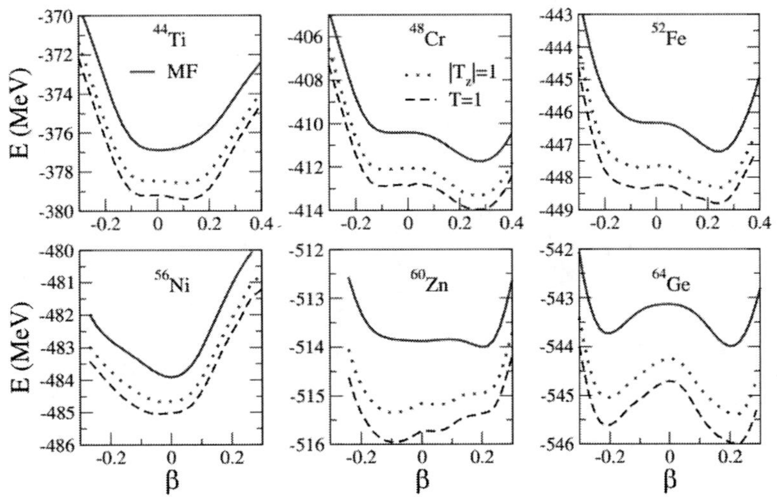

Figure 1: (Color online) Total energy as a function of the deformation parameter calculated in the mean-field (MF) approach (red-full line) is compared with different kind of SM calculations: $|T_z| = 1$ results (red-dotted line) where only particle like pairing is considered and $T = 1$ (black-dashed line) where all the components of the isovector pairing are included.

$(T = 1, S = 0)$ and $(T = 0, S = 1)$ channels. In our scheme, it is possible to identify pairs of time-reversed states that enters into the $T = 1$ channels as well as in the $(T = 0, S_z = 0)$ channel. Due to the specific space symmetry used in the *EV8* code, the interaction matrix elements in $T = 0$, $S_z = \pm 1$ cannot be easily restricted to $J = 0$ and $J = 1$. For this reason, only the $S_z = 0$ component is considered here. Therefore we consider the isovector pair operators:

$$P_1^\dagger(k) = \nu_k^\dagger \nu_{\bar{k}}^\dagger \;\; P_{-1}^\dagger(k) = \pi_k^\dagger \pi_{\bar{k}}^\dagger, \;\; P_0^\dagger(k) = (\nu_k^\dagger \pi_{\bar{k}}^\dagger + \pi_k^\dagger \nu_{\bar{k}}^\dagger)/\sqrt{2} \qquad (2)$$

and the isoscalar one:

$$D_0^\dagger(k) \;\; = \;\; (\nu_k^\dagger \pi_{\bar{k}}^\dagger - \pi_k^\dagger \nu_{\bar{k}}^\dagger)/\sqrt{2}. \qquad (3)$$

Different types of Shell Model (SM) calculations will be performed in order to single out specific effects:

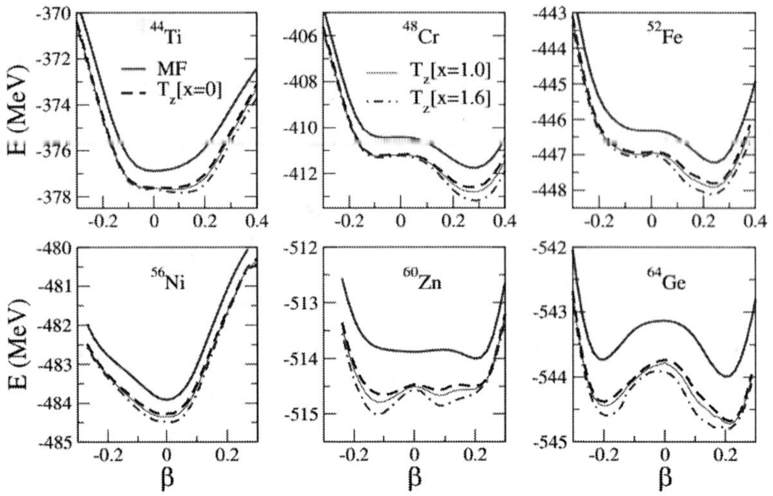

Figure 2: (Color online) Total energy as a function of the deformation obtained using only p-n pairing in the T=1 channel (black-dashed line, $T_z[x = 0]$), including also the isoscalar one with equal strength ,(thin-solid red line, $T_z[x = 1]$) and using the more realistic value $x = 1.6$ (blue dot-dashed line, $T_z[x = 1.6]$). The results obtained in the mean-field (MF) calculations are plotted for comparison.

- particle-like pairing: only n-n and p-p pairing are treated by considering the Hamiltonian

$$H = H_{s.p.} + \sum_{i\neq j, T_z=\pm 1} V_{ij}^{10} P^{\dagger}_{T_z}(i) P_{T_z}(j). \tag{4}$$

  The corresponding SM calculations will be denoted as $|T_z| = 1$;

- full isovector-pairing: The Hamiltonian includes the full $T = 1$ component

$$H = H_{s.p.} + \sum_{i\neq j, T_z=0,\pm 1} V_{ij}^{10} P^{\dagger}_{T_z}(i) P_{T_z}(j) \tag{5}$$

  and the results will be labelled as $T = 1$;

- p-n pairing: this calculation will illustrate qualitatively the competition between isovector and isoscalar p-n pairing. The Hamiltonian:

$$H = H_{s.p.} + \sum_{i \neq j} V_{ij}^{10} P_0^\dagger(i) P_0(j) + \sum_{i \neq j} V_{ij}^{01} D_0^\dagger(i) D_0(j). \quad (6)$$

is employed and the corresponding calculations denoted as $T_z = 0$.

In Fig. 1, the total energy E is shown as a function of the deformation parameter. The results obtained in the $(S = 0, T = 1)$ channel are shown either including all channels or only particle like pairing. In all cases, the results are compared with the energy landscape obtained directly from *EV8* and we see that additional correlation energy is systematically gained compared to the original *EV8* case, due to the restoration of the particle number and isospin symmetries. In the same figure, the SM results obtained considering also the p-n pairing in the $T = 1$ channels (black-dashed lines) are plotted. Compared with the $|T_z| = 1$ results (red-dotted lines), we see that the inclusion of p-n correlations provides some extra correlation energy.

In order to study the competition of the isoscalar and isovector pairing, we consider $v_{is} = x v_{iv}$, x being a constant defining the relative weight between the isovector (v_{iv}) and isoscalar (v_{is}) pairing strengths. In Fig. 2, the potential energy curve obtained including only isovector p-n pairing (black-dashed line), i.e. x=0, is compared with the results where isoscalar correlations also are included, for different x values. It is clear from this figure that the addition of the isoscalar channel gives a gain in correlation. We also see, comparing the results with $x = 1.0$ and $x = 1.6$ that the dependence of the isoscalar correlation energy on the x value is rather weak.

3 Conclusions

In the present work, we investigate particle-like and p-n correlations and their dependence on nuclear deformation in a framework that combines self-consistent mean-field calculations and the diagonalization techniques for the treatment of the pairing Hamiltonian. This framework is here used to systematically investigate fp-shell even-even $N = Z$ nuclei, from ^{44}Ti to ^{64}Ge. We found that deformation plays also an important role, producing a strong quenching of the pairing correlations [11]. The interplay of the isoscalar and isovector correlations, and their quantitative role in the binding energies is analyzed. It is found that the isoscalar p-n pairing is generally much weaker than the isovector contribution. However, in some cases, we observed that large deformation can favor the coexistence of isoscalar and isovector p-n condensates.

Acknowledgments

D.G. acknowledges the support from the Extreme Light Infrastructure Nuclear Physics (ELI-NP) Phase I, a project co-financed by the Romanian Government and European Union through the European Regional Development Fund.

References

[1] Macchiavelli A. O. *et al.* Phys. Rev. **C61** (2000) 041303R

[2] O' Leary C. D. *et al.* Phys. Rev. **C67** (2003) 021301(R)

[3] Cederwall B. *et al.* 2011 Nature (London) **469** (2011) 68

[4] Frauendorf S. and Macchiavelli A. O Prog. Part. Nucl. Phys. **78** (2014) 24

[5] Parikh J. C. Nucl. Phys. **63** (1965) 214

[6] Evans J. A., Dussel G. G., Maqueda E. E., Perazzo R. P. J. Nucl. Phys. **A367** (1981) 77

[7] Engel J., Pittel S., Stoitsov M., Vogel P., Dukelsky J. Phys. Rev. **C55** (1997) 1781

[8] Lerma H. S., Errea B., Dukelsky J., Satula W. Phys. Rev. Lett. **99** (2007) 032501

[9] Bertsch G. F. and Luo Y. Phys. Rev. **C81** (2010) 064320

[10] Bonche P., Flocard H. and Heenen P. H. Comp. Phys. Comm. **171** (2005) 49

[11] Gambacurta D. and Lacroix D. Phys. Rev. **C91** (2015) 014308

EPJ Web of Conferences **117**, 04006 (2016) DOI: 10.1051/epjconf/201611704006
NN2015

Low-lying 1^- and 2^+ states in ^{124}Sn via inelastic scattering of ^{17}O

L. Pellegri[1,2*], A. Bracco[1,2], F.C.L. Crespi[1,2]
and the AGATA Collaboration

[1] Dipartimento di Fisica dell'Università degli Studi di Milano, Milano, Italy
[2] INFN, Sezione di Milano, Milano, Italy

Abstract

The γ decay of low-lying 1^- and 2^+ states up to the neutron separation energy in ^{124}Sn populate by the inelastic scattering of ^{17}O was measured. The Angular distributions were measured both for the γ rays and the scattered ^{17}O ions. The results are presented.

The study of the Pygmy Dipole Resonance (PDR), the low-energy part of the electric dipole response in nuclei, is particularly relevant in the nuclear structure investigations and photo-disintegration reaction rates of astrophysical importance [1–3]. Its description, within the hydrodynamical model, corresponds to a vibration of the neutron skin against a isospin-saturated core [4, 5]. In recent years, the study of the PDR has attracted particular attention since its microscopic structure is presently under discussion. A dipper understanding of its nature can be achieved through its excitation using different probes [5]. Indeed, recent works comparing photon and α scattering experiments show that these reactions behave differently when populating these states [6, 7]. While a set of states at lower energy is excited by both reactions, another set of states at higher energies is not populated by α scattering. This interesting finding has motivated further work using another probe with strong isoscalar character as ^{17}O [8–10].

*Current affiliation: University of the Witwatersrand and iThemba LABS, South Africa

DOI: 10.1051/epjconf/201611704006

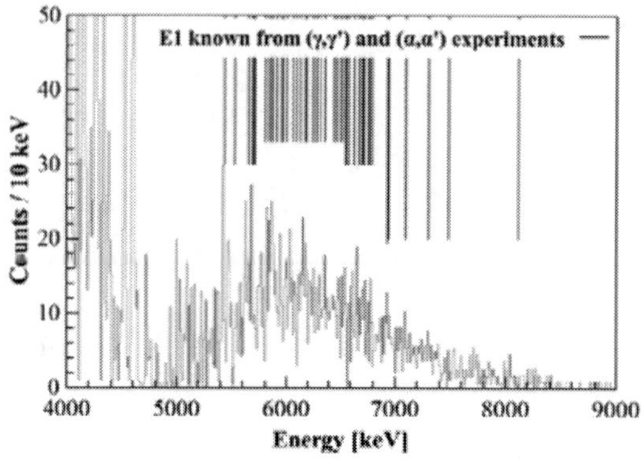

Figure 1: AGATA energy spectrum of the ^{124}Sn nucleus in the PDR region with the gating conditions as described in the text. Adapted from [10].

Building on this, a detailed investigation of the nature of the low-lying E2 strength in neutron-rich nuclei is required to understand if a neutron skin affects the excitation of other multipolarities [11, 12].

To study the low-lying 1^- and 2^+ states in ^{124}Sn, inelastic scattering of ^{17}O at 20 MeV/u was used as a probe since it preferentially enhanced the excitation of the isoscalar component of populated states [5]. The experiment was performed at Legnaro National Laboratories (LNL, Italy). The experimental setup consisted of the AGATA Demonstrator [13, 14] and an array of 9 large volume LaBr3:Ce scintillation detectors [15], for the detection of high-energy γ rays. The scattered ions were detected by the of TRACE detector system consisting of two ΔE-E silicon telescopes [14, 16]. The combination of the silicon telescopes and the AGATA Demonstrator enabled the identification of different decay paths by imposing a gate on the two-dimensional ^{17}O-γ coincidence matrix. Fig. 1 shows the γ-ray spectrum measured with the AGATA demonstrator array in the energy region above 4 MeV where the γ decay of the pygmy states is expected to be. Since the dominant decay of these pygmy states is to the ground state, this spectrum was obtained by imposing the constraint E_x=E_γ after selecting the inelastically scattered ^{17}O events. Since these states had a life-time of the order of the femtoseconds, a Doppler correction for the recoil motion was also needed.

The high position sensitivity of the silicon detectors and the AGATA Demonstrator allows clear identification of the multipolarity of the decaying transitions. The angular distributions of the gamma rays of the pygmy states exhibit an electric dipole character, as expected [10]. A distorted wave Born

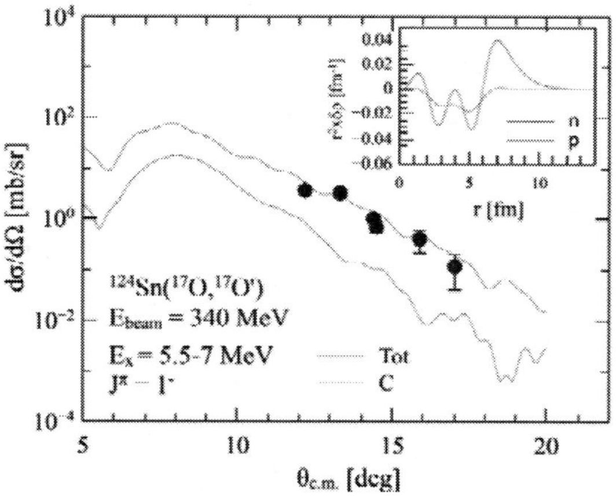

Figure 2: Differential cross sections for the 1^- states between 5.5 - 7 MeV. The blue line shows the DWBA calculation with only the Coulomb contribution while the red line uses the total microscopic form factor based on the transition densities shown in the inset of the figure. Adapted from [10].

approximation (DWBA) analysis was performed to gain a more detailed understanding of the nature of these states. The DWBA calculations were performed using the computer code FRESCO [17]. A normalization factor deduced from the elastic scattering was used to account for the combined contribution of the beam current, target thickness and dead time. This normalization factor was kept the same for all the inelastic scattering cross sections. The optical model parameters of the Woods-Saxon potentials used for these calculations are reported in Ref. [10]. A microscopic form factor based on the transition density associated with these pygmy states was used. Fig. 2 shows the cross sections measured in our experiment together with the DWBA calculations. The use of the microscopic form factor proved crucial to reproduce the data, leading on to conclude that the main contribution to the excitation comes from the isoscalar nuclear contribution. It is clear from the transition densities that these 1^- states have a strong isospin mixing character. Furthermore, the only contribution on the surface comes from the neutrons of the skin giving rise to a isovector and isoscalar components of the same intensity.

A recent theoretical study [12] discusses the possible influence of the neutron skin on excitations of other multipolarities. The use of the silicon detectors allowed us to also investigate the low-lying 2^+ states since they measured a wide range of excitation energies. A number of 2^+ states were observed between 3 and 5 MeV. For several states the multipolaries were not

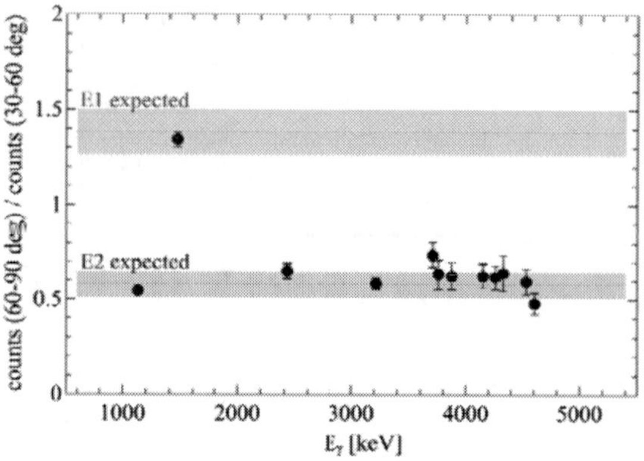

Figure 3: Ratio between the number of counts in the 60° - 90° angular interval over the number of counts in the 30° - 60° angular interval measured for different transitions of ^{124}Sn. The horizontal red and blue bands correspond to the expected ratio for E1 and E2 transitions, respectively. Adapted from [11].

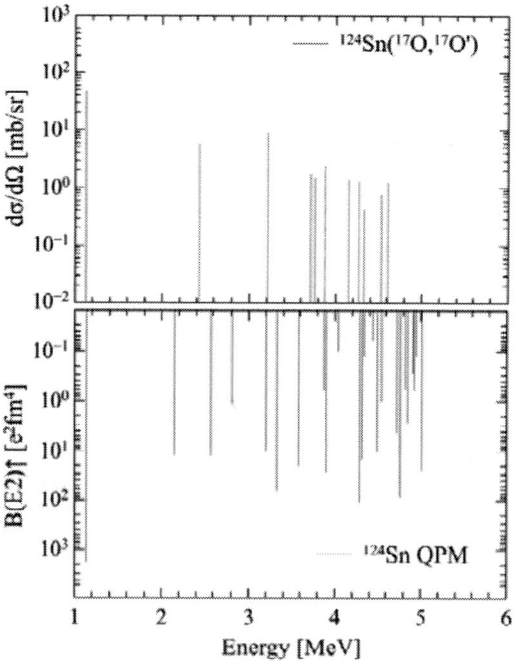

Figure 4: Comparison between the transition probabilities theoretically calculated [12] (bottom panel), and the cross sections measured in our experiment (top panel). Adapted from [11].

previously known. Through the γ-rays angular distributions extracted from our experimental data, it was possible to assign the multipolarity to these states, as can be seen from Fig. 3. The cross sections measured for the 2^+ states were extracted and compared with the Quasi-particle Phonon Model (QPM) calculations. This is shown in Fig. 4. The presence of a number of 2^+ states grouped together in the energy region 3-5 MeV supports the prediction of the HFB+QPM model. This quadrupole strength clustering appears to be similar to the known Pygmy Dipole Resonance at 5-7 MeV. Moreover, the microscopic analysis of these 2^+ states reveal that they have a unique structure closely connected with the excitation of the neutron skin. It will be important to obtain information on the transition densities of these states, the B(E2)s and to study other isotopes to learn more about the quadrupole degrees of freedom of the neutron skin.

References

[1] Goriely S *et al.* 2004 *Nucl. Phys. A* **739** 331

[2] Carbone A *et al.* 2010 *Phys. Rev. C* **81** 041301(R)

[3] Piekarewicz J *et al.* 2012 *Phys. Rev. C* **85** 041302

[4] Savran D *et al.* 2013 *Prog. Part. Nucl. Phys.* **70** 210

[5] A Bracco F C L C and Lanza E G 2015 *Eur. Phys. J. A* **51** 99

[6] Govaert K *et al.* 1998 *Phys. Rev. C* **57** 2229

[7] Endres J *et al.* 2010 *Phys. Rev. Lett.* **105** 212503

[8] Crespi F C L *et al.* 2014 *Phys. Rev. Lett.* **113** 012501

[9] Crespi F C L *et al.* 2015 *Phys. Rev. C* **91** 024323

[10] Pellegri L *et al.* 2014 *Phys. Lett. B* **738** 519

[11] Pellegri L *et al.* 2015 *Phys. Rev. C* **92** 014330

[12] Tsoneva N and Lenske H 2011 *Phys. Lett. B* **695** 174

[13] Akkoyun S *et al.* 2012 *Nucl. Instr. and Meth. A* **668** 26

[14] Gadea A *et al.* 2011 *Nucl. Instr. and Meth. A* **654** 88

[15] Giaz A *et al.* 2013 *Nucl. Instr. and Meth. A* **729** 910

[16] Mengoni D *et al.* 2014 *Nucl. Instr. and Meth. A* **764** 241

[17] Thompson I J *et al.* 1988 *Computer Physics Reports* **7** 167

EPJ Web of Conferences **117**, 04007 (2016) DOI: 10.1051/epjconf/201611704007
NN2015

g-factor measurements of isomeric states in ^{174}W

M. Rocchini[1,2], A. Nannini[2], G. Benzoni[3], B. Melon[2], P.R. John[4,5], C.A. Ur[5], R. Avigo[3,6], D. Bazzacco[5], N. Blasi[3], G. Bocchi[3,6], S. Bottoni[3,6], A. Bracco[3,6], F. Camera[3,6], S. Ceruti[3,6], F.C.L. Crespi[3,6], G. Georgiev[7] A. Giaz[3,6], A. Gottardo[8], S. Leoni[3,6], R. Menegazzo[4,5], D. Mengoni[4,5], C. Michelagnoli[9] B. Million[3], V. Modamio[10], A.I. Morales[3,6], D.R. Napoli[10], M. Ottanelli[2], L. Pellegri[3,6], A. Perego[1,2], J.J. Valiente-Dobon[10] and O. Wieland[3]

[1] Università degli Studi di Firenze, Firenze, Italy
[2] INFN, Sezione di Firenze, Firenze, Italy
[3] INFN, Sezione di Milano, Milano, Italy
[4] Università degli Studi di Padova, Padova, Italy
[5] INFN, Sezione di Padova, Padova, Italy
[6] Università degli Studi di Milano, Milano, Italy
[7] CSNSM, Orsay, France
[8] CNRS, Paris, France
[9] GANIL, Caen, France
[10] INFN - Laboratori Nazionali di Legnaro, Legnaro (PD), Italy

Abstract

The experimental setup GAMIPE used for gyromagnetic factor measurements at Laboratori Nazionali di Legnaro and a recent experimental work regarding K-isomers in ^{174}W are described. Aim of the experiment is to study the detailed structure of the isomeric states wave functions, by the measurement of the magnetic dipole moments. This piece of information can provide interesting hints for theoretical models. Preliminary results concerning the population of the isomers of interest and half-lives are presented.

1 Introduction

The study of the high-K states in the region of mass A $\sim$ 180 has attracted in recent years much experimental and theoretical work, aiming at obtaining insight on specific nuclear structure aspects [1]. The knowledge of the properties of these long-living states may be useful also for practical applications in medicine and for energy storage methods [1, 2].

In axially deformed symmetric systems, away from closed shell, the most common kind isomers are the so called K-isomers. The name refers to the projection K of the nuclear angular momentum on the symmetry axis. The existence of this kind of isomers is due to the approximate conservation of K: the transition from a state characterized by K_i to another state with K_f is governed by the selection rule $\Delta K = |K_f - K_i| \leq \lambda$ being λ the transition order. Transitions that violate this rule have been observed and are called K-forbidden. Such transitions are strongly hindered, resulting in long-lived isomeric states. In the study of these transitions it is useful to define the degree of forbiddenness $\nu = \Delta K - \lambda$ and the hindrance factor

$$F = \frac{T_{1/2}^{exp}}{T_{1/2}^{W}} \tag{1}$$

which is the half-life of the level respect to the corresponding Weisskopf estimate.

K-isomers have been observed in many regions throughout the Segrè chart and have been extensively studied in past years. However many important properties of such states are not well understood yet, in particular those related to their de-excitation towards levels with lower K values. One of the model used to describe the K-hindered decays for the deformed nuclei (characterized by the parameters of Hill-Wheeler γ, β) is the γ-tunneling, in which the transition from an initial state to a final state is described like a tunneling of the nucleus through a barrier of γ degree of freedom, along a line of constant β deformation [3].

The levels structure of ^{174}W has been recently studied and twenty rotational bands have been identified, two of them built on isomeric levels of spin 12^+ and 8^- with half-lives of 128 ns and 158 ns, respectively [4, 5]. An interesting open issue in the decay pattern of ^{174}W is the decay mechanism of the level 12^+, which decays mainly directly to the 10^+ level of the ground state band with an E2 transition at 1879 keV. This transition is characterized by high degree of forbiddenness, ν = 10, but the hindrance factor comparable to those of other similar states in mass region A $\sim$ 180, could not be described by the above-mentioned γ-tunneling model. A satisfactory explanation of this discrepancy remains to be found, in particular other contributions can be considered, as the K-mixing mechanism. The K-hindrance in gamma decay by using quasi-continuum gamma spectroscopy techniques, was recently studied [6]. The results show that a K-mixing process due to temperature effects plays an important role already also at rather low excitation energy and could be the basic, general phenomenon, responsible for the K-hindrance to the gamma decay, both for discrete excited rotational bands and specific isomeric states with large K-values. Therefore, the knowledge of the detailed structure of the know high-K isomers in ^{174}W, given by the evaluation of the static magnetic dipole moment, can give useful hints for the theoretical modelling of their decays.

In this contribution we report about the preliminary results about a recent measurement of the gyromagnetic factors of the isomeric levels in ^{174}W, with the goal of assign their quasi-particle configuration. Attempts to give a quasi-particle configuration assignment were already performed by Tandel *et al.* [4] on the basis of experimental in-band branching ratios, from which $|g_K - g_R|/Q_0$ were derived. These analysis are, however, dependent on the assumption concerning the quadrupole deformation and the effects associated with the reduction of the pairing correlations on both the rotational g factor g_R and the alignment. In this respect the knowledge of the static magnetic dipole moment is of crucial importance, as it provides independent information on the composition of the wave-functions.

2 Experimental details

The isomeric states were populated in the ^{162}Dy(^{16}O, 4n)^{174}W reaction (81.1% of the total cross section estimated by CASCADE code [7] and also experimentally confirmed [5]) using an 84 MeV ^{16}O beam delivered by the XTU-Tandem, at the Laboratori Nazionali di Legnaro. The beam has been pulsed into packets of 2 ns duration and repetition time of 800 ns to register the complete decay of the isomers of interest. The target consisted in a layer of ^{162}Dy with a thickness of 500 μg/cm^2 and a backing of Pb with a thickness of 50 mg/cm^2 to implant the ^{174}W nuclei and let them decay at rest. In order to reduce the dealignment effects due to the interaction with radiation-induced defects, the target was heated to a temperature of 400 K in a special oven. This temperature was monitored with an automatic control system that can provide a constant temperature with an error of about 0.5 K. For the detection of gamma rays we have used the GAMIPE apparatus that consists of 4 GASP-type HPGe detectors, placed at 12 cm from the target and positioned at $\pm 45°$ and $\pm 135°$ respect to the beam axis between the yokes of an a electromagnet that can provide magnetic field of the order of 10 kG.

The half-lives of the levels of interest are suitable for the application of the Time Differential Perturbed Angular Distribution (TDPAD) technique. This technique consists in the observation of the anisotropy of gamma emission of an ensemble of oriented nuclei located in an external magnetic field, in function of the time. In this way it is possible to obtain the Larmor frequency and so the gyromagnetic factor. The target was located in the gap of the electromagnet, set to a magnetic field $B = 14.65$ kG applied perpendicular to the beam-detector plane, for observing 2-3 complete oscillations in the range of 3 $T_{1/2}$. Data have been acquired by using commercial digital electronics adapted for the TDPAD measurements. A new structure supporting the HPGe detectors has been utilized during this experiment, now every single detector can be positioned individually.

3 Preliminary results

Energy and time signals, recorded as difference with the beam radiofrequency, for each HPGe detector, were recorded in list-mode and analysed offline. Data were sorted in two kinds of correlation matrices: time signals of each detector were correlated to the measured γ-ray energies and then coincidences between emitted γ-rays were registered in γ-γ 2D matrices. Using the first kind of matrices it's possible to obtain time-gated energy spectra and energy-gated time spectra, useful to obtain the Larmor frequency and to measure half-lives. The γ-γ matrices were used to obtain coincidence spectra, useful for the identification of the transitions in the populated channels.

The delayed energy spectrum obtained from a γ-γ matrix by imposing the coincidence with the transition $12^+ \longrightarrow 10^+$ of 1879 keV and a time condition in-between the beam bursts is reported in *fig.* 1. The transitions belonging to states of the yrast line are clearly visible. As mentioned above isomeric 12^+ state mainly de-excites towards the ground state band.

From our data it was possible to give a confirmation of the half-life of the 12^+ isomer. The decay spectrum of the $12^+ \longrightarrow 10^+$ gamma transition are reported in *fig.* 2 as a function of the time, measured with respect to the pulsed beam. The measured value of the half-life is $T^{exp}_{1/2} = 124(21)$ ns and is compatible with the adopted value $T^{ado}_{1/2} = 128(8)$ ns [4]. The higher error

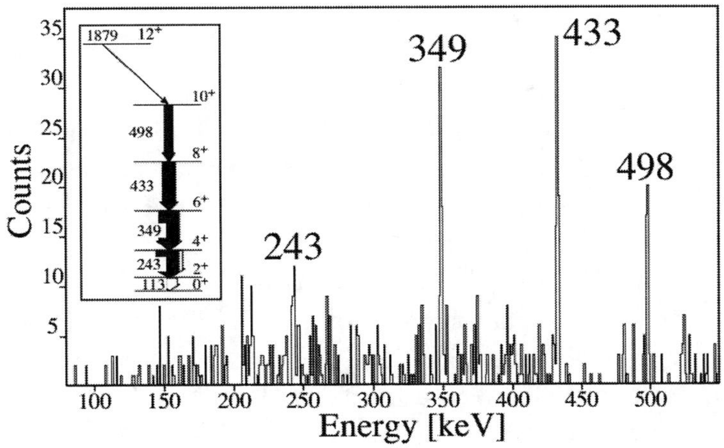

Figure 1: Delayed energy spectrum obtained from γ-γ matrix by imposing an energy condition on the $12^+ \longrightarrow 10^+$ transition and a time condition in-between the beam bursts, background subtraction has been performed (in the box on top left is shown the decay pattern of the isomeric level 12^+).

obtained in this work is due to the fact that the apparatus is not optimized for half-life measurements.

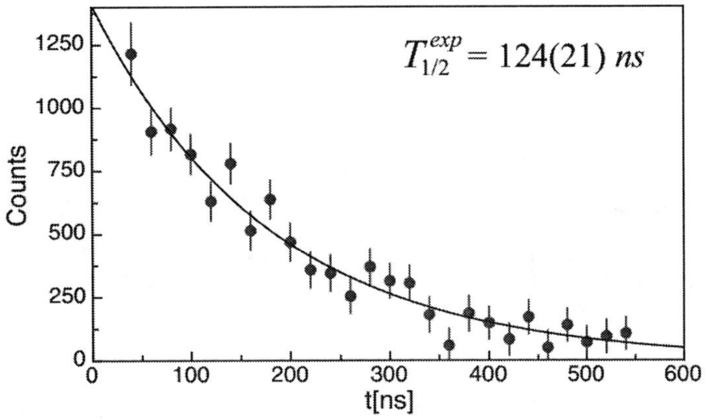

Figure 2: Exponential decay of the isomeric level 12^+ observed with respect to the pulsed beam (on top right is shown the obtained half-life).

4 Summary and perspectives

Preliminary results, obtained in a g-factor measurement with TDPAD technique in ^{174}W performed at the Laboratori Nazionali di Legnaro, have been presented. The first results of the analysis showed a clear evidence of the population of the isomers of interest. Also the value of the half-life of the 12^+ isomeric level is confirmed. For the determination of the g-factor of this level the analysis is still in a preliminary phase due to the low population rate of the 12^+ level and to the presence of a high background contribution. We expect to reach a 3% precision in the g-factor value which is usually achieved in similar experiments (see for example [8]).

The knowledge of the wave-function composition of the isomeric states in ^{174}W, obtainable with the experimental results of this experiment, could be very useful for a better understanding of the decay of K-isomers, in particular for the evaluation of the K-mixing contribution.

References

[1] P. M. Walker and G. D. Dracoulis, *Nature*, **399** (1999) 35.

[2] P. M. Walker and J. J. Carrol, *Phys. Today*, **58** (2005) 39.

[3] K. Narimatsu et al., *Nucl. Phys. A*, **601** (1996) 69.

[4] S. K. Tandel et al., *Phys. Rev. C*, **73** (2006) 044306.

[5] G. D. Dracoulis, P. M. Walker and A. Johnston, *J. Phys. G*, **4** (1978) 713.

[6] V. Vandone et al., *Phys. Rev. C*, **88** (2013) 034312.

[7] F. Puhlhofer, *Nucl. Phys. A*, **280** (1977) 267.

[8] M. Ionescu-Bujor et al., *Phys. Lett. B*, **541** (2002) 219.

EPJ Web of Conferences **117**, 04008 (2016) DOI: 10.1051/epjconf/201611704008
NN2015

Theoretical studies of possible toroidal high-spin isomers in the light-mass region

Andrzej Staszczak[1], Cheuk-Yin Wong[2]

[1]Institute of Physics, Maria Curie-Skłodowska University, Lublin, Poland
[2]Physics Division, Oak Ridge National Laboratory, Oak Ridge, TN USA

Abstract

We review our theoretical knowledge of possible toroidal high-spin isomers in the light mass region in $28 \leq A \leq 52$ obtained previously in cranked Skyrme-Hartree-Fock calculations. We report additional toroidal high-spin isomers in ^{56}Ni with I=114$\hbar$ and 140$\hbar$, which follow the same (multi-particle)–(multi-hole) systematics as other toroidal high-spin isomers. We examine the production of these exotic nuclei by fusion of various projectiles on ^{20}Ne or ^{28}Si as an active target in time-projection-chamber (TPC) experiments.

1 Introduction

Wheeler suggested that under appropriate conditions the nuclear fluid may assume a toroidal shape [1]. If toroidal nuclei could be made, there would be a new family tree for the investigation of the nuclear species.

The rotating liquid-drop model is useful as a qualitative guide to point out essential energy balances leading to possible toroidal figures of equilibrium [2]. A quantitative assessment relies on microscopic descriptions that include both the bulk properties of the nucleus and the single-particle shell effects in self-consistent mean-field theories, such as the Skyrme-Hartree-Fock (SHF) approach [3]. The non-collective rotation with an angular momentum about the symmetry axis is permissible quantum mechanically for an axially symmetric toroid by making particle-hole excitations and aligning the angular momenta of the constituents along the symmetry axis [4]. As a consequence, only a discrete, quantized set of total angular momentum $I{=}I_z$ states is allowed.

In our recent works [5], we showed by using a cranked SHF approach that toroidal high-spin isomeric states have general occurrences in $28 \leq A \leq 52$ for even-even $N{=}Z$ and $N{\neq}Z$ nuclei. Toroidal high-spin isomers have also been found theoretically in similar HF calculations in this mass region [6].

We would like to review the systematics of these nuclei and suggest ways how these nuclei may be produced.

2 Light-mass toroidal high-spin isomers

We have located the toroidal high-spin isomers at their energy minima using the cranked SHF approach [5]. We find that the ratio of the torus major radius R to the torus minor radius d, R/d, increases with angular momentum and approximately linearly with the mass number while the minor radius d remains essentially unchanged (see Ref. [5]). It is useful to classify these nuclei according to the n-particle n-hole nature of the isomer, relative to the toroidal nucleus at $I = 0$. One finds that all np-nh families follow a regular well-behaved pattern as shown in Fig. 1, where we plot the total energy $E^{\rm tot}(I)$ of the isomer ${}^{A}Z^{t}(I)$ as a function of R/d for different toroidal isomers with various aligned angular momenta I.

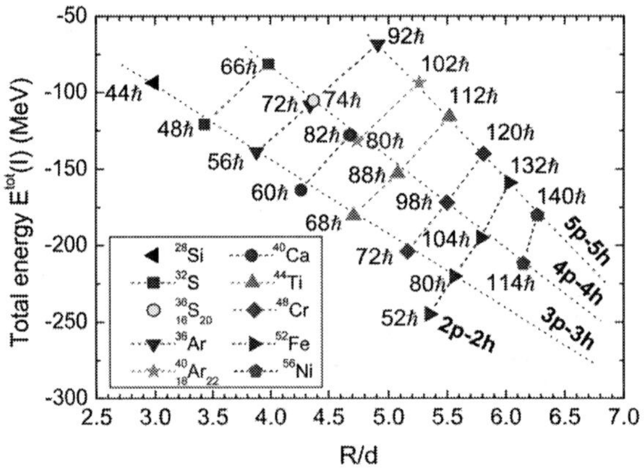

Figure 1: (Color online.) The total energies of the isomeric toroidal states of ${}^{28}_{14}$Si, ${}^{32}_{16}$S, ${}^{36}_{16}$S, ${}^{36}_{18}$Ar, ${}^{40}_{18}$Ar, ${}^{40}_{20}$Ca, ${}^{44}_{22}$Ti, ${}^{48}_{24}$Cr, ${}^{52}_{26}$Fe, and ${}^{56}_{28}$Ni and their associated angular momenta $I{=}I_z$ values along the symmetry axis, as a function of R/d. The np-nh configurations relative to the $I{=}0$ states are also indicated.

We collect the properties of all known 21 toroidal high-spin isomers up to ^{52}Fe obtained previously in [5] in Fig. 1. These systematics predict the possible presence of np-nh toroidal high-spin isomers for ^{56}Ni at $R/d \sim 6.0$. Indeed, we found energy minima for ^{56}Ni with I=114$\hbar$ and 140$\hbar$ at R/d=6.15 and 6.27, respectively, in subsequent cranked SHF calculations.

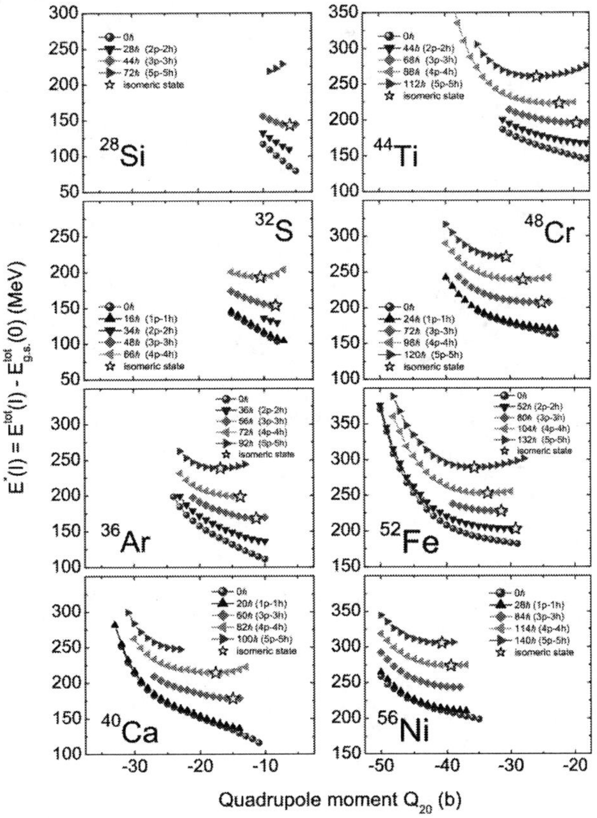

Figure 2: (Color online.) The excitation energies $E^*(I)$ of high-spin toroidal states of ^{28}Si, ^{32}S, ^{36}Ar, ^{40}Ca, ^{44}Ti, ^{48}Cr, ^{52}Fe, and ^{56}Ni as a function of the quadrupole moment Q_{20} for different angular momenta along the symmetry axis, I=I_z.

In Fig. 2 the excitation energies $E^*(I)$ of even-even N=Z toroidal states relative to the energy of the ground states are presented as a function of the quadrupole moment Q_{20} for different I values along the symmetry axis. They include those from our earlier work in [5] and the newly found ^{56}Ni toroidal isomers. The locations of energy minima $E^*(^AZ^t(I))$ representing toroidal high-spin isomeric states are indicated by star symbols.

3 Production of light-mass toroidal isomers

As the question of the mean half-life of these isomers remains unresolved, one can design experiments that could detect the existence of the exotic toroidal high-spin nuclei of Figs. 1 and 2 at appropriate energies, if the mean half-lives are longer than $\hbar$/MeV or 200 fm/c. One way is to search for these isomers as resonances or metastable nuclei by bombarding projectile nucleus ${}^{A_p}Z_p$ on an active-target nucleus ${}^{A_T}Z_T$ for the production of the toroidal high-spin isomer ${}^{A}Z^t(I)$ with angular momentum $I{=}I_z$,

$$ {}^{A_p}Z_p + {}^{A_T}Z_T \to {}^{A}Z^t(I). \tag{1} $$

The active target can be, for example, ^{20}Ne, ^{36,38,40}Ar, or ^{28}Si. We shall consider the cases of ^{20}Ne and ^{28}Si as active-targets.

In recent years, TPC chambers have been used to study the nuclear spectroscopy of metastable nuclei [7, 8]. The idea is to use a chamber of noble gas under a high voltage so that the gas itself or an embedded solid layer serves as the target, and the nuclear trajectories show up as tracks. The production of a composite nucleus with a long half-life would show up as a single track with the mass and charge arising from the fusion of the projectile and target nuclei. The production of binary products indicates a two→two reaction from which one can examine the elastic and inelastic channels and study the excitation function and angular distribution to search for various meta-stable states. Previously, many metastable states formed by colliding various projectile nuclei with an active He target have been found by such a technique [7].

The cross section for producing a toroidal isomer at the correct energy and angular momentum is [9] (p. 517)

$$ \sigma_{\rm res}(E, {}^{A}Z^t(I)) = \frac{4\pi}{k^2}(2I+1)\frac{\Gamma^2/4}{[E - E_{\rm res}({}^{A}Z^t(I))]^2 + \Gamma^2/4} B_{\rm in} B_{\rm out}, \tag{2} $$

where E is the c.m. energy, $E_{\rm res}({}^{A}Z^t(I))$ is the c.m. resonance energy for the toroidal high-spin isomer with spin I, k is the c.m. momentum in the initial state, and Γ is the full width at half maximum height of the resonance. The quantities $B_{\rm in}$ and $B_{\rm out}$ are the branching fractions for the resonance into the initial-state and finial-state channel, respectively. Here, the width Γ and the branching fractions $B_{\rm in}$ and $B_{\rm out}$ may need to be determined experimentally. The resonance energy $E_{\rm res}$ (in the c.m. system) is given by

$$ E_{\rm res}({}^{A}Z^t(I)) = M({}^{A}Z^t(I)) - M({}^{A_T}Z_T) - M({}^{A_p}Z_p), \tag{3} $$

where $M({}^{A}Z^t(I))$, $M({}^{A_T}Z_T)$, and $M({}^{A_p}Z_p)$ are nuclear masses of ${}^{A}Z^t(I)$, ${}^{A_T}Z_T$, and ${}^{A_p}Z_p$, respectively. In terms of the binding energies $B({}^{A}Z)$, $B({}^{A_T}Z_T)$, and $B({}^{A_p}Z_p)$, and excitation energy $E^*({}^{A}Z^t(I))$, we have

$$ E_{\rm res}({}^{A}Z^t(I)) = E^*({}^{A}Z^t(I)) - B({}^{A}Z) + B({}^{A_T}Z_T) + B({}^{A_p}Z_p). \tag{4} $$

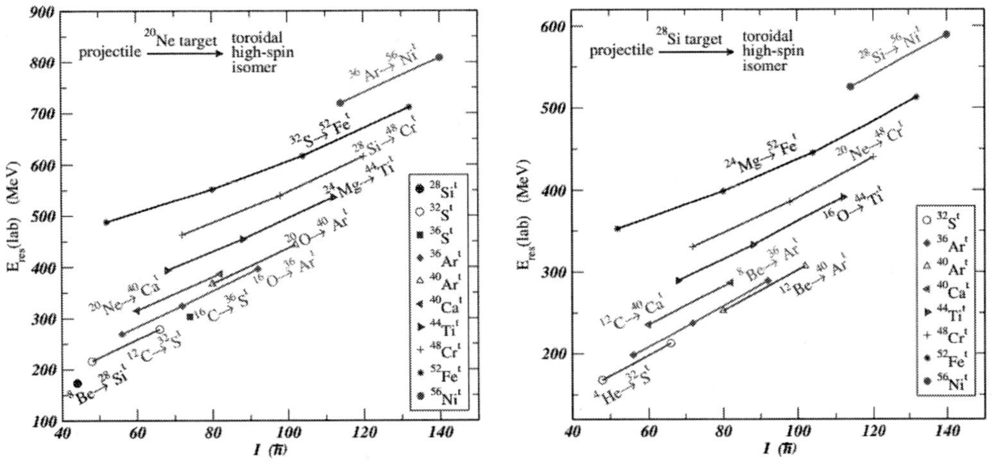

Figure 3: (Color online) The resonance energies $E_{\rm res}(^AZ^t(I))$(lab) in the laboratory system in the bombardment of various projectiles $^{A_p}Z_p$ on ^{20}Ne (left panel) or ^{28}Si (right panel) as an active target for the production of toroidal isomers $^AZ^t(I)$ with different angular momenta I.

The resonance energy $E_{\rm res}(^AZ^t(I))$(lab) in the laboratory system is given by $E_{\rm res}(^AZ^t(I)) \times (A_p + A_T)/A_T$. For the production of toroidal high-spin isomers by collision on ^{20}Ne or ^{28}Si as an active target, the resonance energies are shown in Fig. 3. The knowledge of the predicted values of $E_{\rm res}(^AZ^t(I))$(lab) will facilitate the search of toroidal high-spin isomers.

4 Conclusions and discussion

Under (multi-particle)–(multi-hole) excitation involving large orbital angular momentum orbitals, non-collective rotations of many light nuclei lead to equilibrium configurations whose densities may assume the shape of a torus. The np-nh systematics of toroidal high-spin isomers fit a regular pattern which can be used to predict possible presence of toroidal high-spin isomer in ^{56}Ni. We found additional equilibrium energy minima for ^{56}Ni with I=114$\hbar$ and 140$\hbar$ in subsequent cranked SHF calculations.

We examine the production of light-mass toroidal high-spin isomers by fusion of various projectile nuclei with an active target, in which the trajectories of the reaction products can be examined as a function of the collision energies. Resonance energies for the production of toroidal high-spin isomers have been calculated for ^{20}Ne or ^{28}Si as an active target, based on the Skyrme-Hartree-Fock energies obtained for the isomers.

The technology of building a TPC using 90% of ^{20}Ne as its dominant ingredient has been developed by the ALICE collaboration [10]. The utilization of a similar TPC detector with ^{20}Ne or ^{28}Si as an active target for nuclear spectroscopy may prove useful for the search of toroidal isomers.

Acknowledgements

The authors wish to thank Profs. J. Natowitz and K. Read for helpful discussions. This work was supported in part by the Division of Nuclear Physics, U.S. Department of Energy, Contract No. DE-AC05-00OR22725.

References

[1] See a reference to J. A. Wheeler's toroidal nucleus in Gamow G. 1961 *Biography of Physics* (New York: Harper & Brothers Publishers) pp. 297.

[2] Wong C.Y., *Ann. Phys.* **77** (1973) 279; *Phys. Rev. C* **17** (1978) 331.

[3] Vautherin D. and Brink D.M., *Phys. Rev. C* **5** (1972) 626; Engel Y.M., Brink D.M., Goeke K., Krieger S., and Vautherin D., *Nucl. Phys.* **A 249** (1975) 215.

[4] Bohr A. and Mottelson B.R., *Nucl. Phys.* **A 354** (1981) 303c.

[5] Staszczak A. and Wong C.Y., *Phys. Lett. B* **738** (2014) 401; *Acta Phys. Pol. B* **46** (2015) 675, arXiv:1504.07646; *Phys. Scripta* (2015) in press, arXiv:1412.0050.

[6] Ichikawa T., Maruhn J.A., Itagaki N., Matsuyanagi K., Reinhard P.-G., Ohkubo S., *Phys. Rev. Lett.* **109** (2012) 232503; Ichikawa T., Matsuyanagi K., Maruhn J.A., and Itagaki N., *Phys. Rev. C* **89** (2014) 011305(R); *Phys. Rev. C* **90** (2014) 034314.

[7] Suzuki D. *et al.*, *Nucl. Instr. Meth. A* **691** (2012) 39; Suzuki D. *et al.*, *Phys. Rev. C* **87** (2013) 054301.

[8] Roger T. *et al.*, *Nucl. Instr. Meth. A* **638** (2011) 134, arXiv:1012.3560.

[9] Olive K.A. *et al.*, (Particle Data Group), *Chin. Phys. C* **38** (2014) 1.

[10] B. Ketzer for the GEM-TPC, ALICE TPC Collaborations, *Nucl. Instr. Meth. A* **732** (2013) 237, arxiv:1303.6694.

^{16}O + ^{16}O molecular structures of positive- and negative-parity superdeformed bands in ^{34}S

YASUTAKA TANIGUCHI

Department of Medical and General Sciences,
Nihon Institute of Medical Science,
1276 Shimogawara, Moroyama-machi, Iruma-gun, Saitama 350-0435, Japan

Abstract

The structures of excited states in ^{34}S are investigated using the antisymmetrized molecular dynamics and generator coordinate method (GCM). The GCM basis wave functions are calculated via energy variation with a constraint on the quadrupole deformation parameter β. By applying the GCM after parity and angular momentum projections, the coexistence of two positive- and one negative-parity superdeformed (SD) bands are predicted, and low-lying states and other deformed bands are obtained. The SD bands have structures of ^{16}O + ^{16}O + two valence neutrons in molecular orbitals around the two ^{16}O cores in a cluster picture. The configurations of the two valence neutrons are δ^2 and π^2 for the positive-parity SD bands and $\pi^1\delta^1$ for the negative-parity SD band.

S isotopes ($Z = 16$) are suitable nuclei for studying deformation and clustering caused by excitation. S isotopes are expected to be favorable for the formation of SD bands because $Z = 16$ is considered a magic number of superdeformation. In terms of clustering, S isotopes are key nuclei in the *sd* shell. S isotopes are analogs of Be isotopes because those isotopes can form systems consisting of two doubly closed shell nuclei (^{16}O and α for S and Be isotopes, respectively) and valence neutrons. In Be isotopes, structures consisting of $\alpha + \alpha +$ valence neutrons in molecular orbitals are thought to develop in low-lying states, with the valence neutrons in molecular orbitals around two α cores [1]. The SD states in ^{32}S are predicted to contain a large amount of ^{16}O + ^{16}O cluster structure components [2, 3]. They suggest the existence of SD states that have ^{16}O + ^{16}O + valence neutrons in the molecular orbital structure in S isotopes. By a γ spectroscopy experiment, many states up to the $J^\pi = 10^+$ and (9^-) are observed in ^{34}S. But the structures with multi-particle-multi-hole configurations have never been clarified. This paper aims to clarify the structures of SD states in ^{34}S theoretically using antisymmetrized molecular dynamics (AMD) and the generator coordinate method (GCM). The coexistence of positive- and negative-parity SD bands and their structures are discussed, focusing on ^{16}O + ^{16}O + valence neutrons in molecular orbitals around the two ^{16}O cores. Details of this study was reported in Ref. [4].

The wave functions in low-lying states are obtained using parity projection and angular momentum projection (AMP) and the GCM with deformed-basis AMD wave functions. A deformed-basis AMD wave function is a Slater determinant of Gaussian wave packets. The basis wave functions of the GCM are obtained by energy variation with a constraint on quadrupole deformation parameter β after projection onto eigenstates of parity. The optimized wave functions are superposed after parity projection and AMP by employing the quadrupole deformation parameter β. Then we obtain the energy spectra and the corresponding wave functions. Gogny D1S force is used as the effective interaction.

Figures 1 and 2 show the energy surfaces as functions of the quadrupole deformation parameter β for the positive- and negative-parity states, respectively, obtained by energy variation with a constraint on β after parity projection. The energies projected onto the $J^\pi = 0^+$, 3^-, and 4^- states are also shown. In the positive-parity energy surface (Fig. 1), three excited local minima or shoulders exist around $\beta = 0.4$, 0.6, and 0.8 as well as the minimum at $\beta = 0.25$, which suggest the existence of three excited deformed bands in the positive-parity states. In the negative-parity energy surface (Fig. 2), three local minima exist around $\beta = 0.2$, 0.5, and 0.7.

Figures 3 and 4 show the single-particle energies as functions of the quadrupole deformation β for positive- and negative-parity states, respectively. The quanta $[Nn_z\Omega]$ in the Nilsson picture are also shown for the two highest orbits of neutrons in $\beta > 0.6$ region, which have [202] and [321] as shown in the figures. The single-particle energies of [202] and [321] are flat for the quadrupole deformation parameter β, and they resemble each other. Left and center panels in Fig. 5 show the density distributions of the

DOI: 10.1051/epjconf/201611704009

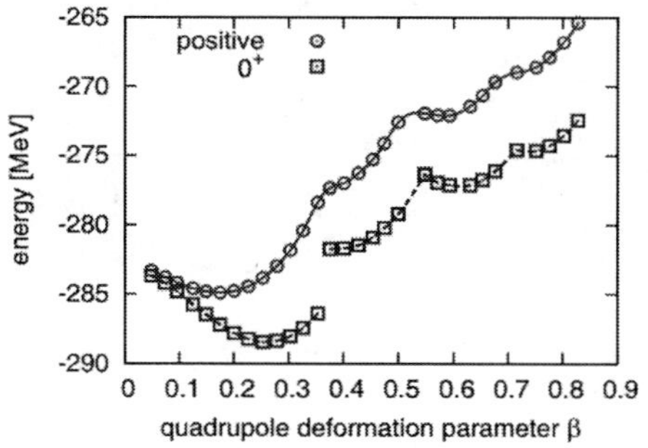

Figure 1: Energy surfaces as functions of quadrupole deformation parameter β for positive-parity states. Circles and squares show energies projected onto positive-parity and $J^{\pi} = 0^{+}$ states, respectively. This figure is taken form Ref. [4].

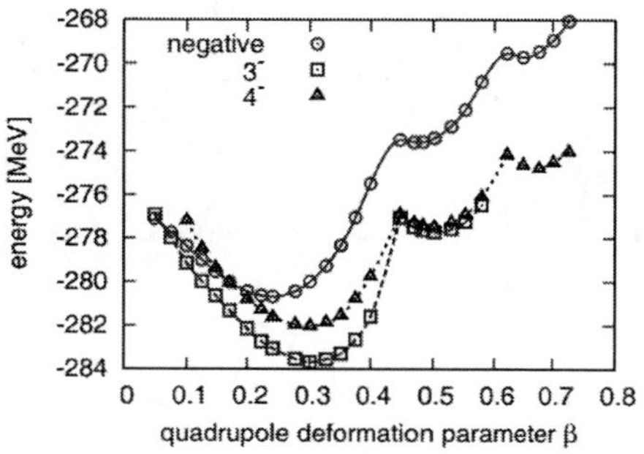

Figure 2: Energy surfaces as functions of quadrupole deformation parameter β for negative-parity states. Circles, squares, and triangles show energies projected onto positive-parity, $J^{\pi} = 3^{-}$, and $J^{\pi} = 4^{-}$ states, respectively. This figure is taken form Ref. [4].

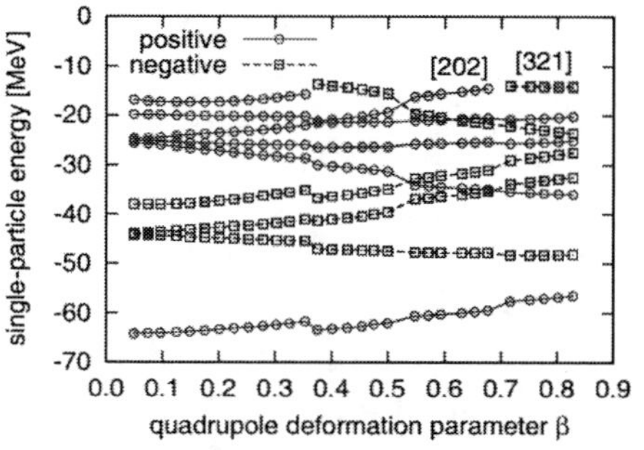

Figure 3: Single-particle energies of neutrons as functions of quadrupole deformation parameter β for positive-parity states. Circles and squares show positive- and negative-parity orbits, respectively. Numbers in brackets show the Nilsson quanta for the two highest orbits of neutrons (see text). This figure is taken form Ref. [4].

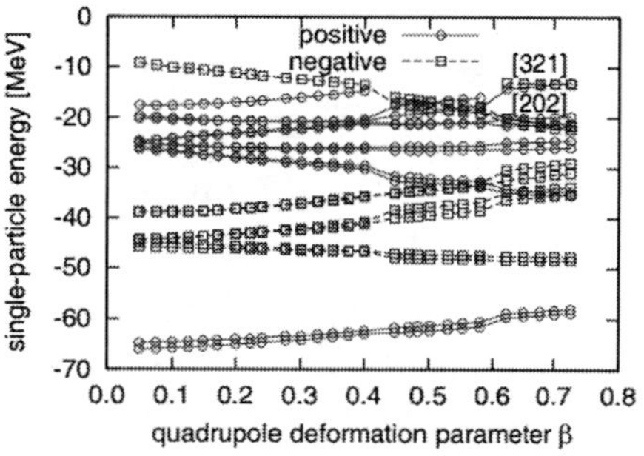

Figure 4: Same as Fig. 3 but for negative-parity states. This figure is taken form Ref. [4].

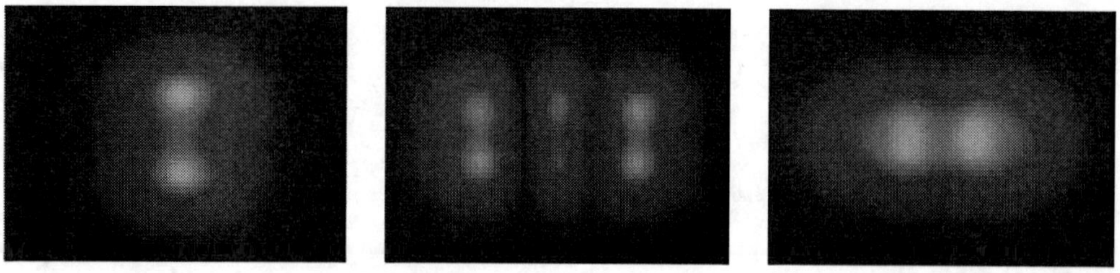

Figure 5: Density distributions of [202] (left) and [321] (center) neutron orbits and that of lower 16 protons and 16 neutrons in $\beta > 0.6$ region.

[202] and [321] orbits, respectively. The ^{32}S core part have neck structures in $\beta > 0.6$ region as shown in the right panel in Fig. 5, and the ^{32}S core part have $4\hbar\omega$ excited configuration, which is similar to SD states in ^{32}S. On the energy surface, ^{32}S (SD) core + $[202]_\nu^2$ and $[321]_\nu^2$ configurations are obtained in positive-parity states, and ^{32}S (SD) core + $[202]_\nu^1[321]_\nu^1$ configurations are obtained in negative-parity states.

Figure 6 shows the level scheme of the positive- and negative-parity states. Various rotational bands, called $K^\pi = 0^+_{\rm ND}$, $0^+_{\rm SD1}$, and $0^+_{\rm SD2}$ for positive parity and $K^\pi = 2^-_{\rm ND}$, $3^-_{\rm ND}$, and $4^-_{\rm SD}$ for negative parity, are obtained, as well as low-lying $0\hbar\omega$ and $1\hbar\omega$ states. The dominant components of the $K^\pi = 0^+_{\rm SD1}$ and $0^+_{\rm SD2}$ bands have ^{32}S (SD) core + $[202]_\nu^2$ and $[321]_\nu^2$ configurations, respectively, and those of the $K^\pi = 3^-_{\rm ND}$ and $2^-_{\rm ND}$ bands have ^{32}S (SD) core + $[202]_\nu^1[321]_\nu^1$ configurations.

The [202] and [321] Nilsson orbits correspond to δ and π molecular orbitals around the ^{16}O + ^{16}O cores. The lowest orbit around an ^{16}O core is a $0d_{5/2}$ orbit. By linear combination of $0d_{5/2}$ orbits around two ^{16}O cores, molecular orbitals around the two ^{16}O cores are formed. Figure 7 shows schematic illustrations of the formation of molecular orbitals around two spherical ^{16}O cores. The $0d_{5/2}$ orbits around the ^{16}O cores form δ and π orbitals, which are formed from two $(l, |l_z|) = (2, 2)$ and $(2, 1)$ orbits, respectively, around the left and right ^{16}O cores. For the δ orbital, the $(l, |l_z|) = (2, 2)$ orbits have no node in the z direction, as shown in the left column of Fig. 7(a). Therefore, the δ orbital also has no node in the z direction, as shown in the right column of Fig. 7(a), and the numbers of nodes of the π orbitals are two in the direction of the z axis, as shown in Fig. 7(b). In the radial direction, they have no node because a $0d$ orbit has no node in the radial direction. This shows that the quanta of the δ and π orbitals are [202] and [321], respectively, in the Nilsson picture.

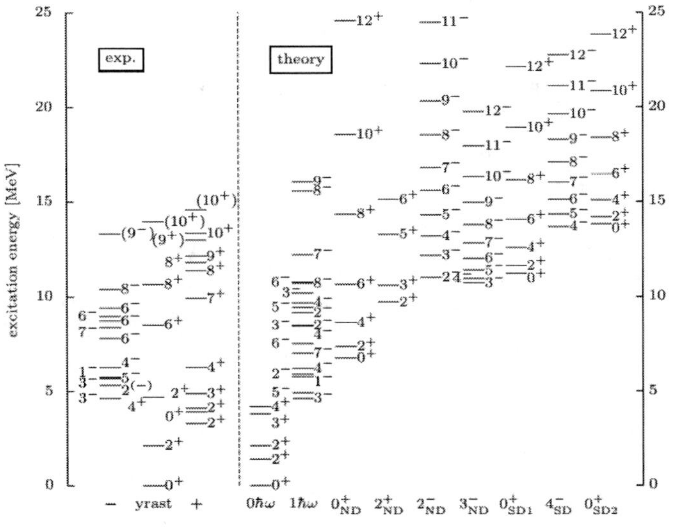

Figure 6: Left and right parts show experimental and theoretical level scheme, respectively, in ^{34}S. The "yrast" and "+" in the experimental part show yrast states and other positive-parity states, respectively, and the "−" shows negative-parity states. In the theoretical part, dominant components and labels of rotational bands are shown (see text). This figure is taken form Ref. [4].

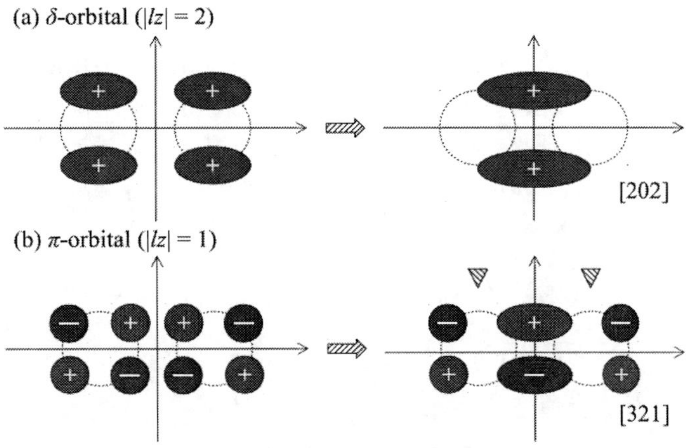

Figure 7: Schematic illustrations of molecular orbitals generated from $0d$ orbits around two ^{16}O cores for (a) δ and (b) π orbitals. Left and right columns show phases of $0d$ orbits around two ^{16}O cores and molecular orbitals, respectively. Dotted circles show two ^{16}O cores located on the z axis. Inverse triangles show locations of nodes in molecular orbitals in the z direction. Numbers in brackets show Nilsson quanta (see text). This figure is taken form Ref. [4].

The GCM calculation yielded three SD bands, called the $K^\pi = 0^+_{\rm SD1}$, $0^+_{\rm SD2}$, and $4^-_{\rm SD}$ bands, the structures of which are interpreted as ^{16}O + ^{16}O + two valence neutrons that have δ^2, π^2, and $\delta^1\pi^1$ configurations, respectively, in a cluster picture. The ^{32}S core part of the $K^\pi = 0^+_{\rm SD1}$, $0^+_{\rm SD2}$, and $4^-_{\rm SD}$ band members are same as SD states in ^{32}S, which contains a large amount of ^{16}O + ^{16}O cluster structure components [3]. Therefore, the three SD bands have structures of ^{16}O + ^{16}O + valence neutrons in a cluster picture. The configurations of the valence neutrons of the $K^\pi = 0^+_{\rm SD1}$, $0^+_{\rm SD2}$, and $4^-_{\rm SD}$ bands are $[202]^2$, $[321]^2$, and $[202]^1[321]^2$, respectively. The [202] and [321] orbits correspond to the δ and π molecular orbitals, respectively, around the two ^{16}O cores. Therefore, the configurations of the valence neutrons of the $K^\pi = 0^+_{\rm SD1}$, $0^+_{\rm SD2}$, and $4^-_{\rm SD}$ bands are interpreted as δ^2, π^2, and $\delta^1\pi^1$, respectively.

In conclusions, the structure of the SD states in ^{34}S were investigated using the AMD and GCM. By superposing the AMD wave functions calculated via energy variation with a constraint on the quadrupole deformation parameter β, the coexistence of two positive- and one negative-parity SD bands is predicted. The SD states have mp-mh configurations, and they are interpreted as a structure consisting of ^{16}O + ^{16}O + valence neutrons in molecular orbitals around ^{16}O + ^{16}O cores in a cluster picture. Both of clustering and deformation are important to understand structures in $A \gtrsim 30$ nuclei.

This work was supported by JSPS KAKENHI Grant Number 25800124. The numerical calculations for this work were conducted under the Interdisciplinary Computational Science Program of the Center for Computational Sciences, University of Tsukuba.

References

[1] Kanada-En'yo Y. Kimura M. and Ono A., Prog. Theor. Exp. Phys., **2012** (2012) 01A202, and references therein.

[2] Ohkubo S. and Yamashita K., Phys. Rev. C, **66** (2002) 021301.

[3] Kimura M. and Horiuchi H., Phys. Rev. C, **69** (2004) 051304.

[4] Taniguchi Y., Phys. Rev. C, **90** (2014) 054308.

EPJ Web of Conferences **117**, 04010 (2016) DOI: 10.1051/epjconf/201611704010
NN2015

Sub-barrier backward quasielastic scattering: A probe of the hexadecapole deformation

H.Q. Zhang, C.J. Lin, H.M. Jia, L. Yang, X.X. Xu, F. Yang, Z.D. Wu, Z.H. Liu, N.R. Ma, L.J. Sun and D.X. Wang

China Institute of Atomic Energy, P. O. Box 275(10), Beijing 102413, P. R. China

Abstract

The quasielastic scattering excitation functions for ^{16}O + ^{152}Sm, ^{170}Er and ^{174}Yb were measured with high precision at a backward angle with small energy intervals at energies near the Coulomb barriers. The hexadecapole deformation (β_4) of the target nuclei was extracted by using the lower-energy data and the obtained values agree with the available results reasonably well. This offers a sensitive method to extract β_4. This method is especially meaningful for the radioactive nuclei considering the low beam intensities.

1 Introduction

The role played by nuclear structure in near-barrier fusion reactions with heavy-ions has been observed several decades ago [1] and has been described well within the coupled-channels (CC) theoretical framework. Beside the quadrupole deformation (β_2) effect, the possible role of the hexadecapole deformation (β_4) in sub-barrier fusion has been frequently addressed [2, 3]. The shape of the experimental fusion barrier distribution is equivalent to the theoretical result obtained by including classically random orientation of the deformed nucleus [4] and β_4 is expected to have a significant effect on fusion. The experimental fusion barrier distributions [5] mean that fusion reactions are very sensitive not only to β_2 but also to small β_4.

Up to now, the methods of sub-barrier α scattering [6], above-barrier α excitation [6,7], electron scattering [8], and muonic x rays [9] have been used to determine the shape of the deformed nucleus experimentally. Usually, β_4 is difficult to extract experimentally and the obtained results are model-dependent and quite different with big errors. Theoretically, the macroscopic-microscopic method to calculate the ground-state deformation has been given in Ref. [10].

Based on the two complementary processes of fusion and backward quasielastic (QEL) scattering, we try to extract β_4 by using the lower-energy QEL data where the coupling effect to nuclear deformation is expected to be dominant. Also QEL is easier to be measured experimentally at lower-energy region. To this end, the spherical ^{16}O was selected as projectile. The nuclei of ^{152}Sm, ^{170}Er and ^{174}Yb, which can be seen as good rotors with similar β_2 but very different β_4 for the ground-state rotational band, were selected as target nuclei.

2 Experimental procedure and data analysis

The experiment was performed based on the HI-13 tandem accelerator at China Institute of Atomic Energy (CIAE). Thin targets (50-100 μg/cm^2 thick) of ^{152}Sm, ^{170}Er and ^{174}Yb with a diameter of 3 mm evaporated onto 20-25 μg/cm^2 carbon foils were used. The beam energy range is E_{Lab} = 40-80 MeV. The QEL particles were measured by using four silicon surface barrier detectors placed at the backward angle of $\theta_{\text{Lab}} = 175°$. Another group of four silicon surface barrier detectors were placed symmetrically at the forward angle $\theta_{\text{Lab}} = 41°$ relative to the beam direction as monitor detectors and also for data normalization. More details can be found in [11].

The energy window for QEL was defined to include the elastic peak and the tail corresponding to all other peripheral processes. For all reactions, the QEL counts were divided by the sum of the elastic scattering counts in the four monitor detectors, giving a quantity proportional to the ratio of the QEL cross sections to Rutherford scattering $d\sigma_{\text{QEL}}/d\sigma_{\text{Ru}}$, where 'Ru' denotes Rutherford scattering. The measured experimental QEL excitation functions for ^{16}O + ^{152}Sm, ^{170}Er and ^{174}Yb are shown in fig. 1 (left), where only statistical errors are given and the typical errors are smaller than 1%.

An effective energy E_{eff} was used by considering the minor centrifugal energy E_{cent} correction, that is $E_{\text{eff}} = E_{\text{c.m.}} - E_{\text{cent}}$, in order to compare the barrier distribution $D_{\text{QEL}}(E_{\text{c.m.}}, 175°)$ with $D_{\text{QEL}}(E_{\text{eff}}, 180°)$. The corresponding barrier distributions are shown in fig. 1 (right). Considering the

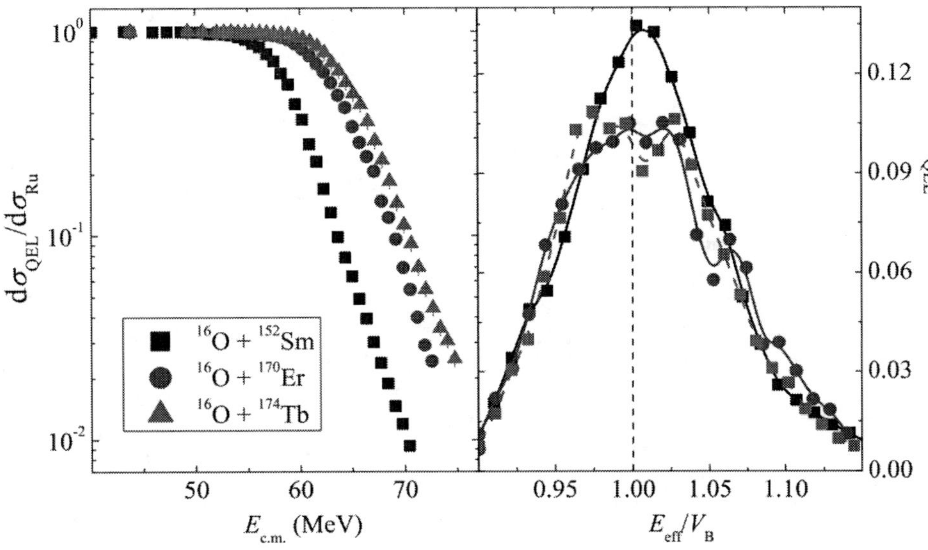

Figure 1: Experimental QEL excitation functions (left) for ^{16}O + ^{152}Sm, ^{170}Er and ^{174}Yb measured at $\theta_{Lab} = 175°$ and the extracted barrier distributions (right).

prevailing sensitivity of QEL to nuclear deformation at lower energies and to simplify the problem further by reducing the possible effect of other reaction channel couplings on the extracted β_4, only the data of $d\sigma_{QEL}/d\sigma_{Ru} > 0.7$ was used in the following analysis somewhat arbitrarily.

The CC calculations were performed with a modified version of the code CCFULL [12]. A Woods-Saxon (WS) potential with $r_0 = 1.20$ fm and $a = 0.65$ fm was used, the potential depth V was varied to reproduce the barrier heights. As usual, a short range imaginary potential with $W = 30$ MeV, $r_{0W} = 1.0$ fm, and $a_W = 0.4$ fm was used to simulate the compound nucleus formation. The used Coulomb radius parameter is $r_{0C} = 1.1$ fm, which has little influence on the calculated cross sections. The calculated results are insensitive to the parameters of the imaginary part of the potential as long as it is strong enough and well localized inside the Coulomb barrier.

Considering the coupling effect, the excitation energies of ^{16}O are high compared with the barrier curvature, so only produces an adiabatic potential renormalization, without affecting the structure of the barrier distribution [13]. This effect can be included in the potential and will not be considered explicitly in the CC calculations. Excitation states up to 10^+ in the rotational band of the target nuclei were included.

For ^{16}O + ^{152}Sm, both fusion and backward QEL have been studied

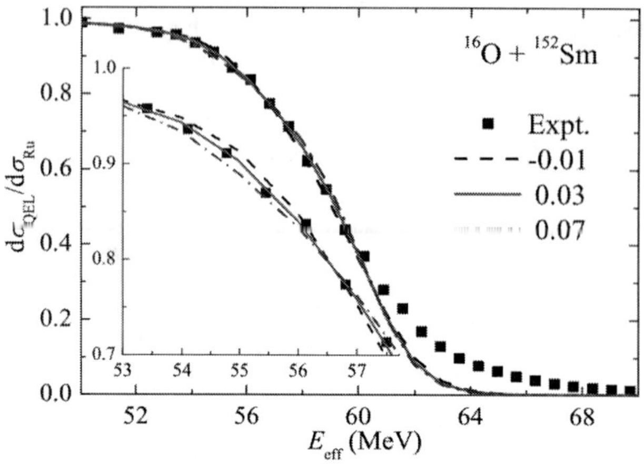

Figure 2: Experimental QEL excitation function for ^{16}O + ^{152}Sm compared with the CC calculations by using different β_4.

widely [14]. Figure 2 shows the present-work backward QEL excitation function (solid squares). The lines are the CC calculations which take the permanent deformation of ^{152}Sm into account. $\beta_2 = 0.31$ [15] is used and fixed for ^{152}Sm. Different β_4-values were used in the CC calculations for getting the optimal value by comparison with the experimental data. The parameter which gives the best reproduction for the lower-energy data well is $\beta_4 = +0.037$ by using a chi-square analysis, the corresponding error is ± 0.006 by making $\chi^2_{min}/N_{pt} = 1.1$ for the optimal fit increased by 1, where 'N_{pt}' denotes the number of data points used in the fitting procedure. This means that the lower-energy backward QEL is indeed sensitive to higher order deformation β_4.

The extracted β_4-value for ^{152}Sm here is similar with both $\beta_4^N = 0.038 \pm 0.007$ extracted from 14-18 MeV α scattering [16] and 0.048 obtained from 50 MeV α scattering [7], but smaller than both the charge deformation parameter of $\beta_4^C = 0.08$ [17] and $\beta_4 = 0.07$ ($\beta_2 = 0.287$) [8] obtained from electron scattering.

Similar analysis was also performed for ^{16}O + ^{170}Er and ^{16}O + ^{174}Yb. The β_4 extracted in this work are shown in fig. 3 (solid circles). Figure 3 also represents some available results for Sm, Er and Yb isotopes, which include the charge deformation β_4^C (α^C) extracted by means of the Coulomb excitation experiments [6, 17–19], nuclear deformation β_4^N (α^N) by using the interference effects [16] and inelastic scattering of α particles at energies well above the barrier [6, 7]. Also included in this plot are the β_4 of $^{152,154}Sm$,

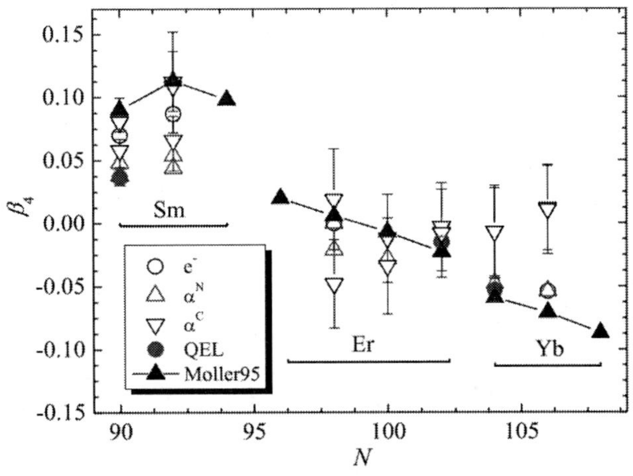

Figure 3: The variation of β_4 with neutron number N for Sm, Er and Yb isotopes by using different experimental methods and Möller's theoretical prediction [10].

^{166}Er and ^{176}Yb determined by electron scattering (e^-) [8]. Deformations of the nuclear ground states calculated with a macroscopic-microscopic model (Möller95) [10] are also shown in fig. 3. It can be seen that the extracted β_4-values in this work qualitatively agree with this overall trend.

3 Summary

In summary, we have measured the QEL scattering excitation functions with high precision at a backward angle and extracted the barrier distributions for ^{16}O + ^{152}Sm, ^{170}Er and ^{174}Yb. The extracted β_4-values show a transition from positive for ^{152}Sm and near-zero for ^{170}Er, to negative for ^{174}Yb, which is consistent with the available trend. This means that the backward QEL at the lower-energy region, like fusion excitation function, is indeed sensitive to the higher-order deformation and especially the sign of β_4. Of course, more precise data and more systems are needed to further check the feasibility of this method. Comprehensive data analysis is also needed in order to get a more reliable quantitative result.

Acknowledgements

This presentation has been supported by the National Key Basic Research Development Program of China under Grant No. 2013CB834404, and the National Natural Science Foundation of China under Grants Nos. U1432246, U1432127, and 11505293.

References

[1] Stokstad R.G. et al., *Phys. Rev. Lett.*, **41** (1978) 465.

[2] Rhoades-Brown M.J. and Oberacker V.E., *Phys. Rev. Lett.*, **50** (1983) 1435.

[3] Dasgupta M. et al., *Annu. Rev. Nucl. Part. Sci.*, **48** (1998) 401.

[4] Nagarajan M.A., Balantekin A.B. and Takigawa N., *Phys. Rev. C*, **34** (1986) 894.

[5] Lemmon R.C. et al., *Phys. Lett. B*, **316** (1993) 32.

[6] Lee I.Y. et al., *Phys. Rev. C*, **12** (1975) 1483.

[7] Hendrie D.L. et al., *Phys. Lett. B*, **26** (1968) 127.

[8] Cooper T. et al., *Phys. Rev. C*, **13** (1976) 1083.

[9] Powers R.J. et al., *Phys. Rev. Lett.*, **34** (1975) 492.

[10] Möller P., *At. Data Nucl. Data Tables*, **59** (1995) 185.

[11] Jia H.M. et al., *Phys. Rev. C*, **90** (2014) 031601(R).

[12] Hagino K., Rowley N. and Kruppa A.T., *Comput. Phys. Commun.*, **123** (1999) 143.

[13] Hagino K. et al., *Phys. Rev. Lett.*, **79** (1997) 2014.

[14] Zhang Huanqiao et al., *Phys. Rev. C*, **57** (1998) R1047.

[15] Raman S., Nestor C.W. Jr., and Tikkanen P., *At. Data Nucl. Data Tables*, **78** (2001) 1.

[16] Brückner W. et al., *Phys. Rev. Lett.*, **30** (1973) 57.

[17] Fischer H. et al., *Phys. Rev. C*, **15** (1977) 921.

[18] Erb K.A. et al., *Phys. Rev. Lett.*, **29** (1972) 1010.

[19] Wollersheim H.J., Wilcke W. and Elze T.W., *Phys. Rev. C*, **11** (1975) 2008.

EPJ Web of Conferences **117**, 04011 (2016) DOI: 10.1051/epjconf/201611704011
NN2015

Cluster rotational bands in ^{11}B

A.N. Danilov[1], A.S. Demyanova[1], A.A. Ogloblin[1],
T.L. Belyaeva[2], S.A. Goncharov[3] and W. Trzaska[4]

[1]National Research Center "Kurchatov Institute" Moscow, 123182, Russia
[2]Universidad Autonoma del Estado de Mexico, Toluca, 50000, Mexico
[3]Lomonosov Moscow State University, Moscow, 119991, Russia
[4]JYFL, Jyväskylä, FIN-40500, Finland

Abstract

Differential cross-sections of ^{11}B+α inelastic scattering at $E(\alpha)$ = 65 MeV leading to most of the known ^{11}B states at excitation energies up to 14 MeV were measured [1]. The data analysis was done using Modified diffraction model (MDM) [2] allowing determining radii of excited states. Radii of the states with excitation energies less than ~ 7 MeV coincide with the radius of the ground state with an accuracy not less than 0.1 - 0.15 fm. This result is consistent with traditional view on shell structure of low-lying states in ^{11}B. Most of the observed high-energy excited states are distributed among four rotational bands. Moments of inertia of band states are close to the moment of inertia of the Hoyle state of ^{12}C. The calculated radii, related to these bands, are 0.7 - 1.0 fm larger than the radius of the ground state, and are close to the Hoyle state radius. These results are in agreement with existing predictions about various cluster structure of ^{11}B at high excitation energies.

1 Introduction

During long time ^{11}B nucleus was considered as a good example of shell effects in light nuclei. Up to excitation energies ~ 7 MeV ^{11}B states were described by different variants of shell models. Recently, however, a number of theoretical and experimental works appeared [3–5] with predictions about cluster configurations of various types co-existed in ^{11}B.

Particular attention was drawn to the idea that there may be states in ^{11}B, which are analogs of the famous 0_2^+ state in ^{12}C nucleus (the so-called Hoyle state). The Hoyle state consists of three weakly interacting alpha clusters and its properties were crucial for verification of the alpha - particle condensation theory [6]. One of the main suggestions of this theory is abnormally large radius of the Hoyle state. Accordingly, Hoyle state analogs in ^{11}B must also have increased size.

It was assumed [3] that the Hoyle state analog in ^{11}B is the state $3/2^-$ with excitation energy 8.56 MeV, which is not described by any variant of the shell model. The radius of 8.56 MeV state was considered to be abnormally large, and it was predicted that this state is a base for rotational band.

There are a lot of experimental studies of ^{11}B (see, e.g., [7] and references therein), but they did not affect the excitation energy region of interest for the problem. Due to the fact that many questions about ^{11}B states remained open, we have undertaken a new study of inelastic $^{11}\mathrm{B} + \alpha$ scattering at $E(\alpha) = 65$ MeV [1]. Experimental results were analyzed using Modified diffraction model (MDM). In this article we discuss results for high-energy excited states and possible cluster rotational bands formed from them.

2 Radii and moments of inertia of high-lying rotational bands in ^{11}B

The following rotational bands were predicted [3,5] in high excitation energy region in ^{11}B. Most of these states were observed in our experiment:
$K = 3/2^-$: 8.56 ($3/2^-$) - 10.34 ($5/2^-$) - 11.60 - 13.14 ($9/2^-$) MeV,
$K = 1/2^+$: 6.79 ($1/2^+$) - 9.88 ($3/2^+$) 11.60 ($5/2^+$) 13.16 ($7/2^+$) MeV,
$K = 3/2^+$: 7.98 ($3/2^+$) - 9.27 ($5/2^+$) - 10.60 ($7/2^+$) - 12.63 ($9/2^+$) MeV,
$K = 5/2^+$: 7.29 ($5/2^+$) - 9.19 ($7/2^+$) - 11.27 ($9/2^+$) MeV.

These rotational bands are shown in Fig. 1 together with the band in ^{12}C, based on the Hoyle state. Data on angular momentum transfer with excitation states belonging to the specified bands, received from our experiment, are in agreement with known spin-parities of ^{11}B states. However, for 6.79, 9.88, 10.34, 13.14 13.16 MeV states it could not determined unambiguously due to insufficient energy resolution. Several special features in $J(J+1)$ dependence of excitation energies can be seen in Fig. 1. Firstly, moments of inertia of the band states are very high and comparable. The largest of them ($2I/\hbar^2 \sim 4.0$, by the energy difference between the excitation

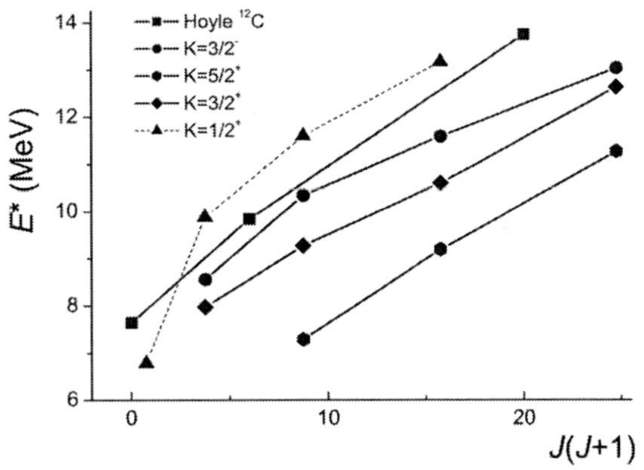

Figure 1: Predicted [3,5] rotational bands in ^{11}B at excitation energies $E^* > 7$ MeV. For comparison, rotational band [8], based on the Hoyle state (0_2^+, 7.65 MeV) of ^{12}C, is shown.

energies 11.60 and 10.34 MeV) are observed for higher members of the rotational band $K = 3/2^-$, for which cluster structure $2\alpha + t$ is predicted. It is interesting that it is much larger than the moment of inertia of its analog - the Hoyle state, for which $2I/\hbar^2 = 2.7$. Secondly, there is a clear correlation between the moments of inertia and the values of radii obtained from scattering data using MDM. Low-lying states of ^{11}B have "normal" radii and "reduced" moments of inertia about $2I/\hbar^2 \sim 1.1$. These values are close to values for the first excited state of ^{12}C, 4.44 MeV. "Big" moments of inertia correspond to increased radii. Summary of the radii of the states measured using MDM is given in the Fig. 2.

As seen from Fig. 2, increased radii were found, at least, for one of the members of each band, and in most cases they are about 0.7 - 1.0 fm larger than the radius of the ground state of ^{11}B. This leads to the conclusion that all states belonging to the bands under consideration, have abnormal size. Theoretical works [3–5] suggest a significant deformation of the rotational states of ^{11}B with $E^* > 7$ MeV and it allows the increase of their radii. The radii and moments of inertia of these states are close to the corresponding values of the Hoyle state in ^{12}C nucleus and the rotational band based on it. These facts indicate probable cluster nature of the ^{11}B states discussed. In particular, the 8.56 MeV state can be considered as an analog of the Hoyle state. But some open questions remain regarding the rotational band that is based on the 6.79 MeV state, including the very existence of band [1].

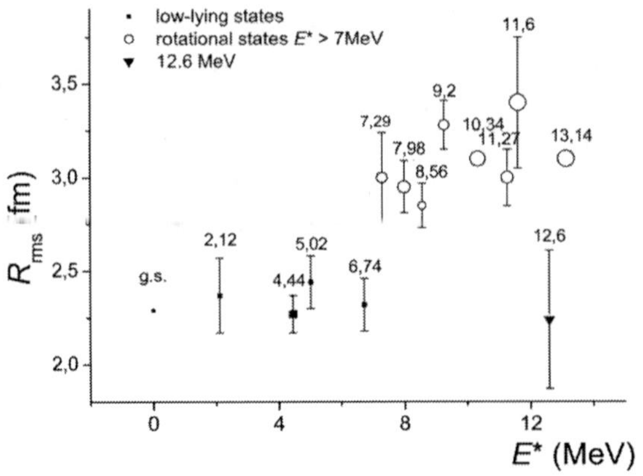

Figure 2: Dependence of R_{rms} (root-mean-square radii, determined by MDM) on the excitation energy for states of ^{11}B. The sizes of the plotted points are proportional to the moments of inertia of the states.

Acknowledgements

The work was supported in part by grants of Russian Science Foundation 14-12-00079, Russian Foundation for Basic Researches 14-02-00560 and 15-02-01503 and the mobility grant from the Academy of Finland

References

[1] A.N. Danilov, A.S. Demyanova et al., Physics of Atomic Nuclei 78, No. 6, 777 (2015)

[2] A.N. Danilov et al., Phys.Rev. C 80, 054603 (2009)

[3] Y. Kanada-Enyo, Phys. Rev. C 75, 024302 (2007)

[4] T. Yamada and Y. Funaki, Phys. Rev. C 82, 064315 (2010)

[5] H. Yamaguchi et al., Phys. Rev. C83, 034306 (2011)

[6] 6. A. Tohsaki, H. Horiuchi, P. Schuck, and G. Ropke, Phys. Rev. Lett. 87, 192501 (2001)

[7] N. Burtebaev et al., Physics of Atomic Nuclei68, 1303 (2005)

[8] A.A. Ogloblin et al., EPJ Web of Conferences, 66, 02074, (2014)

EPJ Web of Conferences **117**, 04012 (2016) DOI: 10.1051/epjconf/201611704012
NN2015

States of ^{13}C with abnormal radii

A.S. Demyanova[1], A.A. Ogloblin[1], A.N. Danilov[1], S.A. Goncharov[2], T.L. Belyaeva[3], Yu.G. Sobolev[4], S.V. Khlebnikov[5], N. Burtebaev[6], W. Trzaska[7], P. Heikkinen[7], G.P. Tyurin[7], D. Janseitov[8] and Yu.B. Gurov[9]

[1]National Research Center "Kurchatov Institute" Moscow, 123182, Russia
[2]Lomonosov Moscow State University, Moscow, 119991, Russia
[3]Universidad Autonoma del Estado de Mexico, Toluca, 50000, Mexico
[4]JINR, Dubna, Moscow Region, 141700, Russia
[5]V.G. Khlopin Radium Institute St.-Petersburg, 194021, Russia
[6]Nuclear Phys. Ins. Almaty, 050032, Kazakhstan
[7]JYFL, Jyvaskyla, FIN-40500, Finland
[8]L.N. Gumilyov Eurasian National University, Astana, 010008, Kazakhstan
[9]MEPHI, Moscow, 115409, Russia

Abstract

Differential cross-sections of the elastic and inelastic ^{13}C + α scattering were measured at $E(\alpha) = 90$ MeV. The root mean-square radii ($< R_{rms} >$) of ^{13}C nucleus in the states: 8.86 ($1/2^-$), 3.09 ($1/2^+$) and 9.90 ($3/2^-$) MeV were determined by the Modified diffraction model (MDM). The radii of the first two levels are enhanced compared to that of the ground state of ^{13}C, confirming the suggestion that the 8.86 MeV state is an analogue of the Hoyle state in ^{12}C and the 3.09 MeV state has a neutron halo. Some indications to the abnormally small size of the 9.90 MeV state were obtained.

1 Introduction

During recent years evidence of the existence of nuclear excited states with abnormally large radii was obtained. In our previous experiment on the

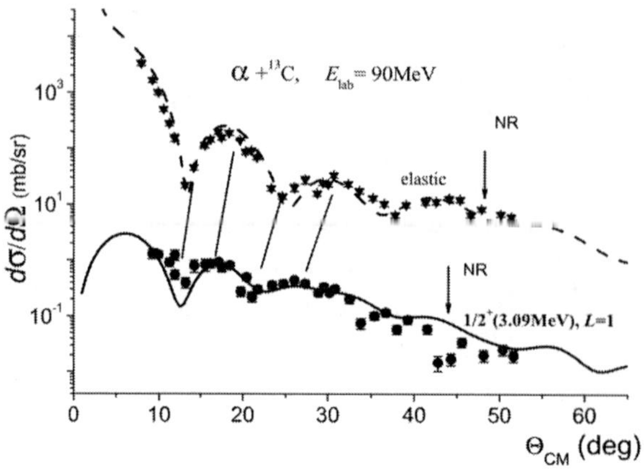

Figure 1: Differential cross sections of the $^{13}C + \alpha$ elastic scattering and inelastic one leading to the 3.09 MeV state in ^{13}C at $E(\alpha) = 90$ MeV. Results of the optical model calculations are shown by the dashed curve. The solid curve corresponds to the DWBA calculations ($L = 1$). The positions of the rainbow minima are denoted by the vertical arrows.

inelastic scattering $^{13}C\ (\alpha, \alpha')$ at $E(\alpha) = 65$ MeV [1] we observed three excited states whose radii differed from that of the ground state: 3.09 MeV $(1/2^+)$, 8.86 MeV $(1/2^-)$ and 9.90 MeV $(3/2^-)$. Because of importance of the obtained result new measurements of the inelastic scattering were performed at $E(\alpha)$=90 MeV.

2 Results and discussion

Differential cross sections of the elastic and inelastic $^{13}C + \alpha$ scattering were measured at the Jyvaskyla University cyclotron, Finland. The data for elastic scattering and inelastic one leading to the 3.09 MeV state are presented in Figure 1. The extremes (joined by the straight lines) should be out of phase if the elastic and inelastic diffraction radii are equal. The observed shift of the extremes of the inelastic scattering cross section to smaller angles indicates the enhancement of the radius for the 3.09 MeV state in comparison with the ground state one. Evaluation of the radius was performed by three independent methods: Modified diffraction model (MDM) [2], Nuclear rainbow method (NRM) [3, 4] and method using the asymptotic normalization coefficients (ANC) [5, 6]. All three approaches gave similar values verifying the validity of the used methods (Table 1) and confirm existence of neutron halo in this states [1, 5].

Differential cross-sections of the inelastic ^{13}C + α scattering leading to the 8.86 MeV state at $E(\alpha)$ = 90 and 65 MeV are shown in Figure 2. Observed identity of the forward angles (up to 45 deg) structure confirms the diffraction origin of the oscillations and similarity of the diffraction radii measured at different energies. The minima at $q \approx 2.5$ fm^{-1} in 65 MeV data and $q \approx 2.0$ fm^{-1} in 90 MeV data were identified as the rainbow (Airy) one. Application of MDM and NRM to the 8.86 MeV state showed that the latter also has an enhanced radius (Table 1) close to that of the Hoyle state (7.65 MeV, 0^+) of ^{12}C [2].

Table 1: $< R_{rms} >$ for the different states in ^{13}C nucleus.

E^*, MeV, J^π	0.00, $1/2^-$	3.09, $1/2^+$	8.86, $1/2^-$	9.90, $3/2^-$
Table data	2.33			
MDM, 65MeV		2.98±0.09 *	2.68±0.12 [1]	2.02±0.14 [1]
NRM, 65MeV		$\geqslant$ 2.7 [1]	$\geqslant$ 2.5*	
ANC,65MeV		2.62±0.10 [5] 2.72±0.02 [6] 2.68 [7], theory		
MDM,90MeV		2.88±0.19*	2.63±0.16*	1.76±0.23 [8]
NRM,90MeV		$\geqslant$ 2.6*	$\geqslant$ 2.5*	
Structure	Shell model	Neutron halo	Dilute cluster	Compact cluster

* this work

Estimates done by MDM method gave the value of the radius of the 9.90 MeV state which is less than that of the ground state (Table 1). This conclusion is confirmed by comparison of different inelastic scattering cross-sections with angular momentum transfer L = 2 (Figure 3). Diffraction structure of the cross-section for the 9.90 MeV state is shifted to the larger angles indicating its smaller radius. This result contradicts to the predictions [9] according to which the 9.90 MeV state should have the radius comparable with that of the Hoyle state. Physical reason of such reduction of the nucleus size is not clear.

Thus, diluted states at 3.09 and 8.86 MeV and a compact one at 9.90 MeV co-exist in ^{13}C together with a number of states with normal radii.

DOI: 10.1051/epjconf/201611704012

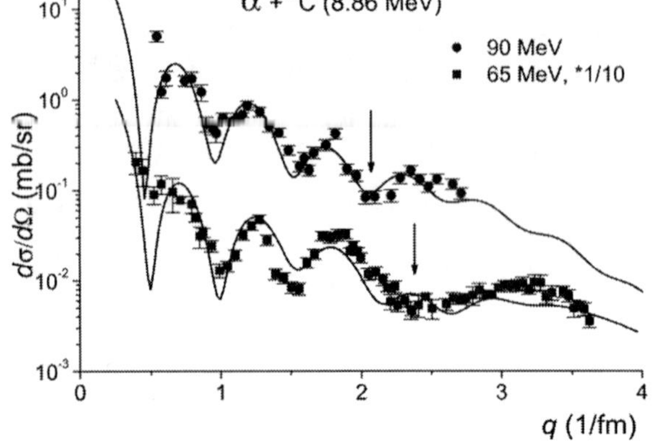

Figure 2: Differential cross-sections of the inelastic ^{13}C + α scattering leading to the 8.86 MeV state at 90 and 65 MeV as a function of the linear momentum transfer q. The positions of the rainbow minima are denoted by arrows.

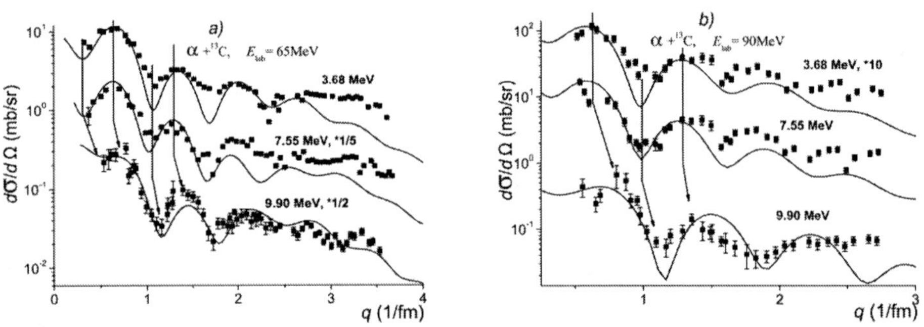

Figure 3: Differential cross sections of the inelastic ^{13}C$(\alpha,\alpha)^{13}$C* scattering corresponding to $L = 2$ transfer for the states of ^{13}C nucleus: 3.68 MeV, 7.55 MeV and 9.9 MeV as a function of the linear momentum transfer q. Vertical lines are drawn through the extremes of the differential cross sections for the 3.68 and 7.55 MeV. The corresponding extremes of the cross sections for the 9.9 MeV state are denoted by arrows. Solid curves correspond to the DWBA calculations ($L = 2$). The figure was taken from the publication [8].

The work was supported in part by grants of Russian Foundation for Basic Researches 14-02-00560 and 15-02-01503 and the mobility grant from the Academy of Finland.

References

[1] A.S. Demyanova et al., EPJ Web of Conferences 66, 02027 (2014)

[2] A.N. Danilov et al., Phys. Rev. C 80, 054603 (2009)

[3] S. Ohkubo and Y. Hirabayashi, Phys. Rev. C 75, 044609 (2007)

[4] A.S. Demyanova et al., Int. J. Mod. Phys. E 20, No 4, 915 (2011)

[5] Z.H. Liu et al., Phys. Rev. C 64, 034312 (2001)

[6] T.L. Belyaeva et al., Phys.Rev. C 90, 064610-1 (2014)

[7] T. Yamada and Y. Funaki, Int. J. Mod. Phys. E 17, 2101 (2008)

[8] A.A. Ogoblin et al. JETP Letters, V. 102, No 4, 199 (2015)

[9] N. Furutachi, M. Kimura, Phys. Rev. C 83, 021303 (2011)

EPJ Web of Conferences **117**, 04013 (2016) DOI: 10.1051/epjconf/201611704013
NN2015

Effect of deformations and orientations in ^{100}Sn daughter radioactivity

Gudveen Sawhney[1], Kanishka Sharma[2], Manoj K. Sharma[2] and Raj K. Gupta[1]

[1] Department of Physics, Panjab University, Chandigarh - 160014, India
[2] School of Physics and Materials Science, Thapar University, Patiala - 147004, India

Abstract

Based on the preformed cluster model (PCM), we have extended our earlier study to investigate the effects of nuclear deformations and orientations of nuclei in context of ground-state de-excitation of Xe to Gd parents, resulting in a doubly closed shell ^{100}Sn daughter and the complementary clusters. The comparison is also made with spherical choice of fragments to extract exclusive picture of the dynamics involved. Since PCM is based on collective clusterization picture, the preformation and penetration probabilities are shown to get modified considerably by inclusion of the quadrupole deformations (β_{2i}) alone, which in turn affects the decay half-lives of the clusters.

The phenomenon of cluster radioactivity is now well established on experimental as well as theoretical front, besides the familiar α, β and γ-emissions. The whole family of such decays consisting of light to heavy (^{14}C to ^{34}Si) clusters have been observed from various actinides (^{221}Fr to ^{242}Cm) in trans-lead region. In all these cases, the daughter nucleus is always ^{208}Pb (a doubly closed shell nucleus) or its neighboring nucleus and hence this phenomena has been termed in some literature as "lead radioactivity". In order to explore the region of possible Sn radioactivity, Gupta and collaborators [1] studied the ground-state decays of neutron-deficient

$^{108-116}$Xe, $^{112-120}$Ba, $^{116-124}$Ce, $^{120-124}$Nd, $^{124-128}$Sm, and $^{128-132}$Gd nuclei, and showed a clear preference for A=4n, α-nuclei, like ^{4}He, ^{8}Be, ^{12}C, etc. However, in their later work [2] carried out for neutron-rich isotopes, it was observed that non-α like clusters are preferred for decays into ^{132}Sn-daughter nucleus, though less probable than A=4n, α-nuclei clusters for the case of ^{100}Sn-daughter radioactivity. Since the shape of a nucleus provides an intuitive understanding of spatial density distribution, therefore, together with shell effects, it is expected that nuclear deformations and orientations should play an important role in the context of Sn radioactivity.

In the present work, we take up this study on the basis of preformed cluster model (PCM) [3] to analyze the influence of quadrupole deformations (β_{2i}) alongwith 'optimum' orientations (θ_i^{opt}) [4] on the behavior of possible fragmentation of Xe to Gd parents. The study is confined to only N=Z parent nuclei, referring to ^{100}Sn as the daughter product. It is relevant to mention here that a majority of parents ^{108}Xe, ^{112}Ba, ^{116}Ce, ^{120}Nd, ^{124}Sm, and ^{128}Gd and their respective emitted clusters ^{8}Be, ^{12}C, ^{16}O, ^{20}Ne, ^{24}Mg and ^{28}Si considered here are deformed and hence the deformation effects seem indispensable, which were not analyzed explicitly in the earlier works [1, 2]. Because the PCM treats the fragmentation process in a collective clusterization approach, it is of great interest to see, in what way the deformation and orientation effects of the decay fragments influence the potential energy surfaces (PES) of these neutron-deficient nuclei.

The decay constant λ or the decay half-life $T_{1/2}$ in PCM is defined as

$$\lambda = \nu_0 P_0 P, \qquad T_{1/2} = \frac{\ln 2}{\lambda}. \tag{1}$$

The clusters are assumed to be preborn in the parent nucleus with preformation probability P_0, hit the barrier with impinging frequency ν_0, and penetrate it with penetrability P. The structure information of the decaying nucleus is contained in P_0 via the fragmentation potential, defined as

$$V_R(\eta) = -\sum_{i=1}^{2} \Big[B_i(A_i, Z_i)\Big] + V_C(R, Z_i, \beta_{\lambda i}, \theta_i) + V_P(R, A_i, \beta_{\lambda i}, \theta_i) \tag{2}$$

with B_i (i=1,2) as binding energies of two fragments and, V_C and V_P, respectively, as the Coulomb and nuclear proximity potentials for deformed and oriented nuclei. For details, we refer to ref. [3].

Figure 1(a) illustrates the mass fragmentation potential, minimized in charge coordinate, as a function of the cluster mass for the considered parents. The effects of deformations up to quadrupole (β_{2i}), in reference to

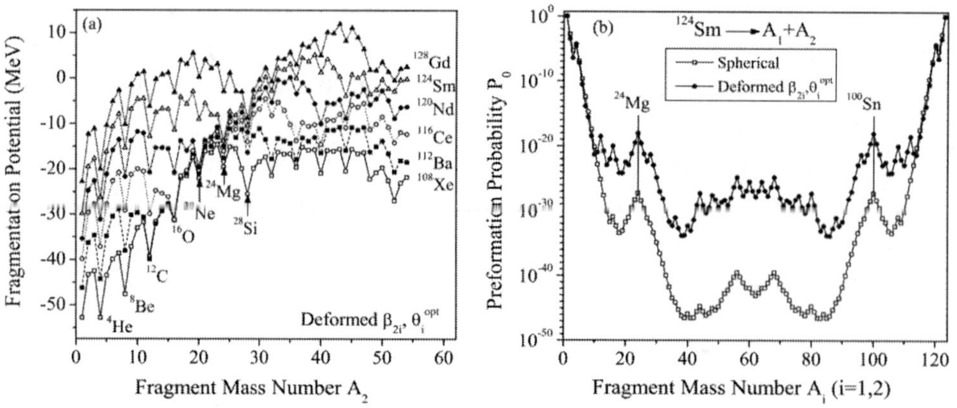

Figure 1: (a) Fragmentation potential for different N=Z parents. (b) Preformation probability P_0 for decay of ^{124}Sm with spherical and deformed configurations.

"optimum" orientation of the cold decay process [4], are duly incorporated in the calculations. For a comprehensive analysis, the calculations have been performed at the touching configuration to estimate the possible existence of the most probable cluster emitted with doubly magic daughter ^{100}Sn. We notice from fig. 1(a) that, even after inclusion of the deformation and orientation effects, potential energy minima occur only at A_2=$4n$, which are alpha-like clusters ^{4}He to ^{28}Si, although the magnitudes of the fragmentation potential are quite different for different N=Z parents. This observation is in contrast to $A_2 \neq 4n$, non-alpha like clusters emitted for the case of cluster decays of radioactive nuclei with Pb or its neighboring nuclei as the daughter product.

Since, both deformations and orientations are found [3] to have significant effect on decay half-lives investigated in trans-lead region, we explore the same here in the trans-tin region. Figure 1(b) presents our calculations of preformation probability P_0 for the illustrative decay of ^{124}Sm, using spherical as well as β_2-deformed choice of fragments. We find that inclusion of deformation and orientation effects of decay fragments changes the relative preformation probability P_0, quite significantly. Interestingly, the emergence of ^{24}Mg cluster in fig. 1(b), corresponding to doubly closed shell ^{100}Sn daughter, seems more probable for deformed configuration in the fragmentation of ^{124}Sm. Similarly, the barrier position as well as its height (not shown here) also get modified with inclusion of deformations, thereby affecting the barrier penetrability P accordingly.

Figure 2 shows the calculated preformation probabilities, penetrabilities and logarithms of half-life times for clusters referring to minima in fragmen-

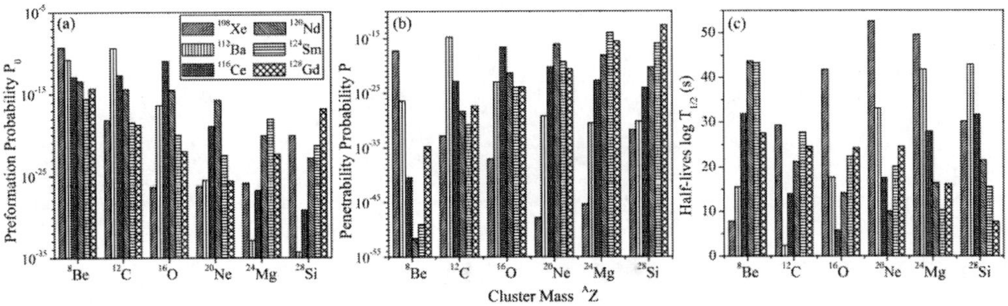

Figure 2: Calculated (a) P_0, (b) P, and (c) $\log T_{1/2}$, for ^{8}Be to ^{28}Si clusters emitted from different N=Z parents. The deformation effects are included upto β_2.

tation potentials of all the considered N=Z parents. Knowing that, in PCM, $T_{1/2}$ exhibits a combined effect of both P_0 and P (ν_0 being almost constant, $\sim 10^{21}$ s^{-1}), we notice in fig. 2 that the clusters referring to doubly magic daughter nucleus ^{100}Sn have the minimal decay half-lives. This is due to the fact that, both P_0 and P have larger magnitudes for each of the most probable (minimal $T_{1/2}$) cluster emission. Apparently, the shortest half-life time is obtained for, say, ^{12}C decay of ^{112}Ba. In other words, the calculated decay half-lives indicate the possible decay of ^{8}Be from ^{108}Xe, ^{12}C from ^{112}Ba, ^{16}O from ^{116}Ce, etc., giving rise to the doubly magic ^{100}Sn (Z=N=50) daughter, and hence are predicted to be most probable cases for measurements.

Summarizing, proper understanding of nuclear shapes along with the relative orientations is essential to make concrete and explicit predictions of cluster dynamics in trans-tin region. As an extension of this work, it will be of further interest to analyse the possible role of deformations in ^{132}Sn radioactivity predicted in the intermediate mass region of the periodic table.

Acknowledgments

Work supported by University Grants Commission (UGC), New Delhi under Dr. D. S. Kothari program.

References

[1] Kumar S. et al., *Phys. Rev. C* **51** (1995) 1762.

[2] Kumar S. et al., *J. Phys. G: Nucl. Part. Phys.* **22** (1996) 215.

[3] Arun S. K. et al., *Phys. Rev. C* **79** (2009) 064616.

[4] Gupta R. K. et al., *J. Phys. G: Nucl. Part. Phys.* **31** (2005) 631.

EPJ Web of Conferences **117**, 04014 (2016) DOI: 10.1051/epjconf/201611704014
NN2015

Triaxiality of the ground states in the ^{174}W

TU YA[1*], Y.S.CHEN[2], L.LIU[1], Z.C.GAO[2]

[1] College of Physics science and Technology, ShenYang Normal University, ShenYang 110034, P.R. China
[2] China Institute of Atomic Energy, Beijing 102413, P.R. China

Abstract

We have performed calculations for the ground states in ^{174}W by using the projected total energy surface (PTES) calculations. Both the ground state (g.s.) band and its γ band reproduce the experimental data. Further discussion about the triaxiality in ^{174}W has been made by transition quardrupole moment (Q_t) and comparing between the PTES and TRS methods.

1 Introduction

The triaxiality of highly rotating nuclei has been a substantial subject of study for several decades. Nuclear triaxiality associated with the breaking of axial symmetry of the quadrupole deformation, brings up many interesting collective motions, e.g., wobbling motion, chiral and γ bands. And the role of the beyond-mean-field-effect on the nuclear triaxiality has become another hot topic.

several theoretical approaches have been developed for studying nuclear triaxiality [1,2]. However, much less research focus had been paid to the axial asymmetry of the nuclear ground state. Recently, the first global calculation across the nuclear chart of axial symmetry breaking was carried out by using the macroscopic-microscopic finite-range liquid-drop model (FRLDM) [3,4].

*tuya_sy@126.com

Many nuclear ground states have been predicted to be triaxiality shaped or γ-soft by the FRLDM calculations and, however, some of which were previously predicted to be axial symmetric by the total routhian surface (TRS) method. TRS method has been extensively used to describe the shape of highly rotating nuclei. Despite the many success of the TRS in the description of nuclear shapes, the disadvantages are clear in the fact that the angular momentum is not a good quantum number, but instead the rotational frequency is employed. And the assumption of the fixed rotational axis in the TRS method is unreasonable in the description of the triaxial nuclear system. The different results of the FRLDM from the TRS method come from their different parameterizations although the two methods are belong to the same type of macroscopic-microscopic approach.

Recently, the PTES method has been developed based on the triaxial projected shell model (TPSM) hybridized with the macroscopic–microscopic method [5], which describes the total energy surface of a triaxially deformed nucleus with the good angular momentum quantum number. To overcome the shortcomings of the TRS method the classical rotational term is replaced by the full quantum mechanics rotational energy in the PTES. The total energy of an atomic nucleus is decomposed into the macroscopic, microscopic and rotational terms. The macroscopic and microscopic parts are described with the liquid drop model and Strutinsky method, respectively, and the rotational energy is given by the TPSM, as the beyond mean field term. In the present work, we will discussed the triaxiality in ^{174}W by performing PTES method.

2 Projected Total Energy Surface method

In the PTES method, the total energy of the nuclear system includes three terms and is expressed as (see Ref [5] for details)

$$E^{tot} = E_{LD} + E_{corr} + E_{rot}. \tag{1}$$

Where E_{LD} is the liquid-drop model energy [6]. E_{corr} is the quantum effect correction to the energy, which is given by the Strutinsky method [7,8]. E_{rot} is the rotational energy obtained by the TPSM calculation. All the terms in Eq.(1), of course, depend on the neutron and proton numbers (N, Z), the deformation parameters, ε_2, γ and ε_4, which are not written explicitly.

The TPSM wave function can be written as

$$|\Psi_{IM}\rangle = \sum_{K\kappa} F^I_{\kappa,K} \hat{P}^I_{MK} |\Phi_\kappa\rangle \ , \tag{2}$$

in which the projected multi-q.p. states span the shell model space. Where, $|\Phi_\kappa\rangle$ represents the set of multi-quasiparticle states labeled by κ , and for even-even nuclei it includes the 2- and 4-q.p, states associated with the triaxially deformed q.p. vacuum $|0\rangle$, $\alpha^\dagger_{\nu_1}\alpha^\dagger_{\nu_2}|0\rangle$, $\alpha^\dagger_{\pi_1}\alpha^\dagger_{\pi_2}|0\rangle$, $\alpha^\dagger_{\nu_1}\alpha^\dagger_{\nu_2}\alpha^\dagger_{\pi_1}\alpha^\dagger_{\pi_2}|0\rangle$.

The rotational energies together with the wave functions, the coefficients $F^I_{\kappa,K}$, are obtained by solving the eigenvalue equation,

$$\sum_{K\kappa} F^I_{\kappa K}\left(\langle\Phi_{\kappa'}|HP^I_{K'K}|\Phi_\kappa\rangle - E^I\langle\Phi_{\kappa'}|P^I_{K'K}|\Phi_\kappa\rangle\right) = 0\,. \quad (3)$$

The Hamiltonian in the TPSM [see [] for details] is written as

$$H = H_0 - \frac{1}{2}\sum_{\lambda=2}^{4}\chi_\lambda \sum_{\mu=-\lambda}^{\lambda} Q^\dagger_{\lambda\mu}Q_{\lambda\mu} - G_0P^\dagger_{00}P_{00} - G_2\sum_{\mu=-2}^{2} P^\dagger_{2\mu}P_{2\mu}\,. \quad (4)$$

The nuclear equilibrium deformation can be obtained for each spin by minimizing the total energy, calculated by Eq.(1), in function of spin with respect to the deformation parameters, ε_2 and γ. The minimization procedure can be performed through the calculation of the total energy surface in the considered deformation plan.

3 Results of calculations and discussion

Three major shells of N=4,5,6 for neutrons and N=3,4,5 for protons are included to calculate the Nilsson single particle states in the TPSM. The used Nilsson potential parameters for each main shells are taken from the literature [9]. In the present calculation, the hexadecapole deformation is not considered as the variable but taken to be $\varepsilon_4 = 0$ for simplicity. The monopole pairing strength parameter G_0 is calculated with the standard parameters $g_1 = 20.12$ MeV and $g_2 = 13.13$ MeV [10], which approximately reproduces the observed odd-even mass differences in the mass region. The quadrupole pairing strength parameter G_2 is calculated from $G_2 = fG_0$ with the proportional coefficient $f = 0.16$ as usual.

The projected total energy surface is calculated for an given angular momentum as well as a given parity, this may provides an opportunity to make a direct comparison between the theory and the experiment. Figure 1 shows the PTES for the g.s. $I^\pi = 0^+$ of ^{174}W, which has a local minimum representing the equilibrium deformations of ($\varepsilon_2 = 0.24, \gamma = 18°$). The results imply that the g.s. of the nucleus is well deformed, and it is particularly

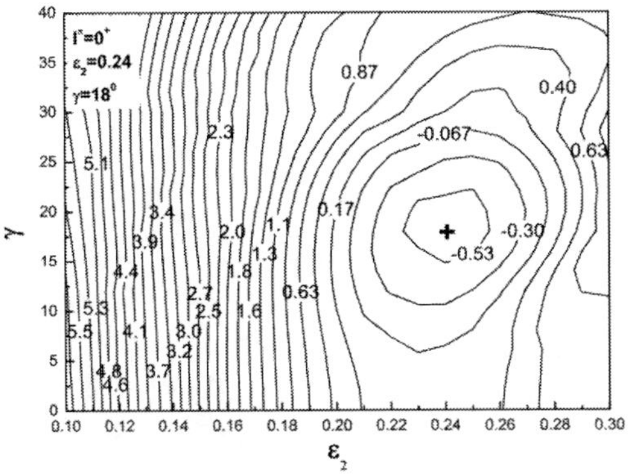

Figure 1: The contour plot of total energy in units of MeV for the ground state in ^{174}W, the minimum is marked by "+".

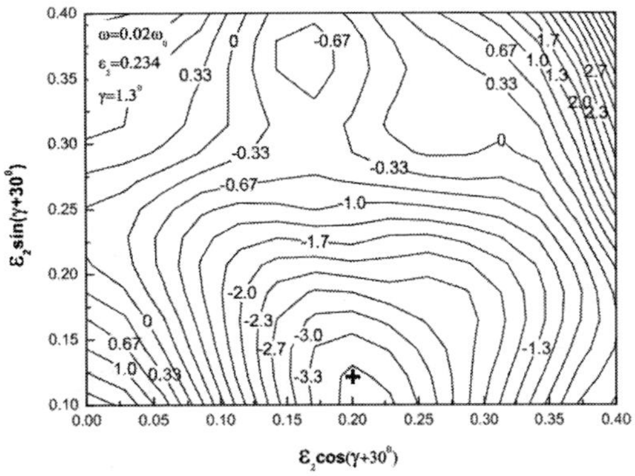

Figure 2: The contour plot of total routhian in units of MeV for ^{174}W, calculated at the rotational frequency $\hbar\omega = 0.02\hbar\omega_0$, the minimum is marked by "+".

noteworthy to have the axial asymmetry with $\gamma = 18°$. The TRS calculation has been intensively employed to study highly rotating nuclei [1],especially, in the rare earth mass region, thus it may be instructive to make a comparison between the PTES and TRS results. Because of the semi-classical nature of the TRS method it is impossible to strictly describe the g.s. which has a good angular momentum $I^\pi = 0^+$. However, the TRS calculated at a small rotational frequency may describe the property of the g.s. as a good approximation, and thus a comparison between the PTES and TRS for the g.s. has the certain significance. Figure 2 shows the TRS calculated with the same set of the Nilsson potential parameters for the ground band state of ^{174}W, at the rotational frequency of $\hbar\omega = 0.02\hbar\omega_0$, which reports a local minimum at the deformations of $(\varepsilon_2 = 0.234, \gamma \sim 0°)$,which consistent with the result of FRLDM. The three methods give the same equilibrium elongation deformation but a considerably different triaxiality for the g.s. of ^{174}W. The γ-deformation of 18 ° is not a large triaxiality but an indication of the axial symmetry breaking in the g.s. of the nucleus. In fact, the axial asymmetry of the g.s. of ^{174}W manifests the appearance of the characteristic γ band observed experimentally at 0.977 MeV above the g.s.. The calculations of the PTES for nonzero spins of g.s. band and its γ band of ^{174}W, have been performed in a similar way to the g.s. calculation. Figure 3 presents the calculated g.s. band and its γ band for ^{174}W, from the PTES minima, compared with the experimental data [11]. By comparison, it is seen that the beyond mean field effects incorporated in the PTES approach may give rise to the axial asymmetry in some nuclei which have been predicted to

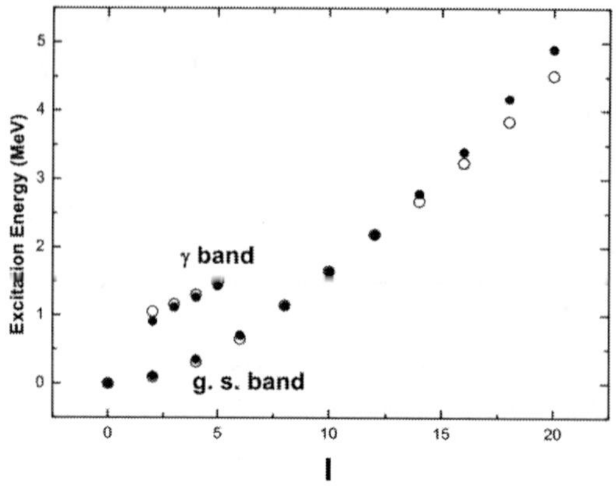

Figure 3: Calculated g.s. band and its γ band (open circle), the energy versus spin, for ^{174}W, from the PTES minima, compared with the experimental data [11] (solid circle).

have an axial symmetric shape by the TRS calculations and FRLDM calculation. In another way, the transition quadrupole moment (Q_t) is one of best physical quantities to identify the intrinsic deformed shape of the nucleus. The calculated Q_t with the deformations determined by minimizing the projected total energies for different spin, well reproduce the available experimental data [12, 13]within the error bars for W isotopes (see [5] and next work). Unfortunately, there are no more clear experimental data for ^{174}W to compare with the calculations.

4 Summary

In summary, we have studied the triaxiality of the ground states in ^{174}W by using PTES method. Our PTES calculations and the comparison with TRS method shows that beyond-mean-field-effect may play an important role on the nuclear triaxial deformation.

Acknowledgements

The work is supported by National Nature Science Foundation of China under Grant Nos.11047171, 11301508, 11175258, 11021504, 11275067,11575290, and by the Knowledge Innovation Project of the Chinese Academy of Sciences under Grant No.KJCX2-SW-N02, the key project of Science and Technology Research of Education Ministry of China under grant number 209053.

References

[1] Tu Ya and Y.S.Chen, et al., Nucl. Phys. A848, 260(2010).

[2] Z.C. Gao, Y.S.Chen, Y.Sun, Phys. Lett. B634, 195(2006).

[3] P. Möller and R. Bengtsson et al., Phys. Rev. Lett.97, 162502(2006).

[4] P.Möller, A.J. Sierk, R. Bengtsson, H.Sagawa and T. Ichikawa, Atomic Data and Nuclear Data Tables 98, 149(2012).

[5] Tu Ya and Y.S.Chen et al.,Sci. China Ser G-Phys. Mech. Astron. 57(11), 2054(2014).

[6] W.D.Myers and W. Swiatecki, Ark. Fys.361,343(1967).

[7] V.M.Strutinsky, Nucl. Phys. A122 , 1(1968).

[8] V.M.Strutinsky, Nucl. Phys. A85, 420(1967).

[9] T. Bengtsson and I.Ragnarsson, Nucl. Phys. A436, 14(1985).

[10] K.Hara and Y.Sun, Int. J. Mod. Phys. E4, 637(1995).

[11] T.Kibédi et al., Nucl. Phys. A688, 669(2001).

[12] F.K.Mcgowan and N.R.Johnson et al., Nucl. Phys. A530, 490(1991).

[13] F.K.Mcgowan and N.R.Johnson et al., Nucl. Phys. A580, 335(1994).

EPJ Web of Conferences **117**, 05001 (2016) DOI: 10.1051/epjconf/201611705001
NN2015

Nuclear physics and particle therapy

G. Battistoni

INFN, Sezione di Milano, Milano, Italy

Abstract

The use of charged particles and nuclei in cancer therapy is one of the most successful cases of application of nuclear physics to medicine. The physical advantages in terms of precision and selectivity, combined with the biological properties of densely ionizing radiation, make charged particle approach an elective choice in a number of cases. Hadron therapy is in continuous development and nuclear physicists can give important contributions to this discipline. In this work some of the relevant aspects in nuclear physics will be reviewed, summarizing the most important directions of research and development.

1 Introduction and basic principles

Charged particle therapy (CPT in the following), or hadron therapy, is an innovative cancer radiotherapy based on nuclear particles (protons, neutrons and light ions) for treatment of early and advanced tumors. In 1946, Robert Wilson proposed the therapeutic use of protons for treating cancer [1]. This is considered as the start of hadron therapy with charged particle. Proton therapy has now become an advanced clinical modality, and CPT with heavier ions (generally ^{12}C) is now becoming more and more attractive. The clinical interest in hadron therapy resides in the fact that it delivers precision treatment of tumors, exploiting the characteristic shape of the Bragg curve for charged hadrons, i.e. dose deposition as a function of depth of traversed matter. As compared to the standard X-ray radiotherapy, accurate and efficient irradiation of the tumor can be obtained reducing the dose to the surrounding healthy tissue, thus achieving less complication probability. Especially for heavy ions, an increased relative biological effectiveness (RBE) in killing cancer cells can also be obtained, making this approach very interesting in a number of cases and in particular for radio-resistant tumors. After a rather long period in which hadron treatments were exclusively delivered in research laboratories, the first hospital-based treatment centre to be established was that of Loma Linda (USA) in 1990. Today proton therapy has grown into an advanced, cutting-edge clinical modality. According to recent statistics [2], more than 110,000 patients worldwide have been now treated with charged hadrons (about 10% with carbon ions), and the number of clinical centers dedicated to CPT is now rapidly increasing.

Hadron therapy represents a paradigmatic case of a topic in between research and actual clinical practice. It is a discipline in evolution and the contribution coming from nuclear physics has still a fundamental role to help CPT to reach in practice the high level of precision which would be in principle attainable. Several R&D projects are in progress with two main goals: to reduce the costs of infrastructures and treatments and reduce existing uncertainties, such those connected to radiobiology, the knowledge of particle range inside the patient and the monitoring of actual dose delivery during the treatment.

In order to reach these goals, nuclear physics can contribute to different items: i) optimization of treatment procedures and planning; ii) improvement of the present comprehension of radiobiological effects and of the related calculable models; iii) development of new detectors and imaging techniques to achieve specific and optimized monitoring; iv) development of accelerators and of related components; v) improving the knowledge of relevant nuclear processes, like fragmentation, and their modeling; vi) contribute to the development of new therapeutic beams using alternative ion species, like ^{4}He and ^{16}O.

The Nuclear Physics European Collaboration Committee has dedicate its 2014 report to the contribution of nuclear physics to medicine [3] where a comprehensive review of the key issues in CPT can be found. Here we limit ourselves to a few selected issues. In Section 2 the main aspects of nuclear physics relevant for CPT will be summarized. The topic of particle range uncertainties and the development of a specific imaging approaches will be presented in Section 3, while Section 4 will be dedicated to real time monitoring techniques based on the exploiting of nuclear interactions.

2 Nuclear Physics: what really matters

Several nuclear processes are relevant in hadron therapy. Inelastic interactions are responsible of beam attenuation along the longitudinal profile, while elastic scattering, especially in the case of proton therapy, contributes to the transversal profile of dose distribution. Fragmentation of both projectile and target is probably one of the most relevant processes to be studied in detail, since it affects the attenuation of primary beam and the biological effect. Compared to the radiation field of the primary ions, secondary fragments lead to an altered spatial dose distribution due to differing ranges and angular distributions of the fragments and to a modification of the linear-energy transfer (LET) spectra which results in a difference of RBE for the same delivered dose.

Nuclear reactions experienced by the primary and its possible fragments are also responsible of radioactive isotope production, which may be used for monitoring purposes, as mentioned in Section 4, in case of β^+ decaying radionuclides. Last, but not least, nuclear evaporation and de-excitation have to be considered, mostly for the production of lowest energy nucleons in the target. Prompt photon production from gamma–decay is a particularly interesting case, since they can also be used for monitoring. Of great interest, for the same reason, is the production of secondary fast charged particles.

Accurate modeling of all the mentioned processes is one of the most important contributions of nuclear physicists to the discipline of CPT, since an accurate knowledge of ion beam fragmentation is important for a precise description of biological effects inside and outside the treatment volume. Secondly, the precise prediction of nuclear particle interactions and resulting residual nuclei distributions are needed for imaging techniques which aim at in vivo dose monitoring. Important developments in Monte Carlo models have been produced in the last years but still there exist relevant uncertainties. Nuclear reaction models are continually evolving, but the amount of available experimental data and their limited precision (expecially in the case of nucleus–nucleus interaction in the relevant range of energies and masses) is not enough to provide a complete benchmarking. Devoted experiments aimed to measure nuclear cross-section are described in ref. [4, 5]. The relevance of existing discrepancies in numerical models (see for example ref. [6, 7]) should be discussed in the context of clinical applications. The detailed quantitative evaluation of biological dose distributions would require the simulation of a carbon ion treatment field including radiobiological modeling. Still, a simplified calculation for a single Bragg curve can give an estimate the degree of impact on the biological dose. It has been estimated that, in the case of carbon ions, at maximum around 40% of the dose in the region in front of the Bragg peak is delivered by fragments and that a large fraction of this dose, about 15%, is delivered by protons. A discrepancy in predicted proton fluences of about 30% translates into a difference of 5% in dose delivered by protons. When taking generic values for the RBE of 1.1, 3.0 and 2.0 for protons, primary carbon ions and all other fragments, respectively, and assuming that the 5%dose which is not delivered by protons is delivered by primary carbon ions instead, one predicts an over-estimation of the biological dose of 4%.

3 Imaging for hadron therapy: proton Computed Tomography

The spatial precision of the dose delivered to the volume to be treated is one of the major specific advantages of CPT. However (see [3]) the precise knowledge of the position where particles deliver the maximum dose in the patient (*i.e.* the particle range), remains one of the most important uncertainties. In order to achieve the desired precision goals, the Stopping Power (SP) map of the patient should be reconstructed before the treatment to set up a detailed plan. At present the SP maps are extracted from 3-dimensional images obtained by X-ray Computed Tomography; here the photon attenuation coefficients are translated into SP using conversion tables. This intermediate step introduces an intrinsic uncertainty resulting into an error in the proton range calculation that can be of several millimeters [8]. To mitigate this effect the SP map could be directly determined using a proton beam with kinetic energy larger than the one used for treatment and reduced intensity. Such an imaging system should be able to determine the 3-dimensional SP map with a position resolution less than 1 mm and a density resolution of the order of 1%. A system which aims to fulfill these requirements should be able, at least, to partially overcome the problems introduced by Multiple Coulomb Scattering (MCS) on charged particles crossing matter. In fact typical objects under study could be as thick as 20 cm water equivalent: in this case a 200 MeV kinetic energy proton undergoes a r.m.s. MCS angle of the order of 40 mrad, which corresponds to an r.m.s. projected displacement of about 3.2 mm, considering a 8 cm minimum distance between the target and the detecting plane. The concept is to perform an imaging based on the use of proton beams with energies above 200 MeV.

An example of the research and development work for a prototype of proton Computed Tomography (pCT) is described in ref. [9, 10]. The scanner

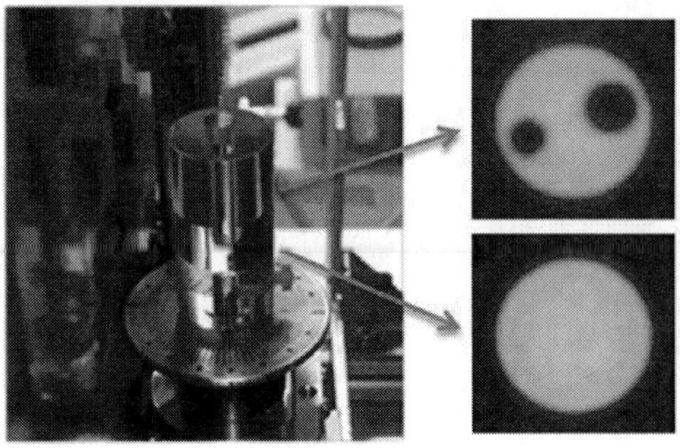

Figure 1: Example of pCT test results for inhomogeneous phantom [10].

is based on a tracker and a calorimeter to measure single protons trajectories and their residual energy. The tracker is composed of four planes of silicon microstrip detectors to measure proton entry and exit positions and angles. Residual energy is measured by a calorimeter composed of YAG:Ce scintillating crystals. A first prototype of this pCT scanner, with an active area of about 55 cm^2, has been constructed and characterized with 60 MeV protons at the INFN Laboratori Nazionali del Sud. A first test to reconstruct the tomographic image of a 2 cm radius, 4 cm high PMMA cylinder has been performed. The phantom, mounted on a rotating platform, has been installed in the middle of the tracker (Fig.1) of the pCT prototype. A total of 36 data sets, each of them containing on average 9.5 10^5 events, have been acquired. The phantom has been rotated by 10° each run. The 36 profiles have been used as input to a "Filtered Back Projection" tomographic reconstruction algorithm. Fig.1 displays two tomographic sections of the phantom: 4 and 6 mm diameter holes (top) and uniform part (bottom). The reconstruction algorithm has to be verified in a more clinical-like setting (*i.e.* higher proton energy and larger object thickness).

4 Real time monitoring

Beyond the question of Stopping Power determination, the need for a real time monitoring, *i.e.* during the treatments, of actual particle range in the patient is another important goal addressed in present research activity. The uncertainty on the position of the dose release in CPT treatments can be due to different factors, such as quality and calibration of the Computed Tomography (CT) images or possible morphologic changes occurring between

 DOI: 10.1051/epjconf/201611705001

CT and each of the several irradiation sessions, operated in different days, that compose a treatment in CPT. Finally, also patient mis-positioning and organ motion during the treatment itself can be sources of uncertainty. All these effects can add up to give an overall uncertainty of the order of few millimeters [11]. A real time monitoring procedure can therefore increase the quality assurance of a CPT treatment. The techniques proposed for in-vivo range monitoring aim to exploit the secondary production coming from the hadronic interactions of the therapeutic beams. In ref. [12, 13] a discussion of range verification methods and of related physics can be found. Three are the nuclear processes that can yield a radiation suited for this purpose: production of β^+ emitters nuclei, excitation of nuclei and charged particle production in inelastic interactions. In order to make use of these processes, the comparison of measured and pre-calculated distributions of secondary particles is needed (see the discussion in Section 2 about Monte Carlo models).

4.1 In-beam PET

Nuclear β^+ decays produce positrons that can be traced exploiting their annihilation with electrons yielding almost back-to-back 511 keV photon pairs. Since organic tissue is mostly constituted of carbon, hydrogen and oxygen, the most likely β^+ emitting isotopes that can be formed are ^{10}C, ^{11}C, ^{15}O and ^{13}N. Positron Emission Tomography has proven to be the most mature technique for range verification. Original discussion and first investigations are reported in ref. [14–16]. An example of recent research and development activity to design a device to be used in real time during clinical operation is reported in ref. [17]. The proposed system consists of two planar detector heads [18] each having an area of 10×10 cm^2 based on the use of segmented scintillating LYSO crystal matrix. Measured activity results were compared with predictions from the FLUKA MC generator [19,20]. For PMMA phantoms the difference between the MC prediction and the data, considering the distance between the 50% rise and 50% fall-off position of the activity distribution is less than 1 mm [21]. The system showed to be able to detect the presence of inhomogeneities in the phantom.

4.2 Prompt photon detection

Nuclear interactions of the beam in the patient can also excite nuclei along particle path. De-excitation photons which are emitted in a very short (< 1 ns) decay times interval and have a useful energy range extending up to

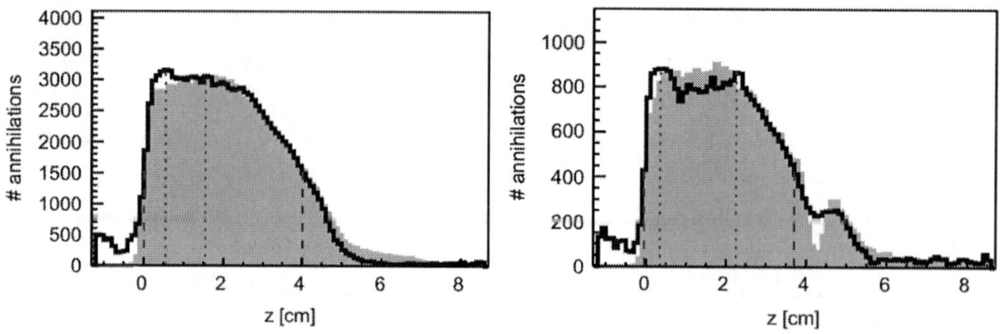

Figure 2: Experimental 1-D activity profiles along the beam direction at 180 s from the beginning of the irradiation, for a homogeneous PMMA phantom (left) and for the same phantom with an air cavity (right). The experimental data, black line, are superimposed to Monte Carlo simulation profile, light brown area. The profiles are normalized. The vertical segmented lines represent: in red the 50% positions for the activity rise and the local maximum of the considered interval; in blue the local maximum of the fall-off used interval and the 50% fall-off positions [18].

about 10 MeV. A review of results on the yields of photons emitted during irradiation with proton and Carbon beams can be found in [22]. At phantom entrance, the average number of detected prompt γ's was found to be of order 10^{-4} per incident carbon ion and 10^{-5} per incident proton. The energies of these prompt photons are too high for standard single gamma detection devices, like SPECT, to be efficient, and dedicated detectors are needed. There are different system at present under investigation. Among them the Compton Camera and the Collimated Camera approaches [23]. A Knife-edge-shaped slit camera [24] is now being developed for clinical application. Promising measurements with a collimated slit-camera prototype, tested with clinical proton beams, have recently been presented in ref. [25]. A precision on single spot range determination of 2 mm is achieved.

4.3 Detection of charged particles

The detection of charged particles can be a promising alternative approach, mostly in the case of CPT performed with ions heavier than protons. Recent studies have been focused on the possibility of exploiting particle production (and in particular the high penetrating proton component) for monitoring purposes, since it can be used to estimate the position of the distal edge of the dose profile. Experimental studies of carbon ion collisions with water reported in refs. [26–29] showed that fragmentation products are peaked in the forward region and mostly contained within few degrees of the beam axis, apart from protons, that represent the largest sample and show tails at large emission angles and energies. Other measurements performed at small angle [30, 31] suggested that, using solid state tracking devices at 30° with respect to the beam direction, the distal edge of the beam could be estimated with an accuracy of 1.3 mm. In addition, variations in the beam width could be measured with a precision of 0.9 mm. In principle, due to obvious geometrical considerations, production at large angle is the most interesting case. The quality of the reconstruction of the trajectory of the single charged particle compensates for the reduced statistics expected at large angle.

Experimental data about charged particle emission under irradiation by a therapeutic ion beam can be taken by ref. [32], in which a a $20\times5\times5$ cm^3 PMMA target was irradiated with a 220 MeV/u ^{12}C beam at GSI. Charged particles were identified with tracks reconstructed in a Drift Chamber and a LYSO crystal detector providing Time-of-Flight and scintillation light output information, so to allow particle identification. Measurements were performed at 90° and at 60° with respect to the beam direction. The resulting flux of secondary p per incident ion, measured for the 90° experimental configuration, is

$$\Phi_p(\theta = 90^\circ) \quad = \quad (1.83 \pm 0.02_{\text{stat}} \pm 0.14_{\text{sys}}) \times 10^{-3}\ sr^{-1}$$

The production region of charged secondary particles has been studied extrapolating backwards the reconstructed tracks, demonstrating the possibility of correlating the spatial profile of emitted particles with the longitudinal dose profile. New preliminary results using ^{4}He and ^{16}O primary ions have been reported at this conference [33].

5 Conclusions

The contribution of nuclear physics to the development of CPT is still fundamental and this review could cover only a very limited part of the research items relevant in this discipline. Several groups in the world, in collaboration with physicians, biologists and medical physicists, are in scientific competition in this area, searching for solutions which can be actually implemented in the clinical practice. An example is the multi-modal system for real time monitoring developed by the INSIDE collaboration [34].

The author is indebted to the members of INFN-RDH Collaboration, CERN and to the teams of CNAO, Trento and HIT for many useful discussions and suggestions.

 DOI: 10.1051/epjconf/201611705001

References

[1] R.R. Wilson, *Radiology*, vol. 47, pp. 487491 (1946).

[2] http://ptcog.web.psi.ch/ *Particle Therapy Co-Operative Group (PTCOG)*

[3] NuPECC, *Nuclear Physics for Medicine*, NuPECC Report 2013, ESF.

[4] R. Pleskac et al., *Nucl. Instr. Meth. in Phys. Res. A* **678** (2012) pp. 130–138.

[5] C. Agodi et al., *Journal of Physics: Conference Series*, **420** (2013) 012061.

[6] T.T. Böhlen et al., *Phys. Med. Biol.*, **55** no. 19 (2010) pp. 5833-5847.

[7] C. Robert et al., *Phys. Med. Biol.*, **58** (2013) pp. 2879–2899.

[8] B. Schaffner and E. Pedroni, *Phys. Med. Biol.* **43** (1998) 1579.

[9] C. Civinini et al., *Nucl. Instr. Meth. in Phys. Res. A* **732** (2013) pp. 573–576.

[10] M. Scaringella and et al., *Nucl. Instr. Meth. in Phys. Res. A* **730** (2014) pp. 178–183.

[11] A. Knopf and A. Lomax, *Phys. Med. Biol.* **58** (2013) pp. 131–160/

[12] A.C. Kraan, *Frontiers in Oncology*, **5** (2015) 150.

[13] R. Faccini and V. Patera, *Modern Physics Letters*, **30** no. 17 (2015) 1540023.

[14] . K. Parodi, *On the Feasibility of Dose Quantification with In-Beam PET Data in Radiotherapy with ^{12}C and Proton Beams.* Ph.D. thesis, Dresden University of Technology, Dresden (2004).

[15] W. Enghardt, et al. *Nucl. Instr. Meth. in Phys. Res. A* **525** (2004) pp. 284288

[16] P. Crespo et al., *Phys. Med. Biol.* **51** no. 9 (2006) pp. 2143-2163.

[17] A.C. Kraan and et al., *Nucl. Instr. Meth. in Phys. Res. A*, **786** (2015) pp. 120–126.

[18] V. Rosso and et al., *Journal of Instrumentation*, **8** (2013) C03021.

[19] A. Ferrari et al., CERN 2005-10, INFN/TC_05/11, SLAC-R-773L (2005).

[20] T.T. Böhlen *et al,*, *Nuclear Data Sheets*, **120** (2014) pp. 211–214.

[21] A.C. Kraan and et al., *Physica Medica*, **30** no. 5 (2014) pp. 559–569.

[22] M. Pinto et al., *Phys. Med. Biol.* **60** (2015) pp. 565–594.

[23] J. Smeets et al. *Phys. Med. Biol.* **57** (2012) pp. 3371–3405.

[24] V. Bom *et. al.*, *Phys. Med. Biol.* **57** (2012) pp. 297–308.

[25] I. Perali et al. *Phys. Med. Biol.* **59** no. 19 (2014) pp. 5849–5871.

[26] N. Matsufuji et al., *Phys. Med. Biol.*, **50** (2005) pp. 3393–403.

[27] E. Haettner, H. Iwase, and D. Schardt, *Radiation protection dosimetry*, **122** (2006) pp. 485–487.

[28] K. Gunzert-Marx, H. Iwase, D. Schardt, and R. S. Simon, *New J. of Phys.*, **10** (2008) 075003.

[29] B. Braunn et al., *Nucl. Instr. Meth. in Phys. Res. B*, **269** (2011) pp. 2676–2684.

[30] P. Henriquet et al., *Phys. Med. Biol.* **57** (2012) pp. 4655–4669.

[31] K. Gwosch et al., *Phys. Med. Biol.* **58** (2013) p. 3755.

[32] L. Piersanti and et al., *Phys. Med. Biol.*, **59** no. 7 (2014) pp. 1857–1872.

[33] G. Battistoni et al., *Measurement of secondary particle production induced by particle therapy ion beams impinging on a PMMA target*, these proceedings.

[34] M. Marafini et al., *Proc. of the II Symposium on positron emission Tomography, Krakow (2014)*, *ACTA PHYSICA POLONICA A*, **127** no. 5 (2015) pp. 1465–1467.

EPJ Web of Conferences **117**, 05002 (2016) DOI: 10.1051/epjconf/201611705002
NN2015

High intensity X/γ photon beams for nuclear physics and photonics

L. Serafini[1], D. Alesini[2], A. Bacci[1], N. Bliss[5], K. Cassou[3], C. Curatolo[1], I. Drebot[1], K. Dupraz[3], A. Giribono, V. Petrillo[1], L. Palumbo[4], C. Vaccarezza[2], A. Variola[2], F. Zomer[3].

[1] INFN, Sezione di Milano and Università degli Studi di Milano, Milano, Italy
[2] LNF-INFN, Frascati (RM), Italy
[3] LAL-Orsay, Orsay, France
[4]Università la Sapienza, Roma, Italy
[5] STFC Daresbury Laboratory, Warrington, UK

Abstract

In this manuscript we review the challenges of Compton back-scattering sources in advancing photon beam performances in the $1 - 20\ MeV$ energy range, underlining the design criteria bringing to maximum spectral luminosity and briefly describing the main achievements in conceiving and developing new devices (multi-bunch RF cavities and Laser recirculators) for the case of ELI-NP Gamma Beam System (ELI-NP-GBS).

A new generation of high brilliance gamma ray photon beams is under development at several laboratories world-wide, based on Compton back-scattering Sources run as mini-colliders of high brightness electron beams and high power laser beams, with the aim to generate photon beams in the 1-20 MeV energy range characterized by unprecedented performances in terms of mono-chromaticity, brilliance, spectral density, tunability and polarization.

Such a new generation of gamma photon beams are aimed at opening the era of Nuclear Photonics stepping up by orders of magnitude w.r.t. present generation machines, therefore enabling to use the Nuclear Resonance Fluorescence technique: nuclear waste remote sensing and diagnosis, special nuclear material recognition for national security and non-proliferation related issues, isotope sensitive imaging for medical and cultural heritage investigations are among the most appealing and high social impact applications that will be made feasible by such advanced gamma photon beams [1]. Not to mention several research fields in nuclear physics dealing with fundamental nuclear structure studies about giant resonances and astro-physical open problems in star nucleo-synthesis that will greatly benefit from the availability of these new probes.

One of the main projects in progress in this field is the ELI-NP Gamma Beam System [2], designed and being constructed by the European EuroGammaS collaboration under a contract with IFIN-HH, the Romanian Institution in charge of developing the ELI-NuclearPhysics Romanian Pillar of E.L.I. (Extreme Light Infrastructure) in Bucharest.

The availability in the near future of gamma ray photon beams of high mono-chromaticity (less than 0.5%), high peak brilliance (larger than $10^{21} N_{ph}/(s.0.1\%.mm^2 mrad^2)$), large tunability, fully controlled polarization, spectral density larger than $10^4 N_{ph}/(s.eV)$ and focused down to micron-scale spot sizes, open several further horizons in the photon-photon and photon-particle collider scenarios.

Quite often in the literature radiation sources based on back-scattering of optical photons carried by laser pulses by relativistic (MeV to GeV energy) electron beams are referred as either Thomson or Compton back-scattering Sources (hereafter referred as TCS). The reason for such an ambiguity is the particular regime of photon back-scattering that is applied, either with (Compton) or without (Thomson) electron recoil. As a matter of fact, the frequency ν of the back-scattered photon, for the simplest case of head-on collision between the laser optical photon (frequency ν_L) and the electron (relativistic factor γ) is given by:

$$\nu \simeq \nu_L \frac{2}{\frac{1}{2\gamma^2} + \frac{2h\nu_L}{\gamma mc^2}} \tag{1}$$

where the first term in the denominator is responsible for the typical $4\gamma^2$ scaling of Thomson Sources (and recalls the $2\gamma^2$ scaling for FEL's based on a classical view of relativistic Doppler effect combined to Lorenz boost of the incoming e.m. wave which induces dipole radiation emission onto the electron), while the second term in the denominator accounts for quantum

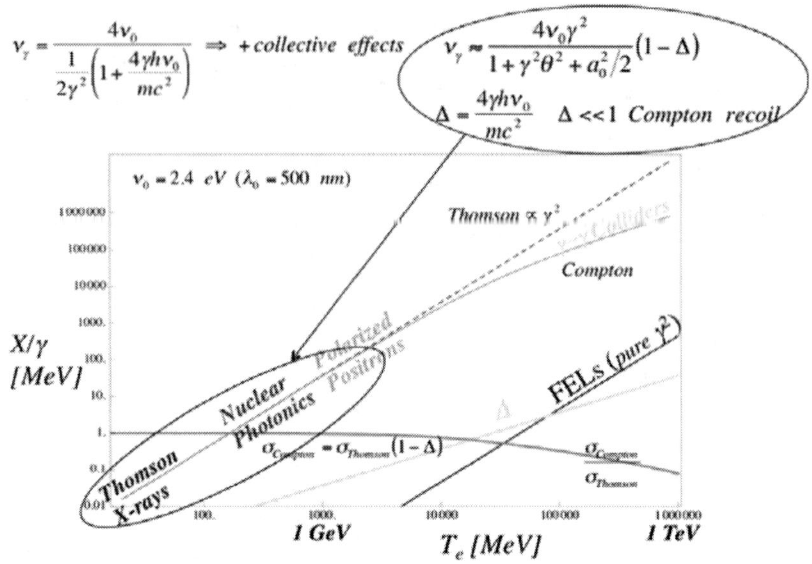

Figure 1: Operating Diagram for Thomson/Compton Sources, showing the elastic (no-recoil) Thomson regime for X-rays and the small recoil regime typical of Compton Sources for Nuclear Photonics. The deep Compton regime, dominated by recoil effects and typical of $\gamma-\gamma$ collider is also shown, together with the behavior of the recoil parameter Δ and the Compton cross section decrease together with increasing Δ. The enhancement of the formula for the frequency of the back-scattered photon is also shown, taking into account the red-shift effects due to scattering angle θ and non-linear multi-photon effects generated by laser field intensity (represented by the laser parameter α_0).

effects, namely the electron recoil, and it is responsible for the correct energy and momentum conservation in the scattering reaction. This second term is negligible for FEL's and X-ray Thomson Sources, while it is small but not negligible for Compton Sources of Nuclear Physics and Photonics, and becomes the dominant term for deep Compton Sources as those planned for $\gamma-\gamma$ colliders in TeV range. The scaling of eq.1 for the typical case of a green light laser ($\lambda_L = 500$ nm) back-scattered by an electron beam of variable energy (from 10 MeV up to 1 TeV) is illustrated in Fig.1.

There are at least two nice properties of the Thomson and quasi-Thomson regions, where the electron recoil is either negligible or quite small (the region enclosed inside an oval in Fig. 1), which are of great interest for the design of an optimized TCS: the total cross section is very close to the maximum value achievable of $0.64barn$ at zero recoil, and the X/γ beam is almost fully polarized even using unpolarized electrons [3] as shown in Fig. 2 where distribution of the Stokes parameters as the color-map of intensity distribution is plotted.

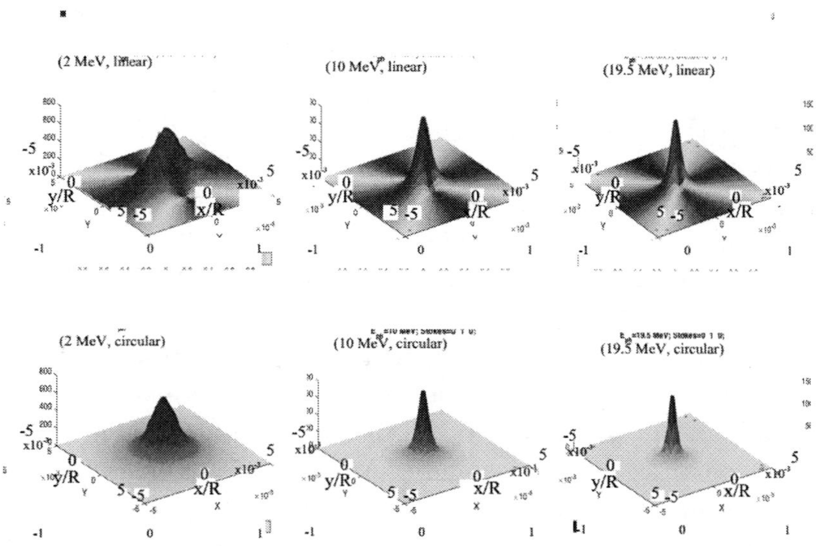

Figure 2: Intensity-polarization graphs for photons at 2, 10 and 19.5 MeV, with linear and circular laser polarization.

The Klein-Nishina cross section for Compton scattering σ_C is in fact almost equal to the Thomson cross section $\sigma_T = 8\pi r_e^2/3 = 0.64$ barn (where r_e is the classical electron radius) whenever the recoil parameter $\Delta \ll 1$, and its value is given approximately by $\sigma_C = \sigma_T(1 - \Delta)$ under this approximation, as plotted in fig. 1 [4]. Furthermore, since the back-scattering in Thomson or quasi-Thomson region is basically elastic, the incident laser photon is bouncing on the electron as if the electron were an ideal mirror moving at relativistic speed in the laboratory frame: this guarantees the preservation of the photon polarization as far as the scattering angle θ is small. As a consequence the back-scattered photon beam collimated within a small angle of scattering $\theta < 1/\gamma$, i.e. corresponding to a small bandwidth bdw $bdw \approx \theta^2\gamma^2 \ll 1$, will keep the same polarization of the laser pulse colliding with the electron beam: as discussed extensively in [3]. This leads to very high polarized X/γ beams which are strategic and of great interest for Nuclear Physics and Photonics applications.

New generation TCS machines are characterized by narrow bandwidths, below 0.5%, combined to large spectral flux, i.e. number of photons/sec per unit bandwidth. In order to achieve these performances the luminosity L_{TSC} of the TCS must be larger than $10^{35}\ s^{-1}cm^{-2}$, as specified by

$$L_{TSC} = \frac{N_L N_e}{4\pi\sigma_x^2} f \tag{2}$$

where N_L are the photons carried by the laser pulse at collision, N_e the electrons carried in the bunch, σ_x the spot size at collision IP (Interaction Point) and f the repetition rate of the collisions (assuming ideal overlap in space and time of the two colliding pulses, as well as negligible diffraction of the two beams over the interaction distance). Using ELI-NP-GBS nominal values for the 2 colliding beams (0.4 J for the laser pulse energy, 250 pC the electron bunch charge, focused down to about 20 μm with an effective rep. rate of 3.2 kHz) we find $L_{TSC} = 1.2\ 10^{35}\ s^{-1}cm^{-2}$, a quite impressive value, similar to the high-luminosity upgrade of LHC.

The total number of back-scattered photons per second, all over the spectrum and solid angle, would be $N_{all} = L_{TSC}\sigma_T$, i.e. $N_{all} = 7\ 10^{10}\ photons/s$. As discussed in details in ref. [2]- [4], any TCS is a polychromatic source of back-scattered photons: most of energetic photons are confined in a narrow solid angle (semi-aperture $1/\gamma$) with a large frequency spread (at $\theta = 1/\gamma$ the photons have half the energy than that given by eq. 1, which determines the maximum energy of photons back-scattered on axis at $\theta = 0$). In order to produce a mono-chromatic photon beam one needs to select a narrow angle around the electron beam propagation axis quite smaller than $1/\gamma$ - this is achieved by means of special collimators, which become quite challenging at photon energies above 1 MeV [2], [5].

So what really matters for experiments and applications, of the collimated photon beams, is the number of photons N_γ^{bw} carried by the back-scattered radiation pulse within such a small angle, and the rms bandwidth associated with it - that is the concept of spectral luminosity and spectral density, which are the real figure of merit for Nuclear Physics and Photonics. These are approximately given by

$$N_\gamma^{bw} = 7\ 10^8 \frac{U_L[J]Q[pC]}{h\nu_L[eV]\sigma_x^2[\mu m^2]}\gamma^2\theta^2 \tag{3}$$

in terms of the laser pulse energy U_L, the electron bunch charge Q, σ_x the spot size at collision IP, and the laser photon energy $h\nu_L$ [2], [4]. The rms bandwidth $\Delta\nu/\nu$ of the photon beam collimated within the acceptance angle $\Psi = \gamma\theta_{max}$ is approximately given by:

$$\frac{\Delta\nu}{\nu} \cong \sqrt{\left(\frac{\Psi^2}{\sqrt{12}} + 2\frac{\epsilon_n^2}{\sigma_x^2}\right)^2 + 4\left(\frac{\Delta\gamma}{\gamma}\right)^2 + \left(\frac{\Delta\nu_L}{\nu_L}\right)^2 + \left(\frac{M^2\lambda_L}{2\pi W_0}\right)^2 + \left(\frac{a_0^2/3}{1+a_0^2/2}\right)^2} \tag{4}$$

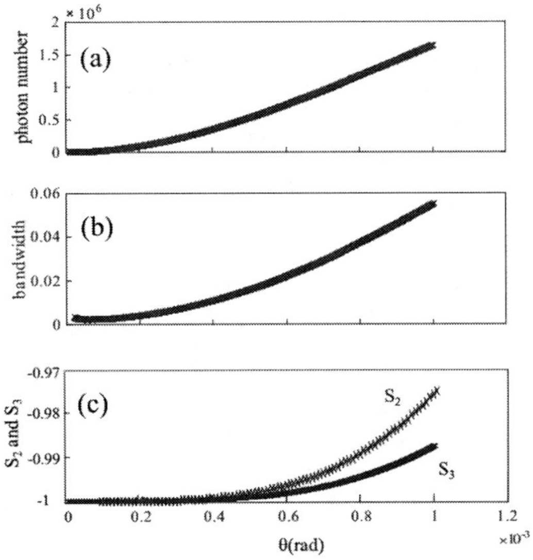

Figure 3: Number of scattering photons (a), bandwidth (b) and Stokes parameters (c) for the case of linear (S_3) and circular (S_2) polarizations as function of the acceptance angle. The red dot marks the bandwidth 0.5%.

in terms of the collimation angle θ, the electron beam phase space density parameters (rms energy spread $\Delta\gamma/\gamma$ and transverse total normalized emittance ε_n and the laser quality parameters, rms spectral bandwidth $\Delta\nu_L/\nu_L$ and mode purity M^2. The last term represents the non-linear effects due to the laser intensity: when the laser parameter a_0 is not much smaller than 1, multi-photon absorption start being effective and the radiation spectrum is significantly modified in shape and frequency distribution [6]. Narrow bandwidth TCS, as those for Nuclear Physics and Photonics, must minimize these non-linear effects by using laser pulses characterized by small value of $a_0 = 4.3\frac{\lambda_L}{W_0}\sqrt{\frac{U_L[J]}{\sigma_t[ps]}}$ (dimensionless amplitude of the vector potential associated to the laser e.m. wave), where W_O is the laser waist size at the IP and σ_t its rms pulse length. In case of ELI-NP-GBS we have, at maximum, $a_0 = 0.04$. In fig. 3 we present the dependence of the number of scattered photons, bandwidth and Stokes parameters as functions of the acceptance angle.

An important figure of merit is the Spectral Density, defined as $SPD = N_\gamma^{bw}/\sqrt{2\pi}h\Delta\nu$, typically expressed in units of $photons/s.eV$. Various generations of γ-ray sources have improved this figure of merit, from values of the order of 1 for bremsstrahlung sources, to about 100 for the present $Hi\gamma S$ facility [7], [8], toward the 10^4 range which is the goal of ELI-NP-GBS.

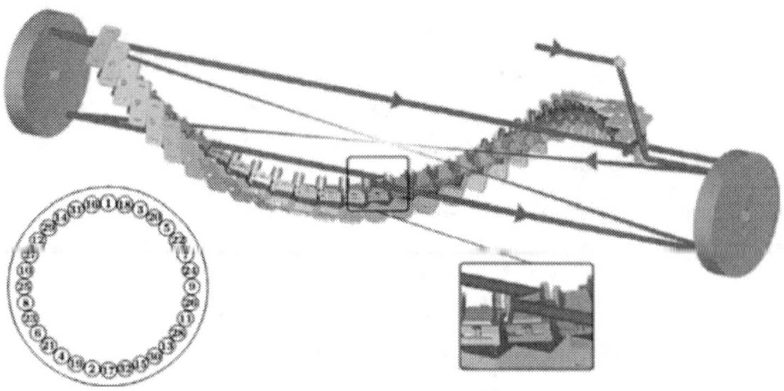

Figure 4: Lay-out of the laser recirculator under construction for ELI-NP-GBS, made of two confocal parabolic mirrors and 32 plane mirror pairs to focus, deflect, collimate and re-circulate the J-class green light $Yb : Yag$ laser pulse for collision at IP (at the center of the device) with each of the 32 electron bunches accelerated to the IP by the multi-bunch RF Linac.

In order to achieve such an impressive upgrade in the performances, the strategy of our TCS is to adopt a multi-bunch operation mode for the electron Linac and to re-circulate the laser pulse as many times as possible at the Interaction Point. In this way we will combine the capability of a RF Linac driven by high brightness photo-injector to provide outstanding peak electron beam quality in the single bunch (as per FEL Linac drivers) with a higher repetition rate than the one typical of such Linacs (100 Hz). As a matter of fact eq. 3 and eq.4 foresee about $N^b w_\gamma = 7\ 10^7\ s^{-1}$ with $\frac{\Delta\nu}{\nu} = 0.4\%$, which means about $SPD \approx 700$ for a single bunch operation at 100 Hz repetition rate, for a TCS radiating 10 MeV γ-rays (530 MeV electrons). Since the laser pulse carries about 10^{18} photons at the IP, but only 10^7 photons maximum are back-scattered at each collision (in other words the electron beam is almost transparent to the laser pulse), then we can conceive to re-use the laser pulse and bring it back to a new collision at the same IP with a new fresh incoming electron bunch. This requires a Linac able to operate in multi-bunch mode and an optical device able to re-circulate the laser pulse. The laser recirculator for 32 pulses was conceived as shown in fig.4. For a full description of this new optical device see ref. [2], [9].

Such an advanced and innovative laser re-circulator is under test by the EuroGammaS collaboration and will be integrated in the ELI-NP-GBS installation phase starting in 2016. Due to several constraints to fulfil in its design, concerning mirror quality, damage threshold, alignment and synchronization of the 2 parabolic mirrors and the 32 plane mirror pairs (a challenging 20 μm and 20 μrad alignment tolerance is required), the fo-

Table 1: ELI-NP-GBS expected performances: ranges for electron beam, laser pulse and γ-ray photon beam parameters are listed.

Electron beam Parameters, 32 bunches per RF pulse, 16 ns separation bunch-to-bunch, 100 Hz rep rate	
Electrons $[MeV]$	$75 - 750$
Bunch charge $[pC]$	$25 - 250$
Bunch rms length $[\mu m]$	$200 - 300$
Total projected rms transverse emittance $\varepsilon_n [mm\ mrad]$	$0.3 - 0.5$
Total projected rms energy spread [%]	$0.05 - 0.1$
Focal spot size σ_x, σ_x $[\mu m]$	$15 - 18$
Laser Parameters, 100 Hz rep rate	
Laser pulse energy $[J]$	$0.2 - 0.4$
Laser pulse length $[psec]$	1.5
Laser focal spot size W_0 $[\mu m]$	28
Laser rms bandwidth [%]	< 0.1
Laser M^2	< 1.2
Laser parameter a_0	$0.02 - 0.04$
Collision angel α_0 $[deg]$	8
Synchronization to RF $[ps]$	< 1
Pulse energy stability [%]	< 1
γ-ray Photon beam Parameters	
Photon Energy $[MeV]$	$0.2 - 19.5$
Spectral Density $[ph/s.eV]$	$0.5 - 2\ 10^4$
Bandwidth rms [%]	$0.2 - 0.5$
Collimation angle $[\mu rad]$	$60 - 400$
# photons per shot within collimation angle	$1.1 - 3.5\ 10^5$
# photons/sec within collimation angle	$0.3 - 1.1\ 10^9$
Source rms size $[\mu m]$	$10 - 15$
Peak Brilliance $[N_ph/s.mm^2.mrad^2.0.1\%]$	$2\ 10^{19} - 4\ 10^{21}$
Radiation pulse length $[ps]$	$0.7 - 1.0$

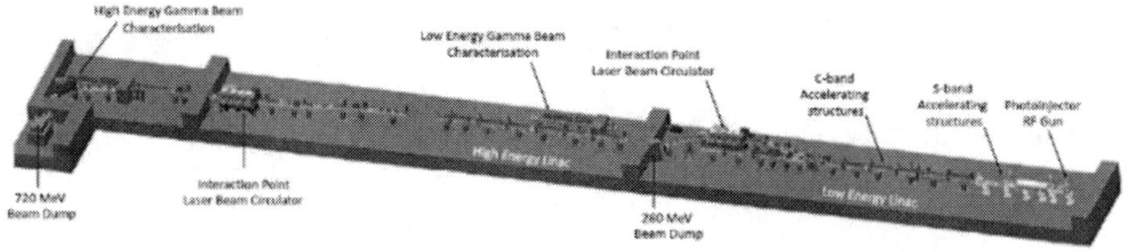

Figure 5: Lay-out of ELI-NP-Gamma Beam System as designed by EuroGammaS (CAD provided by STFC members of EuroGammaS).

Figure 6: High-Order Mode damped C-band accelerating RF cavities, designed and built by INFN-LNF for EuroGammaS, under full power test.

cal length of the parabolic mirrors was set at 1.2 m, implying a round-trip time for the laser pulse of about 16 ns. A total of 32 round-trips have been designed and carefully simulated with a physical laser transport simulation code in order to asses the quality of the laser pulse at the IP at any round-trip. The expected performance is quite close to 100% in the total accumulated spectral density over the 32 pulses as far as the 32 electron bunches will collide at the IP with same constant beam quality over the multi-bunch train. The time structure of the generated γ-ray pulses will be therefore made of 32 long trains repeating at 100 Hz (10 $msec$ far apart), consisting of 32 micro-pulses separated by 16 ns (each micro-pulse with rms pulse length a bit shorter than 1 ps).

In order to generate the high brightness electron beam requested, ELI-NP-GBS adopts an advanced photo-injector coupled to a C-band high gradient RF Linac capable to bring the electron beam up to a maximum energy of 750 MeV [10], [11] with outstanding beam quality. The lay-out of the machine is shown in Fig.5.

An innovative C-band High Order Mode damped RF cavity has to be conceived and designed in order to avoid emittance and energy spread degradation due to the Beam Break-Up instability along the Linac over the 32 multi-bunch train. The low level RF tests successfully conducted at INFN-LNF were followed by successful high power tests with achievement of the nominal 33 MeV/m accelerating gradient in the accelerating structures with the requested rejection level of high order RF modes which are potentially dangerous for BBU [12] as shown in Fig. 6.

Thanks to these advanced components and to many other devices developed specifically for the ELI-NP-GBS (like the collimator-characterization stage for the γ photon beam), the expected performances are at least 2 orders of magnitude higher than the present state of the art, in terms of bandwidth, brilliance and spectral density. These are listed Table 1.

Conclusions

The future availability of advanced X/γ ray photon beams in the energy range $1-20$ MeV range will open many opportunities of addressing strategic applications in Nuclear Photonics.The EuroGammaS collaboration is building the most advanced source in this field that will start operation in 2018.

References

[1] Proceedings of NPNSNP 2014, Ricotti, Tokai-mura, Japan

[2] ELI-NP-GBS T. D. R.,http://arxiv.org/abs/1407.3669, (2014)

[3] Petrillo V. and al., submitted to PRSTAB, (2015)

[4] Petrillo V. and al., NIM A, 693 (2012)

[5] Cardarelli P. and al., NIM B, 355 (2015)

[6] Maroli C. and al., Phys. Rev. STAB, 16 (2013) 030706

[7] http://www.tunl.duke.edu/web.tunl.2011a.higs.php

[8] Sun C. and Wu Y. K., Phys. Rev. ST Acc. Beams, 14 (2011) 044701

[9] K. Dupraz et al., Phys. Rev. STAB 17, 033501 (2014)

[10] Bacci A. and al., J. Appl. Phys., 113 (2013) 194508

[11] Vaccarezza C. and al., Proc. IPAC2014, Dresden, Germany, (2014)

[12] Alesini D. and al., THPRI042, Proceedings of IPAC2014, (2014)

EPJ Web of Conferences **117**, 05003 (2016) DOI: 10.1051/epjconf/201611705003
NN2015

The Muon Portal Project: A large-area tracking detector for muon tomography

F. RIGGI for the MUON PORTAL COLLABORATION

Department of Physics and Astronomy,
University of Catania, Catania, Italy

Abstract

The Muon Portal Project [1] is a joint initiative between research and industrial partners, aimed at the construction of a real size detector protoype to search for hidden high-Z fissile materials inside containers by the muon scattering technique. The detector is based on a set of 48 detection modules (1 m $\times$ 3 m), so as to provide four X-Y detection planes, two placed above and two below the container to be inspected. After a research and development phase, which led to the choice and test of the individual components, the construction of the full size detector has already started and will be completed in a few months.

1 Introduction

An estimate of about 200 million containers are traveling each year through the custom borders of many countries. They are in principle potential sources of small quantities of hidden nuclear material, such as fissile elements. As an alternative to traditional systems based on X-ray inspection it has been long suggested [2] to employ the scattering process of the secondary cosmic muons, which strongly depends on the atomic number of the traversed material, hence particularly sensitive to high-Z fissile elements. For a detection system employing such technique (muon tomography) a large area muon detector is required, to reconstruct the muon tracks with good angular resolution before and after traversing the container volume.

Figure 1: Schematic view of the detector prototype for inspection of real size containers by the muon tomography technique. Four X-Y detection planes are located above and below the container, to provide track reconstruction of the cosmic muons before and after traversing the volume to be inspected.

Reconstruction and visualization algorithms are then applied to produce a tomographic image of the container volume, able to signal the presence of hidden, high-Z materials against a large background originating from low- and medium-Z objects.

Different Projects have been exploited over the past years, aiming at building prototype detectors for muon tomography. They differ in the sensitive area and in the details of the detection technique employed. A new Project was recently started by the Muon Portal Collaboration [1] with the goal to build a real size detector (18 m^2 sensitive area for each plane) with all potential features to be used in a real situation to probe the interior of a standard 20' container. The main parameters of the design of such setup, together with the instrumentation required, are described in Sect. 2. A considerable amount of simulations and experimental tests on the individual components (scintillators, WLS fibres, photosensors, front-end electronics,...) have been already done. Different reconstruction and visualization algorithms have been implemented and have demonstrated the feasibility of this Project, with several papers reporting specific aspects [3-8].

2 The Muon Portal Project

2.1 Geometrical and mechanical structure of the Muon Portal

The detection setup is based on four XY position-sensitive detection planes, two placed above and two below the container volume to be inspected (Fig.1).

The overall size of the detector fits that of a real TEU (Twenty-foot Equivalent Units) container, namely 6 m $\times$ 3 m $\times$ 3 m. Each plane is made by 6 modules (1 m $\times$ 3 m each) in a proper geometry, such as to cover both the X- and the Y-coordinates by the same type of modules, without leaving any dead area between close modules. The detector will be constituted by a total of 48 modules. An appropriate mechanical structure was built to move and hold the detection planes in their planned Z-positions.

2.2 The detection modules

Each of the 48 detection modules is segmented into 100 strips of extruded plastic scintillator ($1\times1 \times 300$ cm^3), with two wavelength-shifting (WLS) fibres embedded in each strip, to transport the photons to the photosensors placed at one of the fibre ends. The overall number of channels (fibres and photosensors) is then 9600.

After a series of experimental tests carried out on several prototypes of scintillator strips and WLS fibres from different suppliers, the final design of the detector makes use of scintillator strips from Amcrys with two embedded WLS Y11 1 mm Kuraray fibres.

Extensive tests of the individual strips and WLS fibres have been done and are reported in a previous paper [7].

A spatial resolution in the order of a few mm is required for all planes, in order to provide good tracking capabilities for the incoming charged particles, reconstruct their trajectories above and below the container and evaluate the amount of scattering suffered by each track.

2.3 Characterization of the photosensors

The photosensors employed are Silicon Photomultipliers (SiPM). These are compact devices, with a high gain and a low operating voltage, which have been used for several years in most demanding applications. The total number of SiPMs needed for the whole detector is 9600. The silicon photomultiplier (SiPM) prototype designed for the *Muon Portal* Project had to maximize the photon detection efficiency and the cell fill factor, as well as to ensure a low cross-talk and dark count rate. Different SiPM prototypes (1 mm^2 area), both with the *p-on-n* and *n-on-p* technologies, have been produced by STMicroelectronics for this project before the mass production.

The layout of the final chip is based on *n-on-p* technology and embeds 4 independent round shaped SiPMs: two MUON60 type with 548 cells and 60 μm pitch, and two MUON75 type with 320 cells and 75 μm pitch.

A custom procedure has been implemented for the characterization of all the devices. The setup is made of a black box where the device under test is placed inside a proper socket; the device is biased by a Keithley picoammeter/voltage source, that is also used to read the current from the device and transmit data to a computer. A LabVIEW program was developed to measure the complete I-V curve in an automatic way, and find the Breakdown Voltage BV value for each SiPM. The breakdown voltage distribution from a sample of about 4500 SiPMs is shown in Fig.2. Detailed results on this procedure and results are reported in Ref. [8].

2.4 Electronic readout and data acquisition

Taking into account the number of modules and the required granularity to achieve a reasonable space and angular resolution, a large number of channels is required for such detector. A compression technique within each module (100 strips) is envisaged in our Project, in order to reduce such number to a suitable level for the electronics and data acquisition. This is achieved by the use of two WLS fibres running along the same strip (for

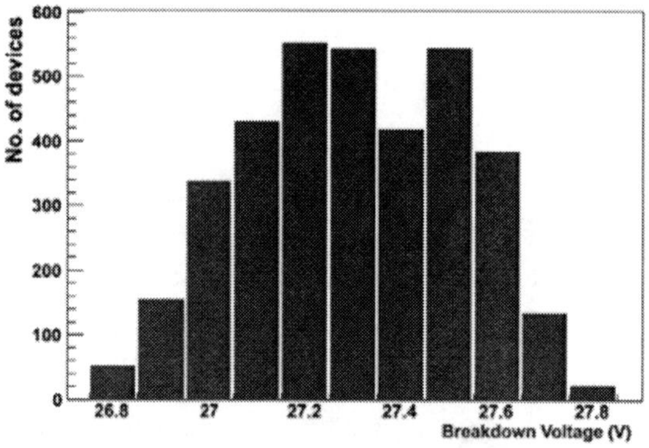

Figure 2: Breakdown voltage distribution, measured over a sample of about 4500 devices.

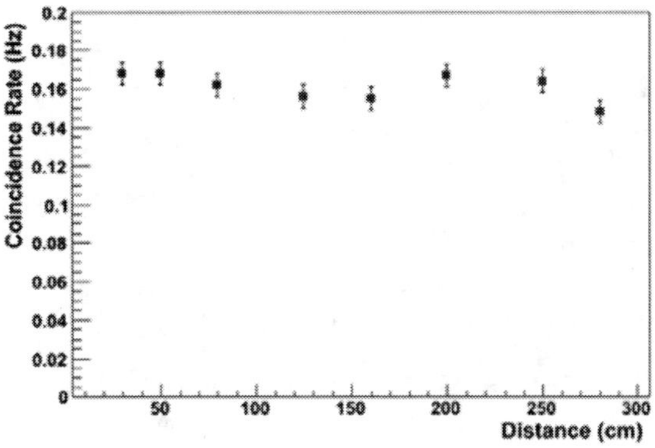

Figure 3: Coincidence rate between one of the scintillator strips and an external trigger scintillator, as a function of its distance with respect to the photosensor.

a total of 9600 WLS fibres) and going to an equal number of SiPMs, with a proper combination of the signals originating from ten channels grouped together.

2.5 Reconstruction and imaging algorithms

Identification of hits and clusters in each detection plane allows the tracks in the upper and lower part of the detector to be reconstructed and the scattering angle between the two tracks estimated.

Several algorithms have been tested for the reconstruction of the muon scattering process. The simplest of them is based on the Poca (Point-of-Closest-Approach) method, from which a spatial distribution of the scattering centres is derived, with a weight proportional to some power of the scattering angle.

Such method, although of easy implementation, neglects the multiple scattering through the volume material and therefore has the drawback of providing poor-resolution images. This motivated the implementation of alternative algorithms (log-likelihood, clustering algorithms, ...), with the Friend-Of-Friend (FOF) producing the best results.

All these reconstruction algorithms were tested under different scenarios, simulating by GEANT the presence of dense objects embedded in low or medium Z materials within the volume to be inspected.

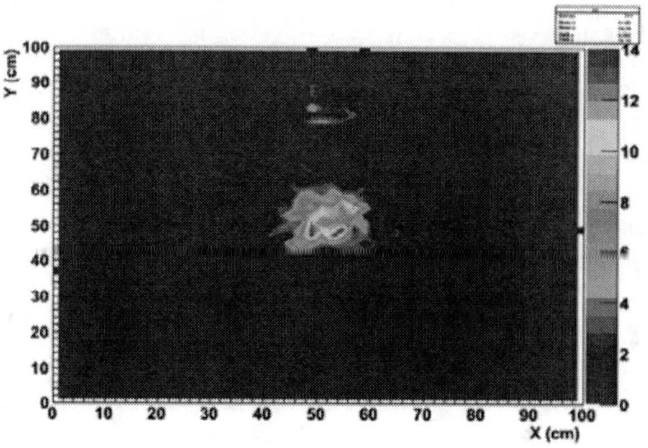

Figure 4: Two-dimensional map obtained by the detection of cosmics passing through the X- and Y-planes and triggered by a small area external scintillation detector.

3 Construction and tests of the detection modules with cosmic rays

Each detection module is built as a sandwich which incorporates 100 extruded scintillator strips and 200 WLS fibres in a compact, light-tight case. Presently, all the 48 detection units have been assembled. Working tests have been performed on some of them, especially at the beginning of the construction activity, to ensure that the assembly procedure was reliable. Two complete detection planes (24 modules) are already installed and tested with cosmic rays.

An important test was to check that the detection efficiency stays almost constant along the length of the strip (3 meters long). This was verified by moving a small area external scintillator in various positions above the module and measuring the coincidence rate. As an example, Fig.3 shows the result obtained with a threshold corresponding to about 2.5 p.e, as a function of the distance along the strip. Similar measurements were done on various strips and at different values of the SiPM bias voltage and discriminator level.

After installing the detection modules in their final positions, threefold coincidence measurements between the external scintillator and the X- and Y-planes were carried out, moving the scintillator over the surface of the plane, so as to explore a large part of the sensitive area of the module. Fig.4 shows one of the maps obtained in such conditions. First coincidences between the top and bottom planes were also observed recently, and an

Figure 5: An exhibition box devoted to outreach activities concerning various aspects of this Project.

angular distribution of the incoming cosmic muons extracted. Complete tests of the overall detection setup will be performed after installing the two additional detection planes.

4 Conclusions

After an intensive phase of research and development to test and choose the individual components for the construction of the detector prototype, as well as to design the mechanical structure and the electronics and readout architecture, the construction of the entire Muon Portal detector has been almost completed. This paper reports on some details concerning the construction and testing of the detection modules, as well as the characterization of the individual photosensors. At the time of writing all the detection modules have been already fully prepared and half of them installed in the mechanical structure of the detector and tested with cosmic rays. The overall completion of the Muon Portal detector requires in the next months the installation of the remaining detection modules, together with the associated front-end electronics and data routing, in order to test the overall functionality of the prototype and carry out the first commissioning tests.

Finally, a proper value will be given in our project to outreach activities concerned also with applications of physics in the everyday life, addressed to students and citizens. For this purpose an exhibition box was specifically built close to our Department (Fig.5).

Acknowledgments

The author warmly thanks all the colleagues and young people in the Muon Portal Collaboration for the fuitful work carried out together during the last years.

References

[1] For further details visit our Web site: *http://muoni.oact.inaf.it*

[2] Borozdin K R 2003 *Nature* **422** 277.

[3] Riggi S et al 2010 *Nuclear Instruments and Methods in Physics Research* **A624** 583.

[4] Riggi S et al., *Nuclear Instruments and Methods in Physics Research* **A728**(2013) 59.

[5] La Rocca P, et al., 2014 *Journal of Instrumentation* **9** C01056.

[6] Pugliatti C, et al, 2014 *Journal of Instrumentation* **9** C05029.

[7] Russo G V, et al, 2014 *Journal of Instrumentation* **9** P11008.

[8] La Rocca P, et al., 2015 *Nuclear Instruments and Methods in Physics Research*, **A787**(2015)236.

EPJ Web of Conferences **117**, 05004 (2016) DOI: 10.1051/epjconf/201611705004
NN2015

Laser-based acceleration for nuclear physics experiments at ELI-NP

O. Tesileanu, Th. Asavei, I. Dancus, S. Gales, F. Negoita, I.C.E. Turcu, D. Ursescu and N.V. Zamfir

Extreme Light Infrastructure - Nuclear Physics, "Horia Hulubei" National R&D Institute for Physics and Nuclear Engineering (IFIN-HH), 30 Reactorului street, 077125 Măgurele, Romania

Abstract

As part of the Extreme Light pan-European research infrastructure, Extreme Light Infrastructure – Nuclear Physics (ELI-NP) in Romania will focus on topics in Nuclear Physics, fundamental Physics and applications, based on very intense photon beams. Laser-based acceleration of electrons, protons and heavy ions is a prerequisite for a multitude of laser-driven nuclear physics experiments already proposed by the international research community. A total of six outputs of the dual-amplification chain laser system, two of 100TW, two of 1PW and two of 10PW will be employed in 5 experimental areas, with the possibility to use long and short focal lengths, gas and solid targets, reaching the whole range of laser acceleration processes. We describe the main techniques and expectations regarding the acceleration of electrons, protons and heavy nuclei at ELI-NP, and some physics cases for which these techniques play an important role in the experiments.

1 Introduction

Extreme Light Infrastructure (ELI) is a distributed European research infrastructure dedicated to the generation and use in various experiments of extremely intense photon beams. The project initiated a decade ago with the idea of reaching 100PW laser powers by the coherent combination of several beams with powers of tens of PW, gained recognition by being adoped in the ESFRI (European Strategy Forum for Research Infrastructures) list of high-priority research projects in Europe in 2006 and then it was defined by a Preparatory Phase (ELI-PP, 2007-2010) funded by FP7.

ELI PP reached the conclusion that, given the dimensions and extremely large scientific scope of the newly proposed research infrastructure, the best implementation strategy would be to adopt a distributed model, with the first three pillars to be hosted in three European countries: the Czech Republic, Hungary and Romania. These will focus on laser development up to the 10PW level and the related scientific applications. A fourth pillar would be decided later on, aiming, based on the experience of the first three pillars, to reach the hundreds of Petawatts power level.

The Romanian pillar of ELI is ELI-NP, devoted to experiments in laser science, nuclear physics and related areas, including applications, employing two 10PW laser beams and an extremely intense gamma photons beam. This unique combination of beams that are at their turn unique worldwide assures for ELI-NP the possibility to tackle areas at the frontiers of present-day knowledge and advance science in many areas, from the fundamental studies to society-relevant developments. The block diagram of ELI-NP presented in Fig.1 shows the large research equipment and the experimental areas with the corresponding science topics.

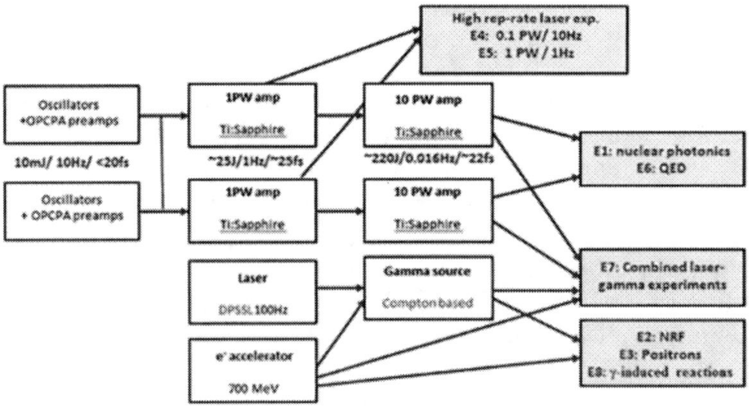

Figure 1: Block diagram of the ELI-NP infrastructure [16].

Technical Design Reports (TDR) for the various types of experiments envisaged at ELI-NP were developed and then reviewed by an international scientific advisory board. At the time of writing the experimental setups are entering the implementation phase.

2 The ELI-NP high-power laser system (HPLS)

The ELI-NP HPLS is a complex ultra-short pulse laser system, having two parallel amplification chains up to 10PW power, with intermediate outputs at 100TW and 1PW levels. Starting from the same front-end, pulses are stretched and amplified to a final energy in the range of 220J, and then compressed to pulse lengths of 22fs.

HPLS will be constructed by an association between Thales Optronique SA France and Thales Romania. It consists of a chirped pulse amplification system at ~820nm central wavelength, with a dual front-end architecture (to minimize down-time for the laser facility). Each of the two parallel chains includes Ti:Sapphire amplifiers. The repetition rate for the two 10PW outputs is one shot per minute. The general layout of the HPLS laser system is similar with the one under development in France for CILEX-APOLLON facility [1].

The additional outputs at 0.1PW and 1PW along the two amplification chains, will be installed with the corresponding optical compressors. The pulse repetition rates for these outputs are 10Hz and 1Hz, respectively. The pre-pulse temporal contrast for all the outputs is specified to be better than 10^{12}:1 both at picosecond and nanosecond delays and the requested Strehl ratio is 0.9. The intensity to be attained is $\sim 10^{23}\mathrm{W/cm}^2$, securing a properly defined laser pulse in both space and time for interaction with thin targets. The energy stability is expected to reach 5% RMS and the pointing stability shall be less than 5 microrad RMS for both axes. In order to create the conditions to attain the pointing stability specifications the building has a vibration-damped floor across the entire laser and experiments areas.

3 Laser-based acceleration of particles

The acceleration of particles with the help of lasers is one of the research areas of great interest nowadays due to a multitude of factors, such as advances in theory, the continuous technological progress in laser pulse power and intensity, more reliable numerical simulations and availability of ever increasing computational resources for these. The final goal is fundamental research and applied research with society-relevant applications, therefore the need for facility grade, laser driven particle accelerators that are cheaper, much more compact, and sometime more suited to the purpose than their "traditional" counterparts.

3.1 Acceleration of electrons

The acceleration of electrons is in a rapid progress, experiments proving higher electron energies with similar laser pulse powers. Recently an experiment at the BELLA laser in Berkeley ([9]) reported 4.2 GeV electron bunches of 6 pC with 6% energy spread, very well collimated (0.3 mrad RMS divergence).

Laser Wake Field Acceleration (LWFA) in a gas-filled capillary discharge waveguide is the main technique applied for attaining the highest energies. Energies up to 40 GeV are expected from simulations [13] using the parameter of ELI-NP laser and capillaries up to 50 cm. Focusing parabola with focal length from 10 m up to 40 m, corresponding to F/20 and possibly upgraded to F/80 in full beam aperture, are considered in the implementation, together with adequate electron magnetic spectrometers and beam dump.

3.2 Acceleration of protons and heavy ions

The laser-based acceleration of protons and heavy ions has the potential of producing beams with some major advantages with respect to classical accelerators when it comes to medical applications, such as the much lower secondary (prompt) ionizing radiation and activation of facility components.

Various ion acceleration mechanisms will be studied at ELI-NP in the new regimes of laser focus intensities approaching and hopefully even going beyond 10^{23} W/cm^2. Proton energies up to 500 MeV are expected and taken as term source in beam dump and shielding calculations. However some of the physics cases require rather high fluxes and bunch densities at moderate energies, in the range of tens of MeV/nucleon for heavy ions.

The Radiation Pressure Acceleration (RPA) [8] mechanism is of highest interesting since it promises acceleration at solid state density which is many orders of magnitude above the density of ion bunches produced by any classical acceleration and opens therefore perspective for new studies. Other acceleration mechanisms such as Break Out Afterburner acceleration (BOA) [7] and Collisionless Shock Acceleration (CSA) in near critical density gas targets are proposed to be studied. CSA has been demonstrated experimentally using CO_2 lasers [5,14] and according to simulations, should enhance the efficiency of laser acceleration of protons and ions compared to TNSA (Target Normal Sheath Acceleration) as well as ion energy scaling [6].

Acceleration of heavy ions from solid targets is the only one giving the access to full range of elements and nuclear reactions of interest. The methods for energy selection and angular spread reduction (refocusing) are relevant for many applications and also proposed for study at ELI-NP.

4 ELI-NP experiments using accelerated particles

4.1 Experimental areas and expected accelerated particles

Out of the eight experimental areas at ELI-NP, four are devoted to laser-based experiments at the three power levels available, and one is devoted to experiments with combined laser and gamma/electron beams.

The E1 and E6 experimental areas feature both 10PW beams. In E1 it is foreseen the acceleration by RPA of protons and heavy ions, up to 500MeV (protons) or 50MeV/A (heavy ions), and their use in laser-driven Nuclear Physics experiments. Accelerated proton bunches should have more than 10^{10} p/pulse, energy FWHM of 10% and divergence better than 5°. E6 [15] is dedicated to fundamental physics research in High Field Physics and QED Experiments, aiming the study of Radiation Reaction and High Field Quantum Electrodynamics mainly through the Non-linear Compton Scattering and Multiphoton Breit-Wheeler Process. The experiments will use for the first time two such powerful laser beams in a pump-probe configuration: first 10PW laser beam (pump) focused on a gas or solid target accelerating electrons to multi-GeV, relativistic, energies; the second 10PW laser beam (probe) is tightly focused on the relativistic electron bunch generating the QED processes [10].

The E7 experimental area is devoted to experiments in fundamental physics with combined laser and gamma or electron beams. The laser-accelerated electron beams envisaged to be used in radiation reaction, pair creation and vacuum birefringence experiements will have energies up to 5 GeV and low divergence (less than $\sim 0.3\,\mathrm{deg}$) and energy spread. These beams will be used for the generation of gamma photons in the GeV range by Compton backscattering effect.

The experiments at E4 and E5 employing 0.1 and 1PW laser beams at higher repetition rates are covering a wide range of topics, from materials science to the quest for dark matter candidates. Protons up to 60MeV in large bunches of $\sim 10^{10} - 10^{12}$, and electrons in the range of $50 - 2,000$ MeV and intensities of up to a few 10^{10} / pulse are foreseen to be produced. Thermal electrons will also be generated at intensities of $\sim 10^{7}$ / pulse and moderate divergence of a few degrees. In E5 the study of materials behaviour in extreme environments will be a central topic, with a direct application to the development of accelerator components and societal applications like the understanding of structural materials degradation in next generation fusion and fission reactors or the shielding of equipment and human missions in outer space.

4.2 Research proposals

More than 100 researchers contributed to the ELI-NP Whitebook [3] proposing the first ideas for the experimental program at the new facility. The next step was detailing the categories of experiments in the TDRs. From the multitude of proposals employing the laser acceleration of particles few are presented below, in their general lines.

Enhanced decay of 26Al in hot plasma environments

The ^{26}Al nucleus was the first radioisotope detected in the interstellar medium, by the observation of the characteristic 1809 keV γ-emission associated with the decay of its ground state [12]. There is still a debate regarding the production sites of this isotope, that is a marker of ongoing nucleosynthesis. Moreover, in stellar conditions (temperatures up to $\sim$0.4GK) it is predicted through theoretical calculations [2] that the effective life time of ^{26}Al would be reduced by many orders of magnitude due to a variety of physical processes.

At ELI-NP there is the possibility to produce plasma conditions similar to the ones in various astrophysical setups, and thus the population of the states of interest will provide a disintegration rate closer to the true astrophysical one [17]. The fact that the radiation pulses are ultrashort in time and synchronous will provide good conditions for the resolution of ps time scales.

Production of extremely neutron-rich isotopes

The waiting point around N=126 is extremely important for the *r*-process nucleosynthesis, and presently very little is known about the nuclei in this area. The proposal [4] aims producing these nuclei by fissioning a dense thorium ion bunch with about 7 MeV/u in a thick thorium target (covered by a thin carbon layer), where the light fission fragments of the beam fuse with the light fission fragments of the target. The accelerated ^{232}Th bunches of solid state density are produced via the RPA process with a high intensity laser pulse (in the range of 10^{23}W/cm^2) produced by one of the 10PW HPLS arms. These bunches pass through a thin carbon layer and desintegrate into light and heavy fission fragments. In addition light ions (H,C) from the CH2 backing of the first (thin) Th target will be accelerated as well, inducing the fission process of ^{232}Th in the second, thick Th foil.

Electron screening in astrophysical plasmas

The measurement of rates and cross-sections (S-factors) in conditions close to the astrophysical ones (extremely low energies and including plasma effects) in the laboratory has been a great challenge for the experimental nuclear physics for many decades. A setup is proposed based on two high power laser beams generating, by interaction with a solid and a gas target, two colliding plasma bunches. The study of the role played by free and bounded electrons on the coulombian screening is one of the first topics to be investigated, but also other subjects are proposed. The TNSA acceleration regime demonstrated in previous experiments [11] will be acting for the solid thin target, while the other target will be a gas jet. Two regimes for the secondary (gas) target are envisaged.

RA5 - electron acceleration

At Research Activity 5 at ELI-NP, experiments with combined laser and gamma / electron beams are envisaged. Laser acceleration of electrons is needed for the proposed experiments in two energy ranges, one up to a few tens of MeV and one of 2.5 – 5 GeV. In the first case, a high intensity laser pulse will generate dense accelerated bunches of electrons that will be used for Bremsstrahlung radiation, which will interact with a secondary target and promote some isotopes of interest to isomeric states. In the second case, electrons are accelerated to 2.5 – 5 GeV by one 1PW laser pulse, with the aim to produce high energy (up to GeV level) secondary gama beams after their interaction with another laser pulse. Studies of radiation reaction, gamma polarization and vacuum birefringence are proposed.

5 Conclusions

The acceleration of particles, electrons and protons/heavy ions, is a research and development area in extremely fast evolution. The possibility to obtain extremely high acceleration gradients and consequently accelerators of much reduced dimensions and costs is for sure a powerful driver for a very active research community. ELI-NP will focus on the study and perfectioning of these methods of acceleration, and on fundamental science employing the accelerated particles. Apart from laser science, important results in nuclear physics, astrophysics, life sciences, materials science and fundamental physics are expected from the new research infrastructure.

Acknowledgments

The authors acknowledge the support from the Extreme Light Infrastructure Nuclear Physics (ELI-NP) Phase I, a project co-financed by the Romanian Government and European Union through the European Regional Development Fund.

References

[1] Cheriaux G., Giambruno F., Freneaux A., Leconte F. *et al*, *AIP Conference Proceedings on Light at Extreme Intensities*, **1462** (2012) 78

[2] Coc A. *et al.*, *Phys. Rev. C*, **61** (1999) 015801

[3] ELI-NP White Book, http://www.eli-np.ro/documents/ELI-NP-WhiteBook.pdf

[4] Habs D. *et al.*, *Applied Phys. B* **103** (2011) 485

[5] Harberger D. *et al.*, *Nature Physics* **8** (2012) 95

[6] d'Humières E. *et al.*, *Phys. of Plasmas* **20** (2013) 023103

[7] Jung D. *et al.*, *New Journal of Physics* **15** (2013) 023007

[8] Kar S. *et al.*, *Phys. Rev. Lett.* **109** (2012) 185006

[9] Leemans W.P. *et al*, *Phys. Rev. Letters*, **113** (2014) 245002

[10] Luo W. *et al.*, *Phys. of Plasmas* **22** (2015) 063112

[11] Macchi A. *et al.*, *Rev of Mod. Phys.* **85** (2013) 751

[12] Mahoney W.A. *et al.*, *Astrophysical Journal*, **286** (1984) 578

[13] Martins S.F., Fonseca R. A., Lu W., Mori W. B., Silva L. O., *Nature Physics* **6** (2010) 311

[14] Palmer C. *et al.*, *Phys. Rev. Lett.* **106** (2011) 014801

[15] Turcu I.C.E., Balascuta S., Negoita F., Jaroszynski D., McKenna P., *Proc. CSSP14*, *AIP Conf. Procs.* **1645** (2015) 416

[16] Ursescu D.,Tesileanu O., Cernaianu M., Gales S., Zamfir N.V., *Review of Laser Engineering*, **42** (2014) 123

[17] Utsunomiya H. *et al.*, *Nucl. Phys. A* **777** (2006) 459

EPJ Web of Conferences **117**, 05005 (2016) DOI: 10.1051/epjconf/201611705005
NN2015

A Compton camera prototype for prompt gamma medical imaging

P.G. Thirolf[1], S. Aldawood[1,2], M. Böhmer[3], J. Bortfeldt[1], I. Castelhano[1,4], G. Dedes[1], F. Fiedler[6], R. Gernhäuser[3], C. Golnik[5], S. Helmbrecht[6], F. Hueso-González[6], H. v.d. Kolff[1,7], T. Kormoll[5], C. Lang[1], S. Liprandi[1], R. Lutter[1], T. Marinšek[1], L. Maier[3], G. Pausch[5], J. Petzoldt[5], K. Römer[6], D. Schaart[7] and K. Parodi[1]

[1] Ludwig-Maximilians-Universität München, 85748 Garching, Germany
[2] King Saud University, Riyadh, Saudi Arabia,
[3] Technische Universität München, 85748 Garching, Germany
[4] University of Lisbon, Lisbon, Portugal
[5] OncoRay - National Center for Radiation Research in Oncology and Technische Universität Dresden, Germany
[6] Helmholtz-Zentrum Dresden Rossendorf, Dresden, Germany
[7] Delft University of Technology, Delft, The Netherlands

Abstract

A Compton camera prototype for a position-sensitive detection of prompt γ rays from proton-induced nuclear reactions is being developed in Garching. The detector system allows to track the Compton-scattered electrons. The camera consists of a monolithic $LaBr_3$:Ce scintillation absorber crystal, read out by a multi-anode PMT, preceded by a stacked array of 6 double-sided silicon strip detectors acting as scatterers. The $LaBr_3$:Ce crystal has been characterized with radioactive sources. Online commissioning measurements were performed with a pulsed deuteron beam at the Garching Tandem accelerator and with a clinical proton beam at the OncoRay facility in Dresden. The determination of the interaction point of the photons in the monolithic crystal was investigated.

1 Introduction

In order to be able to fully exploit the beneficial properties of tumour treatment by particle beams (protons, ions), provided by the highly conformal dose deposition within the Bragg peak, precise knowledge of the ion beam stopping range is mandatory. Our approach towards this goal is to detect the prompt photon range in tissue emitted during nuclear interactions between the hadron beam and the organic target as surrogate signal of the stopping range. The Compton scattering process can be exploited to reconstruct the γ incidence direction from the Compton scattering kinematics of the primary photon. Perspectives of prompt-γ imaging have been extensively studied both theoretically and experimentally [1–4], and Compton camera systems have been studied by several groups [5–10]. Fig. 1 displays the principle of a Compton camera, consisting of a scatterer and absorber component: in 'γ-tracking mode' (left panel), the position dependent detection of the scattering and absorption interaction of the impinging γ ray (with full absorption of the photon energy in the absorber) allows to reconstruct the photon source position on the surface of the 'Compton cone'. Intersecting reconstructed Compton cones from various events finally reveals the source position, as visible from the bottom row of Fig. 1, where this is shown for one (N_γ =1) and 10 events (N_γ =10), respectively [9,10]. In case

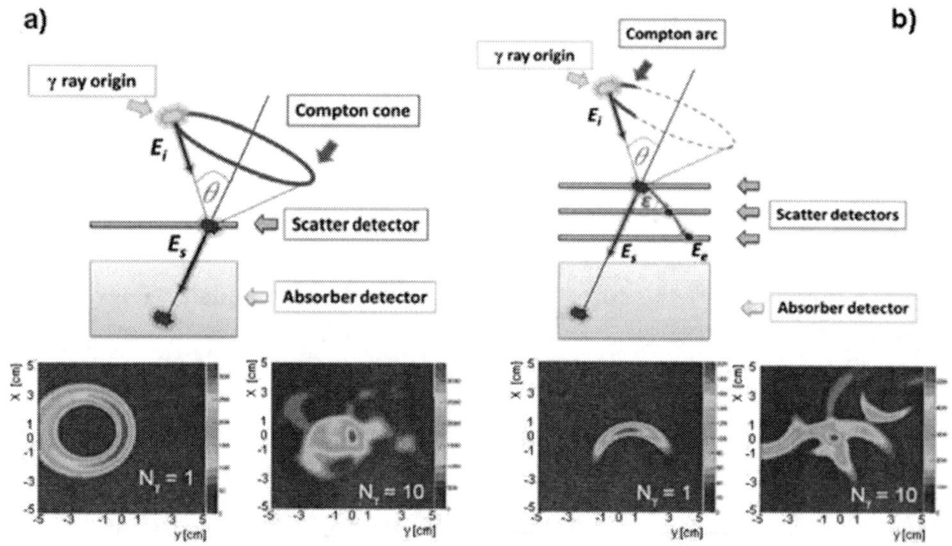

Figure 1: Principle of a Compton camera: a) with γ tracking only, b) with additional electron tracking capability.

of sufficiently high photon energies (few MeV), a layered structure of the scatterer detector (e.g. by an array of position-sensitive silicon strip detectors) enables a tracking of the Compton electron. This not only reduces the Compton cone to an arc segment (see bottom row), but allows to include also incompletely absorbed events into the source reconstruction, thus increasing the reconstruction efficiency [10, 11].

2 Compton camera layout

The Compton camera prototype, presently being under commissioning in Garching, is composed of a scatterer/tracker array consisting of 6 double-sided silicon strip detectors (DSSSD, area: $50\times50\,\text{mm}^2$, thickness: 0.5 mm, 128 strips/side) and a monolithic $LaBr_3$:Ce scintillation absorber crystal ($50\times50\times30\,\text{mm}^3$), read out by a multi-anode PMT (Hamamatsu H9500, 16×16 segments with 3×3 mm^2 pixel size). The strip detectors are read out

Figure 2: Left: Front of the Compton camera scatter/tracker array: double-sided Si-strip detector with ASIC-based signal processing electronics. Right: Absorbing $LaBr_3$:Ce scintillator.

via a compact ASIC-based electronics [12], while the signals from PMT segments of the absorbing scintillator are processed via individual channels of spectroscopy electronics. Energy signals are digitized in VME-based charge-to-digital converters (QDC, mesytec MQDC-32), while timing signals can be acquired via time-to-digital converters (mesytec MTDC-32). Fig. 2 shows the front sight of the scatterer array with its surrounding, directly attached signal processing boards (left), while the $LaBr_3$:Ce scintillator with its 256 signal outputs can be seen on the right-hand side. Design simulation studies

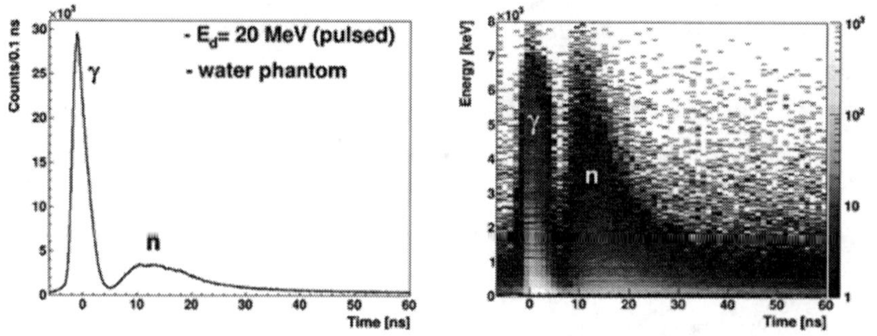

Figure 3: Time spectrum acquired with the $LaBr_3$:Ce scintillator of the Compton camera from a pulsed deuteron beam (E_d = 20 MeV) hitting a water phantom. Right panel: photon detection time (relative to RF trigger signal) vs. photon energy, left panel: projection on time axis.

with GEANT4 predict an angular resolution of $\theta \approx 2^o$ and an image reconstruction efficiency of 10^{-3} - 10^{-5} (E_γ = 2-6 MeV) for a camera arragment with 50 mm distance between the first scatterer detector and the photon source (representing a small animal irradiation scenario) [10, 11].

3 Characterization of camera components

The $LaBr_3$:Ce detector crystal (with reflective side surface coating) has been characterized in the laboratory with calibration sources. Using a collimated 2D irradiation of the front surface with 662 keV from ^{137}Cs, an average relative energy resolution of $\Delta E_\gamma/E_\gamma$ = 3.8% (FWHM) was obtained, independent of the irradiation position. Relative to a fast plastic (BC-418) reference detector, a time resolution of 270 ps was obtained for the coincidence of photons from ^{60}Co.
Using a 20 MeV pulsed deuteron beam (pulse distance 200 ns, pulse width 6 ns) from the Garching Tandem accelerator hitting a water phantom, the time spectrum (relative to the RF trigger signal) shown in Fig. 3 was acquired with the scintillator, showing the well-separated prompt photon and delayed neutron component (right: γ arrival time vs. photon energy, left: projection on time axis).

Using the clinical proton beam of the research area at the Universitäts Protonen Therapie Dresden (UPTD), water and polymethyl methacrylate (PMMA) phantoms, respectively, were bombarded with 100, 160 and 225 MeV protons (pulse distance 9.4 ns, according to the cyclotron frequency of 106 MHz). Fig. 4 displays the resulting time spectrum versus the photon energy for E_p = 225 MeV hitting the water phantom, indicating the pulse

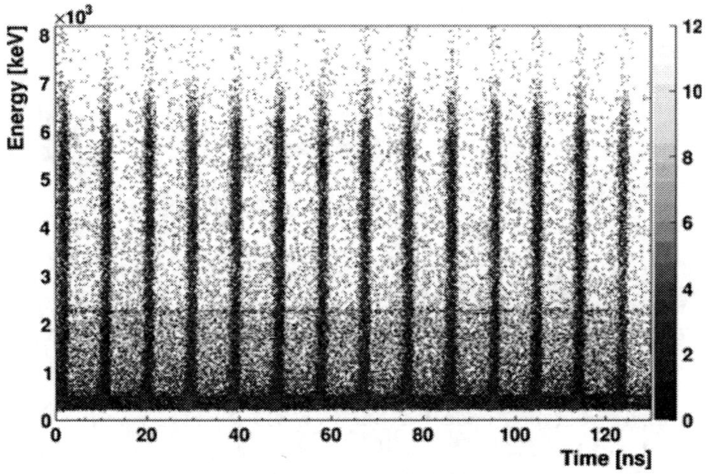

Figure 4: Timing spectrum for 225 MeV protons hitting a water phantom with a pulse period of 9.4 ns, as measured with the $LaBr_3$:Ce scintillator of the Compton camera.

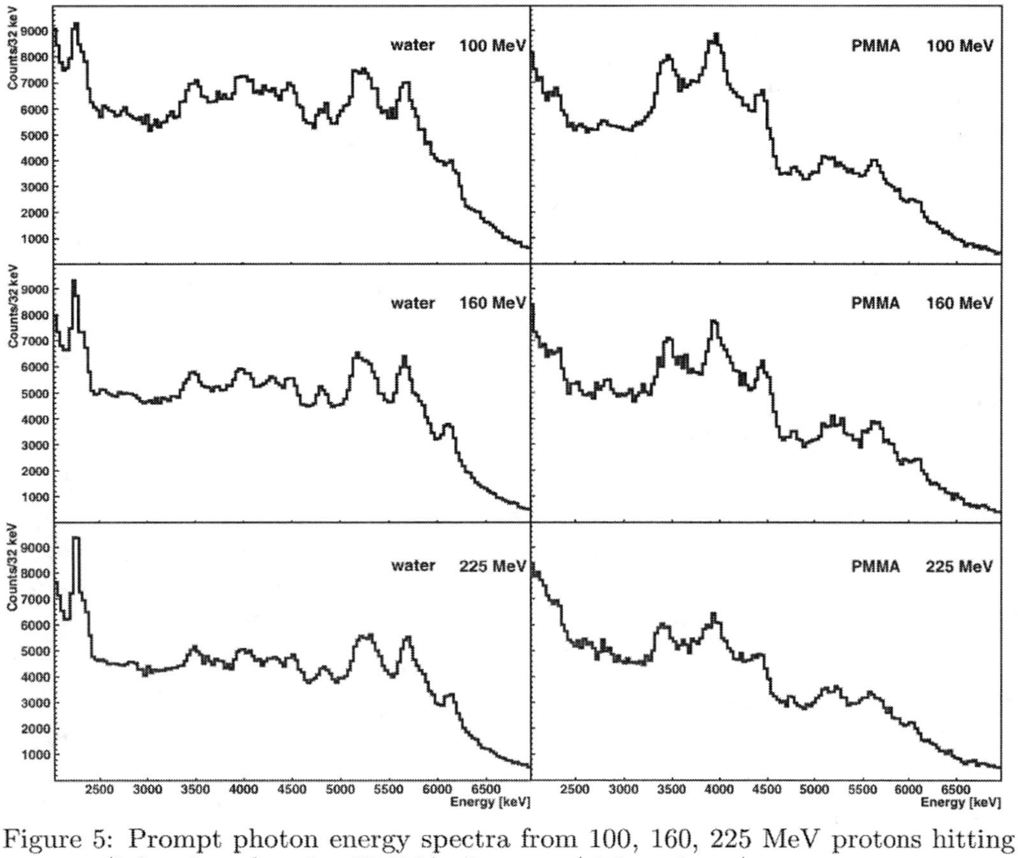

Figure 5: Prompt photon energy spectra from 100, 160, 225 MeV protons hitting a water (left column) and a PMMA phantom (right column).

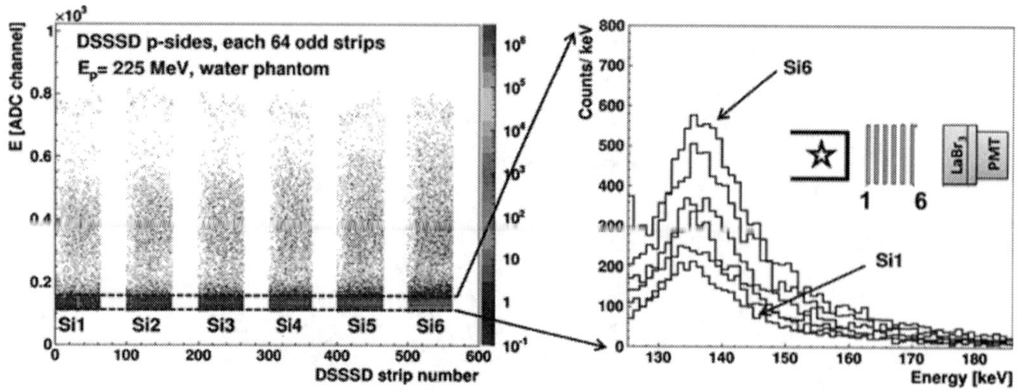

Figure 6: Energy loss spectra measured with the DSSSD array for 225 MeV protons and the water phantom. Exemplarily shown are the digitized energies from the odd 64 p-side strips of detector 1 (close to the phantom) to detector 6 (in front of the scintillator). The zoomed region displayed in the right part highlights the energy deposit by Compton scattered electrons successively accumulating in intensity across the scatterer array towards the scintillator.

structure. The horizontal line at 2.2 MeV represents the capture of a neutron, releasing the deuteron binding energy. Obviously, no strong neutron background component is visible.

The prompt photon energy spectra measured with the $LaBr_3$:Ce scintillator both for the water (left column) and PMMA phantom (right column) are shown in Fig. 5 for the three beam energies of 100, 160 and 225 MeV, respectively. The energy spectra from the water phantom show as most prominent feature the deexcitation from the first excited state of ^{16}O at 6.13 MeV plus the corresponding single- and double-escape peaks, while for the PMMA phantom the contribution from excited ^{12}C (4.44 MeV and its escape peaks) can be observed much stronger compared to the water phantom.

Data from part of the scatterer array (for 225 MeV protons hitting the water phantom) are shown in Fig. 6 (left part). Energy loss signals are displayed exemplarily for the 64 odd strips of the p sides of the 6 DSSSDs ‘Si 1’ to ‘Si 6’. The structure around 130 keV reflects the energy deposition of Compton scattered electrons as a zoomed projection of the spectra shown on the right side onto the energy axis. The intensity contributed by Compton electron successively accumulates along the path of primary photons through the DSSSD stack, therefore exhibiting increasing intensity with increasing proximity to the absorbing scintillator placed behind DSSSD ‘Si 6’.

A mandatory prerequisite for photon source reconstruction using the Compton camera is the determination of the primary interaction position

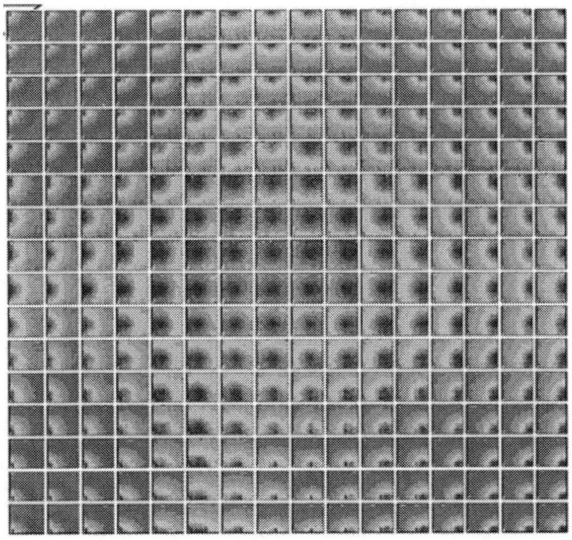

Figure 7: Grid scan of 16×16 irradiation positions of the monolithic $LaBr_3$:Ce detector, using a 1 mm collimated ^{137}Cs source and a grid step size of 3 mm in x and y direction. All 256 segments of the multi-anode PMT are individually read out. The resulting 2D (16×16) light amplitude distribution of each irradiation position clearly indicates the source position movement from the upper left corner to the lower right corner.

of photons hitting the absorbing scintillator. The 'k-Nearest Neighbour' algorithm, developed at TU Delft [13], is used to derive spatial information from the monolithic scintillator. It is based on a reference library of 2D light amplitude distributions, obtained from a 2D scan of the detector front surface with a narrowly collimated calibration source. In Fig. 7, we show the resulting 2D light amplitude map obtained from a grid scan with 16×16 irradiations positions (step size 3 mm in 2 dimensions), using a 1 mm collimated ^{137}Cs source (82 MBq). The correlation of the displayed 256 light amplitude distributions with their respective irradiation position, continuously moving from the upper left corner of the crystal to the lower right corner, is clearly visible. A similar scan with 102×102 irradiation positions (0.5 mm grid step size, 20 seconds measurement time per position) was performed to create the reference library needed for the k-NN algorithm. In order to derive the spatial resolution, irradiation position coordinates of k= 2000 closest matching reference light amplitude distributions, derived from a calculation of the Euclidean distance of the distribution of an 'unknown' event to every distribution from the reference set (with 400 events analyzed for each of the ca. 10^4 irradiation positions), were accumulated into a 2D

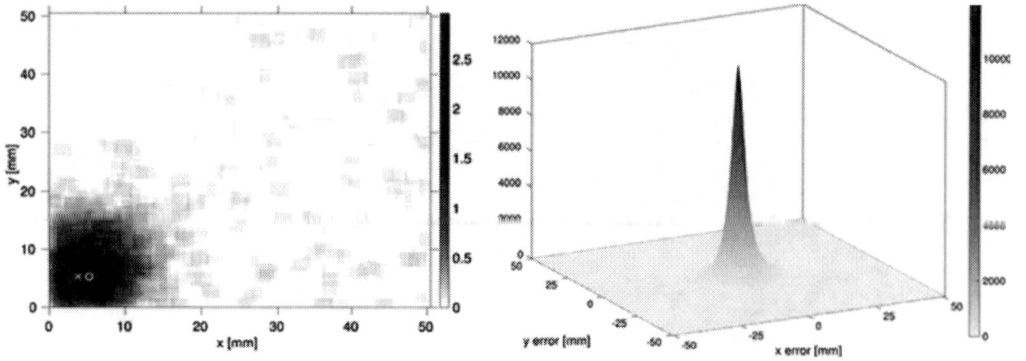

Figure 8: Left: Smoothed histogram of 2000 closest matching 2D light distributions for each 'unknown' irradiation event selected by the k-NN algorithm. The cross (open circle) denotes the actual (calculated) irradiation position. Right: Error histogram of the k-NN algorithm containing the distances between actual and calculated irradiation position for each of the events in the reference library. The corresponding FWHM indicates the spatial resolution.

histogram. After smoothing, the maximum of this distribution (displayed in the left part of Fig. 8) was identified with the calculated source position.

The spatial resolution can be quantified by performing this procedure for each of the reference events, treated as 'unknown'. The differences between actual and reconstructed source position are filled into an 'error histogram' (see right panel of Fig. 8) and the FWHM of this distribution reflects the corresponding spatial resolution. So far, a spatial resolution of 5.5(5) mm has been achieved, with room for further improvement, e.g., by using a tighter collimator (0.5 mm).

In conclusion, the detector components of a Compton camera, designed for prompt-γ ray based imaging of the particle beam range in hadron therapy and consisting of a stack of 6 doubled-sided Si strip detectors as scatterers (allowing also for Compton electron tracking) and a $LaBr_3$:Ce scintillator, have been successfully commissioned off- and online. All prerequisites have been achieved that enable to study the photon source reconstruction capabilities of the Compton camera prototype as the next step on the road towards an online ion beam range verification device.

This work is supported by the DFG Cluster of Excellence MAP (Munich-Centre for Advanced Photonics).

References

[1] C.H. Min et al., Applied Phys. Lett. **89**, 183517 (2006).
[2] E. Testa et al., Appl. Phys. Lett. **93**, 093506 (2008).
[3] F. Fiedler et al., Nucl. Sci. Symp. Conf. Rec. (NSS/MIC), 2011 IEEE.
[4] J.C. Polf et al., Phys. Med. Biol. **59**, 2325 (2014).
[5] F. Roellinghoff et al., Nucl. Instr. Meth. **A648**(2011) S20.
[6] X. Lojacono et al., Nucl. Sci. Symp. Conf. Rec., 2011 IEEE **1**, 3518.
[7] T. Kormoll et al., Nucl. Instr. Meth. **A626** (2011) 114.
[8] M.-H. Richard et al., IEEE Trans. Nucl. Sci. **58** (2011).
[9] P.G. Thirolf et al., EPJ Web of Conferences 66 (2014) 11036.
[10] C. Lang et al., Journal of Instrumentation 9 (2014) P01008.
[11] C. Lang, PhD thesis, LMU Munich (2015).
[12] W. Beusch et al., Internal Note, CERN-ECP/94-17 (1994).
[13] H.T. van Dam et al., IEEE Trans. Nucl. Sci. 58 (2011) 2139.

EPJ Web of Conferences **117**, 05006 (2016) DOI: 10.1051/epjconf/201611705006
NN2015

Nuclear contribution into single-event upset in 3D on-board electronics at moderate energy cosmic proton impact

N. G. Chechenin, T. V. Chuvilskaya and A. A. Shirokova

Skobeltsyn Institute of Nuclear Physics, Moscow State University, Moscow, 119991 Russia

Abstract

In continuation and development of our previous works where nuclear reactions of moderate energy (10 – 400 MeV) protons with Si, Al and W have been investigated, the results of reactions with Cu are reported in this paper. Cu is a most important component in composition of materials in contact pads and pathways of modern and perspective ultra large-scale integration circuitry, especially in 3D topology.

It is impossible to imagine a modern spacecraft without electronic integrated schemes with connecting and contact paths and areas, inter-layer conducting connections, and other metallic components. Such interior components normally have a high concentration of Al, Fe, Ni, Cu, Mo, Ta, W, Pt, Au, and other chemical elements.

The kinematical effect of the energy transfer in a head-on collision from incident proton to a target nucleus is weaker for heavier target nucleus, but the resulting recoil energy can be sufficient to cause a dangerous upset of the electronics in the case of a high energy of incident proton. For example, 1 GeV proton encountering a head-on collision with a gold nucleus loses only about 2% of its energy but transfers an energy of about 20 MeV to the target gold nucleus. Head-on collisions are quite rare, elastic collisions occur predominantly at nonzero impact parameters, with the result that the kinetic energies of recoil nuclei form a spectrum associated with a specific angular distribution of recoil nuclei [1, 2].

In addition to processes of elastic collisions between protons and target nuclei, it is necessary to consider the contribution of nuclear reactions, which result in the formation of the mass spectrum of recoil nuclei arising as reaction products, which extends from the atomic mass of heavy target nuclei down to helium and hydrogen. Each nuclear-reaction product has its own kinetic-energy spectrum and its own angular distribution. Thus, the picture that arises during the propagation of high-energy cosmic ray particles is quite complicated, so that, in order to estimate quantitatively radiative effects in the onboard electronics, it is necessary to invoke present day nuclear data, as well as modern models and code packages implementing them, which make it possible to compensate for missing nuclear data.

Unfortunately, the presence of heavy elements is disregarded in the majority of the calculations, and this of course impairs the reliability of the respective forecasts. The results of our present calculations make it possible to fill this gap.

Special features of silicon fragmentation induced by high-energy cosmic-ray protons were explored in previous studies of our group [3–6]. In [5], it was shown that the results of the calculations performed on the basis of the reaction model according to the EMPIRE code are sensitive to the choice of values for the parameters of the optical-model potential. In the present article, the results obtained by calculating for the ^{28}Si + p, ^{27}Al + p and ^{183}W + p nuclear reactions, the cross-sections for elastic and inelastic scattering and the charge, mass, and energy distributions of heavy reaction products according to the TALYS [7] codes are supplemented with the results of the calculations for the same features of the ^{63}Cu + p reaction.

The idea to create computer codes that would provide a simultaneous description of many nuclear reaction channels as precise as the most detailed description of only one or several reaction channels is not new. Such codes known to date include the GNASH [8], ALICE [9], STAPRE [10] and EMPIRE [11] ones.

The TALYS code [7] is a modern software facility for implementing the modern nuclear-reaction models. It includes the coupled-channel method featuring a number of software procedures. We will now highlight the most important special features of the TALYS code package. First of all, this is a tool for precise calculations of nuclear-reaction mechanisms such as direct, pre-equilibrium and fission ones. The high-precision code makes it possible to construct a reliable description of various reaction mechanisms over a broad region of projectile-particle energies and mass numbers (0.001-200 MeV and 12< A <239, respectively).

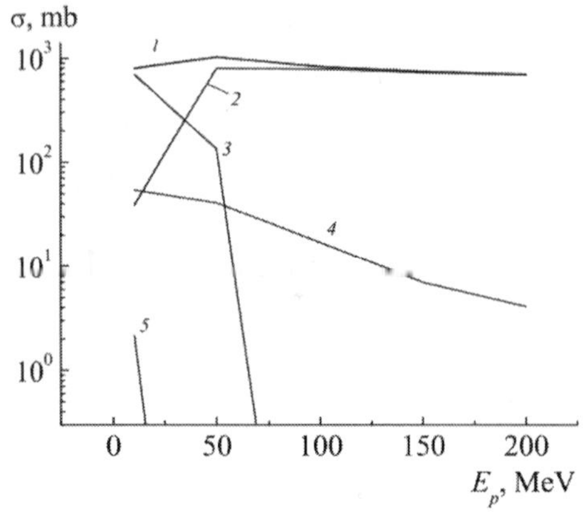

Figure 1: Cross-sections for nuclear processes, $\sigma(E_p)$, in the interaction of protons with a ^{63}Cu nucleus. The curves represent the contributions of (1) inelastic nuclear processes, (2) pre-equilibrium processes, (3) processes leading to compound-nucleus formation, (4) direct nuclear reactions and (5) elastic nuclear processes.

Elastic scattering of protons from a target nucleus with a transfer of a recoil energy to this nucleus is not the most probable mechanism of formation of recoil nuclei. Fig. 1 shows the calculated contributions of the dominant nuclear mechanisms, $\sigma(E_p)$, in the p+^{63}Cu reaction at proton energies between 10 and 200 MeV. The flux of space protons is strongest in this energy region [12]. The TALYS code package was used in this calculation. Without going into details of the formalism used, we note only that elastic proton scattering from ^{63}Cu nucleus is severely suppressed by competing inelastic nuclear processes (curve 1). The degree of suppression is such that we can disregard elastic scattering. The probability of direct nuclear reactions is somewhat higher (curve 4). They include the direct knockout and pickup of nucleons from the target nucleus without excitation of the residual nucleus. Fig. 1 shows that, in the region of proton energies below 100 MeV, a significant contribution comes from reactions involving the production of a compound nucleus (curve 3); at energies below 50 MeV, these reactions are dominant. As the proton energy becomes higher, the compound-nucleus mechanism gives way to the competing mechanism of pre-equilibrium processes (curve 2).

It is necessary to consider that the emission of a light particle cascade proceeds within a very short time interval, so that the time of the whole process is less than 10^{-16} s. Within this time, the nucleus involved travels a short distance (smaller than one interatomic spacing in a target material) without causing any shift and ionization of neighboring atoms. We note that

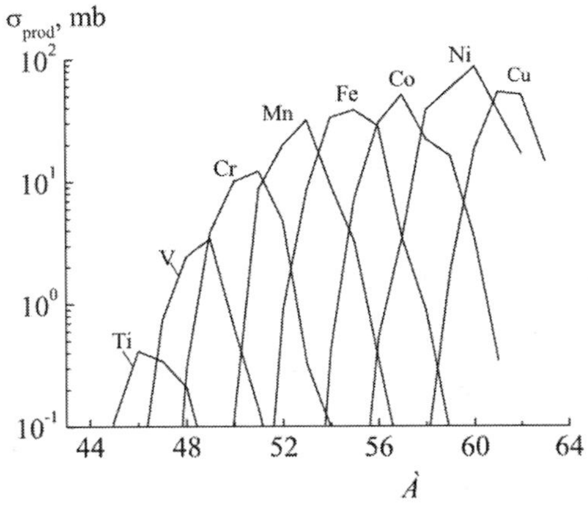

Figure 2: Cross-section for the production of residual nuclei, $\sigma_{prod}(A)$ (mass spectrum), in the interaction of protons with a ^{63}Cu nucleus at the proton energy of $E_p = 150$ MeV.

the production of electron-hole pairs upon medium-atom ionization induced by nuclear fragments is of many orders of magnitude more vigorous than that in the case of primary protons.

The excitation energy of the nucleus involved in the nuclear reaction may be exhausted either by the emission of one particle or by the emission of several particles. As a result, a collision between a proton and a target nucleus leads to the formation of a mass spectrum of residual nuclei. The cross-sections calculated for the production of residual nuclei, $\sigma_{prod}(A)$, on the basis of modern nuclear-physics models used in the TALYS code package are given in Fig. 2 for isotopes of eight residual nuclei. It can be seen that the product of the reaction ^{63}Cu (p, 2p2n) ^{60}Ni is formed with the grate probability. In this reaction, the incident proton leaves part of its energy in the excited nucleus of primary copper. The excitation energy is sufficient for the emission of several protons and neutrons. Isotope production accompanied by the emission of one to ten neutrons has the largest cross-sections ranging from several tens of μb to above 100 mb. Ni isotopes of mass number A between 56 and 64 are produced with the largest cross-section. Isotopes of Cu are produced with a moderately small cross-section below 50 mb in (p,pxn) reaction i.e. via neutron emission and without additional proton emission. As for Co isotopes they are produced in the reaction (p,3pxn) with emission of tree protons. Isotopes of Fe, which is the element that precedes Co in the periodic table of elements, are produced with a rather large cross-section of 40 mb or more in the reaction involving the emission of four protons and

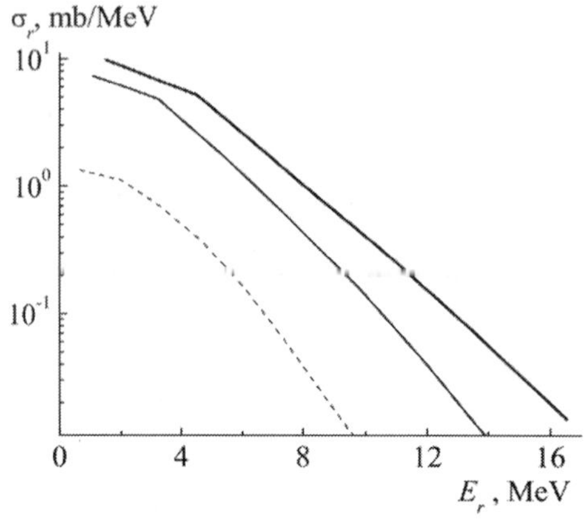

Figure 3: Recoil-energy spectra $\sigma_r(E_r)$ of the isotope ^{53}Mn produced in the ^{63}Cu(p, 5p6n)^{53}Mn reaction versus the product energy at the proton-energy values of E_p = 100 (dashed curves), 150 (thin solid curves), and 240 MeV (thick solid curves).

several neutrons. The lower isotope Z-value, the larger number of protons is to be emitted and, hence, the lower cross-section, because of lower penetrability of the Coulomb barrier with de-excitation of the emitting nucleus. The neutron emission becomes dominating in the pre-equilibrium processes. Thus, V and Ti isotopes are produced with sharply smaller cross-sections.

The kinetic energy distribution of residual nuclei is also of great importance for possible radiation effects in the onboard spacecraft electronics. In Fig. 3 the energy spectra $\sigma_r(E_r)$ are shown of ^{53}Mn residual nucleus produced in the p+^{63}Cu nuclear reaction for the incident proton energy E_p = 100 (dashed curve), 150 (thin solid curve), and 240 MeV (thick solid curve). This figure shows that the region of kinetic energies of residual nucleus depends both on the incident proton energy and on the mass of the target nucleus. The maximum kinetic energy is observed for the residual nucleus ^{53}Mn at E_p = 240 MeV. Our calculations have revealed that the recoil energy of residual nuclei may be as high as 1 to 3 MeV, depending on the projectile-proton energy.

The results obtained in the present work and outlined partially in this paper seem to be of significance in the estimates of reliability of spacecraft apparatus.

References

[1] Pickel J. C., *IEEE Trans. Nucl. Sci.*, **43** (1996) 483.

[2] Chumakov A. I., *Effect of space radiation on integral circuitry* (in Russian), (Radio I Svyaz, Moscow, 2004).

[3] Chuvilskaya T. V., Shirokova A. A., Kadmenskii A. G., Chechenin N. G., *Phys. Atom. Nucl.*, **71** (2008) 1293.

[4] Chechenin N. G., Chuvilskaya T. V., Shirokova A. A., Kadmenskii A. G., *Phys. Atom. Nucl.*, **72** (2009) 182.

[5] Chechenin N. G., Chuvilskaya T. V., Shirokova A. A., Kadmenskii A. G., *Bull. Russ. Acad. Sci. Phys.*, **74** (2010) 1595.

[6] Chechenin N.G., Chuvilskaya T. V., Shirokova A. A., Kadmenskii A. G., *Phys. Atom. Nucl.*, **78** (2015) 159.

[7] Koning A. J., Hilaire S. et al., *TALYS-1.2: A nuclear reaction program* (User manual, Netherlands, NRG, 2009).

[8] Young P. G., Arthur E. D. and Chadwick M. B., in *Proceedings of the Workshop on Computation and Analysis of Nuclear Data Relevant to Nuclear Energy and Safety*, edited by Mehta M. K. and Schmidt J. J., Trieste, Italy, 1993 p. 662.

[9] Blann M., in *Proceedings of the Workshop on Computation and Analysis of Nuclear Data Relevant to Nuclear Energy and Safety*, edited by Mehta M. K. and Schmidt J. J., Trieste, Italy, 1993 p. 622.

[10] Uhl M. and Strohmaier B., IRK report No. 76/01 Vienna, 1976.

[11] Herman M., in *Proceedings of the Workshop on Nuclear Reaction Data and Nuclear Reactors: Physics, Design and Safety*, edited by Paver N., Herman M. and Gandini A., Triest, Italy, 2001 p. 137.

[12] Chechenin N. G., Kadmenskii A. G., Motavekh X. A., Panasyuk M. I., *Journal of Surface Investigation. X-ray, Synchrotron and Neutron Techniques*, **4** (2012) 17.

EPJ Web of Conferences **117**, 05007 (2016) DOI: 10.1051/epjconf/201611705007
NN2015

Measurement of secondary particle production induced by particle therapy ion beams impinging on a PMMA target

M. Toppi[1], G. Battistoni[2], F. Bellini[3,4], F. Collamati[3,4], E. De Lucia[1], M. Durante[5], R. Faccini[3,4], P.M. Fralliciardi[6], M. Marafini[3,4], I. Mattei[2], S. Morganti[3], S. Muraro[2], R. Paramatti[3], V. Patera[3,6,7], D. Pinci[3], L. Piersanti[1,7], A. Rucinski[3,7], A. Russomando[3,4,8], A. Sarti[1,6,7], A. Sciubba[3,6,7], M. Senzacqua[3,7], E. Solfaroli Camillocci[8], G. Traini[3,4] and C. Voena[3,4]

[1] INFN - Laboratori Nazionali di Frascati, Frascati (RM), Italy
[2] INFN, Sezione di Milano, Milano, Italy
[3] INFN, Sezione di Roma, Roma, Italy
[4] Dipartimento di Fisica, Sapienza Università di Roma, Roma, Italy
[5] GSI, Darmstadt, Germany
[6] Museo Storico della Fisica, Centro "E. Fermi", Roma, Italy
[7] Dipartimento di Scienze di Base e Applicate per Ingegneria, Sapienza Università di Roma, Roma, Italy
[8] Center for Life Nano Science@Sapienza, IIT, Roma, Italy

Abstract

Particle therapy is a technique that uses accelerated charged ions for cancer treatment and combines a high irradiation precision with a high biological effectiveness in killing tumor cells [1]. Informations about the secondary particles emitted in the interaction of an ion beam with the patient during a treatment can be of great interest in order to monitor the dose deposition. For this purpose an experiment at the HIT (Heidelberg Ion-Beam Therapy Center) beam facility has been performed in order to measure fluxes and emission profiles of secondary particles produced in the interaction of therapeutic beams with a PMMA target. In this contribution some preliminary results about the emission profiles and the energy spectra of the detected secondaries will be presented.

1 Introduction

In particle therapy accelerated light ions (Z $\leq$ 10), expecially p and ^{12}C, are used for the treatment of tumors [1]. The greatest advantage of the particle therapy with respect to the conventional therapy with photons (radiotherapy) is a highly localized dose deposition, due to the typical profile of the dose released by a charged particle ending with the Bragg Peak, and hence a higher biological effectiveness in killing tumor cells. For such reasons this technique can be used for the treatment of highly radiation resistent tumors, sparing surrounding organs at risk (OAR) [2].

Due to the improved therapy spatial selectiveness, a novel monitoring technique, capable of providing a high precision "in-treatment" feedback on the dose release position, is required in order to act on the beam control and to prevent damages to OARs. Since the primary beam is fully stopped inside the patient body, it is necessary to exploit the secondary particles emitted in the nuclear interaction processes between the ion beam and the target nuclei, in order to reconstruct "online" the dose release [3–6]. Beyond p and ^{12}C there is a growing interest in other ion beams like ^{4}He and ^{16}O: an improved characterization of the secondaries production for these beams is becoming crucial for their deployment in treatment centers. The precise knowledge of the secondary particles production, and in particular of their angular and energy distributions, is a key ingredient also for Treatment Planning Software (TPS) development, allowing to improve the MC description of the ion beams interaction with the patient body [9].

In this contribution we report about the data collected at the HIT beam facility (Heidelberg Ion-Beam Therapy Center), measuring the secondary particles produced in the interaction of ^{4}He, ^{12}C and ^{16}O beams, with therapeutical energies, against a beam stopping PMMA target. Three type of secondaries have been studied: charged fragments (from both projectile and target fragmentation), PET γ (from β^{+} emitters fragments) and prompt γ (from nuclear de-excitation), aiming to correlate the emission point with the dose release and to compute the production fluxes for each type of secondaries. In this contribution we present a brief description of the HIT experiment apparatus and the preliminary results about secondaries emission profiles and energy spectra.

2 The HIT experimental setup

The HIT experimental apparatus (Fig. 1, left) is composed of several subdetectors, each optimized for the detection of one or more type of secondaries

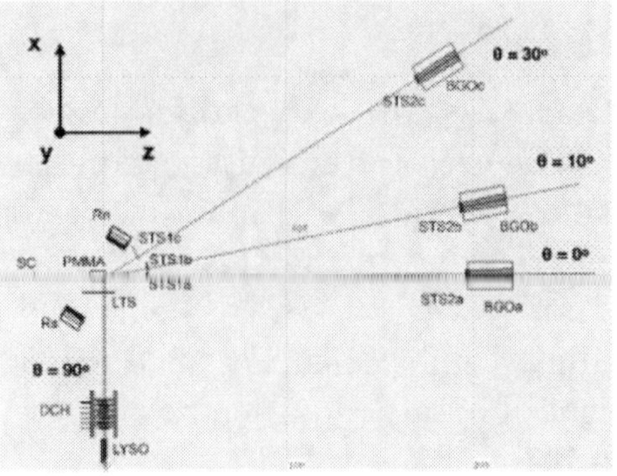

Beam	Energy (MeV/u)	PMMA (cm)	N_b (x 10^8) θ_{LYSO}=90°	N_b (x 10^8) θ_{LYSO}=60°
^{4}He	102	7.65	missing	14.4
	125	10	7.7	12.4
	145	12.65	7.7	11.0
^{12}C	120	12.65	13.2	missing
	160	12.65	4.4	
	180	12.65	2.9	
	220	12.65	2.4	
^{16}O	210	7.65	4.7	4.1
	260	10	3.4	3.3
	300	12.65	2.4	2.4

Figure 1: Left: HIT experimental setup (the scale along the three axis is in cm). Right: Table summarizing the different set up configurations, for each ion beam and for each beam energy. The number of primaries N_b, provided by the start counter detector (SC), is different for the two LYSO position ($\theta_{LYSO} = 60°, 90°$)

produced by the interaction of the ion beam with the PMMA target. For each ion beam different beam energies and geometrical configurations have been studied, as summarized in the table of Fig. 1 (right). The PMMA thickness has been changed, depending on the beam and on its energy, in order to keep the Bragg Peak always inside the target. Before interacting with the target the ion beam goes through the start counter (SC), a thin scintillator that provides the number of primary ions impinging on the PMMA target and that provides the triggers of the experiment in combination with other subdetectors.

The forward emitted charged fragments are detected by a system of three BGO crystals (truncated pyramids 24 cm thick and with square bases 4.7×4.7 cm^2 and 6.0×6.0 cm^2 respectively, read by EMI 98117B PMTs) and three pairs of thin plastic scintillators (STS1 - 0.2×4.0×4.0 cm^3 and STS2 1.0×4.7×4.7 cm^3, read by H10580 PMTs) providing energy and ToF measurements. Each BGO and STS pair combination is an independent coaxial system positioned at different angular position ($\theta \sim 0°$, 10°, 30°, with θ emission angle respect to the beam direction z, as shown in Fig 1).

The charged fragments emitted at large angle, while having a reduced yield, can still be of interest for dose monitoring purpose because of their improved spatial resolution σ_z on the fragment emission point ($\sigma_z \sim 1/sin\theta$, see [6]). The study of such large angle events is performed using a coaxial system composed of a thin plastic scintillator (LTS - 0.2×5.0×17.0 cm^3, read by a H10580 PMT), a drift chamber (DCH) and a LYSO crystals detector, already described in [6]. Data were acquired with this system placed at

$\theta \sim 60°$ and 90° in order to measure the energy, the ToF and the emission angle θ of the charged fragments. The same LYSO scintillator has also been used to measure the prompt γ energy spectra and the integrated production fluxes. Since no collimator was used during the data acquisition, the profile of the emission position of the prompt γ could not be measured.

In order to study the β^+ emission, gaining access both to the position and to the flux, a device able to measure the energy and the ToF of the two back-to-back 511 keV γ in time coincidence was set up. Two pairs of PET heads, shown as Rn and Rs in Fig. 1, composed of a matrix of 23×23 LYSO crystals each have been used.

A MC simulation of the full setup has been developed with FLUKA [7,8] for efficiencies and angular acceptances studies.

3 Preliminary results

Fig. 2 (left) shows the measured energy versus the ToF distribution for detected fragments ($\theta = 10°$) originating from the interactions of an ^{4}He beam of 125 MeV/u energy with the PMMA. The three bands, that are clearly visible, are related to the hydrogen ions (^{1}H, ^{2}H and ^{3}H). The heavier fragments are not visible due to their smaller production cross section and their lower range. A similar behavior is observed for all the exploited ion beams.

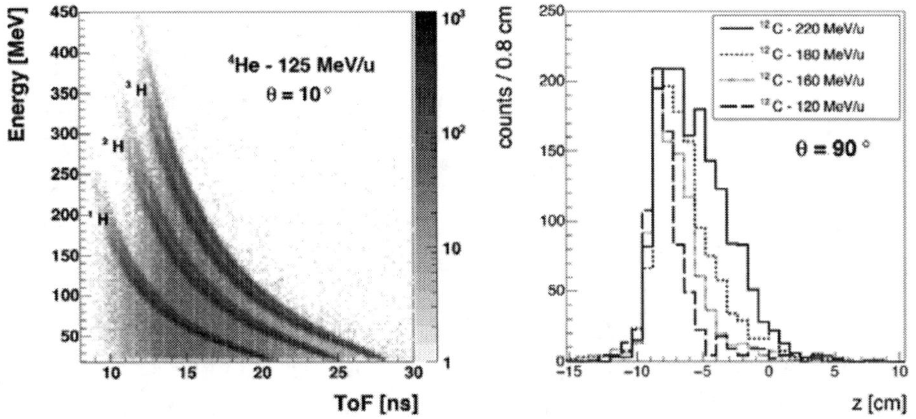

Figure 2: Left: Energy measured by the BGO vs the ToF measured by the STS of the setup at 10° for the charged fragments emitted in the case of the 125 MeV/u^{4}He beam. The bands corresponding at the ^{1}H, ^{2}H and ^{3}H fragments are clearly visible. Right: Emission profile of the charged fragments in the case of the ^{12}C beam at different energies with LYSO detector at $\theta = 90$ °.

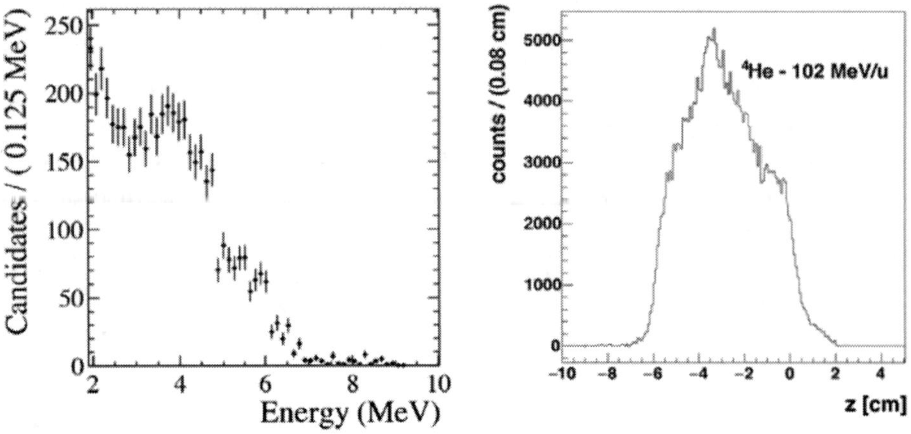

Figure 3: Left: prompt γ energy spectra in the case of the ^{12}C beam. Right: β^+ emission profile in the case of the ^{4}He beam at 102 MeV/u.

The fragments yield at large angles is lower than the yield at small angle as expected from fragmentation cross sections measurements [10]. However a non negligible production of hydrogen ions at large angles is observed for all beam types. The large angle hydrogen ions emission shape can be correlated to the beam entrance window and the Bragg Peak position exploiting the LYSO detector, as already measured with ^{12}C and discussed in [6]. Fig. 2 (right) shows the hydrogen ions reconstructed emission positions, along the z axis, measured at $\theta = 90°$, for carbon beam of different energies.

The prompt γ produced by the nuclear de-excitation of ions and target nuclei along the beam path have been studied with the LYSO detector (measuring Energy and ToF), the SC detector (providing the trigger) and the LTS (used to veto the charged fragments). The energy spectra of the emitted γ have been measured, for all the different beam types and energies, using the LYSO detector to identify the prompt γ using the Energy vs ToF informations as described in [11]. In Fig. 3 (left) the measured energy spectra of the prompt γ emitted in the interaction of the ^{12}C beam with the PMMA target in the case of the LYSO detector at $\theta = 90$ ° is shown. Similar spectra have been obtained for all the beams and angular configurations ($\theta = 60°$, $90°$). These raw energy spectra have not yet been corrected for efficiencies or resolution effects in order to obtain the production energy spectra.

The PET heads have been used to study the ^{4}He beam only. Due to the huge background coming from prompt γ and neutrons during the beam spill, the PET γ events have been selected in the time window with the beam off between two consequent spills (offspill) requiring the double time coincidence of the Rn and Rs heads. In Fig. 3 (right) the β^+ emitters profile has been reconstructed for the ^{4}He beam at 102 MeV/u. The raw emission profile has yet to be corrected for the detector efficiencies and for the angular acceptance.

4 Conclusions

In this contribution the preliminary results for the secondaries emission spectra produced in the interaction of different ion beams (^{4}He, ^{12}C and ^{16}O) with a thick PMMA target have been presented.

References

[1] M. Durante and J. S. Loeffler, *Nat. Rev. Clin. Oncol.*, **7**, 37–43, (2010).

[2] D. Schardt *et al.*, *Rev.Mod.Phys*, **82**, No. 1, 383–425, (2010).

[3] E. Testa *et al.*, *Nucl. Instrum. Meth. B*, **267**, 993, (2009).

[4] K. Parodi *et al.*, *Phys. Med. Biol.*, **47**, 21, (2002)

[5] C. Agodi *et al.*, *Phys. Med. Biol.*, **57**, 5667, (2012)

[6] L. Piersanti *et al.*, *Phys. Med. Biol.*, **59**, 1857–1872, (2014).

[7] A. Ferrari *et al.*, CERN 2005-10, INFN/TC_05/11, (2005).

[8] T.T. Böhlen *et al.*, *Nuclear Data Sheets*, **120** (2014) pp. 211–214.

[9] T.T. Böhlen *et al.*, *Phys. Med. Biol.*, **55**, 5833–47, (2010).

[10] J. Dudouet *et al.*, *Physical Review C*, **88**, 024606, (2013).

[11] C. Agodi *et al.*, *Journal of Instrumentation*, vol. **7**, P03001, (2012).

EPJ Web of Conferences **117**, 05008 (2016) DOI: 10.1051/epjconf/201611705008
NN2015

Study of nuclear reactions in laser plasmas at future ELI-NP facility

G. Lanzalone[1,2], C. Altana[1,3], A. Anzalone[1], F. Cappuzzello[1,3], M. Cavallaro[1], L. A. Gizzi[5,6], L. Labate[5,6], L. Lamia[1,3], D. Mascali[1], A. Muoio[1,4], F. Negoita[7], F. Odorici[8], H. Petrascu[7], A. Trifirò[1,9], M. Trimarchi[1,9] and S. Tudisco[1]

[1]INFN - Laboratori Nazionali del Sud, Catania, Italy
[2]Univ. degli Studi di Enna "Kore", Italy
[3]Dip. di Fisica e Astronomia, Univ. di Catania, Italy
[4]Dip. di Fisica e Scienze della Terra, Univ. di Messina, Italy
[5] Istituto Nazionale di Ottica - CNR Pisa, Italy
[6]INFN, Sezione di Pisa, Pisa, Italy
[7]IFIN - Bucharest Măgurele, Romania
[8]INFN, Sezione di Bologna, Bologna, Italy
[9]INFN, Sezione di Catania, Catania, Italy

Abstract

In this contribution we will present the future activities that our collaboration will carry out at ELI-NP(Extreme Light Infrastructure Nuclear Physics), the new multi peta-watt Laser facility, currently under construction at Bucharest (Romania). The activities concerns the study of nuclear reactions in laser plasmas. In this framework we proposed the construction of a new, general-purpose experimental set-up able to detect and identify neutrons and charged particles.

1 Introduction

Plasma is the most common form of ordinary matter known in the universe. One of the crucial aspects is the possible appearance of new phenomena in nuclear physics (nuclear reactions, nuclear structure or nuclear properties) in plasmas. The discovery, for example, that, in the nuclear reactions, at very low energies the electrons of the target atoms partially screen the Coulomb barrier between the projectile and the target nuclei, with a resulting enhancement of the fusion cross section, opens new and important perspective on astrophysics and cosmology. Such an effect, called "electron screening", affects our knowledge of nuclear processes in stellar plasmas. Getting information about this process wold be extremely important. A further important issue concerns the influence of nuclear excited states on fusion cross-section. Nuclear reactions, relevant for astrophysical studies have been performed in the past at other laboratories, with both target and projectile in their ground state. However, in high temperatures plasmas ($> 10^8 K$), an important role can also be played by the excited states, as already deeply discussed in the pioneering theoretical work of Bahcall and Fowler [1]. In that case, the authors studied the influence of low lying excited ^{19}F states on the final ^{19}F(p, alpha) reaction, predicting an increase of reaction rate of a factor of about 3 at temperatures of about 10^9 K. Thus, determining the appropriate experimental conditions that allow the role of the excited states in the stellar environment to be evaluated could strongly contribute to the development of nuclear astrophysics. The direct measurement of reaction rates in plasma offers this chance. In addition, other new topics can be conveniently explored by studying nuclear reactions in plasmas, such as, for instance, three body fusion reactions (as those predicted by Hoyle [2]), lifetime changes of unstable elements [3] or nuclear and atomic levels [4] in different plasma environments and fundamental physics issues like non-extensive statistical thermodynamics [5]. This last topic can be investigated in order to validate/confute the general assumption of local thermal equilibrium that is traditionally adopted for plasmas [6].

2 Physics case

In 2013 our collaboration submitted a scientific project, which aims to perform nuclear reactions studies in Laser plasmas at the future ELI-NP facility. Among the various nuclear reactions which have attracted relevant attention also for astrophysical or cosmological reasons, the collaboration selected as the first case study the ^{13}C(^{4}He,n)^{16}O and ^{7}Li(d,n)^{4}He-^{4}He reactions; the first for its relevance in the frame of stellar nucleosynthesis, while the second for the role played in Big Bang primordial nucleosynthesis. We aim to produce plasmas containing mixtures of ^{13}C + ^{4}He and ^{7}Li + d in order to investigate inner-plasma thermo-nuclear reactions. In relation to the studies of weakly bound nuclear states, the collaboration selected also, as a first case of study, the $^{11}B(^{3}He, d)^{12}C^{*}$ reaction. Nucleonic matter displays a quantum-liquid structure, but in some cases finite nuclei behave like molecules consisting of clusters of protons and neutrons. Clustering is a recurrent feature in light nuclei, from beryllium to nickel. Cluster structures are typically observed as excited states close to the corresponding decay threshold; the origin of this phenomenon lies in the effective nuclear interaction, but the detailed mechanism of clustering in nuclei has not yet been fully understood. It is extremely interesting to study such aspects in a laser produced plasma.

3 Experimental set-up

The project will take advantage from the excellent and unique performance of the ELI-NP facility to create two colliding plasmas using two separate laser beams. The use of colliding plasma plumes suitable for nuclear physics studies was proposed few years ago [7] and recently adopted to achieve such goal [8]. The idea is the following: a first laser pulse impinging on a ^{13}C, ^{7}Li or ^{11}B solid thin target (few micro-meters) produces, through the TNSA (Target Normal Sheath Acceleration) [9] acceleration scheme, boron, carbon or lithium plasma. In view of this, an extensive experimental investigation programme is in progress, aiming at the optimization of targets and interaction configuration [10,11] The rapidly streaming plasma impacts on a secondary plasma, prepared through the interaction of a second laser pulse on a gas jet target (made by ^{4}He, D_2 or ^{3}He).
The proposed activity requires also the construction of a highly segmented detection system for neutrons and charged particles. The segmentation is required for the reconstruction of the reactions kinematic. The ideal neutron detection module for these studies must exhibit: high efficiency, good discrimination of gammas from neutrons, good timing performance for ToF(Time of Flight) neutron velocity reconstruction. In addition, it must be able to work in hard environmental conditions, like the ones established in the laser-matter interaction area. All these aspects may be met by configuration based on 50x50x50 mm^3 PPO-Plastic scintillator plus a SiPM [12] read-out and a totally digital acquisition of the multi-hit signals. Concerning the charged particle detector, an R&D activity was funded by INFN on Silicon Carbide. In this frame work we aim to realize a wall of detectors. The SiC detectors have been proven recently to have excellent properties [13-14]:

high energy and time resolution, resistance to radiation, low sensitivity to visible light, etc. The use of segmented SiC detectors would be very helpful for the study of nuclear reactions where only the position and energy measurement of light charged particles can give access to the desired information. In conclusion, we present in this contribution the research project that our collaboration will conduct in the next years @ELI-NP. More details about these activities are reported in reference [15].

References

[1] N. A. Bahcall and W.A. Fowler, Astrophysical Journal 157 (1969) 645; doi: 10.1086/150103
[2] F. Hoyle, Monthly Notices of the Royal Astronomical Society 106 (5): 343-383 (1946); doi: 10.1093/mnras/106.5.343
[3] B. N. Limata, Eur. Phys. J. A 27, s01, 193-196 (2006); doi: 10.1140/epja/i2006-08-029-2
[4] F. Hannachi et al., 2007 Plasma Phys. Control. Fusion 49 B79; doi: 10.1088/0741-3335/49/12B/S06
[5] C. Tsallis, Introduction to Nonextensive Statistical Mechanics, Springer 2009; ISBN 978-0-387-85359-8
[6] A. Semerok et al., Applied Surface Science, vol.138-139,311-314(1999); doi: 10.1016/S0169-4332(98)00411-5
[7] D. Mascali et al., Rad. Eff. and Def. in Sol. 165, Issue 6-10 (2010) 730-736; doi:10.1080/10420151003729847
[8] C. Labaune et al., Nature Communications 4, art.2506 (2013) ; doi: 10.1038/ncomms3506
[9] R. A. Snavely et al., Phys. Rev. Lett. 85, 2945 (2000); doi: 10.1103/PhysRevLett.85.2945
[10] S. Tudisco et al. Rev. Sci. Instrum. 87, 02A909 (2015); doi: 10.1063/1.493469.
[11] L. A. Gizzi et al., Phys. Review ST Acc. Beams, 14, 011301 (2011); doi: 10.1103/PhysRevSTAB.14.011301
[12] S. Privitera et al., Sensors 2008, 8(8) 4636-4655; doi: 10.3390/s8084636
[13] M. Moll, NIM in Physics Research A 511 (2003) 97-105; doi: 10.1016/S0168-9002(03)01772-8
[14] K. Sandeep et al., NIM A 728 (2013) 97-101; doi: 10.1016/j.nima.2013.06.076
[15] TDR@ELI-NP in press on Romanian Reports in Physics; ISSN 1221-1451 43 822

EPJ Web of Conferences **117**, 05009 (2016) DOI: 10.1051/epjconf/201611705009
NN2015

Determination of integral cross sections of ^{3}H in Al foils monitors irradiated by protons with energies ranging from 40 to 2600 MeV

Yu.E. Titarenko[1], V.F. Batyaev[1], M.V. Chauzova[1], I.A. Kashirin[1], S.V. Malinovskiy[1], K.V. Pavlov[1], V.I. Rogov[1], A.Yu. Titarenko[1], V.M. Zhivun[1,2], S.G. Mashnik[3] and A.Yu. Stankovskiy[4]

[1]NRC Kurchatov Institute - Institute for Theoretical and Experimental Physics, Moscow, Russia
[2]NRNU MEPhI (Moscow Engineering Physics Institute), Moscow, Russia
[3]Los Alamos National Laboratory, NM 87545, USA
[4]SCK·CEN, Boeretang, Mol, Belgium

Abstract

The results of ^3H production in Al foil monitors ($\sim$ 59 mg/cm^2 thickness) are presented. These foils have been irradiated in 15×15 mm polyethylene bags of $\sim$ 14 mg/cm^2 thickness together with foils of Cr ($\sim$ 395 mg/cm^2 thickness) and ^{56}Fe ($\sim$ 332 mg/cm^2 thickness) by protons of different energies in a range of 0.04 – 2.6 GeV. The diameters of all the foils were 10.5 mm. The irradiations were carried out at the ITEP accelerator U−10 under the ISTC Project # 3266 in 2006−2009. ^3H has been extracted from Al foils using an A307 Sample Oxidizer. An ultra low level liquid scintillation spectrometer Quantulus1220 was used to measure the ^3H β−spectra and the SpectraDec software package was applied for spectra processing, deconvolution and ^3H activity determination. The values of the $Al(p,x)^3H$ reaction cross sections obtained in these experiments are compared with data measured at other labs and with results of simulations by the MCNP6 radiation transport code using the CEM03.03 event generator.

1 Introduction

Tritium is a gaseous product of nuclear reactions, whose formation, in addition to the issues of radiation production in specific parts of nuclear installations, causes additional environmental problems, associated with its high migration ability. This stimulates an interest in the study of the production cross-sections of this nuclide in different structural materials of nuclear installations. A compilation of the experimental values of the cross-sections of the reaction $Al(\text{p,x})^3H$, taken from data libraries [1,2] and together with values obtained in this work in comparison with the simulated data both in [3] and in this work are presented in Fig. 1.

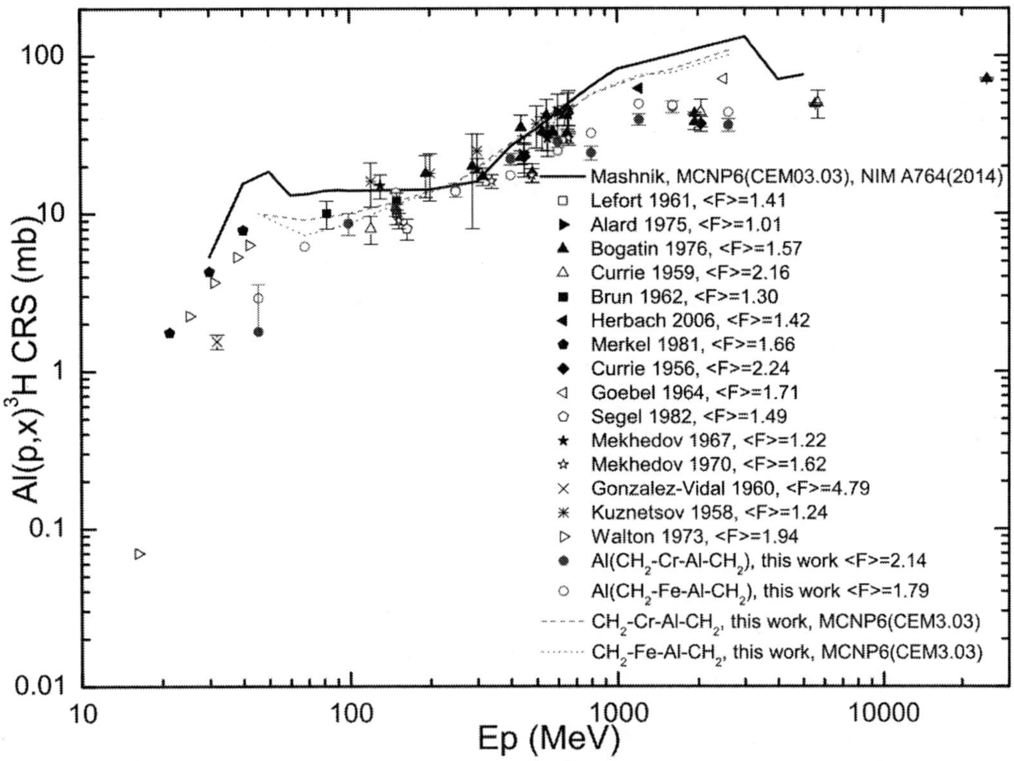

Figure 1: $Al(\text{p,x})^3H$ reaction cross-sections measured in [1,2], simulated [3] as well as obtained in this work.

A comparison of the simulated excitation functions of the $Al(\text{p,x})^3H$ reaction with various experimental data sets reveals that the mean squared deviation factor $< F >$, often used to assess the predictive power of codes [4], is within the range from ~ 1 to ~ 5. In addition to the above, the spread of the experimental data is much wider than the experimental uncertainties.

2 Justification of the experiments, simulation and results

During 2006 - 2009, under the framework of the ISTC Project # 3266, the ITEP synchrotron was used to determine cross sections of residual nuclei production in Cr, ^{56}Fe, and other thin samples induced by 0.04 – 2.6 GeV protons. All the samples were placed inside polyethylene bags together with Al foils to monitor proton flux. Both the samples and foils were thin enough ($\ll 1g/cm^2$) to get a negligible degradation of proton energy within each "sample-foil" sandwich. That is why the losses of residual nuclei heavier ^7Be could be neglected. The main project results obtained by γ–spectrometry of the irradiated samples can be found in [4, 5].

Subsequently, the idea arose to measure tritium inside the irradiated samples and foils. The event generator CEM03.03 inside the MCNP6.1.1 transport code was used to simulate the spectra of tritium nuclei produced by the protons and showed that a significant part of produced tritium has a range above the foils thickness (see Fig. 2).

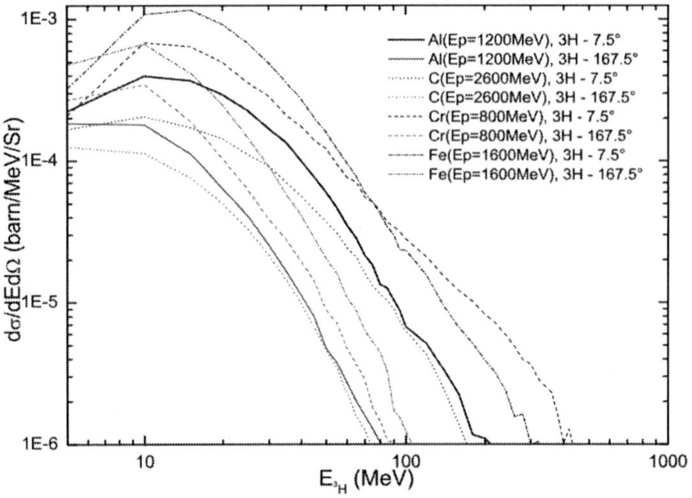

Figure 2: Energy spectra at different angles of ^3H nuclei released from the Al, C, Cr and ^{56}Fe nuclei irradiated by 1200, 2600, 800 and 1600 MeV protons, accordingly.

This fact leads to some redistribution of tritium produced in the polyethylene, samples, and monitor layers through the exchange of tritium nuclei. Therefore, the cross section results obtained in this work are not quite identical to the cross section for tritium production in Al presented in [1, 2]; we identify our results as "integral" cross sections. Nevertheless, as shown in this work, the obtained results can certainly be used to verify the nuclear models of radiation transport codes.

^3H activity in Al foil was determined using a low-background spectrometric α-, β-radiometer, Quantilus 1220, and a system of automatic sample preparation, A307 Sample Oxidizer. An analysis of the measured β-spectra was carried out with the SpectraDec code [6]. The formula for determining the reaction rate and cross-section production of ^3H are presented in [5].

All results were simulated using the CEM03.03 event-generator incorporated in the radiation transport code MCNP6.1.1 [7]. The real parameters (thickness, diameter, density) of irradiated samples (Al, Cr, ^{56}Fe) were modelled, including the thickness of polyethylene bags in the appropriate layers-cells.

3 Conclusion

Simulated values of independent cross-sections of ^3H production in Al [5], together with the calculations of the integral cross-sections and measured values, are shown in Fig. 1. The values of $< F >$ for Al from pair $Cr \leftrightarrow Al$ are equal to 2.14 and 1.79 from pair $^{56}Fe \leftrightarrow Al$ (see details in [5]).

Acknowledgements

The authors very much appreciate the support received from the ISTC projects, as well as from the current pilot project of the National Research Center "Kurchatov Institute". Part of the work performed at LANL was carried out under the auspices of the National Nuclear Security Administration of the U.S. Department of Energy. We thank Dr. Roger L. Martz for a very careful reading of the manuscript and useful suggestions on its improvement.

References

[1] *Experimental Nuclear Reaction Data (EXFOR) database*, http://www-nds.iaea.org/exfor/exfor.htm
[2] *Production of Radionuclides at Intermediate Energies*, Ed. H.Schopper. Springer Verlag, Landolt-Bernstei.
[3] S.G. Mashnik, L.M.Kerby, *NIM A*, 764 (2014) 5981.
[4] Yu.E. Titarenko, et. al., *Physical Review. C*, 84 (2011), 064612-12011.
[5] Yu.E. Titarenko, et. al., *IAEA Nuclear Data Section, INDC (CCP)-0453*, VIC, A-1400 Vienna, Austria, October 2011.
[6] I.A. Kashirin, et al., *Applied Radiation and Isotopes*, 53, 303308 (2000).
[7] M.R. James, et. al., *LA-CP-14-00745*, Los Alamos (2014).

EPJ Web of Conferences **117**, 06001 (2016) DOI: 10.1051/epjconf/201611706001
NN2015

Continuum spectroscopy of light nuclei

R. J. Charity for the HiRA Collaboration

Department of Chemistry, Washington University,
St. Louis MO 63130, USA

Abstract

Resonance spectroscopy of light nuclei is discussed with emphasis on the invariant-mass measurements performed with the HiRA detector. For three-body exit channels, we consider the exact conditions necessary such that the decay can be described as either sequential or prompt. However experimentally, we find some cases where the decay is intermediate between these two limits. Finally, two-proton decay from isobaric analog states is discussed.

Light nuclei near the drip lines have very few particle-bound levels, with most of their states being in the continuum. Of course those beyond the drip lines have all their levels in the continuum including their ground states. Thus the study of continuum spectroscopy is quite important for these light nuclei. There are many techniques to study these continuum states, however, in this work we will consider just the invariant-mass method and discuss results obtained with the HiRA array [1] at the the National Superconducting Cyclotron Laboratory at Michigan State University and the Texas A&M University cyclotron.

It is often assumed that continuum states which are associated with binary exit channels are dominant. However, three-body and high-order exit channels are quite common in light nuclei. For example in the ^{9}C+^{9}Be reaction at E/A=70 MeV studied with the HiRA detectors [17], we have identified 42 resonances in the continuum which decay into exit channels containing charged particle. These resonances are continuum states in ^{5,6,7}Li, ^{6,7,8}Be, ^{7,8,9}B, and ^{8,9,10}C nuclei. Of these, only half were associated with

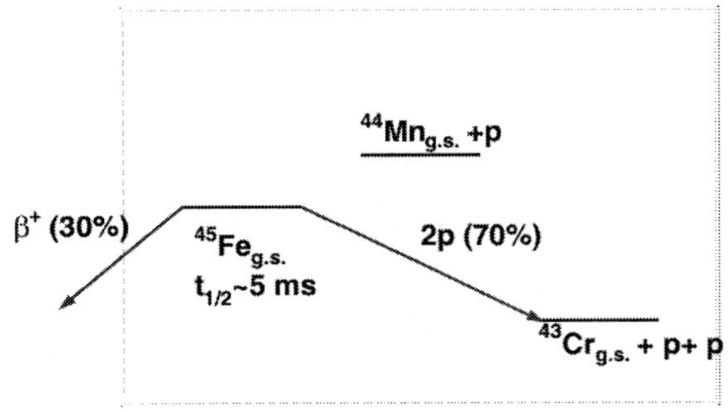

Figure 1: Level scheme for the decay of ^{45}Fe.

binary exit channels with the remainder from three, four, and five-body decays. Thus these higher-order decays are quite common and will be the subject of the rest of this paper.

These higher-order decay channels can be associated with either a prompt decay scenario where all of the exit-channel particles are emitted at the same time or the disintegration could be sequential where the decay proceeds through a binary decay channel producing one or more intermediate states which are themselves resonances and subsequently decay.

As an example of three-body decay, Goldansky considered prompt two-proton emission from ground states in 1960 [3]. He envisaged this process occurring for systems where no intermediate state in the one-proton daughter is energetically available, so the system is forced to decay by a prompt two-proton emission. As an example, the level scheme for the decay of $^{45}\mathrm{Fe}_{g.s.}$ is shown in fig 1. Here the $^{44}\mathrm{Mn}_{g.s.}$ intermediate is not available for one-proton decay so the only allowed particle-decay mode is a two-proton emission. This state also has a smaller β-decay branch.

For odd-Z parents, the pairing interactions pushes the mass of the even-Z one-proton daughter down always making it energetically accessible in these light systems. Thus "Goldansky" type ground-state two-proton emitters must have even-Z. Also the Goldansky's scheme assumes that if a one-proton decay channel is available, the level will chose that branch rather than a prompt two-proton channel for decay.

Light nuclei beyond the proton drip line are more complicated than the Goldansky scheme as the small Coulomb barriers can give rise to substantial widths for the intermediate states. For a true sequential three-body decay, the intermediate state should be narrow (long lived) so that there is a time

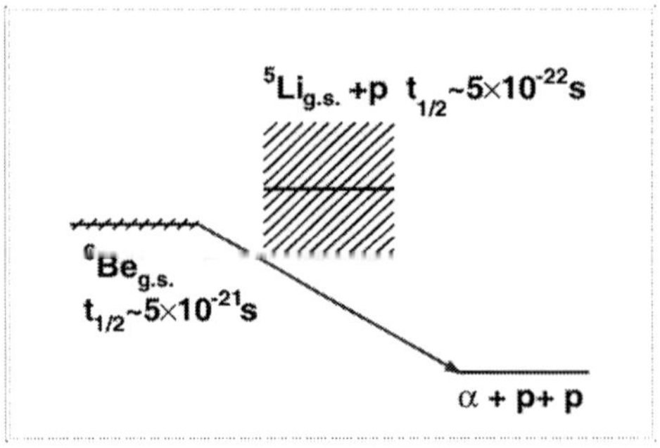

Figure 2: Level scheme for the decay of ^{6}Be.

separation in the two binary emission steps allowing no interactions between the particles produced in these two steps. In the limit of a very large width for this intermediate state, the decay will turn into a prompt decay. For potential sequential two-proton emitters where the energy of the first proton is of the same magnitude as the width of the intermediate state, the decay is called democratic [4] and considered prompt as the first proton will not have traveled any distance before the second proton is emitted. One should be careful of the term democratic decay as it has also been used to describe decays where the momentum correlations between the final fragments follow phase-space considerations which is quite different from the present usage.

The level scheme for the decay of ^{6}Be, the lightest two-proton ground-state emitter, is shown in fig. 2. One can see that most of the strength of the $^5Li_{g.s.}$ intermediate state is energetically inaccessible so it has been considered as a Goldansky decay by some people. However as this intermediate state is very wide, decays though its low-energy tail are still permissible. But even in this case, the width of this state is of the same order as the energies of the allowed single-proton decays, so this state must be considered as democratic and prompt. The pairing effect is also important for ground-state democratic decay. The valence protons for odd-Z systems sit higher up in the Coulomb well than their more-bound even-Z neighbors and thus have much larger widths, Therefore even-Z parents are required in order to have wide odd-Z intermediate states.

Figure 3 shows a portion of the chart of nuclides for light systems highlighting the known ground-state two-proton and two-neutron emitters. It is not surprising from the above arguments that all two-proton emitters are even-Z. From a similar logic, one expects the ground-state two-neutron

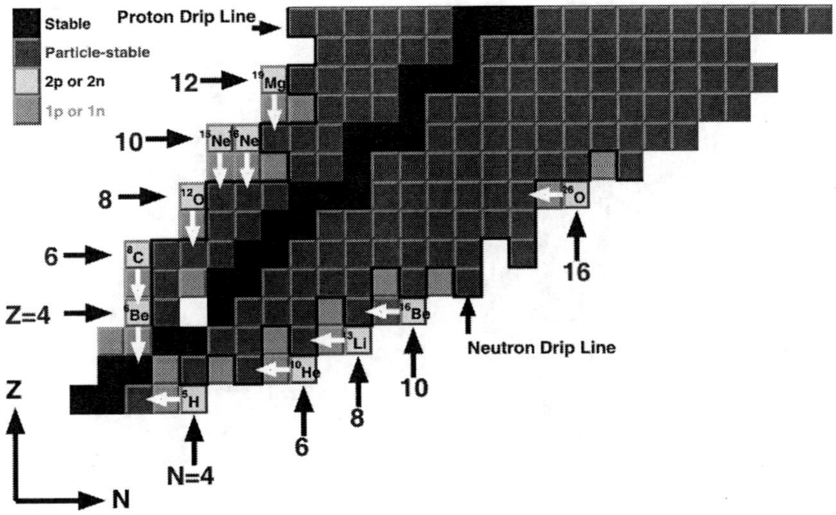

Figure 3: A portion of the chart of nuclides showing the location of known ground-state two-proton and two-neutron emitters in light systems. Each two-nucleon emitter is labeled and a white arrow indicates it decay.

emitters on the other side of the chart of nuclides to be even N. Important information about these higher-order decays is contained in the momentum correlations between the final products. In three-body decay, with three momentum components for each fragment, there are nine degrees of freedom in total. However, three degrees of freedom are constrained by momentum conservation and a further one from energy conservation. In addition any overall rotation of the three momentum vectors is not of interest, reducing the problem by a further three degrees of freedom associated with the three Euler angles of rotation. This leaves us finally with two degrees of freedom. Thus all three-body decays have a two-dimensional distribution describing their momentum correlations. One is of course free to choose which two parameters to use.

One choice of parameters is obtained from the Jacobi Y hyperspherical coordinates. Here we define an energy parameter E_x/E_T and angle parameter θ_k where E_x is the relative energy between the core and a proton, E_T is the total decay energy, and θ_k is the angle between the proton-core relative velocity and the center-of-mass velocity of the other proton. Figure 4 compares measured two-dimensional correlations for (a) $^{6}Be_{g.s.}$ [5] a prompt two-proton emitter and (b) the 5.3-MeV excited state of ^{10}C [6], a sequential two-proton emitter.

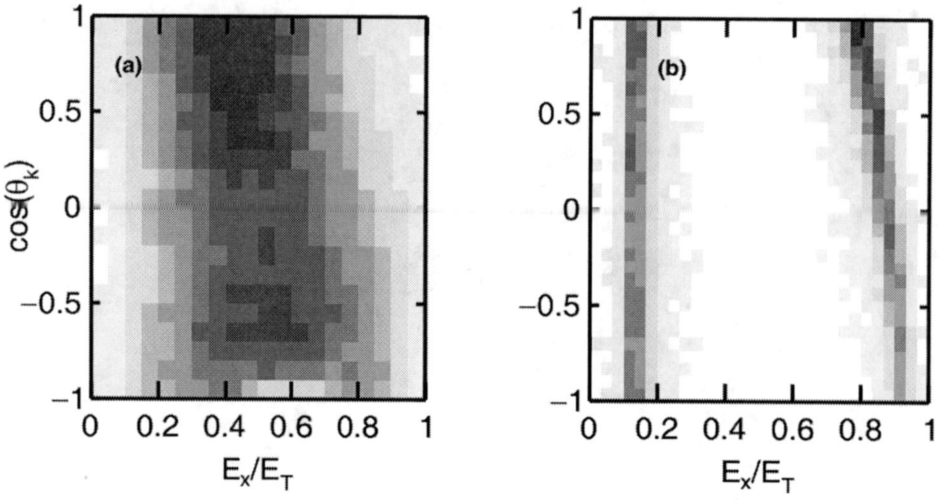

Figure 4: Jacobi Y correlation plots measured for the two-proton decay of (a) ^{6}Be$_{g.s.}$ and (b) the 5.3-MeV excited state of ^{10}C.

One can see clear differences between the correlations for prompt and sequential decay. The sequential decay of fig. 4(b), which is easiest to understand, consists of two narrow ridges that correspond to the two emitted protons. The low-energy ridge, which is independent of the angle θ_k is the second emitted proton. For the high-energy ridge (first proton), E_x (the relative p-core energy) depends on θ_k due to the recoil imparted by the second proton. In contrast, the prompt decay of fig. 4(a) has strength in a single, more-spread-out region of correlation space.

Systematics of the projection of the correlations on the energy axis are shown in fig. 5 for three measured prompt-two-proton emitters: ^{6}Be$_{g.s.}$ [5], ^{16}Ne$_{g.s.}$ [7], and ^{45}Fe$_{g.s.}$ [8]. The results for ^{45}Fe are not from the HiRA group, but were obtained with an optical time projection chamber [8]. All distributions have similar shape with a peak at $E_x/E_T \sim 0.5$, i.e., equal energy protons. This result was originally predicted by Goldansky [3] who indicated that the two-proton emission probability must be proportional to the product of two barrier penetration factors $P(E)$, i.e.

$$P_{2p} \propto P(E_x)P(E_T - E_x) \tag{1}$$

if the two protons are emitted simultaneously. This product peaks for equal energy protons. As one becomes more sub-barrier, the penetration factors increase more rapidly with energy and thus the E_x/E_T distribution should become narrower. This is clearly demonstrated in fig. 5 where the E_x/E_T width decreases quite substantially with increasing Z (larger Coulomb bar-

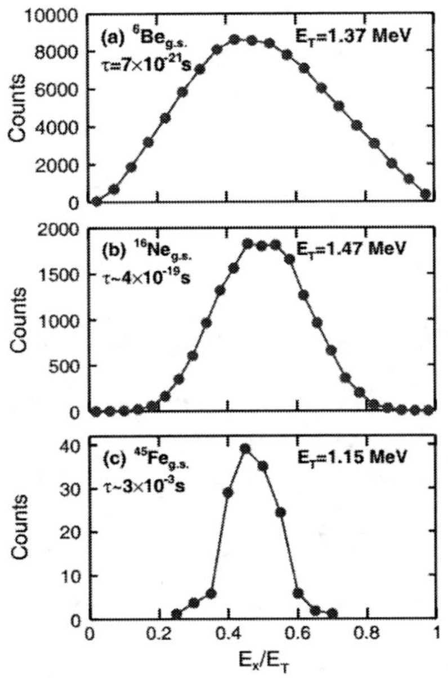

Figure 5: Measured distributions of E_x/E_T for three ground-state two-proton emitters.

riers for the proton emission). Note the total decay energy of the three systems in fig. 5 are all similar and thus it is only the change in the height of the Coulomb barriers that are affecting the degree of sub-barrier emission in these three cases. ^{45}Fe with the largest Coulomb barrier is most subbarrier and has the narrowest E_x/E_T distribution and the longest lifetime.

Theoretical predictions of these momentum correlations must include the initial-state correlation and the final-state interactions as the fragments move apart. The latter has contributions from the spin-singlet *p*-*p* interactions and the long-range three-body Coulomb interaction. In the three-body cluster model developed by Grigorenko et al [9], the wavefunction of the three-body system is solved in hyperspherical coordinates and matched to approximate outgoing waves associated with the Coulomb interaction. Further refinements have been made in more recent years [7]. This model reproduces the measured correlations in the ground and first-excited ($J = 2^+$) states of ^{6}Be [5] very well. More recently it has been applied to the two-proton decay of the ground state of ^{16}Ne where the long-range Coulomb final-state interactions are much stronger [7]. The calculations were able to reproduce the experimental data quite well as long as the three-body Coulomb interactions are considered out to distances of ~100000 fm. This results validates these calculations for use with heavier two-proton emitters.

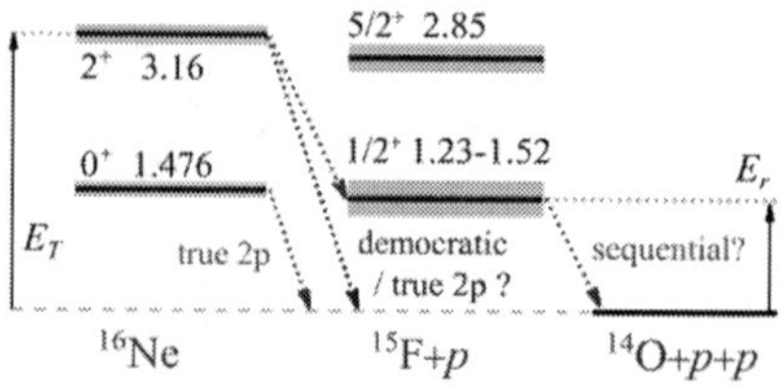

Figure 6: Level scheme of ^{16}Ne and the states associated with its decay.

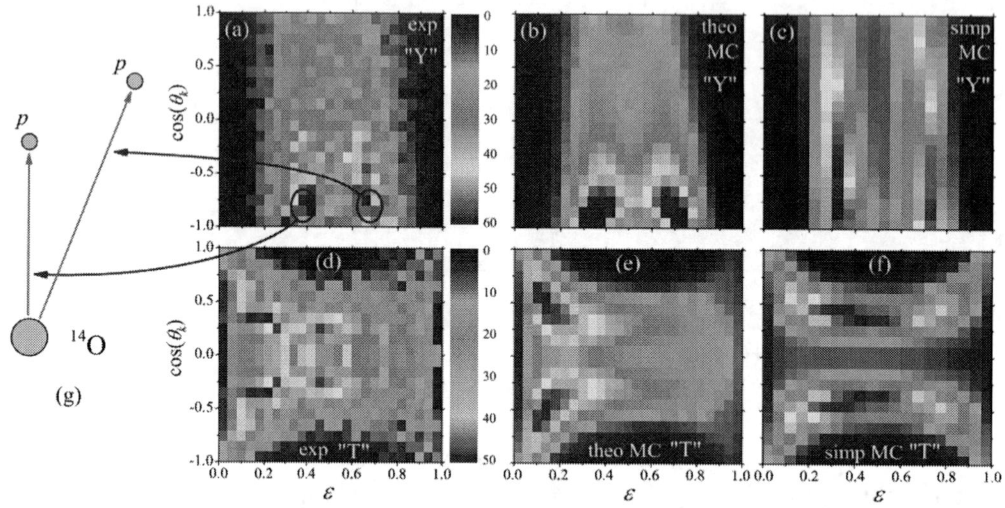

Figure 7: Momentum correlations for the decay of the first excited state of ^{16}Ne. The correlations (a-c) are in the Jacobi Y representation while (d-f) are in the Jacobi T representation. The experimental results (a,d) can be compared to the prediction of the three-body cluster model (b,e), and a sequential calculation (c,d). The prominent configuration indicated by the circles in (a) is shown in (g). Here $\varepsilon = E_x/E_T$.

Although the ground states of ^{6}Be and ^{16}Ne are prompt two-proton emitters, it is interesting to determine the nature of their excited states. The level scheme of ^{16}Ne and its potential ^{15}F intermediate states are shown in fig. 6. For the $J = 2^+$ first excited state, neither the Goldansky or the democratic conditions apply. The ground state of ^{15}F is energetically accessible by one-proton decay and the kinetic energy of such a proton is large compared to the width of this state. So does this states decay by sequential two-proton decay? The correlations measured for this state are shown in fig. 7(a,d) [10]. In addition to the Jacobi Y representation, we also show the Jacobi T representation where here $\varepsilon = E_x/E_T$ and E_x is now the relative energy between the two protons. In the Jacobi T representation,

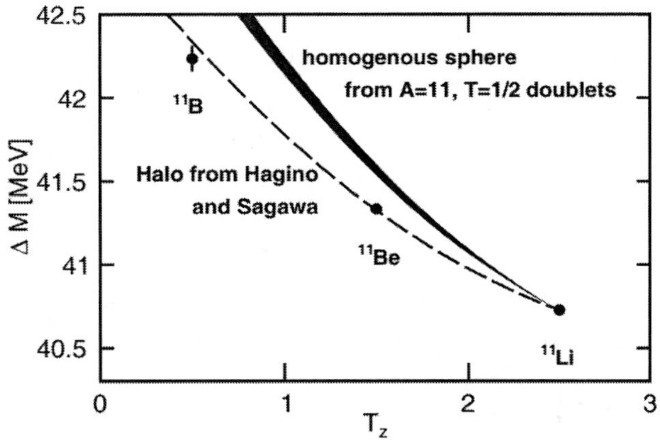

Figure 8: Mass excesses of the three known members of the A=11 sextet plotted as a function of isospin projection. The solid band shows the prediction where the Coulomb energy is determined with a homogeneous-sphere assumption with the radius extracted from the T=1/2 A=11 doublet. The dashed curve shows the prediction using the halo wavefunction of Hagino and Sagawa [16].

the angle θ_k is now the angle between the relative velocity vector between the proton pair and the center-of-mass velocity of the core.

In comparison, fig. 7(c,f) shows the predicted pattern expected for a sequential decay which displays a two-ridge pattern similar to that obtained for ^{10}C* in fig. 4(b). However, the experimental distribution only shows the two-ridge structure for $cos(\theta_k)$ ~-1 in the Jacobi Y representation of fig. 7(a), i.e., where the relative angle between the two protons is small and thus large *p-p* final-state interactions are expected. This behavior is reproduced by the three-body cluster-model prediction in fig. 7(b,e). In this model one finds the existence of both prompt and sequential decay paths that interfere and produce the strange correlation pattern. A similar behavior is found in the ^{6}Be continuum at excitations of 6 to 9 MeV [5]. Clearly the exact conditions for pure sequential decay have yet to be determined.

So far we have only considered two-proton emission from ground and low-lying excited states. Two-proton emission has also been observed from isobaric analogs of ground-state two-proton emitters. These include the isobaric analog of $^8C_{g.s.}$ in ^{8}B [11, 12] and the isobaric analog of $^{12}O_{g.s.}$ in ^{12}N [13]. Both of these two cases can be considered as extensions of the Goldansky argument, i.e, single-proton emission violates either energy or isospin conservation leaving two-proton decay as the only channel satifiying both conservation laws.

Finally, we have also observed two-proton emission in an analog of the two-neutron-halo state $^{11}Li_{g.s.}$. A peak associated with the double isobaric analog of $^{11}Li_{g.s.}$ in $^{11}B(T{=}5/2)$ was found in the $2p{+}^9Li$ exit channel created with the $^{12}Be{+}^9Be$ reaction [14]. To the extent that isospin is conserved, this state should have the same structure as $^{11}Li_{g.s.}$ except that two of the neutrons have been turned into protons and therefore there are components of the wavefunction where the halo contains one or two protons. Figure 8 shows the evolution of the mass excess for the $T{=}5/2$ sextet associated with these states. In addition to the new $T_z{=}1/2$ ^{11}B member, the $T_z{=}3/2$ ^{11}Be member is also known [15]. The increase in the mass excess with decreasing T_z is due to the larger Coulomb energies of these more proton-rich analogs. The predicted evolution when the Coulomb energy is obtained from a homogeneous-sphere assumption produces a much larger T_z dependence than observed. If however one allows for a halo structure and uses the halo wavefunction predicted by Hagino and Sagawa [16], then the mass excesses can be reproduced (dashed curve). Of course putting protons in an extended halo structure give rise to smaller Coulomb energies and masses. Thus the extent of the halo structure of ^{11}Li is also contained in the masses of its analogs.

In summary, the field of continuum spectroscopy is quite exciting for these light nuclei. Momentum correlations have proved to be a powerful tool in determining the sequential and prompt aspects of three-body decay. However the exact nuclear structure necessary for sequential and prompt decays have yet to be determined. While this contribution has focused on two-proton emission, there are many other 3-body and 4-four decays observed in light nuclei [17]. In addition the binary decays are also of interest, for example in astrophysics [2].

This work is supported by the U.S. Department of Energy, Office of Science, Office of Nuclear Physics under Award No. DE-FG02-87ER-40316. The HiRA group acknowledges the considerable theoretical support of Leonid Grigorenko.

References

[1] M. Wallace et al., *Nuclear Instruments and Methods A* **583** (2007) 302.

[2] L.G. Sobotka et al., *Phys. Rev.C* **87** (2013) 054329.

[3] V.I. Goldansky, *Nucl. Phys.* **19** (1960) 482.

[4] O.V. Bochkarev, L.V. Chulhov, A.A. Koreninniicov, E.A. Kuz'min, I.G. Mukha, and G.B. Yankov, *Nucl. Phys. A* **505** (1989) 215.

[5] I.A. Egorova, et al., *Phys. Rev. Lett.* **109** (2012) 202502.

[6] R.J. Charity et al., *Phys. Rev. C* **80** (2009) 024306.

[7] K.W. Brown et al., *Phys. Rev. Lett.* **113** (2014) 232501.

[8] K. Miernik et al., *Phys. Rev. Lett.* **99** (2007) 192501.

[9] L.V. Grigorenko et al., *Phts. Rev. Lett* **88** (2002) 042502.

[10] K.W. Brown et al., *Phys. Rev. C* **92** (2015) 034329.

[11] R.J. Charity et al., *Phys. Rev. C* **82** (2010) 041304.

[12] K.W. Brown et al., *Phys. Rev. C* **90** (2014) 027304.

[13] M.F. Jager et al., *Phys. Rev. C* **86** (2012) 011304.

[14] R.J. Charity et al., *Phys. Rev. C* **86** (2012) 041307.

[15] T. Teranishi et al., *Phys. Lett. B* **407** (1997) 110.

[16] K. Hagino and H. Sagawa, *Phys. Rev. C* **72** (2005) 044321.

[17] R.J. Charity et al., *Phys. Rev. C* **84** (2011) 014320.

EPJ Web of Conferences **117**, 06002 (2016) DOI: 10.1051/epjconf/201611706002
NN2015

Reaction theory: Status and perspectives

A.M. Moro[1] and J. Gómez Camacho[1,2]

[1] Departamento de FAMN, Universidad de Sevilla, Apartado 1065, 41080 Sevilla, Spain
[2] CNA (U. Sevilla, J. Andalucía, CSIC), Av. Tomas A. Edison 7, 41092 Sevilla, Spain

Abstract

The current status of the reaction theory of nuclear collisions involving weakly-bound exotic nuclei is presented. The problem is addressed within the Continuum Discretized Coupled Channel (CDCC) framework, recalling its foundations and applications, as well as its connection with the Faddeev formalism. Recent developments and improvements of the method, such as core and target excitations and the extension to three-body projectiles, are presented. The use of the CDCC wave function in the calculation of inclusive breakup reactions is also introduced.

1 Introduction

Ongoing and planned radioactive beam facilities will produce in near future accurate reaction data covering unexplored areas of the nuclear chart. The extraction of meaningful information from these experiments will demand new advances in nuclear structure and reaction theories.

The description of the collision between composite nuclei is a very complicated quantum mechanical many-body problem which, in general, cannot be solved exactly. Although, ideally, the problem could be solved from a many-body viewpoint, starting with some effective nucleon-nucleon interaction, in practice one needs to resort to additional approximations, aimed at making the problem numerically tractable. These approximate models try to emphasize specific degrees of freedom, those which are expected to dominate the reaction dynamics, or those in which one is interested, and recast the problem in terms of an effective Hamiltonian in which these selected degrees of freedom appear explicitly.

Within the Coupled Channels (CC) method, the Hamiltonian that describes a scattering problem, in which some internal degrees of freedom (from projectile, target or both) can be excited, is:

$$H = H_{\text{int}}(\xi) + \hat{T}_{\mathbf{R}} + U(\mathbf{R}, \xi), \tag{1}$$

with H_{int} as the internal Hamiltonian, depending on the internal coordinates ξ, $\hat{T}_{\mathbf{R}}$ the kinetic energy operator, and U the projectile-target interaction, depending both on their relative coordinate $\mathbf{R}$ and ξ. In CC methods [1], the scattering wave function is written as an expansion of the form

$$\Psi^{\text{CC}}(\xi, \mathbf{R}) = \sum_{\beta} \Phi_{\beta}(\xi)\chi_{\beta}(\mathbf{R}), \tag{2}$$

where $\Phi_{\beta}(\xi)$ are eigenstates of H_{int}. The index β represents as set of quantum numbers characterizing a given projectile or target state. The functions $\chi_{\beta}(\mathbf{R})$ describe the projectile-target relative motion and are obtained by inserting the expansion (2) into the Schrödinger equation, $[H - E]\Psi^{\text{CC}} = 0$. This gives rise to a set of coupled differential equations which, solved with the proper boundary conditions, produce the scattering observables. In practice, the CC method is applied using a partial wave decomposition of the total wave function as explained elsewhere [1, 2].

In the following section, we will adapt the general CC method to relevant cases in exotic nuclei scattering. We will identify the relevant internal degrees of freedom ξ, and specify the particular expression of the internal wave function $\Phi_{\beta}(\xi)$.

2 The Continuum Discretized Coupled Channels (CDCC) method

When dealing with the scattering of a two-body weakly-bound nucleus on a well bound target, an important degree of freedom that must be taken into account is the dissociation (i.e., breakup) of the projectile into its constituents. Within the CC scheme, this problem can be accurately treated using the CDCC method. This method was originally introduced by G. Rawitscher [3] and later refined by the Pittsburgh-Kyushu collaboration [2, 4]

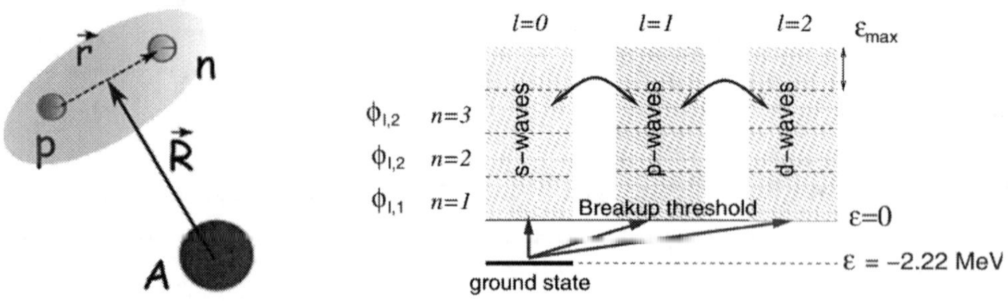

Figure 1: Left: Relevant coordinates for a deuteron-target scattering problem. Right: Illustration of continuum discretization for the same problem.

to describe the effect of the breakup channels on the elastic scattering of deuterons. Denoting the reaction by $a + A$, with $a = c + v$ (referred hereafter as the *core* and *valence* particles, respectively), the method assumes the effective three-body Hamiltonian

$$H = H_{\text{proj}} + \hat{T}_{\mathbf{R}} + U_{cA}(\mathbf{r}_{cA}) + U_{vA}(\mathbf{r}_{vA}), \tag{3}$$

with $H_{\text{proj}} = T_{\mathbf{r}} + V_{vc}$ the projectile internal Hamiltonian, $\hat{T}_{\mathbf{r}}$ and $\hat{T}_{\mathbf{R}}$ are kinetic energy operators, V_{cv} the intercluster interaction and U_{cA} and U_{vA} are the core-target and valence-target optical potentials evaluated at the corresponding incident energy per nucleon. In the CDCC method the three-body wave function of the system is expanded in terms of the eigenstates of the Hamiltonian H_{proj} including both bound and unbound states. Since the latter form a continuum, a procedure of discretization is applied, consisting in representing this continuum by a finite and discrete set of square-integrable functions. So, in this case, $\xi \equiv \mathbf{r}$, and the internal wave functions in Eq. (2) take the form

$$\Phi_{\beta}(\xi) \equiv \phi^{p}_{n,\ell,j\pi}(k_n, \mathbf{r}), \tag{4}$$

where $\phi^{p}_{n,\ell,j\pi}(k_n, \mathbf{r})$ is a discrete continuum wave function representing a region of the continuum with orbital (total) angular momentum ℓ (j) and nominal wave number k_n.

The standard CDCC method relies on a strict three-body model of the reaction ($c + v + A$), and has proven to be rather successful to describe elastic and exclusive breakup cross sections of deuterons and other weakly bound two-body nuclei (like ^{6,7}Li) [2]. However, it has limitations. The assumption of inert bodies is not always justified, since excitations of the projectile constituents (c and v) and of the target (A) may take place in addition to the projectile dissociation. Furthermore, the two-body picture may be inadequate for some nuclei as, for example, in the case of the Borromean systems, i.e., three-body systems in which the binary sub-systems are unbound, such as ^{6}He and ^{11}Li. Some extensions of the CDCC method to deal with the these situations are outlined in the following.

2.1 Inclusion of target excitations

The explicit inclusion of target excitation was done by the Kyushu group in the early days of the CDCC method [4]. Target excitation arises from the non-central part of the valence-target and core-target interactions and, consequently, the effective Hamiltonian, Eq. (3), must be generalized to:

$$H = H_{\rm proj}(\mathbf{r}) + H_{\rm tar}(\xi_t) + \hat{T}_{\mathbf{R}} + U_{cA}(\mathbf{r}_{cA}, \xi_t) + U_{vA}(\mathbf{r}_{vA}, \xi_t), \tag{5}$$

where $H_{\rm tar}(\xi_t)$ is the target Hamiltonian, depending on the target degrees of freedom (denoted as ξ_t). Note that $c - A$ and $v - A$ interactions depend now, in addition to the corresponding relative coordinate, on ξ_t. So, the relevant degrees of freedom are $\xi \equiv \{\mathbf{r}, \xi_t\}$ and the internal states become

$$\Phi_\beta(\xi) \equiv \left[\phi^p_{n,\ell,j\pi}(k_n, \mathbf{r}) \otimes \phi^t_m(\xi_t)\right]_{JM}, \tag{6}$$

which are just products of the target eigenstates times the continuum discretized projectile wave function. They can be coupled to a total internal angular momentum J, M.

In the calculations performed in Ref. [4], the target excitation was treated within the vibrational model and the $c-A$ and $v-A$ interactions represented by deformed optical potentials. They performed calculations for the d+^{58}Ni reaction at $E_d = 22$ and 80 MeV, including s-wave breakup of the deuteron and excitation of the target to its first excited state (2^+). In this case, deuteron breakup was found to dominate over target excitation.

Recently, the problem has been also addressed by P. Chau [5] and applied also to deuteron elastic and inelastic scattering, but the method does not seem to have been implemented in publicly available codes.

2.2 Inclusion of core excitations

Similarly to the case of target excitation, excitations of the projectile constituents (c and v in our case) are also possible. For example, in the important case of the scattering of two-body halo nuclei, collective excitations of the core can be important. These core excitations will affect not only the reaction dynamics, but also the structure of the projectile. In the inert-core picture, the projectile states will correspond to pure single-particle or cluster states. On the other hand, if the core is allowed to excite, the projectile states will contain in general admixtures of these core excited components. These two effects (dynamic and structure effects) have been recently investigated within extended versions of the DWBA and CDCC methods [6–9]. To this end, the following effective three-body Hamiltonian has been used:

$$H = H_{\text{proj}}(\mathbf{r}, \xi_c) + \hat{T}_{\mathbf{R}} + U_{cA}(\mathbf{r}_{cA}, \xi_c) + U_{vA}(\mathbf{r}_{vA}). \tag{7}$$

Note that the core degrees of freedom (ξ_c) appear in the projectile Hamiltonian (structure effect) as well as in the core-target interaction (dynamical effect). In the weak coupling limit, the basis functions can be expressed as

$$\Phi_\beta(\xi) \equiv \sum_\alpha [\varphi^p_\alpha(\mathbf{r}) \otimes \Phi_I(\xi_c)]_{JM}, \tag{8}$$

where $\xi \equiv \{\xi_c, \mathbf{r}\}$. The label α denotes the set of quantum numbers $\{\ell, s, j, I\}$, with I the core intrinsic spin, $\vec{j} = \vec{\ell} + \vec{s}$ and $\vec{J} = \vec{j} + \vec{I}$. The functions $\Phi_I(\xi_c)$ and $\varphi^p_\alpha(\mathbf{r})$ describe, respectively, the core states and the valence–core relative motion. For continuum states, a procedure of continuum discretization is used, similar to the case of the standard CDCC.

Once the projectile states (8) have been calculated, the three-body wave function is written as an expansion in terms of these states, according to Eq. (2). Calculations using this extended CDCC method (XCDCC) were first performed by Summers *et al.* [8,10] for ^{11}Be and ^{17}C on ^{9}Be and ^{11}Be+p. Later on, De Diego *et al.* [9] performed calculations for ^{11}Be on ^{1}H, ^{64}Zn and ^{208}Pb. These calculations suggest that, for the proton target case, dynamic core excitations may increase significantly the breakup cross sections whereas, for the heavier targets, the dynamic core excitation mechanism is small but its effect on the structure of the projectile is still important. As an example, we show in Fig. 2 the XCDCC calculations for the reaction ^{11}Be+p at 63.7 MeV/nucleon (adapted from Ref. [9]).

A no-recoil DWBA version of this approach has been also proposed in Refs. [6,7]. Calculations performed using this formalism have confirmed the conclusions of the more elaborated XCDCC calculations and have also served to evidence that the interplay between the single-particle and dynamic core excitation mechanisms produce a distinctive effect on the interference pattern of the resonant breakup angular distributions [11].

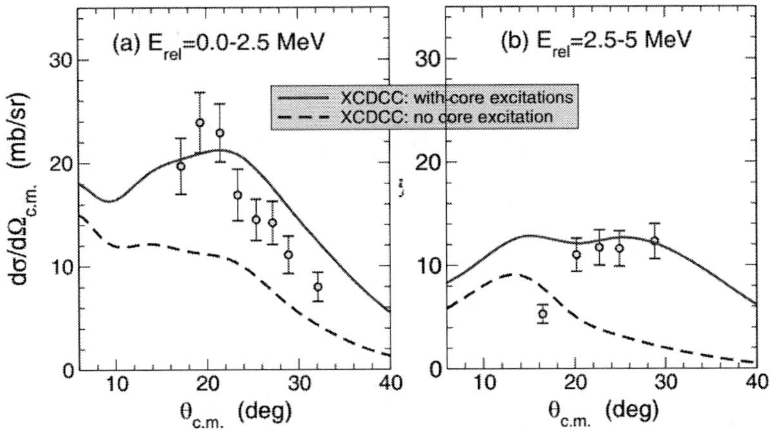

Figure 2: (Color online) Differential breakup cross sections, with respect to the outgoing ^{11}Be* c.m. scattering angle, for the breakup of ^{11}Be on protons at 63.7 MeV/nucleon. Left and right panels correspond to the neutron-core relative energy intervals $E_{\rm rel}$=0–2.5 MeV and $E_{\rm rel}$=2.5–5 MeV, respectively. Adapted from [9].

2.3 Extension to three-body projectiles

To study the scattering of three-body projectiles, such as Borromean nuclei, the Hamiltonian (3) must be generalized in order to take into account the three-body structure of the projectile. For example, for a Borromean system with a structure of the form $c + v_1 + v_2$ one may use the Hamiltonian

$$H = H_{\rm proj}(\mathbf{r}_1, \mathbf{r}_2) + \hat{T}_{\mathbf{R}} + U_{cA} + U_{v_1A} + U_{v_2A}, \tag{9}$$

where $H_{\rm proj}(\mathbf{r}_1, \mathbf{r}_2)$ is the projectile (three-body) Hamiltonian (depending now on two relative coordinates), and $U_{v_i}(\mathbf{r}_{v_iA})$ the valence-target effective interactions. Clearly, the calculation of the projectile states will be much more involved than in the two-body case. The basis functions can be expressed in several ways. For example, using independent coordinates,

$$\Phi_\beta(\xi) = \sum_\alpha \left[\varphi_\alpha(\mathbf{r}_1, \mathbf{r}_2)\right]_{JM}, \tag{10}$$

where $\xi \equiv \{\mathbf{r}_1, \mathbf{r}_2\}$, which are the internal coordinates of the three-body system. The label α denotes the set of quantum numbers $\{\ell_1, \ell_2, K, ...\}$, required to characterize the three-body state, which may vary depending on the three-body approach used. Note that $\Phi_\beta(\xi)$ is an (approximate) eigenstate of the projectile three-body Hamiltonian, with a given energy and angular momentum.

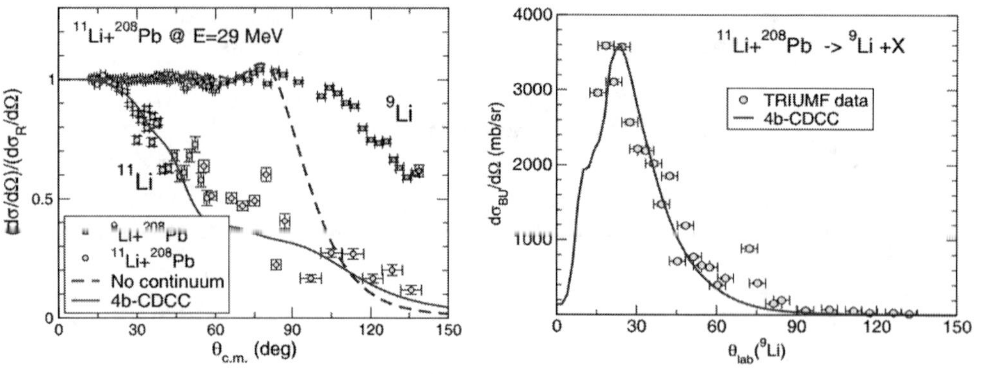

Figure 3: Left: Differential elastic cross section, relative to Rutherford, for the ^{11}Li+^{208}Pb reaction at an incident energy of 29 MeV. The solid (dashed) line is the four-body CDCC calculation including (excluding) ^{11}Li continuum states. The ^{9}Li elastic scattering data are also included for comparison. Right: Angular distribution of ^{9}Li fragments resulting from the same reaction, compared with the four-body CDCC calculation. The data and calculations are from Refs. [14, 15].

The formulation and application of this four-body CDCC method was first done by the groups at Kyushu [12] and Seville [13]. As an application of the method, we show in Fig. 3 the elastic (left) and breakup (right) of ^{11}Li on ^{208}Pb at 29 MeV [14, 15]. In both panels, the full four-body CDCC calculations are represented by the solid line.

3 Comparison with the Faddeev formalism

The CDCC method was originally devised as a physically sound and numerically appealing *ansatz* for the three-body wave function, rather than as a formally rigorous solution of the three-body problem. The latter was provided by Faddeev in the 60s [16], who showed that the exact solution of this problem can be obtained by solving a system of coupled-differential equations, named the Faddeev equations after him. The numerical solution of the Faddeev equations is very involved but, recently, it has been possible to solve these equations for a number of situations [17, 18], thus providing a valuable benchmark to more approximate models. These comparative studies have shown that the elastic and breakup observables calculated with the CDCC method agree in general very well with the Faddeev solution (see Fig. 4). However, there are also kinematical situations [18] in which the agreement is not good, for example, when the incident energy is too low. This result calls for additional studies and, possibly, for extensions or improvements of the CDCC formalism.

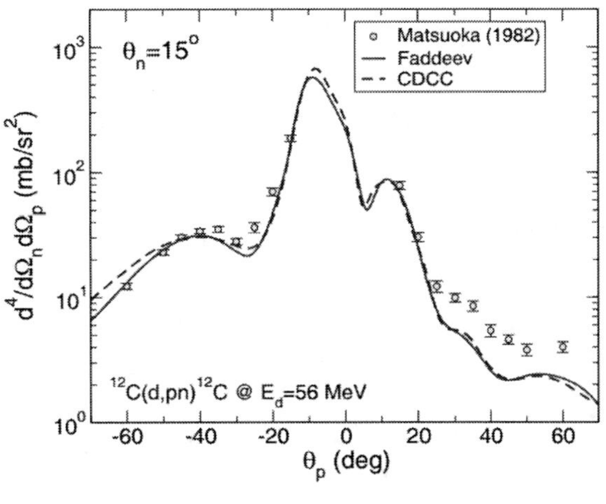

Figure 4: Breakup differential cross section for $^{12}C(p,pn)^{12}C$ at $E_d = 56$ MeV, as a function of the proton scattering angle and for a fixed neutron detection angle of 15°. The Faddeev and CDCC calculations are given by the solid and dashed lines, respectively. The circles are the data from Ref. [19].

4 The problem of inclusive breakup

In previous sections, we have considered only exclusive breakup processes in which the three fragments (c, v and A) survive after the collision and are observed in a definite internal state. In particular, when all fragments end up in their ground state, the process is called *elastic breakup* (EBU). A qualitatively different scenario occurs when the final state of one or more fragments is not specified in the final state. For example, this is the case of reactions of the form $A(a,c)X$, in which only one of the projectile constituents (the core in this case) is observed. The angular/energy distribution of the c fragments will contain contribution from all possible final states of the $v + A$ system. This includes the EBU channel, in which v and A remain in their ground state, but also v transfer, breakup accompanied by excitations of A, and v+A fusion [named *incomplete fusion* (ICF)]. These non-elastic breakup components (NEB) must be added to the EBU component to give the total inclusive breakup. Whereas the calculation of EBU can be accurately done within the CDCC method and other approaches, the calculation of NEB is more involved due to the large number of accessible states. The problem was addressed in the 1980s by several groups, which developed formal techniques to reduce the sum over final states to a closed form. We cite here the theory developed by Ichimura, Austern and Vincent (IAV) [2,20,21], in which the double differential cross section for NEB with respect to the angle and energy of the c fragments is given by

$$\left.\frac{d^2\sigma}{dE_c d\Omega_c}\right|_{\text{NEB}} = -\frac{2}{\hbar v_i}\rho_c(E_c)\langle\varphi_v(\vec{k}_c)|\text{Im}[U_{vA}]|\varphi_v(\vec{k}_c)\rangle, \quad (11)$$

where $\rho_c(E_c) = k_c\mu_c/((2\pi)^3\hbar^2)$, U_{vA} is the optical potential describing $v + A$ elastic scattering, and $\varphi_v(\vec{k}_c, \mathbf{r}_{vA})$ is the wave function describing the evolution of the v particle after dissociating from the projectile, when the core is scattered with momentum $\vec{k}_c$ and the target remains in its ground state. This function is obtained as a solution of the inhomogeneous equation

$$(E_v^+ - K_v - U_{vA})\varphi_v(\vec{k}_c, \mathbf{r}_{vA}) = (\chi_c^{(-)}(\vec{k}_c)|V_{\text{post}}|\Psi^{3b}\rangle, \quad (12)$$

where $E_v = E - E_c$, $\chi_c^{(-)}(\vec{k}_c, \mathbf{r}_{cB})$ is the distorted-wave describing the scattering of the outgoing c fragment with respect to the $B \equiv v + A$ system, obtained with some optical potential U_{cB}, and $V_{\text{post}} \equiv V_{vc} + U_{cA} - U_{cB}$ is the post-form transition operator. This equation is to be solved with outgoing boundary conditions.

Although IAV suggest approximating the three-body wave function appearing in the source term of Eq. (11), Ψ^{3b}, by the CDCC one, a simpler choice is to use the DWBA approximation $\psi_v^{3b} \approx \chi_a^{(+)}(\mathbf{R})\phi_{cv}(\mathbf{r})$, where $\chi_a^{(+)}$ is a distorted wave describing $a + A$ elastic scattering, obtained with some optical potential, and ϕ_{cv} is the projectile ground state wave function.

This DWBA version of the IAV model has been recently revisited by some groups, and applied to several deuteron [22, 23] and ^{6}Li induced reactions [24], showing in most cases a very promising agreement with the existing inclusive data. Further applications and developments are in progress.

5 Conclusions

The nuclear reaction theory community has developed in recent years quantum mechanical methods, based on the coupled channels approach, which allow to interpret the exclusive scattering experiments of exotic nuclei. These methods have been compared, in some cases, with exact three-body Faddeev calculations. Preliminary extensions of these methods to the calculation of inclusive breakup cross sections is in progress show very encouraging results.

References

[1] I.J. Thompson, F.M. Nunes, Nuclear Reactions for Astrophysics, Cambridge, UK: Cambridge University Press, 2009 **1** (2009)

[2] N. Austern et al., Phys. Rep. **154**, 125 (1987)

[3] G.H. Rawitscher, Phys. Rev. C **9**, 2210 (1974)

[4] M. Yahiro, Y. Iseri, H. Kameyama, M. Kamimura, M. Kawai, Prog. Theor. Phys. Suppl. **89**, 32 (1986)

[5] P. Chau Huu-Tai, Jour. of Phys.: Conf. Series **312**, 082018 (2011)

[6] R. Crespo, A. Deltuva, A.M. Moro, Phys. Rev. C **83**, 044622 (2011)

[7] A.M. Moro, R. Crespo, Phys. Rev. C **85**, 054613 (2012)

[8] N.C. Summers et al., Phys. Rev. C **74**, 014606 (2006)

[9] R. de Diego et al., Phys.Rev. C **89**, 064609 (2014)

[10] N.C. Summers, F.M. Nunes, Phys. Rev. C **76**, 014611 (2007)

[11] A.M. Moro, J.A. Lay, Phys. Rev. Lett. **109**, 232502 (2012)

[12] T. Matsumoto et al., Nucl. Phys. A **738**, 471 (2004)

[13] M. Rodríguez-Gallardo et al., Phys. Rev. C **77**, 064609 (2008)

[14] M. Cubero et al., Phys. Rev. Lett. **109**, 262701 (2012)

[15] J.P. Fernández-García et al., Phys. Rev. Lett. **110**, 142701 (2013)

[16] L.D. Faddeev, Zh. Eksp. Teor. Fiz. **39**, 1459 (1960), [Sov. Phys. JETP **12**, 1014 (1961)]

[17] A. Deltuva et al., Phys. Rev. C **76**, 064602 (2007)

[18] N.J. Upadhyay, A. Deltuva, F.M. Nunes, Phys. Rev. C **85**, 054621 (2012)

[19] N. Matsuoka et al., Nucl. Phys. A **391**, 357 (1982)

[20] N. Austern, C.M. Vincent, Phys. Rev. C **23**, 1847 (1981)

[21] M. Ichimura, N. Austern, C.M. Vincent, Phys. Rev. C **32**, 431 (1985)

[22] G. Potel, F. Nunes, I. Thompson, to be published in Phys. Rev. C, arXiv:1508.04822 (2015)

[23] B.V. Carlson, R. Capote, M. Sin, arXiv:1508.01466 (2015)

[24] J. Lei, A.M. Moro, submitted to Phys. Rev. C (2015)

EPJ Web of Conferences **117**, 06003 (2016) DOI: 10.1051/epjconf/201611706003
NN2015

Reaction studies with low-energy weakly-bound beams at INFN-LNS

A. Di Pietro[1], J.P. Fernandez-Garcia[1], P. Figuera[1], M. Fisichella[1], M. Lattuada[1,2], D. Torresi[1,2] and M. Zadro [3]

[1] INFN-Laboratori Nazionali del Sud, Catania, Italy
[2] Dipartimento di Fisica e Astronomia, Università di Catania, Catania, Italy
[3] Ruđer Bošković Institute, Bijenička cesta 54,HR-10000 Zagreb, Croatia

Abstract

The reaction dynamics of collisions involving halo or weakly bound nuclei, at energies around the Coulomb barrier, can be strongly affected by the structure of such nuclei. Very strong entrance channel effects have been observed on various reaction pocess such as, elastic scattering, fusion and direct reactions when comparing collision induced by the ^{6}He and ^{11}Be halo nuclei with the ones induced by their cores ^{4}He and ^{10}Be. Collisions induced by the stable weakly bound nuclei ^{6}Li, ^{7}Li show also some peculiarities in comparison to the ones induced by well bound nuclei; coupling with the break-up channel is in fact very important in reproducing low energy data. In this contribution an overview of our present understanding of the discussed topic will be given along with the discussion of some new preliminary results.

1 Introduction

The study of nuclear collisions around the Coulomb barrier induced by halo and, more in general, weakly bound nuclei such as the stable ^{6}Li, ^{7}Li, ^{9}Be, has been widely investigated over the last twenty years. Results have shown that the very low breakup threshold of such nuclei, combined with the halo or cluster structure of the ground state, will strongly affect the reaction

dynamics around the Coulomb barrier. A recent overview can be found in [1]. A complete theoretical description of such collisions requires a reaction theory which includes all of the structure and dynamic effects. This is partially done in Continuum Discretized Coupled Channels (CDCC) calculations. However such calculations include only coupling with elastic break-up and other effects such as coupling to inelastic breakup or transfer must be included with a different approach. Recently, coupling with excited states of the core have also been included in the CDCC in the case of collisions induced by the 1n-halo ^{11}Be [2]. For the 2n-halo nuclei the inclusion of coupling with core excitation is still beyond reach.
Fusion excitation functions may also be affected both by static and/or dynamic effects being due respectively to the diffuse surface of the halo and to the coupling effects also to the continuum. Moreover, fusion reactions in collisions induced by weakly bound nuclei are complicated by the fact that, owing to the large breakup probability, in addition to complete fusion (CF) one can have a non negligible contribution of incomplete fusion (ICF) following breakup of the projectile or transfer see e.g. [3, 4, 33].

2 Elastic scattering

Elastic scattering with halo nuclei, ^{6}He, ^{11}Be and ^{11}Li, at energies around the Coulomb barrier has been measured by different authors mostly on heavy targets [6–13]. In collisions induced on heavy nuclei the data showed a suppression of the elastic cross-section in the region of the Coulomb-nuclear interference peak when compared with the scattering induced by the corresponding stable isotope at similar c.m. energies. Optical Model (OM) analysis allowed to reproduce the data only using very large imaginary diffuseness, indicating the presence of long range absorption [8–10, 13]. Recently, the ^{11}Li+^{208}Pb elastic scattering around the Coulomb barrier was reported and compared with the scattering of ^{9}Li on the same target [13]. A huge suppression of the ^{11}Li elastic cross-section was found with respect to the ^{9}Li one, much larger than the one observed in the case of ^{6}He on similar mass targets [8–10]. CDCC calculations show that coupling to continuum via the Coulomb-dipole interaction gives an important contribution to the disappearance of the Coulomb-nuclear interference peak. Results are somehow different on medium mass and light targets.
On medium mass targets very few experiments have been performed. The measurement of elastic scattering of ^{6}He+^{64}Zn was reported in [6, 7]. Although the elastic cross-section is overall suppressed when compared to the

one for ^{4}He+^{64}Zn, the Coulomb-nuclear interference pattern is not as much reduced as it was observed in the case of heavy targets. More recently, the ^{4,6}He+^{64}Zn measurements were extended to higher c.m. energies $E_{c.m.}$=13.5 and 16.5 MeV and compared with CDCC calculation [14] . These calculations underestimate the experimental data (see figure 1 of [14]), therefore, other effects than coupling to the break-up channel must be into play in the energy range explored in that study. To investigate on that, Coupled Reaction Channel (CRC) calculations were performed in order to include coupling to the one-neutron transfer to bound and unbound states of the target. Such calculations described the elastic, break-up and fusion cross section [15] of the ^{6}He+^{64}Zn reaction at lower energy $E_{c.m.}$=12.4 MeV [6,7]. It is found that coupling to the one-neutron transfer reproduces better the data than when coupling to the two-neutron transfer to bound and unbound states of the target is considered.
At the REX-ISOLDE facility of CERN, scattering and direct reactions for collisions induced by the 1n halo nucleus ^{11}Be and its core ^{10}Be at the same center of mass energy $E_{c.m.}$ = 24.5 MeV [11, 12] were measured. As shown in figure 1, in the ^{11}Be case it was found a suppression of the elastic angular distribution in the region of the Coulomb-nuclear interference peak which is not present in the ^{6}He case on the same target at similar $E_{c.m.}/V_C$. The Coulomb-nuclear interference seems to be constructive in the case of ^{6}He and distructive in the case of ^{11}Be. The difference observed could be the result of the lower binding energy of ^{11}Be compared to the one of ^{6}He and the consequent more extended halo, in addition the effect of Coulomb-dipole coupling is weaker in the ^{6}He case.
The ^{11}Be elastic scattering angular distribution has been reproduced within the OM, using as bare potential the one that reproduces ^{10}Be scattering, and adding a phenomenological surface Dynamic-Polarization-Potential (DPP). A good fit was obtained with a DPP having a large diffuseness of the order of 3.5 fm in agreement with [16]. The ^{11}Be data were also reproduced within the CDCC frame, showing that the suppression of the Coulomb nuclear interference peak is due to the combined effect of Coulomb and nuclear couplings to continuum.

Core excitation effects have been found to be important in the scattering and break-up of ^{11}Be on the heavy Au target [17].
Although scattering data in collision induced by halo nuclei around the Coulomb barrier start to show some common features, most of those data have been obtained with ^{6}He beams and, in some cases, the results are limited by the poor beam quality. Therefore, additional data, including the ones obtained with p-halo nuclei, are needed to build a wider systematics

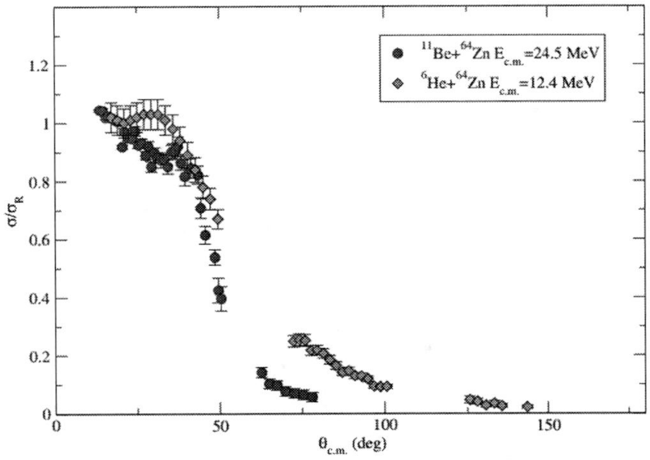

Figure 1: (Color online) Comparison of scattering angular distribution for ^{11}Be (red circles) and ^{6}He (green diamonds) on a ^{64}Zn target at similar energies with respect to the Coulomb barrier.

and allow a deeper understanding of the different effects involved.
The study of collisions induced by stable weakly bound nuclei such as ^{6}Li, ^{7}Li and ^{9}Be, which do not have a halo structure, has also a considerable interest to investigate on the problems related to the presence of the continuum in the vicinity of the ground state. When compared with halo nuclei, they have larger breakup thresholds and do not show the same extended mass distribution. However, coupling to continuum effects are still expected to be important and, since beams of such stable nuclei are available with high currents, high quality data can be gathered.
For weakly bound nuclei the projectile breakup channel is expected to be important even at energies below the Coulomb barrier. In fact, breakup is an open reaction channel also at energies much below the barrier. Coupling to the breakup produces a repulsive polarization potential (e.g. [18]) which is responsible for the disappearance the usual threshold anomaly (TA) in the OP.
At Laboratori Nazionali del Sud in Catania elastic scattering angular distributions were measured at several energies around and below the barrier for the systems ^{6,7}Li+^{64}Zn [19,20]; clear absence of the usual threshold anomaly has been observed in the case of ^{6}Li+^{64}Zn. Absence of the usual threshold anomaly in the OP, in collisions induced by weakly bound nuclei both stable and radioactive has been observed also for other systems (e.g. [21–25]). The ^{6,7}Li+^{64}Zn elastic scattering data have been also analised within the CDCC approach. In figure 2 it is shown a comparison between the ^{6}Li+^{64}Zn data

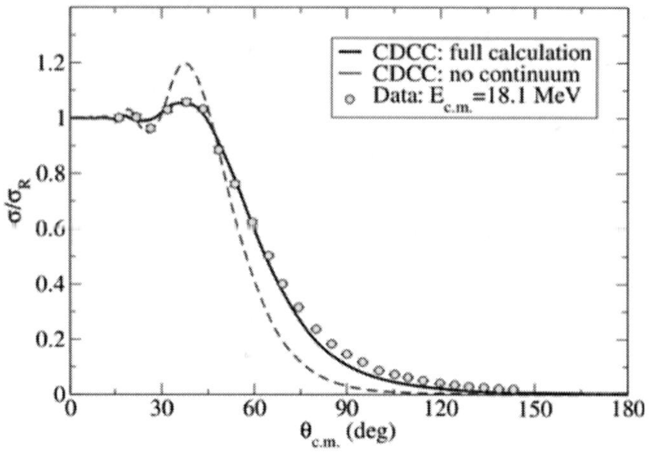

Figure 2: (Color online) Elastic scattering angular distribution for $^{6}Li+^{64}Zn$ at $E_{c.m.}$=18.1 MeV. Symbols are the experimental data, the full line represent the CDCC calculation with full coupling while the dashed line represent the calculation with no-continuum coupling.

and the CDCC calculation both, with and without coupling with the continuum; it can be seen that continuum coupling is essential to reproduce the data. In addition, coupling to resonances in the continuum is also found to be important in collisions induced by ^{6}Li at the barrier energies [20].

3 Fusion reactions

Recently, a complete review on different fusion data for collisions induced by halo and weakly bound nuclei has been published [1]. Systematic studies suggests that, for collisions induced by halo nuclei, once static effects are removed, the residual effect of coupling to continuum or to transfer is a suppression of fusion above barrier and, possibly, a small enhancement of fusion below the barrier. Unfortunately most of presently available data do not explore the sub-barrier region with reasonable errors and new data would be necessary for a complete understanding of this topic.
Fusion excitation functions in collisions induced by stable weakly bound nuclei have been measured for several systems in a wide target mass range [e.g. [3, 26–32]. For fusion-evaporation reactions induced on heavy targets (e.g. [3, 26–28]), where the emission of charged particles is hindered by the presence of a strong Coulomb barrier, complete (CF) and incomplete fusion (ICF) contributions can usually be clearly separated. The main finding for

such systems is that CF is suppressed above the barrier by about 30% with respect to the prediction of Single Barrier Penetration calculations or Coupled Channel calculations not including coupling to continuum effects.
For fusion of weakly bound nuclei on medium mass or light targets (e.g. [29–32]), the compound nucleus can evaporate charged particles; therefore, the same residues can be populated by the CF and ICF mechanisms, making their separation much more difficult. For this reason, most of the fusion data for light and medium mass systems refer to total fusion (TF) cross section. However, also the single nucleon transfer could populate the same heavy fragments and therefore the measured cross-section could contain contributions of direct processes. In the case of collision on medium mass targets no effects on TF are observed at energies above barrier, whereas below the barrier not much data are available.
Recently, in Catania, the excitation functions for heavy residue (H.R.) production was measured, in the collisions ^{6,7}Li+^{64}Zn around the Coulomb barrier, with the aim to extract some information on the competition between CF, ICF and other reaction mechanisms in the H.R. production [33]. It was found that, at energies below the barrier, the H.R. cross-section measured in the ^{6}Li collision is larger than for ^{7}Li, in agreement with [31]. Since ^{6}Li and ^{7}Li have weakly bound cluster structures of the ground state, ICF reactions, where only an alpha particle, a deuteron or a tritium are captured by the target, are expected to contribute to the H.R.. Moreover, as it has been shown in [34, 35], transfer triggered breakup is an important reaction channel in ^{6,7}Li induced reaction, in some cases even more important than direct breakup. Those type of processes would contribute to the production of the same H.R. as for CF and ICF. In [33] it is shown that at energies above the barrier CF is the dominant mechanism; however, at energies below the barrier the H.R. experimental relative yield is rather different than the one expected for CF, showing that CF is no longer the dominant reaction mechanism while other processes such as ICF, direct cluster transfer (DCT) or single nucleon transfer become very important. The present results suggest that, in order to study fusion below the barrier in collisions induced by weakly bound nuclei on medium mass targets, the simple integration of the H.R. yield is not giving the TF cross-sections since direct processes may give an important contribution to the cross-section. Therefore, one needs to find experimental techniques allowing to properly separate the different reaction mechanisms contributing to the H.R. yield.

4 Summary and conclusions

In this paper experimental results concerning the study of elastic scattering and reactions for halo and stable weakly bound nuclei around the Coulomb barrier have been summarized . We have shown that, in collisions induced by halo nuclei, a strong suppression of the elastic scattering cross-section and, correspondingly, a large enhancement of the total reaction cross-section has been observed. Core excitation effects and coupling with the 1n-transfer channel are also found to be important in the elastic scattering with halo nuclei.

In elastic scattering with weakly bound stable nuclei coupling to continuum effects are found to be very important.

Fusion excitation functions in collisions induced by halo nuclei suggest the presence of a fusion cross-section enhancement around the barrier which is due mainly (but not only) to static effects. Fusion with stable weakly bound nuclei on heavy targets shows at energies above the barrier a well established suppression of CF due to ICF following projectile breakup. In collisions with medium mass or light targets CF and ICF cannot be easily separated and most of available data refers to TF. ^{6,7}Li+^{64}Zn fusion data suggest that above barrier H.R. production is dominated by CF. Below the barrier, on the contrary, other processes such as ICF but also DCT or single neutron transfer are the dominating processes for H.R. production.

References

[1] L.F. Canto et al Phys. Rep. **596**, (2015) 1-86

[2] R. de Diego et al. Phys.Rev. C **89**, 064609 (2014)

[3] M. Dasgupta et al. Phys. Rev. C **70**, (2004) 024606

[4] V. Scuderi et al. Phys. Rev. C **84**, (2011) 064604

[5] A. Di Pietro et al. Phys. Rev. C **87**, (2013) 064614

[6] A. Di Pietro et al. Europhys. Lett. **64** (2003) 309

[7] A. Di Pietro et al. Phys. Rev. C **69** (2004) 044613

[8] L. Acosta et al. Phys. Rev. C **84** (2011) 044604

[9] A. M. Sanchez Benitez et al. Nucl. Phys. A **803** (2008) 30

[10] O. R. Kakuee et al. Nucl. Phys. A **765** (2006) 294

[11] A. Di Pietro et al. Phys. Rev. Lett. **105** (2010) 022701

[12] A. Di Pietro et al. Phys. Rev. C **85** (2012) 054607

[13] M. Cubero et al. Phys. Rev. Lett. **109** (2012) 262701

[14] J.P. Fernandez-Garcia et al. *Contribution to these proceedings*

[15] A. M. Moro et al. EPJ Web of Conferences **17** (2011) 08001

[16] A. Bonaccorso et al. Nucl. Phys. A **706** (2002) 322

[17] V. Pesudo et al. Act. Phys. Pol. B **45** (2014) 375

[18] Y. Sakuragi et al. Phys. Rev. C **35** (1987) 2161

[19] M. Zadro et al. Phys. Rev. C **80** (2009) 064610

[20] J.P. Fernandez-Garcia et al. *Accepted for publication Phys. Rev. C*

[21] A.M.M. Maciel et al. 1999 Phys. Rev. C 59 2103

[22] A. Pakou et al. 2004 Phys. Rev. C 69 054602

[23] H. Kumawat et al. 2008 Phys. Rev. C 78 044617

[24] J.M. Figueira et al. 2010 Phys. Rev. C 81 024613

[25] N.N. Deshmukh et al. 2011 Phys. Rev. C 83 024607

[26] P.R.S. Gomes et al. Phys. Rev. C **73** (2006) 064606

[27] Y.W. Wu et al. Phys. Rev. C **68** (2003) 044605

[28] Pradhan M K et al. Phys. Rev. C **83** (2011) 064606

[29] P.R.S. Gomes et al. Phys. Rev. C **71** (2005) 034608

[30] R.M. Anjos et al. Phys. Lett. B **534** (2002) 45

[31] M. Sinha et al. Eur. Phys. J. A **44** (2010) 403

[32] C. Beck et al. Phys. Rev. C **67** (2003) 054602

[33] A. Di Pietro et al. Phys. Rev. C **87** (2013) 064614

[34] D.H. Luong et al. Phys. Lett. B **695** (2011) 105

[35] D.H. Luong et al. Phys. Rev. C **88** (2013) 034609

EPJ Web of Conferences **117**, 06004 (2016) DOI: 10.1051/epjconf/201611706004
NN2015

Self-consistent description of deformed nuclei at the proton drip line

LIDIA S. FERREIRA[1], ENRICO MAGLIONE[2] and PETER RING[3]

[1] Centro de Física e Engenharia de Materiais CeFEMA and
Departmento de Física, Instituto Superior Técnico,
Universidade de Lisboa, Av Rovisco Pais, 1049 001, Lisboa, Portugal
[2] Dipartimento di Fisica e Astronomia "G. Galilei",
Via Marzolo 8, I-35131 Padova, Italy
and INFN, Sezione di Padova, Italy
[3] Physik-Department der Technischen Universität München,
D-85748 Garching, Germany

Abstract

Proton radioactivity from deformed nuclei is described for the first time by a fully self–consistent calculation based on covariant relativistic density functionals derived from meson exchange and point coupling models.

1 Introduction

The production and detection of nuclei towards the limits of stability saw a considerable development in the last decade, due to a better capability to achieve fusion evaporation and multi fragmentation reactions [1] involving very small production cross sections, and producing exotic nuclei with extremely short half lives, below a microsecond. The position of the extremes of bound matter has been confirmed on the neutron side up to Magnesium, and on proton side for charges up to Z=83, with only few exceptions of nuclei yet to be found. The observation of one proton radioactivity, mapped the drip-line between Tin and Bismuth.

Nuclei at the proton drip-line, provide a challenging domain for spectroscopic studies far from stability [2]. Recent advances in tagging techniques lead to the discovery of many excited states [3], resulting in a significant progress on the knowledge of the nuclear structure properties of proton rich nuclei, and providing an important input towards the understanding of fundamental symmetries and of the nucleosynthesis of the elements. In fact, the path of the rapid proton capture reactions for the production of nuclei, can involve nuclei at the proton drip-line [4], but the reaction rates will depend strongly on details of the structure of nuclei along the path, and in particular, on the knowledge of resonances, since it is mainly dominated by resonant proton capture. Penning traps, have also allowed to isolate isomeric states, and to determine high precision mass measurements [5–7], that lead to a clear definition of proton separation energies, which are very import to establish the drip lines and for astrophysical calculations of the abundances.

In spite of the experimental advances in the area, a great part of the nuclear structure knowledge of nuclei at the limits of stability, still relies on theoretical extrapolations. In this context, the need to have a solid theoretical ground, with great independence from parameterizations, and with the capability to interpret the data and help the search of new emitters, is crucial.

Theoretical studies on proton radioactivity, were able to interpret the experimental half-lives and branching ratios measured in the observation of decay, and to predict the properties of the decaying state and shape of the emitter. Non-relativistic calculations using simple semi-classical WKB methods [8], were used for spherical emitters, while deformed proton emitters were studied more rigorously by identifying that the decay proceeds from a single-particle resonances [9, 10]. Microscopic studies of deformed proton emitters allow to test the details of the nuclear wave function to which the half-lives are quite sensitive. The most consistent non-relativistic theoretical approach for proton emission is the non-adiabatic quasiparticle approach which is very successful [11] in bringing out several interesting features of deformed odd-even proton emitters including the triaxially deformed ones [12]. Its extension to odd-odd nuclei [13], suggested that it is possible to test the residual neutron proton interaction and to determine the neutron Nilsson level even if the neutron does not participate actively in the decay.

The above non relativistic microscopic models, rely on the various parameterizations of the nuclear mean field available in the literature, described in terms of phenomenological Wood-Saxon type forms, with parameters fitted to global nuclear properties, or by semiclassical considerations based on the liquid drop model and the proximity force. There is no interactions yet from first principles.

Covariant density functional theory (CDFT) [14–16], might present some advantages. It is derived from relativistic quantum field theory, fulfils Lorentz invariance allowing to describe consistently the spin-orbit coupling, thus restricting the number of parameters of the functionals to be adjusted in order to reproduce bulk properties of spherical and deformed nuclei. The empirical pseudospin symmetry is also explained, and the model is consistent with the non linear realization of chiral symmetry. The functionals can be considered universal since they are valid all over the periodic table, where mean field theory is applicable.

A density dependence is introduced to allow for a quantitative description of the nuclear surface [17]. This can be achieved either by using non-linear coupling terms or by considering coupling constants dependent on the density of the exchanged mesons. Examples of these interactions are the non-linear meson exchange model the NL3 [18] and the density dependent point coupling models the DD-PC1 [19], and the PC-PK1 [20], which were able to describe ground-state properties and collective excitations of β stable and exotic nuclei at the extremes of stability [21]. Ground-state quadrupole deformations have been predicted for proton rich nuclei, and proton separation energies have been calculated, allowing the proton drip-line to be mapped with high precision [21, 22], and the single particle configurations and spectroscopic factors to be derived very successfully.

Relativistic mean-field calculations based on CDFT, have been used to interpret proton radioactivity from spherical nuclei [23, 24]. In our study of Ref. [23], we have performed a fully self-consistent relativistic mean field calculations for spherical nuclei, using the NL3 and DD-PC1 interactions. The decay process is described there within the exact formalism of quantum mechanics for decay of a resonance, and no additional parameters were introduced, besides the ones implicit in the definition of the NL3 and DD-PC1 interactions. The calculation was very successful in reproducing the experimental data, and provided a clear evidence for a mixing of configurations. The calculation of Ref. [24], interprets proton radioactivity, similarly to α decay, and not decay from a single particle resonance. The proton escape is simulated by a WKB tunneling process, where the potential barrier is extracted within the renormalization group approach. Therefore, it is not a self consistent calculation.

We have extended our spherical calculation in order to account for deformation [25], and in the present work, we discuss a fully self consistent calculation of proton emission from ^{131}Eu, a well deformed nucleus.

1.1 Covariant density function theory for decay of deformed proton emitters

In CDFT, the nucleons are described by the Dirac spinors ψ, interacting in an effective Lagrangian through the exchange of the isoscalar scalar σ, isoscalar vector ω, and isovector vector ρ mesons, and the electromagnetic field. This simple model, with interaction terms linear in the meson fields, does, however, not provide a quantitative description of the nuclear surface properties, thus the need for an effective density dependence, as mentioned above. From the Lagrangian density the classical variation principle leads to the equations of motion, which are the Dirac equation for the nucleons, equivalent to the Kohn-Sham equations in non-relativistic density functional theory, and the Klein-Gordon equations for the meson fields. Pairing correlations are taken into account in the constant gap approximation. This set of equations is non-linear and is solved by iteration starting with an initial guess for the potentials, until self-consistency is achieved, and the final mean field interaction obtained.

For the calculation of proton emission from ^{131}Eu, we have used the relativistic point-coupling models, DD-PC1 and PC-PK1 of the Peking group, both successful in describing nuclear properties, but with some differences between them [26, 27], regarding their structure and fitting procedure. The functional PC-PK1 has phenomenological parameters fitted to the binding energies and radii of a large number of spherical semi-magic nuclei all over the periodic table. The functional DD-PC1, was instead fitted to masses of 64 heavy deformed nuclei and a few nuclear matter properties. Therefore, both interactions describe spherical and deformed nuclei differently, being the masses of deformed nuclei better reproduced by the DD-PC1, whereas for spherical nuclei, the masses are over-bound.

The scalar and vector potentials that describe the deformed proton emitter, were obtained in a fully self-consistent way, by solving the relativistic mean field equations for the even-even daughter nucleus including pairing correlations in the BCS approximation. The resulting self-consistent axial potentials $S(\mathbf{r})$ and $V(\mathbf{r})$ were then expanded in spherical harmonics with different L-values, obtaining the corresponding potentials $S_L(r)$ and $V_L(r)$ on a fine mesh in r-space. The latter interactions, were then used to calculate the proton Nilsson resonances, which are the solutions of the Dirac equations imposing outgoing wave boundary conditions. This was achieved by solving a full coupled channel problem in coordinate space, leading to a precise description of the energy, width and wave function of the proton resonances.

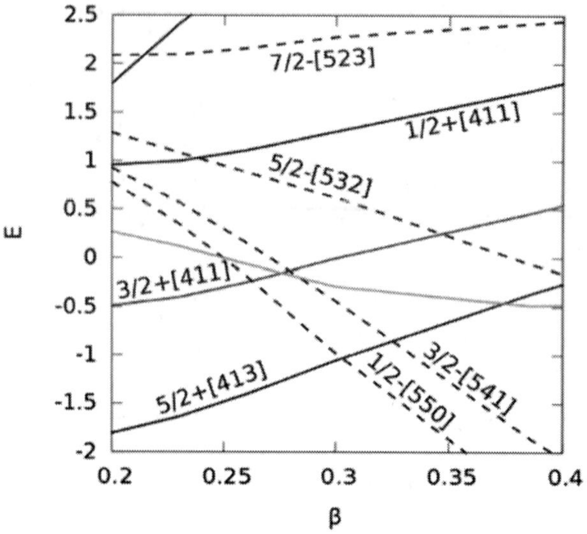

Figure 1: Proton Nilsson resonances in ^{131}Eu as a function of quadrupole deformation (β_2) using the DD-PC1 field. The Fermi energy is represented by the green line.

This is a complex problem, solved previously only for K=$1/2^+$ resonant states by the Beijing group [28], and in [29] as an extension of the complex scaling method. We have solved the full relativistic coupled channel problem for any value of the angular momentum in coordinate space [25], using similar techniques as in the non-relativistic case [30]. The Nilsson diagram for resonances in ^{131}Eu is presented in Fig. 1 for the self-consistent density dependent point coupling DD-PC1 interaction. Radioactive decay of ^{131}Eu, was observed from the ground state to the ground and first excited 2^+ states of the daughter nucleus ^{130}Sm [31,32]. The nucleus is highly deformed, and if there is no change of deformation during the decay, ^{131}Eu, should have a similar deformation to the one expected for ^{130}Sm, i.e. a value $\beta_2 \approx 0.34$, deduced from the empirical relation of Ref. [33]. As it can be seen from Fig.1, the $3/2^+[411]$ state is close to the Fermi surface for a large quadrupole deformation, so it is the candidate to be the decaying state, similarly to what was predicted by the non-relativistic calculation [34].

We have calculated the half-life for decay from the various states close to the Fermi surface and the branching ratios for the decay of these states, to the first excited 2^+ state of the daughter ^{130}Sm. For details, see Ref. [25].

The half-life depends strongly on the quantum numbers of the decaying state, and also on details of the interaction like for example the nuclear radius, and strength of the spin-orbit force. In contrast, the branching ratios

Table 1: Branching ratio for decay from different J^+ states of ^{131}Eu to the first excited 2^+ state of the daughter ^{130}Sm, with a deformation $\beta_2 = 0.34$, and for the DD-PC1 model. The experimental branching ratio is 0.24(5) [31].

J^π	Branching
$3/2^+$ [411]	0.191
$1/2^+$[411]	0.024
$5/2^+$[413]	0.031
$5/2^+$[402]	0.023
$7/2^+$[404]	0.029

are quite stable regarding these quantities, but are very sensitive instead to details of the wave function components with different angular momentum, since their calculation involves the sum over all allowed partial decay widths for decay, divided by the total width, so they are a ratio between very different amplitudes. Therefore, the branching ratio depends on a balance between details of the wave function components, and for this reason an excellent probe of nuclear structure properties.

We present in Table 1 the calculation, of the branching ratio for decay from the Nilsson single particle states of ^{131}Eu, close to the Fermi energy, to the 2^+ state of the daughter nucleus, assuming they have a quadrupole deformation predicted by the Grodzins relation, $\beta_2 = 0.34$. As it can be seen, only the $3/2^+$[411] state interprets perfectly the experimental value of 0.24(5) [31], in agreement with what was found in the non-relativistic calculation [34], and also the predictions based on microscopic-macroscopic mass calculations [35]. The calculation was done with the DD-PC1 model, but similar conclusions can be drawn if the PC-PK1 is used, since only minor differences in the behaviour of the half-lives was found.

This is the first fully self-consistent calculation of proton emission from deformed nuclei, with interactions derived from covariant density functionals. It was very successful to reproduced the experimental data, and has provided a new test of the relativistic point-coupling functionals, the DD-PC1 and the PC-PK1.

2 Conclusions

In conclusion, theoretical calculations within relativistic mean field models present some advantages with respect to the non-relativistic ones, concerning

DOI: 10.1051/epjconf/201611706004

the parameterization of the nuclear interaction, but a unique parameterization for the Lagrangian, which is able to describe properties of nuclei from light to very heavy, and from the proton drip-line to the neutron one, is still not available. Therefore, it is desirable to provide a new test of the interactions derived from the relativistic density functionals, and observe their performance. Proton radioactivity provides this possibility. We have presented the first self-consistent calculation in the literature for decay of deformed emitters, using interactions derived from covariant density functional theory, and determined exactly the Nilsson resonances and the decay observables. We have used two density dependent point coupling models, the DD-PC1 [19] and the PC-PK1 [20], and concluded that they could also describe proton radioactive nuclei, and predict their properties.

From the above discussion, one can conclude that a very solid theory exists that can describe proton radioactivity, interpret the data, and predict features of the structure of nuclei at the extremes of stability, within covariant relativistic mean field theory.

References

[1] M. Thoennessen, AIP Conf.Proc. **8**02 (2005) 250.

[2] B. Blank and M. J. G. Borge, Prog. Part. Nucl. Phys. **6**0 (2008) 403.

[3] P.T. Wady, *et al.* Phys. Lett. **B 740** (2015) 243.

[4] M. Arnould, S. Goriely, Phys. Reports **3**84 (2003) 1.

[5] C. Rauth *et al.*, Phys. Rev. Lett. **10**0 (2008) 012501.

[6] H. J. Kluge, Int. Journal of Mass Spectrometry **3**49 (2013) 26.

[7] M. Kankainen, *et al.*, Proceedings of Science, NIC XII **0**29 (2013).

[8] S.Åberg, P. B. Semmes, W. Nazarewicz, Phys. Rev. **C** 56 (1997) 1762.

[9] L. S. Ferreira, E. Maglione, Phys. Rev. Lett. **8**6 (2001) 1721.

[10] C. N. Davids and H. Esbensen, Phys. Rev. C **6**9 (2004) 034314.

[11] G. Fiorin, E. Maglione, L. S. Ferreira, Phys. Rev. **C** 67 (2003) 054302.

[12] P. Arumugam, L. S. Ferreira, E. Maglione, Phys. Lett. **B**680 (2009) 443.

[13] M. Patial, P. Arumugam, L. S. Ferreira, E. Maglione, Phys. Rev. **C80** (2013) 004300.

[14] M. H. Johnson and E. Teller, Phys. Rev. **98** (1955) 783.

[15] H.-P. Dürr, Phys. Rev. **103** (1956) 469.

[16] J. D. Walecka, Ann. Phys. (N.Y.) **83** (1974) 491.

[17] J. Boguta and A. R. Bodmer, Nucl. Phys. **A 292** (1977) 413.

[18] G. A. Lalazissis, J. König, P. Ring, Phys. Rev. **C55** (1997) 540.

[19] T. Nikšić, D. Vretenar, P. Ring, Phys. Rev.**C78** (2008) 034318.

[20] P. W. Zhao, *et al*, Phys. Rev. **C82** (2010) 054319.

[21] G. A. Lalazissis, *et al.* Nucl. Phys. A **650** (1999) 133.

[22] G. A. Lalazissis, D. Vretenar, P. Ring, Phys. Rev. **C69** (2004) 017301.

[23] L. S. Ferreira, E. Maglione, P. Ring, Phys. Lett. **B701** (2011) 508.

[24] Q. Zhao, *et al*, Phys.Rev. **C90** (2014) 054326.

[25] L. S. Ferreira, E. Maglione, P. Ring, Phys. Lett. B, in press.

[26] S. E. Agbemava, *et al*, Phys. Rev. **C89** (2014) 054320.

[27] Q. S. Zhang, *et al*, Front. Phys. **9** (2014) 529.

[28] Z. P. Li, *et al*, Phys. Rev. **C81** (2010) 034311.

[29] M. Shi, *et al*, Phys. Rev. **C90** (2014) 034319.

[30] L. S. Ferreira, E. Maglione, R. Liotta, Phys. Rev. Lett. **78** (1997) 1640.

[31] A. A. Sonzogni, *et al*, Phys. Rev. Lett. **83**(1999) 1116.

[32] C. N. Davids, *et al.* Phys. Rev. Lett. **80** (1998) 1849.

[33] L. Grodzins, Phys. Lett. **2** (1962) 88.

[34] E. Maglione, L. S. Ferreira, Phys. Rev. **C61**(2000) 047307.

[35] P. Möller, *et al*, At. Data Nucl. Data Tables **66** (1997) 131.

EPJ Web of Conferences **117**, 06005 (2016) DOI: 10.1051/epjconf/201611706005
NN2015

A consistent four-body CDCC model of low-energy reactions: Application to ^{9}Be+^{208}Pb

M. S. Hussein[1,2,3], P. Descouvemont[4] and L. F. Canto[5,6]

[1] Departamento de Física Matemática, Instituto de Física, Universidade de São Paulo, C.P. 66318, 05314-970, São Paulo, SP, Brazil
[2] Instituto de Estudos Avançados, Universidade de São Paulo, C.P. 72012, 05508-970, São Paulo, SP, Brazil
[3] Departamento de Física, Instituto Tecnológico de Aeronáutica, CTA, São José dos Campos, São Paulo, SP, Brazil
[4] Physique Nucléaire Théorique et Physique Mathématique, C.P. 229, Université Libre de Bruxelles (ULB), B 1050 Brussels, Belgium
[5] Instituto de Física, Universidade Federal do Rio de Janeiro, C.P. 68528, 21941-972 Rio de Janeiro, RJ, Brazil
[6] Instituto de Física, Universidade Federal Fluminense, Av. Gal. Milton Tavares de Souza s/n, Niterói, RJ, Brazil

Abstract

We investigate the ^{9}Be+^{208}Pb elastic scattering, breakup and fusion at energies around the Coulomb barrier. The three processes are described simultaneously, with identical conditions of calculations. The ^{9}Be nucleus is defined in an $\alpha+\alpha+n$ three-body model, using the hyperspherical coordinate method. We first analyze spectroscopic properties of ^{9}Be, and show that the model provides a fairly good description of the low-lying states. The scattering with ^{208}Pb is then studied with the Continuum Discretized Coupled Channel (CDCC) method, where the $\alpha+\alpha+n$ continuum is approximated by a discrete number of pseudostates. Optical potentials for the α+^{208}Pb and n+^{208}Pb systems are taken from the literature. We present elastic-scattering and fusion cross sections at different energies.

1 Introduction

Many experiments have been performed with the ^{9}Be nucleus, used as a target or as a projectile [1]. Although ^{9}Be is stable, it presents a Borromean structure, as the well known halo nucleus ^{6}He. None of the two-body subsystems $\alpha + n$ or $\alpha + \alpha$ is bound in ^{9}Be, which has important consequences on the theoretical description of this nucleus. Precise wave functions must include the three-body nature of ^{9}Be. The hyperspherical formalism [2] is an ideal tool to describe three-body Borromean systems, as it does not assume a specific two-body structure, and considers the three particles $\alpha + \alpha + n$ on an equal footing.

In the present work [3], we aim at investigating ^{9}Be scattering and fusion on a heavy target. The reaction framework is the Continuum Discretized Coupled Channel (CDCC) method (see Ref. [4] for a recent review), which is well adapted to weakly bound projectiles since it allows to include breakup channels. Going from two-body projectiles (such as d=p+n or ^{7}Li=α+t) to three-body projectiles, however, strongly increases the complexity of the calculations, even if both options eventually end up with a standard coupled-channel system.

Many data have been obtained for ^{9}Be+^{208}Pb elastic scattering [5, 6] and fusion [7, 8]. These experimental data provide a good opportunity to test ^{9}Be wave functions. No assumption should be made about the cluster structure, and α+^{208}Pb as well as n+^{208}Pb optical potentials are available in the literature.

2 Three-body model of ^{9}Be

The determination of the ^{9}Be wave functions is the first step for the ^{9}Be+^{208}Pb CDCC calculation. For a three-body system, the Hamiltonian is given by

$$H_0 = \sum_{i=1}^{3} \frac{\boldsymbol{p}_i^2}{2m_i} + \sum_{i<j=1}^{3} V_{ij}(\boldsymbol{r}_i - \boldsymbol{r}_j), \tag{1}$$

where $\boldsymbol{r}_i$ and $\boldsymbol{p}_i$ are the space and momentum coordinates of the three particles with masses m_i, and V_{ij} a potential between nuclei i and j. For the $\alpha + \alpha$ interaction, we use the deep potential of Buck *et al.* [9]. The α+n interaction is taken from Kanada *et al.* [10]. Both (real) potentials accurately reproduce the $\alpha + \alpha$ and α+nucleon phase shifts over a wide energy range.

With these potentials, the ^{9}Be ground state is too bound (-3.12 MeV, while the experimental value is -1.57 MeV with respect to the $\alpha+\alpha+n$ threshold). We have therefore introduced a phenomenological three-body force which reproduces the experimental binding energy of the $3/2^-$ ground state. The r.m.s. radius, the quadrupole moment, and the magnetic moment are $\sqrt{r^2} = 2.36$ fm, $Q(3/2^-) = 4.96$ e.fm^2, and $\mu = -1.33$ μ_N, respectively. These values are in fair agreement with experiment (2.45 ± 0.01 fm, 5.29 ± 0.04 e.fm^2 and -1.18 μ_N, respectively)

3 The CDCC method

We present here a brief outline of the CDCC method, and we refer to Ref. [11] for specificities of three-body projectiles. The CDCC method is based on approximate solutions of the projectile Hamiltonian (1)

$$H_0\Phi_k^{jm\pi} = E_{0,k}^{j\pi}\Phi_k^{jm\pi}, \tag{2}$$

where k are the excitation levels in partial wave $j\pi$. Solutions with $E_{0,k}^{j\pi} < 0$ correspond to bound states of the projectile, whereas $E_{0,k}^{j\pi} > 0$ correspond to narrow resonances or to approximations of the three-body continuum. These states cannot be associated with physical states, and are referred to as pseudostates. The ^{9}Be spectrum used in the present model is shown in Fig. 1.

The Hamiltonian of the projectile + target system is written as

$$H(\boldsymbol{R},\boldsymbol{x},\boldsymbol{y}) = H_0(\boldsymbol{x},\boldsymbol{y}) - \frac{\hbar^2}{2\mu_{PT}}\Delta_R + V(\boldsymbol{R},\boldsymbol{x},\boldsymbol{y}) \tag{3}$$

where μ_{PT} is the reduced mass of the system, and $\boldsymbol{R}$ the relative coordinate. The Jacobi coordinates $\boldsymbol{x}$ and $\boldsymbol{y}$ are proportional to $\boldsymbol{r}_{\text{Be-n}}$ and $\boldsymbol{r}_{\alpha-\alpha}$, respectively. The potential term reads

$$V(\boldsymbol{R},\boldsymbol{x},\boldsymbol{y}) = V_{1t}(\boldsymbol{R},\boldsymbol{y}) + V_{2t}(\boldsymbol{R},\boldsymbol{x},\boldsymbol{y}) + V_{3t}(\boldsymbol{R},\boldsymbol{x},\boldsymbol{y}) \tag{4}$$

where the three components V_{it} are optical potentials between fragment i and the target.

In order to solve the Schrödinger equation associated with (3), the total wave function with angular momentum J and parity Π is expanded as

$$\Psi_{(\omega)}^{JM\Pi}(\boldsymbol{R},\boldsymbol{x},\boldsymbol{y}) = \sum_{j\pi kL} \varphi_{j\pi kL}^{JM\Pi}(\Omega_R,\boldsymbol{x},\boldsymbol{y})g_{jkL(\omega)}^{J\Pi}(R), \tag{5}$$

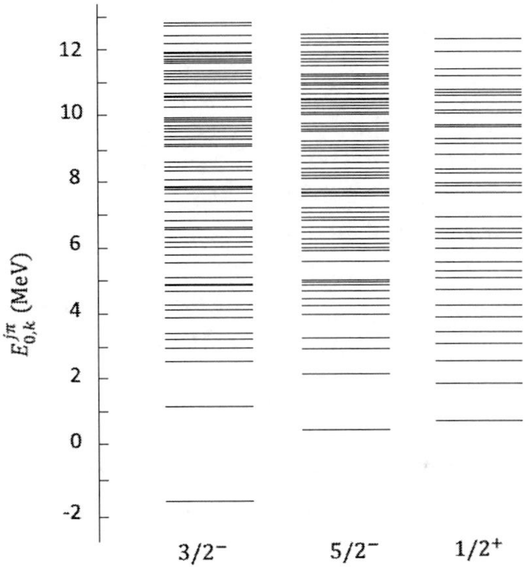

Figure 1: Pseudostate energies (defined from the $\alpha+\alpha+n$ threshold) for the dominant partial waves $j = 3/2^-, 5/2^-, 1/2^+$.

where ω is the entrance channel, and where the channel wave functions are given by

$$\varphi_{j\pi kL}^{JM\Pi}(\Omega_R, \boldsymbol{x}, \boldsymbol{y}) = i^L \Big[\Phi_k^{j\pi}(\boldsymbol{x}, \boldsymbol{y}) \otimes Y_L(\Omega_R)\Big]^{JM}. \tag{6}$$

The radial functions $g_c^{J\Pi}(R)$ (we use the notation $c = (j, \pi, k, L)$) are obtained from the system

$$\left[-\frac{\hbar^2}{2\mu_{PT}}\frac{d^2}{dR^2} + E_c - E\right] g_{c(\omega)}^{J\Pi}(R) + \sum_{c'} V_{c,c'}^{J\Pi}(R) g_{c'(\omega)}^{J\Pi}(R) = 0, \tag{7}$$

where E is the c.m. energy, and where the coupling potentials are given by matrix elements

$$V_{c,c'}^{J\Pi}(R) = \langle \varphi_c^{JM\pi} | V | \varphi_{c'}^{JM\pi} \rangle + \frac{\hbar^2}{2\mu_{PT}} \frac{L(L+1)}{R^2} \delta_{cc'}. \tag{8}$$

As the fragment-target potentials (4) are optical potentials, matrix elements (8) contain a real and an imaginary parts (see Ref. [3] for detail). We solve Eq. (7) with the R-matrix theory [12].

4 Application to the ^{9}Be+^{208}Pb system

4.1 Elastic scattering

The calculations are performed with the $j^\pi = 3/2^\pm, 5/2^\pm, 1/2^\pm$ partial waves on ^{9}Be, and the cutoff energy is 12.5 MeV. Figure 2 shows ^{9}Be+^{208}Pb elastic cross sections at two typical energies (the data are taken from Refs. [5,13]). In contrast with optical-model calculations [5], no renormalization of the potential is needed. The absorption is simulated by the imaginary parts of the $\alpha-^{208}$Pb and n$-^{208}$Pb interactions, and by the $\alpha + \alpha + n$ discretized continuum. Calculations involving only the $3/2^-$ ground state are shown as dashed lines, and the full calculations as solid lines. From Fig. 2, it appears that continuum couplings are more important at low energies. At $E = 38$ MeV, the single-channel calculation, limited to the ^{9}Be ground state is significantly different from the data. This confirms the conclusion of a previous work [14] which suggests that continuum couplings are more important at low energies.

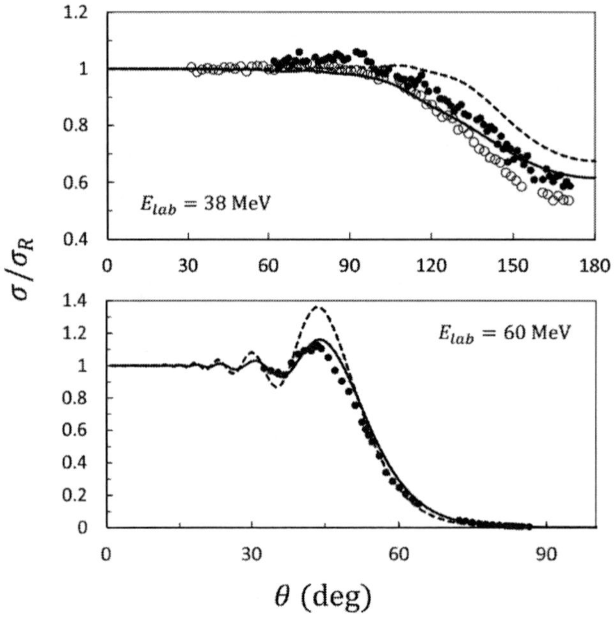

Figure 2: ^{9}Be+^{208}Pb elastic cross sections (divided by the Rutherford cross section) at two ^{9}Be laboratory energies, and for different sets of ^{9}Be partial waves. The experimental data are taken from Ref. [5] (filled circles) and Ref. [13] (open circles).

4.2 Fusion

The $^9Be+^{208}Pb$ fusion reaction has been studied theoretically by several groups (see Ref. [15] and references therein). Data about the various mechanisms (total TF, complete CF and incomplete ICF fusion) are also available in the literature [7, 8, 16].

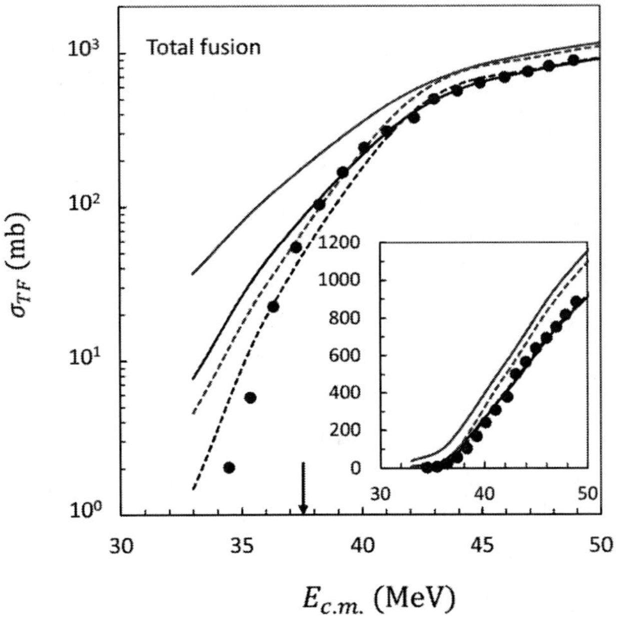

Figure 3: Total fusion cross sections with (black lines) and without (red lines) removal of the neutron capture. The dashed lines represent the single-channel calculations, and the solid lines represent the full calculations. The Coulomb barrier energy V_B is indicated by a vertical arrow.

The total fusion cross section essentially involves two contributions: the complete fusion, where the whole projectile is absorbed, and the incomplete fusion, where only a part of the projectile charge is absorbed by the target. As far as the total fusion is concerned, the cross section can be obtained by using the scattering matrices. The CDCC total fusion cross sections obtained in this way are shown in Fig. 3 as red lines. Above the Coulomb barrier, the present calculation overestimates the data by about a factor two. This overestimation is due to the different definitions adopted for fusion in the experiment and in the calculation. In other words, the theory defines fusion as the absorption of the whole projectile or any of its fragments, including the neutron. On the other hand, the experimental cross section only includes contributions from the absorption of charge [7]. Thus, the two cross sections differ by the neutron absorption which, owing to the lack of a Coulomb barrier, may be quite large.

The CDCC calculation can be adapted to the experimental conditions by using an equivalent definition of the fusion cross section, and by removing the neutron-target imaginary potential. The results obtained in this way are shown in Fig. 3 as black lines. By excluding the neutron capture, we obtain an excellent agreement with experiment above $E_{\rm cm} \approx 40$ MeV. This is more clearly shown in the inset, drawn in a linear scale. We confirm previous CDCC calculations [15, 21] which conclude that the influence of breakup channels increases at sub-barrier energies. Below the Coulomb barrier, the single-channel calculation is in better agreement with experiment, and the full calculation overestimates the data. This discrepancy is common to most CDCC fusion calculations (see, for example, Fig. 1 of Ref. [15]), and deserves further investigation.

5 Conclusion

We have applied the CDCC formalism to the ^{9}Be+^{208}Pb system. In a first step, we have computed ^{9}Be wave functions in a three-body $\alpha+\alpha+n$ model. These calculations were performed for low-lying states, but also for pseudostates, which represent positive-energy approximations of the continuum. The main advantage of the hyperspherical approach is that it treats the three-body continuum without any approximation concerning possible $^8\text{Be}+n$ or $^5\text{He}+\alpha$ cluster structures, which are in fact strictly equivalent.

The three-body model of ^{9}Be relies on $\alpha+\alpha$ and $\alpha+n$ (real) interactions, which reproduce very well the elastic phase shifts. With these bare interactions, the ^{9}Be ground state is slightly too bound. We therefore introduce a phenomenological three-body force to reproduce the experimental ground-state energy. The spectroscopic properties of low-lying states is in fair agreement with experiment.

We used first the ^{9}Be wave functions in a CDCC calculation of the ^{9}Be+^{208}Pb elastic scattering at energies close to the Coulomb barrier. As expected, including continuum channels improves the theoretical cross sections. A fair agreement with the data is obtained.

We also applied the four-body CDCC to the ^{9}Be+^{208}Pb fusion, by using the same potentials as for elastic scattering. The total fusion cross section is found in good agreement with experiment, except at very low energies, where we overestimate the data.

References

[1] N. Keeley et al., Prog. Part. Nucl. Phys. **63**, 396 (2009)

[2] M.V. Zhukov et al., Phys. Rep. **231**, 151 (1993)

[3] P. Descouvemont et al., Phys. Rev. C **91**, 024606 (2015)

[4] M. Yahiro et al., Prog. Theor. Phys. Supp. **196**, 87 (2012)

[5] R.J. Woolliscroft et al., Phys. Rev. C **69**, 044612 (2004)

[6] S.K. Pandit et al., Phys. Rev. C **84**, 031601 (2011)

[7] M. Dasgupta et al., Phys. Rev. Lett. **82**, 1395 (1999)

[8] M. Dasgupta et al., Phys. Rev. C **70**, 024606 (2004)

[9] B. Buck, H. Friedrich, C. Wheatley, Nucl. Phys. A **275**, 246 (1977)

[10] H. Kanada et al., Prog. Theor. Phys. **61**, 1327 (1979)

[11] M. Rodríguez-Gallardo et al., Phys. Rev. C **77**, 064609 (2008)

[12] P. Descouvemont, D. Baye, Rep. Prog. Phys. **73**, 036301 (2010)

[13] N. Yu et al., J. Phys. G **37**, 075108 (2010)

[14] T. Druet, P. Descouvemont, Eur. Phys. J. A **48**, 147 (2012)

[15] V. Jha, V.V. Parkar, S. Kailas, Phys. Rev. C **89**, 034605 (2014)

[16] M. Dasgupta et al., Phys. Rev. C **81**, 024608 (2010)

[17] L.F. Canto et al., Phys. Rep. **424**, 1 (2006); Phys. Rep. **596**, 1 (2015)

[18] K. Hagino, N. Takigawa, Prog. Theor. Phys. **128**, 1001 (2012)

[19] L.F. Canto, M.S. Hussein, *Scattering Theory of Molecules, Atoms and Nuclei* (World Scientific Publishing, Singapore, 2013)

[20] A. Diaz-Torres, I.J. Thompson, Phys. Rev. C **65**, 024606 (2002)

[21] A. Diaz-Torres, I.J. Thompson, C. Beck, Phys. Rev. C **68**, 044607 (2003)

EPJ Web of Conferences **117**, 06006 (2016) DOI: 10.1051/epjconf/201611706006
NN2015

^{7}Be- and ^{8}B-reaction dynamics at Coulomb barrier energies

M. Mazzocco[1,2], A. Boiano[3], C. Boiano[4], M. La Commara[3,5], C. Manea[2], C. Parascandolo[3], D. Pierroutsakou[3], C. Signorini[6], E. Strano[1,2], D. Torresi[1,2], H. Yamaguchi[7], D. Kahl[7], L. Acosta[8,9], P. Di Meo[3], J.P. Fernandez-Garcia[9], T. Glodariu[10], J. Grebosz[11], A. Guglielmetti[4,12], N. Imai[7,13], Y. Hirayama[13], H. Ishiyama[13], N. Iwasa[14], S.C. Jeong[13,15], H.M. Jia[16], N. Keeley[17], Y.H. Kim[13], S. Kimura[13], S. Kubono[7,18], J.A. Lay[1,2], C.J. Lin[16], G. Marquinez-Duran[8], I. Martel[8], H. Miyatake[13], M. Mukai[13], T. Nakao[19], M. Nicoletto[2], A. Pakou[20], K. Rusek[21], Y. Sakaguchi[7], A.M. Sánchez-Benítez[8], T. Sava[10], O. Sgouros[20], C. Stefanini[1,2], F. Soramel[1,2], V. Soukeras[20], E. Stiliaris[22], L. Stroe[10], T. Teranishi[23], N. Toniolo[6], Y. Wakabayashi[18], Y.X. Watanabe[13], L. Yang[16] and Y.Y. Yang[24]

[1] Dipartimento di Fisica e Astronomia, Università di Padova, Padova, Italy
[2] INFN, Sezione di Padova, Padova, Italy
[3] INFN, Sezione di Napoli, Napoli, Italy
[4] INFN, Sezione di Milano, Milano, Italy
[5] Dipartimento di Fisica, Università di Napoli "Federico II", Napoli, Italy
[6] INFN - Laboratori Nazionali di Legnaro, Legnaro (PD), Italy
[7] CNS - The University of Tokyo, RIKEN campus, Wako, Saitama, Japan
[8] Departamento de Fìsica Aplicada, Universidad de Huelva, Huelva, Spain
[9] INFN, Sezione di Catania, Catania, Italy
[10] NIPNE, Măgurele, Romania
[11] IFJ PAN, Kraków, Poland
[12] Dipartimento di Fisica, Università di Milano, Milano, Italy
[13] KEK, Tsukuba, Ibaraki, Japan
[14] Department of Physics, Sendai, Miyagi, Japan
[15] Institute for Basic Science, Daejeon, Korea
[16] China Institute of Atomic Energy, Beijing, China
[17] National Centre for Nuclear Research, Otwock, Poland
[18] RIKEN Nishina Center, Wako, Saitama, Japan
[19] Advanced Science Research Center, JAEA, Tokai, Ibaraki, Japan
[20] Department of Physics, University of Ioannina, Ioannina, Greece
[21] Heavy Ion Laboratory, University of Warsaw, Warsaw, Poland
[22] Department of Physics, University of Athens, Athens, Greece
[23] Department of Physics, Kyushu University, Hakozaki, Fukuoka, Japan
[24] Institute of Modern Physics, Chinese Academy of Sciences, Lanzhou, China

Abstract

We investigated the reaction dynamics induced by the Radioactive Ion Beams ^{7}Be and ^{8}B on a ^{208}Pb target at energies around the Coulomb barrier. The two measurements are strongly interconnected, being ^{7}Be ($S_\alpha = 1.586$ MeV) the loosely bound core of the even more exotic ^{8}B ($S_p = 0.1375$ MeV) nucleus. Here we summarize the present status of the data analysis for the measurement of the elastic scattering process for both reactions and the preliminary results for the optical model analysis of the collected data.

1 Introduction

The reaction dynamics induced by light weakly-bound Radioactive Ion Beams (RIBs) at near-barrier energies has attracted the interest of the Nuclear Physics community for the last 20 years at least. The exotic features of several light RIBs together with their extremely weak binding energies (S_p, S_n and S_α typically lower than 1.0 MeV) make these nuclei an ideal playground for the nuclear reaction theory. From an experimental point-of-view, a large enhancement of the reaction cross section at sub-barrier energies has been observed. This enhancement is mostly due to direct processes, such as the 2n-transfer for the 2n-halo nucleus ^{6}He [1–3] and the neutron skin nucleus ^{8}He [4, 5] and, apparently, the breakup process for ^{8}B- [6] and ^{11}Li-induced [7, 8] reactions. Contrary to initial expectations, only a marginal enhancement, when recorded, of the fusion cross section at sub-barrier energies has been reported [4, 5]. Several review papers have been published on this topic over the last decade [9–15]

In this contribution we describe the present status of the data analysis for our measurements of the elastic scattering process for the systems ^{7}Be + ^{208}Pb and ^{8}B + ^{208}Pb at near-barrier energies.

2 The system ^{7}Be + ^{208}Pb

This reaction was studied at the Laboratori Nazionali di Legnaro (LNL, Italy) of the Istituto Nazionale di Fisica Nucleare (INFN), where a ^{7}Be RIB is routinely delivered by the facility EXOTIC [16, 17]. The ^{7}Be beam for this experiment was obtained starting from a 48.8 MeV ^{7}Li^{3+} primary beam impinging on a 5-cm long gas target filled with 1 bar of H_2 gas cooled down to liquid nitrogen temperature ($\sim$ 90 K). In these conditions, a 40.5-MeV ^{7}Be secondary beam with an average intensity 2-3$\times 10^5$ pps was produced. A second ^{7}Be energy of 37.6 MeV was obtained by means of an approximately 12.5 μm thick Al degrader inserted along the beam line right before the main scattering chamber. Finally, a third energy of 42.4 MeV was achieved at the end of the experiment after warming up the target station at room temperature. In all cases the secondary beam purity was as good as 99%.

Charged particles originated by the interaction with a 1 mg/cm^2 ^{208}Pb target were detected in the angular ranges θ_{lab} = [55°,84°], [96°,125°] and [138°,165°] by means of 6 ΔE-E_{res} telescopes of the newly developed detector array EXPADES [18]. The telescopes were displaced in a cylindrical configuration around the target at an average distance of about 111.5 mm.

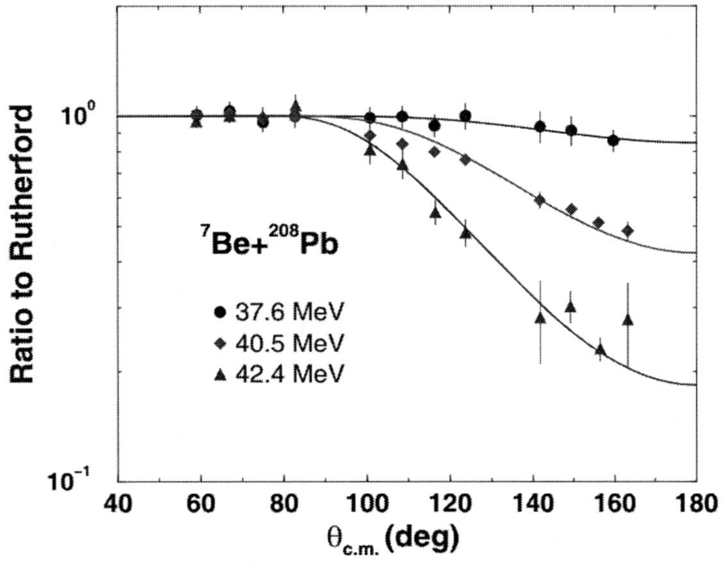

Figure 1: Elastic scattering angular distributions for the system ^{7}Be + ^{208}Pb at three bombarding energies. Continuous lines represent the results of an optical model best-fit analysis of the experimental data.

Three modules were located in the left hemisphere and the other three in symmetrical locations of the right hemisphere. Each telescope consisted of a 43-57 and 300 μm thick Double Sided Single Strip Detector (DSSDS) as inner and outer layer, respectively. Each DSSSD had an active area of 62.3 mm $\times$ 62.3 mm and each side was segmented into 32 strips, determining a position resolution of about 2 mm $\times$ 2 mm.

Fig. 1 shows a preliminary evaluation of the elastic scattering angular distributions for the system ^{7}Be + ^{208}Pb. We can directly appreciate the drop of the differential cross section at backward angles due to the nuclear absorption. To contain statistical fluctuations, at this stage of the analysis, the data were grouped into bins of 8 vertical strips. A better angular resolution should be achieved after completing the sorting procedure of the collected data.

Continuous lines in Fig. 1 are the result of an optical model best fit analysis of the experimental data within the framework of the optical model. The analysis was performed with the SFRESCO subroutine of the main code FRESCO [19] and allowed the extraction of the reaction cross section. The comparison of the reaction cross sections for the system $6,7$Li, ^{7}Be and ^{8}B + ^{208}Pb will be presented in the last section of this contribution.

3 The system ^{8}B + ^{208}Pb

The study of this reaction was carried out at the CRIB (CNS Radioactive Ion Beam) [20,21] facility located in the RIKEN campus at Wako (Japan). A 50 MeV ^{8}B RIB was produced starting from a 66 MeV $^6Li^{3+}$ primary beam impinging on a 8-cm long gas target inflated with ^{3}He gas at a pressure of 1 bar. The outcoming ^{8}B beam had an intensity of 2×10^4 pps and a purity of about 20%, being the ^{6}Li scattered beam, ^{3}He ions recoiling out from the gas target and ^{7}Be the main contaminant beams. However, all these beams had different times of flight (ToF) through the CRIB separator and, by using proper gates in the ToF spectra, it will be possible to discriminate events originating from different components of the cocktail beam.

Charged reaction products arising from the interaction with a 2.2 mg/cm^2 ^{208}Pb (evaporated on a 1.5 μm thick Mylar backing foil) were detected by means of the six ΔE-E_{res} modules of the already mentioned detector array EXPADES. In this experiment a slightly asymmetric configuration of the telescopes was employed in order to ensure the coverage of the following angular ranges: θ_{lab} = [16°,41°], [55°,84°], [96°,125°] and [139°,164°]. The average distance between the detectors and the target central position was about 113 mm.

Fig. 2 shows a very preliminary evaluation of the angular distribution for the elastic scattering process in the system ^{8}B + ^{208}Pb. Only four point were plotted, each one corresponding to the mean location of the four angular ranges covered by the EXPADES array. The experimental data were normalized by using a Monte-Carlo simulation, which took into account the geometry of the detector array and the kinematics of the elastic scattering process. Simulated events were generated according to a pure Rutherford cross section. The ratio between experimental and simulated data was finally set equal to unity in the angular range covered by the detector located at the forward-most angles. In the next months we expect to improve the angular resolution of the elastic scattering differential cross section by performing a detailed strip-by-strip analysis of the gathered data and, if allowed by the collected statistics, to proceed also with a pixel-by-pixel analysis.

Continuous line in Fig. 2 represent the result of an optical model best fit analysis performed with the code FRESCO. This procedure allowed us to extract a first (extremely preliminary) evaluation of the reaction cross section for the system ^{8}B + ^{208}Pb at 50 MeV beam energy.

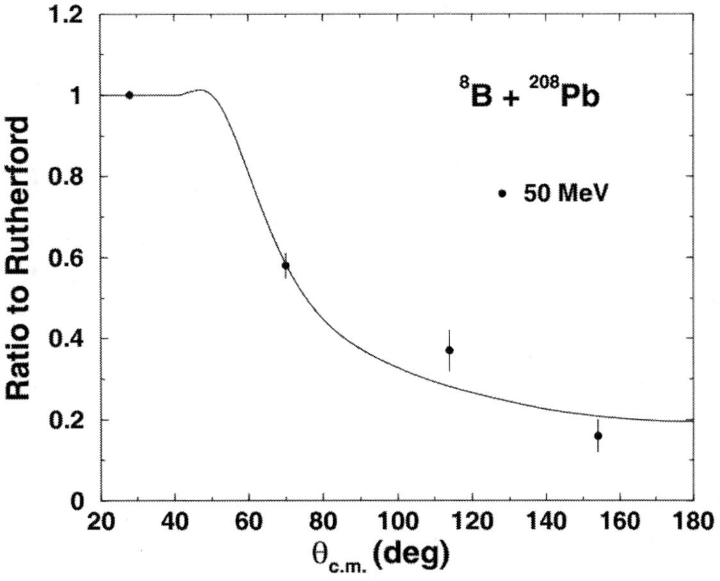

Figure 2: Preliminary elastic scattering angular distribution for the system ^{8}B + ^{208}Pb at 50 MeV beam energy. The continuous line is the result of the optical model analysis of the collected data.

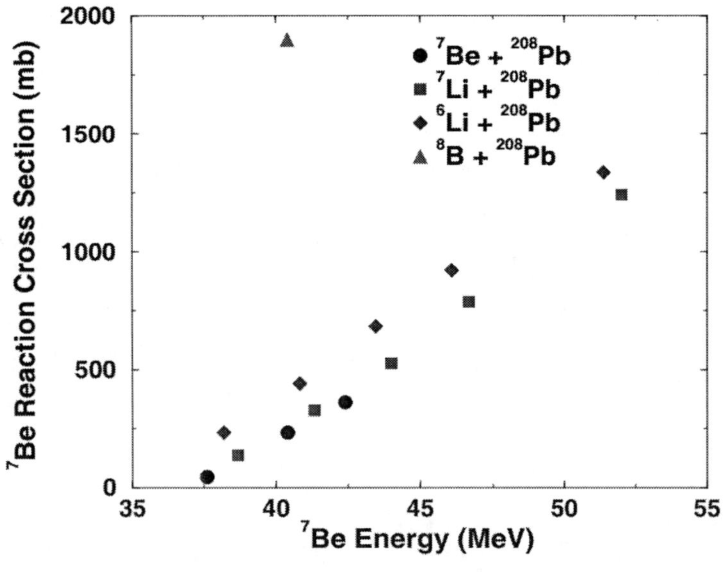

Figure 3: Reaction cross sections for the systems ^{6,7}Li, ^{7}Be and ^{8}B + ^{208}Pb. The data were reduced according to the procedure recommended in Ref. [23] and then normalized to the system ^{7}Be + ^{208}Pb.

4 Discussion

Fig. 3 illustrates the reaction cross sections for the systems ^{6,7}Li [22], ^{7}Be and ^{8}B + ^{208}Pb in the energy range around the Coulomb barrier. The data for the different systems were reduced according to the procedure described in Ref. [23] and then multiplied by the normalization factor of the reaction ^{7}Be + ^{208}Pb. It is interesting to observed that, according to our preliminary analysis, the ^{7}Be ($S_\alpha = 1.586$ MeV) nucleus exhibits a reactivity rather similar to that of its more bound mirror nucleus ^{7}Li ($S_{alpha} = 2.467$ MeV), rather than to that of the similarly weakly bound nucleus ^{6}Li ($S_{alpha} = 1.474$ MeV).

On the other side, it is particularly evident that, even at this very preliminary stage of the data analysis, the reaction induced by the ^{8}B p-halo nucleus ($S_p = 0.1375$ MeV) on the ^{208}Pb target has a reaction cross section much larger those measured for reactions induced by the other weakly-bound projectiles on the same target. This scenario is rather consistent with that observed for reactions induced by the same projectiles on the medium-mass target ^{58}Ni [6]. In our case, in an ultimate stage of the data analysis, it should be possible to established whether the reaction cross section enhancement for the system ^{8}B + ^{208}Pb is mainly due to direct channels, such as breakup and/or transfer, or to the fusion process.

Acknowledgments

This work was partially supported by the Italian Minister for Education, University and Research (MIUR) within the project RBFR08P1W2 (FIRB 2008) and the National Science Centre of Poland under contract No. UMO-2014 14 M ST2 00738 (COPIN-INFN Collaboration). The research leading to these results has also received funding from the European Commission, Seventh Framework Programme (FP7 2007-2013) under Grant Agreement n. 600376. J.A.L. is a Marie Curie Piscopia fellow at the University of Padova.

References

[1] A. Di Pietro *et al.*, Phys. Rev. C **69**, 044613 (2004).

[2] A. Navin *et al.*, Phys. Rev. C **70**, 044601 (2004).

[3] R. Raabe *et al.*, Nature **431**, 823 (2004).

[4] A. Lemasson *et al.*, Phys. Rev. Lett. **103**, 232701 (2009).

[5] A. Lemasson *et al.*, Phys. Rev. C **82**, 044617 (2010).

[6] E.F. Aguilera *et al.*, Phys. Rev. C. **79**, 021601(R) (2009).

[7] M. Cubero *et al.*, Phys. Rev. Lett. **109**, 262701 (2012).

[8] J.P. Fernandez-Garcia *et al.*, Phys. Rev. Lett. **110**, 142701 (2013).

[9] L.F. Canto, P.R.S. Gomes, R. Donangelo and M.S. Hussein, Phys. Rep. **424**, 1 (2006).

[10] J.F. Liang, and C. Signorini, Int. J. Mod. Phys. E **14**, 1121 (2005).

[11] N. Keeley, R. Raabe, N. Alamanos, and J.L. Sida, Prog. Part. Nucl. Phys. **59**, 579 (2007).

[12] N. Keeley, N. Alamanos, K.W. Kemper, and K.Rusek, Prog. Part. Nucl. Phys. **63**, 396 (2009).

[13] M. Mazzocco, Int. J. Mod. Phys. E **19**, 977 (2010).

[14] N. Keeley, K.W. Kemper and K. Rusek, Eur. Phys. J. A **50**, 145 (2014).

[15] L.F. Canto, P.R.S. Gomes, R. Donangelo, J. Lubian and M.S. Hussein, Phys. Rep. **596**, 1 (2006).

[16] F. Farinon *et al.*, Nucl. Instrum. Meth. B **266**, 4097 (2008).

[17] M. Mazzocco *et al.*, Nucl. Instrum. Meth. B **317**, 223 (2013).

[18] E. Strano *et al.*, Nucl. Instr. and Meth. B **317** (2013) 657.

[19] I.J. Thompson, Comput. Phys. Rep. **2**, 167 (1988).

[20] Y. Yanagisawa *et al.*, Nucl. Instr. and Meth. A **539** (2005) 74.

[21] H. Yamaguchi *et al.*, Nucl. Instr. and Meth. A **589** (2008) 150.

[22] N. Keeley *et al.*, Nucl. Phys. A **571** (1994) 326.

[23] P.R.S. Gomes, J. Lubian, I. Padron, R.M. Anjos, Phys. Rev. C **71** (2005) 017601.

EPJ Web of Conferences **117**, 06007 (2016) DOI: 10.1051/epjconf/201611706007
NN2015

Probing nucleon-nucleon correlations in heavy-ion transfer reactions

S. Szilner

Ruđer Bošković Institute, Zagreb, Croatia

Abstract

The γ-particle coincident measurements, performed by coupling of the PRISMA spectrometer to the large γ arrays (CLARA and AGATA), demonstrate a strong interplay between single-particle and collective degrees of freedom that is pertinent to the reaction dynamics. By using the unique PRISMA performance in terms of both resolution and efficiency, measurements at very low bombarding energies have been performed. Via transfer of nucleon pairs, valuable information on the component responsible for particle correlations has been derived.

1 Introduction

The recent revival of transfer reaction studies [1] greatly benefited from the construction of the new generation instrumentation that reached an unprecedented efficiency and selectivity. In the region close to the Coulomb barrier, these reactions reflect a strong interplay between single-particle and collective degrees of freedom and the reaction dynamics [2–4].

Significant progress has been recently achieved by performing studies far below the Coulomb barrier with the PRISMA spectrometer. Transfer cross sections obtained from excitation functions for the closed shell ^{40}Ca+^{96}Zr [5] and superfluid ^{60}Ni+^{116}Sn [6] systems have been measured from the Coulomb barrier energy to deep below it. The transfer probabilities have been compared with microscopic calculations that incorporate nucleon-nucleon correlations, essential for the population pattern of the single particle levels around the Fermi energy. For the first time in a heavy ion collision, one was able to provide a consistent description of one and two neutron transfer reactions by incorporating, in the reaction mechanism, all known structure information of entrance and exit channels nuclei.

In this work I will focus on the main outcome of the recent studies, critically addressing the new achievements, the present problems and new challenges.

2 Transfer reaction mechanism

Transfer reactions have an important impact in the understanding of correlations in the nuclear medium, and play a very important role for the study of the evolution from the quasi-elastic to the deep-inelastic regime [1]. In heavy-ion induced transfer reactions, the constituents of the collision may exchange many nucleons, thus providing information on the contribution of single particle and correlated particle transfers, and on the contribution of surface vibrations (bosons) and their coupling with single particles (fermions). In this sense it is very important to determine the absolute cross sections which can be compared with the state-of-art calculations. From the comparison between data and theory it is possible to evaluate the importance of the elementary modes of excitation in the reaction model.

Total angle and Q-value integrated cross sections for multi-neutron and multi-proton channels have been investigated with spectrometers in various systems close to the Coulomb barrier (^{58}Ni+^{208}Pb [7], ^{40}Ca+^{208}Pb [8], ^{40}Ca+^{96}Zr, ^{90}Zr+^{208}Pb [9, 10], and ^{40}Ar+^{208}Pb [11]). In these systems one finds that most nuclei produced in transfer reactions have N/Z ratio smaller than one of the compound nucleus, implying the dominance of a direct mechanism in the population of different fragments. For the massive proton transfer channels the isotopic distributions drift toward lower masses, a clear indication that these distributions are affected by evaporation processes. The cross sections for the neutron pick-up drop by almost a constant factor for each transferred neutron, as an independent particle mechanism would suggest.

In all mentioned systems, it has been noticed that the pure proton cross sections display a population of the $(-2p)$ channel as strong as the $(-1p)$ one, suggesting the contribution of processes involving the transfer of proton pairs in addition to the successive transfer of single protons. This apparent proton and neutron asymmetric behavior is due to the fact that the one-neutron transfer cross section is almost one order of magnitude larger than the one-proton transfer masking the contribution of a pair-transfer mode in the neutron sector by the successive mechanism. As the very short-range pairing interaction redistributes the strength around the pure configurations, it is very important to study the yields distribution of the individual states. This subject will be addressed in next section.

The newly accumulating data are also attracting a renewed interest by theorists. Microscopic models have been used to compute transfer cross sections in light ion induced reactions [12]. Time Dependent Hartree Fock theory has been employed without [13, 14] and with [15] approximate inclusion of correlations. Calculations based on Langevin [16] and diffusion equations [17] have been recently performed. Calculations for multinucleon transfer processes have been extensively and successfully performed within semiclassical theories [1,9], in particular by using the GRAZING model [18].

2.1 Particle vibration couplings

In the description of the experimental yields, it has been established [1] that the important degrees of freedom in the transfer dynamics are the internal degrees of freedom of the two ions in terms of elementary modes, surface vibrations and single particles. In fact, it is through the excitation of these elementary modes that energy and angular momentum are transferred from the relative motion to the intrinsic degrees of freedom and that mass and charge are exchanged among the two partners. Heavy-ion transfer reactions, thus, should provide the ideal tool for the studies of the particle-vibration coupling scheme. Even if such couplings are essential for the description of many basic states in the vicinity of closed shells, their effects are largely unexplored.

The Ar (neutron transfer channels) [2] and Cl [19] (one proton stripping channels) isotopes have been populated in the ^{40}Ar+^{208}Pb reaction. Their γ spectra display strong transitions which can be connected with the single-particle or single-hole states. In addition, through the whole isotopic chain also states that involve combinations of a single particle or hole with a collective boson have been populated. This is illustrated for the Ar isotopic chain in fig. 1. For example, in ^{41}Ar, the one-neutron transfer channel, beside the low lying states with a pronounced single-particle character, the decay of the $11/2^-$ state has been observed. This $11/2^-$ state can be understood as a coupling of a collective boson to single-particle states (i.e. $|2^+, (f_{7/2})^1>$) giving a $11/2^-$ stretched configuration. The properties of such states are closely connected with the properties of the vibration quanta, allowing one

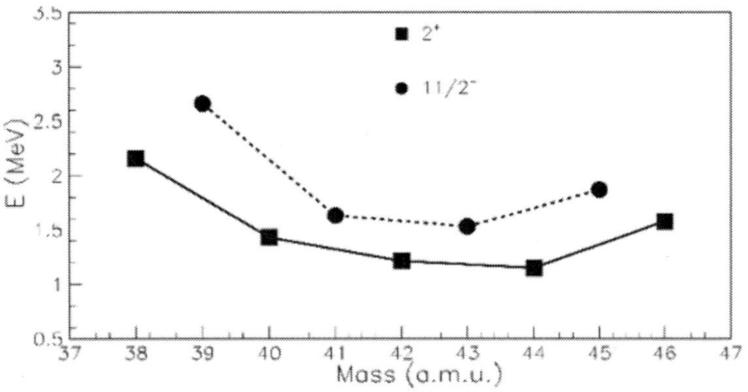

Figure 1: Energies of the 2^+ (squares), and $11/2^-$ (circles) states of argon isotopes with N = 20-28 populated in ^{40}Ar+^{208}Pb at E_{lab} =255 MeV. The energies of $11/2^-$ in ^{41}Ar and ^{43}Ar are from refs. [2,20]. In the most recent compilation of the ^{45}Ar level properties, a level at 1911(5) keV is a good candidate for the $11/2^-$ state, and was added for completeness.

to follow the development of collectivity in odd isotopic chain, a phenomenon widely discussed in even-even isotopes.

2.2 Nucleon-nucleon correlations

With heavy ions multiple transfer of nucleons becomes available in the reaction, giving the possibility to study the relative role of single particle and pair transfer modes. In the sub-barrier domain the two ions probe their mutual interaction only at very large distances, where the nuclear couplings are dominated by transfer processes. The number of open reaction channels is reduced, and the Q-value distributions get narrower. These conditions diminish the complexity of coupled channel calculations. Even if the cross sections becomes smaller, the coming into operation of large acceptance magnetic spectrometers made it possible to perform sub-barrier measurements with good ion identification. Transfer yield could be detected at the level of 10^{-4} with respect to the elastic channel.

In very recently measured systems one successfully demonstrated the powerful method of using PRISMA [9, 21] for such studies, exploiting its unique performance in terms of both resolution and efficiency. Making use of inverse kinematics, target recoils have been detected in multinucleon transfer reactions for the systems ^{96}Zr+^{40}Ca [5] and ^{116}Sn+^{60}Ni [6]. In both cases the excitation functions at several bombarding energies have been obtained from the Coulomb barrier to 20-25% below, reaching about 15.5 fm of distance of closest approach. The results of the ^{116}Sn+^{60}Ni measurement are presented,

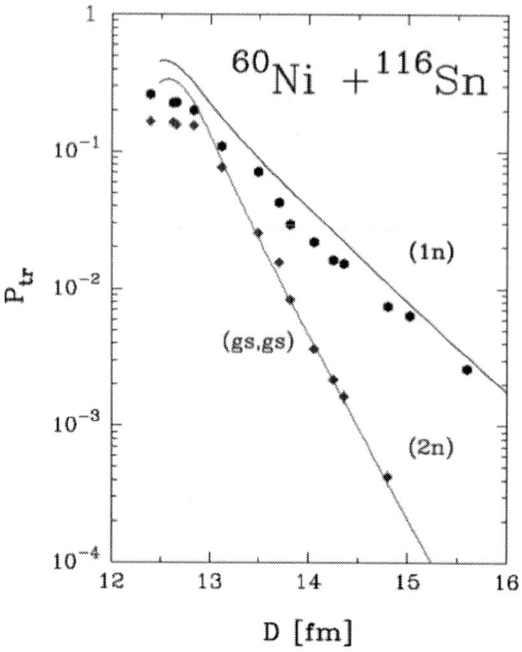

Figure 2: Experimental (points) and microscopically calculated (lines) transfer probabilities (P_{tr} defined as the ratio of the transfer cross section to the corresponding Rutherford cross section) as a function of the distance of closest approach D for one- (^{61}Ni) and two-neutron (^{62}Ni) transfers in the ^{116}Sn+^{60}Ni system (from ref. [6]).

together with the calculations, in fig. 2 for the $(1n)$ and $(2n)$ neutron transfer channels.

To compute the inclusive $(1n)$ cross section (full line) one calculated the transfer probability for a given single particle transition and one obtained the total transfer probability by summing over all possible transitions that can be constructed from the single particle states in projectile and target (see ref. [6]), and by weighting each transition with the corresponding spectroscopic factor. The calculations well reproduce the experimental slope and the absolute values. To calculate the $(2n)$ channel one has to solve the well-known system of semiclassical coupled equations up to the second-order Born approximation. The amplitude consists of three terms, the simultaneous transfer of the pair of nucleons, the nonorthogonality term, and the term which represents the successive process via an intermediate channel. In an independent particle description the first and second term cancel exactly. For the calculation of the $(2n)$ channel the ground to ground state transition has been included following the formalism of ref. [22]. In this

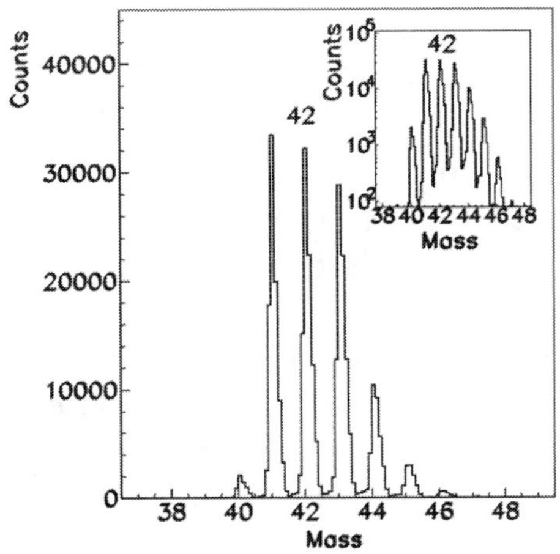

Figure 3: Mass distribution of the K isotopes obtained in the ^{40}Ar+^{208}Pb reaction [11] measured at $\theta_{lab} = 54°$ and E_{lab} =255 MeV. The insert displays the same mass distribution in logarithmic scale.

formalism one can employ the same single particle form factors used for the $(1n)$ channel. The ground states of ^{62}Ni and ^{114}Sn have been described in the BCS approximation.

It is very important to stress that the shown results are not coming from best fit procedure, but that the employed microscopic theory, that incorporates nucleon-nucleon correlations, essential for the population pattern of the single particle levels around the Fermi energy, very well reproduces the experimental data in the whole energy range, in particular, the transfer probability for two neutrons is very well reproduced, in magnitude and slope, by considering solely the ground-ground state transition.

2.3 Proton-neutron correlations

In this context, it is also important to investigate the role played by proton-neutron correlations. In order to study these correlations one has to use systems where the population of the $(\pm pn)$ channels is allowed by the transfer mechanism (Q-value consideration). These conditions were fulfilled in the ^{40}Ar+^{208}Pb system [11], where the angular distributions were studied by PRISMA in a wide angular range (3 different PRISMA angular settings). The total cross sections, obtained from integrating the angular and Q-value distributions for the proton striping and neutron pick-up channels concede with the behaviour of previously studied and already discussed systems. The $(+pn)$ channel behaves differently, with the cross section as strong as the (+1*p*) cross section (see fig. 3). Such behaviour indicates that some additional degrees of freedom have to be considered for that channel.

In heavy-ion reactions the large energy losses may occur, thus we very carefully studied the total kinetic energy loss distributions and the transition from a regime of direct reactions (quasi-elastic) to the more complicated regime of deep inelastic collisions. Even with careful extraction of the quasi-elastic components only, the obtained cross sections for the (+*pn*) channel turned out to be enhanced. The strength and presence of the correlation may emerge from a direct comparison of the measured and the microscopically calculated cross sections, where all essential ingredients for accurate description of the nuclear structure and dynamics are employed.

3 Summary and outlook

The advent of the large solid angle magnetic spectrometers, coupled to large γ arrays, ensured significant advances in the field of multinucleon transfer reactions at energies close to the Coulomb barrier. The detailed studies of the fragments yield distributions and in-beam γ emission are important for the understanding of the reaction mechanism and nuclear structure. Via transfer of multiple pairs valuable information on nucleon-nucleon correlations can also be derived. In the comparison between data and theory, the importance of the elementary degrees of freedom, i.e. single particle, phonon and pair modes, can be deduced. It is very important to continue these studies, and to compare the experimental observables with microscopic calculations where nucleon-nucleon correlations are treated together with the relevant structure and dynamics information.

Acknowledgements

The material presented in this paper is the result of the work of many people which I would like to thank, in particular L. Corradi, T. Mijatović, G. Pollarolo, D. Montanari, E. Fioretto, A. M. Stefanini, J. J. Valiente-Dobón, G. Montagnoli, F. Scarlassara, C. A. Ur, and A. Goasduff. This work was partly supported by the EC FP7 ENSAR (262010), by Croatian Science Foundation (project 7194), and by the Scientific center of excellence for advance materials and sensors.

References

[1] L. Corradi, G. Pollarolo, and S. Szilner, J. of Phys. G **36** (2009) 113101.

[2] S. Szilner *et al.*, Phys. Rev. C **84** (2011) 014325.

[3] S. Lunardi *et al.*, Phys. Rev. C **76** (2007) 034303.

[4] J. J. Valiente Dobón, this proceeding.

[5] L. Corradi *et al.*, Phys. Rev. C **84** (2011) 034603.

[6] D. Montanari *et al.*, Phys. Rev. Lett **113** (2014) 052501.

[7] L. Corradi *et al.*, Phys. Rev. C **66** (2002) 024606.

[8] S. Szilner *et al.*, Phys. Rev. C **71** (2005) 044610.

[9] S. Szilner *et al.*, Phys. Rev. C **74** (2007) 024604.

[10] M. Varga Pajtler *et al.*, Nucl. Phys. A **941** (2015) 273.

[11] T. Mijatović, Ph.D. thesis, Faculty of Science, University of Zagreb (2015).

[12] G. Potel, F. Barranco, E. Vigezzi, and R. A. Broglia, Phys. Rev. Lett. **105** (2010) 172502.

[13] C. Simenel, Phys. Rev. Lett. **105** (2010) 192701.

[14] K. Sekizawa, and K. Yabana, Phys. Rev. C **88** (2013) 014614.

[15] G. Scamps, and D. Lacroix, Phys. Rev. C **87** (2013) 014605.

[16] V. Zagrebaev, and W. Greiner, Phys. Rev. Lett. **101** (2008) 122701.

[17] V. V. Sargsyan *et al.*, Phys. Rev. C **88** (2013) 064601.

[18] A. Winther, Nucl. Phys. A **572**, 191 (1994); Nucl. Phys. A **594** (1995) 203.

[19] S. Szilner *et al.*, Phys. Rev. C **87** (2013) 054322.

[20] D. Mengoni *et al.*, Phys. Rev. C **82** (2010) 024308.

[21] D. Montanari *et al.*, Eur. Phys. J. A **47** (2011) 4.

[22] J. H. Sorensen and A. Winther, Nucl. Phys. A **550** (1992) 306.

EPJ Web of Conferences **117**, 06008 (2016) DOI: 10.1051/epjconf/201611706008
NN2015

Using CHIMERA detector at LNS for gamma-particle coincidences

G. Cardella[1], L. Acosta[9], L. Auditore[4], M.B. Chatterjiee[8], A. Castoldi[7], E. De Filippo[1], D. Dell'Aquila[6], S. De Luca[4], B. Gnoffo[1], C. Guazzoni[7], L. Francalanza[6], G. Lanzalone[2,5], I. Lombardo[6], N. Martorana[2,3], S. Norella[4], A. Pagano[1], E.V. Pagano[2,3], M. Papa[1], S. Pirrone[1], G. Politi[1,3], L. Quattrocchi[4], F. Rizzo[2,3], P. Russotto[1], A. Trifirò[4], M. Trimarchi[4], G. Verde[1] and M. Vigilante[6]

[1] INFN, Sezione di Catania, Via S. Sofia, 95123 Catania, Italy
[2] INFN - Laboratori Nazionali del Sud, Via S. Sofia, Catania, Italy
[3] Dip. di Fisica e Astronomia, Università di Catania, Via S. Sofia, Catania, Italy
[4] INFN - Gruppo collegato di Messina, Dip. di Fisica, Università di Messina, Italy
[5] Facoltà di Ingegneria e Architettura, Università Kore, Enna, Italy
[6] INFN, Sezione di Napoli, Dipartimento di Fisica, Università Federico II, Napoli, Italy
[7] Politecnico di Milano, Dipartimento di Elettronica, Informazione e Bioingegneria and INFN, Sezione di Milano, Italy
[8] Saha Institute for Nuclear Physics, Kolkata, India
[9] Instituto de Física, Universidad Nacional Autonoma de Mexico, Mexico D.F. 0100, Mexico

Abstract

We have recently evaluated the quality of γ-ray angular distributions that can be extracted in particle-gamma coincidence measurements using the CHIMERA detector at LNS. γ-rays have been detected using the CsI(Tl) detectors of the spherical part of the CHIMERA array. Very clean γ-rays angular distributions were extracted in reactions induced by different stable beams impinging on ^{12}C thin targets. The results evidenced an effect of projectile spin flip on the γ-rays angular distributions. γ-particle coincidence measurements were also performed in reactions induced by neutron rich exotic beams produced through in-flight fragmentation at LNS. In recent experiments also the Farcos array was used to improve energy and angular resolution measurements of the detected charged particles. Results obtained with both stable and radioactive beams are reported.

1 Introduction

Relatively large yields of various neutron rich and poor exotic beams, of energy ranging from 20 to 50 MeV/A, are available at LNS [1], and in particular we used ^{6,8}He, ^{8,9}Li, ^{10,11}Be, ^{13}B, ^{16,17}C, ^{68}Ni. In order to benefit of such beams with CHIMERA detector [2, 3] a tagging system [4] was developed to identify event by event the isotopic composition of the produced exotic particles. This system was recently updated to allow trajectory measurement with a PPAC detector. To improve the energy and angular resolution measurements of particles produced in reactions induced by the exotic beams we used a prototype of the new FARCOS detection system under construction in Catania [5]. Moreover, as recently shown [6], it was possible to extract further information on reaction mechanisms also by looking to γ-ray energy and angular distribution by using the CsI(Tl) detectors of the CHIMERA sphere (from 30° to 174°). This latter information will be of fundamental interest in one of the next experiments planned to be performed in CHIMERA, i.e. the study of isoscalar excitation of the PIGMY resonance on ^{68}Ni [7]. In this paper we shortly describe in the next paragraph the production and identification of the fragmentation beam. In the last paragraph we summarize some of the recently published performances of our CsI(Tl) crystals when they are used as γ-ray detectors and some results concerning the coincidences between particles and γ-rays, together with future perspectives.

2 Beam production, and tagging system

The beam is produced by fragmentation reactions, using the beams accelerated by the superconducting Cyclotron (CS) of the Laboratori Nazionali del Sud in Catania. A Beryllium target is placed, close to the CS output extraction channel. The standard primary beam intensity, up to now used, is of the order of 100 Watts. The beam selection is obtained with appropriate setting of two 45° dipoles of the main beam transport line of the CS.

DOI: 10.1051/epjconf/201611706008

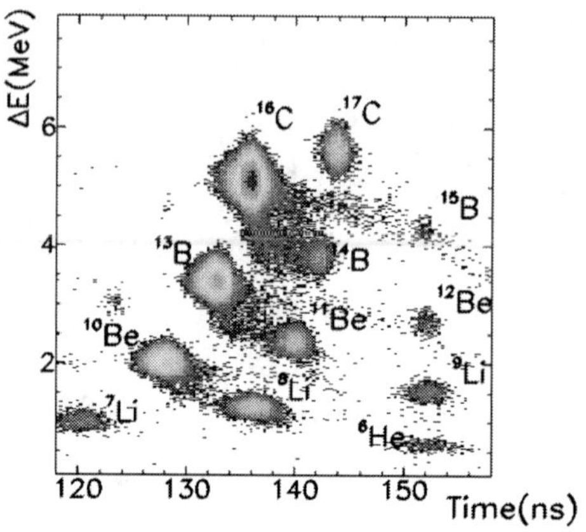

Figure 1: Identification scatter plot for the production of $^{16}C/^{10,11}Be$ beams.

The tagging system is based on the use of a 32x32 strips double side silicon strip detector (DSSSD) measuring the energy loss of the impinging beam particles. The identification of the particles is achieved by combining this energy loss with the time of flight information (TOF). The start of this TOF measurement is provided by a large size micro-channel plate (MCP) [4] mounted 13 m away from the DSSSD. The stop is given by the signal delivered by the front strips of the DSSSD.

An example of the obtained identification scatter plot is shown in fig.1 for the production of ^{16}C and $^{10,11}Be$ by the fragmentation of ^{18}O 55 AMeV primary beam. To measure trajectories we have installed, at the entrance of the CHIMERA sphere, a PPAC position sensitive detector. The position information provided by DSSSD and PPAC allows to extract the beam trajectory. This is a basic measurement to sketch the beam profile at the target position during the beam commissioning phase.

3 Gamma ray detection

In reactions with radioactive beams, due to the low beam intensity, it is today mandatory to perform measurements as complete as possible. It is for instance very important to be able to detect the γ-rays in coincidence with particles emitted in the reaction. With CHIMERA we detect and identify γ-rays in all CsI(Tl) detectors. However we have a better signal to noise ratio in the detectors of the spherical part due to kinematical focusing, and

 DOI: 10.1051/epjconf/201611706008

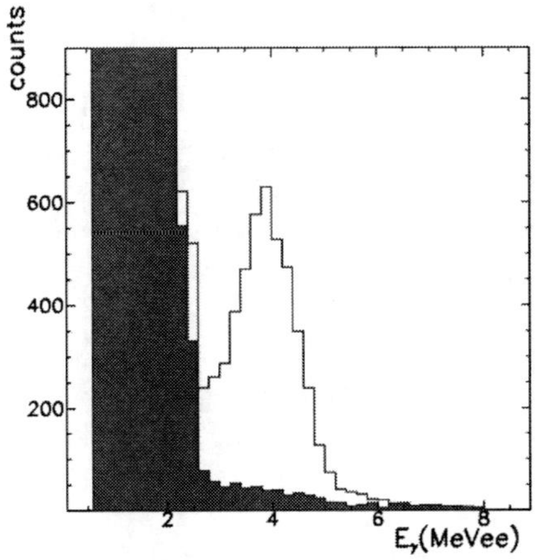

Figure 2: γ-ray spectrum detected in CHIMERA CsI(Tl) in coincidence with proton from elastic scattering on carbon target (filled spectrum), and in coincidence with proton from inelastic scattering leaving the carbon excited at the 4.44 MeV level (empty spectrum).

to the larger solid angle of the detectors. We measured the CsI(Tl) response function to γ-rays, looking to the detection of the well know 4.44 MeV γ-ray, emitted by the decay of the first excited level of ^{12}C. This state can be simply excited using proton scattering. In such kind of reactions one can look for the detection of this γ-ray triggering the event with the inelastic scattered proton. This is a way to measure also the detector efficiency. In fig.2 we show the spectrum collected in one ring of the sphere, in coincidence with such inelastic scattering events (empty spectrum). Background can be evaluated collecting the same number of events, triggered by the detection of protons from elastic scattering (full spectrum). One can see that the peak to noise ratio is rather good. Resolution is of the order of 0.5 MeV, as can be deduced by the missing observation of the first and second escape peaks. From comparison with GEANT-IV simulations, one understand that the energy response function is centered on the first escape peak at about 3.9 MeVee. We have also verified that we can simply extract angular distributions from such events. They are very useful to understand the multipolarity of the emitted γ-ray, so obtaining information about the spin of the observed resonance. It is useful to obtain information about the multipolarity of the emitted γ-ray, in such a way the spin of the observed resonance can be evaluated. For instance, in fig.3 the CM angular distributions of γ-ray obtained at different proton beam energy from the decay

DOI: 10.1051/epjconf/201611706008

of the 4.44 MeV level is reported. The angular distribution is described by the formula $sin^2(2\theta)$+cost as it is expected, for quadrupole decay, in case of proton spin-flip events [6]. Similar results were obtained for all investigated energies, more details can be found in ref. [6]. A pure $sin^2(2\theta)$ behavior is obtained using a zero spin projectile as ^{16}O.

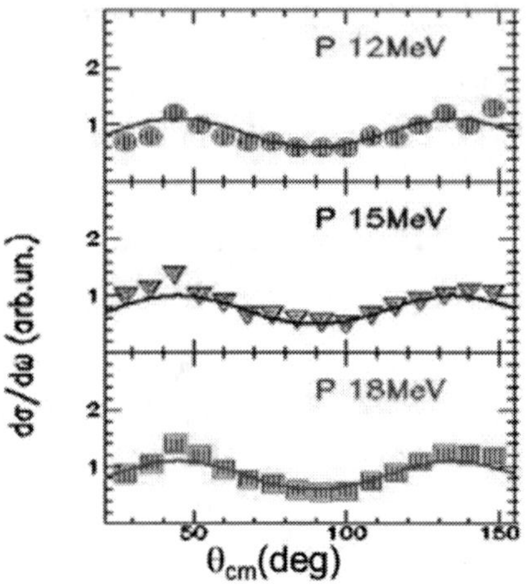

Figure 3: Center of mass angular distributions of γ-ray from the decay of $^{12}C^{4.44}$ populated at different proton beam energies.

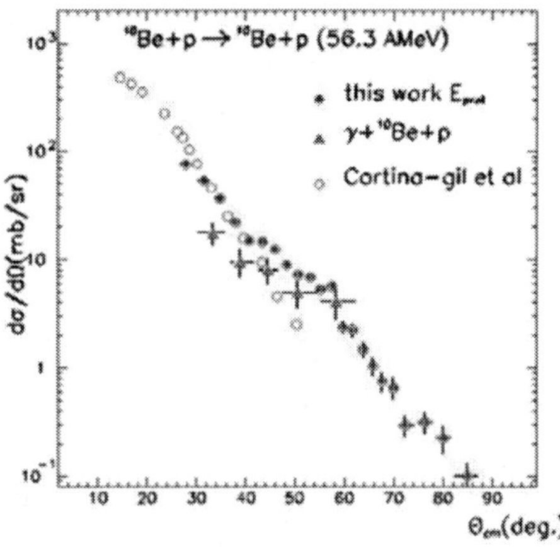

Figure 4: Center of mass angular distributions measured for the reaction ^{10}Be+p, see text.

It is well known that γ-rays can be used as a signature of inelastic channels. In fig. 4 we show how this effect is observed with CHIMERA. In this figure we plot as filled dots the elastic scattering angular distribution measured for the reaction ^{10}Be+p at 56 A.MeV. This angular distribution was extracted with the kinematical coincidence method that we have recently developed [8]. Comparing such angular distribution with the one measured in the past by Cortina-Gil [9] with SPEG (open dots), we note that our data have the same behavior of those obtained with spectrometer up to about 35°. At larger angles we observe a change in the slope of the angular distribution. This is due to the contribution of inelastic channels that cannot be discriminated by the elastic one in our data, due to the resolution of our CsI(Tl) detector. This effect is shown by the inelastic angular distribution (blue triangles) obtained measuring in coincidence with γ-ray the couples of proton and ^{10}Be.

Further work is in progress to improve the energy resolution of CsI(Tl) detectors by using the new GET electronics [10]. The new electronics will also add new trigger capabilities for the CsI(Tl) stage that at present time is not available with standard DAC of CHIMERA, where the main trigger is given by electronic signals delivered by silicon detectors.

References

[1] see http://fribs.lns.infn.it/upgrade-results.html

[2] A.Pagano et al Nucl.Phys. A 734 (2004) 504 and ref. therein

[3] A.Pagano, Nuclear Physics News International, 22:1(2012)25

[4] I. Lombardo et al Nucl.Phys.B (Proc.Suppl.) 215(2011)272

[5] G.Verde et al Journ. of Phys.: Conf. Ser. 420 (2013) 012158; E.Pagano et al proceedings of this conference; Technical design report -TDR/INFN for the farcos project (https://192.84.151.50/joomla/)

[6] G.Cardella et al NIMA 799(2015)64

[7] E. G. Lanza, A. Vitturi, M. V. Andres, F. Catara and D. Gambacurta, Phys. Rev. C 84, 064602 (2011)

[8] L.Acosta et al NIM A 715 (2013) 56

[9] Cortina-Gil et al Phys.Lett.b 401(1997)9

[10] E.Pollacco et al Physics Procedia, Volume 37, 2012, Pages 1799

EPJ Web of Conferences **117**, 06009 (2016) DOI: 10.1051/epjconf/201611706009
NN2015

Preliminary study of the ^{10}Li nucleus via one-neutron transfer

M. Cavallaro[1], M. De Napoli[2], F. Cappuzzello[1,3], C. Agodi[1], M. Bondí[1,3], D. Carbone[1], A. Cunsolo[1], B. Davids[4], T. Davinson[5], A. Foti[2,3], N. Galinski[4], R. Kanungo[6], H. Lenske[7], S.E.A. Orrigo[8], C. Ruiz[4] and A. Sanetullaev[4]

[1]INFN - Laboratori Nazionali del Sud, I-95123 Catania, Italy
[2]INFN, Sezione di Catania, I-95123 Catania, Italy
[3]Dipartimento di Fisica e Astronomia, Università di Catania, I-95123 Catania, Italy
[4]TRIUMF, Vancouver, British Columbia V6T2A3, Canada
[5]School of Physics and Astronomy, University of Edinburgh, EH9 3JZ, Edinburgh, UK
[6]Astronomy and Physics Department, Saint Mary's University, Halifax, Nova Scotia B3H 3C3, Canada
[7]Institut für Theoretische Physik, Universität Giessen, Germany
[8]Instituto de Física Corpuscular, CSIC-Universidad de Valencia, Spain

Abstract

The structure of the ^{10}Li unbound nucleus is a subject of large interest and its description is nowadays a matter of debate. We have investigated this system using the d(^{9}Li,p)^{10}Li one-neutron transfer reaction at 100 MeV in inverse kinematics. The experiment was performed at the ISACII facility at TRIUMF laboratory. The excitation energy spectrum has been reconstructed by measuring the emitted protons at backward angles and the ^{9}Li at forward angles.

1 Introduction

The exploration of the ^{10}Li nucleus is an interesting topic since it is an exotic system, unbound in the ground state. New phenomena are expected and/or observed in the continuum, showing an evolution of nuclear structure from closed to open systems beyond a mean-field description [1], [2]. In addition the knowledge of the ^{10}Li structure is a key ingredient in the description of the two-neutron halo nucleus ^{11}Li [3], [4]. Structure calculations for the weakly bound system ^{11}Li = ^{9}Li + 2n require in fact the interaction between one neutron and the ^{9}Li core as an important input quantity. This can be deduced directly from the binding energy and excitation energies of the states of ^{10}Li.

The properties of the ^{10}Li continuum remain still unclear and even the energy and the spin-parity assignment of the ground state are still controversial. The systematics of the N = 7 isotones suggest that the energy difference between the 2s1/2 and 1p1/2 energy levels is small and that the order of the two levels could be inverted, with the 2s1/2 level lower than the 1p1/2 level. The energy and ordering of the 2s1/2, 1p1/2 and 1d5/2 orbitals can be represented as in Fig. 1, according to the excitation energy of the known single particle states. In the case of ^{15}O, ^{14}N, ^{13}C and ^{12}B ground states, it is well known that the main component of the 1d5/2 shell lies above the 2s1/2, which is in turn above the 1p1/2. In the ^{11}Be (Z = 4) ground state a shell inversion between 2s1/2 and 1p1/2 is known to appear [5], [6]. If this anomaly is maintained also in the ^{10}Li (Z = 3) case, the lowest energy state in ^{10}Li is expected to have a dominant configuration with one neutron in the 2s1/2. On the other hand, if one supposes that the presence of shell inversion in ^{11}Be is related to the 2-alpha cluster structure of Be isotopes, such a structure is not possible for ^{10}Li and the configuration of the ^{10}Li ground state is expected to be a ^{9}Li core + one neutron in the 1p1/2.

Transfer reactions are essential tools to probe selected components of the nuclear wave functions [7], [8]. Thanks to that they have been used in the past and are still crucial for nuclear spectroscopy purposes. With the advent of radioactive beam facilities, the new opportunity to explore nuclear phenomena far from the stability valley has driven a renewed interest to transfer reactions, to be studied in inverse kinematics. On one hand there is a specific attention to understand how the transfer mechanism is influenced by the reduction of the binding energy of the projectile. On the other hand, the measured spectra could reveal unexpected features due to nucleon correlations which are beyond the mean field description of nuclear structure.

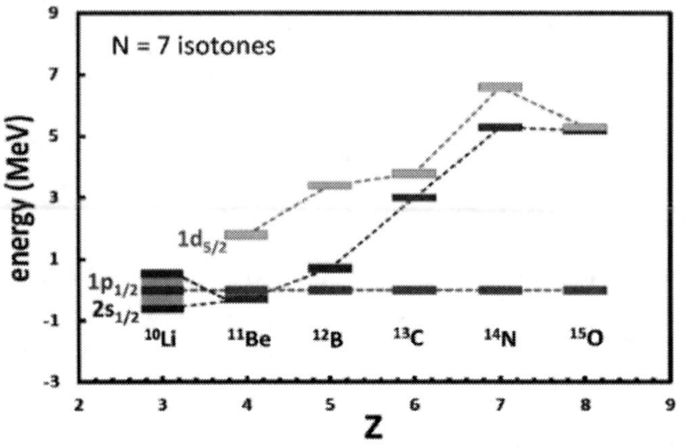

Figure 1: Relative energy of the 1d5/2, 2s1/2 and 1p1/2 shells for the N = 7 isotones in the region of ^{10}Li.

Recently ^{10}Li has been the subject of different theoretical works and challenging experimental studies [9], [10], [11], [12] including two attempts to explore the resonant energy spectrum by the d(^{9}Li,p)^{10}Li transfer reaction at 2.35 AMeV incident energy at REX-ISOLDE [10] and at 20 AMeV at NSCL-MSU, respectively [9]. However, due to the poor statistics, not much was added to our understanding of ^{10}Li states.

In a recent theoretical study of the ^{10}Li low energy resonances [13], a state dependent treatment of the pairing interaction, beyond the usual BCS approximation, has proven to be necessary to reproduce the energy spectra measured in the REX-ISOLDE experiment. In addition it was shown that, despite the complications due to the resonant structure of the final states, the cross sections can be accurately described within the DWBA. Nevertheless, due to the limitations of the existing data, many of the details predicted by the theory were not tested. For example only a measurement of the angular distribution at forward angles in the centre of mass would disentangle contributions from s, p and d orbitals in different portions of the energy spectrum. Also the behaviour of the observed s1/2 and p1/2 orbitals should be better addressed, since in both cases the coupling with the p3/2 proton generates a doublet of ^{10}Li states, namely 1^-, 2^- for the s1/2 and 1^+, 2^+ for the p1/2, while experimentally there is no indication of such doublets. In addition, according to the theory, p3/2 and d5/2 orbitals should generate resonances above 1 MeV, but these have not been observed experimentally.

DOI: 10.1051/epjconf/201611706009

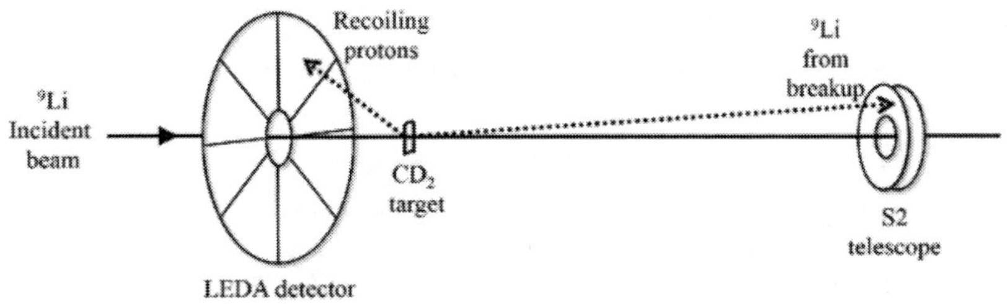

Figure 2: Schematic layout of the experiment.

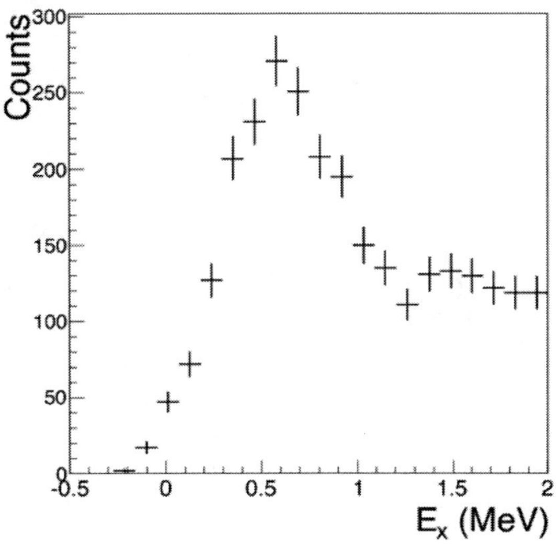

Figure 3: ^{10}Li excitation energy spectrum reconstructed from the d(^{9}Li,p)^{10}Li reaction at 100 MeV incident energy in the angular region $\theta_{c.m.}$ = 8.3° - 16.2°.

2 Experiment and results

We studied the d(^{9}Li,p)^{10}Li reaction at 11 AMeV incident energy at the TRIUMF laboratory (Canada). At this energy the recoiling protons can be detected at backward laboratory angles, thus allowing the exploration of the crucial region at forward angles in the centre of mass.

A ^{9}Li beam, produced by the ISAC-II facility with an average current of 10^6 pps. The beam charge was collected by a Faraday cup located downstream of the target. In order to check the reliability of the Faraday cup measurement, the absolute cross-section of elastic scattering data was compared to Rutherford scattering at forward angles. The beam impinged on a

CD2 target, 126 g/cm^2 thick. The target thickness was chosen to improve the energy resolution maintaining an acceptable count rate. The recoiling protons were detected at backward angles θ_{lab} = 127° - 161° by the LEDA array of silicon strip detectors [14], thus allowing the study of the ^{10}Li emitted at forward angles. Protons were detected in coincidence with the ^{9}Li fragments produced from the break-up of the corresponding ^{10}Li. ^{9}Li fragments were detected and identified by a $\Delta E - E$ telescope of annular Double Sided Silicon Detectors located downstream the target. A schematic layout of the experimental setup is shown in Fig. 2.

The ^{10}Li excitation energy was reconstructed with significant statistics allowing to explore its level structure at low excitation energy. Fig. 3 shows the excitation energy spectrum obtained up to 2 MeV in the angular region $\theta_{c.m.}$ = 8.3° - 16.2°. The centroid (E = 0.6 MeV) and full width at half maximum (FWHM = 0.6 MeV) of the unbound ground state were extracted by a best fir procedure. A second peak, weaker than the first, appears at around 1.4 MeV. The sharp fall down of the tail of the first peak toward zero excitation energy shows that there is no evidence in the present data of the existence of a s1/2 low lying virtual state as supposed by [10]. The data analysis of the region at higher excitation energy is in progress and will be published elsewhere.

The highly segmented detection system also allowed to measure the angular distributions of the observed resonances at forward angles. The analysis of the angular distribution, which is still in progress, would give more indication about the controversial question of the spin-parity of the ^{10}Li unbound ground state.

References

[1] F. Cappuzzello, et al., Phys. Lett. B **711**, 347 (2012).

[2] S.E.A. Orrigo, et al., Phys. Lett. B **633**, 469 (2006).

[3] I. Tanihata, et al., Phys. Rev. Lett. **55**, 2676(1985).

[4] T. Kobayashi, et al., Phys. Rev. Lett. **60**, 2599 (1988).

[5] F. Cappuzzello et al., Phys. Lett. B **516**, 21 (2001).

[6] F. Cappuzzello et al., Nucl. Phys. A **739**, 30 (2004).

[7] G. R. Satchler, Direct Nuclear Reactions, Oxford University Press, NY, 1983.

[8] M. Cavallaro, et a., Phys. Rev. C **88**, 054601 (2013).

[9] P. Santi et al., Phys. Rev. C **67**, 024606 (2007).

[10] H. B. Jeppesen et al., Phys. Lett. B **642**, 449 (2006).

[11] H. G. Bohlen, Z. Phys. A **344**, 381 (1993).

[12] H. G. Bohlen , Nucl. Phys. A **616**, 254c (1997).

[13] S. E. A. Orrigo and H. Lenske, Phys. Lett. B **677**, 214 (2009).

[14] T. Davinson et al., Nucl. Instrum. Methods Phys. Res. A **454**, 350 (2000).

Projectile deformation effects in the breakup of ^{37}Mg

SHUBHCHINTAK[1], R. CHATTERJEE[2] and R. SHYAM[3,2]

[1]Department of Physics and Astronomy, Texas A&M University Commerce, TX-75428, USA
[2]Department of Physics, Indian Institute of Technology Roorkee, 247667, India
[3] Theory Group, Saha Institute of Nuclear Physics 1/AF Bidhannagar, Kolkata 700064, India

Abstract

We study the breakup of ^{37}Mg on Pb at 244MeV/u with the recently developed extended theory of Coulomb breakup within the post-form finite range distorted wave Born approximation that includes deformation of the projectile. Comparing our calculated cross section with the available Coulomb breakup data we determine the possible ground state configuration of ^{37}Mg.

1 Introduction and formalism

Coulomb breakup of nuclei away from the valley of stability has been one of the most successful probes to unravel their structure. However, it is only recently that one is venturing into medium mass nuclei like ^{23}O [1] and ^{31}Ne [2], especially in and around the so called "island of inversion". This is a very new and exciting development which has expanded the field of light exotic nuclei to the deformed medium mass region.

We consider the elastic breakup of a two body composite 'deformed' projectile a in the Coulomb field of a target t, $a + t \to b + c + t$, where projectile a breaks up into fragments b (charged) and c (uncharged). The reduced transition amplitude, $\beta_{\ell m}$, for this reaction is given by [3]

$$\begin{aligned}\beta_{\ell m} &= \left\langle e^{i(\gamma \mathbf{q_c} - \alpha \mathbf{K}).\mathbf{r_1}} \left| V_{bc}(\mathbf{r}_1) \right| \phi_a^{\ell m}(\mathbf{r}_1) \right\rangle \\ &\times \left\langle \chi_b^{(-)}(\mathbf{q_b}, \mathbf{r_i}) e^{i\delta \mathbf{q_c}.\mathbf{r_i}} | \chi_a^{(+)}(\mathbf{q}_a, \mathbf{r_i}) \right\rangle . \end{aligned} \tag{1}$$

The ground state wave function of the projectile $\phi_a^{lm}(\mathbf{r}_1)$ appears in the first term (vertex function), while the second term that describes the dynamics of the reaction, contains the Coulomb distorted waves $\chi^{(\pm)}$. This can be expressed in terms of the bremsstrahlung integral. α, γ and δ are the mass factors pertaining to the three-body Jacobi coordinate system (see Fig. 1 of Ref. [2]). In Eq. 1, $\mathbf{K}$ is an effective local momentum appropriate to the core-target relative system and $\mathbf{q}_i$'s ($i = a, b, c$) are the Jacobi wave vectors of the respective particles.

$V_{bc}(\mathbf{r}_1)$ [in Eq. (1)] is the interaction between b and c, in the initial channel. We introduce an axially symmetric quadrupole-deformed potential, as

$$\begin{aligned} V_{bc}(\mathbf{r}_1) &= \frac{V_{ws}}{1 + exp(\frac{r_1 - R}{a})} \\ &\quad - \beta_2 R V_{ws} \frac{df(r_1)}{dr_1} Y_2^0(\hat{\mathbf{r}}_1), \end{aligned} \tag{2}$$

where V_{ws} is the depth of the spherical Woods-Saxon potential, β_2 is the quadrupole deformation parameter. The first part of the Eq. (2) is the spherical Woods-Saxon potential $V_s(r_1)$ with radius $R = r_0 A^{1/3}$. r_0 and a being the radius and diffuseness parameters, respectively. To preserve the analyticity of our method, we calculate the radial part of the ground state wave function of the projectile using the undeformed Woods-Saxon potential (radius and diffuseness parameters taken as 1.24 fm and 0.62 fm respectively, which reproduce the ground state binding energy). We emphasize that the deformation parameter (β_2) has already entered into the theory via V_{bc} in Eq. (1). For more details on the formalism we refer to Ref. [4].

2 Results and discussions

The nucleus ^{37}Mg has a large uncertainty in its one-neutron separation energy (0.162±0.686 MeV [7]) and has controversies regarding its ground state spin-parity. Recently measured large breakup cross section [5] and reaction cross section [6] seems to suggest a halo structure in ^{37}Mg.

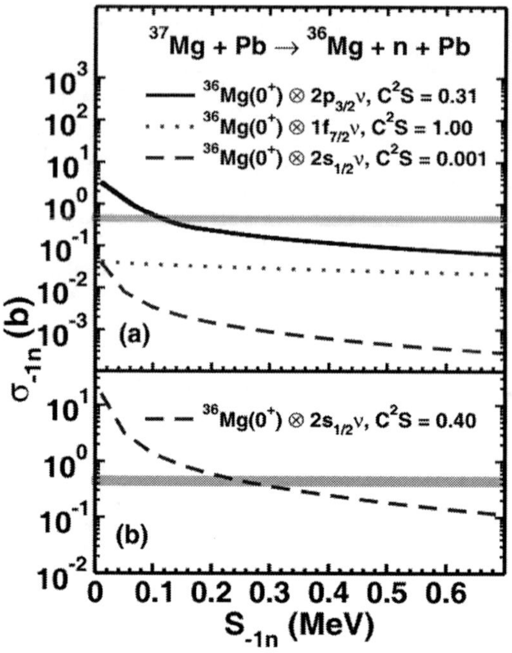

Figure 1: (a) Pure Coulomb total one-neutron removal cross section, σ_{-1n}, in the breakup reaction of ^{37}Mg on a Pb target at 244 MeV/nucleon beam energy as a function of one-neutron separation energy S_{-1n} obtained with configurations ^{36}Mg$(0^+) \otimes 2p_{3/2}\nu$ (solid line), ^{36}Mg$(0^+) \otimes 2s_{1/2}\nu$ (dashed line) and ^{36}Mg$(0^+) \otimes 1f_{7/2}\nu$ (dotted line) for ^{37}Mg$_{gs}$ using the shell model spectroscopic factors (C^2S) as indicated in each case. The experimental cross section (taken from Ref. [5]) is shown by the shaded band. (b) Same reaction as in (a) for the ^{36}Mg$(0^+) \otimes 2s_{1/2}\nu$ configuration using the C^2S value of 0.40 as deduced in Ref. [5].

The nuclei in island of inversion are expected to have significant components of $2p - 2h$ $[\nu(sd)^{-2}(fp)^2]$ neutron intruder configurations. Indeed, in Ref. [5], it has been argued that the valence neutron in ^{37}Mg$_{gs}$ is most likely to have a spin parity (J^π) of $3/2^-$ that corresponds to the $2p_{3/2}$ orbital. For the sake of completeness, in this work we have considered neutron removal from $2p_{3/2}$, $2s_{1/2}$ and $1f_{7/2}$ orbitals.

In Fig. 1(a), we show the results of our calculations for the pure Coulomb σ_{-1n} in the breakup reaction of ^{37}Mg on a Pb target at the beam energy of 244 MeV/nucleon as a function of S_{-1n} corresponding to one-neutron removal from the $2p_{3/2}$, $2s_{1/2}$ and $1f_{7/2}$ orbitals. For C^2S we have used the shell model values as given in Ref. [5], which are 0.31 and 0.001 for the ^{36}Mg$(0^+) \otimes 2p_{3/2}\nu$, and ^{36}Mg$(0^+) \otimes 2s_{1/2}\nu$ configurations, respectively. However, for the ^{36}Mg$(0^+) \otimes 1f_{7/2}\nu$ configuration the SM C^2S is not given in this reference. Therefore, we have assumed a C^2S of 1.0 for this case. The shaded band in this figure shows the corresponding measured cross section taken from Ref. [5] with its width representing the experimental uncertainty. We note that calculated cross sections obtained with the ^{36}Mg$(0^+) \otimes 2p_{3/2}\nu$ configuration (solid line in Fig. 1), overlap with the experimental band in the S_{-1n} region of 0.10 ± 0.02. Theoretical cross sections for the $2p_{1/2}$ case are almost identical to those of the $2p_{3/2}$ case. On the other hand, for the ^{36}Mg$(0^+) \otimes 2s_{1/2}\nu$ and ^{36}Mg$(0^+) \otimes 1f_{7/2}\nu$ configurations there is no overlap between calculated cross sections and the data band. Therefore, our results support a $J^\pi = 3/2^-$ ground state for ^{37}Mg with a one-neutron separation energy of 0.10±0.02. The S_{-1n} deduced in our work is closer to the evaluated value of 0.16 ± 0.68 [7], with lesser uncertainty.

Nevertheless, with C^2S for the ^{36}Mg$(0^+) \otimes 2s_{1/2}\nu$ configuration as extracted in Ref. [5] (0.40), the cross section curve for this case will also overlap with the experimental data band as shown in Fig. 1(b). However, the extracted S_{-1n} in our case is 0.26±0.04 MeV, instead of $0.40^{+0.19}_{-0.13}$ MeV deduced in Ref. [5]. In fact, with the combination of the mean values of S_{-1n} and C^2S (0.40 MeV and 0.40) for this configuration deduced in Ref. [5], there would be no overlap between the theoretical cross section and the experimental data band in our calculations.

In Fig. 2(a), we show our results for σ_{-1n} as a function of β_2 for the ^{36}Mg$(0^+) \otimes 2p_{3/2}\nu$ configuration of ^{36}Mg$_{gs}$ with C^2S values of 0.42 and 0.31 and taking a S_{-1n} of 0.22 MeV (as in Ref. [5]) in both the cases. For $\beta_2 = 0$, the σ_{-1n} in each case is the same as that shown in Fig. 1, which is below the experimental data. With increasing β_2, the cross sections in both the cases increase, and the overlaps between calculations and the data take place in ranges $0.35 < \beta_2 < 0.68$ and $0.62 < \beta_2 < 0.94$ for C^2S of 0.42 and 0.31, respectively. We add that if for $C^2S = 0.31$ the S_{-1n} as deduced in Fig. 1 is taken then calculated cross section would overlap with the data band at much lower values of β_2.

In Fig. 2(b) we show the same results for the ^{36}Mg$(0^+) \otimes 2s_{1/2}\nu$ configuration with C^2S and S_{-1n} values of 0.40 and 0.40 MeV, respectively, (which is the mean value of these quantities as deduced in Ref. [5]). In this case, we see that in contrast to the results in Fig. 2(a), there is no overlap between calculated cross sections and the data band for any value of β_2 in the range of 0−1.0. We have checked that the situation remains the same for $\beta_2 > 1.0$. Moreover, the contribution of the deformation term to the cross section is substantially low for the s-wave configuration, which results in almost constant σ_{-1n} as a function of β_2 as seen in Fig. 2(b). Therefore, in our calculation, even with deformation a s-wave configuration is ruled out

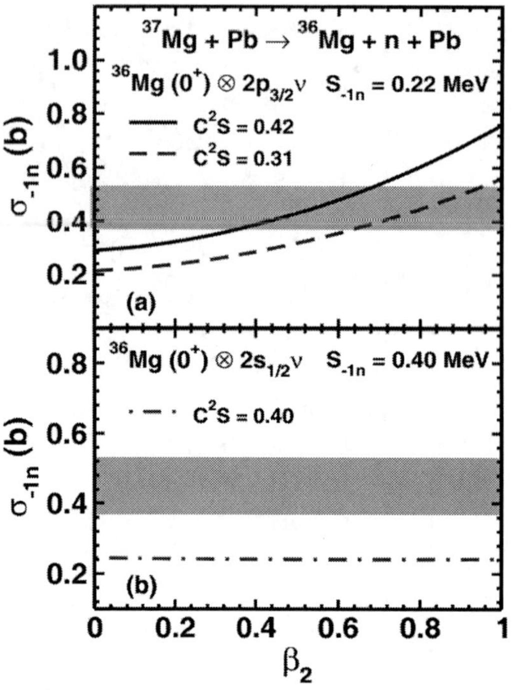

Figure 2: (a) σ_{-1n} as a function of the deformation parameter β_2 in the Coulomb breakup of ^{37}Mg on a Pb target at the beam energy of 244 MeV/nucleon with the configuration ^{36}Mg(0^+) $\otimes$ $2p_{3/2}\nu$ for ^{37}Mg$_{gs}$. The S$_{-1n}$ is taken to be 0.22 MeV with C^2S values being 0.42 (solid line) and 0.31 (dashed line). (b) Same as in Fig. 2(a) for ^{36}Mg(0^+) $\otimes$ $2s_{1/2}\nu$ configuration with C^2S and S_{-1n} of 0.40 and 0.40 MeV, respectively. In both (a) and (b) the experimental data (shown by the shaded region) are taken from Ref. [5].

for ^{37}Mg$_{gs}$ for the C^2S and S_{-1n} combination deduced in Ref. [5].

3 Conclusion

In this paper we have studied the Coulomb breakup reaction ^{37}Mg + Pb $\rightarrow$ ^{36}Mg + n + Pb at the beam energy of 244 MeV/nucleon, within the framework of the post form finite range distorted wave Born approximation theory that is extended to include the projectile deformation effects. In this formalism the transition amplitude is factorized into two parts - one containing the dynamics of the reaction and the another the projectile structure informations such as the fragment-fragment interaction and the corresponding wave function in its ground state. Analytic expressions can be written for both parts. This formalism opens up a route to perform realistic quantum mechanical calculations for the breakup of neutron-drip line nuclei in the medium mass region that can be deformed.

We calculated the total one-neutron removal cross sections (σ_{-1n}) in this reaction and compared our results with the corresponding data reported in a recent publication [5]. Our calculations seem to favor a $J^\pi = 3/2^-$ spin assignment to the ^{37}Mg ground state with one-neutron separation energy (S_{-1n}) of 0.10 ± 0.02 MeV, if the spectroscopic factor (C^2S) for this state is taken to be the corresponding shell model value of 0.31. However, the deduced S_{-1n} depends on the chosen value of C^2S. Our study shows that S_{-1n} rises steadily with increasing C^2S. Indeed, due to the uncertainty in the C^2S value for the ^{36}Mg$(0^+) \otimes 2s_{1/2}\nu$ configuration for the ^{37}Mg ground state, the $J^\pi = 1/2^+$ spin assignment to it can not be fully excluded based on the present data.

In order to gain more insight in the ground state structure of ^{37}Mg, we studied the effect of the projectile deformation on σ_{-1n}. We find that for the configuration ^{36}Mg$(0^+) \otimes 2p_{3/2}\nu$ for the ^{37}Mg ground state, the calculated σ_{-1n} overlaps with the experimental data band in certain range (that depends on the value of C^2S) of the quadrupole deformation parameter (β_2). However, with the ^{36}Mg$(0^+) \otimes 2s_{1/2}\nu$ configuration, the overlap does not occur between calculated and measured σ_{-1n} for any reasonable combination of β_2 and C^2S values. This supports the $J^\pi = 3/2^-$ spin assignment for the ^{37}Mg ground state.

However, for unambiguous confirmation one also needs to calculate [4] more exclusive observables such as the core-valence neutron relative energy spectra, the energy-angle and the angular distributions of the emitted neutron and the parallel momentum distribution of the core fragment. The position of the peak as well as the magnitude of the cross section near the peak of the core-valence neutron relative energy spectra would to be dependent on the configuration of the projectile ground state as well as on its deformation.

Our study is expected to provide motivation for future experiments on breakup reactions of the neutron rich medium mass nuclei.

Acknowledgments

This work is supported by the Council of Scientific and Industrial Research (CSIR) and the Department of Science and Technology(SR/S2/HEP-040/2012), Govt. Of India .

References

[1] R. Chatterjee, R. Shyam, K. Tsushima, A. W. Thomas, Nucl. Phys. A **913**, 116 (2013).

[2] Shubhchintak and R. Chatterjee, Nucl. Phys. A **922**, 99 (2014).

[3] R. Chatterjee, P. Banerjee and R. Shyam Nucl. Phys. A **675**, 477 (2000).

[4] Shubhchintak, Neelam, R. Shyam and K. Tsushima, Nucl. Phys. A **939**, 101 (2015).

[5] N. Kobayashi *et al.*, Phys. Rev. Lett. **112** (2014) 242501.

[6] M. Takechi *et al.*, EPJ Web of Conferences. **66**, 02101 (2014).

[7] M. Wang *et al.*, Chin. Phys. C **36** (2012) 1603.

EPJ Web of Conferences **117**, 06011 (2016) DOI: 10.1051/epjconf/201611706011
NN2015

Study of cluster structures in ^{10}Be and ^{16}C neutron-rich nuclei via break-up reactions

D. Dell'Aquila[4*], L. Acosta[11], F. Amorini[2], R. Andolina[3], L. Auditore[5], I. Berceanu[8], G. Cardella[1], M.B. Chatterjiee[9], E. De Filippo[1], L. Francalanza[4], B. Gnoffo[1], A. Grzeszczuk[10], G. Lanzalone[2,6], I. Lombardo[4], N. Martorana[2], T. Minniti[5], A. Pagano[1], E.V. Pagano[2,3], M. Papa[1], S. Pirrone[1], G. Politi[1,3], A. Pop[8], F. Porto[2], L. Quattrocchi[5], F. Rizzo[2,3], E. Rosato[4†], P. Russotto[1], A. Trifirò[5], M. Trimarchi[5], G. Verde[1,7] and M. Vigilante[4]

[1] INFN, Sezione di Catania, Catania, Italy
[2] INFN-Laboratori Nazionali del Sud, Catania, Italy
[3] Dipartimento di Fisica e Astronomia, Università di Catania, Catania, Italy
[4] Dip. Fisica, Univ. di Napoli Federico II and INFN, Sezione di Napoli, Napoli, Italy
[5] INFN, Gr. Coll. Messina and Dip. Fis, Univ. di Messina, Italy
[6] Facoltà di Ingegneria e Architettura, Università Kore, Enna, Italy
[7] Institute de Physique Nucléaire d'Orsay, Orsay, France
[8] Institute for Physics and Nuclear Engineering, Bucharest, Romania
[9] Saha Institute for Nuclear Physics, Kolkata, India
[10] Institut of Physics, University of Silesia, Katowice, Poland
[11] National Autonomous University of Mexico, Mexico City, Mexico

*e-mail: dellaquila@na.infn.it
†Deceased.

Abstract

Projectile break-up reactions induced on polyethylene (CH_2) target are used in order to study the spectroscopy of ^{10}Be and ^{16}C nuclei. For the present experiment we used ^{10}Be and ^{16}C beams delivered by the FRIBs facility at INFN-LNS, and the CHIMERA 4π multi-detector. ^{10}Be and ^{16}C structures are studied via a relative energy analysis of break-up fragments. The ^{4}He+^{6}He break-up channel allowed us to study the spectroscopy of ^{10}Be; in particular we find evidence of a new state in ^{10}Be at 13.5 MeV excitation energy. The ^{16}C nucleus is studied via ^{6}He-^{10}Be correlation; we find the fingerprint of a possible state at about 20.6 MeV.

1 Introduction

The investigation of cluster structures in neutron-rich nuclei is an important tool to explore the behaviour of nuclear forces in few body interacting systems [1]. In presence of extra neutrons clustering effects could assume very different features from the α clustering in self-conjugated nuclei. These neutrons can assume the role of covalent particles acting as a glue-like between the α cluster centers and leading to the so called nuclear molecules [1].

In this complex scenario, beryllium and carbon neutron-rich isotopes assume a very important role thanks to their possible, respectively, dymeric and trimeric molecular structures [1].

An interesting neutron-rich beryllium isotope is the ^{10}Be, for which the spectroscopy is still not fully understood. This nucleus is characterized by a deformed ground state [2] on which a positive parity rotational band is built. While the 2^+ member of the $0^+_{g.s.}$ band is known, the subsequent members assignment is still missing. For example, the 4^+ member, predicted at about 11–12 MeV, was initially claimed by Bohlen et al. [3] and after contradicted by Suzuki et al. in a very recent paper [4]. A negative parity rotational band with a 1^- state at 5.96 MeV as band-head is also reported in [1]. A molecular band is also built on the 0^+ state at 6.18 MeV, strongly characterized by a $\alpha : 2n : \alpha$ configuration [2]. For this band, the 2^+ state has been found at 7.54 MeV and the subsequent 4^+ member was assigned in Ref. [5] and in recent ^{6}He+^{4}He inverse kinematic resonant elastic scattering experiments [4].

The carbon isotopic chain has recently been object of several studies [6–10]. Three centers molecular structures could be present in ^{16}C, as predicted in a very recent theoretical calculations [11], where some linear chains, also with parity-asymmetric nature (i.e. ^{10}Be+α+2n like configurations), and

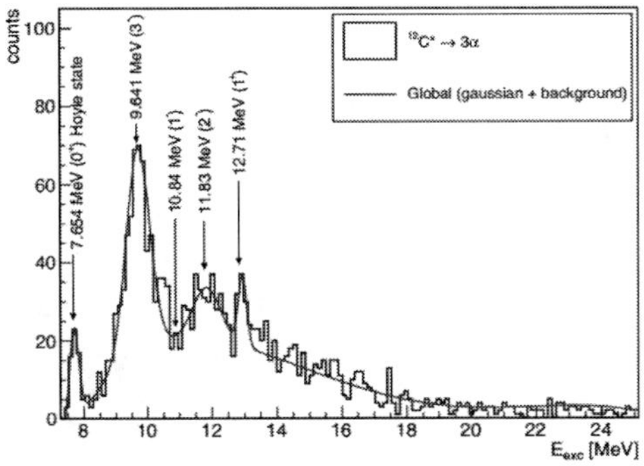

Figure 1: (color online) 3α relative energy spectrum (blue line). The red line is the result of a multiple parameters fit of the data. The fit is obtained by considering the states indicated by arrows.

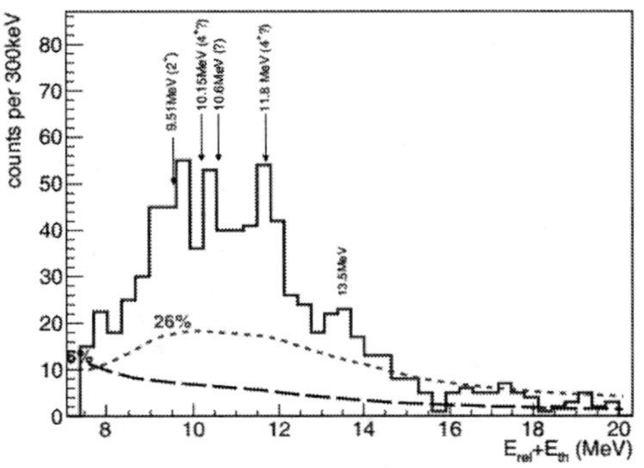

Figure 2: (color online) ^{10}Be relative energy spectrum for the ^{4}He+^{6}He break-up channel. The red and the black dashed lines represent, respectively, the simulated detection efficiency for inelastic scattering on proton and carbon

triangular configurations are pointed out above the helium disintegration threshold. However, the spectroscopy of ^{16}C is extremely poorly known and an experimental evidence of these states is still missing. For this reason, ^{16}C has recently attracted a large interest [9, 10], but very few data have been reported and no definitive conclusions can be drawn.

In the present paper we report new results on the spectroscopy of ^{10}Be and ^{16}C excited states above the cluster emission thresholds, investigated via sequential projectile break-up reactions. Reaction products have been detected by using the Chimera array. From a relative energy analysis of break-up fragments we found indications of a possible state at about 13.5 MeV in ^{10}Be. For the ^{16}C nucleus, the ^{6}He+^{10}Be correlations suggest the presence of a new state at about 20.6 MeV.

2 Experimental details

The experiment was performed at the FRIBs facility of the INFN-Laboratori Nazionali del Sud (LNS). A fragmentation beam, mainly constituted by ^{16}C, ^{13}B and ^{10}Be at 56 MeV/u, was produced starting from a 55 MeV/u ^{18}O primary beam impinging on a ^{9}Be production target. A tagging system [12], made by a Micro Channel Plate (MCP) ad a Double Side Silicon Strip Detector (DSSSD) was used in order to provide a particle by particle identification of the fragmentation cocktail.

Break-up reactions have been induced on a polyethylene CH_2 target. For the present experiment we have selected reaction products from the $^1H,^2H,^{12}C(^{10}Be,^4He^6He)$ and $^1H,^2H,^{12}C(^{16}C,^6He^{10}Be)$ reactions. To obtain spectroscopic information on ^{10}Be and ^{16}C nuclei we have measured masses, energies and flight directions of the corresponding break-up fragments with the CHIMERA 4π multi detector [13]. Particle identification was carried out via the ΔE-E identification technique, by using the energy loss of the particle inside the first detection stage of the Chimera device (Si, 300μm) and the energy residual in the last stage (CsI(Tl) scintillator, $\approx$ 12 cm).

3 Data analysis

The ^{10}Be and ^{16}C structure has been studied via a relative energy analysis of break-up fragments. The excitation energy of the projectile nucleus before decaying has been reconstructed thanks to the relation $E_x = E_{rel} + E_{thr}$, where E_{rel} is the relative energy of the break-up fragments, obtained from the experimental data, and E_{thr} is the energy decay threshold of the selected break-up channel.

In order to check our experimental technique, we have studied 2α and 3α kinematical correlations. In the first case we have found evidence of a prominent peak corresponding to a relative energy of about 91 keV, compatible with the emission of α particles from the decay of ^8Be ground state, and a large bump at about 3 MeV relative energy, produced by the first excited state of ^8Be. In Fig. 1, is reported the 3α relative energy spectrum obtained by including the data from all the available beams. In this case we have found evidence of the population of Hoyle state (7.654 MeV) and also some other excited states in ^{12}C, indicated by the arrows. The red line is the result of a global fit, obtained by including the populated excited states and a polynomial smooth background. As clearly visible from Fig. 1 the fit well reproduces the behaviour of experimental data, providing a further confirmation of the validity of the present analysis.

3.1 ^4He-^6He correlations: the structure of ^{10}Be

The ^4He-^6He correlations have been used to study the spectroscopy of ^{10}Be. In this case we found the relative energy $(E_{rel} + E_{thr})$ spectrum reported in Fig. 2. Despite the limited resolution and statistics, the spectrum exhibits peaks corresponding to excited states known in literature and indicated by

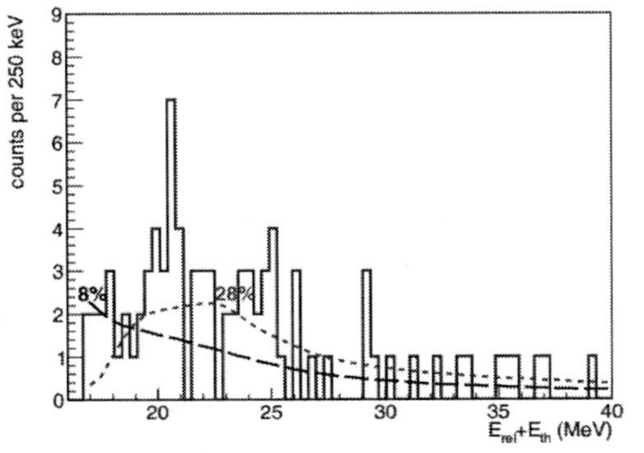

Figure 3: (color online) ^{16}C E_{rel} spectrum for the ^{10}Be+^{6}He break-up channel. The red and the black dashed lines represent, respectively, the simulated detection efficiency for inelastic scattering on proton and carbon

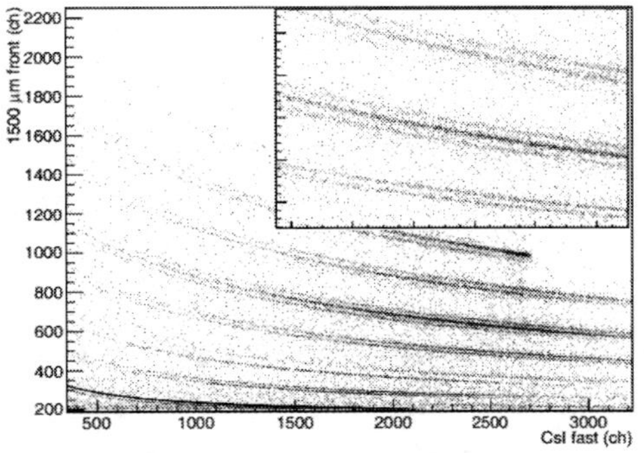

Figure 4: (color online) ΔE-E identification matrix obtained from a calibration run ^{16}O+CH$_2$ of the CLIR experiment. The insert shows a zoom of the boron, carbon and oxygen line.

arrows. It is noticeable the presence of a bump corresponding to an excitation energy of about 13.5 MeV, not previously observed in ^{10}Be, being the possible evidence of a new state. To check if the observed peak can be really ascribed to an excited state in ^{10}Be we evaluated the detection efficiency. This contribution was calculated via a MonteCarlo simulation. We performed two different simulation taking into account the different possible target nuclei. The obtained efficiency curves are represented by the dashed lines in Fig. 2 for the excitation on hydrogen (red line peaking at 26%) and carbon (black line peaking at 6%). In both cases the contribution is smooth, confirming the possibility to attribute the 13.5 MeV peak to the possible existence of a new state in ^{10}Be.

3.2 ^{10}Be-^{6}He correlations: the structure of ^{16}C

The ^{16}C nucleus has been investigated via the ^{6}He-^{10}Be correlations. The corresponding relative energy spectrum is shown in Fig. 3. The spectrum, with a very low statistics, shows the appearance of an yield enhancement at about 20.6 MeV excitation energy. Also in this case the simulated efficiency doesn't show any sharp peak, ant therefore it could be the evidence of a new state in ^{16}C, energetically compatible with a liner state predicted in [11]. Interestingly, a similar yield enhancement, but with lower statistics, has also been reported in both the previous works [9, 10].

4 Conclusions and future perspectives: the CLIR experiment

In conclusion, the structure of ^{10}Be and ^{16}C is investigated via break-up reactions induced by radioactive beams produced at the FRIBs facility (LNS).

We find the possible evidence of a new state in ^{10}Be at about 13.5 MeV in the ^{4}He+^{6}He disintegration channel. The ^{16}C is studied via the ^{6}He-^{10}Be correlations finding also evidence of a possible new state at about 20.6 MeV, even with an extremely low statistics.

Very recently, we have performed a new experiment (CLIR at LNS) with the aim to obtain more information on ^{16}C structure following the suggestions of the present work. The new experiment was carried out by coupling the CHIMERA device to the FARCOS array, covering the most forward directions. As an example of the capabilities of this new device, we report, in Fig. 4, a typical ΔE-E plot obtained from the CLIR data. As clearly visible, the identification is present in the whole Z domain, confirming the possibility to improve the present results.

References

[1] W. Von Oertzen, M. Freer, and Y. Kanada-En'yo, *Phys. Rep.* **432** (2006) 43.

[2] Y. Kanada-En'yo, *J. Phys. G* **24** (1998) 1499.

[3] H.G. Bohlen *et al.*, *Phys. Rev.* **C 75** (2007) 054604.

[4] D. Suzuki *et al.*, *Phys. Rev.* **C 87** (2013) 054301.

[5] M. Freer *et al.*, *Phys. Rev. Lett.* **96** (2006) 042501.

[6] M. Freer *et al.*, *Phys. Rev.* **C 84** (2011) 034317.

[7] I. Lombardo *et al.*, *Nucl. Instrum. Meth. Phys. Res.* **B 302** (2013) 19.

[8] M. Freer *et al.*, *Phys. Rev.* **C 90** (2014) 054324.

[9] P.J. Leask, *Jour. Phys. G: Nucl. Part. Phys.* **27** (2001) B9.

[10] N. I. Ashwood *et al.*, *Phys. Rev.* **C 70** (2004) 0644607.

[11] T. Baba, Y. Chiba and M. Kimura, *Phys. Rev.* **C 90** (2014) 064319.

[12] I. Lombardo *et al.*, *Nuc. Phys.* **B 215** (2011) 272.

[13] A. Pagano *et al.*, *Nucl. Phys. News* **22** (2012) 25.

EPJ Web of Conferences **117**, 06012 (2016) DOI: 10.1051/epjconf/201611706012
NN2015

Effects of coupling to breakup channels in reactions induced by weakly bound and halo nuclei

J. P. Fernández-García[1], A. Di Pietro[1], P. Figuera[1], M. Fisichella[1], M. Lattuada[1,2], A. M. Moro[3], A. Musumarra[1,2], M. G. Pellegriti[1,2], V. Scuderi[1], D. Torresi[1] and M. Zadro[4]

[1] INFN - Laboratori Nazionali del Sud, via S. Sofia 62, 1-95123 Catania, Italy
[2] Dipartamento di Fisica e Astronomia, via S. Sofia 64, I-95123 Catania, Italy
[3] Departamento de FAMN, Universidad de Sevilla, Apartado 1065, E-41080 Seville, Spain
[4] Ruder Bošković Institute, Bijenička cesta 54, HR-10000 Zagreb, Croatia

Abstract

The experimental elastic scattering angular distributions of the collisions involving the weakly bound nuclei ^{6,7}Li and the halo nucleus ^{6}He on the same ^{64}Zn target at several energies around the Coulomb barrier were measured at the Laboratori Nazionali del Sud (LNS, Italy), and the Cyclotron Research Center, Louvain La Neuve (Belgium), respectively. The experimental elastic scattering angular distributions of the ^{6,7}Li + ^{64}Zn systems have been analyzed within the continuum-discretized coupled-channels method, while the ^{6}He + ^{64}Zn data have been compared with both continuum-discretized coupled-channels and coupled-reaction-channels calculations.

1 Introduction

In the last decade, the study of collisions involving halo and weakly bound nuclei had a notable interest. In fact, due to the coupling to breakup channels, strong effects on the elastic scattering and fusion have been observed [1–4].

In the case of the stable lithium isotopes, ^{6,7}Li, elastic scattering angular distributions on different targets have been extensively measured and analyzed within the Continuum-Discretized Coupled-Channel (CDCC) formalism, showing that the couplings to the continuum states have a significant effect on the elastic scattering involving the ^{6}Li nucleus [5].

One of the most studied exotic nuclei in the literature is ^{6}He (e.g. Refs. [6–11]), since such beam is available in different Radioactive Ions Beams (RIB) facilities with high intensity. The ^{6}He nucleus has a bound core (α particle) and two loosely bound neutrons (S_{2n}=0.97 MeV). These two neutrons have a large probability of being far away from the core, creating the so-called nuclear halo. The ^{6}He nucleus constitutes an archetype of Borromean nucleus, since the subsystems α-neutron and neutron-neutron are unbound.

Recently, a systematic study of ^{6,7}Li + ^{64}Zn reactions at energies around the Coulomb barrier was performed at LNS. The elastic scattering angular distributions of both systems were measured at different energies [12, 13]. We present results of the elastic scattering for the ^{6,7}Li+^{64}Zn reactions at the same center of mass energies. In addition, new preliminary experimental data for the ^{6}He+^{64}Zn collision are presented and compared with the three-body CDCC and Coupled-Reaction-Channels (CRC) calculations, assuming a simple di-neutron model of the projectile ^{6}He.

2 Experimental setups and results

The ^{6,7}Li beams were generated by the SMP Tandem Van de Graaff accelerator and transported to the CT2000 scattering chamber at center of mass energies of 11.7, 12.4, 13.5, 15.0, 16.3 and 18.1 MeV. The experimental setup consisted in five silicon telescopes positioned on a rotating plate, in order to cover a wide angular range. The telescope systems were composed by a 10 μm detector plus a 200 μm detector. For the ^{6}Li+^{64}Zn reactions a ^{64}Zn target of 400 μg/cm^2 was used [12], while a 70 μg/cm^2 target for the ^{7}Li+^{64}Zn collisions.

On the other hand, the experiment corresponding to the ^{6}He + ^{64}Zn reaction was carried out at the Cyclotron Research Center of Louvain La Neuve, Belgium. The reactions ^{6,4}He+^{64}Zn at center of mass energies of 13.5 and 16.5 MeV were measured to continue old measurements performed by our group on the same systems at different energies [3,14]. The experimental setup was composed by an array of 4 LEDA detectors [15] covering the forward angles $5° < \theta_{lab} < 12°$. Another array of 3 LEDA detectors, placed at 45° with respect to the beam axis, covered the angular range $22° < \theta_{lab} < 65°$. Finally, two Double Sided Silicon Strips Detectors (DSSSD) were used to cover the backward angles, between 67° and 120°. For additional details see Refs. [3, 10].

The preliminary elastic scattering angular distributions of the ^{6,7}Li and ^{6}He on ^{64}Zn systems at the same center of mass energy ($E_{c.m.}$=13.5 MeV) are presented in Fig. 1.

3 Theoretical calculations

In the last years, the CDCC method has been successfully applied to describe reactions induced by halo and weakly bound nuclei, such as ^{6,7,11}Li, ^{11}Be and ^{6}He [1, 2, 7, 14, 16–18]. To solve the scattering problem, this formalism uses the coupled channel method. Due to the low breakup threshold, it is important to consider the continuum states of the weakly bound and halo nuclei. These states are included and discretized by means of the so-called *binning* method. The calculations were performed using the code FRESCO [19], in which both nuclear and Coulomb effects were considered.

The CDCC method is based on the elastic direct breakup mechanism, where the breakup process is considered as an inelastic excitation of the projectile to its continuum states. We would like to emphasize that this calculation do not consider other reaction mechanisms, such as neutron transfer or non-elastic breakup.

In this framework, we have analyzed the reactions induced by ^{6,7}Li and ^{6}He on the same ^{64}Zn target.

The ^{6}Li nucleus was described as an α+d structure, with a break-up threshold at 1.47 MeV above the ground state. The resonances 3^+, 2^+, 1^+ and the non-resonant states were included in the calculations. On the other hand, the ^{7}Li isotope was described as an α+t system, with a break-up threshold at 2.47 MeV. The excited state $1/2^-$, the resonances $7/2^-$, $5/2^-$ and the non-resonant states were considered.

In Fig. 1, the solid red lines represent the CDCC calculations, while the dotted blue lines correspond to the calculations without couplings to the continuum states of the projectile. A good agreement between the predictions of the CDCC calculations and the experimental data was observed. In the case of the ^{6}Li induced collisions, it was necessary to include the couplings to the breakup channels to reproduce the experimental data, while for the ^{7}Li reactions these couplings were less important. This behavior could be due to the fact that the breakup threshold of ^{6}Li (S_{α}=1.47 MeV) is smaller than that of ^{7}Li (S_{α} =2.47 MeV).

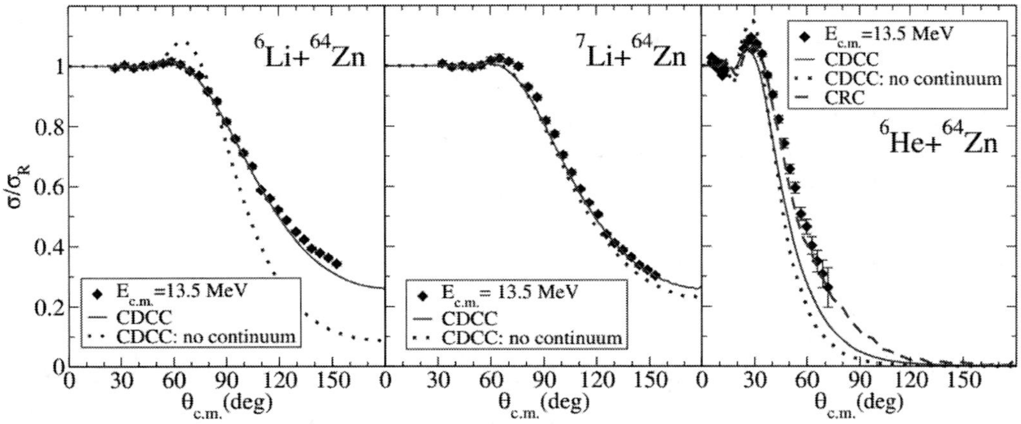

Figure 1: Preliminary elastic scattering angular distributions for the ^{6,7}Li and ^{6}He + ^{64}Zn collisions at the same center of mass energy ($E_{c.m.}$=13.5 MeV). The CDCC calculations are represented by solid red lines, while the same calculation without the continuum couplings are shown by dotted blue lines. In the case of the ^{6}He + ^{64}Zn reaction, the CRC calculation is represented by the dashed green line.

The ^{6}He+^{64}Zn reaction has been analyzed within the three-body (α+2n+^{64}Zn) CDCC method. The improved di-neutron model for ^{6}He (α+2n) of Ref. [20] was considered. This model, which assumes a two-neutron cluster with spin zero bound to the α core with an effective separation energy of 1.6 MeV, was applied to describe several reactions involving ^{6}He nucleus [20].

In Fig. 1, the CDCC calculation of the elastic scattering angular distribution for the ^{6}He+^{64}Zn system is represented by the solid blue line. To study the effects of the coupling to the continuum states of the projectile, we have included the calculations without such couplings, represented by the dotted red line. Comparing both calculations we observe a strong effect of the couplings to breakup channels on the elastic cross section and a reduction of the Coulomb-nuclear interference peak.

However, the CDCC calculation underestimates the experimental data of the ^{6}He+^{64}Zn reaction. Therefore, additionally to this calculation, a Coupled Reaction Channel (CRC) calculation considering the one-neutron transfer to bound and unbound states of the target has been performed, since this calculation described the elastic cross section of the ^{6}He+^{64}Zn reaction at $E_{c.m.}$=12.4 MeV [21]. In this reference, the one-neutron transfer mechanism was found to be a more appropriate description than that considering two-neutron transfer to bound and unbound states of the target. In addition, the CRC calculations, performed in [21], were useful to describe the experimental fusion excitation function reported in Ref. [14].

The calculated elastic scattering angular distribution is represented by the green dashed line in Fig. 1. Unlike to the CDCC result, the CRC calculation improves the agreement with the experimental data.

The obtained result suggests that the elastic direct breakup process (CDCC calculation), where the core and the target are considered as inert particles, is not a good approach to describe the elastic scattering of the ^{6}He+^{64}Zn reaction at $E_{c.m.}$=13.5 MeV. On the other hand, the CRC calculation, based on a breakup process in which one neutron is transferred to bound and unbound states of the target, reproduces the experimental data. This result is different from that obtained in Ref. [21], where both CDCC and CRC calculations were in reasonable agreement with the experimental data at $E_{c.m.}$=12.4 MeV, which suggests a possible energy dependence of the different breakup mechanisms.

4 Conclusions

Experimental results for elastic scattering of the weakly bound nuclei ^{6,7}Li and the halo nucleus ^{6}He on ^{64}Zn at the same center of mass energy have been presented.

CDCC calculations for the ^{6,7}Li+^{64}Zn collisions have been performed, showing a good agreement with the experimental data. The couplings to the breakup channels have an important effect in the case of the ^{6}Li scattering, whereas minor effects have been found for the ^{7}Li case. This behavior could be linked to the smaller binding energy of the ^{6}Li (S_{α}=1.47 MeV) nucleus compared with ^{7}Li (S_{α}=2.47 MeV).

In the ^{6}He+^{64}Zn reaction, we have observed a strong effect of the coupling to breakup channels. Two different calculations, CDCC and CRC calculations, including couplings to the breakup channels, have been performed within different breakup mechanism. The CDCC calculation fails to describe the experimental elastic scattering angular distribution. However, we have observed that the CRC method provides a more suitable approach to describe the elastic scattering angular distribution, where the ^{6}He breakup is treated within one-neutron transfer to the continuum picture.

References

[1] Di Pietro A. et al. *Phys. Rev. Lett.*, **105** (2010) 022701.

[2] Cubero M. et al. *Phys. Rev. Lett.*, **109** (2012) 262701.

[3] Scuderi V. et al. *Phys. Rev. C*, **84** (2011) 064604.

[4] Canto L. F. et al. *Nucl. Phys. A*, **821** (2009) 51.

[5] Canto L. F., Gomes P. R. S., Donangelo R., and Hussein M. S. *Phys. Rep.* **424** (2006) 1.

[6] Sánchez-Benítez A. M. et al. *Nucl. Phys. A*, **803** (2008) 30.

[7] Acosta L. et al. *Phys. Rev. C*, **84** (2011) 044604.

[8] Aguilera E. F. et al. *Phys. Rev. C*, **63** (2001) 061603.

[9] Kakuee O. R. et al. *Nucl. Phys. A*, **728** (2003) 339.

[10] Di Pietro A. et al. *Europhys. Lett.*, **64** (2003) 309.

[11] Fernández-García J. P., Rodríguez-Gallardo M. , Alvarez M. A. G., and Moro A. M. *Nucl. Phys. A*, **840** (2010) 19.

[12] Zadro M. et al. *Phys. Rev. C*, **80** (2009) 064610.

[13] Zadro M. et al. *Phys. Rev. C*, **87** (2013) 054606.

[14] Di Pietro A. et al. *Phys. Rev. C*, **69** (2004) 044613.

[15] Davinson T. et al. *Nucl. Inst. and Meth. A*, **454** (2000) 350.

[16] Fernández-García J. P. et al. *Phys. Rev. Lett.*, **110** (2013) 142701.

[17] Sakuragi Y. *Phys. Rev. C*, **35** (1987) 2161.

[18] Beck C., Keeley N., and Diaz-Torres A. *Phys. Rev. C*, **75** (2007) 054605.

[19] Thompson I. J. *Comp. Phys. Rep.* **7** (1988) 167.

[20] Moro A. M., Rusek K., Arias J. M., Gómez-Camacho J. G. and Rodríguez-Gallardo M. *Phys. Rev. C*, 75:064607, 2007.

[21] Moro A. M., Fernández-García J. P., Alvarez M. A. G., and Rodríguez-Gallardo M. *EPJ Web of Conferences*, **17** (2011) 08001.

EPJ Web of Conferences **117**, 06013 (2016) DOI: 10.1051/epjconf/201611706013
NN2015

Measuring fusion excitation functions with RIBs using the stacked target technique: Problems and possible solutions

M. Fisichella[1], A. C. Shotter[2], A. Di Pietro[1], P. Figuera[1], M. Lattuada[1,3], C. Marchetta[1], V. Privitera[4], L. Romano[3,4], C. Ruiz[5] and M. Zadro[6]

[1]INFN - Laboratori Nazionali del Sud, via S. Sofia 62, I-95123 Catania, Italy
[2]School of Physics and Astronomy, University of Edinburgh, JCMB, Mayfield Road, Edinburgh EH9 3JZ, UK
[3]Dipartimento di Fisica ed Astronomia, via S. Sofia 64, I-95125 Catania, Italy
[4]CNR-IMM MATIS, via S. Sofia 64, Catania, Italy
[5]TRIUMF, 4004 Wesbrook Mall, Vancouver, BC V6T 2A3, Canada
[6]Ruđer Bošković Institute, Bijenička cesta 54, HR-10000 Zagreb, Croatia

Abstract

For measuring fusion excitation functions, the activation technique with a stack of targets offers the considerable advantage that several reaction energies may be simultaneously measured by using one beam energy. However, its main drawback is the degradation of the beam quality as it passes through the stack due to statistical nature of energy loss processes and any non-uniformity of the stacked targets. If not taken properly into account, this degradation can lead to a wrong determination of the fusion excitation function. In this contribution some results of the investigation of this problem are reported.

1 Introduction

In the last decades, many fusion reactions induced by low intensity RIBs have been studied by irradiating stacks of several thick targets and measuring on-line or off-line the radiation emitted in the decay of the evaporation residues [1–7].
In these activation experiments, the fusion excitation function has been determined by associating the cross-section σ_{mean}(E), measured in each target, either with the energy of the beam in the middle of the considered target $\bar{E}$ or to the effective energy E_{eff}, defined as follows:

$$E_{eff} = \frac{\int_0^\infty E\sigma(E)D(E,t_0)dE}{\int_0^\infty \sigma(E)D(E,t_0)dE} \tag{1}$$

In the last definition, the value of the energy is obtained by an averaging procedure which weights the beam energy with energy distribution $D(E,t_0)$ inside the considered target of thickness t_0 together with the trend of the fusion cross sections. The $D(E,t_0)$ distribution represents the probability that a beam particle will have energy E inside the considered target of thickness t_0. The physical factors that influence the shape of the energy distribution function $D(E,t_0)$ are the shape of the beam energy distribution entering the target, the beam energy loss in the target, and the energy straggling of the beam. The $D(E,t_0)$ distribution taking esplicitly into account all the above effects, can be determined by using the SRIM code [9] as discussed in [8].
The present authors found that neither $\bar{E}$ nor E_{eff} lead to the correct reconstruction of the fusion excitation function. In particular, it has been found that in the case of an uniform target $\bar{E}$ is an underestimation and E_{eff} is an overestimation, respectively, of the real value E_r, defined as the reaction energy such that $\sigma_{mean} = \sigma(E_r)$, where $\sigma(E)$ is the true cross section function. As an example it was supposed to know the behaviour of the ^{9}Li+^{120}Sn fusion excitation function and it was supposed that at sub-barrier energies it has an exponential trend with a slope α=1.72 MeV^{-1}, which is within the range of typical slopes observed in sub-barrier fusion experimental data. We considered an incoming energy beam of 17 MeV and calculated analitically the values of E_r, $\bar{E}$ and E_{eff} as a function of the energy losses in the target (i.e., targets of different thicknesses). The results are reported in figure 1 in the case of a delta incoming beam energy distribution (Fig.1a) and Gaussian incoming beam energy distribution with FWHM = 2 MeV (Fig.1b). Only for the ideal case of a mono-energetic beam and for thin targets will E_r , $\bar{E}$, and E_{eff} coincide. In this example uniform

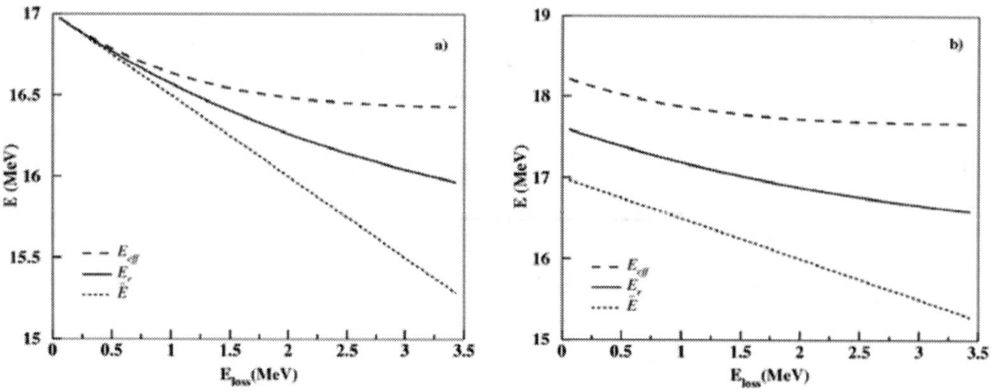

Figure 1: Calculations for the ^{9}Li+^{120}Sn system. $\bar{E}$, E_{eff} and E_r are plotted as a function of the energy loss in the target for an average incident beam energy is 17 MeV. a) Delta beam energy distribution; b) Gaussian beam energy distribution with FWHM=2 MeV. See text for details.

targets have been considered, but it has to be noticed that the presence of non-uniformities in the target will further modify this behaviour, as it will be shown in the following section.

The present authors developed an alternative deconvolution procedure to determine the correct fusion excitation function of the studied system. The idea is to determine an analytical expression for σ(E) such that for the $D(E, t_0)$ associated to each given target one would obtaion by using Eq. 3 the measured mean cross section. Details about this new method are reported in [8]. Application of this procedure requires a detailed knowledge of $D(E, t_0)$. The best way to determine $D(E, t_0)$ is to characterise target properties, as well as performing numerical beam energy loss simulations. Of course, these effects are amplified when a stack of targets is used. For this reason, an accurate analysis has been done concerning the the target non-uniformity and the effect that this could have on the determination the fusion excitation function. It has been developed an experimental technique based on simple measurements of residual energy spectra of α emitted by an ^{231}Am source from which it is possible to determine in a very accurate way the target thickness distribution [8].

2 ^{9}Li+^{120}Sn simulation

Generally, in the previous articles where these kind of experiments are discussed, very few information are given about target non-uniformity, straggling effects and incident beam quality. However, as discussed previously,

since these factors can significantly influence the deduced excitation functions the present authors performed simulations in order to investigate in details how they might influence the determination of the fusion excitation function and try to quantify the magnitudes of these effects.
Numerical simulations of the $^{9}Li+^{120}Sn$ fusion reaction have been performed. It has been supposed that a 28 MeV ^{9}Li beam impinged on a stack of five targets each composed of one ^{120}Sn foil with an average thickness of 5 mg/cm^2, followed by an uniform ^{93}Nb catcher/degrader foil 1.5 mg/cm^2 thick. The results of the simulations were studied for the case of uniform targets and degrader foils, and for the situation where both targets and foils are non-uniform. The $^{9}Li+^{120}Sn$ fusion excitation function behaviour has been supposed to be known and to have a Wong-like behaviour:

$$\sigma(E) = \frac{A}{E} ln(1 + e^{B(E-C)}) \tag{2}$$

with A=1936 mb·MeV, B=1.83 MeV^{-1}, C=18.95 MeV. The chosen parameters in the Eq.(2) are consistent with the cross-section slopes of other similar reactions. In this way, for each target, the mean cross-section that would be measured in it has been calculated by using the following relation:

$$\sigma_{mean} = \frac{\int_0^\infty \sigma(E) D(E, t_0) dE}{\int_0^\infty D(E, t_0) dE} \tag{3}$$

As an example, in Fig. 2a the $D(E, t_0)$ distributions of the fitht target obtained by the SRIM simulation for the case of uniform (continuos line) and non-uniform (dotted line) stack foils are reported. As it is possible to see, not only the shapes but also the explored energy ranges of the distributions are different in the two cases.
In Fig. 2b the σ_{mean} obtained by applying Eq. (3) for the uniform and non-uniform target cases, is plotted versus the mean energy $\bar{E}$ (open symbols), and versus the effective energy E_{eff}(closed symbols). The circles and triangles correspond to case of uniform and non-uniform target, respectively. It has to be put in evidence that the cross-section values that would be measured for the two cases are slightly different due to the different energy distributions inside the targets. The continuous line represents the cross-section of Eq. (2) assumed to represent the real behavior of the fusion excitation function in the simulated experiment. As one can see, at energies below the Coulomb barrier, the real excitation function is not reproduced by any of the two representations. The differences are bigger in the case of non-uniform targets with respect to the uniform case, as one can expect due to the wider energy distributions $D(E, t_0)$. Plotting the measured

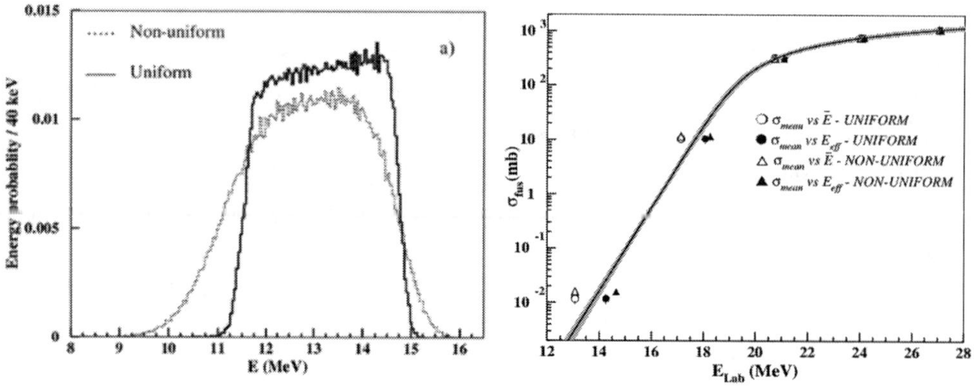

Figure 2: a) Energy distribution of the beam inside the fifth target of the stack considering uniform (continuous line), and non uniform (dotted line) foils. b) Results of the simulations for the case of uniform (circles) and for non-uniform (triangles) targets. The continuous black line shows the *real* fusion excitation function assumed for the collision ^{9}Li+^{120}Sn. The open and closed symbols are the experimental results that would be obtained plotting the measured σ_{mean} versus $\bar{E}$ and E_{eff}. In this figure the results of the deconvolution procedure cannot be distinguished from the black line. The blue shaded area represent the uncertainty associated with the deconvolution procedure. See text for details.

cross-sections versus $\bar{E}$ or E_{eff} generates differences up to a factor of 5 with respect to the real fusion excitation function.

The deconvolution procedure, developed by the present authors, has also been applied to these simulated data. For applying this procedure a typical error on the measured mean cross-section, σ_{mean}, in each of the targets, has been supposed to be 15% . This error would originate from experimental uncertainties associated with detector efficiencies, mean target thickness, target thickness distribution etc. The results obtained with this procedure cannot be distinguished from the continuum line representing the assumed real cross-section. The shaded blue area represents the error band associated with the procedure. This error band has been calculated by applying the standard error propagation formula to Eq. 2, considering the three parameters (A,B,C) as free variables.
According to the present results the best way of determining the fusion excitation function is to use the deconvolution procedure.

3 Summary and conclusion

From the present study it is clear that that the traditional way to represent the fusion cross section as function of the energy in the middle of the target, or as a function of an effective energy based on a weighted average which takes into account both the beam energy distribution and the energy dependence of the cross section, may lead to a wrong determination of the fusion excitation function, especially when the energy distribution within a target is large, and the cross-section is rapidly changing with the energy. The present authors suggest as a possible solution to infer the true excitation function, to apply a deconvolution procedure as discussed in previous paragraph. In order properly apply this procedure a detailed knowledge of the the target thickness distribution and of the incoming beam distribution beam must be available. Therefore the authors recommend that for future publications where such data is presented, enough information should be included concerning the used foil non-uniformities, beam energy distribution and their effects on the deduced fusion excitation functions.

References

[1] J. J. Kolata, et al., Phys. Rev. Lett. 81, 4580 (1998).

[2] P. A. DeYoung, et al, Phys. Rev. C 58, 34423444 (1998).

[3] Y. Penionzhkevich, et al., The European Physical Journal A 31, 185194 (2007).

[4] A. Lemasson, et al., Phys. Rev. Lett. 103, 232701 (2009).

[5] Y. E. Penionzhkevich, et al, Phys. Rev. Lett. 96, 162701 (2006).

[6] C. Signorini, et al, Nuclear Physics A 735, 329 344 (2004).

[7] R. Wolski, et al., The European Physical Journal A 47, 111 (2011).

[8] M. Fisichella et al., Accepted for publication in Phys. Rev. C

[9] J. Ziegler, Srim code - http://www.srim.org .

[10] C. Y. Wong, Phys. Rev. Lett. 31, 766769 (1973).

EPJ Web of Conferences **117**, 06014 (2016) DOI: 10.1051/epjconf/201611706014
NN2015

Cluster Thomas-Ehrman Shift in ^{10}Be–^{10}C

Makoto Ito

Department of Pure and Applied Physics, Kansai University, 3-3-35
Yamate-cho, Suita, Osaka 564-8680, Japan

Abstract

Coulomb shift of the ^{10}Be and ^{10}C nuclei, which can be nicely described by the $\alpha + \alpha + N + N$ four-body degrees of freedom, is investigated by the microscopic cluster model. The Coulomb shift is prominent for the non S-wave configurations, while the shift is suppressed for the S-wave dominant configurations, which have the well-developed α cluster structure. This result can be interpreted in terms of the so-called Thomas-Ehrman shift, which is originally discussed in the Coulomb shift of the ^{17}O–^{17}F systems.

1 Introduction

In light neutron-excess systems, many kinds of molecular structures are discussed from the viewpoint of the clustering phenomena. In particular, much attention has been concentrated on Be isotopes from the theoretical and experimental approaches. The Be isotopes can be considered as typical examples of two-center superdeformed systems which build on an $\alpha + \alpha$ rotor of ^{8}Be. Various cluster structures are generated by the coupling the two α degrees of freedom and the single particle motion of the excess neutrons, and the structures are analogous to the so-called chemical-bonding structure in molecular systems.

The low-lying states can be nicely understood by the molecular orbital (MO) picture, in which the excess neutrons perform the single-particle motion in the π^- or σ^+ orbitals around two α cores [1]. The σ^+ orbit, which has a large mixture of the S-wave component, enhances the α–α clustering.

On the contrary, in their highly-excited states, recent experiments revealed the existence of the interesting resonant states, which dominantly decay into the ^{4,6,8}He fragments [2,3]. These resonances are interpreted in terms of the ionic or atomic configurations of ^{x}He + ^{y}He, in which neutrons are trapped in one of the α cores and generates the exchange force [4, 5].

The systematic studies of the chemical-bonding structure identified in Be isotopes is very interesting. One of the systematic studies is the analysis of the mirror systems, and the Coulomb shift in the proton-rich side is an interesting research subject. For example, a typical example of the Coulomb phenomena is known as the so-called Thomas-Ehrman shift (TES), which was originally discussed in the mirror systems, such as the ^{17}O and ^{17}F nuclei [6]. In the ^{16}O core plus one nucleon model, the Coulomb shift in ^{17}F $= {}^{16}\mathrm{O} + P$ is exceptionally suppressed for the S-wave orbit although the Coulomb shift occurs uniformly for other non S-wave orbits. The S-wave proton can escape from the Coulomb repulsion because of no centrifugal barrier, which tends to confine the wave function inside of the Coulomb barrier.

One of the characteristic features in cluster structure is the spatial extension, which involves the S-wave dominance. In the covalent configurations, for example, the σ^+ orbit with a large mixture of the S-wave component enhance the α clustering, while two nuclei dominantly perform the S-wave relative motion in the atomic or ionic configurations. These cluster picture strongly suggests that the development of the clustering involves the suppression of the Coulomb shift. In this report, the Coulomb shift of ^{10}Be–^{10}C is investigated, which are typical example of the nuclear chemical-bonding systems [5].

We apply the generalized two-center cluster model (GTCM) [4, 5] to ^{10}Be ($= \alpha + \alpha + 2N$) and ^{10}C ($= \alpha + \alpha + 2P$). This model is available to handle the formations of the covalent MOs and the ionic or atomic structures in a unified manner. Therefore, the Coulomb shift can be discussed in connection to the structural changes from the low-lying covalent structures to the highly-excited atomic structure.

2 Framework

The detailed explanation of GTCM has already been published in Refs. [4,5], and we briefly explain the formulation of GTCM in the following. In this model, the basis functions of $^{10}\mathrm{Be} = \alpha + \alpha + 2N$ and $^{10}\mathrm{C} = \alpha + \alpha + 2P$

are given by the $\{\Phi_{\mathbf{m}}^{J^\pi K}(S)\}$, where,

$$\Phi_{\mathbf{m}}^{J^\pi K}(S) = \hat{P}_K^{J^\pi} \mathcal{A} \left\{ \psi_L(\alpha)\psi_R(\alpha) \prod_{j=1}^{2} \varphi_j(m_j) \right\}_S . \qquad (1)$$

The α-clusters at the left (L) and right (R) side center are shown by $\psi_i(\alpha)$ ($i = L, R$), and they are placed with a relative distance parameter of S, which is the generator coordinate [7]. The internal structure of the α clusters are expressed by the $(0s)^4$ configurations of the harmonic oscillator (HO). The single-particle wave function for the valence nucleons localized around the α clusters is given by an atomic orbital (AO) $\varphi(m_j)$, which is labeled by the AO indices of $m_j = (p_k, i, \tau)$. Here, p_k ($k = x, y, z$) means the $0p$-orbitals around i-th core ($i = L$ or R) with the spin τ ($= \uparrow$ or $\downarrow$). In Eq. (1), $\mathbf{m}$ represents a set of AOs for the valence nucleons, $\mathbf{m}$=(m_1, m_2). The intrinsic basis functions with the full anti-symmetrization $\mathcal{A}$ are projected to the eigenstate of the total spin J, its intrinsic angular projection K, and the total parity π by the projection operator $\hat{P}_K^{J^\pi}$.

The total wave function is finally given by taking the superposition over S, $\mathbf{m}$ and K as

$$\hat{\Psi}_\nu^{J^\pi} = \int dS \sum_{\mathbf{m},K} C_{\mathbf{m}K}^{\nu}(S)\, \Phi_{\mathbf{m}}^{J^\pi K}(S) . \qquad (2)$$

The coefficients for the ν-th eigenstate, $C_{\mathbf{m}K}^{\nu}(S)$, are determined by solving a coupled-channel GCM (Generator Coordinate Method) equation [7]. In the present calculation, we consider all the possible $0p$ atomic-orbit configurations, which lseads to the complete formation of the covalent, ionic, and atomic configurations around two α cores.

As for the nucleon-nucleon (NN) interaction, we use the Volkov No.2 [8] and the G3RS [9] for the central and spin-orbit parts, respectively. In ^{10}Be, the parameters in the interactions and the size parameter of HO are the same as those applied in the analysis of ^{10,12}Be [4, 5]. On the contrary, the NN interactions are slightly modified in ^{10}C so as to reproduce the α + ^{6}Be threshold energy. The majorana parameter (M) in ^{10}C is increased by about 20 % in comparison to the parameter in ^{10}Be.

In the present calculation, all of the excited states are calculated by the bound state approximation although the treatment of the scattering problem is also available [4, 5]. The analysis of the scattering states will be shown in future studies.

3 Results

Figure 1 shows the energy levels of ^{10}Be (left levels) and ^{10}C (right levels). In the individual spectra, the lowest four 0^+ levels are plotted. The dominant components of the 0_1^+ and 0_2^+ states are the molecular orbit configuration of $(\pi_{3/2}^-)^2$ and $(\sigma_{1/2}^+)^2$, respectively, in which two valence nucleons performed the single particle motion around two α cores. On the contrary, the unbound 0_3^+ and 0_4^+ states have the atomic configurations of the α + ^{6}He$_{g.s.}$ and α + ^{6}He(2^+), respectively.

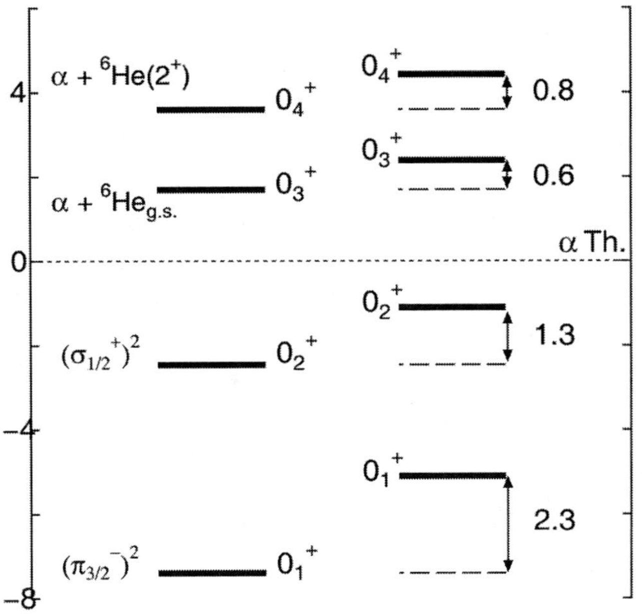

Figure 1: Energy levels of the ^{10}Be (left levels) and ^{10}C (right levels) nuclei. The α threshold is set to be zero energy. In the ^{10}C spectra, the corresponding levels to ^{10}Be are plotted by the dashed line. The numbers attached to the levels show the energies of the Coulomb shifts. The Coulomb shifts are shown in units of MeV.

In the ^{10}C spectra, the Coulomb shifts to the higher energy region are confirmed, and the magnitude of the shifts depend on the energy levels. The shift of the 0_1^+ state is 2.3 MeV, which is the largest of all the energy levels. This prominent shift is due to the spatially compact structure of the $(\pi_{3/2}^-)^2$ configuration. The $\pi_{3/2}^-$ orbit, which has a main component of the P-wave orbit, strongly reduce the α-α distance, and the 0_1^+ state has a spatially compact structure similar to the shell model-like structure. Therefore, the effect of the Coulomb repulsion is prominent for the compact 0_1^+ state in the proton-rich side, the ^{10}C nucleus.

On the contrary, the Coulomb shift of the 0_2^+ state is 1.3 MeV, which is smaller than the shift of the ground 0_1^+ state, 2.3 MeV. The reduction of the shift for 0_2^+ is due to the spatially extended structure of the $\sigma_{1/2}^+$ orbit, which has a large mixture of the S-wave component. The large mixing of the S-wave component in the valence orbit leads to the enhancement of the α-α clustering. Since the repulsive effect of the Coulomb interaction is reduced in the spatially extended configuration, the Coulomb shift for the 0_2^+ state is suppressed in comparison to the shift for the 0_1^+ state.

Moreover, the Coulomb shifts for the 0_3^+ and 0_4^+ states are suppressed much more, which are 0.6 MeV and 0.8 MeV, respectively. This reduction is generated by the well-developed binary cluster structures of α + ^{6}He, which makes it possible to escape from the Coulomb repulsion. In the 0_3^+ state, the α and $^6\mathrm{He}_{g.s.}$ perform the S-wave relative motion, while the α + ^{6}He(2^+) with the D-wave relative motion is a main component in the 0_4^+ state. Since the S-wave component is predominant in 0_3^+, its Coulomb shift is suppressed in comparison to the 0_4^+ state.

In all the spectra, the Coulomb shift is reduced for the S-wave dominant states (0_2^+ and 0_3^+), while the shift is enhanced for the respective non S-wave states (0_1^+ and 0_4^+). Therefore, these suppressed Coulomb shift for the S-wave states are basically same as the Thomas-Ehrman Shift, which was originally discussed in the ^{17}O and ^{17}F systems [6]. The Coulomb suppression in ^{10}Be and ^{10}C, which are shown in Fig. 1, correspond to the extension of the Thomas-Ehrman Shift to the cluster degrees of freedom, which is induced by the $\alpha + \alpha + 2N$ structure.

4 Summary and discussion

We have applied the generalized two-center cluster model to the mirror systems, such as $^{10}\mathrm{Be} = \alpha + \alpha + 2N$ and $^{10}\mathrm{C} = \alpha + \alpha + 2P$, and discussed the Coulomb shift in the lowest four 0^+ states. The present calculation predicts the suppression of the Coulomb shift for the S-wave dominant state, which is generated by the development of the α clustering. Thus, the suppressed Coulomb shift presented here should be called the "Cluster Thomas-Ehrman Shift", which corresponds to the extension of the traditional Thomas-Ehrman Shift on the single particle motion.

In the present calculation, we have employed the bound state approximation for the unbound states above the α decay threshold. In our previous analysis on the continuum state [4, 5], the 0_3^+ state in ^{10}Be, which has a developed $\alpha + {}^6\mathrm{He}_{g.s.}$ structure, is not a sharp resonant states but a broad continuum-like state, while the 0_4^+ state is a resonance. This result is consistent to the recent experiment of the ^{6}He scattering by the α target [10]. Therefore, the Coulomb shift in the continuum states should be carefully discussed by employing the scattering boundary condition. The calculation with the scattering boundary condition is now underway.

Furthermore, the present calculation also predicts the asymmetry in the monopole transition from 0_1^+ to 0_2^+; specifically, the $0_1^+ \to 0_2^+$ strength in ^{10}C is about twice as large as the respective strength in ^{10}Be. Due to the suppression of the Coulomb shift for the 0_2^+ state, the excitation energy of 0_2^+ in ^{10}C is reduced in comparison to the respective excitation energy in ^{10}Be. This reduction leads to the enhancement of the monopole transition in ^{10}C. The detailed analysis of the monopole transitions will be shown in forthcoming papers.

References

[1] N. Itagaki and S. Okabe, Phys. Rev. C **61**, 044306 (2000), and references therein.

[2] A. Saito *et al.*, Mod. Phys. Lett. A **25** 1858 (2010), and references therein.

[3] Z. H. Yang, *et al.*, Phys. Rev. Lett. **112**, 162501 (2014), and references therein.

[4] M. Ito, Phys. Rev. C **85**, 044308 (2012); M. Ito and K. Ikeda, Rep. Prog. Phys. **77**,096301 (2014).

[5] M. Ito, Phys. Lett. B**636**, 293 (2006)

[6] Bohr A and Mottelson B R 1969 Nuclear Structure vol I (New York: Benjamin) p 320.

[7] H. Horiuchi, Suppl. Prog. Theor. Phys. **62**, 90 (1977).

[8] A. B. Volkov, Nucl. Phys. **74**, 33 (1965).

[9] N. Yamaguchi, T. Kasahara, S. Nagata, and Y. Akaishi, Prog. Theor. Phys. **62**, 1018 (1979).

[10] D. Suzuki et al., Phys. Rev. C **87**, 054301 (2013).

EPJ Web of Conferences **117**, 06015 (2016) DOI: 10.1051/epjconf/201611706015
NN2015

Coupled-rearrangement-channels calculation of the three-body system under the absorbing boundary condition

M. Iwasaki[1], R. Otani[1], M. Ito[1] and M. Kamimura[2]

[1]Department of Pure and Applied Physics, Kansai University, 3-3-35 Yamatecho Suita, Osaka, Japan
[2]RIKEN Nishina Center for Accelerator-based Science, RIKEN, Wako, Saitama, Japan

Abstract

We formulate the method of the absorbing boundary condition (ABC) in the coupled-rearrangement-channels variational method (CRCMV) for the three-body problem. In the present study, we handle the simple three-boson system, and the absorbing potential is introduced in the Jacobi coordinate in the individual rearrangement channels. The resonance parameters and the strength of the monopole breakup are compared with the complex scaling method (CSM). We have found that the CRCVM + ABC method nicely works in the three-body problem with the rearrangement channels.

1 Introduction

One of the current issues in nuclear physics is the structure of the borromean three-body systems, which mainly appear around Neutron- or Proton-drip lines and its dynamics of continuum states above the particle decay threshold [1]. The borromean systems around the drip line are weak binding systems and hence, they are easily excited to unbound continuum states. Therefore, it is very important to describe the structure of the borromean systems as well as its reaction dynamics in a consistent manner.

In a naive treatment of the unbound states, a scattering boundary condition should be imposed among the interacting particles, but imposing the scattering boundary condition for three particles is complicated in general. In order to avoid the complexity of the three-body scattering boundary condition, non-Hermitian transformation is often introduced. There are two representative methods of the non-Hermitian transformations: the complex scaling method (CSM) [2] and the method of the absorbing boundary condition (ABC) [3,4].

In CSM, the coordinates and momenta in the Hamiltonian of the total system is simply rotated in the complex plane, while, in ABC, the negative imaginary potential is placed at the outer region of the system. These two procedures commonly give rise to the damping of the wave function at the asymptotic region. Therefore, the three-body scattering problem can be reduced to the bound-state-like problem, in which the variational calculation using the basis-expansion technique is available.

The CSM calculations are greatly successful in describing the Coulomb breakup into the three-body continuum [2]. In the CSM calculations, the rearrangement of the Jacobi coordinates among the three particles, is completely taken into account. On the contrary, in the ABC method, no coordinate rearrangements are considered in the three-body calculations [5,6].

The variational calculations with the rearrangement channels is called the coupled-rearrangement channels variational method (CRCVM), and the CRCVM treatment is essential to obtain the rapid convergence of the binding energy of the three-body system [7]. The CSM + CRCVM calculation in the three-body system has already been established, but the formulation of the CRCVM + ABC still remains unclear. Since the application of CSM is difficult in some special problems [4], it is important to develop ABC in the CRCVM treatment. In the present report, we apply ABC to CRCVM in the three-body problem, and the formulation of CRCVM + ABC is discussed.

2 Framework

In the present calculation, we handle the identical three-boson system, which has the following total Hamiltonian,

$$\hat{H} = -\frac{\hbar^2}{2\mu_{\mathbf{r}}}\nabla_{\mathbf{r}}^2 - \frac{\hbar^2}{2\mu_{\mathbf{R}}}\nabla_{\mathbf{R}}^2 + \sum_{i>j} V(s_{ij}), \tag{1}$$

where, the first two terms represent the kinetic energies with the Jacobi coordinate of $\mathbf{r}$ and $\mathbf{R}$, which denote the relative coordinate of the first and second particles (1–2) and the coordinate between the center of mass of two particle and the third particle ((12)–3), respectively. $\mu_{\mathbf{r}}$ and $\mu_{\mathbf{R}}$ denote the reduced masses respective to each of the Jacobi coordinates. In the present calculations, a kinematic factor is set to be $\hbar^2/m = 1/2$ (MeV·fm^2) for simplicity.

Thc last term denotes the two body potential between the ith particle and the jth particle with relative distance s_{ij}, which is assumed to be the Gaussian potential [5], such as

$$V(s) = -2e^{-0.16s^2} + e^{-0.04s^2}. \tag{2}$$

In the ABC method, total Hamiltonian is modified to include the negative imaginary potential $-i\hat{W}$ with a strength η, such as

$$\hat{H} \rightarrow \hat{H}^\eta = \hat{H} - i\eta\hat{W}, \quad \hat{W} = \sum_{c=1,2,3} (r_c^4 + R_c^4). \tag{3}$$

Here, the absorbing potential $\hat{W}$, which is given by the polynomial function with the fourth power [3,4], are introduced along the Jacobi coordinate in the c-th rearrangement channels. We solve the Schrödinger equation for $\hat{H}^\eta$

$$\hat{H}^\eta \Psi_n^\eta(r,R) = E_n(\eta)\Psi_n^\eta(r,R). \tag{4}$$

The wave function $\Psi_n^\eta(r,R)$ and energy eigenvalue $E_n(\eta)$ for the n-th eigenstate depend on the strength of the absorber η. The S-wave relative motion is assumed in $\Psi(r,R)$ and hence, it depends on the magnitudes of the Jacobi coordinates, r and R. Eq. (4) is solved by the basis-expansion method, in which the oscillating Gaussian basis is employed [7]. The boson symmetry is exactly considered by introducing the rearrangement channels.

We also calculate the strength function of the three-body breakup into the continuum. A general definition of the strength function is given by

$$\mathcal{S}(E) = \sum_\nu \left|\left\langle \Psi_\nu | \hat{O} | \Psi_i \right\rangle\right|^2 \delta(E - E_\nu), \tag{5}$$

where Ψ_ν is the ν-th final state having the energy eigenvalue E_ν in continuum. In Eq. (5), $\hat{O}$ denotes the operator of an external field, and the monopole field is considered, which is the squared particle coordinate measured from the center of mass for the total system.

The calculation of the strength function in Eq. (5) can be achieved by implementing the extended completeness relation (ECR) [8]. The detailed explanation of ECR for the ABC solutions is shown in Ref. [8]. We have also applied the complex scaling method (CSM) on the calculation of the energy eigenvalues and the strength function. The ABC calculation is checked by comparing with the CSM calculation.

3 Results

Figure 1 shows the distribution of the energy eigenvalues obtained by solving Eq. (4) for the identical three-bosons system under the absorbing boundary condition. If the absorbing potential is switched off, the eigenvalues are located along the real (horizontal) axis. The energy eigenvalues are distributed in the complex energy plane, when the absorbing potential is switched on. In the negative real energy region, there is a bound state (open square), which is almost invariant under the absorbing boundary. In the positive energy region, the two resonant poles (solid circle and solid square) are separated from the continuum spectra (open circles).

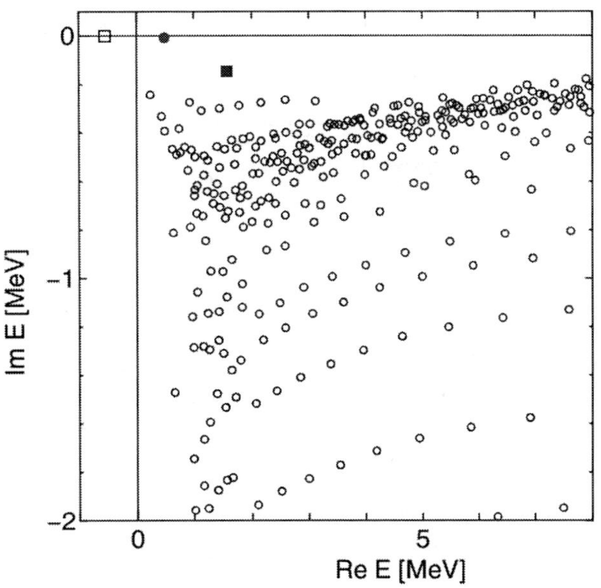

Figure 1: Distribution of the energy eigenvalues in the complex energy plane. The open square and the open circles show the bound state and the non-resonant continuum states, respectively. The solid circle and the solid square show the first and second resonances, respectively.

When the absorber strength η increases, the imaginary energies of the continuum states monotonically increase. The energies of the resonant states $E_{res}(\eta)$ are almost invariant with respect to the variation of η, but it slightly depends on η. The strength η is fixed so as to minimize the expectation value of the absorber, which corresponds to the numerical error, arising from the existence of the absorber and the incomplete basis-expansion [3, 4].

The resonance parameters of E_R and Γ_R are calculated at the optimal strength of η_o according to the relation of $E_{res}(\eta_o) = E_R - i\Gamma_R/2$. The resonance parameters for the first resonance are summarized in table 1. The optimal strength η_o is quite weak ($\sim 10^{-6}$) and hence, the expectation value of $< W(\eta_o) >$ is also small ($\sim 10^{-5}$). The parameters of E_R and Γ_R in ABC (the second row) is almost the same as the CSM results (the bottom row). We have confirmed that the ABC results on the second resonance also

Table 1: Resonance parameters for the first resonance calculated at the optimal strength η_o. The resonance energy (E_R) and the resonance width (Γ_R) are shown in unit of MeV. $< W(\eta_o) >$ represent the expectation value of the absorber at η_o. In the second row, the result of the ABC method is shown, while the results of CSM is shown in the lowest row.

Method	E_R	$\Gamma_R/2$	$< W(\eta_o) >$	η_o
ABC	0.424	5.402×10^{-3}	5.717×10^{-5}	1.22×10^{-6}
CSM	0.424	5.260×10^{-3}	X	X

reproduce the CSM result.

Finally, we discuss the application of the ABC method to the strength function of the three-body breakup. Figure 2 shows the strength function from the ground state to the three-body continua, which is induced by the monopole external field. The solid curve shows the result of the ABC method, while the dashed curve shows the CSM result. The ABC calculation follows the curve obtained by the CSM calculation over all energy region although a small oscillation arise in the higher energy region of $E \geq$ 3 MeV, where the magnitude of the strength is smaller by about 10^{-6} than the peak magnitude at $E \sim 0.5$ MeV.

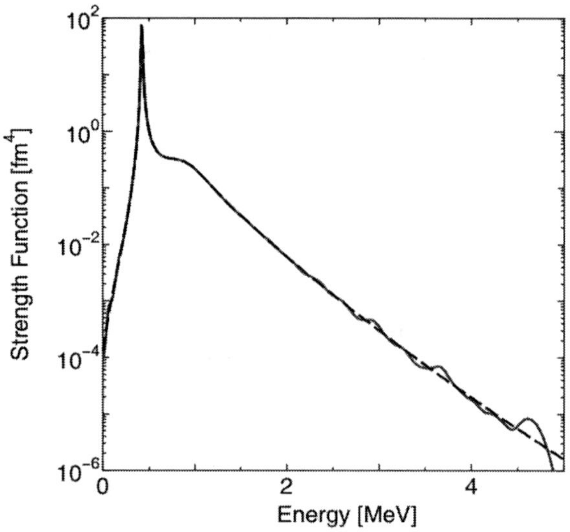

Figure 2: Monopole strength functions calculated from the ABC and CSM results. The solid and dashed curves show the ABC and CSM results, respectively.

The sharp peak at $E \sim 0.5$ MeV is generated by the contribution from the first resonance, while the broad shoulder at $E \sim 1$ MeV is due to the existence of the second resonance. The individual contributions, such as the resonance and the non-resonant continuum, in the strength function is common between ABC and CSM. Therefore, the strength function in the ABC calculation is consistent to the CSM treatment.

4 Summary

In the present study, we have formulated the absorbing boundary condition (ABC) method in the three-body problem with the coupled-rearrangement channel variational method (CRCVM). The CRCVM + ABC has been applied to the schematic three-boson system, and the results are compared to the CRCVM with the complex scaling method (CSM). The resonance parameters and the strength function calculated in ABC is completely consistent to the CSM calculation. Therefore, we can conclude that the formulation of the CRCVM + ABC method is successful in the calculation of the three-body system with the rearrangement channels.

In the future subject, we should apply the ABC method to the more realistic system such as the triple α system, in which its reaction rate is often discussed [9]. The realistic application of CRCVM + ABC is now under progress.

References

[1] I. Tanihata, H. Savajols, and R. Kanungo, Prog. Part. Nucl. Phys. **68**, 215 (2013).

[2] S. Aoyama, T. Myo, K. Katõ, and K. Ikeda, Prog. Theor. Phys. **116**, 1-35 (2006), and references therein.

[3] U. V. Riss and H-D. Meyer, Jour. Phys. **B**26,4503 (1993).

[4] Y. Takenaka, et al., Prog. Exp. Theor. Phys. **2014**, 113D04 (2014), and references therein.

[5] H. Masui and Y. K. Ho, Phys. Rev. C **65**, 054305 (2002).

[6] M. Ueda et al., Phys. Rev. C **67**, 014606 (2003).

[7] E. Hiyama et al., Prog. Part. Nucl. Phys. **51**, 223 (2003).

[8] M. Iwasaki, et al., Prog. Exp. Theor. Phys. **2015** 023D01, (2015).

[9] K. Ogata, M. Kan, and M. Kamimura, Prog. Theor. Phys. Vol. **122**, 1055 (2099).

EPJ Web of Conferences **117**, 06016 (2016) DOI: 10.1051/epjconf/201611706016
NN2015

Evaluation of inclusive breakup cross sections in reactions induced by weakly-bound nuclei within a three-body model

JIN LEI, ANTONIO M. MORO

Departamento de FAMN, Universidad de Sevilla, Apartado 1065, 41080 Sevilla, Spain

Abstract

In 1980s Ichimura, Austern and Vincent [Phys.Rev.C 32, 431 (1985)] proposed a quantum mechanical theory to study the problem of inclusive breakup cross sections in the nuclear reactions. Using the DWBA version of this model, applications to deuteron and ^{6}Li inclusive breakup reactions are presented and compared with some available data.

1 Introduction

The breakup of a nucleus into two fragments is an important mechanism in nuclear reactions, especially when one of the colliding nuclei is weakly bound. For example, $a + A \to b + x + A$, where $a = b + x$. If both fragments (b and x) are detected, the reaction is said to be exclusive. In addition, when all the fragments and target are emitted in their ground state, it is called Elastic Breakup (EBU). On the other hand, if only one of the fragments (say,b) is detected, the reaction is inclusive. This includes the EBU, and also the target (A) excitation, the fusion of x with A, and the exchange of particles between x and A, which are globally referred to as Nonelastic Breakup (NEB).

The problem of calculating inclusive breakup cross sections was studied in 1980s by Pampus *et al.* [1], with the works of Udagawa and Tamura [2] and Austern and Vincent [3] appearing later. In [1–3], the authors considered the reactions in which the projectile is broken up into two fragments. These processes can be represented as $a + A \to b + B^*$, where B^* is any possible state of the $x + A$ system. This includes the elastic scattering of $x + A$, which corresponds to the EBU, and non-elastic $x + A$ scattering, which gives rise to the NEB processes defined above. In [2], Udagawa and Tamura described this problem in the prior form DWBA formalism, whereas Austern and Vincent used the post form [3]. The later was refined by Kasano and Ichimura [4], who found a formal separation between EBU and NEB contributions. These results were carefully reviewed by Ichimura, Austern and Vincent [5] and the model was subsequently referred to as IAV formalism.

In this contribution, we have applied the finite-range DWBA version of IAV formalism [5,6] to different inclusive breakup reaction systems and compared with the available data.

2 Theoretical formulation

Following Ichimura, Asutern and Vincent [5]. We write the process under study as,

$$a(= b + x) + A \to b + B^*, \tag{1}$$

in which a projectile a colliding with a target A breaks into two fragments, b and x. We assume that only b is observed in the experiment and therefore the corresponding experimental cross section will correspond to a sum over all possible final states of the $x + A$ system. This process will be described with the Hamiltonian

$$H = K + V_{bx} + U_{bA}(\vec{r}_{bA}) + H_A(\xi) + V_{xA}(\xi, \vec{r}_x), \tag{2}$$

where K is the total kinetic energy operator, V_{bx} is the interaction between the clusters b and x in the projectile a, $H_A(\xi)$ is the Hamiltonian of the target nucleus (with ξ denoting its internal coordinates) and V_{xA} and U_{bA} are the fragment–target interactions.

Note that we make a distinction between the two cluster constituents; the interaction of the fragment b, the one that is assumed to be detected in the experiment, is described with an optical potential. Non-elastic processes arising from this interaction (e.g. target excitation), are included through U_{bA}. The particle b is said to act as *spectator*. On the other hand, the interaction of the particle x with the target retains the dependence of the target degrees of freedom (ξ).

Using the post form DWBA, the inclusive breakup differential cross section, as a function of the angle and energy of the fragment b, is given by

$$\frac{d^2\sigma}{dE_b d\Omega_b} = \frac{2\pi}{\hbar v_a}\rho(E_b)\sum_c |\langle \chi_b^{(-)}\Psi_{xA}^c|V_{post}|\chi_a^{(+)}\phi_a\rangle|^2\delta(E-E_b-E^c) \quad (3)$$

where $V_{post} = V_{bx} + U_{bA} - U_{bB}$ and c labels the states of x-A system. In the theory of IAV, they used Sokhotski–Plemelj theorem [7] and the Feshbach optical reduction, leading to the NEB differential cross section:

$$\left.\frac{d^2\sigma}{dE_b d\Omega_b}\right|_{\mathrm{NEB}} = -\frac{2}{\hbar v_i}\rho_b(E_b)\langle\psi_x|W_x|\psi_x\rangle \quad (4)$$

where $\rho_b(E_b) = k_b\mu_b/((2\pi)^3\hbar^2)$ is the density of states for the particle b, W_x is the imaginary part of the optical potential U_x which describes $x + A$ elastic scattering. The function ψ_x, which represents the $x - A$ relative motion when the target in the ground state, verifies the equation

$$(E_x^+ - K_x - U_x)\psi_x = (\chi_b^{(-)}|V_{post}|\chi_a\phi_a\rangle, \quad (5)$$

where $E_x = E - E_b$, $\chi_b^{(-)}$ is the distorted-wave describing the scattering of b in the final channel with respect to B^*.

3 Calculations

In this section, we present calculations for reactions induced by deuteron and ^{6}Li projectiles, and compare with the available experimental data. In all cases, we compute the separate contributions for the elastic (EBU) and non-elastic (NEB) breakup cross sections. For the former, we use the CDCC formalism [8], using the coupled-channel code FRESCO [9]. For the NEB part, we use Eq. (4). The intrinsic spins are ignored in the NEB Calculations.

First we consider the d+^{118}Sn reaction at E_d = 56 MeV and compare with the data from Ref. [10]. These data were already analyzed in Ref. [11], using the zero-range version of the post-form DWBA formula.

In the CDCC calculations the deuteron breakup is treated as inelastic excitations to the p-n continuum. These p-n continuum states are truncated in excitation energy and discretized in energy bins. For the p-n interaction, we consider the simple Gaussian form of Ref. [8]. The proton-target and neutron-target interactions are adopted from the global parametrization of

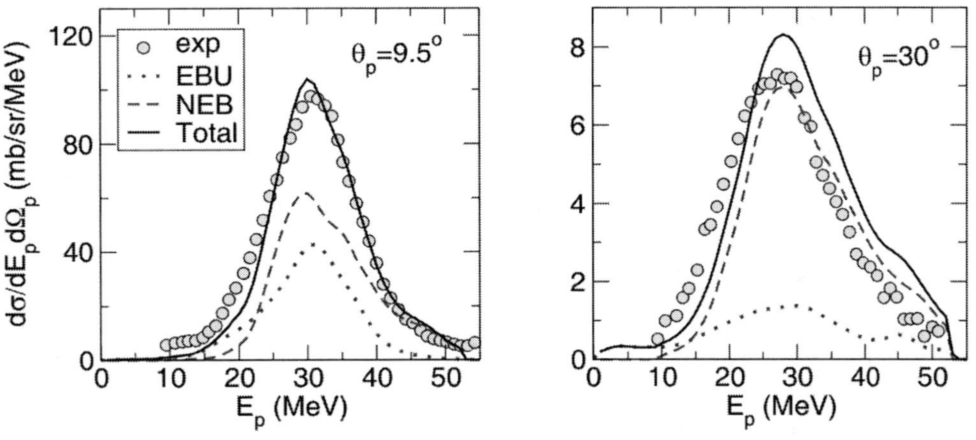

Figure 1: Experimental and calculated double differential cross section, as a function of the outgoing proton energy, for the protons emitted in the ^{118}Sn(d,pX) reaction with angles of 9.5^o (left panel) and 30° (right panel) at incident energy of 56 MeV.

Koning and Delaroche (KD) [12], omitting the spin-orbit term, and evaluated at half of the deuteron incident energy. For the NEB calculations, we use also the KD parametrization for the proton-target and neutron-target interactions, but evaluated at the corresponding proton (E_p) and neutron (E_n) energies. In DWBA, one needs also the incoming channel optical potential (d+^{118}Sn), which is taken from Ref. [13].

In Fig. 1 we compare the experimental and calculated inclusive double differential cross section, $\mathrm{d}^2\sigma/\mathrm{d}E_p\mathrm{d}\Omega_p$, corresponding to different proton angles as a function of outgoing proton energy in the laboratory frame. The dotted line is the EBU calculation, which is found to underestimate the data. The dashed line is the calculation for the NEB part. The solid line is the sum of the EBU and NEB contributions. It is seen that for the smaller angle ($\theta_p = 9.5^o$), the contribution of EBU is comparable with NEB, whereas for the larger angle ($\theta_p = 30^o$), the inclusive breakup cross section is largely dominated by the NEB contribution. The summed EBU and NEB contributions reproduce reassembly well the magnitude and shape of the data.

As a second example, we calculate the inclusive α cross sections for the reaction ^{6}Li+^{159}Tb at bombarding energies of 27 MeV and 35 MeV and compare with the data from [14]. The ^{6}Li nucleus is treated within a two-cluster model ($\alpha + d$). The EBU contribution is obtained from the CDCC calculations. The $\alpha - d$ interaction is taken from [15]. The $d-^{209}$Bi and

DOI: 10.1051/epjconf/201611706016

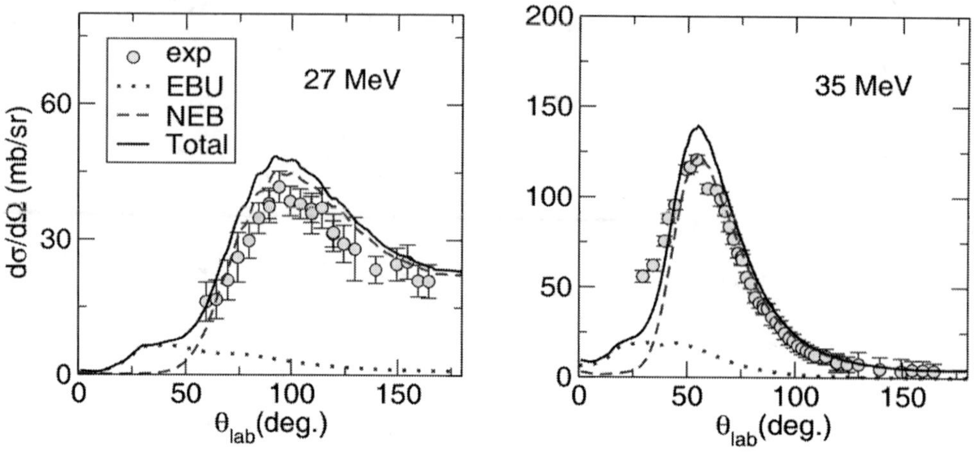

Figure 2: Experimental and calculated differential cross section angular distribution for the ^{159}Tb(^{6}Li,αX) reaction at energies of 27 MeV (left panel) and 35 MeV (right panel).

$\alpha-^{159}$Tb optical potentials are taken from Refs. [13] and [16], respectively. For the NEB calculation we adopt the same optical potential of $\alpha/d+^{159}$Tb as used in the CDCC calculations. For the ^{6}Li+^{159}Tb optical potential we adopted the parametrization of Cook [17].

In Fig. 2, we compare the calculated and experimental angular distributions of α particles, for the two considered energies of ^{6}Li. The dotted line is the calculation of EBU, which only contributes significantly at the very forward angles, whereas the dashed line is the result of NEB, which dominates the whole range of α production except the smallest angles. The summed EBU + NEB cross sections (thick solid lines) reproduce fairly well the shape and magnitude of the data.

4 Conclusion

In summary, we have investigated the problem of inclusive breakup in reactions induced by weakly bound nuclei. For the EBU part, we have considered the CDCC method, whereas for the NEB part, we revisited the model proposed by Ichimura, Austern and Vincent. We have performed calculations for ^{118}Sn(d,pX) and ^{169}Tb(^{6}Li,αX), finding in both cases a good agreement with existing data and a dominance of the NEB contribution. Calculations for other systems are underway.

Acknowledgements

This work has been partially supported by the Spanish Ministerio de Economía y Competitividad, under grant FIS2013-41994-P, by the Spanish Consolider-Ingenio 2010 Programme CPAN (CSD2007-00042) and by Junta de Andalucía (FQM160, P07-FQM-02894). J.L. is partially supported by a grant funded by the China Scholarship Council.

References

[1] J. Pampus, et al, Nucl. Phys. **A311**, 141 (1978)

[2] T. Udagawa, T. Tamura, Phys. Rev. C **24**, 1348 (1981)

[3] N. Austern, C.M. Vincent, Phys. Rev. C **23**(5), 1847 (1981)

[4] A. Kasano, M. Ichimura, Phys. Lett. **115B**, 81 (1982)

[5] M. Ichimura, N. Austern, C.M. Vincent, Phys. Rev. C **32**, 431 (1985)

[6] J. Lei, A. Moro, Phys. Rev. C **submitted** (2015)

[7] Y. Sokhotski, Saint Petersburg **Doctor thesis** (1873)

[8] N. Austern, et al, Phys. Rep. **154**, 125 (1987)

[9] I.J. Thompson, Comp. Phys. Rep. **7**, 167 (1988)

[10] N. Matsuoka, et al, Nucl.Phys. **A345**, 1 (1980)

[11] B.V. Carlson, R. Capote, M. Sin, (2015)

[12] A. Koning, J. Delaroche, Nucl. Phys. A **713**(34), 231 (2003)

[13] Y. Han, Y. Shi, Q. Shen, Phys. Rev. C **74**, 044615 (2006)

[14] M.K. Pradhan, et al, Phys. Rev. C **88**, 064603 (2013)

[15] H. Nishioka, et al, Nucl. Phys. A **415**, 230 (1984)

[16] J. Huizenga, G. Igo, Nuclear Physics **29**, 462 (1962)

[17] J. Cook, Nucl. Phys. A **388**, 153 (1982)

EPJ Web of Conferences **117**, 06017 (2016) DOI: 10.1051/epjconf/201611706017
NN2015

Microscopic nuclear form factor for the pygmy dipole resonance

E.G. Lanza[1,2], A. Vitturi[3,4], M.V. Andrés[5] and F. Catara[1,2]

[1]INFN, Sezione di Catania, Catania, Italy
[2]Dipartimento di Fisica e Astronomia, Università di Catania, Italy
[3]Dipartimento di Fisica e Astronomia, Università di Padova, Italy
[4]INFN, Sezione di Padova, Padova, Italy
[5]Departamento de FAMN, Universidad de Sevilla, Spain

Abstract

Pygmy dipole resonance can be populated via isoscalar probes due to the strong mixing of isoscalar and isovector components. The physical quantities are usually extracted by the use of reaction theory calculations. There, the most important ingredient is the form factor which contains all the relevant properties of the states we are dealing with. We show that the use of microscopic form factor is required for a better analysis of experimental results.

1 Introduction

Among the most interesting new phenomena related to the nuclei with neutron excess, the low lying dipole states [1–3], also known as Pygmy Dipole Resonance (PDR), acquired a special role also for its connections with other branch of physics. The PDR states are clearly connected to the neutron skin of atomic nuclei [4]. Its isovector E1 strength is related to the isovector parameters in the equation of state of nuclear matter [5]. Furthermore, it is connected to the reaction rates in the astrophysical r-process which is responsible of the formation of the elements heavier than iron [6].

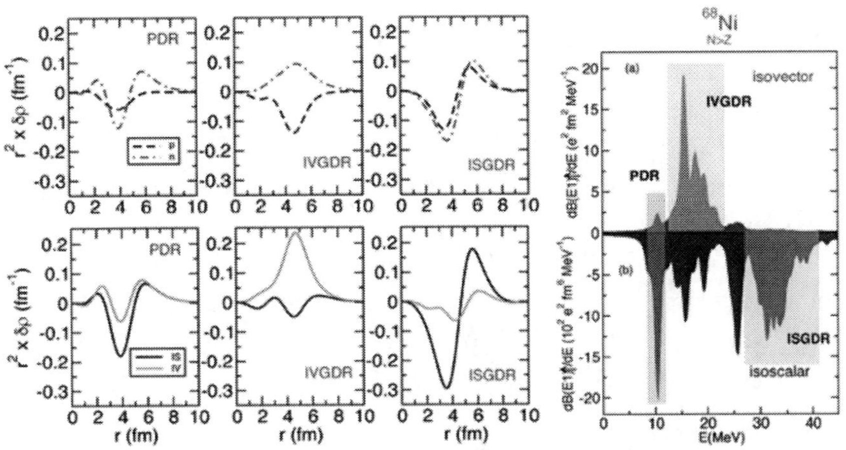

Figure 1: (Color online) RPA transition densities for three different dipole states for the nucleus ^{68}Ni. In the frame at the right, the RPA isovector (upper part, red) and isoscalar (lower part, blue) response for the same nucleus (see text).

PDR states are better described within microscopic self consistent many-body approaches or on the relativistic meson-exchange Lagrangian [1]. All of these calculations show similar behaviors for the transition densities: the neutron and proton ones are in phase inside the nucleus while at the surface only the neutron part contributes significantly. This feature generates a strong mixing of isospin character and therefore even an isoscalar probe can be used to explore these low lying dipole states.

From the experimental point of view, these modes have been investigated using several methods on stable as well as on nuclei far from the stability line. The first evidence in unstable nuclei was found with a relativistic Coulomb excitation at GSI for the ^{130,134}Sn [7] and later on also for the ^{68}Ni [8]. Systematic studies on stable nuclei with the method of nuclear resonance fluorescence (NRF) have also been carried out on different mass regions [2]. Recently, the investigation of the nature of the PDR has been pursued via the use of isoscalar probes, namely via $(\alpha, \alpha'\gamma)$ [2, 9–12] or (^{17}O, ^{17}O' γ) [3, 13–15]. Most of the physical information obtained by using isoscalar probes depend on how well the nuclear properties of the state under investigation are described within the radial form factors. Here, we present a comparison among different ways to calculate the radial nuclear form factor. Full details can be found in ref. [16].

2 Transition densities and nuclear form factors

Inelastic nuclear excitation is a useful method to extract information about nuclear structure. This is often done using different approaches

DOI: 10.1051/epjconf/201611706017

like DWBA or a more complete description like the coupled channel models or their semiclassical versions especially developed for the studies of heavy ion collisions. The construction of the radial nuclear form factor is very important in all these approaches because it includes all the relevant information about the properties of the state under consideration. An extremely useful way to build up the form factors is through the well-known double folding procedure [17] which requires the knowledge of the transition density of the state under study. The transition densities can be calculated in a macroscopic or microscopic way according to the various models describing the different excitation modes.

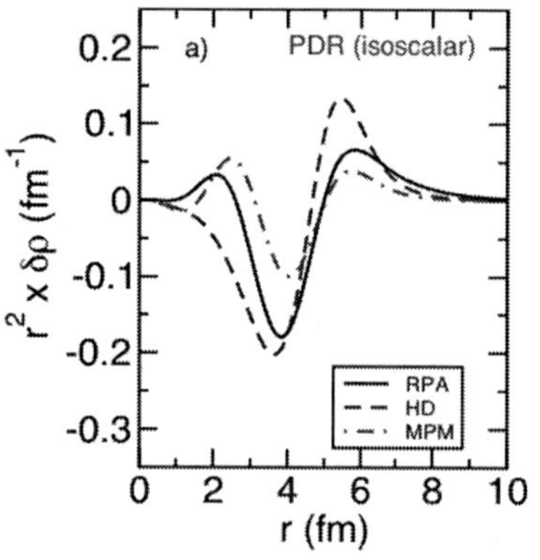

Figure 2: (Color online) Comparison among macroscopic isoscalar transition densities with the RPA based microscopic one (see text).

Microscopic transition densities can be obtained from several many body models calculations. As an example we have performed Hartree-Fock (HF) plus Random Phase Approximation (RPA) calculation with a SGII interaction [18] for the nucleus ^{68}Ni. In the right part of fig. 1 the isovector (upper part, red) and isoscalar (lower part, blue) dipole response are shown with the three region of interest: the isovector giant dipole resonance (IVGDR) in the isovector part, the pure isoscalar giant dipole resonance at very high energy and the peak corresponding to the PDR centered at around 10 MeV. In the six frames on the left, the transition densities for the three states: the neutron (dot-dashed, red) and proton (dashed, black) transition densities in the upper part and the isovector (green) and isoscalar (black) one in the lower one. The IVGDR transition densities exhibit the well known behaviour with protons and neutrons oscillating out of phase giving rise to a strong isovector character of the mode; for the ISGDR, the protons and neutrons are oscillating in phase producing an isoscalar transition density with a node at the nuclear surface, typical of the compressional modes. For the PDR, neutrons and protons are in phase inside the nucleus while only the neutrons give a contribution at the surface. As a consequence, the isoscalar and isovector transition densities have the same strength at the surface which is the region of the nucleus explored in the nuclear reactions used to investigate the PDR states. These important features should be taken into account in the construction of the nuclear form factors.

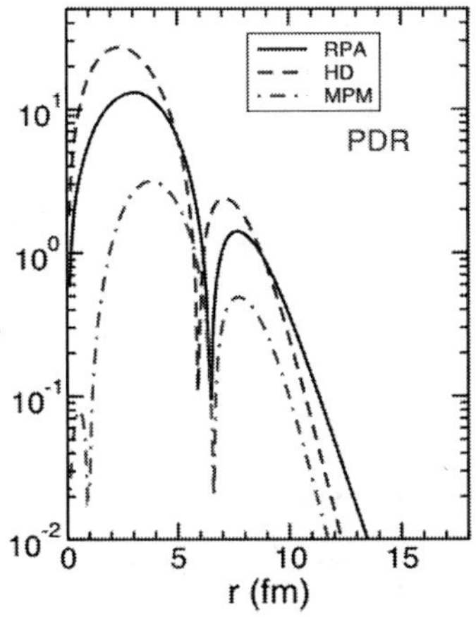

Figure 3: (Color online) Comparison among nuclear radial form factors calculated with the transition densities of fig.2 (see text).

For the very well known collective states the transition densities can be also calculated, with a good accuracy, by means of macroscopic models. We quote the well-known Goldhaber-Teller and Steinwedel-Jensen models which describe the IVGDR as due to an out-of-phase oscillation of neutrons against protons. Also for the ISGDR, which is a $3\hbar\omega$ nuclear transition, it is possible a description in terms of a macroscopic model. A macroscopic collective transition densities has been deduced in ref. [19, 20] by assuming that the isoscalar dipole energy weighted sum rule is exhausted by a single collective state. The isoscalar transition density deduced with such conditions has the typical shape of the compressional modes.

There is another way to calculate a transition density based on the common picture that one has in mind about the nature of the PDR mode. It is a widespread belief, in the literature, that the low lying dipole states are generate by an out-of-phase oscillation of the neutrons on the nuclear surface against a proton-neutron core. Assuming this picture, we have calculated the neutron and proton transition densities and the corresponding isoscalar and isovector ones [16]. The comparison among these macroscopic isoscalar transition densities and the RPA-based microscopic one is shown in fig. 2. The shapes of the three isoscalar transition densities are similar but they are quite different in their magnitude especially at the nuclear surface region which is the region explored by the nuclear reaction considered here. In particular, the one calculated by Harakeh and Dieperink [20], dashed blue curve (HD), are bigger with respect to the RPA one, both inside the nucleus and at the surface. This is not surprising because this transition density has been constructed for a pure isoscalar mode as it is the ISGDR. Indeed for this state the comparison with the microscopic RPA transition density is very good being the two almost undistinguishable [16]. The isoscalar transition density calculated according to the macroscopic pygmy model described in ref. [16], denoted as MPM in the figure with a dot-dashed red line, follows closely the RPA one but it is smaller in the important surface region. All the transition densities have been calculated from the HF ground state density used in the RPA calculations.

In constructing the formfactor with the double folding procedure we use the M3Y [21] nucleon-nucleon interaction together with the HF ground state density and the transition densities shown in fig. 2. As an example we have chosen the system ^{68}Ni $+^{12}$C and their nuclear form factors are shown in fig. 3. The legend indicate which of the three transition densities shown in fig. 2 have been used for the form factor calculations. The three curves are different in all the radial interval; at the surface they differ in magnitude as well as in the slope, the RPA having a slightly larger radial extension. We recall the fact that the excitation processes we are dealing with are mainly active on the surface and therefore the most important parts of the form factor are the ones around 8 fm for the system under consideration. In this region the three form factors are quite different and as a consequence one can expect that the corresponding inelastic cross section can be different. Indeed, the almost factor three in the form factor gives rise to an almost factor 9 in the cross section shown in fig. 4. These were obtained by DWBA calculations for the system ^{68}Ni $+^{12}$C at 10 Mev/u with the DWUCK4 code [22].

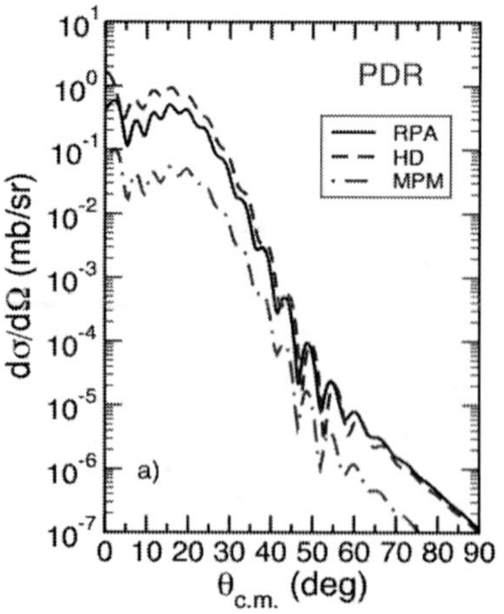

Figure 4: (Color online) Comparison among angular distributions calculated with the form factors of fig.3 (see text).

3 Conclusion

The study of the PDR via isoscalar mode is a well established method [23,24]. The information extracted form the experiment depends on the radial form factors used in the data analysis. Here we have shown that there are strong differences when macroscopic transition densities are used in comparison with the microscopic ones. On the other hand, only the microscopic form factor includes, by construction, the important isospin mixing of the PDR states. Our microscopic form factors have already been successfully employed in the analysis of various experiments performed with ^{17}O [13–15].

References

[1] N. Paar, D. Vretenar, E. Khan and G. Colò, *Rep. Prog. Phys.* **70** (2007) 691.

[2] D. Savran, T. Aumann and A. Zilges, *Prog. Part. Nucl.Phys.* **70** (2013) 210.

[3] A. Bracco, F. C. L. Crespi and E. G. Lanza, *Eur. Phys. J. A* **99** (2015)

[4] J. Piekarewicz, *Phys. Rev. C* **73** 044325 (2006).

[5] J. Piekarewicz et al., *Phys. Rev. C* **85** 041302 (2012).

[6] E. Litvinova et al., *Nucl. Phys.* **823** 26 (2009).

[7] P. Adrich et al., *Phys. Rev. Lett.* **95** 132501 (2005).

[8] O. Wieland et al., *Phys. Rev. Lett.* **102** 092502 (2009).

[9] D. Savran et al., *Phys. Rev. Lett.* **97** 172502 (2006).

[10] J. Endres et al., *Phys. Rev. C* **80** 034302 (2009).

[11] J. Endres et al., *Phys. Rev. Lett.* **105** 212503 (2010).

[12] J. Endres et al., *Phys. Rev. C* **85** 064331 (2012).

[13] F. C. L. Crespi et al., *Phys. Rev. Lett.* **113** 012501 (2014).

[14] L. Pellegri et al., *Phys. Lett. B* **738** 519 (2014).

[15] F. C. L. Crespi et al., *Phys. Rev. C* **91** 024323 (2015).

[16] E.G. Lanza, A. Vitturi, M.V. Andrés, *Phys. Rev. C* **91** (2015) 054607.

[17] G. R. Satchler and W. G. Love, *Phys. Rep.* **55** 183 (1979).

[18] N. Van Giai and H. Sagawa, *Phys. Lett. B* **106** 379 (1981).

[19] T. J. Deal, *Nucl. Phys. A* **217** 210 (1973).

[20] M. N. Harakeh and A. E. L. Dieperink, *Phys. Rev. C* **23** 2329 (1981).

[21] G. Bertsch et al., *Nucl. Phys. A* **284** 399 (1977).

[22] P. D. Kunz, http://spot.colorado.edu/?kunz/DWBA.html

[23] E. G. Lanza, A. Vitturi, M. V. Andrés, F. Catara and D. Gambacurta: *Phys. Rev. C* **84** (2011) 064602.

[24] E. G. Lanza, A. Vitturi, E. Litvinova and D. Savran: *Phys. Rev. C* **89** (2014) 041601(R).

EPJ Web of Conferences **117**, 06018 (2016) DOI: 10.1051/epjconf/201611706018
NN2015

Importance of the single-particle continuum in BCS pairing with a pseudostate basis

J. A. LAY[1], C. E. ALONSO[2], L. FORTUNATO[1], and A. VITTURI[1]

[1]Dipartimento di Fisica e Astronomia "Galileo Galilei", Università di Padova, and INFN, Sezione di Padova, via Marzolo, 8, I-35131 Padova, Italy
[2]Departamento de Física Atómica, Molecular y Nuclear, Facultad de Física, Universidad de Sevilla, Apartado 1065, 41080 Sevilla, Spain

Abstract

In a recent work [arXiv:1510.03185] the use of the Transformed Harmonic Oscillator (THO) basis for the discretization of the single-particle continuum into a Generalized Bardeen-Cooper-Schrieffer (BCS) formalism was proposed for the description of weakly bound nuclei. We make use of the flexibility of this formalism to study the evolution of the pairing when the nucleus becomes more and more weakly bound. Specifically we focus on the evolution of the occupation of the different partial waves in ^{22}O when the Fermi level approaches zero.

Present developments in nuclear facilities allow to study new exotic nuclei, approaching the proton and neutron driplines for both light and medium mass regions of the Segrè Chart. Nuclei close to the dripline are weakly bound and, therefore, affected by the continuum, situations where the last nucleon or nucleons are no longer bound and can explore distances far from the rest of the nucleus.

The effects of the continuum have been traditionally explored for light nuclei where the proximity of the driplines provides a rich variety of weakly bound nuclei and experimental data. In the medium mass region, several attempts for generalizing mean field approaches have been proposed with different results [1,2]. Regarding the Bardeen-Cooper-Schrieffer (BCS) approximation, the single-particle continuum is first discretized and then treated normally splitting resonant and background contributions or including a continuum single-particle level density [2].

In a recent work [3] we studied the possibility of treating background and resonant parts of the continuum on the same footing. In order to do so, the single-particle continuum is discretized using a square-integrable basis. The pseudostates or eigenstates obtained through the diagonalization of the matrix elements of the single-particle Hamiltonian are expected to give a good representation of the continuum. Then, the pairing Hamiltonian is written in terms of the pair creator and annihilator operators built on top of these pseudostates. The strength of the pairing is determined by a density-dependent delta interaction and matrix elements are calculated separately for each couple of pairs. Once we set a maximum energy, it is possible to solve the BCS equations.

The basis chosen for this purpose is the Transformed Harmonic Oscillator (THO) basis. It is constructed by applying a Locale Scale Tranformation to the Harmonic Oscillator basis of the form [4]:

$$s(r) = \left[\frac{1}{\left(\frac{1}{r}\right)^m + \left(\frac{1}{\gamma\sqrt{r}}\right)^m} \right]^{\frac{1}{m}}, \qquad (1)$$

that depends on the parameters m and γ. Even though we increase the number of parameters with respect to the simpler HO, two main advantages are obtained: the asymptotic behaviour of the functions is now exponential, rather than gaussian as in the HO case, and γ can be used to tailor the basis to the different cases. This is of great help when assessing the importance of the low-energy continuum since, by reducing the value of γ, the pseudostates tend to concentrate at low energies, thus optimizing the calculation.

Already in [3], an interesting increase of the occupation of the non-resonant low-energy continuum was found for extremely weakly bound situations. In this work, we will concentrate on the evolution of the total occupation when we vary the strength of the single-particle continuum. Remember that, in the BCS theory, the quantities v^2 and u^2 stand for the occupation and deoccupation of the different level. We will choose the same case and parameters chosen in [3], so that we will look to the occupations in ^{22}O when varying the strength of the central single-particle potential V.

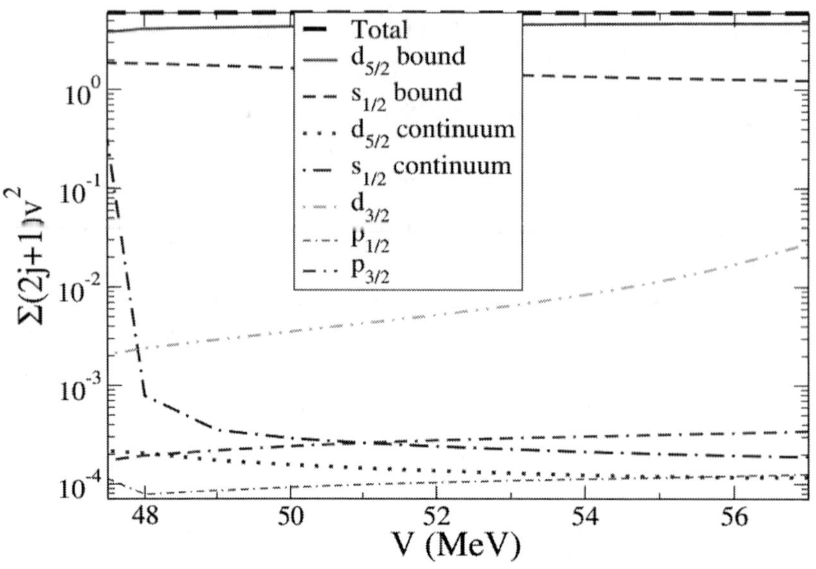

Figure 1: Occupation corresponding to the six valence neutrons considered in ^{22}O for the different partial waves. We separate the $d_{5/2}$ (solid line) and $s_{1/2}$ (dashed line) bound contribution from the corresponding $d_{5/2}$ (dotted line) and $s_{1/2}$ (dotted dashed line) continuum contribution. The thick dashed line corresponds to the total occupation as the sum of the different contributions considered which, by construction, is always the number of valence neutrons.

In [3], the pairing interaction between the different pairs is calculated by using a density dependent delta interaction (DDDI) fitted to reproduced the neutron separation energy in ^{18}O, and then renormalized along the isotopic chain following a $1/A$ reduction.

In fig. 1 we show the occupations in ^{22}O for all the partial waves considered, calculated for different depths of the central single-particle potential V. We have also considered separately the bound and continuum parts of the occupation in the $d_{5/2}-$ and $s_{1/2}-$waves. It is commonly stated that the contribution of the continuum is mainly governed by the resonances present in each case. In our case, this fact should be reflected in the dominance of the $d_{3/2}$ contribution where a low-lying resonance can be found. This is true for a well-bound ^{22}O, which in fig. 1 corresponds to the right part of the figure around $V = 57$ MeV. It is interesting to see how the occupation decreases when the strength of the potential is reduced. However, the occupation of the resonant part of the continuum is still larger by more than one order of magnitude than the non-resonant part of the continuum.

This situation changes for really weakly-bound situations. A sudden increase of the $s_{1/2}$ continuum contribution is found, even though only for an extreme situation. One has to keep in mind than the Fermi level for $V = 47.5$ MeV is practically zero. On the other hand, the bound part of the $s_{1/2}$ contribution is more or less constant with a slight increase.

If we look at the remaining contributions, only the $d_{5/2}$ continuum part is also constantly increasing when reducing the strength V. However, this represents a modest variation if we compare with the behaviour of the $s_{1/2}$ waves. Nevertheless, one has to remind that we are using a DDDI, which can overestimate the relative importance of the $s-$wave in detritment of contributions with larger orbital angular momentum. A more realistic treatment of the pairing interaction could lead to a larger increase of the occupation of the $d_{5/2}$ continuum, at least in comparison with the $s_{1/2}$ contribution.

The fact that only those continuum components with a bound state of the same orbital and total angular momentum show such an increase is telling us that this effect is not related with the extension of the wave functions as in the case of the formation of a halo nucleus. In such a case, $p_{3/2}$ occupation should show an increase since the extension of $p-$waves also goes to infinity when the binding energy goes to zero. Instead, they decrease as we can see in fig.1.

In conclusion, non-resonant continuum contributions to the ground state of a nucleus cannot be neglected when dealing with weakly bound nuclei. The importance of the resonant configuration decreases consistently when the system becomes more and more weakly bound until a situation is reached where is no longer the dominant part of the single-particle continuum. However, we also show that for not so extreme situations, it is the resonant part that carries almost all the importance of the continuum.

Acknowledgments

This work has been supported by MIUR research fund PRIN 2009TWL3MX, by the Spanish Ministerio de Economía y Competitividad under Project FIS2014-53448-C2-1-P by Junta de Andalucía under group number FQM-160 and Project P11-FQM-7632, and by the Consolider-Ingenio 2010 Program CPAN (CSD2007-00042). The research leading to these results has received funding from the European Commission, Seventh Framework Programme (FP7/2007-2013) under Grant Agreement n° 600376. J.A.L. is a Marie Curie Piscopia fellow at the University of Padova.

References

[1] J. Dobaczewski *et al.*, *Phys. Rev. C*, **53** (1996) 2809.

[2] R. Id Betan, *Nucl. Phys. A*, **879** (2012) 14.

[3] J. A. Lay *et al.*, arXiv:1510.03185 Preprint, 2015.

[4] J. A. Lay *et al.*, *Phys. Rev. C*, **82** (2010) 024605.

Competition of β-delayed protons and β-delayed γ rays in ^{56}Zn and the exotic β-delayed γ-proton decay

S. E. A. ORRIGO[1*], B. RUBIO[1], Y. FUJITA[2], B. BLANK[3], W. GELLETLY[4], J. AGRAMUNT[1], A. ALGORA[1,5], P. ASCHER[3], B. BILGIER[6], L. CÁCERES[7], R. B. CAKIRLI[6], H. FUJITA[8], E. GANIOĞLU[6], M. GERBAUX[3], J. GIOVINAZZO[3], S. GRÉVY[3], O. KAMALOU[7], H. C. KOZER[6], L. KUCUK[6], T. KURTUKIAN-NIETO[3], F. MOLINA[1,9], L. POPESCU[10], A. M. ROGERS[11], G. SUSOY[6], C. STODEL[7], T. SUZUKI[8], A. TAMII[8] and J. C. THOMAS[7]

[1]Instituto de Física Corpuscular, CSIC-Universidad de Valencia, E-46071 Valencia, Spain
[2]Department of Physics, Osaka University, Toyonaka, Osaka 560-0043, Japan
[3]Centre d'Etudes Nucléaires de Bordeaux Gradignan, CNRS/IN2P3 - Université Bordeaux 1, 33175 Gradignan Cedex, France
[4]Department of Physics, University of Surrey, Guildford GU2 7XH, Surrey, UK
[5]Inst. of Nuclear Research of the Hung. Acad. of Sciences, Debrecen, H-4026, Hungary
[6]Department of Physics, Istanbul University, Istanbul, 34134, Turkey
[7]Grand Accélérateur National d'Ions Lourds, BP 55027, F-14076 Caen, France
[8]Research Center for Nuclear Physics, Osaka University, Ibaraki, Osaka 567-0047, Japan
[9]Comisión Chilena de Energía Nuclear, Casilla 188-D, Santiago, Chile
[10]SCK.CEN, Boeretang 200, 2400 Mol, Belgium
[11]Physics Division, Argonne National Laboratory, Argonne, IL 60439, USA

Abstract

Remarkable results have been published recently on the β decay of ^{56}Zn. In particular, the rare and exotic β-delayed γ-proton emission has been detected for the first time in the fp shell. Here we focus the

*e-mail: sonja.orrigo@ific.uv.es

discussion on this exotic decay mode and on the observed competition between β-delayed protons and β-delayed γ rays from the Isobaric Analogue State.

1 Introduction

Decay spectroscopy is a powerful tool for exploring the structure of nuclei at the drip-lines. β-decay studies, in particular, provide direct access to the absolute values of the Fermi and Gamow-Teller transition strengths, B(F) and B(GT), respectively.

The proton-rich ^{56}Zn nucleus was observed for the first time at GANIL in 1999 [1]. ^{56}Zn is a weakly-bound nucleus lying very close to the proton drip-line. It has a quite small proton separation energy, S_p = 560(140) keV [2], and third component of the isospin quantum number T_z = -2.

The first study of the β decay of ^{56}Zn was reported in *ref.* [3]. More recently, some interesting results on ^{56}Zn decay have been reported in *ref.* [4]. Among them the discovery of a rare and exotic decay mode, β-delayed γ-proton decay, which has been seen for the first time in the fp shell. The consequences of this rare decay sequence for the determination of the Gamow-Teller (GT) strength have also been analyzed.

2 The experiment

The experimental study of ^{56}Zn decay was performed at GANIL in 2010. The experiment used a primary beam of ^{58}Ni^{26+} to produce ^{56}Zn. The ^{58}Ni beam, of 3.7 eμA and accelerated to 74.5 MeV/nucleon, was fragmented on a natural Ni target, 200 μm thick. The fragments were selected by the LISE3 separator and implanted into a Double-Sided Silicon Strip Detector (DSSSD). The detection set-up comprised the aforementioned DSSSD detector, 300 μm thick, a silicon ΔE detector located 28 cm upstream, and four EXOGAM Ge clovers surrounding the DSSSD.

The EXOGAM clovers were used to detect β-delayed γ rays. The purpose of the DSSSD was the detection of both the implanted fragments and the subsequent charged-particle decays, *i.e.*, β particles and β-delayed protons. An implantation event was defined by simultaneous signals in both the ΔE and DSSSD detectors. A decay event was defined by a signal above threshold (50-90 keV) in the DSSSD and no coincident signal in the ΔE.

The implanted ions were identified and selected by putting a gate in a two-dimensional identification matrix, obtained by combining the energy loss signal from the ΔE detector and the Time-of-Flight. The latter was defined as the time difference between the cyclotron radio-frequency and ΔE signal.

3 Results on the β decay of ^{56}Zn

The results on the β decay of ^{56}Zn [4] are summarized in the decay scheme in *fig.* 1 and in table 1, and discussed below.

A half-life of $T_{1/2} = 32.9(8)$ ms was obtained for ^{56}Zn, in agreement with *ref.* [3]. To determine $T_{1/2}$, a decay-time spectrum has been constructed from the time correlations between a decay event in a given pixel of the DSSSD (with a total of 256 pixels) and any implantation signal that occurred before and after it in the same pixel, satisfying the identification condition required to select ^{56}Zn.

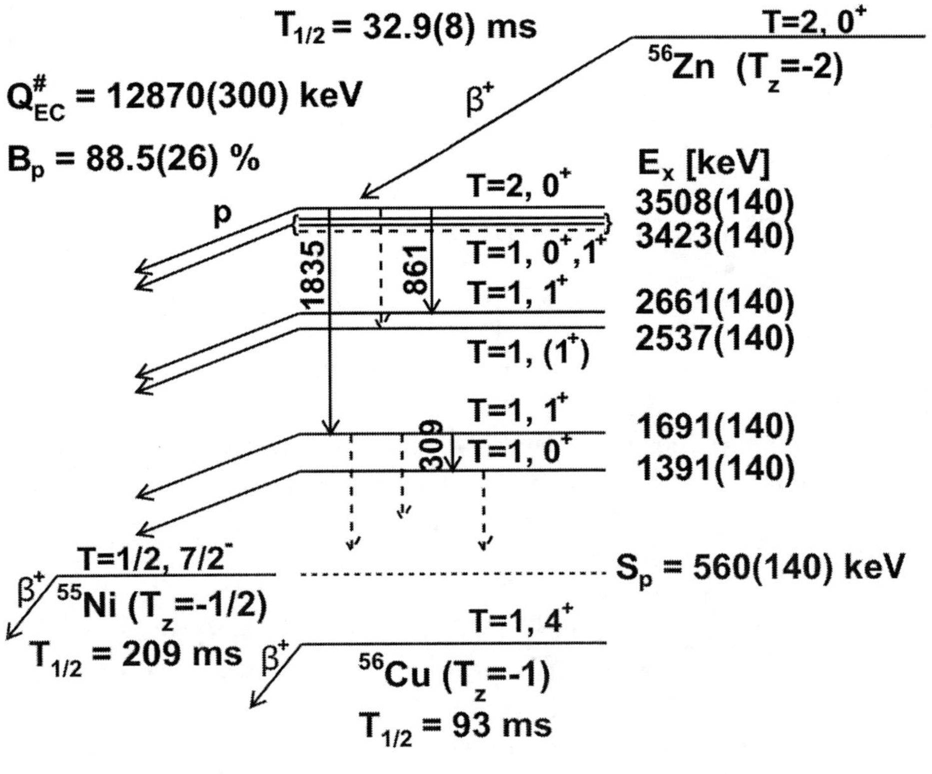

Figure 1: Scheme of the β decay of ^{56}Zn. The solid lines indicate observed proton or γ transitions, while the dashed lines correspond to transitions observed in the mirror ^{56}Co nucleus.

DOI: 10.1051/epjconf/201611706019

The analysis of the charged-particle spectrum measured in the DSSSD has provided new spectroscopic information on the energy levels populated in the ^{56}Cu nucleus, the β-daughter of ^{56}Zn. These levels are shown in *fig.* 1. The comparison of this level spectrum with that of the mirror ^{56}Co, obtained by the ^{56}Fe(^{3}He,t) charge exchange reaction [5], has been very fruitful.

The analysis of the γ spectrum measured in the EXOGAM clovers and γ-proton coincidences have identified three γ rays at 309, 861 and 1835 keV.

Absolute B(F) and B(GT) strengths have been determined (table 1).

Table 1: β feedings, Fermi and Gamow Teller transition strengths to the ^{56}Cu levels populated in the β^{+} decay of ^{56}Zn.

E_X(keV)	I_β(%)	B(F)	B(GT)
3508(140)*	43(5)	2.7(5)	
3423(140)	21(1)	1.3(5)	≤0.32
2661(140)	14(1)		0.34(6)
2537(140)	0		0
1691(140)	22(6)		0.30(9)
1391(140)	0		0

*Main component of the IAS.

3.1 Competition of β-delayed protons and β-delayed γ rays

In the first study of the ^{56}Zn β decay [3], the emission of β-delayed protons was observed but no β-delayed γ rays were seen. This was not a surprise because, in general, in proton-rich nuclei the proton decay is expected to dominate for states well above (>1 MeV) the proton separation energy S_p. The consequence is that normally the β feeding is directly inferred from the measured intensities of the proton peaks. However, cases where there is a competition between β-delayed proton emission and β-delayed γ de-excitation have also been observed, *e.g.*, in *refs.* [3,6].

In the $T_z = -2 \rightarrow -1, \beta^{+}$ decay of ^{56}Zn to ^{56}Cu, the ^{56}Zn ground state decays with a Fermi transition to its Isobaric Analogue State (IAS) in ^{56}Cu. It should be noted that the de-excitation of this $T = 2$, $J^{\pi} = 0^{+}$ IAS via proton decay to the ground state of ^{55}Ni ($T = 1/2$, $J^{\pi} = 7/2^{-}$) is isospin forbidden. Therefore the proton emission that we observe can only happen through a $T = 1$ isospin impurity present in the IAS. Moreover in general, when the proton emission is isospin forbidden, the competitive emission of de-exciting γ rays from the IAS also becomes possible and can be observed even from IAS lying at an excitation energy well above S_p [3,6].

The competition between β-delayed protons and γ rays has indeed been observed in ^{56}Zn. The γ decays represent 56(6)% of the total decays from the 3508 keV IAS. Thus one has to take into account the intensities of both the proton and γ peaks to determine the Fermi strength correctly.

We have also found evidence for the fragmentation of B(F) due to a strong isospin mixing with a 0^+ state at 3423 keV [4], which is important in terms of the mass evaluation [7]. The isospin impurity in the ^{56}Cu IAS, $\alpha^2 = 33(10)\%$ (defined as in *ref.* [5]), and the off-diagonal matrix element of the charge-dependent part of the Hamiltonian, $\langle H_c \rangle = 40(23)$ keV, which is responsible for the isospin mixing of the 3508 keV IAS ($T = 2$, $J^\pi = 0^+$) and the 0^+ part of the 3423 keV level ($T = 1$), are similar to the values obtained in the mirror ^{56}Co nucleus [5].

Thus, the proton decay of the IAS proceeds thanks to the $T = 1$ component. However, considering the quite large isospin mixing in ^{56}Cu, the much faster proton decay ($t_{1/2} \sim 10^{-18}$ s) should dominate on the γ de-excitation ($t_{1/2} \sim 10^{-14}$ s in the mirror). This is not the case since we are still observing the γ decay of the IAS in competition with it.

The knowledge on the nuclear structure of the three nuclei involved in the decay, *i.e.*, ^{56}Zn, ^{56}Cu and ^{55}Ni, can provide us with a possible explanation for the hindrance of the proton decay. Shell model calculations are in progress to clarify this point.

3.2 The β-delayed γ-proton decay

Besides the competition between β-delayed proton emission and γ decay, the exotic sequence of β-delayed γ-proton decay has been detected. Indeed ^{56}Zn does β decay to its IAS in ^{56}Cu and from there we observe the emission of two γ rays of 861 and 1835 keV, populating the ^{56}Cu levels at 2661 and 1691 keV, respectively. Due to the low S_p, these levels are still proton-unbound and thereafter they decay by proton emission. Consequently the rare and exotic β-delayed γ-proton decay has been observed. In addition to these two branches, there is a third case. The 1691 keV level emits a γ ray of 309 keV, going to the level at 1391 keV that is again proton-unbound and then it de-excites by proton emission.

The β-delayed γ-proton decay has been observed here for the first time in the *fp* shell. This rare decay mode was seen only once before, in the *sd* shell in ^{32}Ar [6], but the consequences for the determination of B(GT) were not addressed in *ref* [6].

The observation of this special decay mode is very important because it does affect the conventional way to determine B(GT) near the proton drip-line. For a proper determination of $B(\mathrm{GT})$, indeed, it is crucial to correct the intensity of the proton transitions for the amount of indirect feeding coming from the γ de-excitation. This finding indicates that it is important to employ γ detectors in such studies. This decay mode is expected to be significant in heavier proton-rich nuclei with $T_z \leq -3/2$ under study at RIKEN.

Acknowledgements

This work was supported by the Spanish MICINN grants FPA2008-06419-C02-01, FPA2011-24553; Centro de Excelencia Severo Ochoa del IFIC SEV-2014-0398; CPAN Consolider-Ingenio 2010 Programme CSD2007-00042; *Junta para la Ampliación de Estudios* Programme (CSIC JAE-Doc contract) co-financed by FSE; MEXT, Japan 18540270 and 22540310; Japan-Spain coll. program of JSPS and CSIC; Istanbul University Scientific Research Projects, Num. 5808; UK Science and Technology Facilities Council (STFC) Grant No. ST/F012012/1; Region of Aquitaine. R.B.C. acknowledges support by the Alexander von Humboldt foundation and the Max-Planck-Partner Group. We acknowledge the EXOGAM collaboration for the use of their clover detectors.

References

[1] J. Giovinazzo et al., *Eur. Phys. J. A*, **11** (2001) 247.

[2] G. Audi et al., *Nucl. Phys. A*, **729** (2003) 1.

[3] C. Dossat et al., *Nucl. Phys. A*, **792** (2007) 18.

[4] S.E.A. Orrigo et al., *Phys. Rev. Lett.*, **112** (2014) 222501.

[5] H. Fujita et al., *Phys. Rev. C*, **88** (2013) 054329.

[6] M. Bhattacharya et al., *Phys. Rev. C*, **77** (2008) 065503.

[7] M. MacCormick and G. Audi, *Nucl. Phys. A*, **925** (2014) 61.

EPJ Web of Conferences **117**, 06020 (2016) DOI: 10.1051/epjconf/201611706020
NN2015

α cluster structures in unbound states of ^{19}Ne

R. Otani, M. Iwasaki and M. Ito

Department of Pure and Applied Physics, Kansai University, 3-3-35 Yamatecho, Suita, Osaka 564-8680, Japan

Abstract

Cluster structures in ^{19}Ne are studied by the microscopic and macroscopic cluster models. In the microscopic calculation, the coupled-channels problem of (α+^{15}O) + (^{3}He+^{16}O) are solved, and the calculated energy spectra nicely reproduce the low-lying levels. In the macroscopic approach, the α + ^{15}O potential model is applied. The calculation of the potential model predicts the existence of the resonances above the α threshold, which has a weak-coupling structure of the α particle plus one hole inside of the ^{16}O nucleus. The coupling effect of the $5p$–$2h$ configuration to ^{19}Ne is briefly discussed.

1 Introduction

The α cluster structures have been extensively studied for the so-called $4N$ nuclei with $N = Z$, such as ^{8}Be = 2α, ^{12}C = 3α, ^{16}O = α + ^{12}C, and ^{20}Ne = α + ^{16}O [1]. In current studies, the importance of the cluster degrees of freedom has been extended to the neutron-rich ($N > Z$) systems, which are obtained by adding extra neutrons to the $4N$ cluster systems. In the neutron-rich systems, various chemical-bonding-like structures are generated by the coupling of the cluster relative motion and the single particle motion of extra neutrons [2–4].

On the contrary, cluster structures in the $4N$ systems with a hole in a cluster core, are also interesting research subjects [5, 6]. Pioneering work of the hole system is the study of the ^{19}F nucleus [5, 6]. In the α + ^{15}N system, a proton hole inside of the ^{16}O core weakly couples to the relative motion of the binary cluster cores. The weak coupling state of the hole in ^{16}O and the α particle is also expected in the neutron-deficient system, α + ^{15}O in ^{19}Ne [5, 6]. However, detailed analysis of the energy levels over a wide energy region, which covers the unbound continuum states, have not been undertaken.

Recently, we have applied the potential model to the α + ^{15}O system [7] and predicted the existence of the resonances above the α threshold and the excitation function of the α resonant scattering. The potential model is useful to understand the resonant states, in which the α clustering is well developed. However, the microscopic cluster model on the basis of the nucleon degrees of freedom is very important in analyzing the energy levels from the low-lying states to the highly-excited states in a unified manner. In the present report, we apply the microscopic cluster model of (^{3}He+^{15}O) + (α+^{16}O) and mainly analyze the low-lying states in ^{19}Ne.

There is another reason why we focus on the α + ^{15}O structure in the ^{19}Ne nucleus. The ^{15}O(α,γ)^{19}Ne reaction is known to play a crucial role in the advanced stages of astrophysical hydrogen burning [8]. The most crucial resonance is known to arise through the $3/2^{+}$ resonance level at 504 keV with respect to the α + ^{15}O threshold (E_x = 4.03 MeV). The analyses in Ref. [9] have pointed out that the intrinsic structure of the resonance at 504 keV in ^{19}Ne is the five particle–two hole ($5p$–$2h$) configuration with the ^{14}O$_{g.s.}$ core. The $5p$–$2h$ configuration has a large overlap with the shell model limit of the ^{5}He + ^{14}O configuration. Thus, the coupling effect of ^{5}He + ^{14}O on the $J^{\pi} = 3/2^{+}$ is also discussed.

2 Framework

We apply the generalized two-center cluster model (GTCM) for the calculations of the low-lying levels. The GTCM is the extended model of the microscopic cluster model on the basis of the generator coordinate method (GCM) [10]. The application of GTCM to Be isotopes has already been published in Ref. [2], and we briefly explain the formulation of GTCM in the ^{19}Ne nucleus. The basis function for ^{19}Ne is given by

$$\Phi^{J^{\pi}K}(S) \;=\; \hat{P}^{J^{\pi}}_{K}\mathcal{A}\left\{\psi(^{3}\mathrm{He})\psi(^{14}\mathrm{O})\prod_{j=1}^{2}\varphi_j\left(m_j\right)\right\}_S . \tag{1}$$

The basis function is constructed on the basis of the ^{3}He + ^{14}O cores, and the two neutrons are treated as the active valence-nucleons. The $\psi(^{3}$He) and $\psi(^{14}\mathrm{O})$ represent the core wave functions of the ^{3}He and ^{14}O nuclei, respectively. The former part is described by the $(0\mathrm{s})^3$ configuration in the harmonic oscillator (HO) potential, while the latter part is constructed by the $(0\mathrm{s})^4(0\mathrm{p})^{10}$ configuration. Two core nuclei are placed with the relative distance parameter S. The single-particle wave function for the jth valence neutrons ($j = 1, 2$) localized around the ^{3}He or ^{14}O cluster is given by an atomic orbital (AO) $\varphi(m_j)$, which is classified by a set of the AO quanta $m_j = (k, C, \tau)$ with the labels of the Brink orbitals k, the core C, and the neutron spin τ ($= \uparrow$ or $\downarrow$). The Brink orbitals around the core C ($C = {}^3$He or ^{14}O) are $k = 0s$ for $C = {}^3$He and $k = 0p$ for $C = {}^{14}$O. The basis functions with the full anti-symmetrization $\mathcal{A}$ are projected to the eigenstate of the total spin J, its intrinsic angular projection K, and the total parity π by the projection operator $\hat{P}_K^{J\pi}$.

In the present calculation, we solve the coupled-channels problem of (^{3}He+^{16}O) + (α+^{15}O) at a fixed S and calculate the adiabatic energy curves, which are the sequence of the energy eigenvalues as a function of the distance parameter S [2]. We search the energy minimum in the energy curves calculated for the individual spin-parity states. The distance parameter S is the variational parameter, which is called the generator coordinate [10]. The superposition of S is possible but we fix S to the minimum point in evaluating the energy levels. As for the nucleon-nucleon (NN) interaction, we use the Volkov No.2 [11] and the G3RS [12] for the central and spin-orbit parts, respectively. The parameters in the NN interactions and the size parameter of HO are tuned to reproduce the ^{3}He and ^{4}He decay thresholds in ^{19}Ne as much as possible.

The macroscopic potential model is also applied to calculate the unbound resonant states [7]. The Woods Saxon (WS) potential is assumed for the nuclear potential in α + ^{15}O, and its parameter set is fixed so as to reproduce the angular distribution of the α + ^{15}N elastic scattering. The resonant levels are calculated from the α + ^{15}N potential by adding the Coulomb force. Above the α decay threshold, the absorbing boundary condition (ABC) [13] is applied to identify the resonant levels. The detailed explanations of the potential model plus ABC is shown in Ref. [7].

3 Results

In Fig. 1, the calculated spectra (middle and right levels) are compared with the experimental data (left levels). The results of the negative parity states are shown in this figure. The middle levels is obtained by the microscopic model of (^{3}He+^{16}O) + (α+^{15}O), while the right levels is the result of the potential model. The microscopic calculation reproduces the level ordering of the low-lying bound spectra in the experiment although the level spacing of the theoretical calculation is a little different from the experimental levels.

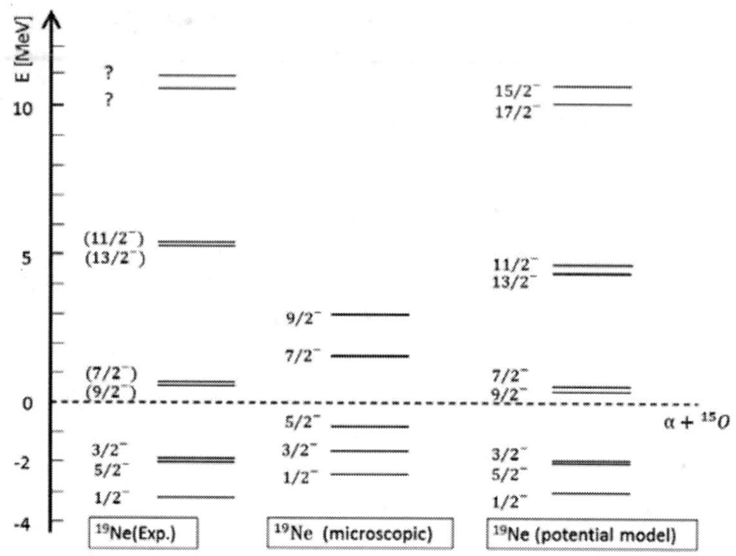

Figure 1: Energy levels of the negative parity states in the ^{19}Ne nucleus. The left levels are the experimental data, while the middle and right levels are the results of the microscopic and macroscopic models, respectively. The dotted line represents the α threshold energy, which is set to the zero energy.

In the potential model calculation, the low-lying bound levels are nicely reproduced. Furthermore, the potential model under the absorbing boundary condition predicts the resonant levels above the α threshold. Unfortunately, the experimental information is still insufficient in the unbound region. Thus, the experimental investigation of the highly-excited states are strongly desired.

The result of the positive parity states are shown in Fig. 2. The microscopic calculation with (^{3}He+^{16}O) + (α+^{15}O) is shown at the center, while the experimental levels and the result of the potential model are plotted at the left and right positions, respectively. The relative wave function of the α+^{15}O cluster configuration in the microscopic calculation (middle levels) corresponds to the HO states with the total oscillator quantum of $N =$ 7, while the resonant levels obtained by the potential model (right levels) belong to the one higher nodal state, $N = 9$. The microscopic calculation reproduces the level ordering of the experiment in the bound region. The macroscopic calculation predicts the existence of the resonant states with

 DOI: 10.1051/epjconf/201611706020

the higher spins of $J^\pi \geq 11/2^+$ and hence, the experimental investigation of these resonances is interesting.

Both the microscopic and macroscopic calculations do not reproduce the $J^\pi = 3/2^+$ state existing around the α threshold. This failure is the same as the results of the previous coupled-channel OCM calculation in $^{19}\mathrm{F} = (^3\mathrm{H}+^{16}\mathrm{O}) + (\alpha+^{15}\mathrm{N})$ [5]. In order to improve the calculation for the $3/2^+$ state, we have performed the extended microscopic calculation, which is the coupled-channels of $(^3\mathrm{He}+^{16}\mathrm{O}) + (\alpha+^{16}\mathrm{O}) + (^5\mathrm{He}+^{14}\mathrm{O})$, to include the $5p$-$2h$ configuration, which is the shell model limit ($S = 0$) of $^5\mathrm{He} + {}^{14}\mathrm{O}$.

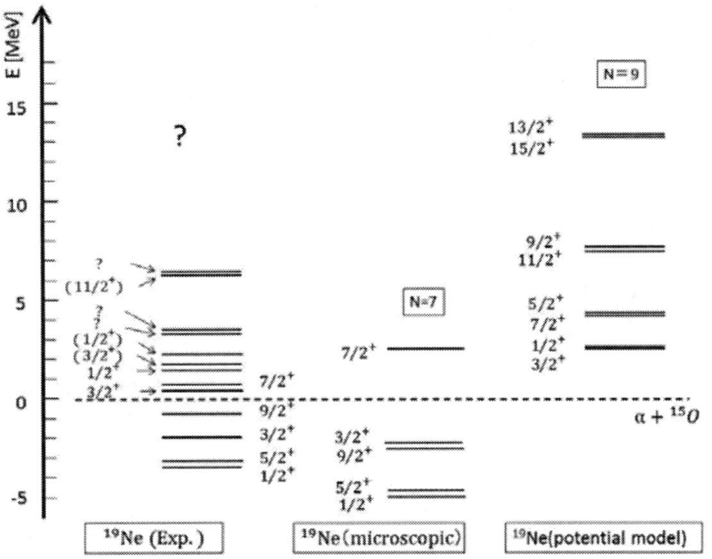

Figure 2: Same as Fig. 1 except for the positive parity states in ^{19}Ne.

A new levels with a dominant component of $^5\mathrm{He} + {}^{14}\mathrm{O}$ appear in this extended calculation. However, the energy of this $^5\mathrm{He}+^{14}\mathrm{O}$ is higher by about 10 MeV than the experimental $3/2^+$ state around the α threshold. This is because the theoretical threshold energy of the $^5\mathrm{He} + {}^{14}\mathrm{O}$ channel is also higher by about 10 MeV than the experimental threshold. Therefore, there is a possibility that a new $3/2^+$ state appear around the α threshold if the theoretical ^{5}He threshold is reproduced correctly. The reproduction of the ^{5}He threshold is important in the future calculation.

4 Summary and discussion

In sum, we have applied the microscopic and macroscopic cluster models to the level analysis of the ^{19}Ne nucleus. The microscopic model is successful in reproducing the low-lying bound spectra, while the macroscopic model predicts the existence of the several resonant states in continuum region. The experimental identification of the predicted levels is interesting in future studies.

The present calculations do not reproduce the $3/2^+$ state, which is important in the astrophysical phenomena. The extended microscopic calculation suggests the formation of a new $3/2^+$ state, which corresponds to the ^{5}He+^{14}O configuration, but its excitation energy is much higher than the experimental energy. Since the failure of the $3/2^+$ energy is due to the incorrect reproduction of the theoretical ^{5}He threshold, the parameters of the NN interaction must be revised to reproduce the ^{5}He threshold. The improved calculations is now underway.

References

[1] H. Horiuchi et al., Suppl. Prog. Theor. Phys. **192**, 1 (2012), and references therein.

[2] M. Ito and K. Ikeda, Rep. Prog. Phys. **77**, 096301 (2014).

[3] N. Itagaki, T. Otsuka, K. Ikeda and S. Okabe, Phys. Rev. Lett. **92**, 142501 (2004).

[4] Masaaki Kimura, Phys. Rev. C**75**, 034312 (2007).

[5] T. Sakuda and F. Nemoto, Prog. Theor. Phys. **62**, 1274 (1979).

[6] P. Descouvemont and D. Baye, Nucl. Phys. A**463**, 629, (1986).

[7] R. Otani et al., Phys. Rev. C**90**, 034306 (2014).

[8] K. Langanke et al., Astrophys. Jour. **301**, 629 (1986).

[9] Z. Q. Mao et al., Phys. Rev. Lett. **74**, 3760 (1995).

[10] H. Horiuchi, Suppl. Prog. Theor. Phys. **62**, 90 (1977).

[11] A. B. Volkov, Nucl. Phys. **74**, 33 (1965).

[12] N. Yamaguchi, T. Kasahara, S. Nagata, and Y. Akaishi, Prog. Theor. Phys. **62**, 1018 (1979).

[13] Y. Takenaka et al., Prog. Exp. Theor. **2014**, 113D04 (2014).

EPJ Web of Conferences **117**, 06021 (2016) DOI: 10.1051/epjconf/201611706021
NN2015

Three-body breakup of ^{22}C

E. C. Pinilla[1] and P. Descouvemont[2]

[1] Universidad Nacional de Colombia, Sede Bogotá, Facultad de Ciencias, Departamento de Física, Grupo de Física Nuclear, Carrera 45 N° 26-85, Edificio Uriel Gutiérrez, Bogotá D.C. C. P. 111321, Colombia
[2]Physique Nucléaire Théorique et Physique Mathématique, C.P. 229, Université Libre de Bruxelles (ULB), B 1050 Brussels, Belgium

Abstract

Coulomb breakup excitation functions of ^{22}C are calculated as a function of its ground state energy. We consider three ground states that fall into the recent experimental limit $S_{2n} < 0.3$ MeV. The ground states are constructed by varying the scattering length value of a $1s_{1/2}$ virtual state in ^{21}C. We find enhanced breakup cross sections sensitive to the weakly bound character of the ground state.

1 Introduction

The breakthrough experiment of Tanaka et al. [1] gave evidenced for a large matter radius for ^{22}C of 5.4 ± 0.9 fm, characterizing this nucleus as the heaviest halo nucleus known so far. Data on high-energy two-neutron removal from ^{22}C suggest that this nucleus is Borromean with both neutrons in a $s-$orbit [2]. A recent mass measurement places a limit in the two neutron separation energy of $S_{2n} < 0.3$ MeV [3]. Mosby et al. [4] predict an experimental scattering length limit of $|a_0| < 2.8$ fm and, from this value and a zero-range three-body model, suggest a limit $S_{2n} < 70$ keV.

Halo nuclei are seen as made of two or three bodies [5]. In particular, the four-body coupled-channels adiabatic and the adiabatic plus eikonal dynamical approaches has been introduced in Ref. [6] to compute reaction cross sections around 300 AMeV of ^{22}C. Four-body models make use of core$-n$ and $n-n$ interactions to describe the three-body projectile, and core-target and target$-n$ interactions are needed into the reaction framework. A four-body study of the Coulomb breakup mechanism of ^{22}C is particularly difficult since neither the ^{20}C$-n$ potential nor the ^{22}C ground state energy is exactly known. A theoretical description of the ^{22}C breakup should clarify many aspects of the very weakly bound character of this nucleus.

In this work we apply a Coulomb corrected eikonal model [7,8] to predict breakup excitation functions of ^{22}C on ^{208}Pb at 240 AMeV as a function of the ground state energy of ^{22}C. We use three ground state energies that fall into the limit $S_{2n} < 0.3$ MeV. These ground states are computed by using ^{20}C$-n$ potentials constructed to give different scattering lengths of a ^{21}C virtual state.

The paper is organized as follows. Section 2 briefly review the eikonal model. In section 3 we exhibit the ^{22}C breakup cross section for different ground state energies. Some conclusions are shown in Section 4.

2 Four-body eikonal

In this section we briefly introduce the four-body Coulomb corrected eikonal model of Ref. [7]. For details we refer the reader to this reference.

For a three-body projectile impinging on a non-composite target the time-independent four-body Schrödinger equation is written as

$$\left[-\frac{\hbar^2}{2\mu_{PT}}\Delta_R + H_0(\xi) + V_{PT}(\boldsymbol{R},\xi)\right]\Phi(\boldsymbol{R},\xi) = E_T\Phi(\boldsymbol{R},\xi), \qquad (1)$$

where μ_{PT} is the projectile-target reduced mass, ξ represents the set of internal coordinates of the projectile and $H_0(\xi)$ is its internal Hamiltonian. In Eq. (1) the center of mass coordinate has been removed, and $\boldsymbol{R} = (\boldsymbol{b}, Z)$ is the relative coordinate between the center of masses of the projectile and the target, with transverse component $\boldsymbol{b}$. The projectile-target interaction $V_{PT}(\boldsymbol{R},\xi)$ is the sum of three terms corresponding to the interaction between each body of the projectile with the target. The total energy in the center of mass system is $E_T = E_0 + \frac{\hbar^2 K^2}{2\mu_{PT}}$, with E_0 being the ground state energy of the projectile.

In eikonal models one assumes that the collision energy is much higher than the Coulomb barrier to make the following ansatz for the solution of Eq. (1)

$$\Phi(\boldsymbol{R},\xi) = \frac{1}{(2\pi)^{3/2}}e^{iKZ}\hat{\Phi}(\boldsymbol{R},\xi), \qquad (2)$$

where $\boldsymbol{K}$ is the initial relative projectile-target wave vector defined along Z. In Eq. (2) $\hat{\Phi}(\boldsymbol{R},\xi)$ is a slow varying function over $\boldsymbol{R}$, which means

$$\left|\Delta_R\hat{\Phi}(\boldsymbol{R},\xi)\right| \ll \left|K\frac{\partial}{\partial Z}\hat{\Phi}(\boldsymbol{R},\xi)\right|. \tag{3}$$

Inserting Eq. (2) into the Schrödinger equation (1), using Eq. (3) and performing the adiabatic approximation which consists in replacing H_0 by E_0 [9], we find the so-called eikonal wave function

$$\Phi_{\rm eik}(\boldsymbol{R},\xi) = \frac{1}{(2\pi)^{3/2}}\exp\left[i\left(KZ - \frac{\mu_{PT}}{\hbar^2 K}\int_{-\infty}^{Z} dZ' V_{PT}(\boldsymbol{b},Z',\xi)\right)\right]\Psi_{J_0M_0\pi_0}(\xi), \tag{4}$$

where $\Psi_{J_0M_0\pi_0}(\xi)$ is the three-body projectile ground state with total angular momentum, projection over the quantization axis and parity J_0, M_0 and π_0, respectively. This state is determined here from a three-body model. For the internal coordinates ξ, we use hyperspherical coordinates (for details we refer the reader to Refs. [10,11]).

Breakup cross sections are proportional to the T-matrix

$$T_{\mathbf{K}',\mathbf{K}} = \langle \phi_{\mathbf{K}'}|V_{PT}|\Phi_{\mathbf{K}}^{\rm eik.}\rangle \tag{5}$$

with the definition $\phi_{\mathbf{K}'}(\mathbf{R},\xi) = \frac{1}{(2\pi)^{3/2}}e^{i\mathbf{K}'\cdot\mathbf{R}}\Psi^{(+)}_{k_\xi,\xi}(E,\xi)$, where $\boldsymbol{K'}$ is the final relative projectile-target wave vector and $\Psi^{(+)}_{k_\xi,\xi}(E,\xi)$ is the three-body continuum state of the projectile. In this work, we use the three-body R-matrix method in hyperspherical coordinates to find continuum states with the appropriate boundary conditions [12]. The set of wave vectors associated to the internal coordinates of the projectile is represented by k_ξ. The projectile excitation energy E is defined from the three-body breakup threshold. The adiabatic approximation introduces a logarithmic divergence for Coulomb interactions. This problem is overcome by the Coulomb corrected eikonal model [7,13,14].

3 Breakup cross sections

Aiming at showing the sensitivity of the Coulomb breakup cross section of ^{22}C with its ground state energy, we consider three sets of ^{20}C$-n$ potentials. The $n-n$ potential for all cases is a central Minnesota interaction with exchange parameter $u = 1$ [15]. We take the ^{20}C$-n$ potential from Ref. [6] with $a = 0.65$ fm and $R_c = 3.393$ fm. The spin-orbit depth is modified to

$V_{ls} = 35$ MeV, which is close to the values given in Ref. [16]. We assume the existence of a $1s_{1/2}$ virtual state which scattering length is varied by modifying the $V_{l=0}$ depth. Table 1 lists the scattering lengths for three $V_{l=0}$ values and their respective three-body ground state energies and $r_{\rm rms}$ radii.

Table 1: Depths $V_0^{l=0}$ for the determined ^{20}C-n potential sets and the corresponding scattering lengths a_0 of the $1s_{1/2}$ virtual state computed with the technique of Ref. [17]. E_0 and $r_{\rm rms}$ are the ^{20}C+$n+n$ ground state energies and $r_{\rm rms}$ radii associated with each two-body potential.

	$V_0^{l=0}$ (MeV)	a_0 (fm)	E_0 (MeV)	$r_{\rm rms}$ (fm)
Set 1	33.47	-490.73	-0.23	3.95
Set 2	33.00	-47.58	-0.14	4.22
Set 3	30.50	-4.57	-	-

From Table 1 we observe that for the Set 1 and Set 2, the ground state energy falls into the experimental limit $S_{2n} < 0.3$ MeV. The scattering length $a_0 = -4.57$ fm, close to the experimental limit $|a_0| < 2.8$ fm [4] which is link to the prediction $S_{2n} < 70$ keV, from a zero-range three-body model, does not provide a bound state. However, we should note that our prediction is more accurate since we use finite-range interactions.

The breakup cross sections are computed with the ground states that reproduce $E_0 = -0.23$ MeV and $E_0 = -0.14$ MeV. As the core$-n$ potential Set 3 in addition to the $n-n$ interaction do not give a bound state, we use a three-body phenomenological potential [18] with $V_{3B_0} = -3$ MeV. This is with the aim of getting a more weakly bound ground state that the previously mentioned, with $E_0 = -0.07$ MeV.

Figure 1 shows the total Coulomb breakup cross section of ^{22}C (right) and its partial wave decomposition (left) as a function of its ground state energy. For each total breakup cross section, the bound and continuum partial waves are computed by using the same ground state Hamiltonian. We use the 1^-, 0^+ and 2^+ partial continuum states which provide the prominent contributions, and are truncated in $K_{\rm max}$ =25, 30 and 20, respectively. The ^{20}C$-^{208}$Pb and $n-^{208}$Pb potentials are taken from Ref. [19] and Ref. [20], respectively. The core-target potential uses charged and matter densities from Ref. [21]. In Fig. 1 we observe a similar behavior of the partial breakup contributions. The total breakup cross section is shifted to low energies as the more weakly bound state is used.

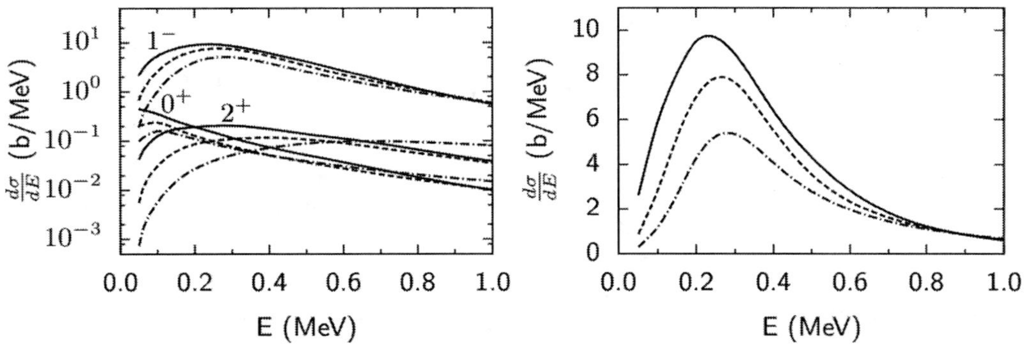

Figure 1: Dominant partial breakup cross sections (left) and total contribution (right) computed by using three-different ground state energies. The solid, dashed and dashed-dotted lines correspond to E_0=-0.07, -0.14, -0.23 MeV, respectively.

4 Conclusions

We have shown breakup cross sections of ^{22}C on ^{208}Pb at 240 AMeV as a function of the ground state energy of ^{22}C. The predictions show that if a weakly bound state exists, it should show up from the very peaked experimental breakup cross section at low excitation energies. Further analysis should be done to understand if the peaked behavior is related to a resonance or if it is related with the very weak binding of the ground state. The role of the core-target potential should be investigated too.

Acknowledgments

E.C.P. is financed by the Fondo Nacional de Financiamiento para la Ciencia, la Tecnologia y la Innovación Francisco José de Caldas and the Universidad Nacional de Colombia, Colombia. P.D. acknowledges the support of F.R.S.-FNRS, Belgium.

References

[1] K. Tanaka et al., Phys. Rev. Lett. **104**, 062701 (2010).

[2] N. Kobayashi et al., Phys. Rev. C **86**, 054604 (2012).

[3] L. Gaudefroy et al., Phys. Rev. Lett. **109**, 202503 (2012).

[4] S. Mosby et al. Nucl. Phys. A **909**, 69 (2013).

[5] J. Al-Khalili, Lect. Notes Phys. **651**, 77 (2004).

[6] Y. Kucuk and J. A. Tostevin, Phys. Rev. C **89**, 034607 (2014).

[7] D. Baye et al., Phys. Rev. C **79**, 024607 (2009).

[8] E. C. Pinilla, P. Descouvemont and D. Baye, Phys. Rev. C **85**, 054610 (2012).

[9] Y. Suzuki et al., *Structure and Reactions of Light Exotic Nuclei* (Taylor & Francis, London, 2003).

[10] P. Descouvemont, C. Daniel and D. Baye, Phys. Rev. C **044309**, 67 (2003).

[11] M. V. Zhukov et al., Phys. Rep. **231**, 151 (1993).

[12] P. Descouvemont, E. M. Tursunov and D. Baye, Nucl. Phys. A **765**, 370 (2006).

[13] B. Abu-Ibrahim and Y. Suzuki, Prog. Theor. Phys. **112**, 1013 (2004).

[14] J. Margueron, A. Bonaccorso and D. M. Brink, Nucl. Phys. A **720** 337 (2003).

[15] D. R. Thompson, M. LeMere and Y. C. Tang, Nucl. Phys. A **286**, 53 (1977).

[16] W. Horiuchi and Y. Suzuki, Phys. Rev. C **74**, 034311 (2006).

[17] O. L. Ramrez Surez and J.-M. Sparenberg, Phys. Rev. C **88**, 014601 (2013).

[18] B. V. Danilin et al., Nucl. Phys. A **632**, 383 (1998).

[19] C. A. Bertulani, M. S. Hussein and G. Münzenberg, *Physics of radioactive beams* (Nova Science, New York, 2001).

[20] A. J. Koning and J. P. Delaroche, Nucl. Phys. A **713** 231, (2003).

[21] L. C. Chamon et al., Phys. Rev. C **014610**, 66 (2002).

EPJ Web of Conferences **117**, 06022 (2016) DOI: 10.1051/epjconf/201611706022
NN2015

Elastic, inelastic and inclusive alpha cross sections in $^{6}Li+^{112}Sn$ system

D. Chattopadhyay[1,2], S. Santra[1,2], A. Pal[1,2], A. Kundu[1], B. K. Nayak[1,2], K. Mahata[1], K. Ramachandran[1], R. Tripathi[3], V. V. Parkar[1], S. Sodaye[3], D. Sarkar[1], B. Pandey[4] and G. Kaur[5]

[1]Nuclear Physics Division, Bhabha Atomic Research Centre, Mumbai 400085, India
[2]Homi Bhabha National Institute, Anushaktinagar, Mumbai 400094, India
[3]Radio Chemistry Division, Bhabha Atomic Research Centre, Mumbai 400085, India
[4]Fusion Neutronics Laboratory, Institute For Plasma Research, Gandhinagar 382428, India
[5]Department of Physics, Panjab University, Chandigarh 160014, India

Abstract

Elastic, inelastic and inclusive alpha cross sections have been measured. Coupled-channels calculations by including the couplings of projectile breakup channels and target excitations explain the measured elastic and inelastic data. The inclusive alpha particle production is found to be more than two-third of the total reaction cross section. From coupled-channels calculations, the contribution of α from the non-capture α-d breakup of the ^{6}Li projectile is calculated to be very small. It suggests the possibility of many transfer induced breakup channels, having α as one of the breakup fragments, to be responsible for such a large α particle production.

1 Introduction

Study of nuclear reactions involving weakly bound projectiles is very interesting because of the observation of several unusual features compared to the case of strongly bound projectiles. Suppression of complete fusion, breakup threshold anomaly in optical potential describing elastic scattering, large production of alpha particles in the reactions and larger peak to valley ratio of the fission fragment mass distribution at sub/near barrier energies are some of those interesting observations [1–4]. Projectile breakup in the field of a target nucleus is known to play an important role in the manifestation of all the above features. To understand the underlying reaction mechanism, the experimental data on projectile breakup cross section is thus very important to compare with the coupled-channels calculations that include the breakup channels. In order to constrain the values of coupling parameters and the potentials in the coupled-channels calculations it is also important to reproduce simultaneously the experimental data for as many reaction channels as possible for the same target+projectile system.

With the above motivation it was planned to measure the elastic, inelastic, transfer and breakup cross sections simultaneously for a system involving weakly bound projectile ^{6}Li and a target ^{112}Sn at a near-barrier energy. Choosing ^{112}Sn as a target has two advantages: i) the inelastic states of the target are well separated from its ground state compared to other isotopes of Sn and ii) the Q-value of the transfer induced breakup cross sections are favorable. Although the experimental data on elastic scattering for the above system at several energies are available in the literature but there is no data on inelastic and other reaction channels at those energies available.

2 Experimental details

The angular distributions for elastic, inelastic and inclusive alpha cross sections for ^{6}Li+^{112}Sn system have been measured at a bombarding energy of 30 MeV using BARC-TIFR Pelletron facility at Mumbai. Four telescopes of single Si detectors, placed at 10° apart, on a rotatable arm inside a 1.5 m diameter scattering chamber were used to detect the light charged particles in the angular range of $\theta_{\rm lab} = 40^o - 140^o$. Each telescope consists of a ΔE detector of thickness $\sim$ 50μm and a E-detector of $\sim$ 1500 $-$ 2000μm. A typical 2-dimensional (ΔE-E) spectrum acquired using a single telescope at $\theta_{\rm lab}$=100° is shown in Fig. 1(a). The one dimensional projection spectra for ^{6}Li, α and deuteron particles are shown in Fig. 1(b), Fig. 1(c) and Fig. 1(d) respectively. Along with the elastic peak, the yields of two inelastic states corresponding to 1st two excited states of target i.e., 2^+ (1.25 MeV) and 3^- (2.35 MeV) states are dominant. For the alpha and deuteron yield, broad spectra of width $\sim$6.4 MeV and $\sim$3.4 MeV peaking at energies corresponding to the beam velocity are observed implying a major contribution could possibly coming from projectile breakup channels.

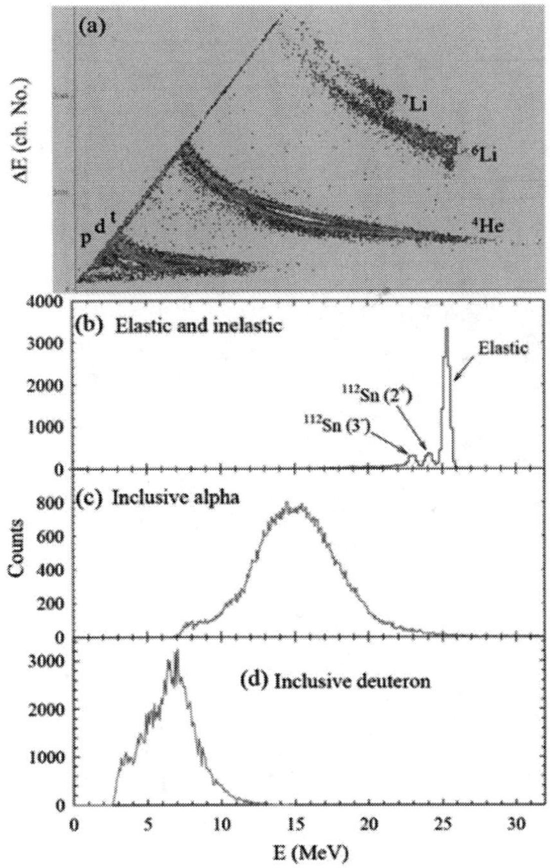

Figure 1: Typical 2-dimensional (ΔE-E) spectrum acquired using a single telescope at θ_{lab}=100° for ^{6}Li+^{112}Sn reaction at E_{lab}=30 MeV is shown in upper panel (a). One dimensional projections of ^{6}Li, alpha and deuteron particle spectra are shown in panel (b), (c) and (d) respectively.

3 Elastic and inelastic scattering

Differential cross sections for the elastic scattering angular distributions normalized to the Rutherford cross sections are shown in Fig. 2(a). The inelastic cross sections corresponding to ^{112}Sn(2$^+$,1.256 MeV) and ^{112}Sn(3$^-$,2.355 MeV) are shown in Fig. 2(b) and (c) respectively. Optical model (OM) analysis using SNOOPY code has been made to fit the elastic scattering data and a total reaction cross section of 914 mb is obtained. To include the effect of breakup coupling, continuum discretized coupled channels (CDCC) calculations using the code FRESCO have been performed using cluster-folded (CF) potential. The potential parameters for fragment target interactions i.e., α+^{112}Sn (d+^{112}Sn) are taken from Ref. [5] ([6]). An overall factor of

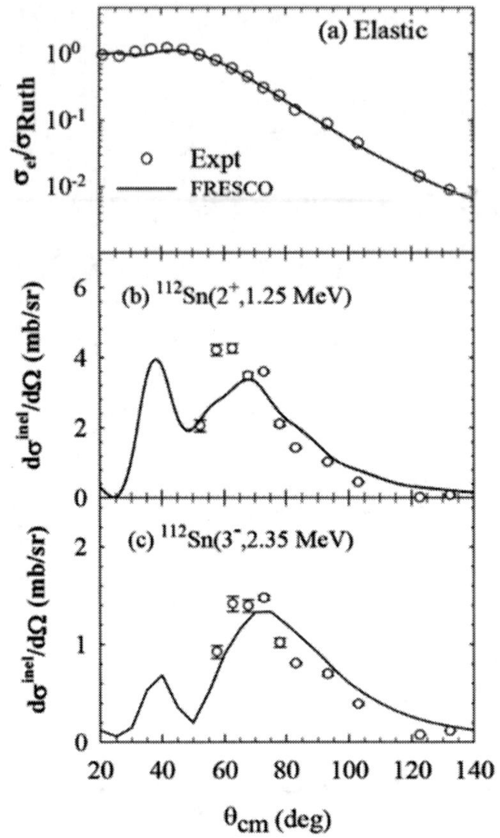

Figure 2: Experimental differential cross sections for (a) elastic scattering and (b,c) inelastic scattering corresponding to (2+, 1.256 MeV) and (3-, 2.35 MeV) excited states of ^{112}Sn respectively. Solid lines correspond to the results of coupled-channels calculations using FRESCO.

0.8 has been multiplied to the CF potential in order to reproduce the elastic scattering angular distribution. The projectile excitation details and the binding potentials of $\alpha - d$ cluster are taken to be same as in Ref. [1]. Since projectile breakup states and target inelastic excitations cannot be coupled together, the calculations are done in two steps. First CDCC calculations are performed. Next, CDCC generated polarization potential is added to CF potential to get an effective potential which in turn is used in the coupled reaction channels (CRC) calculations by FRESCO including the elastic and the target inelastic states only. The coupling parameters available in the literature [7,8] have been used for the two inelastic states. The results of the FRESCO calculations along with the measured data are shown in Fig. 2 as solid lines.

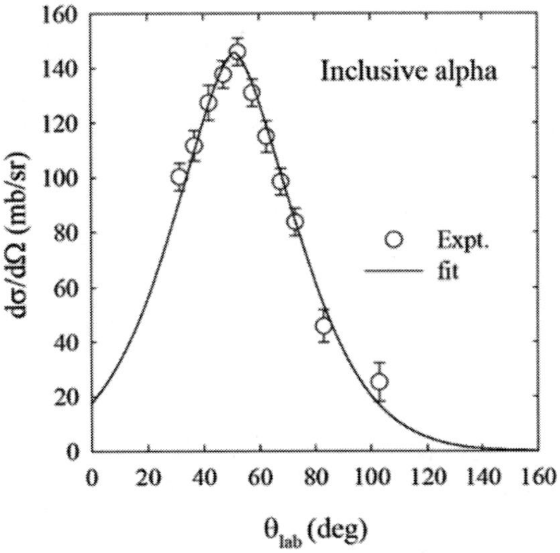

Figure 3: Angular distribution of inclusive alpha cross section. Solid line represents the fit to the measured data to obtain angle-integrated cross section.

4 Inclusive alpha

The measurement of inclusive alpha particle production in the present reaction is another interesting area. As mentioned earlier and shown in Fig. 1, a large alpha yield with velocity equal to that of the incident beam is observed compared to its complementary breakup fragment 'd'. The angular distribution of the inclusive alpha cross section is extracted and shown in Fig. 3. An arbitrary fit that describes the measured data well (shown as a solid line) is applied and the angle-integrated inclusive alpha cross section of 647$\pm$35 mb is obtained. This is more than two-third of the total reaction cross section (914 mb). From the CDCC calculations, the non-capture breakup cross section of ^{6}Li$\rightarrow$ α+d is found to be only $\sim$60 mb. The rest of the inclusive alpha cross section could be originated from several other reactions [3] e.g., (i) ^{6}Li $(-1n) \rightarrow$ ^{5}Li$\rightarrow$ α+p (ii) ^{6}Li $(+d) \rightarrow$ ^{8}Be$\rightarrow$ $\alpha + \alpha$ (iii) ^{6}Li$\rightarrow$ α+d followed by d capture, etc.

5 Summary

The differential cross sections for elastic, inelastic and inclusive alpha have been measured for ^{6}Li+^{112}Sn system at a beam energy of 30 MeV. Coupled-channels calculations are performed to include the effect of projectile breakup and target excitations. The normalized cluster-folded potential that explains simultaneously the elastic and two inelastic states are used to calculate the projectile breakup cross sections. The calculated non-capture breakup cross section of ^{6}Li$\rightarrow$ α+d is found to be very small compared to the inclusive alpha yield suggesting possible α contributions from various transfer induced breakup channels. Detailed calculations and measurements of some of these channels are underway to understand the various reaction mechanisms that are responsible for such a large inclusive alpha cross section.

References

[1] S. Santra, V. Parkar, K. Ramachandran, U. Pal, A. Shrivastava, B. Roy, B. Nayak, A. Chatterjee, R. Choudhury and S. Kailas, Phys.Lett.B **677**, 139 (2009) and references therein.

[2] S. Santra, S. Kailas, K. Ramachandran, V. V. Parkar, V. Jha, B. J. Roy and P. Shukla, Phys.Rev.C **83**,034616 (2011).

[3] S. Santra, S. Kailas, V. V. Parkar, K. Ramachandran, V. Jha, A. Chatterjee, P. K. Rath and A. Parihari, Phys.Rev.C **85**,014612 (2012).

[4] S. Santra, A. Pal, P. K. Rath, B. K. Nayak, N. L. Singh, D. Chattopadhyay, B. R. Behera, Varinderjit Singh, A. Jhingan, P. Sugathan, K. S. Golda, S. Sodaye, S. Appannababu, E. Prasad, and S. Kailas, Phys.Rev.C **90**,064620 (2014).

[5] M. Avrigeanu, W. von Oertzen, A.J.M. Plompen, V. Avrigeanu, Nucl. Phys. **A 723**, 104 (2003).

[6] J. D. Childs, W. W. Daehnick, and M. J. Spisak, Phys.Rev.C **10**, 217 (1974).

[7] S. Raman, C. W. Nestor, Jr., and P. Tikkanen, At. Data Nucl. Data Tables **78**, 1 (2001)

[8] N.-G.Jonsson, A.Backlin, J.Kantele, R.Julin, M.Luontama and A.Passoja, Nucl.Phys. **A371**, 333 (1981)

[9] G. Bruge, J. C. Faivre, I-L Faragg and A. Bussiere, Nucl. Phys. **A146** (1970) 597

EPJ Web of Conferences **117**, 06023 (2016) DOI: 10.1051/epjconf/201611706023
NN2015

^{9}Be scattering within a four-body continuum-discretized coupled-channels framework

J. Casal, M. Rodríguez-Gallardo and J. M. Arias

Departamento de Física Atómica, Molecular y Nuclear, Facultad de Física, Universidad de Sevilla, Apto. 1065, E-41080 Sevilla, Spain

Abstract

The scattering of ^{9}Be on ^{208}Pb is addressed within a four-body Continuum-Discretized Coupled-Channels (CDCC) framework, considering a three-body $\alpha + \alpha + n$ projectile plus a structureless target. The projectile states are generated using the analytical Transformed Harmonic Oscillator (THO) basis in hyperspherical coordinates. Both the elastic and breakup channels are described on the same footing. We find a good agreement between our calculations and the experimental data at beam energies around and below the Coulomb barrier.

1 Introduction

CDCC calculations [1] rely on a good theoretical description of the projectile states and on the knowledge, either experimentally or theoretically, of the interaction potential between the projectile fragments and the target. The continuum discretization for two-body projectiles can be achieved using the standard *binning* procedure [1]. This method was extended for three-body projectiles such as the halo nuclei ^{6}He [2] or ^{11}Li [3], comprising a core and two valence neutrons.

An alternative to the *binning* procedure are the pseudo-state (PS) methods, which consist in representing the continuum spectrum of the projectile by the eigenstates of its Hamiltonian in a discrete basis. This method has been successfully applied to reactions induced by two-body [4] and three-body projectiles [5]. PS methods can also be applied to reactions induced by three-body projectiles with more than one charged particle [6, 7], for which the calculation of actual continuum states is a non-trivial problem [8].

In this work we describe the scattering of ^{9}Be by a ^{208}Pb target within a four-body CDCC framework. We discretize the ^{9}Be spectrum using PS, diagonalizing the three-body Hamiltonian in an analytical THO basis in hyperspherical coordinates (See Refs. [10, 11] for details). We calculate the elastic scattering and breakup angular distributions, as well as the total breakup cross section, at different incident energies.

2 ^{9}Be projectile

Although ^{9}Be is a stable nucleus, it has a small binding energy with respect to the two-body and three-body thresholds [9]. It shows a Borromean structure $\alpha+\alpha+n$, since none of its binary subsystems ^{8}Be or ^{5}He is bound. Two-body models for ^{9}Be need to assume a $^{8}\text{Be}+n$ or $^{5}\text{He}+\alpha$ cluster structure, while both configurations are naturally included in a three-body model. For that reason, we study reactions induced by ^{9}Be within a four-body CDCC framework, considering a three-body projectile plus a structureless target.

3 Results

Calculations are performed including ^{9}Be states $j^{\pi} = 3/2^{\pm}, 5/2^{\pm}, 1/2^{\pm}$, solving the coupled equations with multipole couplings to all orders. The cutoff energy and number of projectile states are fixed to provide converged results. Details are in Ref. [7].

3.1 Elastic scattering

In Fig. 1 we show the elastic cross sections relative to Rutherford at 44 and 38 MeV, i.e. around and below the Coulomb barrier, respectively. Dashed lines are calculations including only the ^{9}Be ground state and solid lines represent full CDCC calculations. The data are from Refs. [12] (circles) and [13] (squares). We show that the agreement with the data is improved when including the coupling to the continuum, as previously reported in the literature [6]. Continuum effects are important even at energies well below the Coulomb barrier. Calculations underestimate the experimental data in the nuclear-Coulomb interference region, especially at 44 MeV. This

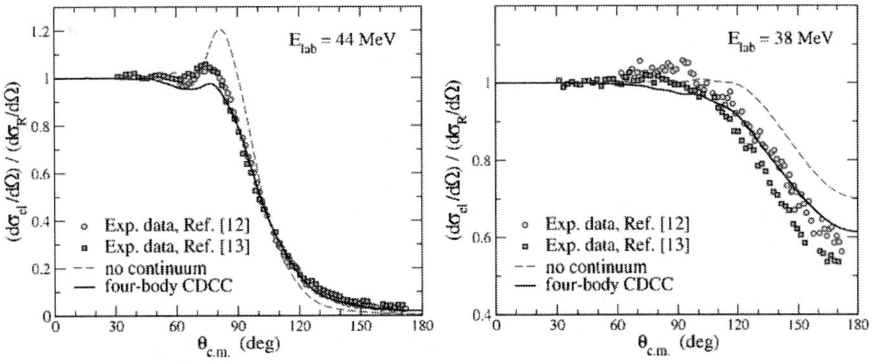

Figure 1: ^{9}Be + ^{208}Pb elastic cross section at 44 MeV (left panel) and 38 MeV (right panel).

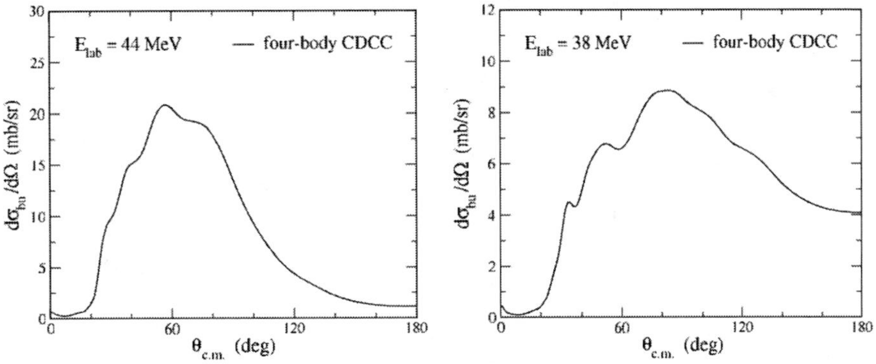

Figure 2: ^{9}Be + ^{208}Pb breakup angular distribution at 44 MeV (left panel) and 38 MeV (right panel).

difference may arise from experimental or theoretical problems, and needs a more exhaustive analysis.

3.2 Breakup

In Fig. 2 we show the breakup angular distributions in the center of mass frame at 44 and 38 MeV. The large breakup probability in the nuclear-Coulomb interference region is associated with the reduction of the elastic cross section at the corresponding angles. No data are available in the literature on this quantity. However it could be measured by performing an exclusive Coulomb dissociation experiment. This would allow to asses the validity of the model to describe the elastic and breakup processes on the same footing. The integrated breakup cross sections at 44 and 38 MeV are 112.0 and 80.5 mb, respectively. This values are in fair agreement with the experimentally reported values of 92.1 and 84.1 mb [14]. As expected, the total breakup increases with the incident energy. Breakup effects are important even below the Coulomb barrier, as already shown for the elastic scattering (see Fig. 1).

Acknowledgements

This work has been supported by the Spanish Ministerio de Economía y Competitividad under FIS2014-53448-c2-1-P and FIS2014-51941-P, and by Junta de Andalucía under group number FQM-160 and Project P11-FQM-7632. J. Casal acknowledges a FPU grant from the Ministerio de Educación, Cultura y Deporte, AP2010-3124. M. Rodríguez-Gallardo acknowledges a postdoctoral contract by the VPPI of the Universidad de Sevilla.

References

[1] Austern N. et al., *Phys. Rep.* **154** (1987) 125.

[2] Rodríguez-Gallardo M. et al., *Phys. Rev. C* **80** (2009) 051601(R).

[3] Fernández-García J. P. et al., *Phys. Rev. Lett.* **110** (2013) 142701.

[4] Moro A. M. et al., *Phys. Rev. C* **80** (2009) 054605.

[5] Rodríguez-Gallardo M. et al., *Phys. Rev. C* **77** (2008) 064609.

[6] Descouvemont P. et al., *Phys. Rev. C* **91** (2015) 024606.

[7] Casal J., Rodríguez-Gallardo M. and Arias J. M., submitted to *Phys. Rev. C* (2015).

[8] Nguyen N. B. et al., *Phys. Rev. Lett.* **109** (2012) 141101.

[9] Tilley D. R. et al., *Nucl. Phys. A* **745** (2004) 155.

[10] Casal J., Rodríguez-Gallardo M. and Arias J. M., *Phys. Rev. C* **88** (2013) 014327.

[11] Casal J. et al., *Phys. Rev. C* **90** (2014) 044304.

[12] Woolliscroft R. J. et al., *Phys. Rev. C* **69** (2004) 044612.

[13] Yu N. et al., *J. Phys. G: Nucl. Part. Phys.* **37** (2010) 075108.

[14] Woolliscroft R. J. et al., *Phys. Rev. C* **68** (2003) 014611.

EPJ Web of Conferences **117**, 06024 (2016) DOI: 10.1051/epjconf/201611706024
NN2015

Transfer to the continuum calculations of quasifree (p,pn) and (p,2p) reactions

M. Gomez-Ramos and A.M. Moro

Departamento de FAMN, Facultad de Física, Universidad de Sevilla,Apdo. 1065, E-41080 Sevilla, Spain

Abstract

Nucleon removal (p, pn) and $(p, 2p)$ reactions at intermediate energies have gained renewed attention in recent years as a tool to extract information from exotic nuclei. The information obtained from these experiments is expected to be sensitive to deeper portions of the wave function of the removed nucleon than knockout reactions with heavier targets. In this contribution, we present calculations for $(p, 2p)$ and (p, pn) reactions performed within the so-called *transfer to the continuum* method (TR*). Results for stable and unstable nuclei are presented, and compared with experimental data, when available.

1 Introduction

Quasi-free nucleon removal reactions with proton projectiles, or (p, pN) reactions, have been extensively used to obtain information on the single-particle observables of nuclei. In these reactions, a proton beam of high-energy (> 100 MeV) collides with a nucleus with A nucleons, and events are selected in which one nucleon is extracted and a residual nucleus of $A - 1$ nucleons is detected, in either its ground or an excited state. In this regime of energies, the proton has a maximum mean free path in the nucleus so it can be assumed to interact only with the extracted nucleon, hence the name "quasi-free".

Thanks to the new facilities which produce beams with unstable nuclei, (p, pN) reactions have been extended to study unstable nuclei, employing reactions in inverse kinematics. These reactions are currently being measured and/or planned in facilities such as GSI or RIKEN, and their results are expected to be complementary to those of knockout reactions with heavier targets, since they are expected to be sensitive to more interior parts of the wavefunction of the extracted nucleon.

In this contribution we present calculations performed with the transfer to the continuum (TR*) formalism, which is more exhaustively described in [1] but whose main features will be indicated in the following section.

2 Theoretical formalism

The TR* formalism is very similar to the continuum-discretized coupled channels formalism (CDCC). In the former, we study the reaction $p + A \rightarrow p + N + C$ through the transition operator:

$$\mathcal{T}_{if} = \left\langle \Psi_{3b}^{(-)} | V_{pN} + U_{pC} - U_{pA} | \phi_{NC} \chi_{pA}^{(+)}(\vec{K}_{pA}, \vec{R}) \right\rangle , \tag{1}$$

in which the following approximations have been made:

1. The nucleus A is modelled as a system formed by the extracted nucleon N and a core C. As such, we use an optical potential U_{pC}, and the final wavefunction is approximated by a 3-body wavefunction $\Psi_{3b}^{(-)}$.

2. The incoming wavefunction between p and A is distorted by the potential U_{pA}. It must be noted that when $\Psi^{(-)}$ is the exact wavefunction of the effective 3-body problem, the matrix element is independent on U_{pA}.

The final 3-body wavefunction is expanded in states of a subsystem of two bodies with a defined energy and angular momentum. In the CDCC formalism, the subsystem chosen is $N - C$. In TR*, the selected one is the $p - N$ subsystem. Therefore, the calculations are very similar to those of transfer reactions, which gives the formalism its name.

At the energies considered, the CDCC formalism becomes impractical due to the large values of the $N-C$ angular momentum which must included in the calculations. On the contrary, TR* calculations have been shown to converge quite rapidly with the angular momentum of the $p-N$ subsystem [2] and they include naturally the (p,d) channel, which cannot be included in CDCC calculations.

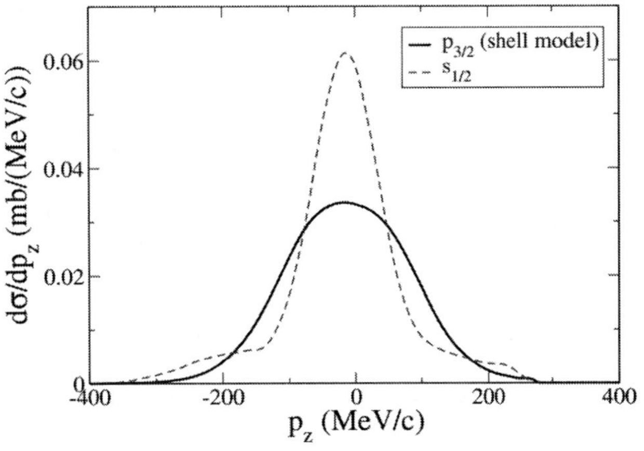

Figure 1: Longitudinal momentum distribution for $^{11}C(p,pn)^{10}C$ at 325 MeV/A, with spectroscopic factor equal to 1, assuming different ground state single-particle configurations for ^{11}C: $p_{3/2}$ (solid line), the single-particle state of the least bound nucleon of ^{11}C, according to mean field calculations and a fictitious $s_{1/2}$ state (dashed line).

3 Results

3.1 $^{11}C(p,pn)^{10}C$ at 325 MeV/A

First, we present calculations for the (p,pn) reaction on nuclei of ^{11}C performed at 325 MeV/A. For the V_{pn} potential we have used the Reid93 interaction [3] while the U_{pC}, U_{nC} and U_{pA} interactions have been calculated through a folding of the Paris-Hamburg g-matrix effective NN interaction [4] with the nuclear density of the nucleus, obtained from a Hartree-Fock calculation with the Skyrme NN interaction.

In Fig. 1 we present the longitudinal momentum distribution assuming different single-particle configurations for the ground state of ^{11}C. The solid line corresponds to a neutron extracted from a $p_{3/2}$ orbital, the least-bound neutron according to mean field calculations. We also present calculations in which the nucleon is extracted from a $s_{1/2}$ orbital, with the same separation energy. As can be seen in Fig. 2, the momentum distribution is very sensitive to the angular momentum of the extracted nucleon.

3.2 $^{18}C(p,pn)^{17}C^*$ at 81 MeV/A

We have performed calculations for the reaction $^{18}C(p,pn)^{17}C^*$ leading to the ^{17}C in its excited $5/2^+$ state. This reaction was measured by Kondo *et al.* [5], and was compared to a CDCC calculation [5]. We present a TR* calculation: the Reid93 interaction was used for V_{pn} while for $U_{p^{18}C}$, $U_{p^{17}C}$ and $U_{n^{17}C}$ a folding of the JLM interaction [6] with the ground state densities of the corresponding nuclei was employed.

Calculations are rescaled in order to reproduce the experimental data, and the required factor is compared with the theoretical spectroscopic factor

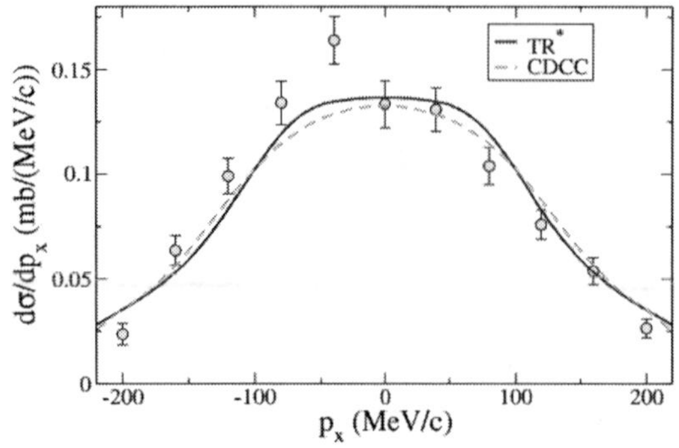

	TR*	CDCC	Theor.
S_F	2.11	2.13	2.80

Figure 2: Transversal momentum distribution for $^{18}C(p,pn)^{17}C^*$ at 81 MeV/A. TR* and CDCC calculations are presented. (CDCC and experimental data are extracted from [5]). Spectroscopic factors (S_F) are presented for both calculations.

(S_F), obtained with a shell-model calculation with the WBP interaction [7]. We find a very good agreement between both calculations and the experimental data, which validates the use of TR* in this regime of energies.

4 Conclusions

Preliminary calculations indicate that the TR* provides a suitable framework to study (p,pN) reactions at energies of $\sim$ 100 MeV/A. We expect the method to be adequate at higher energies, of around hundreds of MeV/A, where the CDCC method becomes unfeasible.

References

[1] A. M. Moro and F. Nunes, *Nucl. Phys. A*, **767** (2006) 138; A. M. Moro, *To be published in Phys. Rev. C*

[2] A. Deltuva et al., *Phys. Rev. C*, **76** (2007) 064602

[3] V. G. J. Stoks et al., *Phys. Rev. C*, **49** (1994) 2950

[4] L. Rikus et al.: *Nucl. Phys. A*, **414** (1984) 413

[5] Y. Kondo et al.: *Phys. Rev. C*, **79** (2009) 014602

[6] J. P. Jeukenne et al., *Phys. Rev. C*, **16** (1977) 80

[7] E. K. Warburton and B. A. Brown: *Phys. Rev. C*, **46** (1992) 923

EPJ Web of Conferences **117**, 06025 (2016) DOI: 10.1051/epjconf/201611706025
NN2015

Analysis of reaction size by method of scattering radius

M. Ito[1,2], R. Otani[1], M. Iwasaki[1], M. Tomita[1]

[1]Department of Pure and Applied Physics, Kansai University, Yamatecho, 3-3-35, Suita, Japan
[2]Research center for Nuclear Phtsics (RCNP), Osaka University, Mihogaoka 10-1, Suita 567-0047, Japan

Abstract

We formulate the method of a scattering radius, which characterizes a spatial size of the scattering area, from partial wave decompositions in the coupled-channel problem. The method of the scattering radius is applied to the p + ^{12}C scattering in the range of $E_p = 29.95$ MeV to 65 MeV, and the scattering radii for the elastic scattering and the various inelastic channels are derived. We have found that the scattering radii for the inelastic channels with a well developed 3α structure are strongly enhanced in comparison to the non-3α channels.

1 Introduction

Clustering phenomena, in which a nucleus is decomposed into some subunits, are well known to appear in the excited states of the lighter mass system [1]. In ^{12}C, for instance, three α particles are strongly overlapped to each other in its ground 0_1^+ state, while they are weakly coupled in the excited 0_2^+ state, called Hoyle state, at $E_x = 7.65$ MeV [2]. One of characteristic features in cluster states is an extension of a matter radius. Previous theoretical calculations suggest that a radius of a cluster state is larger by about 50 % than a radius of a ground state [1,2]. Unfortunately, a size of a cluster state is difficult to measure directly by experiments because a cluster state has a very short life-time. In a naive expectation, however, such a spatial extension may be reflected in observables of scattering phenomena. In previous studies, some enhancement factors originated from a spatial extension of a cluster structure are studied especially in the 3α state in ^{12}C [3,4].

Recently, we have proposed the method of the gscattering radiush on the basis of a partial wave decomposition of a scattering cross section [5]. In this report, the method of the scattering radius is applied to the multi-channel problem in the $p + {}^{12}$C scattering, going to the various inelastic channels, and the results of the scattering radii are discussed in connection to the intrinsic structures of ^{12}C.

2 Framework

We solve a set of the coupled-channel equations for the proton plus ^{12}C system, which is given in a symbolic form

$$(T_f(\mathbf{R}) - V_{f,f}(\mathbf{R}) - E_f)\,\chi_f(\mathbf{R}) = -\sum_{i\neq f} V_{f,i}(\mathbf{R})\chi_i(\mathbf{R}) \ . \qquad (1)$$

Here the subscripts of f and i design a channel. $T_f(\mathbf{R})$ represents the kinetic energy of the relative motion of the $p + {}^{12}$C system with a relative coordinate $\mathbf{R}$, while $V_{f,i}(\mathbf{R})$ denotes the coupling potential for the transition from the channel i to the channel f. The total energy in the channel f, E_f, is given by the relation of $E_f = E - \epsilon_f$ with a proton incident energy E and an internal energy of ^{12}C, ϵ_f. $\chi_f(R)$ is the proton–^{12}C relative wave function for the channel f, which should be solved in the coupled-channel equations. In the present calculation, we include all the discrete states of ^{12}C in addition to the ground 0_1^+ state; the rotational state of 2_1^+ , the vibrational 3_1^- state, the 3α cluster states of 0_2^+ and 2_2^+.

The coupling potential ($V_{f,i}(\mathbf{R})$) has a component of the real and the phenomenological imaginary potential. The real potential is calculated by the folding model [6]. We employ the DDM3Y (Density-Dependent Michigan 3-range Yukawa) [7] effective nucleon-nucleon force and the ^{12}C transition density calculated by the microscopic cluster model [2]. The imaginary potential, which has a Woods-Saxon form factor, is introduced in the diagonal transition ($f = i$) [5,8], in order to reproduce the observed differential cross sections as much as possible. In the present calculation, the proton is treated as the spin-less particle.

In order to characterize a size of scattering area, we simply define the effective orbital spin $\bar{L}$ [5] from the partial cross section $\sigma(L)$ for the incident orbital spin L like

$$\bar{L} = \sqrt{\frac{\sum_L \left\{\sqrt{L(L+1)}\right\}^4 \sigma(L)}{\sum_L \left\{\sqrt{L(L+1)}\right\}^2 \sigma(L)}} \ . \tag{2}$$

The scattering radius R_{SC} corresponding to $\bar{L}$ can be simply obtained from the relation of $\bar{L} = kR_{SC}$, where k denotes the wave number of the incident channel, measured in the laboratory system.

3 Results

We calculate the differential cross sections of the elastic and inelastic scattering in the energy range of $E_p = 29.95$ MeV to 65 MeV. The coupled-channels calculation with the 3α RGM + DDM3Y nicely reproduces all of the differential cross sections in the angular range of $\theta_{c.m.} = 30°$ to $120°$ [8]. From the partial cross sections obtained by the coupled-channels, the effective orbital spins $\bar{L}$ and the scattering radii, R_{SC} are calculated for all the channels. The $\bar{L}$ and R_{SC} derived for $E_p = 65$ MeV are summarized in table 1. In this table, the matter radii of all the states ($\bar{r}$) [2] are also shown in the bottom row for comparison.

Table 1: The effective orbital spin $\bar{L}$ and the scattering radius R_{SC} at $E_p = 65$ MeV. The respective matter radii $\bar{r}$ are also shown in the bottom column [2]. R_{SC} and $\bar{r}$ are shown in units of fm.

	0_1^+	0_2^+	2_1^+	2_2^+	3_1^-
$\bar{L}$	4.6	6.4	5.7	6.8	6.0
R_{SC} (fm)	2.6	3.6	3.2	3.9	3.4
$\bar{r}$ (fm)	2.40	3.5	2.4	4.0	2.8

From table 1, we can see the enhancement of R_{SC} for the the 0_2^+ and 2_2^+ channels, which are the resonant states just above the 3α threshold, in comparison to the scattering radii of the 0_1^+ and 2_1^+ channels. The enhancement of the scattering radii in the 0_2^+ and 2_2^+ channels can be attributed to the spatial extension of the 3α structures of these states [2]. There is a considerable enhancement of the scattering radius in the 3_1^- channel. This enhancement may be due to the extended feature of the resonant 3_1^- states.

4 Summary and discussion

In summary, we have introduced a scattering radius from the partial wave decomposition in the coupled-channels problem. The method of the scattering radius has been applied to the $p + {}^{12}$C scattering. We have found a clear enhancement of the scattering radius for the final 3α channels in comparison to the elastic and non-3α channels. This enhanced scattering radius for the cluster channels is attributed to the spatial extension of the 3α structure.

The scattering radius does not necessarily mean the matter radius of a target nucleus itself but our scattering radius is consistent to the magnitude correlation of the matter radius, which is predicted by the structure calculation [2]. Thus, the scattering radius can be used for the measure of the spatial size of the individual reaction channels. The systematic studies are very important to establish the validity of the method of the scattering radius. In particular, a recent study in Ref. [9] suggests that the 0_6^+ state in ^{16}O corresponds to the 4α dilute state, and the state is expected to have the extended matter radius. Systematic studies are now underway.

References

[1] K. Ikeda et al., Suppl. Prog. Theor. Phys. 68, 1 (1980), and references therein.

[2] M. Kamimura, Nucl. Phys. A**351**,456 (1981).

[3] A. N. Danilov et al., Phys. Rev. C**80**, 054603 (2009).

[4] K. Iida, S. Koide, A. Kohama, K. Oyamatsu, Mod. Phys. Lett. A**27**, 1250020 (2012).

[5] M. Tomita, M. Iwasaki, R. Otani, and M. Ito, Phys. Rev. C**89**, 034619 (2014).

[6] M. Ito, Y. Hirabayashi, and Y. Sakuragi, Phys. Rev. C**66**, 034307 (2002).

[7] M. E. Brandan, G. R. Sarchler, Phys. Rep. **285**, 143 (1997).

[8] M. Tomita, M. Iwasaki, R. Otani, K. Horio, and M. Ito, Phys. Rev. C **92**,024609 (2015).

[9] Y. Funaki et al., Phys. Rev. Lett. 101, 082502 (2008).

EPJ Web of Conferences **117**, 06026 (2016) DOI: 10.1051/epjconf/201611706026
NN2015

^{8}Be and ^{9}B nuclei in dissociation of relativistic ^{10}C and ^{11}C nuclei

D. A. Artemenkov[1,2], V. Bradnova[1], E. Firu[3], N. K. Kornegrutsa[1], M. Haiduc[3], K. Z. Mamatkulov[1], R. R. Kattabekov[1], A. Neagu[3], P. A. Rukoyatkin[1], V. V. Rusakova[1], R. Stanoeva[4], A. A. Zaitsev[1,5], P. I. Zarubin[1,5]* and I. G. Zarubina[1,5]

[1]Joint Institute for Nuclear Research, Dubna, Russia
[2]National Research Nuclear University MEPhI (Moscow Engineering Physics Institute), Moscow, Russia
[3]Institute of Space Science, Măgurele, Romania
[4]South-Western University, Blagoevgrad, Bulgaria
[5] P. N. Lebedev Physical Institute of the Russian Academy of Sciences, Moscow, Russia

Abstract

Progress in the study of nuclear clustering in the relativistic ^{10}C and ^{11}C nuclei dissociation in nuclear track emulsion is presented. The contribution of the unbound ^{8}Be and ^{9}B nuclei to their structure is determined on the basis of measurements of the emission angles of relativistic He and H fragments.

Nuclear track emulsion (NTE) is exposed at the JINR Nuclotron to relativistic nuclei ^{7,9}Be, ^{8,10}B, ^{10,11}C and ^{12}N for the experimental study of the evolution of the cluster structure of light nuclei [reviewed in ref. [1], [2]]. Virtual cluster configurations in an incident nucleus are completely captured in coherent dissociation in which the target nucleus is not destroyed visibly ("white"stars). Therefore, the contributions of cluster states to the structure

*e-mail: zarubin@lhe.jinr.ru

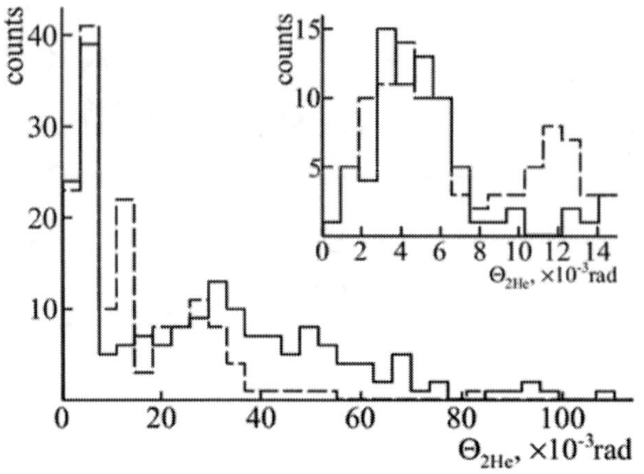

Figure 1: Distributions over the opening angle Θ_{2He} of α-particle pairs in "white"stars ^{10}C $\rightarrow$ 2He + 2H (solid histogram) and ^{9}Be $\rightarrow$ 2He (dashed histogram) at 1.2 A GeV; insertion part of the distribution in an interval Θ_n.

of the nucleus in question can be estimated by the probability of appearance of respective fragment ensembles. The nuclei ^{7}Be, ^{8}Be in the ground (^{8}Be$_{g.s.}$) and first excited states (^{8}Be$_{2+}$), as well as the ^{9}B may serve as a basis for the neighboring nuclei contributing to them with particular probabilities. Verification of this concept can be carried out on the sequence of the nuclei ^{9}Be, ^{10,11}C and ^{12}N.

Reconstruction of the decays of relativistic ^{8}Be and ^{9}B nuclei is possible by the energy variable Q = M* - M, where $M^{*2} = (\Sigma P_j)^2 = \Sigma(P_i \cdot P_k)$, M the total mass of fragments, and their $P_{i,k}$ 4-momenta defined under the assumption of conservation of an initial momentum per nucleon by fragments. For the "white"stars of ^{9}Be and ^{10}C nuclei the assumption that He fragments correspond to ^{4}He, and H ones in ^{10}C – ^{1}He is justified. Then ^{8}Be and ^{9}B identification is reduced to measurements the opening angles of between the directions of fragment emission.

Distributions over the opening angle Θ_{2He} for pairs of He fragments of "white"stars ^{9}Be$\rightarrow$2He and ^{10}C $\rightarrow$ 2He + 2H (82% of the ^{10}C statistics) produced at energy of 1.2 A GeV are presented in Fig.1. In both cases the values of Θ_{2He} of 75-80% of the pairs are distributed about equally in the intervals of $0 < \Theta_{n(arrow)} < 10.5$ mrad and $15.0 < \Theta_{w(ide)} < 45.0$ mrad. The remaining pairs are attributed to a "medium"$10.5 < \Theta_m < 15.0$ and "widest"of $15.0 < \Theta_{vw} < 45.0$ intervals. The Q distribution directly

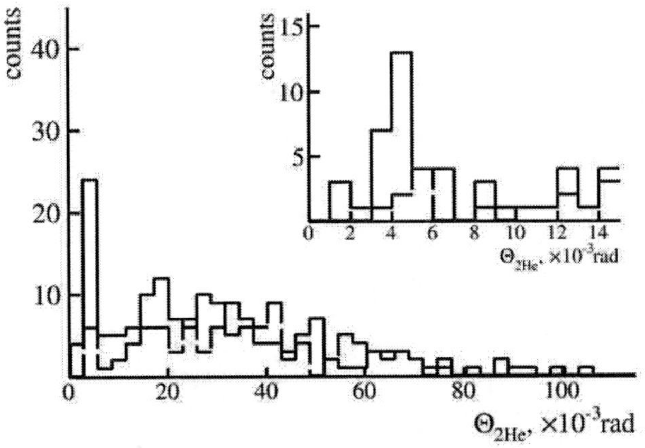

Figure 2: Distribution over the opening angle Θ_{2He} of α-particle pairs in "white" stars $^{11}C \rightarrow 2He + 2H$ (solid histogram), $^{11}C \rightarrow 3He$ (dashed histogram) at 1.2 A GeV at 1 A GeV; insertion part of the distribution in interval Θ_n.

connected with Θ_{2He} point out that "narrow" pairs of Θ_n are produced via $^8Be_{g.s.}$ while pairs Θ_w via $^8Be_{2+}$. There is a peak in the interval Θ_m reflecting the level 5/2- (2.43 MeV) of ^{9}Be. Fractions of events in the intervals Θ_n and Θ_w are equal to 0.56 ± 0.04 and 0.44 ± 0.04 for ^{9}Be, while for ^{10}C 0.49 ± 0.06 and 0.51 ± 0.06, i. e. they practically coincide. They indicate to a simultaneous presence of virtual $^8Be_{g.s.}$ and $^8Be_{2+}$ states in the ground states of the ^{9}Be and ^{10}C nuclei. Earlier, basing on the $Q_{2\alpha}$ energy distribution of the triples $2\alpha + p$ from the "white" stars $^{10}C \rightarrow 2\alpha + 2p$ it was concluded that in the structure of the ^{10}C nucleus the core ^{9}B is manifested with a probability of around $(30 \pm 4)\%$, and the $^8Be_{g.s.}$ decays are arise only through the ^{9}B decays. These conclusions allow one to interpret a significant fraction of "white" stars produced by ^{11}C nuclei only on the basis of angular measurements.

It is already established that 144 "white" stars produced by the ^{11}C in NTE are distributed over the charge channels in the following way: 2He + 2H (50%), 3He (17%), ^{7}Be + He (13%), He + 4H (11%), B+H (5%), Li + He + H (3%), 6H (2%). The distributions of He fragments over the opening angle Θ_{2He} (Fig.2) show that $^8Be_{g.s.}$ decays are presented in 21% 2He + 2H and 19% in the 3He events. These distributions allow one to assume a strong contribution of $^8Be_{2+}$ decays but it is a subject of future consideration.

The ^{9}B nucleus can exist in ^{11}C as an independent virtual component. Decays $^9B_{g.s.}$ in "white" stars $^{10}C \rightarrow 2He + 2H$ are identified in accordance with a limitation on the opening angle between directions of $^8Be_{g.s.}$ and each

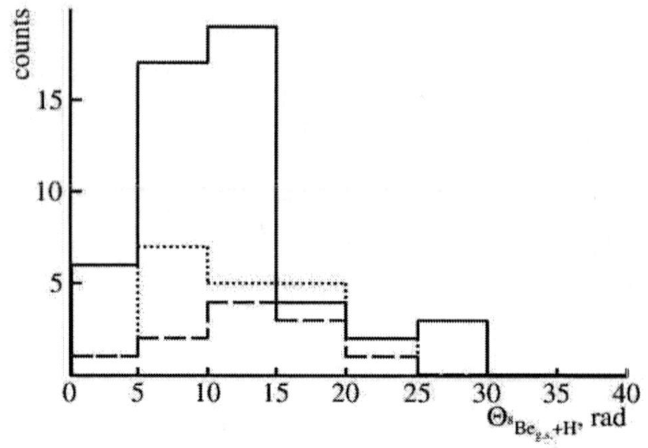

Figure 3: Distributions over the opening angle $\Theta(^{8}\mathrm{Be}_{g.s.} + \mathrm{H})$ in "white"stars ^{10}C $\rightarrow$ $^{8}\mathrm{Be}_{g.s.}$ + 2H (solid histogram), ^{11}C $\rightarrow$ $^{8}\mathrm{Be}_{g.s.}$ + 2H (dashed histogram) all found stars ^{11}C $\rightarrow$ $^{8}\mathrm{Be}_{g.s.}$ + 2H (dotted histogram).

H fragments $\Theta(^{8}\mathrm{Be}_{g.s.} + \mathrm{H}) < 40$ mrad (Fig.3) [3]. Application of such a condition the "white"stars ^{11}C $\rightarrow$ 2He + 2H allows one to identify 20 $^{9}\mathrm{B}_{g.s.}$ decays (Fig.3) constituting 30% of events in this charge channel or 18% of the ^{11}C "white"stars.

The authors are grateful to A. I. Malakhov (JINR), N. G. Polukhina and S. P. Kharlamov (LPI) for their support and critical discussion of the results. This work was supported by the grant from the Russian Foundation for Basic Research 12-02-00067 and 16-02-00062 and grant of plenipotentiary representatives of the government of Bulgaria, Egypt, Romania and the Czech Republic at JINR.

References

[1] P. I. Zarubin Lecture Notes in Physics, Vol. **875**, Clusters in Nuclei, Volume 3. Springer Int. Publ., 51 (2013) [arXiv:1309.4881].

[2] D. A. Artemenkov et al., Phys. At. Nucl. **78**, 794(2015) [arXiv:11411.5806].

[3] T. Toshito et al., Phys. Rev. C **78**, 067602(2008).

Cluster dynamics and symmetries in light nuclei

Martin Freer

School of Physics and Astronomy, University of Birmingham, Birmingham, B15 2TT, UK

Abstract

Many light nuclei display behaviour that indicates that, rather than behaving as an A-body system of individual nucleons, the degrees of freedom are those of clusters. The appearance of α-particle clustering is most widespread. In the present proceedings the symmetries and dynamics of the nuclei ^{8}Be, ^{12}C and ^{16}O are examined together with some recent experimental measurements.

1 Introduction

Symmetries in nature are ubiquitous, be it in natural, biological, systems, through to the atomic scale. On the nuclear scale, symmetries may be traced to the appearance of many nuclear phenomena, from those that underpin the shell model described by Elliot's SU(3) approach, to the appearance of deformed structures [1]. For an atomic system constrained to two dimensions, such as the carbon atoms in a graphite lattice, then hexagonal-like patterns are favoured. Similarly, in nuclei the arrangement of alpha-particles in two dimensions are predicted by the alpha-cluster model to form hexagonal arrangements [23]. Such symmetries are a manifestation of the minimisation of the energy of the system and change as the internal energy increases.

In the ground-state of nuclei, for most part, the structure is best described in terms of the motion of the constituent nucleons; so-called single particle behaviour. However, as the energy of the nucleus is increased, then it is possible for pairs of protons and neutrons to condense into alpha-particles.

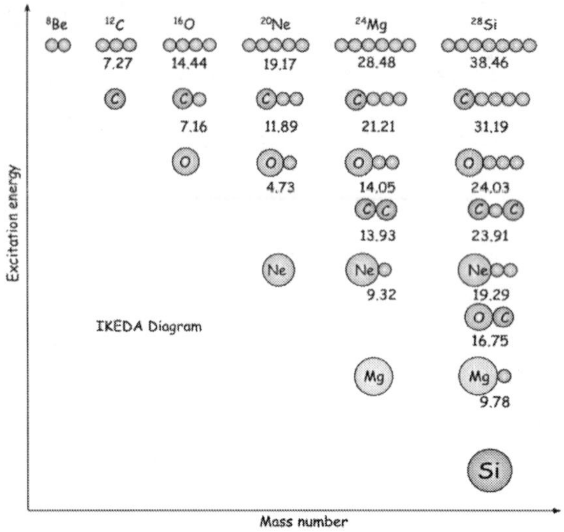

Figure 1: The Ikeda diagram [3]. This approach reveals that cluster structures, rather than appearing in the ground-state, should appear close to the cluster decay thresholds. At this point there is the possibility to convert the internal excitation energy into the mass of the cluster constituents.

There is a threshold energy for this to occur, which is close to the energy to form the alpha-particle. This naturally coincides with the alpha-particle decay threshold. Similarly, if a system is to condense into $N\alpha$-particles then an energy corresponding to the $N\alpha$-decay threshold is required. This has been encapsulated in the Ikeda diagram [3] (Fig. 1). Here as the internal energy of the nucleus is increased to the point an α-decay threshold is encountered then the nucleus is able to dissipate the internal energy into the mass of the clusters. For example, at an excitation energy of 7.27 MeV the 3α decay threshold in ^{12}C is encountered. The Ikeda picture would suggest that at this point that one possible structural mode for ^{12}C is that of 3 α-clusters. The ^{8}Be nucleus is, however, itself unbound to decay to 2α-particles by 93 keV. So, by the same token, ^{8}Be would contain a 2α-cluster structure. Excited states located just above such decay thresholds would be strongly influenced by such clusters. Famously, the 7.65 MeV Hoyle state has been linked to 3α-cluster structure (as outlined in Ref. [4] and references therein), which has been linked, in turn, to the triple α-process [5, 6] responsible for the synthesis of carbon-12.

The existence of 0^+ cluster states close to the α-decay thresholds of ^{8}Be and ^{12}C is mirrored in other light nuclei, which are α-conjugate systems. In ^{16}O the α-decay threshold is 7.16 MeV and the first excited 0^+ state lies at 6.05 MeV. This latter state has been identified with a $^{12}C+\alpha$ cluster structure [8]. Again, the Ikeda picture is reproduced. However, why is it that nature arranges for such cluster states to be close to the decay thresholds? Weakly bound nuclei close to decay thresholds can be thought of as open quantum systems where the properties of the unbound states influence, or couple to, the bound states below the threshold. The link between the appearance of cluster states at threshold and the role of the continuum has recently been explored in Ref. [9].

The transition from a shell-model ground-state, where the degrees of freedom are those of a strongly correlated single-particle system, to a clustered excited state, in the present oral presentation were mirrored to the transitions that take place in water. Water is a many body system where the interplay between hydrogen bonds and van der Waals interactions not only describe the bulk fluid but the dynamics of the formation of water clusters [10, 11]. As heat is added to water, the excess energy results in the formation of steam or water vapour. Here these droplets can be viewed as liberated water clusters. In the nuclear analogue, as energy is added, α-particles can be evaporated - when the decay threshold is reached. If clusters can exist in the bulk water state, one could ask if the formation of nuclear clusters is also possible below the α-decay threshold, i.e. in the nuclear ground-state?

Fig. 2 shows a modified version of the Ikeda diagram (Fig. 1) which shows the water-line linked with the explicit precipitation of clusters above, and the shell-model like structures below. The two densities shown in Fig. 2 correspond to Antisymmetrized Molecular Dynamics (AMD) calculation for ^{12}C [12]. The densities show the resulting intrinsic structures for the ground-state, 0_1^+, and second 0^+ state; 0_2^+. Both reveal the three α-structure, with it being more explicit in the case of the excited state, which is linked to the ^{12}C Hoyle-state. However, it is clear that the triangular, 3-fold, symmetry is present also in the ground-state. The energy levels of ^{12}C calculated using the AMD approach are shown in Fig. 3. There is seen to be excellent agreement with experiment, a feature which is not found in the case of mean-field inspired models such as the no core shell model (NCSM) [15]. It is well-known that the strength of the AMD approach is its ability to capture simultaneously shell-model and cluster-like states. This makes it ideal for revealing the appearance of the degree of clustering in the ground-state. An experimental verification of the structure of ^{12}C linked to these two intrinsic configurations is explored below.

DOI: 10.1051/epjconf/201611707001

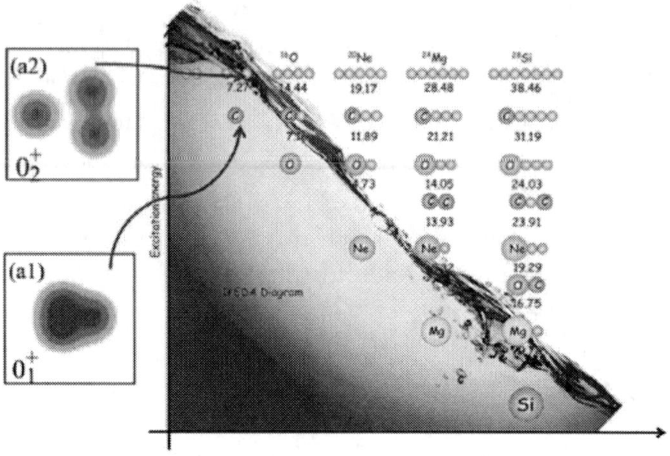

Figure 2: An adapted version of the Ikeda diagram, Fig. 1, delineating the change in structure that occurs above and below the α-decay threshold. The AMD calculations [12] for ^{12}C show that above the threshold (0_2^+) the 3α cluster structure is evident. For the ground state, 0_1^+, the cluster structure is still apparent, but the clusters are in a more compact configuration.

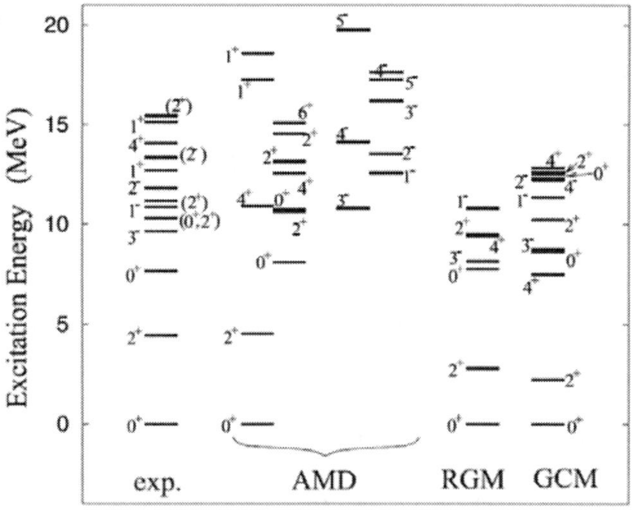

Figure 3: Energy levels for ^{12}C predicted by the AMD approach [12], resonanting group method (RGM) [13] and generator coordinate method (GCM) [14], from [12].

2 Experimental signatures for the 3α system

2.1 Excited states linked to the Hoyle-state; above the water line

If the structure of the excited Hoyle-state were close to that of the ground state, then the coupling of rotational modes would produce a corresponding 2^+ state at 4.4 MeV (the first excited state in ^{12}C is at 4.4 MeV) above 7.65 MeV, i.e. 12.05 MeV. There is no known 2^+ state at this energy. The closest state which has been reported with $J^\pi = 2^+$ is at 11.16 MeV [16]. This state was originally observed in the ^{11}B(^{3}He,d)^{12}C reaction, however has not been observed in measurements subsequently. A re-measurement of this reaction using the K600 spectrometer at iThemba in South Africa demonstrates that the earlier observation of a state at 11.16 MeV was most probably an experimental artifact and no such state exists [17].

If the Hoyle-state is more extended than the ground-state (as suggested in Fig. 2) then the 2^+ state would be lower in energy. Alternatively, the Hoyle-state could possess no collective excitations. Studies of the ^{12}C(α,α') and ^{12}C(p,p') [18] reactions indicated the presence of a 2^+ state close to 9.6-9.7 MeV with a width of 0.5 to 1 MeV. The state is only weakly populated in these reactions, presumably due to its underlying cluster structure, and is broad. Consequently, its distinction from other broad-states and dominant collective excitations (e.g. the 9.6 MeV, 3^- state) made its unambiguous identification in these measurements challenging. Further, and perhaps definitive, evidence for such an excitation comes from measurements of the ^{12}C(γ,3α) reaction performed at the HIGS facility, TUNL [19]. Here a measurable cross section for this process was observed in the same region of 9-10 MeV which cannot be attributed to other known states in this region. Furthermore, the angular distributions of the α-particles are consistent with an L=2 pattern, demonstrating a dominant 2^+ component.

Based on a rather simple description of this state in terms of three α-particles with radii given by the experimental charge radius, it is possible to use the 2 MeV separation between the Hoyle-state and the proposed 2^+ excitation to draw some conclusions as to the arrangements of the clusters [18]. A linear arrange of the 3α-particles, in which the separation is close to that of the ^{8}Be ground state, would give a separation of between the Hoyle-state and the 2^+ state of close to 1 MeV - as opposed to the 2 MeV that is observed experimentally. The data would then indicate that rather than a linear arrangement of the three clusters, a more appropriate description would be a loose arrangement of the α-particles in something approaching a triangular structure. Alternatively, the states may be represneted by the motion of an α-particle around the symmetry axis of the ^{8}Be nucleus.

A natural extension of the rotational model is that there should also be a collective 4^+ state. Using the simple $J(J+1)$ scaling, a 4^+ excitation close to $E_x(^{12}\mathrm{C}) = 14$ MeV would be expected. Measurements of the two reactions ^{9}Be(α,3α)n and ^{12}C(α,3α)^{4}He have been performed [20]. These measurements indicate a candidate state close to 13.3 MeV with a width estimated to be 1.7 MeV. It is believed that this is not a contaminant and is observed with similar properties in all spectra. Angular correlation measurements made using the ^{12}C target are not definitive, but indicate a 4^+ assignment.

2.2 Dynamical excitations of the ground state; below the water line

In 1938 Hafstad and Teller [21] identified for α-particle systems that there should be a set of dynamical symmetries. In the case of ^{12}C the symmetries of 3α-system correspond to a spinning top with a triangular point symmetry (D_{3h}). The rotational properties of these states are given by

$$E_{J,K} = \frac{\hbar^2 J(J+1)}{2\mathcal{I}_{Be}} - \frac{\hbar^2 K^2}{4\mathcal{I}_{Be}} \qquad (1)$$

where $\mathcal{I}_{Be}$ is the moment of inertia corresponding to two touching α-particles, which can be determined from the ^{8}Be ground-state rotational band [21]. K is the projection of the angular momentum onto the symmetry axis of the 3α system.

One would expect that based on this structure there should be a number of rotational bands with different values of K. For K^π=0^+, the rotations will be around an axis which lies in the plane of the three α-particles, generating a series of states 0^+, 2^+, 4^+ .... The next set of rotations correspond to the rotation around an axis perpendicular to the plane of the triangle, with each α-particle having one unit on angular momentum - giving $L = 3 \times 1\hbar$; K^π=3^-. Rotations around this axis and that parallel to the plane combine to give a series of states 3^-, 4^-, 5^-....

As indicated in Fig. 2 the triangular structure seen in the AMD calculations for the Hoyle-state is also found in the ground-state. Here the α-particles exist in a more compact configuration and at such separations the internal fermionic degrees of freedom become important and the Pauli principle plays a role in perturbing the cluster structure. It is well-known that the ground-state has a significant overlap with shell-model like structures, whereas approaches the NCSM [15] fail to reproduce the properties of the Hoyle-state.

Nevertheless, the triangular symmetry that is observed in the calculations may also be found in the experimental spectroscopic fingerprint. The rotations corresponding to $K^{\pi}=0^{+}$ have long been known to correspond to the 0^{+} ground state, 2^{+} 4.4 MeV and 4^{+} 14.1 MeV excited states. The next collective rotation occurs when each α-particle is provided with one unit of angular momentum around the 3-fold symmetry axis that passes through the centre of the triangle. These three units of angular momenta are linked to $K^{\pi}=3^{-}$. The collective excitations are then 3^{-}, 4^{-}, 5^{-}..... A precision measurement of the width of the 9.64 MeV, 3^{-}, state has recently been published [22] indicating that α-clustering may play a non-negligible role in the structure of this state. A collective 4^{-} excitation of this state has been suggested at 13.35 MeV [23, 24], with a 5^{-} member at 22.4 MeV [25]. The rotational behaviour of the ground-state rotational band and the $K^{\pi}=3^{-}$ band (associated with the 9.64 MeV state) are indicative of a D_{3h} dynamical symmetry reflecting the underlying triangular 3α-structure of ^{12}C [25].

3 The 4α system ^{16}O

The work by Hafstad and Teller [21] indicates the collective properties of the 4α system should be described by the tetrahedral symmetry group; T_d. Here the properties are those of a spherical top, with equal moments of inertia. If one assumes the separation of the α-particles is that which is associated with the ^{8}Be ground state, $\mathcal{I}_{Be}$, then the rotational energies are given by

$$E_J = \hbar^2 \frac{J(J+1)}{4\mathcal{I}_{Be}} \tag{2}$$

The rotations of the tetrahedral structure corresponds to the equivalent rotation of two ^{8}Be nuclei around their symmetry axis and hence the $4\mathcal{I}_{Be}$ in the denominator. The symmetry then dictates that all values of J are permitted except J=1, 2 and 5; states with J= 0, 4 and 8 have even parity and J=3, 7 and 11 have negative parity. A key feature of this structure would be degenerate 6^{+} and 6^{-} states. A similar conclusion can be found in the recent work of Bijker and Iachello [26]. The experimentally observed states at 6.130 MeV, 3^{-}; 10.356, 4^{+} and 21.052 MeV 6^{+} have been linked in this latter work to the collective excitations of the tetrahedral structure. These same calculations predicted states at 6.132, 10.220 and 21.462 MeV and electromagnetic transition strengths $B(E3)$ and $B(E4)$ of 181 and 338 e^2fm^{2L} compared with experimental values of 205(10) and 378(133) e^2fm^{2L}. The comparison between experiment and theory is compelling.

An alternative theoretical approach is provided by the Alpha Cluster Model (ACM) calculations of Bauhoff, Shultheis and Shultheis [8]. These calculations identify a number of cluster structures, including a tetrahedral arrangement of the four α-particles in the ground-state. In addition, a planar arrangement of α-particles is found for the first excited 0^+ state. These structures gives rise to rotational bands. The main difference between the ACM and Algebraic Cluster Model (ACM') of Ref. [26] is evident in the assignment of the 10.356 MeV 4^+ state. The ACM assigns it to the planar rotational structure, whereas the ACM' links it to the tetrahedral ground-state. What is clear from measurements of the α-decay branching ratios for decay to the ^{12}C ground state and first excited states is that the states in the ACM planar band, above the alpha-decay threshold, all have very similar decay properties - they predominantly decay to the ^{12}C ground state [27]. This similar structure conflicts with the tetrahedral interpretation and indicates a collective excitation built around a $^{12}C{+}\alpha$ cluster structure where the total angular momentum of the state is generated by the orbital motion of the α-particle around the ^{12}C core. To arrive at a better understanding of the cluster symmetries of ^{16}O further electromagnetic transition strengths need to be determined. These include states above the α-decay threshold, where small branching ratios ($< 10^{-5}$) make such studies very challenging.

A further outstanding puzzle is the question over the existence of the 4α-particle chain state. This dates back to some ground-breaking measurements performed by Chevallier [28] of the ^{12}C(^{4}He,^{8}Be)^{8}Be reaction which showed a number of resonances that decayed into two ^{8}Be nuclei. Given the structure of the final state and that the resonances appeared to lie on a $J(J+1)$ trajectory with a moment of inertia consistent with a 4α linear arrangement, a chain state was conjectured. These data have stood without proper test for many decades. The theoretical study using a cranked Hartree-Fock approach [29] indicated that, unlike the experimental data, the chain-like structure should *not* appear at low spins, but is rather stabilised by the rotations and hence wold exist at higher spins. These are spins beyond when the experimental data indicated sharp resonances. A subsequent experimental study, repeating the measurement of the same reaction used by Chevallier came to a different conclusion regarding the nature, J^π, of the resonances [30]. As such, there remains no definitive evidence for the existence of chain-like states in ^{16}O. A study at higher spins would help in this exploration.

4 Summary

One of the original predictions of the structure of light nuclei [21] suggested the ground state of nuclei such as ^{8}Be, ^{12}C and ^{16}O are composed of geometric arrangements of α-particles. Though this idea has largely fallen from favour, it is now clear that the two nuclei ^{8}Be and ^{12}C have a spectroscopic finger print that reveals this underlying cluster structure. In these systems the cluster symmetries remain, even if the clusters themselves are not completely intact. The 4α system ^{16}O presents the next challenge to our understanding. Within a model which exploits the dynamical symmetries of the tetrahedral system, T_d, there appears to be a reasonable basis for the description of the experimental states. However, there remain inconsistencies from the experimental perspective, which mean that there remains experimental challenges to provide data that can definitively constrain the symmetries and structure.

Acknowledgements:
The author would like to acknowledge the contribution from his colleagues at Birmingham who have contributed significantly through the experimental programme that underpins the present contribution.

References

[1] Arima A., J. Phys. G, **25** (1999) 581

[2] Freer M., Rep. on Prog. in Phys., **70** (2007) 2149

[3] Ikeda K., *et al.*, Prog. Theor. Phys. Suppl., Extra Numbers (1968) 464

[4] Freer M. and Fynbo H.O.U., Prog. in Part. and Nucl. Phys. **78** (2014) 1

[5] Öpik E., Proceedings of the Royal Irish Academy **A 54** (1951) 49

[6] Salpeter E.E., Astrophysical Journal **115** (1952) 326

[7] Cook C.W., Fowler W.A., Lauritsen C.C. and Lane T., Phys. Rev. **107** (1957) 508

[8] Bauhoff W., *et al.*, Phys. Rev. C, **29** (1984) 1046

[9] Okołowicz J., *et al.*, Fortschritte der Physik, **61** (2013) 66 (2013)

[10] Ludwig R., Angew. Chem. Int. Ed., **40** (2001) 1808

[11] Keutsch F.N. and Saykally R.J., PNAS **98** (2001) 10533

[12] Kanada En'yo Y., Prog. Theor. Phys., **117** (2007) 655

[13] Marumori T., J. Phys. Soc. Jpn. **44** (1978) 225; Kamimura M., Nucl. Phys. **A351** (1981) 456

[14] Uegaki E., Okabe S., Abe Y. and Tanaka H., Prog. Theor. Phys. **57** (1977) 1262

[15] Navrátil P., *et al.*, Phys. Rev. Lett. **84** (2000) 5728

[16] Reynolds G.M., Rundquist D.E. and Poichar R., Phys. Rev. C **3** (1971) 442

[17] Smit F.D., *et al.*, Phys. Rev. C, **86**(2012) 037301

[18] Itoh M. *et al.*; Nucl. Phys., **A738** (2004) 268; Phys. Rev. C, **84** (2011) 054308; Freer M. *et al.* Phys. Rev. C **80** (2009) 041303(R); Zimmerman W.R. *et al.* Phys. Rev. C **84** (2011) 027304(BR)

[19] Zimmerman W.R. *et al.*; Phys. Rev. Lett. **110** (2013) 152502

[20] Freer M. *et al.*; Phys. Rev. C **83** (2011) 034314

[21] Hafstad L.R., Teller E., Phys. Rev. **54** (1938) 681

[22] Kokalova Tz., *et al.* Phys. Rev. C **87** (2013) 057307

[23] Freer M. *et al.*, Phys. Rev. C **76** (2007) 034320

[24] Kirsebom O.S. *et al.*, Phys. Rev. C **81** (2010) 064313

[25] Marín-Lambarri D.J., *et al.*, Phys. Rev. Lett. **113** (2014) 012502

[26] Bijker R. and Iachello F., Phys. Rev. Lett. **112** (2014) 152501

[27] Wheldon C., *et al.*, Phys. Rev. C **83** (2011) 064324

[28] Chevallier P. *et al.*, Phys. Rev. **160** (1967) 827

[29] Ichikawa T., Maruhn J.A., Itagaki N., and Ohkubo S., Phys. Rev. Lett. **107** (2011) 112501

[30] Curtis N., *et al.* Phys. Rev. C **88** (2013) 064309

EPJ Web of Conferences **117**, 07002 (2016) DOI: 10.1051/epjconf/201611707002
NN2015

Pairing effects in nuclear dynamic

Denis Lacroix[1], Guillaume Scamps[2] and Yusuke Tanimura[1]

[1]Institut de Physique Nucléaire, IN2P3-CNRS, Université Paris-Sud, F-91406 Orsay Cedex, France
[2] Department of Physics, Tohoku University, Sendai 980-8578, Japan.

Abstract

In recent years, efforts have been made to account for super-fluidity in time-dependent mean-field description of nuclear dynamic [1–5]. Inclusion of pairing is important to achieve a realistic description of static properties of nuclei. Here,we show that pairing can also affect the nuclear motion. State of the art TDHF approach can describe from small to large amplitude collective motion as well as the collision between nuclei. Very recently, this microscopic approach has been improved to include pairing either in the BCS or HFB framework. Recent applications of the 3D TDHF + BCS (TDHF+BCS) model introduced in [4] will be presented. The role of super-fluidity on collective motion [6, 7], on one- and two-particle transfer [8] and on fission [9, 10] will be illustrated.

1 Introduction

Nuclear time-dependent mean-field based on the energy density functional approach is experiencing nowadays a renewal of interest [11–16]. In particular, it allows one to describe a wide variety of dynamical processes ranging from small to large amplitude collective motions, including nuclear reactions. Recently, the inclusion of pairing correlations in Time-Dependent Energy Density Functional Theory (TD-EDF) has opened new perspectives [1–5]. The possibility to describe pairing in dynamical models is an important step in the field since (i) it allows to describe dynamical processes involving superfluid nuclei, that are the majority of nuclei in the nuclear chart; (ii) it opens the perspective to uncover expected or new dynamical phenomena emerging from pairing dynamics. It is indeed well known, especially in condensed matter that the occurrence of Cooper pairs can lead to new macroscopic effects like the occurrence of Josephson effect or vortex dynamic [17].

In this proceedings, we highlight recent progresses in the field and illustrate through applications how superfluidity can affect the nuclear motion either at the level of one nucleus or when two nuclei are involved like during Heavy-Ions collisions.

2 TD-EDF with pairing

The TD-EDF with pairing generalizes the usual TD-EDF by propagating in time both the normal and anomalous densities ρ and κ. These two densities are defined through

$$\rho_{ij}(t) \;=\; \langle \hat{a}^\dagger_j \hat{a}_i \rangle, \quad \kappa_{ij}(t) = \langle \hat{a}_j \hat{a}_i \rangle,$$

where $(\hat{a}^\dagger_i, \hat{a}_i)$ correspond to creation/annihilation operators of a complete set of single-particle states. The TD-EDF equation of motion are similar to the TDHFB ones and can be written as (see for instance [18, 19]):

$$i\hbar \frac{d}{dt}\rho \;=\; [h(\rho), \rho] + \kappa\Delta^* - \Delta\kappa^*, \tag{1}$$

$$i\hbar \frac{d}{dt}\kappa \;=\; h(\rho)\kappa + \kappa h^*(\rho) - \rho\Delta - \Delta\rho^* + \Delta. \tag{2}$$

Here $h(\rho)$ and Δ are respectively the mean-field and pairing field matrix. These operators can be generically written as:

$$h(\rho)_{ij} \;=\; \left(\frac{p^2}{2m}\right)_{ij} + U_{ij}(\rho) = \left(\frac{p^2}{2m}\right)_{ij} + \sum_{kl} v^M_{ikjl}\rho_{lk} \tag{3}$$

$$\Delta_{ij} \;=\; \frac{1}{2}\sum_{kl} v^P_{ijkl}\kappa_{kl}. \tag{4}$$

where U is the mean-field potential. Here v^M and v^P denote effective vertex respectively in the particle-hole and particle-particle channels and can be directly defined as functional derivative of the energy. In the nuclear context, the difference between TDHFB and TD-EDF stems from the flexibility in choosing the effective interactions.

Pairing will affect both the evolution of one-body macroscopic variables and their fluctuations in the following way:

- **Pairing effect on one-body evolution.** Let us consider a collective one-body degree of freedom $\hat{Q} = \sum_{ij} \langle i|\hat{Q}|j\rangle \hat{a}_i^\dagger \hat{a}_j$. Its evolution is obtained through:

$$\begin{aligned} i\hbar \frac{d\langle \hat{Q}\rangle}{dt} &= i\hbar \mathrm{Tr}\left(Q \frac{d\rho}{dt}\right) \\ &= \mathrm{Tr}\left(Q[h(\rho),\rho]\right) + \mathrm{Tr}\left(Q[\kappa\Delta^* - \Delta\kappa^*]\right). \\ &= \mathrm{Tr}\left(Q[h(\rho),\rho]\right) \end{aligned}$$

The fact that the pairing field disappears in the last line has been proved in [10] for local pairing interaction. This does not however mean that pairing does not affect one-body motion. Indeed, pairing directly change the one-body density matrix property, and ultimately the expectation value $\langle \hat{Q}\rangle = \mathrm{Tr}\left(Q\rho(t)\right)$.

- **Pairing effect and correlations.** A more direct evidence of pairing effects is on two-body correlations. Indeed, considering the fluctuation of the observable $\hat{Q}$, we have:

$$\begin{aligned} \sigma_Q^2 &= \langle \hat{Q}\hat{Q}\rangle - \langle \hat{Q}\rangle\langle \hat{Q}\rangle = \sum_{ijkl} \langle i|\hat{Q}|j\rangle \langle k|\hat{Q}|l\rangle \left(\langle a_i^\dagger a_j a_k^\dagger a_l\rangle - \langle a_i^\dagger a_j\rangle \langle a_k^\dagger a_l\rangle \right) \\ &= \mathrm{Tr}(Q\rho Q(1-\rho)) + \sum_{ijkl} \langle i|\hat{Q}|j\rangle \langle k|\hat{Q}|l\rangle \kappa_{ki}^* \kappa_{lj} \end{aligned} \tag{5}$$

where the first term is the standard expectation value in Fermi system when pairing is neglected, while the second term is directly due to pairing. Due to pairing correlation two-body correlation are expected to be enhanced.

In the following, we will illustrate either the influence of pairing on one-body evolution (on vibration and fission) or directly on correlations (two-particle transfer).

3 Small amplitude vibrations

The inclusion of pairing correlations significantly extends the domain of applicability of TD-EDF. First, it allows to describe ground state properties of many nuclei in the nuclear chart rather reasonably. In particular, deformation properties of the ground state is better accounted for compared to the EDF theory without pairing. The TD-EDF with BCS approximation for pairing offers a rather versatile alternative approach to QRPA especially to

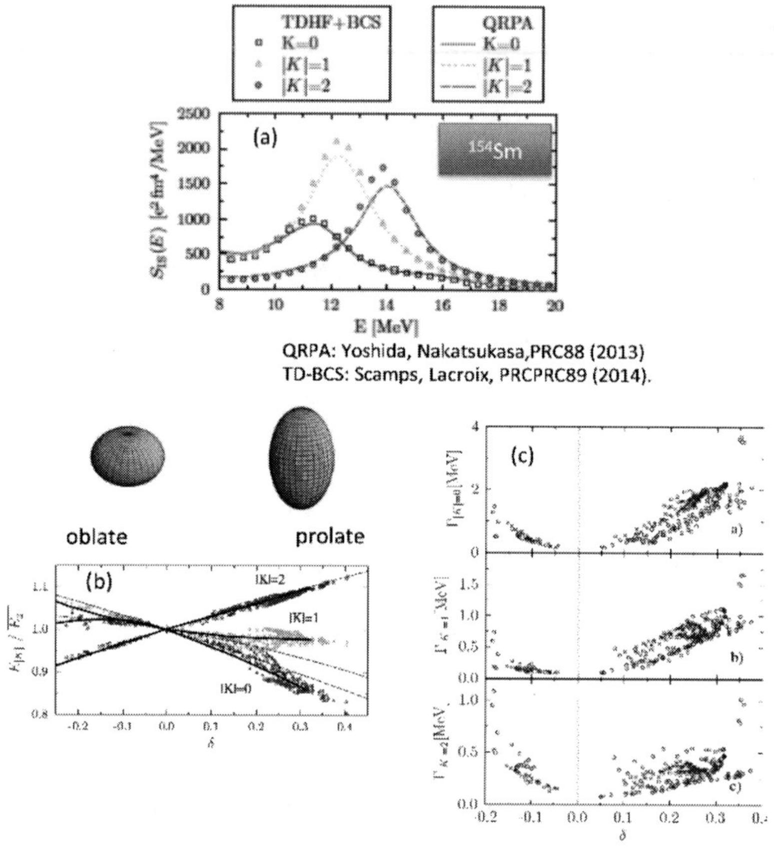

Figure 1: Giant quadrupole collective motion described with TD-EDF including pairing. (a) Comparison between the time-dependent calculation (called in the figure TDHF-BCS) and the QRPA response of ref. [20]. (b) Systematic study of the giant quadrupole energy splitting due to deformation. Each point in this figure corresponds to one nucleus with a given value of the deformation parameter δ (for more details see [6]). The lines correspond to the result of a hydrodynamical model [21] including deformation and coupling to the Giant Monopole Vibration. (c) Width of the three quadrupole response functions depending on the quadrupole moment projection of the excitation operator. Again each point corresponds to a nucleus with a given deformation δ.

study axially deformed nuclei or nuclei with triaxial deformation. Recently, the isoscalar (IS) and isovector (IV) Giant Quadrupole Response (GQR) of more than 700 nuclei has been systematically investigated in Refs. [6, 7]. Important information can be inferred from these studies:

- First, the TD-EDF gives results that are consistent with QRPA even if the BCS approximation is used in the former while the full Hartree-Fock Bogolyubov theory is implemented in the latter (see panel (a) of Fig. 1). With symmetry unrestricted codes, nuclei without specific axis of symmetry that still challenge state of the art QRPA codes, can also be studied without additional difficulty compared to axially symmetric nuclei.

- The effect of deformation on the splitting of isoscalar GQR has been specifically studied. This splitting is well reproduced by the hydrodynamical model of Ref. [21] and essentially depends on the initial deformation as well as the coupling to the isoscalar giant monopole motion (see panel (b) of Fig. 1).

- The TD-EDF as well as the QRPA theories are globally able to describe the energy of the first low-lying 2^+ state but fails to reproduce its collectivity.

- The TD-EDF approach is unable to describe the life-time (width of the response) in spherical nuclei, even if pairing is included. As it is well known, damping mechanisms require to include the coupling to complex internal degrees of freedom. As a matter of fact, see panel (c) of Fig. 1, the width of the response significantly increases as the deformation increases. Interestingly enough, at large deformation, the fragmentation obtained in strongly deformed nuclei seems compatible with the experimental width. Therefore, as far as the damping width is concerned, a transition seems to occur between spherical nuclei, where correlations dominate, to deformed nuclei, where one-body dissipation seems to dominate.

4 Fission dynamic

One obvious advantage of the time-dependent simulation is that, contrary to the linear response theory, it can be applied to situations with strong external perturbations or when the nucleus encounters very large deformations. Larger amplitude motion in heavy systems like the fission process

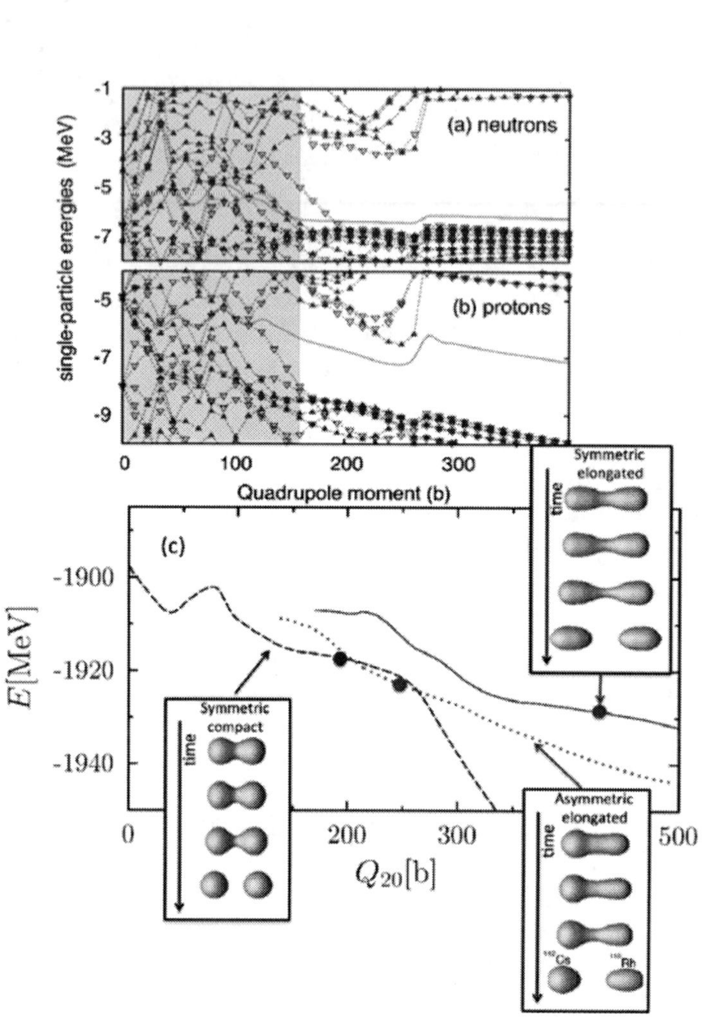

Figure 2: (Color online) Simulation of different fission paths in ^{258}Fm with TD-EDF including pairing. Evolution of neutron (a) and proton (b) single-particle energies along the adiabatic path (symmetric compact case). The red closed and blue open triangles correspond respectively to positive and negative parity states. The thick green lines represent the Fermi energies. (c) Adiabatic surfaces corresponding respectively to three different paths: symmetric compact (blue dashed line), symmetric elongated (red solid line) and asymmetric elongated (green dotted line). The three insets correspond respectively to three non-adiabatic evolutions obtained using TD-EDF. In the three cases, the starting point corresponds to the filled circles indicated in the corresponding adiabatic curve. See Ref. [9] for details.

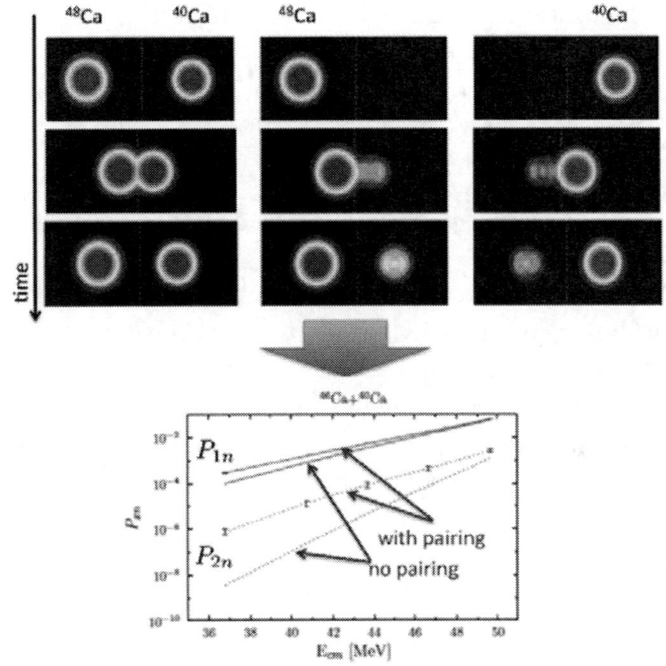

Figure 3: Illustration of the transfer process for a central collision ^{40}Ca + ^{48}Ca with energy below the Coulomb barrier. Top: snapshot of the density at different time of the reaction. Top-left: the density of the two nuclei are shown. Top-Middle (resp. Top-Right): the density of nucleons initially in the ^{48}Ca (resp. ^{40}Ca) and eventually transmitted to the other nucleus are shown. Bottom: probability to transfer one P_{1n} or 2 nucleons P_{2n} as a function of the center of mass energy for the reaction ^{40}Ca + ^{46}Ca, the latter nucleus being superfluid.

can be studied using the TD-EDF. However, as pointed out soon after the introduction of this approach in nuclear physics [22, 23], this problem is difficult to address in a mean-field theory. This stems from the coexistence of quantum effects in both single-particle and collective space (see for instance the recent review [24]). The description of fission passes through the proper treatment of quantum tunneling in many-body system, spontaneous symmetry breaking, non-adiabatic effects,... In spite of this complexity, this problem has recently been revisited in a recent series of works [9, 10, 25–27]. An illustration of different fission paths in ^{258}Fm is shown in Fig. 2 (adapted from Refs [9,10]). A clear advantage of quantal microscopic transport theory is that they consistently include nuclear structure and dynamical effects. In particular, they do not rely on the assumption that the motion is adiabatic. It was shown recently, that experimental observation are compatible with a slow and nearly adiabatic motion before scission and a mean-field+pairing like dynamics around the scission point [9]. Work is actually underway to connect microscopic mean-field theory and macroscopic transport properties [10].

5 Binary collisions below the Coulomb barrier

One interest of time-dependent EDF approach is that it allows to treat a single nucleus or binary systems in a unified framework. We summarize below some recent results obtained when describing collisions of superfluid nuclei. More precisely, central collisions below the Coulomb barrier have been considered. This study was motivated by the recent experimental observation of the 1 particle, 2 particles and 3 particles transfer probabilities, denoted hereafter by resp. P_{1n}, P_{2n} and P_{3n} [28].

In Ref. [8], the increase of two particles transfer due to pairing correlation has been carefully analyzed. An illustration of results obtained for the reaction ^{40}Ca + ^{46}Ca at various beam energy is given in panel (b) of Fig. 3. In this reaction, the nucleus ^{46}Ca has initially non-zero pairing while ^{40}Ca is in a normal phase. To extract multi-nucleon transfer probabilities, specific method based on particle projection should be used [8, 29]. In this figure, we see that the inclusion of pairing or not affects both one-particle and two-particle transfer. Indeed, the one-particle transfer is affected by pairing in particular due to non-zero occupation of single-particle states above the Fermi energy. In the two-particle transfer channel, a strong enhancement is observed due to pairing, as expected. In ref. [8] it was shown that the enhancement is directly proportional to the initial pairing gap in the superfluid nucleus. Application of the TD-BCS approach to the reaction considered experimentally in Ref. [28] was also made. Two important conclusions were drawn (i) The theoretical results obtained with the dynamical theory are compatible with those reported in Ref. [28] where a static BCS approach coupled to a semi-classical model for reaction has been used. It is worth mentioning that, the TD-BCS theory treats both structure and reaction on the same footing and does not a priori make truncation on possible important excited states during the transfer. (ii) A second important conclusion is that the inclusion of pairing is not sufficient to explain the enhancement of 2 nucleon transfer and other effects should be invoked.

6 Conclusion

We presented here most recent applications of TD-EDF including pairing correlation. In recent years, it was demonstrated that pairing can be accurately incorporated in quantum transport theories. This has opened the possibility to describe realistically the evolution of superfluid nuclei all along the nuclear chart including the heaviest ones. When the BCS approximation is made, a rather versatile and computational friendly approach is obtained leading to a unified description of many processes in nuclei from small to large amplitude collective motion.

Acknowledgments

We thank P2IO for Y. T. postdoctoral grant. G.S. acknowledges the Japan Society for the Promotion of Science for the JSPS postdoctoral fellowship for foreign researchers. This work was supported by Grant-in-Aid for JSPS Fellows No. 14F04769.

References

[1] B. Avez, C. Simenel, and Ph. Chomaz, Phys. Rev. C 78, 044318 (2008).

[2] S. Ebata, T. Nakatsukasa, T. Inakura, K. Yoshida, Y. Hashimoto, and K. Yabana, Phys. Rev. C 82, 034306 (2010).

[3] I. Stetcu, A. Bulgac, P. Magierski, and K. J. Roche, Phys. Rev. C 84, 051309 (2011).

[4] G. Scamps, D. Lacroix, G. F. Bertsch, and K. Washiyama, Phys. Rev. C 85, 034328 (2012).

[5] Y. Hashimoto, Eur. Phys. J. A 48, 1 (2012).

[6] G. Scamps and D. Lacroix, Phys. Rev. C 88, 044310 (2013).

[7] G. Scamps and D. Lacroix, Phys. Rev. C 89, 034314 (2014).

[8] G. Scamps and D. Lacroix, Phys. Rev. C 87, 014605 (2013).

[9] G. Scamps, C. Simenel and D. Lacroix, Phys. Rev. C 92, 011602(R) (2015).

[10] Y. Tanimura, D. Lacroix and G. Scamps, Phys. Rev. C 92, 034601 (2015).

[11] K.-H. Kim, T. Otsuka, and P. Bonche, J. Phys. **G23**, 1267 (1997).

[12] C. Simenel, Ph. Chomaz, and G. de France, Phys. Rev. Lett. **86**, 2971 (2001).

[13] T. Nakatsukasa and K. Yabana, Phys. Rev. **C 71**, 024301 (2005).

[14] J. A. Maruhn, P.-G. Reinhard, P. D. Stevenson, J. R. Stone, and M. R. Strayer, Phys. Rev. **C 71**, 064328 (2005).

[15] A. S. Umar and V. E. Oberacker, Phys. Rev. **C 71**, 034314 (2005).

[16] K. Washiyama and D. Lacroix, Phys. Rev. **C 78**, 024610 (2008).

[17] P.-G. de Gennes, *Superconductivity of Metals and Alloys*, Westview Press (1999).

[18] J.P. Blaizot, G. Ripka, *Quantum Theory of Finite Systems*, MIT Press, Cambridge, Massachusetts, (1986).

[19] C. Simenel, D. Lacroix, and B. Avez, *Quantum Many-Body Dynamics: Applications to Nuclear Reactions* (VDM Verlag, Sarrebruck, Germany, 2010).

[20] K. Yoshida and T. Nakatsukasa, Phys. Rev. **C88**, 034309 (2013).

[21] S. Nishizaki and K. Ando, Prog. Theor. Phys. **73**, 889 (1985).

[22] J.W. Negele, Rev. Mod. Phys. **54**, 913 (1982).

[23] J.W. Negele, Nucl. Phys. **A502**, 371c (1989).

[24] D. Lacroix, arXiv:1504.01499.

[25] C. Simenel and A. S. Umar, Phys. Rev. C **89**, 031601 (2014).

[26] Ph. M. Goddard, *A Microscopic Study of Nuclear Fission using the Time-Dependent Hartree-Fock Method* PhD thesis, (University of Surrey, 2014).

[27] P. M. Goddard, P. D. Stevenson, A. Rios, arXiv:1504.00919.

[28] L. Corradi et al., Phys. Rev. **C 84**, 034603 (2011).

[29] C. Simenel, Phys. Rev. Lett. **105**, 192701 (2010).

EPJ Web of Conferences **117**, 07003 (2016) DOI: 10.1051/epjconf/201611707003
NN2015

Cluster production within transport theory

Akira Ono

Department of Physics, Tohoku University, Sendai 980-8578, Japan

Abstract

Clusters are quite important at various situations in heavy-ion collisions and in excited low-density nuclear matter. Antisymmetrized molecular dynamics has been improved to take into account the correlations to form light clusters such as deuterons and α particles. The present status of this approach for the description of multifragmentation is reported. Calculated results demonstrate strong impacts of clusters in various observables including those usually regarded as probes of the density dependence of symmetry energy.

1 Introduction

Fragments and light clusters are copiously produced in various situations in heavy-ion collisions. For example, in Xe + Sn central collisions at 50 MeV/nucleon, the INDRA data [1] show that only 10% of total protons in the system are emitted as free protons and all the other protons are bound in light clusters, such as α particles, and in heavier fragment nuclei. Even though the incident energy is raised, the free-proton ratio increases only gradually [2]. A similar feature can be found in the data of Ref. [3] which shows the cluster mass fractions as functions of the excitation energy of the *vaporized* ^{36}Ar projectile nucleus. The cluster composition in excited low-density nuclear matter can also be an important factor to describe core-collapse supernovae [4].

Recently the heavy-ion data from Texas A&M have been analyzed to extract the cluster composition and other properties of nuclear matter as functions of the evolution of the temperature and the density [5]. Comparison of the extracted equilibrium constants for the cluster composition with various equation-of-state (EOS) models suggests necessity of a cluster suppression mechanism such as medium effects [6]. Nevertheless, in general, the link between heavy-ion collisions and EOS should be carefully checked because of the dynamical nature of collisions.

The property of a cluster in medium is typically calculated with an inmedium Schrödinger equation as in Ref. [7]. A cluster can exist as a bound state, depending on the density of the medium and the momentum of the cluster relative to the medium. Then, in actual systems, many clusters may coexist and change the nuclear matter properties and the collision dynamics.

It is therefore demanded that the transport models should solve the time evolution of collisions by correctly treating cluster correlations. However, transport models have been developed based on the single-nucleon motions, and only few models treat cluster correlations explicitly. In a version of BUU by Danielewicz *et al.* [8], clusters up to ^{3}H and ^{3}He have been introduced as new particle species and the equations for the distribution functions are coupled by the collision terms representing various reactions of clusters and nucleons. A recent calculation demonstrates that collision dynamics is influenced by the clusters [9]. On the other hand, in the antisymmetrized molecular dynamics (AMD) approach, we have introduced cluster correlations by improving the two-nucleon collision procedure [10]. The aims of the rest of this article are to report the present status of this approach and to demonstrate the importance of clusters in the studies of heavy-ion collisions.

2 AMD with clusters

In AMD, a many-body state is described by a Slater determinant of Gaussian wave packets. The time evolution of the wave-packet centroids is determined by the motion in the mean field, derived from the time-dependent variational principle, and the stochastic two-nucleon collision process. The latter is a quantum mechanical transition from an AMD state to another AMD state with a transition rate

$$W_{i\to f} = \frac{2\pi}{\hbar}|\langle\Psi_f|V|\Psi_i\rangle|^2\delta(E_f - E_i). \tag{1}$$

The most basic version is described in Ref. [11], where only the states of the colliding nucleons are changed by a two-nucleon collision.

It has been known that this basic version of AMD does not well reproduce the fragmentation in collisions of heavy nuclei such as Xe + Sn central collisions at 50 MeV/nucleon, with serious underproduction of α particles and overproduction of nucleons (see Ref. [10]). It should be noted that the AMD wave function is suitable for describing cluster correlations in the ground state and low-lying excited states of nuclei [12]. The problem is the too small probability that clusterized states are realized during the evolution of reactions. In fact, after a two-nucleon collision, one (or both) of the scattered nucleons may accidentally form a cluster with surrounding nucleon(s). The probability of this kind of cluster formation should be calculated as a transition to a specific quantum state because it does not necessarily agree with the classical phase space to form a cluster.

Therefore, in the new version of AMD with cluster correlations, the transitions to the clusterized final states are explicitly taken into account, by including those states in the set of final states $\{|\Psi_f\rangle\}$. When two nucleons N_1 and N_2 collide, we consider the cluster-forming reaction $N_1+B_1+N_2+B_2 \rightarrow C_1 + C_2$ in the most general case. Here B_1 (B_2) is a nucleon or a set of nucleons which is forming a cluster C_1 (C_2) with the nucleon N_1 (N_2). The transition rates are still given by Eq. (1). See Ref. [10] for more details. We consider the formation of clusters up to α clusters as C_1 and C_2. A cluster is represented by placing the Gaussian wave packets of nucleons at the same phase-space point, so that the overlap probabilities from the matrix element in Eq. (1) are expressed analytically. The interaction V in Eq. (1) is that between N_1 and N_2, and therefore we need the two-nucleon matrix elements which are essentially given once we have chosen a parametrizattion of two-nucleon collision cross sections which may be modified in medium. For example, in the calculations in the next section, we have chosen a so-called screened cross section [9]

$$\sigma_{\mathrm{NN}} = \sigma_0 \tanh(\sigma_{\mathrm{free}}/\sigma_0), \quad \sigma_0 = y\rho^{-2/3}, \quad y = 0.85. \tag{2}$$

No parameters have been introduced to adjust individual probabilities of different cluster formation reactions, but the overall cluster production probability is suppressed for low momentum-transfer collisions by a factor $1-\exp[-\mathbf{q}^2/(64\ \mathrm{MeV}/c)^2]$. There are many possible final states $\{|\Psi_f\rangle\}$ corresponding to different ways to form clusters. The non-orthogonality of these states should be carefully handled [10]. Finally, it should be commented that the formed cluster (C_1 or C_2) is treated in the following time evolution just as several nucleons which are accidentally at the same point (except for the extension in the next paragraph). A cluster may be broken by the mean field or by a two-nucleon collision between a nucleon in it and

DOI: 10.1051/epjconf/201611707003

another nucleon in the system like a process $d + X \to n + p + X'$. It is also possible that, in the final state of this two-nucleon collision, the same cluster is formed again like an elastic process $d+X \to d+X'$. Thus, without introducing many parameters, various kinds of cluster reactions have been introduced.

Many of light nuclei (Li, Be *etc.*) have only one or a few bound states which may be regarded as bound states of internal clusters. Then the same issue exists about the mismatch of classical phase space and the quantum-mechanical transition. Therefore, for a better description, inter-cluster correlation is introduced as a stochastic process of binding clusters. The basic idea is to replace the relative momentum between clusters by zero if moderately separated clusters ($1 < R_{\text{rel}} < 5$ fm) are moving away from each other with a small relative momentum ($\mathbf{R}_{\text{rel}} \cdot \mathbf{P}_{\text{rel}} < 0$ and $P_{\text{rel}}^2/2\mu < E_{\text{rel}}^{\text{max}}$). It is allowed that more than two clusters are linked by this condition and to be bound simultaneously. The energy conservation should be achieved by scaling the relative momentum between the center-of-mass of the bound clusters (CC system) and a third particle chosen with a suitable criterion. The numerical results depend on the adopted conditions for this cluster-cluster binding process. In the following calculations, we have chosen $E_{\text{rel}}^{\text{max}} = 8$ MeV, but it is reduced by a factor $e^{-(A_{\text{CC}}/35)^2}$ when many clusters (with total mass number A_{CC} of the CC system) are to be bound simultaneously. The binding is not considered if $10 \le A_{\text{CC}} \le 18$, in order to compensate the issue of the overbinding of the nuclei in this mass region with the adopted effective interaction.

3 Basic characters of multifragmentation

In this section, the calculated fragment charge distributions for multifragmentation in some central collisions are shown in comparison with experimental data. All the calculations were performed with the AMD with cluster correlations with the same model parameters as described in the previous section. The Skyrme SLy4 interaction was adopted with the spin-orbit term neglected. The decays of excited fragments at the end of the AMD calculation were calculated by a statistical decay code.

Figure 1 shows the comparison for Xe + Sn collisions at 50 and 32 MeV/nucleon. The light particle multiplicities (multiplied by the particle mass number) are also shown in the inset. The overall distribution of the fragments ($Z \geq 3$) and the individual multiplicities of light charged particles are all reproduced very well. The deviations in the yields around

 DOI: 10.1051/epjconf/201611707003

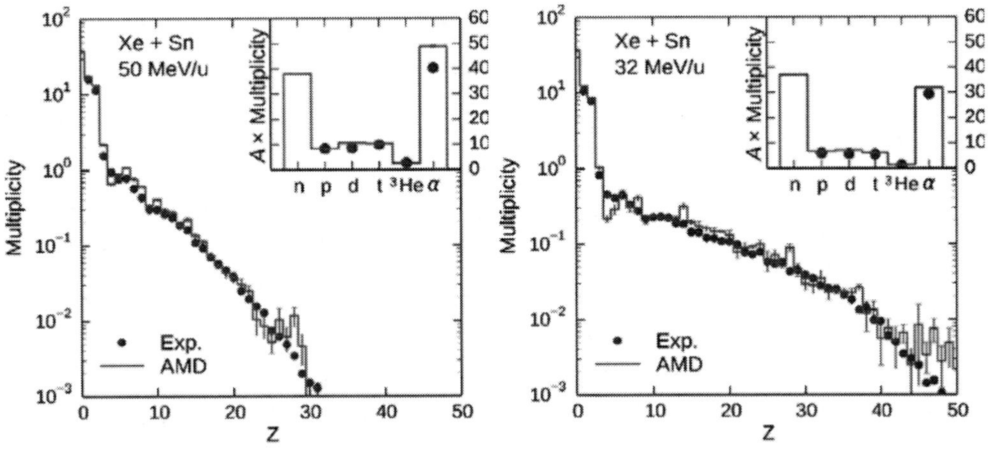

Figure 1: Fragment charge distribution in central Xe + Sn collisions ($0 < b < 2$ fm) at the incident energies of 50 (left) and 35 (right) MeV/nucleon, calculated by AMD with cluster correlations. The inset shows the multiplicities of light particles multiplied by the mass number. The INDRA experimental data are taken from Ref. [1].

$Z = 4$ at both energies have not been well understood. As well as the cluster correlations (to form clusters up to α particles), the cluster-cluster correlations are quite important to reproduce the data at both incident energies simultaneously.

The fragment multiplicities were calculated by the same model for violent collisions of ^{40}Ca + ^{40}Ca at 35 MeV/nucleon (left part of Fig. 2) and central collisions of Au + Au at 250 MeV/nucleon (right part of Fig. 2). The reproduction of data is almost satisfactory if we consider the wide difference of the system sizes and the incident energies. The α particles in Ca + Ca at 35 MeV/nucleon are underproduced in contrast to the overproduction in Xe + Sn at 50 MeV/nucleon. However, the reproductions of these two systems are much more consistent than with any other previous versions of AMD. The yield of relatively heavy fragments in the Au + Au collision may depend on the details of the cluster-cluster binding process.

The yields of clusters and fragments in these various reaction systems represent the essential information on how many-body spatial correlations develop in expanding systems with various expansion speeds and/or thermal excitations. A consistent description of this feature is an important milestone to enable precise studies of other features, as in the following sections.

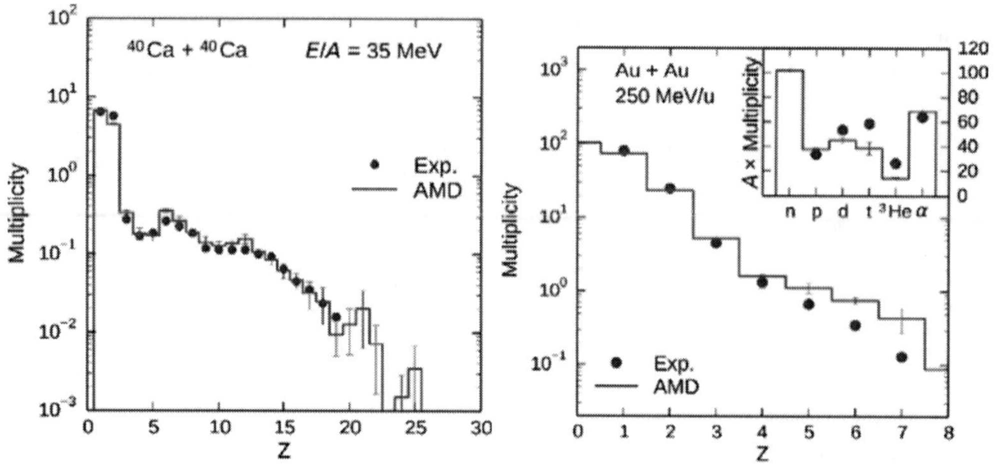

Figure 2: Left: Fragment charge distribution in violent events in ^{40}Ca + ^{40}Ca collisions at 35 MeV/nucleon. A filter for the experimental setup has been applied to the calculated events, and the violent events are selected by the same condition as for the experimental data [13]. Right: Fragment charge distribution in central Au+Au collisions at 250 MeV/nucleon. The calculated result by AMD with cluster correlations is compared to the FOPI experimental data [2].

4 Clusters at 50 MeV/nucleon

Emission of light particles, in systems of various neutron-proton asymmetries, is expected to be a probe of nuclear matter properties such as the density dependence of the symmetry energy and the neutron-proton effective mass difference (see Ref. [14] and references therein). It is evidently important to have a good theoretical understanding of emission of clusters.

The left part of Fig. 3 shows the calculated distributions of the emitted light particles as functions of the energy per nucleon $\varepsilon = E/A$ in the center-of-mass system in the central collisions of neutron-rich nuclei ^{124}Sn + ^{124}Sn at 50 MeV/nucleon. The distributions are weighted by the mass number of the particles. It is clearly seen that a large part of nucleons are bound in α particles in the low energy part. The neutron-proton ratio is $(N/Z)_{\text{gas}} \approx 2$ for the sum of these particles ($A \leq 4$). This is significantly larger than the ratio $(N/Z)_{\text{system}} = 1.48$ for the total system, which is understood as the effect of neutron fractionation. Since the same numbers of neutrons and protons are bound in many α particles (and deuterons), only a small number of protons and ^{3}He are left. Consequently the free n/p ratio and the $t/^3$He ratio can be very large, $(N_{\text{gas}} - 2n_\alpha)/(Z_{\text{gas}} - 2n_\alpha) \gg (N/Z)_{\text{gas}}$. In the right part of Fig. 3, the free n/p ratio is compared with the experimental values

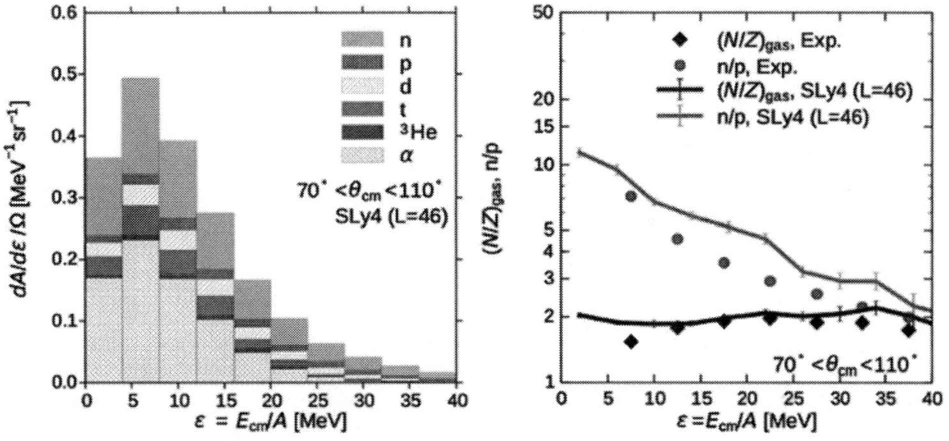

Figure 3: Left: The mass-weighted distributions of the energies $\varepsilon = E/A$ of light particles emitted to the angles $70° < \theta < 110°$ in the center-of-mass system of $^{124}\mathrm{Sn} + {}^{124}\mathrm{Sn}$ central collisions at 50 MeV/nucleon. Right: The spectral n/p ratio of free neutrons and protons and the $(N/Z)_{\mathrm{gas}}$ ratio of the sum of light particles.

from Ref. [14]. The n/p ratio can be very large (~ 10 at low energies) due to the above-mentioned effect of the α-particle formation. The exact value of the n/p ratio is probably quite sensitive to the details of the emission of α particles. On the other hand, the $(N/Z)_{\mathrm{gas}}$ ratio for the sum of these light particles takes a moderate value (~ 2) without a strong dependence on the energy ε. It seems that the present calculation with the Skyrme SLy4 interaction (corresponding to $L = 46$ MeV) is slightly overestimating $(N/Z)_{\mathrm{gas}}$ compared to the data.

5 Nuetron-proton dynamics at 300 MeV/nucleon

Heavy-ion collisions at several hundred MeV/nucleon provide us an opportunity to explore compressed nuclear matter. In the AMD calculation with clusters, the maximum density slightly higher than $2\rho_0$ is reached at about $t = 20$ fm/c, in $^{132}\mathrm{Sn} + {}^{124}\mathrm{Sn}$ central collisions at 300 MeV/nucleon. In such neutron rich systems, the neutron-proton ratio in the compressed part of the system is sensitive to the value of the symmetry energy at high densities [15]. The left part of Fig. 4 shows the neutron-proton density difference at $t = 25$ fm/c defined by $\Delta_{np}(r) = 4\pi r^2[(A/N)\rho_n(r) - (A/Z)\rho_p(r)]$ for the angle-averaged densities $\rho_n(r)$ and $\rho_p(r)$. The coefficients are defined by using A, N and Z of the reaction system so that we have $\int_0^\infty \Delta_{np}(r)dr = 0$. The results are shown by two lines for the effective interactions corresponding to

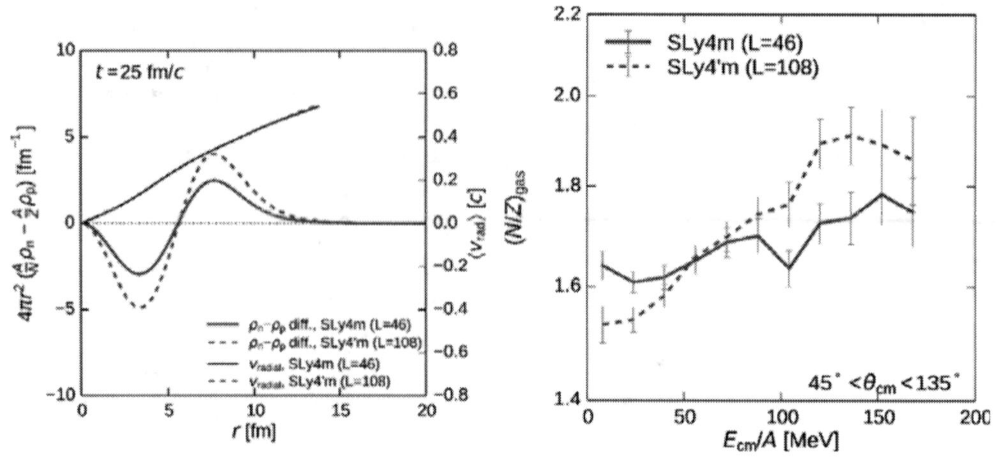

Figure 4: Left: The difference between neutron and protons densities defined as in the text as a function of the distance from the center of the system at the time $t = 25$ fm/c in $^{132}\mathrm{Sn} + {}^{124}\mathrm{Sn}$ central collisions at 300 MeV/nucleon. The collective radial velocity is also shown. Right: The $(N/Z)_{\mathrm{gas}}$ ratio for the sum of emitted particles $1 \le A \le 4$ for the final state of the same reaction system.

$L = 46$ MeV (solid) and $L = 108$ MeV (dashed). As expected, the central high-density part becomes less neutron rich than the exterior low-density part, and the effect is stronger for the stiffer symmetry energy.

The AMD calculations here were performed with the free nucleon-nucleon cross sections, with a parameter set for cluster correlations different from Sec. 2. For the stiffer symmetry energy, the density dependent term of the Skyrme SLy4 interaction $v_\rho^{(L=46)} = \frac{1}{6}t_3(1+x_3P_\sigma)\rho(\mathbf{r}_1)^\alpha\delta(\mathbf{r}_1-\mathbf{r}_2)$ was replaced by $v_\rho^{(L=108)} = \frac{1}{6}t_3(1+x_3'P_\sigma)\delta(\mathbf{r}_1-\mathbf{r}_2)\rho(\mathbf{r}_1)^\alpha + \frac{1}{6}t_3(x_3-x_3')\rho_0^\alpha P_\sigma\delta(\mathbf{r}_1-\mathbf{r}_2)$ with $x_3' = -0.5$. The momentum dependence of the energy functional was modified for high energy collisions in a similar way to Ref. [16].

An important question is, of course, how this effect at an early stage can be observed in experiments. The charged pion ratio can be a good probe [15]. We here explore another complementary possibility of using spectra of nucleons and clusters. A hint is found in the left part of Fig. 4 in which the radial expansion velocity is also shown. The almost linear dependence on the distance from the center suggests that the system is very simply expanding in this high energy collision. Therefore the exterior part, which is neutron rich, will contribute to the high velocity particles in the final state, while the neutron-poor interior part will be observed as low velocity particles. In fact, this very simple picture seems correct in the calculation as in the right part of Fig. 4 which shows the $(N/Z)_{\mathrm{gas}}$ spectral ratio for the sum of light particles $1 \leq A \leq 4$. The relation between the two lines for the different density dependences of the symmetry energy in the $(N/Z)_{\rm gas}$ spectral ratio is similar to $\Delta_{np}(r)$ in the left part.

It should be noted that in the literature the n/p spectral ratio has been found to be insensitive to the density dependence of the symmetry energy, while it is sensitive to the neutron-proton effective mass difference [17], based on the stochastic mean field (SMF) calculation without cluster correlations. On the other hand, the present AMD calculation takes into account the cluster correlations. In fact, when we turn off the cluster correlations in AMD, we find the sensitivity to the symmetry energy is lost in the $(N/Z)_{\rm gas}$ spectral ratio. This suggests that the expansion dynamics is largely influenced by cluster correlations, which is understandable because the number of particles (namely clusters and nucleons) is smaller in the calculation with cluster correlations than without them. The interaction between particles ceases at an earlier time with cluster correlations, while the interaction between nucleons continues longer in the calculations without clusters so that the final neutron and proton spectra are modified from those at early times. This effect is similar to what we found in a different context in the comparison of AMD and SMF [18].

6 Summary

Clusters are important in heavy-ion collisions, not only simply because they are emitted, but also because formation and existence of light clusters influence very much the global reaction dynamics and the bulk nuclear matter properties. AMD has been extended to include cluster correlations in the final states of two-nucleon collisions. The binding of several clusters to form a nucleus should also be considered. The consistent reproduction of basic character of multifragmentation is much more satisfactory than before. Some observables, such as the n/p spectral ratio, are sensitive to the α-particle formation. If cluster correlation is strong, the expansion is simple in collsions at 300 MeV/nucleon so that the high-density effect of the symmetry energy is reflected almost directly in the $(N/Z)_{\rm gas}$ spectral ratio.

Acknowledgments

This work was supported by JSPS KAKENHI Grant Numbers 21540253 and 24105008. The numerical calculations were partly carried out on SR16000 at YITP in Kyoto University.

References

[1] S. Hudan *et al.*, Phys. Rev. C **67** (2003) 064613.

[2] W. Reisdorf *et al.*, Nucl. Phys. A **848** (2010) 366.

[3] B. Borderie *et al.*, Eur. Phys. J. A **6** (1999) 107.

[4] K. Sumiyoshi and G. Röpke, Phys. Rev. C **77** (2008) 055804.

[5] L. Qin *et al.*,, Phys. Rev. Lett. **108** (2012) 172701.

[6] M. Hempel, K. Hagel, J. Natowitz, G. Röpke and S. Typel, Phys. Rev. C **91** (2015) 045805.

[7] G. Röpke, Nucl. Phys. A **867** (2011) 66.

[8] P. Danielewicz and G.F. Beartsch, Nucl. Phys. A **533** (1991) 712.

[9] D.D.S. Coupland, W.G. Lynch, M.B. Tsang, P. Danielewicz and Y. Zhang, Phys. Rev. C **84** (2011) 054603.

[10] A. Ono, Journal of Physics: Conference Series **420** (2013) 012103.

[11] A. Ono, H. Horiuchi, T. Maruyama and A. Ohnishi, Prog. Theor. Phys. **87** (1992) 1185.

[12] Y. Kanada-En'yo, M. Kimura and A. Ono, Prog. Theor. Exp. Phys. **2012** (2012) 01A202.

[13] K. Hagel *et al.*, Phys. Rev. C **50** (1994) 2017.

[14] D.D.S. Coupland *et al.*, arXiv:1006.4546 [nucl-ex] (2014).

[15] B. A. Li, Phys. Rev. Lett. **88** (2002) 192701.

[16] C. Gale, G. Bertsch, and S. Das Gupta, Phys. Rev. C **35** (1987) 1666.

[17] V. Giordano, M. Colonna, M. Di Toro, V. Greco and J. Rizzo, Phys. rev. C **81** (2010) 044611.

[18] M. Colonna, A. Ono and J. Rizzo, Phys. Rev. C **82** (2010) 054613.

AUTHOR INDEX

AUTHOR INDEX

AUTHOR INDEX

AUTHOR INDEX

AUTHOR INDEX

AUTHOR INDEX

AUTHOR INDEX

AUTHOR INDEX

AUTHOR INDEX

12th International Conference on Nucleus-Nucleus Collisions 2015

EPJ Web of Conferences Volume 117 (2016)

Catania, Italy
21 - 26 June 2015

Part 2 of 2

Editors:

V. Greco
M. La Cognata
S. Pirrone
F. Rizzo
C. Spitaleri

ISBN: 978-1-5108-2405-8

Printed from e-media with permission by:

Curran Associates, Inc.
57 Morehouse Lane
Red Hook, NY 12571

Some format issues inherent in the e-media version may also appear in this print version.

Printed by Curran Associates, Inc. (2016)

For additional information, please contact EDP Sciences – Web of Conferences at the address below.

EDP Sciences – Web of Conferences
17, Avenue du Hoggar
Parc d'Activité de Courtabœuf
BP 112
F-91944 Les Ulis Cedex A
France

Phone: +33 (0) 1 69 18 75 75
Fax: +33 (0) 1 69 28 84 91

contact-edps@webofconferences.org

Additional copies of this publication are available from:

Curran Associates, Inc.
57 Morehouse Lane
Red Hook, NY 12571 USA
Phone: 845-758-0400
Fax: 845-758-2633
Email: curran@proceedings.com
Web: www.proceedings.com

TABLE OF CONTENTS

PART 1

PART 2

EPJ Web of Conferences **117**, 07004 (2016) DOI: 10.1051/epjconf/201611707004
NN2015

Constraints on the asymmetric equation of state from heavy-ion collisions

SHERRY YENNELLO, ALAN MCINTOSH and LAUREN HEILBORN

Department of Chemistry & Cyclotron Institute, Texas A&M University, College Station, TX, USA

Abstract

Nuclear matter is one of the most fascinating materials that exists. Therefore elucidating the equation-of-state of nuclear matter is a fundamentally interesting question. Additionally, the nuclear equation-of-state has impacts on astrophysical observables. One important means of constraining the nuclear equation-of-state is through studying heavy-ion collisions. Nuclear material has two components - neutrons and protons - the ratio of which impacts the equation-of-state. Measurements of fragments emitted from reactions of nuclei with different ratios of neutrons and protons - and comparison to simulations based on various underlying interactions - have placed constraints on both the symmetric and asymmetric parts of the equation of state.

1 Introduction

The equation-of-state for nuclear matter (nEOS) is important for many astrophysical processes and vital to answer the question of how the elements are formed. At the heart of the question is what is the isospin dependent nucleon-nucleon interaction, and how does it get modified when it is in the presence of other nucleons. This interaction affects the neutron star mass-radius relationship and the cooling rate of neutron stars. [1, 2] However the underlying interaction is not directly input to the astrophysical calculations, but rather the nEOS - which is derived from the comparison of observables to models with various interactions - is the input to the astrophysical calculations. Recent measurements of 2 solar mass neutron stars have challenged many of the previously accepted nEOS. [3] Additionally, the recent conjecture that the radius of neutron stars around 9 km requires a much denser core of the neutron star and opens the discussion of deconfined nuclear matter or strangeness that could soften the nEOS at these high densities.

Nuclear matter is one of the most fascinating materials that exists. The constituent nucleons are fermions with a size on the order of a femtometer. The nucleon-nucleon interaction is attractive, but has a hard core repulsion resulting in an energy minimum at a specific distance leading to a preferred, or saturation, density. As the system is moved away from the saturation density an energy cost must be paid. This energy cost can be quantified as the incompressibility of nuclear matter. The strong interaction manifests itself in nuclei in terms of the neutron skin, the ground state masses and the saturation density. Additionally it controls nuclear reaction mechanisms such as multifragmentation, transverse collective flow and transport leading to equilibration. Nuclear matter has often been described as a quantum liquid. At high temperatures there is a transition from a liquid to a gas. At very high temperatures there is a further transition to a deconfined quark gluon plasma. At very low densities clustering occurs. Over a range of densities various interesting shapes of so called nuclear pasta are preferred. These various behaviors are a result of the nEOS, therefore elucidating that nEOS is a fundamentally interesting question. Yet whether we are describing the neutron skin thickness on a single nucleus or the density in a neutron star - which vary in size by many orders of magnitude - one needs a common nEOS. An important mechanism for constraining the nEOS is through studying femtonovae in the lab.

Giant resonance measurements have given us the best constraints on the nEOS for "normal" nuclear material, ie nuclei that are near the valley of stability, relatively cold and not far from saturation density. The incompressibility of nuclear matter, K_{NM}, is related to the excitation of the GMR of finite nuclei, but one must account for surface, Coulomb and asymmetry contributions. Measurements taken over a variety of stable nuclei have constrained the incompressibility near the valley of stability. In 1999, measurements of the ISGMRs [4] were compared to HF-based RPA calculations which used the Gogny interaction and took into account pairing and anharmonicity corrections, and a value of $K_{NM} \sim 230$ MeV was obtained. The measurement of the GMR for nuclei with increasing asymmetry can help refine the K_{NM}. [5] Moving away from saturation density, the data from the measurement of transverse collective flow has been used to constrain

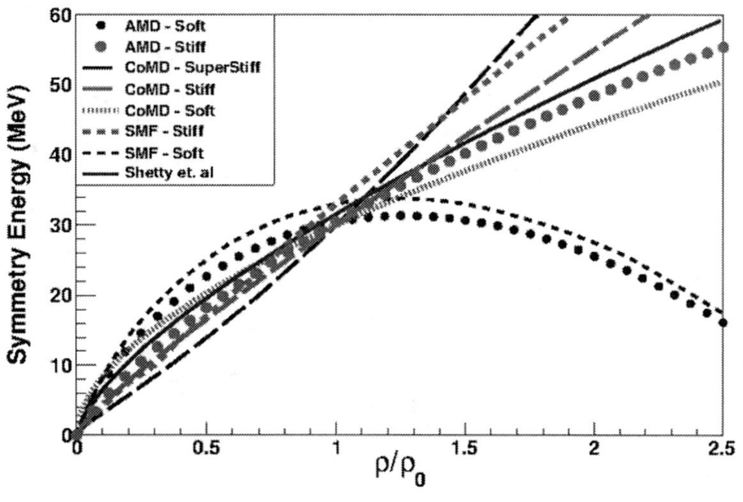

Figure 1: Various forms of the density dependence of the symmetry energy as implemented in the AMD, CoMD and SMF models. Additionally shown is the experimental data from Shetty. [7]

the nEOS at up to $\rho/\rho_0 = 1.5$. [6] However this data is unable to address what happens away from the valley of stability. As one moves away from the valley of stability to more neutron-rich systems the symmetry energy must be taken into account in the nEOS. The magnitude and shape of the symmetry energy depends on the underlying interaction. Many different curves result from different model calculation as shown in figure 1. These are often broadly classified into "soft", where the symmetry energy versus density curve is convex and turns over at high density, and "stiff", where the symmetry energy continues to increase as a function of density. Since these are relative terms it is more accurate to specify value, slope and curvature at saturation density. While the symmetry energy is often characterized by the behavior near saturation density it must be recognized that there could be a softening at higher density or a flattening of this curve at lower density. This paper will discuss the progress that has been made on constraining the asymmetric part of the nEOS away from saturation density using heavy-ion collisions.

2 Constraints on nEOS from isoscaling

The nEOS directly impacts the fragmentation of hot nuclei. When examining the breakup of two equilibrated nuclear systems that are nearly identical in all aspects except for the neutron-to-proton ratio one can extract a measure of the nuclear symmetry energy through isoscaling. The work of Tsang [8] demonstrated that scaling appears when you compare the ratios of fragment yields between a neutron-rich and a neutron-poor system. Ono used this technique to analyze the fragments produced from AMD calculations for two underlying interactions and proposed the relationship as the first observable that could constrain the nEOS. [9] This calculation was compared with experimental data to put forth the first constraints on the symmetry energy of the nEOS from heavy-ion collisions. [10]

Such indirect measurements depend on comparison to theoretical calculations in order to extract the symmetry energy. This technique was used with experimental data as well as statistical calculations and dynamical calculations to map how the symmetry energy changes as a function of excitation energy. A decrease in symmetry energy as a function of excitation energy has been linked with a decrease in the density. [11] In order to extract a meaningful value of symmetry energy from isoscaling one must accurately know the $(N-Z)/A$ of the sources and the temperature of the system. Recent work by McIntosh [12–15] has demonstrated that there is an $(N-Z)/A$ dependence in the nuclear caloric curve. One also must take into account any secondary cooling or experimental filters that are needed to map the observed fragments to primary fragments. In order to better define the source Wuenschel [16] and Galanopolous [17] developed the technique of source reconstruction of the fragmenting quasiprojectile. Another technique for connecting the fragment yield ratios to the symmetry energy is through the Landau expansion of the free energy near the critical point using mirror nuclei. This work extracts a symmetry energy versus excitation energy that is consistent with that extracted from isoscaling. [18, 19] Isoscaling of data from many experiments and theoretical simulations has provided initial constraints on the symmetry energy of the nEOS. These constraints have been augmented by measurements of other observables, such as equilibration and fragment flow.

3 Nucleon transport as measured by isospin equilibration

The amount of mixing of nucleons between two reacting systems is a result of the underlying nucleon transport, which is dependent on the symmetry energy of the nEOS. Early measurements using isotopically identified fragments from collisions of isobaric nuclei were used to measure the onset of isospin non-equilibration. [20, 21] At higher energies this non-equilibration was quantified through development of the isospin tracer method. [22] The underlying nucleon transport was broken down into drift and diffusion by Shi [23]and the diffusion coefficient was connected to the symmetry potential. Tsang used the tracer method to measure isospin diffusion and through comparison to theoretical calculations determined the slope of the symmetry energy near saturation density to be L=88±25 MeV.

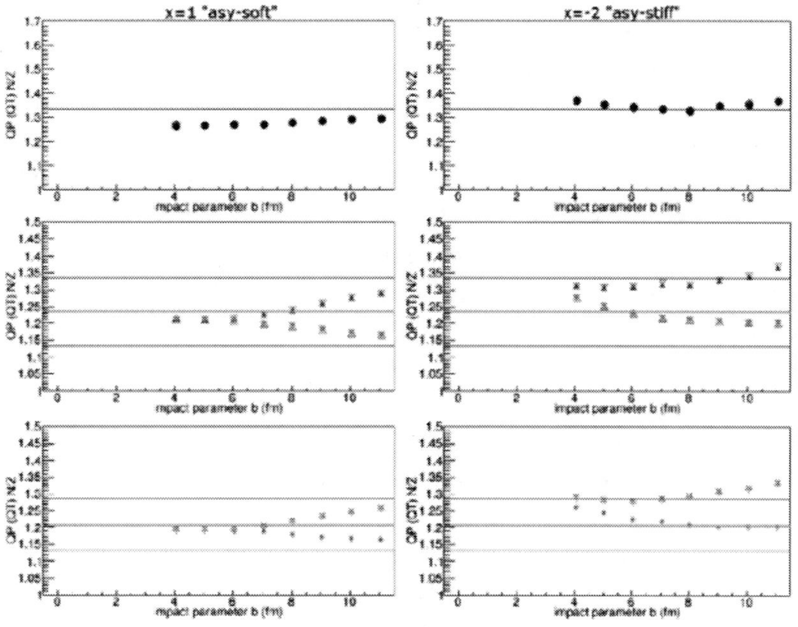

Figure 2: QP (solid symbol) and QT (open symbol) N/Z values as a function of impact parameter. The left and right side correspond to the "asy-soft" and "asy-stiff" parameterizations of the symmetry energy, respectively. Top row: ^{70}Zn+^{70}Zn (black circles) QP and QT N/Z values compared to the projectile/ target/composite system N/Z value (black line). Middle row: ^{70}Zn+^{64}Zn (red triangles) and ^{64}Zn+^{70}Zn (green inverted triangles) QP and QT N/Z values compared to the ^{70}Zn (black line), ^{64}Zn (dark blue line) and composite system (light green thick line) N/Z values. Bottom row: ^{64}Zn+^{64}Ni (pink stars) and ^{64}Ni+^{64}Zn (light blue diamonds) QP and QT N/Z values compared to the ^{64}Ni (brown line), ^{64}Zn (blue line) and composite system (green thick line) N/Z values.

Recent calculations [24, 25] show that equilibration as a function of impact parameter is achieved faster with an "asy-soft" nEOS than an "asy-stiff" nEOS. For example figure 2 shows the asymmetry (N/Z) of the quasiprojectile (QP) and the quasitarget (QT) from a series of iBUU calculations using either an "asy-soft" or an "asy-stiff" EOS. The top panels are for the symmetric ^{70}Zn+^{70}Zn reaction, so the results for the QP and QT are nearly identical. The middle panel shows the equilibration that takes place in the charge symmetric ^{70}Zn+^{64}Zn (and ^{64}Zn+^{70}Zn) systems. The bottom row represents the mass symmetric systems of ^{64}Zn+^{64}Ni and ^{64}Ni+^{64}Zn. As the impact parameter changes from a very peripheral interaction, with small contact time, to mid impact parameter collisions, representing reactions with increasing contact time between the QP and QT, the nucleon exchange results in the difference in the asymmetry of the reacting partners decreasing. This can been seen for both the asy-soft and the asy-stiff calculations, but the equilibration is much faster for the asy-soft interactions. The asy-stiff interaction also results in a more neutron rich QT and QP. The asy-soft interaction reaches equilibration at a value of N/Z that is just slightly lower than that of the composite system, while the asy-stiff interaction approaches an equilibration point that is between the composite system and the N/Z of the more neutron-rich reaction partner. Additionally it has been seen that equilibration is faster for the lower beam energies and for momentum independent interaction. [25]

Various measurements [25–30] have demonstrated migration of neutrons to the neck region and an enhancement of the asymmetry of light clusters at mid rapidity due to nucleon exchange. In general when these measurements have been compared to models they tend to agree better with asy-stiff parametrizations of the nEOS. While there were various measurements of asymmetry of the charged particles that came from the QP, and reconstruction of the QP by measurement of all the charged particles coming from the QP, Kohley made a direct measure of the QP using the MoNA-Lisa-Sweeper setup. In comparison with CoMD calculations he was able to say the data agreed best with an L=78. [31] In a study of the N/Z effects in competition between binary reactions and complete fusion a higher probablility for fusion was found using neutron-rich systems. This same data in comparison with CoMD agrees better with a "stiff" symmetry energy. [32]

Equilibration of the N/Z asymmetry can also be measured by the breakup of a deformed quasiprojectile as done by Brown. [33] The average asymmetry as a function of the rotation angle allows one to measure the equilibration as a function of time. There is a clear decrease in the asymmetry of the light fragment as a function of time. [33, 34] Recent analysis was performed on data from the ^{70}Zn+^{70}Zn reaction at 35 MeV/nucleon using the NIMROD array. Events were selected where at least two mass identified fragments were detected. The events were sorted based on proton number followed by atomic number, with the heaviest fragment designated as Z_H and the second heaviest as Z_L. In order to ensure the fragments cor-

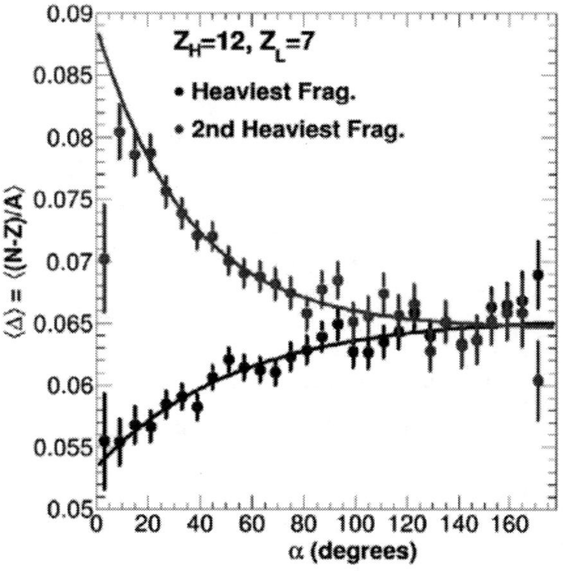

Figure 3: The (N-Z)/A of the heaviest fragment and the second heaviest fragment from an excited QP as a function of the degree of rotation of the system, a measure of its equilibration time.

responded to a large fraction of the initial projectile, events were required to have Z≥7 for Z_H and Z≥3 for Z_L. A total Z cut of minimum Z=21 (70% of beam) and maximum Z=32 was also included. The angular distributions for pairing of Z_H=12 and Z_L=7 are presented in figure 3. A large difference between the N/Z of the Z_H and the asymmetry of the Z_L at low rotation angle is consistent with previous findings, indicating a timescale of dynamical binary splitting of the excited projectile-like fragment much shorter than its rotational period. As the rotation angle increases there is clear equilibration of the asymmetry between the heavy fragment and the light fragment. This data is currently being analyzed to extract the rate constant for equilibration. Since calculations have demonstrated that the equilibration time is sensitive to the symmetry energy of the nEOS and now we have a mechanism to sensitively measure the equilibration as a function of time future work on isospin equilibration may help to further constrain the symmetry energy of the nEOS.

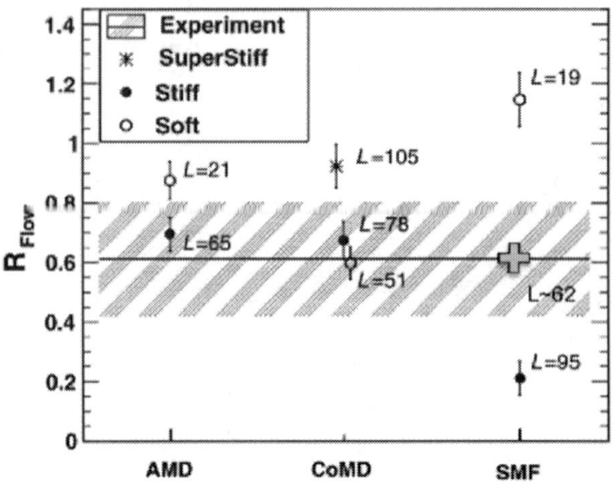

Figure 4: R_{flow} value from mid-peripheral reactions for various symmetry energy parameterizations of the AMD, CoMD, and SMF models. The corresponding symmetry energy slope (L) is shown next to each calculation. The experimental value is represented by the solid blue line with the statistical uncertainty shown as the hatched blue area.

4 Intermediate mass fragment flow

Another method of assessing the symmetry energy in the nEOS is measurement of the transverse collective flow of light charged particles and intermediate mass fragments. Work by Kohley compared the transverse collective flow of IMFs for mid-peripheral reactions in the 35 MeV/nucleon $^{70}Zn+^{70}Zn$, $^{64}Zn+^{64}Zn$, and $^{64}Ni+^{64}Ni$ systems to AMD, CoMD and SMF model predictions with various parameterizations of the nEOS. Using the mid impact parameter events - when the flow signatures are largest - a comparison was performed to AMD, CoMD and SMF as shown in figure 4. For all calculations agreement was better with an asy-stiff parametrization rather than an asy-soft one. All calculations agree with an L of about 60. [35] This same data showed a better agreement with the AMD asy-stiff parametrization when compared over a variety of impact parameters. [36]

5 Future directions

There have been many proposed observables and various experiments and theoretical simulations used to explore the symmetry energy in the nEOS. We need to understand the differences in the various model predictions and continue to explore observables or combinations of observables with increased power to discriminate between various theoretical predictions. Higher order correlation of variables can provide more robust signatures. [37] Additionally, higher energy collisions can provide constraints at higher densities. [38–40] While this has all been done in the realm of neutron and protons as the relevant elemental particles, recent theoretical work by Di-Toro [41] suggests that for very isospin asymmetric matter we can expect a much earlier transition to the quark phase. This may help to explain the recent neutron star data that requires a softening of the nEOS for the core in order to explain the measurements of neutron star radii.

Acknowledgements

This work was supported in part by the US Department of Energy under grant DE-FG02-93ER40773 and the Robert A. Welch Foundation under grant A-1266.

References

[1] J.M. Lattimer, M. Prakash, Science **304**, 536 (2004)

[2] J.M. Lattimer, Ann. Rev. Nucl. Part. Sci. **62**, 485 (2012)

[3] P. Demorest et al., Nature **467**, 1081 (2010)

[4] D.H. Youngblood et al., Phys. Rev. Lett. **82**, 691 (1999)

[5] E. Khan et al., Phys. Rev. Lett. **109**, 092502 (2012)

[6] P. Danielewicz, R. Lacey, W.G. Lynch, Science **298**, 1592 (2002)

[7] S.J.Y. D. V. Shetty, G.A. Souliotis, Phys. Rev. C **75**, 034602 (2007)

[8] M.B. Tsang et al., Phys. Rev. Lett. **92**, 062701 (2004)

[9] A. Ono et al., Phys. Rev. C **68**, 051601(R) (2003)

[10] D.V. Shetty et al., Phys. Rev. C **70**, 011601(R) (2004)

[11] D.V. Shetty, S.J. Yennello, G.A. Souliotis, Phys. Rev. C **76**, 024606 (2007)

[12] A.B. McIntosh et al., Phys. Lett. B **719**, 337 (2013)

[13] A.B. McIntosh et al., Phys. Rev. C **87**, 034617 (2013)

[14] A.B. McIntosh et al., EPJA **50**, 35 (2014)

[15] A.B. McIntosh et al., *Progress in Constraining the Asymmetry Dependence of the Nuclear Caloric Curve*, in *Proceedings of the NN2015 Conference* (2015)

[16] S. Wuenschel et al., Phys. Rev. C **79**, 061602(R) (2009)

[17] S. Galanopoulos et al., Nucl. Phys. A **837**, 145 (2010)

[18] R. Tripathi et al., Phys. Rev. C **83**, 054609) (2011)

[19] J. Mabiala et al., Int. J. Mod. Phys. E **22**, 1350090 (2013)

[20] S. Yennello et al., Phys. Lett B. **15**, 321 (1994)

[21] H. Johnston et al., Phys. Lett. B **371**, 187 (1996)

[22] F. Rami et al., Phys. Rev. Lett. **84**, 1116 (2000)

[23] L. Shi, P. Danielewicz, Phys. Rev. C **68**, 064604 (2003)

[24] L. May, Ph.D. thesis, Texas A&M Univ. (2015)

[25] Rizzo et al., Nucl. Phys. A **806**, 79 (2008)

[26] D. Theriault et al., Phys. Rev. C **74**, 051602(R) (2006)

[27] E.D. Filippo et al., Acta Physica Polonica B **37**, 199 (2006)

[28] P. Russotto et al., Procs of IWM 05 **105**, 13 (2006)

[29] E.D. Filippo et al., Phys. Rev. C **86**, 014610 (2012)

[30] Galichet et al., Phys. Rev. C (2010)

[31] Z. Kohley et al., Phys. Rev. C **88**, 041601 (2013)

[32] G. Cardella et al., Phys. Rev. C **85**, 084609 (2012)

[33] K. Brown et al., Phys. Rev. C **87**, 061601 (2013)

[34] K. Stiefel et al., Phys. Rev. C **90**, 061605 (2014)

[35] Z. Kohley et al., Phys. Rev. C **85**, 064605 (2012)

[36] Z. Kohley et al., Phys. Rev. C **86**, 044605 (2012)

[37] P. Cammarata et al., Nucl. Instr. Meth. A **761**, 1 (2014)

[38] P. Russotto et al., JoP conference series **420**, 012092 (2013)

[39] G. Jhang et al., ARIS conference proceedings (2014)

[40] R. Shane et al., Nucl. Instr. Meth. A **784**, 512 (2015)

[41] W. DiToro et al., Phys. Rev. C **83**, 014911 (2011)

EPJ Web of Conferences **117**, 07005 (2016) DOI: 10.1051/epjconf/201611707005
NN2015

Neutron stars: Cosmic laboratories for matter under extreme conditions

Ignazio Bombaci[1,2] and Domenico Logoteta[2]

[1]Dipartimento di Fisica, Università di Pisa, Pisa, Italy
[2]INFN, Sezione di Pisa, Pisa, Italy

Abstract

The true nature and the internal constitution of the compact stars known as neutron stars (NSs) is one of the most fascinating enigma in modern astrophysics. We discuss some of the present models for the internal structure of NSs and the connection with the properties of ultra dense hadronic matter. In particular, we discuss the role of strangeness on the equation of state and the implications of the measurement of 2 solar mass NSs in PSR J1614-2230 and PSR J0348+0432.

Neutron stars (NSs) are incomparable natural laboratories that allow us to investigate the fundamental constituents of matter and their interactions under extreme conditions that cannot be reproduced in terrestrial laboratories. The bulk properties and the internal constitution of NSs primarily depend on the equation of state (EoS) of strong interacting matter [1], *i.e.* on the thermodynamical relation between the matter pressure, energy density and temperature. Determining the correct EoS model describing NS is a fundamental problem of nuclear physics and astrophysics, and a major effort has been made during the last few decades to solve it by measuring different NS properties using the data collected by various generations of X-ray and γ-ray satellites and by ground-based radio telescopes. The rather recent accurate measurement of the masses, $M = 1.97 \pm 0.04\, M_{sun}$ [2] and $M = 2.01 \pm 0.04\, M_{sun}$ [3], of the neutron stars in PSR J1614-2230 and PSR J0348+0432 respectively, has ruled out all the EoS models which cannot support such high values of stellar masses.

Due to the large values of the stellar central density, various particle species and phases of dense matter are expected in NS interiors. Thus different types of "neutron stars" are hypothesized to exist. In the simplest and conservative picture the core of a NS is modeled as an uncharged uniform fluid of neutrons, protons, electrons and muons in equilibrium with respect to the weak interactions (β-stable nuclear matter). These are the so-called nucleon stars. Even in this simplified picture, the microscopic determination of the EoS from the underling nuclear interactions remains a formidable theoretical problem. In fact, one has to determine the EoS to extreme conditions of high density and high neutron-proton asymmetry, i.e. in a regime where the EoS is poorly constrained by nuclear data and experiments. The nuclear symmetry energy is thus one of the most relevant quantities to control the composition, and the pressure of β-stable nuclear matter [4], and therefore many NS attributes such as the radius, moment of inertia, and crustal properties [5].

Another important issue is related to the role of three-nucleon interactions (TNIs) on the EoS at high density. In fact, it is well known that TNIs are essential to reproduce the experimental binding energy of few-nucleon (A = 3, 4) systems and the empirical saturation point (n_0 = 0.16 fm^{-3}, $(E/A)_0 = -16$ MeV) of symmetric nuclear matter [6]. As shown by various microscopic calculations [7–10] of the EoS of β-stable nuclear matter based on realistic nucleon-nucleon (NN) interactions supplemented with TNI, it is possible to obtain NS sequences with maximum mass $M_{max} > 2\ M_{sun}$ and thus in agreement with presently measured masses. However, the value of M_{max} strongly depends on the strength of the TNI at high density and in addition the central densities for the maximum mass configurations are in the range $n_c(M_{max}) = (6-8)\rho_0$.

At these high densities hyperons are expected among the stellar constituents. The reason for hyperons formation is very simple, and it is mainly due to the fermionic nature of nucleons, which makes the neutron and proton chemical potentials very rapidly increasing functions of the density. As soon as the chemical potential μ_n of neutrons becomes sufficiently large, the most energetic neutrons (i.e. those on the Fermi surface) can decay via the weak interactions into Λ hyperons and form a new Fermi sea for this hadronic species with $\mu_\Lambda = \mu_n$. The Σ^- can be produced via the process $e^- + n \to \Sigma^- + \nu_e$ when the Σ^- chemical potential fulfill the condition $\mu_{\Sigma^-} = \mu_n + \mu_e$ (we consider neutrino-free matter). Other hyperons can be formed with similar weak processes.

To study the influence of hyperons on NS structure, we have performed a Brueckner-Hartree-Fock calculation of the EoS of hyperonic matter using the following interactions: the Argonne v18 (Av18) NN interaction [11]; the TNI used in [8] to reproduce the empirical nuclear matter saturation point; the the Nijmegen ESC08b potential [12] to describe the hyperon-nucleon (NY) interaction. No three-body interactions of the type nucleon-nucleon-hyperon (NNY) and NYY, YYY have been considered.

The chemical potentials μ_i for the different stellar constituents calculated in this scheme are shown in Fig. 1 (upper panel). The onset of Λ (Σ^-) occurs at $n = 0.35$ fm^{-3} (0.64 fm^{-3}), thus at a density well below the central density $n_c^{max} = 1.02$ fm^{-3} for the M_{max} pure nucleonic star calculated within the same approach and the Av18+TNI interaction. The composition of β-stable hyperonic matter is reported in Fig. 1 (lower panel). Notice that at $n = 5\,n_0$ hyperons represent about 43% of the total number of baryons. The effect of hyperons on the EoS is shown in Fig. 2 (upper panels), where we compare the EoS for β-stable pure nucleonic matter (curves Av18+TNI) with that of β-stable hyperonic matter (curves Av18+TNI+ESC08b). As we can see the presence of hyperons produces a significant reduction of the pressure of the system (upper right panel in Fig. 2). As a consequence, solving the relativistic stellar structure equations, we find an impressive decrease of the stellar maximum mass from $M_{max} = 2.28$ M$_{\rm sun}$ to $M_{max} = 1.38$ M$_{\rm sun}$ when hyerons are included among the stellar constituents. The prediction of a value for $M_{max} < 2\ M_{sun}$ is a common feature of many present calculations

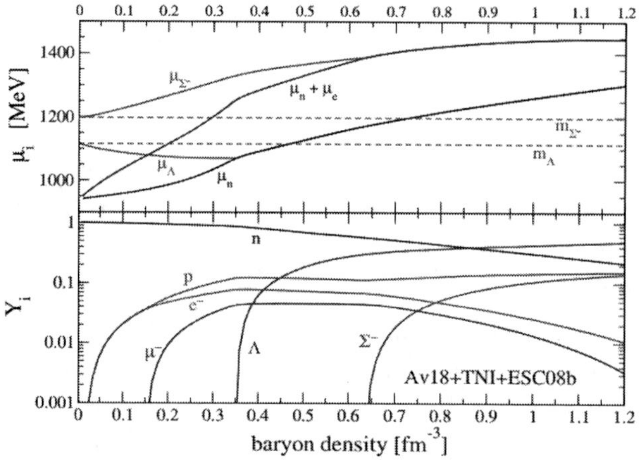

Figure 1: Chemical potentials (upper panel) and concentrations (lower panel) of the different stellar constituents in β-stable hyperonic matter as a function of the total baryon density.

[10, 13–15] based on microscopic hyperonic matter EoSs.

Thus, on the one hand the presence of hyperons seems unavoidable, on the other hand their presence results in a stellar maximum mass not compatible with measured NS masses. This baffling problem is known as the "hyperon puzzle" in NSs. Clearly, one should try to trace back the origin of this problem to the underlying NY and YY two-body interactions or to the possible repulsive NNY, NYY and YYY three-body interactions. Presently, this is a very active research field. Thus, the use of microscopic EoS of hyperonic matter in the contest of NS physics is of fundamental

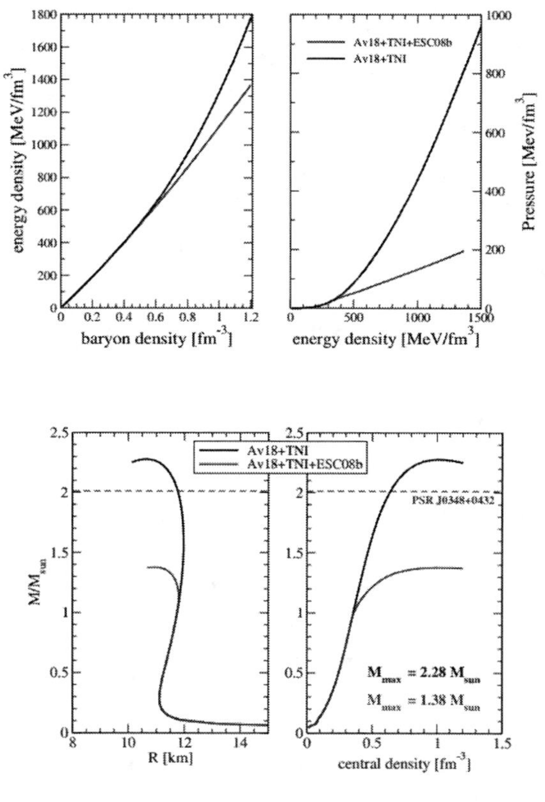

Figure 2: Upper panels: EoS of β-stable matter as a function of the total baryon density. The upper (lower) curves refer to the case of nuclear (hyperonic) matter. Lower panels: gravitational mass as a function of the stellar radius (lower left panel) and of the central baryon density (lower right panel) in the case of nucleon stars (upper curves) and hyperon stars (lower curves). The dashed horizontal line represents the measured mass $M = 2.01 \pm 0.04\, M_{sun}$ [3] of PSR J0348+0432. The mass of the Sun is denoted as $M_{sun} = 1.99 \times 10^{33}$ g.

importance for the understanding of strong interactions involving hyperons, and to learn how these interactions behave in a dense many-body system.

The core of a massive NS is one of the best candidates in the Universe where a transition from a phase where quarks are confined within baryons and mesons ("hadronic matter") to a quark deconfined phase ("quark matter") could occur. In the low temperature T and high baryon chemical potential region of the QCD phase diagram (which is the one relevant for NS physics) several QCD inspired models suggest the deconfinement transition to be a first-order phase transition [16]. As it is well known, all first order phase transitions are triggered by the nucleation of a critical size drop of the new (stable) phase in a metastable mother phase. This is a very common phenomenon in nature (*e.g.* fog or dew formation in supersaturated vapor, ice formation in supercooled water).

One of the most exciting astrophysical consequences of the nucleation process of quark matter (QM) in the core of massive hadronic stars (HSs) (*i.e.* NSs in which no fraction of QM is present) is that above a threshold value of their mass, HSs are metastable [17–19] to the "decay" (conversion) to quark stars (QSs) (*i.e.* to hybrid stars or to strange stars). This stellar conversion process liberates a huge amount of energy (a few 10^{53} erg) and it could be the energy source of Gamma Ray Bursts (GRBs) [20]. In addition, within this scenario, one has two coexisting families of compact stars in the Universe: pure hadronic stars and quark stars [17, 18]. The members of these two families could have similar values for their gravitational masses but different values for their radii [18].

The metastability of HSs originates from the finite size effects (which represents the driving "force" of first order phase transitions) in the formation process of the first QM drop in the hadronic environment.

In cold ($T = 0$) bulk matter the deconfinement transition takes place at *static transition point* defined by the Gibbs' criterion for phase equilibrium

$$\mu_H = \mu_Q \equiv \mu_0\,, \qquad P_H(\mu_0) = P_Q(\mu_0) \equiv P_0 \tag{1}$$

where $\mu_H = (\varepsilon_H + P_H)/n_H$ and $\mu_Q = (\varepsilon_Q + P_Q)/n_Q$ are the Gibbs energies per baryon (average chemical potentials) for the hadron and quark phase respectively, ε_H (ε_Q), P_H (P_Q) and n_H (n_Q) denote respectively the total (*i.e.*, including leptonic contributions) energy density, the total pressure and baryon number density for the hadron (quark) phase.

Consider now the more realistic situation in which one takes into account the energy cost due to finite size effects in creating a drop of deconfined QM in the hadronic environment. As a consequence of these effects, the formation of a critical-size drop of QM is not immediate and it is necessary to have an overpressure $\Delta P = P - P_0$ with respect to the static transition point. Thus, above P_0, hadronic matter is in a metastable state, and the formation of a real drop of QM occurs via a quantum nucleation mechanism. Thus, a HS having a central pressure larger than P_0 is metastable with respect to the conversion to a QS. These metastable HSs have a *mean-life time* which is related to the nucleation time to form the first critical-size drop of deconfined matter in their interior [1]. The *critical mass* M_{cr} of the metastable HS is defined [17, 18] as the value of the gravitational mass for which the nucleation time is equal to one year: $M_{cr} \equiv M_{HS}(\tau{=}1\,\mathrm{yr})$. Pure hadronic stars with $M_H > M_{cr}$ are very unlikely to be observed. M_{cr} plays the role of an *effective maximum mass* for the hadronic branch of compact stars [18]. While the Oppenheimer–Volkov maximum mass $M_{HS,max}$ is determined by the overall stiffness of the EOS for hadronic matter, the value of M_{cr} will depend in addition on the bulk properties of the EOS for quark matter and on the properties at the interface between the confined and deconfined phases of matter (*e.g.*, the surface tension σ).

These findings are exemplified in Fig. 3, where we show the mass-radius (MR) curve for hadronic stars (HS) and that for quark stars (QS). The configuration marked with an asterisk on the hadronic MR curve represents the HS for which the central pressure is equal to P_0 and thus the nucleation time $\tau_q = \infty$. The full circle on the HS sequence represents the critical mass configuration M_{cr}, whereas the full circle on the QS curve represents the QS which is formed from the conversion of the HS with $M^{HS} = M_{cr}$. As we can see, for the EoS parametrizations used in the calculations reported in Fig. 3 (see figure caption for informations on the EoS parametrizations), PSR B1913+16 (which has a mass $M = 1.4398 \pm 0.0002\, M_{sun}$) can be interpreted as a pure HS, whereas PSR J1614-2230 is more likely a QS.

We assume [20] that during the stellar conversion process the total number of baryons in the star (or in other words the stellar baryonic mass M_B) is conserved. Thus the total energy liberated in the stellar conversion is given by [20] the difference between the critical mass HS (M_{cr}) and that of the final QS (M_{fin}) configuration with the same baryonic mass: $E_{conv} = (M_{in} - M_{fin})c^2$. It has been shown [17, 18, 20] that $E_{conv} = 0.5-4.0 \times 10^{53}$ erg. This huge amount of released energy will cause a powerful neutrino burst, likely accompanied by intense gravitational waves emission,

[1] The actual *mean-life time* of the HS will depend on the mass accretion or on the spin-down rate which modifies the nucleation time via an explicit time dependence of the stellar central pressure.

and conceivably it could cause a second delayed (with respect to the supernova explosion) explosion. Under favorable physical conditions this second explosion could be the energy source of a powerful GRB [17, 20]. Thus this scenario is able to explain a "delayed" connection between supernova explosions and GRBs.

The stellar conversion process, described so far, will start to populate the new branch of quark stars, *i.e.* the part of the QS sequence above the full circle (see Fig. 3). Long term accretion on the QS can next produce stars with masses up to the maximum mass M^{QS}_{max} for the quark star configurations. Thus within this scenario one has two coexisting families of compact stars: HSs and QSs [18]. The quark star branch is occasionally referred to as the "third family" of compact stars, considering white dwarfs as the first family and HSs as the second family. Notice also that there is a range of values of stellar gravitational mass (see Fig. 3) where HSs and QSs with the

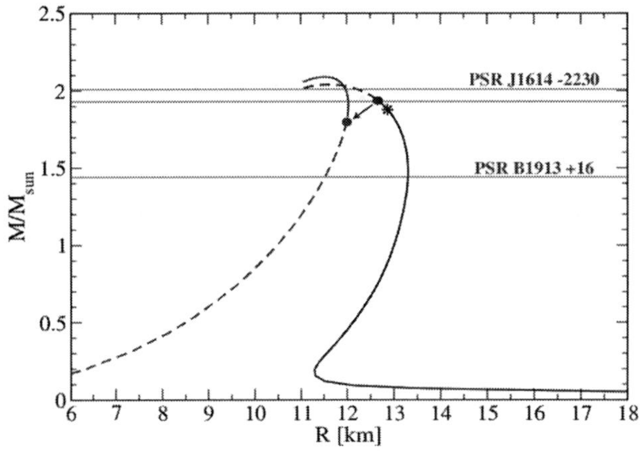

Figure 3: Mass-radius relation for hadronic star (HS) and quark stat configurations (QS). The configuration marked with an asterisk represents the HS for which the nucleation time $\tau_q = \infty$ (*i.e.* $P_c = P_0$). The conversion process of the HS, with a gravitational mass equal to the critical mass M_{cr}, into the final QS is denoted by the full circles connected by an arrow. Results are relative to the the GM1 model [21] of hyperonic matter with hyperon-σ meson coupling $x_\sigma = 0.7$, and for strange star configurations with the extended bag model EOS of Ref. [22] with $B_{eff} = 47.2$ MeV/fm^3 and $a_4 = 0.7$. The values of the critical gravitational (baryonic) mass and of the final QS mass are calculated for a surface tension $\sigma = 10$ MeV/fm^2. The lower horizontal line represents the mass $M = 1.4398 \pm 0.0002\, M_{sun}$ of the pulsar PSR B1913+16, whereas the higher horizontal lines represent the mass $M = 1.97 \pm 0.04\, M_{sun}$ of PSR J1614-2230 [2].

same gravitational mass can exist ("twin stars").

References

[1] M. Prakash et al., Phys. Rep. **280** (1997) 1.

[2] P. Demorest, et al., Nature, **467** (2010) 1081.

[3] J. Antoniadis et al., Science, **340** (2013) 1233232.

[4] I. Bombaci and U. Lombardo, Phys. Rev. C **44**, (1991) 1892.

[5] J.M. Lattimer and A.W. Steiner, Eur. Phys, J. A **50** (2014) 40.

[6] D. Logoteta, et al., Phys. Rev. C **91** (2015) 064001.

[7] R.B. Wiringa, V. Fiks, and A. Fabrocini, Phys. Rev. C **38** (1988) 1010.

[8] M. Baldo et al., Astron. Astrophys. **328** (1997) 274.

[9] A. Akmal et al., Phys. Rev. C **58**, (1998) 1804.

[10] Z.H. Li and H.-J. Schulze, Phys. Rev. C **78** (2008) 028801.

[11] R.B. Wiringa et al., Phys. Rev. C **51**, (1995) 38.

[12] Th.A. Rijken et al., Nucl. Phys. A **835** (2010) 160.

[13] M. Baldo et al., Phys. Rev. C **61** (2000) 055801.

[14] I. Vidaña et al., Phys. Rev. C **62** (2000) 035801.

[15] H.-J. Schulze and Th. A. Rijken, Phys. Rev. C **84** (2011) 035801.

[16] S.D.H. Hsu, M. Schwetz, Phys. Lett. **B 432** (1998) 203; Z. Fodor, S.D. Katz, JHEP **04** (2004) 050.

[17] Z. Berezhiani et al., Astrophys. Jour. **586** (2003) 1250.

[18] I. Bombaci, I. Parenti, I. Vidaña, Astrophys. Jour. **614** (2004) 314.

[19] A. Drago, A. Lavagno and G. Pagliara, Phys. Rev. D **69** (2004) 057505.

[20] I. Bombaci and B. Datta, Astrophys. Jour. Lett. **530** (2000) L69.

[21] N.K. Glendenning, S.A. Moszkowski, Phys. Rev. Lett. **67** (1991) 2414.

[22] E. Fraga et al., Phys. Rev. D **63** (2001) 121702(R); S. Weissenborn et al., Astrophys. Jour. **740** (2011) L14.

EPJ Web of Conferences **117**, 07006 (2016) DOI: 10.1051/epjconf/201611707006
NN2015

Excited nuclear matter at Fermi energies: From transport properties to the equation of state

O. Lopez, D. Durand and G. Lehaut
(INDRA Collaboration)

Laboratoire de Physique Corpusculaire, ENSICAEN, Université de Caen Basse Normandie, CNRS/IN2P3, F-14050 Caen Cedex, France

Abstract

Properties of excited nuclear matter are one of the main subject of investigation in Nuclear Physics. Indeed, the response of nuclear matter under extreme conditions encountered in heavy-ion induced reactions (large compression, thermal and collective excitations, isopin diffusion) around the Fermi energy is strongly needed when studying the nuclear equation of state and the underlying in-medium properties concerning the nuclear interaction. In this contribution, we will present some experimental results concerning the transport properties of nuclear matter, focusing specifically on the determination of in-medium quantities such as mean free pathes and nucleon-nucleon cross sections around the Fermi energy. We will see that, in this specific energy range, energy and isospin dissipations exhibit very peculiar features, such as the crossover between 1-body to 2-body dissipation regimes corresponding to the transition between the nuclear response from Mean-Field to the nucleonic response through the appearance of nucleon-nucleon collisions.

1 Introduction

The study of transport phenomena are of primary importance for understanding the fundamental properties of nuclear matter [1]. They are critical in the description of the supernova collapse and the formation of a neutron star [2], hence they could provide important insights for the determination of the nuclear equation of state. Transport properties are also one of the basic ingredients for microscopic models [3, 4]. In this study, we are looking at the global energy dissipation achieved in heavy-ion induced reactions in the Fermi energy domain. We are using the large experimental dataset available in this energy range for symmetric systems recorded with the 4π array *INDRA* [5]. We are specifically looking at central collisions, *i.e.* collisions corresponding to the maximal overlap and thus leading to the maximal dissipation. Doing so, we will extract information concerning the stopping encountered in such collisions and relate it to the nucleon mean free path and cross section in the nuclear medium.

2 Stopping for central collisions

In this work, we are studying *INDRA* data coming from central collisions, selected through the total multiplicity of charged particle as proposed in [6]. The data span a large body of symmetric systems, with total mass between 70 and 500 mass units, and incident energies covering the full range of the Fermi domain $E_{inc}/A = 10 - 100\ MeV$. To evaluate the degree of stopping achieved in central collisions, we are using the energy isotropy ratio R_E as defined in [6]. This ratio is taking positive values, with 1 for an isotropic emission. We only consider $Z = 1$ particles (p, d, t) in this study since we are mainly interested in the properties for nucleon-nucleon collisions. We find here higher R_E values than those of [6]. Indeed, the isotropy ratio in [6] was computed by using all charged products, including fragments. It has be shown in [7] that clusterization effects on nucleon phase space lower the isotropy ratio, mainly below $100A\ MeV$. We display in Fig. 1 the mean isotropy ratio $< R_E >$ calculated for central collisions as a function of incident energy. It is worthwhile to note that in case of asymmetric systems, the incident energy should be conveniently replaced by the total available cm energy.

The different curves on Fig. 1 correspond to several analytical calculations associated to the entrance channel; we consider here 2 Fermi spheres separated by the relative momentum $(1-\alpha)p_{rel}$ corresponding to the relative energy of the entrance channel. α is a free parameter which represents the level of dissipation observed in the collision; $\alpha = 1$ (red curve) means full dissipation while $\alpha = 0$ (thick black curve) corresponds to the *sudden* approximation without any dissipation. The displayed error bars correspond

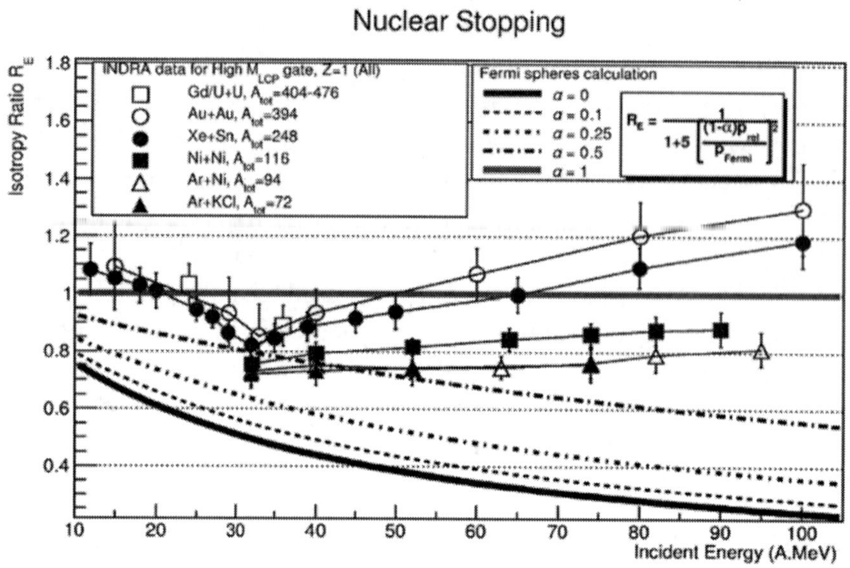

Figure 1: Mean Isotropy ratio $< R_E >$ as a function of incident energy for central collisions in symmetric systems recorded with *INDRA* (symbols). The error bars correspond to statistical + systematic errors.

to statistical supplemented by an estimation of systematic errors coming from the determination of the mean value [6]. We observe that the closest distance to the *no-dissipation* scenario ($\alpha = 0$) corresponds to an incident energy slightly below the Fermi energy, around $30A\ MeV$. Above the Fermi energy, in the nucleonic regime, we can observe a nice mass hierarchy concerning the stopping: the heavier the system, the larger the stopping is. This is fully consistent with a *Glauber* picture where the degree of stopping is closely related to the number of participants. To get further on this point, we have performed Monte Carlo simulations of 2 Fermi spheres, taking into account nucleon-nucleon collisions through the definition of the nucleon mean free path in the nuclear medium λ_{NN}. A simple relationship between the degree of stopping and the nucleon mean free path can then be obtained [8].

3 Nucleon mean free path in nuclear medium

By applying this to the experimental data, we obtain the results displayed by Fig. 2. We note that all curves exhibit the same behaviour whatever is the total mass. We observe a maximum at incident energy corresponding to minimal stopping at $E_{inc} \approx 30A\ MeV$ as can be expected from Fig. 1. The non-physical decrease at low incident energy (below $30A\ MeV$) can be attributed to the Mean-Field dissipation which is well active in this energy domain, and consequently affects the entrance channel values; in this domain, we should adjust the α parameter to better describe the energy dissipation in the entrance channel. In the limited scope of the present study, we are mainly interested on the nucleonic regime, *i.e.* incident energy larger than the Fermi energy. Focusing on this incident energy domain, we observe a decrease from $\lambda_{NN} \approx 10\ fm$ to $\lambda_{NN} = 5\ fm$ at $E_{inc}/A \approx 100\ MeV$. It is worthwhile to mention that the value at high incident energy is in agreement with recent microscopic *Dirac-Brueckner-Hartree-Fock* calculations using realistic interactions [9] and hence gives credit to our rather crude determination of the mean free path.

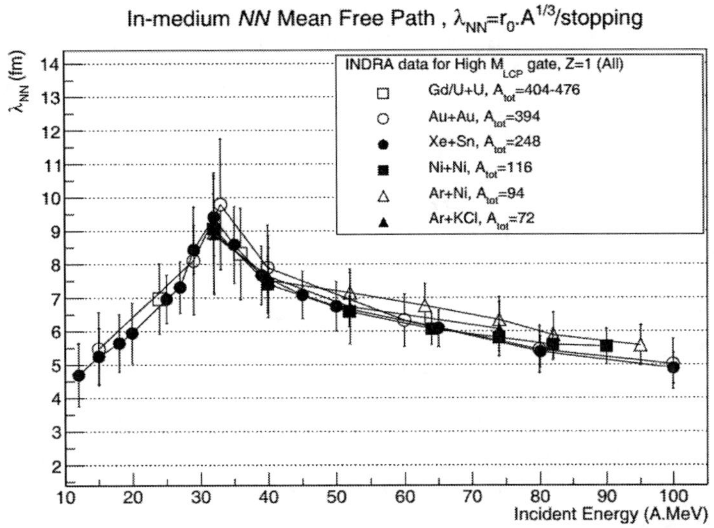

Figure 2: Nucleon mean free path $< \lambda_{NN} >$ as a function of incident energy extracted from experimental data. The errors bars correspond to the statistical+systematic errors.

4 In-medium effects for nucleon-nucleon cross section

From the mean free path, we can estimate the nucleon-nucleon cross section by taking the well-known formula: $\sigma_{NN} = 1/(\rho\lambda_{NN})$ where ρ is the nuclear density which varies linearly from $\rho = \rho_0$ at $E_{inc}/A = 30\ MeV$ to $\rho = 1.5\rho_0$

at $E_{inc}/A = 100\ MeV$ in order to take into account compression effects [1]. To get the *real* nucleon-nucleon cross section, we have then to correct from the *Pauli* blocking as a trivial (but essential) 2-body effect in the medium. To do so, we use the prescription of ref. [10]. More sophisticated approaches concerning the *Pauli* blocking estimation can be found in the litterature [11] but should not change significantly the results. Then, we compute the quenching factor $F = \sigma_{NN}^{in-medium}/\sigma_{NN}^{free}$ by taking the energy-dependent free nucleon-nucleon cross sections σ_{NN}^{free} available in the litterature [12].

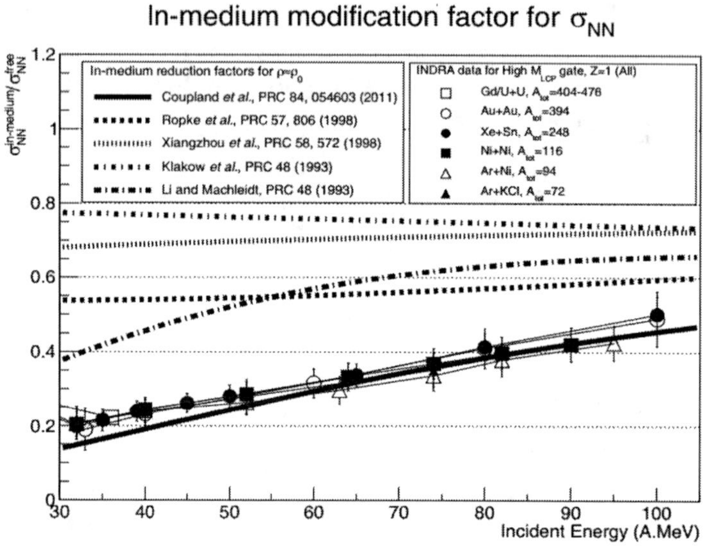

Figure 3: Mean in-medium quenching factor $< F >= \sigma_{NN}^{in-medium}/\sigma_{NN}^{free}$ as a function of incident energy for experimental data (symbols) and various theoretical prescriptions (curves). The errors bars for data correspond to the statistical+systematic errors.

The results for the quenching factor F are displayed on Fig. 3 for the incident energy range above $E_{inc}/A = 30\ MeV$. The 'experimental' values (symbols) are compared to standard theoretical prescriptions derived from various works and currently used in microscopic transport models [13–17]. We have taken the same linear density dependence as mentioned before to account for compressional effects. We can see that the medium effects, out-of-*Pauli* blocking, suppress strongly nucleon-nucleon collisions at low energy with $F \approx 15\%$. We obtain values close to $F = 50\%$ at the highest incident energy, here at $100A\ MeV$. It is interesting to note that all parametrizations and the data seem to converge toward a common value around $60 - 70\%$, indicating that in-medium effects should be quite correctly taken into account above $100A\ MeV$ incident energy. For the low energy domain ($30A - 100A\ MeV$), the parametrizations give quite different results. The best prescription is the one of ref. [17]. It is worthwhile to mention that this parametrization is rather strongly density-dependent and produces large reduction factors compared to other prescriptions [17].

5 Conclusions

In this study, we have studied the degree of stopping connected to the energy dissipation in heavy-ion reactions. We have found a minimal stopping around $E_{inc} = 30A\ MeV$ connected to the crossover between 1-Body to 2-Body dissipation regime. For the latter, we have estimated the nucleon mean free path in the nuclear medium from the degree of stopping achieved in central collisions. The mean free path decreases from $\lambda_{NN} \approx 10\ fm$ at $E_{inc}/A = 30\ MeV$ to $\lambda_{NN} = 5\ fm$ at $E_{inc}/A = 100\ MeV$. These values are in agreement with recent theoretical findings using microscopic approaches. The large value relative to the nuclear size ($\lambda_{NN} > R$) around the Fermi energy suggests that *full thermalization* is not achieved in such central collisions [6]. In-medium effects, namely *Pauli* blocking and *high-order correlations*, have also been evaluated and are found to be large in the Fermi energy range; it is clear that this energy/density dependence of the nucleon-nucleon cross section has to be properly taken into account in any microscopic transport model used in the Fermi energy range. The best agreement for the in-medium factor F is the one proposed in [17], based on simple geometrical arguments. More detailed analyses of experimental data and comparisons with transport models on stopping properties are indeed welcome to confirm these results. As a perpective, the advent of new radioactive beam facilities will indeed provide the opportunity to investigate the isovector properties of the nucleon-nucleon interaction by scanning the isospin degree of freedom concerning mean free pathes and cross-sections.

References

[1] D. Durand, B. Tamain and E. Suraud, *Nuclear Dynamics in the nucleonic regime*, Institute Of Physics, New York (2001) and refs. therein

[2] J.M. Lattimer and M. Prakash, *The Physics of Neutron Stars*, Science **304** (2004) 536

[3] C. Fuchs and H.H. Wolter, Eur. Phys. J. A **30** (2006) 5-21 and refs. therein

[4] A. Andronic *et al.*, Eur. Phys. J. A **30** (2006) 31 and refs. therein

[5] J. Pouthas *et al.*, Nucl. Inst. and Meth. A **357** (1995) 418-442

[6] G. Lehaut *et al.* (INDRA collaboration), Phys. Rev. Lett. **104** (2010) 232701

[7] G. Q. Zhang *et al.*, Physical Review C 84 (2011) 034612

[8] O. Lopez *et al.* (INDRA collaboration), Phys. Rev. C **90** (2014) 064602

[9] A. Rios and V. Som, Phys. Rev. Lett. **108** (2012) 012501

[10] K. Kikuchi and M. Kawai, *Nuclear matter and Nuclear Collisions*, North Holland, New York (1968)

[11] B. Chen, F. Sammarruca and C.A. Bertulani, [nucl-th] arXiv:1304.6096v1

[12] N. Metropolis *et al.*, Phys. Rev. **110** (1958) 204-220

[13] D. Klakow, G. Welke, and W. Bauer, Phys. Rev. C **48** (1993) 1982-1987

[14] G. Q. Li and R. Machleidt, Phys. Rev. C **48** (1993) 1702

[15] A. Schnell, G. Rpke, U. Lombardo, and H.J.Schulze, Phys. Rev. C **57** (1998) 806-810

[16] C. Xiangzhou *et al.*, Phys. Rev. C **58** (1998) 572-575

[17] D. Coupland *et al.*, Phys. Rev. C **84** (2011) 054603

EPJ Web of Conferences **117**, 07007 (2016) DOI: 10.1051/epjconf/201611707007
NN2015

Clusters in neutron-rich light nuclei

D. Jelavić Malenica[1], M. Milin[2], A. Di Pietro[3], P. Figuera[3], M. Lattuada[3], D. Miljanić[1], A. Musumarra[3], M.G. Pellegriti[3], L. Prepolec[1], V. Scuderi[3], N. Skukan[1], N. Soić[1], D. Torresi[3] and M. Uroić[1]

[1] Ruder Bošković Institute, Zagreb, Croatia
[2] Faculty of Science, University of Zagreb, Zagreb, Croatia
[3] INFN - Laboratori Nazionali del Sud and Sezione di Catania, Catania, Italy

Abstract

Due to their high selectivity, transfer and sequential decay reactions are powerful tools for studies of both single particle (nucleon) and cluster states in light nuclei. Their use is particularly simple for investigations of α-particle clustering (because α-particle has $J^{\pi}=0^{+}$, which simplifies spin and parity assignments to observed cluster states), but they are also easily applicable to other types of clustering. Recent results on clustering in neutron-rich isotopes of beryllium, boron and carbon obtained measuring the $^{10}B+^{10}B$ reactions (at 50 and 72 MeV) are presented. The highly efficient and segmented detector systems used, built from 4 Double Sided Silicon Strip Detectors (DSSSD) allowed detection of double and multiple coincidences and, in that way, studies of states populated in transfer reactions, as well as their sequential decay.

Clustering was recognized [1] as an essential ingredient in description of light nuclei already in the early days of nuclear physics, even before the actual formulation of nuclear shell model. A systematic picture of the phenomenon for the α-conjugate nuclei (with $N=Z$) emerged in the late sixties, resulting in insights like *threshold rule* (depicted in Ikeda diagram, [2]) or parity splitting of rotational bands of asymmetric cluster states [3]. Next decades of both theoretical and experimental research established the picture, though number of questions remained open (see *e.g.* [4,5]). In the new millennium the appearance of clustering was theoretically obtained even within *ab initio* calculations - *e.g.* the classical example of clustering, description of the ^{8}Be ground state as two touching α-particles, was nicely reproduced within the Green function Monte Carlo ab initio calculations [6].

Isotopes of light nuclei having additional neutrons (compared to the $N=Z$ ones) show a different type of clustering, with extra neutrons acting as valence particles in molecule-like structures. The best example of such clustering is found among states of beryllium isotopes $^{9-12}$Be [4, 5] - neutrons fill different "molecular orbits" around the existing two-centre (α - α) structure and corresponding states have been identified [7]. The simplest molecular orbits are σ-orbits which possess an increased density along the axis connecting the two cluster centers, and π-orbits in which neutrons occupy a ring orbit perpendicular to that axis and centered on it. Delocalization of neutrons in the two-centered orbits reduces their kinetic energies giving thus a strong contribution to the stability of the structure. The Pauli repulsion effect of neutrons in σ-orbits can furthermore push away two cores to larger distances; such states are predicted [4] to be among the strongest deformed nuclear configurations. For this reason σ-orbits play an important role in stabilization of all linear chain structures. States with neutrons in π- and σ-orbits are successfully theoretically modeled for ^{10}Be within the framework of antisymmetrized molecular dynamics.

This simplified picture is complicated with additional configurations - *e.g.* Ito and Itagaki [8, 9] distinguish between "ionic", "covalent" and "atomic" states in ^{10}Be and ^{12}Be, and further types of states in ^{10}Be are proposed in refs. [10,11]. Recently the molecular states in ^{10}Be are also reproduced within the framework of energy density functionals, without any presumption of clustering [12].

Experimental search for nuclear molecules is ongoing for more than 20 years (*e.g.* [13]), with associate states finally clearly established in ^{10}Be [14, 15] as a rotational band starting with the 0_2^+ state at E_x= 6.18 MeV. The band is confirmed and extended in subsequent experiments [16, 17]. Molecular states are predicted to exist in a number of other nuclei (see ref [4] for a recent overview), but all of them have in common the fact that the valence nucleons are *neutrons*. Due to the Coulomb repulsion, *protons* as valence particles are expected to make the molecules less stable, especially while filling the σ-orbits. The effect is not expected to prohibit the existence of proton molecular states, although in some cases it may make them rather broad and thus hard to detect. The well-known example is the $1/2^+$ state in ^{9}Be (E_x= 1.684 MeV), for which the ^{9}B analog is still not clearly identified.

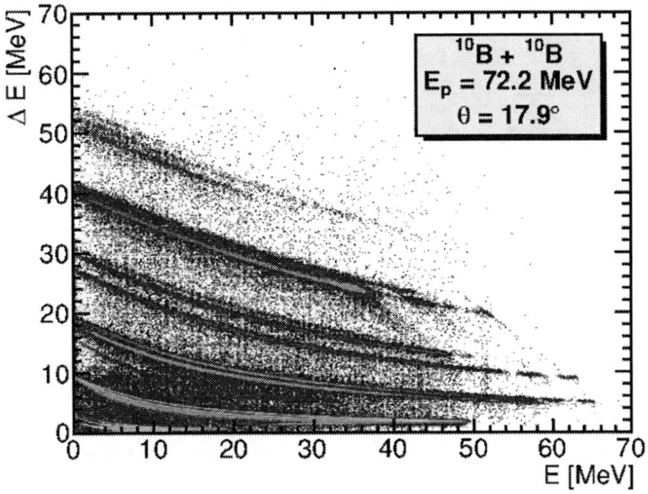

Figure 1: Particle identification spectrum for one of the quadrants of a forward angle telescope. Loci corresponding to different hydrogen, helium, lithium, beryllium, boron and carbon isotopes can be clearly seen.

In order to further study the structure of nuclei in the A=10 mass region, the ^{10}B+^{10}B reactions were measured at beam energies of 50 and 72.2 MeV. The experiment was performed at INFN-LNS, using the SMP Tandem accelerator and targets enriched in ^{10}B up to 99.8%. Reaction products were detected with a highly segmented detector setup covering a large solid angle and allowing for the detection of single events as well as two- and three-particle coincidences. The selectivity of the ^{10}B+^{10}B reactions in populating different states of neighbouring nuclei was studied, together with a sequential decay of states in question.

The complex structure of the low-lying states of the ^{10}B nucleus, which can be described as a mixture of shell model and cluster configurations of the type ^{6}Li$_{gs}$+α or ^{6}Li(0_2^+, 1)+α, together with a high spin of the ground state J$^\pi$ = 3^+, enables population of a range of different high-spin states at high excitation energies. Among excited states of ^{10}B one should also

DOI: 10.1051/epjconf/201611707007

look for molecular states analog to the ^{10}Be ones, but with the neutron and proton pair as valence particles, however due to rather different structure of these states with respect to the ^{10}B ground state, they are not expected to be populated strongly in this experiment.

Detector setup consisted of four ΔE-E silicon telescopes, each composed of thin ΔE detector (57-67 μm), divided into 4 quadrants and thick E DSSSD detector (500 or 1000 μm), divided into 16 strips in both front and back sides.

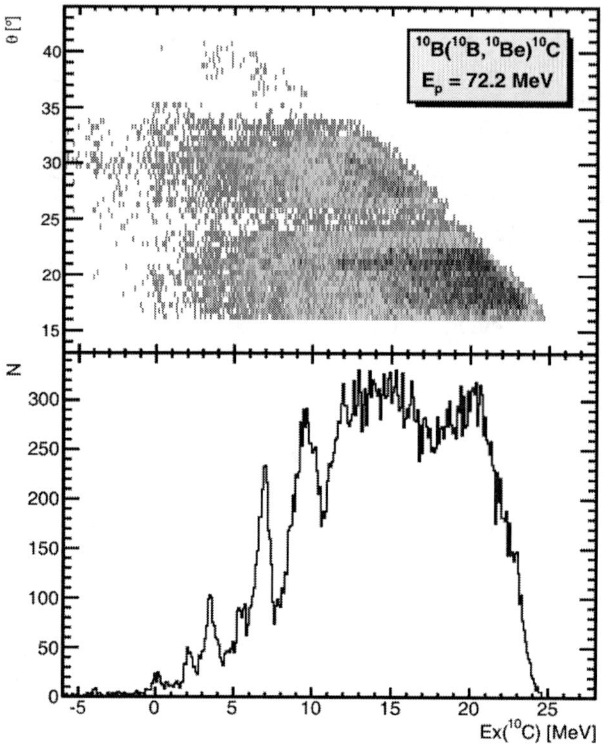

Figure 2: Inclusive spectra of the ^{10}C excitation energy (below) and its angular dependence (above).

Due to the large size of all detectors (50×50 mm^2) and their fine segmentation, telescopes covered rather large solid angle with good angular resolution. This same segmentation makes the analysis of obtained experimental data rather demanding - complications associated with dead layers [18] or interstrip gaps [19,20] are only part of the problem, the very basic one being a calibration of a large number of apparently independent detector channels.

Several novel calibration improvements described in detail in ref. [21] were used in the analysis, as well as a new set of tools that allows an easy and accurate calibration of DSSSD detectors [22].

On both beam energies (50 and 72 MeV) the number of detected α-particles was remarkably higher than number of any other detected nuclei, as can be seen in an example of the particle identification spectra in Fig. 1. This effect was even more pronounced in coincidence measurements, where even triple α-particle coincidences were detected with rather large statistics.

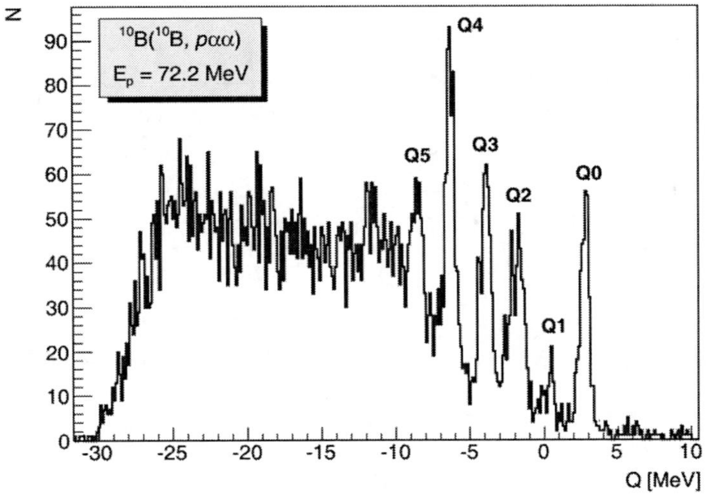

Figure 3: Q-spectrum reconstructed from the triple coincidences α+α+p (see text for details).

Applicability of the used detector set-up (with highly segmented strip detectors) for studies of multiple coincidences is nicely illustrated through the spectrum given in Fig. 3. It shows a Q-spectrum reconstructed from the triple coincidences α+α+p with several additional assumptions: (i) two α-particles have relative energy corresponding to the ground state of ^{8}Be; (ii) the reconstructed ^{8}Be and proton have relative motion corresponding to the ground state of ^{9}B (no other ^{9}B state is clearly seen in the data). With such cuts the spectrum actually shows the ^{11}B states populated in the reaction ^{10}B(^{10}B,^{9}B$_{g.s.}$)^{11}B, which was of course never measured before due to the fact that ^{9}B is particle unbound. The selectively excited ^{11}B states shown in Fig. 3 are the mirror states of the ^{11}C ones seen in the ^{10}B(^{10}B,^{9}Be)^{11}C reaction [22]. In these spectra several states not yet reported in the nucleon transfer reactions on ^{10}B are seen at higher excitation energies.

A number of other double and triple coincidence spectra is obtained in the experiment, yielding new results for ^{10}B, ^{11}B, ^{11}C, ^{12}C, ^{13}C and ^{14}N. As expected, preferably populated are the states at high excitation energies and spins (details will be given in subsequent publications).

Several inclusive spectra from the experiment also show interesting results - one of them is given in Fig. 2 for the ^{10}B(^{10}B,^{10}Be)^{10}C reaction. Several peaks correspond to the known ^{10}C (and ^{10}Be) states, but the peak seen at E_x= 9.45 MeV does not. The energy of 9.45 MeV can be obtained combining the 3.35 excitation in ^{10}C with the ≈6 MeV excitation in ^{10}Be, but since there is no significant peak at ≈6 MeV in the spectrum in Fig. 2, this contribution is probably weak, and the peak mainly corresponds to a ^{10}C state at E_x= 9.45 MeV. The state is not yet reported in the TUNL compilation [23] or later experiments related to ^{10}C (see *e.g.* [24]); it is most likely the ^{10}C isospin analog of the 2^+ state in ^{10}Be at E_x= 9.56 MeV.

Table 1: Analog states in ^{10}Be, ^{10}B and ^{10}C. The numbers in brackets give the excitation energy of the states relative to the lowest T=1 state in ^{10}B (at 1.74MeV).

J^π	E_x (^{10}Be) (MeV)	E_x (^{10}B) (MeV)	E_x (^{10}C) (MeV)
0^+	0.00	1.74 (0.00)	0.00
2^+	3.37	5.16 (3.42)	3.35
2^+	5.96	7.46 (5.81)	
1^-	5.96	7.43 (5.69)	
0^+	6.18	7.56 (5.82)	5.38
2^-	6.26	7.48 (5.74)	
3^-	7.37	8.89 (7.15)	
2^+	7.54	8.89 (7.15)	6.58
2^+	9.56		9.45

The list of the known analog states in A=10 nuclei is given in the table 1. One can see that the states with compact geometry (like *e.g.* ground state) have rather similar excitation energies in all three isotopes, while the well deformed (clusterized) states (like *e.g.* 0_2^+) have energies which are very different. The ^{10}C state at E_x= 9.45 MeV obviously falls in the first category, as expected from the reaction mechanism. Further work and experiments are needed to establish the analogs and fully understand the existence of nuclear molecules in the A=10 nuclei.

To conclude, the ^{10}B+^{10}B reactions (at 50 and 72 MeV) were measured yielding new results on clustering at high excitations of several isotopes in mass region around A=10. These results are important for completing the spectroscopy of light nuclei and picture of different structures that appear in this mass region. Even today, after many decades of intensive research, a number of subjects related to light atomic nuclei still offer surprises and interesting new phenomena.

References

[1] Hafstad L. R. and Teller E., *Phys. Rev.* **54** (1938) 681.

[2] Ikeda K., Tagikawa N. and Horiuchi H., *Prog. Theor. Phys. Extra Suppl.* (1968) 464.

[3] Horiuchi H. and Ikeda K., *Prog. Theor. Phys.* **40** (1968) 277.

[4] von Oertzen W., Freer M. and Kanada-En'yo Y., *Phys. Rep.* **432** (2006) 43.

[5] Freer M., *Rep. Prog. Phys.* **70** (2007) 2149.

[6] Wiringa R. B. et al., *Phys. Rev. C* **62** (2000) 014001.

[7] von Oertzen W., *Z. Phys. A* **354** (1996) 37; **357** (1997) 355.

[8] Ito M., Itagaki N., Sakurai H. and Ikeda K., *Phys. Rev. Lett.* **100** (2008) 182502.

[9] Ito M. and Itagaki N., *Phys. Rev. C* **78,** (2008) 011602(R).

[10] Itagaki N. et al., *Phys. Rev. C* **77** (2008) 067301.

[11] Kobayashi F. and Kanada-En'yo Y., *Phys. Rev. C* **86** (2012) 064303.

[12] Ebran J. -P., Khan E., Nikšić and Vretenar D., *Phys. Rev. C* **90** (2014) 054329.

[13] Soić N. et al., *Europhys. Lett.* **34** (1996) 7.

[14] Milin M. et al., *Nucl. Phys.* **A 753** (2005) 263.

[15] Freer M. et al., *Phys. Rev. Lett.* **96** (2006) 042501.

[16] Rogachev G. et al., *Prog Theor. Phys Suppl.* **196** (2012) 184.

[17] Suzuki D. et al., *Phys. Rev. C* **87** (2013) 054301.

[18] Tengblad O., Bergmann U. C., Fraile L. M., Fynbo H. O. U. and Walsh S., *Nucl. Instrum. Meth. Phys. Res. A* **525** (2004) 458.

[19] Torresi D. et al., *Nucl. Instrum. Meth. Phys. Res. A* **713** (2013) 11.

[20] Grassi L. et al., *Nucl. Instrum. Meth. Phys. Res. A* **767** (2014) 99.

[21] Uroić M. et al., *Eur. Phys. J.* **A 51** (2015) 93.

[22] Jelavić Malenica D., Ph.D. thesis, University of Zagreb, 2015.

[23] Tilley D. R. et al., *Nucl. Phys.* **A 745** (2004) 155.

[24] Charity R. J. et al., *Phys. Rev. C* **80** (2009) 024306.

EPJ Web of Conferences **117**, 07008 (2016) DOI: 10.1051/epjconf/201611707008
NN2015

Dynamical description of heavy-ion collisions at Fermi energies

P. Napolitani[1] and M. Colonna[2]

[1] IPN, CNRS/IN2P3, Université Paris-Sud 11, 91406 Orsay cedex, France
[2] INFN-LNS, Laboratori Nazionali del Sud, 95123 Catania, Italy

Abstract

Descriptions of heavy-ion collisions at Fermi energies require to take into account in-medium dissipation and phase-space fluctuations. The interplay of these correlations with the one-body collective behaviour determines the properties (kinematics and fragment production) and the variety of mechanisms (from fusion to neck formation and multi-fragmentation) of the exit channel. Starting from fundamental concepts tested on nuclear matter, we build up a microscopic description which addresses finite systems and applies to experimental observables.

Heavy-ion collisions at Fermi energies are open systems which require a non–equilibrium dynamical description when the process should be followed from the first instants. The evolution of the heated system produced in such collisions is determined by the nuclear mean-field potential, as well as by the explicit action of two-body correlations and fluctuations. At large beam energies, when exceeding few hundred MeV per nucleon, those beyond-mean-field contributions become prominent, while the Fermi-energy domain is characterised by the interplay of nucleon-nucleon (N-N) collisions and one-body collective behaviour. In order to exploit both mean-field and beyond-mean-field contributions, the one-body description can be extended in terms of BBGKY hierarchy to include N-N collisions and, as their natural corollaries, isoscalar and isovector fluctuations. N-N collisions affect flow and stopping [1] while fluctuations produce an ensemble of mean field trajectories which reflect in a variety of exit channels and induce fragment formation. This representation is the principle of stochastic TDHF which, in the semiclassical context, is analogous to the Boltzmann-Langevin (BL) equation [2], written in terms of the one-body distribution function $f(\mathbf{r},\mathbf{p},t)$

$$\partial_t f - \{H[f], f\} = \bar{I}[f] + \delta I[f] \; . \tag{1}$$

The left-hand side gives the Vlasov evolution for f in its own self-consistent mean field; the right-hand side contains the average Boltzmann hard two-body collision integral $\bar{I}[f]$ and the fluctuating term $\delta I[f]$, both written in terms of the one-body distribution function [3]. Numerical solutions of the BL equation have been worked out in different frameworks. In the regime of small-amplitude fluctuations, a stochastic definition of the initial states is sufficient, while Fermi energies are mostly related to large-amplitude fluctuations. In this latter case, fluctuations can be continuously generated through the collision term: they can be projected on a suited subspace like in the SMF approach [4] or, more efficiently, they can be let develop spontaneously in full phase space from agitating extended portions of the phase space in each single scattering event like in the BLOB approach [5]. The BLOB model constrains the fluctuating term $\delta I[f]$ to act on phase-space volumes with the correct occupation variance so that the Pauli blocking is never violated. As a consequence, when tested for unstable nuclear matter [6], the growth rate of the corresponding (spinodal) unstable modes are connected to the form of the mean-field potential according to the dispersion relation [7]. The propagation of the one-body distribution function is described through the test-particle method; a Skyrme-like (SKM^*) effective interaction [8] is employed, defined according to a soft isoscalar equation of state (of compressibility $K = 200$ MeV) and a linear (stiff) density dependence of the potential symmetry energy per nucleon (see [9] for details).

In order to have some general examples, we use thereafter the BLOB model to simulate collisions of ^{130}Xe nuclei. This system is chosen because it recalls widely investigated systems in the region from Sn to Xe [10]; in order to rely on simpler entrance-channel properties, the system is chosen symmetric and along β stability. We investigate firstly central collisions, then we focus on semiperipheral collisions and finally we draw some general prescriptions.

Head-on collisions are investigated in fig. 1, where density distributions are analysed as a function of time for some events related to different incident energies. Each event samples the most probable mechanism leading to intermediate-mass fragments (IMFs) in the exit channel for the selected incident energy. The event at 32 AMeV shows the arising at early times (from 100 to 200 fm/c) of a pattern of several almost-equal-size inhomogeneities in the density landscape. These undulations reflect a spinodal behaviour [11,12], i.e. a condition of mechanical instability where the size of the emerging blobs reflects the leading instability mode of the dispersion relation. At later times, if the radial expansion is not sufficient, the equal-size inhomogeneities may reaggregate producing a less symmetric pattern and a small multiplicity of fragments. In such chaotic process, the competition between these two antagonist tendencies, the disintegration into several pieces driven by spinodal instability on the one hand, and the action of the attractive nuclear force which tends to bond fragments together, imposes that different exit channels are favoured depending on the bombarding energy. The event at 23 AMeV shows for instance a highly frustrated fragmentation resulting into an almost complete re-aggregation, so that the final exit channel appears as a very asymmetric binary split. It was argued that a similar mechanism also appears in spallation reactions induced by protons and deuterons on heavy nuclei in the 1AGeV range [9]. The spinodal signal,

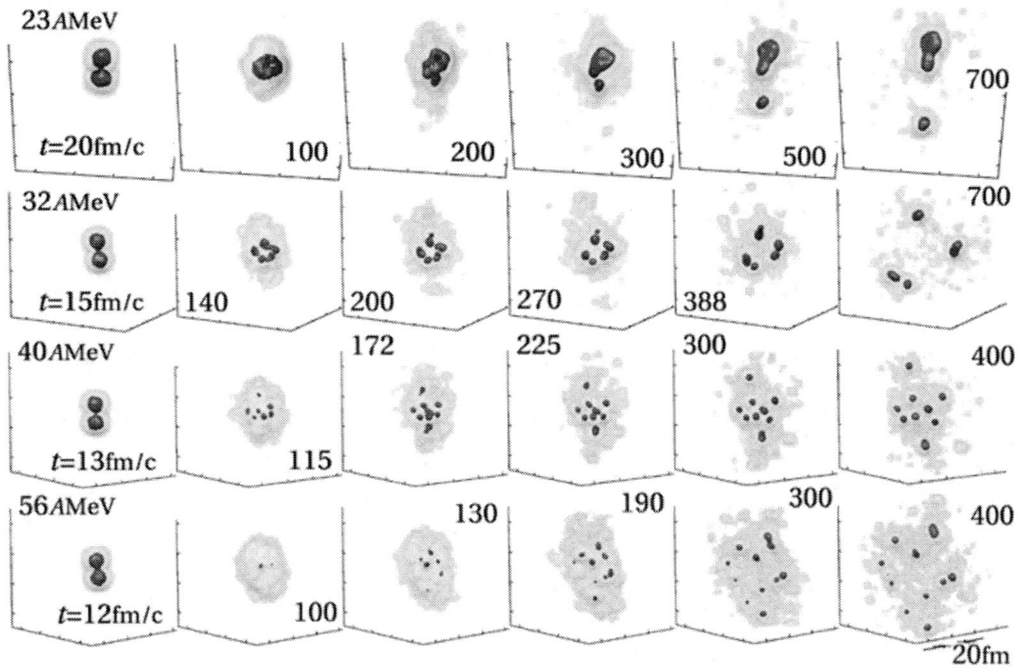

Figure 1: ^{130}Xe+^{130}Xe, examples of most probable mechanisms leading to IMFs in the exit channel for central impact parameters as a function of the incident energy.

imparting fragment-size symmetry, becomes prominent in the event at 40 *A*MeV and fades at larger bombarding energies (i.e. the event at 56 *A*MeV). Within this same approach, the analysis of a statistics of exit channels connected to a given set of initial conditions could reveal phase-transition signals in central collisions [5].

Fig. 2 investigates semi-peripheral collisions sampling the most probable mechanisms of IMF production for the selected incident energy. At 15 *A*MeV the typical mechanism where IMFs are produced is the formation and separation of a neck region in a rather long process. At larger bombarding energies the neck gradually transforms into a diluted midrapidity region where more than one blob can form. At 23 *A*MeV the process is still long and the separation of more than one IMF is too rare. At 32 *A*MeV neck fragmentation producing two or more IMFs becomes a favoured mechanism. In less peripheral collisions (i.e. b = 7 fm) IMFs arise close to the centre of the midrapidity region and they are repelled at large angles with respect to the collision axis, while in more peripheral collisions (i.e. b = 8 fm) at

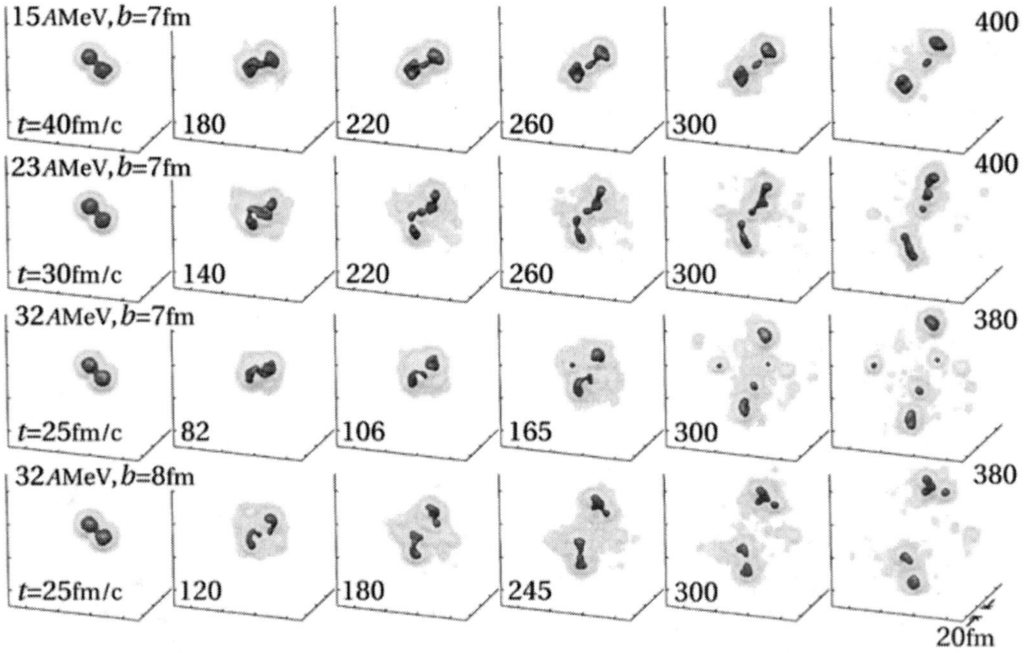

Figure 2: $^{130}Xe+^{130}Xe$, examples of most probable mechanisms leading to IMFs in the exit channel for semiperipheral impact parameters ($b \approx 7$ fm) as a function of the incident energy and the impact parameter.

the same bombarding energy two IMFs tend to form in the proximity of the quasiprojectile (QP) and the quasitarget (QT), respectively; in this case, the IMFs may orbit around the QP or QT and be eventually pulled outside of the collision axis with forward angles. These exotic mechanisms were suggested in refs. [13]. The neck process has been widely investigated especially for its connection to the isospin migration mechanism [14]: this process, driven by density gradients, induces neutron currents from the QP/QT regions towards the diluted midrapidity region where the neck forms. As illustrated in fig. 3 for an event at 15 AMeV, the consequence is the prominent neutron enrichment of the neck fragment with respect to the QP and QT, which were chosen along β-stability; in this situation, the hot neck fragment neither succeeds in de-exciting towards the residue corridor, nor it can approach β-stability.

Samples for the most probable fragment configurations at 300 fm/c for the system ^{130}Xe+^{130}Xe as a function of incident energy and impact parameter are collected in fig. 4, where all transitions between the different mechanisms discussed above can be followed. More quantitatively, the corresponding map of mean multiplicity $\langle M_{\rm frag}(Z>4)\rangle$ of fragments with $Z>4$ as a function of the impact parameter and incident energy is shown in fig. 5 (top left), from analysing a statistics of about 3500 events: projections on the two coordinates are also shown in the lower the panels. To build the map of fig. 5 the multiplicity of fragments was extracted at the time when its mean value stops growing. The right panel illustrates for some systems that such saturation time can be extracted from analysing the time evolution of the mean multiplicity $\langle M_{\rm frag}\rangle$ of fragments with $A>1$.

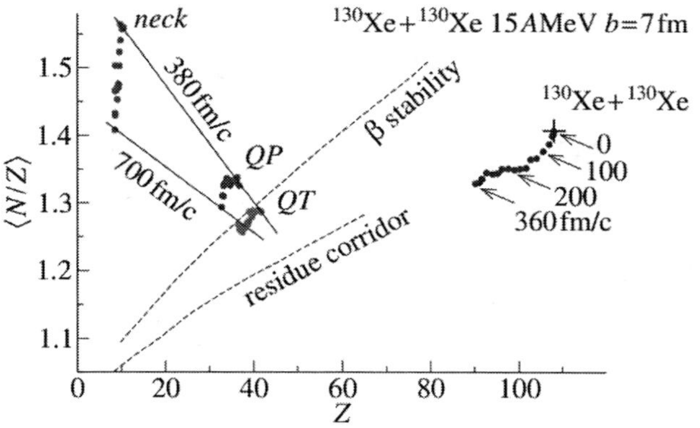

Figure 3: Neutron enrichment (N/Z) of different components of the system ^{130}Xe+^{130}Xe at successive times.

The maps of figs. 4 and 5 indicate in particular the regions where transitions between different mechanisms can arise, and they adapt without significant changes to collisions at Fermi energies of nuclei in the region of Sn and Xe. Some of those transitions may be pointed out. The fusion cross section fades in favour of asymmetric binary splits above 20 *A*MeV in central collisions and, when approaching 30 *A*MeV, binary splits gradually change into the regular pattern of almost equal-size IMFs, indicating the onset of spinodal multifragmentation. Still for central collisions, such symmetric break-up pattern persists till around 45 *A*MeV, and further increasing of bombarding energy brings the system outside of the spinodal region. Below 20 *A*MeV, along the impact-parameter coordinate *b*, fusion

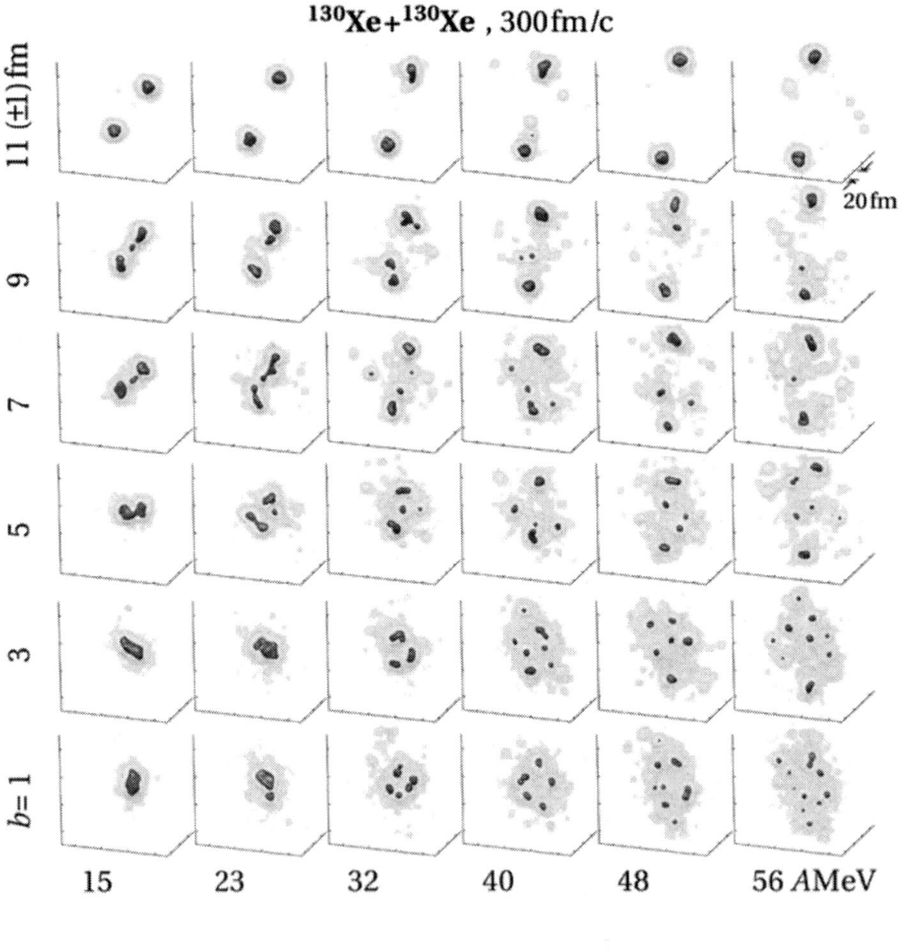

Figure 4: Examples of most probable configurations with IMFs at 300 fm/c.

changes into a binary mechanism. In the interval of about $5 < b < 9$ fm, semiperipheral collisions lead to neck formation. A transition from ternary channels to events with more than one IMF at mid-rapidity appears when moving to larger bombarding energies (above about 30 AMeV) and smaller impact parameters (below about 8 fm).

In conclusion, we find that a stochastic one-body approach can efficiently apply to dissipative collisions in the Fermi-energy domain, giving hints on the fragment production, the thresholds between different mechanisms, the variety of exit channels and the related isospin properties. In particular, we may signal the mechanism of frustrated fragmentation arising from a competition between spinodal instability and mean-field resilience: this process can describe the low-energy threshold of multifragmentation and its possible

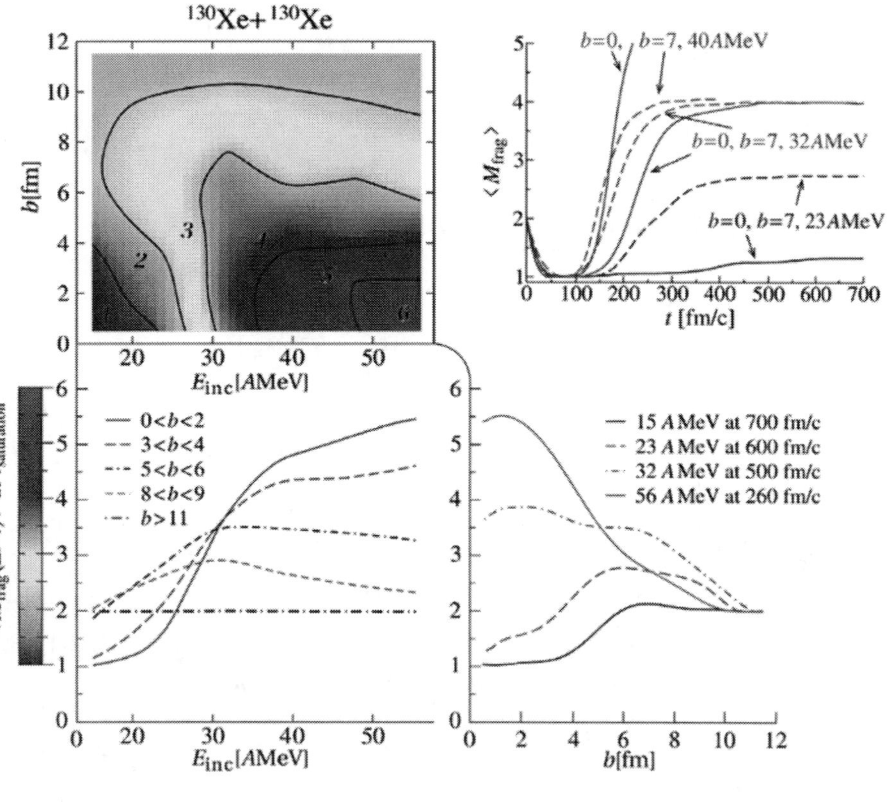

Figure 5: Top-left and bottom. Mean multiplicity of fragments with $Z > 4$ as a function of impact-parameter b and incident energy $E_{\rm inc}$ for the system ^{130}Xe+^{130}Xe at multiplicity-saturation time; projections on b and $E_{\rm inc}$ are shown in the lower the panels. Top-right. Time evolution of the mean multiplicity of fragments with $A > 1$ in central and peripheral collisions.

association to asymmetric splits in two or few fragments. We also signal the richness of the process of neck fragmentation, which may result in various unusual patterns to be further investigated in forthcoming theoretical and experimental works.

References

[1] Lopez. O. et al., *Phys. Rev. C*, **90** (2014) 064602.

[2] Reinhard P.-G. and Suraud E., *Annals Phys.*, **216** (1992) 98.

[3] Ayik S. and Grégoire C., *Nucl. Phys. A* **513** (1990) 187.

[4] Colonna M. et al., *Nucl. Phys. A* **642** (1998) 449.

[5] Napolitani P. and Colonna M., *Phys. Lett. B* **726** (2013) 382; *EPJ Web Conf.* **31** (2012) 00027.

[6] Rizzo J., Chomaz Ph. and Colonna M., *Nucl. Phys. A* **806** (2008) 40; Napolitani P. et al., *EPJ Web Conf.* **88** (2015) 00003.

[7] Colonna M. and Chomaz Ph., *Phys. Rev. C* **49** (1994) 1908.

[8] Guarnera A., Colonna M. and Chomaz Ph., *Phys. Lett. B* **373** (1996) 297.

[9] Napolitani P. and Colonna M., *Phys. Rev. C*, **92** (2015) 034607.

[10] Kohley Z. and Yennello S.J., *Eur. Phys. J. A* **50** (2014) 31; De Filippo E. and Pagano A., *Eur. Phys. J. A* **50** (2014) 32; Ademard G. et al., *Eur. Phys. J. A* **50** (2014) 33.

[11] Chomaz Ph., Colonna M. and Randrup J., *Phys. Rep.* **389** (2004) 263.

[12] Borderie B., Rivet M.F., *Progr. in Part. and Nucl. Phys.* **61 (Book Series)** (2008) 551; Borderie B. et al., *Phys. Rev. Lett.* **86** (2001) 3252; Tabacaru G. et al., *Eur. Phys. J. A* **18** (2003) 103.

[13] Baran V. et al., Phys. Rev. C **85** (2012) 054611; Rizzo C. et al., *Phys. Rev. C* **90** (2014) 054618.

[14] Colonna M. et al., *Eur.Phys.J. A* **50** (2014) 30; Lionti R. et al., Phys. Lett. B **625** (2005) 33; De Filippo E. et al., Phys. Rev. C **86** (2012) 014610; Di Toro M., Olmi A. and Roy R., Eur. Phys. J. A **30** (2006) 65; Rizzo J. et al., *Nucl. Phys. A* **806** (2008) 79.

EPJ Web of Conferences **117**, 07009 (2016) DOI: 10.1051/epjconf/201611707009
NN2015

Properties of nuclear matter and finite nuclei with finite range simple effective interaction

T. R. Routray[1*], X. Viñas[2], M. Centelles[2], L. Robledo[3], S. P. Pattnaik[1] and B. Behera[1]

[1]School of Physics, Sambalpur University, Jyotivihar-768 019, India.
[2]Departament d'Estructura i Constituents de la Matèria and Institut de Ciències del Cosmos (ICC), Facultat de Física, Universitat de Barcelona, Diagonal 645, E-08028 Barcelona, Spain
[3]Departamento de Física Teórica (Módulo15), Universidad Autónoma de Madrid, E-28049, Spain

Abstract

The nuclear mean field as a function of momentum and density is the quantity of crucial relevance in the study of nuclear matter and the equation of state can be obtained from it at momentum equals to Fermi momentum. In order to simulate the momentum dependence of the mean field, the effective interaction is constructed in the simplest form and the parameters are fixed with proper care on the momentum dependent aspects of the mean fields in different kinds of nuclear matter. By fixing the parameters of the interaction, as discussed in the work, microscopic trends of nuclear matter properties and predictions in finite nuclei, not less than any other conventional interaction, could be obtained.

*E-mail: trr1@rediffmail.com (corresponding author)

1 Introduction

The study of nuclear matter (NM) and finite nuclei in a given model is a subject of contemporary interest in the area of nuclear research. The most fundamental *ab initio* calculations of Dirac-Brueckner-Hartree-Fock (DBHF), Brueckner-Hartree-Fock (BHF) and variational types [1–6] start from a Hamiltonian which is adjusted to reproduce the nucleon-nucleon (N-N) scattering phase shifts and properties of few nucleon bound systems. The predictions in the regime of nuclear matter of this kind of *ab initio* calculations are usually considered as a standard. However, the extension to finite nuclei of microscopic calculations has severe constraints due to the much involved theoretical and computational procedures. In general, effective mean field models, either in the relativistic [7–9] and non-relativistic [10–12] frameworks are widely used for the studies in finite nuclei. However, a global description of NM and finite nuclei with a single set of interaction parameters is not available in the literature. In this work, we have made an attempt to study starting from NM to finite nuclei using the finite range simple effective interaction (SEI) [13, 14] without readjustment of any of its parameters while going from NM to finite nuclei. Ten of its total twelve parameters are fixed from the NM studies. The finite nuclei results may be considered as the predictions and shall justify the procedure adopted in the adjustment of parameters in NM.

In section 2 we have outlined the formulation of NM with SEI and discussed the fixation of parameters required for the studies of *isospin* and *spin* asymmetric NMs. We have also examined the predictions of SEI in NM under extreme conditions in this section. In section 3 we have discussed the predictions of finite nuclei properties by fitting the two remaining parameters. The last section 4 contains the conclusions.

2 Formalism

The neutron and proton mean fields, $u^n(k, \rho_n, \rho_p)$, $u^p(k, \rho_n, \rho_p)$ which are functions of momentum k, neutron and proton densities ρ_n and ρ_p, are the quantities of crucial importance in the studies of *isospin* asymmetric nuclear matter (ANM). The studies of these nucleonic mean fields are performed in terms of isoscalar $u(k, \rho)$ and isovector $u_\tau(k, \rho)$ parts. The isoscalar part is the avarege of the neutron and proton mean fields, where as, the isovector part is their difference averaged over the isospin. It has been shown that the momentum dependent parts of the isoscalar and isovector parts of the mean fields can be separated out and the resulting expressions become [15],

$$u(k,\rho) = e(\rho) + \rho\frac{de(\rho)}{d\rho} - \frac{\hbar^2 k_f^2}{2m} + u^{ex}(k,\rho) \tag{1}$$

$$u^{ex}(k,\rho) = \frac{\rho}{2}\int [j_0(kr) - j_0(k_f r)]\frac{3j_1(k_f r)}{(k_f r)}\left[v_{ex}^l + v_{ex}^{ul}\right] d^3r \tag{2}$$

$$u_\tau(k,\rho) = 2E_s(\rho) - \frac{\hbar^2 k_f^2}{3m^*(k=k_f,\rho)} + u_\tau^{ex}(k,\rho) \tag{3}$$

$$u_\tau^{ex}(k,\rho) = \frac{\rho}{2}\int [j_0(kr) - j_0(k_f r)]\left[v_{ex}^l - v_{ex}^{ul}\right] j_0(k_f) d^3r \tag{4}$$

where, $e(\rho) = H(\rho)/\rho$ is the energy per particle in symmetric nuclear matter (SNM), $H(\rho)$ being the energy density, $\rho = \rho_n + \rho_p$ is the total density, $k_f = \frac{(3\pi^2\rho)^{1/3}}{2}$ is the Ferrmi momentum, j_0 is the spherical Bessel function of zeroth order, m^* is the effective mass in SNM and $E_s(\rho)$ is the symmetry energy. v_{ex}^l and v_{ex}^{ul} are the exchange interactions acting between pairs of like (l) and unlike (ul) nucleons. The functionals $u^{ex}(k,\rho)$ and $u_\tau^{ex}(k,\rho)$ in eqs.(2) and (4) are the momentum dependent contributions to the isoscalar and isovector parts of the nucleonic mean field, which vanishes at the Fermi momentum k=k_f and connection to the nuclear matter equation of state (EOS) can be obtained from eqs.(1) and (3). Further, these functionals, given by eqs.(2) and (4), have non-vanishing contributions only for finite range exchange interactions, while zero-range interactions do not contribute to the momentum dependence of the mean fields. So, in the simplest form one can consider a single finite range term of any of the conventional form *Yukawa/ Gaussian/ exponential* in the interction and examine how far it can account for the mean field properties. The finite range effective interaction (SEI) thus constructed in the simplest form containing a single finite range term is given as,

$$\begin{aligned} v_{eff}(r) &= t_0(1+x_0P_\sigma)\delta(r) + \frac{t_3}{6}(1+x_3P_\sigma)\left(\frac{\rho(\mathbf{R})}{1+b\rho(\mathbf{R})}\right)^\gamma \delta(r) \\ &\quad + (W + BP_\sigma - HP_\tau - MP_\sigma P_\tau) f(r), \end{aligned} \tag{5}$$

where, $f(r)$ is the functional form of the finite range interaction containing the single range parameter α and can be of any conventional form *Yukawa/*

Gaussian/exponential. Here, we have made explicit studies for the Gaussian and Yukawa forms.The SEI in eq.(5) contains of 11-parameters, namely, t_0, x_0, t_3, x_3, b, W, B, H, M, γ and α (the spin-orbit strength parameter W_0 will enter in the formulation of finite nuclei). The energy density of ANM, $H(\rho_n, \rho_p)$, and neutron (proton) mean field $u^{n(p)}(k, \rho_n, \rho_p)$ for the Gaussian form of $f(r)$ are given in Refs. [16, 17] and for Yukawa form in Refs. [18, 19].

The complete study of ANM requires the knowledge of altogether nine parameters, namely, b, γ, α, ε^{l}_{ex}, ε^{ul}_{ex}, ε^{l}_{γ}, $\varepsilon^{ul}_{\gamma}$, ε^{l}_{0} and ε^{ul}_{0}. For the sake of simplicity, the formulation has been based on the fact that the range between a pair of like or unlike nucleons is the same but they differ in their strengths. The connection between the parameters of ANM and the interaction parameters is given in the earlier works [16, 18]. The knowledge of the nine parameters that characterize the ANM can be obtained from the independent studies of spin saturated SNM and pure neutron matter (PNM). The parameters γ, b and α along with the combinations of the strength parameters $(\frac{\varepsilon^{l}_{0}+\varepsilon^{ul}_{0}}{2}) = \varepsilon_0$, $(\frac{\varepsilon^{l}_{\gamma}+\varepsilon^{ul}_{\gamma}}{2}) = \varepsilon_\gamma$ and $(\frac{\varepsilon^{l}_{ex}+\varepsilon^{ul}_{ex}}{2}) = \varepsilon_{ex}$ are the six parameters needed for a complete description of the mean field properties and the EOS in SNM. In determining these parameters, we have first constrained the range α and finite range strength parameter ε_{ex} those governs the momentum dependence of the mean field in SNM. These two parameters are determined by adopting a simultaneous optimization procedure with the constraint that the attractive optical potential changes sign for a kinetic energy $\frac{\hbar^2 k^2}{2m}$=300 MeV of the incident nucleon (see Ref. [13] for details), a result extracted from the analysis of nucleon-nucleus scattering data in intermediate energy range [20, 21]. The values thus obtained are ε_{ex}=-96.24 MeV, α=0.7596 fm for the Gaussian and ε_{ex}=-129.25 MeV, α=0.4044 fm for the Yukawa form, where we have used as input only the standard values of the nucleon mass m=939 MeV, Fermi kinetic energy $\hbar^2 k^2_{f_0}/2m$=37 MeV (that corresponds to the saturation density ρ_0=0.161 fm^{-3}) and energy per particle at saturation in SNM $e(\rho_0)$= -16.0 MeV. The momentum dependence of the nuclear mean field in SNM, $u^{ex}(k, \rho)$, computed using this SEI with Gaussian form at three different densities, ρ=0.1, 0.3, and 0.5 fm^{-3}, is shown in figure 1 of Ref [16] and agree very well over a wide range of momentum with the prediction of the realistic interaction UV14+UVII [22]. The same comparison is shown in the lower panel of the same figure in Ref. [16] for the Gogny D1 and D1S sets with the microscopic results [22]. The conclusion regarding the comparison in case of Yukawa form with the microscopic prediction is the same.

With the knowledge of these two parameters, range α and exchange

Table 1: Values of the effective mass and mean field in SNM at normal nuclear matter density for various limiting values of the momentum k togather with the rearrangement energy for Gaussian and Yukawa forms of SEI.

Properties	Yukawa	Gaussian
$\frac{m^*}{m}(k = k_{f_0}, \rho_0)$	0.686	0.709
$\frac{m^*}{m}(k = 0, \rho_0)$	0.609	0.660
$u(k = \infty, \rho_0)$ MeV	35.89	12.68
$u(k = 0, \rho_0)$ MeV	-73.1	-70.05
$u_R(\rho_0)$ MeV	15.93	16.91

strength ε_{ex}, we can make a study of mean field, $u(k, \rho_0)$ and effective mass, m^*/m, properties in SNM at saturation density ρ_0. The predictions of m^*/m and $u(k, \rho_0)$ at different limiting values of k, and the rearrangement energy $u_R(\rho_0)$ are given in Table 1 for both Yukawa and Gaussian forms. However, the calculation of the mean field properties at densities other than saturation density and EOS of SNM, requires the knowledge of the rest four parameters b, ε_γ, ε_0, and γ. The parameter b is fixed to avoid the supraluminous behaviour in SNM [23]. The condition reads $b\rho_0 = \left[\left(\frac{mc^2}{T_{f_0}/5 - e(\rho_0)}\right)^{\frac{1}{(\gamma+1)}} - 1\right]^{-1}$, with $T_{f_0} = \frac{\hbar^2 k_{f_0}^2}{2m}$. Its calculation requires again the knowledge of the NM values ρ_0, $e(\rho_0)$ and the parameter γ only. The two parameters, ε_γ and ε_0, are obtained from the saturation conditions, that is, from the values of $e(\rho_0)$ and ρ_0. The stiffness parameter γ is kept open and its admissible values are constrained by the condition that the pressure-density curve lies within the region extracted from the analysis of flow data in heavy-ion collisions (HIC) at intermediate and high energies [24]. With the knowledge of the two parameters α and ε_{ex} determined from the momentum dependence of isoscalar mean field at normal density ρ_0, the complete study of SNM can be performed for a given γ if one assumes only the standard NM values of ρ_0, $e(\rho_0)$ and m. To extend the study to ANM, one needs to know how ε_{ex}, ε_γ and ε_0 split into like and unlike isospin channels. The splitting of ε_{ex} into ε_{ex}^{l} and ε_{ex}^{ul} is decided using the physical constraint resulting from the studies of the thermal evolution of NM properties [19]. This study predicts a critical value of the splitting of ε_{ex} for which the thermal evolution of NM properties as well as the entropy per particle in PNM does not exceed that of SNM. The resulting critical value of the splitting is $\varepsilon_{ex}^{l} = \frac{2}{3}\varepsilon_{ex}$. The n-p effective mass splitting predicted with this choice of ε_{ex}^{l} nicely coincides with the results of DBHF calculations [2] as has been shown in the previous work [16]. The splitting of the remaining two parameters, namely ε_γ and ε_0, is obtained by assuming a standard value of symmetry energy $E_s(\rho_0)$ at saturation and a value of its derivative $E_s^{'}(\rho_0) = \rho_0 \frac{dE_s(\rho_0)}{d\rho_0}$ for which the asymmetric contribution of the nucleonic part of the energy density in charge neutral beta-stable $n+p+e+\mu$ matter becomes maximum. This choice predicts a density dependence of the symmetry energy which is neither very stiff nor soft and does not allow the direct URCA process to occur in neutron stars, in agreement with the predictions of the population synthesis models [25] based on cooling calculations [26]. The SEI with the parameters obtained in this way is able to reproduce the microscopic trends of the density dependence of the EOS and the momentum dependence of the mean fields in ANM [18,27]. With the knowledge of all nine parameters of ANM, the neutron star matter EOS has been calculated by solving the charge neutral beta-stability conditions and neutron star bulk properties are obtained from the solution of the Tolman-Oppenheimer-Volkov (TOV) equation. In the Figures 1, we show the mass-radius relations predicted for different sets of SEI corresponding to five different values of γ.

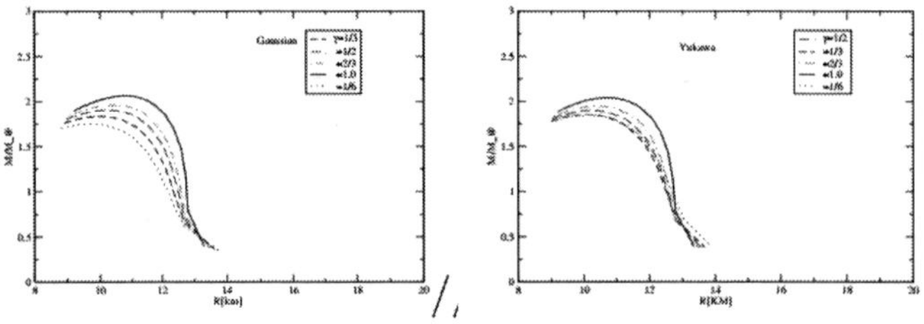

Figure 1: Mass-radius relations predicted for Gaussian (left panel) and Yukawa (right panel) for the EOSs having γ=1/6, 1/3, 1/2, 2/3 and 1.0.

We have still two parameters open which we have taken to be t_0 and x_0. These two parameters, along with the spin-orbit strength W_0, were determined by a simultaneous fit to the experimental binding energies of ^{40}Ca and ^{208}Pb and to the splitting of the $1p$ single-particle levels in ^{16}O in our work on finite nucleus in Ref. [16]. However, in a subsequent work [17], it is shown that the x_0 and W_0 parameters are, actually, correlated and different combinations of them leave the *rms* deviations of binding energies and charge radii practically unchanged but predict very different behavior in spin polarized matter, which is sensitive to the value of x_0. To determine x_0 we consider the particular case of spin polarized neutron matter. Its description requires to know how the strength parameters ε_0^l, ε_γ^l and ε_{ex}^l of spin saturated neutron matter splits into the two channels of like (parallel, "l,l") and unlike (anti-parallel, "l,ul") spin orientations subject to the constraints, ε_0^l=($\varepsilon_0^{l,l} + \varepsilon_0^{l,ul}$)/2, ε_γ^l=($\varepsilon_\gamma^{l,l} + \varepsilon_\gamma^{l,ul}$)/2 and ε_{ex}^l=($\varepsilon_{ex}^{l,l} + \varepsilon_{ex}^{l,ul}$)/2. The parameters $\varepsilon_{ex}^{l,l}$ and $\varepsilon_{ex}^{l,ul}$ determine the momentum dependent aspects of the mean field in polarized neutron matter. From a fit to the DBHF predictions on effective mass splitting between spin-up and spin-down neutrons in spin polarized neutron matter [28], it is found that the SEI predictions agrees well with the DBHF ones for $\varepsilon_{ex}^{l,l} = \varepsilon_{ex}^l/3$, as shown in Ref. [17]. This consideration allows to determine x_0 in a unique way as,

$$x_0 = 1 - \frac{2\varepsilon_0^l - \varepsilon_{ex}^l}{\rho_0 t_0}, \tag{6}$$

subject to the condition that t_0 is known. The two only open parameters, t_0 and W_0, were fitted to reproduce the binding energies of ^{40}Ca and ^{208}Pb, respectively.

3 Finite nuclei

It is clear from the discussions of the last section that the study from infinite NM to finite nuclei requires only the NM parameters $e(\rho_0)$, ρ_0 and $E_s(\rho_0)$ for a given stiffness γ of SNM. The values of these NM parameters are not uniquely defined and their emperical values lies with in certain ranges. The values of these three NM parameters used in different model calculations broadly cover the ranges, $-15.8MeV \leq e(\rho_0) \leq -16.02MeV$, $0.14fm^{-3} \leq \rho_0 \leq 0.18fm^{-3}$ and $27MeV \leq E_s(\rho_0) \leq 38MeV$. In the last section the studies have been made for the values of these parameters $e(\rho_0)$=-16 MeV, ρ_0=0.161fm^{-3} and $E_s(\rho_0)$=30 MeV. But computation of finite nuclei with a set of values of these three NM parameters chosen arbitrarily does not ensures the best description of binding energies and charge radii for interaction with a given γ.In order to verify this point we have calculated the binding energies and charge radii of 161 even-even spherical nuclei using the EOSs corresponding to γ=1/6, 1/3, 1/2 and 2/3 but having same $e(\rho_0)$=-16 MeV, ρ_0=0.161fm^{-3} and $E_s(\rho_0)$=33 MeV. Finite nuclei calculations with the SEI have been performed in the framework of the Density Functional Theory by constructing a quasi-local energy density functional obtained from a Thomas-Fermi approach of the density matrix [29,30] plus a BCS treatment of pairing correlations,as discussed in details in Ref. [16].

The results of the deviations of the predicted charge radii, $\delta r_{ch} = {r_{ch}}^{th} - {r_{ch}}^{expt}$, of 86 even-even spherical nuclei whose values are experimentally measured are shown in Figure 2. It can be seen that for γ=1/6, the calculated results are overestimated, whereas, it is underestimated for γ=1/2 and 2/3. These results are found to be very weakly sensitive to the

Table 2: Critical values of the NM parameters $e(\rho_0)$, ρ_0 and $E_s(\rho_0)$ for the four sets of EOSs of SEI corresponding to γ=1/6, 1/3, 1/2 and 2/3. The *rms* deviations of energies, δE_{rms}, of 161 and radii, δr_{rms} of 86 spherical nuclei for the four sets of EOSs are given. The corresponding *rms* values for the same set of nuclei obtained for SLy4 Skyrme set [31] and D1S and NL3 RMF set are given for comparison.

γ	T_{f0} MeV	ρ_0 fm^{-3}	$E_s(\rho_0)$ MeV	δr_{rms} fm	δE_{rms} MeV
1/6	37.2	0.1623	36.0	0.0189	1.6993
1/3	36.8	0.1597	35.5	0.0170	1.6754
1/2	36.4	0.1571	35	0.0155	1.8518
2/3	36.1	0.1552	35	0.0152	1.8297
SLy4				0.024	1.71
D1S				0.020	2.41
NL3				0.020	3.58

variations of $e(\rho_0)$ and $E_s(\rho_0)$ with in their ranges, and depend very strongly on the value of ρ_0. The results of energies of the 161-nuclei for these four EOSs show that the energies are overestimated for γ=1/6 and underestimated for γ=2/3 and the results are sensitive to the value of $E_s(\rho_0)$ rather than to the variations of $e(\rho_0)$ and ρ_0. So we have fixed the value of $e(\rho_0)$=16 MeV and varied ρ_0 and $E_s(\rho_0)$ with in their ranges and found out the characteristic values of these two NM parameters for each γ for which the charge radii of the 86 and energies of 161 even-even spherical nuclei are reproduced with minimum *rms* deviations. The vaules are given in Table 2.

It can be seen from these values that as NM incompressibilty increases from 206 MeV (corresponding to γ=1/6) to 265 MeV (γ=2/3), the saturation density decreases from 0.1623 fm^{-3} to 0.1552 fm^{-3}, and symmetry energy decreases from 36 MeV to 35 MeV, respectively. The finite nuclei calculations of spherical even-even nuclei over the nuclear chart have performed for the four EOSs having γ=1/6, 1/3, 1/2 and 2/3 with their charecterstic values of the NM parameters, in Refs. [16,17]. The SEI predictions of binding energies, radii, single particle spectra, isotopic shift in ^{208}Pb, etc. of the spherical nuclei are found to be no less in quality than that of any other conventional interaction.The description of finite nuclei has been extended by performing full Hartree-Fock-Bogoliubov (HFB) calculations including deformation degrees of freedom using the SEI. The HFB formulation is restricted to axially symmetric geometry. The pairing interaction is the same as in Refs. [17]. The solution of the HFB equations has been recast as a minimization procedure of the energy density functional, where the HFB wave function of the Bogoliubov transformation is chosen to minimize the

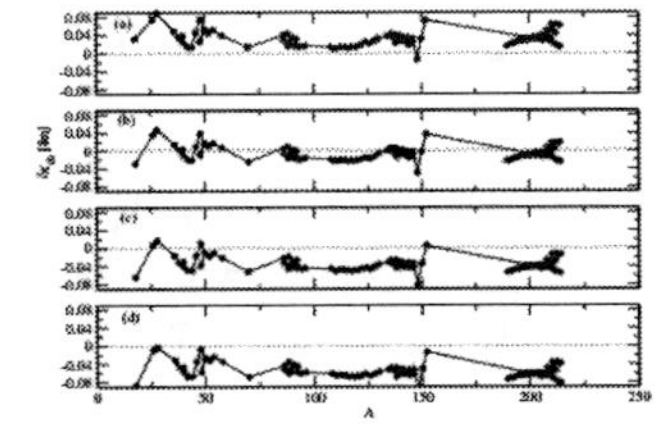

Figure 2: δr_{ch} as a function of mass number A for the 86 even-even spherical nuclei

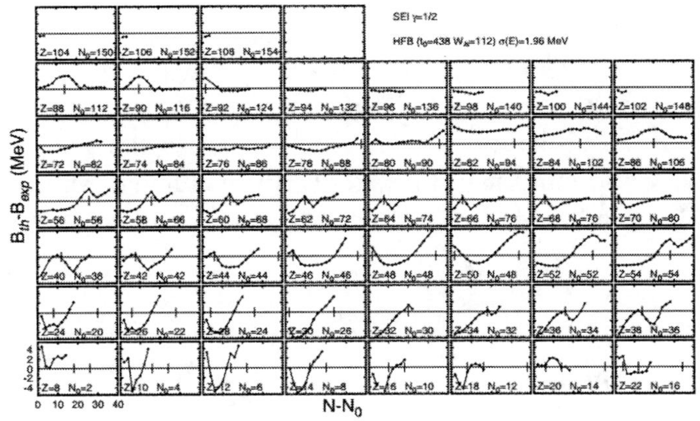

Figure 3: HFB results of $(B_{th} - B_{exp})$ of 620 even-even nuclei as a function of mass number A for the EOS of SEI having γ=1/2.

energy. We have minimized the binding energy *rms* deviation $\sigma(E)$ for the 620 even-even (spherical and deformed) nuclei with known experimental masses taken from Ref. [32]. The values of t_0 and W_0 corresponding to the minimum *rms* deviations $\sigma(E)$ turn out to be the same as found in the quasi-local energy density formulation in Ref. [17]. The energies of 620 nuclei could be reproduced with *rms* $\sigma(E)$=1.873 MeV and 1.958 MeV in the HFB calculation for the EOS having γ=1/3 and γ=1/2, respectively. The results of γ=1/2 are shown in Figure 3. In this figure, the binding energy difference $\Delta B = B_{th} - B_{exp}$, is plotted as a function of the neutron number N, shifted by N_0, which indicates the origin of the horizontal axis for the different isotopic chains displayed in this figure.

4 Conclusions

A global study from nuclear matter to finite nuclei is performed using the finite range simple effective interaction.Ten of the twelve parameters of the interaction are fixed in the regime of nuclear matter. Emphasis has been given to fix the parameters responsible for the momentum dependence of the mean fields in ANM and neutron matter from appropriate considerations.While going for the study of finite nuclei by adjusting the two open parameters t_0 and W_0 from the fits to the binding energies of two magic nuclei, the predictions with regards to the momentum dependence of the mean fields and EOSs in different types of nuclear matters remain unchanged.

It has been found that different properties of nuclear matter similar in trends as given by the microscopic calculations and results in finite nuclei similar in quality to the ones of the conventional interactions could be reproduced. The finite nuclei results can be considered as the predictions of the interaction, as is evident from the procedure of parameters determination. So it can be said that the momentum dependence of the mean field in nuclear matter is a crucial aspect and its appropriate consideration is an essential feature while making a unified study of nuclear matter and finite nuclei in a given model.

References

[1] Haar B ter and Malfliet R 1987 *Phys. Rep.* **149** 207

[2] Sammarruca F 2010 *Int. J. Mod. Phys. E* **19** 1259

[3] Brockmann R and Machleidt R 1990 *Phys. Rev. C* **42** 1965

[4] Van Dalen E N E, Fuchs C and Faessler A 2005 *Phys. Rev. Lett.* **95** 022302

[5] Akmal A, Pandharipande V R and Ravenhall D G 1998 *Phys. Rev C* **58** 1804

[6] Wiringa R B, Fiks V, Fabrocini A 1988 *Phys. Rev. C* **38** 1010

[7] Ring P 1996 *Prog. Part. Nucl. Phys.* **37** 193

[8] Roca-Maza X, Viñas X, Centelles M, Ring P and Schuck P 2011 *Phys. Rev. C* **84** 054309.

[9] Afanasjev A V, Agbemava S E, Ray D and Ring P 2013 *Phys. Lett. B* **726** 680.

[10] Decharge J and Gogny D 1980 *Phys. Rev. C* **21** 1568.

[11] Blaizot J P, Berger J F, Decharge J and Girod M 1995 *Nucl. Phys. A* **591** 435.

[12] Than H S, Khoa Dao T, and Giai Van N 2009 *Phys. Rev. C* **80**, 064312.

[13] B. Behera, T. R. Routray and R. K. Satpathy, J. Phys. G: Nucl. Part. Phys. **24**, 2073 (1998).

[14] Behera B, Routray T R, Sahoo B and Satpathy R K 2002 *Nucl. Phys. A* **699** 770.

[15] Behera B, Routray T R and Pradhan A 2005 *Mod. Phys. Lett. A***20** 2639.

[16] Behera B, Viñas X, Bhuyan M, Routray T R, Sharma B K and Patra S K 2013 *J. Phys G: Nucl. Part. Phys.* **40** 095105

[17] Behera B, Viñas X, Routray T R and Centelles M 2015 *J. Phys G: Nucl. Part. Phys.* **42** 045103.

[18] Behera B, Routray T R, Pradhan A, Patra S K and Sahu P K 2007 *Nucl. Phys. A* **794**, 132.

[19] Behera B, Routray T R and Tripathy S K 2011 *J. Phys. G: Nucl. Part. Phys.* **38**, 115104.

[20] Bertsch G F, Das Gupta S 1988 *Phys. Rep.* **160** 189.

[21] Gale C, Welke G M, Prakash M, Lee S J, Das Gupta S, 1990 *Phys. Rev. C* **41** 1545.

[22] R. B. Wiringa, Phys. Rev. C **38**, 2967 (1988).

[23] Behera B, Routray T R and Satpathy R K 1997 *J. Phys G: Nucl. Part Phys.* **23** 445.

[24] Danielewicz P, Lacey R and Lynch W G 2002 *Science* **298** 1592.

[25] Popov S, Grigorian H, Turolla R and Blaschke D 2006 *Astron. Astrophys.* **448** 327.

[26] Blaschke D, Grigorian H and Voskrenensky D 2004 *Astron. Astrophys.* **424** 979.

[27] Behera B, Routray T R and Tripathy S K 2009 *J. Phys. G: Nucl. Part. Phys.* **36**, 125105.

[28] Sammarruca F and Krastev P G 2007 *Phys. Rev. C* **75** 034315.

[29] Soubbotin V B and Viñas X 2000 *Nucl. Phys. A* **665** 291

[30] Soubbotin V B, Tselyaev V I and Viñas X 2003 *Phys. Rev. C* **67** 014324

[31] Dutra M et *al.* 2012 *Phys. Rev. C* **85**, 035201.

[32] M.Wang, G. Audi, A.H. Wapstra, F.G. Kondev, M. Mac- Cormick, X. Xu, and B. Pfeiffer, Chinese Phys. **C 36**, 1603 (2012).

The ASY-EOS Experiment at GSI

P. Russotto[1], M. Chartier[2], M.D. Cozma[3], E. De Filippo[1], A. Le Fèvre[4], S. Gannon[2], I. Gašparić[5,6], M. Kiš[4,5], S. Kupny[7], Y. Leifels[4], R.C. Lemmon[8], Q. Li[9], J. Łukasik[10], P. Marini[11,12], P. Pawłowski[10], W. Trautmann[4], L. Acosta[13], M. Adamczyk[7], A. Al-Ajlan[14], M. Al-Garawi[15], S. Al-Homaidhi[14], F. Amorini[13], L. Auditore[16,17], T. Aumann[6], Y. Ayyad[18], V. Baran[13,19], Z. Basrak[5], R. Bassini[20], J. Benlliure[18], C. Boiano[20], M. Boisjoli[12], K. Boretzky[4], J. Brzychczyk[7], A. Budzanowski[10], G. Cardella[1], P. Cammarata[21], Z. Chajecki[22], A. Chbihi[12], M. Colonna[13], B. Czech[10], M. Di Toro[13,23], M. Famiano[24], V. Greco[13,23], L. Grassi[5], C. Guazzoni[20,25], P. Guazzoni[20,26], M. Heil[4], L. Heilborn[21], R. Introzzi[27], T. Isobe[28], K. Kezzar[15], A. Krasznahorkay[29], N. Kurz[4], E. La Guidara[1], G. Lanzalone[13,30], P. Lasko[7], I. Lombardo[31,32], W.G. Lynch[22], Z. Matthews[2], L. May[21], T. Minniti[1], M. Mostazo[18], A. Pagano[1], M. Papa[1], S. Pirrone[1], R. Pleskac[4], G. Politi[1,23], F. Porto[13,23], R. Reifarth[4], W. Reisdorf[4], F. Riccio[20,25], F. Rizzo[13,23], E. Rosato[31,32], D. Rossi[4,22], S. Santoro[16,17], H. Simon[4] I. Skwirczynska[10], Z. Sosin[7], L. Stuhl[29], A. Trifirò[16,17], M. Trimarchi[16,17], M.B. Tsang[22], G. Verde[1], M. Veselsky[33], M. Vigilante[31,32], A. Wieloch[7], P. Wigg[2], H.H. Wolter[34], P. Wu[2], S. Yennello[21], P. Zambon[20,25], L. Zetta[20,26] and M. Zoric[5]

[1] INFN, Sezione di Catania, Catania, Italy
[2]University of Liverpool, Liverpool, UK
[3] IFIN-HH, Măgurele-Bucharest, Romania
[4] GSI Helmholtzzentrum, Darmstadt, Germany
[5] Ruder Boškoviċ Institute, Zagreb, Croatia
[6] Technische Universität, Darmstadt, Germany
[7] Jagiellonian University, Kraków, Poland
[8] STFC Laboratory, Daresbury, UK
[9] Huzhou Teachers College, China
[10] IFJ-PAN, Kraków, Poland
[11] CENBGn Université de Bordeaux, CNRS/IN2P3,Gradignan, France
[12] GANIL, Caen, France
[13] INFN-Laboratori Nazionali del Sud, Catania, Italy
[14] KACST Riyadh, Riyadh, Saudi Arabia
[15] King Saud University, Riyadh, Saudi Arabia
[16] INFN-Gruppo Collegato di Messina, Messina, Italy
[17] Università di Messina, Messina, Italy
[18] University of Santiago de Compostela, Santiago de Compostela, Spain
[19] University of Bucharest, Bucharest, Romania
[20] INFN-Sezione di Milano, Milano, Italy
[21]Texas A&M University, College Station, TX, USA
[22] NSCL Michigan State University, East Lansing, MI, USA
[23] Università di Catania, Catania, Italy
[24] Western Michigan University, MI, USA
[25] Politecnico di Milano, Milano, Italy
[26] Università di Milano, Milano, Italy
[27] INFN, Politecnico di Torino, Torino, Italy
[28] RIKEN, Wako, Japan
[29] Institute of Nuclear Research, Debrecen, Hungary
[30] Università Kore, Enna, Italy
[31] INFN, Sezione di Napoli, Napoli, Italy
[32] Università di Napoli, Napoli, Italy
[33] Institute of Physics, Slovak Academy of Sciences, Bratislava, Slovakia
[34] LMU, München, Germany

Abstract

The elliptic-flow ratio of neutrons with respect to protons or light complex particles in reactions of heavy ions at pre-relativistic energies has been proposed as an observable sensitive to the strength of the symmetry term of the nuclear equation of state at supra-saturation densities. In the ASY-EOS experiment at the GSI laboratory, flows of neutrons and light charged particles were measured for ^{197}Au+^{197}Au collisions at 400 MeV/nucleon. Flow results obtained for the Au+Au system, in comparison with predictions of the UrQMD transport model, confirm the moderately soft to linear density dependence of the symmetry energy deduced from the earlier FOPI-LAND data.

1 Introduction

The neutron-proton elliptic flow ratio and difference have been proposed as effective probes of the high-density behavior of the nuclear symmetry energy [1]. The comparison of existing data from the FOPI-LAND experiment [1] with calculations performed with the UrQMD transport model [2] suggests a moderately soft to linear symmetry term. The result suffers, however, from the considerable statistical uncertainty of the experimental data [1]. The same data set was also compared to calculations performed with the Tübingen QMD model and a constraint compatible with the UrQMD result was obtained [3]. At the same time, a thorough study of the parameter dependence of the model predictions was performed in order to devise a route towards a model independent constraint of the the high-density symmetry energy. It was, in particular, found that different parameterizations of the isovector part of the equation of state, the Gogny inspired (momentum dependent) vs. the power law (momentum independent) potential, lead to very similar results. In order to improve the statistical accuracy of the experimental flow parameters for the Au+Au reaction and to extend the flow measurements to other systems, the symmetric collision systems ^{197}Au+^{197}Au, ^{96}Zr+^{96}Zr and ^{96}Ru+^{96}Ru at 400 MeV/nucleon incident energies have been chosen for the ASY-EOS experiment, conducted at the GSI laboratory in May 2011 (S394 experiment).

2 Results of the ASY-EOS experiment

The Large Area Neutron Detector (LAND) [4] was positioned to cover laboratory angles around 45° with respect to the beam direction, at a distance of about 5 m from the target. A veto-wall of plastic scintillators in front of LAND allowed discriminating neutrons and charged particles. In this configuration, it was possible to measure the direct and elliptic flows of neutrons and charged particles at mid-rapidity with high precision in the same angular acceptance. In addition, the Krakòw Triple Telescope Array, KRATTA [5], was installed to measure the energy, emission angles and isotopic composition of light charged reaction products. The 35 modules of KRATTA were placed opposite to LAND at a distance of 40 cm from the target, at polar angles between 24° and 68°. Very good isotopic resolution

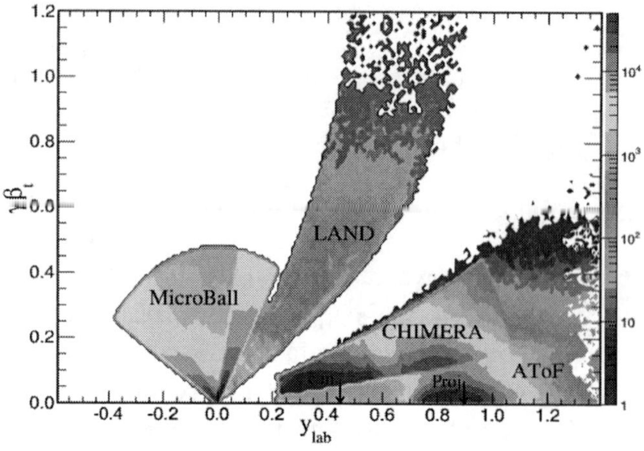

Figure 1: Measured invariant hit distribution for $^{197}Au +^{197} Au$ collisions at 400 MeV/nucleon incident energy in the transverse-velocity vs. rapidity plane for charged particles detected with the three systems Microball, CHIMERA, and AToF Wall with full azimuthal coverage and for neutrons detected with LAND. The velocity of particles detected with the Microball is not measured and shown here with an assumed distribution.

has been obtained in the whole dynamic range up to $Z \sim 6$. The determination of the impact parameter and the orientation of the reaction plane required the use of several devices: i) the ALADIN Time-of-Flight (AToF) wall [6] was used to detect charged particles at polar angles smaller than 7°; two walls (front and rear) of 2.5*100 cm^2 plastic scintillators, read by two photo-multipliers at both ends, gave information on emission angle, atomic number and velocity of forward-emitted ions; ii) 50 thin (between 3.6 and 5.6 mm) CsI(Tl) elements, read out by photo-diodes and arranged in 4 rings, of the Washington-University Microball array [7], surrounded the target at polar angles between 60° and 147° with the aim of measuring the multiplicty and angular distribution of backward emitted particles in order to discriminate against background reactions on non-target material; iii) 352 CsI(Tl) scintillators, 12 cm thick, of the CHIMERA multidetector [8], arranged in 8 rings with 2π azimuthal coverage around the beam axis and covering polar angles between 7° and 20°, were used to detect light charged particles from mid-rapidity emissions.

The kinematic coverage achieved with this assembly of detection systems is illustrated in Fig. 1. In particular, the enhanced particle yields in the kinematic regimes of participant and spectator emissions are clearly visible. The product yields from the decay of the projectile spectator seen with

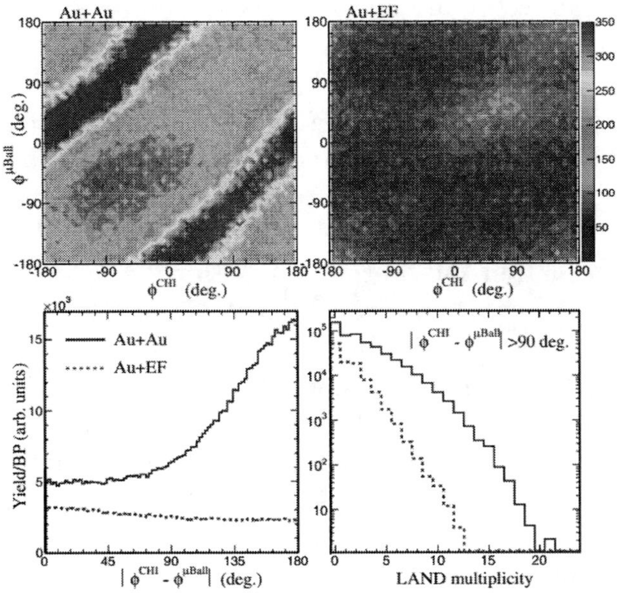

Figure 2: Top row: Correlation between the Q-vector orientations determined with CHIMERA (abscissa) and with the Microball (ordinate) for data sets taken with (left panel) and without (right panel) a target foil in place (EF = Empty Frame). Bottom row: difference of the Q-vector orientations for Au+Au and for Au+EF data (left panel), normalized with respect to the integrated beam intensity (BP stands for beam particles), and the raw hit multiplicities (right panel) registered with LAND for Au+Au (full line) and for Au+EF data sets (dotted).

CHIMERA and the AToF Wall do not exactly match because the AToF efficiency for hydrogen isotopes in this energy range is lower than that of the CHIMERA modules.

With beam intensities of about 10^5 pps and targets of 1-2% interaction probability, about $5 * 10^6$ events were collected for each system. Special runs were performed with and without target, in order to measure the background from interaction of projectile ions with air, and with iron shadow bars in front of LAND or the neutron background measurement.

For rejecting background reactions due to the interaction of Au projectiles with non-target material (mainly air), the correlation of the two reaction plane orientations given by CHIMERA and by the Microball detectors, was used, as shown in fig. 2. From the Microball data, the reaction plane orientation has been estimated by averaging over the azimuthal directions of the recorded hits, i.e. by calculating a $\vec{Q}$ vector (see Ref. [9] for definition) as $\vec{Q} = \sum_{i=1}^{N} \hat{r}_t^i$, where $\hat{r}_t^i$ is the azimuthal unit vector in the direction of the location of the detector module that recorded the ith hit and $N \geq 2$. For CHIMERA, the $\vec{Q}$ vector has been calculated as $\vec{Q} = \sum_{i=1}^{N} Z^i \beta_t^i \gamma^i$, with $N \geq 4$, taking only the forward emitted particles with rapidity in the center of mass reference frame $y_i^{CM} > 0.1$. Figure 2 shows the correlation between CHIMERA and Microball reaction plane orientations for Au+Au reactions (top left panel) and Au+Empty Frame (top right panel) data, relative to each other according to the integrated beam intensity. In top left panel, the strong anti-correlation for on-target reactions is evident, as expected because CHIMERA covers the forward and the Microball mainly the backward hemispheres in the c.m. frame. In runs with empty target frames, the recorded yields are low and only a weak positive correlation is observed. The difference between the two reaction plane orientations, normalized with respect to the integrated beam intensity, is presented in bottom left panel. In order to minimize the contributions of non-target collisions in the data analysis, an anti-correlation of the CHIMERA and Microball reaction-plane orientations was required. The condition $|\Phi_{RP}^{CHI} - \Phi_{RP}^{\mu Ball}| > 90°$ was applied, resulting in a relative weight of background reactions of less than 20%. This clearly shows the relevance of the Microball data in rejecting background reactions. Bottom right panel of Fig. 2 shows the LAND raw multiplicity (number of fired paddles) normalized with respect to the integrated beam intensity for Au+Au and Au+no target data, after applying the indicated CHIMERA-Microball anticorrelation condition. The contribution from non-target backgrounds in the kinematic region of LAND is apparently rather weak.

For the impact parameter selection global variables were constructed from the CHIMERA and AToF data. Strong correlations between the used global variables and a good agreement between experimental data and UrQMD calculations was found. The reaction plane orientation has been obtained by merging CHIMERA and AToF data Using the method of [10] we have also estimated the correction factors for the Fourier parameters of the azimuthal distribution, i.e. v_1 and v_2 attenuation factors. Azimuthal distributions of neutrons and light charged particles measured with LAND with respect to the reaction plane given by CHIMERA have been extracted for Au+Au reactions from runs with and without the shadow bar. After subtracting the background, the obtained distributions have been fitted with formula (1) of Ref. [1] in order to determine the direct (v_1) and elliptic (v_2) flow coefficients. Due to insufficient energy resolution in the LAND veto-wall, charge identification with the ΔE-vs.-time-of-flight technique has not been possible. Therefore, the results for neutrons and for all recorded charged particles (Ch) are presented. The v_1 and v_2 parameters for Ch and neutrons

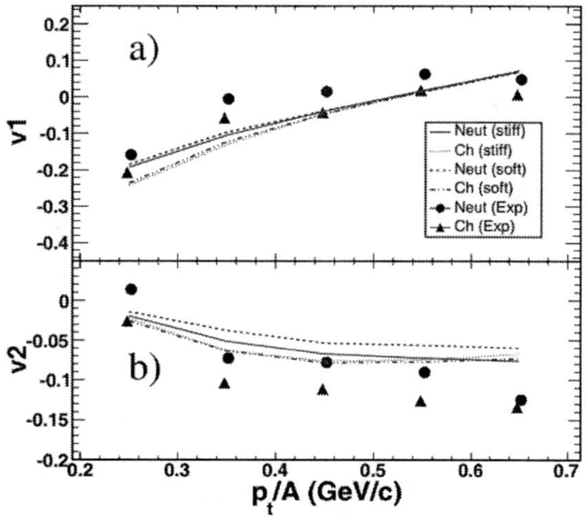

Figure 3: Measured flow parameters v_1 (top) and v_2 (bottom) for impact parameter b < 7.5 fm in Au+Au collisions at 400 MeV/nucleon for neutrons (black full dots) and charged particles (black full triangles) as a function of the transverse momentum p_t/A. The UrQMD predictions for neutrons and charged particles obtained with a stiff ($\gamma = 1.5$, red full and dotted lines, respectively) and a soft ($\gamma = 0.5$, blue dashed and dash-dotted lines, respectively) density dependence of the symmetry term have been filtered to correspond to the geometrical acceptance of the experiment.

as functions of the transverse momentum p_t/A are shown in Fig. 3, respectively. In the same figure, also the UrQMD predictions for a soft (γ=0.5) and a stiff (γ=1.5) symmetry energy potential [1] are given. Constraints for the symmetry energy were determined as in Ref. [1] by comparing the ratios of the elliptic flows of neutrons and Ch, v_2^n/v_2^{Ch}, with the corresponding UrQMD predictions for the soft and stiff cases. The results obtained as functions of the transverse momentum p_t/A are shown in Fig. 4. A preliminary value for the power-law coefficient γ, deduced by linearly interpolating between the predictions, is $\gamma = 0.75 \pm 0.10$. The new constraint is compatible with the previous one presented in Ref. [1] by using FOPI-LAND data and the same UrQMD model but the statistical error has been significantly reduced by a factor of ~ 2. The determination of systematic uncertainties is still on-going; a preliminary analysis indicates that they are substantial and may lead to an overall final uncertainty of the order of $\Delta\gamma \approx \pm 0.20$. Comparisons of the experimental data with other theoretical transport models will be useful in order to pursue the route towards a model-independent constraint of the high-density symmetry energy initiated in Ref. [3]. The

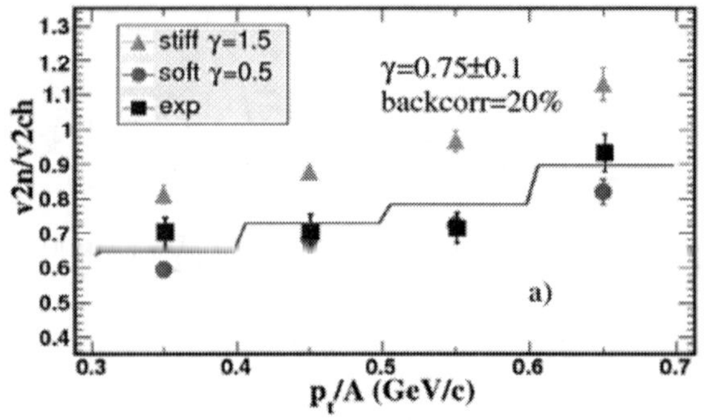

Figure 4: Elliptic flow ratio of neutrons and charged particles for moderately central ($b < 7.5$ fm) collisions as a function of the transverse momentum per nucleon p_t/A. The full squares represent the experimental data, the triangles and dots represent the UrQMD predictions for stiff (γ=1.5) and soft (γ=0.5) power-law exponent of the potential term. The full line is the result of a linear interpolation between the predictions, leading to the indicated γ=0.75± 0.10.

promising preliminary results of the present experiment may also be seen as an encouragement for extending the measurement of neutron and charged particle flows to other reaction systems and energies. Future experiments may thereby benefit from the unique possibilities offered by the NeuLAND detector presently constructed as part of the R^3B experimental set-up and from the availability of radioactive ion beams for reaction studies at FAIR.

References

[1] P. Russotto et al., Phys. Lett. **B697** (2011) 471; M.D. Cozma, Phys. Lett. **B700** (2011) 139; P. Russotto et al., Eur. Phys. Jour. **A50** (2014) 38; Y. Leifels et al., Phys. Rev. Lett. **71** (1993) 963.

[2] Q. Li et al., J. Phys **G** 31 (2005) 1359.

[3] M.D. Cozma et al., Phys. Rev. **C88** (2013) 044912.

[4] Th. Blaich et al., NIM **A314** (1992) 136.

[5] J. Łukasik et al., Nucl. Instr. and Meth. **A709** (2013) 120.

[6] A. Schüttauf et al., Nucl. Phys. **A607** (1996) 457.

[7] D.G. Sarantites et al., Nucl. Instr. and Meth. **381** (1996) 418.

[8] A. Pagano et al., Nucl. Phys. **A734** (2004) 504.

[9] P. Danielewicz et al., Phys. Lett. **B157** (1985) 146.

[10] J.-Y. Ollitrault, preprint nucl-ex/9711003.

EPJ Web of Conferences **117**, 07011 (2016) DOI: 10.1051/epjconf/201611707011
NN2015

Impact of Fock terms on the isospin properties of nuclear matter

Bao Yuan Sun, Qian Zhao and Wen Hui Long

[1]School of Nuclear Science and Technology, Lanzhou University, Lanzhou 730000, People's Republic of China
[2]Key Laboratory of Special Function Materials and Structure Design of Ministry of Education, Lanzhou 730000, People's Republic of China

Abstract

Several topics on the isospin properties of nuclear matter studied within the density-dependent relativistic Hartree-Fock theory are summarized. In detail, the effects of the Fock terms on the nuclear symmetry energy are listed, including the extra enhancement from the Fock terms of the isoscalar meson-nucleon coupling channels, the extra hyperon-induced suppression effect originating from the Fock channel, self-consistent tensor effects embedded automatically in the Fock diagrams, the enhanced density-dependent isospin-triplet potential part of the symmetry energy at high densities, a reduced kinetic symmetry energy at supranuclear density and so on. The results demonstrate the importance of the Fork diagram, especially from the isoscalar meson-nucleon coupling channels, on the isospin properties of the in-medium nuclear force.

The isospin properties of nuclear matter, such as the nuclear symmetry energy, the nucleon isovector (symmetry) potential, and the neutron-proton effective mass splitting, play essential roles in studying several topics of nuclear structure, nuclear reactions and nuclear astrophysics [1–7]. Among them of great contemporary interest is the nuclear symmetry energy E_S [8], which has been investigated extensively within both phenomenological approaches and microscopic many-body theories using almost all available effective and/or realistic nuclear interactions, see, e.g., Refs. [9–14]. Due to our limited knowledge about the in-medium effects of nucleon-nucleon (NN) interactions and correlations along with technical difficulties to treat quantum many-body systems accurately, distinct deviations of predicted E_S values appear beyond the nuclear saturation density ρ_0.

For the description of the structural properties of nuclear systems, the covariant density functional (CDF) theory has achieved great success during recent decades [15–23]. One of the most popular representatives is the relativistic Hartree approach with the no-sea approximation, namely, the relativistic mean field (RMF) theory [15,24–26]. However, due to the limit of the approach itself, significant system degrees of freedom are missing in RMF, such as the one-pion exchange. The non-local effects of nuclear forces and the momentum dependence of self-energies are also hardly to be treated. With the growth of computational facilities and the development of new methods, such defects can be eliminated with the inclusion of exchange (Fock) terms, which generates a new CDF model — the relativistic Hartree-Fock (RHF) theory with density-dependent meson-nucleon couplings (i.e., the DDRHF theory) [27]. Substantial improvements are gained by DDRHF in the self-consistent description of the ground state properties [27–38], the excitation modes [39–42] of finite nuclei and nuclear astrophysics as well [43–46].

It has been shown from several studies that the inclusion of the Fock terms in the CDF theory could improve descriptions of the isospin-related nuclear structure. In fact, non-locality of the mean field induced by the Fock terms gives rise to the momentum dependence of the RHF self-energies, improving the isospin and energy dependence for the nucleon effective mass [27]. Besides the Fock terms in the isovector channel, those in the isoscalar channel are found also to play a dominant role in the evolution of the shell structure and the spin-orbit splitting [30]. Moreover, it has been demonstrated that the isoscalar Fock terms are essential for self-consistent description of the spin-isospin resonances like the GT and SD ones within RPA [39, 41] and the prediction of neutron star properties [43].

In this proceeding, we will give a short review about the impact of Fock terms on the isospin properties of nuclear matter within the DDRHF theory, in particular, its influence on the symmetry energy. The detailed discussion on these topics refers to Refs. [27, 43, 44, 46, 47].

The general formulae of the DDRHF theory in nuclear matter can be found in Refs. [27, 43, 46–48]. In RHF theory, the energy density functional (EDF) is obtained by taking the expectation value of the Hamiltonian with respect to the Hartree-Fock ground state, which consists of three parts: the kinetic EDF ε_k, the direct (Hartree) ε_ϕ^D and the exchange (Fock) ε_ϕ^E terms of the potential EDF. The meson-NN interaction vertices consist of various meson-nucleon coupling channels, namely, the Lorentz scalar (σ-S), vectors (ω-V, ρ-V), vector-tensor (ρ-VT), tensor (ρ-T) and pseudovector (π-PV) couplings. In DDRHF, the relativistic mass-energy relation has the form $E^{*2} = M^{*2} + \boldsymbol{p}^{*2}$, where the starred quantities are defined as

$$M^* = M + \Sigma_S(p), \qquad \boldsymbol{p}^* = \boldsymbol{p} + \hat{\boldsymbol{p}}\Sigma_V(p), \qquad E^* = E - \Sigma_0(p), \quad (1)$$

where Σ_S is the scalar self-energy, and Σ_0 and Σ_V correspond to the time and space components of the vector self-energy, respectively. M^* is also called the scalar mass or Dirac mass in several literatures. As a result, the binding energy per nucleon E_b in nuclear matter for a given baryonic density ρ_b and isospin asymmetry $\delta \equiv (\rho_n - \rho_p)/\rho_b$ can be obtained, and correspondingly the density-dependent symmetry energy is deduced from the second order coefficient of its Taylor expansion as

$$E_S(\rho_b) = \frac{1}{2}\frac{\partial^2 E_b(\rho_b,\delta)}{\partial\delta^2}\Big|_{\delta=0} = E_{S,k} + E_S^D + E_S^E, \quad (2)$$

including the kinetic part $E_{S,k}$, the Hartree E_S^D and Fock E_S^E potential parts.

The isospin dependence of the effective masses has been studied by defining its non-relativistic (M^*_{NR}) and relativistic (M^*_R) versions in the DDRHF theory [27]. As the Fock terms bring non-local effects into the RHF self-energies, the isospin splitting of the nucleon effective mass in RHF calculations is significantly different from those in RMF. It is revealed that for the effective masses M^*_{NR} at the corresponding Fermi energy E_F, the trend in RHF is that $M^*_{NR,n}(E_{F,n}) > M^*_{NR,p}(E_{F,p})$ at lower density but reversed when approaching to the saturation density.

The DDRHF theory has then been utilized to study the symmetry energy and neutron star properties. The inclusion of the Fock terms strongly affects the density-dependence of the symmetry energy at high densities, and in turn the radius and cooling process of neutron stars [43,44,46,47]. Especially, it is found that not only the isovector mesons but also the isoscalar mesons can contribute to the symmetry energy with the inclusion of the Fock diagrams of meson-nucleon couplings [43]. In fact, the isoscalar mesons via the Fock terms provide more significant contributions to the symmetry energy than the isovector ones in the high-density region.

Motivated by recent observational data, the equations of state with the inclusion of Λ hyperons and the corresponding properties of neutron stars are studied based on the DDRHF theory as well [44]. Because of the extra suppression effect originating from the Fock channel, large reductions induced by the occurrence of Λ hyperons on the symmetry energy are predicted at high densities, resulting in a relatively small predicted value of the neutron-star radius.

In addition, with the possibility to extract the nuclear tensor interaction directly from the Fock diagrams of various meson-nucleon couplings [38], distinct tensor effects are illustrated in the exploration of the isospin properties of nuclear matter and neutron star structures [46]. The naturally involved tensor force components in the Fock diagrams soften the density-dependent behavior of the symmetry energy and consequently lead neutron stars to be more compact. Then the threshold density for the direct Urca process that cools the neutron star rapidly is raised by the nuclear tensor force.

To clarify the isospin structure of in-medium NN interaction, the isospin coupling-channel decomposition of the symmetry energy is carried out within the DDRHF theory [47]. When the Fock terms is involved, it is found that both isospin-singlet and isospin-triplet components of the potential energy play an important role in determining the symmetry energy. A strong density-dependent behavior is revealed in the isospin-triplet potential part of the symmetry energy at high densities. The results demonstrate again the importance of the Fork diagram, especially from the isoscalar-meson coupling channels, on the isospin properties of the in-medium nuclear force.

Recently, the kinetic EDF ε_k has been identified as a particularly good indicator of the short range correlations due to the tensor force in the nuclear ground state [49–52]. It is found that the contribution from ε_k to the symmetry energy is strongly reduced and even becomes negative due to such correlations when compared to the non-interacting case [50–52]. In CDF theory, the kinetic EDF ε_k, which contains the self-energies, gets the in-medium NN interaction involved as well [43].

$$\varepsilon_k = \frac{1}{\pi^2}\sum_{i=n,p}\int_0^{k_{F,i}} M\hat{M}p^2dp + \frac{1}{\pi^2}\sum_{i=n,p}\int_0^{k_{F,i}} p\hat{P}p^2dp \equiv \varepsilon_k^M + \varepsilon_k^P, \quad (3)$$

where $\hat{M} = M^*/E^*$, $\hat{P} = \mathbf{p}^*/E^*$, and ε_k^M and ε_k^P denote the contributions from the nucleon mass and momentum, respectively. The hatted quantity $\hat{M}$ could be further separated into

$$\hat{M} = \frac{M}{E^*} + \frac{\Sigma_S^D}{E^*} + \sum_\phi \frac{\Sigma_S^{E,\phi}}{E^*}. \quad (4)$$

Thus, the in-medium effects hidden in ε_k could be investigated in detail within the CDF theory, in particular the contribution due to the short range correlations [47]. It is found that the inclusion of the Fock terms in DDRHF causes a sizable reduction on the kinetic part $E_{S,k}$ of the symmetry energy, as compared to the RMF results [47]. At supranuclear density region, the DDRHF results even become negative, as seen in the left panel of Fig. 1.

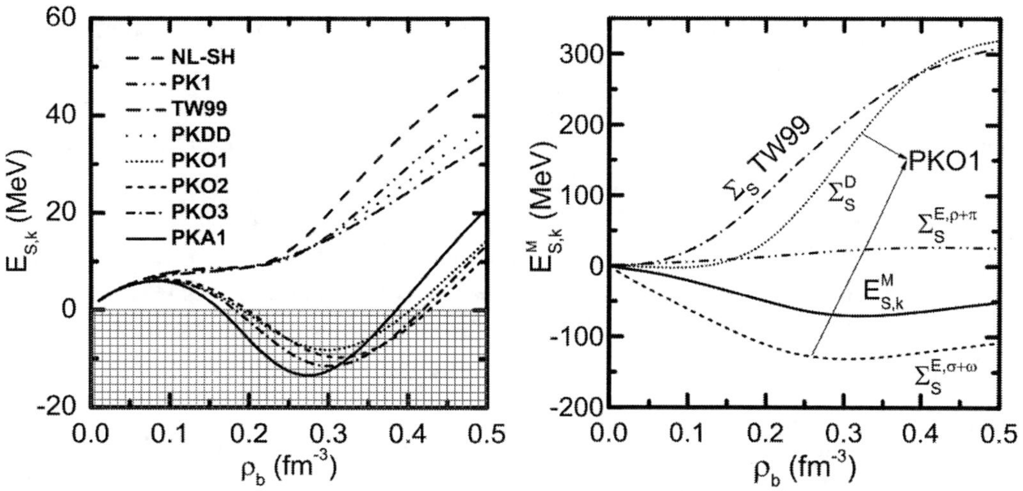

Figure 1: Left panel: the kinetic energy part $E_{S,k}$ of the nuclear symmetry energy as a function of baryonic density ρ_b. The results are calculated with RHF models PKO1, PKO2, PKO3 and PKA1, in comparison with RMF ones NL-SH, PK1, TW99 and PKDD. Right panel: the mass-related term $E^M_{S,k}$ of the kinetic symmetry energy is separated according to various components of the self-energy, see Eq.(4) for details. The results are calculated with RHF model PKO1, in comparison with RMF one TW99.

To identify the physical mechanism, $E_{S,k}$ is further decomposed into the mass- and momentum-related parts according to Eq. (3), namely, $E^M_{S,k}$ and $E^P_{S,k}$, and the deviations between RMF and RHF results of $E_{S,k}$ are actually dominated by the mass-related term $E^M_{S,k}$ [47]. Such model deviations could be further clarified in the right panel of Fig. 1, where the contributions to $E^M_{S,k}$ from different components of the scalar self-energy Σ_S are extracted according to Eq. (4). It is found that the direct terms of the scalar self-energy Σ^D_S in both RMF and RHF present similar contributions to $E^M_{S,k}$. For the Fock terms in PKO1, i.e., $\Sigma^{E,\sigma+\omega}_S$ and $\Sigma^{E,\rho+\pi}_S$, remarkably negative contribution to $E^M_{S,k}$ is found from the isoscalar-meson coupling channel, while relatively weak but still considerable contribution results from the isovector-meson coupling channel, eventually leading to a significant difference in the sign and magnitude of $E_{S,k}$ between RMF and RHF results. In fact, it has been demonstrated very recently that the nuclear tensor interactions are naturally embedded in both the isoscalar and the isovector meson-nucleon coupling channels with the presence of the Fock diagrams in the RHF EDF [38,46]. In consequence, the extra contributions from the Fock terms of Σ_S to the kinetic symmetry energy could be regarded partly as the effects of the nuclear tensor-force components therein, see Refs. [46,47].

In short, present status of DDRHF studies on the isospin properties of nuclear matter is summarized, especially the effects of the Fock terms on the nuclear symmetry energy such as those from the hyperon and the nuclear tensor force. The reduction of the kinetic symmetry energy at supranuclear density in DDRHF models is discussed as well, which is related to the fact of a remarkably negative contribution from the isoscalar-meson coupling channel via Fock terms. The results manifest the importance of the Fork diagram in the CDF theory, essentially from the isoscalar-meson coupling channels, on the isospin properties of the in-medium nuclear force.

Acknowledgements

This work is partly supported by the National Natural Science Foundation of China under Grants No. 11205075 and No. 11375076, and the Specialized Research Fund for the Doctoral Program of Higher Education under Grants No. 20120211120002 and No. 20130211110005.

References

[1] Brown B. A., Phys. Rev. Lett., **85** (2000) 5296.

[2] Lattimer J. M. and Prakash M., Phys. Rep., **333** (2000) 121.

[3] Baran V. et al., Phys. Rep., **410** (2005) 335.

[4] Steiner A. W. et al., Phys. Rep., **411** (2005) 325.

[5] Li B. A., Chen L. W. and Ko C. M., Phys. Rep., **464** (2008) 113.

[6] Tamii A. et al., Phys. Rev. Lett., **107** (2011) 062502.

[7] Newton W. G. et al., Astrophys. J. Lett., **779** (2013) L4.

[8] Li B. A. et al., Eur. Phys. J. A, **50** (2014) 9.

[9] Chen L. W., Ko C. M. and Li B. A., Phys. Rev. C, **72** (2005) 064309.

[10] Chen L. W., Ko C. M. and Li B. A., Phys. Rev. C, **76** (2007) 054316.

[11] Akmal A. et al., Phys. Rev. C, **58** (1998) 1804.

[12] Dalen V. E., Fuchs C. and Faessler A., Phys. Rev. C, **72** (2005) 065803.

[13] Taranto G. et al., Phys. Rev. C, **87** (2013) 045803.

[14] Lopes L. L. and Menezes D. P., Phys. Rev. C, **89** (2014) 025805.

[15] Serot B. D. and Walecka J. D., Adv. Nucl. Phys., **16** (1986) 1.

[16] Reinhard P. G., Rep. Prog. Phys., **52** (1989) 439.

[17] Ring P., Prog. Part. Nucl. Phys., **37** (1996) 193.

[18] Bender M. et al., Rev. Mod. Phys., **75** (2003) 121.

[19] Vretenar D. et al., Phys. Rep., **409** (2005) 101.

[20] Meng J. et al., Prog. Part. Nucl. Phys., **57** (2006) 470.

[21] Paar N. et al., Rep. Prog. Phys., **70** (2007) 691.

[22] Liang H. Z., Meng J. and Zhou S. G., Phys. Rep., **570** (2015) 1.

[23] Meng J. and Zhou S. G., J. Phys. G, **42** (2015) 093101.

[24] Miller L. D. and Green A. E. S., Phys. Rev. C, **5** (1972) 241.

[25] Walecka J. D., Ann. Phys. (NY), **83** (1974) 491.

[26] Boguta J. and Bodmer A., Nucl. Phys. A, **292** (1977) 413.

[27] Long W. H., Van Giai N. and Meng J., Phys. Lett. B, **640** (2006) 150.

[28] Long W. H. et al., Phys. Rev. C, **76** (2007) 034314.

[29] Long W. H. et al., EPL, **82** (2008) 12001.

[30] Long W. H. et al., Phys. Lett. B, **680** (2009) 428.

[31] Long W. H. et al., Phys. Rev. C, **81** (2010) 024308.

[32] Long W. H. et al., Phys. Rev. C, **81** (2010) 031302(R).

[33] Ebran J. P. et al., Phys. Rev. C, **83** (2011) 064323.

[34] Lu X. L., Sun B. Y. and Long W. H., Phys. Rev. C, **87** (2013) 034311.

[35] Niu Z. M. et al., Phys. Rev. C, **87** (2013) 037301.

[36] Wang L. J. et al., Phys. Rev. C, **87** (2013) 047301.

[37] Wang L. J. et al., Phys. Rev. C, **87** (2013) 054331.

[38] Jiang L. J. et al., Phys. Rev. C, **91** (2015) 034326.

[39] Liang H. Z. et al., Phys. Rev. Lett., **101** (2008) 122502.

[40] Liang H. Z., Van Giai N. and Meng J., Phys. Rev. C, **79** (2009) 064316.

[41] Liang H. Z., Zhao P. W. and Meng J., Phys. Rev. C, **85** (2012) 064302.

[42] Liang H. Z. et al., Phys. Rev. C, **86** (2012) 021302(R).

[43] Sun B. Y. et al., Phys. Rev. C, **78** (2008) 065805.

[44] Long W. H. et al., Phys. Rev. C, **85** (2012) 025806.

[45] Niu Z. M. et al., Phys. Lett. B, **723** (2013) 172.

[46] Jiang L. J. et al., Phys. Rev. C, **91** (2015) 025802.

[47] Zhao Q., Sun B. Y. and Long W. H., J. Phys. G, **42** (2015) 095101.

[48] Bouyssy A. et al., Phys. Rev. C, **36** (1987) 380.

[49] Vidaña I. et al., Phys. Rev. C, **84** (2011) 062801(R).

[50] Carbone A., Polls A. and Rios A., EPL, **97** (2012) 22001.

[51] Xu C., Li A. and Li B. A., J. Phys. Conf. Ser., **420** (2013) 012090.

[52] Hen O. et al., Phys. Rev. C, **91** (2015) 025803.

EPJ Web of Conferences **117**, 07012 (2016) DOI: 10.1051/epjconf/201611707012
NN2015

Three-body force effect on the properties of neutron-rich nuclear matter

Wei Zuo[1], Ignazio Bombarci[2] and Umberto Lombardo[3]

[1] Institute of Modern Physics, Chinese Academy of Sciences, Lanzhou 730000, China
and State Key Laboratory of Theoretical Physics, Institute of Theoretical Physics, Chinese Academy of Sciences, Beijing 100190, China
[2] Dipartimento di Fisica "E. Fermi", Università di Pisa, and INFN, Sezione di Pisa, Largo B. Pontecorvo, 3 I-56127 Pisa, Italy
[3] Università di Catania and INFN - Laboratori Nazionali del Sud, Catania 95123, Italy

Abstract

We review our research work in recent years on the properties of neutron-rich nuclear matter within the framework of the Brueckner theory extended to include a microscopic three-body force (TBF). We discuss specially the TBF effect on the equation of state and the single-particle (s.p.) potentials. The TBF turns out to be crucial for describing the saturation properties of nuclear matter in nonrelativistic microscopic framework. The TBF effect on the EOS of neutron-rich nuclear matter is to result in a significant stiffening of its isovector part at supra-saturation densities. Within the Brueckner approach, the TBF may lead to a rearrangement contribution to the s.p. potentials, which enhances strongly the repulsion and momentum-dependence of the s.p. potentials at high densities and high momenta. Our results are also compared with the predictions by other *ab initio* approaches.

The equation of state (EOS) and single-particle properties of neutron-rich nuclear matter play an important role in many issues of nuclear physics and nuclear astrophysics, including the structure and collective properties of heavy nuclei and neutron-rich nuclei [1–7]. The isovector part of the EOS of neutron-rich nuclear matter is determined by nuclear symmetry energy. At densities around and below nuclear saturation density, symmetry energy is not only closely related to the structure of neutron-rich nuclei away from the nuclear stability line [8], it is also shown to be decisive for explaining the experimentally observed enhancement of the stability against α-decay with increasing the mass number along an isotope chain for the synthesized superheavy nuclei not around shell closures [9]. At high-densities well above the normal density, the density dependence of symmetry energy is expected to be extremely important for modeling the structure and cooling properties of neutron stars.

Theoretically, the properties of neutron-rich nuclear matter can be predicted within various nuclear many-body frameworks, including both phenomenological and microscopic approaches. In the phenomenological methods, the parameters of the effective interactions are essentially obtained by reproducing the ground state properties of stable nuclei and/or the saturation properties of nuclear matter, and the predicted high-density behavior using different parameter sets may be largely different and even be opposite [10]. Within the *ab initio* approaches, the properties of neutron-rich nuclear matter have been investigated extensively in recent years. Almost all the *ab initio* predictions indicate an monotonically increasing symmetry energy as a function of density, whereas the stiffness of the density dependence predicted by adopting different approaches and/or different NN interactions turns out to be largely different at high densities. In our investigation, we use the Breuckner-Hartree-Fock (BHF) method based on the BBG theory [11]. The extension of the BBG scheme to asymmetric nuclear matter and to include microscopic three-body forces can be found in Refs. [12, 13] and Refs. [14–16], respectively.

As well known, a nonrelativistic many-body approach of rigid nucleons interacting via realistic two-body forces fitting in-vacuum nucleon-nucleon scattering data can not reproduce the empirical saturation properties of nuclear mattery [17]. To solve this problem, one has to introduce three-body forces or to adopt a relativistic theory, such as Dirac-BHF method. Up to now, several different kinds of TBF model have been applied in the BHF calculations. One is the semiphenomenological TBF model, such as the Urbana TBF, in which the adjustable parameters are usually determined by fitting the observed triton binding energies and/or the empirical saturation properties of symmetric nuclear matter. Another kind of TBF model adopted in the Brueckner theory is the microscopic one [15, 16, 18] based on the meson exchange theory for NN interactions. There is no adjustable parameters in the sense that the meson parameters are essentially

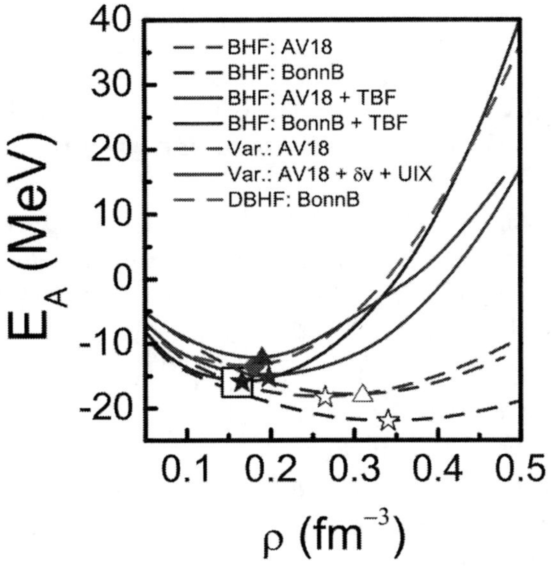

Figure 1: EOS of symmetric nuclear matter predicted by different microscopic approaches. The results of the BHF approach are taken from Refs. [16] and [18]. The Dirac-BHF prediction is taken from Ref. [19]. The results by the variational approach are from Ref. [20].

determined self-consistently by the corresponding two-body force. The extension to include the σ and ω exchanges as well as the associated virtual nucleon-antinucleon pair excitations and nucleon resonances have been done by Grange *et al.* [15]. Further improvement and development of the model have been achieved in Refs. [14, 16, 18]. In our BHF calculation, the microscopic TBFs have been adopted and the TBF contribution is included by reducing the TBF to an equivalently effective two-body interaction according to the standard and extensively adopted scheme [15].

In Fig. 1, we show the TBF effect on the EOS of symmetric nuclear matter. In the figure, the predictions from different *ab initio* theoretical frameworks are compared, including the BHF approach [16, 18], the many-body variational method [20], and the Dirac-BHF theory [19]. It is clear that the TBF provides a repulsive contribution to the nuclear EOS in both the nonrelativistic BHF and variational approaches, and the TBF repulsion increases monotonically as a function of density. Inclusion of the TBF contribution turns out to be necessary for reproducing the empirical saturation property of nuclear matter in a non-relativistic microscopic framework. Within the BHF framework, the predicted saturation densities are approximately 0.265fm^{-3} and 0.33fm^{-3} by using the AV18 and BonnB two-body interactions, respectively, which are much larger than the empirical value and the Dirac-BHF prediction of roughly 0.18fm^{-3}. By including the corresponding TBFs, the saturation densities are improved remarkably to about 0.167fm^{-3} and 0.19fm^{-3}.

In comparison with the Dirac-BHF theory, it has been shown [16, 18] quantitatively that the main relativistic correction to the EOS of nuclear matter can be reproduced within the nonrelativistic BHF framework by the TBF component involving the virtual excitations of nucleon-antinucleon pairs due to the 2σ-meson exchange.

As for neutron-rich nuclear matter, the energy per nucleon has been shown to fulfil satisfactorily a linear dependence on β^2 in the whole asymmetry range of $0 \leq \beta \leq 1$ by various *ab initio* investigations by adopting different realistic NN interactions, i.e., $E_A(\rho, \beta) = E_A(\rho, 0) + E_{sym}(\rho)\beta^2$. Inclusion of the TBF effect does not destroy the linear dependence of $E_A(\rho, \beta)$ on β^2 [16]. The simple β^2-law implies that the isospin dependence of the EOS of neutron-rich nuclear matter is governed completely by the symmetry energy. Due to the linear dependence of $E_A(\rho, \beta)$ on β^2, the symmetry energy can be calculated directly by the difference between the EOS of pure neutron matter and that of symmetric nuclear matter, i.e., $E_{sym}(\rho) = E_A(\rho, 1) - E_A(\rho, 0)$. Another important consequence of the β^2-law is that the difference of the neutron and proton chemical potentials in neutron star matter can be related to the symmetry energy in a explicitly way: $\mu_n - \mu_p = 4\beta E_{sym}$.

To see clearly the TBF effect on the EOS of neutron-rich nuclear matter, we display, in Fig. 2, the density dependence of symmetry energy predicted within various *ab initio* frameworks, including the BHF approach [16, 18], the variational methed [20], and the Dirac-BHF theory [21]. At low densities around and below the saturation density, the TBF effect turns out to be reasonably weak, and the difference among the different *ab initio* predictions is seen to be quite small. It is worth noticing that the symmetry energy predicted by the three different *ab initio* approaches and/or using different realistic NN interactions increases monotonically as a function of density. However, the density dependence of symmetry energy from different approaches may become quite different at high densities. The TBF effect on symmetry energy is repulsive. The TBF repulsion increases rapidly at supra-saturation density, and leads to a stiffening of the density dependence of symmetry energy at high densities. It is worth noticing that the TBF effect on symmetry energy is much stronger at high densities within the

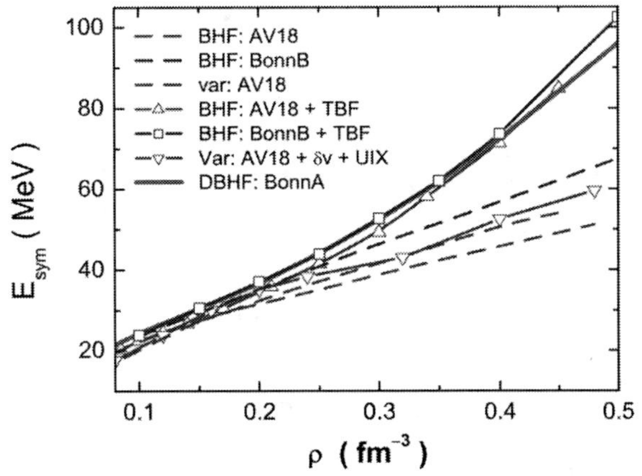

Figure 2: Density dependence of symmetry energy predicted within three different *ab initio* theoretical frameworks. The results of the BHF approach are from Refs. [16] and [18]. The Dirac-BHF prediction is taken from Ref. [21]. The results by the variational approach are from Ref. [20].

BHF framework than that within the variational framework. Inclusion of the TBFs in the BHF calculations makes the predicted symmetry energy almost completely coincide with the Dirac-BHF prediction up to $\rho = 0.5\text{fm}^{-3}$. Whereas, the symmetry energy obtained within the variational approach using the AV18 interaction plus the UIX TBF and the boost correction remains far away from the Dirac-BHF result at high densities well above the saturation density.

At the lowest mean field approximation, there exist two problems in the BHF approach for predicting nuclear s.p. properties. First, at densities around the saturation density, the predicted optical potential depth is too deep as compared to the empirical value, and it destroy the Hugenholtz-Van Hove (HVH) theorem. This problem essentially by including the effect of ground state correlations in calculating the s.p. properties. Second, at high densities, the predicted potential is too attractive and its momentum dependence turns out to be too weak at the lowest-order BHF approach for describing the experimental elliptic flow data [22]. Concerning these two problems, we have improved the BHF calculation of the s.p. properties inn two aspects. One is to extend the calculation of the effect of ground state correlations to neutron-rich nuclear matter. The second is to take into account the TBF-induced rearrangement contribution for calculating the s.p. properties within the Brueckner theory.

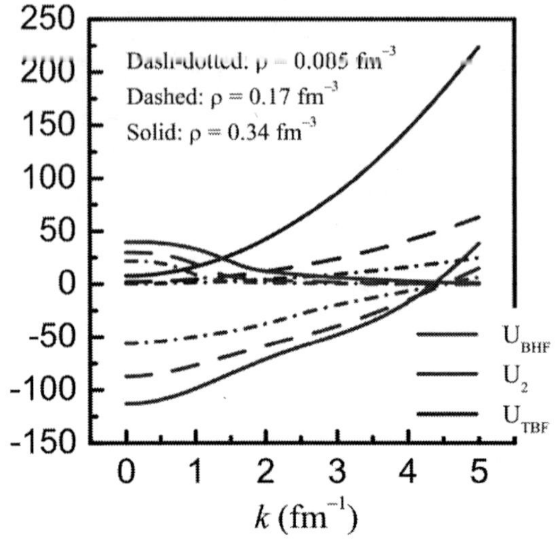

Figure 3: The three different contributions $U_{\rm BHF}$, U_2, and $U_{\rm TBF}$ in symmetric nuclear matter at three densities. Taken from Ref. [23].

In Fig. 3, we show separately the three different contributions (i.e., the BHF s.p. potential $U_{\rm BHF}$, the Pauli rearrangement term U_2, and the TBF rearrangement contribution $U_{\rm TBF}$) in symmetric nuclear matter at there typical densities of $\rho = 0.085$, 0.17, and 0.34 fm^{-3}. The BHF s.p. potential is strongly attractive at low momenta and its attraction decreases as a function of momentum. The g.s. correlations affect the s.p. potential mainly at low momenta, and lead to a repulsive contribution U_2. which is considerably smaller in magnitude than the lowest-order BHF contribution $U_{\rm BHF}$. The contribution of g.s. correlations decreases rapidly around the Fermi mumentum and vanishes at high momenta well above the Fermi momentum. The TBF induces a strongly repulsive and momentum-dependent rearrangement contribution to the nucleon s.p. potential, at high densities and momenta in nuclear medium, which has been shown [24,25] to be crucial for improving the *large-density* and *high-momentum* behavior of the BHF s.p. potential. For a detailed discussion about the isospin-dependence of the three different contributions in neutron-rich nuclear matter, we refer to our review paper [1].

In summary, We have reported briefly our research work on the properties of neutron-rich nuclear matter within the framework of the Brueckner theory. Special attention has been paid to the discussion of the TBF effect and the comparison of our results with the predictions by other *ab initio* approaches. It is shown that the TBF effect provides a repulsive contribution to the EOS of nuclear matter and improves remarkably the predicted saturation properties within a nonrelativistic *ab initio* approach. Inclusion of the TBF does not destroy the empirical parabolic law fulfilled by the EOS of neutron-rich nuclear matter. The TBF may lead to a strong enhancement of the stiffness of symmetry energy at high densities. In predicting the s.p. properties in neutron-rich nuclear matter, we have improved the Brueckner calculation in two aspects. One is to extend the calculation of the g.s. correlation effect to neutron-rich nuclear matter; the second is to include the rearrangement contribution induced by the TBF. The TBF-induced rearrangement contribution turns out to be strongly repulsive and momentum-dependent at high densities and momenta, which is necessary for improving the large-density and high-momentum behavior of the s.p. potentials.

The work is partly supported by the National Natural Science Foundation of China (11435014,11175219), the 973 Program of China (Grant No. 2007CB815004), and the Knowledge Innovation Project(KJCX2-EW-N01) of the Chinese Academy of Sciences.

References

[1] Wei Zuo, Ignazio Bombaci, and Umberto Lombardo, Eur. Phys. J. A **50** (2014) 12 and reference therein.

[2] B. A. Li, L. C. Chen and C. M. Ko, Phys. Rep. **464**, (2008) 113

[3] P. Danielewicz, R. Lacey and W. G. Lynch, Science **298**, (2002) 1592.

[4] W. Zuo, A. Li, Z. H. Li and U. Lombardo, Phys. Rev. C **70**, (2004) 055802.

[5] M. Baldo and C. Maieron, J. Phys. G **34**, (2007) R243.

[6] C. Fuchs and H. H. Wolter, Eur. Phys. J. A **30**, (2006) 5.

[7] V. Barana, M. Colonna, V. Grecoc and M. Di Toro, Phys. Rep. **410**, (2005) 335.

[8] J. M. Dong, W. Zuo, J. Z. Gu, and U. Lombardo, Phys. Rev. C **85**, (2012) 034308.

[9] J. M. Dong. W. Zuo and W. Scheid, Phys. Rev. Lett. **107**, (2011) 012501.

[10] L. W. Chen, C. M. Ko and B. A. Li, Phys. Rev C **72**, (2005) 064309.

[11] B. D. Day, Rev. Mod. Phys. **39**, (1967) 719.

[12] I. Bombaci, U. Lombardo, Phys. Rev. **C 44**, (1991) 1892.

[13] W. Zuo, I. Bombaci, and U. Lombardo, Phys. Rev. C **60**, (1999) 024605.

[14] W. Zuo, A. Lejeune, U. Lombardo and J. F. Mathiot, Eur. Phys. J. A **14**, (2002) 469.

[15] P. Grangé, A. Lejeune, M. Martzolff, and J. F. Mathiot, Phys. Rev. C **40**, (1989) 1040.

[16] W. Zuo, A. Lejeune, U. Lombardo and J. F. Mathiot, Nucl. Phys. A **706**, (2002) 418.

[17] Z. H. Li, U. Lombardo, H. J. Schulze, W. Zuo, L. W. Chen and H. R. Ma, Phys. Rev. C **74**, (2006) 047304.

[18] Z. H. Li, U. Lombardo, H. J. Schulze and W. Zuo, Phys. Rev. C **77**, (2008) 034316.

[19] R. Brockmann and R. Machleidt, Phys. Rev. C **42**, (1990) 1965.

[20] A. Akmal, V. R. Pandharipande, and D. G. Ravenhall, Phys. Rev. C **58**, (1998) 1804.

[21] E. N. E. van Dalen, C. Fuchs and A. Faessler, Nucl.Phys. A **744**, (2004) 227; Phys. Rev. C **72**, (2005) 065803.

[22] P. Danielewicz, Nucl. Phys. A **673**, (2000) 375.

[23] W. Zuo, Nucl. Phys. A **834**, (2010) 574c.

[24] W. Zuo, L. G. Cao, B. A. Li, U. Lombardo and C. W. Shen Phys. Rev. C **72** (2005) 014005.

[25] W. Zuo, U. Lombardo, H. J. Schulze and Z. H. Li, Phys. Rev. C **74**, (2006) 014317.

EPJ Web of Conferences **117**, 07013 (2016) DOI: 10.1051/epjconf/201611707013
NN2015

Study of ^{12}C excited states decaying into three α particles using the thick target inverse kinematic technique

M. Barbui[1], K. Hagel[1], J. Gauthier[1], S. Wuenschel[1], R. T. deSouza[2], S. Hudan[2], D. Fang[3], V.Z. Goldberg[1], H. Zheng[1], G. Giuliani[1], G. Rapisarda[1], E-J. Kim[1], X. Liu[1,4] and J.B. Natowitz[1]

[1] Cyclotron Institute, Texas A&M University, College Station, TX, USA
[2] Indiana University, Bloomington, IN, USA
[3] Shanghai Institute of Applied Physics (SINAP), Chinese Academy of Sciences, Shanghai, China
[4] Institute of modern physics, Chinese Academy of Sciences, Lanzhou, China

Abstract

We will show that the Thick Target Inverse Kinematics (TTIK) technique can be used to investigate the breakup of excited self-conjugate nuclei into many alpha particles. Two test runs were performed at Cyclotron Institute of Texas A&M University to study the reaction ^{20}Ne+α at maximum beam energies of 10 and 12 AMeV. Due to the limited statistics, only events with alpha multiplicity up to three were analyzed. The analysis of the three α-particle emission data allowed the identification of the Hoyle state and other ^{12}C excited states decaying into three alpha particles. The results will be shown and compared with other data available in the literature.

Searching for alpha cluster states analogous to the ^{12}C Hoyle state in heavier alpha-conjugate nuclei can provide tests of the existence of alpha condensates in nuclear matter. Such states are predicted for ^{16}O, ^{20}Ne, ^{24}Mg, ^{28}Si etc. at excitation energies slightly above the multi-α particle

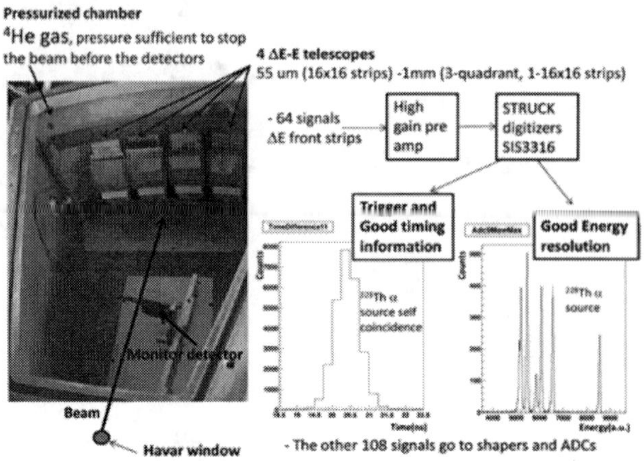

Figure 1: Experimental setup and scheme of the electronics. Good energy and time resolution are obtained by using the STRUCK digitizers SIS3316

decay threshold [1-3]. The Thick Target Inverse Kinematics (TTIK) [4] technique can be successfully used to study the breakup of excited self-conjugate nuclei into many alpha particles. This technique is suited for this purpose because it allows the exploration of a large range of incident energies in the same experiment. In inverse kinematics, the reaction products are focused at forward angles and can be detected with detectors covering a relatively small portion of the solid angle in the forward direction. Since we stop the beam in the gas target volume we can detect the particles emitted at zero degrees. A first test run was performed at Cyclotron Institute at Texas A&M University to study the reaction ^{20}Ne+α. A ^{20}Ne beam at 11 AMeV was provided by the TAMU K150 Cyclotron. After the Havar entrance window the energy of the beam was 10 AMeV. The pressure of the ^{4}He gas was 3800 mbar, sufficient to stop the beam before reaching the detectors. Here the TTIK method was used to study both single α-particle emission and multiple α-particle decays. Due to the limited statistics, only events with alpha multiplicity up to three were analyzed. A description of the experimental setup used in this run and the results regarding the events with α multiplicity 1 and 2 are reported in ref [5]. The analysis of the three α-particle emission data allowed the identification of the Hoyle state and other ^{12}C excited states decaying into three alpha particles. These results will be presented in the following.

We recently improved the experimental setup increasing the solid angle coverage and the detector granularity. Figure 1 shows a picture of the new detector arrangement and some details. Four DeltaE-E telescopes placed at

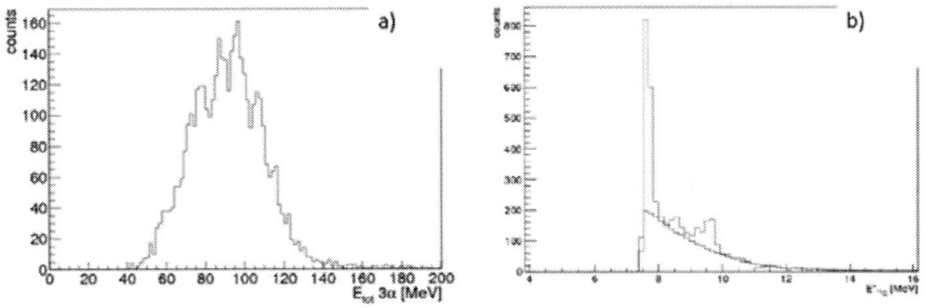

Figure 2: Panel a): Sum of the energy of the 3 alpha particles. Panel b): Excitation energy of ^{12}C. The red line shows the total events the blue line shows the uncorrelated events obtained by mixing α particles from random events

the end of the pressurized chamber are used to detect the reaction products. The signals from the front strips of the DeltaE detectors are processed by high gain pre-amplifiers from Indiana University and digitized using Struck SIS1366 Flash ADCs. Those digitizers provide the energy and time information relative to the cyclotron radio frequency. As before the particle identification is obtained from the two dimensional scatter plots Time-Energy and DeltaE-E. The new setup was tested on December 2014 using a ^{20}Ne beam at 13 AMeV provided by the TAMU K150 Cyclotron on ^{4}He gas at a pressure of 4964 mbar. The overall statistics collected during this test run was 1/3 of that collected in the previous run. However the results are consistent with those obtained in the first test run.
The analysis of the events with α multiplicity three was carried out in two steps. First we analyzed the events where three alpha particles were detected in the central telescope; second we improved the statistics considering the events where two alpha particles were detected in the central telescope and the third could be elsewhere. In both cases we selected the events where the 3 α particles arrived to the detectors within a time window of 15 ns.

Fig. 2-a shows the sum of the energies of the three alpha particles detected in the central telescope. This spectrum is obtained from the measured energies without any assumption or energy loss correction. It is interesting to note that the total energy spectrum shows a series of peaks. Since the detected alpha particles are correlated in time and position we can reasonably assume that they are coming from an excited ^{12}C decaying into three α-particles. If we assume that an excited state in ^{24}Mg decayed into two ^{12}C, one in the ground state, the other with enough excitation energy to split in three α-particles, we can use a recursive procedure to reconstruct

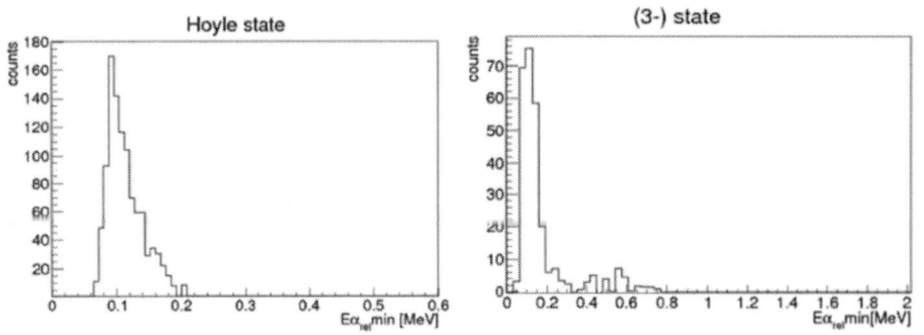

Figure 3: Left panel: minimum relative energy of two alpha particles for the Hoyle state. Right panel: minimum relative energy of two alpha particles for the (3-) state.

the position of the interaction point, based the reaction kinematics and the energy and momentum conservation. The excitation energy of the ^{12}C splitting into three α-particles is obtained from the sum of the kinetic energies of the three α-particles in the center of mass of the ^{12}C and the Q value. It is important to note that the error on the determination of the ^{12}C excitation energy due to an incorrect determination of the interaction point is minimal. In fact, the three detected α-particles have similar and quite large energy in the laboratory system and travel similar flight-paths so that energy loss correction almost cancels out when we calculate ^{12}C excitation energy. The ^{12}C excitation function is shown in Fig. 2-b, together with the spectrum of the uncorrelated events obtained by randomly mixing three alpha particle energies from different events. The spectrum in Fig. 2-b shows two peaks, one at 7.65 MeV, corresponding to the energy of the Hoyle state, and one at 9.64 MeV, corresponding to a (3-) state. In order to determine if the ^{12}C decay is proceeding through the ground state of ^{8}Be we calculated event by event the relative energy of the three possible couples of alpha particles. The minimum of these relative energies is plotted in Fig.3 for the Hoyle state and the (3-) state. In both cases this relative energy peaks around 92 keV, showing that the decay proceeded through the ^{8}Be ground state.

To improve the statistics and allow the detection of higher excitation energy states we extended the event selection to those events with two alpha particles in the central telescope and the third elsewhere. Fig. 4 shows the three alpha particles correlation function obtained in this way.

Compared to Fig.2-b, Fig. 4 shows peaks at larger excitation energies such as 11.8 MeV and 12.7 MeV. Those states correspond to known (2-) and (1+) states in ^{12}C. The small spurious peak at 8.6 MeV can be explained

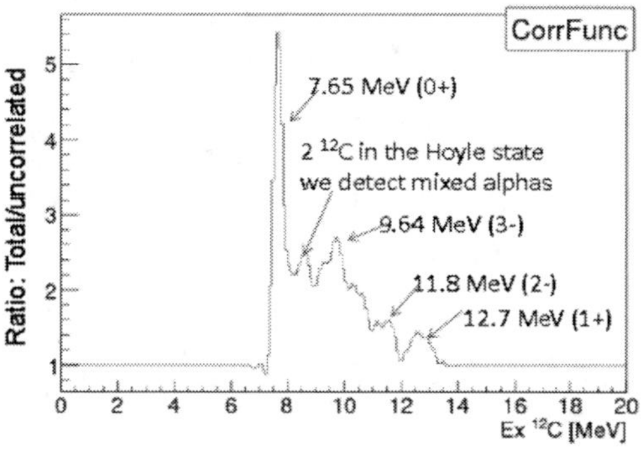

Figure 4: Correlation function obtained form the ratio of the measured spectrum and the uncorrelated events spectrum. The events with two α-particles in the central telescope and the third everywhere are selected

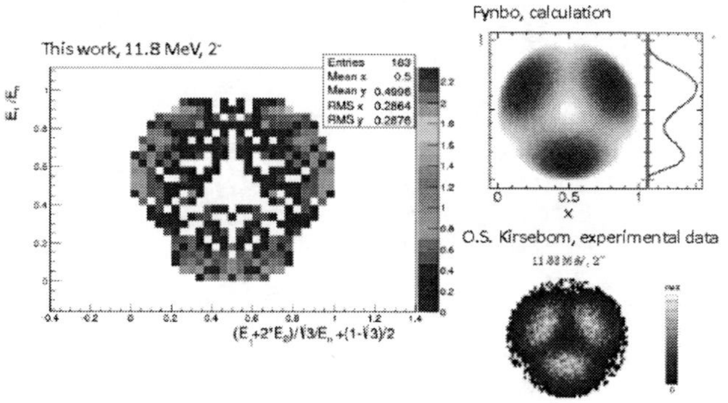

Figure 5: Dalitz plot for the 2^- state at 11.8 MeV compared with other experimental data and calculation

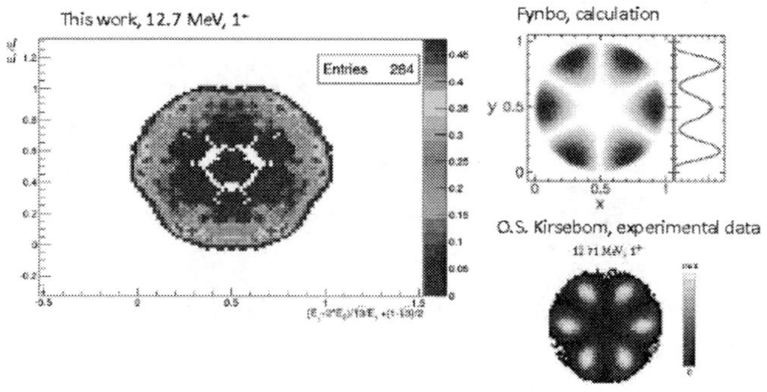

Figure 6: Dalitz plot for the 1^+ state at 12.7 MeV compared with other experimental data and calculation

using a Monte Carlo simulation. It is due to the mixing of alpha particles coming from two ^{12}C nuclei in the Hoyle state. The events corresponding to the Hoyle state, (3-), (2-) and (1+) states were analyzed in more details. The relative energies of the three possible couples of alpha particle energies were calculated event by event, in order to determine if the decay proceeds through the ground state of ^{8}Be or not. The Hoyle state and the (3-) were found to decay through the ground state of ^{8}Be, while the (2-) and (1+) states, were not. Dalitz plots were produced for each state and compared with other available experimental data and calculations [6,7]. Fig.5 and Fig.6 show the results for the (2-) and (1+) states respectively. Even though the statistics is quite low the results are in agreement with other data available in the literature. In conclusion we have shown that it is possible to use the TTIK technique to study excited states of ^{12}C decaying into 3 α-particles, including the Hoyle state. In the future we plan to use the same method and the improved detector system shown in Fig.1 to look for a state analogous to the Hoyle state in ^{16}O. Calculations suggest a candidate 0^+ state at excitation energy of 15.1 MeV.
This work was supported by US DOE and Robert A. Welch Foundation Grant A0330.

References

[1] K. Ikeda, N. Takigawa, and H. Horiuchi, Prog. Theor, Phys. Suppl. Extra Number, 464 (1968);

[2] W. von Oertzen, M. Freer, and Y. Kanada-Enyo, Phys. Rep. 432, 43 (2006);

[3] C. Beck Conference Series 436 (2013) 012014;

[4] K. Artemov et al., Sov. J. Nucl. Phys. 52, 406 (1990);

[5] M. Barbui et al., EPJ Web of Conferences 66, 03005 (2014);

[6] H. Fynbo et al., Phys. Rev. C,79, 054009 (2009);

[7] O. S. Kirsebom et al, Phys. Rev. C, 81, 064313 (2010).

EPJ Web of Conferences **117**, 07014 (2016) DOI: 10.1051/epjconf/201611707014
NN2015

Alpha-particle clustering in excited expanding self-conjugate nuclei

B. Borderie[1], Ad. R. Raduta[1,2], G. Ademard[1], M. F. Rivet[1†], E. De Filippo[3], E. Geraci[3], N. Le Neindre[1,4], G. Cardella[3], G. Lanzalone[5], I. Lombardo[5], O. Lopez[4], C. Maiolino[5], A. Pagano[3], S. Pirrone[3], G. Politi[3], F. Rizzo[5,3] and P. Russotto[5,3]

[1] Institut de Physique Nucléaire, CNRS/IN2P3, Université Paris-Sud 11, 91406 Orsay, France
[2] National Institute for Physics and Nuclear Engineering, Bucharest-Măgurele, Romania
[3] INFN, Sezione di Catania and Dipartimento di Fisica e Astronomia, Università di Catania, Italy
[4] LPC, CNRS/IN2P3, Ensicaen, Université de Caen, Caen, France
[5] INFN, Laboratori Nazionali del Sud, Catania, Italy

Abstract

The fragmentation of quasi-projectiles from the nuclear reaction ^{40}Ca+^{12}C at 25 MeV/nucleon was used to produce α-emission sources. From a careful selection of these sources provided by a complete detection and from comparisons with models of sequential and simultaneous decays, strong indications in favour of α-particle clustering in excited ^{16}O, ^{20}Ne and ^{24}Mg are reported.

1 Introduction

Clustering is a generic phenomenon which can appear in homogeneous matter when density decreases; the formation of galaxies as well as the disinte-

†Deceased

gration of hot dilute heavy nuclei into lighter nuclei are extreme examples occuring in nature. As far as nuclear physics is concerned, the nucleus viewed as a collection of α-particles was very early discussed and in the last forty years both theoretical and experimental efforts were devoted to clustering phenomena in nuclei. Very recently the formation of α-particle clustering from excited expanding self conjugate nuclei was revealed in two different constrained self consistent mean field calculations [1, 2]. The aim of the present work was to obtain, from the experimental side, some information on α-particle clustering from excited and consequently expanding alpha-conjugate nuclei. The chosen experimental strategy was to use the reaction ^{40}Ca+^{12}C at an incident energy (25 MeV per nucleon) sufficient to possibly produce some hot expanding reaction products, associated with a high granularity-high solid angle particle array (to precisely reconstruct directions of velocity vectors). Then, by selecting the appropriate reaction mechanism and specific events the required information was derived.

2 Experimental details and event selection

The experiment was performed at INFN, Laboratori Nazionali del Sud in Catania, Italy. The beam impinging on a thin carbon target (320 μg/cm^2) was delivered by the Superconducting Cyclotron and the charged reaction products were detected by the CHIMERA 4π multi-detector [3]. The beam intensity was kept around 10^7 ions/s to avoid pile-up events. CHIMERA consists of 1192 telescopes (ΔE silicon detectors 200-300 μm thick and CsI(Tl) stopping detectors) mounted on 35 rings covering 94% of the solid angle, with very high granularity at forward angles. Details on A and Z identifications and on the quality of energy calibrations can be found in refs. [3–5]. Energy resolution was better than 1% for silicon detectors and varies between 1.0 and 2.5% for alpha particles stopped in CsI(Tl) crystals.

As a first step in our event selection procedure, we want to exclude from the data sample poorly-measured events. Without making any hypothesis about the physics of the studied reaction one can measure the total detected charge Z_{tot} (neutrons are not measured). In relation with their cross-sections and with the geometrical efficiency of CHIMERA, the well detected reaction mechanisms correspond to projectile fragmentation (PF) [6–8] with Z_{tot}=19-20 (target fragmentation not detected) and to incomplete/complete fusion with Z_{tot}=21-26. At this stage we can have a first indication on the multiplicity of α-particles, M_α, emitted per event for well detected events ($Z_{tot} \geq$19 - see fig.1). M_α extends up to thirteen, which means a deexcita-

DOI: 10.1051/epjconf/201611707014

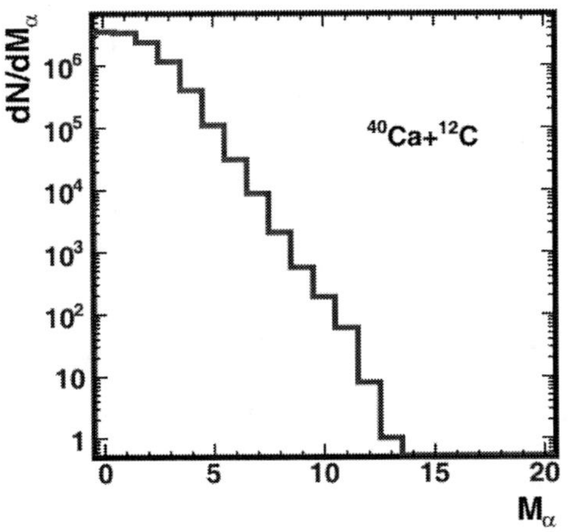

Figure 1: Distribution of α-particle multiplicity, M_{α}, for well detected events ($Z_{tot} \geq 19$)

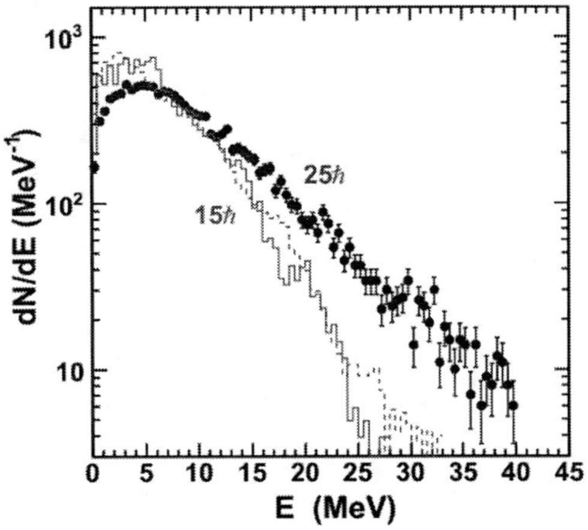

Figure 2: Sequential decay of excited Ca projectiles: energy spectra (in the $N\alpha$=5 system reference frame) of evaporated α-particles associated to a ^{20}Ne evaporation residue . Full points are experimental data and histograms are results of GEMINI simulations (see text)

tion of the total system into α-particles only. Moreover a reasonable number of events exhibit M_{α} values up to about 6.

The goal is now to tentatively isolate, in events, reaction products emitting α-particles only. To do this, knowing that at such incident energy ^{40}Ca as ^{20}Ne PF is dominated by alpha-conjugate products [6], we restrict our selection to completely detected PF events (Z_{tot}=20) composed of one projectile fragment and of four to six α-particles. Charge conservation imposes Z_{frag}=20 - $2M_{\alpha}$.

After this double selection, the question is: from which emission source are the α-particles emitted? Several possibilities have to be considered and complementary selections must be done before restricting our study to alpha-sources emitting exclusively the M_{α} observed (called $N\alpha$ sources in what follows). Possibilities associated to selected PF events are the following:

I) considering the incident energy of the reaction and the strong forward focusing of reaction products, it is important to identify the possible presence of preequilibrium (PE) α-particles in our selected PF events. With the hypothesis that all the α-particles are emitted from their centre-of-mass reference frame, we noted a thermal distribution with the presence of a high energy tail starting at 40 MeV, which signs PE emission. Events in which such PE emission was possibly present were suppressed to prevent errors on alpha emitter properties; an upper energy limit was imposed to the α-particle energy, with a value of 40 MeV irrespective of M_α.

II) α-particles can be emitted from deexcitation of PF events via unbound states of ^{12}C, ^{16}O, ^{20}Ne and not directly from excited expanding $N\alpha$ sources. Multi-particle correlation functions [5, 9] were used to identify unbound states 100% α-particle emitters and suppress a small percentage of events (1.6-3.6%).

III) it must be verified that the fragments associated with M_α are not the evaporation residues of excited Ca projectiles emitting only α-particles.

As far as the two first items are concerned the effect was to suppress from 8.5 (M_α=4) to 12.8% (M_α=6) of previously selected events. The last item will be discussed in the following section.

To conclude on this part, one can also indicate that if excited $N\alpha$ sources have been formed their excitation energy thresholds for total deexcitation into α-particles vary from 20 to 50 MeV when $N\alpha$ moves from 4 to 6. Their mean excitation energy per nucleon is rather constant around 3.3-3.5 MeV which indicates that average lowest densities around 0.7 the normal density may have been reached due to thermal pressure [10, 11].

3 Evidence for α-particle clustering

Before discussing different possible deexcitations of Ca projectiles and $N\alpha$ sources, information on the selected reaction mechanism is needed. Major features of PF events are reproduced by a model of stochastic transfers [8]. For primary events with Z_{tot}=20 excitation energy extends up to about 200 MeV, which allows the large excitation energy domain (20-150 MeV) measured for $N\alpha$ sources when associated to a single fragment; angular momenta extend up to 24 $\hbar$, which gives an upper spin limit for Ca projectiles or $N\alpha$ sources.

Are α-particles emitted sequentially or simultaneously? To answer the question α-energy spectra are compared to simulations. For excited Ca projectiles and $N\alpha$ sources, experimental velocity and excitation energy distributions and distributions for spins are used as inputs. Then, results of simulations are filtered by the multi-detector replica. Simulated spectra are normalized to the area of experimental spectra.

For sequential emission the GEMINI++ code [12] was used. Before discussing $N\alpha$ sources, as said before, we must consider the possible evaporation from Ca projectiles. Excitation energy for projectiles is deduced from

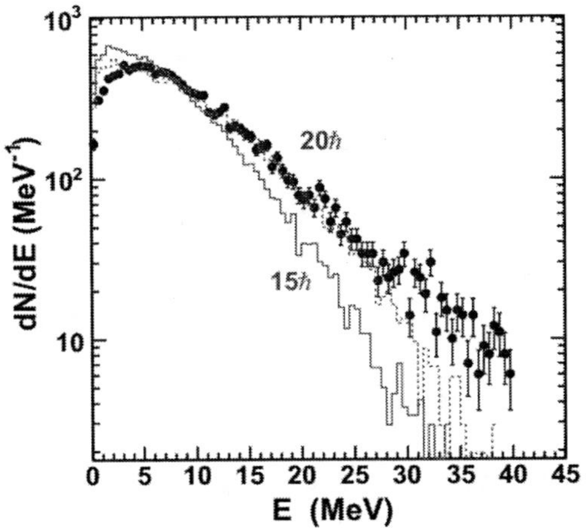

Figure 3: Decay of excited ^{20}Ne ($N\alpha$=5): α-particle energy spectra; full points are experimental data and histograms correspond to GEMINI simulations (see text).

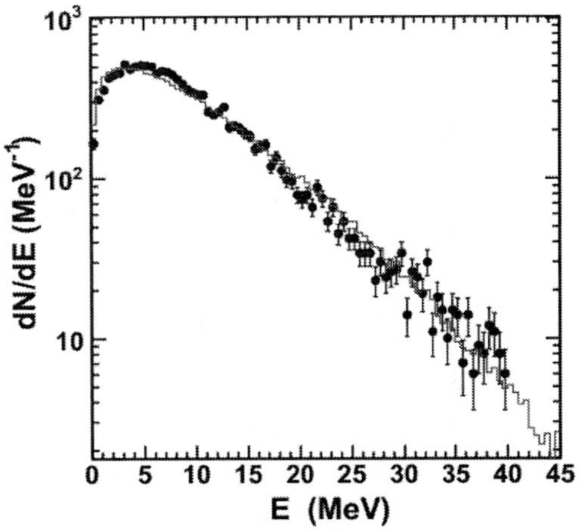

Figure 4: same as fig. 3 but histogram corresponds to a simulation of a simultaneous decay (see text).

E^{*}=$E^{*}(N\alpha)$+E_{rel}+Q. E_{rel} is the relative energy between the $N\alpha$ source and the associated fragment (evaporation residue). Fig. 2 displays results of simulations with reconstructed excitation energy distribution for ^{40}Ca ($< E^{*} >$=88.2 MeV) and gaussian distributions centred at 15 and 25$\hbar$ for spins (RMS=1.5$\hbar$) as inputs; note that no more ^{20}Ne residues are produced for spin distributions centred at values larger than 25$\hbar$. Comparison with experimental data shows a rather poor agreement indicating that such an hypothesis seems not correct. Same kind of results are observed for $N\alpha$ equal 4 and 6.

Considering now excited $N\alpha$ sources only, histograms in fig. 3 are examples of GEMINI simulation results for $N\alpha$=5. Gaussian distributions for spins are used as inputs and the best agreement with data is obtained with RMS=1.5$\hbar$ for spin distributions. The agreement between data and simulations appears poorer and poorer when $N\alpha$ value decreases. Moreover an important disagreement between data and simulations is observed for the percentages of $N\alpha$ sources which deexcite via ^{8}Be emission [13].

For simultaneous emission from $N\alpha$ sources, a simulation was done which mimics a situation in which α clusters are early formed when the source is expanding [1, 2] due to thermal pressure. The $N\alpha$ source is first splitted into α's. Then the remaining available energy ($E^{*}+Q$) is directly randomly shared among the α-particles such as to conserve energy and linear momentum. Histogram in fig. 4 is the result of such a simulation, which shows a good agreement with data. Such agreement is also observed for $N\alpha$ equal 4 and 6. Similar histograms (within a few percents) were also obtained with simulations containing an intermediate freeze-out volume stage and then propagation in the Coulomb field.

4 Conclusions

In conclusion, the reaction ^{40}Ca+^{12}C at 25 MeV/nucleon bombarding energy was used to produce and carefully select minor classes of events from which excited $N\alpha$ sources can be unambiguously identified. Their excitation energy distributions are derived with mean values around 3.4 MeV per nucleon, which indicates that mean densities about 0.7 the normal density may have been reached. Their energetic emission properties were compared with two simulations, one involving sequential decays and a second for simultaneous decay. For excited expanding $N\alpha$ sources composed of 4, 5 and 6 α-particles, for which statistics is good enough for conclusives comparisons with simulations, evidence in favour of simultaneous emission (α-particle clustering) is reported.

References

[1] M. Girod and P. Schuck, *Phys. Rev. Lett.* **111** (2013) 132503.

[2] J. P. Ebran et al., *Phys. Rev.* **C 89** (2014) 031303(R).

[3] A. Pagano et al., *Nucl. Phys.* **A 734** (2004) 504.

[4] N. Le Neindre et al., *Nucl. Instr. and Meth. in Phys. Res.* **A 490** (2002) 251.

[5] Ad. R. Raduta et al., *Phys. Lett.* **B 705** (2011) 65.

[6] M. Morjean et al., *Nucl. Phys.* **A 438** (1985) 547.

[7] B. Borderie et al., *Ann. Phys. Fr.* **15** (1990) 287.

[8] L. Tassan-Got and C. Stephan, *Nucl. Phys.* **A 524** (1991) 121.

[9] R. J. Charity et al., *Phys. Rev.* **C 52** (1995) 3126.

[10] W. A. Friedman, *Phys. Rev.* **C 22** (1990) 667.

[11] B. Borderie, in *Large-Scale Collective Motion of Atomic Nuclei*, edited by G. Giardina, G. Fazio and M. Lattuada (World Scientific, Singapore) 1997, pp. 1-14.

[12] R. J. Charity, *Phys. Rev.* **C 82** (2010) 014610 and references therein.

[13] B. Borderie et al., *Phys. Lett.* **B** to be published.

EPJ Web of Conferences **117**, 07015 (2016) DOI: 10.1051/epjconf/201611707015
NN2015

Impact of pairing on thermodynamical properties of stellar matter

S. Burrello[1,2], F. Aymard[3], M. Colonna[1], F. Gulminelli[3] and Ad. R. Raduta[4]

[1] INFN - Laboratori Nazionali del Sud, 95123 Catania, Italy
[2] Dipartimento di Fisica e Astronomia, 95123 Catania, Italy
[3] CNRS and ENSICAEN, UMR6534, LPC, 14050 Caen cédex, France
[4] NIPNE, Bucharest - Măgurele, POB - MG6, Romania

Abstract

Superfluidity in the crust is a key ingredient for the cooling properties of proto-neutron stars. Investigations on crust superfluidity carried out so far typically assumed that the cluster component was given by a single representative nucleus and did not consider the fact that at finite temperature a wide distribution of nuclei is expected to be populated at a given crust pressure condition. We want to assess the importance of this distribution on the calculation of the heat capacity in the inner crust, in the framework of an extended NSE model. We additionally show that it is very important to consider the temperature evolution of the proton fraction, imposed by the β-equilibrium condition, for a quantitatively reliable estimation of the heat capacity.

Introduction

Superfluidity in the crust is essential to understand many different phenomena in compact star physics, such as the cooling of neutron stars. It is well-known indeed that pairing correlations reduce the crust thermalization time by a large fraction [1]. The specificity of the inner crust is the simultaneous presence of clusters and homogeneous matter. The studies of crust superfluidity at finite temperature have always been done solving Hartree-Fock-Bogoliubov (HFB) equations in the so-called Wigner-Seitz approximation, meaning that the cluster component is given by a single representative quasi-particle configuration, corresponding to a single representative nucleus immersed in a neutron gas. These works do not consider the fact that at finite temperature a wide distribution of nuclei is expected to be populated at a given crust pressure and temperature condition.

Finite temperature Nuclear Statistical Equilibrium (NSE) models [2, 3] take into account the whole distribution of clusters, which is obtained self-consistently under conditions of statistical equilibrium. Our aim is to analyze how the non-homogeneity of crust matter and the associated wide distribution of nuclear species affects the superfluid properties of the crust.

In most HFB calculations for the cooling problem, moreover, it is assumed that the proton fraction does not evolve with the temperature and can be estimated by the value imposed, at each baryonic density, by the condition of neutrinoless chemical equilibrium at zero temperature. As the NSE model is much less numerically demanding than a full HFB calculation, we have released this approximation and imposed the β-equilibrium condition at each finite temperature.

1 The extended NSE model

The complete formalism that we use can be found in [4]. We detail here our extension of the NSE model, that is the inclusion of pairing in the bulk region of the Wigner-Seitz cell.

The energy density of a superfluid and spin saturated nuclear gas at finite temperature T, in the mean field approximation, reads [5] (q=n,p):

$$\epsilon_{HM} = 2\sum_q \int_0^\infty \frac{dp}{2\pi^2\hbar^3} p^2 f_q \frac{p^2}{2m_q^*} + \mathcal{E}_{sky} + \frac{1}{4}\sum_{q=n,p} v_\pi(\rho_{gq})\tilde{\rho}^*_{gq}\tilde{\rho}_{gq} \qquad (1)$$

We use the Sly4 parametrization [6] of the Skyrme energy functional for the local energy density $\mathcal{E}_{sky}$ and the effective nucleon mass m_q^*. In Eq.(1), f_q is the particle occupation number including pairing, $\tilde{\rho}_{gq} = 2\Delta/v_\pi(\rho_{gq})$ denotes the anomalous density while Δ is the temperature dependent pairing gap. The pairing interaction $v_{\pi q}(\rho_{gq})$, as a function of the gas density ρ_{gq}, is given by [7] :

$$v_{\pi q}(\rho_{gq}) = V_\pi \left[1 - \eta\left(\frac{2\rho_{gq}}{\rho_{sat}}\right)^\alpha\right], \qquad (2)$$

 DOI: 10.1051/epjconf/201611707015

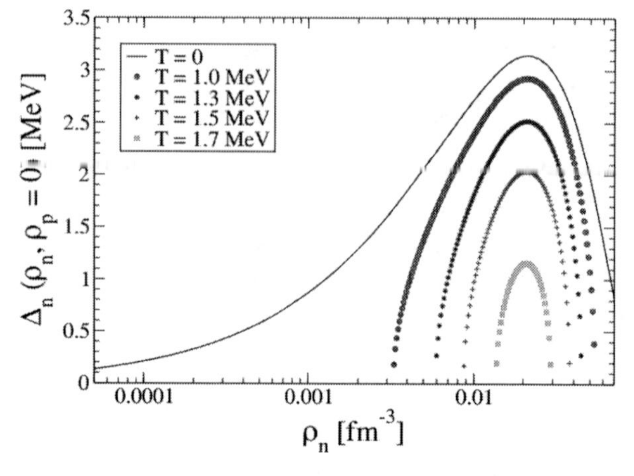

Figure 1: (Color online) $^1S_0$ pairing gap as a function of density for homogeneous neutron matter at zero temperature, as obtained from Brueckner-Hartree Fock calculations (full line from [8]). The figure also shows the energy gap deduced solving the BCS gap equations at finite temperature (symbols from [5]).

where ρ_{sat} is the saturation density and the parameters V_π, η, α are fixed imposing to reproduce the $^1S_0$ pairing gap of pure neutron matter, as obtained in Brueckner-Hartree-Fock calculations [8]. The resulting gap is displayed in Fig.1.

The binding energy of the clusters is evaluated by the analytical liquid drop model (LDM) expression reported in [9], where the different parameters are fitted from numerical calculations which make use of the same Skyrme energy functional employed for the unbound component.

The equilibrium distribution is obtained by minimizing the total free energy corresponding to an arbitrary collection of different Wigner-Seitz cells, subject to the constraint of total baryonic and charge density conservation [4].

2 Main results

In order to facilitate a quantitative comparison with the previous literature, we have chosen ten representative values for the baryonic density which have been proposed in [10]. These values cover the inner crust of the neutron star, approximately from the emergence of the neutron gas close to the drip point (denoted as cell 10) to a density close to the crust-core transition (cell 1).

2.1 Clusters distribution

The distribution of the cluster size obtained by our NSE calculation is displayed in Fig. 2. We can see that at the lowest densities and temperatures

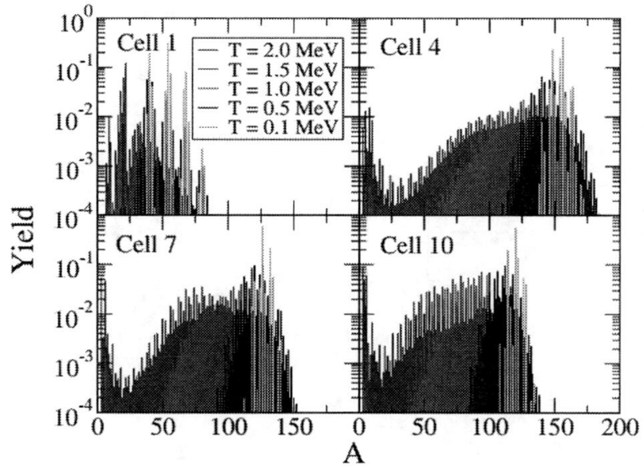

Figure 2: (Color online) Normalized cluster size distribution at different temperatures for four representative values of the total baryonic density: $\rho_B\,(\text{Cell } 1) = 4.8 \times 10^{-2}$ fm^{-3}, $\rho_B\,(\text{Cell } 4) = 5.8 \times 10^{-3}$ fm^{-3}, $\rho_B\,(\text{Cell } 7) = 9.0 \times 10^{-4}$ fm^{-3}, $\rho_B\,(\text{Cell } 10) = 2.8 \times 10^{-4}$ fm^{-3} and under the imposition of the β-equilibrium condition.

the distribution is strongly peaked and can be approximated by a unique nucleus, but increasing the temperature and/or moving towards the inner part of the crust, many different nuclear species can appear with comparable probability. Moreover, at sufficiently high temperatures, light resonances can appear and even become dominant in the composition of matter. Such configuration cannot be addressed in mean-field based formalisms like HFB. Looking at the isotopic distribution of clusters, especially for the lowest density cells (cells 7 and 10), one observes that the β-equilibrium condition leads to a significantly larger cluster asymmetry. This happens especially for light resonances, so at the highest temperatures, where the contribution of the most unbound clusters is increased.

2.2 Energy and heat capacity

The effects of the temperature dependence of the β-equilibrium condition and of the whole distribution of cells can also be appreciated looking at the behavior of the total energy density of the system. We have seen that these effects are very important not only at the highest densities, but also at the lowest ones when the temperature gets higher, whereas the emergence of light clusters occurs. The transition temperature from the superfluid to the normal phase is signalled by a kink in the behavior of the energy density, which will lead to a peak in its derivative with respect to the temperature, that is the associated heat capacity shown in Fig. 3. The temperature derivative was performed numerically following the trajectory of β-equilibrium: this means that only the total baryonic density is constant, but the proton fraction is not. As we can see, the temperature dependence

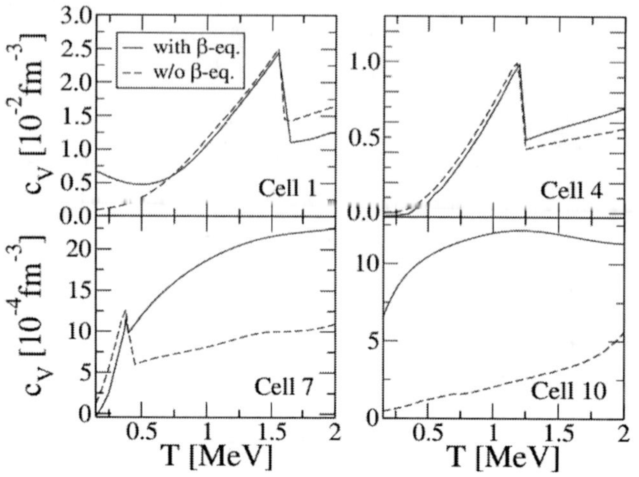

Figure 3: (Color online) Temperature evolution of the heat capacity for the same representative cells as in Fig.2. Full line: complete NSE calculation. Dashed line: as the full line, but the value of the global proton fraction is assumed equal to that one calculated from β-equilibrium at the lowest temperature, that is T=100 keV.

of the β-equilibrium condition has a dramatic effect on the heat capacity. In particular the peak due to the phase transition is strongly smeared out in the outer region of the inner crust, from cell 7 to 10, while on the contrary, at the highest densities (cells 1 to 3) the consideration of the temperature variation of the proton fraction increases the size of the peak. In this case, indeed, it is possible to show that β-equilibrium path favors a discontinuous trend of all thermodynamic quantities at the transition point.

Actually, the quantitative value of the heat capacity, as shown in Fig. 3, cannot be directly compared to the results of previous HFB works [1] because of the different mean-field and/or pairing model. However, we have verified that the temperature location of the heat capacity peak, its height and width are almost identical, if we take the same parameters for the pairing interaction employed in that work.

2.3 The effect of mass functional

In all the calculations presented in the previous sections, we have used the Skyrme-based liquid-drop formula given in [9]. This choice allows a consistent treatment of the bound and unbound matter component within the same energy functional. However, as it happens also in other mean-field model, light clusters are systematically underbound, with respect to heavier ones, by this formula and this effect is even more dramatic for the most neutron rich light resonances, at the limit of nuclear binding, which can appear in our distribution. We have therefore repeated the same calculations, making use of the experimental mass, whenever this value is known [11].

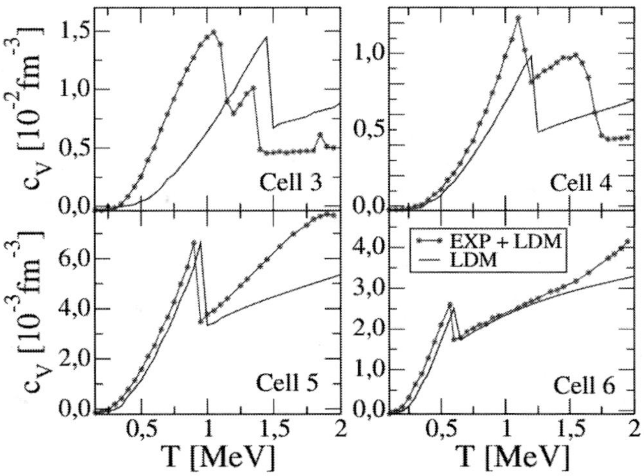

Figure 4: (Color online) Temperature evolution of heat capacity for four intermediate cells. Full lines: complete NSE calculation making use of the analytical expressions reported in [9] for binding energies (labeled as LDM in the figure). Line with symbols: as the full line, but experimental binding energies from [11] are used whenever available (EXP + LDM in the figure).

As we can see in Fig. 4, in the range of temperatures considered, the results are similar to the ones presented in Fig. 3 both for the highest (cells 1-2) and lowest (cells 7 to 10) densities. However, in the cells from 3 to 6 the consideration of the experimental masses has dramatic consequences because it modifies the height of the heat capacity peak and also its position in temperature. Moreover, a second peak appears in the case of cells 3 and 4 whose location, it is possible to show, coincides with the temperature where a sharp transition to the dominance of light clusters occurs. The same features could also appear in cells at lower densities, but at temperature values that are beyond the range considered in the present study.

References

[1] Fortin M. et al., Phys. Rev. C, **82** (2010) 065804.

[2] Hempel M. and Schaffner-Bielich J., Nucl. Phys., **A837** (2010) 210.

[3] Raduta A.R., Gulminelli F., Aymard F., Eur. Phys. J., **A50**, (2014) 24.

[4] Burrello S. et al., Phys. Rev. C., (2015).

[5] Burrello S., Colonna M., Matera F., Phys. Rev. C, **89** (2014) 057604 .

[6] Chabanat E. et al., Nucl. Phys., **A635** (1998) 236.

[7] Chamel N., Phys. Rev. C, **82** (2010) 014313.

[8] Cao L.G., Lombardo U., Schuck P., Phys. Rev. C, **74** (2006) 064301.

[9] Danielewicz P., Lee J., Nucl. Phys., **A818** (2009) 36.

[10] Negele J.W., Vautherin D., Nucl. Phys., **A207** (1973) 298.

[11] Audi G. et al., Chinese Physics, **C36** (2012) 1287.

EPJ Web of Conferences **117**, 07016 (2016) DOI: 10.1051/epjconf/201611707016
NN2015

Impact of the in-medium conservation of energy on the π^-/π^+ multiplicity ratio

M.D. Cozma

IFIN-HH, Reactorului 30, 077125 Măgurele-Bucharest, Romania

Abstract

An upgraded version of the isospin dependent Tübingen QMD transport model, which allows the conservation of the total energy, is presented. This is achieved by including in the energy-balance equations of the density, isospin asymmetry and momentum dependent in-medium baryon potential energies. It leads to an effective modification of particle production thresholds with respect to the vacuum ones. Compatible constraints for the symmetry energy stiffness from π^-/π^+ multiplicity ratio and elliptic flow experimental data of Au+Au collisions at 400 MeV/nucleon can be extracted in this case. However, an important dependence of the π^-/π^+ observable on the strength of the isovector part of the $\Delta(1232)$ isobar potential is also demonstrated. The present lack of information on this quantity prevents a precise extraction of the value for the symmetry energy stiffness employing the mentioned observable alone.

1 Introduction

The π^-/π^+ multiplicity ratio (PMR) has been proposed as a suitable observable to constrain the density dependence of the symmetry energy (SE) above the saturation point [1]. Various attempts in this direction [2–5] have resulted in a confusing picture: constraints on the high density dependence of SE ranging from a very soft to stiff have been determined by employing different transport models and/or parametrizations of the symmetry potential. Additionally, most models lead to a contradiction between the

π^-/π^+ multiplicity ratio and neutron/proton elliptic flow ratio extracted constraints. Attempts to find a solution to this problem by studying the impact of in-medium modification of the pion-nucleon interaction [6], of the kinetic part of the SE term [7], neutron skin thickness [8] or particle production thresholds due to the inclusion of self-energy contributions [9,10] on the PMR value have proven unsuccessful.

Transport models that employ momentum or isospin dependent mean-fields do not conserve the total energy at microscopical level, in two-body collision, resonance decay or meson absorption processes. This fact translates into an energy conservation violation at an event by event basis and also when an average is performed over a large number of events. Consequently, an upgrade of the Tübingen transport model, that alleviates this problem, has been developed. All the relevant details of this upgrade together with the most relevant results to the PMR problem have been presented in detail in Ref. [11]. In this conference proceeding we will present in greater detail the motivation that has lead to the mentioned upgrade, some relevant details of the model and end with a few selected results.

2 The model

The starting point for the model employed in this study has been the QMD transport model developed in Tübingen [12] which has been upgraded to accommodate for density dependent cross-sections, various parametrizations of the optical potential and of the isovector part of the equation of state (EoS) of nuclear matter [13]. For the present study the Gogny inspired parametrization of the SE has been used [14], which allows the adjustment of the stiffness of the isovector part of the EoS by the introduction of a parameter denoted x. Positive and negative values of this parameter correspond to a soft and respectively stiff choice for the stiffness of the SE.

The poorly known in-medium baryonic resonance's potentials are chosen as follows. The isoscalar part is set to be equal to that of the nucleon, a choice common to most transport models. The isovector resonance potential is related to that of the nucleon by making use of the decay branching ratios of each isospin quadruplet component into the possible pion-nucleon pairs [15]. The total resonance potential can thus be written as

$$\begin{aligned} V_{\Delta^-} &= V_N + (3/2)\,V_v \\ V_{\Delta^0} &= V_N + (1/2)\,V_v \\ V_{\Delta^+} &= V_N - (1/2)\,V_v \\ V_{\Delta^{++}} &= V_N - (3/2)\,V_v \end{aligned} \tag{1}$$

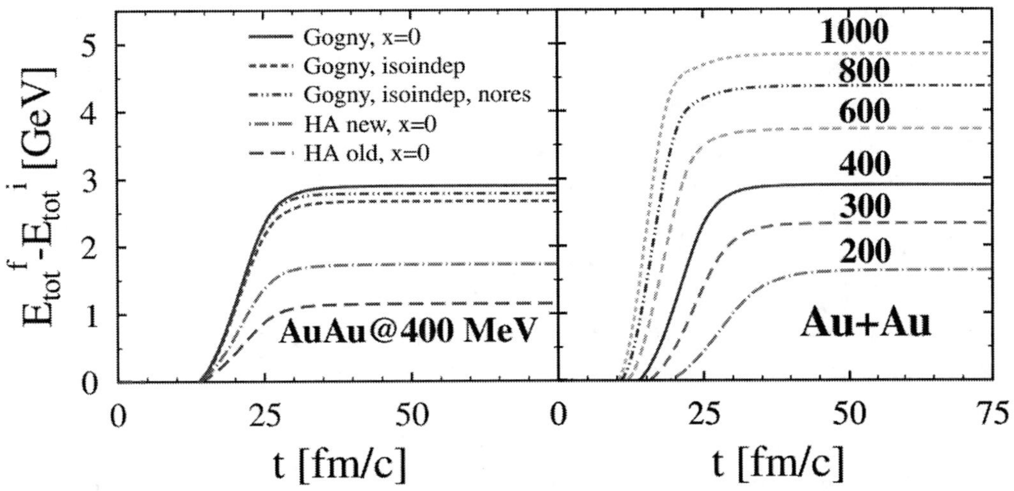

Figure 1: Left Panel: Magnitude of the energy conservation violation in central Au+Au collisions at an impact energy of 400 MeV/nucleon. Results for three different parametrizations of the optical potential are presented: Gogny (full curve), old Hartnack-Aichelin (long dashed-curve) and new Hartnack-Aichelin (dash-dotted curve). Additionally, for the case of the Gogny potential results for the cases when the isovector part of the potential (short dashed curve) and the inelastic channels (dashed double dotted curve) are ignored are presented. Right Panel: Dependence of the magnitude of the energy conservation violation on the impact energy below 1 GeV/nucleon for the Gogny inspired potential.

where V_N is the isoscalar nucleon potential and V_v=δ, with the definition δ=$(1/3)(V_n$-$V_p)$, V_n and V_p being the isovector components of the neutron and proton potential respectively.

To justify the approximations introduced next, the extent to which the conservation law of total energy is obeyed is investigated, the results being presented in Fig. 1. The left panel presents results for three different parametrizations of the optical potential for central Au+Au collisions at an impact energy of 400 MeV/nucleon: the Gogny inspired one of Ref. [14] and the two parametrization (Hartnack-Aichelin) extracted from the analysis of proton-nucleus scattering data in Ref. [16]. It is observed that the amount of energy conservation violation (ECV) dependents strongly upon the choice of the optical momentum but is it in all cases much larger than the rest mass of the pion. The isospin dependent part of the potential and the inelastic channels are seen to impact the magnitude of ECV rather modestly at this impact energy. The right panel of Fig. 1 presents the dependence of the amount of ECV on the kinetic energy of the projectile nucleus. An monotonically increasing behavior on this quantity is observed, however a saturation phenomenon takes place towards higher impact energies, a trend that can be explained by the same behavior of the optical potential as a function of the momentum of the incident particle.

It is thus clear that tranoport modols that fail to conserve the total energy at the level of few GeV are not appropriate for the description of production of particles with rest masses much smaller than this value. The problem can be alleviated by including in the process of the determination of the kinematics of final states of two-body collisions, resonance decay and meson absorption processes of the in-medium meson and baryon potentials. In this processes elementary two-body reactions lose their local character becoming part of a much more complicated N-body process which allows the conservation of total energy via exchanges mediated by the intermediate and long-range part of the nucleon-nucleon interaction. In a Feynman diagrammatic picture, the leading order contributions are expected to arise from initial-state and final-state interactions, while contributions originating from more complicated diagrams (rescattering terms) are assumed to be much smaller and thus are completely neglected. All the pertinent details of the approximations involved, the invariant masses at which elastic and inelastic channels cross-sections are evaluated, the needed modifications to the detailed balance formula, in-medium effects on cross-sections and numerical approximations required to make such calculations feasible time-wise are presented in Ref. [11].

3 Selected results

The inclusion of contributions of the hadronic potentials in the equation of energy conservation leads to threshold shifts for particle production with respect to their vacuum position [9]. In the case of π mesons the effect is much stronger for the negatively charged pion [11] which leads to a total different dependence of the PMR on the SE stiffness compared with the case when total energy is not conserved: a higher PMR value is favored by a stiffer asy-EoS, a behaviour that was also reported by the study in Ref. [10]. This allows, using the prescription for the in-medium Δ potential of Eq. 1, to extract a constraint for the density dependence of SE above the saturation point which is compatible with the one extracted from experimental data of the elliptic flow ratio of neutrons and protons [11]. Additionally, the

 DOI: 10.1051/epjconf/201611707016

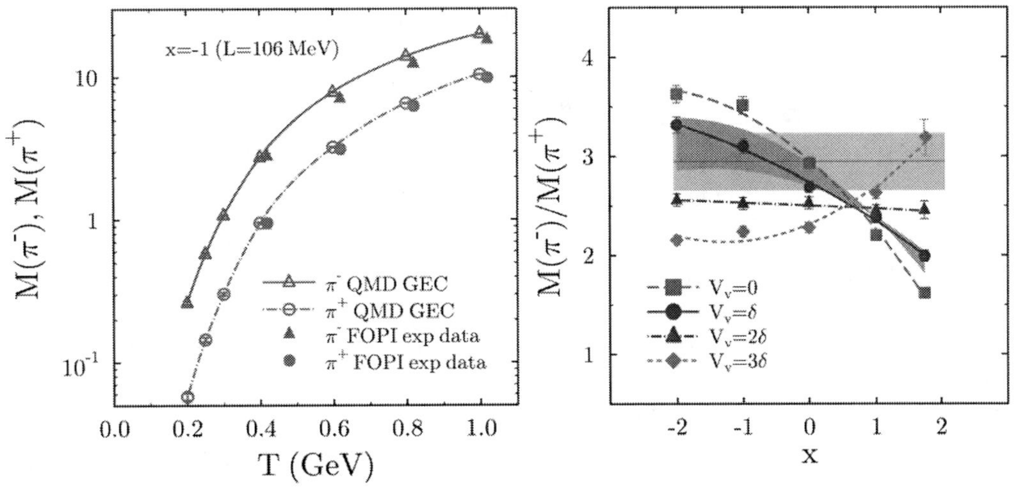

Figure 2: Left Panel: Comparison between theoretical and experimental π^- and π^+ multiplicities for impact energies below 1.0 GeV/nucleon. Theoretical predictions for sub-threshold impact energies are also provided. Right Panel: Sensitivity of the PMR to variations of the strength of the isocalar (band) and isovector (curves) $\Delta(1232)$ potentials at an impact energy of 400 MeV/nucleon as a function of the stiffness parameter x. The experimental value is depicted by the horizontal band. Both panels display results for central ^{197}Au+^{197}Au collisions.

dependence of the PMR on model parameters like compressibility modulus of the isoscalar part of the EoS, parametrization of the optical potential, inclusion or exclusion of medium effects on elastic and inelastic elementary cross-sections is found to be small [11]. A comparison of theoretical and experimental data for the charged pion multiplicities up to values of the impact energy of 1.0 GeV/nucleon shows a good agreement (left panel of Fig. 2). The PMR is however found to be extremely sensitive to the strength of the isovector Δ potential and to a smaller extent to its isoscalar component, as shown in the right panel of Fig. 2. By changing the strength of the isovector Δ potential from 0 to 3 times that of Eq. 1 almost any values for the stiffness parameter x can be extracted from a comparison with the experimental data.

In conclusion, the conservation of the total energy in transport models is a mandatory ingredient if a consistent description of particle production is to be achieved. In particular, the multiplicities and multiplicity ratio of charged pions in heavy-ion collisions close to or slightly above the vacuum pion production threshold are greatly impacted by imposing this constraint consistently. Constraints for the symmetry energy dependence on density compatible with the ones extracted from others heavy-ion observables (most notably flows) can be extracted only within this scenario. However, it was also demonstrated that PMR is extremely sensitive to the strength of the unknown isovector Δ potential, which hinders at present the use of this observable for the purpose of determining the density dependence of the symmetry energy above the saturation point.

References

[1] B. A. Li, G. C. Yong and W. Zuo, *Phys. Rev. C*, **71** (2005) 014608.

[2] Z. Xiao, B. A. Li et al., *Phys. Rev. Lett.*, **102** (2009) 062502.

[3] Z. Q. Feng and G. M. Jin, *Phys. Lett. B*, **683** (2010) 140.

[4] W. J. Xie, J. Su, L. Zhu et al., *Phys. Lett. B*, **718** (2013) 1510.

[5] J. Hong and P. Danielewicz, *Phys. Rev. C*, **90** (2014) 024605.

[6] J. Xu, L. W. Chen, C. M. Ko et al., *Phys. Rev. C*, **87** (2013) 067601.

[7] B. A. Li, W. J. Guo and Z. Shi, *Phys. Rev. C*, **91** (2015) 044601.

[8] G. F. Wei, B. A. Li, J. Xu et al., *Phys. Rev. C*, **90** (2014) 014610.

[9] G. Ferini, T. Gaitanos et al., *Phys. Rev. Lett.*, **97** (2006) 202301.

[10] T. Song and C. M. Ko, *Phys. Rev. C*, **91** (2015) 014901.

[11] M. D. Cozma, arXiv:1409.3110 [nucl-th] Preprint (submitted to Phys. Lett. B), 2014.

[12] D. T. Khoa, N. Ohtsuka et al., *Nucl. Phys. A*, **548** (1992) 102.

[13] M. D. Cozma et al., *Phys. Rev. C*, **88** (2013) 044912.

[14] C. B. Das, S. D. Gupta, C. Gale et al., *Phys. Rev. C*, **67** (2003) 034611.

[15] B. A. Li, *Nucl. Phys. A*, **708** (2002) 365.

[16] C. Hartnack and J. Aichelin, *Phys. Rev. C*, **49** (1994) 2801.

[17] W. Reisdorf et al. [FOPI Collab.], *Nucl. Phys. A*, **848** (2010) 366.

EPJ Web of Conferences **117**, 07017 (2016) DOI: 10.1051/epjconf/201611707017
NN2015

Sensitivity of N/Z ratio in projectile break-up of isobaric systems

E. De Filippo[1], A. Pagano[1], P. Russotto[1], L. Acosta[2], L. Auditore[3], V. Baran[4], T. Cap[5], G. Cardella[1], M. Colonna[2], L. Francalanza[6], B. Gnoffo[1], G. Lanzalone[7,2], I. Lombardo[6], G. Marquínez-Durán[8], C. Maiolino[2], T. Minniti[1], S. Norella[3], , E.V. Pagano[2,9], M. Papa[1], E. Piasecki[5,10], S. Pirrone[1], G. Politi[9,1], F. Porto[9,2], L. Quattrocchi[3], F. Rizzo[9,1], E. Rosato[6], A. Trifirò[3], M. Trimarchi[3], G. Verde[1], M. Vigilante[6], K. Siwek-Wilczyńska[5] and J. Wilczyński[10]

[1] INFN, Sezione di Catania, Catania, Italy
[2] INFN - Laboratori Nazionali del Sud, Catania, Italy
[3] INFN, Gruppo Collegato di Messina and Univ. of Messina, Italy
[4] Physics Faculty, University of Bucharest, Romania
[5] Faculty of Physics and Heavy Ion Laboratory, Univ. of Warsaw, Poland
[6] INFN, Sezione di Napoli and Dip. di Fisica, Univ. of Napoli, Italy
[7] Università "Kore", Enna, Italy
[8] Departamento de Fisica Aplicada, Huelva, Spain
[9] Dipartimento di Fisica e Astronomia, Univ. di Catania, Italy
[10] National Centre for Nuclear Research, Otwock-Świerk, Poland

Abstract

The binary break-up of projectile-like fragments in non central heavy-ion collisions follows different decay patterns, from equilibrated emission towards dynamical (prompt) fission. Recently, comparing two systems with different N/Z in the entrance channel, it has been shown that the dynamical emission cross-section is enhanced for the most neutron rich system while the statistical emission cross-section is independent from the isotopic composition. In order to understand this dependence and disentangle it from the initial size of the nuclei, we have studied the two isobaric systems $^{124}Xe+^{64}Zn$ and $^{124}Xe+^{64}Ni$ at 35 A MeV (InKiIsSy experiment), in comparison with the previous studied reactions ($^{124}Sn+^{64}Ni$ and $^{112}Sn+^{58}Ni$) at the same bombarding energy. We present the first results evidencing a striking similar effect in the dynamical decay as a function of the N/Z of the target for equal size systems.

1 Introduction

In non-central heavy ion collisions at intermediate energies, the reaction dynamics can result in binary products such as excited projectile-like (PLF) and target-like (TLF). An important cross-section is also associated with ternary reactions where the PLF and TLF residues are accompanied by the emission of intermediate mass fragments (IMF) and light charged particles. Light IMFs ($Z < 10$) are produced preferentially in the midrapidity region ranging between the PLF and TLF velocities. An important aspect, in fragments and particles emission, is the fact that they are continuously produced all along the dynamical evolution path of the system, from shortest times (pre-equilibrium, prompt neck rupture emission) to longer time scales in the sequential decay of the PLF or TLF primary fragments [1–5]. The analysis of the reactions $^{124,112}Sn + {}^{64,58}Ni$ at 35 A MeV beam incident energy, have shown a well-defined chronology: light IMFs (Z<9) are likely to be emitted either dynamically on a short time scale (within 50 fm/c) with a prompt neck rupture mechanism or statistically in the sequential decay of the PLF and/or TLF primary fragments. Isotopic properties of the neck fragmentation products have been linked to the symmetry energy term parametrizzation of the nuclear equation of state [5]. In contrast, heavier fragments (Z≥9) are predominantly emitted on a longer time scale (>120 fm/c) and they have been associated either with a massive binary splitting of deformed and excited PLF or TLF ("dynamical fission") or a statistical equilibrated fission decay. These different decay patterns for the PLF break-up are characterized by peculiar angular distributions: an aligned break-up with the recoil velocity of the PLF source in the dynamical emission and an isotropic emission of the fission-like fragments in the PLF reference frame in the sequential statistical decay. It has been shown [6, 7] that while the statistical-sequential emission assumes the same cross section for the two systems as a function of the IMF atomic number Z of the emitted fragments, the dynamical IMF emission cross section is enhanced for the neutron rich system, clearly indicating an isospin effects of the entrance channel sensitive to the density dependence of the symmetry term in the nuclear equation of state. However, a contribution to the observed effect related to the different sizes of the studied systems can not be ruled-out [8]. By using the 4π detector CHIMERA, a new experimental investigation was recently carried out (InKiIsSy, Inverse Kinematic

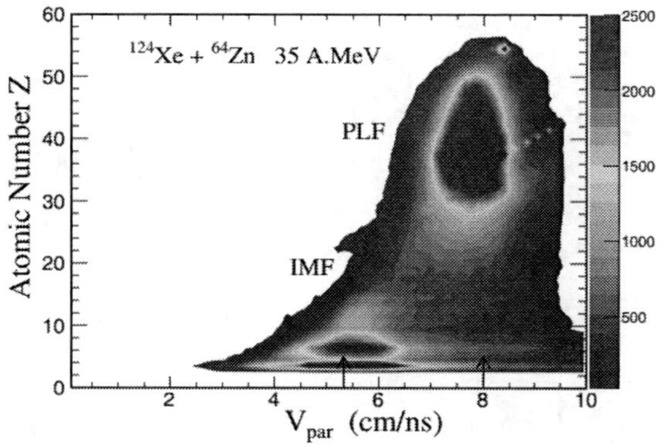

Figure 1: Yield distribution for particles punching-through the Silicon detectors as a function of their atomic number Z and parallel to the beam velocity V_{par}.

Isobaric Systems) at INFN-LNS. Our goal was to extend the previous study, at the same incident energy, using a projectile/target combination having the same mass of the neutron rich system ($^{124}Sn +^{64} Ni$) and a N/Z close to the value of the neutron poor one, in order to disentangle genuine isospin effects by the ones related to the size of the two interacting systems.

2 Results

The experiment was performed at the LNS Cyclotron facility in Catania using the ^{124}Xe beam at 35 A MeV incident on targets of ^{64}Zn (308 $\mu g/cm^2$) and ^{64}Ni (370 $\mu g/cm^2$) by using the CHIMERA 4π detector. A prototype of the FARCOS correlator array (four telescopes) [9], subtending the polar angles of $15^o \leq \theta_{lab} \leq 45^o$ and the azimuthal interval $\Delta\phi \approx 90^o$, was used for the first time to measure with better angular and energy resolution than in CHIMERA, IMF-IMF and proton-proton correlations for events triggered by the CHIMERA array. In the following, data analysis will be limited to particles well identified in charge by using the $\Delta E - E$ technique in the Silicon-CsI(Tl) CHIMERA telescopes and light charged particles isotopically identified in the CsI(Tl). Future progresses in data analysis will permit to add the particles of lower energies stopped in the silicon detectors and identified in charge and/or mass by Time-of-Flight and Pulse-Shape techniques. A 2-dimensional contour plot of the fragments yields for the $^{124}Xe +^{64} Zn$ reaction is shown in Fig. 1. In the figure the parallel velocity along the beam axis is correlated with the atomic number. The figure shows the regions of PLF and IMF for well identified events around or above the c.m. velocity. Binary decay of the PLF produced in peripheral reactions were

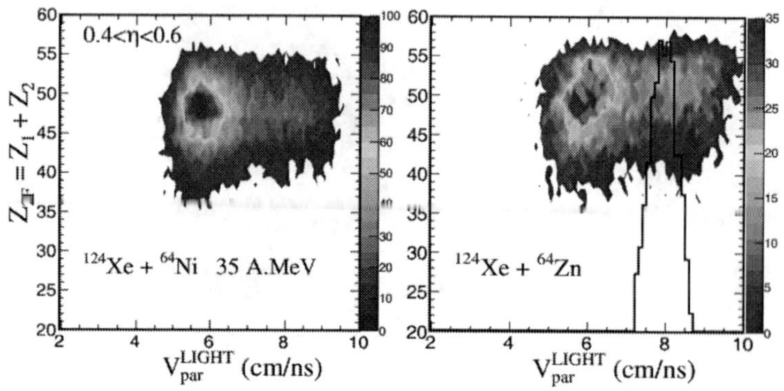

Figure 2: Distribution of the sum of charges of the two heaviest fragments (Z_{2F}) vs. V_{par} of the lighter fragment for ^{64}Ni (left) and ^{64}Zn (right) targets.

selected assuming that the sum of the charge of the two heaviest fragments $Z_{2F} = Z_1 + Z_2$ was larger than the value of 70% of the projectile charge. We show in Fig. 2 the distribution of Z_{2F} vs. the parallel velocity of the lighter fragment for the reactions $^{124}Xe +^{64} Ni$ (left) and $^{124}Xe +^{64} Zn$ (right) for a large asymmetric splitting of quasi-projectiles, $0.4 < \eta < 0.6$, where $\eta = (Z_1 - Z_2)/(Z_1 + Z_2)$ is the asymmetry parameter and for a window of the total kinetic energy of the two fragments E_{2F} >2700 MeV, so removing in this way the more dissipative collisions. The velocity distribution of the heaviest fragment, centered around the projectile velocity, is shown for comparison in the right panel. We observe that the light fragment velocity has two components characteristic of "forward" and "backward" emission in the PLF frame as compared with the projectile heavy residue velocity distribution centered around the beam velocity, consistent with the scenario of the scattering of the PLF followed by its splitting into two fragments [2, 6]. The largest component is the backward one when the lighter fragment is emitted with a velocity smaller that the heavier fragment, around 6 cm/ns. To better understand this aspect we compute the θ_{prox} angle, defined as the angle between the reconstructed PLF flight direction in the c.m. system and the relative velocity between Z_1 and Z_2 fragments.

Fig. 3 (left) shows the $\cos(\theta_{prox})$ distributions for three different values of the asymmetry parameter η. For the most symmetric splitting of the PLF ($\eta < 0.2$) the $\cos(\theta_{prox})$ spectrum is symmetric around $\cos(\theta_{prox}) = 0$ as expected for an equilibrated process of binary PLF splitting. For large asymmetric splittings, the distribution results from the superposition of a symmetric component and a pronounced peak $\cos(\theta_{prox}) \approx 1$ increasing with the charge asymmetry. This yield enhancement corresponds to a back-

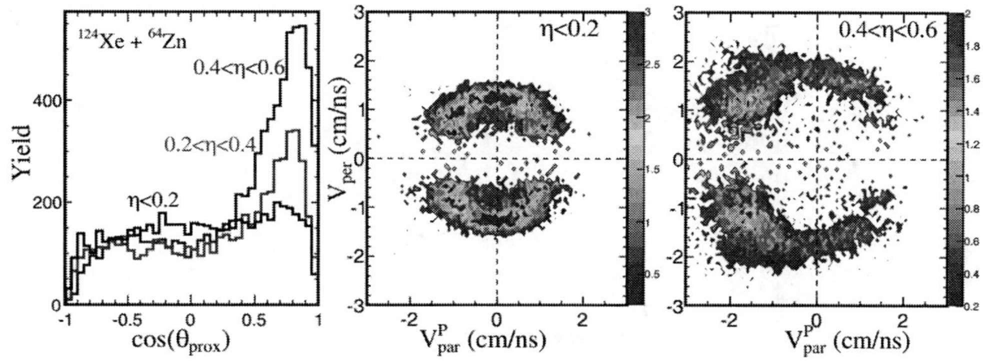

Figure 3: (left) $\cos(\theta_{prox})$ distribution for three different asymmetries η in the binary splitting of the PLF. Invariant cross-sections for the ligher fragment in the reference frame of the PLF for $\eta < 0.2$ (middle) and $0.4 < \eta < 0.6$ (right). Invariant plots are weighted with the fragments transverse momentum $1/p_t$.

ward emission of the fragment, aligned along the direction of the PLF. This anisotropy is one signature of a "dynamical" process [5, 7, 10]. This is confirmed looking at Fig. 3 (right) where the V_{par} vs. the V_{per} Galilean invariant cross-section plot is shown for the lighter fragment in the PLF asymmetric binary splitting in the reference frame of the reconstructed PLF. The forward-backward asymmetry clearly indicates here a component due to a non-sequential emission. This can be compared with the Fig. 3 (middle) plot for symmetric splitting ($\eta < 0.2$) where we observe a typical pattern of a forward-backward symmetry emission around a well defined Coulomb ring expected for an equilibrated break-up. We have used the $\cos(\theta_{prox})$ distributions in order to separate the statistical from the dynamical contribution for the two systems studied here, following the method described in Refs. [6, 7]. In Fig. 4 the ratio of the dynamical component respect to the dynamical+statistical one is shown for the two systems $^{124}Xe +^{64} Ni$ and $^{124}Xe +^{64} Zn$. These preliminary results show clearly that the dynamical component is prevailing for the reaction with a neutron rich target (^{64}Ni). Thus this result indicates independence from the inizial size of the projectile and target.

Extended analysis (including the achievement of data analysis for particles stopping in the Si detectors) and absolute cross sections determination for the statistical and dynamical components, will permit the comparison with results of the previous experiment $^{124}Sn +^{64} Ni$. The dynamical IMF emission is an important mechanism to constraint the density dependence of the symmetry energy but this need calculations that are able to follow the

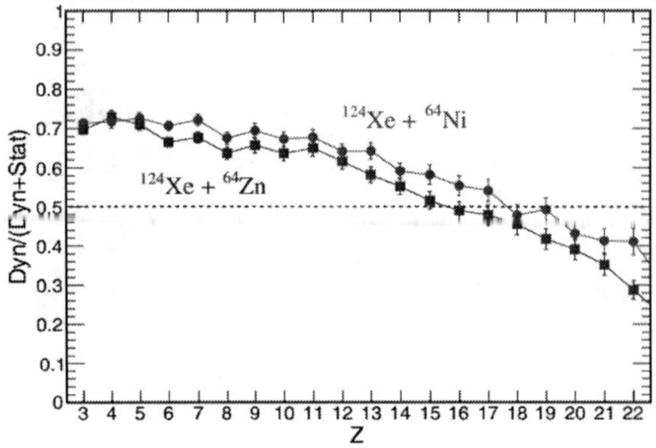

Figure 4: Ratio of the dynamical component vs. the total (statistical+dynamical) as a function of the IMF atomic number for the $^{124}Xe +^{64} Ni$ (red circles) and $^{124}Xe +^{64} Zn$ (blue squares) reactions.

largest possible reaction time-scales of the IMF emission. On this light we have started calculations using the Constrained Molecular Dynamics model CoMD3 [11] in order to simulate the two reaction studied here with promising results. Finally the merging of the CHIMERA and FARCOS data will certainly extend our experimental capabilities, due to Farcos enhanced isotopic resolving power for IMFs and light charged particles.

References

[1] E. De Filippo *et al.*, Phys. Rev. C**71**, 044602 (2005); *ibidem* Phys. Rev. C**71** 064604 (2005).

[2] A.B. McIntosh *et al.*, Phys. Rev. C**81**, 034603 (2010).

[3] B. Davin *et al.*, Phys. Rev. C**65**, 064614 (2002).

[4] S. Piantelli *et al.*, Phys. Rev. C**74**, 034609 (2006).

[5] E. De Filippo *et al.*, Phys. Rev. C**86**, 014610 (2012).

[6] P. Russotto *et al.*, Phys. Rev. C**81**, 064605 (2010).

[7] P. Russotto *et al.*, Phys. Rev. C**91**, 014610 (2015).

[8] G. Giuliani *et al.*, IWM2007 Proc., ed. SIF Bologna, Vol. 95, 311 (2008); M. Papa et al., Phys. Rev. C**75**, 054616 (2007).

[9] E.V. Pagano, proceedings of this conference; Technical Design Report - TDR/INFN for the Farcos project: *https://192.84.151.50/joomla*

[10] S. Hudan *et al.*, Phys. Rev. C**86**, 021603 (2012).

[11] M. Papa, Phys. Rev. C**87**, 014001 (2013).

EPJ Web of Conferences **117**, 07018 (2016) DOI: 10.1051/epjconf/201611707018
NN2015

From femtonova to supernova: Heavy-ion collisions and the supernova equation of state

K. Hagel[1], M. Hempel[2] J. B. Natowitz[1], G. Röpke[3], S. Typel[4], S. Wuenschel[1], R. Wada[1], M. Barbui[1] and K. Schmidt[5]

[1]Cyclotron Institute, Texas A & M University, College Station, Texas, USA
[2]Department of Physics, University of Basel, 4056 Basel, Switzerland
[3]University of Rostock, FB Physik, Rostock, Germany
[4]GSI Helmholtzzentrum fr Schwerionenforschung GmbH, D-64291 Darmstadt, Germany
[5]Institute of Physics, Silesia University, Katowice, Poland

Abstract

Calculations using astrophysical equations of state at low densities comparable to that of the neutrino emission surface in supernovae and accretion disks are confronted with experimental results from heavy ion collisions. An extension of previous work shows that it is important to include all of the measured experimental data to draw conclusions about the astrophysical equation of state. Armed with this information, the calculations of the astrophysical equation of state are significantly constrained. Predictions of temperatures and densities sampled in black hole accretion disks are compared to those sampled in the experimental data.

1 Introduction

The evolution of explosions of massive stars samples densities from $10^{-10} < \rho < 2$ nuc/cm^3 and temperatures between $0 < T < 100$ MeV. Ninety nine percent of the energy released in a core collapse supernova is radiated in neutrinos. The neutrinos interact until a temperature and density is reached where they no longer interact. This region is known as the neutrinosphere and is expected to have temperatures ranging from 5 - 10 MeV and densities that range from 6×10^{-5} - 6×10^{-3} fm^{-3}. Details of neutrino heating, therefore, depend on the properties of low density nuclear matter. Heavy ion collisions offer the possibility of studying low density nuclear matter and as such provide the intriguing possibility of studying stellar evolution and core-collapse supernovae [1, 2].

There are many calculations of core collapse supernovae [3–14] using different astrophysical equations of state (EOS). Calculations by design are performed with different assumptions and different compositions of competing particle species. In order to reduce discrepancies resulting from these different assumptions, we made comparisons of experimental data to the results of calculations using equilibrium constants [2]. The equilibrium constants are independent of particle species in the ideal gas limit of no interaction. When an interaction is present, differences in the equilibrium constants may provide a window into the nature of the interaction.

In reference [2], we compared α-particle equilibrium constants from the experimental data to those from a number of supernovae calculations. The results showed that the experimental data significantly constrain the calculations. This work was extended to include all light particle species and more detail on the interaction, Coulomb effects and particle degrees of freedom [15]. These proceedings are, in part, a summary of that work.

The first extension of the analysis, in addition to that presented in [2], was to quantify the effect on the equilibrium constants extracted from α-particles in [2] on the variation of different parameters in the same calculation. Figure 1 shows the results of varying the parameters of the calculation presented in [7]. The solid diamonds show the experimental data of [2] and the solid circles show the results of the ideal gas calculation whereas the open symbols show the equilibrium constants for the various parameters that are being varied. Panel (a) of figure 1 shows the dependence of the equilibrium constants on the particle composition assumed in the calculation. We note that the value of the equilibrium constants increases with increasing proton fraction, y_p.

Panel (b) of figure 1 shows a slight variation depending on whether Coulomb screening is present in the calculation or not. This is an important point in the comparison as supernova matter is charge neutral but the matter in heavy ion collisions has a net charge.

Finally panel (c) of figure 1 shows the variation with assumed particle degrees of freedom. The calculation shows almost no dependence when A is

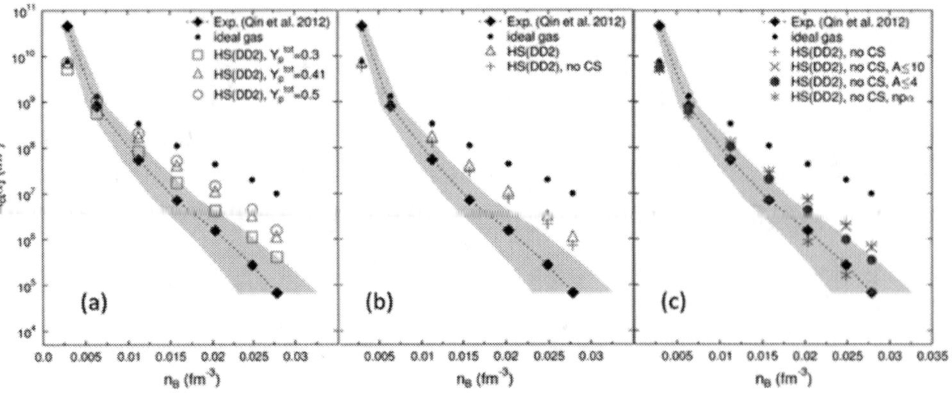

Figure 1: The dependence of α-particle equilibrium constants on the variation of different parameters in the calculations. Panel (a) shows the variation with particle composition, panel (b) shows the variation with coulomb effects and panel (c) shows the variation with particle degrees of freedom.

constrained to be less than 10 as compared to no constraint. In the experiment there is, however, only a very small amount of fragment production of A>4 from the source analyzed. The figure shows that there is a slight dependence when A is constrained to be less than 4 in the calculation.

The star symbols in panel (c) illustrate a very important point in comparing calculations of supernova EOSs to experimental data in general and heavy ion data in particular. Those equilibrium constants result from constraining the particle degrees of freedom to include only n, p and α. Taken by themselves, this implies the best agreement with the experimental data. This is, however, an accidental agreement as the experiment shows significant production of d, t, ^{3}He. The d, t, ^{3}He equilibrium constants from calculations ignoring these species are by definition zero and that assumption will modify the results for the species included.

The rest of the calculations use the constraints that are present in the experiment. i. e., in the following $y_p = 0.41$, Coulomb screening is turned off and the particle species are constrained to A$\leq$4 since those constraints are what was observed in the experiment.

Figure 2 shows the equilibrium constants with the experimental constraints for all of the light particles measured. As mentioned above, it is seen that while the calculation constraining production to n, p and α agrees very well with the α-particle equilibrium constants, this is partially because competing species are ignored. They cannot agree with the equilibrium constants of the other measured light particles. We do observe that the calculations that agree best with the experimental data are those where particle

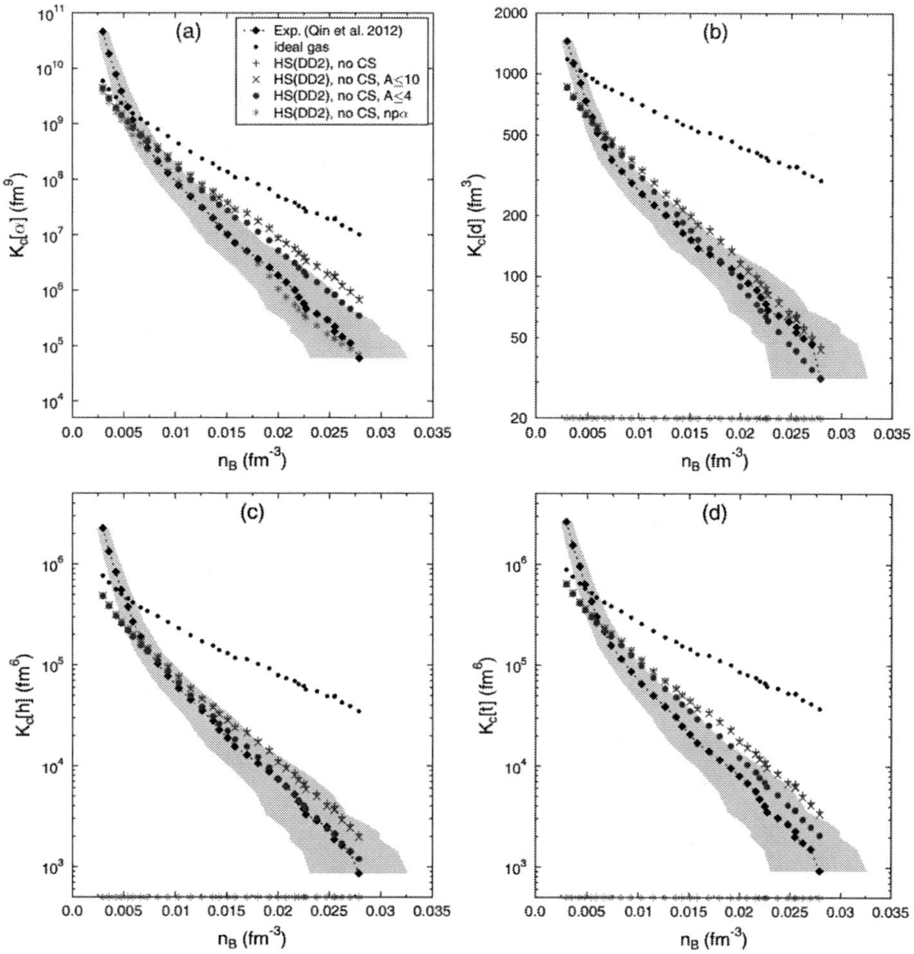

Figure 2: Equilibrium constants for light particles measured in the experiment. The experimental data is shown as solid diamonds, the ideal gas is shown as solid circles and the open symbols show the calculations with variations of different constraints on the particle degrees of freedom. Panel (a) shows the equilibrium constants for α-particles, (b) for deuterons, (c) for tritons and (d) for ^{3}He.

production is constrained to A$\leq$4.

As mentioned in the introduction, there are many calculations of the astrophysical EOS. Comparisons of results of other calculations to the experimental data have been made as well. Where possible the experimental constraints were added into the calculations. Figure 3 shows the results of these calculations. The figure shows two groups of calculations. The first group is the EOS calculations which include only n, p, α (plus a surrogate heavy species) and predict only K(α), STOS [3,4], LS EOS [5], the

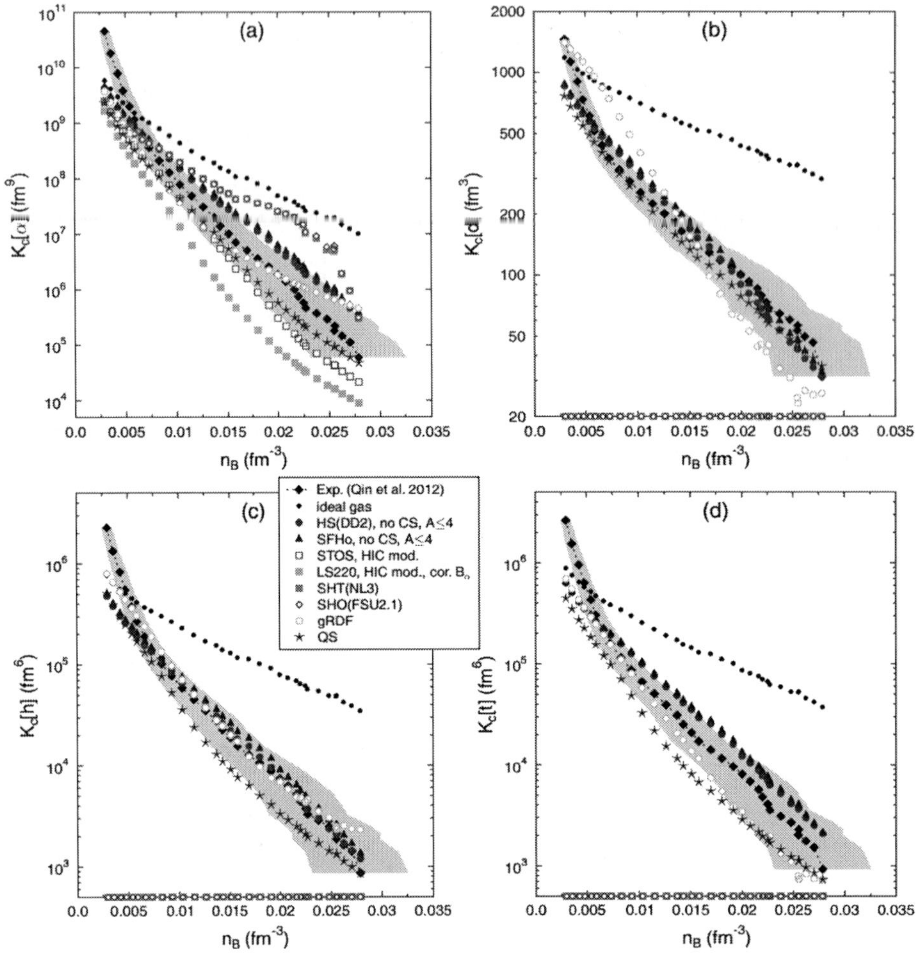

Figure 3: Equilibrium constants from experimental data compared to the predictions of various calculations of the astrophysical EOS. Calculations that treat only n, p, α-particles are shown on the x-axis for particles other than α-particles. See text for details.

NL3 [8] and FSU Gold [9]. The other models shown, HS(DD2) [7], SFHo [16], gRDF [11], and QS [17, 18] comprise a second group and these calculations include all of the particle species measured in the experiment. We observe that the equilibrium constants from all calculations converge to the ideal gas limit at low densities. The models that treat all light particles generally lie within the error bars of the experiment. The convergence at low densities is below the experimental data. This can probably be explained because the low energy particles leading to the equilibrium constants at low densities are emitted during the late stages of the reaction where there is probably con-

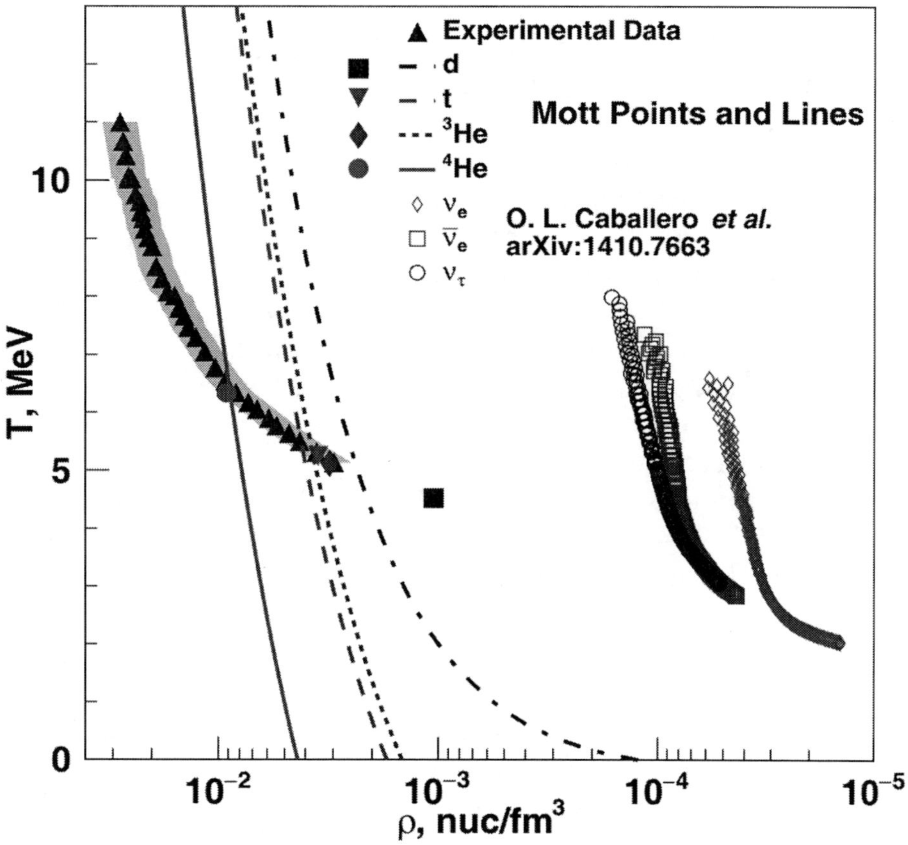

Figure 4: The evolution of T vs ρ. Solid triangles indicate the experimental data. Solid squares, inverted triangles, diamonds and circles indicate Mott points shown in [20]. Open diamonds, squares and circles indicate the temperature density map of neutrinos from black hole accretion disks

tamination by particles from sources other than the coalescing low density gas. Such contributions of low energy particles are very difficult to separate.

We have also explored how the density vs temperature profiles sampled in the neutrinosphere or in accretion disks compares to the density vs temperature profile that is sampled in our experimental data. The solid triangles in figure 4 show the density vs temperature profile from our experimental data [2, 19, 20]. Note the reversed axis in density. The other solid symbols show the Mott points extracted from the experimental data [20] along with theoretical predictions of the temperature vs density profile of the Mott points [7]. The open diamonds, squares and circles show the prediction of the temperature vs density sampling from black hole accretion disks [21,22] for neutrinos, anti-neutrinos, and tau-neutrinos, respectively.

We observe from the figure that heavy ion collisions do indeed sample temperatures and densities in the vicinity of the neutrinosphere. The calculations from black hole accretion disks indicate that similar temperatures are sampled, but that the density sampled in the experiment is higher.

In summary we have significantly constrained calculations of the astrophysical EOS. An extension of the work in [2] has shown that it is important to include all of the measured experimental data in order to draw meaningful conclusions [15] on the astrophysical EOS. We have also compared the evolution of temperature and density that is sampled in heavy in collisions to the predictions of that sampled by neutrinos in black hole accretion disks. The experiment is shown to be sampling nuclear matter in the range of the neutriosphere and black hole accretion disks.

This work was supported by the United States Department of Energy under Grant # DE-FG03- 93ER40773 and by The Robert A. Welch Foundation under Grant # A0330.

References

[1] B. A. Li, L. W. Chen and C. M. Ko, Phys. Rep. **464**, 113 (2008).

[2] L. Qin *et al.*, Phys. Rev. Lett. 108, 172701 (2012).

[3] H. Shen *et al.*, Nucl. Phys. **A637**, 435 (1998); Prog. Theor. Phys. **100**, 1013 (1998).

[4] H. Shen, H. Toki, K. Oyamatsu, and K. Sumiyoshi, Astrophys. J. Suppl. **197**, 20 (2011).

[5] J. M. Lattimer and F. D. Swesty, Nucl. Phys. **A535**, 331 (1991).

[6] C. J. Horowitz and A. Schwenk, Nucl. Phys. **A776**, 55 (2006).

[7] S. Typel, G. Röpke, T. Klahn, D. Blaschke and H. H. Wolter, Phys. Rev. C **81**, 015803 (2010).

[8] G. Shen, *et al.*, Phys. Rev. C **83**, 035802 (2011).

[9] G. Shen *et al.*, Phys. Rev. C **83**, 065808 (2011).

[10] M. Hempel and J. Schaffner-Bielich, Nucl. Phys. **A837** , 210 (2010).

[11] M. D. Voskresenskaya and S. Typel, Nucl. Phys. **A887**, 42 (2012).

[12] G. Röpke, M. Schmidt, L.Münchow, and H. Schulz, Nucl. Phys. **A399**, 587 (1983).

[13] G. Röpke, L. Münchow, and H. Schulz, Nucl. Phys. **A379**, 536 (1982).

[14] G. Röpke, L. Münchow, and H. Schulz, Phys. Lett. **B110**, 21 (1982).

[15] M. Hempel *et al.*, Phys. Rev. C **91**, 045805 (2015).

[16] A. W. Steiner, M. Hempel, and T. Fischer, Astrophys. J. **774**, 17 (2013). (2013).

[17] G. Röpke, Phys. Rev. C 79, 014002 (2009).

[18] G. Röpke, Nucl. Phys. A 867, 66 (2011).

[19] R. Wada *et al.*, Phys. Rev. C **85**, 064618 (2012).

[20] K. Hagel *et al.*, Phys. Rev. Lett. **108**, 062702 (2012).

[21] O. L. Caballero *et al.*, ArXiv:1410.7663 (2014).

[22] O. L. Caballero *et al.*, EPJ Web of Conferences **93**, 03002 (2015).

EPJ Web of Conferences **117**, 07019 (2016) DOI: 10.1051/epjconf/201611707019
NN2015

Progress in constraining the asymmetry dependence of the nuclear caloric curve

Alan B. McIntosh and Sherry J. Yennello

Cyclotron Institute and Department of Chemistry, Texas A&M University, College Station TX, USA

Abstract

The nuclear equation of state is a basic emergent property of nuclear material. Despite its importance in nuclear physics and astrophysics, aspects of it are still poorly constrained. Our research focuses on answering the question: How does the nuclear caloric curve depend on the neutron-proton asymmetry? We briefly describe our initial observation that increasing neutron-richness leads to lower temperatures. We then discuss the status of our recently executed experiment to independently measure the asymmetry dependence of the caloric curve.

The nuclear equation of state (EOS) relates the thermodynamic state variables temperature, pressure, density, internal energy, and neutron-proton asymmetry. It describes the evolution of finite excited nuclear systems created in terrestrial accelerator experiments and finds application in many aspects of nuclear astrophysics as well. Our knowledge of the EOS is well constrained for nuclear systems near beta stability. However, extrapolation to very exotic environments far from beta stability (e.g. neutron stars) represents the largest uncertainty in the EOS. We approach this deficiency in a unique way by seeking to answer the question: How does the caloric curve depend on the neutron-proton asymmetry?

To date, the theoretical approaches [1–6] that have been applied to this question have provided answers that differ not only in magnitude but in sign: some predict a decrease in temperature with increasing neutron content, while others predict just the opposite. This discrepancy stems in part

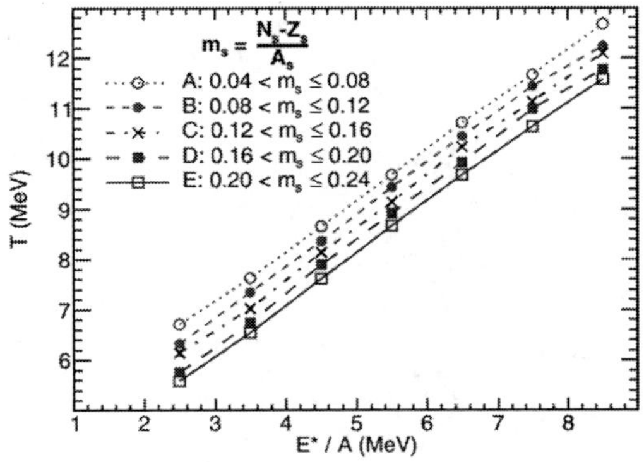

Figure 1: For reconstructed, isotopically-resolved quasi-projectile sources, the caloric curve shifts to lower temperature with increasing asymmetry.

from a dearth of experimental constraints. This paper summarizes our first observation of dependence to the caloric curve, and describes the status of our subsequent measurement for which the analysis is in progress. We hope that these measurements may stimulate and guide theoretical progress.

We have previously reported our observation of the asymmetry dependence of the nuclear caloric curve [7–9]. Charged particles and neutrons produced in reactions of ^{70}Zn+^{70}Zn, ^{64}Zn+^{64}Zn, ^{64}Ni+^{64}Ni @ 35A MeV were measured with the 4π NIMROD-ISiS array at Texas A&M University Cyclotron Institute. Charged particles were isotopically identified using the ΔE-E technique up to at least Z=17 in many detectors. We recognized that it was crucial to reconstruct the quasi-projectile (QP), the primary excited large piece of nuclear material produced near beam velocity in non-central collisions. Such a reconstruction allowed us to select on the neutron-proton asymmetry of the QP rather than the asymmetry of the entrance channel. This provides a much more sensitive measurement of the asymmetry dependence of the caloric curve. Building on the method of Wuenschel et al. [10], we selected events in which the charged particles comprising the QP were all isotopically identified and further required that the reconstructed QP be thermally equilibrated by selecting events which are on average spherically symmetric. Excitation energies were calculated from measured kinetic energies of charged fragments, the neutron multiplicity and average kinetic energy, and the Q-value of the QP breakup.

Temperatures were calculated using three methods: the momentum quadrupole fluctuation thermometer (MQF) [11], the Albergo yield ratio [12], and the slope-temperature method [13]. The first and third of these both make use of the momenta of charged particles, while the second is a

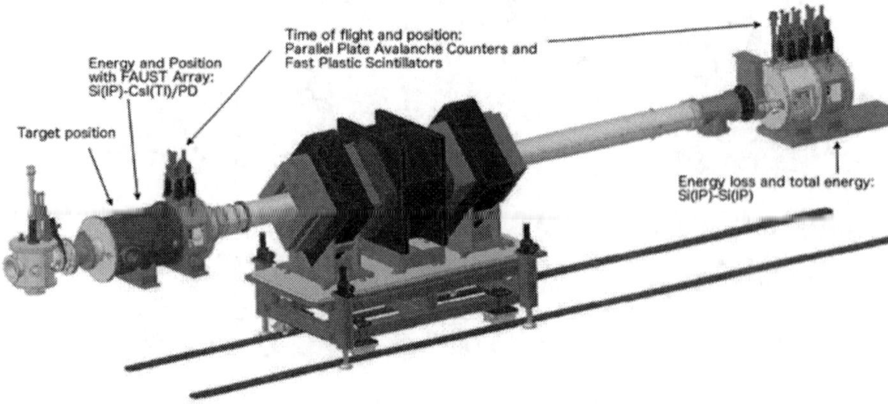

Figure 2: The quadrupole triplet spectrometer at TAMU provides a time-of-flight measurement for heavy ions around zero degrees ($0.9° \leq \theta_{lab} \leq 2.3°$). The FAUST array complements the QTS by providing charged particle measurement over a large angular range.

chemical method. The caloric curve, the correlation between the temperature and excitation energy of the reconstructed QP, is shown in Fig. 1 using the MQF method with protons as the probe particle. As expected, the temperature rises with increasing excitation energy. Due to the relatively light mass of the QP ($48 \leq A_s \leq 52$), the plateau is expected to begin around $E*/A \approx 9$ MeV [14], which is above the energies probed here. The data is broken into 5 ranges of asymmetry ($m_s = \frac{N_s - Z_s}{A_s}$) which span a broad range in asymmetry. For any given excitation energy, as the source becomes more neutron-rich, the temperature decreases. This occurs relatively uniformly for all excitation energies measured. A change from $m_s = 0.06$ to 0.22 for this $A \approx 50$ system corresponds to 4 neutrons, and elicits a decrease in temperature of 1.1 MeV. Other charged particle probes of the temperature using all methods mentioned above provide the same result that a larger neutron content corresponds to a decrease in the temperature. The magnitude of the effect differs from one probe to another (between 0.1 and 2.2 MeV), and in all cases is statistically significant. In addition to verifying this initial observation, future work should attempt to constrain the magnitude of the effect.

We have designed and executed an experiment to independently measure the asymmetry dependence of the caloric curve by studying incomplete fusion reactions of ^{78,86}Kr + ^{12}C @ 15, 25, 35 A MeV. The excitation and composition of the compound nuclei produced are determined by the fraction of the target that fuses with the projectile and does not rely on a free

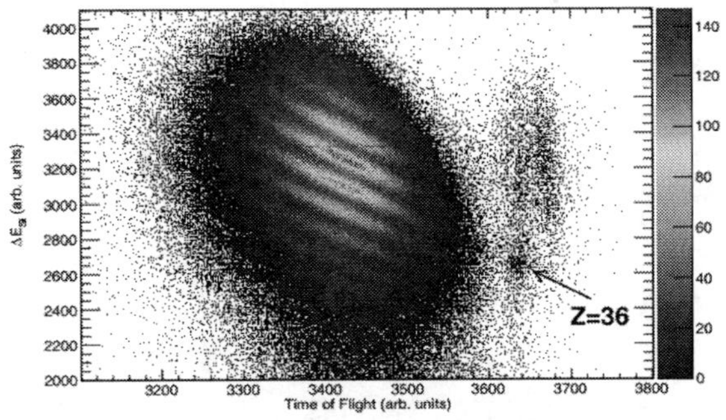

Figure 3: Particles can be identified by atomic number in the QTS by the correlation between energy loss and time of flight. Data shown is measured within $0.9° < \theta_{lab} < 2.3°$ for ^{86}Kr + ^{12}C @ 25A MeV.

Figure 4: The FAUST array provides nearly complete coverage for charged particles in the angular range $2.3° < \theta_{lab} < 34.1°$ with partial coverage for $\theta_{lab} < 45.5$ and $\theta_{lab} > 1.6$.

neutron measurement. The fraction of the targets that fuses with the projectile can be deduced by measuring the velocity of the fusion-evaporation residue event-by-event [15], and applying a correction (< 10%) for the recoil of the residual target-like fragment [16]. The temperature can be obtained by measuring charged particles evaporated from the compound nucleus and then employing the methods discussed above.

The experiment was conducted using the FAUST-QTS detector system at the Texas A&M University Cyclotron Institute in July-August 2015. Fig. 2 illustrates shows a mechanical drawing of the experimental apparatus. The krypton beams entered from the left and collided with an $830\mu g$ carbon target. Light charged particles were measured in the FAUST array (discussed in more detail below). Downstream of FAUST, a Faraday cup collected unreacted and elastically-scattered beam with $\theta_{lab} < 0.9°$. Fragments in the angular range $0.9° < \theta_{lab} < 2.3°$ were focused by the quadrupole triplet spectrometer (QTS), over a $\approx 5.5m$ flight path. On either end of this

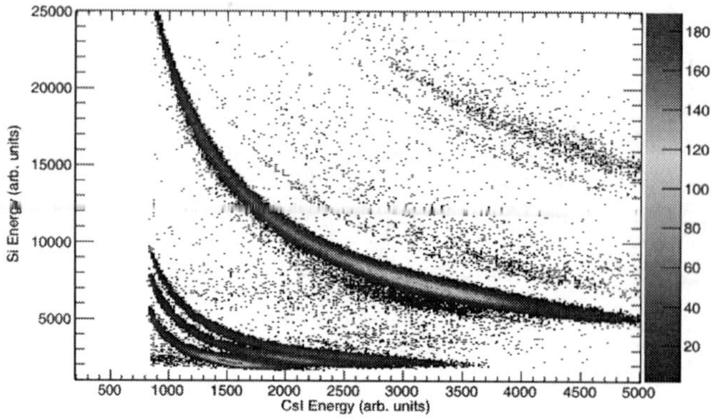

Figure 5: Light charged particles are identified in FAUST using the dE-E technique. Data shown is for ^{86}Kr + ^{12}C @ 25A MeV, FAUST telescope 49.

flight path, parallel plate avalanche detectors and thin plastic scintillator detectors were used to measure the time of flight. At the downstream end, silicon detectors were used to measure energy loss and total energy. Figure 3 shows the correlation between the energy loss and the time of flight for reactions of ^{86}Kr+^{12}C @ 25 A MeV. With the timing resolution in the vicinity of 500ps and energy on the 1% level, determination of the atomic number of the heavy ion is possible with unit resolution. A small contribution of elastically scattered beam can be used to identify the $Z = 36$ band. Further details of the FAUST-QTS detector system may be found in references [17, 18].

Figure 4 shows a GEANT4 model of the 68 Si-CsI(Tl)/PD detector telescopes of the FAUST array. These provide $\approx$ 95% coverage in the angular range $2.3° < \theta_{lab} < 34.1°$ with partial coverage for $\theta_{lab} > 1.6°$ and $\theta_{lab} < 45.5°$. According to the evaporation codes GEMINI++ and PACE4, this range provides kinematic coverage for the vast majority of the phase space for light charged particles in the reactions studied. The silicon detectors are position sensitive to better than $200\mu m$ [19]. Particle Identification in FAUST uses the ΔE-E technique. Figure 5 shows the isotopic resolution obtained for isotopes of hydrogen and helium obtained for detector 49 in reactions of ^{86}Kr + ^{12}C @ 25A MeV. Details of the FAUST array may be found in references [19–21].

With the excellent coverage and resolution of the evaporated particles and the fusion-evaporation residues in these measurements of Kr + C, we expect to provide constraints on the dependence of the nuclear caloric curve on the neutron-proton asymmetry.

Acknowledgement

This work was supported by the Robert A. Welch Foundation (Grant A-1266) and the United States Department of Energy (Grant DE-FG02-93ER40773).

References

[1] J. Besprosvany, S. Levit, Phys. Lett. B **217**, 1 (1989)

[2] Y.J. Zhang et al., Phys. Rev. C **54**, 1137 (1996)

[3] V.M. Kolomietz, A.I. Sanzhur, S. Shlomo, S.A. Firin, Phys. Rev. C **64**, 024315 (2001)

[4] R. Ogul, A.S. Botvina, Phys. Rev. C **66**, 051601(R) (2002)

[5] C. Hoel, L.G. Sobotka, R.J. Charity, Phys. Rev. C **75**, 017601 (2007)

[6] J. Su, F.S. Zhang, Phys. Rev. C **84**, 037601 (2011)

[7] A.B. McIntosh et al., Phys. Lett. B **719**, 337 (2013)

[8] A.B. McIntosh et al., Phys. Rev. C **87**, 034617 (2013)

[9] A.B. McIntosh et al., Euro. Phys. J A **50**, 35 (2014)

[10] S. Wuenschel et al., Phys. Rev. C **79**, 061602(R) (2009)

[11] H. Zheng, A. Bonasera, Phys. Lett. B **696**, 178 (2011)

[12] S. Albergo, S. Costa, E. Costanzo, A. Rubbino, Il Nuovo Cimento **89**, 1 (1985)

[13] R. Yanez et al., Phys. Rev. C **68**, 011602(R) (2003)

[14] J.B. Natowitz et al., Phys. Rev. C **65**, 034618 (2002)

[15] H. Morgenstern et al., Phys. Lett. B **113**, 463 (1982)

[16] W. Bohne et al., Rev. Mod. Phys. **41**, 5 (1990)

[17] P. Cammarata et al., Nucl. Inst. Meth. A **792**, 61 (2015)

[18] A.B. McIntosh et al., Nucl. Inst. Meth. A (2016), to be submitted

[19] S. Soisson et al., Nucl. Inst. Meth. A **613**, 240 (2010)

[20] F. Gimeno-Nogues et al., Nucl. Inst. Meth. A **399**, 94 (1997)

[21] L. Heilborn et al., Nucl. Inst. Meth. A (2016)

EPJ Web of Conferences **117**, 07020 (2016) DOI: 10.1051/epjconf/201611707020
NN2015

Study of two- and multi-particle correlations in $^{12}C+^{24}Mg$ and $^{12}C+^{208}Pb$ reactions at E=35 AMeV

L. Quattrocchi[1,2], L. Acosta[3,13], F. Amorini[5], A. Anzalone[5], L. Auditore[1,2], I. Berceanu [14], G. Cardella[3], A. Chbihi[6], E. De Filippo[3], D. Dell' Aquila[7,10], L. Francalanza[7], B. Gnoffo[3,8], A. Grzeszczuk [15], G. Lanzalone[5,9], I. Lombardo[7,10], I. Martel[4], T. Minniti[8], S. Norella[1,2], A. Pagano[3], E.V. Pagano[5,8], M. Papa[3], S. Pirrone[3], G. Politi[3,8], F. Porto[5,8], F. Rizzo[5,8], E. Rosato [7,10†], P. Russotto[3], A. Trifirò[1,2], M. Trimarchi[1,2], G. Verde[3,11], M. Veselsky[12], and M. Vigilante[7,10]

[1] Università di Messina, Dip. di Fisica e Scienze della Terra, Messina, Italy
[2] INFN, Gruppo Collegato di Messina, Messina, Italy
[3] INFN, Sezione di Catania, Catania, Italy
[4] Departamento de Física Aplicada, Universidad de Huelva, Huelva, Spain
[5] INFN - Laboratori Nazionali del Sud, Catania, Italy
[6] GANIL,CEA-IN2P3-CNRS, Caen, France
[7] INFN, Sezione di Napoli, Napoli, Italy
[8] Università di Catania, Dip. di Fisica e Astronomia, Catania, Italy
[9] Università Kore, Enna, Italy
[10] Università di Napoli Federico II, Dipartimento di Fisica, Napoli, Italy
[11] IPN Orsay, Orsay, France
[12] Slovak Academy of Sciences, Bratislava, Slovakia
[13] Instituto de Física, Universidad Nacional Autónoma de México, México
[14] Institute for Physics and Nuclear Engineering, Bucharest, Romania
[15] Institute of Physics, University of Silesia, Katowice, Poland

†Deceased

Abstract

Two and multi particle correlations from the decay of sources produced in $^{12}C+^{24}Mg$ and $^{12}C+^{208}Pb$ collisions at E=35 $AMeV$ have been studied by using the forward part ($1^o < \theta_{lab} < 30^o$) of the CHIMERA multi-detector. Correlations and invariant mass spectroscopy are used to explore simultaneous and sequential decays of resonances in light isotopes with Z∼3-6, produced in peripheral collisions via the break-up of excited quasi-projectiles. Among them we mention ^{5}Li, ^{6}Li, ^{6}Be, ^{8}Be and the astrophysically important state in ^{12}C decaying into three alpha particles. Results and future perspectives at the INFN-LNS will be presented.

1 Introduction

Heavy-ion collisions at intermediate energy have been extensively studied to produce nuclear systems under extreme conditions of temperature, density and excitation energy and they represent the only terrestrial means to explore the Nuclear Equation of State at low density ($\rho \leq \rho_0 = 0.17fm^3$). The correlations between different particles emitted during a collision provide important information about space-time properties of emitting sources produced and quantitative understanding of reaction dynamics [1]. Multi-particle correlations are also used to study the decay of unbound states produced during nuclear reactions [2, 3]. In this respect invariant-mass spectroscopy has become a tool to explore spectroscopic properties such as spins [4] but also to evaluate the competition between simultaneous and sequential channels in decay of produced resonances [2, 7]. In this contribution we study the three-α correlations from the decay of ^{12}C quasi-projectiles produced in peripheral $^{12}C+^{24}Mg,^{208}Pb$ collisions at 35 $AMeV$. The experiment was performed at INFN-LNS with the aim of exploring decay of resonances in heavy-ion collisions.

2 Experiment and results

The ^{12}C beam at 35 $AMeV$ was delivered by the Superconducting Cyclotron at Laboratori Nazionali del Sud and INFN of Catania and the charged reaction products were detected by the CHIMERA 4π multi-detector [5]. To investigate the reliability of energy calibration, in Fig. 1 the $\alpha - \alpha$ (left panel) and $\alpha - d$ (right panel) coincidence pairs are shown. The plots are reported as a function of excitation energy, $E_{ex} = E_{tot}Q$, where E_{tot} is the total kinetic energy in their pairs center of mass reference frame and Q is the Q-value for the corresponding decay channel i.e., ^{8}Be$\rightarrow \alpha - \alpha$ and ^{6}Li$\rightarrow \alpha - d$, respectively.

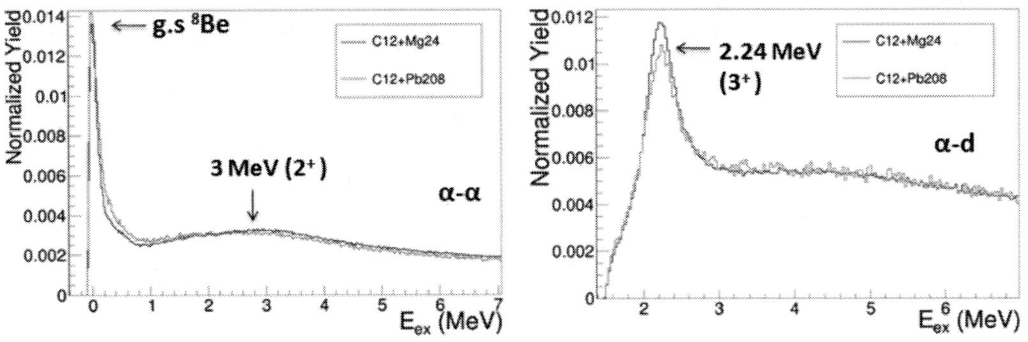

Figure 1: (Colour online) Yield of correlated $\alpha-\alpha$ (left panel) and $\alpha-d$ (right panel) couples as function of excitation energy, for ^{12}C+^{24}Mg (blu line) and ^{12}C+^{208}Pb (pink line) reactions.

The $\alpha - \alpha$ spectrum shows a narrow peak centred at 0 MeV corresponding to the ground state of ^{8}Be (Qvalue =92 keV) and a much broader peak centred around 3 MeV, corresponding to its first excited state at 3.03 MeV (2^{+}). The $\alpha - d$ spectrum, on the other hand, shows a well-shaped peak centred at ≈ 2.2 MeV corresponding to the first excited state of ^{6}Li at 2.18 MeV (3^{+}). In both cases, the low energy levels are accurately determined, thus showing a good quality of energy calibrations and a good angular resolution of the detector array. The FWHM of the peak at 92 keV in $\alpha - \alpha$ correlation amounts ≈ 70 keV, while FWHM of first peak in $\alpha - d$ correlation is about 300 keV. These widths are consistent with those measured in the experiment of Ref. [7] performed with the same experimental setup.

The α particles resulting from the decay of ^{12}C quasi-projectiles, are selected by requiring their parallel velocity, $V_{//}$, to be larger than 80% of beam velocity. This criterion restricts the analysis to the peripheral collisions as it is confirmed by a comparison to simulations performed with HIPSE model calculations [6]. Then information on unstable excited states of ^{12}C may be extracted from the 3α correlation function defined by the following ratio:

$$1 + R(E_{ex}) = \frac{Y_{coinc}(E_{ex})}{Y_{uncorr}(E_{ex})}, \qquad (1)$$

where the coincidence yield spectrum, $Y_{coinc}(E_{ex})$, is obtained from 3α particles detected in the same event (left panel of Fig. 2), while uncorrelated 3α spectrum, $Y_{uncorr}(E_{ex})$, is built by mixing particles from different events.

The 3α correlation function reported in right panel of Fig. 2 shows two peaks; the first one, centered around $E_{ex} = 7.66\ MeV$, corresponds to the Hoyle state ($E_{ex}^{th} = 7.65\ MeV$, $\Gamma = 8.5\ eV$) [8]. The second peak, centered at $E_{ex} = 9.83\ MeV$, arises from the overlap of the states at $E_{ex} = 9.64\ MeV$ (3^-), $E_{ex} = 10.3\ MeV$ (0^+) and possibly at $E_{ex} = 9.7\ MeV$ (2^+). In order to study the decay mechanisms of the observed resonaces, possibly disentangling sequential decays from direct three body ones, we follow a strategy which is similar to that obtained in Ref. [9].

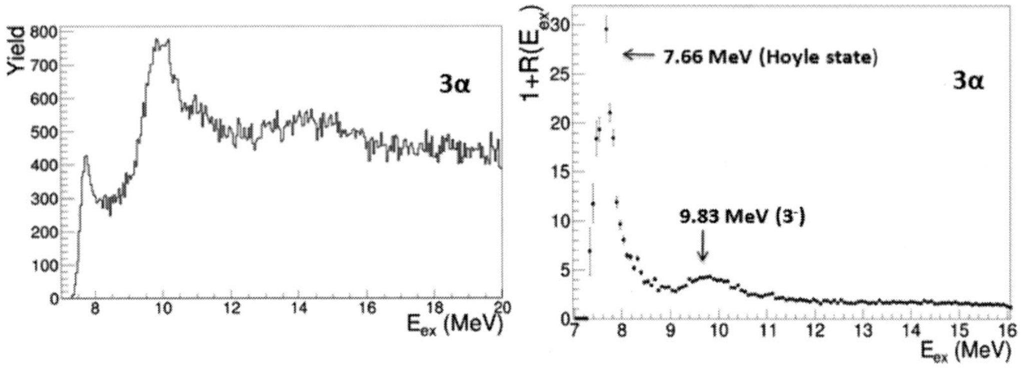

Figure 2: Yields of correlated 3α particles (left panel) and 3α correlation function (right panel) versus excitation energy of ^{12}C obtained in ^{12}C+^{24}Mg reactions at 35 $AMeV$.

In particular we build Dalitz plots with the following coordinates:

$$X = \sqrt{3}(\epsilon_j - \epsilon_k), \qquad Y = 2\epsilon_i - \epsilon_j - \epsilon_k, \tag{2}$$

where $\epsilon_{i,j,k} = E_{i,j,k}/(E_i + E_j + E_k)$ are the energies of particles in the center of mass reference frame, normalized to total energy of three α decay ($E_{i,j,k}$ are selected with $E_i > E_j > E_k$ for simplicity of the symmetry).

The experimental results, are compared to numerical Monte-Carlo simulations filtered through the geometry and detector response of the CHIMERA multi-detector. Panels (a) and (b) of Fig. 3 show the Dalitz plots constructed using simulated events corresponding to sequential ($^{12}\mathrm{C}^* \rightarrow {}^8\mathrm{Be}_{gs} + \alpha \rightarrow 3\alpha$) and direct ($^{12}\mathrm{C}^* \rightarrow 3\alpha$) decays of the Hoyle state, respectively. Panels (c) and (d), on the other hand, show the same Dalitz plots constructed with experimental data collected in ^{12}C+^{24}Mg and ^{12}C+^{208}Pb reactions, respectively; they exhibit a more uniform distribution that does not allow us to exclude any of the two decay mechanisms. For better evaluations, monodimensional ϵ_i distributions are used. ϵ_i is the

DOI: 10.1051/epjconf/201611707020

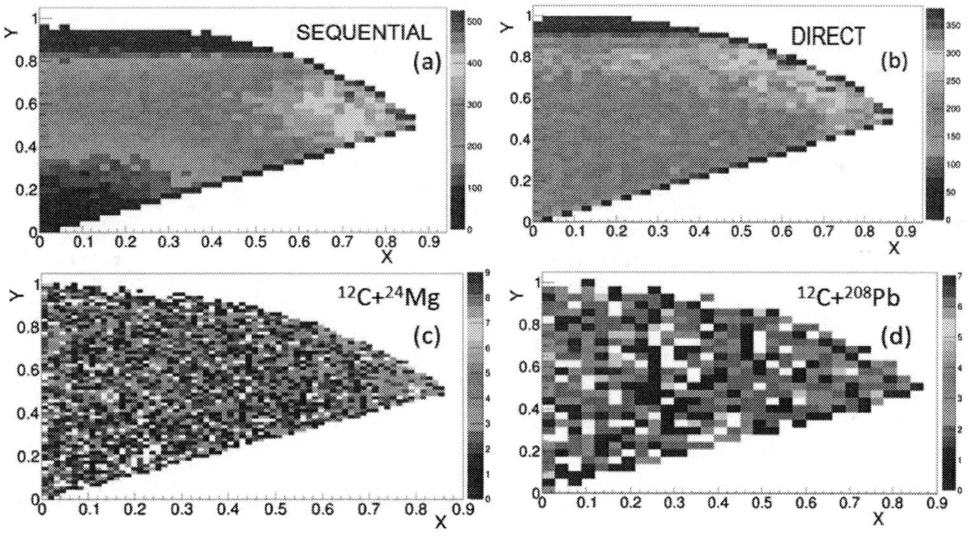

Figure 3: Panels (a) and (b): Dalitz plots of simulated events for sequential ($^{12}C^* \rightarrow ^8 Be_{gs} + \alpha \rightarrow 3\alpha$) and direct ($^{12}C^* \rightarrow 3\alpha$) decay of Hoyle state. Panels (c) and (d): Dalitz plot constructed with experimental data collected in $^{12}C+^{24}Mg$ and $^{12}C + ^{208}Pb$ reactions.

highest normalized energy among those of the 3α particles decaying from the Hoyle state; in the case of sequential mechanism it should have a peak around 0.5 because the first decay α particle takes away two thirds of the released energy. On the other hand, in the direct processes, it should range between 1/3 and 2/3. On Fig. 4 the experimental distributions (pink lines), constructed using the data collected in $^{12}C+^{24}Mg$ and $^{12}C+^{208}Pb$ reactions, are shown on the left and right panel, respectively. They are compared with simulated events for sequential (blue lines) and direct (green lines) decays of Hoyle state. The experimental ϵ_i distributions show a peak typical of a sequential mechanism, but they have also the tails similar to the case of direct processes. These results, in both reactions, suggest a strong contribution of direct decay, a mechanism already observed in in Ref. [7] for $^{40}Ca+^{12}C$ at 25 $AMeV$.

3 Conclusions and perspectives

Multi-particle correlations are under study to explore the decay of excited states in peripheral $^{12}C+^{24}Mg$ and $^{12}C+^{208}Pb$ collisions at 35 $AMeV$. As a first physical system, we focus on α decay mechanisms of Hoyle state by

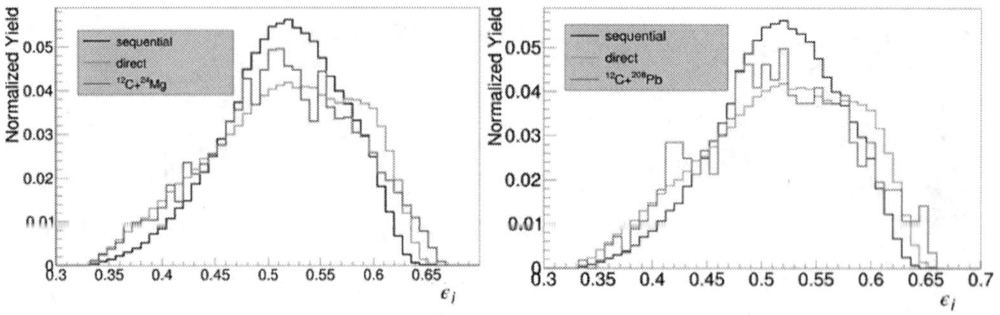

Figure 4: ϵ_i distributions for ^{12}C+^{24}Mg (left panel) and ^{12}C+^{208}Pb (right panel) reactions, compared with simulated data. Blue and green lines refer to simulated events for sequential and direct decays of Hoyle state, respectively, while pink lines represent experimental ϵ_i distributions.

using Dalitz plots. The comparison with Monte-Carlo simulations evidences the contribution of a direct mechanism component, in agreement with results reported in the literature [2,7]. Similar contributions of sequential and direct decays exist for the state at 9.64 MeV in ^{12}C. This and other resonances are under study to understand if differences in decay mechanisms are observed in different reaction systems (^{12}C+^{24}Mg, ^{208}Pb), or in data collected in other experiments with heavy-ion collisions. Among the aim of present study we plan to explore the possible existence of in-medium effects in nuclear structure properties.

References

[1] G. Verde et al., Eur. Phys. J. A**30**, 81-108 (2006).

[2] F. Grenier et al., Nucl. Phys. A **811** 233 (2008).

[3] B. Charity et al., Phys. Rev. C, **52** (1995).

[4] W.P. Tan et al., Phys. Rev. C, **69** 061304.

[5] A. Pagano et al., Nucl. Phys. A **734** 504 (2004).

[6] D. Lacroix et al., Phys. Rev. C **69** 054604 (2004).

[7] Raduta et al., Phys. Lett. B **705** 65 (2011)

[8] F. Hoyle, Astrophys. J. Suppl. Ser., **1** 121 (1954).

[9] M. Itoh et al., Phys. Rev. Lett., **113** 102501 (2014).

EPJ Web of Conferences **117**, 07021 (2016) DOI: 10.1051/epjconf/201611707021
NN2015

Properties of excited A = 40 nuclear systems with varying matter composition

K. Schmidt[1], J.B. Natowitz[2], M. Barbui[2], K. Hagel[2], A. Bonasera[2], G. Giuliani[3], H. Zheng[3], M. Rodrigues[4], R. Wada[2], M. Huang[2,5], C. Botosso[2] and S. Kowalski[1]

[1]Institute of Physics, University of Silesia, Katowice, Poland,
[2]CyclotronInstitute, Texas A&M University, College Station, TX, USA,
[3]INFN - Laboratori Nazionali del Sud, via Santa Sofia 62, 95123 Catania, Italy,
[4]Instituto de Física, Universidade de Sao Paulo, Caixa Postal 66318, CEP 05389-970 Sao Paulo, Brazil,
[5]Institute of Modern Physics, Chinese Academy of Sciences, Lanzhou 730000, China

Abstract

There exists an intriguing problem when bosonic clusters as bound states of fermions are produced in the reaction, and the Bose character of the composite clusters competes with the fermionic properties of their constituents. In the analysis of ^{40}Ca + ^{40}Ca reactions at 35MeV/A we selected classes of projectile-like sources with exit channels consisting of only bosons, only fermions, only even-even nuclei, only odd - odd nuclei, only even - odd nuclei and only alpha-conjugate nuclei, respectively and searched for kinematic characteristics of these systems which might differ depending upon the type of matter selected. The distributions of various observables for the different classes of matter and comparisons between them will be presented and discussed.

1 Introduction

The study of systems composed of mixtures of bosons and fermions stimulates significant theoretical and experimental efforts in different fields of physics. Atomic nuclei are commonly described as systems of strongly interacting fermions, namely neutrons and protons. However, in experiments we observe the existence of phenomena that can be explained by considering nuclei as systems composed of bosonic clusters, the most common being α particles. By considering the nucleus as a mixture of fermions and bosons one may wonder whether the bosonic properties may dominate over the fermionic properties in some instances. Recent theoretical studies indicate, that in heavy ion collisions the produced particles do not follow classical statistics [1] because of the quantum nature, the correct distribution function must be used in the calculations. Protons (p), neutrons (n), tritium (t), etc., follow the Fermi-Dirac statistics [3] [4], while deuterons (d), alphas (α), etc., should follow the Bose-Einstein statistics [5]. In this paper we study the projectile-like sources of different composition and compare the results to the Antissymmetrized Molecular Dynamics Model (AMD) [6] [7] incorporating a statistical decay code GEMINI [8] as a afterburner.

2 Experiment

The experiment was performed in the Cyclotron Institute, Texas A&M University with the multidetector NIMROD-ISiS [9]. The array consists of 14 concentric rings ranging from 3.6^o to 176.0^o in lab. The entire device is housed inside the Texas A&M Neutron Ball, which provides an average neutron multiplicity for each event. Two different detector module configurations were used in the array. The single telescope configuration consisted of a 150μm or 300μm silicon placed in front of the CsI(Tl)-PMT detector. The supertelescope configuration had two silicon detectors, 150μm and 500μm, placed in front of a CsI(Tl)-PMT detector.

3 Event selection

We studied the projectile-like sources consisting of only bosons, only fermions, only even-even nuclei, only odd - odd nuclei, only even - odd nuclei and only alpha-conjugate nuclei respectively. We employed the total α -like mass (d-like mass) of events consisting of α -conjugate (d-conjugate) nuclei

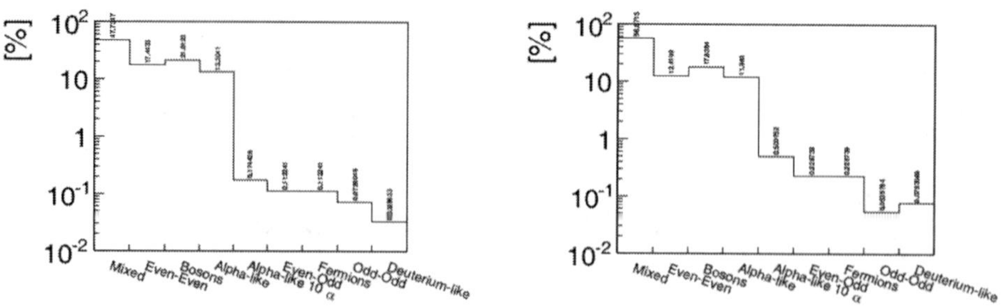

Figure 1: The percentage all events which decayed into a particular type of matter. Left: the experimenta data, right: AMD calculations.

is 40 as well as the total mass A_{TOT} of remaining class of events in order to select the project-like sources. The moving source fit performed on selected class of events indicated, that the light particles (p, d, t) come mainly from the nucleon-nucleon source prior to PLF disassembly and therefore they are not taken into calculations of dynamical variables or PLF excitation energies however they are present in the investigated events. In the Figure 1 the percentage all events which decayed into a particular type of matter is shown, the left panel refers to the experimental data, while the right one to theoretical simulations. It can be seen that almost half of all experimental events are those consisting of Bosons, even-even fragments or α -like fragments and the other types of matter are 2 orders of magnitude less likely. Therefore further results will be presented only for the most populated types of matter. In the AMD calculations the same tendency is also visible, but the difference is not so significant.

4 The number fluctuations

It has been known that fluctuations are large for statistical systems near their critical points, hence the study of multiplicity fluctuations of hadrons produced in high-energy heavy-ion collisions is of importance to study the phase transition [11] [12] [13] [14].

In intermediate energy nuclear collisions the enhancement of multiplicity fluctuations of bosons may be a signal of Bose-Einstein condensation in nuclei [5]. Fermions emitted from a nucleus are subject to the Pauli exclusion principle which may block certain emission channels (Pauli blocking), leading to suppression of the multiplicity fluctuations. In the Figure 2 the

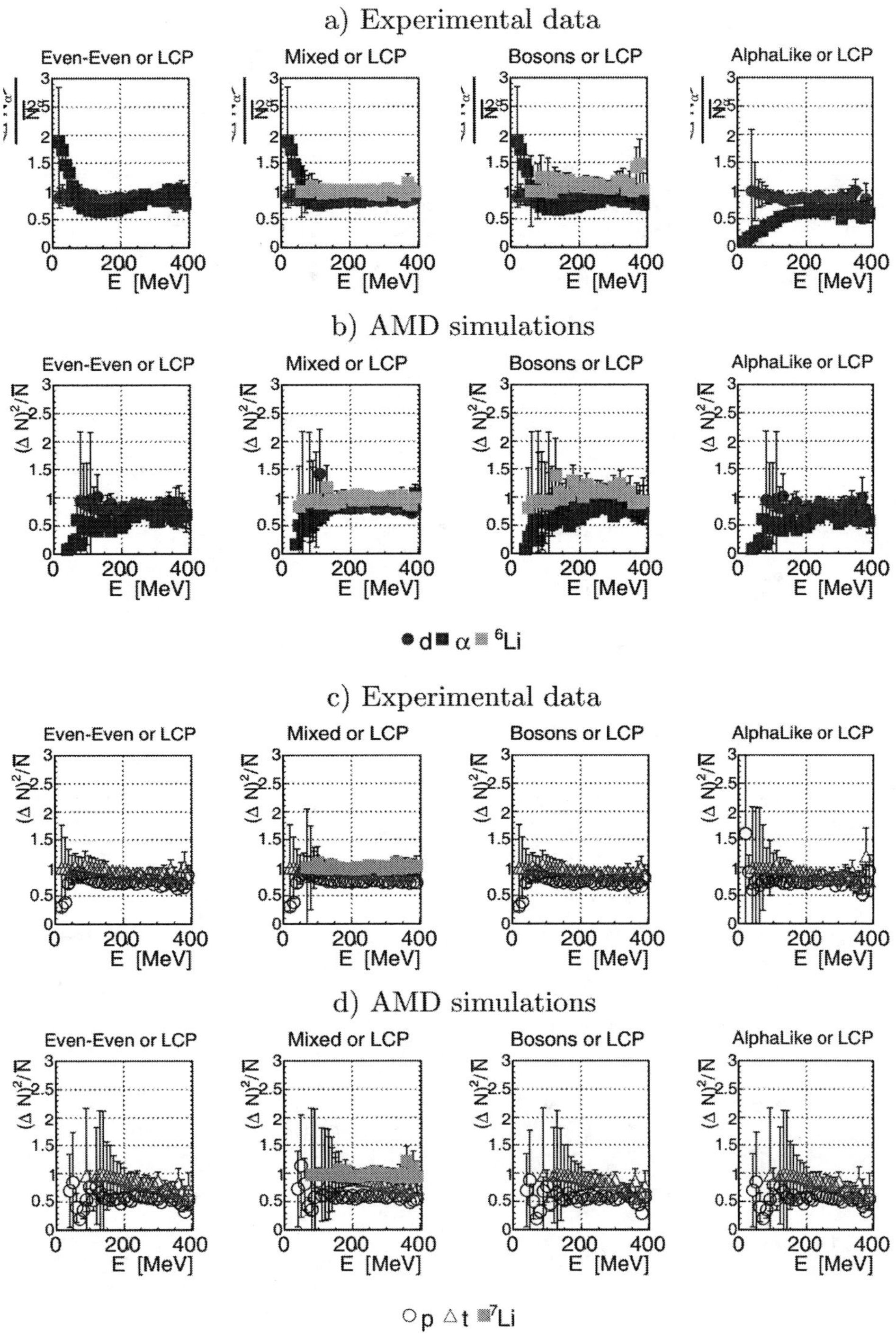

Figure 2: Multiplicity fluctuations of bosons (a,b) and fermions (c,d), experimental data (a,c) and AMD calculations (b,d).

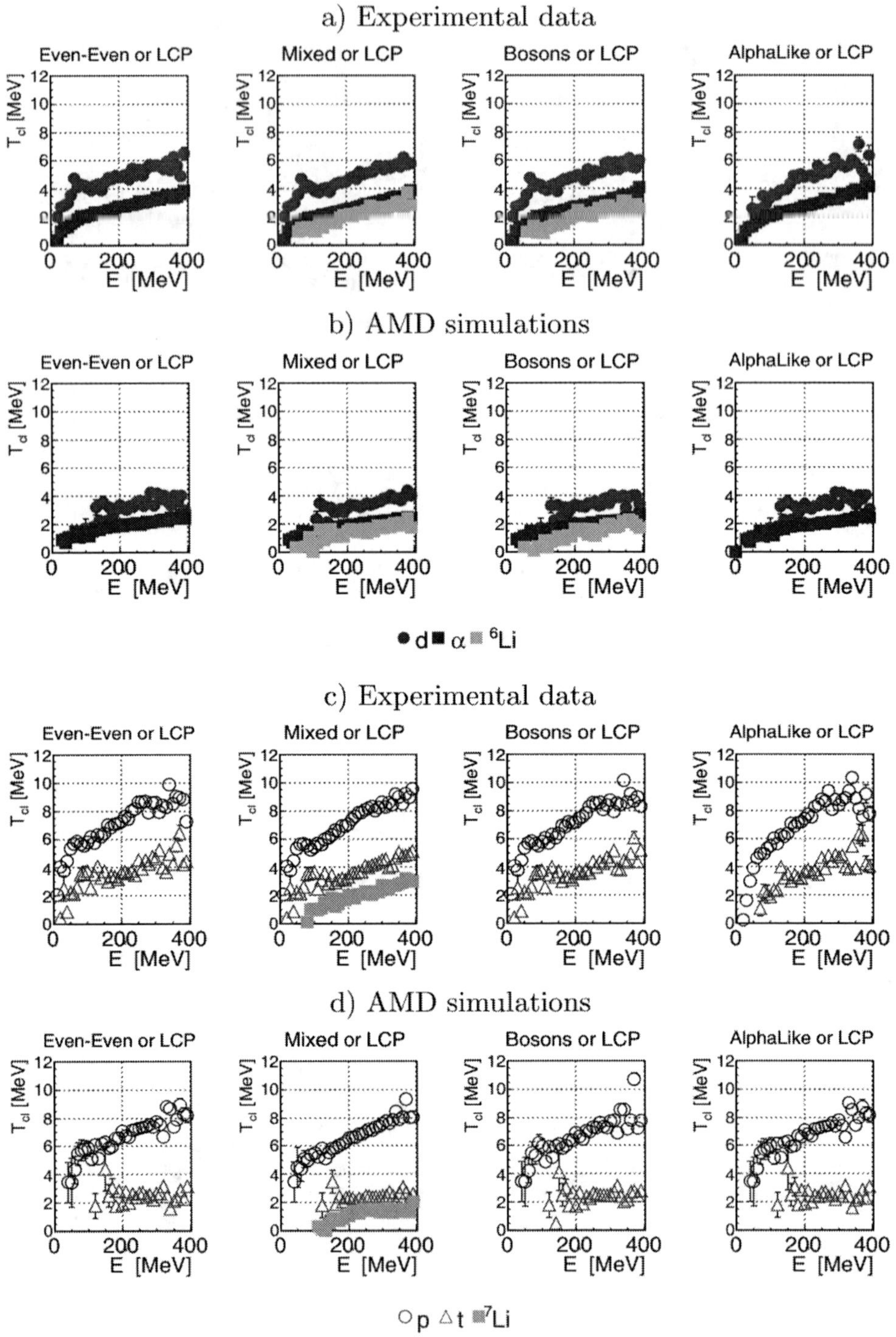

Figure 3: Temperature of bosons (a,b) and fermions (c,d), experimental data (a,c) and AMD calculations (b,d).

multiplicity fluctuations of bosons and fermions for each type of matter as a function of excitation energy E* calculated through calorimetry are presented. Panels a, c refer to experimental data, b,d to AMD calculations. In the experimental data the enhancement of multiplicity fluctuations is observed for low excitation energies for 3 types of matter: even - even, bosons and mixed. Whether those events are the candidates for Bose-Einstein condensates is an open issue. Multiplicity fluctuations higher than the classical limit are not observed for alpha-like matter or for all other types of matter in the AMD calculations, where we can observe the suppression of multiplicity fluctuations.

5 Momentum fluctuations

A method for measuring the temperature based on momentum quadruple fluctuations of detected particles was proposed in [15]. A quadruple moment $Q_{xy} = p_x^2 - p_y^2$ is defined in a direction transverse to the beam axis (z-axis) to minimize non-equilibrium effects [3] [4] [5]. The average Q_{xy} is zero for a given particle type in the center of mass of the equilibrated emitting source. In the Figure 3 the temperatures extracted for bosons and fermions using the method described in [15] are presented as a function of excitation energy. It can be noticed that the temperature of bosons is lower than the fermions and the temperature of bosons extracted from experimental data is higher than extracted from AMD simulations. It is interesting, that the temperature extracted for tritons from AMD data does not show a dependence on the excitation energy. The temperatures of protons from both the experiment and from the simulation are very similar.

6 Summary

The multiplicity fluctuations and the momentum quadrupole fluctuations for bosons and fermions for different classes of events, selected on the basis of exit channel spins, have been studied. Differences between the experimental results and predictions of theoretical calculations have been noticed in both observables. It is highly recommended to compare the experimental results with a model that takes the quantum nature of emitted fragments into account. In future analyses we plan to incorporate the quantum and Coulomb corrections to density and temperature determinations as proposed in [2].

We acknowledge support from the United States Department of Energy under Grant No. DE-FG03-93ER40773 and the Robert A. Welch Foundation under Grant No. A0330

References

[1] Zheng H., Giuliani G., Bonasera A. Nuclear Science and Techniques **24** (2013) 050512

[2] Zheng H., Giuliani G., Bonasera A. Phys Rev C **88** (2013) 024607

[3] Zheng H. and Bonasera A. Phys Lett B **696** (2011) 178181.

[4] Zheng H. and Bonasera A. Phys Rev C **86** (2012) 027602

[5] Zheng H., Giuliani G., Bonasera A. Nucl Phys A **892**(2012) 4357.

[6] Ono A. et al Phys. Rev. C **53** (1996) 2958

[7] Ono A. Phys. Rev. C **59** (1999) 853

[8] Charity R. J. et al Nucl. Phys. A **483** (1988) 371

[9] Schmidt K. et al J. Phys.: Conference Series **420** (2013) 012088

[10] Tarnowsky T J. J. Phys. G: Nucl. Part. Phys. **38** (2011) 124054

[11] Hwa R C, Yang C B. Phys. Rev. C **85** (2012) 044914

[12] Friman B, Karsch F, Redlich K et al. Eur. Phys. J. C (2011) 1694

[13] Gorenstein M I. Phys. Part. Nucl. **39** (2008) 1102-1109

[14] Konchakovski V P, Gorenstein M I, Bratkovskaya E L. Indian J. Phys.**85** (2011) 1-4

[15] Wuenschel S, Bonasera A, May L W, et al. Nucl Phys A **843** (2010) 113.

EPJ Web of Conferences **117**, 07022 (2016) DOI: 10.1051/epjconf/201611707022
NN2015

Excited states of ^{26}Al studied via the reaction ^{27}Al(d,t)

Vishal Srivastava[1,*], C. Bhattacharya[1], T. K. Rana[1], S. Manna[1], S. Kundu[1], S. Bhattacharya[1], K. Banerjee[1], P. Roy[1], R. Pandey[1], G. Mukherjee[1], T. K. Ghosh[1], J. K. Meena[1], T. Roy[1], A. Chaudhuri[1], M. Sinha[1], A. K. Saha[1], Md. A. Asgar[1], A. Dey [1], Subinit Roy[2] and Md. Moin Shaikh[2]

[1]Variable Energy Cyclotron Centre, 1/AF, Bidhan Nagar, Kolkata 700064, India
[2]Saha Institute of Nuclear Physics, 1/AF, Bidhan Nagar, Kolkata 700064, India

Abstract

The reaction ^{27}Al(d,t) at 25 MeV was utilized to study the excited states of ^{26}Al. The angular distributions of the observed excited states of ^{26}Al were analyzed with zero range distorted wave Born approximation as well as by incorporating finite range correction parameters to extract spectroscopic factors. The two sets of extracted spectroscopic factors were compared with each other to see the effect of using finite range correction in the transfer form factor.

The nucleus ^{26}Al is the first cosmic radioactivity detected in the interstellar medium and its half life (7.2×10^5 years) is much shorter than the time for galactic evolution ($\sim10^{10}$ years) and its detection directly indicates that nucleosynthesis is currently active in our galaxy. The origin and importance of ^{26}Al has been very much studied in Refs. [1–5]. To study the structure of the nucleus ^{26}Al, excitation energies, spin and parity and spectroscopic factors etc. are required. Single nucleon transfer reactions are important

*Email: vishalphy@vecc.gov.in

tools to extract spectroscopic information about the nuclei of interest. Very recently, we studied the reaction ^{27}Al(d,t) and extracted neutron spectroscopic factors for the observed excited states of ^{26}Al in Ref. [6]. We also compared the results of our study of one neutron pick-up reactions with the previously reported values obtained using different reaction probes and the comparison has been given in Ref. [6].

In the present study, we extracted the spectroscopic factors for the observed states of ^{26}Al produced through ^{27}Al(d,t) reaction using zero range distorted wave Born approximation (ZR-DWBA) and also attempted the ZR-DWBA calculation by incorporating finite range correction parameters in ZR-DWBA computer code DWUCK4 [7]. So, to examine the effect of using finite range correction parameter in ZR-DWBA on spectroscopic factors was the main motivation of the present paper.

The experiment was performed at the Variable Energy Cyclotron Centre, Kolkata, India to study the reaction ^{27}Al(d,t). The details of the experiment have been given in Ref. [6].

Table 1: The best fit OMPs used in DWUCK4 code for the reaction ^{27}Al(d,t).

Parameters	^{a}d+^{27}Al	^{a}t+^{26}Al	n+^{26}Al
V_R(MeV)	90.301	161.91	$V^{b)}$
R_R(fm)	1.055	1.20	1.20
a_R(fm)	0.675	0.72	0.65
W (MeV)		39.99	
W_D(MeV)	2.407		
R_I(fm)	1.400	1.40	
a_I(fm)	0.850	0.84	
V_{ls}(MeV)	9	2.50	
r_{ls}(fm)	1.055	1.20	
a_{ls}(fm)	0.780	0.72	
R_c (fm)	1.25	1.30	

^{a}taken from Ref. [6]
^{b}Adjusted to give the required separation energy transferred particle.

In this study, we used only one set of optical model potential parameters (OMP) from the Ref. [6] to extract spectroscopic factors of the different observed states of ^{26}Al which are given in Table 1. The angular distributions of the observed states of ^{26}Al were fitted with theoretical predictions from

ZR-DWBA calculations to extract the spectroscopic factors. The fitted angular distributions of the observed states of ^{26}Al with and without using finite range correction parameter and nonlocal parameters were shown in Figs. 1 and 2. The value of the transferred angular momentum (l_{tr}) was estimated from the relation given in [8]. To extract spectroscopic factors, we used the following relation between experimental and theoretical cross sections as used in Refs. [6] and [9];

$$\left(\frac{d\sigma}{d\Omega}\right)_{exp.} = \frac{NC^2S}{2J+1}\left(\frac{d\sigma}{d\Omega}\right)_{DWBA} \tag{1}$$

where $(\frac{d\sigma}{d\Omega})_{exp.}$ is the experimental differential cross-section and $(\frac{d\sigma}{d\Omega})_{DWBA}$ is the cross-section predicted by the DWUCK4 code. C^2 is the isospin Clebsch-Gordon coefficient, S the spectroscopic factor and J is the total angular momentum of that orbital from which the neutron was picked up.

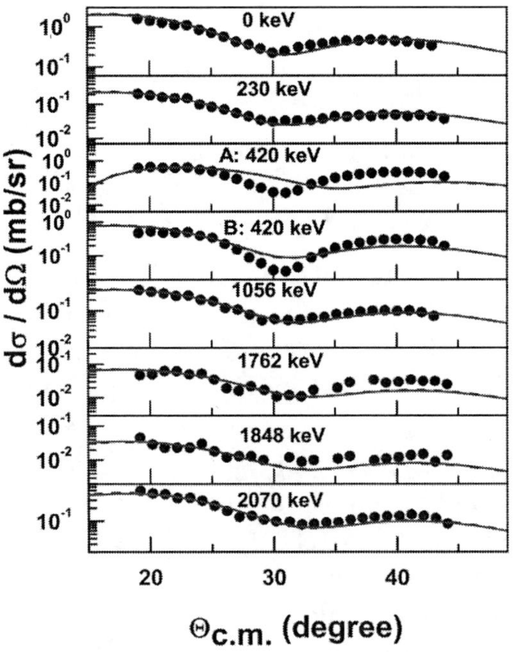

Figure 1: Fitted angular distribution of 0.0, 230, 420, 1056, 1746, 1848 and 2070 keV states of ^{26}Al. The filled circles represent experimental data points, solid lines represents ZR-DWBA predictions without finite range parameter and dash-dash line represent ZR-DWBA predictions using finite range parameter 1.36 fm^{-1}.

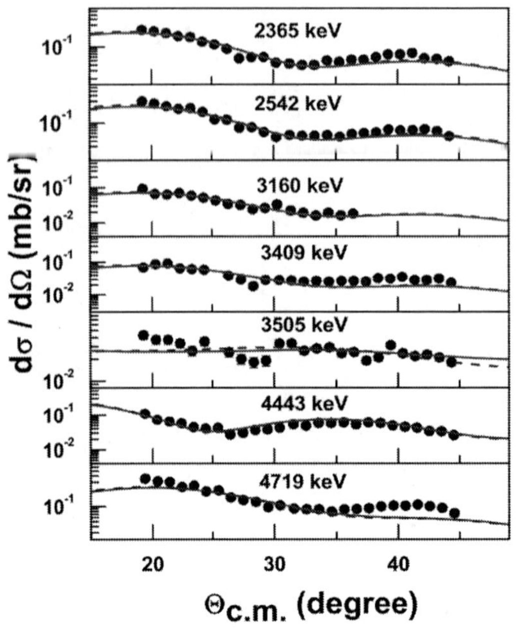

Figure 2: Fitted angular distributions of 2365, 2542, 3160, 3409, 3505, 4443 and 4719 keV states of ^{26}Al (same notation as in Fig. 1).

In the present work, we extracted spectroscopic factors for 14 states of ^{26}Al populated through the reaction ^{27}Al(d,t) which are listed in Table 2. The ground, 230, 1056, 1762, 1848, 2070, 2365, 2542, 3160, 3409 and 4719 keV states of ^{26}Al were studied by assuming pick up from $0d_{5/2}$ single particle orbital. The excited states at 3505 and 4443 keV were studied by assuming pick up from $0g_{9/2}$and $0p_{1/2}$ single particle orbitals respectively. The analysis of 420 keV state was performed for both $1s_{1/2}$ (shown by A:420 keV in Fig. 1) and $0d_{5/2}$ (shown by B:420 keV in Fig. 1) single particle orbitals separately. We performed ZR-DWBA calculation using the finite range correction value and nonlocal parameters 0.54, 0.25 and 0.85 for deuteron, tritium and neutron respectively. First a ZR-DWBA calculation was performed by using normalization $N = 3.33$ taken from Ref. [7] without finite range parameter and non local parameters. Then the ZR-DWBA calculation was again performed using a normalization constant $N = 2.54$ as used in [9] with a finite range parameter of 1.36 fm^{-1} for (d,t) [10]. It

Table 2: Extracted values of C^2S for different excited states of ^{26}Al.

Ex(KeV)	J^{π}	l	$^{a)}C^2S$	$^{b)}C^2S$
0	5^+	2	0.83	0.69
230	0^+	2	0.10	0.08
420	3^+	0	0.08	0.06
		2	0.36	0.30
1056	1^+	2	0.20	0.16
1762	2^+	2	0.043	0.035
1848	1^+	2	0.022	0.018
2070	2^+	2	0.29	0.24
2365	3^+	2	0.15	0.12
2542	3^+	2	0.17	0.15
3160	2^+	2	0.06	0.05
3409	5^+	2	0.07	0.06
3505	6^+	4	0.08	0.06
4443	2^-	1	0.25	0.22
4719	4^+	2	0.31	0.25

^{a}Extracted from pure ZR-DWBA calculation.
^{b}Extracted from ZR-DWBA calculation by incorporating finite range parameters.

was observed that the theoretical predictions, with and without using finite range correction parameters in ZR-DWBA calculation, were found to fit the experimental data in the same way. The extracted spectroscopic factor values were found to be reduced approximately by 10% to 25% for using finite range correction parameters 1.36 fm^{-1} and $N = 2.54$ as compared with those without using finite range correction parameter in ZR-DWBA calculation. The deviation estimated may be more for the states with poor statistics in data. The extracted C^2S values using the above two calculations were tabulated in Table 2 for comparison.

To sum up, the reaction ^{27}Al(d,t)^{26}Al at $E_d = 25$ MeV was used to study ground as well as excited states of ^{26}Al. A total of 14 states of ^{26}Al were studied using ZR-DWBA calculation as well as by incorporating finite range correction parameters in DWUCK4 code. The theoretical predictions were found be in fair agreement with the data. Two different calculations were used to examine the variation in C^2S values of the observed states of ^{26}Al and approximately 10% to 25% reduction in C^2S values was estimated using ZR-DWBA calculation with finite range effect comparative to ZR-DWBA calculation without finite range parameters.

The authors thank the cyclotron operating staff for their cooperation during the experiments. One of the authors (S.B.) acknowledges with thanks the financial support received as Raja Ramanna Fellow from the Department of Atomic Energy, Government of India. One of the authors (A.D.) acknowledges with thanks the financial support provided by the Science and Engineering Research Board, Department of Science and Technology, Government of India.

References

[1] R. Diehl *et al.*, Nature(London) **439**, 45 (2006).

[2] N. Prantzos *et al.*, Phys. Rep. **267**, 1 (1996).

[3] C. Fitoussi *et al.*, Phys. Rev. C **78**, 044613 (2008).

[4] P. Finlay *et al.*, Phys. Rev. Lett. **106**, 032501(2011).

[5] W. Satula, J. Dobaczewski, W. Nazarewicz and R. Rafalski *al.*, Phys. Rev. Lett. **106**, 132502(2011).

[6] Vishal Srivastava *et al.*, Phys. Rev. C **91**, 054611 (2015).

[7] http://spot.colorado.edu/∼kunz/DWBA.html

[8] Direct Nuclear Reactions By G. R. Satchler, Clarendon Press-Oxford,Oxford University Press - New York,1983.

[9] R. E. Tribble *et al.*, Nucl. Phys. A **282**, 269(1977).

[10] W. R. Hering *et al.*, Nucl. Phys. A **151**, 33(1970).

EPJ Web of Conferences **117**, 07023 (2016) DOI: 10.1051/epjconf/201611707023
NN2015

Charge-changing cross section measurement of neutron-rich carbon isotopes at 50 AMeV

D. T. Tran[1,2], T. T. Nguyen[2,3], I. Tanihata[1,4], H. J. Ong[1], M. Fukuda[5], N. Aoi[1], Y. Ayyad[1], H. Sakaguchi[1], J. Tanaka[1], P.Y. Chan[1], T. H. Hoang[1,2], T. Hashimoto[6], E. Ideguchi[1], A. Inoue[1], T. Kawabata[7], L. H. Khiem[2], K. Matsuta[5], M. Mihara[5], S. Momota[8], D. Nagae[9], A. Ozawa[9], P.P. Ren[10], S. Terashima[4], R. Wada[10], W.P. Lin[10] and T. Yamamoto[1]

[1] Research Center for Nuclear Physics, Osaka University, Japan
[2] Institute Of Physics, Vietnam Academy of Science and Technology, Hanoi, Vietnam
[3] Faculty of Physics and Engineering Physics, VNU-HCMUS, Vietnam
[4] Beihang University, Beijing, China
[5] Department of Physics, Osaka University, Osaka, Japan
[6] Institute for Basic Science, Daejon, Korea
[7] Department of Physics, Kyoto University, Kyoto, Japan
[8] Kochi University of Technology, Kochi, Japan
[9] Institute Of Physics, University of Tsukuba, Tsukuba, Japan
[10] Institute of Modern Physics, Lanzhou, China

Abstract

Charge Changing Cross Sections (CCCS or σ_{CC}) of neutron-rich carbon isotopes on carbon target were measured at low energy ($50A$ MeV) for the first time. The consistency between Glauber calculation and experimental σ_{CC} of ^{12}C isotope at low energy region shows that proton distribution radii can be derived from CCCS at low energy.

1 Introduction

The world of atomic nuclei is governed by nuclear physics which can be described by nuclear structure models. The proton and neutron distribution radii of nuclei are good observables for testing nuclear structure models. The rms radii of proton distribution for some of the light and stable nuclei have been determined by isotope shift measurements [1–4]. However, applications of this method to unstable nuclei from B to Ne isotopes are extremely challenging due to the uncertainty in atomic physics calculation. The recent progress in Glauber model analysis enables the extraction of the matter distribution radii from interaction/reaction cross section (σ_{R}) [5]. The Glauber model can be extended to extract proton distribution radii from Charge-Changing Cross Sections (CCCS or σ_{CC}) on a known target. Eq.(1) shows the formula to calculate σ_{CC} in the Glauber model framework.

$$\sigma_{\mathrm{CC}} = \int d\mathbf{b}\,[1 - T(\mathbf{b})] \tag{1}$$

where $T(\mathbf{b})$ is transmission function:

$$T(\mathbf{b}) = \exp\left\{-\int d\mathbf{r}\rho_{\mathrm{Pp}}(\mathbf{r}-\mathbf{b})\left[\sigma_{\mathrm{pp}}\rho_{\mathrm{Tp}}(\mathbf{r}) + \sigma_{\mathrm{pn}}\rho_{\mathrm{Tn}}(\mathbf{r})\right]\right\}.$$

The formula includes five physical quantities: the projectile proton density $\rho_{\mathrm{Pp}}(r)$, the target one $\rho_{\mathrm{Tp}}(r)$, the target neutron density $\rho_{\mathrm{Tn}}(r)$, and the precisely determined bare proton-proton (σ_{pp}) and proton-neutron (σ_{pn}) total cross sections.

This extension has been shown to work well for stable nuclei and Li isotopes at high energy [6] and has been applied to Be [7], B [8] and C [9] isotopes at high energy region ($E > 200A$ MeV). Experimental results on proton distribution radii showed that increasing number of neutron in nuclei leads to increases in the proton distribution radii in He [1,3], Li [2], Be [4] and B [8] isotopes, especially in neutron-halo nuclei. The experimental results are consistent with the predictions of Antisymmetrized Molecular Dynamics (AMD) model [10]. However, the AMD calculation predicted that proton densities (radii) of C isotopes are almost constant. The weak N dependence of the proton radius indicates the insensitivity of the proton distribution to that of the neutrons.

However, the agreement between the zero-range Glauber calculation with the experimental values of reaction-cross section is not so good at intermediate energies even for stable nuclei, especially, the discrepancy at a few ten MeV/nucleon reaches almost 20%. To reduce such discrepancy, therefore, we used the finite-range optical limit approximation Glauber type calculation.

DOI: 10.1051/epjconf/201611707023

2 Experiment

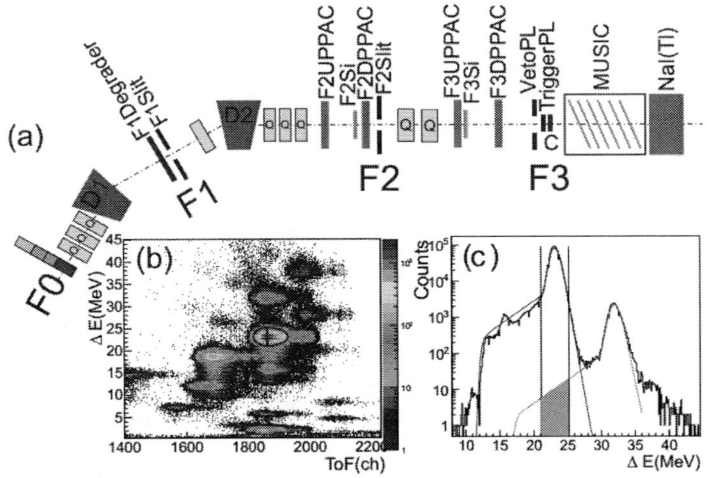

Figure 1: (a) Schematic view of the experiment setup. Identification spectrum of (b) ^{12}C before reaction target and (c) contamination estimation.

The experiment was performed at the EN course [11], RCNP, Osaka University using transmission method. In this method, the σ_{CC} is calculated as $\sigma_{\mathrm{CC}} = \frac{1}{t}\ln\left[\frac{1}{\gamma}\right]$, in which t is target thickness, γ is the ratio of the number of outgoing nuclei, which have the same atomic number with incoming nuclei (N_{sameZ}) to the number of incoming beam (N_{inc}). The detector setup is shown in fig. 1(a). Carbon-isotope beams were produced in separate runs by fragmentation of the ^{22}Ne primary beam at 80A MeV interacting with thick Be target ($1-5$ mm for different isotopes) located at F0. The interested isotope was selected in flight by setting the appropriate magnetic rigidity (Bρ) using dipole magnets D1 and D2. The beam energy was tuned by using combination of Be target, F1Degrader, F1Slit and determined using the Bρ values. The selected secondary beam was focused at F2 focal plane; contaminants were eliminated using a set of collimators at F2. The incoming beam (N_{inc}) was identified and counted by using the information of energy loss in a 320μm-thick silicon detector (F3Si) and time of flight (TOF) between the production target and a 100μm-thick plastic scintillator, which was also used as the trigger for data taking. The secondary beams were focused on a 450 mg/cm^2 carbon target placed at F3. The track and the spot of beam was provided by four Parallel Plate Avalanche Counters (PPAC) [12] located at F2 and F3. The particles that scattered at large angles after the last PPAC was rejected using a plastic veto and trigger counter, which have the same size and were located in front of the carbon target. The outgoing particles went through a MUlti Sampling Ionization Chamber (MUSIC) [13] before being stopped in a 7 cm-thick NaI(Tl) scintillator. The ΔE$-$E method was

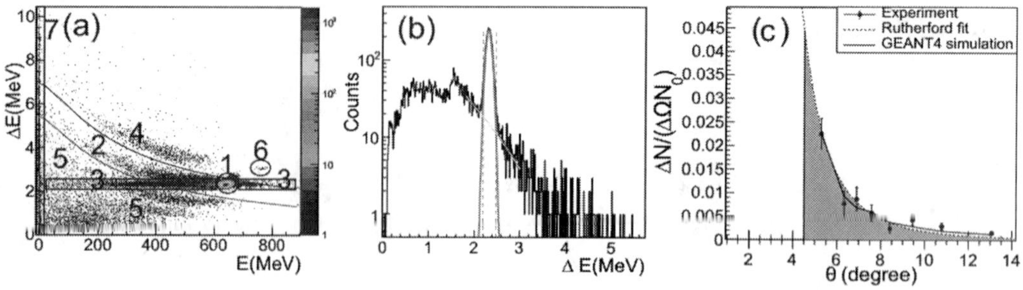

Figure 2: (color online). (a) Scattered particles identification, (b) background estimation of region 5 in region 3, (c) acceptance factor calculation.

used to identify and count the scattered particles (N_{sameZ}).

3 Results and discussion

The incoming particle was identified and counted by ΔE−TOF method (fig. 1(b)) with less than 0.1% of admixing of other nuclides due mainly to channelling effect in silicon crystal. We use MUSIC instead of Si detectors after the reaction target to avoid complication due to the channeling effect in silicon. The contaminants was estimated using the distribution obtained by selecting the contaminant events detected by the detector after the target. The contaminants contribute to cross section ranging from 0.1 to 3.5 mb depending on isotopes.

Particle identification of scattered particles is more complicated at low-energy region than at high energy. The outgoing particles were identified and categorized into 7 regions (fig. 2(a)): (1) beam-like particles, (2) elastic and inelastically scattered beam-like particles, (3) particles reacted in NaI(Tl), (4) proton-removed particles, (5) proton-picked-up particles, (6) contaminants, and (7) out-of-acceptance particles, which we did not detect. The number of events with the same Z as the incoming beam was calculated by summing the events in region 1, 2 and 3. To estimate the number of counts of light particles in region 3, the a Gaussian plus exponential background function was used to fit the experimental data (fig. 2(b)). The systematic uncertainties attributed to the background was bellow 1mb in final σ_{CC}.

The main systematic uncertainty, about 20% of the final σ_{CC}, is attributed to the uncertainty in the estimation of the "*unreacted*" particles (about 2%) that failed to reach the NaI(Tl) detector. To consider this effect, we introduced the acceptance factor (P) and the final σ_{CC} was deduced

DOI: 10.1051/epjconf/201611707023

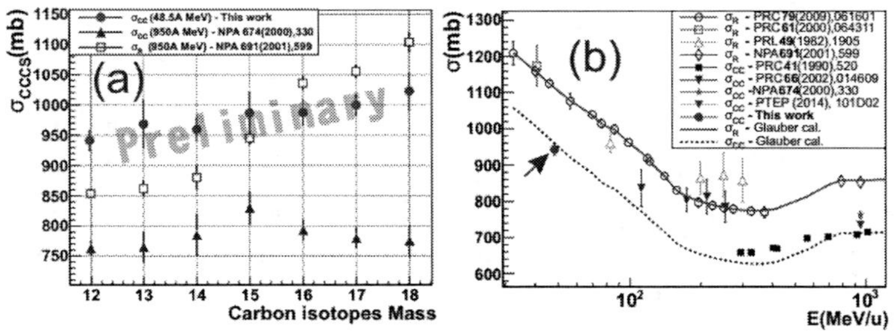

Figure 3: (a) The CCCS of carbon isotopes. (b) Reaction cross section and CCCS of ^{12}C.

as shown in eq. 2.

$$\sigma_{\rm CC} = \frac{1}{t} \ln \left[\frac{\gamma_{out}(1 - P_{in})}{\gamma_{in}(1 - P_{out})} \right] \tag{2}$$

where *in* and *out* indicate measurements with and without target respectively. This is the main difference in data analysis from the high energy experiment. The P_i $(i : in, out)$ was calculated by considering Rutherford scattering of the incident beam in the target. The difference in solid angle $(\Delta\Omega)$ covered by a particular layer (MUSIC electrode) but not by the next layer (MUSIC electrode or NaI(Tl)), was calculated. The number of beam with signal in one layer of MUSIC but without signal in next layer was considered as the number of particles scattered into the solid angle $\Delta\Omega$. The ratio $\Delta N/(\Delta\Omega N_{\rm inc})$ is proportional to the differential cross sections of scattered particles and was fitted with Rutherford distribution. The area below distribution curve from the maximum angle covered by NaI to 180 degree is the P_i (the shadowed in fig. 2(c)). We note that the assumed Rutherford distribution is consistent with the experiment data, as well as with a Monte Carlo simulation using the GEANT4 code (fig. 2(c)). The calculated P_{in} value varies from 0.003 to 0.004, resulting in 2 to 10% uncertainty, and the P_{out} value varies from 0.0005 to 0.0012, resulting in 5 to 15% uncertainty. These uncertainties result in systematic uncertainty of 6 - 10 mb for the $\sigma_{\rm CC}$ for difference isotopes. The determined CCCS's for $^{12-18}$C isotopes are shown in fig. 3(a). The error bar includes the above-mentioned systematic uncertainty, the statistical uncertainty as well as the uncertainty in the target thickness ($< 0.43\%$). For comparison, $\sigma_{\rm CC}$'s and $\sigma_{\rm R}$'s for the same isotopes at 950*A* MeV are also shown.

For $^{12-18}$C, our $\sigma_{\rm CC}$'s are almost constant in contrast with $\sigma_{\rm R}$'s. This trend is consistent with the AMD predictions [10]. Our results also qualitatively agree with the ones at 950*A* MeV for $^{12-15}$C; both results show different systematics for $^{16-18}$C.

It is important to note that we have applied the Glauber model adopting the finite-range optical limit approximation. To eliminate possible systematic error due to the energy dependence, we have fine-tuned the Glauber model so as to reproduce σ_R for ^{12}C on ^{12}C target at various energies as well as for σ_{CC} at 50A MeV. Fig. 3(b) shows σ_R and σ_{CC} for ^{12}C on ^{12}C target at energies up to 950A MeV. Except the points between 300A – 500A MeV, σ_{CC}'s at higher energies are also reproduced relatively well by our Glauber model calculations. It will interesting to compare our results with σ_{CC}'s for $^{12-18}$C at 800A MeV [14].

4 Summary

We have measured the Charge-Changing Cross Sections (CCCS's) of $^{12-18}$C on a ^{12}C target at 50A MeV at the RCNP EN Course in Osaka University. We have shown, for the first time, the possibility to extend the Glauber model to the σ_{CC} measurements to determine the proton distribution radii of nuclei.

References

[1] L. -B Wang *et al.*, *Phys. Rev. Lett.*, **93** (2004) 142501.

[2] R. Sanchez *et al.*, *Phys. Rev. Lett.*, **96** (2006) 033002.

[3] P. Mueller, *Phys. Rev. Lett.*, **99** (2007) 252501.

[4] W. Nortershauser *et al.*, *Phys. Rev. Lett.*, **102** (2009) 062503.

[5] A. Ozawa *et al.*, *Nucl. Phys. A*, **691** (2001) 599.

[6] I. Tanihata *et al.*, *Prog. Part. Nucl. Phys.*, **68** (2013) 215.

[7] S. Terashima *et al.*, *Prog. Theor. Exp. Phys.* (2014) 101D02,.

[8] A. Estrade *et al.*, *Phys. Rev. Lett.*, **113** (2014) 132501.

[9] T. Yamaguchi *et al.*, *Phys. Rev. Lett.*, **107** (2011) 032502.

[10] Y. Kanada-En'yo, *Phys. Rev. C*, **91** (2015) 014315.

[11] T. Shimoda *et al.*, *Nucl. Instr. Methods Phys. Res. B*, **70** (1992) 320.

[12] H. Kumagai *et al.*, *Nucl. Instr. Methods Phys. Res. A*, **470** (2001) 562.

[13] K. Kimura *et al.*, *Nucl. Instr. Methods Phys. Res. A*, **538** (2005) 608.

[14] R. Kanungo *et al.*, *GSI-S395 experiment proposal* (2010).

EPJ Web of Conferences **117**, 07024 (2016) DOI: 10.1051/epjconf/201611707024
NN2015

High-spin molecular resonances in $^{12}C + ^{12}C$

E. UEGAKI[1] and Y. ABE[2]

[1]Dept. of Mathematical Science, Akita University, Akita 010-8502, Japan
[2]RCNP, Osaka University, Ibaraki 567-0047, Japan

Abstract

Resonances observed in the $^{12}C + ^{12}C$ collisions are studied with a molecular model. At high spins $J = 10$–18, a stable dinuclear configuration is found to be an equator-equator touching one. Firstly, normal modes have been solved around the equilibrium, with spin J and K-quantum number being specified for rotation of the whole system. Secondly, with respect to large centrifugal energy, Coriolis coupling has been diagonalized among low-lying 11 states of normal-mode excitations, which brings K-mixing. The analyses of decay widths and excitation functions have been done. The molecular ground state exhibits alignments of the orbital angular momentum and the ^{12}C spins, while some of the molecular excited states exhibit disalignments with small widths. Those results are surprisingly in good agreement with the experiments, which will light up a new physical picture of the high-spin $^{12}C + ^{12}C$ resonances.

1 Introduction

Well above the Coulomb barrier of the $^{12}C+^{12}C$ system, series of resonances have been found with high spins over $10\hbar$, which exhibit prominent peaks in the elastic and inelastic 2^+ channels [1]. Band Crossing Model (BCM), based on the double resonance mechanism, has successfully explained resonance mechanism with the aligned configurations of the orbital angular momentum and the spins of the excited states of ^{12}C nuclei [2].

On the other hand, for high-spin resonances observed in ^{24}Mg + ^{24}Mg and ^{28}Si + ^{28}Si, we have developed a new molecular model [3, 4]. The model has an advantage to describe various geometrical configurations of two constituent deformed nuclei with the method of normal mode around the stable energy minimum under strong nucleus-nucleus interaction, which gives intuitive understanding on nuclear structure of the molecular resonances. For the oblate-oblate systems such as ^{28}Si $-$ ^{28}Si and ^{12}C $-$ ^{12}C, the stable configurations with axial asymmetry exhibit a series of low-lying excited states with $K = 2$ or 4, in which spins of the constituent nuclei incline to the molecular z'-axis. Thus, those states are expected to have an interesting property of spin disalignments with the orbital angular momentum in contrast to the alignments in BCM. How to resolve an apparent contradiction?

In rotating systems, Coriolis coupling plays an important role in optimization of the centrifugal energy. Actually, a diagonalization of the coupling has brought a result that the molecular ground state with the $K = 1$ mixed configuration has aligned components similar to those of BCM [5]. In addition, a chain coupling of Coriolis is expected to connect the $K = 2$ and $K = 4$ states with the elastic channel, which gives rise to intermediate structures comparable to the observation.

2 Dynamical aspect and normal modes

Assuming a constant deformation and axial symmetry of the constituent nuclei, we have seven degrees of freedom $(q_i) = (\theta_1, \theta_2, \theta_3, R, \alpha, \beta_1, \beta_2)$, as illustrated in fig. 1. The relative vector (R, θ_2, θ_1) of two ^{12}C nuclei defines the molecular z'-axis. Orientations of the symmetry axes of ^{12}C as internal degrees of freedom are described with Euler angles (α_i, β_i) referring to the

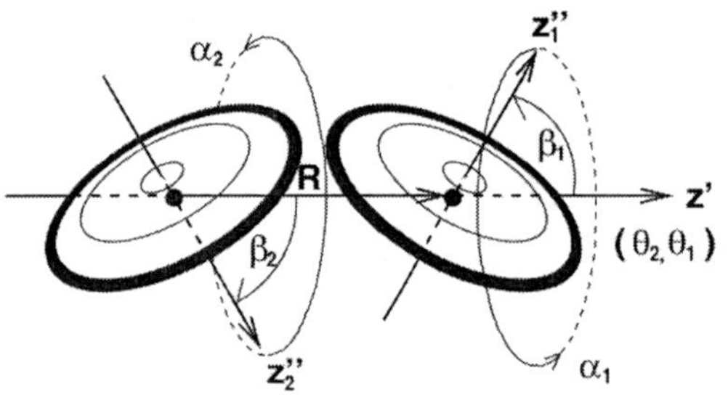

Figure 1: Dinuclear molecular coordinates; seven degrees of freedom of ^{12}C + ^{12}C (oblate-oblate system) are described with $(q_i) = (\theta_1, \theta_2, \theta_3, R, \alpha, \beta_1, \beta_2)$.

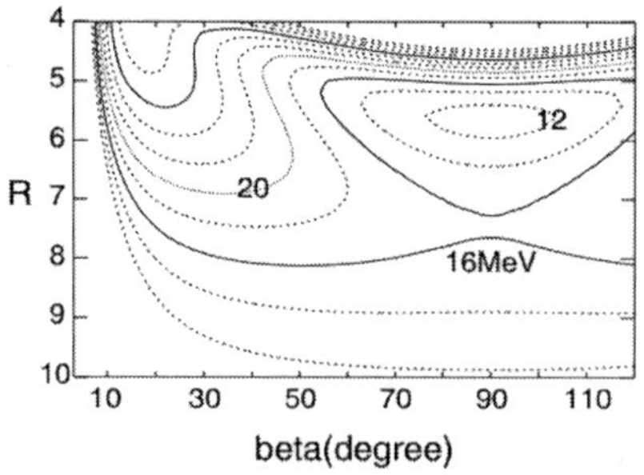

Figure 2: Contour map of the effective potential energy V_{JK} with $J = 14$ and $K = 0$ for the $R - \beta(\beta_1 = \beta_2)$ degrees at $\alpha = \pi/2$. Contours are in 2 MeV.

molecular axes. With α_1 and α_2, $\theta_3 = (\alpha_1 + \alpha_2)/2$ of the total K-rotation around z'-axis and $\alpha = (\alpha_1 - \alpha_2)/2$ of twisting motion are defined.

The kinetic energy is given classically in terms of angular velocities at first, and then the quantization is done to describe the whole rotating system. The resultant kinetic energy is classified into three parts, as $\hat{T} = \hat{T}_{\rm rot} + \hat{T}_{\rm int} + \hat{T}_{\rm C}$, where $\hat{T}_{\rm rot}$ denotes the rotational energy with the total angular momentum operator $\hat{\boldsymbol{J}}'$, $\hat{T}_{\rm int}$ the energy of internal motions, and finally $\hat{T}_{\rm C}$ the Coriolis coupling. The nucleus-nucleus interaction is given by a folding potential with nucleon-nucleon interaction DDM3Y. Expecting a stable configuration, we describe the interaction potential by the internal degrees of freedom $(R, \alpha, \beta_1, \beta_2)$, which specify geometrical configurations. Consistently with the coordinate system, at first we introduce a rotation-vibration type wave function specified with K-quantum number as basis one, $\Psi_\lambda \sim D^J_{MK}(\theta_i)\chi_K(R, \alpha, \beta_1, \beta_2)$, where χ_K describes the internal motions.

In order to know dynamical aspects of multidimensional internal motions, we calculate the effective potential for given J and K,

$$V_{JK}(R, \alpha, \beta_1, \beta_2) = V_{\rm interaction}(R, \alpha, \beta_1, \beta_2) + T'_{\rm rot}(J, K; R, \alpha, \beta_1, \beta_2), \quad (1)$$

where $T'_{\rm rot}$ denotes a multidimensional centrifugal potential with specified J and K, reduced from the rotational kinetic energy operator $\hat{T}_{\rm rot}$. In fig. 2, a multidimensional energy surface is displayed for $J = 14$. A local minimum is obtained around $\beta = 90°$ and $R = 5.7$ fm, which is called as *equator-equator* (E-E) configuration. The barrier height is 4.8 MeV located at $R = 7.5$ fm.

Couplings among various molecular configurations are taken into account by the method of normal mode around the equilibrium at $\beta_i = 90°$, which gives rise to the molecular modes of excitation. We expand the effective potential $V_{JK}(R, \alpha, \beta_1, \beta_2)$ into a quadratic form to describe normal modes. Combining the kinetic energy operator $\hat{T}_{\rm int}$, we obtain hamiltonian H_0, which is almost separated into each internal degree of freedom, as $H_0 = H(R) + H(\alpha) + H(\Delta\beta_1) + H(\Delta\beta_2) + H_{\rm coupl}(\alpha, \beta_1, \beta_2)$, where $\Delta\beta_i$ denote $\beta_i - \pi/2$. This means the internal degrees of freedom $(R, \alpha, \Delta\beta_1, \Delta\beta_2)$ are the normal modes themselves. Thus we use basis wave functions, $\Psi_{JMK}(\theta_i; R, \alpha, \beta_1, \beta_2) \sim D^J_{MK}(\theta_i) f_n(R)\phi_\nu(\alpha)\varphi_{n_1}(\Delta\beta_1)\varphi_{n_2}(\Delta\beta_2)$, where f_n, φ_{n_1} and φ_{n_2} are H.O. functions, while ϕ_ν is a rotational function $e^{i\nu\alpha}$ as the confinement is very weak in α-degree. The energy of the basis state is given as $E_{JK}(n, \nu, n_1, n_2) = E_0(R_{\rm e}) + \{J(J+1) - K^2 - 1\}\hbar^2/2\mu R_{\rm e}^2 + \{K^2 - 2\}\hbar^2/4I + \nu^2\hbar^2/4I + (n + 1/2)\hbar\omega_R + (n_1 + n_2 + 1)\hbar\omega_\beta$, where $\hbar\omega_\beta \sim 5$ MeV. Additional $H_{\rm coupl}$ includes coupling among β_i and α such as $\Delta\beta_1\Delta\beta_2\cos 2\alpha$ as well as corrections for weak K-dependence of confinements on β_i-motions.

H_{coupl} and $H(\alpha)$ for the twisting motion have been diagonalized. In fig. 3, the spectrum of the normal modes with $J = 12$ is displayed, classified with quantum numbers $K = 0$–4. The axially asymmetric shape of the whole composite system allows states with $K = 2m$, m being integer, in low excitation. With $1\hbar\omega_\beta$ excitation around the equilibrium, the system has states with $K = 2m + 1$. The molecular quantum numbers also indicate rotational motions of two constituent ^{12}C nuclei by $\{n_1, \mu_1 = (K + \nu)/2\}$ and $\{n_2, \mu_2 = (K - \nu)/2\}$, respectively. A set of numbers $I_i = n_i + |\mu_i|$ is related to the set of spins of ^{12}C nuclei, like a channel assignment. In fig. 3, single 2^+ and mutual 2^+ assignments are depicted with connected lines.

3 Role of Coriolis coupling

With the normal-mode hamiltonian, the total hamiltonian is written as

$$H = H_0 + \hat{T}_{\rm C} + (higher\ order). \quad (2)$$

Generally in the rotating system, we have Coriolis coupling $\hat{T}_{\rm C}$, which can be described with a simple product $-\boldsymbol{J}' \cdot (\boldsymbol{S}'_1 + \boldsymbol{S}'_2)/\mu R^2$ of the total angular momentum operator $\boldsymbol{J}'$ and spins of ^{12}C $\boldsymbol{S}'_i$. Note that they refer to the molecular axes as designated with prime. In the model, it is given with many terms; $\boldsymbol{J}'$ is consisted with differential operators of (θ_i), which are combined into lowering and raising operators of K, $J'_\pm = J'_x \pm iJ'_y$, respectively. As for spins $\boldsymbol{S}'_i$, they are described usually with differential operators of Euler angles β_i, *i.e.*, given with those of the internal variables $(\alpha, \beta_1, \beta_2)$.

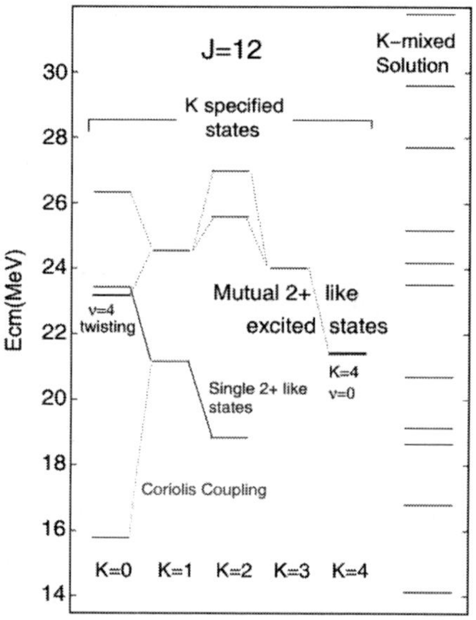

Figure 3: Energy spectra of molecular normal modes of ^{12}C+^{12}C for $J = 12$, specified with $K = 0$ to 4 from the *l.h.s.* Diagonalization of the Coriolis coupling gives a new spectrum shown in the rightmost column.

We consider the Coriolis terms as residual coupling, which causes K-mixing with $\Delta K = 1$. The mixing induces aligned components. Actually, in the molecular ground state, a $(K = 1, \nu = 1)$ mixed component appears to bring strong alignments in the single 2^+ channel [5]. In the rightmost column of fig. 3, levels of the K-mixed solutions caused by the Coriolis coupling are displayed for $J = 12$.

4 Molecular spectrum and excitation functions

We have found the stable E-E configuration for high spins from $J = 10$ to $J = 18$. For lower spins than $J = 10$, stable configurations are not found in the E-E one, but belly-belly one, *etc.* in closer contact of ^{12}C nuclei. In fig. 4, the molecular bands with the E-E structure are shown. Of course, there are many possible molecular bands for lower spins, which may correspond to those observed around the Coulomb barrier region.

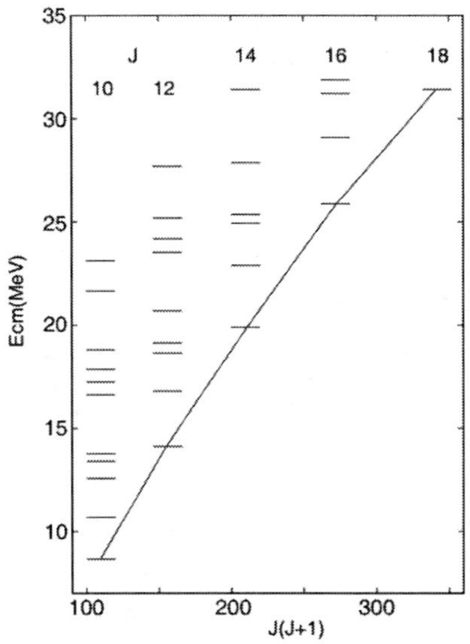

Figure 4: Molecular band structure for J = 10–18. The ground band is connected for eye-guide. The levels with widths larger than 4 MeV are omitted.

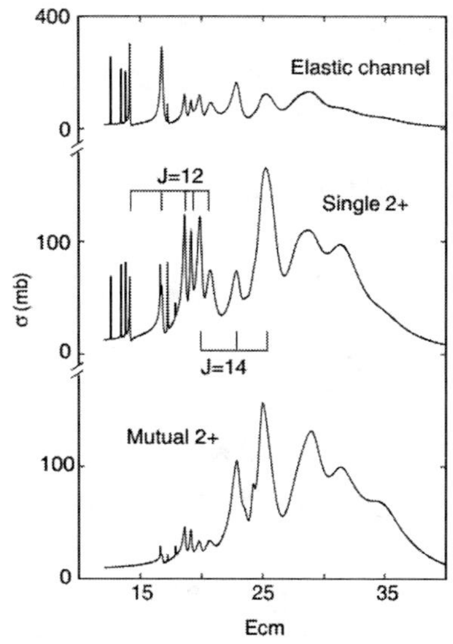

Figure 5: Theoretical excitation functions for the elastic, single and mutual 2^+ channels. The levels of J = 10–18 are included in. Spins are indicated for some resonance peaks (see text).

We have obtained many molecular excited states with small widths for the same spin, the number of which are comparable with the experimental observations. To obtain such a wide variety of the excited states, we introduced 11 bases of the molecular-mode states with low excitation energies, corresponding to single 2^+ and mutual 2^+ excitations. Important are low-lying states with high μ_i excitations, such as $\nu = 4$ or $K = 4$, which contains components of spin disalignments due to rotations of constituent ^{12}C nuclei around the molecular z'-axis. Surprisingly, widths of those states are not large due to very small amounts of the elastic components. The elastic components are almost exhausted in the molecular ground state and the first excited state.

In fig. 5, calculated excitation functions are displayed. Ledoux et al. assigned spins of three resonance states around $E_{\rm cm} = 20$ MeV as $J = 12$ for lower two states (18.4, 19.3 MeV) and $J = 14$ for the 20.3 MeV state [6]. This spin assignment is very important, because spins of all three resonances are suggested to be $J = 12$ by Cormier [1]. In the present results, the molecular ground state with $J = 14$ appears at 19.9 MeV in good agreement with the Ledoux's data. As for $J = 12$, the ground state is obtained at 14 MeV, while the second and third excited states appear around 19 MeV (see fig. 5). Thus three close-by resonances obtained around $E_{\mathrm{cm}} = 19$ MeV are in good correspondence with three intermediate peaks on the gross peak observed in the single 2^+ channel. It is noted that the excitation functions of both single and mutual 2^+ excitations are in good agreement with experiment [1].

5 Conclusions

Dinuclear molecular model has been applied to high-spin resonances of the $^{12}C + ^{12}C$ system. The model is capable to describe various angular momentum couplings, by the role of the Coriolis coupling. The resultant molecular ground state has its own characteristics as a token of aligned configuration, consistent with BCM, while in the excited states high K-rotational states appear relatively low in energy as non-aligned configurations. The latter multi-excited states are expected to weakly couple to the elastic channel and to be observed as resonances with small intermediate widths. As a result, characteristic features of excitation functions in the single 2^+ and mutual 2^+ channels, *i.e.*, the prominent gross peaks topped with the intermediate structures have been very well reproduced over a wide energy range. This gives rise to a novel visage to the known molecular resonances. Experiments to explore nature of those resonance states are strongly called for.

References

[1] Cormier T. M. et al., *Phys. Rev. Lett.* **40** (1978) 924.

[2] Abe Y., Kondō Y. and. Matsuse T., *Prog. Theor. Phys. Suppl.* No. **68** (1980) 303, and *references therein.*

[3] Uegaki E. and Abe Y., *Prog. Theor. Phys.* **90** (1993) 615.

[4] Uegaki E. and Abe Y., *Prog. Theor. Phys.* **127** (2012) 831, *ibid.* 877.

[5] Uegaki E. and Abe Y., *J. Phys.: Conf. Series* **569** (2014) 012091.

[6] Ledoux R. J. et al., *Phys. Rev. C* **27** (1983) 1103.

EPJ Web of Conferences **117**, 07025 (2016) DOI: 10.1051/epjconf/201611707025
NN2015

Influence of Skyrme-type interaction on HICs observables

Yingxun Zhang[1], M.B.Tsang[2], Zhuxia Li[1] and Hang Liu[3]

[1] China Institute of Atomic Energy, Beijing 102413, P.R.China
[2] National Superconducting Cyclotron Laboratory, Michigan State University, East Lansing, MI 48824, USA
[3] Texas Advanced Computing Center, University of Texas, Austin, TX 78758, USA

Abstract

A new version of the improved quantum molecular dynamics model has been developed by including Skyrme type momentum dependent interaction. 12 Skyrme like parameter sets $\{K_0, S_0, L, m_s^*, m_v^*\}$ are adopted in the transport model code to calculate the isospin diffusion, single and double ratios of transverse emitted nucleons, neutron proton isoscaling ratios. The calculations and correlation analysis evidence that isospin diffusion observable at lower beam energy is sensitive to the slope of symmetry energy and m_s^*. The high energy neutrons and protons yield ratios from reactions at different incident energies provide a sensitive observable to study the nucleon effective mass splitting, at higher beam energy.

The properties of isospin asymmetric nuclear equation of state is of fundamental importance in our understanding of nature's asymmetric objects including neutron stars as well as heavy nuclei with very different number of neutrons and protons. Currently, there exist, lot of effective interaction parameter sets [1, 2], for fitting the ground-state properties of stable nuclei, symmetric and asymmetric nuclear matter. It also leads very different neutron Equation of state which may have different nuclear matter (NM) parameters, such as incompressibility K_0, isoscalar effective mass m_s^*, symmetry energy coefficient S_0, slope of symmetry energy L and isovector effective mass m_v^* [3,4], or different sign or magnitude of effective mass splitting. Thus, constraining isospin asymmetric nuclear equation of state become one of hot topics in nuclear physics, especially for heavy ion collisions.

Some properties of symmetric nuclear matter, such as $K_0 = 230\pm30$MeV [1,5,6] and $m_s^*/m = 0.65-0.99$ [1,5,7,8], have been extracted from isoscalar collective vibrations, giant quadrupole resonance and heavy ion collisions measurements. Recently, a consistent set of constraints on the symmetry energy near saturation density, i.e. $S_0 \sim 31MeV \pm \Delta S_0$ and its slope $L \sim 50-60MeV \pm \Delta L$, has been obtained from observables measured in both nuclear structure and nuclear reaction experiments [9–12]. Up to now, the challenge of symmetry energy constraints around normal density is how to reduce the uncertainties of the constraints on symmetry energy, i.e. how to reduce ΔL and ΔS_0 or obtain precise and accurate symmetry energy constraints. As a first step to reach this goal, one has to quantitatively understand where the uncertainties of symmetry energy come from, and then to identify what experimental observables are crucial for better constraining the interested physical quantities in the theoretical models [13]. In this work, we firstly investigate the correlation between symmetry energy and effective mass, effective mass splitting within the framework of Skyrme-Hartree-Fock approaches, and then we quantify the sensitivity of HIC observables to L, m_s^* and m_v^* by analyzing its correlation coefficient by using the transport code, ImQMD-Sky [14], and a strategy to refine the symmetry energy constraints is proposed.

Within the framework of Skyrme-Hartree-Fock approaches, the density dependent of symmetry energy is expressed as,

$$\begin{aligned} S(\rho) &= \frac{1}{3}\epsilon_F\rho^{2/3} + A_{sym}\rho + B_{sym}\rho^{\sigma+1} + C_{sym}(m_s^*, m_v^*)\rho^{5/3} \qquad (1) \\ &= S_0 + \frac{L}{3}(\frac{\rho-\rho_0}{\rho_0}) + \frac{K_{sym}}{18}(\frac{\rho-\rho_0}{\rho_0})^2 + ... \end{aligned}$$

This formula shows the density dependence of symmetry energy also depends on isoscalar effective mass (m_s^*) and isovector effective mass (m_v^*) in the mean field level. In Table 1, we list the correlation coefficients C_{AB} [15] between the pair of K_0, S_0, L, m_s^* and m_v^* which are obtained from 120 Skyrme parameter sets. $C_{AB} = \pm1$ means there is a linear dependence between A and B, and $C_{AB} = 0$ means no correlations. As shown in Table 1, the slope of symmetry energy also strongly correlate to m_s^* and m_v^* (or effective mass splitting), because the effective mass splitting, $(m_n^* - m_p^*)/m$, is related to m_s^* and m_v^* based on the following relationship. Thus, reducing the uncertainties of symmetry energy constraints obtained from heavy ion collision data needs us to reduce the uncertainties of constraints on m_s^*, m_v^* or $\Delta m^*/m$ since all of them are related to the symmetry energy.

$$\begin{aligned}\frac{\Delta m*}{m} &= \frac{m_n^* - m_p^*}{m} = -\frac{\delta}{m}\frac{2m_s^* m_v^*(m_v^* - m_s^*)}{m_v^{*2} - \delta^2(m_v^* - m_s^*)^2} \\ &= -2\frac{m_s^*}{m}[\frac{m_v^* - m_s^*}{m_v^*}\delta + (\frac{m_v^* - m_s^*}{m_v^*})^2\delta^2 + \cdots\end{aligned} \quad (2)$$

where $\delta = (\rho_n - \rho_p)/(\rho_n + \rho_p)$.

Table 1: Correlation coefficient C_{AB} between the pair of K_0, S_0, L, m_s^*, m_v^* from 120 Skyrme parameters.

C_{AB}	K_0	S_0	L	m_s^*/m	m_v^*/m
K_0	1	0.003	0.161	0.131	0.295
S_0	0.003	1	**0.764**	0.397	0.228
L	0.161	**0.764**	1	**0.460**	0.212
m_s^*/m	0.131	0.397	**0.460**	1	**0.715**
m_v^*/m	0.295	0.228	0.212	**0.715**	1

In order to disentangle the impacts from different K_0, S_0, L, m_s^* and m_v^*, we analysis the correlation coefficients C_{AB} between five observables B=CI-$R_2(n/p)$, CI-$DR(n/p)$, CI-$R_{21}(n/n)$, CI-$R_{21}(p/p)$ and R_i, and five NM parameters K_0, S_0, L, m_s^* and f_I in the reaction of ^{124,112}Sn+^{124,112}Sn [14,16]. Here, we use $f_I = \frac{1}{2\delta}(\frac{m}{m_n^*} - \frac{m}{m_p^*}) = \frac{m}{m_s^*} - \frac{m}{m_v^*}$ to replace m_v^* or the effective mass splitting $\Delta m^*/m$, because this value is related to the effective mass splitting, and independent of the isospin asymmetry of system and can be easily incorporate into the transport code. All the nucleon yield observables in the simulations are obtained for ^{124}Sn+^{124}Sn and ^{112}Sn+^{112}Sn at E_{beam} = 50 and 120 MeV/u at impact parameter 2fm, and for emitted nucleons with $E_k > 40MeV$. For isospin diffusion R_{diff} (or R_i), an additional mixed reaction ^{124}Sn+^{112}Sn is included and the calculations are performed with mid-peripheral impact parameter of 6 fm.

Figure 1 shows the correlation coefficient C_{AB} for Sn+Sn reactions at E_{beam}=50 MeV/u (upper panels) and at E_{beam}=120 MeV/u (lower panels) [16]. Red color bars represent positive correlations which mean observable B increases with parameter A, and blue color bars show negative correlations which mean observable B decreases with increasing parameter A. To focus our search for the best experimental observables which are most sensitive to the model parameters, we will discuss only C_{AB} with values greater than 0.6, which were represented as solid blue and red bars in Figure 1. With this criterion, S_0 and K_0 are not very sensitive to any of the observables studied here.

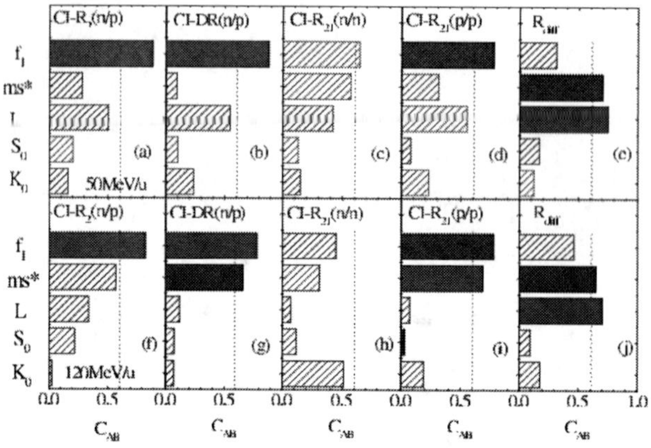

Figure 1: (Color online) Correlations of five obervables, CI-$R_2(n/p)$ (a), CI-$DR(n/p)$ (b), CI-$R_{21}(n/n)$ (c), CI-$R_{21}(p/p)$ (d), R_{diff} (e) with five force parameter, K_0, S_0, L, m_s^* and f_I. Up panels are the results for 50 MeV/u, and bottom panels are for 120 MeV/u.

In the case of L, the slope of the symmetry energy at saturation density, $C_{L,Rdiff} \sim 0.7$ meets this criterion. f_I shows the strongest correlations to the experimental coalescence invariant nucleon observables, the single ($R_2(n/p)$), double n/p ($DR(n/p)$) and isoscaling p/p ($R_{21}(p/p)$) yield ratios at both incident energies. As shown in panels (a), (f) and (b), (g) of Figure 1, the correlation between the single and double n/p yield ratios and f_I is larger than 0.8. Thus, one should be able to extract constraints for both f_I and m_s^* from the single and double n/p ratios as well as the p/p ratios at 120MeV/u incident energy. Alternately, f_I can be extracted using the value of $m_s^*/m = 0.65 - 0.9$ obtained in ref. [1,5,7,8]. Our analysis also shows negative correlations of R_{diff} with m_s^*, $C_{m_s^*,R_{diff}} \sim -0.64$ to -0.70. The observation that R_{diff} is also sensitive to m_s^* is consistent with the results from BUU calculations [17,18]. Strong correlation of single and double n/p yield ratios with m_s^* are also observed mainly at high energy, because the more violent nucleon-nucleon collisions at 120MeV/u cause larger momentum transfer and lead to the momentum dependent interaction (which can be characterized by the effective mass) to play more important roles.

In summary, by varying the interaction parameter sets in the ImQMD-Sky code, we study the influence of different Skyrme parameter sets on the coalescence invariant neutron and proton yield ratios at high energy region, for ^{124}Sn+^{124}Sn and ^{112}Sn+^{112}Sn at 50MeV/u and 120MeV/u incident energy. Sensitivities to S_0 and K_0 are relatively small, $C_{AB} < 0.5$. At incident energy of 120MeV/u, strong correlations are observed between observables constructed from coalescence invariant nucleon spectra with $E_{c.m.}/A > 40$ MeV, such as CI-$R_2(n/p)$, CI-DR(n/p), and CI-$R_{21}(p/p)$, and the effective mass splitting as well as the isoscalar effective mass, important input parameters to the transport models. The calculations also confirm the sensitivity of L to the isospin diffusion observable. Since the same observable is also sensitive to isoscalar effective mass, this should allow one to extract the constraints of m_s^* and L with reasonable uncertainties. Similarly the opposite correlations of the nucleon yield ratios, such as CI-DR(n/p) and CI-R$_2$(p/p) ratios to f_I and m_s^*, at 120 MeV/u reactions should allow one to disentangle the effects of the effective nucleon mass splitting and the isoscalar effective mass.

Acknowledgements This work has been supported by the National Key Basic Research Development Program of China under Grant No. 2013CB834404, the National Natural Science Foundation of China under Grants No.(11475262, 11375062,11275072). MBT acknowledges support from the USA National Science Foundation Grants No. PHY-1102511 and travel support from CUSTIPEN (China-US Theory Institute for Physics with Exotic Nuclei) under the US Department of Energy Grant No. DE-FG02-13ER42025.

References

[1] M.Dutra, O. Lourenco, J.S. Sa Martins, and A. Delfino, J.R.Stone, P.D.Stevenson, Phys.Rev.C 85, 035201(2012).

[2] M.Dutra, O. Lourenco, S.S.Avancini, B.V.Carlson, A. Delfino, D.P.Menezes, C.Providencia, S.Typel, J.R.Stone, Phys.Rev.C 90, 055203(2014).

[3] B. A. Brown, Phys. Rev. Lett. 85, 5296 (2000).

[4] B. A. Li, L.W.Chen, C.M.Ko, Phys. Rep 464, 113 (2008).

[5] J. P. Blaizot, J. F. Berger, J. Decharge, and M. Girod, Nucl.Phys. A 591, 435 (1995).

[6] D.H. Youngblood, H.L. Clark, Y.-W. Lui, Phys. Rev. Lett. 82 (1999) 691.

[7] P. Klupfel, P.-G. Reinhard, T. J. Burvenich, and J. A. Maruhn, Phys.Rev.C79,034310(2009).

[8] Pawel Danielewicz,Nucl.Phys.A673(2000)375.

[9] M. B. Tsang, Yingxun Zhang, P. Danielewicz, M. Famiano, Zhuxia Li, W. G. Lynch and A.W. Steiner, Phys.Rev.Lett. 102, 122701 (2009).

[10] M. B. Tsang, J. R. Stone, F. Camera,et al.,Phys. Rev. C 86, 015803(2012).

[11] J.M.Lattimer, A.W.Steiner, Eur.Phys.J. A50, 40(2014).

[12] C. J. Horowitz, E. F. Brown, Y. Kim, W. G. Lynch, R.Michaels, A. Ono, J. Piekarewicz, M. B. Tsang, and H. H. Wolter, J. Phys. G: Nucl. Part. Phys. 41 (2014) 093001.

[13] D.G. Ireland, W. Nazarewicz, J. Phys. G: Nucl. Part. Phys. 42, 030301(2015).

[14] Yingxun Zhang, M.B.Tsang, Zhuxia Li, Hang Liu, Phys.Lett.B 732, 186(2014).

[15] Siegmund Brandt, Data Analysis-Statistical and Computational Methods for Scientists and Engineers, Fourth Edition, Springer.

[16] Yingxun Zhang, M.B.Tsang, Zhuxia Li, Phys.Lett.B749, 262(2015).

[17] L.W.Chen, Che Ming Ko, Bao-An Li, Phys.Rev.Lett.94, 032701(2005).

[18] J.Rizzo, M.Colonna, V.Baran, M.Di Toro, H.H.Wolter, M.Zielinska-Pfabe, Nucl.Phys.A806,79(2008).

EPJ Web of Conferences **117**, 07026 (2016) DOI: 10.1051/epjconf/201611707026
NN2015

Effects of symmetry energy and momentum dependent interaction on low-energy reaction mechanisms

H. Zheng[1], M. Colonna[1], V. Baran[2], S. Burrello[1,3]

[1] INFN-LNS, Laboratori Nazionali del Sud, 95123 Catania, Italy,
[2] Faculty of Physics, University of Bucharest, Măgurele, Romania,
[3]Physics and Astronomy Dept., University of Catania, Catania, Italy.

Abstract

We study the dipole response associated with the Pygmy Dipole Resonance (PDR) and the Isovector Giant Dipole Resonance (IVGDR), in connection with specific properties of the nuclear effective interaction (symmetry energy and momentum dependence), in the neutron-rich systems ^{68}Ni, ^{132}Sn and ^{208}Pb. We perform our investigation within a microscopic transport model based on the Landau-Vlasov kinetic equation.We observe that the peak energies of PDR and IVGDR are shifted to higher values when employing momentum dependent interactions, with respect to the results obtained neglecting momentum dependence. The calculated energies are close to the experimental values and similar to the results obtained in Hartree-Fock (HF) with Random Phase Approximation (RPA) calculations.

1 Introduction

The electric dipole (E1) response of nuclei, including also exotic systems, was intensively explored during the last decade through a variety of experimental methods and probes [1]. From these data, the experimental evidence of an additional dipole strength, located below the IVGDR and dubbed as PDR, has emerged for several systems. The position of the PDR peak changes from around 7.35 MeV for ^{208}Pb [2,3] to 8-10 MeV for Sn isotopes [4,5] and 9.55-11 MeV in the case of ^{68}Ni [6,7].

In refs. [8,9], a microscopic transport model based on the Landau-Vlasov kinetic equation was used to systematically study the correlation between the PDR peak position and the mass number of the nucleus considered, and the correlation between the Energy Weighted Sum Rule (EWSR) attributed to the PDR mode and the neutron skin thickness. The role of the isovector sector of the nuclear effective interaction was also explored, by considering three different parametrizations of the symmetry energy, but momentum dependent effects were neglected. In these calculations the peak energies, for the different exotic systems considered, were below the experimental values, especially in the case of the IVGDR mode. In the present work, we implement the standard Skyrme interaction, including momentum dependence, into the microscopic transport model. Three parametrizations, the SAMi27, SAMi31 and SAMi35 [10] are considered, in order to explore also the impact of the symmetry energy on the properties of PDR and IVGDR.

2 Theoretical framework

The energy density associated with standard Skyrme interactions can be expressed in terms of the isoscalar, $\rho = \rho_n + \rho_p$, and isovector, $\rho_3 = \rho_n - \rho_p$, densities and kinetic energy densities ($\tau = \tau_n + \tau_p, \tau_3 = \tau_n - \tau_p$) as [11]:

$$\mathscr{E} = \frac{\hbar^2}{2m}\tau + C_0\rho^2 + D_0\rho_3^2 + C_3\rho^{\alpha+2} + D_3\rho^{\alpha}\rho_3^2 + C_{eff}\rho\tau + D_{eff}\rho_3\tau_3 + \\ + C_{surf}(\bigtriangledown\rho)^2 + D_{surf}(\bigtriangledown\rho_3)^2, \quad (1)$$

where C_i, D_i are combinations of traditional Skyrme parameters. The terms with coefficients C_{eff} and D_{eff} are the momentum dependent contributions to the EoS.

Having as main ingredients the fermionic nature of the constituents and the self-consistent mean-field, the Vlasov equation represents the semi-classical limit of Time-Dependent Hartree-Fock (TDHF) and, for small-oscillations, of the RPA equations. While the model is unable to account for effects associated with the shell structure, our self-consistent approach is suitable to describe robust quantum modes, of zero-sound type, in both nuclear matter and finite nuclei. The two coupled Landau-Vlasov kinetic equations for neutrons and protons:

$$\frac{\partial f_q}{\partial t} + \frac{\partial \epsilon_q}{\partial \mathbf{p}}\frac{\partial f_q}{\partial \mathbf{r}} - \frac{\partial \epsilon_q}{\partial \mathbf{r}}\frac{\partial f_q}{\partial \mathbf{p}} = I_{coll}[f_n, f_p], \quad (2)$$

determine the time evolution of the one-body distribution functions $f_q(\vec{r},\vec{p},t)$, with $q = n,p$ [12]. In the equation above, ϵ_q represents the single particle energy, which can be deduced from the expression of the energy density eq. (1). Since we are interested in collective motion at zero temperature, we shall switch-off the collision integral in the following. The parameters of the Skyrme effective interactions are fixed in order to reproduce the saturation properties of symmetric nuclear matter, $\rho_0 = 0.16 fm^{-3}$, $E/A = -16$ MeV and a compressibility modulus $K \approx 200 - 240$ MeV [12]. Moreover, some specific properties of finite nuclei (radii, binding energies) can be fitted as well [10]. In the case when momentum dependent terms are neglected ($C_{eff} = 0$, $D_{eff} = 0$), we consider, for the symmetry energy coefficient $C(\rho)$ [8] the following expressions: $C(\rho) = \rho_0(482 - 1638\rho)$ MeV (asysoft EoS), $C(\rho) = 36$ MeV (asystiff EoS) and $C(\rho) = 36\frac{2\rho}{(\rho+\rho_0)}$ MeV (asysuperstiff EoS). The corresponding symmetry energy is plotted in fig. 1 (lines). One can see that the symmetry energy at saturation density is almost the same for the three EoS. In this study, we also employed three effective interactions including momentum dependence: SAMi27, SAMi31 and SAMi35 [10]. Since the SAMi EoS are obtained by fitting the main properties of finite nuclei, the symmetry energy is not normalized at saturation density but around $0.65\rho_0$, see fig. 1.

The integration of the transport equations is based on the test-particle (or pseudo-particle) method, with a number of 1500 t.p. per nucleon in all the cases, ensuring in this way a good spanning of the phase-space. From the one-body distribution functions one obtains the local densities:

$$\rho_q(\vec{r},t) = \frac{2}{(2\pi\hbar)^3}\int d^3\mathbf{p} f_q(\vec{r},\vec{p},t), \tag{3}$$

as well as the center of mass of neutron and proton distributions, which is needed to evaluate the dipole moment:

$$\mathbf{R}_q = \frac{1}{N_q}\int \vec{r}\rho_q(\vec{r},t)d^3\mathbf{r}. \tag{4}$$

3 Results

It is known from quantum mechanics that the E1 response of nuclei is [9,13]

$$S(E) = \sum_{n>0} |\langle n|\hat{D}|0\rangle|^2 \delta(E - (E_n - E_0)), \tag{5}$$

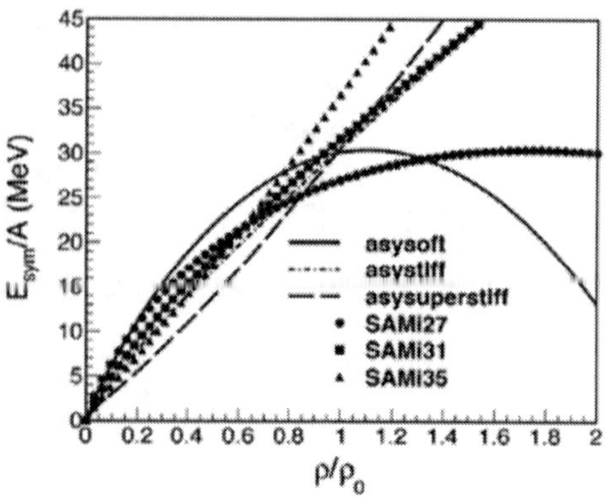

Figure 1: The symmetry energy versus density for the EoS with (symbols) and without (lines) momentum dependent terms.

where $\hat{D}$ is the dipole operator, E_0 is the energy of the ground state $|0\rangle$, while the energies E_n correspond to the excited states $|n\rangle$, eigenstates of the Hamiltonian of the system. The corresponding strength function in the semiclassical framework can be obtained by the Fourier transformation of the perturbation observable

$$S(E) = |\int_0^{\infty} dt\ \delta\mathbf{D}(t) e^{i\frac{Et}{\hbar}}|. \tag{6}$$

To study the isovector dipole response, we perturbed the ground state of the nucleus by boosting the neutrons against protons in coordinate space at the beginning (but keeping the system center of mass at rest), then following the time evolution of the relative distance between the center of mass of protons and the center of mass of neutrons up to 1800 fm/c. In our transport approach:

$$\delta\mathbf{D}(t) = \frac{NZ}{A}\left(\mathbf{R}_p(t) - \mathbf{R}_n(t)\right), \tag{7}$$

where $\mathbf{R}_p(t)$ and $\mathbf{R}_n(t)$ have been given in eq. (4).

We performed the calculations for the three neutron-rich nuclei ^{68}Ni, ^{132}Sn and ^{208}Pb, employing the Skyrme effective interactions, with and without momentum dependent terms, introduced above. The results obtained for ^{132}Sn are shown in fig. 2. As one can see, the peak energies of the IVGDR show a strong dependence on the EoS, when momentum dependent terms are neglected. Indeed, owing to surface effects, the IVGDR energies are mostly sensitive to the symmetry energy below saturation density ($\rho \approx 0.65\rho_0$), which takes different values for the three parametrizations

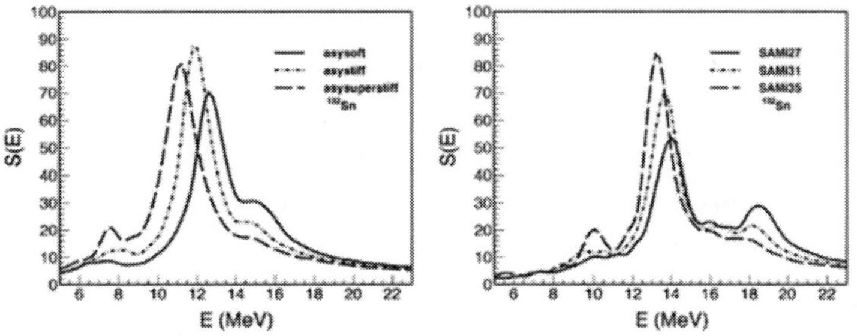

Figure 2: The isovector dipole strength obtained for ^{132}Sn employing effective interactions without (left) and with (right) momentum dependent terms.

considered (see the lines of fig.1). For the EoS including momentum dependence, the sensitivity of the IVGDR centroid to the symmetry energy decreases, because the three parametrizations employed cross below saturation density (see fig.1). We also notice that there is a second peak above the main IVGDR peak, especially in the case of the EoS including momentum dependence, similarly to HF-RPA calculations [10]. In the PDR region, the strength of the isovector response appears sensitive to the symmetry energy, both for momentum dependent and momentum independent interactions. Indeed, as pointed out in ref. [8], the PDR strength can be connected to the extension of the neutron skin, which depends on the symmetry energy parametrization, being larger in the stiffer case. On the other hand, the energy peak of the PDR seems to locate at the same position, independently of the parametrization employed. This indicates that the PDR is mostly of isoscalar nature [8]. Thus, to better understand the PDR features, it is worth investigating also the isoscalar dipole modes. From fig. 2, one can see that both the PDR and the IVGDR peak energies are shifted to higher values when the momentum dependent terms are considered, close to the experimental values (especially for the IVGDR) and similar to the results of HF-RPA calculations [10]. Similar conclusions can be extracted from the study of the systems ^{68}Ni and ^{208}Pb.

4 Conclusion

Employing several effective interactions, with and without momentum dependent terms, in a semi-classical transport model, we investigate the isovector dipole response in the neutron-rich systems ^{68}Ni, ^{132}Sn and ^{208}Pb. Our calculations indicate a clear dependence of the peak energy of the IVGDR on the behavior of the symmetry energy below saturation density. On the other hand, the peak energies of the PDR do not depend on the symmetry energy, however the strength located in this energy region is sensitive to it. These findings are in agreement with the results of ref. [8]. The peak energies of the PDR and the IVGDR are shifted to higher values when momentum dependent terms are included. The latter are close to the experimental values and similar to the results of HF-RPA calculations.

References

[1] Savran D. et al., *Prog. Part. Nucl. Phys.* **70** (2013) 210.

[2] Ryezayava N. et al., *Phys. Rev. Lett.* **89** (2002) 272502.

[3] Tamii A. et al., *Phys. Rev. Lett.* **107** (2011) 062502.

[4] Adrich P. et al., *Phys. Rev. Lett.* **95** (2005) 132501.

[5] Toft H. K. et al., *Phys. Rev. C* **83** (2011) 044320.

[6] Wieland O. et al., *Phys. Rev. Lett.* **102** (2010) 092102.

[7] Rossi D. M. et al., *Phys. Rev. Lett.* **111** (2013) 242503.

[8] Baran V. et al., *Phys. Rev. C* **88** (2013) 044610.

[9] Baran V. et al., *AIP Conference Proceedings* **1645** (2015) 267.

[10] Roca-Maza X. et al., *Phys. Rev. C* **85** (2012) 024601.

[11] Raduta Ad. R. et al., *Eur. Phys. J. A* **50** (2014) 24.

[12] Baran V. et al., *Phys. Rep.* **410** (2005) 335.

[13] Urban M., *Phys. Rev. C* **85** (2012) 034322.

EPJ Web of Conferences **117**, 07027 (2016) DOI: 10.1051/epjconf/201611707027
NN2015

Elastic scattering of ^{9}Be+^{51}V near the Coulomb barrier

J. C. Morales-Rivera[1,2], E. Martinez-Quiroz[1*], T. L. Belyaeva[2], E. F. Aguilera[1], D. Lizcano[1] and P. Amador-Valenzuela[1]

[1]Departamento de Aceleradores, Instituto Nacional de Investigaciones Nucleares, A. P. 18-1027, C. P. 11801, México, D. F. México.
[2]Universidad Autónoma del Estado de México, C. P. 50000, Toluca, México.

Abstract

Elastic scattering angular distributions for the ^{9}Be+^{51}V system were measured at three near Coulomb barrier energies, E_{lab} = 16.35, 17.44 and 18.53 MeV. The data were analyzed by using a Semimicroscopic Optical Model. This combines a microscopic calculation of the mean-field double folding potential and a phenomenological construction of the dynamical polarization potential. The calculations reproduced the data very well and the total reaction cross sections were also calculated.

1 Introduction

Nuclear reactions induced by weakly bound projectiles have attracted much attention in the last few decades. The ^{9}Be nucleus, with a separation energy of 1.57 MeV, is an interesting case study. In the present work, angular distributions for the ^{9}Be+^{51}V system are measured and analyzed. A Semimicroscopic Optical Model, which combines microscopic calculations of the mean-field double folding potential (DFP) and a phenomenological construction of the dynamical polarization potential [1, 2], is used in the

*E-mail: enrique.martinez@inin.gob.mx (corresponding author)

analysis. A comparison with recent data for ^{7}Be+^{58}Ni [3, 4] and other ^{9}Be systems is made [5–7].

2 Description of the experiment and analysis

The beam was produced by the 6 MV Tandem Van de Graaff Accelerator at ININ. The experimental setup consisted of six surface barrier detectors placed on a turntable that allowed measurement of particles scattered in a large angular range (35-165 degrees). The measurements were performed in two stages, measuring first the scattering in an initial configuration and then rotating the turntable by 30 degrees to complement the angular range.

Fig. 1 shows the elastic scattering angular distributions measured for the ^{9}Be+^{51}V system at energies E_{lab} = 16.35, 17.44 and 18.53 MeV with OM-model fits discussed below.

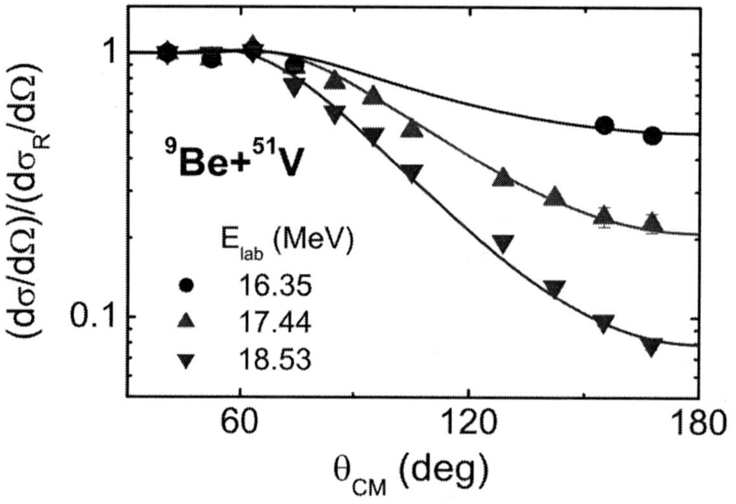

Figure 1: Elastic scattering angular distributions for ^{9}Be+^{51}V system at energies E_{lab} = 16.35, 17.44 and 18.53 MeV.

The data were analyzed with an optical model potential of the form $U = V_F + V_P + iW + V_C$, Ref. [1,2], where V_F is the double folding potential, V_C is the Coulomb potential, and V_P, W are given by

$$V_P(r, E) = \alpha(E)W(E)f(x_W) + \beta(E) \cdot 4W_D(E)\frac{df(x_D)}{dx_D}, \tag{1}$$

$$iW = -iW(E)f(x_W) + i \cdot 4W_D(E)\frac{df(x_D)}{dx_D}. \tag{2}$$

A fit to the data was made by using the computational codes Foldeg v2007 and Fresco, see Table 1. The results are shown with the curves in Fig. 1. Table 1 also shows the total reaction cross sections and the volume integrals obtained in the analysis.

Table 1: Parameters of the Optical Model Potential for the ^{9}Be+^{51}V system. The energy and depths of the potential are measured in MeV, $r_{W,D}$ y $a_{W,D}$ in fm. Total reaction cross sections (mb) and volume integrals (MeV· fm^3) also are included.

E_{lab}	α	β	W	r_W	a_W	W_D	r_D	a_D	σ_R	J_F
16.35	-0.62	-0.001	8.60	1.55	0.40	1.33	1.45	0.40	242	403.30
17.44	-0.53	-0.84	10.28	1.31	0.61	2.10	1.30	0.53	350	403.00
18.53	-0.83	-0.41	12.0	1.22	0.70	4.16	1.15	0.71	542	402.80

2.1 Comparison with other systems

In Fig. 2, our data are compared to total reaction cross sections reported previously for the ^{7}Be+^{58}Ni system [3], such comparison is made in reduced units as suggested in ref [8, 9] and the solid line corresponds to a Wong function fit [10].

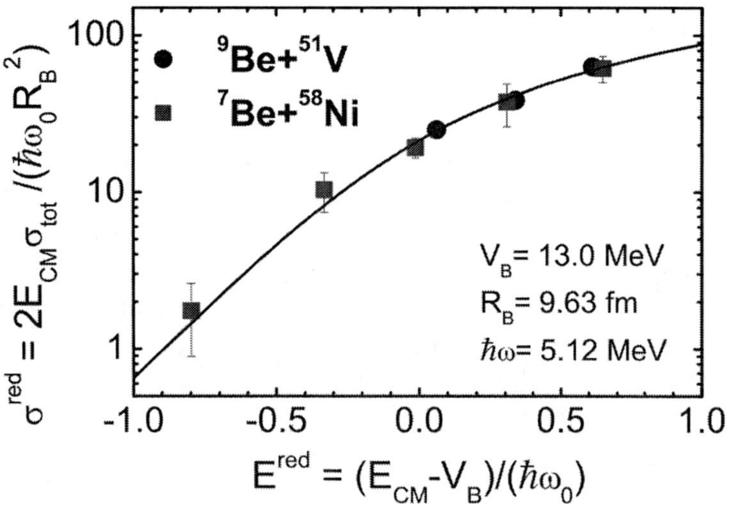

Figure 2: Comparison of the reaction cross sections for ^{9}Be+^{51}V and ^{7}Be+^{58}Ni [3] at different energies.

Fig. 3 shows a comparison with several systems having the same projectile, ^{9}Be. Previously reported data were taken from the literature [5–7]. All experimental points seem to be clustered in a specific region, regardless of the mass number of the target.

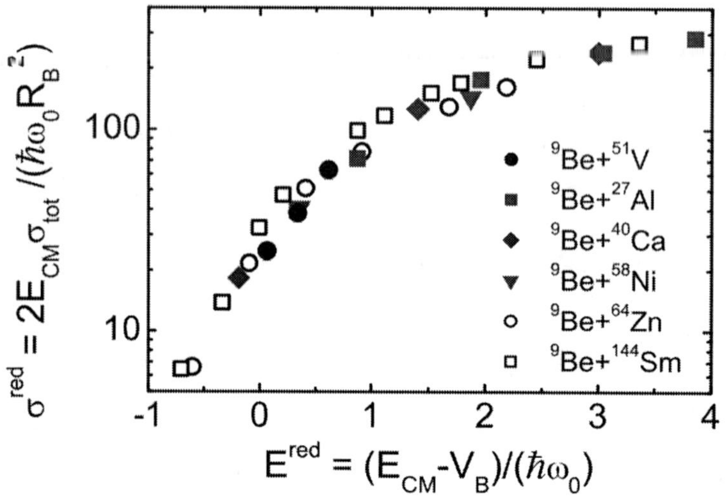

Figure 3: Comparison of the reduced reaction cross sections for ^{9}Be+^{51}V with data for different systems reported in the literature [5–7].

3 Summary

Elastic scattering angular distributions for ^{9}Be+^{51}V system were measured at bombarding energies E_{lab} = 16.35, 17.44 and 18.53 MeV. By doing a proper optical model analysis, the corresponding total reaction cross sections were obtained and compared with experimental results previously reported for other systems. The reduced total reaction cross sections obtained for different systems having the same projectile show perfect consistency. On the other hand, our results are consistent with the reduced total reaction cross sections obtained for the other system, which has the ^{7}Be isotope as a projectile.

Acknowledgments

This work was partially supported by CONACYT, Mexico

References

[1] Goncharov S. A. and Izadpanah A., *Phys. At. Nucl.* **70**, 1 (2007) 18.

[2] Goncharov S. A. and Izadpanah A., *Phys. At. Nucl.* **70**, 9 (2007) 1491.

[3] Aguilera E. F. *et al.*, *Phys. Rev. C* **79** (2009) 021601(R).

[4] Martinez-Quiroz E. *et al.*, *Phys. Rev. C* **90** (2014) 014616.

[5] Satchler G. R. and Love W. G., *Phys. Rep.* **55**, 3 (1973) 183.

[6] Gomes P. R. S. *et al.*, *Nucl. Phys. A* **828** (2009) 3.

[7] Gomes P. R. S. *et. al.*, *Nucl. Phys. A* **828** (2009) 233.

[8] Canto L. F. *et al.*, *J. Phys. G* **36** (2009) 015109.

[9] Canto L. F. *et al.*, *Nucl. Phys. A* **821** (2009) 51.

[10] Wong C. Y., *Phys. Rev. Lett.* **31** (1973) 766.

EPJ Web of Conferences **117**, 08001 (2016) DOI: 10.1051/epjconf/201611708001
NN2015

Dynamical effects in fission investigated at high excitation energy

J. Benlliure

Universidad de Santiago de Compostela, E-15782 Santiago de Compostela, Spain

Abstract

The experimental techniques used for the investigation of nuclear fission have progressed considerably during the last decade. Most of this progress is based on the use of the inverse kinematics technique allowing for the first time the complete isotopic and kinematic characterization of both fission fragments. These measurements make possible to characterize the fissioning system at saddle and at scission, and can be used to benchmark fission model calculations. One of the important ingredients in transport models describing the dynamics of the process is the dissipation parameter, governing the coupling between intrinsic and collective degrees of freedom. Recent experiments got access to the magnitude of this parameter and could also investigate its dependence in temperature and deformation.

1 Introduction

More than 75 years after its discovery, fission still represents one of the most fascinating processes in nuclear physics, but also an important challenge for physicists. Indeed, nuclear fission has implications in many different domains such as the investigation of shell effects at large deformation [1] or the dynamics of highly excited nuclear matter [2], but also in the understanding of the abundances of elements produced in the stellar nucleosynthesis r process [3]. Moreover, fission also has a relevant societal impact because of its role in energy generation and radioisotope production for medical imaging.

DOI: 10.1051/epjconf/201611708001

Despite the fact that a global understanding of nuclear fission was proposed soon after its discovery [4], we are still far from a complete microscopic description. The reason being that fission is a conceptually complicated process where collective and single-particle degrees of freedom interplay. The first attempt to describe the dynamics of fission was put forward by Kramers [5] assuming that the evolution of the deformation degree of freedom can be viewed as a Brownian motion of this collective degree of freedom with a heat bath defined by the single-particle degrees of freedom. The time-dependent formulation of this model using a Fokker-Planck equation with a dissipative term describing the coupling between the collective and single-particle degrees of freedom was introduced by Grangé and collaborators [6]. An analogue formulation is the one based on the Langevin equation [2] or as a random walk, in the limit of strong dissipative coupling [7], on a multidimensional macroscopic-microscopic potential energy surface [8].

Accurate microscopic descriptions of the potential energy surface are nowadays obtained with modern implementations of the density functional theory [9]. However, a microscopic description of the full fission dynamics still represents a challenge for theorists. In this respect, approaches based on the time-dependent Hartree-Fock theory have provided relevant progress in the description of the fission dynamics by including nonadiabatic effects close to the scission point [10].

From an experimental viewpoint, the complete characterization of the fission fragments has been a challenge because of the low energy and high Z values of the heavy fission fragments. The use of inverse kinematics and state-of-the-art detection setups made possible to overcome this situation very recently. On another hand, the dynamics of the fission process has been mostly addressed by using low-energy fission reactions. However, it has been demonstrated that transient effects governing the ground-to-saddle dynamics only manifest in fast fission reactions induced at high energies. In this work we will briefly review the main advances in the investigation of the fission dynamics and we will then concentrate of recent results concerning these investigations at high excitation energy.

2 Investigation of dynamical effects in fission

The time scale and energy dissipation during the fission path from saddle to scission have been investigated at low energies using observables giving access to the the scission-point configuration. For example, total kinetic energies of the fragments give access to the elongation at scission of the fissioning system [11] The sawtoothlike behavior of the prompt-neutron yields was generally attributed to the contribution of the deformation energy to the excitation energy of the fragments [12]. Finally, the odd-even effect in fission-fragment element distributions is another observable used to characterize the energy dissipated between saddle and scission in low-energy fission [13]. However, the interpretation of some of these observable has been recently revisited. K.-H. Schmidt and collaborators explain the sawtoohlike pattern of the prompt-neutron yields, and in particular its energy dependence, and the general trends observed in the odd-even staggering, as due to the energy sorting between both fission fragments according to their size [14, 15].

The ground-to-saddle dynamics can be investigated using observables sensitive to the corresponding time. The emission of prescission gamma rays and particles, fission cross sections and the width of the charge distribution of the fission fragments have shown that typical ground-to-saddle times are longer than expected from pure statistical considerations [16]. This delay of the fission time is understood as the transient time needed for the coupling between intrinsic and collective degrees of freedom until a stationary flow of the fission width over the barrier is reached. That time can be quantified in terms of a nuclear viscosity parameter in transport equations [6]. Most of the measurements are compatible with a value of the transient time around 1-5 10^{-21} s. Those values are negligible when compared to typical statistical times for low-energy fission, therefore reactions inducing fission at high-excitation energies are better suited [17]. Indeed, Grangé and collaborators [6] established that the best conditions for the manifestation of transient effects are the use of reactions leading to fissioning systems not only with large values of excitation energy but also with low values of angular momentum and spherical shapes.

3 Experimental approaches

Most of the information we have on the saddle-to-scission dynamics at low energy was obtained using neutron-induced fission [12]. This approach allows a precise definition of the initial compound nuclei but can only be applied with a limited number of fissionable target materials. This limitation was overcome using fusion-fission reactions [18] or beta-delayed fission [19]. However, these reactions do not allow for a complete identification of the fission fragments because the charge and velocity of the heavy-fission fragments do not allow to correlate any energy loss with their atomic number.

The real step forward in the investigation of fission was brought by the use of the inverse kinematics. Under those conditions, the boosted fission fragments can be fully identified in atomic and mass number from magnetic rigidity, time of flight and energy-loss measurements [20]. Moreover, one can take advantage of the secondary beams produced in-flight using fragmentation reactions to investigate Coulomb-induced fission for a large number of actinides and pre-actinides [21]. This technique has been also used to investigate the saddle-to-scission dynamics in transfer-induced fission reactions [22]. However, the complete identification of both fission fragments was just achieved recently by combining the inverse kinematics with a large acceptance detection setup [23].

The inverse kinematics also allows to induce fission using fragmentation or spallation reactions [24–29]. Those reactions produce fissioning nuclei with large values of excitation energy giving access to the short time delay induced by dissipation until the fission flow over the barrier reaches a stationary regime. Moreover, spallation and fragmentation reactions introduce small shape distortions and low angular momenta. Therefore, these reactions are considered as an optimum tool for the investigation of the ground-to-saddle dynamics [30, 31].

4 Ground-to-saddle dynamics

The investigation of the dynamics between the initial state of the fissioning compound nucleus and its configuration at saddle should provide relevant information on the energy transfer between intrinsic and collective degrees of freedom in fission that can be quantified in terms of a dissipation parameter [6]. Those investigations have been mostly carried out using fusion-fission reactions [17, 32–38]. However, those works provided rather unclear conclusions concerning the absolute value of the dissipation parameter and its possible dependences on temperature and deformation. Some reasons could be the strong influence of angular momentum and deformation in the fission probability, and the limited values of excitation energies reached in these reactions [30, 31, 39].

An interesting alternative to fusion-fission is the use of spallation reactions induced by energetic protons [40, 41] or anti-protons [42], and fragmentation reactions [30, 43]. These reactions are expected to fulfill the conditions for the manifestation of dissipative effects in fission [6] because they can produced fissioning nuclei with large values of excitation energy but low angular momenta. Moreover, the investigation of these reactions in inverse kinematics facilitates the identification of the fission fragments, giving access to new observables to better constraint dissipative effects in fission [44].

The ground-to-saddle time, characterizing the fission dynamics at small deformation, can only be deduced from indirect measurements. Some of those are based on measurements of the multiplicity of light-charged particles emitted pre-scission [45]. Unfortunately, those measurements do not only depend on the presaddle transient time but also on the saddle-to-scission time. Fission probabilities [33,37,39,46,47] or the complementary cross sections of evaporation residues [36,41,48] seem to be better suited to constraint presaddle effects. However, fission or evaporation residue cross sections depend on factors other than dissipation such as the fission barriers or the level densities [37,49,50]. Indeed, some authors claim that cross sections of fission reactions induced at relatively high energies can be described using purely statistical models, provided that level densities at saddle and ground configuration are adjusted to specific values [51,52].

In order to overcome this confusing situation it was proposed the combined measurement of several independent observables constraining the main parameters defining the fission probability. The first observable proposed, together with the fission cross sections, was the width of the charge distribution of the final fission fragments [40]. The choice of this observable was based on the fact that the width of the charge or mass distribution of the fission fragments at their initial formation close to the saddle point is determined by statistical fluctuations and therefore by temperature [53]. Moreover, if one takes into account that fission fragments deexcite mainly emitting neutrons, then we conclude that the charge distribution of the final fragments should be very close to the same distribution at saddle and should provide an indirect measurement of the energy dissipated between ground and saddle.

A step further was proposed in Refs. [43,44] where the charge distribution of the final fission fragments was sorted as a function of the sum of the atomic numbers of the two fragments. In this case, the sum of the atomic numbers of the two fragments represents the size of the fissioning nuclei. In spallation and fragmentation reactions the size of the compound nuclei is correlated with the impact parameter and then, with the energy gained during the collision phase. In these works it was also proposed to correlate the fission cross sections with the size of the fissioning nucleus. In Fig. 1 we depict an example of both graphs obtained for the reaction ^{208}Pb+p at 500 A MeV taken from [54]. As can be seen in the upper panel, fission cross sections

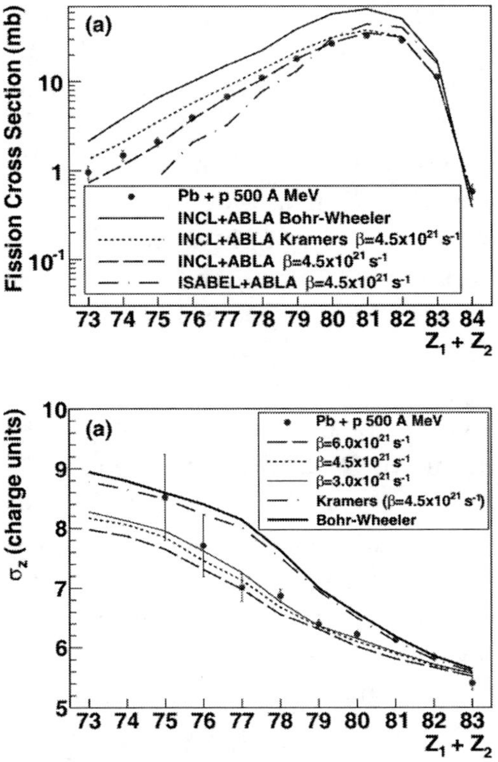

Figure 1: Upper panel: partial fission cross sections (dots) as a function of the size of the fissioning nuclei obtained for the reaction $^{208}Pb + p$ at 500 A MeV compared to different model calculations performed with the intranuclear cascade codes INCL and ISABEL coupled to deexcitation code ABLA. Lower panel: correlation between the width of the charge distribution of the fission fragments and the size of the fissioning nuclei compared with different model calculations.

decreases with the size of the fissioning nuclei as expected for nuclei with lower fissility. An exception are the fissioning nuclei with $Z_1+Z_2=83$ and $Z_1+Z_2=84$ which are produced in charge-exchange reactions with a very small probability [55]. In the lower panel we observe a clear increase of the width of the charge distribution of the final fragments when decreasing the size of the fissioning nuclei. This behavior is interpreted as an increase of the temperature at saddle with the initial excitation energy. In this figure the measurements are used to benchmark different model calculations in order to quantify the magnitude of the dissipation coefficient. In particular we couple two intranuclear cascade codes describing the first stage of the collision, INCL4.6 [56] and ISABEL [57], to the deexcitation code ABLA07 [58].

In intranuclear cascade codes, the collision of a high-energy projectile within a target nucleus initiates a succession of binary nucleon-nucleon collisions that may lead to the emission of pre-equilibrium nucleons. In INCL and ISABEL, nucleons are described as point-like particles, and as a continuous medium perturbed by the collisions induced by cascade particles, respectively. Nucleons in INCL are endowed within a realistic phase-space density (Woods-Saxon in configuration space, hard Fermi sphere in momentum space), while in ISABEL, both projectile and particle follow a folded Yukawa density distribution. In INCL, the excitation energy and the angular momentum induced by the intranuclear cascade are computed according to particle-hole excitations. The cascade stops self-consistently when the prefragment is assumed to be thermalized. In contrast, in ISABEL the cascade stops when the energy is below a specific cutoff energy. In both codes, the remaining excitation energy is distributed among all the nucleons.

The ABLA code describes the deexcitation of a thermalized compound nucleus by fission or by the emission of light charged particles, γ-rays, neutrons and intermediate-mass fragments (IMF) following Weisskopfs model [59]. The fission probability is computed according to a time-dependent fission width following the analytical description of the solutions of the corresponding Fokker-Plank equation proposed in Refs. [30,60,61]. The code also allows to evaluate the fission width according to the transition-state model of Bohr and Wheeler [4] or the time-independent formulation of Kramers [5]. Important parameters in the code are the level density parameters that are calculated according to the parametrization proposed by Ignatyuk [62] and angular momentum dependent fission barriers which are taken from Sierks finite-range liquid drop model [63].

The benchmarking of these model calculations presented in Fig. 1 shows the sensitivity of the partial fission cross sections and the width of the charge

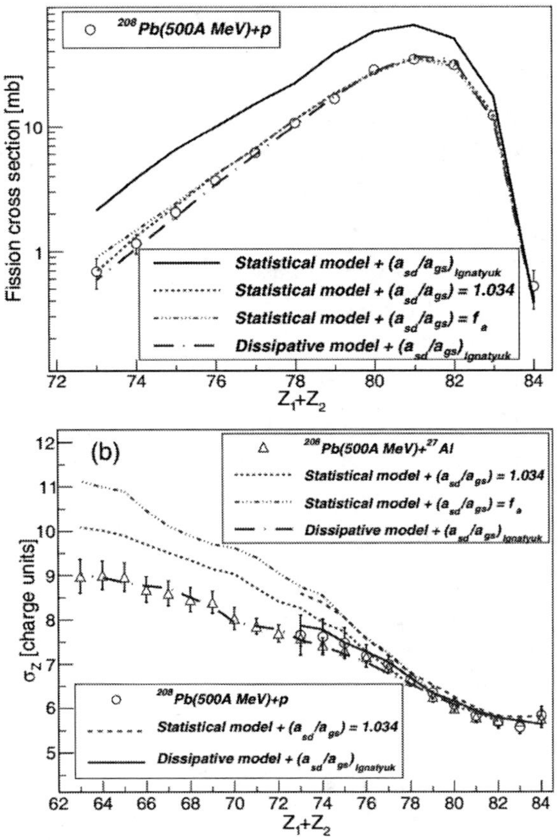

Figure 2: Upper panel: Fission cross section of the reaction ^{208}Pb +p at 500A MeV as a function of the atomic number of the fissioning nuclei (open circles). The lines represent model calculations with different descriptions of the fission decay width and the level density parameter (see text for details). Lower panel: Width of the atomic-number distribution of the final fission fragments measured in the reaction ^{208}Pb+p at 500A MeV as a function of the atomic number of the fissioning nuclei compared to the same model calculations shown in the upper panel.

distribution of the final fragments to different intranuclear cascade codes as well as to the different descriptions of the fission decay width. It is clear that purely statistical calculations overestimate the partial fission cross sections and the width of the charge distributions. Only model calculations including a time-dependent description of the fission width with a viscosity parameter around 4.5 10^{21} s^{-1} describe correctly both observables. This magnitude of the dissipation parameter corresponds to a ground-to-saddle fission transient time of 1-5 10^{-21} s. This result is in good agreement with others previously obtained using fusion-fission reactions [37–39, 49]. Moreover, the fact that we are able to describe the data over a broad range of Z_1+Z_2 values with the same dissipation parameter would indicate that, contrary to some previous results, this parameter does not depend on temperature.

These observables have also been used in Ref. [50] to investigate the role of the description of the level densities at ground and saddle deformation in the fission process. The upper panel in Fig. 2 shows that different descriptions of the fission decay width can reproduce fission cross sections provided that specific values of the ratio of the level density parameter at ground and saddle deformation are chosen. This argument was used by the authors of Refs. [51, 52] to conclude that purely statistical descriptions of the fission decay width could describe the cross sections of fission reactions induced at high energies by using appropriate values for the level densities. We can then conclude that the theoretical description of the ground-to-saddle fission dynamics is a multiparametric problem that can only be solved by using several independent observables measured simultaneously.

5 Saddle-to-scission dynamics

Dissipative effects at large deformation has been traditionally investigated by measuring the prescission emission of neutrons [18, 32, 33], GDR γ-rays [17, 34–36] or light-charged particles [38, 45]. In many of these works fission cross sections measured in coincidence were used to determine the ground-to-saddle dissipation parameter. These results are in good agreement with the values obtained in spallation and fragmentation reactions discussed in the previous section $\beta_{ground-saddle} \approx$ 4-5 10^{21} s^{-1}. Most of the experiments providing values of the dissipation parameter at large deformation using prescission GDR γ-rays [35, 36] and neutrons [32, 33] obtained values around $\beta_{saddle-scission} \approx$ 20-30 10^{21} s^{-1}. However, model calculations based on the Langevin equation obtained values for the dissipation parameter at large deformation around $\beta_{saddle-scission} \approx 6$ 10^{21} s^{-1} [64].

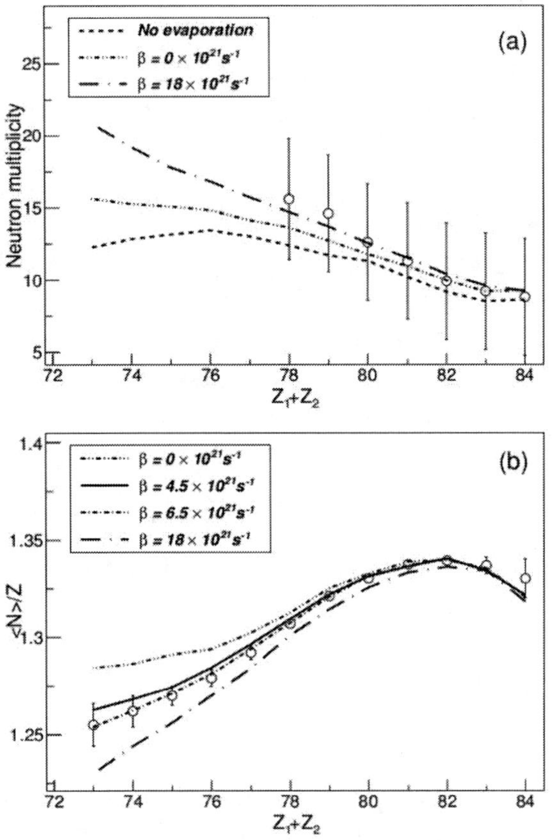

Figure 3: Upper panel: Prescission neutron multiplicities obtained for the reaction ^{208}Pb + p at 500A MeV as a function of the atomic number of the fissioning nuclei (open circles). The lines represent model calculations with different dissipation factors between saddle and scission. Lower panel: Neutron excess of the final fission fragments as a function of the atomic number of the fissioning nuclei also compared with model calculations with different dissipation factors between saddle and scission.

Recently, an experiment to investigate dissipative effects at large deformation using the reaction ^{208}Pb+p at 500A MeV in inverse kinematics was conducted at GSI. The combination of a large acceptance dipole magnet with accurate tracking and time-of-flight detectors, and performant ionization chambers made it possible for the first time to identify in atomic and mass numbers both fission fragments and determine their velocities with high accuracy [23]. These measurements provided a precise characterization of the fissioning systems at scission and allowed to determine the prescission neutron multiplicities from the isotopic composition of the final fragments and their velocities using momentum conservation and the the kinematic model for fission from Ref. [65]. The results are shown in the upper panel of Fig. 3 as a function of the size of the fissioning nuclei.

The figure also depicts the result of several model calculations using different assumptions to describe the saddle-to-scission time. In particular calculations were run for the cases where no particle evaporation between saddle and scission is considered, the saddle-to-scission time is calculated from statistical considerations or this time is obtained with a dynamical calculation assuming a value of the dissipation parameter of $\beta_{saddle-scission}$ = 18 10^{21} s^{-1} according to Ref. [66]. As can be observe, the rather large error bars associate to the multiplicity of prescission neutrons caused by the accuracy in the measurement of the velocity of the fission fragments does not allow to really benchmark those calculations.

An alternative was to use the neutron excess of the final fission fragments. Although the final value of the neutron excess of the fragments is defined by the neutron evaporation pre- and postscission, dissipative effects will only affect the prescision fraction. Moreover, this is an observable that it is determined with high accuracy in the mentioned experiment. In the lower panel of Fig. 3 we depict the neutron excess of the final fission fragments measured in the reaction ^{208}Pb+p at 500A MeV as a function of the size of the fissioning nuclei. As can be seen, the error bars are relatively smaller than in the case of the prescission neutron multiplicity shown in the upper panel of the figure.

In the lower panel of Fig. 3 we also show the results of the model calculations described above. This benchmarking clearly shows that the new observable is more sensitive to the different model calculations. Moreover, the evolution of the neutron excess over the complete range of size of the fissioning nuclei can be nicely describe with a value of the dissipation coefficient of $\beta_{saddle-scission} = 6.5 \pm 2.0\ 10^{21}\ s^{-1}$. The difference with respect to previous measurements could be due to the low efficiency affecting prescission neutron and γ-ray measurements, and the limited sensitivity of these observables to the model calculations unless extremely accurate measurements are done.

Considering the final accuracy in the dissipation parameter at large deformation and the one obtained at small deformation $\beta_{ground-saddle} = 4.5 \pm 1.0\ 10^{21}\ s^{-1}$ it was concluded that both values are compatible, and no dependence of the dissipation parameter on deformation is then observed.

6 Conclusions

The investigation of the fission dynamics still represents a real challenge for physicists however, the last decade brought impressive progress in experimental and theoretical studies. From the experimental side one can highlight the use of inverse kinematics to better identify the fission fragments. Indeed, very recently it was possible for the first time to fully identify in atomic and mass number both fission fragments using inverse kinematics and a state-of-the-art detection setup.

The progress in the experimental techniques has also allowed to use fragmentation or spallation reactions to investigate fission at high excitation energy, low angular momenta and small shape distortion in the fissioning nuclei, fulfilling the optimal conditions for the study of the dynamics through dissipative effects in fission. These advanced experiments made also possible the use of new observables. Some of these observables could be measured simultaneously in the same experiment providing stringent constraints to the different parameters affecting the theoretical description of the fission process such as level densities at ground and saddle deformation and the dissipation parameter governing the coupling between intrinsic and collective degrees of freedom. Indeed, the magnitude and possible dependences of this parameter on deformation and temperature have been a subject of vivid debate.

Partial fission cross sections and the width of the charge distribution of the final fission fragments sorted as a function of the sum of the atomic numbers of the two final fragments, have been proven to be optimum observables to constraint the nuclear viscosity at small deformation. All studies suggest a value of the dissipation parameter at small deformation of $\beta_{ground-saddle} = 4.5 \pm 1.0\ 10^{21}\ s^{-1}$.

The recent isotopic identification of the final fission fragments gave access to a new and accurate observable to constraint the dissipation parameter at large deformation, the neutron excess of the final fragments. The description of this observable with model calculations is compatible with a value of the dissipation parameter from saddle to scission of $\beta_{saddle-scission} = 6.5 \pm 2.0$ 10^{21} s^{-1}. Considering the accuracy in both dissipation parameters it was concluded that they are compatible and then no evidence for a deformation dependence of this parameter is observed. Moreover, all observables are described over a broad range of sizes of the final fragments, very much correlated to the initial excitation energy, with the same values, therefore no dependence on temperature was observed.

References

[1] K.-H. Schmidt, J. Benlliure and A. Junghans, Nucl. Phys. A **693**, 169 (2001).

[2] P. Fröbrich and I.I. Gontchar, Phys. Rep. **292**, 131 (1998).

[3] S. Goriely, Eur. Phys. J. A **51**, 22 (2015).

[4] N. Bohr and J.A. Wheeler, Phys. Rev. **56**, 426 (1939).

[5] H.A. Kramers, Physica **7**, 284 (1940).

[6] P. Grangé, L. Jun-Qing, and H. A. Weidenmüller, Phys. Rev. C **27**, 2063 (1983).

[7] P. Möller and J. Randrup, Phys. Rev. C **91**, 044316.

[8] P. Möller et al., Phys. Rev. C **91**, 024310 (2015).

[9] J.D. MacDonnell et al., Phys. Rev. C **90**, 021302 (2014).

[10] C. Simenel and A.S. Umar, Phys. Rev. C **89**, 031601 (2014).

[11] C. Böckstiegel et al., Phys. Lett. B **398**, 259 (1997).

[12] A.A Naqvi et al., Phys. Rev. C **34**, 218 (1986).

[13] H. Nifenecker et al., Zeit für Phys. A **308**, 39 (1982).

[14] K.-H. Schmidt and B. Jurado, Phys. Rev. lett. **104**, 212501 (2010).

[15] B. Jurado and K.-H. Schmidt, J. Phys. G **42**, 5 (2015).

[16] D. Jacquet and M. Morjean, Prog. Part. Phys. **63**, 155 (2009).

[17] M. Thoennessen and G.F. Bertsch, Phys. Rev. Lett. **71**, 4303 (1993).

[18] D.J. Hinde et al., Phys. Rev. C **45**, 1229 (1992).

[19] A.N. Andreyev et al., Phys. Rev. lett. **105**, 252502 (2010).

[20] M. Bernas et al., Phys. Lett. B **331**, 19 (1994).

[21] K.-H. Schmidt et al., **665**, 221 (2000).

[22] M. Caamaño et al., Phys. Rev. C **92**, 034606 (2015).

[23] J.L. Rodriguez-Sánchez et al., Phys. Rev. C **91**, 064616 (2015).

[24] J. Benlliure et al., Nucl. Phys. A **683**, 513 (2001).

[25] T. Enqvist et al., Nucl. Phys. A **686**, 481 (2001).

[26] T. Enqvist et al., Nucl. Phys. A **703**, 435 (2002).

[27] M. Bernas et al., Nucl. Phys. A **725**, 213 (2003).

[28] T. Enqvist et al., Nucl. Phys. A **658**, 47 (1999).

[29] J. Pereira et al., Phys. Rev. C **75**, 014602 (2007).

[30] B. Jurado et al., Nucl. Phys. A **757**, 329 (2005).

[31] W. Ye and J. Tian, Phys. Rev. C **91**, 064603 (2015).

[32] D. Hilscher and H. Rossner, Ann. Phys. **17**, 471 (1992).

[33] P. Fröbrich et al., Nucl. Phys. A **556**, 281 (1993).

[34] D. J. Hofman et al., Phys. Rev. C **51**, 2597 (1995).

[35] N.P. Shaw et al., Phys. Rev. C **61**, 044612 (2000).

[36] I. Diosdegi et al., Phys. Rev. C **63**, 014611 (2001).

[37] J.P. Lestone and S.G. McCalla, Phys. Rev. C **79**, 044611 (2009).

[38] E. Vardaci et al., Phys. Rev. C **92**, 034610 (2015).

[39] K.X. Jing et al., Phys. Lett. B **518**, 221 (2001).

[40] J. Benlliure et al., Nucl. Phys. A **700**, 469 (2002).

[41] J. Benlliure et al., Phys. Rev. C **74**, 014609 (2006).

[42] B. Lott et al., Phys. Rev. C **63**, 034616 (2001).

[43] C. Schmitt et al., Phys. Rev. C **81**, 064602 (2010).

[44] B. Jurado et al., Phys. Rev. Lett. **93**, 072501 (2004).

[45] J.P. Lestone et al., Phys. Rev. Lett. **67**, 1078 (1990).

[46] Y. Ayyad et al., Phys. Rev. C **89**, 054610 (2014).

[47] J.L. Rodriguez-Sánchez et al., Phys. Rev. C **90**, 064606 (2014).

[48] W. Ye, Phys. Rev. C **83**, 044611 (2011).

[49] W. Ye and N. Wang, Phys. Rev. C **87**, 014610 (2013).

[50] J.L. Rodriguez-Sánchez et al., Phys. Rev. C **92**, 044612 (2015).

[51] D. Mancusi et al., Phys. Rev. C **82**, 044610 (2010).

[52] S. Lo Meo et al., Nucl. Phys. A **933**, 43 (2015).

[53] M.G. Itkis et al., Sov. J. Part. Nucl. **19**, 301 (1988).

[54] Y. Ayyad et al., Phys. Rev. C **91**, 034601 (2015).

[55] A. Kelic et al., Phys. Rev. C **70**, 064608 (2004).

[56] A. Boudard et al., Phys. Rev. C **87**, 014606 (2013).

[57] Y. Yariv and Z. Fraenkel, Phys. Rev. C **20**, 2227 (1979).

[58] A. Kelic, M. V. Ricciardi, and K.-H. Schmidt, in Proceeding of Joint ICTP-IAEA Advanced Workshop on Model Codes for Spallation Reactions, ICTP Trieste, Italy, 48 February, 2008, edited by D. Filges, S. Leray, Y. Yariv, A. Mengoni, A. Stanculescu, and G. Mank (IAEA, Vienna, 2008), INDC(NDS)- 530, pp. 181221.

[59] V. Weisskopf, Phys. Rev. **52**, 295 (1937).

[60] B. Jurado, K.-H. Schmidt, and J. Benlliure, Phys. Lett. B **533**, 186 (2003).

[61] B. Jurado, et al., Nucl. Phys. A **747**, 14 (2005).

[62] A. V. Ignatyuk et al., Sov. J. Nucl. Phys. **21**, 612 (1975).

[63] A. J. Sierk, Phys. Rev. C **33**, 2039 (1986).

[64] A.V. Karpov et al., Phys. Rev. C **63**, 054610 (2001).

[65] B.D. Wilkins et al, Phys. Rev. C **14**, 1832 (1976).

[66] H. Hofmann and J.R. Nix, Phys. Lett. B **122**, 117 (1983).

EPJ Web of Conferences **117**, 08002 (2016) DOI: 10.1051/epjconf/201611708002
NN2015

Low-energy fusion dynamics of weakly bound nuclei: A time dependent perspective

A. Diaz-Torres and M. Boselli

ECT*, Strada delle Tabarelle 286, 38123 Villazzano, Trento, Italy

Abstract

Recent dynamical fusion models for weakly bound nuclei at low incident energies, based on a time-dependent perspective, are briefly presented. The main features of both the PLATYPUS model and a new quantum approach are highlighted. In contrast to existing time-dependent quantum models, the present quantum approach separates the complete and incomplete fusion from the total fusion. Calculations performed within a toy model for ^{6}Li + ^{209}Bi at near-barrier energies show that converged excitation functions for total, complete and incomplete fusion can be determined with the time-dependent wave-packet dynamics.

1 Introduction

Nuclear reactions constitute the engine of stars and control nucleosynthesis. They are the key tool for extracting the properties of exotic nuclei at rare-isotope beam facilities. The physics of low-energy nuclear reactions is exciting because it integrates many-body nuclear structure, reaction dynamics and mechanisms as well as quantum mechanics.

The present contribution addresses the fusion dynamics of weakly bound nuclei from a time-dependent perspective which is interesting, as it provides an intuitive picture of the reaction dynamics and how reaction observables are set during a nuclear collision. The question is how two complex nuclei

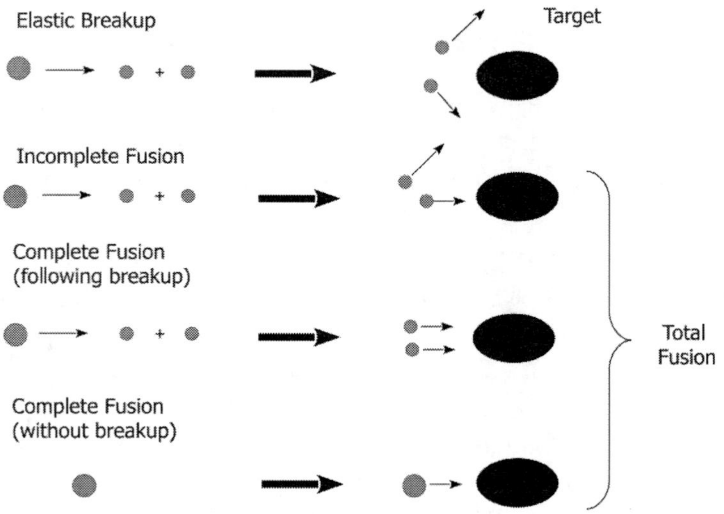

Figure 1: Some reaction processes of a weakly-bound two-body projectile.

collide and the reaction observables emerge when the nuclei pass each other slowly. There is spectacular experimental evidence that nuclear structure and reaction dynamics are strongly coupled, the coupling determining the reaction outcomes [1].

Most of the low-lying states of weakly-bound nuclei have a few-body cluster structure. A few states are bound and most of the spectrum lies in the continuum of scattering states. The strong coupling to that continuum is essential for understanding the reaction dynamics. These fragile structures dissociate when they collide with stable targets, as illustrated in Fig. 1. Many reaction paths emerge, such as elastic breakup and various fusion processes. In the incomplete fusion (ICF) only a part of the projectile is captured by the target, while in the complete fusion (CF) all the projectile fragments are captured. The sum of ICF and CF defines the total fusion (TF). It is a challenge to describe all these reaction processes in a unified picture as well as to measure all the cross sections unambiguously. To separate the fusion processes is a great problem as well. Different approaches have been used to investigate these issues [2]. A classical and a quantum mechanical model, which are based on a time-dependent perspective, are discussed below.

2 Selected dynamical fusion models

2.1 The PLATYPUS model

The PLATYPUS model uses two notions: classical trajectory and Monte-Carlo sampling of probability densities. This approach has been developed in close collaboration with experimenters [3, 4]. The model is implemented in the PLATYPUS code [5].

For a given impact parameter and bombarding energy, a weakly bound two-body projectile describes an orbit on which the projectile can be dissociated by the projectile-target interaction. A crucial input of this model is an experimentally determined breakup function, constructed as follows. For instance, if a ^{9}Be(=2α+n) dissociates on a stable target, the alpha particles can be detected in coincidence for a given scattering angle of ^{9}Be. The breakup cross section is divided by the Rutherford scattering cross section, defining a breakup probability as a function of the distance of closest approach R_{min}. This probability is interpreted as a total breakup probability, which results from the integration of a local probability density over the projectile-target distances [4]. The observed breakup function of weakly-bound stable nuclei (^{9}Be, ^{6,7}Li) appears to be a linear function of R_{min} in logarithmic scale [6, 7], whose slope is the same as that of the local probability density. Other probability densities are employed in the PLATYPUS model, which can be found in Refs. [3,4]. The final destination of the projectile fragments (captured or non-captured by the target) is determined solving classical equations of motion. Based on those trajectories, probabilities and (integrated and differential) cross sections are constructed. The PLATYPUS model connects the sub-Coulomb breakup process to the above-barrier CF and ICF processes [3, 4].

Figure 2 shows the above-barrier CF excitation function for ^{7}Li + ^{198}Pt (solid curve). It is observed that the sequential fusion component (dashed curve) becomes important as energy increases. This CF component (CF following breakup in Fig. 1) is usually neglected in quantum mechanical models, but should be included for describing the CF precisely. For this system, the ICF has also been studied with the PLATYPUS model, as reported in Refs. [8,9]. PLATYPUS provides useful information (angular momentum and excitation energy distributions) on the primary ICF products along with differential cross sections of the surviving breakup fragment [9]. The interaction between the fusing fragment and the survival fragment is very important, as this interaction affects both the angular momentum and kinetic energy brought by the captured fragment into the target. These

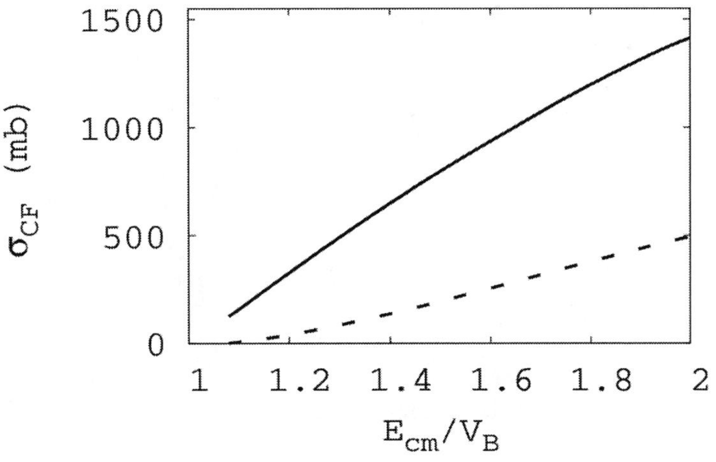

Figure 2: The above-barrier CF excitation function (solid curve) for ^{7}Li + ^{198}Pt from the PLATYPUS model. The sequential fusion component (dashed curve) becomes important as energy increases.

observables are useful for planning and interpreting particle-gamma coincidence measurements [8]. In Ref. [8], the spin and excitation energy distributions from PLATYPUS are used in combination with the evaporation code PACE for calculating the production of specific evaporation residues. Some evaporation residue cross sections can be explained by the breakup-fusion mechanism implemented in the PLATYPUS model.

Dedicated sub-Coulomb breakup measurements, involving the ^{6,7}Li and ^{9}Be projectiles, have revealed that the projectile fragmentation is triggered by transfer dominantly [7, 10]. This process enriches the fusion scenario of weakly bound nuclei, motivating the development of more comprehensive fusion models. As an example, Fig. 3 presents a preliminary PLATYPUS calculation of ICF for ^{7}Li + ^{209}Bi at $E_{cm}/V_B = 1.2$. It shows the contribution of the observed fragmentation processes of ^{7}Li to the total ICF cross section. The contribution of the ^{8}Be mediated breakup yield on ICF is much larger than the contribution of the direct ^{7}Li breakup into alpha and triton. Some projectile-like nuclei (^{5}Li, ^{8}Be) are unbound in their ground states, so these resonant states may change the breakup location of these nuclei with respect to the target position. Only prompt breakup near the target may affect the CF and ICF cross sections. The understanding of the role of the breakup-triggered-by-transfer process in fusion seems to be important for

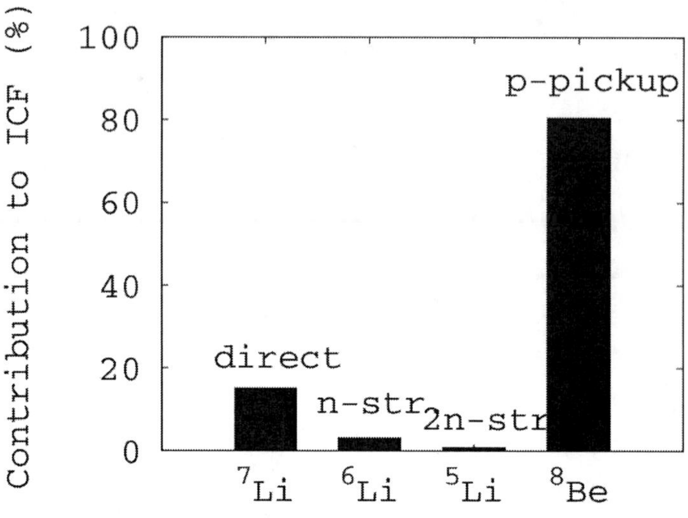

Figure 3: The contribution of various ^{7}Li fragmentation processes to the ICF for ^{7}Li + ^{209}Bi at $E_{cm}/V_B = 1.2$, calculated with the PLATYPUS model.

making further progress in this field.

2.2 Time-dependent wave-packet dynamics

A quantum dynamical approach is being developed [9] using the time-dependent wave-packet (TDWP) method. This model can deal with quantum tunneling, making it possible to investigate the interplay of breakup and fusion at sub-Coulomb energies. This perspective has been already employed in the past, see e.g. Ref. [11]. The method is based on three key steps [12]:

(i) Preparation of the initial three-body wave function, $\Psi(t = 0)$,

(ii) Time propagation, $\Psi(t) = \exp(-i\hat{\mathcal{H}}t/\hbar)\,\Psi(0)$, where $\hat{\mathcal{H}}$ is the total three-body Hamiltonian,

(iii) Analysis of the long-time propagated wave function, $\Psi(t_f)$, to calculate energy-resolved cross sections.

2.2.1 A one-dimensional toy model with two degrees of freedom

As a starting point, we have studied the ^{6}Li+^{209}Bi fusion within a one-dimensional model with two degrees of freedom, where the ^{209}Bi target and the ^{6}Li fragments (^{4}He and ^{2}H) are always on a line [9]. X_{CM} denotes the distance between the target and the center of mass of the projectile, whereas ξ gives the distance between the projectile fragments. M is the mass of the target nucleus, while m_1 and m_2 are the masses of the projectile fragments. The Hamiltonian of the system in terms of these coordinates reads:

$$\hat{\mathcal{H}} = \frac{\hat{P}^2_{X_{CM}}}{2\mu_{TP}} + \frac{\hat{p}^2_\xi}{2\mu_{12}} + U_{12}(\xi) + V_{T1}(X_{CM} - \alpha\,\xi) + V_{T2}(X_{CM} + \beta\,\xi) \quad (1)$$

where $\mu_{TP} = M(m_1 + m_2)/(M + m_1 + m_2)$, $\mu_{12} = m_1 m_2/(m_1 + m_2)$, $\alpha = m_2/(m_1 + m_2)$, $\beta = m_1/(m_1 + m_2)$. U_{12} represents the interaction between the fragments. V_{T1} and V_{T2} describe the interaction between target and the two projectile fragments, respectively, which depend on the relative distances $x_1 = X_{CM} - \alpha\,\xi$ and $x_2 = X_{CM} + \beta\,\xi$. In the calculations below, Woods-Saxon nuclear potentials are used for the binary systems, while for their Coulomb interaction the potential of a uniformly charged sphere is employed.

To describe fusion of the projectile fragments with the target, the Hamiltonian (1) is augmented with two strong imaginary potentials, $iW_{T1}(x_1)$ and $iW_{T2}(x_2)$, which operate in the interior of the individual Coulomb barriers between the target and the projectile fragments. This is usually employed in the coupled-channels model of fusion [1]. The imaginary potentials have a Woods-Saxon shape, centered at the minimum of the individual potential pockets.

Initial wave function. Since at the initial time the projectile is considered to be far apart from the target, V_{T1} and V_{T2} in Eq. (1) can be neglected and the Hamiltonian becomes separable. Consequently, the initial wave function can be factorized as

$$\Psi(\xi, X_{CM}, t = 0) = \Phi_0(X_{CM})\,\chi_0(\xi), \quad (2)$$

where χ_0 describes the ground state of the projectile, while $\Phi_0(X_{CM})$ is chosen to be a Gaussian wave-packet centered in X_0, with spatial width σ and average energy E_0. The ^{6}Li(= ^{4}He + ^{2}H) wave function is calculated using a nuclear Woods-Saxon potential between the alpha particle and the deuteron, which provides a 1s-state with a separation energy of -1.47 MeV. Please note that the X_{CM} and ξ coordinates are strongly coupled when the ^{6}Li and ^{209}Bi nuclei come together, so a product state (2) is only justified asymptotically.

Time propagation. The time evolution operator is represented as a convergent series of modified Chebyshev polynomials [12,13]. In the absence of the imaginary potentials, the norm of the wave function is preserved with an accuracy of $\sim 10^{-14}$.

Analysis. A limitation of most fusion models of weakly bound nuclei is the lack of an unambiguous calculation of σ_{CF} and σ_{ICF} [9]. We think that a key idea to overcome this problem is to examine the location of each fragment with respect to the position of the individual Coulomb barriers, irrespective of the internal excitation of the ^{6}Li projectile: if a fragment is located inside the radius of its Coulomb barrier it is considered captured by the target. Otherwise, it is not captured. We identify a CF event when both fragments are located inside their individual barrier radii, while an ICF event occurs when just one of the fragment is inside its barrier radius. This idea is realized by means of position projection operators, as explained in detail in Ref. [9].

The energy-resolved fusion cross section can be expressed in terms of the energy-resolved transmission coefficient through the Coulomb barrier:

$$\sigma(E) = \frac{\pi\hbar^2}{2\mu E}\mathcal{T}(E), \tag{3}$$

where each fusion process (TF, ICF and CF) has its own transmission coefficient $\mathcal{T}(E)$.

The energy-resolved transmission coefficients can be determined using the window operator method [14]. In the following discussion, Ψ refers to a wave function which has any asymptotic contribution at large X_{CM}, such as the initial and final *total* wave functions as well as the parts of the total wave function describing the ICF and the scattering process [9]. The main idea is, for instance, to calculate the energy spectrum, $\mathcal{P}(E_k)$, of the initial and final wave functions. E_k is the centroid of a *total energy bin* of width 2ϵ. A vector of reflection coefficients, $\mathcal{R}(E_k)$, is determined by the ratio [11,12]:

$$\mathcal{R}(E_k) = \frac{\mathcal{P}^{final}(E_k)}{\mathcal{P}^{initial}(E_k)}. \tag{4}$$

The transmission coefficients are:

$$\mathcal{T}(E_k) = 1 - \mathcal{R}(E_k). \tag{5}$$

The energy spectrum $\mathcal{P}(E_k) = \langle\Psi|\hat{\Delta}|\Psi\rangle$, where $\hat{\Delta}$ is the window operator [14]:

$$\hat{\Delta}(E_k, n, \epsilon) \equiv \frac{\epsilon^{2^n}}{(\hat{\mathcal{H}}_{asym} - E_k)^{2^n} + \epsilon^{2^n}}, \tag{6}$$

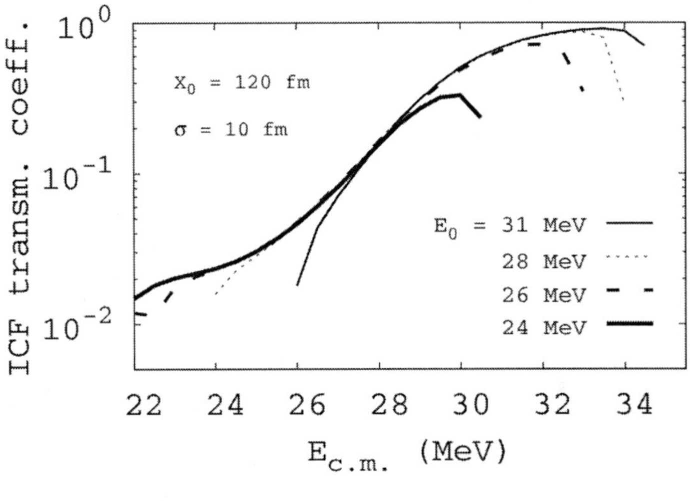

Figure 4: Energy-resolved ICF excitation function from the TDWP method. A converged curve can be constructed.

$\hat{\mathcal{H}}_{asym}$ is the asymptotic part of the system Hamiltonian $\hat{\mathcal{H}}$, and n determines the shape of the window function. As n is increased, this shape rapidly becomes rectangular with very little overlap between adjacent energy bins [14], the bin width remaining constant at 2ϵ. The spectrum is constructed for a set of E_k where $E_{k+1} = E_k + 2\epsilon$. Thus, scattering information over a range of incident energies can be extracted from a time-dependent wave function that has been calculated on a grid. In this work, $n = 2$ and $\epsilon = 0.25$ MeV. Solving two successive linear equations for the vector $|\chi\rangle$:

$$(\hat{\mathcal{H}}_{asym} - E_k + \sqrt{i}\,\epsilon)(\hat{\mathcal{H}}_{asym} - E_k - \sqrt{i}\,\epsilon)\,|\chi\rangle = |\Psi\rangle, \tag{7}$$

yields $\mathcal{P}(E_k) = \epsilon^4 \langle\chi|\chi\rangle$.

Calculations of energy-resolved transmission coefficients (*fusion probabilities*) have been performed using a Fourier grid in the (ξ, X_{CM}) coordinates [12]: $\xi = 0 - 30\,\text{fm}$ and $X_{CM} = 0 - 200\,\text{fm}$ with 128 and 256 evenly spaced points, respectively.

As an example, Fig. 4 shows the energy-resolved ICF probability and its dependence on the average energy E_0 of the initial wave-packet. We have checked [15] the convergence of the calculations regarding the X_0 and σ parameters of the initial wave packet for a broad range of values around those presented in Fig. 4. A few wave-packet propagations allows one to calculate the converged ICF excitation function in a broad range of incident energies. In contrast to the usual CDCC fusion calculations [17–19], this ICF excitation function includes ICF from the projectile bound state(s) [20].

3 Summary

The time dependent perspective is useful for understanding and quantifying low-energy reaction dynamics of exotic nuclei. It finds realization in the classical and quantum mechanical realms. PLATYPUS is a powerful tool for planning and interpreting (fusion and breakup) measurements, which allow its fine tuning. A quantum approach to low-energy reaction dynamics of weakly-bound few-body nuclei, based on the time-dependent wave-packet method, is being developed. It incorporates novel features into existing models based on this perspective, such as the separation of the complete and incomplete fusion from the total fusion. This type of approach is powerful because (i) it provides an intuitive description of the reaction dynamics, and (ii) all the continuum couplings are included. We have implemented this approach in a one-dimensional reaction model with two degrees of freedom. A method for an unambiguous calculation of complete and incomplete fusion cross sections is used, which includes incomplete fusion from bound states as well as complete fusion from unbound states of the weakly bound projectile. Model calculations show that converged reliable fusion excitation functions can be obtained with the time-dependent wave-packet method. This approach is being developed further, using a three-dimensional model. Additional cross sections, such as elastic, breakup and transfer cross sections, can also be calculated within this approach. Understanding the breakup mechanisms and their impact on unambiguously separated complete and incomplete fusion processes could make for further progress in the field. The video of this presentation can be seen in Ref. [16].

References

[1] Back B.B., Esbensen H., Jiang C.L. and Rehm K.E., *Rev. Mod. Phys.*, **86** (2014) 317.

[2] Canto L.F., Gomes P.R.S., Donangelo R., Lubian J. and Hussein M.S., *Phys. Rep.*, (2015) in press.

[3] Diaz-Torres A., Hinde D.J., Tostevin J.A., Dasgupta M. and Gasques L.R., *Phys. Rev. Lett.*, **98** (2007) 152701.

[4] Diaz-Torres A., *J. Phys. G*, **37** (2010) 075109.

[5] Diaz-Torres A., *Comp. Phys. Comm.*, **182** (2011) 1100.

[6] Ramin R. *et al.*, *Phys. Rev. C*, **81** (2010) 024601.

[7] Luong D.H. *et al.*, *Phys. Rev. C*, **88** (2013) 034609.

[8] Shrivastava A. *et al.*, *Phys. Lett. B*, **718** (2013) 931.

[9] Boselli M. and Diaz-Torres A., *J. Phys. G*, **41** (2014) 094001.

[10] Luong D.H. *et al.*, *Phys. Lett. B*, **695** (2011) 105.

[11] Yabana K., *Prog. Theor. Phys.*, **97** (1997) 437.

[12] Tannor D.J., in *Introduction to Quantum Mechanics: A Time-Dependent Perspective* (University Science Books, Saulito, California) 2007, pp. 273-331.

[13] Mandelshtam V.A. and Taylor H.S, *J. Chem. Phys.*, **103** (1995) 2903.

[14] Schafer K.J. and Kulander K.C., *Phys. Rev. A*, **42** (1990) 5794.

[15] Boselli M. and Diaz-Torres A., *in preparation.*

[16] https://www.youtube.com/watch?v=gLe-RxC8hZU.

[17] Hagino K., Vitturi A., Dasso C.H. and Lenzi S.M., *Phys. Rev. C*, **61** (2000) 037602.

[18] Diaz-Torres A. and Thompson I.J., *Phys. Rev. C*, **65** (2002) 024606.

[19] Diaz-Torres A., Thompson I.J. and Beck C., *Phys. Rev. C*, **68** (2003) 044607.

[20] Hashimoto S., Ogata K., Chiba S. and Yahiro M, *Prog. Theor. Phys.*, **122** (2009) 1291.

EPJ Web of Conferences **117**, 08003 (2016) DOI: 10.1051/epjconf/201611708003
NN2015

Present status of coupled-channels calculations for heavy-ion subbarrier fusion reactions

K. Hagino[1,2,3] and J.M. Yao[1,4]

[1] Department of Physics, Tohoku University, Sendai 980-8578, Japan
[2] Research Center for Electron Photon Science, Tohoku University, Sendai 982-0826, Japan
[3] National Astronomical Observatory of Japan, 2-21-1 Osawa, Mitaka, Tokyo 181-8588, Japan
[4] School of Physical Science and Technology, Southwest University, Chongqing 400715, China

Abstract

The coupled-channels method has been a standard tool in analyzing heavy-ion fusion reactions at energies around the Coulomb barrier. We investigate three simplifications usually adopted in the coupled-channels calculations. These are i) the exclusion of non-collective excitations, ii) the assumption of coordinate independent coupling strengths, and iii) the harmonic oscillator approximation for multi-phonon excitations. In connection to the last point, we propose a novel microscopic method based on the beyond-mean-field approach in order to take into account the anharmonic effects of collective vibrations.

1 Introduction

The field of heavy-ion subbarrier fusion reactions started in the late '70s, when a large enhancement of fusion cross sections was experimentally discovered with respect to the prediction of a simple potential model [1]. Even though the potential model works well for light systems, such as ^{14}N+^{12}C, it has turned out that it largely underestimates fusion cross sections for heavier systems, such as $^{16}O+^{154}Sm$, at energies below the Coulomb barrier. It has been well recognized by now that this large enhancement of subbarrier fusion cross sections is caused by the couplings of the relative motion between the colliding nuclei to several nuclear intrinsic degrees of freedom, such as low-lying collective excitations in the colliding nuclei as well as several nucleon transfer processes [1–4].

Naturally a standard framework for heavy-ion subbarrier fusion has thus been the coupled-channels method [5] by including relevant degrees of freedom. This method has not only successfully accounted for the subbarrier enhancement of fusion cross sections for many systems but has also provided a natural interpretation of the so called fusion barrier distributions [1, 2].

In the coupled-channels approach for heavy-ion fusion reactions, the following simplifications are usually employed. Firstly, the model space is restricted only to low-lying collective excitations, excluding non-collective excitations and giant resonances. Secondly, the coupling strength and the excitation energy for each state are taken to be the same as those in an isolated nucleus and are assumed to be unaltered during the whole process of fusion reaction. Thirdly, a simple harmonic oscillator or a rigid rotor is assumed when multiple excitations to higher collective states are involved. In this contribution, we shall investigate the validity of each of these assumptions.

2 Role of non-collective excitations

Let us start with the first assumption, that is, the role of non-collective excitations, which are usually not included in coupled-channels calculations.

Low-lying collective motions are strongly coupled to the ground state, and also have a strong mass number and atomic number dependences. They play a major role in heavy-ion subbarrier fusion reactions, and are explicitly taken into account in coupled-channels calculations. In addition to the low-lying collective excitations, there are many other modes of excitations in atomic nuclei. Among them, non-collective excitations couple only weakly to the ground state and usually they do not affect in a significant way heavy-ion fusion reactions, even though the number of non-collective states is large [6]. Couplings to giant resonances are relatively strong due to their collective character. However, since their excitation energies are relatively large and also are smooth functions of mass number, their effects can be effectively incorporated in a choice of internuclear potential through the

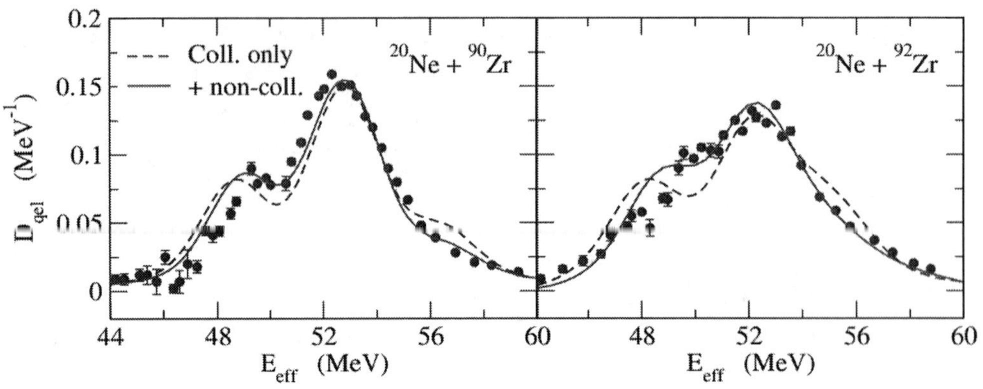

Figure 1: The quasi-elastic barrier distributions for the ^{20}Ne+^{90}Zr (the left panel) and the ^{20}Ne+^{92}Zr (the right panel) systems. Here, the quasi-elastic barrier distribution is defined as $D_{\rm qel} = -d[d\sigma_{\rm qel}/d\sigma_{\rm R}]/dE$, where $\sigma_{\rm qel}$ and $\sigma_{\rm R}$ are the quasi-elastic and the Rutherford cross sections, respectively. These barrier distributions are evaluated at the scattering angle of $\theta_{\rm lab} = 150^\circ$ and are plotted as a function of effective energy defined by $E_{\rm eff} = 2E\sin(\theta_{\rm c.m.}/2)/(1+\sin(\theta_{\rm c.m.}/2))$. The dashed lines show the results of the coupled-channels calculations with the collective excitations in the projectile and the target nuclei, while the solid lines take in addition the non-collective excitations in the target nuclei into account with a random matrix model. The experimental data are taken from Ref. [7].

adiabatic potential normalization [1].

Although in most of cases, the non-collective excitations do not play a significant role unless the system is not very heavy, there are certain cases in which they play a role. An example is the ^{20}Ne + ^{92}Zr system, for which the quasi-elastic scattering at backward angles has been measured experimentally [7]. Here, the quasi-elastic scattering refers to the sum of elastic, inelastic, and transfer cross sections, and is a counter part of fusion reactions [8]. Figure 1 shows a comparison between the quasi-elastic barrier distribution, defined as $D_{\rm qel} = -d[d\sigma_{\rm qel}/d\sigma_{\rm R}]/dE$, where $\sigma_{\rm qel}$ and $\sigma_{\rm R}$ are the quasi-elastic and the Rutherford cross sections, respectively, for the ^{20}Ne + ^{90}Zr system and that for the ^{20}Ne + ^{92}Zr systems [9]. One striking feature is that the experimental quasi-elastic barrier distribution for the ^{20}Ne + ^{92}Zr system is much more smeared than that for the ^{20}Ne + ^{90}Zr system [7]. The dashed lines in the figure show the results of the coupled-channels calculations that include the rotational excitations in ^{20}Ne as well as the collective phonon excitations in ^{90,92}Zr. This calculation reproduces the experimental data for the ^{20}Ne + ^{90}Zr system but not for the ^{20}Ne + ^{92}Zr system. The solid lines, on the other hand, take into account also the non-collective excitations in ^{90,92}Zr with a random matrix model [9]. One can see that the smearing of quasi-elastic barrier distribution for the ^{20}Ne + ^{92}Zr system is now well reproduced by the non-collective excitations of the ^{92}Zr nucleus, whose level density is much larger than that of ^{90}Zr due to the two extra neutrons outside the $N = 50$ shell closure.

3 Coordinate dependent coupling strength and deep subbarrier fusion hindrance

The coupled-channels approach expands the total wave function with the basis of isolated nuclei. Important inputs for coupled-channels calculations are, together with an internuclear potential, the excitation energy and the coupling strength for each excitations. Usually the experimental data are available for the excitation energy, and the coupling strength can be estimated from a measured electric transition probability [1]. These values are usually employed in coupled-channels calculations assuming that they are not altered during the reaction process. This assumption has been examined recently by Ichikawa and Matsuyanagi [10]. They have carried out random-phase approximation (RPA) calculations with a two-center shell model potential for the ^{16}O+^{16}O, ^{40}Ca+^{40}Ca, and ^{16}O+^{208}Pb systems, and have demonstrated that the coupling strengths are indeed constant at large distances but they decrease appreciably in the vicinity of the touching point.

This implies that the assumption of the constant coupling strength is reasonable for fusion reactions at energies around the Coulomb barrier. However, at deep subbarrier energies, the inner turning point is close to or even inner the touching point [11], and it is important to take into account the effect of variation of the coupling strength. Notice that these are the energies at which the deep subbarrier fusion hindrance phenomenon has been observed [4]. In fact, the finding of Ichikawa and Matsuyanagi provides a microscopic justification for the damping factor introduced phenomenologically in the adiabatic model for the deep subbarrier fusion hindrance phenomenon [12].

4 Semi-microscopic modeling of heavy-ion fusion reactions with a beyond-mean field method

In heavy-ion fusion reactions at energies around the Coulomb barrier, multiple excitations to higher members of collective states, such as multi-phonon states and high-spin states in the ground state rotational band, often play an

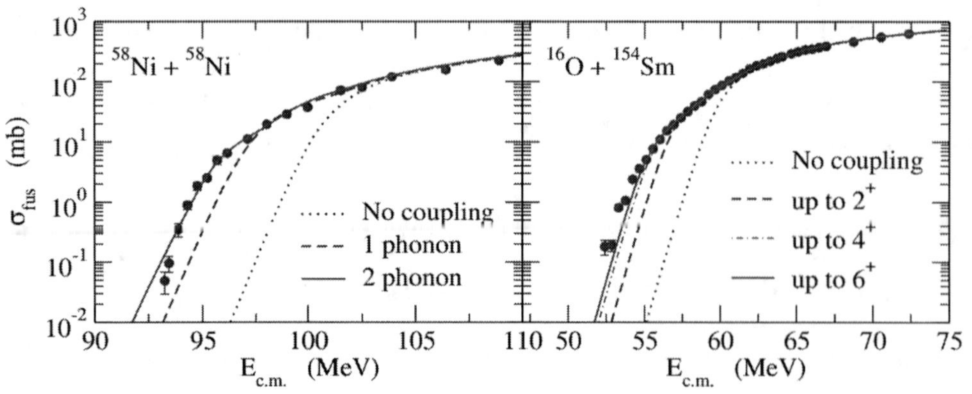

Figure 2: The results of coupled-channels calculations for the fusion cross sections for the ^{58}Ni+^{58}Ni (the left panel) and the ^{16}O+^{154}Sm (the right panel) systems. For the ^{58}Ni+^{58}Ni system, the vibrational coupling to multi-quadrupole-phonon states are considered in the harmonic oscillator approximation. On the other hand, for the ^{16}O+^{154}Sm system, the rotational coupling within the ground state rotational band is taken into account. The experimental data are taken from Refs. [13, 14].

important role. Figure 2 shows typical examples for this. The left and the right panels show fusion cross sections for the ^{58}Ni+^{58}Ni and the ^{16}O+^{154}Sm systems obtained with the coupled-channels calculations. For the former system, the vibrational coupling to quadrupole phonon states are considered, while the rotational coupling within the ground state rotational band is taken into account for the latter system. One can see that for both the systems the coupling to the first excited state is insufficient and the couplings to the higher members are necessary in order to account for the subbarrier enhancement of fusion cross sections. This feature has been demonstrated beautifully also through the analyses of fusion barrier distributions [2,14,15].

In order to take into account those multiple excitations in coupled-channels calculations, one usually uses the rigid rotor model for deformed nuclei and the harmonic oscillator model for vibrational nuclei [1]. In reality, however, most nuclei have neither a pure harmonic oscillator spectrum nor a pure rigid body rotational band, although the rigid rotor approximation is reasonable for medium-heavy and heavy deformed nuclei. For example, the ^{58}Ni nucleus, which has usually been considered to be a typical vibrational nucleus, does not exhibit a level spectrum characteristic to the harmonic vibration, *e.g.*, the degeneracy of the two-phonon triplet is considerably broken. Moreover, a recent theoretical calculation also indicates that the $B(E2)$ strengths among the collective levels in ^{58}Ni deviate largely from the harmonic oscillator limit [16].

DOI: 10.1051/epjconf/201611708003

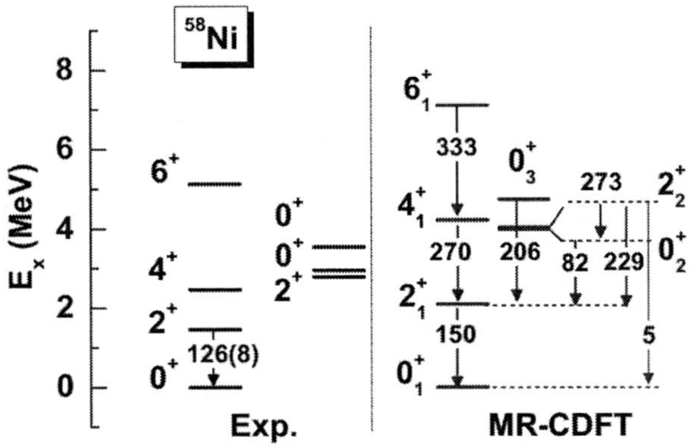

Figure 3: The low-lying energy spectra of ^{58}Ni obtained with the multi-reference covariant density functional theory (MR-CDFT) method with the PC-PK1 force. The arrows indicate the $E2$ transition strengths, given in units of $e^2\text{fm}^4$. The experimental data are taken from Refs. [21, 22].

There are several theoretical ways to describe an anharmonic vibration. Among them, a multi-reference density-functional theory (MR-DFT) has been rapidly developed for the past decade [17–19]. This method is based on the so called beyond-mean-field approximation, which incorporates the angular momentum and particle number projections as well as the quantum fluctuation of the mean-field wave function described by the generator coordinate method (GCM).

Figure 3 shows the result of the MR-DFT calculation for the ^{58}Ni nucleus. The calculation employs the covariant density functional theory (CDFT) with PC-PK1 interaction [23], and we thus call it MR-CDFT. One can see that the main feature of the energy spectrum, as well as the $E2$ transition strength from 2_1^+ to 0_1^+, are reproduced rather well. It is interesting to notice that the overall pattern of $B(E2)$ values is quite different from what would be expected for a harmonic vibrator, even though the excitation energies of the 4_1^+, 2_2^+, and 0_2^+ states are about twice the energy of the 2_1^+ state. In particular, the $E2$ transition from the 0_2^+ to the 2_1^+ states is much smaller than that from the 4_1^+ and the 2_2^+ states to the 2_1^+ state. Instead, the 0_2^+ state has a strong transition from the 2_2^+ state, which clearly indicates that the 0_2^+ state is not a member of the two-phonon triplet. Compared to the 0_2^+ state, the $E2$ transition strength from the 0_3^+ to the 2_1^+ states is much larger and is comparable to that from the 4_1^+ and the 2_2^+ states to the 2_1^+ state. This fact makes the 0_3^+ state a better candidate for a member of the two-phonon triplets, even though the excitation energy is a little bit large.

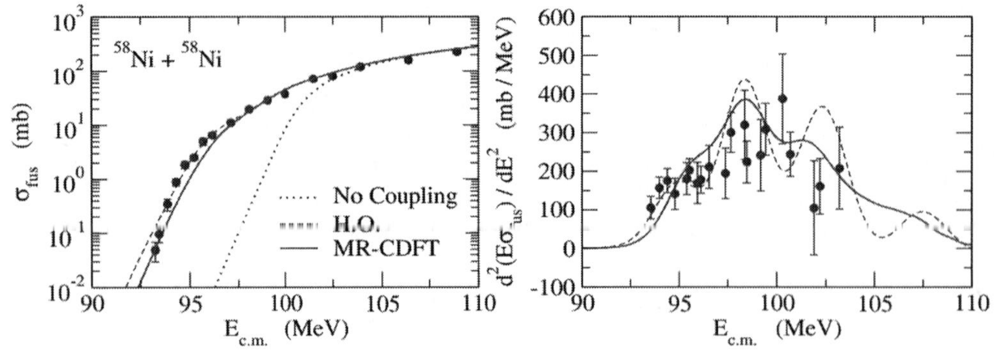

Figure 4: The fusion cross sections (the left panel) and the fusion barrier distributions (the right panel) for the ^{58}Ni+^{58}Ni system. The dashed line is the result of the coupled-channels calculations including the double quadrupole phonon excitations in each ^{58}Ni nucleus in the harmonic oscillator limit, while the solid line is obtained with the multi-reference covariant density functional theory (MR-CDFT) method by including the 0_1^+, 2_1^+, 0_3^+, 2_2^+, and 4_1^+ states. The dotted line in the left panel denotes the result in the absence of the channel couplings. The experimental data are taken from Ref. [13] for the fusion cross sections and from Ref. [15] for the fusion barrier distribution.

Notice that, in the harmonic oscillator limit, the $B(E2)$ value from any of the two-phonon triplet states to the 2_1^+ state is exactly twice the $B(E2)$ value from the 2_1^+ state to the ground state. The calculated $B(E2)$ values shown in Fig. 3, together with the strong transition from the 2_2^+ to the 0_2^+ states, indicate the presence of large anharmonicity in the quadrupole vibrations in ^{58}Ni. That is, the calculated $B(E2)$ values are significantly quenched from the values in the harmonic limit.

We can now ask ourselves how the deviation of the spectrum from the harmonic limit affects the subbarrier fusion reactions of Ni isotopes. Figure 4 shows the fusion cross section $\sigma_{\rm fus}(E)$ and the fusion barrier distribution $D_{\rm fus}(E) = d^2(E\sigma_{\rm fus})/dE^2$ [2, 24] for the ^{58}Ni+^{58}Ni reaction. The dashed line shows the result of the coupled-channels calculations including up to the double phonon states in the harmonic oscillator limit. All the mutual excitations between the projectile and the target nuclei are included. On the other hand, the solid line in the figure is obtained with the coupling strengths calculated with the MR-CDFT method while adopting a phenomenological Woods-Saxon potential. To this end, we include the 0_1^+, 2_1^+, 0_3^+, 2_2^+, and 4_1^+ states in the coupled-channels calculations. Again, all the mutual excitation channels are taken into account. See Ref. [20] for the details of the calculations. For a comparison, the figure also shows the result of no-coupling limit by the dotted line. One can see that the calculations in the harmonic limit overestimate fusion cross sections at the two lowest energies, while the MR-CDFT calculations underpredict fusion cross sections around 95 MeV. For the energy dependence of fusion cross sections, shown in terms of fusion barrier distribution in the right panel of the figure, the MR-CDFT calculation leads to a minor improvement by considerably smearing each peak.

We have carried out similar calculations also for the ^{40}Ca + ^{58}Ni system [20]. We have observed again also for this system that the anharmonicity effect in ^{58}Ni smears the fusion barrier distribution, leading to a better agreement with the experimental fusion barrier distribution as compared to the results in the harmonic oscillator limit.

5 Summary

The coupled-channels approach has been a standard tool in the field of heavy-ion subbarrier fusion reactions. We have discussed three assumptions used in usual coupled-channels calculations.

We have first discussed the role of non-collective excitations. Usually, coupled-channels calculations take into account a few selected low-lying collective states, neglecting the couplings to non-collective states. We have demonstrated that even though the non-collective excitations can indeed be neglected in many cases, there are certain systems which show those effects. One example is ^{20}Ne + ^{92}Zr system, which shows a considerably smeared barrier distribution as compared to the ^{20}Ne + ^{90}Zr system. By explicitly including many non-collective excitations, we have demonstrated that they indeed smear the barrier distribution for ^{20}Ne + ^{92}Zr system while the effect is much smaller for the ^{20}Ne + ^{90}Zr system.

We have next discussed two important input quantities in coupled-channels calculations, that is, the coupling strength and the excitation energy for each state. In most of coupled-channels calculations, these quantities are assumed to be unchanged during the reaction process and are taken to be constants. The recent RPA calculations have indicated that this is the case at large distances but they decreases considerably around the touching point. We have argued that this leads to important consequences for the deep subbarrier hindrance of fusion cross sections.

We have then proposed a semi-microscopic approach to heavy-ion subbarrier fusion reactions. The basic idea of this approach is to combine a multi-reference density functional theory (MR-DFT) to a coupled-channels calculation. The MR-DFT provides transition strengths among collective states without resorting to the harmonic oscillator model or the rigid rotor model. The advantages of this approach include i) deviations from the harmonic limit as well as the rigid rotor limit can be taken into account, ii) it can therefore be applied also to transitional nuclei, which show neither the vibrational nor the rotational characters, and iii) a natural truncation is introduced in the coupling schemes. We have applied this approach to the ^{58}Ni+^{58}Ni and ^{40}Ca+^{58}Ni fusion reactions, and have found that the anharmonicities smear the fusion barrier distributions, somewhat improving the agreement with the experimental data.

One of the important current issues in nuclear reaction theory is to develop a microscopic framework starting from the nucleon degree of freedom. It has however been extremely challenging to construct a fully microscopic theory which is applicable to heavy-ion subbarrier fusion reactions, thus to many-particle quantum tunneling. We believe that the semi-microscopic approach presented in this paper provides an important step towards this direction.

Acknowledgements

We thank N. Rowley, T. Ichikawa, and S. Yusa for useful discussions. This work was partially supported by the National Natural Science Foundation of China under Grant Nos. 11305134, 11105111, and the Fundamental Research Funds for the Central University (XDJK2013C028).

References

[1] Hagino K. and Takigawa N., Prog. Theor. Phys. **128** (2012) 1001.

[2] Dasgupta M., Hinde D.J., Rowley N. and Stefanini A.M., Annu. Rev. Nucl. Part. Sci. **48** (1998) 401.

[3] Balantekin A.B. and Takigawa N., Rev. Mod. Phys. **70** (1998) 77.

[4] Back B.B., Esbensen H., Jiang C.L. and Rehm K.E., Rev. Mod. Phys. **86** (2014) 317.

[5] Hagino K., Rowley N. and Kruppa A.T., Comp. Phys. Comm. **123** (1999) 143.

[6] Yusa S., Hagino K. and Rowley N., Phys. Rev. C **88** (2013) 044620.

[7] Piasecki E. *et al.*, Phys. Rev. C**80** (2009) 054613.

[8] Hagino K. and Rowley N., Phys. Rev. C**69** (2004) 054610.

[9] Yusa S., Hagino K. and Rowley N., Phys. Rev. C **88** (2013) 054621.

[10] Ichikawa T. and Matsuyanagi K., Phys. Rev. C**88** (2013) 011602 (R); arXiv: 1506.07963.

[11] Ichikawa T., Hagino K. and Iwamoto A., Phys. Rev. C**75** (2007) 057603; Phys. Rev. C**75** (2007) 064612.

[12] Ichikawa T., Hagino K. and Iwamoto A., Phys. Rev. Lett. **103** (2009) 202701.

[13] Beckerman M. *et al.*, Phys. Rev. C**23** (1981) 1581.

[14] Leigh J.R. *et al.*, Phys. Rev. C**52** (1995) 3151.

[15] Stefanini A.M. *et al.*, Phys. Rev. Lett. **74** (1995) 864.

[16] Yao J.M., Bender M. and Heenen P.-H., Phys. Rev. C**91** (2015) 024301.

[17] Bender M., Heenen P.-H. and Reinhard P.-G., Rev. Mod. Phys. **75** (2003) 121.

[18] Yao J.M., Hagino K., Li Z.P., Meng J. and Ring P., Phys. Rev. C**89** (2014) 054306.

[19] Vretenar D., the contribution in this conference.

[20] Hagno K. and Yao J.M., Phys. Rev. C**91** (2015) 064606.

[21] National Nuclear Data Center, http://www.nndc.bnl.gov/.

[22] Allmond J.M. *et al.*, Phys. Rev. C**90** (2014) 034309.

[23] Zhao P.W., Li Z.P., Yao J.M. and Meng J., Phys. Rev. C**82** (2010) 054319.

[24] Rowley N., Satchler G.R. and Stelson P.H., Phys. Lett. **B254** (1991) 25.

EPJ Web of Conferences **117**, 08004 (2016) DOI: 10.1051/epjconf/201611708004
NN2015

Sub- and near-barrier fusion reactions experimental results

G. Montagnoli

Dipartimento di Fisica e Astronomia, Università di Padova, Padova, Italy
and INFN, Sezione di Padova, Padova, Italy

Abstract

Early data of sub-barrier fusion teached us that cross sections may strongly depend on the structure of colliding nuclei and on couplings to transfer channels. The influence of transfer is clearly indicated in the excitation functions of different nickel isotopes and various Ca+Zr systems. Fusion barrier distributions often yield the fingerprint of the relevant inelastic and transfer couplings. At lower energies, far below the barrier the slope of the excitation function keeps increasing in many cases, so that the cross sections are strongly over-predicted by standard coupled-channels (CC) calculations; this was named a hindrance effect.

Furthermore, light heavy-ion systems show cross section oscillations above the Coulomb barrier. Recent experiments have been performed on the fusion of ^{28,30}Si+^{28,30}Si systems where all phenomena cited above show up. In particular regular oscillations that have been revealed above the barrier for ^{28}Si+^{28}Si and have been interpreted as the consequence of the strong channel couplings and/or the oblate deformation of ^{28}Si.

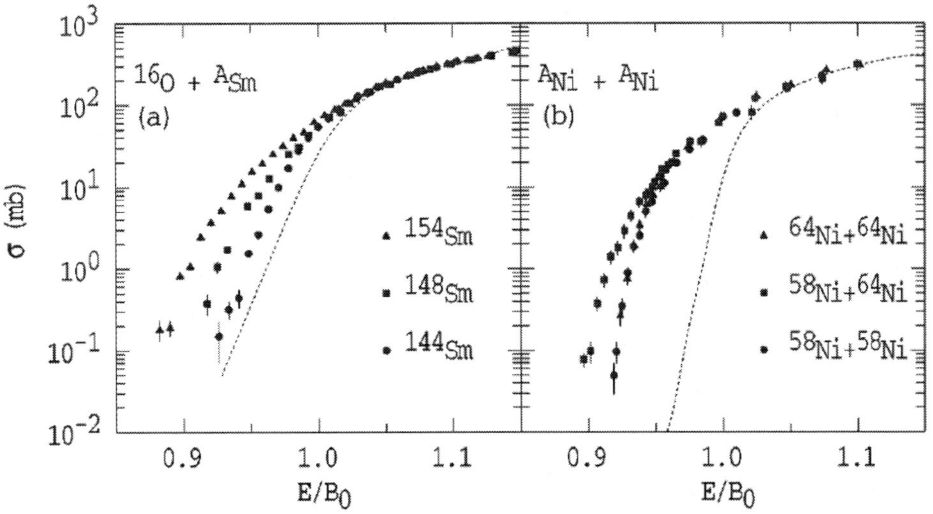

Figure 1: (a) Fusion excitation functions for ^{16}O on Sm isotopes [3] show a marked increase of cross section with increasing mass number and deformation. (b) The difference in the energy dependence of the cross sections [4] for the ^{58}Ni + ^{58}Ni, ^{58}Ni + ^{64}Ni, and ^{64}Ni + ^{64}Ni reactions indicates that transfer of nucleons affects the fusion process. The dashed lines show calculated cross sections within the one dimensional model.

1 The various features of near- and sub-barrier heavy-ion fusion

A wide range of interesting features have been shown by the studies of heavy-ion fusion reactions [1,2]. Indeed it was already observed in the experiments concerning ^{16}O on various Sm isotopes [3] and Ni+Ni systems [4] that there is a strong interplay of dynamics and nuclear structure near and below the barrier as shown in Fig. 1.

The effect of coupling to transfer channels was revealed in several cases, and one of the most significant evidences was observed in the trend of the Ca + Zr systems which show very different behaviours as reported in Fig. 2 (left panel). Here the fusion excitation functions have been plotted in a appropriate reduced energy scale to take into account the different Coulomb barriers and the high energy octupole vibrations of ^{40}Ca and ^{48}Ca [5,6]. In this way, we highlight the effect of the neutron pick-up transfer channels that have positive Q-values only for ^{40}Ca + ^{94,96}Zr.

The concept of barrier distribution introduced in Ref. [7] was often successfully exploited to identify the nature of the coupled channels. In the

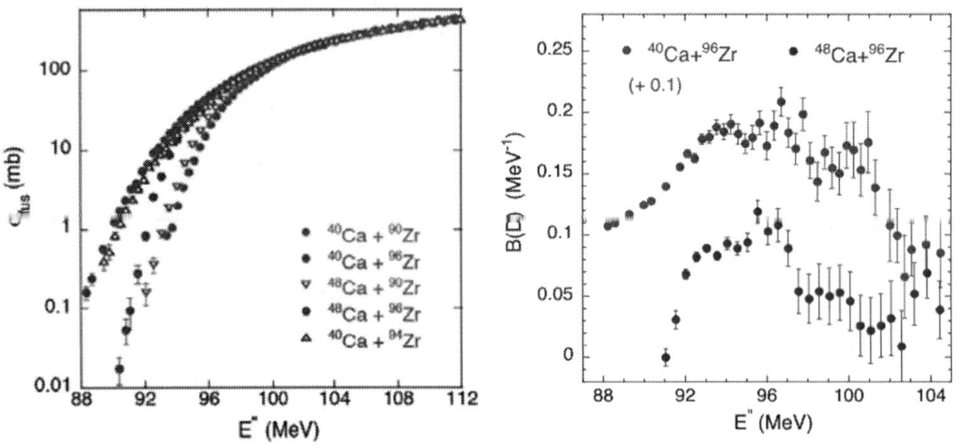

Figure 2: Left: Excitation functions for several Ca+Zr systems. Right: Barrier distributions for the two cases^{40,48}Ca+^{96}Zr.

examples reported in Fig. 2 (right panel) one can see that the effect of transfer channels is to produce a wider distribution with a tail towards low energies.

2 Hindrance vs. enhancement

In the energy region far below the Coulomb barrier fusion cross sections result from a balance between the enhancement produced by couplings which are dominant at energies close to the barrier, and the more recently observed hindrance phenomenon [8]. The left panel of Fig. 3 shows the excitation functions of the two systems ^{48}Ti + ^{58}Fe [9] and ^{58}Ni + ^{54}Fe [10]. One notices the very large enhancement with respect to the no coupling limit and a fast change of slope for ^{58}Ni + ^{54}Fe indicating the appearance of hindrance already at relatively large cross sections (energies), because these two nuclei are rather stiff. ^{48}Ti and ^{58}Fe are softer and no indication of hindrance is seen down to $\sim$1 μb.

A convenient way to show the trend of the low-energy excitation function is to plot the logarithmic derivative of the energy-weighted cross section L(E) (right panel of Fig. 3). Indeed the onset of hindrance is conventionally located at the energy where L(E) reaches the L_{CS} value which is expected for a constant astrophysical S factor [11]. The S-factor, originally introduced in nuclear astrophysics as a convenient representation for light-ion fusion reactions, shows a maximum at that energy. This is observed for ^{58}Ni + ^{54}Fe, but not for ^{48}Ti + ^{58}Fe.

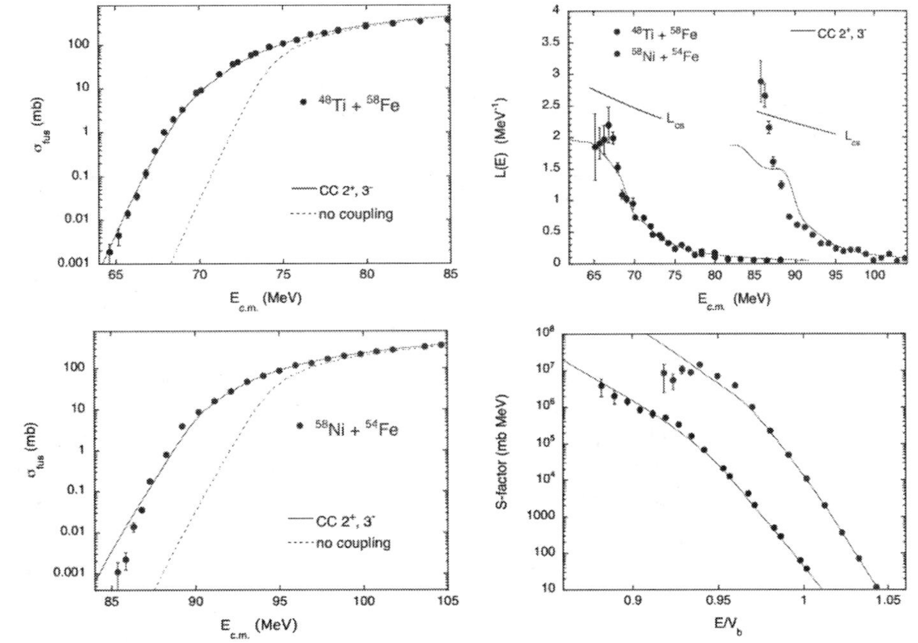

Figure 3: Excitation function of ^{48}Ti + ^{58}Fe and ^{58}Ni + ^{54}Fe (left panels), logarithmic derivatives and S-factors (right panels). Standard CC calculations do not reproduce the low energy ^{58}Ni + ^{54}Fe data indicating the presence of hindrance.

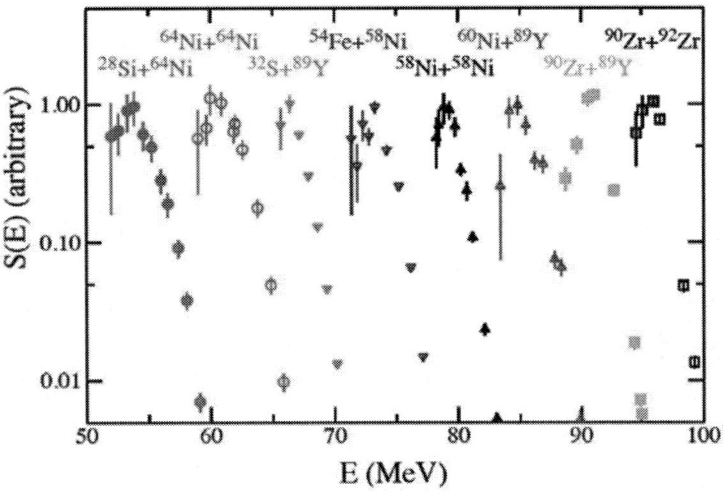

Figure 4: $S(E)$ for the systems whose cross sections at low energies are dominated by fusion evaporation and where maxima have been experimentally observed [12].

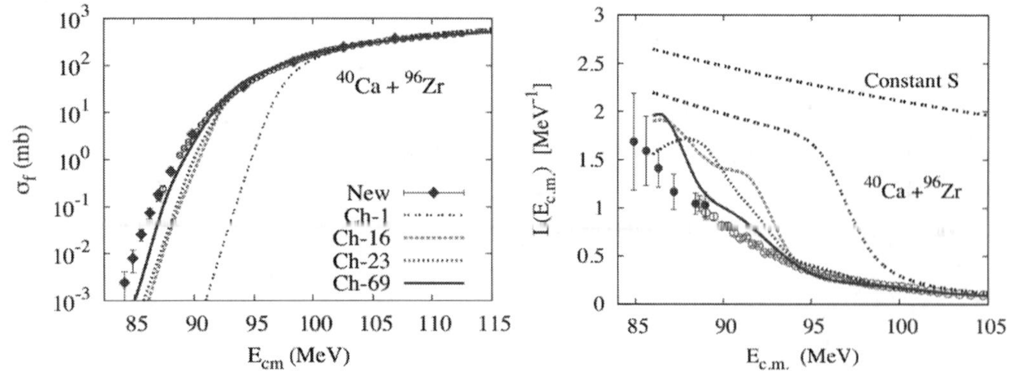

Figure 5: (Colour online) Left: Excitation function of ^{40}Ca+^{96}Zr in a logarithmic scale. The red and black symbols are the cross sections measured in Refs. [5, 13], respectively. The lines are the results of the CC calculations described in the text. Right: Slope L(E) of the fusion excitation function compared to the CC results.

Clear S-factor maxima have been detected in various cases (see Fig. 4 [12]). For other systems the hindrance is no so strong to produce a S-factor maximum and it is necessary to refer to a calculation in order to recognise that effect.

3 Coupling to transfer channels

Let us consider the special case of ^{40}Ca +^{96}Zr [5, 13] where many neutron transfer channels with positive Q-values are available. Detailed CC calculations were performed [14] but even using a standard Woods-Saxon potential the fusion data at low energies are strongly underestimated and no indication of hindrance appears yet. The low-lying 2^+ and 3^- states (Ch-28 in Fig. 5) produce strong effects. Couplings to Q>0 one- and two-nucleon transfer channels (Ch-84) bring further significant enhancements, even at the level of a few μb, Ch-1 is the no-coupling limit.

This is confirmed when we plot the slope L(E) (right panel of Fig. 5), which remains very low with respect to the value expected for a constant S factor. In lighter cases as the various Ca+Ca systems, analogous systematic trends have been observed. The effect of transfer is strong and clear for ^{40}Ca+^{48}Ca [15, 16], whose cross sections exceed the ^{48}Ca+^{48}Ca data at low energies and are suppressed compared to the ^{40}Ca+^{40}Ca data at high energies. In spite of the large enhancement in the sub-barrier fusion of ^{40}Ca+^{48}Ca, a hindrance does eventually occur but the onset is pushed down to very low energies, where L(E) increases rapidly.

 DOI: 10.1051/epjconf/201611708004

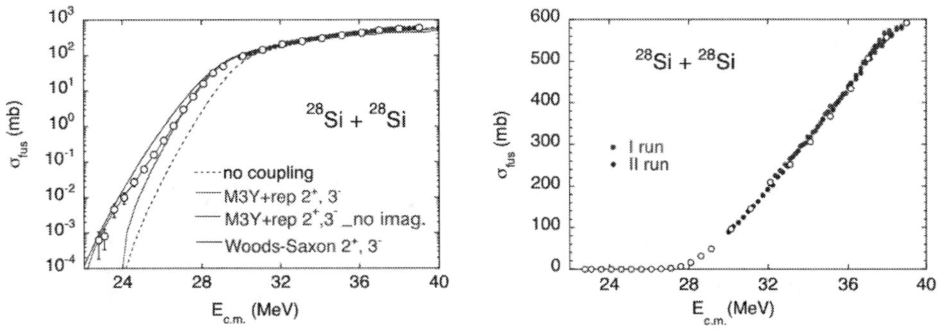

Figure 6: Fusion excitation function of ^{28}Si $+^{28}$Si in logarithmic (left) and linear (right) scales [22, 23]. The results of the CC calculations discussed in the text are also shown in the left panel.

4 The case of the ^{28,30}Si + ^{28,30}Si systems

The excitation functions of various light heavy-ion systems indicate that oscillatory structures appear above the barrier, probably due to the penetration of successive centrifugal barriers [17]. A good example for this is the case of ^{16}O $+^{16}$O [18,19]. A shallow potential is needed to fit the data above (as well as below) the barrier [20]. More recently TDHF calculations have also been performed to reproduce those oscillations [21].

It is very interesting to investigate heavier systems, where the sub-barrier hindrance phenomenon is stronger and better established. A convincing observation of such oscillations may provide useful information on the ion-ion potential in a wide energy range. Therefore we recently measured the fusion excitation function of ^{28}Si $+^{28}$Si from below to above the barrier with very small energy steps in the higher energy region, please see Fig. 6 [22,23]. In the left panel of this figure the results of CC calculations are shown, including the low-lying 2^+ and 3^-states, and their mutual excitations. Using a standard Woods-Saxon potential, the calculation overpredicts the data just below the barrier but the fit improves at the lowest energies.

On the other hand by employing the M3Y shallow potential (see the green line) the data are clearly underestimated below the barrier. To reproduce the experimental data a weak and short-ranged imaginary potential (W_0=5 MeV, a=0.2 fm) is required at low energies, that possibly simulates the effect of the strong oblate deformation of ^{28}Si.

Using the first derivative of the excitation function $d(E\sigma)/dE$ makes it easier to observe the oscillations [20]. The first derivative obtained from the detailed cross section measurement above the barrier shows distinct

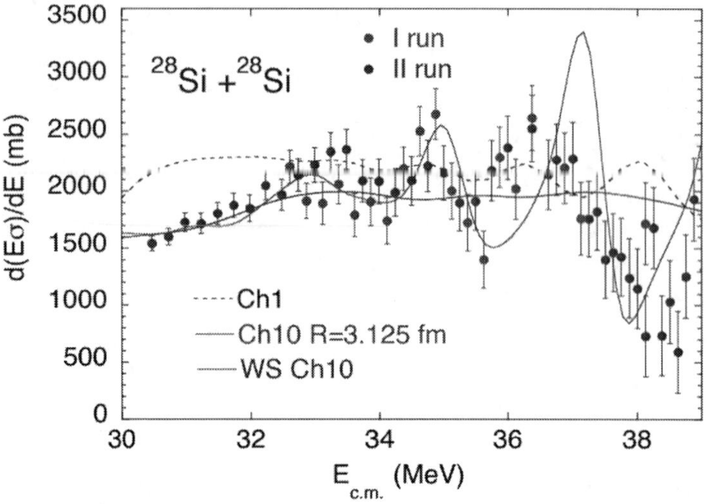

Figure 7: The derivative of the energy-weighted fusion excitation function $d(E\sigma)/dE$ of ^{28}Si $+^{28}$Si in an energy range above the barrier. The derivative is obtained as the incremental ratio between successive points, with an energy step $\Delta E_{cm} = 0.75$ MeV. The lines are the results of the theoretical calculations discussed in the text.

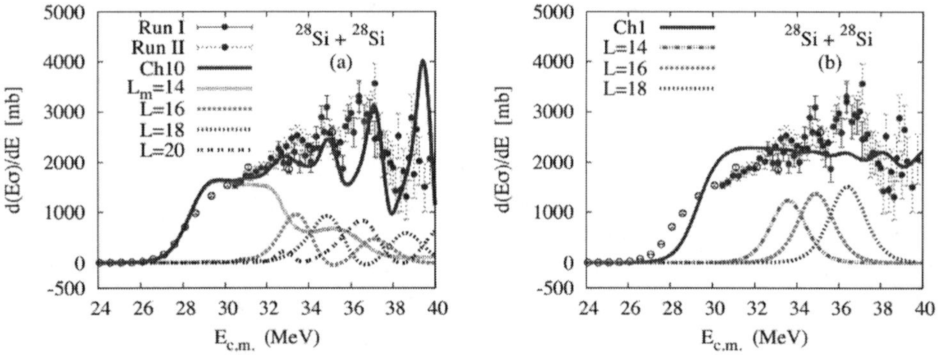

Figure 8: (Colour online) (a) The first derivative of the excitation function is compared to the results of Ch10 calculations. The red (magenta) curve is the result using a maximum angular momentum for fusion $L_m = 38$ (L_m=14). The individual contributions for L = 16-20 are also reported. (b) The results of Ch1 calculations are shown. The experimental peaks at 33, 35, and 37 MeV are associated with L = 14, 16, and 18, while they are primarily related to L = 16, 18, and 20 in Ch10 calculations shown in panel (a).

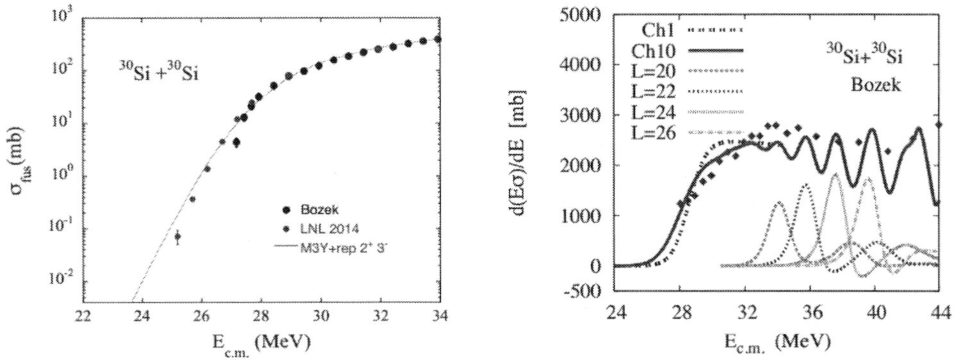

Figure 9: (Colour online) Left: Excitation function of ^{30}Si+^{30}Si as measured by Bozek et al. [24], with the few additional (preliminary) points recently measured at LNL. The red line is the early CC calculation cited in the text. Right: The derivative obtained from the data of Bozek et al. (blue points) is shown together with the CC results (red line) predicting well defined oscillations.

oscillations although the uncertainties are rather large, as reported in Fig. 7. This behaviour has been confirmed in two different series of measurements (I run and II run in Fig. 7). In that figure the comparison with the CC calculations using M3Y+repulsion potential, have been shown. The Ch1 is the no-coupling limit and Ch10 is the result when the 2^+ and 3^-states are considered. One can notice that only by including the channel couplings, a reasonable fit of the data is obtained.

The calculation has also been performed employing the Woods-Saxon potential but the result is not equally satisfying (WS Ch10 in the Fig. 7) because very weak oscillations show up, providing a poor fit to the data.

Although the channel couplings are essential to reproduce the data, they split the contribution of a single L (see Fig. 8(a)) so that the correspondence of the experimental peaks to the penetration of specific centrifugal barriers is lost. Fig. 8(b) reports the results of no-coupling calculations (Ch1) where the oscillations are smoothed out because the contributions to the single L-values (reflecting the individual centrifugal barriers) produce wider peaks.

We have undertaken very recently the study of ^{30}Si+^{30}Si because this nucleus has a spherical shape. Therefore it can be a good reference point to better understand the role of the oblate deformation of ^{28}Si. The Fig. 9 (left) shows the excitation function of ^{30}Si+^{30}Si near and below the barrier as measured by Bozek et al. [24] and, very recently, at LNL. The result of preliminary CC calculations is also shown, but the data have to be completed with more detailed measurements.

An experiment aiming at the measurement of oscillations in this system will be performed at LNL very soon, where oscillations have been predicted (see Fig. 9 (right)). The coupling to the low-lying states is weaker than in the case of ^{28}Si+ ^{28}Si, so that the oscillatory structures, if confirmed, could be associated to the individual centrifugal barriers as for lighter systems. The derivative extracted from the existing data [24] is reported in the figure but the very large errors (not shown here) do not allow any conclusion.

5 Summary

Sub-barrier fusion cross sections are strongly enhanced by couplings of the relative motion to nuclear shape deformations and vibrations [1, 2]. Analogous effects due to nucleon transfer degrees of freedom have been searched for, following the early experiments for the various Ni+Ni systems [4], but the role of nucleon transfer has been unambiguously identified only in few other cases. The discovery of the hindrance phenomenon [8] at very low energies, has made the situation more interesting but even more complex. Indeed, fusion cross sections are determined by the concurring contributions of hindrance and enhancement in that energy range where transfer couplings are expected to be still effective.

The appearance of oscillations and the trend of sub-barrier cross sections in ^{28}Si+^{28}Si have been reproduced within the same theoretical frame, i.e., the CC model using the shallow M3Y+repulsion potential. Within that model the existence of oscillations is tightly bound to channel couplings, while in lighter cases (e.g. ^{16}O+^{16}O) the oscillations are related to the overcoming of successive centrifugal barriers well spaced in energy. As a consequence, the one-to-one relation between each peak and the height of a centrifugal barrier is lost in ^{28}Si+^{28}Si. The nature of these oscillations and the behaviour of the sub-barrier excitation function might be clarified by analogous measurements for ^{30}Si+^{30}Si because ^{30}Si is a spherical nucleus while ^{28}Si has a permanent oblate deformation. Such experiments on ^{30}Si+^{30}Si are in progress at LNL.

My grateful thanks go to all my colleagues who made possible to perform the research and to obtain the results reported in this paper.

References

[1] M.Dasgupta, D.J.Hinde, N.Rowley and A.M.Stefanini, Annu. Rev. Nucl. Part. Sci. **48**, 401 (1998).

[2] B.B.Back, H.Esbensen, C.L.Jiang and K.E.Rehm, Rev. Mod. Phys. **86**, 317 (2014).

[3] W.Reisdorf et al., Nucl. Phys. A **438**, 212 (1985).

[4] M.Beckerman et al., Phys. Rev. Lett. **45**, 1472 (1980).

[5] A.M.Stefanini et al., Phys. Lett. B **728**, 639 (2014).

[6] A.M.Stefanini et al., Phys. Rev. C **76**, 014610 (2007).

[7] N.Rowley, G.R.Satchler and P.H.Stelson, Phys. Lett. B **254**, 25 (1991).

[8] C. L. Jiang *et al.*, Phys. Rev.Lett. **89**, 052701 (2002).

[9] A.M.Stefanini et al., EPJ Web of Conf. **86**, 00057 (2015).

[10] A.M.Stefanini et al., Phys. Rev. C **81**, 037601 (2010).

[11] C.L.Jiang et al., Phys. Rev. Lett. **93**, 012701 (2004).

[12] C.L.Jiang, private communication.

[13] H.Timmers et al. Nucl. Phys. A **633**, 421 (1998).

[14] H.Esbensen and A.M. Stefanini, Phys. Rev. C **89**,044616 (2014).

[15] G. Montagnoli et al., Phys. Rev. C **85**, 024607 (2012).

[16] C. L. Jiang *et al.*, Phys. Rev. C **82**, 041601(R) (2010).

[17] N.Poffé, N.Rowley and R. Lindsay, Nucl. Phys. A **410**, 498 (1983).

[18] I.Tserruya et al., Phys. Rev. C **18**, 1688 (1978).

[19] J.Thomas et al., Phys. Rev. C **33**, 1679 (1986).

[20] H.Esbensen, Phys. Rev. C **85**, 064611 (2012).

[21] C.Simenel et al., Phys. Rev. C **88**, 024617 (2013).

[22] G.Montagnoli et al, Phys. Rev. C **90**, 044608 (2014).

[23] G.Montagnoli et al., Phys. Lett. B **746**, 300 (2015).

[24] E. Bozek et al., Nucl. Phys. A **451**, 171 (1986).

EPJ Web of Conferences **117**, 08005 (2016) DOI: 10.1051/epjconf/201611708005
NN2015

Breakup locations: Intertwining effects of nuclear structure and reaction dynamics

M. Dasgupta*, E. C. Simpson, D. H. Luong, Sunil Kalkal, K. J. Cook, I. P. Carter, D. J. Hinde, and E. Williams

Department of Nuclear Physics, Research School of Physics and Engineering, Australian National University, ACT 2601, Australia

Abstract

Studies at the Australian National University aim to distinguish breakup of the projectile like-nucleus that occurs when approaching the target from that when receding from the target. Helped by breakup simulations, observables have been found that are sensitive to the breakup location, and thus to the mean-lives of unbound states; sensitivity to even sub-zeptosecond lifetime is found. These results provide insights to understand the reaction dynamics of weakly bound nuclei at near barrier energies.

Understanding the interactions of weakly bound nuclei, and their reaction outcomes, is a key challenge in nuclear reactions research [1]. For well-bound nuclei, collisions near the barrier are reasonably well-described by including couplings to quantum states of the approaching nuclei. For weakly bound nuclei, the reaction dynamics has added complexity due the presence of low-lying particle unbound states. The situation is made even more challenging by the fact that weakly bound nuclei can breakup not only following direct excitation of continuum states (direct breakup), but also following nucleon transfer. The latter occurs when the neighbouring nuclei, formed by transfer, are themselves weakly bound or unbound. Modelling this is a major theoretical challenge, and currently there is no quantum

*E-mail: mahananda.dasgupta@anu.edu.au

model that describes all breakup modes and their effect on other reaction channels simultaneously.

One of the first unambiguous experimental demonstrations of the effects of weak binding on other reaction channels came from using the concept of fusion barrier distributions [2]. Using the experimentally determined average barrier energy, it was shown that above-barrier complete fusion cross-sections of the weakly-bound nucleus ^{9}Be + ^{208}Pb is suppressed [2] compared with the expectations of well bound nuclei. Complete fusion suppression has since been seen in many reactions with weakly bound stable nuclei [3–9]. However, studies on medium-mass and light target nuclei are sparse as experimentally it is difficult to separate complete fusion (experimentally defined as capture of the full charge of the projectile) from incomplete fusion (only part of the projectile charge is captured). In the case of radioactive nuclei too, the majority of experimental observations agree that suppression is found above the barrier [10].

It was natural to think that the above-barrier suppression was due to breakup of weakly bound nuclei reducing the flux of intact nuclei at the barrier radius. Indeed, the relationship [8] between observed suppression, quantified by $1 - F_{CF}$ with F_{CF} being the fraction of complete fusion suppression, and the energy required for direct breakup, shown in Fig. 1, seemed to support this idea. However, many works (*e.g.* [11–14]) have indicated that the picture is not so simple, as transfer-triggered breakup, which can be important, is neglected. In ^{6}Li, ^{7}Li and ^{9}Be induced reactions for example, these channels are highly probable. This is demonstrated in Fig. 1(b), which shows coincident charged particles that are detected following collisions of ^{7}Li with ^{208}Pb. Only events following breakup close to the target are shown due to their relevance to fusion, as will be discussed later. It is clear from Fig. 1(b) that apart from $\alpha - t$ pairs from direct breakup of ^{7}Li, $\alpha - d$, and $\alpha - p$ and $\alpha - \alpha$ coincidences, arising from breakup following stripping of n, $2n$ and pickup of p respectively, are significant contributors [13, 14]. Therefore, in this manuscript, the term breakup will include both direct and transfer-triggered modes.

In the absence of a quantum model, a classical approach [15] was developed to relate measured probabilities for various breakup modes at below barrier energies to above barrier fusion suppression. The classical model requires knowledge of probability of breakup in the incoming trajectory as a function of centre-to-centre separation of the two colliding nuclei. This information has been obtained by making measurements at a range of below barrier energies, where fragment absorption is minimal and thus interpretation of breakup is unambiguous [13, 14, 16]. This gives an exponential

EPJ Web of Conferences **117**, 08005 (2016) DOI: 10.1051/epjconf/201611708005
NN2015

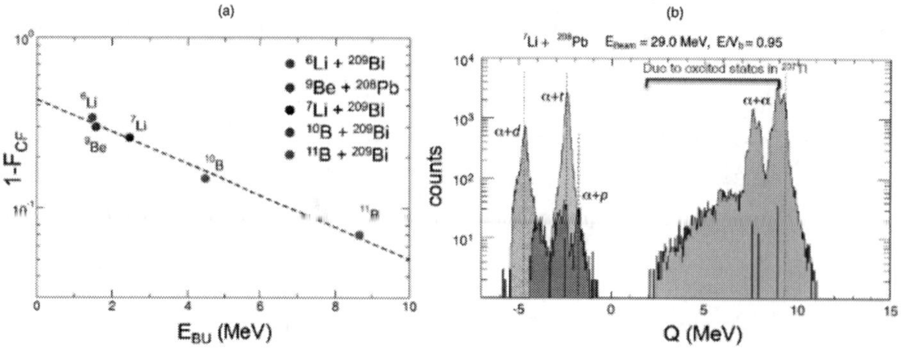

Figure 1: (Color online) (a) The experimentally determined suppression of complete fusion [2,5,8] as a function of the energy required for direct breakup of the projectile for the reactions indicated. The quantity F_{CF} is the observed fraction of complete fusion. Whilst this strong correlation has led to the notion that complete fusion suppression is linked to direct breakup, it masks the fact that a large proportion of breakup occurs via nucleon-transfer as shown in panel (b) for near-target breakup events (see text) that are observed following interactions of ^{7}Li with ^{208}Pb [13, 14].

fall-off with radial separation, and allows extrapolation to above-barrier energies. The investigations of breakup that followed focussed attention on the position at which breakup occurs due to its importance to fusion suppression. Complete fusion is expected to be suppressed if breakup of the projectile-like nucleus occurs whilst approaching the target, before reaching the fusion barrier radius. However, complete fusion will hardly be affected if the projectile-like nucleus passes inside the fusion barrier before disintegrating. This paper discusses recent work at the Australian National University aimed at distinguishing these two scenarios.

1 Identifying breakup modes

The experiments were performed using pulsed beams from the Heavy Ion Accelerator Facility at the Australian National University. Recent studies have focussed on a range of targets to study the dependence of the breakup modes on target mass (A) and charge (Z). The charged breakup fragments were detected in coincidence using an array of four double sided Silicon strip detectors with a large coverage in both scattering angle θ and azimuthal angle ϕ. The signals in intersecting arcs and sectors provide information on (θ, ϕ) of the detected particles. This information for each of the coincident fragments, along with their measured energy allows kinematic reconstruction of events. From the measured energies, angles and known beam energy, the

Q-value for each event can be calculated. The Q-value spectrum [14] for events that arise from near-target breakup in collisions of ^{7}Li with ^{208}Pb is shown in Fig. 1(b). It is clear that in addition to direct breakup into α and t clusters, breakup via other modes is significant [14]. The transfer-induced breakup modes arise as the nuclei that are populated following nucleon transfer are themselves weakly bound or unstable *e.g.*, ^{8}Be formed following proton pickup breaks up into $\alpha - \alpha$, and ^{5}Li formed following $2n$ stripping breaks into $\alpha - p$. These experiments, and those that followed using lighter targets, have all shown the dominance of transfer triggered breakup. The component of direct breakup is negligible for interactions with Ni and lower Z targets. However, as discussed in the introduction, knowing the breakup modes alone is not sufficient to understand fusion suppression. A knowledge of the breakup location is crucially important.

2 Relative energies of fragments as an indicator of breakup locations

The relative energy of two coincident breakup fragments has been used in recent works [13, 14, 16] to get information on the location of breakup. The concept is discussed here using the example of ^{8}Be, which is populated in reactions involving ^{7}Li following p pickup or in ^{9}Be induced reactions following n-stripping. The ground state of ^{8}Be is unstable against α-decay by 92 keV and has a mean-life of 10^{-16} s. The mean-life is very long compared with the typical reaction time scale of 10^{-21} s. The first excited state of ^{8}Be is much shorter lived, with a mean-life close to the reaction time scale. If ^{8}Be is populated in its ground state then, at the sub-barrier energies at which the measurements are made, it will recede far from the target before it breaks into two α-particles. The relative energy of the two breakup α-particles is therefore expected to be equal to 92 keV. If however ^{8}Be is populated in its excited state, its breakup can occur soon after its formation, while it is still in close proximity to the target. The Coulomb interaction between the target and the fragments will alter their relative velocity and thus the relative energy. The experimentally measured relative energy spectra [14], shown in Fig. 2, follows this expectation: at low energy there is a narrow peak which arises from breakup of the ^{8}Be ground state. The higher relative energy events arise due to breakup from the second resonant state of ^{8}Be and higher excitations. This realization means that a distinction between breakup near the target and far from the target (asymptotic) can be made experimentally. It is also clear that for fusion suppression, our focus must be

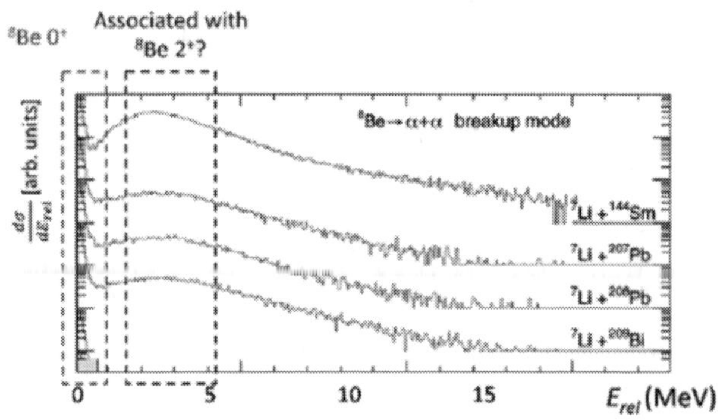

Figure 2: (Color online) Relative energy spectra for the two coincident α-particles from the breakup of ^{8}Be formed via p-pickup by ^{7}Li incident on various targets. The narrow peak at relative energy of 92 keV is due to breakup of ^{8}Be ground state. The higher relative energy components arise due to breakup of excited states of ^{8}Be (see text).

on near target breakup processes (called prompt breakup in Refs. [13,14,16]).

3 New observables for better sensitivity to breakup locations

Having identified near-target breakup, an even finer distinction must be made which is of relevance to fusion suppression – whether breakup of the projectile-like nucleus occurs on the incoming trajectory or when receding from the target. In previous studies [13, 14, 16], it has been assumed that of events with high relative energy, half arise from breakup on the incoming trajectory and half from the outgoing trajectory. This assumption is valid provided breakup (direct or following transfer) is instantaneous, since the probability of both direct excitation and transfer is expected to be symmetric around the turning point. However, if the unbound state of the nucleus (formed following direct excitation or nucleon transfer) has a lifetime, then it may continue on its trajectory before it breaks-up. Depending on the mean-life, the unbound state populated on the incoming trajectory may remain intact until it begins to recede from the target. It is therefore necessary [17] to have a more nuanced experimental observable than the fragment relative energy described in the previous section.

In the case of break-up close to the target, the fragments will be scattered to large angles due to their interaction with the target, resulting in a large angle θ_{12} between the two coincident fragments as shown in the inset of Fig.

3 (a). On the other hand, if breakup occurs on the receding trajectory, far from the target, then the opening angle will on average be smaller. How different the opening angles will be in the two situations will depend on the target Z and on the excitation energy of the nucleus that breaks-up. With heavy targets the strong Coulomb interaction will force the fragments to backward angles, which is likely to wash-out the distinction. Thus the distinction in opening angles for the incoming and outgoing breakup may be expected to be more visible for lighter targets. The distinction is clearly observed [17] in the reaction of ^{7}Li with ^{58}Ni by plotting the measured θ_{12} against the angle β between the direction of the unbroken ^{8}Be and the deduced breakup axis, as shown in Fig. 3 (a). The data lie mainly in two distinct bands – one shows small θ_{12} for all values of β with a tight correlation; the other has much larger θ_{12} with a larger spread for a given β. That such bands emerge from the fairly featureless (except for the peak at 92 keV) E_{rel} spectrum shown in Fig. 2 is very interesting, and point to a common physical origin for events in each band. The superimposed lines in Fig. 3(a) show the expectations for asymptotic breakup of the ground state of ^{8}Be (dashed line) and 2^+ state of ^{8}Be (full line); as expected, there is a one-to-one correlation between β and θ_{12}, as effects of the Coulomb field of the target are negligible for asymptotic breakup. We see that the breakup of the ground state of ^{8}Be (that can be unambiguously identified from the E_{rel} value of 92 keV) follows the expectations beautifully. Clearly this state breaks up far from the target, consistent with its mean-life of 10^{-16} s.

The other events, which include breakup from the 2^+ state of ^{8}Be with mean-life 0.4 zs, are not fully consistent with expectations of breakup far from the target (full line). As discussed earlier, large values of θ_{12} (strongly correlated with E_{rel}) are indicative of breakup close to the target. Classical calculations that include the Coulomb effects due to proximity of the fragments to the target were done using the code PLATYPUS [15] to quantify the deviations from the asymptotic expectations. In order to illustrate the sensitivity of θ_{12} to the location of breakup, calculations were performed by varying the location of breakup with respect to the distance of closest approach. Representative calculations for a fixed excitation energy of ^{8}Be and fixed angular momentum are shown in Fig. 3(b) by lines that are superimposed on the experimental data. Each line corresponds to breakup at the indicated distance Δr from the turning point on the outgoing trajectory. Variation in excitation energy and angular momentum will lead to some smearing around the lines shown. Calculations for breakup on the incoming trajectory (negative Δr) produce even larger θ_{12}, which the current experimental arrangement was not sensitive to. However, the data and

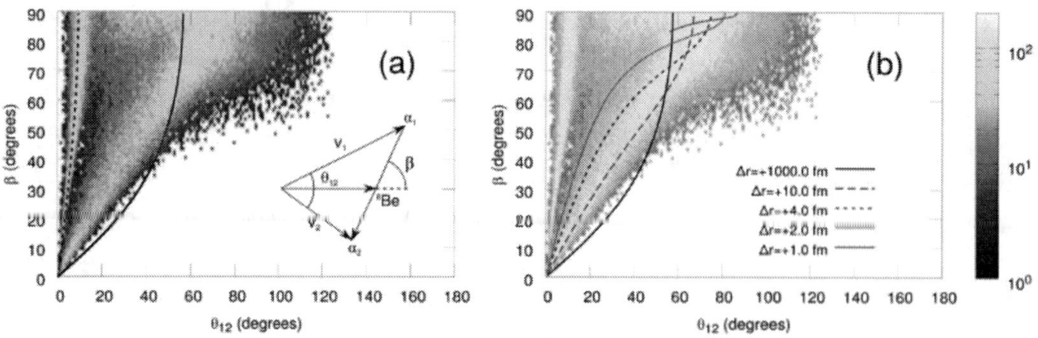

Figure 3: (Color online) The angle β between the unbroken ^{8}Be and one of the breakup fragments is plotted against the angle between the two coincident fragments θ_{12}. The angles β and θ_{12} are illustrated in the inset in panel (a). The data for breakup of ^{8}Be formed in collisions of ^{7}Li with ^{58}Ni are shown in in both panels. Results of classical calculations are shown by the lines. Calculations in panel (a) show the expectations for asymptotic breakup of the ground state of ^{8}Be (dashed line) and the 2^+ state of ^{8}Be (full line). Calculations presented in panel (b) show the effect of changing the breakup position with respect to the target; a Δr value of 1000 fm corresponds to asymptotic breakup, and decreasing values of Δr represent reducing distance from the turning point.

the simulations do illustrate the fact that the θ_{12} vs. β plots are sensitive to breakup location [17], and thus can, in principle, be used to make a distinction between breakup on the incoming and outgoing trajectories.

4 Summary

Recent studies to understand the breakup of weakly bound light nuclei and its relationship to suppression of complete fusion have led to many new insights. It has been shown that direct breakup is not the only mode, and that breakup following nucleon transfer is extremely significant if transfer populates nuclei that are weakly bound. A more nuanced understanding of breakup locations can be obtained from measurements with detectors that cover a wide angular range. Comparison with classical simulations show that the mean-lives of the unbound states will play a role in determining the location of breakup. Experimental variables have been identified that allow us to probe this aspect and sensitivity to even sub-zeptosecond mean-life can be expected. These results provide insights into breakup dynamics near the barrier, and should help advance models to understand breakup and its effect on fusion.

Acknowledgments

The authors acknowledge operations support for the ANU Heavy Ion Accelerator Facility from NCRIS. Financial support from ARC grants DP130101569, DE140100784, and FL110100098 is acknowledged.

References

[1] L.F. Canto *et al.*, *Phys. Rep.* **596** (2015) 1, and references therein.

[2] M. Dasgupta *et al.*, *Phys. Rev. Lett* **82** (1999) 1395.

[3] V. Tripathi *et al.*, *Phys. Rev. Lett.* **88** (2002) 172701.

[4] Y.W. Wu *et al.*, *Phys. Rev. C* **68** (2003) 044605.

[5] M. Dasgupta *et al.*, *Phys. Rev. C* **70** (2004) 024606.

[6] A. Mukherjee *et al.*, *Phys. Lett. B* **636** (2006) 91.

[7] P.K. Rath *et al.*, *Phys. Rev. C* **79** (2009) 051601.

[8] L.R. Gasques *et al.*, *Phys. Rev. C* **79** (2009) 034605.

[9] M. Dasgupta *et al.*, *Phys. Rev. C* **81** (2010) 024608.

[10] N. Keeley *et al.*, *Prog. Part. Nucl. Phys.* **59** (2007) 579.

[11] K.O. Pfeiffer *et al.*, *Nucl. Phys. A* **206** (1973) 545.

[12] A. Shrivastava *et al.*, *Phys. Lett. B* **633** (2006) 463.

[13] D.H. Luong *et al.*, *Phys. Lett. B* **695** (2011) 105.

[14] D.H. Luong *et al.*, *Phys. Rev. C* **88** (2013) 034609.

[15] A. Diaz-Torres *et al.*, *Phys. Rev. Lett.* **98** (2007) 152701.

[16] R. Rafiei *et al.*, *Phys. Rev. C* **81** (2010) 024601.

[17] E.C. Simpson *et al.*, submitted to *Phys. Rev. C* (2015).

EPJ Web of Conferences **117**, 08006 (2016) DOI: 10.1051/epjconf/201611708006
NN2015

Systematic study of quasifission characteristics and timescales in heavy element formation reactions

D.J. Hinde[1], E. Williams, G. Mohanto[2], C. Simenel, M. Dasgupta, A. Wakhle[3], I.P. Carter, K.J. Cook, D.Y. Jeung, D.H. Luong, C.S. Palshetkar[4], E. Prasad, D.C. Rafferty, R. du Rietz[5] and E.C. Simpson

Department of Nuclear Physics, Research School of Physics and Engineering, Australian National University, ACT 2601, Australia

Abstract

Superheavy elements can only be created in the laboratory by the fusion of two massive nuclei. Mass-angle distributions give the most direct information on the characteristics and time scales of quasifission, the major competitor to fusion in these reactions. The systematics of 42 mass-angle distributions provide information on the global characteristics of quasifission. Deviations from the systematics reveal the major role played by the nuclear structure of the two colliding nuclei in determining the reaction outcome, and in hindering or favouring heavy element production.

To form very heavy and superheavy elements (SHE), heavy-ion fusion reactions are used. Their cross sections can be significantly suppressed [1] by quasifission [2]. This dynamical non-equilibrium process results when the

[1]email: david.hinde@anu.edu.au
[2]Current address: TIFR, Mumbai, India
[3]Current address: National Superconducting Cyclotron Laboratory, Michigan State University, East Lansing, Michigan 48824, USA
[4]Current address: TIFR, Mumbai, India
[5]Current address: Malmö University, Faculty of Technology and Society, 205 06 Malmö, Sweden

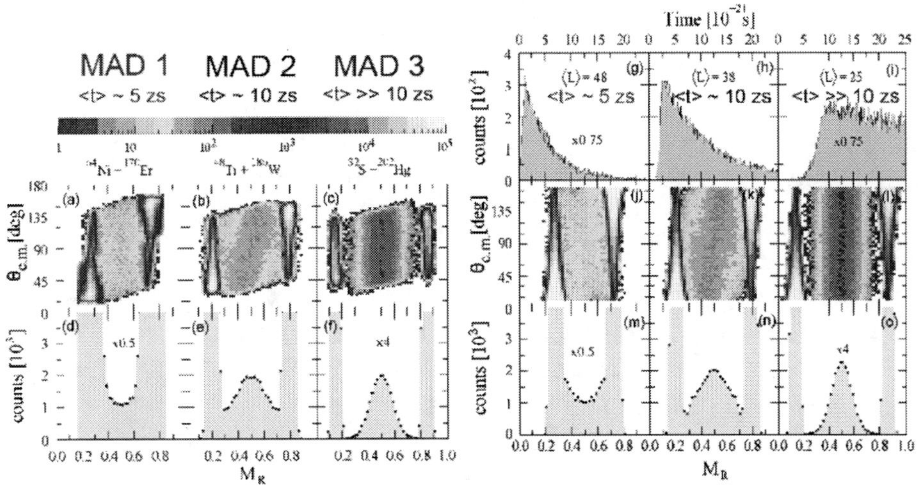

Figure 1: (Color online) Panels (a)-(c) show the experimental MADs, with (d)-(f) showing the projections onto M_R, for the reactions ^{64}Ni+^{170}Er, ^{48}Ti+^{186}W and ^{32}S+^{202}Hg, which all form ^{234}Cm. The multiplicative factors shown scale the y-axis. The simulated MADs for same reactions and beam energies are shown in panels (g),(h),(i), with the corresponding M_R spectra in panels (m),(n),(o), and the sticking time distributions used for the simulations in (j),(k),(l). The assigned MAD category for each reaction, and associated approximate mean sticking times are also shown (adapted from Ref. [3]).

combined system formed after capture separates into two (fission-like) fragments in times $\sim 10^{-20}$s, before a compact compound nucleus is formed. The probability of quasifission (P_{QF}) can be very large, with the corresponding probability of compound nucleus formation (P_{CN} = 1 - P_{QF}) being lower than 10^{-3} in unfavourable cases. Understanding the competition between quasifission and fusion is thus very important in predicting the best fusion reactions to use to form new isotopes of heavy and super-heavy elements.

A key quantity characterizing quasifission is its "sticking time" between capture of the two nuclei inside the entrance channel potential barrier [4] and breakup (scission). Quasifission mass-angle distributions (MAD) first measured at GSI in the 1980s [2,5] showed that quasifission timescales could often be shorter than the rotation time of $\sim 10^{-20}$s. However, subsequently only a few measurements [6,7] were made until recent years, when an extensive series of experiments (using the Australian National University Heavy Ion Accelerator Facility and CUBE spectrometer) were carried out [3,8–16]. The kinematic coincidence technique used in the measurements [2,3,17] provides direct information on the mass-ratio of the fragments at scission; thus, the data are represented in terms of mass ratio M_R, rather than pre- or post-evaporation masses of the fragments themselves.

Examples of MAD and deduced quasifission sticking time distributions are shown in Fig.1, for reactions forming the compound nucleus ^{234}Cm [3]. According to the characteristics of the MAD (minimum mass yield at symmetry, mass-angle correlation with peak yield at symmetry, and no significant mass angle correlation), they are assigned as type MAD1, MAD2 and MAD3 respectively [3]. There is a clear correlation between the MAD class and the entrance channel charge product. Other entrance channel characteristics are important in determining the sticking times and MAD characteristics, including neutron richness [14, 18], and shell structure including static deformation [11] and magic numbers [14].

To improve our quantitative understanding of the role of shell structure in the dynamics of quasifission, we make an analogy with the liquid drop model approach to nuclear masses, in which localized shell effects can be quantified when the underlying smooth (liquid drop) trends are well defined. To define the smooth trends in quasifission, a large number of MAD measurements have been selected, for beam energies somewhat above the capture barrier (typically by $\sim 6\%$). Here the known effects of deformation alignment [10, 11, 17, 19] and shell structure observed in measurements at below-barrier energies [14, 20] are much reduced [11, 21, 22]. However the beam energies should not be too far above the capture barriers, otherwise high angular momenta would be introduced in the collisions, which would then not be representative of heavy element formation reactions.

Now we map the reaction outcomes (defined in terms of the MAD classes discussed above) against variables that reflect the balance between nuclear and Coulomb forces during the collision, expected to determine the dynamics of reactions forming very heavy elements, as discussed below.

Ref. [23] proposed that there should be scaling behavior between reactions, on the basis of the schematic "chaotic regime dynamics" model of fusion of heavy nuclei. This is the same model [24] that predicted the "extra push" and "extra-extra-push" kinetic energies needed to overcome the conditional and unconditional saddle-point energies respectively. We have ordered the experimental MAD outcomes as a function of the two scaling parameters expected to most strongly define the reaction outcome: (i) the effective fissility parameter of the entrance channel, x_{eff}, and (ii) the fissility parameter of the compound nucleus, x_{CN}. The former is most relevant to necked shapes close to the (generally) mass-asymmetric contact configuration in the entrance channel, whilst the latter applies to shapes without a constricted neck [24], and where the mass-asymmetry degree of freedom is not constrained.

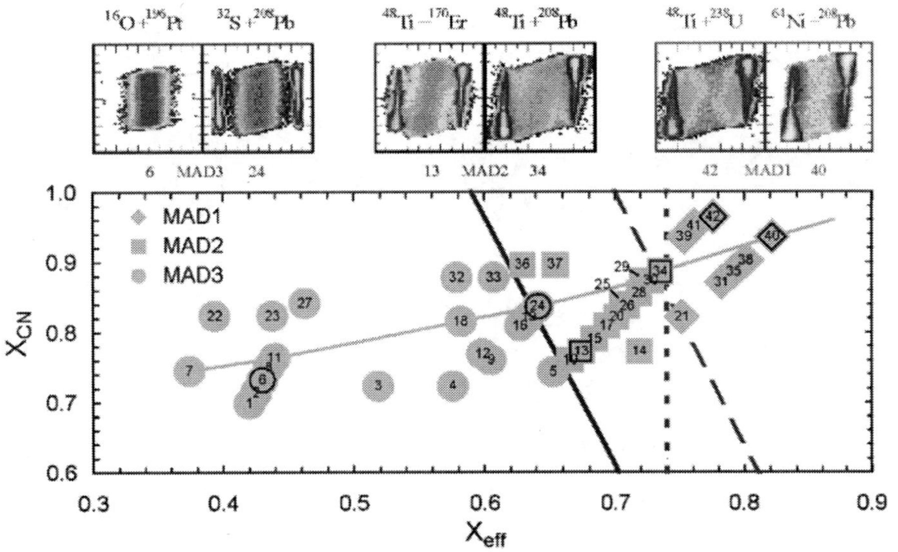

Figure 2: (Color online) The numbers (inside a symbol indicating the classification of MAD observed) refer to the reaction number in Ref. [3], and are plotted as a function of the effective fissility parameter in the entrance channel x_{eff} and the compound nucleus fissility parameter x_{CN} (see text). The diagonal full line represents the empirical boundary between reactions with no mass-angle correlation (left) and those that have (right). The dashed and dotted lines indicate the uncertainty in the boundary of reactions which no longer exhibit a peak at symmetry in the angle-integrated fission mass distribution. The thin line represents the locus of reactions with ^{208}Pb. Examples of MAD are shown in the panels above, identified by their reaction number.

Fig. 2 shows the experimental MAD class as a function of these two variables. As can be seen, the MAD classes are clustered into groups. The boundary across which a mass-angle correlation becomes significant (between MAD classes 3 and 2) shows a dependence on both variables, with a stronger dependence on the entrance-channel than on the compound nucleus fissility. The full blue line is our estimate of this boundary based on the current data. Mass-angle distributions for reactions on this line should show similar mass-angle correlations, associated with similar reaction trajectories and timescales. The same should be true for reactions on nearby parallel lines. The equation of this boundary line is $0.75x_{eff} + 0.25x_{CN} = 0.68$, giving x_{eff} three times the weight of x_{CN}. In terms of the MAD, this defines a mean fissility parameter x_m which seems to give a good characterization of the quasifission observed, at the higher beam and excitation energies chosen here. The thin full line almost orthogonal to the others represents the locus of reactions with the doubly-magic ^{208}Pb. At the above-barrier energies

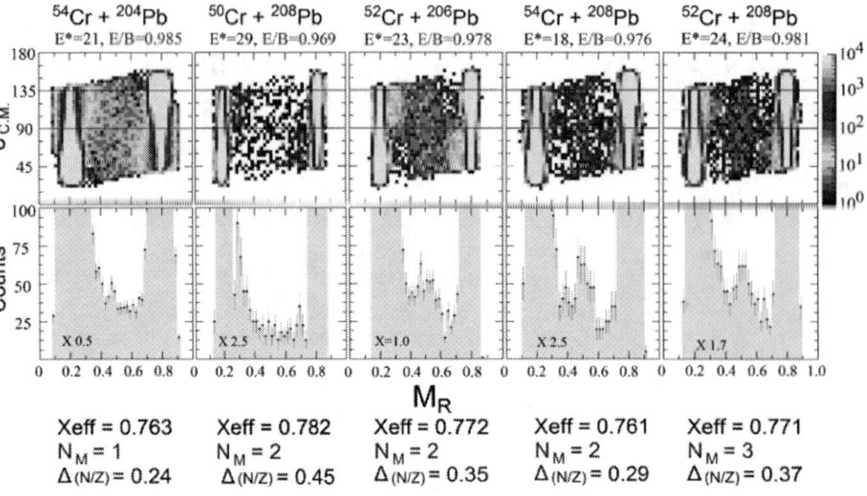

Figure 3: (Color online) MAD and projected mass-ratio distributions for backward angles from 90° to 135° (as indicated in the MAD plots), for reactions of ^{50,52,54}Cr isotopes with ^{204,206,208}Pb. Sub-barrier beam energies (denoted by E/B), resulted in the low excitation energies E*. The entrance channel fissility x_{eff}, the number of entrance-channel magic numbers, and the mismatch in N/Z values are indicated for each reaction (see text). The reaction outcome changes from a minimum in yield at mass-symmetry (left) to a narrow peak at symmetry, having no evidence of a mass-angle correlation.

chosen, no significant departure from systematic behaviour is seen.

The story is very different at sub-barrier energies. Here it is well-known that collisions with the tips of deformed nuclei are those that lead to capture [4], and that fission angular distributions [17], mass distributions [10,13,25] and MAD [11,16] point to the changing nature (shorter sticking time) of quasifission under these circumstances. Microscopic TDHF calculations of quasifission masses and angles give a good match [16] to experimental results in reaction ^{40}Ca+^{238}U, across the transition from sub-barrier to above-barrier energies (at which all collision orientations contribute). In sub-barrier reactions with closed-shell nuclei, it has been shown [14] that increasing the "magicity" in the entrance channel results in narrower mass distributions, correlated with a reduced mass-angle correlation, indicative of longer sticking times. Indeed, for the ^{48}Ca+^{208}Pb reaction, measured mass distributions [20] appear consistent with a fusion-fission mechanism. These results led to the conclusion that "magicity" plays its strongest role when the N/Z values of the projectile and target nuclei are matched. When this is the case, transfer reactions that destroy entrance-channel magicity (as in the

^{40}Ca+^{208}Pb reaction [14]) are minimized, preserving the closed shell nature of the collision partners as long as possible during the merging of the two nuclei [14].

Very recent ANU results, for the reaction of isotopes of Cr with Pb, are shown in Fig. 3. These measurements were at sub-barrier energies, resulting in the low excitation energies above the ground-state as indicated. They support the picture from the Ca+Pb data. The left three reactions all form the compound nucleus ^{258}Sg. The difference between the N/Z values of the target and projectile nuclei is denoted by $\Delta_{N/Z}$ in the figure. The panels are ordered from left to right first by the number of magic numbers in the entrance channel (N_M), and then by $\Delta_{N/Z}$. The left-most reaction has only a single magic number in the entrance channel, and shows a U-shaped mass distribution, consistent with MAD class 1. With two magic numbers, the reactions with $\Delta_{N/Z}$ closer to zero show a peak at mass-symmetry, associated with an angle-independent ridge in the MAD. With three magic numbers, but less favourable $\Delta_{N/Z}$, similar result is seen. This mass-symmetric peak becomes a smaller and smaller fraction of the total fission yield for $0.3<M_R<0.7$, as the beam energy is increased. These data show a very similar behaviour to the ^{40,44,48}Ca+^{204,208}Pb reactions; however, the transition from a U-shaped mass distribution to a narrow mass-symmetric mass distribution is in some ways an even more drastic change in reaction outcome. The very sudden change in outcome, depending on neutron number and magicity, will be a severe challenge for models of quasifission to reproduce. And yet it is this level of sensitivity that models must strive to reproduce, to optimise experimental opportunities to create new isotopes of superheavy elements.

Acknowledgments

The authors acknowledge operations support for the ANU Heavy Ion Accelerator Facility from NCRIS, and support from Dr. N. Lobanov and Dr. T. Kibedi and the ANU Heavy Ion Accelerator Facility staff in operating the Linac. Financial support from ARC grants DP130101569, DP140101337, FL110100098, FT120100760 and DE140100784 is acknowledged.

References

[1] D.J. Hinde, M. Dasgupta, A. Mukherjee, Phys. Rev. Lett. **89**, 282701 (2002)

[2] J. Tõke, B. Bock, G.X. Dai, A. Gobbi, S. Gralla, K.D. Hildenbrand, J. Kuzminski, W. Müller, A. Olmi, H. Stelzer, Nucl. Phys. A **440**, 327 (1985)

[3] R. du Rietz, E. Williams, D.J. Hinde, M. Dasgupta, M. Evers, C.J. Lin, D.H. Luong, C. Simenel, A. Wakhle, Phys. Rev. C **88**, 054618 (2013)

[4] M. Dasgupta, D.J. Hinde, N. Rowley, A.M. Stefanini, Annu. Rev. Nucl. Part. Sci. **48**, 401 (1998)

[5] W.Q. Shen, J. Albinski, A. Gobbi, S. Gralla, K.D. Hildenbrand, N. Herrmann, J. Kuzminski, W.F.J. Müller, H. Stelzer, J. Tõke et al., Phys. Rev. C **36**, 115 (1987)

[6] J. Velkovska, C.R. Morton, R.L. McGrath, P. Chung, I. Diószegi, Phys. Rev. C **59**, 1506 (1999)

[7] B.B. Back, P.B. Fernandez, B.G. Glagola, D. Henderson, S. Kaufman, J.G. Keller, S.J. Sanders, F. Videbæk, T.F. Wang, B.D. Wilkins, Phys. Rev. C **53**, 1734 (1996)

[8] R.G. Thomas, D.J. Hinde, D. Duniec, F. Zenke, M. Dasgupta, M.L. Brown, M. Evers, L.R. Gasques, M.D. Rodriguez, A. Diaz-Torres, Phys. Rev. C **77**, 034610 (2008)

[9] R. Rafiei, R.G. Thomas, D.J. Hinde, M. Dasgupta, C.R. Morton, L.R. Gasques, M.L. Brown, M.D. Rodriguez, Phys. Rev. C **77**, 024606 (2008)

[10] D.J. Hinde, R. du Rietz, M. Dasgupta, R.G. Thomas, L.R. Gasques, Phys. Rev. Lett. **101**, 092701 (2008)

[11] D.J. Hinde, R.G. Thomas, R. du Rietz, A. Diaz-Torres, M. Dasgupta, M.L. Brown, M. Evers, L.R. Gasques, R. Rafiei, M.D. Rodriguez, Phys. Rev. Lett. **100**, 202701 (2008)

[12] R. du Rietz, D.J. Hinde, M. Dasgupta, R.G. Thomas, L.R. Gasques, M. Evers, N. Lobanov, A. Wakhle, Phys. Rev. Lett. **106**, 052701 (2011)

[13] C.J. Lin, R. du Rietz, D.J. Hinde, M. Dasgupta, R.G. Thomas, M.L. Brown, M. Evers, L.R. Gasques, M.D. Rodriguez, Phys. Rev. C **85**, 014611 (2012)

[14] C. Simenel, D.J. Hinde, R. du Rietz, M. Dasgupta, M. Evers, C.J. Lin, D.H. Luong, A. Wakhle, Phys. Lett. B **710**, 607 (2012)

[15] E. Williams, D.J. Hinde, M. Dasgupta, R. du Rietz, I.P. Carter, M. Evers, D.H. Luong, S.D. McNeil, D.C. Rafferty, K. Ramachandran et al., Phys. Rev. C **88**, 034611 (2013)

[16] A. Wakhle, C. Simenel, D.J. Hinde, M. Dasgupta, M. Evers, D.H. Luong, R. du Rietz, E. Williams, Phys. Rev. Lett. **113**, 182502 (2014)

[17] D.J. Hinde, M. Dasgupta, J.R. Leigh, J.C. Mein, C.R. Morton, J.O. Newton, H. Timmers, Phys. Rev. C **53**, 1290 (1996)

[18] K. Hammerton, Z. Kohley, D.J. Hinde, M. Dasgupta, A. Wakhle, E. Williams, V.E. Oberacker, A.S. Umar, I.P. Carter, K.J. Cook et al., Phys. Rev. C **91**, 041602 (2015)

[19] D.J. Hinde, M. Dasgupta, J.R. Leigh, J.P. Lestone, J.C. Mein, C.R. Morton, J.O. Newton, H. Timmers, Phys. Rev. Lett. **74**, 1295 (1995)

[20] E.V. Prokhorova, A.A. Bogachev, M.G. Itkis, I.M. Itkis, G. Knyazheva, N.A. Kondratiev, E.M. Kozulin, L. Krupa, Y.T. Oganessian, I.V. Pokrovsky et al., Nucl. Phys. A **802**, 45 (2008)

[21] K. Nishio, H. Ikezoe, S. Mitsuoka, I. Nishinaka, Y. Nagame, Y. Watanabe, T. Ohtsuki, K. Hirose, S. Hofmann, Phys. Rev. C **77**, 064607 (2008)

[22] I.M. Itkis, E.M. Kozulin, M.G. Itkis, G.N. Knyazheva, A.A. Bogachev, E.V. Chernysheva, L. Krupa, Y.T. Oganessian, V.I. Zagrebaev, A.Y. Rusanov et al., Phys. Rev. C **83**, 064613 (2011)

[23] J. Blocki, H. Feldmeier, W. Swiatecki, Nucl. Phys. A **459**, 145 (1996)

[24] S. Bjornholm, W. Swiatecki, Nucl. Phys. A **391**(2), 471 (1982)

[25] K. Nishio, H. Ikezoe, I. Nishinaka, S. Mitsuoka, K. Hirose, T. Ohtsuki, Y. Watanabe, Y. Aritomo, S. Hofmann, Phys. Rev. C **82**, 044604 (2010)

EPJ Web of Conferences **117**, 08007 (2016) DOI: 10.1051/epjconf/201611708007
NN2015

Description of the fusion-fission reactions in the framework of dinuclear system conception

Sh. A. Kalandarov[1,2], G. G. Adamian[1], N. V. Antonenko[1] and J.P. Wieleczko[3]

[1] Joint Institute for Nuclear Research, 141980 Dubna, Russia
[2] Institute of Nuclear Physics,702132 Tashkent, Uzbekistan
[3] GANIL, CEA et IN2P3-CNRS, B.P. 55027, F-14076, Caen Cedex, France

Abstract

Within the dinuclear system model fusion-fission reactions ^{78}Kr+^{40}Ca and ^{86}Kr+^{48}Ca is investigated. The charge distributions of the decay products are predicted at bombarding energy 10 MeV/nucleon. The competition is treated between complete fusion followed by the decay of compound nucleus and quasifission channels. The possible explanation of the odd-even staggering in the yield of the final reaction products at high excitation energies is discussed.

1 Introduction

The experimentally measured charge, mass, and isotopic distributions of reaction products carry important information about both the reaction mechanism and the interfacing structure of nuclei and reaction dynamics [1–7]. One of such interesting experimental findings in the binary heavy-ion reactions is the persistence of the odd-even staggering in the charge distributions of reaction products at relatively high bombarding energies [3,5,6]. The odd-even staggering effect was found to be smaller for neutron-rich system than for the neutron-deficient one [2,5]. One of the possible explanations for the odd-even staggering at high excitation energies is that the fragments with

$5 \leq Z \leq 16$ mainly come out at the first step of de-excitation cascade and these nuclei preserve their structure during the formation stage and, thus, the primary charge distribution has strong odd-even staggering. The further de-excitation of these fragments slightly decreases the amplitude of the odd-even staggering.

The purpose of the present work is to demonstrate the influence of isospin of the system on the reaction mechanism and, correspondingly, on the charge, mass, isotopic distributions of the reaction products and on the odd-even staggering effect.

The detailed theoretical study will be carried out within the dinuclear system (DNS) model [7]. In this model the cluster emission is treated under the assumption that the complex fragments are produced by a collective motion of the nuclear system in the charge (mass) asymmetry coordinate with further thermal escape over the Coulomb barrier. The emission barriers for complex fragments are calculated within the DNS model by using the double-folding procedure (with the Skyrme-type density-depending effective nucleon-nucleon interaction) for the nuclear part of the nucleus-nucleus interaction potential. Both the evaporation and binary decay are treated in the same way. The correct definition of the emission barriers and of their dependence on the angular momentum allows us to calculate the charge, mass, and kinetic energy distributions of the emitted complex fragments. The main ingredients of our description is the sophisticated potential energy as a function of angular momentum. In our case the maximum possible angular momentum in the system is not adjustable parameter. It is calculated within our model by using the nucleus-nucleus interaction potential [7]. The dynamics plays a role at high angular momenta when the quasifission becomes important contributing to the yields of fission-like fragments. Note that the high angular momenta have larger contribution to the cross section.

The DNS model was applied to predict the charge and mass distributions of the products in the reactions ^{78,82}Kr+^{40}Ca at the incident energy 5.5 MeV/nucleon [2]. Our results were in a good agreement with experimental data slightly underestimating the cross sections near the entrance channel, where the contributions from the deep-inelastic reaction channel may be present in the experimental data. Here we apply the well tested model to predicts the yields of products in the binary reactions ^{78}Kr+^{40}Ca and ^{86}Kr+^{48}Ca at bombarding energy 10 MeV/nucleon. For these reactions, the ISODEC experiments have been done at the INFN-Laboratori Nazionali del Sud by using the CHIMERA device. Some preliminary experimental results were published in Ref. [4].

2 DNS model

The DNS model [7,8] describes an evolution of the charge and mass asymmetry degrees of freedom, which are defined here by the charge and mass (neutron) numbers Z and A ($N = A - Z$) of light nucleus of the DNS. According to this description, there are nucleon drift and nucleon diffusion between the DNS nuclei and eventually either the CN is formed (the complete fusion) or the DNS with given Z and A is formed and decays (quasifission). After the formation, the excited CN decays by various channels including again the formation of certain DNS and their decay. The CN formation and its consequent decay are not necessarily the ultimate results of the evolution of the initial DNS. In addition to the contributions from the CN decay, the binary decay component is related to the quasifission mechanism. The competition between the complete fusion and quasifission depends on the value of maximum angular momentum deposited in the system. The quasifission and CN decays are hardly distinguished in the experiments because in both cases two fragments are produced by the decay of the DNS formed during the diffusion process in the mass (charge) asymmetry coordinate with and without stage of the CN formation.

The production cross section of nucleus with charge Z and mass A numbers is calculated as follows [7]

$$\sigma_{Z,A}(E_{\rm c.m.}) = \sum_{J=0}^{J^*_{max}} \sigma_{Z,A}(E_{\rm c.m.}, J) = \sum_{J=0}^{J^*_{max}} \sigma_{cap}(E_{\rm c.m.}, J) W_{Z,A}(E_{\rm c.m.}, J), \quad (1)$$

where σ_{cap} is the partial capture cross section which defines the transition of the colliding nuclei over the Coulomb barrier and the formation of the initial DNS and $E_{\rm c.m.}$ is the kinetic energy. This transition probability is calculated with the Hill-Wheeler formula. The value of $W_{Z,A}$ is the formation-decay probability of the DNS with a given asymmetries Z and A. The probability of the DNS formation is calculated statistically by using the stationary solution of the master equation with respect to the charge and mass asymmetries and depends on the potential energy of the DNS configurations at touching distance and on the thermodynamical temperature. The probability of the DNS decay in the relative distance R is calculated by using the transition state method. This probability depends on the difference between the potential energies of the DNS configurations at the touching distance and at the barrier position. The maximum angular momentum $J^*_{max} = J_{max} + \delta J_{max}$ which leads to the capture without the pre-equilibrium decay is defined here by taking into account the effect of pre-equilibrium emission. Here, the value of J_{max} is limited by either the kinematical angular momentum $J_{kin} = [2\mu(E_{\rm c.m.} - V(R_b))/\hbar^2]^{1/2}R_b$ (R_b is the position of the Coulomb barrier with the height $V(R_b)$ and μ is the reduced mass) or by the calculated critical angular momentum J_{cr} depending on which one is smaller, $J_{max} = \min[J_{kin}, J_{cr}]$, and δJ_{max} is the average angular momentum which is taken away from the system by pre-equilibrium particles (Table 1). For the case of partial wave J [$J > \delta J$], we calculate the partial formation-decay probability $W_{Z,A}$ and the partial capture cross section at $J - \delta J$ and J, respectively. It should be noted that the calculated capture cross sections with this method are in a good agreement with those obtained with the dynamical model [9]. The excitation energy of the system after the pre-equilibrium emission is decreased by the δE value according to the Table 1. Finally, after each pre-equilibrium emission we calculate the evolution in mass asymmetry degree of freedom and decay of the DNS. The details of calculations of σ_{cap}, $W_{Z,A}$, and, correspondingly, $\sigma_{Z,A}(E_{\rm c.m.})$ are given in Ref. [7, 8].

3 Results and discussions

We calculate the charge distributions of the products in the ^{86}Kr+^{48}Ca reaction at bombarding energy $E_{\rm lab}$=10 MeV/nucleon and compared them with the charge and mass distributions in the neutron-deficient ^{78}Kr+^{40}Ca reaction at the same bombarding energy. The calculated maximum angular momenta and CN excitation energies (at J=0) involved in the ^{86}Kr+^{48}Ca and ^{78}Kr+^{40}Ca reactions are J_{max}=90, E^*_{CN}=270 MeV and J_{max}=73, E^*_{CN}=217 MeV, respectively. The driving potentials (normalized to the energies of the corresponding rotating CN), which are responsible for the formation of different DNS configurations, are shown in Fig. 1 for both reactions. Comparing these driving potentials, one can conclude that the odd-even staggering decreases with increasing N/Z ratio in the system. The main reason for this is that the pairing energy decreases with increasing mass number of the nucleus. The pairing energies and shell corrections of the DNS nuclei are included into the driving potential through their binding energies. So, the staggering effects in the yields of final products are related to the corresponding structure effects in the nuclear binding energies [10].

One can see that for the neutron-deficient system the potential energy is lower than for neutron-rich system at the same value of angular momentum (the value of $B_1 + B_2 - B_{12}$ increases faster than $V(R_m, Z, A, J)$ decreases with increasing total neutron number). This means that the formation of

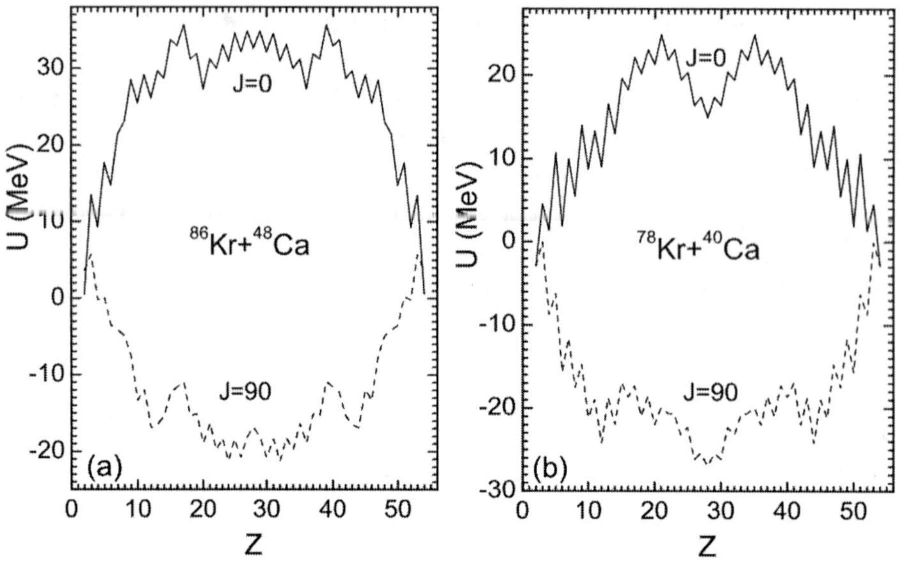

Figure 1: The driving potentials at angular momenta J=0 and 90 for the systems ^{86}Kr+^{48}Ca (a) and ^{78}Kr+^{40}Ca (b). The value of U is normalized to the energy of the rotating CN. The value A is related to Z to supply the minimum of U.

the DNS configurations is more probable for the neutron-deficient system. So, one can expect the larger yield of fission-like fragments in the neutron-deficient system than in the neutron-rich one. With increasing angular momentum the potential energy for symmetric DNS configurations becomes negative at $J > J_0$=45 for ^{78}Kr+^{40}Ca and at $J > J_0$=70 for ^{86}Kr+^{48}Ca. At these large angular momenta the calculated driving potentials show global minimum at the symmetric DNS but not at the CN configuration and the charge (mass) drift pushes the system toward symmetry. The complete fusion becomes energetically denied and higher J waves do not lead to fusion, but to the quasifission resulting in the fragments produced as the binary decay products of the transient DNS originated from the target-projectile DNS. This implies that at higher partial waves, most of the heavy complex fragments are produced by quasifission. At $J \approx J_0$ the CN and symmetric DNS potential energies coincide, and one can observe the coexistence of the quasifission and fusion-decay events. Thus, at the value of J_0 the reaction mechanisms become less clear-cut. At $J < J_0$ the driving potential is positive, the CN configuration is energetically favorable, and the decay products are mainly from the decay of the excited CN.

In Fig. 2, we compare the charge distributions of the emitted complex fragments in the reactions ^{78}Kr(10 MeV/nucleon)+^{40}Ca and ^{86}Kr(10

MeV/nucleon)+^{48}Ca. In both cases, the odd-even effects are present in the production cross sections

$$\sigma_Z(E_{\rm c.m.}) = \sum_A \sigma_{Z,A}(E_{\rm c.m.}) \qquad (2)$$

of the fragments with Z=2–12. For heavier fragments, these effects are not visible because the odd-even structures of the charge distributions are washed out due to the sequential evaporation. So, the amplitude of the staggering depends on the deposited energy. The primary decay products with $Z \leq 12$ are less excited and emit mainly α-particles. This retains the relatively strong odd-even staggering effect in the charge distributions for these nuclei. Thus, the reason why the odd-even staggering effect persists even at the intermediate energies [5, 6] becomes clear. The comparison of the predicted charge distributions with the preliminary experimental results for the reactions considered has already made in Ref. [4].

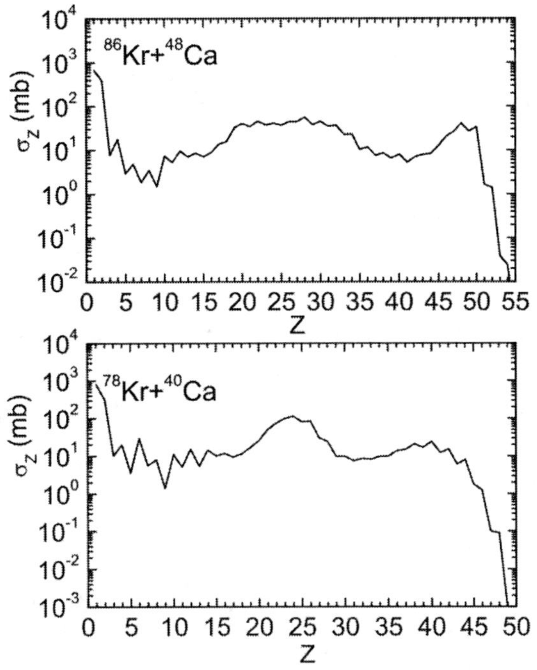

Figure 2: The calculated charge distributions of the products in the reactions ^{86}Kr+^{48}Ca and ^{78}Kr+^{40}Ca at bombarding energy 10 MeV/nucleon.

As found in Ref. [11], the staggering depends in a more complicated way on the history of the evaporation chain than the schematic interpretation suggested in Ref. [10]. In Ref. [10], the staggering is presented in the isotopic yields at the last but one evaporation step through the pairing effect on nucleon separation energies.

In Fig. 2, the odd-even staggering is much smaller for the neutron-rich system comparing to the neutron deficient one. The maxima of the yields of symmetric fragments are located at the same position for both system, but for the neutron-rich system the dependence on Z is more flat. The evaporation residue charge distributions is shifted to smaller Z values for the neutron-deficient system. It is interesting to see that the yields of the emitted light complex fragments are much smaller in the ^{86}Kr+^{48}Ca reaction. For example, the cross section for carbon emission is about 6 times smaller than in the ^{78}Kr+^{40}Ca reaction. This is mainly because of smaller formation probability of corresponding DNS configurations in the ^{86}Kr+^{48}Ca reaction comparing to the ^{78}Kr+^{40}Ca reaction.

4 Conclusions

We investigated the reactions ^{78}Kr+^{40}Ca and ^{86}Kr+^{48}Ca at bombarding energy 10 MeV/nucleon within the DNS model. Our aim was to provide the predictions for these reactions. We found the persistence of odd-even staggering effects in the charge and mass distributions of the reaction products that is in agreement with the experimental observations of Refs. [5,6]. The charge distributions for the neutron-rich system are more flat as compared with the neutron-deficient system.

To elucidate the influence of neutron-excess of the initial system on the formation of final decay fragments, it would be better to perform the experiments at the incident energies about 4 MeV/nucleon for the reactions considered.

Acknowledgements

This work was partially supported by RFBR and IN2P3–JINR Cooperation program.

References

[1] W.-U. Schröder and J.R. Huizenga, in *Treatise on Heavy-Ion Science*, edited by D.A. Bromley, Vol. 2 (Plenum Press, New York, 1984) p.115;

DOI: 10.1051/epjconf/201611708007

V.V. Volkov, in *Treatise on Heavy-Ion Science*, edited D.A. Bromley, Vol. 8 (Plenum Press, New York, 1989) p.101.

[2] G. Ademard *et al.*, Phys. Rev. C **83**, 054619 (2011).

[3] I. Lombardo *et al.*, Phys. Rev. C **84**, 024613 (2011).

[4] S. Pirrone *et al.*, Journal of Physics: Conf. Ser. **515**, 012018 (2014).

[5] S. Piantelli *et al.*, Phys. Rev. C **88**, 064607 (2013).

[6] G. Casini *et al.*, Phys. Rev. C **86**, 011602(R) (2012).

[7] Sh. A. Kalandarov, G.G. Adamian, N.V. Antonenko, and W.Scheid, Phys. Rev. C **82**, 044603 (2010); **83**, 054611 (2011).

[8] V.V. Volkov, Izv. AN SSSR ser. fiz. **50**, 1879 (1986).

[9] V.V. Sargsyan, G.G. Adamian, N.V. Antonenko, and W. Scheid, Eur. Phys. J. A **45**, 125 (2010); V.V. Sargsyan, G.G. Adamian, N.V. Antonenko, W. Scheid, and H.Q. Zhang, Phys. Rev. C **84**, 064614 (2011); **85**, 024616 (2012); **91**, 014613 (2015).

[10] M.V. Ricciardi, A.V. Ignatyuk, A. Kelic, P. Napolitani, F. Rejmund, K.-H. Schmidt, and O. Yordanov, Nucl. Phys. **A733**, 299 (2004).

[11] M. D'Agostino *et al.*, Nucl. Phys. **A861**, 47 (2011).

The influence of the 2-neutron elastic transfer on the fusion of ^{42}Ca + ^{40}Ca

A.M. STEFANINI[1], G. MONTAGNOLI[2], L. CORRADI[1], E. FIORETTO[1], A. GOASDUFF[1,3], J. GREBOSZ[4], F. HAAS[5], M. MAZZOCCO[2], F. SCARLASSARA[2], E. STRANO[2]

[1] INFN, Laboratori Nazionali di Legnaro, I-35020 Legnaro, Padova, Italy
[2] Dip. di Fisica e Astron., Univ. di Padova and INFN, Sezione di Padova, Padova, Italy
[3] CSNSM, CNRS/IN2P3 and Univ. Paris-Sud, F-91405 Orsay, France
[4] Institute of Nuclear Physics, PL 31-342 Cracow, Poland
[5] IPHC, CNRS-IN2P3, Univ. de Strasbourg, F-67037 Strasbourg, France

Abstract

Strong coupling to a single channel with zero Q-value is predicted to produce a characteristic fusion barrier distribution with two peaks, one on each side of the original uncoupled Coulomb barrier. In practical cases, only coupling to an elastic transfer channel may produce such a distribution which, however, has never been observed sofar, probably because low-lying surface vibrations usually have a dominant role, and this may obscure the two-peak structure. The case of the two-neutron (2n) elastic transfer in ^{42}Ca + ^{40}Ca is particularly attractive, because of the relatively rigid nature of the two nuclei.

We have measured the fusion excitation function of this system using the ^{42}Ca beam of the XTU Tandem of LNL on a thin ^{40}Ca target enriched to 99.96% in mass 40. Cross sections have been measured down to $\leq$1 mb. The extracted barrier distribution shows clearly two main peaks. We have performed preliminary CC calculations where the 2^+ coupling strengths have been taken from the literature and the schematic 2n pair transfer form factor has been used, with a deformation length σ_t= 0.39 fm. The excitation function is well reproduced by

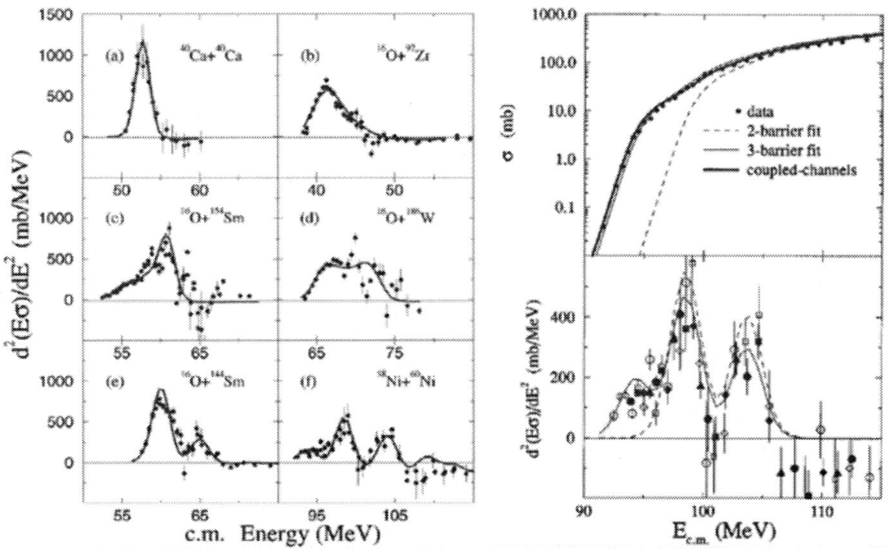

Figure 1: Left: Experimental fusion barrier distributions for six different systems, showing the sensitivity to the structure of the colliding nuclei, and to the nature of couplings involved. Right: Top panel: The measured fusion excitation function for ^{58}Ni + ^{60}Ni is compared with phenomenological fits and with CC calculations. Bottom panel: The barrier distribution is compared with the phenomenological fits.

the calculation including the 2n transfer channel. However, including the octupole excitations destroys the agreement.

1 The fusion barrier distributions

In the near- and sub-barrier fusion of heavy ions, several measurements exploited the concept of a fusion barrier distribution (BD) to identify the nature of couplings responsible for cross section enhancements. Different shapes of barrier distribution are predicted for the coupling to low-energy surface vibrations and for the transfer of one or more nucleons (see Fig. 1, left panel for some well-known examples [1]). Multi-phonon excitations have been shown [2] to become dominant for medium-heavy nuclei and produce complex fusion barrier distributions, in some cases with discrete structures [3,4]. Moreover, below the lower energy limit of such distributions, it has been discovered in recent years [5] that fusion excitation functions show a sharp decrease (named "hindrance") with decreasing energy, well below the expectations based on standard coupled-channels (CC) calculations.

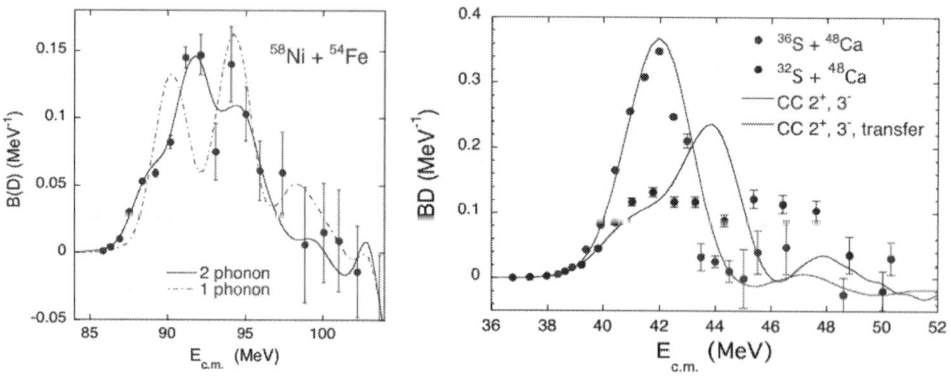

Figure 2: Left: The barrier distribution of ^{58}Ni + ^{54}Fe. Right: Fusion barrier distributions of the two systems ^{36,32}S + ^{48}Ca as derived from the experimental data and calculated within the CC model (see text for comments).

However, identifying the effect of coupling to transfer channels has often been elusive, when deduced from comparing with calculations. One might say that transfer couplings are clear only in the cases where the experimental evidence is conclusive in itself.

2 Coupling to Q=0 channels

A striking fusion barrier distribution is predicted for strong coupling to a single channel with zero Q-value. Irrespective of the type of coupling (phonon or transfer channel) one should obtain in this case a roughly symmetric distribution possessing two peaks, one on each side of the original uncoupled Coulomb barrier.

In practical cases, only coupling to an elastic transfer channel (of one or two nucleons) may produce such a distribution which, however, has never been observed sofar, because low-lying surface vibrations usually have a dominant role. This complicates and may totally obscure the two-peak structure. One should also keep in mind the model calculations of Ref. [6], where it was shown that one obtains a two-peak BD for Q=0 only in the case of *simultaneous* transfer of any number of nucleons, while the number of peaks equals the number of channels when the transfer is sequential.

For the fusion of combinations of Ni isotopes both phonon and transfer channels may play an important role. Several years ago, we decided, therefore, to perform a precision experiment on the system ^{58}Ni + ^{60}Ni [4], so to investigate the possible role of the 2-neutron elastic transfer channel.

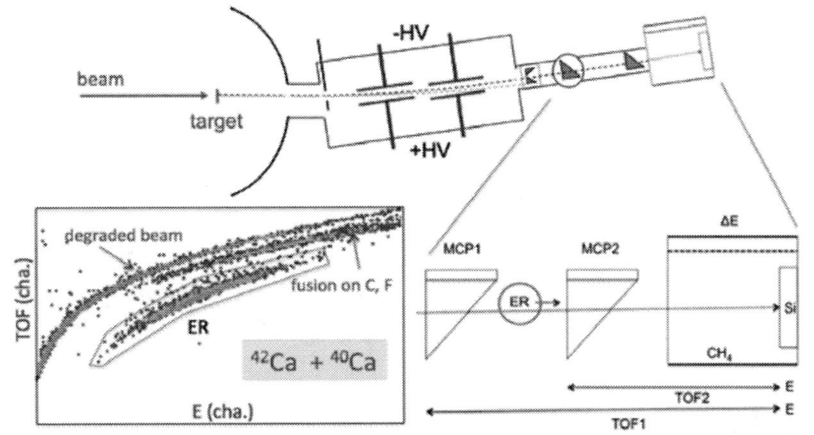

Figure 3: Scheme of the set-up used for the detection of fusion-evaporation residues in the experiment on ^{42}Ca + ^{40}Ca.

The phonon excitation energies and vibrational deformation parameters are very similar for those two nuclei, and the result of the experiment was that multi-phonon surface vibrations determine the shape of the barrier distribution to a large extent, thus preventing to observe the possibly underlying symmetric distribution expected for the 2n elastic transfer. The right panel of Fig. 1 is taken from the original paper on ^{58}Ni + ^{60}Ni [4] and illustrates this situation.

A similar situation was found for ^{58}Ni + ^{54}Fe a few years later [7], where the complex structure of the barrier distribution closely resembles the BD of ^{58}Ni + ^{60}Ni, and it is nicely reproduced by CC calculations, see Fig. 2 (left). The fusion dynamics is dominated by low-energy surface modes in this case too, and little space is left for the possible influence of the α-elastic transfer.

By the way, strong transfer couplings tend to produce wide and flat barrier distributions, even if Q is not zero. A recent example is given by the measurement on ^{32}S + ^{48}Ca [8], whose BD is reported in Fig. 2 (right) in comparison with the case of ^{36}S + ^{48}Ca where transfer does not play any role.

3 This experiment

The case of ^{42}Ca + ^{40}Ca seemed to us particularly attractive, because:

- ^{40}Ca is a magic nucleus with double shell closure, having a strong

DOI: 10.1051/epjconf/201611708008

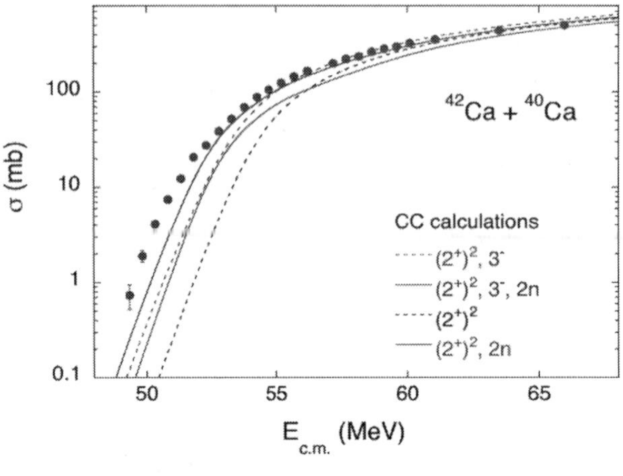

	I^{π}	E_x(MeV)	β_C	β_N
^{40}Ca	2+	3.905	0.12	0.11
	3-	3.747	0.41	0.28
^{42}Ca	2+	1.525	0.25	0.19
	3-	3.447	0.30	0.16

Figure 4: Measured fusion excitation function of ^{42}Ca + ^{40}Ca compared to CC calculations. The structure information on the two nuclei used in the calculations is shown in the Table on the right (see text).

octupole vibration at high energy (3.737 MeV). Its effect on sub-barrier fusion is expected to be mainly a potential renormalization "rigidly" shifting the barrier distribution to lower energies;

- ^{42}Ca, with two neutrons in the $1f_{7/2}$ shell, is rather rigid with a weak 2^+ quadrupole excitation at 1.524 MeV. This makes it plausible that the effect of elastic 2-neutron pair transfer can be recognized in the barrier distribution of this system.

We report here on the measurement of the near- and sub-barrier excitation function of ^{42}Ca + ^{40}Ca, where no data have been obtained sofar. In recent years we have investigated the fusion of various Ca+Ca systems down to very low cross sections [9–11], and the 2-neutron pick-up channel has been found to play an important role in the fusion of the asymmetric system ^{40}Ca + ^{48}Ca. The CC analysis of those data gives us an idea of the coupling strength to be associated to the 2n transfer channel.

The ^{42}Ca beam from the XTU Tandem accelerator of LNL has been used, on thin $^{40}CaF_2$ targets $\sim$ 50μg/cm^2, so to minimize beam energy corrections and straggling effects. The ^{40}Ca target isotopic enrichment was very high (99.96%). The evaporation residues (ER) have been detected by the electrostatic separator set-up (Fig. 3), already used for several sub-barrier fusion measurements at LNL.

The fusion excitation function has been measured in a wide energy range around the Coulomb barrier. It is shown in Fig. 4, together with the results

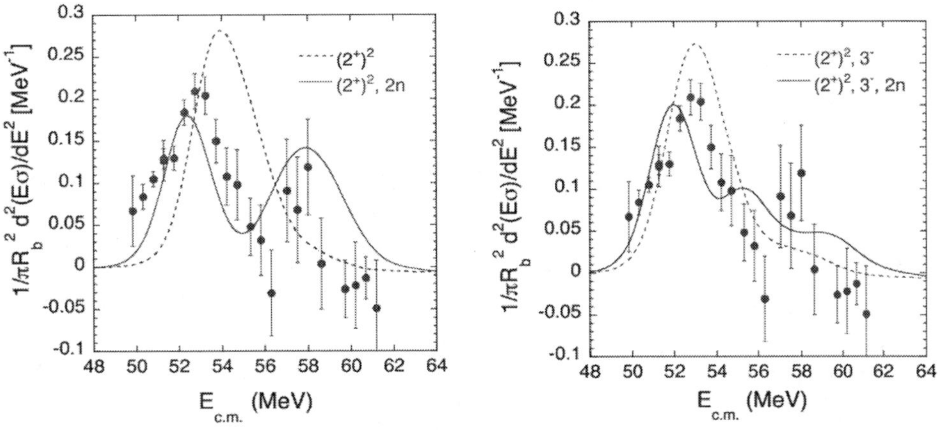

Figure 5: Extracted barrier distributions and CC calculations.

of CC calculations described in the next Section. The energy step and the statistical errors in the measured points were small enough, and this has allowed us to extract the barrier distribution with good accuracy. The BD is shown in Fig. 5, and two nice peaks can be clearly seen.

4 Results and CC calculations

We have performed CC calculations, using the code CCFULL [12], whose results are shown in Fig. 4. The 2^+ and 3^- coupling strengths have been taken from the literature. Two quadrupole phonons have been included and the Akyüz-Winther potential [13] $U(r)$ has been slightly modified to better fit the cross sections in the barrier region. The 2n pair transfer is schematically described (simulated) by the form factor [14]

$$V_t = -\sigma_t dU(r)/dr$$

In the present calculations σ_t has been given the value 0.39 fm best fitting the data on $^{40}Ca + ^{48}Ca$ [9].

One sees that adding both the octupole modes and the 2n transfer to the quadrupole excitations, produces a large effect on the cross sections. The full calculation still underestimates the sub-barrier excitation function. But let us see what happens when the barrier distribution is considered.

When only quadrupole excitations are included (left panel), the additional transfer mode produces a nice two-peak structure closely resembling

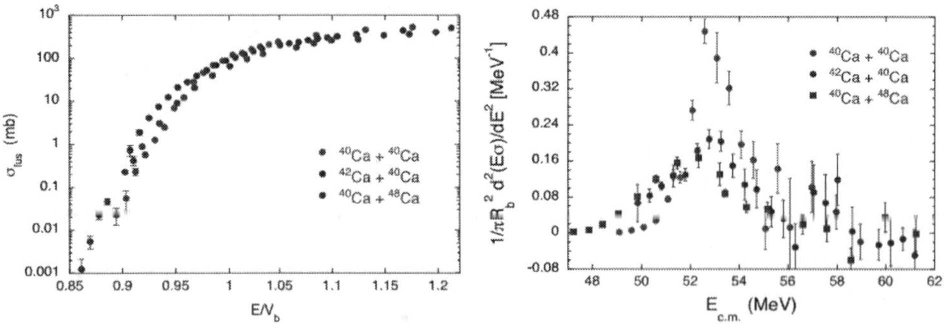

Figure 6: Left: The three excitation functions of the systems ^{42}Ca + ^{40}Ca, ^{40}Ca + ^{40}Ca and ^{40}Ca + ^{48}Ca, compared in an energy scale relative to the Coulomb barrier. Right: Barrier distributions for the same three systems.

the experimental BD. However, by including also the octupole excitations that good agreement is lost, even if a couple of oscillations can be seen above the main peak of the BD. The present comparison with the CC results would tell us that, while the effect of transfer is surely large, the evidence of an *elastic* transfer coupling is not clear at all in ^{42}Ca + ^{40}Ca. We should not forget that the expression of the 2n form factor is very rough, and its coupling strength is arbitrary in our calculations to a large extent.

A final remark: when comparing the BD of the three systems ^{42}Ca + ^{40}Ca, ^{40}Ca + ^{40}Ca [9] and ^{40}Ca + ^{48}Ca [10] (Fig. 6, right panel), the two asymmetric cases are not so different from each other. These are the two cases when transfer couplings play an important role, no matter whether the Q-values are zero or not. The corresponding excitation functions are reported in the left panel of the same Figure.

5 Summary

We have measured the near- and sub-barrier fusion excitation function of ^{42}Ca + ^{40}Ca, where no previous data on the fusion cross sections existed sofar. The energy step and the statistical errors of the measurements were small enough to allow extracting the fusion barrier distribution with good accuracy. This barrier distribution clearly shows a double-peak structure which is tempting to associate with the elastic 2n-transfer, since it is known that any coupling to a zero Q-value channel should produce a symmetric barrier distribution with two equal peaks on either side of the uncoupled barrier.

When comparing the present data with simple CC calculations, we find that the octupole vibrations (very strong in ^{40}Ca, but rather high in energy) do not essentially influence the shape of the barrier distribution if no transfer coupling is considered. However, the simple two-peak structure is lost in the more complete calculations where the quadrupole modes, the octupole vibrations *and* the 2n transfer are included. The CC predictions presented in this contribution are based only on a schematic (approximate) formulation of the 2n transfer form-factor. While keeping this in mind, we may conclude that transfer couplings are important in ^{42}Ca + ^{40}Ca, but the evidence of an elastic two-neutron transfer is sofar marginal.

References

[1] M.Dasgupta, D.J.Hinde, N.Rowley and A.M.Stefanini, Annu. Rev. Nucl. Part. Sci. **48**, 401 (1998).

[2] H.Esbensen, Phys. Rev. C **72**, 054607 (2005).

[3] N.Rowley, G.R.Satchler and P.H.Stelson, Phys. Lett. B **254**, 25 (1991).

[4] A.M.Stefanini et al., Phys. Rev. Lett. **74**, 864 (1995).

[5] C. L. Jiang et al., Phys. Rev. Lett. **89**, 052701 (2002).

[6] N. Rowley, I. J. Thompson, and M. A. Nagarajan, Phys. Lett. B **282**, 276 (1992).

[7] A.M.Stefanini et al., Phys. Rev. C **81**, 037601 (2010).

[8] G. Montagnoli *et al.*, Phys. Rev. C **87**, 014611 (2013).

[9] G. Montagnoli *et al.*, Phys. Rev. C **85**, 024607 (2012).

[10] C.L.Jiang et al., Phys. Rev. C **82**, 041601(R) (2010).

[11] A.M. Stefanini et al., Phys. Lett. B **679**, 95 (2009).

[12] K.Hagino, N.Rowley and A.T.Kruppa, Comput. Phys. Commun. **123**, 143 (1999).

[13] Ö.Akyüz and Å.Winther, in Nuclear Structure and Heavy-Ion Physics, Proc. Int. School of Physics "Enrico Fermi", Course LXXVII, Varenna, eds. R.A. Broglia and R.A. Ricci (North Holland, Amsterdam, 1981).

[14] H.Esbensen, S.H.Fricke and S.Landowne, Phys. Rev. C **40**, 2046 (1989).

EPJ Web of Conferences **117**, 08009 (2016) DOI: 10.1051/epjconf/201611708009
NN2015

Study of the ^{20,22}Ne+^{20,22}Ne and ^{10,12,13,14,15}C+^{12}C fusion reactions with MUSIC

M. L. Avila[1], K. E. Rehm[1], S. Almaraz-Calderon[1], P. F. F. Carnelli[2], B. DiGiovine[1], H. Esbensen[1], C. R. Hoffman[1], C. L. Jiang[1], B. P. Kay[1], J. Lai[3], O. Nusair[1], R. C. Pardo[1], D. Santiago-Gonzalez[1,3], R. Talwar[1] and C. Ugalde[1]

[1] Physics Division, Argonne National Laboratory, Argonne IL 60439, USA
[2] Laboratorio Tandar, B1650KNA San Martín, Buenos Aires, Argentina
[3] Department of Physics and Astronomy, Louisiana State University, Baton Rouge, LA 70803, USA

Abstract

A highly efficient MUlti-Sampling Ionization Chamber (MUSIC) detector has been developed for measurements of fusion reactions. A study of fusion cross sections in the ^{10,12,13,14,15}C+^{12}C and ^{20,22}Ne+^{20,22}Ne systems has been performed at ATLAS. Experimental results and comparison with theoretical predictions are presented. Furthermore, results of direct measurements of the $^{17}O(\alpha,n)^{20}$Ne, $^{23}Ne(\alpha,p)^{26}$Mg and $^{23}Ne(\alpha,n)^{26}$Al reactions will be discussed.

1 Introduction

Fusion cross sections play an important role for both nuclear structure and nuclear astrophysics. In nuclear astrophysics the prime example is the ^{12}C+^{12}C fusion reaction, which is the primary route for the production of heavier elements. This is a key reaction in Type Ia supernovae and X-ray bursts and has been the subject of intense experimental efforts. Also, fusion reactions between light neutron-rich nuclei have been proposed to be a possible energy source in X-ray superbursts that originate in the crust of an accreting neutron star [1–3]. The reactions included in the calculations of superburst range from ^{12}C+^{12}C to neutron-rich reactions such as ^{24}C+^{24}C and ^{28}Ne+^{28}Ne. Many of these reactions are outside of our experimental capabilities. However, performing experimental studies in accessible systems will be instrumental to test the predictive power of theoretical models.

For nuclear structure studies, it is important to understand the reaction mechanism of fusion reactions, which according to studies of heavy-ion induced reactions is more complicated than previously suggested. Moreover, there are reported oscillations at energies above the Coulomb barrier in the fusion cross sections of light symmetric systems such as ^{12}C+^{12}C, ^{12}C+^{16}O and ^{16}O+^{16}O (see Refs. [4, 5] and references therein). This phenomenon has been suggested to be cause by the successive penetration of centrifugal barriers with different angular momenta [4]. However, the exact origin is not completely understood and more experimental data are needed in order to gain a better understanding.

The development of the MUlti-Sampling Ionization Chamber (MUSIC) detector [6] opened new possibilities for fusion reaction studies. The high efficiency and flexibility to measure the excitation function in a large energy range in a single measurement make the MUSIC detector an ideal tool for performing measurements of fusion cross sections. Moreover, MUSIC is a self normalizing detector and no additional monitor detectors are needed for the absolute normalization of the beam. This work shows a systematic study of the excitation function of the ^{10,12,13,14,15}C+^{12}C and ^{20,22}Ne+^{20,22}Ne systems and first data of (α, p) and (α, n) reactions with MUSIC.

2 The ^{10,12,13,14,15}C+^{12}C experiments

In a series of recent experiment using MUSIC, measurements of fusion reactions between ^{12}C and ^{10,12,13,14,15}C were obtained for the first time [7]. The energy of the excitation functions in the center-of-mass (c.m.) ranged from 8-18 MeV. The ^{12,13}C+^{12}C fusion cross sections which were previously measured in Refs. [8,9] were measured in order to verify the technique. The beams from the ATLAS accelerator were delivered to the MUSIC detector which was filled with 200 mbar of CH_4 gas. Fig. 1 (left) shows that the detector was able to separate events from the ^{15}C beam and for evaporation residues produced by ^{15}C particles in e.g. strip 4 and by the contaminant ^{15}N beam in e.g. strip 10 due to the difference in the ΔE signals.

The extracted S factors of the ^{10,12,13,14,15}C+^{12}C reactions are shown by

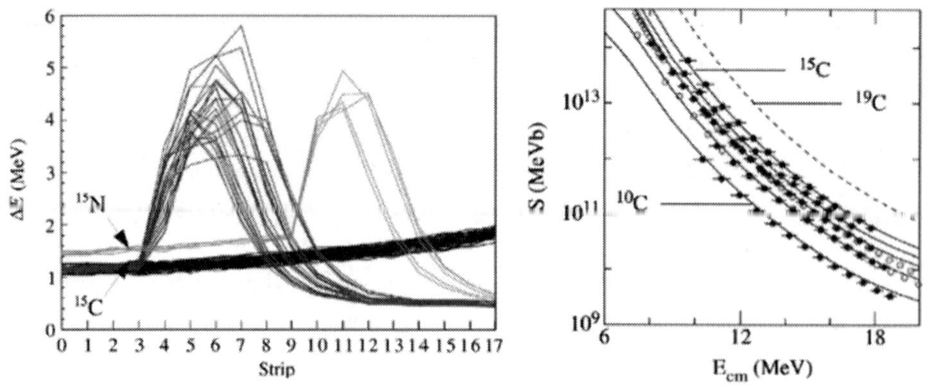

Figure 1: Left: Events measured in MUSIC for the ^{15}C beam, and for evaporation residues produced by ^{15}C particles in strip 4 and by ^{15}N particles in strip 10. Right: S factors of the ^{10,12,13,14,15}C+^{12}C systems measured with MUSIC [7] (solid circles) and the ^{12,13}C+^{12}C systems measured in Refs. [8, 9] (open circles) compared with calculations using the Sao Paulo potential from Ref. [2].

the solid circles in Fig. 1 (right). The solid lines are calculations performed from Ref. [2] using the Sao Paulo potential and the open circles are data for the ^{12,13}C+^{12}C reactions previously obtained in Refs. [8, 9]. As can be seen in this figure there is a very good agreement with earlier measurements between the MUSIC data and with the theoretical predictions.

3 The ^{20,22}Ne+^{20,22}Ne experiments

It has been suggested that fusion reactions between neutron-rich nuclei other than carbon, such as oxygen or neon also play a role in the crust of neutron stars. Moreover, it is also of interest to investigate oscillations in Ne+Ne systems, since oscillations have been seen before in light symmetric spin-0 systems, such as carbon and oxygen, and thus, oscillations are also expected to be found in Ne+Ne systems. Therefore, we have started a series of measurement involving Ne isotopes at ATLAS using the MUSIC detector.

These experiments were carried out with ^{20,22}Ne beams of 120 MeV from the ATLAS accelerator and enriched ^{20}Ne and ^{22}Ne gases in MUSIC. As a test the ^{12}C+^{20}Ne excitation function which had been previously measured in Refs. [10] was also studied. Preliminary results of the fusion cross sections of ^{12}C+^{20}Ne measured with MUSIC are shown by squares in Fig. 2 (left) compared with the cross sections previously measured in Ref. [10] (open circles). As seen in the figure there is a good agreement between the two data sets. The cross sections presented in Fig. 2 (right) shows hints of oscillations

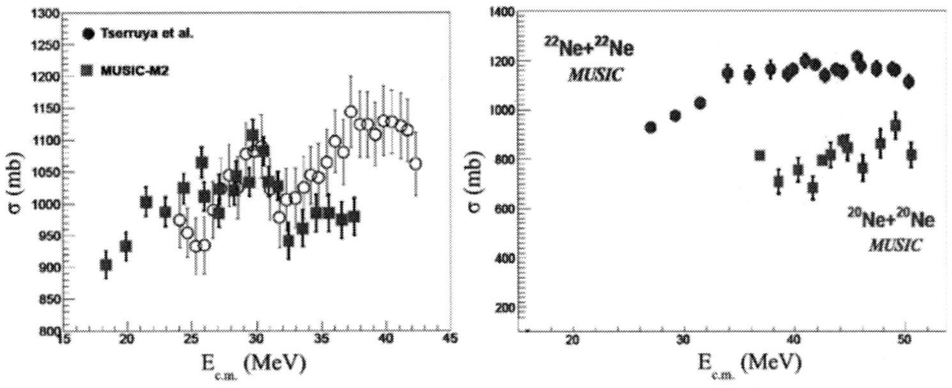

Figure 2: Left: Fusion cross section of ^{12}C+^{20}Ne from Ref. [10] (open circles) and the MUSIC detector (squares). Right: Fusion cross sections of the ^{22}Ne+^{22}Ne (circles) and ^{20}Ne+^{20}Ne (squares) systems measured with MUSIC.

for both the ^{20}Ne+^{20}Ne (squares) and ^{22}Ne+^{22}Ne (circles) systems, but further measurements are needed in order to confirm this claim.

4 (α, n) and (α, p) measurements

(α, p) and (α, n) reactions are known to play an important role in nuclear astrophysics. The capabilities of MUSIC for measuring (α, n) reactions were demonstrated using the ^{17}O$(\alpha, n)^{20}$Ne reaction as a benchmark. Fig. 3 (left) shows that it is possible to separate events of the (α, n) reaction from the incoming beam. The cross sections obtained from this measurement are in good agreement with the cross section from Ref. [11]. This is shown in Fig. 3 (right), where the cross sections of the ^{17}O$(\alpha, n)^{20}$Ne reaction as a function of c.m. energy extracted from the MUSIC data are compared with the cross sections obtained from Ref. [11]. To compare both data sets the data from Ref. [11] were averaged over a wider energy steps (solid curve in Fig. 3), since it was measured in smaller energy steps.

Recently, we have been successful in measuring the ^{23}Na$(\alpha, p)^{26}$Mg and ^{23}Na$(\alpha, n)^{26}$Al reactions using this technique. The (α, p) reaction has been identified in a sensitivity study in Ref. [12] as a key process for the understanding of ^{26}Al nucleosynthesis in massive stars. Figure 4 (left) shows that MUSIC allows the separation of the traces from (α, p) and (α, n) reactions. The cross sections shown in Fig. 4 (right) for the ^{23}Na$(\alpha, p)^{26}$Mg and ^{23}Na$(\alpha, n)^{26}$Al reactions were measured simultaneously with two different beam energies 57.5 MeV and 52 MeV labeled as High and Low energy, re-

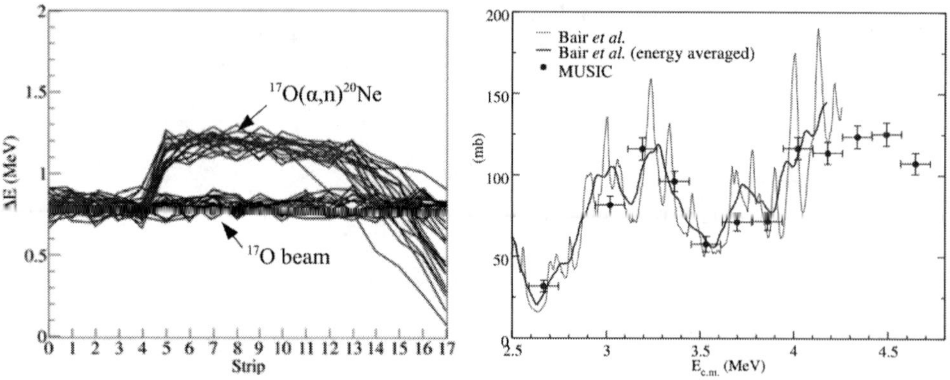

Figure 3: Left: Events of the $^{17}O(\alpha, n)^{20}Ne$ reaction occurring in strip 5 and of the ^{17}O beam. Right: Cross section of the $^{17}O(\alpha, n)^{20}Ne$ reaction, obtained from Ref. [11] (dashed line) and its energy-averaged cross section (solid line) compared with the MUSIC data (solid circles).

spectively. This preliminary analysis shows that our data are in good agreement with the statistical model calculations performed in Ref. [13] (dashed lines). These are important results because MUSIC can now be used to study (α, p) and (α, n) reactions directly, offering a unique opportunity of study for α-capture reactions with both stable and radioactive beams.

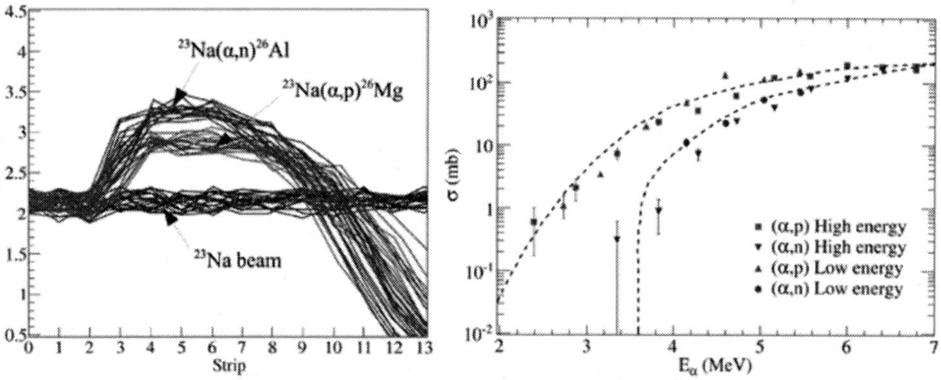

Figure 4: Left: Events for reactions occurring in strip 3 for (α, p), (α, n) and the ^{23}Na beam in the MUSIC detector. Right: Cross section of the $^{23}Ne(\alpha, p)^{26}Mg$ and $^{23}Na(\alpha, n)^{26}Al$ reactions.

5 Summary

A highly efficient active target detector system has been developed. The fusion cross sections of $^{10,12,13,14,15}C+^{12}C$ were measured and compared with theoretical predictions. For experiments with radioactive $^{10,14,15}C$ beams a good agreement between our results and theoretical calculations was found. Measurements of the $^{12,13}C+^{12}C$ systems are in good agreement with the cross sections measured by standard techniques in Refs. [8, 9]. Preliminary results of the fusion cross sections of $^{20,22}Ne+^{20,22}Ne$ are also presented. Along with that, MUSIC has been successfully used to measure the $^{17}O(\alpha, n)^{20}Ne$, $^{23}Na(\alpha, p)^{26}Mg$ and $^{23}Na(\alpha, n)^{26}Al$ reactions, opening new possibilities for the study of α-induced reactions.

This material is based upon work supported by the U.S. Department of Energy, Office of Science, Office of Nuclear Physics, under contract number DE-AC02-06CH11357. This research used resources of ANL's ATLAS facility, which is DOE Office of Science User Facility.

References

[1] A. S. Umar et al., Phys. Rev. C **85**, 055801 (2012).

[2] D. G. Yakovlev et al., Phys. Rev. C. **82**, 044609 (2010).

[3] M. Beard et al., ADNDT **96**, 541 (2010).

[4] H. Esbensen, Phys. Rev. C **85**, 064611 (2012).

[5] N. Rowley and K. Hagino, Phys. Rev. C **91**, 044617 (2015).

[6] P. F. F. Carnelli et al., Nucl. Inst. Method A, **799**, 197 (2015).

[7] P. F. F. Carnelli et al., Phys. Rev. Lett, **112**, 192701 (2014).

[8] D. G. Kovar et al., Phys. Rev. C **20**, 1305 (1979).

[9] R.A. Dayras et al., Nucl. Phys. A **265**, 153 (1976).

[10] I. Tserruya et al., Phys. Rev. C **21**, 1864 (1980).

[11] J. K. Bair and F. X. Hass, Phys. Rev. C **7**, 1356 (1973).

[12] C. Illiadis et al., ApJS **193**, 16 (2011).

[13] P. Mohr, EPJA **51**, 56 (2015).

EPJ Web of Conferences **117**, 08010 (2016) DOI: 10.1051/epjconf/201611708010
NN2015

Sub-barrier fusion and transfers in the ^{40}Ca + ^{58,64}Ni systems

D. Bourgin[1,2], S. Courtin[1,2,3], F. Haas[1,2], A. Goasduff[4], A.M. Stefanini[4], G. Montagnoli[5], D. Montanari[1,2,3], L. Corradi[4], J. Huiming[4], F. Scarlassara[5], E. Fioretto[4], C. Simenel[6], N. Rowley[7], S. Szilner[8] and T. Mijatović[8]

[1]IPHC, Université de Strasbourg, F-67037 Strasbourg, France
[2] CNRS, UMR7178, F-67037 Strasbourg, France
[3] USIAS, Université de Strasbourg, F-67083 Strasbourg, France
[4] INFN - Laboratori Nazionali di Legnaro, I-35020 Legnaro (Padova), Italy
[5] Dipartimento di Fisica, Università di Padova, I-35131 Padova, Italy
[6] School of Physics and Engineering, Australian National University, Canberra, Australia
[7] IPN, Université Paris-Sud, F-91406 Orsay, France
[8] Ruder Bošković Institute, HR-10002 Zagreb, Croatia

Abstract

Fusion cross sections have been measured in the ^{40}Ca + ^{58}Ni and ^{40}Ca + ^{64}Ni systems at energies around and below the Coulomb barrier. The ^{40}Ca beam was delivered by the XTU Tandem accelerator of the Laboratori Nazionali di Legnaro and evaporation residues were measured at very forward angles with the LNL electrostatic beam deflector. Coupled-channels calculations were performed which highlight possible strong effects of neutron transfers on the fusion below the barrier in the ^{40}Ca + ^{64}Ni system. Microscopic time-dependent Hartree-Fock calculations have also been performed for both systems. Preliminary results are shown.

1 Introduction

The fusion process accounts for the major part of the reaction cross section in collisions of heavy ions at above the barrier beam energies. Some of the most striking recent results on the fusion process have nevertheless been obtained for reactions below the Coulomb barrier (CB). At energies around the CB, the process is driven by couplings of the relative motion of the colliding nuclei to their low energy surface vibrations and/or stable deformations. Coupled-channels (CC) calculations taking into account these couplings have proven a very powerful tool to reproduce fusion cross sections at moderate sub-barrier energies. The CC description is discussed in details in a recent review article [1]. Distributions of barriers, extracted from high precision fusion cross section data are also very good tools to test the couplings to collective degrees of freedom of the target and projectile [2]. In the last decade, fusion cross sections have been measured down to the ~ 1 μb level, below effects of the lowest barrier of the distributions of barriers. There, it was found that the measured fusion cross sections decrease sharply, well below predictions of usual CC calculations based on standard Woods-Saxon potentials. This fusion hindrance phenomenon has been shown to depend on the parameters included in the CC calculations and the logarithmic derivative of the energy-weighted cross section is an interesting representation to explore the fusion hindrance phenomenon [3]. The present contribution focusses on fusion measurements at and below the CB for the two ^{40}Ca + ^{58}Ni and ^{40}Ca + ^{64}Ni systems.

The starting point of this experimental study is the recent exploration of Ca+Ca systems and also results on Ni+Ni systems. All systems, except ^{40}Ca + ^{48}Ca have negative Q values for fusion. Fusion excitation functions have been measured for these systems down to very low energies and very low cross sections [4–9].

The ^{58}Ni, ^{60}Ni and ^{64}Ni are vibrational nuclei with well established 1 and 2 phonon states. Most of the Ni+Ni systems have been measured to show hindrance at low energies. The ^{64}Ni + ^{64}Ni is a textbook case of fusion enhancement and hindrance. For this system, vibrational couplings around the barrier lead to an enhancement of the fusion cross section. These coupling effects get weaker at lower energies and large hindrance is observed. Below the CB, it has been measured that the slope of the fusion excitation function is less steep for the asymmetric ^{58}Ni + ^{64}Ni system, as regards to the ^{58}Ni + ^{58}Ni and ^{64}Ni + ^{64}Ni systems. This was interpreted by the authors by possibly strong positive Q value multiple transfer channels.

In the ^{40,48}Ca+^{40,48}Ca cases, a fusion hindrance was reported at deep sub-barrier energies for ^{40}Ca + ^{48}Ca. The effect is different for ^{40}Ca + ^{40}Ca and ^{40}Ca + ^{48}Ca, because of the coupling to the strong 3^- state in the soft ^{40}Ca nucleus with regard to the ^{48}Ca stiff nucleus.
Moreover besides including all couplings to low-lying excited states of the target and projectile in the CC calculations, it has been shown that it is necessary to take into account transfer channel effects to best describe the ^{40}Ca + ^{48}Ca data. In fact, Q values for the +2n, +3n and +4n transfer channels (from ^{48}Ca towards ^{40}Ca) are positive as well as Q values for the -1p, -2p, -3p and -4p transfer reactions. Other possible particle tranfers have negative Q values. In the symmetric ^{40}Ca + ^{40}Ca and ^{48}Ca + ^{48}Ca systems, all nucleon transfer channels have negative Q values.
To put it in a nutshell, strong couplings to octupole vibrational states are observed in the fusion of the Ca+Ca systems at CB and below, fusion hindrance is measured at very deep sub-barrier energies in the asymmetric ^{40}Ca + ^{48}Ca collision as well as effects of neutron transfer channels. The Ni+Ni systems show coupling effects, different from Ca+Ca, but typical of collisions between vibrational nuclei, and presumably the influence of transfer channels in the asymmetric ^{58}Ni + ^{60}Ni system and even more in the ^{58}Ni + ^{64}Ni system. Based on the good knoweledge of the Ca+Ca and Ni+Ni systems, and effects of couplings under control, we have decided to explore the ^{40}Ca + ^{58,64}Ni systems at energies around the CB and below to look for eventual transfer effects in the ^{40}Ca + ^{64}Ni system.

2 Fusion of ^{40}Ca + ^{58,64}Ni

For the ^{40}Ca+^{64}Ni system, Q values are positive for the +2n to +4n transfer channels. The situation is much less favourable for nucleon transfers for the ^{40}Ca + ^{58}Ni system: all neutron and proton transfer channels have negative Q values.

A 9 pnA intensity ^{40}Ca$^{9+,10+}$ beam was provided by the LNL XTU Tandem with energies ranging from E_{lab} = 104.75 MeV to 153.5 MeV in steps of 1.25 MeV below the Coulomb barrier and 2.5 MeV above the Coulomb barrier for both ^{40}Ca + ^{58}Ni and ^{40}Ca + ^{64}Ni reactions. The ^{40}Ca beam was impinging on 50 μg.cm^{-2} ^{58}Ni and ^{64}Ni targets deposited on a 20 μg.cm^{-2} ^{12}C backing. The LNL electrostatic beam deflector was used to deflect the incident beam with a rejection factor of 10^7, and evaporation residues were measured at 2° with a detector set-up composed of 2 micro-channel plate

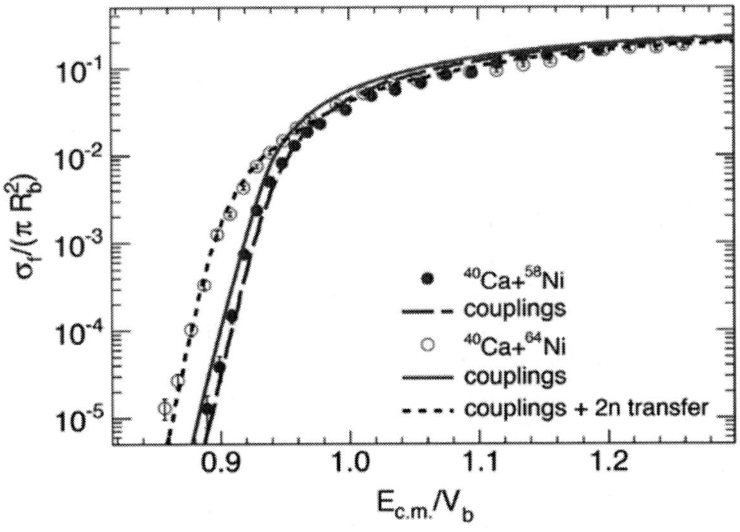

Figure 1: Reduced fusion excitation functions measured in this work for the ^{40}Ca + ^{58}Ni and ^{40}Ca + ^{64}Ni systems. The solid and dashed lines are results of CC calculation (see text for details).

detectors, a transverse field ionization chamber and a large silicon detector (600 mm^2). Details on the set-up can be found in [10].
Very accurate fusion cross sections were measured in an energy range from 11 % (resp. 14%) below the CB to 19% (resp. 26%) above the CB for the ^{40}Ca + ^{58}Ni (resp .^{40}Ca + ^{64}Ni) system. Fig. 1 shows these fusion excitation functions.
Coupled-channels calculations have been performed for both systems using the CCFULL code [11]. These calculations use a standard O. Akyüz and A. Winther [12, 13] parametrization of the ion-ion potential and relevant couplings to low-lying states of the target and projectile. Parameters of the calculations are given in Table. 1. Interestingly enough, for ^{40}Ca + ^{58}Ni, good description of the cross sections is found including only the octupole 3_1^- state of ^{40}Ca and 2_1^+ state of ^{58}Ni (dashed blue line of Fig. 1). The same type of calculation including couplings to the relevant states, was thus performed for ^{40}Ca + ^{64}Ni. This is shown by the red solid line of Fig. 1. This underestimates the cross sections at sub-barrier energies and it was found that good reproduction of the data is obtained if a 2n transfer channel is included in the CC calculations in some schematical way (black dashed

curve). Neutron transfer channels may thus play an important role in the ^{40}Ca + ^{64}Ni reaction. More details can be found in [14].
Coupled-channels calculation results are of course sensitive to the number of channels that one take into account. In the next section, we shall present

Table 1: ^{40}Ca, ^{58}Ni and ^{64}Ni spectroscopic data used as parameters of the CC calculations. The transition strengths S and the quadrupole and octupole deformation parameters $\beta_{2,3}$ were determined from the γ-ray energies E_γ of the gs transitions and the mean lifetimes of the states given in Ref. [15].

Nucleus	J^π	E_γ [keV]	S [W.u.]	$\beta_{2,3}$
^{40}Ca	3_1^-	3736	25.95	0.40
^{58}Ni	2_1^+	1454	9.78	0.18
^{64}Ni	2_1^+	1346	7.57	0.16

some preliminary results of alternative calculations based on microscopic methods.

3 Preliminary microscopic TDHF calculations of ^{40}Ca + ^{58}Ni and ^{40}Ca + ^{64}Ni reactions

The time-dependant Hartree-Fock (TDHF) method has the advantage that all the couplings are naturally existing at the mean-field level. These can be couplings to surface excitations as well as transfer channels. The method also allows excitations to evolve during the reaction, while the nuclei overlap. Such calculations, using the Sly4 Skyrme energy density functionnal, have been performed for both ^{40}Ca + ^{58}Ni and ^{40}Ca + ^{64}Ni reactions. Details on the method can be found in [16]. Preliminary results are presented in Fig. 2 which shows the evolution of the nuclear density as a function of time. The corresponding center-of-mass energy is the fusion barrier calculated with the TDHF method, E_{fus} = 71.70 MeV and 69.06 MeV for ^{40}Ca + ^{58}Ni and ^{40}Ca + ^{64}Ni, respectively. The octupole (pear-shaped) deformation of ^{40}Ca during the fusion process is clearly seen on graphs on the 3rd line of the figure (from top, t = 1.425 zs). These calculations do not take into account the sub-barrier tunneling. The next step of this study will thus be to use TDHF results such as the frozen potential in CC calculations. This will be the object of a further article.

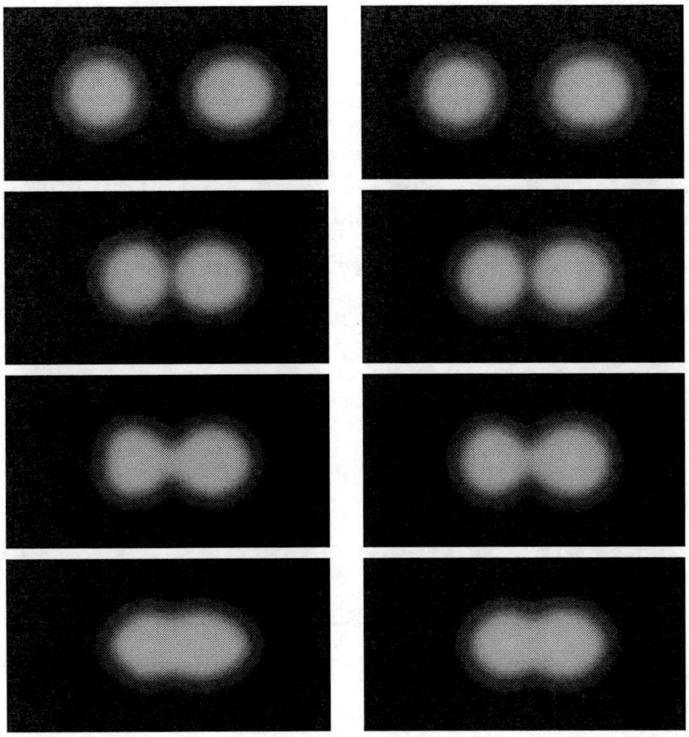

Figure 2: TDHF calculations of the evolution of the nuclear density at the fusion barrier for $^{40}Ca + ^{58}Ni$ (graphs at the left) and $^{40}Ca + ^{64}Ni$ (graphs at the right). From top to bottom, the time is t = 0.05 zs, 0.975 zs, 1.425 zs and 3.15 zs.

4 Conclusions and perspectives

Fusion cross sections have been measured in the $^{40}Ca + ^{58}Ni$ and $^{40}Ca + ^{64}Ni$ systems at beam energies from 11 % (resp. 14%) below the CB to 19 % (resp. 26%) above the CB for the $^{40}Ca + ^{58}Ni$ (resp. $^{40}Ca + ^{64}Ni$) system. CC calculations have been performed to describe these fusion cross sections. Hints of strong effects of neutron transfer channels in the $^{40}Ca + ^{64}Ni$ system were found. Preliminary TDHF calculations have been performed as well. These have the advantage of taking into account all entrance channel effects during the fusion process.
Following this work, a recent experiment has been performed by our team to measure nucleon transfer cross sections for these systems using the large acceptance magnetic spectrometer PRISMA at LNL. Data is under analysis and should shed light on the importance of transfer in $^{40}Ca + ^{64}Ni$ at energies at and below the CB.

5 Acknowledgments

We would like to thank the professional work of the XTU Tandem and target LNL laboratory staff. The research leading to these results has received funding from the European Union Seventh Framework Programme under Grant Agreement No. 262010 ENSAR.

References

[1] B. B. Back, H. Esbensen, C. L. Jiang, and K. E. Rehm, Rev. Mod. Phys. **86** 317 (2014).

[2] N. Rowley, G.R. Satchler, P.H. Stelson, Phys. Lett. **B 254** 25 (1991).

[3] C.L. Jiang *et al.*, Phys. Rev. Lett. **89** 052701 (2002).

[4] G. Montagnoli *et al.* Phys. Rev. **C 85** 024607 (2012) and references therein.

[5] M. Beckerman *et al.*, Phys. Rev. **C 23** 1581 (1981).

[6] M. Beckerman *et al.*, Phys. Rev. Lett. 1472 (1980).

[7] D. Ackermann *et al.*, Nucl. Phys. **A 609** 91 (1996).

[8] A.M. Stefanini *et al.*, Phys. Rev. Lett. **74** 864 (1995).

[9] C.L. Jiang *et al.*, Nucl. Phys. **A834** 615c (2010) and references therein.

[10] A.M. Stefanini *et al.*, Phys. Rev. **C 81** 037601 (2010).

[11] K. Hagino *et al.*, Comput. Phys. Commun. **C 123** 143 (1999).

[12] O. Akyüz, A. Winther, in Nuclear Structure and Heavy-Ion Physics, Proc. Int. School of Physics, Enrico Fermi, Varenna, North-Holland, Amsterdam (1981) 492.

[13] A. Winther, Nucl. Phys. A **594** 203 (1995).

[14] D. Bourgin *et al.*, Phys. Rev. **C 90** 044610 (2014).

[15] National Nuclear Data Center, www.nndc.bnl.gov/nudat2.

[16] C. Simenel, Eur. Phys. J. A **48** 152 (2012) and references therein.

EPJ Web of Conferences **117**, 08011 (2016) DOI: 10.1051/epjconf/201611708011
NN2015

Fusion reactions of ^{58,64}Ni+^{124}Sn

F. Galtarossa[1,2], C. L. Jiang[3], A. M. Stefanini[1], H. Esbensen[3], K. E. Rehm[3], S. Almaraz Calderon[3], M. L. Avila[3], B. B. Back[3], D. Bourgin[4], L. Corradi[1], S. Courtin[4], E. Fioretto[1], A. Goasduff[8], F. Haas[4], M. Mazzocco[5], D. Montanari[4], G. Montagnoli[5], T. Mijiatovic[6], R. Sagaidak[7], D. Santiago-Gonzalez[3], F. Scarlassara[5], E. Strano[5] and S. Szilner[6]

[1]INFN - Laboratori Nazionali di Legnaro, I-35020 Legnaro (Padova), Italy
[2]Dipartimento di Fisica e Scienze della Terra, Università di Ferrara, I-44121 Ferrara, Italy
[3]Physics Division, Argonne National Laboratory, Argonne, IL 60439, USA
[4]IPHC, CNRS/IN2P3, Université de Strasbourg, F-67037 Strasbourg Cedex 2, France
[5]Dipartimento di Fisica e Astronomia, Università di Padova, INFN, Sez. di Padova, I-35131 Padova, Italy
[6]Ruder Bošković Institute, HR-10002 Zagreb, Croatia
[7]Joint Institute for Nuclear Research, RU-141980 Dubna, Russia
[8]CSNSM, CNRS/IN2P3 and Université Paris-Sud, F-91405 Orsay Campus, France

Abstract

In order to better understand the influence of transfer in sub-barrier nuclear reactions, cross sections for the system ^{58,64}Ni+^{124}Sn have been measured down to 0.5-1 μb and compared to detailed coupled-channel calculations. In agreement with a phenomenological Q-value systematics, calculations show the importance of including the coupling to the transfer channel for these heavy systems. No clear evidence of fusion hindrance is observed, probably due to the fact that the cross sections measured in this experiment are not low enough for the appearance of that phenomenon.

In the last years many theoretical and experimental works on heavy-ion fusion reactions have been devoted to the study of the influence of nuclear structure on the fusion process and the main results have been summarized in several review articles [1–3].

It has been clear since the 80's that the structure of colliding nuclei strongly affects the fusion probability: the most evident effect is the large enhancement of the experimental fusion cross section with respect to the predictions of the one-dimensional barrier model. This effect can be explained by means of coupled-channels calculations, where couplings to the low-lying surface modes of the colliding nuclei are taken into account.

In more recent years, thanks to the capability of reaching cross sections at the nb level, the phenomenon of the fusion hindrance was discovered [4] and explained through the saturation properties of nuclear matter [5] or the damping of the coupling strength during the adiabatic fusion process [6].

A still open question for a more complete understanding of fusion reactions mechanisms is the influence of nucleon transfer between the colliding nuclei. Recent measurements on Ca+Ca [5–7] and Ca+Zr [8] systems have showed evidence for the effect of positive Q-value transfer coupling at sub-barrier energies but no unambiguous conclusion concerning heavier systems has yet been drawn. As a matter of fact in heavy systems involving soft vibrational nuclei multi-phonon excitations become dominant and, as a consequence, it is more difficult to disentangle a possibile influence of transfer in the trend of the excitation function.

Fusion in the Ni+Sn system, widely studied in the past down to cross sections of about 0.1-1 mb [9–12], can give important information in this sense. Two recent publications, concerning the influence of transfer in the fusion process, came to different conclusions. In the first one [13], where the excitation function of ^{58}Ni+^{132}Sn could only be measured down to the 0.1-1 mb region due to the low intensity of the secondary ^{132}Sn beam, it was claimed that the influence of transfer for Ni+Sn is negligible compared to lighter systems; in the other one [14] it was concluded that the fusion enhancement due to transfer couplings should be present in heavier systems as well.

In the latter an attempt of extracting a systematics from several cases is presented. The Wong formula [15] is used to explore the influence of transfer on fusion in a phenomenological way. This formula is a good approximation of the fusion cross section in the region from 1000 to around 1 mb by means of only three parameters, as we can see in Eq. 1:

$$\sigma = \frac{R_C^2}{2E}\,\hbar\omega\, ln\left[1 + e^{\frac{2\pi}{\hbar\omega}(E - V_C)}\right], \tag{1}$$

where R_C, V_C and $\hbar\omega$ correspond to the radius, the height and the curvature of the fusion barrier, respectively.

The results clearly show that large values of $\hbar\omega$ (corresponding to thin barriers and then in favour to a fusion enhancement) are always associated with positive Q values for neutron transfer, in both medium (Ca+Ca, Si+Ni, Ni+Ni) and heavy systems (Ca+Zr, Ca+Sn, Ni+Sn). In particular, excitation functions have a shallower slope for systems where a neutron-poor projectile collides on a neutron-rich target [14].

In Figure 1 excitation functions for different Ni+Sn isotopes are reported and fitted with the Wong formula, which of course fails to reproduce the experimental trend in the deep sub-barrier energy region. For ^{58}Ni+^{124}Sn and ^{64}Ni+^{124}Sn the new experimental data are also reported but they are irrelevant for this part of the discussion.

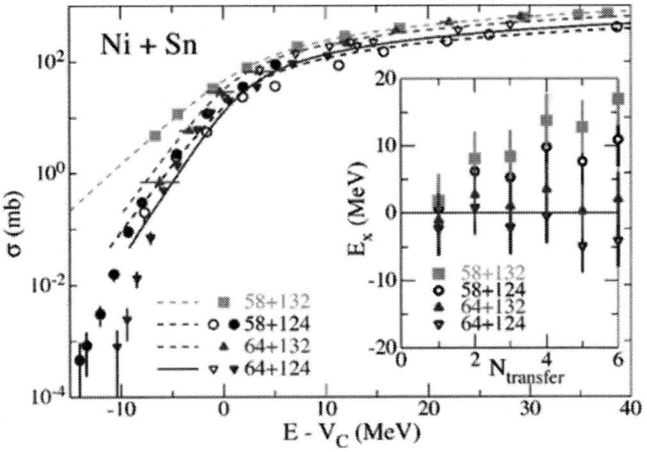

Figure 1: Fusion cross sections for different Ni+Sn systems plotted versus the difference between the centre-of-mass energy and the Coulomb barrier extracted from the fit with the Wong formula (see Equation 1). Insert: average excitation energies of neutron pickup reactions calculated from the Q-value systematics for the same systems. Colours in the online version.

The most remarkable feature is, as already pointed out, the close connection between transfer channels with positive Q value (see the insert in Figure 1 and Table 1) and the shallower slope of the associated excitation function.

This is not sufficient to draw the conclusion that the coupling to transfer

Table 1: Q-values for one- and two-neutron pickup by ^{58}Ni and ^{64}Ni from the two Sn isotopes of interest. For the combination neutron-poor projectile with neutron-rich target (or viceversa) the Q values are always positive and larger with respect to the more symmetric combinations.

	^{58}Ni+^{124}Sn	^{64}Ni+^{124}Sn	^{58}Ni+^{132}Sn	^{64}Ni+^{124}Sn
Q(+1n)	0.510 MeV	-2.391 MeV	1.050 MeV	-1.245 MeV
Q(+2n)	5.952 MeV	0.615 MeV	7.833 MeV	8.497 MeV

channels plays an important role also in the fusion process of heavier systems. A possible confirmation needs an extension of the experimental data to lower cross sections and a comparison with detailed coupled-channels calculations.

To this end an experiment was performed at the XTU Tandem accelerator of Laboratori Nazionali of Legnaro, Italy, using doubly-stripped beams of ^{58,64}Ni with a current of 1-3 pnA on a ^{124}Sn target with a thickness of 50 μg/cm^2, evaporated on a 20 μg/cm^2 carbon backing. The evaporation residues were detected with the LNL electrostatic separator in its upgraded configuration [16], consisting on two micro-channel plate detectors, one ionization chamber and a silicon detector. More details about the experiment can be found in Ref. [17].

In Figure 2 the experimental results are presented together with the theoretical calculations.

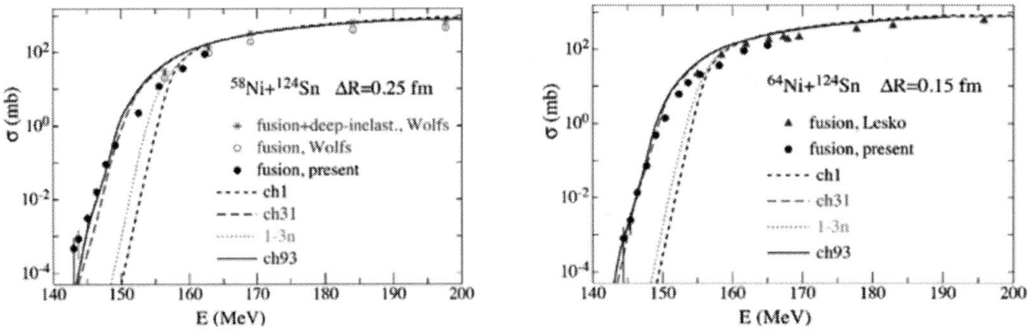

Figure 2: Experimental excitation functions from the present experiment and comparison with CC calculations for ^{58}Ni+^{124}Sn (on the left) and ^{64}Ni+^{124}Sn (on the right). The parameter ΔR represents the radius shift to be added to the fusion radius of the reactions in order to obtain a good reproduction of the experimental data at low energies. Colours in the online version.

The fusion excitation functions already include the fusion-fission contribution, measured by Lesko *et al.* [10] and Wolfs *et al.* [11]. In the vicinity of the Coulomb barrier the CC calculations still overestimate the fusion cross section, maybe because the deep-inelastic yield is not included, as it was measured only for ^{58}Ni and only at relatively high energies.

Figure 3 shows a zoom of Figure 2 in the low-energy region.

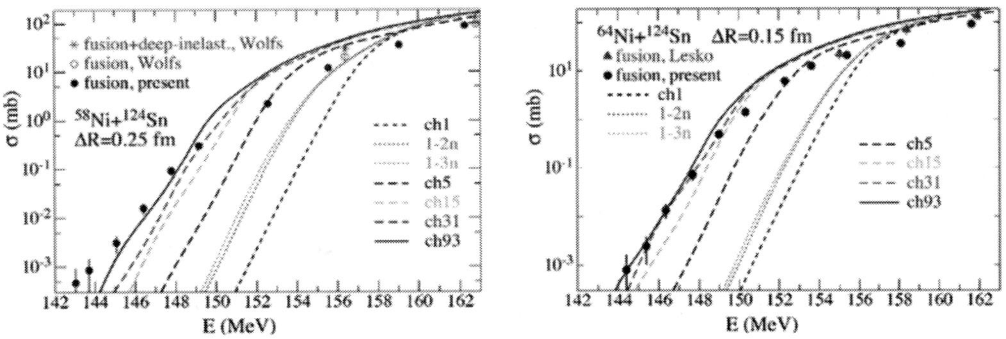

Figure 3: A zoom of Figure 2 in the low-energy region. For details on the CC calculations see text. Colours in the online version.

As no indication of fusion hindrance was observed, a standard Woods-Saxon potential has been used to describe the ion-ion interactions. In Figure 2 and 3 the black dot-dashed line (indicated with ch1) is the result of the calculation where no coupling is taken into account. The dotted magenta and green lines include the coupling to one- and two- (1-2n) and to one-, two- and three-neutron transfer (1-3n), respectively. The dashed curves represent the effect of including inelastic excitations. The black and light-blue lines show the effect of including one- (ch5) and two-phonon (ch15) excitations; in the latter all one- and two-phonon states and mutual excitations of the low-lying 2^+ and 3^- states in the projectile and/or the target are included. The red dashed curve shows the results of a three-phonon calculation where the three mutual one-phonon excitations and the mutual excitations of one- and two-phonon states are included whereas the three-phonon excitations of the same mode are excluded, resulting in 31 channels (ch31).

When we include in the calculations the couplings of both inelastic (one-, two- and three-phonon) and transfer (one- and two-neutron) reactions we obtain a total of 93 channels (ch93), represented by the solid blue line.

Parameters for inelastic excitations were taken from the literature: in the case of ^{58}Ni from previous coupled-channel calculations for multi-neutron transfer reactions in ^{58}Ni+^{124}Sn collisions around the Coulomb barrier [18]; in the case of ^{64}Ni from the analysis of the fusion of ^{64}Ni+^{64}Ni [19]. These parameters can be found, together with spectroscopic factors for the strongest one-neutron pickup channel by ^{58}Ni and ^{64}Ni and parameters of the Woods-

Saxon potentials used in the CC calculations, in Table II, III and IV of Ref. [17].

A comparison between the two excitation functions exhibits several interesting features.

- The largest enhancement of the fusion cross section is due to the coupling of inelastic excitations, as expected for these heavy systems.

- The system ^{58}Ni+^{124}Sn shows a larger influence from the coupling to transfer reactions, as expected from the Q-value systematics previously described. For the system ^{64}Ni+^{124}Sn the contribution from transfer is weaker, due to the smaller Q values, but not negligible.

- For both systems the influence of transfer shows up mainly in the cross section region below 1 mb. This may partially explain why in the work by Kohley *et al.* [13] no contribution from transfer was identified.

- Both systems show a reduction of the experimental cross section compared to the CC calculations around 151 MeV. The reason is not yet fully understood, although it could be ascribed, as pointed out before, to the contribution from the deep-inelastic scattering.

- There is no evidence for fusion hindrance. From systematics involving both soft (open-shell) and stiff (closed-shell) systems, the appearance of an S-factor maximum, which would be an evidence for hindrance, is expected at center-of-mass energies lower than 140 MeV and at cross sections at the nb level. Therefore to see the hindrance behaviour for these systems an improvement of the experimental sensitivity by a factor $\sim$100 is needed.

References

[1] M. D. Dasgupta, D. J. Hinde, N. Rowley and A. M. Stefanini, Annu. Rev. Nucl. Part. Sci. **48**, 401 (1998).

[2] A. B. Balantekin and N. Takigawa, Rev. Mod. Phys. **70**, 77 (1998).

[3] B. B. Back, H. Esbensen, C. L. Jiang and K. E. Rehm, Rev. Mod. Phys. **86**, 317 (2014).

[4] C. L. Jiang *et al.*, Phys. Rev. Lett. **89**, 052701 (2002).

[5] G. Montagnoli *et al.*, Phys. Rev. C **85**, 024607 (2012).

[6] C. L. Jiang *et al.*, Phys. Rev. C **82**, 041601 (2010).

[7] A. M. Stefanini *et al.*, Phys Lett. B **679**, 95 (2009).

[8] A. M. Stefanini *et al.*, Phys Lett. B **728**, 639 (2014).

[9] W. S. Freeman *et al.*, Phys. Rev. Lett. **50**, 1563 (1983).

[10] K. T. Lesko *et al.*, Phys. Rev. Lett. **55**, 803 (1985);
K. T. Lesko *et al.*, Phys. Rev. C **34**, 2155 (1986).

[11] F. L. H. Wolfs *et al.*, Phys Lett. B **196**, 113 (1987);
F. L. H. Wolfs, Phys. Rev. C **36**, 1379 (1987).

[12] J. F. Liang *et al.*, Phys. Rev. C **78**, 047601 (2008).

[13] Z. Kohley *et al.*, Phys. Rev. Lett. **107**, 202701 (2011).

[14] C. L. Jiang *et al.*, Phys. Rev. C **89**, 054602 (2014).

[15] C. Y. Wong, Phys. Rev. Lett. **31**, 766 (1973).

[16] A. M. Stefanini *et al.*, Phys Rev. C **82**, 014614 (2010).

[17] C. L. Jiang *et al.*, Phys. Rev. C **91**, 044602 (2015).

[18] H. Esbensen, C. L. Jiang and K. E. Rehm, Phys Rev. C **57**, 2401 (1998).

[19] C. L. Jiang *et al.*, Phys. Rev. Lett. **93**, 012701 (2004).

EPJ Web of Conferences **117**, 08012 (2016) DOI: 10.1051/epjconf/201611708012
NN2015

Isospin influence on the decay modes of the systems produced in the $^{78,86}Kr + ^{40,48}Ca$ reactions at 10 AMeV

B. Gnoffo[1], S. Pirrone[1], G. Politi[1,2], M. La Commara[3,4], J.P. Wieleczko[5], E. De Filippo[1], P. Russotto[1], M. Trimarchi[7,8], M. Vigilante[3,4], G. Ademard[11], F. Amorini[6], L. Auditore[7,8], C. Beck[9], I. Bercenau[10], E. Bonnet[5], B. Borderie[11], G. Cardella[1], A. Chibihi[5], M. Colonna[6], A. D'Onofrio[4,12], J.D. Frankland[5], E. Geraci[1,2], E. Henry[13], E. La Guidara[1,14], G. Lanzalone[15,6], P. Lautesse[16], D. Lebhertz[5], N. LeNeidre[17], I. Lombardo[3], K. Mazurek[5], S. Norella[7,8], A. Pagano[1], E.V. Pagano[2,6], M. Papa[1], E. Piasecki[5,10], F. Porto[9,1], L. Quattrocchi[7,8], M. Quinlann[13], F. Rizzo[1,6], U. Shoroeder[13], G. Spadaccini[3,4], A. Trifirò[7,8], J. Toke[13], G. Verde[1].

[1] INFN, Sezione di Catania, Catania, Italy
[2] Dipartimento di Fisica e Astronomia, Univ. di Catania, Italy
[3] Dip. di Fisica, Univ. Federico II Napoli, Italy
[4] INFN, Sezione di Napoli, Napoli, Italy
[5] Ganil, CAEN, France
[6] INFN - Laboratori Nazionali del Sud, Catania, Italy
[7] Dip. di Fisica, Univ. Messina, Italy
[8] INFN, Gruppo coll. di Messina, Messina, Italy
[9] IN2P3-IPHC Strasbourg, France
[10] IPNE, Bucarest, Romania
[11] IN2P3-IPN Orsay, France
[12] Dip. di Matematica e Fisica, Seconda Univ. di Napoli, Caserta, Italy
[13] University of Rochester, Rochester, NY, USA
[14] Centro Siciliano Fisica Nucleare e Struttura della Materia, Catania, Italy
[15] Università "Kore", Enna, Italy
[16] IN2P3-IPN Lyon, France
[17] IN2P3-LPC CAEN, France
[18] Faculty of Physics and Heavy Ion Laboratory, Univ. of Warsaw, Poland
[19] INFN - Laboratori Nazionali di Legnaro, Legnaro (PD), Italy

Abstract

The results of the analysis of the reactions $^{78,86}Kr +^{40,48} Ca$ at 10 AMeV are presented. The experiment was performed at the INFN Laboratori Nazionali del Sud (LNS) in Catania by using the 4π multidetector CHIMERA, with beams delivered by the Superconductive Cyclotron. The competition among the various disintegration paths and in particular the isospin effects on the decay modes of the produced composite systems are investigated; this provides information about fundamental nuclear quantities such as level density, fission barrier and viscosity. Different isotopic composition and relative richness are observed among the reaction products of the two systems. An odd-even staggering effect is present in the charge distributions, in particular for the light fragments produced by the neutron-poor system. The kinematical characteristics of the IMF seem to indicate a high degree of the relaxation of the formed system. Besides, global features analysis seems to show some differences in the contribution arising from the various reaction mechanisms for the two reactions.

1 Introduction

Nuclear reactions between medium-mass nuclei at low energy domain are characterized by the competition between binary processes and compound nucleus de-excitations. The isospin plays a crucial role in the competition between these decay channels, thus novel information on fundamental nuclear quantities such as level density, fission barrier and viscosity could be extracted across the long isotopic chains of compound nuclei, extending from the neutron-rich to neutron-poor side. The neutron richness influence on the formation and decay modes of compound nuclei ^{118}Ba and ^{134}Ba, produced respectively in the $^{78}Kr +^{40} Ca$ and $^{86}Kr +^{48} Ca$ reactions at 10 AMeV are investigated [1, 2]. The experiment was performed by using the 4π multidetector, for charge particles, CHIMERA [3], allowing precise measurements of velocity, angular distributions and a good identification [4] of the reaction products with low energy thresholds.

2 Experimental results

2.1 Fission Fragments characteristics

The decay modes populate the whole mass/charge domain from evaporated light charged particles (LCP) up to the Evaporation Residues (ER) with the symmetric Fission Fragments (FF) and the Intermediate Mass Fragments (IMF) in between. The IMF cross section production exhibits for both systems a strong even-odd effect, the staggering, due to a preferential production of fragments with an even value of the atomic number. In agreement with other examples in literature [5–7] the staggering is more pronounced for the n-poor system respect to the n rich one, in particular for Z<10 reaction products as it is shown in fig. 1. This effect persists for higher Z, but with a smaller amplitude.

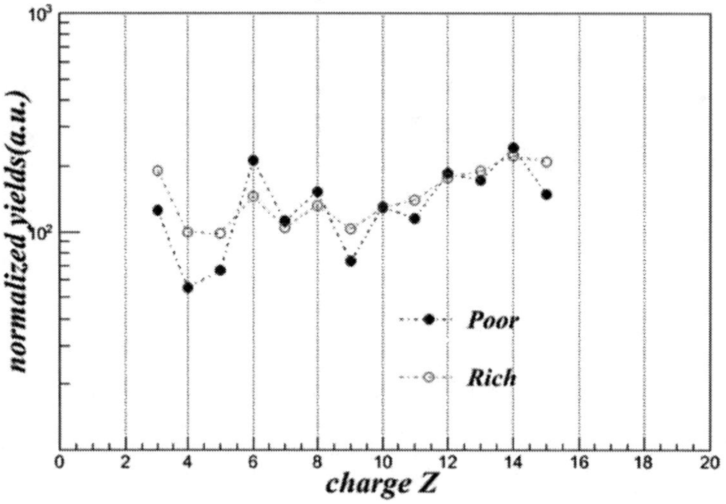

Figure 1: IMF Charge distributions, in red for the $^{86}Kr + ^{48}Ca$ reaction, in black for the $^{78}Kr + ^{40}Ca$ reaction (color on line)

The analysis of the kinematical features of the IMF indicates a high relaxation of the degrees of freedom, suggesting a production via a long lived system. In fact the mean values of the velocity of the IMF in the center of mass frame (left hand panel of figure 2), that are independent of the emission angle, show a quasilinear decrease increasing the atomic number and they are well reproduced by the values predicted by the systematic of Viola, with the Hinde's correction for the asymmetric fission [8]. Besides the angular distributions of the IMF and the other FFs, in the center of mass frame, follow a $1/sin\theta_{c.m.}$ behavior as it is shown in right hand panel of the figure 2.

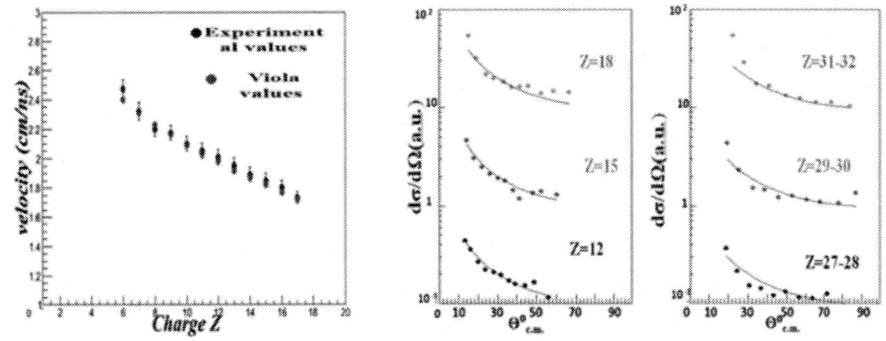

Figure 2: For the $^{78}Kr + ^{40}Ca$ reaction. Comparison of the mean values of the velocity, averaged over all angles, with the values of Viola (Left). FFs angular distributions, in the center of mass frame, the solid line is $1/sin\theta_{c.m.}$ (Right) (color on line)

2.2 Evaporation Residues features

In the figure 3 are reported the angular distributions of the heavier reaction products in the laboratory frame; these distributions are very strongly forward peaked, as it is expected for the evaporation residues. Besides the extracted centroids of the velocity spectra at each angle θ, in the laboratory frame, lie along the solid line, that represents the velocity of the Compound Nucleus (CN) multiplied by the $cos\theta$. This is a signal that these products have the same velocity of the Compound Nucleus.

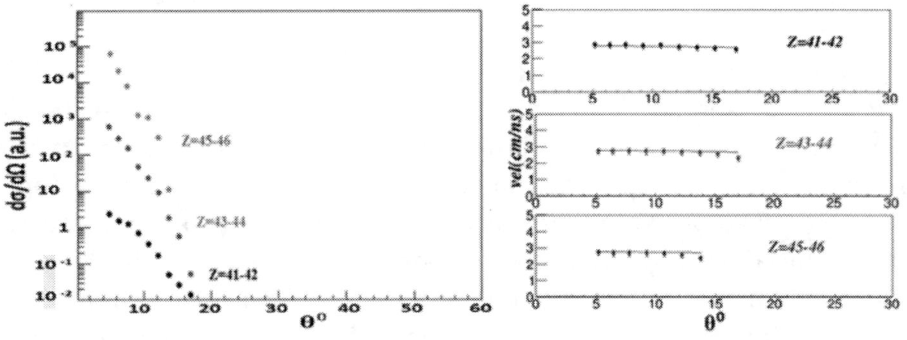

Figure 3: Left hand panel: the angular distributions in the laboratory frame for ER. Right hand panel: values of the mean velocity in the laboratory frame for Er at different angles θ.

2.3 Global decay patterns

Complete events are selected by imposing the conditions: detected mass M_{tot} >90 and momentum $P_{tot} > 0.7P_{beam}$. By looking at the plot mass versus parallel velocity for each reaction product, shown in fig 4, some qualitative information about the competition among the different decay channels could be extracted. In fact in this plot the different reaction products, fragments coming from the Fusion-Fission and Quasi-Fission or Evaporation Residues, can be easily localized. The analysis shows that the ER/FF ratio is higher for the n-poor system compared to n-rich one.

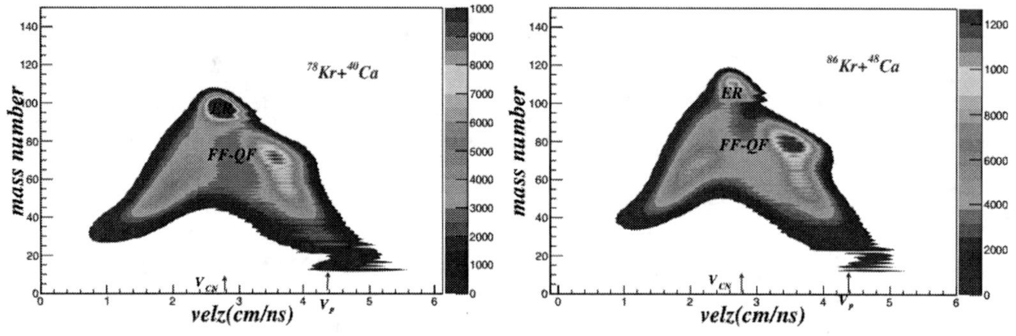

Figure 4: mass vs parallel velocity plot for the n-poor system (Left) and for the n-rich one (Right)

3 Conclusions

Some results of the analysis of the reactions $^{78}Kr +^{40} Ca$ and $^{86}Kr +^{48} Ca$ at 10 AMeV are presented. The exit channel retains some memories of the entrance channel, in fact the reaction products are neutron richer in the reaction $^{86}Kr +^{48} Ca$ [9]. As a consequence, deep differences in the isotopic composition and relative abundance in the production cross sections of the fragments produced are observed. An even-odd staggering effect is observed in the IMF charge distributions and it is stronger for the n-poor system. The angular distributions and the features of the velocity of the IMF suggest a high degree of relaxation of the degrees of freedom of the system formed in the collisions before the breakup, as in a typical fusion-fission like reaction mechanism. The analysis of the kinematical characteristics of the ER suggests that these products have the same velocity of the CN. The analysis of global features seems to show differences in the contribution arising from the various reaction mechanisms like fusion-fission, quasi-fission and fusion-evaporation. Data analysis are in progress. Important information could be obtained from the study of the LCP-FF, LCP-IMF and LCP-ER coincidences. To confirm the preliminary qualitative results, it is necessary to complete the calculations of the cross sections. Precise comparisons with statistical and dynamical models are necessary to provide indications of the isospin influence on the competition between different decays modes of the Compound Nuclei. In order to study the evolution, with isospin asymmetry, of the reaction dynamics and to investigate the interplay between the nuclear structure and reaction mechanism a LOI was presented at the "Second SPES International Workshop at INFN-LNL" to study the reactions: $^{92}Kr +^{40,48} Ca$ at 10 AMeV by using neutron rich radioactive beams.

References

[1] S. Pirrone *et al.*, Journ. Phys. Conf. **515**, 012018 (2014).

[2] G. Politi *et al.*, JPS: Conf. Proc. **6**, 030082 (2015).

[3] A. Pagano *et al.*, Nuc. Phys. news **22**, 25 (2012).

[4] G. Politi *et al.*, IEEE NSS: Conf. Rec., N28-5 (2005).

[5] G. Ademard *et al.*, Phys. Rev. C**83**, 054619 (2011).

[6] I. Lombardo *et al.*, Phys. Rev. C**84**, 024613 (2011).

[7] G. Casini *et al.*, Phys. Rev. C**86**, 011602 (2012).

[8] D. J. Hinde *et al.*, Nuc. Phys. A**472**, 318 (1987).

[9] E. De Filippo and A. Pagano, EPJ A**50**, 32 (2014).

EPJ Web of Conferences **117**, 08013 (2016) DOI: 10.1051/epjconf/201611708013
NN2015

Effect of breakup and transfer on complete and incomplete fusion in ^{6}Li+^{209}Bi reaction in multi-body classical molecular dynamics calculation

Mitul R. Morker; Subodh S. Godre

Department of Physics, Veer Narmad South Gujarat University,
Surat 395007, India

Abstract

The effect of breakup and transfer in ^{6}Li+^{209}Bi reaction is studied in a multi-body classical molecular dynamics approach in which the weakly-bound projectile ^{6}Li is constructed as a 2-body cluster of ^{4}He and ^{2}H in a configuration corresponding to the observed breakup energy. This 3-body system with their individual nucleon configuration in their ground state is dynamically evolved with given initial conditions using Classical Rigid Body Dynamics (CRBD) approach up to distances close to the barrier when the rigid-body constraint on the target, inter-fragment distance, and ^{2}H itself are relaxed, allowing for possible breakup of ^{2}H which may result in incomplete fusion following the transfer of the n or p. Relative probabilities of the possible events such as scattering with and without breakup, DCF, SCF, ICF(x) where x may be ^{4}He, ^{2}H, ^{4}He+n, ^{4}He+p, n, p are calculated. Comparison of the calculated event-probabilities, complete, and incomplete fusion cross sections with the calculation in which ^{2}H is kept rigid demonstrates the effect of the transfer reactions on complete and incomplete fusion in the 4-body reaction. Events ICF(^{4}He+n) corresponding to n-stripping followed by breakup of the resultant ^{5}Li to ^{4}He+p are found to contribute significantly in the fusion process in agreement with a recent experimental observation of direct reaction processes in breakup of weakly-bound projectiles.

DOI: 10.1051/epjconf/201611708013

1 Introduction

Fusion reactions involving weakly-bound projectiles are complicated by the breakup of the projectile into two fragments. Complete fusion (CF) involves capture of both the fragments directly (DCF) or sequentially (SCF); incomplete fusion (ICF) involves capture of only a part of the projectile [1]. If one of the fragment itself is a weakly-bound nucleus like ^{2}H, then its own breakup in the approach phase is also possible, resulting in nucleon transfer processes and breakup of the remaining unstable projectile residue. In a recent experiment it is observed that breakup of projectiles like ^{6}Li is predominantly triggered by nucleon transfer such as n-stripping [2].

Such reactions are usually studied in CDCC formalism [1], a semi-classical coupled channel approximation [3] and, a classical trajectory model [4]. However, none of these approaches account for CF, ICF, and ICF following nucleon transfer, within the same model calculation. The multi-body, 3-Stage Classical Molecular Dynamics (3S-CMD) model [5], apart from demonstrating CF and ICF events is also able to account for a process equivalent to a direct reaction leading to ICF in the same model calculation.

The details of the present model calculation are given in section-2. Using this model, the effect of breakup and transfer on CF and ICF in ^{6}Li+^{209}Bi collisions is studied by calculating various event-probabilities which are discussed in section-3. Calculated CF, ICF and total fusion (TF) cross section are presented in section-4. Finally conclusions are given in section-5.

2 Calculation details

The ^{6}Li+^{209}Bi collision is studied in a multi-body, 3S-CMD model [5] in which ^{6}Li is constructed as a cluster of ^{4}He(α) and ^{2}H(d) nuclei held together in a configuration corresponding to the observed breakup energy equal to 1.467 MeV. The projectile fragments and the target are first generated with a variational potential minimization code with a soft-core Gaussian NN-potential and approximately reproduced ground state properties of the nuclei as in ref [5]. Collision simulation is carried out in the three stages:(1)The projectile and the target nuclei are initially brought along their Rutherford trajectories, (2) This system is then dynamically evolved using Classical Rigid Body Dynamics (CRBD) up to distances close to the barrier, followed by (3) CMD evolution of the entire many-body system. If one or both the projectile fragments are further constrained to be rigid then these nuclei are dynamically evolved as in the CRBD calculation even in the stage-3.

The rigid-body constraint on the bond between ^{2}H and ^{4}He in ^{6}Li, as well as on the target ^{209}Bi are relaxed in the stage-3 for R_{cm} <13 fm. One of the projectile fragments (^{4}He) is always kept rigid. By allowing ^{2}H in the projectile to be non-rigid, thereby allowing the possibility of its own breakup, and comparing the results with the calculation in which it is kept rigid even in the stage-3 near or inside the barrier, demonstrates the effect of direct reaction process in this reaction.

3 Event probabilities

Dynamical simulations of ^{6}Li+^{209}Bi collisions with different impact parameters, energies and initial orientations may result in events such as scattering with (NCBU) and without breakup (NBUS), DCF, SCF, ICF(x) where x is either ^{4}He or ^{2}H which is captured. When ^{2}H is also allowed to breakup near the target, x may also be ^{4}He+n, ^{4}He+p, n or p.

Events fractions defined as $F(b)=(\boldsymbol{N}_{events}/\boldsymbol{N}_{total})$ are calculated, where $\boldsymbol{N}_{total}$ is total no of initially random orientations for given E_{cm} and b, and $\boldsymbol{N}_{events}$ is no of trajectories analyzed as DCF, SCF, ICF etc. events. Trajectories for b=0 fm to b_{max}=8.6, 7.0, 5.0 fm for E_{cm}=50, 36, 29 MeV, respectively, in steps of 0.2 fm are analyzed. For each value of b, $\boldsymbol{N}_{total}$=500 for E_{cm}=50 and 36 MeV, and 2000 for 29 MeV. Calculated $F(b)$ when ^{2}H is kept rigid are shown in figure-1(a) for E_{cm}=50 MeV. Events following breakup (ICF+NCBU) increases with b but decreases again for larger values of b. DCF is the major component of CF with a few SCF events also. Events ICF(^{4}He) which are negligible at low b, rises to a peak value at higher b.

Figure-1(b) shows the results when ^{2}H is kept non-rigid in stage-3 for E_{cm}=50 MeV. This figure shows events ICF(^{4}He+n) which are equivalent to n-stripping followed by breakup of the resultant unstable $^5\text{Li} \rightarrow {}^4\text{He}+p$. Distribution of ICF(^{4}He+n) is much broader and larger compared to ICF(^{4}He) in figure-1(a). This is in qualitative agreement with the experiment [2] which shows the importance of direct reaction processes. Similar calculations for E_{cm}= 36 and 29 MeV shows reduction in events following breakup.

Integration of $F(b)$ over $b \leq b_{max}$ for a given E_{cm} gives event-probabilities, shown as staked bar-charts in figure-2. It can be noted from figure-2(a), for ^{2}H-rigid case, that DCF, SCF and CF events increase with E_{cm}. Figure-2(b) shows ICF(^{4}He+n) events which are seen even at the lowest energy. Its relative importance increases as E_{cm} increases, contributing significantly in the fusion process. The relative strength of ICF(^{4}He+p) events corresponding to p-stripping is much smaller compared to ICF(^{4}He+n) events.

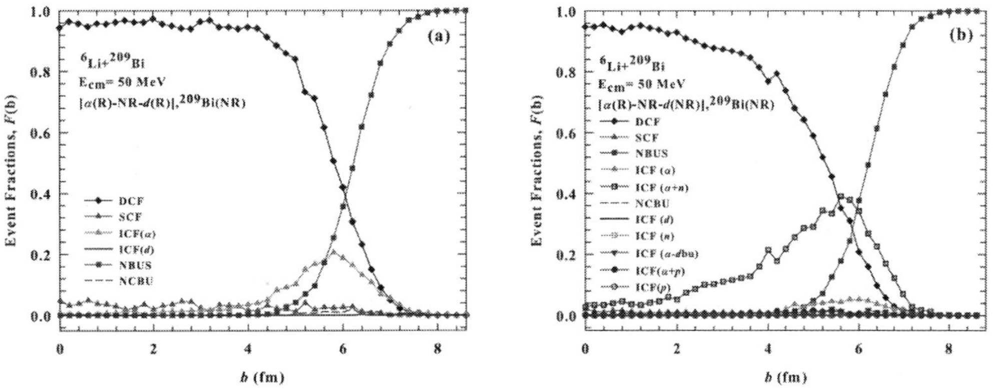

Figure 1: Event fractions for (a) ^{2}H(d)-rigid(R), (b) ^{2}H(d)- non-rigid(NR)

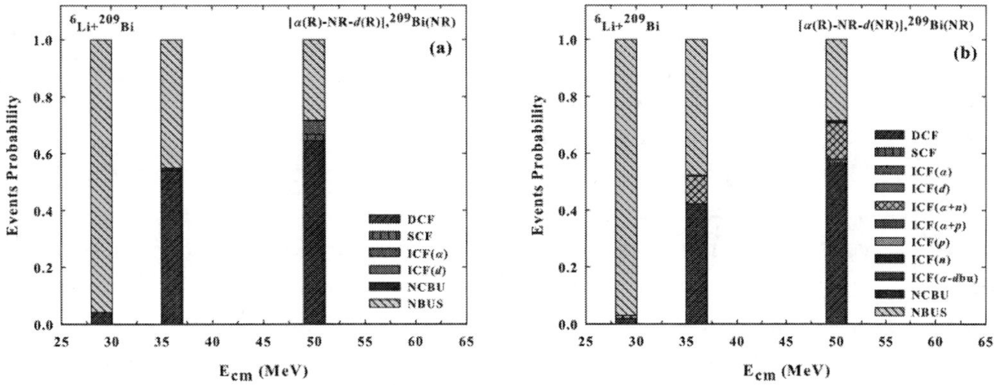

Figure 2: Event probabilities for (a) ^{2}H(d)-rigid(R), (b) ^{2}H(d)- non-rigid(NR)

4 Fusion cross sections

CF is defined as an event in which ^{2}H and ^{4}He both are captured by ^{209}Bi for a long interval of time. ICF is defined as an event in which only one of the projectile fragments or a part of the projectile is captured by ^{209}Bi after their break-up. In a sharp cut-off approximation, it is assumed that all the trajectories with $b<b_{cr-CF}$, critical impact parameter for CF, are fused and those for $b>b_{cr-CF}$ are scattered or lead to ICF events up to $b<b_{cr-TF}$. Scattering with or without breakup occur for $b>b_{cr-TF}$.

For given E_{cm}, a large number of Monte-Carlo sampled initial orientations (as in section-3) are considered and b_{cr-CF} and b_{cr-TF} are determined for every orientation. Ion-ion potential is obtained as a function of the separation between the centre of masses of the target and all the projectile

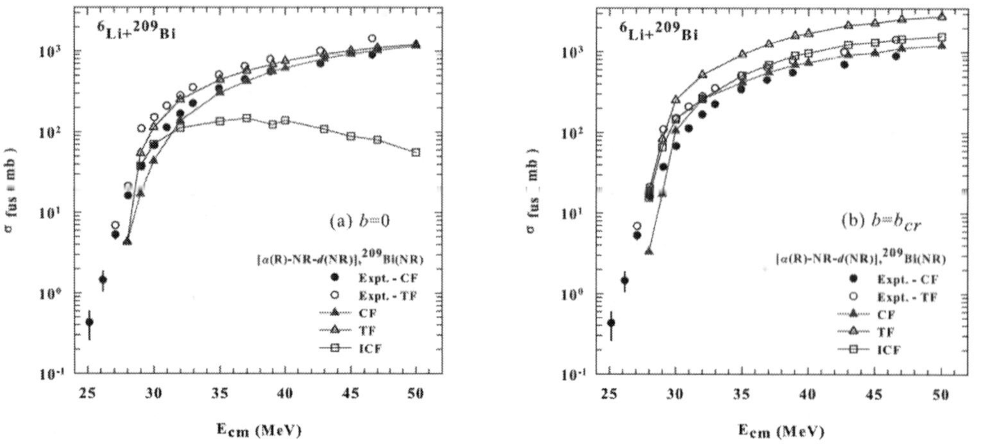

Figure 3: CF, TF and ICF cross sections for (a) b=0 and (b) b=b_{cr}

fragments for b=b_{cr-CF} or the the centre of masses of the target and the projectile-fragment that is captured for b=b_{cr-TF}.

Barrier parameters obtained from the ion-ion potential corresponding to b_{cr-CF} or b_{cr-TF} are used in the Wong's formula [6] to calculate $\sigma_{\rm CF}$ or $\sigma_{\rm TF}$ respectively for a given orientation. An average over different orientations gives $\sigma_{\rm CF}$ and $\sigma_{\rm TF}$ for a given collision energy. ICF cross section are calculated from $\sigma_{\rm TF}$=$\sigma_{\rm CF}$+$\sigma_{\rm ICF}$.

Conventionally, one uses the Wong's formula with b=0 approximation. Calculated $\sigma_{\rm CF}$ and $\sigma_{\rm TF}$ for ^{6}Li+^{209}Bi reaction for b=0 are shown in figure-3(a) and compared with the experimental $\sigma_{\rm CF}$ and $\sigma_{\rm TF}$ [7]. Experimental $\sigma_{\rm TF}$ shown in figure-3 are obtained as a sum of the $\sigma_{\rm CF}$ and $\sigma_{\rm ICF}$ of ref[7]. The calculated cross sections with b=0 seems to match well with the experimental $\sigma_{\rm CF}$ and $\sigma_{\rm TF}$ respectively, except at very low energies.

The b=0 approximation in the Wong formula is, however, justified at low energies only. Higher partial wave may contribute significantly at higher energies [8]. We still use the Wong formula but, with the barrier parameters corresponding to the critical impact parameter as in ref [5, 9]. Calculated $\sigma_{\rm CF}$ at b_{cr-CF} from ref [5] are shown in figure-3(b). Moreover, since the contribution of ICF increases at higher values of b, reaches a maximum and diminishes again at grazing impact parameters. Therefore, to account for the effect of increased number of ICF events, the cross sections must include the contributions from trajectories with higher b also. Therefore, $\sigma_{\rm TF}$ is calculated for b=b_{cr-TF} and the results for CF and TF for central and non-central collisions are compared in figure-3.

The $\sigma_{\rm CF}$ and $\sigma_{\rm TF}$ calculated using the Wong formula with barrier parameters corresponding to b_{cr-CF} or b_{cr-TF} are shown in figure-3(b) and compared with the corresponding experimental data. The differences in the $\sigma_{\rm CF}$ and $\sigma_{\rm TF}$ correspond to $\sigma_{\rm ICF}$ which are also shown in figure-3. Calculated $\sigma_{\rm ICF}$ in figure 3(b) corresponding to b=b_{cr-TF} is much larger at higher energies compared to the calculated $\sigma_{\rm ICF}$ in figure-3(a) for b=0 case. This large difference arises because of the increased number of ICF events at b>0 when ^{2}H is non-rigid. The $\sigma_{\rm CF}$ and $\sigma_{\rm ICF}$ in figure-3(b) are of the same order of magnitude, although $\sigma_{\rm CF}$ in figure-3(b) are enhanced compared to those in figure-3(a). The $\sigma_{\rm TF}$ in figure-3(b) are also enhanced due to the enhanced values of $\sigma_{\rm ICF}$ and are overestimated compared to the experimental values.

5 Conclusions

The effect of breakup and transfer in ^{6}Li+^{209}Bi reaction is studied in a multi-body classical molecular dynamics approach. Comparison of the calculated event probabilities, $\sigma_{\rm CF}$ and $\sigma_{\rm TF}$ between, a 4-body simulation in which ^{2}H is allowed to breakup near the barrier, and the 3-body calculation in which ^{2}H is kept rigid, clearly demonstrates the effect of the transfer reactions on complete and incomplete fusion. It is found that ICF(^{4}He+n) corresponding to n-stripping contributes significantly in the fusion process. This observation is in agreement with the experimental observation of the importance of direct reaction processes in breakup of weakly-bound projectiles. The present approach is able to account for DCF, SCF, ICF and a process equivalent to a direct reaction leading to ICF process in the same model calculation.

References

[1] L. F. Canto *et al*, Phys. Rep. **424**, 1 (2006); *and references therein.*
[2] D. H. Luong *et al*, Phys. Rev. C **88**, 034609 (2013).
[3] H. D. Marta *et al*, Phys. Rev. C **89**, 034625 (2014).
[4] A. Diaz-Torres *et al*, Phys. Rev. Lett. **98**, 152701 (2007).
[5] M. R. Morker, S. S. Godre, EPJ Web of Conf. **86**, 28 (2015).
[6] C. Y. Wong, Phys. Rev. Lett. **31**, 766 (1973).
[7] M. Dasgupta *et al*, Phys. Rev. C **70**, 024606 (2004).
[8] N. Rowley and K. Hagino, Phys. Rev. C **91**, 044617 (2015).
[9] P. R. Desai, S. S. Godre, Proc. Symp. on Nucl. Phys. **55**, 400 (2010).

EPJ Web of Conferences **117**, 08014 (2016) DOI: 10.1051/epjconf/201611708014
NN2015

Studies for the effect of the positive Q-value neutron transfers on sub-barrier heavy-ion fusion

H.M. Jia, C.J. Lin, L. Yang, X.X. Xu, N.R. Ma, L.J. Sun, F. Yang, Z.D. Wu, H.Q. Zhang, Z.H. Liu and D.X. Wang

China Institute of Atomic Energy, P. O. Box 275(10), Beijing 102413, P. R. China

Abstract

This contribution reports the recent experimental studies for the coupling effect of the positive Q-value neutron transfer (PQNT) channels on near-barrier fusion with heavy-ions, measured by an electrostatic deflector setup. This effect is expected to be important in some situations. In this presentation, the experimental studies for the near-barrier fusion of ^{32}S + ^{90,94,96}Zr, ^{16}O + ^{76}Ge, ^{18}O + ^{74}Ge, and ^{16,18}O + ^{58}Ni are reported. Also a method, based on the CCFULL calculations with suitable inelastic couplings, to sort out the PQNT effect is presented.

1 Introduction

Near-barrier fusion with heavy-ions offers a good platform to study the two basic mechanisms of tunneling and coupling [1,2] in the quantum world. The involvement of the couplings may lead to strong sub-barrier fusion enhancement in some situations. So far, couplings to these collective excitation states can be described well in near-barrier fusion by using the coupled-channels (CC) theory [2]. While couplings to nucleon transfers are even troublesome and have not yet been well understood.

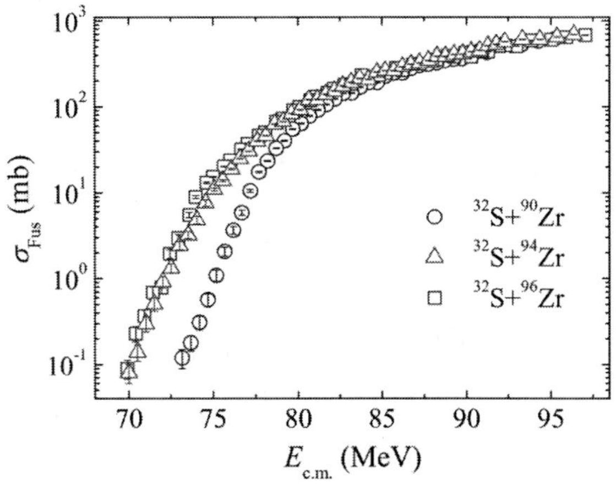

Figure 1: Experimental near-barrier fusion excitation functions of $^{32}S + ^{90,94,96}Zr$.

For the transfer reactions, neutron transfers might be more important than proton transfers especially at sub-barrier energy region. The possible effect of positive Q-value neutron transfers (PQNT) on near-barrier fusion was first discovered by Beckerman *et al.* [3] in the experimental comparison of the fusion excitation functions of $^{58,64}Ni + ^{58,64}Ni$ and, explained firstly by Broglia *et al.* [4] by considering the gained kinematic energy for the (virtual) intermediate states as an important doorway to fusion [5]. Afterwards, many studies [1, 6] have been devoted to PQNT effect. But it is still not clarified for the relevant reaction dynamics up to now.

This contribution reports the recent experimental studies for the coupling effect of PQNT on near-barrier fusion with heavy-ions measured by an electrostatic deflector setup. Also a method, based on the CCFULL [7] calculations with suitable inelastic couplings, to sort out the PQNT effect is presented.

2 Recent near-barrier fusion experiments at CIAE

The experiments were performed at the HI-13 tandem accelerator of CIAE, Beijing. Fusion evaporation residual (ER) cross sections have been measured by using an electrostatic deflector (see Ref. [8] for more details).

2.1 $^{32}S + ^{90,94,96}Zr$

The first studied case is $^{32}S + ^{90,94,96}Zr$ [9]. The idea is to further search for a correlation of sub-barrier fusion enhancement with PQNT channels.

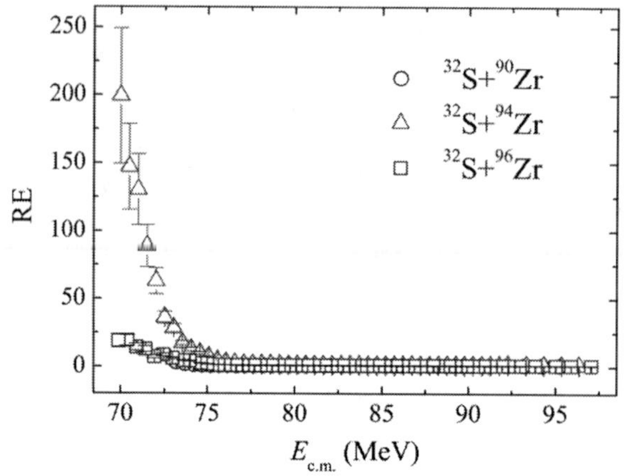

Figure 2: Residual enhancement for near-barrier fusion of ^{32}S + ^{90,94,96}Zr. The original experimental fusion data are taken from Refs. [9].

The fusion excitation functions measured for ^{32}S + ^{90,94,96}Zr are shown in fig. 1, where the energy scale $E_{\mathrm{c.m.}}$ (energy in the center-of-mass system) was corrected for the target thickness and the error bars represent purely the statistical uncertainties. It can be seen that the experimental fusion of ^{32}S + ^{94,96}Zr really shows a strong relation of sub-barrier fusion enhancement to the PQNT channels.

Fusion cross section enhancement reflects the whole effect of couplings to all the relevant channels. For understanding the PQNT effect, the key is to exclude the impact of the collective excitation states. For sorting out the effect related to PQNT, residual enhancement (RE) is defined as the ratio of the experimental fusion data (σ_{Exp}) to the CC calculation result (σ_{CC}) [7] including the major inelastic couplings. That is RE $= \sigma_{\mathrm{Exp}}/\sigma_{\mathrm{CC}}$, by excluding the inelastic coupling effect considering the good description of the CC theory for the inelastic couplings [2].

The corresponding RE for ^{32}S + ^{90,94,96}Zr is shown in fig. 2. It can be seen that the experimental fusion excitation function of ^{32}S + ^{90}Zr is reproduced well. While RE shows deviation from unity with the decreasing energy for ^{32}S + ^{94,96}Zr. Notably, bigger RE shows for ^{32}S + ^{94}Zr than ^{32}S + ^{96}Zr at the same energy, although of the smaller neutron transfer Q_{gg}-values for the former system and larger neutron separation energies for ^{94}Zr. The obtained bigger RE for ^{32}S + ^{94}Zr is reasonable considering the similar experimental fusion cross sections for the two systems but the larger inelastic (3^-) coupling effect for ^{96}Zr. This suggests that more should be considered for a full clarification of the relevant dynamic reaction processes.

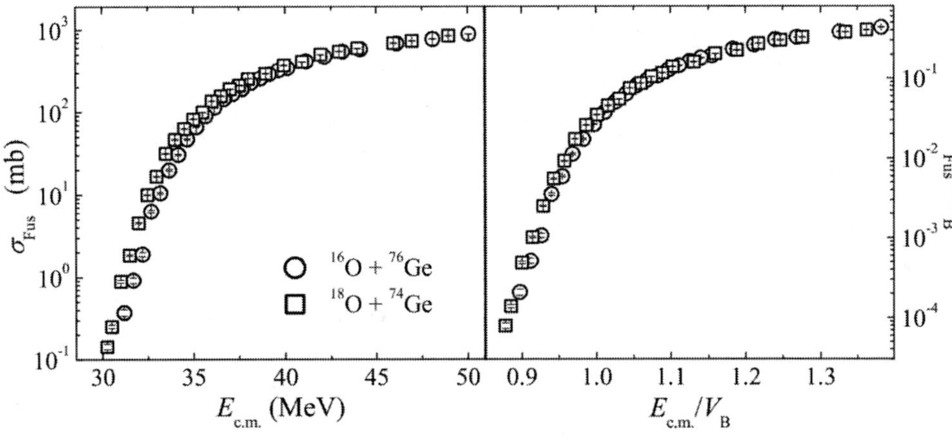

Figure 3: Experimental fusion excitation functions of ^{16}O + ^{76}Ge and ^{18}O + ^{74}Ge (circles and squares, respectively) in absolute (left) and reduced (right) scales.

The role of the neutron orbital populations, such as shell closure of ^{90}Zr and sub-shell closure of ^{96}Zr, should be considered theoretically. Certainly, experimental measurements for the transfer reactions of the ^{94,96}Zr-involved systems should give a meaningful clue for this quantitative correlation.

2.2 ^{16}O + ^{76}Ge and ^{18}O + ^{74}Ge

For simplifying the problem, the systems with only one positive $Q_{-2\mathrm{n}}$-value neutron transfer channel were selected to study. At first, the fusion of ^{18}O + ^{74}Ge with $Q_{-2\mathrm{n}} = +3.75$ MeV and lower inelastic coupling ($Z_\mathrm{p}Z_\mathrm{t}$) effect was studied experimentally [6].

Fusion of ^{16}O + ^{76}Ge was also measured as a reference. The results of the experimental fusion excitation functions are shown in fig. 3 (left). Usually, a method is used in order to remove the trivial geometric effect for a comparison of the coupling effect between different systems. The right panel shows the reduced fusion excitation functions for the two systems in a reduced energy scale, where V_B is the Coulomb barrier energy and R_B is the fusion barrier radius. One can see that the two reduced fusion excitation functions almost overlap at the whole energy region. This means no visible sub-barrier fusion enhancement related to the positive $Q_{-2\mathrm{n}}$-value neutron transfer channel for ^{18}O + ^{74}Ge at the measured energy region. This conclusion is consistent with the result that obtained from the experimental studies of ^{36}S + ^{58}Ni [10] and ^{18}O + ^{A}Sn [11] with positive $Q_{-2\mathrm{n}}$-value neutron stripping channels.

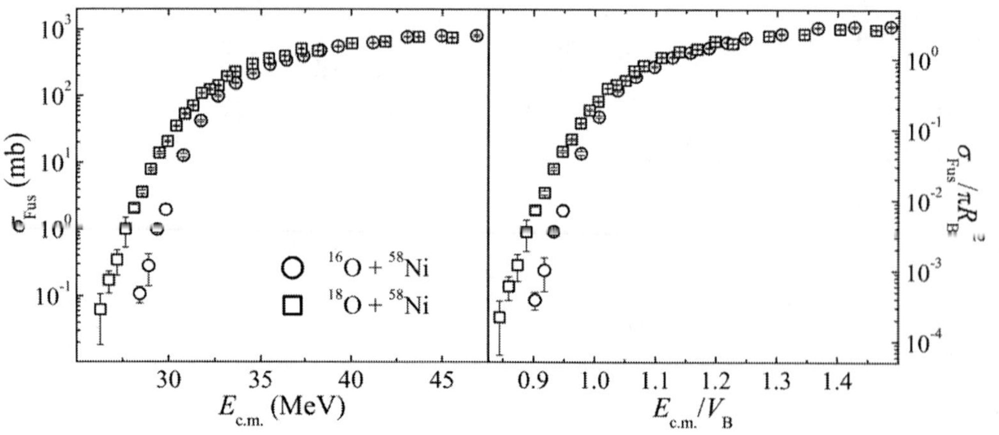

Figure 4: Preliminary experimental fusion excitation functions of $^{16,18}O$ + ^{58}Ni (circles and squares) in absolute (left) and reduced (right) scales.

2.3 $^{16,18}O + ^{58}Ni$

For further studying the effect of the neutron transfer channels with positive Q_{-2n}-values, another ideal system ^{18}O + ^{58}Ni with bigger Q_{-2n}-value and smaller Z_pZ_t was studied. The fusion excitation function of ^{18}O + ^{58}Ni was once measured [12] but only at near-barrier energies and with lower data quality. Fusion of ^{18}O + ^{58}Ni is interesting considering the bigger Q_{-2n}/V_B and therefore has attracted many studies, but precise fusion data down to sub-barrier energy region is still absent.

For the moment, we have tried to complete the near-barrier experimental fusion of ^{18}O + ^{58}Ni with rather good accuracy by extending the measurement to the sub-barrier energy region. The fusion of ^{16}O + ^{58}Ni was also measured as a reference system. The Preliminary results of the experimental fusion excitation functions of $^{16,18}O$ + ^{58}Ni are shown in fig. 4 (left), where the energy scale $E_{c.m.}$ is corrected for the carbon backing (faced to the beam) and the target thickness. The right panel is what obtained by a representation of the fusion excitation functions in a reduced energy scale. It can be seen that the two reduced fusion excitation functions deviate from each other at near- and sub-barrier energy region. This means the appearance of the PQNT effect on fusion considering the minor coupling effect to the collective inelastic excitation states for the two lighter systems.

3 Summary

The PQNT effect has been studied experimentally. These experimental results further show a complicated PQNT effect, which should be further studied for a full understanding of the underlying physics in the nuclear reactions with heavy-ions.

Acknowledgements

This work has been supported by the National Key Basic Research Development Program of China under Grant No. 2013CB834404, and the National Natural Science Foundation of China under Grants Nos. 11475263 and 11375268.

References

[1] Back B. B. et al., *Rev. Mod. Phys.*, **86** (2014) 317.

[2] Hagino K. and Takigawa N., *Prog. Theor. Phys.*, **128** (2012) 1001.

[3] Beckerman M. et al., *Phys. Rev. Lett.*, **45** (1980) 1472.

[4] Broglia R. A., Dasso C. H., Landowne S., and Winther A., *Phys. Rev. C*, **27** (1983) R2433.

[5] Henning W. et al., *Phys. Rev. Lett.*, **58** (1987) 318.

[6] Jia H. M. et al., *Phys. Rev. C*, **86** (2012) 044621.

[7] Hagino K., Rowley N., and Kruppa A. T., *Comput. Phys. Commun.*, **123** (1999) 143.

[8] Zhang Huan-Qiao et al., *Chin. Phys. C*, **34** (2010) 1628.

[9] Jia H. M. et al., *Phys. Rev. C*, **89** (2014) 064605.

[10] Stefanini A.M. et al., *Nucl. Phys. A*, **456** (1986) 509.

[11] Jacobs P. et al., *Phys. Lett. B*, **175** (1986) 271.

[12] Borges A. M. et al., *Phys. Rev. C*, **46** (1992) 2360.

EPJ Web of Conferences **117**, 08015 (2016) DOI: 10.1051/epjconf/201611708015
NN2015

Dissipation strength of the tilting degree of freedom in fusion-fission reactions

P. N. Nadtochy, D. V. Vanin, A. V. Cheredov, S. V. Fedorov, E. G. Ryabov and G. D. Adeev

Omsk State University, Physics Department, Mira prospekt 55-A, Omsk, 644077, Russia

Abstract

The four-dimensional Langevin model was applied to calculate a wide set of experimental observables for compound nuclei, formed in heavy-ion fusion-fission reactions. A modified one-body mechanism for nuclear dissipation with a reduction coefficient k_s of the contribution from a "wall" formula was used for shapes parameters. Different possibilities of deformation-dependent dissipation coefficient for the K coordinate (γ_K) were investigated. Presented results demonstrate that the influence of the k_s and γ_K parameters on the calculated quantities can be selectively probed. It was found that it is possible to describe experimental data with the deformation-dependent γ_K coefficient. One of the possibility is to use large values of $\gamma_K \simeq 0.2$ (MeV zs)$^{-1/2}$ for compact shapes featuring no neck and small values of $\gamma_K \simeq 0.0077$ (MeV zs)$^{-1/2}$ for elongated shapes.

Fission still is one of the most interesting and challenging topics in nuclear physics providing a perfect opportunity to investigate the large scale evolution of initial compound nucleus into fission products. During the past two decades stochastic approach based on multidimensional Langevin equations has been extensively and rather successfully used to elucidate many problems of collective nuclear dynamics in fusion-fission reactions at high excitation energies [1, 2]. A reasonable choice of collective degrees of freedom for modeling shape evolution and considering particle evaporation allow modeling the complex interplay between static and dynamical effects in fission and succeeding in explaining a wide range of experimental data [3, 4].

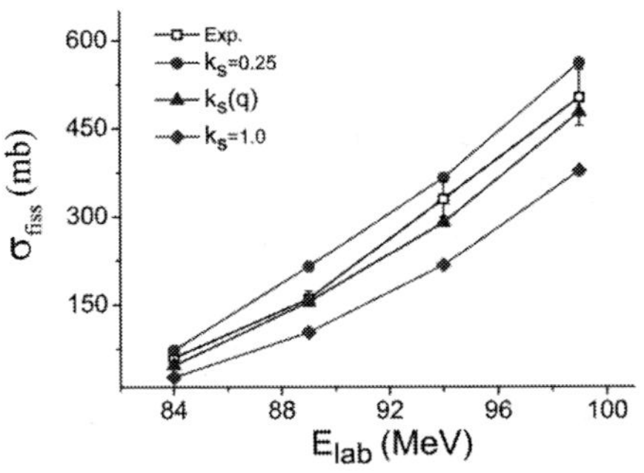

Figure 1: (Color online) The fusion-fission cross section for ^{204}Po. The open squares are experimental data. The filled symbols are results of 4D calculations: circles with k_s=0.25, triangles with $k_s(\mathbf{q})$, and diamonds k_s=1.

The significance of orientation degree of freedom (K coordinate), which is the projection of the total angular momentum onto the symmetry axis of fissioning nucleus, was demonstrated and an overdamped Langevin equation for K coordinate were introduced [5, 6].

In the present study we used recently developed 4D dynamical model for the description of fissioning nucleus shape evolution based on a stochastic approach [1, 2, 4, 7, 8]. The detailed description of the model could be found in Refs. [2, 4] and here we give only a short summary. Three collective shape coordinates and the orientation degree of freedom (K coordinate) were considered dynamically from the ground state deformation to the scission into fission fragments. A modified one-body mechanism of nuclear dissipation [9, 10] was employed to determine the dissipative part of the driving forces with reduction coefficient from the "wall" formula k_s. The value $k_s = 1.0$ corresponds to the "wall" and "wall-plus-window" formulas, whereas values $0.2 < k_s < 0.5$ allow to reproduce different features of the experimental fission fragment MED and particle multiplicities in multidimensional Langevin calculations [2, 8, 11].

In Refs. [12–14] it was argued that chaos-theory related ideas [14] can be used to calculate the value of the reduction coefficient k_s as a function of deformation of the fissioning nucleus. The applications of this approach to calculate the coefficient $k_s(\mathbf{q})$ for studying different fission characteristics were rather successful, and have shown that such calculations yield almost the same results as those using the constant k_s coefficient from the interval $0.25 < k_s < 0.5$ [6, 12, 15].

We use three different options to model the deformation dependence of $\gamma_K(\mathbf{q})$ coefficient as described in Ref. [16]. As the first one we used

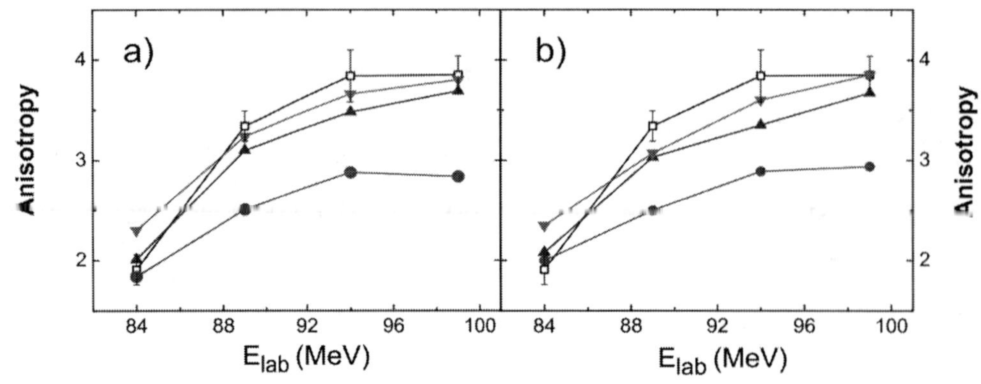

Figure 2: (Color online) The anisotropy of fission fragment angular distribution at $k_s(\mathbf{q})$ (a) and k_s=0.5 (b) for ^{204}Po. The open squares are experimental data. The filled triangles correspond to 4D calculations with $\gamma_K^{(1)}$=0.077 (MeV zs)$^{-1/2}$; the inverted triangles with $\gamma_K^{(2)}$ (γ_K^{const} = 0.2 (MeV zs)$^{-1/2}$); and the circles with $\gamma_K^{(3)}$.

$\gamma_K^{(1)} = const$. The second option is determined by the following equation

$$\gamma_K^{(2)}(\mathbf{q}) = \begin{cases} \gamma_K^{\text{const}}, \text{ for shapes without a neck;} \\ \gamma_K^{\text{neck}}(\mathbf{q}), \text{ for shapes with a neck,} \end{cases} \tag{1}$$

where γ_K^{const} is a variable parameter independent of nuclear deformation. After appearance of the neck in nuclear shape the γ_K^{const} value is joined smoothly with $\gamma_K^{\text{neck}}(\mathbf{q})$, which was obtained in Refs. [5, 16]. The third one is given by equation

$$\gamma_K^{(3)}(\mathbf{q}) = \begin{cases} \gamma_K^{\text{cshape}}(\mathbf{q}), \text{ for shapes without a neck;} \\ \gamma_K^{\text{neck}}(\mathbf{q}), \text{ for shapes with a neck,} \end{cases} \tag{2}$$

where $\gamma_K^{\text{cshape}}(\mathbf{q})$ is the extrapolation of $\gamma_K^{\text{neck}}(\mathbf{q})$ to the mononuclear shapes, featuring no neck. The detailed description of these three prescriptions could be found in Ref. [16].

In the present calculations we investigate the influence of k_s and deformation-dependent $\gamma_K(\mathbf{q})$ parameters on the fusion-fission (σ_{fis}), evaporation residues (σ_{ER}) cross sections, and anisotropy of fission fragment angular distribution in 4D dynamical model. We performed calculations for the reactions $^{16}\text{O} + ^{188}\text{Os} \rightarrow ^{204}\text{Po}$ (E_{lab} = 84, 89, 94, and 99 MeV) [17–19] and $^{16}\text{O} + ^{184}\text{Pt} \rightarrow ^{200}\text{Pb}$ (E_{lab} = 91.6, 102.5, and 107.9 MeV) [20–23]. The calculations were performed at different values of k_s and γ_K. In figures we present the options which allow good reproduction of experimental data. In fig. 1 one can see that $k_s = 0.25$ or $k_s(\mathbf{q})$ obtained from chaos-weighted wall formula allows well reproduce the fission cross section for the fission of ^{204}Po. The calculated fission cross section does not depend neither on the γ_K option used nor on the absolute γ_K values, so we did not show it in fig. 1 in order not to complicate figure. Our investigations [16] show that only the anisotropy of fission fragment angular distribution is dependent on both k_s and γ_K values, as it was found in previous investigation with heavy compound nuclei [6]. The calculated anisotropy of fission fragment angular distribution is presented in fig 2. One can see that at both k_s values ($k_s = 0.5$ and $k_s(\mathbf{q})$) it is possible to reproduce the experimental data on anisotropy at constant $\gamma_K = 0.077$ (MeV zs)$^{-1/2}$ or with the deformation-dependent γ_K coefficient: to use large values of $\gamma_K \simeq 0.2$ (MeV zs)$^{-1/2}$ for compact shapes featuring no neck and small values of $\gamma_K(\mathbf{q})$ for dinuclear shapes. The 4D calculations with $\gamma_K^{(3)}(\mathbf{q})$ option underestimate the experimental anisotropy of fission fragment angular distribution.

The comparison of theoretical 4D calculations with experimental data for the ^{200}Pb is presented in fig. 3. In figure 3(a) the calculated fusion-fission (σ_{fis}) and evaporation residues (σ_{ER}) cross sections are compared with the experimental data at different excitation energies and k_s values. We found that γ_K does not influence the cross sections, so we did not present in fig. 3(a) the results obtained with different γ_K values. One can see that 4D calculations reproduce worse the cross sections for the ^{200}Pb compound nucleus than for the ^{204}Po compound nucleus. The same is true also for the calculated results for the anisotropy of fission fragment angular distribution presented in fig. 3(b). In the present calculations we needed to increase the k_s and/or γ_K values, which generate large values of the anisotropy. However, even very large values of γ_K (the option $\gamma_K^{(2)}(\mathbf{q})$ with the large values of $\gamma_K^{const} = 0.4$ - 0.6 (MeV zs)$^{-1/2}$ for compact shapes) and large k_s values does not allow to reproduce the experimental values of the anisotropy of fission fragment angular distribution. Moreover, the anisotropy reaches the saturation at $\gamma_K^{const} = 0.4$ (MeV zs)$^{-1/2}$, and does not grow up after the further increase of γ_K^{const} independently on k_s value used. The results of present 4D calculations with k_s=1.0 and $\gamma_K^{(2)}(\mathbf{q})$ with the large values of $\gamma_K^{const} = 0.4$ - 0.6 (MeV zs)$^{-1/2}$ for compact shapes predicts the anisotropy about 15% less than experimental data. However, we should mention that there is significant difference in experimental data on fusion cross section for the ^{200}Pb compound nucleus between old [23] and recent [20] experimental data. Thus, possible revision of experimental data [23] on anisotropy of angular distribution could influence on the conclusions from comparison

DOI: 10.1051/epjconf/201611708015

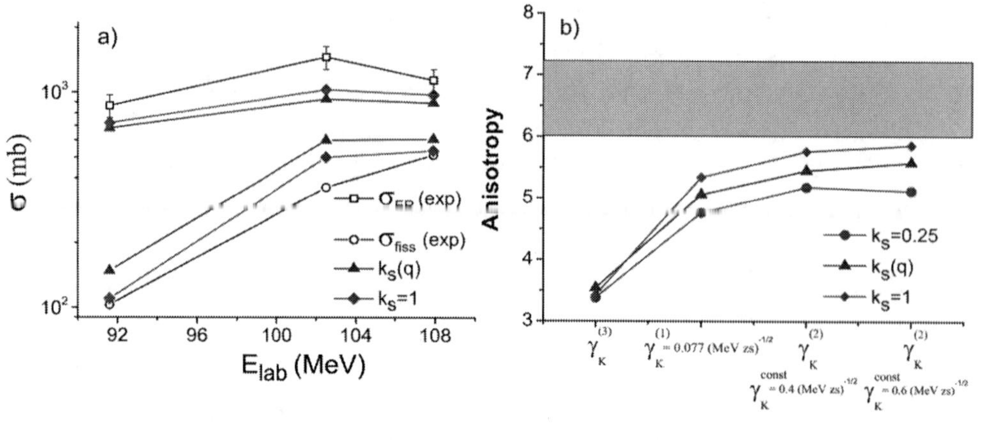

Figure 3: (Color online) The σ_{fis} and σ_{ER} (a) and the anisotropy of fission fragment angular distribution (b) for ^{200}Pb at different k_s values and γ_K options. The panel (b) demonstrates results for the E_{lab} = 107.9 MeV. The open symbols are experimental data. The hatched area at panel (b) determines experimental data with error bars. The filled symbols are results of 4D calculations: circles with k_s=0.25; triangles with $k_s(\mathbf{q})$; and diamonds with k_s=1.0.

between theoretical and experimental data.

Concluding we can state that it was found that in the 4D Langevin calculations it is possible to describe experimental data with the deformation-dependent γ_K coefficient. One of the possibility is to use large values of $\gamma_K \simeq 0.2$ (MeV zs)$^{-1/2}$ for compact shapes featuring no neck and small values of $\gamma_K \simeq 0.0077$ (MeV zs)$^{-1/2}$ for elongated shapes. Using such a different value of γ_K for deformations before saddle and along the saddle-to-scission descent tends the dynamics of the K coordinate to the prediction of the transition state model at saddle point.

This study was partially supported by the Russian Foundation for Basic Research, Research Project No. 13-02-00168 (Russia).

References

[1] Abe Y., Ayik S., Reinhard P-G., Suraud E., *Phys. Rep.,* **275** (1996) 49.

[2] Adeev G. D.*et al., Fiz. Elem. Chast. At. Yadra,* **36** (2005) 732.

[3] Vardaci E. *et al., Phys. Rev. C,* **92** (2015) 034610.

[4] Nadtochy P. N., Ryabov E. G., Gegechkori A. E., Anischenko Yu. A., Adeev G. D., *Phys. Rev. C,* **85** (2012) 064619.

[5] Lestone J. P. and McCalla S. G., *Phys. Rev. C,* **79** (2009) 044611.

[6] Nadtochy P. N., Ryabov E. G., Gegechkori A. E., Anischenko Yu. A., Adeev G. D., *Phys. Rev. C,* **89** (2014) 014616.

[7] Kramers H. A., *Physica,* **7** (1940) 284.

[8] Nadtochy P. N., Adeev G. D., Karpov A. V., *Phys. Rev. C,* **65** (2002) 064615.

[9] Randrup J. and Swiatecki W. J., *Nucl. Phys. A,* **429** (1984) 105.

[10] Nix J. R. and Sierk A. J., in *Proceedings of the 6th Adriatic Conference on Nuclear Physics: Frontiers of Heavy Ion Physics, Dubrovnik, Yugoslavia, 1987* ed N. Cindro, R. Caplar, W. Greiner (World Sci., Singapore) 1990, pp. 333–340.

[11] Ryabov E. G., Karpov A. V., Nadtochy P. N., Adeev G. D., *Phys. Rev. C,* **78** (2008) 044614.

[12] Pal S. and Mukhopadhyay T., *Phys. Rev. C,* **54** (1996) 1333.

[13] Blocki J., Brut F., Srokowski T., Swiatecki W. J., *Nucl. Phys. A,* **545** (1992) 511.

[14] Swiatecki W. J., *Nucl. Phys. A,* **488** (1988) 375c.

[15] Schmitt C., Mazurek K., Nadtochy P. N., *Phys. Lett. B,* **737** (2014) 289.

[16] Nadtochy P. N., Ryabov E. G., Adeev G. D., *J. Phys. G,* **42** (2015) 045107.

[17] J. van der Plicht *et al.*, *Phys. Rev. C,* **28** (1983) 2022.

[18] Rafiei R. *et al.*, *Phys. Rev. C,* **77** (2008) 024606.

[19] Tripathi R. *et al.*, *Phys. Rev. C,* **79** (2009) 064607.

[20] Shidling P. D. *et al.*, *Phys. Rev. C,* **74** (2006) 064603.

[21] Leigh J. R., Bokhorst J. J. M., Hinde D. J., Newton J. O., *J. Phys. G,* **14** (2015) L55.

[22] Hinde D. J. *et al.*, *Nucl. Phys. A,* **385** (1982) 109.

[23] Forster J. S. *et al.*, *Nucl. Phys. A,* **464** (1987) 497.

EPJ Web of Conferences **117**, 08016 (2016) DOI: 10.1051/epjconf/201611708016
NN2015

Dynamical Dipole mode in the ^{40,48}Ca $+^{152,144}$Sm fusion reactions at 11 MeV/nucleon

C. Parascandolo[1], D. Pierroutsakou[1], R. Alba[2], A. Del Zoppo[2], C. Maiolino[2], D. Santonocito[2], C. Agodi[2], V. Baran[3,2], A. Boiano[1], M. Colonna[2], R. Coniglione[2], E. De Filippo[4], M. Di Toro[2,5], U. Emanuele[6], F. Farinon[7], A. Guglielmetti[8], M. La Commara[9,1], B. Martin[9,1], C. Mazzocchi[8], M. Mazzocco[10], C. Rizzo[2,5], M. Romoli[1], C. Signorini[10], R. Silvestri[9,1], F. Soramel[10], E. Strano[10], D. Torresi[10], A. Trifirò[6] and M. Trimarchi[6]

[1]INFN - Napoli, Napoli, Italy
[2]INFN - Laboratori Nazionali del Sud, Catania, Italy
[3]University of Bucharest, Bucharest and NIPNE-HH, Măgurele, Romania
[4]INFN, Sezione di Catania, Catania, Italy
[5]Dip. di Fisica e Astronomia, Università di Catania, Catania, Italy
[6]INFN, Gr. Col. di Messina and Università di Messina, Messina, Italy
[7]GSI, Darmstadt, Germany
[8]Dip. di Fisica, Università di Milano and INFN, Sezione di Milano, Milano, Italy
[9]Dip. di Scienze Fisiche, Università di Napoli, Napoli, Italy
[10]Dip. di Fisica e Astronomia, Università di Padova and INFN, Sezione di Padova, Padova, Italy

Abstract

The excitation of the dynamical dipole mode along the fusion path was investigated in the formation of a heavy compound nucleus in the A=190 mass region. To form the compound nucleus, the ^{40}Ca + ^{152}Sm and ^{48}Ca + ^{144}Sm reactions were employed at E_{lab}=11 and 10.1 MeV/nucleon, respectively. Both fusion–evaporation and fission events were studied simultaneously for the first time. Our results for evaporation and fission events (preliminary) show that the dynamical dipole mode survives in reactions involving heavier nuclei than those studied previously.

1 The physical problem

The "Dynamical Dipole mode" (DD throughout the text) is a collective dipole oscillation that can be excited in N/Z asymmetric heavy-ion collisions. It develops along the symmetry axis of the deformed composite system, the dinucleus, and decays emitting prompt γ-rays in addition to those coming from the Giant Dipole Resonance (GDR) thermally excited in the hot compound nucleus (CN) [1–3]. The DD radiation is characterized by i) a centroid energy lower than that of the CN GDR in the same mass region due to the high deformation of the emitting source [2, 3] ii) an anisotropic angular distribution with respect to the beam axis since the oscillation is confined in the reaction plane [4] and iii) a γ yield that should depend on both the reaction dynamics and the symmetry term of the EOS [3].

The existence of the DD mode has been probed in deep inelastic and fusion-evaporation heavy-ion collisions [5–9]. In these measurements, an excess of γ-rays was observed in the GDR energy region for a charge asymmetric reaction, with respect to that of a more charge symmetric one forming the same CN at identical conditions [6–8] or with respect to statistical model calculations [9]. This γ excess was attributed to the decay of the predicted DD. The emission of DD γ-rays decreases the excitation energy and hence the initial temperature of the nucleus reaching the statistical phase. This cooling mechanism could be suitable to favour the production of super-heavy elements in hot fusion processes. TDHF calculations [2] showed that the DD γ yield decreases as the mass of colliding ions increases since the reactions with small nuclei are less damped than those involving more nucleons. In order to understand the behavior of the DD in heavier systems than those studied before and to test its usefulness in super-heavy element production, we decided to study the DD in a composite system in the mass region A=190.

2 ^{40,48}Ca+^{152,144}Sm at 11 MeV/nucleon

The experiment was performed by using the ^{40}Ca (^{48}Ca) pulsed beam provided by the Superconducting Cyclotron of the Laboratori Nazionali del Sud (LNS, Italy), impinging on a 1 mg/cm^2 thick self-supporting ^{152}Sm (^{144}Sm) target enriched to 98.4%(93.8%) in ^{152}Sm (^{144}Sm) at E_{lab} = 440 (485) MeV. Both entrance channels populate the same CN through a quite different initial dipole moment, 30.6 fm for the ^{40}Ca + ^{152}Sm charge asymmetric reaction and 5.3 fm for the ^{48}Ca + ^{144}Sm more charge symmetric one. The mass asymmetry of the two entrance channels is very similar, namely 0.22(0.18) for the ^{40}Ca + ^{152}Sm (^{48}Ca + ^{144}Sm) system. Furthermore, the formed CN had identical excitation energy in both reactions, as explained in the following, and identical spin distribution: $L_{max} = 74\hbar$ for fusion and $L_{max} = 42\hbar$ for fusion-evaporation, according to PACE2 calculations [10] with a level density parameter a = A/9.5 MeV^{-1}, A being the CN mass. The γ-rays and the light particles were detected by using the MEDEA BaF_2 sphere [12], with a full azimuthal coverage in the polar angular range between 42° and 170°. They were identified by combining a pulse-shape analysis of the scintillator signal with a time of flight measurement (obtained with respect to the radiofrequency signal of the cyclotron). The fusion-evaporation residues were detected by four position sensitive Parallel Plate Avalanche Counters (PPACs) placed symmetrically around the beam direction at 70 cm from the target at $\theta = 7°$ and subtending 7° in θ. The fission events were selected by detecting the two kinematically coincident fission fragments with position sensitive PPACs, centered at $\theta = 52.5°$ symmetrically around the beam axis at 16 cm from the target covering 22° in both θ and ϕ and allowing the study of γ-ray - fragment angular correlations. Down-scaled single events and coincidence events between at least one fired BaF_2 and a PPAC (two PPACs) for evaporation (fission) events were collected during the experiment. The coincidence request eliminated any cosmic-ray contamination of the γ-ray spectra. Moreover, the trigger avoids to use any normalization factors in the γ-ray spectra as the double differential γ multiplicity is obtained from the ratio of the number of coincidences between γ-rays and evaporation residues (fission fragments) and the number of single events of evaporation (fission).

2.1 Results and discussion

The average excitation energy, the average mass and the average charge of the composite system after pre-equilibrium particle emission were evaluated by studying the energy spectra of the light charged particles (p, α) in coincidence with evaporation residues, while the pre-equilibrium neutron emission was estimated from our proton data and from systematics. These spectra were analyzed with a moving source fit where the particles were assumed to be emitted isotropically from two moving sources: a *slow* source simulating the statistical evaporation from the hot CN and an *intermediate-*

velocity (between the CN and the projectile velocity) source related to the pre-equilibrium particles emitted by the composite system before thermalization. The analysis demonstrated that the two reactions lead to the formation of a CN with the same average mass and charge at the same average excitation energy. Hence, as in our previous works [7, 8], the only different parameter in the two reactions is the initial charge asymmetry. Therefore any difference in the γ-ray emission between the two reactions can be related to this quantity.

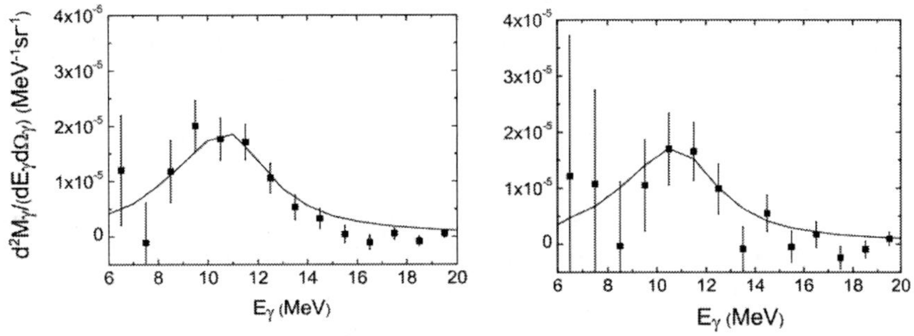

Figure 1: Difference between the charge asymmetric and charge symmetric reaction bremsstrahlung-subtracted center-of-mass γ-ray spectra for fusion-evaporation (left) and fission (right) events. The lines are described in the text.

By comparing the bremsstrahlung-subtracted center-of-mass γ-ray spectra, an excess of γ-rays in the more charge asymmetric reaction was observed, concentrated in the energy range E_γ = 8-15 MeV. This can be seen in Figure 1 where the difference between the spectra of the two reactions is displayed for evaporation (fission) events in the left (right) side. This excess is related to a pre-equilibrium effect caused by the large charge asymmetry of the $^{40}Ca+^{152}Sm$ reaction, namely to the DD γ decay, and can be reproduced by means of a lorentzian curve folded by the experimental apparatus response function [13] (line in the figure). The DD centroid energy E_{DD} and width Γ_{DD} used to reproduce our data are E_{DD} = 11 MeV and Γ_{DD} = 3.5 MeV, in both exit channels. It is interesting to note that E_{DD} is lower than the centroid energy of the GDR built on the ground state of a nucleus of similar mass, E_{GDR} = 14 MeV. This result confirms the high deformation of the emitting source at the moment of the DD γ emission, in agreement with expectations [2, 3] and with our previous works [7, 8].

A clear confirmation of the pre-equilibrium nature of the DD comes from the γ-rays angular distribution for evaporation events. This is related to the interplay between the rotation angular velocity of the dinuclear system

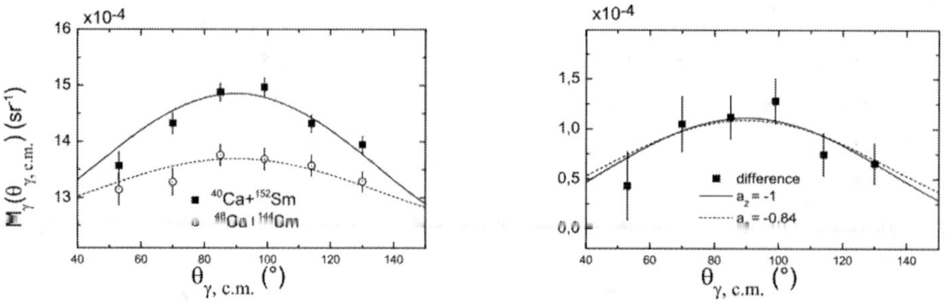

Figure 2: Center-of mass angular distribution of the evaporation γ-rays for the two reactions (left) and of their difference (right) in the energy range 9-15 MeV corrected by the experimental setup efficiency. The lines are described in the text.

during the DD emission and the instant when this emission takes place [4]. Figure 2 displays the center-of-mass angular distribution with respect to the beam direction of the fusion-evaporation γ-rays for the ^{40,48}Ca+^{152,144}Sm reactions (left) and for their difference (right), integrated over energy from 9 to 15 MeV and after the subtraction of (nn)-bremsstrahlung component. The angular distribution was corrected by the experimental setup efficiency. The lines in both panels of figure 2 describe the angular distribution of the emitted γ-rays given by the Legendre polynomial expansion $M_\gamma(\theta_\gamma) = M_0[1 + Q_2 a_2 P_2 cos(\theta_\gamma)]$, where a_2 is the anisotropy coefficient and Q_2 is an attenuation factor for the finite γ-ray counter [14] (0.98 in our case). From a best fit to the data, shown with a solid (dashed) line for the ^{40}Ca+^{152}Sm (^{48}Ca+^{144}Sm) reaction, we found a_2 = -0.13 $\pm$ 0.03 for the ^{40}Ca+^{152}Sm reaction and a_2 = -0.06 $\pm$ 0.02 for the ^{48}Ca+^{144}Sm one. The charge asymmetric reaction (squares) displays a more anisotropic angular distribution around 90° than the charge symmetric one (circles). Since we have the same CN, with the same excitation energy and spin distribution, such a difference is related to entrance channel effects. Consequently, the experimental angular distribution of the difference (shown in the right-hand side of the figure 2) is very anisotropic around 90°. The data can be reproduced well with a_2 = -1 (solid line) that results in an angular distribution of the $sin^2(\theta_\gamma)$ form of emission from a dipole oscillation along the beam axis. The dashed line in the figure gives the DD angular distribution with a_2 = -0.84 obtained within BNV calculations for impact parameters integrated up to 2 fm (evaporation events). The above a_2 values indicate a preferential oscillation axis of the DD, triggered at the early stage of the fusion path, along an axis that has not rotated much on the reaction plane during the DD lifetime.

Our data therefore suggest that the DD γ-emission time scale is confined at the beginning of the reaction, in agreement with our previous results [8] for evaporation events and with theoretical expectations [4].

By taking into account the an DD γ-ray angular distribution (a_2 = -1) for evaporation events and the response function of the experimental setup, the DD yield, integrated over energy and over angle, is $(1.2 \pm 0.2)*10^{-3}$. The analysis for the DD angular distribution for fission events is under way.

The experimental results on the DD in $^{40}Ca+^{152}Sm$ reaction were compared with calculations performed within the BNV transport model and based on a collective bremsstrahlung approach of the entrance channel reaction dynamics [3]. These calculations give centroid energy, width and angular distribution of the DD in good agreement with those of the experiment. However, the theoretical γ yield for evaporation events overestimates the data. This aspect should be further investigated.

References

[1] P. Chomaz, M. Di Toro and A. Smerzi *Nuc. Phys.* A **563** (1993) 509

[2] C. Simenel, P. Chomaz and G. de France *Phys. Rev. Let.* **86** (2001) 2971

[3] V. Baran *et al. Phys. Rev. Let.* **86** (2001) 182501 and *Nuc. Phys.* A **679** (2001) 373

[4] V. Baran *et al. Phys. Rev.* C **79** (2009) 021603 (R)

[5] D. Pierroutsakou *et al. Eur. Phys. Jour.* A **16** (2003) 423

[6] S. Flibotte *et al. Phys. Rev. Let.* **77** (1996) 1448; F. Amorini *et al. Phys. Rev.* C **69** (2004) 014608

[7] D. Pierroutsakou *et al. Eur. Phys. Jour.* A **17** (2003) 71,*Phys. Rev.* C **71** (2005) 054605

[8] B. Martin *et al. Phys. Let.* B **664** (2008) 47; D. Pierroutsakou *et al. Phys. Rev.* C **80** (2009) 024612

[9] A. Corsi *et al. Phys. Let.* B **679** (2009) 197; A. Giaz *et al. Phys. Rev.* C **90** (2014) 014609

[10] A. Gavron *Phys. Rev.* C **21** (1980) 230

[11] M. P. Kelly *et al. Phys. Rev. Let.* **82** (1999) 3404

[12] E. Migneco *et al. Nuc. Inst. Meth.* A **314** (1992) 31

[13] G. Bellia *et al. Nuc. Inst. Meth.* A **329** (1993) 173

[14] M. E. Rose *Phys. Rev.* **91** (1953) 610

Dissipative effects in fission investigated with proton-on-lead reactions

J. L. Rodríguez-Sánchez[1*], J. Benlliure[1], J. Taïeb[2], H. Álvarez-Pol[1], L. Audouin[3], Y. Ayyad[1], G. Bélier[2], G. Boutoux[2], E. Casarejos[4], A. Chatillon[2], D. Cortina-Gil[1], T. Gorbinet[2], A. Heinz[5], A. Kelić-Heil[6], B. Laurent[2], J.-F. Martin[2], C. Paradela[1], E. Pellereau[2], B. Pietras[1], D. Ramos[1], C. Rodríguez-Tajes[7], D. M. Rossi[6], H. Simon[6], J. Vargas[1] and B. Voss[6]

[1] Universidad de Santiago de Compostela, E-15782 Santiago de Compostela, Spain
[2] CEA, DAM, DIF, F-91297 Arpajon, France
[3]Institut de Physique Nucléaire d'Orsay, F-91406 Orsay, France
[4] Universidad de Vigo, E-36200 Vigo, Spain
[5] Chalmers University of Technology, SE-412 96 Gothenburg, Sweden
[6] GSI Helmholtzzentrum für Schwerionenforschung GmbH, D-64291 Darmstadt, Germany
[7] Grand Accélérateur National d'Ions Lourds, F-14076 Caen Cedex 5, France

Abstract

The complete kinematic measurement of the two fission fragments permitted us to investigate dissipative effects at large deformations, between the saddle-point and the corresponding scission configurations. Up to now, this kind of study has only been performed with fusion-fission reactions using a limited number of observables, such as the mass distribution of the fission fragments or the neutron multiplicities. However, the large angular momenta gained by the compound nucleus

*joseluis.rodriguez.sanchez.phy@gmail.com

could affect the conclusions drawn from such experiments. In this work, the use of spallation reactions, where the fissioning systems are produced with low angular momentum, small deformations and high excitation energies, favors the study of dissipation, and allowed us to define new observables, such as postscission neutron multiplicities and the neutron excess of the final fission fragments as a function of the atomic number of the fissioning system. These new observables are used to investigate the dissipation at large deformations.

1 Introduction

It is known that the gross features of a fissioning system can be described in terms of a small number of parameters, the so-called collective degrees of freedom, where the dissipation represents the average effect of the interactions between the collective and intrinsic degrees of freedom. In this picture, the interaction between these degrees of freedom results in a the fluctuating force acting on the collective dynamics, which in effect causes the diffusion of the dynamical coordinates (elongation, mass asymmetry, etc.) [1].

However, other questions such as the onset of dissipation and the temperature or deformation dependences of the dissipation parameter are still under debate [2]. A systematic study was carried out by Thoennessen et al. [3] to find the threshold excitation energy at which the statistical model starts losing its validity. Their work opened up the problem of understanding the properties of nuclear dissipation and its dependence on the excitation energy. It was found that the ratio of the threshold temperature ($T_{threshold}$) and the temperature-dependent fission barrier (B_f) is independent of the mass of the fissioning nucleus and equal to 0.26. The authors suggested that this could reflect an onset of dissipation with increasing temperature. According to this conclusion, Hofman and collaborators [4] found an analytical expression for the temperature dependence of dissipation from the study of the γ-ray multiplicities associated with several fusion-fission and quasifission reactions. However, the authors of Ref. [5] showed that this systematic behaviour could be explained without assuming an onset of dissipation effects or a temperature dependence of it. This work also pointed out that the reason why the statistical model fails to reproduce the experimental data for $T_{threshold}/B_f > 0.26$ is that at that point the dynamical delay time starts to be larger than the decay time predicted by the statistical model. Similar conclusions were also found by Lestone et al. [6,7], who were able to describe evaporation and fission cross sections as well as light-charged particles multiplicities of fusion reactions considering a temperature independence of the dissipation parameter.

Recently, Jurado et al. [8], Schmitt et al. [9], and Ayyad et al. [10,11] investigated dissipative effects in fission by using spallation and fragmentation reactions. In these works, both fission fragments were measured with a novel experimental approach [12], similar to the one used in this work [13], which allowed to investigate observables sensitive to dissipation at small deformations, such as the total and partial fission cross sections, and the width of the charge distribution of the fission fragments. The authors demonstrated that dissipative and transient time effects are needed to describe the data, leading to a constant value of the reduced dissipation parameter at small deformations of $\beta = 4.5 \times 10^{21}$ s^{-1}. Moreover, the overall good description of the data also validated the conclusions on the temperature independence of the dissipation parameter. These conclusions, obtained with respect to the magnitude and temperature independence of the reduced dissipation parameter at small deformations, were also confirmed by us investigating the same observables [14, 15].

On the other hand, recent works concerning the study of γ-ray multiplicities in fusion-fission reactions were interpreted assuming a deformation dependence of dissipation instead of a temperature dependence. Using this hypothesis, Shaw and collaborators [16] were able to reproduce the γ-ray multiplicity emitted in the fission process of ^{240}Cf at different energies by using a value of $\gamma_{ground-saddle} = 2$ inside the saddle point, and of $\gamma_{saddle-scission} = 5 - 10$ beyond the saddle point [1]. Diószegi and coworkers [17] were able to investigate dissipation at small and large deformations by separating the presaddle and the saddle-to-scission γ-ray components. They used evaporation-residue cross sections of the reaction ^{9}F + ^{181}Ta at 161 and 181 MeV to determine the dissipative parameter up to the saddle point, $\gamma_{ground-saddle} = 3$. Fixing this value of the dissipation coefficient at the saddle-point deformation, they varied the dissipation coefficient from saddle to scission to reproduce the γ-ray multiplicities obtaining $\gamma_{saddle-scission} = 10$. Such a deformation dependence with very low values of dissipation for compact shapes up to the saddle point seems to be compatible with the results obtained by Fröbrich and collaborators [18]. These authors analyzed the prescission neutron multiplicities and fission probabilities with one-dimensional Langevin calculations, coupled to a statistical evaporation code. They found that the experimental observables could only

[1] Note that γ_i (where the subscript i denotes ground-saddle or saddle-scission) is related with the reduced dissipation parameter according to $\gamma_i = \beta_i/2\omega$, being ω the frequency of the inverted oscillator potential at the saddle point.

be described with a deformation-dependent dissipative parameter by considering a constant value $\beta_{ground-saddle} = 2 \times 10^{21}$ s^{-1} at small deformations, and a linear increase with deformation when the fissioning system descends from the saddle to the scission point.

In order to contribute to this discussion, we take advantage of the complete characterization of both fission fragments produced in proton-induced fission reactions on ^{208}Pb at 500A MeV. The measurement was performed by using the SOFIA setup [19,20] mounted at the GSI facilities. The inverse kinematics used in this experiment gave us, for the first time, access to the determination of the velocities, atomic and mass numbers of both fission fragments [13] simultaneously. With these measurements, we can define observables sensitive to dissipation, such as the neutron excess of the fission fragments, constraining the value of the dissipation parameter at large deformations by using model calculations.

2 Model calculations

The calculations were done by using the code INCL4.6 [21] to describe the first stage of the reaction induced by energetic protons according to the intranuclear cascade approach. The deexcitation of the resulting remnants was described by employing the code ABLA07 [22]. This statistical code describes the emission of γ-rays, neutrons, light-charged particles and intermediate-mass fragments (IMFs) according to the Weisskopf formalism [23], which provides a good description of the evaporation residues produced in spallation and fragmentation reactions of nuclei from iron to uranium [21, 24–26]. The fission decay width is described by an analytical approximation of the solution of the Fokker-Planck equation, describing fission as a diffusion process across the fission barrier [27]. Fission barriers are calculated according to the finite-range liquid-drop model of Sierk [28]. Finally, if fission is reached, the mass- and atomic-number distributions of the fission fragments are described by the semiempirical model proposed in Ref. [29].

In order to include the fission dynamics beyond the saddle point, the saddle-to-scission time, needed by the fissioning system to descend from the saddle to the scission point, can be compared with the statistical evaporation time obtained from the neutron decay width. If the saddle-to-scission time of the fissioning system is longer than the statistical evaporation time, the fissioning system could emit neutrons during that phase, increasing the prescission neutron multiplicities. With the aim of introducing these dynamical effects in our calculations, we have used the pioneering theoretical model proposed by Hofmann and Nix [30], who found an analytical solution of the average saddle-to-scission time based on the dynamical picture of Kramers [31]. Following this formalism, the average saddle-to-scission dissipative time (τ_{ssc}) can be calculated according to:

$$\tau_{ssc} = \left(\left[1 + \left(\frac{\beta}{2\omega_0} \right)^2 \right]^{1/2} + \frac{\beta}{2\omega_0} \right) \tau_{ssc}^0 \tag{1}$$

where ω_0 is the frecuency of the inverted oscillator potencial at the saddle point, β is the reduced dissipation parameter, and τ_{ssc}^0 is the saddle-to-scission statistical time.

3 Results

The complete kinematic measurement of the two fission fragments was used to reconstruct the postscission neutron multiplicities using an indirect method based on the Wilkins model [32]. The resulting postscission neutron multiplicities are displayed in Fig. 1 as a function of the atomic number of the fissioning system. This observable is compared with different model calculations: assuming no evaporation of particles between the saddle and scission points (dotted lines), calculations based on the saddle-to-scission statistical time τ_{ssc}^0 (long-dashed line), and calculations based on Eq. 1 with a dissipation parameter of $\beta = 18 \times 10^{21}$ s^{-1} (dot-long-dashed line). The latter calculation is displayed to show the sensitivity of this observable to dissipative effects.

From the comparison of the neutron multiplicities with our calculations, one can see that, as expected, the saddle-to-scission dissipative effects do not affect the postscission neutron multiplicities because the multiplicities seem to be very similar for all the fissioning systems. Thus, these results indicate us that the neutron excess of the two final fission fragments could be as sensitive to dissipation at large deformation as the prescission neutron multiplicities. Therefore, we propose using this observable as a function of the atomic number of the fissioning system. As demonstrated in Ref. [13], the neutron excess as a function of the atomic number of the fission fragments provides important information on the partition of the excitation energy between the two fission fragments at scission. However, the neutron excess as a function of the atomic number of the fissioning system should help us to obtain information on the saddle-to-scission dynamics.

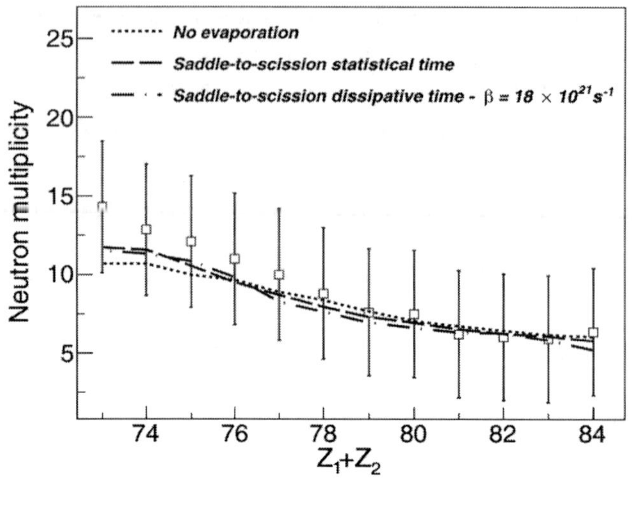

Figure 1: (Color online) The postscission neutron multiplicities are compared with different model calculations (lines).

The average neutron excess of the fission fragments, obtained in Ref. [13], is displayed as a function of the atomic number of the fissioning nuclei in Fig. 2 for the reaction ^{208}Pb + p at 500A MeV. In the figure, we also compare the neutron excess with our previous model calculations. First, the average neutron excess is compared with a calculation assuming the saddle-to-scission statistical time τ^0_{ssc} (long-dashed line). As can be observed, this calculation clearly overestimates the neutron excess for the lightest fissioning systems. This overestimation indicates the need for dissipation at high excitation energies where the statistical evaporation time is comparable to the time needed by the fissioning nucleus to descend from the saddle point to scission. Therefore, we also compare the data with dissipative calculations based on Eq. 1, assuming a reduced dissipation parameter β of 4.5 $\times$ 10^{21} s^{-1} (short-dashed line), 6.5 $\times$ 10^{21} s^{-1} (solid line), and 18 $\times$ 10^{21} (dot-long-dashed line) s^{-1}. As can be seen in the figure, the calculation considering a reduced dissipation parameter of 4.5 $\times$ 10^{21} s^{-1} (short-dashed line) or 6.5 $\times$ 10^{21} s^{-1} (solid line) can describe the average neutron excess for all the fissioning systems. These results could indicate that the dissipation parameter does not depend on deformation because, in our previous works, we have determined a dissipation parameter of $(4.5 \pm 1.0) \times 10^{21}$ s^{-1} at small deformations [14, 15].

Our results seem to contradict most of the works performed with fusion-fission reactions [16–18, 33–35], where a deformation-dependent dissipation parameter was claimed to describe the prescission particle multiplicities.

However, the conclusions obtained using fusion-fission reactions could be affected by the initial conditions of the compound nucleus and by the fact that several independent observables are needed for constraining the dissipation parameter accurately.

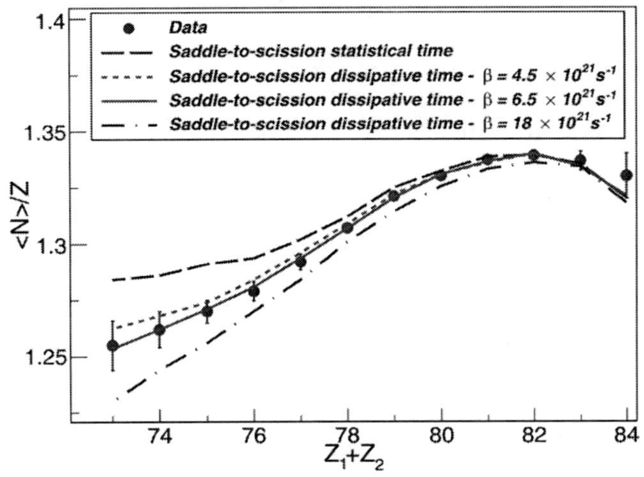

Figure 2: (Color online) The average neutron excess of the final fission fragments produced in the reaction ^{208}Pb + p at 500A MeV (solid circles) is compared to different model calculations (lines).

4 Conclusions

The complete kinematics measurement of the fission fragments produced in the reaction ^{208}Pb(500A MeV) + p was used to reconstruct the average postscission neutron multiplicities. This observable, together with the average neutron excess of the final fission fragments, were displayed as a function of the atomic number of the fissioning nucleus to investigate dissipative effects at large deformations. The comparison of different model calculations with these observables shows that the neutron excess is more sensitive to dissipative effects than the average postscission neutron multiplicities. The higher accuracy of the neutron excess allowed us to determine the value of the dissipation parameter at large deformations, obtaining a value of the reduced dissipation parameter of $\beta_{saddle-scission} = (6.5 \pm 2.5)\times\ 10^{21}\ s^{-1}$. Taking its uncertainty into account, this value is in agreement with the value of the reduced dissipation parameter obtained at small deformations,

$\beta_{ground-saddle} = (4.5 \pm 1.0) \times 10^{21}$ s^{-1}, and therefore we can conclude that no dependence on deformation is observed.

References

[1] Y. Abe et al., Phys. Rep. **275**, 49 (1996).

[2] D. Jacquet and M. Morjean, Prog. Part. Nucl. Phys. **63**, 155 (2009).

[3] M. Thoennessen et al., Phys. Rev. Lett. **71**, 4303 (1993).

[4] D. J. Hofman et al., Phys. Rev. C **51**, 2597 (1995).

[5] H. van der Ploeg et al., Phys. Rev. Lett. **75**, 970 (1995).

[6] J. P. Lestone, Phys. Rev. Lett. **70**, 2245 (1993).

[7] J. P. Lestone and S. G. McCalla, Phys. Rev. C **79**, 044611 (2009).

[8] B. Jurado et al., Phys. Rev. Lett. **93**, 072501 (2004).

[9] C. Schmitt et al., Phys. Rev. C **81**, 064602 (2010).

[10] Y. Ayyad et al., Phys. Rev. C **89**, 054610 (2014).

[11] Y. Ayyad et al., Phys. Rev. C **91**, 034601 (2015).

[12] K.-H. Schmidt et al., Nucl. Phys. A **665**, 221 (2000).

[13] J. L. Rodríguez-Sánchez et al., Phys. Rev. C **91**, 064616 (2015).

[14] J. L. Rodríguez-Sánchez et al., Phys. Rev. C **90**, 064606 (2014).

[15] J. L. Rodríguez-Sánchez et al., accepted in Phys. Rev. C (2015).

[16] N. P. Shaw et al., Phys. Rev. C **61**, 044612 (2000).

[17] I. Diószegi et al., Phys. Rev. C **63**, 014611 (2001).

[18] P. Fröbrich et al., Nucl. Phys. A **556**, 281 (1993).

[19] G. Boutoux et al., Phys. Procedia **47**, 166 (2013).

[20] T. Gorbinet et al., Phys. Procedia **64**, 101 (2015).

[21] A. Boudard, J. Cugnon, J.-C. David, S. Leray, and D. Mancusi, Phys. Rev. C **87**, 014606 (2013).

[22] A. Kelić, M. V. Ricciardi, and K.-H. Schmidt, Proceedings of Joint ICTP-IAEA Advanced Workshop on Model Codes for Spallation Reactions, ICTP Trieste, Italy, 4-8 February 2008, edited by D. Filges, S. Leray, Y. Yariv, A. Mengoni, A. Stanculescu, and G. Mank (IAEA INDC(NDS)-530, Vienna, 2008), pp. 181-221.

[23] V. F. Weisskopf and D. H. Ewing, Phys. Rev. **57**, 472 (1940).

[24] M. V. Ricciardi et al., Phys. Rev. C **73**, 014607 (2006).

[25] C. Villagrasa-Canton et al., Phys. Rev. C **75**, 044603 (2007).

[26] J. Alcántara-Núñez et al., Phys. Rev. C **92**, 024607 (2015).

[27] B. Jurado, K.-H. Schmidt, and J. Benlliure, Phys. Lett. B **553**, 186 (2003).

[28] A. J. Sierk, Phys. Rev. C **33**, 2039 (1986).

[29] J. Benlliure et al., Nucl. Phys. A **628**, 458 (1998).

[30] H. Hofmann and J. R. Nix, Phys. Lett. B **122**, 117 (1983).

[31] H. A. Kramers, Physica **7**, 284 (1940).

[32] B. D. Wilkins, E. P. Steinberg, and R. R. Chasman, Phys. Rev. C **14**, 1832 (1976).

[33] P. Paul and M. Thoennessen, Ann. Rev. Nucl. Part. Sci. **44**, 65 (1994).

[34] G. Chaudhuri and S. Pal, Phys. Rev. C **63**, 064603 (2001).

[35] P. Fröbrich, Prog. Theor. Phys. **154**, 279 (2004).

EPJ Web of Conferences **117**, 08018 (2016) DOI: 10.1051/epjconf/201611708018
NN2015

Fusion and neutron transfer reactions with weakly bound nuclei within time-dependent and coupled channel approaches

V. V. SAMARIN

Flerov Laboratory of Nuclear Reactions, JINR, Dubna, Moscow Region, Russia

Abstract

The time-dependent Schrödinger equation and the coupled channel approach based on the method of perturbed stationary two-center states are used to describe nucleon transfers and fusion in low-energy nuclear reactions. Results of the cross sections calculation for the formation of the ^{198}Au and fusion in the ^{6}He+^{197}Au reaction and for the formation of the ^{65}Zn in ^{6}He+^{64}Zn reaction agree satisfactorily with the experimental data near the barrier. The Feynman's continual integrals calculations for a few-body systems were used for the proposal of the new form of the shell model mean field for helium isotopes.

1 Time-dependent model for neutron transfer

It is a well-established fact that ^{6}He nucleus consists of the ^{4}He core and the two-neutron cluster, see *e.g.* Refs. [1–4]. The probability density for the ground state of the three-body systems ^{6}He $(\alpha + n + n)$ was calculated in Ref. [5] using Feynman's continual integrals method [6, 7]. The ground state probability densities for few-body systems of ^{4,5,6}He (see Fig. 1a for ^{6}He) were used for the proposal of the new forms $V_1(r)$ for mean field in the shell model of helium isotopes (see Fig. 1c). For ^{4}He properties of the mean field are similar to these determined in Ref. [8].

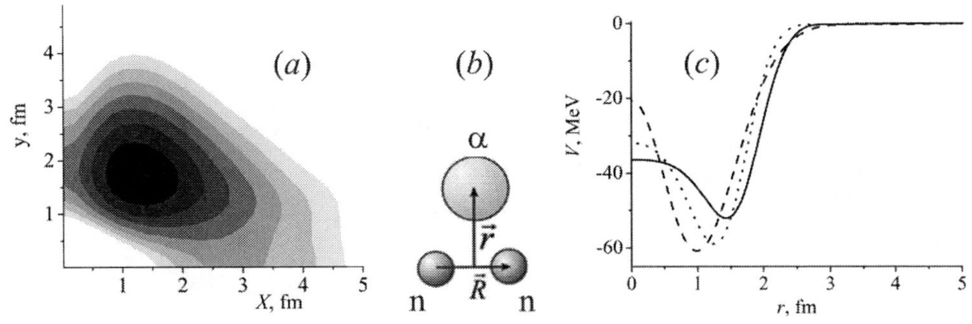

Figure 1: The probability density (a) for neutron-position symmetric configurations $\alpha + n + n$ (b) of ^{6}He calculated by Feynman's continual integrals method; (c) the mean field $V_1(r)$ for neutrons in the shell model of ^{6}He (solid line), ^{5}He (dotted line) and ^{4}He (dashed line).

The approximation of the independent external (valence) neutrons was used for the description of transfer reactions and the first (capture) stage of fusion reactions [9]. The two component spinor wave function of the neutron with the mass m and the radius vector $\mathbf{r}$ changes according to the time-dependent Schrödinger equation (TDSE) with the spin-orbital interaction [10–12]

$$i\hbar\frac{\partial}{\partial t}\Psi(\mathbf{r},t) = \{-\frac{\hbar^2}{2m}\Delta + W(\mathbf{r},t) + \hat{W}_{LS}(\mathbf{r},t)\}\Psi(\mathbf{r},t), \tag{1}$$

$$W(\mathbf{r},t) = V_1(\mathbf{r} - \mathbf{R}_1(t)) + V_2(\mathbf{r} - \mathbf{R}_2(t)). \tag{2}$$

Here $\mathbf{R}_{1,2}(t)$ are the radius vectors of the center of mass of the He projectile and the heavy target nuclei moving along classical trajectories. Potential $V_2(r)$ is the energy of the neutron interaction with the heavy target nucleus in the Wood-Saxon form [13]. The operator of the spin-orbit interaction is

$$\hat{W}_{LS} = -\frac{b_1}{2\hbar}\sigma\left[(\nabla V_1)\hat{\mathbf{p}}\right] - \frac{b_2}{2\hbar}\sigma\left[(\nabla V_2)\hat{\mathbf{p}}\right]. \tag{3}$$

Here $\hat{\mathbf{p}}$ is the momentum operator, σ are Pauli matrices, the factors $b_{1,2}$ are

$$b_i = R_0^2\lambda_i\frac{\hbar^2}{2m^2R_0^2c^2} = 0.022R_0^2\lambda_i, \tag{4}$$

c is the speed of light, $R_0 = 1$ fm and λ_i are phenomenological dimensionless constants. In this study value $\lambda_2 \approx 35$ [13] for heavy nuclei and value $\lambda_1 \approx 10$

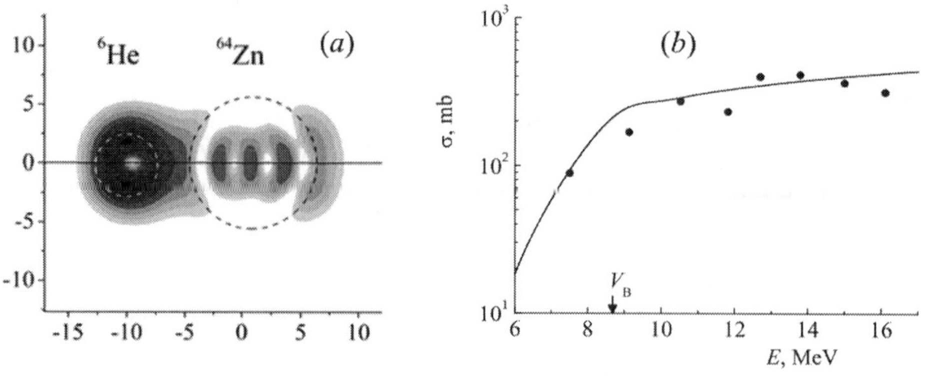

Figure 2: (a) The probability density of the valence neutrons of the ^{6}He nucleus during a frontal collision with ^{64}Zn at energy in the center of mass system E =8 MeV, a scale value is 1 fm, the radii of the circumferences equal to the radii of the nuclei. (b) The excitation function for the formation of the ^{65}Zn isotope in the reaction ^{6}He+^{64}Zn. Experimental data (solid circles) is from [15] . The theoretical curve is calculated within the TDSE approach; V_B is the Coulomb barrier.

for helium nuclei are used. The evolution of the probability density of the valence $1p_{3/2}$ neutrons of the ^{6}He nucleus during the collision with the ^{197}Au nucleus was studied in Ref. [12]. The probability density during the frontal collision of ^{6}He+^{64}Zn is shown in Fig. 2. The used mesh spacing in the TDSE method was not large then 0.2 fm, which is substantially smaller than 0.8 fm in a typical time-dependent Hartree-Fock (TDHF) calculation [14]. The colliding nuclei are enclosed in a box of typical dimensions $50 \times 25 \times 25$ fm^3 for central collisions and $50 \times 35 \times 25$ fm^3 for grazing collisions.

The presence of a stable structure in Fig. 2a indicates the formation of two-center (molecular) states of external neutrons. The neutron cloud in this case covers both nuclei, and an corresponding integral can serve as a measure of the probability w of the neutron transfer from one nucleus to another. Results of the cross section calculation using integration over the impact parameter $b \geq b_0$

$$\sigma = 2\pi \int_{b_0}^{\infty} w(b) b db. \tag{5}$$

for the formation of the ^{65}Zn isotope in the ^{6}He+^{64}Zn reaction agree satisfactorily with the experimental data [15] near the barrier (Fig. 2b). The impact parameter $b = b_0$ in Eq. (5) is corresponded to nuclei touching.

2 Coupled channel model for neutron transfer and fusion reactions

The semiclassical model with few independent quantum light particles (valence neutrons) and two classical heavy cores describes experimental neutron transfer cross sections satisfactory (see Fig. 3a and Ref. [12]). For simultaneous description of neutron transfer and fusion reactions we may use coupled channel (CC) approach (see *e.g.* Refs. [16,17]). The program CCFULL [16] includes pair-transfer coupling between the ground states only. It uses the macroscopic coupling form factor and the value of the coupling strength $F_t = \text{const}$ is adjusted manually, which makes its choice unobvious. Microscopic approach proposed in Refs. [5,10,18] takes into account shell structure of colliding nuclei. In this approach TDSE and CC solutions are used simultaneously. For neutron rearrangement CC equations based on the method of perturbed stationary two-center states are

$$y''_{L,\mu} - \frac{L(L+1)}{R^2} y_{L,\mu} + \frac{2M}{\hbar^2}\left[E_\mu(R) - U(R)\right] y_{L,\mu} = F_t \sum_\nu T_{\mu\nu}(R) y_{L,\nu}. \quad (6)$$

Here M is the reduced mass, $U(R)$ is the nucleus-nucleus potential, L is the relative orbital momentum, $E_\mu = E_{\text{c.m.}} + Q_\mu(R)$, $E_{\text{c.m.}} = \hbar^2 k_0^2 / 2M$, $Q_\mu(R) = \varepsilon_0 - \varepsilon_\mu(R)$ is the distance-dependent Q-value, ε_0 is the energy of the initial neutron state in the distant nucleus, $\varepsilon_\mu(R)$ is the two-center (molecular) energy level and $T_{\mu\nu}(R)$ is the reduced kinetic energy coupling matrix

$$T_{\mu\nu}(R) = \int \frac{\partial}{\partial R}\phi^*_\mu(\mathbf{r}; R) \frac{\partial}{\partial R}\phi_\nu(\mathbf{r}; R)\, d\mathbf{r}. \quad (7)$$

Here $\phi_\nu(\mathbf{r}; R)$ are wave functions of the stationary two-center states calculated using series expansion of Bessel functions [10]. The effective coupling strength F_t in Eq. (6) is used to compensate the deletion of some complicated expressions in the exact equations of the perturbed stationary states method [5, 10]. The value of the coupling strength $F_t \approx 10$ was determined from the requirement of the correlation between the values of the channel functions $|y_{L,\nu}|^2$ and the coefficients $|a_\nu(R)|^2$ of the series expansion

$$|y_{0,\nu}(R)|^2 \sim |a_\nu(R(t))|^2 \quad (8)$$

$$a_\nu(R(t)) = \int \phi^*_\nu(\mathbf{r}; R(t))\, \Psi(\mathbf{r}, t) d\mathbf{r} \quad (9)$$

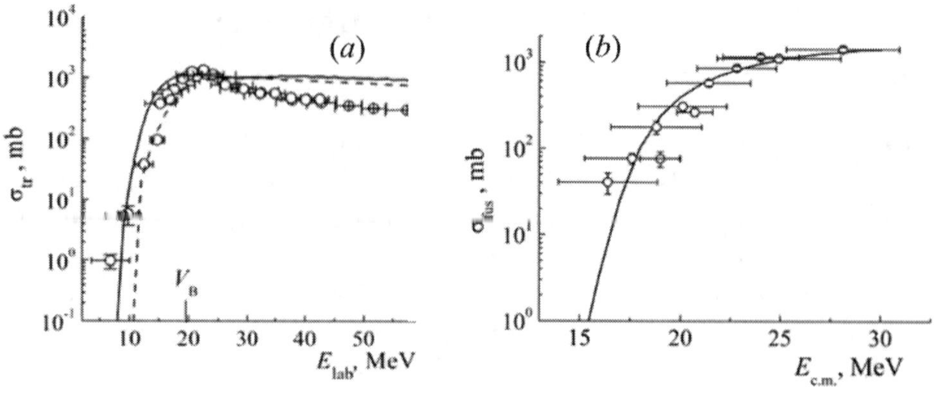

Figure 3: The excitation functions for the formation of the ^{198}Au isotope (a) and fusion (b) in the reaction ^{6}He+^{197}Au. Experimental data (circles) is from Refs. [19,20]. Theoretical curves were calculated within the coupled channel (solid lines) and the TDSE (dashed line) approaches in Ref. [21]; V_B is the Coulomb barrier.

The transfer cross section $\sigma_{\rm tr}$ and the fusion cross section $\sigma_{\rm fus}$ are calculated by summation of probability density flux in the infinity and in the region of the nuclear surfaces contact $R = R_c$

$$\sigma_{\rm tr} = \frac{\pi k_0^2}{j_0} \sum_{L=0}^{\infty} (2L+1) \sum_{\nu} |j_{L,\nu}(R \to \infty)|. \qquad (10)$$

$$\sigma_{\rm fus} = \frac{\pi k_0^2}{j_0} \sum_{L=0}^{\infty} (2L+1) \sum_{\nu} |j_{L,\nu}(R \to R_c)|. \qquad (11)$$

Here $j_0 = \hbar k_0/M$ is the flux incoming from the infinity. In the simplest approximation coupling matrix elements have nonzero values only for the states localized in different nuclei in the limit $R \to \infty$. Results of calculations of the transfer and fusion cross section plotted in Fig. 3 agree satisfactorily with the experimental data near the barrier.

References

[1] Yu. E. Penionzhkevich, Phys. Atom. Nucl. **74** 1615 (2011)

[2] Yu. Ts.Oganessian, V. I. Zagrebaev and J. S. Vaagen, Phys. Rev. C **60** 044605 (1999)

[3] V. I. Kukulin *et al.*, Nucl. Phys. A **453** 365 (1986)

[4] M. V. Zhukov *et al.*, Phys. Rep. **231** 151 (1993)

[5] V. V. Samarin, Phys. Atom. Nucl. **78** 861 (2015)

[6] R. P. Feynman and A. R. Hibbs, *Quantum Mechanics and Path Integrals* (McGraw-Hill, New York, 1965)

[7] E. V. Shuryak, Sov. Phys. Usp. **27** 448 (1984)

[8] H. Kanada *et al.*, Prog. Theor. Phys. **61** 1327 (1979)

[9] V. I. Zagrebaev *et al.*, Phys. Rev. C **75** 035809 (2007)

[10] V. V. Samarin, Phys. Atom. Nucl. **78** 128 (2015)

[11] V. V. Samarin, EPJ Web Conf. **66** 03075 (2014)

[12] V. V.Samarin, EPJ Web Conf. **86** 00040 (2015)

[13] S. Cwiok, *et al.*, Comput. Phys. Commun. **46** 379 (1987)

[14] C. Golabek and C. Simenel, Phys. Rev. Lett. **103** 042701 (2009)

[15] V. Scuderi *et al.*, Phys. Rev. C **84** 064604 (2011)

[16] K. Hagino *et al.*, Comput. Phys. Commun. **123** 143 (1999)

[17] V. V. Samarin and V. I. Zagrebaev, Nucl. Phys. A **734** E9 (2004)

[18] V. V.Samarin, EPJ Web Conf. **86** 00039 (2015)

[19] Yu. E. Penionzhkevich *et al.*, Eur. Phys. J. A **31**, 185 (2007)

[20] A. Kulko *et al.*, J. Phys. G **34**, 2297 (2007)

[21] V. V. Samarin *et al. Proceedings of the International Symposium on Exotic Nuclei (Kaliningrad, 2014)*, (World Scientific, Singapore, 2015) 115

EPJ Web of Conferences **117**, 08019 (2016) DOI: 10.1051/epjconf/201611708019
NN2015

Microscopic description of ^{258}Fm fission dynamic with pairing

GUILLAUME SCAMPS[1]*, CÉDRIC SIMENEL[2] and DENIS LACROIX[3]

[1] Department of Physics, Tohoku University, Sendai 980-8578, Japan
[2] Department of Nuclear Physics, Research School of Physics and Engineering
Australian National University, Canberra, Australian Capital Territory 2601, Australia
[3] Institut de Physique Nucléaire, IN2P3-CNRS, Université Paris-Sud, F-91406 Orsay Cedex, France

Abstract

Fission dynamic remains a challenge for nuclear microscopic theories. In order to understand the dynamic of the last stage of the fission process, the time-dependent Hartree-Fock approach with BCS pairing is applied to the describe the fission of the ^{258}Fm. A good agreement is found for the one-body observables: the total kinetic energy and the average mass asymmetry. The non-physical dependence of two-body observables with the initial shape is discussed.

1 Introduction

The fission process is an ideal phenomenon to test the predictive power of dynamical theories. Indeed, this process incorporates many aspects of nuclear dynamic, dissipation, superfluidity, tunneling, as well as a large number of degrees of freedom. The fission dynamic has been the object of several approaches: Brownian motion on the potential energy sur-

*E-mail: scamps@nucl.phys.tohoku.ac.jp

face [1], time-dependent generator coordinate method (TDGCM) [2], time-dependent Hartree-Fock (TDHF) [3] and adiabatic models that suppose that the two fragments choose the path that minimizes the energy from the initial well to the scission point.

The adiabatic assumption is often assumed along the fission process, where the friction is important and so the evolution is slow enough to follow the path that minimizes the energy. Nevertheless, close to scission, non-adiabatic effects play an important role [4]. A smooth transition occurs between the adiabatic motion when the two fragments are in contact and the fast evolution due to the Coulomb repulsion when the neck is broken. This transition involves many degrees of freedom: distance between the fragments, deformations of the fragments, neck configuration ... The TDHF theory is an ideal tool to study this process as it makes no restriction on the shape of the one-body density. Then the results do not depend of an arbitrary choice of parametrization of the nuclear shape during the scission.

In order to take into account pairing in the TDHF framework, the natural candidate is the time-dependent Hartree-Fock-Bogoliubov theory (TD-HFB). The TDHFB theory is numerically demanding [5]. In consequence, we choose to adopt the TD-BCS approximation. In several studies, it has been shown that the BCS approximation is a good compromise to take into account pairing in the mean-field framework. In particular, comparisons to QRPA calculations show a good agreement with the BCS approximation for the small amplitude motion [6, 7].

2 The fission process

In order to study the fission process, the system is first initialized after the barrier on the adiabatic path. This initialization is done using the Constraint Hartree-Fock with BCS approximation (CHF+BCS). A modified version of the code EV8 [8] has been used, where two of the reflection symmetry have been removed allowing the octupole deformation necessary to study asymmetric fission. As a test case, we study the ^{258}Fm fission where experimental data are available [9]. In the literature, three modes are considered: a symmetric fission with compact fragments, a symmetric fission with elongated fragments and an asymmetric fission. The three modes exhibit different dynamical behaviors due to different structure effects and Coulomb energy at the scission point.

Using the CHF+BCS, the evolution of the potential energy is shown on fig. 1 (right). Each mode is associated with a valley in the 3 dimensional space (Q_{20}, Q_{30}, Q_{40}) and is shown here as a function of the distance between the two fragments. Similar results for the potential energy as a function of the deformation for the three modes have been found using different interactions or methods in ref. [10, 11]. Starting from the three valleys, three corresponding TD-BCS evolutions have been done. The evolutions until complete scission are shown on the fig. 1 (right) . The details of the calculation can be found in ref. [12].

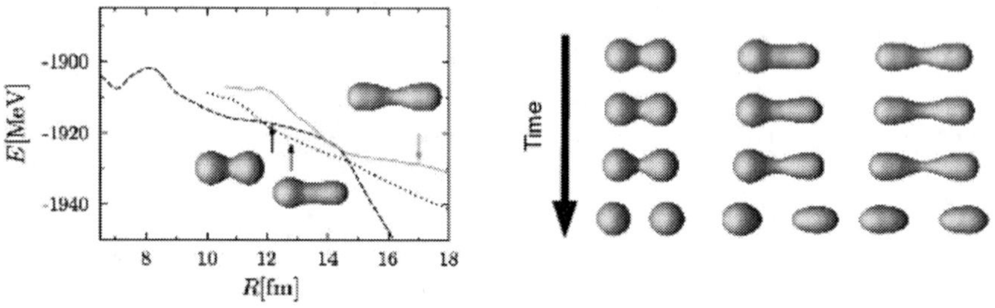

Figure 1: Left: Energy as a function of the distance between the two fragments for the three modes. The arrows represent the starting configuration of the dynamical calculation. Right : Isosurface density as a function of time for the three modes. The time between two pictures is different for the three modes, from left to right, $\Delta t = 0.675$ zs, 1.8 zs and 1.08 zs.

2.1 One-body observables

Several observables can be extracted from this simulation of the fission process. In particular, the total kinetic energy (TKE) is calculated after the complete separation of the two fragments by adding the Coulomb energy to the kinetic energy. A comparison between the TD-BCS and the experimental data is shown in fig. 2. The TD-BCS can predict neither the population of each of the mode nor the fluctuation of the TKE. Nevertheless, it predicts an average TKE and mass distribution for each mode in good agreement with the experimental distributions. For the elongated mode, it is found that this mode is compatible with the tail of the TKE distribution.

We can worry about the dependence of those observables with the starting point of the dynamic. In a naive picture, starting with a more compact shape increases the total energy and should then increase the TKE. In reality, due to the strong friction forces, all this additional energy is dissipated during the dynamic. For example, for initial distance of the fragment between 9.5 and 12.5 fm (both well before scission), the TKE is changed by less

DOI: 10.1051/epjconf/201611708019

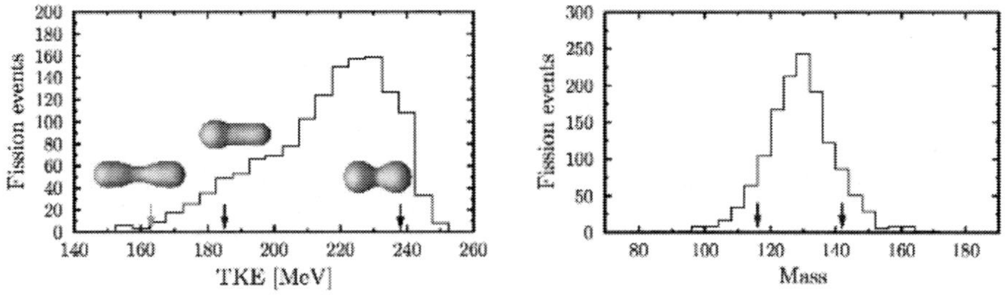

Figure 2: Left panel: Comparison between the TKE obtained for the three mode to the experimental TKE distribution. Right panel: Comparison between the average mass in the asymmetric mode to the experimental mass distribution (only events with TKE < 220 MeV are shown).

than 1 MeV for the symmetric compact mode. The fact that this observable is independent of the starting point confirms that the evolution between $R=$ 9.5 fm and 12.5 fm is adiabatic.

2.2 Two-body observables

The two-body observables, like the odd-even effects or the fluctuations of the mass distribution show a different behavior. To study this effect, we display on fig. 3 (left), the neutron distribution with different initial distances between the fragments. The distribution is obtained using the projection technique developed for TDHF in ref. [13] and extended to the case with pairing in ref. [14].

Starting from a configuration close to the scission point (R =12.5 fm), the neutron distribution shows an odd-even effects due to the pairing correlations. When more compact initial shapes are chosen, the odd-even effects are smoothed out. This effect can be understood simply by the fact that the initial energy is larger for compact shape. Then more energy is dissipated during the descent of the potential. This dissipated energy break the pair correlation and so the odd-even effects are reduced.

We next investigate the impact of the initial configuration on the width of the fragment mass distribution in fig. 3 (right). The fluctuations of the mass distribution is obtained without pairing with the TDHF theory and is compared to fluctuation obtained with the time-Dependent Random Phase Approximation (TDRPA) theory [15, 16]. The mass distribution is shown in fig. 3 (right) assuming a gaussian shape. The TDRPA goes beyond the TDHF approach with a variational approach not only for the one-body observables but also for the fluctuations. The TDRPA result is then closer to the experimental data. Nevertheless, for the two theories a dependence of the results with the initial distance is found.

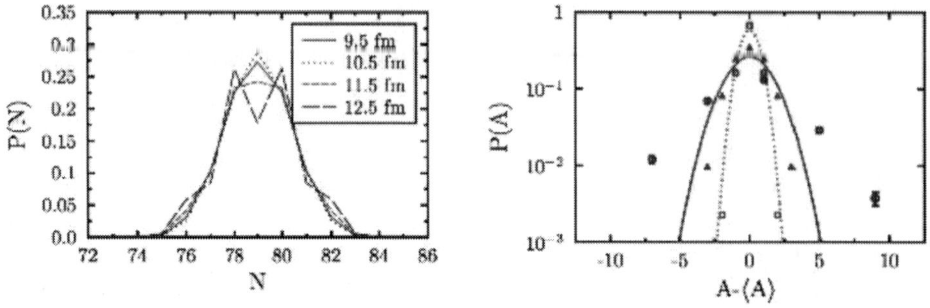

Figure 3: Left panel: Neutron distribution of the symmetric compact mode for different initial distances between the fragments. Right panel: Mass distribution obtained with TDHF (for ^{264}Fm) with initial distance R =10 fm (triangles) and R = 12 fm (squares). The TDRPA results are shown by solid line for R =10 fm and dotted line for R =12 fm assuming a Gaussian distribution. The experimental mass distribution (for ^{258}Fm) is also shown with blue circles.

This is expected as the fluctuations accumulate along the entire fission path from the compound nucleus to scission. We can expect that a dynamical theory starting from the initial well would reproduce the experimental fluctuation. Nevertheless as shown in ref. [17,18], within mean-field dynamics theories, the fission process does not occur for too compact shape.

3 Conclusion

The fission process has been study with the TD-BCS theory. It is shown that the one-body observables does not depends of the initial adiabatic configuration and reproduce the experimental data. For the two-body observables, a dependence is found. The latter could be due to the non-consideration of the initial excitation energy in the dynamical calculations. This dependence shows the necessity to go beyond the present approach with finite temperature calculations.

Acknowledgments

G.S. acknowledges the Japan Society for the Promotion of Science for the JSPS postdoctoral fellowship for foreign researchers. This work was supported by Grant-in-Aid for JSPS Fellows No. 14F04769. This work has been supported by the Australian Research Council Grants No. FT120100760.

References

[1] J. Randrup and P. Möller, Phys. Rev. Lett. **106** (2011) 132503.

[2] H. Goutte, J. F. Berger, P. Casoli, and D. Gogny, Phys. Rev. C **71** (2005) 024316.

[3] J. W. Negele, S. E. Koonin, P. M oller, J. R. Nix, and A. J. Sierk, Phys. Rev. C **17** (1978) 1098.

[4] C. Simenel and A. S. Umar, Phys. Rev. C **89** (2014) 031601.

[5] B. Avez, C. Simenel, and Ph. Chomaz, Phys. Rev. C **78** (2008) 044318.

[6] G. Scamps and D. Lacroix, Phys. Rev. C **88** (2013) 044310.

[7] S. Ebata, T. Nakatsukasa, T. Inakura, et al., Phys. Rev. C **82** (2010) 034306.

[8] P. Bonche, H. Flocard, and P. H. Heenen, Comp. Phys. Com. **171** (2005) 49.

[9] E. K. Hulet, J. F. Wild, R. J. Dougan, et al., Phys. Rev. C **40** (1989) 770.

[10] L. Bonneau, Phys. Rev. C, **74** (2006) 014301.

[11] A. Staszczak, A. Baran, J. Dobaczewski, and W. Nazarewicz, Phys. Rev. C **80** (2009) 014309.

[12] G. Scamps, C. Simenel, and D. Lacroix, Phys. Rev. C **92** (2015) 011602.

[13] C. Simenel, Phys. Rev. Lett., **105** (2010) 192701.

[14] G. Scamps and D. Lacroix, Phys. Rev. C **87** (2013) 014605.

[15] R. Balian and M. Vénéroni, Phys. Lett. B **136** (1984) 301.

[16] C. Simenel, Phys. Rev. Lett. **106** (2011) 112502.

[17] P. M. Goddard, P. D. Stevenson, and Arnau Rios, arXiv:1504.00919, 2015.

[18] Y. Tanimura, D. Lacroix, and G. Scamps, Phys. Rev. C **92** (2015) 034601 .

Measurement of fusion excitation function for ^{7}Li+^{64}Ni near the barrier

Md. Moin Shaikh[1], Subinit Roy[1], S. Rajbanshi[1], A. Mukherjee[1], M. K. Pradhan[1], P. Basu[1], S. Pal[2], V. Nanal[2], A. Shrivastava[3], S. Saha[2] and R. G. Pillay[2]

[1]Saha Institute of Nuclear Physics, 1/AF, Bidhan Nagar, Kolkata-700 064, India,
[2]Department of Nuclear and Atomic Physics, Tata Institute of Fundamental Research, Mumbai-400 005, India and
[3]Nuclear Physics Division, Bhabha Atomic Research Centre, Mumbai-400 085, India

Abstract

Total fusion (TF) excitation function has been measured for the system ^{7}Li + ^{64}Ni at the energies near the Coulomb barrier of the system. The evaporation residue (ER) cross sections have been estimated through the online detection of characteristic γ-rays of the ERs. The summed ER cross sections yielding the experimental TF cross section have been compared with the theoretical one dimensional barrier penetration model (1DBPM) prediction. The measured and the model cross sections are very close to each other at above barrier energies. However, an enhancement of the experimental TF cross section with respect to the 1DBPM prediction is observed at below barrier energies. Coupled channels (CC) calculation with inelastic excitations alone could not explain the enhancement. The origin of the enhancement is identified as due to the enhanced population of the αxn channels.

A complete understanding of the effect of direct reaction channels like, breakup or transfer followed by breakup on fusion of weakly bound systems at near barrier energies is of primary importance [1, 2]. In case of fusion of weakly bound stable projectiles like, ^{6}Li (S_α = 1.47 MeV), ^{7}Li (S_α =

2.47 MeV) and ^{9}Be (S_n =1.67 MeV), two principal observations are: the suppression of complete fusion (CF) cross section with respect to one dimensional barrier penetration model (1DBPM) prediction at the above barrier energies and the enhancement of fusion cross section at below barrier energies [3–12]. While, the suppression at above barrier energies received maximum attention in the study of fusion of weakly bound systems, the sub-barrier enhancement of fusion cross section of these systems is also equally important to understand the interplay of reaction mechanisms in this energy regime.

Though the measurements of fusion cross sections for these projectiles show consistent behavior of suppression at above barrier energies for heavy mass targets, the evolution of the behavior as the target mass decreases needs to be clearly understood. The problem in the collision with lower medium target is to experimentally distinguish the CF cross sections from other reaction processes producing the same residues. Hence the observable cross section is known as total fusion (TF) cross section [10,13–15]. However, very recently our group has reported the extraction of CF excitation function from the measured TF excitation function for the system ^{6}Li+ ^{64}Ni where the target has six additional neutrons over the N = 28 closed sub-shell [12].

To further explore the method, we have measured the fusion excitation function of the other weakly bound Li-isotope, ^{7}Li (S_α= 2.47 MeV), with the target ^{64}Ni at near-barrier energies. We present here the experiment, analysis and interpretation of only the below barrier behavior of the fusion excitation function.

The experiment was performed at BARC-TIFR Pelletron Facility in Mumbai, India. An enriched metallic (~99%) and 507 μg/cm^2 thick ^{64}Ni target, procured from Oak Ridge National Laboratory, USA, was used for the present experiment. The target thickness was verified using α-energy loss method with 3-line α source (^{239}Pu, ^{241}Am and ^{244}Cm). The estimated uncertainty in the thickness was ~2%. The target was bombarded with ^{7}Li beam with energies varying from 12 to 28 MeV. The online characteristic γ-ray detection method was used for the measurement of fusion cross section [16]. The γ-rays were detected by a HPGe detector placed at 135° and a clover detector placed at 45° with respect to the beam direction. The detector resolutions were 2.8 keV and 5.0 keV, respectively for HPGe and clover detectors for 1408 keV γ-line of ^{152}Eu. The data were recorded using the data acquisition system LAMPS [17]. Fig. 1 shows a representative γ-ray spectrum recorded at E_{lab} = 26 MeV with some of the observed γ-lines.

The compound nucleus (CN) ^{71}Ga, produced in the collision of the system ^{7}Li+^{64}Ni, predominantly decays through the $2n$ and $3n$ evaporation

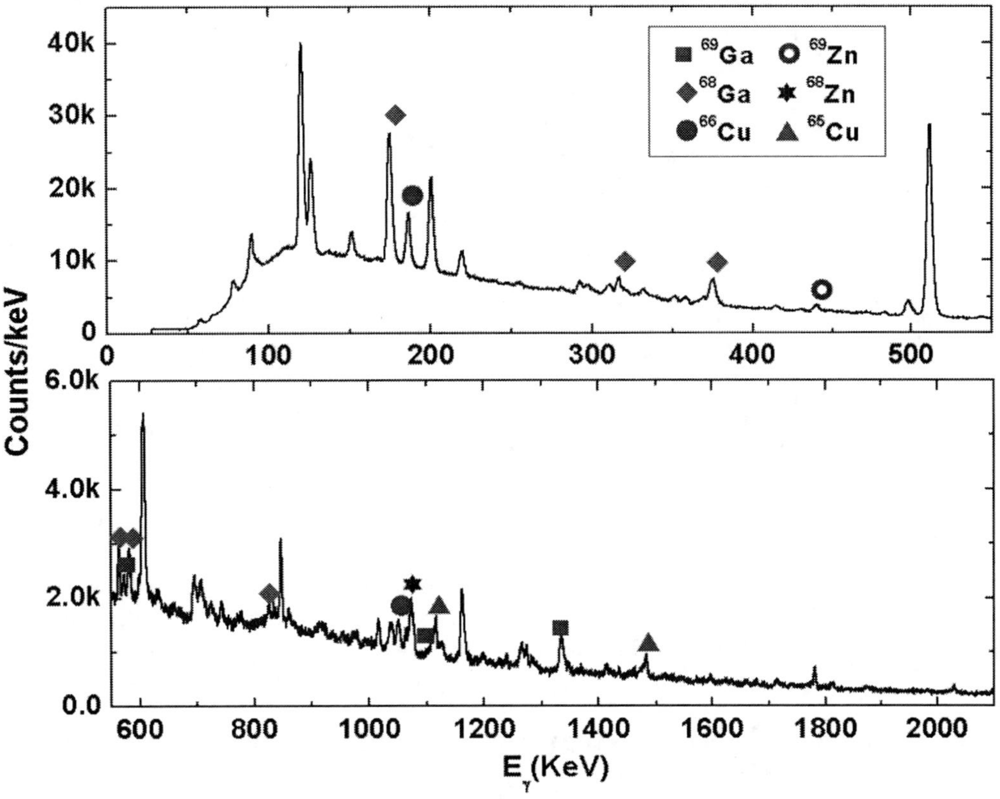

Figure 1: Representative characteristic γ-ray spectrum obtained at E_{lab}= 26 MeV from the collision of the projectile ^{7}Li with the target ^{64}Ni.

channels. The evaporation residues from decay channels pn, $p2n$ or t, αn and α2n were also observed. The residues produced through $2n$, $3n$ and pn evaporation channels are associated only with CF process. The other channels are produced from CF as well as from the other reaction processes like breakup or transfer-breakup reactions. As an example the residue produced from CF process through αn channel, can also be produced from t-ICF process with subsequent evaporation of a neutron. The direct cluster transfer (DCT) to unbound state of ^{67}Cu can also populate the channel.

Each evaporation residue channel cross section (σ_{chn}^{exp}) was obtained by summing over the measured cross sections of the observed γ-transitions directly to the ground state of the residue. The experimental fusion cross section (σ_{fus}^{exp}) at each energy was obtained from,

$$\sigma_{fus}^{exp} = \sum_{chn} \sigma_{chn}^{exp}. \tag{1}$$

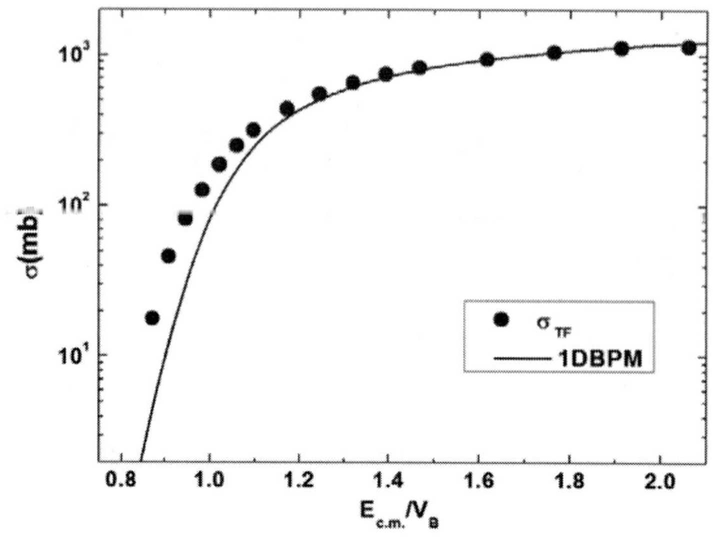

Figure 2: Experimental and theoretical fusion excitation functions for the system $^{7}Li+^{64}Ni$ at near barrier energies. The data points are represented by solid bullets and the 1DBPM predictions is denoted by solid line.

Since, the evaporation residues produced from the ICF processes were experimentally indistinguishable from the residues produced in the CF process, the measured fusion cross section provides only the total fusion cross section.

The measured TF excitation function has been plotted in Fig. 2. The data have been compared with the theoretical 1DBPM prediction calculated with the code CCFULL [18]. The potential parameters used in the calculation are: strength V_0 = 48.0 MeV, radius parameter r_0 = 1.12 fm and diffuseness $a_0 = 0.66$ fm and the corresponding barrier parameters are: 12.20 MeV, 9.19 fm and 3.41 MeV, respectively. It can be observed from Fig. 2 that the 1DBPM reproduced the above barrier fusion excitation function quite well. However, as the energy is decreased towards the barrier the experimental TF cross sections start to increase and at the sub-barrier energies they are significantly larger compared to the theoretical values.

To probe the possible admixture from ICF and/or transfer channels with CF processes at below barrier region, we have constructed the summed cross sections of the measured xn and αxn channels. The experimental $\sum xn$ and $\sum \alpha xn$ cross sections are shown in Fig. 3. In the figure, the experimental cross sections are compared with the statistical model predictions of $\sum xn$ and $\sum \alpha xn$ channel cross sections by PACE4 [19]. Remarkably good reproduction of the experimental $\sum xn$ cross section is obtained with the statisti-

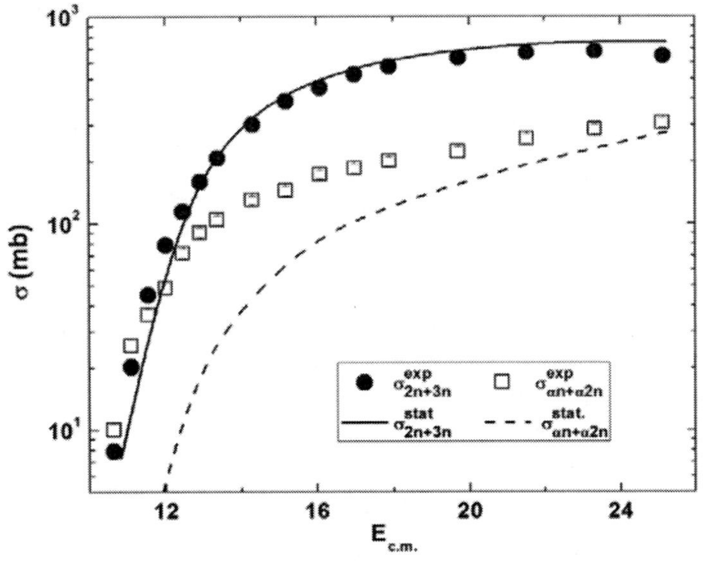

Figure 3: Experimental excitation functions for the sum of pure CF evaporation channels and the sum of mixed channels in comparison with their corresponding statistical model predictions.

cal model indicating the purity of their CF origin. The small enhancement at the subbarrier energies clearly shows the effect of channel coupling in CF cross sections. On the other hand, large underprediction of experimental $\sum \alpha xn$ cross sections by the statistical model highlights the strong presence of non-compound processes in $\sum \alpha xn$ cross sections. It can also be noticed from Fig. 3 that at energies below the barrier $\sum \alpha xn$ cross sections are higher than $\sum xn$ cross sections coming only from CF evaporation. Thus at these energies the non-compound process has comparable cross section relative to the CF process.

The possible non-compound processes that can contribute are the incomplete fusion of triton, or t-ICF, followed by neutron evaporation; the transfer of clusters like 't' or 'd' to unbound state of the residues. Clear estimation of the contributions of these reaction mechanisms is however limited by the setup used in the present experiment.

To summarize, in the present work we have measured the TF excitation function for the weakly bound projectile ^{7}Li with medium mass target ^{64}Ni at near barrier energies. The TF cross sections have been obtained from the measured residue cross sections identified from the characteristic γ rays. There is no suppression in TF cross sections in above barrier region with respect to the theoretical predictions. This observation agrees with the earlier observations from all the weakly bound stable projectiles, that the TF cross sections are unaffected by the breakup of projectile irrespective of target mass [4–7,10,12–15]. But the observed subbarrier enhancement of TF cross sections is not understood clearly. The admixture of cluster transfer cross section may be the reason behind this large enhancement. To settle the issue, exclusive measurements with particle- γ coincident is required.

We sincerely thank the Pelletron staff for providing a steady ^{7}Li beam during the experiment. One of the authors, Md. Moin Shaikh, would like to thank the Council of Scientific & Industrial Research, Govt. of India for financial support vide File No. 09/489(0084)/2010-EMR-I.

References

[1] L.F. Canto, *et al.*, Phys. Rep. **424**, 1 (2006).

[2] B.B. Back,*et al.*, Rev. Mod. Phys. **86**, 317 (2014).

[3] D. J. Hinde, *et al.*, Phys. Rev. Lett. **89**, 272701 (2002).

[4] M. Dasgupta, *et al.*, Phys. Rev. C **66**, 041602(R) (2002).

[5] A. Diaz-Torres, *et al.*, Phys. Rev. C **68**, 044607 (2003).

[6] M. Dasgupta, *et al.*, Phys. Rev. C **70**, 024606 (2004).

[7] A. Mukherjee, *et al.*, Phys. Lett. B **636**, 91 (2006).

[8] P.R.S. Gomes *et al.*, Phys. Rev. C **84**, 014615 (2011).

[9] P. K. Rath, *et al.*, Phys. Rev. C **88**, 044617 (2013).

[10] A. Di Pietro *et al.*, Phys. Rev. C **87**, 064614 (2013).

[11] C.S. Palshetkar, *et al.*, Phys. Rev. C **89**, 024607 (2014).

[12] Md. Moin Shaikh, *et al.*, Phys. Rev. C **90**, 024615 (2014).

[13] C. Beck, *et al.*, Phys. Rev. C **67**, 054602 (2003).

[14] P.R.S. Gomes, *et al.*, Phys. Lett. B **601**, 20 (2004).

[15] P. R. S. Gomes, *et al.*, Phys. Rev. C **71**, 034608 (2005).

[16] P.R.S. Gomes, *et al.*, Nucl. Instrum. Meth. Phys. Res. A **280**, 395 (1989).

[17] Computer code LAMPS (Linux Advanced Multiparameter System), www.tifr.res.in/~pell/lamps.html.

[18] K. Hagino, *et al.*, Comput. Phys. Commun. **123**, 143 (1999).

[19] A. Gavron, Phys. Rev. C **21**, 230 (1980).

EPJ Web of Conferences **117**, 08021 (2016) DOI: 10.1051/epjconf/201611708021
NN2015

Exploring dissipative processes at high angular momentum in ^{58}Ni+^{60}Ni reactions

E. Williams, D.J. Hinde, M. Dasgupta, I.P. Carter, K.J. Cook, D.Y. Jeung, D.H. Luong, S.D. McNeil, C.S. Palshetkar[1], D.C. Rafferty, K. Ramachandran[2], C. Simenel, E.C. Simpson and A. Wakhle[3]

Department of Nuclear Physics, Research School of Physics and Engineering, Australian National University, Acton, ACT 2601 Australia

Abstract

Current coupled channels (CC) models treat fusion as a coherent quantum-mechanical process, in which coupling between the collective states of the colliding nuclei influences the probability of fusion in near-barrier reactions. While CC models have been used to successfully describe many experimental fusion barrier distribution (BD) measurements, the CC approach has failed in the notable case of ^{16}O+^{208}Pb. The reason for this is poorly understood; however, it has been postulated that dissipative processes may play a role. Traditional BD experiments can only probe the physics of fusion for collisions at the top of the Coulomb barrier ($L = 0\hbar$). In this work, we will present results using a novel method of probing dissipative processes inside the Coulomb barrier. The method exploits the predicted sharp onset of fission at $L \sim 60\hbar$ for reactions forming compound nuclei with $A < 160$. Using the ANU's 14UD tandem accelerator and CUBE spectrometer, reaction outcomes have been measured for the ^{58}Ni+^{60}Ni reaction at a range of energies, in order to explore dissipative processes at high angular momentum. In this reaction, deep inelastic processes have been

---

[1]Present address: Department of Nuclear and Atomic Physics, Tata Institute of Fundamental Research, Mumbai 400005, India

[2]Present address: Bhabha Atomic Research Centre, Mumbai, India 400085, India

[3]Present address: National Superconducting Cyclotron Laboratory, Michigan State University, MI 48824, USA

found to set in before the onset fission at high angular momentum following fusion. The results will be discussed in relation to the need for a dynamical model of fusion.

1 Introduction

In the study of nuclear reactions, one of the most elusive goals of the field has been to develop a comprehensive model of nuclear reactions: one which is capable of predicting the probabilities of the full range of possible reaction outcomes, for the complete range of nuclear collisions. The challenge in creating such a model arises from the complex dynamics of the nuclear reactions process. For a given reaction, many degrees of freedom are at play, including nuclear shell structure, collective excitations, reaction kinematics, energy damping, and angular momentum. Experimentally, we cannot vary one degree of freedom while holding all others fixed. At best, we can minimise the importance of one or two variables through clever tricks, or compare data with models reliant on different assumptions in an effort to tease out the physics behind any disagreement.

Unfortunately, no single answer exists for most experiment-model disagreements. One example of this can be found in the body of work surrounding ^{16}O+^{208}Pb fusion cross sections. High precision cross section data below and well above the Coulomb barrier cannot be consistently reproduced with a coupled channels model of fusion, unless one is willing to use different nuclear potential parameters to reproduce the data below and above the Coulomb barrier [1]. Different methods of addressing this discrepancy have been put forth, ranging from altering the ion-ion potential [2,3] to a realistic representation of energy dissipation within the model [4]. For now, though, all explanations amount to adding another parameter or two into an already parameter-heavy model, in an effort to make the model fit the data.

In this work, we aim to investigate the role energy dissipation plays in fusion, because it is an essential component of the fusion process but is not explicitly included in the coupled channels model of fusion. In the coupled channels approach [5,6], fusion is rendered irreversible via the use of either an imaginary potential or the Incoming Wave Boundary Condition, both of which set in at an internuclear separation r inside the fusion barrier. This onset of the irreversibility of fusion is not angular momentum dependent; the condition is either there or it is not for a given r. This sharp onset of irreversibility can have very different effects on fusion outcomes as a function of angular momentum.

Experimentally, partial fusion cross sections σ_L cannot be measured independently, making it very difficult to get a sense of whether the means by which the coupled channels approach mocks up fusion irreversibility is a reasonable approximation. For reactions forming a compound nucleus of mass $A_{cn} < 160$, the fission barrier becomes sufficiently low for fusion with angular momenta $L > 60\hbar$. Fission can be used as a high angular momentum fusion tag, providing us with a unique opportunity to study high angular momentum fusion outcomes.

In this paper, we present the first study to test this high angular momentum separation method, for the ^{58}Ni+^{60}Ni reaction. We will outline the preliminary findings, and discuss the possible implications of our findings on the ^{16}O+^{208}Pb problem and the coupled channels approach to fusion.

2 Experimental methods

The ^{58}Ni+^{60}Ni experiment was performed at Australian National University's Heavy Ion Accelerator Facility, using the 14UD tandem accelerator and CUBE two-body fission spectrometer [7]. The ^{58}Ni beam impinged upon the 60 μg/cm^2-thick ^{60}Ni target for 22 separate beam energies ranging from E_{lab} =198-266 MeV ($E/V_B \sim 0.97 - 1.35$). The CUBE spectrometer, consisting of two large-area multiwire proportional counters, was placed at forward angles 45° relative to the beam axis and a distance of 22.24 cm from the target. The detectors provided energy loss, time of flight, and (x, y) position information with a resolution of 1 mm.

Two calibration runs with a ^{58}Ni beam at 160 MeV in the lab frame bombarding ^{58,60}Ni targets of thicknesses 50 and 60 μg/cm^2, respectively were also carried out. The Mott scattering patterns for the former and Rutherford scattering data for the latter were both used to calibrate the geometry of the setup, define the resolution of the CUBE detector system, and provide a solid angle calibration for cross section determinations.

Using the position and time of flight information provided by the CUBE detector, a full kinematic reconstruction of each fission event was performed using the kinematic coincidence method [7,8], providing total kinetic energy, mass ratio,

$$M_R = \frac{m_1}{m_1 + m_2}, \tag{1}$$

and scattering angle information.

The total angular coverage for this experiment can be seen in Fig. 1, where the lab angle for one detector versus the folding angle ($\theta_1 + \theta_2$ is shown for the highest energy run in this experiment, at 135.19 MeV.

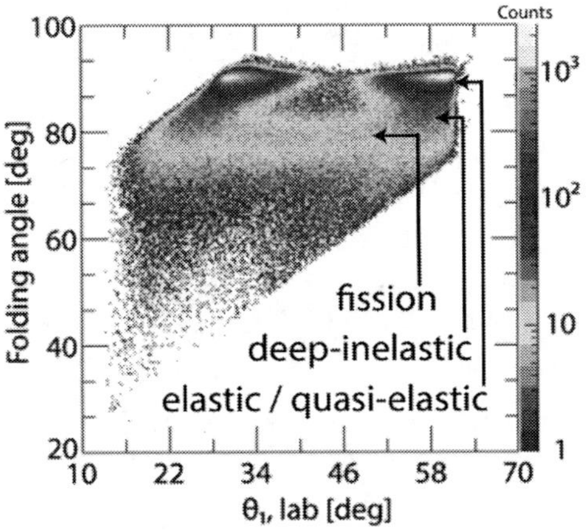

Figure 1: Lab angle for one detector versus the folding angle ($\theta_1 + \theta_2$) for the ^{58}Ni+^{60}Ni reaction at 135.19 MeV.

In this work, our interest in studying the influence of dissipation on fusion reactions at high angular momentum led to a focus on the observed mass ratio versus total kinetic energy. Before producing the final mass versus total kinetic energy plots, gates were placed on (1) the detector positions and times to exclude unphysical events, (2) the folding angle to exclude heavy target impurities, and (3) on the out-of-plane velocity of the fission fragments, in order to exclude random coincidences. The TKE results are presented below, and full discussion of the analysis will be given in a forthcoming publication.

3 Results and discussion

The results of this work are summarized by the mass ratio versus total kinetic energy (TKE) plots shown in Fig. 2 for three representative above-barrier runs.

The data show a distinct evolution as a function of energy. Just above the barrier in Fig. 2 (a), elastic and quasielastic scattering dominate the results, and the projectile-like and target-like species can be distinguished in mass and appear at a total kinetic energy consistent with that of elastic scattering. As the energy increases (Fig. 2 (b)), a lower TKE component appears below the original elastic scattering component, and no clear distinction between the projectile-like and target-like species is observed. This lower TKE group is consistent with deep inelastic scattering (energy is dissipated, some mass

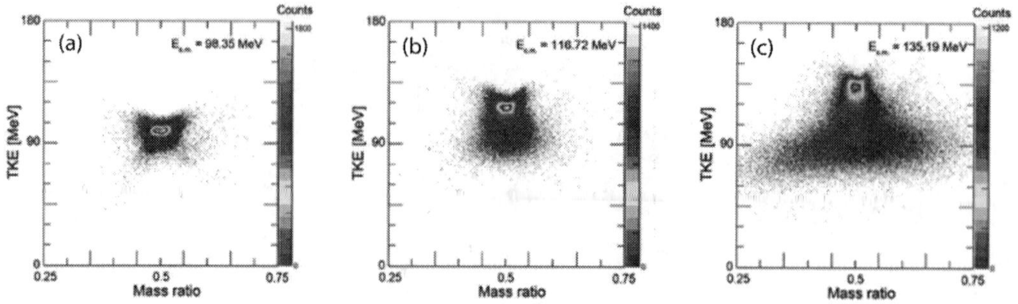

Figure 2: Mass ratio versus total kinetic energy for three representative above-barrier runs: (a) 98.35 MeV, (b) 116.72 MeV, and (c) 135.19 MeV. Note the excellent mass resolution of the CUBE detector; in the lowest energy dataset shown—the projectile-like and target-like species are distinguishable.

equilibration, very little change in the mass of the two fragments). While the intermediate energies are not shown here, the progression from (a) to (b) is smooth, with the onset of deep inelastic scattering appearing at progressively lower TKE values above ~105 MeV. From Fig. 2 (b) to (c), a third class of reaction outcomes appears with a threshold of ~125 MeV, corresponding to events with a wide spread in mass and full energy damping, consistent with expectations for fusion-fission.

The appearance of deep inelastic scattering before fission was an unexpected outcome in this experiment. Traditionally, one might expect deep inelastic scattering to set in at high angular momentum, higher than fusion-fission. The early onset of deep inelastic scattering provides firm evidence of the role dissipation plays even at lower angular momenta.

4 Conclusions

In this work, two-body reaction outcomes were used to search for evidence of dissipative processes in high angular momentum ^{58}Ni + ^{60}Ni reactions. Deep inelastic scattering was observed to play a significant role in this reaction, with its onset (~ 105 MeV) preceding that of fusion-fission (~ 125 MeV). While these results are preliminary, they demonstrate that deep-inelastic scattering competes with fusion, and should be taken into account in fusion models. The inclusion of deep inelastic scattering as a process that competes with fusion may offer a potential solution for experiment-model disagreements below and above the barrier, e.g., for the case of ^{16}O+^{208}Pb, though this cannot be confirmed in the current experiment. This work does, however, provide further evidence [9,10] for the need for a dynamical model of reactions, in which all possible reaction outcomes can be calculated within a single consistent framework.

Acknowledgments

The authors thank the Heavy Ion Accelerator Facility accelerator and technical staff for their invaluable help with this project, and NCRIS for support for accelerator operations. This work was supported by the Australian Research Council, grant numbers DP130101569, DP140101337, FL110100098, FT120100760 and DE140100784.

References

[1] C. R. Morton *et al. Phys. Rev. C*, **60** (1999) 044608.

[2] H. Esbensen and M. Serban, *Phys. Rev. C*, **76** (2007) 054609.

[3] A. S. Umar, C. Simenel, and V. E. Oberacker, *Phys. Rev. C*, **89** (2014) 034611.

[4] M. Dasgupta *et al. Phys. Rev. Lett.*, **99** (2007) 192701.

[5] C. H. Dasso, S. Landowne, and A. Winther, *Nucl. Phys.*, **A405** (1983) 381.

[6] K. Hagino, N. Rowley, and A. T. Kruppa, *Comp. Phys. Comm.*, **129** (1999) 143.

[7] D. J. Hinde *et al.*, *Phys. Rev. C*, **53** (1996) 1290.

[8] R. Rafiei *et al.*, *Phys. Rev. C*, **77** (2008) 024606.

[9] E. Williams *et al.*, *Phys. Rev. C*, **88** (2013) 034611.

[10] R. du Rietz *et al.*, *Phys. Rev. C*, **88** (2013) 054618.

Systematics for low energy incomplete fusion: Still a puzzle?

Abhishek Yadav[1], Mohd Shuaib[2], Abhay V. Aggarwal[2], Vijay R. Sharma[2], Indu Bala[1], D. P. Singh[3], P. P. Singh[4], Unnati[5], M. K. Sharma[6], R. Kumar[1], R. P. Singh[1], S. Muralithar[1], B. P. Singh[2] and R. Prasad[2]

[1]NP Group: Inter-University Accelerator Centre, New Delhi-110 067, India
[2]Phys. Dept., Aligarh Muslim University, Aligarh-202 002, U. P., India
[3]Phys. Dept., Univ. of Petro. & Energy Studies, Dehradoon-248007, India
[4]Phys. Dept., Indian Institute of Technology Ropar, Ropar-140 001, India
[5]Dept. of Phys. & Astrophys., University of Delhi, Delhi-110007, India
[6]Phys. Dept., S. V. College, Aligarh-202 001, U. P., India

Abstract

In order to have a better and clear picture of incomplete fusion reactions at energies ≈4-7MeV/nucleon, the excitation function measurements have been performed for ^{18}O+^{159}Tb system. The experimental data have been analyzed within the framework of compound nucleus decay. The cross-section for xn/pxn-channels are found to be well reproduced by PACE4 predictions, which suggest their production via complete fusion process. However, a significant enhancement in the excitation functions of α-emitting channels has been observed over the theoretical ones, which has been attributed due to the incomplete fusion processes. The incomplete fusion fractions have been deduced at each studied energy and compared with other nearby systems for better insight into the underlying dynamics. The incomplete fusion fraction has been found to be sensitive to the projectile's energy and α-Q-value.

1 Introduction

Fusion reactions, induced by heavy-ions (HIs), play an important role in nuclear physics since they enable to study the production & properties of nuclei away from the valley of stability [1]. For this reason the understanding of HI-reaction mechanism has always been an active area of study. During the last couple of decades, with the observation of incomplete fusion (ICF) reactions at energies in the vicinity of Coulomb barrier (V_b), considerable efforts are being employed by both experimental and theoretical physicists to understand the presence of ICF-reactions at such low energies, where complete fusion (CF) is supposed to be the sole contributor to the total fusion cross section [2–6]. The presence of ICF-reactions at such low energies triggered the resurgent interest to understand & find out the general systematics for low enery ICF reactions. It is not out of place to mention that, presently, no theoretical model is available which could reproduce the low energy ICF data satisfactorily.

The CF-reactions correspond to the complete amalgamation of interacting nuclei for input angular momentum $\ell < \ell_{crit}$, leading to the formation of completely fused excited composite system, which may decay via particle and/or γ-emission. However, in case of ICF-reactions, for partial waves $\ell > \ell_{crit}$ (as per sharp-cut off model), the projectile may break-up into its constituent to provide the sustainable input angular momentum. One of the fragments may fuse with the target nucleus forming the reduced excited composite system with relatively less mass, charge and excitation energy. While, the remnant flows in the forward direction with almost beam velocity. Since the observation of ICF reactions, several theoretical models/theories have been proposed to explain these reactions. In general, all the proposed models are found to fit the experimental data at projectile energies $\geq$ 10 MeV/nucleon to a large extent [6]. But, at low beam energies ($\approx$3-7 MeV/nucleon), the ICF-reactions are not well understood. In addition to this, the effect of various entrance channel parameters such as the beam energy [5], the angular momentum [7], the entrance channel mass asymmetry [8], the projectile structure in terms of alpha-Q-value and/or binding energy of the projectile [6] have been studied and contradicting dependence of ICF-reactions on these parameters have been reported [6]. Morgenstern *et al.* [9]correlated the ICF fraction with entrance channel mass asymmetry (μ_A). Recently, Singh *et al.* [5] introduced the importance of projectile structure in ProMass-systematics. Hence, in order to explore the low-energy incomplete fusion and to find a consistent general systematics for low energy ICF reactions, measurements of excitation function for $^{18}O+^{159}Tb$ system at energies ≈ 4-7MeV/A have been performed and compared with nearby systems.

2 Experiments

The experiments have been performed at the Inter-University Accelerator Center (IUAC), New Delhi to measure the excitation functions of radio-nuclides populated during ^{18}O+^{159}Tb interactions. Here, brief experimental details are given for the ready reference; however, the detailed description is given in ref [6]. Isotopically pure ^{159}Tb targets and Al-catcher/energy degrader foils, thicknesses ranging ≈ 1.5-2.5 mg/cm^2, were prepared using rolling technique. An *Al*-catcher foil of sufficient thickness has been placed behind each target foil so that the recoiling products during the irradiations may be trapped in the catcher foil thickness. The stacked-foil (to cover wide energy range in an irradiation) activation technique followed by off-line γ-ray spectroscopy has been used. The induced activities in the irradiated samples were recorded by counting each target-catcher foil assembly, using a pre-calibrated HPGe γ-ray spectrometer. A 50Hz pulser was used to determine the dead time of the detector. The efficiency calibration of the detector in the specified geometry was carried out using a standard ^{152}Eu source. The characteristic γ-lines have been used to identify the reaction products. Further, in order to make sure the identification of the reaction products, the decay curves of the identified reaction products have also been analyzed. Nuclear data of radio-nuclides, such as the characteristic gamma-lines, their abundances, half-lives, etc. were taken from ref [10]. After the identification and confirmation of the residues, the production cross sections of the reaction products have been determined using the standard formulation [5].

3 Results and discussion

In order to understand the formation mechanism of the residues produced during ^{18}O+^{159}Tb interactions, the experimentally measured excitation functions have been analyzed within the framework of statistical model code PACE4 [11], which is based on equilibrated CN-decay of Hauser-Feshbach theory. It may, however, be pointed out that the ICF and pre-equilibrium-emission (PEE) are not taken into consideration in this code. In this code, level density parameter (a=A/K) is an important input parameter which affects the CF cross-sections and where K may be varied to match the experimental cross-sections. As a representative case, in Fig.1 (a) the EFs for 3n-channel have been compared with corresponding PACE4-predictions

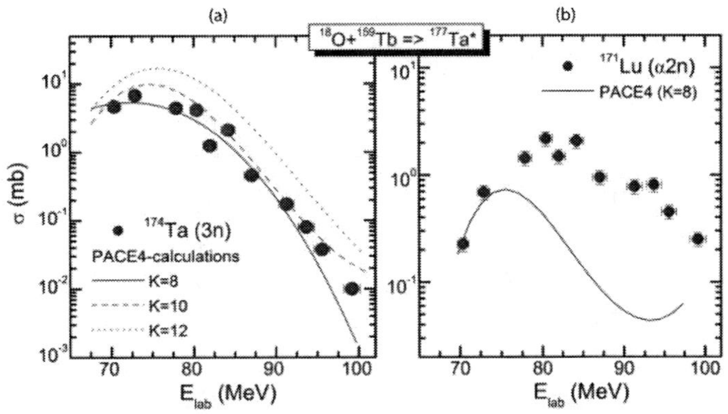

Figure 1: (Color online) Experimental EFs of ^{174}Ta($3n$) (a), ^{171}Lu($\alpha 2n$) (b) have been compared with that predicted by PACE4 for different values of level density parameter (a = A/K MeV^{-1}, where K = 812).

for three different values of level density parameter and "K=8" has been found to reproduce satisfactorily the experimental data, which shows the production of this residue via CF process. Similar observations have been observed for other xn & pxn-channels, showing their production via CF processes. The same set of parameters has been used to check the production mechanism of α-emitting channels, also. As shown in Fig.1 (b), the measured cross-sections for ^{171}Lu ($\alpha 2n$) residue are found to be significantly enhanced than theoretical predictions. It has already been mentioned that PACE4 do not take ICF, PEE into account and hence, this enhancement may be attributed as the contribution due to ICF-reaction mechanism.

It is evident from the analysis that ICF-reactions contribute significantly to the production cross-section of α-emitting channels at studied energy range. Further, the ICF-contribution for individual channels has been deduced by subtracting CF cross-sections (σ_{CF}) from the experimentally measured total fusion cross sections (σ_{TF}) at each studied energy and plotted in Fig.2(a). As can be seen from Fig.2(a) that the ICF contribution is increasing with beam energy. It is not out of place to mention that the σ_{TF} has been corrected for the missing channels (which could not be measured experimentally) by their PACE4 values. Hence, the σ_{ICF} may be taken at least as the lower limit of ICF-contribution.

3.1 α-Q-value systematics

In order to understand the effect of projectile on ICF-reactions, the F_{ICF}, the fraction of ICF to total fusion, have been deduced for ^{18,16}O+^{159}Tb systems and plotted in Fig.2(b). This comparison of F_{ICF} for different projectiles on same target reveals a strong projectile dependence of low-energy ICF

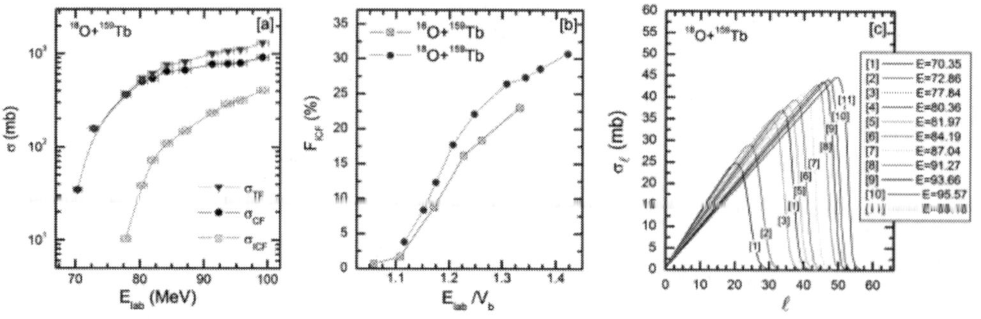

Figure 2: (color online) (*a*) The incomplete fusion cross-section along with total and complete fusion cross-section. (*b*) The comparison of F_{ICF} for $^{18,16}O$ projectiles on same target. (*c*) The fusion ℓ-distribution to understand the population of ℓ-bin at each studied energy.

reactions. It is clear from this figure, that the F_{ICF} is larger for ^{18}O than ^{16}O as projectile on the same target ^{159}Tb, which can be understood by recently proposed alpha-Q-value systematics [5]. The more-negative Q_α-value for ^{16}O translates into the smaller breakup probability into constituent α clusters, resulting in a smaller ICF-fraction than for ^{18}O induced reactions. The present work strengthens the recently observed alpha-Q-value systematics [5] for strongly bound projectiles.

3.2 Observation of ICF at $\ell < \ell_{crit}$

In order to study the diffuseness in ℓ-distribution, the critical angular momentum ℓ_{crit} for the present system at which the pocket in the entrance channel potential vanishes has been calculated [12]. The calculations give ℓ_{crit} as $63\hbar$. The fusion ℓ-distributions for $^{18}O+^{159}Tb$ interactions over a broad energy range $\approx$ 70-99 MeV have been calculated using the code CCFULL [13], and are plotted in Fig.2(c). The values of ℓ_{max} at studied energies, are less than the ℓ_{crit} ($63\hbar$) for fusion for this system, in general. It is evident from this figure that there is no significant cross-section above ℓ_{crit}, even at the highest beam energy, and hence, suggest that a significant number of ℓ-bins below ℓ_{crit} may contribute to the ICF. The present observations clearly indicate a diffused boundary for ℓ-values, contrary to the sharp cut-off model, that may penetrate close to the barrier.

3.3 Remark on general systematics for low energy ICF

The mass-asymmetry systematics proposed by Morgentern *et al* [9] states that ICF contribution increases with mass-asymmetry of the interacting partners (but at some what higher energies > 10MeV/nucleon). Later, Singh *et al.* [5]introduced the importance of projectile structure on ICF reactions at low energies in his ProMass systematics. Recently, the α-Q-value systematics [6] for strongly bound projectiles has been proposed, which gives α-Q-value as a parameter to understand the projectile strucute effect on low energy ICF reactions. Further, no systematics is available which can include both the effect of projectile and target into account. Hence, some general systematics for low energy ICF reactions, which can deal both projectile-target effect into account, is still lacking and some more conclusive measurements are required.

Acknowledgements

The authors thank to the Director, IUAC, New Delhi, India, for providing all the necessary facilities to carry out this work. One of authors A.Y. thanks the DST for providing support through Young Scientist Scheme under start-up research grant ref: SB/FTP/PS-194/2013.

References

[1] S. Hofmann et al., Eur. Phys. J. A **32**, 251 (2007); **14**, 147 (2002).

[2] M. Dasgupta *et al.*, Nucl. Phys. A **787**, 144-149 (2007).

[3] A. Diaz-Torres and I. J. Thompson, Phys. Rev. Lett. **98**, 152701 (2007).

[4] P. R. S. Gomes *et al*, Phys. Rev. C **73**, 064606 (2006); Phys. Lett. B **601**, 20 (2004).

[5] Pushpendra P. Singh *et al.*, Phys. Rev. C **77**, 014607 (2008); Euro. Phys. J. A **34**, 29-39 (2007).

[6] Abhishek Yadav *et al.*, Phys. Rev. C **85**, 034614 (2012); *ibid* **85**, 064617 (2012); *ibid* **86**, 014603 (2012) and references therein.

[7] D. P. Singh *et al.*, Phys. Rev. C **81**, 054607 (2010).

[8] V R Sharma *et al.*, Phys. Rev. C **89**, 024608 (2014).

[9] H. M. Morgenstern *et al.*, Phys. Rev. Lett. **52**, 1104 (1984).

[10] E. Browne and R. B. Firestone, Table of Radioactive Isotopes (Wiley, New York, 1986).

[11] A. Gavron, Phys. Rev. C **21**, 230 (1980).

[12] J. Wilczynski *et al.*, Phys. Rev. Lett. **45**, 606 (1980); Nucl. Phys. A **373**, 109 (1982).

[13] K. Hagino et al., Computer Physics Communications **123**, 143 (1999).

EPJ Web of Conferences **117**, 08023 (2016) DOI: 10.1051/epjconf/201611708023
NN2015

Study of fusion-fission dynamics in $^{19}F+^{238}U$ reaction

R. Dubey[1], P. Sugathan[1], A. Jhingan[1], Gurpreet Kaur[2], Ish Mukul[3], Davinder Siwal[4], N. Saneesh[1], Tathagata Banerjee[1], Abhishek Yadav[1], Meenu Thakur[2], Ruchi Mahajan[2] and M.B. Chaterjee[5]

[1]Inter University Accelerator Centre, Aruna Asaf Ali Marg, New Delhi - 110067, India,
[2]Dept of Physics, Panjab University, Chandigarh - 160014, India,
[3]Dept of Particle Physics and Astrophysics, Weizmann Institute of Science, Rehovot 76100, Israel,
[4]Dept of Chemistry, 2401 Milo B. Sampson Ln.,Bloomington, IN 47405, USA,
and[5]Saha Institute of Nuclear Physics, Kolkata, West Bengal 700064, India

Abstract

Mass angle distribution measurements for $^{19}F+^{238}U$ reaction were carried out around the sub barrier energies. Mass angle correlation has not been observed at above and below the fusion barrier in present reaction. This infer the minimal presence of non compound like events at these bombarding energies range.

1 Introduction

Over the years, a large number of experiments have been performed to study the dynamics of fusion-fission mechanism in heavy ion induced reactions. These studies are important for better understanding of the reaction dynamics involved in the formation of super heavy elements in laboratory [1]. Apart from pure fusion-fission, the existence of pre-equilibrium fission and

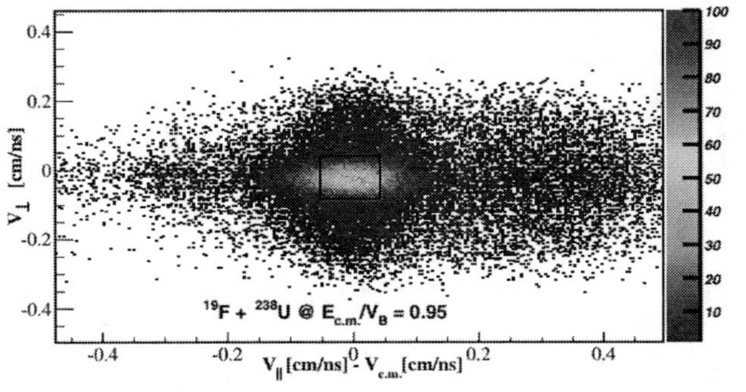

Figure 1: Measured distribution of FF velocity components at 5 % below the fusion barrier. Full momentum transfer fission events are shown inside black rectangular box.

quasi-fission add further complexity to the reaction dynamics [2–4] . They are found to be the competing reaction channels for light ion induced reactions with actinide targets. Though many studies have been performed in this mass region, there are still inconsistencies in the results of many of these reactions. For example, the angular anisotropies of fission fragments (FF) measured in reactions ^{19}F, ^{16}O, ^{12}C + ^{232}Th, ^{238}U have been found inconsistent with the statistical saddle point model (SSPM) predictions [5–7]. Hence for understanding the anomaly of this mass region, we performed the measurement of mass angle distribution (MAD) of FF in the reaction ^{19}F+^{238}U.

2 Experimental details

Measurements of fission fragments from reaction ^{19}F + ^{238}U were carried out at the 15UD Pelletron accelerator facility of Inter University Accelerator center (IUAC), New Delhi. ^{19}F pulsed beam with energy 96 MeV and 112 MeV was used to bombard the ^{238}U target. The thickness of enriched target ^{238}U was about $\sim$ 110 μg/cm^2. It was sandwiched between the carbon layers of thickness of thickness $\sim$ 20 μg/cm^2. The FF were detected in to large area multi-wire proportional counters (MWPCs) [8]. Front detector was placed at 35 cm from target centre covers the scattering angle from 25°-60°. The back detector was placed at a distance of 27 cm away from the target centre covers the scattering angle from 105°-135°. Two silicon detectors were placed at ±10° with respect to the beam axis to monitor the beam position on the target during the experiment.

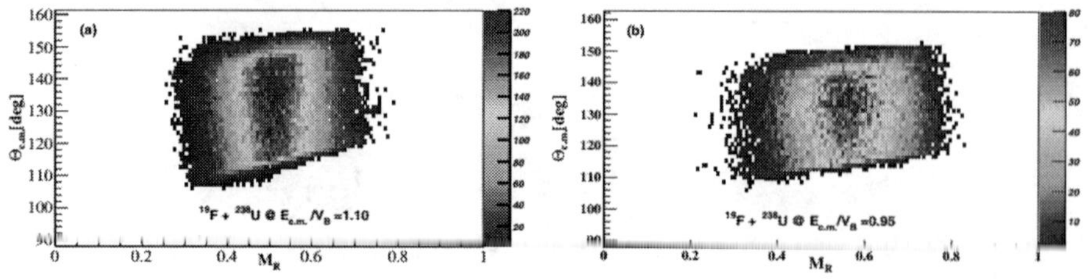

Figure 2: FFs mass angle correlations for reactions $^{19}F+^{238}U$ at a) above the fusion barrier ($E_{c.m.}/V_B$=1.1) and b) below the fusion barrier ($E_{c.m.}/V_B$=0.95) are shown. No mass angle correlation for $^{19}F+^{238}U$ are observed around this barrier energies.

The velocity component $V_\perp$, in the plane perpendicular to the beam axis was determined from the projection of the fragment velocities onto the azimuthal plane. FF velocities in laboratory frame were calculated by using polar angles and Viola systematics [9]. Parallel components of velocity ($V_{||}$) were adjusted such that it became equal to velocity of compound nucleus in the centre of mass frame ($V_{c.m.}$) which is pre-condition for full momentum transfer in fission process, i.e, (($V_{||}$ -$V_{c.m.}$, $V_\perp$)=(0, 0)) which has been as shown in inside the rectangular gate of Figure 1. Only the events within the black rectangle marked as FF were used to gate the mass spectrum (Fig 1).

3 Result

The dependence of mass ratio on emission angle in center of mass frame ($\theta_{c.m.}$), known as mass-angle-correlation, is a probe to disentangle the compound and non-compound nucleus fission processes. The mass ratio yield is independent of the emission angle of FFs coming from compound nucleus process, however, the yield for FF coming from non-compound processes exhibit the dependence on $\theta_{c.m.}$. In the present work, the mass-angle-correlation have been studied for $^{19}F+^{238}U$ at near and above barrier energies. As a representative case, the mass-angle-correlation for this reaction around barrier energies are shown in Fig. 2. Here, the correlation angles ($\theta_{c.m.}$) are lying $\approx$ 100° to 160°. It is evident from figures that mass ratio yield is independent of emission angle. Thus the absence of correlation between mass ratio yield and $\theta_{c.m.}$ in present nuclei concluded the minimal presence of non compound nucleus processes around the barrier energies. Similar observations for $^{12}C+^{238}U$, $^{16}O+^{238}U$ [10, 11] have been observed.

4 Conclusion and summary

MAD for ^{19}F+^{238}U reaction were measured at the 5% below and 10% above the fusion barrier energies. The reaction products, mainly fission fragments were detected by two large area MWPCs. The absence of correlation between mass ratio yield and $\theta_{c.m.}$ has been observed for present reaction. The characteristics of mass angle correlation for ^{19}F+^{238}U exhibit similar to the ^{12}C+^{238}U and ^{16}O+^{238}U reactions. This infer the fission mechanism of ^{12}C, ^{16}O, ^{19}F +^{238}U reactions indicate the minimal yield of non compound like events around sub barrier energies range.

Acknowledgments

One of the authors (RD) acknowledges the financial support from Council of Scientific Industrial Research, Govt. of India.

References

[1] Yu. Ts. Oganessian, *et al.*, Phys. Rev. C 74 (2006) 044602 .

[2] V. S. Ramamurthy,*et al.*, Phys. Rev. Lett. 54 (1985) 178.

[3] B. B. Back, *et al.*, Phys. Rev. Lett.50 (1983) 818.

[4] D. J. Hinde *et al.*, Phys. Rev. Lett. 74 (1995)1295 .

[5] J.C. Mein, *et al.*, Phys. Rev. C 55 (1997) R 995 .

[6] J. P. Lestone, *et al.*, Phys. Rev. C 56 (1997) R2907.

[7] V. S. Ramamurthy, *et al.*, Phys. Rev. Lett. 65 (1990) 25.

[8] A. Jhingan, *et al.*, Rev. Sci. Instrum. 80 (2009) 123502.

[9] D. J. Hinde, *et al.*, Phys. Rev. C 53 (1996)1290 .

[10] C.Yadav, *et al.*, Phys. Rev. C 86 (2012) 034606 .

[11] R. du Rietz, *et al.*, Phys. Rev. C 88 (2013) 054618

EPJ Web of Conferences **117**, 08024 (2016) DOI: 10.1051/epjconf/201611708024
NN2015

Semiclassical approach to sequential fission in peripheral heavy-ion collisions

Andrea Strazzeri[1] and Antonio Italiano[2]

[1]Dipartimento di Fisica e Astronomia dell'Università and INFN, Sezione di Catania, Catania, Italy
[2] INFN, Gruppo collegato di Messina, Messina, Italy

Abstract

A closed-form theoretical approach describing in a single picture both the evaporation component and the fast nonequilibrium component of the sequential fission of projectilelike fragments in a semiperipheral heavy-ion collision is derived and then applied to the dynamical fission observed in the $^{124}Sn+^{64}Ni$ semiperipheral collision at 35A MeV. Information on opposite polarization effects of the fissioning projectilelike fragments and on their "formation-to-fast fission lifetimes" are obtained.

1 Introduction.

In the past years several studies of sequential fission of the heavier product observed in peripheral heavy-ion collisions have revealed that the measured in-plane angular correlations between the fissioning projectilelike (PL) nucleus and the heavier fission fragment are sharply forward peaked and not symmetric with respect to the direction of the PL fragment with marked differences between distributions for positive and negative angles with a clear indication of a fast nonequilibrium component besides the equilibrated evaporative component ([1, 2] and references therein). We here show that many of the features observed can be explained by means of a closed-form which combines in a single picture both the evaporation component and the fast nonequilibrium component of the sequential fission of the PL fragments following a peripheral heavy-ion collision. This approach is applied to the dynamical fission of PL fragments in the ^{124}Sn+^{64}Ni collision at 35A MeV [2]. Only an outline of the analytic expressions is given here; detailed calculations have been presented in an earlier brief paper [3].

2 Theory.

The theoretical background of our semiclassical approach is described in Ref. [4]. Firstly, let us consider a three-body sequential process such as $A+a\rightarrow B+b\rightarrow C+c+b$, where C and c indicate respectively the heavier and the lighter fragments from the fission of the heavier product of the peripheral collision $a(A,B)b$, the PL fragment B. From now on, $\vec{k}_X, \varepsilon^*_X, J_X$ and m_X indicate the momentum, the excitation energy, the spin and its z-component, respectively, of the nucleus X. The spherical polar angles (ϑ_B, φ_B) of $\vec{k}_B$ are defined in the $(A+a)$ centre-of-mass (c.m.) system while $\vec{k}_C$ has polar angles (ϑ, φ) defined in the recoil-centre-of-mass (r.c.m.) system (rest frame of the nucleus B). We assume that the PL fragment B is highly excited to levels that lie in the continuum region, within an energy interval Δ centered at the energy ε^*_B, and the fission fragments C and c are left in any of a large number of states contained within intervals $\Delta' < \Delta$ about the energy $\varepsilon^*_C < \varepsilon^*_B$ and $\Delta'' < \Delta$ about the energy $\varepsilon^*_c < \varepsilon^*_B$.

To get the average value of the B–C angular correlation over the energy interval Δ centered at energy ε^*_B let us split the $\mathcal{S}$ matrix, which describes the process $B\rightarrow C$+c, as $\mathcal{S} = \mathcal{S}_E + \mathcal{S}_{NE}$ with $\mathcal{S}_{NE} \equiv \langle \mathcal{S} \rangle$ and $\mathcal{S}_E \equiv (\mathcal{S} - \langle \mathcal{S} \rangle)$ [3,4] Then we can express the energy-averaged $B-C$ angular correlation as [3]

$$\langle d^2\sigma/d\omega_B d\omega \rangle = (d^2\sigma/d\omega_B d\omega)_E + (d^2\sigma/d\omega_B d\omega)_{NE}, \qquad (1)$$

$$(d^2\sigma/d\omega_B d\omega)_E = \sum_{m_C}\sum_{\ell} w_\ell \, \langle | S_\ell - \langle S_\ell \rangle |^2 \rangle \left| \sum_{m_B} q_\ell(m_B, m_C; \omega_B) Y_\ell^m(\omega) \right|^2, \qquad (2)$$

where $m = m_B - m_C$, the amplitude q_ℓ determines the population of the substates $(J_B m_B)$ in the PL fragment B and the quantity w_ℓ describes the probability of orbital momentum ℓ transferred in the decay $(B; \varepsilon^*_B J_B) \rightarrow (Cc; \varepsilon^*_C \varepsilon^*_c J_C \ell)$.

$$(d^2\sigma/d\omega_B d\omega)_{NE} = \sum_{m_C} \left| \sum_{\ell} \langle S_\ell \rangle \sum_{m_B} q_\ell(m_B, m_C; \omega_B) Y_\ell^m(\omega) \right|^2, \qquad (3)$$

We choose coordinate axes having the z-axis along $\vec{k}_B \times \vec{k}_A$ (normal to the reaction plane) and the x-axis along $\vec{k}_B$, so that the direction φ=0 corresponds to the fragment C moving along the PL fragment flight direction, and we suppose that the emission of the fragment C does not change the sign of initial polarization of $\vec{J}_B$ at the moment of the fission of B.

Besides we consider J_B and J_C large as compared to ℓ so that the Clebsch-Gordan coefficient in Eq.(4) can be approximated by a reduced rotation matrix $d^{\ell}_{m\kappa}(\zeta)$ with $\kappa = J_B - J_C$ and $\cos\zeta = m_B/J_B$ [5].

We consider a coordinate rotation to a body-fixed system where the new quantization axis is oriented along $\vec{J}_B$ at a certain angle ζ with respect to the z-axis and lies in a plane perpendicular to the reaction plane and to the unit vector $\hat{k}_0$, oriented along the average momentum transferred in the collision $a(A,B)b$, corresponding to the angle φ_0 with respect to the x-axis [3]. Then, if we define as (Θ, Φ) the polar angles of $\vec{k}_C$ with respect to the body-fixed $(\hat{k}_0, \hat{J}_B \times \hat{k}_0, \hat{J}_B)$ system, we have, for $\vartheta = 90°$ [6]

$$\cos\Theta = -\sin\zeta \sin(\varphi - \varphi_0), \quad \cot\Phi = \cos\zeta \cot(\varphi - \varphi_0). \tag{4}$$

Moreover, in Eq.(2) we get the parametrization $\langle |\,(\mathcal{S}_E)_\ell\,|^2\rangle = (T_\ell/G)$, with T_ℓ the optical model transmission coefficient and G all decay modes energetically open for the decay $B \to C + c$, and in Eq.(3) we assume that the fast decay described by $\langle \mathcal{S}\rangle$ can be associated with a situation where the fragment C is promptly emitted from a peripheral region of the nucleus B.

For the differential multiplicity, in units (sr^{-1}), from Eqs.(1)-(4) we get [3]

$$M(\vartheta,\varphi,\zeta) = [M(\vartheta,\varphi,\zeta)]_E + [M(\vartheta,\varphi,\zeta)]_{NE}, \tag{5}$$

$$[M(\vartheta,\varphi,\zeta)]_E \sim \sum_{\ell,\kappa} w_\ell (T_\ell/G)(Y^{\kappa}_{\ell}(\Theta,0))^2, \tag{6}$$

$$[M(\vartheta,\varphi,\zeta)]_{NE} \sim \left|\sum_\ell \langle S_\ell\rangle_+ \exp(+i\ell\Phi)\right|^2 + H_0 \left|\sum_\ell \langle S_\ell\rangle_- \exp(-i\ell\Phi)\right|^2. \tag{7}$$

In Eq.(7) $\langle S_\ell\rangle_+$ and $\langle S_\ell\rangle_-$ describe the splitting of the PL fragment with $m_B \equiv +m_0$ (positive polarization of B) and $m_B \equiv -m_0$ (negative polarization of B), respectively, and H_0 depends on the ratio between negative and positive polarization of the nucleus B.

The final expression of the differential multiplicity is given by ($\cos\zeta_0 = m_0/J_B$)

$$M(\vartheta,\varphi) = \left(\int d\zeta \, \exp(-(\zeta-\zeta_0)^2/2\xi_0^2)\right)^{-1} \int d\zeta \, \exp(-(\zeta-\zeta_0)^2/2\xi_0^2)\, M(\vartheta,\varphi,\zeta), \tag{8}$$

Here, for M_E we get the semiclassical approach of Ref. [7]

$$[M(\vartheta,\varphi,\zeta)]_E = C_E \exp(\gamma \sin^2 \Theta), \tag{9}$$

where C_E is a constant independent of ϑ and φ, and γ is the so-called anisotropy coefficient.

If we admit than only a relatively narrow ℓ-window centered at a certain ℓ_0 contributes to the fast process described by $\langle S \rangle$, for M_{NE} we obtain [3]

$$[M(\vartheta,\varphi,\zeta)]_{NE} = C_{NE}[\exp(-(\Phi+\chi_0)^2/\Gamma_+^2) + H_0 \exp(-(\Phi-\chi_0)^2/\Gamma_-^2)], \tag{10}$$

where C_{NE} is a constant independent of ϑ and φ, χ_0 is the quantal deflection and Γ is the total angular dispersion. In the hypothesis that the PL fragment B during the "formation-to-fast fission lifetime" τ_0 rotates with rotational frequency $\omega_0 = \hbar\ell_0/\mathcal{I}$ ($\mathcal{I}$ is the moment of inertia) we assume [3]

$$\chi_0 = \omega_0\tau_0 = (\hbar\ell_0/\mathcal{I})\tau_0. \tag{11}$$

Table 1: List of the parameters obtained in the analysis of the experimental in-plane angular correlations shown in Fig.1 by the total multiplicity formula for three ranges of the total kinetic energy of the two heaviest fragments E_{2F}, in units MeV, and for three heavy-light mass ratios (A_H/A_L).

E_{2F}	(A_H/A_L)	C_E^a	C_{NE}	γ^a	ζ_0^a	ξ_0^a	φ_0^a	H_0	Γ_+	Γ_-	χ_0
	2.6-4.6	0.3	4.8	3.0	6°	36°	16°	0.20	0.62	0.11	15°
3200-3650	1.6-2.6	0.9	3.8	2.9	6°	45°	10°	0.52	0.30	0.14	18°
	1.0-1.6	1.2	3.0	2.8	6°	39°	8°	0.54	0.21	0.18	21°
	2.6-4.6	0.3	6.0	3.0	6°	36°	23°	0.28	0.56	0.14	13°
2850-3200	1.6-2.6	0.7	3.6	2.9	6°	45°	20°	0.48	0.56	0.20	17°
	1.0-1.6	1.2	2.4	2.8	6°	39°	17°	0.58	0.16	0.22	19°
	2.6-4.6	0.2	5.6	2.9	6°	36°	22°	0.30	0.58	0.16	10°
2500-2850	1.6-2.6	0.6	4.4	2.7	6°	45°	20°	0.48	0.56	0.16	12°
	1.0-1.6	1.6	1.6	2.6	6°	39°	18°	0.62	0.18	0.70	14°

[a]Quantities obtained by fitting the in-plane data in the region $|\varphi| \geq 100°$ by the evaporative component.

3 Application.

Now, we briefly quote the application of the semiclassical approach above derived to the in-plane angular distributions of the heavier fragment from

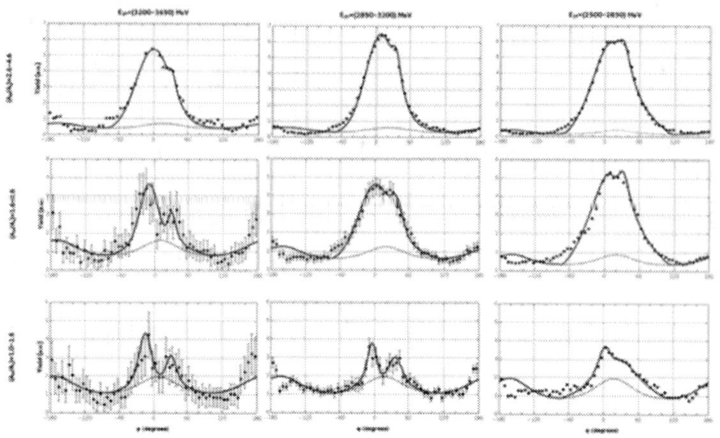

Figure 1: (Color online) In-plane angular distributions of the heavier fragment from the sequential fission of the projectilelike nucleus in the ^{124}Sn(35A MeV)+^{64}Ni semiperipheral collision for three ranges of the total kinetic energy of the two heaviest fragments E_{2F}, and for three heavy-light mass ratios (A_H/A_L) [2]. The dashed and solid lines correspond to the equilibrium M_E and the total M_E+M_{NE} multiplicities, respectively (see text).

the fission of the PL nucleus in the ^{124}Sn +^{64}Ni semiperipheral collision at 35A MeV reported in Fig. 5 of Ref. [2].

The parameters (C_E,γ,ζ_0,ξ_0,φ_0) are obtained by fitting the experimental data in the angular region $|\varphi|\geq 100°$ (where the NE emission is negligible) by the equilibrium multiplicity $[M(\varphi)]_E$, given by Eqs.(4),(8),(9). Then, the parameters ($C_{NE}, H_0, \Gamma_+, \Gamma_-, \chi_0$) are obtained by fitting the experimental data in the complete angular region $|\varphi|\leq$ 180° by the total multiplicity $[M(\varphi)]_E$+$[M(\varphi)]_{NE}$, given by Eqs.(4),(8)-(10), where the above determined values of the parameters (C_E,γ,ζ_0,ξ_0,φ_0) were inserted. Best χ^2 values, estimated within 5%, are listed in Table 1.

The in-plane differential multiplicities corresponding to the parameters listed in Table 1 are represented in Fig.1. The dashed and solid lines correspond to the equilibrium M_E and total M_E+M_{NE} multiplicities, respectively. The above semiclassical approach seems to be able to reproduce many of the observed features of the experimental in-plane angular correlations. The two *angular bumps* are associated to opposite polarizations of the fissioning PL fragment, the higher one with positive polarization and the lower one with negative polarization [3].

For each range of the total energy E_{2F}, for larger (A_H/A_L) the quantal deflection χ_0 is smaller and consequently the two *angular bumps* merge

Table 2: Estimates of the in-plane integrated M_E and M_{NE} multiplicities, the positive alignment p_+ and the "formation-to-fast fission lifetime" τ_0, in units (fm/c), for three ranges of the total kinetic energy of the two heaviest fragments E_{2F}, in units MeV, and for three heavy-light mass ratios (A_H/A_L).

E_{2F}	(A_H/A_L)	M_E	M_{NE}	p_+	τ_0
	2.6-4.6	0.2	5.3	0.83	352
3200-3650	1.6-2.6	0.6	2.5	0.66	369
	1.0-1.6	0.7	1.6	0.65	401
	2.6-4.6	0.2	6.4	0.78	305
2850-3200	1.6-2.6	0.4	4.2	0.67	349
	1.0-1.6	0.7	1.2	0.63	363
	2.6-4.6	0.1	6.2	0.77	235
2500-2850	1.6-2.6	0.4	5.0	0.66	246
	1.0-1.6	1.0	1.7	0.62	267

together (see Fig.1) whereas the parameter φ_0, which is the angular gap between $\vec{k}_0$ and $\vec{k}_B$ (corresponding to φ=0°), is larger.

From Eqs.(9) and (10) one can obtain a rough estimate of the in-plane E and NE multiplicities integrated in the angular region $|\varphi| \leq 180°$. The values of integrated multiplicities M_E and M_{NE} coming from the parameters (C_E,C_{NE},Γ_+,Γ_-,H_0) of Table 1, estimated within 30%, are listed in Table 2.

We can define a positive alignment parameter p_+ of the fissioning PL fragment, on a quantization axis perpendicular to the reaction plane [3], whose values, corresponding to the parameters H_0 given in Table 1, estimated within 10%, are listed in Table 2.

From Eq.(11) with the values of χ_0 given in Table 1 we can estimate the "formation-to-fast fission lifetime" τ_0. We used $\ell_0 = 30\hbar$ [2] and we calculated the moment of inertia of the fissioning nucleus $B(\equiv C + c)$ assuming a rigid rotation of two hard spheres, at contact [3]. The values of τ_0, in units (fm/c), estimated within 15%, for the three mass ratios (A_H/A_L), are listed in Table 2.

The decrease of the lifetimes τ_0 with increasing mass ratio (A_H/A_L), for each range of the total energy E_{2F}, is in clear agreement with results reported in Ref. [8].

References

[1] P. Russotto *et al.*, *Phys. Rev. C* **91** (2015) 014610.

[2] E. De Filippo *et al.*, *Phys. Rev. C* **71** (2005) 064604.

[3] A. Strazzeri and A. Italiano, *Int. J. Mod. Phys. E* **23**(2014)1450081.

[4] A. Strazzeri, *Nuovo Cimento A* **52** (1979) 323.

[5] D.M. Brink and G.R. Satchler, *Angular Momentum* (Oxford University Press, Oxford, 1971).

[6] R.M. Green, *Spherical Astronomy* (Cambridge University Press, Cambridge, 1985).

[7] I. Halpern, *Bull. Amer. Phys. Soc. II* **5** (1960) 510.

[8] G. Casini *et al.*, *Phys. Rev. Lett.* **71** (1993) 2567.

EPJ Web of Conferences **117**, 08025 (2016)
NN2015
DOI: 10.1051/epjconf/201611708025

Barrier distribution from $^{28}Si+^{154}Sm$ quasielastic scattering: Coupling effects in the fusion process

Gurpreet Kaur[1], B.R. Behera[1], A. Jhingan[2], B.K. Nayak[3], R. Dubey[2], Priya Sharma[1], Meenu Thakur[1], Ruchi Mahajan[1], N. Saneesh[2], Tathagata Banerjee[2], Khushboo[4], A. Kumar[1], S. Mandal[4], A. Saxena[3], P. Sugathan[2], and N. Rowley[5]

[1]Department of Physics, Panjab University, Chandigarh-160014, India
[2]Inter University Accelerator Centre, New Delhi-110067, India
[3]Nuclear Physics Division, Bhabha Atomic Research Centre, Mumbai-400085, India
[4]Department of Physics and Astrophysics, Delhi University, New Delhi-110067, India and
[5]Institut de Physique Nucléaire, F91406-Orsay Cedex, France

Abstract

Barrier distribution for the $^{28}Si+^{154}Sm$ system has been extracted from large angle quasielastic scattering measurement to investigate the role of various channel couplings on fusion dynamics. The coupled channel calculations, including the collective excitation of the target and projectile, are observed to reproduce the experimental BD rather well. It seems that the role of neutron transfer, relative to collective excitation, is in fact weak in the $^{28}Si+^{154}Sm$ system even though it has positive Q-value for neutron transfer channels.

1 Introduction

Interplay between nuclear structure and reaction dynamics is well established while explaining the enhancement of sub-barrier fusion cross sections, compared to one-dimensional barrier predictions. However, the role of neutron transfer in fusion dynamics is not fully understood. Recently, using diffusion model, Sargsyan *et al.* [1] reported for positive Q-value neutron transfer channels, fusion enhancement is significant only if the transfer leads to increase in the deformation of the colliding nuclei. In their study, for ^{28}Si + ^{154}Sm system, no fusion enhancement has been observed due to transfer coupling as the deformation of both projectile and target decreases after neutron transfer. On the contrary, reduced fusion cross section calculations by Shorto *et al.* [2] showed a significant effect of neutron transfer on fusion enhancement for the ^{28}Si + ^{154}Sm system. Such a contradictory result can limit the validity of the conclusion by Sargsyan *et al.* [1].

Moreover, both the above mentioned theoretical models have investigated the neutron transfer effect on fusion process through fusion excitation function. However, the barrier distribution (BD) provides deeper understanding about the intrinsic degrees of freedom involved in the fusion than the fusion excitation function alone. From the existing experimental fusion excitation function [3], we have extracted the BD and studied the coupling effect using `CCFULL` program [4], incorporating $\sim$ 10% error in the data. But the uncertainty in the data forbids us from resolving the influence of various coupling channels on fusion enhancement as reported earlier [5].

In the present article, we have reported a precise measurement of quasielastic (QE) excitation function for the ^{28}Si+^{154}Sm system at large backward angles. From the measured data, BD has been extracted and coupled channel calculations have been performed to probe the effects of coupling on fusion dynamics through BD.

2 Experimental details

The experiment has been performed in the General Purpose Scattering Chamber (GPSC) facility at IUAC, New Delhi using ^{28}Si beam, from 15UD Pelletron, on ^{154}Sm target (typical thickness $\approx$ 180 $\mu g/cm^2$). Beam energy has been varied in steps of 2 MeV ranging from 90.0 MeV (25% below barrier) to 135.0 MeV (11% above barrier). The particle identification has been obtained using hybrid telescope detectors with gas ionization chamber as dE detector and PIPS detector as E detector. Four telescope detectors, two of them in plane and other two out of plane, each at an angle of 170° have been arranged in a symmetrical cone geometry to minimize the uncertainty due to beam misalignment and also to attain good statistics in small time. To check the consistency of the BD obtained from measured QE scattering events, one more telescope has been placed at an angle of 140°. Two 300 μm thick silicon detectors have been placed at $\pm$10° with respect to the beam direction for beam monitoring and normalization purpose.

DOI: 10.1051/epjconf/201611708025

3 Data analysis and results

The counts of elastic, inelastic and transfer events, from particle identification spectra, have been considered for the QE events as per definition. After centrifugal correction, the obtained QE cross sections, σ_{qe}(E, θ), normalized to rutherford cross section, *i.e.*, $d\sigma_{qe}/d\sigma_R(E)$ for θ_{lab}=170° and 140°, have been shown in *fig. 1(left)*. The statistical error is found to be less than 1% at lower energies and around 2% at higher energies. The BD, $D_{qe}(E)$, as shown in *fig. 1(right)*, has been extracted from QE excitation function using the method given in *ref*. [6]. Similarity in BD structure from QE measurements at two different angles give a check on the consistency of the extracted experimental BD, as shown in *fig. 1(right)*.

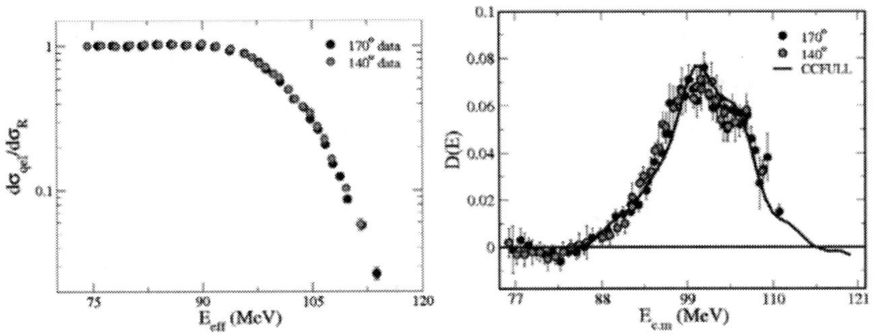

Figure 1: Left: Quasielastic excitation function obtained from two large backward scattering angles, *i.e.*, θ_{lab} = 170° and 140° for the system ^{28}Si+^{154}Sm. Right: Experimental barrier distribution (dots) for the ^{28}Si+^{154}Sm system with solid line representing the best fitted result of coupled channel calculation.

To quantitatively understand the experimental BD, quantum coupled channel (QCC) calculations have been performed using `CCFULL` code [4]. This program provides fusion excitation function from which BD has been extracted using the method mentioned in *ref*. [7]. The Woods-Saxon type interaction potential has been used with the parameters as given in *ref*. [5]. The various coupling parameters have also been taken from *ref*. [5]. At first, the rotational excitation of permanently deformed target, ^{154}Sm, has been considered in the calculation as reported for the ^{16}O+^{154}Sm system in the *ref*. [8]. It has been observed that only rotational coupling of ^{154}Sm is not able to reproduce the BD structure for the ^{28}Si+^{154}Sm system. Moreover, the BD for the ^{28}Si+^{154}Sm system seems to be wider as compared to the ^{16}O+^{154}Sm system. As coupling strength is proportional to the product of charges of the interacting nuclei. So, higher charge of projectile ^{28}Si, causing an increase in the coupling strength, may be attributed to the broader width for the ^{28}Si+^{154}Sm system. Furthermore, the vibrational excitation of projectile ^{28}Si has been considered in the coupled channel calculations in addition to rotational excitation of the target. The best fit to experimentally observed BD has been obtained with coupling involving three rotational states of target along with its octupole vibration and double phonons quadrupole vibrational excitation of projectile, as shown in *fig. 1* (*right*) with solid line. As the inelastic collective excitation alone could explain the experimentally observed BD, hence the CCFULL calculations do not reveal any significant influence of neutron transfer coupling on fusion enhancement.

4 Summary and conclusion

For ^{28}Si+^{154}Sm system, the projectile excitation, in addition to excitation of permanently deformed target, play a significant role in the fusion process. Furthermore, the coupled channel calculations predict that the inelastic excitation of target and projectile alone is sufficient to explain the experimental BD. In other words, it reveals weak influence of positive Q-value neutron transfer channel on fusion enhancement for the ^{28}Si+^{154}Sm system. Hence, our results show that increase in deformation parameter, as suggested by Sargsyan *et al.* [1], is a sufficient condition for fusion enhancement due to the presence of positive Q-value neutron transfer channels.

References

[1] V.V. Sargsyan *et al.*, Phys. Rev. C **91**, 014613 (2015).

[2] J.M.B. Shorto *et al.*, Phys. Rev. C **81**, 044601 (2010).

[3] S. Gil *et al.*, Phys. Rev. Lett. **65**, 3100 (1990).

[4] K. Hagino *et al.*, Comput. Phys. Commun. **123**, 143 (1999).

[5] Gurpreet Kaur *et al.*, EPJ Web of Conferences **86**, 00018 (2015).

[6] H. Timmers *et al.*, Nucl. Phys. A **584**, 190 (1995).

[7] M. Dasgupta *et al.*, Annu. Rev. Nucl. Part. Sci. **48**, 401 (1998).

[8] J.X. Wei *et al.*, Phys. Rev. Lett. **67**, 3368 (1991).

EPJ Web of Conferences **117**, 08026 (2016) DOI: 10.1051/epjconf/201611708026
NN2015

Probing dynamics of fusion reactions through cross-section and spin distribution measurement

Maninder Kaur[1,2] *, B.R. Behera[1], Gulzar Singh[1], Varinderjit Singh[1], N. Madhavan[3], S. Muralithar[3], S. Nath[3], J. Gehlot[3], G. Mohanto[3], Ish Mukul[3], D. Siwal[4], M. Thakur[1], K. Kapoor[1], P. Sharma[1], T. Banerjee[3], A. Jhingan[3], T. Varughese[3], Indu Bala[3], B.K. Nayak[5], A. Saxena[5], M.B. Chatterjee[5] and P.D. Stevenson[6]

[1] Department of Physics, Panjab University, Chandigarh - 160014, India
[2] Department of Physics, SGTB Khalsa College, Anandpur Sahib - 140118, Punjab, India
[3] Inter University Accelerator Centre, Aruna Asif Ali Marg, New Delhi - 110067, India
[4] Department of Physics and Astrophysics, University of Delhi - 110007, India
[5] Nuclear Physics Division, Bhabha Atomic Research Centre, Mumbai - 400085, India
[6] Department of Physics, University of Surrey, Guildford, Surrey, GU2 7XH, UK

Abstract

Present work aims to explicate the effect of entrance channel mass asymmetry on fusion dynamics for the Compound Nucleus ^{80}Sr populated through two different channels, ^{16}O+^{64}Zn and ^{32}S+^{48}Ti, using cross-section and spin distribution measurements as probes. The evaporation spectra studies for these systems, reported earlier indicate the presence of dynamical effects for mass symmetric ^{32}S+^{48}Ti system.

*E-mail: manisaini153@gmail.com

The CCDEF and TDHF calculations have been performed for both the systems and an attempt has been made to explain the reported deviations in the α-particle spectrum for the mass symmetric system.

1 Introduction

The Compound nucleus (CN) populated at high excitation energy and angular momentum can undergo decay through fission or emission of different light particles. The study of these light particles and their analysis in terms of statistical model gives important information about the properties of the CN. Though a large amount of such data exists in the literature, but many of these studies show anomalous results for the mass symmetric systems [1–3] that could not be explained in terms of the statistical model predictions. These deviations from the statistical model predictions were explained through the modification of important ingredients of the statistical model codes such as spin dependent level density, lowering the level density parameter or using lower values of l contributing to fusion. However the simultaneous analysis of other variables such as spin distribution, cross-section measurement, angular distribution measurements etc. should also be taken into account to have a better understanding of fusion dynamics. With this motivation we have performed the spin distribution and the cross-section measurements for two systems ^{16}O + ^{64}Zn (mass asymmetric) and ^{32}S + ^{48}Ti (mass symmetric) populating the same CN (^{80}Sr) for which the evaporation spectra studies have already been performed [1, 2]. These studies also report the existence of deviations for the symmetric system. Present measurements have been performed using 15 UD Pelletron accelerator and Heavy Ion Reaction Analyzer (HIRA) facility at Inter University Accelerator Centre (IUAC), New Delhi. The details of these measurements and analysis procedure have been described elsewhere [4].

2 Analysis and results

The evaporation residues (ERs) populated during the reaction were separated from beam-like background using HIRA. For this, a two-dimensional spectrum was generated using energy loss and Time of Flight (TOF) of ERs. The value of transmission efficiency of different ERs through HIRA was estimated by γ-ray co-incidence method and using the TERS code. The experimental cross-sections thus obtained were compared with CCDEF and TDHF [5] calculations. For both the systems the experimental cross-sections

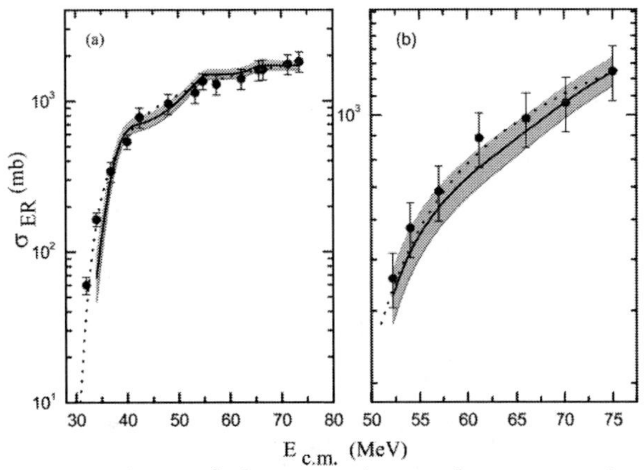

Figure 1: The comparison of the experimental cross-sections (solid circles) with the theoretical calculations for (a) $^{16}O + ^{64}Zn$, (b) for $^{32}S + ^{48}Ti$. The solid line corresponds to the TDHF calculations and shaded area represents the uncertainty in cross-sections obtained from the TDHF calculations. The dotted lines corresponds to the CCDEF calculations

were in agreement with these calculations as shown in fig. 1. However, the fusion time scales obtained using the TDHF calculations, at highest energies, were found to approach the decay times showing that the fusion time scales are elongated that may lead to pre-equilibrium emission resulting in the modification of the CN spin distribution.

So to have a better picture of the contribution of various partial waves to the fusion process, we have also performed the ER-gated spin distribution measurements. Van der Werf's prescription was used for unfolding the γ-multiplicity distributions from γ-fold distributions. γ-multiplicity distributions were converted to spin distribution by assuming that average spin carried away by each non-statistical γ-ray is 1.6 and incorporating the corrections from the spin carried away by the light particles. Using the experimental cross-sections, the CN spin distributions were obtained. These were compared with CCDEF calculations. It was observed that for the symmetric system, the contribution of the higher partial waves towards fusion goes on decreasing with the increase in energy (fig. 2(a)). These observations are in agreement with the calculations of fusion time scales extracted using TDHF calculations. These observations were not reflected in the cross-section measurements as the cross-sections correspond to the zeroth moment of the spin distribution, so there can be many shapes of the spin distribution, reflecting very different physical processes involved, but may yield the same fusion cross-section. The experimental spin distributions, for $^{32}S + ^{48}Ti$ at E_{lab} = 125 MeV, were give as an input to the PACE code and the generated α-particle spectrum was found to be in excellent agreement with experimental spectrum (fig. 2(b)), thus explaining the deviations reported earlier.

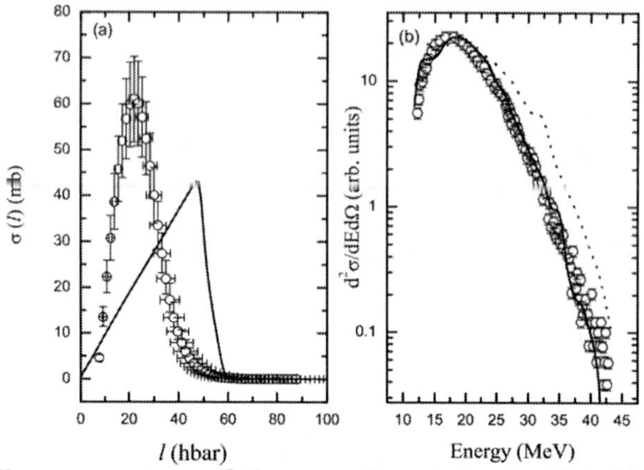

Figure 2: (a) The comparison of the experimental spin distributions (circles) with CCDEF calculations (solid line) for $^{32}S + ^{48}Ti$ at $E_{lab} = 125$ MeV. (b) The explanation of experimental α-particle spectrum (circles) by the spectrum generated (solid line) using experimental spin distribution in PACE. The dotted line corresponds to default calculations using PACE.

3 Conclusions

The measured experimental cross-sections were well explained by the CCDEF and TDHF calculations. However, the comparison of experimental spin distributions with the the CCDEF calculations indicate the effect of entrance channel mass asymmetry on fusion dynamics and the deviations in the experimental α-particle spectrum for the system $^{32}S+^{48}Ti$ were reasonably explained. Therefore, we conclude that the shape of the spin distribution can act as a sensitive tool for the investigation of fusion dynamics.

Acknowledgements

We thank the accelerator crew of IUAC for providing beams of excellent quality throughout the experiments. One of the authors (M.K.) would like to thank University Grants Commission (UGC) for the financial assistance.

References

[1] J. Kaur et al., Phys. Rev. C **70**, (2004) 017601.

[2] Ajay Kumar et al., Phys. Rev. C **68**, (2003) 034603.

[3] M. Kaur et al., Phys.Rev. C **89**, (2014) 034621.

[4] M. Kaur et al., DAE Symp. Nucl. Phys. **58**, (2013) 434; **59**, (2014) 350.

[5] J. A. Maruhn et al., Comput. Phys. Commun. **185**,(2014) 2195.

EPJ Web of Conferences **117**, 08027 (2016) DOI: 10.1051/epjconf/201611708027
NN2015

Elastic scattering of $^{17}O+^{208}Pb$ at energies near the Coulomb barrier.

D. Torresi[1,2], E. Strano[3,4], M. Mazzocco[3,4], A. Boiano[5], C. Boiano[6], P. Di Meo[5], M. La Commara[5,7], C. Manea[4], M. Nicoletto[4], J. Grebosz[8], A. Guglielmetti[6,9], P. Molini[3,4], C. Parascandolo[5], D. Pierroutsakou[5], C. Signorini[10], F. Soramel[3,4], N. Toniolo[10], D. Filipescu[11], A. Gheorghe[11], T. Glodariu[11], S. Jeong[12], Y.H. Kim[12], J. A. Lay[3,4], H. Miyatake[12], A. Pakou[13], O. Sgouros[13], V. Soukeras[13], L. Stroe[11], A. Vitturi[3,4], Y. Watanabe[12] and K. Zerva[13]

[1]Dipartimento di fisica e astronomia, Università di Catania, Italy
[2]INFN - Laboratori Nazionali del Sud, Catania, Italy
[3]Dipartimento di fisica e astronomia, Università di Padova, Italy
[4]INFN, Sezione di Padova, Padova, Italy
[5]INFN, Sezione di Napoli, Napoli, Italy
[6]INFN, Sezione di Milano, Milano, Italy
[7]Dipartimento di Fisica, Università di Napoli, Italy
[8]IFJ PAN, Krakow, Poland
[9]Dipartimento di Fisica, Università di Milano, Italy
[10]INFN - Laboratori Nazionali di Legnaro, Legnaro (PD), Italy
[11]NIPNE, Bucharest, Romania
[12]KEK, Tsukuba, Japan
[13]Physics Department, University of Ioannina, Greece

Abstract

Within the frame of the commissioning of a new experimental apparatus EXPADES we undertook the measurement of the elastic scattering angular distribution for the system $^{17}O+^{208}Pb$ at energy around the Coulomb barrier. The reaction dynamics induced by loosely bound

Radioactive Ion Beams is currently being extensively studied [4]. In particular the study of the elastic scattering process allows to obtain direct information on the total reaction cross section of the exotic nuclei. In order to understand the effect of the low binding energy on the reaction mechanism it is important to compare radioactive weakly bound nuclei with stable strongly-bound nuclei. In this framework the study of the ^{17}O+^{208}Pb elastic scattering can be considered to be complementary to a previous measurement of the total reaction cross section for the system ^{17}F+^{208}Pb at energies of 86, 90.4 MeV [5, 6]. The data will be compared with those obtained for the neighboring systems ^{16,18}O+^{208}Pb and others available in literature.

1 Introduction

In the last decades the study of the reaction mechanisms at energies around the Coulomb barrier for halo and weakly bound nuclei has attracted a large interest. The peculiar characteristics of these nuclei can affect deeply their reaction dynamics increasing or enhancing the total reaction cross section and/or specific reaction channels (for instance, breakup, transfer, fusion). Thus, it is quite important to have a large and reliable systematics of reaction cross section measurements for stable nuclei interacting with light, medium and heavy targets in order to compare the behavior of exotic nuclei with that of stable and strongly bound nuclei. We measured at the Laboratori Nazionali di Legnaro the elastic scattering process for the system ^{17}O+^{208}Pb in the energy range 80 - 87 MeV. This measurement, performed within the commissioning of the detection system EXPADES, is particularly interesting since, being ^{17}O the mirror nucleus of ^{17}F, the two projectiles have very similar nuclear structure and essentially differ by the binding energy of the valence nucleon (S_p =0.6 MeV for ^{17}F and S_n=4.3 MeV for ^{17}O).

2 The experimental setup

The elastic scattering angular distribution was measured by taking advantage of the detector array EXPADES (EXotic PArticle DEtection System) [1]. It is an apparatus aimed at studying the properties of exotic nuclei far from stability, since it has to deal with very small beam intensities requires high-efficiency detection set-ups and high granularity in order to obtain high angular resolution and to measure coincidences to characterize the final state of reactions. The array consist of 16 squared DSSSD arranged in 8 ΔE-E telescopes. Each detectors is 64$\times$64 mm^2 wide and each side is

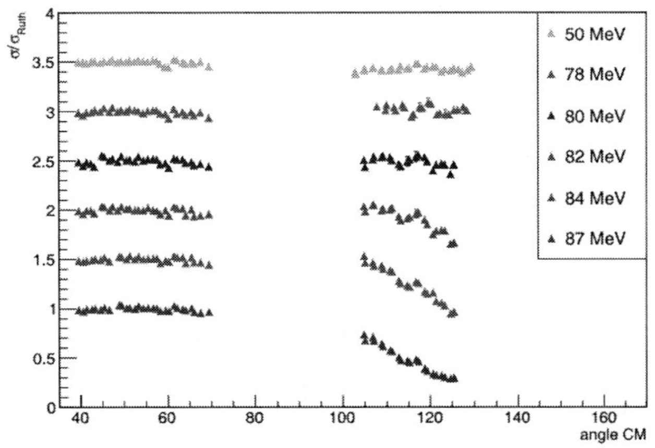

Figure 1: Elastic scattering angular distribution for system ^{17}O+^{208}Pb at different energies.

segmented in 32 strips per each side, being the strip of one side orthogonal to the ones of the opposite side. The ΔE thickness is 40 μm while the E thickness is 300 μm. Ionization chambers can be used as possible alternative ΔE stages. The detector signals are read by using standard electronic chains (developed in Milano and Napoli) both for the silicon ΔE layers and the ionization chambers, while 32-channel ASIC chips manufactured by IDEAS-GM are employed for the E stages. For the present experiment two single stage 300 μm thick DSSSD were placed at 50° and a DSSSD telescope was placed at 110°.

3 Results and conclusions

A preliminary optical model analysis of the experimental data has been performed. The fits were performed with the SFRESCO subroutine of the main coupled-channel code FRESCO [2]. We assumed all events as originating from a pure elastic scattering process, even if the experimental energy resolution did not allow to separate the events leading to the excitation of the ^{17}O first excited state at E_x=0.871 MeV.

The scattering process of the ^{17}O on a ^{208}Pb target has been measured in the energy range around the Coulomb barrier. The data have been analysed within the framework of the optical model to extract the reaction cross section and to investigate the relevance of direct reaction channels at near-barrier energies. Our measurements confirm that the trend of the two sys-

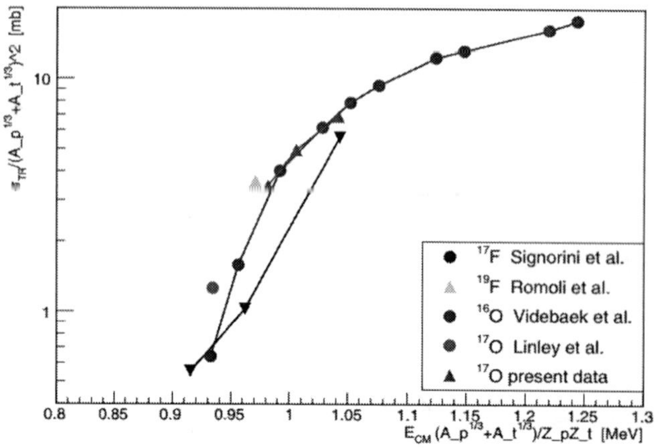

Figure 2: Total reaction cross section for different systems.

tems $^{17}O+^{208}Pb$ and $^{16}O+^{208}Pb$, after the reduction that eliminate differences arising from system size and barrier height [3], do not show substantial differences. It also preliminary confirms that ^{17}F should have a reactivity very similar to those of the stable isotopes ^{16}O and ^{17}O, as stated in [3, 4] This outcome suggests that the low binding energy of the additional proton in ^{17}F has a rather small influence on the reaction dynamics at near-barrier energies. Coupled channel calculations are presently being performed to investigate the relevance of inelastic excitations and 1n-stripping for the reaction $^{17}O+^{208}Pb$.

References

[1] E. Strano et al., Nucl. Instr. and Meth. B 317, 657 (2013).

[2] I.J. Thompson, Comput. Phys. Rep. 2, 167 (1998).

[3] N. Keeley, K.W. Kemper and K. Rusek, Eur. Phys. J. A 50, 145 (2014) and references therein.

[4] P.R.S. Gomes, et al., Phys. Lett. B 601 (2004) 20.

[5] M. Romoli et al., Phys. Rev. C 69, 064614 (2004).

[6] C. Signorini et al. Eur. Phys. J. A 44, 63 (2010).

EPJ Web of Conferences **117**, 09001 (2016) DOI: 10.1051/epjconf/201611709001
NN2015

Reaction models in nuclear astrophysics

PIERRE DESCOUVEMONT

Physique Nucléaire Théorique et Physique Mathématique, C.P. 229,
Université Libre de Bruxelles (ULB), B 1050 Brussels, Belgium

Abstract

We present different reaction models commonly used in nuclear astrophysics, in particular for the nucleosynthesis of light elements. Pioneering works were performed within the potential model, where the internal structure of the colliding nuclei is completely ignored. Significant advances in microscopic cluster models provided the first microscopic description of the $^3\mathrm{He}(\alpha,\gamma)^7\mathrm{Be}$ reaction more than thirty years ago. In this approach, the calculations are based on an effective nucleon-nucleon interaction, but the cluster approximation should be made to simplify the calculations. Nowadays, modern microscopic calculations are able to go beyond the cluster approximation, and aim at finding exact solutions of the Schrödinger equation with realistic nucleon-nucleon interactions. We discuss recent examples on the $d+d$ reactions at low energies.

1 Introduction

Nuclear reactions determine the nucleosynthesis in stars, and produce the energy released to compensate their gravitational contraction [1]. Stellar models are in general based on large reaction networks, involving many reaction rates. In the Big Bang nucleosynthesis, only a few reactions are important, producing elements up to $A = 8$. In stellar nucleosynthesis, the reaction networks depend on the physical conditions of the star (mass, temperature, density, etc.). At low temperatures, the stellar evolution is mainly determined by the pp chain and by the CNO cycle. Both processes convert hydrogen in Helium. Advanced stages of stellar evolution involve He burning, followed by reactions involving heavier elements. At high temperature, neutron capture (s and p processes), as well as explosive burning determine the star evolution.

The calculation of the reaction rates relies on the cross sections [2]. There are in general two main problems in nuclear astrophysics: (i) the stellar energies being much smaller than the Coulomb barrier, the relevant cross sections between charged particles are too small to be measured in the laboratory; (ii) explosive burning involves short-live nuclei which, even if they can be produced with modern technologies, are available with weak intensities only. Consequently a theoretical support is necessary, either to extrapolate the cross sections down to astrophysical energies, or to predict unknown cross sections.

Theoretical models used in nuclear astrophysics can be roughly classified in three categories [3,4]:
(*i*) Models involving adjustable parameters, such as the R-matrix method [5,6]; parameters are fitted to the available experimental data and the cross sections are extrapolated down to astrophysical energies. These fitting procedures of course require the knowledge of data, which are sometimes too scarce for a reliable extrapolation.
(*ii*) "Non fitting" models, where the cross sections are determined from the wave functions of the system. The potential model [7], the Distorted Wave Born Approximation (DWBA) [8], and microscopic models [9] are, in principle, independent of experimental data. More realistically, these models depend on some physical parameters, such as a nucleus-nucleus or a nucleon-nucleon interaction which can be reasonably determined from experiment only.
(*iii*) Models (*i*) and (*ii*) can be used for low level-density nuclei only. This condition is fulfilled in most of the reactions involving light nuclei ($A \leq 20$). However when the level density near threshold is large (i.e. more than a few levels per MeV), statistical models, using averaged optical transmission coefficients, are in general more suitable [10]. In some specific applications, shell-model theories can provide the gamma widths of relevant states.

2 Theoretical models

As mentioned before, reaction models are essential in nuclear astrophysics. Many models have been used to describe low-energy reactions. Here we give a brief overview of various approaches, commonly used in the field.

2.1 The potential model

Solving the Schrödinger equation associated with a many-body system is a difficult problem, which does not have an exact solution when the nucleon number is larger than three. The potential model is fairly simple to use, and has been applied to many reactions in low-energy nuclear physics [7,11]. The basic assumptions of the potential model are: (*i*) the nucleon-nucleon interaction is replaced by a nucleus-nucleus force $V(\rho)$, which depends on the relative coordinate ρ only; (*ii*) the wave functions of the unified nucleus can be described by a cluster structure with $A_1 + A_2$ nucleons; (*iii*) the internal structure of the nuclei does not play any role. Since we are dealing with low energies, the potential is in general real. The extension to higher energies, which requires complex potentials to simulate absorption channels, is known as the optical model. A generalization to coupled-channel problems is also possible, but seldom used in nuclear astrophysics.

The radial function $g_{\ell J}^{J\pi}(\rho)$ for bound and scattering states is deduced from the equation

$$\left(-\frac{\hbar^2}{2\mu m_N}\left(\frac{d^2}{dr^2} - \frac{\ell(\ell+1)}{r^2}\right) + V(\rho)\right) g_{\ell J}^{J\pi}(\rho) = E\, g_{\ell J}^{J\pi}(\rho), \tag{1}$$

where E is the relative energy ($E > 0$ for scattering states and $E < 0$ for bound states). Let us notice that the potential may depend on ℓ and J. In nuclear physics, the nucleus-nucleus potential involves a Coulomb term $V_C(\rho)$ and a nuclear term $V_N(\rho)$. According to the application, the choice of the nuclear contribution is guided by experimental constraints. In radiative-capture calculations it is crucial to reproduce the final-state energy. If phase shifts are available, they can be used to determine the initial potential.

In this simple model, the capture cross sections are deduced from integrals involving scattering functions $g_{\ell_i I}^{J_i\pi_i}(\rho)$ at energy E, and bound-state wave functions $g_{\ell_f I}^{J_f\pi_f}(\rho)$ (see Ref. [11] for detail).

2.2 The phenomenological R-matrix method

The R-matrix method is well known in atomic and nuclear physics [6]. The basis idea is to divide the space in two regions: the internal region (with radius a), where the nuclear force is important, and the external region, where the interaction between the nuclei is governed by the Coulomb force only. Although the R-matrix parameters do depend on the channel radius a, the sensitivity of the cross section with respect to its choice is quite weak. In the

R-matrix method, the energy dependence of the cross sections is obtained from Coulomb functions, as expected from the Schrödinger equation.

The physics of the internal region is determined by a number N of poles, which are characterized by energy E_λ and reduced widths $\gamma_{\lambda i}$. In a multi-channel problem, the R-matrix at energy E is defined as

$$R_{if}(E) = \sum_{\lambda=1}^{N} \frac{\gamma_{\lambda i}\gamma_{\lambda f}}{E_\lambda - E}, \tag{2}$$

which must be given for each partial wave J (not written for the sake of clarity). Indices i and f refer to the initial and final channels. The pole properties are associated with the physical energy and width of resonances, but not strictly equal. This is known as the difference between "formal" and "observed" parameters, deduced from experiment. In a general case, involving more than one pole, the link between those two sets is not straight-forward (see Refs. [12, 13] for a general formulation of this problem). The collision matrix, and therefore the cross sections, are directly determined from the R matrices in the different partial waves (see Refs. [5, 6]).

The R-matrix method can be applied to transfer as well as to capture reactions. It is usually used to investigate resonant reactions but is also suited to describe non-resonant processes. In the latter case, the non-resonant behavior is simulated by a high-energy pole, referred to as the background contribution, which makes the R-matrix nearly energy independent.

2.3 Microscopic cluster theories

Microscopic models are based on fundamental principles of quantum mechanics, such as the treatment of all nucleons, with exact antisymmetrization of the wave functions. Neglecting three-body forces, the Hamiltonian of a A-nucleon system is written as

$$H = \sum_{i=1}^{A} T_i + \sum_{i<j=1}^{A} V_{ij}, \tag{3}$$

where T_i is the kinetic energy and V_{ij} a nucleon-nucleon interaction [9].

The Schrödinger equation associated with this Hamiltonian can not be solved exactly when $A > 3$. For very light systems ($A \sim 4 - 5$) efficient methods [14] exist, even for continuum states [15]. However most reactions relevant in nuclear astrophysics involve heavier nuclei essentially with nucleon or α projectiles. Recent developments of *ab initio* models (see for example Refs. [16–18]) are quite successful for spectroscopic properties of low-lying states. These models make use of realistic interactions, including three-body forces, and fitted on many properties of the nucleon-nucleon system. Recent works succeeded in computing the ^{3}He$(\alpha,\gamma)^7$Be [17] and ^{2}H$(d,\gamma)^4$He [19] cross sections from realistic interactions. However, a consistent description of bound and scattering states of an A-body problem remains a very difficult task [15], in particular for transfer reactions.

In the cluster approximation, it is assumed that the nucleons are grouped in clusters [9, 20]. The internal wave functions of the clusters are denoted as $\phi_i^{I_i\pi_i\nu_i}(\boldsymbol{\xi_i})$, where I_i and π_i are the spin and parity of cluster i, and $\boldsymbol{\xi_i}$ represents a set of their internal coordinates. In a two-cluster system, a channel function is defined as

$$\varphi_{\ell I}^{JM\pi}(\Omega_\rho,\boldsymbol{\xi_1},\boldsymbol{\xi_2}) = \Big[Y_\ell(\Omega_\rho)\otimes[\phi_1^{I_1\pi_1}(\boldsymbol{\xi_1})\otimes\phi_2^{I_2\pi_2}(\boldsymbol{\xi_2})]^I\Big]^{JM}, \qquad (4)$$

where different quantum numbers show up: the channel spin I, the relative angular momentum ℓ, the total spin J and the total parity $\pi=\pi_1\pi_2(-)^\ell$.

The total wave function of the A-nucleon system is written as

$$\Psi^{JM\pi} = \sum_{\alpha\ell I}\Psi_{\alpha\ell I}^{JM\pi} = \sum_{\alpha\ell I}\mathcal{A}\, g_{\alpha\ell I}^{J\pi}(\rho)\,\varphi_{\alpha\ell I}^{JM\pi}(\Omega_\rho,\boldsymbol{\xi_1},\boldsymbol{\xi_2}), \qquad (5)$$

which corresponds to the Resonating Group (RGM) definition [20,21]. Index α refers to different two-cluster arrangements, and $\mathcal{A}$ is the antisymetrization operator. In most applications, the internal cluster wave functions $\phi_i^{I_i\pi_i\nu_i}$ are defined in the shell model. Accordingly, the nucleon-nucleon interaction must be adapted to this choice, which leads to effective forces, such as the Volkov [22] or the Minnesota [23] interactions. The relative wave functions $g_{\alpha\ell I}^{J\pi}(\rho)$ are to be determined from the Schrödinger equation, which is transformed into a integro-differential equation involving a non-local potential [21]. In most applications, this relative function is expanded over Gaussian functions [9, 24], which corresponds to the Generator Coordinate Method (GCM).

The main advantage of cluster models with respect to other microscopic theories is their ability to deal with reactions, as well as with nuclear spectroscopy. The first applications were done for reactions involving light nuclei, such as d, ^{3}He or α particles [25, 26]. More recently, much work has been devoted to the improvement of the internal wave functions: multicluster descriptions [27], large-basis shell model extensions [28], or monopolar distortion [29].

2.4 Ab-initio models: the ^{2}H$(d,\gamma)^4$He, ^{2}H$(d,p)^3$H and ^{2}H$(d,n)^3$He cross sections

The knowledge of the reaction cross sections at astrophysical energies is of great interest not only for establishing imprints of the properties of nuclei in the universe but also for understanding an interplay between the structure and reactions of these nuclei based on a nucleon-nucleon (NN) interaction. Aside from the astrophysical interest, the ^{2}H$(d,\gamma)^4$He capture reaction is extremely important from the nuclear physics viewpoint because its cross section at low energies (below 0.3 MeV) is expected to be dominated by D-wave components in the α particle. Hence it should be very sensitive to the tensor force in the NN interaction [30].

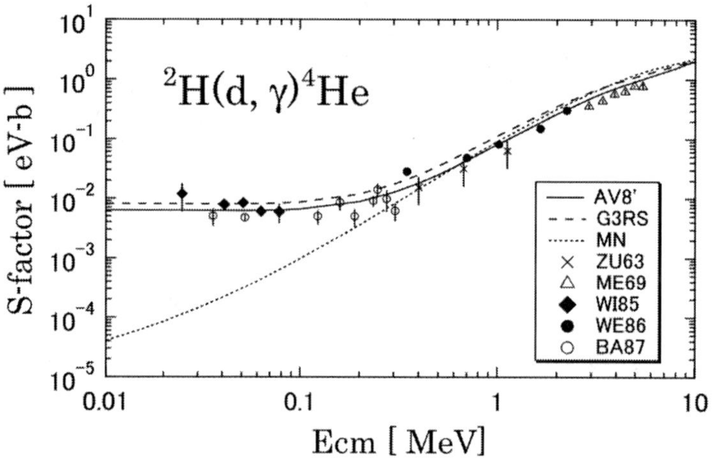

Figure 1: The astrophysical S-factor of the ^{2}H(d, γ)^{4}He reaction (taken from Ref. [19]). Results calculated with the realistic (AV8′, G3RS) and the effective (MN) potentials are compared to experiment (see Ref. [34]).

We have applied an ab initio model to study the phase shifts of the p+^{3}He [31] and $d+d$, p+^{3}H, n+^{3}He [19] systems. For the two-body NN interaction V_{ij}, we use two different realistic potentials, AV8′ [32] and G3RS [33], that consist of central, tensor, and spin-orbit components. Because our main aim is to clarify the role of the tensor force, it is useful to compare results obtained with the realistic interactions with that of an effective NN interaction that contains no tensor force. Using such effective interaction is reasonable because only s-shell nuclei participate to the reactions. We adopt the MN central potential [23] with a standard value for the admixture parameter $u = 1$.

Figure 1 displays the calculated astrophysical S-factor for the ^{2}H$(d,\gamma)^4$He reaction. Results with both AV8′ (solid line) and G3RS (dashed line) potentials reproduce very well the experimental data, especially its flat behavior at low energies (typical of an initial s wave), whereas the MN potential (dotted curve) shows a rapidly decreasing pattern as $E_{\rm cm}$ decreases. Without the tensor force, we cannot reproduce the ^{2}H$(d,\gamma)^4$He astrophysical S-factor below 0.3 MeV.

Similar conclusions can be drawn for the ^{2}H$(d,p)^3$H and ^{2}H$(d,n)^3$He transfer reactions at low energies. The S factors are presented in Fig. 2. They mainly occur from the transitions of the $d+d\ ^5S_2$ channel to the D-wave continuum of p+^{3}H or n+^{3}He, which is also due to the tensor force. Without the tensor force, these cross sections cannot be reproduced.

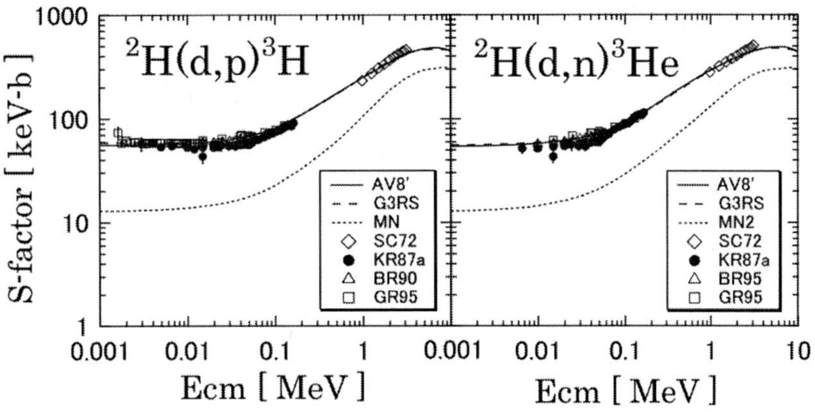

Figure 2: ^{2}H$(d,p)^3$H and ^{2}H$(d,n)^3$He astrophysical S-factors calculated with the realistic (AV8′, G3RS) and the effective (MN) potentials (taken from Ref. [19]). See Ref. [34] for the experimental data.

The ^{2}H$(d,p)^3$H and ^{2}H$(d,n)^3$He reactions play an important role in Bib-Bang nucleosynthesis. As the observed D/H ratio is currently known with 1.5% accuracy, a high precision is required for the reaction rates. The cross sections have been measured by several groups [35, 36], but the extrapolations down to low energies are still uncertain. A compilation of the latest data has been undertaken to reduce the extrapolation uncertainties [37]. In order to disentangle among the different data sets, the ratio $(d,n)/(d,p)$ provides a stronger constrain than the individual cross sections. At low energies, this ratio essentially depends on the Coulomb interaction, and is therefore expected to be very reliable in an ab initio calculation. This ratio is shown in Fig. 3, with values deduced from recent experiments.

As expected this ratio is close to unity since the entrance channel is

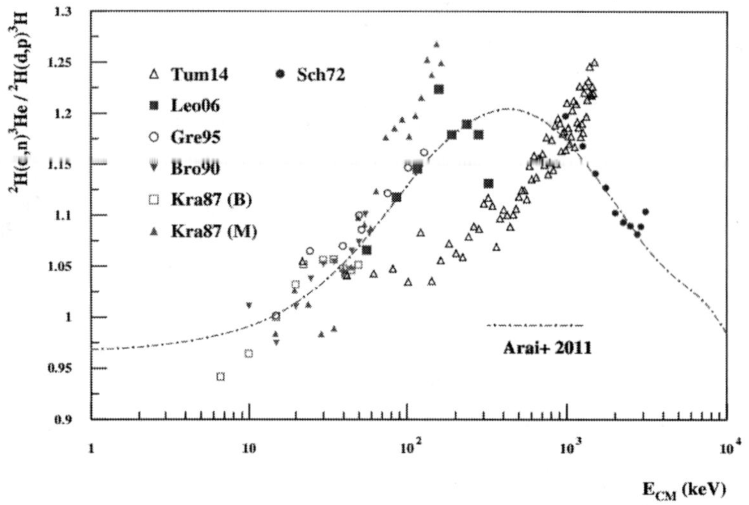

Figure 3: Ratios of ^{2}H$(d,n)^3$He to ^{2}H$(d,p)^3$H experimental and theoretical [19] S-factors.

common to both reactions. The ab initio calculation predicts a smooth increase, with a maximum near 0.6 MeV. This behaviour is supported by previous experimental data, in particular by the data of Ref. [35]. However the indirect data of Ref. [36], obtained in the framework of the Trojan Horse method, are probably affected by a systematic error, which shifts this ratio to higher energies.

3 Conclusion

Nuclear astrophysics is a broad field, where many nuclear inputs are necessary. In particular, charged-particle cross sections are quite important, and difficult to measure, owing to the low energies and cross sections. Cluster models are well adapted to these reactions, since in the low-mass region, the number of open channels is fairly small. The assumption of a cluster structure is in general realistic and allows to find approximate solutions of then A-body Schrödinger equation. Many applications have been performed so far in nuclear astrophysics. One of the future challenges for these models is the use of more realistic nucleon-nucleon interactions, and their extension to higher mass systems.

We have discussed different theoretical models often used in nuclear astrophysics. The potential model and the R-matrix method are widely applied in this field; they are fairly simple and well adapted to low-energy reactions. On the other hand microscopic cluster models have a stronger predictive power, since the only rely on a nucleon-nucleon interaction, and on the assumption of a cluster structure for the nucleus. An analysis of the $^2H(d,n)^3He$ and $^2H(d,p)^3H$ cross sections suggests that the Trojan Horse data of Tumino *et al.* [36] are probably affected by a systematic error.

References

[1] C. Iliadis, *Nuclear Physics of Stars* (Wiley-VCH Verlag GmbH, 2007)

[2] P. Descouvemont, M.S. Hussein, A. Lépine-Szily, *Nuclear reactions in stars: theoretical and experimental aspects* (Nova Science, New-York, 2013)

[3] P. Descouvemont, *Theoretical models for nuclear astrophysics* (Nova Science, New York, 2003)

[4] P. Descouvemont, T. Rauscher, Nucl. Phys. A **777**, 137 (2006)

[5] A.M. Lane, R.G. Thomas, Rev. Mod. Phys. **30**, 257 (1958)

[6] P. Descouvemont, D. Baye, Rep. Prog. Phys. **73**, 036301 (2010)

[7] T.A. Tombrello, Nucl. Phys. **71**, 459 (1965)

[8] I. Thompson, F. Nunes, *Nuclear Reactions for Astrophysics: Principles, Calculation and Applications of Low-Energy Reactions* (Cambridge University Press, 2009)

[9] K. Wildermuth, Y.C. Tang, *A Unified Theory of the Nucleus* (Vieweg, Braunschweig, 1977)

[10] W. Hauser, H. Feshbach, Phys. Rev. **87**, 366 (1952)

[11] C. Bertulani, Comput. Phys. Commun. **156**, 123 (2003)

[12] C. Angulo, P. Descouvemont, Phys. Rev. C **61**, 064611 (2000)

[13] C.R. Brune, Phys. Rev. C **66**, 044611 (2002)

[14] A. Kievsky *et al.*, J. Phys. G **35**, 063101 (2008)

[15] P. Navrátil, R. Roth, S. Quaglioni, Phys. Rev. C **82**, 034609 (2010)

[16] E. Caurier, P. Navrátil, Phys. Rev. C **73**, 021302 (2006)

[17] T. Neff, Phys. Rev. Lett. **106**, 042502 (2011)

[18] P. Navrátil, S. Quaglioni, Phys. Rev. C **83**, 044609 (2011)

[19] K. Arai, S. Aoyama, Y. Suzuki, P. Descouvemont, D. Baye, Phys. Rev. Lett. **107**, 132502 (2011)

[20] P. Descouvemont, M. Dufour, *Clusters in Nuclei*, Vol. 2 (Springer, 2012)

[21] H. Horiuchi, Prog. Theor. Phys. Suppl. **62**, 90 (1977)

[22] A.B. Volkov, Nucl. Phys. **74**, 33 (1965)

[23] D.R. Thompson, M. LeMere, Y.C. Tang, Nucl. Phys. A **286**, 53 (1977)

[24] M. Dufour, P. Descouvemont, Phys. Lett. B **696**, 237 (2011)

[25] Q.K.K. Liu, H. Kanada, Y.C. Tang, Phys. Rev. C **23**, 645 (1981)

[26] H.M. Hofmann, G.M. Hale, Nucl. Phys. A **613**, 69 (1997)

[27] P. Descouvemont, D. Baye, Nucl. Phys. A **573**, 28 (1994)

[28] P. Descouvemont, Nucl. Phys. A **596**, 285 (1996)

[29] D. Baye, M. Kruglanski, Phys. Rev. C **45**, 1321 (1992)

[30] K. Sabourov et al., Phys. Rev. C **70**, 064601 (2004)

[31] K. Arai, S. Aoyama, Y. Suzuki, Phys. Rev. C **81**, 037301 (2010)

[32] B.S. Pudliner *et al.*, Phys. Rev. C **56**, 1720 (1997)

[33] R. Tamagaki, Prog. Theor. Phys. **39**, 91 (1968)

[34] C. Angulo et al., Nucl. Phys. A **656**, 3 (1999)

[35] D.S. Leonard *et al.*, Phys. Rev. C **73**, 045801 (2006)

[36] A. Tumino et al., Astrophys. J. **785**, 96 (2014)

[37] A. Coc et al., to be published (2015)

EPJ Web of Conferences **117**, 09002 (2016) DOI: 10.1051/epjconf/201611709002
NN2015

LUNA: Present status and future prospects

A. Caciolli (for the LUNA Collaboration)

University and INFN of Padua, Padua, Italy

Abstract

One of the main ingredients of nuclear astrophysics is the knowledge of the thermonuclear reactions responsible for powering the stellar engine and for the synthesis of the chemical elements. At astrophysical energies the cross section of nuclear processes is extremely reduced by the effect of the Coulomb barrier and often extrapolations are needed. The Laboratory for Underground Nuclear Astrophysics (LUNA) is placed under the Gran Sasso mountain. Thanks to the environmental background reduction provided by its position many reactions involved in hydrogen burning has been measured directly at astrophysical energies. Based on this progress, currently there are efforts in several countries to construct new underground accelerators. The exciting science that can be probed with these new facilities will be highlighted.

1 Introduction

Nuclear processes generate the energy that makes stars shine. Moreover they are responsible of the synthesis of the elements (and isotopes) in stars. As a matter of fact, hydrogen, helium and all isotopes until lithium and beryllium are synthesized during the Big Bang Nucleosynthesis (BBN). All other nuclei are produced during the different characteristic phases of the star evolution [1].

The understanding of these nuclear processes is the goal of nuclear astrophysics and, in particular, the knowledge of the nuclear cross-sections involved in that processes. At astrophysical energies the cross section is highly reduced by the effect of the Coulomb repulsion and the nuclear reactions can occur only via tunnel effect.

Due to these small cross section values, the rate of the reactions, characterized by a typical energy release of a few MeV, is too low, down to a few events per year, in order to stand out from the laboratory background. In many cases it is not possible even to reach energy values close to the Gamow peak and extrapolations are needed, leading to substantial uncertainties. A way to handle that background problem is to go in an underground environment. As a matter of fact the natural shielding provided by an underground site will guarantee a reduction of the cosmic flux of orders of magnitude leading to the success of experimental nuclear physics. LUNA [2] is placed under the Gran Sasso National Laboratories of INFN. Two accelerators were used during years. First a 50 kV accelerator (hereafter LUNA1) [3] and then a 400kV accelerator (hereafter LUNA2) [4].

Under the Gran Sasso Laboratory the muon flux is reduced by a factor 10^6 and the neutron flux by a factor of 1000 [5, 6]. Further background reduction in the region below 3 MeV in the gamma spectrum can be achieved by implementing a shielding made by copper and lead [7]. A review of the results achieved by the LUNA collaboration will be presented in this paper combined with a discussion on the future projects for nuclear astrophysics in underground with a MV accelerator.

2 Solar hydrogen burning

Hydrogen burning in the Sun proceeds mainly by the proton-proton chain, with a 0.8% contribution from the carbon-nitrogen-oxygen cycle (CNO cycle) [8]. The basic processes are by now well understood, leading to the so-called standard solar model [9] that explains both helioseismological data and neutrino observations. The main uncertainty affecting this model, the solar neutrino puzzle, has been spectacularly solved by large neutrino detectors [10–12, e.g.] showing that the missing solar neutrinos have undergone flavour oscillation.

LUNA started its work studying the ^{3}He(^{3}He,2p)^{4}He reaction since there was the discussion on a possible resonance at the Gamow peak energy [13, 14]. This study was done By using the LUNA1 50 kV accelerator this reaction was studied directly at the energy of the Gamow peak ruling out the resonance existence [15]. The ^{2}H(p,γ)^{3}He reaction, responsible for the production of ^{3}He, was also studied at LUNA1 [16].

The neutrino fluxes emitted by the Sun are strictly correlated with the nuclear processes involved in the hydrogen burning. The LUNA2 program was focused on these processes achieving important results. The ^{3}He(α,γ)^{7}Be reaction was studied by using both the gamma prompt detection [17] and the activation techniques [18] finding a perfect agreement within the two methods. This result was important not only to reduce the systematics, but also to solve the discrepancy previously shown in experiments based on the two different techniques.

The CN neutrino fluxes are governed by the ^{14}N(p,γ)^{15}O reaction. It is the bottleneck of the first CNO cycle and therefore the ^{13}N and ^{15}O neutrinos are controlled by this reaction. LUNA deeply studied this reaction finding that the S-factor was a factor of two lower than what reported in the NACRE database [19]. This results are shown in [20] and reference therein.

A new precise knowledge of the ^{14}N(p,γ)^{15}O cross section has been raised to solve the so called Solar Composition Problem [8]: the conflict between helioseismology and the new metal abundances (i.e. the amount of elements different from hydrogen and helium) that emerged from improved modelling of the photosphere [21]. As a matter of fact, the CNO neutrino flux is decreased by about 30% in going from the high to the low metallicity scenario. This way it will be possible to test whether the early Sun was chemically homogeneous [22], a key assumption of the standard Solar Model. In order to reduce the nuclear uncertainties in the solar model a new measurement was performed reaching the final value of $S_{1,14}(0)=1.57\pm0.13$ keV barn [23, 24].

2.1 Second, third CNO cycles, and the Mg-Al

In recent years the LUNA collaboration focused its attention on several reactions involved in hydrogen burning Nova explosion. The first reaction studied in this program was the ^{15}N(p,γ)^{16}O [25]. This is the link from the first to the second CNO cycle and it was studied intensively at the LUNA accelerator. Two new experiments were performed by using ^{15}N enriched solid targets ([26] and reference therein). The LUNA measurements cover totally the Gamow peak for Nova explosion where the ^{15}N(p,γ)^{16}O is important and the cross section was found to be lower than a factor of 2 with respect what reported in the NACRE database [19]. This leads to a reduction of ^{16}O produced by Novae of a 40% [26].

The ^{17}O(p,γ)^{18}F reaction was investigated from 2011 to 2013. In particular, the ratio between the reaction rates of ^{17}O(p,α)^{14}N (Q = 1.2 MeV) and ^{17}O(p,γ)^{18}F (Q = 5.6 MeV) channels is one of the most important parameters for the galactic synthesis of ^{17}O, the stellar production of radioactive

^{18}F, and for predicted O isotopic ratios in presolar grains [27, 28].

Since the ^{18}F is a radio-nuclide with a half life of $\approx$110 min, the cross section has been derived both by detecting the prompt gamma rays and by counting the 511 keV γs emitted by the ^{18}F decay. The results are perfectly in agreement reducing considerably the systematic uncertainties [29]. The LUNA results affect not only the direct capture evaluation, but the 183.3keV resonance strength was also measured with a value of $\omega\gamma = (1.67 \pm 0.12) \mu$eV. As a matter of fact, the LUNA measurements cover the whole Gamow peak referred to Nova scenarios reducing by a factor of 4 the uncertainty on this reaction in stellar models and in particular on the oxygen and fluorine isotopes produced in Nova explosions [29, 30]. The very low uncertainty obtained in this experiment was possible thank to an intensive study of the target, realised and tested directly by the LUNA group with IBA and SIMS technique [31].

To measure also the ^{17}O(p,α)^{14}N reaction, a new chamber has been constructed which allows to place 8 silicons detectors in backward directions. The setup is described in [32] in details.

The study of the CNO cycles is the natural precursor for the hydrogen burning in Ne-Na and Mg-Al cycles. One of the reaction more interesting of the Ne-Na cycle is the ^{22}Ne(p,γ)^{23}Na. This is one of the slowest reactions of the entire cycle and dominates the uncertainty for the production of many isotopes from neon to aluminum in different stellar scenarios. This is due to the presence of many resonances at stellar energies where only upper limits were given. With a setup characterized by two high purity germanium detectors fully shielded [33] this reaction has been studied and several resonance strengths measured reducing drastically the uncertainties on the reaction rate [34].

The problem of the ^{26}Al production is one of the most interesting cases [35]. LUNA measured precisely several resonances for ^{24,25}Mg(p,γ)^{26,27}Al in order to reduce the uncertainties on those reactions as requested in the astrophysical models [36,37]. The impact of the LUNA results is discussed in details in a recent work [38]. It is worth to mention that the ^{26}Al uncertainty in C/Ne explosive burning is dominated by the ^{25}Mg(α,n)^{26}Al reaction as discussed in [39] and reference therein.

3 Big Bang nucleosynthesis

The ^{3}He(α,γ)^{7}Be reaction has an important role in solving the problem of the Spite plateau [40]. LUNA measured this reaction in the Gamow peak for Big Bang Nucleosynthesis reducing the uncertainties to 3% overall. Another problem concerning lithium isotopes as been raised recently: the ^{6}Li has been measured to be 3 order of magnitude higher than what expected in BBN [41, 42]. For ^{6}Li production in the Big Bang, the main nuclear physics unknown is the ^{2}H(α,γ)^{6}Li reaction rate. The setup used to study this reaction is already described in details [43]. A long and detailed study of this background was required in order to perform the analysis of the data [43, 44]. This way a possible nuclear solution of the ^{6}Li problem has been ruled out by LUNA.

4 Science case for a future higher-energy accelerator underground

Recent advances in astronomy and astrophysics require nuclear data at energies that are higher than the high-energy limit of LUNA2.

Most notably, the ^{12}C(α,γ)^{16}O reaction still eludes experimental and theoretical efforts to pin down its precise rate. This reaction, together with the triple-α reaction, determines the ratio of carbon to oxygen at the end of helium burning, a value that has wide-ranging impacts on the nucleosynthesis of heavier elements.

Whereas a direct ^{12}C(α,γ)^{16}O study at the relevant energy of 300 keV is impossible due to the forbiddingly-low absolute yield, a study in a low-background environment such as LUNA at higher energies can help improve necessary extrapolations by providing constraints at energies where there are currently no data. The study of this reaction is based on a precise knowledge of the targets, since the background induced by the parasitic (α,γ) reaction on ^{13}C can overwhelm the signal of the ^{12}C(α,γ)^{16}O reaction if the isotopic ratio ^{12}C/^{13}C is less than 10^5 (at least three order of magnitude higher than in natural carbon). The LUNA collaboration has started a deep investigation on ^{12}C enriched targets and their stability, by performing analysis tests on different backings and cleaning techniques and to understand the behaviour of the produced targets against irradiated charge. Those tests are performed at the Laboratori Nazionali di Legnaro and they will continue also in 2015 in order to have a complete understanding of the targets and in order to keep better under control the production techniques used in their creation. This work is essential for the success of the ^{12}C(α,γ)^{16}O cross section measurements.

Another important open issue of nuclear astrophysics is the neutron source reactions. In particular the ^{13}C(α,n)^{16}O and ^{22}Ne(α,n)^{25}Mg reactions. They are responsible for the production on neutrons involved in the slow neutron capture process, called the astrophysical s-process. Whereas those reactions are the subject of intensive experimental study, so far the reactions actually producing the neutrons have not yet been measured in the relevant energy range since they should be addressed by an underground accelerator.

A third topic is to complement some of the proton- and α-capture reactions studied at the LUNA2 accelerator at higher energy. Such a continuation is particularly important for the Big Bang reactions ^{3}He(α,γ)^{7}Be and ^{2}H(α,γ)^{6}Li, where the present LUNA2 400 kV machine can only cover the lower part of the relevant energy region.

5 Future underground accelerator facilities

Based on the successes of the LUNA collaboration, around the world several efforts are underway to install high-current, stable-beam accelerators in underground sites. LUNA-MV project was started in order to install a 3MV machine in the underground laboratories of Gran Sasso. The new accelerator has already been financed by the Italian government and it should be installed in the next years at the Gran Sasso Laboratory. The synergy between the existing LUNA2 and the new LUNA-MV accelerator will allow to perform reaction studies in a wide range of energies with complete understanding of the setups involved. The planned DIANA facility at the Deep Underground Science Laboratory DUSEL in the Unites States also includes a megavolt and a lower-energy machine. Another project is under discussion at the Canfranc underground laboratory in the Pyrenees, Spain. As part of a staged approach, even an accelerator laboratory in a shallow-underground facility such as Felsenkeller (Dresden, Germany) is under consideration.

At present, the existing 400 kV LUNA2 machine continues its scientific program outlined here. The next few years will show where this highly successful approach will eventually be complemented by one or more higher-energy accelerators underground. The technique is sufficiently mature to address not only the data needs of the astrophysics community, but it has the potential to benefit also the astroparticle and other communities.

Acknowledgments

The work at LUNA is carried out in the framework of the LUNA collaboration; see e.g. Ref. [43] for an author list. Financial support by OTKA

K101328, NN83261, DFG (BE 4100/2-1), and NAVI is also gratefully acknowledged.

References

[1] C. Iliadis, *Nuclear Physics of Stars*, Wiley-VCH, 2007.

[2] C. Broggini, D. Bemmerer, A. Guglielmetti, and R. Menegazzo, Annu. Rev. Nucl. Part. Sci. **60**, 53 (2010).

[3] U. Greife et al., Nucl.Instr.Meth. A 350 (1994) 327.

[4] A. Formicola et al., Nucl.Instr.Meth. A 507 (2003) 609.

[5] D. Bemmerer, et al., *Eur. Phys. J. A* **24**, 313–319 (2005).

[6] T. Szücs, et al., *Eur. Phys. J. A* **44**, 513 (2010).

[7] A. Caciolli, et al., *Eur. Phys. J. A* **39**, 179–186 (2009).

[8] A. M. Serenelli, S. Basu, J. W. Ferguson, and M. Asplund, *Astrophys. J. Lett.* **705**, L123–L127 (2009), 0909.2668.

[9] J. N. Bahcall, and M. H. Pinsonneault, *Phys. Rev. Lett.* **92**, 121301 (2004).

[10] J. Hosaka, et al., *Phys. Rev. D* **73**, 112001–+ (2006).

[11] I. Ahmad, J. P. Greene, E. F. Moore, S. Ghelberg, A. Ofan, M. Paul, and W. Kutschera, *Phys. Rev. C* **74**, 065803–+ (2006).

[12] C. Arpesella, et al., *Phys. Rev. Lett.* **101** (2008).

[13] W.A. Fowler, Nature 238 (1972) 24

[14] V.N. Fetysov and Y.S. Kopysov, Phys. Lett. B 40 (1972) 602

[15] R. Bonetti, et al., *Phys. Rev. Lett.* **82**, 5205–5208 (1999).

[16] C. Casella, et al., *Nucl. Phys. A* **706**, 203–216 (2002).

[17] D. Bemmerer, et al., *Phys. Rev. Lett.* **97**, 122502 (2006).

[18] F. Confortola, et al., *Phys. Rev. C* **75**, 065803–+ (2007).

[19] C. Angulo et al. (NACRE Coll.), Nucl. Phys. A 656 (1999) 3.

[20] D. Bemmerer, et al., *Nucl. Phys. A* **779**, 297–317 (2006).

[21] C. Pena-Garay and A.M. Serenelli, arXiv: 0811.2424 (2008.)

[22] W.C. Haxton and A.M. Serenelli, arXiv:0902.0036 (2009).

[23] M. Marta, et al., *Phys. Rev. C* **78**, 022802(R) (2008).

[24] M. Marta, et al., *Phys. Rev. C* **83**, 045804 (2011).

[25] D. Bemmerer, et al., *J. Phys. G* **36**, 045202 (2009).

[26] A. Caciolli, et al., *A&A* **533**, A66 (2011).

[27] C. Fox *et al.*, Phys. Rev. C 71 (2005) 055801.

[28] J.R. Newton *et al.*, Phys. Rev. C 81 (2010) 045801.

[29] D. A. Scott, et al., *Phys. Rev. Lett.* **109**, 208001 (2011).

[30] A. Di Leva, et al., *Phys. Rev. C* **89**, 015803 (2014).

[31] A. Caciolli, et al., *Eur. Phys. J. A* **48**, 144 (2012).

[32] C. G. Bruno, et al., *Eur. Phys. J. A* **51**, 94 (2015).

[33] F. Cavanna, et al., *Eur. Phys. J. A* **50** 179 (2014).

[34] F. Cavanna, et al., submitted to PRL.

[35] C. Iliadis, et al., *Astrophys. J. Supp. Series* **193**, 16 (2011).

[36] B. Limata et al., *Phys. Rev. C* 82 (2010) 015801

[37] F. Strieder et al., *Phys. Lett. B* **707**, 60 (2012).

[38] O. Straniero et al., *Astrophys. J.* **763**, 100 (2012).

[39] A. Caciolli, et al. *Eur. Phys. J. A* **50**, 147 (2014).

[40] M. Spite, and F. Spite, *Nature* **297**, 483–485 (1982).

[41] V. Smith, D. Lambert, and P. Nissen, *Astrophys. J.* **408**, 262 (1993).

[42] M. Asplund, et al., *Astrophys. J.* **644**, 229–259 (2006).

[43] M. Anders et al., *Eur. Phys. J. A* **49**, 28 (2013).

[44] M. Anders et al., *Phys. Rev. Lett.* **113**, 042501 (2014).

Indirect techniques for astrophysical reaction rates determinations

F. Hammache[1], N. Oulebsir[1,2], S. Benamara[3,1], N. de Séréville[1], A. Coc[4], A. Laird[5], I. Stefan[1], P. Roussel[1] and collaborators

[1] IPN, IN2P3-CNRS, Université Paris sud 11, 91406 Orsay, France
[2] LPT, Université A. Mira, 06000 Béjaia, Algérie
[3] LPCQ, Université Mouloud Mammeri, Tizi-Ouzou, Algérie
[4] CSNSM, IN2P3-CNRS, Université Paris Sud 11, 91405 Orsay, France
[5] Department of Physics, University of York, York, UK

Abstract

Direct measurements of nuclear reactions of astrophysical interest can be challenging. Alternative experimental techniques such as transfer reactions and inelastic scattering reactions offer the possibility to study these reactions by using stable beams. In this context, I will present recent results that were obtained in Orsay using indirect techniques. The examples will concern various astrophysical sites, from the Big-Bang nucleosynthesis to the production of radioisotopes in massive stars.

The main characteristics of the nuclear reactions involved in stellar nucleosynthesis is the low energy where they generally occur, between few keV to few MeV and at these energies the cross sections of these reactions are very small ranging from fbarn to hundreds of pbarn especially when it involves charged particles. This makes the direct measurements at stellar energies very difficult and often impossible. Hence direct measurement are usually performed at higher energies and then extrapolated down to stellar energies where the reactions occur. However the extrapolations can lead sometimes to wrong results when a possible very low energy resonance or the tail of a possible sub-threshold resonance are not taken into account.

The effect of these resonances may change the extrapolated S-factor by a huge factor. The other problem with direct measurements is related to the radioactive nature of the nuclei involved in reactions occurring for instance in novae, r-process,... The radioactive beam intensities are usually low, they rarely exceed 10^5 -10^6 pps and for nuclei which have relatively long half lives, making targets with enough at/cm^2 is quite difficult.

To bypass all these difficulties, related to extrapolation and/or radioactive nuclei, indirect techniques such as transfer reactions, ANC, inelastic scattering and charge exchange-reactions may help to determine the most important ingredients to evaluate the cross section and reaction rates of resonant reactions: the energy level, spin-parity and partial decay widths of the state of interest in the formed compound nucleus.

In the following, three cases of astrophysical interest that were studied in Orsay using indirect techniques will be presented.

1 Study of $^{12}C(\alpha,\gamma)^{16}O$ through α-transfer reaction

The importance of the challenging $^{12}C(\alpha,\gamma)^{16}O$ reaction in energetics, nucleosynthesis and final fate of the massive stars is notorious [1]. Despite the various experiments [2] which studied this reaction, its cross section remains highly uncertain. At the Gamow peak of 300 keV where the reaction occurs during the He burning stage, the expected cross section is about 10^{-7} nbarn which means impossible to measure directly. Hence direct measurements were performed down to 900 keV in center of mass energy and then extrapolated down to 300 keV using R-matrix formalism. But the extrapolation in this case is very complicated. Indeed, the capture cross section at 300 keV which correspond to the excitation energy of 7.46 MeV in ^{16}O is dominated by the overlap of several contributions, the most important ones being the E1 and E2 transitions to the ground state via the low energy tail of the 1^- broad state at 9.6 MeV of ^{16}O and the high energy tails of the 1^- and 2^+ sub-threshold states at 6.92 and 7.12 MeV of ^{16}O. The effect of these two states was badly known because their alpha-spectroscopic factors and so their alpha-reduced widths were spread over a large range of values.

In view of the large uncertainties surrounding the alpha-spectroscopic factors of the two sub-threshold states, a new measurement of these parameters was performed through the α transfer reaction $^{12}C(^7Li,t))^{16}O$ [3].

The experiment [3] was performed at the Alto facility of Orsay using the magnetic Split-Pole spectrometer. A ^{7}Li beam was sent on a self-supporting enriched ^{12}C target. The reaction products were momentum analyzed by the Split-Pole and the tritons were detected in the focal plane first by a position sensitive gas chamber and then by a ΔE proportional counter. The angular distribution measurements were performed at two incident energies, 34 and 28 MeV , and at angles up to 45° in cm in order to check the direct character of the transfer mechanism.

In Fig. 1 are displayed the measured differential cross sections for 4 populated states of ^{16}O at the two incident energies together with the FR-DWBA calculations performed with FRESCO code in dotted curves and HF calculations in dotted-dashed curves. The solid curves are the incoherent sum of the two contributions.

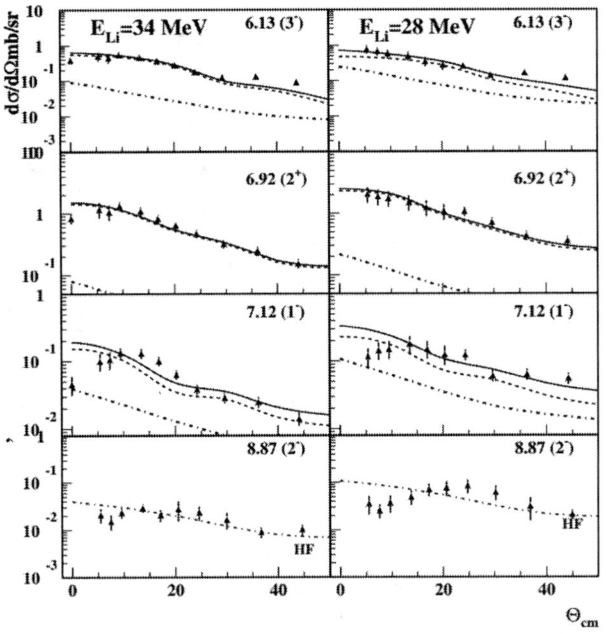

Figure 1: Experimental differential cross sections of the ^{12}C(^{7}Li,t)^{16}O reaction obtained at 34 MeV (left column) and 28 MeV (right column) for the 6.13, 6.92, 7.12 and 8.87 MeV states of ^{16}O [3].

One can see, that except for the non-natural parity state at 8.87 MeV, the data are quite well described by the FR-DWBA calculation at the two incident energies and this indicates that the direct transfer mechanism is the dominant one in this transfer measurement.

The S_α of the two sub-threshold states were extracted from the normalization of the FR-DWBA calculations to the data. The values obtained for

DOI: 10.1051/epjconf/201611709003

the 6.92 MeV and 7.12 MeV states, 27±10 keV and 8±3 keV respectively, were then used to calculate their α-reduced widths at the radius of 6.5 fm where the radial part of the α-^{12}C cluster wave function reaches its asymptotic behavior. The assymptotic normalization coefficients (ANC) of the two states of interest were also deduced, C^2(6.92 MeV)=(2.07±0.80)$\times10^{10}$ fm^{-1} and C^2(7.12 MeV)=(4.00±1.38)$\times10^{28}$ fm^{-1} and were found to be in agreement with the results of ANC experiments of Brune et al [4] and the most recent ones of Avila et al [5].

The obtained α-reduced widths of the 2^+ and 1^- sub-threshold states were included in an R-matrix calculation of the E1 and E2 component of ^{12}C(α,γ)^{16}O using Descouvemont code. The details of the R-matrix calculations and the fitting procedure of the E2 and E1 measured phase shifts and astrophysical S-factors is given in [3]. The deduced S-factors at 300 keV for the E2 and E1 components, 50±19 keV-b and 100±28 keV respectively, were found in excellent agreement with those deduced in the ANC experiment of Brune et al. [4] where the alpha-reduced widths of the two sub-threshold states were also fixed in the R-matrix fitting procedure. The obtained values are also in agreement within the error bars with the values recommended in the very recent updated Nacre compilation (NACRE II) [6]; S_{E2}(300 keV)=61±19 keV-b and S_{E1}(300 keV)=80±18 keV-b.

2 The ^{7}Li cosmological problem

The primordial nucleosynthesis of light elements, D, ^{3,4}He and ^{7}Li, is one of the three observational pillars of the Big-Bang model. The observed primordial abundances for D and ^{3,4}He agree well with the predictions of BBN model together with the precise WMAP cosmic baryon density while ^{7}Li abundance observations in metal poor halo stars are three times smaller than BBN+WMAP predictions [7]. This discrepancy is what is known as the ^{7}Li problem. Many ideas and explanations were addressed to explain the ^{7}Li deficit but none of them were satisfactory.

^{7}Li is produced mainly by ^{7}Be EC decay and the most important reactions which produces and destroys ^{7}Be are ^{3}He(^{4}He,γ)^{7}Be and ^{7}Be(n,p)^{7}Li(p,α)α respectively. The cross section of ^{3}He(^{4}He,γ)^{7}Be reaction was measured by several groups using different techniques and its knowledge is nowadays better than 8% [8]. The cross sections of ^{7}Be(p,n)^{7}Li and ^{7}Li(p,α)α reactions are also well known [9]. Recently some authors [10–13] proposed the idea that some missed resonances in secondary destruction channels of ^{7}Li and ^{7}Be could be at the origin of the ^{7}Li deficit. Three promising candidates came out from their investigation: ^{7}Be+d, ^{7}Be+^{3}He and ^{7}Be+^{4}He. The ^{7}Be+d reaction channel was investigated by various works ([14] and references therein) and it was dismissed. Hence we took care about the two other candidates in Orsay for which a presence of a narrow 1^- or 2^- state close to 15 MeV in the compound nucleus ^{10}C or a presence of a state close to 7.8 MeV in the compound nucleus ^{11}C may reconcile the ^{7}Li predictions with the observations.

The search for the missing resonant states has been performed at the split-pole (SP) spectrometer of the ALTO facility. ^{10}C and ^{11}C nuclei were populated with the (^{3}He,t) charge exchange reaction on ^{10}B and natural Boron targets at 35 MeV and the emitted tritons were detected at four SP angles [15].

In Fig. 2 and Fig. 3 are displayed the measured Bρ position spectra for ^{11}C and ^{10}C. The well populated peaks observed in ^{11}C spectrum are the already well known ^{11}C levels and no new peak is seen in the energy region of interest between 7.79 and 7.9 MeV. Thanks to the very large signal to background ratio observed in the spectrum, one can assert that it is very unlikely that a new state in ^{11}C exists in the energy region of investigation. Moreover , all known ^{11}B states below Ex=9 MeV have their counterpart in ^{11}C mirror nucleus.

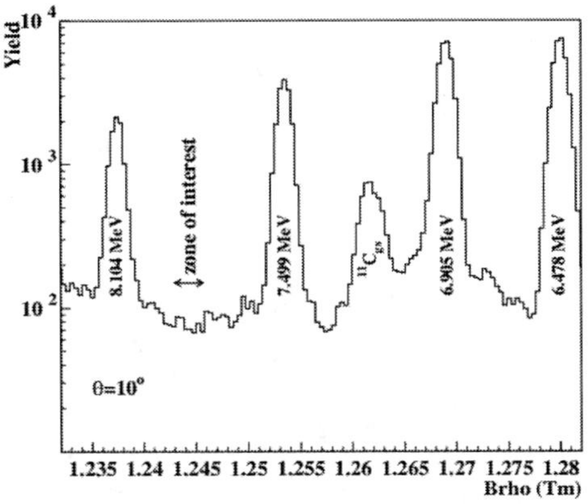

Figure 2: ^{11}C Bρ position spectrum measured at 10°.

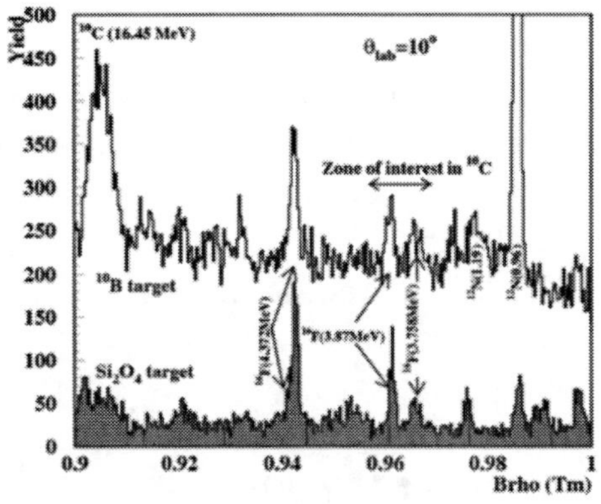

Figure 3: Bρ position spectra measured at 10° on ^{10}B and Si_2O_4 targets.

No additionnal states in ^{10}C are observed in Fig. 3 around 15 MeV neither, the only states we observe are those of ^{16}F populated via (^{3}He,t) on the ^{16}O contamination of the target as one can see from the grey filled histogram obtained with Si_2O_4 target. The same results were obtained at the other measured angles. But since the background in this case is very important and the signal to background ratio is 10 times less than in the case of ^{11}C, one can not exclude that a state may be hidden in the background. However, from a χ^2 study [15] of a simulated state close to 15 MeV with various widths and populated cross sections on the top of the measured background, we could conclude that any 1^- or 2^- state of ^{10}C in the excitation energy region of interest has very likely, if present, a total width larger than 590 keV to escape our detection. With this limit and even a three times lower one, the calculated ^{7}Be+^{3}He reaction rates [15] do not solve the ^{7}Li problem. Finally, from our results we conclude that the ^{7}Li problem is not due to resonant states in ^{10}C and ^{11}C. With this conclusion and those of previous works concerning other reactions, we may even say that the solution has very likely to be found outside of nuclear physics.

3 ^{26}Al nucleosynthesis in massive stars

Massive stars are large producers of the ^{26}Al radioisotope which is a β^+ emitter decaying to ^{26}Mg g.s. after emitting a characteristic γ-ray line at 1.809 MeV. ^{26}Al has been observed since the late 70s in the galactic plane and its emission has been correlated to massive star population. It is also observed through excesses of its daughter nucleus ^{26}Mg in meteorites inclusions.

The interpretation of both these observations need precise yields of ^{26}Al. In massive stars, ^{26}Al is mostly produced during explosive burning and it was shown that the ^{26}Al yield depends crucially on the ^{26}Al(n,p) and ^{26}Al(n,α) reactions rates [16]. Unfortunately, the latters are badly known because of the lack of spectroscopic information in the ^{27}Al compound nucleus (CN) above the neutron threshold.

For typical temperatures of 2 billion degrees achieved during explosive burning, resonances of interest in the CN ^{27}Al are located from neutron threshold (at 13 MeV) to about 500 keV above neutron threshold. Several direct and time reverse measurements of these reactions were performed but the covered energy range did not extend up to 500 keV above neutron threshold in the center of mass. Moreover, the experimental ^{26}Al(n,p) reaction rates show a disagreement between the different sets of measured data [17]. Hence an improvement of the knowledge of ^{27}Al spectroscopy is really needed.

The excited states of ^{27}Al were populated at the Alto facility of Orsay via ^{27}Al(p,p') reaction [18] at 18 MeV incident energy and using Split-Pole spectrometer for the detection of the protons. In total, more than 30 new levels were observed above the neutron threshold (see Fig.4 and Fig.5 in ref

[18]). The Orsay Split-Pole results were confirmed by the recent ^{27}Al(p,p') measurements performed at the Q3D of MLL Munich where a better energy resolution (7 keV) and a better energy precision (2 keV) could be achieved. New states were observed in the Q3D experiment due to the improvement in the energy resolution as one can see in Fig. 4 where a small part of the measured spectra with the Split-Pole and the Q3D spectrometers are displayed.

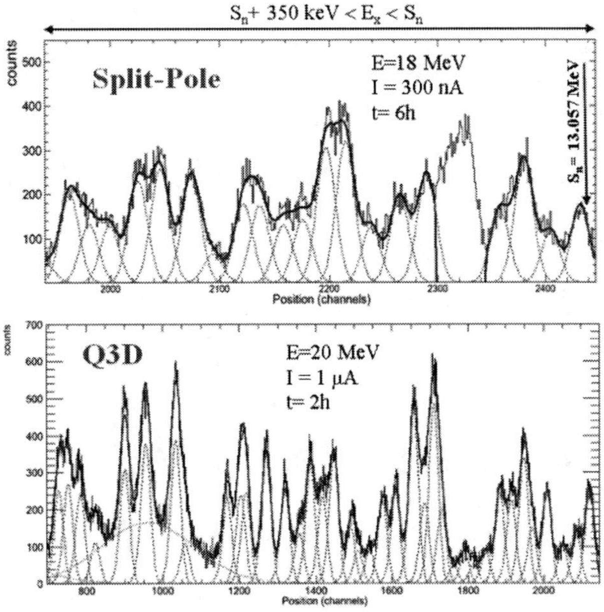

Figure 4: Bρ position spectra of ^{27}Al measured with Split-Pole in Orsay (top) and with Q3D in MLL Munich (bottom). The dotted curves are fit of the spectra after background substraction.

A coincidence measurement coupling the SP spectrometer with three DSSSDs placed around the target in a close geometry was performed recently in order to study the decay modes of the measured new resonances and try to extract the decay branching ratios needed for reaction rates calculations. The analysis of the data is still in progress.

Acknowledgments

I would like to thank all my collaborators from IPNO, CSNSM, York, Ganil, ASCR, Algeria,Barcelona, Huelva, Santiago de Compostella and Saha Institut who participate in the experiments described in this contribution.

References

[1] T.A. Weaver and S.E. Woosley, Phys. Reports 227 (1993) 65.

[2] A. Belhout et al., Nucl. Phys. A793 (2007) 178 and references therein.

[3] N. Oulebsir, F. Hammache et al., Phys. Rev. C 85 (2012) 035804.

[4] C. R. Brune, W. H. Geist, R. W. Kavanagh and K. D. Veal, Phys. Rev. Lett. 83 (1999) 4025.

[5] M.L.Avila, G.V.Rogachev et al., Phys.Rev.Lett. 114 (2015) 071101.

[6] Y. Xu, K. Takahashi, S.Goriely et al., Nucl. Phys. A 918 (2013) 61

[7] A. Coc, E. Vangioni, J. Cosmology. Astropart. Phys. 10 (2014) 050.

[8] E. G. Adelberger et al., Rev. Mod. Phys. 83 (2011) 195.

[9] P. Descouvemont, A. Adahchour, C. Angulo, A. Coc and E. Vangioni-Flam, At. Data Nucl. Data Tables 88 (2004) 203.

[10] N. Chakraborty, B. D. Fields, K. Olive, Phys. Rev. D 83 (2011) 063006.

[11] C. Broggini, L. Canton, G. Fiorentini, F.L. Villante, J. Cosmology Astropart. Phys. 06 (2012) 030.

[12] O. Civitarese, M.E. Mosquera, Nucl. Phys. A 898 (2013) 1

[13] R.H. Cyburt, M. Pospelov, Int. J. Mod. Phys. E 21 (2012) 1250004.

[14] O. S. Kirsebom, B. Davids, Phys. Rev. C 84 (2011) 058801 and references therein.

[15] F. Hammache, A. Coc et al., Phys. Rev. C 88 (2013) 062802(R).

[16] C. Iliadis, A. Champagne, A. Chieffi, and M. Limongi, Astrophys. J. Suppl. 193 (2011) 16.

[17] B. M. Oginni, C. Iliadis et al. Phys.Rev.C 83,025802 (2011)

[18] S. Benamara, N.de Sereville, et al., Phys.Rev. C 89 (2014) 065805.

EPJ Web of Conferences **117**, 09004 (2016) DOI: 10.1051/epjconf/201611709004
NN2015

The ^{12}C(^{12}C,α)^{20}Ne and ^{12}C(^{12}C,p)^{23}Na reactions at the Gamow peak via the Trojan Horse Method

A. Tumino[1,2], C. Spitaleri[2,3], S. Cherubini[2,3], L. Guardo[2,3], M. Gulino[2,3], I. Indelicato[2,3], M. La Cognata[2], L. Lamia[3], R.G. Pizzone[2], S.M.R. Puglia[2,3], G.G. Rapisarda[2,3], S. Romano[2,3], M.L. Sergi[2,3] and R. Spartá[2,3]

[1]Facoltà di Ingegneria e Architettura, Università degli Studi di Enna "Kore", Enna, Italy
[2] INFN - Laboratori Nazionali del Sud, Catania, Italy
[3]Dipartimento di Fisica e Astronomia, Università degli Studi di Catania, Catania, Italy

Abstract

A measurement of the ^{12}C(^{14}N,α^{20}Ne)^{2}H and ^{12}C(^{14}N,p^{23}Na)^{2}H reactions has been performed at a ^{14}N beam energy of 30.0 MeV. The experiment aims to explore the extent to which contributing ^{24}Mg excited states can be populated in the quasi-free reaction off the deuteron in ^{14}N. In particular, the ^{24}Mg excitation region explored in the measurement plays a key role in stellar carbon burning whose cross section is commonly determined by extrapolating high-energy fusion data. From preliminary results, α and proton channels are clearly identified. In particular, ground and first excited states of ^{20}Ne and ^{23}Na play a major role.

1 Introduction

After more than four decades, carbon burning stands out as one of the most interesting subjects to study because of several reasons. ^{12}C +^{12}C is the system with more evidence for quasimolecular structure, derived mainly from excitation function measurements for many of the possible reaction channels [1–7]. The vast amount of data accumulated in this respect still awaits a sound and comprehensive theoretical model capable of explaining all the observed details.

Starting from intermediate mass stars (8–10 $M_\odot$) and up with more than 10 $M_\odot$, star evolution and nucleosynthesis is strongly influenced by the carbon fusion process [8]. Superbursts from accreting neutron stars are also important phenomena that have been related to carbon burning, for which accurate reaction rates are needed to constrain superburst models with neutron and strange stars [9]. The temperatures at which carbon burns range from 0.8 to 1.2 GK, corresponding to center-of-mass energies from 1 to 3 MeV. The Coulomb barrier height for the ^{12}C + ^{12}C system is around 6.3 MeV, much higher than the energies of interest. In that region, the cross section falls rapidly below one nanobarn. This is the reason why the measurement of the cross section at astrophysical energies remains a difficult task. The excitation energy of the compound nucleus ^{24}Mg is high enough to decay by particle emission. Alpha, proton and neutron are the dominant evaporation channels, leading respectively to ^{20}Ne, ^{23}Na and ^{23}Mg, which can also be produced in excited bound states. Below a center of mass energy E_{cm} of 2.5 MeV there is not enough energy to feed ^{23}Mg even in its ground state and α and p channel are the only relevant ones at low energies. Considerable efforts have been devoted to measure the ^{12}C + ^{12}C cross section at astrophysical energies, involving both, charged particle [11–13] and gamma ray spectroscopy [14–19]. Nevertheless, it has only been previously measured down to E_{cm} = 2.5 MeV, still at the beginning of the region of astrophysical interest. However, below E_{cm} = 3.0 MeV the reported cross sections disagree and are rather uncertain, because at these energies the presence of ^1H and ^2H contamination in the C targets hampered the measurement of the $^{12}C+^{12}C$ process both in particle and gamma ray studies. In a more recent study [20], the astrophysical S(E) factor exhibits new resonances at $E_{cm} \leq$ 3.0 MeV, in particular, a strong resonance at E_{cm} = 2.14 MeV, which lies at the high-energy tail of the Gamow peak. This resonance, if confirmed, would increase the present nonresonant reaction rate of the alpha channel by a factor of 5 near T = 0.8 GK. On the other hand, it has been proposed that a sub-barrier fusion hindrance effect might drastically reduce the reaction rate at astrophysical energies. As known, measurements at lower energies are extremely difficult. Moreover, in the present case the extrapolation procedure from current data to the ultra-low energies is complicated by the presence of possible resonant structures even in the low-energy part of the excitation function. Thus, further measurements extending down to at least 1 MeV would be extremely important. In this paper, we are going to discuss the indirect study of the $^{12}C(^{12}C,\alpha)^{20}Ne$ and $^{12}C(^{12}C,p)^{23}Na$ reactions via the Trojan Horse Method (THM) [21–23] applied to the $^{12}C(^{14}N,\alpha^{20}Ne)^{2}H$ and $^{12}C(^{14}N,p^{23}Na)^{2}H$ three-body processes in the quasi-free (QF) kinematics regime, where ^{2}H from the ^{14}N TH nucleus is spectator to the ^{12}C + ^{12}C two-body processes. There is a number of works providing evidence of direct ^{12}C transfer in the $^{12}C(^{14}N,d)^{24}Mg^{*}$ reaction at 30 MeV of beam energy and up [24, 25].

2 The Trojan Horse Method

The Trojan Horse Method (THM) [21–23] is a powerful indirect technique to determine the astrophysical factor for rearrangement reactions. It has already been applied several times to reactions of relevance for astrophysics ([26–35] and references therein). The purpose of the THM is to determine the cross section for the binary reaction (2 particles → 2 particles) $A+x \to C+c$, from the measured cross section for the TH reaction 2 particles → 3 particles $A+a \to C+c+s$ in the quasifree (QF) kinematics regime, where the "Trojan Horse" (TH) particle, $a = (x\,s)$, has a strong $(s\,x)$ cluster structure. The reaction $A+a \to C+c+s$ can proceed through different reaction mechanisms. The TH reaction mechanism, gives the dominant contribution to the cross section in a restricted region of the three-body phase space where the relative momentum of the fragments x and s, p_{sx}, fulfill the QF conditions. For a $x-s$ relative motion in s-wave, p_{sx} is zero in the QF conditions or small compared to the bound state $(s\,x)$ wave number as given below. Since the transferred particle x in the TH reaction $A+a \to C+c+s$ is virtual, its energy and momentum are not related by the on-energy-shell (OES) equation $E_x = k_x^2/(2\,m_x)$. The energy for the $A+a$ relative motion is chosen to be above the Coulomb barrier in such a way that the two body interaction can be considered as taking place inside the nuclear field, without experiencing either Coulomb suppression or electron screening effects. The $A+a$ relative motion is compensated for by the $x-s$ binding energy, determining the so called "quasi-free two-body energy" given by

$$E_{QF} = E_{Aa} - B_{x-s} \tag{1}$$

where E_{Aa} represents the beam energy in the center-of-mass system and $B_{x\,s}$ is the binding energy for the $x-s$ system. In Eq.1, E_{QF} is not changed varying the projectile energy. What is done is to keep the beam energy at a fixed value and to vary the relative momentum p_{sx} from its QF value to an upper limit given by $\kappa_{xs} = \sqrt{2\,\mu_{xs}\,B_{xs}}$ that represents the on-energy-shell (OES) $a = (x\,s)$ bound state wave number [36]. This upper limit is usually few tens of MeV/c. This little variation is linked to the $x - s$ inter-cluster motion inside a and it is taken on the p_s variable, the momentum of the spectator particle (in the laboratory system $\vec{p}_{xs} = \vec{p}_x = -\vec{p}_s$) and/or on its emission angle, both measured or easily reconstructed. This brings to the following formula:

$$E_{QF} = \frac{m_x}{m_x + m_A} E_A - \frac{p_s^2}{2\,\mu_{sB}} + \frac{\vec{p}_s \cdot \vec{p}_A}{m_x + m_A} - B_{x\,s}. \tag{2}$$

with μ_{sB} reduced mass of the $s - B$ system $(B = A + x = C + c)$. Thus, the TH cross section can be used to determine the energy dependence of the astrophysical factor, $S(E)$, of the binary process, $A + x \to C + c$, down to zero relative kinetic energy of particles A and x without distortion due to electron screening [37]. The absolute value of $S(E)$ is obtained normalizing to direct measurements at higher energies. At low energies where electron screening becomes important, comparison of the THM astrophysical factor to the direct result provides a determination of the screening potential. The THM has been applied successfully to many direct and resonant processes, usually involving light particles as spectators, such as d, ^{3}He, ^{6}Li. Here, we address the first application of the THM to heavy ion reactions, using ^{14}N as spectator.

3 The experiment

The ^{12}C(^{14}N,α^{20}Ne)^{2}H and ^{12}C(^{14}N,p^{23}Na)^{2}H reactions were investigated at the Laboratori Nazionali del Sud - INFN, by using a ^{14}N beam at 30 MeV accelerated by the SMP TANDEM and delivered onto a 100 μg/cm^2 C target. Special care was taken in beam collimation in order to produce a beam spot on the target smaller than 1.5 mm. The experimental setups consisted of four telescopes (38 μm silicon detector as ΔE- and 1000 μm position sensitive detector (PSD) as E-detector) placed on both sides with respect to the beam direction in symmetric configuration (two on each side), covering angles from 7° to 30° and from 47° to 68°. We have detected the ejectile of the two-body reactions (either α or p) in coincidence with the spectator d particle. The heavy counterparts in the two-body reactions have quite low energy and, if detected, this would imply a high detection threshold. In order to fulfill the QF requirement for the spectator d particle to be essentially part of the beam, this particle was detected at forward angles. The trigger for the event acquisition was given by the coincidences between one of the two most forward telescopes and any one of the two telescopes on the opposite side with respect to the beam direction. The angular regions covered by the detectors were optimized for measuring the break-up process of interest in QF kinematics, and the investigated range of deuteron momentum values was feasible to check the existence of the QF mechanism.

4 Data analysis and preliminary results

Energy calibration of each PSD was performed by means of a standard eight-peak α source and data from elastic scattering of ^{14}N on ^{1,2}H at beam energies of about 13, 22 and 30 MeV. with an overall energy resolution of about 1.5%. Angular calibration was achieved by using grids with equally spaced slits placed in front of each PSD during preliminary runs of the experiment, with an angular resolution of about 0.3°. Several steps are involved in the data analysis before the two-body cross section of astrophysical relevance can be extracted. First, it is necessary to identify the events due to the three-body reactions of interest and select those experiencing the QF mechanism. Channel selection begins with the separation of the d and p/α locus in the ΔE-E 2D plot by means of graphical cuts. Selected events were then used to reconstruct the experimental Q-values.

If one considers a two-dimensional spectrum with a kinematical variable, such as the energy or the angle of any one of the involved particles as a function of the Q-value, coincidence events of interest should lie on a vertical line that cuts the Q-value axis at the expected value. Typical spectra for the present experiment are reported in Figs.1 (^{12}C(^{14}N,α^{20}Ne)^{2}H) and 2 ^{12}C(^{14}N,p^{23}Na)^{2}H where the deuteron detection angle is shown as a function of the Q-value. Two dominant sharp vertical lines show up in each figure, crossing the Q-value axis at about -5.7 and -7.3 MeV in Fig.1, corresponding to the ground and first excited state of ^{20}Ne, and at about -8 and -8.4 MeV in Fig.2, referring to the ground and first excited state of ^{22}Na. The spectra makes us confident of the quality of the calibration and of the possibility to identify the channels of interest. Events within these vertical region will be selected for further analysis.

 DOI: 10.1051/epjconf/201611709004

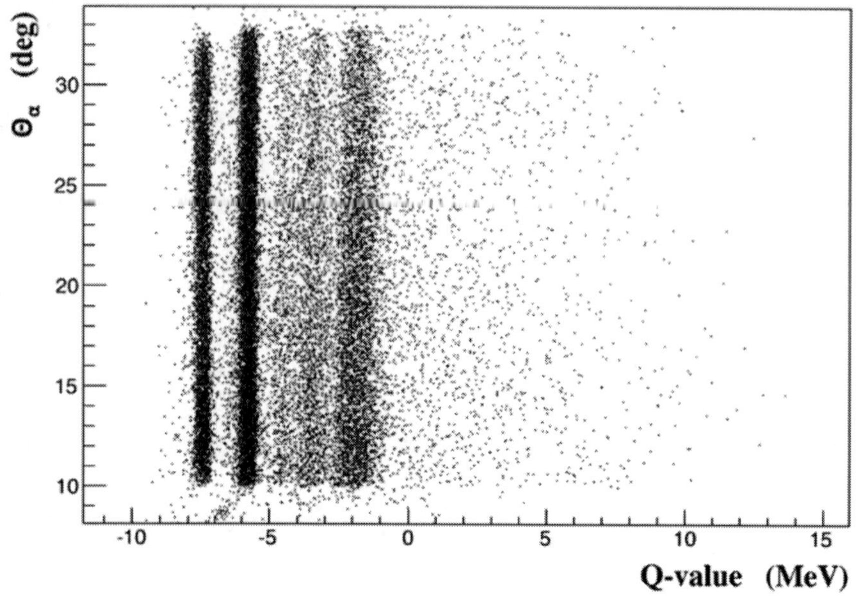

Figure 1: Two-dimensional spectrum showing the deuteron detection angle vs the Q-value. The sharp vertical regions at about -5.7 and -7.3 MeV of the Q-value, corresponds to the ground and first excited state of ^{20}Ne.

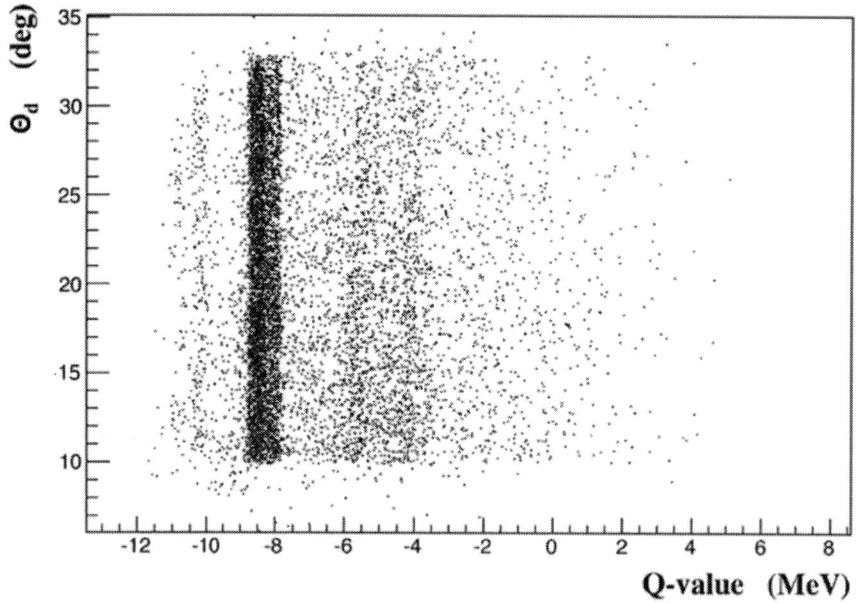

Figure 2: Two-dimensional spectrum showing the deuteron detection angle vs the Q-value. The sharp vertical regions at about -8 and -8.4 MeV of the Q-value, corresponds to the ground and first excited state of ^{22}Na.

References

[1] E. Almquist, D. A. Bromley, and J. A. Kuehner, Phys. Rev. Lett. 4, 515 (1960)

[2] W. Greiner, J. Y. Park, and W. Scheid, Nuclear Molecules (World Scientific, Singapore, 1995)

[3] M. Aliotta *et* al., Zeit. Phys. A, 355, (1996) 119

[4] S. Cherubini *et* al., Zeit. Phys. A, 357, (1997) 291

[5] M. Lattuada *et* al., Nuovo Cimento, 110 (1997) 1007

[6] E. Costanzo *et* al., Eur. Phys. Journ. A, 5 (1999) 69

[7] A. Tumino *et* al., Eur. Phys. Journ. A, 12 (2001) 327

[8] E. Garcia-Berro *et* al., Astrophys. J. 286, (1997) 765

[9] R.L. Cooper*et* al., Astrophys. J. 702, (2009) 660

[10] D. Page, A. Cumming, Astrophys. J. 635 (2005) L157

[11] M.G. Mazarakis, W.E. Stephens, Phys. Rev. C 7 (1973) 1280

[12] J.R. Patterson, H. Winkler, C.S. Zaidins, Astrophys. J. 157 (1969) 367

[13] H.W. Becker, K.U. Kettner, C. Rolfs, H.P. Trautvetter, Z. Phys. A 303 (1981) 305

[14] M.D. High, B. Cujec, Nucl. Phys. A 282 (1977) 181

[15] K.U. Kettner, H. Lorenz-Wirzba, C. Rolfs, Z. Phys. A 298 (1980) 65

[16] P. Rosales *et* al., Rev. Mex. Phys. 49 S4 (2003) 88

[17] L. Barrón-Palos *et* al., Nucl. Phys. A 779 (2006) 318

[18] L. Barrón-Palos *et* al. Eur. Phys. J. A Direct (2005), doi:10.1140/epjad/i2005-06-008-2

[19] E.F. Aguilera *et* al., Phys. Rev. C 73 (2006) 064501

[20] T. Spillane *et* al., Phys. Rev. Lett. 98, (2007) 122501

[21] C. Spitaleri *et* al., Phys. At. Nucl., **74**, (2011) 1763

[22] A. Tumino et al., *Few Body Syst.*, **54**, Issue 5-6 (2013) 745

[23] R. Tribble et al., 77, Issue: 10 (2014) 106901

[24] K.P. Artemov et al., Phys. Lett., 149 B (1984) 325

[25] R.W. Zurmuehle et al., Phys. Rev. C 49, (1994) 2549

[26] RG Pizzone *et al.*, *A&A* **438** (2005) 779

[27] S. Romano *et al.* , *Eur. Phys. J.* A **27** (2006) 221

[28] M La Cognata *et al.*, *Phys. Rev. Lett.* **101** (2008) 152501

[29] L Lamia *et al.*, *Il Nuovo Cimento* **31** N.4 (2008) 423

[30] A Tumino *et al.* , *Phys. Rev. Lett.* **98** (2007) 252502

[31] A Tumino *et al.*, *Phys. Rev.* C **78** (2008) 064001

[32] A Tumino *et al.*, *Phys. Lett.* B **700** (2011) 111

[33] A Tumino *et al. Phys. Lett.* B **705** (2011) 546

[34] A Tumino *et al. Astrophysical J.* **785** (2014) 96

[35] Li Chengbo *et al. Phys. Rev.* C **92** (2015) 025805

[36] Shapiro, I.S.*et al.*, 1965, *Nucl. Phys.* A, **61**, 353

[37] H. Assenbaum,K.L. Langanke and C. Rolfs, *Z. Phys.* A **327**, 461 (1987)

EPJ Web of Conferences **117**, 09005 (2016) DOI: 10.1051/epjconf/201611709005
NN2015

Studying astrophysical reactions with low-energy RI beams at CRIB

H. Yamaguchi[1], D. Kahl[1], S. Hayakawa[1,2], Y. Sakaguchi[1], Y. Wakabayashi[3,1], T. Hashimoto[4,1], S. Cherubini[2,5], M. Gulino[2,6], C. Spitaleri[2,5], G.G. Rapisarda[2,5], M. La Cognata[2], L. Lamia[5], S. Romano[2,5], S. Kubono[3,1], N. Iwasa[7], T. Teranishi[8], T. Kawabata[9], Y.K. Kwon[4], D.N. Binh[10,1], L.H. Khiem[11], N.N. Duy[12], S. Kato[13], T. Komatsubara[4], A. Coc[14], N. de Sereville[15], F. Hammache[15], G. Kiss[16], and S. Bishop[2,17]

[1] Center for Nuclear Study (CNS), University of Tokyo, Tokio, Japan
[2] INFN - Laboratori Nazionali del Sud, Catania, Italy
[3] RIKEN Nishina Center, Wako, Saitama, Japan
[4] Institute for Basic Science, Daejeon, Korea
[5] Dipartimento di Fisica e Astronomia, Università di Catania, Catania, Italy
[6] Università "Kore", Enna, Italy
[7] Department of Physics, Tohoku University, Tohoku, Japan
[8] Department of Physics, Kyushu University, Kyushu, Japan
[9] Department of Physics, Kyoto University, Kyoto, Japan
[10] 30 MeV Cyclotron Center, Tran Hung Dao Hospital, Hanoi, Vietnam
[11] Institute of Physics, Vietnam Academy of Science and Technology, Hanoi, Vietnam
[12] Dong Nai University, Dang Nai, Vietnam
[13] Department of Physics, Yamagata University, Yamagata, Japan
[14] Centre de Spectrometrie Nucleaire et de Spectrometrie de Masse, Orsay, France
[15] Institut de Physique Nucleaire, Orsay, France
[16] Institute for Nuclear Research (MTA-Atomiki), Debrecen, Hungary
[17] Technische Universität München, München, Germany

Abstract

Studies on nuclear astrophysics, nuclear structure, and other interests have been performed using the radioactive-isotope (RI) beams at the low-energy RI beam separator CRIB, operated by Center for Nuclear Study (CNS), the University of Tokyo. A typical measurement performed at CRIB is the elastic resonant scattering with the inverse kinematics. One recent experiment was on the α resonant scattering with ^{7}Li and ^{7}Be beams. This study is related to the astrophysical ^{7}Li/^{7}Be(α,γ) reactions, important at hot *p*-*p* chain and νp-process in supernovae. There have also been measurements based on other experimental methods. The first THM measurement using an RI beam has been performed at CRIB, to study the ^{18}F(p, α)^{15}O reaction at astrophysical energies via the three body reaction ^{2}H(^{18}F, α^{15}O)n. The ^{18}F(p, α)^{15}O reaction rate is crucial to understand the 511-keV γ-ray production in nova explosion phenomena, and we successfully evaluated the reaction cross section at novae temperature and below experimentally for the first time.

1 Introduction

CRIB [1,2] is a radioactive-isotope (RI) beam separator operated by Center for Nuclear Study (CNS), the University of Tokyo, installed at the RIBF facility of RIKEN Nishina Center. CRIB can produce low-energy ($<$ 10 MeV/u) RI beams by the in-flight technique, using primary heavy-ion beams accelerated at the AVF cyclotron of RIKEN (K=70). Most of the RI beams are produced via direct reactions such as (p, n), (d, p) and (^{3}He, n), taking place at an 8-cm-long gas target with a maximum pressure of 760 Torr. A cryogenic target system, in which the target gas can be cooled down to about 90 K, is currently available, and an intense ^{7}Be beam of 2×10^8 pps was produced using the system [3]. One main feature of the target system is the forced circulation of the target gas. We have found that the circulation of the target gas at a rate of 55 standard liters per minute (slm) was effective in eliminating the density reduction, caused by heat deposition of the beam. Solid targets such as beryllium foils can also be used as the production target. The secondary beam is purified with a magnetic analysis using dipole magnets, and with a Wien filter, which separates the beams according to their velocities. For relatively light RI beams such as ^{7}Be, we obtained a purity close to 100% after the Wien filter. The Wien filter is operated with high voltages of $\pm$50–100 kV, supplied for a pair of 1.5-m long electrodes with a gap of 8 cm. For a stable operation at a higher voltage, we are making improvements on the insulators and other parts of the system.

A list of typical parameters of RI beams produced at CRIB is found in [4].

The low-energy RI beams at CRIB are particularly suitable for studies on astrophysical reactions and nuclear resonant structure. An experimental method extensively used is the thick-target method in inverse kinematics [5]. In that method, the beam energy is degraded in a thick reaction target, and reactions occur at various center-of-mass energies. We detect light particles emitted after reactions, and reconstruct the kinematics. This method has several advantages, namely, (a) using inverse kinematics, we can study reactions with short-lived RI which cannot be used as the target, (b) we can perform simultaneous measurements of cross sections at various excitation energies without varying the incoming RI beam energy, and (c) when the beam is stopped in the target, we can perform measurements at 180° in center-of-mass angle, where the Coulomb scattering is minimal. Many fruitful results have been obtained at CRIB with this method [6–16].

A recent major topic of our interest is the measurement of alpha-induced reactions. Several (α, p) reactions, such as ^{14}O(α, p) [17], ^{11}C(α, p), ^{21}Na(α, p), ^{18}Ne(α, p), ^{30}S(α, p), and ^{22}Mg(α, p), have been studied at CRIB. For some of the recent measurements, an active target, referred to as "GEM-MSTPC" [18], has been used. Measurements on the elastic resonant scatterings of a helium target and heavy-ion beams such as ^{14}O, ^{21}Na, ^{7}Li [12] and ^{7}Be have also been performed. These measurements provide information on astrophysical (α, γ) reaction rates, and are also very suitable for studying nuclear cluster structure of the compound nuclei.

There are other projects for the determination of the astrophysical reactions using indirect methods with RI beams. The indirect measurement of the ^{12}N(p, γ) reaction, which is a key reaction to synthesize nuclei heavier than carbon, was performed by measuring ^{12}N(d, n) reaction in inverse kinematics. Using the asymptotic normalization coefficient (ANC) method, the reaction rate was reevaluated [19]. We have performed an indirect reaction measurement using the Trojan horse method (THM) [20–22], as discussed later.

2 ^{7}Be+α elastic resonant scattering

A typical example of the resonant scattering experiment is the one on the ^{7}Be+α scattering measurement. The measurement of the ^{7}Be+α scattering allows us to evaluate the rate of the ^{7}Be(α, γ) reaction, which is considered to play an important role in the hot p-p chain and related reaction sequences [23]. Several reaction sequences including the ^{7}Be(α, γ) reaction should take place in some high-temperature environments at $T_9 > 0.2$, where T_9 is the temperature in GK. In the νp-process in core-collapse supernovae [24], the ^{7}Be(α, γ) reaction may contribute as much as the triple-α process to the synthesis of elements heavier than boron at the relevant temperature of T_9 = 1.5–3, according to a theoretical calculation [25]. The Gamow energy window at the highest temperature T_9 = 3 corresponds to the excitation energy E_{ex}=8.2–9.6 MeV in ^{11}C. By our study, the resonant reaction rate should be evaluated more precisely by determining α widths for the resonances at high temperatures.

In 2010, we performed the measurement of the ^{7}Be+α resonant elastic and inelastic scatterings with the thick-target method in inverse kinematics at CRIB [26]. A low-energy ^{7}Be beam at 14.7 MeV was produced using a 2.3-mg/cm^2-thick hydrogen gas target and a primary ^{7}Li beam at 5.0 MeV/u. The typical ^{7}Be beam intensity used in the measurement was 1–2 $\times$ 10^5 per second at the secondary target, and the main measurement using a thick helium-gas target was performed for 4 days, injecting 2.9 $\times 10^{10}$ ^{7}Be particles into the target.

correspond to the resonance structure in ^{11}C, We obtained an excitation function of the elastic scattering with several peaks, corresponding to the resonance structure in ^{11}C. The obtained excitation is shown in the left panel of Figure 1. cross section. The curve is by an R-matrix calculation. An R-matrix analysis was performed to deduce the parameters of the resonances, as the calculated curve also shown in the figure. A similar measurement was independently carried out by M. Freer *et al.* at other facilities [27], but our measurement included γ-ray detection to identify inelastic scattering events, and several differences were found in the obtained spectra [26].

The resonances observed in the present work might contribute to the astrophysical ^{7}Be(α, γ)^{11}C reaction rate at high temperature, $T_9 > 1.5$. We calculated the resonant reaction rate and compare it with the total reaction rate evaluated in NACRE [28, 29]. In the evaluation reported in NACRE, only 2 resonances at 8.1045 and 8.420 MeV are included. These two resonances dominate the reaction rate $N_A\langle\sigma v\rangle$ up to the temperature $T_9 \sim 3$, and a Hauser-Feshbach calculation was used to provide the reaction rate at higher temperatures. The resonant reaction rates were calculated for three resonances using analytical formula described in [28], and plotted in Fig. 1.

r resonances, but none of In conclusion, the resonances at 8.90 MeV and 9.20 MeV have a possibility to give significant contributions to the reaction rate for T_9 = 1.5–3, although they are unlikely to be more than the contribution of the 8.420-MeV resonance, which dominates the reaction

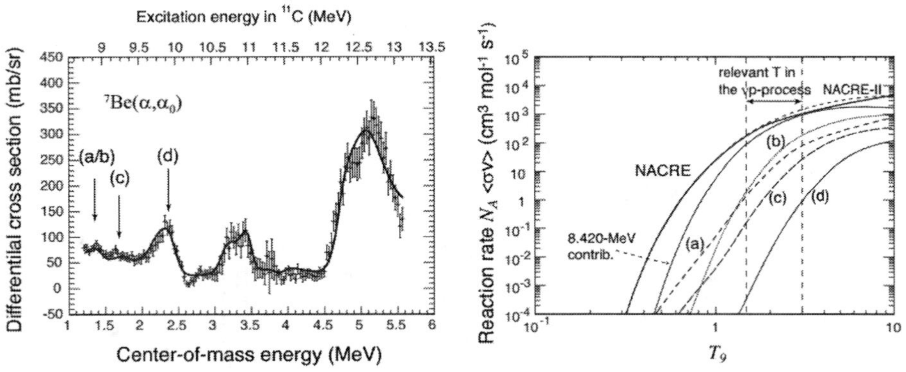

Figure 1: Excitation function of ^{7}Be+α elastic scattering with an R-matrix fit curve (left panel) and resonant reaction rate of ^{7}Be(α, γ) for the 8.90, 9.20, and 9.97-MeV resonances, calculated by the analytical formula. The evaluation by NACRE and NACRE-II are shown for comparison. The contribution by the 8.420-MeV resonance, included in NACRE, is also shown.

rate.

3 Study with the Trojan horse method and an RI beam

The first measurement using the THM with an RI beam has recently been performed at CRIB [30]. The measurement was to study the ^{18}F(p, α)^{15}O reaction at low energies relevant to astrophysics via the three body reaction ^{2}H(^{18}F, α^{15}O)n. The ^{18}F(p, α)^{15}O reaction rate is particularly responsible for the 511-keV γ ray emission in nova explosion phenomena. The relevant temperature in novae is corresponding to the Gamow energy of 100–400 keV. A direct measurement of the ^{18}F(p, α) reaction was recently performed at TRIUMF [31], reaching to thee energy as low as 250 keV, but the reaction rate below that energy has been totally unknown experimentally.

The measurement at CRIB was performed with a ^{18}F RI beam, of which maximum intensity was 2 $\times 10^6$ pps. A CD_2 target with a thickness of 150 μg/cm^2 was irradiated by the beam, and 3-body reaction events were detected by a silicon detector array, referred to as ASTRHO, with additional double-sided silicon strip detectors. The analysis was performed by following a standard scheme of the Trojan horse method. 3-body events were selected according to the kinematical relationship of the detected charged particles, and converted into a Q-value spectrum. The momentum distri-

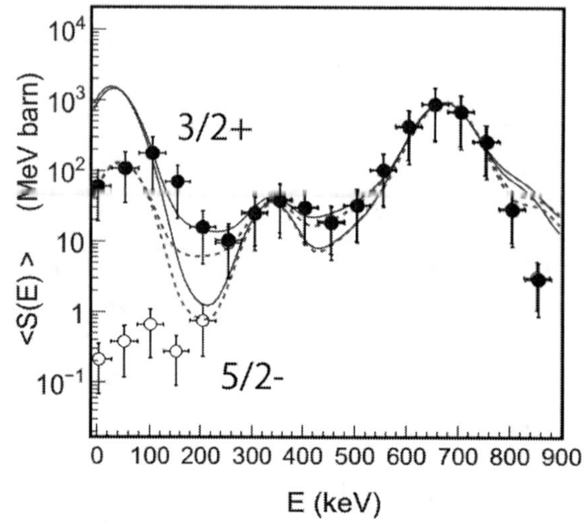

Figure 2: The astrophysical S-factor of the $^{18}F(p, \alpha)^{15}O$ reaction [30]. The filled and open circles show the S-factor obtained by the THM experiment at CRIB assuming $J^{\pi} = 3/2^{+}$ and $5/2^{-}$ for the 6460-keV resonance, respectively. The solid and dashed curves are calculations discussed in [31].

bution for the p-n intercluster motion was in agreement with a Hulthén function, which proves the dominance of the quasi-free mechanism. Then the three-body cross section was converted into the two-body cross section, using the momentum distribution of the p-n intercluster motion and a kinematical factor (see [30] for further details). We successfully evaluated the reaction cross section at the novae temperature and even below experimentally for the first time, as shown in Fig. 2. Because of the limited statistics due to the RI beam experiment, we could not uniquely determine the spin and parity (J^{π}) of each level from the angular distribution. As a result, we had to assume J^{π} for some of the resonances to obtain the total reaction cross section. The S-factor greatly deviated according to the J^{π} assignment for the 6460-keV resonance, which indicates the importance of the determination of J^{π}. Thereby we are planning to perform another measurement with greater statistics in 2015.

Acknowledgements

The experiments were performed at RI Beam Factory operated by RIKEN Nishina Center and CNS, the University of Tokyo. We are grateful to the

RIKEN and CNS accelerator staff for their help. This work was partly supported by JSPS KAKENHI (No. 21340053,25800125) from the Ministry of Education, Culture, Sports, Science and Technology (MEXT) of Japan.

References

[1] S. Kubono, et al., Eur. Phys. J. A13 (2002) 217.

[2] Y. Yanagisawa, et al., Nucl. Instrum. Meth. Phys. Res., Sect. A 539 (2005) 74–83.

[3] H. Yamaguchi, et al., Nucl. Instrum. Meth. Phys. Res., Sect. A 589 (2008) 150–156.

[4] D. Kahl, et al., AIP Conference Proceedings 1594 (2014) 163–170.

[5] K. P. Artemov, et al., Sov. J. Nucl. Phys 52 (1990) 408.

[6] M. Notani, et al., Nucl. Phys. A 746 (2004) 113c–117c.

[7] T. Teranishi, et al., Phys. Lett. B 556 (2003) 27–32.

[8] T. Teranishi, et al., Phys. Lett. B 650 (2007) 129–134.

[9] J. J. He, et al., Phys. Rev. C 76 (2007) 055802.

[10] H. Yamaguchi, et al., Phys. Lett. B 672 (2009) 230–234.

[11] J. J. He, et al., Phys. Rev. C 80 (2009) 015801.

[12] H. Yamaguchi, et al., Phys. Rev. C 83 (3) (2011) 034306.

[13] H. S. Jung, et al., Phys. Rev. C 85 (2012) 045802.

[14] J. J. He, et al., Phys. Rev. C 88 (2013) 012801.

[15] S. J. Jin, et al., Phys. Rev. C 88 (2013) 035801.

[16] L. Y. Zhang, et al., Phys. Rev. C 89 (2014) 015804.

[17] A. Kim, et al., Phys. Rev. C 92 (2015) 035801.

[18] T. Hashimoto, et al., Nuclear Instruments and Methods in Physics Research Section A 556 (1) (2006) 339 – 349.

[19] B. Guo, et al., Phys. Rev. C 87 (2013) 015803.

[20] G. Baur, Physics Letters B 178 (2–3) (1986) 135–138.

[21] S. Cherubini, et al., Astrophys. J. 457 (1996) 855.

[22] C. Spitaleri, et al., Phys. Rev. C 69 (2004) 055806.

[23] M. Wiescher, et al., Astrophys. J. 343 (1989) 352–364.

[24] C. Fröhlich, et al., Astrophys. J. 637 (2006) 415.

[25] S. Wanajo, et al., Astrophys. J. 729 (2011) 46.

[26] H. Yamaguchi, et al., Phys. Rev. C 87 (3) (2013) 034303.

[27] M. Freer, et al., Phys. Rev. C 85 (2012) 014304.

[28] C. Angulo *et al.*, Nucl. Phys. A 656 (1999) 3–187.

[29] Y. Xu, et al., Nucl. Phys. A 918 (2013) 61–169.

[30] S. Cherubini, et al., Phys. Rev. C 92 (2015) 015805.

[31] C. E. Beer, et al., Phys. Rev. C 83 (2011) 042801.

Neutron star structure from a quark-model baryon-baryon interaction

K. Fukukawa,[1] M. Baldo,[2] G. F. Burgio,[2] and H.-J. Schulze[2]

[1] Research Center for Nuclear Physics, Osaka University, 10-1, Mihogaoka, Osaka, 567-0047, Japan
[2] INFN, Sezione di Catania, Via Santa Sofia 64, I-95123, Catania, Italy

Abstract

We derive the equation of state (EOS) of nuclear matter from a realistic constituent quark model for the nucleon-nucleon interaction. We use the Brueckner-Bethe-Goldstone approach with the inclusion of the three hole-line contribution. We find that the resulting EOS reproduces correctly the saturation point, moreover the symmetry energy at low density, its slope, and the incompressibility turn out to be compatible with phenomenology. We calculate the mass-radius relation for neutron stars, and find maximum values close to two solar masses, in agreement with recent observational data.

1 Introduction

One of the most fundamental issues in theoretical nuclear physics is the understanding of nuclear many-body systems starting from realistic nucleon-nucleon (NN) and three-nucleon ($3N$) interactions. In particular, it is essential to devise an elementary interaction among nucleons which is able to describe consistently both few-body systems and nuclear matter properties, such as the experimental binding energy of light nuclei and the correct saturation point.

Quark-Model (QM) baryon-baryon (BB) interactions have been modeled to describe the short-range repulsion with the resonating group method for

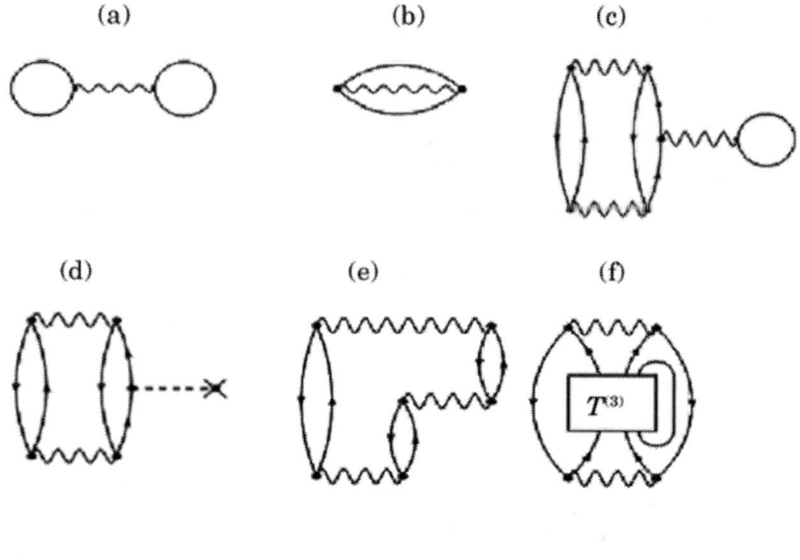

Figure 1: Different Goldstone diagrams contributing to the nuclear matter EOS. The wavy line indicates the Brueckner G-matrix. For details, please see the text.

two $(0s)^3$ three-quark clusters. One of the most developed version of the QM BB interaction, called fss2 [1], has been successfully applied to the three- and four-nucleon systems without resorting to $3N$ interactions [2]. In this model, the role of $3N$ interactions may be not so large as usually considered.

In the following we will discuss the EOS of nuclear matter [3] derived with the Bethe-Brueckner-Goldstone (BBG) approach, where the NN interaction is obtained using the fss2. In §2, we explain the theoretical framework and show our numerical results both for the equation of state and the structure of neutron stars. The final section is devoted to the conclusions.

2 Nuclear matter EOS and neutron star structure

We analyze the EOS within the many-body microscopic approach based on the BBG expansion up to three hole-line (THL) level. We show in Fig. 1 the Goldstone diagrams whose contribution has been calculated following Refs. [4,5]. Diagrams (a) and (b) correspond to the Brueckner-Hartree-Fock (BHF) calculations, which include the two hole-line contribution, whereas the sum of the other diagrams gives the THL contribution. In Fig. 2 we display for symmetric matter each contribution to the EOS obtained in the so-called continuous choice (CC) for the single-particle (s.p.) potential. For details concerning the theoretical approach, the reader is referred to

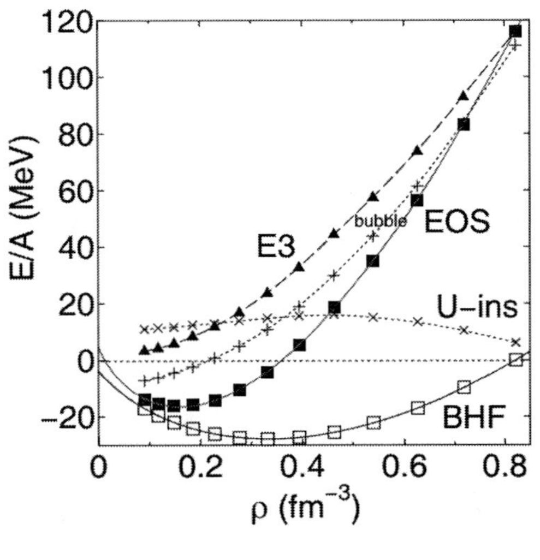

Figure 2: The binding energy as a function of nucleon density is shown for symmetric nuclear matter in the continuous choice. The contribution from the BHF calculation [open squares], the bubble diagram [Fig. 1(c), '+' symbols], the corresponding U-insertion diagram [Fig. 1(d), crosses], the total contribution from the THL diagrams [Figs. 1(c)-(f), black triangles], and the resultant EOS [black squares], are plotted separately. The contributions from the ring diagram [Fig. 1(e)], and the higher-order term [Fig. 1(f)] are not shown, because they are small.

Refs. [3, 5]. The peculiar feature is that THL give large contributions at high densities in the fss2 case, and this is at variance with the results found in the case of the Argonne v_{14} or v_{18} NN potentials, where the THL are small [6]. We found that the THL contribution is important to predict the correct saturation point [$\rho_0 \approx 0.157\,\mathrm{fm}^{-3}$, $(E/A)(\rho_0) \approx -16.3\,\mathrm{MeV}$], and an incompressibility $K = 219\,\mathrm{MeV}$ compatible with phenomenological data. Around saturation density, the contributions from the U-insertion diagram are appreciable, whereas at high densities the contributions from the bubble diagram rapidly increase. Since the THL contribution is larger than the BHF contribution at high densities, the BBG expansion becomes questionable beyond a density range where it can be expected that strange baryonic matter and/or quark deconfinement may appear.

In Fig. 3 we show the fss2 EOS for symmetric (SNM) and pure neutron matter (PNM), both compared with other theoretical predictions. In particular, we show results obtained with the well-known APR EOS (dot-dashed curve) [7], and with the BHF approach (dotted curve) [8]. In those cases the Argonne v_{18} NN potential is used along with the Urbana model for $3N$ forces [9]. The other curves show results obtained with BHF and

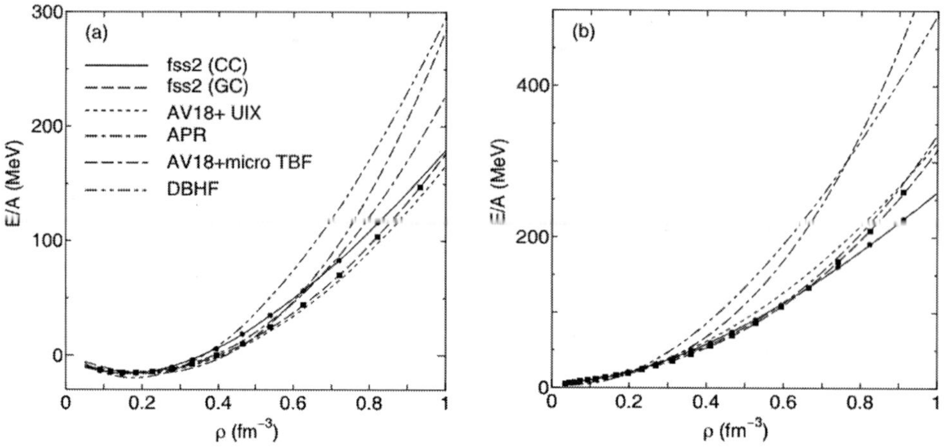

Figure 3: EOS of (a) symmetric nuclear matter and (b) pure neutron matter. The calculations in the continuous and gap choice are denoted by (CC) and (GC), respectively.

microscopic $3N$ forces [10], and the relativistic DBHF [11]. We observe that the fss2 EOS is quite soft at high densities, and that all EOS agree well in symmetric matter up to nucleonic density $\rho \sim 0.5$ fm^{-3}. The fss2 EOS in the continuous and the gap choice agree well with each other up to nucleonic density $\rho \sim 0.7$ fm^{-3}, so that the final EOS turns out to be independent of the choice of the s.p. potential.

For studying the neutron star structure, we first calculate the composition $\{\rho_i(\rho)\}$ and the EOS for asymmetric, neutrally charged and beta-stable matter. We adopt a conventional description of stellar matter as composed of neutrons, protons, electrons, and muons. The EOS reads

$$P(\rho) = \rho^2 \frac{d}{d\rho} \frac{\varepsilon(\{\rho_i(\rho)\})}{\rho} = \rho \frac{d\varepsilon}{d\rho} - \varepsilon \,, \tag{1}$$

being ϵ the energy density. For non-rotating and spherically-symmetric neutron stars, the Tolman-Oppenheimer-Volkoff equations [12] must be solved

$$\begin{aligned} \frac{dP}{dr} &= -\frac{Gm\varepsilon}{r^2} \frac{(1+p/\varepsilon)(1+4\pi r^3 p/m)}{1-2Gm/r} \,, \\ \frac{dm}{dr} &= 4\pi r^2 \varepsilon \,, \end{aligned} \tag{2}$$

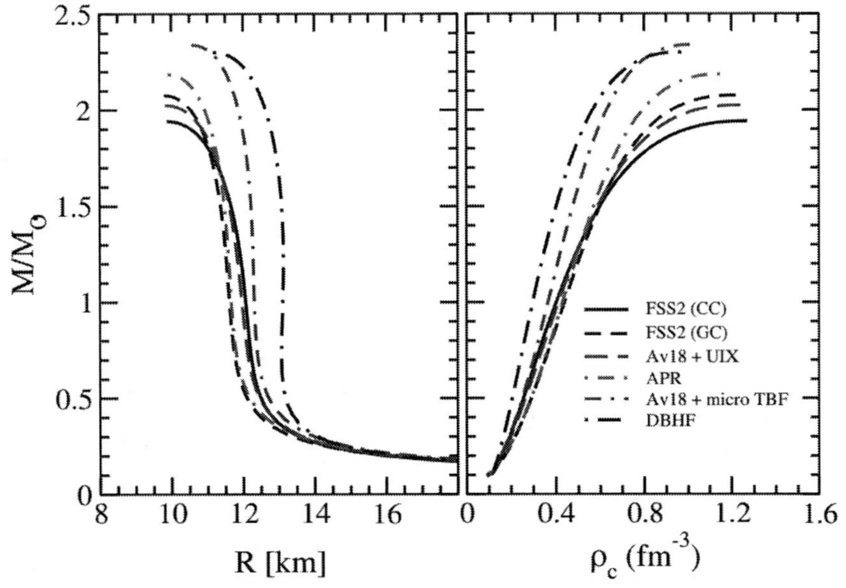

Figure 4: Neutron star mass as a function of radius (left panel) and central baryon density (right panel) for several EOS.

where G is the gravitational constant. We integrate Eqs. (2) from the center to the surface $r = R$ and obtain the gravitational mass $M_G = m(R)$. In Fig. 4, we show the mass-radius relation (left panel) and the mass-central density relation (right panel) for the cases discussed above. The fss2 EOS gives slightly different maximum masses for the GC and CC approximations, in line with their different stiffness at high density. However, the range of the maximum mass predicted by fss2 turns out to be compatible with the largest mass observed up to now, which is $(2.01 \pm 0.04)M_\odot$ [13].

3 Summary

We have derived the fss2 EOS within the BBG approach up to the THL level of approximation. It provides a saturation point compatible with experimental constraints and reproduces the binding energy of three- and four-nucleon systems without introducing $3N$ interactions. We found that the $3N$ correlations are important to predict the saturation point. The obtained neutron star maximum mass is about 2 solar masses.

Acknowledgement

The authors wish to acknowledge the "Newcompstar" COST Action MP1304.

References

[1] Y. Fujiwara et al., Prog. Part. Nucl. Phys. **58**, 439 (2007).

[2] Y. Fujiwara, Y. Suzuki, M. Kohno, and K. Miyagawa, Phys. Rev. C **77**, 027001 (2008); Y. Fujiwara and K. Fukukawa, Few-Body Systems **54**, 2357 (2013); Y. Fujiwara, Few-Body Systems **55**, 993 (2014).

[3] M. Baldo, and K. Fukukawa, Phys. Rev. Lett. **113**, 242501 (2014); K. Fukukawa et al., arXiv:1507.07288 [nucl-th] (2015).

[4] R. Rajaraman and H. Bethe, Rev. Mod. Phys. **39**, 745 (1967); B. D. Day, Phys. Rev. C **24**, 1203 (1981), Phys. Rev. Lett. **47**, 226 (1981).

[5] M. Baldo, in *Nuclear Methods and the Nuclear Equation of State*, International Review of Nuclear Physics Vol. **8**, pp. 1 (World Scientific, 1999).

[6] H. Q. Song, M. Baldo, G. Giansiracusa, and U. Lombardo, Phys. Rev. Lett. **81**, 1584 (1998).

[7] A. Akmal, V. R. Pandharipande, and D. G. Ravenhall, Phys. Rev. C **58**, 1804 (1998).

[8] G. Taranto, M. Baldo, and G. F. Burgio, Phys. Rev. C **87**, 045803 (2013).

[9] J. Carlson, V. R. Pandharipande, and R. B. Wiringa, Nucl. Phys. **A401**, 59 (1983).

[10] W. Zuo, A. Lejeune, U. Lombardo, and J.-F. Mathiot, Nucl. Phys. **A706**, 418 (2002); Eur. Phys. J. **A14**, 469 (2002).

[11] T. Gross-Boelting, C. Fuchs, and A. Faessler, Nucl. Phys. **A648**, 105 (1999).

[12] S. L. Shapiro and S. A. Teukolsky, *Black Holes, White Dwarfs, and Neutron Stars* (John Wiley & Sons, New York, 1983).

[13] J. Antoniadis et al., Science **340**, 1233232 (2013).

EPJ Web of Conferences **117**, 09007 (2016) DOI: 10.1051/epjconf/201611709007
NN2015

Nuclear magics at explosive magnetization

V. N. Kondratyev

Physics Department, Taras Shevchenko National University of Kiev,
03022-UA Kiev, Ukraine
Bogolubov LTP, JINR, 141980, Dubna, Russia

Abstract

Effects of ultra-strong magnetization in creation of iron group nuclides are considered by employing arguments of nuclear statistical equilibrium. Nuclear magnetic reactivity is demonstrated to enhance the portion of titanium product due to magnetic modification of nuclear structure. The results are corroborated with an excess of ^{44}Ti revealed from the Integral mission data.

1 Introduction

Ultrastrong magnetic fields exceeding *teratesla* (TT) arise at core-collapse supernovae (SNe) [1, 2] and in heavy ion collisions [3]. Nuclides produced in such processes contain an information on matter structure and explosion mechanisms. In this contribution we analyze possibilities for using radionuclides to probe internal regions of respective sites. Magnetization of hot dense matter makes plausible explosion mechanism and can leave its trace at nucleosynthesis [1, 2]. Such fields (i.e., larger than 0.1 TT) can modify the structure and properties of atomic nuclei (see. [1, 2, 4–6] and refs. therein), that requires to consider a possible influence of magnetism on structure, transformation and transmutation of radionuclides. Employing relevant data in analyses of nucleosynthesis and nuclear reaction chains may provide more detailed information as on dynamics and mechanism of supernova (SN) explosions as on heavy ion collisions.

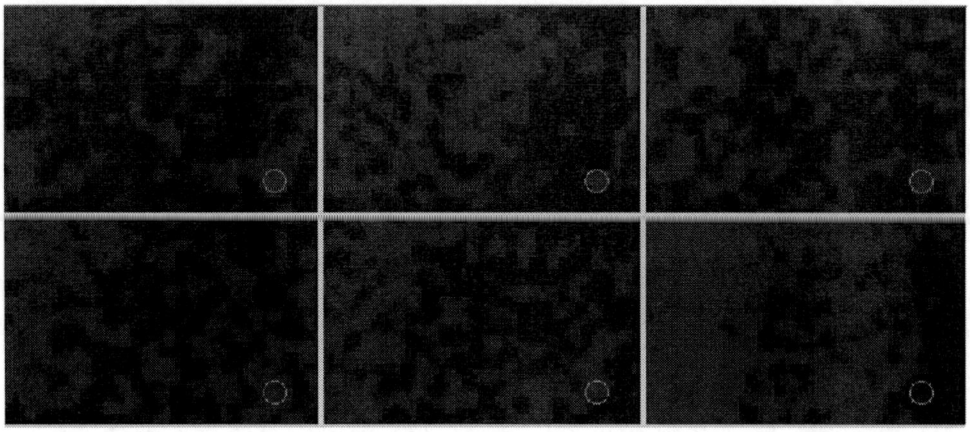

Figure 1: Direction (pixel number) dependence of the registered gamma-ray flux at different energy ranges; top: left - 20 – 50 keV, middle - 50 – 67 keV, right - 67 – 70 keV, bottom: left - 70 – 77 keV; middle –77 – 82 keV, right - 82 –– 100 keV; for the Cassiopeia region. SNR CAS A, (J2000) R.A. 350.86^o, decl. 58.81^o, indicated by circle.

In present study we demonstrate that the magnetic effects in the nuclear binding energy lead to an increase of the titanium portion in a synthesis of nuclides of mass numbers close to the iron "peak". Consequently, the characteristic lines of respective nuclei in spectra of astrophysical objects are considerably strengthened and become noticeable allowing for an analysis of synthesized elements. The radioactive decay chain ^{44}Ti→ ^{44}Sc→ ^{44}Ca gives rise to an emission of lines with energies of 67.9 keV and 78.4 keV (from ^{44}Sc*) and 1157 keV (from ^{44}Ca*) of approximately equal intensity. The respective image-mosaics from Integral data (e.g., [1, 2]) for the Cassiopeia region in various energy ranges of registered photons are presented in Fig. 1. The color (brightness) is proportional to the gamma-quanta flux: as larger the flux as lighter (brighter) the color of a pixel. As is seen the SNR CAS A gives the brightest spot for energies matching ^{44}Sc lines,

Then ^{44}Ti half-life, about 60 years, allows to determine this isotope initial mass in SN remnants. Table. 1 shows the observational results for the total mass of ^{44}Ti nuclide synthesized in SN explosions [6]. These values are significantly larger as compared to model predictions, see. [7], giving the mass of initially synthesized ^{44}Ti, $M_{\rm Ti} \sim 10^{-5} M_{\rm Sun}$ (in solar masses $M_{\rm Sun}$) in an absence of magnetic effects.

Table 1: Volume M_{Ti} of nuclides ^{44}Ti (in solar masses M_{Sun}) initially synthesized in young SNe, CAS A and SN1987A, see [6] and refs. therein.

SN	$M_{\mathrm{Ti}}/10^{-5}M_{\mathrm{Sun}}$
CAS A	$3.3^{+0.9}_{-0.7}$
SN1987A	3.1 ± 0.8

2 Nucleosynthesis in strong magnetic field

The nuclear statistical equilibrium (NSE) approach is used very successfully for a description of abundances of iron group and nearby nuclides for over half a century, cf. [1,2,6,7]. We briefly recall that at NSE conditions abundance of i-th nuclear particle Y_i (e.g., nucleons, nuclei, electrons) at a temperature T is determined by the respective chemical potential from the condition of entropy S extremum. At considered parameters of magnetized matter, i.e. magnetic field strengths H <100 TT, the yield Y_i of atomic nucleus i is mainly determined by corresponding binding energy B_i as: $Y_i \approx \exp\{-B_i/kT\}$, [6,7], for more extended discussion of statistical models see [8]. The magnetization brings, however, a sensitivity of nuclide creation processes to a projection of respective magnetic moments on the field direction. When accounting for thermal fluctuations of the spin projection on the field vector in corresponding magnetic components of the partition function G, the dependence of the relative output for nucleosynthesis products ($y = Y(H)/Y(0)$) on the magnetic field strength H with statistical accuracy can be written as

$$y \simeq G_{\mathrm{n}}^{-N} G_{\mathrm{p}}^{-Z} G_i \exp\{\Delta B/kT\}. \quad (1)$$

Here the spin-magnetic part in the partition function is represented by $G_i = \sum_M \exp\{g_i M \omega_{\mathrm{L}}/kT\}/(2I_i+1)$ and is determined by the energy for an interaction of the magnetic field with magnetic moments of atomic nuclei $i =^A_N Z$ with spin I_i, free N neutrons and Z protons, $g_{n(p,i)}$ gives the g-factor of a neutron (proton, atomic nucleus), and $\omega_{\mathrm{L}} = \mu_{\mathrm{N}} H$ with nucleon magneton μ_{N}. At conditions $\omega_{\mathrm{L}} \gg kT$ Eq. (1) is reduced to simplified exponential limit [1,2].

Since magnetic effects are determined by nuclear shell structure [1,2,4–6] we identify magnetic field dependence of binding energy $B(H)$ with a change of shell correction energy C (i.e. $\Delta B(H) = C(H) - C(0)$). Employing the Nilsson model with harmonic oscillator (HO) spectrum of a frequency $\omega_0 = 41/A^{1/3}$ MeV and using Eq. (1) we examine features of nuclide composition in ultra-magnetized matter. As is seen in Fig. 2 the nuclide relative

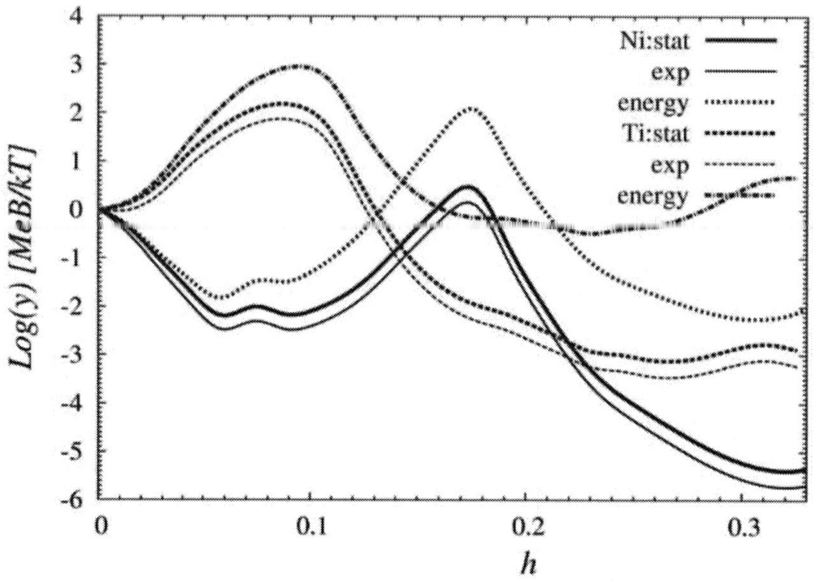

Figure 2: Magnetic field dependence ($h = \omega_L/\omega_0$) of relative yield for ^{56}Ni and ^{44}Ti. Statistical accuracy given by Eq. (1) is indicated by notation *stat*, *exp* denotes exponential limit, and *energy* corresponds to an effect of binding energy.

yields display oscillations as a function of field strength. The production of Ti grows and Ni is reduced with increasing magnetization at small field strengths where various approximations give very similar results.

3 Conclusion

Creation of atomic nuclei at conditions of nuclear statistical equilibrium is investigated for the ultra-magnetized matter. For iron group nuclides the magnetic modification of nuclear structure shifts a maximum of nucleosynthesis products towards smaller mass numbers approaching titanium. Magnetic effects in nucleosynthesis are favorably compared to observational Integral data. However, detail comparison requires accounting for realistic field geometry [6]. Thermal effects can also affect structural and collective properties of atomic nuclei at high temperatures, see [9]. Finally, we notice that similar magnetic effects in atomic clusters [10] will also lead to a shift of electronic magic numbers and change the mass distribution.

This work is supported in part by scientific data center of Integral mission and SCOPES IZ73Z0-152485. The author (V.N.K.) thanks JINR (Dubna, Russia) for warm hospitality and financial support.

DOI: 10.1051/epjconf/201611709007

References

[1] Kondratyev V.N. *Phys. Atom. Nucl.,* **75** (2012) 1368.

[2] Kondratyev V.N. *Eur. Phys. J.* A **50** (2014) 7..

[3] Voronyuk V. et al *Phys. Rev.* C **83** (2011) 054911..

[4] Kondratyev V.N. *Phys. Rev. Lett.* **88** (2002) 221101.

[5] Kondratyev V.N. *Phys. Rev.* C **69** (2004) 038801.

[6] Kondratyev V.N., Korovina Yu.V. *JETP Letters* **102** (2015) 131..

[7] Woosley S.E. et al *Rev. Mod. Phys.* **74** (2002) 1015.

[8] Kondratyev V.N. et al *J. Chem. Phys.* **106** (1997) 7766.

[9] Kondratyev V.N., Di Toro M. *Phys. Rev.* C **53** (1996) 2176.

[10] Kondratyev V.N., Lutz H.O. *Eur. Phys. J.* D **9** (1999) 483.

EPJ Web of Conferences **117**, 09008 (2016) DOI: 10.1051/epjconf/201611709008
NN2015

The Trojan Horse Method as a tool for investigating astrophysically relevant fusion reactions

L. Lamia[1], C. Spitaleri[1,2], E. Tognelli[3,4], S. Degl'Innocenti[3,5], R.G. Pizzone[2], P.G. Prada Moroni[3,5]

[1] Dipartimento di Fisica e Astronomia, Università di Catania, Catania, Italy
[2] INFN - Laboratori Nazionali del Sud, Catania, Italy
[3] INFN Sezione di Pisa, Pisa, Italy
[4] Dipartimento di Fisica, Università di Roma Tor Vergata, Roma, Italy
[5] Dipartimento di Fisica, Università di Pisa, Pisa, Italy

Abstract

The Trojan Horse Method (THM) has been largely adopted for investigating astrophysically relevant charged-particle induced reactions at Gamow energies. Indeed, THM allows one to by pass extrapolation procedures, thus overcoming this source of uncertainty. Here, the recent THM results and their impact in astrophysics are going to be discussed.

1 Introduction

Among the deduced cosmic elemental abundances, the ones relative to the light element lithium, beryllium and boron (LiBeB) play a crucial role for understanding and constraining cosmology or stellar physics, as firstly highlighted in the B^2FH paper [1]. In the framework of stellar nucleosynthesis and models, the combined study of LiBeB gives an unique opportunity for understanding stellar structure and mixing phenomena, because of their different fragility against (p,α) reactions at different stellar depths [2]. Being

Table 1: Some of the available values for electron screening potential U_e, for both theoretical and experimental determinations. The third column gives the ratio between experimental and theoretical values to highlight the strong discrepancy still present.

Reaction	$U_{theor.}$ (eV)	$U_{exp.}$ (eV)	Ratio	Reference
^{2}H(d,p)^{3}H	14	13.2±4.3	~ 0.95	[6]
^{6}Li(p,α)^{3}He	186	440±150	~2.36	[4]
^{6}Li(d,α)^{4}He	186	330±120	~1.77	[4]
^{1}H(^{7}Li,α)^{4}He	186	300±160	~1.74	[4]
^{2}H(^{3}He,p)^{4}He	65	109±9	~1.68	[8]
^{3}He(^{2}H,p)^{4}He	120	219±7	~1.82	[8]
^{1}H(^{9}Be,α)^{6}Li	240	900±50	~3.75	[9]
^{1}H(^{11}B,α)^{8}Be	340	430±80	~1.26	[10]
^{1}H(^{17}O,α)^{14}N	594	1356±1037	~2.28	[7]

these reactions ignited at temperatures of few 10^6 K, experimental nuclear astrophysics has to use often extrapolation procedures to access the corresponding Gamow energy peak. Extrapolations are in turn inevitably affected by several uncertainties such as the currently not-well-understood *electron screening effects* [3]. These effects have been largely studied in the past, as in [4], or more recently in [5], even if they appear far to be completely understood since the current available theoretical dynamics models (i.e. adiabatic approximation) largely underestimate the electron screening potential values with respect those measured in terrestrial laboratories, as clearly visible from the values given in Table 1.

To overcome such difficulties related to both the very low signal-to-noise ratio (as due to Coloumb penetrability in the entrance channel) and the electron screening phenomenon, the Trojan Horse Method (THM) has been largely applied for measuring the bare nucleus S(E)-factor for astrophysically relevant reactions, being its power the capability of accessing to the bare-nucleus S(E)-factor measurement without any kind of extrapolation [11–13]. THM allows one to extract the bare-nucleus cross-section of a charged-particle induced reaction a+x→c+C at astrophysical energies free of Coulomb suppression, by properly selecting the quasi-free (QF) contribution of an appropriate reaction a+A→c+C+s, performed at energies well above the Coulomb barrier, where the nucleus A has a dominant x⊕s cluster configuration. Usually, deuteron has been used in THM application because its obvious p-n structure mainly occurring in *s*-wave, even if a small component of *d*-wave could influence THM data (see [14] for instance).

DOI: 10.1051/epjconf/201611709008

The main advantage of the method is the possibility of accessing the a(x,c)C two-body reaction cross section measurement by performing a devoted experiment aimed at studying the quasi-free three-body reaction a+A→c+C+s. By using the Plane Wave Impulse Approximation (PWIA), the link between the cross sections of the two processes is given by

$$\frac{d^3\sigma}{dE_c C\Omega_c C\Omega_C} \propto KF \cdot \mid \Phi(\vec{p_s}) \mid^2 \cdot \left(\frac{d\sigma}{d\Omega}\right)\Big|_{a-x}^{HOES} \quad (1)$$

where KF represents the kinematical factor, $\mid \Phi(\vec{p_s}) \mid^2$ is the square of the momentum distribution for the $x - s$ relative motion inside the TH-nucleus A, and $\frac{d\sigma}{d\Omega}\Big|_{a-x}^{HOES}$ the half-off energy shell cross section. This last quantity represents the "bare-nucleus" cross section of interest for astrophysics, once it has been corrected for the penetrability through the Coulomb barrier and normalized to the available high-energy direct data. A complete review of the experimental approach to THM can be found in [15, 16] and references therein.

2 Astrophysical impact

In last years, THM helped in understanding several unresolved issues in astrophysics, as in the case of the stellar physics case discussed in [17]. In the last work of [18], the impact of the THM measurements for the (p,α) cross sections involving ^{9}Be and ^{10}B has been evaluated. In particular, the ^{9}Be(p,α)^{6}Li THM measurement has been firstly investigated in [19] while an improved measurement is given in [20]. The latter one has been used to firstly extract the bare-nucleus S_b(E)-factor and then for calculating the corresponding reaction rate, by means of standard formulas (see [18] for details). The THM S_b(E)-factor measurements are shown in the left panel of Fig.1, where the experimental points are shown as black-circles with their errors. The full line represents the corresponding fit leading to a value of S(0)=21.8±0.8 (MeV b) while the dashed line gives its enhancing due to an electron screening potential of U_e=676±86 eV [20]. The fit of the THM S(E)-factor has been then used in [18] for calculating the corresponding reaction rate through the standard formula

$$N_A < \sigma v >= \left(\frac{8}{\pi\mu}\right)^{\frac{1}{2}} \frac{N_A}{kT_9^{\frac{3}{2}}} \int_0^\infty S_b(E) e^{-2\pi\eta - \frac{E}{kT}} dE \ \ (\mathrm{cm^3\,mol^{-1}\,s^{-1}}) \quad (2)$$

where the temperature T_9 is expressed in units of 10^9 K and center of mass energy E in MeV. In Eq. 2 the bare-nucleus S(E)-factor, S_b(E), is the one

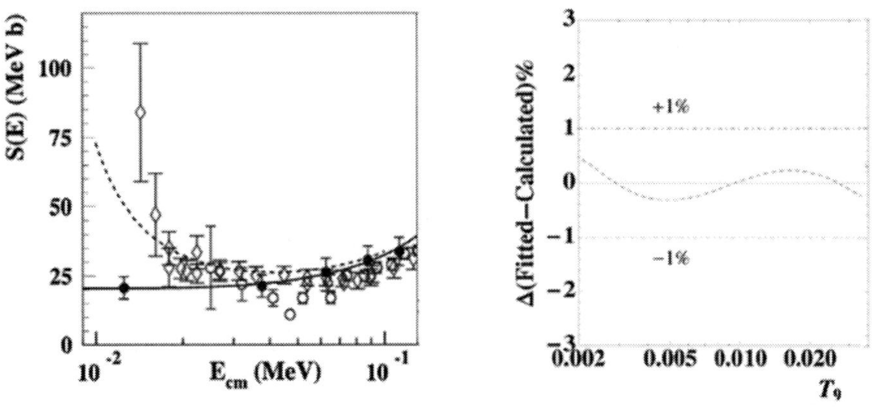

Figure 1: Left panel: The ^{9}Be(p,α)^{6}Li S(E)-factor as given by the THM measurements [20] (black points) compared with the direct measurements available in the NACRE compilation [21]. The full line represents the bare-nucleus THM S(E)-factor while the dashed one refers to its enhancing with the extracted U_e=676±86 eV [20]. Right panel: Discrepancy between the calculated and the parametrized THM reaction rate, as given in [18].

measured by THM. The THM reaction rate has been then parametrized in terms of the temperature T_9 as given in [18]. The differences between the THM calculation of the reaction rate and its parametrization is significantly lower than 1%, as shown in the right panel of Fig.1. However, the comparison between the THM reaction rate and the one proposed in the widely used NACRE compilation [21] leads to a ~20% of discrepancy, thus calling for an evaluation of the impact of the THM reaction rate in astrophysical environments.

Besides the role played by deuteron, lithium-7 and lithium-6 abundances already investigated in [25–28], the THM ^{9}Be(p,α)^{6}Li S(E)-factor measurements have been then used in [18] for evaluating the fate of ^{9}Be in pre-main sequence (PMS) stars by means of the PROSECCO stellar code derived from the well tested FRANEC one [22, 23]. The calculations have been performed in a wide stellar mass range, starting from 0.08 $M_\odot$ up to 0.7 $M_\odot$ for different stellar ages and metallicity. The data show a remarkable difference in the depletion result of beryllium at different stellar masses, being this the combined effect of both stellar structure and metallicity of the adopted models. As matter of example, models of solar metallicity with masses of about $0.08 \leq M/M_\odot \leq 0.5$ with effective temperatures 2600-3600 K show a maximum difference of the ^{9}Be logarithmic abundance of more than 1 dex when using the THM ^{9}Be(p,α)^{6}Li reaction rate instead of the NACRE one [18].

Once again, THM has been proved to be an useful tool for nuclear astrophysics allowing us to access, without any extrapolation, the Gamow energy region at which astrophysically relevant reactions take place. Further studies are still under investigations, either for induced reactions with stable beam or unstable beams [29].

Acknowledgments

This work has been partially supported by the Italian Ministry of University MIUR under the grants RFBR082838 (FIRB2008) and 'LNS-Astrofisica Nucleare (fondi premiali)'.

References

[1] Burbidge E.M., Burbidge G.R., Fowler W.A., and Hoyle F. 1957 *Rev. of Mod. Phys.* **29** 4

[2] Boesgaard, A.M. *Origin and Evolution of the Elements*, Carnegie Observ. Astrophys. Ser. Vol.4, ed. Mc William

[3] Rolfs, C. & Rodney 1988, W., *Cauldrons in the Cosmos* (The Univ. of Chicago, Chicago)

[4] Engstler, S. et al., 1992, Z. Phys. A, **342**, 471

[5] Vesic, J. et al., 2014, Eur. Phys., Journ. A, **50**, 153-1

[6] Chengbo Li et al., 2015, Phys. Rev. C, **92**, 025805

[7] Sergi, M.L. et al., 2015, Phys. Rev. C., **91**, 065803

[8] Aliotta, M.L. et al., 2001, Nucl. Phys. A, **690**, 790

[9] Zahnow, D. et al., 1997, Z. Phys. A, **359**, 211

[10] Angulo, C. et al., 1993, Z. Phys. A, **187**, 345

[11] Baur, G. et al., 1986, Nucl. Phys. A, **458**, 188

[12] Spitaleri, C. et al. 1990, *Problems of Fundamental Modern Physics*, 211, World Scientific-New York

[13] Spitaleri C et al. 2004, Phys.Rev.C, **69**, 055806

[14] Lamia, L. et al. 2012, Phys. Rev. C, **85**, 025805

[15] Spitaleri, C. et al. 2011, Phys. of Atom. Nucl., **74**

[16] Tribble R., et al. 2014, Rep. Prog. Phys., **77**, 106901

[17] Palmerini, S. et al., 2013, The Astrophysical Journal, **764**, 128

[18] Lamia, L. et al. 2015, The Astrophysical Journal, **811**, 69

[19] Romano, S. et al. 2006, Eur. Phys. J., **A 27**, s01, 221-225

[20] Wen, Q.G. et al. 2008, Phys. Rev. C, **78**, 035805

[21] Angulo, C. et al. 1999, Nucl. Phys. A, **656**, 3

[22] Degl'Innocenti, S. et al. 2008, Ap&SS, **316**, 25

[23] Dell'Omodarme, M. et al., 2012, A&A, **540**, A26

[24] Pizzone, R.G. et al. 2005, A&A. **438**, 779

[25] Pizzone, R.G. et al., 2013, The Astrophysical Journal, **87**, 025805

[26] Tumino, A. et al. 2014, Astr. Phys. Journ., **785**, 96

[27] Lamia, L. et al. 2012, Astron. & Astrophys., **541**, A158

[28] Lamia, L. et al. 2013, The Astrophysical Journal, **768**, 65

[29] Cherubini, S. et al. 2015, Phys. Rev. C, **768**, 015805

EPJ Web of Conferences **117**, 09009 (2016)
DOI: 10.1051/epjconf/201611709009
NN2015

New measurement of the $^{10}B(p,\alpha_0)^7Be$ reaction cross section at low energies and the structure of ^{11}C

I. Lombardo[1,2*], D. Dell'Aquila[1,2], F. Conte[1], L. Francalanza[2], M. La Cognata[3], L. Lamia[3,4], R. La Torre[1,2], G. Spadaccini[1,2], C. Spitaleri[3,4] and M. Vigilante[1,2].

[1] Dipartimento di Fisica "Ettore Pancini", Università di Napoli Federico II, Via Cintia, Napoli, Italy and INFN, Sezione di Napoli, Napoli, Italy
[2] INFN, Sezione di Napoli, Via Cintia, Napoli, Italy
[3] INFN - Laboratori Nazionali del Sud, Via Santa Sofia, Catania, Italy
[4] Dipartimento di Fisica e Astronomia, Università di Catania, Via Santa Sofia, Catania, Italy

Abstract

We report preliminary results of a new measurement of the $^{10}B(p,\alpha_0)^7Be$ cross section in the 0.6 - 1.0 MeV bombarding energy domain. In this region very few data have been reported in the literature. Excitation functions at both forward and backward angles have been obtained. Angular distributions testify the contribution due to several excited states in the ^{11}C compound nucleus. The experimental S-factor is about 30% lower than the one reported in the literature and based on activation methods.

1 Introduction

The $^{10}B(p,\alpha)^7Be$ reaction is important in various fields of Nuclear Physics. In fact, (1) it allows to study the spectroscopy of the non self-conjugated

*e-mail address: ivlombardo@na.infn.it

^{11}C compound nucleus, with particular regard to the appearance of α-cluster states (see, e.g. [1–3]); (2) it is important to accurately describe the destruction of the ^{10}B isotope in stars [5–7]; (3) it is involved in radio-protection issues in Nuclear Fusion technology [4].

For example, from the Nuclear Structure point of view, it is important to have available angular distributions and excitation functions of several reaction channels leading to the population (or coming from the decay) of ^{11}C excited states. In this case, a R-matrix analysis of experimental data allows to determine the partial width and the branching ratios of the various channels. In particular, the analysis of states near and above the α emission threshold is an useful tool to unveil the possible existence of molecular rotational bands [2, 3] in ^{11}C, that can constitute a fingerprint of α-clustering in non self-conjugated nuclei.

^{10}B(p,α) reaction is important also for Applied Physics purposes, and in particular in Nuclear Fusion technology [4]. In fact, the ^{11}B(p,α)$\alpha\alpha$ reaction has been proposed as a possible candidate for laser-induced fusion reactions without neutron production (*aneutronic fusion*) [8,9]. In this way, it would be possible to overcome radio-protection hazards due to the large neutron flux emitted by the usual $d+d$ or $d+t$ fusion reactions. Unfortunately, the natural boron is composed both of ^{11}B (80.1% natural abundance) and ^{10}B (19.9% natural abundance). Reactions induced by protons on ^{10}B contaminants can lead to the emission of α particles and of long-lived ^{7}Be radioactive nuclei ($\tau \simeq 52.7$ days) that can implant on the reactor vessel, leading to radiation safety issues. For this reason, it is necessary to isotopically enrich the ^{11}B target, but a correct estimate of the enrichment level needed to avoid radioactive hazard necessarily relies on an accurate knowledge of the ^{10}B(p,α)^{7}Be cross section at low energies (< 0.3 MeV) [10].

Unfortunately, direct data reported in the literature at low bombarding energies, especially in the $E \approx 1-2$ MeV region, are few [11–13], being often characterized by large error bars. Recent measurements performed with the Trojan Horse method allowed to explore the S-factor down to zero energy [5]. These data are usually obtained in arbitrary units, and therefore they need to be normalized to higher energy direct data. Considering all these reasons, we decided to perform a new experiment aimed at measuring the ^{10}B(p,α)^{7}Be reaction in the E_p=1.028-0.630 MeV bombarding energy domain. Some experimental details and preliminary results of this analysis are reported in the following sections.

2 Experimental methods

Proton beams at bombarding energies E_p=0.630-1.028 MeV were delivered by the TTT3 tandem accelerator of the Laboratorio dell'Acceleratore (LdA) at Federico II University of Naples [14]. The energy step here used was about 40 keV. To reduce pile-up effects, in this experiment the beam intensity did not overcome 1 nA. The beam energy calibration was determined by analysing the γ yield of the ^{19}F(p,$\alpha\gamma$)^{16}O reaction around the 0.872 MeV resonance and by investigating several elastic scattering resonances induced by proton and α-particles on several light targets. The maximum beam energy spread was of the order of 0.2%, and the diameter of the beam on target was smaller than 2 mm. The beam intensity was measured by means of a Faraday-cup. The vacuum in the scattering chamber was of the order of 10^{-6} mbar.

A boron foil (38 μg/cm^2 thick) was used as reaction target. Because of its extreme fragility, a small amount of polyvinyl formal ($C_3H_6O_2$, some μg/cm^2 thick) has been used to strengthen it. The target surface was orthogonal to the beam direction. Because of the manufacturing process, the target was affected by the presence of several contaminants. They have been characterized with elastic backscattering analysis, with proton beams of about 1 MeV. The main observed contaminants were Li, C, O, Al, Cl, Cu, Ba. Because of kinematic reasons, peaks due to the heavier of these contaminants are expected to overlap with the α_0 signal at backward angles, leading to strong difficulties in the extraction of α_0 yields from the energy spectrum of ejectiles.

For this reason, to measure the α_0 yields at backward angles ($\theta_{lab} = 120° - 160°$) we used a method based on the simultaneous measurements of two ejectile spectra at the same angle (in opposite directions with respect to the beam axis), the first one taken with an unshielded silicon detector, the second one taken by a silicon detector covered by a thin Al foil (3μm thick) that is able to stop (or highly slow down) the α_0 particles, allowing the elastically scattered protons to punch through. Then, the α_0 yields have been derived by subtraction, taking into account the energy loss and straggling of protons in the aluminium foil. Details on this experimental technique will be described in a forthcoming paper.

The experimental setup consisted of an array of 10 silicon detectors. Four detectors were placed at forward angles (θ_{lab} =30°, 40°, 50°, 60°), while the remaining six detectors were placed at $\theta_{lab} = \pm 120°, 140°, 160°$. In this way, we obtained angular distributions at 7 angles, covering the range from θ_{lab}=30° to 160°. α_0 yields have been transformed into absolute cross

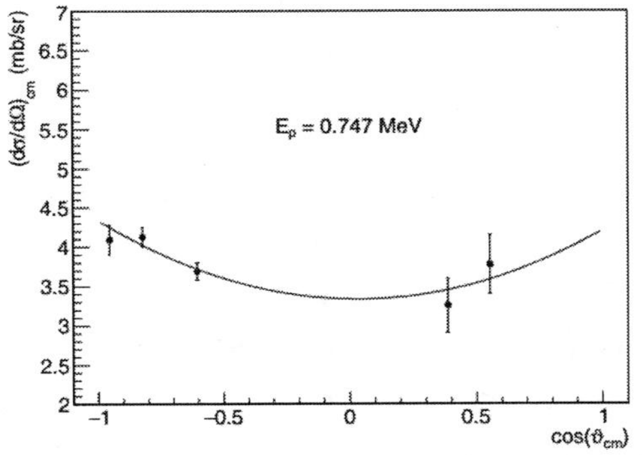

Figure 1: (color online) Angular distribution of the ^{10}B(p,α_0)^{7}Be reaction at E_p=0.747 MeV. The red line indicates Legendre polynomial fit of data (truncated to the 2nd order).

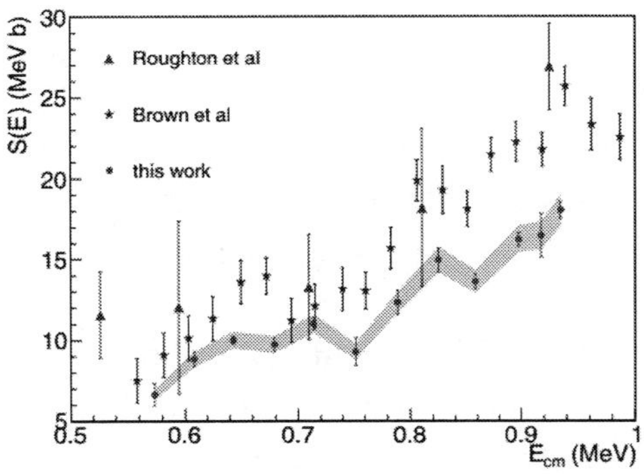

Figure 2: (color online) Preliminary S-factor of the ^{10}B(p,α_0)^{7}Be reaction (green dots). The grey band represents non statistical errors. Blue stars: data by Brown et al [18]; red triangles: data by Roughton et al [11] as reported by [12, 13].

sections by internal normalization to the ^{10}B(p,p$_0$) elastic scattering signal, taking into account the ^{10}B(p,p$_0$) elastic scattering differential cross sections reported in Refs. [15, 16].

3 Preliminary experimental results

An example of preliminary angular distribution obtained at E_p=0.747 MeV is reported in Figure 1. The red line represents its fit by using Legendre Polynomials up to the 2nd degree, corresponding to ℓ_{max}=1 in the p+^{10}B channel. This angular distribution might be slightly anisotropic, testifying the possible contribution given by $\ell \neq 0$ partial waves. The preliminary analysis of experimental data shows a continuous evolution of angular distribution shapes as a function of energy. In particular, in the E_p = 0.630-0.785 MeV energy range, angular distributions show quite small anisotropies, similar to the one reported in Figure 1, while at higher energies (E_p = 0.827-1.028 MeV) the angular distributions are more anisotropic. R-matrix calculations of angular distributions are currently in progress; they will help us to improve the spectroscopy of ^{11}C states in an excitation energy region $E_x \approx$9-11 MeV,) where several ambiguities in the J$^\pi$ assignments exist [17].

A preliminary estimate of the angle integrated reaction cross section has been obtained from the present angular distributions; consequently, our preliminary ^{10}B(p,α_0)^{7}Be S-factor is shown in Figure 2, as green dots. Vertical bars represent statistical errors, while the grey band indicates non statistical ones, derived from the internal normalization procedure. We compared our experimental data with results taken from the NACRE and NACRE2 compilations [12, 13], and derived from a differential cross section reported by Brown et al [18] at θ_{lab}=137.8° (blue stars) and from an activation experiment reported by Roughton et al [11] (red triangles; these measurements refer to the $\alpha_0 + \alpha_1$ channels; the α_1 channel is expected to have cross section well lower than 10% of the α_0 channel at energies E_p <1 MeV [19]). Our data are ≈30% lower than the estimates based on Roughton et al data, even if they agree within the error bars. Similar conclusions can be drawn by considering the Brown et al data.

The next step in the data analysis will be the related to the definitive estimate of the S-factor from the present experimental data. This analysis is currently in progress.

4 Conclusions

In these proceedings we report some preliminary results concerning a new measurement of the ^{10}B(p,α)^{7}Be reaction at low energies (E_p=0.630-1.028 MeV). The experiment has been performed by using the TTT3 tandem accelerator in Naples. Angular distributions have been obtained at several angles, both in the forward and backward hemisphere; their shape changes as a function of energy, indicating the contribution of several excited states of ^{11}C to the excitation function. A preliminary estimate of the S-factor has been compared with results reported in the literature in the same energy region.

References

[1] M. Wiescher et al, Phys. Rev. **C 28** (1983) 1431

[2] M. Freer et al, Phys. Rev. **C 85** (2012) 014304

[3] H. Yamaguchi et al, Phys. Rev. **C 87** (2013) 034303

[4] R.J. Peterson et al, Ann. Phys. Energy **2** (1975) 503

[5] C. Spitaleri et al, Phys. Rev. **C 90** (2014) 035801

[6] S.M.R. Puglia et al, Mem. S.A.It. Suppl. **14** (2010) 43

[7] L. Lamia et al, Nucl. Phys. **A 787** (2007) 309

[8] V.S. Belyaev et al, Phys. Rev. **E 72** (2005) 026406

[9] M.C. Spraker et al, J. Fusion Energy **31** (2012) 357

[10] A. Kafkarkou et al, Nucl. Instrum. Meth. Phys. Res. **B 316** (2013) 48

[11] N.A. Roughton et al, At. Data Nucl. Data Tables **23** (1979) 177

[12] C. Angulo et al, Nucl. Phys. **A 656** (1999) 3

[13] Y. Xu et al, Nucl. Phys. **A 918** (2013) 61

[14] L. Campajola et al, Nucl. Instrum. Meth. **B 29** (1987) 129, I. Lombardo et al, Nucl. Instrum. Meth. **B 302** (2013) 19, I. Lombardo et al, J. Phys. G 40 (2013) 125102, I. Lombardo J. Phys. Conf. Ser. 569 (2014) 012068

[15] M. Chiari, L. Giuntini, P.A. Mandò, N. Taccetti, Nucl. Instrum. Meth. Phys. Res. **B 184** (2001) 309

[16] M. Chiari, M. Bianconi, I. Bogdanović Radović, M. Mayer, Nucl. Instrum. Meth. Phys. Res. **B 343** (2015) 70

[17] J.H. Kelley et al, Nucl. Phys. **A 880** (2012) 88

[18] A.B. Brown et al, Phys. Rev. **82** (1951) 159

[19] J.W. Cronin et al, Phys. Rev. **56** (1956) 298

EPJ Web of Conferences **117**, 09010 (2016) DOI: 10.1051/epjconf/201611709010
NN2015

Primordial nucleosynthesis revisited via Trojan Horse Results

R.G. Pizzone[1], R. Spartá[1,2], C. Bertulani[3], C. Spitaleri[1,2], M. La Cognata[1], L. Lamia[2], A. Mukhamedzhanov[4] and A. Tumino[1,5]

[1]INFN - Laboratori Nazionali del Sud, Catania, Italy
[2]Dipartimento di Fisica e Astronomia, Università degli Studi di Catania, Catania, Italy
[3]Department of Physics and Astronomy, Texas A&M University, Commerce, TX 75025, USA
[4] Cyclotron Institute, Texas A& M University, College Station, TX, USA
[5] Facoltà di Ingegneria e Architettura, Università Kore, Enna, Italy

Abstract

Big Bang Nucleosynthesis (BBN) requires several nuclear physics inputs and nuclear reaction rates. An up-to-date compilation of direct cross sections of d(d,p)t, d(d,n)^{3}He and ^{3}He(d,p)^{4}He reactions is given, being these ones among the most uncertain bare-nucleus cross sections. An intense experimental effort has been carried on in the last decade to apply the Trojan Horse Method (THM) to study reactions of relevance for the BBN and measure their astrophysical S(E)-factor. The reaction rates and the relative error for the four reactions of interest are then numerically calculated in the temperature ranges of relevance for BBN ($0.01<T_9<10$). Their value were therefore used as input physics for primordial nucleosynthesis calculations in order to evaluate their impact on the calculated primordial abundances of D, ^{3,4}He and ^{7}Li. These were compared with the observational primordial abundance estimates in different astrophysical sites. A comparison was also performed with calculations using other reaction rates compilations available in literature.

DOI: 10.1051/epjconf/201611709010

1 Introduction

One of the foundation stones of the Big Bang model, together with the Hubble expansion and the Cosmic Microwave Background (CMB) radiation [1] is the Big Bang Nucleosynthesis (BBN). BBN probes the Universe to the earliest times, the so called radiation dominated era, from a fraction of second to few minutes. It involves events that occurred at temperatures below 1 MeV, and naturally plays a key role in forging the connection between cosmology and nuclear physics [2]. Focusing only on the products of the BBN, according to the Standard Big Bang Nucleosynthesis model (SBBN), only the formation of light nuclei (^{2}H,^{3,4}He,^{7}Li) is predicted in observable quantities, starting from protons and neutrons. Today, with the only exception of ^{3}He and lithium, the abundances of these isotopes in the appropriate astrophysical environments are rather consistent with SBBN predictions [3]. A comparison between the primordial abundances from WMAP observations and the calculated ones constrains the baryon-to-photon ratio, η, which is the only free parameter in the presently accepted model of the SBBN. A recent observation yields $\eta = 6.16 \pm 0.15 \times 10^{-10}$ [4], which is the value that we adopt in our calculations.

BBN nucleosynthesis requires several nuclear physics inputs and, among them, an important role is played by nuclear reaction rates. Due to the relatively small amount of key nuclear species involved in the BBN nuclear reaction network, only 12 reactions play a major role [5]. The reaction rates are calculated from the available low-energy cross sections for reactions which are also a fundamental information for a number of other still unsolved astrophysical problems, e.g. the so called "lithium depletion" either in the Sun or in other galactic stars [6, 7]. Cross sections should be measured in the astrophysically relevant Gamow window [8], of the order of few hundreds keV. In the last decades these reactions have been widely studied and, in particular, great efforts have been devoted to their study by means of direct measurements at the relevant astrophysical energies, sometimes in underground laboratories [9, 10]. However, for many of the relevant reactions, no direct experiments exist at astrophysical energies (mostly because of difficulties connected with the presence of the Coulomb barrier in charged particle induced reactions) and the cross section within the Gamow window has to be extrapolated from higher energy measurements. Alternative and challenging ways to obtain σ_b for charged-particles at sub-Coulomb energies have been provided by indirect methods such as the Coulomb dissociation method [11, 12] and the ANC (Asymptotic Normalization Coefficient) [13]. Among them, the Trojan-horse Method (THM) [14] is particularly suited to investigate binary reactions induced at astrophysical energies by neutrons or charged particles by using appropriate three-body reactions. It allows one to avoid both Coulomb barrier suppression and electron screening effects, thus preventing the use of unreliable extrapolations. In the next sections we will show the calculations of the reaction rates based also on the THM measurements of the cross sections σ_b. For recent reviews on the THM see [15, 16]). Thus, the method can be regarded as a powerful indirect technique to get information on bare nucleus cross section for reactions of astrophysical interest, which leads to new reaction rates determination. Some of the reactions relevant for the SBBN, i.e. ^{7}Li(p,α)^{4}He, ^{2}H(d,p)^{3}H, ^{2}H(d,n)^{3}He, ^{3}He(d,p)^{4}He, were studied by means of the THM in the energy range of interest and their measurements were performed in an experimental campaign which took place in the last decade [17–24].

1.1 Reaction rates with TH data

The reaction rates for the the four reactions mentioned above (from a compilation of direct and THM data, as reported in [25]), have been calculated numerically. Then, we fitted the rates with the parametrization displayed in Equation 1. This is the common procedure adopted in previous works (see, e.g., [26–28]). For the 4 reactions of interest, we have fully included the experimental errors from measurements, allowing us to evaluate the respective errors in the reaction rates. The numerical results are then fitted with the expression

$$N_A \langle \sigma v \rangle = \exp[a_1 + a_2 \ln T_9 + \frac{a_3}{T_9} + a_4 T_9^{-1/3} + a_5 T_9^{1/3} + a_6 T_9^{2/3} + a_7 T_9 + a_8 T_9^{4/3} + a_9 T_9^{5/3}] \quad (1)$$

which incorporates the relevant temperature dependence of the reaction rates during the BBN. The a_i coefficients for the ^{2}H(d,p)^{3}H, ^{2}H(d,n)^{3}He, ^{3}He(d,p)^{4}He and ^{7}Li(p,α)^{4}He reaction rate expression are given in [25]. The direct data were considered from the compilation described in [25] for energies above 100 keV for ^{3}He(d,p)^{4}He and ^{7}Li(p,α)^{4}He and for energies above 10 keV for ^{2}H(d,p)^{3}H and ^{2}H(d,n)^{3}He, in order to avoid the enhancement due to the electron screening in the direct data.

For all the cases we noticed that deviations of up to 20% are obtained from previous compilations.

Table 1: BBN predictions using different set of data (see text) compared with observations. (a) The mass fraction for ^{4}He, Y_p is from Ref. [36]. (b) The mean deuterium abundance is the mean average from [37]. (c) The ^{3}He abundances are adopted from Ref. [38] as a lower limit to the primordial abundance. (d) The lithium abundance arises from observations of stars which provide a sample of the "lithium plateau" [39]. D/H is in units of 10^{-5}, ^{3}He/H in 10^{-6} and Li/H in 10^{-10}.

Yields	present work	Observed
Y_p	$0.2485^{+0.001}_{-0.002}$	$0.256 \pm 0.006^{(a)}$
D/H	$2.692^{+0.177}_{-0.070}$	$2.82 \pm 0.26^{(b)}$
^{3}He/H	$9.441^{+0.511}_{-0.466}$	$\geq 11. \pm 2.^{(c)}$
^{7}Li/H	$4.683^{+0.335}_{-0.292}$	$1.58 \pm 0.31^{(d)}$

2 Discussion and conclusion

The reaction rates of four of the main reactions of the BBN network in the temperature range ($0.001<T_9<10$), namely, ^{2}H(d,p)^{3}H, d(d,n)^{3}He, ^{3}He(d,p)^{4}He, ^{7}Li(p,α)^{4}He, have been calculated numerically including the recent THM measurements [29–35]. The uncertainties of experimental data for direct and THM data have been fully included for the above reactions. The extension of the same methodology to the other reactions forming the BBN reaction network will be examined in a forthcoming paper. The parameters of each reaction rates as given in Eq. 1 are reported in [25]. The obtained reaction rates are compared with some of the most commonly used compilations found in the literature and used to calculate the BBN abundance for ^{3,4}He, D and ^{7}Li. The obtained abundances are in agreement, within the experimental errors, with those obtained using the compilation of direct reaction rates. Moreover, a comparison of our predictions with the observations for primordial abundance of ^{3,4}He, D and ^{7}Li show an agreement for ^{3,4}He and D, while showing a relevant discrepancy for ^{7}Li as reported in table 1. The present results show the power of THM as a tool for exploring charged particle induced reactions at the energies typical of BBN.

This work was supported by the Italian Ministry of the University under Grant No. RBFR082838 and by the Italian Ministry of University MIUR under the grant "LNS Astrofisica Nucleare (fondi premiali)". This work was partially supported by the US-DOE grants DE-FG02-08ER41533 and DE-FG02-10ER41706. A.M. acknowledges the support by US Department of Energy under Grant Nos. DE-FG02-93ER40773, DE-FG52- 09NA29467, and DE-SC0004958.

References

[1] G. Steigman 2007, Ann. Rev. Nucl. Part. Sci. 57, 463

[2] B. D. Fields and S. Sarkar 2006, J. Phys. G33, 220

[3] G. Israelian 2012, Nature 489, 37

[4] E. Komatsu et al. 2011, Ap. J. Sup. 192, 18

[5] E.W. Kolb and M.S. Turner 1990, "The Early Universe", Addison-Wesley

[6] R. Weymann and E. Moore 1963, Ap. J. 137, 552

[7] D. Ezer and A.G.W. Cameron 1963, Icarus 1, 422

[8] C. Iliadis 2007, "Nuclear Physics of Stars", Wiley

[9] R. Bonetti et al. 1999, Phys. Rev. Lett. 82, 5205

[10] C. Casella et al. 2002, Nucl. Phys. A706, 203

[11] G. Baur, C.A. Bertulani & H Rebel 1986, Nucl. Phys. A458, 188

[12] C.A. Bertulani & A. Gade 2010, Phys. Rep. 485, 195

[13] A. Mukhamezhanov et al. 2008, Phys. Rev. C 78, 0158042008

[14] C. Spitaleri et al. 2003, Nucl. Phys. A 719, 99c

[15] C. Spitaleri et al. 2011, Physics of At. Nucleus 74, 1725

[16] R.E. Tribble et al. 2014, Rep. Progr. Phys. 77, 10,106901

[17] A. Tumino et al. 2008, Phys. Rev. C 78, 064001

[18] R.G. Pizzone et al. 2005, A. & A. 438, 779

[19] L. Lamia et al., 2013, Ap. J. 768, 65

[20] M. La Cognata et al. 2011, Ap. J. L. 739, L54

[21] L. Lamia, M. La Cognata, C. Spitaleri, B. Irgaziev, R.G. Pizzone 2012, Phys. Rev. C 85, 025805

[22] S. Romano et al. 2006, Eur. Phys. J. A 27, 221

[23] Q. Wen et al. 2008, Phys. Rev. C 78, 035805

[24] S. Cherubini et al., 2015, Phys. Rev. C, 92, 015805

[25] R.G. Pizzone et al. 2014, Ap. J.,786, 112

[26] M.S. Smith, L.H. Kawano and R.A. Malaney 1993, Ap. J. 85, 219

[27] R.H. Cyburt 2004, Phys. Rev. D 70, 023505

[28] A. Coc, S. Goriely, Y. Xu, M. Saimpert, and E. Vangioni 2012, Astrophys. J. 744, 18

[29] L. Lamia et al. 2012, A. & A. 541, 158

[30] M. Lattuada et al. 2001, Ap. J. 562, 1076

[31] A. Tumino et al. 2011, Phys. Lett. B 700 (2), 111

[32] A. Tumino et al. 2014, Ap. J., 785, 96

[33] M. La Cognata et al. 2005, Phys. Rev. C 72, 065802

[34] R.G. Pizzone et al. 2013, Phys. Rev. C 87, 025805

[35] Li C. et al., 2015, Phys. Rev. C, 92, 025805

[36] Y.I.Izotov and T.X.Thuan 2010, Ap. J. Lett. 710, L67

[37] J.M. O'Meara, S. Burles, J.X. Prochaska, and G.E. Prochter 2006, Ap. J. 649, L61

[38] T. M. Bania, R. T. Rood, D. S. Balser 2002, Nature 415, 54

[39] L. Sbordone, et al. 2010, Astron. Astro. 522, 26

EPJ Web of Conferences **117**, 09011 (2016) DOI: 10.1051/epjconf/201611709011
NN2015

Fusion measurements of $^{12}C+^{12}C$ at energies of astrophysical interest

D. Santiago-Gonzalez[1,2], C. L. Jiang[2], K. E. Rehm[2], M. Alcorta[3], S. Almaraz-Calderon[2,*], M. L. Avila[2], A. D. Ayangeakaa[2], B. B. Back[2], D. Bourgin[4], B. Bucher[5], M. P. Carpenter[2], S. Courtin[4], H. M. David[2], C. M. Deibel[1], C. Dickerson[2], B. DiGiovine[2], X. Fang[6], J. P. Greene[2], F. Haas[4], D. J. Henderson[2], R. V. F. Janssens[2], D. Jenkins[7], J. Lai[1], T. Lauritsen[2], A. Lefebvre-Schuhl[8], D. Montanari[4], R. C. Pardo[2], M. Paul[9], D. Seweryniak[2], X. D. Tang[10], C. Ugalde[2] and S. Zhu[2]

[1] Department of Physics and Astronomy, Louisiana State University, Baton Rouge, LA 70803, USA
[2] Physics Division, Argonne National Laboratory, Argonne, IL 60439, USA
[3] TRIUMF, Vancouver, BC V6T 2A3, Canada
[4] IPHC and CNRS, Université de Strasbourg, F-67037 Strasbourg, France
[5] Lawrence Livermore National Laboratory, Livermore, CA 94551, USA
[6] University of Notre Dame, Notre Dame, IN 46556, USA
[7] Department of Physics, University of York, Heslington, York YO10 5DD, UK
[8] IN2P3-CNRS and University of Paris Sud, F-91405 Orsay-campus, France
[9] Racah Institute of Physics, Hebrew University, Jerusalem 91904, Israel
[10] Institute of Modern Physics, Lanzhou, China
* Present address: Department of Physics, Florida State University, Tallahassee, FL 32306, USA

Abstract

The cross section of the $^{12}C+^{12}C$ fusion reaction at low energies is of paramount importance for models of stellar nucleosynthesis in different astrophysical scenarios, such as Type Ia supernovae and X-ray superbursts, where this reaction is a primary route for the production of heavier elements. In a series of experiments performed at Argonne National Laboratory, using Gammasphere and an array of Silicon detectors, measurements of the fusion cross section of $^{12}C+^{12}C$ were successfully carried out with the γ and charged-particle coincidence technique in the center-of-mass energy range of 3-5 MeV. These were the first background-free fusion cross section measurements for $^{12}C+^{12}C$ at energies of astrophysical interest. Our results are consistent with previous measurements in the high-energy region; however, our lowest energy measurement indicates a fusion cross section slightly lower than those obtained with other techniques.

In the interior of highly developed stars the fusion of $^{12}C+^{12}C$, known as carbon burning, drives the nucleosynthesis of heavier elements [1]. Because of its importance in nuclear astrophysics, many measurements of the $^{12}C+^{12}C$ reaction have been performed in the past [2–9]. However, the uncertainty in the cross section measured at and near the Gamow window is too large to put constraints on computations of nuclear abundances and reaction rates. At the moment, one has to rely on phenomenological extrapolations and/or model calculations in order to obtain the appropriate astrophysical reaction rates for use in stellar models. Even in explosive scenarios, such as Type Ia supernovae and X-ray superbursts, where the temperature is high compared to many stellar environments, the corresponding Gamow energies for the $^{12}C+^{12}C$ reaction are still very low (roughly between 1 and 3 MeV in the center of mass). This results in extremely small cross sections and, at this moment, direct measurements at low center-of-mass energies ($E_{\mathrm{c.m.}} < 3$ MeV) are extremely challenging.

The $^{12}C+^{12}C$ fusion cross section can be measured in the laboratory by impinging a beam of ^{12}C ions onto a carbon target. The two ^{12}C nuclei fuse to form a highly excited nucleus of ^{24}Mg, which then decays primarily by emitting protons or α particles. The resulting evaporation residue is often left in an excited state, which then decays to the ground state via γ emission. This is schematically depicted in Fig. 1. There are numerous $^{12}C+^{12}C$ fusion cross section measurements and in general they agree for $E_{\mathrm{c.m.}} > 3$ MeV, with few exceptions. However, there are very large discrepancies between the different experimental results of Refs. [2–9] in the low-energy region that can be attributed to the difficulties of measuring low cross sections in combination with various known and yet-unknown backgrounds. In the

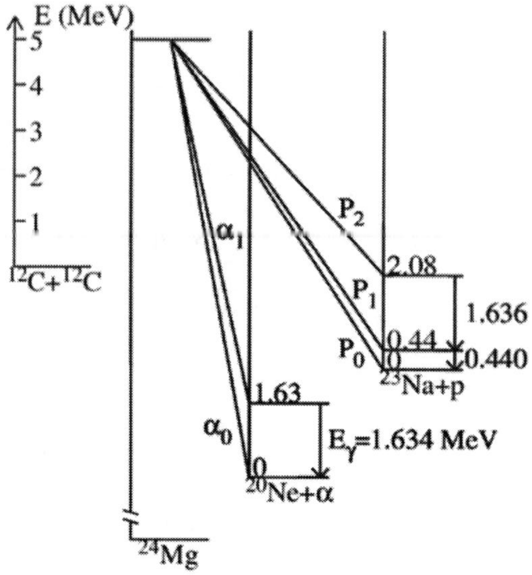

Figure 1: Most favorable decay channels of ^{24}Mg following the fusion of ^{12}C+^{12}C at low center-of-mass energies.

experiments mentioned above, only a single signal from the decay cascade is measured. Although serious attempts were made to reduce the backgrounds, such as those arising from hydrogen and deuterium contaminants in the target, the background is not completely eliminated, especially at the lowest energies.

In order to circumvent these experimental difficulties we have developed a coincidence technique that effectively suppresses the background in ^{12}C+^{12}C fusion measurements [10]. For this, we utilize thin carbon targets (thickness $\sim$ 50 μg/cm^2) and a compact array of silicon detectors placed at the center of Gammasphere, which is an array of over 100 high-purity germanium detectors. Figure 2 shows a hemisphere of Gammasphere with the charged-particle detector chamber located at the center. A schematic picture of the array of silicon detectors inside the detector chamber is presented in Fig. 3. By detecting γ rays and charged particles from the decays of excited ^{24}Mg states in coincidence, the background has been substantially reduced. Furthermore, any measured event from ^{12}C+^{12}C fusion identified by energy signals from a charged particle and from a γ ray can be clearly separated from the background events. Thus, a cross section measurement of the ^{12}C+^{12}C fusion reaction of interest is possible, even with low statistics, as a result of such well-characterized events.

A series of experiments have been performed since 2010 [11] at Ar-

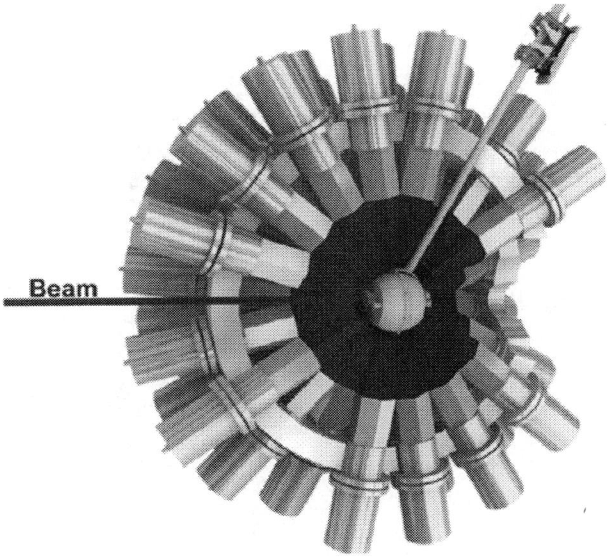

Figure 2: Hemisphere of Gammasphere with the chamber for Si detectors and (thin) targets located at the center.

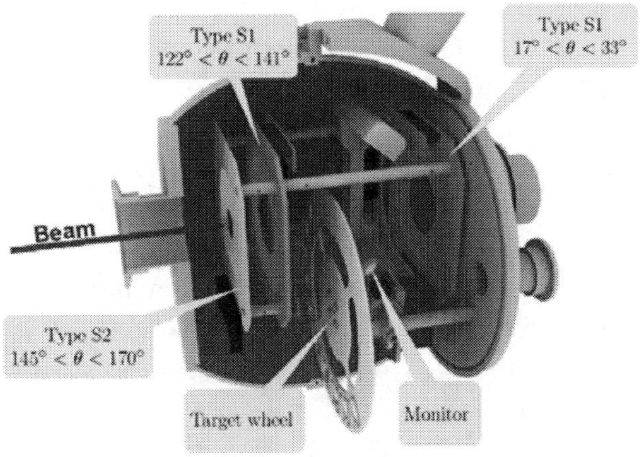

Figure 3: Target and Si detector chamber cut away showing the experimental setup. For the Si detectors, the labels show the angles covered in the laboratory reference frame. The detector labeled "Monitor" and a Faraday cup (not shown) located downstream of the chamber were used for normalization.

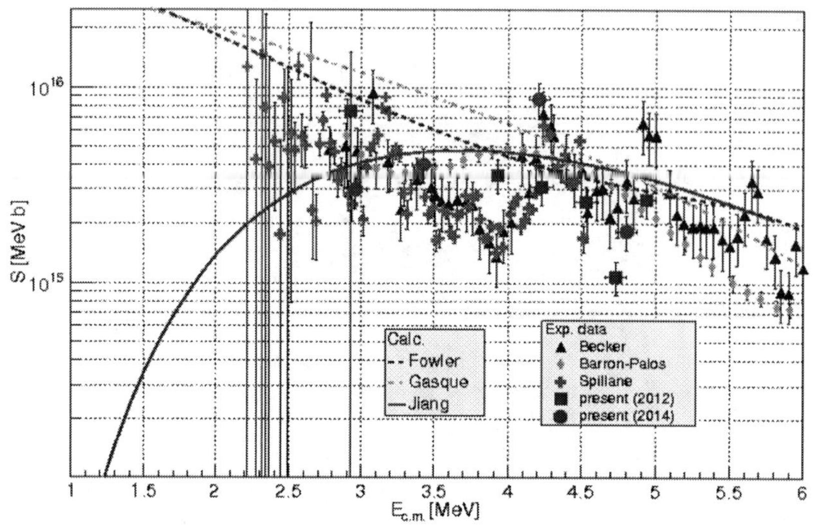

Figure 4: S-factor for the $^{12}C+^{12}C$ fusion reaction as a function of center-of-mass energy.

gonne National Laboratory (ANL) using the coincidence technique described above. Background-free measurements of the fusion cross section of $^{12}C+^{12}C$ were successfully carried out with the γ and charged-particle coincidence technique in the center-of-mass energy range of 3-5 MeV. In order to compensate for the strong Coulomb effect we present the extracted S factor ($S(E) = \sigma E e^{2\pi\eta}$) in Fig. 4 instead of the cross section. Two of our data sets are presented, one from the most recent experiment in 2014 and another one from an experiment performed in 2012 [12]. The lowest energy point in the 2012 data set was obtained from one count only and is therefore only an upper limit. Data points from Refs. [5, 6, 8] are presented as well. In general, our results are consistent with previous measurements in the high-energy region, especially our latest results from 2014 (evidence that the coincidence technique has matured from its earlier implementations). However, our lowest energy measurement from 2014 indicates a fusion cross section slightly lower than those obtained with other techniques. The calculations with the barrier penetration model [13, 14] and empirical extrapolations with fusion hindrance [15] result in opposite trends with decreasing energy. New measurements at lower energies will be carried out at ANL in the near future. The new data may help to discern between the extrapolations shown in Fig. 4. The ramifications of using the fusion hindrance extrapolation method on simulations of the late stages of giant star evolution are substantial [16] and, in the astrophysical community, there is a keen interest to settle this question on the basis of solid experimental data.

This material is based upon work supported by the U.S. Department of Energy, Office of Nuclear Physics, under contract No. DE-AC02-06CH11357 and No. DE-FG02-96ER40978. This research used resources of ANL's ATLAS facility, which is a DOE Office of Science User Facility.

References

[1] Rolfs C. E. and Rodney W. S., in *Cauldrons in the Cosmos*, The University of Chicago press, Chicago, 1988.

[2] Patterson L. J. *et al.,* Astrophys. J. **157**, 367 (1969).

[3] Mazarakis M. *et al.,* Phys. Rev. **C7**, 1280 (1973).

[4] High M. D. and C̃ujec B., Nucl. Phys. **A282**, 181 (1977).

[5] Becker H. W. *et al.,* Phys. Rev. Lett. **38**, 337 (1977).

[6] Barron-Palos L. R. *et al.,* Nucl. Phys. **A779**, 218 (2006).

[7] Aguilera E. F. *et al.,* Phys. Rev. **C73**, 064601 (2006)

[8] Spillane T. *et al.,* Phys. Rev. Lett. **98**, 122501 (2007)

[9] Zickefoose J., Thesis: "^{12}C+^{12}C Fusion: Measurement and Advances Toward the Gamow Energy", U. Connecticut, 2010.

[10] Jiang C. L. *et al.,* Nucl. Instru. Meth. **A682**, 12 (2012).

[11] Jiang C.L. EPJ Web of Conferences **17** 01002 (2011).

[12] Jiang C. L. *et al.,* Jour. of Phys. Conference Series **420**, 012120 (2013).

[13] Fowler W., *et al.,*, Annu. Rev. Astrophys. **13**, 69 (1975).

[14] Gasques L. R. *et al.,* Phys. Rev. **C72**, 025806 (2005).

[15] Jiang C. L. *et al.,* Phys. Rev. **C75**, 015803 (2007).

[16] Gasques L. R. *et al.,* Phys. Rev. **C76**, 035806 (2007).

EPJ Web of Conferences **117**, 09012 (2016) DOI: 10.1051/epjconf/201611709012
NN2015

Large electron screening effect in different environments

Aleksandra Cvetinović, Matej Lipoglavsek, Sabina Markelj and Jelena Vesić

Jožef Stefan Institute, Jamova cesta 39, Ljubljana, Slovenia

Abstract

Electron screening effect was studied on different hydrogen containing targets with the ^{7}Li, ^{11}B and ^{19}F ion beams. Results show large electron screening potentials strongly dependent on the proton number Z of the projectile. The largest ever measured screening potential with the value about a factor of 50 above the calculations from the model in adiabatic limit was observed in the graphite target containing hydrogen as an impurity.

1 Introduction

In reactions between nuclei at very low energies, when the energy of the incident beam in the center of mass system is far below the Coulomb barrier, tunneling is the only way the fusion process can happen. But, since the probability for a reaction drops steeply with decreasing energy, the cross sections at thermal energies become extremely difficult to measure. Nowadays, the reactions at these energies can be investigated in underground laboratories with high-current low energy accelerators (such as LUNA [1] and in future DIANA [2]). However, even when astrophysical energies are reached, the measurements do not give the nuclear cross section, since the reaction rate at these energies is strongly influenced by the electrons that surround the reacting nuclei and increase the tunneling probability through the Coulomb barrier.

The tunneling probability at low energies can be approximated with the Gamow factor $G{=}e^{-2\pi\eta}$ [3] which describes the s-wave penetration through the Coulomb barrier of point like charges. Since G strongly depends on the projectile energy, in order to separate the strong energy dependence from effects of pure nuclear interaction, the cross section $\sigma(E)$ is usually presented by introducing the astrophysical S-factor, which in the case of non-resonant reactions varies smoothly with energy [3], *i.e.*:

$$\sigma(E) = \frac{S(E)}{E} e^{-2\pi\eta}. \tag{1}$$

Here $\eta = Z_1 Z_2 e^2 / 4\pi\varepsilon_0 \hbar\sqrt{2E/\mu}$ represents the Sommerfeld parameter (Z_1 and Z_2 are the charge numbers of the interacting nuclei; E is the center of mass energy and μ is the reduced mass). When the reaction rate is very low, it is sensitive to electronic properties of the target material and the electrons surrounding the reacting nuclei can increase the tunneling probability through the Coulomb barrier, enhancing the cross sections compared to the cross section in the case of the bare nuclei. In order to take into account this effect, an enhancement factor defined as the ratio of screened and bare cross sections [4] is introduced:

$$f(E) = \frac{\sigma(E + U_e)}{\sigma(E)}, \tag{2}$$

where U_e represents the electron screening potential. There are many experimental results published by different groups [5-13] that are inconsistent with the theoretical models (*i.e.* model in adiabatic limit [4]). Also linear dependence of U_e on target Z number was suggested in ref. [10], while findings of our group show a dependence of Ue on target preparation [12, 13].

2 Experiment

To investigate the electron screening potential, we performed experiments with the ^{1}H(^{7}Li,α)^{4}He, ^{1}H(^{11}B,$\alpha\alpha$)^{4}He and ^{1}H(^{19}F,$\alpha\gamma$)^{16}O reactions in inverse kinematics on physically the same targets. Measurements were performed on five thick proton containing targets. The 50 μm thick polyimide (Kapton, $C_{22}H_{10}N_2O_5$) foils were used as a reference hydrogen target. The 250 μm thick graphite foil was fabricated by Chempur. This target originally contained about 6 % of hydrogen. A small additional

amount of hydrogen was implanted using our Tectra IonEtch ion gun. The titanium hydride target was pressed from TiH_2 powder into a 1 mm deep hole in a Cu backing which helped to cool the target after heating with the ion beam. The palladium target was first loaded and unloaded with hydrogen more than 10 times. The cycling was done by leaving the Pd foil in hydrogen gas at 1 bar for 24 hours and then heating in vacuum to 300° C. After that, the foil was radiation damaged with $45.3{\cdot}10^{15}$ ^{19}F ions per cm^2 at 7.8 MeV in order to create lattice defects to which hydrogen could be trapped [14] and then again left in hydrogen gas at 1 bar for another 2 hours. The tungsten target was treated in a similar way. The foil was produced by Plansee and it was mechanically polished and outgassed at 1200 K for stress relief. Electron microscopy showed large grains (5-20 μm) [15]. The foil prepared in this way was pre-damaged with 20 MeV ^{186}W ions up to radiation damage of 0.45 displacements per atom. Additionally, the foil was implanted with 5 keV protons from the ion gun with a beam intensity of 400 μA/cm^2 for four hours in steps of half an hour. After each step we performed a reference measurement in order to check for hydrogen concentration. The maximum hydrogen concentration was achieved already after the first half hour. All targets were checked for hydrogen loss by repeatedly measuring yields at the same beam energy before and after each measurement at a given energy. It was found that only the graphite and Kapton targets lost some hydrogen during all measurements, while the W target lost some hydrogen only in the experiment with ^{19}F ion beam. In order to correct for the hydrogen loss, the detected yields for a given beam energy were normalized to the average of the two control measurements.
The α-particle yield from the ^{1}H(^{7}Li,α)^{4}He reaction, according to the definition of the cross section $\sigma(E)$ in the case of a thick target, was calculated as [16]:

$$N_\alpha = 2N_{Li}n\frac{\rho N_A}{M}\int_{E_0}^{0}\varepsilon\omega\frac{\sigma(E)}{dE_{Li}/dx}dE_{Li}. \tag{3}$$

Here ε is the efficiency of the detector, ω is the angular distribution factor of emitted α-particles (taken from ref. [17]) and n is the number of hydrogen atoms per crystal lattice atom. The stopping power was calculated using the SRIM code [18]. Electron screening was observed by fitting eq. 3 to experimental data with U_e and n as free parameters. The cross sections for bare nuclei $\sigma(E)$, were calculated using equation 2, with the S-factor taken from ref. [5]. The U_e and n values resulting from least-squares fits to our data together with the predicted values of the electron screening potential

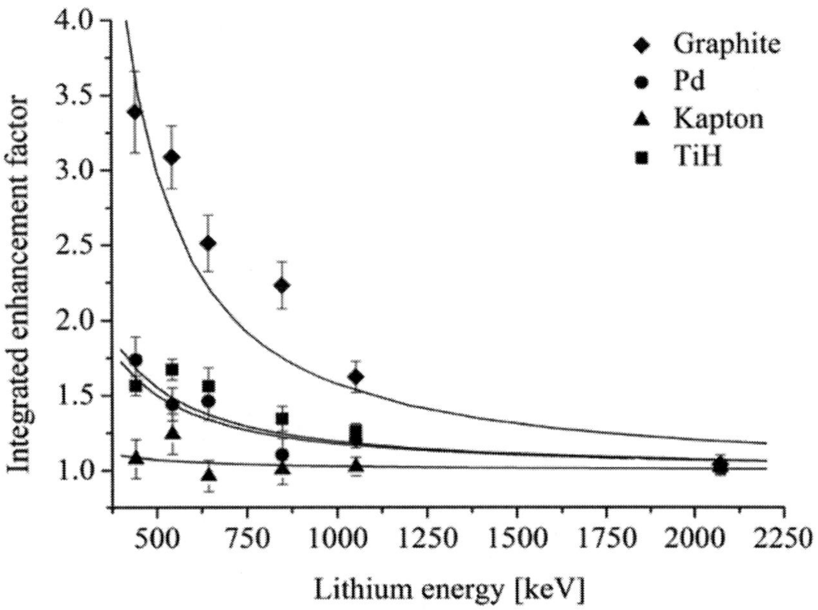

Figure 1: Integrated enhancement factors as a function of lithium beam energy for the ^{1}H(^{7}Li,α)^{4}He reaction in Kapton, Pd, graphite and TiH targets. The solid lines represent least-squares fits using eq. 3.

calculated from the approximation in adiabatic limit U_{ad} are summarized in table 1. Fig. 1 shows obtained enhancement factors as a function of Li beam energy in Kapton, Pd, graphite and TiH targets.

The ^{1}H(^{11}B,$\alpha\alpha$)^{4}He and ^{1}H(^{19}F,$\alpha\gamma$)^{16}O reactions were studied by measuring the narrow resonances at proton energies of 163 keV [19] and 340.46 keV [20], respectively. Since the beams lose all kinetic energy inside of the target region implanted with hydrogen, we observed step-like shaped resonances which are described by the infinitely thick target yield [3]:

$$Y_{\Delta E \to \infty}(E) = \frac{\lambda^2}{2\pi} \frac{\omega\gamma}{\varepsilon_r^{eff}} \left[arctan \left(\frac{E - E_r}{\Gamma/2} \right) + \frac{\pi}{2} \right]. \quad (4)$$

Here ε_r^{eff} is the effective stopping power at the resonance energy E_r, calculated by the SRIM [18] code. Measured α-particle and γ-ray yields we fitted to eq. 4, with E_r, $\omega\gamma$ and Γ as free parameters. Since $\sigma(E)$ is proportional to $\omega\gamma$, *i.e.*:

$$\sigma(E) = \frac{\lambda^2}{\pi} \frac{\omega\gamma}{\Gamma} \quad (5)$$

Table 1: The obtained electron screening potentials in reactions with lithium, boron and fluorine ion beam incident on implanted hydrogen in different targets together with bulk target hydrogen concentrations. The calculated values of the electron screening potentials in the adiabatic limit U_{ad} [2] are also given.

Target	n	$U_e(^7\text{Li})$ [keV]	$U_e(^{11}\text{B})$ [keV]	$U_e(^{19}\text{F})$ [keV]
U_{ad}		0.24	0.68	2.19
TiH	1.03 ± 0.04	3.9 ± 0.4	6.7 ± 1.8	62 ± 6
Pd	0.21 ± 0.03	3.6 ± 0.7	8.0 ± 1.9	63 ± 6
W	$(4.2 \pm 0.2)10^{-2}$	5.9 ± 0.9	–	74 ± 15
C	$(5.9 \pm 0.3)10^{-2}$	10.3 ± 0.4	32 ± 4	115 ± 8

and assuming $U_e{}^{Kapton} = 0$, the screening potential was calculated according to eq. 2 from the ratio $(\omega\gamma)_{metal}/(\omega\gamma)_{Kapton}$. The hydrogen concentrations we took from the experiment with the ^{7}Li beam since the same targets were used with all three ion beams and since the experiments were performed immediately after each other. For the W target we observed a surface hydrogen peak in experiments with ^{19}F (see fig. 2). Since this peak did not result from surface contamination, but from a larger number of hydrogen trapping sites at the surface [21], the W target was treated as a sum of a thin and a thick target. We assumed that the screening potential on the surface equals the one in the bulk and we found that the surface contains about 4 times higher hydrogen concentration than the bulk. The calculated values of the electron screening potential for the ^{1}H(^{11}B,$\alpha\alpha$)^{4}He and ^{1}H(^{19}F,$\alpha\gamma$)^{16}O reactions are given in table 1.

As it can be seen from table 1, large electron screening potential was observed in all targets with the highest value observed in graphite target. These results can be interpreted in the following way. The presence of a hydrogen impurity atom in the hexagonal graphite lattice creates a lattice distortion [22] placing the proton always closer to one carbon atom than to the others. This is different from cubic lattice of our metallic targets, where in the undamaged lattice the hydrogen finds an equilibrium position equidistant from the metallic atoms. However, the palladium and tungsten targets were radiation damaged and protons were trapped into crystal lattice vacancies [14], where they are again closer to one metallic atom than to the others. But we observed lower electron screening potentials in Pd

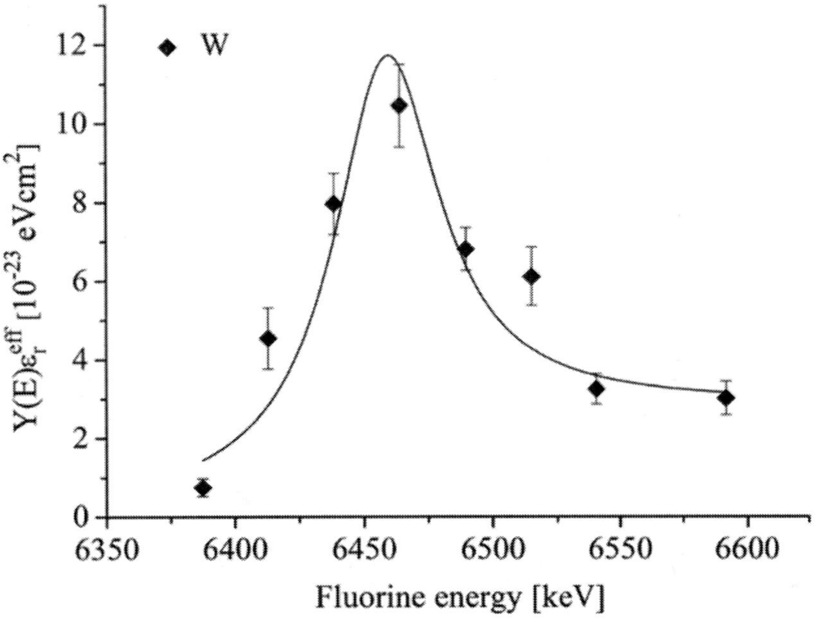

Figure 2: The resonance observed from the ^{1}H(^{19}F,$\alpha\gamma$)^{16}O reaction for W target fitted to a sum of a thin and an infinitely thick target yields. The γ-ray yield multiplied by the effective stopping power is shown as a function of beam energy.

and W targets than in the graphite target. The reason for this is that in polycrystalline metals, hydrogen can also be trapped at grain boundaries and voids [14], where we assume lower electron screening, making the effective electron screening for the two kinds of trapping sites lower. Next, the observed electron screening potential in the Pd target was lower than in the W target. A possible explanation for this is that our palladium sample was cold rolled, while the tungsten one was annealed. It is known that the cold rolling process increases the number of grain boundary defects, but when the foil is annealed the number of these defects is reduced. Due to this, the resulting effective screening potential in the Pd target is lower than in W. The TiH target did not have the full TiH_2 stoichiometry, since the powder has been stored in air for several years. The measured stoichiometry of 1.03 means that $TiH_{1.03}$ is a mixture of fcc and tetragonal fct lattices [23]. As we have shown in earlier experiments with Pd targets [12], the hydrogen on regular interstitial sites in a fcc lattice does not produce a large electron screening effect. Only when the protons are pulled away from their fcc equilibrium positions, a large screening effect occurs. This is the case in the fct lattice in TiH. Finally, from the presented results we calculated that the dependence of the electron screening potential on the proton number Z of the projectile seems to be higher than Z^2.

3 Summary

The electron screening effect on thick hydrogen containing graphite, titanium hydride, Pd, and W targets was investigated by studying the ^{1}H(^{7}Li,α)^{4}He, ^{1}H(^{11}B,$\alpha\alpha$)^{4}He and ^{1}H(^{19}F,$\alpha\gamma$)^{16}O reactions in inverse kinematics. Our measurements are the first ones that studied reactions with different charge numbers using physically the same targets. Contrary to expectations, large electron screening potentials were found in the graphite target: with ^{7}Li beam we observed $U_e(Li) = 10.3(4)$ keV, with ^{11}B beam $U_e(B) = 32(4)$ keV and $U_e(F) = 115(8)$ keV was deduced from the experiment with the ^{19}F beam. Large electron screening of $U_e(Li) = 3.9(4)$ keV, $U_e(B) = 6.7(18)$ keV and $U_e(F) = 62(6)$ keV was found in titanium hydride. Similarly, high values of Ue were found also in the tungsten target, *i.e.* $U_e(Li) = 5.9(9)$ keV and $U_e(F) = 74(15)$ keV and in palladium: $U_e(Li)$ $= 3.6(7)$ keV, $U_e(B) = 8.0(19)$ keV and $U_e(F) = 63(6)$ keV.
Also, these results clearly show that the dependence of the electron screening potential on the proton number Z of the projectile is not linear as it was previously supposed, but higher than Z_2.

References

[1] D. Bemmerer et al., Eur. Phys. J. A 24, 313 (2005).
[2] D. Leitner et al., Proc. 2011 Particle Accelerator Conf., New York, USA, 2552 (2011).
[3] C. Iliadis, Nuclear Physics of Stars, Wiley-VCH, Weinheim (2007).
[4] H. J. Assenbaum et al., Z. Phys. A 327, 461 (1987).
[5] L. Lamia et al., Astron. Astrophys. A188, 541 (2012).
[6] A. Huke et al., Phys.Rev. C 78, 015803 (2008).
[7] K. Czerski et al., J. Phys. G 35, 014012 (2008).
[8] F. Raiola et al., Eur.Phys.J. A 19, 283 (2004).
[9] J. Kasagi, Prog. Theor. Phys. Suppl. 154, 365 (2004).
[10] J. Cruz et al., J. Phys. G 35, 014004 (2008).
[11] F. Raiola et al., J. Phys. G 31, 1141 (2005).
[12] M. Lipoglavsek et al., Eur. Phys. J. A 44, 71 (2010).

[13] J. Vesic et al., Eur. Phys. J. A 50, 153 (2014).
[14] Y. Fukai, The Metal Hydrogen System, Springer, Berlin, Germany, (2005) Chapter 5.
[15] O. Ogorodnikova et al., J. Nucl. Mat. 451, 379 (2014).
[16] C. Rolfs and R. W. Kavanagh, Nucl. Phys. A 455, 179 (1986).
[17] S. Engstler et al., Z. Phys. A 342, 471 (1992).
[18] J. F. Ziegler et al., The Stopping and Range of Ions in Matter, Lulu Press Co., Morrisville, NC, (2008), www.srim.org.
[19] Y.S. Chen et al., Nucl. Phys. A 106, 1 (1967).
[20] K. Spyrou et al., Z. Phys. A 357, 283 (1997).
[21] S. Markelj et al., Appl. Surf. Sci. 282, 478 (2013).
[22] C. Herrero and R. Ramirez, J. Phys. D 43, 255402 (2010).
[23] C.P. Liang and H.R. Gong, Mater. Lett. 115, 252 (2014).

Indirect study of the ^{16}O+^{16}O fusion reaction toward stellar energies by the Trojan Horse Method

S. HAYAKAWA[1], C. SPITALERI[2,3], N. BURTEBAYEV[4], A. AIMAGANBETOV[5], P. FIGUERA[2] M. FISICHELLA[2] G.L. GUARDO[2], S. IGAMOV[6], I. INDELICATO[2], G. KISS[7], S. KLICZEWSKI[8], M. LA COGNATA[2], L. LAMIA[3], M. LATTUADA[2], E. PIASECKI[9], G.G. RAPISARDA[2], S. ROMANO[2,3], S.B. SAKUTA[10], R. SIUDAK[8], A. TRZCIŃSKA[9], A. TUMINO[2,11] and A. URKINBAYEV[5]

[1] Center for Nuclear Study, University of Tokyo, Wako, Japan
[2] INFN - Laboratori Nazionali del Sud, Catania, Italy
[3] Department of Physics and Astronomy, University of Catania, Catania, Italy
[4] Institute of Nuclear Physics of National Nuclear Center, Almaty Kazakhstan
[5] Gumilyov Eurasian National University, Astana, Kazakhstan
[6] Uzbek. Acad. Sci., Inst. Nucl. Phys., Tashkent, Uzbekistan
[7] MTA-Atomiki, Debrecen, Hungary
[8] The H. Niewodniczański Institute of Nuclear Physics PAN, Kraków, Poland
[9] Heavy Ion Laboratory, University of Warsaw, Warsaw, Poland
[10] National Research Center "Kurchatov Institute", Moscow, Russia
[11] Kore University of Enna, Enna, Italy

Abstract

The ^{16}O+^{16}O fusion reaction is important in terms of the explosive oxygen burning process during late evolution stage of massive stars as well as understanding of the mechanism of low-energy heavy-ion fusion reactions. We aim to determine the excitation function for the most major exit channels, α+^{28}Si and p+^{31}P, toward stellar energies indirectly by the Trojan Horse Method via the ^{16}O(^{20}Ne, α^{28}Si)α and ^{16}O(^{20}Ne, p^{31}P)α three-body reactions. We report preliminary results involving reaction identification, and determination of the momentum distribution of α-^{16}O intercluster motion in the projectile ^{20}Ne nucleus.

The ^{16}O+^{16}O fusion reaction is important in terms of the explosive oxygen burning process during late evolution stage of massive stars as well as understanding of the mechanism of heavy-ion fusion reactions at low energies. The astrophysical S-factor of such a heavy-ion fusion strongly depends on energy at corresponding stellar temperatures far below the Coulomb barrier. There are large discrepancies among different experiments [1–4], and among theoretical predictions [5,6], and is a lack of data below $E_{\mathrm{cm}} = 7$ MeV. We aim to determined the excitation function of the most major products, α+^{28}Si and p+^{31}P, of the ^{16}O+^{16}O reaction at stellar energies by the Trojan Horse Method (THM) [7].

We have performed THM measurements via the ^{16}O(^{20}Ne, α^{28}Si)α and ^{16}O(^{20}Ne, p^{31}P)α three-body reactions at $E_{20\mathrm{Ne}} = 45$ MeV at the Heavy Ion Laboratory, Warsaw, Poland, covering center-of-mass energy ranges of 8–15 MeV. In these three-body reactions, the α particles in the exit channels may act as the "spectator" through the quasi-free mechanism, where the momentum transfer of α decaying from the possible α cluster state in the projectile ^{20}Ne is sufficiently small. The momentum of the spectator is defined by masses and momenta of α and ^{20}Ne; $\boldsymbol{p}_s \equiv \boldsymbol{p}_\alpha - m_\alpha/m_{20\mathrm{Ne}} \times \boldsymbol{p}_{20\mathrm{Ne}}$. To guarantee quasi-free mechanism, the two-cluster α-^{16}O system in the nucleus ^{20}Ne should preferably be in s state, so that the momentum distribution of the spectator α is single-peaked at $p_s = 0$. Here we report preliminary p_s distribution investigated for the first time, which is crucial to determine the two-body reaction cross section by THM.

The experimental setup is illustrated in Fig. 1.

The ^{20}Ne^{3+} beam was provided at 45 MeV from the $K = 160$ cyclotron with a typical intensity around 20 enA on target, and the production run was performed for about 180 hours in total. For the beam collimator, a ϕ6-, a ϕ3- and a ϕ2-mm hole are laid straight on the beam axis within a distance of 380 mm from the upstream, respectively. We used WO_3 evaporated onto Au backing as solid oxygen target with a typical thickness of 116 mg/cm^2 for

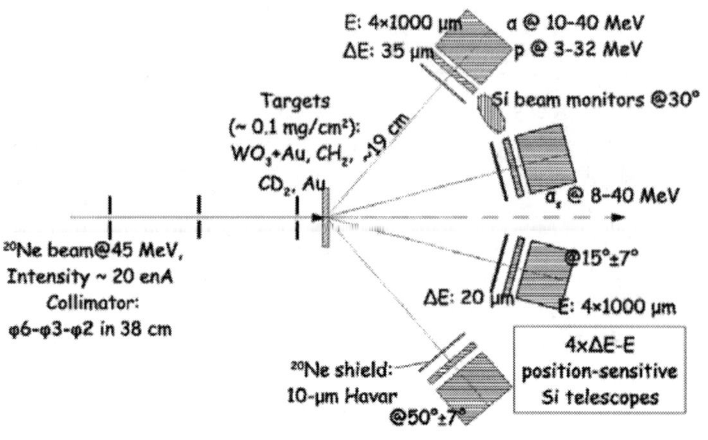

Figure 1: Schematic view of the experimental setup.

WO_3 and 193 mg/cm^2 for Au. Three silicon beam monitoring detectors were installed at 30°. For the reaction product measurement, four ΔE-E silicon telescopes were mounted symmetrically with respect to the beam axis at 15° and 50°. The thickness of each ΔE layer at 15° was 20 μm in order

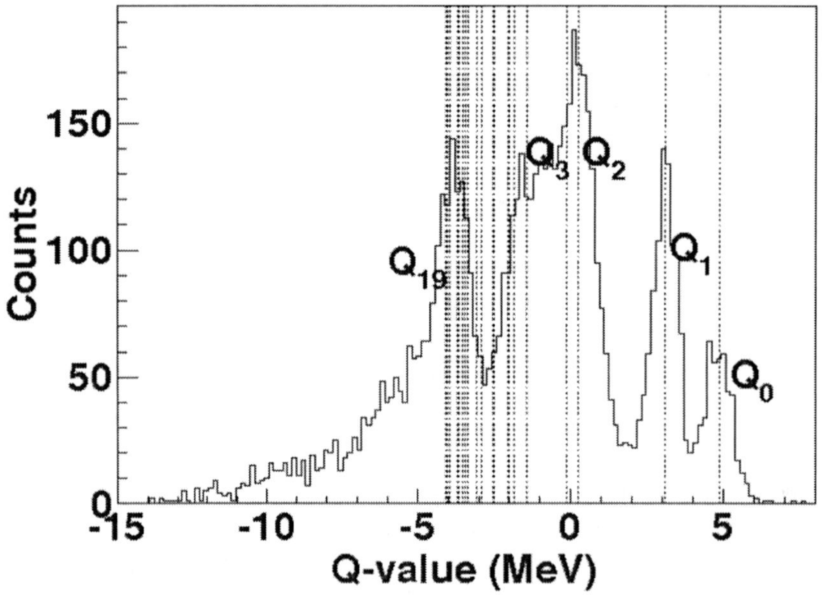

Figure 2: Q-value spectrum of the $^{16}O(^{20}Ne, \alpha^{28}Si)\alpha$ channel. The dotted lines corresponds to the excited states of ^{28}Si.

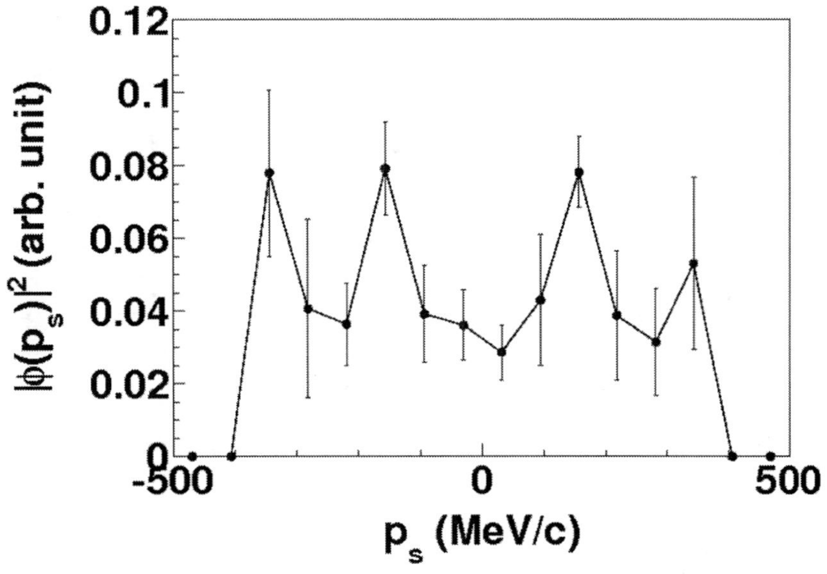

Figure 3: Preliminary momentum distribution of α in ^{20}Ne.

to measure low-energy spectator α, while that at 50° was 35 mm focusing on higher energy up to 40 MeV of α of the coincidence pair. Each E layer consisted of a stack of four 1-mm-thick silicon detectors for high-energy proton up to 32 MeV. The first E layer was position-sensitive by charge division, and the distances from the target were typically 190 mm. We put a 10-mm Havar foil right in front of each ΔE layer in order to prevent the detectors from plenty of beam scattering on W and Au in the target. During the production run with the WO_3 target, we mostly observed protons and α particles in the ΔE-E telescopes.

By selecting only α-particle data, we confirmed that the peaks found in the Q-value spectrum which is defined by $Q = E_{28\mathrm{Si}} - E_{20\mathrm{Ne}} + E_{\alpha 1} + E_{\alpha 2}$ correspond well to the excited energy of ^{28}Si nucleus as shown in Fig. 2, which evinces the $^{16}\mathrm{O}(^{20}\mathrm{Ne}, \alpha^{28}\mathrm{Si})\alpha$ reaction.

The preliminary momentum distribution is show in Fig. 3, assuming energy and angular distribution of the differential cross section of the two-body reaction $^{16}\mathrm{O}(^{16}\mathrm{O}, \alpha)^{4}\mathrm{He}$. The fact that the momentum distribution does not have the maximum value around $p_s = 0$ suggests that the three-body reactions $^{16}\mathrm{O}(^{20}\mathrm{Ne}, \alpha^{28}\mathrm{Si})\alpha$ and $^{16}\mathrm{O}(^{20}\mathrm{Ne}, p^{31}\mathrm{P})\alpha$ might not proceed through the 0^+ ground state of ^{20}Ne dominantly but the 2^+ first excited state. Further data analysis to determine the two-body cross section of interest is ongoing, also for the $^{16}\mathrm{O}(^{20}\mathrm{Ne}, p^{31}\mathrm{P})\alpha$ channel.

References

[1] H. Spinka and H. Winkler, Nucl. Phys. **233** (1974) 456.

[2] G. Hulke *et al.*, Zeitschrift für Physik A, Atoms and Nuclei, **297** (1980) 161.

[3] S. C. Wu and C. A. Barnes, Nucl. Phys. A, **422** (1984) 373.

[4] J. Thomas *et al.*, Phys. Rev. C, **33** (1986) 1679.

[5] C. L. Jiang *et al.*, Phys. Rev. C, **75** (2007) 1.

[6] A. Diaztorres, L. Gasques, and M. Wiescher, Phys. Lett. B, **652** (2007) 255.

[7] C. Spitaleri *et al.*, Phys. of Atomic Nuclei, **74** (2011) 1725.

[8] G. F. Chew and G. C. Wick, Phys. Rev. **85** (1952) 636.

EPJ Web of Conferences **117**, 09014 (2016) DOI: 10.1051/epjconf/201611709014
NN2015

Statistics of magnetar crusts magnetoemission

V. N. Kondratyev[1], Yu. V. Korovina[2]

[1]Physics Department, Taras Shevchenko National University of Kiev, 03022-UA Ukraine, and Bogolubov LTP, JINR, 141980, Dubna, Russia
[2]Moscow Institute of Open Education, 125167-RU Moscow, Russia

Abstract

Soft repeating gamma-ray (SGR) bursts are considered as magnetoemission of crusts of magnetars (ultranamagnetized neutron stars). It is shown that all the SGR burst observations can be described and systematized within randomly jumping interacting moments model including quantum fluctuations and internuclear magnetic interaction in an inhomogeneous crusty nuclear matter.

1 Introduction

The pioneering evidence for ultramagnetized astrophysical objects (magnetars) is associated with the March 5, 1979 event from SGR 0526-66 [1]. The magnetar concept, see [2,3], is strongly corroborated by further observations of soft gamma repeaters (SGRs) indicating significant multipole magnetic fields being substantially larger than the dipole component $B_{\rm dip} \sim 10^{15}$ G. For a noticeable contribution of magnetic pressure in a balance of crust forces in neutron star (NS) we write

$$dB^2_{\rm crust}/dR \approx 8\pi {\rm G} M n(R)/R^2, \qquad (1)$$

where the gravitational constant G, and the star mass $M(R)$ within radius R is related to the matter density $n(R)$ as $4\pi R^2 n(R) = dM/dR$. Substituting this relation into Eq. (1) and integrating over NS crust area we obtain the field strength $B \sim 10^{1.5}$ *teratesla* $(M/M_o)(10 \text{ km}/R)^2$, with the solar mass M_o. Thus field torroidal components can reach tens *teratesla* (TT) that is consistent with estimates [4, 5] based on the supernova explosion energy. Such fields can affect the structure and properties of atomic nuclei [3–5]. In this contribution the randomly jumping interacting moments (RJIM) model [3] is further extended for an analysis of SGR bursts.

2 Magnetodynamics of crusty nuclear matter

The RJIM model for magnetodynamics simulations in NS crusts has already been described in [3]. We briefly remind that in simulations of demagnetization dynamics we use a very general form for magnetic moments m of atomic nuclei $m = \mu \sum_n \nu_n \theta(b - b_n) = \mu I g$ with the nucleon magneton μ, nuclear g-factor g and spin I, as well as step function $\theta(x)$. Atomic nuclei occupied a volume V_{D} contribute to the magnetization $P = m/V_{\mathrm{D}}$. Taking g=3 for nuclear component of magnetic induction we get

$$P \;=\; 1.5 \text{ TG } I \; n/(10^{13}\text{g/cm}^3). \tag{2}$$

In a case of comparable sizes for nucleus and occupied volume V_{D} (i.e., $n \sim 10^{13,5}$ g/cm^3) internuclear interaction is ferromagnetic [3, 6]. Taking for magnetic coupling strength between nearest-neighbor (nn) elements the value J total Hamiltonian H for atomic nuclei array in a field H can be expressed as follows $\mathrm{H} = -\sum_i b_i m_i$ through an interaction of magnetic moment m_i with a local field $b_i = H(t) + J \sum_{j=\mathrm{nn}} + h_i$. Here the sum runs over nn elements and random fields h_i with Gaussian distribution of a width R called the disorder [3, 6–8], which allow to account for inhomogeneity, disorder and fluctuations.

3 RJIM model for SGR-burst activity

Let us consider adiabatically changing in time crust magnetic field H. When the local field value b_i of a certain NS crust domain becomes less than certain value b_n magnetization is changing stepwise. Due to the ferromagnetic interaction moment hopping may be triggered for nearest neighbors, which in turn may cause some discontinuity point for their neighbors, and so forth, producing, thereby, avalanches. Such avalanches lead to erratic jumps in magnetization curves and sharp release of magnetic energy estimated as

$$E = \; H \; \Delta P \; V_a \;=\; 10^{41}\text{Ergs}(H/\text{TT}) \; (\Delta I \; n/10^{13}\text{g/cm}^3) \; (V_a/10^6\text{m}^3). \tag{3}$$

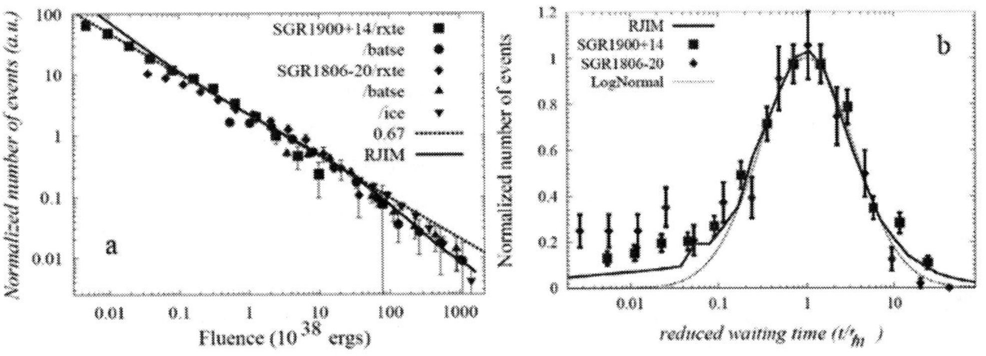

Figure 1: a) Normalized cumulative energy distributions of SGR-bursts are compared with the results of RJIM for the cubic lattice of a size $(150)^3$ represented by the solid line. The data of the RXTE and BATSE observations for SGR 1900+14 from [9] are shown by squares and circles, respectively. RXTE (diamonds), BATSE (up-triangles), and ICE (down-triangles) data for SGR 1806-20 are from [10]. The dashed line denotes the power law distribution. b) The reduced waiting time distribution between the successive RXTE/PCA bursts from SGR 1900+14 (squares) [9] and SGR1806-20 (diamonds) [10] are compared with the waiting time distribution between avalanches (solid curve). The dashed line represents the fit to the lognormal distribution of the width 3.6.

For a field strength $H \sim 3$ TT, typical magnitar crust density $n \sim 10^{13,5}\text{g/cm}^3$, and avalanche linear size of order of outer NS crust thickness, 100 m (i.e., V_a about 10^6 m^3) the amount of energy obtained from Eq. (3) is consistent with an energy of soft gamma-ray bursts.

Figure 1a represents the cumulative distributions of detected burst energy, i.e. the burst number with an energy exceeding certain value. The RJIM results are in a good agreement with observations for 7 periods of energy. The obtained event number dependence is well fitted by the power law with an exponent 0.67, which corresponds to the value 1.67 for the differential distribution and provides a signal of self-organized criticality in the burst statistics.

Within RJIM model a time-scale in a period of burst activity is determined by the ratio of the disorder parameter R and the field change rate $\tau = R/(dB/dt)$. Therefore, the scaling with respective time leads to an universal function as is shown in Fig. 1b for normalized waiting time, i.e., inter-burst time interval, distribution.

4 Conclusion

We considered magnetodynamics of inhomogeneous crusty nuclear matter. Accounting for inter-nuclide magnetic coupling we show that quantum fluctuations in nuclear magnetic reactivity give rise to erratic jumps in magnetotransport of neutron star crusts. The resulting sharp energy releases lead to gamma-ray bursts. The properties of such a noise are favorably compared with burst statistics of Soft Gamma Repeaters. As is shown the predicted by RJIM model scaling properties for, e.g., the burst intensity and waiting time distributions, are in a good agreement with SGR observations supporting thereby the credibility of RJIM model. Thermal effects can also affect structural and collective nuclear matter properties, see [8].

This work is supported in part by scientific data center of Integral mission and SCOPES IZ73Z0-152485. One of us (V.N.K.) thanks JINR (Dubna, Russia) for warm hospitality and financial support.

References

[1] Mazets E.P. et.al. *Nature* (London) **282** (1979) 587.

[2] Svinkin D.S. et.al. *Mon. Not. R. Astron. Soc.* **447** (2015) 1028.

[3] Kondratyev V.N. *Phys. Rev. Lett.* **88** (2002) 221101.

[4] Kondratyev V.N. *Phys. Atom. Nucl.,* **75** (2012) 1368.

[5] Kondratyev V.N. *Eur. Phys. J.* A **50** (2014) 7..

[6] Kondratyev V.N., Lutz H.O. *Eur. Phys. J.* D **9** (1999) 483.

[7] Kondratyev V.N. *Phys. Lett.* A **179** (1993) 209; *Phys. Lett.* A **190** (1994) 465.

[8] Kondratyev V.N., Di Toro M. *Phys. Rev.* C **53** (1996) 2176.

[9] Gogus E. et.al. *Ap.J Letters* **526** (1999) L93.

[10] Gogus E. et.al. *Ap.J Letters* **532** (2000) L121.

EPJ Web of Conferences **117**, 09015 (2016) DOI: 10.1051/epjconf/201611709015
NN2015

Preliminary studies for three experiments at Treiman-Yang criterion

I.V. Kres[1], V.N. Kondratyev[1], S. Cherubini[2] and C. Spitaleri[2]

[1] Taras Shevchenko National University of Kyiv, Ukraine
[2] Dipartimento di Fisica e Astronomia, Università di Catania e INFN - Laboratori Nazionali del Sud, Catania, Italy

Abstract

Nuclear reactions with three bodies in their final state may proceed through different reaction mechanisms. The Feynman graph technique has been widely used to describe such reactions. However, it is very difficult in general to select the graphs that dominate in given process. The Treiman-Yang criterion is one of the most powerful experimental tests for verifying the pole approximation prediction when describing a quasi-free reaction mechanism. We propose the theoretical study of the $H^2(B^{10}, \alpha\ Be^7)n_s$, $H^1(B^{11}, \alpha_1\alpha_2)\alpha_s$, $He^3(Be^9, \alpha_1\alpha_2)\alpha_s$ reactions at different energies. The preliminary study helps to check the existence of a QF channel by using the TY creterion.

The Trojah Horse Method (THM) is a powerful indirect technique that allows to determine the astrophysical factor for rearrangement reactions. The THM involves obtaining the cross section of the binary process 0+T→1+2 at astrophysical energies by measuring the Trojan Horse (TH) reaction (the two-body to three-body process) 0+N→1+2+s in the quasi-free (QF) kinematics mode, where the Trojan Horse particle, N = (s+T), which has a dominant cluster structure, is accelerated at energies above the Coulomb barrier.

The advantage of the THM is that the extracted cross section of the binary subprocess does not contain the Coulomb barrier suppresion factor and electron screening does not play a role in the cross section. After penetrating the barrier, the TH-nucleus N undergoes breakup leaving particle T

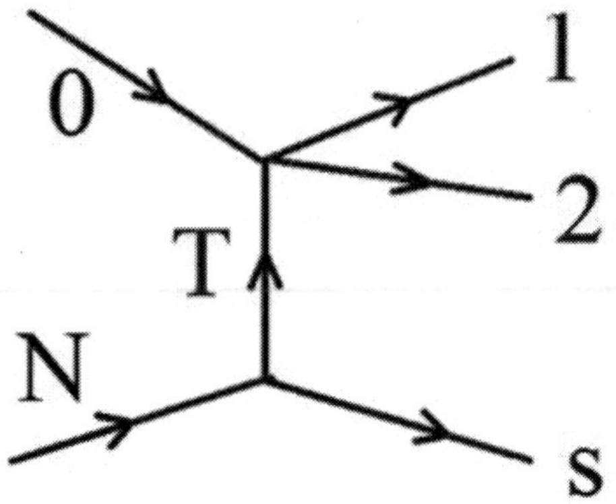

Figure 1: The diagram describing the TH reaction 0+N→1+2+s in the QF kinematics

to interact with particle 0 while particle s, also called spectator, flies away. From the measured cross section of TH reaction, the energy dependence of the binary sub-process is determined. The TH reaction mechanism shown schematically in fig.1, gives the dominant contribution to the cross section in a restricted region of the three-body phase space when the relative momentum of the fragments s and T is zero (the QF kinematical condition, when l=0). [1]

Nuclear reactions with three bodies in their final state may proceed throught a variety of reaction mechanisms. In general the calculation of the reaction cross-section should include terms from quasi-free processes, expecially when appropriate regions of the phase-space are selected. Many models have been developed to describe such direct reactions. Among these the Feynman graph technique has been widely used but unfortunately it is very difficult in general to select the particular graphs that are dominant for a given process. In particular the role of the pole mechanism has been investigated in detail and attempts have been made to find sensitive criteria able to establish its relative importance. One of these criteria is the study of the distribution with respect to the Treiman-Yang angle. This criterion shows that under given conditions such a distribution should be isotropic, provided the pole machanism is dominant in reaction. The Treiman-Yang criterion is a necessary criterion but not a sufficient condition for dominance of the pole graph, due to some triangular graph that may fulfil this criterion, it remains strongly indicative of the pole character of the process under consideration.

The Treiman-Yang (TY) criterion establishes that the amplitude of the

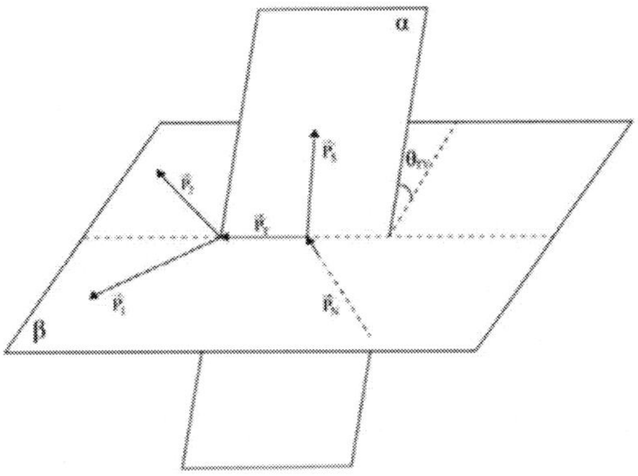

Figure 2: TY criterion as the rotation of plane α with respect to plane β.

three-body reaction should be invariant under rotation of the ($\vec{p_1}$,$\vec{p_2}$) plane around the sum of these momenta, in a reference frame in which the projectile is at rest [2]. The quantities of three Galilean invariants do not change if in the anti-laboratory system (that is the system for which $\vec{p_0}$=0) the momenta of the particles 1 and 2 are rotated with respect to their sum. In these condition the differential cross section must remain constant under this rotation.

The application of the Treiman-Yang criterion is related to the comparison of the differential cross-sections measured at different detection conditions corresponding to the same value of the Galilean invariants but with differents angles Θ_{TY}. In particular a Treiman-Yang distribution can be visualized as the rotation of the plane α with respect to plane β as shown in fig.2.

The plane α is defined by $\overrightarrow{p'_T}$ and $\overrightarrow{p'_N}$, the plane β by $\overrightarrow{p'_T}$ and $\overrightarrow{p'_1}$:

$$\cos\Theta_{TY} = \frac{(\overrightarrow{p'_T}\times\overrightarrow{p'_N})\bullet(\overrightarrow{p'_T}\times\overrightarrow{p'_1})}{|\overrightarrow{p'_T}\times\overrightarrow{p'_N}|\bullet|\overrightarrow{p'_T}\times\overrightarrow{p'_1}|}$$

And by taking into account the identity:

$$(\vec{A}\times\vec{B})\bullet(\vec{C}\times\vec{D}) = (\vec{A}\bullet\vec{C})\bullet(\vec{B}\bullet\vec{D}) - (\vec{A}\bullet\vec{D})\bullet(\vec{B}\bullet\vec{C})$$

It can be shown that

$$\cos\Theta_{TY} = \frac{{p'_T}^2 p'_{1z} + p'_{Tz}(p'_{Tx}p'_{1x} + p'_{Ty}p'_{1y} + p'_{Tz}p'_{1z})}{\sqrt{{p'_{Ty}}^2 + {p'_{Tx}}^2}\cdot\sqrt{(p'_{Ty}p'_{1z} - p'_{Tz}p'_{1y})^2 + (p'_{Tz}p'_{1x} - p'_{Tx}p'_{1z})^2 + (p'_{Tx}p'_{1y} - p'_{Ty}p'_{1x})^2}}$$

Where all the momenta are in the antilaboratory system. The transformation from laboratory to the antilaboratory system is:

$$\overrightarrow{p'_k} = \overrightarrow{p_k} - \frac{m_k}{m_0}\overrightarrow{p_0}$$

In the case of the present work, we want to extend and experimentally proof the pole approximation validity by means of the TY criterion applied

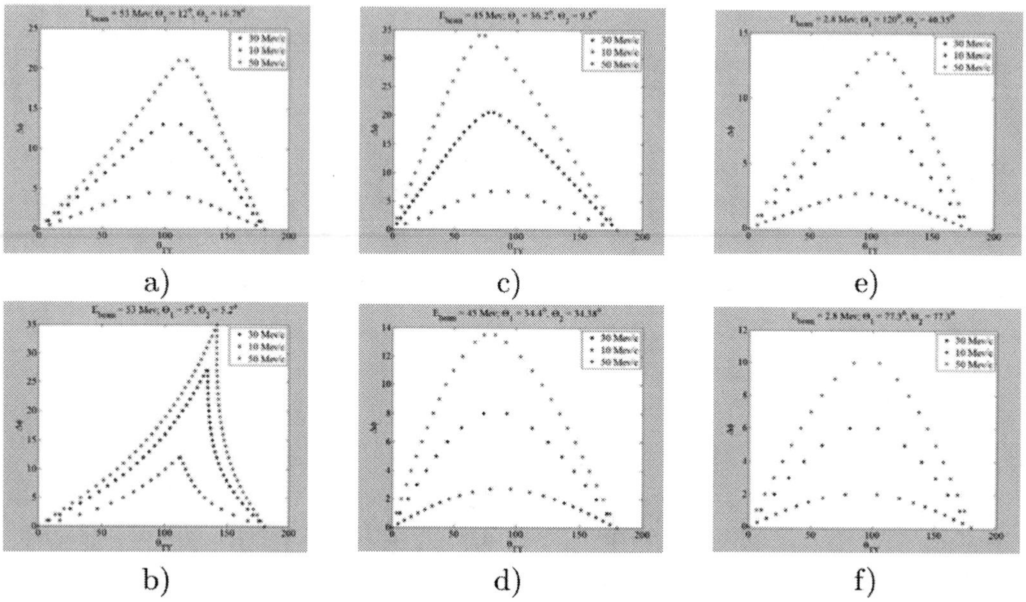

Figure 3: Correlation between ϕ_2 and Θ_{TY} angles for various fixed value of the spectator momentum modulus

to three cases:

•$H^2(B^{10}, \alpha\ Be^7)n_s$ reaction induced at a boron beam energy of 53 MeV in the laboratory(fig.3.a, fig.3.b). The plan is to use the $p + n_s$ cluster configuration of the target H^2 nucleus and detect the emitted Be^7 and α particles in the binary virtual reaction $H^1(B^{10}, Be^7)\alpha$.

•$H^1(B^{11}, \alpha_1\alpha_2)\alpha_s$ reaction induced at a boron beam energy of 45 MeV in the laboratory(fig.3.c, fig 3.d). The plan is to use the $\alpha_s + Li^7$ cluster configuration of the projectile B^{11} nucleus and detect the emitted α_1 and α_2 particles in the binary virtual reaction $H^1(Li^7, \alpha_1)\alpha_2$.

•$He^3(Be^9, \alpha_1\alpha_2)\alpha_s$ reaction induced at a helium beam energy of 2.8 MeV in the laboratory(fig.3.e, fig.3.f). The plan is to use the $He^5 + \alpha_s$ cluster configuration of the target Be^9 nucleus and detect the emitted α_1 and α_2 particles in the binary virtual reaction $He^3(He^5, \alpha_1)\alpha_2$.

References

[1] R.Tribble, C. Bertulani, M. La Cognata, A. Mukhamedzhanov, C. Spitaleri, Indirect techniques in nuclear astrophysic, Rep. Prog. Phys. 77 (2014) 106901 (49pp).

[2] S. B. Treiman and C. N. Yang, Phys. Rev. Lett. 8 (1962) 140.

EPJ Web of Conferences **117**, 09016 (2016) DOI: 10.1051/epjconf/201611709016
NN2015

Resonance strength measurement at astrophysical energies: The $^{17}O(p,\alpha)^{14}N$ reaction studied via THM

M.L. Sergi[1], C. Spitaleri[1,2], M. La Cognata[1], L. Lamia[2], R.G. Pizzone[1], G.G. Rapisarda[2] A. Mukhamedzhanov[3], B. Irgaziev[4], X.D. Tang[5], M. Wischer[5], J. Mrazek [6], V. Kroha[6]

[1] INFN - Laboratori Nazionali del Sud, Catania, Italy
[2] Dipartimento di Fisica e Astronomia, Università di Catania, Catania, Italy
[3] Cyclotron Institute, Texas A&M University, College Station, TX 77843, USA
[4] GIK Institute of Engineering Sciences and Technology, Topi, Districti Swabi, K.P. Pakistan
[5] Nuclear Physics Institute of ASCR, Rez, Czech Republic

Abstract

In recent years, the Trojan Horse Method (THM) has been used to investigate the low-energy cross sections of proton-induced reactions on ^{17}O nuclei, overcoming extrapolation procedures and enhancement effects due to electron screening. We will report on the indirect study of the $^{17}O(p,\alpha)^{14}N$ reaction via the Trojan Horse Method by applying the approach developed for extracting the resonance strength of narrow resonance in the ultralow energy region. The mean value of the strengths obtained in the two measurements was calculated and compared with the direct data available in literature.

1 Introduction

The ^{17}O+p reactions are of paramount importance for the nucleosynthesis in a number of stellar sites, including red giants (RG), asymptotic giant branch (AGB) stars, massive stars and classical novae [1]. At temperatures typical of the above mentioned astrophysical scenario, T=0.01-0.1 GK for RG, AGB [2] and massive stars and T=0.1-0.4 GK for classical nova explosion, the^{17}O(p,α)^{14}N reaction cross section is dominated by two resonances: one at about $E^R_{c.m.}$ =65 keV above the ^{18}F proton threshold energy, corresponding to the E_X=5.673 MeV level in ^{18}F, and another one at $E^R_{c.m.}$=183 keV (E_X=5.786 MeV). In the last few years, several measurements [3–7] of the $E^R_{c.m.}$=183 keV resonance both in the (p,γ) and (p,α) channels have drastically reduced the uncertainties on both ^{17}O(p,α)^{14}N and ^{17}O(p,γ)^{18}F rates in the context of explosive H burning, whereas only one direct measurement [8] of the $E^R_{c.m.}$=65 keV resonance was performed in the (p,α) channel. However, the screening of the nuclear Coulomb field by atomic electrons was not taken into account in [8].

In this paper, we report on the indirect measurement of the ^{17}O(p,α)^{14}N reaction at energies below 300 keV by means of the Trojan Horse Method (THM). In particular, by using the THM formalism developed for the resonant reaction, [9,10], both the 65 and 183 keV resonances were observed and the resonance strength of the 65 keV resonance has been deduced. Here, we apply the THM to measure the cross section of the ^{17}O(p,α)^{14}N reaction by selecting the QF contribution to the ^{2}H(^{17}O,^{14}Nα)n reaction. The proton is brought inside the nuclear field of ^{17}O, while the neutron acts as a spectator to the ^{17}O(p,α)^{14}N QF reaction. Deuteron was used as the Trojan Horse nucleus because of its $p-n$ structure and its relative low binding energy ($\sim$ 2.2 MeV) [9,10].

2 The measurements and the results

The study of the ^{17}O(p,α)^{14}N via the THM application was made via two experiments: the first one at the Laboratori Nazionali del Sud (LNS) in Catania (Italy) and the second one at the Nuclear Science Laboratory (NSL) of the University of Notre Dame (USA). The experimental setup of the two experimental measurements, the data analysis and the results have been already described step by step in [10–12], here only the resulting 65 keV resonance strengths for the LNS and NSL experiments are reported. In particular, by using the formalism reported in [10], one gets the strength of the

DOI: 10.1051/epjconf/201611709016

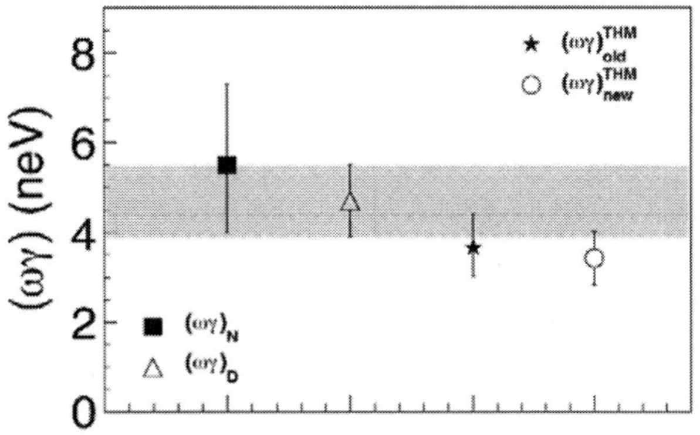

Figure 1: Comparison between the different values of the 65 keV resonance strength discussed in the text. The gray band represents the recommended range deduced by [3].

resonance at 65 keV for the LNS, $(\omega\gamma)_1^{THM} = (3.72\pm0.78)\times10^{-9}$ eV, and for the NSL experiment, namely $(\omega\gamma)_1^{THM} = (3.16\pm0.68)\times10^{-9}$ eV. The total error (∼21% for the LNS experiment and ∼22% for the NSL experiment) on the two $(\omega\gamma)_1^{THM}$ values is the sum in quadrature of the independent uncertainties (20.6% for LNS experiment and 21.4% for NSL experiment) due to the statistical error, the combinatorial background subtraction and peak value correlation (see Ref. [10]), and of the common uncertainty due to the normalization procedure (4.2% for both LNS and NSL experiments). For both LNS and NSL experiments, the normalization has been performed by scaling the strength of the 65 keV resonance, $(\omega\gamma)_1$, to the 183 keV one, $(\omega\gamma)_2$, which is well known from the literature. In more detail, the adopted value for the strength of the 183 keV resonance is $(\omega\gamma)_2 = (1.67\pm0.07)\times10^{-3}$ eV, obtained by the weighted average of the four strength values reported in literature [3–5, 13]. Considering the upper and lower limits, the resulting weighted average $(\omega\gamma)_{new}^{THM} = (3.42\pm0.60)\times10^{-9}$ eV is in good agreement with the strength given by NACRE, $(\omega\gamma)_N = (5.5^{+1.8}_{-1.5})\times10^{-9}$ eV [14] and with the direct value, $(\omega\gamma)_D$=$(4.7\pm0.8)\times10^{-9}$ eV calculated by using the same Γ_p and Γ_α reported in [3, 13], namely Γ_α=130 eV [15] and Γ_p=19± 3 neV [16,17]. The $(\omega\gamma)_{new}^{THM}$ value is, at the end, in agreement with the value $(\omega\gamma)_{old}^{THM}$=$(3.66^{+0.76}_{-0.64})\times$ 10^{-9} eV measured in the previous THM analysis, that has been instead deduced considering the only three 183 keV resonance strength values reported in literature [3–5].

Acknowledgement

This work has been partially supported by the Italian Ministry of University MIUR under the grants RFBR082838 (FIRB2008) and 'LNS-Astrofisica Nucleare (fondi premiali)'. It has been partially supported also by the grant LC 07050 of the Czech MŜMT, grant M10480902 of the Czech Academy of Science, grant LH1101 of the AMVIS project. A. M. M. acknowledges the support by U.S. Department of Energy under Grants No. DE-FG52-09NA29467, No. DE-FG02-93ER40773, and No. DE-SC0004958 and by NSF under Grant No. PHY-0852653.

References

[1] M. Hernanz et al. 1999, *Astrophys. J.* **526** L97-L100

[2] S. Palmerini et al. 2013, *Astrophys. J.***764** 128

[3] A. Chafa et al. 2007, *Phys. Rev. C***75** 035810

[4] B. Moazen et al. 2007, *Phys. Rev. C***75** 065801

[5] J. Newton et al. 2007, *Phys. Rev. C***75** 055808

[6] D. Scott et al. 2012, *Phys. Rev. Lett.***109** 202501

[7] A. Di Leva et al. 2014, *Phys. Rev. C***89** 015803

[8] J. Blackmon et al. 1995, *Phys. Rev. Lett.***74** 2642

[9] R. Tribble et al. 2014, *Rep. Prog. Phys.***77** 106901

[10] M. L. Sergi et al. 2015, *Phys. Rev. C***92** 065803

[11] M. L. Sergi et al. 2010, *Phys. Rev. C***82** 032801(R)

[12] M.L. Sergi 2010, *Il nuovo Cimento* **125 B** N.4, p. 457

[13] C. Iliadis et al. 2010, *Nucl. Phys. A* **841** 251

[14] C. Angulo et al. (NACRE), *Nucl. Phys. A***656** 3

[15] H. Mak et al. 1980, *Nucl. Phys. A***343** 79

[16] C. Fox et al. 2005, *Phys. Rev. C***71** 055801

[17] M. Hannam and W. Thompson 1999, *Nucl. Instrum. Methods A***431** 239

EPJ Web of Conferences **117**, 09017 (2016) DOI: 10.1051/epjconf/201611709017
NN2015

Exchange effects in the radiative capture reactions ^{3}H$(\alpha,\gamma)^7$Li and ^{3}He$(\alpha,\gamma)^7$Be

A.S. Solovyev[1], S.Yu. Igashov[1] and Yu.M. Tchuvil'sky[2]

[1]All-Russia Research Institute of Automatics, Moscow, 127055 Russia
[2]Skobeltsyn Institute of Nuclear Physics, Lomonosov Moscow State University, Moscow, 119991 Russia

Abstract

The mirror ^{3}H$(\alpha,\gamma)^7$Li and ^{3}He$(\alpha,\gamma)^7$Be reactions have been considered using the algebraic versions of the resonating group model and of the orthogonality conditions model. Exchange effects in interaction of the colliding nuclei and influence of the corresponding exchange terms in matrix elements of the interaction potential on calculated astrophysical S-factors for the reactions have been studied.

At the present time one of the most important aims of nuclear reaction theory is to build so called *ab initio* approaches to description of nuclear reactions, which use realistic NN-potentials and exactly account for all exchange effects caused by the antisymmetry of a wave function of nuclear system. Such approaches are in continuous progress but, nevertheless, turn out to be very complicated procedures, which are, in fact, unrealizable for sufficiently heavy colliding nuclei. Though, there are alternative approaches suitable for heavier systems. They are based on the resonating group model (RGM) [1] or its algebraic version (AVRGM) [2] with effective NN-potentials and, as above mentioned approaches, rigorously allow for the exchange effects. However, complexities rapidly grow even in such approaches at increasing system mass. Thus, a development of reliable approximations within the existing microscopic approaches is rather topical problem of nuclear reaction theory. Such approximations will allow to simplify the calculations significantly and even to make ones realizable for sufficiently complicated systems.

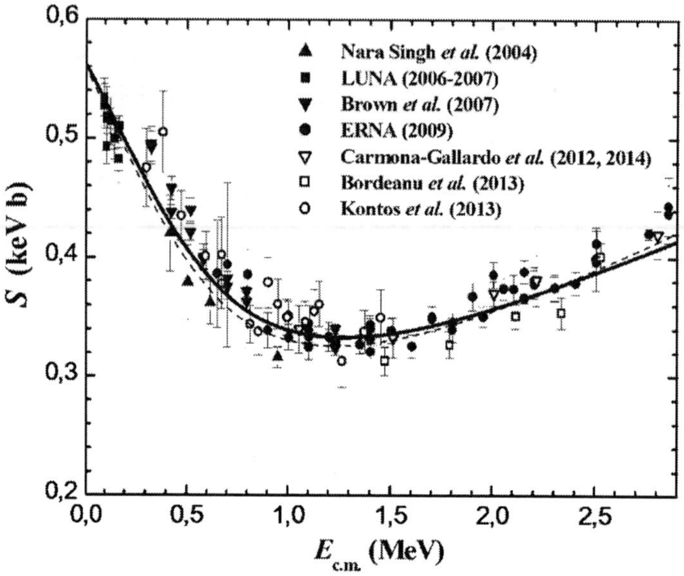

Figure 1: Astrophysical S-factor for the $^3He(\alpha,\gamma)^7Be$ reaction.

One of approximations originating from the RGM consists in the exchange term neglect in the matrix elements of Coulomb and nuclear potentials of microscopic Hamiltonian. The approximation based on such simplified treatment of the exchange effects is known as the orthogonality conditions model (OCM) [3]. The same assumption done in the framework of the AVRGM leads to the algebraic version of the OCM (AVOCM) [4]. So, the main aim of the present work is to study the role of the discussed exchange effects in the radiative capture reactions $^3H(\alpha,\gamma)^7Li$ and $^3He(\alpha,\gamma)^7Be$ using the AVRGM and the AVOCM, as well as to consider a possibility of the exchange term neglect at the radiative capture description.

It should be noted, that radiative capture reactions at low energies play very significant role in nuclear astrophysics [5]. It is a reason for permanent interest to this type of reactions, in general, and to the $^3H(\alpha,\gamma)^7Li$ and $^3He(\alpha,\gamma)^7Be$ reactions, in particular.

Availability of the AVRGM to description of energy dependence of astrophysical S-factors for the radiative capture reactions as well as its high efficiency were demonstrated in [6,7]. All details of the model, which are necessary for the calculations, can be found in [2,6,7]. Details of the AV-OCM are presented in [4]. It should be emphasized, that in the AVOCM exchange terms connected only with the permutations of nucleons between different clusters are neglected in the matrix elements of interaction poten-

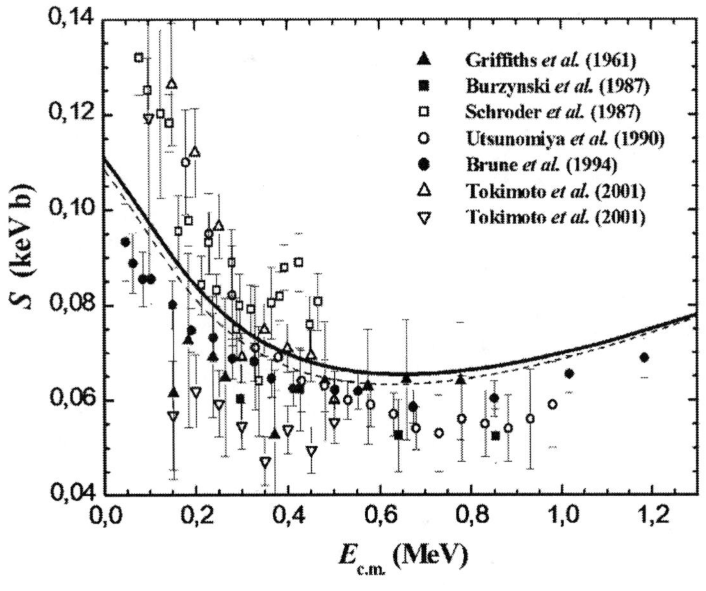

Figure 2: Astrophysical S-factor for the $^3\mathrm{H}(\alpha,\gamma)^7\mathrm{Li}$ reaction.

tial only. At the same time, all exchange terms are taken into consideration in matrix elements of all other operators (exchange normalization kernel, kinetic energy, *etc*). This character feature distinguishes the AVOCM from the potential cluster model that excludes all exchange terms.

Astrophysical S-factor for reaction induced by charged particles is related to cross section σ in the following conventional way

$$S(E_{\mathrm{c.m.}}) = E_{\mathrm{c.m.}}\, \sigma(E_{\mathrm{c.m.}}) \exp{(2\pi\eta)}, \tag{1}$$

where $E_{\mathrm{c.m.}}$ is the relative motion energy of colliding nuclei in center-of-mass system, η is Coulomb parameter. At sub-barrier energies the behavior of astrophysical S-factor looks smoother than cross section and, therefore, is more suitable for the analysis. Astrophysical S-factors for the mirror $^3\mathrm{He}(\alpha,\gamma)^7\mathrm{Be}$ and $^3\mathrm{H}(\alpha,\gamma)^7\mathrm{Li}$ reactions are shown in figs. 1 and 2 respectively. Solid lines are the AVRGM-calculation including all exchange terms. Curves obtained by the AVRGM-calculation without the Coulomb exchange terms practically coincide with the solid curves in the figs. Dashed lines are the AVRGM-calculation without the nuclear exchange terms. Curves calculated within approach combining the AVRGM and the AVOCM, *i.e.* without the exchange terms, simultaneously, in matrix elements of the Coulomb and nuclear interactions between colliding nuclei, almost do not differ from dashed lines in the figs. Only two adjustable parameters were used in the calculations. The first one is intensity of the central Majorana forces g_c entering into the modified Hasegawa–Nagata potential [8] employed to describe nuclear interaction. The second one is oscillator radius r_0 involved in the internal wave functions of the colliding clusters, which are chosen in form of the translationally invariant oscillator shell model lowest states eigenfunctions compatible with the Pauli exclusion principle. All the calculations are performed with the unified set of the parameter values: $r_0 = 1.22$ fm, $g_c = 1.035$. As it can be seen from the figs. 1 and 2, all curves are in good agreement with the experimental data (see refs. cited in [6, 7]).

Thus, the nuclear exchange terms affect the energy dependence of the astrophysical S-factors stronger than the Coulomb ones. For example, the nuclear exchange term neglect slightly decreases values of the astrophysical S-factors for the ^{3}H$(\alpha,\gamma)^7$Li and ^{3}He$(\alpha,\gamma)^7$Be reactions at low energies. But influence of the exchange matrix elements of the Coulomb and nuclear interactions between the colliding nuclei on the form of the curves, as a whole, turns out to be unessential enough. Consequently, approximations typical for the AVOCM allow to describe the radiative capture reactions and it makes opportunities for further applications of this model to perform calculations in the case of more complicated systems, when fully microscopic approaches become unrealizable at all.

References

[1] Wildermuth K. and Tang Y.C., *A Unified Theory of the Nucleus* (Vieweg, Braunschweig) 1977.

[2] Filippov G.F. and Okhrimenko I.P., *Phys. Atom. Nucl.*, **32** (1980) 480.

[3] Saito S., *Prog. Theor. Phys.*, **41** (1969) 705.

[4] Igashov S.Yu. and Tchuvil'sky Yu.M., *Eur. Phys. J. Web Conf.*, **38** (2012) 16002.

[5] Xu Y. et al., *Nucl. Phys. A*, **918** (2013) 61.

[6] Solovyev A.S., Igashov S.Yu., and Tchuvil'sky Yu.M., *Eur. Phys. J. Web Conf.*, **86** (2015) 00054.

[7] Solovyev A.S., Igashov S.Yu., and Tchuvil'sky Yu.M., *J. Phys. Conf. Ser.*, **569** (2014) 012020.

[8] Kanada H. et al., *Prog. Theor. Phys.*, **61** (1979) 1327.

DOI: 10.1051/epjconf/201611710001

Extreme Light Infrastructure Nuclear Physics (ELI–NP): Present status and perspectives

N.V. Zamfir

ELI–NP, "Horia Hulubei" National Institute for Physics and Nuclear Engineering (IFIN–HH), 30 Reactorului Street, Bucharest-Măgurele, 077125 Romania

Abstract

Extreme Light Infrastructure – Nuclear Physics (ELI–NP), a new Research Center under construction, will use extreme electromagnetic fields for nuclear physics research and will be operational in 2018. The status of the Project implementation will be presented. At ELI–NP, a high power laser system together with a very brilliant gamma beam are the two main research tools. Their targeted operational parameters will be described. The related experimental set-ups will be presented, together with the main directions of the research envisioned.

Introduction

The Extreme Light Infrastructure (ELI), listed on the ESFRI (European Strategic Forum for Research Infrastructures) 2006 road-map, is distributed among three European countries (the Czech Republic, Hungary and Romania) and funded by European Commission through Structural Funds. The Extreme Light Infrastructure – Nuclear Physics (ELI–NP) pillar [1], located near Bucharest in the Magurele Physics research campus, is midway the implementation phase. The project was approved by European Commission in September 2012 and will be operational in 2018. The civil construction of the new center, consisting of laser building, accelerator and experiments building, laboratories building, offices, guest house and canteen started in June 2013 and will be finished in the beginning of 2016.

Figure 1 shows the status of the construction.

Figure 1: The civil construction stage of ELI–NP in May 2015.

The total cost of the facility will be 300 million euro, without VAT.

At the time of entering operation, ELI–NP will be the most advanced research infrastructure in the world focused on photonuclear physics studies and applications. The facility will cover frontier fundamental physics, new nuclear physics and astrophysics as well as applications in nuclear materials and radioactive waste management, materials science and life sciences [2–4].

2 Scientific equipment

Two large, cutting-edge pieces of equipment will promote ELI–NP to the forefront of scientific research at international level: the High Power Laser System (HPLS) and the high intensity Gamma Beam System (GBS).

The High Power Laser System [5] consists of 2 arms of 10 PW each based on Optical Parametric Chirped Pulse Amplification (OPCPA) system [6, 7] at about 820 nm central wavelength, which includes Ti: Sapphire amplifiers to bring the final output energy to the level of a few hundreds of Joule. Subsequently, the pulses are compressed to around a 20 fs pulse duration. The repetition rate is 1 shot per min for each of the two arms. Along the two amplification chains, additional outputs with corresponding optical compressors will be installed. Their corresponding power levels are 0.1 PW and 1 PW at repetition rates of 10 Hz and 1 Hz, respectively. The high-power laser system of ELI–NP will be constructed by an association between Thales Optronique France and Thales Romania.

The Gamma Beam System will produce brilliant Gamma ray beam, with energy tuneable up to 20 MeV, obtained by back-scattering of optical photons on electrons from a warm LINAC with energies up to 720 MeV [8, 9]. The very low cross section for Compton scattering will be compensated in obtaining high brilliance gamma beam by very intense high repetition rate photon beam, very intense low emittance electron beam and a very small and precise interaction volume.

The output specifications for the Gamma beam (see Table 1) were established in a series of workshops and meetings organized with the scientific community interested in ELI–NP, in order to satisfy the needs of the scientists for the progress of their research but also to have realistic expectancies that are technically feasible within the timeframe of project implementation. The main characteristics of the gamma beam that will make it unique will be the very narrow relative bandwidth (of the order of 10^{-3}), one order of magnitude better than present-day machines, and the extremely high brilliance – at least two orders of magnitude better than present state-of-the-art machines. Consequently, very good energy level selection capabilities and reduced irradiation times will be possible in experiments, making possible experiments that were not feasible before.

Table 1: Output parameters for the gamma beam system

Type	**Range**
Photon energy [MeV]	0.2 – 19.5
Divergence [μRad]	25 – 200
Average Bandwidth	$\leq 5 \cdot 10^{-3}$
Nr. Photon/s within FWHM bdw [1/(s eV)]	$< 2.6 \cdot 10^{5}$
Peak Brilliance [N_{ph}/s·mm^2·mrad^2·0.1%]	10^{20} - 10^{23}
Minimum Frequency γ Ray Macropulses [Hz]	≥ 100

The Gamma Beam System will have two gamma-ray outputs available for use independently, one for the lower part of the energy range (up to approximately 3.5 MeV) obtained using the first half of the accelerator, and a second one for higher energies (up to approximately 19.5 MeV).

The gamma beam system will be constructed by EuroGammas, a European consortium led by INFN Italy and composed of leading research institutions and private companies in accelerator and laser technologies from 8 European countries.

3 Experimental areas

The scientific case for ELI–NP was elaborated by an international collaboration of more than 100 scientists from 30 countries and published as the ELI–NP White Book [2]. It is based on the unique features of the high–power laser and gamma beams. The main research topics of interest are: laser driven nuclear physics experiments, characterization of the laser–target interaction by the means of nuclear physics methods, photonuclear reactions, exotic nuclear physics and astrophysics. In addition to fundamental themes, applications of HPLS and GBS are under study. Radiation induced damage and gamma induced nuclear reactions are major research area in nuclear engineering. Their applications extend from nuclear power plants to medicine and from space science to material science.

The ELI–NP team together with their collaborators from the international scientific community shaped the future scientific program of ELI–NP is a series of workshops and defined ten development directions for the facility. The Technical Design Reports (TDRs) are finalized and approved by ELI–NP International Scientific Advisory Board in June 2015.

ELI–NP Center will have eight experimental rooms:

- E1: laser induced nuclear reactions;
- E2: nuclear resonance fluorescence and applications;
- E3: positrons source;
- E4 / E5: accelerated particle beams induced by HPLS (0.1/1 PW) at high repetition rates;
- E6: intense electron and gamma beams induced by high power laser beams;
- E7: experiments with combined laser and gamma beams;
- E8: nuclear reactions induced by high energy gamma beams.

4 Experiments

4.1 Laser-based experiments

The powerful lasers will be used to obtain accelerated ions. The acceleration of particles using lasers can occur following several mechanisms, depending on the characteristics of the incident beam and target: Wake-Field acceleration [10] for gas targets, Target Normal Sheath Acceleration (TNSA) [11] for solid targets or Radiation Pressure Acceleration (RPA) [12] for much thinner (nanometers) foil targets. The acceleration of particles using laser beams is of great interest due to many applications it may trigger in a number of fields, from medicine (compact and more versatile accelerators for ion-therapy) to management of nuclear waste.

The study of the interaction of radiation with matter by means of nuclear physics techniques is one of the new areas where ELI–NP will constitute a big step forward with respect to existing facilities. Interaction of the high power (PW) laser radiation with the solid state matter produces specific effects, not completely known and/or understood. The detailed knowledge of these effects has a fundamental interest for understanding the behaviour of various materials in extreme conditions of irradiation. The study of the irradiation effects significantly helps to optimization of the materials and components operating in high radiation fields.

One of the experiments proposed aims to the production of neutron-rich nuclei around the $\mathrm{N} = 126$ waiting point of the r-process via the fission-fusion reaction mechanism [13].

4.2 Experiments based on the gamma beam

The state-of-the-art characteristics of the gamma beam will make feasibly the study of detailed structure of dipole states in nuclei via Nuclear Resonance Fluorescence [14], including the fine structure of Giant Dipole Resonances (GDR) and Pigmy Dipole Resonances. The fine structure of fission barrier will be also studied. Compared to former gamma-ray facilities, the much improved bandwidth is decisive for this new facility, because an important factor in this type of experiments is the energy bandwidth of the beam, which should be smaller than the normal energy gap between nuclear levels. Several experiments, like the parity violation experiment, only become possible also due to this much better bandwidth.

Laboratory astrophysics experiments aiming at explaining the nucleosynthesis processes will be possible, through direct or inverse reactions, i.e. through (γ,n), (γ,p) or (γ,α) reactions. Reactions relevant for the p- and r-processes will be investigated, to advance the explanation of the formation of a large part of the known elements in the Universe.

The gamma beam can be used to study new techniques to map the isotope distributions of nuclear materials or radioactive waste remotely via Nuclear Resonance Fluorescence (NRF) measurements. At lower energies, around 100 keV the high resolution of the beam is very important for protein structural analysis. In addition it will be produced low energy, brilliant, intense neutron and positron beams, which will open new fields in materials science and life sciences.

The new production schemes of medical isotopes [15] (e.g., ^{99}Mo – currently used in therapies, ^{195m}Pt – nuclear imaging to determine efficiency of chemotherapy, ^{117m}Sn – emitter of low energy Auger electrons for tumour therapy) via (γ,n) processes may also reach socio-economical relevance.

Applied research and development will be enabled by the very high degree of collimation and the intensity of the gamma beam, allowing studies in materials science. A relevant example is the investigation of methods new methods to produce thermal neutrons, through photonuclear reactions (γ,n). Another example would be the creation of an intense positron sources by means of the (γ,e$^+$e$^-$) reaction [16].

4.3 Experiments employing lasers and gamma beam

The possibility to study fundamental processes with very brilliant optical and gamma beams will be unique and shall advance science much faster. The proposals for future experiments include two main topics with staged developments [17]. The first one is linked to the possibility to reach at ELI–NP conditions similar to the interior of stars and production and photoexcitations of isomers can be studied. The second topic is probing the photon-photon interactions at low energy (below MeV), which have not been explored to date, by using the laser-laser,laser-gamma, and gamma-gamma collisions. By these various combinations can be tested nonlinear QED effects. It also includes the pair production via the tunnelling process in extremely high-intensity laser fields. The laser-electron experiments aims at the generation of polarized gamma-rays from sub-GeV to GeV as probes for the later stage experiments to explore the vacuum birefringence under high-intensity laser fields.

5 Conclusions

The ELI–NP project is the result of a pan-European collaborative effort that extended beyond the borders of the European Union. Research communities in more than 20 countries actively contributed to the definition of the project. The present report is a short, updated synthesis of these efforts.

ELI–NP will host two cutting edge infrastructures, namely 2×10 PW laser systems and a brilliant gamma beam. They will open new directions in fundamental research. They will allow also applied research and development in unique conditions, relevant also for companies acting in the fields of medicine (radionuclides and hadron-therapy), telecommunications (materials in high intensity radiation fields), engineering industry (non-destructive testing), security (scanners based on nuclear resonance fluorescence of sensitive nuclear materials).

Acknowledgments

The ELI–NP project is the result of an international collaborative effort of more than 100 scientists from 20 countries and their contribution to the definition and implementation of the project scientific program is gratefully acknowledged. The tremendous work of the ELI–NP scientific team and the essential contribution of my enthusiastic and tenacious colleagues from the Management team in the implementation of the project are also deeply acknowledged. The ELI–NP project is co-financed by the Romanian Government and European Union through the European Regional Development Fund.

References

[1] www.eli-np.ro

[2] ELI-NP White Book eds. Habs D. et al. (2010), http://www.eli-np.ro/documents/ELI-NP-WhiteBook.pdf

[3] Zamfir N.V., *EPJ Web of Conferences* **66** (2014) 11043

[4] Zamfir N.V., *EPJ Special Topics* **223** (2014) 1221

[5] Ursescu D. et al., *Proc. of SPIE*, 87801H-1 (2013)

[6] Strickland D., Mourou G., *Opt.Comm.* **56** (1985) 219

[7] Ross I.N. et al., *Opt.Commun.* **144** (1997) 125

[8] Adriani O. et al., Technical Design Report; EuroGammaS proposal for the ELI-NP Gamma beam System, arXiv:1407.3669v1 [physics.acc-ph].

[9] Ur C. A. et al., *Acta Physica Polonica* **B46** (2015) 743

[10] Tajima T. and Downson J.M., *Physical Review Letters* **43** (1979) 267

[11] Wilks S.C. et al., *Phys. Plasma* **8** (2001) 542

[12] Henig A. et al. , *Phys. Rev. Lett.* **103** (2009) 245003

[13] Thirolf P.G. et al., *EPJ Web of Conferences* **38** (2012) 08001

[14] Ulrich Kneissl, Norbert Pietralla, Andreas Zilges, *J. Phys. G: Nucl. Part. Phys.* **32** (2006) R217

[15] Habs D., Köster U., *Applied Physics* **B103** (2011) 501

[16] Mills A.P. and Gullikson E.M., *Appl. Phys. Lett.* **49** (1986) 1121

[17] TDR for combined laser gamma experiments at ELI-NP, eds. Kensuke Homma, Keita Sato, Ovidiu Tesileanu (2015)

EPJ Web of Conferences **117**, 10002 (2016) DOI: 10.1051/epjconf/201611710002
NN2015

Highlights of the ISOLDE facility and the HIE-ISOLDE Project

M.J.G Borge

PH Department, CERN, CH-1211 Geneva-23, Switzerland
Instituto de Estructura de la Materia, CSIC, Serrano 113bis, 28006-Madrid, Spain

Abstract

The ISOLDE Radioactive Beam Facility is the dedicated CERN installation for the production and acceleration of radioactive nuclei far from stability. Exotic nuclei of most chemical elements are available for the study of nuclear structure, nuclear astrophysics, fundamental symmetries and atomic physics, as well as for applications in condensed-matter and life sciences. Since more than a decade it offers the largest variety of post-accelerated radioactive beams in the world today. In order to broaden the scientific opportunities beyond the reach of the present facility, the on-going HIE-ISOLDE (High Intensity and Energy) project will provide major improvements in energy range, beam intensity and beam quality. Post-accelerated beams will be available already this year boosting the beam energy of the current REX LINAC to 4.3 MeV/u reaching 5.5 MeV/u next spring. In this new energy regime the Coulomb excitation cross sections are strongly increased with respect to the previous energy of bearly 3 MeV/u and many transfer reaction channels will open. The second stage of the energy upgrade will allow energies of the beam up to 10 MeV/u for the worst scenario of A/q = 4.5. The funds are already secured and it is expected to be completed in 2017. In this contribution the present status of the ISOLDE facility will be discussed, some highlights will be briefly described to illustrate the advances of the facility. The HIE-ISOLDE project will be described together with a panorama of the physics cases to be addressed in the near future with emphasis in the day-one experiment to be done this year.

1 The ISOLDE facility

The ISOLDE Radioactive Beam Facility is the CERN experiment for the production and acceleration of radioactive nuclei. Isotopes from a variety of chemical elements are produced in a target directly connected to the ion-source of an isotope separator to minimise the time delay between production of a nucleus and its arrival at the experimental setup.
This ISOL-type facility which started in 1964 at the 600 MeV proton Synchrocyclotron is located since 1992 at the Proton-Synchrotron Booster (PSB) of the European Organization for Nuclear Research, CERN. Pulses of about 3 10^{13} protons are sent to the ISOLDE target every 1.2 seconds or multiples of this period resulting in an average beam intensity of 2 particle μA [1]. The use of high-energy protons such as the ones delivered by the CERN PSB injector of 1.4 GeV has been recognized to be optimum for the production of radioactive nuclei via spallation, fission and fragmentation reactions on thick targets. As many as twenty different solid and liquid target materials are used to produce a wide spectrum of radioactive isotopes covering essentially the whole nuclear chart below uranium ($Z = 92$). The reaction products are thermalized in the bulk of the target material, and thereafter transported from the target container to an ion source via diffusion and effusion processes. The speed of the diffusion and effusion process is governed by the physical and chemical properties of the radioactive elements of interest and of the specific materials that the target container, transfer line and ion source are made of. Chemical selectivity is obtained by the right combination of target-ion sources giving rise to a selective production of more than 1300 isotopes of 73 different chemical elements, being Boron the last element incorporated this year to this long list. ISOLDE's decades of experience in target-ion source development makes the whole process, from isotope production in target to beam delivery to experiments, efficient, fast and selective giving rise to a wide spectrum of beams. Fig. 1 overlays on the nuclear chart the ISOLDE beams with their production yields. The list is not fully updated. Radioactive beams has been produced with intensities reaching the nA level (e.g. ^{213}Fr with 8.4 10^9 particles per second) and of very short-lived nuclei with a few milliseconds half life (e.g. ^{14}Be, $T_{1/2}$=4.45 ms). These beams are used for experiments in the field of nuclear physics (taking $\sim$ 67% of the beam time), weak interaction studies ($\sim$ 3 %), solid-state physics studies ($\sim$ 19%), nuclear astrophysics ($\sim$4 %) and, biology and medicine ($\sim$8 %). Statistics given for 2014.

The ions are subsequently extracted from the ion source, accelerated to 60 keV, mass separated according to mass-over-charge ratio using a dipole

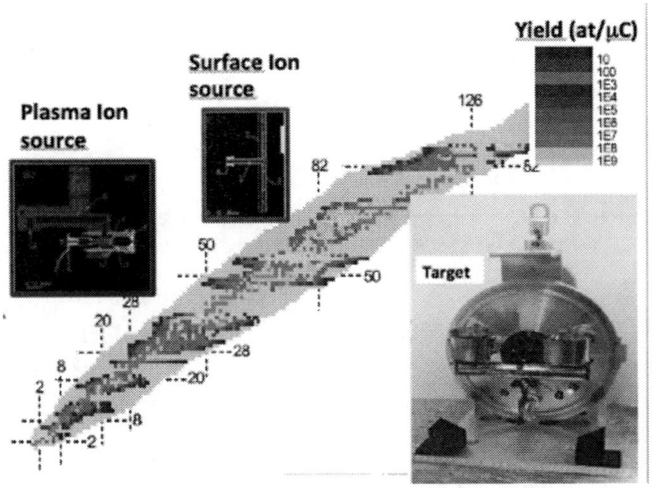

Figure 1: Chart of nuclide colour code follows the ISOLDE yields, scale given in the upper right corner. The gaps with no beams available correspond to refractory-type elements. Off diagonal to the left the surface and plasma ion source designs are shown, and to the right a photo of the target unit.

magnet and send to the different experimental set-ups. Two mass separator units exit with different mass resolution power, the GPS with M/δM = 1000 and the HRS with M/δM = 5000. A radiofrequency cooler-buncher was installed in 2006 after the HRS allowing to deliver short bunches of radioactive ion beams to experiments without a substantial loss in intensity. This development has resulted in an important increase in the peak-to-background ratio for some experiments e.g. the collinear laser spectroscopy measurements [2], and a even better beam waist for all of them. A general description of the ISOLDE facility is given in [1].

The success of ISOLDE is due to the continuous development of new radioactive ion beams and to the improvement of the experimental conditions. Decades of RIB research has shown that different purification steps have to be applied one after the other as in most cases none of the purification steps result in a 100% pure beams. Different ways are implemented to purify ISOLDE's beams. The most straightforward way is increasing the wanted isotope over its isobaric contaminants using the high-resolution mass separator [1]. Using the high-resolution mass separator, however, often induces losses in intensity. Other ways to purify the beam is making use of the physical and chemical properties of the element. The most elegant and effective way to purify the ion beam is the use of the so-called 'Resonant Ionisation Laser Ion Source (RILIS)' [3]. Resonant laser ionisation is an element selective process where atoms are irradiated by two or three lasers exciting the electron in two or three resonant steps into the continuum. This allows the production of very pure beams as only atoms of the element of interest are ionized.

RILIS has become extensibly used at ISOLDE [4] because of its ability to efficiently ionize a range of 35 elements with an unmatched degree of element selectivity. In recent years, the RILIS team has contributed to a highly successful collaboration with several experimental groups at ISOLDE [5]. This collaboration has performed extensive laser and nuclear-spectroscopic studies of shape evolution/coexistence in the lead region using the technique of in-source laser spectroscopy. This is the most sensitive laser spectroscopy method available at ISOL facilities, and has enabled the measurement of isotopes with a yield as low as 0.1 ions/second. In particular, long chains of Pb (Z=82) [6,7], Po (Z=84) [8,9], Tl (Z=81) [10–12], isotopes have been studied. This year At (Z=85), Au (Z=79) and Hg (Z=80) have been studied employing a combination of particle detection techniques: ($\alpha/\beta/\gamma$) with the Windmill setup from Leuven (Belgium); direct ion counting with the ISOLTRAP multi-reflection time of flight mass spectrometer (MR-TOF MS); and ion beam current measurements using the ISOLDE faraday cups. The combination of all these methods has allowed to determine the hyperfine structure measured by RILIS of the neutron defficient Hg isotopes from A=177 to 181. The mean square radius shows absence of the odd-even staggering observed in the $^{181-184}$Hg isotopes. This confirms the predictions than well below the mid-shell neutron number N=104 the structure of the Hg-isotopes becomes spherical [13]. Using the same techniques the radii of the n-defficient Au isotopes from A = 176-183 were measured. The results show that the deformed Au isotopes are symmetric around the mid-shell N=104 and return to sphericity for neutron defficient isotopes lighter than ^{179}Au.

In order to boost the energy of the existing radioactive ion beams of ISOLDE from typically 60 keV to a few MeV/u a new concept was proposed in which the existing singly-charged ion beams from ISOLDE were accelerated in a universal, fast, efficient and cost-effective way [14,15]. The concept was based on ion beam cooling and bunching in the buffer gas of a Penning trap [16], charge-state breeding in an Electron Beam Ion Source (EBIS) [17] and post-acceleration in a room-temperature linear accelerator [18]. The 60 keV radioactive ion beams from ISOLDE are retarded and continuously injected into a Penning trap situated at a high voltage platform. In the trap the ions are accumulated and at the same time cooled via buffer gas collisions. After a fixed time, the ion bunch is extracted from the Penning trap and transferred to the EBIS. The typical duration of the ion

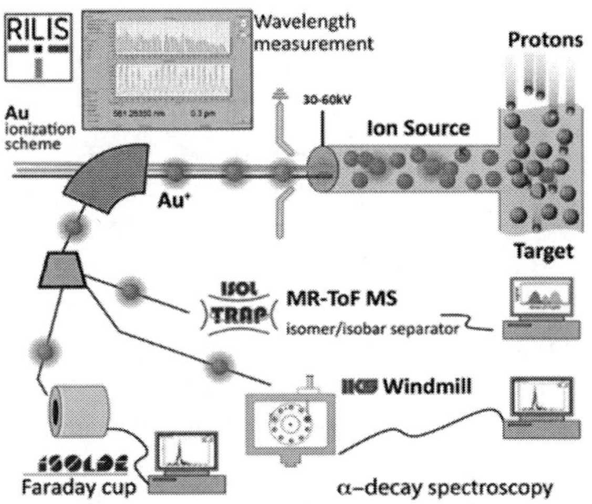

Figure 2: Picture of the RILIS ion source selection system combine with the different devices used used for $\alpha/\beta/\gamma$ detection by the Leuven windmill device, ion selection and counting using the ISOLTRAP MR-ToF mass spectrometer or ion beam current in the ISOLDE Faraday cup.

bunches extracted from the Penning trap is 10 μs. After charge breeding to the required mass-to-charge ratio (A/q) between 2.5 - 4.5, the ion bunch is extracted from the EBIS, mass separated and injected in the radiofrequency quadrupole (RFQ) accelerator. The charge breeding time depends on the required charge-to-mass ratio and thus on the isotope of interest. It varies from typically 20 ms for light mass elements (A<40) to over 200 ms for the heavy elements in the lead region (A$\simeq$200). The room temperature accelerator is typically running with a 10% duty cycle. The RFQ accelerator is followed by an interdigital H-type (IH) structure with a final energy between 1.1 and 1.2 MeV/u, three seven-gap resonators that create an energy range from 0.8 to 2.2 MeV/u and finally by a nine-gap resonator boosting the beam energy to a maximum of about 3 MeV/u.

While the original goals of the REX-ISOLDE project were limited to energies up to 2 MeV/u and masses below A=50, the universal concept proved to be very successful and meanwhile beams with A/q ratio < 4.5 and with masses up to 224 have been accelerated up to 3 MeV/u, with efficiencies with varying from $\sim$2 to 10 % . The experimental program with post-accelerated beams focussed on Coulomb excitation, few-nucleon transfer reaction studies and fusion evaporation studies using radioactive beams. Dedicated instrumentation for γ-ray and charge particle detection was developed especially tailored to the particularities of radioactive ion beam research: low beam intensities, low-γ-ray multiplicity and strong background radiation from the radioactivity of the beam itself. For γ-ray detection, an array of six-fold segmented, individually encapsulated and tapered germanium crystals was constructed [19]. The so-called Miniball array is optimized with respect to full-energy peak efficiency and angular resolution rather then resolving power.

Miniball consists of eight cryostats, each containing three individual germanium crystals, that are mounted in close geometry around the target position. In its standard configuration for Coulomb excitation experiments, the typical target to detector distance is about 11 cm resulting in a solid-angle coverage of $\sim$ 60% of 4π. The photo-peak efficiency for the $1.3MeV$ reaches up to about 8%. In case of Coulomb excitation experiments the spherical detection chamber is equipped with a double sided silicon strip detector (DSSSD) of the CD type [20]. The DSSSD covers an angular range typically from 16° to 53° laboratory angle in the forward direction. This range can be changed by changing the target to detector distance. To study few-nucleon transfer reactions using radioactive ion beams in inverse kinematics, two detection set-ups were developed. The first one mainly used for transfer reactions in the light mass range consisted out of one or more [21] Si double-sided strip detectors with a specially developed [22] thin dead layer. The second set-up, called T-REX (the T stands for transfer), is mainly used for transfer reactions using heavier mass beams and combines particle detection with gamma detection [23, 24]. It consists of the Miniball array to detect γ-rays in coincidence with particles detected by an array of segmented Si detectors. The array comprises two so-called CD detectors, in forward and backward direction, and a barrel of eight planar DSSSD's around 90°. A good description of the REX-ISOLDE experiments done during its first decade of existence can be found in [25]

The heaviest accelerated REX-ISOLDE beams of Radon and Radium nuclei were employed to investigate shape asymmetric configurations. Strong octupole correlations leading to pear shapes can arise when nucleons near the Fermi surface occuping states of opposite parity with orbital and total angular momentum differing by 3 units. Pear shape nuclei have enhanced E1 and E3 transitions connecting rotational states of opposite parity. The E1 moments are small and dominated by single-particle and cancellation effects, while the E3 transition moments are collective in behaviour and insensitive to single-particle behaviour. Octupole correlations were studied by the determination of the electric octupole transition strengths in ^{220}Rn and ^{224}Ra. The E3 moments are an observable that should provide direct evidence for enhanced octupole correlations for deformed nuclei. The measured E1, E2

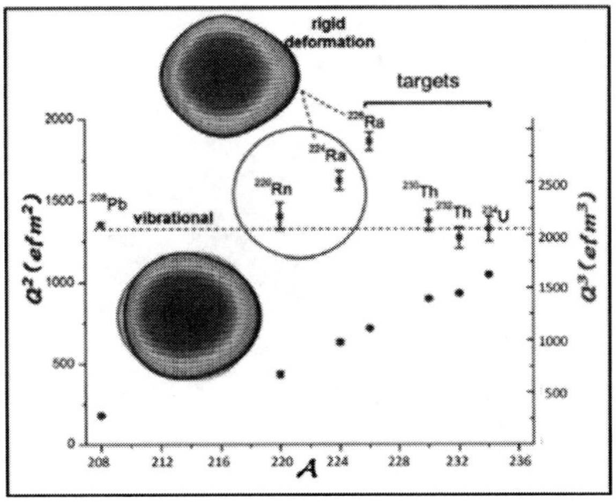

Figure 3: Systematics of the quadrupole and octupole moments for heavy nuclei with A = 208-234 including the new cases measured at ISOLDE ^{220}Rn and ^{224}Ra. The present results indicate a clear evidence for strong octupole deformation in ^{224}Ra [26].

or E3 transitions for ^{220}Rn and ^{224}Ra allowed the determination of the reduced matrix elements and the intrinsic moments. As shown in Fig. 3, the E2 moment increases a factor of 6 from ^{208}Pb to ^{334}U while the E3 moment only varies up to 50% in the same region. The large values of Q3 measured for ^{220}Rn and ^{224}Ra point to an enhancement of octupole collectivity indicating the onset of octupole deformation in this region. While ^{220}Ra has Q3 values typical of an octupole vibrator, the Q3 values obtained for ^{224}Ra gives convincing evidence that this nucleus is of quadrupole-octupole shape in its ground state. These relevant results for nuclear structure and of importance for EDM searchers are summarized in a recent Nature paper [26].

2 The HIE-ISOLDE project

The HIE-ISOLDE upgrade (HIE stands for High Intensity and Energy), intends to improve the experimental capabilities at ISOLDE over a wide front [27]. The main features are to boost the energy of the beams, going in steps from currently 3 MeV via 5.5 MeV to finally 10 MeV per nucleon, and to increase roughly sixfold the production due to an increase in intensity and energy of the proton injector. In addition improvements in several aspects of the secondary beam properties such as purity, ionisation efficiency and optical quality are addressed in the project, see Fig. 4.

Major project components include a new superconducting (SC) linear ac-

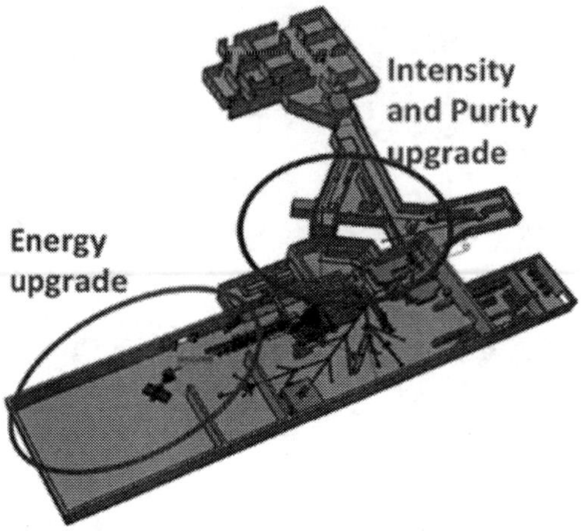

Figure 4: Layout of the ISOLDE facility showing the target-ion source unit connected to each mass separator as well as the the post accelerator line at the experimental hall. The expected upgrades of the facility are indicated: Increase in intensity and energy of the injector in 2020; increase of the post-accelerator beam from 4.3 MeV/u this year to 10 MeV/u in 2017. Improvements of the beam purity and quality are a continuos effort of the technical team.

celerator (LINAC) based on Quarter Wave resonators (QWRs) for the post-acceleration and the necessary 4.5 K cryogenic station for helium. The decision to keep the existing experimental hall has imposed severe constraints on the LINAC, so it has been necessary to design and build accelerating cavities with a very high voltage gradient of 6 MV/m and low heat dissipation below 10 W. The superconducting accelerator part dedicated to the increase of energy is based on a quart wave resonator (QWR) geometry with twenty high-β cavities cooled by helium and installed in four cryo-modules. The transverse focusing is achieved using four superconducting solenoids housed inside the cryo-modules maximising the transverse acceptance. Each cryo-module contains five cavities based on 101.28 MHz niobium-sputtered copper (Nb/Cu) Quarter Wave Resonator and one solenoid, see inset of Fig. 5. The first two cryo-modules with five high-β cavities each will permit to increase the energy to 5.5 MeV/u for A/q = 4.5 and constitute the first phase of the project. The first cryo-module was installed at ISOLDE in May. Hardware commissioning of the first cryo-module of the HIE-ISOLDE superconducting LINAC has been now ongoing for the last 3 months, with a start planned for mid-October 2015. Two beam lines are already operative for physics and a third beam line will be built next year.

The first call for proposals was made in October 2012. So far thirty four experiments have been approved with more than six hundred 8h shifts allocated. The physics cases approved expand over the wide range of post-accelerated beams available at ISOLDE, where the increase in energy of the radioactive beams will enhance the cross section in most of the cases and the accessibility to detail nuclear structure information at higher excitation energy.

In the light nuclear region, reaction studies of astrophysical interest such as the search for high-excited states in ^{8}Be to address the cosmological ^{7}Li problem. Nuclear structure studies are planned to characterise cluster structure in ^{10}Be by transfer reaction. Characterisation of resonant states in the proton-rich nucleus ^{21}Al will be determined by resonance elastic and inelastic scattering using the active target MAYA in order to check isospin conservation beyond the drip line. For middle mass nuclei, the validity of a shell model description around ^{78}Ni will be studied as well as shape coexistence in the region A = 70-80 will be determined with high precision. Statistical properties of warm nuclei will be investigated by the low-energy enhancement of the gamma strength function of neutron rich nuclei. For heavier mass nuclei, quadrupole and octupole collectivity will be addressed in the neutron rich Te, Xe and Ba isotopes by Coulomb excitation, lifetime measurements and magnetic moment determination. Collective effects around the double magic ^{132}Sn will also be studied. For heavy nuclei, shape coexistence in the light Pb isotopes will be explored. Measurements of octupole collectivity in the Rn and Ra nuclei using Coulomb excitation will continue. In the quest of super-heavies, it is proposed to investigate the influence of the predicted shell closures at Z = 120 and N = 184 by probing the height of the fission barrier in the compound nucleus. This will be achieved by exploring the contributions of quasi-fission and fusion-fission reactions; in particular by the use of the reaction of the deformed ^{95}Rb beam on a ^{209}Bi target is expected to permit the study of these features.

The proposed studies will be realised with the existing workhorses Miniball [19] and T-REX [28] plus new instrumentation for transfer reaction studies such as the active targets MAYA [29] and the future ACTAR, a new general purpose scattering chamber, the two arms CORSET setup from GSI,...etc. The Miniball array will be complemented with an electron spectrometer, SPEDE, to realise Coulomb excitation studies of odd-heavy nuclei. While writing these pages the first beams of carbon and rubidium have been accelerated preparing the machine to start with radioactive Zn beam for the 10th of October 2015. The nuclei approved for physics experiments at HIE-ISOLDE are shown in red in the chart of the nuclides displayed in Fig. 5. We will start with the study of the evolution of the nuclear structure

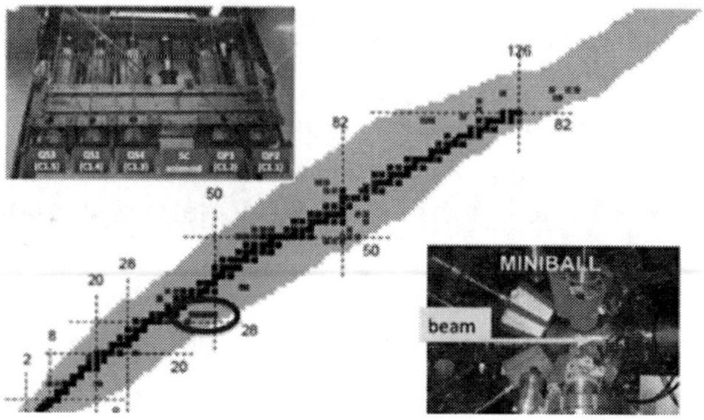

Figure 5: In the diagonal the nuclear chart shows in red the cases already approved for nuclear studies at HIE-ISOLDE. Indicated with a circle the first physics cases that will be studied this year, the n-rich Zn isotopes studied with Miniball. On the left a photo of the elements inside the cryo-module vessel and thermal shield are shown, the five cavities are clearly visible. On the right the Miniball setup that will be used in 2015 is shown

along the Zinc isotopic chain close to the doubly magic nucleus ^{78}Ni. This study will probe recent shell-model calculations in this area of the nuclear chart. Excitation energies and connecting B(E2) values will be measured with Miniball through a multiple Coulomb excitation experiment with laser ionised purified beams of $^{74-80}$Zn.

3 Summary and outlook

The future of ISOLDE is bright. ISOLDE restarted with the low energy physics program the first of August 2014. The knowledge accumulated over decades on how to construct targets and ion sources tailored to release pure beams of specific elements are one of ISOLDE strong points. With more than 50 years since approval ISOLDE remains as the pioneer ISOL-installation both at the level of designing new devices and production of frontier physics. The first phase of HIE-ISOLDE will start for physics this autumn. The physics cases approved expand over the wide range of post-accelerated beams available at ISOLDE with more than six hundred shifts approved for physics. Our first experiment will explore the evolution of nuclear structure near ^{78}Ni by performing multi-step Coulomb excitation in n-rich Zn isotopes. We expect to complete the increase of energy of post accelerated beams up to 10 MeV/u in 2017.

I acknowledge the support of the ISOLDE collaboration and technical ISOLDE and HIE-ISOLDE teams.

References

[1] E. Kugler E, Hyperfine Interactions **129** (2000) 23.

[2] E. Mané, *et al.*, Eur. Phys. J. **A42** (2009) 503.

[3] V. I. Mishin, *et al.*, Nucl. Instrum. Methods B **73** (1993) 550.

[4] V. Fedoseyev, *et al.*, Hyperfine Interactions, **127** (2000) 409.

[5] B. A. Marsh, *et al.*, Nuclear Instruments and Methods in Physics Research Section B: Beam Interactions with Materials and Atoms, **317** (2013) 550.

[6] H. De Witte, *et al.*,Phys. Rev. Lett. **98** (2007) 112502.

[7] M. Seliverstov, *et al.*, Eur. Phys. J., **A41** (2009) 315.

[8] T. E. Cocolios, *et al.*, Phys. Rev. Lett. **106** (2011) 052203.

[9] M. Seliverstov, *et al.*, Phys. Rev. C **89** (2014) 034323.

[10] J. Elseviers, *et al.*, Phys. Rev. C **84**, 034307 (2011);

[11] A. N. Andreyev, *et al.*, Phys. Rev. C **87**, 054311 (2013);

[12] C. van Beveren, *et al.*, submitted to Phys. Rev. C (2015).

[13] B. A. Marsh, *et al.*, *in preparation*

[14] D. Habs, *et al.*, Hyperfine Interactions **129** (2000) 43.

[15] D. Voulot, *et al.*, Nucl. Instrum. Methods B **266** (2008) 4103.

[16] F Ames, *et al.*, Nucl. Instrum. Methods A **538** (2005) 17.

[17] B Wolf, *et al.*, Rev. Scien. Instr. **73** (2002) 682.

[18] T. Siebert, *et al.*, Nucl. Phys. **A701** (2002) 656.

[19] J Eberth, *et al.*, Progr. Part. Nucl. Phys. **46** (2001) 389.

[20] A. Ostrowski, *et al.*, Nucl. Instrum. Methods A **480** (2002) 448.

[21] H B Jeppesen, *et al.*, 2005 Nucl. Phys. A **748** (2005) 374.

[22] O. Tengblad, *et al.*, Nucl. Instrum. Methods A **525** (2004) 458.

[23] V. Bildstein, *et al.*, Progr. Part. Nucl. Phys. **59** (2007) 386.

[24] Th. Kröll it et al., Procedings of "Nuclear Structure and Dynamics '09 *AIP Conf. Proceedings* **1165** (2009) 363.

[25] P. Van Duppen and K. Riisager, J. Phys. G: Nucl. Part. Phys. **38** (2011) 024005.

[26] L. P. Gaffney, *et al.*, Nature **497** (2013) 199.

[27] HIE-ISOLDE, the scientific opportunities ed. By K. Riisager, P. Butler, M. Huyse and R. Krücken, CERN Report, CERN-2007-008.

[28] V. Bildstein, *et al.*, Eur. Phys. J. A bf 48 (2012) 85.

[29] C. E. Demonchy, *et al.*, Nuc. Instrum. Methods in Physics Research A **583** (2007) 341.

EPJ Web of Conferences **117**, 10003 (2016) DOI: 10.1051/epjconf/201611710003
NN2015

The nuclear matrix elements of $0\nu\beta\beta$ decay and the NUMEN project at INFN-LNS

F. Cappuzzello[1,2], C. Agodi[1], F. Balestra[3,4], R. Bijker[5], D. Bonanno[6], D. Bongiovanni[6], V. Branchina[2,6], S. Calabrese[1,2], L. Calabretta[1], A. Calanna[1], D. Calvo[3], D. Carbone[1], M. Cavallaro[1], M. Colonna[1], S. Ferrero[3,4], A. Foti[2,6], P. Finocchiaro[1], G. Giraudo[3,4], V. Greco[1,2], F. Iazzi[3,4], R. Introzzi[3,4], G. Lanzalone [1], A. Lavagno[3,4], D. Lo Presti[2,6], F. Longhitano[6], A. Muoio[1], L. Pandola[1], D. Rifuggiato[1], M.V. Ruslan[5], E. Santopinto[5], L. Scaltrito[3,4], S. Tudisco[1], V. Zagatto[1]

[1]INFN - Laboratori Nazionali del Sud, Catania, Italy
[2]Dipartimento di Fisica e Astronomia, Università di Catania, Italy
[3]INFN, Sezione di Torino, Torino, Italy
[4]Politecnico di Torino, Torino, Italy
[5]INFN, Sezione di Genova, Genova, Italy
[6]INFN, Sezione di Catania, I-95123 Catania, Italy

Abstract

An innovative technique to access the nuclear matrix elements entering the expression of the life time of the double beta decay by relevant cross sections measurements of double charge exchange reactions is proposed. A key aspect of the project is the use of the MAGNEX large acceptance magnetic spectrometer, for the detection of the ejectiles, and of the LNS K800 Superconducting Cyclotron (CS), for the acceleration of the required high resolution and low emittance heavy-ion beams, already in operation at INFN Laboratory Nazionali del Sud in Catania (Italy).

1 Introduction

Neutrinoless double beta decay, $0\nu\beta\beta$, is at the present time strongly pursued both experimentally and theoretically [1]. Its observation will determine whether the neutrino is a Dirac or Majorana particle and will provide a measurement of the average neutrino mass, which is nowadays one of the most fundamental problems in physics. An innovative technique to access the Nuclear Matrix Elements (NME) entering the expression of the life time of the $0\nu\beta\beta$ by relevant cross sections of double charge exchange (DCE) reactions is proposed. The basic point is the coincidence of the initial and final state wave-functions in the two classes of processes and the similarity of the transition operators, which in both cases present a superposition of Fermi, Gamow-Teller and rank-two tensor components with a relevant implicit momentum transfer. First pioneering experimental results obtained at the INFN-LNS laboratory for the ^{40}Ca(^{18}O,^{18}Ne)^{40}Ar reaction at 270 MeV, give encouraging indication on the capability of the proposed technique to access relevant quantitative information.

First experimental results, obtained at the INFN-LNS laboratory in Catania, for the ^{40}Ca(^{18}O,^{18}Ne)^{40}Ar reaction at 270 MeV, in a wide range of transferred momenta, give encouraging indication on the capability to access quantitative information towards the determination of the NME for $0\nu\beta\beta$ decay [2]. On the basis of the above mentined ground-breaking achievement, we propose an ambitious project, NUMEN, with the aim to go deep inshight in the HI-DCE studies on nuclei of interest in $0\nu\beta\beta$ decay[3]. There are a number of important similarities among DCE and $0\nu\beta\beta$ decay processes, despite they are mediated by different interactions, also the description of NMEs extracted from DCE and $0\nu\beta\beta$ presents the same degree of complexity, with the advantage for DCE to be accessible in laboratory. However a simple relation between DCE cross sections and $\beta\beta$-decay half-lives is not trivial and needs to be explored.

2 The project

The availability of the MAGNEX spectrometer [4] for high resolution measurements of very suppressed reaction channels was essential for the first pilot experiment. Moreover the measurement of DCE high resolution energy spectra and accurate cross sections at very forward angles are key points to identify the transitions of interest [5]. The concurrent measurement of the other relevant reaction channels allows to isolate the direct DCE mechanism

Figure 1: A view of MAGNEX spectrometer at Laboratori Nazionali del Sud in Catania.

Figure 2: A view of K800 Superconducting Cyclotron at Laboratori Nazionali del Sud in Catania.

from the competing transfer processes. These are at least of 4th-order and can be effectively minimized by the choice of the proper projectile-target system and incident energy [6].

The present limit of low beam current we have experienced both for the CS accelerator and for the MAGNEX focal plane detector must be sensibly overcome. For a systematic study of the many hot cases of $\beta\beta$ decays an upgraded set-up, able to work with two orders of magnitude more current than the present, is thus necessary. This goal can be achieved by a substantial change in the technologies used in the beam extraction and in the detection of the ejectiles. For the accelerator the use of a stripper induced extraction is an adequate choice. However with the present set-up it is difficult to suitably extend this research to the hot cases, where $\beta\beta$ decay studies are and will be concentrated. For the spectrometer the main foreseen upgrades are:

1. The substitution of the present Focal Plane Detector (FPD) [7-8]

technology with tracker based on a micro-pattern electron amplifier;

2. The substitution of the wall of silicon pad stopping detectors with a wall of telescopes based on SiC-CsI detectors;

3. The enhancement of the maximum magnetic rigidity;

4. The introduction of an array of detectors for measuring the coincident γ-rays.

In this framework we propose four phases in the NUMEN project, looking forward to do, in the same time, both the experimental and the up-grade activity, as indicated in the following Phases of the project.

3 The facility upgrade

Once all the building block for the upgrade of the accelerator and spectrometer facility will be ready at the LNS a Phase3, connected to the disassembling of the old set-up and re-assembling of the new will start. An estimate of about 18-24 months is considered. Phase4: the experimental campaign The Phase4 will consist of a series of experimental campaigns at high beam intensities (some pμA) and long experimental runs in order to reach in each experiment integrated charge of hundreds of mC up to C, for the experiments in coincidences, spanning all the variety of candidate isotopes for $0\nu\beta\beta$ decay, like: ^{48}Ca, ^{82}Se, ^{96}Zr, ^{100}Mo, ^{110}Pd, ^{124}Sn, ^{128}Te, ^{130}Te, ^{136}Xe, ^{148}Nd, ^{150}Nd, ^{154}Sm, ^{160}Gd, ^{198}Pt.

4 Perspectives

Once selected the optimal experimental condition for the different cases in the Phase2, with the upgrades, both of CS and the MAGNEX array, the Phase4 will be devoted to collect data addressed to give, with an accurate analysis, a rigorous determination of the absolute cross sections values and their uncertainties for all the system of interest, to the challenging determination of the $0\nu\beta\beta$ decay nuclear matrix elements, that is the ambitious goal of NUMEN Project.

References

[1] Report to the Nuclear Science Advisory Committee, Neutrinoless Double Beta Decay, 2014.

[2] F. Cappuzzello, et al., Heavy-ion double charge-exchange reactions: a tool towards $0\nu\beta\beta$ nuclear matrix elements, Eur. Phys. J. A (2015) accepted.

[3] C. Agodi et al., Nuclear and Particle Physics Proceedings Vol 265-266, pag.28-30.

[4] F. Cappuzzello et al.,MAGNEX: an innovative large acceptance spectrometer for nuclear reaction studies in: Magnets: Types, Uses and Safety, Nova Publisher Inc., New York, 2011, pp 1-63.

[5] H. Matsubara, et al., Few-Body Systems 54 (2013) 1433.

[6] D. M. Brink, Phys. Lett. B 40 (1972) 37.

[7] M. Cavallaro, et al., Eur. Phys. J. A (2012) 48:59.

[8] D. Carbone, et al., Eur. Phys. J. A (2012) 48:60.

DOI: 10.1051/epjconf/201611710004

The performance of AGATA: From the LNL demonstrator to the GANIL setup

Caterina Michelagnoli

GANIL CEA/DSM-CNRS/IN2P3, Caen, France

Abstract

The performance of the *Advanced–GAmma–Tracking–Array* AGATA is demonstrated through selected examples from the different recent experimental campaigns.

1 Introduction

In order to perform nuclear structure studies in extreme conditions of neutron/proton asymmetry and angular momentum, in the last decade, the challenge of the realization of the *gamma–ray spectroscopy dream* has been taken on by two international collaborations, AGATA (Advanced–GAmma–Tracking–Array) [1] in Europe and GRETA (Gamma–Ray Energy Tracking Array) in the US [2]. The *dream* consists in covering the whole 4 π solid angle by germanium material and track the path of the gamma rays inside the detector medium, corresponding to maximum efficiency and "infinite" position resolution, in addition to an excellent energy resolution.

On this purpose, the digitally recorded waveforms from highly–segmented HPGe detectors are treated with Pulse–Shape–Analysis (PSA) techniques in order to extract the position of the interaction points in the detector with a 4 mm position resolution (at 1 MeV) [3,4]. These interaction points (*hits*) are grouped in events on the basis of the *timestamp*. Different gamma rays in the same event and their ordered path are reconstructed via *tracking* algorithms. An efficiency four times larger than standard arrays (also for high gamma–ray multiplicities) is expected, together with the capability to stand high count–rate (up to hundreds of kHz) thanks to the use of digital electronics. For the first time an almost continuous angular distribution of gamma interaction points will be available, allowing for a "perfect" Doppler correction and new degree of sensitivity in the determination of nuclear structure observables.

This contribution focuses on the AGATA array. It run its Demonstrator phase in 2009–2010 [5] and first physics campaign with stable beams at the Legnaro National Laboratories, consisting mainly in the spectroscopy and lifetime measurements in neutron–rich nuclei produced in deep–inelastic reactions at the PRISMA magnetic spectrometer. It then moved to GSI for the PreSpec campaign at the FRS separator and used for the first measurements with fast moving radioactive ion beams (beta up to 70%). It is now running its first experimental campaign at GANIL (France) coupled to the large acceptance magnetic spectrometer VAMOS [6]. Details on the AGATA project can be found in the reference paper [1]. It the following, the focus will be on particular aspects, that bring to new physics opportunities already with a sub–system of the whole array.

2 High count–rate capabilities

The use of digital electronics to preprocess the detector preamplifier signals allow to preserve a good efficiency and energy resolution even for count–rates larger than 10 kHz (typical limitation when using analogue electronics). In Fig. 1 an example of the output of the detector preamplifier on the core contact of an AGATA detector in response to a count rate of 30 kHz is shown (in red), together with the corresponding amplitudes reconstructed via the trapezoidal filter, corresponding to two different risetimes of the trapezoid (in blue and black). The proper choice of the risetime (and of the filter parameters in general), together with the baseline restoration, allow for extracting the amplitudes from the signals and avoiding pileup. A detailed study of the efficiency and resolution of an AGATA detector as a function of the count rate is reported in [7].

3 Correction for neutron damage

The operation of the germanium detectors in position–sensitive mode allows for the correction for the effects of the damage of the germanium crystalline

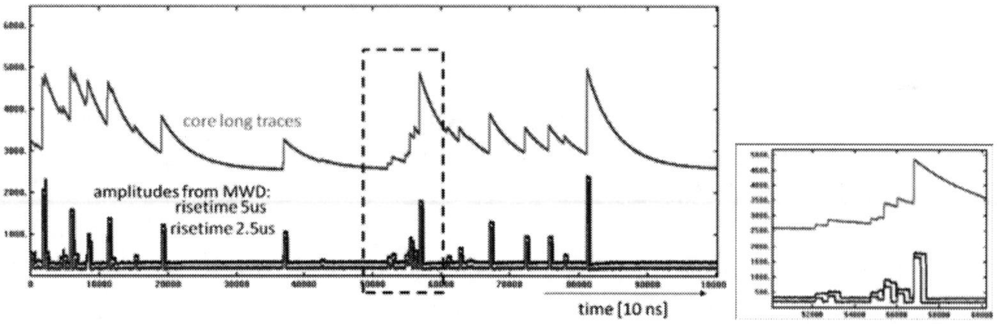

Figure 1: Response of the core preamplifier of an AGATA HPGe detector to an average 30 kHz count rate, during an in–beam experiment (in red). The amplitudes extracted from the signals when using a trapezoidal filter with two different risetimes are reported in black and blue. In the right panel, a zoom of the region indicated by the dashed rectangle is reported. The shorter risetime acts as a pile–up rejector.

structure due to the interaction of fast neutrons. Such an effect results in an incomplete charge collection causing a worsening of the energy resolution.

Since the charge loss due to neutron damage is proportional to the length of the path to the electrodes, the interaction points resulting from PSA can be used together with a modeling of charge trapping to correct for such an effect [8]. In Fig. 2 the energy spectra corresponding to the 1.3 MeV line (^{60}Co calibration source) before and after the correction are compared. This procedure is available and used for the detectors that suffers from neutron damage and applied in the data flow (or offline processing) after the PSA, before the data are passed to the global level stage, where event building and tracking are performed.

4 Some benefits from position resolution

The position resolution of AGATA can be used for the determination of nuclear structure observables with a new degree of sensitivity, giving access to new physics opportunities. Few selected examples from the Legnaro and GANIL campaigns will be given in the following.

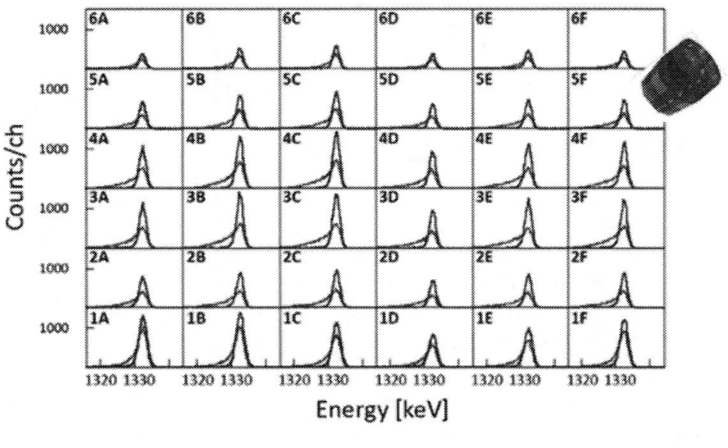

Figure 2: Segment spectra corresponding to the 1.3 MeV line (^{60}Co calibration source) –see schematic representation of an AGATA detector in the inset for segment labeling. The left tail consequent to inefficiency in charge collection due to neutron damage is evident in the spectra in red (no correction applied) and it is significantly reduced in the spectra in blue, obtained after correction for neutron damage (resolution for the sum of all segments from 5.9 keV to 2.9 keV after correction. The data are taken during the Legnaro campaign (picture taken from [9]).

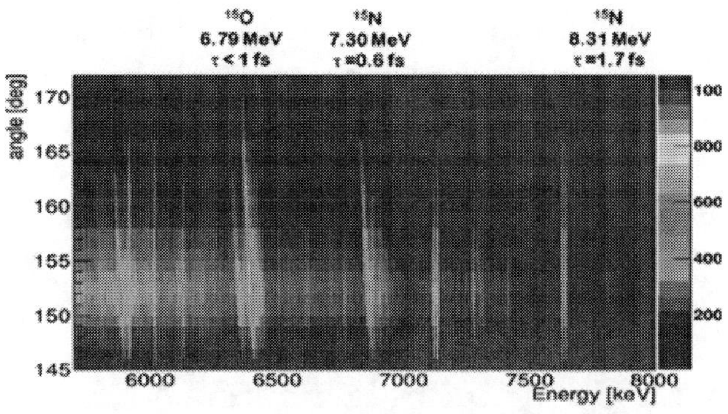

Figure 3: Tracked gamma–ray energy sorted according to the angle of the reconstructed first interaction point, from the experiment for the measurement of the lifetime of the 6.79 MeV state in ^{15}O. The "straight" lines correspond to the emission at rest by a radioactive source (present while beam–on–target for monitoring of gain instabilities). The tilted broad lines correspond to gamma rays emitted from short ($\approx$ fs) lived states. The lifetimes are obtained with Doppler Shift Attenuation technique over an almost continuous distribution of angles.

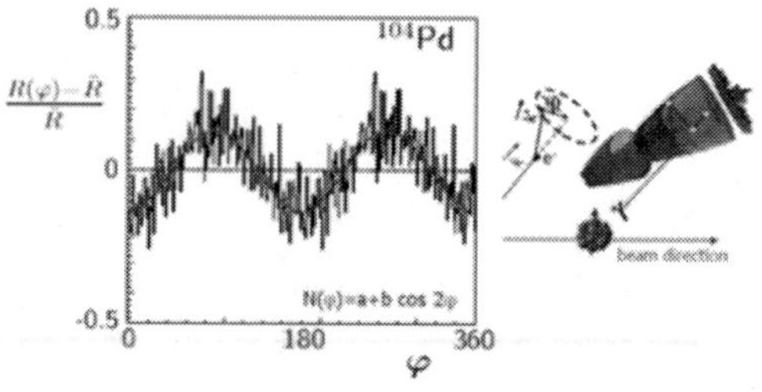

Figure 4: (taken from [11]) Experimental asymmetry as a function of the azimuthal angle of Compton scattering measured for the deexcitation of the first 2^+ state in ^{104}Pd populated by Coulomb excitation. The gamma linear polarization is obtained with a fit to the experimental data [11].

4.1 Sub–femstosecond lifetime of interest in nuclear astrophysics

The position resolution of AGATA can be used to measure nuclear level lifetimes via Doppler Shift techniques exploiting a continuous distribution of angles. It is well known that especially in the case of the Doppler Shift Attenuation Method (DSAM) the contribution of the detector response function, namely the finite angular coverage of the detectors, is one of the main factors that limits the sensitivity. A new sub–femtosecond upper limit has been recently obtained for the 6.79 MeV state in ^{15}O [10]. The experiment was performed during the Demonstrator campaign in Legnaro. The reaction of ^{14}N beam on a ^{2}H target has been used to populate the state of interest. The effect of fs lifetimes can be observed by analysing the line–shape of the gamma rays, emitted by the reaction products slowing down in a gold layer, as a function of the observation angle. In Fig. 3 the energy of the reconstructed gamma rays sorted according to the angle of the first interaction point is reported with an unprecedented angular resolution of 2 degrees (see figure caption for details).

4.2 AGATA as a Compton polarimeter

The second point of interaction of a gamma ray in its attenuation path in the germanium medium can be used to determine the linear polarization of gamma rays emitted by aligned nuclear states. The asymmetry in the distribution of the Compton scattered gamma rays as a function of the azimuthal angle can be used to determine the linear polarization. A test of AGATA modules as Compton polarimeters has been performed in the Legnaro campaign [11], using the Coulomb excitation of the first excited

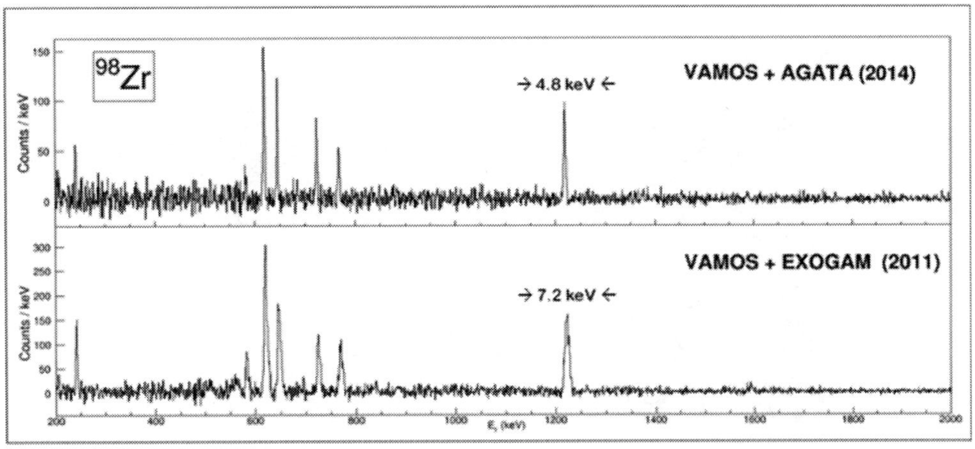

Figure 5: Gamma–ray spectra corresponding to the ^{98}Zr nucleus produced in the fission of a ^{238}U beam on a ^{9}Be target at 6.6 MeV/u and identified in the magnetic spectrometer VAMOS. The comparison of performance in gamma–ray detection between the EXOGAM (lower panel, data taken from [13]) and AGATA (upper panel, data taken from December 2014 commissioning runs) arrays is shown. The FWHM at 1.2 MeV is reported.

states in ^{104,106}Pd targets by a ^{12}C beam. As evident from Fig. 4 (taken from [11]), with respect to standard clover detection system, this asymmetry can be analysed as a continuous function of the angle. This results in a larger analysing power, that allows for the measurement of gamma linear polarization in cases of weaker nuclear alignments or, in general, for rarer phenomena. This is the case of the study of the *photon entanglement* through the measurement of the relative polarization of the 511 keV gamma rays emitted by the singlet positronium [12].

4.3 Doppler correction capabilities

In Fig. 5 the advantages of using PSA and tracking techniques for the detection of gamma rays with respect to traditional arrays are shown, by using the comparison of spectra obtained in the same experimental conditions, exception made for the gamma array. The ^{98}Zr nucleus was populated in the fission process in inverse kinematics of ^{238}U beam on a ^{9}Be target at 6.6 MeV/u. The same reaction has been recently used in GANIL for the study of isotopically–identified neutron–rich fission fragments at the VAMOS++ magnetic spectrometer [13]. Here a spectrum obtained from those data is used for comparison. The gain of almost a factor of two in energy resolution is evident.

(a) AGATA at GANIL (eight triple clusters).

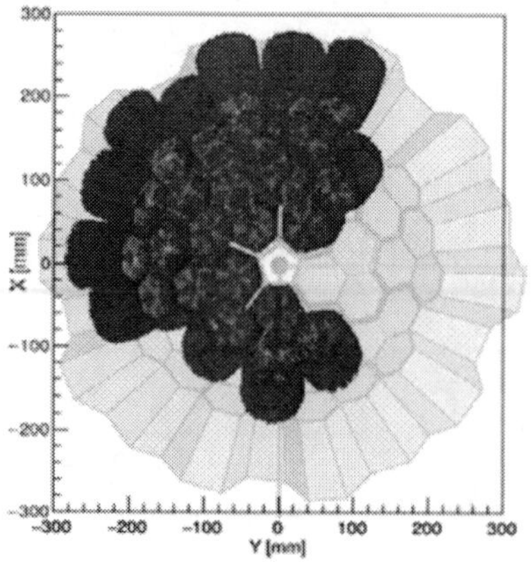

(b) Hit pattern from 24 crystals.

Figure 6: The AGATA array installed for the first campaign at the VAMOS spectrometer at GANIL (spring/summer 2015). Eight triple clusters are mounted at backward angles with respect to the axis of the VAMOS spectrometer. The corresponding hits reconstructed via PSA are also shown (in the XY plane) together with the simulation of the 1 π array foreseen for the forthcoming experimental campaign at GANIL.

5 The AGATA+VAMOS campaign at GANIL

The commissioning runs in GANIL took place in November–December 2014 and the Doppler correction capabilities and physics correlations between the gamma and magnetic spectrometers were verified and tested for different range of velocities and A and Z of the emitting nuclei [14].
Nowadays 8 triple clusters of the AGATA array are available, Fig. 6. The AGATA array positioned at backward angles with respect to the beam line coupled to the magnetic spectrometer VAMOS++ were used in the first campaign with stable beams at GANIL, in March–July 2015. The sensitivity of this unique combination was used to measure lifetimes and other spectroscopic properties of neutron rich nuclei produced by fission and deep–inelastic reactions in different region of the nuclear chart.

References

[1] S. Akkoyun et al., Nuclear Instruments and Methods in Physics Research Section A **668**, 26 (2012)

[2] I. Lee et al., Nuclear Physics A **746**, 255 (2004)

[3] F. Recchia et al., Nuclear Instruments and Methods in Physics Research Section A **604**, 555 (2009)

[4] P.A. Söderstrom et al., Nuclear Instruments and Methods in Physics Research Section A **638**, 96 (2011)

[5] A. Gadea et al., Nuclear Instruments and Methods in Physics Research Section A **654**(1), 88 (2011)

[6] M. Rejmund et al., Nuclear Instruments and Methods in Physics Research Section A **646**(1), 184 (2011)

[7] F. Recchia et al., LNL Annual Report 2010 p. 60 (2011)

[8] B. Bruyneel et al., The European Physical Journal A **49** (2013)

[9] C. Michelagnoli, Ph.D. thesis, University of Padova (2013)

[10] C. Michelagnoli et al., submitted to PRL

[11] P. Bizzeti et al., The European Physical Journal A **51**(4) (2015)

[12] P. Bizzeti et al., private communication

[13] A. Navin et al., Physics Letters B **728**, 136 (2014), ISSN 0370-2693

[14] C. Michelagnoli et al., in preparation

EPJ Web of Conferences **117**, 10005 (2016) DOI: 10.1051/epjconf/201611710005
NN2015

FAZIA applications

S. Piantelli[1], G. Casini[1], P.R. Maurenzig[1,2], A. Olmi[1],
S. Barlini[1,2], M. Bini[1,2], G. Pasquali[1,2], G. Pastore[1,2],
G. Poggi[1,2], A.A. Stefanini[1,2], S. Valdrè[1,2], G. Ademard[5],
L. Auger[3], R. Bougault[3], E. Bonnet[4], B. Borderie[5], A. Chbihi[4],
J.D. Frankland[4], D. Gruyer[1], O. Lopez[3], N. LeNeindre[3],
M. Parlog[3,6], M.F. Rivet[5,†], E. Vient[3], E. Rosato[7†],
G. Spadaccini[7], M. Vigilante[7], M. Bruno[8], T. Marchi[9],
L. Morelli[8], M. Cinausero[9], M. Degerlier[10], F. Gramegna[9],
A. Kordyasz[11],T. Kozik[12], T. Twarog[12], R. Alba[13],
C. Maiolino[13], D. Santonocito[13] and E. Galichet[5,14]
FAZIA Collaboration

[1]INFN, Sezione di Firenze, Italy
[2] Dipartimento di Fisica, Univ. di Firenze, Italy
[3]LPC, IN2P3-CNRS, ENSICAEN et Université de Caen, Caen, France
[4]GANIL, CEA/DSM-CNRS/IN2P3, Caen, France
[5]IPN, CNRS/IN2P3, Université Paris-Sud 11, Orsay, France
[6]Horia Hulubei, Nat. Inst. of Phys. and Nucl. Eng., Măgurele, Romania
[7]INFN, Sezione di Napoli and Dip. di Fisica, Univ. di Napoli, Italy
[8]INFN, Sezione di Bologna and Dip. di Fisica, Univ. di Bologna, Italy
[9]INFN - Laboratori Nazionali di Legnaro, Legnaro (PD), Italy
[10]Nevsehir Univ. Science and Art Faculty, Phys. Dep., Nevschir, Turkey
Heavy Ion Laboratory, University of Warsaw, Warsaw, Poland
[12]Jagiellonian University, Institute of Physics, Krakow, Poland
[13]INFN Laboratori Nazionali del Sud, Catania, Italy
[14]Conservatoire National des Arts et Métiers, F-75141 Paris Cedex 03, France

[†]Deceased

DOI: 10.1051/epjconf/201611710005

Abstract

The present status and the perspectives of the FAZIA project are presented. The main achievements in terms of identification thresholds and isotopic resolution are discussed, together with the adopted technical solutions. The detector is particularly well suited for the investigation of isospin transport phenomena at intermediate beam energies; perspectives to reduce the identification thresholds to cope with lower energy ISOL beams are briefly introduced. Some experimental results concerning isospin transport effects obtained with a test telescope are presented. The study of isospin transport phenomena can give information on the symmetry energy term of the nuclear equation of state by comparing the experimental results on isospin related observables with the predictions of transport codes.

1 Introduction

The aim of the FAZIA (Four-π A and Z Identification Array) project [1] is the design and construction of a 4π detector with high resolution and low thresholds, which should be used both with stable beams (at GANIL (Caen, France), INFN-LNL (Legnaro, Italy), INFN-LNS (Catania, Italy)) and with radioactive beams (at SPIRAL2, SPES, EURISOL) in the range 10-50AMeV. Several European institutions are involved in the project and the R&D phase (concerning detectors, electronics and identification techniques) started in 2006.

The detector consists of blocks, each one comprising 16 three-layer telescopes. Each telescope, with an active area of $20x20mm^2$, includes a first $300\mu m$-thick Silicon (Si) layer, a second $500\mu m$-thick Si layer, followed by a $10cm$-thick CsI(Tl) crystal read out by a photodiode. Each telescope is fully equipped with digital electronics and the Si detectors are reverse mounted in order to improve the Pulse Shape Analysis (PSA). Concerning identification techniques, the Collaboration uses the standard ΔE-E method for particles punching through the first Si layer, obtaining isotopic resolution up to $Z \sim 23$. Particles stopped in the first layer are, on the contrary, identified by means of the PSA, exploiting either the "Energy (E) vs. Rise time of the Charge signal (Q_{trise})" correlation or the "E vs. Maximum of the Current signal (I_{max})" correlation; if the range of the particles in Si is greater than a given threshold, with a minimum value of $30\mu m$ and increasing with the charge of the ion, they can be identified in charge, up to that of the projectile. Concerning isotopic resolution, this is possible up to $Z \sim 15$ if the range of the particles in Si is greater than a certain threshold, starting from a minimum value of $150\mu m$ and increasing with Z.

During the R&D phase in the period 2007-2012 many tests of prototypes of the telescope were performed under beam at LNL-INFN , LNS-INFN and GANIL, obtaining very good performances in terms of identification capabilities (mass and charge) and thresholds. In December 2014 there was the commissioning of the first complete block and in June 2015 there was the first physics experiment with 4 blocks; data analysis is in progress. In December 2015 the second physics experiment will take place and, starting from 2016, 12 blocks of FAZIA (the so-called FAZIETTO) will be moved to GANIL, to be coupled with INDRA [2]. After 2020 FAZIETTO will be probably moved to LNL in order to exploit the first beams of SPES.

2 The FAZIA recipe

The main results obtained during the R&D phase are described in many technical papers (see [1] and references therein, with ref [41] now substituted by [3]).

Among the main prescriptions used by the Collaboration we can cite that:

- n-Transmutation Doped Si detectors with good doping uniformity ($< 3\%$) are used. Detector uniformity has strong influence on the identification capability: for example, in fig. 5 of [4] $E - Q_{trise}$ correlations for detectors with different homogeneity are presented, clearly showing the improvement on the charge resolution when detectors have homogeneities below 1%.

- The thickness uniformity of Si is within $\sim 1\mu m$; this is important to improve the quality of the $\Delta E - E$ correlation.

- Si detectors are obtained from wafers cut at $7°$ off the $< 100 >$ axis, in order to minimize channeling; this prescription improves significantly the isotopic resolution both on the $\Delta E - E$ correlation and in PSA, as it is shown in fig. 2 and fig. 3 of [4]: when particles enter along a crystallographic axis or plane, the isotopic resolution considerably worsens.

- Si detectors are reverse mounted; this is fundamental for the PSA. In fact the differences of signal shapes for light and heavy ions is maximized for rear injection (i.e. entrance from the low electric field side), as shown in fig. 2 of [5]; as a consequence, for the same detector used both in rear and in front configuration, the charge identification threshold significantly decreases for the reverse mounting, as it is clearly shown in fig. 10 of [5].

- a thin Al layer is deposited on both sides of the Si in order to reduce sheet resistance to few Ω; this prescription allows to get better timing properties, thus improving the quality of the signal rise time measurement and -as a consequence- of the PSA.

- The Front End Electronics (FEE) and the preamplifiers, all purposely developed by the collaboration, are located inside the vacuum scattering chamber, in order to reduce the noise pickup; the connection between the vacuum chamber and the external world takes place only through a 3Gb/s optical fiber.

- The collaboration has also developed proper algorithms for digital signal shaping, whose implementation is mandatory to improve the quality of the charge and mass resolution. For example, we can cite the use of a cubic interpolation on four consecutive samples to extract the maximum of the current signal (left side of fig. 1), to be compared with the result obtained taking directly the sample with the maximum value of the current (right side of fig. 1): the isotopic resolution associated with the former technique is significantly better than in the latter case.

- Finally, the voltage drop across the detector is kept constant, by online monitoring the reverse current and consequently adjusting the applied voltage.

The quality of the obtained identification is very good: as an example, fig. 2 presents the Particle Identification (PI) spectrum obtained from the $\Delta E - E$ correlation for particles punching through the first Si layer and stopped in the second Si layer (from [6]). Only the regions with Z=11-14 and Z=22-27 are shown. Isotopic resolution is clearly seen up to $Z \sim 23$. Concerning PSA, the results obtained for the charge identification are shown in fig. 10, 11, 13, 15 of [6], while Ref. [3] shows the extremely good performances of a not completely depleted detector.

3 FAZIA applications and results

The physics which should best exploit the very good capabilities of FAZIA in terms of isotopic resolution and low thresholds concerns the study of the

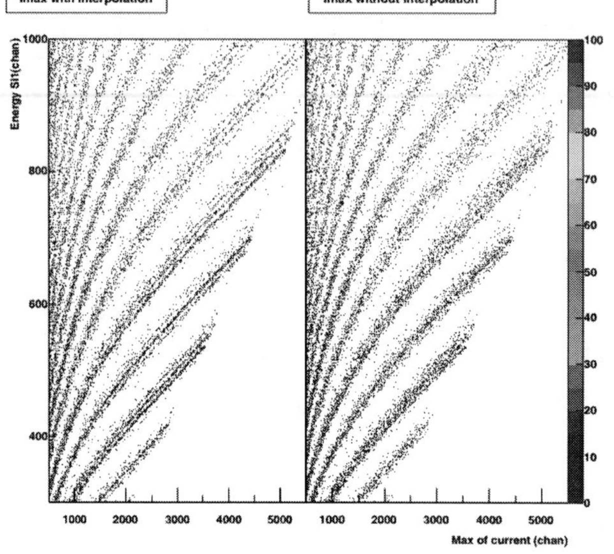

Figure 1: E vs. I_{max}. Left side: I_{max} obtained via cubic interpolation on four consecutive samples. Right side: I_{max} without interpolation.

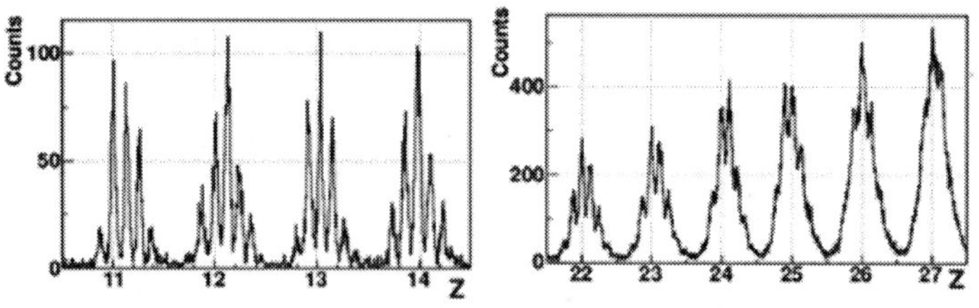

Figure 2: PI spectrum from $\Delta E - E$ correlation for particles punching through the first Si layer and stopped in the second Si layer (with Z=11-14 and Z=22-27) (from [6])

isospin of the ejectiles at medium-low beam energies; this kind of study is important because it can give information on isospin transport phenomena, which are strictly connected to the density dependence of the symmetry energy term of the nuclear equation of state, not well known far from normal conditions. In particular, we expect effects of isospin diffusion, sensitive to the symmetry energy itself, driven by the isospin gradient between target and projectile, and effects of isospin drift, sensitive to the derivative of the symmetry energy and driven by the density gradient between regions at normal density (QuasiTarget (QT) and QuasiProjectile (QP) zone) and the more diluted neck zone [7,8]. Information on the symmetry energy behaviour can be obtained comparing experimental results on isospin related observables to the prediction of theoretical model, such as for example transport models like SMF [9] or AMD [10], including different recipes for the density dependence of the symmetry energy.

Experimental evidence of isospin transport phenomena have been already obtained also by the FAZIA Collaboration thanks to inclusive data taken in 2011 during a test experiment performed at LNS-INFN with one prototype of the FAZIA telescope [11]. In particular, we investigated the reactions $^{84}Kr+^{112,124}Sn$ at 35AMeV and the angular coverage of the detector was such that it mainly detected ejectiles coming from the QP phase space and the neck zone. We found evidence of isospin diffusion by comparing the average isospin ($< N/Z >$) of the ejectiles as a function of their charge for the two different reactions: the $< N/Z >$ is sistematically higher when the target is the n-rich ^{124}Sn, thus indicating a isospin transport from target to projectile (see fig. 5 of [11]). Moreover, we found evidence of isospin drift looking at $< N/Z >$ as a function of the lab velocity for light products: light fragments emitted closer to the centre of mass (i.e. possibly coming from the neck source, a dilute zone) show a neutron enrichment with respect to those emitted closer to the QP velocity (i.e. coming from a zone at normal density), see, for example, fig. 3, adapted from fig. 6 of [11].

On the same set of data, the N and Z staggering was investigated too [12]. This phenomenon might affect the comparison of experimental data and theoretical models if not properly taken into account by the afterburner codes.

In June 2015 a first physics experiment with 4 complete block was done at INFN-LNS (see left part of fig. 4), on the system $^{80}Kr +^{48,40} Ca$ at 35AMeV; at present data analysis is in progress. The experimental counts in polar representation are shown in fig. 4 right part. Note that this was the first time that more than one block was acquired in coincidence. The main goal of the experiment is to extend the study of the isospin transport

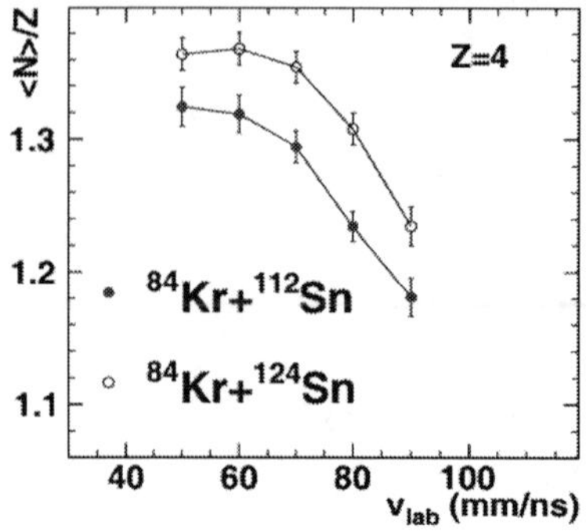

Figure 3: Average isospin as a function of the lab velocity for Z=4. Adapted from fig. 6 of [11]

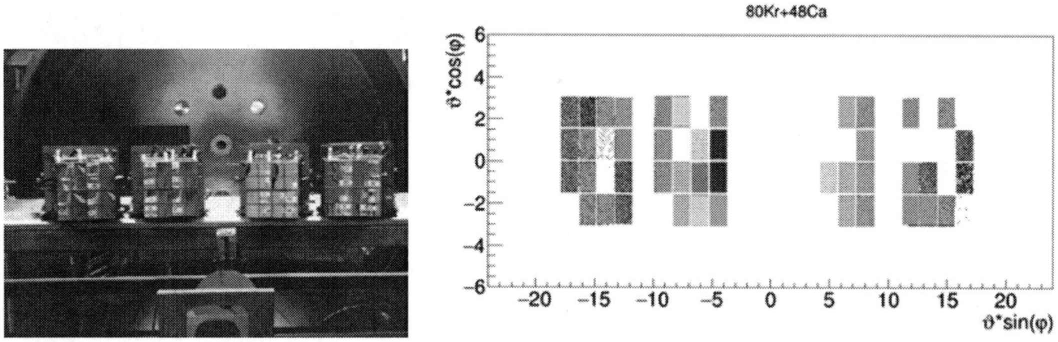

Figure 4: Left part: Four complete blocks of FAZIA inside the Ciclope scattering chamber of INFN-LNS during the first physics measurement in June 2015. Right part: Experimental raw counts in polar representation

process already undertaken in the test experiment of 2011. In particular, it will be possible to classify the events in different bins of centrality. Moreover the isotopic composition of both fission fragments coming from the QP will be measured in coincidence.

Starting from the end of 2016, 12 blocks of FAZIA will be moved to GANIL, in order to be coupled to INDRA after removing the first rings. This composite setup will have innovative characteristics; in fact in the forward direction it will provide performances comparable to a spectrometer for medium-light nuclei, but it will also allow the simultaneous measurement of more fragments in coincidence (for example, this point is extremely important for fission events), togheter with a 4π coverage.

In 2020 low energy ($\leq 10AMeV$) radioactive beams from SPES will become available and FAZIETTO will be moved back to LNL. This opportunity represents a challenge from the point of view of the identification thresholds. Among the possible solutions we can cite the use of the time of flight (strongly dependent on the available flight path) and the use of very thin Si detectors as first identification layer. Concerning this latter possibility, in 2012 the FAZIA Collaboration performed a test under beam using a standard FAZIA telescope in which the first Si layer was $21\mu m$ thick. The obtained results were very encouraging [13] from the point of view of the charge identification by means of the $\Delta E - E$ technique; some hints on the mass reconstruction might be obtained by means of energy loss calculations. PSA was of course not possible because the Si thickness was smaller than the minimum threshold for the charge resolution ($30\mu m$ for light ions).

All these results demonstrate the very good capabilities of FAZIA in terms of charge and mass identification with low thresholds; as a consequence we can expect that this detector will contribute in a substantial way to the study of isospin related phenomena.

References

[1] Bougault R. et al., *Eur.Phys.J.A*, **50** (2014) 47.

[2] Pouthas J. et al., *Nucl.Instr.MethA*, **357** (1995) 418.

[3] Pasquali G. et al., *Eur.Phys.J.A*, **50** (2014) 86.

[4] Bardelli L. et al., *Nucl.Instr.Meth.A*, **654** (2011) 272.

[5] LeNeindre N. et al., *Nucl.Instr.Meth.A*, **701** (2013) 145.

[6] Carboni S. et al., *Nucl.Instr.Meth.A*, **664** (2012) 251.

[7] Baran V. et al., *Phys.Rep.*, **410** (2005) 335.

[8] Di Toro M. et al., *J.Phys.G*, **37** (2010) 083101.

[9] Baran V. et al., *Nucl.Phys.A*, **730** (2004) 329.

[10] Ono A., *Phys.Rev.C*, **853** (1999) 853.

[11] Barlini S. et al., *Phys.Rev.C*, **87** (2013) 054607.

[12] Piantelli S. et al., *Phys.Rev.C*, **88** (2013) 064607.

[13] Kordyasz A.J. et al., *Eur.Phys.J.A*, **51** (2015) 15.

Silicon carbide detectors study for NUMEN project

A. Muoio[1,2], C. Agodi[1], D.L. Bonanno[3], D.G. Bongiovanni[3], S. Calabrese[1,4], L. Calcagno[3,4], D. Carbone[1], M. Cavallaro[1], F. Cappuzzello[1,4], A. Foti[3,4]. P. Finocchiaro[1], G. Lanzalone[1,5], F. La Via[6], F. Longhitano[3], D. Lo Presti[3,4], L. Pandola[1], S. Privitera[6] and S. Tudisco[1]

[1]INFN - Laboratori Nazionali del Sud, Catania, Italy
[2]Dip. di Fisica e Scienze della Terra, Univ. di Messina, Messina, Italy
[3]INFN, Sezione di Catania, Catania, Italy
[4]Dip. di Fisica e Astronomia, Univ. di Catania, Catania, Italy
[5]Univ. degli Studi di Enna Kore, Enna, Italy
[6]CNR-IMM, Catania, Italy

Abstract

In this contribution, we will illustrate the main results of the R&D activities related to the Silicon Carbide detectors associated with NUMEN project.

1 NUMEN Physics

The NUMEN project[1-4] proposes an innovative technique to access the nuclear matrix elements entering the expression of the half-life of the double beta decay by relevant cross sections measurements of heavy-ion induced double charge exchange (DCE) reactions, $(^{18}O,^{18}Ne)$, $(^{20}Ne,^{20}O)$ and $(^{12}C,^{12}Be)$. DCE reactions will be investigated at the INFN-LNS laboratory with incident energies ranging from 10 to 60 MeV/A. A key aspect of the project is the use of the MAGNEX [5-6] large acceptance magnetic spectrometer to detect the ejectiles produced in the nuclear collisions[7]. With the present set-up it is difficult to suitably extend this research to the hot cases, where $\beta\beta$ decay studies are and will be concentrated. The present limitations concern mainly the low beam current from the L.N.S. superconducting cyclotron CS accelerator and the low counting rate tolerated (< 5 kHz)[8-9] by the MAGNEX Focal Plane Detector (FPD) . In the view of a systematic study of the many hot cases of $\beta\beta$ decays, an upgraded set-up, able to work with at least two orders of magnitude more current than the present, is thus mandatory. This goal can be achieved by a substantial change in the technologies used in the beam extraction and transport and in the detection of the ejectiles. For the spectrometer, the main foreseen upgrade is the substitution of the present FPD. In particular the existing wall of stopping detectors based on 50 x 70 mm^2, 0.5 mm thick silicon pad detectors needs to be upgraded in view of the higher detection rate. Many requirements must be taken into account for the design of the new stopping wall telescopes for particle identification. The most relevant aspects are: a) energy resolution better than 1% to maintain the performances in the particle identification reached with the present set-up of the FPD (mass resolution 1/160) [10]; b) unambiguous and fast particle identification procedure; c) time resolution better than $1\div2$ ns in order to guarantee an accurate measurement of the drift time in the gas chamber used to reconstruct the vertical track of the ejectiles; d) a high granularity(modules of 1 cm^2) is required in order to limit double-hit events probability below 10% in the whole FPD; e) the thickness of the telescope must be chosen in order to permit the detection of the ejectiles in the wide dynamical range of incident energies (from 10 up to 60 MeV/A). In this context, the choice of solid state detectors, such as SiC, is the most appropriate.

2 Silicon carbide

Radiation damage can affect various properties of a detector. Phenomena connected with the alteration of these properties are the increase of the leakage current, the decrease of charge collection efficiency (CCE) and the removal of free carriers from the conductive regions of the device. Radiation hardness is the inertness of these device parameters to high doses of particle irradiation [11]. SiC, due to its wide gap and strength of its chemical bonds, has been seriously considered as a valid alternative to Si for the production of radiation hard ionizing particle detectors. The leakage current of a p-n junction consists of diffusion current from the quasi-neutral areas and generation current from depletion area [12]. The second term depends essentially on the temperature T and on the energy gap E_g. Increasing the E_g from 1.1 eV (Si) to 3.2 eV (4H-SiC) determines a reduction of the diffusion term of the leakage current of about 17 orders of magnitude at room temperature($\sim 300K$). Silicon Carbide technology offers then an ideal response to such challenges, since it gives the opportunity to cope the excellent properties of silicon detectors (resolution, efficiency, linearity compactness) with a much bigger radiation hardness (up to five orders of magnitude for heavy ions), thermal stability and insensitivity to visible light. R&D activity is mandatory to develop innovative processes, which allows a massive production of thick (>100 μm) and large area (about 1 cm^2) SiC detectors. Following these indications we started the research work with a Monte Carlo simulation with the aim to study ΔE - E configuration (with respective thicknesses 100μm and 1000μm) from the point of view of defects dislocation generated by an^{18}O at 25 MeV/A. For this work we used the SRIM [13] software; it is a collection of packages which calculate many features of the transport of ions in matter. The following figure 1.a shows the number of simulated vacancies as function of protons (at 25 MeV and 1 GeV) and ^{18}O (25 MeV/A) ions fluency. One can notice that the number defects create in the same stage (ΔE or E) from protons beam at 25 MeV/A is about two orders of magnitude smaller then the corresponding number coming from heavy oxygen ions of 25 MeV/A. In Figure1.b, we show the results of such calculation, we can be notice as several order of magnitude of leakage current increased are expected at high doses irradiation/fluency. Such increase could be acceptable for SiC detectors, they have a leakage current five order of magnitude less than Silicon detectors.(Where the maximum of acceptable fluency is about 10^9 heavy ions/cm^3).

We have studied, moreover, the effects of the irradiation comparing SiC and Si detectors. These detectors were irradiated with C^+ ions at 740 MeV and at different doses. For the experiment we used a standard Si detector of 1 cm^2 of active area and 300 μm thickness and SiC detector which was built growing a 4H-SiC n-type epitaxial layers with a carrier concentration of $1.0 \cdot 10^{-14} cm^{-3}$ and a thickness of 80 μm on a highly doped n-type ($\sim 7.0 \cdot 10^{-18} cm^{-3}$) substrate. The following Table 1 shows the results of the electrical static characterization before and after the irradiation. Can be observe as Si detectors are acceptable to work at the irradiation 94kGy dose; while SiC are able to survive at 218 kGy dose.

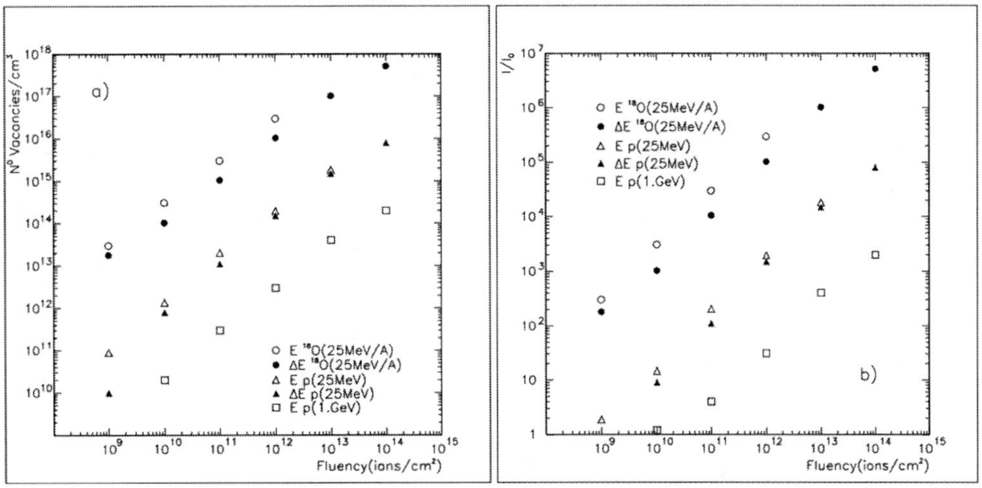

Figure 1: The figure 1.a shows the comparison of Number of silicon vacancies%Fluency on ΔE and E in SiC stages; while the figure 1.b shows the comparison of Increment of current leakage%Fluency in SiC for thickness of ΔE and E. In both figures the solid symbols indicate the first stage of the detector (with a thickness of 100μm), while the empty symbols the second stage (with a thickness of 1000μm). The following energies has been simulated: oxygen ions at 25 MeV/A (circle), protons ions at 25 MeV/A(triangle) and protons ions at 1 GeV/A(square).

Table 1: Comparison, with several doses, between SiC and Si detectors

	SiC reverse current[A/cm^3]			**Si reverse current[A/cm^3]**		
Voltage[V]	**SiCFresh**	**94kGy**	**218kGy**	**SiFresh**	**94kGy**	**21kGy**
-190	1.9 10^{-6}	8.6 10^{-6}	4.5 10^{-4}	8.3 10^{-5}	3.9 10^{-4}	8.3 10^{-5}
-150	8.0 10^{-7}	5.4 10^{-6}	2.9 10^{-4}	6.5 10^{-6}	3.2 10^{-4}	8.3 10^{-5}
-110	3.4 10^{-7}	3.2 10^{-6}	1.4 10^{-4}	4.8 10^{-6}	2.5 10^{-4}	8.3 10^{-5}
-70	1.4 10^{-7}	1.8 10^{-6}	3.5 10^{-5}	2.6 10^{-6}	7.8 10^{-5}	3.0 10^{-5}
-30	7.0 10^{-8}	6.1 10^{-7}	3.9 10^{-7}	7.9 10^{-7}	6.1 10^{-5}	2.2 10^{-5}
-10	4.7 10^{-8}	2.8 10^{-7}	5.1 10^{-8}	1.9 10^{-7}	3.5 10^{-5}	1.7 10^{-5}

3 Conclusions

Silicon Carbide is one of the promising radiation hard materials. Our goal is to realize a two stages detector increasing, beyond the state of the art, thickness and active area. The results of our simulations and test demonstrate that SiC is a good compromise between several detectors materials. Due to its wide band-gap it has a very low leakage current, good radiation resistance and reasonable sensing capability for charges created during the ionization processes.

References

[1]F. Cappuzzello et al. Jour. of Phys. Conf. Ser. 630 (2015) 012018
[2] C. Agodi et al. Nucl. and Part. Phys. Proc. 265266 (2015) 28-30
[3] F. Cappuzzello et al. Eur. Phys. Jour. A,
[4] F. Cappuzzello, C. Agodi, F. Balestra et al. This Conference Proceeding
[5] F. Cappuzzello et al., MAGNEX: an innovative large acceptance spectrometer for nuclear reaction studies, in Magnets: Types, Uses and Safety (Nova Publisher Inc., New York, 2011) pp. 163.
[6] A. Lazzaro, et al.Nucle. Instr. and Meth. A 591 (2008) 394-405
[7]M. Cavallaro. et al. Eur. Phys. J. A (2012) 48- 59.
[8]M.Cavallaro et al. Eur. Phys. Jour. A 48 (2012)59
[9]D. Carbone, F. Cappuzzello and M. Cavallaro Eur. Phys. Jour. A 48 (2012)60
[10]F. Capuzzello. et al. NIM A 638 (2011) 74-82.
[11]A.A. Lebedev et al. Fiz. Tekh. Poluprovodnikov 36. 1354 (2002).
[12]S.M. Sze. Physics of Semiconductor Devices (Ed. Wiley). 1969).
[13]J.F.Ziegler et al. The stopping and range of ions in matter(Ed. Wiley).(1969).

Detector developments for the Super-FRS

C. Nociforo[1], F. Schirru[1], M. Kiš[1] and S. Schlemme[1,2]

[1] GSI Helmholtzzentrum fr Schwerionenforschung, Darmstadt, Germany
[2] TU Darmstadt, Darmstadt, Germany

Abstract

Some developments of Super-FRS focal plane detector prototypes and results of recent in-beam tests performed at GSI and LNS-INFN with diamond detectors are presented.

1 Introduction

The magnetic in-flight separator Super-FRS [1] under construction at FAIR consists of 2-degrader stages, the pre-separator made mainly of normal conducting magnets and the main separator made of superconducting magnets. As shown in Fig. 1, it can deliver high energy radioactive beams (RIBs) starting from 20 Tm to the high-energy branch, slowed down RIBs to the low-energy branch (the beam can be even stopped at the end of the energy buncher) and pulsed RIBs for in-ring experiments. The right combination of beam energies and target/degrader thickness will typically let to optimize the delivery of in-flight highly-charged well separated fragments. Even though a strong reduction of contaminants is ensured at the end of the three branches, the particle rate in the first half of the main separator can still be high (1-10 MHz). Various combinations of the magnetic sections of the Super-FRS operated in dispersive, achromatic or dispersion-matched spectrometer modes will allow measurements of momentum distributions of secondary reaction products with high precision [2]. Due to the different experiments and operation modes a large variety of detectors is foreseen. All

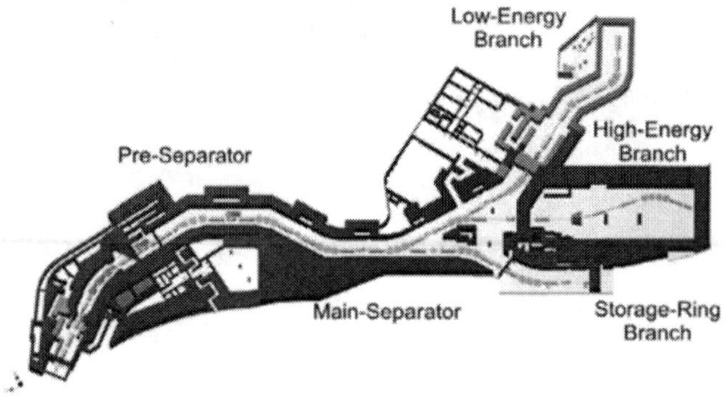

Figure 1: Layout of the Super-FRS.

this has strong implication on the design of the Super-FRS detecting system, which has to cope with two main technical challenges: high intensity and resolution.

2 Super-FRS radiation-hard detectors

2.1 Tracking detectors

The identification of different isotopes at the Super-FRS requires a combined event-by-event analysis of the magnetic rigidity (Bρ), time-of-flight (ToF) and energy deposition [3]. Toward the higher beam intensity we plan to use a set of Time Projection Chambers (TPC) with GEM (Gas Electron Multiplier) amplification. In addition to high resolution and adjustable gain over a wide range (from protons to Z=92), these detectors have the advantage of being able to perform particle tracking on an event-by-event basis exposing very little additional material to the ion beam. Considerable progress has been made in the past years in the development of GEM-TPC tracking detector prototypes [4] for the Super-FRS. Even though gas detectors are very robust against beam bombardment, they are too slow for timing measurements.

2.2 Time-of-flight detectors

The use of radiation hard silicon and diamond detectors for fast timing at Super-FRS has been recently exploited. A few experiments have been carried out in order to assess the ToF performances of radiation detectors

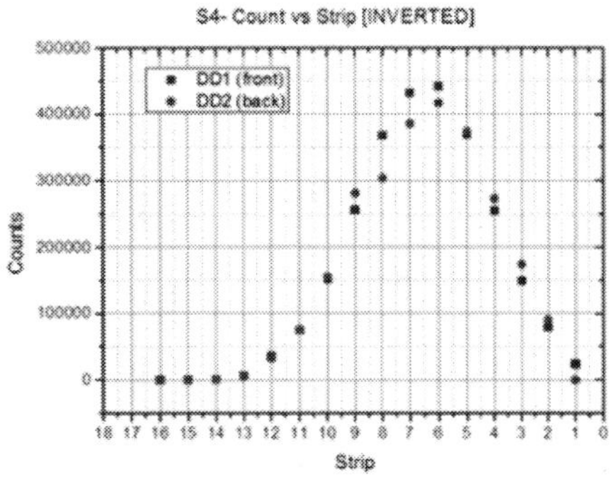

Figure 2: Left panel: PADI7 board with strip diamond detectors (front and back) Right panel: Horizontal beam profile reconstructed on diamonds.

based on diamond material. We have developed position-sensitive diamond detectors made of polycrystalline CVD (chemical vapour deposition) material of (20x20x0.3) mm^3 size provided by Element Six. Their electrodes were fabricated in house in 16-strips geometry by depositing on both sides of the diamond samples Cr/Au layers of thickness 50/100 nm respectively. The vertical geometry was chosen to match the horizontal dispersion of the separator. The device was mounted on a new board (see Fig. 2) having bias voltages, threshold and integrated amplification stage and able to provide LVDS output signals.

3 Beam tests measurements

3.1 ToF measurements

In August 2014, a test experiment was performed at the FRS [5] in GSI with ^{197}Au primary beam using a diamond pair as start and stop detectors, in experimental condition similar to those expected at the Super-FRS. A printed circuit board equipped with four PADI7 chips [6] and fabricated in house was used to readout each single diamond strip. The digital output signals from PADI after splitting were sent to the FPGA TDC VFTXs [7] synchronized with a 200 MHz external clock [1]. An example of beam profile reconstructed with two diamond detectors DD1 and DD2 located at the end of the FRS is shown in Fig. 2. A ToF resolution σ_{ToF} of about 40 ps has been measured between two strips of the front and back diamonds (see Fig. 3). A ToF resolution of about 50 ps was measured

[1]PADI and VFTX electronics had an intrinsic resolution below 15 ps.

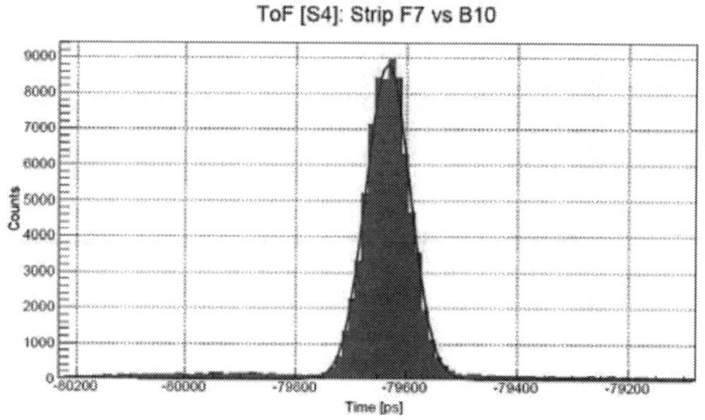

Figure 3: Diamond ToF measurement.

after positioning the DD2 detector at the mid-focal plane of the separator, about 35 m apart. The intrinsic time resolution of each diamond could be extracted taking advantage of the presence of the FRS TPC detectors [9] and plastics scintillator mounted along the beam line. After applying a TPC position correction, an intrinsic resolution σ_t=$\sigma_{ToF}/\sqrt{2}$ $\approx$25 ps is achieved.

3.2 Irradiation tests

Damage studies on the same diamond material have been also performed. At the LNS-INFN laboratory, a polycrystalline CVD diamond detector (pcCVD-DD) 0.3 mm thick has been irradiated via series of long irradiations of 10^7 ^{12}C/mm^2/s at 62 MeV/u. The detector has four squared metalized faces Q_{1-4} (10x10) mm^2 each. Each of them was connected to a broadband DBA amplifier [8]. The time resolution and the charge collection efficiency of the pcCVD-DD have been monitored via digital scope during low intensity runs taking place at the end of each series. Digital waveforms have been registered also for a single-crystal diamond detector (scCVD-DD) 0.09 mm thick, which was moved in during each low intensity run. The diamond analogue signals were split and sent to the channels of a oscilloscope (10 GS/s) and to a scaler after leading-edge discriminator for counting. All detectors worked inside a vacuum chamber at a pressure of 10^{-7} mbar. A collimator with 2.6 mm diameter was used at the entrance of the chamber. The beam current was measured by using an ionization chamber located behind the diamonds. The pcCVD-DD was kept constantly biased at 300 V since no change in the leakage current was observed. The number of ions counted by each squared section summed up during the long irradiation se-

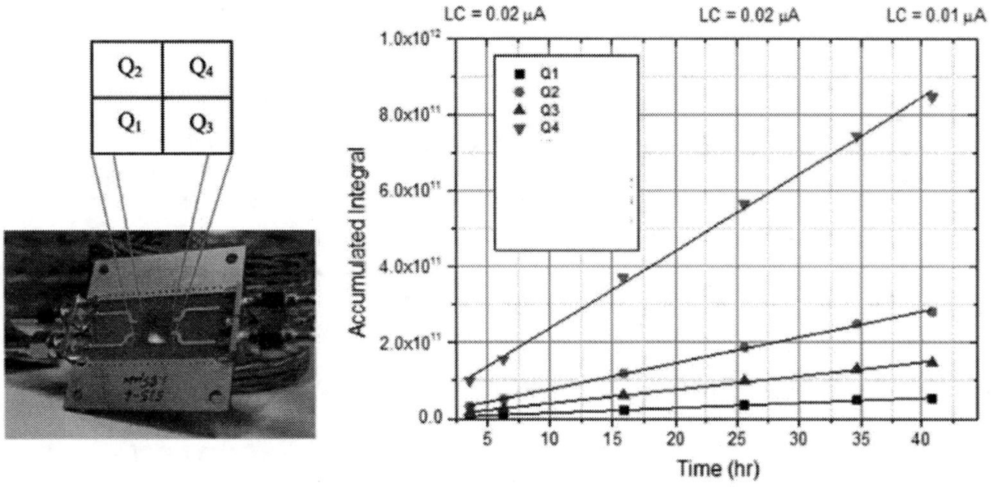

Figure 4: Left panel: pcCVD-DD with four squared faces Q_{1-4}. Right panel: Count numbers measured by each face of the pcCVD-DD during the irradiation measurement.

ries is shown in Fig. 4. Their sum provides a minimum accumulated dose of about 1.7 MGy corresponding to 4.3x10^9 ^{238}U/mm^2 s at 350 MeV/u ($\approx$400 days operation of Super-FRS). Detector efficiency corrections will be taken into account to estimate the final accumulated dose. In Fig. 5 the slope (left side) and the amplitude (right side) distributions of Q_4 signals measured after 1 hour and 40 hours of irradiation are shown. Pulses below 30 mV have been discarded. The slope distribution was obtained by a linear fit of the sampled waveform at 10-90% of the maximum amplitude. From the present results, no degradation of the pcCVD-DD signal is observed within the 500 ps resolution (due to the DBA use). Preliminary results on the ToF measured between the scCVD-DD and Q_4 pcCVD-DD showing no change in the time resolution are also encouraging.

4 Future developments

The active area of the Super-FRS ToF detectors ranges from 10 up to 10^4 mm^2, depending on the focal planes. Not only material choice is important. Additional experimental requirements like multi-channel electronics and data acquisition need to be considered in relation to the increasing number of channel and amount of digitised data. In the next future not

DOI: 10.1051/epjconf/201611710007

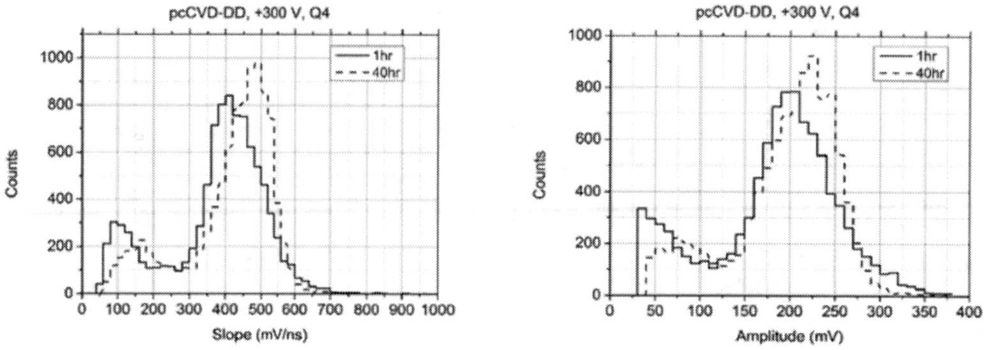

Figure 5: Pulse slope (left panel) and amplitude (right panel) of pcCVD diamond detector after 1 hour and 40 hours of irradiation.

only further irradiation tests with uranium ions will be performed but time measurements based via high precision time distribution and time stamps.

Acknowledgments

The authors wish to acknowledge the support of: M. Ciobanu, J. Frühauf and N. Kurz in providing the PADI and VFTX electronics, M. Träger and R. Visinka for detector processing, A. Kratz, P. Figuera, A. Musumarra and S. Salamone during the experiment at LNS-INFN.

References

[1] M. Winkler et al., Nucl. Instr. and Meth. B 266 (2008) 4183.

[2] J. Äystö et al., JPS Conf. Proc. 6, 020035 (2015).

[3] C. Nociforo, 2014 JINST 9 C01022.

[4] F. Garcia et al., GSI Report 2014, in press.

[5] H. Geissel et al., Nucl. Instr. and Meth. B 70 (1992) 286.

[6] M. Ciobanu et al., IEEE T-NS 61 (2014) 1015.

[7] E. Bayer and M. Traxler, IEEE T-NS 58 (2011) 1547.

[8] P. Moritz et al., Diamond and Related Materials 10 (2001) 1765.

[9] R. Janik et al., Nucl. Instr. and Meth. A 640 (2011) 54.

EPJ Web of Conferences **117**, 10008 (2016) DOI: 10.1051/epjconf/201611710008
NN2015

Status and perspective of FARCOS: A new correlator array for nuclear reaction studies

E.V. Pagano[1,2], L. Acosta[3,4], L. Auditore[5], C. Boiano[6], G. Cardella[3], A. Castoldi[7], M. D'Andrea[3], D. Dell'aquila[8], E. De Filippo[3], S. De Luca[5], F. Fichera[3], L. Francalanza[8], N. Giudice[1,3], B. Gnoffo[3], A. Grimaldi[3], C. Guazzoni[6,7], G. Lanzalone[2,9], I. Lombardo[8], T. Minniti[11,12], S. Norella[5], A. Pagano[3], M. Papa[3], S. Pirrone[3], G. Politi[3], F. Porto[1,2], L. Quattrocchi[5], F. Rizzo[1,2], P. Russotto[3], G. Saccá[3], A. Trifirò[5], M. Trimarchi[5], G. Verde[3,10] and M. Vigilante[8]

[1]Università di Catania, Catania, Italy
[2]INFN - Laboratori Nazionli del Sud, Catania, Italy
[3]INFN, Sezione di Catania, Catania, Italy
[4]Istituto de Física, Universidad Nacional Autónoma de México, México City, México
[5]Dipartimento di Fisica e Scienze della Terra, Università di Messina, INFN, gr. collegato di Messina, Messina, Italy
[6]INFN, Sezione di Milano, Milano, Italy
[7]Politecnico di Milano, Dipartimento di Elettronica, Informazione e Bioingegneria, Milano, Italy
[8]INFN, Sezione di Napoli, Dipartimento di Fisica, Università di Napoli Federico II, Napoli, Italy
[9]Università di Enna Kore, Enna, Italy
[10]Institute de Physique Nucléaire Orsay, Orsay, France
[11]CNR - Istituto per I Processi Chimico-Fisici, Messina, Italy
[12]ISIS Facility, Science and Technology Facilities Council, Rutherford Appleton Laboratory, Oxfordshire, UK

Abstract

The experimental investigation of Heavy Ion reactions at Fermi energies requires an accurate measurement of observables, such as linear momentum and energy of the detected particles. In order to address this problem, dedicated and flexible correlator arrays are useful tools to be coupled with 4π detectors. One of these arrays is FARCOS, presently under construction at the INFN Sezione di Catania and Laboratori Nazionali del Sud (LNS).

The study of dynamics of nuclear reactions in heavy ion (HI) collisions at Fermi energy (20 MeV/u < E/A< 200 MeV/u) is an active area in nuclear science [1]. In this energy region, an impressive range of physical phenomena occurs within large time-scales (from 10 to 1000fm/c) which are revealed through imprints on space-time and particle-particle correlations [2]. In particular, nuclear matter can experiences variation with respect to the saturation density ρ_0 ($\approx$ 0.17 nucleon/fm^3) [3]. Exotic nuclei can be also produced and studied in close connection with nuclear astrophysics [4]. With the availability of new-generation radioactive ion-beam facilities (RIB), like for instance SPES at LNL, SPIRAL II at GANIL, FAIR at GSI or FRIB at MSU or the possible upgrade of the heavy ion CS facility in Catania, such studies assume renewed interest. Experimentally, large efforts has been devoted in the last two decades in order to achieve high energy and angular resolution of dedicated detection systems [5]. In this paper we briefly describe the next-generation correlator FARCOS (Femtoscope ARray for COrrelation and Spectroscopy) for reaction and spectroscopic studies [6]. FARCOS is a modular array of telescopes arranged in a single cluster, each one consisting of three detection stages. The first two stages are two Double Sided Silicon Strip Detectors (DSSSD) of active area 6.4×6.4 cm^2 with 32 vertical strips in the front side and 32 horizontal strips in the back side and are 300 μm and 1500 μm thick respectively. The last stage consists of four CsI(Tl) crystals (3.2×3.2 cm^2) of 6 cm of thickness read by a photodiode (18×18 mm^2). We show in Fig. 1 a typical close assembling of one cluster.

Particle identification is performed by ΔE/E, pulse shape(PSD) and Time of Flight (TOF) techniques. Portability and modularity are important characteristics making feasible the easy coupling with 4π detectors, other correlators and/or magnetic spectrometers. Digital Signal Acquisition (DSA) is also envisaged. We used, preliminary, in the electronic front-end, a preamplifier (PAC) derived from an improved versions of CHIMERA-type PACs developed to minimize the electronics noise and to optimize TOF performances [7]. So, the read-out of silicon strips was performed by using a compact housing (86x80x10 mm^3 (NPA-16FL), 98x80x15 mm^3 (NPA-

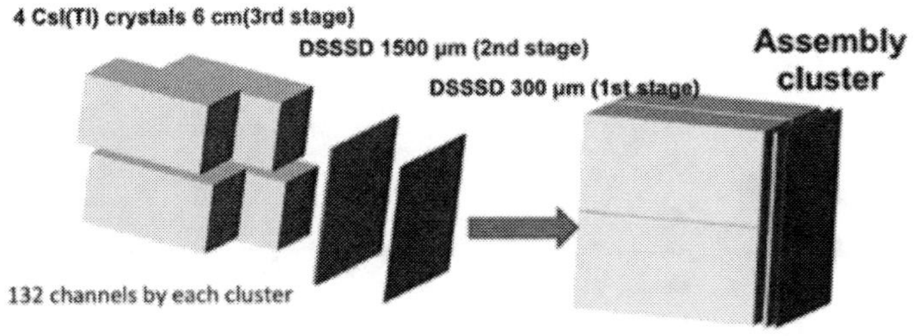

Figure 1: schematic view of a cluster of FARCOS

Figure 2: The INKIISSY experimental setup.^{124}Xe+^{64}Zn @ 35MeV/A was the studied reaction.

16FE)) allocating 32-channels PACs with low power consumption (<900 mW for 32 channels). FARCOS prototype was tested and coupled for the first time with the 4π CHIMERA [8] in the INKIISSY (INverse KInematic ISobaric System) experiment [9] in Catania on April 2013 with beams delivered by INFN-LNS facility. FARCOS covered the range of polar and azimuthal opening, $\theta_{lab} \approx 16^\circ \div 44^\circ$ and $\Delta\phi \approx 75^\circ$, respectively, corresponding to a solid angle coverage of $\approx 262msr$ (see Fig. 2). As a consequence, a portion of CHIMERA was shielded by FARCOS telescopes, thus giving the opportunity to investigate about the detection of neutrons [10].

In reference [6], the ΔE/E identification procedures in the first two DSSSD stages (see Fig. 3) have been already discussed and a good isotopic identification was shown for IMFs up to atomic number Z=10. Good results have been also obtained by applying PSD method in CsI(Tl) for light charged particles up to lithium and ^{8}Be (see Fig. 4).

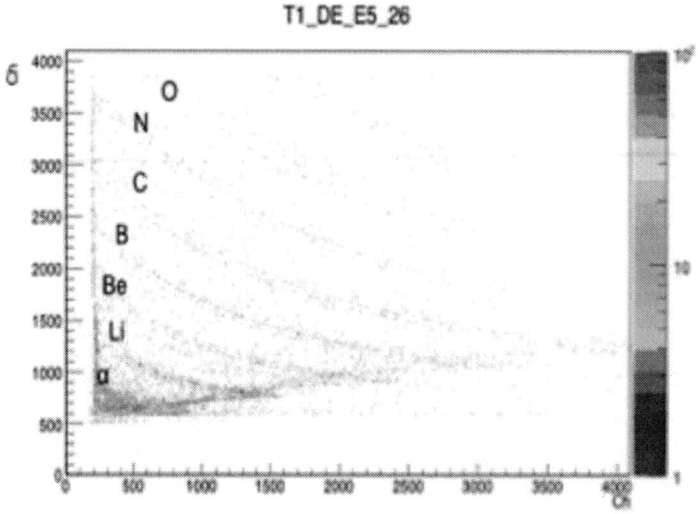

Figure 3: ΔE/E identification matrix of two strips between the first and the second DSSSD stages (INKIISSY experiment).

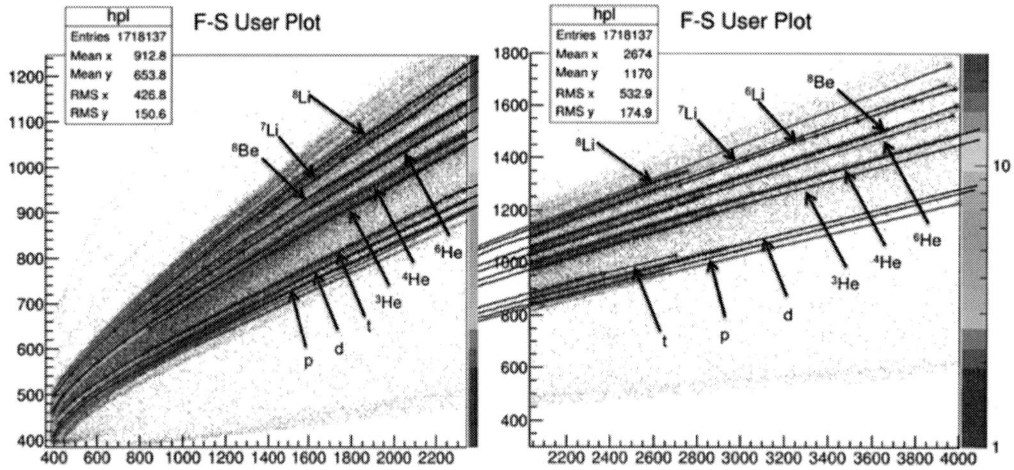

Figure 4: Typically PSD analysis identification in a CsI(Tl) of FARCOS (INKIISSY experiment).

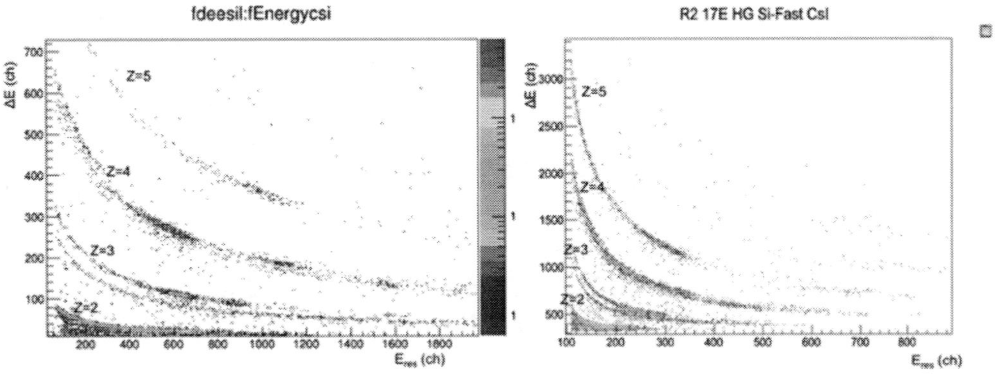

Figure 5: comparison between two $\Delta E/E$ identification matrix obtained in the reaction $^{16}C+^{12}C$ @ 40 MeV/A (CLIR experiment) from two telescopes of CHIMERA detector of the ring two. Left: signals processed by GET. Right: signals processed by standard CHIMERA electronic

The GET (Generic Electronic for TPC) electronics for read-out and DAQ, was also tested. GET is an compact system designed to handle a very large number of independent electronic channels (more than 30kch). The innovative performances such as high versatility and compactness make the GET project a good candidate for FARCOS array (132 channels for each cluster allocated in a small volume, about 7x7x10 cm^3). GET is based on one chip, AGET, processing 64 independent channels. Each channel consists of a preamplifier, a filter, a discriminator and an array of 512 capacitors having sampling frequency ranging from 1 to 100 MHz. Four AGET chips are allocated on a ASAD board for the digitalization of 256 channels. A block of four ASADs (1024 chs) can be read-out by a single CoBo. Up to 10 CoBo boards can be house in the μTCA crate. In this same crate, a MUTANT card is used for triggering, synchronizing different CoBos and for managing inter-communications among different devices [11]. Silicon detectors and CsI(Tl) crystals were tested by comparing the CHIMERA and the GET DAQ systems (see Fig. 5). The result of this comparison is quite satisfactory.

Milestones of FARCOS construction are briefly summarized in the following. The test of GET electronics will be completed within the second semester of year 2015, while in the first semester of 2016 it is expected to test the new ASIC preamplifiers designed by INFN and Politecnico of Milan. In 2016 six new clusters will be assembled; in 2017 and in 2018 four new clusters for each year are planned to be added to the array. The final configuration of 20 clusters of FARCOS array is planned to be in operation on 2019.

In conclusion, we have tested performances of four clusters of the FARCOS prototype. The obtained results are quite satisfactory and fit the requirements for new generation correlators for charged light particles and fragments, in HI physics at intermediate energies. On 2015 year the FARCOS prototype will be used in two experiments at LNS, i.e., CLIR and SIKO. The use of the detector is also envisaged in various Letters of Intent submitted to the new radioactive Ion beam facility, SPES, that is under construction at INFN-Laboratori Nazionali di Legnaro.

References

[1] Proceedings of the IWM-EC (2014), Catania, (Italy), Vol.108, pub. by the Italian Physical Society, Bologna, Eds. B. Borderie, R. Bougault, A. Pagano, S. Pirrone and G. Politi. www.epj-conferences.org/articles/epjconf/abs/2015/07/contents/contents.html.

[2] S. Koonin, Phys. Lett. B 70 43 (1977)

[3] Bao-An Li et al., Phys. Rev. C 71044604 (2005)

[4] V. Baran, M. Colonna, V. Greco, M. Di Toro, Physics Reports, Volume 410, Issues 56, May 2005, Pages 335-466

[5] R.T. De Souza, N. Le Neindre, A. Pagano and K. H. Schmidt, Eur. Phys. J. A 3, 275-291 (2006).

[6] E. V. Pagano et al., EPJ Web of Conferences **88**, 00013 (2015), and contribution in Proceedings ISTROS 2015 International Conference (in press).

[7] C. Boiano, et al., IEEE Trans. On Nucl. Science 51 1931 (2004)

[8] E. De Filippo, A.Pagano, European Physical Journal **A50** (2014) 32, and reference therein.

[9] P. Russotto et al., Phys. Rev. C **91**, 014610 (2015).

[10] L. Auditore, et al., EPJ Web of Conferences **88**, 01001 (2015).

[11] E.Pollacco et al., Phys. Procedia 37 (2012) 1799-1804.

EPJ Web of Conferences **117**, 10009 (2016) DOI: 10.1051/epjconf/201611710009
NN2015

Effects of the interstrip gap on the efficiency and response of Double Sided Silicon Strip Detectors

D. Torresi[1], J. Forneris[2], L. Grassi[3], L. Acosta[1], A. Di Pietro[1], P. Figuera[1], M. Fisichella[1], V. Grilj[2], M. Jakic[3], M. Lattuada[1,4], T. Mijatovic[3], M. Milin[5], L. Prepolec[3], N. Skukan[3], N. Soic[3], D. Stanko[5], V. Tokic[3], M. Uroic[3], M. Zadro[3]

[1]INFN - Laboratori Nazionali del Sud, Catania, Italy
[2]Ruder Bošković Institute, Zagreb, Croatia
[3]Physics Department and NIS Interdepartmental Centre, University of Torino; INFN, Sezione di Torino, CNISM, Sezione di Torino, Italy
[4]Dipartimento di Fisica e Astronomia, Università di Catania, Catania, Italy
[5]Department of Physics, University of Zagreb, Zagreb, Croatia.

Abstract

In this work the effects of the segmentation of the electrodes of Double Sided Silicon Strip Detectors (DSSSDs) are investigated. In order to characterize the response of the DSSSDs we perform a first experiment by using tandem beams of different energies directly sent on the detector and a second experiment by mean of a proton micro-beam. Results show that the effective width of the inter-strip region and the efficiency for full energy detection, varies with both detected energy and bias voltage. The experimental results are qualitatively reproduced by a simplified model based on the Shockley-Ramo-Gunn framework.

1 Introduction

Highly segmented double sided silicon detectors (DSSSD) are widely used in nuclear physics to perform accurate measurements of angular distributions, or to study reactions where coincidences of different particles are required to fully characterize the final state of the interaction process. The DSSSD under study in this work is the model W1, manufactured by Micron Semiconductor Ltd. It has an active area of 50×50 mm^2, each electrode is divided into 16 strips, vertical for the front side and horizontal for the back side. Each strip is 3 mm wide and they are separated by a SiO_2 layer 1 μm wide that in the following is called interstrip. it is known that that particles entering the detector through the SiO_2 insulating interstrip region, can produce signals with an amplitude which is different than the full energy and also signals with an opposite polarity [1–6]. For this reason it is very important to be able to reject interstrip events by selecting only the events producing the correct full energy signals. Therefore, with the aim to characterize the DSSSD we performed [1] a systematic study to investigate the efffects of the different incident ion, energy, and polarization voltage on the efficiency for the full energy detection. Moreover we tried to identify an appropriate selection procedure of events which allows to maximize the efficiency for the full energy detection.

2 Measurements of full energy detection efficiency

A first experiment was performed at the Laboratori Nazionali del Sud, Catania, with the aim to study the dependence of the interstrip effects on the beam energy, the ion species and polarization voltages. ^{7}Li and ^{16}O Tandem beams at energies between 10 and 50 MeV, with intensity of about 100 pps, were directly sent onto a 1000 μm thick DSSSD placed along the beam direction.

From the energy spectra collected is observed that some particle is detected with an energy lower than the full one. In fact particles whose trajectory crosses an interstrip region give rise to signals in the two adjacent strip and has a lower amplitude and sometimes opposite polarity [5,6]. In particular for particles crossing a back interstrip region (back inter-strip events) summing the signals from the two adjacent strips we obtain a full energy signal. Also for front inter-strip events two adjacent front strips give signals, but in this case one of them can have an opposite polarity and the sum of the two signals does not give a full energy signal.

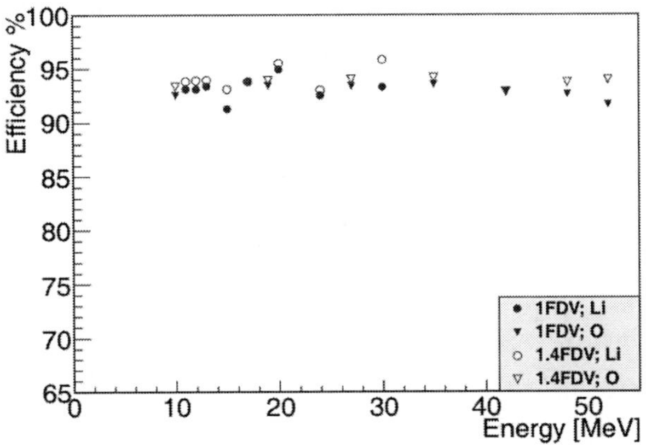

Figure 1: Efficiency for full energy detection as a function of the beam energy, for two different bias voltages applied to the detector. In the left panel is shown the efficiency obtained imposing the condition $E_{front} \approx E_{back}$, the continuous line corresponds to the efficiency for the full energy detection calculated using the geometrical interstrip width; in the right panel is shown the efficiency obtained recovering the back interstrip events. See the text for more details.

Therefore for data gathered with DSSSD, it is necessary a procedure to select the correct full energy events rejecting the others. A possible procedure for selecting the full energy events is to compare the energy measured by the front side E_{front} with the one measured by the back side E_{back} by imposing $E_{front} \approx E_{back}$, within a given tolerance. Using this selection procedure we obtained an efficiency that is slightly dependent on the energy and on the bias applied on the detector as shown in figure 1 We proposed another events selection where the energy measured by the front strips is compared to the sum of the energy of two adjacent back strips. In this way all the back inter-strip events are recovered increasing the efficiency and removing the dependence on energy and on bias as shown in right panel of figure 1 [1]. The average value of the efficiency for full energy detection is ~94%.

3 Measurements of the effective inter-strip width

The efficiency for full energy detection is directly related with the effective inter-strip width, therefore we performed a second experiment with the aim to have a direct measurements of the effective interstrip width for different operating conditions of the DSSSD. The experiment was performed at the

 DOI: 10.1051/epjconf/201611710009

Ruder Bošković Institute in Zagreb using the local microbeam facility. The use of a microbeam allows to know the position of the impinging particle with a precision of few μm, in this way it is possible to correlate the presence of interstrip effects with the impinging position of the particle on the DSSSD. Proton microbeams of 1.7 and 6 MeV at low intensity ($\sim$ 100 pps) were used to irradiate the DSSSD. For each beam energy the measurements were repeated for bias voltages equal to 0.5, 1 and 1.5 times the full depletion voltage (FDV).

In the left panel of figure 2 is shown the measured energy as a function of

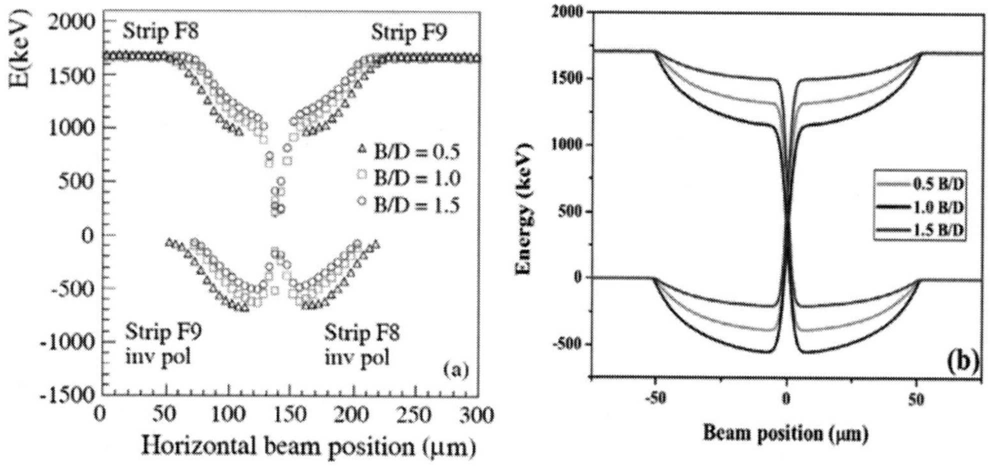

Figure 2: Energy as function of position for proton of 1.7 MeV for three different bias voltages. On the left panel are shown experimental data refering to a front interstrip region, on the right panel are shown data coming from simulations (fro more information see the text). B/D indicates the ratio between the applied bias and the full depletion voltage.

the impinging position of the incident proton. From the plot it is possible to distinguish clearly the interstrip regions corresponding to the region where the amplitude of the energy signal drops. In order to provide a qualitative interpretation of the experimental findings a simplified simulation based on the Shockley-Ramo-Gunn framework was performed. As suggested in [6] it was assumed a build up of positive charge at the oxide interface in the front inter-strip and of negative charge at the oxide interface in the back inter-strip. In the right side of figure 2 we show the result of the simulation, a satisfactory qualitative reproduction of all the observed inter-strip effects has been obtained [2].

4 Summary and conclusions

The response of Double Sided Silicon Strip Detectors and its dependence on the incident ion type, energy, and polarization voltage was studied in two different experiments. A first study was performed by mean of ^{7}Li and ^{16}O tandem beams at different energies. It showed that the efficiency for full energy detection depends on the energy and type of the detected ion and on the applied bias voltage. A selection procedure of events that maximize the efficiency for the full energy reconstruction and remove the cited dependences has been proposed. In a second study, a proton micro-beams was used to perform a direct measurements of the effective interstrip width at different energies and for different detector bias. The results confirmed that both front and back effective inter-strip widths, which are directly related to the efficency for full energy detection, can be much larger than the nominal geometric width of the SiO_2 zone and depend on the operating conditions. Finally the experimental observations were qualitatively interpreted and described by performing a simplified simulations based on the Shockley-Ramo-Gunn framework. Concluding, for those experiments aiming at an accurate measurements of absolute cross-sections, especially if two or more DSSSD are used in coincidence, efficiency for full energy detection cannot be deduced from the geometric inter-strip width but a characterization of the used DSSSDs is required.

Acknowledgements

This work has been partly supported by the European Community as an Integrating Activity "Support of Public and Industrial Research Using Ion Beam Technology - SPIRIT" (EC Contract no. 227012) by the European Community FP7 REGPOT project Particle Detectors (EC Contract no. 256783) and by the Italian Ministry of Education and Research (PRIN 2010-2011, project 2010TP-SCSP).

References

[1] Torresi et al., Nucl. Instr. Meth. A 713 (2013) 11.

[2] Grassi et al. Nucl. Instr. Meth. A 767 (2014) 99.

[3] Poehlsen, et al., Nucl. Instr. Meth.A 721 (2013) 26.

[4] Takeda et al., Nucl. Instrum. Meth. Phys. Res. A579 859, (2007).

[5] Blumenfeld et al., Nucl. Instrum. Meth. Phys. Res. A421, 471 (1999).

[6] Yorkstone et al., Nucl. Instrum. Meth. Phys. Res. A262, 353 (1987).

EPJ Web of Conferences **117**, 10010 (2016) DOI: 10.1051/epjconf/201611710010
NN2015

Recent applications of nuclear track emulsion

K. Z. Mamatkulov[1], I. Ambřozová[2], D. A. Artemenkov[1,3], V. Bradnova[1], D. V. Kamanin[1], R. R. Kattabekov[1], L. Majling[2], A. Marey[4], O. Ploc[2], V. V. Rusakova[1], R. Stanoeva[5], K. Turek[2], A. A. Zaitsev[1,6], P. I. Zarubin[1,6*], I. G. Zarubina[1,6]

[1] Joint Institute for Nuclear Research, Joliot-Curie 6, Dubna, Moscow region,141980, Russia
[2] Nuclear Physics Institute, of the ASCR, v.v.i., Řež 130, 250 68 Řež, Czech Republic
[3] National Research Nuclear University MEPhI (Moscow Engineering Physics Institute), Kashirskoe Shosse 31,Moscow, 115409, Russia
[4] Minoufia University, Tala't street, EL azhar territory, Menouf city, Menofia, Egypt
[5] South-West University, Ivan Michailov st. 66, 2700 Blagoevgrad, Bulgaria
[6] P. N. Lebedev Physical Institute of the Russian Academy of Sciences, Leninskij Prospekt 53, Moscow, 119991, Russia

Abstract

Application of the nuclear track emulsion technique (NTE) in radioactivity and nuclear fission studies is discussed. It is suggested to use a HSP-1000 automated microscope for searching for a collinear cluster tri-partition of heavy nuclei implanted in NTE. Calibrations of α-particles and ion ranges in a novel NTE are carried out. Surface exposures of NTE samples to a Cf-252 source started. Planar events containing fragments and long-range α-particles as well as fragment triples only are studied. NTE samples are calibrated by ions Kr and Xe of energy of 1.2 and 3 A MeV. Use of the image recognition program "ImageJ" for obtaining characteristics of individual events and for events from the large scan area is presented.

*E-mail: zarubin@lhe.jinr.ru

1 Introduction

Nuclear track emulsion (NTE) retains the status of a universal and inexpensive detector. With unsurpassed spatial resolution NE provides complete observation of tracks starting from fission fragments and down to relativistic particles [1-3]. Unique opportunities of NTE deserve further use in fundamental and applied research in state-of-art accelerators and reactors, as well as with sources of radioactivity, including natural ones. Application of NTE is especially justified in those pioneering experiments in which nuclear particle tracks cannot be reconstructed with the help of electronic detectors.

The NTE technique is based on intelligence, vision and performance of researchers using traditional microscopes. Despite widespread interest, its labor consumption causes limited samplings of hundreds of measured tracks which present as a rule only tiny fractions of the available statistics. Implementation of computerized and fully automated microscopes in the NTE analysis allows one to bridge this gap. These are complicated and expensive devices of collective or even remote usage allow one to describe unprecedented statistics of short nuclear tracks.

To make such a development purposeful it is necessary to focus on such a topical issue of nuclear physics the solution of which can be reduced to simple tasks of recognition and measurement of tracks in NTE to be solved with the aid of already developed programs. One of the suggested problems is a search for the possibility of a collinear cluster tri-partition [4]. The existence of this phenomenon could be established in observations of such a type of ternary fission of heavy nuclei in which a lightest fragment is emitted in the direction of one of the heavy fragments.

Despite distinct observability of fission fragments they can not be fully identified in NTE. However, NTE is valuable due to combination of the best angular resolution and maximum sensitivity. Besides, it is possible to measure the lengths and thicknesses of tracks, and, thus, to classify the fragments. As an initial stage, to provide statistics of ternary fissions it is suggested to analyze a sufficient NTE area exposed to ^{252}Cf source with an appropriate density of tracks of α-particles and spontaneous fission fragments. Such an approach will be developed by a NTE with an admixture of the ^{252}Cf isotope [5,6]. Another option is exposure of NTE manufactured with a ^{235}U isotope addition by thermal neutrons.

A large-scale NTE scanning is suggested to be performed on the microscope HSP-1000 [7] of the Department of radiation dosimetry (DRD) of Nuclear Physics Institute of the ASCR, v. v. i.. The use of the NTE resolution will be full if the microscope will be adapted to operate with lenses of the highest magnification. Development of algorithms for automatic search and analysis of short tracks of heavy ions in NTE will be required. On the experimental side, ion ranges in NTE must be calibrated in the α-decay and fission energy scale[8]. Progress of the preparatory phase of the proposed study is summarized below.

2 Exposure to Cf source

Surface exposures of NTE samples in DRD were performed by a manually moving ^{252}Cf source. Most likely, the ^{252}Cf isotope decays by emission of α-particles of energy of 5-6 MeV, the tracks of which mainly populate an exposed sample. This isotope also undergoes a spontaneous fission to a pair or even triple of fragments with probabilities of 3%, and about 0.1%, respectively.

In the surface exposure should not be observed more than two ternary fission fragments as the third one is emitted in the contacting source side. The sign of a ^{252}Cf exposure consists in presence of α-particle tracks from ternary fission events whose ranges significantly exceed the decay α-particle ranges. This channel dominates in the ^{252}Cf ternary fission having a 90% probability.

Search for heavy ion tracks on surface of the NTE samples exposed to the ^{252}Cf source is carried out on the KSM microscope with a 15$\times$ objective. Usually MBI-9 microscopes are used for this stage. Using of KSM to search for very rare fission events has eased immediate transitions to their precise measurements with a 90$\times$ objective. Planar triples are found consisting of a pair of fragments and a long-range α-particle as well as of fragments only. It is worth to emphasize a remarkable fact of the observation of triples in NTE but not only pairs of fragments. For such a full observation of triples their vertices should be dipped to a depth not less than a typical track thickness. Fig. 1a shows the distribution of vertices of Cf fissions into three fragments over NTE layer depth which has an average value of 2.7 ± 0.3 μm (RMS equal to 2.2 μm). Perhaps this effect is due to the binding of Cf atoms of AgBr micro crystals and their drift. Apparently, the source surface protection with initial thickness of the 50 μg/cm^2 gold deposition (according to the source passport) does not prevent such a penetration.

In 45 events of a true ternary fission, i. e., not containing α-particles the ranges of all fragments are measured. Effective criteria for a fission into three heavy fragments is their amount range (Fig. 1b), which has an average value of 15.5 ± 0.9 μm when RMS 6.2 μm. In addition, the opening angles between the fragments are measured in these events (Fig. 1c). Their

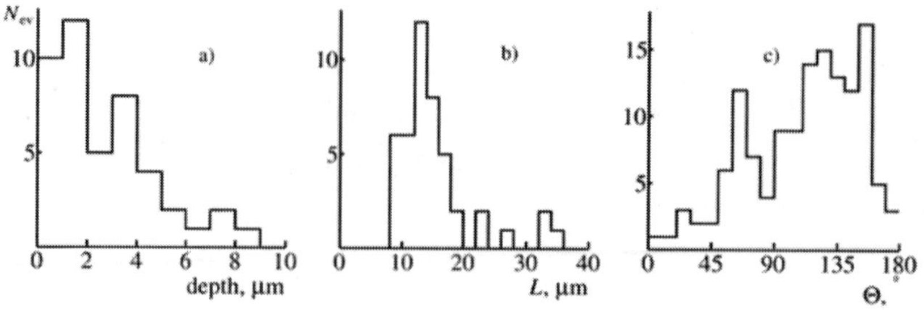

Figure 1: Distributions for the ^{252}Cf fissions into three fragments over depth in NTE layer (a), amount ranges of three fragments (b) and opening angles between fragments (c).

distribution is characterized by a mean value 111 ± 3° when RMS 39°.

3 Experience of automatic measurements

The initial experience of a computer analysis of heavy ion tracks in NTE is obtained using the ImageJ program [9], available online. Steps and results of image recognition of such event analysis are presented in [8]. Recently, the microscope HSP-1000 with a lens 20× was used to scan a large area of NTE exposed to 3 A MeV ^{84}Kr ions at a 30^0 inclination. An analysis by orders of magnitude beyond a man possibility is carried out on an array of 8 frames of 3000 × 2000 pixels. The program ImageJ has found 571 tracks on an area of 0.74 mm^2 and determined their lengths and planar angles ^{84}Kr in an ellipse approximation (Fig. 2). The distribution of lengths is described by a Gaussian function with a parameter of 1.4 μm corresponding to the manual measurements. Statistics ^{84}Kr is divided into two groups differing of 11^0 in the average values. These groups ^{84}Kr are described by Gaussian functions with parameters of 1.8^0 and 3.1^0. Apparently, ^{84}Kr splitting is caused by a change of a magnetic rigidity of approximately 20% fraction of ions extracted from the accelerator U400M as a result of an electron pick-up in the residual magnetic field. The noticed effect can be used to analyze an ion beam composition. Currently, similar approaches to computer analysis are developed for the NTE exposures discussed above.

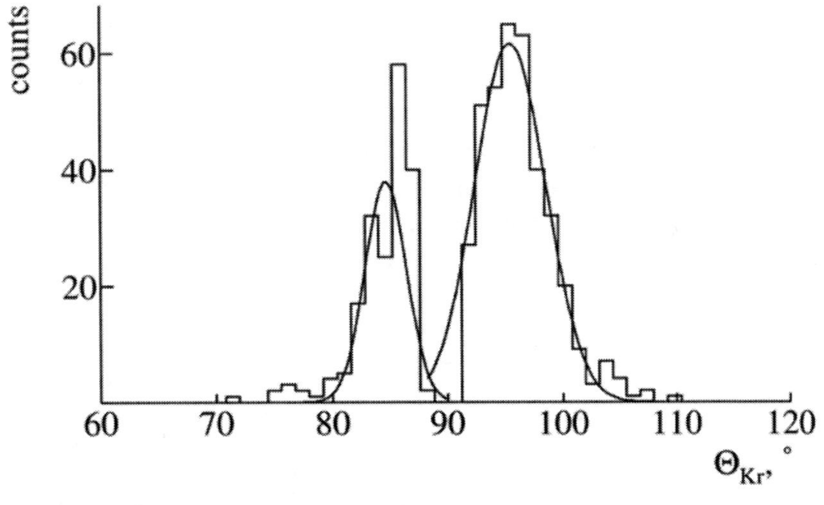

Figure 2: Distribution over planar angles of 3 A MeV ^{84}Kr ions.

4 Conclusions

Computer analysis of images will allow one to select the decays for a perfect manual analysis. Automation of search for ternary fission events will drastically reduce the most time-demanding stage and will help to focus the manual analysis on already found events. Thus, manual and automatic analyses complement each other. When the time of the NTE exposure is controlled the computer analysis can be applied in a desired scale and the diversity of tracks with estimation of ion energy, both to ion beam profilometry and to α-dosimetry with random track directions.

In general, the synergy of modern radioactive sources, NTE proven metrology and advanced microscopy seems to be a promising prospect for α-radioactivity and nuclear fission research. It can be assumed that ions of transfermium elements will be implanted some when in NTE. Their bright decays can be found as common vertices of few α-particles and fission fragments. This perspective emphasizes the fundamental value of preservation and modernization of the NTE technique.Thus, the present study focused on the NTE return in practice of nuclear experiment will serve as a prototype of solution of an impressive variety of problems. Macro photos of the discussed exposures and videos based on them are available on the BECQUEREL project website [10].

Acknowledgements

The authors express gratitude to Profs. T. Hussein and M. Ghoneim (Cairo University), H. El-Samman of the Minoufia University, Yu. V. Pyatkov (Moscow Engineering Physics Institute), A. I. Malakhov (JINR) and N. G. Polukhina (Lebedev Physics Institute FIAN) for discussions and support. This work was supported by grants of plenipotentiary representatives of governments of Bulgaria, Czech Republic, Egypt and Romania at JINR.

References

[1] Powell C F, Fowler P H and Perkins D H 1959 *The Study of Elementary Particles by the Photographic Method.* London - New York: Pergamon Press.

[2] Barkas W H 1963 *Nuclear Research Emulsions.* New York - London: Academic Press.

[3] Goldschmidt-Cremont Y 1953 *Annu. Rev. Nucl. Sci.* 141.

[4] Kamanin D V and Pyakov YV 2014. Clusterization in ternary fission. *Clusters in Nuclei.* Lecture Notes in Physics. **875** 183.

[5] Titterton E W and Brinkley T A 1960 Rare Modes in the Spontaneous Fission of Californium-252. *Nature.* **187** 228.

[6] Muga ML, Bowman H R and Thompson SG 1961 Tripartition in the Spontaneous-Fission Decay of Cf^{252}. *Phys. Rev.* **121** 270.

[7] WEB site: *http://www.odz.ujf.cas.cz/home/resources/microscope-hsp-1000.*

[8] Mamatkulov K Z *et al* 2015 Proceedings of the Conference of Fundamental Research and Particle Physics. 18-20 February 2015. Moscow, Russian Federation. arXiv: 1508.02707.

[9] "Image processing and analyses in Java". WEB site: *http://rsb.info.nih.gov/ij/.*

[10] "The BECQUEREL Project". WEB site: *http://becquerel.jinr.ru/.*

AUTHOR INDEX

AUTHOR INDEX

AUTHOR INDEX

AUTHOR INDEX

AUTHOR INDEX

AUTHOR INDEX

AUTHOR INDEX

AUTHOR INDEX

AUTHOR INDEX

About us

DNV

The purpose of DNV is to safeguard life, property, and the environment. DNV offer world-renowned testing, certification and technical advisory services to the energy value chain including renewables, oil and gas, and energy management.

For more than 20 years the company has carried out R&D activities within all their business areas.

GreenPowerMonitor
a DNV company

GreenPowerMonitor, founded in 2008, is an international provider of leading monitoring, control and data management services for renewable power plants.

GreenPowerMonitor is operating in 90 countries and 6000 plants worldwide, ranging from 2kW to 1.3 GW.

DNV

40th European Photovoltaic Solar Energy Conference and Exhibition

PV System Modelling and Degradation

WHAT?

- **Machine learning** model to estimate **long-term degradation** of PV plants at **system level**.

WHY?

- Long-term degradation is an **important** topic for **PV industry** with **significant financial** implications.
- Most methodologies assume a degradation behavior and require detailed **metadata**.

HOW?

- ✓ A **neural network** (NN) models power production as a function of **time**.
- ✓ The degradation rate is calculated as the **derivative** of power w.r.t. time.
- ✓ Main techniques: set of neural networks (**SNN**), exponentially weighted moving average (**EWMA**) applied to NN's weights, and **automatic differentiation**.
- ✓ Implemented in **PyTorch**.

DNV

Data Requirements

- Irradiance on the plane of the array (**I_{POA}**)
- Ambient temperature (**T_{AMB}**)
- PV plant power production (**P_{AC}**)
- Unix timestamps (**UTS**)

Time elapsed since plant started operating (**T_{OPE}**):

$$T_{OPE} = \frac{UTS_t - UTS_{init}}{8760 \cdot 3600}$$

UTS_t: Unix timestamp at time t
UTS_{init}: first recorded Unix timestamp.

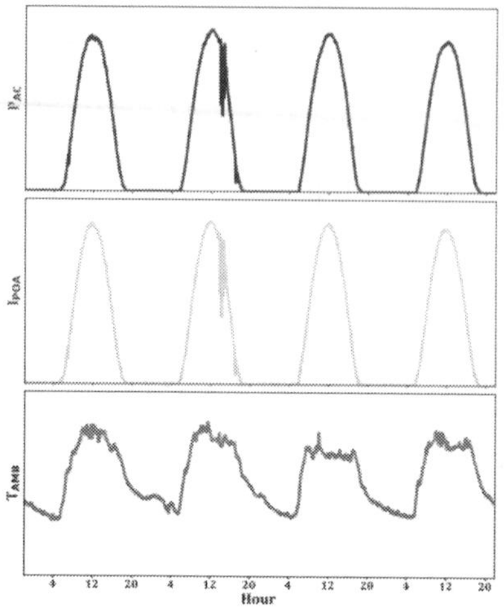

DNV

020278-005

Data Cleaning[1]

1. Remove **missing** data
2. Remove **duplicated** time stamps
3. Remove **out-of-range** data
4. Clean data along **axes** (I_{POA} vs P_{AC})
5. Remove **low-power** periods
6. Discard **erroneous** data (I_{POA}, T_{AMB}, P_{AC})
7. Remove I_{POA} **lower** than **500 W/m²**

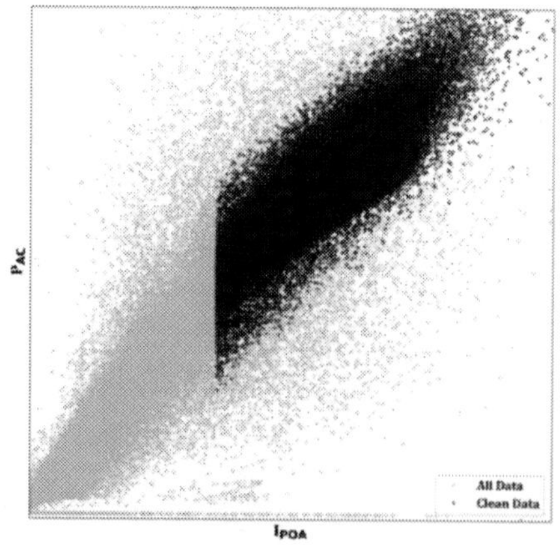

[1]G. Guerra, P. Mercade Ruiz, L. Landberg, A data driven model for solar inverters, 37th EU PVSEC, (2020).

DNV

020278-006

Model Structure and Training

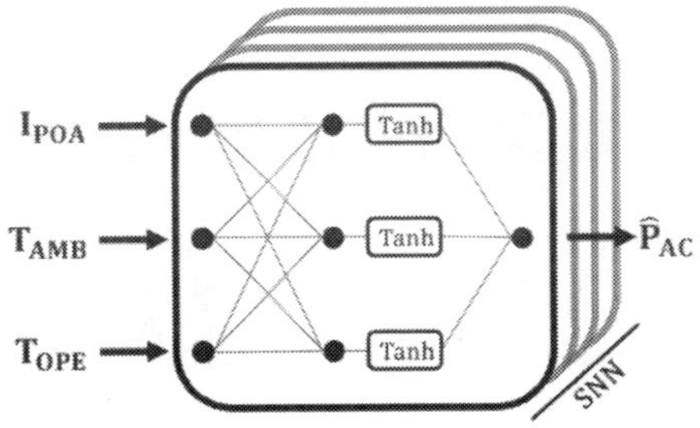

$$\widehat{P}_{AC} = \frac{1}{N}\sum_{n=1}^{N} NN_n(I_{POA}, T_{AMB}, T_{OPE})$$

Algorithm 1

```
Choose number of networks in SNN
Initialise SNN with random weights
Copy SNN in SNN_t
For epoch in total epochs do
  For batch in training data do
    Select NN at random from SNN
    Update NN weights using gradient descent
    Update NN weights SNN_t in using EWMA
  end For
end For
Calculate RMSE and MAE for every NN in SNN_t
Calculate third quartile RMSE (Q75_RMSE)
Calculate third quartile MAE (Q75_MAE)
For NN in SNN_t do
  If (RMSE_NN > Q75_RMSE) v (MAE_NN > Q75_MAE) then
    Remove NN from SNN_t
  end If
end For
```


DNV

020278-007

Degradation Rate

- **Mean** I_{POA} and T_{AMB}
- T_{OPE} covers **operational life**
- **Derivative** of $\widehat{P}_{AC}$ w.r.t. T_{OPE} using ***autograd***[2]
- **Normalised** by the mean $\widehat{P}_{AC}$ of first year
- **Final** rate is the **mean** of all normalised derivatives

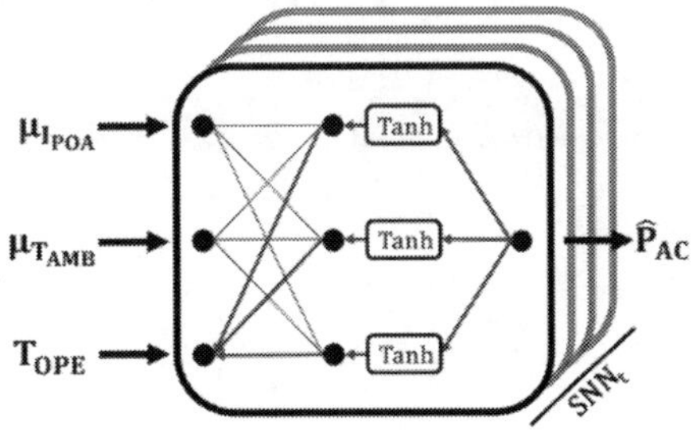

[2] M. Raissi, et. al., Physics Informed Deep Learning (Part I): Data-driven Solutions of Nonlinear Partial Differential Equations, (2017).

DNV

020278-008

Case Study 1 – rdtools notebook data[3,4]

DC rated power [kW]	Temperature coefficient [1/C]	Cell reference temperature [C]	Data sampling period [minute]
1000	-0.005	25	1

Training results

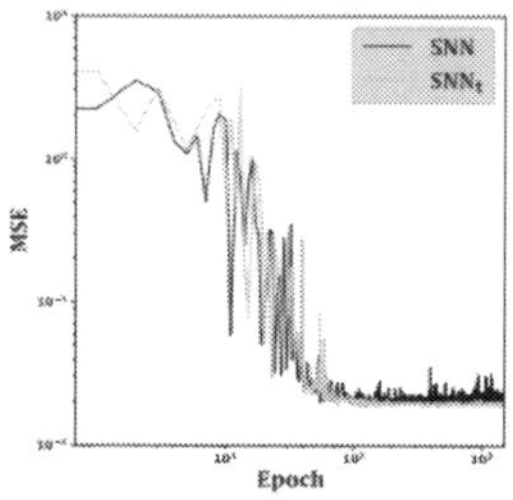

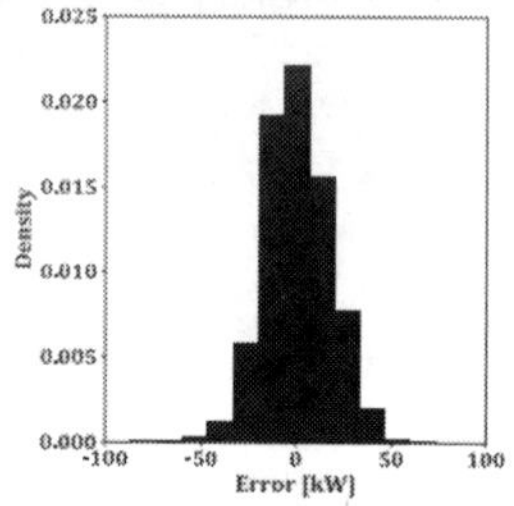

Degradation analysis

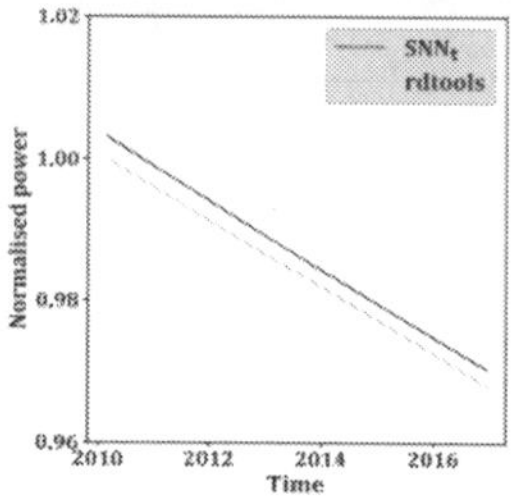

Degradation rate [%/year]	
rdtools	-0.47
SNN_t	-0.48

[3]https://github.com/NREL/rdtools/blob/master/docs/degradation_and_soiling_example_pvdaq_4.ipynb
[4]M.G. Deceglie, et. al., RdTools, version 2.0.5, Computer Software.

DNV

40th European Photovoltaic Solar Energy Conference and Exhibition

Case Study 2 – Internal database

DC rated power [kW]	Temperature coefficient [1/C]*	Cell reference temperature [C]*	Data sampling period [minute]
106.26	-0.005	25	5

Training results

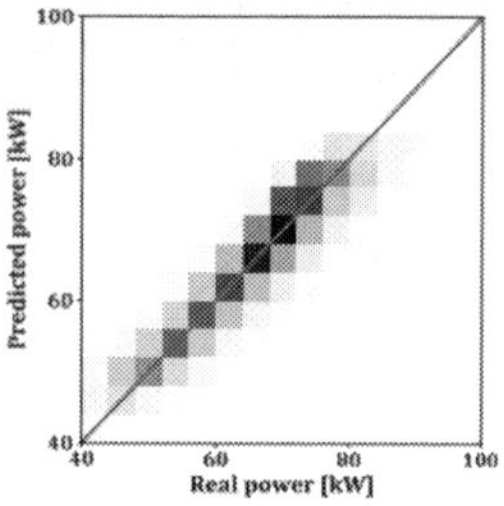

Degradation analysis

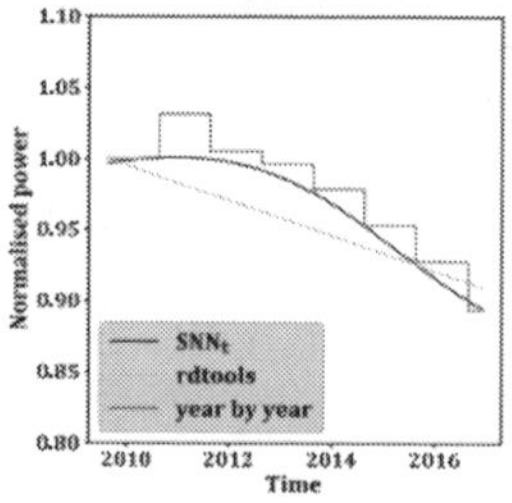

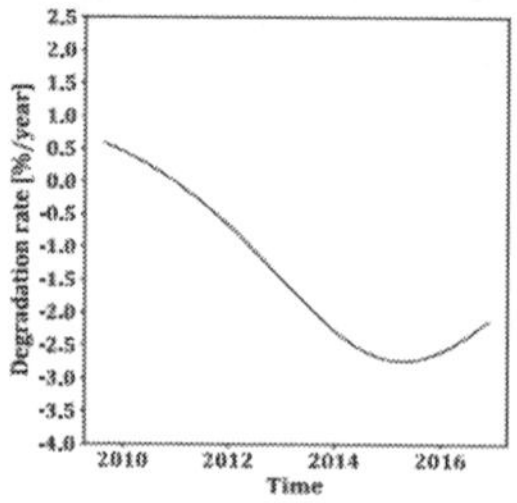

Degradation rate [%/year]	
rdtools	-1.24
SNN_t	-1.41

*Taken from Case Study 1

DNV

Conclusions

- No overfitting
- Comparison to rdtools
- SNN and EWMA
- Automatic differentiation
- Data-driven nature
- Additional work
- Future work

DNV

40th European Photovoltaic Solar Energy Conference and Exhibition

WHEN TRUST MATTERS

Thank you and visit our posters!

Session number: 4CV.1.5, 4CV.1.31

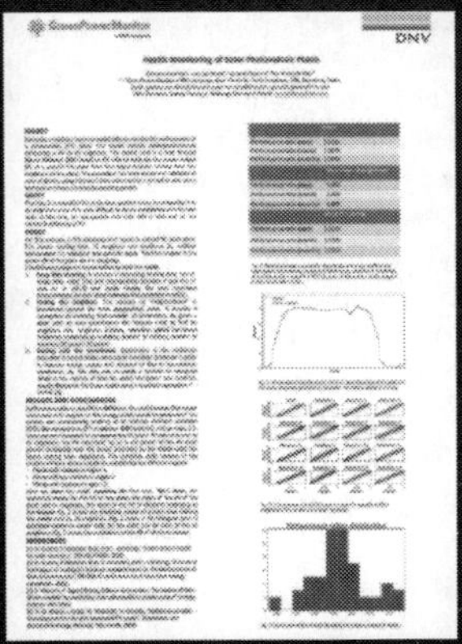

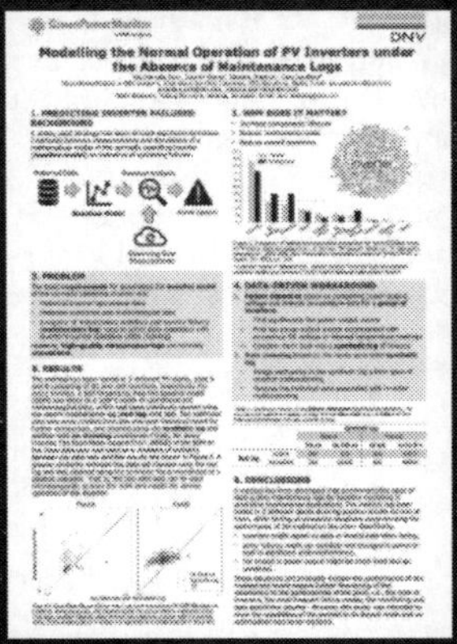

www.dnv.com

DNV

This presentation was selected by the Sc. Committee of the EU PVSEC 2023 for submission of a full paper to one of the EU PVSEC's collaborating peer-reviewed journals.

TUNING PV TEMPERATURE ESTIMATION MODELS USING NORMALIZED EFFICIENCY

Anastasios Kladas, Bert Herteleer, Jan Cappelle
KU Leuven Research Group ELECTA Ghent,
Gebroeders De Smetstraat 1, 9000 Ghent, Belgium
Keywords: PV temperature, Normalized efficiency, Thermal model calibration

ABSTRACT: PV temperature (T_{PV}) is critical for analyzing the behavior of PV systems. Several models have been proposed for T_{PV} estimation based on weather measurements. Those models are usually built following physics concepts and adjusted based on PV measurement data. According to the authors, the inputs of the models have to be assessed following a specific measurement protocol for accurate predictions. Nevertheless, in real-world measurement set-ups some of those guidelines could be overpassed, particularly in cases where they are increasing the complexity of the measurement system. On the other hand, special installation characteristics or types could have a significant impact on PV thermal behavior. Taking the above into consideration, calibration of the models is suggested for accurate estimations. To the best of our knowledge, the calibration of T_{PV} models relies on T_{PV} measurements. Those measurements are very challenging by nature, while they hold an uncertainty portraying the equivalent T_{PV} (the temperature that accounts for the T_{PV} losses), especially when they are assessed from the back sheet of the PV.
In this work, a new methodology for calibrating the T_{PV} models is presented. This method is based on the normalized efficiency (nN) decomposition for isolating the T_{PV} from the rest losses (soiling, degradation, etc.). The decomposition consists of filtering and optimization steps. Based on the extracted temperature losses, equivalent system temperature can be estimated which is used as the reference for the calibration of the examined models.
The results showed that the adjusted coefficients, improved the accuracy of the models especially when the given parameters did not achieve their maximum potential under the examined measurement set-up, while the loses extracted from the optimization step, improved significantly the produced power (PAC) estimations.

1 Introduction

A critical factor impacting photovoltaic (PV) module efficiency is PV temperature (T_{PV}). For evaluating the effectiveness of PV systems, accurate T_{PV} assesment is vital. Several models have been proposed for T_{PV} estimation based on weather measurements [1]–[4]. These models are constructed based on fundamental physical principles and are adjusted using actual PV measurement data. The guidelines for assessing the required weather measurements are provided by the authors, but in real-world scenarios, these guidelines are often not fully adhered to due to the complexity they introduce to the measurement setup. Additionally, the specific characteristics of PV module installations can significantly influence thermal behavior.
Given these challenges, model calibration is suggested to improve the precision of T_{PV} estimations. However, traditional calibration methods rely on T_{PV} measurements, which are inherently challenging and subject to uncertainties, especially when measured from the back sheet of PV modules. This paper presents a novel methodology for calibrating T_{PV} models based on the normalized efficiency (n_N) decomposition, which separates T_{PV} losses from other losses like soiling and degradation. The decomposition process involves filtering and optimization steps, ultimately leading to the estimation of the equivalent system temperature, which serves as the reference for calibrating the models.

2 Method

The n_N is a performance indicator calculated as the ratio of the normalized incident irradiance on the PV array plane (G) to the normalized produced energy (P), as shown in (1) [5]. It serves as a comprehensive performance metric encompassing T_{PV} losses, degradation, soiling, and other factors.

$$n_{N\,(meas)} = \frac{P/P_{STC}}{G/1000} \quad (1)$$

For a single-day analysis, soiling and degradation losses are considered constant, and the remaining n_N comprises T_{PV} losses and partial shading effects. To estimate the n_N without T_{PV} measurements, (2) is utilized, where L_{rest} represents all non-T_{PV} losses as a bias correction element, and γ is the power temperature coefficient provided by module manufacturers.

$$n_{N\,(est)} = 1 + \gamma \times (T_{PV} - 25) - L_{rest} \quad (2)$$

Typically, T_{PV} models depend on plane-of-array irradiance (G), ambient temperature (Tamb), wind speed (WS), and their associated coefficients (a, b, c). If a T_{PV} model is represented as T_{PV}(G, Tamb, WS, a, b, c), Equation (2) can be rewritten as n_N(G, Tamb, WS, a, b, c, L_{rest}).
The goal is to filter out outliers and estimate optimal coefficients (a, b, c) to minimize the function in Equation (3):

$$\min\left(\left|n_{N\,(meas)} - n_{N\,(est)}(G, T_{amb}, WS, a, b, c, L_{rest})\right|\right) \quad (3)$$

Subject to the constraint:

$$Coeff_{given} - 0.9 \times Coeff_{given} < Coeff < Coeff_{given} + 1.2 \times Coeff_{given} \quad (4)$$

Where, $Coeff_{given}$ represents the coefficient proposed by the original author.
The calibration process involves:

1. Calculating n_N as per (1), excluding clipping events and low irradiance conditions.
2. Applying a filtering step to reduce noise in n_N by performing linear regression with weather measurements (G, Tamb, WS) and removing data points that deviate beyond a threshold from the estimations of linear regression.
3. Optimizing the coefficients (a, b, c) based on daily analysis, with daily coefficients and losses

10.4229/EUPVSEC2023/4AO.8.5
020279-001

stored in a data frame. The final coefficients for the thermal models are computed as the average of daily values.

3 Data used

The data utilized for this study, are originated from the NIST ground dataset, which is described in [6], comprising measurements from PV systems, including G, Tamb, WS, and power output (P) collected over a 1-year time period (2018). These measurements were used to calculate the n_N and assess T_{PV} parameter and extra losses in the calibration process. Additional data from T_{PV} sensors and power output measurements were used for validation and comparison. The resolution of the data is one minute.

4 Results and Discussion

Table 1 provides an overview of the examined T_{PV} models, including their respective equations and references.

Table 1 Examined models for the parameter estimation

Name	Equation	Ref
Ross	$T_{PV} = T_{amb} + a * G_T$	[2]
Faiman	$T_{PV} = T_{amb} + \frac{G_T}{(a + b * WS)}$	[3]
King	$T_{PV} = G_T * (e^{a+b*WS}) + T_{amb} + \frac{G_T}{G_{STC}} * 3$	[1]
Skoplaki	$T_{PV} = T_{amb} + \omega * \left(\frac{a}{b + c * \frac{WS}{0.67}}\right) * G_T$	[4]

Table 2 presents both the given and estimated coefficients (a, b, c) for the examined T_{PV} models, including Faiman, King, Ross, and Skoplaki. These estimated coefficients are compared to the values provided by the original authors. The results are further evaluated in Figure 1, which illustrates the T_{PV} model performance with both the given and estimated coefficients concerning T_{PV} sensor measurements (above) and power output estimations (below).

Model	Coefficients	a	b	c
Faiman	Given	25.00	6.84	
	Estimated	32.77	4.81	
King	Given	-3.56	-0.75	
	Estimated	-3.70	-0.15	
Ross	Given	0.02		
	Estimated	0.03		
Skoplaki	Given	0.32	8.91	2.00
	Estimated	0.60	18.82	1.59

Figure 1 Given and estimated parameters for the examined models

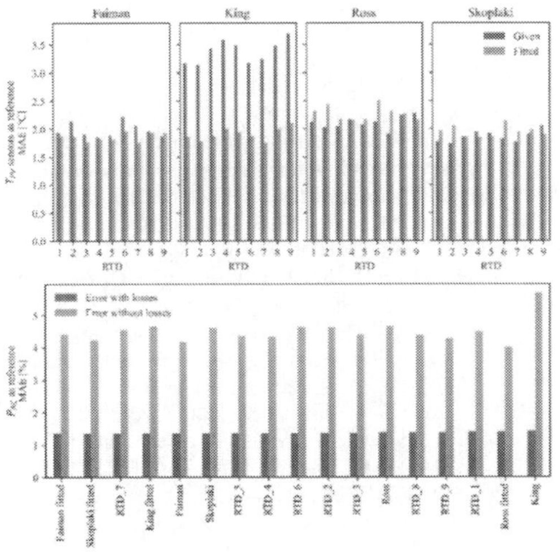

Figure 2 Evaluation of the models with TPV (above) and PAC (below) as reference.

The new estimated coefficients are compared to the coefficients provided by the original authors. Notably, the estimated coefficients are based on equivalent temperature derived from the normalized efficiency (n_N) of the entire PV system, while the sensors measure temperature on the rear side of multiple modules. Therefore, some deviations are expected. The King's model exhibits the most significant improvement, as its given coefficients do not accurately align with the specific measurement setup. Additionally, including the estimated daily losses from the optimization procedure in power estimation results in a significant enhancement in accuracy. The Faiman and Skoplaki models perform slightly better in power output estimation.

5 Conclusions

In this study, we've introduced a novel methodology for calibrating T_{PV} models using normalized efficiency decomposition. This approach proves valuable when traditional T_{PV} measurements are challenging or unavailable. By estimating T_{PV} parameters through n_N decomposition, we've demonstrated notable improvements in model accuracy, particularly for scenarios where strict measurement protocols may be impractical.

It's important to note that this method is recommended when reference temperature measurements for model calibration are lacking. In cases where such measurements can be reliably obtained, the traditional calibration approach remains preferable. Our findings highlight the versatility of the n_N-based calibration method as a valuable alternative, providing a pathway for enhanced T_{PV} model accuracy in real-world photovoltaic applications.

6 References

[1] D. L. King, W. E. Boyson, and J. A. Kratochvil, "PHOTOVOLTAIC ARRAY PERFORMANCE MODEL," 2003.

[2] R. Ross Jr, "Interface design considerations for terrestrial solar cell modules," 1976.

10.4229/EUPVSEC2023/4AO.8.5
020279-002

[3] D. Faiman, "Assessing the outdoor operating temperature of photovoltaic modules," *Progress in Photovoltaics: Research and Applications*, vol. 16, no. 4, pp. 307–315, Jun. 2008, doi: 10.1002/pip.813.
[4] E. Skoplaki, A. G. Boudouvis, and J. A. Palyvos, "A simple correlation for the operating temperature of photovoltaic modules of arbitrary mounting," *Solar Energy Materials and Solar Cells*, vol. 92, no. 11, pp. 1393–1402, Nov. 2008, doi: 10.1016/j.solmat.2008.05.016.
[5] B. Herteleer, B. Huyck, F. Catthoor, J. Driesen, and J. Cappelle, "Normalised efficiency of photovoltaic systems: Going beyond the performance ratio," *Solar Energy*, vol. 157, pp. 408–418, Nov. 2017, doi: 10.1016/j.solener.2017.08.037.
[6] M. Boyd, "Performance data from the NIST photovoltaic arrays and weather station," *J Res Natl Inst Stand Technol*, vol. 122, 2017, doi: 10.6028/JRES.122.040.

This presentation was selected by the Sc. Committee of the EU PVSEC 2023 for submission of a full paper to one of the EU PVSEC's collaborating peer-reviewed journals.

Photovoltaic inverter-based quantification of snow conditions and power loss

Emma C. Cooper, Laurie Burnham, and Jennifer L. Braid
Sandia National Laboratories, Albuquerque, NM, 87123, USA

Abstract—**Snow is a significant challenge for photovoltaic (PV) systems at northern latitudes, where the pace of deployment is rapid but snow-related power losses can exceed 30% of annual production. Accurate snow-related power loss estimation methods for utility-scale sites are needed to validate snow mitigation strategies. In this study, we introduce and demonstrate a snow loss estimation approach for time-series inverter data that is based on the identification of physical snow conditions. We implement this method to compare performance in snowy conditions between three utility-scale sites differing in tilt angle. Results clearly show that utility-scale systems at higher tilt angles consistently shed snow more quickly/completely and lose less power to snowfall than their low-tilt counterparts. These results are consistent with the findings of previous studies. The success of this replication implies that, with this method as our analysis approach, we may be able to extract insights on snow-related performance from any number of the numerous monitoring data sets collected by utility-scale systems around the world.**

I. Introduction

Many studies have demonstrated that snow significantly compromises photovoltaic (PV) output during winter [1]–[3], often a period of high energy demand in snowy regions, with PV power loss as high as 90% - 100% during winter months for some systems [1], [4], [5].

Large-scale PV systems are particularly vulnerable to snow losses, as the labor requirements of mechanical snow clearing increase with system scale past the point of financial feasibility. Despite these challenges, the solar industry continues to expand northward with installed utility-scale (defined as > 1 MW) PV capacity above 40° latitude increasing by over 700% since 2015 to reach 18.8 GW in 2021 [6]. As renewable penetration on the electric grid increases, the winter performance of these utility-scale systems will become of increasing importance to the communities that they power. To contribute to year-round reliability, systems must be designed to facilitate passive snow shedding and minimize the use of mechanical clearing. This optimization requires quantitative comparison of the individual and combined impacts of design components on utility-scale system performance [7]. Snow-related power loss will be a key metric in such comparisons. Additionally, snow losses will be an increasingly important component of resource adequacy assessments as demand on the grid is increasingly met by weather-dependent generators.

Sandia National Laboratories is a multimission laboratory managed and operated by National Technology & Engineering Solutions of Sandia, LLC, a wholly owned subsidiary of Honeywell International Inc., for the U.S. Department of Energy's National Nuclear Security Administration under contract DE-NA0003525. This work was supported by the U.S. Department of Energy's Office of Energy Efficiency and Renewable Energy (EERE) under the Solar Energy Technologies Office under Award 38527.

The development of methods for estimating snow losses has, until this point, not been targeted for utility-scale application. A variety of snow loss identification/ quantification approaches have been published over the last twenty years [3], [8]–[11]. While some have been employed in studies to reduce error in generation predictions during winter months [1], many either explicitly require panel or ground snow depth measurements [1], [10] or have been shown to significantly underpredict snow losses if they do not reference panel snow cover conditions [1], [4], [8], [9], [12]. Standard utility-scale monitoring systems are typically limited to inverter-level DC and AC-side monitoring, plane-of-array (POA) irradiance, and back-of-module (BOM) temperature. A reliance on high temporal resolution snow depth measurements or other site-specific snow condition measurements precludes the use of these models to determine snow losses for existing utility-scale systems using historical data.

Beyond requiring specialized instruments that are not typically installed at utility-scale sites, many established snow loss models have only been validated on smaller test systems [4], [9], [12]. Andrews *et al.* [8] validated a direct-loss model on a utility-scale site, and cited inverter clipping behavior as a possible cause of the poor fit of the model to the measured data. In addition to these limitations, approaches where power loss is calculated by estimating snow cover assume snow cover to be opaque [11], [13] or opaque and uniform [9]. However these models do not allow for power production by fully snow-covered PV systems, which our research has recently shown occurs at significant frequencies [14]. Our recent work [14] identified four distinct snow-loss modes corresponding to decreases in voltage and/or current in inverter data.

In this article, we introduce and implement an approach to quantifying snow losses in utility-scale inverter data sets that builds our previous work on snow detection and power loss mode classification in field data. Power losses are calculated for three utility-scale sites using a performance model, and data is classified into snow-loss modes based on ratios of measured vs. modeled operating current and voltage. We compare losses and modal frequencies across sites, and find that increases in tilt angle enable systems to shed snow earlier and more quickly. We find that higher tilt systems experience significantly lower snow losses on both a monthly and seasonal scale even after cross-site differences in cumulative snowfall are accounted for. Both of these results are consistent with previous findings [3]; the results in it and of themselves are not novel. However, the replication of previous findings serves as a demonstration of the validity of this method. Additionally, the identification of physical snow conditions presented in this paper serves as evidence of the untapped potential of standard monitoring data.

TABLE I

Data types	Metadata
DC voltage $[V]$	Location
DC current $[A]$	Panel tilt $[°]$ and azimuth $[°]$
AC power $[W]$	Panel make/model
POA irradiance $[W/m^2]$	Inverter make/model (MPPT voltage range $[V]$, maximum AC power output $[W]$)
BOM temperature $[°C]$	Number of strings connected in parallel to each combiner, number of modules connected in series to each string

II. Methods

A. Data

Fifteen-minute resolution data sets for three monofacial fixed-tilt utility sites (> 1 MW installed DC capacity) located within thirty miles of one another in the northeastern United States were provided by an electric utility. The sites are referred to as S10, S20, and S35 to indicate the system tilt angle in degrees. Data and metadata types collected from all sites are listed in Table I. All data collected is readily available from typical utility-scale monitoring systems and are recorded as standard by the inverters.

B. Performance modeling

Module output was modeled using *pvlib_python* to solve the single diode model [15] and obtain with module nameplate parameters retrieved from the California Energy Commission (CEC) module database. Modeled voltage was scaled up to the measured resolution by the number of modules per string; modeled current was scaled by the number of strings per combiner. Periods where the system operated outside of its nameplate limits (i.e. clipping, shading) were excluded from analysis or corrected.

C. Mode definition and categorization

Data was categorized into five distinct snow-related modes based on measured voltage, effective transmission, and a voltage PI based on the effective transmission. The effective transmission of light-blocking matter on the array, T_{eff}, is estimated using the Sandia Array Performance Model (SAPM) [16] with measured current, irradiance, and BOM temperature data:

$$I_{mp} = N_{strings} \times I_{mp0}(C_0 E_e T_{eff} + C_1 (E_e T_{eff})^2) \times (1 + \alpha_{Imp} (T_{cell} - T_0)) \tag{1}$$

where C is a vector of coeffients specific to the module type, E_e is effective irradiance, T_{cell} is cell temperature, T_0 is 25 °C, I_{mp0} is the nameplate I_{mp} of the module, and $N_{strings}$ is the number of strings connected in parallel to the combiner measurement system. Inverter voltage given T_{eff} is modeled using the SAPM,

$$V_{mp} = N_{modules} \times [V_{mp0} + C_2 N_s \delta \ln(E_e T_{eff}) + C_3 N_s (\delta \ln(E_e T_{eff}))^2 + \beta_{Vmp} (T_{cell} - 25)] \tag{2}$$

using T_{eff} and sensor data, where N_s is the number of cells connected in series per module, V_{mp0} is the nameplate V_{mp} of the module, and $N_{modules}$ is the number of modules connected in parallel per string. V_{ratio}, which is equivalent to the average snow-free fraction of the array, is defined as the PI of measured voltage in relation to V_{mp}.

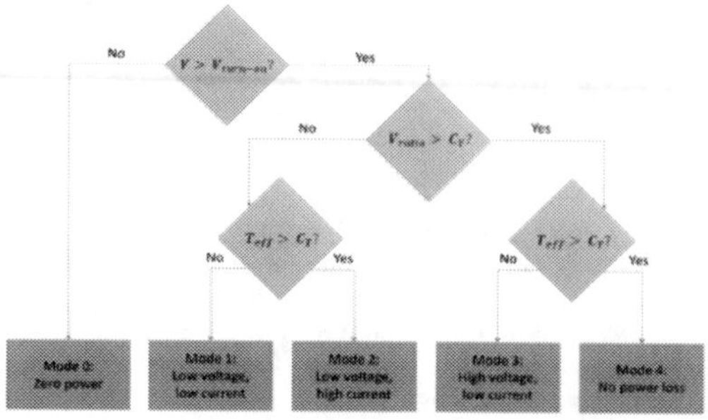

Fig. 1. Mode-based framework for snow identification. Modes with low voltage correspond to conditions with partial coverage by opaque snow; modes with low current correspond to full or partial coverage by light-transmissive snow.

Figure 1 shows the mode-based framework developed by [14]. Mode 0 refers to outages caused by either full or extensive partial opaque snow coverage such that the system is unable to reach the minimum turn-on voltage. Mode 1 corresponds to lower-than-predicted current and voltage, which we attribute to non-uniform snow conditions with partial coverage by both opaque and transparent snow. Mode 1 is hypothesized to occur when partially shed or melted snow is covered by a second, newer layer of snow. Modes 2 and 3 correspond to physical conditions where there is partial coverage by opaque snow (lower-than-predicted voltage, as-predicted current) and partial to full coverage by transparent snow (as-predicted voltage, lower-than-predicted current), respectively. Mode 2 captures periods of shedding while mode 3 describes periods of snow melt or very light snow cover. Mode 4 refers to business-as-usual operations, where there may be small losses due to inefficiencies in an inverter's MPPT algorithm or model inaccuracy.

III. Results

A. Power losses and modal frequencies

Seasonal comparisons of normalized energy loss $[\%/in]$ (Figure 2) revealed that while all sites experienced greater power losses during periods of snowfall, S20 and S10 consistently experienced significantly higher rates of snow-related losses than S35. Energy losses shown in Figure 2 excluded minor losses incurred while systems were operating in mode 4.

Modal frequencies during winter months (November - March) showed that S10 and S20 had significantly higher rates of complete outages (mode 0 occurrences) than S35 (Figure 3), indicating a negative correlation between tilt angle and outage duration. Tilt angle was also negatively correlated with mode 1 occurrences, but the margin of difference in mode 1 occurrences between the sites was decidedly slimmer than the differences in mode 0 frequencies. Given that mode 1 occurs when partially shed opaque snow is covered by a light-transmissive layer of snow, the higher frequencies of mode 1 seen at lower tilt sites indicate that snow sheds slower from lower tilt angle systems.

10.4229/EUPVSEC2023/4AO.9.1
020281-002

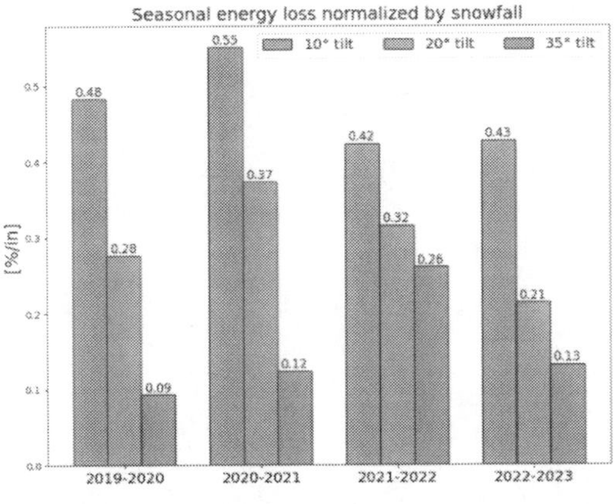

Fig. 2. Seasonally aggregated energy losses from modes 0 - 3 [%] normalized by seasonal snowfall [in]

Frequencies of mode 2, which indicate the presence of partially shed opaque snow, were low across sites and overlapped within the margin of error. The low frequencies can likely be attributed to the systems' portrait module orientation. As snow sheds downward, snow cover is roughly uniform across the bottom edge of a module. For a portrait-orientation module, this means that all three of its substrings remain partially covered as snow sheds off of the top half of a module, as opposed to a landscape-orientation module which would reveal substrings sequentially. If snow cover is opaque, this would translate into a decreased likelihood that the portrait-orientation system turns on until shedding is nearly complete (light reaches all rows of cells).

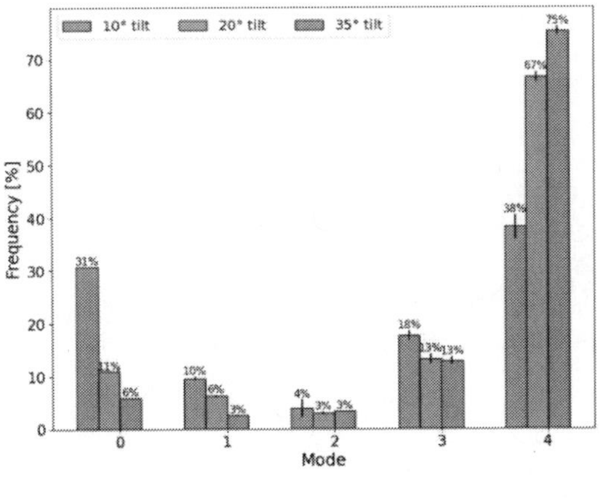

Fig. 3. Modal frequencies during the months of November - March. Error bars were assigned using the spread in modal frequencies generated by varying modal threshold values by their calculated standard deviations.

IV. Conclusions

Snowfall-normalized utility-scale losses are quantified over a four year period for three systems with different modules, inverters, and tilts using standard monitoring data. Seasonal comparisons of normalized energy loss reveal that, for the systems surveyed in this study, lower tilt systems tend to lose more energy than higher tilt systems under the same snowfall conditions. Modal frequency analyses show that tilt angle enables both earlier shedding and shorter shedding durations. These results replicate previous findings and are an example of the type of nuanced information that can be gleaned from inverter data. Future studies should continue to seek performance insights using monitoring even when system instrumentation is as minimal as that of the systems surveyed in this study.

The methods described in this study are location-agnostic and can easily be adapted to different data resolutions and system configurations. Potential research applications of this work include validation for advanced snow shedding models and predictive snow loss models. There is also a substantial audience for whom this approach and future improvements to it have immediate industry applications; as is such, implementing this approach in an industry-friendly tool is of high priority.

References

[1] R. E. Pawluk, Y. Chen, and Y. She, "Photovoltaic electricity generation loss due to snow – A literature review on influence factors, estimation, and mitigation," *Renewable and Sustainable Energy Reviews*, vol. 107, pp. 171–182, Jun. 2019. [Online]. Available: https://www.sciencedirect.com/science/article/pii/S1364032118308268

[2] E. Andenæs, B. P. Jelle, K. Ramlo, T. Kolås, J. Selj, and S. E. Foss, "The influence of snow and ice coverage on the energy generation from photovoltaic solar cells," *Solar Energy*, vol. 159, pp. 318–328, Jan. 2018. [Online]. Available: https://www.sciencedirect.com/science/article/pii/S0038092X17309581

[3] N. Heidari, J. Gwamuri, T. Townsend, and J. M. Pearce, "Impact of Snow and Ground Interference on Photovoltaic Electric System Performance," *IEEE Journal of Photovoltaics*, vol. 5, no. 6, pp. 1680–1685, Nov. 2015, conference Name: IEEE Journal of Photovoltaics.

[4] L. Powers, J. Newmiller, and T. Townsend, "Measuring and modeling the effect of snow on photovoltaic system performance," in *2010 35th IEEE Photovoltaic Specialists Conference*, Jun. 2010, pp. 000 973–000 978, iSSN: 0160-8371.

[5] T. Townsend and L. Powers, "Photovoltaics and snow: An update from two winters of measurements in the SIERRA," in *2011 37th IEEE Photovoltaic Specialists Conference*, Jun. 2011, pp. 003 231–003 236, iSSN: 0160-8371.

[6] U. E. I. Administration, "Form EIA-860 detailed data with previous form data (EIA-860A/860B)." [Online]. Available: https://www.eia.gov/electricity/data/eia860/

[7] J. Braid, D. Riley, and L. Burnham, "Design Considerations for Photovoltaic Systems Deployed in Snowy Climates," *37th European Photovoltaic Solar Energy Conference and Exhibition*, pp. 1626–1631, Oct. 2020, iSBN: 9783936338737 Publisher: WIP. [Online]. Available: http://www.eupvsec-proceedings.com/proceedings?paper=49817

[8] R. W. Andrews and J. M. Pearce, "Prediction of energy effects on photovoltaic systems due to snowfall events," in *2012 38th IEEE Photovoltaic Specialists Conference*, Jun. 2012, pp. 003 386–003 391, iSSN: 0160-8371.

[9] B. Marion, R. Schaefer, H. Caine, and G. Sanchez, "Measured and modeled photovoltaic system energy losses from snow for Colorado and Wisconsin locations," *Solar Energy*, vol. 97, pp. 112–121, Nov. 2013. [Online]. Available: https://www.sciencedirect.com/science/article/pii/S0038092X13003034

[10] L. B. Bosman and S. B. Darling, "Performance modeling and valuation of snow-covered PV systems: examination of a simplified approach to decrease forecasting error," *Environ Sci Pollut Res Int*, vol. 25, no. 16, pp. 15 484–15 491, Jun. 2018.

[11] J. L. Braid, D. Riley, J. M. Pearce, and L. Burnham, "Image Analysis Method for Quantifying Snow Losses on PV Systems," in *2020 47th IEEE Photovoltaic Specialists Conference (PVSC)*, Jun. 2020, pp. 1510–1516.

10.4229/EUPVSEC2023/4AO.9.1
020281-003

[12] C. Baldus-Jeursen, A. L. Petsiuk, S.-A. Rheault, S. Pelland, A. Côté, Y. Poissant, and J. M. Pearce, "Snow Losses for Photovoltaic Systems: Validating the Marion and Townsend Models," *IEEE Journal of Photovoltaics*, vol. 13, no. 4, pp. 610–620, Jul. 2023, conference Name: IEEE Journal of Photovoltaics.

[13] A. Dhyani, C. Pike, J. L. Braid, E. Whitney, L. Burnham, and A. Tuteja, "Facilitating Large-Scale Snow Shedding from In-Field Solar Arrays using Icephobic Surfaces with Low-Interfacial Toughness," *Advanced Materials Technologies*, vol. 7, no. 5, p. 2101032, 2022. [Online]. Available: https://onlinelibrary.wiley.com/doi/abs/10.1002/admt.202101032

[14] E. Cooper, J. L. Braid, and L. Burnham, "Identifying the electrical signature of snow in photovoltaic inverter data," San Juan, PR, USA, 2023.

[15] W. F. Holmgren, C. W. Hansen, and M. A. Mikofski, "pvlib python: a python package for modeling solar energy systems," *JOSS*, vol. 3, no. 29, p. 884, Sep. 2018. [Online]. Available: http://joss.theoj.org/papers/10.21105/joss.00884

[16] D. L. King, W. E. Boyson, and J. A. Kratochvill, "Photovoltaic Array Performance Model."

SOLARGIS

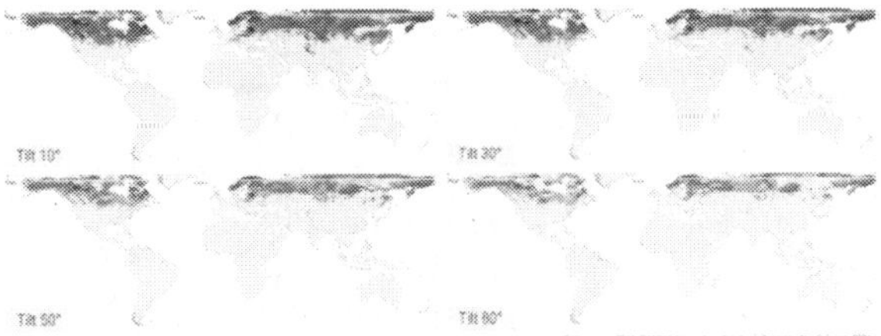

Average expected monthly snow losses in PV systems with different tilts 10, 30, 50 and 60 degrees, example of December 2022

Assessing PV energy loss due to snow with Meteorological Models

Artur Skoczek, Oliver Osvald, Branislav Schnierer, Lubos Helienek, Tatiana Harcinikova
Solargis, Slovakia

EU PVSEC

18 — 22 September 2023

Objectives

Estimate snowfall related losses using meteorological model data and satellite model irradiation data

Meteorological models:
- Historical reanalysis ERA5 from ECMWF
- Forecast IFS from ECMWF
- Solar irradiation data from Solargis Satellite model

Meteorological parameters used:
- Fresh snowfall water equivalent
- Snow density
- Irradiation on tilted surface (Solargis satellite model)
- Air temperature

Derived parameters (Solargis PV simulation software)
- PV power production which considers basic PV system parameters

Estimation of snowfall related losses in PV systems.

Test datasets (power production data-sets)

- NREL data from PV from multiple PV systems in the USA (hourly time-series, various time-periods, 2012 – 2018)
- MyLight Systems (partner in Serendi PV) residential rooftops PV , selected systems in snowy regions of France (15 minute time-series 2021 – 2022)
- Multiple 1 MWp power plants in Central Europe (15-minute time-series 2012 – 2021)

Methodology

1) Human assisted data flagging using Solargis Analyst software package, real PV production and simulated PV production – very time-consuming process!
   Flags
- total power loss,
- partial power loss,
- snow melting / sliding

2) Automatic flagging using meteorological model as an input and model inspired by the SunPower dynamic snow losses model *

* Dynamic Snow Loss Model in PVSim: Modeling Impact of Snow on PV Production, Defne Gun, et. al SunPower Corporation, Richmond, CA 94804, USA

020282-003

40th European Photovoltaic Solar Energy Conference and Exhibition

Distribution of test sites

NREL data from PV from multiple PV systems in the USA

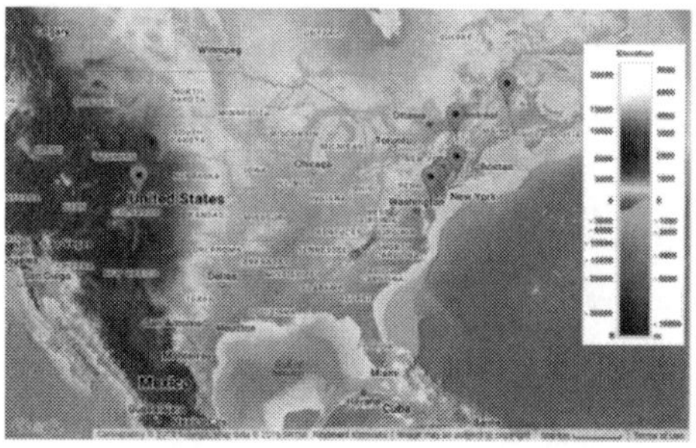

MyLight residential rooftops in France and Switzerland

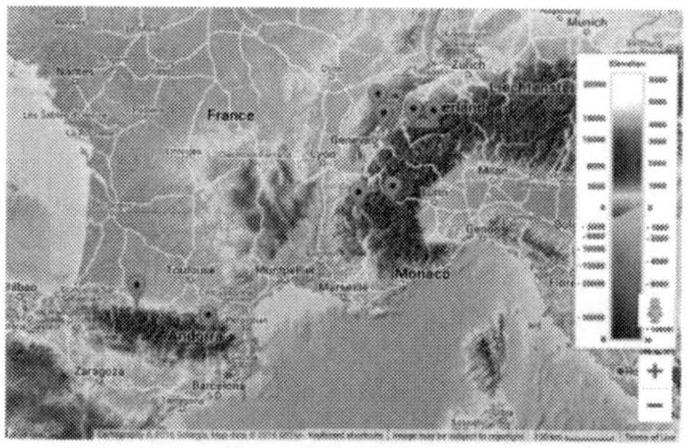

Multiple 1 MWp power plants in Central Europe

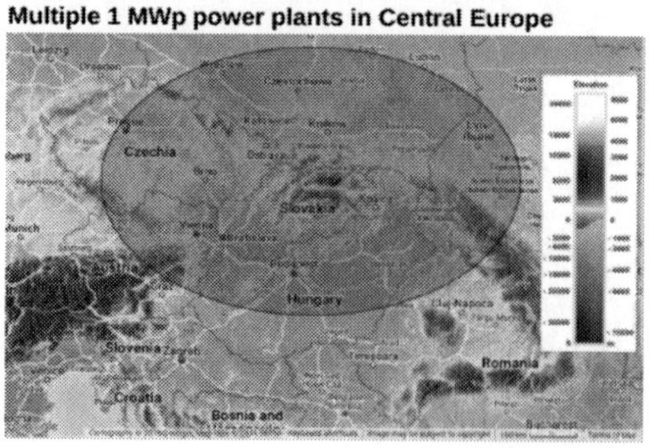

020282-004

Estimation of snowfall related losses in PV systems. Flagging snow events in Solargis Analyst

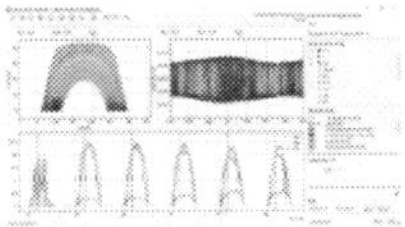

Flagging in Interactive QC window based on difference between real and simulated PV production, fresh snow data and air temperature

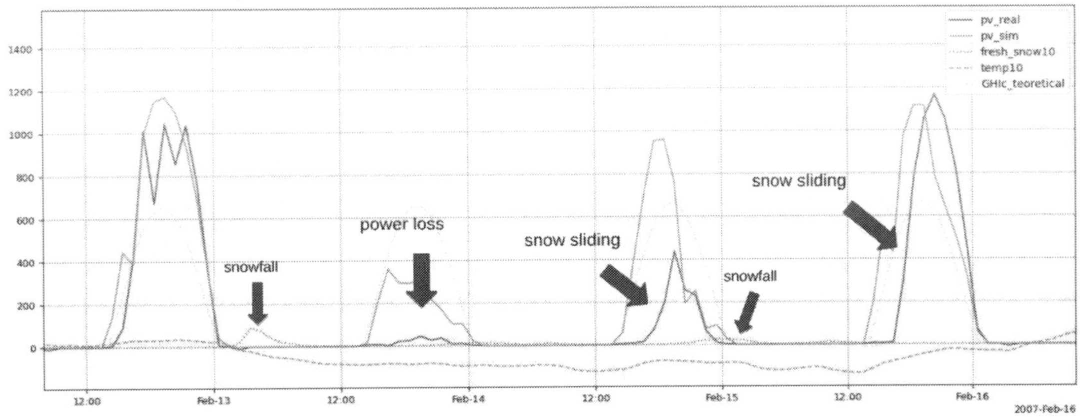

020282-005

40th European Photovoltaic Solar Energy Conference and Exhibition

Snow sliding probability vs module temperature

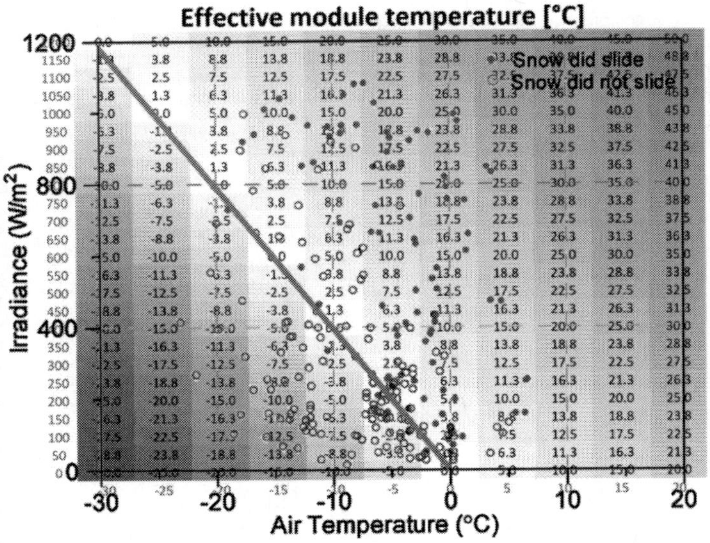

Information on the coefficient of friction of snow on PV modules is not available, but work by the ski industry provides some insight ☺

For a waxed wood ski on wet snow the static coefficient of friction is 0.14
This corresponds to a minimum tilt angle of 8 for wet snow to slide ($atan^{-1}(0.14) = 8$).

Another authors reported that for module tilt angles of 10 or less, snow slides less frequently and snow removal by melting is more likely. This is consistent with a minimum static coefficient of friction of snow on PV modules, presumably when melting, of about 0.14, similar to a waxed wood ski on wet snow.

Measured and modeled photovoltaic system energy losses from snow for Colorado and Wisconsin locations
NREL, **Bill Marion**, Robert Schaefer, Holden Caine, Gonzalo Sanchez

SOLARGIS

020282-006

Dynamic Snow Loss Model

Sunpower model

$\Delta SD_{PVaccum} = \Delta SD_G \times \cos\beta \; (\Delta SD_G > 0)$

$\Delta SD_{PVmelt} = \Delta SD_G \times \cos\theta / \cos\theta z \; (\Delta SD_G < 0)$

$SD_{PV} = SD_{PVprevious} + \Delta SD_{PVaccum} + \Delta SD_{PVmelt}$

SD_G - snow accumulated on the ground
SD_{PV} - snow accumulated on the PV modules
FS – fresh snow
β – tilt/inclination of the panels
θ - the incidence angle normal to the PV modules
θz - the solar zenith angle

Proposed modification

$\Delta SD_{PVaccum} = SF \times \cos\beta \times a$

$ST = Tair + GTI \times tc$
$\Delta SD_{PVmelt} = tmc \times ST \; (ST > 0)$

$SD_{PV} = SD_{PVprevious} + \Delta SD_{PVaccum} + \Delta SD_{PVmelt}$

a – model specific snow "settling" coefficient
ST – surface (snow/module system) temperature
SF – hourly / 15-minute time-series of fresh snow
tmc – thermal melting coefficient
tc – thermal GTI coefficient
Tair – air temperature

 7 SOLARGIS

020282-007

40th European Photovoltaic Solar Energy Conference and Exhibition

Model evaluation methodology

Flags developed by experts

total power loss - loss = 100%
partial power loss - loss = 50%
snow melting / sliding - loss = 50%

Automatic flags based on NWP

total power loss - loss = 100% for SDpv > T1
partial power loss - loss = 50% for T1 > SDpv > T2

$$\text{Power loss} = \sum (GTI_{sat} * \text{loss}) / \sum GTI_{sat}$$

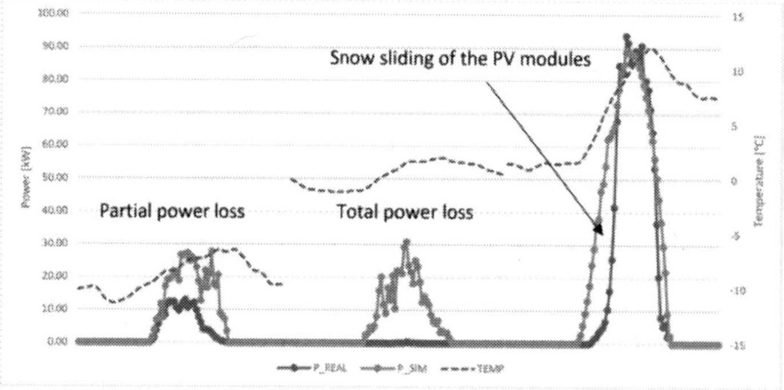

 SOLARGIS

020282-008

Estimation of snowfall losses in PV systems, preliminary results

NREL USA

SiteID	Flag type	Jan	Feb	Mar	Apr	May	Jun	Jul	Aug	Sep	Oct	Nov	Dec
u_1201	ExpertFlag	22.3	20.3	10.9	0.2	0	0	0	0	0	0	0	9.1
u_1201	AutoFlagERA5	17.6	9.3	4.2	0	0	0	0	0	0	0	0.6	5.4
u_57	ExpertFlag	7.2	11.1	5.7	4.3	1.4	0	0	0	0.1	1.2	2.9	5.6
u_57	AutoFlagERA5	11.7	10.0	8.1	6.0	1	0	0	0	0	2.1	1.9	21.3
u_1201	ExpertFlag	22.3	20.3	10.9	0.2	0	0	0	0	0	0	0	9.1
u_1201	AutoFlagERA5	17.6	6.3	4.2	0	0	0	0	0	0	0	0.6	5.4
u_1208	ExpertFlag	12	7.9	4.4	0	0	0	0	0	0	0	0	3.5
u_1208	AutoFlagERA5	15.4	7	3.9	0	0	0	0	0	0	0	0.2	2.7
u_1234	ExpertFlag	22.4	30	14.5	0.3	0	0	0	0	0	0	2.1	11.3
u_1234	AutoFlagERA5	18.6	15.9	5.4	0.1	0	0	0	0	0	0.2	1	3.7
u_1236	ExpertFlag	18.3	38.6	5.7	0.6	0	0	0	0	0	3.3	1.6	18.1
u_1236	AutoFlagERA5	18	21.2	1.2	0	0	0	0	0	0	0.2	1.4	3.8
u_1237	ExpertFlag	26.6	18.7	7.6	0	0	0	0	0	0	2.2	7.2	15.6
u_1237	AutoFlagERA5	14.8	35.1	2.8	0	0	0	0	0	0	0.6	1.3	10.1
u_1239	ExpertFlag	88.3	90.1	98	19.5	0.6	0	0	0	0	0.2	21.5	79.9
u_1239	AutoFlagERA5	100	100	80.1	12	0.1	0	0	0	0	0.1	10.6	81.3
u_1283	ExpertFlag	18.7	36.8	15.3	14.4	5.7	0	0	0	0	4.7	9	25
u_1283	AutoFlagERA5	16.3	28	10	7	2.1	0	0	0	0	2	9.9	27.3
u_1422	ExpertFlag	99.1	56.3	19.2	6.6	0	0	0	0	0	3.1	9.7	39.6
u_1422	AutoFlagERA5	87.4	59.2	15.9	2	0	0	0	0	0	0	7.2	85.6
u_1431	ExpertFlag	9.1	11.7	2.9	1.9	0.9	0	0	0	0	0.9	2.9	11.5
u_1431	AutoFlagERA5	14.0	25.5	8.1	5.5	1.0	0	0	0	0	2.5	8.8	25.1

Central Europe

SiteID	Flag type	Jan	Feb	Mar	Apr	May	Jun	Jul	Aug	Sep	Oct	Nov	Dec
g	ExpertFlag	32.7	19.4	2.3	0.5	0	0	0	0	0	0	5.1	27.2
g	AutoFlagERA5	13.5	7.1	2.3	0	0	0	0	0	0	0.1	0.4	8.2
u	ExpertFlag	10.7	10.1	1.9	0.1	0	0	0	0	0	0	2.9	15.7
u	AutoFlagERA5	9.1	5	1.6	0	0	0	0	0	0	0	0.3	8.9
b	ExpertFlag	19	8.1	1.3	0	0	0	0	0	0	0	1.9	9
b	AutoFlagERA5	9.5	3	0.8	0	0	0	0	0	0	0	0.1	5
m	ExpertFlag	19.7	6.9	1.2	0.1	0	0	0	0	0	0	2.8	14.8
m	AutoFlagERA5	21.2	7.7	2.1	0	0	0	0	0	0	0	0.8	5.9
r	ExpertFlag	16.7	7.2	1.1	0	0	0	0	0	0	0	1.8	6.1
r	AutoFlagERA5	8.9	3.5	0.5	0.5	0	0	0	0	0	0.1	0.4	5.7

MLS France

SiteID	Flag type	Jan	Feb	Mar	Apr	May	Jun	Jul	Aug	Sep	Oct	Nov	Dec
f_179	ExpertFlag	25.3	22.9	7.8	0.7	0	0	0	0	0	0	0	51.6
f_179	AutoFlagERA5	41.3	9.1	13.6	0.5	0	0	0	0	0	0	0	32.4
f_268	ExpertFlag	22.1	12.4	10	1.7	2.2	0	0	0	0.7	1.2	0.5	37.2
f_268	AutoFlagERA5	54.8	21.1	19.6	3.6	9.2	0	0	0	1	6.6	0.1	70.4
f_304	ExpertFlag	28.4	0	0	0	0	0	0	0	0.5	0.6	0	46.9
f_304	AutoFlagERA5	100	9.7	17.8	2.8	0.1	0	0	0	0.2	0.7	0	66.1
f_314	ExpertFlag	9	1.7	2.5	0.5	0.1	0	0	0	0.9	2.7	0.2	18.4
f_314	AutoFlagERA5	8.8	0.1	2.8	0	0	0	0	0	0	0.4	0	14.9
f_335	ExpertFlag	39.2	10	2.2	0.6	0	0	0	0	0	0.1	0	23.6
f_335	AutoFlagERA5	14.8	6.2	5.4	1	0	0	0	0	0	0	0	9.8
f_348	ExpertFlag	25.4	10.4	6.1	0.8	0	0	0	0	0	2	0.3	63.9
f_348	AutoFlagERA5	48.9	86.9	48	4.3	8.6	0	0	0	4.3	21.2	2.7	97.7
f_356	ExpertFlag	35.6	10.2	5.1	2	0.2	0	0	0	0.4	0.3	0	51.1
f_356	AutoFlagERA5	22.6	7.4	6.3	0.7	0	0	0	0	0	0.1	0	29.8
f_371	ExpertFlag	0	0	2.9	0	0	0	0	0	0	0	0	0
f_371	AutoFlagERA5	50	65.8	37	5.6	5	0	0	0	3.8	8.5	1.7	87
f_407	ExpertFlag	7.8	5.8	1.1	0.1	0	0	0	0	0	0.3	0	16.7
f_407	AutoFlagERA5	23.6	5.3	5.2	0.4	0	0	0	0	0	0	0	33.9
f_471	ExpertFlag	9.6	2.4	3.8	0.9	0.1	0	0	0	0	1.3	0.1	24.3
f_471	AutoFlagERA5	25.7	10.4	8.1	2.1	1.2	0.5	0	0	0.9	4.5	5.9	98
f_47	ExpertFlag	13.2	4.6	2.9	0.6	0	0	0	0	0.9	0.2	0	22
f_47	AutoFlagERA5	96.2	6.9	5	1.4	0	0	0	0	0	0.2	0	92.3

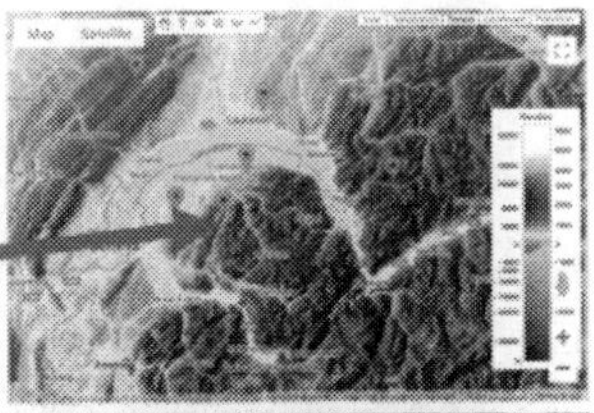

Estimation of snowfall losses in PV systems, preliminary results

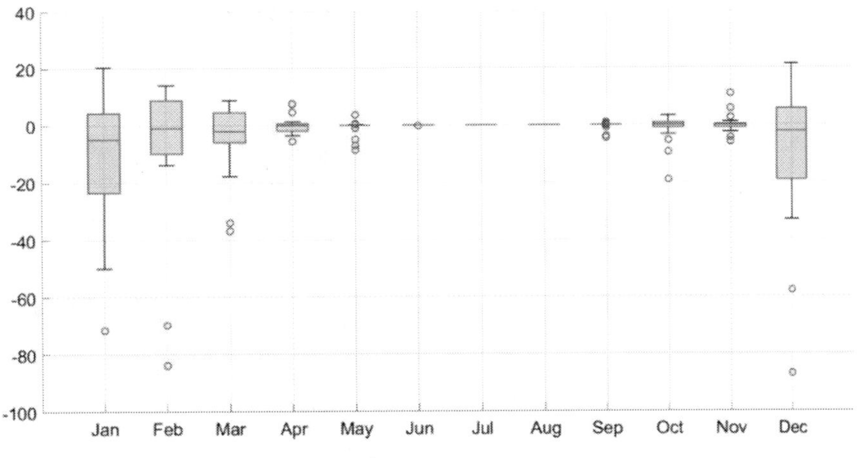

The differences in the power loss between the manual and automatic flags (calculated as manual - automatic)

It can be seen that the medians (mid-line in the boxplot) for all months are close to zero, indicating a good averaged performance, however, in the snowy months there are a number of outliers in the negative part of the graph, indicating that the automatic flagging may overestimate the snow loss in some cases.
These are in general sites located in complex terrain where the spatial resolution of the ERA5 data is insufficient to model local microclimate

Comparison of ERA5 reanalysis and IFS forecast model with higher resolution NOAA NGSA model

The utilization of elaborated gridded analysis products for Technical Description and Evaluation for the forecast and reanalysis verification like the NOAA's National Gridded Snowfall Analysis Continental US-wide (CONUS) with the resolution of 4 km

National Snowfall Analysis: accumulation from 2021-09-30 to 2022-01-26

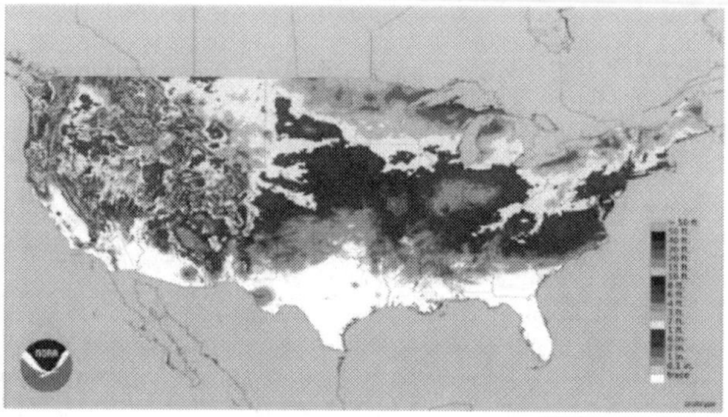

This analysis synthesizes several NWS data sources to produce a unified CONUS-wide gridded snowfall analysis. These sources include the Stage IV mosaic of quantitative precipitation estimates (QPE) generated operationally at NWS River Forecast Centers, precipitation type estimates from short-term numerical weather prediction (NWP) model forecasts, and climatological snowfall-to-liquid ratio (SLR) estimates generated from the Global Historical Climate Network (GHCN) dataset. The analysis reconciles these products with real-time snowfall observations from NWS Cooperative observers, NWS spotters, CoCoRaHS observers, and other observing networks.

The National Snowfall Analysis combines an unbiased background, or first guess, with a two-pass ordinary kriging (OK) process that is parameterized based upon the real-time spatial correlation among differences between observed and background snowfall. Version 2.1 was placed into production for the 2018-2019 winter season. It is accompanied by a ten-year (October 2009-September 2018) retrospective analysis, and the combined 11-year dataset constitutes the first period of record with which future analyses can be compared.

020282-011

Comparison of the 24-hours fresh snowfall NOAA (4 km) vs ERA5 (25 km) vs IFS (10 km)

Snow24h_20200226

Snow24h_20200227

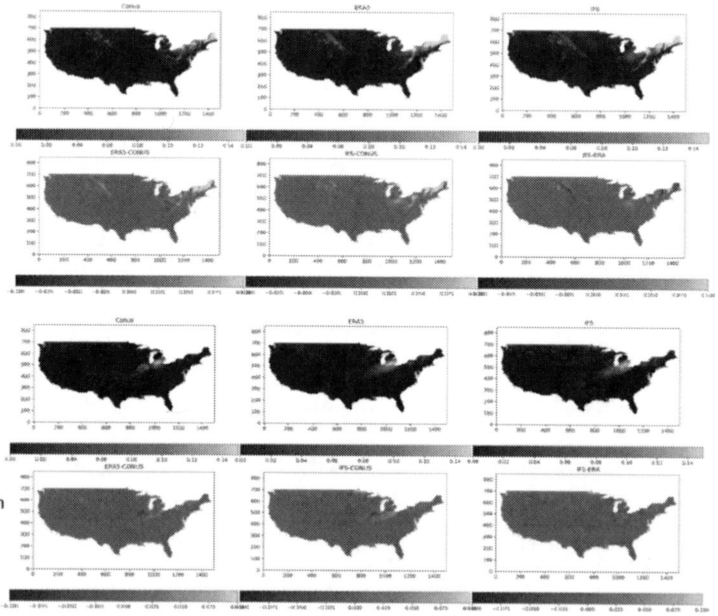

Reasonably good match between the datasets considering different spatial resolution

020282-012

Comparison of the 24-hours fresh snowfall NOAA NGSD (4 km) vs ERA5 (25 km) vs IFS (10 km)

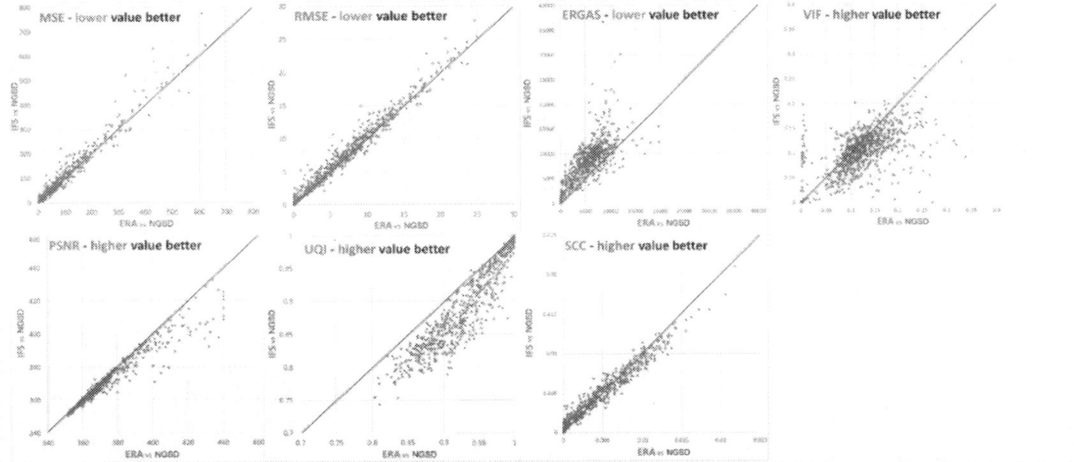

RMSE	Root Mean Squared Error	Is the standard deviation of the residuals (prediction errors). Residuals are a measure of how far from the regression line data points are; RMSE is a measure of how spread out these residuals are. In other words, it tells you how concentrated the data is around the line of best fit.
MSE	Mean Squared Error	MSE is measured in units that are the square of the target variable, while RMSE is measured in the same units as the target variable. Due to its formulation, MSE, just like the squared loss function that it derives from, effectively penalizes larger errors more severely.
PSNR	Peak Signal-To-Noise Ratio	The term peak signal-to-noise ratio (PSNR) is an expression for the ratio between the maximum possible value (power) of a signal and the power of distorting noise that affects the quality of its representation.
SCC	Spatial Correlation Coefficient	reflects the indirect correlation based on the spatial contiguity between any two geographical entities. Pearson's correlation coefficient minus the spatial cross-correlation coefficient leaves the direct correlation coefficient
UQI	Universal Image Quality Index	Instead of using traditional error summation methods, the proposed index is designed by modeling any image distortion as a combination of three factors: loss of correlation, luminance distortion, and contrast distortion. Although the new index is mathematically defined and no human visual system model is explicitly employed, experiments on various image distortion types show that it exhibits surprising consistency with subjective quality measurement. It performs significantly better than the widely used distortion metric mean squared error.
VIF	Pixel Based Visual Information Fidelity	It is deployed in the core of the Netflix VMAF video quality monitoring system, which controls the picture quality of all encoded videos streamed by Netflix.
ERGAS	Error Relative Global Dimensionless Synthesis	A more advanced image quality index than RMSE. ERGAS is a global statistic expressing the quality of the enhanced resolution image. ERGAS measures the transition between spatial and spectral information Relative dimensionless global error synthesis (ERGAS) is used to calculate the accuracy of Pan sharpened image considering normalized average error of each band of the result image

Estimation of snowfall losses, monthly time series and monthly averages. Preliminary results

USA, East coast, New Jersey, North Brunswick

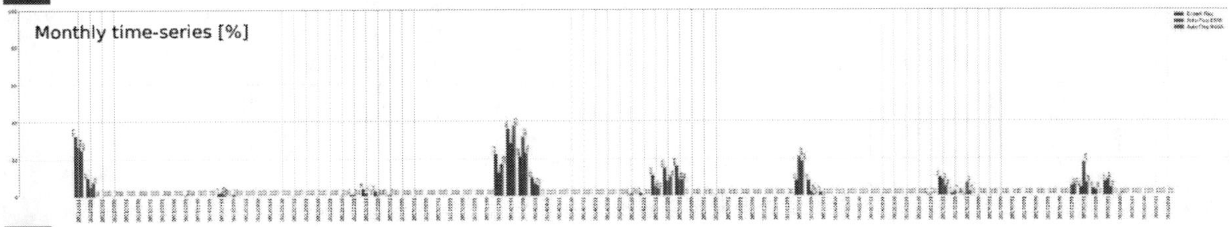

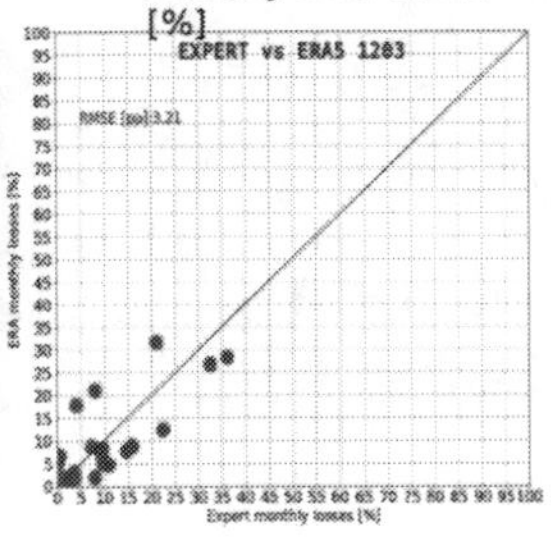

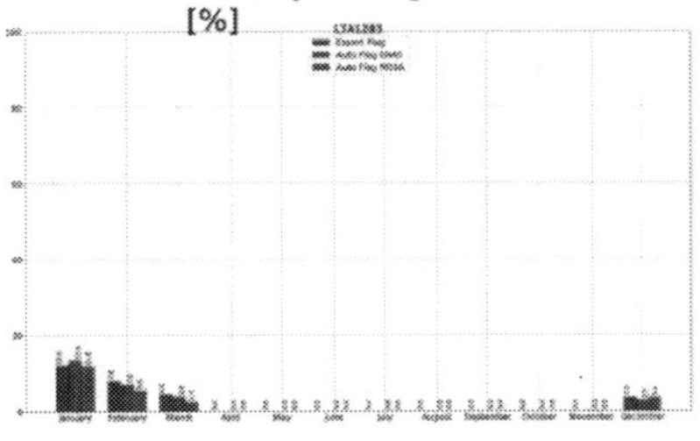

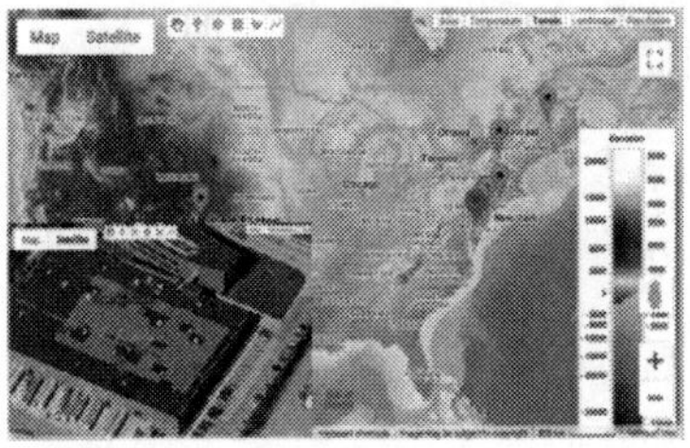

Estimation of snowfall losses, monthly time-series and monthly averages. Preliminary results

USA, NREL, Research Support Facility, Golden

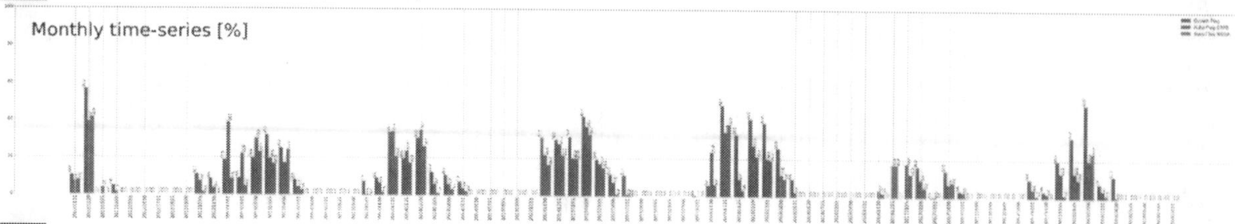

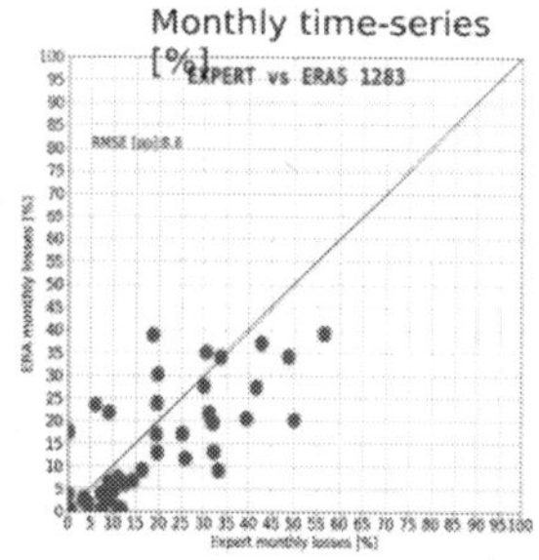

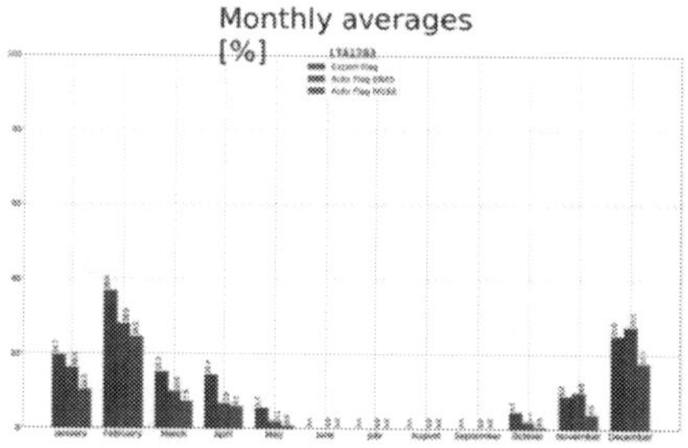

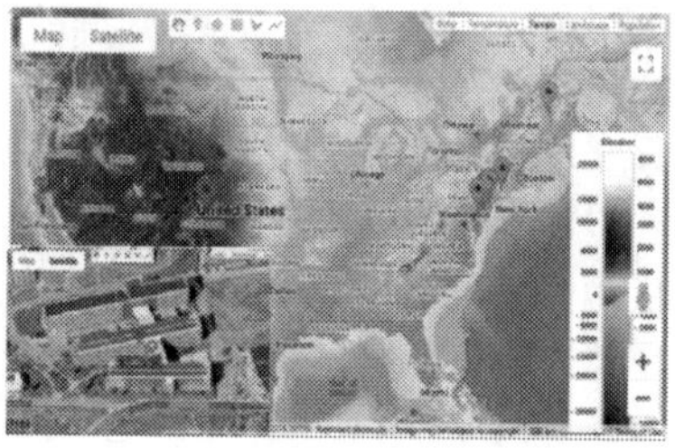

Estimation of snowfall losses, monthly time series and monthly averages. Preliminary results

Canton of Valais, Switzerland

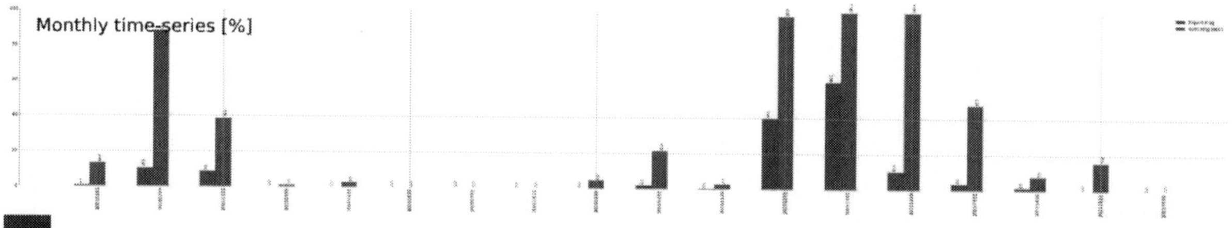

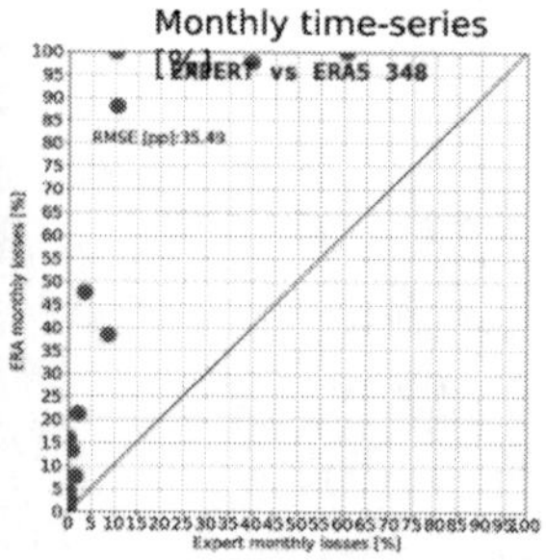

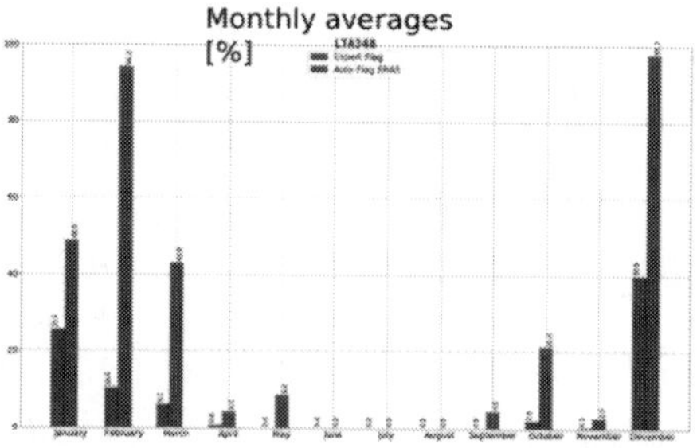

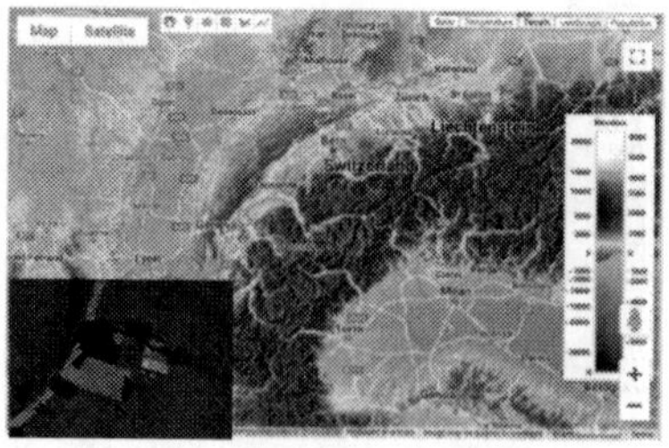

Estimation of snowfall losses, monthly time series and monthly averages. Preliminary results

Foothill of French Pyrenees

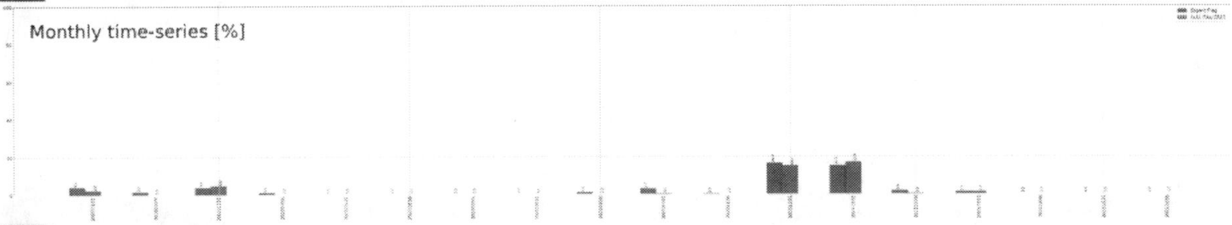

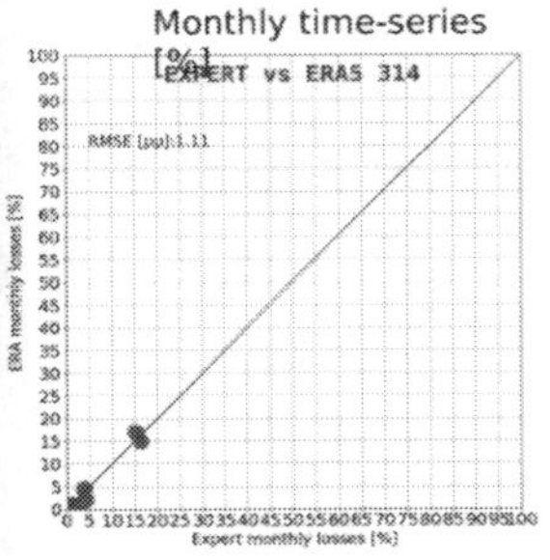

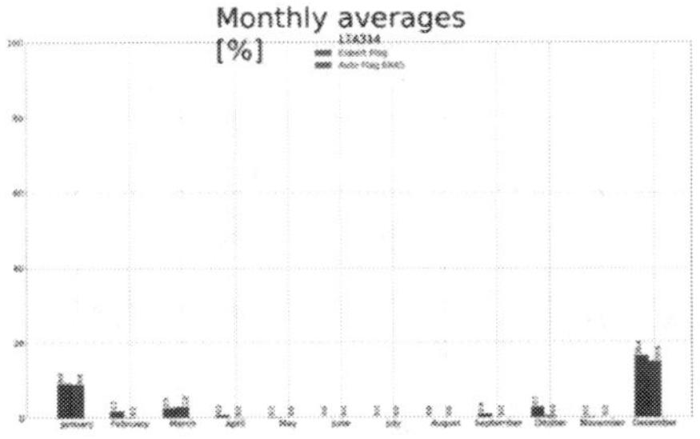

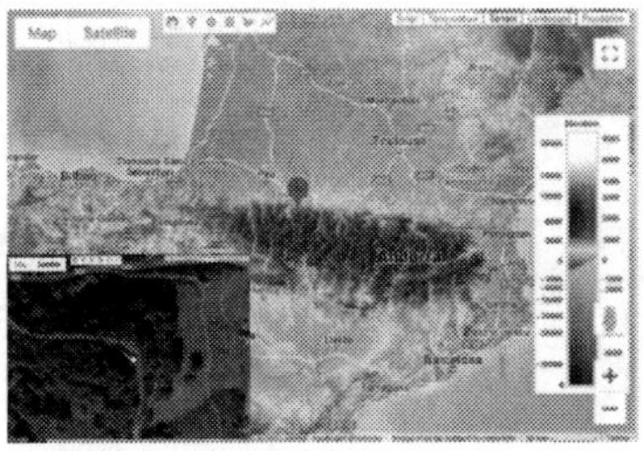

Raster version of monthly energy losses due to snow

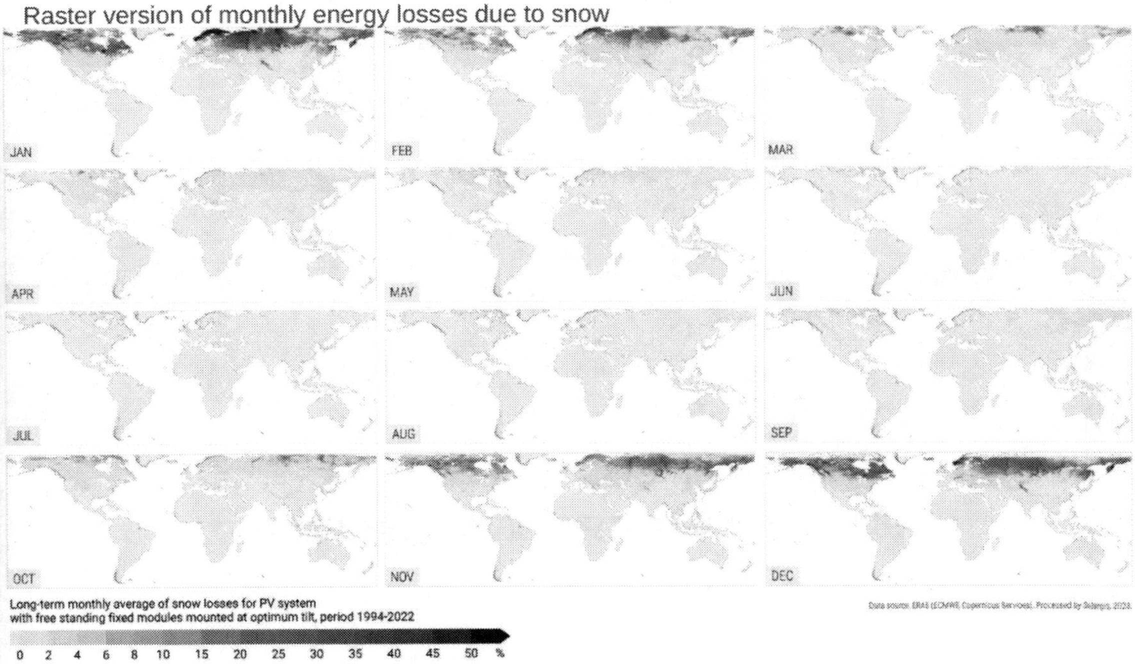

Long-term monthly average of snow losses for PV system with free standing fixed modules mounted at optimum tilt, period 1994-2022

Data source: ERA5 (ECMWF Copernicus Services). Processed by Solargis, 2023.

Raster version of monthly snow losses – interannual variability (basic statistical indicators)

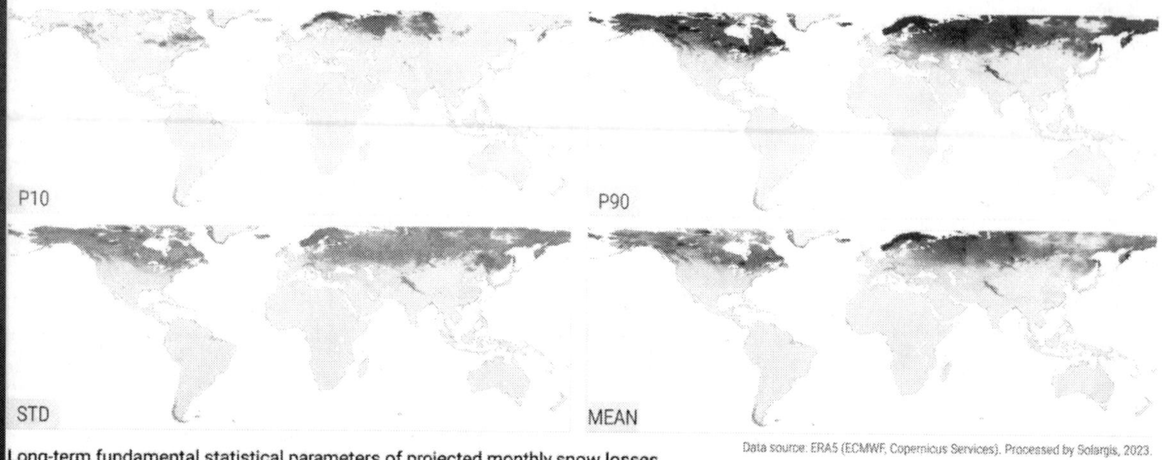

Long-term fundamental statistical parameters of projected monthly snow losses, example of the month January: Minimum P10, maximum P90, standard deviation, and average values

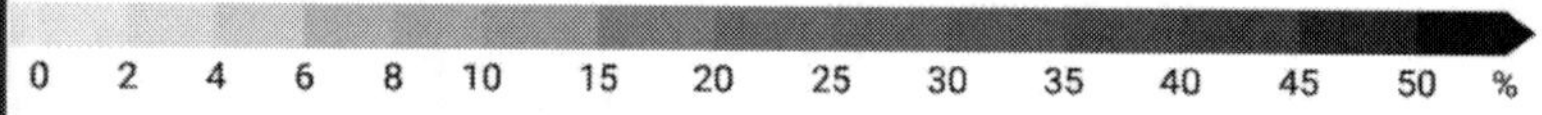

020282-019

Raster version of monthly snow losses - dependence on PV modules tilt

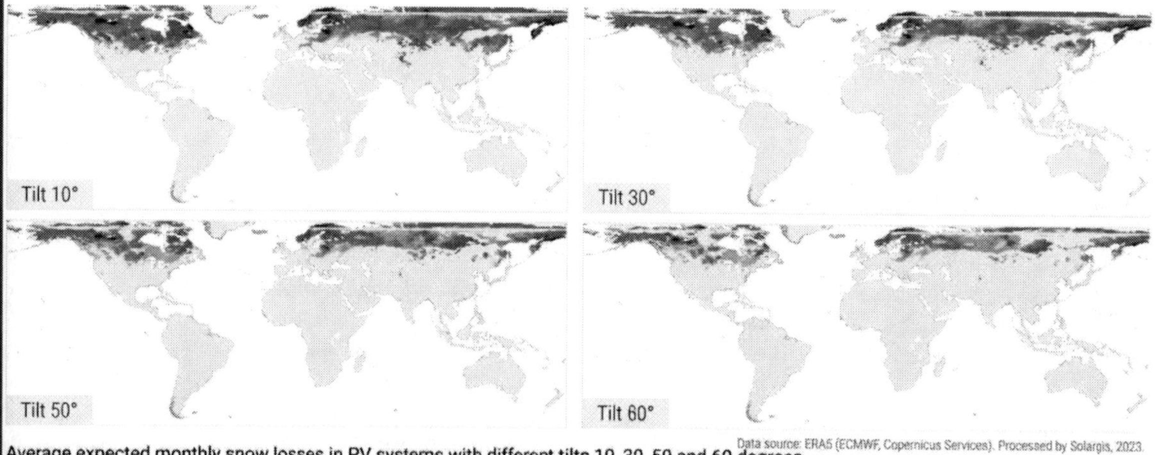

Average expected monthly snow losses in PV systems with different tilts 10, 30, 50 and 60 degrees, example of December 2022

020282-020

Modern approach, snow detection with deep learning

- **Recurrent neural network** model - **LSTM** (Long Short Term Memory)
  - LSTM is a type of a recurrent network cell which enables it to better retain information from earlier time steps of a sequence (compared to plain RNN)
  - Proven model of choice for time series prediction tasks
- **Supervised classification task formulation**
  - Using manually labelled data - partial, sliding, full snow flags
  - Model output is currently a binary flag for presence of snow
  - Initial training completed on ~50,000 5-day sequences

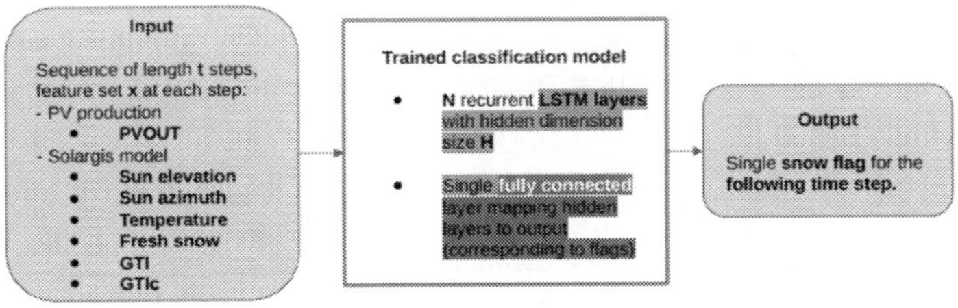

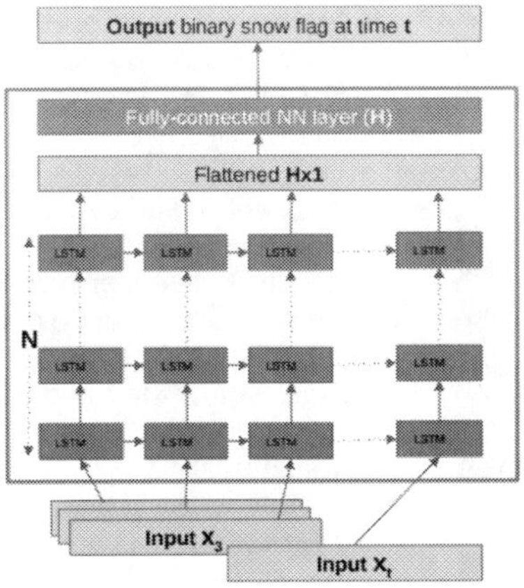

Modern approach, snow detection with deep learning

- Results preview

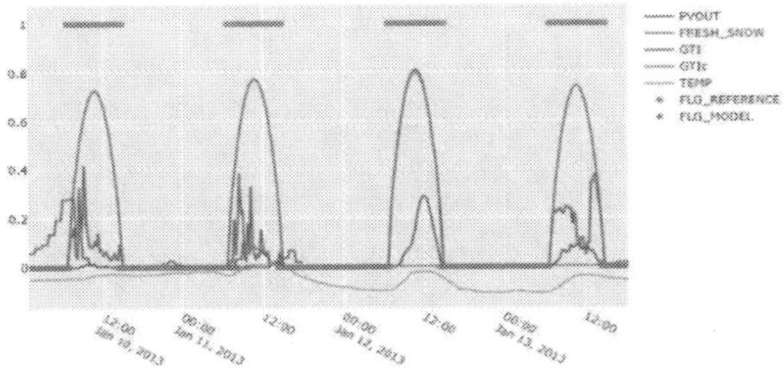

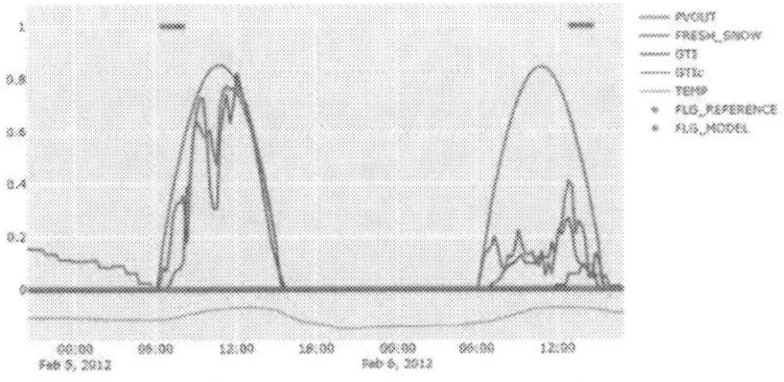

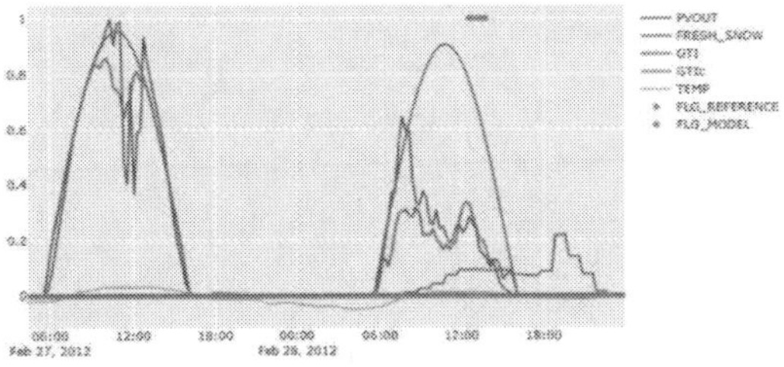

- Limitations
  - Overlap with other issues / site features (e.g. shading, inverter issues etc.)
  - The model reflects the training data - further efforts to be carried out to ensure high veracity of labelled data and diversify the training set

Conclusions and the main challenges

1) The inherent features of the complex physical modeling of the weather
- coarse resolution ~25 km for ERA5 reanalysis and ~10 km for IFS forecast
- false positives (model predicts snow which doesn't occurs in reality)
- false negatives (model doesn't predict snowfall which occurs in reality)
- mismatch in the timing of snow events
- high uncertainty in the magnitude of snowfall event (quantitative)

2) Behavior of the snowpack on the panels
- difficulty in the prediction of the snow sliding moments
- difficulty in the prediction of snowpack distribution on the PV panels and snowpack transparency (partial power loss)

3) Problems in the "expert" manual flagging of snow losses
- Time consuming
- Requires the experienced operator, often quite ambiguous
- Snow losses occurs mainly in a wintertime when the Sun is low over horizon and there is a thick cloud cover (similar profile of snowfall losses and the inaccuracy in the simulation of the cloud cover)

4) Data availability
- Never enough of high quality data from snow affected PV power plants (PV power production, webcam video footage) which is often proprietary

020282-023

THE IMPACT OF SOILING IN EUROPE: ESTIMATION AND ERROR INDUCED BY TYPICAL LOSS ASSUMPTIONS

Álvaro Fernández Solas[1], Eduardo F. Fernández[1], Leonardo Micheli[2], Nicholas Riedel-Lyngskær[3,4]
[1]Advances in Photovoltaic Technology (AdPVTech), University of Jaén, 23071 Jaén, Spain
Email: afsolas@ujaen.es
[2]Sapienza University of Rome, Rome, Italy
[3]Technical University of Denmark, Roskilde, Denmark
[4]European Energy A/S, Søborg, Denmark

ABSTRACT: Soiling in photovoltaic (PV) systems is a significant factor that decreases their efficiency, resulting in lower yields and reduced revenues. However, many PV plants lack specialized equipment to monitor soiling losses. As a result, yield assessors often rely on constant annual loss assumptions to estimate yield losses caused by soiling. These assumptions can lead to inaccurate estimates of the losses and unprofitable cleaning schedules. This study analyses the reliability of different technical soiling loss assumptions by comparing their results against measured soiling data from various sites across Europe. The results demonstrate that all the assumptions tend to overestimate the soiling losses in the Nordic sites. In fact, in one Danish site, the actual soiling losses were less than half of what the least conservative assumption (1% annual soiling loss) predicted. Also, this work evaluates the impact of electricity price fluctuations on revenue losses. The results reveal that significant increases in electricity prices over time can triple the revenue losses associated with soiling in PV systems.
Keywords: Soiling Losses, PV Yield Assessment, Electricity price, Economic analysis.

1 INTRODUCTION

The optimization of the operation and maintenance (O&M) tasks in photovoltaic (PV) systems is a topic that has garnered significant interest due to its impact on economic revenues [1]. One of these tasks is the cleaning optimization of PV modules. The cleaning activities should be only conducted when it is economically profitable [2,3]. This implies that the cleaning cost has to be lower than the economic value of the energy losses due to soiling. This latter value depends both on the amount of the losses and on the electricity price, which is continuously fluctuating in spot markets.

Therefore, having precise knowledge of soiling losses is crucial for maximizing profits. Presently, many PV yield assessors rely on typical soiling loss assumptions [4,5], which involve a constant annual value. However, this does not correspond to the actual loss experienced by the PV systems. Indeed, in the majority of sites in Europe, soiling presents a seasonal pattern with higher losses during the dry summer season and negligible losses during the months with frequent precipitation. Also, soiling losses are strongly dependent on the surroundings and specific climate conditions of the site. Consequently, these assumptions can result in substantial errors when estimating actual losses, thus leading to suboptimal cleaning schedules. This highlights the importance of having soiling data that are either measured at the site or specifically estimated for its characteristics and conditions.

This study evaluates the impact of soiling in different European locations. The monitored soiling losses from optical sensors have been analyzed to obtain the typical soiling profiles for each site. Also, the revenue losses have been calculated by combining these profiles with wholesale electricity prices. The results obtained with measured soiling data have been compared with the typical fixed soiling loss assumptions to assess the reliability of these.

2 METHODOLOGY

2.1 Experimental locations

Measured soiling data from 6 different locations in Europe have been used in this study to assess the variability in the impact of soiling across the continent and to check the accuracy of technical loss assumptions. In Figure 1, the experimental sites are marked with red circles within a Europe map. The three Danish sites are located in a rural environment, surrounded by forests and vegetation. The location in Norway is surrounded by a farm, which presents an intense activity during the spring and the summer. In Spain, the southernmost site is located in a desert environment, whereas the other one is situated close to a park with large green areas.

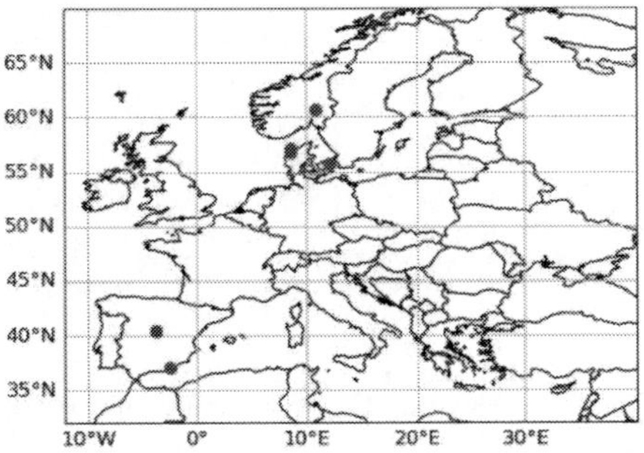

Figure 1: Map of Europe with the locations that have been analyzed in this study: 2 sites in Spain, 1 site in Norway and 3 sites in Denmark.

2.2 Soiling monitoring equipment

The soiling losses in the aforementioned locations were measured using DustIQ optical sensors from Kipp&Zonen [6]. A DustIQ sensor employs a photodiode to measure the light emitted from an LED and reflected by the dust accumulated on a glass cover's surface. The integrated software of the sensor performs internal calculations to provide the soiling loss as output. To ensure representativeness, the sensors were installed vertically in

10.4229/EUPVSEC2023/4AO.9.3

all locations, mirroring the orientation and tilt angle of the PV modules within the respective PV plants.

The sensors operate at a resolution of 1 minute. However, for this study, daily average values of soiling losses were utilized.

2.3 Soiling profile and revenue losses calculation

The methodology employed to extract the soiling energy and the economic losses at each site is summarized in the following steps:

i. Identification of the soiling profile. This was done by extracting the relative daily soiling loss [%], which was calculated as the simple average of the DustIQ data within 2 h of the solar noon, after filtering out the data points affected by dew, frost or precipitation.
ii. Calculation of the lossless PV yield. The theoretical PV energy generated per kWp of installed capacity was calculated using the PVWatts DC power model [7]. This model was run with the global-in-plane irradiance and the ambient temperature data from PVGIS [8]. It should be noted that the actual configuration (orientation and tilt) of the DustIQ sensors was used to download the PVGIS data.
iii. Evaluation of the energy losses. The daily energy loss due to soiling was calculated by multiplying the relative daily soiling loss by the lossless daily PV yield.
iv. Assessment of the revenue losses. The daily revenue loss due to soiling was the product of the multiplication of the daily energy loss by the baseload daily electricity price. The data sources of the electricity prices were: OMIE (Spanish electricity market operator) [9] and Nord Pool (Scandinavian and Baltic countries market operator) [10].

2.4 Technical soiling loss assumptions

One of the focus of this work is to assess the reliability of soiling loss assumptions, which are typically employed by yield assessors. Table I indicates the soiling loss assumptions evaluated in this study.

Table I: Overview of the typical soiling loss assumptions.

Technical assumption [% annual soiling loss]	Reference
1.0%	IEA-PVPS Task 13 [4]
1.5%	IEA-PVPS Task 13 [4]
3.5%	ESMAP [5]

The estimated energy and economic losses resulting from these assumptions have been compared to the actual values of losses.

3 RESULTS

3.1 Actual soiling losses

The soiling profiles of the different sites under study are shown in Figure 2-top. The markedly seasonal pattern of the soiling losses can be clearly appreciated in the Danish (DK1 to 3) and Norwegian (NO) sites where the losses are negligible during the fall and winter months, but there are some soiling accumulation periods during the spring and the summer that usually coincide with dry spells. The high soiling loss (~6%) registered in the Norwegian site in late April was due to the intense agricultural activity near the PV plant. In the case of the DK2 site, the pollination of nearby trees in April might be the reason of the high soiling rate in that period. The sites in Spain (ES) present higher soiling losses (up to 6% in the southernmost site), especially during the summer when long dry spells occur.

The bottom plot of Figure 2 indicates the annual average soiling loss. None of the sites in Denmark or Norway reach the least conservative assumption of 1% soiling loss. Notably at the DK1 site, the average soiling loss is below 0.5%. For the two sites in Spain (ES), average annual soiling losses of ~1.5% are reported. This plot suggests that the typical soiling assumptions will result in inaccurate energy and economic modeling. The extent of these inaccuracies are established with the results shown in Fig. 3

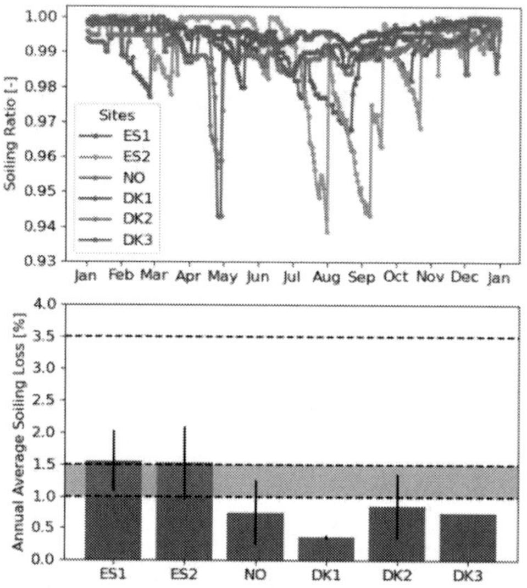

Figure 2: Top: daily soiling profiles of the different sites. The y-axis represents the soiling ratio, which is 1 – soiling loss. Bottom: annual mean soiling loss in the different sites. The dashed lines represent the technical assumptions analyzed in this work. The vertical black lines represent the standard deviation in the sites with at least 2 years of soiling data.

3.2 Validation of typical soiling assumptions

The annual yield losses due to soiling are shown in the top plot of **Figure** *3*. The values range between 5.1 kWh/kWp and 24.7 kWh/kWp. Contrary to the results shown in Figure 2, these values also take into account the effect of solar radiation, and thus, although if all the sites had had the same soiling profile, the absolute soiling losses would have been higher in those sites with greater solar radiation. The annual revenue losses in the sites are represented in Fig. 3-bottom. In this case, the values vary between 0.38 €/kWp and 2.97 €/kWp. The most conservative assumption (3.5% soiling loss) significantly overestimates the energy and economic losses in all of the sites. In the case of the DK1 site, this assumption returns modeled revenue losses that are 7.8x higher than the actual ones. In all of the Scandinavian sites, with the exception of DK2, the other two assumptions also overestimate the losses. Only the 1% assumption for the two Spanish locations and for the DK2 site slightly underestimates the

10.4229/EUPVSEC2023/4AO.9.3

actual losses (17% or relative error for the ES1 and the DK2 sites respectively and 26% for the ES2 location).

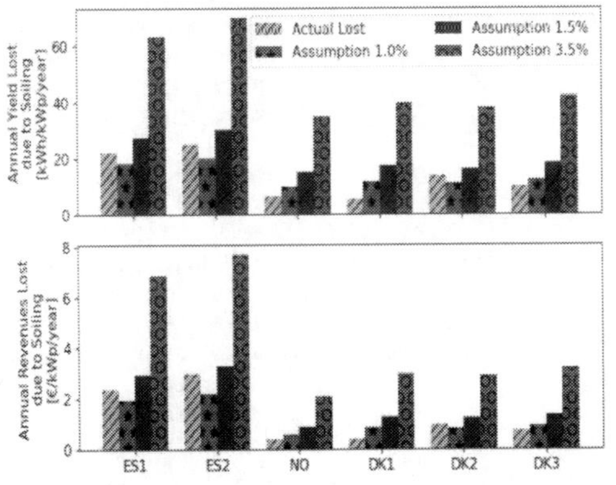

Figure 3: Top plot: annual yield lost due to soiling in the different sites. Bottom plot: annual revenue lost associated to the yield lost due to soiling. The results obtained when considering the different technical assumptions are also represented.

3.3 Assessment of electricity price fluctuations

The electricity prices in most European countries undergo constant fluctuations. This is primarily because a significant portion of electricity is traded on daily markets [11]. In these markets, the price of electricity for each hour of the following day is determined by considering the available energy sources and the expected demand. Due to the seasonal nature of certain energy sources like wind and PV, as well as variations in demand, the electricity price also exhibits a seasonal pattern. Additionally, there are other factors to consider, such as oil and gas prices, which can lead to short-term deviations from the normal trend. Moreover, unforeseen events like pandemics [12] and international conflicts [13], can greatly influence the electricity prices.

Figure 4 shows a comparison of electricity prices recorded in Spain and Denmark during the years 2021 and 2022. Throughout the first nine months of 2022, electricity prices in Europe notably surged compared to 2021, primarily due to the energy crisis resulting from the Ukrainian conflict. For instance, in March, the average electricity price in 2022 surpassed the average price in 2021 by more than fourfold in both countries. Then, the prices returned to normal values or even lower in Spain.

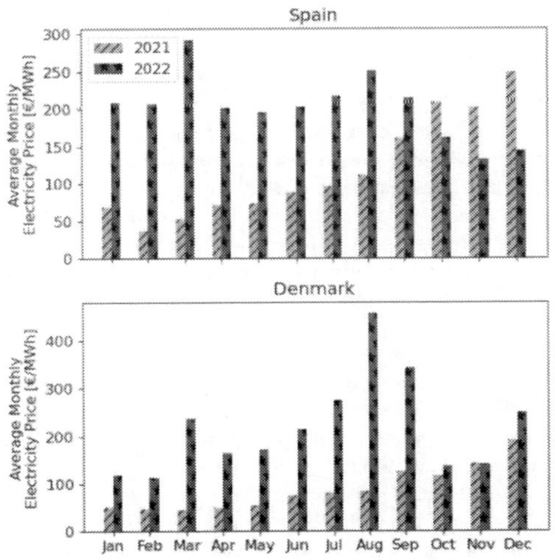

Figure 4: Monthly average electricity price in Spain and in Denmark. Comparison between 2021 and 2022 prices.

The impact of variations in electricity prices on revenue losses caused by soiling can be observed in **Figure 5**. In the Spanish sites (ES1 and ES2), the revenue losses in 2022 are twice as high as those in 2021, with the southernmost site experiencing a maximum loss of 5.76 €/kWp. Similarly, in the Scandinavian sites (NO and DK1 to DK3), the losses in 2022 are approximately three times greater than the previous year. These substantial differences in economic losses are expected to heavily influence the scheduling of cleaning activities, leading to an increased number of cleanings aimed at maximizing energy production and, consequently, reducing revenue losses.

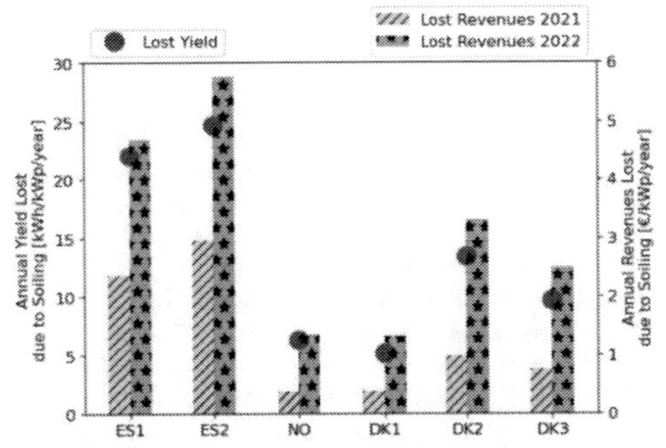

Figure 5: Impact of different electricity prices on revenue losses due to soiling. Left: the circle red markers indicate the annual energy soiling loss. Right: the bars represent the annual economic losses that the energy losses cause in two years with different electricity prices.

4 SUMMARY

This work evaluates the varying impact of soiling on the performance of PV systems installed in different regions of Europe, each characterized by distinct surroundings and climate conditions. The findings reveal that the distribution of soiling losses is not uniform throughout the year. This trend is particularly prominent in the Spanish sites, where the losses are more pronounced during the summer months (reaching maximum daily values of approximately 6%), primarily due to the scarcity of precipitation during this season. Building upon this observation, we demonstrated the potential errors that arise from assuming constant annual values for soiling loss. Notably, even the least conservative assumption (1% annual soiling loss) significantly overestimates the losses for three out of the four Nordic sites (NO, DK1, and DK3).

Furthermore, we assessed the influence of the electricity price fluctuations on the revenue losses due to soiling. The results show that the increases in the electricity prices in consecutive years lead to economic losses that can be more than 3 times higher. Future studies should evaluate how these changes in revenue losses can impact the optimization of cleaning schedules.

ACKNOWLEDGEMENTS

Alvaro Fernández-Solas is supported by the Spanish Ministry of Science, Innovation and Universities under the program “Ayudas para la formación de profesorado universitario (FPU), 2018 (Ref. FPU18/01460)”. The authors thank Mari Øgaar, Jesús Polo and Natalie Hanrieder from IFE, CIEMAT and DLR, respectively, for providing the experimental soiling data for this work.

REFERENCES

[1] Livera A, Theristis M, Micheli L, Fernandez EF, Stein JS, Georghiou GE. Operation and maintenance decision support system for photovoltaic systems. IEEE Access 2022:1–1. https://doi.org/10.1109/access.2022.3168140.

[2] Micheli L, Fernandez EF, Almonacid F. Tracking Soiling Losses and Cleaning Profits Trends. 48th IEEE Photovolt. Spec. Conf., 2021, p. 0144–6. https://doi.org/10.1109/pvsc43889.2021.9518866.

[3] Mithhu MMH, Rima TA, Khan MR. Global analysis of optimal cleaning cycle and profit of soiling affected solar panels. Appl Energy 2021;285:116436. https://doi.org/10.1016/j.apenergy.2021.116436.

[4] IEA-PVPS Task 13. Technical Assumptions in Financial PV Models: Review of Current Practices and Recommendations. 2017.

[5] Energy Sector Management Assistance Program (ESMAP). Global Photovoltaic Power Potential by Country. Washington, D.C.: World Bank, Washington, DC; 2020. https://doi.org/10.1596/34102.

[6] Korevaar M, Mes J, Nepal P, Snijders G, van Mechelen X. Novel soiling detection system for solar panels. 33rd Eur. Photovolt. Sol. Energy Conf., Amsterdam: 2017.

[7] Dobos AP. PVWatts Version 5 Manual 2014.

[8] European Commision. Photovoltaic Geographical Information System (PVGIS) 2016. https://re.jrc.ec.europa.eu/pvg_tools/en/tools.html#MR (accessed December 28, 2022).

[9] OMIE. Precio final medio mercado diario n.d. https://www.omie.es/es/market-results/monthly/average-final-prices/prices-spanish-market?scope=monthly&year=2022&month=12 (accessed January 8, 2023).

[10] Nord Pool. Nord Pool Day-ahead prices n.d. https://www.nordpoolgroup.com/en/Market-data1/Dayahead/Area-Prices/ (accessed January 8, 2023).

[11] Micheli L, Theristis M, Talavera DL, Nofuentes G, Stein JS, Almonacid F, et al. The economic value of photovoltaic performance loss mitigation in electricity spot markets. Renew Energy 2022;199:486–97. https://doi.org/10.1016/j.renene.2022.08.149.

[12] Santiago I, Moreno-Munoz A, Quintero-Jiménez P, Garcia-Torres F, Gonzalez-Redondo MJ. Electricity demand during pandemic times: The case of the COVID-19 in Spain. Energy Policy 2021;148. https://doi.org/10.1016/j.enpol.2020.111964.

[13] Feveile Adolfsen, Jakob Kuik F, Magdalena Lis E, Schuler T. The impact of the war in Ukraine on euro area energy markets. Eur Cent Bank 2023. https://www.ecb.europa.eu/pub/economic-bulletin/focus/2022/html/ecb.ebbox202204_01~68ef3c3dc6.en.html (accessed July 12, 2023).

40th European Photovoltaic Solar Energy Conference and Exhibition

This presentation was selected by the Sc. Committee of the EU PVSEC 2023 for submission of a full paper to one of the EU PVSEC's collaborating peer-reviewed journals.

DETECTION OF SHADING FOR SOLAR POWER FORECASTING USING MACHINE LEARNING TECHNIQUES

Tim Kappler, Anna Sina Starosta, Nina Munzke, Bernhard Schwarz, Marc Hiller

Karlsruher Institut for Technology (KIT)

76344 Eggenstein-Leopoldshafen , Hermann-von-Helmholtz-Platz 1

ABSTRACT: This paper presents a machine learning based solar power forecast methode that can take into account partial shading related fluctuations. Generated PV power is difficult to predict because there are various fluctuations. Such fluctuations can be weather related when a cloud passes over the array. But they can also occur due to shading caused by stationary obstacles. In this work an approach is presented that improves the forecast under such fluctuations caused by partial shading. This adaptation is necessary, because partial shading is usually not detected directly. Such shading occurs after the growth of trees or later built buildings. The presented algorithm can detect such effects itself and thus works self-learning. A correction of the prediction of a forecast model could successfully reduce forecast error due to partial shading. The model is evaluated on the basis of two months of recorded shading data in which shading was caused by a tree infront a PV array. The correction uses internal inverter data and irradiance values of the previous day to perform the correction and was able to reduce the RMSE of a 10 kWp under shading and thus improve the prediction accuracy by up to 40% depending on how strong the shading is. The model can detect how intense the shading is and correct the forecast by itself.

Keywords: shading detection, solar power forecasting, machine learning

1 Introduction

The number of photovoltaic systems installed worldwide and the associated installed capacity rose by 23.6 % to more than 1100 GW in 2022. Further increase is expected to meet CO_2 emission targets in the future [1]. Since the generated power fluctuates as PV systems are subject to cloud movements, rain or changes in irradiation, problems of grid stability are more and more in focus [2]. Accurate forecasts of generated power and energy are necessary to maintain and guarantee stability and availability. Especially machine learning methods have gained popularity in this context, since their advantage is to have a good generalization capability and are therefore able to adapt quickly to new situations [3]. The disadvantage of these methods is the large amount of data that has to be collected to achieve high accuracy. Typically, exogenous data such as irradiation data, wind speed or air temperature are used for solar power forecasting [4–8]. In addition, it is difficult to predict newly occurring situations that are not considered in the recorded data set because no information about them is available. For example, incoming shade from trees that have grown taller or buildings that were constructed later can reduce the PV system's output. Other possible effects are pollution [9] by dust and leaves or degradation of the PV modules [10]. With information about irradiation, wind speed, ambient temperature or sun angle, it is not possible to represent such caused drops in power. The forecast error becomes larger when shading is present because the trained models are not able to handle the shading on their own. There is already a wide range of papers dealing in general with the prediction of the generated solar power [11,8] as well as with the effects of shading [12, 13] and how to model and analyze shading or solar systems under yield reducing effects [14,15]. Shading was already taken into account in power predictions, where the yield loss over the period of one year [16] was considered. Also, in daily forecasts as in [17], shading was included by reducing the effective irradiance and thus by preprocessing. In this paper, the focus is on a correction of the already existing prediction value with a model that has seen training data over years and thus could develop a good generalization capability. The consideration is therefore done by a postprocessing. In addition, loss effects such as shading and soiling are quantified and allow later condition monitoring approaches in an extension. To the authors' knowledge, there is no research that combines quantification of shading losses based only on PV yield data with solar power forecasting. Thus, no extra soiling or shading sensors are needed. This is investigated due to the significant impact of shading on solar cells. Long Short-Term Memory (LSTM) networks are used as a basis for the prediction model. The model is validated with four shading configurations with PV data recorded over several months. The RMSE of the shaded PV systems, which have different orientations and inclinations, is compared. In addition, soiling is taken into account. Additionally, soiling is considered. In this paper, one shading scenario is presented.

The paper is structured as follows. First, the basic method of the procedure to consider shading for PV forecasts is described. For this purpose, three individual submodels are described, which are required for the understanding of the method. Then, the presented method is validated using data from a array shaded by a tree. Finally, the results are discussed and an outlook is given.

2 Methodology

The following section explains how the whole approach works. Three sub-models are described, with a basic introduction of the models used for the forecasting part. The shading is taken into account by means of a correction value. For this purpose, the forecast values of the prediction model are corrected to take the effect of shading into account. First, the forecast model itself is described in this section. This method is first presented in a fundamentals section. Afterwards it is explained how the forecast model is trained to predict the generated PV power one day ahead. Then, a PV array model is used, which simulates the power that an unshaded PV array would provide based on irradiation and temperature values. Finally, it is discussed how the shading power is determined from the simulated PV power and the actual measured values. This refers to the power that is lost due to the shadows that occurs.

The principle of the approach is visualized in Figure 1. The forecast model provides a predicted value, which represents the solar power one day ahead. The PV array model calculates the power that an unshaded array will deliver under the same conditions. Together with the calculated and actual measured value, the shading power can be determined and the actual predicted value can be corrected afterwards so that the prediction can take the shading into account.

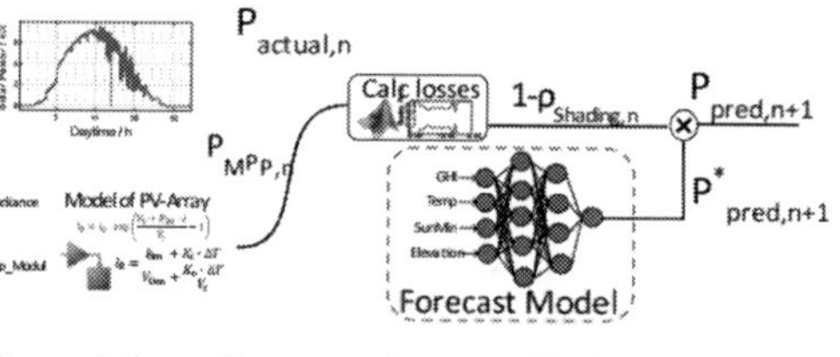

Figure 1 Overall approach to considering shading for solar performance predictions. The approach is divided into three submodels with the prediction model, the PV array model and the separation of losses.

2.1 Fundamentals

In the following, the used forecast model is introduced. The focus lies on how the forecast models work and which parameters need to be trained or optimized using collected data (which will be explained more in detail in section 2.3).

2.1.1 Long Short-Term Memory Networks (LSTM)

The basic unit of each neural network is the perceptron. It maps a weighted sum of the input data x_i with the edge weights w_{ij} and transfer function ψ to the output h_j.

Layers can be built up based on this unit. The output values of each neuron are finally calculated according to Equation 1.

$$h_j = \Psi(\sum_{i=1}^{n} w_{ij} \cdot x_i) \quad (1)$$

If several of these layers are linked together, this arrangement is called a neural network as displayed in Figure 2. These are called feed-forward networks (FFN) because the layers with p_n neurons in the k-th layer mesh in the forward direction from input to output with the transfer function Ψ [18].

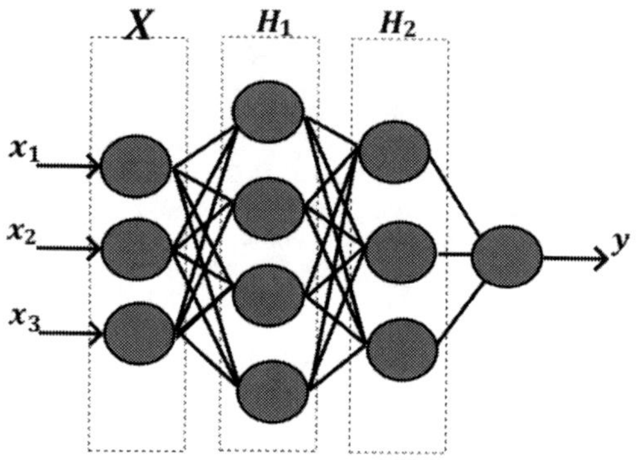

Figure 2 Structure of a two-layer neural network with three input features, two hidden layers and one output

Finally, the data are transformed from the input layer to the output y_k.

Such FFNs have been able to demonstrate good prediction capabilities in many publications in the past [19, 20]. In more recent publications, however, recurrent neural network architectures are used, which have shown an improvement in prediction accuracy several times [21,22]. In particular, LSTM networks have been able to achieve more and more popularity in scientific publications due to their memory capability [22,23] and their ability to deal with the exploding and vanishing gradient problem [24]. LSTM networks consist of a large number of gates that store knowledge about the previous state. These data are either written to, stored in or read from a cell that serves as a type of memory. When the cell reads, writes or deletes information using the input and forget gates, it makes a decision about whether to store the data. Based on the signals received, they become active and use their own weighted filters to decide whether to forward or suppress the information based on its importance and strength. These weights are similar to those that adapt during the training phase of the network to modulate the input and hidden states [25].

2.2. Data

The data are classified into endogenous and exogenous data.

10.4229/EUPVSEC2023/4AO.9.6
020285-002

Exogenous data: Since 1981 the German Weather Service (DWD) offers historical and forecast weather data from selected weather stations on its publicly accessible website [29,30]. Weather data includes data such as air temperature, irradiation, cloud cover, humidity and many others. In this work, weather data are used as input features to train and validate the forecast model. An LSTM network is used as the forecast model, as described in Section 2.1.1.

Endogenous data: Data from the solar park at KIT Campus North are used. It is located at 49.1° north and 8.44° east. There are a total of 102 PV arrays with an installed capacity of around 10 kWp each. The arrays have inclinations between 2° and 60° and an orientation between 60° west and 60° east. The data of the solar park is recorded since 2014. Power data of the inverters (string voltage, string currents, string power) as well as irradiance and module temperatures of selected arrays are available. In addition, the geographical location is used to determine the solar angles with the help of the library pvLib [31] for the description of the solar trajectory.

The hourly mean values of the data are used for all investigations and calculations. Since data can be corrupted, the data are filtered beforehand. These outliers can be explained by sensor errors or communication problems during transmission of the data to the database. From the outset negative values and values that are measured significantly above the Standard Test Conditions (STC) can be considered corrupt. However, since this concerns only a small amount of data, the corrupted values can be easily compensated by interpolation of the neighboring values.

2.2.1 Shading structures

For the evaluation of the methodology, one of the 102 PV arrays is selected and is marked in Figure 3. Later, the calculated power losses are used to subsequently correct the forecast values of the model.

Figure 3. Solar park at the North Campus of the KIT with shaded array. The array has an inclination of 15° and orientation of 30° west.

2.2.1 Forecast model

An LSTM network is used to forecast the power one day ahead. As a data-driven algorithm, the endogenous and exogenous data described earlier are used to determine the parameters of the neural network. For this purpose, a total of seven years of recorded data are split into two data sets. A training data set to optimize the parameters and a validation data set to test the model on unknown data. Six years are used for training and one year for validation.

2.4 PV array model

To determine the power of an unshaded array, a model of a PV array is needed. This model uses irradiance and temperature data to determine the power of a PV array that does not face any shading. In general, it returns the power of an ideal PV array without yield degrading effects.

This model can be built from the 1-diode model [35] and thus can be parameterized completely from existing data sheets of the used modules. A perturbation and observation (PO) algorithm is used to operate the array at the maximum power point (MPP).

2.5 Separation of losses

Now the ideal power can be described using the PV array model and the predictions can be calculated using the forecast model. The recorded data and the ideal PV model can be used to estimate the shading losses similar to [36]. A ratio ρ is defined, which puts the ideal (P_{ideal}) and actual power (P_{DC}) into a direct relation, as in Equation 2.

$$\rho = \frac{P_{\text{ideal}}}{P_{\text{DC}}} \tag{2}$$

Since soiling contributes to a reduction in output in addition to shading, a soiling ratio $\rho_{soiling}$ can be calculated as the average value of the ratio at midday hours (ρ_{Noon}), as in Equation 3.

$$\rho_{\text{Soiling}} = \overline{\rho_{\text{Noon}}} \tag{3}$$

In general, the average ρ_{Noon} should be calculated at unshaded times. The power dissipation on the soiling and shading are then calculated according to Equation 4-5.

$$P_{\text{Soiling}} = \rho_{\text{Soiling}} \cdot P_{\text{ideal}} \tag{4}$$

$$\rho_{\text{Shading}} = \rho - \rho_{\text{Soiling}} \tag{5}$$

The shading $P_{Shading}$ losses are calculated through Equation 6.

$$P_{\text{Shading}} = \rho_{\text{Shading}} \cdot P_{\text{ideal}} \qquad (6)$$

3. Results

In the following section the results are presented. In the result part, an explaination is given how shading data are recorded and evaluated and how the entire approach is validated. Furthermore, the limitations of the method are discussed.

3.1. Separation of losses

Data has been recorded over two months. This allows the manipulation of power due to shading to be recreated and observed in real terms using the inverter data. Using the strongly shaded array as an example, the power that falls on the shading can be visualized well in Figure 4.

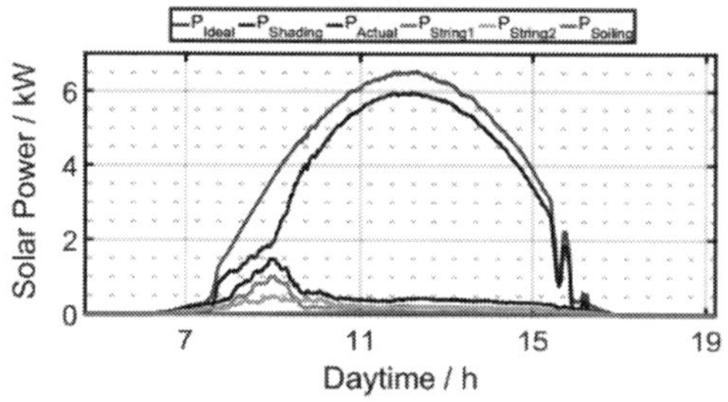

Figure 4 The calculated portion of the power due to shading is shown in black. The theoretical maximum power is reduced by this amount.

3.2. Validation

With the help of the losses from the previous day and from the shading ratio $\rho_{Shading}$ a correction can now take place by multiplication with the forecast value. It is also conceivable to subtract the power loss from the forecast value. But the correction via the shading ratio makes more sense here, since the losses proportional to the occurring shading always depend on the actual irradiation and thus the power.

It makes sense to multiply the forecast values by $1 - \rho_{Shading}$ because the shading remains constant over the days if it comes from a stationary obstacle. At the same time, the forecast model that was trained with the historical data provide forecasts in the shading period. In Figure 1 the scheme of the correction is shown as well as the change of the *RMSE* over the hours of the day when the model is applied. However, since the correction model detects the shading power at these times, it can make the correction and adjust the power values downward. So, you can also see in Figure 5 the change of the *RMSE* by the correction model. In principle, other forms of correction are also possible with the method shown. Figure 5 also shows that a correction using the shading power of the previous day ("lag correction") can already result in a significant improvement of the forecast error. A perfect correction would be if the power losses of the previous day are exactly equal to those of the current day. This illustrates the maximum potential of the presented method. In the period from February to March there was a 40% improvement in the RMSE, while in the April to May the improvement was 12.5%. The difference is in the intensity of the shading. Due to the low sun elevation in spring, there is much more shading of the array. Thus, the uncorrected LSTM model is more inaccurate than with a smaller amount of shading.

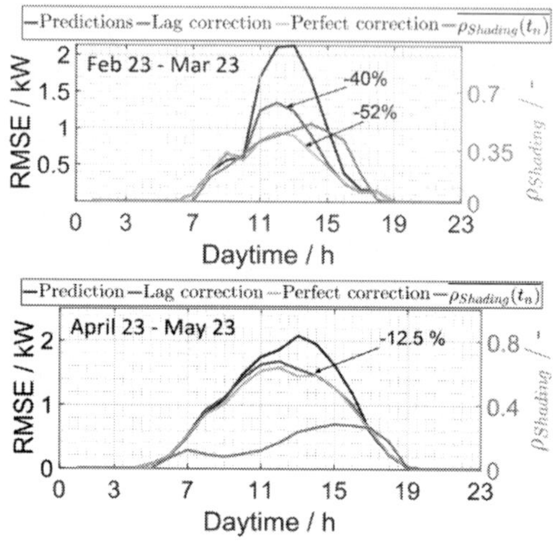

Figure 5 Decrease in RMSE due to the correction process

4. Discussion

Although the correction method in the previous validation has consistently contributed to a noticeable improvement in the prediction error under shading, the method also has limitations. It is in the nature of forecast models to over- or underestimate the true value. If the prediction values are now reduced, although the true value is underestimated, an improvement in the prediction error is not guaranteed.

In general, the power loss due to shading must be greater than the bias of the prediction model to improve the forecast error. The validation dataset showed that the true measured power value is underestimated, especially in the early morning hours and in general during the summer months. A correction in the case of very weak shading does not always lead to an improvement in the RMSE due to the negative Bias.

5. Conclusion

In this work, a forecast model was trained and a PV model was parameterized using endogenous and exogenous data. The PV model uses a PO algorithm to guarantee the MPP. Based on the power values of the PV model and the actual measured data, it was possible to calculate how much power is lost due to shading. This could finally be used to subsequently correct the prediction value of the trained prediction model and thus achieve an improvement in the prediction error under shading.

10.4229/EUPVSEC2023/4AO.9.6
020285-004

In summary the subsequent correction and therefore postprocessing of day-ahead PV power forecasts can help improve forecast error. The presented procedure was able to contribute to an improvement of the forecast error for a shaded array based on two months of recorded shading data. Particularly large shadings can be recognized and corrected in the forecast model. An improvement of the RMSE by up to 40 % could be achieved depending on the extend of shading.

As an outlook, the aproach should be extended. It would make sense to extend the forecast correction by including soiling. In principle, the following methodology could also be extended to other inverter faults. The error would then be detected at time *n* and the forecast would be corrected at time $n + 1$ day depending on the forecast horizon. In addition, the limitations of the methods were worked out. At the same time, the presented method is interesting for solar systems in general, since the quantified shading power can determine whether solar power optimizers can sensibly retrofit their performance. This application would lead to a reduction in the levelized cost of electricity (LCOE).

Acknowledgement

This work contributes to the research performed at KIT Battery Technology Center. The results were generated within the "Solarpark 2.0" project (funding code 03EE1135A) funded by the Federal Ministry for Economic Affairs and Climate Action (BMWK). The authors thank the project management organization Jülich (PTJ) and the BMWK.

Authors contribution

The development and evaluation of the correction model, the construction of the shading setup, and the writing of the manuscript were performed by Tim Kappler. Anna Sina Starosta contributed significantly to the development of the forecast procedures. Nina Munzke, Bernhard Schwarz, Anna Sina Starosta, and Marc Hiller also contributed significantly to the development with suggestions and intellectual input.

References

1. *Solar pv*, Renewable electricity – Renewables 2022 – Analysis - IEA: 2023-09-21
2. A.S. Saidi, *Impact of grid-tied photovoltaic systems on voltage stability of tunisian distribution networks using dynamic reactive power control* (2022), Vol. 13, p. 101537
3. H. Ye, B. Yang, Y. Han, N. Chen, *State-Of-The-Art Solar Energy Forecasting Approaches: Critical Potentials and Challenges* (2022), Vol. 10
4. J. Zeng, W. Qiao, *Short-term solar power prediction using an RBF neural network*, in *2011 IEEE Power and Energy Society General Meeting* (2011), pp. 1–8
5. A. Fentis, L. Bahatti, M. Mestari, B. Chouri, *Short-term solar power forecasting using Support Vector Regression and feed-forward NN*, in *2017 15th IEEE International New Circuits and Systems Conference (NEWCAS)* (2017), pp. 405–408
6. P. Bacher, H. Madsen, H. Nielsen, *Online Short-term Solar Power Forecasting* (2011), Vol. 83
7. F. Almonacid, P. P´erez-Higueras, E.F. Ferna´ndez, L. Hontoria, *A methodology based on dynamic artificial neural network for short-term forecasting of the power output of a PV generator* (2014), Vol. 85, pp. 389–398
8. A. Starosta, K. Kaushik, P. Jhaveri, N. Munzke, M. Hiller,*A Comparative Analysis of Forecasting Methods for Photovoltaic Power and Energy Generation with and without Exogenous Inputs*, in *38th European Photovoltaic Solar Energy Conference and Exhibition : proceedings of the International Conference : 06 September - 10 September 2021. Ed.: J.M. Almeida Serra* (WIP-Renewable Energies (WIP), 2021), pp. 938 – 945, ISBN 3-936338-78-7, 37.12.02;
LK 01
9. M.R. Maghami, H. Hizam, C. Gomes, M.A. Radzi, M.I.Rezadad, S. Hajighorbani, *Power loss due to soiling on solar panel: A review* (2016), Vol. 59, pp. 1307–1316
10. G.G. Kim, W. Lee, B.G. Bhang, J.H. Choi, H.K. Ahn,
Fault Detection for Photovoltaic Systems Using Multivariate Analysis With Electrical and Environmental Variables (2021), Vol. 11, pp. 202–212
11. A. Alcan˜iz, D. Grzebyk, H. Ziar, O. Isabella, Energy Reports **9**, 447 (2023)
12. Y. Chaibi, M. Malvoni, A. Chouder, M. Boussetta,M. Salhi, Energy Conversion and Management **196**, 330 (2019)
13. A. Dolara, G.C. Lazaroiu, S. Leva, G. Manzolini, Energy**55**, 466 (2013)
14. R. Ahmad, A.F. Murtaza, H. Ahmed Sher, U. TabrezShami, S. Olalekan, Renewable and Sustainable Energy Reviews **74**, 721 (2017)
15. A. Babatunde, S. Abbasoglu, M. Senol, Renewable andSustainable Energy Reviews **90**, 1017 (2018)
16. S. Pareek, R. Dahiya, Energy **95**, 561 (2016)
17. M.D.L.F.L.A.P.R.L.L.G.M.F.F.R.P.M.S.R. Lucaferri, V.; Radicioni, *An indirect approach to forecast produced power on photovoltaic plants under uneven shading condition* (2022), https://iris.uniroma1.it/handle/11573/1675678
18. C.C. Aggarwal, *Neural Networks and Deep Learning* (Springer Cham, 2018)
19. M. Rana, A. Rahman, *Multiple steps ahead solar photovoltaic power forecasting based on univariate machine learning models and data re-sampling* (2020), Vol. 21, p. 100286
20. H. Malki, N. Karayiannis, M. Balasubramanian, *Shortterm electric power load forecasting using feedforward neural networks* (2004), Vol. 21, pp. 157 – 167

21. A. Yona, T. Senjyu, T. Funabashi, *Application of Recurrent Neural Network to Short-Term-Ahead Generating Power Forecasting for Photovoltaic System*, in *2007 IEEE Power Engineering Society General Meeting* (2007), pp. 1–6
22. S. Srivastava, S. Lessmann, *A comparative study of LSTM neural networks in forecasting day-ahead global horizontal irradiance with satellite data* (2018), Vol. 162, pp. 232–247
23. M. Gao, J. Li, F. Hong, D. Long, *Day-ahead power forecasting in a large-scale photovoltaic plant based on weather classification using LSTM* (2019), Vol. 187, p. 115838
24. F. Harrou, F. Kadri, Y. Sun, *Forecasting of Photovoltaic Solar Power Production Using LSTM Approach* (2020), ISBN 978-1-83880-091-8
25. C.H. Liu, J.C. Gu, M.T. Yang, *A Simplified LSTM Neural Networks for One Day-Ahead Solar Power Forecasting* (2021), Vol. 9, pp. 17174–17195
26. M. Awad, R. Khanna, *Support Vector Regression* (Apress, Berkeley, CA, 2015), pp. 67–80, ISBN 978-1-4302-5990-9, https://doi.org/10.1007/978-1-4302-5990-9$_4$
27. A. Anghel, N. Papandreou, T. Parnell, A. Palma, H. Pozidis, *Benchmarking and optimization of gradient boosted decision tree algorithms* (2018)
28. V.K. Ayyadevara, *Gradient Boosting Machine* (Apress, Berkeley, CA, 2018), pp. 117–134, ISBN 978-1-4842-3564-5, https://doi.org/10.1007/978-1-4842-3564-5$_6$
29. DWD, https://opendata.dwd.de/climate$_e$*nvironment/CDC/*
30. DWD, https://opendata.dwd.de/weather/local$_f$*orecasts/*
31. W. Holmgren, C. Hansen, M. Mikofski, *pvlib python: a python package for modeling solar energy systems* (2018), Vol. 3, p. 884
32. H. Chen, X. Chang, *Photovoltaic power prediction of LSTM model based on Pearson feature selection* (2021), Vol. 7, pp. 1047–1054, 2021 International Conference on Energy Engineering and Power Systems
33. M.F.N. Tanvir Ahmad, Sharmin Sobhan, *Comparative Analysis between Single Diode and Double Diode Model of PV Cell: Concentrate Different Parameters Effect on Its Efficiency* (2016)
34. M.H. Qais, H.M. Hasanien, S. Alghuwainem, K. Loo,M. Elgendy, R.A. Turky, *Accurate Three-Diode model estimation of Photovoltaic modules using a novel circle search algorithm* (2022), Vol. 13, p. 101824
35. E. Batzelis, G. Anagnostou, C. Chakraborty, B. Pal, *Computation of the Lambert W function in photovoltaic modeling* (2019)
36. S. Ghosh, J. Roy, C. Chakraborty, *A model to determine soiling, shading and thermal losses from PV yield data* (2022)
37. T. Selmi, M. Abdul-Niby, L. Devis, A. Davis, *PO MPPT implementation using MATLAB/Simulink*, in *2014 Ninth International Conference on Ecological Vehicles and Renewable Energies (EVER)* (2014), pp. 1–4

10.4229/EUPVSEC2023/4AO.9.6
020285-006

VALIDATION OF THREE MODULE TEMPERATURE MODELS (FAIMAN, SANDIA AND ROSS) USING MEASUREMENT DATA FROM MULTIPLE OUTDOOR RESEARCH FACILITIES

Authors: Taras Sheremet[1]*, FH-Prof. Ing. Momir Tabakovic PhD. MSc[1], Dr. ir. R.M.E. (Roland) Valckenborg[2]
Company / Institutes:
[1] University of Applied Sciences Technikum Wien (Fachhochschule Technikum Wien, FHTW), Höchstädtplatz 6, 1200, Vienna, Austria,
[2] TNO – Energy Transition, High Tech Campus 21, 5656 AE Eindhoven, The Netherlands
*Corresponding author: ee20m045@technikum-wien.at; Tel: +49 1717840684

ABSTRACT: In this work three steady-state photovoltaic (PV) module temperature models were validated in the context of building-integrated photovoltaic (BIPV) Modules. The first one is Faiman model, which has ambient temperature, irradiance, and wind speed as input and has a wide variety of applications including international standards, the second one is Sandia National Laboratories (Sandia) model, which has the same input as Faiman model, but uses another function type for the prediction of temperatures, the third one is Ross model which is more simple and has only ambient temperature and irradiance as input. The ability of all the named thermal models to accurately predict the temperatures and energy generation of the BIPV modules was assessed.
The models were validated using experimental data measured on mock-ups that imitate the installation of a BIPV module as a part of a ventilated building facade. These test sites were located mostly in Europe and one of them was located in South Korea. On each test site, the necessary for thermal models measurements as well as modules' temperatures and power output were monitored during different periods. For each site, theoretical temperatures and theoretical generated energies were calculated and compared with the measured temperatures and energies.
In general, all the mentioned thermal models showed high to acceptable accuracy in predicting temperatures of the modules and neither model showed significantly higher accuracy than the other two.
Keywords: BIPV, Temperature, Thermal Model Validation, Façade, PV Module

1 INTRODUCTION

1.1 Goal and Approach

For any type of PV module, including BIPV modules, temperature is the second (after irradiance) most important factor influencing module performance. Steady-state temperature models are a low resource and time-consuming method to quickly estimate the module temperature based on irradiance and environmental conditions. These models have shown a high level of accuracy in assessing the temperature of PV modules that are optimally tilted and ventilated. BIPV modules have different mounting and ventilation conditions, so it is unclear whether steady-state temperature models can be used to assess the temperature of BIPV modules.

The current research aims to compare the temperatures, modelled by the several steady-state thermal models with the measured temperatures of the BIPV modules installed on the mock-up test stands.

The IEA-PVPS Task 15 focuses on international networking and joint research in the area of BIPV and is subdivided into 5 sub-working groups. Within Task 15, a group of over 60 experts from 16 different countries collaborated to create an Enabling Framework for the Acceleration of BIPV [1]. As part of this collaboration, a joint action was initiated to compare available outdoor tests facilities in a "Round Robin" test. A goal of Task 15 is to create insight in the thermal and energy performance of BIPV elements mounted as ventilated façade elements in different climatic conditions [2].

Data gathered during the "Round Robin" test is used in this work the following way: measured climate conditions are used as input of the analyzed thermal models, then measured module temperatures and power output are compared with the modelling results.

1.2 Contributing Sites

At the beginning of the activity, the total number of contributing test sites was 9 [3]. Due to reasons such as storm damaging or absence of one or more strictly necessary measurements, data from 5 sites is analyzed in this work. The original idea was to put the test sites in evenly distributed all over the world locations [3], but finally (in context of data suitable for temperature analysis in this work) 4 sites are located in Europe and 1 site is located South Korea.

The locations of the test sites and the description of the climate zones in which test sites are located are presented in Figure 1 and Table 1.

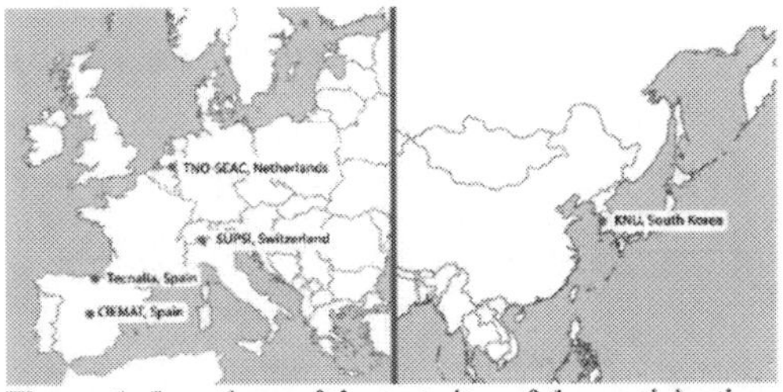

Figure 1: Locations of the test sites of the participating institutions on the map (map images source: worldmapblank.com).

Table 1: Participating institutions.

Participant	Country	Latitude /Longitude	Module Number
CIEMAT	Spain (ESP)	40.45° / -3.73°	M17-04276
KNU	S-Korea (KOR)	36.81° / 127.11°	M17-04275
SUPSI	Switzerland (CHE)	46.027° / 8.916°	M17-02281
Tecnalia	Spain (ESP)	43.293° / -2.862°	M17-02286
TNO-SEAC	Netherlands (NLD)	51.446° / 5.485°	M17-02288

10.4229/EUPVSEC2023/4BO.5.2
020295-001

1.3 BIPV-module and mock-up construction

The characteristics of the BIPV-module and mock-up installation have been extensively described in preceding studies [2] [3] [4]. Only the module characteristics will be shortly recalled now.

Module construction parameters:

- Size: 1.100 mm × 1.600 mm × 11.5 mm
- Type: glass/glass BIPV module of the type VSG-L (Ertex-solar)
- Configuration: 3 strings of 18 (9 x 2) poly-c Si-cells (type Poly TT; 156 × 156mm) cells, each string with by-pass diode.
- Distance between the cells: 5 mm
- Transparency: 25.33%

Module electrical parameters:

- Nominal rated power (Pmp, STC): 227.3 Wp (+/- 5%); Temperature coefficient: -0.37 %/K
- Short circuit current (ISC): 8.34 A; 0.05 %/K
- Open circuit voltage (VOC): 34.07 V; -0.32 %/K

The constructive boundaries of the mock-up installations at all the sites were agreed by all participating institutions and designed to imitate the installation of the BIPV module on a ventilated cold façade and must be identical. This makes the comparison of the energy performance of the modules as accurate as possible [3]. The detailed construction of the mock-ups is presented in [4].

1.4 Data Monitoring Method

The common data monitoring methodology was also defined by all participants. On each test site minimum monitoring requirement was to record data from morning till evening when plane of array irradiance is higher or equal to 20 W/m^2 [4].

On each site, following parameters were measured and monitored:

- I_{mp} (A) – current at maximum power point,
- V_{mp} (V) – voltage at maximum power point,
- T_{mod} (°C) - module temperature,
- Gi (W/m^2) – plane of array irradiance,
- T_{amb} (°C) – ambient temperature at test location,
- WS (m/s) - wind speed at test location.

The list of deployed sensors as well as full description of the monitoring method can be found in [4].

2 METHODOLOGY

The experimental data was prepared using different data analysis software. Then, the prepared measurement datasets were used to calculate the theoretical module temperatures and theoretical generated energies using thermal models. Finally, the theoretical temperatures and energies were compared with the measured ones.

The validation methodology is based on the comparison of the modelled BIPV module temperatures with the measured ones and on the comparison of the calculated theoretical generated by the module energy with the measured during the experiment energy.

Climate data, temperatures, and power output of the BIPV mock-ups are measured during the experiment. Theoretical temperatures can then be modelled by the thermal models using the measured climate data and compared with the measured temperatures of the BIPV modules from the mock-ups. In the next step, the theoretical energy produced by the BIPV module is calculated from the measured climate data and modelled module temperatures using a method presented in [5] and compared with the measured during the experiment energy (see Figure 2).

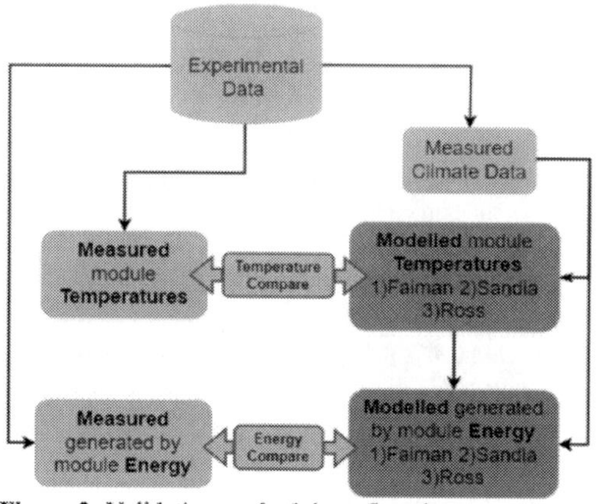

Figure 2: Validation methodology flowchart

3 ANALYZED THERMAL MODELS

In this work, Faiman, Sandia National Laboratories (SNL) and Ross temperature calculation models are analyzed. The Faiman model has ambient temperature, plane-of-array irradiance and wind speed as input variables and is currently used by the i.e., Photovoltaic Geographical Information System [6] and in the IEC 61853 standard [7]. The SNL model has the same three input variables as the Faiman model. This model is widely used by the industry [8], and is one of the models used by NREL's System Advisor Model [9]. The Ross model is a simpler model, that has two input variables: ambient temperature and plane-of-array irradiance. This model is used by i.e. Homers' PV software [10].

The Faiman and SNL thermal models consider the influence of wind speed on the temperature of the PV module. These two models estimate the PV module's temperature using the following climate data as input:

- Tamb - ambient temperature (°C),
- Gi - plane of array irradiance (W/m2),
- WS - wind speed (m/s).

The Faiman Model [11] allows to assess the module temperature using simple heat transfer concepts. The model formula:

$$\mathrm{Tmod} = \mathrm{Tamb} + \frac{\mathrm{Gi}}{\mathrm{U0} + \mathrm{U1} \cdot \mathrm{WS}}$$

where:

- T_{mod} - module temperature (°C),
- T_{amb} (°C), Gi (W/m^2), WS(m/s) – climate measurements, as described before,

10.4229/EUPVSEC2023/4BO.5.2

- U0 – constant heat transfer component ($\frac{W}{m^2 \cdot K}$)
- U1 – convective heat transfer component ($\frac{W}{m^3 \cdot s \cdot K}$)

The SNL model presented by David King [12] also allows to assess the module temperature having meteorological data as an input:

$$Tmod = Tamb + Gi \cdot \left(e^{a+b \cdot WS}\right)$$

where:

- T_{mod} - module temperature (°C),
- T_{amb} (°C), Gi (W/m^2), WS(m/s) – climate measurements, as described before,
- a, b – parameters, that are dependent from module materials, construction, and mounting configuration.

In the Ross Model [13] a linear proportionality of the difference between PV module and ambient temperatures ($T_{mod} - T_{amb}$) and plane of array irradiance (Gi) is assumed. In this model, irradiance and ambient temperature are input variables:

$$Tmod = Tamb + k \cdot Gi$$

where:

- T_{mod} - module temperature (°C),
- T_{amb} (°C), Gi (W/m^2) – climate measurements, as described before,
- k – parameter, dependent from module's materials, construction, and mounting configuration.

The accuracy of the described models is limited when used for the prediction of temperatures for short periods (comparable to days). But it is hypothesized that accuracy of the models is high enough when applied to long periods of at least 6 months. In this work periods of different lengths are used to validate the accuracy of the models.

Generated energy is the most important characteristic of a BIPV system, that is why validation of the temperature models is performed not only by assessing the temperature prediction accuracy, but also by comparing the theoretical energy, calculated using modelled temperatures with the measured energy output of the BIPV modules on the test sites.

4 DATA HANDLING

Firstly, the data is extracted and prepared for analysis. After that, the parameters of the calculation methods are extracted from the datasets with the help of regression analysis. Then, modelled module temperature timelines are generated using different thermal models. These temperatures are compared with the measured during the experiment module temperatures. As the next aspect of validation, theoretical energies generated by the modules are calculated using temperatures, provided by different thermal models and climate data, and then compared with the measured energies.

On the step of *basic primary filtering* the following data was rejected:

- Failed Gi measurements (points, where P_{mp} is significantly higher than 0 W and Gi = 0 W/m^2)
- Gi below 5 W/m^2,
- WS speed below 0.25 m/s.

For the energy comparison calculation also following data points were rejected:

- Failed Pmp measurements (points, where Pmp = 0 W and Gi is significantly higher than 0 W/m2).

The threshold of 5 W/m^2 was chosen in order to reduce measurement noise during sunrise and sunset. This threshold level should be chosen with care i.e. must not be too high, because the energy, produced at Gi < 100 W/m^2 makes a significant contribution to the annual energy production [3]. Datapoints with measured wind speed lower than 0.25 m/s were rejected in accordance with measurement procedure, described in [7]. Points that did not have all the data necessary for the analysis of the temperature models were also filtered during the *basic primary filtering*.

Additionally to *basic primary filtering,* in accordance to [7] *standard primary filtering* (2) performed.

For the Nominal Operating Cell Temperature (NOCT) case, during *standard primary filtering* datapoints measured under the following conditions were rejected:

- Irradiance below 400 W/m^2,
- In a 10 min interval after the irradiance varied by more than 10% from the maximum value to the minimum value recorded during that 10 min period,
- Wind speed outside the range 1 m/s ± 0.75 m/s,
- Ambient temperature outside the range 20 C ± 15 C.

For the Nominal Module Operation Temperature (NMOT) case, during *standard primary filtering* data taken under the following conditions were rejected:

- Irradiance below 400 W/m^2,
- In a 10 min interval after the irradiance varied by more than 10% from the maximum value to the minimum value during the preceding 10 min period,
- In a 10 min interval from a deviation of the instantaneous wind speed to below 0.25 m/s or gusts larger than + 200% from a 5 min running average,
- When the 5 min running average was less than 1 m/s or greater than 8 m/s.

The *secondary filtering* step deals with all the measurement anomalies, biases and all types of failed measurements that were not filtered during *basic primary filtering* or *standard primary filtering.*

The accuracy of the models, built using the following combinations filtering procedures:

1. *basic primary filtering + secondary filtering*
2. *standard primary filtering + secondary filtering*

where compared and no significant difference in accuracy was observed.

5 RESULTS AND DISCUSSION

2.1 Modeled Temperatures Evaluation

The parameters of the thermal models were fitted for a 3-month measurement period (see Table 2) and then used to predict temperatures for the next 6 months.

Table 2: MAE, RMSE and maximum temperature differences, recorded by two sensors on one module.

	Faiman		Sandia		Ross
	U0	U1	a	b	k
CIEMAT	25.7	1.24	-3.25	-0.045	0.036
KNU	21.4	1.05	-3.07	-0.045	0.044
SUPSI	21	0.63	-3.04	-0.028	0.046
Tecnalia	23.1	1.41	-3.14	-0.056	0.038
TNO-SAEC	25.8	1.67	-3.26	-0.057	0.034

Such lengths of the parameters extraction (fitting) period and of the prediction period were chosen because some of the datasets are only 9 months long. The results are presented in Figures 3,4,5, and Table 4:

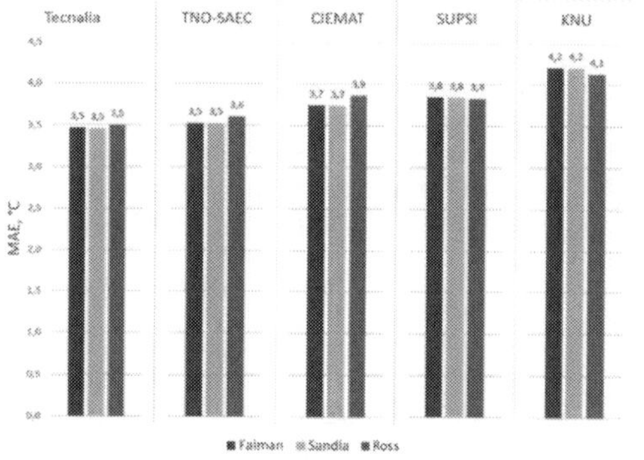

Figure 3: Mean Absolute Error (MAE), °C

MAE has values in between: 3.47°C and 4.21°C.

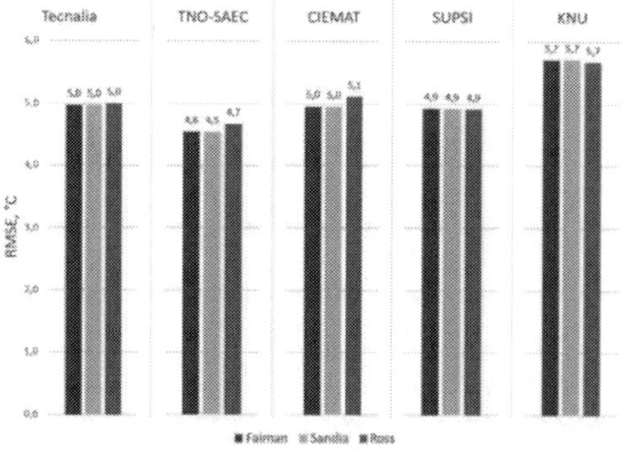

Figure 4: Root Mean Square Error (RMSE), °C

RMSE has values in between: 4.55°C and 5.71°C.

These MAE and RMSE values are comparable to the temperature differences measured by two sensors on a tested BIPV module. Measurements from two sensors were available at the Tecnalia, KNU and TNO-SAEC test sites. The MAE and RMSE as well as maximum temperature differences recorded at the mentioned test sites are presented in Table 3.

Table 3: MAE, RMSE and maximum temperature differences, recorded by two sensors on one module.

	Tecnalia	KNU	TNO-SAEC
MAE, °C	1.179	1.453	0.827
RMSE, °C	1.98	1.875	1.627
max ΔT, °C	9.988	6.051	22

The values in Table 2 match the average difference of temperatures between two cells in one module described in the literature, which is ± 2°C [5].

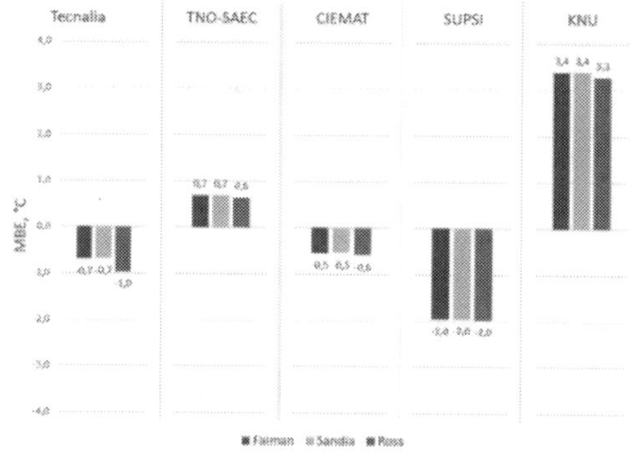

Figure 5: Mean Bias Error (MBE), °C

Table 4: Coefficient of determination of linear regression (R^2)

	R^2		
	Faiman	Sandia	Ross
CIEMAT	0.86	0.86	0.85
KNU	0.87	0.87	0.87
SAEC	0.91	0.91	0.90
SUPSI	0.88	0.88	0.88
Tecnalia	0.87	0.87	0.87

These figures and tables allow us to compare the accuracy of the thermal models for different test sites and data sets and to compare the difference in accuracy between the models within each data set.

A high R^2 between 0.85 and 0.91 (see Table 4) as well as MAE and RMSE values calculated in this work, indicate a high to acceptable prediction accuracy.

As can be seen from the diagrams on Figures 3 and 4 neither thermal model demonstrates a definitely higher accuracy than the other.

MBE calculated for KNU have the highest absolute values. MAE and RMSE values calculated for this dataset are also the highest. This indicates a bias in the measurement of either the climate data or the module's temperatures.

2.2 Calculated Energies Evaluation

The energy prediction accuracy assessment method

10.4229/EUPVSEC2023/4BO.5.2
020295-004

used in this work was presented and described in [5]. This method allows to calculate a theoretical generated by the PV module energy using the temperatures, modeled by the thermal models. Theoretical energies can then be compared with the measured ones:

$$|\varepsilon_E|(\%) = \frac{\left|E_{PV}{}^{exp} - E_{PV}{}^{T_{mod}{}^{th}}\right|}{E_{PV}{}^{exp}}$$

where:

- $E_{PV}{}^{T_{mod}{}^{th}}$ (W·h) – theoretical energy, calculated using on the measured module temperatures,
- $E_{PV}{}^{exp}$ (W·h) – measured experimental energy.

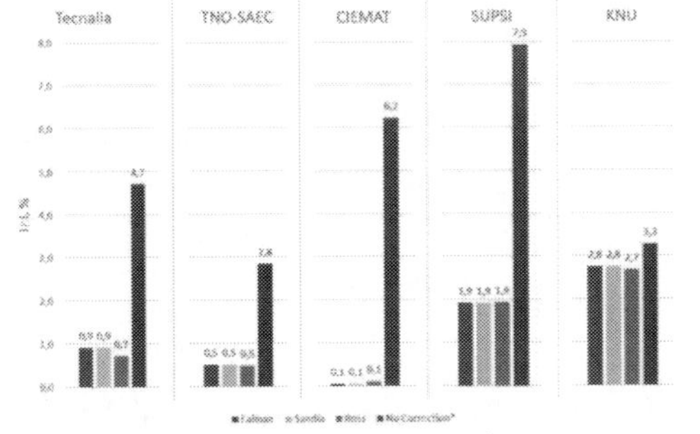

Figure 5: Absolute percentual difference between measured and modeled energies, %

On Figure 5, absolute percentual difference between measured and calculated energies, including energy, calculated without temperature correction, i.e., energy, calculated using the same method, but assuming the temperature of the BIPV module to be a constant 25 °C (red bar on Figure 5).

If the temperature correction is used, the absolute percentual difference between measured and modeled energies varies from 0.05% to 2.8 %, without the temperature correction, energy prediction accuracy varies from 2.8% to 7.9%.

In all cases, except for the dataset, in which measurement bias was detected, temperature correction using thermal models significantly improves the energy prediction accuracy.

2.3 Optimal Parameters Fitting Period

Based on the data from the four datasets, in which measurement bias has not been detected it has been noticed, that two months of measurements are enough to develop temperature models which have the same accuracy level as the models, which are built on the parameters, extracted from the full measurement period.

In Figure 6, MAE values received using different thermal models and different parameters fitting periods for the CIEMAT dataset are displayed. It can be observed that expanding the parameters extraction period beyond 2 months does not improve the temperature prediction accuracy. Similar results were obtained for the other datasets.

It has also been noticed that the accuracy of the models, built on the parameters extracted from short periods and different seasons is very close to the models, which used the parameters, extracted from the full measurement period. Models, built on the parameters, fitted for the period, which consisted of the 5-day periods, taken from three different seasons, two of which were summer and winter (and the third one was spring or autumn) demonstrated a very close level of accuracy to the models, built using the parameters extracted from the full measurement period (see Figure 7).

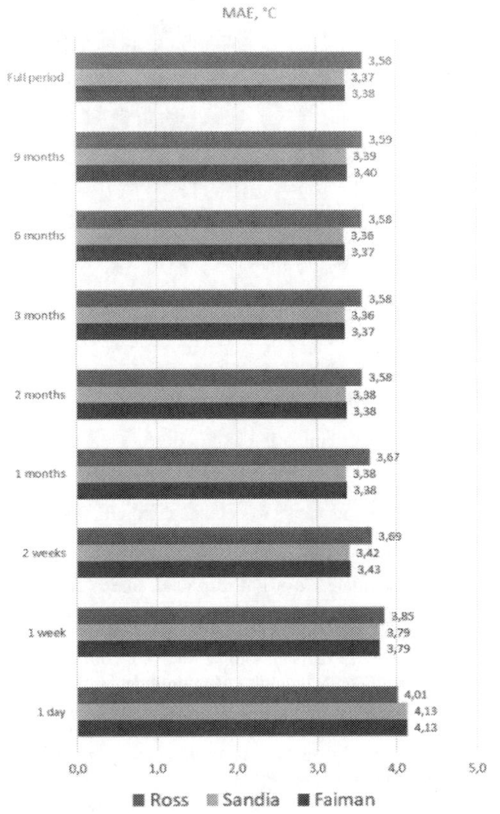

Figure 6: MAE as a function of the parameters fitting period, part 1. CIEMAT dataset.

10.4229/EUPVSEC2023/4BO.5.2
020295-005

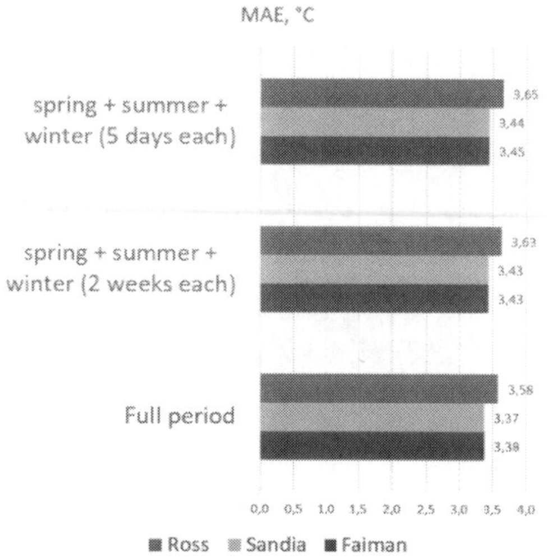

Figure 7: MAE as a function of the parameters fitting period, part 2. CIEMAT dataset.

6 CONCLUSION AND OUTLOOK

In this contribution the ability of the Faiman, Sandia, and Ross thermal models to predict the temperatures of the ventilated façade integrated BIPV modules was studied. The used method allows to validate the thermal models in two different ways: (i) comparing theoretical and measured temperatures and (ii) comparing theoretical energy, calculated using theoretical temperatures and measured climate data with measured energy.

When used to predict temperatures, all the analyzed models showed accuracy comparable with common cell to cell temperature difference in the module. Neither model showed a significantly higher level of accuracy. This makes all the models suitable for the purposes of thermal simulations and assessment of temperatures for building systems regulations purposes.

Ambient temperatures, used for the fitting of the models in this work were the temperatures of air near the mock-up installation. As it was described in [5], measuring the ambient air temperature inside the air gap behind the module and using this temperature for the fitting of the parameters can improve the accuracy of the solid-state temperature models. But even based on ambient temperatures near the mock-up installation, models showed a high level of accuracy.

During the analysis of the models' ability to predict the energy generated by the module, all the models showed high level of accuracy in predicting energy generation by the BIPV module. Temperature correction was proved to significantly improve the energy prediction accuracy: from 3 - 8% down to 0 - 2% (using data in which o measurement bias was detected).

Based on the measurements from the 4 analyzed test facilities it can be concluded that 2 months of measurements before the prediction period or 3 short measurement periods during different seasons before the prediction period are enough for the extraction of the parameters of the thermal models, which will be accurate enough for the thermal simulations as well as for the prediction of energy generation of the BIPV system.

The simplest Ross model, which has irradiance and ambient temperature as input didn't demonstrate lower accuracy, than Faiman and Sandia models, which additionally have wind speed as input. It can be concluded that wind speed measurement is not necessary for the accurate enough forecasting and now-casting of the BIPV modules temperature. However, as long as all the mock-ups were located on low altitudes above the ground and relatively far from open sea, wind speed can have a higher impact on the BIPV module's temperature in the higher altitudes or coastal areas, where wind is usually stronger. Therefore, it would be useful to analyze more measurements of the BIPV modules parameters in more windy locations.

7 REFERENCES

[1] IEA-PVPS, *Enabling Framework for the Development of BIPV - IEA-PVPS.* [Online]. Available: https://iea-pvps.org/research-tasks/enabling-framework-for-the-development-of-bipv/ (accessed: Mar. 27 2023).

[2] L. Gaisberger *et al.*, "BIPV Round Robin Action of IEA PVPS Task 15," *Proceedings of 36th European Photovoltaic Solar Energy Conference and Exhibition*, pp. 1811–1816, 2019, doi: 10.4229/EUPVSEC20192019-6BV.4.7.

[3] R. Valckenborg *et al.*, "IEC61853-Matrix Analysis of PVPS Task 15 BIPV Round-Robin for More Than One Year at Seven Test Sites over the World. 2020," in *35th European Photovoltaic Solar Energy Conference and Exhibition: Proceedings of the international conference held in Brussels, Belgium, 24 September - 28 September 2018*, Brussels, Belgium, op. 2018, pp. 1719–1725. [Online]. Available: https://repository.supsi.ch/12138/

[4] *Comparative Performance Measurements of Identical BIPV-Elements in Different Climatic Environments-A Round Robin Action within the IEA PVPS Task ...* [Online]. Available: https://opus.technikum-wien.at/frontdoor/index/index/docid/2611

[5] N. Martín-Chivelet, J. Polo, C. Sanz-Saiz, L. T. Núñez Benítez, M. Alonso-Abella, and J. Cuenca, "Assessment of PV Module Temperature Models for Building-Integrated Photovoltaics (BIPV)," *Sustainability*, vol. 14, no. 3, p. 1500, 2022, doi: 10.3390/su14031500.

[6] European Commission, *PVGIS data sources & calculation methods.* [Online]. Available: https://joint-research-centre.ec.europa.eu/pvgis-online-tool/getting-started-pvgis/pvgis-data-sources-calculation-methods_en#references (accessed: Jan. 29 2023).

[7] *Part 2: Spectral responsivity, incidence angle and module operating temperature measurements*, IEC 61853-2:2016, Photovoltaic (PV) module performance testing and energy rating. [Online]. Available: https://webstore.iec.ch/publication/25811#additionalinfo

[8] J. Oh, A. Pavgi, and G. Tamizhmani, "Determination of Empirical Coefficients and ΔT for Sandia Thermal Model: Dependence on Backsheet Type," in *2018 IEEE 7th World*

10.4229/EUPVSEC2023/4BO.5.2
020295-006

Conference on Photovoltaic Energy Conversion (WCPEC) (A Joint Conference of 45th IEEE PVSC, 28th PVSEC & 34th EU PVSEC), 2018, pp. 442–446.

[9] *Temperature estimation of an unconventional PV array using the sandia module temperature model*, 2015. [Online]. Available: https://www.researchgate.net/profile/anna-bruce-3/publication/290226135_temperature_estimation_of_an_unconventional_pv_array_using_the_sandia_module_temperature_model/links/5695b96e08ae820ff074e24b/temperature-estimation-of-an-unconventional-pv-array-using-the-sandia-module-temperature-model.pdf

[10] T. A. Olukan and M. Emziane, "A Comparative Analysis of PV Module Temperature Models," *Energy Procedia*, vol. 62, pp. 694–703, 2014, doi: 10.1016/j.egypro.2014.12.433.

[11] D. Faiman, "Assessing the outdoor operating temperature of photovoltaic modules," *Prog. Photovolt: Res. Appl.*, vol. 16, no. 4, pp. 307–315, 2008, doi: 10.1002/pip.813.

[12] J. Kratochvil, W. Boyson, and D. King, *Photovoltaic array performance model*: Office of Scientific and Technical Information (OSTI), 2004.

[13] Ross, R. G., Jr., "Interface design considerations for terrestrial solar cell modules," *12th Photovoltaic Specialists Conference*, pp. 801–806, 1976.

ASSESSMENT OF THERMALLY INDUCED STRESS IN BIPV MODULES USING FEM SIMULATIONS

Andreas J. Beinert[1], Achour Mahfoudi[1], Frank Ensslen[1], Christof Erban[2], Pascal Romer[1]
[1]Fraunhofer Institute for Solar Energy Systems ISE, Heidenhofstraße 2, 79110 Freiburg, Germany
[2]SUNOVATION, Glanzstoffstraße 21, 63820 Elsenfeld, Germany
Corresponding author: Andreas J. Beinert | Phone: +49 (0)761 4588 5630 | E-mail: Andreas.Beinert@ise.fraunhofer.de

ABSTRACT: Environmental conditions like irradiation and temperature cause thermally induced stresses in building-integrated PV modules. This study investigates the influences of different design parameters and additionally environmental conditions on the thermally induced stresses within the glass of BIPV modules. A thermomechanical 3D FEM model of a glass-glass BIPV module in an example ventilated façade is used to perform parameter sensitivity studies. Among the investigated variations, partial shading has the strongest impact. All simulated variations result in maximum stresses above the design edge resistance of annealed glass. Therefore, the use of annealed glass is risky, while the design edge resistance of thermally toughened glass was not exceeded in any variation.
From the detailed 3D FEM model, a simplified 2D FEM model was derived for the purpose of a quick and easy estimation of the thermally induced stress.
The approach "Determination of thermomechanical resistance to partial shading" was proposed for discussion by the standardization committee responsible for revising EN 50583:2016.
Keywords: FEM simulation, Thermal stress, Digital prototyping,

1 INTRODUCTION

For the mechanical design of building-integrated PV modules (BIPV) in façades and roofs, the loads due to self-weight, wind and snow have been sufficiently well studied and are taken into account in engineering practice. However, these modules are also exposed to thermally induced loads from direct solar radiation. The standards and guidelines available to date, both at the German and the European level, contain only simplified information and specifications for calculating the thermally induced stresses. Within the scope of the research project with the acronym "Thermobruch" [1,2], building-integrated glass-glass PV modules, e.g. as components of a ventilated façade, are investigated by means of the finite element method (FEM) using long-term German meteorological data. The aim is to reduce or prevent the occurrence of thermally induced glass breakage by taking appropriate measures already in the planning phase by applying the developed simulation approaches.

2 METHOD

A 3D thermomechanical FEM model of a typical glass-glass BIPV module in a ventilated façade was created as an example, based on an FEM model of conventional PV modules [3–6]. Each glass sheet has a thickness of 4 mm and a silicone casting resin is used as the cell encapsulation. The PV module has four strings of eight solar cells, each with dimensions of 156.75×156.75×0.18 mm³, and a PV module area of 1.2×0.6 m². A square hollow profile frame is attached adhesively to the back surface of the PV module, near its perimeter. Please note that the frame does not enclose the edge of the glass, as is the case with conventional PV module frames. A thermal FEM simulation determines the inhomogeneous temperature distribution inside the BIPV module due to partial shading for different shading geometries and ambient conditions (outdoor temperature and irradiation). Further thermal effects, such as hotspot formation, are not taken into account. We consider thermal radiation and convection as well as absorption within the single layers. To simplify the calculation, it was assumed that the solar reflectance was zero. For the ambient conditions, real and simulated weather data over 11 years are used. Particularly critical data sets are considered, which were identified in advance in another work package within the research project [2]. The coupled mechanical FEM simulation determines the thermally induced mechanical stress. The temperature T and the first principal stress σ_1 (tensile stress) in the front glass cover are evaluated in each case. The values in the back glass pane are only slightly different to those in the front glass cover. As an example for assessment criteria, the simulated stress is compared with the typical design edge resistances of 13.1 MPa and 75 MPa, as determined by Schwind *et al.* for annealed and thermally toughened glass, respectively [7]. If the simulated stress at the edge exceeds this limit, the design is assessed to be at risk of breakage.

The thermal model was successfully validated on full-size modules in a climatic chamber with a solar simulator. For this purpose, fully and partially covered PV modules were equipped with thermocouples at five positions over the module area of one PV module, as indicated in Figure 1, whereby two thermocouples were mounted at each such position, one on the back surface of the solar cell and one on the back surface of the back glass pane. Two such PV modules were mounted above each other, so that a total of ten thermocouple positions were investigated. The FEM model was adapted to the exact conditions of the experimental setup. The PV modules are mounted using conventional PV module clamps in front of a wooden panel, which represents the front of the façade. The climate chamber is set to a constant temperature of +10 °C. The irradiation varies between 900 W/m² to 1000 W/m² with a mean value of 920 W/m², which is the value used in the FEM simulation. Combined heat transfer coefficients, which are adapted to the measured temperature and wind speed are used. Due to the temperature inhomogeneity of the climate chamber, different values are used, as documented in Table 1. For each PV module design, a measurement sequence covering four different scenarios is performed, preceded by a stabilizing sequence:

1. Unshaded at maximum power point (MPP)
2. Unshaded under open circuit conditions (OC)
3. Horizontally shaded (using a metal sheet) in OC

Vertically shaded (using a metal sheet) in OC

10.4229/EUPVSEC2023/4BO.5.3
020296-001

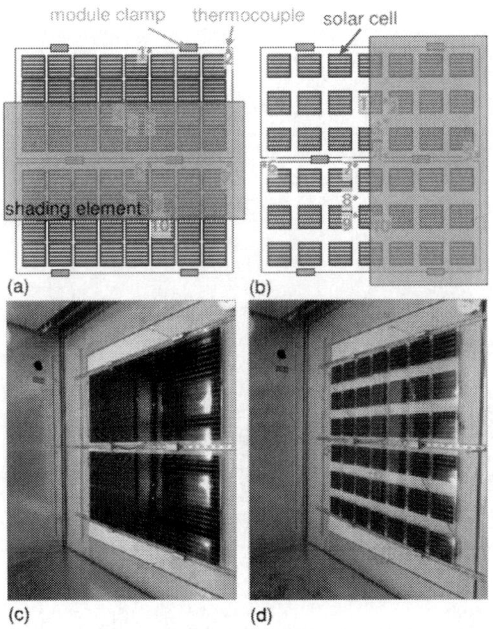

Figure 1: Schematic configuration of the PV modules used for experimental validation with (a) complete cell coverage and a horizontal shading element and (b) partial cell coverage and a vertical shading element. The orange numbers indicate the lateral positions of the thermocouples. For better visibility, the shading elements are illustrated semi-transparently. Corresponding experimental setup in the climate chamber with the PV modules with (c) complete and (d) partial cell coverage, without shading elements.

Table 1: Measured temperature and wind speed, as well as the corresponding calculated heat transfer coefficients calculated according to EN ISO 6946 for the upper and lower PV modules.

	Temperature [°C]	**Wind speed [m/s]**	**Heat transfer coefficients [W/(m²K)]**		
			Irradiation	Convection	Combined
	Upper PV module				
Front	11	4	4.7	20	24.7
Back	12.7	2	4.8	12	16.8
	Lower PV module				
Front	13.2	2	4.8	12	16.8
back	15.6	0.2	4.9	4.8	9.7

Applying the validated FEM model, a sensitivity study is performed to investigate the influence of the following parameters:

1. Environmental conditions (irradiance and temperature, depending on orientation and location).
2. Frame
3. Operating conditions (MPP and OC)
4. Solar cell coverage
5. Edge distance from the cells to the glass edge
6. Glass thickness
7. Shading configuration
8. Crystalline silicon compared to thin film technology

Unless otherwise stated, all variations were carried out for the temporal profile of irradiation and temperature for a representative cold day shown in Figure 2, since higher stresses occur here than on a warm day (see 3.1 Environmental conditions).

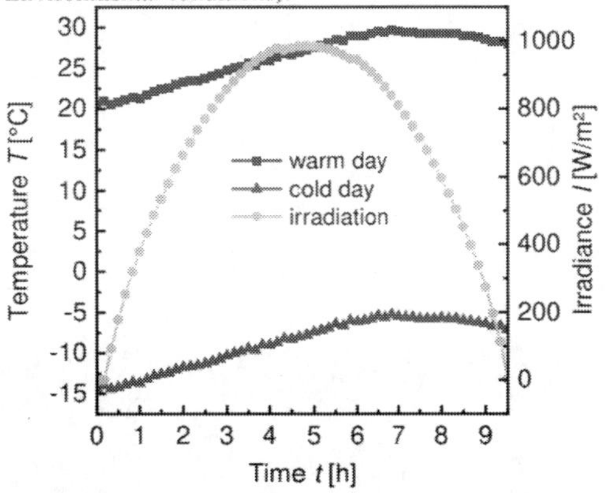

Figure 2: time course of temperature (left axis) of a representative warm (red squares) and cold day (blue triangles) as well as irradiance (yellow circles).

The results of the parameter sensitivity study were used to derive a simplified 2D FEM model, requiring less computational capacity. The 2D FEM model can be used for a standardized calculation procedure, for example in the planning phase of a BIPV façade. The approach "Determination of thermomechanical resistance to partial shading" was proposed for discussion by the standardization committee responsible for revising EN 50583:2016.

3 VALIDATION

Due to the page limitation, we present the results from only two representative sensor positions of the fully covered PV module in Figure 3. The full validation data can be found in the project report [2]. Except for the initial heating up, which was modelled in the FEM simulation with a pre-study, the simulation matches the experimental temperature curve well. Under maximum power point conditions and no shading, the deviation between FEM simulation and experiment is negligible. In open circuit without shading, the FEM simulation slightly overestimates the temperature of the experimental result by <1 K. In the case of shading and the OC state, the deviation becomes much larger, up to 5 K, with the FEM simulation resulting in lower temperatures. We suspect that the semi-shaded area around the metal sheet edges was not modelled correctly in the FEM simulation. Here a fading out of the shade was assumed. However, the unshaded temperature is simulated with good agreement to the measurements. Therefore, the thermal FEM model has been successfully validated.

4 PARAMETER VARIATIONS

4.1 Environmental conditions

In order to identify critical environmental conditions, a number of different conditions for different sky orientations were investigated in advance in a separate working package. As an example, Figure 4 shows the

10.4229/EUPVSEC2023/4BO.5.3
020296-002

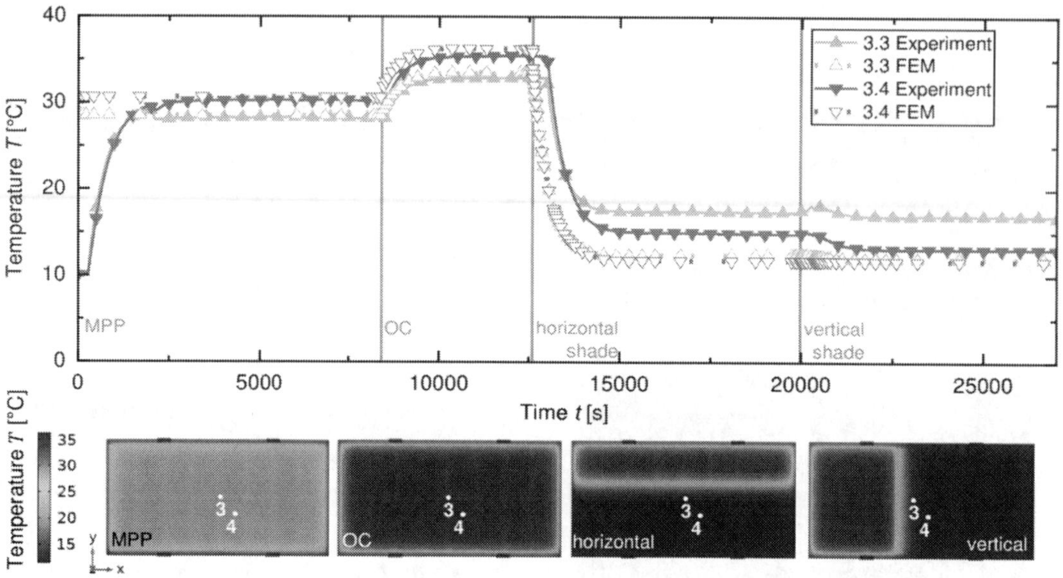

Figure 3: Experimental (closed symbols) and simulated (open symbols) temperatures T of two selected positions on the back surface of a solar cell of the upper fully covered PV module. The 2D color plots show the complete temperature distribution over the module area at the end of each sequence.

temperature distribution and resulting mechanical stress for a warm day with temperatures between +20 °C and +30 °C and a cold day with temperatures between -10 °C and +5 °C (Figure 2) for homogeneously illuminated PV modules. The maximum irradiance is 1000 W/m² in each case and the modules are oriented to the south.

The temperature difference of 38 K between maximum and minimum temperature is 2 K greater on the cold day than on the warm day, which is also reflected in the slightly higher maximum stress in the front glass cover of 18.3 MPa (Figure 4). On both days, the maximum stresses exceed the design edge resistance of annealed glass, which means that even under homogeneous irradiation, the use of annealed glass must be considered to be risky. On the other hand, the maximum stresses are significantly below the design edge resistance of thermally toughened glass and are therefore not critical.

In summary, cold days with high irradiance levels were identified as the most critical environmental conditions. Consequently, all further investigations were carried out for these conditions.

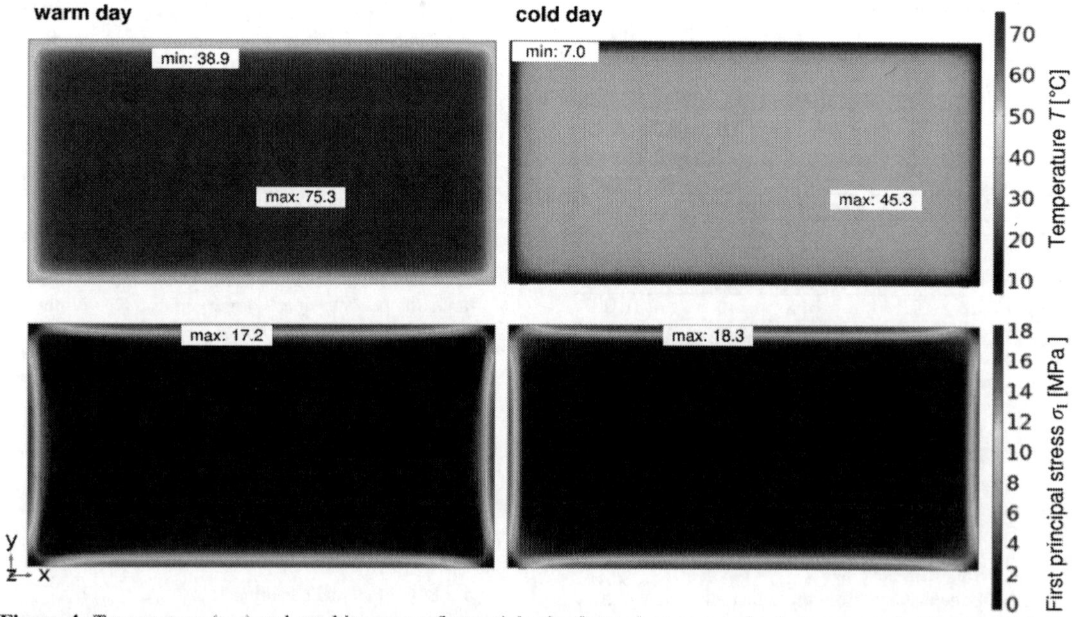

Figure 4: Temperature (top) and resulting stress (bottom) in the front glass cover of a homogeneously illuminated (up to 1000 W/m²) photovoltaic module on a warm (left) and cold day (right).

4.2 Frame

Three variants were simulated to investigate the influence of the module frame:

1. with frame and with absorption in the adhesive
2. with frame and without absorption in the adhesive
3. without frame

Figure 5 shows the resulting temperature distribution in the module cross-section at the position of the frame and the associated maximum first principal stress (tensile stress).

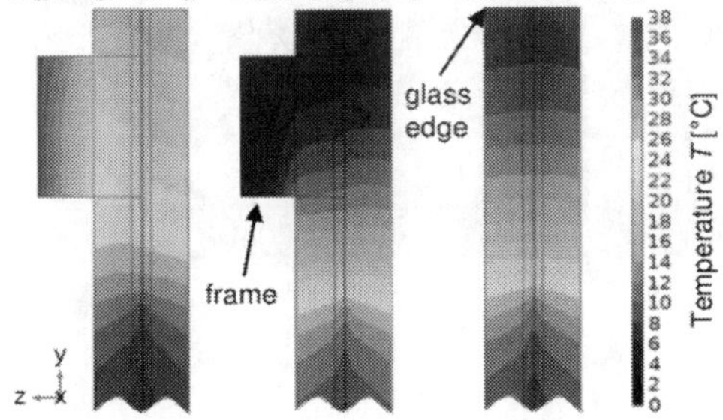

Figure 5: Temperature distribution in the module cross-section at the position of the frame for simulations with frame and absorption in the adhesive joint (left), with frame without absorption in the adhesive joint (center) and without frame (right). The resulting maximum stress at the edge of the glass is specified above the plots.

It can be seen that the frame has a significant influence on the temperature and thus the stress. On the one hand, the high absorption in the adhesive bond causes the module and frame to heat up, which is why a higher edge temperature occurs in the simulation with frame and absorption in the adhesive. On the other hand, the good heat conduction of the aluminum frame contributes to better heat dissipation to the environment, which in turn leads to a lower edge temperature compared to a frameless module. Accordingly, the frameless module's stress level is between that of the framed modules with and without absorption in the adhesive joint. All stresses at the glass edge exceed the design edge resistance of annealed glass, which is why annealed glass is to be considered risky. However, the stress is significantly below the design edge resistance of thermally toughened glass and hence is non-critical in this case. High absorption by the adhesive is advantageous, in order to reduce the mechanical stresses in the glass. This can be achieved, for example, by black adhesives.

4.3 Operating conditions

To investigate the influence of the operating conditions, the absorptance of the solar cell is adjusted. For the open circuit voltage, it is assumed that the incident radiation (after absorption by the glass and encapsulant) is absorbed completely by the solar cells. In the case of the maximum operating point, the absorptance is reduced by the solar cell efficiency of 22 %, resulting in a solar cell absorptance of 73 %. 5 % of the radiation is absorbed by the front glass cover and encapsulant. Figure 6 shows that the lower absorptance of the solar cell at the MPP leads to an approx. 10 K lower temperature in the area of the solar cells, while the temperature at the edge of the module remains almost unchanged. This results in an approx. 4.5 MPa lower stress in the glass. In the MPP state, the maximum stress at the edge of the glass corresponds to the design edge resistance of annealed glass, so that in both operating conditions annealed glass is considered risky, while thermally toughened glass is not.

In summary, OC operation is the more critical operating condition than MPP.

4.4 Solar cell coverage

We investigate three different cell coverages:

1. Fully covered: 4×8 solar cells
2. Partially covered: 3×7 solar cells
3. Lightly covered: 2×6 solar cells

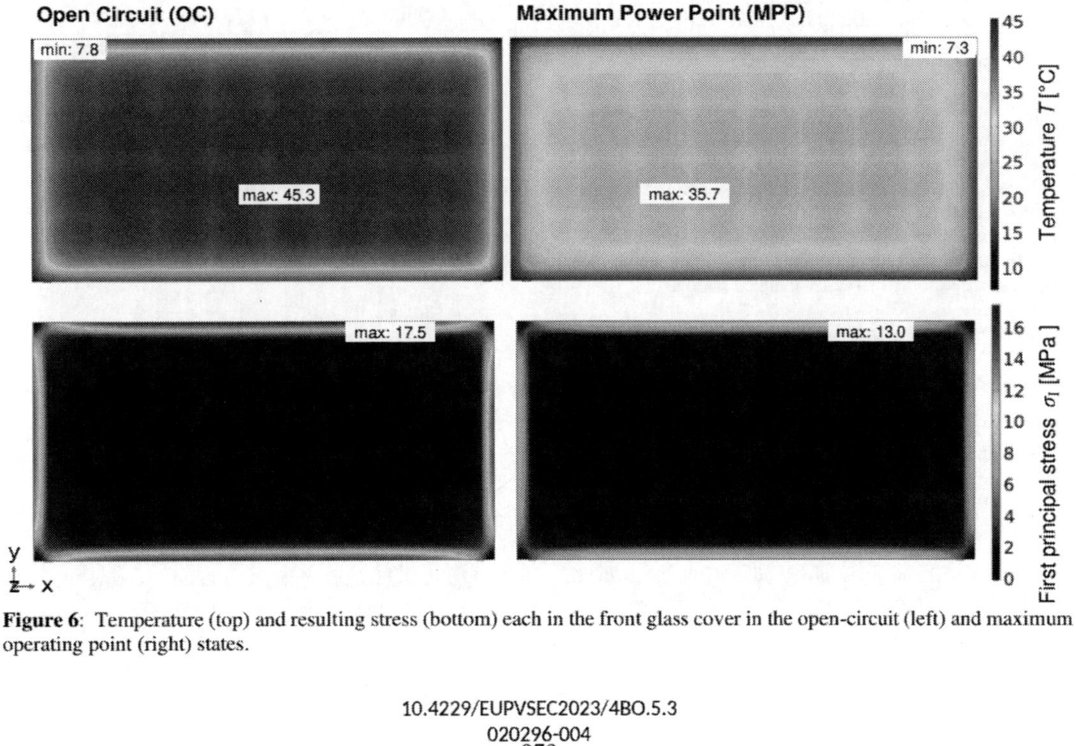

Figure 6: Temperature (top) and resulting stress (bottom) each in the front glass cover in the open-circuit (left) and maximum operating point (right) states.

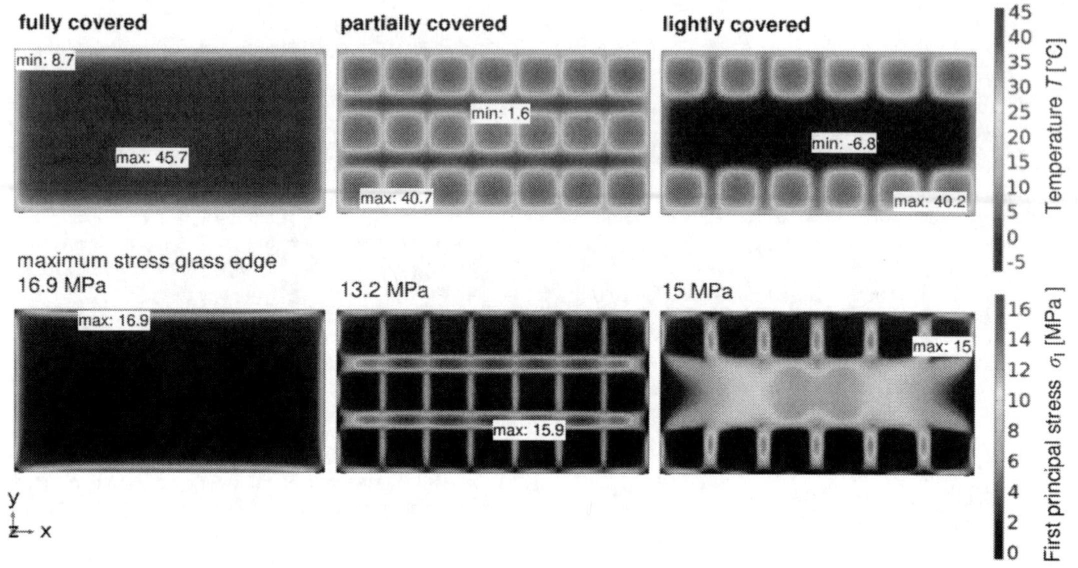

Figure 7: Temperature (top) and resulting stress (bottom) in each case in the front glass cover for fully covered (left), partially covered (middle) and lightly covered (right) PV modules with homogeneous irradiation. In addition to the maximum values, the maximum stress at the edge of the glass is indicated

Figure 7 shows the simulated temperature and mechanical stress. Due to the stronger heat generation, the fully covered PV module shows the highest maximum temperature of 45.8 °C. The difference between partially and lightly covered is not significant, with maximum temperatures of 40.7 °C and 40.2 °C, respectively. This shows that even a slight increase in cell spacing can reduce the temperature of the solar cell by up to 5 K. The coverage ratio has a much stronger effect on the minimum temperature, which drops from 8.7 °C with full coverage to -6.8 °C with light coverage. This significantly increases the temperature difference from 37 K to 47 K. However, the temperature gradient is weakened by lighter coverage, which is why the highest maximum stress of 16.9 MPa occurs in fully covered modules. Moreover, this occurs at the glass edge, which typically has the lowest fracture stress, whereas for partial coverage, the maximum stress at the glass edge is only 13.2 MPa, and is 15 MPa for light coverage. All coverage ratios lead to stresses that exceed the design edge resistance of annealed glass, which is why the use of annealed glass should be considered risky. However, they are well below the edge design resistance of thermally toughened glass and should be considered non-critical here.

In summary, fully covered PV modules are slightly more susceptible to thermally induced stresses.

4.5 Edge distance from the cells to the glass edge

We investigate two different perimeter width: a normal edge distance (long side: 28.8 mm, short side: 18.8 mm) and a very large edge distance (long side: 108.9 mm, short side: 99.9 mm). Figure 9 shows that the maximum temperature does not change significantly, while the minimum temperature drops from 8.7 °C to 2 °C with a large edge distance. Thus, the temperature difference increases from 37 K to 47.6 K. The resulting maximum stress increases from 16.9 MPa to 20.6 MPa, with the maximum stress at the edge of the glass being 17.5 MPa, which is only slightly higher but still above the design edge resistance of annealed glass, which is why the use of annealed glass must be considered risky. However, the design edge resistance of thermally toughened glass is significantly higher, so both edge distances can be rated as non-critical for thermally toughened glass.

In summary, the edge distance has a minor influence on thermally induced stresses.

4.6 Glass thickness

We vary glass thickness in 2 mm steps between 2 mm and 10 mm. The front and back glass panes have the same thickness. Figure 8 shows that the glass thickness has only a slight influence on the maximum stress in the glass. For 10 mm glass, the stress is reduced by 0.6 MPa compared to 2 mm. Due to the larger thermal mass, the time of maximum stress is shifted by about half an hour. Consequently, a greater glass thickness does not provide a significant advantage in reducing thermally induced stresses.

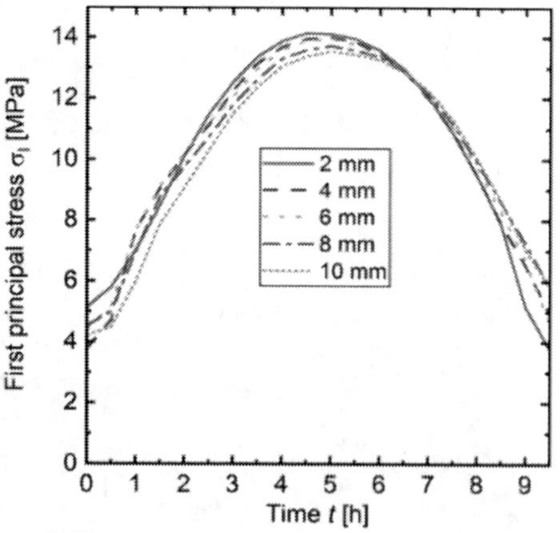

Figure 8: First principal stress in the glass over the course of the day for different glass thicknesses.

10.4229/EUPVSEC2023/4BO.5.3

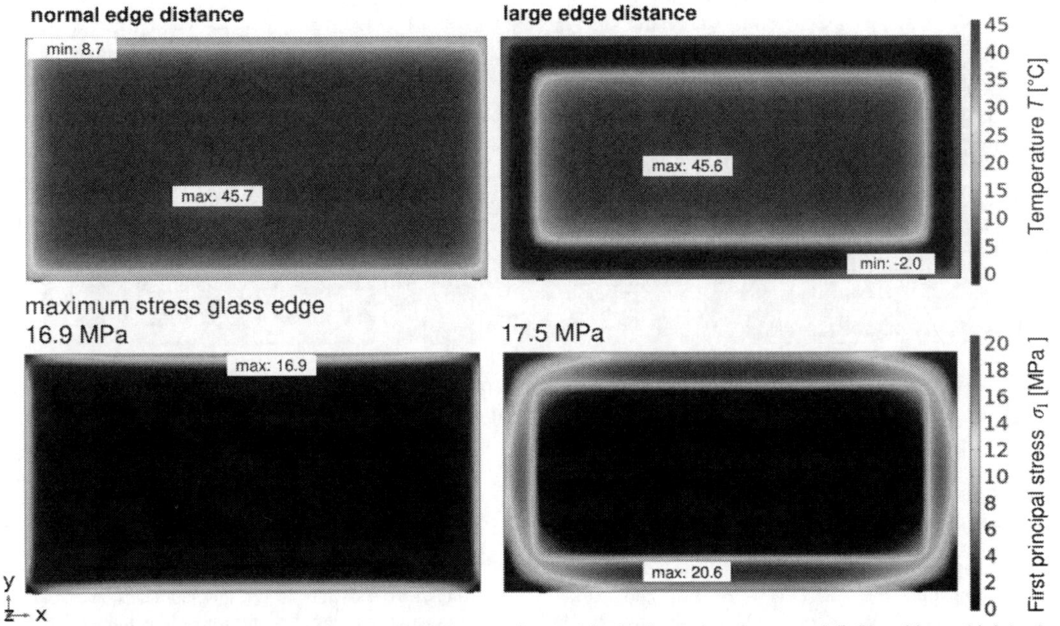

Figure 9: Temperature (top) and resulting mechanical stress (bottom) in the front glass for normal (left) and large (right) edge distance. In addition to the maximum values, the maximum stress at the edge of the glass is indicated.

4.7 Shading

Firstly, horizontal and vertical shading with 5 %, 10 % and 50 % degrees of shading with respect to the modules width or length were investigated. The simulated temperatures are shown in Figure 10. The maximum temperature decreases only at 50 % horizontal shading from 45.3 °C to 44.6 °C. The minimum temperature, however, is strongly dependent on the shadow width and ranges from -0.6 °C for 5 % horizontal shading to -8.0 °C for 50 % vertical shading. The resulting stress is shown in Figure 11. All shadows lead to significantly higher maximum stresses compared to unshaded PV modules, which all occur at the edge of the glass. The values range from 21.1 MPa (10 % horizontal and 5 % vertical) to 25.3 MPa (50 % vertical). There is no clear correlation between the surface area of the shadow and stress.

Secondly, a double diagonal shadow was considered. This starts in the middle of both long edges of the module and has its vertex at the middle of the short edge of the module. According to the "Guide for the Design, Construction and Control of Buildings with Structural Glass Elements" CNR-DT 210/2013 [8], this shading geometry is considered as "most dangerous" for structural glass elements. Figure 12 shows the simulated temperature and resulting stress. The maximum temperature is 45.3 °C, with -7.4 °C as the minimum temperature. Due to the very large temperature gradient at the top of the shadow, the highest maximum stress of 32.4 MPa is reached here. Compared with the unshaded PV module, the stress is almost twice as high, hence such shadows significantly influence thermally induced stress. For annealed glass in particular, the maximum stress is almost three times as high as the design edge resistance, which is why the use of annealed glass is assessed as being very risky. For fully toughened glass, the design edge resistance is more than twice as high as the maximum stress, so the risk of glass breakage purely due to thermal stresses is assessed as low. In a real installation case, however, thermally induced

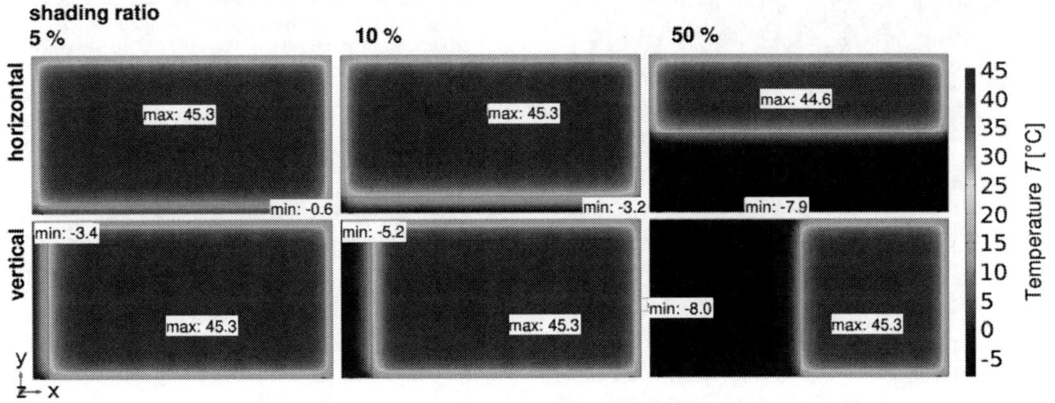

Figure 10: Temperature in the front glass cover for different shading configurations: top: horizontal shading; bottom: vertical shading, each with different shading ratios: 5 % (left) 10 % (center) and 50 % (right).

10.4229/EUPVSEC2023/4BO.5.3
020296-006

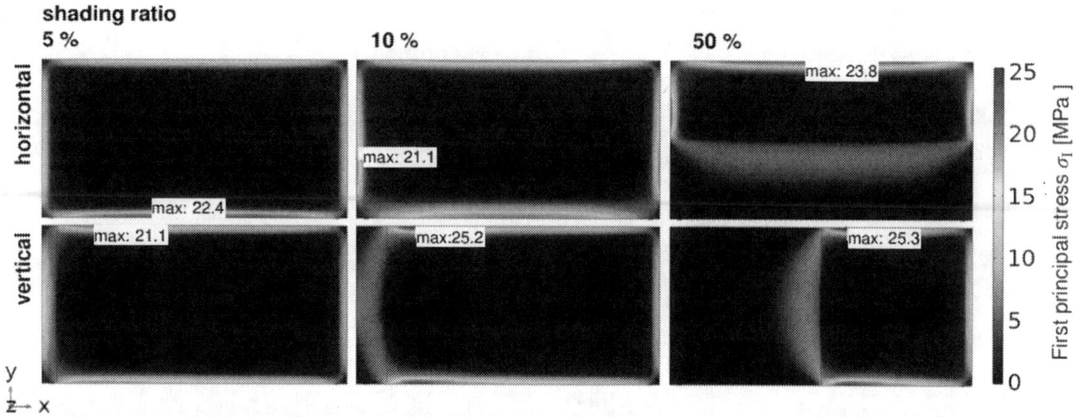

Figure 11: First principal stress in the front glass cover for different shading orientations and ratios: top: horizontal shading; bottom: vertical shading, each with different shading ratios: 5 % (left), 10 % (center) and 50 % (right).

stress is often superimposed with mechanically induced stress. This can result, for example, from wind load, snow load, self-weight or unplanned installation errors. It can be assumed that wind has a reducing effect on the thermally induced stress due to the increased convection and the associated cooling effect. With snow, on the other hand, a distinction must be made between full and partial coverage. In the case of full coverage, no stresses are caused by the shading and only the mechanical loads have an effect. In the case of partial coverage, the snow causes shading (and at the same time cooling of the module), which is why the thermally induced stress is superimposed on the snow load. Likewise, stress due to self-weight (with glass edges under tensile stress) and possible assembly errors in combination with thermally induced stress can lead to glass breakage.

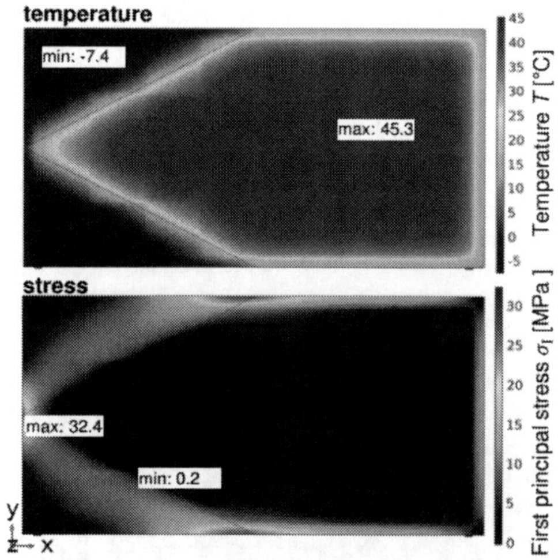

Figure 12: Temperature (top) and resulting first principal stress (bottom) in the front glass cover for a double diagonal shadow.

4.8 Difference between crystalline silicon and thin film cell technology

For the simulation of a thin film PV module, the solar cells were removed and replaced by an absorber layer positioned on the surface of the front glass cover facing away from the sun. Since the simulation was performed for the OC state, the different efficiencies did not have any influence. To simplify the calculation, it was assumed that the solar absorptance of both technologies is equal, which might not be the case depending on the thin film technology.

Figure 13 shows that due to the smaller absorber layer thickness and thus lower thermal mass, the maximum temperature in the thin-film module is almost 1 K lower. However, since the edge temperature is also slightly lower and the gradient from maximum to minimum temperature is larger, the thin-film module has a 1.5 MPa higher maximum stress. Consequently, the simulated thin film technology is slightly more susceptible to thermally induced glass breakage. However, it should be noted that thin-film modules generally use only annealed glass as the back glass, even when they contain drilled holes for cabling. This is risky because the design edge resistance of annealed glass is exceeded. By contrast, crystalline modules usually use thermally toughened glass with a higher design edge resistance that is not exceeded, and therefore is not critical.

5 SIMPLIFIED FEM MODEL

For the development of a standardized method, the 3D FEM model was simplified, in order to save computational cost and enable broad application. For this purpose, the combination with the greatest influence on thermally induced stresses was selected from the parameters investigated above. This resulted in a frameless, partially covered PV module with a double diagonal shadow, which is shown schematically in Figure 14. As a simplification, a 2D approach was chosen. The basic idea is to represent the PV module as a glass pane with the total thickness of the PV module. The glass pane has the material properties of the glass used, with adjusted absorptance coefficients at the positions of the solar cells and in the shaded area. These are shown in Figure 14. In addition, combined effective heat transfer coefficients are used: 10.5 W/(m²K) for the main module surfaces and 4 W/(m²K) for the edges. A constant 10 °C is chosen as the outdoor temperature. The same simplified boundary conditions are used in the 3D FEM model for comparison. Mechanically, the module is fixed at one corner point and a sliding bearing is attached to the neighboring corner point along the long edge.

10.4229/EUPVSEC2023/4BO.5.3

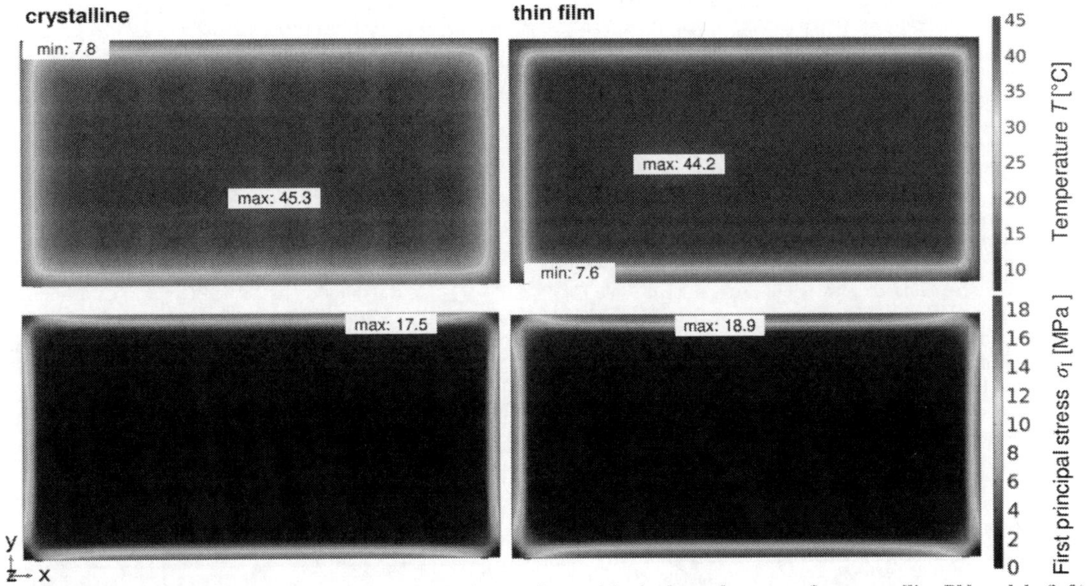

Figure 13: Temperature (top) and resulting mechanical stress (bottom) in the front glass cover for a crystalline PV module (left) and a thin-film PV module (right).

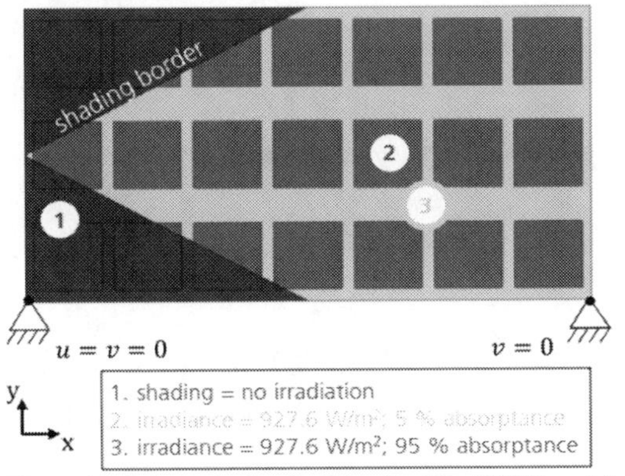

Figure 14: Schematic sketch of the simplified FEM model with division into the different absorptance and irradiance zones as well as the mechanical boundary conditions.

Figure 15 compares the simulation results for the temperature and Figure 16 the resulting mechanical stresses of the 2D simulation and the 3D simulation. Since in the 2D simulation the heat is homogeneously generated in the highly absorbing glass body itself, instead of being conducted from the solar cells to the glass, there is no temperature gradient between the absorbing solar cell and the glass surface. Hence, this leads to a higher temperature level compared to the 3D simulation. In particular, the maximum temperature of 46.5 °C is 5.1 K higher. However, the qualitative temperature distribution agrees well. For stress, the qualitative distribution also agrees very well and the higher temperature level leads to higher stress. The 2D FEM model has a stress of 34.7 MPa, which is 5.5 MPa higher than the 3D FEM model. Thus, the simplified FEM model offers the possibility to quickly estimate thermally induced mechanical stresses.

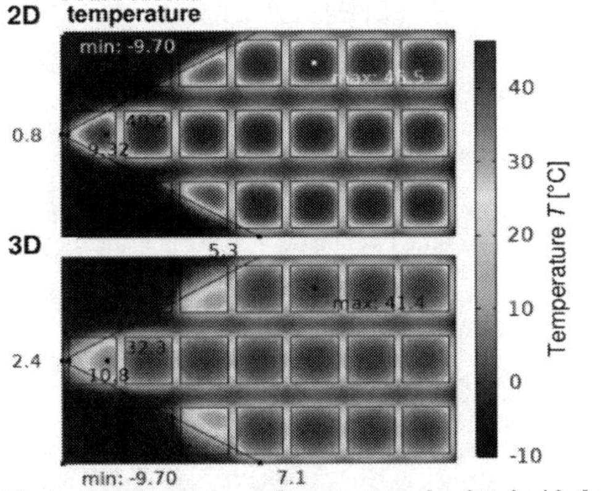

Figure 15: Comparison of temperature simulated with the simplified 2D model (top) and with the full 3D model (bottom). Values at characteristic points are marked.

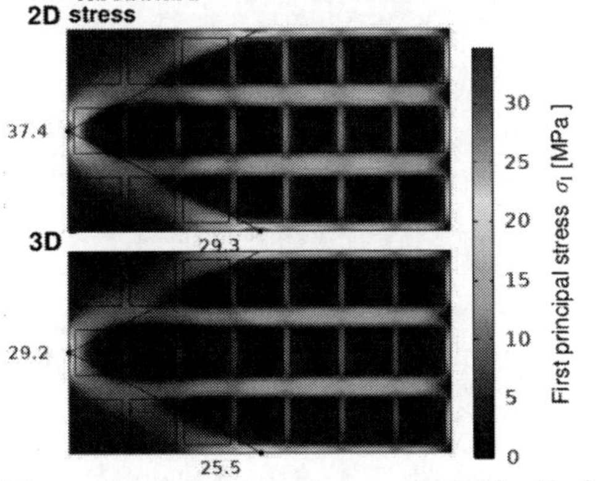

Figure 16: Comparison of stress simulated with the simplified 2D model (top) and with the full 3D model (bottom). Values at characteristic points are marked.

6 CONCLUSION

10.4229/EUPVSEC2023/4BO.5.3
020296-008

Both a very detailed thermomechanical 3D FEM model of a glass-glass BIPV module in a ventilated façade and a simplified 2D FEM model were developed. The 3D thermal model was successfully validated, displaying good agreement with the experimental data. The performed parameter variations have shown that even for homogeneous irradiation, the glass stress can exceed the design limits for annealed glass edges. Partial shading can lead to a significant increase of the stress and should therefore be considered in the design of a PV system, especially if thermally toughened glass is not used. Overall, for the simulated glass-glass BIPV module, all variations resulted in maximum stresses exceeding the reference design edge resistance of annealed glass determined by Schwind *et al.* [7]. Consequently, the use of annealed glass must be considered to be risky. Also, the use of thin glass, which can only be partly thermally toughened, may become risky. However, the edge resistance of fully thermally toughened glass was not exceeded in any variation, so the use of fully thermally toughened is rated as being non-critical. The other results are:

1. During the course of a year, cold days with high irradiation induce the highest stresses.
2. Open Circuit (OC) conditions induce higher stress than maximum power point (MPP).
3. Shading has a significant impact on the stress. In particular, a double diagonal shadow shape can double the stress.
4. Fully covered PV modules have slightly higher mechanical stress than partially covered ones.
5. The edge distance from solar cell to glass edge has a very small effect on the maximum stress at the glass edge.
6. The glass thickness has a minimal influence.
7. Thin-film PV modules have a slightly higher stress than crystalline silicon PV modules.
8. A frame adhesive with high absorption reduces stress in the glass.

7 ACKNOWLEDGEMENTS

The authors would like to thank Markus Heck for the careful planning and execution of the validation experiments at Fraunhofer ISE. Also, the authors would like to thank the German Federal Ministry for Economic Affairs and Climate Action for the financial support within the project: "Thermobruch" (03TN0007C).

8 REFERENCES

[1] F. Ensslen, G. Schwind, J. Schneider, A. Beinert, A. Mahfoudi, *et al.*, Challenging Glass Conference Proceedings (2022), 10.47982/cgc.8.433.

[2] F. Ensslen, A.J. Beinert, A. Mahfoudi, W. Herzberg, E. Lorenz, *et al.*, Normentwurf zur Ermittlung der thermischen Beanspruchung von Glas und Glas-PV-Modulen (BIPV) im Bauwesen (Thermobruch) - Schlussbericht zum WIPANO-Projekt (FKZ: 03TN0007A-D), (2023).

[3] A.J. Beinert, P. Romer, M. Heinrich, J. Aktaa, D.H. Neuhaus, Prog. Photovoltaics Res. Appl. (Progress in Photovoltaics: Research and Applications), (2022), 1–13, 10.1002/pip.3624.

[4] A.J. Beinert, P. Romer, M. Heinrich, M. Mittag, J. Aktaa, *et al.*, IEEE J. Photovoltaics 10, (2020), 70–77, 10.1109/JPHOTOV.2019.2949875.

[5] P. Romer, K.B. Pethani, A.J. Beinert, 8th World Conference on Photovoltaic Energy Conversion: Joint Conference of 39th European Photovoltaic Solar Energy Conference and Exhibition, 50th IEEE Photovoltaic Specialists Conference, 32th International PV Science and Engineering Conference. Conference Record (2022), 924–929, 10.4229/WCPEC-82022-3DV.3.27.

[6] P. Romer, A.J. Beinert, Proceedings of the 38th European Photovoltaic Solar Energy Conference and Exhibition (EU PVSEC) (2021), 602–606, 10.4229/EUPVSEC20212021-4BO.4.5.

[7] G. Schwind, F. Paschke, J. Schneider, Challenging Glass Conference Proceedings – Volume 8 (2022), 1–37, 10.47982/cgc.8.388.

[8] National Research Council of Italy – Advisory Committee on Technical Recommendations for Construction CNR-DT 210/2013, Guide for the Design, Construction and Control of Buildings with Structural Glass Elements (2013).

BIFACIAL PV SYSTEMS AT HIGH LATITUDE: MODELING AND VALIDATION WITH MONITORING DATA

Hugo E Huerta[1], Samuli Ranta[1], Shuo Wang[1], Aleksi Heinonen[1], Sami Jouttijärvi[2], Anton Driesse[3], Kati Miettunen[2]
1. New Energy Research Group, Turku University of Applied Sciences, Joukahaisenkatu 7, 20520 Turku
2. Department of Mechanical and Materials Engineering, University of Turku, Vesilinnantie 5, 20500 Turku, Finland
3. PV Performance Labs, Emmy-Noether-Str. 2, Freiburg, Germany
Corresponding author: Hugo Huerta, hugo.huertamedina@turkuamk.fi

ABSTRACT: In this paper, we model and evaluate the performance of two different Bifacial Photovoltaic implementations at high latitude locations with latitudes close to 60° N, consisting of a set of seaside PV canopies with 100 kWp power and a roof-mounted PV system with 140 kWp power. The modelling is done by using the backward ray-tracing technique in combination with detailed 3D models to achieve a more accurate evaluation of the rear-side irradiance and energy yield prediction. An improved irradiance sampling method was developed to simulate an accurate irradiance map on each cell of the modules with high-speed asynchronous execution. The detailed irradiance power output of the canopy system was simulated over one year and compared with monitored data for an analysis of energy gain and loss. Results show that 7% of the total energy comes from the rear irradiance, in which 1% of the total energy is from the enhancement of snow. However, snow coverage on the front side could cause a loss of 2.8% of the total energy, which indicates that an appropriate snow clean-up measure might be important to further optimize the yield of PV systems in high latitude regions. The methodology was implemented preliminarily for the second configuration, roof-mounted PV to prove its replicability. Preliminary results show only 2% of the energy is produced by the rear side. The modelling methodology was proven to be reliable and useful when evaluating different bifacial PV configurations.

Keywords: Bifacial, Ray-Tracing, High-Latitudes, Modelling.

1 INTRODUCTION

Bifacial Photovoltaic (BPV) modules have emerged as a promising technology in the field of solar energy generation, offering the potential to significantly enhance the efficiency and energy yield of solar installations. Unlike conventional PV modules, which capture sunlight only from their front surfaces, bifacial modules can also harvest energy from the rear side by capturing the light reflected off by the nearby surfaces such as the ground, buildings, and other objects. This unique characteristic has increased the interest in bifacial technology, particularly in applications where maximizing energy production from limited space is crucial, such as installations on flat roofs and solar farms. However, the accurate modeling and simulation of bifacial PV modules present substantial challenges, primarily from the difficulty in accurately estimating the incident irradiance at the rear side of the modules. The incident irradiance, or the amount of sunlight received by the rear surface, plays a critical role in determining the overall performance and energy generation of bifacial modules. Consequently, limitations in accurately characterizing this parameter can significantly impact the reliability of predictive models and the design of efficient bifacial solar systems.

Within the PV community, two primary methods have been used [1], namely geometric factors and ray-tracing methods. The View factor model [2] employs a simplified approach to compute bifacial irradiance. It utilizes view factors for 2D geometric representations of PV arrays. This method is particularly advantageous for estimating the energy production of big PV arrays due to its computational efficiency, completing annual hourly simulations in just a few seconds. The main characteristic in this method is that it considers infinite PV rows, in consequence ignoring the edge effects as well as the lack of light behavior physics behind the model. Ray-tracing methods [3] simulate the paths of individual rays of sunlight as they interact with the module and its surroundings. By tracing the rays through a three-dimensional representation of the PV module, ground, and nearby objects, ray-tracing models can provide a more detailed and accurate estimation of rear-side irradiance. Nonetheless, they require substantial computational resources and can be complex to implement.

In this study, we model and validate results with a pilot site by implementing the ray-tracing method. By using Radiance [4] software which excels in simulating the behavior of light in complex environments to estimate how light interacts with surfaces, materials, and objects when considering factors such as reflections, refractions, and shadows. Additionally, we integrated the use of Bifacial-Radiance [5] tool which is a software extension built upon the Radiance platform and is designed to facilitate the modeling and analysis of bifacial PV systems. These tools are combined to model and evaluate the performance of two different Bifacial PV implementations at high latitude locations, where the seasonal changes, long days during the summer and short and dark days with/without snow during the winter play an important role in the solar energy production. To complete this study, two different BPV system configurations in different locations are considered. The first system, used for validation, hereafter referred to as the 'Canopy-PV array,' consists of a 100 kWp bifacial system integrated into seaside canopies. This system, located in Naantali, Finland, has been in operation since June 2022 and includes six canopies, each equipped with a varying number of bifacial modules (36, 48, or 60) rated at 350 Wp. The modules are tilted at an angle of 15°, with the bottom edge of the modules positioned 3.5 meters above the ground. Different materials integrate the surfaces under the Canopies like gravel, wood and even water, all having different albedo conditions which range from 30 – 55%. The canopies' roof orientations vary slightly (+-10°) from the South.

The second system, modeled for replicability and with preliminary results, hereafter referred to as the 'Flat-Roof

PV array', consists of a 140 kWp bifacial system scheduled for installation in autumn 2023 on the roof of a new student housing unit in Turku, Finland. These bifacial modules, rated at 360 Wp, were designed and manufactured by a local company. They will be installed on the building's roof with a tilt angle of 15° and oriented towards the South; The building's roof is covered by membrane with an albedo value around 14%.

2 METHODOLOGY

2.1 Monitoring data collection

Data collection to validate the modeling of the Canopy-PV array commenced in mid-summer 2022 by monitoring DC voltage and current for two PV strings with a 1-second time resolution. On-site meteorological data is gathered using a ventilated pyranometer to measure global tilt irradiance (GTI) and a weather transmitter to record air temperature, wind speed, and other relevant variables. Additionally, six reference cells were strategically positioned, 5 of them facing downward along the rear-side of the canopy enabling the measurement of rear-side plane of array (POA); The sixth cell was placed at the same modules tilt angle for measuring POA irradiance. The study collected irradiance and power output data from the Canopy-PV system for a one-year period, spanning from July 2nd, 2022, to July 1st, 2023. Throughout this timeframe, a total of 530 hours of monitoring data were unavailable due to blackouts or system updates. Thus, there were in total 8230 hours (343 days) of effective monitoring data that were available for validation and analysis. A comprehensive description of the monitoring system is shown in our previous work [6].

For the Flat-Roof PV array, the simulations utilized the same meteorological data from the Canopy-PV due to delays in the installation of the system.

2.2 System modelling

Both PV arrays underwent 3D modeling to create realistic representations of the buildings and mounting systems **Figure 1**. In the case of smaller BPV systems, such as those studied here, the inclusion of more details becomes essential. These details include accounting for mounting systems, understructures, and even the spacing between modules and cells. At this scale, the inhomogeneity of rear-side irradiance becomes more pronounced due to stronger edge effects leading to power mismatches among cells and modules at various positions.

To address this challenge, a complex and computationally intensive method for detailed irradiance mapping, known as reverse ray-tracing with Radiance, was employed. This method involves tracing a large number of photon paths using the Monte Carlo method and simulating object interference within the modeled system. It allows for optical simulations by incorporating properties such as albedo for various ground/roof conditions as well as specularity, and reflectance of the structures and mounting systems into the model.

This study focused primarily on three columns of PV modules (18 in total) from Canopy-PV, constituting a single string with 6.3 kWp. To evaluate the electrical properties of the PV modules, a two-diode model was employed. The model incorporated average values derived from flashing tests of 50 modules.

Canopy-PV array (top), 3D models rendering (bottom)

Flat-Roof PV- building (top), 3D model rendering (bottom)

Figure 1:PV systems under study: 3D models were built to cell-scale level and used in Radiance software for modelling rear-side irradiance. Rendering using acceleradRT [6]

10.4229/EUPVSEC2023/4BO.5.4
020297-002

The initial model, as described in a previous work [6], underwent expansion to incorporate electrical calculations for the entire system. The integration of the SUNPOWER PVMismatch module enabled the computation of both power output and mismatch loss.

An enhanced sampling method was introduced to facilitate irradiance calculations for individual cells within the PV modules of interest in conjunction with a high-level interface method for asynchronously executing the simulations, resulting in a significant reduction in simulation time. Bifacial-Radiance tool in combination with the measured data and the detailed irradiance maps were used for the thermal and electrical modelling, resulting in the BPV output, the whole process is detailed in **Figure 2**.

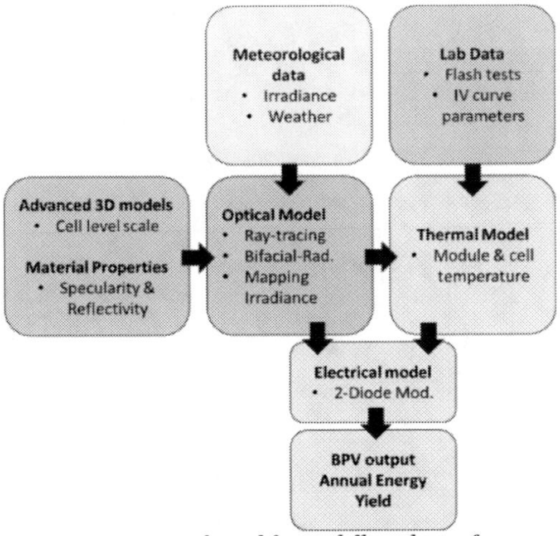

Figure 2: Framework used for modelling the performance of bifacial PV systems

To account for the influence of snow, which is critical in this region, snow depth data was obtained from the Finnish Meteorological Institute at Turku Artukainen station, located approximately 8 km from the Canopy-PV array system and 5 km from the Flat-roof PV array system. Timestamps with snow depth greater than zero were classified as "snow conditions". While acknowledging some variation between the weather station's snow condition data and the canopy system's location, this approach sufficed for distinguishing ground conditions with and without snow. The work as described is intended to establish a practical and accurate methodology using the ray-tracing technique for bifacial systems at high latitudes validating it with on-site measured data, tackling at the same time the previously mentioned challenges.

3 RESULTS

3.1 Canopy-PV

1. Model validation

a) Rear irradiance

The rear irradiance detected by the sensors was used to validate the albedo parameters. The reference albedo of ground objects in the model was optimized by the method as described in our last paper [7]. **Figure 3** shows the comparison between simulated and measured irradiance at five sensors on the backside and the average value over the five sensors for a sunny day in August 2022. In general, the simulations fit with the measured irradiance well, only with slight deviations for sensor 1 in the afternoon.

With the reference albedo, the detailed irradiance on both sides of modules were simulated by an improved sampling method. This method enables fast and accurate sampling on every cell on the module. **Figure** 4 illustrates the irradiance map of the whole canopy from the back side for a sunny day on 25 August. Note that the target string is the left three columns in the figure.

The hourly averaged bifacial gain of the target module string can be evaluated from the irradiance map and shown in **Figure** 5. Since the target string is in the eastern part of the system, in sunny days it gets more light on the backside in the morning, which results in higher bifacial gain up to over 30% during that period. With the Sun moving to south, the bifacial gain decreases to the lowest gain around 9% in the afternoon. In cloudy days like 21 and 24 August, the bifacial gain tends to be constant (around 13%) over the whole day.

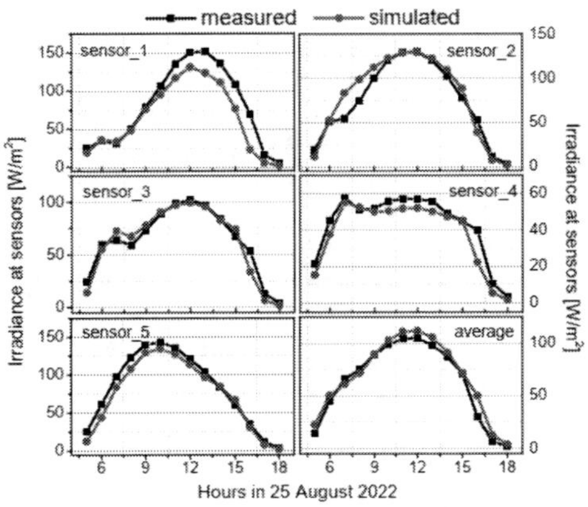

Figure 3: Measured and simulated hourly irradiance on back sensors and their average value over a sunny on 25 August 2022.

b) Power output and mismatch

The power output of the target string of the same week was simulated by the electrical model and shown in **Figure** 6. The sampled irradiance at each cell in the module was used to calculate the power with mismatch loss (P_{mm}), while the average irradiance was used to calculate the power output without mismatch (P_{ave}). The simulations show good coherence with the measured data, although with slight overestimation. The power simulated with mismatch is slightly lower than the power without mismatch, which is due to the mismatch loss. When the power is integrated over the sunny day on the 25th, the simulated power with mismatch is about 4% higher than the measured power. This difference may be partly due to possible losses that we lack in our model, such as PV module variation, soiling effect, wire connection loss, and MPPT loss.

The hourly mismatch loss of the same week is analyzed in detail in **Figure 7.** The mismatch loss (MM) is defined as:

$$MM = (P_{mm} - P_{ave})/P_{ave} \qquad (1)$$

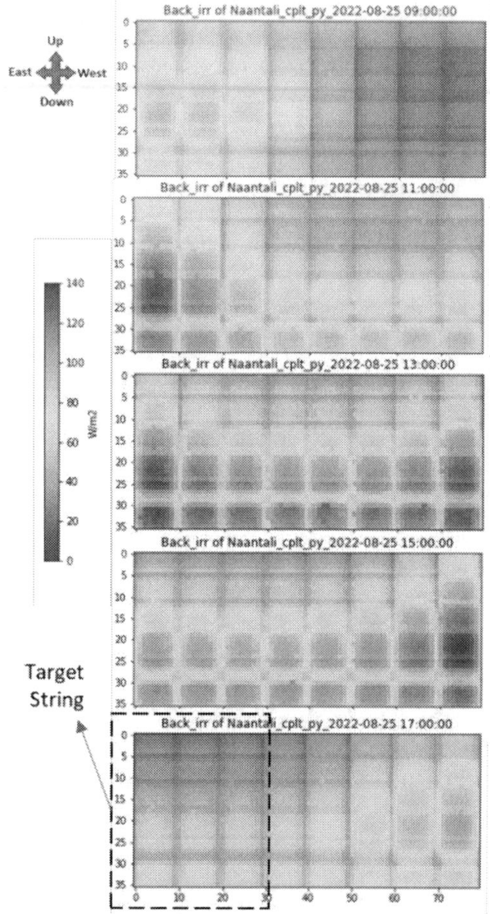

Figure 4: Rear irradiance mapof the system for a sunny day on 25 August 2022. The x- and y-axis indicate the number of cells. The perspective in the figure is from north to south. The three columns of modules in the eastern part were taken as the target string for detailed investigation.

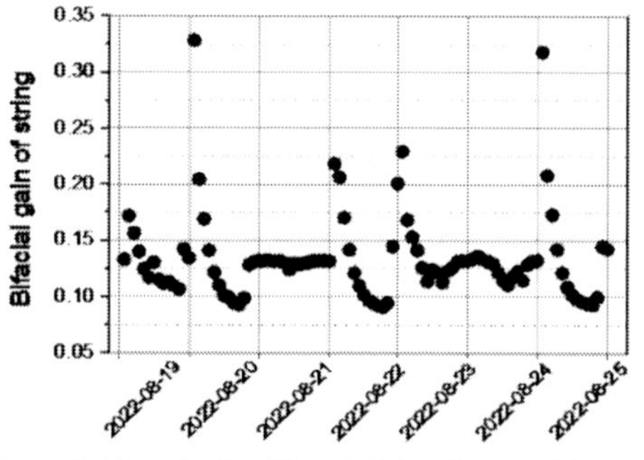

Figure 5: The calculated hourly bifacial gain of the target string in the week from 19 to 25 in August 2022

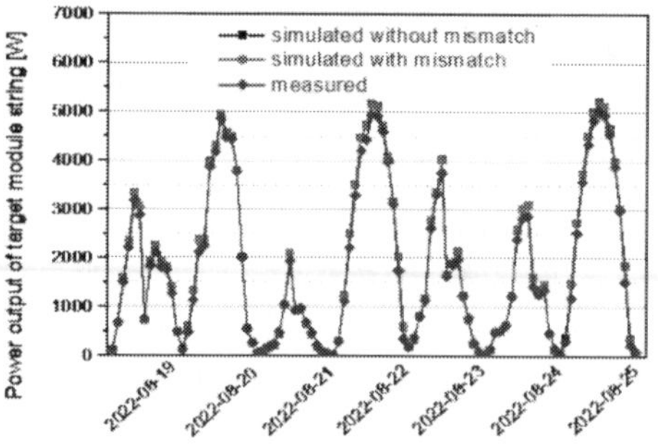

Figure 6: The simulated and measured hourly power output of the target string in one week in August 2022.

The mismatch loss is dependent on the time of the day, which is larger in the morning and late afternoon. The highest loss can be over -1%, while for most time in the day it is around -0.4%.

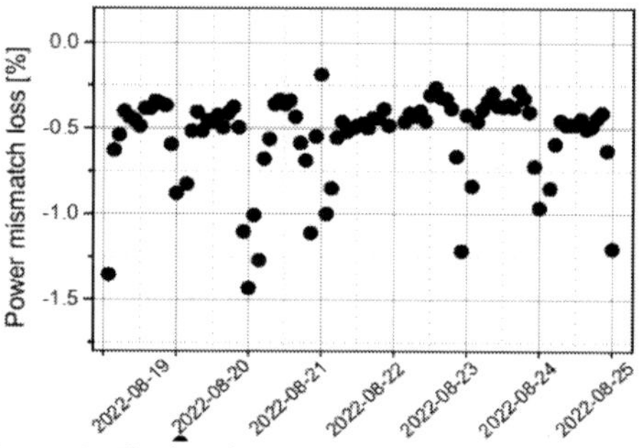

Figure 7: The hourly power mismatch loss of the target string in one week in August 2022.

2. Evaluation over the whole year

After validation and initial analysis of the model, simulations throughout the year were performed to understand the performance of the system and the energy losses. As the albedo of the ground can be dramatically changed by snow, the discussion was divided into two parts, the days with and without snow on the ground.

a) Days without snow

When comparing the simulated irradiance on the rear sensors with the measured data over a longer period, it was found that a single set of albedo parameters was insufficient to fit all the conditions well. The ground conditions keep changing over time, such as alterations in wood color (aging) and its differences in color between wet and dry conditions. As shown in **Figure 8**, the irradiance on rear sensors was aggregated into days, and the ratio of measured to simulated irradiance was calculated. The timestamps with front irradiance lower than 200 W/m^2 were excluded to reduce the influence of low light conditions. The daily fluctuations can be observed, and the ratio tends to decrease over time. When integrating the irradiance on sensors over all the non-snow days, the simulated irradiance is approximately 15% greater than the measured one.

This overestimation may arise due to fluctuating

ground conditions that cannot be easily corrected using a single set of albedo parameters. A possible approach to solve this problem is to calculate albedos based on the measured rear irradiance for each timestamp. However, it faces significant challenges in implementing this method with the time-consuming ray-tracing simulation. What it used to mitigate the impact of this overestimation on power evaluation is to minimize the difference between simulated and measured irradiance averages across all days by using a set of decreased albedo parameters. With well-tuned albedo parameters, the difference between simulated and measured irradiance averages on sensors can be lowered to 0.87%. When applying this to the irradiance sampling on cells, the optical average bifacial gain of 9.7% can be evaluated across all non-snow days.

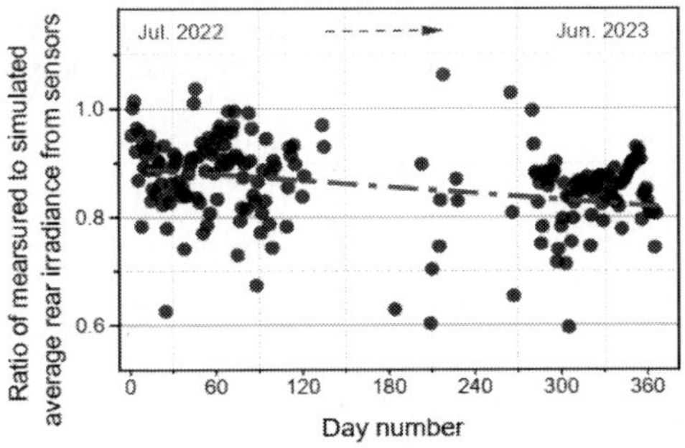

Figure 8 The ratio of measured to simulated daily averaged rear irradiance of days without snow on the ground. The red dash line is guide to eyes.

The hourly outputs with and without mismatch were simulated for all the non-snow days and then aggregated into days to calculate the ratio of measured to simulated daily energy outputs as shown in **Figure 9**. Clear sky filter is used to exclude the influence of low irradiance conditions. As **Figure 9** shows, the ratio is around 0.96 in July 2022 and then decreases to around 0.91 in June 2023. We attribute this reduction to the possible soiling and degradation effects on the system during the first year of installation. The annual degradation rate can be evaluated from the slope of the reduction as 5.6%. The value is very high compared to expected degradation of typical PV systems during the lifetime of the panels. However, this value shows only the first year of operation and it can be expected to slow down later.

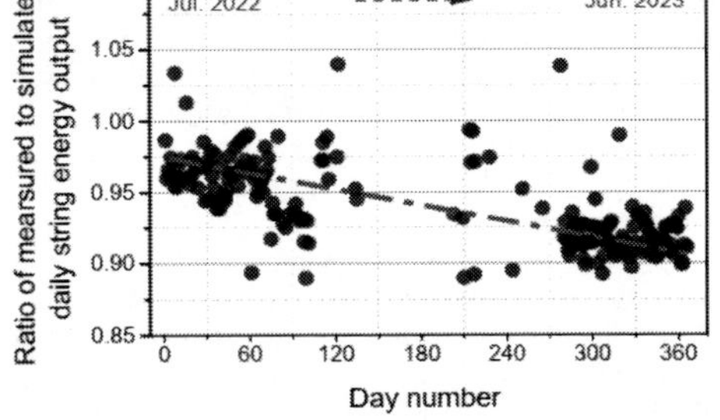

Figure 9 The ratio of measured to simulated daily averaged string power output of days without snow on the ground. The red dash line is guide to eyes.

It should be also noted that, as indicated in **Figure 8**, since the same albedo was used in all the simulations, the ratio of actual to simulated rear irradiance on the string will also decrease with time. This reduction should also contribute to the decrease in power output and cannot be easily distinguished from the electrical degradation unless hourly optimized albedo is used.

The dependence of mismatch loss on total irradiance was investigated. Two indicators of the standard deviation (STD, σ) and the mean absolute difference (MAD, Δ) were calculated as follows:

$$\sigma = \frac{1}{\bar{G}_{total}} \sqrt{\frac{\sum\left(G_{total,i} - \bar{G}_{total}\right)^2}{n}} \quad (2)$$

$$\Delta = \frac{1}{n^2 \bar{G}_{total}} \sqrt{\frac{\sum_{i=1}^{n} \sum_{j=1}^{n} \left|G_{total,i} - G_{total,j}\right|}{n}} \quad (3)$$

The relations between mismatch loss and irradiance variation STD and MAD are shown in **Figure 10**. The fitting proposed by Deline [8] is also drawn in **Figure 10** for comparison. For both indicators, the mismatch losses show a similar relationship between our case and Deline fitting in the low range of variation. However, when the variation is higher than 2%, the mismatch drops at a much steeper rate than the Deline fitting. This trend indicates that dependence could be case sensitive. The overall average power mismatch loss for all non-snow days was evaluated and found to be -0.37%.

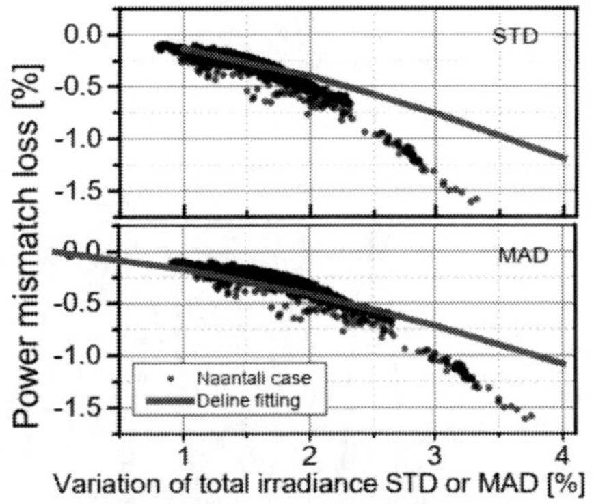

Figure 10 The dependence of power mismatch loss on the irradiance variation of STD or MAD of days without snow on the ground.

When analyzing the energy output across all non-snow days, the simulated energy output can be taken as the ideal energy output without any losses, which is 6.7 MWh for the 6.3 kWp target string, accounting for 87.5% of the total annual energy output. Compared to the measured energy output of 6.26 MWh, even considering the mismatch losses, there is still a gap of 0.41MWh, which accounts for about 6.1% of total energy production in non-snow days. Given that the annual degradation rate is 5.6%, the annual energy loss from degradation can be estimated to cover 2.8%-units by assuming a linear degradation. The remaining losses, approximately 3.3%-units, is then attributed to other factors.

10.4229/EUPVSEC2023/4BO.5.4
020297-005

(2) Days with snow

When the ground is covered by snow, the albedo is dramatically increased. **Figure** 11 shows the irradiance from back sensors for one snow week in March 2023. When the reference albedo for clear ground is used, the simulated average irradiance on sensors, which is shown as the red dot-line, is significantly lower than the measured values. For sunny days like 9 and 10 March, the rear irradiance is enhanced by almost twice at noon time.

The measured rear sensor irradiance can be well replicated with appropriately tuned parameters, as demonstrated by the black dot-line in **Figure** 11. The power output was then calculated and shown in
Figure 12. The difference between the red and black lines indicates the increased power output resulting from enhanced rear irradiance due to snow. However, snow may stick to the front side of the PV modules and lower the power output, particularly in the morning of the 9th and 11th.

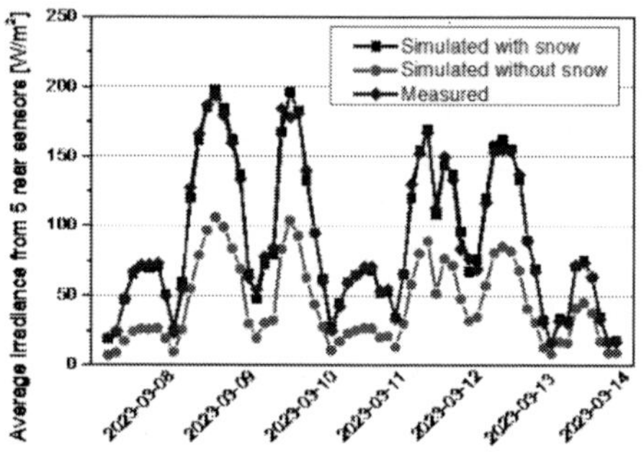

Figure 11 Measured and simulated hourly average irradiance of five rear sensors over one week with snow on the ground in March 2023.

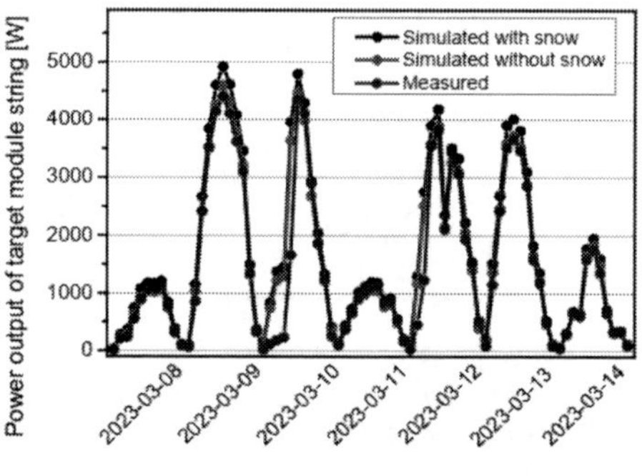

Figure 12 Measured and simulated hourly power output over one week with snow on the ground in March 2023.

Nevertheless, the problem remains that adjusting the albedo for every hour in all the snow days is unfeasible. The ground albedo changes gradually when the snow melts. Therefore, for the following simulations, we utilized a set of albedo parameters that produce an average irradiance on sensors which is comparable to the measured irradiance. With simulations for irradiance on cell and power output, the average bifacial gain during days with snow is 23.4%, which is 141% higher than for the days without snow. Assuming similar degradation and loss percentages as in non-snow days, we estimate that the snow coverage results in a loss of approximately 215 kWh, which accounts for 22.8% of the potential energy output during snow days.

(3) Energy analysis for the whole year

Here we consider both the days with and without snow to analyze the performance of the target string in one-year duration. The energy sources and losses are listed in Table 1. The energy output without mismatch is considered as the ideal energy output without any losses and is used as the reference to calculate the percentage of other components. Approximately 7% of the total energy output comes from the backside of the modules, with roughly 1% benefiting from snow. The mismatch loss accounts for around 0.5% of total energy production, which is a minor loss in our case, while the loss from snow coverage is relatively high at 2.8%. The findings suggest that, in our case, snow tends to have stronger negative effects than the positive one to the energy output. The proper estimation of snow-induced losses and the preventative measures for snow accumulation are important topics for the implementation of PV systems in high latitude areas.

Items			Energy [kWh]	Percentage [%]
Source	Front side		7105	92.97
	Rear side	Normal refelction	451	5.91
		Snow befeficial	86	1.13
Ideal total output			7642	100
Loss	Mismatch		37	0.48
	Snow coverage		215	2.81
	Degradation		214	2.80
	Others		257	3.36
Real total output			6920	90.55
Quasi-annual yield (for 343 days)			1098.45 hr	

Table 1 The Energy analysis of the target string in Naantali canopy system

3.2 Flat-Roof PV

1. Preliminary results

With the validated methodology, a preliminary simulation was performed for the Flat-Roof PV by using the meteorological data available from the Canopy-PV. As the system will be installed during autumn 2023, no DC power measurements are available for comparing the simulations, however preliminary inferences can be obtained for evaluating the benefits of having bifacial systems in this type of configurations.

a) Rear irradiance

For this evaluation, 3 modules were model in which was considered that the rear irradiance was the most affected. **Figure 13** shows the rear irradiance map at cell level for the 3 modules simulated where the edge effects are well defined. The 3rd module located at the end of the row (At the right) receives almost twice the light during some hours of the day compared to the modules that have

10.4229/EUPVSEC2023/4BO.5.4
020297-006

other modules on both sides, therefore a higher mismatch effects are expected.

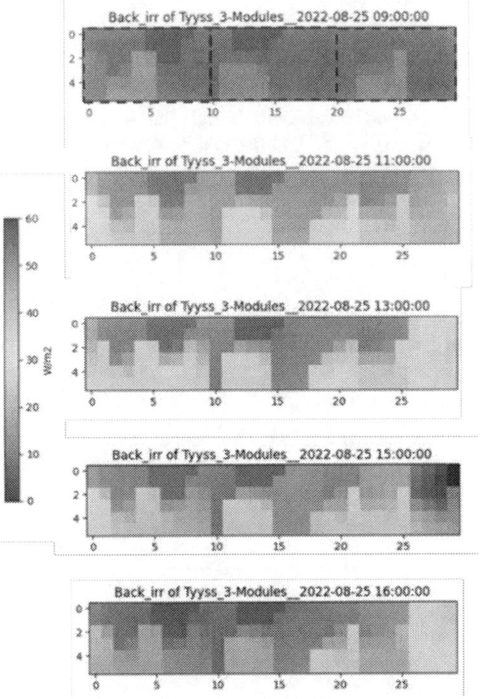

Figure 13: Rear irradiance map of the Flat-Roof array for 3 modules in a sunny day on 25 August 2022. The x- and y-axis indicate the number of cells. The perspective in the figure is from north to south. The three modules were chosen from the most representative array.

Optical bifacial gain from the Flat-Roof PV is shown in **Figure 14**. The estimated value during sunny days is around 2.5%, which is low due to the tilt angle and the clearance of less than 15 cm from the roof to the bottom edge of the modules. Average irradiance at the rear of the modules across the year was found to be 10 W/m^2 with maximum values up to 90 W/m^2.

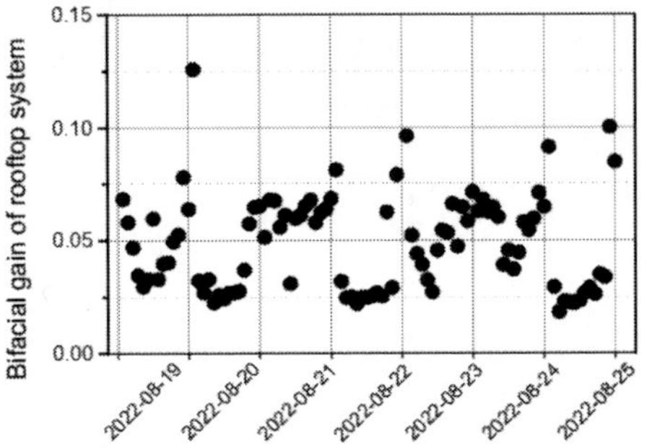

Figure 14: Hourly bifacial gain for the 3 modules, week from 19 to 25 in August 2022

b) Mismatch

The mismatch was calculated for this configuration as well **Figure 15**, the results show that a higher mismatch must be expected for this type of arrays, due to the racking systems and due to the amount of light limitations when the modules are in the middle of other modules compared to the ones at the end of the rows. At first glance seems that the mismatch is higher (estimated 4% for the whole year) for this configuration compared with the Canopy-PV, however a complete string of modules must be simulated, to make a proper comparison.

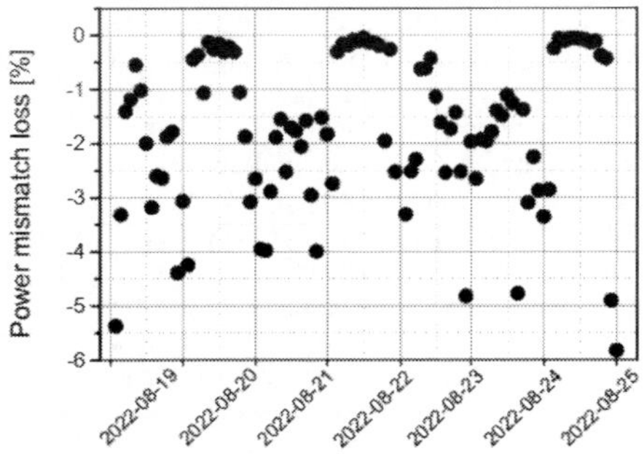

Figure 15. Hourly power mismatch for the Flat-Roof PV, from 19 to 25 in August 2022

c) Snow effects

Finally, a week of data was simulated with high albedo using the same snow conditions as presented for the Canopy-PV. Preliminary results show a 1.5 % increase in the power output of the systems when there is fresh now with an albedo value of 80%. It was found that the average irradiance was double at the rear of the modules reaching 11.5 W/m^2, under perfect snow conditions, i.e., fresh snow fully covering the roof. All these results to be evaluated once field data has been gathered.

4 CONCLUSIONS

Two BPV systems at high latitude sites were simulated using backward ray-tracing technique in combination with detailed 3D models and on-site meteorological data. An improved irradiance sampling method was developed to simulate an accurate irradiance map on each cell of the modules with high-speed asynchronous execution. This enabled detailed and accurate simulation of BPV system in a relatively short time. The goal was to evaluate their energy generation efficiency and losses, particularly in high-latitude locations with variable seasonal conditions, including snow.

The study validated the model's accuracy by comparing simulated and measured rear irradiance and DC power data. The power output of the Canopy-PV system was simulated over one year and compared with monitored data for an analysis of energy gain and loss. It showed that, with an annual bifacial gain of 7.5%, 7% of the total energy comes from the irradiance on the backside of the PV modules. The power mismatch loss due to irradiance variation on the backside was evaluated as 0.5%, which was not a major loss in our case. Snow, which is common during winter at high latitudes, had two opposing effects. On sunny days, it can double the irradiance on backside. The energy output resulted from this enhancement accounts for about 1% of the total energy. On the other hand, snow coverage on the front side could block the front irradiance and leads to a loss of 2.8% of the total energy. Therefore, appropriate snow clean-up measures might be important to further optimize the yield of PV systems in high latitude regions.

Preliminary results for the Flat-Roof PV array revealed that rear irradiance is significantly influenced by the module's position within the array, with edge modules receiving nearly twice the light. Bifacial gain during sunny days was estimated at approximately 2.5%, affected by factors such as the array's tilt angle and clearance. Additionally, mismatch calculations indicated a potential for higher mismatch in this configuration compared to the Canopy-PV system due to the presence of racking systems and shading from neighboring modules. Furthermore, simulations suggested a 1.7% increase in power output during snowy conditions, with rear irradiance peaking at 11.5 W/m^2 under ideal snow conditions.

Overall the study demonstrated the effectiveness of ray-tracing modeling for BPV systems, allowing for accurate predictions of rear irradiance, bifacial gain, power output, and mismatch. It highlighted the significance of variable ground conditions, snow effects, and module-level variations in BPV system performance. These findings emphasize the importance of considering these factors when designing and implementing BPV systems, especially in high-latitude regions with seasonal variations.

5 ACKNOWLEDGMENTS

The project has received funding from the European Union's Horizon 2020 research and innovation program under Grant Agreements 957751 (RESPONSE) and 957769 (TIGON) as well as Finnish Cultural Foundation.

6 REFERENCES

[1] B. Marion et al., "A Practical Irradiance Model for Bifacial PV Modules," 2017 IEEE 44th Photovoltaic Specialist Conference (PVSC), Washington, DC, USA, 2017, pp. 1537-1542, doi: 10.1109/PVSC.2017.8366263
[2] Anoma, M., Jacob, D., Bourne, B.C., Scholl, J.A., Riley, D.M. and Hansen, C.W., 2017. View Factor Model and Validation for Bifacial PV and Diffuse Shade on Single-Axis Trackers. In 44th IEEE Photovoltaic Specialist Conference.
[3] Ward, Gregory J., Francis M. Rubinstein, Robert D. Clear, (1988). A Ray Tracing Solution for Diffuse Interreflection, Computer Graphics, Volume 22.
[4] Ward, G. J. (1994). The RADIANCE lighting simulation and rendering system. In 21st Annual Conference on Computer Graphics and Interactive Techniques (pp. 459–472). doi:10.1145/192161.192286
[5] Ayala Pelaez et al., (2020). Bifacial_radiance: a python package for modeling bifacial solar photovoltaic systems. Journal of Open-Source Software, 5(50), 1865, https://doi.org/10.21105/joss.01865
[6] Ranta S. et al., (2022). Bifacial PV Canopy System in High Latitude, Model Development and Validation with First Months of Monitoring Data. 8th World Conference on Photovoltaic Energy Conversion, doi: 10.4229/WCPEC-82022-4DO.5.6
[7] Nathaniel L. Jones & Christoph F. Reinhart (2019) Effects of real-time simulation feedback on design for visual comfort, Journal of Building Performance Simulation, 12:3, 343-361, DOI: 10.1080/19401493.2018.1449889
[8] Liang, Tian Shen, Pravettoni, Mauro, Deline, Christopher A., Stein, Joshua S., Kopecek, Radovan, Singh, Jai Prakash, Luo, Wei, Wang, Yan, Aberle, Armin G., and Khoo, Yong Shen. A Review of Crystalline Silicon Bifacial Photovoltaic Performance Characterization and Simulation. United States: N. p., 2018. doi:10.1039/C8EE02184H

ESTIMATION OF NEXT-DAY PV POWER AND PREDICTION ERROR RANGE USING PHYSICAL MODEL AND SHORT-TERM LEARNING BY GRADIENT BOOST DECISION TREE

Masaki Sato[1], Kenji Utsunomiya[2], Jun Sasaki[2], Maki Okada[2], Shigeyuki Yoshikawa[2], Koji Yamaguchi[2], Yuzuru Ueda[1]
[1]Tokyo University of Science, 1-3, Kagurazaka, Shinjuku, Tokyo 162-8601, Japan
[2]Japan Weather Association, 55F Sunshine 60, 3-1-1 Higashi-Ikebukuro, Toshima-ku, Tokyo 170-6055, Japan
4322519@ed.tus.ac.jp

ABSTRACT: With the rapid global expansion of photovoltaic(PV) power plant, PV power generation prediction is an essential technology. Therefore, we developed PV power generation predicting technologies that uses forecasted or estimated weather data. Proposing method calculates not only PV power generation, but also the prediction error range. This method is divided into two major steps. The first step is called 'physical model' based on the proven JIS method and statistical processing. The next step is a more accurate improvement of the PV power generation using machine learning. Losses that could not be removed in the first step, such as shadows on the modules, are calculated by machine learning using past data from the last 30 days. Here, we used a machine learning called Gradient Boost Decision Tree (GBDT). At the same time as the loss is calculated, the prediction error range is calculated using quantile regression. We tested the accuracy of this method using weather data calculated from our short-term weather forecasting technology. This method showed a certain improvement in accuracy compared to the only physical model. In addition, a comparison with measured PV power generation showed that the prediction error range could also be estimated appropriately.
Keywords : estimation, prediction, machine learning, gradient boost decision tree

1. INTRODUCTION

1.1 Background

Photovoltaic(PV) power generation is becoming increasingly popular to address future energy problems. However, PV power generation is uncertain because the energy source is solar radiation. This is because it is obtained intermittently and is non-deterministic depending on weather, date and time. Therefore, in order to use PV as a core energy source, it is necessary to develop technology which can forecast PV power generation. In this study, we have developed a PV power generation predicting technology using forecasted or estimated weather data.

1.2 Aim and Approach

We are developing a method for predicting PV power generation with the goal of establishing PV power as a core energy source. In particular, the aim of this study is to develop technology that can be applied to any PV power plant with weather forecast data and site installation data.

To meet this aim, our approach is to combine a physical model with a machine learning model to form a composite model. And this model estimates the prediction error range at the same time as the PV power generation. This allows the validity of the prediction to be measured on the day of the prediction.

The physical model is based on the proven Japanese Industry Standard(JIS) method[2] and can calculate PV power generation taking into account multiple losses from weather forecast values and installation site information. Calculating PV power generation with this model first helps to build a robust prediction model that maintains a certain level of accuracy. It also helps to reduce the black-boxing of machine learning by highlighting losses that are not eliminated and helps with variable selection.

The machine learning model, with 30 days of short-term past data, empirically calculates the losses not eliminated by physical model. There are two reasons for limiting the past data to 30 days: one is to be able to introduce this method with a small number of past data, and the other is to collect train data that is close to the most recent situation without the influence of seasonality.

The prediction error range is calculated by quantile regression. The purpose of the range calculation is to determine whether the previous day's prediction should be corrected or not. A same-day re-prediction allows for building a more flexible predicting model that is robust to sudden condition changes, such as downpour, snow cover or cell failure.

2. METHOD

2.1 Physical Model

The first major step is to calculate the estimated PV power generation of the physical model. The ideal PV power generation obtained from the rated output and the solar irradiation. To calculate irradiation on the tilted surface(H_{Ag}) from the global horizontal irradiance(GHI), we use the general method[3][4] of separating direct and diffuse insolation on a horizontal surface. However, when the height angle of the sun is small, the value of the direct reaching irradiation may diverge in the calculation model. Therefore, cases where the height angle is less than 3 deg are treated as 3 deg.

Then, we consider several calculable losses. The efficiency loss caused by each of these losses is calculated. The estimated value from the physical model is multiplied by these coefficients. The three losses we can calculate are a) Temperature loss, b) Incident angle loss, and c) System power reduction loss.

a) Temperature Loss

To account for changes in solar cell characteristics due to temperature changes, we estimate the module temperature T_{PA} and calculate the temperature loss coefficient K_T . First, the module temperature T_{PA} is estimated using Equation(1)[2].

$$T_{\mathrm{PA}} = T_{\mathrm{A}} + \left(\frac{A}{BV^{0.8}+1} + 2\right) \times G_{\mathrm{A}} - 2 \qquad (1)$$

where G_{A} is the solar irradiation on the tilted surface, T_{A} is the air temperature, and V is the wind speed. A and B are coefficients depending on the solar array installation method, and the coefficients for the stand-mounted type ($A = 46$, $B = 0.41$) are used in this study.

And then, following Equation(2) is used to calculate the temperature loss coefficient K_{T}.

$$K_{\mathrm{T}} = 1 + \alpha_{\mathrm{Pmax}}(T_{\mathrm{PA}} - T_0) \quad (2)$$

where T_0 is the module temperature under STC, and α_{Pmax} is the temperature coefficient of maximum output.

b) Incident Angle Loss

Ideally, the incidence of solar radiation on PV modules should be always reach vertically to the module surface. However, any PV system is affected by changes in the direct irradiance due to the ever-changing position of the sun, diffusing in the atmosphere, and reflections from the ground surface. Therefore, the incident angle loss coefficient K_{I} is calculated from Equation(3), considering the configuration and thickness of the PV module.

$$K_{\mathrm{I}} = \frac{1 - \frac{r_{\mathrm{b}(\theta)} \cdot G_{\mathrm{Ab}} + r_{\mathrm{d}(\theta\mathrm{ed})} \cdot G_{\mathrm{Ad}} + r_{\mathrm{r}(\theta\mathrm{er})} \cdot G_{\mathrm{Ar}}}{G_{\mathrm{Ag}}}}{\tau_0} \quad (3)$$

where G_{Ab} is the solar beam irradiation intensity on the tilted surface, G_{Ad} is the diffuse irradiation intensity on the tilted surface, and G_{Ar} is the reflected irradiation intensity on the tilted surface. $r_{\mathrm{b}(\theta)}$ is the reflectance of direct component at incident angle θ, $r_{\mathrm{d}(\theta\mathrm{ed})}$ and $r_{\mathrm{r}(\theta\mathrm{er})}$ are the reflectance of diffuse and reflected components at inclination angle β, and θ_{ed} and θ_{er} are the equivalent incident angles of diffuse and reflected irradiation.

c) System Power Reduction Loss

The main causes of the system power reduction loss of the PV system are series-parallel imbalance, module surface soiling, and aging. To estimate these factors, the system power reduction loss is calculated as a steady-state error during the evaluation period as follows.

Firstly, Equation(4) is used to calculate the ratio a_{SPL} of the measured value E_{mea}, the calculated temperature loss K_{T} and incident angle loss K_{I}.

$$a_{\mathrm{SPL}} = \frac{1}{K_{\mathrm{T}} \cdot K_{\mathrm{I}}} \cdot \frac{E_{\mathrm{mea}}}{P_{\mathrm{ACap}}} \cdot \frac{G_{\mathrm{S}}}{H_{\mathrm{Ag}}} \quad (4)$$

where P_{ACap} is the PV array rated power and G_s is the standard solar irradiation intensity(= 1000 W/m²).

Secondly, from all a_{SPL}, we extract those with "tilted surface solar irradiation intensity greater than 200 W/m²" and " diffuse ratio D_{g} less than 0.8". The diffuse ratio D_{g} is calculated from the horizontal diffuse irradiation H_{d} and the global horizontal irradiation H_{g} using Equation(5).

$$D_{\mathrm{g}} = \frac{H_{\mathrm{d}}}{H_{\mathrm{g}}} \quad (5)$$

Finally, we create a histogram of the extracted a_{SPL} and obtain the maximum class from it. The average value of a_{SPL} for the class is then used as the system power reduction loss coefficient K_{SPL}.

2.2 Short-Term learning

2.2.1 Other Losses

The second major step is to correct the physical model by machine learning in short-term. In addition to the losses removed by the physical model, there are “other losses” that are difficult to calculate using only data that can be forecasted and measured, such as the effects of shading, surrounding environment change and errors in forecast weather data. We calculate these losses using machine learning called Gradient Boost Decision Tree(GBDT).

When we enter weather data, we calculate the error (Δ_{oth}) derived from “other losses”. From the above, we obtain the estimated PV power generation E_{est} from Equation(6).

$$E_{\mathrm{est}} = \left(P_{\mathrm{ACap}} \times \frac{H_{\mathrm{Ag}}}{G_s} \times K_{\mathrm{T}} \times K_{\mathrm{I}} \times K_{\mathrm{SPL}}\right) - \Delta_{\mathrm{oth}} \quad (6)$$

2.2.2 Gradient Boost Decision Tree(GBDT)

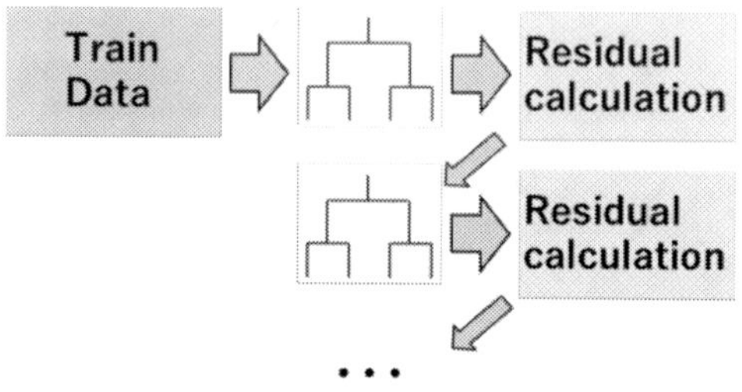

Fig.1 Conceptual figure of the GBDT

In the GBDT, to generalize the relation between the response and explanatory variables, we also define a "loss function," which is the mean of the errors between the estimated and actual values in all data. In this paper, for the loss function, the mean squared error is used.

The GBDT is unique in that it minimizes the loss function by gradient descent. Fig. 1 shows a conceptual figure of the GBDT. As we start this learning, train data are inserted into the decision tree to obtain residuals. From the residuals, the next decision tree brings the next residual. This is repeated many times to minimize it. This process is done using a gradient to rationalize a series of steps.

The gradient is computed at each data point and compared to other data points, moving in the direction with the greatest value. This move is completed when the difference between iterations converges, or when the maximum number such of iterations is reached. The loss function at this point is used as the optimal function when creating the estimate model.

GBDT has both advantages and disadvantages. The advantages are good performance even with small data sets and the ability to generate flexible predicting models by hyper-parameter properties. On the other hand, the disadvantage is that it is susceptible to outliers. Thus, outliers must be removed from the train data to improve prediction accuracy.

The main reason why the GBDT was chosen for this method is because it is an algorithm that can make accurate predictions with a small amount of data. Compared to other algorithms, such as neural networks, LSTM and so on, GBDT requires less data, making it a suitable algorithm for this method which has a limit of the amount of past data.

2.2.3 Train Data Set

The train data set is unique in this method. This is because of the 30-day limit on past data. As mentioned above, this is for two reasons: to make it possible to run with a small number of past data, and to form train data that is closer to the target date of the prediction.

10.4229/EUPVSEC2023/4BO.6.1
020299-002

Specifically, the following shows how the train data is acquired. The time of the prediction target is set to "t".

First, the same time-period data for the past 30 days at "t" is acquired. In this paper, the past 1-day data consists of 25 data points including the 60 minutes(12 time-stamps) before and after "t". This data is collected for the past 30 days. In other words, 750 points of past data are collected for all prediction. Fig. 2 shows an example of data collection when 12:00 is set as "t".

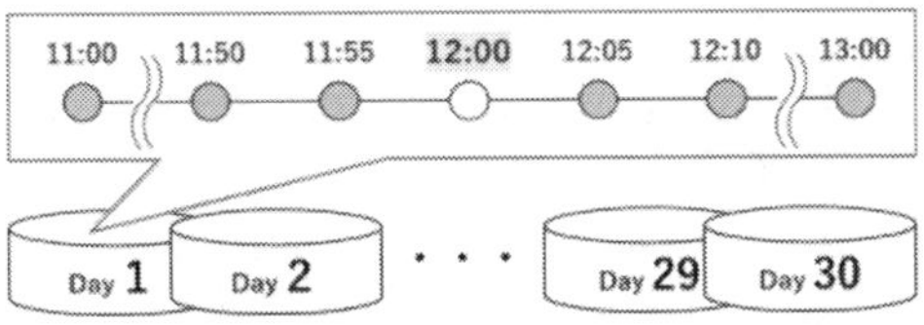

Fig.2 Example of data collection

Next, outlier elimination is applied. The elimination is performed by a process called Tukey's method. The following is a description of this method.

(i) Calculate the difference between "first quartile Q_1" and "third quartile Q_3" of data sets. This is defined as Δ_{qua}.

$$\Delta_{\text{qua}} = Q_3 - Q_1 \tag{7}$$

(ii) Form a range using Δ_{qua} as in Equation(8).

$$X < Q_1 - \Delta_{\text{qua}} \times 1.5, \quad Q_3 + \Delta_{\text{qua}} \times 1.5 < X \tag{8}$$

(iii) Determine outliers outside this range as X.

Applying this to both explanatory and response variables. This process helps refine the train data.

2.2.4 Response and Explanatory Variables

In the past 30 days of data, the response and explanatory variables used for learning are normalized.

The response variable is the error between actual measured value and the physical model estimated value, and four explanatory variables are introduced: GHI, temperature, sun azimuth angle, and sun height angle. The GHI is introduced to account for changes in solar radiation and weather, temperature to account for changes in the ambient environment, and the sun azimuth and height angles to account for seasonality, such as shadow.

2.3 Prediction Error Range

The prediction error range is used to identify errors or anomalies. If the value of a new sample is within this range, then it is an expected result. If it is outside the range, the data is likely to be irregular. When identifying anomalies, a numerical value, a threshold, is needed as a reference. The prediction error range can be used as the basis.

In the machine learning of the GBDT described above, the prediction error range is also calculated at the same time. We use python to calculate the prediction error range by quantile regression. Quantile Regression is an extension of the loss function in linear regression. Ordinary regression analysis estimates the mean value of the response variable given an explanatory variable. In contrast, quantile regression estimates the distribution location of the conditional quantiles(median, quartiles, deciles, etc.). This calculation does not require the creation of a special data set and is performed from all given past data. It allows the influence of the explanatory variable on the distribution of the response variable to be evaluated at different positions of the distribution.

The specific calculation means that instead of calculating the squared error and optimizing the mean, the loss function is optimized using a predefined quantile as τ. The loss function $Loss$ at data y is shown in Equation(9) below. Loss is the weighted sum of the difference between the correct and predicted values, τ when it is positive and $(1-\tau)$ when it is negative.

$$Loss = (\tau - 1)\int_{-\infty}^{\hat{y}} (y - \hat{y}) f(y) \mathrm{d}y + \tau \int_{-\infty}^{\hat{y}} (y - \hat{y}) f(y) \mathrm{d}y \tag{9}$$

The prediction error range is calculated using this algorithm. In this paper, 80% prediction error range is calculated by estimating the upper 10% and lower 10% distributions of the response variable, and this calculated result is used. The following Fig.3 shows the illustration.

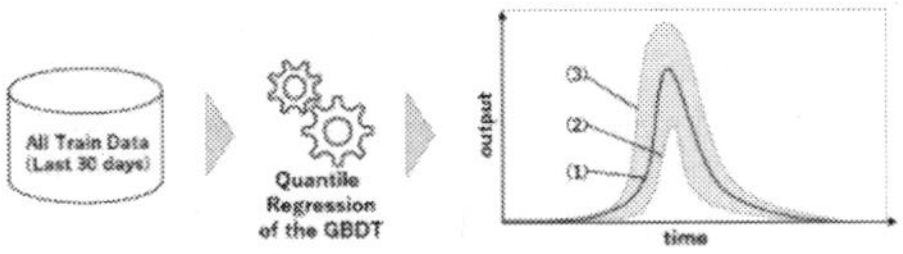

Fig.3 Application of quantile regression.
(1) : Ordinary regression result
(2) : Lower 10% distributions
(3) : Upper 10% distributions

2.4 Flowchart of Learning and Range Calculation

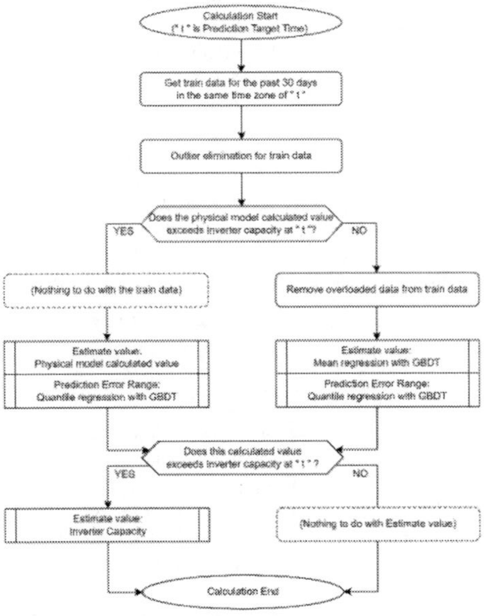

Fig.4 Flowchart of learning and range calculation

The sequence of learning and range calculation is shown in Fig.4.

If the physical model determines that the power generation estimated exceeds the inverter capacity, no learning is performed to estimate the power generation. This is because it is not "other losses" that are assumed to be compensated by learning. The estimated value is then checked to ensure that it doesn't exceed the inverter capacity again.

10.4229/EUPVSEC2023/4BO.6.1
020299-003

2.5 Error Indexes

In the performance evaluation, we used the following 3 error index. Coverage is an index that calculates how much of the actual measured value is within the calculated range. Inv_{cap} is the rated capacity of the inverter.

$$MAPE_{\mathrm{Inv}} = \frac{1}{N}\sum_{i=1}^{N}\left|\frac{E_{\mathrm{mea}\,i} - E_{\mathrm{est}\,i}}{Inv_{\mathrm{cap}}}\right| \times 100\ [\%] \qquad (10)$$

$$RMSE_{\mathrm{Inv}} = \frac{\sqrt{\frac{1}{N}\sum_{i=1}^{N}(E_{\mathrm{mea}\,i} - E_{\mathrm{est}\,i})^2}}{Inv_{\mathrm{cap}}} \times 100\ [\%] \qquad (11)$$

$$coverage = \frac{N - N_x}{N} \times 100\ [\%] \qquad (12)$$

3. RESULTS

3.1 Data

Used datasets are as follow: Measured PV power generation, Predicted GHI, Measured temperature, and Measured wind speed. These all are 5-minuite value.

- Predicted GHI:
  Calculated values from our short-term predicting model. Since there is no on-site data, we use the calculated values from the model created by Japan Weather Association(JWA).
- Measured temperature and wind speed:
  Measurements from the nearest meteorological office.
- Measured PV power generation:
  Actual data of mega-solar in eastern-Japan, where several MW is installed. There are four seasons, with an annual mean precipitation of more than 1500 mm and an average temperature of about 15 °C.

3.2 One-Week Estimated Result

The one-week estimated result of prediction accuracy verification are shown in Fig.5. This figure is an estimated result for one-week, from 1 to 7 August.

Here, the vertical axis is set in units of "ratio of inverter rated capacity" as "1", and the horizontal axis is set as "0h" at 0:00 on the 1st day and "168h" at 24:00 on the 7th day.

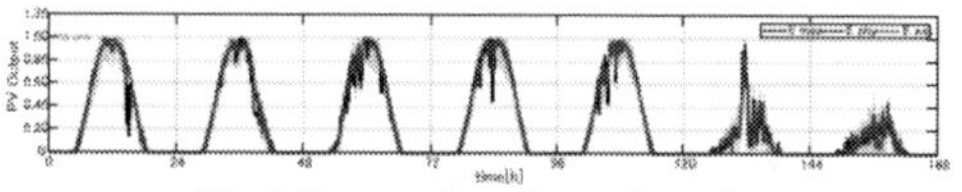

Fig.5 One-week estimated result

Black line : Measured value E_{mea}
Blue line : Physical model estimated E_{phy}
Red line : Result after machine learning E_{est}
Pink line : Limited value of Inverter capacity
Shaded Red area : Estimated prediction error range

The results show that the estimated value E_{est} can follow the measured value E_{mea} overall. On the other hand, we can also confirm a part not to correct for underprediction. This could be improved in learning to change the building way of the train dataset or to adjust the hyperparameters more suitable.

The results are evaluated by MAPE, RMSE and coverage as follows.

Table 1 One-week estimated result

		8/1	8/2	8/3	8/4	8/5	8/6	8/7	7days ave.
MAPE	Only Phy.	4.91	5.10	5.05	4.70	4.17	4.18	2.91	4.44
	Phy. + GBDT	4.52	4.02	3.58	2.64	3.28	4.89	3.52	3.78
RMSE	Only Phy.	10.80	7.72	7.36	6.36	5.75	7.14	4.12	7.03
	Phy. + GBDT	10.26	6.29	5.75	4.49	5.74	7.92	5.59	6.58
Coverage		87.50	84.90	89.06	96.88	91.67	93.23	90.63	90.55

On sunny days, the results are improved compared to the estimation based on only physical model.

The result of the coverage is about 90%. From this, it is found that the estimated prediction error range almost covers the measured value.

3.3 One-Year Estimated Result

Similarly, the 2018 results of the estimation using the proposed method are shown.

Table 2 One-year estimated result

		Jan.	Feb.	Mar.	Apr.	May	Jun.	Jul.	Aug.	Sep.	Oct.	Nov.	Dec.	Ave.
MAPE	Only Phy.	6.26	5.40	4.01	4.77	4.94	7.12	8.40	5.52	5.31	5.85	4.58	4.95	5.61
	Phy. + GBDT	5.98	5.01	3.28	4.27	4.80	6.52	7.81	5.04	4.68	4.88	4.34	4.61	5.11
RMSE	Only Phy.	9.90	9.11	6.83	7.68	7.71	10.12	12.02	8.77	8.84	8.61	7.62	8.59	8.81
	Phy. + GBDT	8.62	8.49	5.14	7.14	7.37	9.68	11.80	8.34	7.23	8.28	7.52	8.92	8.13
Coverage		68.92	70.68	65.27	68.71	69.79	73.96	78.04	79.64	78.96	76.53	81.53	88.41	73.74

In Table 2, the machine learning estimates are more accurate than only physical model throughout the year. As for the prediction error range, coverage is around 70% and needs further improvement.

4. CONCLUSIONS AND FUTURE WORK

In this paper, we proposed a method for predicting PV power generation using a physical model and machine learning, based on weather observation data such as GHI. The values obtained from this predicting model were close to the actual measured values. We also proposed a method to calculate the prediction error range by quantile regression, which allows us to determine times that deviate significantly from the previous day's prediction.

As a future work, we will commit 3 tasks. First, to further improve the prediction accuracy. Second, to make the transition from prediction to forecast. Finally, to develop a same-day correction method for forecast generation using the obtained prediction error ranges.

Acknowledgement

The PV power plant data used in this study was provided by NTT Anode Energy Corporation and this study was supported by NEDO "Development of Solar Radiation Forecast Technology for Short-Term Prediction of Power Generation(Research and Development of Technology for Predicting and Evaluating Power Generation)"

Reference

[1] Yuzuru Ueda, "A study on an advanced performance analysis method for PV systems using monitoring data", Tokyo University of Agriculture and Technology, 2007, Ph. D. thesis.
[2] Japanese Industrial Standards Committee, "JIS C 8907"
[3] D.G.Erbs, S.A.Klein and J.A.Duffie, 'Estimation of the Diffuse Radiation Fraction for Hourly, Daily and Monthly-Averag Global Radiation' Solar Energy Vol.28, No.4, pp.293-302
[4] B.Y.H.Liu and R.C.Jordan, "A Rational Procedure for Predicting the Long-term Average Performance of Flat Plate Solar Energy Collectors" Solar Energy, Vol.7, No.2, pp.53-74

10.4229/EUPVSEC2023/4BO.6.1
020299-004

UNDERSTANDING THE CONSEQUENCES OF SWITCHING TO A PREDICTIVE O&M STRATEGY

Eduardo A. Sarquis Filho[1,2], Björn Müller[2], Paulo J. Costa Branco[1]
[1] IDMEC, Instituto Superior Técnico, Universidade de Lisboa, Av. Rovisco Pais, 1049-001 Lisboa, Portugal,
eduardosarquis@tecnico.ulisboa.pt, pbranco@tecnico.ulisboa.pt
[2] Enmova GmbH, Basler Str. 115, 79115 Freiburg, Germany, eduardo.sarquis@enmova.de, bjoern.mueller@enmova.de

ABSTRACT: Amid growing digitalization in the photovoltaic industry, there is a trend towards increasingly complex and comprehensive monitoring solutions. For the technological advancement of O&M practices, it is fundamental to balance the expectations and objectively quantify the potential gains from investing in better monitoring and maintenance practices. In this study, we develop a methodology to estimate the effects a maintenance strategy can have on a portfolio of systems, and we use it to evaluate the potential benefits of adopting a maintenance strategy based mainly on predictive actions, considering the realistic limitations of monitoring based on automatic fault detection tools. The results show that a predictive strategy can reduce the need for urgent interventions and reduce the total number of interventions by 10% while maintaining average performance comparable to current best practices. These benefits depend on the response time between failure detection and maintenance. A shorter response time (e.g., 3 months) can enhance the average performance of the systems, leading to a 5.8% higher energy yield, with a 16% increase in the average number of interventions.

1 INTRODUCTION

In the continuing technological evolution of photovoltaic (PV) systems, there is a trend toward the expansion of sensing and digitalization of operation and production data. This trend reflects in the increasingly complex and comprehensive online monitoring possibilities for PV systems. In recent years, best-practice manuals have included the adoption of a maintenance paradigm based on predictive analysis as a real possibility. In practice, it is about extracting information from collected data that can be used to improve operation and maintenance (O&M) management [1].

In conventional maintenance practices, preventive actions are based on a pre-established calendar of interventions [1]. There is no assessment of the real need for intervention. Systems without problems are inspected only as a routine. On the other hand, systems can remain with unnoticed failures for a long time until the next inspection occurs. In a digitalized world, remotely collected information can help assess the need for intervention. Reduce the frequency of interventions when possible or anticipate them if necessary.

Corrective interventions are usually urgent and therefore have little planning time. This often translates into more expensive and logistically difficult interventions. Information that allows identifying problems at an early stage, before a critical failure occurs, can extend the time available to plan interventions and reduce the need for urgent actions.

Predictive maintenance is expected to reduce maintenance effort without compromising system performance. Or, improve performance without increasing maintenance efforts. The success of the predictive approach depends on the quality of the data interpretation. To decide to postpone or anticipate an inspection or maintenance intervention, high diagnostic reliability is required. It is key that the automatic monitoring system triggers as few false alerts as possible and ensures no critical problem passes unnoticed.

For the technological advancement of O&M practices, it is important to balance the expectations and objectively quantify the potential gains from investing in improved monitoring and maintenance practices. Various of aspects should be considered, such as the frequency with which faults occur, their impact on system performance, the procedures adopted to detect failures, the sensitivity of the tools used for monitoring and inspection, and the reaction time for maintenance. In this study, we collect information on these aspects and propose a methodology to combine them in order to estimate the effects that a maintenance strategy can have on a portfolio of systems. Using this methodology, we evaluate the potential benefits of adopting a maintenance strategy based mainly on predictive actions, taking into account the realistic limitations of a monitoring based on automatic fault detection tools.

The scope of this study is limited to basic technical aspects related to the O&M strategy, excluding other aspects such as the availability of spare parts, qualified personnel, transport logistics, contractual and safety requirements, etc.

2 METHODOLOGY

The designated function of a photovoltaic system is to generate electrical power, and the occurrence of faults will cause a reduction in system performance until the fault is detected and removed. The total energy yield loss will be a consequence of the severity of the fault and the time elapsed between its occurrence and repair. Different O&M strategies may lead to a shorter or longer time to detect and repair, resulting in a lower or higher total energy yield loss. Quantifying this yield loss is fundamentally important for comparing strategies. To do this, we propose a methodology with five main steps:

1. Define the average PV system design
2. Calculate the expected energy yield
3. Generate a sequence of fault events and their respective losses
4. Add inspection and maintenance events according to the O&M strategy
5. Calculate the impact of events on the energy yield

In addition to evaluating the performance of the systems, we measure the maintenance efforts made over the years, characterized by the number of interventions and the number of repaired components.

10.4229/EUPVSEC2023/4BO.6.2
020300-001

3 AVERAGE PV SYSTEM DESIGN AND EXPECTED ENERGY YIELD

PV systems are designed to match local conditions, but despite their individual characteristics, they are built with very similar architecture and components due to constraints imposed by other factors such as standards, regulations, procurement optimization, etc. It is reasonable to assume that a group of systems with such similar characteristics can be fairly represented by a typical system whose specifications are an average design.

In this study, we considered a typical rooftop PV system with the following specifications:

- PV module with nominal power 255 W.
- PV array with two strings, each with 24 PV modules in series, summing up 12.24 kWp.
- PV array orientation towards the south (azimuth 0 degrees south) and 18 degrees tilt.
- String-type inverters with a nominal power of 20 kW and two MPPT inputs.
- PV system with six inverters, each connected to two arrays, totaling 146.88 kWp of installed capacity.
- Expected lifetime 25 years.

The expected energy yield for the typical system can be obtained using a project design tool for PV systems, e.g., PVSyst. The fault events will be added later, thus the simulation of expected energy yield must consider no losses due to soiling or shading, and 100% availability. The performance loss rate was assumed to be 0.4%/year. The simulation provided a time series of energy yield in kWh/kWp per day for 25 years, and the weather conditions reflect the typical meteorological year (TMY) of Freiburg, Germany, as shown in Figure 1.

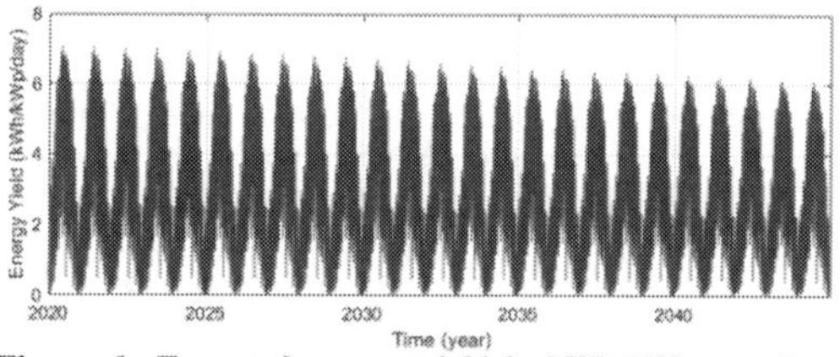

Figure 1: Expected energy yield in kWh/kWp per day simulated using PVSyst for 25 years.

4 STORIES OF FAULT EVENTS AND LOSSES

A component fault is an unexpected event that results in its loss of ability to perform its function as expected. To generate examples of failure events, we rely on reliability engineering theory for modeling failure occurrence.

Two important parameters used to characterize the reliability of a component are *failure rate* (λ) and *mean time to failure* (MTTF). The failure rate is the probability of failure per unit time, given that the component has not yet failed. The MTTF is the average time to the first failure. For most of their operating life, electrical components have a constant failure rate. A constant failure rate leads the time-to-failure $f(t)$ to be an exponential distribution [2] with λ as the rate parameter, as shown in (1). Then the MTTF are defined as in (2).

$$f(t) = \lambda \exp(-\lambda\, t) \tag{1}$$

$$\mathrm{MTTF} = \int t\, f(t)\, dt = 1/\lambda \tag{2}$$

To generate examples of fault events, for each component and fault type, we define continuous random variables that follow an exponential distribution with λ equal to its failure rate. These random variables generates random samples, representing the specific time each component will be affected by each failure type. From a previous study [3], we have a list of the typical failure types for each type of component and their respective failure rates.

Table I shows an example of the sample values for each of the five types of failure that can affect the six inverters. Samples after the end of life (25 years) should be excluded and only the first failure of each inverter is considered. In this way, we assume that faults per component are non-cumulative, i.e., each component can have only one fault occurrence. This assumption is reasonable as long as the maintenance of the failed component does not take long to occur.

Repeating this process for all components, we create a story of failure events over the years of operation.

Table I: Sample values representing the time (in years) that the inverters will be affected by each type of failure.

Fault type	Inverter					
	1	2	3	4	5	6
Inverter capacitor	1.04	**0.50**	10.5	0.46	2.32	11.8
Inverter control com	3.23	1.52	**0.11**	**0.09**	4.67	**0.08**
Inverter cooling fan	**0.19**	3.09	0.95	8.35	3.69	0.38
Inverter IGBT	2.09	0.37	3.78	29.8	**1.46**	0.61
Inverter relay	15.9	11.4	12.2	38.6	17.4	72.7

4.1 Calculate losses

Each failure event causes the total or partial loss of performance of the affected component, which in turn results in reduced performance of the system as a whole. To estimate the impact of faults on system performance, we start by defining how severe the faults are at the component level.

The range of power loss associated with typical failures affecting an inverter or a PV array is shown in Figure 2. For PV array component faults, losses were defined in terms of loss in array performance and were calculated specifically for the PV array configuration defined in Section 3 using the modeling framework defined in [4]. The severity of each fault event is estimated to be somewhere between the worst and the best case.

To calculate the resulting system performance loss, we must know where the faulty components are located and consider the interconnection between the components according to the number of modules per string, strings per array, and arrays per inverter given in the system specifications. To indicate the location of the fault, we identify each component by a number and assign a location number to each fault randomly, according to the type of component.

The total loss in array performance was considered to be the cumulative sum of the loss in its component performances, i.e., losses are added linearly within an array and cannot exceed 100%. The efficiency of a set of arrays connected to the same inverter is the average of their individual efficiencies.

Similarly, the efficiency of an inverter derives from the cumulative sum of the performance losses in that inverter over time. The efficiency of a subsystem is the product of the efficiency of the inverter and the efficiency

10.4229/EUPVSEC2023/4BO.6.2
020300-002

of the set of arrays connected to it. Finally, the efficiency of the system is the average of the efficiencies of the subsystems.

The fault occurrence story, containing the types of faults, their locations and their severities, is then converted into a time series representing the loss of system performance over the years due to the fault events.

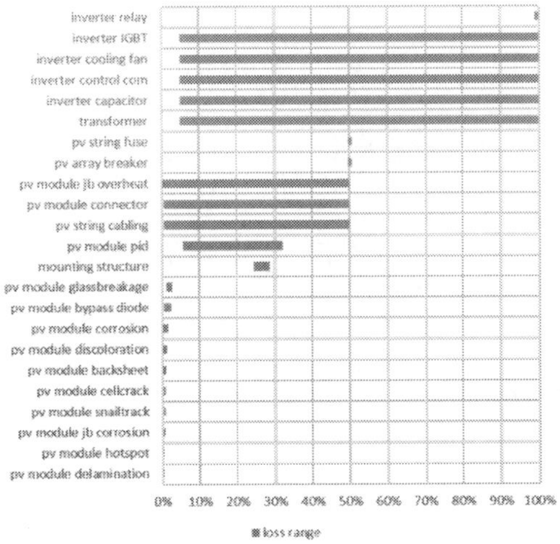

Figure 2: Summary of minimum and maximum losses for each fault type.

5 MAINTENANCE EVENTS

According to the best practice manuals for O&M of PV systems [1], maintenance events are defined using three paradigms: *Preventive*, *Corrective*, and *Predictive*. In this section, we review the best practices associated with each paradigm to understand and model the logic used to define when and how to perform interventions for inspection and maintenance.

5.1 Preventive maintenance events

Traditionally, this is the core of the maintenance of PV systems. According to the best practices manual [1], following a preventive maintenance strategy means defining a schedule of regular inspections for the entire useful life of the system. The activities to be carried out during these inspections can be distinguished between (a) minimal requirements and (b) recommended. The recommended activities can be grouped into (b1) junction box internal inspection, (b2) measurements inspection, and (b3) thermography inspection. Each type of inspection performs a different set of activities, making it sensitive to different types of failures. Table II lists typical faults and which inspections can detect them.

During the inspection, some tests, such as checking the integrity of cables and connectors, cannot be performed on every single component. For these activities, a random subset of components is selected to be tested. The percentage of components checked during an inspection defines its extension, here assumed to be 30% by default.

Although the best practices manual explicitly recommends how often inspections should be performed, this frequency is treated as an input parameter to investigate the difference it makes in the maintenance strategy. Based on the inspection frequency, the complete preventive inspection plan can be created with specific dates for each inspection throughout the system's life.

The execution of a field inspection involves two steps: (a) fault detection and (b) maintenance. In the fault detection step, the existing faults (faults that occurred before the inspection date) are considered detected if they are in the list of fault types possible to detect by the inspection type. For fault types where the inspection extension is limited (< 100%), detection also depends on the fault being within the group of inspected components.

The detected faults are passed to the next step: maintenance. In practice, the faults detected during a preventive inspection are corrected on site on the same day or the following days in case additional resource are needed. To represent this delay, the maintenance day is defined within 10 days after the inspection day.

Defective components are maintained by replacing them with new components of the same type. We assume the maintenance action is perfect, i.e., the performance loss associated with the defective component is fully recovered. Each replaced component is included in a maintenance event list with a component performance recovery equivalent to the additive inverse of its component performance loss, i.e., the number which when added with the original component performance loss gives 0. Thus, when calculating the system performance loss, a recovery equivalent to the replaced components will be observed after the day of maintenance.

The new components are also subject to failures; thus, failure events are generated for them following the logic previously described in Section 4, assuming the maintenance date as the beginning of the life time of these components. These events are added to the system story of events, extending the list of events after inspection.

Table II: A list of typical faults and which inspections can detect them. Inspection types are: (a) minimal requirements, (b1) junction box internal inspection, (b2) measurements inspection, and (b3) thermography inspection.

Fault type	Inspection type			
	(a)	(b1)	(b2)	(b3)
Inverter capacitor	-	-	-	x
Inverter control com	x	-	-	x
Inverter cooling fan	x	-	-	x
Inverter IGBT	-	-	-	x
Inverter relay	x	-	-	x
Mounting structure	x	-	-	-
PV array breaker	x	-	-	-
PV string cabling	x	-	x	-
PV string fuse	x	-	-	-
PV module cellcrack	-	-	x	x
PV module corrosion	-	-	x	x
PV module hotspot	-	-	-	x
PV module pid	-	-	x	x
PV module snailtrack	-	-	x	x
PV module glassbreakage	x	-	x	-
PV module delamination	x	-	x	x
PV module discoloration	x	-	-	x
PV module backsheet	x	-	-	x
PV module junction-box corrosion	-	x	-	x
PV module junction-box overheat	-	x	-	x
PV module bypass diode	-	-	x	x
PV module connector	x	-	-	x

10.4229/EUPVSEC2023/4BO.6.2
020300-003

Figure 3 depicts an example of a 25-year story of fault and maintenance events following preventive maintenance adopting recommended best practices. The resulting system performance loss is shown in Figure 4.

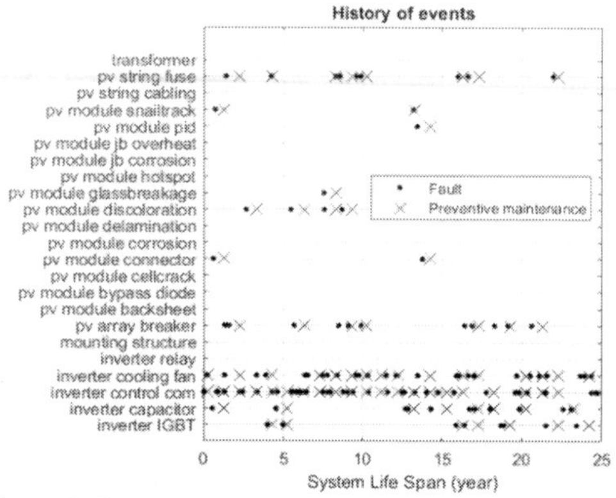

Figure 3: Example of a story of failure and maintenance events following preventive maintenance.

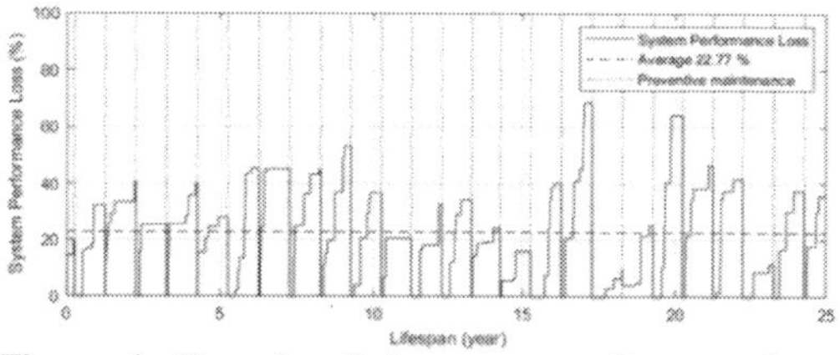

Figure 4: Example of the system performance loss resulting from preventive maintenance.

5.2 Corrective maintenance events

Corrective maintenance is a reaction to serious equipment failure and system breakdowns. It is motivated by safety problems or significant revenue losses. Significant reduction in energy generation can be detected even without a detailed monitoring system. Periodic assessment of system yield can indicate the need for urgent intervention to inspect and correct the causes of poor performance.

Unplanned urgent interventions are more costly and may occur only when losses are high enough to justify logistics. In other words, there might be a certain loss tolerance before triggering a corrective maintenance. In case of such severe events, the intervention must occur within a short time interval. Normally, the maximum response time specified in O&M service contracts for non-safety related issues is 10 days or less.

In this exercise, we have adopted a loss tolerance of 50% and a response time of 10 days. This means that a corrective maintenance intervention is scheduled for 10 days after the system performance loss reaches 50%. In case a preventive inspection is already scheduled within these 10 days, no corrective maintenance is needed. On the other hand, to avoid excessive visits to a system, if preventive maintenance is scheduled within 90 days of a corrective maintenance, it is canceled.

During a corrective maintenance intervention, we assume that all inspection tools are used to detect faults in the system. The fault detection and maintenance steps are performed the same way as described for preventive maintenance. Figure 5 shows an example of a 25-year story of fault and maintenance events following a preventive maintenance combined with corrective maintenance. The resulting system performance loss is shown in Figure 6.

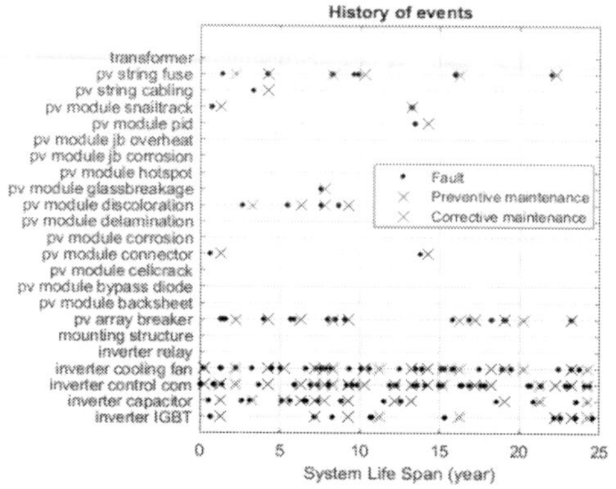

Figure 5: Example of a story of failure and maintenance events following preventive and corrective maintenance combined.

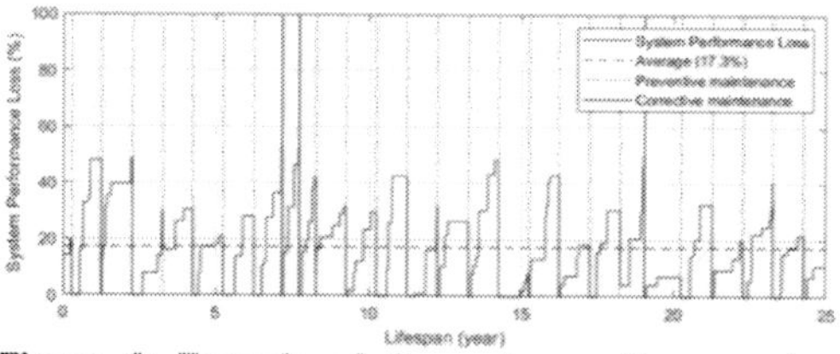

Figure 6: Example of the system performance loss resulting from preventive and corrective maintenance combined.

5.3 Predictive maintenance events

Also known as condition-based maintenance, predictive maintenance is a data-driven strategy that analyzes the monitoring signals from the PV system to identify signals of degradation and detect small failures in their early stages. The extracted information is used to prioritize maintenance activities and resources. To adopt a predictive maintenance strategy, an O&M service provider needs online monitoring combined with advanced analytical tools to detect small performance losses with high sensitivity. Ideally, this advanced monitoring should allow the reduction (or even abandonment) of preventive interventions in favor of intervention planning focused on systems with significant losses.

In the proposed methodology, for a predictive maintenance strategy, the need for maintenance intervention is defined based on the identification of system performance loss above a tolerance, and the maintenance date is scheduled after a predefined time window. This procedure is similar to corrective maintenance, but with two major differences: the tolerance for loss is much smaller, e.g. 10-20%, and the response time is longer, e.g. 90 days. In this way, poor performance is detected earlier and there is more time to plan the intervention, reducing urgency and logistical costs.

Online monitoring combined with algorithms for automated failure detection allows the system's performance to be evaluated daily, with alerts generated in

10.4229/EUPVSEC2023/4BO.6.2
020300-004

case of abnormal behavior. As previously studied in [5], a sensitivity of 99% and a specificity of 93% can be expected from an automatic fault detection system. This means that an alert will be raised for 99% of the days with a fault and 7% of the days without any fault. The upper plot in Figure 7 depicts an example of a fault occurrence (on day 195) and the daily alerts generated by an automatic fault detection with a sensitivity of 99% and a specificity of 93%. Understanding the limited accuracy of automatic alerts, maintenance interventions are scheduled only after a number of alerts within a short period, for example, five alerts within 10 consecutive days. The lower plot in Figure 7 shows the count of alerts within the previous 10 days and a threshold used to decide if a maintenance intervention should be scheduled.

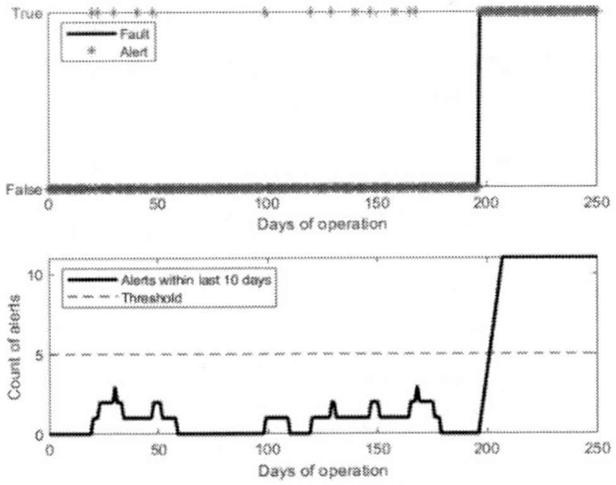

Figure 7: Example of the daily alerts generated by an automatic fault detection and the count of alerts within the previous 10 days to schedule a maintenance intervention.

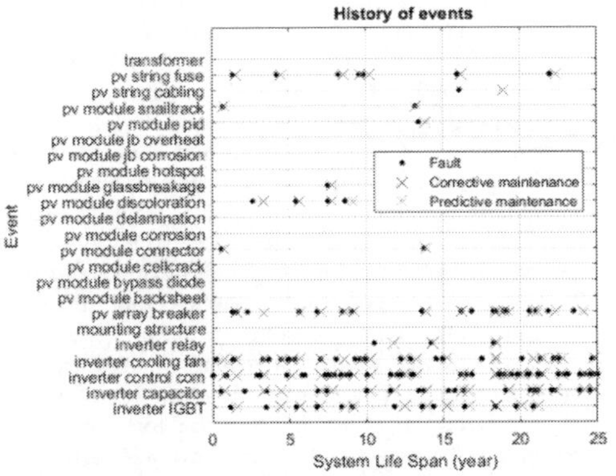

Figure 8: Example of a story of failure and maintenance events following predictive and corrective maintenance combined.

During a predictive maintenance intervention, we have assumed that all inspection tools are used to detect faults in the system. The fault detection and maintenance steps are performed in the same way as described above for preventive maintenance. Figure 8 shows an example of a 25-year story of fault and maintenance events following predictive and corrective maintenance strategies combined. The resulting system performance loss is shown in Figure 9.

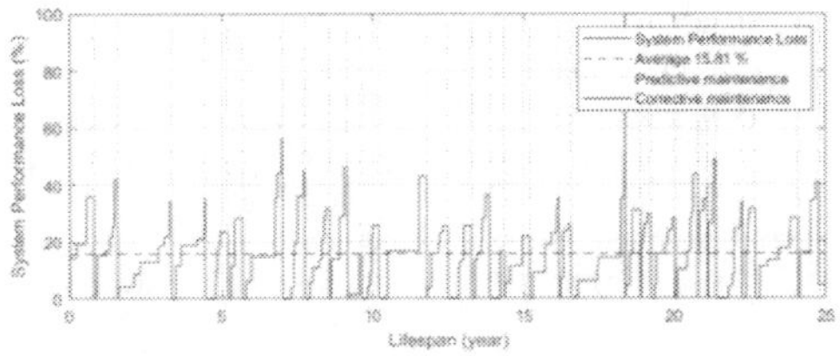

Figure 9: Example of the system performance loss resulting from predictive and corrective maintenance combined.

6 IMPACT ON ENERGY YIELD

The effectiveness of a maintenance strategy can be measured in terms of the performance of maintained PV systems throughout their lifespan. For that, we can simply calculate the average energy yield considering the daily energy yield $Y(t)$, which should consider the impacts of failure and maintenance events. In the proposed methodology, we calculate $Y(t)$ based on (8),

$$Y(t) = Y_{exp}(t) \times (1 - Ploss(t)) \quad (8)$$

where $Y_{exp}(t)$ is the expected energy yield defined in Section 3 and Ploss(t) is the system performance loss calculated as defined in Section 5. These variables are also used to calculate typical performance metrics adopted for PV systems: *Performance Ratio* (PR) and *Energy Performance Index* (EPI).

7 COMPARING MAINTENANCE STRATEGIES

Following the proposed methodology, we tested nine maintenance strategies listed below. For each strategy, 1000 stories of faults and maintenance events are generated and their metrics are averaged for comparison.

- *Corrective only*: 50% loss tolerance, 10 days delay for intervention
- *Minimum Preventive*: minimal inspections scheduled once a year + corrective.
- *Reduced Preventive*: additional inspections scheduled once every two years + minimum preventive.
- *Best Practices Preventive*: all inspection types scheduled once a year + corrective.
- *Predictive (12m)*: 20% loss tolerance, 360 days delay for intervention + corrective.
- *Predictive (9m)*: 20% loss tolerance, 270 days delay for intervention + corrective.
- *Predictive (6m)*: 20% loss tolerance, 180 days delay for intervention + corrective.
- *Predictive (3m)*: 20% loss tolerance, 90 days delay for intervention + corrective.
- *Predictive (1m)*: 20% loss tolerance, 30 days delay for intervention + corrective.

Figure 10 shows the evolution of PR over the years. It shows a steep decrease in efficiency in the first years, followed by a stabilization. The PR shows a decreasing trend due to the degradation rate included in the expected yield. The average PR, EPI, and yield throughout the system's life are shown in Figure 11. The results indicate that current best practices maintain the average system EPI

10.4229/EUPVSEC2023/4BO.6.2

at an 80% level, while predictive maintenance can reach 89%, depending on the reaction time.

Maintenance efforts calculated for each maintenance strategy in terms of the number of interventions and replaced components are shown in Figure 12. Table VI presents the percentage deviation relative to the *Best Preventive Practices* strategy. The results show that using corrective maintenance alone can lead to minimal maintenance efforts, but also results in the lowest system yield.

Performing preventive interventions with only the minimum activities results in a 10% reduction in average performance compared to best practices. It also reduces the total number of replaced components, and increases the need for urgent interventions.

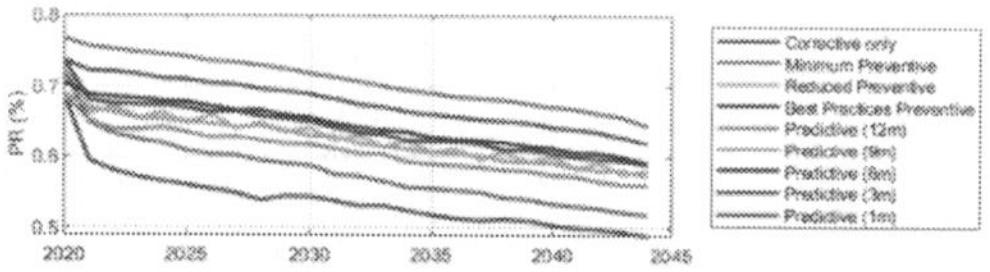

Figure 10: Evolution of PR over the years for each O&M strategy tested.

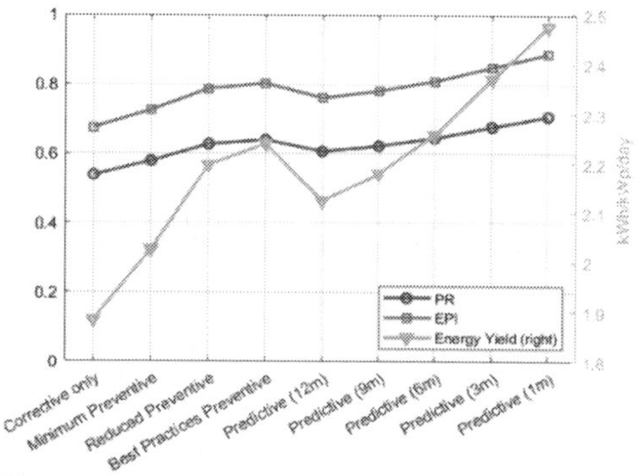

Figure 11: Average PR, EPI, and yield throughout the entire system lifespan for each O&M strategy tested.

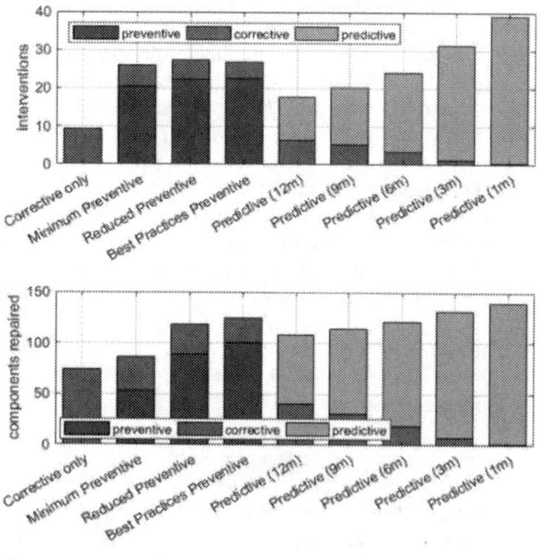

Figure 12: Maintenance efforts calculated for each maintenance strategy.

If complete preventive inspections are performed every two years, as in the *Reduced Preventive* strategy, there will still be a greater need for corrective maintenance, leading to a 2% increase in the total number of interventions, while reducing the average performance by 1.9%, compared to best practices.

The need for corrective interventions can be reduced by adopting predictive maintenance with response times of up to 6 months. For the predictive strategy with a short response time (up to 3 months), the average number of interventions and replacement components would increase by 16% and 5.9%, respectively, but the energy yield would also be 5.8% higher than *Preventive Best Practices*.

A predictive strategy with a response time of 6 months would reduce the total number of interventions and components replaced by 10% and 2.5%, respectively, while maintaining a similar average performance compared to *Preventive Best Practices*.

Table VI: Percentage deviation relative to the *Best Preventive Practices* strategy.

O&M Strategy	Deviation (%)		
	PR	Interventions	Components
Corrective only	-15.9	-65.4	-40.0
Min. Preventive	-9.6	-3.5	-30.5
Red. Preventive	-1.9	2.1	-5.0
Best Pract. Prev.	0.0	0.0	0.0
Predictive (12m)	-5.2	-34.3	-12.9
Predictive (9m)	-2.7	-24.7	-8.2
Predictive (6m)	0.8	-10.0	-2.5
Predictive (3m)	5.8	16.2	5.9
Predictive (1m)	10.5	45.0	12.1

8 CONCLUSION

We combined information from multiple studies to propose a methodology that effectively provides insight into the differences between maintenance strategies.

Results confirm that complete annual inspections, as recommended by current best practices, maintain system performances at a good level. They also show the benefits of moving towards a maintenance strategy mainly based on predictive actions. Namely, the need for urgent interventions is reduced, and the total number of interventions and replaced components can be reduced by 10% and 2.5%, respectively, without compromising the average performance of the systems.

These benefits are very dependent on the response time, which is the time between the detection of a failure and the execution of an intervention for maintenance. Short response times can enhance the average performance of the systems, leading to a 5.8% higher energy yield, with a 16% increase in the average number of interventions.

Attempts to reduce or minimize maintenance efforts should be made with caution, as they can significantly impair system performance and total energy yield. In this regard, the proposed methodology is a promising tool for evaluating O&M strategies. By including financial information, such as the price of electricity and the costs associated with interventions and component repairs, it is possible to carry out techno-economic assessments and quantify the potential impact of changing maintenance practices or investing in better monitoring, for example.

10.4229/EUPVSEC2023/4BO.6.2
020300-006

9 REFERENCES

[1] B. Clarke, A. Ara and W. Hemetsberger (2021). "O&M Best Practice Guidelines. Version 5.0." *SolarPower Europe* (2021).

[2] J. E. Breneman, C. Sahay, E. E. Lewis, "Introduction to reliability engineering." *John Wiley & Sons* (2022).

[3] E. A. Sarquis Filho, A. A. Zúñiga, J. F. Fernandes, P. J. C. Branco, "Failure Rates in Photovoltaic Systems: A Careful Selection of Quantitative Data Available in the Literature." *In Proceedings of the 37th European Photovoltaic Solar Energy Conference and Exhibition* (2020) pp. 1590-1594.

[4] E. A. Sarquis Filho, C. A. F. Fernandes, P. J. C. Branco, "A complete framework for the simulation of photovoltaic arrays under mismatch conditions." *Solar Energy*, 213, (2021) pp.13-26.

[5] E. A. Sarquis Filho, B. Kollosch, B. Müller, D. Melgar, K. Kiefer, C. Reise, P. J. C. Branco "Development and Evaluation of an Automatic Failure Detection System for O&M of PV Portfolios." *In Proceedings of the 8th World Conference on Photovoltaic Energy Conversion* (2022) pp. 1089-1094.

10.4229/EUPVSEC2023/4BO.6.2
020300-007

Using PV Efficiency Models with Monitoring Data to Identify Faults and Degradation

Anton Driesse
PV Performance Labs, Freiburg, Germany

EU PVSEC, Lisbon, Portugal, September 19, 2023

40th European Photovoltaic Solar Energy Conference and Exhibition

PV Performance Labs

a nimble one-person consultancy and research organization
based in Freiburg, Germany

25 years of PV expertise

"we may not be big, but we're small"

Objectives

Convince the audience that fitting a PV efficiency model to field measurements

a) can be done
b) can lead to useful insights into field performance

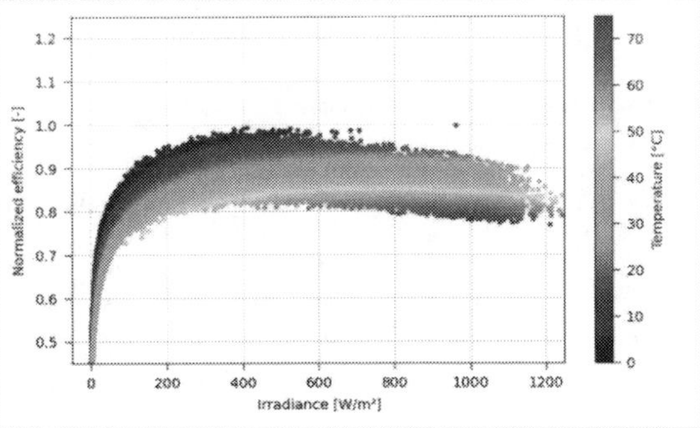

40th European Photovoltaic Solar Energy Conference and Exhibition

Introduction

- Monitoring data is collected to evaluate performance
- Identified faults or degradation are used to trigger corrective action
- Need some kind of comparison of observed and expected performance
- Simple methods:
  - Performance ratio (PR) or similar metrics
  - Few parameters required
- Holy grail:
  - A "digital twin"
  - Many *as-built* parameters required
- Possible middle ground:
  - PV efficiency models

Two types of PV performance models

1. IV-curve models
   - Inputs: irradiance and temperature
   - Model + parameters → IV curve → ***Pmp = Imp * Vmp***
   - Very hard to extract parameters from ***Pmp*** observations

2. Power or efficiency models
   - Same inputs: irradiance and temperature
   - Model + parameters → ***Pmp*** or ***η***
   - Should be much easier to extract parameters from ***Pmp*** observations
   - Frequently used with normalized efficiency:

$$\eta_{rel} = \frac{\eta}{\eta_{STC}}$$

40th European Photovoltaic Solar Energy Conference and Exhibition

Two types of efficiency models

1. Linear
   - Model equation is a weighted sum of predictors variables
   - Example, Huld/PVGIS/6K model:

$$\eta_{rel}(S, dT) = 1 + k_1 \log S + k_2 (\log S)^2 + k_3\, dT + k_4 dT \log S + k_5 dT (\log S)^2 + k_6 dT^2$$

   - The weights (model parameter values) are easily determined by linear regression
   - Most of the published models are linear equations

2. Non-linear
   - Model equation does not fit this form
   - A search algorithm is needed to find the model parameter values

PV efficiency modeling terminology

$$\eta = f(G,T)$$ general form

$$\eta_{rel} = \frac{\eta}{\eta_{STC}} \approx PR_{DC}$$ normalized or relative efficiency [0-1]

$$\eta_{rel} = \frac{P(G,T)/P_{STC}}{G/G_{STC}}$$ any power model is an efficiency model

$$\frac{G}{1000} \rightarrow S$$ normalized irradiance [0-1]

$$(T - 25) \rightarrow dT$$ "normalized" temperature

2023-09-19 Anton Driesse, PV Performance Labs, Germany 7

020302-007

40th European Photovoltaic Solar Energy Conference and Exhibition

Previous work on fitting PV efficiency models

- IEC 61853-1 lab measurements by CFV Labs
- 4 PV module types
- 7 PV efficiency models
- 7 test cases
- Results published in SAND2020-3877

Data are openly available from Sandia Labs
Model and fitting code is openly available from PVPLabs on github

2023-09-19 Anton Driesse, PV Performance Labs, Germany 8

020302-008

Example 1: model fits for a mc-Si module

Using all available measurements (left) vs. only irradiance > 200 W/m² (right)

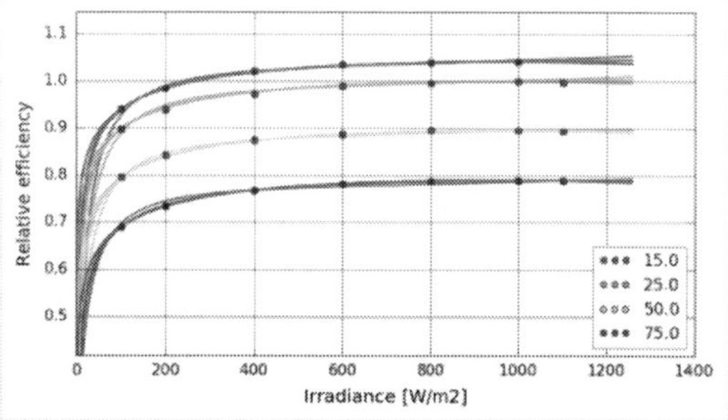

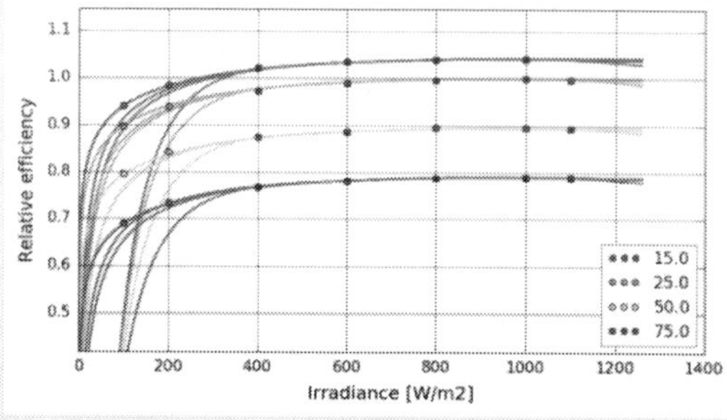

2023-09-19 Anton Driesse, PV Performance Labs, Germany 9

020302-009

40th European Photovoltaic Solar Energy Conference and Exhibition

Example 2: model fits for a CdTe module

Using all available measurements (left) vs. only irradiance > 200 W/m² (right)

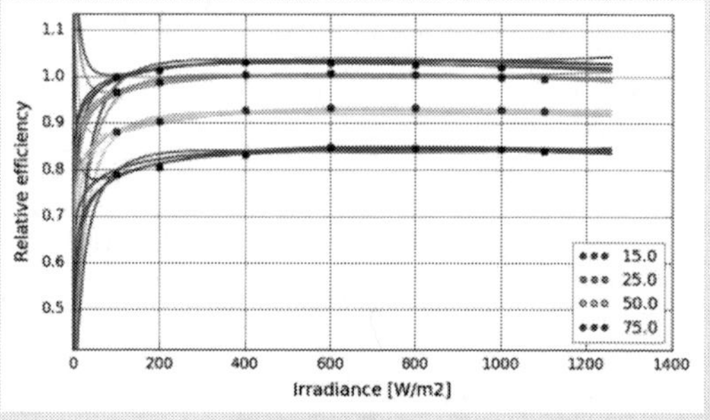

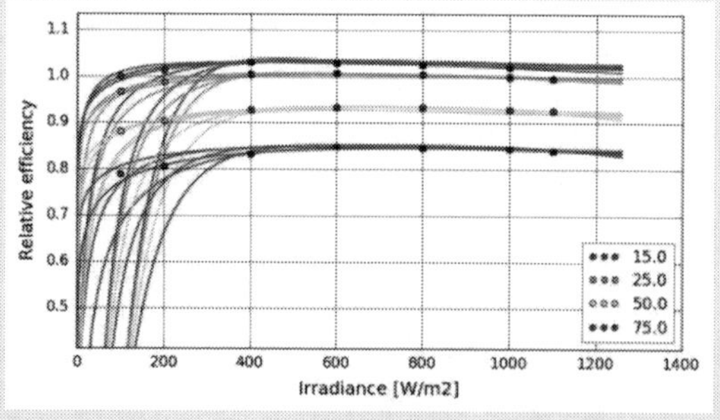

2023-09-19 Anton Driesse, PV Performance Labs, Germany 10

020302-010

Conclusions from fitting to lab measurements

- Models differ substantialy in their ability to fit low irradiance data and to extrapolate from subsets of lab measurements
- The ADR model did well and therefore appears most likely to perform well with imperfect and incomplete field data

Table E1: Qualitative summary and index of test results

#	Fit model to this:	Evaluate this:	Bilinear	HEY	MotherPV	PVGIS	MPM5	MPM6	ADR
1	All points	RMSE all points							
2	Points < 1000 W/m²	RMSE >= 1000 W/m²							
3	Points > 200 W/m²	RMSE <= 200 W/m²							
4	Points in IEC grid	RMSE outside IEC grid							
5	All points	values at G=0		0.0	nan	nan	-inf.	-inf.	0.0
5	All points	values at G=1 mW/m²							
5	All plus G=1 mW/m²	RMSE >= 100 W/m²							
6	All points + 2%offset	RMSE w.r.t. offset points							
7	All points + 1% noise	RMSE w.r.t. no noise							

The ADR PV efficiency model

A non-linear model having five parameters:

k_a: scaling factor, equal to the efficiency at STC ($\eta_{rel,STC}$)

k_d: diode factor, influencing the degree of curvature w.r.t. irradiance

tc_d: temperature coefficient of ***k_d***

k_rs: factor loosely corresponding to the relative magnitude of series resistance losses

k_rsh: factor loosely corresponding to the relative magnitude of shunt resistance losses

In this work:

- Temperature coefficient parameter was converted to a relative value in %
- For field measurement fitting, ***k_rsh*** = 0.0

Full derivation and description published in SAND2020-3877 and subsequent JPV article

Sources of field data

Multi-year outdoor monitoring data

- Open data analyzed in **IEA-PVPS T13-22:2021** report on Performance Loss Ratios (PLR)
  - EURAC, Bolzano
  - FOSS, Cyprus
- Open data used in on-line **RdTools** examples
  - NREL, Colorado, PVDAQ system 4
- Fraunhofer ISE, Freiburg, unnamed system

Five-year synthetic data

- Open data introduced in IEA-PVPS T13-22:2021 report on Performance Loss Ratios
  - EDF, Rennes, 4 systems (not very sunny)
  - Known degradation parameters

2023-09-19 Anton Driesse, PV Performance Labs, Germany 13

020302-013

40th European Photovoltaic Solar Energy Conference and Exhibition

Fitting to field data

Measurements needed:

- Gpoa
- Tmod
- Power (DC preferred)

Calculations:

- $\eta_{rel} = \frac{P/P_{STC}}{G_{POA}/G_{STC}}$

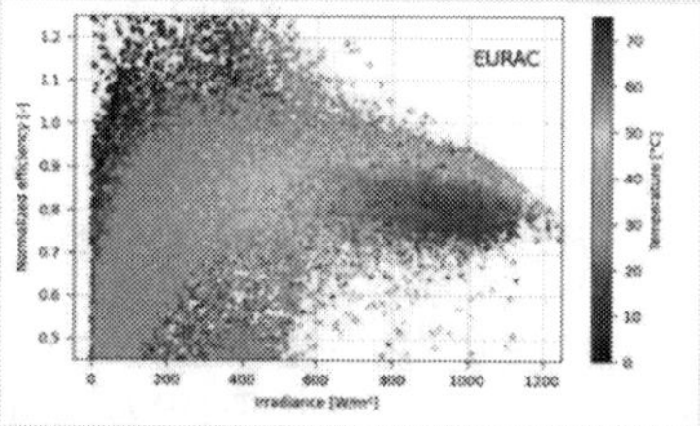

2023-09-19 Anton Driesse, PV Performance Labs, Germany 14

020302-014

Fitting to field data

Measurements needed:
- Gpoa
- Tmod
- Power (DC preferred)

Calculations:
- $\eta_{rel} = \frac{P/P_{STC}}{G_{POA}/G_{STC}}$

Basic filtering:
- sun zenith < 80°
- aoi < 70°
- Gpoa > 50 W/m^2
- P > 5% of P_{STC}
- η_{rel} < 125% of median
- η_{rel} > 50% of median

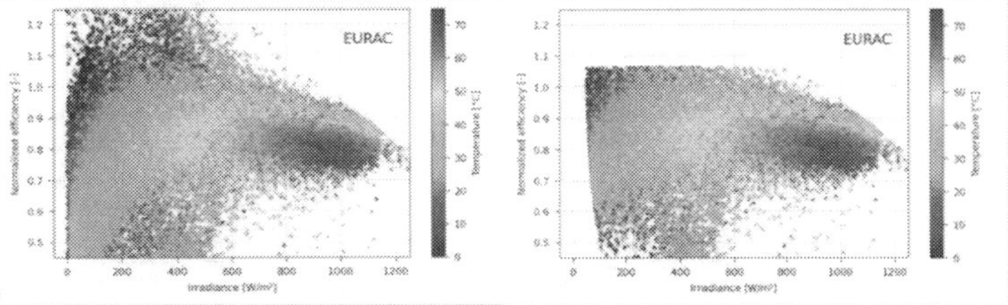

2023-09-19 Anton Driesse, PV Performance Labs, Germany 15

020302-015

40th European Photovoltaic Solar Energy Conference and Exhibition

Fitting to field data

Measurements needed:
- Gpoa
- Tmod
- Power (DC preferred)

Calculations:
- $\eta_{rel} = \frac{P/P_{STC}}{G_{POA}/G_{STC}}$

Basic filtering:
- sun zenith < 80°
- aoi < 70°
- Gpoa > 50 W/m^2
- P > 5% of P_{STC}
- η_{rel} < 125% of median
- η_{rel} > 50% of median

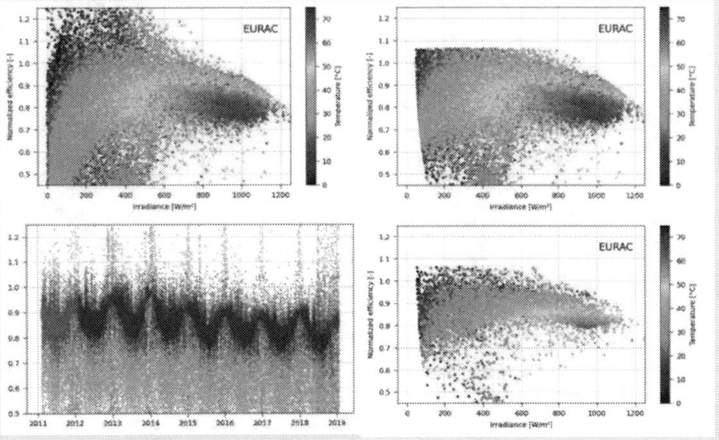

2023-09-19 Anton Driesse, PV Performance Labs, Germany 16

020302-016

Fitting to first year data

- Model fit to filtered data
- As-built temperature and low-irradiance behaviour

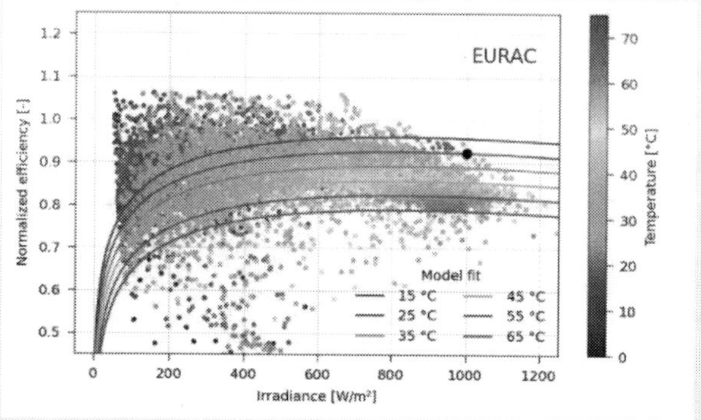

2023-09-19 Anton Driesse, PV Performance Labs, Germany 17

40th European Photovoltaic Solar Energy Conference and Exhibition

Fitting to first year data

- Model fit to filtered data
- Model predictions for all data
- As-built temperature and low-irradiance behaviour

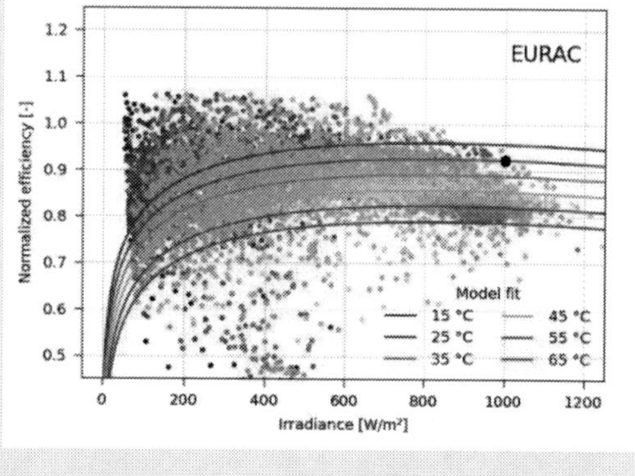

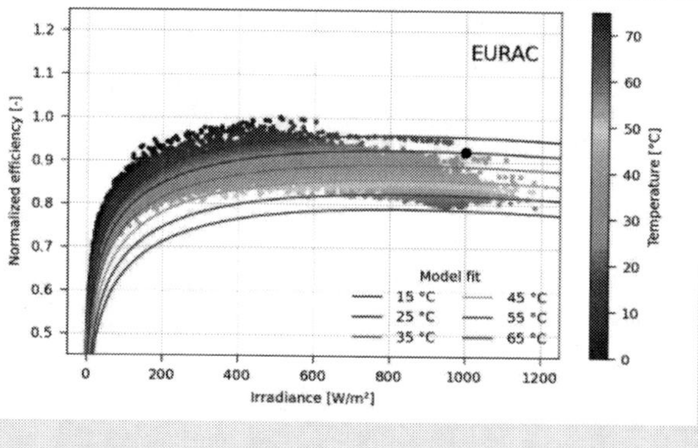

2023-09-19 Anton Driesse, PV Performance Labs, Germany 18

Fitting to first year data

η

- Model fit to filtered data
- But what about the other years?

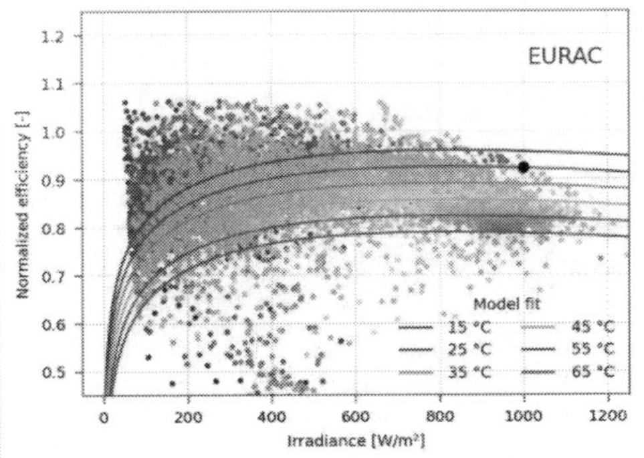

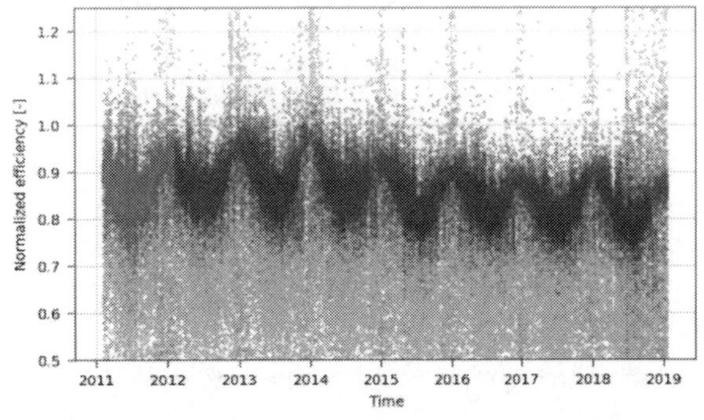

2023-09-19 Anton Driesse, PV Performance Labs, Germany 19

Year 1 vs. Year 7

η

All model parameters fluctuate somewhat over time

Hypothesis: main trends will ***appear*** as photo-current and/or series resistance changes

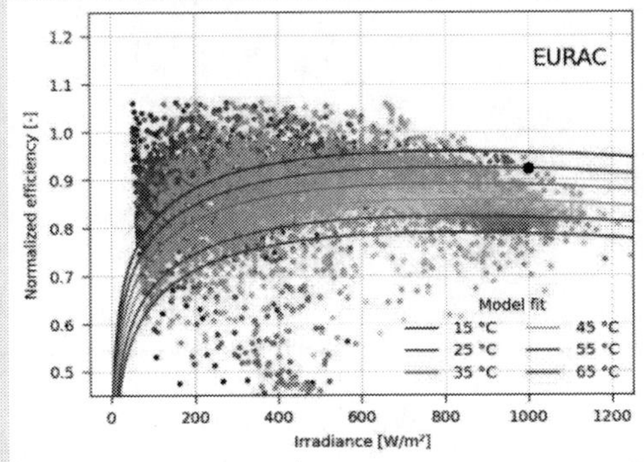

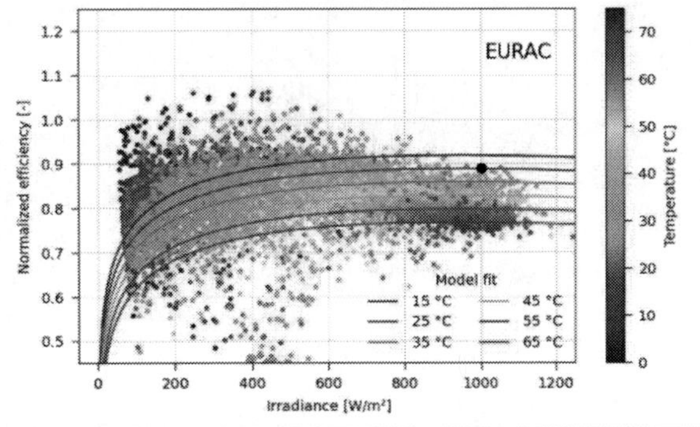

2023-09-19 Anton Driesse, PV Performance Labs, Germany 20

Multi-year fitting to identify linear trends

Make parameters of interest time-dependent:

Photocurrent → parameter ***k_a***

$k_a = k_{a0} \cdot (1 + \Delta k_a \cdot \Delta t)$

Δk_a is the annual rate of change of k_a

Series resistance → parameter ***k_rs***

$k_{rs} = k_{rs0} \cdot (1 + \Delta k_{rs} \cdot \Delta t)$

Δk_{rs} is the annual rate of change of k_{rs}

Δt is elapsed time in years

Now fit the modified ADR model to the complete multi-year time series at once

Single time-varying parameter: ***k_a***

k_a is equal to efficiency at STC

When ***k_a*** varies, efficiency at all irradiance levels varies by the same amount

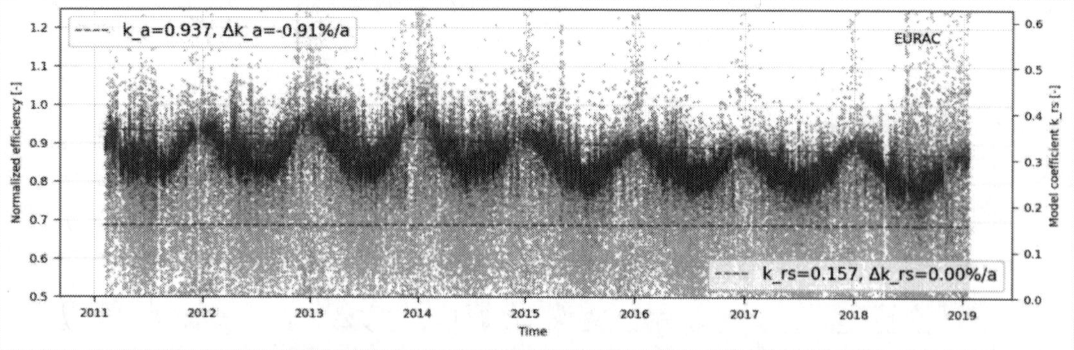

Two time-varying parameters: ***k_a*** and ***k_rs***

η

Increasing ***k_rs*** -> decreasing efficiency at high-irradiance
Decreasing ***k_rs*** -> decreasing efficiency low-irradiance

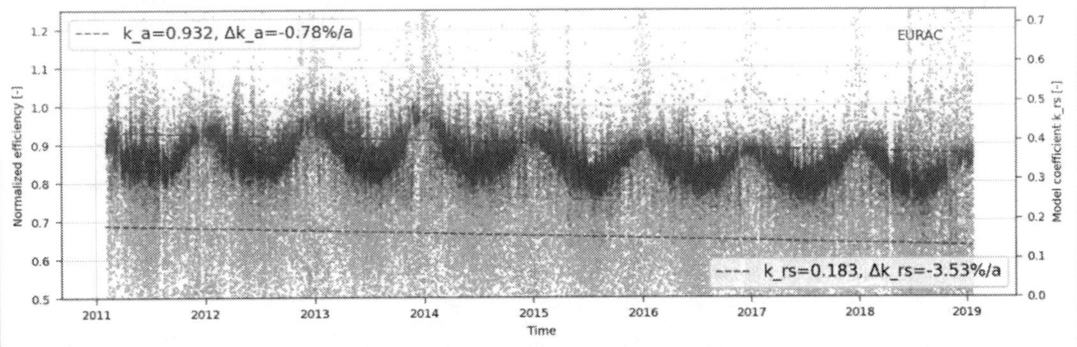

2023-09-19 Anton Driesse, PV Performance Labs, Germany 23

Comparing rates of change with published PLR values

η

Single parameter fit closely matches the mean PLR of other methods

Monitored PV systems	Task 13 Report	RDTools YoY Example		Single param.	Dual parameter fit	
	mean PLR	PLR sensor	PLR clear-sky	Δk_a	Δk_a	Δk_rs
EURAC	-0.85			-0.91	-0.78	-3.50
FOSS	-0.71			-0.69	-0.76	10.00
NREL4		-0.45	-0.51	-0.53	-0.56	2.60
ISE				-0.13	-0.13	-0.20

Dual parameter fit accurately determines loss rate of STC efficiency for simulated systems

Simulated PV systems	Task 13 Report			Single param.	Dual parameter fit	
	mean PLR*	True ener. degr.	True P STC degr.	Δk_a	Δk_a	Δk_rs
No degr. repeated weather	-0.02	0.00	0.00	0.01	0.00	0.30
No degr. different weather	-0.10	0.00	0.00	0.15	0.04	2.82
Strong degr. repeated weather	-4.60	-4.90	-4.41	-4.65	-4.43	-5.21
Strong degr. different weather	-4.75	-4.90	-4.41	-4.58	-4.39	-4.40

*estimated from graphic

2023-09-19 Anton Driesse, PV Performance Labs, Germany 24

Sliding/rolling window fits to detect non-linear trends

A one-year window is shifted in small increments to detect parameter changes
This produces a time-varying model of the system performance

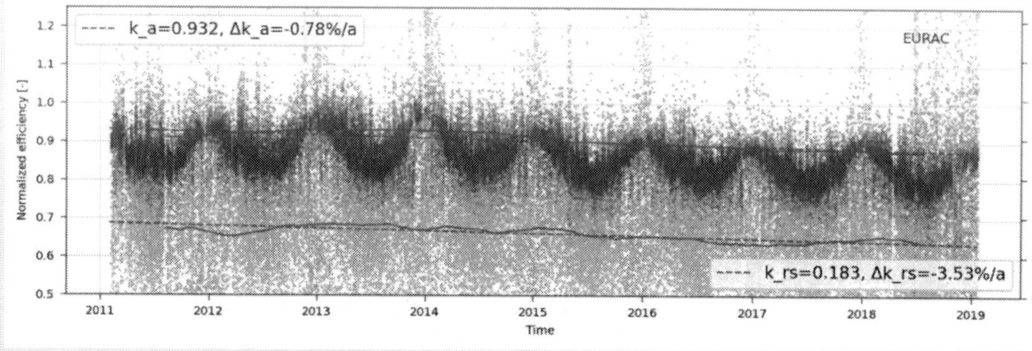

Using the time-varying model to detect faults

Measured vs. modeled power is a stable signal for use in fault detection
Winter-time shading of lower module row is evident (should exclude for fitting)

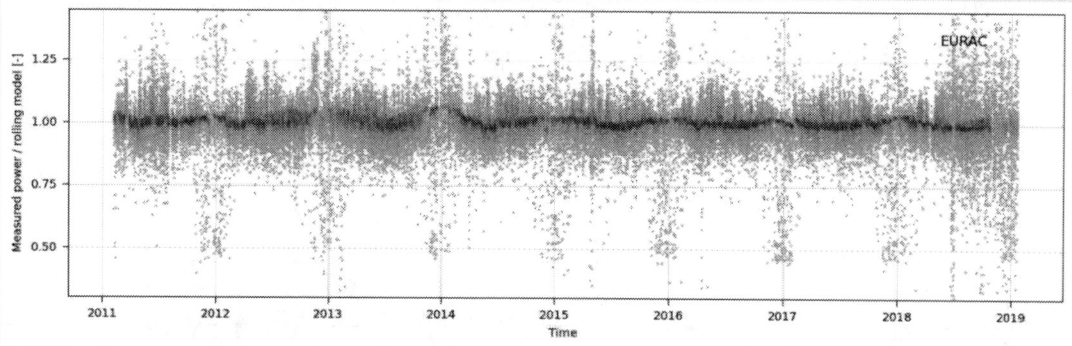

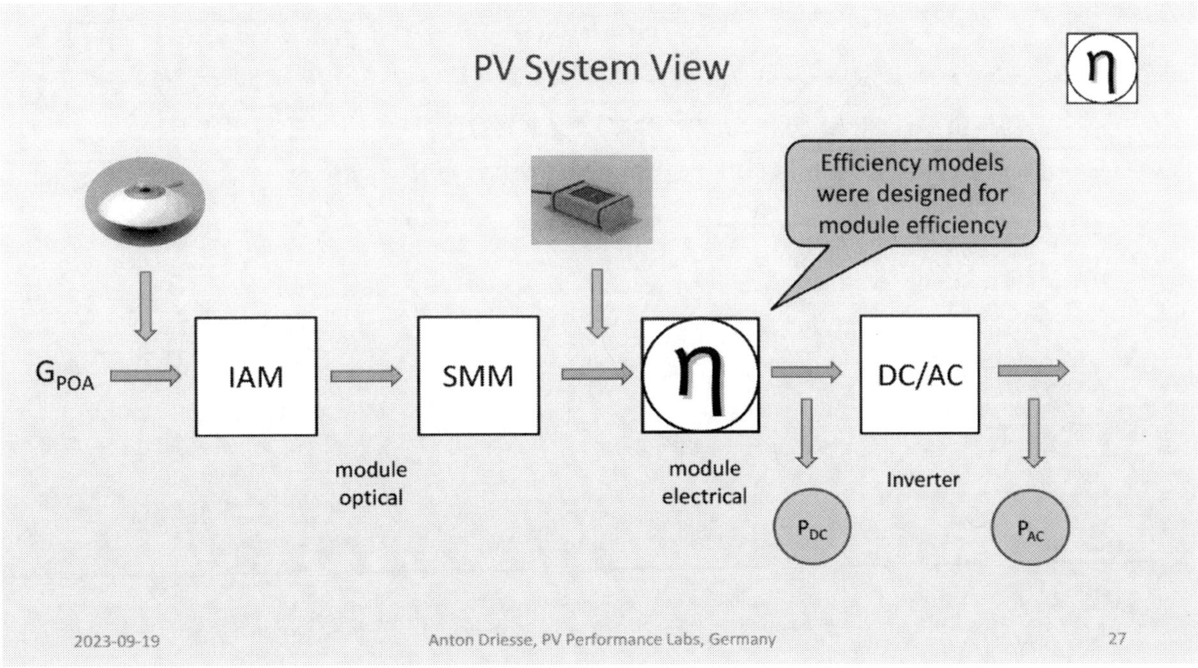

020302-027

40th European Photovoltaic Solar Energy Conference and Exhibition

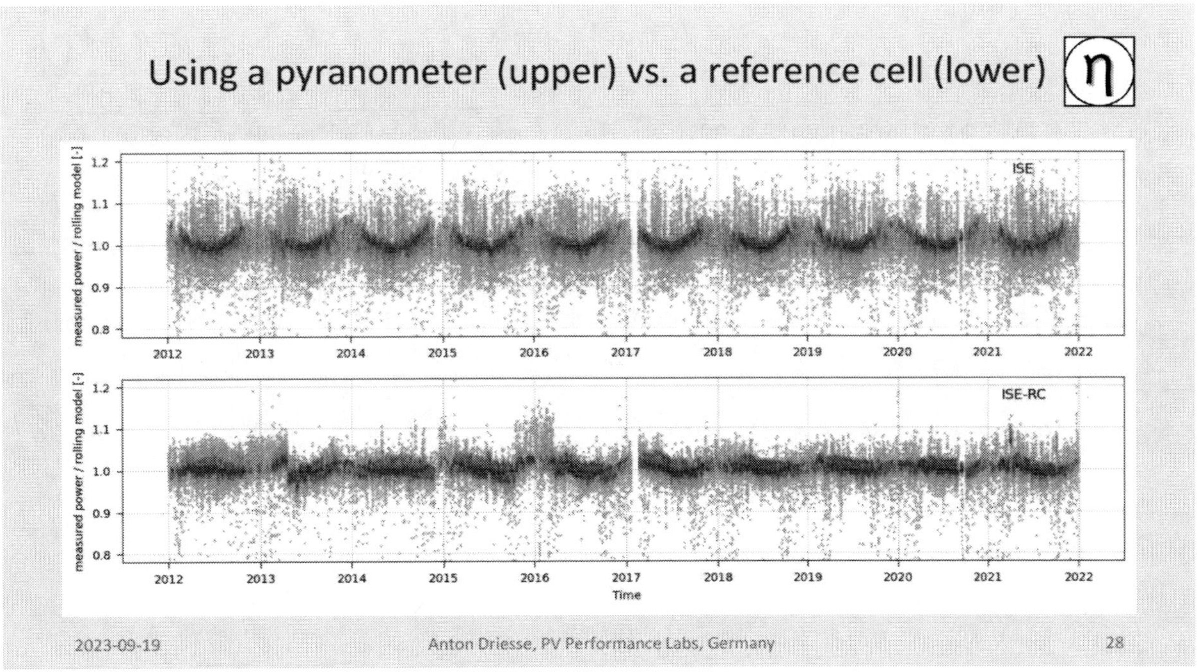

020302-028

Conclusions

- Can we fit an efficiency model to field data? ***Yes!***
- Does this provide useful insights into field performance? ***Yes!***

2023-09-19 Anton Driesse, PV Performance Labs, Germany 29

020302-029

40th European Photovoltaic Solar Energy Conference and Exhibition

Conclusions

- Main advantage of the demonstrated approach is that both low-light performance and temperature dependency are automatically incorporated.
- The linear rates of change of ADR model parameters match published PLR values.
- The dynamically changing ADR model calculates expected output that accounts for degradation, permitting tighter thresholds for fault detection.
- Directional response and spectral mismatch are not properly accounted for when using pyranometer *Gpoa*. Model this separately, or use reference cells.

These were the basics; much more is possible!

2023-09-19 Anton Driesse, PV Performance Labs, Germany 30

020302-030

Acknowledgements

Josh Stein, Marios Theristis, Sandia Labs and DOE
for supporting the lab comparisons and ADR model development

Sascha Lindig, EURAC, FOSS, EDF and IEA Task 13
for open reports and open data

Christian Braun & Fraunhofer ISE
for access to ISE system monitoring data

Useful information

The ADR model, a fitting function, and an example can be found here:
https://github.com/pvlib/pvlib-python

More efficiency models, another fitting function and example can be found here:
https://github.com/adriesse/pvpltools-python

More useful insights into your own system's performance can be found here:
anton.driesse@pvperformancelabs.com

PCS UNIT BASED FAULTY MODULES DETECTION USING POISSON DISTRIBUNTION OF PCS CURRENT FREQUENCY DISTRIBUTION IN A LARGE-SCALE PV POWER PLANTS

Kohei Takino*, Adiyudha Sadono**, Yuzuru Ueda*
Tokyo University of Science*, Looop Inc**
4322528@ed.tus.ac.jp

ABSTRACT: With the increasing number of large-scale photovoltaic power plants (mega solar power plants), fault detection in photovoltaic power generation systems is becoming increasingly important.In this method, we propose a fault detection method based on Poisson distribution, which does not use solar irradiation data and air temperature data, but only use the DC current data of each power conditioning system (PCS) measured per unit.The systems analyzed in this study are PV systems in which DC current and DC voltage are measured per PCS unit, and the case in which there are two sub-string defective modules in a string is considered a target fault level. Faults were detected without using solar irradiance, temperature, or voltage in order to reduce the installation cost of pyranometers, thermometers, and voltmeters.To reduce the influence of solar radiation and temperature, the reference PCS current was determined from the measured PCS current, and each PCS current was normalized by the reference PCS current.For the normalized PCS currents, a frequency distribution diagram was created with the PCS currents on the horizontal axis and the number of occurrences on the vertical axis. By using these data, PCSs containing modules in fault condition were detected by using Poisson distribution for the frequency distribution of PCS currents.

1 INTRODUCTION

The importance of renewable energy has been attracting attention toward achieving carbon neutrality by 2050, and large-scale photovoltaic power plants (mega-solar) have been introduced. In order to ensure long-term operation of photovoltaic power generation systems, it is necessary to detect failures. However, detailed on-site fault detection in large-scale solar power plants requires a lot of cost and time. Therefore, fault detection by remote monitoring of measured data is effective. In this paper, we propose a PV module fault detection method based on Poisson distribution, which does not use solar radiation data and temperature data, but only use the DC current data of distributed PCSs measured per unit. This method is expected to detect PCSs that may have faulty modules and reduce the burden of on-site investigations.

2 PROPOSED METHOD

In this study, fault detection was performed assuming the case of two sub-string faulty modules within a single string in a measured PCS.When a sub-string fault occurs within a module, the voltage value of the faulty module drops because of the operation of the bypass diode.In addition, since each string is connected in parallel, the voltage of the entire string is same.Therefore, the operating point of the faulty string moves to the open sideand the current decreases. In this method, fault detection focuses on the decrease in PCS current in this characteristic.

2.1 Preparation (Creating a simulated fault module)

Using the nameplate values of the PV system, the maximum output operating voltage of the string in the event of the failure assumed in this case is calculated by the following equation.

$$V_f = V_{pm} \times (N_m - 2) + V'_{pm} \times 2 \qquad (1)$$

where V_{pm} is the maximum output operating voltage of the normal module under the standard test conditions (STC) as indicated on the nameplate, V'_{pm} is the value of the maximum output operating voltage of the failed module under STC, and N_m is the number of modules that make up one string. After calculating the value of the maximum output operating voltage of the failed string, V_f, using equation (1), the corresponding PCS current value, I_f, is calculated in the I-V curve of the module under STC. The current values obtained are normalized by the following equation and used as data for the pseudo-faulty PCS.

$$I_{fn} = \frac{I_f}{I_{pm} \times N_m} \qquad (2)$$

where I_{pm} is the maximum output operating current of the module under STC and N_m is the number of modules that make up one string.

2.2 Flow of the method

A. Extraction of analysis time

During periods of low solar irradiation intensity, the output of not only the PCS containing the faulty module but also the normal PCS is reduced, and this makes fault detection difficult. Therefore, the filtering conditions for the current, which is roughly proportional to the solar irradiation intensity, are set as follows.

$$I_{pcs}(n,t) \geq I_{pm} \times N_m \times 0.3 \qquad (3)$$

$$I_{pcs}(n,t) - I_{pcs}(n,t-1) < I_{pm} \times N_m \times 0.05 \qquad (4)$$

The time when more than half of the PCSs satisfy both of these two conditions is used for fault detection. Where I_{pcs} is the PCS current, I_{pm} is the maximum output operating current, and N_m is the number of modules that make up one string.In addition, the time periods with more missing values are eliminated for the time passed by equations (3) and (4).

$$I_{pcs}(n,t) = 0 \qquad (5)$$

Similarly, remove the times when more than half of the PCSs satisfy this.

B. Determination of reference PCS current

Since this method does not correct the measured data for solar irradiation and temperature, it is necessary to select PCSs that are considered to be normal from the measured data. Therefore, the average of the top 10% PCS currents among all PCS measurements at each time is

10.4229/EUPVSEC2023/4BO.6.4
020303-001

determined as the reference PCS current, assuming that there is no failed module in the PCS. This reduces the installation cost of pyranometers and thermometers and makes fault detection more practical.

C. Normalization

The reference PCS determined in B has different currents at each time depending on the solar irradiation intensity and temperature. Therefore, in this method, it is necessary to determine one reference PCS current for each time by normalizing. The following equation is used to normalize the current.

$$I_{nor}(n,t) = \frac{I_{mea}(n,t)}{I_{ref}(t)} \quad (6)$$

where I_{mea} is the measured current of the PCS, I_{ref} is the measured current of the reference PCS, and n is the PCS number.

D. Creation of PCS current frequency distribution chart

Next, the time-series data, PCS currents, are represented as a frequency distribution chart. For the measured PCS currents at all analysis times normalized in the previous C, the graph is created as the horizontal axis is the PCS current and the vertical axis is the number of occurrences. Here, the frequency distribution diagram of PCS currents for one day is shown in Figure 1. By using a frequency distribution of PCS currents, which are time-series data, it is possible to approach fault detection from a broader perspective and is expected to detect PCSs containing faulty modules with higher accuracy.

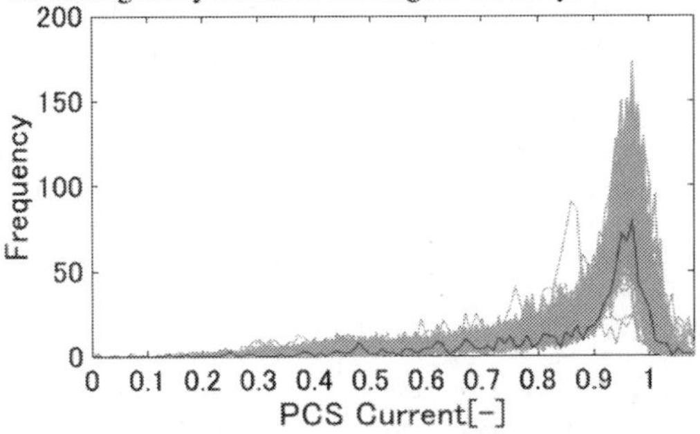

Fig.1 Frequency distribution of PCS currents on a given day

E. Failure determination by Poisson distribution

This method uses the Poisson distribution to determine failures. The Poisson distribution is a distribution of the probability of how many times a rare phenomenon can be expected to occur within a certain period of time. Therefore, we used the Poisson distribution in fault detection because I think that it can be used to detect faults with high accuracy.

From the creation of pseudo-fault data in Preparation, the faulty PCS current is I_{fn} times the normal PCS current when there are two sub-string faulty modules in one string within the measured PCS. Based on this, normal PCS and faulty PCS are created by Poisson distribution and pseudo-fault data.Next, Figure 2 shows the normal and faulty PCSs based on the Poisson distribution.

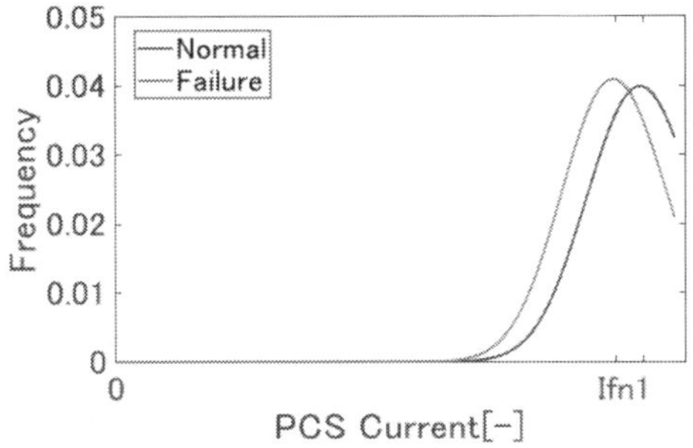

Fig. 2 Poisson distribution

Next, we normalize the vertical axis of the frequency distribution of Normal and Failure between 0 and 1.0 respectively. The diagram is shown in Figure 3.

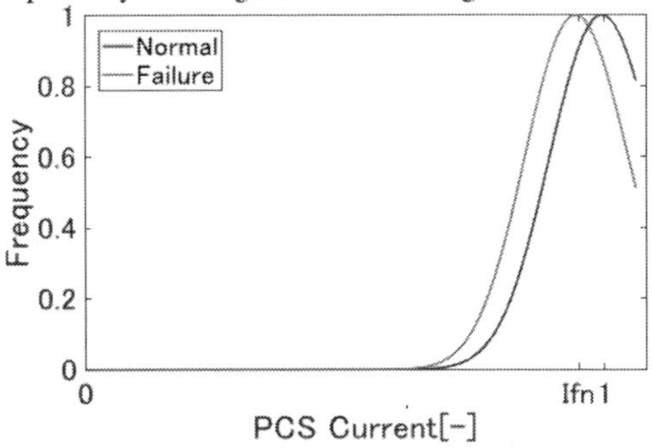

Fig. 3 Poisson distribution after normalization

Next, the frequency distribution of each PCS is normalized between 0 and 1.0 in the vertical peak value. Figure 4 shows the frequency distribution of PCS currents after normalization.

Next, the peak PCS values are determined as follows. First, the peak value is defined as the PCS current value between 0.90 and 1.02.

The peak value is determined as the extreme value that takes 10% or more of the total frequencies.If no peak value is found, the peak value is determined as the extreme value that takes 5% or more of the total frequenciesIf there is no peak value, the PCS is judged as a shadowed PCS.

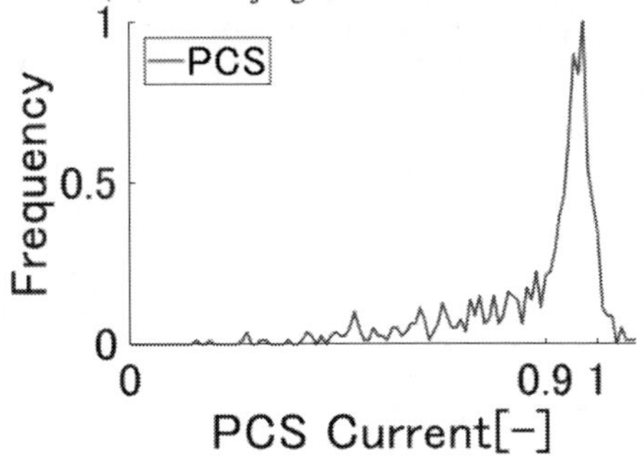

Fig. 4 the frequency distribution of PCS currents after normalization.

Next, each PCS and Normal are compared and the total distance (d_n) for each current value is calculated. Figure 5 shows, as an example, the distance between PCS and Normal when the PCS current is 0.9. In this way, the total distance d_n between PCS and Normal is calculated when the PCS current is between 0 and 1.

10.4229/EUPVSEC2023/4BO.6.4

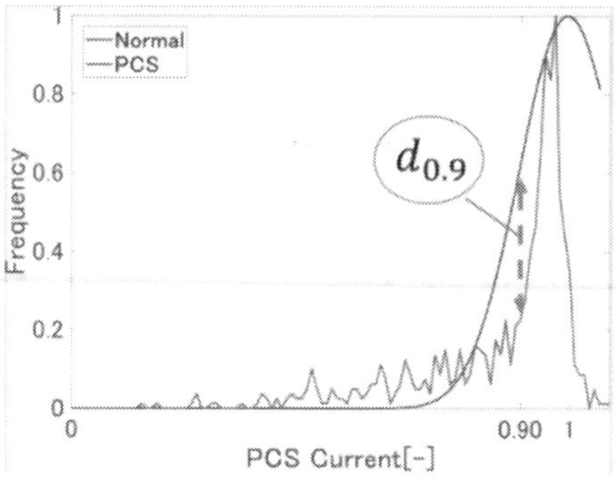

Fig. 5 Distance between PCS and Normal when PCS current is 0.9

Next, compare Failure and Normal and calculate the total distance (d_f) for each current value. Similarly, calculate the total distance df between Failure and Normal for PCS current values from 0 to 1, as shown in Figure 6.

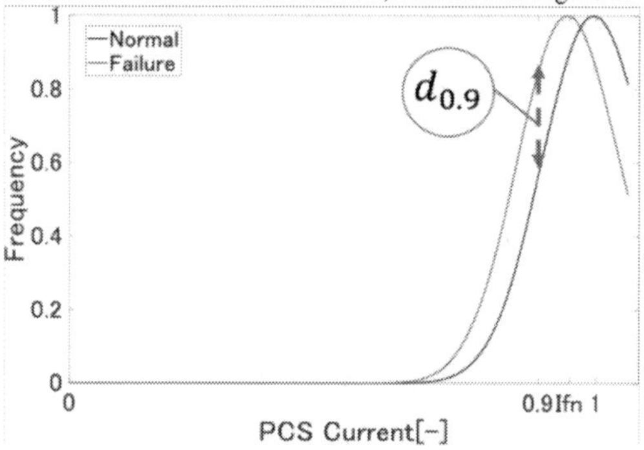

Fig. 6 Distance between PCS and failure when PCS current is 0.9

Finally, when d_n is greater than d_f, the PCS (PCS No. n) is determined to be the faulty PCS.

F. Calculation of failure detection count

The number of times a PCS is determined to be a PCS with a failed module per day is calculated and used as the failure detection result. The vertical and horizontal axes of the graph are the number of detections and the PCS number, respectively.

3 USED DATA

In this paper, a total of seven days from March 27, 2020 to April 2, 2020 are used as the analysis period for one megasolar in Japan. The daily used data is taken from the last 5 days. The target PCSs are 303 PCS units, and the PCS current is measured at one-minute intervals for each PCS. The configuration of this mega solar power plant is shown below.

- PV system: 303PCSs
- PCS: 14 strings
- String: 22 modules
- Module: 60 cells / 3 bypass diodes

4 ANALYSIS RESULT

The results of fault detection by Poisson distribution are shown in Figure 7. In Figure 7, PCSs with 4 or more failures detected are shown in red, PCSs with 1 to 3 failures detected are shown in blue, and PCSs with 0 failures detected are shown in gray. Based on the detection results, the PCSs marked in red are the PCSs with the highest probability of having a failed module. In particular, No. 217 is the PCS with the highest probability of having a failed module, as the number of detected failures was 5.

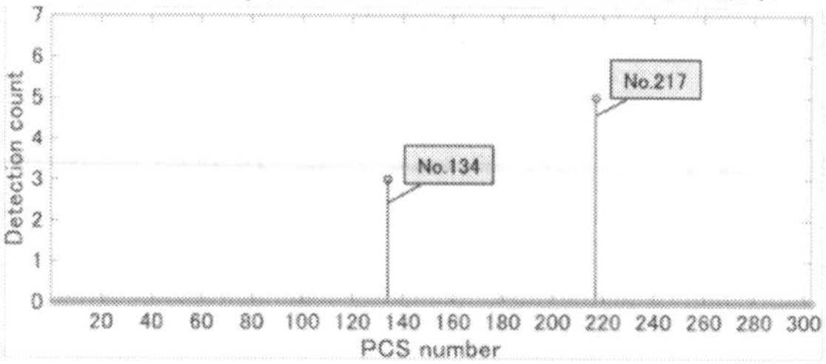

Fig. 7 Failure detection result (from March 27, 2020 to April 2, 2020)

Next, Figure 8 shows the results of the one-year analysis of PCS No. 217 mentioned earlier.

Blue indicates days that are judged normal, red indicates days that are judged faulty, and yellow indicates days that are judged shadowed.

This table shows that during the winter period, many days were judged as either failure or shadow.

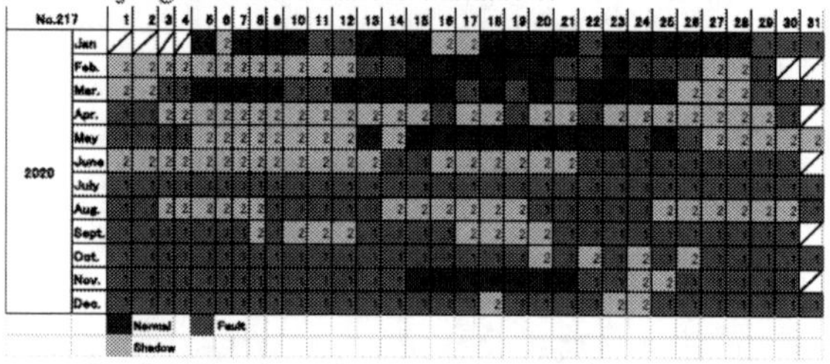

Fig. 8 Failure detection result. (No.217 for one year)

5 CONCLUSION

In this paper, PCSs containing faulty modules were detected by making a frequency distribution diagram of PCS currents and using Poisson distribution, using pseudo-fault data created using only nameplate values, not solar irradiation intensity or temperature. This has enabled us to reduce the installation cost of pyranometers and thermometers, and to develop a highly practical fault detection technology. Future work includes improving the accuracy of fault detection and improving the method so that other fault patterns can also be detected. The power plant data used in this study was provided by Looop Inc.

6 REFERENCE

[1] Kaori Tanina, and Yuzuru Ueda, "PV string fault detection by using density of module operating points for Large-scale PV power plant", PVSEC-31,2021.

[2] Hiroto Koshirae, and Yuzuru Ueda, "Modelingof Cell Performance Variation in One Diode Model for Reference I-V Curve Calculation", Proceeding of 29thPVSEC, pp.2344-2347(2019)

10.4229/EUPVSEC2023/4BO.6.4
020303-003

020304-001

40th European Photovoltaic Solar Energy Conference and Exhibition

Workflow

- Failure appearance in PV plant
- Creation of ticket in SCADA system
- Classification of O&M tickets according to **TRUST PV's Risk Matrix**
- Resolution of O&M tickets
- Statistical assessment of O&M tickets (**TRUST PV Solution Matrix**)
- Economic impact analysis of O&M tickets

TRUSTPV eurac research 2

020304-002

Failure appearance in PV plant
Creation of ticket in SCADA system
Classification of O&M tickets according to TRUST PV's Risk Matrix
Resolution of O&M ticket

TRUST PV Risk Matrix example

TRUST PV RISK MATRIX	O&M TICKETS
Component: Module Subcomponent: Entire Module **Failure: Broken module**	Damaged PV Module Found broken panel Faulty panel Isolated broken panels 2 broken panels found at string X PV panel outage PV Insul KO String isolated due to broken panel Damaged panel Open string Burned panel Burnt panel Damaged module Broken module Faulty module Open circuit module Module broken Smashed module

020304-003

40th European Photovoltaic Solar Energy Conference and Exhibition

TRUST PV Risk Matrix

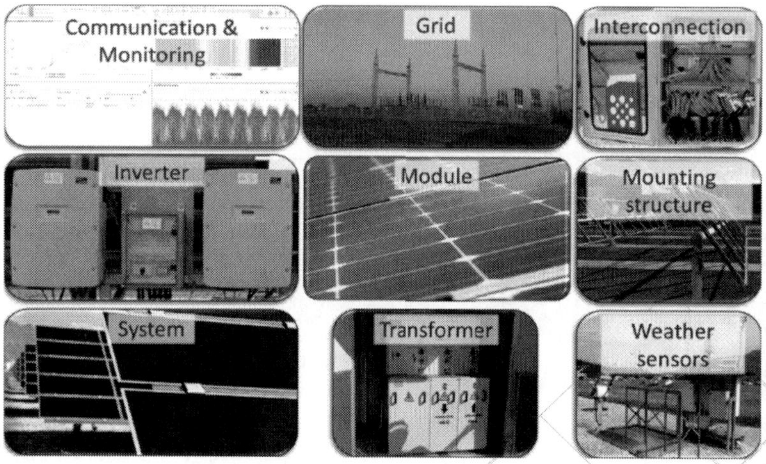

Failure appearance in PV plant
Creation of ticket in SCADA system
Classification of O&M tickets according to TRUST PV's Risk Matrix
Resolution of O&M ticket
Statistical assessment of O&M tickets
Economic impact analysis of O&M tickets (CPN)

https://trust-pv.eu/reports/risk-matrix/

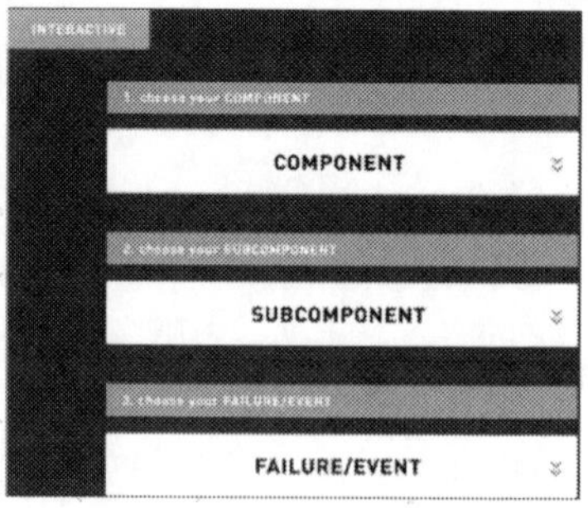

TRUSTPV
eurac research

020304-004

Failure appearance in PV plant
Creation of ticket in SCADA system
Classification of O&M tickets according to TRUST PV's Risk Matrix
Resolution of O&M ticket
Statistical assessment of O&M tickets
Economic impact analysis of O&M tickets

Statistical assessment
~250,000 tickets

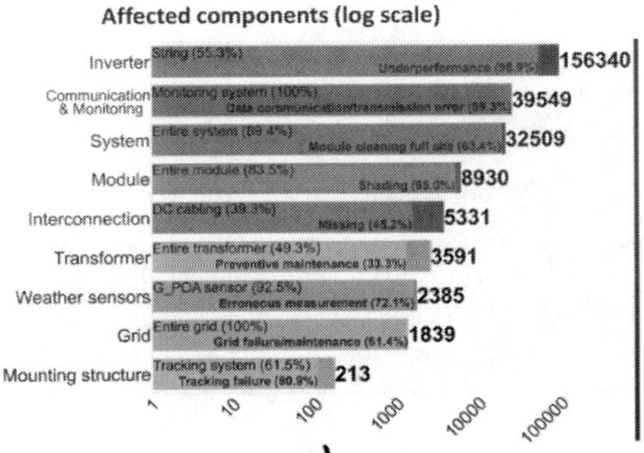

a)

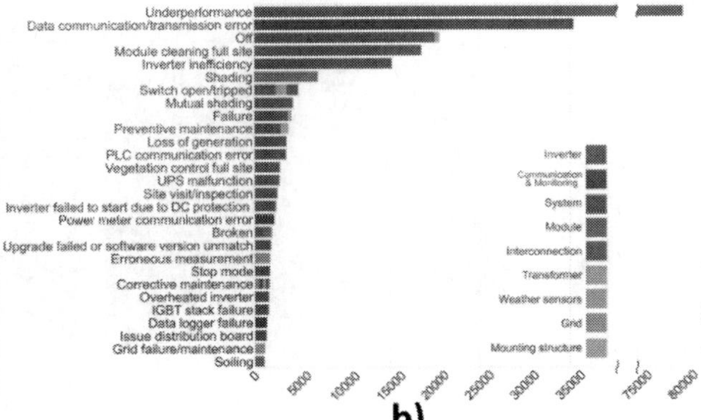

b)

TRUSTPV

eurac research

5

020304-005

Failure appearance in PV plant
Creation of ticket in SCADA system
Classification of O&M tickets according to TRUST PV's Risk Matrix
Resolution of O&M ticket
Statistical assessment of O&M tickets
Economic impact analysis of O&M tickets

Statistical assessment - Solutions
~16,000 tickets

Affected components (log scale)

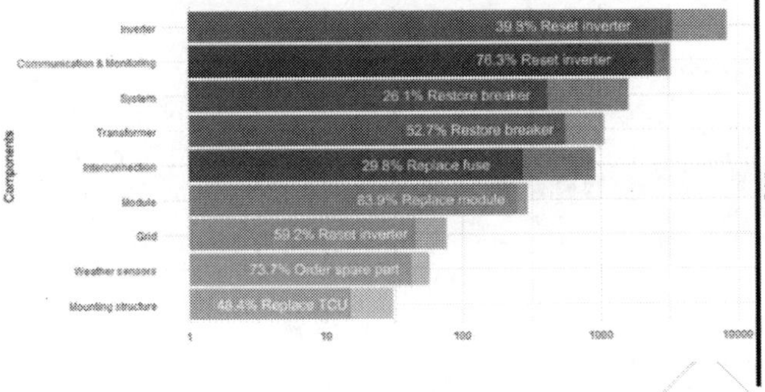

Solution

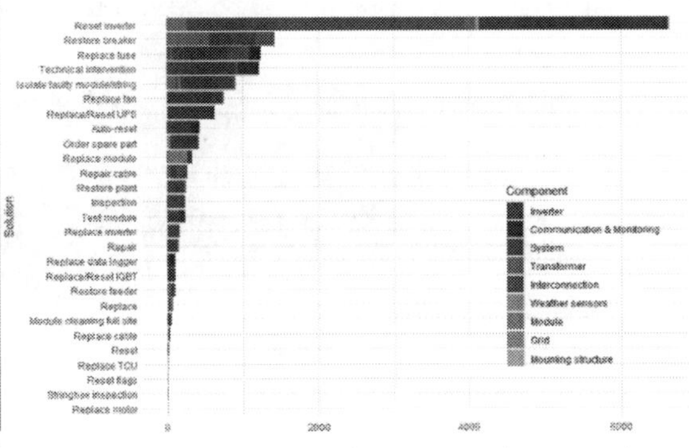

TRUSTPV

eurac research

6

020304-006

Economic evaluation of O&M tickets

Failure appearance in PV plant
Creation of ticket in SCADA system
Classification of O&M tickets according to TRUST PV's Risk Matrix
Resolution of O&M ticket
Statistical assessment of O&M tickets

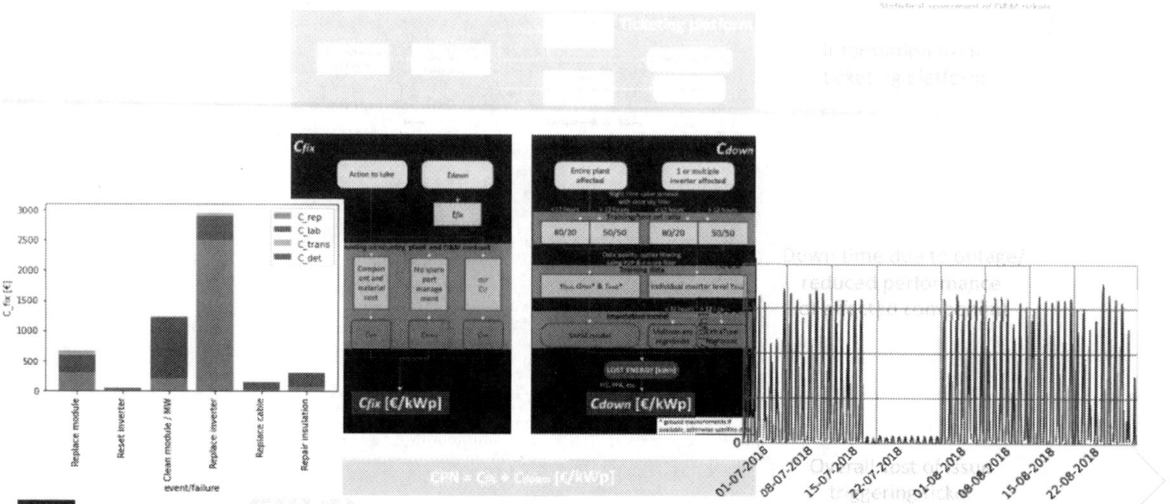

TRUSTPV

eurac research

S. Lindig, et al. (2022). *"Towards the development of an optimized Decision Support System for the PV industry: A comprehensive statistical and economical assessment of over 35,000 O&M tickets"* PiP.

020304-007

40th European Photovoltaic Solar Energy Conference and Exhibition

Risk Matrix

Solution Matrix

Failure appearance in PV plant
Creation of ticket in SCADA system
Classification of O&M tickets according to TRUST PV's Risk Matrix
Resolution of O&M ticket
Statistical assessment of O&M tickets
Economic impact analysis of O&M tickets (CPN²)

Component	Subcomponent	event	$event_{id}$	Solution	t_{fix}	n_P	C_P	C_{rep}	C_{trans}
					h		€/h	€	€
Inverter	Entire inverter	Damaged	inv.15	Replace inverter	3.00	2	50	4350	150

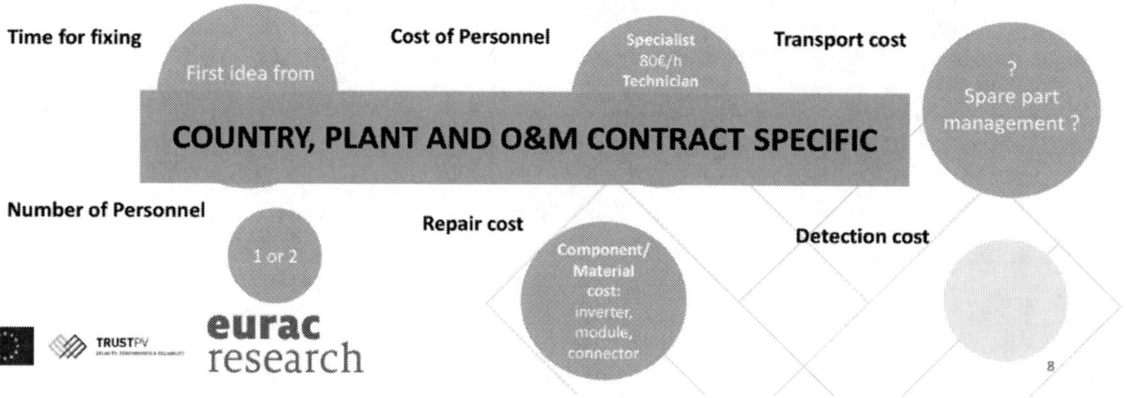

TRUSTPV

eurac research

8

020304-008

Section of Solution Matrix

event_id	Solution	Time for fixing [h]	Number of Personnel	Cost of Personnel [€/h]	Repair cost [€]	Transport cost [€]	Detection
mod.14	Replace module	1.0	2	50	150	10	VI, IR-T, EL, dPL, UV-F
mod.15	Replace module	1.0	2	50	150	10	VI, IR-T, UV-F
mod.16	Further investigation	0.1	1	80	0	0	VI, IR-T, EL, dPL, UV-F
mod.17	Replace connector	0.5	2	50	30	10	VI, IR-T, EL, dPL
mod.18	Replace connector	0.5	2	50	30	10	VI, IR-T, EL, dPL
mod.19	Replace connector	0.5	2	50	30	10	VI, IR-T, EL, dPL
mod.20	Replace connector	0.5	2	50	30	10	VI, IR-T, EL, dPL
mod.21	Replace connector	0.5	2	50	30	10	VI, IR-T, EL, dPL
mod.22	Replace module	1.0	2	50	150	10	VI, IR-T, UV-F
mod.23	Replace module	1.0	2	50	150	10	
mod.24	Replace module	1.0	2	50	150	10	
mod.25	Replace module	1.0	2	50	150	10	
mod.26	Replace module	1.0	2	50	150	10	
mod.27	Replace module	1.0	2	50	150	10	
mod.28	Replace module	1.0	2	50	150	10	
mod.29	Replace module	1.0	2	50	150	10	
mod.30	Replace module	1.0	2	50	150	10	
mod.31	Replace module	1.0	2	50	150	10	
mod.32	Replace module	1.0	2	50	150	10	
mod.33	Replace module	1.0	2	50	150	10	
mod.34	Do maintenance	2.0	2	50	0	0	
mod.35	Replace module	1.0	2	50	150	10	
mod.36	Replace module	0.5	2	50	150	10	IR-T, EL, dPL
mod.37	Remove shading	0.5	1	30	0.08	0	
mod.38	Replace module frame	1.0	2	50	15	5	
mod.39	Replace module frame	1.0	2	50	15	5	
mod.40	Replace module frame	1.0	2	50	15	5	
mod.41	Tighten module frame	0.2	1	50	0	0	
mod.42	Remove the damaged anti-reflective coating and	0.5	1	50	10	0	
mod.43	Replace module	1.0	2	50	150	10	VI, IR-T, EL, dPL, UV-F
mod.44	Replace module	1.0	2	50	150	10	
mod.45	Replace module	1.0	2	50	150	10	
mod.46	Clean module	0.1	1	30	0.28	0	
mod.47	Replace module	1.0	2	50	150	100	
mod.48	Visual inspection Remove shading	1.0	1	50	40	5	
mod.49	[illegible]	1.0	1	50	40	5	
mod.50	Replace junction box	1.0	1	50	40	5	
mod.51	Replace bypass-diode	1.0	1	50	6	5	
mod.52	Replace junction box	1.0	1	50	40	5	
mod.53	Track isolation fault and repair	0.5	1	50	5	0	
mod.54	Fix junction box	0.5	1	50	8	0	
mod.55	Seal junction box	0.5	1	50	5	0	
mod.56	Replace module	0.2	1	50	150	100	
mod.57	De-potting (remove potting material and apply an	1.0	1	50	40	5	
mount.01	Replace clamps	1.0	1	50	2	3	
mount.02	Replace clamps	1.0	1	50	2	3	
mount.03	Replace clamps	1.0	1	50	2	3	
mount.04	Align clamps	0.5	1	50	0	0	
mount.05	Repair mounting structure	0.5	1	50	400	50	
mount.06	Repair mounting structure	0.5	1	50	400	50	
mount.07	Replace profile	1.0	1	50	400	50	
mount.08	Apply anodizing or powder coating	1.0	1	50	2	0.5	
mount.09	Replace profile	1.0	1	50	400	50	

TRUSTPV

eurac research

9

020304-009

40th European Photovoltaic Solar Energy Conference and Exhibition

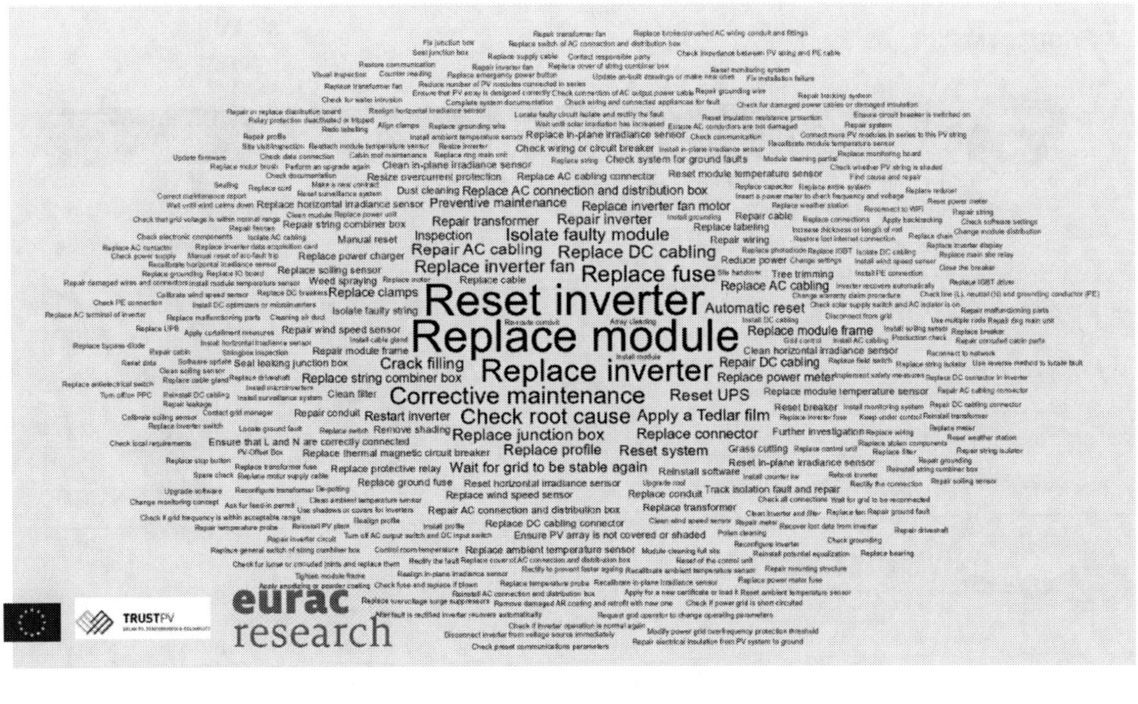

020304-010

Failure appearance in PV plant
Creation of ticket in SCADA system
Classification of O&M tickets according to TRUST PV's Risk Matrix
Resolution of O&M ticket
Statistical assessment of O&M tickets
Economic impact analysis of O&M tickets (CPN¹)

Decision Support System

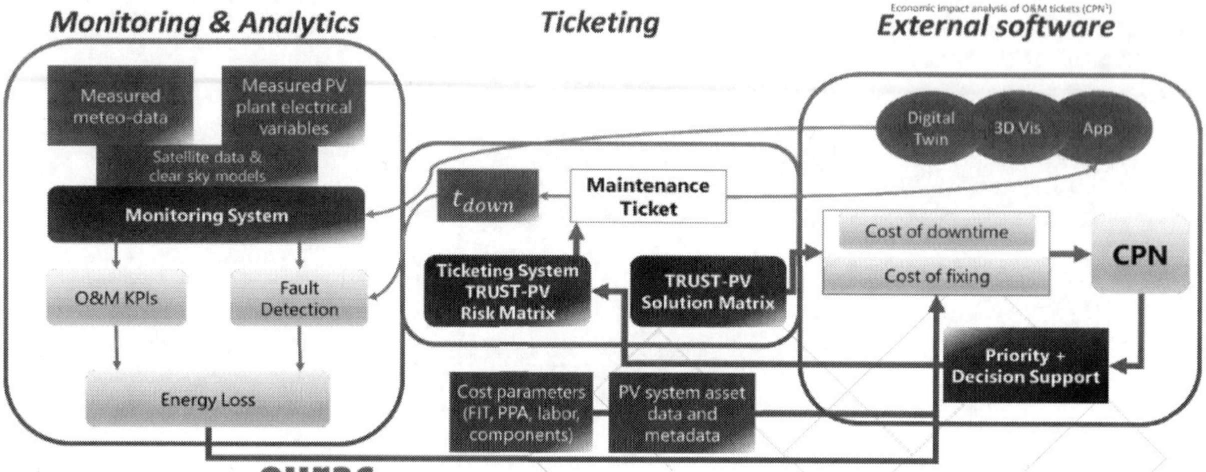

eurac research

11

020304-011

Failure appearance in PV plant

Priority ranking

SAVE 3400€

Ticket ID	Component/ Subcomponent	Event	Outage [days]	Energy Loss [kWh]	C_down [€]	So…on	C_fix [€]	CPN [€]	
194	Inverter/ Entire inverter	Insulation failure	8	81	3	…pair …nsulation	130	133	
3	Inverter/ Entire Inverter	Insulation failure	3	540	1…	Repair insulation	130	149	
622	Inverter/ Entire inverter	Insulation failure	~~19~~ **5**	69258	~~4571~~ **1179**	Repair insulation	130	~~4701~~ **1309**	1
35752	Inverter/ Entire inverter	Loss of generation	21	17530	1125	Replace inverter	2800	3925	2
4	Inverter/ Entire inverter	Loss of generation	21	17773	622	Replace inverter	2800	3422	3
307	Inverter/ Entire inverter	Loss of generation	7	156	5	Replace inverter	2800	2805	

eurac research

12

020304-012

eurac research

Thank you for your attention!

Accepted for publication in Solar RRL:
"Automated Fixing Cost Estimation of PV System Failures for the Creation of a Decision Support System"

S. Gallmetzer, S. Lindig, M. Herz, D. Moser
sandra.gallmetzer@eurac.edu

TRUSTPV

This project has received funding from the European Union's Horizon 2020 research and innovation programme under GA. No. 952957.

020304-013

Inverter Prices

Failure appearance in PV plant
Creation of ticket in SCADA system
Classification of O&M tickets according to TRUST PV's Risk Matrix
Resolution of O&M ticket
Statistical assessment of O&M tickets
Economic impact analysis of O&M tickets (CPN²)

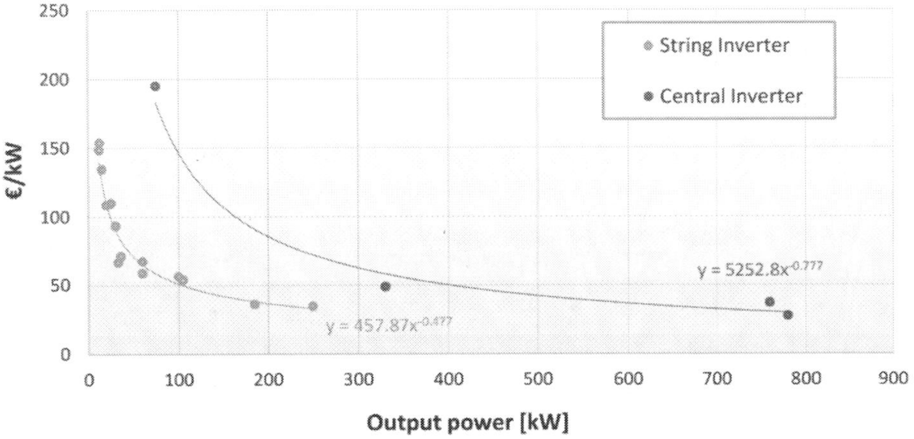

TRUSTPV eurac research

020304-014

Failure appearance in PV plant
Creation of ticket in SCADA system
Classification of O&M tickets according to TRUST PV's Risk Matrix
Resolution of O&M ticket
Statistical assessment of O&M tickets
Economic impact analysis of O&M tickets (CPN[1])

CPN calculation

Table 1: CPN calculation of O&M tickets.

Component	Subcomponent	Event	event$_{ID}$	C_{down} [€/kWp]	Solution	C_{fix} [€/kWp]	CPN [€/kWp]
Inverter	Fan	Failure	inv.48	0.0021	Reset inverter	0.0001	0.0022
					Replace fuse	0.0006	0.0027
					Replace inverter fan	0.0066	0.0087
Inverter	Fan	Failure	inv.48	0.0116	Reset inverter	0.0002	0.0118
					Replace fuse	0.0009	0.0125
					Replace inverter fan	0.0106	0.0222
Inverter	Entire inverter	Inverter failed to start due to DC protection	inv.21	0.0143	Automatic reset	0.0005	0.0148
					Replace DC breakers	0.0071	0.0214
Inverter	Entire inverter	Inverter inefficiency	inv.22	0.0079	Reset inverter	0.0001	0.0080
					Reset UPS	0.0001	0.0080
					Isolate faulty module	0.0006	0.0085
					Replace inverter fan	0.0059	0.0138
					Replace inverter	0.0582	0.0661

TRUSTPV

eurac research

15

020304-015

020305-001

40th European Photovoltaic Solar Energy Conference and Exhibition

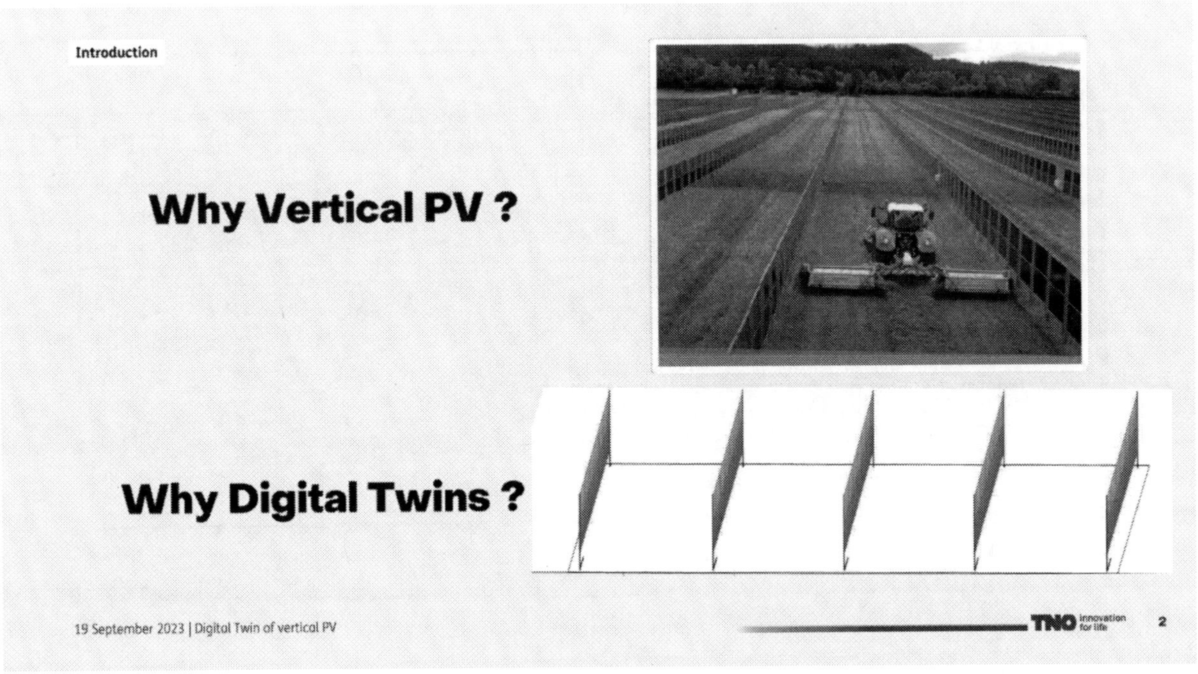

020305-002

System A 22kW peak vertical system

System A

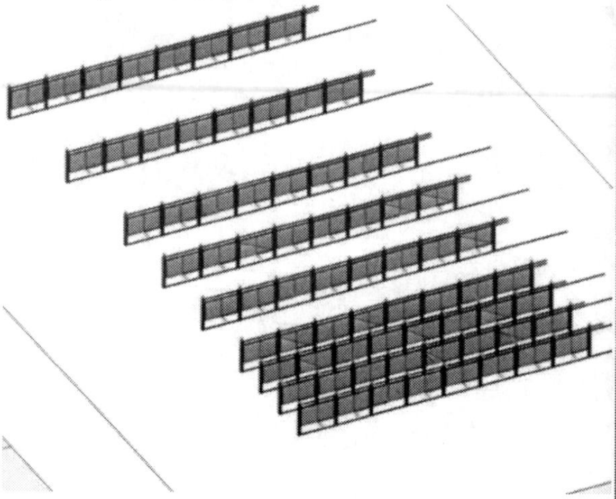

19 September 2023 | Digital Twin of vertical PV

020305-003

40th European Photovoltaic Solar Energy Conference and Exhibition

System A

First model results

Energy measured : 1.9 kWh

Energy modelled : 1.7 kWh

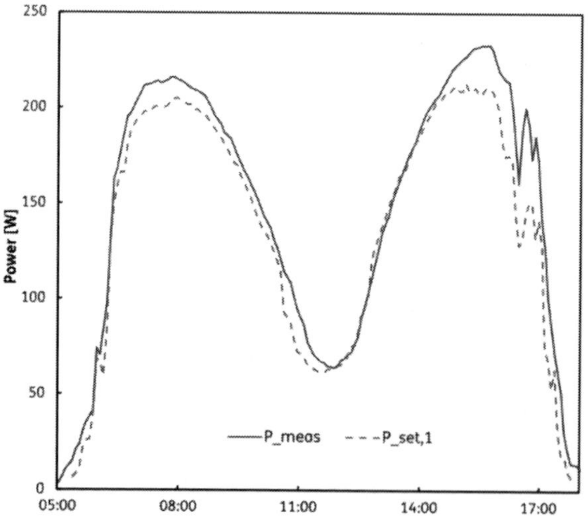

19 September 2023 | Digital Twin of vertical PV

TNO innovation for life 4

020305-004

System A

First model results

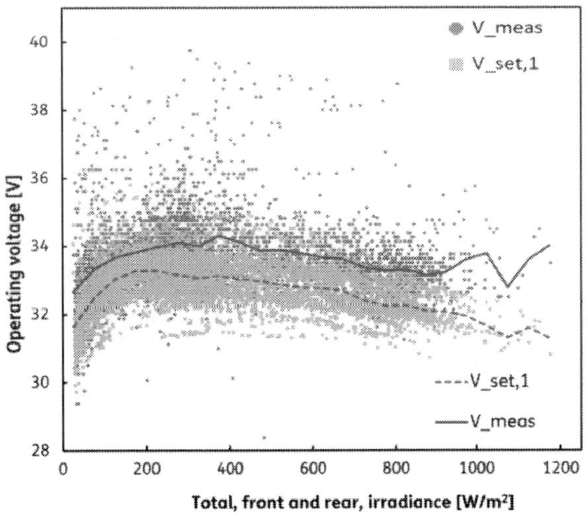

020305-005

40th European Photovoltaic Solar Energy Conference and Exhibition

System A

Additional measurements

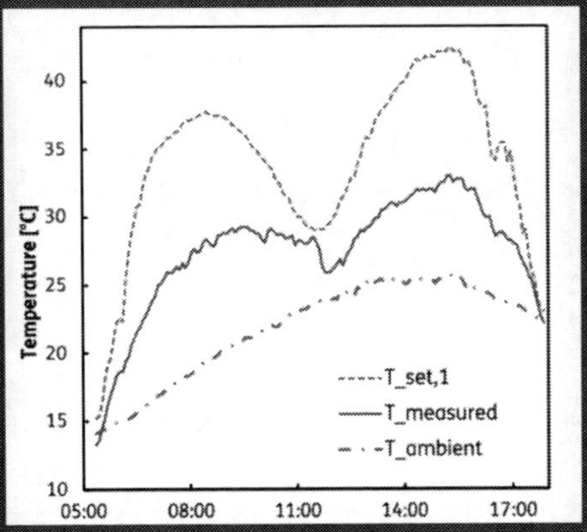

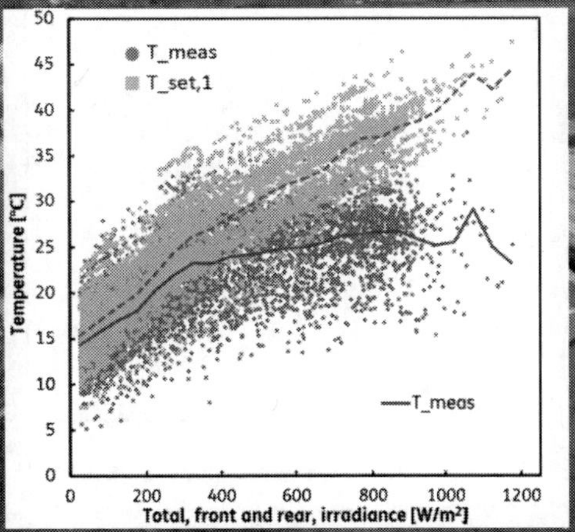

020305-006

System A

Revised model run

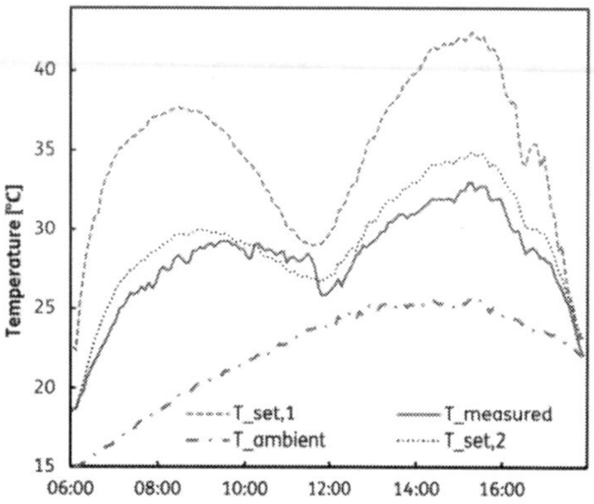

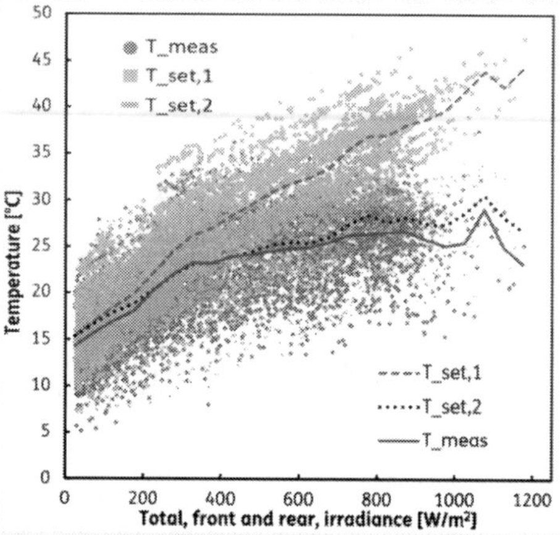

System A

Revised model run

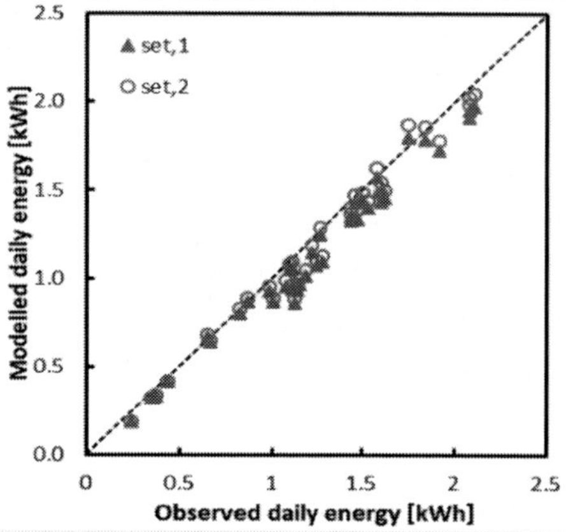

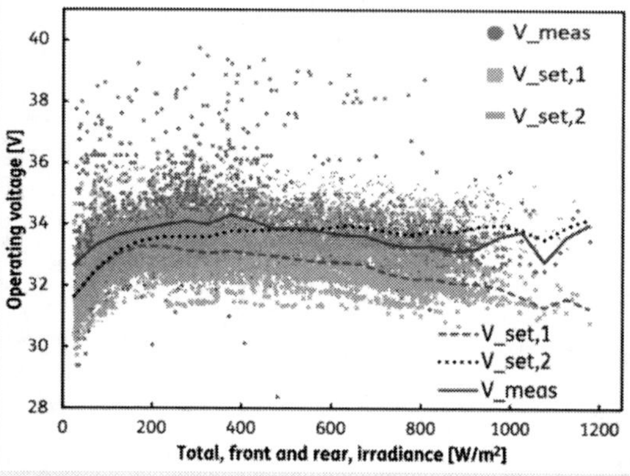

System B

System B

Commercial System

- Netherlands
- Pitch distance is about 6m
- String of modules

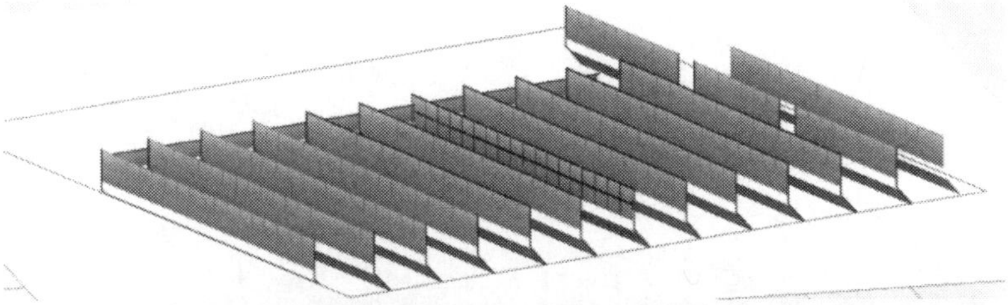

- Updated thermal heat transfer coefficients for vertical PV
  - $U_c = 56\ W/m^2/K$, $U_v = 0\ W/m^2/K/\ m/s$
- Acknowledgement:
  - Vrijstad Energie
  - Farmer

19 September 2023 | Digital Twin of vertical PV

020305-009

40th European Photovoltaic Solar Energy Conference and Exhibition

System B

Simulation results

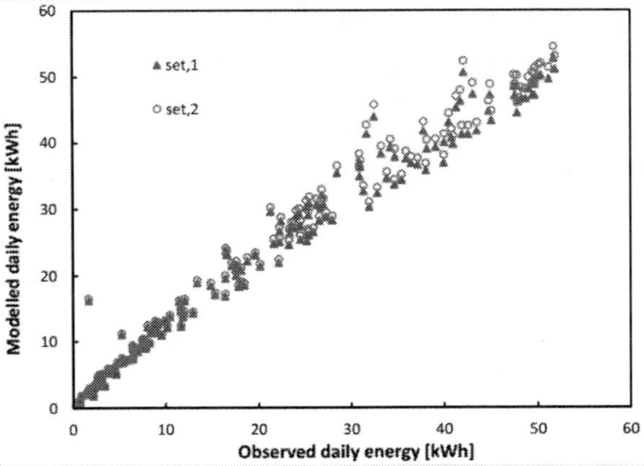

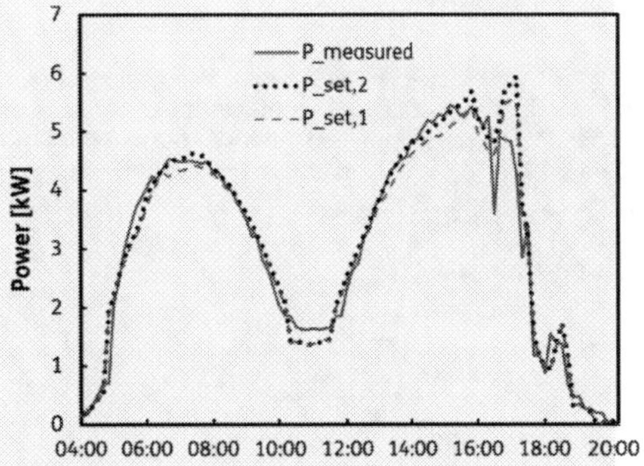

19 September 2023 | Digital Twin of vertical PV

TNO innovation for life 10

020305-010

Takeaway/Conclusion

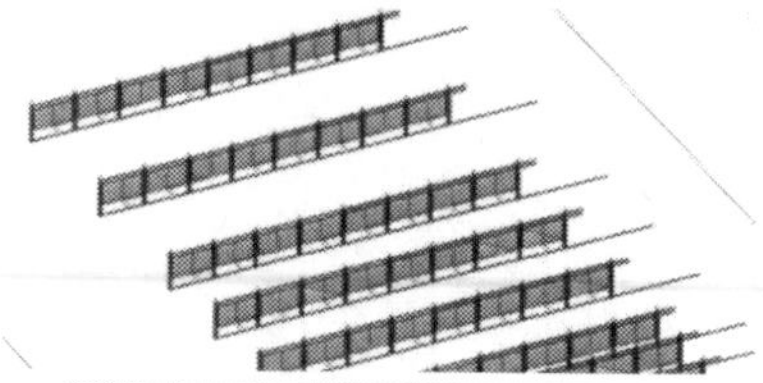

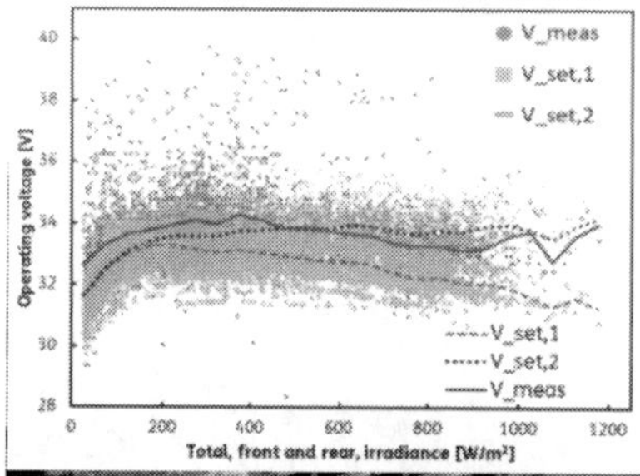

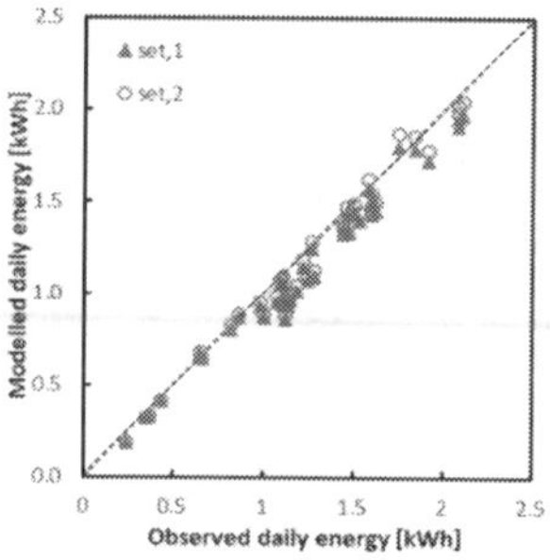

020305-011

40th European Photovoltaic Solar Energy Conference and Exhibition

Takeaway/Conclusion

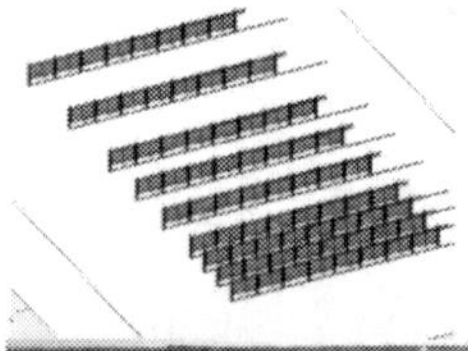

- Digital Twin helps in understanding system

- Default thermal heat transfer coefficient Uc underestimates performance

- Lower module temperature can be observed in higher operating voltage

- Revised Uc is closer to measured yield

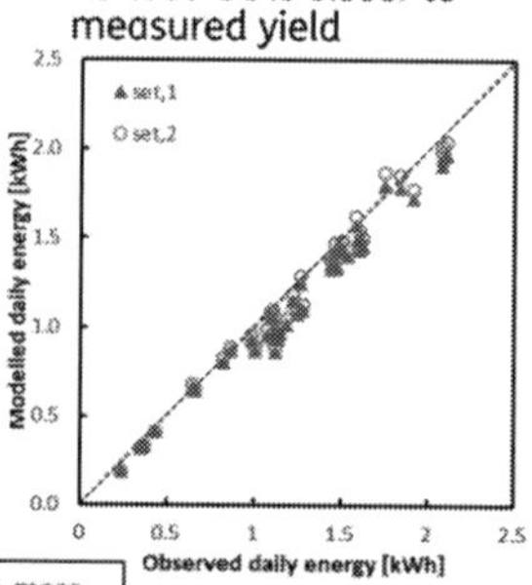

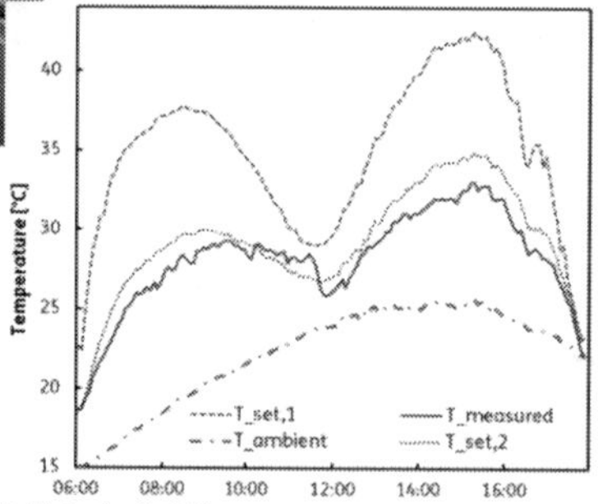

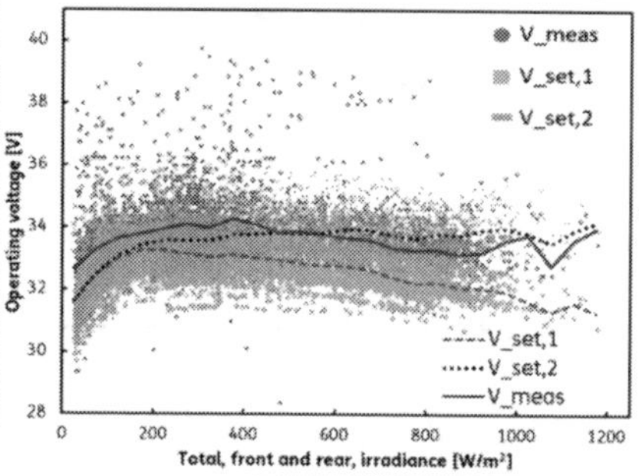

- Vertical PV
- Higher yield
- Lower module temperature

020305-012

40th European Photovoltaic Solar Energy Conference and Exhibition

This presentation was selected by the Sc. Committee of the EU PVSEC 2023 for submission of a full paper to one of the EU PVSEC's collaborating peer-reviewed journals.

EFFECTS OF WIND TURBINE DYNAMIC SHADING ON COMBINED SOLAR AND WIND FARMS

Nico J. Dekker[a], Lenneke H. Slooff[a], Mark Jansen[a], Gertjan de Graaff[a], Jaco Hovius[b], Rudi Jonkman[c], Jesper Zuurbier[d], Jan Pronk[e]

[a] TNO Solar Energy, Westerduinweg 3, 1755 LE Petten, The Netherlands, nico.dekker@tno.nl, lenneke.slooff@tno.nl,
[b] Vattenfall, Hoekenrode 8 1102BR Amsterdam, The Netherlands,
[c] Heliox, De Run 4315, 5503 LP Veldhoven,
[d] Uw-Stroom, Warmenhuizen, The Netherlands
[e] Prosoldiga, Schagen, The Netherlands

ABSTRACT: Combined solar and wind farms offer an interesting solution in reducing the load on the electricity grid. As the energy generation of wind an solar energy are for a large part complementary in time, they could make use of the same grid connection. There is a drawback however, as the static shade of the wind turbine tower, and the dynamic shade of the rotating wind turbine blades can cause serious shadow losses due to reduced irradiance and effects on the electronics of the inverter. This paper studies these effects by millisecond data acquisition of DC and AC signals in non-shaded and shaded strings. By comparing the results, it is concluded that the losses sum up to about 6% for the combined solar and wind farm that was studied.

1 INTRODUCTION

Various energy suppliers see the combination of solar and wind farms as a possible solution[1-2] to reduce the electricity grid congestion problems. [3-7]. It is known that solar and wind energy are to a large extent complementary with respect to the amount of generated energy in time (see Figure 1 below). Overall only during 2-3% of the time do solar and wind operate simultaneously at full capacity.

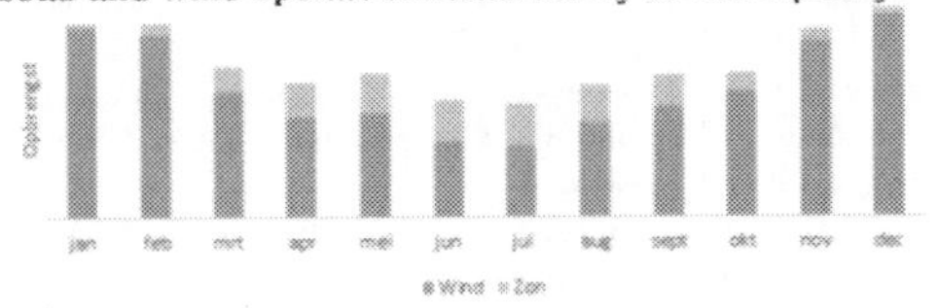

Figure 1. Electricity generated by solar and wind over the year.[8]

But such a combination will always have shade of the wind turbines on the solar farm. To what extend this has an effect depends on the position of the wind turbines with respect to the strings in the solar farm, but also on the way the strings are configured and the type of inverter and its algorithm. There is not much information on the exact nature of the disturbances due to the shade of the wind farm. To get more insight in these effects we have performed a millisecond measurement campaign on two combined solar and wind farms. Part of the results are given in this paper.

2 Experimental setup

2.1 Solar farm 1

The first solar farm was a 30 MW_{AC} solar farm with 115.232 60 cells, 330 Wp solar panels and 163 string inverters. The inverters have 9 maximum power point trackers (MPPT) per inverter with 2-3 strings parallel per MPPT. The farm orientation is South-East. See Fig.2.

2.2 Monitoring system 1

The monitoring system was positioned in two different areas of the farm, one in an area that did not suffer from the shading of the wind turbine, indicated in green in Fig. 3, and one that did suffer from shading, indicated in red in Fig. 3.

Figure 2: Picture of the first combined solar wind farm of this study, showing the shadow of the wind turbine tower and the blades, strings monitored and position of the irradiance sensors (see Fig. 4).

Figure 3: Overview of the first combined solar and wind farm with the areas studied on the left in green, unshaded, and on the right in red, shaded.

The monitoring system was designed to determine:
- the characteristics of the dynamic shade (in relation to the position of the rotor and that of the sun)
- its impact on the PV power electronics (in relation to the hardware and software configuration)
- its impact on the PV system yield (in relation to the design of the PV system).

This was done by measuring the differences in AC and DC yield between the shaded and unshaded areas, as well as the dynamic reaction of the inverter due to the fast moving shades of the turbine blades. This was done by logging DC power, voltage and current and energy of the top string and the parallel connection of top and bottom string, see Fig. 4. Sampling was done every 6 seconds. Additionally, measurements sampled every millisecond were done at the DC input of the inverter, the voltage of the panels at beginning, middle and end side of the string as well as the irradiance near these positions. To complete the measurement system, a camera was overlooking the farm to indicate the shade position at every moment in time and ambient and module temperatures were measured. More details on the setup can be found in ref 9. In this park, the two strings in parallel per MPPT monitored were positioned above each other, thereby suffering from almost identical shading conditions

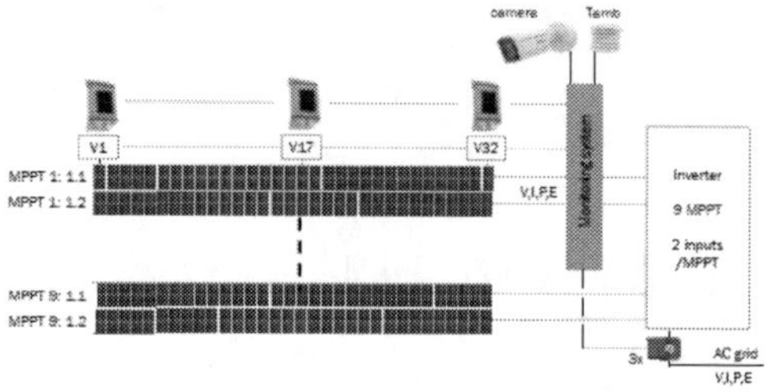

Figure 4: Monitoring system.

2.3 Solar farm 2

The second solar farm was a 13.5 MW_{AC} solar farm with 30.000 144 half cells, and 40 string inverters. See Fig. 5.

Figure 5. Picture of the second combined solar wind farm of this study.

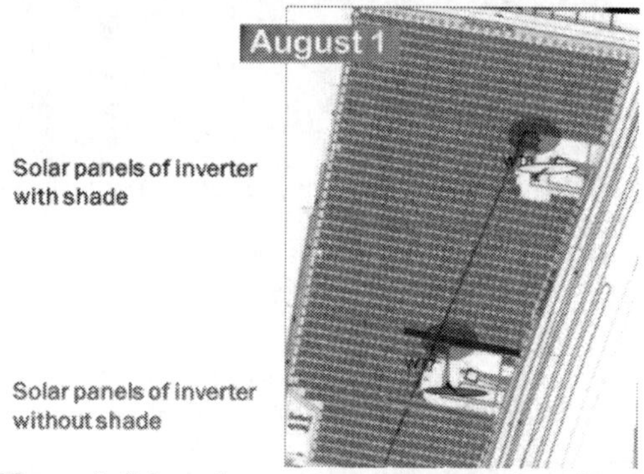

Figure 6. Schematic overview of the second combined solar and wind farm with the areas studied in green, unshaded, and in red, shaded.

The same monitoring system as in the first farm was used, positioned in two different areas of the farm, one in an area that did not suffer from the shading of the wind turbine, indicated in red in Fig. 6, and one that did suffer from shading, indicated in green in Fig. 6. In this park, the two strings in parallel per MPPT monitored were positioned in line of each other, thereby suffering from different shading conditions. See Fig. 7.

Figure 7. String positioning and position of irradiance and voltage sensors at beginning (B), middle (M) and end (E) of string 1.

3 RESULTS AND DISCUSSION

3.1 Results solar park 1

In Fig. 8 the power output of a string is given as function of the in-plane irradiance for the systems without and with shade on a clear sky day. The power output of the string without shade is perfectly proportional with the in-plane irradiance, while the power output of the string with shade is considerably lower.

The lower power is due to a higher and lower voltage (Fig. 9) as well as a lower current (Fig. 10).

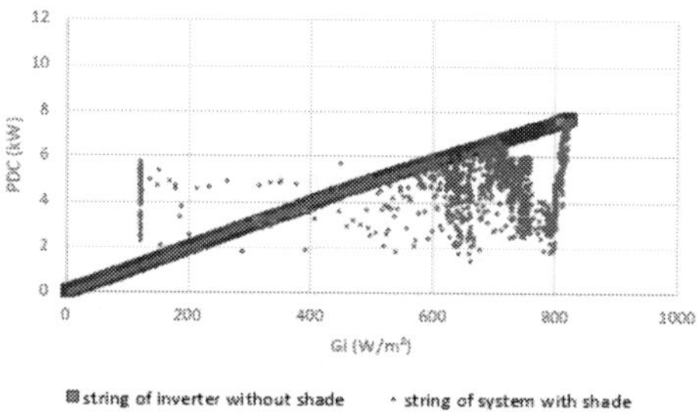

Figure 8. Power of the strings as function of the in-plane irradiance for the system without and with shade.

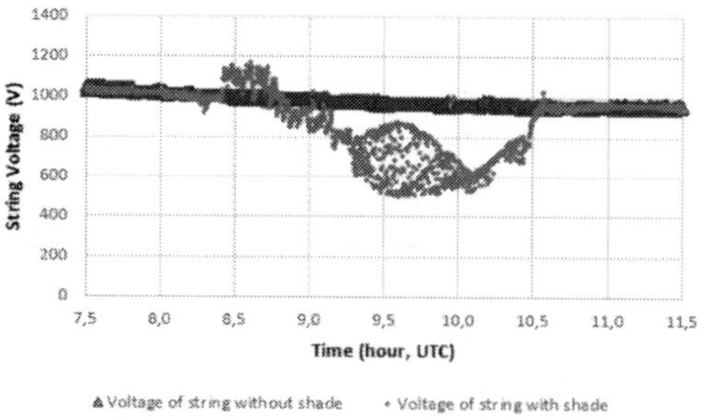

Figure 9. Voltage of a string in time for the system without

10.4229/EUPVSEC2023/4BO.7.4
020306-002

and with shade in time

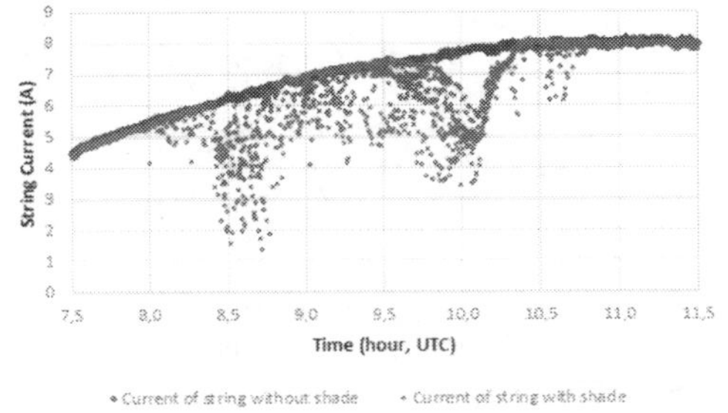

Figure 10. Current of a string in time for the system without and with shade in time

These figures clearly show the effect of the shade on the voltage, current and power of the string as measured by 'slow' data logging (sample per every 6 seconds). A detailed analysis of the fast data of this park can be found in ref. 9, explaining the deviation in MPPT of the inverter.

3.2 Results solar park 2

Fig. 11 shows the fast, ms monitoring results for the irradiance sensors of the second solar park on a sunny afternoon as seen in Fig. 7. At that moment, the shadow of the turbine blades only rotate over the End irradiance sensor of string 1 and reach to almost half of string 2. So both strings are affected by the shade. As can be seen, only the irradiance of the End sensor is dropping to about 1/3 of the unshaded irradiance. The other two sensors are not affected.

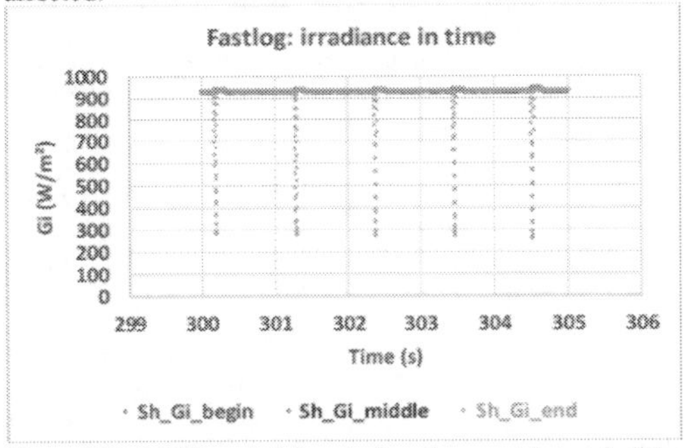

Figure 11. Millisecond drop in irradiance at the end of the string, upon fast passing of the shade of the turbine blade.

The corresponding ms voltage response is shown in Fig. 12. Again, no response is seen for the module voltage of the Middle and Beginning module, but the End module is shortened by the bypass diodes.

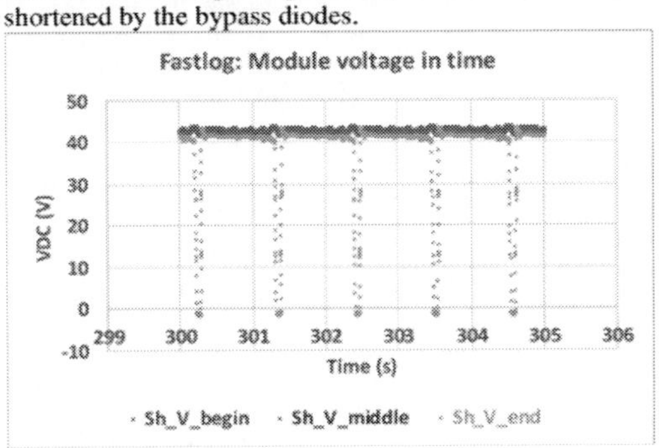

Figure 12. Millisecond module voltage response upon fast passing of the shade of the turbine blade.

After passing of the blade, the module voltage quickly returns back to the voltage at the MPP.

The corresponding DC voltage response of the inverter is seen in Fig. 13.

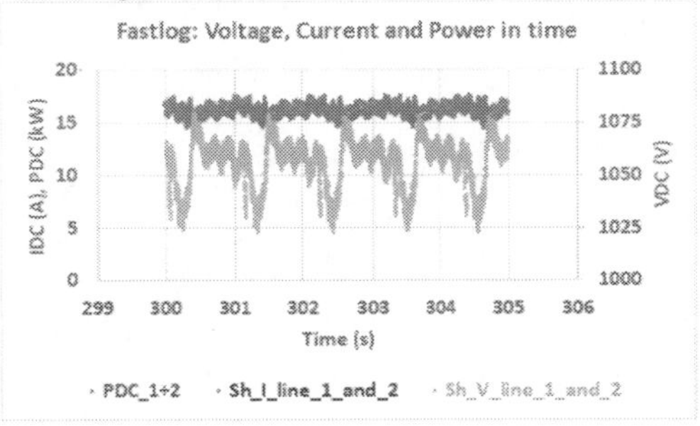

Figure 13. DC voltage of the inverter upon fast passing of the shadow of the wind turbine blade.

The data in Fig. 13 show that the algorithm of the inverter chooses modulates the voltage to keep the current at Impp level. This results in maximum power output of the system.

However, later that day, the situation is different. Around 2PM, the shade of the wind turbine blades affects about ¾ of string 2, but not string 1, as can be seen in Fig. 14.

Figure 14. Position of the shade of the wind turbine blades at 2PM.

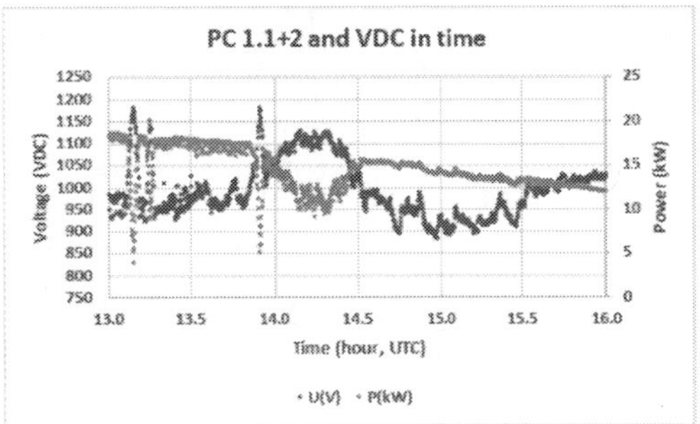

Figure 15. DC voltage of inverter and power output of the inverter around 2PM

10.4229/EUPVSEC2023/4BO.7.4
020306-003

Fig. 15 shows the DC voltage response and the power output around 2PM. The voltage starts around the Vmpp at 950V, but starts to increase to about 1125V. This is a continuous rise, whereas the shadow of the wind turbine blade is causing only a temporal disruption in the irradiance. It is clear that the MPPT is not able to find the most optimal situation, as it did in the previous situation. After roughly half on hour, the MPPT finds the correct settings again. This incorrect behavior results in a drop in power output of the string from 16kW to 9 kW. See Fig. 16.

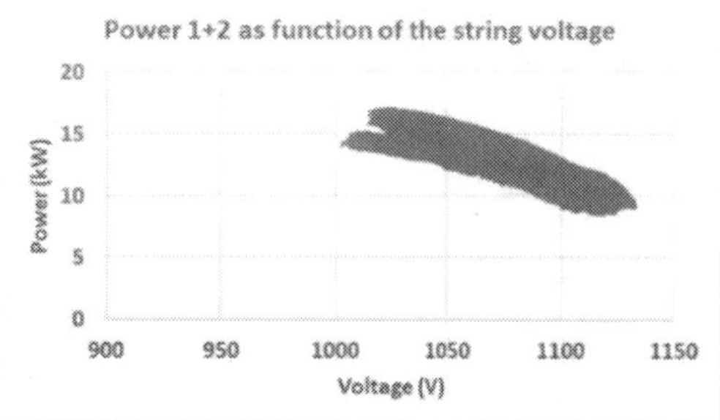

Figure 16. Power of the combined strings around 2PM as a function of the string voltage.

4 CONCLUSIONS

Detailed measurements on second and millisecond sampling rate were performed to study the effect of static and dynamic shadow of wind turbines on two solar farms in a combined solar and wind farm. The data show a significant effect due to both static and dynamic shading. The results are obtained for certain situations for a specific string in a specific solar park and only give an indication what the effect of a fast dynamic shade from a wind turbine blade can be. In general, at a certain moment, only a limited region of the solar park will be affected by the shade and the losses due to the dynamic shadow in the total park can be much lower.

ACKNOWLEDGEMENTS
This work is part of the SolarWind project (reference number TEUE119001) and financed by the Dutch Ministry of Economic Affairs and Climate through RVO.

REFERENCES

[1] S. Mertens, Cleaner Engineering and Technology 6 (2022) 100402.
[2] S.Z.M. Golroodbari, D.F. Vaartjes, J.B.L. Meit, A.P. van Hoeken, M. Eberveld, H. Jonker, W.G.J.H.M. van Sark, Solar Energy, 219 (2021) 65-74.
[3] B. Rausch, P. Stadut, C. Weinhardt, Proceedings of the Tenth ACM International Conference on Future Energy Systems (2019) 443–446.
[4] H. Schermeyer, M. Studer, M. Ruppert, W. Fischer, IFIP Advances in Information and Communication Technology, 495 (2017) 78-89.
[5] N. I. Yusoff, A. A. M. Zin and A. Bin Khairuddin, 2017 3rd International Conference on Power Generation Systems and Renewable Energy Technologies (PGSRET), Malaysia, 2017, 22-27.
[6] S. Nandini, P. Suganya and K. M. Lakshmi, International Journal of Innovative Research in Science Engineering and Technology, vol. 3, no. 1, pp. 682-688, 2014.
[7] C. Hitaj, Location matters, Energy Policy, 86 (2015) 1-16.
[8] https://ponderaconsult.com/meer-duurzame-energie-met-minder-kabels-een-netaansluiting-beter-benutten/
[9] Nico. Dekker, Lenneke Slooff, Mark Jansen, Gertjan de Graaff, Jaco Hovius, Rudi Jonkman, Jesper Zuurbier, Jan Pronk, submitted for publication in Journal of Renewable and Sustainable Energy

40th European Photovoltaic Solar Energy Conference and Exhibition

This presentation was selected by the Sc. Committee of the EU PVSEC 2023 for submission of a full paper to one of the EU PVSEC's collaborating peer-reviewed journals.

THE EARLY DEATH OF PV SYSTEMS IN ASIAN SITES SET IN INDUSTRIAL AREAS

André M. Nobre[1,#], Thornthanut Pakdeepinyo[1], Jason Ward[1], Rohit Jaswal[1], Sai Pranav[1], Shailendra Singh Chouhan[2], Ankush Gaurav[2], Ashish Kumar Sharma[2]
[1]Cleantech Solar, 25 Church Street, #03-04, Capital Square Three, Singapore 049482
[2]B/702, Satellite Garden 1, Gen. AKV Marg, Goregaon East, Mumbai Maharashtra, India 400063
[#] andre.nobre@cleantechsolar.com

ABSTRACT: Photovoltaic (PV) systems utilization in Asia has increased tremendously in the past decade. In the early days of PV, most installations were in the United States, Japan and various European countries. Since the mid-2010s, the deployment of sites picked up in Asia, with China leading the way. As much as technology has advanced in terms of PV systems performance enhancements, and consistently through improved module efficiencies, the reality on the ground where systems are being deployed is different than the perceived "pristine" solar farms deployed across vast expanses. Observed field degradation rates are noticeably higher than those reported by long-standing PV systems present in temperate climates of the world. One of the major culprits identified is advanced soiling found in industrial settings across developing countries in Asia. As Asian markets enter multi-GW territory, systems output must achieve developers' & financiers' expectations, irrespective of the perceived high irradiation/high conventional electricity costs found in these countries. Having that in mind, this investigation intends to assess PV systems that are clearly on a trajectory of "early deaths" – through accelerated degradation rates and other low-performance-inducing mechanisms, which go against investors goals.

Keywords: PV System, System Performance, Long-term Behavior, Degradation, Industrial Pollution.

1 INTRODUCTION

Photovoltaics (PV) have been deployed at rapid speeds for well-over two decades now. From sub 1 GWp levels of installed capacity at the start of the new millennium to ~1.2 TWp by 2022 [1], a compound annual growth rate of ~40% clearly indicates that solar electricity generation is amongst the most successful forms of present energy choices. By the end of 2023, total installed photovoltaics globally will cross the 1.5 TWp threshold [2], with China, Japan and India combined representing circa 50% of all PV systems [3], whereas global penetration rates are still at 6.2% (of electricity demand covered by photovoltaics).

As Asia leads the number of installations worldwide, it would only be natural that systems are well-understood in those parts of the world. That is not the case due to two reasons: firstly, as PV was mostly deployed in Europe, Japan and the United States as pioneering locations, markets are mature in those areas, with several systems having run for more than a decade; and secondly, systems in locations such as Southeast Asia are still very young.

However, as systems are rapidly coming online, a dangerous combination ensues – installations are developed at speeds that do not allow for the proper maturing of local specialized companies, thus workmanship of systems tend to be lower when compared to other global markets. And perhaps most concerning is that given the growth rate of Asian nations being higher than world averages, you have the aggravating aspect of heavy industrial activity. The outcome are systems deployed in areas where air pollution is an issue [4], or systems suffering from various types of loss mechanisms preventing optimum performance, primarily soiling as the most prominent culprit [5].

With that in mind, this paper is targeted at studying low-performing solar photovoltaic assets in Asian locations set in industrial areas, where soiling from site activities acts against optimum PV output. Selected systems for this publication are found in India and Southeast Asia (such as the system at a cement factory shown in Fig. 1, plagued by strong adhesion of dust particles to the front glass surface of the solar panels). The main objective is to bring to light the realities and field experience found in real-world situations so that the scientific community as well as investors & decision-makers at large are made aware of detrimental return on investment conditions presently found in Asia (and potentially elsewhere in the world under similar conditions).

Figure 1: Workers cleaning a solar system deployed at a cement factory in Southeast Asia via "scrapping" the front glass of the solar panels.

2 MOTIVATION

Fig. 2 show us how heavy industrial activity has affected a PV system in India. The fumes from existing industrial activity within the factory compound are sufficient to fully cover the solar panels' glass surface to a point where its output is diminished to more than half of its intended expectations. This system is only one example of several of a fleet of photovoltaic installations across Asia by a single developer holding a portfolio of 500+ assets with little over 0.5 GWp of capacity. Yet, photovoltaics deployed across Asia has crossed 0.5 TWp – one thousand times the volume covered in this study – which could indicate that most systems have performance shortfalls as portrayed here, or also are on the path for an "early death", with loss of output much higher than expectations.

Figure 2: Red metallic steel emissions from chimneys at this rooftop cover the typically blue solar panels to an "unrecognizable look". Power output observed at this system has been cut by more than 60% at the condition shown in the picture.

Photovoltaic systems are now deployed in their majority in the Asian continent. However, most publications in this space are still stemming from installations in the US, Japan and Europe. The work proposed here aims therefore to act as a key contribution to works for Asian systems. Equally important is the ability to showcase the performance records of hundreds of installations, with a considerable volume of sites (versus typical scientific publications which tend to address how a single site has performed with certain characteristics of interest).

The findings presented are not meant as being a deterrent to the progress of photovoltaic systems in developing Asian countries rather to act as guideline to best installation practices. For example, cement factories tend to have major footprints and certain sections of these sites, away from the cement clinker zone (the heart of the factory), may contain suitable space for optimum (or at least less detrimental) PV performance (seen in [5]). Learnings from real-world painful cases can act as best practices for future systems being envisioned, for example the careful design positioning of modules on a roof to avoid certain exhaust chimneys.

3 METHOD

For this investigation, a fleet of 500+ PV systems with a volume of more than 0.5 GWp is evaluated. The footprint of these systems can be seen on the map in Fig. 3, with installations ranging from New Delhi (North India) all the way to Surabaya, Indonesia, across 7 countries. The number of inhabitants of these nations (India, Thailand, Malaysia, Singapore, Cambodia, Indonesia and Vietnam) alone account to nearly 2 billion (or ~24% of the entire global population), highlighting the likelihood of growing electricity needs in these developing areas of the world.

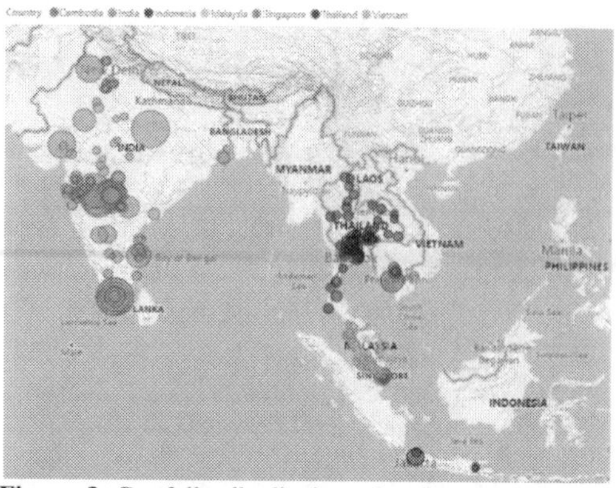

Figure 3: Portfolio distribution of a fleet of 500+ PV systems across more than 0.5 GWp in seven countries across India and Southeast Asia. Bubble sizes are proportional to the size of the PV installation.

The study is subdivided into three areas as follows:

i) Part 1: "Identification of *Sick/Dying* Systems". In this moment, performance ratios (PRs) are calculated for all systems. PRs are the most common metric of a quality of a PV system and are covered in an IEC standard [6]. However, due to challenges in using ground sensors in Asia for PR calculations, the methodology in [7] is repeated, where satellite data sources are used as a much more reliable data point. PRs can be visualized in a yield vs. irradiation plot which then allows for further insights into the pool of facilities. A similar evaluation was used by Reich et al. in [8] to showcase the improvements of PV systems in Germany across time.

ii) Part 2: "Data Scrubbing". All sites have their own monitoring systems, where data from energy meters, inverters and weather sensors (where available) are recorded at either 1-min or 5-min intervals. This type of data is what allows analysis such as in Fig. 4 to be generated, indicating a "sick" PV asset with degradation of -1.8% per annum, considerably higher than similar PV systems for the same location [9].

iii) Part 3: "Deep Dive". In this phase, in situ investigations were launched with the goal of isolating system components. For example, by measuring a single panel's electrical output before and after cleaning such panel, we can gauge how much loss of generation occurs due to less light reaching the solar cells. Once the sites with advanced degradation rates are found, dedicated in-situ investigations are to be performed with the use of IV-tracers, plus thermal and night-time electroluminescence (EL) imaging for mapping of fault mechanisms in PV modules. Such efforts allowed the zoom in to system performance at module level on sites to help understand advanced aging effects.

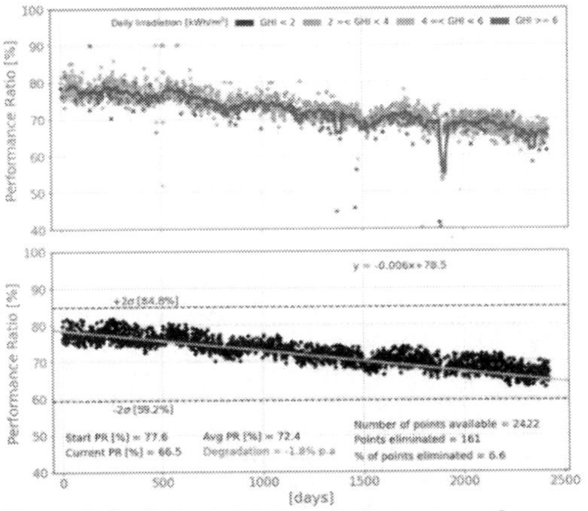

Figure 4: Performance ratio evolution over nearly seven years of a PV system in Singapore. Using the continuous degradation method calculation as per [9], the rate -1.8% p.a. is found which is more than double of what is expected of typical PV systems business models.

Data range for Part 1 included system records for one year, from September 2022 to August 2023. For Part 2, data availability is in general better than 99.5%, recorded with industrial-grade data loggers. All codes in the study were written in Python. The instrumentation used for Part 3 consisted of:

a) IV-curve tracing: Solmetric PV Analyzer/PV-1500T;
b) Thermal imaging: FLIR C2;
c) Electroluminescence (EL) imaging: EL Camera D90 (Nikon).

ELs were performed at nighttime where allowed. For sites where restrictions existed, certain panels were chosen for testing and removed from the roof, brought down to a make-shift laboratory where darkness could be replicated for the images to be generated.

4 RESULTS & DISCUSSIONS

4.1 Part 1: "Identification of *Sick/Dying* PV Systems"

Fig. 5. shows the individual plots for each PV asset in relation to their annual irradiation (in kWh/m^2) versus their annual specific yield (in kWh/kWp). The irradiation source is from a satellite data provider and the selection of the value can be either global horizontal irradiation (GHI) or global tilted irradiation (GTI) depending on certain boundary conditions which were covered in [7]. The country codes for the systems are TH (Thailand), SG (Singapore), KH (Cambodia), MY (Malaysia), VN (Vietnam), ID (Indonesia), with IN (India) being subdivided into two categories namely IN-RT (systems deployed on rooftops) and IN-SF (systems which are solar farms). A green transparent band is present indicating typical zones where profitability of the assets is desired, depending on the age of the installations. The average PV system age of this ~0.5 GWp is approximately two years, so performance ratios of 72-76% are in line with expected investors' returns aspirations. A yellow/red star was calculated as the weighted portfolio average and lands inside this green interval, which is already an indication that the fleet performs within expected ranges of profitability, irrespective of outliers of underperformance.

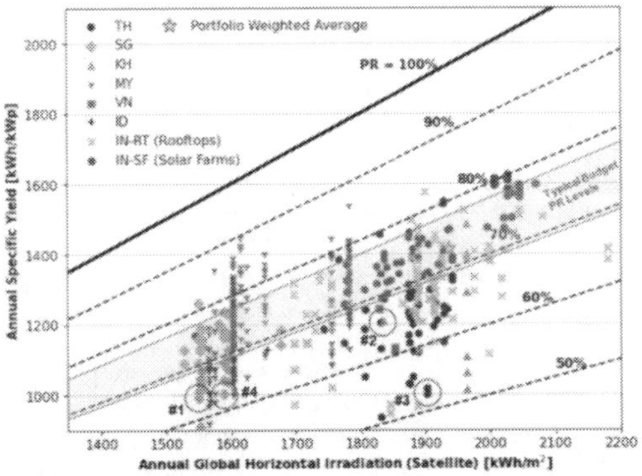

Figure 5: Scatter plot of annual irradiation from satellite sources versus annual specific yield for 500+ PV systems, ~0.5 GWp worth of installations.

Some of the observations we can derive from the figure follow:

a) PRs are much lower than those experienced in temperate climates such as Europe [8]. This is due to much larger thermal losses in PV systems found in tropical climates [10];
b) Some of the sites have the same irradiation data source, as seen from "linear" sequence of MY points. This is because there are several sites sharing the same latitude/longitude satellite data source. They are small systems deployed at petrol stations which are very near to one another (here for Kuala Lumpur). Due to the high satellite data costs, certain simplifications are needed when managing hundreds of small assets as shown in [7];
c) The low PRs found in low profitability zones (65% and below) may not be fully associated with technical problems as previously shown in [5]. Certain systems are not able to export excess solar electricity to the grid due to local grid regulations. The power purchase agreement tariff may already have been adjusted to higher levels to compensate for such undesired losses. Another example may be linked to high shading found in an urban setting such as Singapore. A fleet of systems for a given client may collectively be profitable, but not all sites as individuals.

For this study, four sites were selected for further investigations and marked in Fig. 5. They will be treated for the rest of this work as:

- Case Study #1: "Watermark Soiling"
- Case Study #2: "Aggressive Soiling"
- Case Study #3: "Steel Mill Soiling"
- Case Study #4: "Recoverable Soiling"

Each of the case studies was selected due to having distinct characteristics and being able to serve as examples for the solar community in areas to play close attention to when deploying PV systems in industrial settings. Although some of the findings can be perceived as negative, they are meant to serve as educational material so that future installations do not suffer similar fate. The fourth case study in fact is an example of a solar system which has been recovered and back on its intended generation targets.

4.2 Part 2: "Data Scrubbing"

In Fig. 4, the solar system of Case Study #1 was plotted for its performance ratio evolution across its nearly 7 years of life. Using the methodology applied in [9], an annual degradation of -1.8% was found. In typical business plans for solar PV deployment, these ratios can be anywhere from -0.4% to -0.8%, depending on module technology and aggressive nature of developers. Therefore, the rapid rate of decay here of this solar system is nearly triple that of earlier expectations.

The data can only show a part of the story. Without site visual inspections, one would be left wondering at what causes the unwanted phenomenon. Upon observing these systems in proximity, one can understand the source of the lower generation. It is caused by dust accumulation at the lower end of a solar module, see Fig. 6. This degradation is further compounded from the fact that once a cleaning cycle is performed, the glass surface of the panel is permanently stained, without being able to be cleaned off, even with trials of various types of chemicals. This prevents sunlight from being able to fully hit the solar cells and less power output is obtained.

Figure 6: Installations at low tilt angles such as <7 degrees cause the unwanted accumulation of dirt at lower ends of the panels. Rainwater with panel dust is unable to be drained and the dirt attaches downstream of the panel upon water evaporation.

By using the monitoring systems data from the hundreds of sites, we can isolate case studies and try to understand remotely the reasons for underperformance.

4.3 Part 3: "Deep Dive"

The four cases selected for this study are presented and discussed as follows:

- Case Study #1: "Watermark Soiling"

The glass staining caused by low tilt angles and dust accumulation onto panels could not have been fully explained without further investigations on site. Visually, one can see the dust/stain marks, but it was necessary to validate and quantify the loss mechanism observed. Fig. 7 shows that an electroluminescence (EL) image (top left row) performed on a panel of a Singapore PV system with watermark soiling did not indicate the presence of any issues. Similarly, the thermal images executed on site (bottom left row) also did not indicate any problems with the modules. However, the actual IV curve measurements (on the right of the figure, blue lines), when reconstructed to STC conditions (orange lines), do indicate a drop in power output of the panels against their expected specifications (these points are represented by the datasheet circles in these IV curves). The percentage drop observed, when put into perspective with the age of the asset, was in line with the nearly -2% per annum degradation rates observed.

Although the glass staining seems permanent on site even after strong cleaning pressure and testing of several chemicals, plastic drain clips have been added to the downstream edge of the panels at this site. The intent of such clips is that there is a pathway for rainwater to exit the lowest part of the solar module, thus allowing the dirt also to exit. Even though it is expected that the degradation rate cannot be reversed, we theorize that it likely can be brought back closer to -1% p.a. levels (from current "year 7" PR levels onwards). Due to this, a recommendation for future work could address the performance evolution of sites with and without the addition of the drain clip.

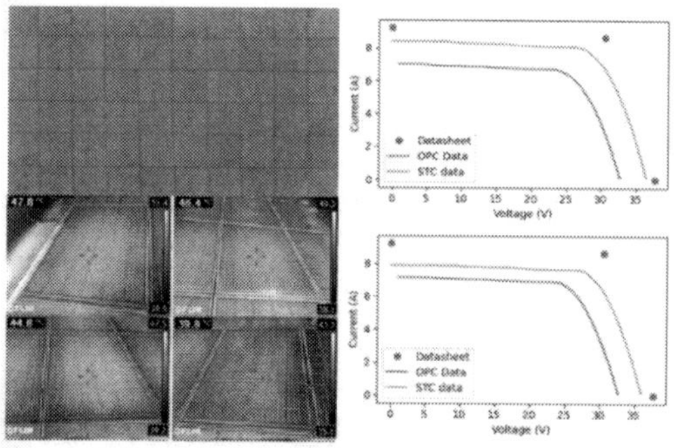

Figure 7: Electroluminescence (EL) image (top left row) performed at panel with watermark soiling did not indicate any issues, so did the thermal images on site (bottom left row). However, the actual IV curves indicated anomalies which could be traced to similar annual degradation rates measured with monitored data.

- Case Study #2: "Aggressive Soiling"

The photo of the workers on a curved roof in Fig. 1 takes place at the site addressed in this case study. As the solar system is installed in a cement factory in Cambodia, the roofs near the core of the facility (clinker zone + storage areas) have cement-like dust adhered to the glass of the solar panels. As locations with hot & humid climates have regular rain showers, it is usually a bad combination together with dust elements – the result being a compound which hardens onto the panel surface.

For this system, several chemical solutions were tested from providers stemming from multiple countries. No specific formula was found that could melt the dirt into a washable level. Without any choice left, it was decided to attempt a scrapping technique, like what is done when removing paint from a wall. This method has worked in allowing the PV system to operate at PRs close to expectations, but at consequences that have yet to be tested in terms of the long-term health of the system. The only confirmed outcome was that a team of ~40 workers must clean the system daily to keep its output close to business targets (thus maintenance expenditure has more than doubled for this asset).

With this investigation, the extra steps needed to further confirm if the scrapping method was damaging the panels were performed. Results, as the ones shown in Fig. 8, corroborate such fears – by using electroluminescence technique, it was confirmed that modules have micro cracks on several of their cells. The IV curves of these panels also indicate abnormal behavior. When inspecting on site cleaning methodology by the maintenance company, it was also confirmed that due to difficult nature of the roof (steep and slippery, due to the cleaning water), that workers tend to anchor themselves between panels and end up adding extra pressure onto the frames of the solar modules. Therefore, beyond the scrapping force applied, there is a constant movement of workers across the PV field which is likely adding to new fractures. Future work on this site may include seasonal execution of new EL images and IV curves, as well as continuous tracking of PR values.

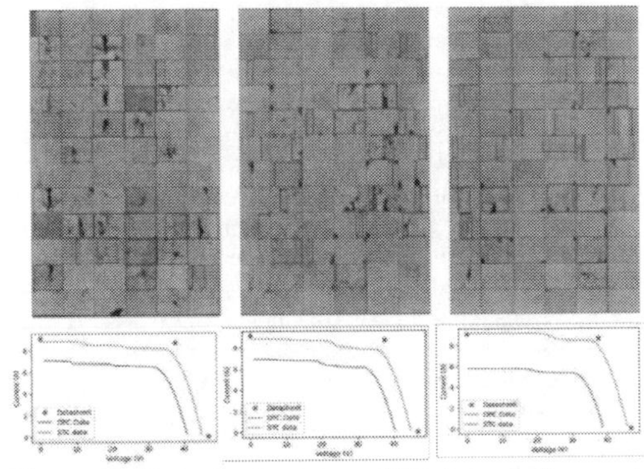

Figure 8: Electroluminescence (EL) images (top row) and their respective IV curves (bottom row) of PV panels installed at a cement factory. The panels are cleaned 40+ times per year with a very aggressive scrapping technique. The ELs clearly indicate breakage of cells (dark spots) while the corresponding IV curves (blue lines) showcasing poor shape, a signal of electrical issues within the module.

- Case Study #3: "Steel Mill Soiling"

The third case study of this work sees the performance evaluation of a PV system at a steel mill factory in Thailand. Fig. 9 (top row) shows the satellite view of the roof before (left) and after (right) PV system addition. Before the solar system installation, it was already possible to see that the roof had stain marks. In fact, this is one major area of discussion in this conference contribution – that roofs ought to be thoroughly evaluated before a solar system is deployed, with certain zones being eliminated from consideration. On the right side of this roof, it can be clearly seen that soiling would have played a role in the optimum performance of the system. It comes as no surprise then by observing the bottom row of Fig. 9 that the solar modules are severely covered in steel mill dust – a stubborn and sticky compound which is hard to clean.

This solar system performance ratio has been hovering at the 60-65% zone for its fifth year of operation, a figure which is about 5-10% PR points short of the performance guarantee the developer has with the client on site. This means, beyond the shortfall in production which hinders the necessary profits to pay back the investment of the system, the developer suffers additional penalties set by the very polluter on site.

Further into the investigation of the health of this asset, the EL image-taking process was complex – as night work was not allowed, a handful of panels had to be lowered down from the roof and tested in a dark room, thus illustrating the complexity to even trying to address sources of the loss mechanics found at certain sites. EL, IV and thermal images came out clear for this system, indicating that issues are simply related to the blockage of light from reaching the solar cells inside the panels.

More recently, the testing of novel chemicals has yielded minor improvements in this asset. But once more, complexity of roof (15 degrees of tilt), minimal client support and challenges in the "reset clean" of these panels stop the PR gains from reaching 70+% levels.

Figure 9: Satellite view of steel mill in Thailand before the installation of the solar PV asset (top left) and after its installation (top right). For the bottom row, the general situation of the site with dust adhesion to panels can be closely seen.

- Case Study #4: "Recoverable Soiling"

One PV system which was investigated in this work had a success story in being recovered back to its intended performance. Fig. 10 (top) shows the asset deployed on a factory rooftop in Singapore. The top of the figure illustrates a system with a brown color soiling. Although classic cleaning using water and mop can help remove this type of sludge-like dirt, it is not fully effective. It also needs to be executed approximately 20 times per year to allow the PV system to operate at its intended levels. All in all, 20% of this 1.6 MWp site suffers from this type of soiling.

Upon execution of the various tests, the EL scans came up clear, indicating the solar cells are healthy. This was a good indication to the team that the issues lied primarily on the front glass top of the panels. Although a few modules showed cell hot spots as seen in Fig. 10 (bottom left), that was only observed in a few small sections of the system (< 2% of the site).

Recently, various chemicals were tested here with the goal of attempting a "reset cleaning" at the site's troubled zone. Fig. 10 (bottom right) shows some panels which were successfully cleaned. The next tests, which are still ongoing, are meant to determine how often cleaning cycles should take place with the objective of not allowing the brown sludge to return and attach itself to the panels.

Figure 10: Top view – solar system deployed at a factory in Singapore with its surface heavily soiled with a sludge-like material. Bottom left – the thermal image at a certain section of this system showed abnormal cell temperatures due to the differing rate of evaporation at a few of the affected panels. Bottom right – ongoing chemical cleaning tests, which have shown successful results, are being rolled out to the rest of the site at affected zones.

Fig. 11 ultimately displays this system and its performance ratio evolution for the past 6+ years. The asset has been mostly profitable for its first half decade of operation but started to face some string/inverter failures, combined with the progressive worsening of the soiling problem. With the fixing of the technical problems with the equipment, the last frontier for the full recovery of this system was the solving of the soiling (sludge-like) problem. Once a successful working formula was discovered, zones of the system started to undergo "reset cleaning" and PRs began climbing back to levels above the threshold of profitability. With the near completion of the chemical cleaning activity, PRs have hovered in the 75-76% zones. There is hope that nearly 80% PR levels can once again be achieved here.

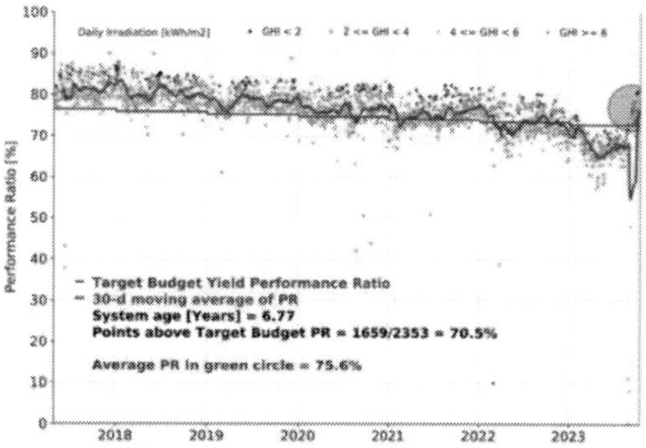

Figure 11: Performance ratio evolution of the site in case study #4, where chemical cleaning of panels affected by chimney fumes were successfully cleaned. PRs have resumed to satisfactory levels after deep cleaning intervention.

4.4 Further Discussions

The drain clip solution which has been added to the system in Case Study #1 has also been applied to other sites of the portfolio with below 7 degrees of tilt in the installation. Additionally, for sites that had just been commissioned or about to be completed, the clip has been either retrofitted or is part of the new bill of materials. This way, it is expected that issues that brought system annual degradation levels to nearly -2% per annum do not occur again.

Also, important to note is the concept of avoidance of recurring errors, or "repeating the same mistake" effect, such as the deployment of PV systems in cement factories. Beyond the system shown in Case Study #2, the developer has deployed a second asset in a cement factory, this time in India, which also has certain system sections fully covered in cement. For the India case, the matter is even worse as the combination cement/dew during winter forms a crust on the glass surface which is even tougher to be cleaned. However, it can be noted that if a cement factory has available land a few km away from the clinker + storage zones of the factory that the deployment of systems is feasible in terms of having optimum power output close to best practice levels.

Continuing, the avoidance of cement factories might sound trivial, an easy step to take, but perhaps the most impactful and achievable action is to avoid deploying modules too close to chimneys, no matter what exhaust types are being expelled. In the example of Case Studies #3 and #4, the assets deployed could have been made smaller (circa 80% of the original size) by avoiding the zones which caused the issues described in this work. There could have been some minor underperformance at a few strings of the system, but any impact would have been constrained to sub 5% levels of the entire system size, versus reported rates of impact >20% here.

Also, paramount to note is that PV systems ought to be classified more and more via their "system degradation" full impact. In some cases, a degradation may be noted to be, for example -1.2% p.a. and lab tests, after removal of dust under controlled conditions can demonstrate that panels have degraded in line with their manufacturing boundary conditions (e.g. < -0.6% p.a.). However, replicating the level of module cleanliness achieved in a factory throughout an entire field of panels in a hot roof in a tropical climate is simply not achievable. This means the observed "system degradation" is in fact as close to the reality expected for the lifetime of the system as one can expect.

A final and more complex discussion revolves around the fact that companies worldwide want to "go green" and reduce the carbon impact to their operations, be it for a stakeholder mandate or simply to save on cost of production (as solar PV is among the cheapest forms of electricity nowadays). Yet, we may reach a situation where solar PV developers (under typical PPA – power purchase agreements – models), upon encountering a "bad roof", may shy away from wanting to deploy an asset in such a factory in the first place. Even if the company eventually opts to deploy a PV system themselves (via an outright capital expenditure), the accelerated aging/degradation of the system due to reasons covered in this study would represent the inability of the target company in recovering its investment. Ways must be identified to have more suitable operational conditions for PV systems in Asia.

10.4229/EUPVSEC2023/4BO.7.5
020307-006

5 CONCLUSIONS

Intense industrial activity plays a role in undermining the performance of PV systems in Asia. Soiling arising from these activities is the main source of losses for PV assets presented in this work (and beyond).

In this paper, hundreds of systems have been investigated in terms of their performance ratio decay. In certain cases, external investigations were needed to find out mechanisms causing accelerated losses (which would likely lead to a de facto "early death" of these systems). Electroluminescence (EL) and IV curve tracing indicated damaged sections for systems which undergo tens of cleaning cycles per year, especially at places where strong forces during cleaning are required (such as cement-like deposits which can only be removed via "scrapping" of panels' glass surface).

As more than half of all PV systems worldwide are currently deployed in Asia, experiences narrated in this work are paramount to be shared. It is hoped that other system owners out there (whether self-owners – under outright purchase, or developers – under PPA models) come forward with their experiences so that the solar PV community can avoid repeat mistakes and can prosper together.

ACKNOWLEDGEMENTS

The authors would like to acknowledge the support of other Cleantech Solar team collaborators across multiple geographies: Bristo Paulose, Vicente Tangcueco, Ng Qi Ying (Singapore), Abhishek Puri, Manav Gupta, Sahaj Mamgain (India), Meas Chansovathdy, Thorn Kimhong (Cambodia) and Sukitti Kraiprasit, Permpat Bumroongsook, Ratwade Dhithijaiyratn and Kamolchat Tappaviboon (Thailand). Also, a further thank you to Dr. Marius Peters (Forschungszentrum Jülich) for his insights into some of the issues plaguing PV systems in rapid degradation mode.

REFERENCES

[1] PV Magazine. *Global solar installations hit 240 GW in 2022*. https://www.pv-magazine.com/2023/04/17/the-world-installed-240-gw-of-solar-in-2020/. Accessed on 27-May-23.

[2] PV-Tech. *Global solar PV installs to reach 270GW in 2023 says Wood Mackenzie*. https://www.pv-tech.org/global-solar-pv-installs-to-reach-270gw-in-2023-says-wood-mackenzie/. Accessed on 22-Aug-23.

[3] International Energy Agency (IEA). Photovoltaic Power Systems Programme. *Snapshot of Global PV Markets 2023*. Online Report, April 2023, 21 pages.

[4] A.M. Nobre, D. Dave, A. Khor, R. Malhotra, S. Karthik, I.M. Peters, T. Reindl. *Advanced analyses of loss mechanisms for PV systems in Delhi, India*. 32nd European Photovoltaic Solar Energy Conference and Exhibition, Munich, Germany (2016).

[5] A.M. Nobre, A. Agarwal, S. Pranav. *Ongoing performance assessment strategies & operational challenges when managing hundreds of distributed PV assets across Asia*. 49th IEEE Photovoltaic Specialists Conference, Philadelphia, USA, 2022.

[6] IEC 61724-1: 2017. *Photovoltaic system performance monitoring – Guidelines for measurement, data exchange and analysis*. Edition 1.0, 2017-03.

[7] A.M. Nobre, S. Karthik, C. Liu, R. Jaswal, R. Baker, R. Malhotra, A. Khor. *Irradiance measurement considerations for system performance assessment when managing fleets of photovoltaic assets across Asia*. 44th IEEE Photovoltaic Specialists Conference, Washington DC, USA, 2017.

[8] N.H. Reich, B. Mueller, A. Armbruster, W.G.J.H.M van Sark, K. Kiefer, C. Reise. Performance ratio revisited: is PR>90% realistic? Progress in Photovoltaics Res. Appl. 20 (6) (2012) 717-726.

[9] A. Nobre, R. Malhotra, C.H. Tang, C. Reise, K. Kiefer, R. Rüther, T. Reindl. *Degradation Analysis of Photovoltaic Systems in a Tropical Environment*. 28th European Photovoltaic Solar Energy Conference and Exhibition, Paris, France, 2013.

[10] A. Nobre, Z. Ye, H. Cheetamun, T. Reindl, J. Luther, C. Reise. *High performing PV systems for tropical regions – optimization of systems performance*. 27th European Photovoltaic Solar Energy Conference and Exhibition, Frankfurt, Germany, 2012.

10.4229/EUPVSEC2023/4BO.7.5
020307-007

This presentation was selected by the Sc. Committee of the EU PVSEC 2023 for submission of a full paper to one of the EU PVSEC's collaborating peer-reviewed journals.

PV GENERATION FORECASTING UTILIZING A CLASSIFICATION-ONLY APPROACH

Spyros Theocharides[1, *], George Makrides[1] and George E. Georghiou[1]

[1]PV Technology Laboratory, FOSS Research Centre for Sustainable Energy, Department of Electrical and Computer Engineering, University of Cyprus, Nicosia, 1678, Cyprus
*corresponding author email: theocharidis.spyros@ucy.ac.cy

ABSTRACT: The growing adoption of photovoltaic (PV) systems in electrical infrastructure introduces new challenges related to reliability, as solar energy supply heavily relies on weather conditions. To address these challenges, improved day-ahead PV production forecasts can be achieved through the application of advanced machine learning techniques and by reducing the uncertainty in solar irradiance predictions through statistical analysis. This study aims to present a methodology for accurately forecasting day-ahead PV production utilizing innovative machine learning techniques and a classification-only forecasting approach. In this classification-based approach, PV generation forecasting is carried out solely through classification. The central element of the proposed method is a classifier model based on the Extreme Gradient Boosting (XGBoost) ensemble algorithm. This model categorizes daily 30-minute profiles of forecasted global horizontal irradiance (GHI), measured incident irradiance (G_i), and AC power (P_{AC}) into a predefined number of classes. The classifier model serves as a reference to assign newly forecasted GHI values to specific classes and subsequently identify the corresponding forecasted P_{AC}. The results of this study indicate that the proposed forecasting solution provides forecasts with a daily normalized root mean square error (nRMSE) of 8.20% and a mean absolute percentage error (MAPE) of 6.91% over a one-year test period.
Keywords: forecasting, machine learning, photovoltaics, power

1 INTRODUCTION

With the decline of traditional energy sources worldwide, photovoltaic (PV) technologies are increasingly emerging as the primary solution to address the growing demand for electricity [1]. The projected increase in renewable energy production capacity by 2024 is expected to reach 50%, with photovoltaic systems contributing to 60% of this expansion [2]. Nonetheless, the incorporation of PV systems into the distribution grid poses distinctive challenges that require adaptable power system solutions to guarantee consistent service amidst abrupt supply and demand fluctuations [3]. Concerns for power system stability are raised by the variability and unpredictability of PV generation, as grid operators must account for the intermittent character of solar-generated electricity in generation planning and dispatch operations [3]-[5]. As the adoption of distributed PV systems grows, utility grids are undergoing a transformation towards modern, digitally-enhanced technologies that facilitate the monitoring and control of distributed energy resources (DERs). This involves actively incorporating intermittent renewable generation into network planning models and optimization procedures. The continuous electrification and decentralization trends within the power sector are driving the adoption of innovative digital tools, aiming to enhance system flexibility and accommodate the increasing penetration of renewable energy sources. [6].

Forecasting PV power generation encompasses a range of methodologies chosen according to data availability, specific application needs, and the desired prediction horizon. Short-term forecasting is essential for efficiently handling power fluctuations, anticipating voltage fluctuations, coordinating control tasks, and optimizing energy dispatch. Furthermore, mid-term forecasting plays a pivotal role in monitoring load consumption, production, and voltage/frequency regulation. This mid-term forecasting encompasses predictions for the current day and the day ahead, enhancing overall operational efficiency.

Initially, the process of PV production forecasting revolved around developing empirically driven models that aimed to achieve optimal performance. This approach demanded a comprehensive understanding of the system's characteristics and behavior. [7]-[9]. Nonetheless, owing to the intricate nature of these systems, accurately forecasting the electricity output from PV systems has posed significant challenges. Current research efforts have shifted towards the development of advanced and adaptable prediction methodologies, harnessing data-driven approaches rooted in machine learning algorithms [10]-[14].

Recent research has shown that PV production forecasts provided by third-party organizations typically depend on a combination of measured resources, including weather data, PV system data, satellite imagery, and sky imagery. These forecasts also make use of numerical weather prediction (NWP) models, which are primarily designed for weather forecasting purposes [15]. While measured data holds greater significance for short-term forecasts, typically covering periods of up to 6 hours, Numerical Weather Prediction (NWP) models come into play for projections extending beyond the 6-hour mark. Moreover, the forecast accuracy varies depending on the geographical area, creating opportunities for both emerging and established providers. Research studies have demonstrated that accurate forecasts can achieve relatively low Root Mean Square Error (RMSE) percentages, as low as 6%, in clear sky conditions. However, for forecasts in different weather conditions, RMSE values tend to be larger, ranging from 20% to as high as 40-80% [16], [17].

In the realm of PV power forecasting, numerous machine learning models have been developed employing a diverse range of methodologies. These models include Multilayer Perceptron Neural Networks (MLPNN), Radial Basis Function Neural Networks (RBFNN), and Recurrent Neural Networks (RNN), all designed to offer power forecasts for the day ahead [18]. Short-term forecasting, on the other hand, employs various models, such as nonlinear autoregressive exogenous neural networks (NARX), a combination of the grey model (GM) with multilayer perceptron neural networks (MLPNN), and adaptive feed-forward back-propagation neural networks (AFFNN). Research has been conducted to test the accuracy of several strategies for regional forecasting. These techniques include Bagging, Random Forest, Boosting, Support Vector Machines (SVM), and Generalised Additive Models (GAM). For the purpose of regional forecasting, Support Vector Regression (SVR) pre-processing, in conjunction with Principal Component Analysis (PCA), has been investigated [19].

While the findings of the studies mentioned above are promising, they have yet to undergo extensive evaluation against large-scale field data, ensuring dependable weather forecasts that remain unaffected by technological advancements or the specific climatic variables of a particular region. To enhance the accuracy of PV power forecasting, it is crucial to employ entirely data-driven approaches capable of capturing system behavior even in the absence of comprehensive system attributes. This becomes particularly vital for decentralized rooftop installations, such as behind-the-meter photovoltaic (PV) systems, where complete system metadata may not always be readily available. In such scenarios, leveraging recent operational datasets can contribute significantly to understanding system behavior and enabling precise PV production estimates. Moreover, implementing strategies to enhance the accuracy of existing photovoltaic production forecasting models is essential for the development of robust and location-independent forecasting models.

This study aimed to introduce a methodology for precise day-ahead PV production forecasting utilizing innovative machine learning techniques centered on a classification-only forecasting approach. In a classification-only approach to PV generation forecasting, machine learning algorithms are employed to categorize a PV system's future power output into predefined categories, rather than predicting the exact power output. This approach is typically chosen when the target variable has a limited range of possible values and is particularly suitable for short-term forecasting, where prediction uncertainty is high. However, the utilization of advanced machine learning methods empowers this specific methodology to provide accurate forecasts up to 24 hours in advance. Notably, the cornerstone of the proposed approach is a classifier model based on an ensemble algorithm known as Extreme Gradient Boosting (XGBoost). This classifier model categorizes the daily 30-minute profiles of forecasted global horizontal irradiance (GHI), measured incident irradiance (G_i), and AC power (P_{AC}) into specific classes. This classifier model serves as a reference to assign newly generated forecasted GHI values to their respective classes and subsequently determine the corresponding forecasted P_{AC}. The performance of this model was assessed using datasets obtained from a test site situated at the University of Cyprus (UCY).

2 OUTDOOR TESTING FACILITY

This study made use of the Outdoor Testing Facility (OTF) at UCY's PV Technology Laboratory. The OTF is equipped with a fixed-plane structure designed for outdoor module and system-level performance assessments. In this investigation, a polycrystalline silicon (poly-c-Si) system was installed in portrait orientation on aluminum mounts, aligned to the optimal annual energy yield plane-of-array (POA) angle of 27.5 degrees, customized for the climate conditions in Cyprus. Furthermore, the PV system was integrated with a data acquisition platform responsible for monitoring and storing meteorological and operational data related to the PV system.

The data acquisition platform includes meteorological and PV operational measurement sensors connected to a central data acquisition system. This system was developed in accordance with the International Electrotechnical Commission (IEC) 61724 standard. [30]. The platform captures various meteorological measurements, including G_i, RH, wind direction (W_a), wind speed (Ws), and ambient temperature (T_{amb}). Additionally, it records operational measurements of the PV system, encompassing current at the maximum power point (I_{mp}), voltage at the maximum power point (V_{mp}), and power at the maximum power point (Pmp), all of which are measured at the DC output of the PV system [31].

The test-bench PV system is composed of five polycrystalline silicon (poly-c-Si) PV modules, each rated at 235 Wp according to the manufacturer's datasheet. These modules are connected in series to create a PV string with a nominal power capacity of 1.175 kWp at the input of a string inverter. The system is installed in an open-field mounting configuration, facing due South, and positioned at the optimal annual energy yield angle for Cyprus, which is 27.5°.

3 METHODOLOGY

The proposed methodology entails creating a data-driven power output classifier model using the XGBoost ensemble algorithm [25]. To build this model, we applied the XGBoost classifier to historical datasets collected from a test-bench PV system and another system situated in Nicosia, Cyprus. The training dataset consisted of two years of data, with one year used for training and the other for testing.

At the heart of our proposed method lies a classifier model that leverages the XGBoost ensemble algorithm to categorize the daily 30-minute profiles of forecasted Global Horizontal Irradiance (GHI), measured G_i, and P_{AC} into distinct classes. This classifier model serves as a reference, assigning the newly generated day-ahead forecasted GHI to a specific class and subsequently identifying the corresponding forecasted PAC. This classification-based approach empowers the model to make accurate PV power generation predictions based on the classification of weather and irradiance conditions.

To evaluate the performance of our day-ahead PV production forecasting model, we utilized a test set consisting of historical measurements collected during the evaluation period. The evaluation process employed predetermined metrics to gauge the accuracy and reliability of the model's forecasts. By comparing the forecasted PV power output to the actual measurements, we could ascertain the forecasting model's accuracy.

In summary, the objective of this research is to improve PV production forecasting by employing a classification-only approach. The methodology entails the creation of a data-driven classifier model for power output based on the XGBoost ensemble algorithm. The model classifies the forecasted GHI and forecasts the corresponding PV power output by utilizing historical data and predefined classifications. Utilizing a test set of acquired measurements and predefined metrics, the forecasting model's performance is evaluated. In addition, the optimal goal of this study is to develop a technology agnostic algorithm that can be replicable to any technology by following the aforementioned steps. Figure 1 demonstrates a visual representation of the proposed methodology.

10.4229/EUPVSEC2023/4BO.7.6
020309-002

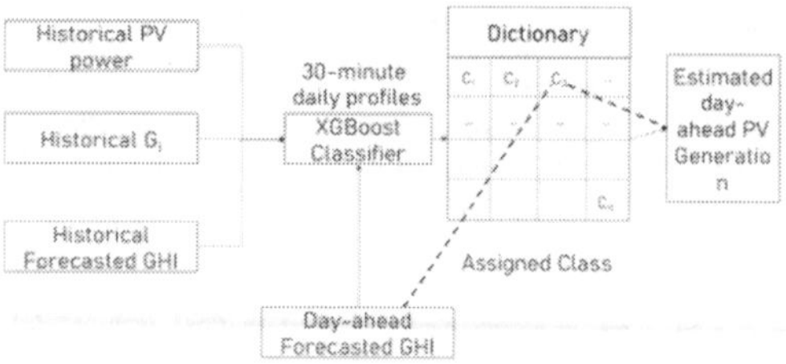

Figure 1: Graphical representation of the proposed methodology.

3.1 Classification with XGBoost

XGBoost is a highly robust and efficient machine learning method, belonging to the gradient boosting family of algorithms. Its versatile architecture allows it to excel in various tasks, including classification, regression, and ranking. At its core, XGBoost relies on the principle of gradient boosting, a process where an ensemble of weak prediction models is iteratively constructed and combined to form a powerful predictive model [25].

The process begins with the initialization of a base learner, often a decision tree, which is then incrementally augmented with more trees during each iteration. XGBoost's objective is to minimize a specific loss function by adding decision trees that rectify the errors of their predecessors. This iterative process is guided by gradient descent, which determines the direction and magnitude of adjustments required at each step.

Mathematically, XGBoost optimizes the objective function by minimizing the sum of the loss function and a regularization term, which helps control the model's complexity. The loss function quantifies the disparity between predicted values and actual labels, while regularization prevents the model from becoming overly specific, enhancing its generalization ability. Depending on the desired level of regularization, regularization terms are derived from L2 norms. XGBoost offers a feature importance measure derived from an analysis of the decision trees within the ensemble. This measure indicates the percentage reduction in the loss function attributed to splits on each feature across all trees. It helps identify the most impactful features and provides deeper insights into underlying data patterns. Additionally, it leverages parallelization techniques to enhance computing performance, enabling parallel processing on multi-core CPUs and distributed computing frameworks. This significantly reduces computation time, making it especially valuable for large datasets and complex models.

Furthermore, it handles missing data seamlessly. During training, it learns to handle missing values by assigning appropriate weights to decision tree branches based on available information. This eliminates the need for imputation techniques and simplifies pre-processing.

XGBoost incorporates early stopping, which monitors model performance on a validation set during training. If no improvement is observed over a set number of iterations, training is halted. This prevents overfitting and conserves computational resources.

3.2 Clearness Index

The clearness index, denoted as k_t, serves as a measure to assess the impact of current weather conditions on forecast reliability. It quantifies the transparency of the atmosphere and is expressed in degrees. Importantly, k_t is a dimensionless value, ranging from 0 to 1, with higher values indicating clear and sunny conditions, and lower values indicating cloudy and overcast conditions. The clear sky index is employed as an indicator, representing a ratio of sky conditions (where $k_t = 0$ signifies an overcast sky, and $k_t = 1$ signifies a clear sky).

$$k_t = \frac{GHI}{GHI_{CS}} \tag{1}$$

where GHI and GHI_{CS} is the observed and clear-sky GHI, respectively.

Specifically, the k_t was separated into three classes. In this study, three classes of the k_t were used as follows:

- Clear Sky: $k_t >= 0.75$;
- Moderate: $k_t < 0.75$ and $k_t > 0.25$;
- Overcast: $k_t <= 0.25$

3.3 Performance Assessment

The prediction performance accuracy was assessed based on several predefined metrics. The mean absolute percentage error ($MAPE$) which is given by [32]:

$$MAPE = \frac{100}{n} \cdot \sum_{i=1}^{n} \left| \frac{y_{\text{observed},i} - y_{\text{forecasted},i}}{y_{\text{actual},i}} \right| \tag{2}$$

The $nRMSE$ which is the relative $RMSE$ normalized to the nominal capacity of the PV system and defined as [32], [33]:

$$nRMSE = \frac{100}{P_{\text{nominal}}} \sqrt{\frac{1}{n} \cdot \sum_{i=1}^{n} (y_{\text{observed},i} - y_{\text{forecasted},i})^2} \tag{3}$$

where P_{nominal} is the maximum installed capacity of the PV system.

For the above performance metrics $y_{\text{observed},i}$ and $y_{\text{forecasted},}$ is the actual and forecasted power, respectively, P_{nominal} is the nominal peak power of the PV system, $RMSE_{\text{forecasted}}$ and $RMSE_{\text{baseline}}$ is the $RMSE$ of the forecasted and baseline models (forecasts of the PM), respectively.

4 RESULTS

The initial validation of the proposed approach took place at the UCY test facility. The primary objective of these initial tests was to assess the performance and operation of the algorithms on a smaller-scale PV system. This crucial step served to confirm the reliability and effectiveness of the methodology before applying the same techniques to a larger utility-scale PV system.

4.1 Identifications of the Classes / Dictionary Implementation

As part of the PV generation forecasting process, the proposed approach was implemented in the study. It involved the creation of 25 distinct classes, which were visually represented through a heat map (see Figure 2). Each class was assigned a unique colour corresponding to the number of days falling into that class.

The heat map served multiple purposes. Firstly, it facilitated the analysis of irradiance levels and ramping rates associated with each class. Secondly, it provided a comprehensive overview of the distribution of days among the different classes. A close examination of the heat map allowed for the identification of recurring patterns and trends in the data. The color-coded representation enabled

10.4229/EUPVSEC2023/4BO.7.6
020309-003

quick identification of groups of days with similar characteristics in terms of photovoltaic energy production.

Additionally, the heat map allowed for an exploration of the relationship between class assignments and associated irradiance levels. This exploration offered insights into how varying weather conditions influenced the PV system's power production. Furthermore, the heat map provided valuable information about the ramping rates present within each class, offering insights into the speed of PV power output changes over time.

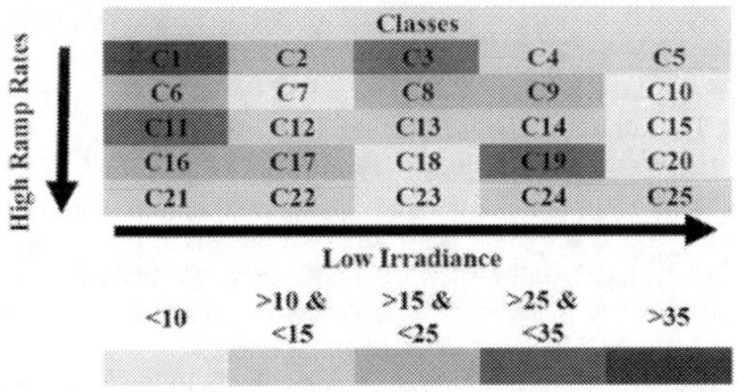

Figure 2: Heatmap representing the number of the identify classes used as a dictionary. The colour code indicates the number days per class.

4.2 Identifications of the Classes / Dictionary Implementation

Furthermore, a daily profile assessment was conducted to evaluate the accuracy of the proposed PV production forecasting model on specific days. This analysis involved comparing the actual power output of the PV system with the two-day forecasted values, providing insights into the model's real-world performance.

On the first selected testing day, which occurred during the winter season (see Figure 3a), the model precisely predicted the PV system's power output trajectory throughout the day. The predicted values closely aligned with the actual power output, signifying a high degree of accuracy, even in the presence of rapid irradiance ramping rates. Overall, the model effectively captured the dynamic variations in power generation.

On the second summer day chosen for testing (see Figure 3b), the model exhibited consistent performance, closely tracking the actual power output throughout the day. The model's ability to predict PV generation patterns was validated by a significant correlation between predicted and observed power output. It accurately accounted for solar irradiance variations, resulting in precise predictions of power output fluctuations. Notably, it correctly forecasted the rapid morning power production increase, followed by a gradual decline as the sun's intensity waned in the late afternoon.

The daily profile evaluation on these selected days demonstrated the proposed day-ahead PV production forecasting model's capacity to accurately predict the system's power output. The alignment between predicted and actual power values showcased the model's ability to capture the nuances of solar irradiance and translate them into precise forecasts. These results provide compelling evidence of the model's performance in real-world conditions, validating its potential as a valuable tool for one-day-ahead PV generation forecasting.

Furthermore, a comprehensive evaluation was conducted to assess errors based on sky conditions, utilizing the k_t index (see Table 1). This evaluation aimed to investigate how the model's performance varied with different weather conditions.

Results revealed that on days with clear skies, the model exhibited the fewest errors, with a nRMSE of 5.30% and a MAPE of 4.10%. These findings highlighted the model's reliability in accurately forecasting PV power output during favourable weather conditions when solar irradiance was high.

On moderate and cloudy days, the error rates were slightly higher. However, they remained within an acceptable range, demonstrating the model's capability to provide reasonably accurate forecasts even under less-than-ideal weather conditions. It's worth noting that the number of days classified as moderate or overcast was significantly lower than those with clear skies, contributing to the observed higher error rates on these days due to more challenging weather conditions.

This analysis of the impact of sky conditions on forecasting errors underscores that the model's performance was most accurate on clear-sky days while still acceptable on moderate or overcast days. These findings demonstrate the model's adaptability to various weather conditions and its ability to generate accurate forecasts for diverse scenarios. Additionally, Table 1 presents the representative classes assigned by the XGBoost classifier for each sky condition, as determined by the k_t index.

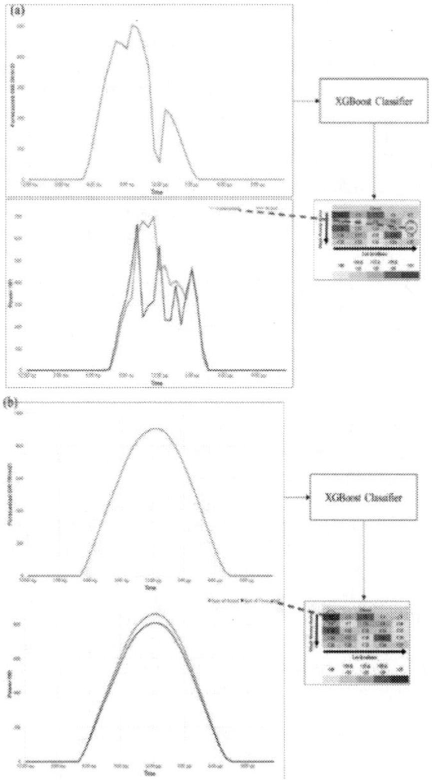

Figure 3: Daily profile assessment of: (a) an overcasted day during the winter period and (b) a clear sky day during summer period.

10.4229/EUPVSEC2023/4BO.7.6

Table I: Correlation coefficients between the input and output parameters (investigated features).

Clearness Index (k_t)	Class	nRMSE (%)	MAPE (%)
Clear Sky	C1 – C5	5.30	4.10
Moderated	C6 – C19	8.99	7.90
Overcasted	C19 – C25	10.30	8.70
Overall		8.20	6.91

5 CONCLUDING REMARKS

The growing adoption of photovoltaic (PV) systems in electricity grids introduces new challenges related to reliability, primarily influenced by weather conditions affecting solar energy supply. To address this challenge, enhanced day-ahead PV production forecasts can be achieved through advanced machine learning techniques, reducing the uncertainty of solar irradiance predictions through statistical processing. Additionally, PV generation forecasting plays a vital role in power grid planning, operation, and optimization. Accurate forecasts of expected solar power generation enable utilities and system operators to schedule dispatchable energy effectively.

This study aimed to introduce a methodology for precise day-ahead PV production forecasting models that leverage innovative machine learning techniques, focusing on a classification-only forecasting approach. In this approach, machine learning algorithms classify a PV system's future power output into predefined categories instead of predicting the exact output. This approach is particularly suitable for short-term forecasting when the target variable has a limited range of possible values. However, by employing advanced machine learning methods, this methodology can provide accurate forecasts up to 24 hours ahead.

Central to this proposed method is a classifier model based on the Extreme Gradient Boosting (XGBoost) ensemble algorithm. This model categorizes daily 30-minute profiles of forecasted global horizontal irradiance (GHI), measured incident irradiance (G_i), and AC power (P_{AC}) into specific classes. The classifier model serves as a reference to assign newly forecasted GHI values to specific classes and, consequently, identify the corresponding forecasted P_{AC}.

The results demonstrated that the proposed forecasting solution provided forecasts with a daily normalized root mean square error (nRMSE) of 8.20% and a mean absolute percentage error (MAPE) of 6.91% over the test set period. Furthermore, the methodology's performance was evaluated across different sky conditions using the clearness index (k_t). The evaluation indicated higher accuracy on clear-sky days, while moderate and overcast days remained within acceptable error bounds.

6 ACKNOWLEDGMENT

This work has received funding from the European Union's Horizon 2020 research and innovation programme under Grant Agreement no 864537, project title Flexible Energy Production, Demand and Storage-based Virtual Power Plants for Electricity Markets and Resilient DSO Operation (FEVER).

7 REFERENCES

[1] IEA, "Electricity security in tomorrow's power systems," 2020. Available: https://www.iea.org/articles/electricity-security-in-tomorrow-s-power-systems.

[2] IEA, "Renewables 2019: Market analysis and forecast from 2019 to 2024," 2020. Available: https://www.iea.org/reports/renewables-2019.

[3] IEA, "Introduction to System Integration of Renewables: Decarbonising while meeting growing demand, 2020," 2020. Available: https://www.iea.org/reports/introduction-to-system-integration-of-renewables.

[4] M. Mahoor, A. Majzoobi and A. Khodaei, "Distribution asset management through coordinated microgrid scheduling," *IET Smart Grid,* 2018. . DOI: 10.1049/iet-stg.2018.0076.

[5] H. Sadeghian and Z. Wang, "A novel impact-assessment framework for distributed PV installations in low-voltage secondary networks," *Renewable Energy,* 2020. . DOI: 10.1016/j.renene.2019.09.117.

[6] -. I. International Renewable Energy Agency, "Innovation landscape for a renewable-powered future, 2019," 2020. Available: https://www.irena.org/publications/2019/Feb/Innovation-landscape-for-a-renewable-powered-future.

[7] H. T. C. Pedro and C. F. M. Coimbra, "Assessment of forecasting techniques for solar power production with no exogenous inputs," *Solar Energy,* 2012. . DOI: 10.1016/j.solener.2012.04.004.

[8] B. Wolff, O. Kramer and D. Heinemann, "Selection of numerical weather forecast features for PV power predictions with random forests," in 2017, . DOI: 10.1007/978-3-319-50947-1_8.

[9] T. Schmidt *et al*, "Short-term solar forecasting based on sky images to enable higher PV generation in remote electricity networks," *Renewable Energy and Environmental Sustainability,* vol. 2, pp. 23, 2017. Available: http://www.rees-journal.org/10.1051/rees/2017028. DOI: 10.1051/rees/2017028.

[10] L. Visser, T. AlSkaif and W. van Sark, "Operational day-ahead solar power forecasting for aggregated PV systems with a varying spatial distribution," *Renewable Energy,* vol. 183, pp. 267-282, 2022. Available: https://linkinghub.elsevier.com/retrieve/pii/S0960148121015688. DOI: 10.1016/j.renene.2021.10.102.

[11] S. Theocharides and G. Makrides, "Day-ahead solar photovoltaic forecasting for EAC supply portfolios," Nicosia, Cyprus, 2022.

[12] S. Pretto *et al*, "A New Probabilistic Ensemble Method for an Enhanced Day-Ahead PV Power Forecast," *IEEE Journal of Photovoltaics,* vol. 12, *(2),* pp. 581-588, 2022. Available: https://ieeexplore.ieee.org/document/9684273/. DOI: 10.1109/JPHOTOV.2021.3138223.

[13] D. Caputo *et al*, "Photovoltaic plants predictive model by means of ANN trained by a hybrid evolutionary algorithm," in 2010, . DOI: 10.1109/IJCNN.2010.5596782.

[14] A. Mellit and A. M. Pavan, "A 24-h forecast of solar irradiance using artificial neural network: Application for performance prediction of a grid-connected PV plant at Trieste, Italy," *Solar Energy,* vol. 84, *(5),* pp. 807-821, 2010. . DOI: 10.1016/j.solener.2010.02.006.

[15] Y. Xue *et al*, "Voltage stability and sensitivity analysis of grid-connected photovoltaic systems," in 2011, . DOI: 10.1109/PES.2011.6039649.
[16] S. Pelland *et al*, "Photovoltaic and Solar Forecasting: State of the Art," *International Energy Agency: Photovoltaic Power Systems Programme, Report IEA PVPS T14,* pp. 1-40, 2013. Available: https://www.researchgate.net/publication/259717376_Photovoltaic_and_Solar_Forecasting_State_of_the_Art. DOI: 978-3-906042-13-8.
[17] W. Glassley *et al*, "California Renewable Energy Forecasting , Resource Data and Mapping Executive Summary," 2010.
[18] A. Yona *et al*, "Application of neural network to 24-hour-ahead generating power forecasting for PV system," in 2008, . DOI: 10.1109/PES.2008.4596295.
[19] J. G. Da Silva Fonseca *et al*, "Forecasting regional photovoltaic power generation - A comparison of strategies to obtain one-day-ahead data," in 2014, . DOI: 10.1016/j.egypro.2014.10.124.
[20] T. Southern *et al*, "Machine Learning Tec Hniques for Short Te Rm Solar Forecasting," pp. 39-43, 2015.

020287-001

40th European Photovoltaic Solar Energy Conference and Exhibition

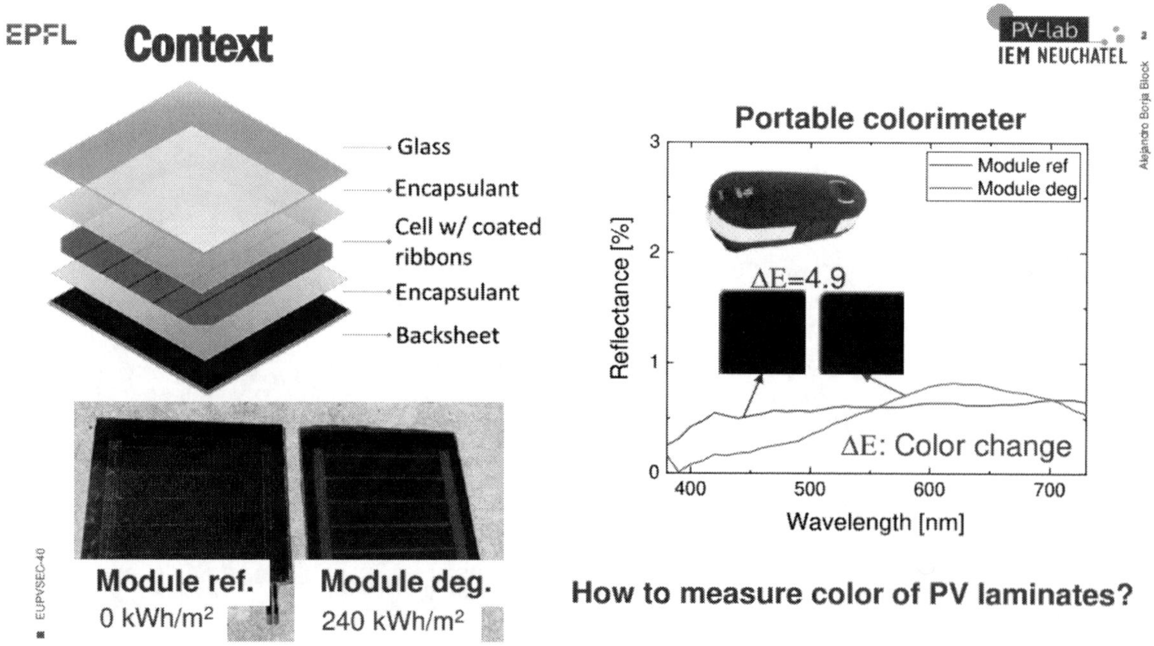

020287-002

EPFL

Motivation

PV-lab IEM NEUCHATEL 3

Alejandro Borja Block

- Assess degradation
- Quality control
- Color design

EUPVSEC-40

020287-003

EPFL

Characterization equipment comparison

PV-lab IEM NEUCHATEL 4

Alejandro Borja Block

- Portable colorimeter
- Integrated sphere spectrometer
- Novel colorimeter

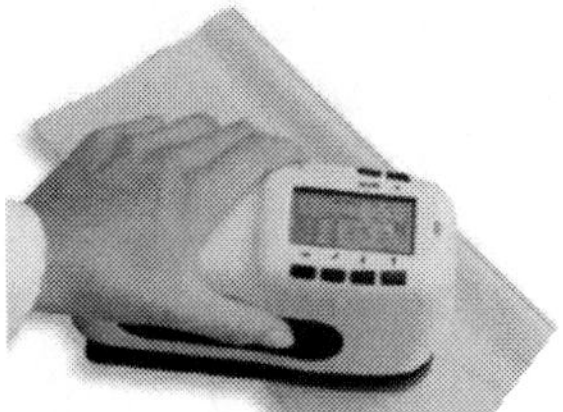

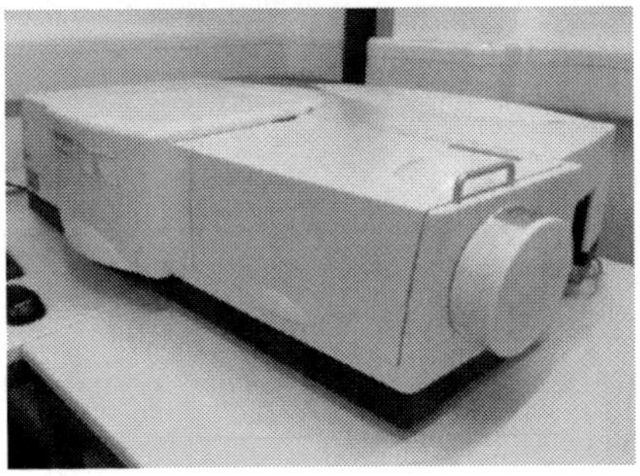

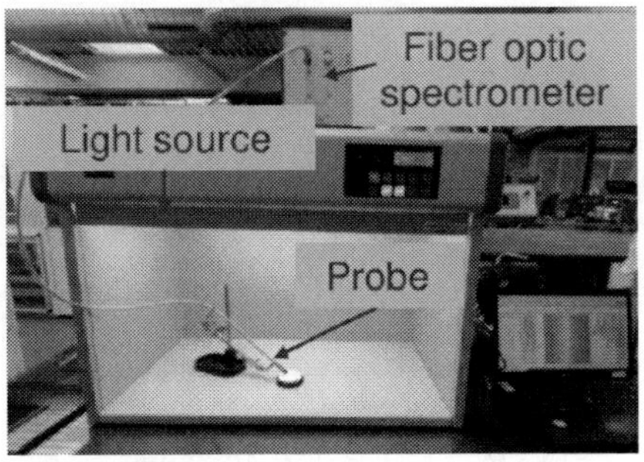

Borja Block et al. "*Accurate color characterization of solar photovoltaic modules for building integration*", submitted for publication

EUPVSEC-40

020287-004

EPFL

Samples

PV-lab IEM NEUCHATEL 5

Alejandro Borja Block

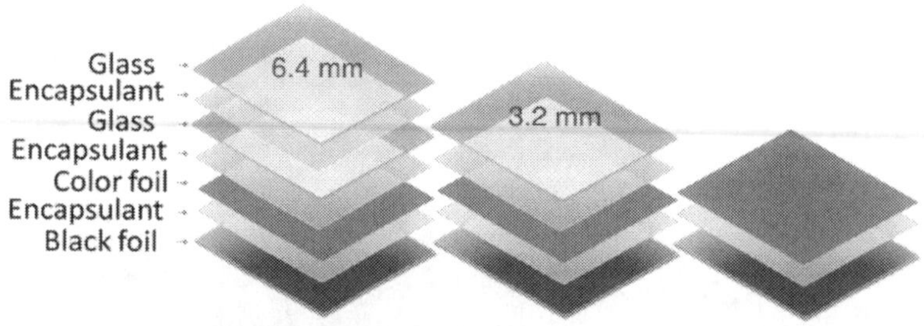

EUPVSEC-40

020287-005

40th European Photovoltaic Solar Energy Conference and Exhibition

EPFL

Reflectance measurements

PV-lab IEM NEUCHATEL 6

Alejandro Borja Block

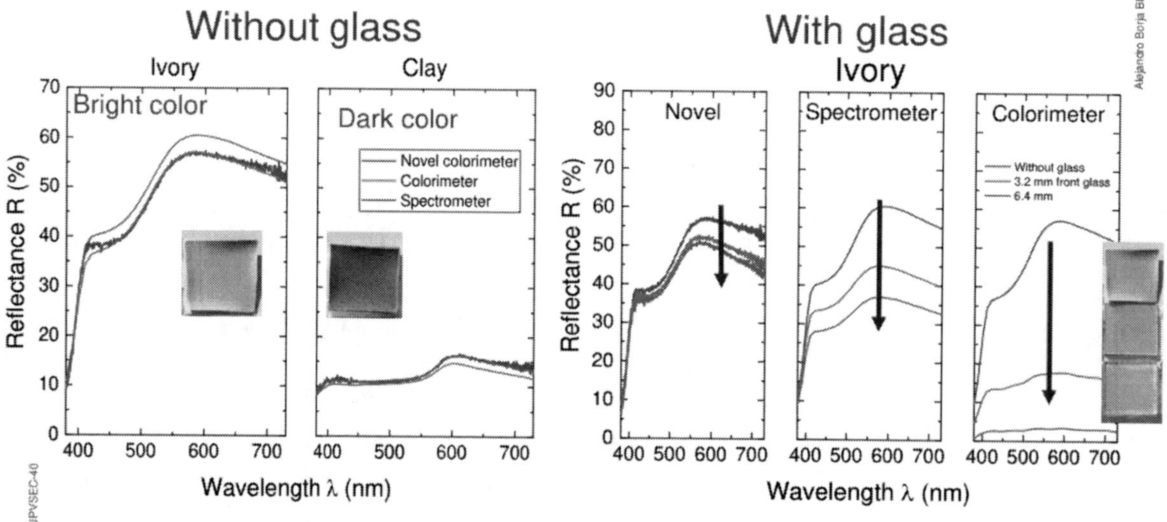

EUPVSEC-40

- Color change: ΔE<2 comparable

020287-006

EPFL

Data processing

PV-lab IEM NEUCHATEL 7

Alejandro Borja Block

Reflectance

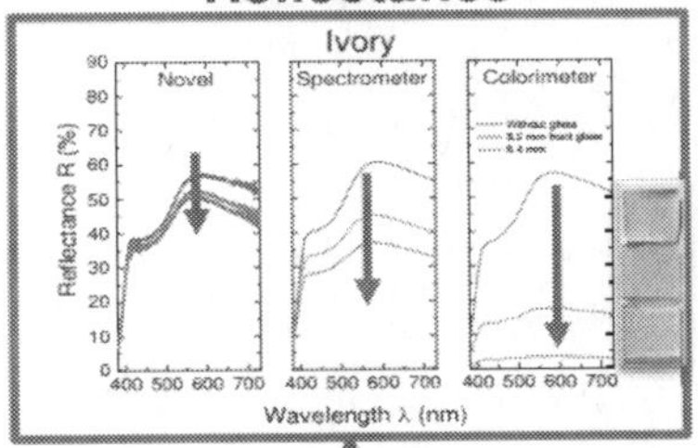

Observer (10°)

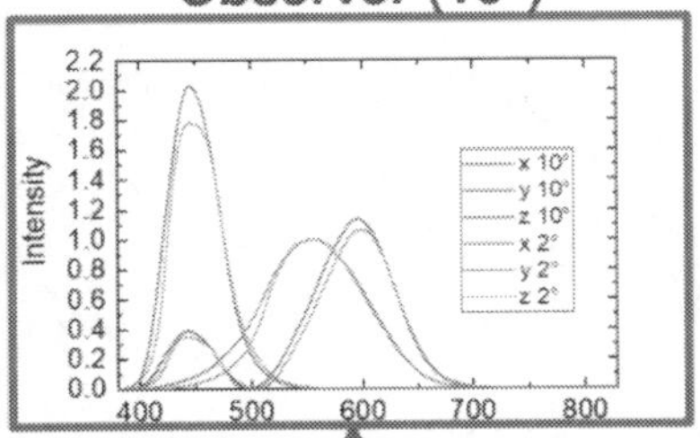

Illuminant (D65)

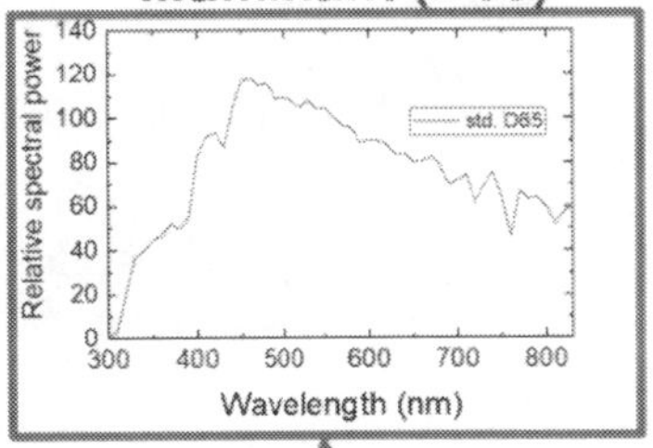

$$X = \frac{1}{N}\int_{\lambda_0}^{\lambda_1} \bar{x}(\lambda)I(\lambda)S(\lambda)d\lambda \qquad Y = \frac{1}{N}\int_{\lambda_0}^{\lambda_1} \bar{y}(\lambda)I(\lambda)S(\lambda)d\lambda$$

$$Z = \frac{1}{N}\int_{\lambda_0}^{\lambda_1} \bar{z}(\lambda)I(\lambda)S(\lambda)d\lambda \qquad N = \int_{\lambda_0}^{\lambda_1} \bar{y}(\lambda)I(\lambda)d\lambda$$

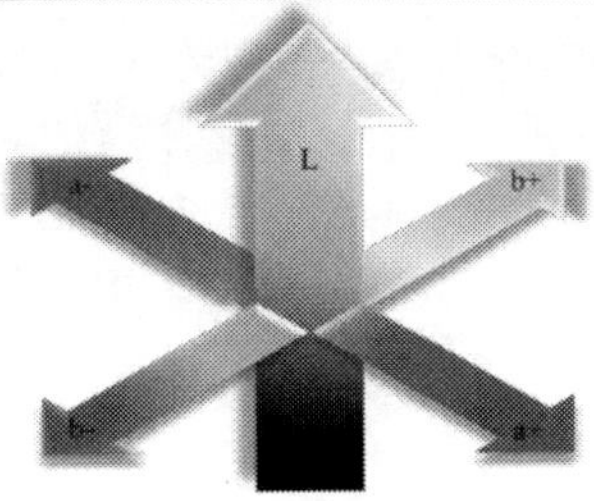

ΔE: Color change

EUPVSEC-40

EPFL

Equipment comparison

PV-lab IEM NEUCHATEL 8

Alejandro Borja Block

Sample ID	Novel colorimeter	Spectrometer	Colorimeter
Ivory 0 mm			
Ivory 3.2 mm			
Ivory 6.4 mm			
ΔE 0 mm vs 3.2 mm	2.5	6.9	25.7
ΔE 0 mm vs 6.4 mm	3.4	11.1	57.4

Why does the signal decrease?

No glass vs 6.4 mm

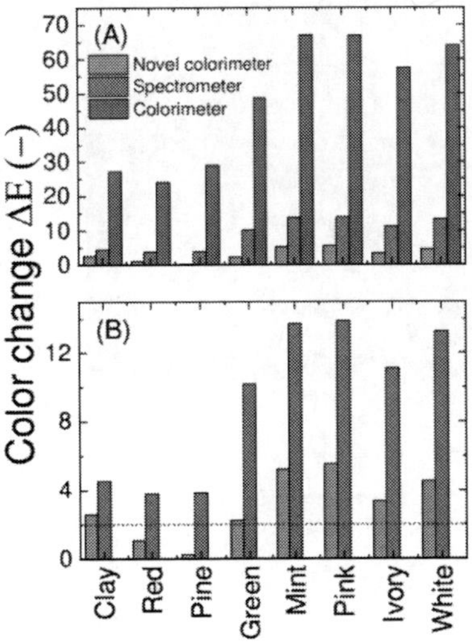

EUPVSEC-40

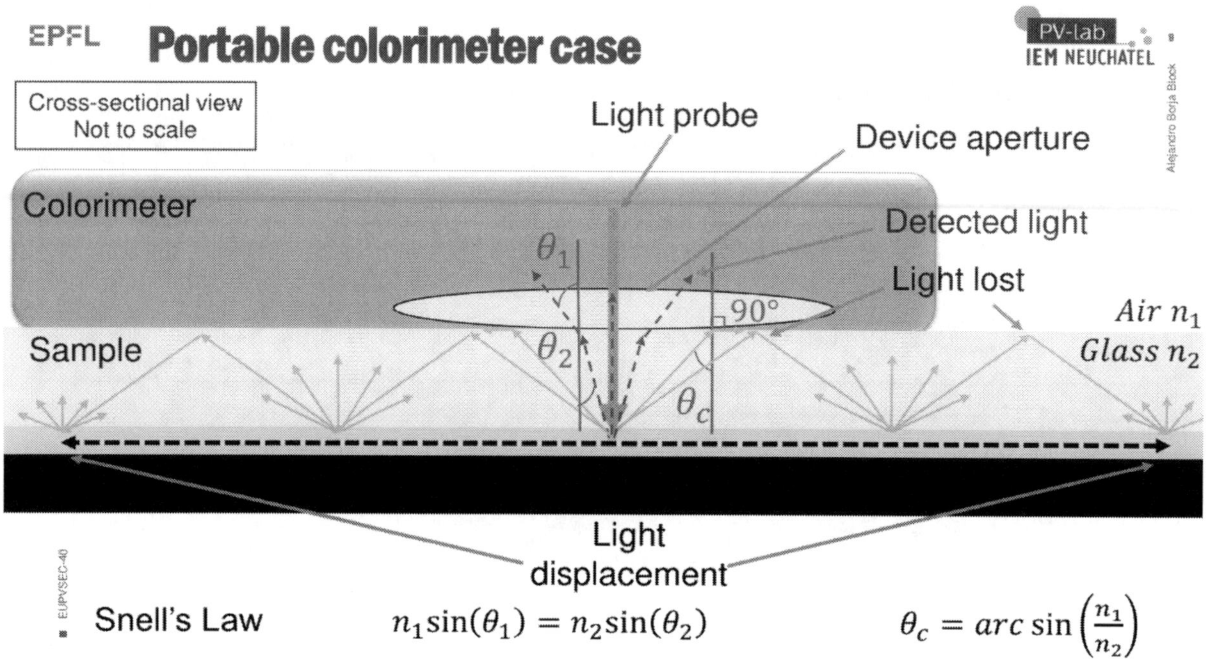

020287-009

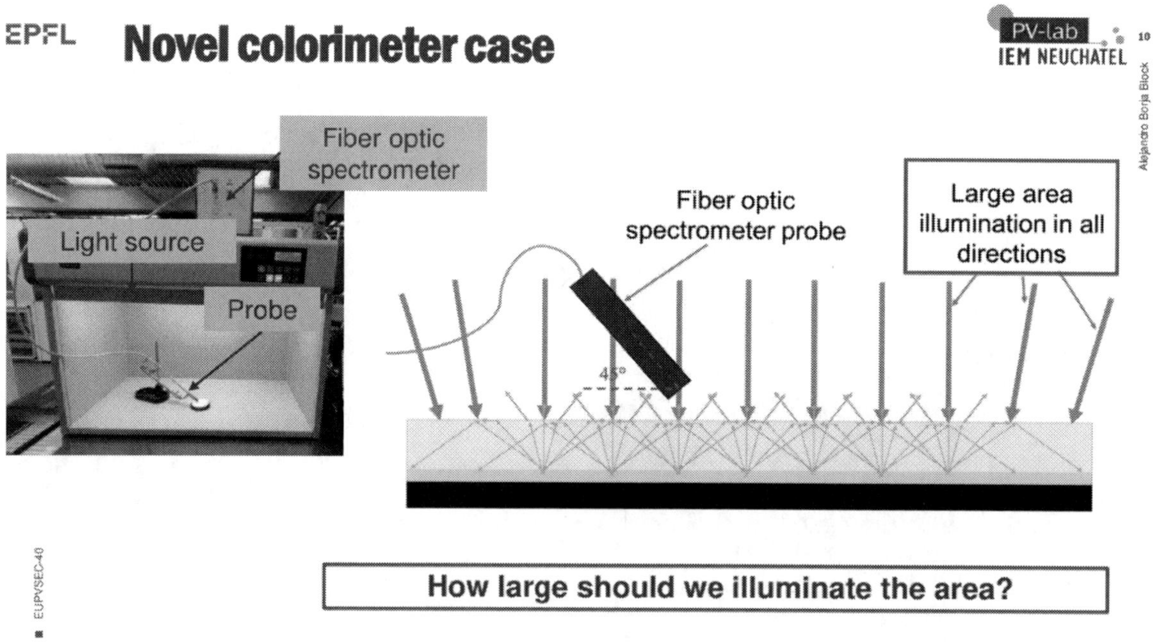

020287-010

EPFL **Large area illumination**

PV-lab IEM NEUCHATEL 11

Alejandro Borja Block

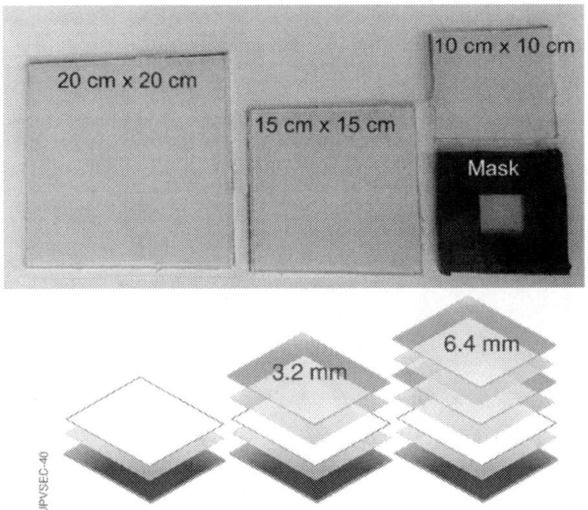

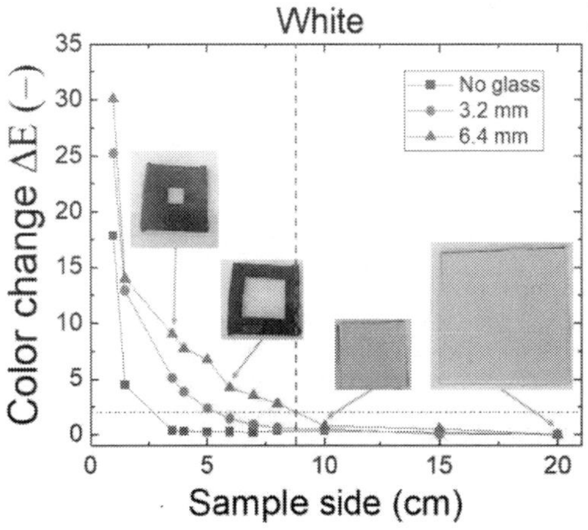

8 cm by 8 cm needed for the 6.4 mm thick white sample

EUPVSEC-40

EPFL **Alternative solutions**

PV-lab IEM NEUCHATEL 12

Alejandro Borja Block

Larger integrated sphere

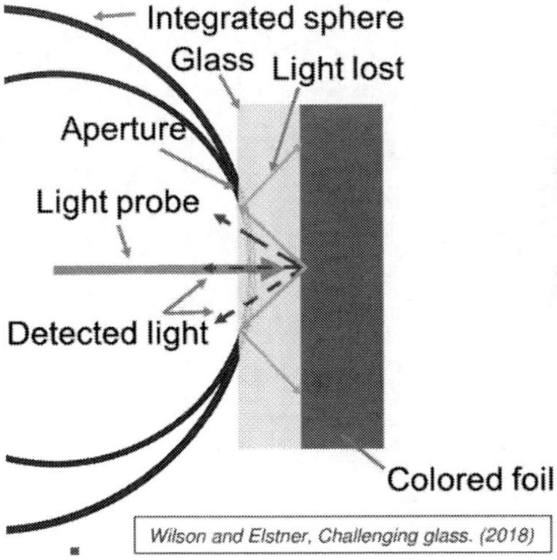

Wilson and Elstner, Challenging glass. (2018)

Reference behind the glass layer

Cross-sectional view
Not to scale

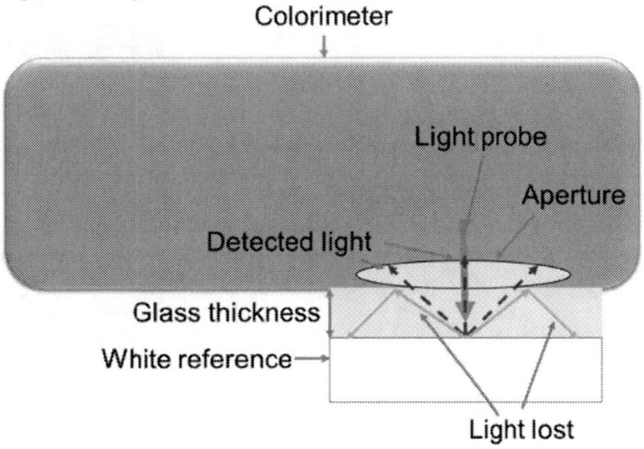

Hergert et al., Johanna Solar Technology. (2010)

EPFL

Conclusions

PV-lab IEM NEUCHATEL 13

Alejandro Borja Block

- **Common color characterization techniques**:
  - Are not suitable for measuring typical PV laminates.
  - High reflectance reduction due to light trapping in the glass and shift of the measuring plane.
- **Novel colorimeter:**
  - The reflectance is only slightly decreased.
  - Large area illumination compensates the light losses.
- **Open activities:**
  - Light sources (stable, spectra)
  - Angle dependent colors
  - Surface effects
  - Calibration
- **Challenge:**
  - Portable novel colorimeter for integrated photovoltaic applications

EUPVSEC-40

020287-013

EPFL

PV-lab IEM NEUCHATEL

14

EUPVSEC-40

020287-014

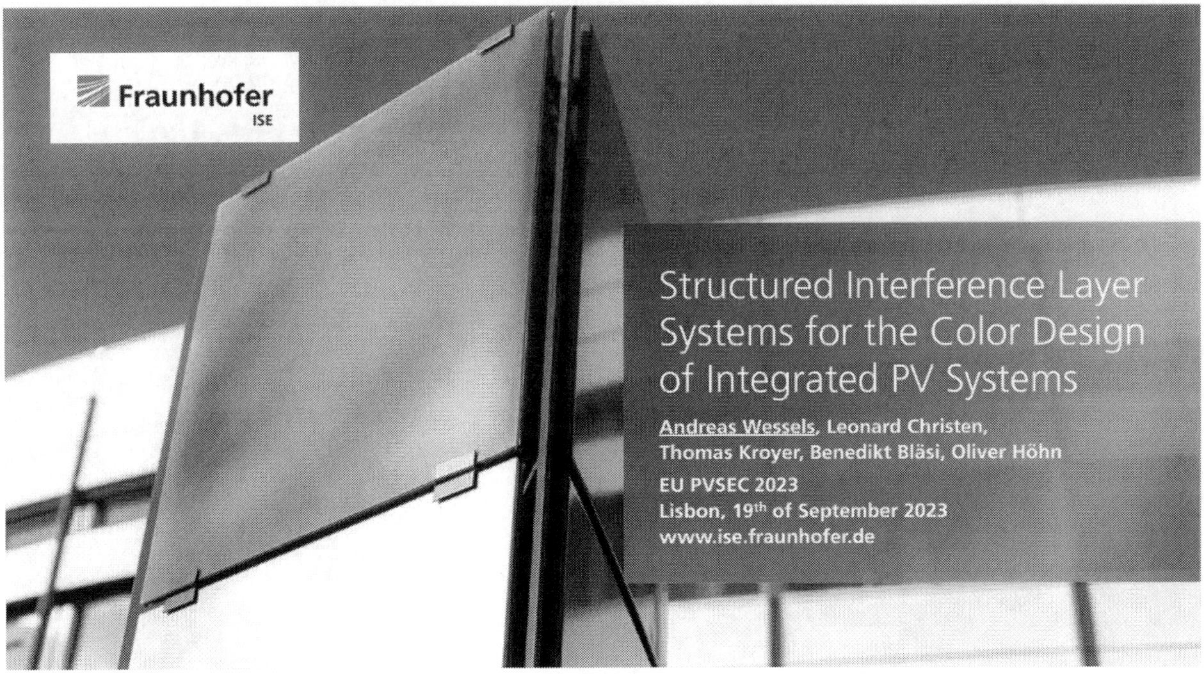

Fraunhofer ISE

Structured Interference Layer Systems for the Color Design of Integrated PV Systems

Andreas Wessels, Leonard Christen, Thomas Kroyer, Benedikt Bläsi, Oliver Höhn

EU PVSEC 2023
Lisbon, 19th of September 2023
www.ise.fraunhofer.de

020288-001

Agenda
Content of This Presentation

1. The MorphoColor® Concept
2. Modeling by Means of Ray Tracing
3. Demo Modules and First Facade Installation

2

Fraunhofer ISE

020288-002

The MorphoColor Concept
Requirements for Colored PV

1. Beautiful saturated colors
2. Low power losses for the photovoltaic
3. Angular stable color appearance

40th European Photovoltaic Solar Energy Conference and Exhibition

The MorphoColor Concept
Basic Principle of Operation

- Standard black PV module
- Coloring by modification of inner glass surface
  - Glass structuring
  - Application of an interference layer system

➢ Narrow reflection in the visible range
➢ High light transmission to the solar cells

➢ Patented technology

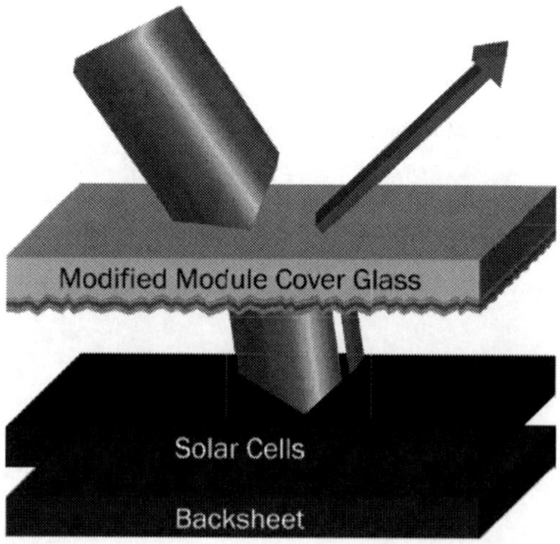

 Granted patent: Höhn et al., Glazing unit, process for their preparation and their use, DE102017203105B4, CN110382974B, EP3757480 B1

The MorphoColor Concept

Structured Interference Layer Systems

- Materials used:
  - Si_3N_4 as low and TiO_2 as high index material
  - SiO_2 spacer layer
- Design optimized for:
  - High color saturation and angular stability
- Using:
  - Higher harmonic reflection order
  - High average refractive index
  - Suitable structured glass

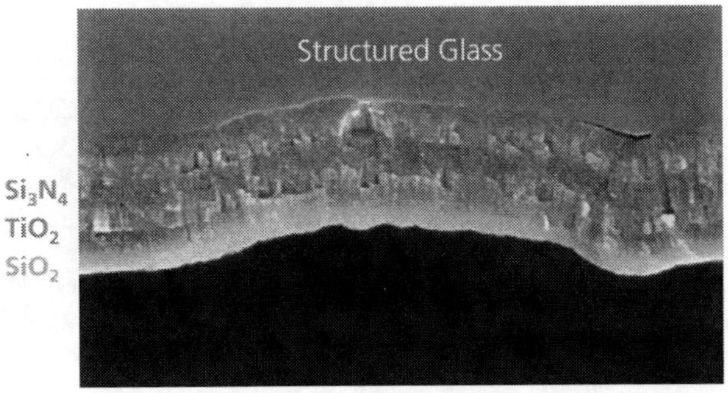

5

Bläsi et al., The MorphoColor Concept for Colored Photovoltaic Modules, 10.1109/JPHOTOV.2021.3090158

Fraunhofer ISE

020288-005

40th European Photovoltaic Solar Energy Conference and Exhibition

Modeling by Means of Ray Tracing

Topographic Characterization and Data Preparation

- 3D laser measuring microscope image
- Representation via triangulation
- Input in ray tracing scenery
  - Implementation of layer stack via TMM
  - Process-related layer thickness variation as a function of the structure angle

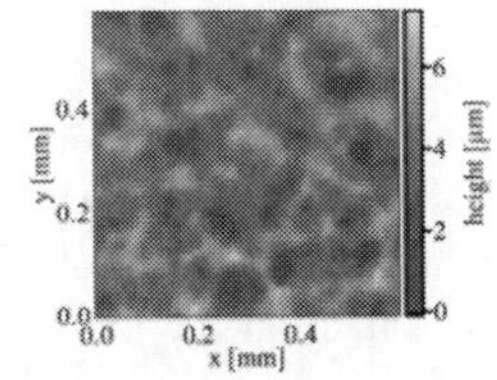

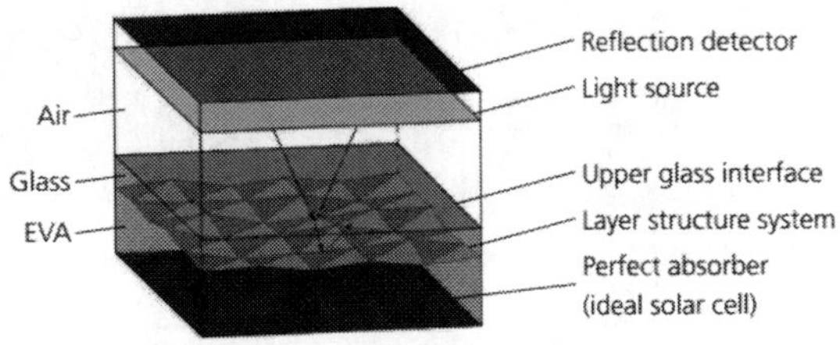

6

Schöttl et al., Raytrace3d by Fraunhofer ISE - Accurate and Efficient Ray Tracing for Concentrator Optics
Wessels et al., Modeling the optical properties of Morpho-inspired thin-film interference filters on structured surfaces, 10.1364/OE.455776
Wessels et al., Modeling of thin-film interference filters on structured substrates: microfacet-based BSDF versus ray tracing, 10.1364/OE.489221

Fraunhofer ISE

020288-006

Modeling by Means of Ray Tracing
Comparison With Measured Data

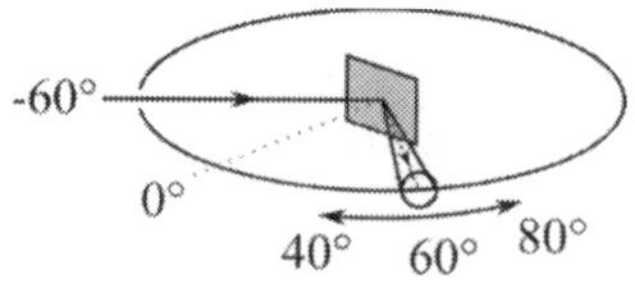

- Incidence and reflection angle dependent spectrometry
- ➢ Good agreement

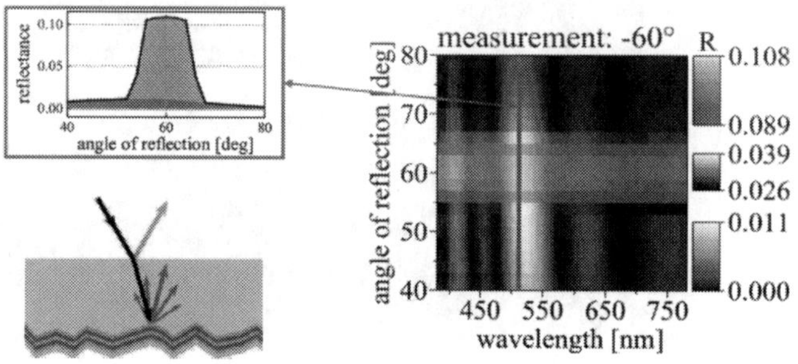

7 Wessels et al., Modeling of thin-film interference filters on structured substrates: microfacet-based BSDF versus ray tracing, 10.1364/OE.489221

Fraunhofer ISE

020288-007

40th European Photovoltaic Solar Energy Conference and Exhibition

Modeling by Means of Ray Tracing
Comparison With Measured Data

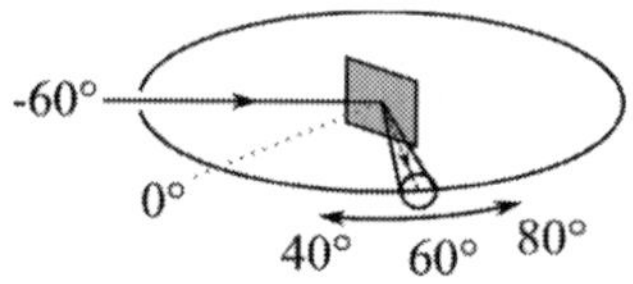

- Incidence and reflection angle dependent spectrometry
- ➢ Good agreement

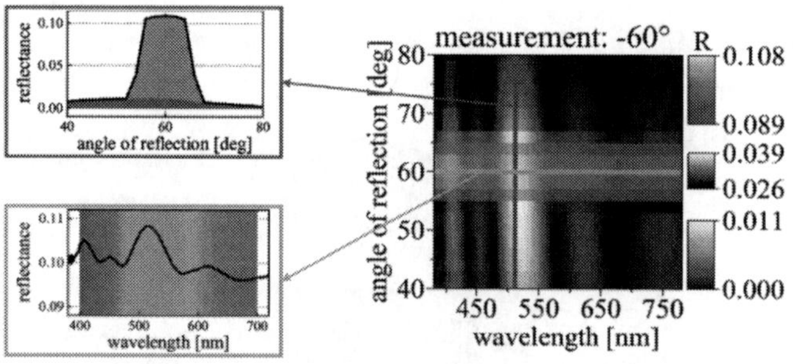

8 Wessels et al., Modeling of thin-film interference filters on structured substrates: microfacet-based BSDF versus ray tracing, 10.1364/OE.489221

Fraunhofer ISE

020288-008

Modeling by Means of Ray Tracing

Comparison With Measured Data

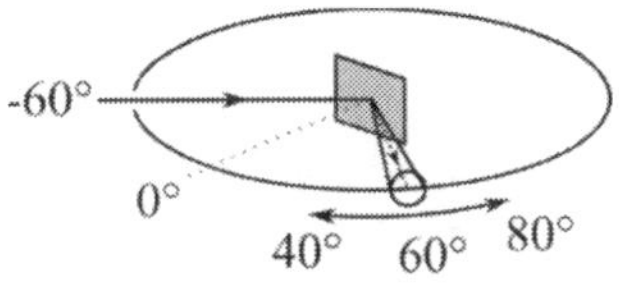

- Incidence and reflection angle dependent spectrometry

➢ Good agreement

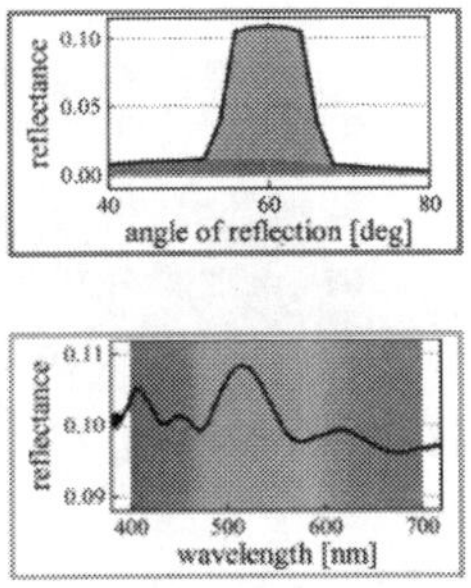

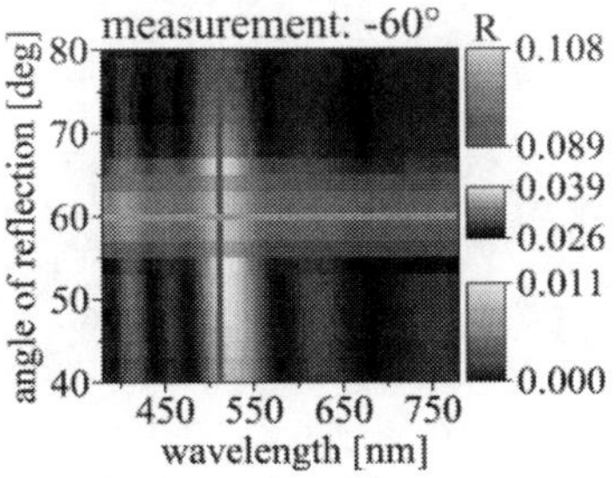

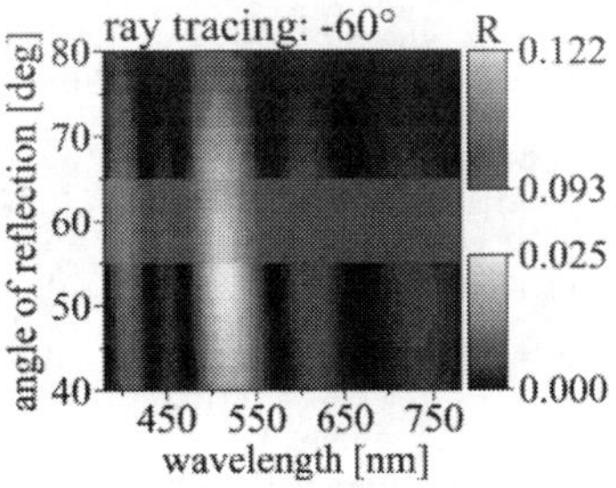

Wessels et al., Modeling of thin-film interference filters on structured substrates: microfacet-based BSDF versus ray tracing, 10.1364/OE.489221

40th European Photovoltaic Solar Energy Conference and Exhibition

Modeling by Means of Ray Tracing

Comparison With Measured Data

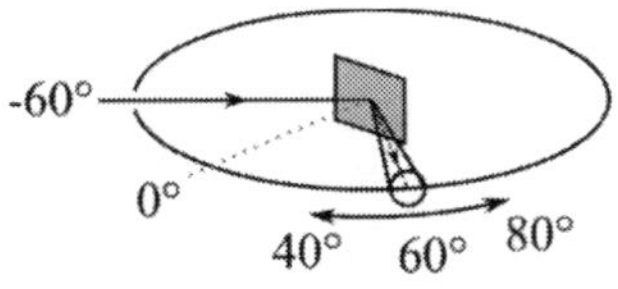

- Incidence and reflection angle dependent spectrometry

➢ Good agreement

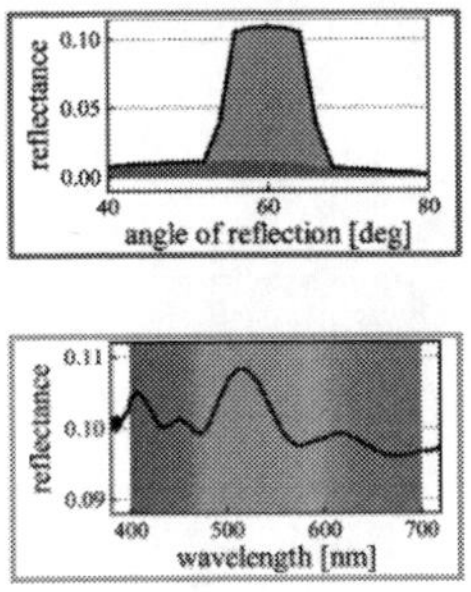

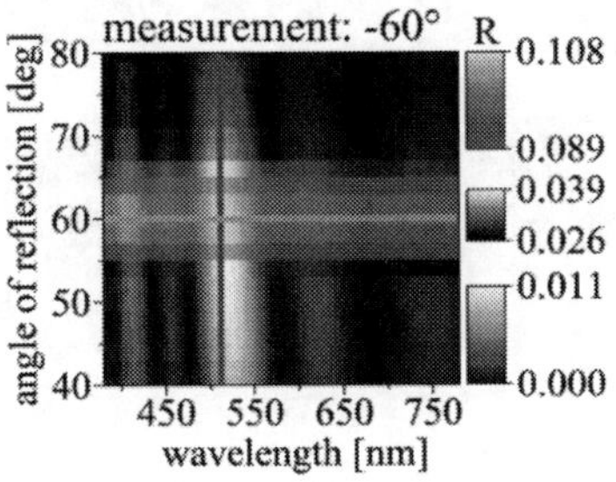

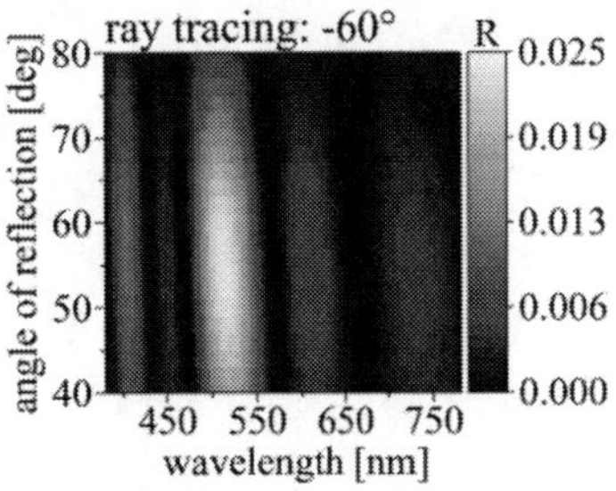

Wessels et al., Modeling of thin-film interference filters on structured substrates: microfacet-based BSDF versus ray tracing, 10.1364/OE.489221

Modeling by Means of Ray Tracing

Comparison of Possible Module Configurations

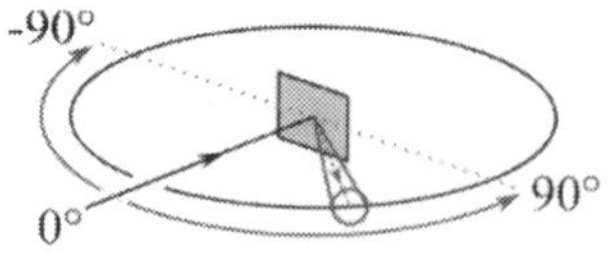

- Structured interference layer system is advantageous in terms of homogeneous appearance
- No significant influence of front side structuring if the layer system is already structured

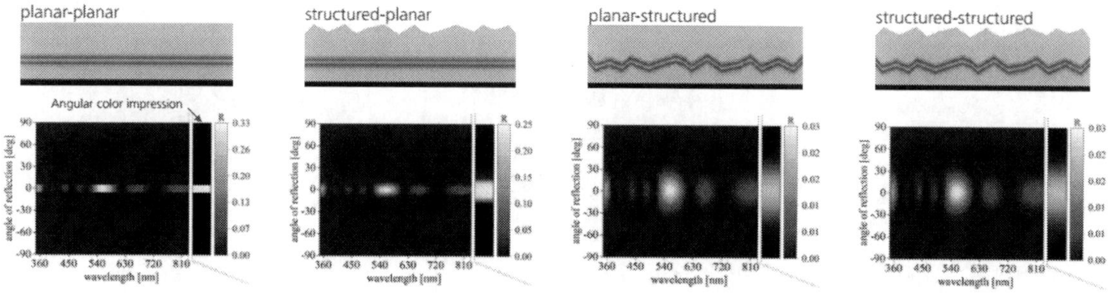

11 Wessels et al., Modeling of thin-film interference filters on structured substrates: microfacet-based BSDF versus ray tracing, 10.1364/OE.489221

Fraunhofer ISE

020288-011

40th European Photovoltaic Solar Energy Conference and Exhibition

Modeling by Means of Ray Tracing

Comparison of Different Structures

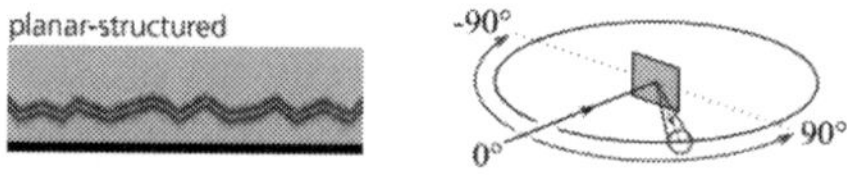

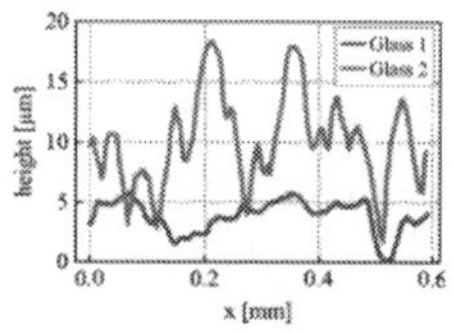

Measured topography

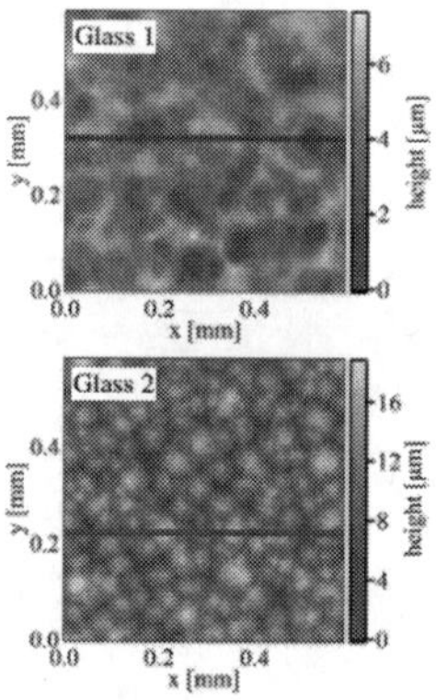

Simulated reflectance and color

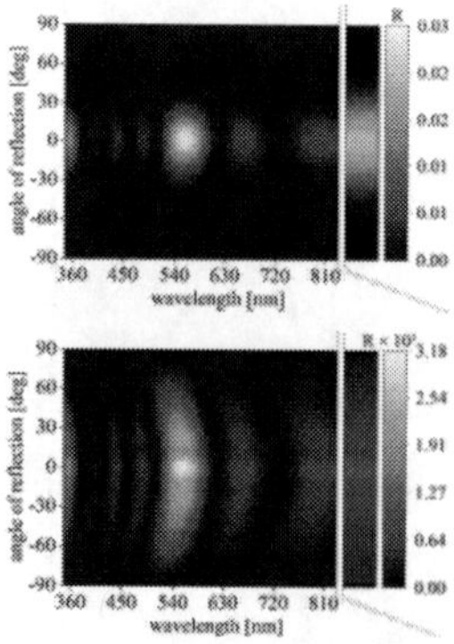

12 Wessels et al., Modeling of thin-film interference filters on structured substrates: microfacet-based BSDF versus ray tracing, 10.1364/OE.489221

Fraunhofer ISE

020288-012

Demo Modules and First Facade Installation

Example for Perfect Angular Stability

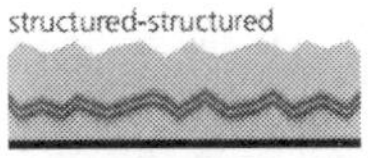

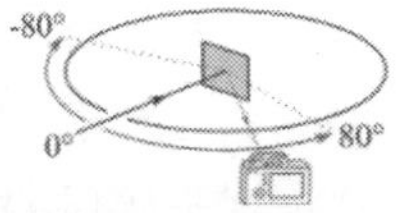

- 1 x 1.6 m² Modules
- Use of shingle matrix modules
  - Serve as homogenous background
- Module efficiency relative to black reference:
  - 94% for red
  - 95% for green
  - 96% for blue

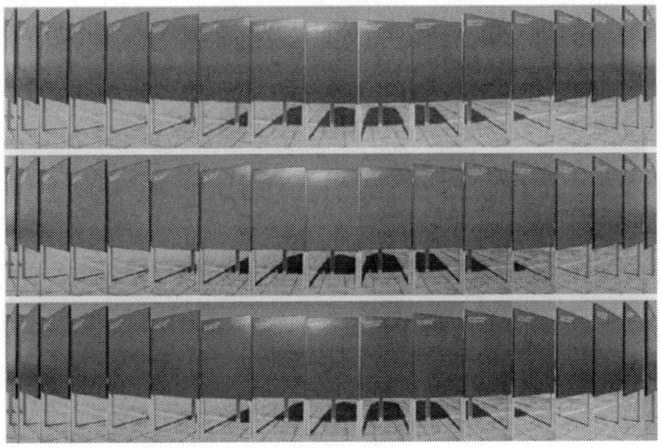

13

Presentation about shingling interconnections using aluminum foil by Jan Paschen (3DO.17.2)

Presentation about the reliability of shingle matrix modules by Najwa Abdel Latifand (3DO.18.1)

Fraunhofer ISE

020288-013

40th European Photovoltaic Solar Energy Conference and Exhibition

Demo Modules and First Facade Installation

Example for a Dynamic Appearance

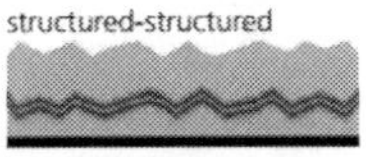

- Center for High Efficiency Solar Cells
- Brightness changes across the width of the installation

14

Fraunhofer ISE

020288-014

Conclusion
Take Away Messages

- The structured interference layer systems can be simulated in good agreement with measurement data
- It is beneficial to applying the interference layer system on a structured surface in terms of angular color distribution
- MorphoColor enables both eye-catching dynamic appearance as well as previously unrealized angular stability

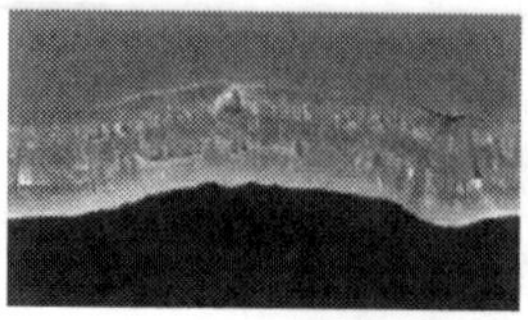

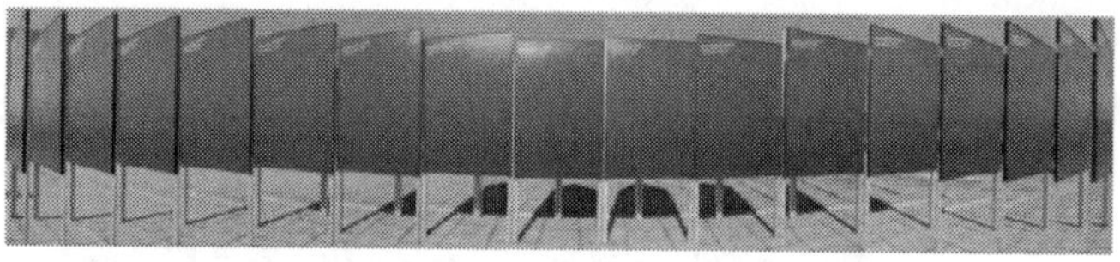

15

Poster presentation about colored PV by Dirk Reinwand 4BV.4.23

Fraunhofer ISE

020288-015

Outlook
Availability and Future Opportunities

- Megasol Energie AG has licensed the technology and offers colored *SOLARCOLOR Morpho* modules
- The MorphoColor technology also offers great potential for example in PV integration in vehicles
  - Here exemplary in the hood of the vehicle

16

Fraunhofer ISE

020288-016

Thank you for your attention

Andreas Wessels
Thermal Systems and Building Technology
andreas.wessels@ise.fraunhofer.de

Fraunhofer ISE
Heidenhofstraße 2
79110 Freiburg
www.ise.fraunhofer.de

Fraunhofer ISE

link to Fraunhofer ISE contributions of the 40th EU PVSEC

This work has received founding from:

German Ministry for Economic Affairs and Climate Action
Projects PV-Hide (03EE1049A) and Farbkollektor (03ETW007A)

Deutsche Bundesstiftung Umwelt

Supported by:
Federal Ministry for Economic Affairs and Climate Action
on the basis of a decision by the German Bundestag

sponsored by
DBU
Deutsche Bundesstiftung Umwelt
www.dbu.de

020288-017

40th European Photovoltaic Solar Energy Conference and Exhibition

The MorphoColor Concept

Impact of the Harmonic on the Peak Width

- Eight layers with thickness $\lambda_0/4n$ resp. $3\lambda_0/4n$
  - λ_0=550nm
  - n_{high}=1,5 and n_{low}=1
- 2nd harmonic is much narrower

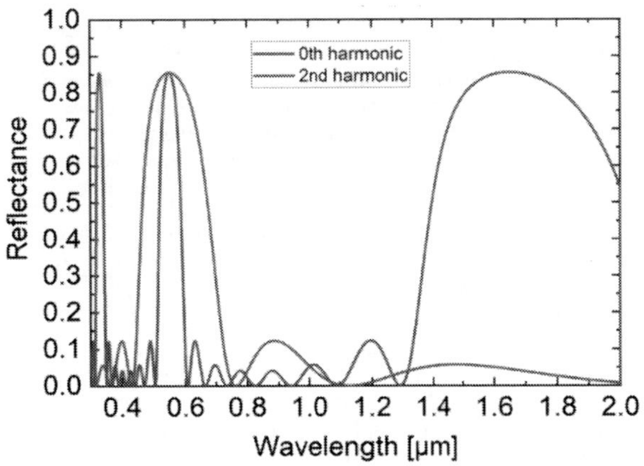

18
© Fraunhofer ISE
FHG-SK: ISE-INTERNAL

[2] Bläsi, The MorphoColor Concept for Colored Photovoltaic Modules, 10.1109/JPHOTOV.2021.3090158

Fraunhofer ISE

020288-018

IAM LOSSES IN COLORED BIPV: FROM THE LAB TO THE FIELD

Anna Bertomeu i Baldé, Markus Babin, Nanna Lysgaard Andersen, Sune Thorsteinsson
Department of Electrical and Photonics Engineering, Technical University of Denmark
Frederiksborgvej 399, 4000 Roskilde, Denmark

ABSTRACT: The aesthetic challenge in deploying building-integrated photovoltaics (BIPV) is addressed by exploring coloration solutions, particularly absorptive materials. These, aside from the direct coloration losses, also introduce additional absorption and reflection losses, which increase with the incidence angle and vary significantly depending on the color. In this study, BIPV modules with colored interlayer foils are investigated, revealing significant differences in incidence angle modifier (IAM) behavior between the five colors studied. The Martin and Ruiz model captures the overall trend but faces difficulties in fitting the IAM of colored samples, especially when combined with satinated or structured glass. In contrast, the polynomial Sandia IAM model achieves a solid fit. Furthermore, this study compares the lab-determined angular optical losses with outdoor measurement data. This involves modeling effective irradiance and power output, accounting for color-dependent module efficiency, temperature effects, and IAM losses. The study also indicates that apart from IAM losses, spectral mismatch variation has a significant effect on the performance of colored BIPV.
Keywords: BIPV, IAM, colored BIPV, spectral variation

1 INTRODUCTION

The growing trend of integrating solar panels into buildings, known as BIPV, is a significant step towards decarbonizing the building stock. To ensure financial viability of these projects, accurately predicting the performance of these systems becomes fundamental. However, the electrical performance of BIPV can deviate from that of traditional PV systems due to their operation under different conditions, like higher operating temperatures, extreme tilts, non-optimal orientations, or the introduction of coloration for greater aesthetics [1].

To customize the appearance of BIPV systems to align with architectural demands, methods include the use of materials that exhibit spectrally selective absorption and reflection, positioned between the glass cover and the PV cell, as well as thin-film coatings.

This paper focuses on a specific aspect: the additional angular reflection losses introduced by absorptive materials in BIPV, which exhibit color-dependent variability. Previously, IAM losses have been estimated to amount to up to 5% extra loss due to the colorant itself [2]. The aforementioned study also identifies the refractive index as a key driving factor behind the fluctuations in IAM associated with color differences. However, IAM losses from colored BIPV have not, according to the authors' knowledge, been shown in the field before.

This study seeks to address this gap by comparing the angular losses of six mini modules measured in the lab with one of six modules placed outdoors, mounted in the curtain wall facing west of a small operational BIPV installation at DTU Risø Campus, using a similar bill of materials.

2 METHODOLOGY

2.1 Mini modules and modules preparation

Both sample types are manually fabricated using monocrystalline silicon solar cells with two busbars encapsulated in a glass-glass laminate. In the façade-mounted modules, the top glass is satinized through acid-etching, while IAM measurements in the lab are performed both on samples with satinated and standard PV glass.

The single-cell mini-modules have dimensions of 20x30 cm, while the full-sized modules contain 48 cells, leading to total module dimensions of 105.5x148.5 cm. Coloration is achieved through colored interlayers laminated between the cover glass and PV cells using additional layers of encapsulant. In total, five interlayer foils made from the same base material but with different colors were studied.

To determine the transmission losses caused by coloration, IV-curves at standard test conditions (STC) were measured on all samples.

2.2 Angular dependent reflection losses

The setup for indoors IAM measurements of the mini modules is the same as in [2] and consists of a rotatory platform with stepper motor where the samples are mounted and illuminated using collimated light from a laser-driven light source.

In the process of fitting the acquired experimental data, two models are employed and evaluated. The first model is based on the Martin and Ruiz formulation [3][4] represented by Eq. (1). The second model is a 5th-order polynomial function, as outlined by Sandia [5] and presented in Eq. (2).

$$IAM_{M\&R} = \frac{1 - \exp[-\cos(AOI)/a_r]}{1 - \exp(-1/a_r)} \quad (1)$$

$$IAM_{Sandia} = b_0 + b_1 AOI + \cdots + b_5 (AOI)^5 \quad (2)$$

The IAMs for the full-size modules are not directly measured, instead the parameter a_r and the coefficient vector b in Eq. (1) and Eq. (2), respectively, are extracted from measurements on the mini-modules with satinated glass were employed to model the performance of the larger modules.

2.3 Outdoors data acquisition and modeling

The studied set of 6 modules is integral to a curtain wall BIPV system located at DTU Risø campus (55.6958°N; 12.1052°E). This system takes the form of an office container featuring three PV arrays integrated into its east, south, and west facades. The focal point of this investigation lies with the west-facing façade, oriented at 275° azimuth, shown in Figure 1.

Each module is connected to a power optimizer, reporting the output power in 5-minute intervals, and outfitted with a back-of-module PT 100 temperature sensor. Irradiance on the façade is measured using an in-plane pyranometer, with irradiance components measured

10.4229/EUPVSEC2023/4BO.16.6
020289-001

at a nearby weather station (< 200 m). For more details on the system and data acquisition, refer to [6].

Figure 1: West-facing BIPV façade featuring five colored modules and one uncolored reference.

The main focus of this study is to explore the color-dependent IAM effect. To achieve this, the study involves creating a model for the DC power output of each module. This model is based on the module's efficiency as measured under Standard Test Condition (STC), temperature, and the in-plane irradiance. Additionally, the IAM effect is integrated into the irradiance model to understand how it impacts the overall performance of the modules. Subsequently, the predicted power output is compared to the measured performance in order to assess the model accuracy.

Equations (1) and (2) are applicable for introducing the IAM to the beam component of the plane-of-array irradiance. However, for increased accuracy, it is essential to extend the IAM effect to also encompass the diffuse and ground reflected components. This can be done by integrating over all angles contributing to the diffuse and reflected irradiance. Assuming isotropic distributions allows for use of an analytical approximation of these integrals based solely on PV module tilt β. Two variations of this approach which can be used to model the effective irradiance have been presented in literature:

The first approach by Martin and Ruiz [7], uses the angular loss coefficient a_r whereas in the second approach, equivalent Angles of Incidence (AOIs) for the diffuse and ground-reflected components are calculated based on work by Brandemuehl and Beckman [8]. Subsequently, these equivalent angles are used in the Sandia model in Eq. (2) to separately determine the irradiance components.

As input to this modelling, DNI and DHI are measured at the nearby weather station and used together with the Hay diffuse sky model for transposition and an albedo value of 0.22 to determine reflected irradiance. Furthermore, to facilitate steady-state temperature modeling, exclusively data originating from clear sky days is used.

2.4 External quantum efficiency

During the analysis phase, it was decided to also measure the external quantum efficiency (EQE) of the 6 mini modules. This is done to uncover potential explanations for observed discrepancies between the modeled and measured power output of the modules in the BIPV system under study.

PV Measurements' QEXL system with a 5 nm resolution is used to conduct the EQE measurements, based on calibration with a reference Si-photodiode for the wavelength range of 300-1100 nm. The IR range is calibrated separately using a Ge-photodiode.

2.5 Solar spectrum measurements

In order to estimate the effect of spectral variations on the differently colored modules during a clear-sky day (June 11, 2023), the EQE measurements described in section 2.4 are combined with measured DNI data. This data is acquired using an EKO MS711DNI spectroradiometer, mounted on a two-axis solar tracker at DTU Risø weather station, located approximately 200 m from the BIPV test installation. Spectral data is recorded in the range of 300 to 1100 nm at 5 minute intervals.

The mismatch factor (MM) is calculated according to Eq. (3), using the standard AM1.5 Direct spectrum as reference (G_{ref}). To allow for easier comparison between the different colors, their mismatch factor is put in relation to that of the uncolored reference according to Eq. (4).

$$MM = \frac{\int EQE \cdot DNI \, d\lambda}{\int EQE \cdot G_{ref} \, d\lambda} \cdot \frac{\int G_{ref} \, d\lambda}{\int DNI \, d\lambda} \quad (3)$$

$$MM_{rel} = \frac{MM(color)}{MM(reference)} \quad (4)$$

3 RESULTS AND DISCUSSION

3.1 Total optical losses

Figure 2 shows the measured short-circuit currents (I_{sc}) for all samples, relative to the uncolored reference. It shows coloration losses of approximately 20 to 40 % under STC conditions, depending on the color. While slight differences can be observed between façade-mounted modules and single-cell samples, both sets show similar trends for coloration losses depending on color.

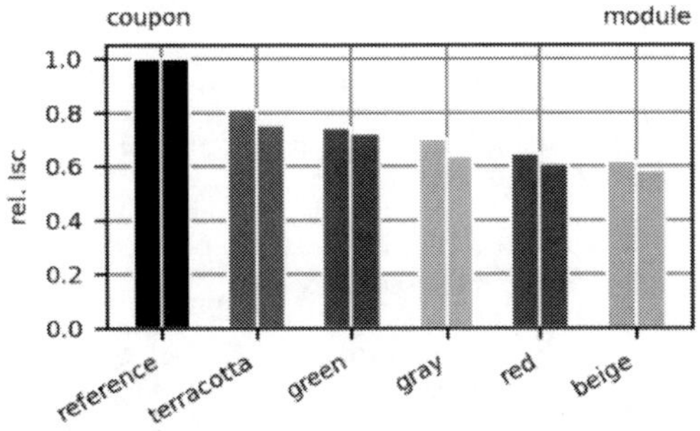

Figure 2: Relative short circuit current of coupon (left) and full-sized modules (right).

3.2 IAM losses

The IAM of the six mini modules with standard PV glass covers, obtained from the measurements in the lab, is presented in Figure 3, together with a fit of the Sandia IAM model. To isolate the effect of coloration from the angular dependent reflections from the glass cover, Figure 4 shows the IAM relative to the reference sample.

It can be observed that the difference in IAM with color is significant, however it is not directly correlated with the coloration loss shown in Figure 2. Comparable findings have also been documented for ceramic printing, demonstrating a lack of direct correlation between trans-mission losses and angular response. Instead, the increase

10.4229/EUPVSEC2023/4BO.16.6
020289-002

in optical losses with incidence angle is correlated with the refractive index of the coloration material, including pigments [2].

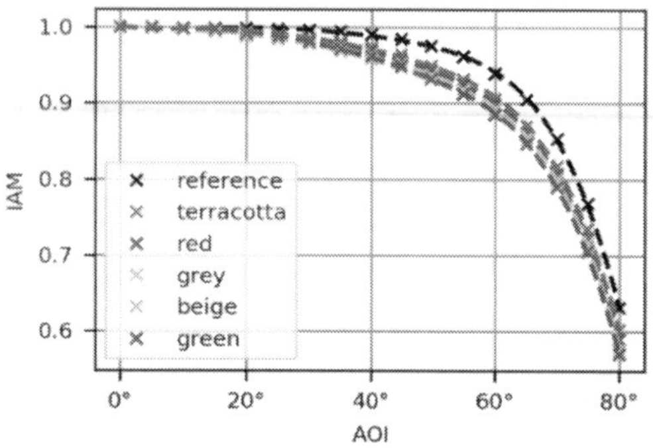

Figure 3: Measured (×) IAMs versus AOI, together with the Sandia fit (--), for each mini module with standard PV glass.

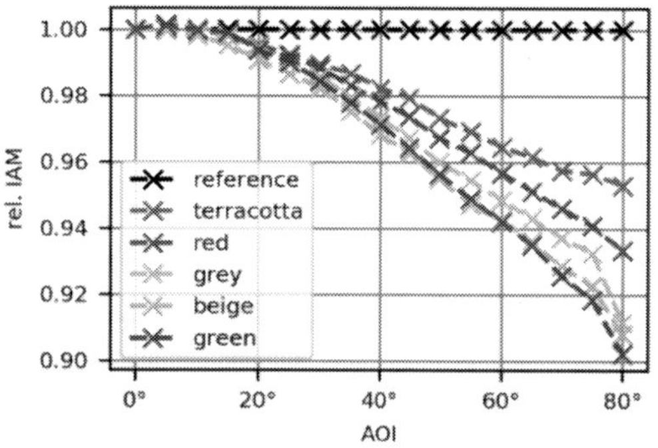

Figure 4: Relative IAM values versus angle of incidence, measured on mini modules with standard PV glass.

To investigate the fitting capabilities of IAM models, Figure 5 and Figure 6 show the measured and fitted IAM of the reference and terracotta-colored samples, with standard PV and satinated glass, respectively.

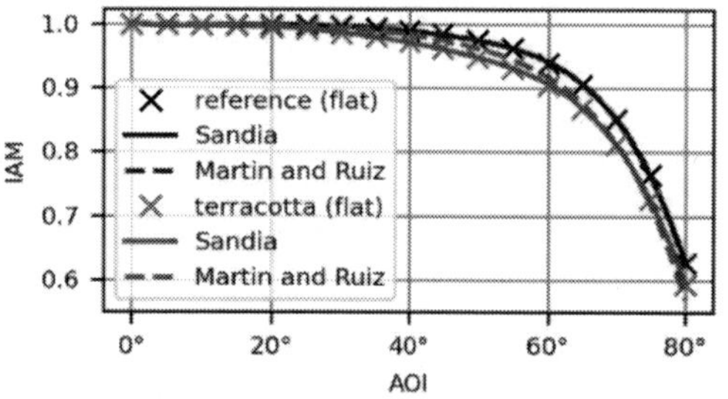

Figure 5: Measured (×) and fitted (-/--) IAM on reference and terracotta-colored samples with standard PV glass.

For samples with standard PV glass, the Martin and Ruiz model does capture the reduction in IAM with coloration, however, the model encounters difficulties in accurately fitting the measured data for incidence angles ranging from 30° to 60°. This error increases significantly when using satinated glass, as the Martin and Ruiz model is unable to capture the changed IAM profile accurately.

On the other hand, the Sandia model achieves a good fit for all samples, regardless of glass cover or coloration.

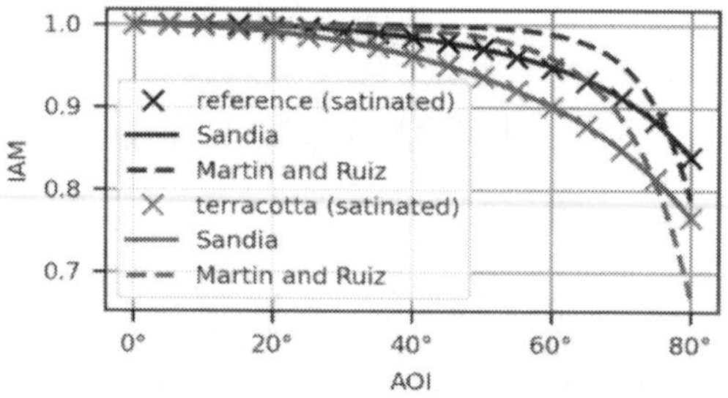

Figure 6: Measured (×) and fitted (-/--) IAM on reference and terracotta-colored samples with satinated glass.

3.3 Effect of IAM in the field

To identify the impact of including IAM models in outdoor performance modelling, Figure 7 shows the measured and modelled power of the terracotta-colored module over a clear-sky day. Solar noon on the shown day occurred at approximately 11:15 (UTC time) at the installation location.

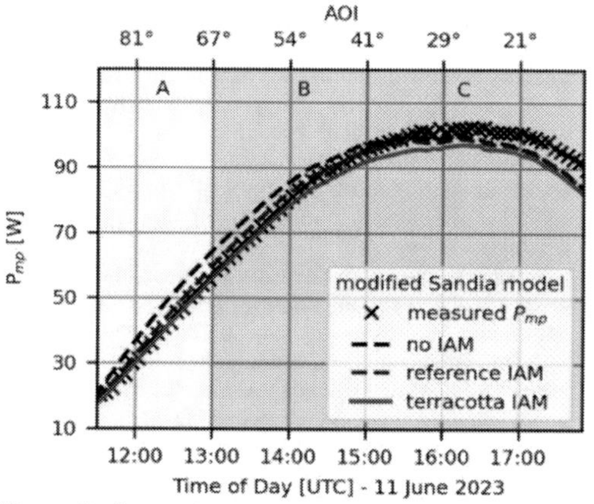

Figure 7: Comparison of P_{mp} values between measured (×) and modeled (-/--) data on a clear sky day for the terracotta-colored BIPV module.

When considering no IAM losses at all, module power output is significantly overestimated in the early afternoon (period A and B). By just considering the angular dependent reflections from the satinated glass surface using the Sandia fit of the reference sample, the fit of the measured data is significantly improved during the early and middle afternoon (period A and B). Considering the measured IAM of the terracotta sample further reduces the modelling error in these time periods, however a significant deviation remains between modelled and measured data in the later afternoon (period C). This discrepancy can be attributed to the fact that the solar spectrum undergoes variations throughout the day, as discussed in next section.

3.4 Intra-daily spectral variation effect

Figure 8 illustrates the impact of introducing a colored interlayer by comparing the external quantum efficiency of colored modules to the reference module. For all colors, the spectral response of the modules are shifted towards longer

10.4229/EUPVSEC2023/4BO.16.6
020289-003

wavelengths, albeit to different degrees.

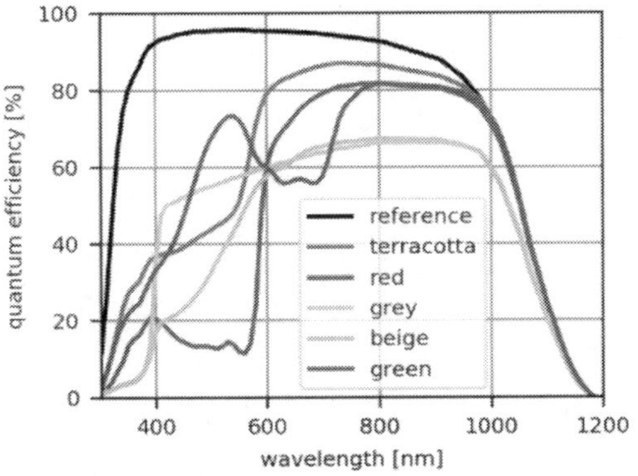

Figure 8: EQE laboratory measurements on the six mini modules.

Based on these measurements it can be hypothesized that such colored modules will perform better at times of day where the solar spectrum similarly shifts to higher wavelengths, i.e. the morning and evening. This is reinforced by the calculated relative mismatch factors shown in Figure 9: Here the red colored module experiences the highest increase in relative efficiency, as its spectral response shows the largest change. As shown by the air mass plotted on the secondary axis, this change in spectral mismatch gain is a direct function of light absorption and scattering in the atmosphere and therefore highly dependent on system location and orientation.

Since the model considered earlier did not account for the daily spectral variation, this provides an explanation for the underestimation of power for hours after 16:00 (UTC), as shown in Figure 7.

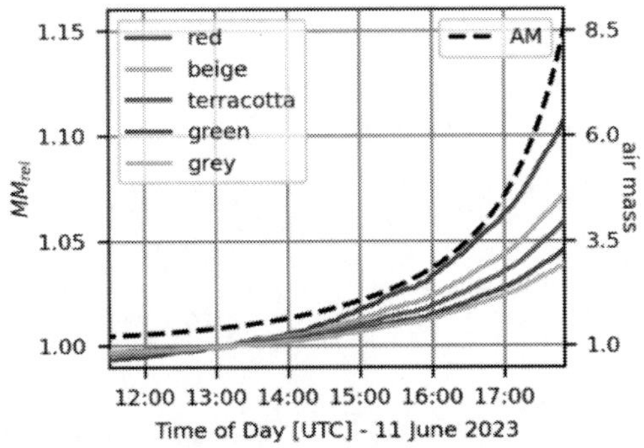

Figure 9: Calculated spectral mismatch for each module color relative to the uncolored reference based on spectral DNI and EQE, geometrical airmass on the right axis.

Implementing this spectral mismatch gain in the power modelling presented in section 3.3 results in Figure 10. It shows significant improvements in the late afternoon (period C). Despite improvements, some deviation remains, which could be associated with different spectral variations of diffuse and/or ground reflected light. Additionally, the isotropy-assumption for diffuse and reflected irradiance may be invalid during the late afternoon, due to the low solar elevation.

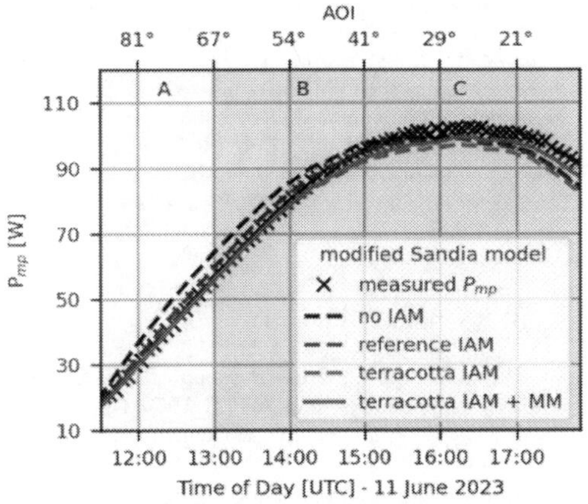

Figure 10: Comparison of P_{mp} values between measured (x) and modeled (-/--) data on a clear sky day for the terracotta-colored BIPV module.

For the specific orientation and location considered in the study, results suggest that it is important to consider daily spectral variation when modeling the effective irradiance, especially for colored modules with EQE curves that exhibit a primary response in the red range (around 600-1000 nm).

The above results are reported for a west facing façade at 56° latitude, where the AOI is at its minimum for high airmass. The airmass variation on a south facing façade is smaller and IAM impact might be larger. Similarly, different results with lower spectral mismatch factors and less daily variation are expected for diffuse lighting conditions.

4 CONCLUSIONS

As determined using indoor IAM measurements, the colored samples exhibit different angular responses compared to the reference sample, indicating distinct reflective properties for each color. While the Martin and Ruiz model effectively captures the general behavior for samples with standard PV glass, it encounteres challenges when fitting data for colored samples in the incidence angle range of 30° to 60°. Notably, it struggles to accurately model samples with satinated glass. On the other hand, the Sandia model provides a good fit based on a polynomial equation.

When bringing this to a colored BIPV installation in the field, it was found that the inclusion of added angular losses associated with glass surface structure, such as satination, as well as coloration-dependent IAM losses in modelling is critically important.

Further modelling errors were observed during the late afternoon on clear sky days. This discrepancy can in part be explained by the temporal evolution of the irradiance spectrum throughout the day. In the presented setup, since the colored modules are facing west, IAM losses are dominating in the early afternoon hours, whereas spectral mismatch losses are dominating in the later afternoon. In south-facing façade or rooftop installation, the strongest impact of these two effects may be seen at the same time, potentially leading to a significantly higher error in performance predictions. On the other hand, since both effects show opposite signs for red-colored modules, they

10.4229/EUPVSEC2023/4BO.16.6
020289-004

may partially compensate for each other.

While only a clear-sky day was investigated in this study, diffuse irradiation can be modeled using equivalent incidence angles greater than 55° [8]. Consequently, it can be expected that the IAM losses due to coloration increase in magnitude on overcast days.

In general, the results show that not accounting for color-dependent IAM losses and/or spectral mismatch can lead to significant errors for the daily energy yield on clear-sky days.

5 ACKNOWLEDGEMENTS

The authors acknowledge the support from EUDP as a funding agency for project “64021-1079 UnitSun”, and Rasmus Nielsen for or his assistance with the EQE experiments.

6 REFERENCES

[1] N. Martín-Chivelet, K. Kapsis, H. R. Wilson, V. Delisle, R. Yang, L. Olivieri, J. Polo, J. Eisenlohr, B. Roy, L. Maturi, G. Otnes, M. Dallapiccola, W.M. Pabasara Upalakshi Wijeratne. “Building-Integrated Photovoltaic (BIPV) products and systems: A review of energy-related behavior”. Energy and Buildings 262 (2022) 111998. https://doi.org/10.1016/j.enbuild.2022.111998

[2] M. Babin, S. Thorsteinsson, A. A. Santamaria Lancia, P. B. Poulsen, A. Thorseth, C. Dam-Hansen, M. L. Jakobsen. “Dependency of IAM Losses in Colored BIPV Products on the Refractive Index of Colorants”. Proceedings of 38th European Photovoltaic Solar Energy Conference (2022) pp. 583-589. https://doi.org/10.4229/EUPVSEC20212021-4BO.4.2

[3] N. Martin and J.M. Ruiz. “Calculation of the PV modules angular losses under field conditions by means of an analytical model”. Solar Energy Materials and Solar Cells 70 (2001) 25–38. https://doi.org/10.1016/S0927-0248(00)00408-6

[4] N. Martín and J. M. Ruiz. “A new model for PV modules angular losses under field conditions”. International Journal of Solar Energy 22 (2002) 19–31. https://doi.org/10.1080/01425910212852.

[5] D. L. King, E. E. Boyson, and J. A. Kratochvil. “Photovoltaic Array Performance Model”. Tech. rep. SAND2004-3535. Sandia National Laboratories, 2004.

[6] A. Bertomeu i Baldé. “Performance model validation of BIPV curtain walls”. Master Thesis. DTU Department of Electrical and Photonics Engineering (2023). https://findit.dtu.dk/en/catalog/64c6fbec5334821f4f64121c

[7] N. Martín and J. M. Ruiz. “Annual angular reflection losses in PV modules”. Progress in Photovoltaics: Research and Applications 13 (2005) 75–84. https://doi.org/10.1002/pip.585

[8] M. Brandemuehl and W. Beckman. "Transmission of diffuse radiation through CPC and flat plate collector glazings". Sol. Energy 24 (1980) 511–513. https://doi:10.1016/0038-092X(80)90320-5

IEA PVPS

Magasin X; Vasakronan - Uppsala/S; White Architects; https://www.besmartproject.eu/magasin-x

CROSS-SECTIONAL ANALYSIS OF BIPV INSTALLATIONS: performance evaluation as building component and energy generator

Frontini Francesco/SUPSI; Wilson Helen Rose/Fraunhofer ISE; Eder Gabriele C./OFI*; Babin Markus/DTU; Thorsteinsson Sune/DTU; Adami Jennifer/EURAC Research, Yang Rebecca/RMIT; Martin Chivelet Nuria/CIEMAT; Boddaert Simon/CSTB

**** Corresponding author: Gabriele.Eder@ofi.at***

Technology Collaboration Programme by iea

MULTI-DIMENSIONAL EVALUATION METHODOLOGY FOR BIPV

Since BIPV systems consist of multifunctional components that provide both power generation and building functions, a multidimensional approach to assessment and comparison is required.

IEA PVPS Task 15 has developed a **multi-dimensional evaluation methodology** for BIPV installations

- With this tool, the multi-functional performance (energy-relevant, economic, environmental and visual appearance) of selected BIPV systems/projects is derived.
- This assessment also allows the comparison of different design drafts for BIPV projects for a specific building (e.g. in a project tender).

→ **The methodology developed by the experts will be presented**

→ **Target groups are architects, developers and other stakeholders in the design process of BIPV projects**

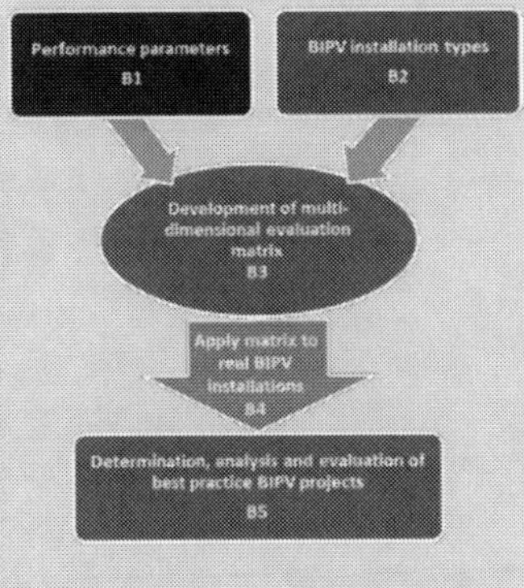

PVPS 2

STEP 1: DEFINITION OF THE PERFORMANCE INDICATORS

The quality of the building integration and the energy efficiency of existing and future BIPV systems were determined and evaluated using performance indicators (PIs) in 4 categories:

- Economic
- Energy-relevant
- Environmental
- Optical / visual appearance

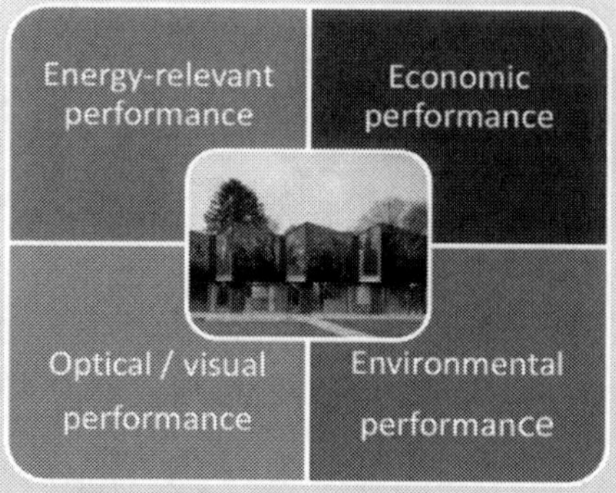

PVPS

3

020290-003

40th European Photovoltaic Solar Energy Conference and Exhibition

PERFORMANCE INDICATORS
for the characterization of the electrical and building functions of BIPV

	Energy-relevant PIs	Economic PIs*	Environmental PIs**	Visual appearance PIs
1	PV array energy yield (DC)	BIPV building element costs (whole BIPV System)	Cumulative energy demand non-renewable	Recognizability
2	Final system yield (AC)	Material replacement value	Climate change	Colour (uniformity)
3	Area-specific AC final system yield (relative to total BIPV system area)	Effective levelized cost of electricity (eLCOE)	Resource use, minerals and metals	Glare (risk)
4	Performance Ratio (annual)	Standard LCOE	Particulate matter	BIPV building in the urban environment
5	Self-sufficiency index	Effective net present value (eNPV)/costs	Acidification	BIPV system in the building
6		Effective discounted payback period (eDPP)	Water use	PV module in the BIPV system

Evaluation: objective fulfilled: y/n

*A review of 45 non-domestic buildings 12 western countries; Weerasinghe, R.P.N.P.; Yang, R.J.; Renewable and Sustainable Energy Reviews 137 (2021) 110622
** Report IEA-PVPS T12-03:2020; Task 12 methodology guidelines, 4th edition (References for PV LCA studies also: TS PEF Pilot PV 2018; European Commission 2017; Fazio et al. 2018; Zampori & Pant 2019).

PVPS

4

020290-004

STEP 2: RATING OF THE PERFORMANCE INDICATORS

As a next step, a numerical rating system (scale 1 → 5) for each PI was established.

Procedure:

- ✓ define possible range; information taken from real installations (min. and max. values);
- ✓ define scale 1 → 5 for each PI => rating 5 4 3 2 1
- ✓ calculate the mean rated PI for each category
- ✓ visualisation of mean value in the middle, but all individual PIs displayed in a "spider" diagram

Energy-relevant PIs (refer to performance of installed BIPV system) rating 1-5

	Energy-relevant PIs	Unit	minimum values	maximum values	value / specific BIPV project	Rating / Energy relevant PIs
1	PV array energy yield (DC)	kWh/kWp	0	1500	1021	3,0
2	AC final system yield	kWh/kWp	0	1350	919	3,0
3	Area-specific AC final system yield	kWh/m²	0	300	204	4,0
4	Performance Ratio (annual)	%	0	100	85	5,0
5	self-sufficiency index	%	0	100	62	4,0
					Mean	3,8

Proposed quantitative rating basis

	Energy-relevant PIs	Unit	rating 1	rating 2	rating 3	rating 4	rating 5
1	PV array energy yield (DC)	kWh/kWp	<550	550 - 825	825 - 1100	1100 - 1375	> 1375
2	AC final system yield	kWh/kWp	< 500	500 - 750	750 - 1000	1000 - 1250	> 1250
3	Area-specific AC final system yield	kWh/m²	<100	100 - 150	150 - 200	200 - 250	> 250
4	Performance Ratio (annual)	%	< 50	50-60	60-70	70-80	>80
5	self-sufficiency index	%	<5	5-20.	20-50	50-80	>80

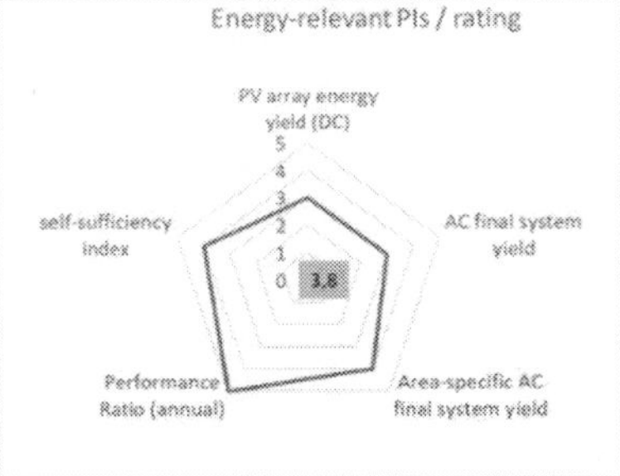

PVPS

5

020290-005

40th European Photovoltaic Solar Energy Conference and Exhibition

VISUALISATION - RATING OF PIs / arbitrary Example

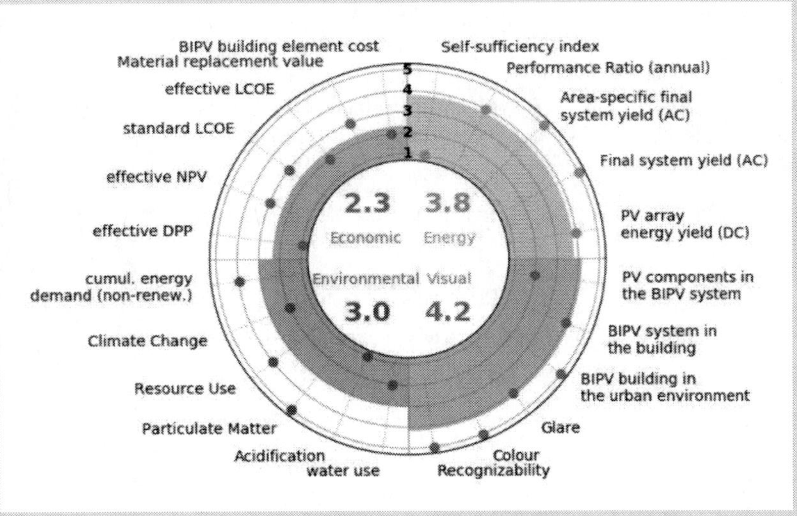

PVPS

6

020290-006

STEP 3: MULTI-DIMENSIONAL EVALUATION TOOL

Data/Information required for assessment of the multifunctional performance of a specific BIPV installation:

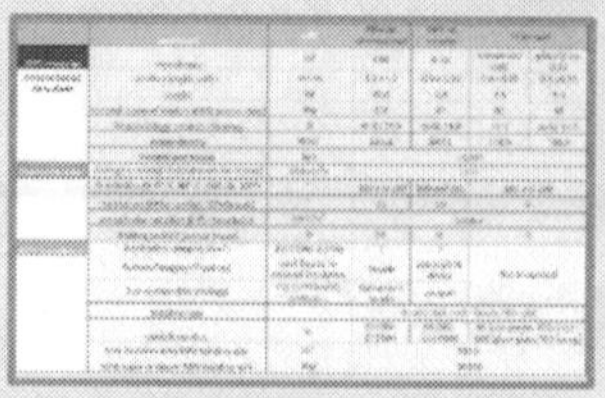

- General information on
  - installation site (geographical, climatic, orientation, inclination, shading);
  - the PV modules (datasheet): technical and construction-related
  - the BIPV installation (application and system category, sub-system, building type, total installed area and total nominal power of the BIPV building skin)

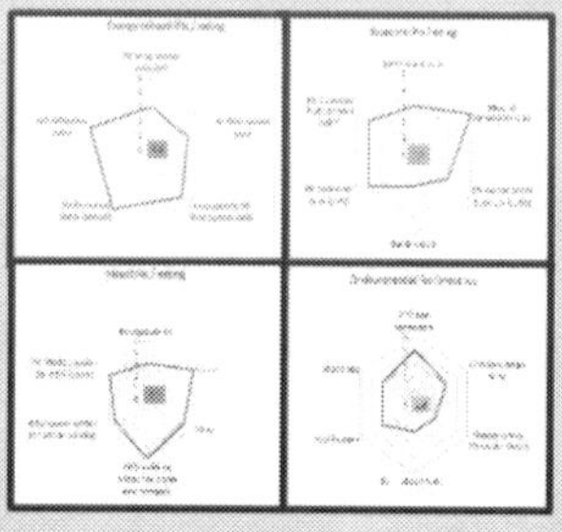

- Rated Performance Indicators (PI) in 4 categories + visualization
  - energy-relevant
  - economic
  - environmental
  - optical/visual performance
- Photos, technical drawings for visual performance assessment

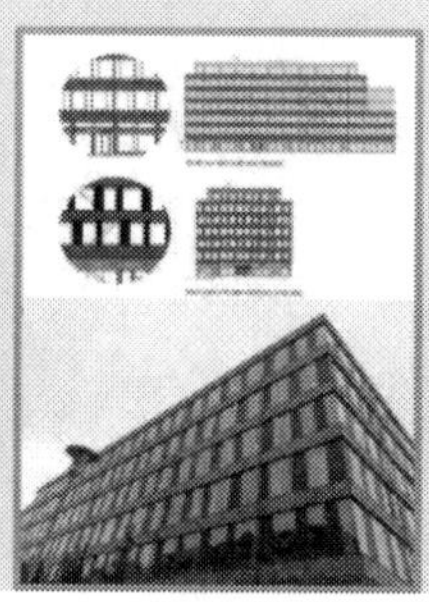

PVPS

7

STEP 4: IDENTIFY REAL BIPV INSTALLATIONS WITH AVAILABLE DATA

To test the tool, several BIPV installations/representative building typologies (different application types*: roof, facade, external device) were identified:

- a multifamily house with a BIPV façade using coloured BIPV panes intended not to be recognizable as PV; architect Karl Viridén

- an office building with BIPV elements in the façade and on the roof by White arkitekter and Vasakronan

- BIPV roofs of residential houses planned and realized by Exasun
- a multifamily building with BIPV parapets and BIPV window balustrades; building retrofit (2021); part of the Horizon 2020 EnergyMatching project

* Report IEA-PVPS T15-12:2021: Categorization of BIPV applications: Breakdown and classification of main individual parts of building skin including BIPV elements; https://iea-pvps.org/wp-content/uploads/2021/09/IEA-PVPS-T15-12_2021_BIPV-categorization_report.pdf

PVPS

ENERGY MATCHING

STEP 5: MULTI-FUNCTIONAL EVALUATION OF BIPV CASES

A preliminary analysis on representative building typologies has been done by a group of experts in order to test and validate the methodology and approaches.

The results of the following examples will be shown in this presentation:

1. Lighthouse project:
   - multifamily house with BIPV facades in grey; renovation project; Switzerland/Karl Viriden
2. Typical new housing estate:
   - BIPV roofs of housing project; newly built housing estate planned and realized by Exasun + housing cooperatives
3. Visual Performance assessment by the T15 expert group:
   - BIPV Facade Hotel Solna; Sweden/ White architects -> two categories (1) optics (colour, glare); (2) visual

Further examples will be discussed and presented in a peer-reviewed publication to be submitted to Buildings & Energy, Special Issue: Photovoltaics in the Built Environment in autumn 2023.

PVPS

9

020290-009

40th European Photovoltaic Solar Energy Conference and Exhibition

PHOTOS & DRAWINGS OF COLOURED BIPV FAÇADE / LIGHTHOUSE PROJECT

Solarchitecture © Viridén & Partners; https://solarchitecture.ch/de/wohngebaeudesanierung/ and https://www.viriden-partner.ch/

PVPS

10

020290-010

CROSS-SECTIONAL ANALYSIS OF BIPV FACADE / PROJECT ZÜRICH (VIRIDEN)

Results of the performance evaluation in 4 categories as a diagram (with mean ratings)

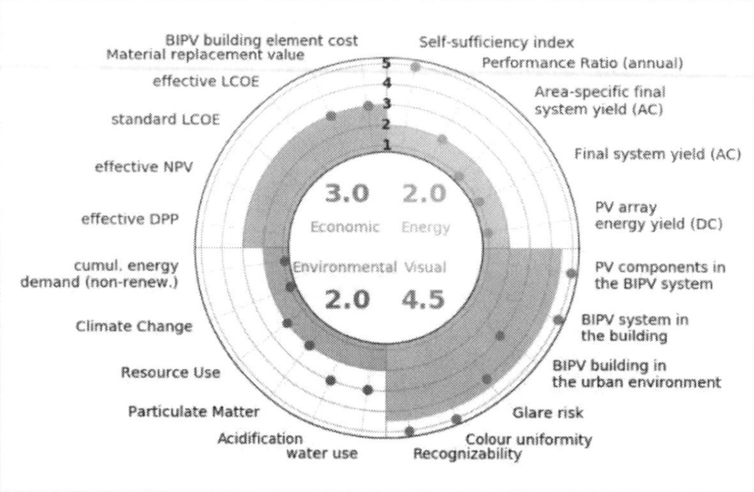

Lighthouse project: renovation project (many restrictions from constructional conditions! Shading; N facade) with innovative coloured modules (> 30% performance loss)

→ learn from data: importance of lighthouse projects for innovation and new product development

→ improve colouring technologies and also constructional details

→ develop customized solution

PVPS

PHOTOS AND DRAWINGS OF BIPV ROOF / HOUSING PROJECT

Exasun © Jan-Jaap van Os; https://exasun.com/

Housing project : Delft

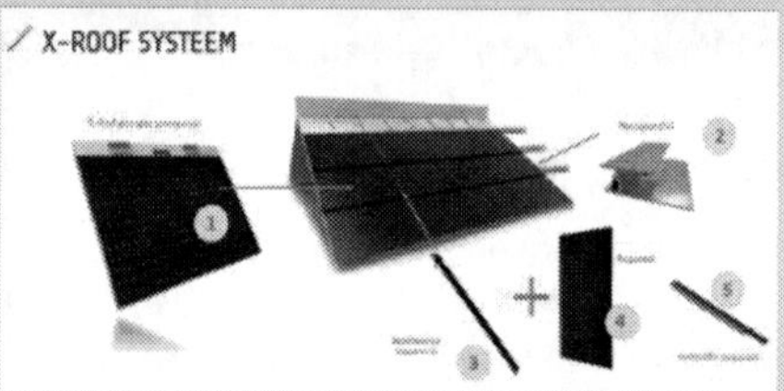

PVPS

12

CROSS-SECTIONAL ANALYSIS OF BIPV ROOF /PROJECT DELFT

Results of the performance evaluation in 4 categories as a diagram (with mean ratings)

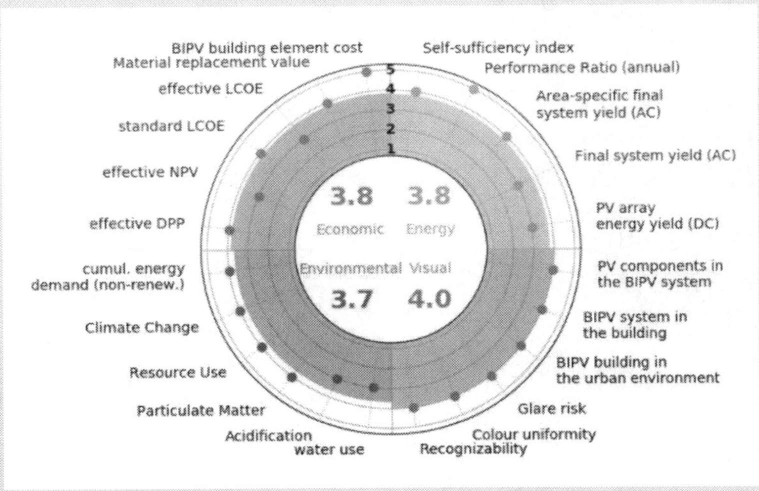

- not individually planned
- standardized products (representative)
- roll-out to several housing estates
- cooperation with housing associations

-> larger market segments
(! Might depend on national conditions!)

PVPS

13

020290-013

COLOUR AND GLARE PIs – POINT-BASED SYSTEM

- Start at base rating of 3
- +/- points for specific features
- Points can be discarded if irrelevant/intentional (if further information is known)
- Designed to be evaluated on-site on sunny days

Colour uniformity

Property	Points
Base rating	3
Hue and/or saturation change within single module/element	-1
Hue change between single modules/elements	-1
Saturation/lightness change over entire façade	-1
Colour-matched to surrounding materials (or no others)	+1
Homogeneous colour over entire façade	+1
Final rating:	1-5

Glare risk

Property	Points
Base rating	3
Satinated or highly structured glass (no specular reflections visible)	+1
Flat glass (specular reflections highly visible)	-1
Facing only low traffic areas	+1
Facing high traffic areas	-1
Small, unconnected areas (module areas < $2m^2$ AND < 25% coverage)	+1
Final rating:	1-5

PVPS

14

020290-014

"LIVE" TRIAL – ASSESSMENT RESULTS – 18 PARTICIPANTS

Comfort Hotel Solna / Stockholm

Design: White Arkitekter

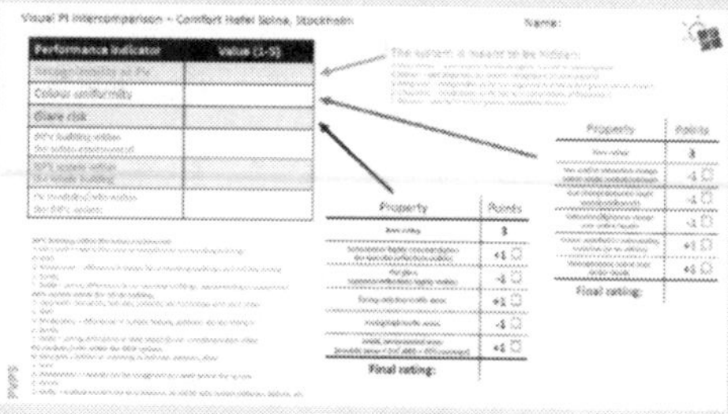

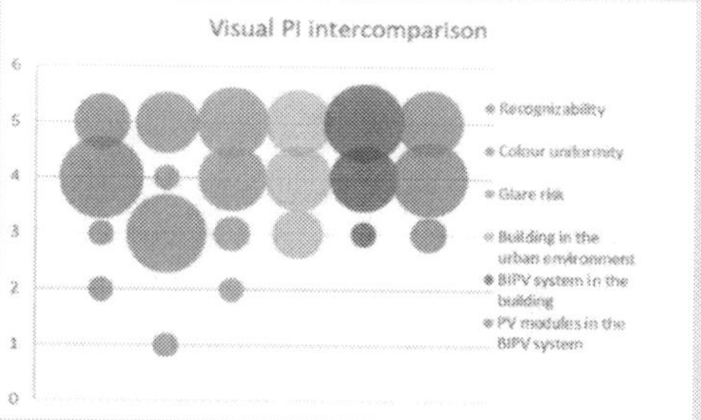

PVPS

15

020290-015

40th European Photovoltaic Solar Energy Conference and Exhibition

STEP 5: MULTI-FUNCTIONAL EVALUATION OF BIPV CASES

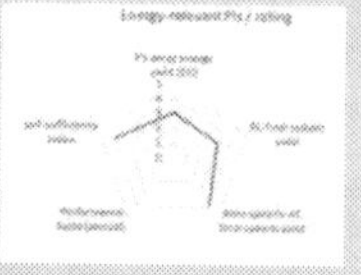

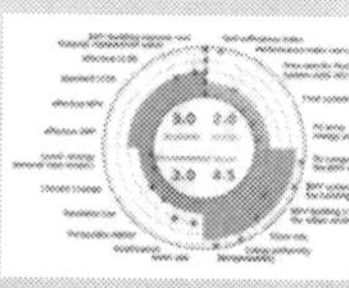

Challenges and difficulties:

1. Data availability and authorization for the use of data
   - Completeness – how to handle missing data?
   - Different methods of data creation for environmental data (→ other reference units)
2. Assessment of visual performance
   - Objectivity in aesthetic assessment (architecture ↔ technology ↔ economy) difficult to achieve → restricted to visual and optical (colour, glare) PI
   - Assessment difficult only with photos; much better "live rating"
3. How to evaluate lighthouse projects, research BIPV installations, ...
   - How to deal with lighthouse projects (disproportionately expensive, development products with possibly lower electricity yield) compared to standardized products?
   - Research or test installations; pioneers of new technologies (coloured BIPV)
   - Implemented projects ↔ planned projects
4. Can we use this assessment method for tenders and competitions for multifunctional evaluation?

PVPS

16

020290-016

CONCLUSIONS AND LESSONS LEARNED.....

- IEA-PVPS Task 15 developed a multidimensional assessment methodology for BIPV systems, starting from BIPV installations.
- After several improvement loops (learning cycles), most of the challenges could be solved. Sometimes solutions to the problems that occurred could be worked out, sometimes a workaround had to be found.
- Several real BIPV installations were identified for testing and optimizing the multifunctional evaluation tool. Results for these examples were shown.
- This methodology and the examples given can help the architects, developers and other stakeholder in the design process of BIPV projects.
- It works well for the major market segments.

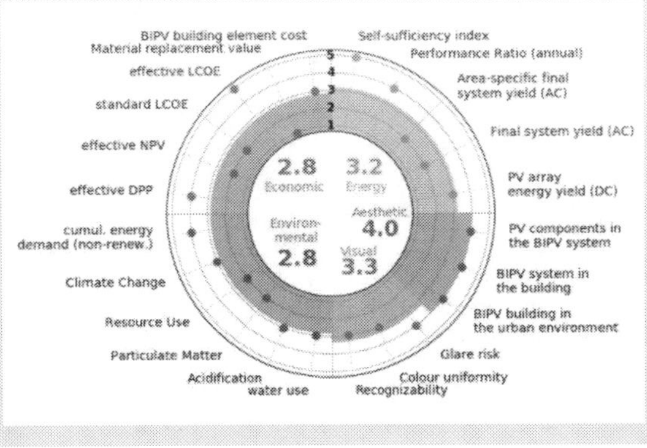

PVPS

eurac research | ofi | FFG | Fraunhofer ISE | Ciemat | CSTB | SUPSI | DTU | RMIT UNIVERSITY | EUDP

All contributors, their institutes and their funding agencies are acknowledged.

Special thanks to
Karl Viridén/ Viridén & Partners, Jan-Jaap van Os/ Exasun and Rickard Nygren/ White Architects
for sharing their project data with us and supporting our work

IEA PVPS

Contact:
gabriele.eder@ofi.at

Technology Collaboration Programme by iea

NEW PHASE IEA PVPS TASK 15 WILL START IN 2024...

Anyone interested in working on the next phase of Task 15,
please contact Francesco Frontini (francesco.frontini@supsi.ch)

These are the topics:

A: Challenges and opportunities of BIPV in a de-carbonised and circular economy

B: BIPV characterization & performance: pre-normative international research

C: BIPV in the digital environment

D: BIPV products, projects and demos: innovation and long-term behaviour

E: BIPV training, dissemination and stakeholders' collaboration

PVPS

19

EX POST ASSESSMENT OF ECONOMIC AND ENERGY PERFORMANCES OF FOUR BIPV RETROFITS CONFIGURATIONS

Elina Bosch[1], Philippe Macé[1], Gaëtan Masson[1]
[1]Becquerel Institute
e.bosch@becquerelinstitute.org

ABSTRACT: This paper assesses the energy and economic performance of BIPV in Europe using real costs and monitoring data from four demonstration sites. First, the energy performance is assessed and shows how the different integration areas and designs affect the BIPV electricity generation. Then, the economic performance is evaluated through the net present value based on an extra cost approach, which allows to determine whether it is more profitable to invest in a BIPV solution compared to a conventional construction solution. Competitiveness levels reached under demonstration site conditions (national retail electricity prices, local irradiation conditions) show that the developed solutions can be competitive with conventional construction solutions even when the BIPV solution design is not primarily performance-driven (e.g., presence of colours and patterns, low power density, suboptimal orientation). The conducted sensitivity analysis to different electricity prices and different irradiation conditions illustrates that the developed BIPV solutions can be competitive with different boundary conditions as well. Eventually, the role of BIPV in nZEBs (nearly Zero Energy Buildings) regulations is highlighted. Results show that the contribution of BIPV in reaching some nZEB objectives is not negligible although it does not constitute a substitution to energy efficiency and thermal insulation measures.
Keywords: BIPV, competitiveness, demonstration, nZEB

1 INTRODUCTION

As part of the BIPVBOOST project, (Horizon 2020 project), various BIPV solutions have been developed and implemented in four demonstration sites distributed in three European countries (Belgium, Italy and Spain). The demonstrated BIPV solutions show a great variety of characteristics in terms of integration (façade, roof, balustrade), in terms of design (patterns, colours, …), in terms of building type (single family house, multifamily house, non-residential building) and are the result of a series of innovations and improvements developed by project partners all along the BIPV value chain (from manufacturing to installation).

The demonstrated BIPV solutions' characteristics are presented more into details in **Table I.**

2 METHODS

2.1 Energy performance

Developed BIPV solutions have been installed at the demonstration sites and their production has been monitored for around one year. In some cases, a full year of monitoring data was not available (typically a few weeks missing). Consequently, the monitoring data (provided by project partner Tecnalia) was complemented by simulation data. The simulated data was obtained through a BIM software (BIM Solar) developed by one of the BIPVBOOST project partners (EnerBIM) as part of the project activities and able to model the 3D designs (building and BIPV solutions) of the demonstration sites and thus to estimate the BIPV electricity production under site-specific orientation, tilt and shading conditions.

2.2 Economic performance

Because BIPV serves multiple functions, its economic performance or competitiveness can be evaluated at different perimeters. First, looking at BIPV as a construction system, its cost (in €/m²) can be compared to that of conventional construction systems such as ceramic tiles roofs, concrete facades, glass facades, … Then, looking at BIPV as an electricity generation unit, the cost of electricity (in €/kWh) can be compared to that of other electricity generating units such as BAPV, or compared to retail grid electricity prices. Eventually, looking at BIPV as an investment, its net present value (NPV in € and normalized in €/m²) can be compared to that of a conventional construction project. Eventually, this holistic competitiveness assessment was selected to help answer the question: "*Is it worth investing in such electricity generating construction material, compared to a conventional building component?*".

In this holistic approach to evaluate the economic performance, all costs and revenues are considered. Typical sources of revenues include the savings on the electricity bill thanks to self-consumption as well as remuneration for injected electricity and potential incentives. Considered costs include the operation and maintenance costs as well as the initial investment. This later has been first obtained based on "as-built" cost (i.e., the costs that were necessary to implement the BIPV solution in the demonstration site including hardware costs and soft costs). Then, the "replicable" cost is determined. Compared to the "as-built" cost, the replicable cost does not include the cost for specific materials and/or interventions to accommodate the demo sites' specificities. For example, in one of the demonstration sites (DS1), a super substructure needed to be applied to the façade for it to be able to welcome the BIPV solution. In another demonstration site, reinforcement works needed to be conducted before the installation of the BIPV balustrade (DS2). Only considering the replicable cost allows to evaluate the competitiveness of a BIPV solution implemented in real conditions (i.e., not hindered by the demonstration site limits and configuration). Eventually, the "extra" cost is estimated (i.e., the additional cost necessary to implement a BIPV solution compared to the cost necessary to implement a conventional construction solution with similar visual and physical properties). This "extra" cost is determined following the methodology laid out in a previous BIPVBOOST deliverable. [1] In summary, the total cost is broken down into different cost items (PV module, inverter, cabling, installation, design, …) and for each cost item it is determined which share is attributable to BIPV and which share would also occur for

Table I: Demonstration sites description

Demo Site	DS1	DS2		DS3	DS4
Country	Aretxabaleta, Spain	Puertollano, Spain		Mons, Belgium	Morbegno, Italy
Type of building	Non-residential building	Non-residential building		Single family house	Multifamily house
Illustration					
BIPV module	Glass-glass cSi	Glass-glass cSi bifacial	Glass-glass back contact	CIGS	Glass-glass cSi
BIPV solution	Ventilated façade	(a) Balustrade	(b) Flat walkable floor	Tilted roof	Opaque cladding
Solutions providers	Onyx Solar (modules) and Tulipps (structure)	Onyx Solar (modules)	Onyx Solar (modules)	Flisom (modules) and Schweizer (structure)	Onyx Solar (modules) and PIZ (cladding)
Orientation	SE	S+E+W	S	E+W	SW+SE+NW+NE
Surface [m²]	152	124	69	150	107
Installed PV capacity [kWp]	21,6	12,8	9,2	9,9	9,8
Specific feature	Red back glass to match the company's logo colours. Integration of modules with different PV cell density to form desired patterns	Bifacial modules	Anti-slip glass	-	Poly crystalline green/gold-coloured cells Integrated insulation

10.4229/EUPVSEC2023/4BO.17.3
020291-002

a conventional construction solution. For example inverter costs are 100% attributable to BIPV while mechanical installation costs are only partly attributable to BIPV. Considering the extra cost allows to directly and easily interpret the NPV. Indeed, by using the "extra" cost, a positive NPV indicates that it is more profitable to invest in the BIPV solution compared to investing in a conventional construction solution. In other words, the "extra" cost of BIPV is compensated by the additional savings and revenues enabled by BIPV.

2.3 BIPV contribution to nZEBs

One of the demonstration site (DS3) is taken as illustrative example of the potential contribution of BIPV in meeting nZEB requirements. In the other three demonstration sites, the demonstrated BIPV solution has a limited size compared to the electricity consumption of the building it is integrated to. Installed capacities range from 10 to 20 kWp while some buildings shelter a very high electricity consumption leading to self-consumption rates around 100%. In DS3, the installed capacity (approx. 10 kWp) is in the order of magnitude of PV system sizes in the residential segment. Details concerning the buildings energy consumption for different energy vectors have been provided by the demonstration site owner (project partner Optimal Computing).

3 RESULTS

3.1 Energy performance

The annual electricity production of the different BIPV solutions is displayed in Figure 1. The bifacial BIPV balustrade (DS2.a) demonstrates the highest yield thanks to good local irradiation conditions and bifaciality. The walkable floor (DS2.b) also demonstrates an attractive yield thanks to the local irradiation conditions. The two BIPV facades (DS1 and DS4), demonstrate moderate energy performances for various reasons. In both cases, the vertical integration leads to relatively low average yearly in-plane irradiation. It can be added that the orientation of the BIPV façade is not optimal. In DS1, the system is oriented south-east while in DS4, the BIPV solution is integrated to all four facades of the building (including north-west and north-east). In DS4, self-shading from some building elements (see **Illustration a**) further deteriorate the energy performance. The BIPV roof shows moderate performances due to lower irradiation conditions (DS3 is located in Belgium compared to the other demonstration sites which are located in Southern Europe) and to a suboptimal orientation (East and West).

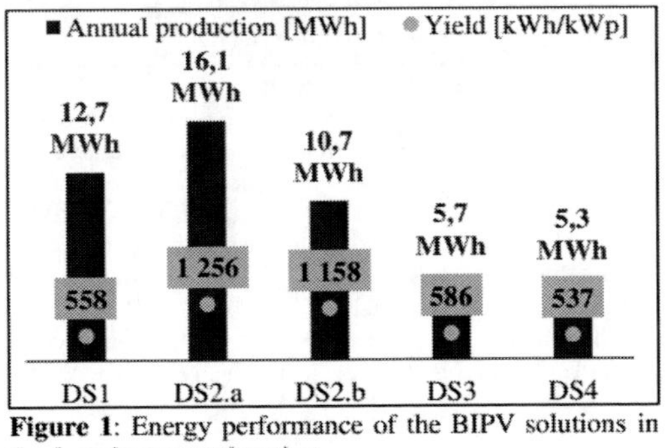

Figure 1: Energy performance of the BIPV solutions in the four demonstration sites.

Illustration a: Self-shading in DS4.

3.2 Economic performance

The economic performance is assessed using a holistic approach based on the replicable extra cost.

i. Costs

The investment costs are not presented here for confidentiality reasons. The annual operation and maintenance cost considered is 2 €/m² for the BIPV roof and 5 €/m² for all other BIPV solutions.

ii. Revenues

In DS1, DS2 and DS4, the self-consumption rate is (very close to) 100%. Consequently, the revenues are only coming from savings on the electricity bill thanks to self-consumption and thus amount to the compensable retail electricity price.

In DS3, the self-consumption rate is close to 50% (a relatively high rate for a 10 kWp PV system on a single family house mainly enabled by the presence of an electric vehicle with at-home charging). Yet, in Wallonia, yearly net-metering is still in place (although soon to be phased-out). This allows the whole electricity production to reduce the electricity bill. Eventually, revenues for each generated kilowatt hour under demonstration site conditions are summarised in **Table II**.

Table II: Revenues (compensable retail electricity price) for each generated kilowatt hour per demonstration site [2]

Demonstration site	DS1	DS2	DS3	DS4
€/kWh	0,21	0,30	0,44	0,34

iii. Other parameters

The annual average inflation rate considered is 2,5%, the cost of equity is assumed to be 2,5% for the residential cases and 4% for the non-residential cases. The NPV is calculated over 30 years which is assumed to be the BIPV system lifetime.

iv. Results

In the following visual representations, the competitiveness of the developed BIPV solutions is assessed. The competitiveness level with demonstration site conditions is represented with a black star.

In order to estimate the competitiveness levels in other locations (different orientation, different irradiation conditions, different self-consumption rate, …) a sensitivity analysis is conducted. On the vertical axis, several yearly average yields are being simulated. On the horizontal axis, several valuation levels of BIPV production are considered. A 150 €/MWh valuation of BIPV production can be interpretated as a 100% self-

10.4229/EUPVSEC2023/4BO.17.3
020291-003

consumption rate with a compensable retail electricity price of 150 €/MWh, as well as a 40% self-consumption rate, with a compensable retail electricity price of 250 €/MWh and a remuneration of injected surplus electricity amounting to 80 €/MWh). This allows to calculate the competitiveness level that would be achieved for different electricity end-users, for different regulatory frameworks for PV, for different countries and regions.

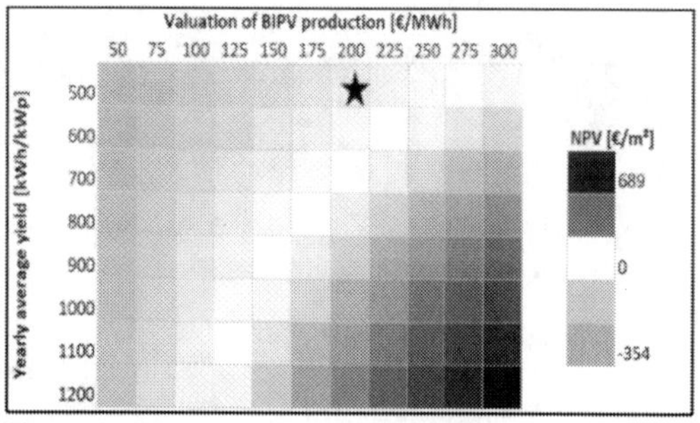

Figure 2: Competitiveness result for DS1.

The NPV at DS1 (represented by a black star on the heatmap) is -60 €/m², thus demonstrating a competitiveness level close to break-even. The case is mainly hindered by the lower retail electricity price (large electricity consumer), the vertical integration combined with a suboptimal orientation as well as a low power density to favour aesthetics. The sensitivity analysis shows how higher valuation of BIPV production (thanks to higher retail electricity price in the case of a residential building) or an optimal orientation could improve the competitiveness.

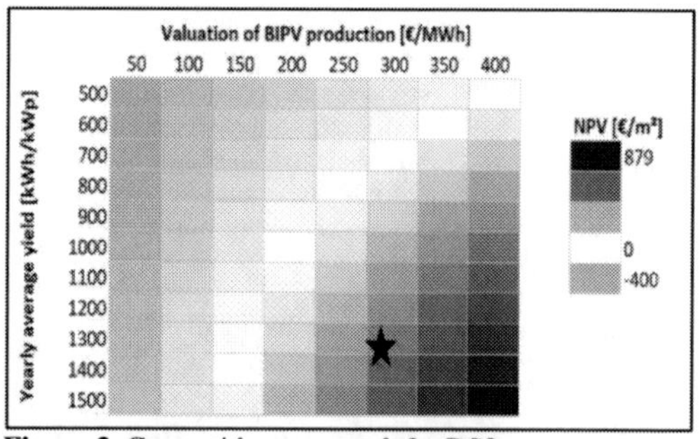

Figure 3: Competitiveness result for DS2a.

The NPV at DS2a (represented by a black star on the heatmap) is 374€/m², thus demonstrating a very attractive competitiveness level. The case mainly benefits from a high yield thanks to the good irradiation conditions and the bifaciality. The sensitivity analysis shows how the BIPV solution can also be competitive under lower irradiation conditions.

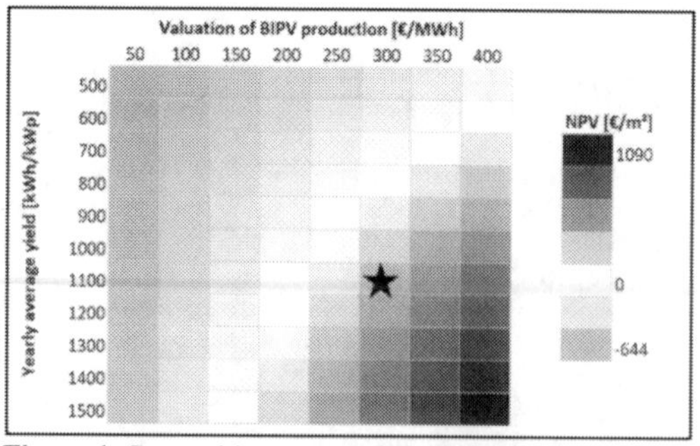

Figure 4: Competitiveness result for DS2b.

The NPV at DS2b (represented by a black star on the heatmap) is 318€/m², thus demonstrating a very attractive competitiveness level. The case mainly benefits from a high yield thanks to the good irradiation conditions. The sensitivity analysis shows how the BIPV solution can also be competitive under lower irradiation conditions.

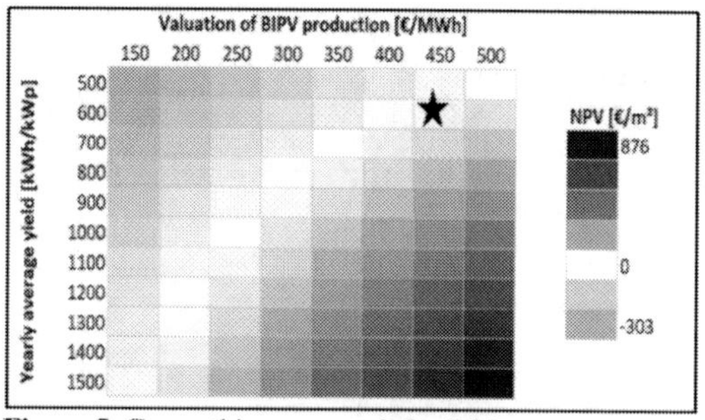

Figure 5: Competitiveness result for DS3.

The NPV at DS3 (represented by a black star on the heatmap) is 10€/m², thus demonstrating a competitive BIPV solution. The case is mainly hindered by the lower yield but benefits from a high retail electricity prices for residential customers in Belgium. The sensitivity analysis shows how the competitiveness would be improved with better irradiation conditions and this even with lower retail electricity prices.

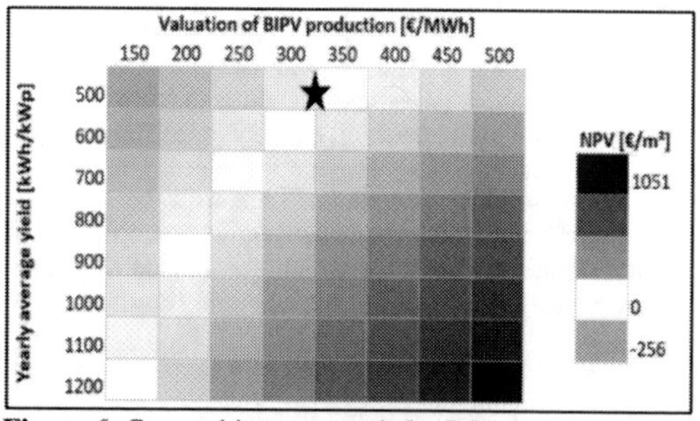

Figure 6: Competitiveness result for DS4.

The NPV at DS4 (represented by a black star on the heatmap) is 5€/m², thus demonstrating a competitive BIPV solution. The case is mainly hindered by the lower yield due to the integration on all four facades of the buildings (including north-east and north west) and the self-shading from building elements. Yet, DS4 shows that a BIPV design giving importance to visual and aesthetical integration to the detriment of energy performance can also lead to competitive projects. The sensitivity analysis

10.4229/EUPVSEC2023/4BO.17.3
020291-004

shows how the competitiveness would be improved with better irradiation conditions.

3.3 BIPV contribution to nZEBs

Requirements for new buildings and major renovations may vary depending on the country and on the type of building (residential, non-residential) and may concern different metrics such as (non-exhaustive): primary energy consumption, thermal insulation (e.g., U-values), share of renewable for certain uses (e.g., for domestic hot water), installed capacity of renewables, CO_2 footprint of construction materials, …

Although these requirements are renewable technology-agnostic, PV and thus BIPV can often contribute to meeting some requirements as presented in **Table III**.

Table III: Possible contributions of BIPV to nZEBs

	Contribution description
Contribution A:	Reducing primary energy balance by deducing primary energy avoided thanks to renewable electricity production. (e.g., a part (typically flat rate or self-consumed part) or the whole (typically through a monthly balance) on-site renewable energy production can be accounted for as avoided primary energy)
Contribution B1:	Reaching renewable energy absolute contribution targets (e.g., a certain installed capacity for renewable electricity generation should be installed). There are often several renewable technologies eligible to meet the target (solar photovoltaic, small wind turbines, …).
Contribution B2:	Reaching renewable energy relative contribution targets (e.g., a certain share of domestic hot water and heating needs should be covered by renewable energy). There are often several renewable technologies eligible to meet the target (heat pumps, solar thermal, solar photovoltaic, small wind turbines, …)

DS3 which is located in Wallonia (Belgium) is taken as an example to illustrate the potential contribution of BIPV to nZEB requirements. In Wallonia, the Decree of the Walloon Government concerning energy performance of buildings of the 28th of November 2013 sets the requirements for new constructions and renovations. According to the regulation, new buildings, building reconstructions and building extensions should comply with primary energy consumption requirements. The primary energy consumption is expressed in kWh/(m².a) where the square meters refer to the heated and/or cooled floor area while the primary energy consumption covers the following uses: heating, cooling, domestic hot water, auxiliary energy and lighting. PV can help meet this primary energy consumption requirement as the whole PV production can be converted into primary energy using a primary energy conversion factor of 2,5 (same as for electricity from the grid) and then deducted in the primary energy calculations. Yet, this can only be done through a monthly balance. [3]

First, the energy consumption of the building is compiled (**Figure 7**). The building is split into a residential and a small office area. Heating is covered by a gas furnace and a wood fire while domestic hot water is covered by the gas furnace only. Air-conditioning and lighting are covered by electricity. The electricity consumption for the electric vehicle charging is not represented here as it is not in the scope of nZEB requirements. It can be added that the building is not a recently renovated nor constructed building.

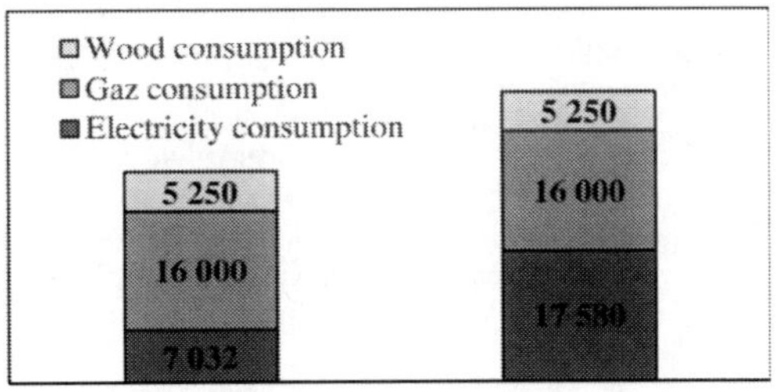

Figure 7: Energy consumption of the DS3 building in kWh and ventilated by energy vector.

Then, following the rule that the deduction of PV production from primary energy consumption can be performed using a monthly balance, around 80% of the total BIPV production (5,7 MWh converted into 14 235 $kWh_{Primary\ Energy}$ using a 2,5 conversion ratio as specified in the regulation) can be deducted from the primary energy consumption.

Eventually, after normalizing the values into kWh/m², the BIPV production allows to reduce the primary energy consumption by 60 kWh_{PE}/m² or from 206 kWh_{PE}/m² to 146 kWh_{PE}/m². Given the mixed use of the building (residential and office), it can be estimated that in case of major renovation, the building would need to reach a primary energy consumption below 85-100 kWh_{PE}/m². Thus, while the BIPV contribution to meeting the primary energy consumption-related requirement is important, it is not a substitution to energy efficiency or thermal insulation measures which remain the primary lever to achieve nZEB regulations' first objective of having "very high energy performing buildings".

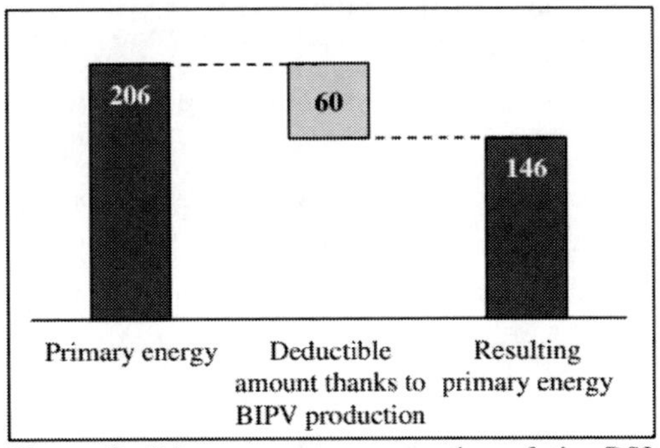

Figure 8: Primary energy consumption of the DS3 building in kWh/m² with and without BIPV contribution.

10.4229/EUPVSEC2023/4BO.17.3
020291-005

4 CONCLUSIONS

The demonstrated BIPV solutions show varied characteristics (different types of integration, coloured solutions, residential and non-residential buildings, …).

The competitiveness of the BIPV solutions has been assessed using the replicable extra cost to reflect to what extent replicating the demonstrated BIPV solution would be profitable.

All BIPV solutions in demonstration site conditions show competitiveness levels above 0, or very close to break even. The conducted sensitivity analysis to retail electricity prices and irradiation conditions shows that the BIPV solutions can be competitive under a wide range of boundary conditions.

Beyond the economic performance of BIPV, the contribution of BIPV in reaching nZEB objectives is not negligible. Yet, it does not constitute an alternative to energy efficiency and thermal insulation measures which remain the primary lever to achieve nZEB regulations' first objective of having "very high energy performing buildings".

5 REFERENCES

[1] 2020, BIPVBOOST, Competitiveness status of BIPV solutions in Europe
[2] Eurostat, Energy statistics, (https://ec.europa.eu/eurostat/web/energy/database)
[3] Arrêté du Gouvernement wallon portant exécution du décret du 28 novembre 2013 relatif à la performance énergétique des bâtiments

6 FUNDING AND ACKNOWLEDGMENT

The work described in this presentation has received funding from the European Union's Horizon 2020 research and innovation programme under grant agreement N° 817991. We thank the BIPVBOOST consortium partners (Tecnalia, SUPSI, Onyx Solar, EnerBim, Mondragon Assembly, Optimal Computing, WIP Renewable Energies, CSTB, EURAC Research, Cycleco, 3E, Tulipps, Flisom, Viridén+Partner, ISFOC, PIZ, Comsa Corporacion and Schweizer). Specifically, we would like to thank the following partners for their help by providing, collecting and processing key data for this work: Andreas Haller (Schweizer), Maria Martinez (ISFOC), Leire Minguez (EURAC Research), Julian Perrenoud (Flisom), Stephane Pierret (Optimal Computing), Merche Polo (Comsa), Elena Rico (Onyx Solar), Claudio Pirotta (PIZ), Marcela Rubio (Comsa), Eneko Setien (Tecnalia), Antonio Valor (Mondragon Assembly), Eugène Widlak (Tulipps).

40th European Photovoltaic Solar Energy Conference and Exhibition

This presentation was selected by the Sc. Committee of the EU PVSEC 2023 for submission of a full paper to one of the EU PVSEC's collaborating peer-reviewed journals.

PHOTOVOLTAIC SYSTEMS ON BUILDING ROOFS: RESEARCH ABOUT FIRE BEHAVIOR RATING

Piergiacomo Cancelliere[a], Giovanni Manzini[b], Giombattista Traina[c], Fabio Parolini[d]

[a] Ministry of Interior, Italian Fire Rescue and Service - CNVVF, Rimini, Italy - piergiacomo.cancelliere@vigilfuoco.it
[b] RSE S.p.A., Milano, Italy - giovanni.manzini@rse-web.it
[c] Istituto Giordano S.p.A., Bellaria (RN), Italy - g.traina@giordano.it
[d] University of Applied Sciences and Arts of Southern Switzerland (SUPSI), Mendrisio, Switzerland - fabio.parolini@supsi.ch

ABSTRACT: The increase in the number of photovoltaic (PV) installations leads to a more significant consideration of the aggravation of fire risk associated with the photovoltaic modules' installation. Given the importance and interest of all stakeholders in fire safety, both internationally and nationally, a process focused on fire risk assessment and mitigation has been started. This paper focuses on photovoltaic systems applied on residential and industrial building rooftops that are not integrated into the roof tiling. In fact, fires triggered directly by the photovoltaic system or in which the photovoltaic system is involved could occur on these installations. The authors, for a proper assessment and mitigation of fire risk on this type of installation, starting from current national and European regulations, the authors outline the experimental program that has been developed. Considering the European specifications for construction products, the research develops new test protocols specifically designed to evaluate the fire behavior of a system consisting of photovoltaic modules applied on the roof of a building in the case of a representative fire scenario. The innovation of these protocols lies in testing the PV module and roofing material as a single unit of construction by considering the effects of fire on the PV module and the roof with which it interacts. The research activity has focused on:

- the analysis of the Italian and European legal regulations and technical standards regarding the PV fire risk and their shortcomings' identification;
- development of appropriate test setup starting from existing test methodologies;
- the utilization of a specific approach focused on roofing samples made jointly by roofing materials and PV samples applied to them;
- the development of new PV roof samples, test protocols and fire rating criteria;
- the utilization of a specific approach focused on roof samples made by roof samples with PV applied.

Among the main results, the new procedures propose a fire behavior rating related to the parameters already used in Europe about the reaction to fire of building products (HRR, THR, FIGRA, etc.).
Keywords: PV roof, Test protocols, Fire behavior, Rating.

1 INTRODUCTION

The current EU (European Union) policy is pushing to increase the amount of energy produced from renewable sources, including solar. According to the European Commission survey, "buildings are responsible for about 40% of the EU's energy consumption and 36% of energy-related greenhouse gas emissions". European policies, therefore, call for an improvement in the number of NZEB (Near Zero Emission Buildings), and PV is part of achieving this goal (1). The growth of photovoltaics has affected all types of installations, particularly those on buildings. In 2022 there were 41.4 GW of new grid-connected solar PV capacity installed across EU27, with a 47 per cent increase over 2021 **Figure *1*** (2).

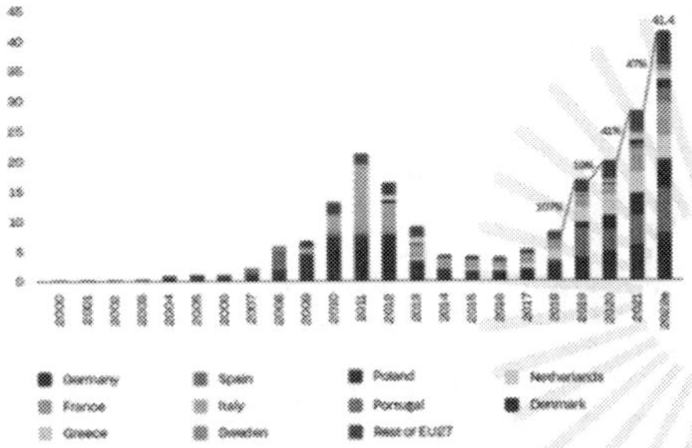

Figure 1 EU27 annual solar PV installed capacity 2000-2022 *(2)*.

The increase in the use of photovoltaics, not only for Italy but for all European countries, has been accompanied by cases of fires developed on photovoltaic systems due to accidental combustion. These events are no longer negligible, but they need specific attention. As expressed by some authors (3), a frequency of occurrence of 0.0293 fires per MW installed through quantitative analysis has been estimated. However, data on the number of fires developed on PV systems is challenging. In Italy and Australia, fire and rescue services have included a specific box in their post-fire assessment reports, but almost no data is available. (In Italy, data provided by the Italian National Fire Rescue and Services - DCPST, Rome, Italy, indicate that in 2012 there were 479 firefighter interventions (4) (5)).

Although fire safety is an international priority in PV projects' implementation, there are still missing gaps and regulatory aspects to be clarified. Many countries directly or indirectly refer to international standards that have parts focused on fire safety (IEC 61730 (6) (7), UL 1703 (8), etc.), but these standards are often not exhaustive.. For example, the European tests about building product ratings are developed for passive and non-active products. Furthermore, these tests do not consider elements in PV modules such as junction boxes, connectors, cables, etc. In the legislative scenario, there is also the variability of local building codes, which constantly leads to complexity in the adoption of standards and difficulties for manufacturers and stakeholders to spread PV on the market (9).

10.4229/EUPVSEC2023/4BO.17.4

1.1 Normative framework

The European approach to PV module and rooftop fire classification is based on the testing and classification of these two products separately, according to each test method. Roofs are classified according to EN 13501-5 (10) based on test protocols with specific fire scenarios that do not represent roofs with PV modules covering an entire roof area. The European standard on photovoltaic products IEC 61730-2: 2016 (7) requires testing the module with a small test flame, according to EN ISO 11925-2 (11) and refers to the regulations of each member state for further fire testing.

Each European country can, therefore, apply its national standard. Manufacturers are thus forced to add tests, requested by each country, to assess their PV modules' fire risk to market them at the European or International level. In Italy, for example, PV modules are tested and classified according to UNI 9177 (12) to obtain class one or two. An Italian guideline (13) assesses the fire risk for PV module applications.

This guideline refers to the fire performance of roofs as specified by (14) (derived from CEN/TS 1187 (15)) in which four test methods T1, T2, T3, and T4 are transcribed. These methods do not indicate a higher or lower fire resistance scale but only the test method used. Instead, performance is classified by letters: from "B_{ROOF}", which indicates the highest possible fire resistance, to "F_{ROOF}", which is equivalent to "no performance".

In Italy, modules that achieve class two can only be installed on roofs with B_{roof} classification, while PV modules with class one can also be installed on class F_{roofs}.

To implement fire safety, standard development processes have been initiated both at the international level (IEC TC 82 "Solar PV energy systems") and at the European level (CENELEC TC 82 "Solar PV energy systems" - WG01 "Wafers, cells and modules ") as well as at the national level, (In Italy: CEI TC 82 "Photovoltaic solar energy conversion systems" - WG11 "The fire risk in photovoltaic systems"[1]).

Specifically, in 2016, CENELEC Technical Report (TR) 50670:2016 (16), was issued at the European level. Following the European TR, the CEI TR 82-89:2021, "Fire hazard of photovoltaic systems - Fire behavior of photovoltaic modules applied on building roofs: test protocols and evaluation criteria" (now withdrawn and replaced by CEI TS 82-89:2023 (17) with the same title) was produced at the Italian national level.

1.2 Research activity

From the authors' experiences and shared understanding among European experts, proper testing of a combined/complete sample (PV module sample with roof sample) can provide a more meaningful understanding of the fire risk of PVs on a roof.

Several tests were conducted to investigate how photovoltaic panels mounted above the roofs modify their fire behavior in the function of different roof materials. Referring to the American literature (18), the first photovoltaic modules and roof coverings were analyzed as system behavior when exposed to fire, using the test bench configurations of the UL 790 (19) standard. Next, the "channeling effect" between the back sheet of the PV module and the roof surface was verified (20). Finally, the specific positions of the PV module (low slope and steep slope) on the roof decks during diffuse flame testing were verified (21).

Research on the topic has also been carried out in Italy, and one such research is the program carried out by RSE S.p.A. as part of the research for the Italian Electric System[2]. The research program included a brief analysis of fire cases of photovoltaic systems (photovoltaics applied on building roofs) and a study of existing legislation and technical standards (22).

The research first led to the publication of CEI TR 82-89:2021 (now withdrawn) and later to the publication of CEI TS 82-89:2023, "Fire risk of photovoltaic systems - Fire behavior of photovoltaic modules applied to building roofs: test protocols and evaluation criteria" (17).

The protocols developed in this standard involve samples made with photovoltaic modules applied to portions of roofs installed inside a test room with some "building roof type" geometries (23) simulating already built case histories.

Currently used under the European Standard for Construction Products (24), the EN 13823:2022 "Single Burning Item" (25), was chosen as a test room. However, the 'equipment described in EN 13823 was used only as a calorimeter to measure the heat release rate of photovoltaic modules applied to different types of roof coverings when subjected to a 30-kW flame attack.

The 30-kW flame was chosen and used starting with the burner described in CLC TR 50670:2016 (16), mounted below the PV module, ensuring a significant value of flame power.

In addition, the mounting of the PV modules and substrate, which represent the roof coverings, differs from the sample mounting scenario required by EN 13823 (25). In the standard, the samples must be mounted vertically, simulating a corner scenario of the room **Figure 2**.

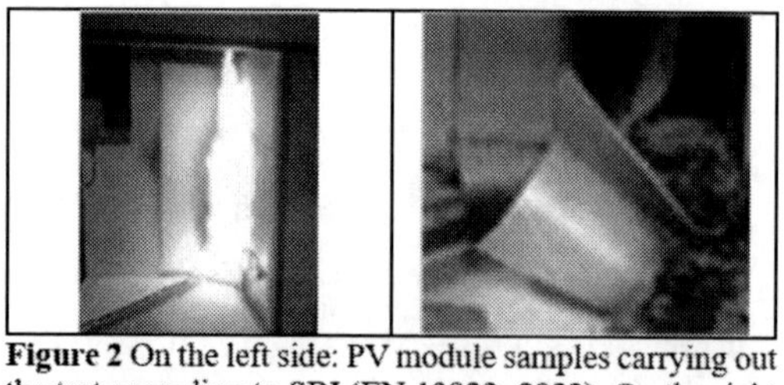

Figure 2 On the left side: PV module samples carrying out the test according to SBI (EN 13823: 2022). On the right side: Original test samples after testing.

[1] The Italian National Fire Rescue and Service has issued a national resolution to test and classify the reaction to fire of PV modules and panels based on the Decree of the Ministry of the Interior published on June 26, 1984. This technical regulation states that reaction-to-fire performance classifications for combustible materials are determined by using the results from a combination of the following tests: UNI 8457, Combustible products which flames can hit on one surface - small flame test; UNI 9174, reaction to fire of products subjected to a flame in the presence of radiant heat and UNI 9177, Classification of reaction to fire of combustible materials. According to test results, flammable materials and products are divided into five categories, 1, 2, 3, 4 and 5, where one (1) represents the best level of performance and five (5) the worst (there is also class 0 for non-combustible materials).

[2] About RSE activity: This work has been financed by the Research Fund for the Italian Electrical System under the Contract Agreement between RSE S.p.A. and the Ministry of Economic Development - General Directorate for the Electricity Market, Renewable Energy and Energy Efficiency, Nuclear Energy in compliance with the Decree of April 16th, 2018.

Otherwise, in the CEI TS 82-89 (17), the sample is laid in the center of the test room. The sample consisting of the roofing specimen over which the PV specimen is applied is shown in **Figure** *3*.

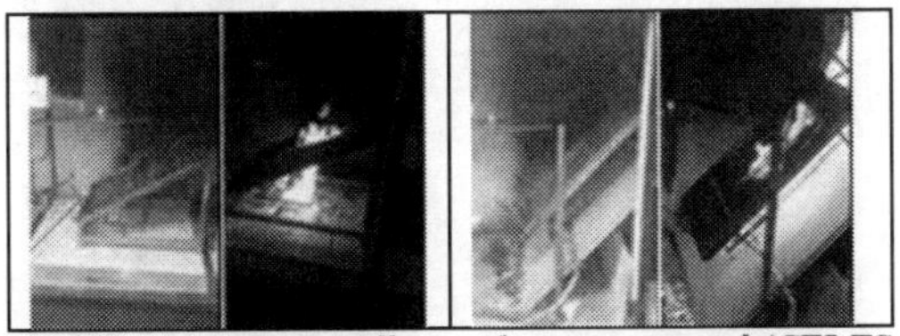

Figure 3 Testing according to the new protocol (CEI TS 82-89:2023) - Sample made by PV and roof. On the left side is the flat roof version, and on the right is the sloped one, both before and during the test.

In addition to conducting the test by pairing the module with a roofing sample, the innovation of the developed test protocol is to express the fire behavior using the same parameters as the ones used within construction products. Thus, the set procedure in CEI TS 82-89 (17) allows the classification of the PV module according to the roofing package used.

There is no such approach in the legislative scenario and published standards at the European level today. To date, testing is focused on the fire spread characteristic of the individual component and never by testing a sample consisting of a portion of the roof and PV module.

Analysis of the complete system, not the individual component, was also carried out in another research program conducted by RISE (Research Institutes of Sweden). The study, reported in PV magazine (26), focused on the space between the solar panels and the roof surfaces, which could play a key role in the fire behavior of the PV system and could contribute to the development of the fires themselves.

2 METHODOLOGY

As remarked by the authors in previous works [12], the research activity was carried out through an experimental program consisting of tests based on the European Specifications for Construction Products. What is set out in EN 13823:2022 (25), and CLC/TR 50670:2016 (16) was considered. To better assess the fire behavior of a PV roof sample, some variations were made from what was expressed in the previously stated standards on the burner, sample placement, frames and structures used.

As expressed above, in the EN 13823 standard (25), samples must be mounted vertically, simulating a room corner scenario. Figure 1 shows the dimensions of the specimen wings to be used according to the EN 13823 standard (25).

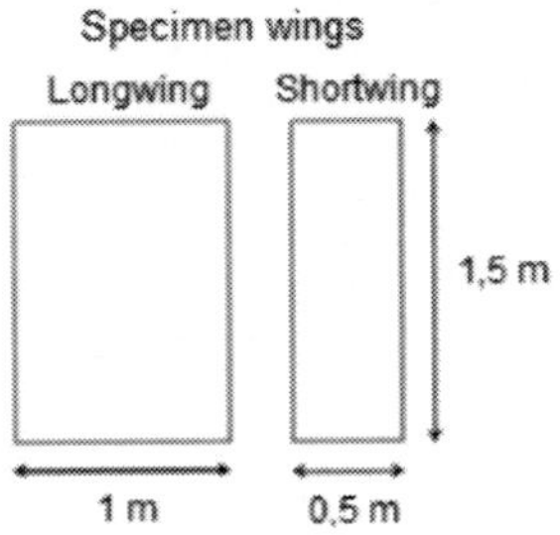

Figure 4 Example of SBI test installation from EN 13823.

The new test bench frame required by CEI TS 89-89 (17) in the modified SBI chamber was made considering representative sizes of the sample that allow for a proper evaluation of the behavior during the test. Figure 8 shows the test bench with the dimensions.

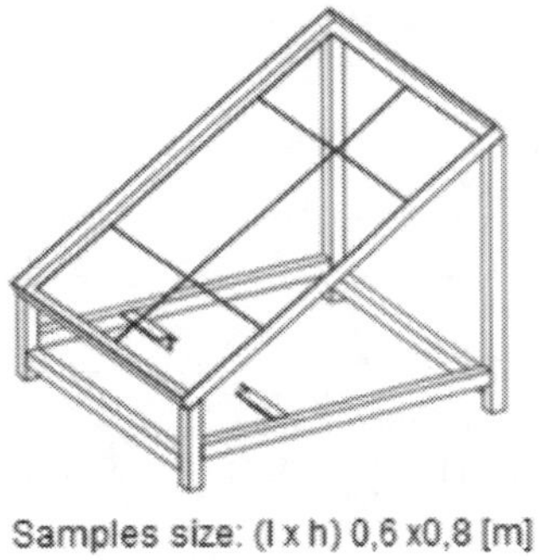

Figure 5 Example of CEI TS 82-89 test installation in modified SBI chamber derived from EN 13823 *(17)*.

It is important to note the impact of the variations also requires minimal resources using the same test chamber, measuring system, and software used in the SBI test (25) as outlined in CEI TS 82-89 (17). This allows the test to be performed easily and without cost burden.

Below, Figure 6 Scenario "A" expressed by CEI TS 82-89.

Figure 7, and

Figure **8**, show the photovoltaic roofing samples in the three different test protocols expressed by CEI TS 82-89 *(17)*, considering the possible different scenarios of the roof and PV module and the burner positioning.

A- Sloped/flat roof test with photovoltaic module attached to the roof and burner superimposed on it.

B- Test on inclined/flat roof with photovoltaic module coplanar to the roof and burner interposed between the roof and photovoltaic module.

C- Test on a flat roof with the PV module inclined to the roof and burner between the roof and the PV module.

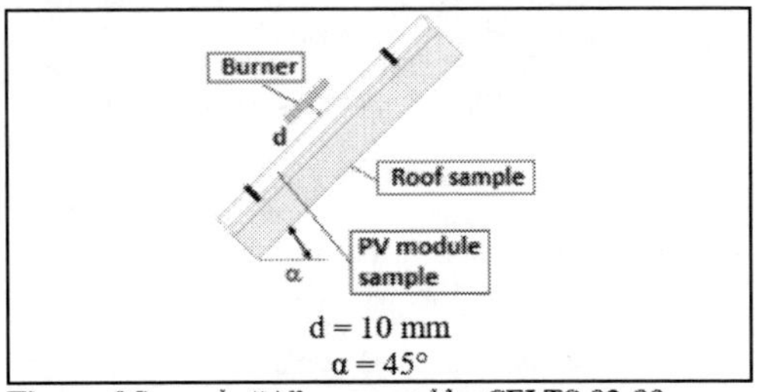

Figure 6 Scenario "A" expressed by CEI TS 82-89.

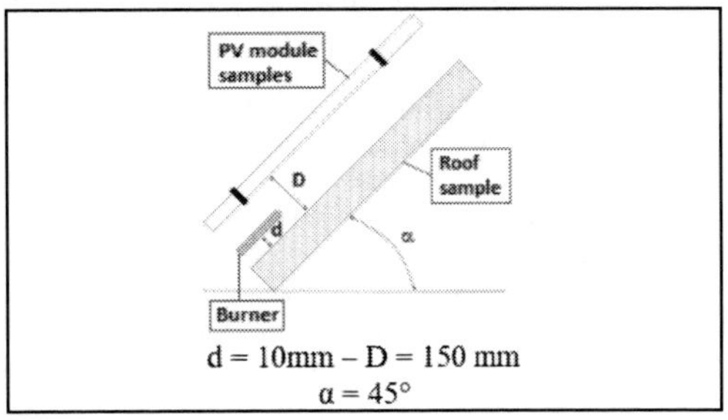

10.4229/EUPVSEC2023/4BO.17.4
020292-003

Figure 7 Scenario “B” expressed by CEI TS 82-89.

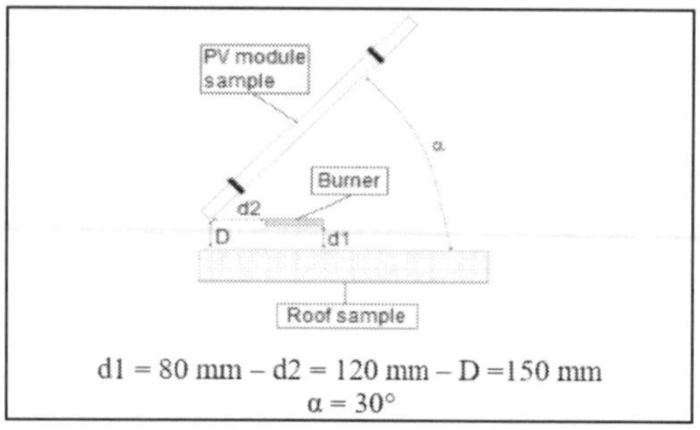

Figure 8 Scenario “C” expressed by CEI TS 82-89.

For better clarity, **Table *1*** shows the main parameters used in the CEI TS 82-89 (17) test protocol compared with those used in EN 13823:2022 (25).

Table 1 Comparison of the main characteristics SBI EN 13823: original - and SBI variant - CEI TS 82-89.

Test Chamber	SBI EN 13823	SBI CEI TS 82-89 derived from EN13823
Sample(s) size	Derived from EN 13823. 0,5x1,5 + 1x1,5 (m)	Defined from CEI TS 82-89. 0,6 x 0,8 (m)
Tilt angle α	90 °	30° or 45° or real tilt angle
L_f [cm]	n.a.	n.a.
HRR_f [kW]	30	30
Test duration Δt_f [s]	1200	1200

Different samples made by roof portions with different waterproof membranes and PV modules were tested. To obtain the test configurations expressed in CEI EN TS 82-89 (17), the requirements of the Italian photovoltaic standard for photovoltaic modules were considered. Samples classified as class 1 and class 2, according to the Italian standard UNI 9177 (12), were used.

The European Construction Products Regulation requirements (24) were met for roof parts, including waterproofing membranes.

The waterproofing membranes, including the thermal insulation layer, were tested according to CEN TS 1187 (T2) (15) and ISO 1716 (27) standards.

The substrates tested and serving as roofs were chosen belonging to the B-roof and F-roof classes (EN 13501-5 (10)) testing according to the test method two (T2) expressed by CEN TS 1187 (15).

In this way, it was possible to find a correlation between the Total Heat Release (THR)[3] by the PV module assembled to the roof system and the characteristics of the roofing layers (membranes and insulation) used during the tests. All the samples thus composed were tested mainly in the configurations indicated in the CEI TS 82-89 (17).

Figure 9 and **Figure 10** show the test according to CEI TS 89-89.

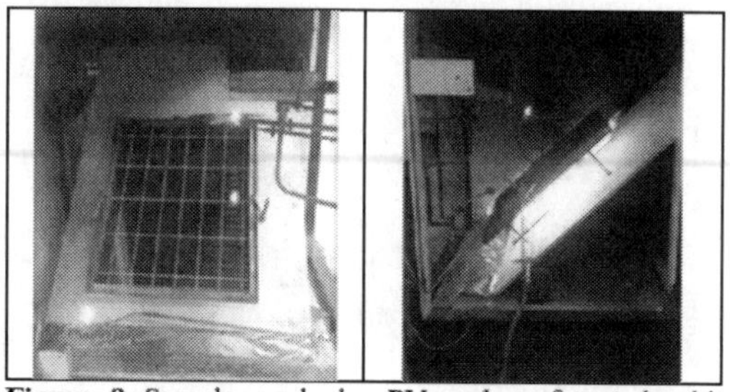

Figure 9 Sample made by PV and roof tested with protocol B according to CEI TS 82-89 during the test. On the left side: frontal view. On the right side: lateral view.

Figure 10 Sample made by PV and roof tested with protocol C according to CEI TS 82-89 during the test. On the left side: frontal view. On the right side: lateral view.

3 MAIN RESULTS

As expressed above, an innovation of the developed protocol is testing a sample consisting of a PV module specimen assembled with a roof portion specimen.

The test methodology developed in CEI TS 82-89 (17) allows the fire behavior classification of the PV module according to the roofing material used for the roof.

Depending on the flame behavior and how much heat and smoke the PV module emits in conjunction with the roofing material, a different fire behavior class will be obtained for the PV module.

Considering the rationale behind developing the new procedure for classifying photovoltaic modules installed in roofing is essential.

Photovoltaic modules are made of combustible plastic polymeric materials. In many cases, the back sheet of PV modules is composed of a multilayer polymeric material consisting of polyester, EVA, Tedlar PVF, or any other material which protects the active photovoltaic part on the back side of the module.

The other polymeric material inside the PV module is the encapsulant.

This material serves the dual purpose of protecting and providing insulation for the photovoltaic cells and keeping the back sheet assembled with the glass.

EVA (Ethylene Vinyl Acetate) or POE (Polyolefin) are commonly used as encapsulant, but other materials are available on the market. From fire test observations, these

[3] THR is defined as the amount of heat energy evolved by the specimen under test during its burning time. It is expressed in terms of energy or as surface area normalized energy (kWh/m^2).

10.4229/EUPVSEC2023/4BO.17.4
020292-004

materials are rapidly combustible and develop gaseous products when thermally degraded.

For this reason, the influence provided by the roofing sample during the fire test appears to be extremely important since it could actively contribute to the flame development depending on the characteristics of the material it is made of.

The products used on roofing can be non-combustible or, as with bituminous sheets, contribute to fire development.

When used in fire tests in combination with PV modules, these roofing materials can significantly influence the final PV fire classes and actual fire development.

To properly discriminate the sample's behavior consisting of the PV module and the roof element (external layer), the CEI TS 82-89 (17) uses the same test variables as those used in conducting the defined SBI test of EN13823 (25).

Starting on these parameters (THR and FIGRA[4]), it was possible to apply new reaction to fire criteria and classes of fire behavior. **Table *2*** summarizes the proposed values.

Table 2 PV roof samples fire rating criteria and classes expressed by CEI TS 82-89.

Classes	$B_{PV(A,B,C)}$	$C_{PV(A,B,C}$	$D_{PV(A,B,C}$	$E_{PV(A,B,C}$
$FIGRA_{0.4MJ}$ [W/s]	≤ 180	≤ 450	≤550	> 550
THR [MJ]	≤ 10	≤ 25	≤ 35	> 35

To obtain an initial validation of the test procedure expressed in CEI TS 82-89 (17), more than 50 samples were tested. All samples consisted of a PV module applied to a roof sample.

The roof's waterproof membranes were chosen from different materials (different chemical compositions) and, consequently, with different Gross Heat of Combustion (PCS). PV modules have been tested according to Italian methods, reaching class fire ratings 1 or 2.

Table *3* Some of the PV roof samples' testing results made according to CEI TS 82-89**Table 3Table *3*** reports the main parameters used during the test according to the CEI TS 82-89 (17) and shows the main features of the tested materials.

Table 3 also shows the PV roof sample class according to the CEI TS 82-89 (17) in the last row.

The parameters and features shown in **Table *3*** are the following:

A: Test protocol described in CEI TS 82-89 (17).

B: Burner HRR [kW].

C: PV module fire rating according to UNI 9177 (12).

D: Waterproof layer typology.

E: Roof membrane class according to EN 13-501-5 (10) and CEN TS 1187 (15).

F: Roof membrane Gross Heat of Combustion value (PCS) [MJ/m^2].

G: Gros heat of combustion value of waterproof membrane layer (PCS) [MJ/kg].

H: Damaged zone from CEN/TS 1187 (15) – T2 [cm].

I: THR in CEI TS 82-89 SBI test [MJ].

L: PV roof sample rating according to CEI TS 82-89 (17).

Table 3 Some of the PV roof samples' testing results made according to CEI TS 82-89.

	Test number					
	26	**27**	**28**	**29**	**30**	**31**
A	C	C	C	C	C	C
B	30	30	30	30	30	30
C	2	2	2	2	2	2
D	SBS	APP	SBS	APP	SBS	TPO
E	B_{roof}	B_{roof}	B_{roof}	F_{roof}	F_{roof}	B_{roof}
F	127,9	158,9	112,2	95,6	80,2	63,8
G	22	26	19	21	23	34
H	43,0	48,0	47,0	57,0	63,0	40,0
I	30,2	25,0	23,3	19,7	14,4	11,0
L	D_{PV}	D_{PV}	C_{PV}	C_{PV}	D_{PV}	C_{PV}
	Test number					
	32	**33**	**34**	**35**	**36**	**37**
A	C	C	C	C	C	C
B	30	30	30	30	30	30
C	2	2	2	2	2	1
D	TPO	PVC	TPO	TPO	TPO+g	Ca-S
E	B_{roof}	B_{roof}	B_{roof}	B_{roof}	B_{roof}*	B_{roof}**
F	54,9	57,9	53,2	47,1	0	0
G	25	24	21	24	0	0
H	38,0	41,0	37,0	33,0	0	0
I	8,3	7,3	6,5	5,1	3,0	1,0
L	B_{PV}	C_{PV}	B_{PV}	B_{PV}	B_{PV}	B_{PV}

* Roof sample with inert layer (EI30) on the upper side.
** Inert panel as roof sample.

Where:
SBS is a Styrene-Butadiene-Styrene membrane.
APP is an Atactic Polypropylene membrane.
TPO is a Thermoplastic Polyolefin membrane.
PVC is Polyvinyl Chloride membrane.
TPO+g is a Thermoplastic Polyolefin membrane with EI30 two Gypsum plasterboard of 12,5 mm on the upper side.
Ca-s is a Calcium Silicate sheet.

Although suitably modified, the SBI test chamber makes it possible to record the parameters that characterize the fire curve and describe the behavior of the photovoltaic sample conjoined to the roof during the test. The following parameters are thus recorded during the test:

- HRR (Heat Release Rate);
- THR (Total Heat Release);
- FIGRA (FIre Growth RAte);
- TSP (Total Smoke Production);

[4] In the SBI test, FIGRA is defined as the growth rate of the burning intensity (HRR), during a test. It is calculated as the maximum value of the function (heat release rate) / (elapsed test time). Its measured unit is W/s.

The Heat Release Rate (HRR) is defined as the fire rate release of energy (the rate at which fire releases energy). Defined also as firepower, its measured unit is Watt (W) or, in the international System unit, Joule per second (J/s).

10.4229/EUPVSEC2023/4BO.17.4
020292-005

- SMOGRA (SMOke Growth RAte).

Figure 11, **Figure 12**, and **Figure 13** report the recorder test graphs of the parameters HRR (30), THR (30) and FIGRA (30) in which it is possible to evaluate the different trends of the curves according to the type of sample tested shown in **Table 3.**

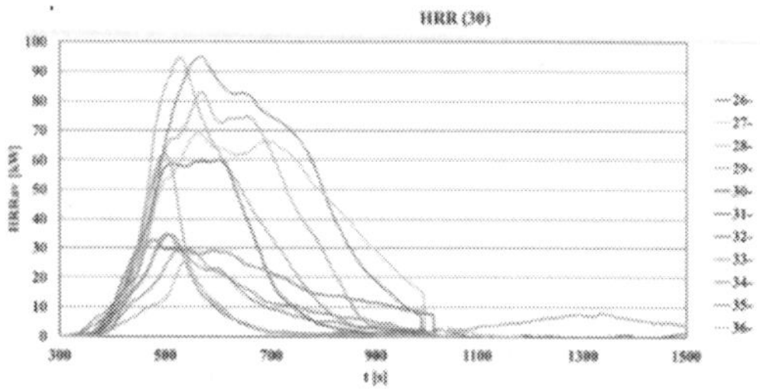

Figure 11 HRR curves recorded during the tests resumed in **Table 3**.

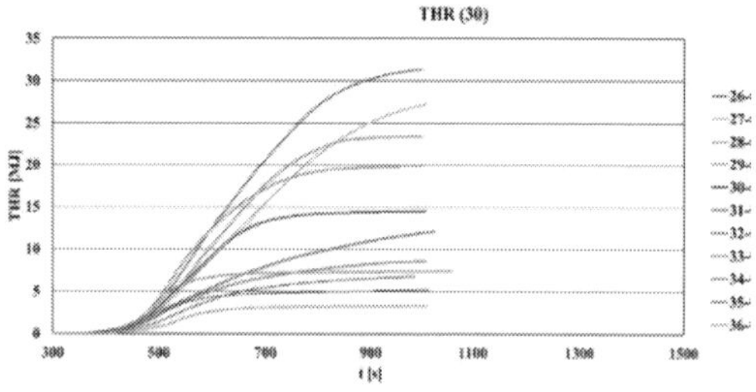

Figure 12 TRR curves recorded during the tests resumed in **Table 3**.

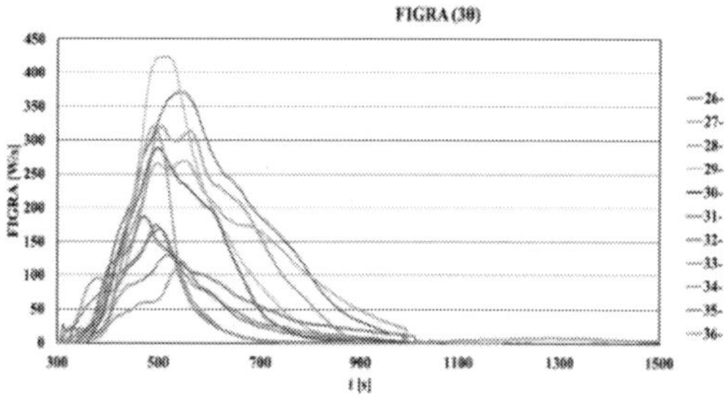

Figure 13 FIGRA curves recorded during the tests resumed in **Table 3**.

3.1 Discussion on results

As already written, a proper test of a construct unit sample (PV module with roof) can provide a more meaningful understanding of the fire risk of PV panels on a roof.

In fact, the materials that make up the roof can actively participate in combustion by altering the behavior obtained from the PV module alone.

In addition to materials, however, the influence of the installation features, such as module inclination and roof distance, must be considered, as well as specific features such as flame ignition, its power and duration, and the initial sample's deterioration.

All these conditions influence the sample's behavior and parameter values used in the test chamber.

Parameters such as FIGRA (growth rate of combustion intensity during a test), THR (total heat release) or parameters expressing smoke release (TSP, SMOGRA) can detect the behavior of the sample as a unit that differs from the individual behavior of the components.

Testing showed a good correlation between the Total Heat Release (THR) parameter of the test and the Gros Heat of Combustion value (PCS) of the waterproof membranes used (roof waterproofing layer), as shown in **Figure 14**.

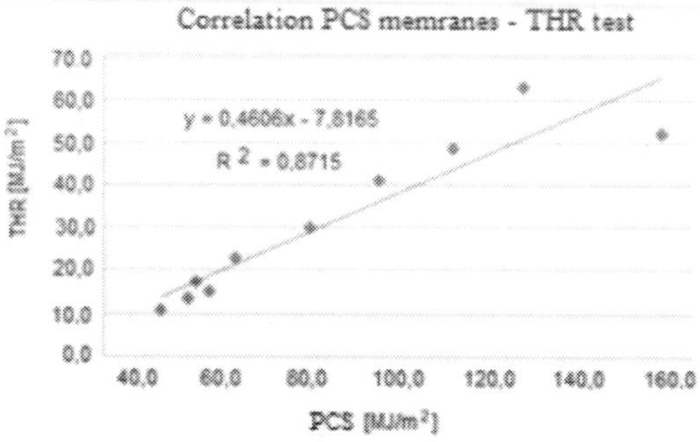

Figure 14 THR of PV roof sample correlated to PCS of the membrane waterproof.

4 CONCLUSIONS

This paper highlights the limits of the current European system for classifying rooftop fire scenarios when a PV installation is present.

Current testing protocols are based on a burning brand ignited on the top layer, to which the effect of wind or radiant heat can be added during the test. However, the application of PV modules changes the originally envisioned fire scenario.

Indeed, some authors agree that it is necessary to use a specific approach that can represent the actual fire scenario (18) (28), and a combined test seems to be the goal of any future work in this field (29).

The test procedure contained in CEI TS 82-89 (17) based on calorimetry was presented to overcome these missing gaps. This procedure allows the classification of the photovoltaic module considering the contribution of the roof part.

The experimental activity used existing test protocols and new tests with variants focused on PV characteristics. This activity made it possible to develop a procedure that allows verifying the behavior of photovoltaic modules in combination with different roofing materials. It helped identify weaknesses in the current European and Italian standards regarding the fire behavior assessment of photovoltaic modules installed on building roofs.

These new test protocols developed partly from those already listed in CLC TR 50670:2016 (16), aim to improve the identification of fire characteristics of PV modules more rigorously by analyzing the PV module and roof material assemblies. This, as also verified by the Solar America Board for Codes and Standards (18), (20) (30), turns out to be an essential characteristic.

As expressed above, the new fire resistance criteria described in CEI TS 82-89 (17) are based on calorimetry. The THR and FIGRA indices help to classify a range of PV modules that exhibit fire behavior when mounted on different roofs. Moreover, the correlation between THR and PCS of the roof waterproofing membrane can help extend the classification to other possible roof

10.4229/EUPVSEC2023/4BO.17.4
020292-006

configurations with a lower PCS in MJ/m^2.

7 ACKNOWLEDGMENTS

RSE S.p.A. carried out the experimental program in collaboration with:

- SUPSI-ISAAC (University of Applied Sciences and Arts of Southern Switzerland – Institute of Applied Sustainability to the Built Environment, Mendrisio. Switzerland), which provided consulting tips about electrical aspects and samples of PV modules;
- Istituto Giordano S.p.A. (Reaction to Fire Lab, Gatteo. Italy), which strongly supported the design and development of the protocol's maintenance and testing activities.

8 REFERENCES

1. **Commission, European.** Energy performance of buildings directive. *An official website of the European Union.* [Online] https://energy.ec.europa.eu/topics/energy-efficiency/energy-efficient-buildings/energy-performance-buildings-directive_en.
2. **Solar Power Europe.** EU Market Outlook for solar power 2022-2026. December 2022.
3. **Nur Aliah Fatin Mohd Nizam Ong, Muhammad Adnan Sadiq, Mohamad Syazarudin Md Said, Grunde Jomaas, Mohd Zahirasri Mohd Tohir, Jens Steemann Kristensen.** Journal of Building Engineering. *Fault tree analysis of fires on rooftops with photovoltaic systems.* 1 April 2022. Vol. 46, 103752.
4. **Piergiacomo, Cancelliere - Ministero dell'Interno, Dipartimento dei Vigili del Fuoco, del Soccorso pubblico e della Difesa Civile - Direzione centrale per la prevenzione e sicurezza tecnica.** Guida per installazione impianti FV. 2012.
5. —. Circolare 04/05/2012. *Soluzioni tecniche per la conciliare la presenza di moduli fotovoltaicie relativi accessori con coperture realizzate in materiali combustibili.* 2012. Ordine degli Ingegneri della Provincia di Vicenza.
6. **(IEC), International Electrotechnical Commission.** IEC 61730-1:2016 . *Photovoltaic (PV) module safety qualification - Part 1: Requirements for construction.* 2016.
7. —. IEC 61730-2:2016. *Photovoltaic (PV) module safety qualification - Part 2: Requirements for testing.* 2016.
8. **(UL), Underwriters Laboratories.** UL 1703. *Standard for Flat-Plate Photovoltaic Modules and Panels (Ed. 3).* 2002.
9. **Rebecca Yang, Yukun Zang, Jiaqi Yang, Ron Wakefield, Kate Nguyen, Long Shi, Bambang Trigunarsyah, Fabio Parolini, Pierluigi Bonomo, Francesco Frontini, Dahai Qi, Yoon Ko, Xin Deng.** Renewable and Sustainable Energy Reviews. *Fire safety requirements for building integrated photovoltaics (BIPV): A cross-country comparison.* 2022.
10. **(CEN), European Committee for Standardization.** EN 13501-5:2016. *Fire classification of construction products and building elements - Part 5: Classification using data from external fire exposure to roofs tests.* 2016.
11. **Standardization, International Organization for.** EN ISO 11925-2:2020. *Reaction to fire tests — Ignitability of products subjected to direct impingement of flame — Part 2: Single-flame source test.* 2020.
12. **(UNI), Ente Nazionale Italiano di Unificazione.** UNI 9177. *Reaction to Fire - Flammable Product Classification.* 2008.
13. **Ministero dell'Interno Dipartimento dei Vigili del Fuoco, del Soccorso Pubblico e della Difesa Civile (VVF).** Testo coordinato della Nota 07 febbraio 2012 . *Guida per l'installazione degli impianti FV – Edizione anno 2012* . 2012.
14. **(UNI), Ente Nazionale Italiano di Unificazione.** UNI CEN/TS 1187:2012. *Metodi di prova per tetti esposti al fuoco dall'esterno.* 2012.
15. **(CEN), European Committee for Standardization.** CEN/TS 1187:2012. *est methods for external fire exposure to roofs.* 2012.
16. **(CENELEC), European Committee for Electrotechnical Standardization.** CLC/TR 50670. *External fire exposure to roofs in combination with photovoltaic (PV) arrays - Test method(s).*
17. **(CEI), Comitato Elettrotecnico Italiano.** CEI TS 82-89:2023. *Rischio d'incendio nei sistemi fotovoltaici - Comportamento all'incendio dei moduli fotovoltaici installati su coperture di edifici: protocolli di prova e criteri di classificazione.* 2023.
18. **Andrew Rosenthal, Larry Sherwood, Bill Brooks, ravinray Ghandi, and Bob Backstrom.** Solar America Board for Codes and Standards Interim Report. *Flammability Testing of Standard Roofing Products In the Presence of Standoff-mounted Photovoltaic Modules.* 2010.
19. **(UL), Underwriters Laboratories.** UL 790. *Standard Test Methods for Fire Tests of Roof Coverings (Ed. 9).* 2022.
20. **Bob Backstrom, Dwayne Sloan.** UL Project Number: – 11CA43479 . *Report of Experiments of Minimum Gap and Flashing for Rack Mounted Photovoltaic Modules Phase 4* . 2012.
21. —. UL Project Number: 10CA49953 . *Effect of Rack Mounted Photovoltaic Modules on the Fire Classification Rating of Roofing Assemblies Phase 2* . 2012.
22. **Giovanni Manzini, Pasqualino Gramazio, Salvatore Guastella, Claudio Liciotti, Gian Luigi Baffoni.** Energy Procedia. *The Fire Risk in Photovoltaic Installations - Test Protocols For Fire Behavior of PV Modules.* s.l. : ELSEVIER, 2015. Vol. 82.
23. **Piergiacomo Cancelliere, Giovanni Manzini, Giombattista Traina, Marco Gabriele Cavriani.** Fire Safety Journal. *PV modules on buildings – Outlines of PV roof samples fire rating assessment.* 2021.
24. **European Commission.** Construction Product Regulation. *Regulation (EU) No 305/2011 of the European Parliament and of the Council of 9 March 2011 laying down harmonised conditions for the marketing of construction products and repealing Council Directive 89/106/EEC Text with EEA relevance.* 2011.
25. **(CEN), European Committee for Standardization.** EN 13823:2020+A1:2022. *Reaction to fire tests for building products - Building products excluding floorings exposed to the thermal attack by a single burning item.* 2023.
26. **Bellini, Emiliano.** PV Magazine. *PV system fires potentially exacerbated by gap between solar panels, rooftops.* December 8, 2022.
27. **(ISO), International Organization for Standardization.** ISO 1716:2018. *Reaction to fire tests*

for products — Determination of the gross heat of combustion (calorific value). 2018.
28. *Experimental study of flame spread underneath photovoltaic (PV) modules.* **Jens Steemann Kristensen, Farah Binte Mohd Faudzi, Grunde Jomaas.** May 2020, Fire safety Journal 120(9):103027.
29. *Experimental Study of the Fire Dynamics in a Semi-enclosure Formed by Photovoltaic (PV) Installations on Flat Roof Constructions.* **Jens Steemann Kristensen, Benjamin Jacobs, Grunde Jomaas,.** 2022, Fire Technology.
30. **Bob Backstrom, Dwayne Sloan, Pravinray Gandhi.** Project Number: – 11CA43479 . *Considerations of Module Position on Roof Deck During Spread of Flame Tests Phase 5* . July 24, 2012.
31. **EOTA.** European Assesment Document. *EAD 090062-00-0404 Kits for external wall claddings mechanically fixied.* 2018.

THE ROLE OF PV WITHIN NEW EU BUILDING TARGETS TO ACHIEVE CLIMATE NEUTRALITY

Delia D'Agostino[1], Carmen Maduta[1], Luca Castellazzi[1], Robert Kenny[1]

[1]Joint Research Centre, European Commission

delia.dagostino@ec.europa.eu; carmen.maduta@ec.europa.eu; luca.castellazzi@ec.europa.eu; robert.kenny@ec.europa.eu

ABSTRACT

Buildings are a key sector to bring down greenhouse gas (GHG) emissions in the European Union (EU). Recent policy developments highlighted the need to accelerate the clean energy transition and boost renewable energy production, especially photovoltaic energy (PV). With this aim, new building targets have been launched in recent EU energy and climate policies. As established in the recast of the Energy Performance of Buildings Directive (EPBD), nearly-zero energy buildings (NZEBs) are the building target since 2021 for new buildings. Furthermore, a new target as of 2030, namely zero-emission buildings (ZEBs), has been proposed at the end of 2021 in the revised EPBD and it is under discussion. Although PV plays a crucial role in both building types, differences can be found depending on methodological issues, established boundaries, and calculation methods. Starting from an overview of the status and progress of the topic, this paper clarifies the role of PV in NZEBs and ZEBs, providing insights into the renewable contribution needed to reach the new building targets in the EU. It offers an analysis of key existing and planned policies and measures that tackle PV in buildings to speed up the green transition and reduce energy dependence. Coordinating energy efficiency upgrades with PV installation reduces building environmental impact in terms of energy use and emissions. The analysis shows how PV is an essential pillar of the EU strategy towards climate neutrality.

Keywords: Energy efficiency; photovoltaic systems; building decarbonization; climate neutrality; energy policy.

1 INTRODUCTION

Buildings account for about 40% of global energy consumption, and 36% of associated CO_2 emissions [1]. In line with the EU Fit for 55 package, the REPowerEU Plan set the objective to raise the renewable energy target to 45% with the objective to cut GHG emissions by 55% by 2030 [2] [3].

An important step towards a decarbonised building stock is defined under the Energy Performance of Buildings Directive 2010/31/EU (EPBD) and focus on setting cost-optimal minimum energy performance requirements, promoting high energy efficient buildings and developing long-term renovation strategies with the view of mobilising energy efficiency investments in the buildings sector [4]. As of 2021, all new buildings are Nearly-Zero Energy Buildings (NZEBs). The broad NZEB definition is given in the EPBD, but NZEB indicators vary widely across the Member States, reflecting national, regional, or local conditions.

NZEBs can be achieved by minimizing the energy needs through energy-efficient measures and adopt renewable sources to fulfil the remaining demand [10]. A common point is the advantage provided in terms of carbon emission reduction, strengthening to be a promising answer to the energy and environmental concerns linked to the sector [5]. However, different terms and building concepts have been globally established over last decade [6].

PV provides convenient and clean energy that can be produced on-site, reducing energy imports to meet the building needs [7]. Moreover, the European Solar Rooftop Initiative proposes a legally binding EU solar rooftop obligation for certain building categories [8]. As proposed in the EPBD revision [9], all new buildings need to optimize their solar energy generation potential. At the same time, the need to electrify energy demand to facilitate GHG emission reductions, and reduce climate change warming impacts, makes it important to increase PV production in NZEBs and ZEBs.

2 METHODS

This paper offers an overview of the recent EU policy developments and examine the provisions related to photovoltaic (PV) technologies and buildings. Both NZEBs and ZEBs are defined as buildings of very high energy performance [10] [11]. However, in NZEBs the nearly zero or very low amount of energy required is covered to a very significant extent by energy from renewable sources produced on-site or nearby, while in ZEBs the very low amount of energy required is fully covered by renewable energy generated on-site, from renewable energy communities, or district heating and cooling systems [12]. Therefore, while renewable energy plays a central role in NZEB, its significance is decisive in achieving the proposed building target of ZEB. Methodological issues, boundaries, and calculation methods are discussed to clarify the PV role within both EU building targets [13]. A particular focus is given to PV in the refurbishment of existing buildings as this represents a priority in the EU. Rooftop PV arrays also coupled with electrical storage serve as a mean for addressing building energy use in a sustainable approach, since roof areas are

10.4229/EUPVSEC2023/4BO.17.5

often unobstructed to solar radiation and freely available for such utilization.
However, the role of PV within the new EU building targets has received limited attention within the scientific community so far [14]. This analysis assembles relevant studies to comprehensively assess the PV contribution in NZEBs and ZEBs in the EU within a future perspective. The research is of utmost importance as buildings, the largest energy-using sector, offer a large untapped potential for PV deployment to reduce carbon emissions and, at the same time, achieve additional benefits.

3 RESULTS

PV plays a crucial role in the new EU building targets aimed at achieving climate neutrality [15]. These targets prioritize the reduction of greenhouse gas emissions and the transition to renewable energy sources, and PV is a key technology for achieving these goals in the building sector.
Here are the main ways through which PV contributes to the attainment of EU building targets:
1. Carbon emissions reduction: The EU building targets aim to reduce carbon emissions from the building sector. PV systems generate electricity from sunlight without any direct carbon emissions, thereby helping to reduce the sector's operational carbon footprint.
2. Renewable energy generation: PV is a proven and widely available renewable energy technology. By installing PV systems on buildings, renewable energy generation can be increased, reducing the reliance on fossil fuel-based electricity and contributing to the overall decarbonisation of the energy sector. Moreover, the energy surplus can be supplied to the grid, allowing consumers to offset emissions produced using grid-supplied electricity. Where the offsetting of PV electricity is high, because the carbon intensity of the electricity grid is high, excess generation creates a large carbon credit that can offset carbon emissions from heat demand and embodied carbon [19]. Where the electricity demand is likely to exceed PV generation, and the electricity grid has a high carbon intensity, this translates into a significant carbon debt which is unlikely to be offset entirely by negative embodied carbon.
3. Energy self-sufficiency: PV systems can enable buildings to produce their own electricity, thereby reducing their dependence on the grid. This helps to create more resilient and sustainable buildings and communities, as well as reducing energy costs for building owners and occupants.
4. Energy efficiency: PV systems are often integrated with energy-efficient building design and technologies, such as improved thermal insulation, LED lighting, heat pumps, and smart energy management systems. This integrated approach optimizes energy use in buildings, enhancing their energy efficiency and further contributing to climate neutrality goals.

5. Job creation and economic growth: The widespread adoption of PV technology in the building sector can lead to increased job opportunities in both the construction and the renewable energy industries. This supports the EU's goal of creating a sustainable and resilient economy while contributing to local economic growth.

To achieve climate neutrality, PV technology needs to be integrated into building design and construction practices. This requires supportive policies, financial incentives, and streamlined administrative procedures to encourage the installation of PV systems on both new and existing buildings. Additionally, research and development efforts are necessary to continue improving the performance, efficiency, and cost-effectiveness of PV technology in order to make it a viable solution for achieving climate neutrality in the building sector.
The EU has implemented several policies and initiatives to promote the installation of PV systems on buildings. Some of EU main policies and initiatives are:
1. Renewable Energy Directive (RED): This directive sets binding targets for EU member states to increase the share of renewable energy in the overall energy consumption. The current target is 32% by 2030, and PV installation on buildings is a crucial element in achieving this goal.
2. Energy Performance of Buildings Directive (EPBD): The EPBD sets energy performance standards for buildings, including requirements for the use of renewable energy. Member states are required to establish measures to promote the use of renewable energy, including PV, in new and existing buildings.
3. Smart Finance for Smart Buildings initiative: This initiative aims to mobilize private investment for energy efficiency and renewable energy in buildings, including PV systems. It provides support through technical assistance, financial instruments, and capacity building to help overcome barriers to financing PV installations.
4. European Regional Development Fund (ERDF): The ERDF provides financial support for sustainable urban development, including investments in renewable energy and energy efficiency in buildings. Many Member States utilize the ERDF to fund PV installations on buildings.
5. European Investment Bank (EIB) initiatives: The EIB provides financing for sustainable energy projects, including PV installations on buildings. The EIB's various initiatives, such as the Energy Efficiency Investment Fund and the InnovFin Energy Demonstration Projects, offer support for PV projects in the building sector.
These policies, along with other national and regional initiatives, aim to create an enabling environment for the installation of PV systems on buildings. They provide regulatory frameworks, financial incentives, and technical support to promote the widespread adoption of PV technology and contribute to the EU's climate and energy goals.
By 15 March 2023, Member States were due to report for the first time in an integrated manner on their progress towards implementing their 2020 National Energy and Climate Plans (NECPs)[1]. This reporting covered, notably, progress towards their objectives, targets and contributions across the five dimensions of the Energy Union as well as on greenhouse gas emissions and removals. It also addressed the implementation or amendment of Member States' policies and measures including their financing. In relation to the promotion of PV on buildings, some key measures identified are:

- Denmark:

Enacted easures to support access to PV energy for poor and vulnerable households, implements actions to remove

[1] Introduced by the by the Governance Regulation (EU)2018/1999.

10.4229/EUPVSEC2023/4BO.17.5
020294-002

barriers to innovation in solar energy deployment and planned a measure to ensure that new buildings are solar ready.

- Italy:

Has in place measures on tax deduction for PV on roofs.

- Cyprus:

Runs a grant scheme for PV charging of electric vehicles at homes and enacted exemptions in obtaining permits for installing rooftop PV.

- Hungary:

Offers non-reimbursable funds for installation of residential PV and implemented a programme for the electrification of heating system in combination with PV system.

- Sweden:

Removed the building permit requirements for solar installations, applies tax reductions and implemented programme to support and guide PV installation.

In terms of most implemented renewable technologies, the Mediterranean climate benefits from higher solar radiation, so that the more convenient sources are solar thermal and PV, while in colder climate countries, e.g., in northern Europe, renewable sources such as geothermal energy and biomass used in heating systems are common. In addition to traditional roof-installed PV, building-integrated photovoltaic systems (BIPV) can be a key player both in existing and new buildings [17]. PV could provide the electricity needed to power a heat pump for heating, cooling and DHW. A secondary generation system, such as a biomass boiler, could support heat pumps in winter.

In relation to the type of energy generation, various options exist concerning renewable energy supply location [18]:

- Generation within the building footprint (e.g. PV and solar thermal installed on the rooftop or on building facades);
- On-site generation from on-site renewables: use of renewable energy sources available at the site (e.g. PV on parking lots);
- On-site generation from off-site renewables: use of renewable energy sources available off site to generate energy on site (e.g. biomass or wood pellets that can be imported);
- Off-site generation: investments in off-site renewable energy production, such as wind turbines;
- Off-site supply: purchase of green energy.

The most used technologies in Member States, apart from PV, are: solar thermal, air- and ground-source heat pumps, geothermal, passive solar, passive cooling, wind power, biomass, biofuel, micro CHP, and heat recovery.

4 CONCLUSIONS

The analysis shows how PV is an essential pillar of the EU strategy towards climate neutrality. The synergy of energy efficiency upgrades and PV installation can significantly reduce the building's operational impact (energy and emissions) while increasing renewable energy self-consumption and additional benefits, such as health and safety, also increases the resilience of EU energy systems. Boosting the roll-out of PV appears of paramount importance to achieve a decarbonised EU building stock by 2050.

Over the past decade, the falling costs of PV and the competing cost of generated electricity made this technology more attractive. The potential rooftop solar PV share in the final electricity consumption varies greatly across European Member States, from over 40% in Malta, Cyprus and Portugal, between 30% and 40% in Italy and Greece, and between 20% and 30% in France, Spain and Germany. In existing buildings, installing PV systems with electrical storage and insulating roofs can provide a cost-effective way to improve the thermal performance, while covering a large portion (55-80%) of annual energy and electrical needs. A joint effort from the policy sector and a massive contribution from buildings is crucial to achieve a decarbonised EU stock by 2050.

ACKNOWLEDGMENTS

The views expressed are purely those of the authors and do not represent any official position of the European Commission. The authors wish to thank Christian Thiel and Maurizio Bavetta (JRC) for supporting this research.

REFERENCES

[1] Economidou M, Todeschi V, Bertoldi P, D'Agostino D, Zangheri P, Castellazzi L. Review of 50 years of EU energy efficiency policies for buildings. Energy and Buildings. 2020;225:110322.

[2] P.R. Shukla, J. Skea, R. Slade, A. Al Khourdajie, R. van Diemen, D. McCollum, et al. IPCC, 2022: Summary for Policymakers. In: Climate Change 2022: Mitigation of Climate Change. Contribution of Working Group III to the Sixth Assessment Report of the Intergovernmental Panel on Climate Change Cambridge University Press, Cambridge, UK and New York, NY, USA. doi: 10.1017/9781009157926.001.

[3] EU 2022, REPowerEU: Joint European Action for more affordable, secure and sustainable energy, COM(2022) 108 final, Available at: https://eur-lex.europa.eu/resource.html?uri=cellar:71767319-9f0a-11ec-83e1-01aa75ed71a1.0001.02/DOC_1&format=PDF.

[4] European Parliament (2010). EPBD recast, Directive 2010/31/EU of the European Parliament and of the Council of 19 May 2010 on the energy performance of buildings (recast). Official Journal of the European Union, 18.6.2010, L 153/13.

[5] D'Agostino D., Mazzarella L. (2019). What is a Nearly zero energy building? Overview, implementation and comparison of definitions, Journal of Building Engineering; 21:200–212, https://doi.org/10.1016/j.jobe.2018.10.019.

[6] D'Agostino D, Tzeiranaki ST, Zangheri P, Bertoldi P. Assessing Nearly Zero Energy Buildings (NZEBs) development in Europe. Energy Strategy Reviews. 2021;36:100680.Delia nzenb

[7] Attia, S., Eleftheriou, P., Xeni, F., Morlot, R., Ménézo, C., Kostopoulos, V., Betsi, M., Kalaitzoglou, I., Pagliano, L., Cellura, M., Almeida, M., Ferreira, M., Baracu, T., Badescu, V., Crutescu, R., Hidalgo-Betanzos, J. M. (2017). Overview and future challenges of nearly zero energy buildings (nZEB) design in Southern Europe. Energy and Buildings, 155, 439–458. https://doi.org/10.1016/J.ENBUILD.2017.09.043

[8] V. Kapsalis, D. Karamanis, On the effect of roof added photovoltaics on building's energy demand, Energy Build. 108 (2015) 195-204. https://doi.org/10.1016/j.enbuild.2015.09.016
[9] Revision of the Directive of the European Parliament and of the Council on the use of energy from renewable sources (recast), L 328/82, 2018/2001. Available at https://eur-lex.europa.eu/legal-content/EN/TXT/PDF/?uri=CELEX:32018L2001&from=EN
[10] D'Agostino Delia, Parker Danny, Melià Paco, Giovanni Dotelli, Optimizing photovoltaic electric generation and roof insulation in existing residential buildings, Energy and Buildings, Volume 255, 2022.
[11] EU 2021, Revision of the Directive of the European Parliament and of the Council on the energy performance of buildings (recast), COM(2021) 802 final. Available at https://ec.europa.eu/energy/sites/default/files/proposal-recast-energy-performance-buildings-directive.pdf
[12] V. Kapsalis, D. Karamanis, On the effect of roof added photovoltaics on building's energy demand, Energy Build. 108 (2015) 195-204. https://doi.org/10.1016/j.enbuild.2015.09.016
[13] D. D'Agostino, D. Parker, I. Epifani, D. Crawley, L. Lawrie, How will future climate impact the design and performance of nearly zero energy buildings (NZEBs)?, Energy, Volume 240, 2022.
[14] D'Agostino, D., Parker, D., Melià, P., Dotelli, G., Data on roof renovation and photovoltaic energy production including energy storage in existing residential buildings , Data in Brief 41,107874
[15] V. Kapsalis, D. Karamanis, On the effect of roof added photovoltaics on building's energy demand, Energy Build. 108 (2015) 195-204. https://doi.org/10.1016/j.enbuild.2015.09.016
[16] Communication from the Commission to the European Parliament, the European Council, the Council, the European Economic and Social Committee and the Committee of the Regions, the European Green Deal, COM(2019) 640, 11.12.2019, https://eur-lex.europa.eu/legal-content/EN/TXT/?uri=COM%3A2019%3A640%3AFIN
[17] Martín-Chivelet N, Kapsis K, Rose Wilson H, Delisle V, Yang R, Olivier L, et al. Building-Integrated Photovoltaic (BIPV) products and systems: A review of energy-related behavior.Energy and Buildings. 2022:111998.
[18] Marszal, A.J., Heiselberg, P., Bourrelle, J.S., Musall, E., Voss, K., Sartori, I., Napolitano, A., 2011. Zero Energy Building – A review of definitions and calculation methodologies. Energy and Buildings 43, 971–979. https://doi.org/10.1016/j.enbuild.2010.12.022
[19] Parkin, A., Herrera, M., Coley, D.A., 2020. Net-zero buildings: when carbon and energy metrics diverge. Buildings and Cities 1, 86–99. https://doi.org/10.5334/bc.27

ANALYSIS OF EFFECT FOR BIFACIAL MODULE IN FLOATING PHOTOVOLTAIC

Jieun Lee, Hyunsik Jo, Jiyeon Im, Hyun Jun Kim
K-water Research
125, Yuseong-daero 1689beon-gil, Yuseong-gu, Daejeon, Republic of Korea
Email: jelee@kwater.or.kr

ABSTRACT:
Bifacial modules have an optical advantage of greater light collection over mono-facial modules. For this reason, there has been a growing interest in bifacial photovoltaic (PV) technology because it promises higher performance yield and lower levelized cost of energy (LCOE) compared to mono-facial PV technology. In this study, the effect of the bifacial module compared with mono-facial module of the floating photovoltaic (FPV) which is located by valley of reservoir nearby Dam in South Korea are described. The system is consisted of mono-facial module, bifacial modules, inverter for checking power of each module and insolation for measuring irradiation and reflection of water. With a result, during 8 months, power generation of bifacial PV are 359.75 kW which are higher 6.69~7.76% than that of mono-facial PV. Bifacial gain had 2% with only water reflection and 4~6% with internal reflection and water reflection.

Keyword: Photovoltaic, Floating PV System, Bifacial PV, Albedo, Water reflection,

1 INTRODUCTION

There are expended to develop the renewable energy according to policy which is reduction of Carbon in South Korea. Photovoltaic which is one of the renewable energy accounts for about 70% [1]. But there is limit to increase of PV system due to topography feature in South Korea, in which FPV(Floating Photovoltaic) is actively used to overcome the topography feature. Additionally, Bifacial photovoltaic technology can be improved the power of module with using front and rear side both. There are result that bifacial module is improved 12.5~14.5% in ground mounted PV system [2,3]. Even more, levelized cost of electricity (LCOE) is lower comparing with mono-facial PV. According to this benefit, bifacial PV installation is increased rapidly from 97 MWp in 2016 to over 5GWp in 2019[4]. The market share of bifacial technology is expected to increase from ~15% in 2019 to ~71% by 2030 [5]. Bifacial PV for floating system was higher value compared with mono-facial [6] and even higher 12.5% with higher supporting height from water and reflection [7]. However, there are less study which is bifacial gain according to water reflect etc. for floating bifacial PV technology. Various altitude of solar depending on the season in South Korea would be have positive effect to improve the bifacial PV due to solar diffusion in water.

In this paper, FPV system consisted with mono-facial module, bifacial module with same power. It will be figured out the effect of bifacial module located in valley which is normal reservoir in South Korea to analyze with bifacial gain and reflection of water for mono-facial and bifacial module, front and rear irradiance and albedo according to time period.

2 EXPERIMENTAL FOR FLOATING PV

2.1 Location

Reservoir located in Chungju Dam, South Korea, was selected for Floating bifacial PV test (Figure 1). The latitude is 36.92°N, and the longitude is 128.06 °W. This test site which is around valley which are general place of floating PV system in South Korea.

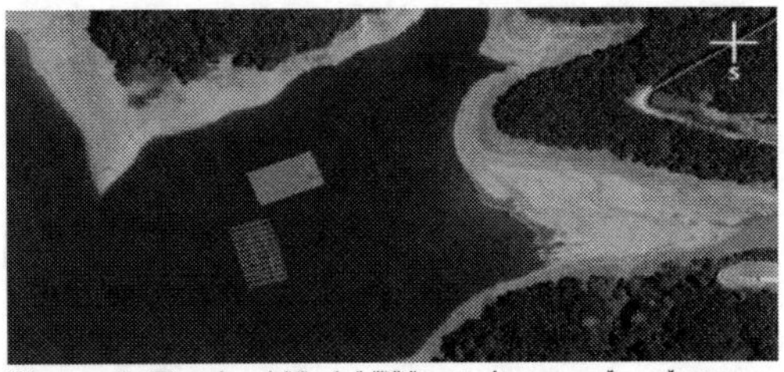

Figure 1: Floating bifacial PV test site was place between valley topography

2.2 Experiment Configuration

For confirming the effect of bifacial in floating, the experiment consists of 1pcs mono-facial module and 3pcs of bifacial modules. Maximum power of modules is same as 445W of mono-facial and bifacial which is only different backside material of module, glass and white backsheet. The layout of test is shown in Figure 2, where the installation angle is 30°. Temperature for module/air and front and rear irradiance with sensor was consisted in system (Figure 3). Water reflection are noticed by Albedo which is defined as the ratio of reflected front incident irradiance which is shown in Figure 3 and can be calculated by the following equation.

$$\text{Albedo} = \frac{\textit{Rear irradiance}}{\textit{Front irradiance}}$$

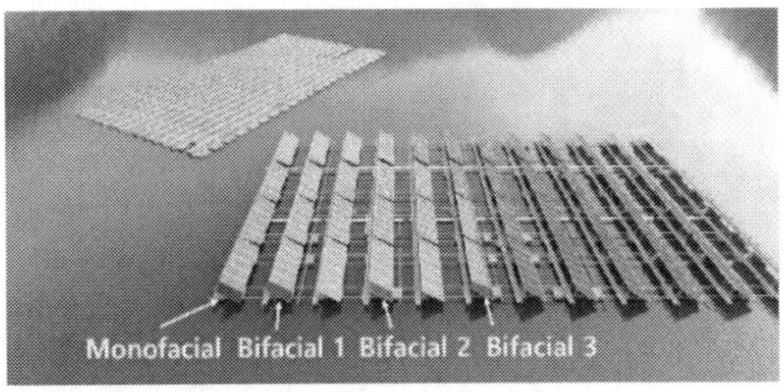

Figure 2: Test layout of Floating bifacial PV. Test modules was side by side in the Floating PV system.

10.4229/EUPVSEC2023/4BV.3.4
020310-001

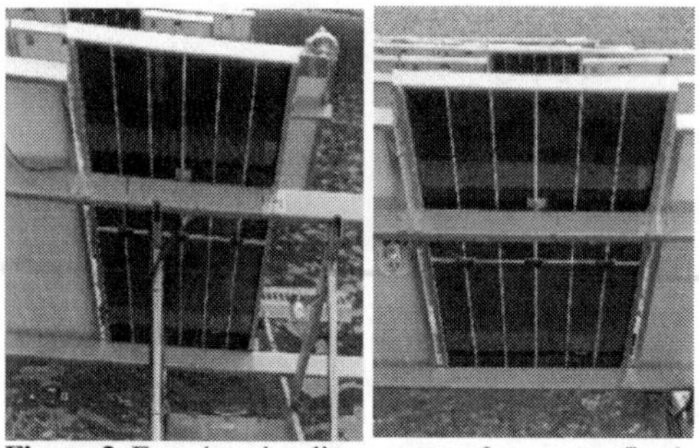

Figure 3: Front/rear irradiance sensor for water reflection and temperature in test system

3 RESULT AND DISCUSSION

3.1 Power generation of Bifacial PV

Figure 4 shows that comparison of cumulative power generation of mono-facial and bifacial PV from '22 Dec. to '23 Aug. Power generation of mono-facial is 333.83 kWh and that of bifacial are 357.08~359.75 kWh. It can be seen the bifacial gains as 6.96% ~7.76%.

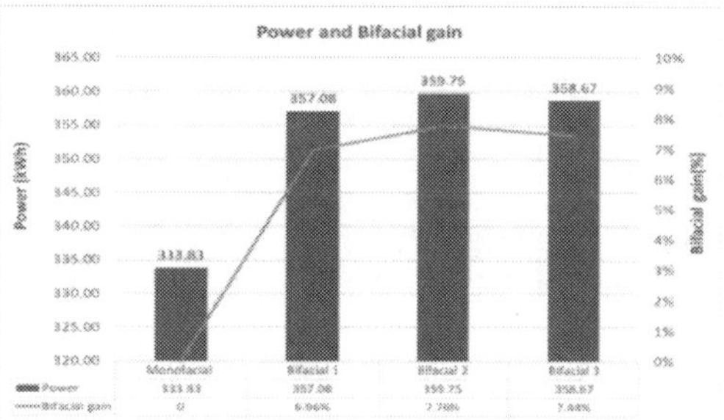

Figure 4: Power generation of bifacial and mono-facial PV from '22 Dec to '23 Aug.

To investigate more detail, power generation of mono-facial/bifacial and bifacial gain are shown with time period from '22 Dec. to '23 Aug (Figure 5). Although it has variation power generation with each day due to weather condition, there are trend that power generation of winter has low and increased to May and then slightly decrease the power generation from June to August. on the other hands, bifacial gain in winter ('22 Dec.~'23 Jan.) has around 8% and that of spring and summer is around 5%.

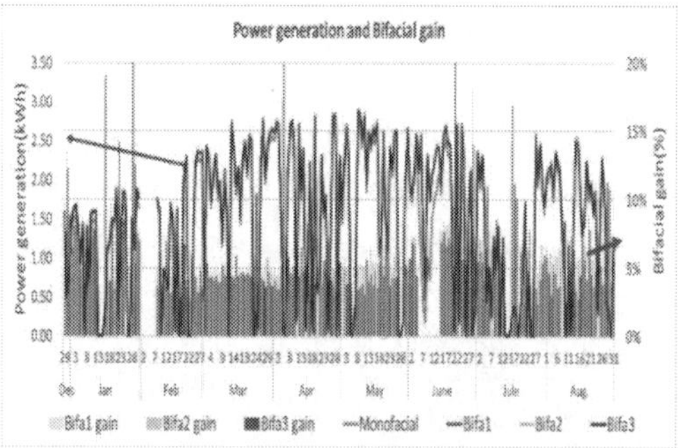

Figure 5: Power generation of mono-facial and bifacial with bifacial gain since '22 Dec with time period

The reason of increase the power generation from Dec. to May are due to increase the latitude of solar and irradiance (Figure 6). In June, on the other hands, the power generation was slightly decrease. It is expected that is low irradiance and higher temperature compared with in May despite high latitude of solar. After then, the rainy season during summer and higher module temp. make decrease the power generation.

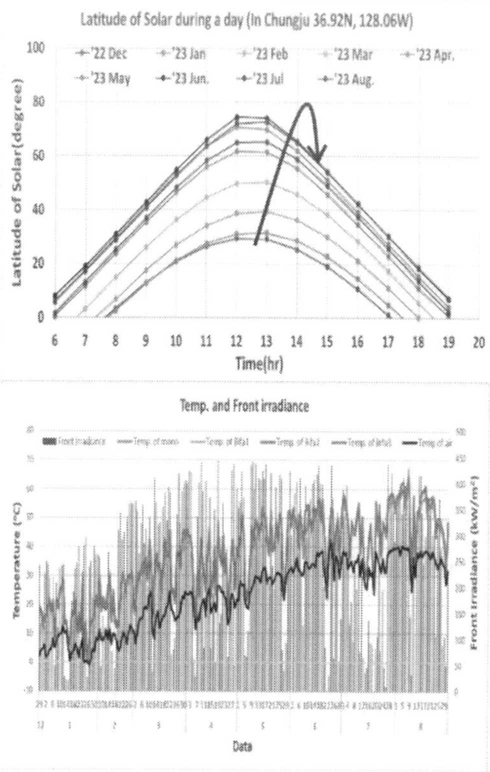

Figure 6: Solar latitude(upward) according to month and module, air temperature and irradiance (downward) with time period

Figure 7 shows the power variation during day and compared each month with mono-facial and bifacial. It can be known the reason that bifacial gain was different with winter and spring /summer. In winter season ('22 Dec.~'23 Jan.), power of mono-facial are lower than usual that expected in the morning (10~11AM) which is difference of 7~9% compared with bifacial power. It needs be had under discussion to figure it out, but one of the reasons was expected due to lower solar latitude (below 30degree in Figure 6 upward) in winter and topography features of the location. The test construction is placed in valley (in Figure 1) which would be made long length of shadow module during sunrise. The mono-facial module would be affected to the shadow. After 11AM and since March which has higher latitude of Solar above 30degree, there are no power degradation of mono-facial module in early morning.

10.4229/EUPVSEC2023/4BV.3.4
020310-002

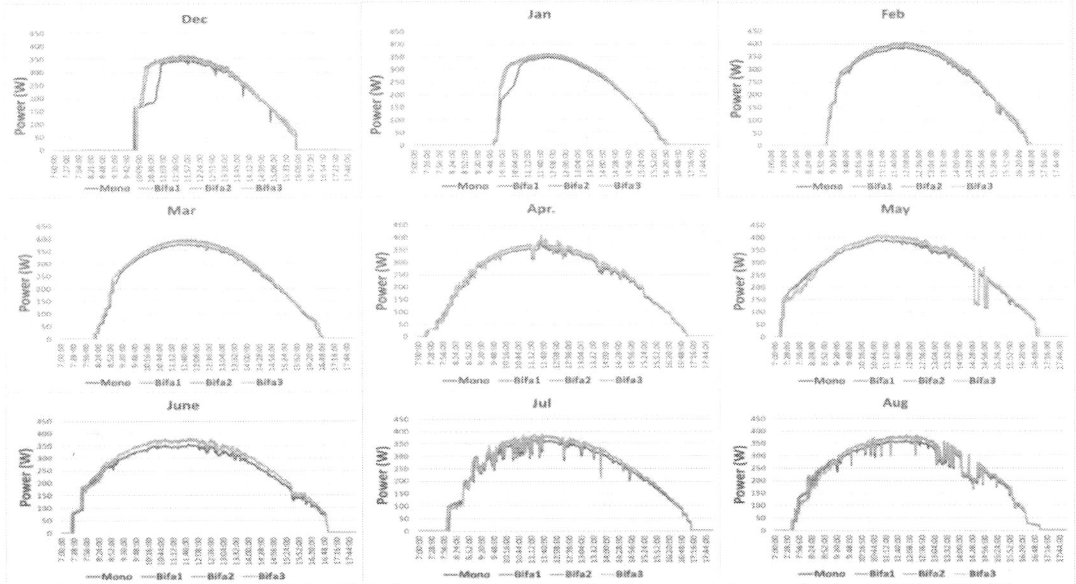

Figure 7: Comparison of mono-facial and bifacial power generation with time period during a day from ´22 Dec. To ´23 Aug.

3.2 Relation of irradiance with bifacial PV power

Figure 8 shows the front irradiance verse power generation of mono-facial and bifacial modules. Power generation of bifacial has higher slope (0.4067~0.4116) compared with that of mono-facial (0.383) as 6.10~7.38%. With this result, it can be known that bifacial effect with same irradiance has 6.10~7.38%.

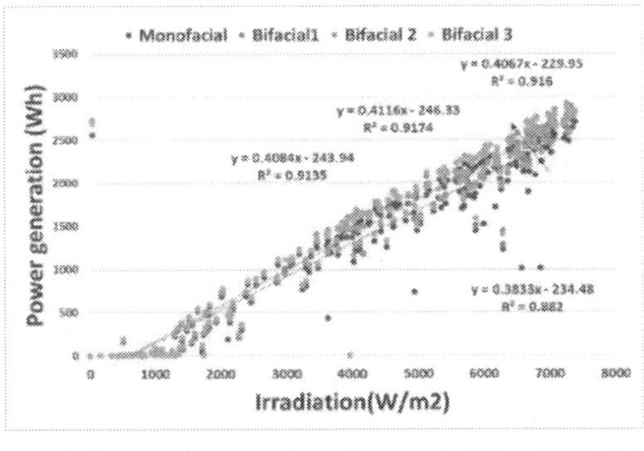

Figure 8: Power generation of mono-facial and bifacial comparison with irradiance

Performance Ratio (PR) is similar result with Figure 8. Performance Ratio calculate with below equation with actual power generation, irradiance, area of PV module and PV efficiency.

$$PR = \frac{\text{Actual Power Generated(kWh)}}{\text{Irradiance} \times \text{Active Arear of PV module(m2)} \times \text{PV module Eff}}$$

Figure 9 is shown PR of mono-facial and bifacial PV with time period from Dec. to Aug. PR bifacial and mono-facial are kept max. 90% and 85% from Dec. to Mar. then, it is decrease around 80% and 75% after April.

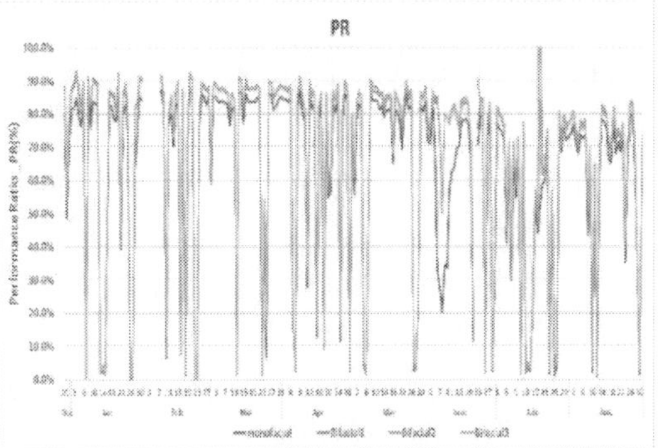

Figure 9: Performance ratio (PR) of mono-facial and bifacial FPV according to day since '22 Dec.

Figure 10 shows the Histogram of PR for comparing more detail. Average PR of bifacial are 69.23~69.91% which is higher result as 5.00~5.68% compared with mono-facial of 64.23%. Even calculate of PR is only use front irradiance, rear side effect would be had higher 5~5.68% in bifacial module in Floating.

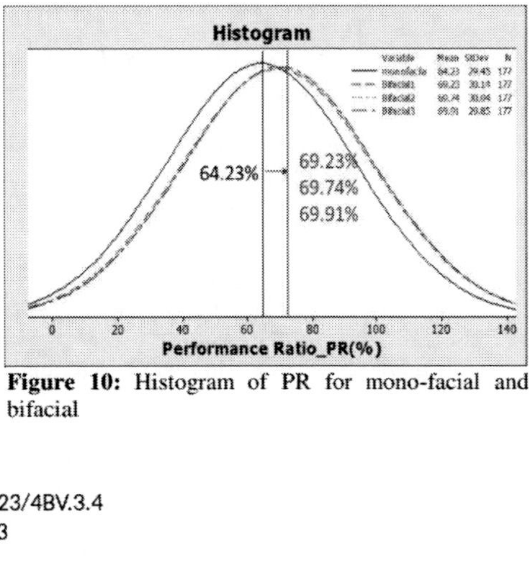

Figure 10: Histogram of PR for mono-facial and bifacial

10.4229/EUPVSEC2023/4BV.3.4
020310-003

3.3 Water reflection effect of bifacial PV

Figure 11 shows the Albedo of water which is combined with rear and front irradiance with month. During early in the morning and lately a day during 2~3hr after sunrise and before sunset, there are higher water Albedo. It is due to low solar latitude which has higher diffusion [8]. Then, it has similar Albedo as 0.04~0.1 during 10:00~15:00 with regardless of the month. When variation of Albedo is compared with bifacial effect in Figure 7 and Figure 11, bifacial effect stands out at noon which has low Albedo.

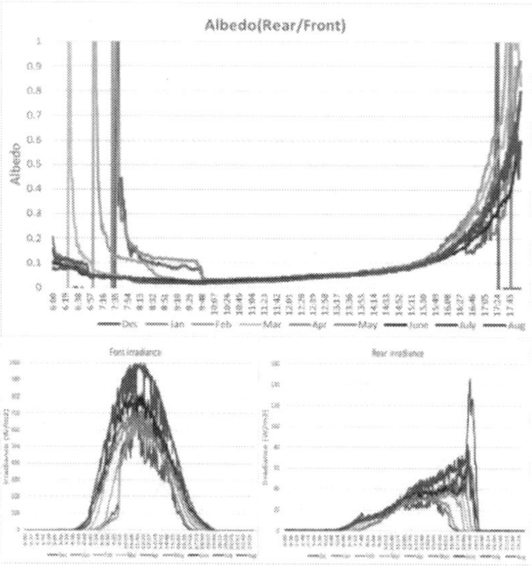

Figure 11: Albedo (upward) and Front(downward-left) and rear (downward-right) irradiance

It is necessary to confirm what effect water reflection to bifacial PV performance. Bifacial gain is included internally diffusion of front and rear irradiance [9]. Because the maximum power of mono-facial during a day is the highest performance with internal diffusion, power which is same time with maximum power of mono-facial are choose to normalize the bifacial power.

Figure12 are shown the comparison of bifacial gain before and after normalized the power during a day. Bifacial gain with including internal reflection and water reflection is 4~6% during 10:00~15:00. With only water reflection effect, bifacial gain is 0~2% during same time. Bifacial effect is independent with month which has same with water albedo.

4 SUMMARY AND CONCLUSION

In this study, it is verified bifacial effect with water reflection in Floating PV system. Albedo of water is 0.04~0.1 during day with regardless of the month. Bifacial gain had 2% with only water reflection and 4~6% with internal reflection and water reflection. Consquentely, power generation of bifacial PV has higher performance as 7% compared with mono-facial during 8monthes ('22 Dec. ~ '23 Aug.). For further study, Applied the back reflector for higher reflection without disturb in environment would be had higher power density in Floating PV.

5 REFERENCE

[1] K. Jiyoung, Korea Energy Economics Institute report (2021-2027)
[2] S. Yeongju, IEEE 46th Photovoltaic Specialists Conference (2019)
[3] P. Dohyun, Current Photovoltaic Research 6(2) 62-67 (2018)
[4] Global Bifacial Module Market Report 2019, 2020, Wood Mackenzie Power & Renewables.
[5] International Technology Roadmap for Photovoltaic (ITRPV) 2019 Results. March (2020)
[6] Z. Hesan, Progress in Photovoltaics 29; 725-743 (2021)
[7] Giuseppe, Applied Energy 281, 116084 (2021)
[8] S. Patel, Southern Africa Sustainable Energy Conference, Nov. (2019)
[9] Report of International Energy Agency Photovoltaic Power Systems Programme(IEA-PVPS) T13-14 (2021)

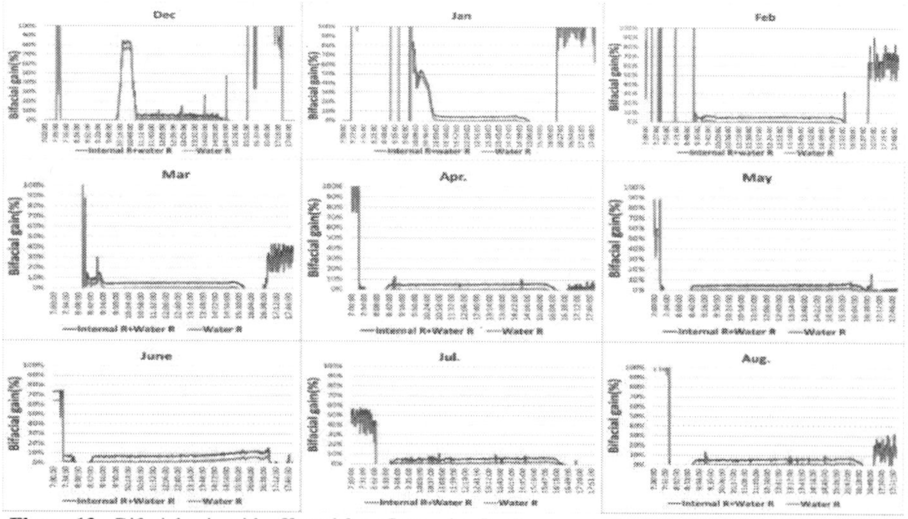

Figure 12: Bifacial gain with affected from Internal reflection + water refection and water reflection only

10.4229/EUPVSEC2023/4BV.3.4
020310-004

Proposal of Solar Cell Modules for Reducing Partial Shading Loss

Kyotaro Nakamura[1], Chikao Okamoto[2], Ryo Ozaki[1], Yoshio Ohshita[1] and Masafumi Yamaguchi[1]
[1] Toyota Technological Institute, Nagoya, Japan
[2] Sharp Corporation, Nara, Japan
phone: +81-52-809-1830, fax: +81-52-809-1879, e-mail: kyotaro@toyota-ti.ac.jp

ABSTRACT: The actual driving range of PV-powered vehicles is shown to be lower than the estimated values due to some losses such as partial shading. This paper presents effects of module interconnection upon power generation under partial shading and proposes new solar cell module interconnection to realize more power generation under partial shading. It is found in this study that new solar cell modules proposed show more power generation under shadow covering. Introduction of bypass diodes is thought to be very effective in the case of horizontal shadowing. In this study, increase in parallel connection of solar cell modules is clarified to be effective for higher generating power under partial shading. The maximum power point tracking (MPPT) control is suggested to be very effective for higher generating power of vehicle integrated PV (VIPV) modules under partial shading. Therefore, it is important to design installation and connection of VIPV modules in order to reduce shadowing loss of the PV-powered vehicles.
Keywords: Design, Module, Simulation, Shading

1. Introduction

Developing a vehicle powered by an onboard photovoltaics (PV) module [1,2] is desirable and essential for reducing CO_2 emissions. In our previous paper, the Toyota Prius demonstration car has shown a 36.6km/day PV-powered driving range at solar irradiation of 6.2kWh/m^2/day [1]. The practical cruise range of PV-powered vehicles is shown to be lower than the estimated values due to some losses such as temperature rise of solar cell modules and partial shading. The effects of partial shading upon VIPV modules have been studied by Tomita et al. [3]. The shading of the vehicle integrated photovoltaics (VIPV) is reported [4,5] to be dependent on the sun's attitude, climate condition and location, shading objects such as buildings, houses, trees, poles and others, driving and parking conditions. Average irradiation losses by partial shading reported to be 0.03 on the high way, 0.07 in the suburbs, 0.23 in the Urban, 0.5 near tall buildings and 0.07-0.57 in the parking [5]. Because driving distance of PV-powered vehicles is thought to be decreased by partial shading, development of solar cell modules with low power loss under partial shading is necessary.

This paper presents effects of module interconnection upon power generation under partial shading and proposes new solar cell module interconnection to realize more power generation under partial shading. It is found in this study that new solar cell modules proposed show more power generated under shadow covering. Therefore, it is important to design installation and connection of VIPV modules in order to reduce shadowing loss of the PV-powered vehicles.

2. Cells and Modules for Partial Shading Test, Simulation and Experiment of Power Generation under Partial Shading

In the partial shading tests, Si hetero-junction (SHJ) solar cells with area of 2 cm x 2 cm and mini-modules with 6 series connections and 6 parallel connections fabricated in our lab. were used. Table 1 shows properties of SHJ solar cell and mini-module with 6 series connection and 6 parallel connection used in the partial shading tests.

Table 1. Properties of SHJ solar cells and modules used in the partial shading tests.

	Efficiency	Voc	Jsc / Isc	FF
Cell	14.2%	0.700V	34.7mA/cm^2	0.585
Module	8.42%	4.20V	0.833A	0.585

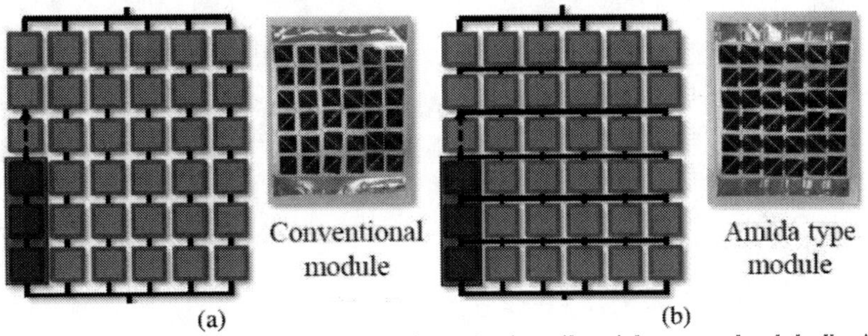

Figure 1. Conventional solar cell module and new (Amida type) solar cell module proposed and shading in the case of vertical shadow.

Figure 1 shows conventional solar cell module with 6 series connection and 6 parallel connection and new solar cell module (we call as Amida-type) proposed in this study. The Amida-type solar cell module is composed of 6 series connections, 6 parallel connections and additionally parallel connections of solar cells with series connection in order to reduce power loss under partial shading.

Simulation was carried out by using the SPICE software. Effects of partial shading upon power loss of solar cell modules were evaluated by using vertical shadow (Shadow A), horizontal shadow (Shadow B) and square-type shadow (Shadow C) as shown in Fig. 1. Figure 1 shows the case of Shadow A (vertical shadow).

3. Effects of Power Generation of Conventional and New Solar Cell Modules under Partial Shading

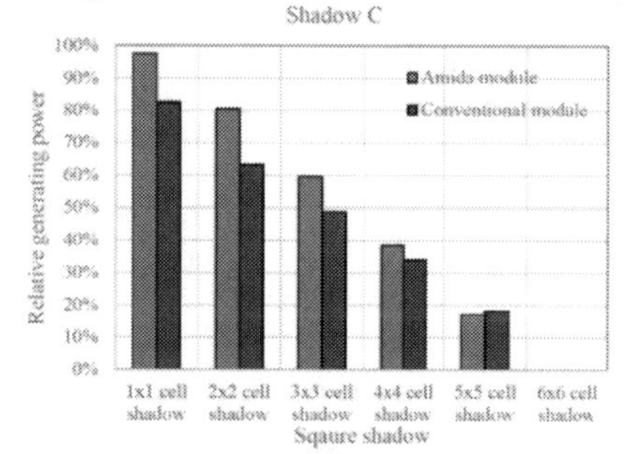

(c)

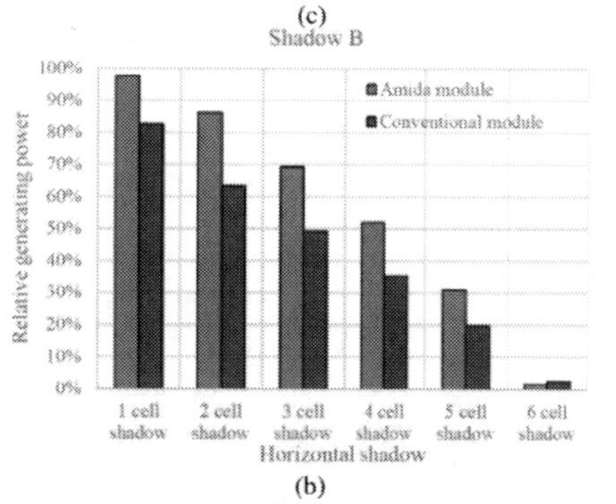

(b)

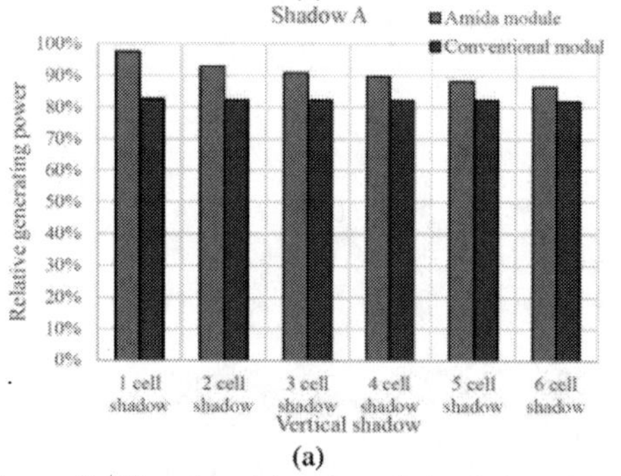

(a)

Figure 2. Experimental values for relative power generation of conventional module and new (Amida-type) module under Shadow A, Shadow B and Shadow C.

Figure 2 shows experimental values for relative power generation of conventional module and new (Amida-type) module under Shadow A, Shadow B and Shadow C. Power loss of new solar cell module under partial shading in the all cases (Shadow A, B and C) is relatively lower compared to the conventional solar cell module.

Figure 3 shows correlation between simulated power ratio and experimental power ratio of new (Amida-type) module under partial shadows (Shadow A, Shadow B and Shadow C). Good correlation between them shows validity of evaluation procedure for partial shading of solar cell modules.

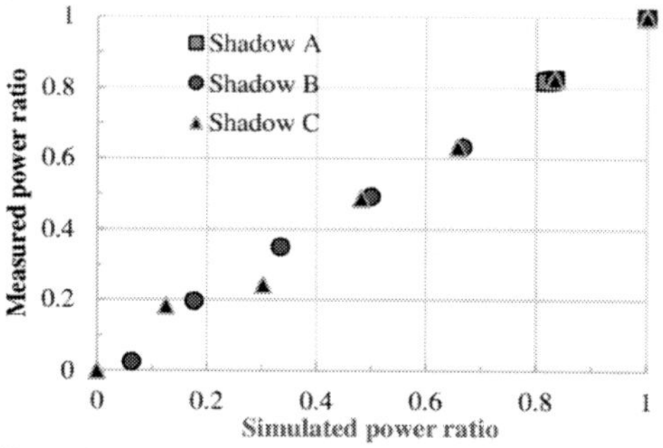

Figure 3. Correlation between simulated power ratio and experimental power ratio of new (Amida-type) module under partial shadows (Shadow A, Shadow B and Shadow C).

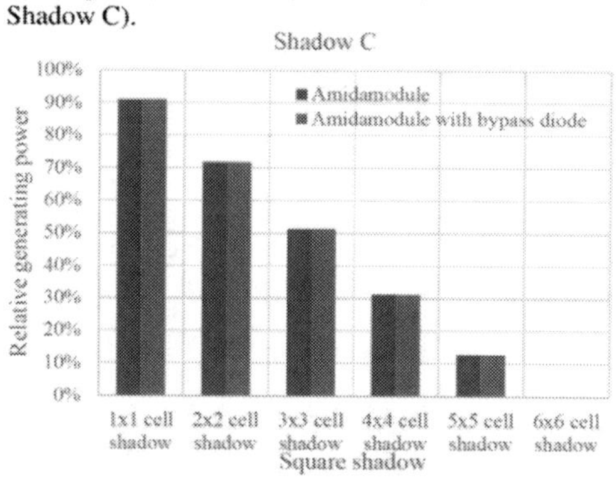

(c)

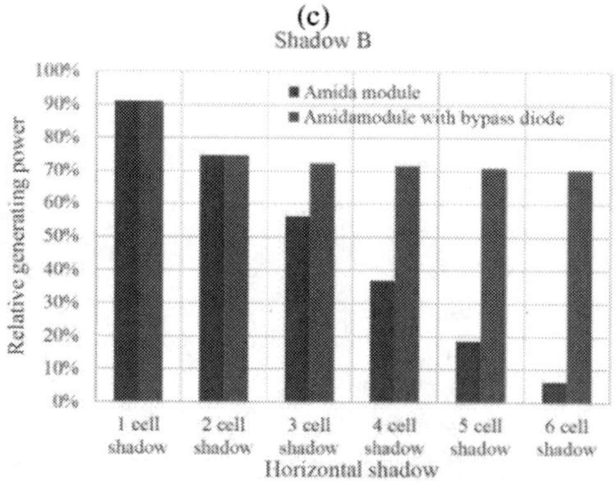

(b)

10.4229/EUPVSEC2023/4BV.3.5
020312-002

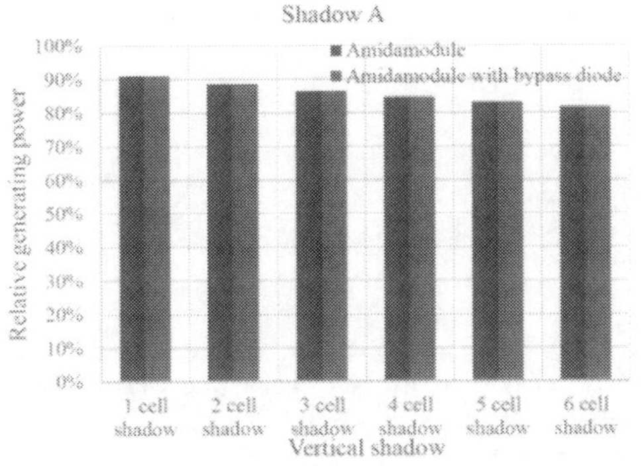

(a)

Figure 4. Effects of bypass diodes upon generating power of new (Amida-type) solar cell modules under partial shading.

Because good correlation between simulated and measured results is shown as shown in Fig.3, effects of bypass diode upon generating power of new solar cell module was calculated. Figure 4 shows effects of bypass diodes upon generating power of new (Amida-type) solar cell modules under partial shading. Bypass diodes equipped each solar cell module can compensate power loss caused by partial shading [6], especially in the case of horizontal shadowing.

4. Discussion about Further Reduction in Power Loss of Solar Cell Modules under Partial Shading

Figure 5 shows calculated results for the effects of parallel connection of solar cell modules from 18-series and 2 parallel to 2-series and 18 parallel upon generating power under partial shading in comparison with experimental results. It is clear that generating power of solar cell modules under partial shading is increased as increase in number of parallel connections [6].

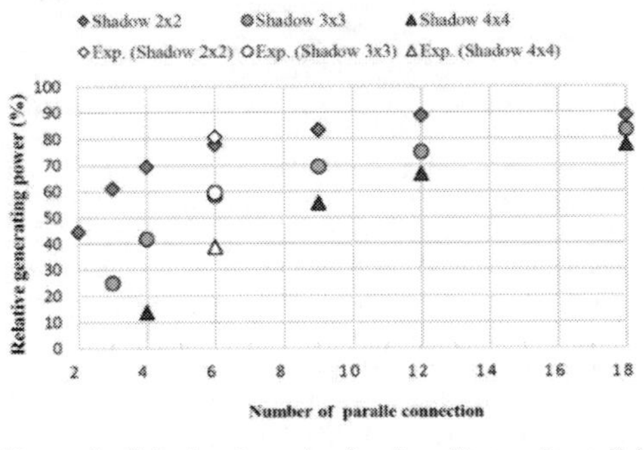

Figure 5. Calculated results for the effects of parallel connection of solar cell modules upon generating power under partial shading in comparison with experimental results.

Figure 6 shows comparison of our (TTI) results and Nissan results [3] and improvement in reduction of partial shading power loss by using maximum power point tracking (MPPT). In addition to increase in parallel connection, MPPT is thought to be very effective for higher generation power under partial shading.

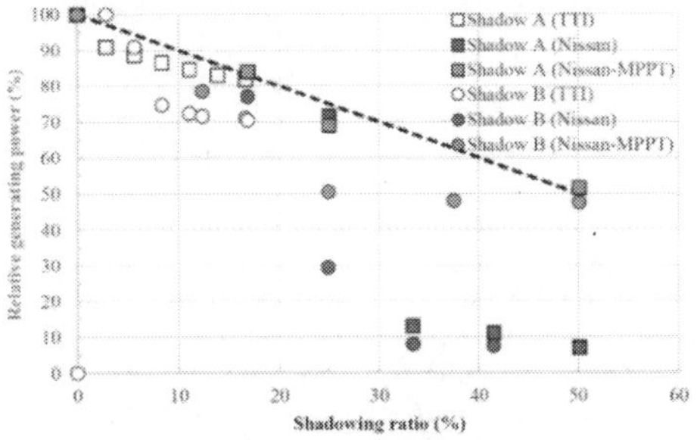

Figure 6. Comparison of our (TTI) results and Nissan results [3] and improvement in reduction of partial shading power loss by using MPPT.

5. Summary

This paper presented effects of module interconnection upon power generation under partial shading and proposed new (we call as Amida-type) solar cell module interconnection to realize more power generation under partial shading. It was found in this study that new solar cell modules proposed showed more power generation under shadow covering in the all case of vertical, horizontal and square shadows. Introduction of bypass diodes was shown to be very effective in the case of horizontal shadowing. In this study, increase in parallel connection of solar cell modules was clarified to be effective for higher generating power under partial shading. The maximum power point tracking (MPPT) control was suggested to be very effective for higher generating power of vehicle integrated PV (VIPV) modules under partial shading. Therefore, it is important to design installation and connection of VIPV modules in order to reduce shadowing loss of the PV-powered vehicles.

Acknowledgments

The authors sincerely thank the New Energy and Industrial Technology Development Organization (NEDO) for supporting R&D and Dr. T. Tanimoto and Dr. Y. Tomita, Nissan Motor, Dr. T. Takamoto, Sharp and Dr. T. Masuda, Toyota Motor for their fruitful discussion.

References

[1] NEDO. Interim Report "PV-Powered Vehicle Strategy Committee". ; 2018. http://www.nedo.go.jp/english/index.html.
[2] M. Yamaguchi et al., Prog. Photovolt. **29**, 684 (2021).
[3] Y. Tomita et al., 5th International Electric Vehicle Technology Conference (EVTech 2021).
[4] Y. Ota et al., IEEE J. Photovolt.**11**, 1299 (2021).
[5] T. Hirota et al., Trans. of the Society of Automotive Engineers of Japan **53**, 784 (2022). (in Japanese)
[6] A. Tsuyoshi et al., Research Report of the Kobe City College of Technology 51, 77 (2013).

10.4229/EUPVSEC2023/4BV.3.5
020312-003

AUTONOMOUS OPTIMIZATION OF AGRIVOLTAIC SYSTEMS WITH VERTICAL BIFACIAL PV PANELS AND THE GRASS CROP TIMOTHY

Erlend Hustad Honningdalsnes*[1,2], Helge Bonesmo[3], Erik Stensrud Marstein[1,2], Steve Völler[4], Heine Nygard Riise[1]
[1]Institute for Energy Technology, Norway
[2]University of Oslo, Norway
[3]Norwegian Institute of Bioeconomy Research, Norway
[4]Norwegian University of Science and Technology, Norway
*Corresponding Author: Erlend.honningdalsnes@ife.no

ABSTRACT: We present an autonomous system that integrates architecture software with Python to optimize both agricultural yield and electrical energy production Our approach employs ray tracing within the Grasshopper environment to analyze ground surface shading and consequent evapotranspiration changes, and PV surface irradiation. These metrics inform a crop model to estimate agricultural yield and a PV model for energy production. An optimization algorithm iteratively refines the system design, evaluating criteria such as land use efficiency or minimum crop yield, to identify the most effective configuration for a given objective. The framework is adaptable to various crops and PV designs, enabling multifaceted optimizations like profitability and land equivalent ratio while meeting regulatory requirements, such as minimum crop yield. As a proof-of-concept, we designed an agrivoltaic setup featuring vertical bifacial PV panels with the grass crop Timothy (Phleum pratense L.) for Norwegian conditions.
Keywords: Agrivoltaics, Optimization, Microclimate, Energy Yield, Crop Yield

1 INTRODUCTION

Agrivoltaics, an emergent discipline that integrates agriculture and photovoltaics (PV), holds significant promise in addressing the pressing issue of land availability. This approach facilitates simultaneous food and energy production on the same land area, thereby achieving efficient land use. While agrivoltaics has gained momentum in regions such as Southern Europe, China, Japan, India, and North America, its applicability in higher-latitude territories such as Norway remains largely unexplored.

This paper addresses this research gap by exploring the viability of vertical bifacial agrivoltaic systems in Norway, a region where grasslands make up 70% of all agricultural areas [1]. We focus on timothy grass, the predominant forage crop, as a prospective pairing for vertical bifacial PV arrays in these settings.

To expedite future design optimizations - which inherently involve a complex interplay of variables and system parameters - we have developed an autonomous optimization workflow within the Grasshopper environment in Rhinoceros 7. This comprehensive methodology comprises five key modules: parametric design, shadow analysis, crop yield assessment, PV energy yield evaluation, and a genetic algorithm-based optimization solver. Our approach can generate multiple optimal designs that cater to diverse objectives, from economic viability to land use efficiency.

The results show that our model identified optimal solutions for the given system and objective functions. They also underscore the potential of the proposed framework in agrivoltaic design workflows and as a promising path for future research and real-world applications.

2 METHODOLOGY

The primary objectives of this study were:

1) Implement a method that allows for autonomous optimization of agrivoltaic systems by mostly depending on existing software.

2) Demonstrate the applicability of the method by optimizing the design of a vertical bifacial PV system in synergy with the agricultural production of the forage grass timothy under Norwegian conditions.

The methodology encompassed:

- A parameterized design framework for vertical PV systems with variable internal row spacing and panel orientation.
- Microclimatic analysis.
- Energy and crop yield modelling.
- Autonomous system evaluation and redesign.

These components were integrated to ensure a comprehensive, holistic approach to agrivoltaic system design and evaluation. The workflow is visually represented in Figure 1.

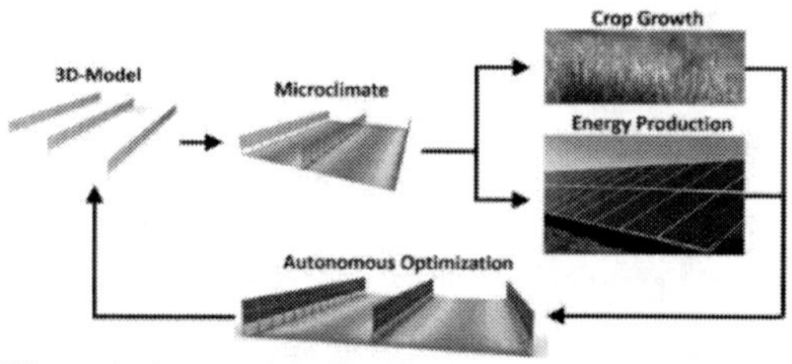

Figure 1: Autonomous loop for system optimization.

2.1 Software Environment - Grasshopper

The entire study was conducted within the Grasshopper environment, an integral part of the Rhinoceros 7 CAD software. Grasshopper is a visual programming environment that enables modelling of complex shapes and systems, often through a drag-and drop interface that does not require traditional coding skills. We opted for it due to its extensive ecosystem that includes a library of prebuilt components for various simulation and modelling tasks, such as for architectural studies. These include popular open-source tools such as Ladybug Tools [2] or commercial ones like ClimateStudio (free for researchers). Additionally, it is possible to write custom applications in languages such as Ironpython, C# and Visual Basic. While some studies on irradiation in

10.4229/EUPVSEC2023/4BV.3.13
020314-001

agrivoltaic systems have been performed in this environment [3, 4], none has to our knowledge used all relevant capabilities it offers like autonomous optimization and system evaluation.

2.2 System Setup

The agrivoltaic system had a set of fixed properties that did not change throughout the optimization procedure. The PV panels were assumed to be CHASER-M6/144P 475W, which is a half-cut 90% bifaciality double glass module of length 2.126 x 1.042 meters and 21.85% efficiency. The lowest point of the bottom panel was placed at a height of 1 meter above the ground, and two panels were placed in height in portrait mode. The supporting structure was made of aluminum and had a diameter of 5 cm. As for the weather data and geographic area used in the study, a TMY (Typical Meteorological Year) file for Fornebu in Oslo was chosen to represent the conditions in the South-Eastern parts of Norway.

2.3 Microclimate Analysis

Microclimatic conditions play a pivotal role in both solar energy production and agricultural yield. Two specific parameters were examined:

- Ground surface irradiation
- Soil-water conditions

For the former, the RadMap module from ClimateStudio (v 1.9.8389.22035) was used for its ray-tracing capabilities and computational efficiency, which proved beneficial for iterative analyses. The TMY file was used as input for the ray tracing procedure and the soil-water calculations.

To model soil-water conditions, we integrated a dual-model system that has been utilized in Norwegian agricultural drought studies [5]. The first part is an empirical regression model that calculates potential evaporation (*Ep*) based on several environmental factors, such as solar irradiation. It was chosen as it has been more accurate for Norwegian conditions than, e.g., Penman-Monteith [6]. The second model computes the actual evaporation (*Ea*), factoring in soil and crop parameters, as well as precipitation. The combined output of these models, *Ea/Ep*, was given as daily input to the crop model to incorporate the effect of panel shading on drought.

2.4 Crop Modelling

We employed the Canadian Timothy Model (CATIMO) to estimate the dry matter yield of timothy grass (*Phleum Pratense* L.) in g/m²/year [7]. This model incorporates various growth drivers such as solar irradiation and temperature, as well as growth limitations like drought to calculate daily biomass growth and crop development. The model was written in IronPython to fit inside the Grasshopper environment.

As mentioned, Timothy was chosen as the crop representing grasslands as it is the most widely used forage grass in Norway. Additionally, timothy is a common forage species in larger areas of Scandinavia and North America.

While a model for timothy was used in this study, the compartmentalized structure of the system allows easy substitution to other crops and growth models.

2.5 Energy Production Modelling

To calculate energy production, we first needed to determine the incoming solar irradiation hitting the vertically mounted PV panels. This was also accomplished using ClimateStudio's RadMap module. While the irradiation is available on an hourly basis, it was aggregated to annual data to investigate annual yield. Subsequently, the irradiation data was fed into a simplified PV model to estimate electrical energy production. The model used the equation:

$$Electrical\ energy = Irradiation * bf * \eta$$

Here, bf refers to the bifaciality factor for the rear side of the PV panels at 0.9, and η is the conversion efficiency of 21.85%. Although quite simple, this model provides an initial approximation and can later be expanded upon to include more variables in future work, as described in section 3.6.

2.6 Autonomous Optimization

For this study's central feature, we implemented an autonomous optimization loop using the Galapagos solver. The solver dynamically adjusts the system's parametric design to optimize a predefined objective function. The parametric design allowed different row distances and system orientations as illustrated in figure 2. The choice of optimizing only row distance and system orientation was due to their high impact on the results, and ease of result interpretation in this study. In practice, more parameters could be included.

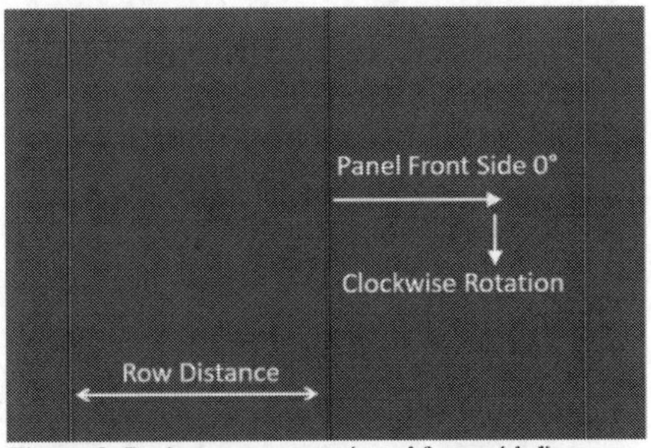

Figure 2: Design parameters viewed from a bird's-eye perspective in the parametric model, where up is North.

One of the pivotal elements in optimization studies is the selection of an appropriate objective function. In the realm of agrivoltaic systems, this function needs to account for both agricultural and electrical yields. A commonly used metric for this purpose is the Land Equivalent Ratio (LER), which measures the land use efficiency of the agrivoltaic system relative to traditional farming and PV systems on an annual basis [8]. Mathematically, it can be expressed as:

$$LER = LER_{crop} + LER_{solar}$$

Where:

$$LER_{crop} = \frac{Yield\ in\ agrivoltaic\ system\ \left(\frac{g}{m^2}\right)}{Yield\ in\ traditional\ farming\ \left(\frac{g}{m^2}\right)}$$

and

$$LER_{solar} = \frac{Yield\ from\ agrivoltaic\ system\ \left(\frac{kWh}{m^2}\right)}{Yield\ from\ standalone\ PV\ \left(\frac{kWh}{m^2}\right)}$$

10.4229/EUPVSEC2023/4BV.3.13
020314-002

An LER coefficient greater than 1 signifies that the agrivoltaic system is utilizing the land more efficiently than traditional methods alone. Note that the annual energy yield is divided by the ground surface area the system occupies, not the panel surface area.

This approach does not capture all challenges posed for such systems. For example, a 10% increase in energy yield from a PV system often has a more significant economic impact than a 10% decrease in food production, particularly for low-value crops like grass. However, this would not present a change in LER, thus warranting adjustments in any optimization model based on this parameter. To protect agricultural production under such pressures, some countries, such as Germany and France, are implementing minimum crop yield requirements in agrivoltaic systems [9].

To capture the multifaceted nature of agrivoltaic system optimization, our study used LER as the objective function. For baseline LER values, the crop yield per surface area were modeled without shading effects, and the energy yield reference was set at 68 kWh/m^2/year, as has been used in other Northern studies [10]. Additionally, some constraints were imposed, namely a minimum crop yield limit of 70% of regular farming and a minimum specific PV energy yield of 750 kWh/kWp/year.

2.7 Computational Considerations

Computational time is a critical factor for the feasibility of real-time or near-real-time optimization. We optimized the ray-tracing procedure and subsequent calculations by reducing the number of sensor points and leveraging system symmetry. This is done by implementing an agrivoltaic unit that represents the majority of the system, assuming a sufficiently large system [11]. The unit is shown in figure 3, and has one sensor per panel surface (front and back), and a ground surface strip with sensors spaced every 10 cm.

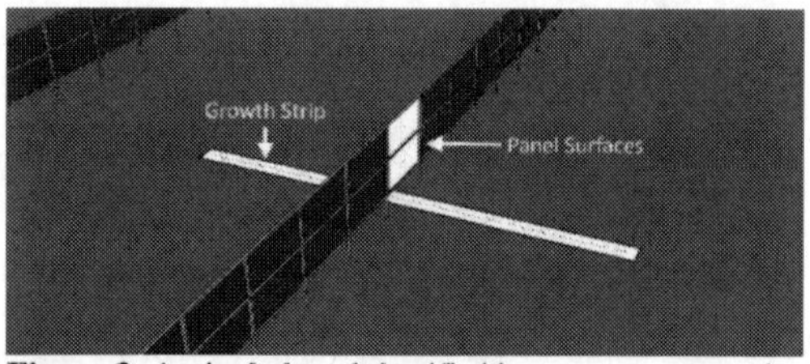

Figure 3: Agrivoltaic unit justified by system symmetries

The hardware used for computation comprises an AMD Ryzen 7 5800-X CPU, an NVIDIA GeForce RTX 3060 Ti GPU, and 32 GB of RAM. With this system we achieved a computational time of 3 seconds per design, of which half was used for ray tracing.

3 RESULTS AND DISCUSSION

This study produced two principal outcomes:

1) A framework that integrates multiple tools for autonomous optimization, and
2) Optimization results for a vertically oriented bifacial PV system in co-existence with the grass crop Timothy.

3.1 Software Environment Performance

The study's computational model successfully interfaced with the Galapagos solver to perform optimization tasks. The Galapagos solver employs a genetic algorithm to efficiently navigate the solution space, typically requiring around 250 iterations for convergence, though this will depend on the implementation of the algorithm and the system design configurations. Each optimization run took approximately 10 minutes, showcasing the feasibility of our framework for real-time or near-real-time optimization. The primary objective function in this study was the Land Equivalent Ratio (LER), a metric that quantifies land-use efficiency.

3.2 Optimization Results

The optimization procedure described in section 2.6 yielded the following system configuration:

- Row Distance: 6 meters.
- Panel Orientation: 124°, meaning the front panel surface facing South-West.

These settings yield the following performance metrics:

- LER = 1.502
- LER_crop = 0.710
- LER_energy = 0.792
- Specific Energy Yield = 774 kWh/kWp

These results indicate a substantial increase in land utilization under Norwegian climatic conditions. It should be noted that this row distance is less then what is typically suggested for such systems (8-12 meters). The reason for this is explained in subsequent sections.

To visually illustrate the validity of the optimized system in terms of LER and the given system limits, the next three subsections provides more details on the crop and energy yields of the different system configurations. More specifically, they show the modelled results for a series of iterations for all the row distances and orientations (10-degree resolution).

3.3 Energy Production

Figure 4 illustrates that the system produces the most energy per installed capacity when the front panel surface is facing a South-Western direction. The grey colored patches signify which system configurations produce less than the minimum limit of 750 kWh/kWp, and thus were not accepted in the optimization procedure. For example, 5-meter row distances for most orientations produce too little energy per panel due to interrow shading. Additionally, the suboptimal orientations are under the limit even at 6-meter row distances.

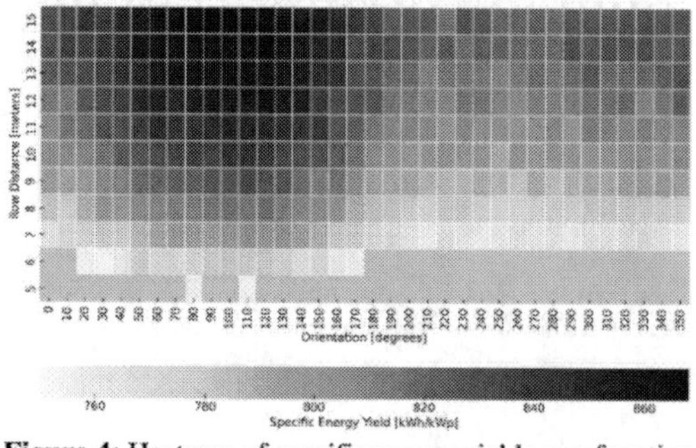

Figure 4: Heatmap of specific energy yields as a function of system row distance and orientation.

As for the energy produced per land surface area, kWh/m^2, as is used in LER calculations, it is clear that it is desirable to place the panels with smaller row distances. E.g., 5 meters vs 15 meters row spacing uses just a third of the land area, while the energy yield difference is only 100 kWh/kWp/year.

Note that an improved PV model may find different results, though the same general trends are expected.

3.4 Crop Yield

Figure 5 illustrates the average crop yield in the system for different system configurations. Unlike the energy yield, the highest average crop yield in the system can be found for roughly 0 and 180 degrees, meaning the panel surfaces are facing East and West. It is also a general pattern that increased row distances produces higher yield. For example, systems with 5-meter row spacing are not even accepted as the crop yield is less than 70% of normal crop yield (992 g/m^2).

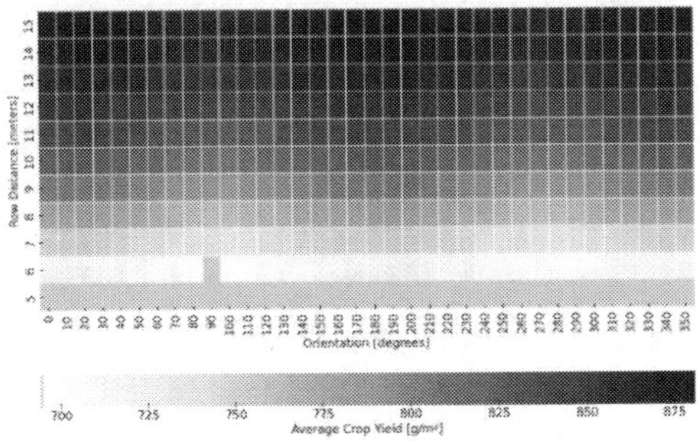

Figure 5: Heatmap of average grass crop yields as a function of system row distance and orientation.

While the average crop yield in the system is often what is used in LER calculations, the interrow distribution of the yield is also of interest. One reason for this may be that all crops will be harvested at the same time, meaning uneven shading may lead to different levels of crop development and optimal harvest times. The crop yield along the growth strip for a system with 10-meter row distance is illustrated in figure 6.

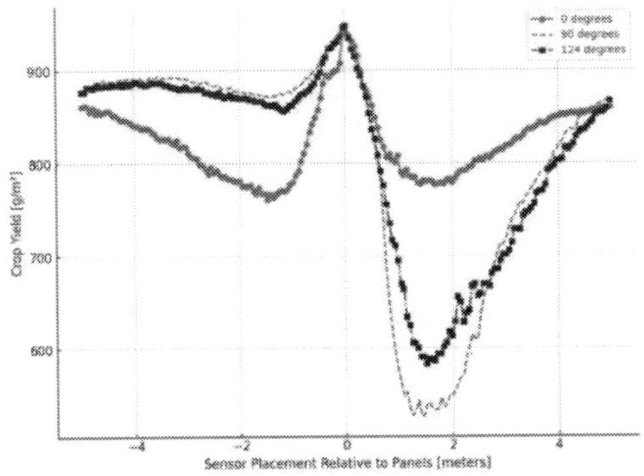

Figure 6: Crop yield in different parts of the system for different orientations and 10-meter row distance.

Orientation has a large effect on crop yield distribution. While East/West-facing panels (0 degrees) allow for more even growth within the system, panels facing South will produce a deep shadow on the Northern side that has a more local negative effect on the crops behind the panels. Furthermore, the modelled crop yield of the system is highly dependent on the weather data that's used. For example, the yield loss may be lower in dry years where reduced evapotranspiration is beneficial.

3.5 LER Optimization

To assess whether the system found the optimal system configuration, a heatmap of LER values with the system configurations that produce sub-limit yields removed is illustrated in figure 7. To more clearly differentiate which configurations exhibit the highest LER, the color values has been subjected to an inverse power-law normalization.

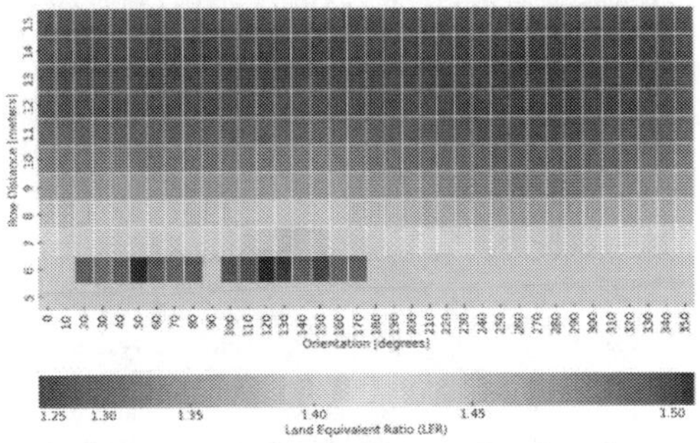

Figure 7: Heatmap of LER values as a function of system row distance and orientation.

We observe that the designs with the highest LER within the acceptable parameter space all have 6-meter row distances. This is because the energy produced per area increases faster than the crop yield decreases. Therefore, the limitations on how close the panels can be placed together in this study are the lower limitations on average crop yield and specific energy yield. This goes to show that LER optimization may not be the best objective function to maximize for optimal agrivoltaic system designs, as it may lead to low crop yields and specific energy yields, both of which are important for economical and practical purposes. Consequently, it may be better to do an economic optimization that includes the cost of, e.g., equipment, land and energy and crop yields, while also including minimum crop yield limits.

As for the orientation of the optimal system, it is in an area of high LER values. The iteration procedure had lower resolution (10 degrees) than the optimization procedure (1 degree), thus making it difficult to visually confirm the exact optimal system configuration.

3.6 Opportunities for further development

There are several avenues for further development of this system. First, the individual components of the model, such as the irradiation model and crop model, need to be validated for agrivoltaic systems like vertical bifacial PV. Secondly, the Galapagos solver can be replaced with other optimization solvers in the Grasshopper environment, such as Wallacei, that allows multi-objective optimization and includes more features. It was not used in this study, as it is not directly compatible with ClimateStudio due to system execution priorities. This can be solved by exchanging ClimateStudio with another ray tracing model, such as Honeybee Radiance (HB) that is part of Ladybug Tools, which is compatible with Wallacei. Both of these programs are freely available for both research and commercial use. While HB was not used in this study due to ClimateStudio being slightly easier to implement and having faster ray tracing, it offers more flexibility and control, in addition to being well integrated with the rest of Ladybug Tools' environment.

It is also possible to significantly expand the complexity of the PV modelling, boosting both the accuracy and usefulness of the optimization procedure. Use cases may include, but are not limited to, local energy price fluctuation matching, mismatch losses due to interrow shading and PV panel temperature efficiency coefficients.

Additionally, it is possible to include more microclimatic factors in the microclimatic analysis. The effect of wind speeds may be of particular interest in vertical bifacial systems, but CFD (Computational Fluid Dynamics) analysis tools, such as Butterfly from Ladybug Tools, is not recommended to include in the optimization loop due to computational time requirements. Instead, preprocessed wind speed changes stored in matrices may be imported and used appropriately.

4 SUMMARY

This study presents an autonomous optimization framework and applies this to an agrivoltaic system design in Norway. The framework merges CAD software (Rhinoceros 7 with Grasshopper) and Python algorithms to simultaneously optimize crop yield and energy production. We focus on pairing vertical bifacial PV panels with Timothy grass, a major forage crop in Norway. Through ray tracing and microclimate analysis, our system informs both a crop model and a simplified PV energy model. An optimization loop, using the Galapagos solver, refines design parameters such as row spacing and PV panel orientation to maximize the Land Equivalent Ratio (LER) while meeting minimum energy and crop yield requirements.

Our optimized design achieves an LER of 1.502 with a row distance of 6 meters and a panel orientation of 124°, exceeding the minimum imposed requirements for both energy and crop yield. In practice, however, this configuration may not be optimal due to practical and economic considerations, thus the finding highlights a weakness of using LER as objective function. The system's computational efficiency, at 3 seconds per design iteration, supports its use for near-real-time applications. The modular and adaptable framework is a robust tool for optimizing a wide range of agrivoltaic systems, with the potential for further refinement in future research.

5 ACKNOWLEDGEMENTS

Parts of this work was performed within the Norwegian Research Center for Sustainable Solar Cell Technology (FME SUSOLTECH, project number 275639/E20). The center is co-sponsored by the Research Council of Norway and its research and industry partners.

The method for system optimization builds on work from the master's thesis of Honningdalsnes, E. H.[12].

6 REFERENCES

[1] AgriAnalyse, "Landbruksbarometeret 2021," 2021.

[2] M. S. Roudsari and M. Pak, "Ladybug: a parametric environmental plugin for grasshopper to help designers create an environmentally-conscious design," 2013.

[3] Y. Gao *et al.*, "Modeling and analyses of energy performances of photovoltaic greenhouses with sun-tracking functionality," *Applied Energy*, vol. 233, pp. 424-442, 2019.

[4] O. A. Katsikogiannis, H. Ziar, and O. Isabella, "Integration of bifacial photovoltaics in agrivoltaic systems: A synergistic design approach," *Applied Energy*, vol. 309, p. 118475, 2022.

[5] H. Riley, "Tørkesommeren 2018-beregninger av hvor mye korn-, potet-og grasavlingene ble påvirket på ulike jordtyper i ulike distrikt," *Jord-og Plantekultur 2019. Forsøk i korn, olje-og proteinvekster, engfrøavl og potet 2018*, 2019.

[6] L. Zotarelli, M. D. Dukes, C. C. Romero, K. W. Migliaccio, and K. T. Morgan, "Step by step calculation of the Penman-Monteith Evapotranspiration (FAO-56 Method)," *Institute of Food and Agricultural Sciences. University of Florida*, vol. 8, 2010.

[7] H. Bonesmo and G. Bélanger, "Timothy yield and nutritive value by the CATIMO model: I. Growth and nitrogen," *Agronomy Journal*, vol. 94, no. 2, pp. 337-345, 2002.

[8] D. Deb, S. Dutta, and R. Erickson, "The robustness of land equivalent ratio as a measure of yield advantage of multi-crop systems over monocultures," *Experimental Results*, vol. 3, p. e2, 2022.

[9] M. Trommsdorff, I. S. Dhal, Ö. E. Özdemir, D. Ketzer, N. Weinberger, and C. Rösch, "Agrivoltaics: Solar power generation and food production," in *Solar Energy Advancements in Agriculture and Food Production Systems*: Elsevier, 2022, pp. 159-210.

[10] P. E. Campana, B. Stridh, S. Amaducci, and M. Colauzzi, "Optimisation of vertically mounted agrivoltaic systems," *Journal of Cleaner Production*, vol. 325, p. 129091, 2021.

[11] D. Wang, Y. Zhang, and Y. Sun, "A criterion of crop selection based on the novel concept of an agrivoltaic unit and M-matrix for agrivoltaic systems," in *2018 IEEE 7th World Conference on Photovoltaic Energy Conversion (WCPEC)(A Joint Conference of 45th IEEE PVSC, 28th PVSEC & 34th EU PVSEC)*, 2018: IEEE, pp. 1491-1496.

[12] E. H. Honningdalsnes, "Autonomous Optimization of Agrivoltaic Systems in Norway," Master's thesis, NTNU, 2022.

10.4229/EUPVSEC2023/4BV.3.13
020314-005

SOLAR@SEA: A CONCEPT FOR PV ON FLEXIBLE FLOATERS FOR OFF-SHORE APPLICATION

Wim Soppe, Marc Koetse, Dorrit Roosen, Arjan vd Brink, Martijn Hoogeland (TNO)
Bastiaan vd Berg, Clemens vd Nat (Bluewater Energy Services)
Alessio di Fino (Endures)
Chrétien Hendriks, Carlo Scheerder (Genap)
Amarante Böttger, Lucas van den Akker (Avans Hogeschool)
William Otto (Marin)

ABSTRACT: This paper presents the results of a Dutch joint industry project in which a concept has been developed for offshore floating PV, based on air-filled flexible floaters combined with flexible thin film PV modules. The concept includes low weight materials that are inflated at the offshore sites to minimize transport and installation costs. The feasibility of the concept was tested by developing a prototype of the floaters to study the hydrodynamic behavior in a wave basin. Thereupon a 20 kWp pilot system is developed and installed on a lake (Oostvoornse Meer - The Netherlands). The system includes a mooring system in which the floaters are individually connected to a pre-laid underwater mooring network, allowing easy access to the floaters for maintenance. Although the significant wave height on that lake is much smaller than for the North Sea, the modelling of the mooring design could be validated. During operation, the PV panels will undergo millions of small deformations per year. To validate suitability of flexible thin film CIGS PV panels, the project included laboratory experiments simulating these frequent deformations and study possible degradation effects. CIGS PV panels turned out to be very forgiving and only small degradation effects were observed. Biofouling of PV panels and floaters is another important issue for offshore floating PV, since cleaning of marine and bird fouling can lead to significant OPEX costs. The project investigated growth of marine organisms on the floaters and PV modules and tested the suitability of several anti-fouling coatings.

Keywords: offshore floating PV; thin film PV; flexible floaters

1 INTRODUCTION

Offshore Floating PV (OFPV) has huge potential, in particular for countries like the Netherlands where land is scarce and large sea areas are available. Especially the areas between the wind turbines in offshore wind parks on the North Sea are envisioned. The environmental conditions at these offshore locations however are harsh, leading to the need of very robust systems as operating and maintenance (O&M) cost are offshore typically much higher than for land-based PV systems[1]. In order to create a perspective on economically viable electricity production by offshore PV systems, the development in the project has a strong focus on all cost elements, but especially on offshore installation costs and the offshore O&M.

The Solar@Sea concept is based on the usage of inflatable floaters and lightweight flexible PV panels. The very small displacement and wave riding properties of the floaters result in smaller mooring forces than for rigid floaters.

The economic viability of OFPV not only depends on the CAPEX costs but also on OPEX costs. It is crucial to achieve high Performance Ratios (PR) and to minimize maintenance efforts. For inland waters the PR of floating PV systems is generally as high or higher than for land based systems[2] but for OFPV this is more challenging as marine and bird fouling is more severe. The project has studied several fouling prevention methods to mitigate this performance impact.

2 DESCRIPTION OF THE SYSTEM

2.1 The floaters

The floaters are made of Double Wall Fabric (DWF), coated with an elastomer. Ballast bags are incorporated along the floater edges to avoid any uplift of the floater due to wind. Various design variations of these ballast bags have been taken into consideration and investigated in hydrodynamical simulations and model tests. Measurements of the mechanical behavior of the floater have been carried out by TNO to determine the ideal pressure (range) for the floater to ensure the flexible for acting as a wave rider, but stiff enough to prevent wrinkling of the floater top surface.

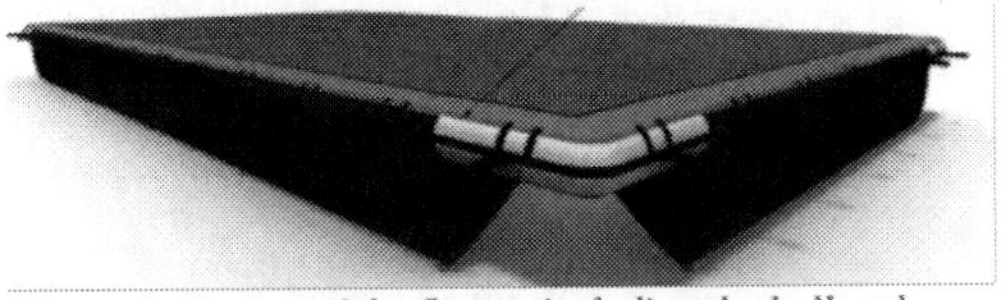

Figure 1: Design of the floater including the ballast bags. Length 13 m; width 7m.

Full scale test were carried out in the wave basin at Marin to investigate the wave riding properties of the floaters and the functionality of the ballast bags.

Figure 2: Wave basin tests at Marin.

2.2 The PV panels and the attachment to the floaters

The project has tested two commercially available flexible CIGS panels. Typical front and back sheet materials of these modules use ETFE and PET respectively. The PV panels were attached to the floaters using special adhesives. Several adhesives were

10.4229/EUPVSEC2023/4BV.3.14
020316-001

investigated in laboratory tests to select the most suitable adhesive, taking into account the harsh offshore conditions. These conditions were also used to investigate the potential degradation of flexible CIGS modules. These tests did especially focus on the panel degradations caused by very frequent but small deformations[3].

2.3 The mooring system

The mooring system is based on a submerged mooring grid where individual floaters are hooked onto. This system avoids the need for complex and fatigue sensitive inter-coupling systems between floaters. The mooring system provides flexibility to allow first order motions of the floaters, while the drift forces are transferred to the submerged mooring grid. This mooring grid is moored to the seabed using a conventional mooring system with small buoys. As the mooring grid is submerged at a several meters depth, the mooring does not hinder small boats to provide access for people and/or drones for O&M purposes.

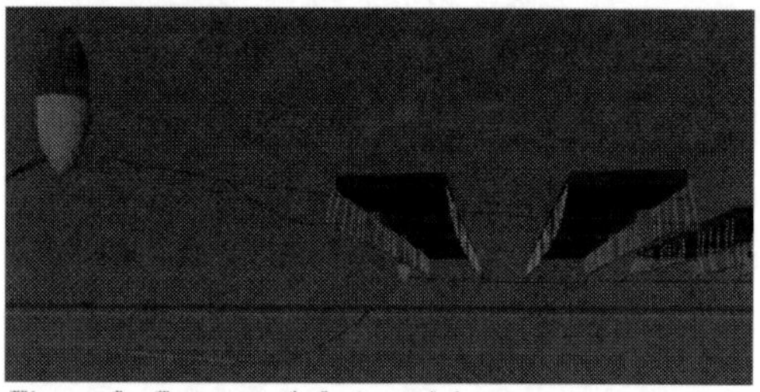

Figure 3: Conceptual design of the mooring system, as seen from an observer at the bottom of the sea.

2.4 The installation concept

For utility scale (offshore) PV projects typically very large surfaces are required that need to be transported and installed at the solar farm. In offshore conditions this activity involves installation vessels and offshore trained crews that are costly. Reducing the transport volume and limiting the offshore installation activities to the minimum is considered by the project key for success. Especially enabling installation in higher wave conditions can limit the impact of weather down times of these activities.

The offshore installation of the system is foreseen to take place by the following steps: 1) onshore preparation of the floaters: attachment of PV modules and longitudinal mooring grid lines; 2) installation of the mooring system and a Technical Equipment Buoy; 3) transport of un-inflated floaters to the location, inflation of the floaters and launching of the floaters (see picture below); 4) connection of the floaters to the mooring system; 5) electrical connection of the PV panels with the inverters on the Technical Equipment Buoy

Figure 4: Envisioned installation of floaters with PV panels at an offshore location.

3 EXPERIMENTAL

3.1 Degradation of PV modules

The PV modules that were used in the pilot system were commercial flexible CIGS modules from two different brands. Prior to the installation of the pilot system, tailor made mini-modules of 40×50 cm^2 with CIGS cells were tested in a mechanical degradation setup. The details of this setup and results are described in ref. [3]. The main finding is that CIGS is a very forgiving solar cell material for deformations but it can degrade when strained. The critical strain limit for CIGS material failure seems to be 2% [4]. The North Sea wave spectrum contains waves which would theoretically lead to strains up to 2% if the PV modules would perfectly follow the water curvature without strain release. In practice however, the cell interconnects, the encapsulants and the adhesion material between the PV modules and the floater will release the strain in the CIGS cells, because the CIGS material is grown on steel substrates.

Our experiments show that the CIGS cells mildly degrade after millions of small deformations that can be expected for deployment on a flexible floater on the North Sea with a performance loss of a few percent.

3.2 Fouling of the PV modules and floaters

Fouling of the system results from the marine environment and organic sources (sea life, mammals, birds). The project has investigated the fouling rate from these sources. In the harbor of Den Helder, Endures tested a mini floater with a PV module, and vertical panels with and without special anti-fouling coatings. .

Figure 5: Endures raft with mini floater at the beginning of the tests.

A number of anti-fouling coatings were tested; both on the floater material and on the front layer of the PV modules.The tests on the floater material were carried out in horizontal and in vertical position. As expected, none

10.4229/EUPVSEC2023/4BV.3.14
020316-002

of the coatings adheared well on the ETFE of the front layers. On the floater material adhesion was much better and several coatings showed good anti-fouling characteristics (see figure below).

Figure 6: Floater material test panels with various antifouling coatings after 157 days of immersion.

The floater material was severely fouled under water during the test period, with algae and barnacles growing on the material. Also an open ballast bag was completely filled with barnacles and mussels, leading to a significant increase of the net weight of the floater. The fouling of the PV module on top of the floater was very modest, despite or thanks to the continous splashing by seawater. Only some minor algae growth was observed after the 60 days testing period, and this fouling could easily wiped off.

Figure 7: Floater and PV module after 60 days.

3.3 Pilot system on the Oostvoornse Meer

In November 2021 a 20 kWp system was installed on a near sea location on the lake of Oostvoorne. The water in this lake is brackish and at the location of the pilot up to 22 m deep. The system consisted of two floaters and contained commercial flexible CIGS modules of two brands. The installation took place in three phases: 1) Installation of the mooring system of the floaters; 2) Mounting of the PV panels on the floater and installation of the floaters, using three different types of adhesives; 3) Installation of the electrical system.

Figure 8: Mounting of the PV modules on the floaters.

The PV modules were connected in series and in four parallel strings. Each string was connected to an inverter on land (see figure below).

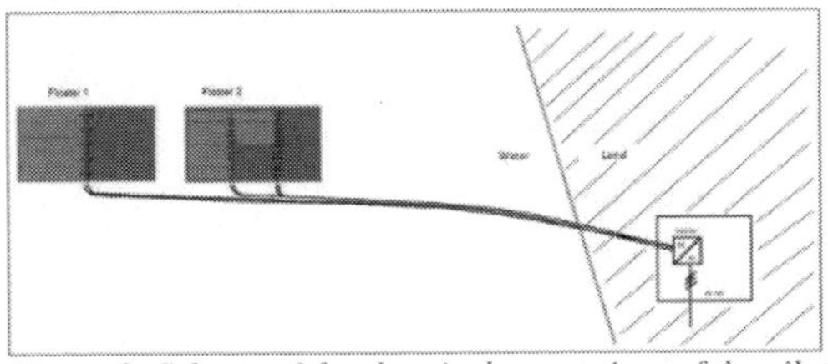

Figure 9: Scheme of the electrical connections of the pilot system.

The system performed well, but a severe serie of storms in February 2022 with strong winds upto 140 km/hr, caused damage to the adhesion of several PV modules. The PV performance test could not be continued.

Using an underwater drone, the underwater fouling of the floaters was inspected in the spring and summer of 2022. These inspection showed a modestly fouling by algae growth on the side walls of the floaters. The bottom of the floater though was free from marine fouling.

Bird droppings on the PV modules almost did not occur in the period November – March. In the spring and summer time the floaters turned out to become a favourite resting place for smaller seagulls, leaving a significant amount of bird droppings (see figure below). Rain fall could wash these droppings from the panel surfaces.

Figure 10: Snapshot of one of the floaters in the summer of 2022.

4 CONCLUSIONS

Offshore Floating PV has a huge potential for large scale production of renewable electricity, but attaining a competitive LCOE is a big challenge. This paper presents a concept that aims to minimize material and installation

10.4229/EUPVSEC2023/4BV.3.14
020316-003

costs. Laboratory tests as well as full scale tests conducted on the flexible CIGS PV panels and the floaters showed the survivability of the system in harsh (offshore) weather conditions. A pilot project on a lake near the sea confirmed these conclusions but also revealed that the laboratory tests conducted on the adhesives, used to mount the PV modules to the floaters, were not fully compliant.

Fouling of the floaters through growth of algae and barnacles is a potential issue for the buoyancy of the floaters but can effectively be suppressed by applying antifouling technologies. The pilot study moreover showed that this type of fouling primarily occurs on the side walls of the floaters and not on the bottom side.

Floaters with flat panels seem to be an attractive resting place for sea birds. Observations elsewhere though, indicate that systems with small angle tilted PV panels seem to be equally attractive for birds. The hydrophobic encapsulant of the flexible PV panels make a non-stick layer for the bird droppings and the PV panels did easily clean by rainfall. To obtain high PV performance, regular cleaning, will be necessary for near shore locations during spring and summer periods.

5 REFERENCES

[1] Challenges and potential for offshore solar (topsectorenergie.nl)

[2] World Bank Group, ESMAP and SERIS (2019) *Where Sun Meets Water: Floating Solar Handbook for Practitioners*, Washington DC: World Bank

[3] Wim Soppe, Aldo Kingma and Dorrit Roosen, *Mechanical Degradation Studies on flexible CIGS cells and modules for floating PV*, 2022 IEEE 49th Photovoltaics Specialists Conference (PVSC)

[4] Chen Yang, Kai Song, Xin Liang Xu, Gang Yao and, ZhenYu Wu, *Strain dependent effect on power degradation of CIGS thin film solar cell,* Solar Energy 195 (2020) 121–128.

COMPARATIVE ANALYSIS OF TWO AGRIVOLTAIC SYSTEMS FOR NIGHTTIME IRRIGATION OF PLAIN VEGETABLE PLOTS

M.A. Muñoz-García[1, *], G.P. Moreda[1], R. Sánchez-Calvo[2], M.B. Nieto-Morone[3, 4], M.C. Alonso-García[4]
[1]Research Group LPF-TAGRALIA. Dep. Ingeniería Agroforestal, ETSIAAB, Universidad Politécnica de Madrid. Av. Puerta de Hierro, 2. 28040 Madrid, Spain. Tel.: +34910670968.
[2] Research Group Hidráulica del Riego. Dep. Ingeniería Agroforestal, ETSIAAB, Universidad Politécnica de Madrid.
[3]PhD student, LPF-TAGRALIA. Dep. Ingeniería Agroforestal, ETSIAAB, Universidad Politécnica de Madrid.
[4]Dep. Energía, Unidad de Energía Solar Fotovoltaica, CIEMAT. Av. Complutense, 40. 28040 Madrid, Spain.
[*]Corresponding author. E-mail: miguelangel.munoz@upm.es

ABSTRACT: Agrivoltaics, the food and solar PV dual land use, is a promising land management strategy. In semi-arid climates, agricultural food production demands crop irrigation. Under these circumstances, an agrivoltaic setup will require a fraction of the PV production to be self-consumed by the irrigation pumping system, while the remainder is injected to the grid to obtain the expected revenue. Here, we assumed that nighttime irrigation would be desirable as it can save water, and this implies some type of PV energy storage. For a flat land and large plot acreage, the cost of erecting a correspondingly large capacity water tank stand might be prohibitive. Then, one possibility would be to resort to hydrogen to store the solar PV energy. Hence, we examined the main parameters to compare two alternative designs, one without the hydrogen conversion pathway and the other with an electrolyzer and a fuel cell. We concluded that the without fuel cell design is simpler and more cost-effective. The with fuel cell design features some merit points, like the possibility of using part of the hydrogen produced to refuel a hydrogen fuel cell tractor, or the absence of grid nighttime consumption.
Keywords: Solar PV irrigation, energy storage, grid-connected, water-energy nexus.

1 INTRODUCTION AND OBJECTIVES

Continuous increase in photovoltaic (PV) installed capacity might eventually cause friction between solar PV developers and farmers due to the latter reluctancy to accept the expropriation deal conditions proposed by the former. Further, although worldwide policies try to promote renewable energy share in the power generation mix, excessive land occupation is a sensitive matter, as it can foster food prices rise. The dual land use [1, 2, 3] known as *agrivoltaics* offers a potential solution to this. The aim of agrivoltaics is to reach a trade-off between agricultural production and solar PV generation, through the coexistence of both activities on the same land and at the same time.

In semi-arid climates, irrigation of vegetables is mandatory to achieve acceptable crop yields and of marketable quality. In these climates, nighttime automated irrigation is recommended as a rule of thumb, as it can save water.

Nighttime irrigation from solar PV requires some kind of energy storage, due to the time offset between electric energy generation and the irrigation operation itself. One possibility is to store the energy in electrochemical batteries. Nonetheless, if the plot features a natural land elevation or promontory, a more cost-effective option could be to install a water tank on the promontory. In this setup, daytime PV pumping replenishes the elevated water tank, conferring gravitational potential energy to the water stored. This head pressure allows for subsequent nighttime gravity i.e. no-pump irrigation. An affordable technology to install circular water tanks on natural elevations uses corrugated galvanized steel sheet tank shell anchored to a reinforced concrete ring beam foundation. The installation process includes some earthworks, basically ground levelling and trench excavation for the ring concrete beam.

If the croplands are plain (flat), yet the possibility of nighttime irrigation based on head pressure of elevated water reservoir remains, at the expense of erecting a tank stand, i.e. a lattice tower supporting structure with a top deck for the tank. However, if the plot acreage is large, PV water pumping to a tank stand becomes expensive, since the tank capacity should be correspondingly large. Hence, the supporting structure should be more robust, which ultimately would raise the cost.

In this work, we outline the main parameters that should be considered in a techno-economic comparison between two agrivoltaic setups for nighttime irrigation in semi-arid plains.

2 MATERIALS AND METHODS

Two alternative designs for nighttime-irrigation of flat fields were compared: One that uses less energy conversion steps and another that includes H_2 as a vector and therefore entails more conversion steps. The systems were termed without-fuel-cell (Fig. 1) and with-fuel-cell (Fig. 2), respectively.

The without fuel cell design (Fig. 1) entails nighttime consumption from the grid for the irrigation pump. The with fuel cell design (Fig. 2) obtains the energy for the irrigation pump via the H_2 pathway, which consists of: electrolyzer, H_2 compressor, fuel cell and inverter (the left-hand inverter in Fig. 2). The nighttime consumption of the solar inverters in both designs (the unique inverter in Fig. 1 and the right-hand inverter in Fig. 2) is considered negligible.

For the comparison, we assume a groundwater irrigation system, a water table deeper than 10-m and thereby a submersible well pump (vertical shaft pumps with motor out of the well are unadvisable for water table depths greater than 10-m, due to shaft warping propensity). For cost-effectiveness, groundwater irrigation systems are sometimes designed without storage tank and therefore they only require one pump, the well pump, which pressurizes the pipeline network. It is a cost-effective design since it saves the water tank itself plus one extra pump. However, installing a water tank brings several operational benefits that overcome the inherent

10.4229/EUPVSEC2023/4BV.3.16
020317-001

drawback of expensiveness.

The water tanks or water reservoirs in both Fig. 1 and Fig. 2 are at ground level, i.e., they are not elevated.

Besides H_2, the electrolyzer (Fig. 2) produces O_2, a gas with multiple uses in metal production, chemical processes, medical, etc. Indeed, following N_2, O_2 is the most used industrial gas [4].

The with fuel cell design (Fig. 2) includes a pathway to recover water, through a Peltier cooler [5]. This is important in semi-arid climates, where even groundwater is scarce and should not be wasted. In this regard, release of fuel cell steam to the atmosphere would be objectionable from an eco-friendly perspective.

Although not represented in Fig. 1 and Fig. 2, the understorey vegetable -e.g. lettuce- crop rows would sit parallel and close to the dripper pipelines depicted as garland blue lines.

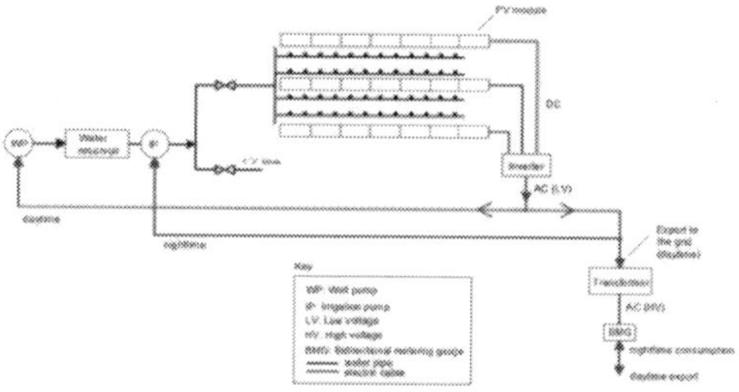

Figure 1: Grid-connected agrivoltaic system for nighttime irrigation, without fuel cell.

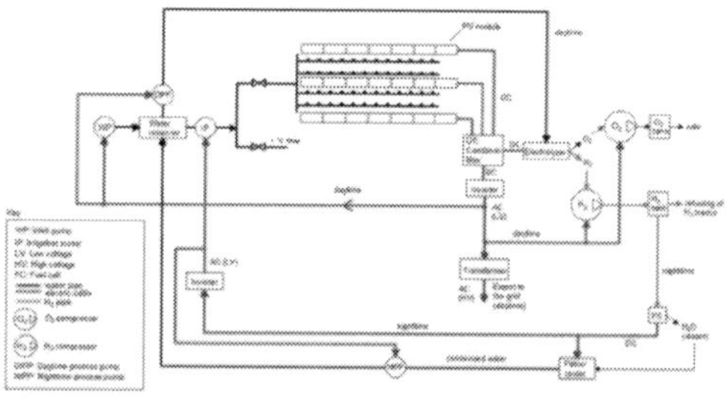

Figure 2: Grid-connected agrivoltaic system for nighttime irrigation including H_2 fuel cell technology.

3 RESULTS AND DISCUSSION

With regards to the parameters shown in Table I, it is important to note that the daily consumption of the well pump in the without-fuel cell design, $DcWP_{woFC}$, is smaller than its counterpart in the with-fuel cell design, $DcWP_{FC}$. The reason is that in the with-fuel cell design the well pump operates more hours, since the daytime process pump (DtPP) withdraws water from the reservoir for the electrolysis, what entails more functioning time of the well pump to replenish the water reservoir.

The yield of the PV array, Y_{PV}, is exactly the same for the two designs, i.e. the without-FC and the with-FC design.

The consumption of both process pumps, namely the daytime process pump (DtPP) and the nighttime process pump (NtPP), is low, since their power is small.

The income associated to the with-fuel cell design, I_{FC}, is given by:

$$I_{FC} = I_{GE} + I_{O_2} = Y_{PV} \cdot p_{ee} - DcWP_{FC} - DcHC - DcDtPP - DcOC + Y_{O_2} \cdot p_{O_2}$$

, where:

I_{GE}: Income obtained from the grid export

I_{O2}: Income obtained from the sale of the O_2 produced

Y_{PV}: Photovoltaic array nominal yield, in kWh

p_{ee} : price of the electricity exported (€/kWh)

$DcWP_{FC}$: Daily consumption of the well pump in the with-fuel cell design

DcHC: Daily consumption of the H_2 compressor

DcDtPP: Daily consumption of the daytime process pump

DcOC: Daily consumption of the O_2 compressor

Y_{O2}: Daily production of O_2

p_{O2}: Sale price of the O_2

With regards to the nighttime consumption for the irrigation pump in the without fuel cell design, it should be noted that it benefits from a lower price of the kilowatt-hour imported in the nighttime hours compared to the conventional price of the kilowatt-hour imported during the daytime hours.

Table I: Incomes and operating expenditures (OPEX) of the two designs explored, without- and with- fuel cell

		Incomes	OPEX
Without fuel cell (woFC)	Daytime	• High export to the grid, GE (equal to the daily yield of the PV array, Y_{PV}, minus the daily consumption of the well pump in the without FC design, $DcWP_{woFC}$)	• Farm tractor diesel fuel
	Nighttime	----	• Grid consumption (irrigation pump)
With fuel cell (FC)	Daytime	• Medium-low GE (equal to Y_{PV} minus $DcWP_{FC}$ minus daily consumption of the H_2 compressor minus daily consumption of the daytime process pump minus daily consumption of the O_2 compressor) • Sale of the O_2 produced	----
	Nighttime	----	----

10.4229/EUPVSEC2023/4BV.3.16

020317-002

With regards to the *with-fuel cell* design, the efficiency of the conversion chain (Fig. 2) that goes from the DC output of the PV array to the AC connection of the irrigation pump (IP) is given by:

$$\eta_{PV\rightarrow IP} = \eta_{electrolyzer} \cdot \eta_{HC} \cdot \eta_{FC} \cdot \eta_{inverter}$$

, where:
η_{HC} : efficiency of the H_2 compressor
η_{FC} : efficiency of the fuel cell
Assuming average efficiency values found in literature:
$\eta_{PV\rightarrow IP} = 0.70 \cdot 0.51 \cdot 0.60 \cdot 0.90 = 19\%$

Table II: Capital expenditures (CAPEX) of the two designs explored, without- and with- fuel cell

	CAPEX
Without FC	$CAPEX_{woFC}$
With FC	$CAPEX_{woFC}$ + purchase cost of: • 2 compressors (H_2 compressor and O_2 compressor): Expensive • 2 process pumps (daytime PP and nighttime PP); small power: Non-expensive • 2 tanks (H_2 tank and O_2 tank): Expensive • 1 electrolyzer: Very expensive[(a)]; 1271 €·kW^{-1} • 1 fuel cell: Very expensive[(a)]; 1119 €·kW^{-1} • 1 Peltier cooler: Non-expensive • 1 additional inverter: Non-expensive

[(a)]Source: [9], 2018 values; currency exchange rate applied: 2018 annual average of 1 € = 1.18 USD

The main shortcoming of the with-fuel-cell alternative is the intricate pathway of conversion steps that decrease overall energy efficiency and raise capital expenditure (Table II). Nonetheless, the with-fuel-cell system features several merit points: First, it is synergistic with agricultural machinery electrification based on H_2 fuel cell tractor [6, 7]. Second, it saves the operating cost of nighttime grid electricity consumption of the without-fuel-cell system. Third, in the case of off-grid agrivoltaic systems, it proves self-sufficient, unlike the without-fuel-cell system (to do the latter system autonomous, batteries should be installed that would increase capital expenditures). The *a priori* theoretical drawback of the water spent to generate the H_2 [8] is partially compensated by the nighttime condensation in the Peltier cooler (Fig. 2).

4 CONCLUSIONS

Agrivoltaics is envisaged as a potential way of harnessing solar PV while maintaining the agricultural use of land. An inalienable objective is to obtain a revenue from the electric energy produced. On the other hand, in semi-arid climates, where crop irrigation is mandatory, a part of the PV energy produced should be self-consumed by the irrigation pumping system.

In this work we assumed that nighttime irrigation is meaningful, as it can save water. Then, two possibilities exist: To use the grid as a backup for nighttime consumption, or to use some type of PV energy storage. With regard to the latter, if we omit batteries and tank stands, another option is to use a H_2 pathway. The merits and drawbacks of two alternative irrigated agrivoltaic designs have been explored. The without fuel cell design is simple and cost-effective, although it does not include energy storage. The other alternative, the with fuel cell design, features an intricate chain of conversion steps (overall efficiency of 19%), and its investment cost or capital expenditure is far more expensive than the counterpart of the without fuel cell design. However, since it includes energy storage in the form of H_2, this could be used to refuel a fleet of H_2FC tractors.

References

[1] A. Goetzberger, A. Zastrow, Int. J. Sol. Energy 1 (1982) 55.
[2] S. Schindele et al. Applied Energy 265 (2020) 114737.
[3] G.P. Moreda et al. Agronomy 11(3) (2021) 593.
[4] Berenschot. Oxygen synergy for hydrogen production. TESN118016-Waterstofversneller. Available at: www.berenschot.nl (2019).
[5] M.A. Muñoz-García et al., Computers and Electronics in Agriculture 93 (2013) 60.
[6] FuelCellsWorks. The first hydrogen tractor in the Netherlands. Available at: https://fuelcellsworks.com/news/the-first-hydrogen-tractor-in-the-netherlands/ . Last accessed: Feb. 5th 2022 (2020).
[7] G.P. Moreda et al. Energy Conversion and Management 115(1) (2016) 117.
[8] M.E. Webber, Environ. Res. Lett., 2 (2007) 034007.
[9] K. Mongird et al., 2020 Grid energy storage technology cost and performance assessment. Technical report. Publication No. DOE/PA-0204 (2020).

10.4229/EUPVSEC2023/4BV.3.16
020317-003

ASSESSMENT AND GUIDELINES FOR AN AGRIVOLTAIC PILOT IN ALENTEJO

Lisa Bunge[a] (lisa.bunge@uevora.pt), Luís Fialho[a,b] (lafialho@uevora.pt), Pedro Horta[a,b] (phorta@uevora.pt)

[a] Renewable Energies Chair, University of Évora. Pólo da Mitra da Universidade de Évora, Edifício Ário Lobo de Azevedo, 7000-083 Nossa Senhora da Tourega, Portugal
[b] Institute of Earth Sciences, University of Évora, Rua Romão Ramalho, 7000-671, Évora, Portugal

ABSTRACT: This paper presents a study of compatibility strategies for the interface of the new PV power plant of the Aquila Capital group at Cercal do Alentejo with agricultural and conservation activities, aiming at minimizing the environmental impact while engaging the local communities in the benefit of such approach. In the first phase, agricultural production and nature conservation compatibility strategies with the PV system were defined. For the agricultural interface, a one-year cultivation cycle of carrots followed by lettuce between the PV rows is proposed, and the integration of areas for sheep grazing and black pig holding. The nature conservation strategies are based on the improvement and recovery of the existing habitats, focusing on the existing floral species and the creation of new habitats through the introduction of native endangered vascular floral species. Furthermore, the incorporation of beehives is proposed such as the installation of shelters and nidification spots for birds. In the second phase, an agrivoltaic pilot was designed, aiming at studying the selected agricultural strategy.

Keywords: Agrivoltaic, Conservation, Environment, Photovoltaic

1 INTRODUCTION

Agrivoltaic presents new opportunities for the coexistence of agriculture with photovoltaic - PV electricity production. Recent research presents agrivoltaic projects and results with good potential when using aloe vera [1], lettuce [2], spinach and root crops (potatoes, beets, and carrots [3]). Other vegetables, like tomatoes [4] [5] and peppers, have also shown promising results. According to [6], all types of crops are generally suited for agrivoltaic application, but different effects on the yield are to be expected due to crop shading. Highly shade-tolerant crops such as leafy greens, stone fruits, berries, soft fruits, and special crops such as wild garlic, asparagus, and hops, show good potential.

Agrivoltaic can have a more pronounced effect in developed countries with arid and semi-arid regions, providing shade for plants and animals, while reducing water consumption [6]. The PV system can provide additional energy, for households, water treatment, food cooling, and processing, with even more benefits in decentralised systems in remote and impoverished areas. Adapted PV structures can also collect rainwater, which could fight desertification and soil degradation in affected countries, promoting ecosystem regeneration. Thus, agrivoltaic could also be used for nature conservation.

Considering the potentials of agrivoltaic, compatibility options of agriculture and nature conservation strategies with the new solar power plant at Cercal do Alentejo will be studied. The aim is to minimize the environmental impact linked to the construction and operation of the PV park and increase the land use potential.

2 CASE STUDY

2.1 Site and climate

Cercal do Alentejo, 37.8033°N 8.6708° W, is a parish located in the Alentejo province, near the Alentejo Coast (*Alentejo Litoral*), in the district of Setúbal and in the municipality of Santiago do Cacém [7], Portugal. Generally, the Alentejo is characterized by hot summers to very hot and fresh to moderate winters, with irregular and relatively low precipitation levels [8], especially in the last years due to climate change.

2.2 Cercal Power Plant

The new solar power plant at Cercal do Alentejo is being developed by the Cercal Power company of the Aquila Capital Group and will be located about 1.0 km from Cercal do Alentejo [9]. The project is still in the planning stage and is envisioned to have a nominal installed power of about 264 MWp [10]. The area that has been studied for the implementation of the PV plant is of almost 816 ha [9]. The area of PV arrays and inter rows has about 312.80 ha, of which only the panels occupy 125.79 ha [11]. The envisaged modules are of bifacial technology, supported by N-S single axis tracking structures fixed to the ground through directly driven metal piles and with an inter row spacing of 10.0 m [10].

2.3 Characterisation of the land

2.3.1 Soil

The implementation area includes some soils with good agricultural suitability integrated into the National Agricultural Reserve (*Reserva Agrícola Nacional* - RAN), although they have little significance [9]. Regarding the Land Use Capacity of the soils, the prevailing class is E followed by class D [12]. According to [13], soils of class E have very severe limitations for grazing, scrubland, and forest exploitation, very high erosion risks, and are not suitable for agricultural purpose, but instead for natural vegetation, protection, or forest restoration. Soils D in turn, have few or moderate limitations, high to very high erosion risk, and are generally not suitable for agriculture.

Figure 1: Landscape at the Cercal Power PV plant implementation area.

2.3.2 Water

The area has five groundwater abstractions, two vertical boreholes and some wells [12]. Through a hydrological study, seven main water lines were identified

10.4229/EUPVSEC2023/4BV.3.19
020318-001

on the project area, which are limited by the hydrographic basins. Most of these watercourses have a markedly torrential regime and remain dry for much of the year.

2.3.3 Biodiversity

2.3.3.1 Flora and Habitats

The designated area for the plant is not situated within areas of high conservation interest classified under the Habitats Directive [12]. However, there are various scattered *montado* areas, which are human created oak forests considered as habitats with protective status. On the site, most *montados* are in poor conservation condition [12]. In the study area, the following habitats were identified: 1) cork oak *montado* units - Habitat 6310 - Evergreen *Quercus spp.*; and 2) units related to riparian vegetation that develops along the banks of existing watercourses: 2a) reedbeds - Habitat 6420 - Mediterranean tall herbaceous *Molinio-Holoschoenion* wet meadows, and 2b) willow stands - Habitat 92A0pt3 - *Psammophilous* wooded *Salix Atrocinerea* stands [12]. Furthermore, some main water lines are associated to the National Ecological Reserve (*Reserva Ecológica Nacional* – REN) [9].

On the site 49 floral species were identified, distributed across 27 families. However, given the size of the study area, it has a limited floral diversity [12]. According to the Red List of Vascular Flora of Continental Portugal [14], the region of Baixo Alentejo has most of the endangered plants.

2.3.3.2 Fauna

During the environmental impact analyses, a total of 138 species of fauna were inventoried in the implementation area [12]. The faunal group with the highest number of identified species is birds, with 96 species, followed by mammals with 23, amphibians with 13, and reptiles with six species. Out of the total inventoried species, 30 were found on the site and 18 are considered concerning from a conservation standpoint.

3 COMPATIBILITY STRATEGIES

3.1 Agricultural strategies

Agriculture compatibility strategies are defined as agricultural activities, such as crop cultivation and livestock grazing, that are compatible with the existing layout of the commercial PV power plant Cercal Power, following the agrivoltaic concept.

3.1.1 Cultivation of crops

The first step was to find suitable crops for the PV plant and site under study. This involved researching various crops, mainly horticultural and herbaceous, that are typically cultivated in Portugal and in Mediterranean climate. The second step consisted in selecting crops with shade tolerant characteristics, once in the PV park the same will experience higher shading levels provided by the modules. As the aim is to identify one or more crops that can be effectively integrated into the PV park, which already has a predefined layout concerning the height of structures and inter row spacing, the third step was selecting crops whose maximal height is lower than the minimum height of the modules. It must be ensured that the selected crops do not create any shading on the PV modules, thus affecting energy production.

After the research, a total of 35 horticultures and 16 herbs typical from Portugal and Mediterranean climate were found. Given that the modules have a minimum height of 0.30 m at their maximum inclination [10], the selected crops should have the same maximum height or less. Among the shade tolerant crops, and considering that many horticultural varieties exceed this height, only carrots and lettuce turn out as suitable options for the PV park, as both have a maximum height of 0.30 m [15]. Consequently, a cycle of carrots and lettuce is suggested.

To guarantee enough space for the machinery, the cultivated area must be between the arrays delimited by the modules' lower edge at their maximum inclination, suggesting a safety distance to the trackers of 1.40 m. Therefore, as the trackers are spaced 10.0 m between rows, the proposed agricultural area between the arrays will be 7.20 m wide.

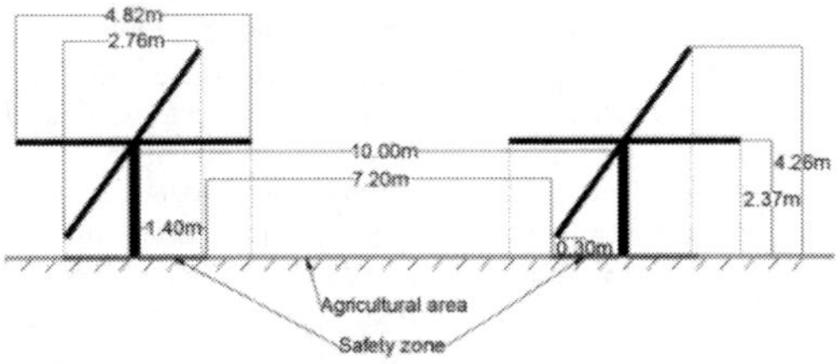

Figure 2: Proposed agricultural area between the PV arrays of the Cercal Power PV power plant.

To facilitate the movement of workers and machinery for maintenance purposes of the PV system, the agricultural area should feature two uncultivated rows. These rows should have a width matching the machinery's tire dimensions and be separated by a distance equivalent to the axle length.

3.1.2 Livestock grazing

During the field visit, landowners, who lease their land to the company for the PV project, were met. Among them, some currently use the land for sheep and cow grazing. According to [16], the PV structure can have a minimum height of 1.3 m for livestock activity, therefore, as the PV plant predicts a structure height of about 2.4 m [10], the same is suitable for livestock. Given the substantial size and weight of cows, it is not advisable to continue their grazing after the installation of the PV park, mitigating the potential damage to the PV infrastructure and to ensure the safety of the animals. However, sheep can safely continue their grazing practices if the PV system fulfils specific safety requirements, such as protecting all cables and electric equipment to avoid any contact with the animals. In this approach, sheep can profit from the modules shade, especially during summer. Additionally, the modules' shade may enhance the natural pasture growth between the rows.

Considering that the project area has several *montado* zones, the option of including black or Alentejo pigs is considered. The black pig/*montado* binomial is the key for a sustainable development of this native species and is important for the equilibrium of the *montado*'s biodiversity [17]. Given that the pigs will primarily remain in the vicinity of the oak trees to consume acorns, this approach would not be categorized as agrivoltaic, as there is little expectation of meaningful interaction and mutual benefit between the PV system and the pigs. Consequently, the black pigs should be held in the zones with higher *montado* density, identified in Figure 3.

10.4229/EUPVSEC2023/4BV.3.19
020318-002

3.2 Nature conservation strategies

Nature conservation compatibility strategies refer to initiatives designed to safeguard and enrich the natural spaces within the PV power plant, while mitigating the environmental impact caused through the construction and operation of the PV park. Moreover, with these strategies it is intended to create new habitats and promote greater biodiversity in the project site. The following approaches are an initial study of possible actions for the PV plant under study, which will be later evaluated and adjusted in collaboration with experts in fauna and flora.

The easiest zones to undertake renaturation efforts are the REN areas, ponds, and the main watercourses. Some of these zones remain untouched during the operation and maintenance of the PV park, what allows establishing small ecosystems functioning as natural sanctuaries. The process of renaturing these areas involves enhancing the existing habitats by promoting the growth of the existing flora and establishing shelters for the associated faunal species. A special focus should be considered on the Habitat 6310 of *Quercus spp.* and the two units of riparian vegetation, Habitat 6420 and Habitat 92A0pt3.

Considering the limited floral diversity and recognising the presence of numerous flora species under conservation status in the Alentejo province, it is worth considering the incorporation of specific endangered species native to the Cercal do Alentejo region within the PV plant. According to the Red List of Vascular Flora [18], the following species could be integrated on the project site:

- *Bupleurum acutifolium*: vulnerable, living in dry and open scrubland and rocky areas only in Cercal do Alentejo [18], 0.30–1.00 m high [19];
- *Chaenorhinum serpyllifolium*: in danger, endemic of the south-west coast of Portugal, living in calcarenite rock outcrops [18], 0.20–0.30 m high [20];
- *Doronicum plantagineum*: vulnerable, endemic of the south Portugal, living on edges and in the undergrowth of evergreen or marcescent scrub and woodland [18], reaches heights up to 0.90 m [21];
- *Klasea algarbiensis*: vulnerable, living in clearings of scrubland on sandy soils [18], floral buds grow close to the ground [22];
- *Scorzonera hispanica*: in danger, herbaceous perennial that inhabits forest clearings [18], 0.60–0.90 m high [23].

A blend of the listed species could be planted across different zones within the project's implementation area. Since each species is linked to a distinct habitat with unique properties, it might be necessary to establish specific environments, e.g., incorporation of calcarenite rocks for *Chaenorhinum serpyllifolium*.

Additionally, given that some studies highlighted increased ground coverage in the partial shaded areas between module rows in PV parks ([24] [25] [26]), it's reasonable to consider implementing such actions in the inter rows of the PV plant. Furthermore, the PV plant's landscape features valleys coincident with some watercourses that, when shaded by the PV arrays, might exhibit higher soil moisture levels that in the surroundings, consequently enhancing the expansion of ground coverage. Hence, valleys within the inter rows of the PV area are potentially favourable sites for the reintroduction of floral species. As for agricultural crops, these species should ideally have a height of less than 0.30, for the same reason as for the latter. Consequently, only *Chaenorhinum serpyllifolium* and *Klasea algarbiensis* from the endangered species listed could be integrated into the PV area. Some potential valleys are highlighted on Figure 3.

Based on a visual impact assessment of the PV park, specific zones require obstruction, which can be created using "green curtains" of indigenous trees, such as the *Arbutus unedo L.* [27]. The *Arbutus unedo L.* serves various purposes, including leather tanning, traditional medicine, and the production of the renowned *medronho* brandy, which makes it economically attractive. Additionally, it thrives on nutrient-poor soils and is fire resistant. Its blossoms provide bees with nectar and pollen, fruits serve as sustenance for birds, and its dense foliage offers shelter to insects and small animals in winter.

Furthermore, integrating beehives across the entire area for pollination could be explored to enhance the natural propagation of the floral species, while preventing the collapse of the bee populations. Additionally, this practice could lead to the production of local honey.

Actions aimed at enhancing floral biodiversity will consequently enhance faunal biodiversity, once a healthy and diverse plant community can provide a wide array of food, shelter, and other resources that the faunal species depend on. Although, it is advised to introduce supplementary wildlife measures, especially for birds, since the same are the predominant faunal group at the project site, having 17 endangered species. Thus, incorporating shelters for birds within the REN and forest areas could be considered. These shelters can include hidden nidification spots on trees or on artificial structures across the project site.

4 AGRIVOLTAIC PILOT DESIGN

In addition to the compatibility strategies, the design of a dedicated agrivoltaic system was caried out. The initial use of the agrivoltaic pilot is for research, such as a selected area with the proposed agriculture compatibility strategy at the PV area. Thus, three experimental plots with the same cultivated crop are suggested, allowing a comparative assessment of the crop development in three distinct scenarios:

1. The agrivoltaic approach;
2. Retrofitting with agrivoltaic of commercial PV plants;
3. The reference situation with full sun exposure.

Furthermore, within the agrivoltaic plot, three different shade patterns are proposed. With the proposed areas and different shading levels in the agrivoltaic pilot, it is intended to answer the following questions:

Q1: Can we integrate agricultural activities in pre-existing commercial PV power plants without modifying their design?

Q2: How does the configuration of the PV modules influence the yield?

Q3: Is agrivoltaic a solution for the Alentejo province?

4.1 Selecting the areas

For implementing an agrivoltaic project, a suitable area for the same had to be identified on the project area. For the later comparison of the results of the pilot with the observations made at the PV arrays and reference areas, the three plots should be relatively close so they experience similar environmental conditions. The selection of the area for implementing these three plots was based on the following criteria: proximity to a water source; gently sloping topography; easy access by

10.4229/EUPVSEC2023/4BV.3.19
020318-003

agricultural machinery; and representative size. Three areas were selected of 25 m x 80 m each (Figure 3), which are near to each other and to a well for the further irrigation of the crops. At the PV park, the area will cover three PV rows, equivalent to two inter row spacings for cultivation.

4.2 Design of the PV system

The following design of the agrivoltaic pilot is conceptual and purely based on engineering considerations.

4.2.1 Structure

Following the recommendations of [28] and [29], the structure should be at least 5 m high to allow agricultural machinery to work safely. Therefore, this height was chosen for the agrivoltaic pilot. Moreover, fixed structures were selected for being economically more attractive, with a module inclination of 33 ° which is the optimal angle for Cercal do Alentejo [30]. Regarding the interrow spacing, 3 m are assumed to be adequate to allow enough light to reach the crop while still achieving satisfactory energy yields [29].

4.2.2 PV module technology

Once it is intended to study the impact of different shading levels or patterns, three different zones are suggested which are provided by different PV technologies and densities:

- High shade: opaque with full density;
- Partial shade: opaque with half density;
- Light shade: semi-transparent with half density.

To reduce the CAPEX expenses associated with the pilot project's higher structures, it is proposed to use second-life opaque modules.

4.3 Design of the agricultural system

In order to enable a meaningful comparison, given that the suggested crops for the PV arrays of the Cercal Power plant are carrots and lettuce, these will also be proposed for the agrivoltaic and reference plots. Moreover, the cultivated area in all three plots needs to be consistent. Given that the PV park and its layout impose restrictions on the available cultivated space, both the reference and agrivoltaic plots have the same cultivated area as defined for the PV plant. Thus, the area effectively cultivated in each plot will be of 2 x 7.2 m x 80 m, equivalent to 1152 m^2. Further, an appropriate irrigation system must be designed to satisfy the water requirements of the crops.

4.4 Monitoring

To understand the impact of the different PV configurations on agricultural conditions and yields, it is necessary to define an appropriate monitoring system to install in each study area. Besides the meteorological conditions measured at the weather station, the following agronomic parameters should be monitored:

- Agricultural yield;
- Evapotranspiration;
- Leaf area index;
- Photosynthesis;
- Photosynthetically Active Radiation - PAR;
- Soil quality through point demonstrations over time;
- Soil moisture and temperature.

4.5 Procedures

For the installation of the agrivoltaic pilot, the following steps were defined:

1. Installation of the modules structure, avoiding soil compaction and land movement, preferably during a dry period and outside the growing season;
2. Installation of the modules and inverter.

The next actions, besides being for the agrivoltaic pilot, are also applicable to the adapted PV area:

3. Installation of all cables at an appropriate and safe height (2 m or more) under the modules and along the mounting structures using additional protection elements [31];
4. Design and installation of the irrigation system according to the hydric needs of the crops;
5. Installation of the monitoring systems;
6. Soil quality analysis;
7. Soil fertilisation based on the crop requirements;

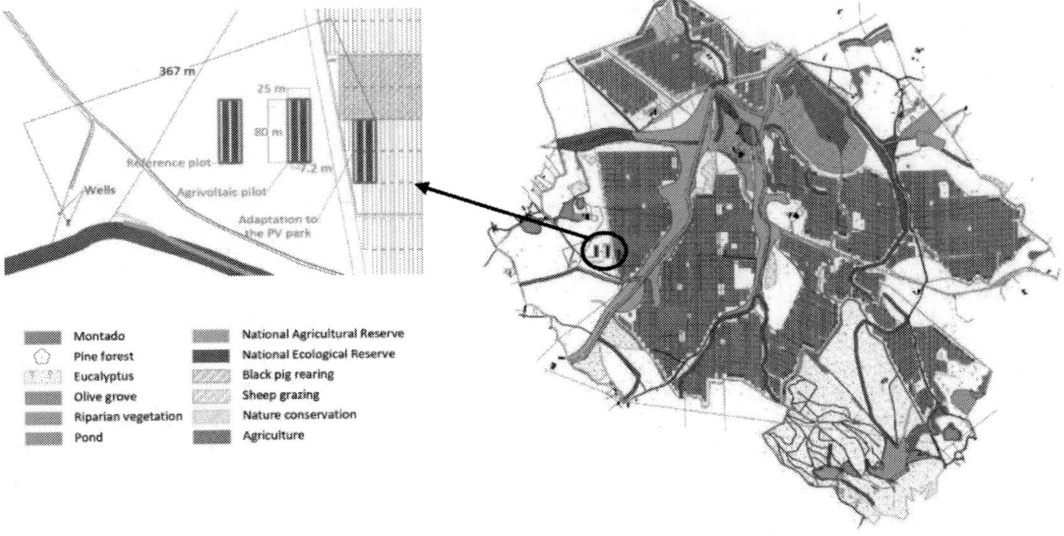

Figure 3: Proposed areas for livestock, nature conservation within valleys with PV arrays, and agrivoltaic experiment plots.

Table I: Characteristic of the experiment plots.

		Agrivoltaic Pilot			*Agrivoltaic adaptation to PV park*	*Reference plot*
		High shade	*Partial shade*	*Light shade*		
Plot dimensions		25 m x 80 m; 2000 m^2				
PV system	***Height***	5.0 m			2.4 m	-
	Interrow spacing	3.0 m			10.0 m	-
	Orientation	N-S			N-S	-
	Flexibility	Fixed, 33 °			Single-axis	-
	Foundation	Driven metal piles			Driven metal piles	-
	Module technology	Opaque 2nd life	Opaque 2nd life	Semi-transparent	Bifacial	-
	Module density	Full density	Half density	Half density	-	-
Agricultural activity		Carrots-Lettuce cycle				
Agricultural area		2 x 7.2 m x 80 m; 1152 m^2				

8. Field works according to the crop requirements. It must be ensured that these works do not coincidence with the maintenance procedures of the PV system;
9. Cultivation of the crop;
10. Module cleaning if any of the previous steps causes elevated soiling levels;
11. Data collection of the different parameters monitored during the crop cycle;
12. Final crop product quality evaluation after harvest;
13. Data analyses and comparison between plots;
14. Assessment of each set up and identification of improvements.

5 CONCLUSIONS

This paper introduces initial agricultural and nature conservation compatibility strategies for the planned PV park at Cercal do Alentejo. The agricultural strategies fall under the concept of agrivoltaic and were defined according to the existing layout of the PV system, which determined the maximum crop height and the available area for cultivation. The proposed compatibility strategies focus on integrating a crop cycle of carrot and lettuce into the predefined layout of the commercial PV park. Additionally, the existing areas for sheep rearing can continue on the PV area where the animals can profit from the shading. Furthermore, besides not being an agrivoltaic approach, rearing of black pigs is proposed on the *montado* areas. Nature conservation strategies are initiatives to enhance the local biodiversity by improving, recovering, and creating habitats for floral and faunal species. These actions include, among others, the introduction of endangered floral species, beehives, and shelters for birds. Furthermore, an agrivoltaic pilot was designed to explore how different PV system configurations influence agronomic parameters. For comparison purposes, it is planned to carry out an experiment with the same crop cycle of lettuce and carrots at the agrivoltaic pilot, a cultivated area within the arrays of the PV park and at a reference plot. It is forecasted to observe variations between the distinct study areas, once different shading conditions and potentially diverse microclimatic conditions are generated by each PV system configuration. It is expected to have an enhanced crop quality in the agrivoltaic and PV areas, since during summer the modules will protect the crops from the intense solar radiation. Additionally, this shelter can help maintaining the moisture levels in the soil. The aim of this pilot and procedure is to understand whether the integration of agricultural activities with PV energy generation, without compromising the respective yields, demands the development of an entirely new system. Simultaneously, the investigation seeks to determine the feasibility of incorporating such activities into existing commercial PV facilities. Should this conjecture hold true, technical guidelines and practices can be defined for the incorporation of agricultural activities at pre-existing PV power plants. Furthermore, this study is the first implemented in the Alentejo region on large scale. If the results are promising, this approach may present a solution to local farmers who are tackling the higher temperatures and less precipitation levels due to climate change, while profiting from cheap and clean energy.

6 FUTURE WORK

In the scope of this project, the following future tasks are planned:

- Procurement of the materials and equipment for the agrivoltaic pilot and monitoring systems;
- Conducting the experimental study on the agrivoltaic pilot, adapted PV area and reference plot;
- Development of guidelines and best practises for the implementation of agrivoltaic systems in Portugal;
- Assessment and adjustment of the nature conservation strategies with experts of the flora and fauna areas;
- Developing a pilot for the nature conservation strategies to study the feasibility of renaturation in PV parks. If this is demonstrated, best practises can be identified which can then be applied to every preexisting commercial PV park.

7 ACKNOLEDGEMENTS

The authors would like to thank the Aquila Capital Group for the opportunity to undertake this study.

8 REFERENCES

[1] S. Ravi, J. Macknick, D. Lobell, C. Field, K. Ganesan, R. Jain, M. Elchinger and B. Stoltenberg, "Colocation opportunities for large solar infrastructures and agriculture in drylands," *Applied Energy, Elsevier*, pp. 383-392, 2016.

[2] H. Marrou, J. Wéry, L. Dufour and C. Dupraz, "Productivity and radiation use efficiency of lettuces grown in the partial shade of photovoltaic panels," *European Journal of Agronomy, Elsevier*, pp. 54-66, 2013.

[3] E. Bellini, "Agrivoltaics works better with leafy greens, root crops," PV Magazine, 8 June 2020.

[Online]. Available: https://www.pv-magazine.com/2020/06/08/agrivoltaics-works-better-with-leafy-greens-root-crops/. [Accessed 17 January 2022].

[4] M. Cossu, L. Murgia, L. Ledda, P. A. Deligios, A. Sirigu, F. Chessa and A. Pazzona, "Solar radiation distribution inside a greenhouse with south-oriented photovoltaic roofs and effects on crop productivity," *Applied Energy - Elsevier*, 2014.

[5] G. A. Barron-Gafford, M. A. Pavao-Zuckerman, R. L. Minor, L. F., I. Barnett-Moreno, D. T. Blackett, M. Thompson, K. Dimond, A. K. Gerlak, G. P. Nabhan and J. E. Macknick, "Agrivoltaics in drylands: Co-location has food, water, and renewable energy benefits".

[6] "Agrivoltaics: Opportunities for Agriculture And The Energy Transition," Fraunhofer Institute for Solar Energy Systems ISE, Freiburg, 2020.

[7] "Caracterização," Cercal do Alentejo, [Online]. Available: https://www.freguesiadecercaldoalentejo.pt/caracterizacao. [Accessed 3 November 2022].

[8] C. M. Faísca, "A Produção Agrícola no Alentejo (1929-2018): Uma Primeira Abordagem," *Revista de Estudios Económicos y Empresariales*, pp. 39-64, 2019.

[9] "Estudo de Impacte Ambiental da Central Fotovoltaica do Cercal e Linha de Muito Alta Tensão Associada Volume 1.1 Relatório Técnico (Capítulos 1 a 5)," Matos, Fonseca & Associados, Estudos e Projetos Lda., 2021.

[10] "Memória Descritiva - Central Fotovoltaica do Cercal," Cercal Power, 2022.

[11] "Nota Técnica - Reformulação do layout da Central Fotovoltaica do Cercal e Demonstração da conformidade da proposta de layout do Projeto," Cercal Power, 2022.

[12] "Estudo de Impacte Ambiental da Central Fotovoltaica do Cercal e Linha de Muito Alta Tensão Associada Volume 1.2 Relatório Técnico (Capítulos 6 a 7)," Matos, Fonseca & Associados, Estudos e Projetos Lda., 2021.

[13] "Nota Explicativa da Carta dos Solos de Portugal e da Carta de Capacidade de Uso do Solo," Direção-Geral de Agricultura e Desenvolvimento Rural, [Online]. Available: https://www.dgadr.gov.pt/nota-explicativa. [Accessed 16 August 2023].

[14] "Lista Vermelha da Flora Vascular de Portugal Continental," [Online]. Available: https://listavermelha-flora.pt/publicacao-final-do-projeto-ja-esta-acessivel-na-sua-versao-digital/. [Accessed 7 November 2022].

[15] D. Almeida, Manual de Culturas Hortícolas Volume 1, Lisboa: Editorial Presença, 2006.

[16] "Linee Guida in materia di Impianti Agrivoltaici," 2022.

[17] A. Oliveira, A. Durão and F. Carvalho, "Biodiversidade da raça suína autóctone portuguesa porco alentejano (Sus ibericus)," *Revista Internacional em Língua Portuguesa*, vol. 35, pp. 151-166, 2019.

[18] A. Carapeto, A. Francisco, P. Pereira and M. Porto, Lista Vermelha da Flora Vascular de Portugal Continental, Lisboa, 2020.

[19] S. Castroviejo, Flora Iberica, Madrid: Real Jardín Botánico, 1986-2012.

[20] "Chaenorhinum serpyllifolium (Lange) Lange subsp. robustum (Loscos) Mateo & Figuerola," Herbario Virtual del Mediterráneo Occidental, [Online]. Available: http://herbarivirtual.uib.es/es/general/1950/especie/chaenorhinum-serpyllifolium-lange-lange-subsp-robustum-loscos-mateo-figuerola. [Accessed 29 August 2023].

[21] "Doronicum plantagineum subsp. tournefortii," Biodiversidade a seus pés, [Online]. Available: http://almargem.org/biodiv/especie/doronicum-plantagineum-subsp-tournefortii/. [Accessed 29 August 2023].

[22] "Sítios de Interesse Botânico de Portugal Continental - Tomo II," Imprensal Nacional , Lisboa, 2021.

[23] J. M. Stephens, "Scorzonera - Scorzonera Hispanica L.," IFAS Extensions - University of Florida, [Online]. Available: https://edis.ifas.ufl.edu/publication/MV131. [Accessed 29 August 2023].

[24] M. Graham, S. Ates, A. P. Melathopoulos, A. R. Moldenke, S. J. DeBano, L. R. Best and C. W. Higgins, "Partial shading by solar panels delays bloom, increases floral abundance dureing the late-season for pollinators in a dryland agrivoltaic ecosystem," *nature portfolio*, vol. 11, 2021.

[25] M. C. Moscatelli, R. Marabottini, L. Massaccesi and S. Marinari, "Soil properties changes after seven years of ground mounted photovoltaic panels in Central Italy coastal area," *Geoderma Regional*, vol. 29, 2022.

[26] E. H. Adeh, J. S. Selker and C. W. Higgins, "Remarkable agrivoltaic influence on soil moisture, micrometeorology and water-use efficiency," *PLos ONE*, vol. 13, no. 11, 2018.

[27] "Medronheiro - Arbutus unedo L.," Brigada da Floresta, [Online]. Available: https://brigadadafloresta.abae.pt/medronheiro/. [Accessed 29 August 2023].

[28] "Agrivoltaics: Opportunities for Agriculture and the Energy Transition," Fraunhofer Institute for Solar Energy Systems ISE, Freiburg, 2022.

[29] A. Weselek, A. Ehmann, S. Zikeli, I. Lewandowski, S. Schindele and P. Högy, "Agrophotovoltaic systems: applications, challenges,and opportunities. A review," *Agronomy for Sustainable Development*, vol. 35, no. 39, 2019.

[30] "Photovoltaic Geographical Information System," European Commission, [Online]. Available: https://re.jrc.ec.europa.eu/pvg_tools/en/.

[31] "Agrisolar Best Practice Guidelines Version 2.0," SolarPower Europe, 2023.

10.4229/EUPVSEC2023/4BV.3.19
020318-006

MPPT TRACKING FOR VEHICLE INTEGRATED PHOTOVOLTAICS APPLICATIONS

Kasagani Naga Bhavya Jyothi, Narendra Shiradkar
Indian Institute of Technology Bombay, Mumbai Maharashtra, India
bhavyajenny18@gmail.com, naren@ee.iitb.ac.in)

ABSTRACT: This paper analysis the effect of rapidly varying irradiation on the solar module performance in vehicle integrated photovoltaics(VIPV) domain due to intermittency in the solar irradiation and the vehicle movement. There could be a huge potential if the solar panels can be mounted on car roofs and maximum power can be extracted efficiently even under irradiation varying conditions. Thus, a simulink model was developed to model different shading patterns for prolonged duration. MPPT control strategy of particle swarm optimization(PSO) is tested with a simulink model developed for a 360W module with three bypass diodes. Two different solar panel topologies – series and series-parallel are considered to analyze the significance of solar cells interconnection in a module for VIPV application. For comparative analysis, a model without MPPT tracking is developed and compared with PSO tracking algorithm results. In this paper, the importance of the tracking speed and accuracy in reaching the module global peak power value are detailed relating to the simulation results.

Keywords: partial shading, varying irradiance, maximum power point tracker, vehicle integrated photovoltaics

1 INTRODUCTION

The emission of greenhouse gasses has led to global warming which has worsened due to the significantly increasing energy demand all over the world. One of the contributors to the global warming is the transportation sector which corresponds to one fourth of the greenhouse gas emissions. Therefore, there is an alarming situation to reduce the greenhouse gas emissions from non-electric vehicles. On the other hand, electric vehicles contribute significantly less to greenhouse gas emissions. Therefore, in future, all the non-electric vehicles will be replaced by electric vehicles. But, this would further enhance the demand for electricity in order to charge these electric vehicles. One of the efficient solutions would be to utilize the natural resources i.e., solar energy which is available in abundance to address the electricity demand of electric vehicles i.e., cars. A solar panel mounted on a car roof would satisfy this energy requirement of the vehicle to some extent which would reduce the burden on the charging stations.

In general, solar panels mounted on rooftop or open areas would face partial shading issues due to surroundings and meteorological conditions. In this condition, maximum power point trackers would ensure the module is operating at its peak power point[1]. However, this might not be an easy solution in case of VIPV as the car itself would be in non-stationary mode leading to varying irradiation. This irradiation would be rapidly changing depending on the car speed and also the module would be under partial shading condition. To utilize the solar panel energy to the fullest, the MPPT tracker in the PV system must be efficient enough to extract the maximum power from the module under rapidly varying irradiation conditions.

In order to extract maximum power from the module, the MPPT tracker must track the peak power point at a significantly faster rate than the varying irradiation speed[2]. This is a challenging task as the car would be moving at around 30 kmph speed. To analyze the capability of the MPPT tracker to extract maximum power from the solar module, a Simulink model is developed to replicate the VIPV scenario. Particle swarm optimization(PSO) algorithm is considered to understand the impact of tracking maximum power and the convergence speed[3]. A shading pattern is considered and the simulation results are studied. For further analysis, Simulink models with different module configurations are also developed and the results are compared with and without including MPPT algorithm.

2 MATLAB Simulink Model for VIPV

A Simulink model is developed in MATLAB to replicate a situation where a solar panel is receiving irradiation that varies in seconds as shown in fig.1.

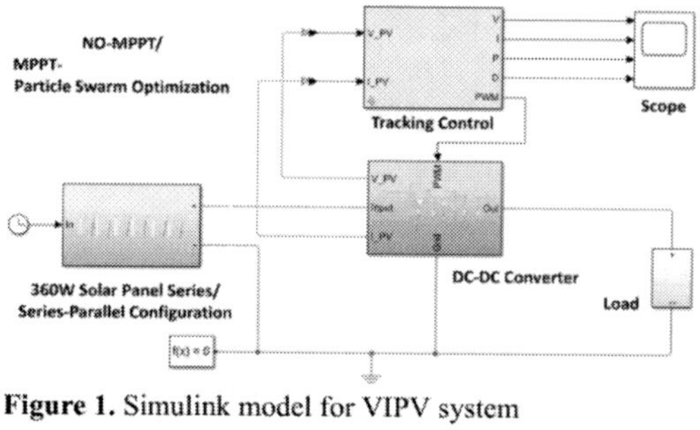

Figure 1. Simulink model for VIPV system

A solar panel of 360W maximum power is connected to a buck-converter designed to deliver maximum power to the load with a tracking control. In a cell series configuration of a module, all the 72 cells are connected in series with three bypass diodes across 24 cells as depicted in Fig.2(a). Whereas, series-parallel configuration has 144 half-cut cells, with 72 cells connected in series and three bypass diodes across 48 cells as illustrated in Fig.2(b).

The DC-DC buck converter is designed to step down the maximum power point voltage(V_{mpp}). An auxiliary battery of 14V would be charged in a car with the power delivered from the solar panel. Therefore, a load is considered to receive 14V in the model. Thus, the buck-converter design would step down the 36V(V_{mpp}) to 14V. Inductor and capacitor values are calculated for a buck-converter switching frequency of 20kHz. Under 1000W/m2, the load would be maintained at 14V and 360W. This would correspond to a load value of around 0.6ohm. The duty cycle of the buck converter switching time would be controlled by an MPPT algorithm to

10.4229/EUPVSEC2023/4BV.3.21
020320-001

deliver maximum power to the load.

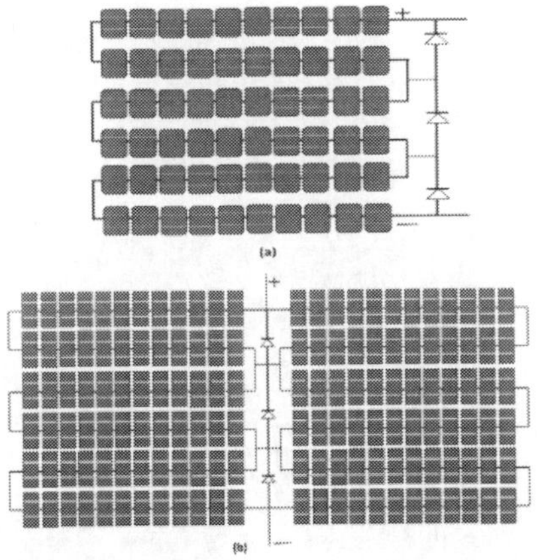

Figure 2. Solar panel topologies (a) series connection (b)series-parallel connection

The developed simulink model is shown in fig.1. In order to analyze the MPPT tracking efficiency, two different scenarios are considered. The following are the scenarios

1. Uniform illumination all over the module
2. Irradiation over one full size cell is varied

In order to identify the effects of the solar panel configuration, series and series-parallel configurations also would be considered for achieving maximum power

3 MAXIMUM POWER POINT TRACKING ALGORITHM

The module mounted on the car roof would be subjected to partial shading which would result in reduced power. Under this condition if maximum power point is not tracked, the output power delivered to the load would further decrease implying the inefficient utilization of the module power. In this case an MPPT algorithm would play a significant role in tracking the maximum power point, thus delivering maximum power generated by the module to the load. There are many MPPT tracking algorithms followed to track the maximum power point. However, under partial shading conditions, most of the MPPT algorithms fail to track the global peak. Among the available MPPT control techniques, particle swarm optimization (PSO) is observed to be an efficient optimization technique to track the global maximum power point (GMPP). This algorithm is based on the scattered group of swarm movement towards a single direction. This replicates to the situation of the module operation point with a global point i.e., GMPP.

The Simulink model has implemented a PSO tracking algorithm that collects the voltage and current values continuously. Firstly, a set of duty cycles would be considered randomly by the algorithm for buck-converter and the corresponding power values would be compared during simulation of the Simulink model. The maximum power among the obtained power values would be taken and new duty cycles would be calculated to track the peak point. Eq(1) would be used to find the delta duty cycle, to track the duty cycle value for GMPP. The delta_duty_cycle value would be added to the current duty cycle set to generate the updated duty cycle set for every iteration.

Delta_duty_cycle=(w*velocity)+(c1*r1)*(pobest-d))+(c2 *r2*(gwbest-d)) (1)

Weight(w), acceleration coefficients(c1,c2) are the values to be specified at the start of the simulation to track the peak point. The tracking efficiency depends on the combination of w, c1 and c2. This combination plays a major role in tracking the GMPP. Also, d is the duty cycle value, pobest is the peak power in the sample set of duty cycles, gwbest is global peak power from the different pobest obtained for the repeated iterations for updated sets of duty cycle values, r1 and r2 are the random values from 0 to 1. At a particular iteration, the duty cycle set would have all the values nearer to each other, implying the duty cycle corresponding to GMPP.

The algorithm time duration, t1 for obtaining the delta_duty_cycle would affect the convergence speed of the algorithm to track GMPP. This time duration would be depending on the components in the PV system and their settling time. If t1 is very low or very high, the algorithm would find it difficult to track the GMPP. This t1, is the perturbation time step for starting a new iteration of the PSO algorithm. This t1 would be one of the limiting factors to track the GMPP. The simulations with the PSO algorithm would be performed for the scenarios mentioned in section 2. In order to understand the significance of the MPPT algorithm, a model without MPPT tracking control is simulated. In this model, the duty cycle for the buck-converter is calculated for module maximum power point voltage of 36V for $1000 W/m^2$ irradiation on the module. Considering the load must be at 14V, duty cycle is calculated by eq(2) i.e., the ratio of the output voltage to the input voltage

$$D = \frac{14}{36} = 0.388 \quad (2)$$

Thus, duty cycle input of 0.388 with a PWM generator frequency of 20kHz is given to the buck converter. The results of the model without MPPT algorithm would be considered with the PSO MPPT model results for comparison.

4 SIMULATION RESULTS

4.1. Uniform illumination all over the module

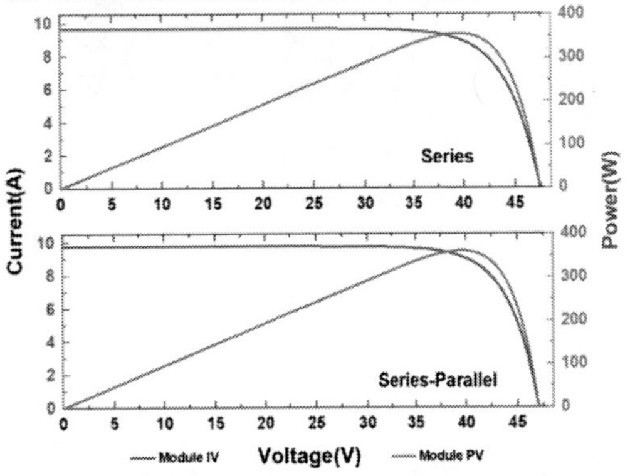

Figure 3. Module IV and PV curves for series and series-parallel topologies for $1000 W/m^2$ uniform illumination all over the module

10.4229/EUPVSEC2023/4BV.3.21
020320-002

The IV and PV characteristics of the series and series-parallel configuration modules for 1000W/m^2 uniform irradiation all over the module are shown in fig. 3. Simulations are performed to track maximum power for uniform irradiation for a time duration of 0.5s. Fig.4 depicts the power tracking time for series and series-parallel configurations. It can be observed that the all the models track the maximum power, however the PSO algorithm is limited by the convergence speed.

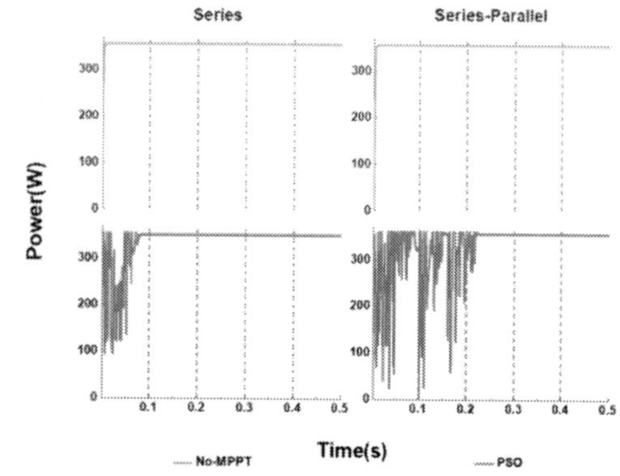

Figure 4. Comparison of power tracking speed of PSO and without MPPT model results considering series and series-parallel module configurations. The module is illuminated uniformly(1000W/m^2)

Overall energy is calculated for 0.5s time duration and shown in fig.5. Model with the PSO algorithm generates lower energy compared to the model without the MPPT algorithm, but the energy is the same for series and series-parallel module configurations. To analyze better, the same simulation procedure is repeated for a scenario where irradiation of one full size cell(6"x6") would be varying for every 0.5s.

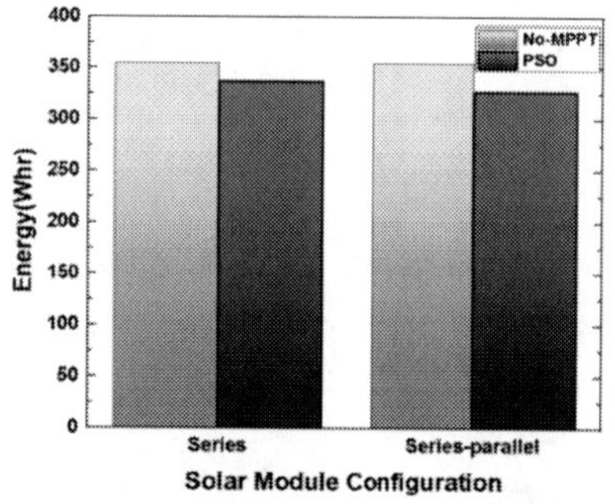

Figure 5. Comparison of overall energy generated for 0.5s by the PSO and without MPPT control strategy. Seri-es a series-parallel configuration are also considered and compared for 1000W/m^2 uniform illumination

4.2. Irradiation over one full size cell is varied

All the cells in the module are maintained at 1000W/m^2 except one full size cell. The shading of a full size cell for both the module topologies is illustrated in fig. 6. Two cells must be shaded in the series-parallel configuration as depicted in fig.6(b) to replicate the same scenario of one full size cell shading in series configuration. The corresponding IV curves are simulated and shown in fig.7. Multiple peaks can be observed in the graph.

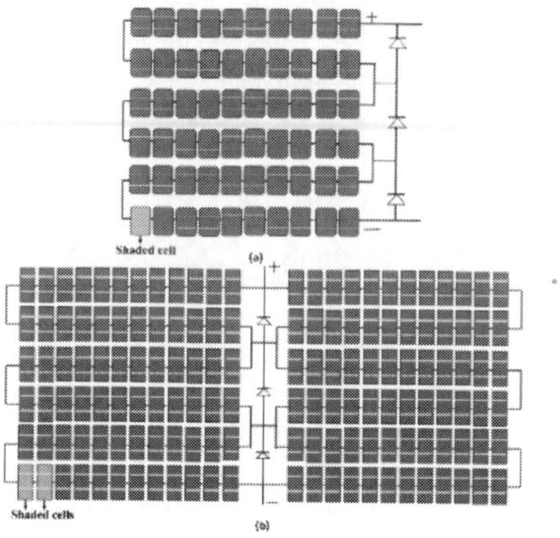

Figure 6. Full size cell shading in both solar panel topologies (a) series connection (b)series-parallel connection

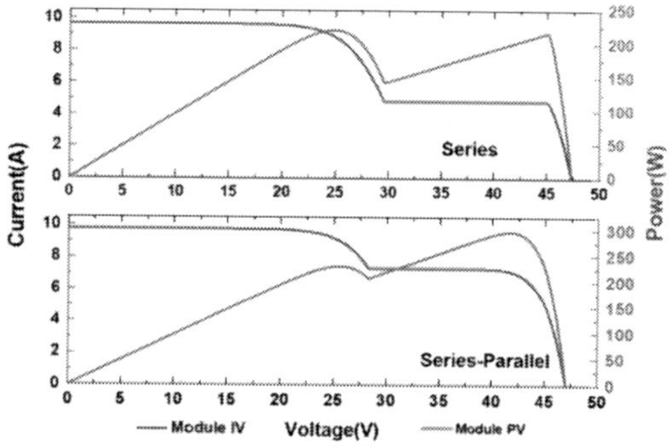

Figure 7. Module IV and PV curves with one full size cell shaded i.e., light intensity-500W/m^2

The models are now tested for tracking the GMPP. Simulations are performed by varying the irradiation on the highlighted cells in fig.6 i.e., irradiation changes from 1000W/m^2 to 500W/m^2 for every 0.5s. The simulation results are shown for a total duration of 2s in fig.8. It can be clearly seen that the model without MPPT tracks faster, but doesn't reach the maximum power. However, PSO tracks the maximum power. Further, the power is high for series-parallel configuration for 500W/m^2 irradiation on full size cell condition. This implies that the cells interconnection in the module plays a vital role in delivering the maximum power to the load under shading conditions.

In order to analyze the significance of the MPPT algorithm and module configuration, overall energy is calculated and depicted in fig.9. The achieved energy is higher for series-parallel configuration by around 10%. The energy for no-MPPT condition is observed to outperform despite not tracking the maximum power. This is because the PSO algorithm takes time to converge at the maximum power point, while the model without MPPT algorithm converges faster. Therefore, a PV

10.4229/EUPVSEC2023/4BV.3.21
020320-003

system in a VIPV application can't provide better results just by using an MPPT algorithm. It is important to note that the algorithm tracking speed must be significantly faster to receive higher energy. This is because the convergence speed of the PSO algorithm limits the total energy for varying irradiation conditions as the algorithm takes time to track the GMPP. Therefore, there is a need for an algorithm whose convergence speed is significantly faster under rapidly varying irradiation conditions.

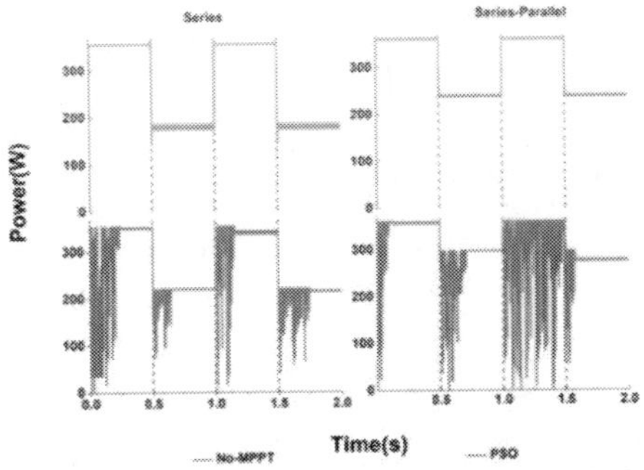

Figure 8. Comparison of power tracking speed of PSO and without MPPT model results considering series and series-parallel module configurations. The module has one full size cell irradiation varying from 1000W/m^2 to 500W/m^2

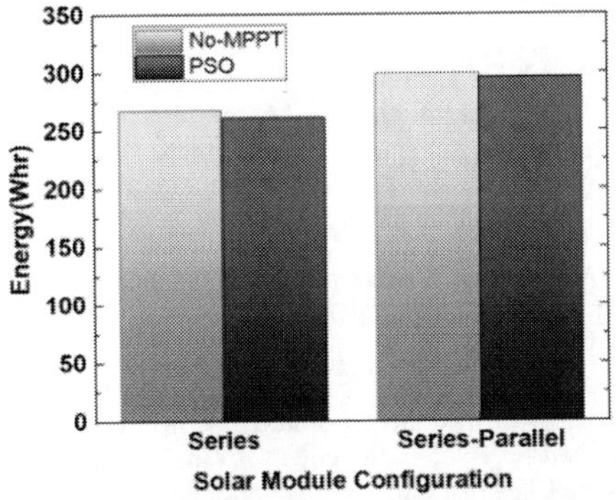

Figure 9. Comparison of overall energy generated for 2s by the PSO and without MPPT control strategy. Series and series-parallel configuration are also considered and compared for varying irradiation on full size cell from 1000W/m^2 to 500W/m^2 with other cells maintained at 1000W/m^2

5 CONCLUSION

Power tracking speed is significantly faster when the PV system has no MPPT control algorithm, but the maximum power can't be extracted under partial shading conditions without an efficient tracking control strategy. Thus, MPPT algorithms play a crucial role during rapidly varying irradiation conditions in tracking maximum power. However, the PSO algorithm is limited by the convergence speed in tracking the maximum power if the irradiance is changing continuously. Solar Module configurations also impact the maximum power generated under partial shading conditions. Series-parallel configuration gave better results compared to series module configuration. Overall energy generated is also observed to be more for series-parallel module configuration which implies that cell interconnection also plays a vital role.

6 REFERENCES

[1] H. Chaieb, A. Sakly, 2015 16th International Conference on Sciences and Techniques of Automatic Control and Computer Engineering (STA) 2015, DOI 10.1109/sta.2015.7505205.

[2] R. Hussan and A. Sarwar, 2018 2nd IEEE International Conference on Power Electronics, Intelligent Control and Energy Systems (ICPEICES), Delhi, India, 2018, pp. 293-298, doi: 10.1109/ICPEICES.2018.8897493.

[3] G. Calvinho, J. Pombo, S. Mariano and M. d. Rosario Calado, "Design and Implementation of MPPT System Based on PSO Algorithm," 2018 International Conference on Intelligent Systems (IS), Funchal, Portugal, 2018, pp. 733-738, doi: 10.1109/IS.2018.8710479

SKY VIEW FACTOR ESTIMATION FROM LIDAR SCENES FOR VEHICLE INTEGRATED PHOTOVOLTAICS

J. Macías, R. Herrero, L. San José, R. Núñez, I. Antón
Instituto de Energía Solar, Universidad Politécnica de Madrid
Madrid, Spain, javier.macias@ies.upm.es

ABSTRACT: The motion of the Electric Vehicle along its trajectory makes the performance of the Vehicle Integrated PhotoVoltaics technology highly dependent on the surrounding elements of the urban streetscape. The direct component of the irradiance is affected by shading effects and the diffuse by shielding effects. The effect of shading on the direct component has already been analysed in previous studies. Regarding the sky diffuse, the shielding effect can be evaluated by the sky view factor, which corresponds to the percentage of visible sky in a hemispheric sky. This factor can be evaluated by images taken from fisheye cameras or by available imaging systems such as GSV. Moreover, they can also be obtained using virtual cameras from LiDAR scenes, as it is a technology that allows to map the cities in 3D and to take sky images in any location/height. In this paper, a LiDAR-based methodology is proposed to obtain the sky view factor from these scenes and its validation is performed by comparing to images taken with a fisheye camera. The calculation of the sky view factor must be related to the photovoltaic surface of the vehicle therefore, the camera should be placed at a given location/height. Thus, the effect of the height of the camera when calculating the sky view factor has also been evaluated. Finally, using LiDAR methodology, the diffuse component of the irradiance has been estimated for a vehicle route in Madrid.
Keywords: Sky View Factor, Diffuse, Modelling, VIPV

1 INTRODUCTION

Vehicle Integrated PhotoVoltaics (VIPV) seeks to increase the autonomy of the Electric Vehicle (EV) from an autonomous and clean way by installing photovoltaic (PV) cells on the different surfaces of the car (roof, bonnet or doors). The defining characteristics of this technology such as the curvature of the vehicle surfaces [1], its motion or dynamic shading, require the development of new studies and tools to determine its viability in terms of solar resource available or energy generation [2] [3].

The motion of the vehicle along its trajectory introduces complexity in modelling the solar resource. The direct and diffuse components of the irradiance on the PV surface will vary not only with the location and orientation of the vehicle but also depending on the urban environment around it.

Objects in the urban environment, in addition to generating shading effects that drastically reduce the direct component of the irradiance, also produce a shielding effect that impacts on the diffuse (sky and reflected).

Scenes based on Light Detection and Ranging (LiDAR) point clouds are useful to represent the urban streetscape in order to analyze the impact of these effects on the available solar resource. Previous works on VIPV have already used LiDAR scenes to represent the urban environment to study the effect of shading on the direct component [4].

Although the contribution of diffuse irradiance to the available solar resource is less significant than the contribution of the direct component, accurate modelling of diffuse irradiance is also necessary to obtain a correct modelling of energy generation, especially considering that urban environments can lead to scenarios of total shading for many hours in the context of VIPV technology, where diffuse irradiance is predominant.

Therefore, the effect of shielding on the diffuse can be analyzed by estimating the diffuse irradiance in the horizontal plane (DHI) and the Sky View Factor [5] seen from the PV surface of the vehicle under evaluation (i.e., at a certain position and height). The Sky View Factor (SVF) corresponds to the percentage of sky visible in a hemispherical image of the sky. This information can be gathered based on images taken by fisheye cameras, by processing Google Street View (GSV) pictures or by using LiDAR scenes.

Therefore, the following work proposes three different methodologies to obtain sky images and validates a methodology based on LiDAR scenes for the calculation of the SVF at the different positions of the EV trajectory and also the effect it has on the available solar resource.

If LiDAR data is available in the region of study, this methodology allows to obtain sky images at any position/height unlike the other proposed methodologies, where a measurement set-up is required (fisheye camera) or where the images are only available at certain positions/heights (GSV).

2 SOLAR RESOURCE MODELLING: EFFECT OF THE URBAN STREETSCAPE

There are several models that estimate the incident irradiance on a surface by combining the direct and diffuse (sky and reflected) components. In the field of VIPV, the inherent characteristics of this technology mean that the solar resource reaching the PV surface is greatly affected by the effects of the urban environment (Figure 1).

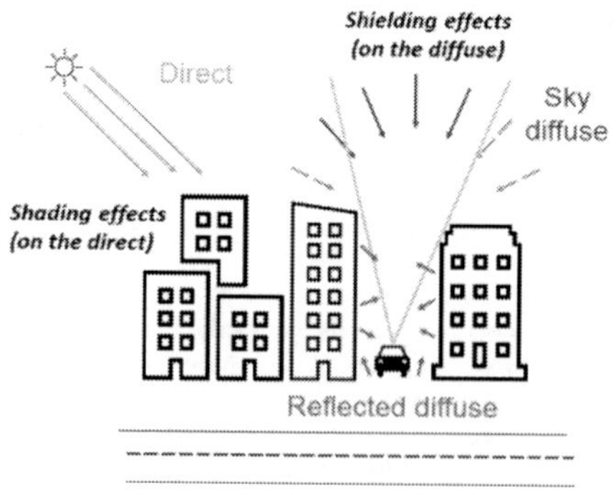

Figure 1: Schema of the impact of the urban landscape on the different components of irradiance.

10.4229/EUPVSEC2023/4BV.3.24
020321-001

2.1 Effective Direct Irradiance

Regarding the direct irradiance, most of the models are based on its estimation as purely geometrical problem from the Direct Normal Irradiance (DNI) and the Angle of Incidence (AOI). The value of this component is strongly affected by shading effects caused by urban landscape objects. To quantify this effect, a *Shading Factor* has been proposed which is 0 when the PV surface is shaded and 1 when it is under shading conditions.

$$Direct = DNI \cdot \cos(AOI) \cdot Shading\ Factor \quad (1)$$

Previous work based on LiDAR scenes has already been carried out with the aim of modelling the effect of shading on the direct component by calculating the shading factor at each point of the PV surface [4].

2.2 Effective Diffuse Irradiance

Models for estimating diffuse component can be classified into two large groups [6]. As an example, we find models such as the isotropic of Liu and Jordan presented below [7] or the anisotropic model of Hay and Davies [8].

These models consider the Diffuse Horizontal Irradiance (DHI) and the elevation of the incident plane (AOI) but they do not consider the effects generated by the surrounding objects of the urban landscape.

As an approximation, assuming that the sky diffuse on the horizontal plane comes from the whole visible hemisphere, it is possible to consider only the proportion of the sky that is clearly visible from the collection plane, discounting the part hidden by obstacles. This is traditionally evaluated with the Sky View Factor [5].

The Sky View Factor (SVF) is defined as the proportion of visible sky in a hemispheric sky image. In other words, the SVF is the ratio of the number of pixels corresponding to the sky with respect to the number of pixels of the total image, considering an equisolid projection. In this projection, the projected pixels have the same solid angle irrespective of where they are.

$$Sky\ View\ Factor = \frac{\sum Sky\ Pixels}{\sum Total\ Image\ Pixels} \quad (2)$$

From the Liu and Jordan model, the value of the sky diffuse irradiance can be calculated using the DHI and the SVF as follows [5]:

$$Sky\ Diffuse = DHI \cdot \left(\frac{1+\cos(AOI)}{2}\right) = DHI \cdot SVF \quad (3)$$

Finally, the reflected component of the diffuse irradiance depends on the position of the sun relative to obstacles and the car. Images of the surroundings may be useful to estimate the value of this component and will be studied in the next steps.

3 DIFFUSE IRRADIANCE MODELLING FROM SKY IMAGES

Diffuse estimation from sky images is interesting especially in the field of VIPV, as vehicles are constantly moving around objects of the urban streetscape.

The high view angle (180°) of the optics required to obtain the sky images may cause distortion of the projected images, leading to errors in the SVF calculation and consequently, in the estimated diffuse irradiance. This can be avoided by a proper calibration process of the camera.

Additionally, regardless of the methodology used to obtain the sky images, it is necessary a pre- or post-processing of the image to obtain them with an equisolid projection, to preserve a constant ratio of solid angles in the sky and pixels in the image. Once the sky image has been obtained with this projection, an additional image processing is necessary to distinguish between sky and objects.

The image processing applied consists of a preprocessing and a segmentation. The preprocessing is composed by a contrast adjustment and an areas homogenization that enhances the contrast between the sky and the objects. The segmentation extracts through iterative algorithms the area of the image that corresponds to the observed sky With the processed image, it is possible to estimate the SVF as explained in the previous section.

There are different methodologies to obtain sky images and three of them have been proposed in this work:

3.1 Fisheye cameras

Fisheye cameras have a high view angle and are useful for taking images of the sky. However, this methodology presents a lack of data availability if thinking about modelling any route in any vehicle as it is the user who has to take each image from a corresponding set-up. In this study, these images are used as a reference to validate the other proposed methodologies.

Once the images have been taken and after further processing, the SVF can be obtained as shown in Figure 2:

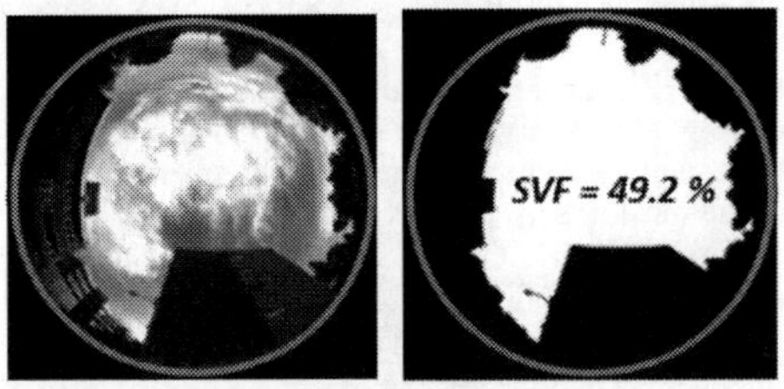

Figure 2: On the left: Hemispheric Sky Image from a Fisheye camera (Mobotix Q26). On the right: Processed image (the sky is shown in white).

3.2 Google Street View

This methodology is based on the processing of images obtained through the Google Street View (GSV) platform, as presented in [9].

One of the strengths of this methodology is that GSV currently has a large number of images available around the world. However, this platform requires payment for a high number of download and its images are restricted to certain locations/altitudes (not necessary at the position of the PV surface of the vehicle).

As an example, a sky image is shown from images available in GSV and its corresponding processing to obtain an equisolid projection (Figure 3).

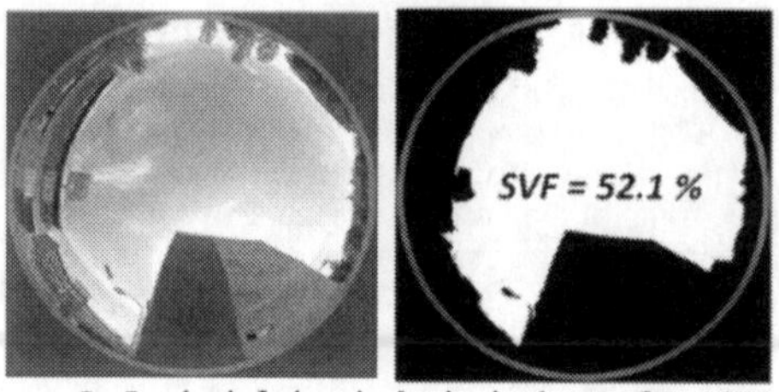

Figure 3: On the left: hemispheric sky image from GSV. On the right: Processed image (the sky is shown in white).

3.3 Hemispheric images from LiDAR Point Clouds

LiDAR point clouds provide a 3D mapping of the topography of cities, representing all the elements of the urban streetscape. From these point clouds, 3D scenes of cities can be obtained by surface triangulation (Figure 4):

Figure 4: On the left: LiDAR Point Cloud, source: Centro Nacional de Información Geográfica [10]. On the right: 3D surface from the Point Cloud.

Using these scenes and 3D visualisation tools, it is possible to take images from a virtual camera to obtain sky images.

From the LiDAR scene of Figure 4, the virtual camera is placed in the position where the PV surface of the vehicle is located and 5 photographs are taken with a field of view of 90° varying the azimuth (θ) and elevation (φ), as shown in Figure 5. These images represent a cube map of the surroundings of the vehicle:

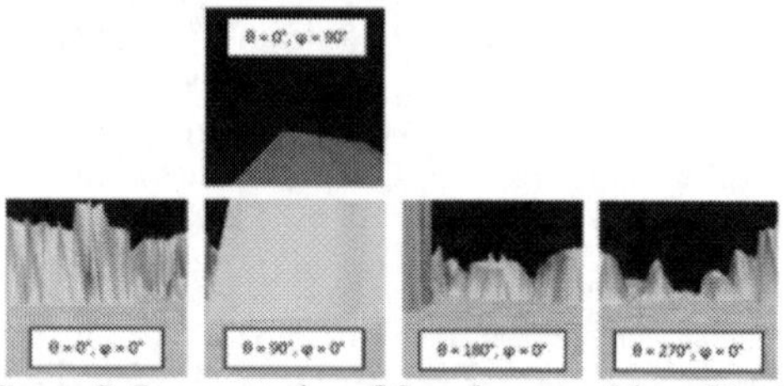

Figure 5: Representation of the cube map of the vehicle's surroundings.

Finally, from the equisolid projection (Figure 6) of the cube map, the SVF is estimated:

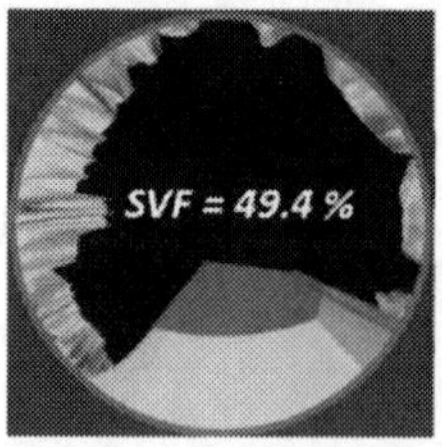

Figure 6: Equisolid projection from the cube map.

4 RESULTS

4.1 Comparative analysis of the SVF using the three proposed methodologies

In this section, a comparative analysis of the SVF has been carried out for three different methodologies (fisheye camera, GSV and LiDAR). The aim of this study is to validate the SVF calculated from the available image acquisition systems (GSV or LiDAR) using the images taken by the fisheye camera.

The images were taken at the same location and altitude (camera height) and correspond to the surroundings of the Instituto de Energía Solar. A table showing the processed images and the corresponding SVF is shown below (Figure 7):

Methodology	Original Picture	Processed Picture	Sky View Factor (%)
Fisheye camera (Mobotix Q26)			49.2
Google Street View – (GSV)	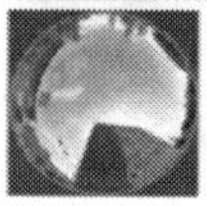		52.1
LiDAR			51.8

Figure 7: Comparative analysis of the SVF for the different methodologies proposed.

Taking the fisheye camera result as a reference (SVF = 49.2%), relative differences of 2.6% are found with respect to the LiDAR methodology and 2.9% to GSV.

Differences from LiDAR may be due to inaccuracies in the surface triangulation of the LiDAR point cloud. Considering the differences with GSV, these may occur due to slight variations in the urban landscape (e.g. trees) or due to variations in the location/camera height/orientation between the two methods, as GSV only offer us images at certain location/height.

4.2 Effect of the camera height on the SVF

In addition to the location (latitude/longitude), the height of the camera can cause variations in the SVF, as objects of the urban streetscape will be at different positions with respect to the reference. Therefore, it is important to take the image in the position where the PV surface is located.

Therefore, in order to evaluate the impact of camera height for the same latitude/longitude, different images were taken at heights from 0 to 3 m using the fisheye camera and LiDAR, obtaining the relative differences between them. (Figure 8).

At this point, it is also necessary to highlight one of the limitations of the GSV methodology as, contrary to fisheye camera or LiDAR, the images available from GSV are only taken at certain locations/heights. For this reason, GSV is excluded from this study.

10.4229/EUPVSEC2023/4BV.3.24
020321-003

Height (m)	Sky View Factor (%)		
	Fisheye Camera	LiDAR	diff
0	45.9	48.1	2.2
1	47.5	50.5	3.0
1.5	49.2	51.8	2.4
2	49.4	53.0	3.7
2.5	50.7	54.2	3.5
3	51.9	55.4	3.51
SVF variation (0 - 3 m)	**6.0%**	**7.3%**	

Figure 8: SVF from fisheye camera and LiDAR for camera heights from 0 to 3 m.

The variations obtained in the SVF calculation for the different heights (0 - 3 m) correspond to 6.0% for the fisheye camera measurements and 7.3% for the LiDAR simulations, so it can be assumed that the camera height has a low impact on the SVF.

Additionally, the differences between fisheye camera and LiDAR (diff column in Figure 8) for images taken at the same location/height remain around 2.2 - 3.5%.

4.3 Diffuse estimation on a route

The objective of this study is to estimate the sky diffuse irradiance reaching the PV surface (car roof) of an EV along its route in Madrid, for a typical day (1 April). To do this, it is necessary to obtain the SVF at the different points along the route.

These simulations provide additional information on the topography of the city, differentiating between parts of the city with a higher presence of urban streetscape objects (low SVF) and areas with clearer surroundings (high SVF).

Obtaining the SVF along a trajectory using a fisheye camera requires mounting a set-up on the roof of the vehicle that takes pictures every few steps.

On the other hand, GSV is an appropriate methodology for this purpose, but the available images are only taken for certain locations/altitudes. In particular, the camera height acquisition is not necessary the same of the height of the PV surface.

Therefore, LiDAR is proposed as the methodology to be used to obtain the hemispheric images and the sky diffuse irradiance for this study. The following figure shows the results of the DHI from the Typical Meteorological Year of PVGIS [11] at the different positions of the EV along its trajectory and the sky diffuse obtained by the SVF calculated using LiDAR scenes (Figure 9).

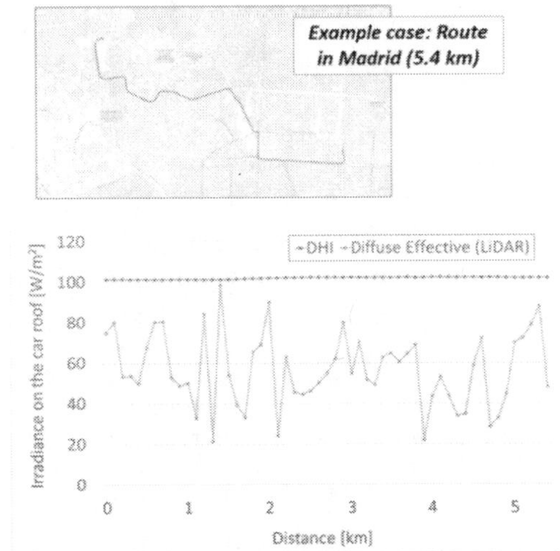

Figure 9: Route of the EV in Madrid (up). Impact of SVF on the sky diffuse irradiance along a route (down).

5 CONCLUSIONS

In this work, three methodologies (fisheye camera, GSV and LiDAR) have been proposed for the estimation of the SVF and the corresponding sky diffuse irradiance from hemispheric sky images.

The fisheye camera methodology requires a set-up on the car roof which makes this procedure more complex. On the other hand, GSV is a payment service where the images are subject to a certain location/height and do not necessarily correspond to the position of the PV surface.

In particular, a methodology based on LiDAR scene imaging has been validated, using real images taken from a fisheye camera at the same location/camera height. The SVF values obtained with LiDAR are 2.6% higher (2.9% in the case of GSV) than those obtained with the fisheye camera.

The effect of camera height on SVF has been also evaluated, showing variations of 6.0% for images (0 to 3 m) taken with the Fisheye camera and 7.3% for images taken with LiDAR. It can be concluded that the camera height has a low impact on the SVF value.

The estimation of the sky diffuse along an EV route has also been carried out in which the effects of the urban landscape on the diffuse irradiance are assessed.

Therefore, if LiDAR data is available in the region of the study, this methodology is considered ideal for this analysis, especially in the field of VIPV.

6 ACKNOWLEDGMENTS

This work has been supported by the project DETEC-PV (reference PID2021-128853OB-I00) financed by the Fondo Europeo de Desarrollo Regional (FEDER) and the Spanish Ministerio de Ciencia e Innovación and Agencia Estatal de Investigación. We acknowledge to open-source libraries PVLIB and PVGis Database for their contribution to this project.

7 REFERENCES

[1] Y. Ota, T. Masuda, K. Araki, and M. Yamaguchi, 'Curve-Correction Factor for Characterization of the Output of a Three-Dimensional Curved Photovoltaic

Module on a Car Roof', *Coatings*, vol. 8, no. 12, p. 432, Nov. 2018, doi: 10.3390/coatings8120432.
[2] K. Araki, L. Ji, G. Kelly, and M. Yamaguchi, 'To Do List for Research and Development and International Standardization to Achieve the Goal of Running a Majority of Electric Vehicles on Solar Energy', *Coatings*, vol. 8, no. 7, p. 251, Jul. 2018, doi: 10.3390/coatings8070251.
[3] B. Commault, T. Duigou, V. Maneval, J. Gaume, F. Chabuel, and E. Voroshazi, 'Overview and Perspectives for Vehicle-Integrated Photovoltaics', *Appl. Sci.*, vol. 11, no. 24, p. 11598, Dec. 2021, doi: 10.3390/app112411598.
[4] J. Macías, R. Herrero, R. Núñez, and I. Antón, 'ON THE PERFORMANCE OF PHOTOVOLTAICS ON CAR ROOFS: CASE STUDY UNDER DIFFERENT STATIC SCENARIOS', p. 4.
[5] I. García, M. de Blas, B. Hernández, C. Sáenz, and J. L. Torres, 'Diffuse irradiance on tilted planes in urban environments: Evaluation of models modified with sky and circumsolar view factors', *Renew. Energy*, vol. 180, pp. 1194–1209, Dec. 2021, doi: 10.1016/j.renene.2021.08.042.
[6] S. A. Khalil and A. M. Shaffie, 'A comparative study of total, direct and diffuse solar irradiance by using different models on horizontal and inclined surfaces for Cairo, Egypt', *Renew. Sustain. Energy Rev.*, vol. 27, pp. 853–863, Nov. 2013, doi: 10.1016/j.rser.2013.06.038.
[7] B. Y. H. Liu and R. C. Jordan, 'The interrelationship and characteristic distribution of direct, diffuse and total solar radiation', *Sol. Energy*, vol. 4, no. 3, pp. 1–19, Jul. 1960, doi: 10.1016/0038-092X(60)90062-1.
[8] J. E. Hay, 'Calculating solar radiation for inclined surfaces: Practical approaches', *Renew. Energy*, vol. 3, no. 4–5, pp. 373–380, Jun. 1993, doi: 10.1016/0960-1481(93)90104-O.
[9] R. Nuñez, A. Muñoz, L. San Jose, J. Macias, R. Herrero, and I. Antón, 'Estimation of Solar Resource for Vehicle-Integrated Photovoltaics in Urban Environments Using Image-Based Shade Detection', *Opt. Express*, Sep. 2023, doi: 10.1364/OE.496856.
[10] 'CNIG - Instituto Geográfico Nacional'. https://www.ign.es/web/qsm-cnig (accessed Sep. 25, 2022).
[11] 'PVGIS Photovoltaic Geographical Information System'. https://joint-research-centre.ec.europa.eu/pvgis-photovoltaic-geographical-information-system_en (accessed Sep. 29, 2022).

GIS-BASED SOLAR IRRADIANCE SIMULATION FOR VIPV APPLICATIONS IN A COMPLEX URBAN ENVIRONMENT

David Pera, Christian Braun, Philippe Pinheiro and Ulrich Leopold
Luxembourg Institute of Science and Technology (LIST))
5, avenue des Hauts-Fourneaux, L-4362 Esch/Alzette,

ABSTRACT: We present two different approaches to assess the spatio-temporal solar potential in complex environments based on geospatial techniques for VIPV/VAPV (vehicle integrated / added photovoltaics) applications. Both were implemented in a LIST developed web-based interoperable GIS platform, iGuess®. One of the methods used a 2.5 dimensional approach, using r.sun from GRASS GIS, to compute real-sky radiation with direct beam, diffuse, and reflected components on a simulated flat-horizontal vehicle's rooftop using a 0.5m Digital Surface Model (DSM). The second approach used high precision CityGML 3D vector models applying a multi-beam projection collision detection method for shadow casting. The simulation outputs were compared with data acquired in a large-scale experimental campaign using fast-response silicon irradiation sensors equipped with GPS and wireless communication installed on vehicles' roofs. Preliminary results show that the simulation model using the 3D approach with CityGML achieves more accurate results. This study also contributes to the identification of the nature of the deviations found between experimental measurements in a real environment and simulations, which could ultimately be used to further improve the computer models.

Keywords: Solar irradiation, modelling, GIS, VIPV, CityGML

1 INTRODUCTION

In parallel with the innovation and optimization of the integration of PV technology in electric vehicles (EV) or non-electric vehicles with the aim of reducing their carbon footprint by supplying a share of the needed energy, it is of the utmost importance to develop mechanisms for the evaluation of the solar resource along the road network, in highly complex environments. Proper computational simulation tools for the solar potential assessment play a fundamental role in the design and operation of the VIPV/VAPV's (vehicle integrated/added photovoltaics) energy management systems, depending on their typology and operation, such as the light-vehicles for daily commute, public transportation, and heavy-duty long-distance transportation, among others.

Recent studies [1][2][3] address the use of Geographical Information Systems (GIS) and geospatial simulation combined with methodologies to perform space-time shadow casting analyses from digital surface models (DSM). Availability of high resolution (0.25-0.5m) digital surface models (DSM) and well-tested solar irradiance algorithms are ubiquitously available [4]. GIS-based approaches have the advantage of providing the required inputs at high resolution and can account for scale changes which help to upscale the approach easily to larger regions, such as complex urban environments.

The simulation of the solar potential at the level of the vehicles requires high-resolution space-time solar irradiance modelling accounting for dynamic cast shadow effects, direct irradiation amplified by multiple reflections from surrounding surfaces, and the impact on the diffuse component by local albedo variations. In this scope, we present two different approaches to assess the spatio-temporal solar potential in the built environment, which were implemented in a LIST developed web-based interoperable GIS platform, iGuess®[5] for visualization and parametrization.

The simulations were performed considering road network of Luxembourg at the level of vehicles to be compared with *in-situ* measurements acquired large-scale experimental campaign using fast-response silicon irradiation sensors equipped with GPS and wireless communication installed on public transportation buses, and heavy-duty transportation trucks and vans operating in the that territory.

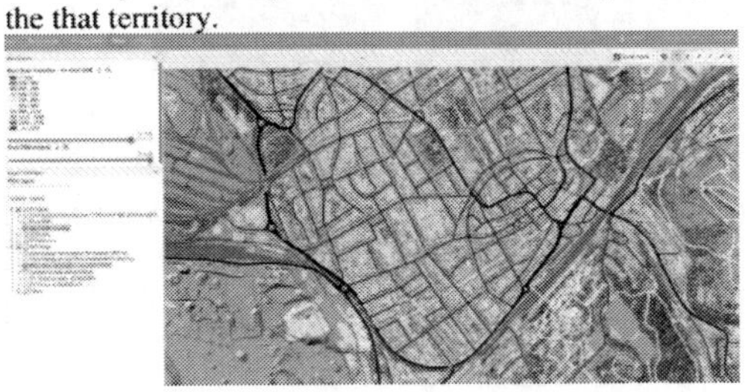

Figure 1: iGuess® web interface showing the solar irradiation at 0.5m resolution for Esch-sur-Alzette, Luxembourg.

2 APPROACH

2.1 2.5D simulation

The first method was applied using the r.sun [6] module of GRASS GIS [7] to compute real-sky radiation with direct beam, diffuse, and reflected components on a simulated vehicle rooftop height along the road framework (Figure 2). The calculation accounts for cast shadows of the built infrastructure as well as vegetation components by making use of a high-resolution DSM (0.5m) from available LIDAR data for Luxembourg. This method doesn't allow to properly represent vertical surfaces of the modelled structures only the volumes extruded based in LiDAR information, therefore is considered a 2.5D approach (Figure 3) and non-exactly tri dimensional. Real-sky radiation is generated by clear-sky radiations reduced by atmospheric factors, and cloudiness, which have been obtained from local long-term meteorological time-series measurements and as well as from the CAMS Radiation Service (CRS) [8][9][10].

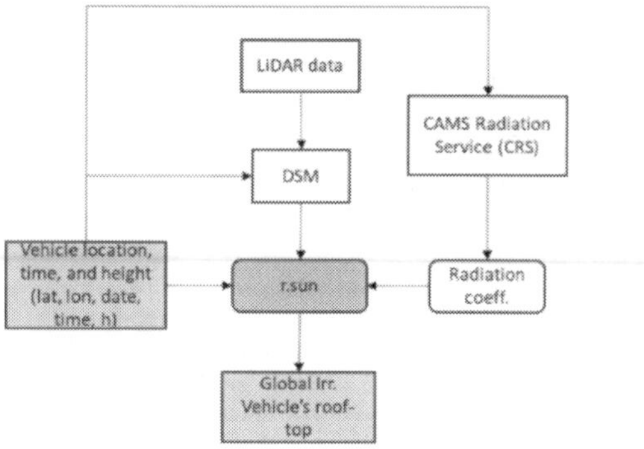

Figure 2: 2.5D model conceptual simulation workflow chart.

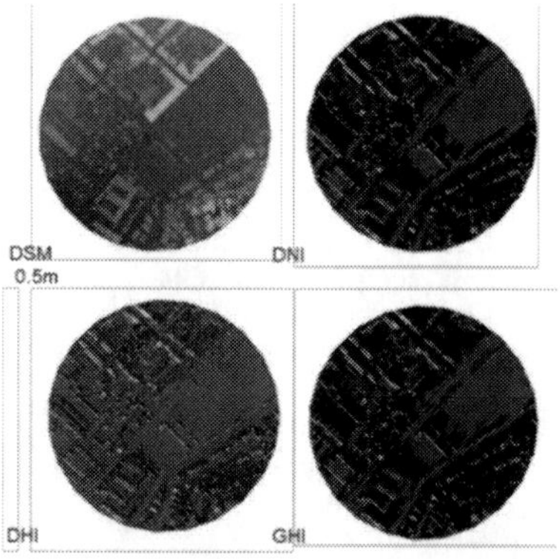

Figure 3: 2.5D model DSM input in the upper left image, and the respective outputs of the DNI, DHI and GHI distributions on built infrastructures in the site under analysis for an arbitrary time.

2.1 3D simulation

The second approach uses CityGML 3D models [11] with a precision of around 10cm applying a multi-beam projection collision detection method for shadow casting taking the solar angles for each time-step of the simulation (Figure 4). CityGML is a widely accepted data standard and exchange format for the representation of 3D virtual geographic landscape and city models. It facilitates processes that can benefit from the integration of digital urban geospatial data as digital twin applications. The models available in the CityGML format are also derived from LiDAR data, but not limited to as a more detailed and accurate modelling of urban infrastructure requires other data sources, such as photogrammetry and computational CAD modelling. For this reason, the existence of the CityGML databases is still limited for many regions, although its availability has been accelerated in recent times. Moreover, CityGML databases are often made available free of charge by local, regional, and national governments, as it is the case for Luxembourg.

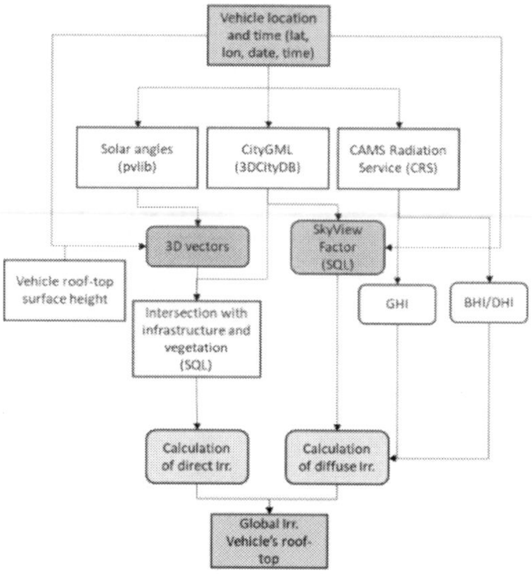

Figure 4: 3D model simulation conceptual workflow chart.

In this approach, the vegetation and other high-definition infrastructure components models, limited to Level of Detail – LoD of the available data, can be added to the 3D simulation environment (Figure 5).

Figure 5: Example of the 3D model approach applied to a specific location in time and space in Luxembourg using CityGML 3D models and 3D geospatial visualisation. (without vegetation: top image; with vegetation: bottom image).

The radiation components were obtained using the Copernicus CAMS Radiation Service (CRS) [8][9][10], for each time step and location in the analysis. The diffuse component is then corrected using the Skyview factor of each location, calculated using the 3D city models.

2.1 Solar irradiation field measurements.

Solar radiation measurements using irradiation sensors installed on vehicles travelling under normal conditions are an important instrument for the enhancement of the quality of solar potential estimation simulations applied to VIPV/VAPV applications. The study of the deviations obtained for the specific radiative envelopes of the

10.4229/EUPVSEC2023/4BV.3.25
020322-002

vehicles surrounding allows to determine the limitations of the models and how to improve them.

For this purpose, fast-response silicon irradiation sensors equipped with GPS and wireless communication (Figure 6) were installed on different types of vehicles' roofs, collecting synchronized measurements of solar radiation, location coordinates and temperature. The acquisition was made during the day-time with rates limited to 30 seconds.

Figure 6: Irradiation sensor used in the experimental campaign (enerjin.com).

3 RESULTS

The results obtained with both approaches show common limitations in resolution of the irradiance models inputs and the accuracy of reflected irradiation simulations. Figures 7 and figure 8, show underpredictions for the 2.5D and 3D simulations compared to the peak irradiance measurements, most likely due to reflective effects from the surrounding. Overestimations may be explained by the lack of representativeness of the radiate source data. Figure 7 shows results obtained using a local solar station, located in the vicinity of the vehicle's performed route on a snowy day. In this case, the radiation components used in the model did not consider the albedo correction of the ground around the pyranometers, which was considerably different from the one affecting the vehicle roof top. The lack of time resolution of the satellite driven data obtained by the CAMS Radiation Service, can also justify some of the under and over-predictions detected with the 3D approach (figure 8), namely under overcast conditions, where CAMS data might be too course in resolution to provide higher spatial accuracy.

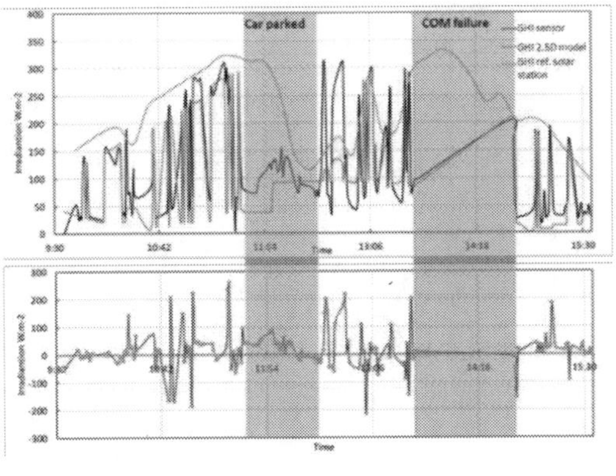

Figure 7: Irradiation profiles (top plot) and respective differences (bottom plot) obtained by experimental measurements using a sensor installed on a vehicle rooftop and with the 2.5D simulation model, taking as radiative reference the radiation components measured by a local solar station (GHI, DHI).

Another common source of error detected in both analysis is due to the GPS positioning inaccuracy that significantly limits the number of valid acquisitions for analysis, whenever the measured coordinates point location of a building, for example, for which the models naturally estimate 0 W/m². This error is clearly expressed by the points over the y-axis in the data the correlations in figure 8. In cases where the mis-positioning of the GPS system is not so drastic, this effect can still significantly affect the estimated irradiance on the vehicle roof by shifting it by a few meters, affecting the reflected component or even eliminating the direct radiation contribution in the case of shading.

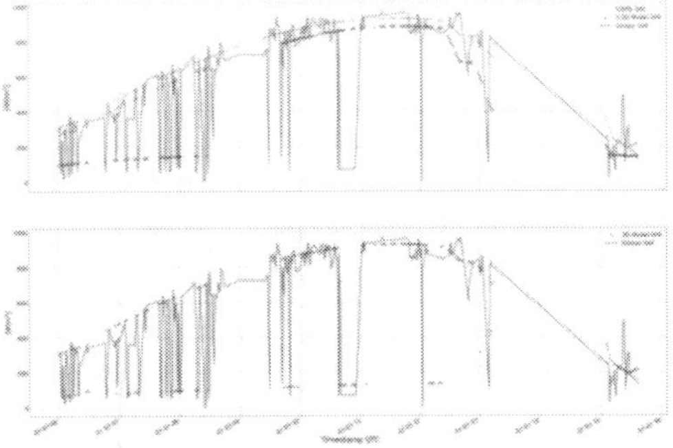

Figure 8: Irradiation profiles obtain by experimental measurements using a sensor installed on a vehicle rooftop, using the 2.5D simulation model (top plot), and the 3D simulation model (bottom plot). Both simulated results were obtained using the CAMS Radiation Service (CRS) as reference for the radiative estimations.

A detailed classification of the nature of the errors obtained in this comparative study will require a more exhaustive study based on measurements over longer periods. The optical characterization of the surroundings of the road network framework will also have to be considered in order improve the radiative models. However, in terms of spatial resolution and representativity, this preliminary study reveals a higher accuracy from the 3D modelling method, which showed a better correlation with the experimental measurements (figure 9) when compared with the 2.5D approach for the same travelled route and using the same radiation data source (CAMS- CRS).

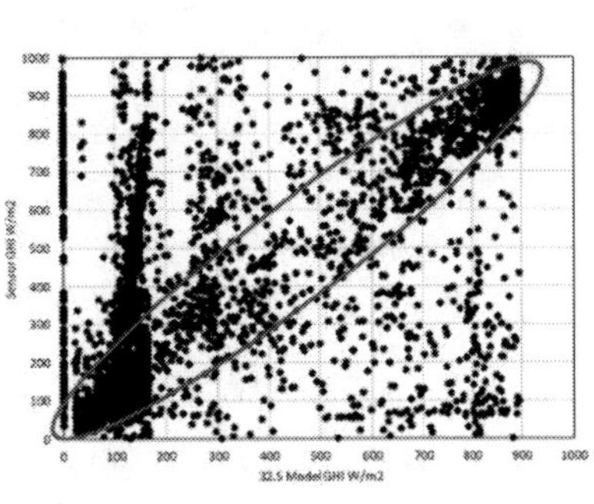

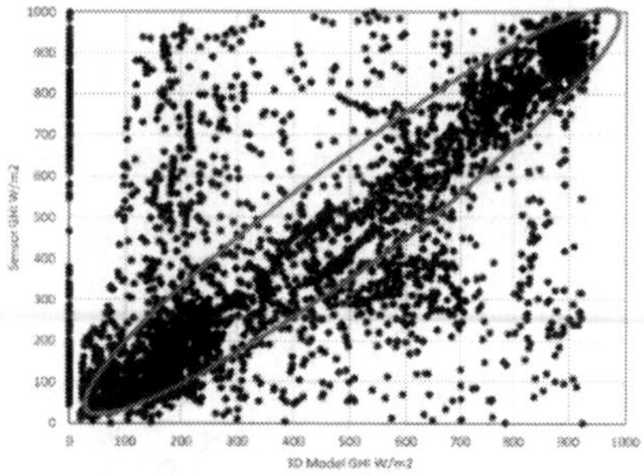

Figure 9: Irradiance data correlations between the measured irradiance and the simulations outputs obtained with the 2.5D (top plot) and 3D simulation models (bottom plot).

4 CONCLUSIONS

The approaches presented for simulating the solar potential of road networks are a first step towards developing more advanced and accurate methods for estimating the energy yield of VIPV/VAPV applications in complex urban environments. The comparison of experimental measurements of solar irradiance sensors installed on vehicles in real operation conditions with the models' outputs revealed the limitations of the resolution and accuracy of the irradiance inputs. Furthermore, it was shown that the accuracy of the geolocation of the measurements can strongly affect impact the quality of the analysis by introducing positional mismatch errors affecting the representativity of the simulation, mainly in the assessment of the shadowing impacts. Underestimations detected using both methods clearly point to the necessity to further evolve the models considering local optical effects, as direct reflections, and the contribution of local albedos, which are particularly relevant in the built environment. Despite several sources of errors of the comparative study, the simulation model using the 3D CityGML model returned consistently better predictions compared to the measured irradiance, which validates the approach for further development.

Future work will be required to further analyze the differences of simulations and field observations.

5 REFERENCES

[1] Brito, M. C., Santos, T., Moura, F., Pera, D., & Rocha, J. (2021). Urban solar potential for vehicle integrated photovoltaics. Transportation Research Part D: Transport and Environment, 94, 102810.
[2] Oh, M., Kim, S. M., & Park, H. D. (2020). Estimation of photovoltaic potential of solar bus in an urban area: Case study in Gwanak, Seoul, Korea. Renewable Energy, 160, 1335-1348.
[3] Sovetkin, Evgenii, et al., Vehicle-integrated photovoltaics irradiation modeling using aerial-based LIDAR data and validation with trip measurements, Solar RRL. 10.1002/solr.202200593
[4] Freitas, S., Catita, C., Redweik, P., Brito, M. (2015) Modelling solar potential in the urban environment: State-of-the-art review. Renewable and Sustainable Energy Reviews, 41, 915 - 931.
[5] Mastrucci, A., Braun, C., Baume, O., Stazi, F., & Leopold, U. (2015) Towards a Comprehensive Approach to Sustainable Urban Planning: Integrated Estimation of Housing Electricity Consumption and Photovoltaic Generation Potential Using the web-based framework iGUESS®. In Renewable Energy in the Service of Mankind Vol I (pp. 791-800). Springer, Cham.
[6] Hofierka, J., Suri, M. (2002) The solar radiation model for Open source GIS: implementation and applications. International GRASS users conference, Trento, Italy, September 2002.
[7] GRASS Development Team (2020) Geographic Resources Analysis Support System (GRASS) Software, Version 7.8. Open Source Geospatial Foundation. https://grass.osgeo.org
[8] Lefèvre, M. et al, 2013. McClear: a new model estimating downwelling solar radiation at ground level in clear-sky conditions. AMT, 6, 2403-2418, doi: 10.5194/amt-6-2403-2013.
[9] Gschwind, B., et al., 2019. Improving the McClear model estimating the downwelling solar radiation atground level in cloud free conditions – McClear-V3., Meteorol. Z., doi:10.1127/metz/2019/0946.
[10] Qu, Z. et al., 2017. Fast radiative transfer parameterisation for assessing the surface solar irradiance: The Heliosat-4 method, Meteorol. Z., 26, 33-57, doi: 10.1127/metz/2016/0781.
[11] https://www.ogc.org/standard/citygml, consulted in September 2023

10.4229/EUPVSEC2023/4BV.3.25
020322-004

MODELLING PERFORMANCE AND AVAILABILITY OF SOLAR STREETLIGHTS IN SHADED ENVIRONMENTS

Luca Morino, Martin Bartholomäus, Sergiu V. Spataru, Peter B. Poulsen
Department of Electrical and Photonics Engineering, Technical University of Denmark
Frederiksborgvej 399, 4000, Roskilde, Denmark

ABSTRACT: In this work, a system performance model of a standalone streetlight was developed, which is comprised of models for the photovoltaic (PV) array, battery, converter, and LED luminaire. This model is paired with an irradiance and shade model, which uses a Digital Surface Model (DSM) and solar position to estimate the irradiance reaching the streetlight when located in environments affected by shade. The system performance model and the component models were developed and validated using test data from DTU's Risø laboratories in Roskilde, Denmark, and measurements from streetlight prototypes present on campus. It is designed to be modular, allowing for the analysis of different systems using the same modelling framework. Combining the system performance model and irradiance and shade model allows for remote assessments of the operation and performance of the lamp in any environment with DSM data available. It can, therefore, be a valuable tool for assisting installation and optimization of the streetlight design. The operation of both models is shown for a case study in the Red Square in Copenhagen, where a proposed installation of eight solar streetlights is analyzed.

Keywords: Streetlight, standalone PV, Product Integrated Photovoltaics, shade modelling, system performance modelling, Digital Surface Model

1 INTRODUCTION

When installing a standalone solar streetlight in a shaded area, it is difficult to predict if the lamp will operate effectively throughout the year. This is because shading caused by surrounding buildings and other urban structures can greatly affect the performance of the solar streetlight. To address this problem, we aim to develop a tool for assisting in the proper positioning and installation of standalone solar streetlights in urban or shaded environments. Such a tool should be able to estimate the actual streetlight operation and availability for a given use load, considering factors such as solar radiation, streetlight orientation, and shading geometry as well as system component characteristics: PV panel electrical and thermal characteristics, converter efficiency, battery size, and roundtrip efficiency.

Furthermore, the tool should also be able to estimate operation over multiple years, taking into consideration the fading of the battery capacity and estimating its lifetime. Last, the tool should also aid in sizing the battery correctly to ensure the lamp has enough power to operate throughout the year.

The tool is then used in a case study to simulate the operation and availability of eight proposed solar streetlight locations in Copenhagen's Red Square. Furthermore, possible system design changes are proposed to increase the operational availability of the streetlights.

There are fewer tools and experience in modelling product-integrated photovoltaic (PIPV) technologies than there are for conventional solar PV systems. Companies producing PIPV products face a significant challenge in ensuring their products will operate as expected in the field. This risk can arise more often when installing in urban environments where shade caused by buildings is present and strong seasonal changes affect available the solar resource. One way to mitigate the risk and increase confidence in PIPV is by using system performance models and implementing shading loss and irradiance modelling [1] using Digital Surface Models (DSM). These tools can help lower the risk of product downtime and support the wider adoption of PIPV technology.

Tools for manual 3D modelling of the surrounding environment are already present in PV- planning software such as NREL's System Advisory Model (SAM) [2] or PVsyst [3], while PV*SOL [4] can import pre-fabricated 3D model files from Sketchup or Blender. However, manual 3D modelling is time-intensive, and this software is designed to work with utility-scale or residential PV applications rather than standalone PIPV.

Previous work on modelling PIPV systems done by Olindo et al. [5] puts focus on modelling PV power generation of the PIPV systems but does not dive deep into modelling complexities of the PIPV operation, for example modelling the other components (battery, converter, LED load profile) and resulting availability for a possible system optimization for the PIPV systems. Other papers based on modelling solar streetlight operation, such as Iwase Takenaka [6], Guo et al. [7] and Singh [8] often do not combine both a model for shading and full system operation of the streetlight. In addition, the modularity of the proposed system model in this study adds flexibility for different kinds of solar streetlight designs.

2 DESCRIPTION OF THE PIPV PRODUCT

The standalone PV lamp (SPVL) being studied is one of four prototypes of SPVL systems that were installed from 2019 to 2021 in DTU's Risø campus to be used for research purposes. **Figure 1** shows a diagram of the SPVL, highlighting the main components: PV panels, a battery energy storage system, a converter that regulates energy flows in the system, and a LED luminaire. Their specifications are detailed in **Table I**.

Table I: - Component specifications of the solar streetlight under study, the SPVL.

Solar Streetlight - SPVL	
PV Modules	9 Modules 32-cell, mono Si, 35Wp each
ECU (Converter)	5 MPPT Inputs 50 W max per input
Battery	LiFePO4 12.8 V, 66 Ah, 845 Wh
Luminaire	LED PMW, Optics fitted 40 W, 8000 lumen
Lamp Height	6-8 m

10.4229/EUPVSEC2023/4BV.3.27
020324-001

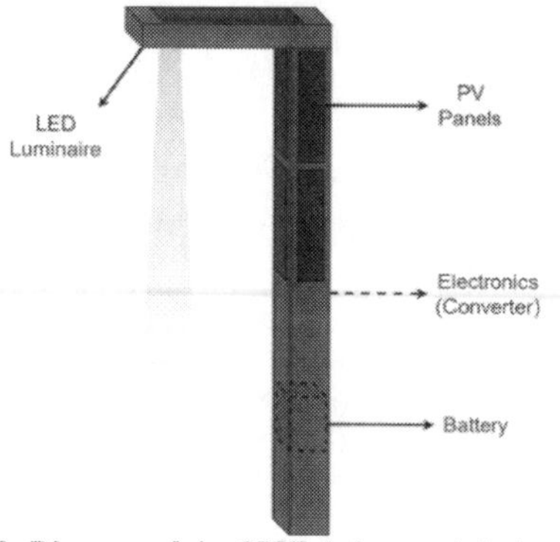

Figure 1: Diagram of the SPVL being modelled and its components. There are four SPVL prototypes present in Risø Campus providing measurement data.

The solar panels transform solar energy into electricity, which is then fed via the converter to the battery storage and accumulated throughout the day. At night, that energy is then used to power the LED luminaire system. The PV arrays of the solar streetlight are mounted to the pole and cover five different orientations: Top, North, East, South, and West. All other sub-components are located inside the streetlight and are therefore not visible from the outside.

Data is obtained from sensors installed on the SPVL prototypes present on campus to continuously measure their performance. This data, along with detailed characterization experiments done in the campus laboratories, is used for the development and validation of the component models.

The intended operation of the SPVL is to be able to operate standalone all throughout the year at a pre-programmed continuous load of $P_{led,max}$ = 40W. However, due to unexpected periods of low irradiance, the battery might not reach enough charge to power the LED luminaire for a full night, first causing it to dim to a lower "minimum power" and then eventually to turn off. These night hours in which the lamp is off would be counted as hours of light not delivered by the streetlight and would impact its reliability in the market.

The availability of the streetlight is therefore measured by the number of hours of light it delivers compared to the number of hours it is requested to deliver at night. The scope of the tool is to aid installers in ensuring maximum availability for the streetlight for a selected location.

$$Availability = \frac{hours\ of\ light\ delivered}{hours\ of\ light\ requested} \quad (1)$$

3 ESTIMATING THE SOLAR RESOURCE IN AN URBAN LOCATION

A case study is undertaken to show the operation of the tool: eight streetlight locations are chosen for possible installation in Superkilen Park, also known as the "Red Square" in Copenhagen, as shown in **Figure 2**. The locations are chosen to provide lighting for the square's main bike path and are therefore placed along it, with an average distance between them of around 25 meters. First, the solar resource reaching the locations is calculated, considering shading from the surrounding buildings. Afterward, the operation of the streetlights is simulated, and the resulting availability is analyzed.

To estimate the solar resource reaching the streetlights, a DSM is used to model nearby shading from buildings and trees in the area of the Red Square. It includes height data in a specific resolution for any buildings and manmade objects as well as for the ground, hills, and trees. In this study, the publicly available DK- DEM data [9] is used as it covers Denmark with an accurate resolution of 0.4 m x 0.4 m [10].

The DSM for the selected area, as well as the specific coordinates for the SPVL, are then used as inputs to GRASS (Geographic Resources Analysis Support System) GIS to perform a terrain analysis, resulting in a horizon line. GRASS uses the *r.horizon* tool to calculate to calculate the horizon line, which gives the maximum elevation angle of obstructions along the horizon for every azimuth angle (with chosen step size) [11].

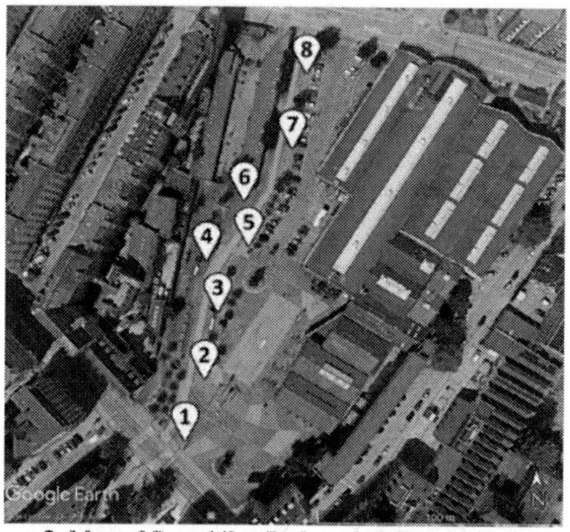

Figure 2: Map of Superkilen Park or the" Red Square" in Norrebrø, Copenhagen, and the proposed streetlight locations. The locations are shown in white, while the bike path crossing the square is shown in green. (Google Earth).

However, GRASS GIS calculates the horizon line from the elevation for the data points already present in the DSM, i.e. from ground level. Because the streetlights have a height of around 6-8 meters, the DSM has to be modified by adding points of added elevation in the locations of interest.This comes in the form of 1x1x6 meter pillars, and the addition of these to the DSM can be seen in **Figure 3**. This addition of these pillars is automated for a specific list of locations. The difference in the calculated horizon lines for ground and "pillar" level for the streetlight in Point 6 is shown in **Figure 4**, while all the horizon lines created for the eight possible streetlight points are shown in **Figure 5**.

10.4229/EUPVSEC2023/4BV.3.27
020324-002

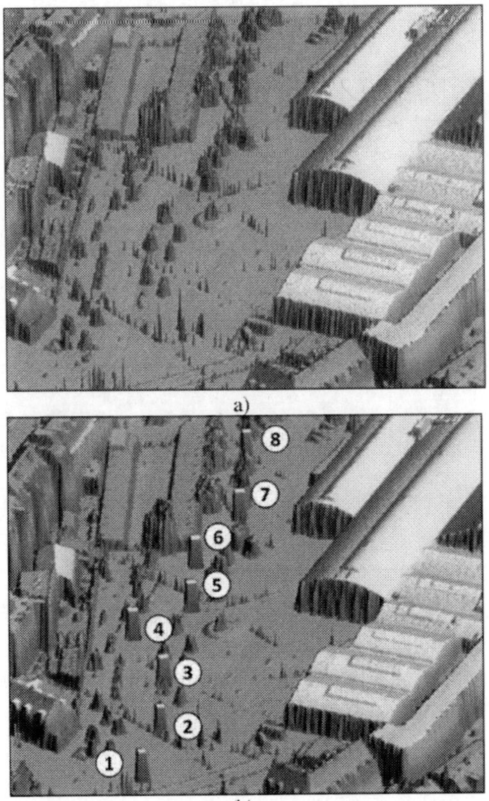

Figure 3: DSM of the Red Square before (a) and after (b) the 3D streetlights "pillars" are placed in their specified locations in the model (screenshots taken from Grass GIS).

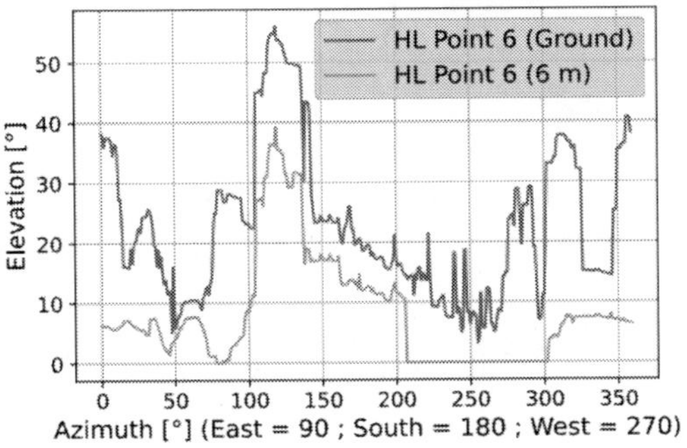

Figure 4: Horizon lines created from r.horizon in GRASS GIS for the streetlight in point 6, at ground level and at 6 meters (height of the streetlight).

The next step is to transpose the horizon lines according to the tilt and orientation of the different sides of the streetlight. This is because the horizon line seen from the vertical North-facing panel is not going to be the same as the horizon line seen from the vertical South-facing panel and so on. Therefore, the horizon line for the different vertical panels is modified to "block" the view of what is behind the panel, e.g. for a vertical South-facing panel, the horizon line is modified to only make visible the horizon 180º around the South (180º azimuth) from East (90º azimuth) to West (270º azimuth), and to block the horizon 180º around the North (0º azimuth): from West to East (270º to 90º azimuth). This is done for all sides, and the results for the streetlight in Point 6 can be seen from **Figure 6**.

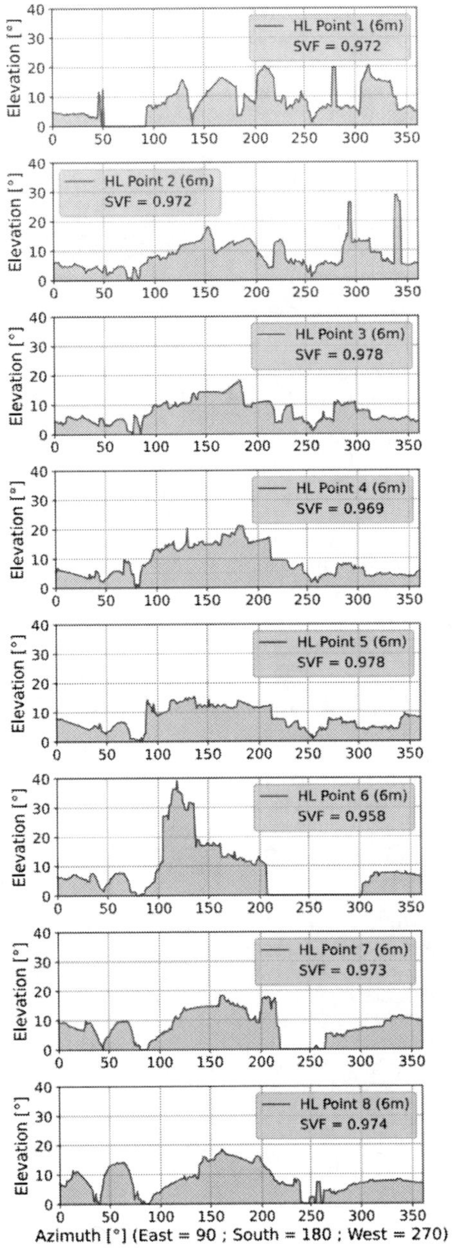

Figure 5: Horizon lines at 6-meter "pillar" height found for all eight possible streetlight points in the Red Square, Copenhagen, created from r.horizon in GRASS GIS. Their resulting sky view factor (SVF) is also shown.

10.4229/EUPVSEC2023/4BV.3.27
020324-003

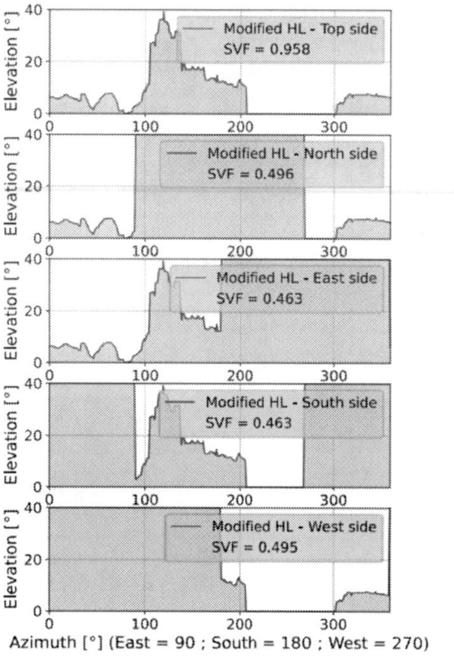

Figure 6: Modified horizon lines are shown for each side of the streetlight in point 6. The Top side is horizontal (tilt = 0º), while the North, East, South, and West sides are vertical (tilt = 90º).

An isotropic sky diffuse model by PVlib [12] is used to find the plane-of-array unshaded irradiance, G_{poa}, using the irradiance components: global horizontal irradiance (GHI), direct normal irradiance (DNI) and diffuse horizontal irradiance (DHI); the sun position, the streetlight orientation (azimuth) of the SPVL system and its individual panel tilts and orientations as inputs. This is then modified to account for shading, resulting in the plane-of-array shaded irradiance, G_{poa}*.

The sun position, modelled using the solar position tool from the PVlib library [13], and the horizon line are used to calculate the shading vector and sky view factor (SVF). The shading vector is applied to the DNI, as in (2), where a calculation is done comparing the sun's position at each hour with the horizon line resulting in irradiance being set to 0 if there is an obstruction and remaining unchanged if not. The SVF, which represents the unobstructed part of the sky seen from the plane of the array, is used to reduce the DHI, as in (3). Reflected irradiance is currently not considered in the modelling (assumed albedo = 0).

$$DNI_{Shaded} = SV \cdot DNI \quad (2)$$

$$DHI_{Shaded} = SVF * DHI \quad (3)$$

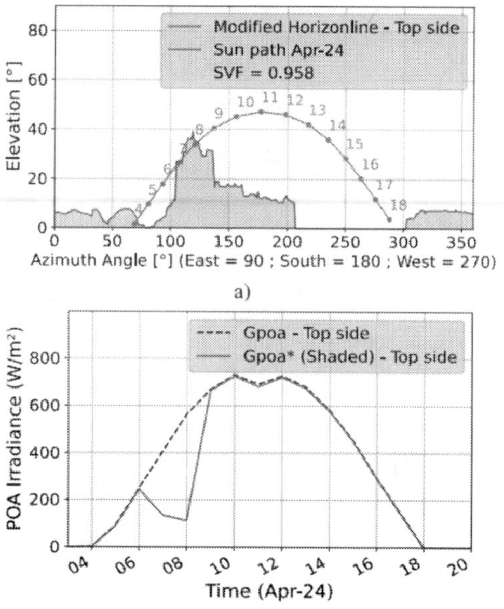

Figure 7: a) Sun-path compared to the modified horizon line seen from the Top side; b) resulting irradiance and shaded irradiance on the plane of array of the Top side for the 24th April for the streetlight in point 6.

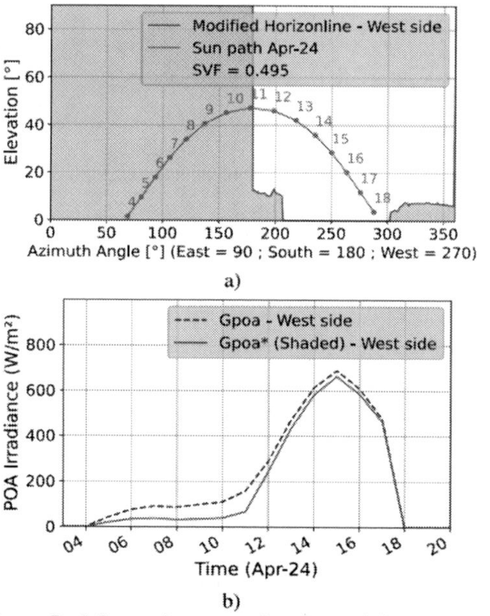

Figure 8: a) Sun-path compared to the modified horizon line seen from the West side; b) resulting irradiance and shaded irradiance on the plane of array of the West side side for the 24th April for the streetlight in point 6.

The sun path for a specific day compared to the horizon line and the resulting calculated irradiance and shaded irradiance on the plane of the array are shown in **Figure 7** and **Figure 8** for two different sides for the streetlight in point 6. It can be seen that when the sun is obstructed by a building (e.g., in hours 7 and 8 on the Top

10.4229/EUPVSEC2023/4BV.3.27
020324-004

side) the shaded irradiance is much lower than the unshaded irradiance. This is because the DNI component of the irradiance is affected by the shading vector. While in periods of non-obstruction (e.g., from hours 12-18), the shaded irradiance is lower than the unshaded irradiance in relation to the SVF, as this reduces the DHI component of irradiance. In the West side (SVF = 0.495), this effect is, therefore, more visible than on the Top side (SVF = 0.958).

4 MODELLING OF THE PIPV PRODUCT

4.1 Component modelling

Component performance models were developed in Python based on datasheet parameters or test data obtained from laboratory characterization tests.

When choosing what models to use for the components, simplicity is valued. It's preferable to have models that can be parameterized using data sheet values rather than those requiring extensive lab or field measurements. Additionally, the modularity of a model is crucial; simpler models with fewer inputs and outputs are more straightforward to integrate into the system model.

PV Panel Model

The PVWatts Model is used to model the operation of the PV panel. It is provided by NREL as a simple way to model PV performance of the panels [14] . It calculates the maximum power from the panel with a specific nameplate rating, P_{mp0}, given a cell temperature, T_c and effective irradiance, E_e, transmitted on the plane of the array. The efficiency of the panel is assumed to decrease linearly with rising cell temperature, based on the temperature coefficient γ. The equation is quite simple and is shown below, where E_0 is the reference solar irradiance (1000 W/m^2), and T_0 is the reference temperature (25°C).

$$P_{mp} = \frac{E_e}{E_0} P_{mp0}[1 + \gamma(T_c - T_0)] \quad (4)$$

This model is based on the latest version of PVWatts, and, therefore, does not include a quadratic correction to reduce power output for irradiances smaller than 125 W/m^2 as this behavior was not observed in modern PV panels.

PV Array Model

The PV Array model uses the shaded G_{poa}* for all sides as an input and applies optical losses through an incidence angle modifier (IAM) model to find the resulting effective shaded irradiance on the plane of the array for all sides: E_{poa}*. The IAM model uses angles of incidence (AOI) found through an AOI model for the various tilts and orientations of the different sides. Both optical loss models are from the PVlib library [12].

The cell temperature of the different PV panels on each side is found first using a module temperature model by Sandia [15], which uses effective irradiance on the plane of array E_{poa} (in this case, after shading, E_{poa}*), wind speed, *WS*, and ambient temperature, T_{amb} as inputs. The cell temperature is then found using Sandia's cell temperature model, using E_{poa}* as an input. The parameters for the module and cell temperature models are assumed to be the typical values for a glass/cell/polymer sheet panel with an insulated back given by [15].

The PV power of the different sides of the array is found through the use of the PV panel models described earlier, using the found effective shaded irradiance, E_{poa}*, and cell temperature, T_c for each side. Because in the SPVL model studied, the sides are connected to five different converter inputs, these sides are considered independent, and so the PV power output of the array is given by the sum of the powers generated by the different sides.

Converter Models

A linear regression model is used both for the PV to the battery (PV2Bat) side of the converter and for the LED to the battery (LED2Bat) side. The parameters of interest of both models are identified through an assessment of converter test data obtained from conversion efficiency experiments carried out in the Risø campus lab. Both the models are one dimensional: the PV2Bat model uses the PV power as input, P_{pv}, and outputs the power given to charge the battery, $P_{bat,ch}$, while the Bat2LED model uses LED load power as an input, P_{load}, and outputs the requested power coming out from the battery $P_{bat,disch}$.

After training on the test data, the regression coefficients for the PV2Bat model are:

$$Y = 9.044 * P_{PV} - 0.062 * P_{PV}^2 \quad (5)$$

And for the Bat2LED model:

$$Y = 7.195 * P_{LOAD} + 0.109 * P_{LOAD}^2 \quad (6)$$

Battery Model

The battery model is based on the General Energy Storage System Model or' General Model' described in [16] This model treats the battery as an" energy buffer" which takes in energy and gives energy out. It does not take into account the voltage of the battery and operating conditions such as temperature and C-rate.

The parameters needed for the model are the maximum power allowed by the battery during charging and discharging $P_{bat,max}$, the total storage energy, E_{bat}, the roundtrip efficiency, η_{RT}, the minimum, SOC_{min}, and maximum, SOC_{max}, allowable values of SOC and a user defined input, SOC_{init}, for the initial SOC of the battery.

The $P_{bat,max}$ = 100 W and E_{bat} = 845 Wh are taken from the datasheet of the battery, while η_{RT} is found as a function of the power going in and out of the battery, P_{bat}, from experiments carried out in the lab. These consist of putting the battery through multiple charge/discharge cycles at different powers. The values for SOC_{min}, SOC_{max}, and SOC_{init} are assumed to be 20%, 90%, and 50%, respectively.

A battery degradation model is also developed based on a pure-lifetime model described in [17]. This modelling approach considers throughput as the stress factor: it counts energy passed through the battery and compare it to a predefined threshold for the battery's End of Life (EOL). This gives an estimate of the battery's remaining useful life (RUL).

$$RUL_{Wh}[years] = \frac{Throughput_{EOL}\,[Wh]}{Throughput_{year}[\frac{Wh}{year}]} \quad (7)$$

The EOL throughput of the battery is determined from a cycles vs depth-of-discharge (DOD) lifetime curve [17] (obtained from manufacturer data through the same methods applied in [18]).

10.4229/EUPVSEC2023/4BV.3.27
020324-005

Energy management system

In addition, an energy management system (EMS) is modelled to improve the energy use of the SPVL and reduce the chances of the battery charge is depleted. In order to have a dynamic operation, the EMS is included in the battery model. It works by monitoring the SOC at every timestep: if the SOC falls below a certain threshold, $SOC_{load,reduced}$, then the EMS simulates a dimming of the light, and reduces the load request to the battery to $P_{load,reduced}$. The basic formula is shown in the equation below for an hourly timestep h, and a battery discharging power $P_{Bat,disch,h}$.

$$if: SOC_{h-1} \leq SOC_{load,reduced} \rightarrow P_{Bat,disch,h} = P_{load,reduced} \quad (8)$$

The value for $SOC_{load,reduced}$ is assumed to be 30%.

LED Load Profile

In the normal operation of the SPVL, the LED luminaire is turned on when sensors in the streetlight detect that the voltage in the PV panels is below a certain threshold, as this indicates that the light source coming from the sun is too low for the panels to produce power. The LED luminaire is set to operate at a predefined power during the night, which is usually lower than the maximum LED power available, especially in the winter months that have low solar resource and long nights. Additionally, a dimming schedule can be set for the LED through a charge controller that regulates the power the LED receives.

A simple night LED load profile is then generated, which is based on GHI input: the LED is set to turn on, and $P_{load} = P_{led,max}$, when GHI is below a certain threshold (e.g. when GHI < 0.01 W/m^2) or equal to 0, as this indicates that there is no solar resource and there is a need of light. An example of the LED load profile generated based on GHI is displayed in **Figure 9.**

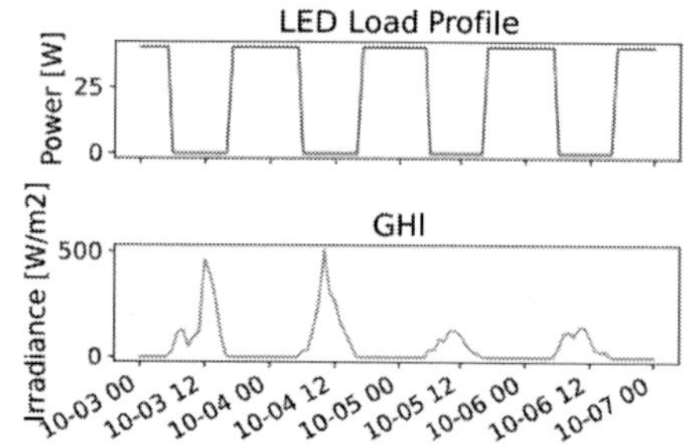

Figure 9: Daily LED load profile based on GHI. When GHI is below 0.01 W/m^2, then $P_{load} = P_{led,max}$.

This simple yearly "normal" load profile is assumed to be the same for the whole year, as can be seen in **Figure 10**.

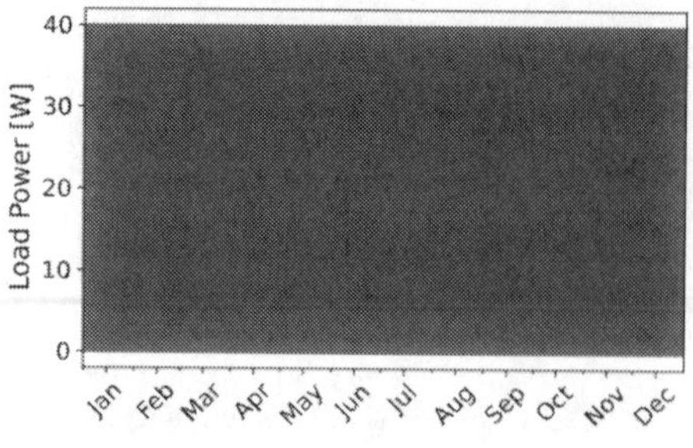

Figure 10: Yearly "normal" LED load profile of the streetlight. It considers $P_{load} = P_{led,max}$ for every night hour during the year.

Availability Model

The availability model can be considered using the $P_{load,request}$ from the LED2Bat converter model, and the $P_{load,delivered}$ from the battery model as well as the $P_{load,reduced}$ from the EMS as inputs. The main output of the model is the number of load hours delivered, not delivered or delivered at a reduced load by the SPVL, and, as a result, the availability of the system.

If the $P_{load,request}$ from the LED2Bat converter model is met, the hours delivered, $h_{Delivered}$, are counted. When the battery SOC drops below the $SOC_{load,reduced}$ the EMS reduces the $P_{load,delivered}$, so that it is equal to $P_{load,reduced} = 0.5 * P_{load}$. The hours in which the load is delivered at a reduced load, $h_{Reducedload}$, are counted. When the battery SOC reaches the SOC_{min}, the $P_{load,delivered}$ becomes equal to 0, and the load is not delivered, and the load hours not delivered $h_{Notdelivered}$ are counted. The total availability and full load availability are then calculated in (9) and (10):

$$Total\ Availability = \frac{h_{Delivered}}{h\_requested} \quad (9)$$

$$Full\ Load\ Avail. = \frac{h_{Delivered}+h_{Delivered\ reduced\ load}}{h_{requested}} \quad (10)$$

4.2 Component model validation

PV Panel and Array Model

The PV array models are validated with measured field data from the SPVL sensors. The plane of array irradiance, G_{poa}, measured by irradiance sensors placed on the different side arrays, is used as an input as the validation of the transposition and optical losses model is not within the scope of the study. The temperature sensors of the SPVL were subject to malfunctions and could not give a reliable output, therefore, the module and cell temperature models from Sandia are used to calculate the cell temperature for each side.

The measured PV power at the input of the converter is used as a reference output; however, this is limited both by the converter and by the battery and does not match the maximum power output of the arrays at all times. Hence, the validation shown in **Figure 11** is for an overcast day with lower PV power output1 (13th of September 2021).

This model seems to overpredict the PV power generated by a small margin quite constantly throughout the day by about 2 W and up to 4 W at PV powers higher than 30 W. This overprediction, however, is relatively less evident at higher powers, where the model shows a relative error of 4-6 % for PV powers higher than 15 W.

At lower PV powers, on the other hand, the overprediction starts to become more evident, and relative errors start to increase dramatically.

Converter Models

Both the PV2Bat and LED2Bat converter models were validated using measured field data from the SPVL sensors.

The input to the PV2Bat model are the measured PV powers PV array, $P_{pv,field}$, while the value that the model is trying to simulate, Y, is considered to be the battery power charging the battery during the day, $P_{bat,ch,field}$. The SPVL converter effectively contains five different converters, one connected to the different side arrays and one connected to the Top panel. Therefore, the converter models are applied five times, with five different sets of inputs coming from the PV array.

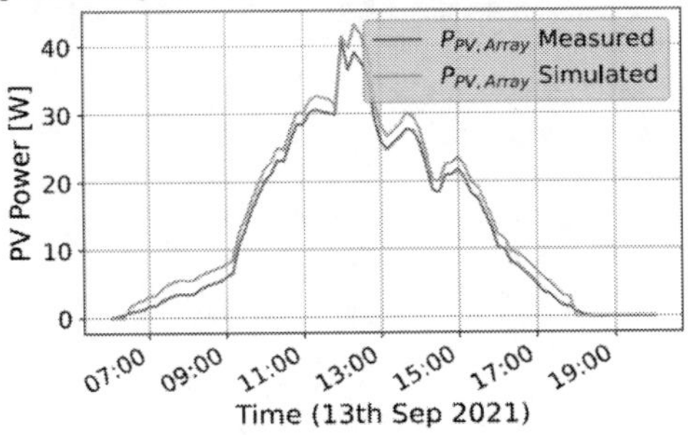

a)

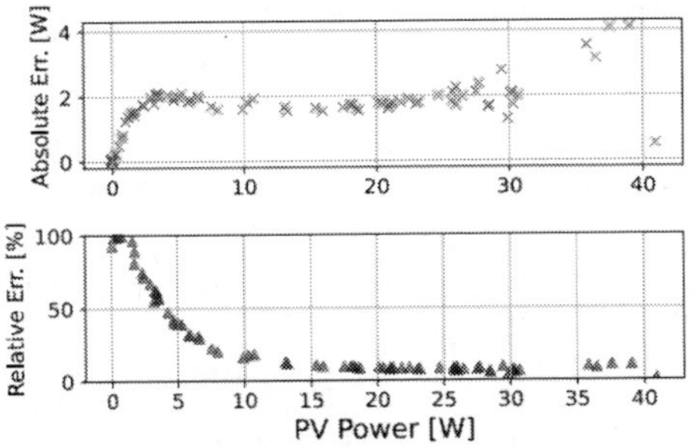

b)

Figure 11: PV array model validation for the 13th of September, using measured irradiance data from the SPVL and module temperatures from the Sandia model as inputs. The simulated PV power from the array is then compared to the measured PV power at the converter of the SPVL in (a) and the resulting relative errors are shown in (b).

The final predicted battery charging power $P_{bat,ch,pred}$ is considered to be the sum of the resulting five battery charge powers predicted from the side arrays.

$$Y_{Top} = LR(P_{PV,TOP}) \tag{11}$$

$$P_{Bat,ch,pred} = Y_{Top} + Y_{North} + Y_{East} + Y_{South} + Y_{West} \tag{12}$$

The field data chosen to be visually compared in **Figure 12** is also from the SPVL daytime measurements of September 13th 2021, as this shows a wide range of PV and battery powers. In addition, the day is chosen because power limitation from the battery is not present, as this could be a cause of potential bias in the results.

The PV2Bat converter model validation results show that the model is fairly accurate, keeping an absolute error lower than 1 W for all PV powers (except one data point). The relative error is more evident for lower PV powers and starts to rise considerably from PV powers of 10 W and under.

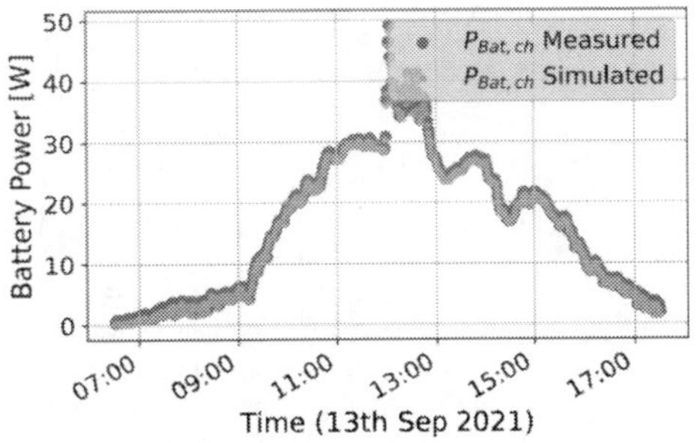

a)

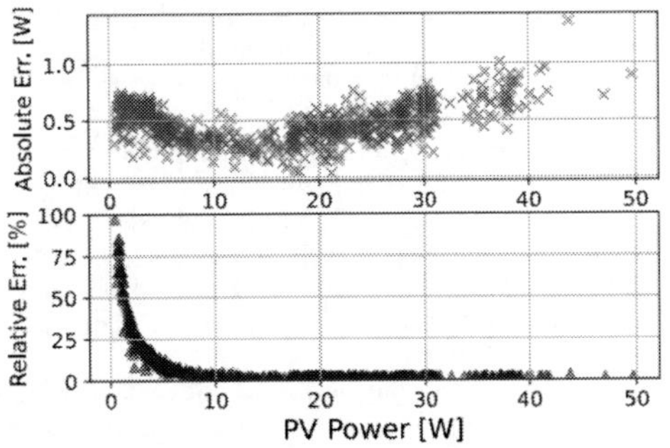

b)

Figure 12: PV2Bat converter model validation for the 13th of September. The simulated PV power from the PV2Bat model is compared with measured PV power data from the SPVL in (a) and the absolute and relative errors are shown in (b)

The LED2Bat model is validated in a similar way to the PV2Bat model. The input to the LED2Bat model, however, is the measured LED power from the LED luminaire, $P_{led,field}$, while the value that the model is trying to simulate, Y, is considered to be the be the battery power discharged from the battery at night, $P_{bat,disch,field}$.

Battery Model

Because the field data of the SPVL systems does not include the battery SOC, the battery model is individually validated with the test data from charge/discharge cycles done in the lab at different conditions.

Furthermore, part of the battery validation comes in the form of the full system model validation explained later, which uses the measured irradiance and measured LED power requested from the lamp as inputs to find daily power and energy going in and out of the battery, and comparing it with the measured daily P_{bat}.

4.3 System modelling

As illustrated in **Figure 13**, the modelling is divided into two main parts: irradiance and shade modelling and system performance modelling. The main inputs are meteorological data and coordinates of the specific location and lamp orientation.

10.4229/EUPVSEC2023/4BV.3.27
020324-007

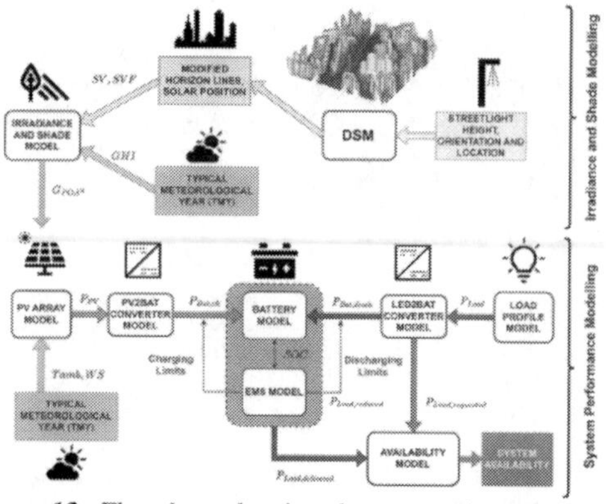

Figure 13: Flowchart showing the two main parts of the modelling: Irradiance and Shade Modelling and System Performance Modelling. Starting from the input of streetlight height, orientation and location into the DSM, using the resulting modified horizon lines and GHI from a TMY file into the irradiance and shade model to obtain the shaded plane of array irradiance, to end with the system availability after considering all the system component models.

The irradiance and shade model requires irradiance data, GHI, as well as horizon lines generated from the DSM for a certain location and calculates the irradiance on each side of the lamp as well as the effect of shade from nearby buildings, giving the shaded plane of array irradiance G_{poa}* as an output. This is then fed to the system performance model.

Starting from the left-hand side of the system performance model, the shaded irradiance on the plane of array G_{poa}*, ambient temperature, and wind speed are given to the PV array model, which is composed of a module temperature model and a PV power model, to get the PV power output, P_{pv}, from the array. This is then fed to the PV2Bat converter model, which converts the PV power to battery charging power, $P_{bat,ch}$, and supplies it to the battery model. The $P_{bat,ch}$ is limited by the EMS if the battery is full.

Moving to the right-hand side, the GHI is used as input to the load profile model to create a load profile for the LED luminaire. The resulting P_{load} is then input to the LED2Bat converter model and converted to the power requested from the LED to the battery, $P_{load,request}$. This, however, is subject to changes by the EMS, based on the state of charge of the battery: if the battery is below a certain SOC, then the power requested to the battery is reduced, $P_{load,reduced}$.

The effective discharging power of the battery, $P_{bat,disch}$, is then either equal to the $P_{load,request}$ or the $P_{load,reduced}$ depending on the conditions set by the EMS. The $P_{bat,disch}$ is limited if the battery has reached minimum SOC. The output of the battery model is the effective power that is delivered to the LED by the battery, $P_{load,delivered}$. If the battery has reached a minimum SOC, this will be equal to 0. The $P_{load,delivered}$ and $P_{load,request}$ are then input to the system availability model, which calculates the number of load hours that are delivered, that are not delivered, or delivered at a reduced load, and calculates the effective light availability of the lamp.

4.4 System model validation and operation

The system was then validated with field measurements coming from sensors installed in the SPVL present on campus. Once the component models were validated, they were combined into a system model using a Python framework, which was then validated using the measured irradiance and measured LED power requested from the lamp as inputs. The resulting simulated power going into (from the PV panels) and out of (from the LED load) the battery is then compared to the measured battery power in **Figure 14**, as well as the simulated and measured daily energy going into and out of the battery, for four days in September 2021.

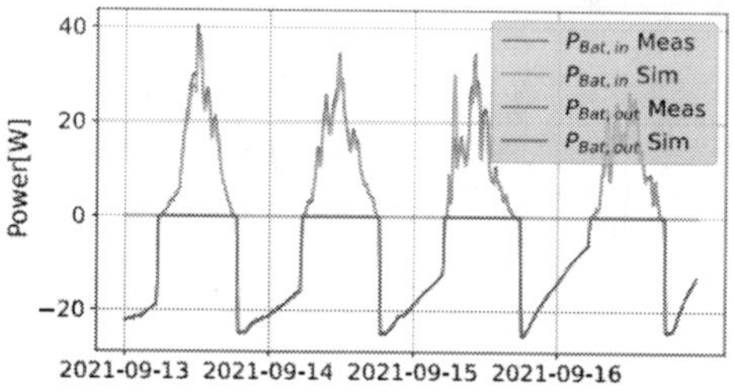

a)

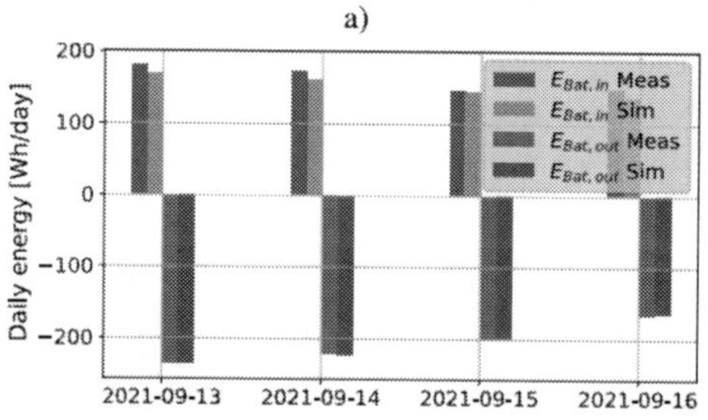

b)

Figure 14: Results from a system performance model validation of the SPVL, showing simulated power (a) and energy (b) going into the battery vs measured power (a) and energy (b) going into and out of the battery, from the 13th to the 17th of September 2021

A simulation of the SPVL system was carried out using coordinates from the Risø campus, considering a position affected by shade from the surrounding buildings, a typical meteorological year (TMY) file obtained from the Photovoltaic Geographic Information System (PVGIS) [19] for Risø and a horizon line created from DSM data of the region as inputs.

In **Figure 15** and **Figure 16**, partial results of the simulation shows how the models operate. The operation of the irradiance and shade model is displayed in **Figure 15**. The irradiance on the different planes of array of the lamp is calculated from the input GHI and orientation of the lamp, as well as the tilt and orientation of the individual panels. The shading model is then applied, reducing the overall irradiance received by the lamp.

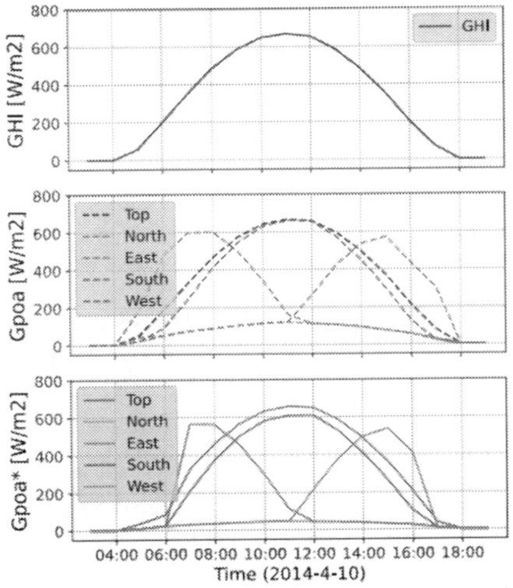

Figure 15: Results from the irradiance and shade model, using GHI, solar position and modified horizon lines as input and giving the shaded G_{poa}*, for all sides as output for a clear-sky day in Risø.

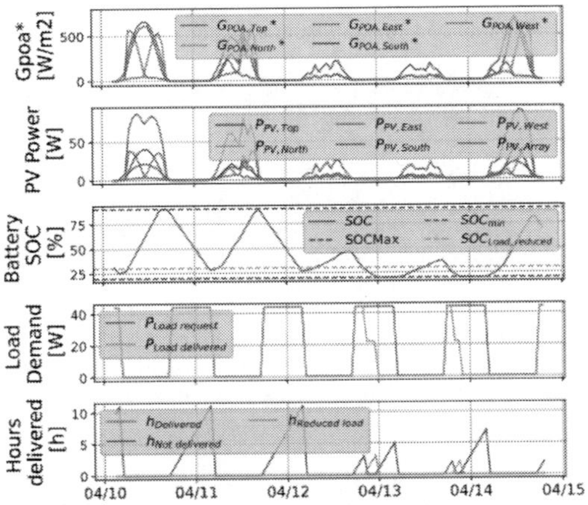

Figure 16: Results from the simulation of the System Performance Model showing five days of operation in Risø. Starting from shaded G_{poa}* as an input from the Irradiance and Shade Model, PV power produced is then obtained, which charges the battery during the day while the load demanded discharges it during the night. The battery SOC limits restrict the actual load delivered and result in the availability calculation.

The power from the different PV modules is then calculated by the system performance model and combined to obtain the total PV power output of the array. As seen in **Figure 16,** this PV power, as well as a pre-established load demand profile, is fed into a battery model which calculates battery SOC and sets power limitations if the battery reaches a minimum or maximum SOC, as well as a reduced demand if the SOC falls below a specified value. The battery SOC rises during the day with the PV power production and falls at night with the load demanded by the LED. The power delivered to the LED depends on the SOC of the battery. If the battery reaches a minimum SOC, SOC_{min}, or falls below the value of $SOC_{load,reduced}$, the availability model counts those load hours as 'not delivered' or 'delivered at a reduced load.'

5 RESULTS AND DISCUSSION

In this section, the results of a one-year simulation run for the streetlights in the eight selected points in the Red Square are evaluated. Possible system design solutions to increase availability are then proposed and analyzed.

5.1 Evaluation of the available solar resource

The total resource reaching the different sides of the streetlights after shading is considered, as shown in **Figure 17**, as well as the resulting losses due to shade. It can be observed that for all streetlights, the solar resource on the top (0º tilt) side is significantly higher than on the other sides, reaching values above 1000 sun hours for all eight points. It is also the side with the least losses in solar resource due to shade. This can be explained by the fact that the other sides are vertical (90º tilt) and have parts of the sky "blocked" from their view, as can be seen in **Figure 6** for the streetlight in point 6, and will therefore be subject to large losses in both the direct normal irradiance component, due to the sun being obstructed, and the diffuse irradiance component of due to the low SVF. As is expected for a country in the latitude of Denmark, the North side has substantially less irradiance than the other orientations.

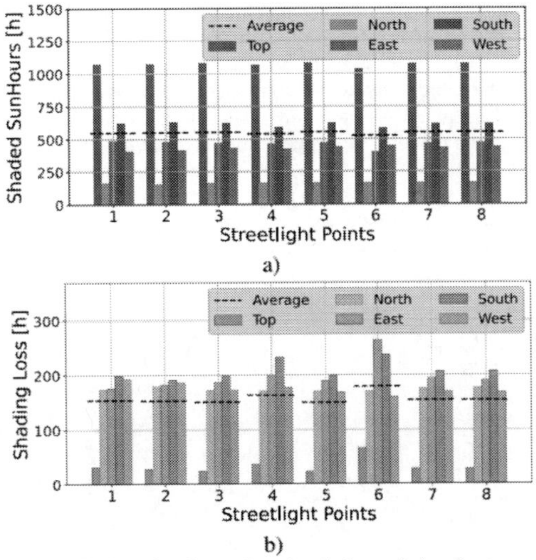

Figure 17: Results from the simulation of the System Performance Model showing five days of operation in Risø. Starting from shaded G_{poa}* as an input from the Irradiance and Shade Model, PV power produced is then obtained, which charges the battery during the day while the load demanded discharges it during the night. The battery SOC limits restrict the actual load delivered and result in the availability calculation.

Losses due to shade in the yearly solar resource for the different vertical sides (North, East, South, and West) are similar for all the streetlights. However, for the streetlight at point 6, the losses from the East side are considerably more pronounced, almost reaching 300 sun hours. This could be explained by the tall obstruction present in point 6's horizon line at around 120º azimuth, as seen in **Figure 7**, which can cause significant obstruction during the year.

10.4229/EUPVSEC2023/4BV.3.27
020324-009

Upon further inspection of the map and DSM shown in **Figure 2** and **Figure 3**, respectively, this obstruction is caused by a tree close to the streetlight. This tree also causes significant losses to the top side and south side of the streetlight in Point 6.

These losses in solar resource are reflected in **Figure 18**, where the average yearly solar resource reaching the streetlight is shown. As expected, the streetlight in point 6 receives the least amount of solar resource in average, 524 sun hours, while the streetlight in point 4 also has noticeably lower solar resource, 540 sun hours, than the other points.

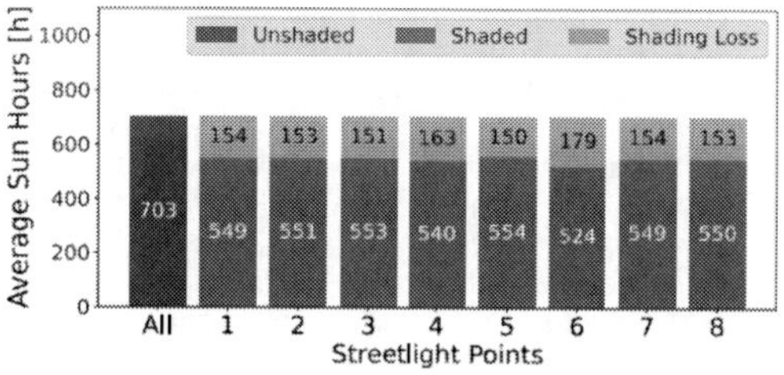

Figure 18: Average yearly solar resource (in sun hours) reaching the streetlights for a one-year simulation in the Red Square. In dark blue is shown the solar resource for all the lamps if the streetlights were not affected by shade. In lighter blue is shown the actual solar resource for the streetlights after shading, and in grey is the corresponding loss in solar resource due to shading.

5.2 Evaluation of the streetlight availability

The resulting total and full load yearly availability for the streetlight in the eight different points is shown in **Figure 19**. As can be observed, all of the streetlights have a very poor yearly performance: the total availability is less than 50 % for all streetlights, and full load availability is less than 40 % for all except streetlight in point 2. This means that all the streetlights have more hours of light not delivered than delivered.

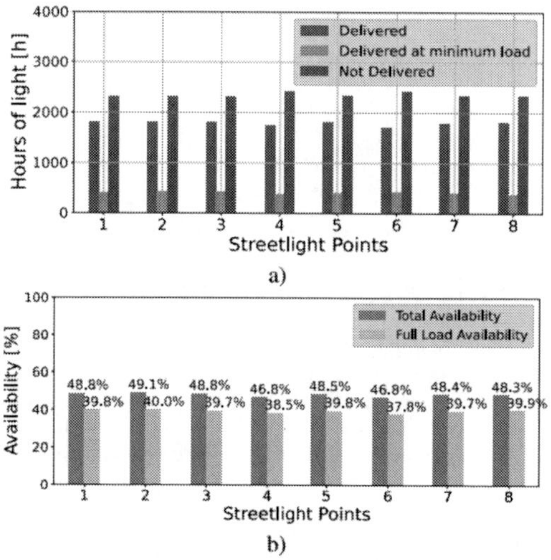

Figure 19: Hours of light delivered from the different streetlights (a) and their resulting total and full load availability (b) for a one year simulation in the Red Square.

When comparing to the total availability for all streetlights, if shading is not considered, **Figure 20**, it can be seen that shading causes from 7-10 % losses in availability for all the streetlight points. The streetlights in points 4 and 6 are the worst performing, as can be expected due to the lower solar resource, but not substantially. Even when shading is not considered, the total yearly availability is not at an acceptable level, 56 %.

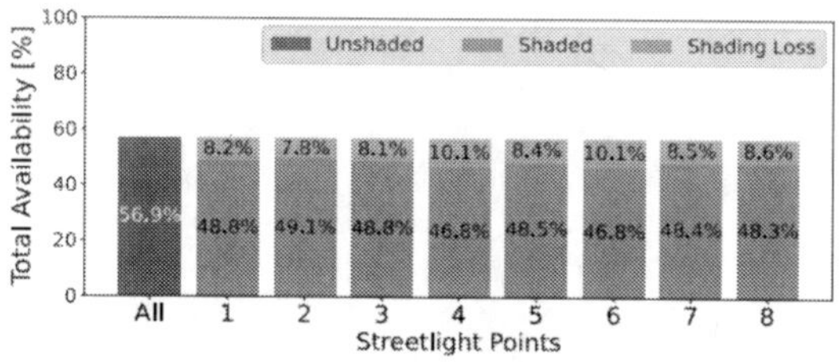

Figure 20: Total availability for the different streetlights after a one-year simulation in the Red Square. Represented in dark green is the availability of all the streetlights if they were not affected by shading, while in lighter green is the availability of the streetlights after considering the shade, and in gray is the corresponding loss in availability due to shade.

The low yearly availability of all the streetlights is explained by the strong seasonal changes in Copenhagen, seen in **Figure 21**, which has winter months with low irradiance and long nights and summers with high irradiance and short nights.

From **Figure 22**, it is evident that the daily energy produced from the PV array is not enough in the winter months to meet the energy requested by the load. Therefore, the battery does not get charged enough and reaches minimum SOC on a nightly basis, limiting the hours of light delivered. The PV energy produced is only enough in the summer months (April to August). In these months, due to the shorter nights, the demand is lower, and the energy produced from the PV panels is higher from the increased irradiance. Because of this, the battery during the summer can reach maximum SOC.

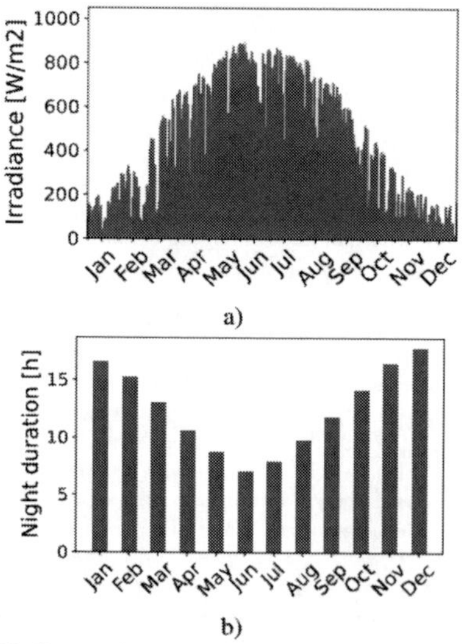

Figure 21: Seasonal variation in Copenhagen: irradiance (a) and length of the nights (b) changes significantly

10.4229/EUPVSEC2023/4BV.3.27
020324-010

during the year. Winter months, having low irradiance, and long nights can be a challenge for the operation of solar streetlights.

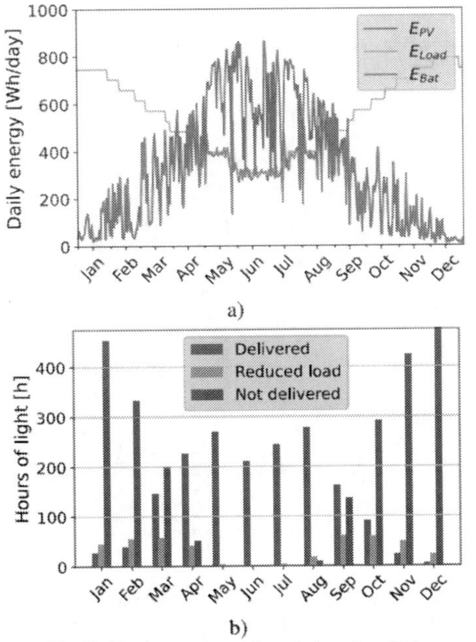

Figure 22: Daily energy produced by the PV array compared to the daily energy requested by the load and that is going into the battery (a) and the resulting hours of light delivered, delivered at a reduced load and not delivered (b) for a one-year simulation of the streetlight at Point 6, considering a "normal" load and a "normal" sized battery.

5.3 Strategies for improving streetlight availability

Seasonal Load

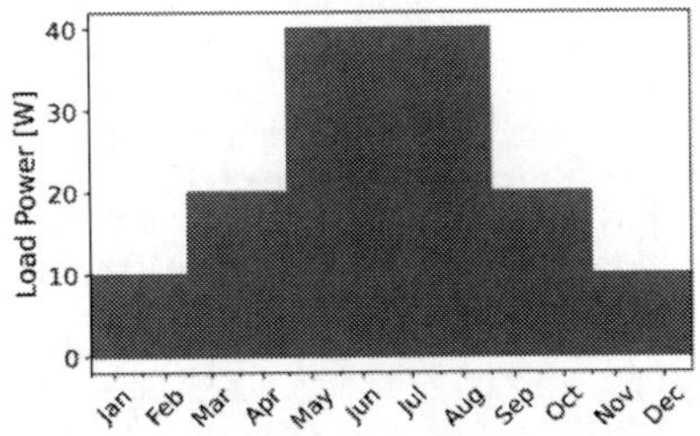

Figure 23: Yearly seasonal LED load profile of the streetlight. It considers $P_{load} = f * P_{led,max}$, where f is a seasonal factor that reduces the load based on the season, reflecting the level of irradiance for the period of the year. For the winter months, $f = 0.25$, for the spring and autumn $f = 0.5$ and for the summer $f = 1$."

One possible way to account for the seasonal changes is to use a seasonal, yearly LED load profile that reflects the periods of low irradiance in the winter. As shown in **Figure 23**, the load uses a seasonal factor that reduces the power requested from the LED in the winter, spring, and autumn months, i.e., simulating a dimming of the luminaire.

The resulting total yearly availability from this change can be observed in **Figure 24**, with all the streetlights showing a great increase, ranging from 27 % to 32 %, and reaching an average total availability of around 78.8 %. This increase can be explained by **Figure 25**; the daily energy demanded by the load in the winter months is now similar to the daily energy produced by the PV array. However, in January and December, the irradiance levels are still too low for the streetlight to produce the required PV power.

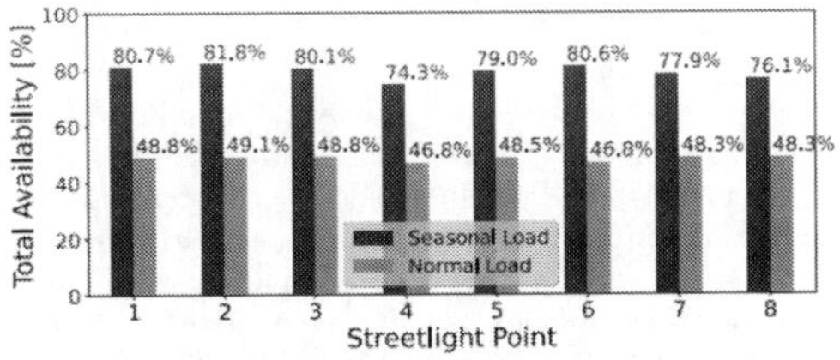

Figure 24: Total availability of the eight streetlights for a one-year simulation in the Red Square. In dark green, the streetlights use a seasonal, yearly load profile of the LED luminaire, and in light green, they use the "normal" constant yearly load profile.

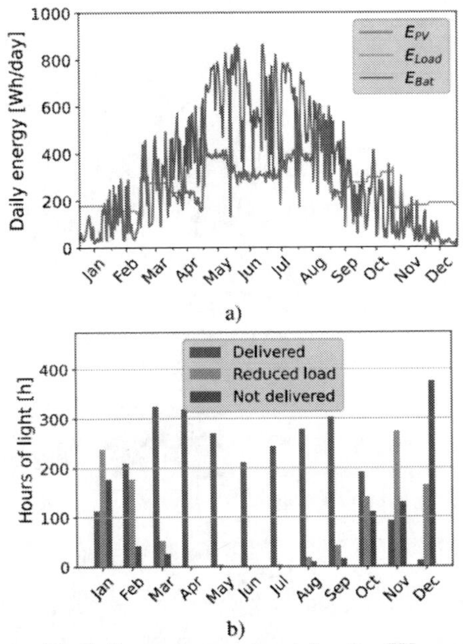

Figure 25: Daily energy produced by the PV array compared to the daily energy requested by the load and that is going into the battery (a) and the resulting hours of light delivered, delivered at a reduced load and not delivered (b) for a one year simulation of the streetlight at Point 6, considering a seasonal load.

Larger Battery

The use of a larger battery (1845 Wh) instead of the "normal" sized battery mentioned in the streetlight specifications (845 Wh) is analyzed as a possible system design change. The resulting total availability for all the streetlights is shown in **Figure 26**.

10.4229/EUPVSEC2023/4BV.3.27
020324-011

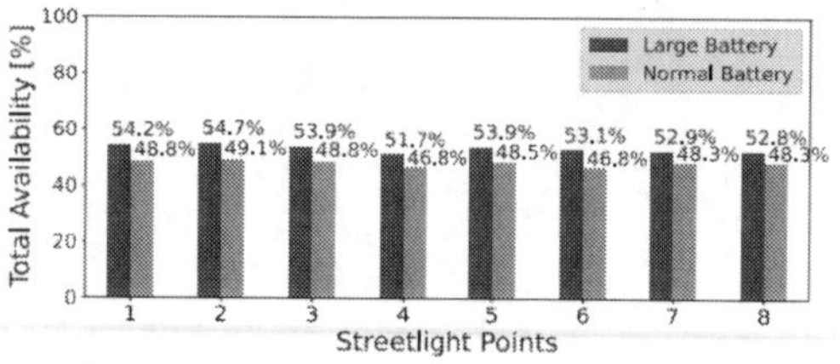

Figure 26: Total availability of the eight streetlights for a one-year simulation in the Red Square, in light green when considering a "normal" sized battery is installed in the streetlights and dark green when considering a "large" battery is installed.

The availability increases slightly for all the streetlights, by around 5 % on average, but clearly not enough to significantly improve the performance to an acceptable level. As can be seen from **Figure 27**, the battery is still getting discharged to the minimum SOC in the winter months and charged to the maximum SOC in the summer, as the energy inputs from the PV power and requests from the LED load have not changed. However, in months like August, the larger battery does allow a bit more "emergency storage" to compensate for sudden days of low irradiance and PV power production. Additionally, the larger capacity allows for more hours of dimmed light due to the EMS still reducing the load at $SOC_{load,reduced}$ = 30%.

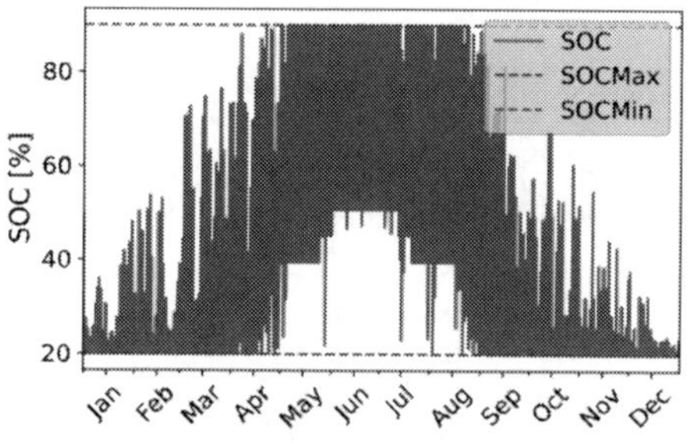

a)

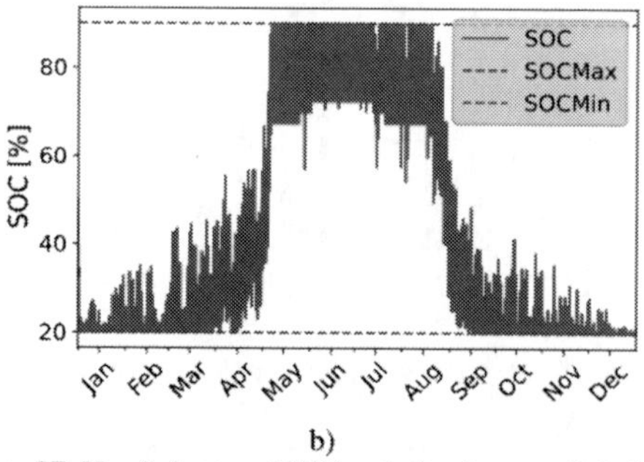

b)

Figure 27: Yearly battery SOC levels for the streetlight in Point 6 for a year simulation, when considering a normal-sized battery installed (a) and a larger-sized battery installed (b). The minimum and maximum SOC levels are shown by the red and green dotted lines, respectively.

One notable change that a larger battery provides is the smaller daily depth of discharge. As can be seen in **Figure 28**, for all the streetlights, the average daily DOD dropped by around half when switching to the larger battery. This leads to the simulated remaining useful lifetime of the batteries being considerably higher in the case of the large batteries, almost three times bigger than the normal-sized battery, due to the fact that the battery is going through fewer cycles and, therefore, less degradation. This could be an important factor when dealing with the projected lifetime costs of the streetlight: the bigger, more expensive battery might end up costing less in the long run than the cheaper, normal-sized battery with a lower lifetime.

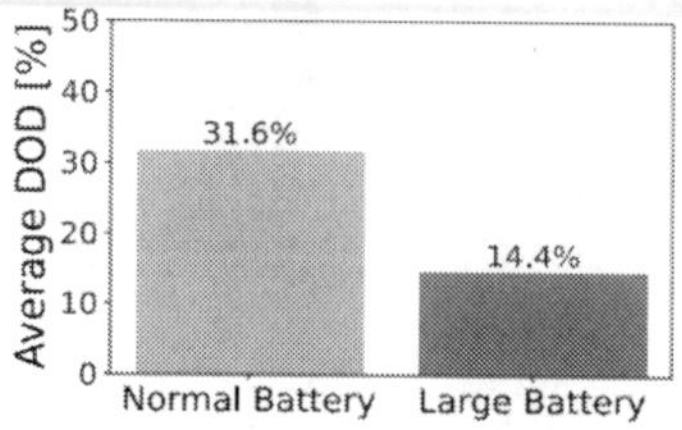

a)

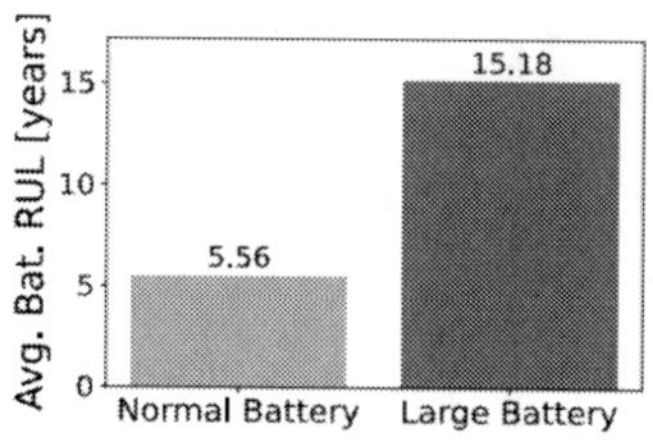

b)

Figure 28: The average daily depth-of-discharge (DOD) of the streetlight batteries for a one-year simulation in the Red Square (a) and their resulting average estimated remaining useful lifetime (RUL) (b) compared in the case the batteries installed are "normal" sized (light orange) and larger sized (dark orange).

6 CONCLUSIONS

A tool has been developed to facilitate the remote estimation of solar resources and system performance for a PIPV product, specifically a solar streetlight, in shaded urban environments. It employs shade and irradiance modeling techniques, drawing data from Digital Surface Models to accurately estimate the solar resource. From this solar resource, the streetlight's availability is found using the system performance model. What makes this tool particularly versatile is its modularity, allowing it to be tailored for different PIPV products in the future. Additionally, the models are designed with scalability in mind, capable of incorporating a vast number of locations in a single simulation.

A case study in Copenhagen's Red Square highlighted the tool's practicality. Results show that the solar streetlights placed in the square would underperform, with an average total availability of 48.2 %. However, by adjusting the load to a seasonal one, which mirrors periods of low irradiance and long nights in the winter time, the availability of streetlights can be enhanced by up to 30%. In contrast, merely increasing the battery size results in a modest availability increase of up to 5%. It's worth noting that larger batteries might have a longer lifespan than standard-sized ones. This is attributed to their lower Depth of Discharge, which means they undergo fewer cycles and consequently experience reduced degradation.

As for future developments, there's potential in accounting for self-shading by employing a precise 3D model of the product.

7 REFERENCES

[1] A. Mira *et al.*, “Improving Site Shading Assessment from Digital Surface Models with Onsite LiDAR Measurements,” 2022, doi: 10.4229/WCPEC-82022-4BV.4.23.

[2] National Renewable Energy Laboratory, “System Advisor Model,” *U.S. Department of Energy*, 2019. https://sam.nrel.gov (accessed Jan. 20, 2023).

[3] “PVsyst - Logiciel Photovoltaïque.” https://www.pvsyst.com/features/ (accessed Jan. 20, 2023).

[4] V. Software, “PV*SOL.” https://pvsol.software/en/features-pricing (accessed Jan. 20, 2023).

[5] O. Isabella *et al.*, “Advanced modelling of E/UIPV systems from location to load,” *2018 IEEE 7th World Conference on Photovoltaic Energy Conversion, WCPEC 2018 - A Joint Conference of 45th IEEE PVSC, 28th PVSEC and 34th EU PVSEC*, pp. 2691–2696, 2018, doi: 10.1109/PVSC.2018.8548039.

[6] O. Iwase and H. Takenaka, “New system designed for monitoring of solar radiation in urban areas,” *Energy Sci Eng*, vol. 10, no. 9, pp. 3471–3492, 2022, doi: 10.1002/ese3.1233.

[7] S. Guo, C. Wu, M. Danner, A. Nobre, A. G. Aberle, and M. Peters, “Modelling of an integrated standalone streetlamp PV system,” *2015 IEEE 42nd Photovoltaic Specialist Conference, PVSC 2015*, 2015, doi: 10.1109/PVSC.2015.7355829.

[8] J. Singh, P. Mahajan, and R. Garg, “Design implementation of solar fed intensity controlled streetlight,” *2018 2nd IEEE International Conference on Power Electronics, Intelligent Control and Energy Systems, ICPEICES 2018*, pp. 671–676, 2018, doi: 10.1109/ICPEICES.2018.8897422.

[9] “SDFE - Kortforsyningen download. Styrelsen for dataforsyning og effektivisering.” https://download.kortforsyningen.dk/ (accessed Jan. 20, 2023).

[10] A. Flatman *et al.*, “Quality assessment of the Danish Elevation Model (DK-DEM),” *Danish Ministry of the Environment - National Survey and Cadastre*, no. 12, 2011.

[11] Huld Thomas et al., “r.horizon manual - GRASS Development Team,” 2007. https://grass.osgeo.org/grass78/manuals/r.horizon.html. (accessed Jan. 20, 2023).

[12] C. W. H. and M. A. Mikofski. William F Holmgren, “pvlib python: A python package for modeling solar energy systems,” *Journal of Open Source Software 3.29*, p. 884., 2018.

[13] I. Reda and A. Andreas, “Corrigendum to ‘Solar position algorithm for solar radiation applications’ [Solar Energy 76 (2004) 577-589] (DOI:10.1016/j.solener.2003.12.003),” *Solar Energy*, vol. 81, no. 6. 2007. doi: 10.1016/j.solener.2007.01.003.

[14] “PVWatts - Sandia National Laboratories - US Department of Energy (2018).”

[15] and D. L. King. Jay A Kratochvil, William Earl Boyson, “Photovoltaic array performance model.”, [Online]. Available: url: https://www.osti.gov/biblio/919131.

[16] D.-I. Stroe, “Modelling of Energy Storage Systems,” 2019.

[17] D. I. Stroe, “Lifetime Models for Lithium-ion Batteries used in Virtual Power Plant Applications,” *Lifetime Models for Lithium-ion Batteries used in Virtual Power Plant Applications*, p. 274, 2014, [Online]. Available: https://www.energy.aau.dk/events/show/phd-defence-by-daniel-ioan-stroe-on-lifetime-models-for-lithium-ion-batteries-used-in-virtual-power-plant-applications.cid131015

[18] K. R. Mallon, F. Assadian, and B. Fu, “Analysis of on-board photovoltaics for a battery electric bus and their impact on battery lifespan,” *Energies (Basel)*, vol. 10, no. 7, 2017, doi: 10.3390/en10070943.

[19] “European Comission - PVGIS Online Tool.” https://joint-research-centre.ec.europa.eu/pvgis-online-tool_en (accessed Jan. 20, 2023).

10.4229/EUPVSEC2023/4BV.3.27
020324-013

LANDSCAPE INTEGRATED PV: A CASE STUDY

Celísio Pires[1], Lisa Bunge[1], Luis Fialho[1,2], Pedro Horta[1,2]
[1] Renewable Energies Chair, University of Évora, Polo da Mitra da Universidade de Évora, Edifício Ário Lobo de Azevedo, 7000-083 Nossa Senhora da Tourega, Portugal
[2] Institute of Earth Sciences, University of Évora, Rua Romão Ramalho, 7000-671 Évora, Portugal

ABSTRACT: This paper presents a comprehensive proposal for a photovoltaic (PV) system aimed at supplying electricity to an ecotourism project at Quinta do Carmo, Estremoz, Portugal. The analysis begins by estimating and analysing electricity consumption for three distinct categories: water pumping, buildings and accommodations, and electric vehicle charging. Based on these estimations, a PV system is sized, and energy production along with self-consumption rates are simulated and calculated. Moreover, a technical-economic analysis of the proposed system and its decarbonization potential is provided. For water pumping, consumption estimates are based on monthly water consumption, area calculations, and elevation differences. The proposed PV system comprises three installations with varying capacities and orientations, strategically placed across different sections of the land. By simulating the consumption and production profiles, the study achieves an 70% self-consumption rate, demonstrating the viability of the PV solution to meet the ecotourism project's energy demands. The proposal also considers innovative solutions like single-axis solar trackers and agrivoltaic integration, showcasing a holistic approach towards sustainability and efficient land use.
Keywords: Photovoltaic System, Ecotourism, Integrated PV, Self-consumption, Sustainability

1 INTRODUCTION

The European Union (EU) is committed to reducing greenhouse gas emission by 55% by 2030 when compared to the 1990 levels and aims to become the first climate neutral continent by 2050. Achieving climate neutrality by 2050 requires an authentic revolution in the energy sector that will only be carried out with local or proximity production. As climate change is becoming more pronounced each year, several solutions must be researched and developed so the goals can be achieved. This will require an effort from all energy sectors, including the tourism sector.

In this paper a study is presented to offer the best solution for the design of a grid-tied PV system for the eco-resort located in Quinta do Carmo, Estremoz, Portugal. This eco-resort is still in the procurement and construction phases and aims to become as sustainable and energetically autonomous as possible.

The design of this PV system comes with some challenges, one of them being the different loads present at the resort, which can be divided into 3 groups: buildings and accommodations; water pumping systems and electric vehicles charging. Beside this challenge, the resort owners want to enhance and keep the natural values of its land by protecting as much as possible the landscape, natural habitats, and biodiversity. This means that all electrical infrastructures must blend with the landscape and remain hidden from guests during their experience. Other challenge is the area of the resort's land, which totals an area of 873.06 ha with several buildings spread, which will require the electrical distribution to be done in medium voltage (MV), thus implying the incorporation of transformers in the electrical distribution infrastructure.

This PV system was designed and optimized to satisfy as much as possible the resort's demand. As no usage data is available in this design phase, demand was estimated using: an average daily guest occupancy estimation for the resort; the average arrival of EVs (both full electric and hybrid); the water requirements of the different agricultural areas and natural areas; and, the electrical consumption of the pumps located in the different water wells, pits, and dams, according to their relative heights.

2 CONSUMPTION CHARACTERIZATION

The electricity consumption for the proposed photovoltaic (PV) system is characterized across three main consumers. First, water pumping consumption is estimated for accommodations, vineyards, gardens, and other areas based on the monthly water usage and elevation differences. Second, electricity consumption for various accommodations and buildings is estimated using forecasted guest numbers. Third, electric vehicle charging consumption is estimated considering the number of EVs, hybrids, and its weekly consumption. Consumption profiles are provided for each category, reflecting seasonal variations linked to irrigation needs and occupancy rates. This data is vital for understanding consumption patterns, aiding in system design and optimization.

Both water pumping and EV charging were treated as exclusively daytime activities, as they can be easily scheduled. Electricity consumption at the accommodations and buildings follows the average daily consumption profile [1] based on the representative profiles given by E-REDES (main electricity distribution grid operator in mainland Portugal) [2].

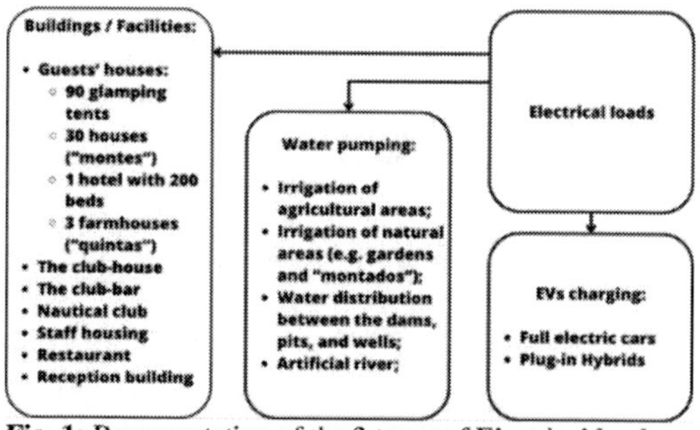

Fig. 1: Representation of the 3 types of Electrical loads.

2.1 Water Pumping

The water pumping consumption entails water supply for various accommodation types, the winery, and irrigation of gardens, agricultural, and forested areas.

10.4229/EUPVSEC2023/4BV.3.28
020325-001

The estimates are based on:

- Monthly water consumption for each sector;
- Water supply pipeline network;
- Pipeline network for inundated soils;
- Maximum elevations (C_{max}) of pipelines near supply areas.

Table I: Annual water consumption volume.

Sector	Water Volume [m³/year]
Vineyard 1,10 ha	25,000.00
Vineyard 2,50 ha	125,000.00
Vineyard 3,60 ha	150,000.00
Cork Forest 1995	120,000.00
Olive Trees	12,500.00
"Oliveiras do Carmo" Garden + Main House + Winery	127,300.00
Glamping + Garden + Natural Pools	156,870.00
Serra da Ossa Hotel	15,792.00
Serra da Ossa Hotel Garden + Valley	60,000.00
"Montes" + Gardens	138,198.00
"Quintas"	3,400.00
Total	**934,060.00**

All measurements of areas, distances, and elevations were determined using Google Earth Pro [3].

For the vineyard, considering the varied elevation across its three distinct areas, water consumption was determined using a specific consumption of 2,500 m³/ha.

The water source was assumed to be the Mesquita Dam at an elevation of 302 m (C_o), four meters below the full reservoir level of 306 m (FSL - Full Storage Level), an estimate for drier periods. ΔH is the height difference between the consumption point elevation and the 302 m elevation of the Mesquita Dam:

$$\Delta H = C_{max} - C_o = C_{max} - 302 \text{ [m]}$$

As the pipelines are still under design, head losses (P) had to be estimated, assuming 0.20 mH_2O for every 100 m of linear pipe distance (D). This value will later be adjusted after the final hydraulic system design. The sum of ΔH and losses gives the manometric height, H_m:

$$H_m = \Delta H + P = \Delta H + 0.2 \times D/100 \text{ [m]}$$

To estimate the energy required for water pumping, the following equation [4] was employed:

$$E_i = (\rho g V_i H_m) / (\eta 3.6 \times 10^6) \text{ [kWh/month]}$$

Where ρ is the water density (998.34 kg/m³ [5]), g the gravitational acceleration (9.81 m/s²), V_i the monthly water volume in m³/month, and η the pump efficiency (assumed as 0.90).

According to the calculations, the consumption for water pumping exhibits pronounced seasonality, with higher electricity consumption during Spring-Summer months, peaking in August. This pattern reflects typical irrigation and occupancy needs in the tourism sector (**Fig. 2**).

2.2 Accommodations and Buildings

In this eco-resort there are several buildings and accommodations that will demand electricity. A summary of this facilities can be found on **Table II**, where the number of units and beds can be observed.

Table II: Summary of the facilities in the eco-resort.

RESORT	Units	Pax/Beds tot
Accommodations	**203**	**714**
Glamping	90	216
Montes	30	272
Hotel	80	200
Quintas	3	26
Buildings	**8**	**60**
Nautical Club	1	0
ETAR	1	0
Equestrian Center	2	0
Staff Housing	1	60
Club-House	1	0
Club-Bar	1	0
Reception Building	1	0
Total	**203**	**774**

The electricity consumption associated to the various types of accommodations, as well as other buildings was estimated based on a forecast of the number of guests per month (**Table III**).

Table III: Forecast of the number of guests per month.

Month	Jan	Feb	Mar	Apr
# Guests	4,508	6,011	9,016	10,519
Month	May	Jun	July	Aug
# Guests	13,524	15,027	18,033	21,038
Month	Sep	Oct	Nov	Dec
# Guests	21,038	9,016	12,022	10,519

Using the data in **Table III** and assuming an average consumption per guest of **23 kWh** [6] [7]**,** the annual consumption curve in **Fig. 2** was obtained.

The estimated consumption for buildings also exhibits seasonal variation, with the higher electricity consumption component occurring during the summer months, peaking in August and September. This pattern reflects the typical needs of hotel establishments as well as the guest occupancy profile in tourism enterprises.

2.3 Electric Vehicles

Electricity consumption associated with electric vehicles charging was estimated based on the forecast of the number of guests per month (**Table III**) and the following assumptions:

- Nights per guest: 7
- Guests per car: 2.5
- Percentage of EVs in 2026:
    - EVs: 30%
    - Hybrids: 33%
- EV consumption for 7 days:
    - EVs: 165 kWh
    - Hybrids: 33 kWh

Based on the above assumptions, the annual consumption curve for electric vehicle charging was obtained, as shown in **Fig. 2**.

As depicted in **Fig. 2**, the estimated consumption for electric vehicle charging also displays seasonal variation, with the primary electricity consumption component occurring during the summer months, peaking in August

10.4229/EUPVSEC2023/4BV.3.28
020325-002

and September. This pattern is directly related to the estimated number of guests.

2.4 Annual consumption curves and daily profile

In the following figures, the annual consumption curves for each type and its combined total, as well the daily consumption profile for each type can be observed.

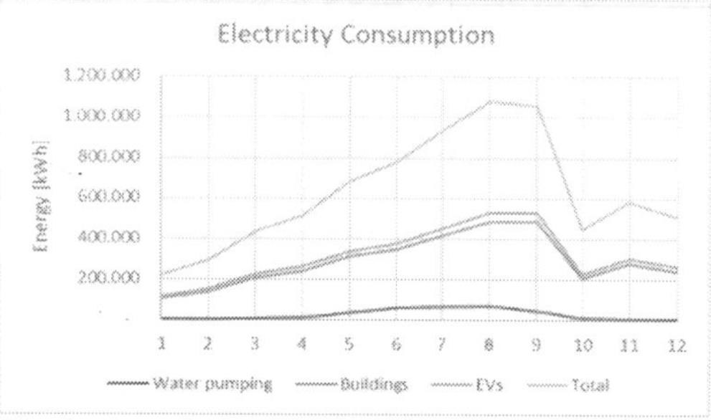

Fig. 2: Annual consumption curves estimated for the 3 types and its combined total.

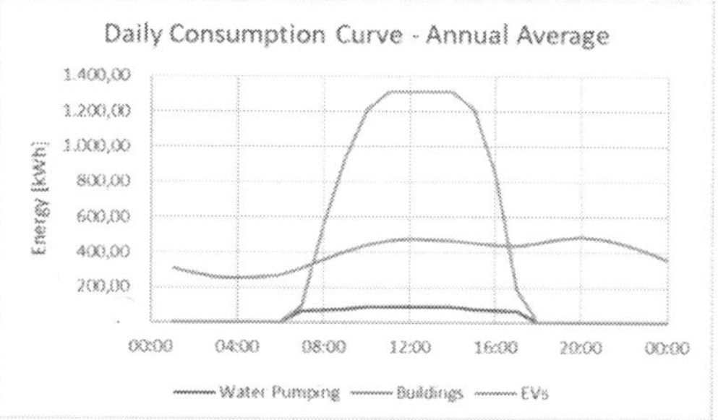

Fig. 3: Daily consumption profile for the 3 types.

As can be seen in **Fig. 3**, 100% of the consumption for water pumping and EVs charging is done during the day, while for the buildings the self-consumption rate reached is 56%.

3 PROPOSED PHOTOVOLTAIC SYSTEM

The sizing of the PV system was optimized for the load profile, maximizing self-consumption of photovoltaic electricity and decarbonization potential, while minimizing surpluses during the autumn and winter months to optimize the technical and economic indicators of this investment. To achieve this, three photovoltaic systems (A, B, and C) are proposed, occupying three distinct plots of the land (I4, I5, I6).

A photovoltaic system (A) is proposed, installed on single-axis horizontal trackers aligned from North to South with 750 kWp of installed photovoltaic power, located on plot I4. Another photovoltaic system (B) is proposed, also installed on single-axis horizontal trackers aligned from North to South with 1,200 kWp of installed photovoltaic power, located on plot I5. Finally, another photovoltaic system is proposed, installed on double East-West CarPort structures with a 15° inclination and 1,800 kWp of installed photovoltaic power, located on plot I6. These structures will serve as shading for the parking area.

The summaries of characteristics for the 3 photovoltaic systems can be observed in **Table IV**.

Table IV: Summary of Photovoltaic Installations.

System	**Structure Type**	**Orientation**	**Tilt (°)**	**Plot**
A	Trackers	N-S	0	I4
B	Trackers	N-S	0	I5
C	Fixed	E-W	15	I6
System	**Available area (m^2)**	**Power to install (kWp)**	**Price (€/Wp)**	**Estimated Installation Cost**
A	18,585	750	1.00 €	750,000 €
B	26,475	1,200	1.00 €	1,200,000 €
C	27,755	1,800	1.10 €	1,980,000 €

The installed systems should have the following general configuration:

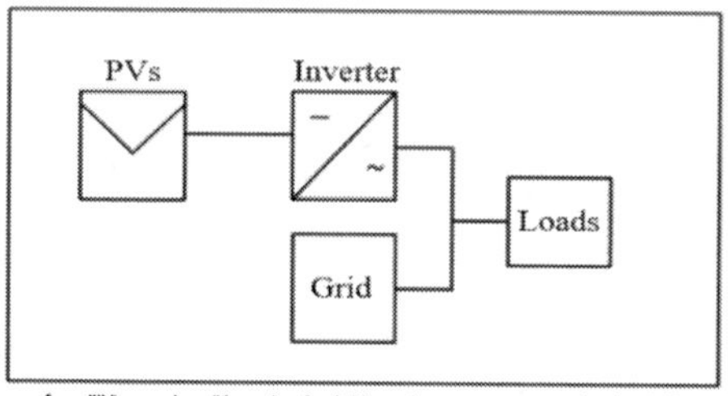

Fig. 4: Electrically hybridized systems: hybridization between PV generators and other energy sources is done in the electrical part of the system.

This configuration involves hybridization between the PV generator and the public utility electricity grid in the electrical part of the system. This system is considered a self-consumption system and is subject to the applicable regulations/legislation in Portugal, under Decree-Law No. 15/2022, dated January 14.

3.1 PV Trackers

A significant observation regarding Quinta do Carmo's electricity consumption (**Fig. 2**) is the notable asymmetry between summer and winter consumption. To address this disparity, the installation of photovoltaic modules on movable support structures capable of tracking the movement of the Sun, rotating around a North-South aligned horizontal axis, has been proposed. This approach of solar trackers proves to be the most effective in terms of maximising the difference in energy production between different seasonal periods.

It is essential to note that the horizontal arrangement of photovoltaic modules plays a pivotal role in the visual context of the landscape. This layout contributes to minimised visual impact, discreetly integrating into the environment.

In summary, the installation of photovoltaic trackers with a single horizontal axis tracking from North to South is a strategic solution for the seasonal asymmetry in electricity consumption at Quinta do Carmo. This approach not only optimises energy production but also offers notable advantages in terms of productivity and visual integration.

This type of tracker can be combined with an agrivoltaic approach. Vegetation control can be addressed through innovative methods such as sheep grazing or mechanical cutting, aimed at keeping the land under the solar trackers productive. Additionally, some areas can be designated for soil recovery measures and promoting native vegetation, contributing to improving the health of

10.4229/EUPVSEC2023/4BV.3.28
020325-003

the local ecosystem.

A prominent highlight in the agrivoltaic approach is the incorporation of horticulture, which focuses on organic farming. To effectively integrate this "farm", the adaptation of a portion of the photovoltaic system is required, raising the axis from approximately 1.75 meters to about 2 meters. This intelligent integration not only demonstrates a holistic approach to space utilization but also exemplifies the potential to combine clean energy production with the production of healthy food.

Furthermore, as part of educational and awareness efforts, a visitor's circuit can be developed in this space. This circuit will allow visitors to closely explore the various sustainability and decarbonization measures implemented at Quinta do Carmo. Through this interactive experience, visitors can gain a deeper understanding of innovative practices and ongoing environmental commitment.

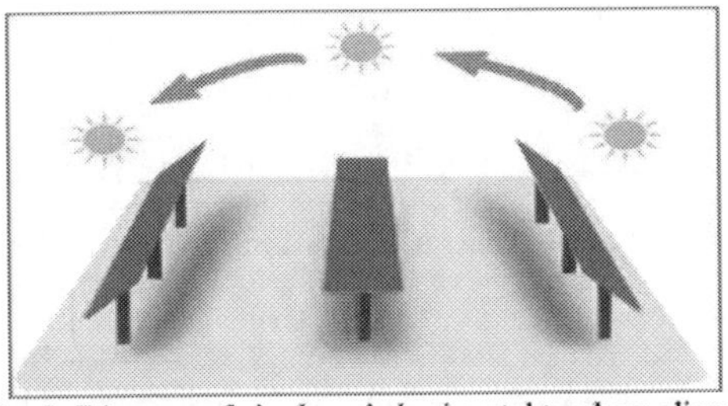

Fig. 5: Diagram of single-axis horizontal trackers aligned from North to South.

3.2 Parking Lot

The effective design and sizing of a shading system for a parking lot involve several important considerations. One approach is to implement a rational space management, opting for a parking layout at a 45° angle. This configuration not only maximises shading capacity but also reduces the width of necessary access lanes. Furthermore, the use of double East-West CarPort structures offers significant benefits, minimising the need for additional posts and ground fixation points.

The integration of electric vehicle charging stations is another crucial consideration in this project. Ideally, these stations would be bidirectional, providing a future-proof approach capable of meeting emerging needs. Strategically positioned next to the central pillars, these charging stations would be accessible from both sides, providing convenience to users.

Additionally, sustainable water management is also contemplated in the design. A practical solution is rainwater collection along the central axis of the system. This simplified approach contributes to operational efficiency and the seamless integration of the shading system with the surrounding environment.

A fundamental consideration in the parking lot design is the compliance with relevant regulations. According to Decree-Law No. 123/97, for parking facilities with a capacity between 25 and 100 spaces, a minimum of 3 spaces must be reserved for people with reduced mobility. In this context, a parking facility with 450 spaces was designed, with 27 spaces for people with reduced mobility, distributed in groups of 3 for every 50 spaces.

Moreover, compliance with safety standards is of utmost importance. Decree-Law No. 66/95 establishes specific regulations for fire safety in covered parking facilities. These guidelines ensure that adequate measures are implemented to ensure the safety of users in case of emergencies.

In summary, the design of the shading system for the parking facility involves careful choices, from spatial layout to the integration of technologies such as bidirectional charging stations. Additionally, the consideration of sustainable elements, such as rainwater management, demonstrates a commitment to environmental efficiency and innovation. It is crucial to also consider compliance with established regulations. Adherence to accessibility, fire safety, and urban planning requirements contributes to creating a functional, safe, and well-integrated environment.

Fig. 6: Example of double East-West CarPort structure.

4 PHOTOVOLTAIC GENERATION

For the purpose of analysing the performance of these systems' energy production, simulations of energy production were conducted, and annual production curves (**Fig. 7**) were generated using the same methodology employed for the load curves of each type of consumption.

The simulations were carried out using PVGIS [8] and SISIFO [9] softwares, both photovoltaic systems simulation tools.

Regarding the environmental parameters of the location, the typical meteorological year (TMY 2005-2020) was used for solar radiation and air temperature data. These data are obtained from the PVGIS - Photovoltaic Geographical Information System database (European Commission).

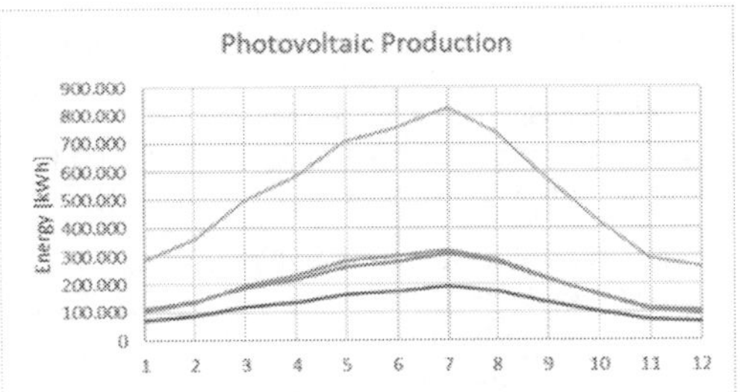

Fig. 7: Annual production curves of the 3 systems and its combined total.

As we can see from the graph, the chosen installation types allow for a significant asymmetry in energy production between the summer and winter months, confirming their suitability for the consumption patterns at Quinta do Carmo's ecotourism facility.

In **Fig. 8** the daily generation profile for the 3 systems can be observed.

10.4229/EUPVSEC2023/4BV.3.28
020325-004

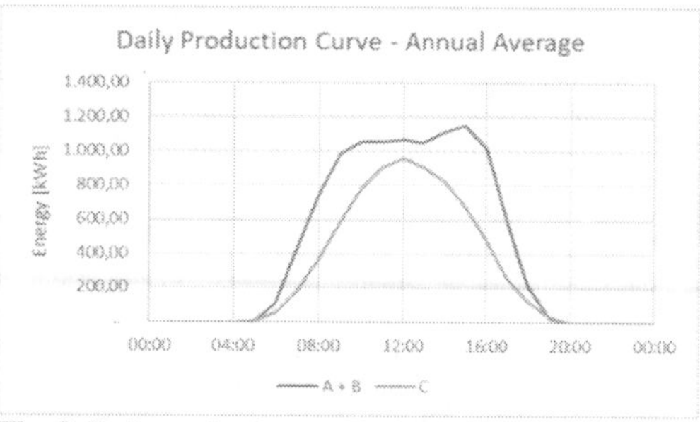

Fig. 8: Daily production profile for the 3 PV systems.

4.1 Production vs. Consumption

After obtaining the consumption profile (**Fig. 2** and **Fig. 3**) and the production profile (**Fig. 7** and **Fig. 8**) for Quinta do Carmo's ecotourism, we overlaid these data sets and generated the following graphs:

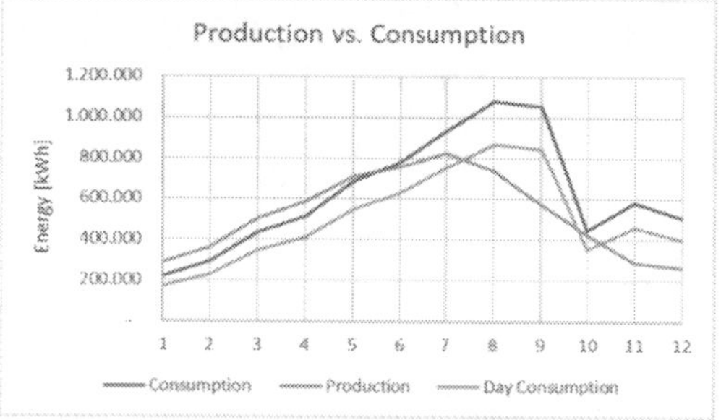

Fig. 9: Comparison of consumption vs. PV production.

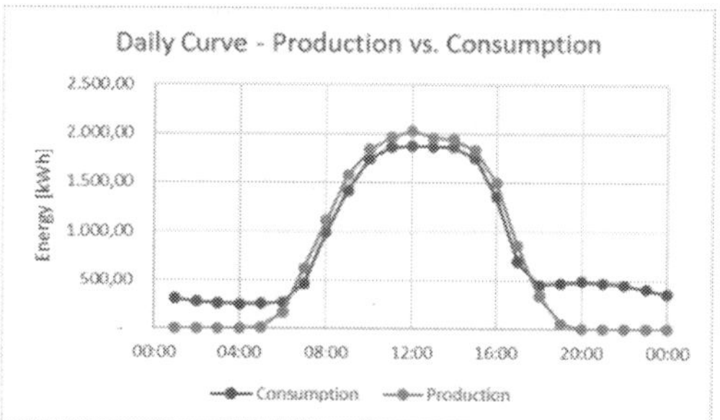

Fig. 10: Daily profiles for consumption vs. PV production.

Upon analysing the graphs, it's evident that the two profiles almost perfectly match. Using these profiles, the data in **Table V** was calculated.

Table V: PV electricity produced, consumed, excess and the self-consumption rate.

PV Electricity – Annual Values (kWh)	
Produced	6,304,515
Consumed	5,295,407
Excess	1,009,108
Self-consumption rate	**70%**

As shown in **Table V**, the proposed photovoltaic system achieves a self-consumption rate of 70% with an initial investment of €3,930,000. This self-consumption rate can be further improved with the installation of an energy storage system capable of storing excess electricity produced by the photovoltaic system.

5 TECHNICAL AND ECONOMIC ANALYSIS

5.1 Methodology

There is no standard or universal method for obtaining the irrefutable value of a company or business, as the elements to be evaluated include, among many others of a material and objective nature, management capacity, the quality of human resources, and market characteristics and expectations. On the contrary, there are accepted formulas to establish parameters of reasonableness to determine the value of a business.

The following general considerations were considered regarding the methodology used for the analysis:

- In this analysis, a conservative approach is assumed in the financial parameters used. This ensures that the results presented represent a conservative scenario.
- Financing costs are not considered in this economic analysis, given their variability related to the source of investment capital selected by Quinta do Carmo (equity, debt, credit, subsidized credit, etc.).
- The electricity price considered is a weighted average defined by the Renewable Energy Chair, using a conservative value.
- A projection period of 25 years was considered, the usual power warranty period for photovoltaic modules.
- The analysis does not include costs for the dismantling of equipment at the end of its life.
- Estimated costs for each photovoltaic system were considered based on the best available knowledge from the Renewable Energy Chair, with a relative margin of safety. However, due to the current European context, the evolution of the international market for raw materials and logistics of goods is particularly unstable, so the costs considered may vary at the time of investment execution.
- The economic analysis does not account for costs related to the medium-voltage to low-voltage distribution network, distribution network transformer substations, parking lot pavement and landscaping, EV chargers, or pumping stations.

4.3 Results

The results of the technical and economic analyses can be observed in **Table VI** and **Fig. 11**.

Table VI: Results of the technical and economic analyses.

Electricity Consumption		
Water Pumping	308,570	kWh
Accomodations and Buildings	3,456,233	kWh
Electric Vehicles Charging	3,756,790	kWh
Total Electricity Consumed	7,521,593	kWh
Photovoltaic Production		
System A	1,485,975	kWh
System B	2,377,560	kWh
System C	2,440,980	kWh
Total PV Electricity Produced	6,304,515	kWh
Consumed PV Electricity	5,295,407	kWh
Excess PV Electricity	1,009,108	kWh
Annual Efficiency Loss	0.45%	
Considered Electricity Price		
Average Electricity Price	0.1200	€/kWh
Financial Assumptions		
Inflation Rate	2.50%	
Discount Rate	5.00%	
Electricity Price Increase Rate	2.00%	

10.4229/EUPVSEC2023/4BV.3.28
020325-005

Economic Indicators		
Net Present Value (NPV)	5.921.082,92 €	
Total Life Cycle Cost (TLCC)	5.305.960,13 €	
Payback Period	6.25	years
Internal Rate of Return (IRR)	10.82 %	
Levelized Cost of Electricity (LCOE)	0.0425	€/kWh
CO2 Emissions		
CO_2 Emissions Factor (EDP 2023)	262.88	g/kWh
Avoided CO_2 Emissions per Year	1,657.33	t

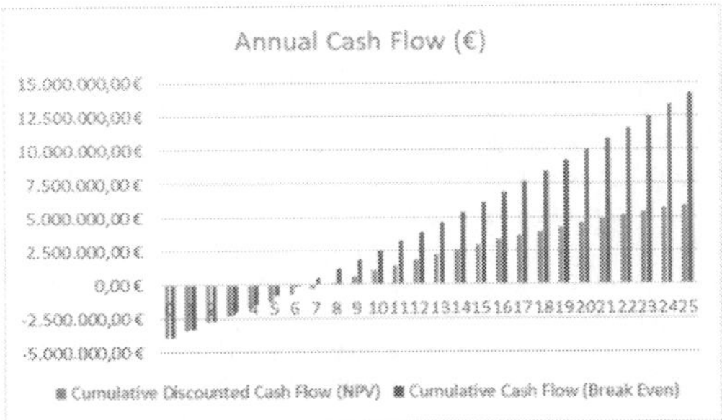

Fig. 11: Annual Cumulative Cash Flow Graph.

5 CONCLUSIONS

In conclusion, this paper has presented a comprehensive analysis of a sustainable energy project aimed at transforming the Quinta do Carmo into an eco-friendly ecotourism destination. By integrating renewable energy sources, the project seeks to minimize its environmental footprint while providing a unique and sustainable experience for visitors.

The analysis began by assessing the different energy consumption profiles on the eco-resort, including buildings, water pumping, and electric vehicles charging. This thorough evaluation allowed us to design and size a photovoltaic system strategically, optimizing energy self-sufficiency during peak demand months while minimizing surpluses in the autumn and winter.

The choice of single-axis solar trackers for the photovoltaic installations proved to be an effective solution, maximizing energy production and ensuring a more even power output throughout the day. Additionally, these solar trackers can be integrated with innovative agrivoltaic practices, promoting sustainable land use and even supporting local organic farming.

The design of the parking lot structures also followed sustainable principles, incorporating shading elements, EVs charging stations, and rainwater collection systems. These elements not only enhance the visitor experience but also contribute to the overall sustainability of the project.

Analysing the technical and economic aspects, the paper presents a conservative approach, ensuring that the results represent a cautious scenario. The analysis includes considerations such as electricity production, consumption, financial parameters, and environmental impacts. Notably, the project achieves an 70% energy self-sufficiency rate, with an initial investment of €3,930,000. This self-sufficiency rate could be further improved with the addition of an energy storage system capable of storing the excess PV electricity.

The economic analysis indicates a positive outlook for the project, with a NPV of €5,921,082.92, a Payback Period of 6.25 years, and a competitive LCOE of 0.0425 €/kWh. Additionally, the project is estimated to save 1,657.33 tons of CO_2 emissions annually.

In financial terms, the project demonstrates its feasibility and potential for long-term sustainability. Furthermore, the Quinta do Carmo's transformation into an eco-friendly destination serves as a model for sustainable development, blending environmental responsibility with a unique and enjoyable visitor experience.

In summary, the Quinta do Carmo sustainable energy project showcases the possibilities of integrating renewable energy solutions within the ecotourism sector. Beyond its economic viability, it exemplifies a commitment to environmental stewardship, sustainability, and innovation, setting a compelling example for similar ventures in the future. This holistic approach to sustainable tourism not only benefits the environment but also contributes to a more resilient and eco-conscious world.

6 ACKNOWLEDGMENTS

The authors acknowledge Bacalhôa and Quinta do Carmo for the trust placed in us to develop this work. This work was also supported by two projects from H2020 (Horizon 2020), POCITYF (grant agreement no 864400) and AURORA (grant agreement no 101036418), through the funding of research grants for two master students in Solar Energy Engineering from the University of Évora.

7 REFERENCES

[1] E-REDES - Distribuição de Eletricidade, S.A., "Perfis de Consumo | E-REDES," [Online]. Available: https://www.e-redes.pt/pt-pt/perfis-de-consumo. [Accessed 13 September 2023].

[2] E-REDES - Distribuição de Eletricidade, S.A., "About us | E-REDES," [Online]. Available: https://www.e-redes.pt/en/about-us. [Accessed 14 September 2023].

[3] Google Earth, [Online]. Available: https://www.google.com/intl/pt-PT/earth/.

[4] "Pumps - Power Calculator," The Engineering Toolbox, [Online]. Available: https://www.engineeringtoolbox.com/pumps-power-d_505.html. [Accessed 4 Agosto 2023].

[5] "Water Density Calculator," Omni Calculator, [Online]. Available: https://www.omnicalculator.com/physics/water-density. [Accessed 4 Agosto 2023].

[6] "Por um turismo sustentável - Eficiência energética no setor hoteleiro," ENA - Agência de Energia e Ambiente da Arrábida, [Online]. Available: http://www.porumturismosustentavel.pt/?cix=1059&lang=1.

[7] D. V. Ventura, "Electrical Energy Management in five-star Hotel in Lisbon," Lisboa, 2018.

[8] "Photovoltaic Geographical Information System," European Commission, [Online]. Available: https://re.jrc.ec.europa.eu/pvg_tools/en/.

[9] "SISIFO - PV Simulation," [Online]. Available: https://www.sisifo.info/en/default. [Accessed 13 September 2023].

10.4229/EUPVSEC2023/4BV.3.28
020325-006

MANAGEMENT PRESSURE IRRIGATION UNITS WITH PHOTOVOLTAIC BOOSTER PUMPING IN THE HYDRANT FOR MAXIMIZING THE PROBABILITY OF MEETING THE CROP WATER NEEDS

Raúl Sánchez[1] (raul.sanchez@upm.es), Miguel A. Muñoz[2*] (miguelangel.munoz@upm.es), Guillermo Moreda[2] (guillermo.moreda@upm.es), Luis Juana[1] (luis.juana@upm.es)
1. Universidad Politécnica de Madrid , Departamento de Ingeniería Agroforestal, Escuela Técnica Superior de Ingeniería Agronómica, Alimentaria y de Biosistemas. Grupo de investigación Hidráulica del Riego.
2. Universidad Politécnica de Madrid , Departamento de Ingeniería Agroforestal, Escuela Técnica Superior de Ingeniería Agronómica, Alimentaria y de Biosistemas, Grupo de investigación LPF-Tagralia.
*. Corresponding author.

ABSTRACT

A proposal is presented which deals with issues related to the necessary power peak of a photovoltaic power plant. The level of assurance provided by the irrigation intensity (gross water irrigation depth Hb divided by irrigation application time tar) to cover the crop water needs is one of the key aspects in the proposal. An irrigation system downstream a hydrant is used to elaborate a procedure for determining the peak power of a photovoltaic system adequate for supplying the irrigation necessities. These last are the given in the water administrative concession and are according to a feasible cropping pattern in the irrigation area. For determining the power generated by the photovoltaic system to install data from a real system placed in the irrigation district is used. A methodology has been proposed to determine the power to be installed, necessary to guarantee the irrigation needs. The power required for booster pumping from a hydrant that supplies irrigation water to the 56 drip irrigation units that cover 14 ha is 15.4 kWp. In the case analyzed, this power is the one necessary to be able to address the irrigation operations during the month of June 2022.

The design of the booster pumping system requires having intra-hourly data of photovoltaic generation.

KEYWORDS: Solar PV irrigation, energy efficiency, booster pumping in pressurized irrigation systems.

1 INTRODUCTION

The purpose of a distribution network is to supply a certain maximum flow in each hydrant, which depends on the area covered by it, with controlled pressure conditions, with a high level of assurance.

A set of irrigation units is fed from a hydrant. Each one of them applies the water to a certain surface, which, except in exceptional cases, is a fraction of the surface irrigated from the hydrant. The operation of the units is not usually simultaneous, but the irrigator must sequence and overlap the operation of several in order not to exceed the maximum available flow. The irrigation units are designed with the aim of applying uniform irrigation whenever the pressure is within certain limits. Sprinkler irrigation is much more demanding than drip irrigation in terms of pressure, because the size of the drops that form varies substantially with pressure and inadequate size results in low irrigation application performance. In all pressurized irrigation, the drained flow rate depends approximately on the square root of the pressure, except in the case of compensating drippers, where the flow rate is constant within a pressure range. Compensating drippers were developed for irrigating land with a steep slope, however, currently their use is widespread since in terms of the manufacturing variation coefficient, they are just as competitive as conventional drippers, thanks to the improvement reached by industrial processes. In any case, below the lower end of the compensating pressure range, compensating drippers behave identically to conventional drippers.

On the other hand, a high uniformity of irrigation application makes it possible to achieve, with an adequate management, a high irrigation water application efficiency, which, in one irrigation unit, indicates the degree of use of the flows applied to the crop. The uniformity is independent of the pressure above a minimum value, specific to each irrigation unit, and which, in drip irrigation, could be around 5 to 7 meters of pressure head. However, in practice, it is intended that all the drippers have a pressure height greater than 10 m, which translates into about 15 m at the head of the drip irrigation unit, which, in turn, with head losses in the rest of pipes and filters, become the usually assigned 25 m head of pressure downstream of the hydrant.

In summary, in general, the discharged flow rate is a function of the pressure and as long as the pressure does not drop below a minimum value, the uniformity of the applications is not reduced. The proposed methodology for the resolution of this objective comprises these two facts. It is based on the measurement of the flow rate of each unit, whose time integral gives rise to the volume, and the management of the booster pumping system so that the pressure does not drop below the minimum pressure. Each unit must be previously characterized by the volume, or corresponding time, that must be applied in a temporary period, for example one day, and its minimum pressure.

The booster pumping system, at a given moment, depending on the photovoltaic power generated and the pressure supplied by the network to the hydrant, through the adaptation of the number of pumps and the rotation speed, will allow the irrigation of a certain number of units and adapt to the pressure provided by the network at the hydrant. On the other hand, at that very moment there could be units that have already applied the desired volume and others that have not. It

10.4229/EUPVSEC2023/4BV.3.31
020327-001

may even be that at the end of the temporary period not all the units have been able to apply the necessary water needs, an issue to consider for the optimal management of irrigation application from the hydrant. Fortunately, the crops do not suffer a reduction in production, because the soil provides storage, unless the restrictions occur repeatedly.

The adequacy of photovoltaic generation to crop water requirements, in principle, seems high, since irrigation needs and photovoltaic generation are governed by a common cause, solar radiation. The integration with pumps with variable rotation speed has been developed for several years, as can be seen in the works of [1] and [2]. However, manufacturers of variable frequency drves are adding new equipment to the market that incorporate, in addition to the inverter with MPPT (maximum power point tracker) function for solar photovoltaic energy, functions closely related to irrigation, as is the case of the [3] range, oriented to the irrigation sector.

Current solar panel technology makes it possible to obtain a peak power (which could be generated under optimal conditions) of about 225 W/m², with a cost that is already less than 0.5 €/W [4]. Solar panel energy efficiency is reduced by several factors, including temperature (at higher temperatures, panel conversion efficiency decreases) and surface particle accumulation, called “soiling”.

In the scientific literature, a considerable number of works can be found about the design of solar photovoltaic power plants for irrigation. This type of work deals with the optimization of the power to be installed based on the hydraulic characteristics of the installation, the crop water requirements, the place and the price of crop yields, such as those of [5], [6], [7], [8], [9], [10], [11], [12], [13], [14], [15], [16], [17], [18], [19], [20], [21], [22] and [23].

It also can be found a considerable number of works on the management and exploitation of solar photovoltaic power plants for irrigation, in which various strategies are addressed to supply the demand for irrigation, taking into account the variability due to changing climatic conditions, such as those by [24], [25], [26], [27] and [28].

Irrigation systems in general, both at the distribution network level and at the farm, plot and unit level, usually have technology for the acquisition and recording of the most relevant operational variables. In any case, the market already offers technology and services with a consolidated level of maturity.

Digitization is an essential "work layer" to build on it energy efficiency strategies in distribution networks and in the water application over irrigated plots. Pressure and flow rate measurements are fundamental, since the analyzes related to the distribution and application of water can be carried out from them. Soil moisture measurements allow analyzing, not without difficulty, the result of an irrigation application, the destinations of the water and the adequacy of soil moisture to the crop water requirements.

Digitization allows addressing strategies to improve water productivity and reduce inputs (mainly water, energy and fertilizers) considering environmental, socioeconomic and legal conditions. This set of strategies within the framework of digitization is what relatively recently was called Precision Irrigation or Smart Irrigation. The integration of photovoltaic generation and precision irrigation has been raised in the work of [29].

This proposal deals with issues related to the necessary power peak of photovoltaic generation. The level of assurance provided by the irrigation intensity (gross water irrigation depth Hb divided by irrigation application time tar) to cover the crop water needs is one of the key aspects in the proposal.

2 MATERIALS AND METHODS

An irrigation system downstream a hydrant is used to elaborate a procedure for determining the peak power of a photovoltaic system adequate for supplying the irrigation necessities. These last are the given in the water administrative concession and are according to a feasible cropping pattern in the irrigation area. For determining the power generated by the photovoltaic system to install data from a real system placed in the irrigation district is used. However, tilt is 30° in this PV system, because it was designed for other purposes different to irrigation, and orientation angle is 176°.

2.1 Irrigation system

The irrigation system used in this study starts from a hydrant that is fed by a low-pressure distribution network. The pressure downstream of the hydrant is variable and depends on the demand from the rest of the hydrants in the network. For the purposes of calculating the necessary power to be installed, it is considered that the hydrant does not have pressure, although there could be hydrants in a network that have a guaranteed minimum pressure, so the elevation height of the booster pumping system would complement this minimum value. The hydrant supplies a surface area of 14 ha, common in pressure irrigation systems in extensive irrigated agriculture. The application of irrigation water from the hydrant is carried out through 56 drip irrigation units, each with a surface area of 0.25 ha, with the following characteristics:

- Drippers with flow rate discharge curve [$q(L/h) = 1.2 \cdot h(m)^{0.52}$] and manufacture variation coefficient CVqm = 0.03.

- Laterals with external/internal diameter 16/13.6 mm, length 50 m, length between drippers sd 0.5 m and descending slope of 0.02.

- Submain pipeline with with external/internal diameter 63/51.4 mm, length 50 m, distance length between laterals sl 2.0 m and horizontal slope.

The uniformity is a function of the pressure and, expressed by the flow rate variation coefficient CVq as a function of the pressure head at the head of the unit h0, is shown in Fig. 1.

10.4229/EUPVSEC2023/4BV.3.31
020327-002

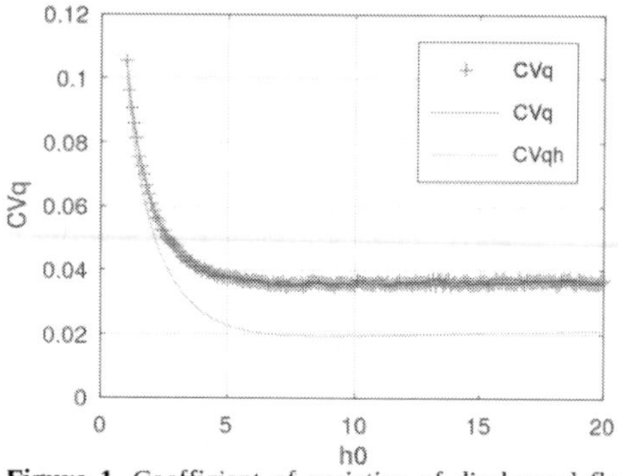

Figure 1. Coefficient of variation of discharged flow rate CVq as a function of pressure head at upstream section h0 in the drip irrigation units

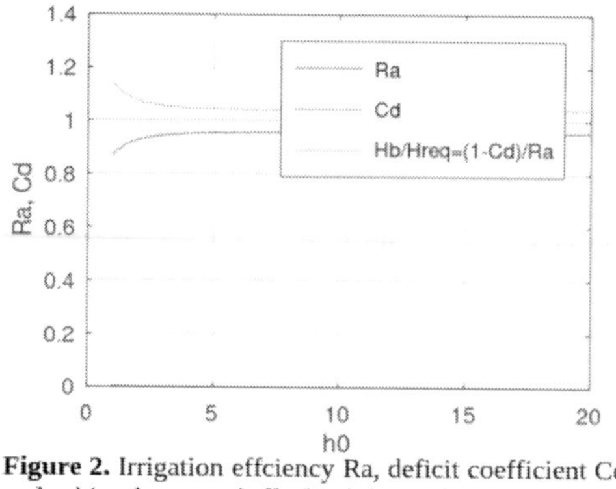

Figure 2. Irrigation effciency Ra, deficit coefficient Cd and raltion between bulk depth irrigation and required depth Hb/Hreq as a function of pressure head h0

Flow rate variation is due to pressure variation within the unit and manufacturing variation. CVqh is the flow variation due exclusively to pressure. The difference between the total flow rate variation CVq and CVqh is due to the flow rate variation due to manufacturing CVqm.

The pressure head at the beggining of the unit h0 should not be less than about 6.5 m so that the irrigation uniformity is not reduced (see Figure 1). This value is increased in order to move away from the uncertainty of the border, and the value of 8.5 m is taken for the average pressure height in the unit. Although above the minimum pressure, the uniformity is not affected, the applied flow depends on the pressure [30]. The corresponding flow according to the discharge curve is 3.65 L/h, and therefore the flow drained by a drip irrigation unit is 9125 L/h = 2.53 L/s. It should be borne in mind that the irrigation uniformity is greater the lower the coefficient of variation of the flow is.

Flow rate in compensating drippers is considered constant, but below the lower end of the compensating pressure range, compensating drippers behave identically to conventional drippers [31].

The uniformity of the irrigation determines the amount of bulk irrigation Hb that has to be applied to satisfy a determined required depth Hreq. This relationship is given through the irrigation application efficiency Ra and the deficit coefficient Cd. In Fig. 2 these coefficients Ra and Cd are shown, as well as the relationship between Hb and Hreq depending on the head pressure at the unit.

For a pressure head at the beginning greater than 8.5 m the Hb/Hreq = 1.0428.

The gross depth Hb (volume applied in the surface unit) is equal to the average flow rate drained by the drippers multiplied by the tar irrigation application time divided by the area corresponding to each dripper (sd·sl = 1 m2). The irrigation application time tar(h) necessary to apply a certain required sheet Hreq(mm) is given by the relationship tar(h)/Hreq(mm) = 0.2856.

From the hydrant, the water is distributed to the 56 drip irrigation units. If the average pressure head takes the minimum value of 8.5 m established so that uniformity is not affected, the flow rate drained by each unit is $2.53 \cdot 10^{-3}$ m^3/s. The distribution network downstream of the hydrant serves the water up to the 56 hydrants and has a filtering system at its head. Having a pressure head downstream of the hydrant based on the flow rate of $15 + 500 \cdot Q(m^3/s)^2$ guarantees that the minimum necessary pressure head is available in all units. This curve is represented with a dashed line in Figure 3.

Downstream of the hydrant, a booster pumping consisting of three pumps in parallel with characteristic curves $H(m) = 22.35 - 8592.9 \cdot Q(m^3/s)^2$ and eff. = $58.824 \cdot Q(m3/s) - 1153.4 \cdot Q(m^3/s)^2$, one of them with a variable frequency drive. The variable frequency drive makes it possible to control a different lifting height value depending on the number of pumps in operation. In Figure 3 it is shown the pressure head supplied by the booster pumping and the power demanded by it, according to the demanded flow and taking into account that the number of pumps in operation must increase as the demanded flow increases. This pumping system has the capacity to supply the necessary flow for 28 of the 56 drip irrigation units assuming a constant flow rate value.

10.4229/EUPVSEC2023/4BV.3.31
020327-003

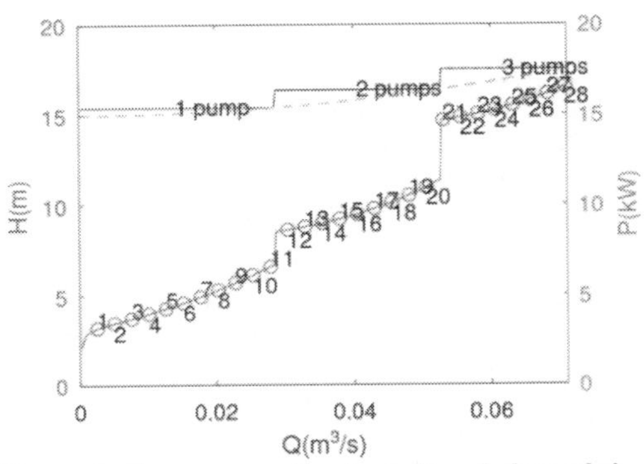

Figure 3. Booster pumping head H and demanded power P as a function of flow rate Q. Numbers over P(Q) curve indicates number of working drip irrigation units

One pump can supply the necessary flow rate for up to 11 drip irrigation units (see Fig. 3), two pumps can supply the flow for up to 20 units and three pumps the flow for up to 28 units. The power required according to the number of drip irrigation units in operation is indicated with a circle and the corresponding number of units over the power curve. The power required for irrigation with one unit is 3.2 kW and for irrigation with 28 units is 16.5 kW.

2.2 Irrigation water supply needings

The cropping plan for the irrigation district area includes the monthly water needs shown in Table 1, expressed in their daily average in mm/day, along with the necessary tar irrigation application time.

Table 1. Crop water requirements Hreq for the crop planning in the irrigation district

month	4th	5th	6th	7th	8th	9th
Hreq (mm/day)	0.48	2.04	5.64	6.54	4.20	1.35
tar(h/day)	0.13	0.56	1.55	1.80	1.15	0.37

The irrigation application time has been calculated with the flow rate corresponding to the average pressure head of 8.5 m as a guarantee that the uniformity is not affected.

2.3 Photovoltaic generation estimation

Hourly generated data during year 2022 of a photovoltaic power plant close to the irrigation district have been used. Tilt of panels in the PV plant is 30º, probably inadequate for irrigation purposes, and orientation is 176º.

In Figure 4 data of June (6th month) and July (7th month) of 2022 are shown. On the left side there are the daily peak hours during the day for all the days of the month, with average (continuous black line) and with average plus and minus standard deviation (dashed black line). On right side are the corresponding cumulative frequency distributions for the whole daily peak hours for all the days of the month.

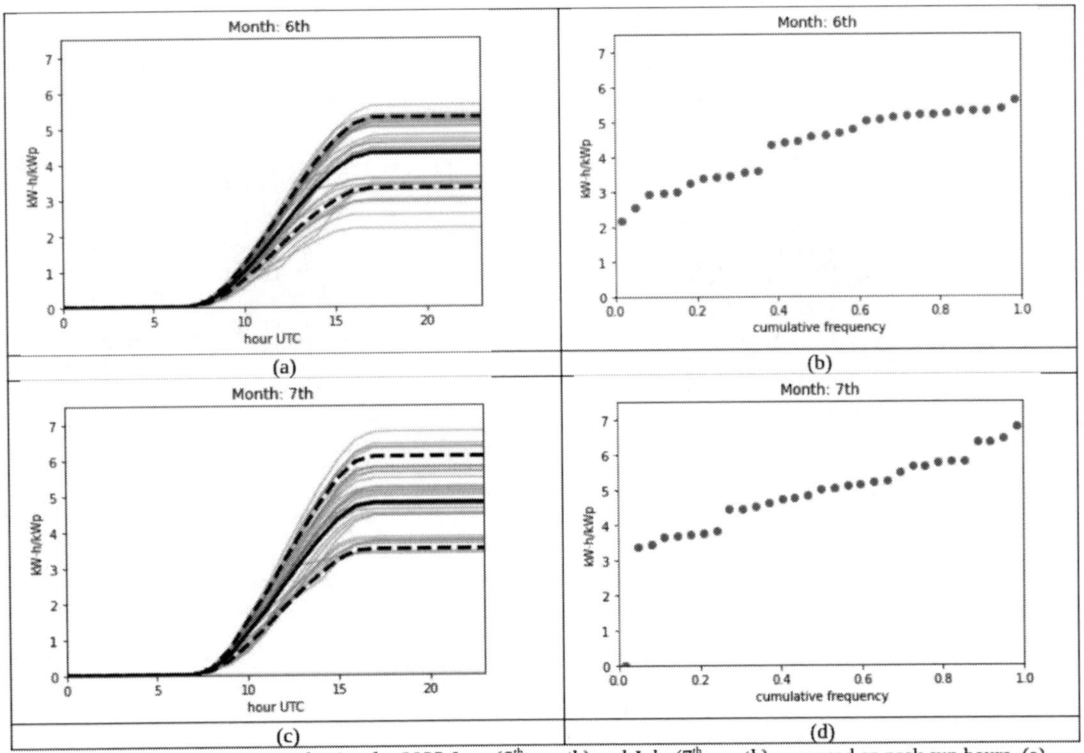

Figure 4. Photovoltaic power production for 2022 June (6th month) and July (7th month), expresed as peak sun hours. (a) and (c) show peak sun hours acumulation for all the days of the month. Continuous black lines represent monthly mean

10.4229/EUPVSEC2023/4BV.3.31
020327-004

cumulative daily power while dashed black lines represents average plus and minus standard deviation. (b) and (d) show the cumulative distribution for the whole daily peak sun hours

3 RESULTS AND DISCUSSION

The achievement of the objective involves determining the peak power to be installed that guarantees sufficient energy to be able to supply irrigation in accordance with the water needs. In this case, the water needs have been considered uniform for every day of the month. This approximation is sustained by the storage capacity provided by the crop soil. However, the procedure could be applied for non-uniform water needs, calculated daily or even hourly. This question has an important agronomic nature component and exceeds the theme of the congress to which this work is presented.

The supply of the daily water needs requires time, which can be seen in Table 1, and the power that the booster pumping will require depends on the number of units irrigate simultaneously (see Figure 3). The maximum water needs occur in July, closely followed by those in the month of June. The maximum generation also occurs in July, while that of June is also somewhat lower.

The calculation of the power of the photovoltaic plant necessary to be able to supply the daily irrigation needs with the average distribution of peak hours is different for each month. In this case, the month of June is the one that requires the highest power, 15.4 kWp, while July would require less, 14.8 kWp. This installation would therefore require a photovoltaic plant with a power of 15.4 kWp. Figures 5 and 7 show the average hourly powers that plants would generate with the necessary powers for each month.

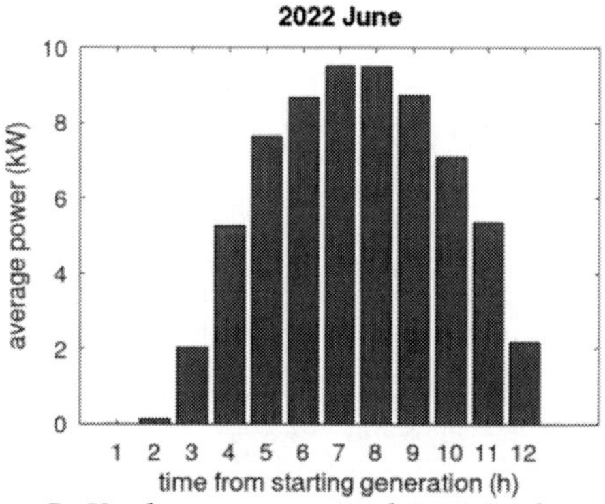

Figure 5. Hourly average power for a 15.4 kWp photovoltaics power plant according to Figure 4a

The power of the photovoltaic generation allows at a moment water supply for a maximum number of irrigation units, in accordance with Figure 3. In June, irrigation from one unit will need 0.2856 h/mm·5.64 mm/day = 1.61 h/day, so for the 56 irrigation units a total of 90.2 units·h/day are needed. While in July are needed 0.2856 h/mm·6.54 mm/day = 1.87 h/day and a total of 104.6 units·h/day for the 56 units.

Considering the generation of the average day, the number of irrigation units that can irrigate simultaneously at a certain time since the moment of the day when the plant begins to produce are shown in Figure 6. In June and for the day with average generation, during the first three hours it is not possible to irrigate any unit, and in the fourth hour the irrigation of some would be completed. It can also be seen that in the last hour, the irrigation of any unit would not be completed either. The reason is that during the mentioned hours the generated power would be less than the minimum value of 3.2 kW required for irrigation with only one unit.

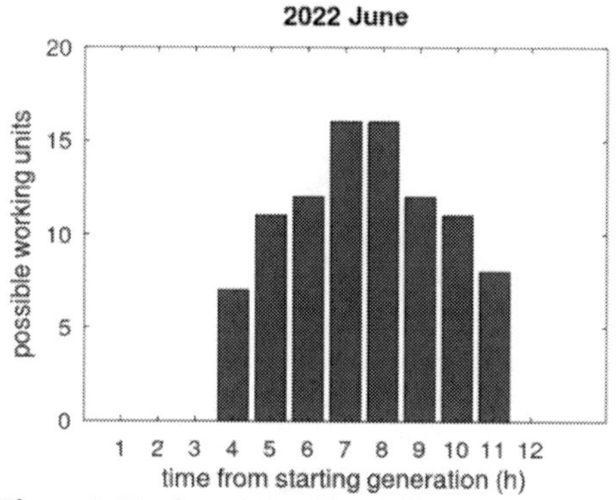

Figure 6. Number of possible working units along day for a 15.4 kWp PV power plant

Figure 6 shows the distribution on the average day of a total of 93 units·h.

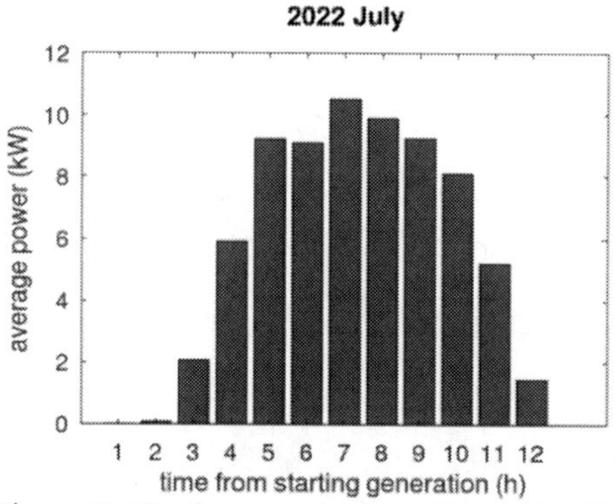

Figure 7. Hourly average power for a 14.8 kWp photovoltaics power plant according to Figure 4c

The number of units with which it could be irrigated at each moment of the day during the month of July is shown in Figure 8. The distribution of a total of 106 units·h on the average day is shown. As was the case for the month of June, and even though the powers analyzed are different, during the first three hours and during the last one there would be no unit in operation.

10.4229/EUPVSEC2023/4BV.3.31
020327-005

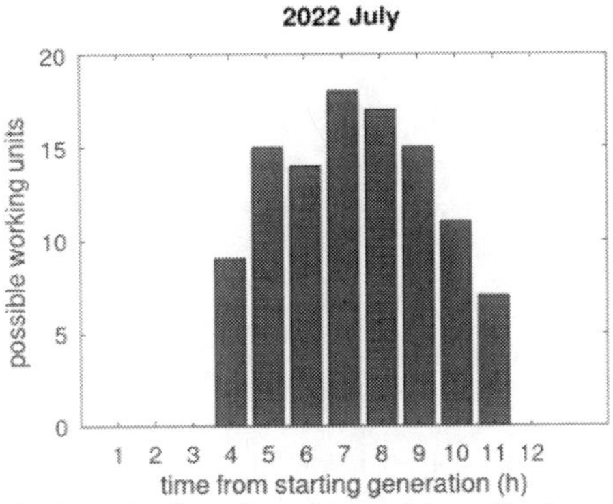

Figure 8. Number of possible working units along day for a 14.8 kWp PV power plant

For both the month of June and July, it can be observed that the powers to be installed studied would allow irrigation during an average day. The number of units that could irrigate simultaneously is less than 28, the maximum capacity of the pumping facility. However, there will be days in which the average needs cannot be supplied because the generation will be less than the average, and that must be compensated with those that the generation is greater than that of the average day. For this reason, the power of the pumping installation must be higher than average. The average peak power in one hour recorded in 2022 was 13.6 kW. Since it is an average value, this value could be exceeded at some point. The booster pumping power is 16.5 kW (see Figure 3) and the analysis of its adequacy cannot be addressed with the available data with hourly resolution.

4 CONCLUSIONS

The general aim is to propose a power needings determination and a management procedure for farm's irrigation units, placed downstream a hydrant where an isolated photovoltaic power plant feeds a booster pump for pressurizing, in order to maximize the probability of meeting the crop water needs. The pressure downstream of the booster pumping should not drop below a minimum value, since the uniformity of the distribution of the irrigation application will be affected. So, in case the photovoltaic power is not enough to reach this pressure, the irrigation operation should be stopped. On the other hand, although above the minimum pressure, the uniformity is not affected, the applied flow depends on the pressure. The registration of the volume that would have been applied in each irrigation unit, will allow a procedure to be elaborated to achieve the mentioned maximization through the irrigation of the most suitable unit.

A methodology has been proposed to determine the power to be installed, necessary to guarantee the irrigation needs. The power required for booster pumping from a hydrant that supplies irrigation water to the 56 drip irrigation units that cover 14 ha is 15.4 kWp. In the case analyzed, this power is the one necessary to be able to address the irrigation operations during the month of June 2022.

The design of the booster pumping system requires having intra-hourly data of photovoltaic generation.

5 REFERENCES

[1] Alonso-Abella, M.; Lorenzo, E.; Chenlo, F. Sistemas de bombeo fotovoltaico. Optimización del uso de convertidores de frecuencia con bombas centrífugas y motores trifásicos. 2002. In Era Solar: Energías Renovables, No 106; SAPT Publicaciones Técnicas S.L.: Madrid, Spain, pp. 14–35.

[2] Abella, M.A.; Lorenzo, E.; Chenlo, F. 2003. PV water pumping systems based on standard frequency converters. Prog. Photovolt. Res. Appl. 11, 179–191.

[3] ABB, 2021, Convertidores para bombeo solar de 0.75 a 22 kW, Catálogo ACQ-80. ABB.

[4] Instituto de Energía Solar (UPM). 2018. Gráficos Significativos Energía Solar Fotovoltaica. p. 61. Available online: https://www.ies.upm.es/sfs/IES/IES-UPM/Portada/2018_PV_Espa%c3%b1a.pdf (accessed on december 4 2021).

[5] Hamidat, A.; Benyoucef, B.; Hartani, T. 2003. Small-scale irrigation with photovoltaic water pumping system in Sahara regions. Renew. Energy, 28, 1081–1096.

[6] Cuadros, F.; López-Rodríguez, F.; Marcos, A.; Coello, J. 2004. A procedure to size solar-powered irrigation (photoirrigation) schemes. Sol. Energy, 76, 465–473.

[7] Ghoneim, A.A. 2006. Design optimization of photovoltaic powered water pumping systems. Energy Convers. Manag., 47, 1449–1463.

[8] Glasnovic, Z.; Margeta, J. 2007. A model for optimal sizing of photovoltaic irrigation water pumping systems. Sol. Energy, 81, 904–916.

[9] Posadillo, R.; López Luque, R. 2008. A sizing method for stand-alone PV installations with variable demand. Renew. Energy, 33, 1049–1055.

[10] Meah, K.; Fletcher, S.; Ula, S. 2008. Solar photovoltaic water pumping for remote locations. Renew. Sustain. Energy Rev. 12, 472–487.

[11] Hamidat, A.; Benyoucef, B. 2009. Systematic procedures for sizing photovoltaic pumping system, using water tank storage. Energy Policy, 37, 1489–1501.

[12] Bouzidi, B. 2013. New sizing method of PV water pumping systems. Sustain. Energy Technol. Assess. 4, 1–10.

[13] Campana, P.E.; Zhu, Y.; Brugiati, E.; Li, H.; Yan, J. 2014. PV water pumping for irrigation equipped with a novel control system for water savings. Energy Procedia, 61, 949–952.

[14] López-Luque, R.; Reca, J.; Martínez, J. 2015. Optimal design of a standalone direct pumping photovoltaic system for deficit irrigation of olive orchards. Appl. Energy, 149, 13–23.

[15] Campana, P.E.; Li, H.; Zhang, J.; Zhang, R.; Liu, J.; Yan, J. 2015. Economic optimization of photovoltaic water pumping systems for irrigation. Energy Convers. Manag. 95, 32–41.

[16] Carroquino, J.; Dufo-López, R.; Bernal-Agustín, J.L. 2015. Sizing of off-grid renewable energy systems for drip irrigation in Mediterranean crops. Renew. Energy, 76, 566–574.

[17] Sontake, V.C.; Kalamkar, V.R. 2016. Solar photovoltaic water pumping system—A

10.4229/EUPVSEC2023/4BV.3.31
020327-006

comprehensive review. Renew. Sustain. Energy Rev. 59, 1038–1067.
[18] Reca-Cardeña, J.; López-Luque, R. 2018. Design Principles of Photovoltaic Irrigation Systems. Adv. Renew. Energies Power Technol. 1, 295–333.
[19] Muhsen, D.H.; Khatib, T.; Abdulabbas, T.E. 2018. Sizing of a standalone photovoltaic water pumping system using hybrid multi-criteria decision making methods. Sol. Energy, 159, 1003–1015.
[20] Ehrmann, S.; Fickert, L.; Nolz, R. 2019. Optimizing the setup of a photovoltaic pumping system for irrigation considering different crop water requirements. JPE, 9, 043104.
[21] Orts-Grau, S.; Gonzalez-Altozano, P.; Gimeno-Sales, F.J.; Balbastre-Peralta, I.; Marquez, C.I.M.; Gasque, M.; Segui-Chilet. 2021. S. Photovoltaic Water Pumping: Comparison between Direct and Lithium Battery Solutions. IEEE Access, 9, 101147–101163.
[22] Gualteros, S.; Rousse, D.R. 2021. Solar water pumping systems: A tool to assist in sizing and optimization. Sol. Energy, 225, 382–398.
[23] Mana, A.; Allouhi, A.; Ouazzani, K.; Jamil, A. 2021. Toward a Sustainable Agriculture in Morocco Based on Standalone PV Pumping Systems: A Comprehensive Approach. Green Energy Technol. 399–433.
[24] Benlarbi, K.; Mokrani, L.; Nait-Said, M.S. 2004. A fuzzy global efficiency optimization of a photovoltaic water pumping system. Sol. Energy, 77, 203–216.
[25] Campana, P.E.; Li, H.; Yan., J. 2013. Dynamic modelling of a PV pumping system with special considerationon water demand. Applied energy, 112, 635-645.
[26] Gao, X.; Liu, J.; Zhang, J.; Yan, J.; Bao, S.; Xu, H.; Qin, T. 2013. Feasibility evaluation of solar photovoltaic pumping irrigation system based on analysis of dynamic variation of groundwater table. Appl. Energy, 105, 182–193.
[27] Reca, J.; Torrente, C.; López-Luque, R.; Martínez, J. 2016. Feasibility analysis of a standalone direct pumping photovoltaic system for irrigation in Mediterranean greenhouses. Renew. Energy, 85, 1143–1154.
[28] Zavala, V.; López-Luque, R.; Reca, J.; Martínez, J.; Lao, M.T. 2020. Optimal management of a multisector standalone direct pumping photovoltaic irrigation system. Appl. Energy, 260, 114261.
[29] Mérida García, A.; Fernández García, I.; Camacho Poyato, E.; Montesinos Barrios, P.; Rodríguez Díaz, J.A. 2018. Coupling irrigation scheduling with solar energy production in a smart irrigation management system. J. Clean. Prod. 175, 670–682.
[30] Chamba D, S Zubelzu y L Juana. 2019a. “Energy, cost and uniformity in the design of drip irrigation systems”. Biosystems Engineering, Elsevier. Vol 178: 200-218pp. https://doi.org./10.1016/j.biosystemseng.2018.11.012.
[31] Chamba D, S Zubelzu y L Juana. 2019b. “Determining Hydraulic Characteristics in Laterals and Drip Irrigation Systems”. Agricultural Water Management 226 (2019) 105791. https://doi.org./10.1016/j.agwat.2019.105791.

EXPERIMENTAL MEASUREMENT AND MODELLING OF SOLAR IRRADIATION ON THE ROOF OF A PV-POWERED URBAN BUS

Michał Jan GĘCA[1], Teresa SANTOS[2], Miguel Centeno BRITO[3]
1 - Lublin University of Technology, ul. Nadbystrzycka 38D, 20-618 Lublin, POLAND
m.geca@pollub.pl
2 - NOVA FCSH, v. de Berna, 26 C, 1069-061 Lisboa, PORTUGAL
teresasantos@fcsh.unl.pt
3 - University of Lisbon, Alameda da Universidade, 1649-004 Lisboa, PORTUGAL
mcbrito@fc.ul.pt

ABSTRACT: This study aims to evaluate the impact of urban shadowing on the potential and range extension of solar-powered vehicles, in particular public transport buses. For this purpose, an experimental measurement campaign was conducted in the city of Lublin, in Poland for over one year, measuring global irradiance on the rooftop of a solar-powered bus. For this analysis, we choose clear sky conditions and the days in which the bus drives the whole day without long stops. To determine the effect of shadowing from neighboring buildings, these irradiation data are compared to a shadow-free fixed pyranometer in the university building. data driving buses lose about 19% of the solar radiation. To assess the impact of urban shadowing, the data is compared to a shadow-free irradiance measurement on a fixed location in a GIS environment estimating solar irradiance along the bus routes. Upon validation, the GIS model is used to estimate the impact of urban shadowing on solar irradiance, solar range extension, and reduction of frequency of charging events.
Keywords: solar radiation, moving bus, model, urban shadowing

1 INTRODUCTION

Solar-powered vehicles are a promising approach to decreasing CO2 emissions and extending the daily range of electric vehicles, reducing the frequency of charging events and thus decoupling, or at least mitigating, the deployment of electric vehicles from the investment in the charging infrastructure.

In the urban environment, multiple and frequent shading events can lead to lower irradiation reaching the vehicle-integrated PV modules thus reducing the solar extended range. This effect is expected to be more relevant at higher latitudes and during the winter due to the lower solar elevation. However, the urban environment is expected to be particularly adequate for the adoption of solar-powered vehicles, and electric vehicles in general, due to the limited daily travelled distances associated with urban commuting. It is therefore of relevance to assess the impact of urban shadowing on the irradiation reaching a solar-powered vehicle in the urban environment.

The study of solar-powered vehicles is emerging as a very relevant issue for the solar and automotive communities, reflected for example by the creation of start-ups for the development of solar-powered vehicles [4], industrial associations [5], institutional reports [6] or a growing scientific literature [7].

Many of the most relevant contributions are towards the development of prototypes for the integration of lightweight but robust and efficient solar modules on vehicles (passenger cars, buses, and trucks) and the assessment of its impact on extension range or CO2 emission reductions.

However, most of these studies consider shadow-free assessments of solar irradiance, lacking the evaluation of the real conditions of complex urban environments with multiple and frequent shadowing events. These urban environments are arguably the most interesting location for the early deployment of solar-powered vehicles, due to the concentration of car and bus users and because electric vehicles, with their limited range, are very adequate to address urban commuting.

To the authors knowledge, there is no publication on the experimental long-term assessment of urban shadowing on the potential of solar-powered vehicles.

2 METHODOLOGY

This study aims to evaluate the impact of urban shadowing on the potential and range extension of solar-powered vehicles, in particular public transport buses.

For this purpose, an experimental measurement campaign was conducted in the city of Lublin, in Poland (latitude 51.2° N) for over one year, measuring global irradiance on the rooftop of a solar-powered bus (fig. 1) travelling in different routes across town. Monitored data includes global irradiance (measured using a second-class pyranometer Hukseflux SR03) and GPS signal to allow for its geographical positioning [1]. For this analysis, we choose clear sky conditions and the days in which the bus drives the whole day without long stops. To determine the effect of shadowing from neighbouring buildings, these irradiation data are compared to a shadow-free fixed pyranometer in the university building

Figure 1: Photovoltaic system in moving bus nr 332.

3 RESULTS

On figure 2 we can see an example of solar radiation from one clear sky day and in figure 3 and figure 4 comparisons between solar radiation reaching the building and 4 buses during different days during the year. According to this measurement data driving buses lose about 19% of the solar radiation.

Solar irradiation is modelled using the Solar Analyst tool for ArcGIS [2] using a Digital Surface Model with 50 cm resolution [3] (figure 5). Model validation with experimental measurements is performed for clear sky days for different seasons. Model results for the full year are then used to estimate the impact of urban shadowing on solar irradiance, solar range extension, and reduction of frequency of charging events.

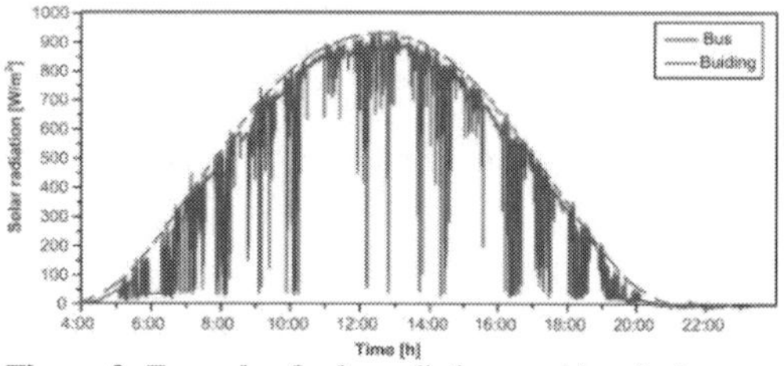

Figure 2: Example of solar radiation reaching the bus driving along a regular route (red) and a roof of a non-shaded building (blue-dashed line).

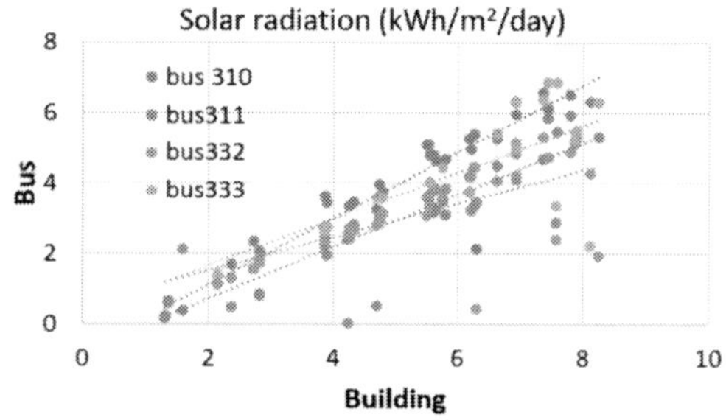

Figure 3: Example of solar radiation reaching 4 buses during clear sky conditions on different days and a roof of a non-shaded building.

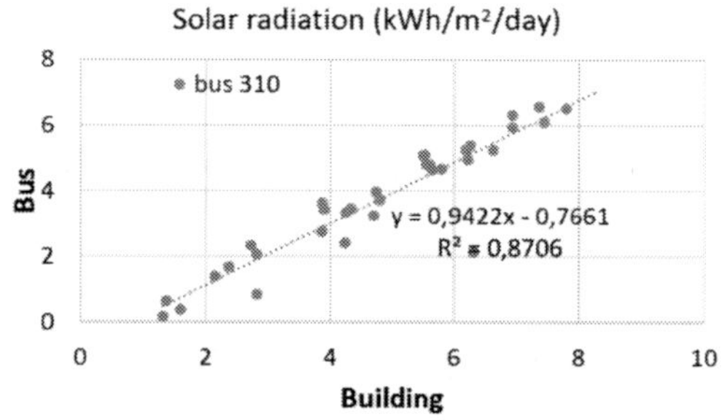

Figure 4: Example of solar radiation reaching bus 2310 during clear sky conditions on different days and a roof of a non-shaded building.

Figure 5: Solar radiation modelling considering shadowing from buildings at 12:00 noon for a summer and a winter day.

10.4229/EUPVSEC2023/4BV.3.35
020328-002

4 REFERENCES

[1] M. Wendeker, M. J. Gęca, Ł. Grabowski, K. Pietrykowski, K. Nanthagopal.: Measurements and analysis of a solar-assisted city bus with a diesel engine, Applied Energy, 309, 2022, 118439, https://doi.org/10.1016/j.apenergy.2021.118439
[2] M. Centeno Brito, T. Santos, F. Moura, D. Pera, J. Rocha: Urban solar potential for vehicle integrated photovoltaics, Transportation Research Part D: Transport and Environment, 94, 2021, 102810, https://doi.org/10.1016/j.trd.2021.102810
[3] https://www.geoportal.gov.pl/dane/numeryczny-model-pokrycia-terenu, accessed February 2022
[4] Sono motors (https://sonomotors.com/) and Lightyear (https://lightyear.one/), accessed February 2022
[5] ASOM – Alliance for Solar Mobility (https://asom.solar/) and Solarunited (https://solar-united.org/), accessed February 2022
[6] IEA PVPS task 17 (https://iea-pvps.org/key-topics/state-of-the-art-and-expected-benefits-of-pv-powered-vehicles/) and NEDO (https://www.nedo.go.jp/content/100885778.pdf), accessed February 2022
[7] Ch. Thiel, A. Gracia Amillo, A. Tansini, A. Tsakalidis, G. Fontaras, E. Dunlop, N. Taylor, A. Jäger-Waldau, K. Araki, K. Nishioka, Y. Ota, M. Yamaguchi: Impact of climatic conditions on prospects for integrated photovoltaics in electric vehicles, Renewable and Sustainable Energy Reviews, 158, 2022, 112109, https://doi.org/10.1016/j.rser.2022.112109

10.4229/EUPVSEC2023/4BV.3.35
020328-003

QUANTIFIED POTENTIAL CONTRIBUTION OF AGRIVOLTAICS TO THE EU PV MARKET

Caroline Plaza[1], André Penas[2], Philippe Macé[2], Gaëtan Masson[2]
[1]Becquerel Institute France, Lyon (France)
[2]Becquerel Institute, Brussels (Belgium)
c.plaza@becquerelinstitute.eu

ABSTRACT: The agricultural land in the EU represented a massive 38% of its total area in 2021. Such area represents an untapped solar PV potential. It can supply the EU's energy needs and ensure its energy security, provided that suitable regulatory frameworks are defined to avoid land usage conflict. Quantifying this market in the EU highlights the economic and energy independence opportunity for Europeans. This has been done for several levels of penetration of PV systems on agricultural land, by making assumptions on the PV applications that could suit agricultural areas in the EU. Results show that less than installing PV on 1% of the selected utilised agricultural areas in the EU could lead to a cumulative capacity of 479 GWp. More than 50% of this potential comes from permanent grassland. It shows that a few % of market penetration in Europe is enough to achieve the PV target of the EU. However, to allow this potential to be realized, a continued setting of regulatory frameworks and intersectoral studies with the agricultural sector are necessary.

1 Context

The EU-27 had around 220 GWp of cumulative installed PV capacity by the end of 2022 and this capacity is expected to keep on growing rapidly [1]. This is demonstrated by the EU goal of 750[1] GWp by 2030, which should be reached easily according to Becquerel Institute's "most probable" market scenario, with a cumulative 777 GWp PV capacity, by the same year [2] [3].

Most of Europe's future installed PV capacity is expected to come from conventional centralized or distributed PV installations (residential, commercial & industrial sectors) [3]. Meanwhile, other non-conventional PV market segments are developing, with agrivoltaics (AgriPV) taking the first steps in Europe with some countries introducing legal frameworks for the development of this innovative market segment. Germany, France and Italy have recently introduced legal frameworks where the conditions for permitting and financial support for the different AgriPV systems are defined. The definition of AgriPV, however, is not consensual among these countries and consequently, their legal frameworks are distinct [4] [5] [6].

The agricultural land use in the EU represented a massive 38% of its total area in 2021 [7]. Such surface area represents an untapped solar PV potential. It is capable of supplying the EU's energy needs and ensuring its energy security, while simultaneously providing a service to farmers in helping them to cope with growing extreme climatic events, securing and stabilizing farmers' cash flow, as well as supporting the EU's food security, provided that suitable regulatory frameworks are defined. There is a need to quantify this market in the EU, to demonstrate that it can represent an economic, energy and food independence opportunity for Europeans, to support lawmakers paving the way to new regulatory frameworks that will lead to further integration of renewable energy capacity.

2 AIM AND APPROACH

The aim of this publication is to quantify the AgriPV market potential at the EU level, following the approach shown below.

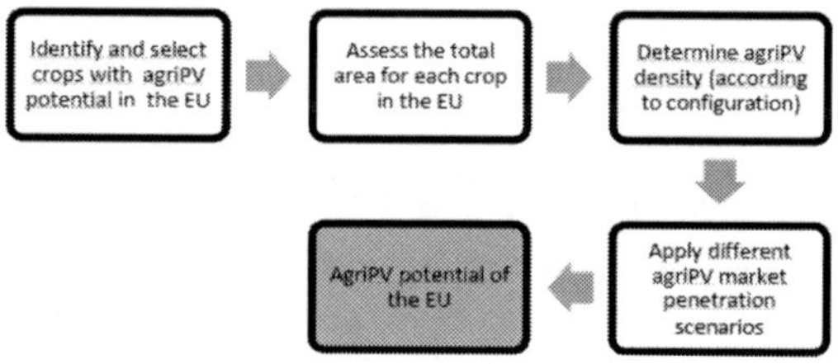

Figure 1: Methodological approach in steps

First of all, the crop areas in the EU that could potentially be suitable for AgriPV will be selected and quantified. Assumptions on the PV configuration used in each of these surfaces will be made. The PV density of each configuration will be estimated, and arbitrary market penetration rates will be applied to each combination of "type of surface" and "PV configuration", which will allow to obtain a potential in the EU.

This potential will be compared to the current and planned installed PV capacities landmarks, to quantitatively illustrate the potential contribution of AgriPV to the energy transition in Europe. The AgriPV market potential will be assessed per country, with visibility on the impact of each selected surface type on the potential.

[1] The defined target proposed by the EU Commission is 600 GW_{ac}. Considering a AC/DC convertion ration of 80% the target translates into 750 GWp.

10.4229/EUPVSEC2023/4BV.3.36
020330-001

3 Main Assumptions

The total area used for agriculture in the EU was obtained from Eurostat with the latest available values (2021) [8].

Table I: Area by type of surface in the EU (2021).

Area type	Surface (1000ha)	% of the EU Area
EU total area	423.300	100,0%
Utilised Agricultural Area	161.024	38,0%
Permanent grassland	51.035	12%
Cereals	51.048	12%
Permanent crops	11.768	3%
Other crops	47.173	11%

The Utilised Agricultural Area (UAA) in the EU represents 38% of the EU's area and is equivalent to around 161 million ha. Permanent grassland is primarily used for grazing, it represents 12% of the EU's area. Another 12% of the EU area is dedicated to cereal production. Permanent crops (e.g., tree fruits, vineyards) represent an area of around 3% of the EU area. Around 11% of the EU area with "Other crops" was not considered due to the non-specification of crop types in our sources. This can be considered as a conservative assumption, but this allows to limit the level of uncertainty of the final results.

The agricultural areas per country can be seen on the map available in Figure 2.

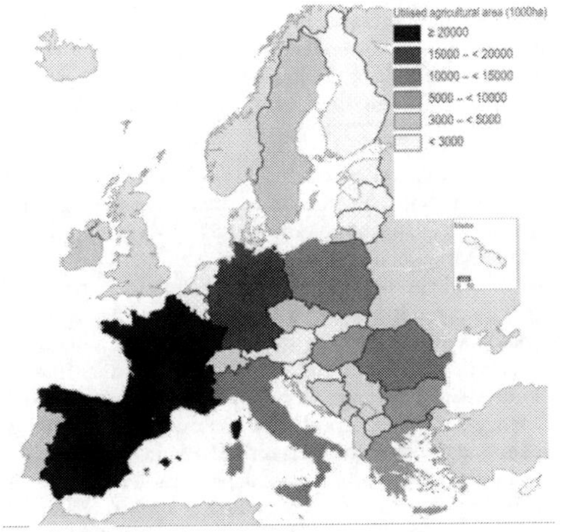

Figure 2: Utilised agricultural area in the EU by country (2021).

The PV configurations considered were:

1. The classical ground-mounted system with a lower density
2. The interrow PV configuration consisting of vertical bifacial PV or conventional PV with well-spaced between rows and same density
3. Elevated PV, which is a PV system above crops. This system can be mobile with tracking or fixed with no tracking.

We assumed PV power densities for each one of these configurations based on existing literature. In the table below, we associate each one of these PV configurations with a type of surface, based on the information available in the literature.

Table II: Assumptions on PV configurations

PV Configuration	Density (MWp/ha)	Surface type	Surface in the EU (ha)
Ground-mounted	0,50	Permanent grassland	51.035.000
Interrow	0,30	Cereals	51.048.000
Elevated	0,60	Permanent crops	11.768.000
Area not considered for AgriPV			47.173.000

4 Results of the simulations

The technical potential is estimated through the multiplication of the selected area of each surface type and the corresponding density of the PV configuration. Then a market penetration rate was applied to this technical potential. The market penetration rates simulate scenarios where a % of the agricultural land area in the EU has PV systems installed. The market penetration rates of 1%, 5% and 10% were arbitrarily chosen. The results of the AgriPV potential in the EU for each market penetration rate are shown in *Table III*.

Table III: Market potential by type of surface and market penetration rate in the EU

	Market penetration (%)	Market Potential EU (GWp)
Permanent grassland	1%	255
	5%	1 276
	10%	2 552
Cereals	1%	153
	5%	766
	10%	1 531
Permanent crops	1%	71
	5%	353
	10%	706

Figure 3: AgriPV market potential distribution per type of agricultural surface in the EU

10.4229/EUPVSEC2023/4BV.3.36

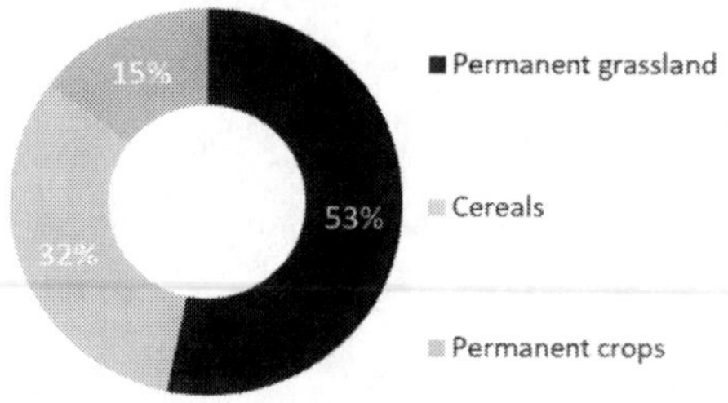

A market penetration rate of 1% on the selected agricultural areas represents an AgriPV technical potential in EU 479 GWp, 5% represents 2 395 GWp and of 10% 4 789 GWp.

The share of each type of agricultural surface in the total agriPV market potential is the same for all market penetration scenarios. More than half of the AgriPV potential in the EU is attributed to "Permanent grassland" surfaces, 32% to "Cereals" and 15% for "Permanent crops".

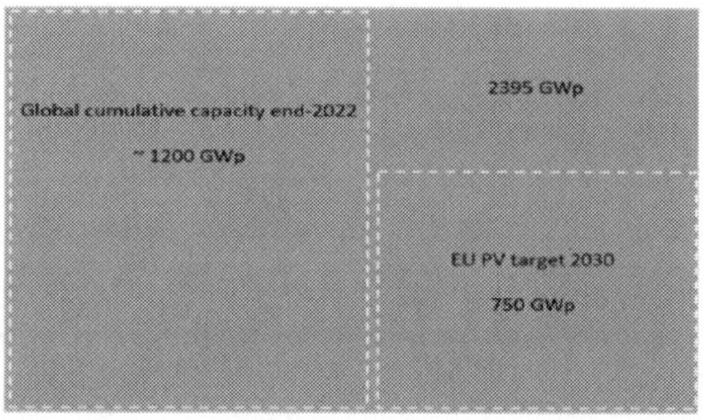

Figure 4: Potential for 5% market penetration in agricultural area in the EU

In Figure 4, the AgriPV technical market potential in case of a 5% market penetration rate is shown, equaling 2 395 GWp. This capacity is around 3-fold the EU PV target for 2030 and around twice as much as the global cumulative capacity installed by the end of 2022.

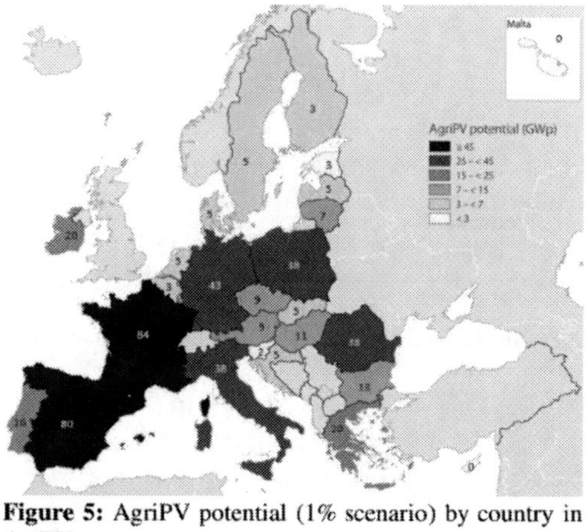

Figure 5: AgriPV potential (1% scenario) by country in the EU

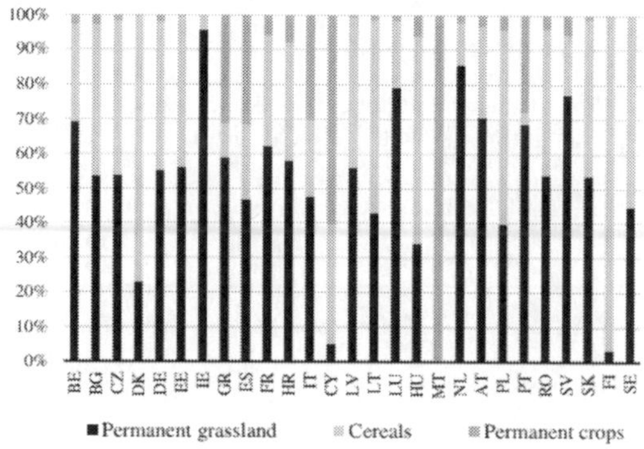

Figure 6: AgriPV potential crop type distribution by EU country

In Figure 5, one can see that Spain (80 GWp) and France (84 GWp) have the highest AgriPV technical potential in the EU. Germany's (43 GWp) and Poland's (38 GWp) potentials are hindered due to less land availability for agriculture and a higher share of cereal crops, for which we hypothesized a lower PV density. Despite the large size of countries such as Sweden (5 GWp) and Finland (3 GWp), the AgriPV potential is limited due to other land uses and the predominance of cereal crops.

5 Conclusions

Across all scenarios, permanent grassland surfaces contribute to more than half (53%) of the AgriPV potential in Europe. In the 1% market penetration scenario, the potential equals 479 GWp in the EU, 2 395 GWp in the 5% scenario and 4 789 GWp in the 10% scenario. This represents a huge installed capacity, that could single-handedly help achieving the REPowerEU PV target. Even if a few % of the total technical AgriPV market potential is activated in the EU would be sufficient to surpass the cumulative installed capacity in the World by the end of 2022 of around 1,2 TWp.

These results can be compared with other recent studies exploring the same questions. Researchers in Denmark obtained a technical potential of 51 TWp in the EU, for a 100% market penetration rate for the totality of the UAA [9]. If we hypothesized a market penetration rate of 100% for the same selected surface types, the calculated potential would have achieved the same order of magnitude, around 48 TWp. The Joint Research Center of the European Commission also explores the potential of AgriPV in its Science for Policy report. They assessed the relative land requirement to reach different EU PV installation targets under different hypotheses of density for PV systems. The AgriPV technical potential in the EU for 1% of the UAA has been estimated at 944 GW for a density of 0,6 MW/ha [10]. The results of our research are consistent with those of the Joint Research Center of the European Commission, but differ slightly. This is due to the application of the 1% scenario to a selection of areas within the UAA, instead of its totality, and to the weighted average of the PV density being 0,42 MWp/ha in our case.

Looking into the potential per country, France and Spain have the highest AgriPV technical potential in the EU,

10.4229/EUPVSEC2023/4BV.3.36
020330-003

with 84 GWp and 80 GWp respectively, for a 1% market penetration scenario.

The AgriPV market in Europe is still in its infancy, as only a few countries defined a regulatory framework. These regulatory frameworks on AgriPV should be as consistent as possible across the Member States to achieve the calculated technical potential. Among the other obstacles to overcome, further research is still needed to identify the optimal set-up for a broad combination of crops, climates and PV configurations. It will bring more confidence in AgriPV to all the stakeholders involved such as governments, agricultural landowners and PV developers. Finally, social acceptance must be worked on as it could hinder AgriPV development, considering that the impact on landscapes could be significant.

6 References

[1] IRENA, "RENEWABLE ENERGY STATISTICS 2023," IRENA, Abu Dhabi, 2023.

[2] European Commission, 05 2022. [Online]. Available: https://energy.ec.europa.eu/topics/renewable-energy/solar-energy_en.

[3] Becquerel Institute, "www.becquerelinstitute.eu/data-hub," Becquerel Institute, 2023. [Online]. Available: www.becquerelinstitute.eu/data-hub. [Accessed 2023].

[4] ADEME, "Characterising solar PV projects on agricultural land and agrivoltaism," ADEME, 2021.

[5] ENEA, Università Cattolica del Sacro Cuore, Italian Standardization Body UNI, REM TEC, 08 2023. [Online]. Available: https://store.uni.com/uni-pdr-148-2023. [Accessed 08 2023].

[6] Fraunhofer ISE, "www.ise.fraunhofer.de," 04 2022. [Online]. Available: https://www.ise.fraunhofer.de/content/dam/ise/en/documents/publications/studies/APV-Guideline.pdf. [Accessed 08 2023].

[7] European Commission, "Eurostat," [Online]. Available: https://ec.europa.eu/eurostat. [Accessed 08 2023].

[8] EU Commission, "Eurostat," [Online]. Available: https://ec.europa.eu/eurostat/databrowser/view/TAG00025/default/table?lang=en. [Accessed 09 2023].

[9] PV Magazine, "www.pv-magazine.com/2023/07/14/," 14 07 2023. [Online]. Available: https://www.pv-magazine.com/2023/07/14/new-research-identifies-potential-for-51-tw-of-agrivoltaics-in-europe/. [Accessed 28 09 2023].

[10] European Commission, "Overview of the Potential and Challenges for Agri-Photovoltaics in the European Union," Publications Office of the European Union, 2023.

10.4229/EUPVSEC2023/4BV.3.36
020330-004

IS A DOUBLE-BUBBLE CURVED SURFACE WITH EXCELLENT AERODYNAMICS SUITABLE FOR VEHICLE-INTEGRATED PHOTOVOLTAIC (VIPV) INSTALLATION?

Shota Matsushita, Kenji Araki, Yasuyuki Ota, Kensuke Nishioka
University of Miyazaki, Japan
1-1 Gakuenkibanadai-Nishi, Miyazaki, Japan, hk19046@student.miyazaki-u.ac.jp

ABSTRACT: A double-bubble roof is one of the roof-shape of the passenger's car known for excellent aerodynamics. The roof of the driver's and front passenger seats is slightly raised to form a hump. In this study, we investigated if this curved shape has the advantages of PV installation compared to conventional curved roof shapes. The calculation was done in the following steps: 1. Capturing the curved surface of the double-bubble roof and interpolation into continuous function using the bi-cubic spline function. 2. The inner products between the unit normal vectors and the ray vectors of entire surface elements were computed for the incident light fraction. 3. The solar radiation was referenced from the solar radiation database. The ray vectors from the sun were converted to the local coordinates of the curved surfaces. Calculations were performed using 2013 solar radiation records for Miyazaki city, Japan, and taking into account the hidden surfaces included in double-bubble curved surfaces. The annual light incident on double-bubble curved surfaces was about 6% less than on surfaces without concave surfaces.
Keywords: VIPV,double-bubble, curved surface, Modeling, Performance

1 INTRODUCTION

Various efforts have been made worldwide to solve global warming in recent years, cited as one of the world's most pressing issues. One of these is the decarbonization of automobiles, and research on solar-powered electric vehicles (VIPV) is underway as a specific initiative [1]. The advantage of VIPVs is that they are charged while the car is in motion and when it is parked, reducing the amount and frequency of charging at charging stations and shortening the waiting time between charging [2]. In addition, the power can be diverted as an emergency power source in a disaster [3].

However, VIPV still faces many challenges [4]. There are many challenges, especially about the amount of electricity generated, which can be divided into two main categories [5].

1) Lack of stability and immediacy in power generation and charging.

Gasoline-powered vehicles can quickly supply the energy the user needs at the required time, but the above cannot be done arbitrarily in solar power generation.

2) Inability to accurately forecast power generation yet [6].

Even with flat panels, predicting the amount of power generated is still challenging because many factors contribute to the amount of power generated, such as shading, weather, module assembly, and cell size [7]. Furthermore, when it comes to vehicle-mounted solar panels, they need to be curved solar panels, not flat panels [8-10]. In addition to the above factors, curved solar panels must consider the so-called self-shadowing factor, where the panel covers the panel itself. In addition, as the car moves, the direction and speed of the car and the shading cast by trees, buildings, and other objects change rapidly, making it even more challenging to predict the amount of electricity that will be generated.

In this study, among the above issues, we addressed the forecasting of power generation, taking weather and self-shadowing into account.

2 RESEARCH OBJECTIVES

Some automobile roofs have a unique shape called a double-bubble roof.

An example of a double-bubble roof is shown in Figure 1.

Figure 1: Peugeot RCZ R

The roof directly above the driver's and front passenger seats is slightly raised in the shape of two humps and is mainly used for sports cars. The same curves are seen in Toyota's 2000GT and Mazda's RX7. The double-bubble type has a lower center of gravity than the roof of an ordinary car, making it more aerodynamic and fuel-efficient. It is predicted that the double-bubble type will be widely used in new energy vehicles, which energy-saving is crucial.

In this study, a calculation software program created a model of a "double-bubble roof" and a "conventional roof" with the exact dimensions but without concavities. The results of this study were based on actual data from Miyazaki City, Japan.

3 RESEARCH METHODS

3.1 Creating a Curved Surface Model

The double-bubble roof surface used in this study was created by calculating the dimensions of the Peugeot RCZ LHD by taking the ratio of the 6i-S EyeSight, a Japanese Subaru car with published sizes (Figures 2 and 3).

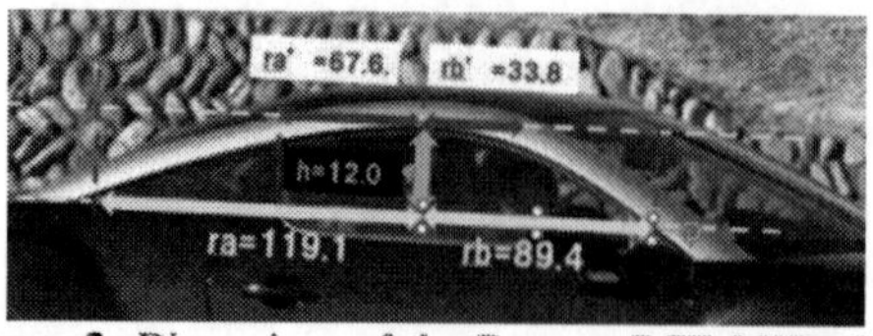

Figure 2: Dimensions of the Peugeot RCZ LHD (Side view)

10.4229/EUPVSEC2023/4BV.3.39
020331-001

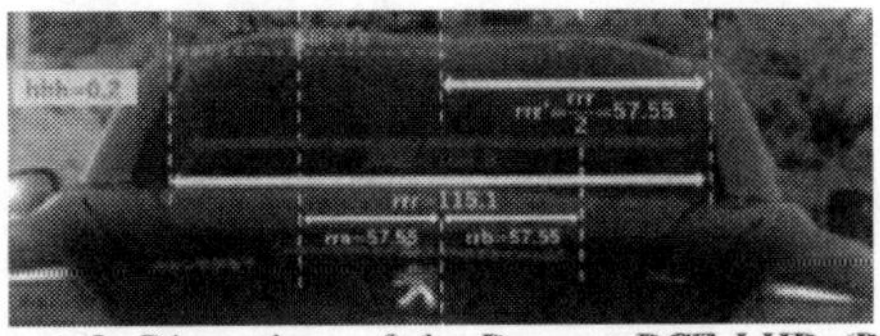

Figure 3: Dimensions of the Peugeot RCZ LHD (Rear view)

Note that since we use a parametric continuous function in this paper, arbitrary double-bubble surfaces can be generated by changing the values.

First, the following approximations were made for the lateral direction.

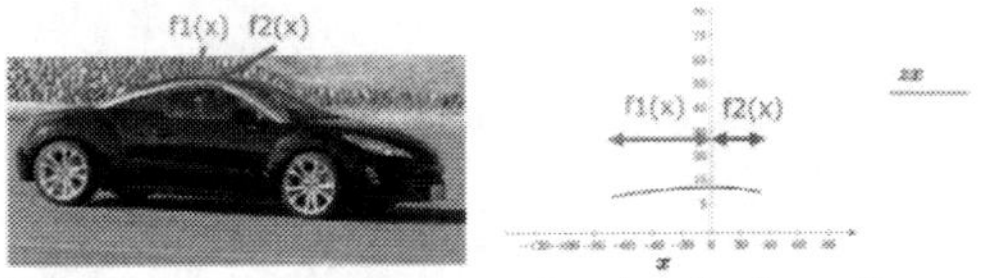

Figure 4: Roof function approximation (horizontal)

$$f1(x)=\frac{-h}{ra^2}(x-ra)(x+ra) \quad (1)$$

$$f2(x)=\frac{-h}{rb^2}(x-rb)(x+rb) \quad (2)$$

The horizontal direction was approximated by two parabolas, $f1(x)$ and $f2(x)$. The above equations are continuous up to the second derivative at the connection point, and the degree of the polynomial was considered to be kept to a minimum. The function connecting these two functions at x=0 is zx, shown on the right in Figure 4. $f2(x)$ was adopted from x=0 to rb', as shown in Figure 3.

We approximate the longitudinal direction.

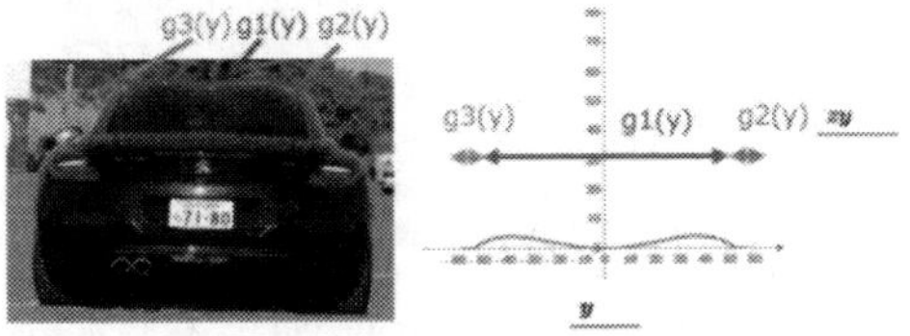

Figure 5: Roof function approximation (longitudinal)

$$g1(y)=\frac{1}{4}a^2y^4-\frac{1}{2}b^2y^2 \quad (3)$$

$$\left(a:\frac{\sqrt{2}b}{rr'} \quad b:\sqrt{\frac{2hh(rr')^2}{((rra^2)-(rra)(rr'))((rra^2)+(rra)(rr'))}}\right)$$

Since rrr' $\neq$ rr'

$$g2(y)=\frac{-hhh}{(rrr'-rr')^2}((y-rrr')-(rrr'-rr'))((y-rrr')+(rrr'-rr')) \quad (4)$$

$$g3(y)=\frac{-hhh}{(rrr'-rr')^2}((y+rrr')-(rrr'-rr'))((y+rrr')+(rrr'-rr')) \quad (5)$$

$g1(y)$ and $g2(y)$ were connected by rr' and $g1(y)$ and $g3(y)$ by $-rr'$. Also, $g2(y)$ was adopted from rr' to rrr' and $g3(y)$ from $-rrr'$ to $-rrr'$. This is zy.

The above zx and zy are interpolated using the Bicubic function (a cubic function with x and y as parameters) to create a curved surface. As shown in Figure 2-9, the interpolation is made to extend zx along zy. The interpolation condition is that each discretized point is always included in the surface. In this case, x and y are given as parameters u and v, respectively.

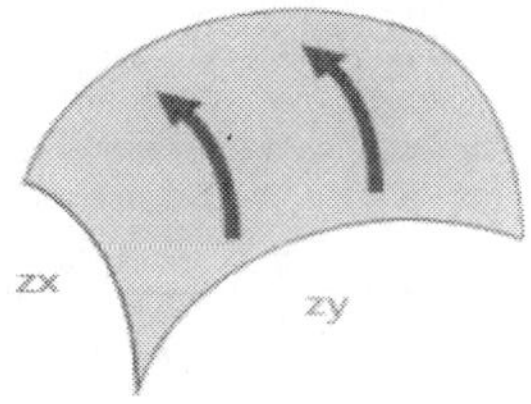

Figure 6: Interpolation Image Diagram

In this case, zy was multiplied by the function $h(x)$ to make the concavity smaller toward the rear of the vehicle body.

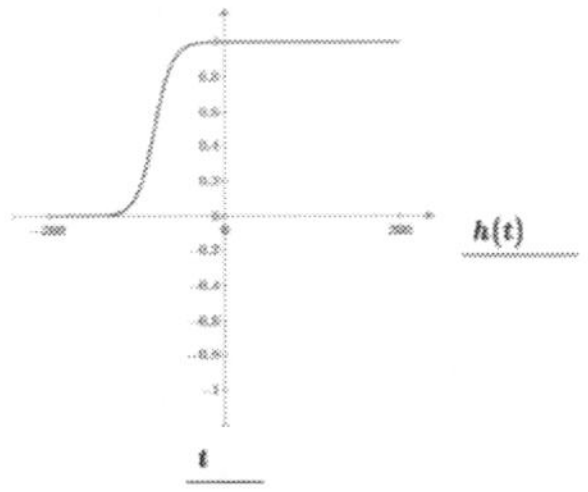

Figure 7: Sizing function $h(x)$

$$h(t)=\frac{y_0L}{y_0+(L-y_0)e^{-k(t+t')}}$$

(k=0.1、L=1、y_0=0.7、t'=70)

The double-bubble surface created by the above procedure is shown in Figure 8. A 42 x 42 areal element was generated from 44 x 44 discretized points.

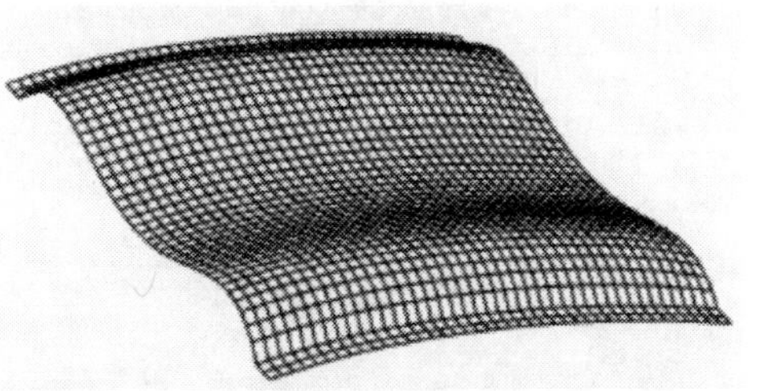

Figure 8: Double-bubble surface used in this study

A conventional roof surface without concavities was created by connecting the maxima of zy with straight lines while leaving zx and $h(x)$ unchanged and interpolating in the same way as when the double-bubble roof was created.

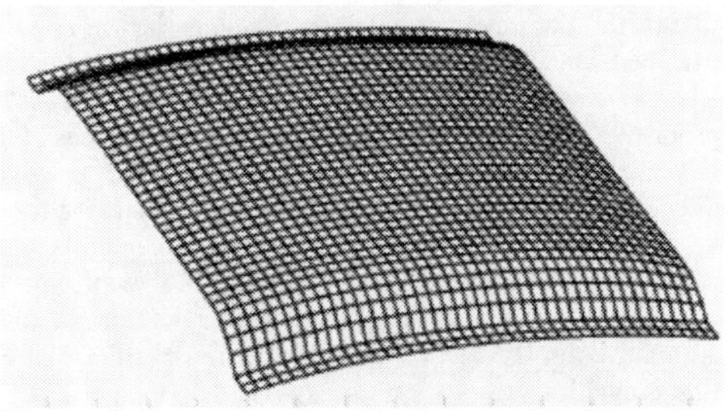

Figure 9: Conventional roof surface used in this study

3.2 Calculate the rays incident on each surface

The direct and scattered light incident on each surface was calculated using cosine loss. In determining the amount of light incident on each of the surfaces, data taken from a solar radiation database called METPV-20 was used as the basis. METPV-20 is a database that contains hourly estimates of average, multi-shine, and oligo-shine years at 835 sites in Japan from 2010 to 2018. Since the angle of the light-receiving surface of the measurement can be set arbitrarily, we set up a situation in which a flat panel was installed horizontally. Since flat panels do not self-shadow, the average direct and scattered light cosine losses can be determined by calculation. Calculate the average cosine loss for double-bubble and conventional surfaces created similarly. However, when calculating the cosine loss of a surface, it is necessary to consider self-shadowing. For the cosine loss calculation considering self-shadowing, the inner product calculation of the ray vector and the unit normal vector of the areal elements were used. A schematic of the calculation is shown in Figure 10.

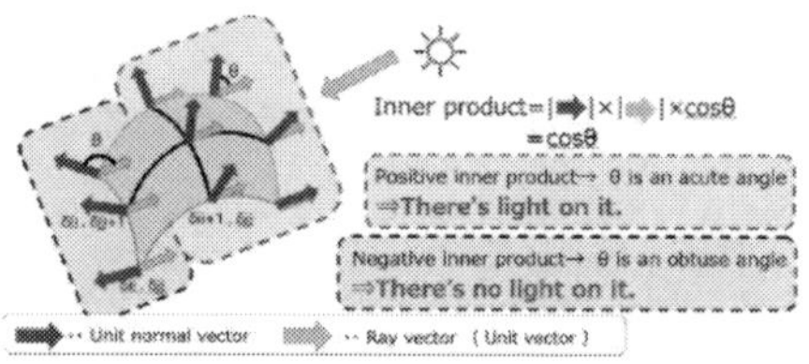

Figure 10: Illustration of the calculation for the cosine loss at surface elements of the curved surface

Since the ray vector and the normal vector of the plane element are unit normal vectors, the result of the inner product calculation is a cosine value. If the cosine value is positive, the angle formed is an acute angle, which means that the ray is incident on that face element. Conversely, if the cosine value is negative, the angle of incidence is obtuse, meaning that no rays are incident on that face element.

However, for surfaces with shapes such as double-bubble surfaces, there are cases where light rays do not enter the surface, even if the inner area is positive due to a hidden surface. Figure 11 shows an example of how the hidden surface is handled.

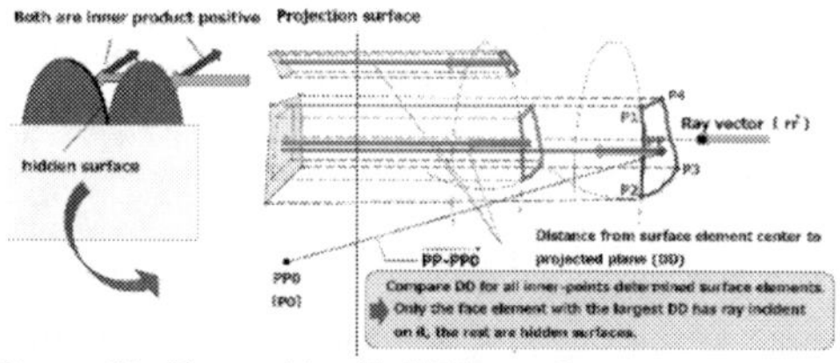

Figure 11: How to identify hidden surfaces

The method of inner point determination is shown. Assume that the face element is infinitely extended and that the face and the ray vector intersection is point "a." The outer product of the vector connecting two adjacent vertices of the face element and the vector connecting the vertex at which the vector starts and point a is calculated. This is done for all vertices of the face element. If the result of the outer product calculation and the positive and negative of the unit normal vector of the face coincide, point a can be determined to be an inner point. It is an outside point if not even one of the four coincides.

By calculating the ratio using the cosine loss average of the flat panel, the METPV-20 data, and the above cosine loss average of the curved surface, the amount of light incident on the curved surface to the curved surface installed in Miyazaki can be calculated for each hour. This paper cites 2013 solar radiation data for Miyazaki City. In other words, we calculated the amount of incident light that would have been available if the curved surface had been installed in 2013.

This operation is performed for direct ray and scattered ray, respectively. The number of direct rays was set to one. The sun's solar altitude and azimuthal angle were used for the incident direction for each date and time in 2013. The scattered ray was set to 10,000. A randomly determined solar altitude and azimuth were used for the incident direction. After applying the above algorithm to direct and scattered light, the cosine loss averages for each curved surface were as follows.

【direct ray】
Double-bubble curved surface: 0.19
Conventional curved surface: 0.55
【scattered ray】
Double-bubble curved surface: 0.19
Traditional curved surface: 0.63

4 RESULTS OF THE RESEARCH

3.1 Cosine loss on each curved surface

When rays of light enter the double-bubble type and conventional type curved surfaces from a solar azimuth of $0° \leq \phi r < 360°$ and a solar altitude of $0° \leq \theta r < 90°$, respectively, the cosine losses are as follows (Figure 12).

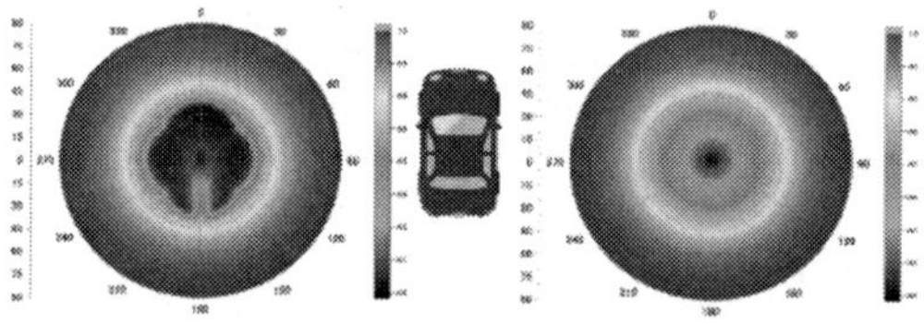

Figure 12: Comparison of cosine loss, double-bubble vs. conventional curved surface
(South: 0° West: 90° North: 180° East: 270°)

The larger the ray incident on the surface, the redder it is, and the more minor the light incident on the surface, the bluer it is. In the figure, the south orientation is 0°, and the car is also assumed to face south. For arbitrary postures of the vehicle (azimuth, pitch angle, roll angle, etc.), the coordinate axes must be transformed and recalculated, so the calculation flow needs to be streamlined in this area. For standard surfaces, the higher the solar altitude, the less loss is independent of the azimuth angle, and the characteristics are similar to those of Lambert surfaces. On the other hand, for the double-bubble surface, it was calculated that at solar altitudes below 30°, the contours are highly distorted, and the values become smaller, decreasing the available solar energy.

3.2 Comparison of incident light intensity for each curved surface

The direct and scattered ray referenced from METPV-20 is shown.

$$ScatCSV = \begin{bmatrix} 0 & 0 & 0 & 0 & 0 & 0 & 0 & 5 & 26 & 49 & 84 & 71 \\ 0 & 0 & 0 & 0 & 0 & 0 & 0 & 12 & 21 & 26 & 34 & 37 \\ 0 & 0 & 0 & 0 & 0 & 0 & 0 & 12 & 32 & 37 & 33 & 37 \\ 0 & 0 & 0 & 0 & 0 & 0 & 0 & 12 & 33 & 36 & 33 & 37 \\ 0 & 0 & 0 & 0 & 0 & 0 & 0 & 3 & 13 & 22 & 43 & 41 \\ 0 & 0 & 0 & 0 & 0 & 0 & 0 & 5 & 32 & 30 & 36 & 71 \\ 0 & 0 & 0 & 0 & 0 & 0 & 0 & 8 & 31 & 31 & 35 & 39 \\ 0 & 0 & 0 & 0 & 0 & 0 & 0 & 3 & 19 & 40 & 59 & 81 \\ 0 & 0 & 0 & 0 & 0 & 0 & 0 & 6 & 29 & 60 & 38 & 37 \\ 0 & 0 & 0 & 0 & 0 & 0 & 0 & 6 & 25 & 30 & 34 & 38 \\ 0 & 0 & 0 & 0 & 0 & 0 & 0 & 7 & 20 & 27 & 33 & 38 \\ 0 & 0 & 0 & 0 & 0 & 0 & 0 & 7 & 27 & 65 & 53 & 90 \\ & & & & & & & & & & & \ddots \end{bmatrix} \quad DirCSV = \begin{bmatrix} 0 & 0 & 0 & 0 & 0 & 0 & 0 & 0 & 0 & 1 & 35 & 163 \\ 0 & 0 & 0 & 0 & 0 & 0 & 0 & 3 & 43 & 97 & 138 & 163 \\ 0 & 0 & 0 & 0 & 0 & 0 & 0 & 0 & 21 & 102 & 144 & 169 \\ 0 & 0 & 0 & 0 & 0 & 0 & 0 & 1 & 13 & 97 & 148 & 171 \\ 0 & 0 & 0 & 0 & 0 & 0 & 0 & 0 & 0 & 0 & 0 & 0 \\ 0 & 0 & 0 & 0 & 0 & 0 & 0 & 0 & 18 & 93 & 132 & 98 \\ 0 & 0 & 0 & 0 & 0 & 0 & 0 & 0 & 20 & 91 & 137 & 150 \\ 0 & 0 & 0 & 0 & 0 & 0 & 0 & 0 & 0 & 0 & 1 & 1 \\ 0 & 0 & 0 & 0 & 0 & 0 & 0 & 3 & 28 & 78 & 129 & 173 \\ 0 & 0 & 0 & 0 & 0 & 0 & 0 & 2 & 34 & 94 & 142 & 172 \\ 0 & 0 & 0 & 0 & 0 & 0 & 0 & 4 & 47 & 103 & 147 & 173 \\ 0 & 0 & 0 & 0 & 0 & 0 & 0 & 1 & 8 & 37 & 0 & 38 \\ & & & & & & & & & & & \ddots \end{bmatrix}$$

10.4229/EUPVSEC2023/4BV.3.39
020331-003

The left matrix represents the scattered ray, and the suitable matrix represents the direct ray values. Both are in a 365 × 24 matrix, showing the amount of ray incident on the flat panel per hour for each date in 2013(0 for January 1 and 364 for December 31.). From this matrix and the cosine loss averages for the flat panels, as well as the cosine loss averages for each curved surface, we can calculate the amount of incident light that would have been expected if a curved panel had been placed instead of a flat panel on any given date in 2013. In the present study, comparisons were made for two cases, one on January 1, a clear day, and the other on June 20, a rainy day. Concerning azimuth, the difference between the actual solar azimuth and the randomly given orientation of the car is taken as the apparent azimuth.

Analysis result (integration in entire surface): January 1, 2013, at noon in sunny Miyazaki City (noon here refers to the total amount of solar radiation from 12:00 to 13:00), and June 20, 2013, at noon in rainy Miyazaki City.
January 1, 2013, at noon in sunny Miyazaki City
The azimuth angle of the panel: 242.439°, sun azimuth angle: 355.153°, sun altitude: 34.956°.
【Double-bubble type curved surface】

$$28.7 \times 10^{-2} [Wh/m^2 \cdot hr]$$

【Conventional type curved surface】

$$18.5 \times 10^{-2} [Wh/m^2 \cdot hr]$$

Double-bubble curved surfaces had 1.55 times more incident light than conventional curved surfaces. Double-bubble curved surfaces had more incident light depending on the sun's relative position.
June 20, 2013, at noon in rainy Miyazaki City
Azimuth angle of the panel: 262.244°, sun azimuth angle: 336.864°, sun altitude: 80.941°
【Double-bubble type curved surface】

$$9.70 \times 10^{-2} [Wh/m^2 \cdot hr]$$

【Conventional type curved surface】

$$11.1 \times 10^{-2} [Wh/m^2 \cdot hr]$$

Conventional curved surfaces had 1.55 times more incident light than double-bubble curved surfaces.

The annual incident ray intensity was compared by adding all the hourly incident ray intensities over the year.
【Double-bubble type curved surface】

$$1040.60 [Wh/m^2 \cdot hr]$$

【Conventional curved surface】

$$1107.11 [Wh/m^2 \cdot hr]$$

The double-bubble curved surface received 6% less than the conventional curved surface.

5 DISCUSSION

Although the double-bubble curved surface seems to have less incident light than a conventional surface because of its concave shape, in some cases, as on January 1, the double-bubble curved surface has more incident light rays than the traditional surface. We elucidated under what circumstances the double-bubble type surface's incident light intensity exceeds the conventional surface's intensity by comparing the cosine loss averages in Figure 10.

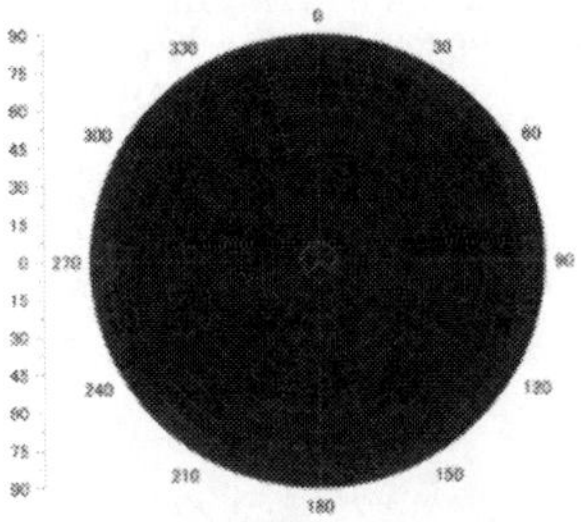

Figure 13: Angular dependence of incident light intensity at each curved surface

Figure 13 shows the angular dependence of the amount of light incident on the curved surface. The area shown in red indicates that the double-bubble curved surface exceeds the incident light intensity of the conventional curved surface. The graph shows that the amount of incident light on the double-bubble surface exceeds that on the traditional surface when the solar altitude is around 7.5°. In other words, when the altitude is 7.5°, the influence of the hidden surface is the smallest, and the concavity allows the light rays to hit the surface well.

6 CONCLUSIONS

This study addressed the prediction of power generation from a double-bubble-roofed vehicle-mounted solar panel that considers self-shadowing and weather conditions. Two curved surface models of a double-bubble roof with and without concavity were created, and the predicted total incident solar radiation was compared for an arbitrary date and time and one year. The solar radiation comparison was conducted at two different times and dates: at noon on January 1, when the weather was clear, and at noon on June 20, when it was raining during the rainy season. The total incident light with and without concavity at any given date and time over a year was added together to obtain a decrease in incident light of less than 10% when concavity was added. As mentioned in the introduction, the double-bubble roof is expected to increase fuel efficiency instead of decreasing power generation; a 10% or even smaller decrease in power generation could be compensated for by an increase in fuel efficiency or even more.

However, this does not accurately represent the double-bubble surface of a Peugeot car because it does not consider mismatching losses, the way the modules are assembled, and the quadratic function used to create the surface only allows for extreme values at the coordinates along the dimensions. It would be possible to create a more accurate surface by subdividing the cut surface into smaller pieces and approximating and connecting them with various functions. Therefore, as the scope of my research, it is necessary to further investigate the above as a future task in this study to predict more accurately the amount of electricity generated.

REFERENCES

[1] C. Tiel et al. Renewable and Sustainable Energy Reviews 158 (2022): 112109.
[2] Y. Ota et al. " IEEE Journal of Photovoltaics 11, no. 5 (2021): 1299-1304.
[3] K. Araki et al. Energies 16, no. 8 (2023): 3580.
[4] K. Araki et al. Coatings 8, no. 7 (2018) 251.

[5] K. Araki et al. Applied Sciences 10, no. 3 (2020): 872.
[6] K. Araki et al. A Energies 16, no. 11 (2023): 4414.
[7] Y. Ota, et al. Solar Energy 184 (2019): 84-90.
[8] Y. Ota et al. Progress in Photovoltaics: Research and Applications 30, no. 2 (2022): 152-163.
[9] Y. Ota, et al. Coatings 8, no. 12 (2018): 432.
[10] Y. Ota, et al. Cleaner Engineering and Technology 7 (2022): 100446.

10.4229/EUPVSEC2023/4BV.3.39
020331-005

3D LAMINATION FOR FLEXIBLE MANUFACTURING OF VARIOUSLY SHAPED VIPV MODULES

Wiebke Wirtz[1], Jens Eilrich[1], Arnaud Morlier[1,2], Sonja Bräunig[1], Susanne Blankemeyer[1], Henning Schulte-Huxel[1]
[1]Institute for Solar Energy Research Hamelin (ISFH), Am Ohrberg 1, 31860 Emmerthal, Germany
[2]now with Interuniversity Microelectronics Center (imec), Imo-Imomec, Thor Park, 3600 Genk, Belgium

ABSTRACT: Integration of photovoltaics (PV) into surfaces that are so far not used for electrical power generation often demands new designs of PV modules. Such new module shapes can include curvatures in three dimensions, which can in general not be handled in standard PV module production processes. In this work, an alternative lamination process using a flexible heat transfer mold (FlexHTM) is presented. The flexible heat transfer mold consists of deformable heat transfer cushions filled with a thermally conductive granulate. The mold supports the heat transfer from a flat heating element to a curved module stack and simultaneously prevents the module shape from deformation during vacuum lamination. The deformable properties enable the adaptation of the flexible heat transfer mold to differently shaped and structured PV modules, which is advantageous over the use of costly massive metallic molds. Several differently shaped PV modules for vehicle integration were manufactured from interdigitated back contact (IBC) solar cells on passenger car hoods with the FlexHTM method, showing its flexibility and ability to produce mechanically stable and aesthetically appealing curved VIPV modules.

Keywords: lamination, curved PV module, VIPV, BIPV, back contact solar cell

1 INTRODUCTION

The developments in the photovoltaics (PV) sector accompanying its continuous growth generate new PV applications such as building-integrated PV (BIPV) and vehicle-integrated PV (VIPV). These applications call for adapted module designs regarding material composition and shape, restricted by the surfaces which shall be solar-activated [1, 2]. A new field of potentially high volume is curved PV modules for integration into complex building skins or vehicle bodies. Curved modules can, in general, not be produced with the same equipment as flat standard PV modules because of their extensiveness in the third dimension. The critical step is the lamination process, particularly the mechanical load and the heat transfer into the module stack during the vacuum process. The mechanical load can be reduced by using a double-membrane laminator instead of a conventional plate-membrane laminator [3], but the heat transfer through the air between flat heating elements and curved module stack remains challenging.

Solutions for single-membrane lamination with metallic or metal-containing molds for mechanical support and improved heat conduction have been suggested before [4]. These solutions have the advantage of a perfect fit to the substrate and a very good heat conductivity, but they lack in adaptability. Every individually shaped substrate needs its individual mold, which generates great effort in design, material and machining, resulting in high costs, especially in the development phase. This disadvantage is eradicated by the method presented in this work, which uses deformable and therefore reusable heat transfer cushions filled with metallic granulate as a flexible heat transfer mold instead of massive metallic molds.

In the following, the flexible heat transfer mold (FlexHTM) method [5] is described in detail and results of curved VIPV modules, manufactured with this method, are presented.

2 METHOD

2.1 3D lamination

We use deformable heat transfer cushions filled with aluminum granulate as a flexible heat transfer mold to support both the mechanical stability of the 3D-shaped module structure and the heat transfer from the heating element to the module stack during vacuum lamination [5]. The outer skin of the heat transfer cushions is either made from heat resistant polymer films or textile fabrics such as linen. The heat transfer cushions are placed on the heating plate of a conventional plate-membrane laminator and brought into shape, such that they roughly match the negative shape of the module's rear side. The substrate, on which the solar cells shall be laminated, is placed on top of the cushions as sketched in Fig. 1. The exact fit of the heat transfer cushions to the structure of the substrate is achieved by applying vacuum to the vacuum chamber of the laminator (or alternatively to a vacuum bag as depicted in Fig. 1) and therefore pressing the substrate into the heat transfer cushions. The adjusted shape of the FlexHTM remains even after venting the vacuum chamber.

Instead of a conventional laminator, the experimental setup in this work features a silicone vacuum bag on a heating plate surrounded by a thermally isolating construction to prevent heat loss to the surroundings (see Fig. 1). The vacuum bag is evacuated with the help of a vacuum pump and the pressure inside the vacuum bag is monitored with a manometer. The module lamination is done sunny side up, i.e. the module stack is built on top of the substrate with the solar cells' front side up and the front cover of the module on the very top. The temperature of the module stack during the lamination process is monitored by multiple thermo couples attached to the rear side of the module substrate at various positions.

Due to the amount of material and its heat capacity, it takes some time, depending on the module size, heating plate, layer stack and curvature, to reach a sufficient

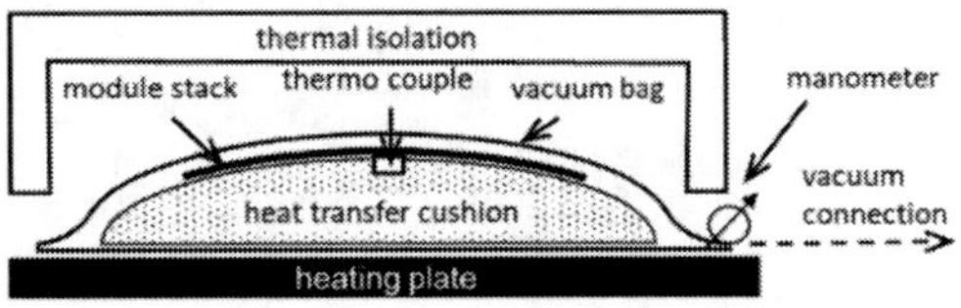

Figure 1: Sketched experimental setup for the 3D lamination process with the help of heat transfer cushions as a flexible heat transfer mold between flat heating plate and curved module stack in a vacuum bag.

10.4229/EUPVSEC2023/4BV.3.41
020332-001

lamination temperature over the whole module area. In order to shorten the heating time, thermoplastic encapsulant materials are used, such that only their melting temperature in the range of 100-120 °C has to be reached and not a higher crosslinking temperature of conventional PV encapsulants in the range of 150 °C.

2.2 Samples and characterization

. Several curved VIPV modules based on passenger car hoods were manufactured with the described lamination process, i.e. sunny side up on heat transfer cushions in a vacuum bag on a flat heating plate. Each of the presented modules is equipped with a matrix of 4x9 half-cut interdigitated back contact (IBC) cells connected in series by soldered ribbons resulting in an active PV area of 0.46 m^2. The proof of concept of the described 3D lamination process was performed by laminating a cutout of a passenger car hood. The result is shown in Fig. 2.

Figure 2: Matrix of IBC cells laminated onto a cutout of a passenger car hood.

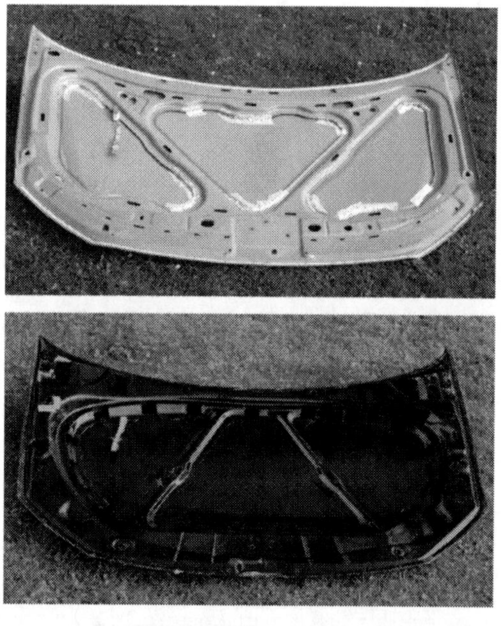

Figure 3: Supporting structures on the rear sides of the two passenger car hoods of two different car models, which were laminated with the help of the same heat transfer cushions (top: gray, bottom: black).

In addition, we fabricate full-size modules, where the cell matrix is extended by eight dummy half-cut IBC cells, which are distributed on the sides of the active cell strings, hiding the cross connectors beneath them. These dummies are not electrically connected to the circuit of the module, but contribute to a more appealing appearance. The two electrical contacts are fed through a hole in the substrate to the rear side of the module (cf. Fig. 3).

After lamination, the *IV* characteristics and therefore the maximum module power of the VIPV modules are measured at standard test conditions (STC). Furthermore, the produced VIPV modules are tested for cracks in the solar cells and interconnection failure by electroluminescence (EL) imaging. The fragile silicon solar cells, which are laminated onto curved substrates, are very likely to break during the lamination process because they are pressed in a bend position and mechanical stress is induced. Potential cell cracks can be detected as dark lines and interconnection failure as dark areas in EL images.

Since VIPV modules are exposed to rougher mechanical impacts during operation than standard PV modules, there are some additional tests recommended in comparison to the licensing requirements for flat standard PV modules [6]. One of these tests is a vibration test imitating the vibrations that occur during rough road driving. The test sequence, which is applied in this work, is based on the random vibration test for equipment mounted on sprung masses (vehicle body) with a power spectral density of acceleration in the frequency range from 10 Hz to 2000 Hz defined in ISO 16750-3 [7]. The vibration test takes 24 hours and is performed at 25 °C ambient temperature. In order to investigate potential mechanical and electrical degradation of the VIPV module under test, visual and EL inspection as well as *IV* measurements are performed before and after the vibration test.

3 RESULTS

3.1 Lamination without cell breakage

As can be seen in Fig. 2, the encapsulant melted evenly over the whole surface of the proof-of-concept cutout. Thus, a uniform heat transfer is achieved even under the critical conditions imposed by the curvature and the complex hollow supporting structure (cf. Fig. 3) on the rear side of the substrate. Furthermore, no deformations of the steel substrate due to the applied pressure during lamination are observed. This implies that, in addition to

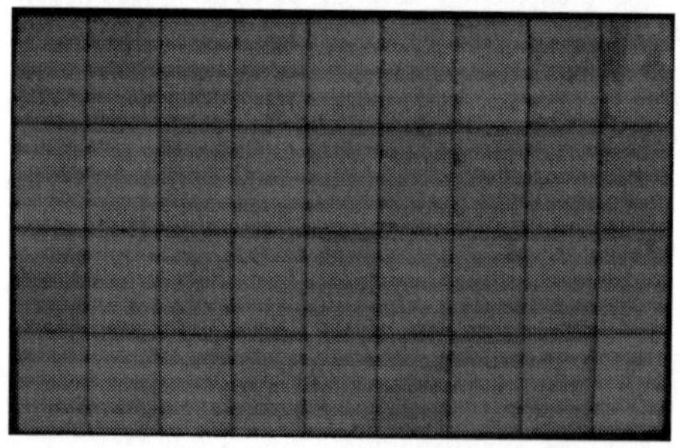

Figure 4: Electroluminescence image of the VIPV module based on IBC cells on a cutout of a passenger car hood showing the absence of cell breakage after lamination with the FlexHTM method.

10.4229/EUPVSEC2023/4BV.3.41
020332-002

Figure 5: Gray VIPV module, which was manufactured with the FlexHTM lamination process, mounted on a passenger car.

Figure 6: Gray full-size VIPV module based on a passenger car hood.

Figure 7: Black full-size VIPV module based on a passenger car hood equipped with half-cut IBC solar cells resulting in a uniform color appearance.

the support of the heat transfer, the mechanical support by the heat transfer cushions works well and enables lamination of 3D-shaped PV modules in a conventional plate-membrane laminator. The EL image in Fig. 4 supports this statement, since no cell breakage in the matrix after lamination can be detected. In total, a power of 96 W is measured for the cutout module with a module area of 0.7 m² at STC.

The preservation of the shape of the module substrate during the vacuum lamination process is also given in the case of laminating a full-size passenger car hood. The car body part still fits in its designated place afterwards, as can be seen in Fig. 5.

3.2 Adaptability to various shapes

Two full-size passenger car hoods (gray and black) of two different car models with different dimensions and supporting structures (see Fig. 3) were laminated with the FlexHTM method. Despite the differing dimensions and structures on the rear side, the lamination was successfully performed with the same heat transfer cushions, which were adapted to the corresponding shape. The encapsulant melted over the whole curved surface of the car hoods of approximately 1.13 m² and a total height difference of 8.8 cm. Furthermore, the shape of the steel substrates was preserved thanks to the mechanical support of the adapted heat transfer cushions during vacuum lamination. Pictures of the two full-size VIPV modules from the front sides are shown in Figs. 6 and 7.

Table I summarizes the *IV* parameters of the gray and black full-size VIPV modules and the proof-of-concept cutout. A good reproducibility of module parameters is achieved with identical materials and process parameters for differently shaped VIPV modules manufactured with the FlexHTM method. Please note that the data are subject to higher measuring uncertainties than usual because of the curvature of the modules, which leads to illumination inhomogeneities across the module area during the test.

This experiment proves the adaptability and ability to reuse the heat transfer cushions for lamination of differently 3D-shaped PV modules with complex structures. This implies that the FlexHTM method is not only applicable to produce VIPV modules but is suitable for manufacturing any 3D-shaped PV module for any application and it is possible to switch between models and structures without the need for new equipment like milled metallic molds. The adaptability of the heat transfer mold drastically reduces the material consumption of the process and enables sustainable development of 3D lamination processes for variously shaped PV modules.

Table I: *IV* parameters of VIPV modules from passenger car hoods manufactured with FlexHTM method.

Module	I_{SC} [A]	V_{OC} [V]	FF [%]	P_{MPP} [W]	η [%]
Cutout	4.9	24.8	79.3	96	20.8
Gray	4.9	24.8	80.1	98	21.3
Black	4.9	24.7	80.6	98	21.3

3.3 Vibration testing of VIPV modules

No visual degradation of the steel substrate, the solar cells or their interconnection could be detected after vibration testing of the gray VIPV module, neither with the eye nor on EL images. This indicates a sound module structure without predetermined breaking points from the vacuum lamination process. There are differences in the *IV* parameters measured before and after the automotive vibration test (see Table II), but it has to be noted that these values feature high measuring uncertainties, which are in the same range as the observed differences. The current measurement is affected by the curvature and the positioning of the VIPV module in the *IV* tester. In addition, the measurements in Table II were performed in a different laboratory without full temperature control, which increases the uncertainty of the voltage measurement. In summary it can be said that if there is a power degradation of the tested VIPV module due to the vibration testing, it can be assumed to be less than 3 $\%_{rel}$.

Table II: *IV* parameters of gray full-size VIPV module before and after automotive vibration test.

	I_{SC} [A]	V_{OC} [V]	FF [%]	P_{MPP} [W]	η [%]
Before	4.9	24.9	80.9	99	21.6
After	4.8	24.6	82.3	97	21.2

10.4229/EUPVSEC2023/4BV.3.41
020332-003

3.4 Uniform color appearance

The increasing call for aesthetics of PV modules, especially PV modules for integration purposes, is met by IBC solar cells with a homogeneous appearance thanks to the absence of metallization on the front side [8]. Simultaneously they allow to generate more power on the limited space due to their high efficiencies.

Taking the advantage of the already dark blue color of IBC solar cells, we laminated IBC solar cells onto a black-painted passenger car hood. The result is shown in Fig. 7. Together with the slightly scattering encapsulant, the black background matches the bluish-black color of the IBC cells, which therefore integrate well into the device. This example shows that inconspicuous VIPV modules can be fabricated without covering the solar cells with an additional coloring layer. Not covering the solar cells also preserves the whole power potential of the module. The presented black VIPV module based on a passenger car hood features a maximum power of 98 W on a limited active area of 0.46 m² thanks to the use of highly efficient IBC solar cells.

4 CONCLUSION AND OUTLOOK

A lamination process for flexible manufacturing of various 3D-shaped PV modules, as they are used for integration into vehicle bodies and building skins, has been developed. This method, featuring a flexible heat transfer mold, overcomes several challenges of 3D lamination at once. First, it provides sufficient heat transfer from a flat heating element to a curved module stack. Secondly, it prevents unwanted permanent deformation of the module shape in a vacuum lamination process. This could easily happen in a plate-membrane laminator without heat transfer cushions by pressing a 3D-shaped module stack into the flat heating plate with the membrane of the laminator. Furthermore, the presented method adds the aspect of adaptability to various structures by using deformable heat transfer cushions instead of massive metallic molds, which is an improvement of the process in terms of sustainability.

With the help of the FlexHTM method it is possible to manufacture curved VIPV modules without cell breakage. Additionally, it was shown that the VIPV modules, which were produced with the FlexHTM method, are able to withstand automotive vibration tests. The unique feature of the FlexHTM method is the adaptability of the heat transfer mold, which was demonstrated by manufacturing two differently structured VIPV modules with the same heat transfer cushions.

It has to be noted that the presented 3D lamination process is not restricted to the manufacturing of VIPV modules but is applicable to the fabrication of 3D-shaped PV modules in general thanks to its geometric flexibility. Another usage example is structured building-integrated photovoltaics (BIPV). For instance, the small BIPV test module shown in Fig. 8 is based on an aluminum square profile, which is used to build façades of industrial buildings. The lamination on the front plane of the square profile with a height of 2 cm is only possible with sufficient heat transfer from the heating plate below the part to the front plane. This is conveniently enabled by the use of heat transfer cushions and in addition they ensure the soundness of the square profile, which is otherwise prone to deformation in a vacuum lamination process.

Figure 8: BIPV test module based on an aluminum square profile. The soundness of the profile after lamination and the heat transfer to the front plane was achieved by the use of heat transfer cushions in a 3D lamination process.

5 ACKNOWLEDGMENTS

This work was funded by the German State of Lower Saxony, the German Federal Ministry for Economic Affairs and Climate Action within the research project AluPV (Contract no. 03EN1069A) and the European Union's Horizon 2020 Programme for research, technological development and demonstration under Grant Agreement no. 857793 (HighLite). The authors appreciate the support. Furthermore, the authors would like to thank Bianca Lim, Kevin Meyer and Christina Hollemann for discussion and proofreading.

6 REFERENCES

[1] Commault et al., Appl. Sci. 2021, 11, 11598, doi: 10.3390/app112411598
[2] Kuhn et al., Energy & Buildings Vol. 214, 2020, doi: 10.1016/j.enbuild.2020.110381
[3] Govaerts et al., WCPEC-8, 2022, doi: 10.4229/WCPEC-82022-3CO.5.3
[4] Neven-du Mont et al., PVinMotion, 2023
[5] "FlexHTM Verfahren zur Aufbringung einer Solarzellenlaminierung auf ein einfach oder mehrfach gekrümmtes Flächenelement", patent number DE 10 2022 104 261 B3
[6] Markert et al., Sustainability 2021, 13, 13341, doi: 10.3390/su132313341
[7] "Road vehicles – Environmental conditions and testing for electrical and electronic equipment – Part 3: Mechanical loads", ISO/FDIS 16750-3:2007
[8] Ortiz Lizcano et al., Progress in Photovoltaics Vol. 30, pp.401-435, 2022

DESIGN AND IMPLEMENTATION OF A PORTABLE IOT SOURCE-SINK METER FOR INDOOR PHOTOVOLTAIC SYSTEM CHARACTERISATION IN REAL ENVIRONMENTS

Zaid Haymoor, Matt Carnie
Swansea University
Swansea University, Bay Campus, Fabian Way, Swansea, Wales, UK
Zaid.haimoor@gmail.com, M.J.Carnie@Swansea.ac.uk

ABSTRACT: This paper presents the design and implementation process of a portable IoT source-sink meter system for real-world indoor photovoltaic (PV) module testing. Specifically tailored for indoor environments and low-light conditions, the meter offers a comprehensive and automated solution for monitoring and testing the performance of various PV module types. It allows the simultaneous scanning of multiple modules under the same conditions, providing long-term detailed data for analysis. The meter measures parameters such as Fill Factor, Maximum Power Point (MPP), humidity, temperature, and irradiance, which reports to a cloud database at scheduled intervals. Our system eliminates manual scans and provides a portable, low-cost solution for accurately characterising indoor PV modules in low-light conditions. The collected performance data is valuable for optimising the efficiency of indoor PV-powered electronic devices.

Additionally, the system provides an overview of energy generation in indoor environments, facilitating the development of power management algorithms considering the consumption, generation yield, and storage element size. Observing long-term historical data enables the understanding of the degradation of PV modules over time, aiding in maintenance and lifespan prediction. Our portable IoT source-sink meter offers an efficient solution for characterising indoor PV modules with its portability, ease of use and setup, and cost-effectiveness. This paper will review our indoor PV tester's PCB design, IT infrastructure, database, communication, testing, and results.

Keywords: Indoor PV, characterisation, Source-sink meter, IoT

1. MOTIVATION

Our journey began when our team set out to design an IoT sensor node powered by a small PV source [1]. This endeavour raised fundamental inquiries: How much power does an indoor PV cell produce? How can we accurately measure this power, especially in low-light conditions? Furthermore, how does room occupancy affect the performance of these cells and modules? These inquiries were fundamental in guiding the development of our self-powered indoor device. Notably, there were no established methods for measuring indoor PV cells [3]. Many studies employed lux values as a metric, which we found unsuitable for real-world scenarios. Measuring individual cells alone provided a misleading indicator of module performance [1]. The voltage at which maximum power is achieved significantly affects the extracted power; the difference between Vmpp and the desired output voltage plays a vital role in system efficiency; the efficiency of power extraction by the electronics can vary widely, ranging from 30% to 90% [12]. Energy transfer occurs twice: first from the cell/module to the storage element and then from the storage element to the load. Loading certain solar modules for specific durations can change their output power. Confining these multifaceted variables was a significant challenge [13].

Consequently, we decided to design a versatile device capable of practically measuring various cells and modules in diverse environments [5]. To achieve this, we designed the portable IoT source-sink meter, which is the outcome of this paper [9]. For verification, a comprehensive range of commercial and developmental indoor PV modules available in the market was presented to derive the results presented in this work [4]..

2. INTRODUCTION

1.1 Low Lightbox Setup

Figure (1) shows a commonly employed low-light controlled box setup used in our laboratory for evaluating indoor PV cells. This setup provides valuable insights for the performance of indoor PV cells under varying lux conditions. The configuration comprises an LED array controlled by a constant current power source. Luminance is precisely managed through the programmable constant current source, with each current value corresponding to a specific luminance level. A linear equation facilitates the determination of the required current for any specified luminance.

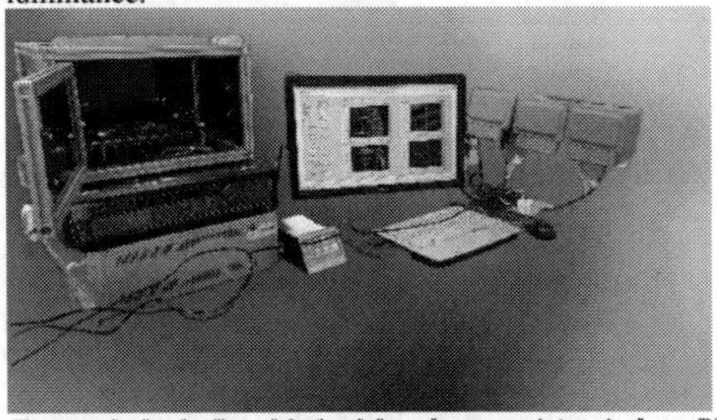

Figure 1: Light Box Method for characterising indoor PV modules

The LED array is positioned at the top within a dark box equipped with a diffuser. Beneath, a test bed is situated, allowing for the mounting of cells/modules, which connects to a stationary source-sink meter for characterisation [9]. Both the LED driver and the source-sink meter necessitate interfacing with a computer via custom-made software.

The software initialises the constant current driver to control the LEDs and set the desired illuminance. Users must configure the software by specifying the expected operating voltage of the cell/module and the voltage stepping for each change during characterisation. These settings are exchanged between the software and the source-sink meter before the characterisation process commences. The source-sink meter then initiates the scanning process and streams an array of voltage and current data to the PC. The PC processes both arrays to compute power and maximum power. Additional parameters, such as Fill Factor (FF), Maximum Power Point Voltage (Vmpp), and possibly cell efficiency, can also be calculated if the light spectrum is known [9].

2.2 Source sink-meter

In characterising indoor PV modules, an essential element is a source-sink meter [2]. This instrument is the cornerstone of precision testing. The source-sink meter operates by dynamically controlling both voltage and current levels. The initialisation cell/module testing begins with communicating with dedicated digital-to-analog converter (DAC) units that adjust the output impedance through specialised amplifiers, establishing the desired output voltage. This precise control over voltage enables the meter to modulate the current flowing through the PV cell/module to any specified value or direction. The characterisation can be programmed to commence testing at the open circuit voltage (Voc) or the current at the module's short-circuit current (Isc). The sweep then systematically varies the current from Isc down to zero amps or vice versa, from Voc down to zero volts. However, direct voltage control can be an easier, more straightforward implementation, so it is usually the design choice for practical installation. Figure (2) shows an overview of the internals of a source measure unit [2].

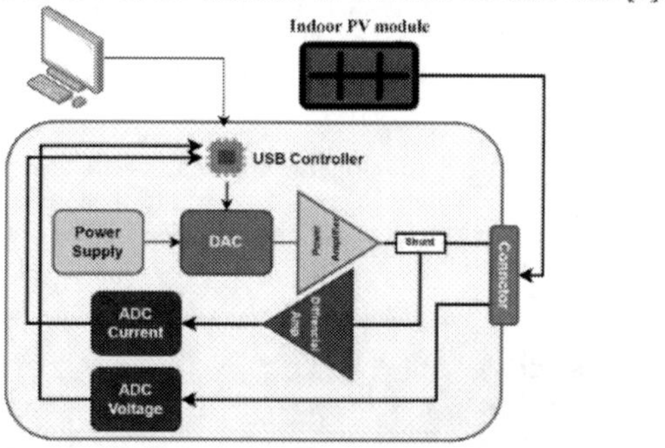

Figure 2: Block Diagram of Source measure unit internals

Simultaneously, the source-sink meter interfaces with high-resolution analog-to-digital converters (ADCs) and current sensing amplifiers to capture voltage and current data throughout the I-V sweep. The resulting dataset, comprising voltage and current arrays, is then stored in a database or commonly used data formats like CSV or Excel for subsequent analysis. This comprehensive approach allows for precisely determining the full I-V curves of modules, including the maximum power point (MPP) and fill factor (FF).

2.3 IoT infrastructure

Figure 3:IoT system Key Blocks

Figure (3) begins on the far left with the things layer; these refer to devices capable of connecting to the internet via a gateway. This gateway is the second key component that takes the form of a Wi-Fi router, Bluetooth, or Lora hub and establishes an internet connection through GSM, Wi-Fi, or cable connections. Data traffic is then transferred to middleware; where one or more servers process and store the data. This data exchange is typically conducted using communication protocols like HTTP or MQTT in most cases for lightweight data transmission. Finally, the application layer known as the user interface, provides a versatile platform for users to access and interact with stored and processed data according to their preferences. Moreover, the interface can also function as an input panel for controlling devices on the first layer. The choice between MQTT and HTTP on the application layer for data communication depends on the type of data and its presentation, with MQTT suitable for real-time actions and HTTP serving various other data scenarios. While these blocks can be further subdivided to illustrate their complex interconnection, this concise overview comprehensively introduces how our device can be integrated within the IoT framework[18].

3. IMPLEMENTATION

1.2 Source measure meter hardware

The implementation presented several challenges. Initially, the project started on a breadboard, but we encountered issues with current measurements due to the impedance of the breadboard. Consequently, we developed several PCB iterations to address these issues, which were impractical to resolve on the breadboard. Figure (4) illustrates the fifth and latest iteration of the PCB, equipped with two current sense amplifiers, DACs, ADCs, amplifiers and a mux capable of accommodating up to four indoor PV cells for simultaneous characterisation within the same environment[9].

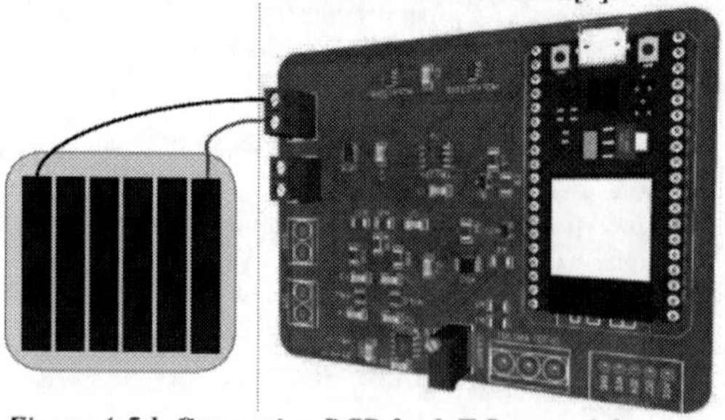

Figure 4:5th Generation PCB for IoT Source sink meter

10.4229/EUPVSEC2023/4BV.3.43
020333-002

At the heart of this board lies an ESP32 driving DACs, cascaded by power-sinking amplifiers and ADCs steaming the current sense amplifier values, enabling the measurement of both voltage and current. The characterisation of a PV module follows a source-sink process. Dynamic load variation influences the current into or out of the connected module. The device operates as a current-driving tester for PV modules, varying the voltage across their terminals while recording the ensuing current. Scans can be initiated from Isc or Voc, and the voltage is gradually adjusted in discrete steps until the desired endpoint is reached [15].

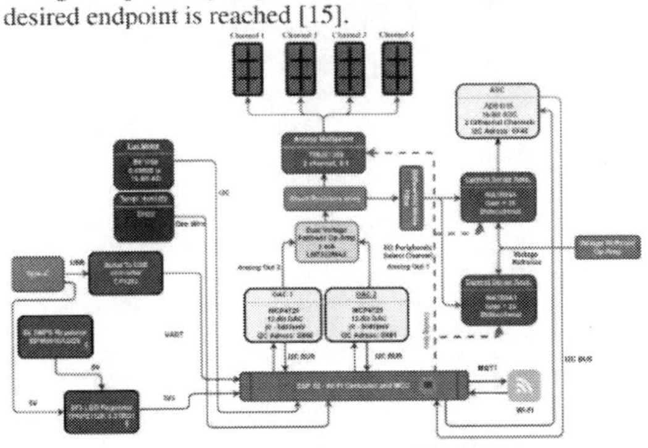

Figure 5: IoT PV charechtersiser Block Diagram

Our system was designed and optimised for low-power PV module testing. The IoT Source-Sink Meter's block diagram is depicted in Figure (1) and encompasses various components, each chosen to fulfil its intended function. This design represents the culmination of five iterative improvements to enhance its performance. Our device can manage up to four modules, each characterised by a voltage ranging from -5 to 5 volts, with a step accuracy of 76 μV. The board integrates multiple current sensing amplifiers with calibrated shunt resistors and filter networks to ensure optimal results. We have incorporated a four-by-one dual-channel bi-directional multiplexer to switch between the four cells/modules even with different voltages and polarities. The ESP32 takes charge of the multiplexer, selecting the desired module when the scan cycle is running. Regarding connectivity between critical components essential for the system's operation, our design heavily relies on the I2C bus protocol.

1.3 Middleware

The middleware operates on an Ubuntu server, which is equipped with essential applications to serve as an IoT server for PV module characterisation. At its core, the Mosquito broker serves as the central application on this server, responsible for establishing and managing MQTT connections across all devices. Additionally, we have integrated MariaDB as our relational database, designed to aggregate and store scans from various devices. Figure (6) provides a comprehensive visual representation of the tables within our database, offering insights into the relationships between these tables and keys, which collectively form an indexed storage system tailored to meet the specific requirements of our system.

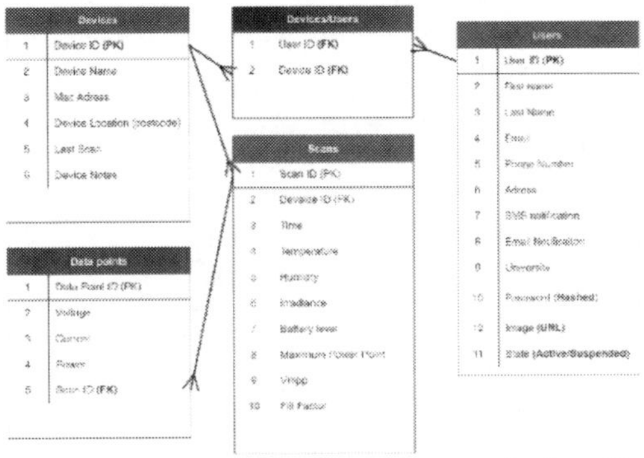

Figure 6: SQL Database system Architecture for Portable PV tester

Furthermore, we have incorporated Node-RED, a NodeJS-based visual programming interface, to orchestrate the communication elements of our project efficiently. Node-RED played a pivotal role in creating the HTTP requests interface, as elaborated upon in Section 3.3. The journey of an HTTP request begins with a user interface-generated GET or POST request. Node-RED adeptly handles the reception and processing of these requests, extracting attached data in JSON format and subsequently effecting the necessary modifications to the database or responding with the requested data. A notable feature worth highlighting is the non-blocking execution of these requests. In practical terms, if a resource-efficient request is received while the system processes a resource-intensive one, it promptly addresses the smaller request before resuming the larger one[18].

1.4 Web interface

The final bit of our Indoor IoT PV testing system is the user interface. This interface was first designed and optimised for friendliness using Adobe XD, and then it was implemented in Flutter; the application mainly consists of 6 interfaces: the dashboard, PV cell connector, user creation, attaching a PV cell into a user and finally, an essential interface that allows the user to observe and download the required data. Figure (7) is a screenshot taken from the dashboard, giving an overview of how the cell performs on the last characterisation.

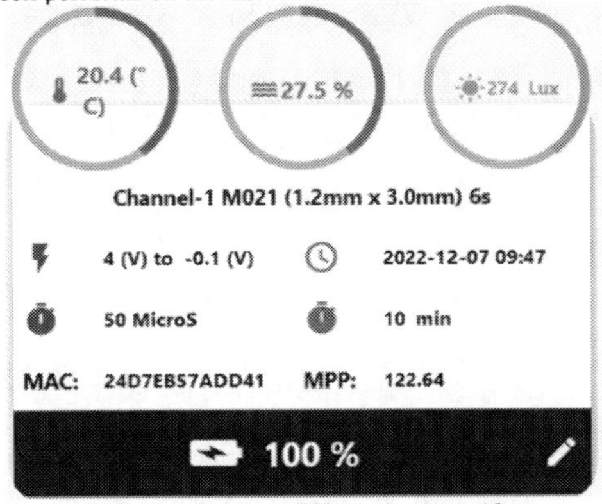

Figure 7: Screenshot from PV monitoring widget

Then there is the language control and the user account control in the top bar of the screen; on the right or left side, the app bar helps navigate between various

10.4229/EUPVSEC2023/4BV.3.43
020333-003

screens. Another important screen is the scans and historical data screen. It gives the option to Navigate the historical data points of Power temperature, humidity and lux values. This data can be downloaded for any further processing or scenario troubleshooting. Figure (8) shows an example of the historical points of the maximum power voltage recorded using our system over five days.

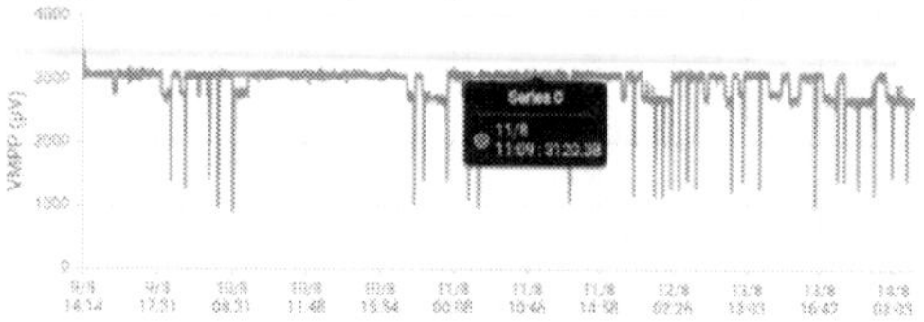

Figure 8: Vmpp Over Five days generated by our Portable PV tester

4. RESULTS

4.1 Long-Term Fulls I-V scans

Numerous tests were conducted using our device in various environmental conditions. However, before proceeding with these tests, a crucial accuracy assessment was performed, comparing our device's performance to that of a fixed test box setup. The results revealed an identical I-V curve across a range of light intensities, including 50, 100, 200, 500, 1000, 2000, and 5000 lux. Notably, calibrating the low lux values presented a challenge, requiring meticulous calculations and procedures to ensure accurate ampere readings under these lighting conditions. The optimal solution emerged by implementing multiple current sensing amplifiers tailored to different illuminance ranges and currents. For further clarity, Figure 9 showcases a screenshot from the dashboard, illustrating one of the I-V characterisations accomplished using our hardware.

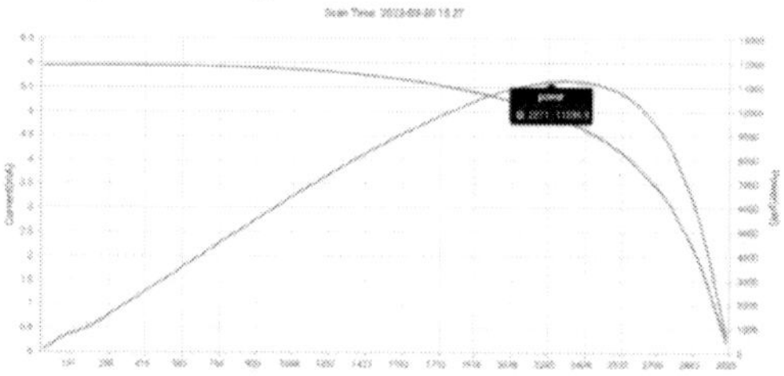

Figure 9: Full I-V and P-V curves for an indoor module

Our system was designed to give an overview of the long-term operation of indoor PV cells. It not only stores the IV scanning data completed on a schedule, it also provides long-term data for the energy generation yield and the fill factor to show how fast a cell is degrading.

4.2 In-situ Stress testing data for various modules

In-situ testing and comparing was done on many cells and modules; an example of that is what was done to compare the quality of fullerene and non-fullerene based cells for one of the promising cells designed by Onnin in brazil, the test compared between fullerene and a non-fullerene cells, all the data including the full I -V scans the maximum power point the fil factor temperature and humidity were all successfully recorded on a scan schedule. The test Rig is shown in Figure (10).

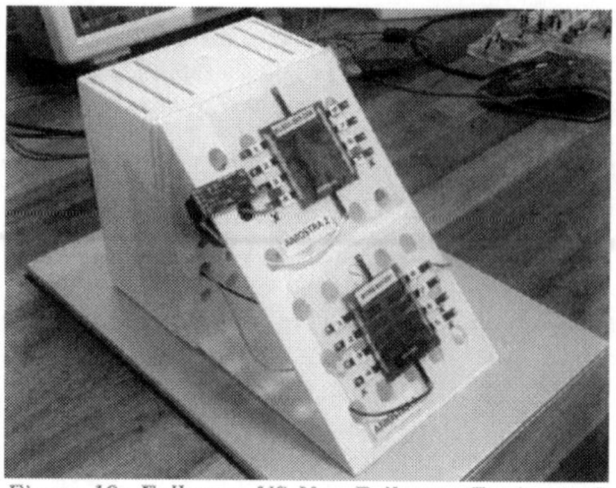

Figure 10: Fullerene VS Non-Fullerene Testing in Real Environments

Figure(11) shows the long-term MPP data of the generated power in microwatt on the left side and illuminance values reflected on the right side.

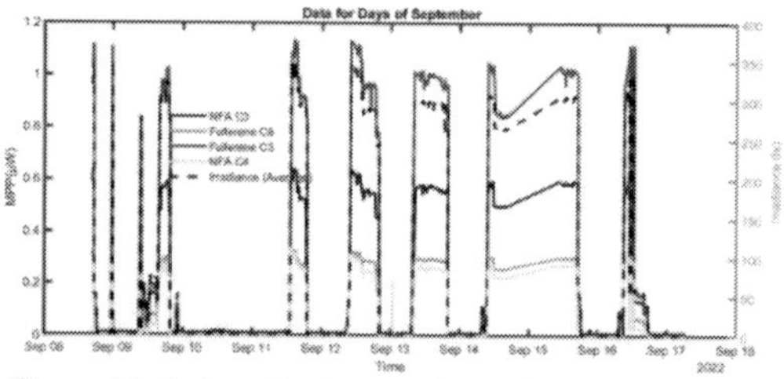

Figure 11: Indoor Testing results of Fullerene and Non-Fullerene Cells By Onnin

By integrating the areas under the Maximum power point curve, we can calculate the energy yield of each cell individually to evaluate which cell should proceed for further development and scaling. Table (1) shows the energy yield of each cell in Micro-Joules. It can be observed that fullerene cells performed better in indoor environments than non-fullerene ones.

Sample	Type	Energy
Amoustra1C3	Fullerene	20.81 µJ
Amoustra1C8	Fullerene	21.99 µJ
Amoustra2C3	Non-Fullerene	20.28 µJ
Amoustra2C4	Non-Fullerene	16.13 µJ

Table 1:Energy Yield Comparison of Fullerene and NFA based Indoor Cells

5. CONCLUSION

The use cases of the generated data can be as simple as building an idea of the cell performance in particular environments, measuring the decay speed of a module over long periods and maybe, This section delves into the rich insights and opportunities from the extensive dataset gathered from our long-term indoor PV cell monitoring. The data, comprising I-V scans, lux readings, and critical parameters such as maximum power point (MPP) and Vmpp, presents a treasure trove of possibilities. Beyond the mere collection of raw data, we explore the actionable steps and analyses that can be derived from this dataset. In later work, we

10.4229/EUPVSEC2023/4BV.3.43
020333-004

discuss, in detail, an energy yield allowance equation that has been meticulously developed and extracted from this data. This equation sheds light on indoor PV systems' performance trends and provides a practical tool for optimising energy efficiency. Furthermore, we examine the methodologies employed, the implications of our findings, and the broader context within which our research contributes to the advancement of indoor photovoltaic technology.

6. References

[1] Buchicchio, E., De Angelis, A., Santoni, F., & Carbone, P. (2022). EasyEIS: A software tool to perform electrochemical impedance spectroscopy using a source measure unit. SoftwareX, 18, 101075.

[2] Cejer, M. A., Director, M., Tucker Sr, J. L., Weng, L., & Instruments, K. (2013). Choosing the optimal source measurement unit instrument for your test and measurement application. Source measure unit.

[3] Nikolov, G., Nikolova, B., Gieva, E., & Ruskova, I. (2017, May). Extraction of solar cell parameters using source measure unit. In 2017 40th International Spring Seminar on Electronics Technology (ISSE) (pp. 1-6). IEEE.

[4] LÓpez-Vargas, A., Fuentes, M., & Vivar, M. (2018, June). On the application of IoT for real-time monitoring of small stand-alone PV systems: Results from a new smart datalogger. In 2018 IEEE 7th World Conference on Photovoltaic Energy Conversion (WCPEC)(A Joint Conference of 45th IEEE PVSC, 28th PVSEC & 34th EU PVSEC) (pp. 605-607). IEEE.

[5] Ali, M., & Paracha, M. K. (2020). An IoT based approach for monitoring solar power consumption with Adafruit Cloud. International Journal of Engineering Applied Sciences and Technology, 4(9), 335-341.

[6] Gupta, V., Sharma, M., Pachauri, R. K., & Babu, K. D. (2021). A low-cost real-time IOT enabled data acquisition system for monitoring of PV system. Energy Sources, Part A: Recovery, Utilisation, and Environmental Effects, 43(20), 2529-2543.

[7] Naskari, V., Doumenis, G., Koutsos, C., & Masklavanos, I. (2022, September). Design and implementation of an indoors light simulator. In 2022 7th South-East Europe Design Automation, Computer Engineering, Computer Networks and Social Media Conference (SEEDA-CECNSM) (pp. 1-6). IEEE.

[8] Torres, D., Dias, J. P., Restivo, A., & Ferreira, H. S. (2020, September). Real-time feedback in node-red for iot development: An empirical study. In 2020 IEEE/ACM 24th International Symposium on Distributed Simulation and Real Time Applications (DS-RT) (pp. 1-8). IEEE.

[9] Kodali, R. K., & Anjum, A. (2018, August). IoT based home automation using node-red. In 2018 Second International Conference on Green Computing and Internet of Things (ICGCIoT) (pp. 386-390). IEEE.

[10] Espressif Systems. (n.d.). ESP32 Datasheet. Retrieved from https://www.espressif.com/sites/default/files/documentation/esp32_datasheet_en.pdf

[11] Renesas Electronics Corporation. (2015). Introduction to Photovoltaic System. Retrieved from https://www.renesas.com/us/en/doc/white-paper/introduction-photovoltaic-system.html

[12] Roy, S., & Tathagata, D. (2016). A Review on Maximum Power Point Tracking Techniques for Photovoltaic Systems. Renewable and Sustainable Energy Reviews, 62, 795-807. doi:10.1016/j.rser.2016.04.051

[13] Syaiful, M., & Ali, R. M. (2019). Design and Implementation of IoT Based Solar Energy Monitoring System Using ESP32. Journal of Physics: Conference Series, 1333, 012139. doi:10.1088/1742-6596/1333/1/012139

[14] S. Qian, Y. Li, X. Zhang, Z. Chen, and J. Li, "A Non-Invasive Method for Indoor Photovoltaic Characterization Using a Single Sensor," Applied Sciences, vol. 9, no. 4, p. 696, 2019.

[15] M. Mukhtar, A. Iqbal, S. Kamal, S. Ali, and A. H. Malik, "Design and Implementation of an IoT-Based Smart Energy Monitoring System for Home Appliances," Sensors, vol. 19, no. 20, p. 4392, 2019.

[16] R. P. Sousa, A. O. A. Cardoso, and J. A. Q. Bressane, "Development of a Remote Monitoring System for Photovoltaic Systems Based on IoT," IEEE Latin America Transactions, vol. 17, no. 1, pp. 10-16, 2019.

[17] J. Zhou, J. Zhang, Y. Xiong, and X. Liu, "IoT-Based Photovoltaic System Monitoring and Performance Evaluation Using Machine Learning," Energies, vol. 14, no. 2, p. 293, 2021.

[18] S. M. Azimi, A. Sharifian, and E. Rokrok, "Design and Implementation of an IoT-Based Solar Energy Monitoring System for Smart Buildings," IET Smart Grid, vol. 1, no. 1, pp. 47-55, 2021.

10.4229/EUPVSEC2023/4BV.3.43
020333-005

DEVELOPMENT OF BUILDING-INTEGRATED PHOTOVOLTAICS (BIPV) FROM 2000 TO 2022 – WHAT CAN WE LEARN FROM NINE BIPV COMPETITIONS?

Gerd Becker[1], Fabian Flade[1], Roland Krippner[2], Bruno Schiebelsberger[1], Walter Weber[1]
[1]Bavarian Association for the Promotion of Solar Energy
Friedrich-List-Str. 88, 81377 München, Germany
Phone: +49 (0) 89/278134-28, info@sev-bayern.de, www.sev-bayern.de
[2] Nuremberg Institute of Technology/Faculty of Architecture, Nuremberg, Germany

ABSTRACT: Starting in the year 2000, the Bavarian Association for the Promotion of Solar Energy has organized the competitions "Building-Integrated Solar Technology". The aim of this award is to make exemplary solutions in qualitatively sophisticated architecture accessible to the greater public. Since the year 2000 nine competitions have been organized, with 783 participants and some 30 prize-winners. The awarded competition entries show the huge potential of coherent integrated solar technology (PV, but also solarthermal) as a part of advanced architectural and energy building concepts. The intension of the work is to illustrate how BIPV has changed within the last 22 years – by means of examples coming from the competitions.
Keywords: Building Integrated PV (BIPV), Façade, Solar Architecture

1 GENERAL DATA BIPV COMPETITIONS

1.1 Why building-integrated photovoltaics (BIPV)

In addition to the generation of electric energy building-integrated photovoltaics fulfil some more classic functions such as thermal insulation, protection against wind and weather or also design functions. Energetic requirements can be met. BIPV provides aesthetically appealing power systems to blend into the urban or rural landscape. Architecturally integrated PV modules in roofs and façades allow aesthetically appealing power generation, they can make a significant contribution to the energy transformation [1]. BIPV is an important component to reduce CO_2 emissions in buildings.

1.2 BIPV competitions

In the time span from the year 2000 to 2022, the "Bavarian Association for the Promotion of Solar Energy" has organized nine competitions "Building-Integrated Solar Technology (BIPV)", with 783 participants in total from all over the world. These nine competitions were organized in the years 2000, 2001, 2005, 2008, 2011, 2014, 2017, 2020 and 2022. They give a good idea of BIPV in the last 22 years [2] [3].

In each of these competitions a considerable prize money was provided, in the last competitions it amounted to 15,000 € for the 1st prize winner and 10,000 € for the 2nd prize winner(s). Commendations and citations (without prize money) were additionally awarded. In total 783 participants from 47 countries formed an excellent data base for this purpose.

Within all the competitions the following criteria had to be fulfilled: Architectural aspects, innovativeness, energy yield/energy efficiency and communicating the system to the general public.

The results of these competitions provide many ideas for future BIPV, as they illustrate the development of BIPV from the year 2000 to 2022.

1.3 Regions of origin

In the 2000 and 2001 competitions, only BIPV projects from Bavaria were considered. In 2005, projects from all over Germany were invited to participate in the competition. In 2008, the group of participants was expanded, as BIPV projects from the whole of Europe participated.

It must be added that from the year 2008 solar thermal systems were admitted to the competition. From the year 2011 all kinds of BIPV projects from all over the world were invited to participate in the competition.

In order to give an idea of characterizing figures, the 2022 competition will be described in detail. In this year, participants from 19 countries have taken part in the competition, with 121 entries, 12 of which came from universities. With one submission each from Monaco, Uganda, and the United Arab Emirates, three more countries joined in 2022. Projects from German-speaking countries nevertheless dominate the current competition, accounting for two-thirds of the total. Roughly 35 % of the schemes come from Germany, ~26 % from Switzerland and ~7 % from Austria. The remaining third come from the other 16 countries, with 13 submissions from the Netherlands alone, which is a considerable proportion compared with the previous competitions and reflects the greatly expanded activities in this field of work in those countries. More projects came from Belgium and Sweden, France and Italy and Norway. It is largely other countries from northern and western Europe that round off the field of participants.

1.4 Modernization or refurbishment

A key challenge for achieving the climate goals is the energy-efficient modernization or refurbishment of the existing building stock. Here, 20 % of the projects submitted involve the retrofitting of old buildings, an increase of one third compared with the competition in 2020. It is noticeable that this proportion - modernization or refurbishment - now makes up around one third of the submissions for apartment buildings.

1.5 Two types of BIPV projects

It can be distinguished between two types of BIPV solutions. First there are exceptional projects - lighthouse projects - where solar technology is integrated into the building skin, sometimes in a technically complex way. Special components are often used in the construction. The costs for this are high. Often lighthouse projects were the first-prize-winners of the competitions. On the other hand, so called scalable projects are considered. In these, standard components are used as far as possible during construction. The costs for these projects should remain within specified limits.

10.4229/EUPVSEC2023/4BV.4.1
020334-001

2 LIGHTHOUSE PROJECTS

2.1 Nikolaus-Fiebiger-Center, Erlangen (GER)

The first BIPV competition was organized in the year 2000. The 1st prize winner - a shadowvoltaic system of a university building - is definitively a lighthouse project.

Figure 1: 1st Prize winner of the year 2000: South orientation and solar awning PV modules plus one axis tracked solar blind

This system of the Nikolaus-Fiebiger-Center (Figure 1) consists of two subsystems, a solar awning and a one axis tracked solar blind. Both systems were designed to achieve a semitransparent effect through the arrangement of the cells and thus to maintain the visual relationship from the interior of the building to the outside. The nominal power of the solar awning is 7.7 kW, the one of the solar blinds is 22 kW. Final yield obtained lies in the range of 700 kWh/kW. The costs were high: 8,000 €/kW for the fixed PV modules and 22,000 €/kW for the one axis tracked solar blinds. This is a very expensive and complex BIPV solution. This façade system has many moving parts and is therefore expensive and prone to failure. It is the question, if such a system would be built up nowadays.

2.2 "Umwelt Arena", Spreitenbach (CH)

The architects of the "Umwelt Arena" in Spreitenbach near Zurich (René Schmid Architekten, Zurich) refer to a "futuristic solar dress". The 1st prize winner of the 2014 BIPV competition shows quite extraordinary forms of the photovoltaic system. The PV system, whose structural and formal implementation provides an equally convincing solution when viewed both from a distance and close up. The southeast/northwest-oriented building succeeds in effectively activating the north-facing PV cells, enabling the roughly 5,400 m^2 of monocrystalline glass-glass modules to achieve a total installed output of 750 kW (Figure 2).

The energy yield (specific in 2014: ≈ 640 kWh/kW) is in total more than double the building's own energy requirements. The costs for the 750-kW-PV systems including panels, inverters etc. amounted to 2,29 Mio CHF or specific roughly 3,400 €/kW.

Figure 2: 750-kW-building-integrated PV System of the Umwelt Arena in Spreitenbach near Zurich (Switzerland)

2.3 Copenhagen International School (DK)

CIS Nordhavn is a new school building (C. F. Møller Architects, København) for Copenhagen International School (Figure 3), which has achieved a commendation in the 2017 competition.

The new school building, prominently located on the waterfront in the Nordhavn harbor area, is characterized by a very differentiated structure of the building mass, reminiscent of stacked shipping containers. Above a partially glazed plinth zone and recessed structures, whose almost square glass panels are characterized by different glass panels create a lively façade image in shades ranging from light blue to turquoise green.

The technically and creatively innovative approach: The specially coated cover glass of the monocrystalline modules moves the photovoltaics into the background and thus enables novel-colored solar façades. It works in such a way that vaporized oxide layers reflect only a certain spectral range of sunlight, see Figure 4.

To accentuate the surface area of 6,048 m^2, the 12,000 modules are mounted at different angles in metal cassettes. They are estimated to produce 300 MWh per year.

Figure 3: Copenhagen International School (CIS) on the waterfront in Nordhavn harbor area (Photos: Adam Mørk)

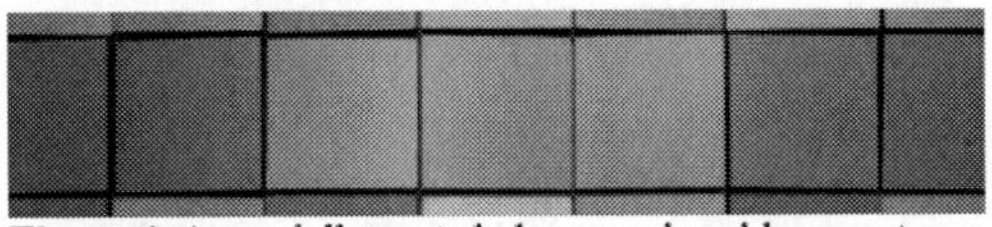

Figure 4: A specially coated glass panel enables new types of colored solar façades.

2.4 "Lycée Technique pour Professions de Santé" in Ettelbrück (LU)

The 1st prize winner of the 2020 BIPV competition (FABECKARCHITECTES, Koerich) is a four-storey building, which is divided into projecting and receding structures (Figure 5). The primary structure is made of timber except for the escape staircases. Asymmetrical gable roofs vary in the inclination of their surfaces and are covered entirely with PV modules of 320 kW_{el}. The crystalline PV generator on the roof surfaces is meticulously realized on a timber substructure with painted flashings and continues familiar approaches from structural engineering practice.

The façade design (Figure 6) posed greater challenges. Solar thermal collectors, to which a 91,000-liter hot water tank is connected, are installed in the southwest and west elevations at the height of the window strips. The solar thermal collectors (150 kW_{th}) are also integrated design-wise so that they align with the window openings and extend the almost square windows along their right-hand reveals.

Based on the energy-plus concept, the new school building vividly combines the use of innovative building services components with the use of ecological materials in a coherent building design with high functionality and design quality.

Figure 5 : „Lycée Technique pour Professions de Santé" in Ettelbrück (LU) (Photo: Christian Aschman)

Figure 6: Solarthermal collectors align with the window openings (Photo: LCGDP).

2.5 "Amt für Umwelt und Energie" (AUE), Basel (CH)

The new building in Basel (Figure 7) is the 1st prize winner of the 2022 BIPV competition (Architect: jessenvollenweider architektur, Basel).

The slender and rectangular eight-storey building fits snugly between existing neighboring buildings.

It is a freestanding structure that responds in height to the surrounding context. The two upper storeys are set back to prevent overshadowing.

The energy concept for the building, which is constructed using a hybrid timber-concrete frame structure, is based on providing district heating and natural night-time cooling, combined with the photovoltaics in the façade.

The façade concept (Figure 8) ultimately realized underwent multiple design iterations. An approach using visible "amber-colored" polycrystalline photovoltaic cells was initially pursued. Not yet satisfied with the result as an "urban solar dress", the architectural practice further developed the concept by using 3D fused glass as the front panel. After experimenting with different textures, they decided on a structure with a square grid and circular depressions in the surface.

Figure 7: "Amt für Umwelt und Energie" in Basel (Photos: Megasol Energie AG)

It was completed in 2021 not far from the market square with its historic town hall. Right from the beginning, the client's goal was to realize a "flagship project for sustainable construction" and a multiplier for other building schemes.

In order to accentuate the slightly translucent effect, especially when viewed from a distance, a PVB film with metallic reflected dots was additionally inserted between the fused glass and the PV cells. These dots are arranged in the surface in three different degrees of density, graded between 5, 7.2 and 9 %. They rise from the first floor, initially each time over two storeys to the seventh floor.

Figure 8: Façade structure with a square grid and circular depressions in the surface

3 SCALABLE PROJECTS

3.1 "Hof 8" in Weikersheim-Schäftersheim (GER)

Buildings in rural areas face numerous challenges. The energetic refurbishment of the building stock using renewable materials is just as relevant as the use of renewable energies for decentralized heat and power generation.

The "Hof 8" project was the 1st prize winner in the competition 2017. The Architect (Architekturbüro Klärle, Bad Mergentheim) has succeeded to make an extraordinary contribution to this complex task, which also has to take into account aspects of inner-city development and demographic change, in a holistic concept. The work provides valuable impulses for the sustainable further development of rural areas and shows that photovoltaics can become a natural building material for building renovation, see Figure 9.

In addition to a coherent mix of uses, the use of regional products and the reuse of materials, photovoltaics (80 kW) are fully integrated into the roof surfaces of the ensemble of a former agricultural farm. In terms of construction, a rather conventional on-roof installation is chosen, which, however, captivates with the carefully detailed treatment of the roof edges of the solar installation, which covers a total area of 550 m^2. The partial residual areas created at the transition of different roofs are harmoniously supplemented with color-matched metal sheets. The combination with existing quarry stone masonry and new wooden facades also illustrates the design potential of commercially available solar technology.

Figure 9: "Hof 8" with PV forming the roof (Photo: Brigida Gonzales)

3.2 Photovoltaic Carport in Rheinfelden (GER)

This BIPV project from the competition in 2022 is a carport parking lot roofing with solar modules, battery storage and e-car charging stations with intelligent charging management, see Figure 10. The PV generator consists of 504 building-integrated glass/glass modules. They are extremely durable; a lifetime > 30 years is expected.

Figure 10: Photovoltaic carport (Photo: Juri Junkov)

A durable and material-efficient carport system for the carport project aluminum construction has been developed, which does not use any sealing rubber and rubber seals. The rainwater is drained through the horizontal main steel beams into the infiltration trough, which means that no additional downpipes are necessary.

In addition, insect-friendly LED lighting has been integrated into the structure. The entire carport construction is very easy to uninstall and all materials used are fully recyclable. The generated electricity is fed to the 14 wallboxes installed under the carport. E-cars charge with sustainably generated electricity. In addition, a battery storage (110 kWh) is charged, so that electric cars can also be supplied in the evening and at night.

At the heart of the system there is an intelligent charging management system that calculates the available energy, controls the charging processes according to demand and thus reduces the demand for power from the grid.

3.3 "Helmholtz-Zentrum" in Berlin-Adlershof (GER)

The Laboratory Building and Testing Hall (Figure 11) has received a building-integrated photovoltaic system with likewise rear-ventilated glass-glass modules with CIGS technology. A total of 360 modules have been installed on the west, south and even the north side of the outer shell. Their concealed suspension enables a frameless design without additional edging at the module edge.

Thus a "real laboratory" has been created, including research on topics such as long-term monitoring and yield analyses in real operation. With more than 120 additional measuring points beyond the standard monitoring, the facades thus become a unique real laboratory for long-term oriented research for building-integrated photovoltaics.

Figure 11: Helmholtz-Zentrum Berlin für Materialien und Energie GmbH (Photo: Andreas Muhs)

3.4 Office building BAURCONSULT in Haßfurt (GER)

The new office building in wood hybrid construction in Haßfurt is based on a classic insulated curtain-type, rear-ventilated facade with glass panels.

On an aluminum substructure of battens and counter battens, which are supported by spacers, aluminum glass mounting points are fixed. On the south-west and east side, the solar panels are hung into these, on the north side normal color-coated glass panels are used. Between the spacers of the aluminum substructure, thermal insulation complementing the wooden structure is applied. The spacers provide rear ventilation of the system with a layer of air between the thermal insulation and the solar panels.

10.4229/EUPVSEC2023/4BV.4.1
020334-004

Figure 12: PV modules on the roof, below a 70 kW PV façade with CIGS glass/glass modules can be seen (Photo: BAURCONSULT Architekten Ingenieure).

The building (Figure 12) can generate almost its own electricity consumption. At the same time, the PV system also produces the electricity to operate the heating and cooling supply. The building is thus self-sufficient and completely emission-free. Only peaks in demand are covered by purchasing green electricity.

There are 8 charging stations for electric vehicles, which are also supplied by the PV system. A further 12 charging columns are pre-equipped. For the growing number of electric vehicles, solar carports will be built in a 2nd construction phase. The necessary cabling and control technology has already been planned and installed. All CO_2 produced during the construction of the building was reforested regionally.

To ensure an ideal PV area utilization of the building, the PV system is installed in two parts on the roof and in the façade. The ventilation center is located in the center of the roof.

The other areas of the roof were designed as a green roof, on which a PV area with 90 modules and almost 30 kW is installed. The main part of the PV façade with 160 modules is installed on the long side of the building in the south direction. On the east and west side of the façade 52 modules each were installed. These complement each other with capturing the morning and evening sun. On sunny days, the PV system can selectively cover the entire electricity demand for the building.

3.5 Children's Surgical Hospital, Entebbe (UGA)

The medical centre (Figures 13 and 14) was built for the Emergency non-government organization that provides free surgical treatment for children from all over Africa. It is slightly oriented to the southwest and consists of two parallel, elongated buildings that are connected by a square building to the east.

The single-storey buildings, which vary in depth, have monolithic walls constructed using rammed earth techniques. To prevent heat gain, a gently sloping butterfly roof with wide overhangs is suspended a clear distance above the building structure. The elegant steel structure functions as a multifunctional roof for providing both weather protection and energy generation. Arranged across a 3,301 m^2 area, 2,352 thin-film modules with an installed PV capacity of 276 kW generated about 245,000 kWh of electricity during the first 12 months.

Children's Surgical Hospital in Entebbe (Architects: Renzo Piano Building Workshop + studio TAMassociati (IT)) provides an excellent example of climate-friendly construction, utilizing local materials in combination with renewable energy sources. The roof is designed as a fifth façade.

Matched to the structural grid, well-proportioned PV surfaces are arranged with a clear structure. The conceptual approach utilizing a "simple" construction method combined with an extremely sensible solution for large-scale energy roofs not only provides an exemplary solution for the African continent, but can also be transferred as a model to other climate regions and construction tasks.

Figure 13: Children's Surgical Hospital (Photos: Emmanuel Museruka)

Figure 14: The roof is designed as a fifth façade.

3.6 New Build Housing in Raalte (NL)

A fine example for scalable BIPV is located in The Netherlands. 28 homes around Dingshof in Raalte have been built. This project in the Salland region of Overijssel province consists of blocks, each with corner houses and terraced houses, see Figure 15.

The installed PV capacity on all the roofs is 119 kW. The modules are facing all directions: 25 kW South East / 15 kW North West / 70 kW South West / 9 kW North East. The actual output of all the PV systems in one year is 96,600 kWh.

Figure 15: 28 homes with BIPV roofs (Photos: Viridian Solar)

The municipality of Raalte set stringent sustainability and energy efficiency requirements resulting in many of the homes being completely energy neutral. The roofs of the homes in this new-build project are equipped with in-roof solar PV systems. Figure 16 illustrates, that a good architectural BIPV solution has been found.

Figure 16: Raalte new homes with PV roof in detail, a good architectural BIPV solution

3.7 "House of Choice" in Solna (SWE)

House of Choice is an 11-storey hotel in Solna designed as a plus-energy building. The solar cells are the most visible sign of the hotel's high energy performance. They are not only contributing to the architecture but also giving an annual surplus of electricity.

The photovoltaic solar system has three parts. Standard modules on the roof, integrated solar cells in glass roofs, and PV panels integrated in the main entrance façade facing southwest. The atrium is covered with glass with integrated solar cells (Figure 17). These cells are distributed in an even grid and an irregular vivid pattern respectively to decrease need of cooling.

Figure 17: The atrium of the hotel is covered with glass with integrated solar cells (Photo: WestblomKlasse).

The façade has three types of panels, two with different textures of the front glass, satinated glass and ribbed, and one with same colour as window frames and screens. Partly to meet a wish for a not too dark façade, but also to go follow a design concept where even and ribbed wall surfaces are varied (Figure 18). The way the different surfaces are reflecting sunlight makes the façade perceived differently. The perception is depending on whether it is sunny or cloudy, the time of the day, or if you are watching it while walking along the street or standing straight in front of it.

Figure 18: Façade "House of Choice" (Photo: Anders Fredriksen)

The solar system is a result of a from the beginning clearly formulated plus-energy goal, which has permeated the good collaboration during project planning, and later implemented by the contractor and their partners.

4 WHAT CAN WE LEARN?

The examples illustrate, that from 2000 to 2022, the appearance and quality of BIPV has changed. Generally, PV modules have become a building material.

There are new ideas - coming from lighthouse projects and scalable projects - which can be applied for new future solutions. However, it can also be seen that some ideas were not expanded.

Nowadays it can be concluded, that tracking systems - like the system at University Erlangen (chapter 2.1) - were a niche within a niche at that time, just as they are today. They were an object-specific solution.

The "Umwelt Arena" in Spreitenbach (2.2) near Zurich is also a lighthouse project. Fixed modules are facing many directions of the sky, as demonstrated by the southeast/northwest-oriented 750-kW-BIPV system, which succeeds in effectively activating the north-facing PV cells. Thus, it can be learnt that southeast/northwest-oriented PV Modules are possible.

Nowadays in an enhanced matter colored and black modules are installed. A technically and creatively innovative approach comes from Copenhagen. A new building for Copenhagen International School applies for its PV system a specially coated cover glass for the monocrystalline modules (2.3). This moves photovoltaics into the background and enables novel-colored solar façades.

PV systems together with solar thermal systems can form a good solution (2.4), as performed for a medical school in Luxembourg. The PV generator on the roof surfaces is meticulously realized on a timber substructure with painted flashings. In the façade there are solar thermal collectors installed, to which a hot water tank is connected.

An interesting energy concept was applied for the building of the Office for Environment and Energy in Basel (2.5), which is constructed using a hybrid timber-concrete frame structure. It is based on providing district heating and natural night-time cooling, combined with the photovoltaics in the façade.

10.4229/EUPVSEC2023/4BV.4.1
020334-006

"Hof 8" in Weikersheim-Schäftersheim illustrates, that energetic refurbishment of the building stock in rural areas - using renewable materials - is possible (3.1).

BIPV can be applied to charge electric cars (3.2). The carport parking lot roofing with solar modules, provides electric power for 14 wallboxes.

A "real laboratory" has been created in Berlin-Adlershof (3.3), including research on topics such as long-term monitoring and yield analyses in real operation. With more than 120 additional measuring points beyond the standard monitoring, the facades thus become a unique real laboratory for long-term oriented research for building-integrated photovoltaics.

The office building in wood hybrid construction in Haßfurt (3.4) can generate almost its own electricity consumption. At the same time, the PV system also produces the electricity to operate the heating and cooling supply. The building is thus self-sufficient and completely emission-free. Only peaks in demand are covered by the by purchasing green electricity.

Children's Surgical Hospital in Entebbe (3.5) is an excellent example of climate-friendly construction, utilizing local materials together with renewable energy sources.

A fine example for scalable BIPV is located in The Netherlands. 28 new-build homes around Dingshof in Raalte (3.6) are equipped with in-roof solar PV systems. This type of PV roof is easy to duplicate.

The 11-storey hotel in Solna in Sweden (3.7) is designed as a plus-energy building, with new ideas. The solar cells are the most visible sign of the hotel's high energy performance. The atrium is covered with glass with integrated solar cells. These cells are distributed in an even grid and an irregular vivid pattern respectively to decrease need of cooling.

5 STATISTICAL EVALUATION

Figure 19 illustrates a data analysis - with 783 BIPV systems over 22 years - on the location of the installation. It is clear to see that the roof is the preferred location for the use of a BIPV system. In all the years - exception 2001 - the share of rooftop installations was always roughly around 50%. BIPV systems in facade and roof and in façade did not play a dominant role. The figures indicate that facade systems are now – in 2022 - gaining in importance.

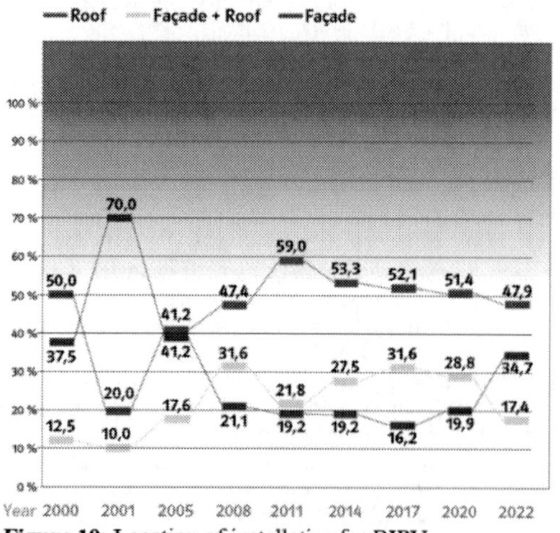

Figure 19: Location of installation for BIPV

Figure 20 indicates the applied technology. For better understanding, it must be considered that at the beginning of the BIPV competitions in 2000 only PV systems were invited to participate, but from 2008 on all types of solar projects. The figure clearly shows the dominance of photovoltaics. The share of PV went up from roughly 60 % to over 90 % in 2022. Pure solar thermal systems never played a significant role. In the years from 2008 onwards, hybrid technology had high numbers of application until 2020; but then dropped to 5%, and PV raised to more than 92 %.

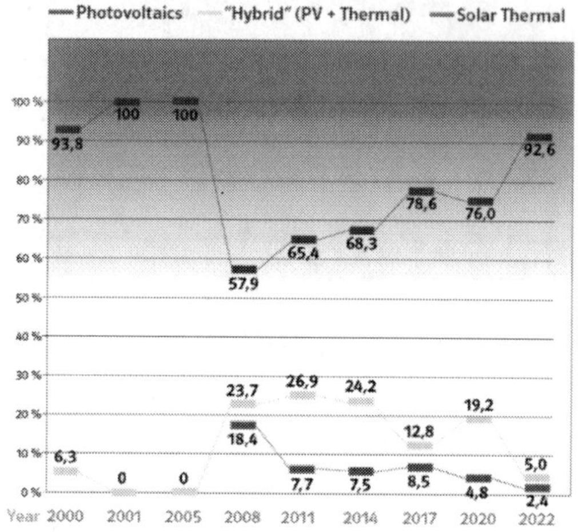

Figure 20: Different types of solar technology

6 CONCLUSION

Nine architectural BIPV competitions confirm the orientation on solar technology in a higher-level overall architectural concept. Today BIPV is a self-evident, unobtrusive application. More and more BIPV systems are combined with modern energy saving technology. PV in façades and on roofs is ready to enable the tremendous amount of new PV capacities also in unobtrusive as in striking architecture.

7 REFERENCES

[1] Roland Krippner (Ed.): Building Integrated Solar Technology; Architectural design with photovoltaics and solar thermal; Detail green books. Munich: DETAIL Business Information GmbH, June/2017)

[2] Krippner, Becker, Flade, Schiebelsberger, Weber: Building Integrated Solar Technology – Evaluating 20 Years of Experience with Solar Buildings from an International Competition. In: ISES Solar World Congress 2021. [virtual], 25 - 29 October 2021. Proceedings. Freiburg: The International Solar Energy Society (ISES) e.V., [July]/2022, S. 140–151

[3] Krippner, Flade: Building-integrated solar technology. Learning from more than 30 years of experience with solar buildings (examples from international competitions). [Session 15 / Sustainable Architectural Design (5)]. In: Waldo Bustamante et al. (Ed.): 36th PLEA 2022. Will Cities Survive? The future of sustainable buildings and urbanism in the age of emergency. Santiago de Chile, November 23 - 25, 2022. Book of Proceedings / Onsite Sessions. Santiago de Chile, November/2022, S. 480–485

10.4229/EUPVSEC2023/4BV.4.1
020334-007

RINGS-BIPV PROJECT: ANALYSIS OF PV SOLUTIONS FOR RETROFITTING BUILDINGS UNDER MEDITERRANEAN CLIMATE CONDITIONS

Nuria Martín-Chivelet[a], Miguel Alonso-Abella[a], Carlos Sanz-Saiz[a], Marina De la Cruz[a], Ana Marcos-Castro[a,b], José Cuenca[a], Jesús Polo[a], Francesca Olivieri[b], Irene Del Hierro López[b], César Bedoya[b], Estefanía Caamaño-Martín[c], Lorenzo Olivieri[b,c]
[a] Solar PV Unit, Department of Energy, CIEMAT, Av. Complutense 40, 28040 Madrid (SPAIN)
[b] Department of Construction and Technology in Architecture, Escuela Técnica Superior de Arquitectura, Universidad Politécnica de Madrid, Av. de Juan de Herrera 4, 28040, Madrid (SPAIN)
[c] Instituto de Energía Solar, Universidad Politécnica de Madrid, Av. Complutense 30, 28040, Madrid (SPAIN)
nuria.martin@ciemat.es; miguel.alonso@ciemat.es; carlos.sanz@ciemat.es; marina.cruz@ciemat.es; cuenca@ciemat.es; jesus.polo@ciemat.es; francesca.olivieri@upm.es; irene.delhierrolop@upm.es; cesar.bedoya@upm.es; estefan@ies.upm.es; lorenzo.olivieri@upm.es

ABSTRACT: The energy renovation of buildings addressing improving their thermal envelope can also integrate photovoltaic systems. Building Integrated Photovoltaics (BIPV) assumes the double functionality of BIPV modules, which are photovoltaic generators and building elements in the role of conventional materials. The ongoing research project *Renovation INnovative Global Solutions with Building Integrated PhotoVoltaics* (RINGS-BIPV), with two multidisciplinary Spanish teams from CIEMAT and the Polytechnic University of Madrid (UPM), respectively, aims to provide efficient energy rehabilitation solutions that include BIPV systems and to improve the modeling and simulation of BIPV systems. The outcomes will help architects and other stakeholders in the decision-making processes to incorporate BIPV solutions at both building and urban scales.
Keywords: BIPV, Energy retrofit, Solar irradiance simulation, PV forecasting, Zero Energy Buildings.

1 INTRODUCTION

Aligned with the objectives established by the European Commission, the Convention on Climate Change, and the UN Sustainable Development Goals, the energy-efficiency renovation of the Spanish building stock, which currently consumes about 30% of the total final energy in Spain, is a priority of the National Energy and Climate Integrated Plan (PNIEC); being the significant increase in renewable energy sharing another key objective, building-integrated photovoltaics (BIPV) appears as the best option for combining the energy retrofitting of buildings with locally produced renewable electrical energy.

However, and despite the potential of BIPV in retrofitting projects, this is still an emerging technology. One main reason is the lack of information and accessible tools to help in the BIPV system design and construction processes: a better understanding of BIPV products and systems and their behavior in rehabilitation projects is needed. The ongoing research project *Renovation INnovative Global Solutions with Building Integrated PhotoVoltaics* (RINGS-BIPV), with two multidisciplinary Spanish groups from the Spanish Centre for Energy, Environmental and Technological Research (Centro de Investigaciones Energéticas, Medioambientales y Tecnológicas, CIEMAT) and the Polytechnic University of Madrid (Universidad Politécnica de Madrid, UPM), respectively, aims to provide efficient energy rehabilitation solutions and to achieve further progress in the modeling and simulation of those BIPV systems. The outcomes will help architects and other stakeholders in the decision-making processes to incorporate BIPV solutions at both building and urban scales.

Building-Integrated Photovoltaics assumes the double functionality of BIPV modules, which become building elements in substitution of conventional materials. Ventilated façades or other construction systems considered when renovating the building envelope can easily include BIPV modules as construction elements. Many examples worldwide illustrate the design and constructive possibilities of BIPV modules and systems [1]. The challenge is that the energy requirements, including the electrical, thermal, and optical ones, should be combined with the constructive ones [2]. Moreover, as the European Standard EN 50583 (parts 1 and 2) details, a BIPV module has to fulfill one or more construction functions according to the Construction Products Regulation. While improving the framework affecting BIPV at European and national scales, products, and systems, there are several reasons for BIPV to stay as a niche market in Spain, as analyzed in the framework of the IEA PVPS Task 15 devoted to BIPV [3].

One of the main weaknesses is the lack of information exchange due to few communication opportunities and low trust among the construction and photovoltaic stakeholders. Aware of that, RINGS-BIPV project will contribute to developing BIPV solutions for building retrofitting but also will improve the BIPV modeling and simulation existing capabilities for a better holistic design, helping in the decision-making process.

2 PROJECT DESCRIPTION

2.1 Team background

CIEMAT team acts as the coordinator of the RINGS-BIPV project. Being experts in testing, characterizing, and modeling photovoltaic (PV) modules and systems, the group is involved in international committees related to PV, BIPV, and solar resource assessment and forecasting, e.g., IEA-PVPS Task 15 and Task 16, IEC TC82, and EERA JP-PV. CIEMAT's main activity in this project is related to the improvement of testing and modeling of BIPV modules and BIPV systems for a holistic simulation of the energy behavior of buildings. Solar radiation incident on the different surfaces available in urban topology is another important part of this project. CIEMAT team has relevant expertise in modeling solar radiation components, transposition models, forecasting solar irradiance, and solar resource assessment, the basis to identify the solar potential on buildings and infrastructure.

The UPM team has been working in recent years on national and international projects aimed at promoting urban regeneration and energy transition in cities. One important research topic of this team is renewable distributed energy generation with photovoltaic solar energy integrated into the built environment.

The UPM founded the project *INTEGRA: Zero energy buildings, an integrative methodology for their design*, with the aim of strengthening the research plans of young researchers by supporting the study of how to design and build zero energy buildings. In the same line, the past year UPM started the project *Transition towards an emission-free university TULE* to support the energy transition of the university by promoting collaborative research between research groups and departments.

2.2 Main objectives

The RINGS-BIPV project intends to deliver further solutions in BIPV systems through research in two main aspects: the design and performance of BIPV solutions and the BIPV modeling, simulation, and forecasting. It aims to contribute to the definition of PV construction technological units and subsystems for renovation projects, adjusting the design to the needs of representative building typologies. The possibilities for flat and sloped roofs, ventilated façades, curtain walls, skylights, and external devices are explored, looking for a good balance between energy performance, ease of use, costs, and architectural quality.

In this area, led by the UPM team, there are two particularly innovative research lines: i) the integration between photovoltaic systems and heritage buildings that require preservation due to their historical, architectural, cultural, or aesthetic value and ii) the combination of photovoltaic systems with Nature-Based metropolitan Solutions (NBS).

Regarding the first line, decision-making concerning BIPV solutions is especially challenging in protected historic urban districts, being in those cases especially important to set the specific requirements for BIPV systems about the constructive, energetic, and visual impact (e.g., texture, grain, color, size).

With regard to the second line, combining photovoltaics and NBS is an interesting area of research, since PV elements and green roofs and façades have traditionally been considered mutually exclusive solutions. However, this combination achieves an ideal symbiotic relationship in which both the promotion of biodiversity and, in general, the ecosystem services provided by vegetation and the production of on-site renewable energy are enhanced, providing the buildings that integrate them with a wide range of environmental, health, aesthetic and economic benefits.

In parallel to the work developed by UPM, CIEMAT focuses on another key aspect of the design of the BIPV system: PV generation forecasting. The energy yield, or energy generated by a PV system over a period, is mostly a function of the received solar irradiation but second-order factors such as module temperature or soiling can play a role, depending on the boundary conditions. BIPV generation forecasting is a more recent activity and entails other difficulties, such as the importance of partial shading and a more complicated determination of the PV module temperature. The adaptability of current forecasting schemes of solar radiation to the BIPV context might play an important role in planning retrofitting and other BIPV actions.

2.3 Work packages and activities

The project is structured through a set of work packages covering the identification of needs, analysis of current standards, assessment of BIPV solutions, characterization of specific systems, and development of new modeling capabilities in both solar cadasters and BIPV systems (Figure 1).

Aware of the lack of dissemination of useful and practical information among BIPV stakeholders, this project dedicates a specific work package to promote the developed BIPV solutions and other outcomes of the project, as well as to serve as a bridge between the construction and the photovoltaic sectors in reference to BIPV. The website https://rings-bipv.com/ will update the main results, while dissemination includes publications, workshops, conferences, and courses.

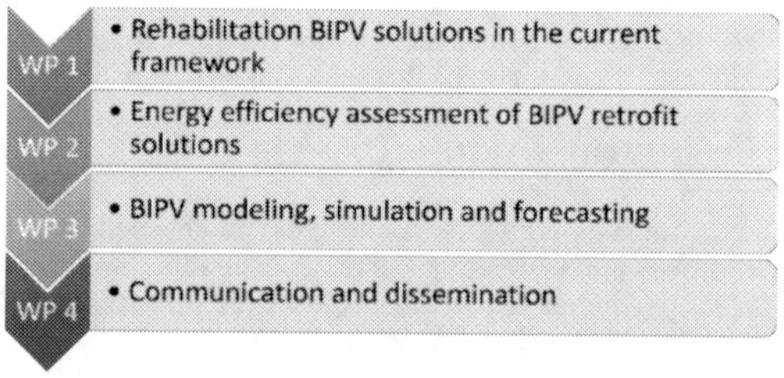

Figure 1: Structure of the project in work packages.

2.4 Experimental work

In order to fulfill the main objectives of the RINGS-BIPV project it is crucial to design specific experimental facilities and test mock-ups to study the performance of the different BIPV systems. The following are some of the experimental facilities of this project:

CIEMAT's Building 42 facility integrates a 27 kW BIPV ventilated façade, having five PV arrays at the top of the south, east and west façades. Three 4 kW PV arrays (each one with 2 parallel strings of 7 modules -7s x 2p-electrical configuration) are on the east façade, another 4 kW array (7s x 2p) on the west one, and one 8 kW array (7s x 4p) on the south façade. The main electrical parameters (i.e., output current, voltage, and power) and the relevant meteorological variables have been monitored since 2016. The in-plane irradiance (*POA*) has been measured at each cardinal point with reference cells.

Two testing cabins are designed to study the performance of semi-transparent PV windows, including their thermal performance and the validation of temperature models (Figure 2).

Figure 2: Testing cabins for BIPV windows and west view of the CIEMAT's BIPV ventilated façade.

The *BIfacial CANopie* (BICAN) testing facility is designed to study the performance of bifacial modules when applied as PV canopies on rooftops. The experimental setup includes ten modules, two of them as references. Reference bifacial modules provide front, rear, and total irradiance by blocking the light to selected strips of cells. In addition, ten reference cells measure front and rear irradiance at different points.

Also, a colored BIPV façade system has been mounted to test the performance of such modules under real operation in a ventilated façade configuration (Figure 3).

Figure 3: Colored ventilated façade developed in the project at CIEMAT.

3 PRELIMINARY RESULTS

3.1 Performance ratio of BIPV systems

Considering the monitored available data from the three BIPV systems of the CIEMAT building, the evolution of the daily performance ratio (*PR*) shows no significant tendency since 2017. *PR* characterizes the efficiency of a PV system when compared to the theoretical performance with the absence of losses. In this case, temperature is the most impacting parameter on the west façade of the building, while partial shading is a challenge on the east one due to nearby deciduous trees. All these effects are analyzed in the project.

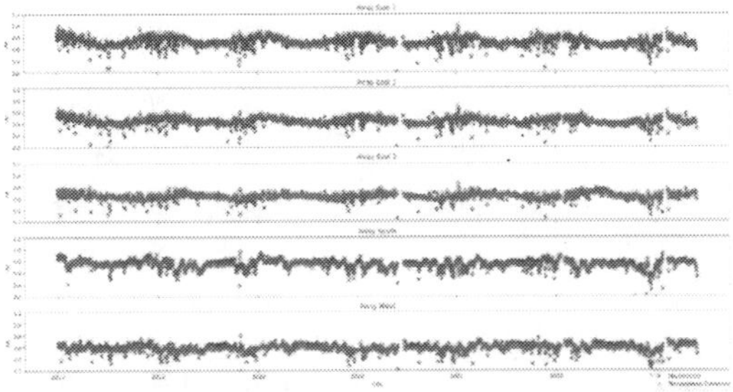

Figure 4: Daily Performance Ratio (*PR*) of the CIEMAT building BIPV arrays since 2017. The temperature-corrected *PR* is also included.

3.2 PV generation forecasting

Photovoltaic generation forecasting directly based on time series data from one of the BIPV vertical façades has been studied using machine learning algorithms based on decision trees. The forecasting scheme employs the *skforecast* library from the *Python* environment, which facilitates the implementation of different schemes for both deterministic and probabilistic forecasting applications [4].

4 REFERENCES

[1] T. Reijenga et al., "Successful Building Integration of Photovoltaics - A Collection of International Projects", IEA, 2020.

[2] N. Martín-Chivelet et al., "Building-Integrated Photovoltaic (BIPV) products and systems: A review of energy-related behavior," Energy Build. 262 (2022).

[3] N. Martín-Chivelet, L. García, E. Martín-Caamaño, and D. Racero. IEA Report IEA-PVPS T15-13:2022 (2022).

[4] J. Polo, N. Martín-Chivelet, N., M. Alonso-Abella, C. Sanz-Saiz, J. Cuenca, M. de la Cruz, "Exploring the PV Power Forecasting at Building Façades Using Gradient Boosting Methods", Energies 16 (2023) 1495.

5 ACKNOWLEDGEMENT

The authors would like to thank the RINGS-BIPV Project (PID2021-124910OB-C31 and – C32), which is funded by the Ministerio de Ciencia e Innovación (Spain). The authors would also like to recognize the efforts, research and contributions of the expert groups of the IEA PVPS Program, in particular those corresponding to Task 15 (BIPV) and Task 16 (Solar Resource), where the authors have an active collaboration.

EXPERIMENTAL PERFORMANCE EVALUATION OF A BIPVT ROOF TILE SYSTEM

L. Clasing[1], C. von Holst[1], J. Münzberg[2], P. Hakenberg[2], C. P. Dick[3], S. Leyer[4], J.-R. Hadji-Minaglou[4], U. Blieske[1]

[1]TH Köln, Faculty of Process Engineering, Energy and Mechanical Systems, Betzdorfer Str. 2, Cologne, Germany
[2]paXos Consulting & Engineering GmbH & Co. KG, Karl-Benz-Str. 9, Langenfeld, Germany
[3]TH Köln, Faculty of Information, Media and Electrical Engineering, Betzdorfer Str. 2, Cologne, Germany
[4]University of Luxembourg, Faculty of Science, Technology and Medicine, Department of Engineering

ABSTRACT: Building-integrated PV can greatly extend the potential installation capacities of building façades and roof covers. However, integrating PV into buildings has a decisive disadvantage regarding the performance of solar cells: Because there is no rear ventilation compared to standard rooftop PV modules, the solar cells show lower performance due to elevated module temperatures. Combining these systems with solar thermal systems results in a hybrid system. In this work, a performance evaluation of an improved hybrid solar roof tile that has the characteristics of a standard roof tile and thus can replace the building envelope is presented. Therefore, the examined system is compared to a standard rooftop PV system by using performance ratio metrics. These systems are installed on two identical roof structures in Cologne, Germany, and measured continuously in 20-second intervals. The system is actively rear-ventilated in order to decrease module temperature and utilize the pre-heated ambient air for a heat pump system. The results of the performance ratio calculation for a period of 9 months show only slightly lower average values of 1.1% for the solar roof tiles. The results show a relative improvement between $9\%_{rel}$ and $14\%_{rel}$ compared to the values of building-integrated systems found in the literature.
Keywords: BIPV, BIPVT, outdoor performance, performance ratio

1 INTRODUCTION

The application of building-integrated photovoltaic (BIPV) systems is gaining increasing attention due to custom design possibilities in recent years [1-3]. One major disadvantage of these systems is the lack of natural rear ventilation, which leads to an increase in PV solar cell operating temperature and thus a reduction in efficiency compared to standard rooftop PV systems. Photovoltaic and thermal (PVT) systems consist of photovoltaic modules that combine the use of solar irradiation to generate electrical and thermal energy. The implementation of PVT systems in buildings (BIPVT) tackles these problems by utilizing the waste heat from the PV solar cells. Additional promising benefits are a decrease in total system costs and primary energy savings, resulting in reduced environmental impact.

Despite the rapidly growing interest, the current state of research on BIPVT lacks not only evaluations on economic and environmental aspects but also studies on the electrical and thermal performance of the systems, as well as comprehensive studies that represent real operating conditions. Therefore, experimental analysis and evaluation of the performance of such systems under real operating conditions are crucial for building owners and investors. In previous work, an improved air-based BIPVT roof tile design has been presented and modeled [4-7]. These studies show promising improvement in coefficient of performance (COP) by connecting to an air-source heat pump (ASHP) system. The objective of this study is to compare the annual performance of the presented improved BIPVT roof tile system with active rear ventilation and a standard rooftop PV system.

2 BIPVT ROOF TILES

The presented improved BIPVT roof tile design features a durable, heat-resistant, accessible, replaceable, efficient, and aesthetic solution that has the characteristics of a standard roof tile and thus can replace the building envelope and be installed by roofers. Fig.1 shows the roof tile design studied in this work. This design consists of two main parts: A custom-designed glass-glass module with monocrystalline silicon PV cells and a sheet metal body as a frame for the module. The active surface area is 0.1 m² with a rated power output of 14.5 Wp. The sheet metal body allows the glass-glass module to be moved for installation and maintenance purposes. Fig. 2 shows the air duct between these main components that is used for active rear ventilation.

Figure 1: BIPVT roof tile with sheet metal body (a) and glass-glass PV module (b).

Figure 2: Cross-sectional profile of the BIPVT roof tile with a rear ventilation air-duct system.

3 EXPERIMENTAL DETAILS

In order to compare the performance of the presented BIPVT roof tile system with a standard rooftop PV system, a measurement campaign was conducted. Therefore, two roof constructions with identical tilt angle as well as orientation (45° Tilt, 172° Azimuth (North = 0°)) are installed at a test site in Cologne, Germany (Fig. 4). The location is classified as oceanic temperate climate (Cfb) according to the Köppen-Geiger climate classification [8]. The experimental test roof is equipped with 192 BIPVT roof tiles, and the electrical system is arranged in two strings (2x1.39 kWp, Fig. 3), resulting in an irradiated surface area of 19.2 m². The modules in the left string are equipped with a standard junction box, while the modules in the right string are optimized by a maximum power point tracker on module level (MPPT). However, only the left string is used for the evaluation to ensure comparability with the rooftop PV system. The air ducts of the roof tiles are vertically connected in 12 parallel systems from the bottom to the top in an open-loop system.

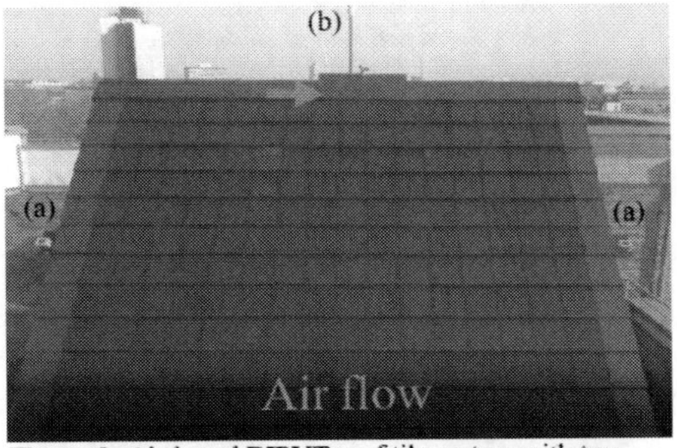

Figure 3: Air-based BIPVT roof tile system with two monocrystalline silicon irradiance sensors (a) and all-in-one weather station (b). The pre-heated ambient air is collected by a horizontal pipe in the ridge of the roof.

A manifold placed horizontally at the top of the roof is connected to each vertical row. Furthermore, the manifold is connected to an ASHP to use the increased temperature levels of the pre-heated ambient air to improve its performance. An additional exhaust fan is used to support the ASHP ventilation system.

Figure 4: Experimental setup with a building-integrated PV and solar thermal roof tile system (a) and a standard rooftop PV system (b) at a test site in Cologne, Germany. A dual-axis solar tracker with pyranometer (c) is used to monitor the irradiance on the system.

The reference roof is covered with conventional concrete roof tiles and a standard rooftop PV system (1.55 kWp, Fig. 4 (b)) on a mounting structure. Both PV systems are grid-connected with identical inverters. The systems are equipped with in-plane global irradiance as well as temperature sensors. In addition, a dual-axis solar tracker with a class A pyranometer and pyrheliometer is used to monitor the direct normal and diffuse horizontal irradiance (Fig. 4 (c)). Likewise, an all-in-one weather station is installed at the test site. Measurements of the electrical output of the PV systems are done at the AC grid feed-in. Data are recorded continuously in 20-second from March to November 2022 in order to account for seasonal impacts on BIPVT system performance. The monitoring system can be classified as a class B system according to the IEC standard 61724 [9].

The performance ratio PR is the quotient of the final AC yield of the system Y_f and its reference yield Y_r at standard test conditions (STC; 1000 Wm^{-2}, AM1.5g, 25°C) and indicates the total effect of losses on the system. The PR is calculated according to the IEC standard 61724 and used in this work to evaluate the electrical performance of the BIPVT system. Accordingly, the PR is calculated by

$$PR = \frac{Y_f}{Y_r} = \frac{E_{out} / P_0}{H_i / G_{i,STC}} = \frac{\sum_k P_{out,k} \cdot t_k}{\sum_k \frac{P_0 \cdot G_{i,k} \cdot t_k}{G_{i,STC}}}$$

where E_{out} is the AC power output of the system, P_0 is the rated power output of the system, G_i is the in-plane global irradiance, t is the time factor of the data and $G_{i,STC}$ is the in-plane global irradiance according to STC. However, - studies have found that the PR is highly dependent on climatic conditions at a particular site [10]. Therefore, the PR_{STC} is calculated with a correction to the STC module temperature (25°C) to reduce the thermal impacts on the system performance. PR_{STC} is calculated by

$$PR_{STC} = \frac{\sum_k P_{out,k} \cdot t_k}{\sum_k \frac{\left(c_{k,STC} \cdot P_0\right) \cdot G_{i,k} \cdot t_k}{G_{i,stc}}}$$

with

$$c_{k,STC} = 1 + \gamma \cdot \left(T_{m,k} - 25\,°C\right)$$

where c_{STC} is the temperature correction factor, γ is the power temperature coefficient of the PV module and T_m is the module operating temperature. Accordingly, measured PV module temperature values of both systems are used to correct the operating temperature effect.

4 RESULTS AND DISCUSSION

Fig. 5 shows the PR calculation results for the 9-month measurement period. The PR is calculated according to the IEC 61724 standard for both systems, the BIPVT roof tile system and the standard rooftop PV system (reference). The average values for a period of 9 months of PR analysis show 85.5% and 86.6% for the BIPVT system and standard PV system, respectively. The largest deviation was observed in August, with around -3%. While the values are lower in the summer season, they are higher in the winter season, which is typical for the oceanic temperate climate conditions in the northern hemisphere. However, the results are limited to the site under consideration due to the climatic effects on PR. The average temperature-corrected PR_{STC} shows 88% for both systems for the same time period. Accordingly, the differences in electrical output can be explained by the increased module temperature of the BIPVT system. Moreover, the similar PR_{STC} averages confirm the comparability of the obtained PR values.

10.4229/EUPVSEC2023/4BV.4.7
020340-002

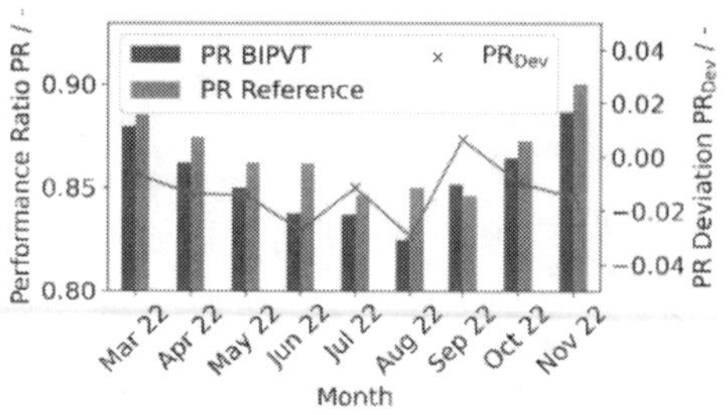

Figure 5: Performance Ratio PR analysis of a BIPVT roof tile system and a standard rooftop PV system (reference) for 9 months of measurement data under north temperate climate conditions (Cfb).

Tab. I summarizes the seasonal and overall average values of the PR calculations for both systems. While the deviation of the average PR from the BIPVT system compared to the reference system is about -1%$_{rel}$ in spring and autumn, it is around -2%$_{rel}$ in summer. The overall averages show a deviation of -1.4%$_{rel}$ for the 9-month period.

Table I: Seasonal and overall average values of the Performance Ratio (PR).

Season	Average PR		
	BIPVT	Reference	Dev.
Spring	86.4%	87.4%	-1.1%$_{rel}$
Summer	83.3%	85.3%	-2.3%$_{rel}$
Autumn	86.7%	87.3%	-0.7%$_{rel}$
Overall	85.5%	86.6%	-1.4%$_{rel}$

Recently published performance studies of BIPV systems without rear ventilation under real operating conditions showed an annual average PR between 74% [11] and 78% [12]. Therefore, a relative improvement between 9%$_{rel}$ and 14%$_{rel}$ in PR can be reported for the presented BIPVT roof tile design in the experimental measurement campaign.

5 CONCLUSIONS AND OUTLOOK

BIPV systems without rear ventilation operate at higher module temperatures compared to standard rooftop systems with natural rear ventilation. These increased operating temperatures lower performance and can cause module deterioration. In this work, an experimental performance evaluation of a BIPVT and conventional rooftop PV system is done for a period of 9 months.

It can be concluded that the presented BIPVT system has a significant performance improvement between 9%$_{rel}$ and 14%$_{rel}$ compared to the values of BIPV systems found in the literature. These improvements indicate higher economic viability and thus facilitate the market introduction of this system. In addition to the application-specific benefits of the improved design, a higher level of acceptance among building owners and architectural professionals is also achieved.

Future work should consider a comparison of different BIPVT roof tile designs. In addition, the thermal energy of the collected air used in the ASHP should be included to determine an overall energy balance of the improved system.

6 ACKNOWLEDGEMENTS

This project has been financed by the European Fonds for Regional Development North Rhine-Westphalia (EFRE.NRW, Code: EFRE-0801561, Reference no. EU-2-2-048B)

7 REFERENCES

[1] N. Sánchez-Pantoja, R. Vidal, and M. C. Pastor, “Aesthetic perception of photovoltaic integration within new proposals for ecological architecture,” Sustainable Cities and Society, vol. 39, pp. 203–214, May 2018.

[2] M. C. M. Probst and C. Roecker, “Criteria and policies to master the visual impact of solar systems in urban environments: The LESO-QSV method,” Solar Energy, vol. 184, pp. 672–687, May 2019.

[3] H. Kaan and T. Reijenga, “Photovoltaics in an architectural context,” Progress in Photovoltaics: Research and Applications, vol. 12, no. 6, Art. no. 6, Sep. 2004.

[4] L. Clasing, C. Dick, U. Blieske, P. Hakenberg, S. Leyer, and J.-R. Hadji-Minaglou, “Analysis of a Novel Improved BIPV/T Roof Tile Design,” in Proc. 15th Conference on sustainable development of energy, water and environment systems (SDEWES), 2020.

[5] L. Clasing, P. H. Yam, C. Ziller, and U. Blieske, “Dynamic Simulation Model of an Air-Based BIPVT Roof Tile System,” in Proc. 8th World Conference on Photovoltaic Energy Conversion, Milan, Italy, 2022, pp. 662–665.

[6] M. Bartkowski, H. R. Saleemi, L. Clasing, C. Brosig, E. Waffenschmidt, U. Blieske, „Development and validation of a performance model for simulation of an air-based BIPVT roof tile system using opensource libraries,“ in Proc. 8th World Conference on Photovoltaic Energy Conversion, Milan, Italy, 2022, pp. 716–723.

[7] C. Brosig, L. Clasing, U. Blieske, and E. Waffenschmidt, “Modeling a novel building-integrated PVT-air-collector system coupled to a heat pump using open source libraries,” in Proc. 16th Conference on sustainable development of energy, water and environment systems (SDEWES), Dubrovnik, Croatia, October 2021.

[8] H. E. Beck, N. E. Zimmermann, T. R. McVicar, N. Vergopolan, A. Berg, and E. F. Wood, “Present and future Köppen-Geiger climate classification maps at 1-km resolution,” Scientific Data, vol. 5, no. 1, Art. no. 1, Oct. 2018.

[9] IEC61724-1, “Photovoltaic system performance - Part 1: Monitoring,” International Electrotechnical Commission, 2021.

[10] N. H. Reich, B. Mueller, A. Armbruster, W. G. J. H. M. van Sark, K. Kiefer, and C. Reise, “Performance ratio revisited: is PR>90% realistic?,” Progress in Photovoltaics: Research and Applications, vol. 20, no. 6, Art. no. 6, Jan. 2012.

[11] N. M. Kumar, M. Samykano, and A. Karthick, “Energy loss analysis of a large scale BIPV system for university buildings in tropical weather conditions: A partial and cumulative performance ratio approach,” Case Studies in Thermal Engineering, vol. 25, p.

100916, Jun. 2021.
[12] A. Alazazmeh et al. "Real-time data-based performance analysis of a large-scale building applied PV system" in Energy Reports Vol. 8, pp. 15408-15420, 2022.

10.4229/EUPVSEC2023/4BV.4.7

FAILURE ROOT CAUSE INVESTIGATION IN CORRUGATED FLEXIBLE THIN FILM MODULES

Remi Aninat[1]*, Aldo Kingma[1], Helena Kirchner[1], Leonard Simeonov[1], Hugo Mallet-Guy[1], Arthur Vandenbroucke[1], Simona Villa[1], Dorrit Roosen-Melsen[1], Monique van den Nieuwenhof and Mirjam Theelen[1]
[1]*TNO, HTC 21, Eindhoven*
**remi.aninat@tno.nl*

ABSTRACT:
$Cu(In,Ga)(Se,S)_2$ (CIGS) - based devices, which, in rigid modules, already have a solid track-record and demonstrated high reliability, are also very promising for the emerging field of curved vehicle- and building-integrated photovoltaics (VIPV and BIPV). However, such new applications imply additional stresses compared to flat, rigid modules, that can be induced for example by bending or coefficient of thermal expansion mismatch. Prior to the present study, we laminated flexible CIGS modules onto corrugated rooftiles and installed them outdoor for power generation for three years. We then unpackaged the modules using an in-house procedure and found two characteristic defects. The focus of this study was on the replication of these defects in the lab, and the search for a realistic root cause. Both defects were found to arise from the sample flexibility, and should therefore not be specific to CIGS. The first defect, which was relatively mild and due to very local front grid delamination, could be replicated in thermal cycling tests. The second one, which severely impacted the performance by causing kinks in IV curves, consisted in local delamination between the Mo back contact and the CIGS absorber layer. We investigated mechanical, thermal and environmental stresses to replicate this last defect. Although thermal cycling and sequential testing failed to replicate it, tensile and hail testing revealed that the Mo/CIGS interface, in the present experiment, was indeed the weakest in the stack and could delaminate when high strain were applied locally and combined with damp-heat conditions.

Keywords: BIPV, thin-film photovoltaics, reliability

1 INTRODUCTION

Although thin film photovoltaic (PV) still represents a small minority of the installed capacity to date, it allows for new applications compared to c-Si modules, especially when applied on curved surfaces as in vehicle- and building-integrated PV (VIPV and BIPV)[1-3]. CIGS, in particular, has been integrated into such curved surfaces and tested in the past [1, 4]. Although rigid CIGS devices have demonstrated high stability with respect to damp-heat and potential-induced degradation [5-7], the impact of mechanical stress in flexible CIGS has been mostly neglected. Yang et al. [3] studied the effect of mechanical strain in flexible CIGS cells, through tensile testing. However, a study of realistic root causes for mechanically induced outdoor degradation in flexible CIGS modules is still missing. Here, we focus on two defects induced in corrugated modules during outdoor deployment, and try to replicate this defect using various standardized test methods on replicas of the modules deployed outdoors.

2 EXPERIMENTAL METHODS

The modules connected and deployed outdoors followed the characterization protocol described in Figure 1.

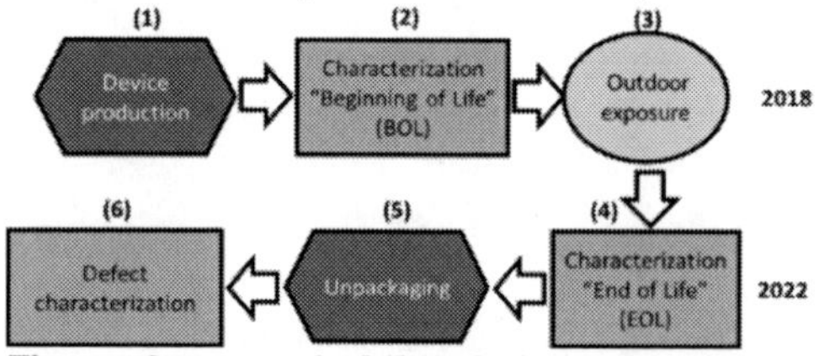

Figure 1: protocol followed for preparation and monitoring of the outdoor modules

Apart from the tensile testing, the modules produced for defect replication all had the same structure as the modules originally exposed to outdoor conditions for three years. A typical corrugated module used in these tests is shown in Figure 2. The module stack is the following: roof tile element (Eternit, Profile 6, made of compressed fiber cement), 160 µm Alu/PET back sheet, 450 µm polyolefin (PO) encapsulant, commercial flexible CIGS solar cells on stainless steel substrate and with a Cu/Ni front grid, 200 µm PO and 125 µm thick PET-based transparent front sheet.

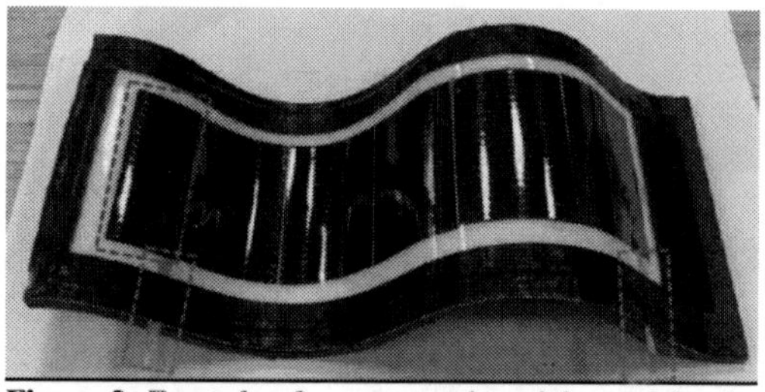

Figure 2: Example of a corrugated module used for the tests. The first cell, boxed in dashed red, is a dummy (not electrically connected), while the tabs are boxed in blue.

For the tensile tests, two products were tested. The first was a simple cell, consisting of its stainless steel substrate and the entire cell stack up to and including the grid-fixation layer, designed to make the front grid adhere to the TCO. The second product was the same cell with, additionally, laminated front and back sheets, in a flat configuration.

For electrical characterization of individual cells, the modules were separated from their supporting tile element, the backsheet was removed and the cells individually contacted from the back, using in-house developed protocols.

The damp-heat testing was carried out in a climate chamber at 85°C and 85 % relative humidity. For UV exposure, the average irradiance in the 280 nm - 400 nm range was 184W/m^2. Humidity freeze was carried out in a PL3J. A humidity-freeze cycle consisted of 20 hours of

Damp Heat exposure followed by 4 hours at -40°C, and humidity settings in accordance with IEC61215. Tensile testing was undertaken in an Instron tensile tester with 10kN loadcell, on 15.5 cm long samples. Finally, the hail tests were carried out using an in-house setup, where 3.5 cm diameter ice balls (made according to the IEC 61215 specifications) were propelled at 27 m/s using 0.6 bar pressure.

3 DEFECTS IN OUTDOOR MODULES

The current-voltage curves at beginning of life (BOL), i.e. before outdoor deployment, and end of life (EOL, i.e. after being taken down from the roof, still packaged) are shown in Figure 3 for a representative module. The corresponding electroluminescence (EL) image is shown in Figure 4. Two features stand out in the EL of the EOL samples, which were not initially present in the at BOL (for more data, see [8]). The first one ("dark interconnects", or **DI**) is a darkening on one side of the interconnect between cells. The second defect ("Sharp border Black Areas, or **SBA**) was characterised by areas without luminescence in EL. Unlike shunts, which are point-like features with a gradual EL signal intensity gradient in their surroundings, defects SBA had an extended area with no luminescence inside, with a sharp border with the luminescent area outside the defect (Figure 4). Scanning Electron Microscopy (SEM) and Energy-dispersive X-ray Spectroscopy (EDS) were carried out on the sample after front sheet removal, showing that the black areas in EL features are due to a delamination between the Mo and the CIGS layer. The edge of the SBA defects is also characterised by a fern-like structure in the encapsulant, most likely due to viscous fingering (not shown).

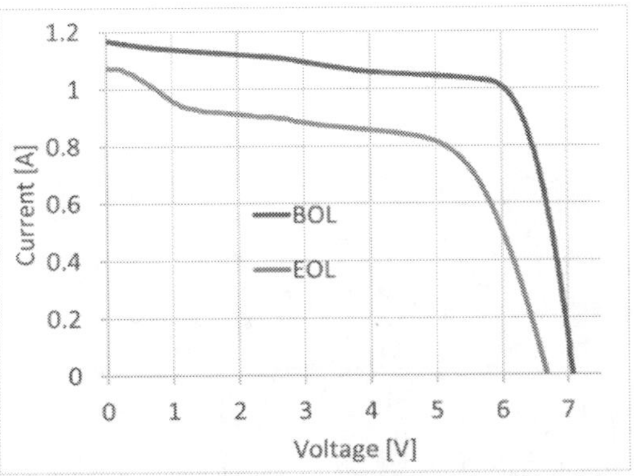

Figure 3: IV characteristic of a module deployed outdoor for three years, showing the defect we try to replicate.

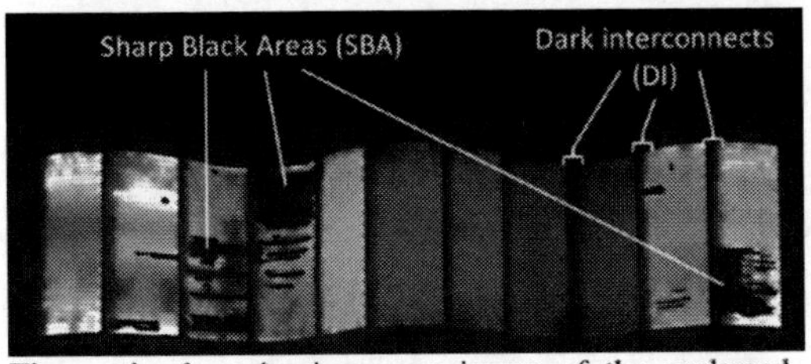

Figure 4: electroluminescence image of the packaged corrugated module shown in Figure 3, showing the two types of defects to replicate

In the following sections, we will describe the different tests made to try and replicate defects DI and SBA.

4. RESULTS OF ACCELERATED STRESS TESTS

4.1 Climatic and thermal cycling stress

The first series of tests aimed to determine whether the thermal expansion/contraction of these flexible modules, in combination with other possible stresses, could have induced defects DI and SBA.

The first test was a thermal cycling experiment, where the samples were exposed to temperature cycles from -40°C to +85°C (3h per cycle) following IEC 61215 specifications. The EL pictures after 100 cycles and 600 cycles are shown in Figure 5. The main observation is that the interconnect area becomes darker in EL after 600 cycles, associated with an increase in series resistance (Rs). The similarities between this defect and defect DI suggest that defect DI can be simply explained by thermal stresses in the laminate. Internal experiments (not shown here) suggest that this Rs increase results from partial delamination of the grid at the interconnect, due to a combination of the interconnect design and the thermal stress induced by thermal cycling. We note that the same thermal cycling tests on an unsupported flat module (i.e., without the corrugated shape and rigid support) yielded a faster DI development. This indicates that the integrated module is less prone to DI than in its support-free form.

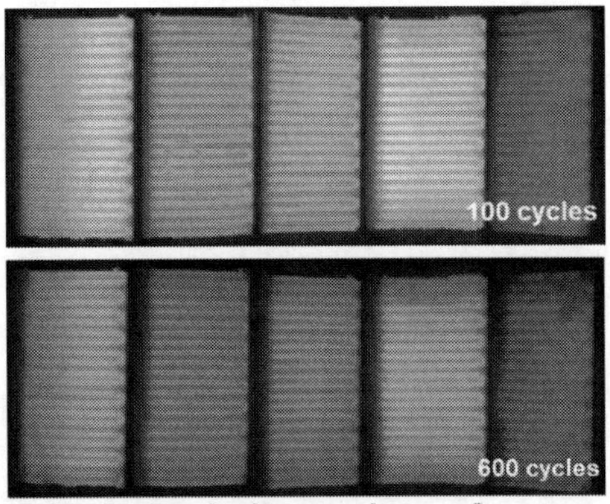

Figure 5: electroluminescence images of a corrugated module after 100 cycles and 600 cycles of thermal cycling.

To test the degradation further, a so-called sequential testing was carried out, whereby a sequence of 200 hours of damp-heat (DH), 60kWh of ultraviolet light exposure and 10 cycles of humidity-freeze (HF) were applied (Figure 6). Although no significant degradation was visible for the first two steps, HF caused a sudden increase in Rs, as well as a slight drop in short-circuit current (Isc), likely due to partial contact degradation in the rightmost cell.

Apart from the Rs increase observed in both experiments, no other damage was found in the devices. In particular, no traces of defect SBA formation were detected.

10.4229/EUPVSEC2023/4BV.4.8
020342-002

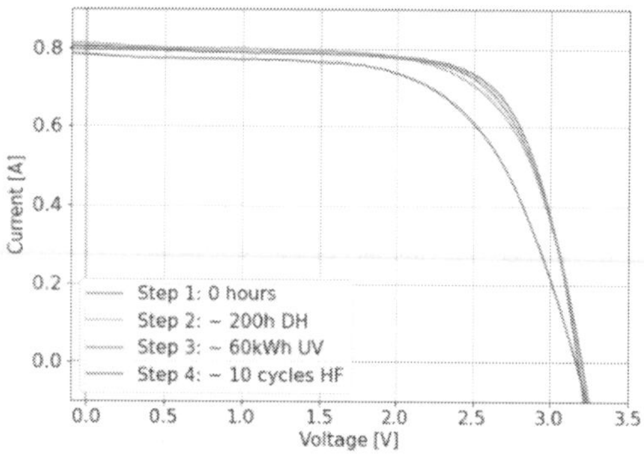

Figure 6: I-V curves of a corrugated module at different steps of the sequential test

4.2 Tensile stress testing

In order to test more extreme tension than what the thermal cycling could produce, and determine the weakest interface in the stack, tensile testing was also carried out (Figure 7).

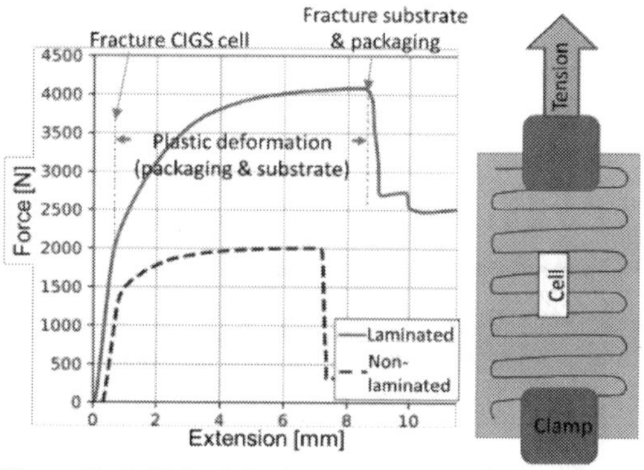

Figure 7: (left) load-displacement curves of a laminated and a non-laminated cell. (Right) schematic of the way in which the sample was clamped and the tension applied

First of all, the laminated cells are shown to withstand higher loads before giving in than unpackaged cells, due to the elastic and plastic properties of the packaging elements. Although both laminated and non-laminated cells show a similar behaviour, let us focus on the laminated (i.e. packaged) case, which is the most similar to the modules deployed outdoor. The first part of the deformation is elastic (i.e. no permanent damage has occurred), characterised by a relatively linear dependency between force applied and extension. At around 2kN, the CIGS material starts giving in and the sample enters a plastic deformation regime, until the packaging itself and the stainless steel substrate break as well. After the experiment, the samples showed clear signs of delamination at the CIGS/Mo interface (Figure 8), similarly to defect SBA but extended to the entire sample area.

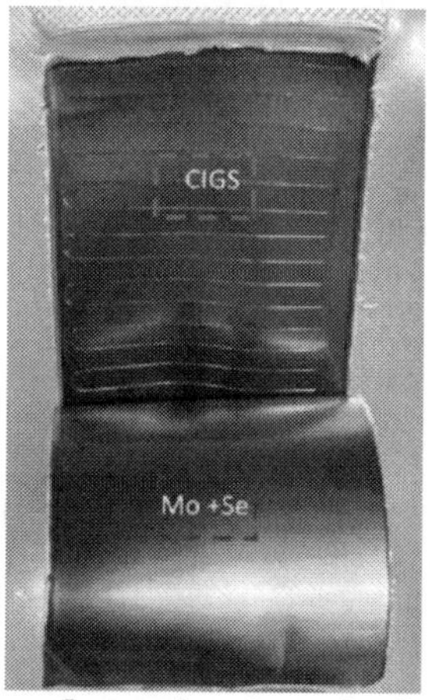

Figure 8: a packaged sample after tensile testing. The dashed boxes indicate the elements detected by EDS on the two sides of the delaminated interface.

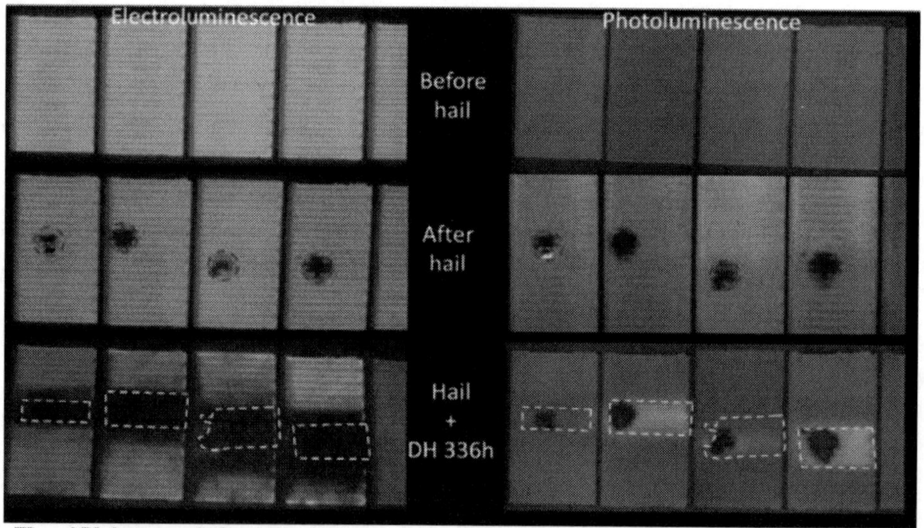

Figure 1: EL and PL images at before hail, after hail and after 336 hours of damp heat. Note that the PL image before hail is missing.

10.4229/EUPVSEC2023/4BV.4.8
020342-003

4.3 Hail testing

At this stage, we have demonstrated that only very high stresses could yield delamination at the CIGS/Mo interface. However, the SBA features observed in the outdoor sample were local features. Furthermore, such a high uniform stretching force is not a realistic scenario in the BIPV application considered here. Therefore, we next investigated how local strain could have potentially led to high, local stress. The most obvious situation where such stress could be induced is hail, where the hail balls transfer a large amount of energy on very small areas. A hail test was therefore applied to a fully packaged corrugated PV rooftile. The EL images are shown in Figure 9.

After hail, dark areas appear (red dashed circles) at the hail impact locations, but only a small feature (blue square) was actually similar to defect SBA. This small SBA region is believed to be the effect of a sharp chip from the hail ball, causing more local and higher gradient strain. The other dark regions circled in red, on the other hand, show the luminescence gradient usually associated with shunting or grid contact loss.

After damp heat, however, SBA-like defects (outlined in yellow) do develop in and around all of the original hail impact locations. IV results (Figure 10) show a dramatic decrease in Isc. However, we note that the strong kink typically associated with defect SBA is not visible, possibly because every cell of the module was similarly affected by the hail test, which was not the case for the original modules exposed outdoor.

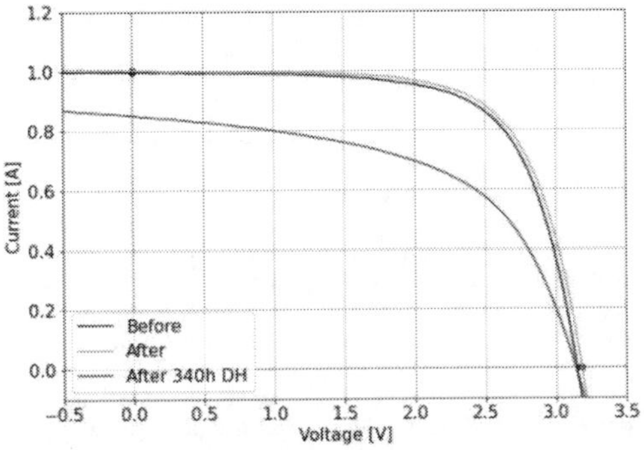

Figure 10: IV curves before and after hail testing, and after subsequent damp heat exposure

3.5 Root cause and mitigation for defect SBA

The hail testing results show that a defect similar to SBA can be generated by hail. This does not necessarily mean that hail is the root cause behind the SBA features found in our outdoor samples. Indeed, it simply supports the hypothesis that a strong local strain is the root cause. This strong local strain can be the result of hail, mechanical impacts during handling, or even of stacking up modules for example, where top weight associated with features from the junction box can initiate a degradation. Another takeaway from the hail test is that the defect is not immediately correlated with performance loss, but only becomes clearly detectable in EL and IV after some time. We would like to emphasize that both defects, DI and SBA, could in principle occur in any other material than CIGS, since the mechanical properties of the PV layers and packaging (and interconnect design in the case of DI) are the only relevant properties.

Possible mitigation measures to prevent SBA formation would therefore include:

- Ensuring good adhesion between the PV active layers, to decrease the risk of stress-induced delamination
- Tuning the front and back sheet material properties to prevent sharp local strain
- Avoiding situations where such sharp local strain might be applied to the PV stack

4 CONCLUSION

We have shown that two specific defects form in our corrugated PV rooftile elements when deployed outdoor. The first one, DI, characterized by dark interconnects in EL, has a relatively minor effect and only affects the series resistance. It could be replicated by thermal cycling. The other defect, SBA, more detrimental to the performance, is characterized by sharp black regions in electroluminescence and causes strong kinks in the IV curve. This defect, due to delamination at the CIGS/Mo interface, could only be replicated in the lab when the module was subjected to extreme tensile stresses (>2kN), or to sharp local strain in hail testing. The second scenario (local strain) is the most likely one, since hail or other local mechanical loads are a frequent occurrence in outdoor deployment.

5 REFERENCES

[1] B.P. Jelle, Building integrated photovoltaics: A concise description of the current state of the art and possible research pathways, Energies, 9 (2015) 21.
[2] A.K. Shukla, K. Sudhakar, P. Baredar, A comprehensive review on design of building integrated photovoltaic system, Energy and Buildings, 128 (2016) 99-110.
[3] C. Yang, K. Song, X. Xu, G. Yao, Z. Wu, Strain dependent effect on power degradation of CIGS thin film solar cell, Solar Energy, 195 (2020) 121-128.
[4] A.K. Shukla, K. Sudhakar, P. Baredar, Recent advancement in BIPV product technologies: A review, Energy and Buildings, 140 (2017) 188-195.
[5] J. Kettle, M. Aghaei, ... & M. J. Theelen, Review of technology specific degradation in crystalline silicon, cadmium telluride, copper indium gallium selenide, dye sensitised, organic and perovskite solar cells in photovoltaic modules: Understanding how reliability improvements in mature technologies can enhance emerging technologies. Progress in Photovoltaics: Research and Applications, 30(12) (2022), 1365-1392.
[6] P. Yilmaz, J. Schmitz & M. Theelen, Potential induced degradation of CIGS PV systems: A literature review. Renewable and Sustainable Energy Reviews, 154 (2022), 111819.
[7] D.C. Jordan et al., Compendium of photovoltaic degradation rates. Progress in Photovoltaics: Research and Applications, 24(7) (2016), 978-989.
[8] A. Kingma, R. Aninat, M. Theelen et al., Post-mortem analysis of building-integrated flexible thin film modules, *under submission*.

10.4229/EUPVSEC2023/4BV.4.8
020342-004

40th European Photovoltaic Solar Energy Conference and Exhibition

This presentation was selected by the Sc. Committee of the EU PVSEC 2023 for submission of a full paper to one of the EU PVSEC's collaborating peer-reviewed journals.

VERTICAL BIFACIAL PV FOR FLAT ROOFTOPS – ENERGY YIELDS FROM PROTOTYPES AND PILOTS IN EUROPE

Sean Erik Foss[1], Simona Petroncini[1], Aria Yadmelat[1], Christoph Seiffert[2], Manuel Auliano[1], Mari B. Øgaard[2], Trygve Mongstad[1]
[1]Over Easy Solar AS, Filipstad Brygge 1, 0252 Oslo, Norway
[2]Institute for Energy Technology, Instituttveien 18, 2007 Kjeller, Norway

ABSTRACT: We present here the main findings from pilot testing of vertical bifacial solar PV installations over the last 18 months. The studies are performed to validate the energy yield and support the commercial development of the concept of the new patent-pending prefabricated vertical bifacial PV Units (VPV Units) developed by Over Easy Solar. From July 2021 to April 2023, Over Easy Solar has commissioned 12 VPV Unit installations, from Tromsø, Norway at 69°N to Alicante, Spain at 38°N. Different azimuths have been used in the different locations, and at one location, two different azimuths have been installed. The sites are equipped with monitoring equipment, or they are located close to a conventional monofacial rooftop installation that can be used for comparison and interpretation of the results.
The preliminary results suggest that installations with VPV Units may give the same or higher specific yield (kWh/kW_p/year) compared to a conventional monofacial rooftop installation. In Norway, they provide high specific yields, comparable to or up to 51 % higher than that of conventional flat rooftop solar installations. The yield depends on several factors, importantly the latitude, amount of clear sky in the local climate, azimuth, albedo and bifaciality. The specific yield is especially good at high latitudes and dry climates, or on rooftops with high albedo. The performance of VPV Units under cloudy conditions is similar to that of monofacial rooftop panels, while VPV Units generally outperform conventional monofacial rooftop PV on sunny days.
Keywords: Vertical PV, Bifacial, Flat roof PV

1 INTRODUCTION

With the introduction of bifacial solar cells and increased availability of heterojunction (HJT) silicon solar cells, highly performant bifacial PV modules have entered the market during the last few years. These have usually been applied to utility scale or commercial solar parks where the bifaciality can be utilized as the modules are not mounted on the surface of a building. In addition, vertically mounted PV modules have been researched and, to a smaller extent, utilized the last few years. This configuration has been used for facades first, then in connection with other infrastructure such as sound barriers, and lately in connection with farmland as a relevant configuration for agri-PV. The advent of bifacial PV modules has been a pre-requisite for the growth of research, pilot and commercial projects utilizing a vertical configuration.

Over Easy Solar was founded in 2021 with the aim of developing a system that would improve on several shortcomings of conventional PV installations for flat rooftops:

- Faster mounting process with prefabricated units
- Lower roof loading with a ballast free system
- Higher energy value with double energy production peaks, especially in markets with price cannibalization
- Access to the rooftop for inspection and repairs
- Easy to combine with existing and new green rooftops

The answer was a system consisting of vertical, bifacial PV panels on a lightweight, pre-mounted frame (vertical PV system - VPV). The idea is to utilize the benefits of vertical mounting which utilizes ground reflection, albedo, better than conventional PV systems as well as generates power earlier in the morning and later in the afternoon than a conventional PV system. In addition, with a vertical mounting of narrow panels the effect of wind load on the system will be smaller than for a conventional low angle system which typically can experience quite high aerodynamic lift. By reducing the wind load, the VPV system can completely avoid ballast, which is a significant factor for roof material wear, adds a significant CO_2 footprint and makes the mounting work less efficient.

The Over Easy VPV unit is shown in Figure 1. It consists of a frame on which 4 narrow panels are fastened. The frame is connected to feet of different heights which can be varied depending on the roofing material. This system also lends itself well to green roofs with varying growth height. The panels are pre-mounted and pre-connected. This enables a 10-fold increase or more in on-roof system mounting speed.

A few similar concepts have been researched previously [1,2], however, there are differences in panel technology and framing in the Over Easy Solar VPV Unit. 12 pilots have been installed using the VPV Units around Europe to gather real-world experience with mounting and performance. This paper shows some of the performance analysis done on the inverter data from these pilots and contains novel data that has never been shown before in other publications, as e.g. yearly energy performance of south-north oriented vertical bifacial panels in a complete array. The results presented here are important to support the advent of rooftop vertical solar installations but may also be of high importance for planning of agrivoltaic and utility scale vertical solar installations.

Figure 1: A rendering of the Over Easy Solar vertical solar PV unit showing the panels, frame and feet. Each panel is 20-25 cm high and contains a single string of 6-8 bifacial cells. The complete height of a unit is <30 cm.

10.4229/EUPVSEC2023/4BV.4.10
020344-001

2 DESCRIPTION OF PILOTS

The pilots are placed as far north as Tromsø, Norway at 69°N and as far south as Alicante, Spain at 38°N. Several different azimuths are chosen as well. The name and geography of each of the pilots are shown in Figure 2. In addition, several different PV panel and cell technologies are tested in these pilots, namely PERC, Topcon and HJT as well as full cell and half-cut. Not surprisingly, the high bifaciality of >90% that can be reached with HJT is strongly beneficial for the performance of VPV Units, but tested cells with both PERC and Topcon also have good performance and bifaciality of >80%.

The pilot installations consist of 1.4-5.0 kW_p of VPV Units connected in series, and to optimizers and an inverter from SolarEdge, or a string inverter from SMA Solar Technology AG. The data used in this study are from the following pilots:

- P2 (Vestby, NO, lat. 59.6°N, azimuth east/west and south/north)
- P3 (Green roof - Oslo, NO, lat. 59.9°N, azimuth SSW/NNE)
- P4 (Tromsø, NO, lat. 69.7°N, azimuth SSW/NNE)
- P8 (Vestby, NO, lat. 59.6°N, azimuth E/W)

These pilots all utilize HJT cells.

Figure 2: Overview of all the pilots with startup date and city.

3 PERFORMANCE ANALYSIS

The data used for the analysis was downloaded through the inverter web portal API for each pilot. Time series resolution varied from 5 min. intervals to 15. min. intervals.

The performance of the selected pilots in terms of monthly specific energy yield is shown in Figure 3. Figure 4 shows the absolute difference of specific yield between the pilot and the corresponding nearby reference. The comparison shows increased or similar specific yield for all pilots and most months. For the northern most pilot (P4) the increase in specific yield for the measured period is 51 % while for the pilots in southern Norway (P2, P3 and P8) the increase in specific yield is around 30 % (28 % for P3 and 31 % for P2 and P8).

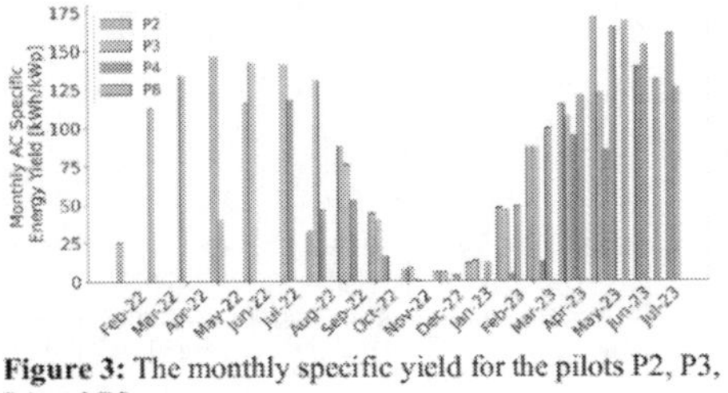

Figure 3: The monthly specific yield for the pilots P2, P3, P4 and P8.

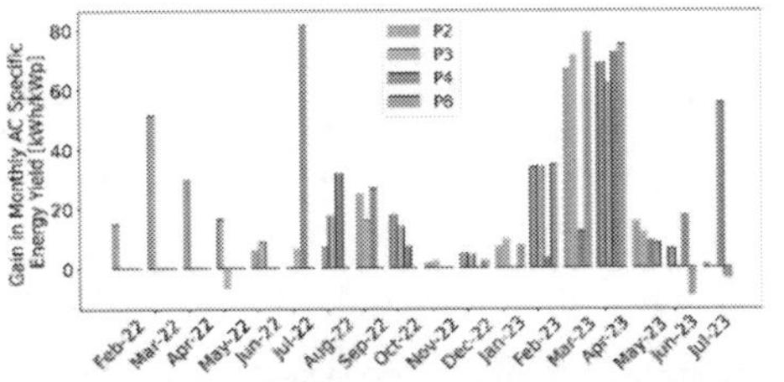

Figure 4: Absolute difference in specific yield between pilot and corresponding nearby reference. In all cases the reference is a conventional flat roof system with 10° tilt and modules facing east-west. P2 is missing data from parts of June and August, as well as all of July, 2022. For most of the months the pilots have a higher specific yield than the reference systems.

Performance of the P2 pilot placed on the roof of a warehouse in Vestby, Norway (59°N) is shown in Figures 5-8. This pilot consists of two nearly identical systems, each with a nominal power of 2.433 kWp. One is placed facing east-west and the other south-north. The conventional reference has a 10° tilt, facing east-west and a size of 214 kWp. The dates are chosen to represent a variation in season.

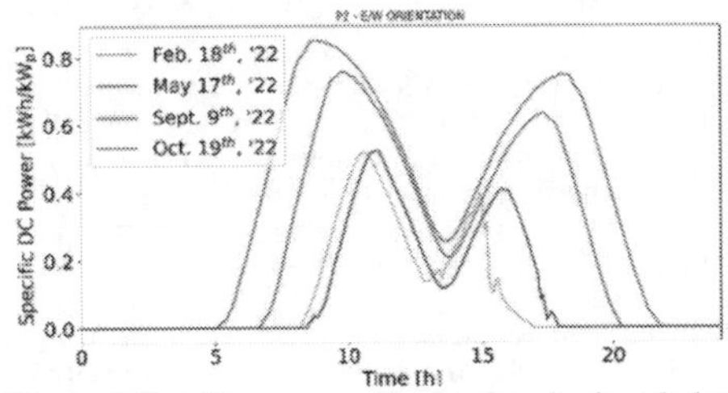

Figure 5: Specific power profiles for clear sky days during each season for the east-west oriented system. Two peaks are present all year. Even during winter, the specific power is significant.

10.4229/EUPVSEC2023/4BV.4.10
020344-002

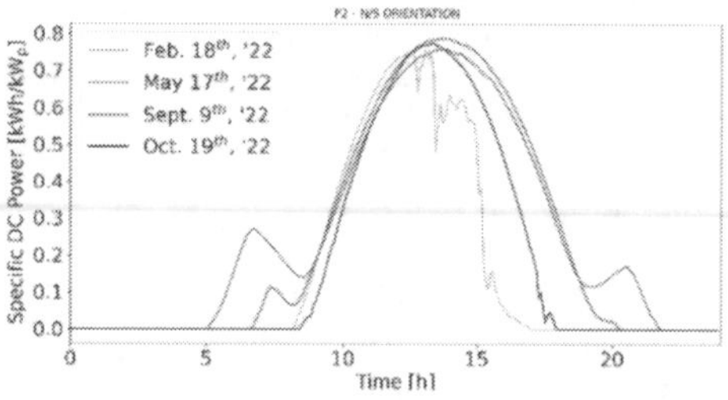

Figure 6: As in Figure 5, but for the north-south oriented system. The significant early morning and late afternoon peaks are due to the sun irradiation reaching the panels from the north between spring and autumnal equinoxes.

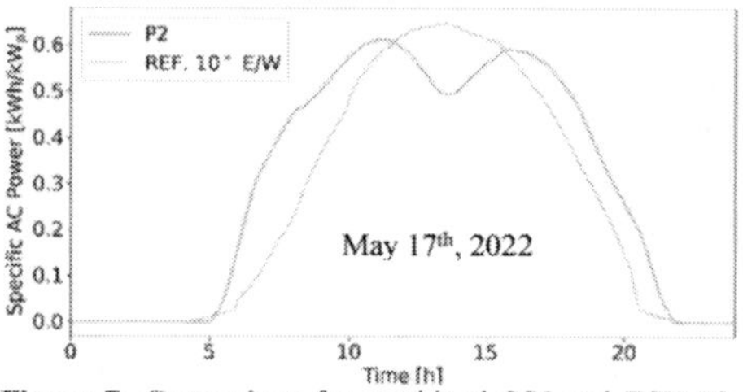

Figure 7: Comparison for combined S/N and E/W P2 systems and reference on the same day. The morning and evening production benefit for P2 is clear. In addition, by combining S/N and E/W VPV systems the profile becomes flatter.

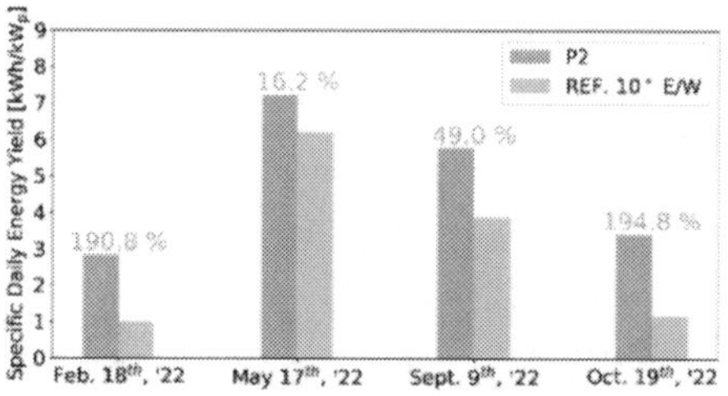

Figure 8: Comparison for the same clear-sky days as for Figures 5 and 6 between combined S/N and E/W P2 systems and reference. Relative difference in yield is shown above the bars. All seasons show a significant increase in specific yield for the pilot with winter and fall showing a large benefit for VPV over 10° tilt systems.

An illustration of the performance during winter is shown in Figure 9 for the pilot P3 which is placed on a green roof in Oslo, Norway. It is clear that the VPV system is more robust against snow covering and will perform even with some snow covering the ground and will also start generating power quicker as it seems snow melts quicker around the system.

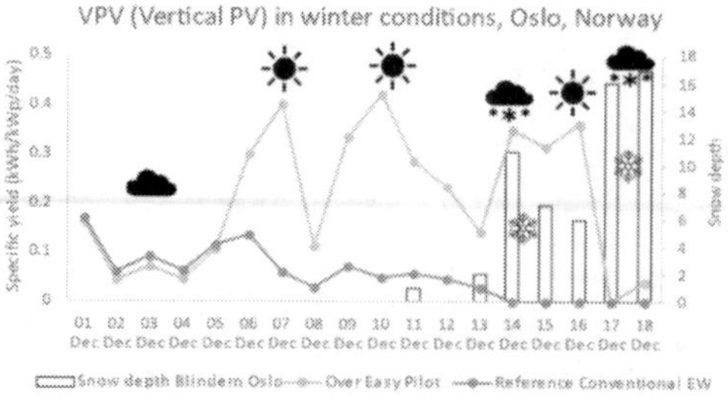

Figure 9: Energy performance from P3 installation in Oslo, Norway, compared to a conventional 10° tilt east-west rooftop installation, during days of December 2022. One can see a generally higher power production during winter days as well as power generation with some snow cover when the conventional system does not generate any power.

4 CONCLUSION

Inverter output time series from the Over Easy Solar pilots P2, P3, P4 and P8, all in Norway, show that the vertical bifacial PV rooftop installations provide high specific yields, comparable to or up to 51 % higher than that of conventional flat rooftop solar installations under varying environmental and solar conditions and at varying latitudes.

5 ACKNOWLEDGEMENTS

The work has been partly funded by The Research Council of Norway under grant number 332198, and Regional Research Fund Oslo under grant number 328867.

REFERENCES

[1] Baumann, T. & Nussbaumer, H. & Klenk, M. & Dreisiebner, A. & Carigiet, F. & Baumgartner, F., *Photovoltaic systems with vertically mounted bifacial PV modules in combination with green roofs*, Solar Energy 190 (2019)

[2] Podlowski, Lars & Wendlandt, Stefan & Cormode, Daniel, *Yield Study on Identical Bifacial Rooftop Systems Installed in the USA and in Germany*, Presentation at the 4th Workshop on Bifacial PV – Konstanz (2017)

10.4229/EUPVSEC2023/4BV.4.10
020344-003

HEATING OF URBAN ENVIRONMENTS BY PHOTOVOLTAIC MODULES

Max Mittag
Fraunhofer Institute for Solar Energy Systems ISE
Heidenhofstr. 2, 79110 Freiburg, Germany

ABSTRACT: The paper answers the question if solar modules contribute to additional heating of urban environment compared to conventional mounting areas on buildings. The effective albedo of a module is calculated that contains the optical reflection as well as the electrical energy that is created by the module. The generated heat of the PV module is compared to a scenario without PV installation.
Energy input on the module as well as on the environment (mounting area, neighboring buildings) is calculated. The module power output is determined using a bottom-up loss channel model that considers optical, electrical, thermal and geometrical aspects ("SmartCalc.Module" by Fraunhofer ISE). Reflection of the module is obtained from an effective measurement as well as from optical modelling. An energy equilibrium model is used to calculate the generated heat. A sensitivity analysis on mounting conditions and ambient albedos shows that a PV module has a higher thermal impact than a "no PV" scenario. An advantage regarding total heat generation of a PV installation is found only for dark mounting areas. The effective albedo of a PV module is calculated to be 0.24.
Keywords: Integrated Phovoltaic, Modelling, Simulation, Heat Generation, Energy Balance, Albedo

1 INTRODUCTION

Building Integrated PV (BIPV) is expected to be a significant factor in future energy systems as it provides many advantages (double use of areas, energy generation close to consumption, investing people in the energy transition etc.). But – as with every application – disadvantages are to be expected and need to be discussed. I provide data to quantify the impact of PV modules on the heat management of cities as they are observed to face temperature rises due to the climate change.

The thermal impact of PV modules on the (urban) environment has not yet been finally discussed and existing work finds both additional cooling and heating of the environment by PV installations which indicates that further work is necessary and detailed analysis is required [1, 2].

In this work I analyze the impact of module design, mounting conditions and mounting area properties on the heat generation of PV modules.

2 APPROACH

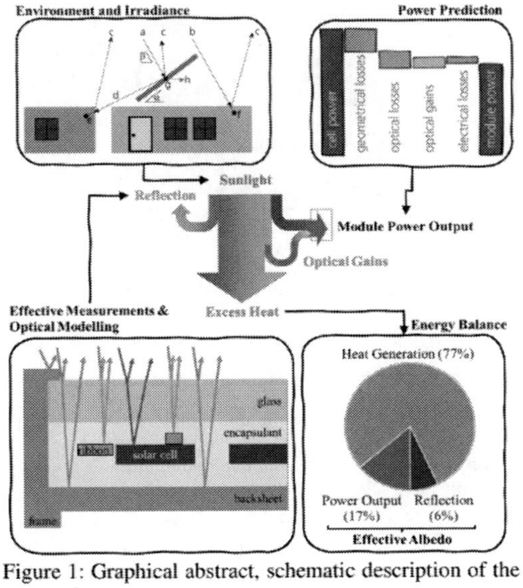

Figure 1: Graphical abstract, schematic description of the approach

I apply the energy conservation law to calculate the total heating of PV module and associated building areas. The heat generated is the difference between the irradiant energy, the reflected energy, and the power production.

The reflection of the module is obtained through the identification of relevant area shares and the measurement of their reflective properties. The area shares of solar cells, frame, backsheet and ribbons are calculated. The measured reflectivity is weighted with the area share of each component area and the sun spectrum to obtain the total module reflection.

The power output is calculated using a bottom-up multi-physics loss channel analysis model of a photovoltaic (PV) module. The cell-to-module (CTM) approach is used for calculation (software: SmartCalc.Module by Fraunhofer ISE) [3, 4].

An effective albedo is calculated, which describes the fraction of energy not converted to excess heat in the energy incident on a module [2].

Energy irradiant on the module and the building areas is calculated using a geometrical model. Mounting area albedo and reflection to other buildings is considered. The shading of the mounting area by the module is calculated. Again, the difference between irradiant energy and reflected energy is used to calculate heat generation in the building surface.

Sensitivity analyses regarding mounting angle, mounting area albedo and sun position are performed. Total heat generation in scenarios with and without PV modules is calculated and compared.

3 DETAILED DESCRIPTION

3.1 Energy Equilibrium and Heat Generation

The generated heat is calculated using the law of energy conservation.

$$E_{module} = R_{module} + P + H_{module}$$

E_{module} = irradiance on the module
R_{module} = reflection of the module
P = module power generation
H_{module} = module heat generation

$$E_{ground} = R_{ground} + H_{ground}$$

$$E_{ground} = I_{ground} \cdot \left(A_{ground} + \left(1 - A_{ground}\right)\right)$$

E_{ground} = irradiance on the mounting area or buildings
R_{ground} = reflection of mounting area or from buildings
H_{ground} = heat generation in the mounting area or in buildings
A_{ground} = albedo of the ground or building

I define the effective albedo $A_{effective}$ as the share of the energy irradiant onto the reference area that is not converted into heat. The effective albedo includes reflection and power output.

$$A_{eff.module} = \frac{R_{module} + P_{module}}{E_{module}} = 1 - \frac{H_{module}}{E_{module}}$$

$A_{eff.module}$ = effective albedo of the PV module considering reflection and power output as a heat sink

3.2 Geometrical Analysis of Area Shares in Solar Modules

I find that a photovoltaic (PV) module can be divided into four segments of different optical properties: the frame area, the visible rear cover (backsheet) area, the active solar cell area and the area of the cell and string interconnection ribbons (Figure 2).

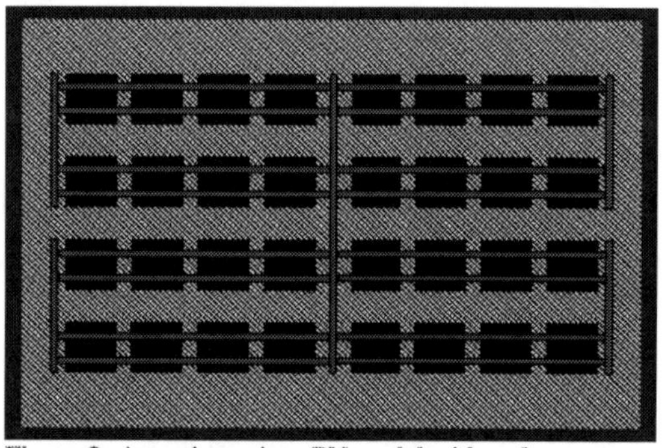

Figure 2: Area shares in a PV module, blue: frame area, red: interconnection ribbon area, black: solar cell area, grey: visible module rear cover area.

Area shares can be calculated using the geometrical properties of the module such as cell format, cell distance, string distance, ribbon width etc. (Figure 3).

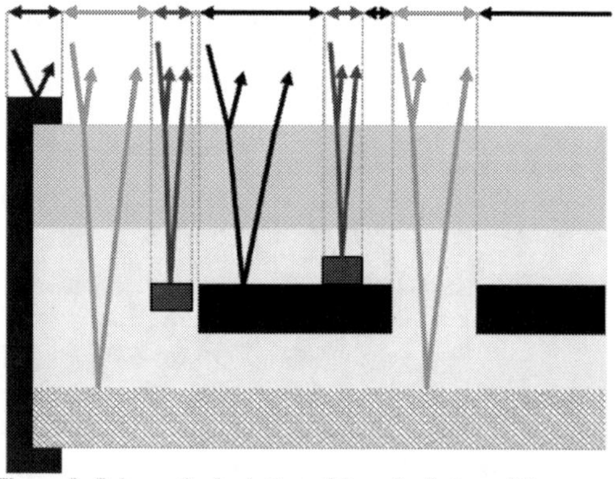

Figure 3: Schematic depiction of the calculation of the area shares from geometrical module and component properties.

I calculate the area shares for commercially available modules based on 144x M10 (182 mm x 91 mm, 247 mm) half cells with total dimensions of 2278 x 1134 mm² as in Table 1. Note that the given area for the ribbons is the optically effective area since internal reflection has to be considered that decreases the effective optical width compared to the geometric footprint. An effective width of 60% is assumed for the cell interconnection ribbons [5–7]. String interconnection ribbons have a 100% effective width. The calculated area shares match previous calculations [8].

Table 1: Area shares in the analyzed PV module.

	Area [m²]	Area share
Solar Cell	2.352	91.0%
Backsheet	0.102	3.8%
Frame	0.088	3.4%
Ribbons	0.067	1.6%
Total	2.583	100.0%

3.3 Optical Characterization of Module Materials

The visible module areas are measured regarding the reflective properties using an integrating Ulbricht sphere. A commercially available black and white backsheet as well as a black and a grey aluminum frame are characterized. The cell used is an industrial PERC cell which is encapsulated into EVA. Ribbon characterization is performed using string interconnector ribbons due to their larger widths. Reflective properties of the string and the cell ribbon are assumed to be the same except using the mentioned effective width of 60% for the area share calculation based on 0.25 mm round wires.

The measurements are performed using a module stack. Therefore, the measurement of the cell, the ribbons and backsheets include the optical properties of the front cover glass and the encapsulant. The frames are measured against air.

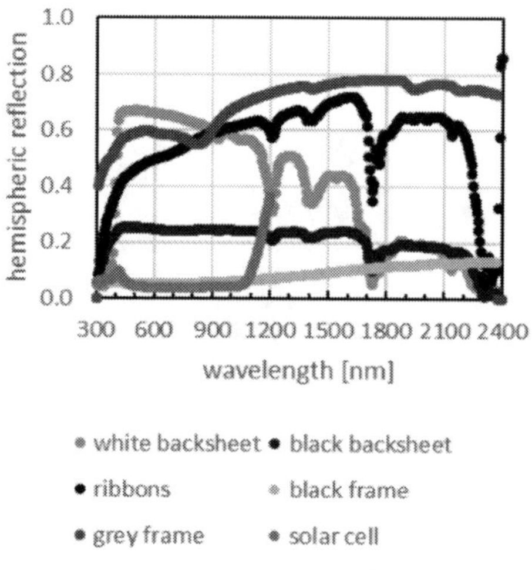

Figure 4: Reflection measurement results of different module area shares.

3.4 Calculation of Area Weighted Module Reflection

I calculate the effective module reflection by weighting the measured reflection with the calculated area share and the AM1.5 solar spectrum.

Table 2: Reflected energy shares, AM1.5 weighted reflection data.

	Per 1 m² module area, area share weighted	Per m²
Solar Cell	43.9 W	48.2 W
White Backsheet	22.0 W	558.4 W
Black Backsheet	9.3 W	237.2 W
Grey Frame	20.7 W	608.5 W
Black Frame	2.0 W	59.1 W
Ribbons	8.4 W	517.1 W

For a module with a white backsheet and a grey frame the reflection is calculated to be 9.5% and for a module with black backsheet and black frame 6.3%, respectively.

No angular measurements for the different module area shares are performed but since the angular resolved module reflection is required for the energy balance calculation, a conversion of the calculated reflectivity is performed using Fresnel's Law for reflection and a refractive index of 1.67 for glass. By doing so, I calculate an initial reflection of 6.3% at perpendicular irradiance which fits to the result obtained from measurement and are weighting. The module rear side has a constant reflectivity of 55.8% for all angles of incidence (AOI) derived from the measurement of the white backsheet.

If all energy not reflected is converted into heat, solar modules generate 90.5 – 93.6% of the absorbed energy into excess heat. Since some of the absorbed energy is converted into electrical power the excess heat is reduced.

3.5 Module Power Estimation

The electrical power generation of the module is estimated using a bottom-up multi-physics loss channel analysis implemented into SmartCalc.Module by Fraunhofer ISE. The software performs a cell-to-module analysis combining geometrical, optical, electrical and thermal calculations to quantify occurring gains and losses inside a PV module from a given solar cell to a module using component specifications and module design information [4, 9–13]. The geometrical data from Table 1 as well as optical data from Figure 4 are used together with additional information i.e. on resistivity of the ribbons, cell metallization patterns, junction box specification etc. Typical values based on industrial components are assumed. The solar cell used has an efficiency of 23.3%.

Module power is calculated to be 540.7 Wp and efficiency to be 20.9%. The module is monofacial.

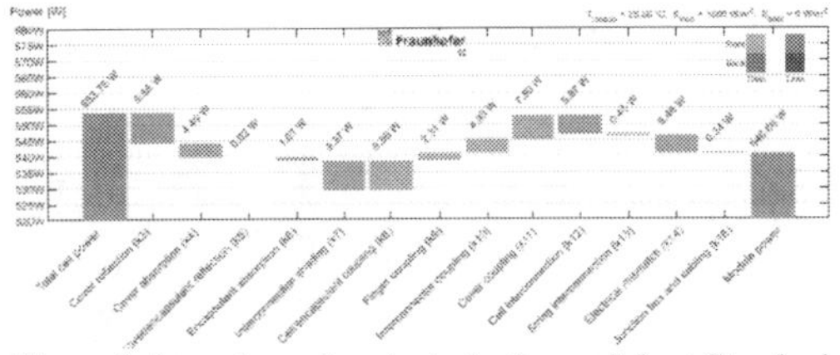

Figure 5: Loss channel analysis for the module at Standard Testing Conditions (STC).

To assess the generated heat of the PV module the CTM analysis is performed for a variation of the angle of incidence (zenith angle 0° - 80°) including a module temperature calculation to assess the reduced module power output in outdoor operation due to increased cell temperature [9]. An outdoor temperature of 30 °C, 1 m/s wind speed and an irradiance of 1000 W/m² (AM1.5) are used. The module is mounted at 40° against the horizontal and the mounting area has an emissivity of 0.8.

The calculated module power for each angle of incidence (Figure 6) is used in the energy balance calculations.

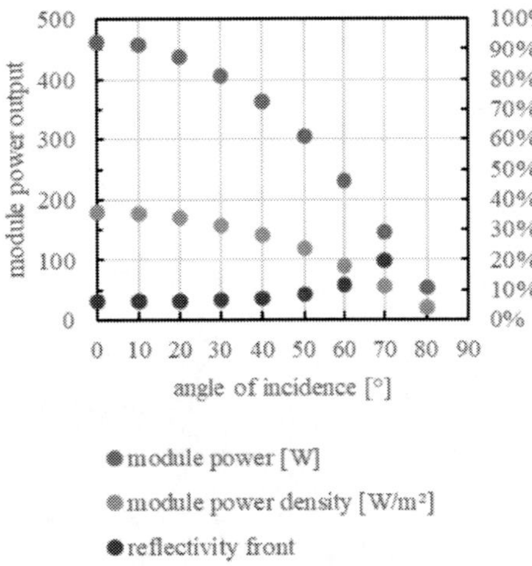

Figure 6: Module power generation considering elevated module temperature and module front side reflectivity for a variation of the angle of incidence.

Using the results from the power calculations and the module reflection, I find the excess heat generation of the PV module to be 757 W/m² at 1000 W/m² irradiance (Figure 7). The module design impacts this energy balance by changing module power and module reflection. Only considering the reflection, the excess heat is reduced to 726 W/m² for a white backsheet, white frame module. With an additional power gain of 1% for a white backsheet compared to a black backsheet due to improved light management inside of the module the module excess heat reduces to 724 W/m².

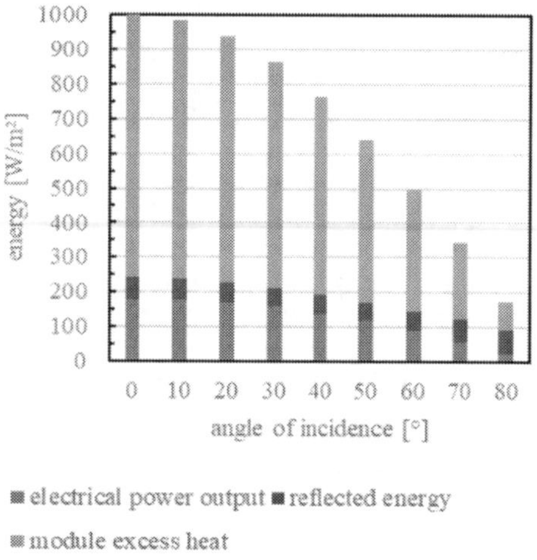

Figure 7: Energy balance of the module for a variation in angle of incidence.

3.6 Irradiance on Module and Surrounding Environment

Irradiance is calculated for the solar module itself, the mounting area and for additional areas (Figure 8). All light reflected from the module is reflected to the additional areas (d) or the sky (c2).

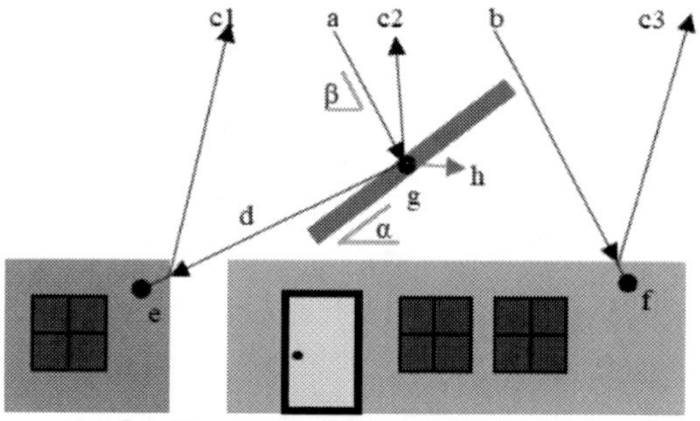

a) Irradiance on the PV module
b) Irradiance on uncovered mounting area
c) Reflection to the sky
d) Reflection on surrounding buildings
e) Absorption in surrounding buildings
f) Absorption in uncovered mounting area
g) Absorption in the PV module
h) Generated electricity

Figure 8: Considered light paths and energy shares.

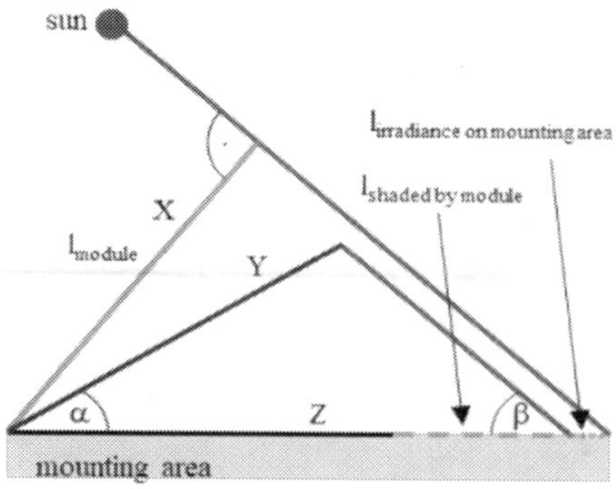

Figure 9: Approach in the calculation of the irradiance on module and mounting area.

The irradiance on the module and the mounting area considers the module tilt angle α and the sun elevation angle β as shown in Figure 9. It is assumed, that a perfectly oriented module X (perpendicular irradiance on the module) would not cause perfect shading of the mounting area. An imperfect orientation Y for a sun inclination β leads to irradiance passing by the module. The irradiance on the mounting area is the difference between the optimal orientation for each sun elevation (X, full shading) and the actual shading caused by Y. For a horizontally mounted module Z this leads to irradiance on the mounting area for all sun elevations ≠ 90°.

The calculation of the module reflection uses the effective reflective index as described in 3.4 and Fresnel's law. Irradiance on the area below the module (mounting area) is calculated using the sun elevation and the module mounting angle.

All reflections by the module front are assumed to be direct. The module rear side reflection is assumed to be isotropic towards sky and mounting area. A view factor model is applied to calculate the respective energy shares for the rear side. The energy reflected by the surrounding buildings is assumed to be directed equally towards the sky and other surrounding buildings where it is being absorbed. Reflection of the mounting area is only directed to the sky (Figure 8, c3). Additional irradiance on the module from mounting area reflection is not considered.

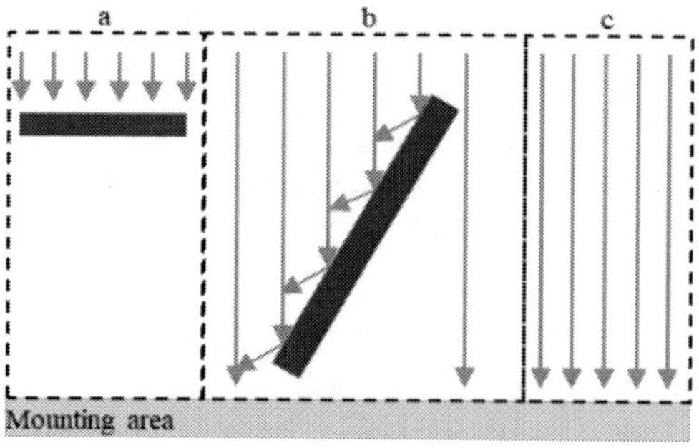

Figure 10: a) no irradiance on the mounting area and full irradiance on the module; b) reduced irradiance on the module and irradiance on the mounting area for tilted modules and irradiance non perpendicular to the mounting area; c) "no PV" scenario with irradiance only on the mounting area.

All calculations assume an irradiation of 1000 W/m² in the horizontal.
An analysis regarding the difference of heat generation with and without the use of solar modules is performed. In the "no PV" scenario all irradiance is onto the mounting area (Figure 10).

3.7 Energy Balance for Changing Sun Elevation Angles

The energy balance analysis combining optical analysis, module power output and heat generation calculation, shows that significant shares of the incident energy are converted into heat (Figure 11). It is found that the module heat generation depends on the sun position and is largest for elevation angles between 50° and 90°. While the module heat generation peaks at a sun inclination of 50°, the total heat generation has a maximum hat larger AOI (90° sun inclination) due to the increased heat generation in the mounting area.

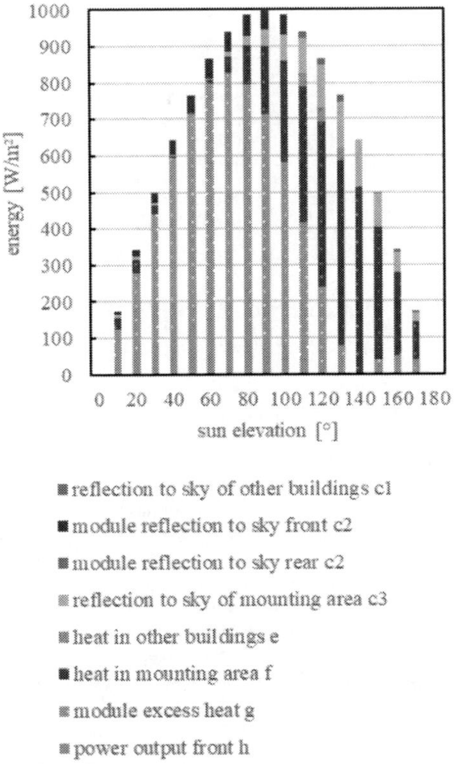

Figure 11: Sensitivity analysis of the sun elevation angle regarding generated heat for a mounting area albedo of 0.2, module tilt angle of 40°.

Reflection and module power output are the effective heat sinks in a PV module. In my analysis the electrical power generation accounts for up to 74% thus making module power generation an important aspect in lowering heat generation.

The heat generation in different surfaces is clearly following the irradiance on the respective surfaces (Figure 12, Figure 13 top).

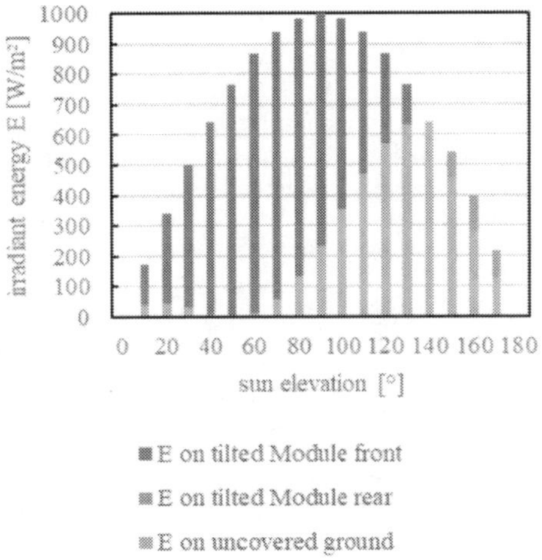

Figure 12: Irradiance on different surfaces with a variation of the sun elevation angle.

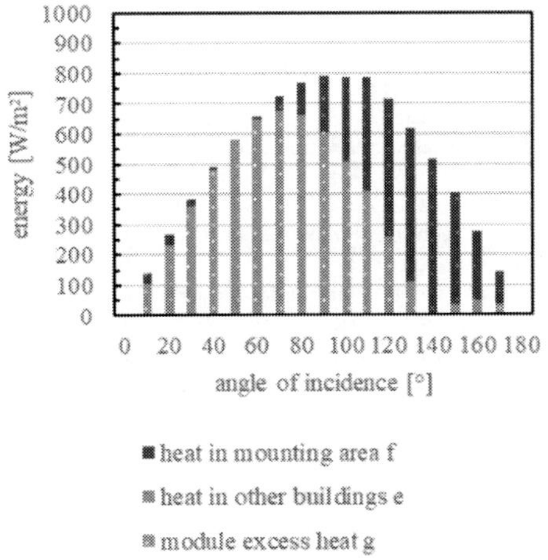

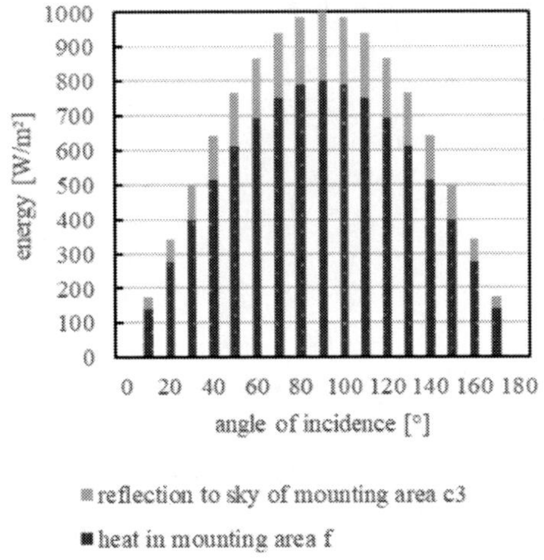

Figure 13: Sensitivity analysis of the sun elevation angle regarding generated heat for a PV installation (top) and a "no PV" scenario (bottom), mounting area albedo of 0.2.

10.4229/EUPVSEC2023/4BV.4.14
020345-005

Comparison of the scenario with a PV installation to a "no PV" scenario (for a mounting area albedo of 0.2) shows no significant difference (Δ ~1%) for the total heat generation (Figure 13).

3.8 Sensitivity Analysis on Module Tilt

The mounting angle of the module is changed from 0° (flat mounted module on horizontal mounting area) to 90° (vertical standing module). The mounting area is kept horizontal for all analyses (flat roof). We analyze the average heat generation for all sun elevation angles (0° - 180°).

As no ground coverage ratio of the modules is a direct input, the assumption is that an imperfect module orientation regarding sun position leads to light passing the module plane and reaching the mounting area (Figure 9). Therefore, even for a flat lying module heat generation is present in the attributed mounting area (Figure 14). Additional absorption in row spacing areas between different modules occurs but is not considered in the comparison.

We find the average module excess heat for the variation in mounting angle to be 280 W/m², the heat generation in other buildings to be 7 W/m² and the heat generation associated to the mounting area to be 188 W/m² which is in total lower (468 W/m²) than for the "no PV" scenario (461 W/m²).

Increasing the mounting area albedo to 0.5 generates an advantage of the "no PV" scenario and shows that light roofing materials effectively reduce heat generation. Mounting area heat generation is reduced by 37% (117 W/m² / 188 W/m²) through the increase in albedo from 0.2 to 0.5. Module power and heat generation are not affected since reflection from the mounting area onto the module is not considered.

An average heat production of 134% (min.: 121%, max.: 143%) compared to the "no PV" scenario is calculated. PV module installations lead to a higher heat generation for low mounting area albedo.

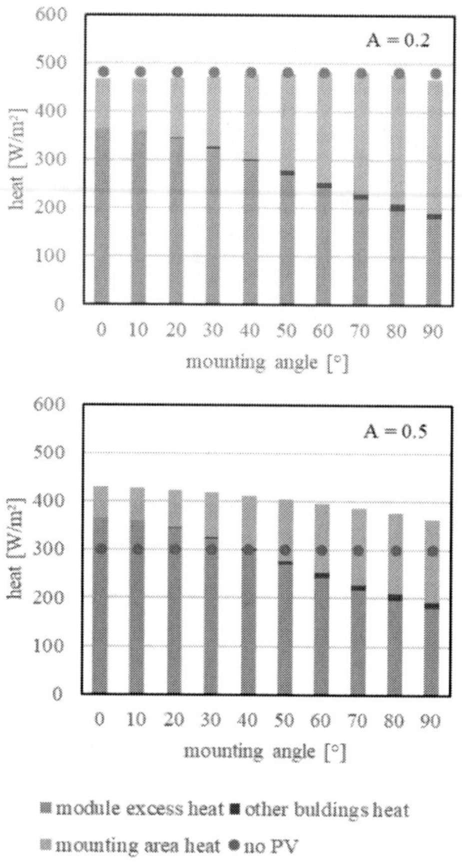

Figure 14: Sensitivity analysis of the module tilt angle regarding generated heat for a mounting area albedo of 0.2 (top) and 0.5 (bottom), average values for sun elevation angles from 0° to 180°.

3.9 Sensitivity Analysis on Mounting Area Albedo

An additional variation of the mounting area albedo is performed for a fixed module mounting (20° - 80°). Every analyzed scenario finds a significantly higher heat generation than the "no PV" scenario (Figure 15). Additional average heat generation is between 123 W/m² (+41%) at 20° and 76 W/m² (+25%) at 80°. Lighter mounting areas increase the disadvantage of PV modules regarding heating.

10.4229/EUPVSEC2023/4BV.4.14

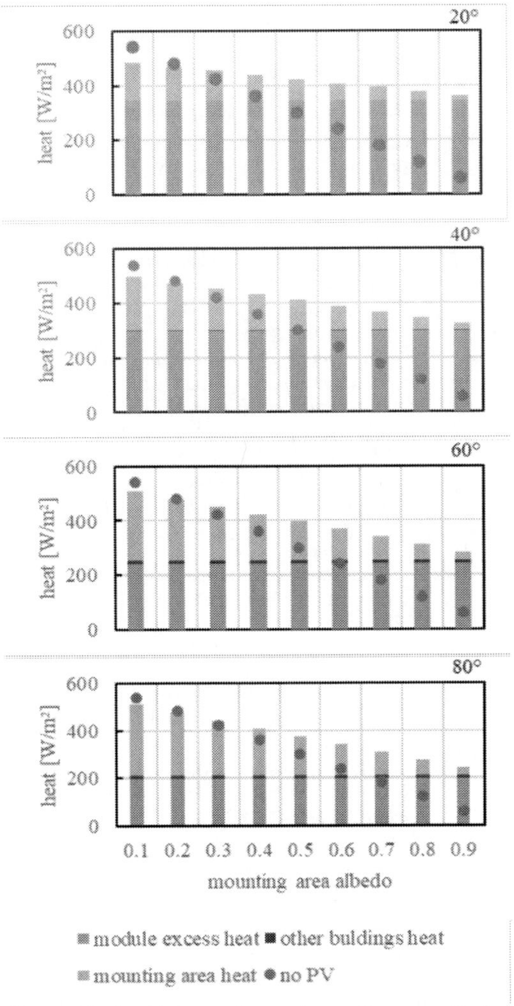

Figure 15: Sensitivity analysis of mounting area albedo regarding generated heat for a module mounting angles between 20° and 80°, average values for sun elevation angles from 0° to 180°.

Figure 16 summarizes all variations in albedo and mounting angle. The advantage of higher mounting area albedo is clearly visible resulting in a lower heat generation. The graph also shows the minor impact of module orientation on heat generation.

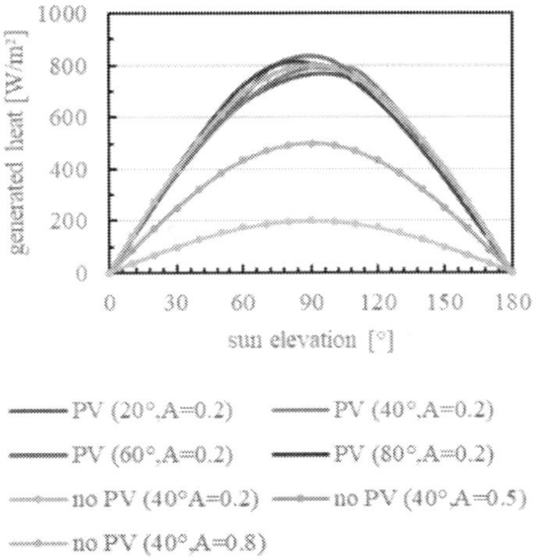

Figure 16: Sensitivity analysis of mounting area albedo and mounting angle regarding generated heat, average values for sun elevation angles from 0° to 180°.

3.10 Effective Albedo

The analyses performed above show that a consideration of the module power output changes the total heat generated by the module. Methods that asses large scale structures like cities only based on optical properties of the infrastructure therefore overestimate the heat generated by PV modules. By applying the effective albedo rather than the purely optical one a correction is possible.

I find the effective albedo of the module to be much higher for conditions where the module rear side is illuminated directly by the sun (Figure 17) which results from the assumption of a white module rear side. Considering only conditions with the sun in front of the module (Figure 17, grey box), I find an average effective albedo of the module of 24.1% for front side irradiance. Including all irradiance conditions, the effective module albedo is found to be 31.1%.

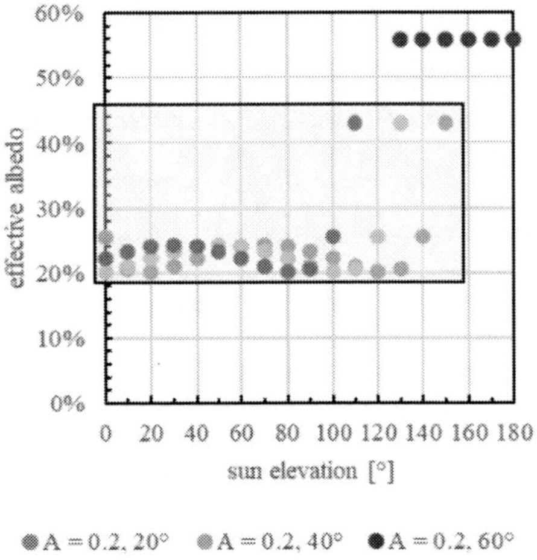

Figure 17: Effective module albedo for a mounting angle variation.

10.4229/EUPVSEC2023/4BV.4.14
020345-007

The effective module albedo is to be applied on the module area. Additional mounting area shares need to be considered separately with their respective albedo.

Average urban albedo are found to be mostly below 0.2 [14] thus showing an advantage of PV modules compared to cities as a whole. On the other hand typical building materials have a higher albedo [15] than solar modules which indicates a disadvantage for PV modules in a more spatially focused study.

4 SUMMARY

An analysis regarding the area shares of different module components and their reflectivity is performed. The solar cells dominate the module area with 90% area share.

I find the total solar reflectivity of a PV module to be between 6.4% to 9.5% depending on the choice of module components (black / grey frame, black / white backsheet). Without significant transmission therefore 90.5% to 93.6% of incident light are absorbed of which some fraction is converted into electrical power. Considering cell-to-module (CTM) power losses and the cell temperature in operation, the module efficiency is calculated to be 17.9%. Therefore ~72.6-75.7% of the total irradiance are converted into heat.

The effective albedo is introduced which combines module reflection and power generation and is the sum of all irradiant energy that is not converted into heat. An effective albedo of 24.1% is found for a PV module.

Module power is confirmed to be an important heat sink accounting for up to 74% of the total energy sinks of a module in operation. Module heat generation (and module power) depends on the choice of module components. "All black" designs with black frame and black backsheet have an increased absorption and therefore higher heat generation.

The irradiance on the module in an urban mounting situation is calculated considering module orientation, sun inclination, mounting area albedo and other building areas. The irradiance is used to calculate the effective heat generation in the module and the building area. I find that the heat generation in the mounting area is significant.

The total heat generation of a module and its mounting area is compared to "no PV" scenarios for variations in mounting area albedo, sun position and module tilt angle. The analyses show that most "no PV" scenarios have a lower total heat generation than scenarios with PV modules. Only for mounting area albedo lower than the effective module albedo an advantage of PV modules is found.

I find the module tilt angle to be of lower importance for the total heat generation while increasing the mounting area albedo leads to a much lower total heat generation.

REFERENCES

[1] A. Khan *et al.*, "On the local warming potential of urban rooftop photovoltaic solar panels in cities," (eng), *Scientific Reports*, vol. 13, no. 1, p. 15623, 2023.

[2] H. Taha, "The potential for air-temperature impact from large-scale deployment of solar photovoltaic arrays in urban areas," *Progress in Solar Energy 1*, vol. 91, pp. 358–367, 2013.

[3] Fraunhofer Institute for Solar Energy Systems ISE, *SmartCalc.Module*.

[4] M. Mittag *et al.*, "Systematic PV module optimization with the cell-to-module (CTM) analysis software," *Photovoltaics International*, no. 36, pp. 97–104, 2017.

[5] R. Witteck *et al.*, "Optimized Interconnection of Passivated Emitter and Rear Cells by Experimentally Verified Modeling," *IEEE Journal of Photovoltaics*, vol. 6, no. 2, pp. 432–439, 2016.

[6] L. C. Rendler *et al.*, "Wave-shaped wires for the interconnection of silicon solar cells without pads or busbars," in *2nd International Conference on Emerging Trends in Engineering*, p. 20007.

[7] R. Witteck *et al.*, "Optimized Interconnection of Passivated Emitter and Rear Cells by Experimentally Verified Modeling," *IEEE Journal of Photovoltaics*, vol. 6, no. 2, pp. 432–439, 2016.

[8] M. Mittag *et al.*, "Impact of solar cell dimensions on module power, efficiency and cell-to-module losses," in *30th PV Solar Energy Conference*, 2020.

[9] M. Mittag *et al.*, "Thermal Modelling of Photovoltaic Modules in Operation and Production," in *36th European Photovoltaic Solar Energy Conference and Exhibition (EUPVSEC) 2019*.

[10] M. Mittag *et al.*, "Cell-to-Module (CTM) analysis for photovoltaic modules with shingled solar cells," in *44th IEEE PV Specialist Conference PVSC*, pp. 1531–1536.

[11] A. Pfreundt *et al.*, "Rapid Calculation of the Backsheet Coupling Gain Using Ray Groups," in *35th European Photovoltaic Solar Energy*.

[12] A. Pfreundt *et al.*, "Cell-To-Module Analysis Beyond Standard Test Conditions," in *47th IEEE Photovoltaic Specialists Conference 2020*.

[13] M. Mittag *et al.*, "Techno-Economic Analysis of Half Cell Modules - The Impact of Half Cells on Module Power and Costs," 2019.

[14] H. Taha, "Urban climates and heat islands: albedo, evapotranspiration, and anthropogenic heat," in *Energy and Buildings*, pp. 99–103.

[15] Racine Tadeu Araújo Prado, Fabiana Lourenço Ferreira, "Measurement of albedo and analysis of its influence the surface temperature of building roof materials," in *Energy and Buildings*, pp. 295–300.

10.4229/EUPVSEC2023/4BV.4.14
020345-008

INDICATIONS OF EARLY CLIMATE CHANGE IMPACTS REVEALED WITHIN WEATHER FILES FOR PREDICTING PHOTOVOLTAIC SYSTEM PERFORMANCE IN EUROPE

Delia D'Agostino[1], Danny Parker[2]

[1]Joint Research Centre, European Commission
[2]FSEC Energy Research Center, University of Central Florida, Cocoa, FL, USA

delia.dagostino@ec.europa.eu; dparker@fsec.ucf.edu

ABSTRACT

In the view of being the first climate-neutral continent by 2050, renewable energy will play a key role in Europe. The need to electrify energy demand to facilitate greenhouse gas (GHG) emission reductions, and reduce climate change warming potentials, makes it important to locate new available areas for renewable energy production and particularly Photovoltaic (PV). PV array energy production can be predicted by the TRNSYS hourly simulation. Weather data is a key driver for realistic simulation of PV system performance that changes over time, seasonally, yearly and stochastically from one year to the next. This paper evaluates PV energy production using different commonly used weather datasets (IWEC2 1994-2011 and TMY 2007-2021) in different European locations (Vienna, Stockholm, Milan, Munich, Lisbon, Rome, Paris. Detected changes reflect the early indicators of future anticipated climate change and are useful to analyse factors that can help control module operating temperatures as well as implications for Nearly Zero Energy Buildings (NZEB).

Keywords: Energy efficiency; photovoltaic systems; building energy simulations; building decarbonization; climate neutrality; nearly zero energy buildings.

1 INTRODUCTION

Photovoltaic (PV) electricity production is a key technology for future renewable energy generation to meet energy needs while addressing greenhouse gas reduction targets around the world [1]. In particular, greater roof-top PV electric production is an integral part of reaching Nearly Zero Energy Building (NZEBs) within changing climate [2][3]. The most modern PV cells and modules are approximately 20% efficient under standard operating conditions. Generally, electric production is proportional to incident solar irradiance on PV arrays. However, the module operating temperatures impact the DC power produced depending on the operative temperature coefficient. With higher temperatures around arrays, the produced electric power is reduced depending on the temperature coefficient and other thermal properties of the simulated crystalline silicon modules [4].

With anticipated climate change, the expectation is that ambient air temperatures will be elevated around the world. However, the geographic distribution of these changes are unknown. The timing of the changes in air temperature is another factor since increases air temperature during evening hours has less impact on potential PV efficiency than increases in daytime.

One predicted aspect of climate change over the next century is to impact the average site related cloud cover will decrease except in the driest and hottest regions. Generally, the expectation in Northern Europe is that cloud cover will decrease with climate change, although with seasonal and geographic variations [5]. This decrease in cloud cover is expected to increase ground level irradiance which will also increase expected PV output. Thus, there are two competing influences expected with climate change: increases to expected array irradiance, but also increase to the air temperature around modules which can be expected to reduce module operating efficiency.

At the same time, older weather files, such as the IWEC2 weather files that are widely used for building simulation have been produced, not only in an older historical time frame, but also before the advent of satellite data sources. These sources have dramatically improved the prediction of ground level solar irradiance as well as the split between direct beam and diffuse radiation that is important to accurate prediction of PV performance[6][7].

Currently, PV array energy production are well predicated by the TRNSYS hourly simulation which shows superior prediction with known irradiance, tilt, PV characteristics and surrounding temperatures [8]. However, weather data is a key driver for realistic simulation of PV system performance. This varies over time, seasonally, yearly and stochastically from one year to the next.

Surprisingly, users sometimes predict PV performance based on weather datasets that may be antiquated. For instance the IWEC2 weather data set (1994-2011) are often used to predict PV performance as it pertains to building energy use within building energy simulations[9]. However, recently more up to date weather data sets have become available such as a TMY2007-2021 data sets from Climate.OneBuilding.org. These data sets, being ten years more recent already show changes in weather data reflecting the early indicators of future anticipated climate change. These files have the advantage of being drawn from real recorded data showing how solar irradiances and prevailing temperatures and how they have change over a twenty period in Europe. This source also uses a sophisticated evaluation method, starting with the ISD (ground station data) and then supplementing with the ERA5 solar radiation data satellite reanalysis data [10][11].

2 METHODS

For this example analysis, we predict how a 10 kW south facing PV array at 30% tilt will produce electricity using the older IWEC2 files against the newer TMY2021 (2007-2021) files. We also show how these files compare to predictions of the finely tuned PVGIS-SARAH2 database [11]. We evaluated both how temperatures and irradiances have changed. For an indicator, we use the Global Horizontal Irradiance (GHI) and average temperature changed from these two files. We also show how TRNSYS predicts electricity output for the two different weather files as well a prediction from PVGIS.

3 RESULTS

Both daytime and nighttime temperatures changes are shown and statistically summarized. Within this analysis, we find, for instance that the average predicted annual PV electricity in Vienna is predicted to be 8-15% greater with the newer TMYs or the PVGIS than when using the older 1994-2011 data, even with higher prevailing ambient temperatures.

Figure 1 shows the monthly average GHI and drybulb temperatures for the IWEC2 vs. the TMY2021 for Vienna-Schwechat International Airport. The averages are superimposed over a 24 hour cycle to give indications on when temperature differences occur against solar irradiance levels.

It can be seen that much of the increased solar irradiance comes in winter and spring. On the daily cycle much of the increased air temperature is in late afternoon and evening when PV efficiency is less affected. Some of the increased PV electric output comes from the satellite-related change in solar irradiance split between diffuse and beam components. Predicted annual electric energy production is 10,533 kWh for the older IWEC2 file against 12,183 for the newer TMY2021.

Our initial analysis considered the eight varied European locations to provide geographic indicators:

- Vienna (Schwechat), Austria
- Stockholm (Arlanda), Sweden
- Milan (Malpensa), Italy
- Munich (AP), Germany
- Lisbon (Portela), Portugal
- Rome, (Fiumincino) Italy
- Paris (Orly), France
- Madrid (Barajas), Spain

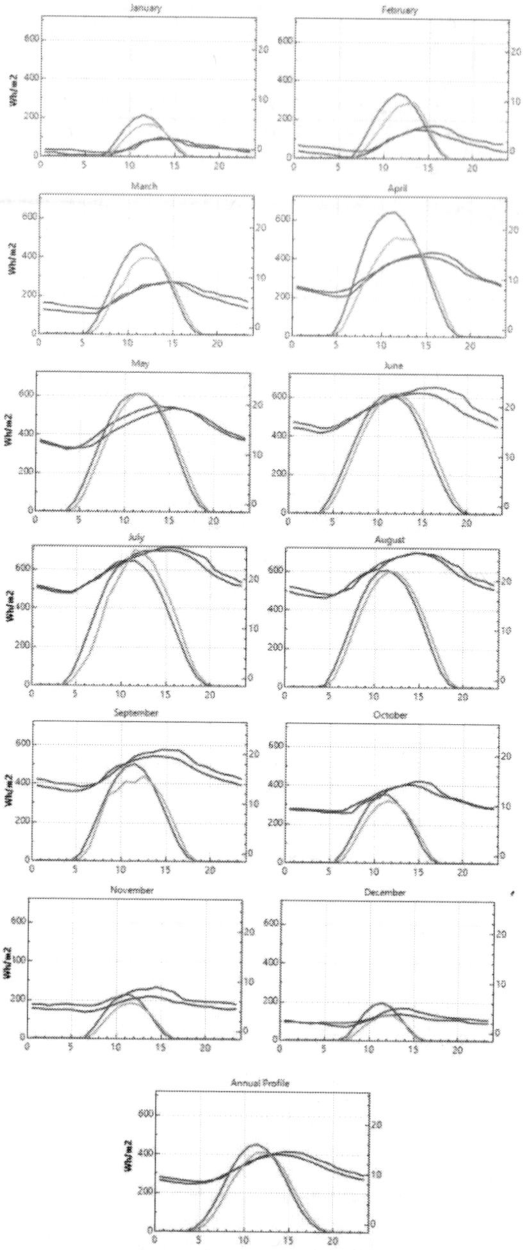

Figure 1: Monthly Global Horizontal Irradiance and drybulb temperatures for IWEC2 file (1994-2011) against newer TMY2021 (2007-2021) for Vienna International Airport. GHI: Blue= TMY2021; Orange= IWEC2; Drybulb ambient temperature: Dark Green (TMY 2021), Dark Red (IWEC2).

10.4229/EUPVSEC2023/4BV.4.19
020346-002

Location of 10 kW PV System	Predicted Annual kWh to Grid	Percent Change (%)
Vienna, Austria IWEC2	10533	
Vienna, Austria TMY 2021	12183	
Change	1650	15.7%
Stockholm, Sweden, IWEC2	10422	
Stockholm, Sweden, TMY2021	10006	
Change	-416	-4.0%
Milan, Italy, IWEC2	12796	
Milan, Italy, TMY2	12852	
Change	56	0.4%
Munich, Germany, IWEC2	10205	
Munich, Germany, TMY2021	11902	
Change	1697	16.6%
Lisbon, Portugal, IWEC2	13382	
Lisbon, Portugal, TMY2021	15217	
Change	1835	13.7%
Rome, Italy, IWEC2	13537	
Rome, Italy, TMY2021	14713	
Change	1176	8.7%
Paris, France, IWEC2	9950	
Paris, France, TMY2021	10806	
Change	856	8.6%
Madrid, ESP, IWEC2	14669	
Madrid, ESP, TMY2021	14428	
Change	-241	-1.6%

Table 1: Energy simulation results and percentage change for different locations and weather datasets.

As shown in Table 1, while there was variation in the impact of more recent weather files from one European site to the next.

Generally, the newer TMY data showed predictions of higher PV energy production, largely due to greater plane or array irradiance. The analysis showed the newer TMY2021s and the PVGIS prediction values were similar although with expected variation. In six of the eight cases, the older IWEC2 sites indicated lower PV electricity production in spite of higher indicated ambient temperatures. Over the eight examined sites, PV electricity to grid increased by about 7% compared with the older weather files.

4 CONCLUSIONS

The newer TMY2021 or PVGIS calculations for predicting PV performance are seen to be more accurate. As average annual temperatures also increased, such increased renewable energy production may be successful to near Zero Energy Homes which will increasingly rely on heat pumps for heating, cooling and water heating. Across the eight sites, the average prediction by the PVGIS-SARAH2 calculations were within 1% of the average prediction using TRNSYS with the TMY2021 weather source.

An essential conclusion is that more recent weather data is desired for better PV electricity predictions. This is particularly true given changing climate with respect to cloud cover and temperature.

In the future, we plan to extend the evaluation will also be made on a single location looking at morphed weather file so the year 2070 for the IECC median and 90% cases where ambient temperatures can be expected to be considerably higher [12]. Implications will be discussed including factors that can help control module operating temperatures as well as suggestions for more successful rooftop PV implementation on NZEBs.

ACKNOWLEDGMENTS

The views expressed are purely those of the authors and do not represent any official position of the European Commission. The authors wish to thank Christian Thiel and Maurizio Bavetta (JRC) for supporting this research.

REFERENCES

[1] D. D'Agostino, D. Parker, I. Epifani, D. Crawley, L. Lawrie, How will future climate impact the design and performance of nearly zero energy buildings (NZEBs)?, Energy, Volume 240, 2022, https://doi.org/10.1016/j.energy.2021.122479.

[2] D'Agostino, D., Tzeiranaki, S.T., Zangheri, P., Bertoldi, P., Assessing Nearly Zero Energy Buildings (NZEBs) development in Europe, Energy Strategy Reviews 36,100680, 10.1016/j.esr.2021.100680

[3] D'Agostino, D., Parker, D., Melià, P., Environmental and economic data on energy efficiency measures for residential buildings, Data in Brief 28,104905, 10.1016/j.dib.2019.104905

[4] Huld T. and Gracia Amillo A. M. Estimating PV Module Performance over Large Geographical Regions: The Role of Irradiance, Air Temperature, Wind Speed and Solar Spectrum. Energies. 2015 8, 5159-5181.

[5] Sfică, L, Beck, C, Nita, A-I, Voiculescu, M, Birsan, M-V, Philipp, A. Cloud cover changes driven by atmospheric circulation in Europe during the last decades. Int J Climatol. 2021; 41 (Suppl. 1): E2211–E2230. https://doi.org/10.1002/joc.6841

[6] González-Peña, D.; García-Ruiz, I.; Díez-Mediavilla, M.; Dieste-Velasco, M.I.; Alonso-Tristán, C. Photovoltaic Prediction Software: Evaluation with Real Data from Northern Spain. Appl. Sci. 2021, 11, 5025. https://doi.org/10.3390/app11115025

[7] D'Agostino Delia, Parker Danny, Melià Paco, Giovanni Dotelli, Optimizing photovoltaic electric generation and roof insulation in existing residential buildings, Energy and Buildings, Volume 255, 2022.

[8] D'Agostino, D., Parker, D., Melià, P., Dotelli, G., Data on roof renovation and photovoltaic energy production including energy storage in existing residential buildings , Data in Brief 41,107874

[9] Y. Joe Huang, ASHRAE Research Project RP-1477, "Development of 3012 Typical Year Weather Files for International Locations," IWEC2, by White Box Technologies, Moraga, California, 2012.

[10] C. C. C. S. ERA5: Fifth generation of ECMWF atmospheric reanalyses of the global climate. Copernicus Climate Change Service Climate Data Store (CDS) (2017).

https://www.ecmwf.int/en/forecasts/datasets/reanalysis-datasets/era5.
[11] Gracia Amillo, A.M.; Taylor, N; Martinez A.M.; Dunlop E.D.; Mavrogiorgios P.; Fahl F.; Arcaro G.; Pinedo I. Adapting PVGIS to Trends in Climate, Technology and User Needs. 38th European Photovoltaic Solar Energy Conference and Exhibition (PVSEC), 2021, 907 - 911.
[12] Faiman, D. Assessing the outdoor operating temperature of photovoltaic modules. Prog. Photovolt. Res. Appl.2008, 16, 307–315. S.E. Belcher, J.N. Hacker and D.S. Powell, "Constructing Design Weather Data for Future Climates, Building Serv. Eng. Res Technol 26, 2005, pp. 49-61.

10.4229/EUPVSEC2023/4BV.4.19
020346-004

EXAMINING THE APPLICABILITY OF END-OF-LIFE (EOL) PHOTOVOLTAIC (PV) PANELS AS A BUILDING MATERIAL

Roshan R Rao* and Monto Mani
Centre for Sustainable Technologies,
Indian Institute of Science, Bangalore, India
*roshanrao@iisc.ac.in *(+91 9538725513), monto@iisc.ac.in

ABSTRACT: Recently, in 2022, global PV installations crossed 1 TW (1000 GW) of solar PV. It is projected that by the end of 2050, a cumulative of 70-80 million tonnes of End-of-Life (EoL-PV) waste is envisaged. Such enormous PV waste heading into landfills severely threatens the environment and human health. A hitherto untried solution is proposed to upcycle EoL-PV panels as a building material to extend its useful life. This work aims to address the challenge of understanding the degradation-induced changes in the optical and thermal transmittance properties of the EoL-PV through extensive field and laboratory measurements. The measured Solar Transmittance of the EoL-PV panels is consistently lower than that of new PV panels by 11 ~ 37.6 %. The encapsulant discolouration in EoL-PV causes a maximum reduction in the Blue to Vis-IRA region of the spectrum, and Infrared B (1400~3000 nm) unaffected. The U-value of the EoL-PV and new PV panels are in the range of 9.85 ~10.58 W/m^{2}K, and the apparent thermal conductivity of the PV panels is in the range of 0.45 ~ 0.85 W/mK. The adoption of EoL-PV may alter thermal performance compared to new PV in buildings.
Keywords: End-of-Life Photovoltaics, Alternative building materials, thermal transmittance, solar transmittance

1 INTRODUCTION

The renewable energy share in the energy mix worldwide is rising sharply to mitigate climate change, with Solar Photovoltaics (PV) carrying a large share of 3.6%. However, the end-of-life (EoL) management of decommissioned PV panels is emerging as a serious concern, globally. It is projected that by the end of 2050, a cumulative 70~80 million tonnes of EoL-PV waste will be generated worldwide, with 4.5~7.5 million tonnes envisaged in India [1]. Such enormous PV waste heading into landfills severely threatens the environment and human health [2], [3]. The current study explores a novel, hitherto untried, approach of upcycling EoL-PV panels as a building material. EoL-PV also carries potential for application in housing, alleviating the pressure on conventional building materials. Further, upcycling PV panels could extend the use-phase by a few decades, buying time till recovery and recycling options become economically viable. This work aims to understand the probability of a particular degradation type for a given climate zone in India and to determine the degradation-induced changes in the properties of EoL-PV. The first objective is to understand the probability of a particular degradation type for a given climate zone in India. The next objective is to determine the degradation-induced changes in the properties of the PV panel (EoL-PV). A probability-based tool has been developed to achieve the first objective. Our previous studies [4], [5], involving visual observations of PV panels from Bangalore (moderate climate zone) showed that 100% of all 33 EoL-PV panels surveyed carried encapsulant discolouration above most cells. The developed tool calculated the probability of degradation for moderate climate zone. The dominant degradation type in this climate zone is encapsulant discolouration which is in concurrence with our observations. The panels carrying degradation were further examined for its applicability in buildings. Extensive field and laboratory measurements were carried out to determine degradation-induced changes in the optical and thermal transmittance properties of the EoL-PV. The approach to measure the Solar Transmittance of an EoL-PV panel of large size is the first of its kind, and the Thermal Transmittance of an EoL-PV panel using the HotBox facility does not have any precedence. This is understandable, as adopting EoL-PV panels as a building material is a new approach introduced in our work.

2 METHODOLOGY

2.1 A Probability based Tool

A probability-based tool has been developed to achieve the first objective. The tool can also calculate each degradation type's impact on the electrical performance which provides an estimate of PV lifetime. The tool is based on survey data of physical degradation observations of PV panels collected all over India during multiple campaigns by NCPRE/NISE in 2013, 2014, 2016 and 2018 [6]. The data of the number of panels surveyed and the fraction of those panels carrying each degradation type is extracted from all the survey years. The probability is calculated based on the collated data. The following block diagram (see Figure *1*) describes the steps involved in the calculation.

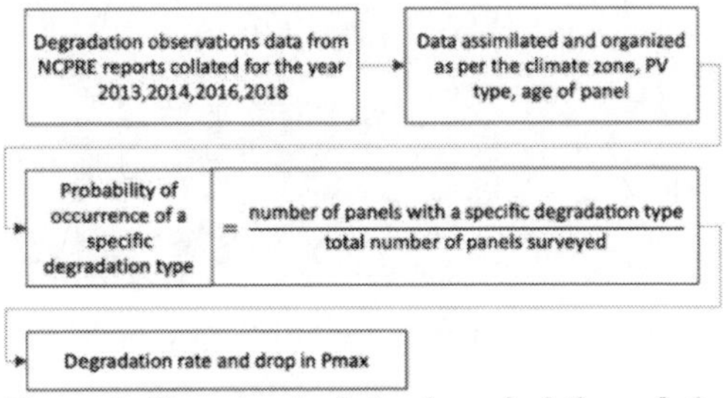

Figure 1 Steps involved in the calculation of the probability of multiple degradation types

2.2 Solar Transmittance Measurement

The solar Transmittance (ST) measurements were performed outdoors using Class 1 pyranometers (to measure irradiance) under the clear-sky sun (see Figure *2*). ST was estimated as the ratio of irradiance below PV panels to irradiance at the plane of the PV panel. This approach closely complies with ASTM E 1084-86:2015 standard [7]. A photo-radiometer has been used to study the spectral band solar transmittance, and a similar procedure was followed to measure spectral band solar transmittance. The EoL-PV and new (unused) PV panels of the exact specification were measured simultaneously.

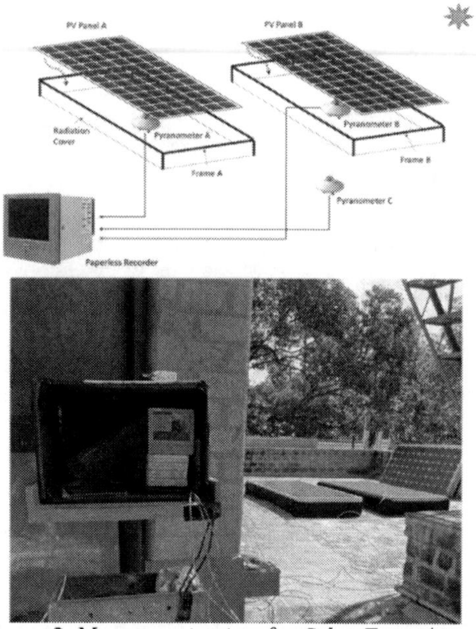

Figure 2 Measurement setup for Solar Transmittance measurement of EoL-PV and new (unused) PV simultaneously.

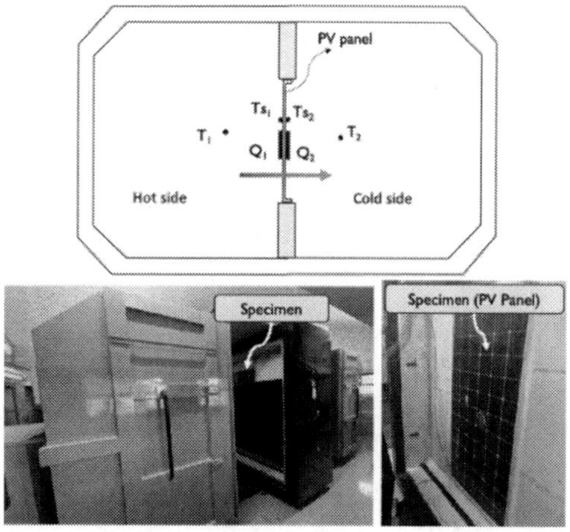

Figure 3 HotBox setup to measure U-value for PV panels (EoL-PV and new (unused) PV panels)

2.3 Thermal Transmittance Measurement

The Thermal Transmittance is an important parameter of a façade/wall/roof. Hence, to understand the EoL-PV's suitability as a building material, thermal Transmittance (U-value) has been determined using a HotBox facility (see Figure 3). The specimen (EoL-PV and new (unused) PV panels) was placed at the centre, and on either side, the chambers were maintained at different air temperatures and this gradient drives the heat flow through the specimen. A heat flux sensor was stuck on the specimen to measure the heat flux, and thermocouples measured the air and specimen surface temperatures. Thermal Transmittance (U) W/m^{2}K (sometimes called as overall coefficient of heat transfer) (ASTM 1363-19[8]) is defined as ,

$$U = \frac{Q}{T_{env,h} - T_{env,c}}$$

Where, Q is the heat flux through PV panel, T_{env} (effective environmental temperature) is the temperature that yields the same net heat exchange in the simple convective mode as the combination of the convective and radiation exchange seen in the test condition. (ASTM 1363-19 [8]) (the sign of the heat transfer coefficients may vary depending on direction of heat flow).

Apparent Thermal Conductivity (k) W/mK (ASTM 1363-19 [8]) is defined as,

$$k = \frac{QL}{T_{PV,h} - T_{PV,c}}$$

L is the thickness of the PV panel. In this study it is 5mm. T_{PV} is the surface temperature of the PV panel. Here, the apparent thermal conductivity of PV panel is the bulk thermal conductivity of the composite laminate.

3 RESULTS AND DISCUSSIONS

3.1 A Probability based Tool

The developed tool has been used to calculate the probability of degradation modes for a moderate climate zone for a PV panel aged 10~20 years. The tool's screenshot (see Figure 4), and the dominant degradation mode is encapsulant discolouration (92 % probability) is highlighted. Metallization discolouration, back sheet degradation, and glass haze are the types of degradation likely to occur following encapsulant discolouration.

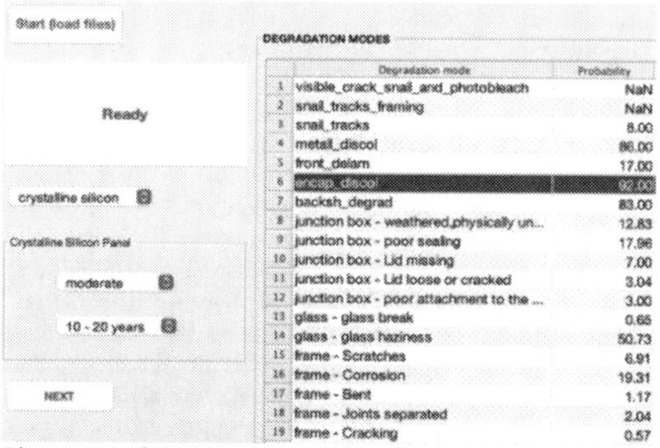

Figure 4 Tool developed on MATLAB based on the collated data from multiple NCPRE/NISE reports.

3.2 Solar Transmittance

The solar Transmittance of the EoL-PV panels is consistently lower than that of new (unused) PV panels by 25 ~ 35 %, as shown in Figure 5 and Figure 6.

The Degradation in EoL-PV causes a maximum reduction in the Blue to Vis-IRA spectrum region, and Infrared B (1400~3000 nm) still passes through the EoL-PV, which is shown in Figure 7 and Figure 8.

The spectral transmittance of the UV, UVA band is zero, in both New(unused) PV and EoL-PVs. The EoL-PVs blue band transmittance reaches as low as less than 0.01 %. The

10.4229/EUPVSEC2023/4BV.4.21
020347-002

IRA band's transmittance is close to 0.1 %. Total broadband transmittance which includes IRB along with the previous bands has a transmittance of 5.2 % ~ 6.2 %. The log of transmittance with wavelength is linear in case of EoL-PVs and New (unused) PV panel.

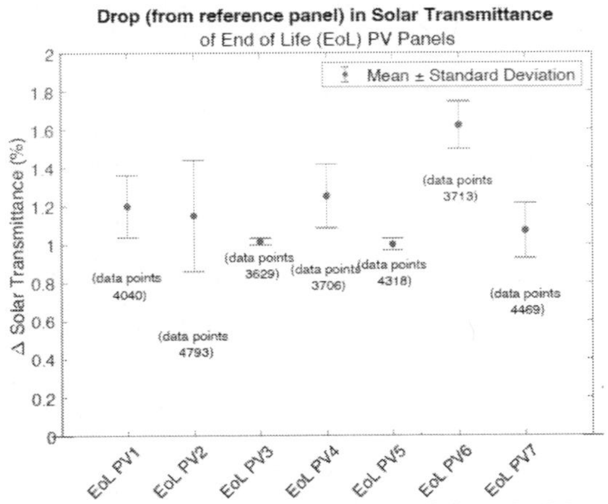

Figure 5 Drop in Solar Transmittance of the EoL-PV compared to reference (new/unused) PV

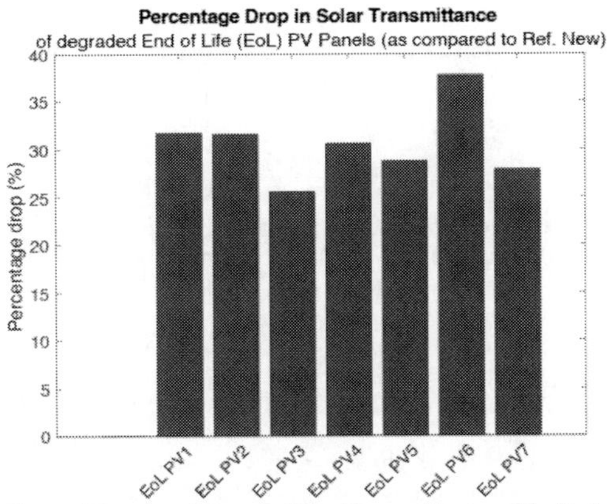

Figure 6 Fractional Drop in Solar Transmittance of the EoL-PV

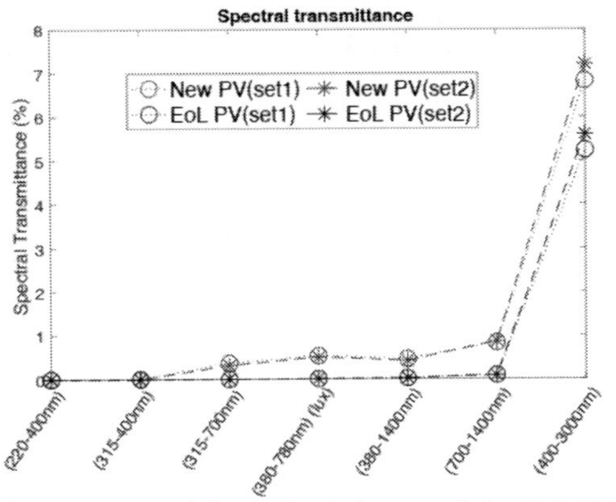

Figure 7 Spectral band Transmittance of the EoL-PV compared to ref. (new/unused) PV panels

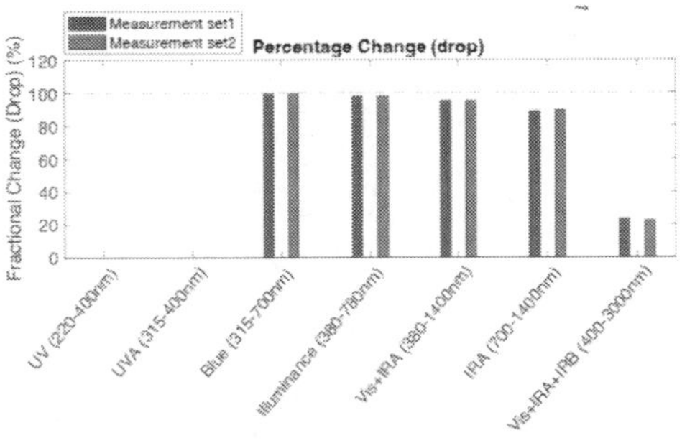

Figure 8 Fractional Drop in Spectral band Solar Transmittance of the EoL-PV

3.3 Thermal Transmittance

A full-scale PV panel (glass- backsheet based crystalline silicon) has been tested for thermal properties in the HotBox. Three EoL-PV panels and two New (unused) PV panels were the specimens examined in the HotBox. There haven't been any such EoL-PV panel measurement found in literature and there are no standards addressing directly for PV panel measurements. The steady state convergence in the HotBox was achieved in around 125 minutes. The air temperature stability (peak to peak variation) is observed to be dependent on the scenario (temperatures maintained in the hot and the cold chamber) and the least stability observed was around ± 0.6 °C where the set temperatures were close to the peak operating condition of the heating/cooling system. The best stability achieved was around ± 0.15 °C for the scenario with the heating/cooling system operating at its moderate capacity. Each PV panel (total 5) was subjected to different temperature conditions in Hot and Cold chambers (coldest being 10 °C and hottest being 45°C). Temperature dependency on thermal Transmittance was not observed. As seen in Figure *10*, apparent thermal conductivity increases with an increase in mean PV surface temperature. No significant changes in Thermal Transmittance are observed between EoL and New (Unused) PV panels (see Figure *9*). The U-value of the EoL-PV and new (unused) PV panels are in the range of 9.85 ~10.58 W/m^{2}K, and the apparent thermal conductivity of the PV panels tested is in the range of 0.45 ~ 0.85 W/mK. The encapsulant discolouration has an impact on the spectral band solar transmittance. Hence, the application of EoL-PV will have changes in thermal performance compared to new (unused) PV in buildings. The quantification of these changes in different climate zone is essential to recommend its suitability as a building material.

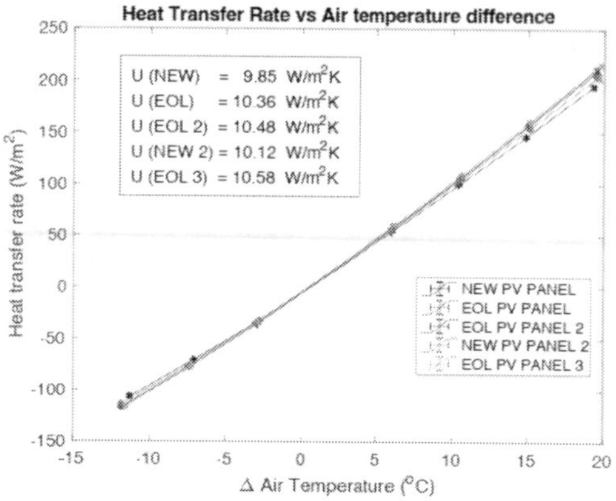

Figure 9 Heat transfer rate for different air temperatures across PV panels.

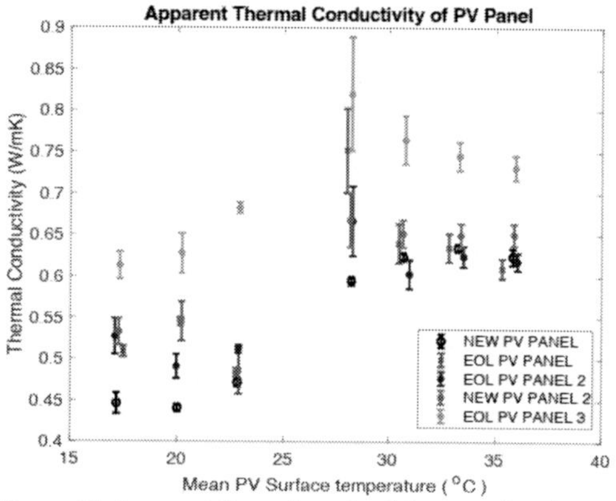

Figure 10 Apparent thermal conductivity as a function of mean PV surface temperature

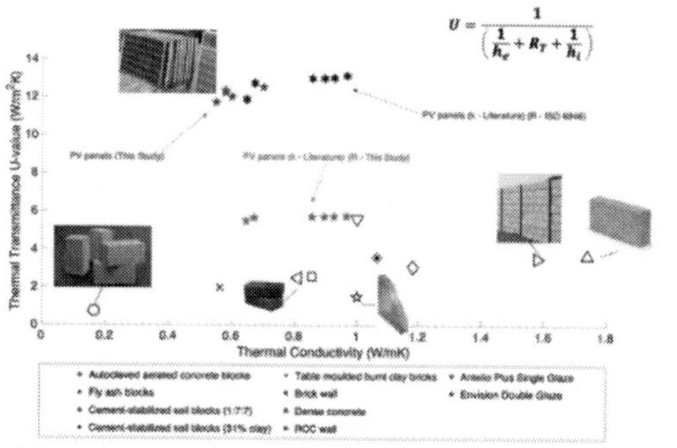

Figure 11 Comparison of the thermal transmittance and thermal conductivity of few typical building materials and the PV panels tested in this study

The thermal transmittance (U-value) of the measured PV panels is found to be very high compared to the conventional building materials currently in use (seeFigure *11*). The high U-value of PV panels compels the need for more insulation to maintain thermal comfort for the occupants. The implication of these properties of EoL-PV panels on the building's indoor thermal performance is understood in the our future studies.

4 CONCLUSIONS

EoL-PV panels were found to carry strong discoloration related degradations. Encapsulation discoloration seems to be the dominant degradation mode in the climate zone considered, followed by metallization discoloration. This gives a direction as to which degradation mode needs to be carefully examined in EoL-PV panels before considering it be adopted in buildings. To investigate the degradation induced changes in the solar and thermal properties, a set of 33 EoL-PV panels and 2 new (unused) panels were taken up for the study . A maiden study has been conducted, measuring under normal sunlight and operating conditions on full-scale PV specimens. The EoL-PV panels carried 11 % (± 4.7%) ~ 37.6 % (± 2.1 %) lower solar transmittance when compared with new PV panels. When integrated as a building envelope, EoL-PVs permit a mere 10 ~ 20 lux of illuminance in comparison to 400 ~ 500 lux observed for new PV panels (96 ~ 97.5% drop in illuminance). The U-value of the EoL-PV and New (unused) PV panels do not show any noticeable change, which discounts the impact of degradation on the thermal transmittance. The U-value of the EoL-PV and New-PV panels are in the range of 11.7 ± 0.17 W/m2K ~ 12.5 ± 0.18 W/m2K. The thermal conductivity of the PV panels was found to be in the range of 0.55 ± 0.02 W/mK ~ 0.7 ± 0.03 W/mK with the slope of the temperature dependency Δk/ΔT (pv mean) (°C) being 0.007 ~ 0.01, with R^2 ranging from 0.3 ~ 0.9. These measurements provide crucial information in examining the climatic response of EoL-PV panels as a building material.

REFERENCES

[1] S. Weckend, A. Wade, and G. Heath, "End-of-Life Management: Solar Photovoltaic Panels," International Renewable Energy Agency and International Energy Agency Photovoltaic Power Systems Programme, 2016.

[2] D. Mathur, R. Gregory, and T. Simmons, "End-of-life Management of Solar PV Panels," Australia, 2020. [Online]. Available: https://www.cdu.edu.au/sites/default/files/the-northern-institute/eolmanagemnetsolarpv_final_e-version.pdf

[3] BIO Intelligence Service, "STUDY ON PHOTOVOLTAIC PANELS SUPPLEMENTING THE IMPACT ASSESSMENT FOR A RECAST OF THE WEEE DIRECTIVE," France, 2011. [Online]. Available: https://ec.europa.eu/environment/pdf/waste/weee/Study on PVs Bio final.pdf

[4] R. R. Rao and M. Mani, "Degradation observation of 9 year old PV modules in Bangalore, India," *Conference Record of the IEEE Photovoltaic Specialists Conference*, pp. 1384–1387, 2019, doi: 10.1109/PVSC40753.2019.8981278.

[5] R. R. Rao and M. Mani, "Case study on degradation of 8-year-old BiPV modules in Bangalore, India," *2018 IEEE 7th World Conference on Photovoltaic Energy Conversion, WCPEC 2018 - A Joint Conference of 45th IEEE PVSC, 28th PVSEC and 34th EU PVSEC*, pp. 758–761, 2018, doi: 10.1109/PVSC.2018.8547411.

[6] NCPRE, "All-India Survey of Photovoltaic Module Reliability," NCPRE IIT BOMBAY, MUMBAI, 2018. Accessed: Feb. 28, 2023. [Online]. Available: http://www.ncpre.iitb.ac.in/ncpre/ncpre_report_download/ncpre_rep.html
[7] "ASTM E1084-86(2015) - Standard Test Method for Solar Transmittance (Terrestrial) of Sheet Materials Using Sunlight." https://webstore.ansi.org/standards/astm/astme1084862015 (accessed Sep. 05, 2023).
[8] ASTM, "ASTM C1363 – 19 Standard Test Method for Thermal Performance of Building Materials and Envelope Assemblies by Means of a Hot Box Apparatus," USA, 2019. doi: 10.1520/C1363-19.2.

STUDYING INTERVENTIONS TO REGULATE INDOOR COMFORT IN BUILDING INTEGRATED WITH END-OF-LIFE (EOL) PV PANELS

Suchi Priyadarshani*, Roshan R Rao, Monto Mani
Centre for Sustainable Technologies, Indian Institute of Science
Bangalore, India

*Corresponding Author email: suchip@iisc.ac.in

ABSTRACT: End-of-life (EoL) Solar Photovoltaic (PV) panels are a concern worldwide, with a projected ~100 million EoL-PV panels by 2035 in India alone. A novel approach of integrating EoL-PV panels as a building material would extend the PV's life and offset the use of conventional building materials. In the present work, we examine the application of this approach. This work focuses on achieving Indoor thermal environments close to that of conventional buildings. A full-scale EoL-BIPV has been constructed at the Indian Institute of Science, Bangalore, to conduct this study. Indoor air temperature, relative humidity, and wall surface temperatures were monitored to validate a whole-building simulation model, which was used to test various wall interventions for improved comfort. It has been found that indoor operative temperature trends were comparable to conventional buildings by integrating EoL-PV with Packaging wood and mud plaster. The bulk thermal transmittance (U-value) of the material assembly changed from 5.60 W/m^{2}K (EoL-PV without intervention) to 0.38 W/m^{2}K (EoL-PV with intervention). This lead to reduction in building cooling load, especially in summer season, thereby promoting comfort. The results suggest that EoL-PV can be used as a wall in buildings with appropriate interventions for indoor comfort comparable to conventional dwellings.

Keywords: End-of-Life Photovoltaics, Alternative Building materials, Thermal Comfort, Building Simulation

1 INTRODUCTION

1.1 Background and Relevance

India is currently at the forefront of harnessing solar energy through Solar Photovoltaic (PV) technology, ranking 4th globally in Solar Power Capacity. PV panels degrade with time, resulting in decommissioning from operation, thereby reaching its End-of-life (EoL). In terms of energy generation, "EoL-PV" is termed so due to a reduction in efficiency and decommissioning due to commercial viability issues. Handling End-of-Life Photovoltaics (EoL-PV) in India will be challenging due to its large numbers (~100 million PV panels by 2035), reducing land resources for dumping, and its adverse environmental consequences[1]. Given the state-of-the-art recycling technologies, there has been a lot of focus on EoL-PV management and use.

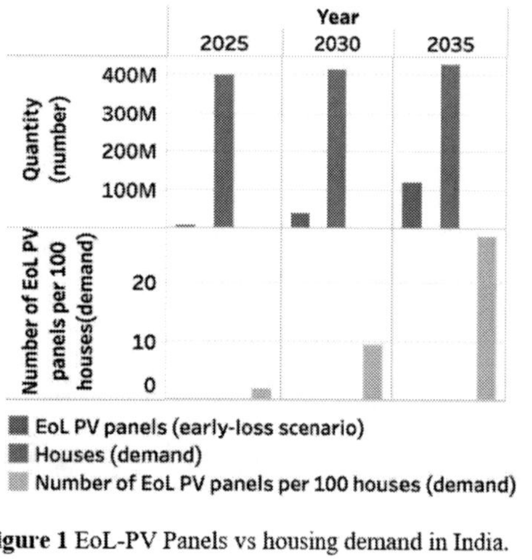

Figure 1 EoL-PV Panels vs housing demand in India.

On the other hand, the growing population in India needs more housing [2]. The use of alternative building materials with minimal cost and low embodied energy is essential and is being promoted by regulatory authorities. Estimation of housing demand in India using Census data [3] and estimates of EoL-PV Projections suggests that India can expect around 25-30 EoL-PV panels per 100 new houses in demand by 2035, as shown in Figure 1. It implies that demand from the building industry can use all the EoL PV generated, both in rural and urban areas due to widespread installation of PV.

Building-Integrated Photovoltaics (BIPV) are predominantly adopted for energy generation in buildings. Looking from an unconventional perspective, we argue that EoL-PV can potentially be used as a building material for the extended life of PV, even if its energy generation is lower than name-plate capacity. Such an approach can provide two-fold benefits of reducing carbon emissions due to conventional building materials and enabling EoL-PV waste management simultaneously [2]. At the same time, the EoL-PV panels can still be used to generate electricity to aid building operation energy requirements.

However, in tropical climatic zones, BIPVs are by themselves unviable for residential applications [4] due to their thermal response to the outdoor climate. Very high thermal transmittance of PV Panels may lead to heated-up indoor environments, increasing cooling loads under normal operations conditions. For its practical application in real-time scenarios, examining pathways for interventions to ensure better thermal performance in these buildings is critical.

1.2 Aim and Approach

The objective of the present study is to present a novel approach to adopt EoL-PV as a building material to extend the life of PV. The study examines pathways for reducing operational energy and enhancing occupant comfort in such buildings. We explore architectural design interventions required in an EoL- BIPV to achieve indoor thermal environments (Ambient Air Temperature, Operative Temperature, and Relative Humidity) comparable to that of conventional buildings using a whole-building simulation approach.

2 MATERIALS AND METHODS

2.1 Full-Scale EoL-BIPV test facility

A full-scale EoL-BIPV (see **Figure** *2*) has been constructed at the Indian Institute of Science, Bangalore, India (Temperate climate zone[5]), to facilitate this study. The panels used for constructing this test facility were decommissioned, 15-year-old PV panels, for which degradations observed experimentally were reported [6], [7] in our previous works. Also, we have discussed the possible theoretical changes in the optical and thermal properties of EoL-PV when compared to a new PV and its implications on operational energy when applied to buildings [8].

Figure 2 Full-Scale EoL-BIPV test facility at Centre for Sustainable Technologies, IISc Bangalore.

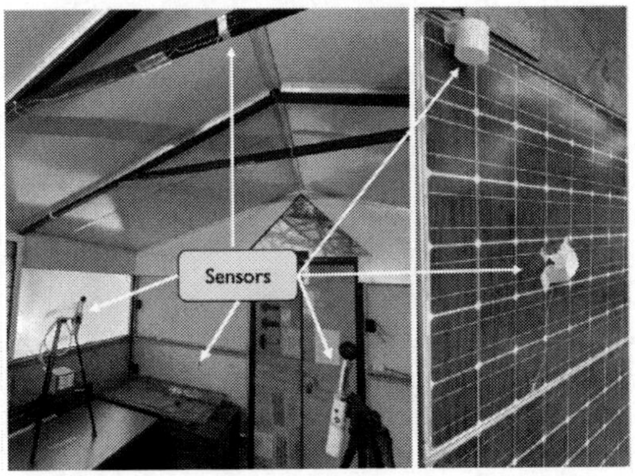

Figure 3 Sensors installed to measure Indoor air parameters and surface temperatures.

The EoL-PV test building was monitored for Indoor environmental parameters (ambient air temperature, globe temperature, wet-bulb globe temperature, relative humidity) with wet-bulb globe thermometer and surface temperatures using thermocouples for nearly three months, with a logging interval of two minutes (see **Figure** *3*). The data obtained was used to examine the response of the EoL PV as a wall and its impact on indoor environmental conditions.

2.2 Whole-building Simulation

A simulation model of the building, with its surroundings, has been created using the Design-Builder tool. The experimentally monitored temperature and relative humidity data was used to validate the building simulation model. The Mean Bias Error (MBE) was less than 2.2%, and the Coefficient of variation of Root Mean Square Error was less than 9.8% for all the data points used for validation.

Figure 4 Simulation model of the EoL-BIPV test facility *(Neighboring structures in gray)*.

Various wall configurations have been tested in the validated simulation model (see Table I). The criterion has been minimal material use (derived from waste), as further addition of material volume would increase cost and embodied energy. The introduced materials for multiple simulations involved EoL-PV with 30mm commercially used Packaging wood, and mud plaster over the EoL-PV panels. The EoL-PV, in this case, had a 50 mm aluminum frame, giving rise to an air gap between the packaging wood and PV. The intent is to achieve an envelope material configuration that would offer indoor thermal environments close to that achieved by replacing wall material with conventionally used wall material (Fired Clay Bricks with Cement Plaster on both interior and exterior surfaces).

Simulation results were used to compare operative temperature (mean indoor air and mean radiant temperature) and relative humidity for all simulation scenarios. Heating cooling loads were computed for indoor operative temperature between 22~24°C for occupant comfort.

Table I: Material assemblies used for simulation

Scenario		Thickness(mm)
EoL-PV Wall		**5**
Conventional Wall	Exterior Cement Plaster	15
	Fired Clay Brick	200
	Interior Cement Plaster	12
		Total 227
Intervention 1	EoL-PV Wall	5
	Air Gap	50
	Plywood	30
		Total 85
Intervention 2	EoL-PV Wall	5
	Air Gap	50
	Plywood (x2)	60
	Mud Plaster	15
		Total 130

10.4229/EUPVSEC2023/4BV.4.22
020349-002

3 RESULTS AND DISCUSSIONS

3.1 EoL-PV as a building wall

EoL-PV as a wall material has many properties significantly different from conventional materials used for wall construction. The thickness of a PV panel is very low, leading to a very low thermal mass. This has resulted in almost equal outdoor and indoor surface temperatures, as shown in **Figure 5**. Also, unlike conventional materials, EoL-PV is not opaque and allows visible light, which can further heat the indoor space, clearly evident with a higher globe temperature close to the wall (see **Figure 6**). During the period of measurements, all the windows in the test facility were kept closed to avoid air exchanges. A significant increase in indoor air temperature was observed when compared to the outdoors due to the absence of any other modes of heat loss through the envelope (see **Figure 7**). The experimental observations indicated that EoL-PV as a wall material does not improve thermal conditions indoors.

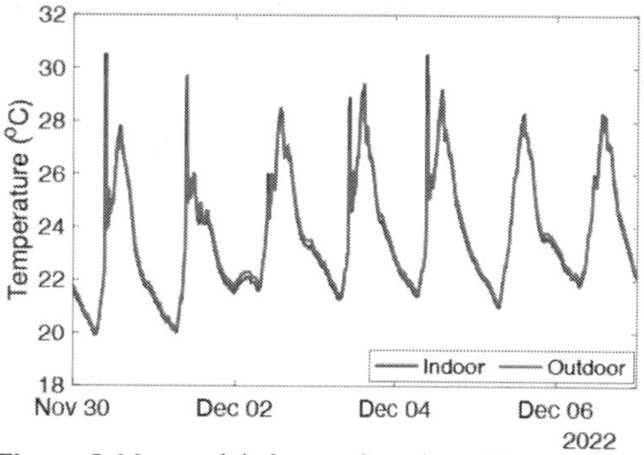

Figure 5 Measured indoor and outdoor Mean surface temperature in EoL-BIPV test facility.

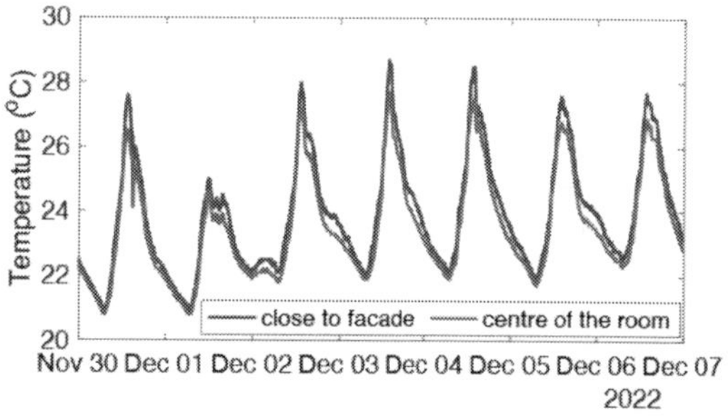

Figure 6 Measured indoor Globe temperature in EoL-BIPV test facility.

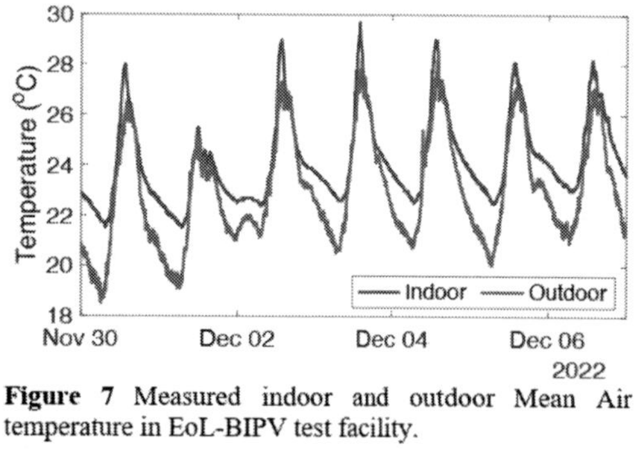

Figure 7 Measured indoor and outdoor Mean Air temperature in EoL-BIPV test facility.

3.2 Interventions to improve indoor comfort

Simulations were conducted to examine various interventions (see **Table I**) to achieve habitable EoL-BIPV indoor environment. These interventions were also compared with thermal environments of conventional dwellings. Since EoL-PV allows visible light through it, it can cause a rise in operative temperatures indoors (mean of Indoor Air temperature and Mean radiant temperature). It has been found that indoor operative temperature, and relative humidity trends in EoL-BIPV with Intervention 2 (EoL-PV+ Air gap + 2 layers of Packaging wood + Mud plaster) are close to that of conventional dwellings (see **Figure 9** and **Figure 10**, respectively). An increase in total thickness of the envelope (Intervention 1 to Intervention 2) intuitively may cause a decrease in the conductive heat transfer, consequently a drop in the indoor temperature. However, contrary to the intuitive approach, adding more layers of wood to Intervention 2 does not significantly change the indoor temperature. Also, only a small drop in U-Value from 0.42 W/m^{2}K (Intervention) 1 to 0.379 W/m^{2}K (Intervention 2) has been observed (see **Figure 8**). However, a drop in wall U-value from 5.607 W/m^{2}K (EoL-PV Wall) to 0.379 W/m^{2}K (Intervention 2), clearly indicates better wall performance with the intervention.

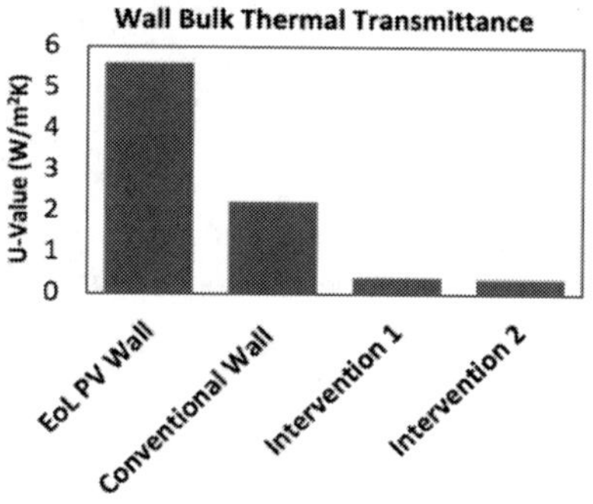

Figure 8 Computed Wall Bulk Thermal Transmittance values for various wall assemblies.

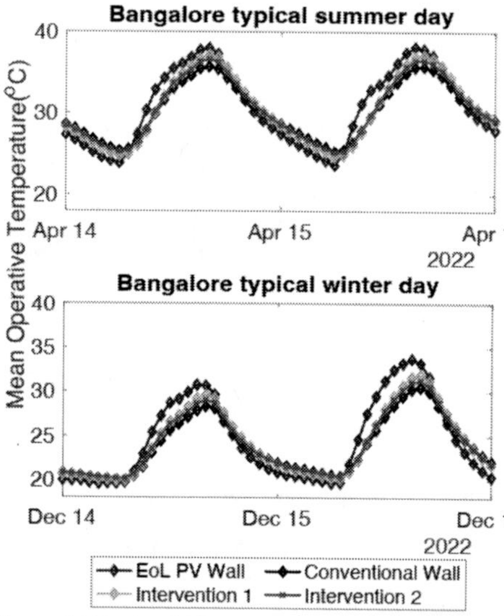

Figure 9 Simulated Mean operative temperature trends for a typical summer and winter day.

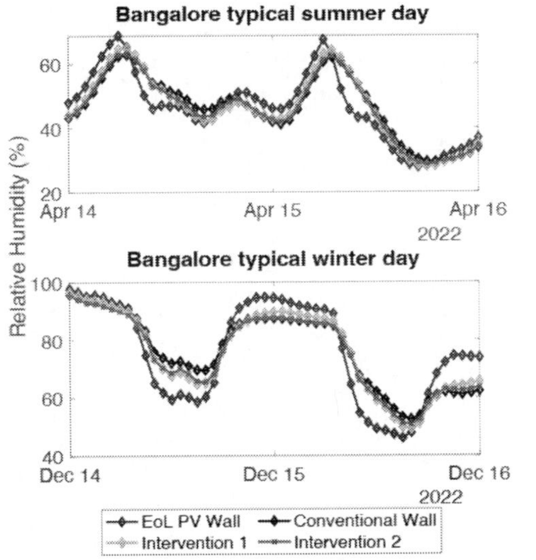

Figure 10 Simulated Relative Humidity trends for a typical summer and winter day.

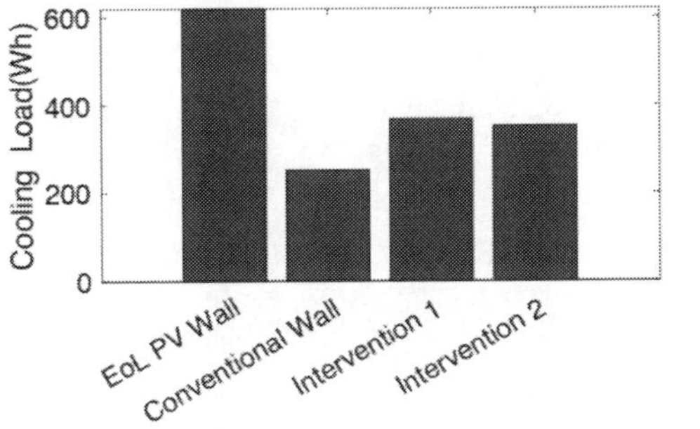

Figure 11 Computed cooling load for a typical winter week.

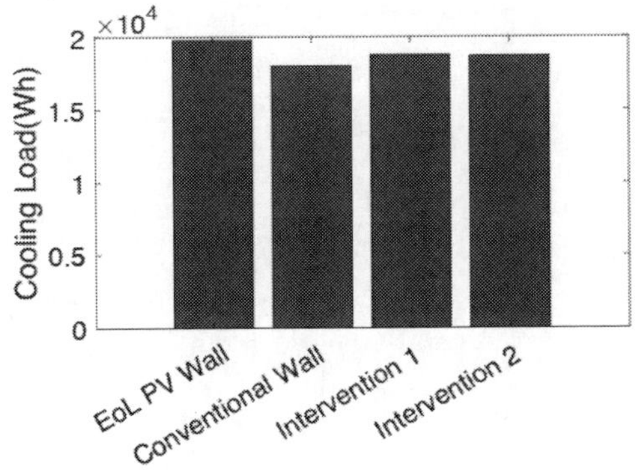

Figure 12 Computed heating and cooling load for a typical summer week.

For the climate of Bangalore, cooling load for typical summer and winter weeks were computed, and has been shown in **Figure 11** and **Figure 12**. The highest operational energy demand was seen in the summer season for cooling. Heating loads were insignificant, as the indoor temperatures always remained greater than 20°C.

The whole-building simulations presented in this article are limited to examining heat transfer through the building envelope. The interventions involve the use of hygroscopic materials like wood, and earth, which are known to moderate the peaks in the indoor relative humidity diurnal cycle[9], [10], thereby decreasing the building cooling load. Detailed hygrothermal investigation could help in examining these aspects for further improvement in indoor comfort and reduction in operational energy. Also, the scope of ventilated cavities in the wall assembly can be explored. Looking forward, a potential investigation of the use of EoL-PV as a building material in different climate zones is pertinent. The results presented in this study are stepping stones towards practically integrating EoL-PV in buildings for human occupation.

4 CONCLUSION

This study explores the application of End-of-Life Photovoltaic (EoL-PV) panels as a building material. Architectural interventions that could aid in improved performance of such buildings for indoor comfort are discussed. The results indicate that EoL-PV, when integrated with packaging wood and mud plaster, can potentially provide indoor thermal environments comparable to that of conventional buildings.

EoL-PV management and alternative building materials (ensuring occupant comfort and wellness) are topics that need urgent addressal, and the present work sets a precedence to examine both hand-in-hand.

5 REFERENCES

[1] P. Nain and A. Kumar, “Ecological and human health risk assessment of metals leached from end-of-life solar photovoltaics,” *Environmental Pollution*, vol. 267, p. 115393, Dec. 2020, doi: 10.1016/J.ENVPOL.2020.115393.

[2] R. R. Rao, S. Priyadarshani, and M. Mani, “Examining the use of End-of-Life (EoL) PV panels in housing and sustainability,” *Solar Energy*, vol. 257, no. June 2023, pp. 210–220, 2023, doi: 10.1016/j.solener.2023.04.033.

[3] “Census of India Website : Office of the Registrar General & Census Commissioner, India.” https://censusindia.gov.in/ (accessed Jan. 12, 2021).

[4] M. Mani, G. Aaditya, and N. C. Balaji, “Appreciating Performance of a Bipv Lab in Bangalore (India),” in *32nd European Photovoltaic Solar Energy Conference and Exhibition*, Munich, Germany, 2016, pp. 2764–2769. doi: 10.1017/CBO9781107415324.004.

[5] N. K. Bansal and G. Minke, “Climatic zones and rural housing in India. Part 1 of the Indo-German project on passive space conditioning.” 1988.

[6] R. R. Rao and M. Mani, “Case study on degradation of 8-year-old BiPV modules in Bangalore, India,” *2018 IEEE 7th World Conference on Photovoltaic Energy Conversion, WCPEC 2018 - A Joint Conference of 45th IEEE PVSC, 28th PVSEC and 34th EU PVSEC*, pp. 758–761, Nov. 2018, doi: 10.1109/PVSC.2018.8547411.

[7] R. R. Rao and M. Mani, “Degradation observation of 9 year old PV modules in Bangalore, India,” *Conference Record of the IEEE Photovoltaic Specialists Conference*, pp. 1384–1387, Jun. 2019, doi: 10.1109/PVSC40753.2019.8981278.

[8] R. R. Rao, S. Priyadarshani, and M. Mani, "An investigation into thermal performance of buildings built using upcycled End-of-Life Photovoltaic panels," in *IBPSA 5th Building Simulation Applications Conference*, Bozen-Bolzano, South Tyrol, Italy, 2022. doi: 10.13124/9788860461919.

[9] S. Priyadarshani, M. Mani, and D. Maskell, "Discerning relative humidity trends in vernacular and conventional building typologies for occupant health," in *Proceedings of the 17th International Healthy Buildings Conference, 21-23 June, 2021*, M.-K. K. and P. G. S. Guangyu Cao, Sverre B. Holøs, Ed., Oslo: SINTEF Proceedings, ISSN: 2387-4295, 2021, pp. 587–598. [Online]. Available: https://www.sintefbok.no/papers/index/38/sintef_proceedings

[10] S. Priyadarshani, M. Mani, and D. Maskell, "Influence of building typology on Indoor humidity regulation," *REHVA Journal*, vol. 6, no. December, pp. 48–52, 2021, [Online]. Available: https://www.rehva.eu/rehva-journal/chapter/influence-of-building-typology-on-indoor-humidity-regulation

[1] M. Smith, A. Miller, Proceedings 17th European Photovoltaic Solar Energy Conference, Vol. I (2002) 903.

[2] K.W. Boer, Solar Cells 16 (1996) 591.

COLOR PROPERTIES AND CELL-TO-MODULE (CTM) LOSSES OF COLORED BUILDING-INTEGRATED PHOTOVOLTAIC MODULES

Dirk Reinwand, Andreas Wessels, Luis Eduardo Alanis, Sebastian Neven-du Mont, Madeleine Antonios, Martin Mattenheimer, Thomas Kroyer, Martin Heinrich
Fraunhofer Institute for Solar Energy Systems ISE, Heidenhofstraße 2, 79110 Freiburg, Germany
Corresponding Author: Dirk Reinwand | +49 (0) 761 4588 2385 | e-mail: dirk.reinwand@ise.fraunhofer.de

ABSTRACT: Aesthetics of the PV modules plays a significant role for their acceptance and usage in BIPV. An important challenge for this concept is the integration of BIPV modules into the building envelope without significant changes of its appearance. Matching the color of BIPV modules to the building envelope is possible by coloring the front side glasses or by using colored encapsulation materials. Therefore, the optical appearance of different coloring techniques is compared in this work. The color properties are determined by using the CIE (Commission Internationale de l'Eclairage) L*a*b* coordinates. The following coloring methods are compared: Screen-printed glass, MorphoColor®-coated glass, colored foils, colored encapsulants and MorphoColor®-coated encapsulants. We determined that the cell-to-module (CTM) losses due to the applied color are in the range of 7 - 85%, depending on the coloring method. For the minimodules with screen printed coloring we observed CTM losses of 84.0% and 84.5% and for MorphoColor® sputtered glass 6.6%.

Keywords: BIPV modules, MorphoColor®, cell-to-module losses, sputtering

1 INTRODUCTION

One advantage of building-integrated photovoltaics (BIPV) is, that no additional land is required for new photovoltaic (PV) installations. This could be a key factor for areas with a high population but limited space for ground-mounted PV [1]. Often, it is required that a BIPV module does not, or only marginally, differs from the optical appearance of the building envelope (façade and roof) materials [2, 3]. Therefore, in addition to power generation, the aesthetic appearance is significant. The objective of this study is to compare different coloring methods in terms of CTM losses caused by the coloring, and the resulting optical appearance for BIPV modules. The color analysis is done by determining the CIE L*a*b* coordinates. With the module power obtained from their current and voltage (IV) characteristics (before and after lamination), cell-to-module (CTM) losses [4] caused by the different coloring methods were compared. The results of this investigation should lead to an optimized solution for colored modules (e.g., for the usage as solar roof tile).

Coloring methods for BIPV modules have been investigated in previous studies [3, 5]. However, in this case the goal is to determine the optimal coloring technology for a new solar roof tile with an appearance that is comparable to that of a regular roof tile.

2 MATERIALS AND METHODS

2.1 Module setup

To investigate the color properties and the CTM losses of colored BIPV modules three separate groups of modules were manufactured. The bill-of-materials (BOM) for the different groups is shown in Figures 1, 2 and 3. In Group A, the glasses were colored with MorphoColor® [2] or with commercially available colored glass (using screen printing technologies).

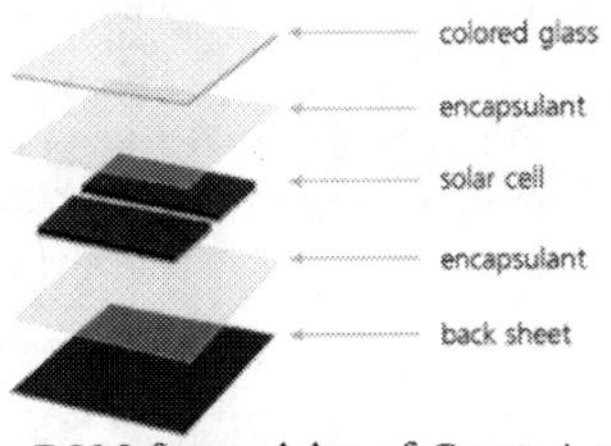

Figure 1: BOM for modules of Group A with colored glass.

All modules were manufactured as glass-backsheet modules in a "sunny-side-down" lamination process, using equivalent processing parameters for all groups. For Group B, the encapsulant itself was colored (with color pigments or sputtered film). Additionally, two different encapsulants, thermoplastic polyolefins (TPO) and ethylene-vinyl acetate (EVA), were used and compared.

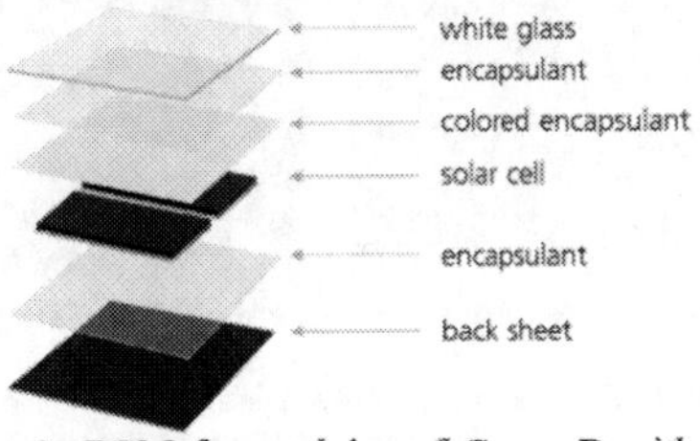

Figure 2: BOM for modules of Group B with colored encapsulants.

For this novel approach, to sputter the coloring film directly on the encapsulant (and not on a glass or on an additional foil), only terracotta colored encapsulants were investigated. For Group C, a MorphoColor®-colored foil was used as an interlayer inside the modules.

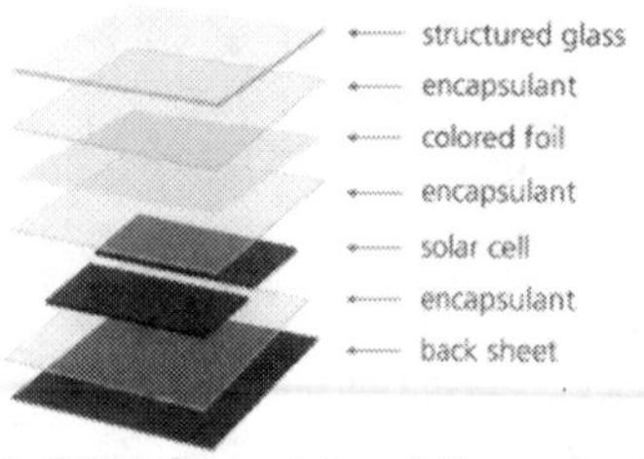

Figure 3: BOM for modules of Group C with colored foils.

This promising approach could offer some advantages for coloring modules e.g., for curved BIPV modules to achieve a homogenous appearance.

2.2 Optical appearance

Besides the electrical performance of BIPV modules, the optical appearance is of high importance when integrating such modules in building envelopes. Therefore, the optical appearance was compared visually for the different modules (for terracotta-colored modules).

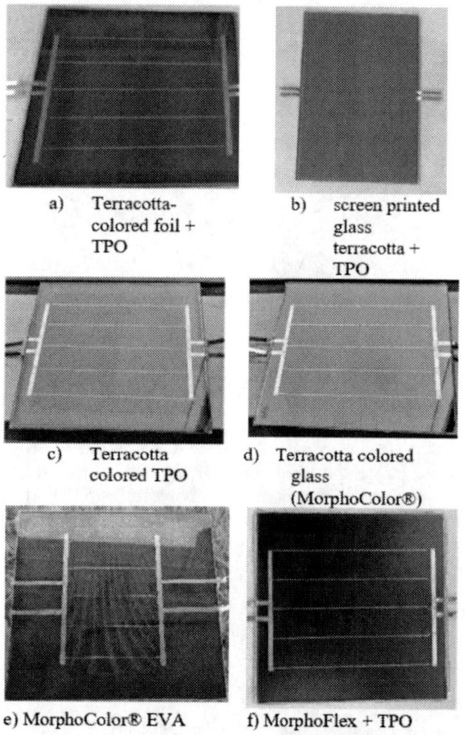

a) Terracotta-colored foil + TPO

b) screen printed glass terracotta + TPO

c) Terracotta colored TPO

d) Terracotta colored glass (MorphoColor®)

e) MorphoColor® EVA

f) MorphoFlex + TPO

Figure 4: Pictures of colored minimodules with colored foils: terracotta (a), screen printed terracotta (b), TPO with color pigments terracotta (c), terracotta colored glass (MorphoColor®) (d), MorphoColor®-colored EVA (e) and terracotta-colored (MorphoColor®) foil (MorphoFlex) (f).

In Figure 4 different minimodules with red-colored glasses, foils or encapsulants are shown. It is visible that the color appearance differs for the different methods, and is highly dependent on the glass (e.g., usage of antireflecting coatings or surface structuring). This shows that not only the thickness of the deposited layers or the amount of color pigments used in foils or encapsulants influence the appearance of the colors, but also the glass itself. One visually appealing result is also shown in Figure 4 (f). For these minimodules, the colored layer was directly sputtered on extra polymeric foil (MorphoFlex) by using the MorphoColor® technology.

2.3 Color properties

The color or optical properties of the different coloring methods were investigated by determining the CIE L*a*b* coordinates [3]. This evaluation of the color space was carried out by the CIE in 1976 and is particularly suitable for the evaluation of color distances with respect to the brightness L*, the red-green color axis a* and the yellow-blue color axis b*. This method allows, in comparison to the RGB table, to evaluate colors with respect to their application (in our case to use it in BIPV modules).

3 RESULTS AND DISCUSSIONS

Subsequently, the results of the visual comparison and the determined color properties are shown. Additionally, the observed CTM losses of diverse groups of minimodules are compared and summarized.

3.1 Optical appearance

For the optical appearance (cf. chapter 2.2) different minimodules with different coloring methods were compared.

Figure 5: Terracotta colored minimodule with MorphoFlex interlayer foil compared to the color of a regular roof tile.

As an example, in Figure 5 one minimodule of Group C with a MorphoFlex foil is compared to the appearance of a regular roof tile.

3.2 Color properties

To derive the CIE L*a*b* coordinates different steps are necessary: First, the spectral reflectance and transmittance was measured by a Fourier spectrometry. For all measurements, a Bruker VERTEX80 Fourier spectrometer equipped with an integrating sphere was used [6]. The measurements were conducted under an angle of 8°, with respect to the orthogonal area to the module surface. As an example, the normal-hemispherical reflectance (R_{nh}) of terracotta and grey colored minimodules (MorphoColor® and TPO) is shown in Figure 6.

10.4229/EUPVSEC2023/4BV.4.23
020351-002

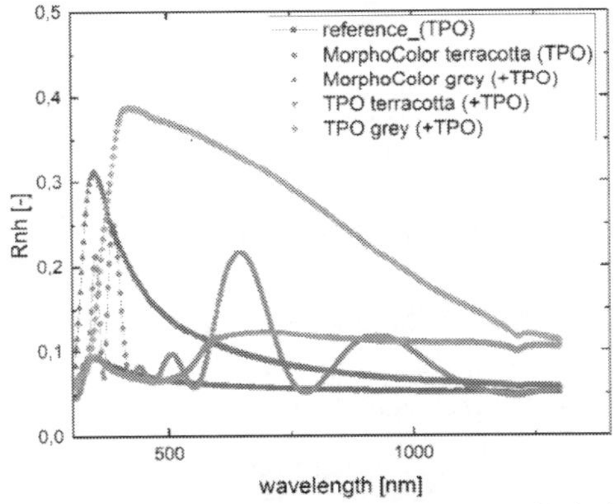

Figure 6: Normal-hemispherical reflectance (R_{nh}) for different grey and terracotta colored minimodules (MorphoColor®, colored encapsulants and reference without coloring).

From the derived spectra, the color values XYZ were calculated [5]. The CIE L*a*b* coordinates were computed according to [7].

Table 1: CIE L*a*b* coordinates for terracotta and grey colored minimodules (MorphoColor®, colored encapsulants and reference modules without coloring).

	Color coordinates L*a*b*		
	L*	a*	b*
Reference module	30.1	0.4	-3.5
MorphoColor® terracotta	37.5	17.3	6.5
TPO (terracotta)	36.2	7.1	8.1
MorphoColor® grey	41.3	1.1	-12.5
TPO (grey)	66.5	-1.1	-3.0

From Table 1 it is visible, that the brightness (L* values) of MorphoColor® terracotta and TPO terracotta is similar. But for the color grey, the brightness of TPO (L * = 66.5) is higher than the MorphoColor®-coated glass (L* = 41.3). One further difference between these two technologies is in the saturation with respect to the green-red axis (a*). The MorphoColor® terracotta (a* = 17.3) corresponds to a red-colored film, whereas TPO terracotta (a* = 7.1) appears more as an orange or golden-brown shade. The CIE L*a*b* coordinates were also determined for reference modules without coloring (cf. Table 1).

3.2 CTM losses

To determine the relative power P_{rel}, IV characterization was performed on all solar cells (pre-lamination), and sample minimodules (post-lamination). To take the overall losses into account (e.g., due to interconnection), reference modules without coloring were built and their post-lamination IV characteristics were compared with their initial cell IV values. In average the CTM losses were between 2 - 4% for all reference modules, independent from the used encapsulant. Subsequently, the results of modules with colored front glass (Group A), colored encapsulants (Group B) and colored foils (Group C) are shown. For each variation of each group five minimodules were manufactured. From the obtained IV data, a deviation (within each group) of ±1% could be determined for all groups.

3.1 Group A – colored front glass

In Group A the relative power of minimodules with screen-printed glasses was compared with sputtered glasses (both in grey and terracotta).

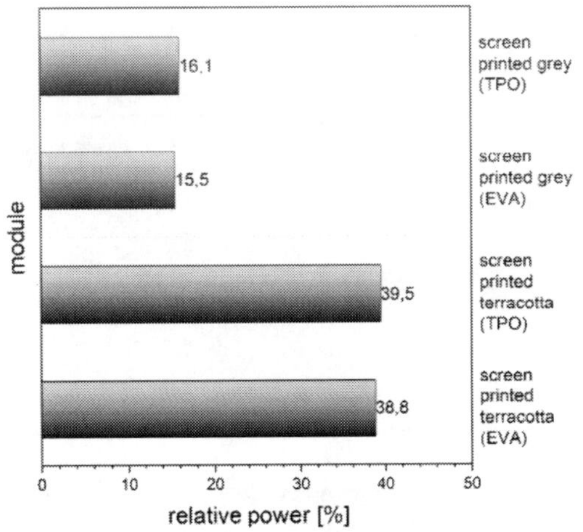

Figure 7: Relative power of minimodules with screen-printed front glasses in grey and terracotta (EVA and TPO).

All minimodules in Figure 7 were manufactured with screen-printed front glasses (in grey and terracotta). Although, the optical appearance of these modules was good, the losses were in a higher range. For the grey colored minimodules, the relative power after lamination was 15.5% and 16.1% and for the modules with terracotta-colored front glass, the relative power was 38.8% and 39.5%.

Another method for coloring the front glasses is the MorphoColor® technology, which was used for the second part of Group A. In Figure 8, the results for mini modules with grey and terracotta sputtered glasses are compared.

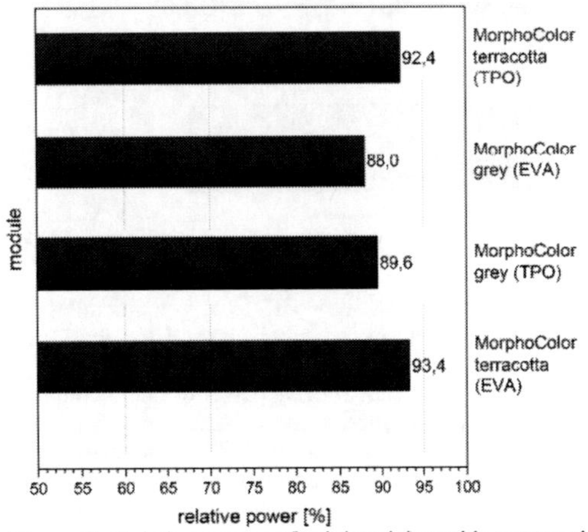

Figure 8: Relative power of minimodules with sputtered front glasses in grey and terracotta (EVA and TPO).

The highest relative Power $P_{rel} = 93.4\ \%$ was measured for the MorphoColor® terracotta minimodules with EVA encapsulants. It is also visible that the relative power of terracotta and grey-colored modules were different for different encapsulants (EVA and TPO). It was also observed that the relative power for the grey-colored modules with EVA was a bit lower than the grey-colored module with TPO, which was not expected. This shall be investigated in further studies.

3.2 Group B – colored encapsulants

In the second group, two different methods of colored encapsulant were investigated. In the first variation an encapsulant with colored pigments was used. Additionally, one reference module with the transparent variant of the same encapsulant (TPO) was built with the same BOM. The results are shown in Figure 9.

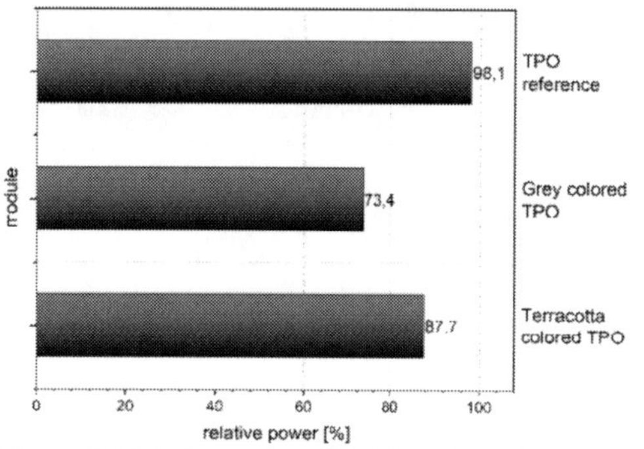

Figure 9: Relative power of minimodules with colored encapsulants (TPO) and one reference module without color.

For the reference modules a relative power of 98.1% was observed. The minimodules with the grey-colored TPO had a relative power of 73.4% and for the terracotta colored encapsulant, the relative power was 87.7% after lamination. In the second variation of group B, the layers were directly sputtered on the surface of the used encapsulants.

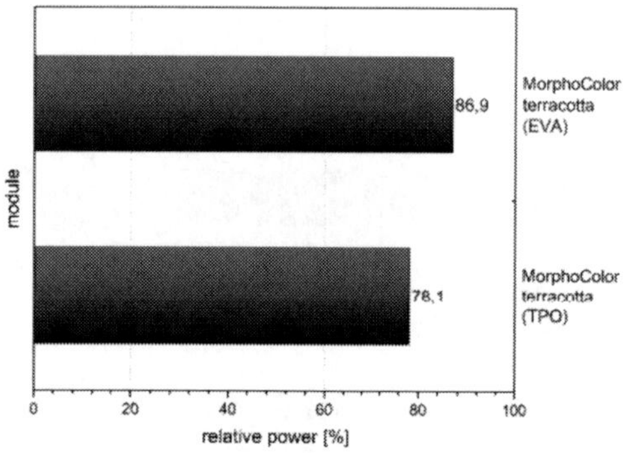

Figure 10: Relative power of minimodules with MorphoColor®-colored encapsulants (EVA and TPO).

In Figure 10 the relative power of minimodules with MorphoColor®-colored encapsulants is shown. For the terracotta-colored modules with TPO, the relative power is 78.1%, whereas for the terracotta-colored modules with EVA the relative power is 86.9%.

3.3 Group C – colored foil

All modules in this group were manufactured with an additional foil to achieve the intended color. Additionally, to improve the performance, structured glasses were used for these minimodules. In the first variation of group C, the modules were manufactured with industrially available grey- and terracotta-colored foils (cf. Figure 11).

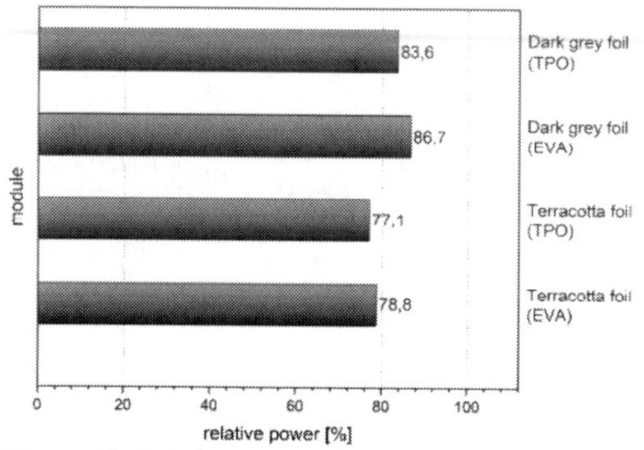

Figure 11: Relative power of minimodules with colored foils in grey and terracotta (and different encapsulants: EVA and TPO).

The relative power for the terracotta-colored modules was between 77.1% and 78.8%. Higher relative power was achieved with the dark, grey-colored modules ($P_{rel} = 83.6\%$ and 86.9%).

Another method for coloring modules, is to use sputtered interlayer foils. In total four minimodules (three in terracotta and one foil without color as reference) were fabricated to investigate the CTM losses in comparison to the above-mentioned methods.

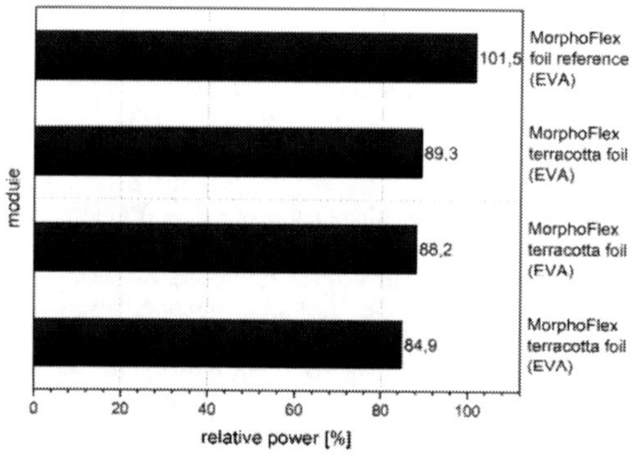

Figure 12: Relative power of minimodules with three different shades of terracotta MorphoFlex foil and structured front glasses.

The relative power $P_{rel} = 101.5\%$ (c.f. Figure 12) for the reference module is due to the usage of structured glass in combination with the MorphoFlex foil (which is also structured). For the shown MorphoFlex terracotta foils (with EVA as encapsulant material) the relative power P_{rel} was between 84.9% and 89.3%. In Figure 13, the relative power for all minimodules of group A, B and C is summarized.

In general, the highest CTM losses (cf. Figure 13) were observed for the screen-printed minimodules (84.0% and 84.5%) and the lowest CTM losses were obtained for MorphoColor® glass (6.6%). The CTM losses for screen-printed glass (terracotta) were approximately 60%, whereas the losses for MorphoColor® terracotta were only in the range of 7 - 8%.

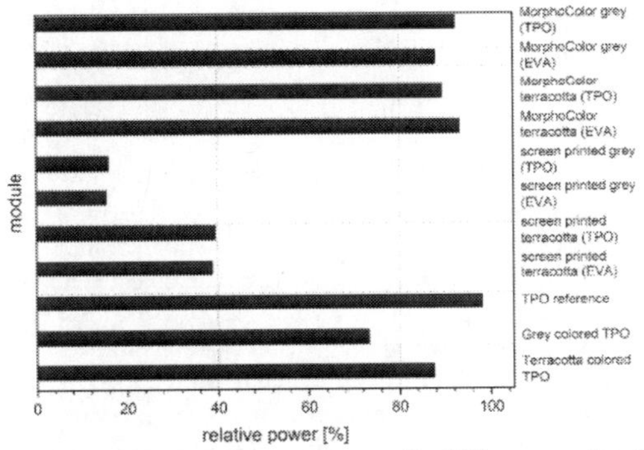

Figure 13: Relative power of different colored minimodules (terracotta) with different encapsulants (EVA and TPO).

Promising results could also be observed for MorphoColor® encapsulants and interlayer foils. E.g., for minimodules with terracotta colored MorphoFlex films the CTM losses were in the range between 10.7% and 15.1% (c.f., Figure 12).

4 CONCLUSION AND OUTLOOK

In conclusion, this study gives an insight on different coloring methods for BIPV modules. Screen-printed glasses, MorphoColor®-sputtered glasses, colored encapsulants, MorphoColor® encapsulants and colored interlayer foils were investigated in terms of optical appearance, color properties and CTM losses.

As one result of the visual inspection of the different colored layers, we successfully demonstrated that with MorphoFlex terracotta interlayer films it is possible to produce modules that are remarkably similar in their appearance to conventional roof tiles. This may also be realized with other technologies shown in this study, however, further optimization would be necessary to achieve comparable results.

Furthermore, the CTM losses of the MorphoColor® technology fall in the range of approximately 6% for terracotta colored glass, while still maintaining a high brightness.

Additionally, determining the color properties of the deposited films allows to optimize the layer characteristics for the desired application.

It must be emphasized that the relatively low losses with MorphoColor-coated glasses, foils or encapsulants could be achieved by various optimization steps specifically taken for this type of BIPV modules, whereas the commercial methods could only be investigated on available samples. It can therefore be assumed that there are still various optimization possibilities for these techniques as well, whereby the losses could be significantly reduced.

5 ACKNOWLEDGEMENTS

This work was supported by the German Federal Ministry of Economics and Climate Protection (BMWK) under the contract number 03EE1115, acronym BALDACHIN and the Ministry of Economics, Labor and Tourism Baden-Württemberg, funding code BW1_2137/03 "MorphoFlex".

6 REFERENCES

[1] T.E. Kuhn, C. Erban, M. Heinrich, J. Eisenlohr, F. Ensslen, D.H. Neuhaus, "Review of Technological Design Options for Building Integrated Photovoltaics (BIPV)", *Energy & Buildings* (2020), doi: https://doi.org/10.1016/j.enbuild.2020.110381.

[2] B. Blaesi, T. Kroyer, O. Höhn, M. Wiese, Claudio Ferrara, U. Eitner, T. E. Kuhn, „Morpho Butterfly Inspired Coloured BIPV Modules", *Proceedings of the 33rd European Photovoltaic Solar Energy Conference and Exhibition,* Amsterdam, 2017, pp. 2630-2634.

[3] C. Kutter, B. Bläsi, H.R. Wilson, T. Kroyer, M. Mittag, O. Höhn, M. Heinrich, "Decorated Building-Integrated Photovoltaic Modules: Power Loss, Color Appearance and Cost Analysis", *Proceedings of the 35th European Photovoltaic Solar Energy Conference and Exhibition,* Belgium, 2018, pp. 1488–1492, doi: https://doi.org/10.4229/35thEUPVSEC20182018-6AO.8.6

[4] Haedrich, I., Eitner, U., Wiese, M., & Wirth, H. (2014). Unified methodology for determining CTM ratios: Systematic prediction of module power. Solar Energy Materials and Solar Cells, 131, 14-23.

[5] Arne Røyset, Tore Kolås, Bjørn Petter Jelle, Coloured building integrated photovoltaics: Influence on energy efficiency, Energy and Buildings, Volume 208 (2020), 109623, ISSN 0378-7788, https://doi.org/10.1016/j.enbuild.2019.109623.

[6] A. Wessels, A. Callies, B. Bläsi, T. Kroyer, and O. Höhn, "Modeling the optical properties of Morpho-inspired thin-film interference filters on structured surfaces," Opt. Express 30, 14586-14599 (2022).

[7] Deutsches Institut für Normung e. V., Farbmetrik — Teil 4: CIE 1976 L*a*b* Farbenraum (ISO/CIE 11664-4:2019)
(11664-4:2019) (Beuth Verlag GmbH, 2020).

ENERGY FLOW ALGORITHM TO THE OPTIMIZATION OF ON-GRID PV BUILDINGS WITH OR WITHOUT BACKUP STORAGE

Kiane Alves e Silva, Luis Miguel Carrasco, Antonio Luis Mata
Instituto de Energía Solar, Universidad Politécnica de Madrid
kiane.asilva@alumnos.upm.es, luismiguel.carrasco@upm.es, a.matag@alumnos.upm.es

ABSTRACT: This article presents a new approach to analyze energy flow rates in self-consumption PV systems with or without backup storage running in a building based on historical hourly energy consumption data. An energy flow analysis tool has been developed, programmed in MATLAB, using the mathematical modeling of the PV system, the meteorological data and the annual energy consumption by a certain consumer. The main objectives of the tool are i) to estimate the energy balance of the PV system in accordance with the PV generator and battery sizes; ii) to optimize the photovoltaic generator and battery size according to the annual energy consumption profile; iii) to characterize the energy flows considering different parameters, e.g. the PV arrays orientations or the energy consumption patterns. The analysis of energy use has been performed using the Gini coefficient as an index to measure the level of inequality between energy consumption and effective solar irradiance. The tool has been applied to three real energy consumption profiles of residential, public administrations and industries, simulating with different PV support structures. The results illustrate the usefulness of this tool in decision making when designing and sizing self-consumption photovoltaic systems.
Keywords: photovoltaic systems, self-consumption, characterization of consumption, energy use.

1 INTRODUCTION

Photovoltaic (PV) self-consumption systems have become increasingly attractive in several countries in recent years, mainly due to the increase in electricity tariffs and reduction of costs associated with their implementation [1]. Just in Spain more than 2.5 GWp of new PV systems for self-consumption has been installed in 2022, a figure 108% higher than that of 2021 [2]. The design and sizing of such systems is commonly based on methods that do not take optimization criteria into account but are limited to considering the available surface on the roof or the user's preferences for whether or not to integrate a battery, without considering aspects such as the self-consumption rate (SCR) or the self-sufficiency rate (SSR) [3]. This practice can lead to the generation of excess energy surpluses, making these investments less profitable and affecting the electrical system. This leads to considering the importance of developing adequate tools for the correct sizing of self-consumption PV systems based on technical-economic optimization.

A literature review shows the development of different tools to size PV systems based on economic optimization seeking to maximize profit from the sale of energy surpluses [4]–[9]. Few studies have been done to optimize the sizing of the system with a focus on better energy use, analyzing how the consumption characteristics influence the energy balance[10], [11] point to the importance of using a real energy consumption data as input for a PV sizing, since the use of aggregated data tends to underestimate costs and is not recommended for self-consumption analysis. Moreover, identifying and understanding the consumption profile's characteristics has a direct impact on overall cost savings, making it possible to predict future energy demand, plan appropriate energy management actions and size a system that responds to the needs of the consumer [12].

Often, energy consumption is obtained through a bottom-up method that may be performed using two different approaches. The first is the consumption estimation of individual households or end-use appliances based on home activity and domestic habits, as shown in [12]–[14]. The second relies on historical consumption data, that allows more accurate results but requires an extensive database which makes it less common due to the difficulty of obtaining such a data, as reported on [15].

Although the approach employing historical data has a lower level of detail about the use of specific equipment, it allows the characterization of the pattern of consumption covering all the variability of use during the monitoring period. For this characterization it is important to establish the correlation between the consumption pattern and the PV generation pattern of the self-consumption system, enabling a better optimization in terms of self-consumption and self-sufficiency [16].

In this regard, this paper presents a new approach to estimate energy flow rates in self-consumption PV systems with or without backup storage based on real hourly energy consumption and meteorological data. A tool has been developed using MATLAB [17] based on the mathematical modeling of the PV system, the meteorological data and the real hourly energy consumption profile of buildings. The main objectives of the algorithm are: i) to estimate the energy balance of the system depending on PV generator and battery sizes; ii) to optimize the PV generator and battery sizes in accordance with the yearly energy consumption profile; and iii) to analyze and compare the energy flows between the PV system, the battery, the building consumption and the grid, considering different parameters, as the location, the type of support structure for the PV generator or the energy consumption pattern.

The energy flow analysis is carried out using SCR, SSR and the so-called Inequality Index (*II*). The *II* is based on the Gini Coefficient (GC), an index commonly used in statistical analysis to measure the level of income inequality in a population. In this work the CG is adapted to measure the level of inequality between energy consumption and effective solar irradiance. Thus, the *II* can be read as an index of the coincidence between the cumulative proportion of the distributions of consumed

10.4229/EUPVSEC2023/4BV.4.24
020352-001

energy and the effective radiation over a year. These three parameters (SCR, SSR and *II*) have been calculated from three real energy consumption buildings (household, public administration and industry) in order to compare the efficiency of the energy use of systems, considering different types of PV support structures (equator-oriented, east-west oriented, or solar tracker) and different energy consumption patterns.

2 ENERGY FLOW ASSESSMENT TOOL

2.1 Energy Flow Algorithm

The methodology to develop the algorithm (Figure 1) has been based on the hourly energy flow between PV production, consumption, backup storage and the grid [18]. This approach enables the comprehensive characterization of energy flows according to the consumption patterns, solar radiation levels, as well as the dimensions of the PV generator and backup storage systems.

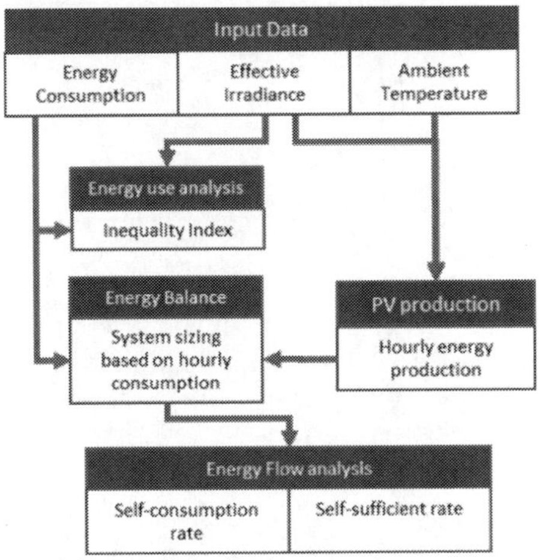

Figure 1: Energy flow algorithm

The input data has an hourly timestep. Effective irradiance depends on the location, type of support structure, orientation and tilt angle of the proposed PV system, given by the building rooftop characteristics. Energy consumption data is obtained by monitoring the electricity consumption of a building over a full year.

Based on the input data, the energy balance between PV, storage, grid and load is performed. The energy balance tool focuses on the energy exchanges and not on the investment profitability. Therefore, the implemented algorithm has the following premises [18]:

- Whenever possible, the energy consumed must be provided by the PV or storage system.
- When photovoltaic production is greater than consumption the battery must be charged, if necessary.
- Excess energy is sold to the grid.
- Battery is only charged from the PV system and never from the grid.

The results are represented in a chart that shows how the energy flow varies according to the size of the PV generator or backup storage. The energy flow charts also allow evaluating how the system sizing influences the SCR, SSR, the participation of the storage system and the purchase (consumption from the grid) and sale (injection of surpluses into the grid) of energy. Considering that the PV system is composed of a PV generator with a backup storage, the SCR indicates the fraction of energy produced by the PV system that is directly consumed by the building and the SSR is the fraction of energy consumed by the building that is supplied by the PV system [19]. The storage consumption ratio ($Storage_{rate}$) indicates the fraction of the energy consumed by the building that is provided by backup storage. The grid purchase rate ($Grid_{purchase}$) pertains to the supply of electricity from the grid to meet the demand, whereas the sell rate ($Grid_{sale}$) pertains to the sale of excess PV energy to the grid. Mathematically, these rates can be defined as:

$$SCR = \frac{E_{PV}^{C}+E_{ST}^{C}}{E_{PV}} \quad (1)$$

$$SSR = \frac{E_{PV}^{C}+E_{ST}^{C}}{E^{C}} \quad (2)$$

$$Storage_{rate} = \frac{E_{ST}^{C}}{E^{C}} \quad (3)$$

$$Grid_{purchase} = \frac{E_{grid}^{C}}{E^{C}} \quad (4)$$

$$Grid_{sale} = \frac{E_{PV}^{Grid}}{E^{C}} \quad (5)$$

where E_{PV}^{C}, E_{ST}^{C}, and E_{grid}^{C} are the consumed energy supplied by the PV generator, backup storage, and grid respectively for a full year. E_{PV}^{Grid} is the PV energy sold to the grid. E^{C} is the annual energy consumed by the building and E_{PV} is the PV energy produced during the same period.

Then, a technical analysis is carried out, considering the SCR and SSR. This analysis is performed considering that the size of the PV generator (C_{PV}) and storage capacity (C_{ST}) are variables, evaluating how these variables are interrelated to serve in the decision making on what should be the optimal size of the system that meet the customer's energy needs. The C_{PV} and C_{ST} can be defined as:

$$C_{PV} = \frac{E_{PV}}{E^{C}} \quad (6)$$

$$C_{ST} = \frac{E_{ST}}{E^{C}} \quad (7)$$

where E_{ST} is the energy capacity of the backup storage.

When the PV energy produced during a full year is equal to that consumed, then $C_{PV} = 1$. Likewise, $C_{ST} = 1$ means that the backup storage capacity is equal to the annual average of the energy consumed daily in the building.

Finally, the inequality index (*II*) analysis is obtained through the elaboration of the Lorenz curve and the Gini coefficient (GC) calculation [20], [21]. The Lorenz Curve is obtained representing the cumulative proportion of the consumption and effective irradiation distributions, as shown in Figure 2. *II* is an indicator of the coincidence between the energy consumed and effective irradiance and is defined by the normalized area between the Lorenz curve and the line of equality. It is calculated as:

$$II = \frac{A}{A+B} \quad (8)$$

where A is the area above Lorenz curve and B is the area below Lorenz curve.

II values close to 1 indicate a high inequality of distribution, while values close to 0 means less inequality, approaching the ideal distribution. Since the effective irradiation varies depending on the PV structure used, this indicator makes it possible to assess energy consumption by comparing different PV support structures and different energy consumption profiles. Analysis II was performed

10.4229/EUPVSEC2023/4BV.4.24
020352-002

for an annual approach (II_a) based on monthly averages and for a daily approach (II_d) based on hourly averages.

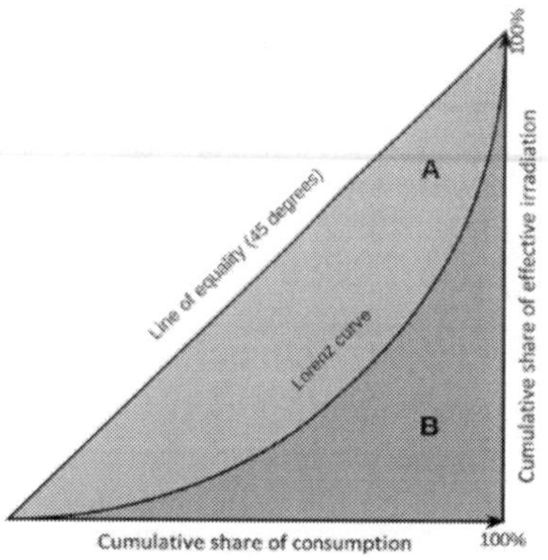

Figure 2: Lorenz Curve Graph example

2.2 Characterization of Energy Consumption

When analyzing hourly energy consumption for a full year, it can be useful to characterize the distribution over time of such consumption. Three segmentation criteria have been used, as shown in Table 1, considering the daily, weekly and yearly energy consumption profiles. Energy consumption has been analyzed according to each criterion individually.

Table 1: Consumption characterization criteria

Criteria	Categories	Reference
Hourly	Morning (8am to 2pm) Afternoon (2pm to 10pm) Nigh (10pm to 8am) Daytime (8am to 10pm) All day	[22], [23]
Daily	Weekday Weekend All week	[23]
Monthly	Summer (June to September) Winter (October to may) All year	[24]

The characterization of consumption is defined for each criterion according to the category with the highest accumulated value. In the case of categories that incorporate other categories (e.g., all week category incorporates weekday and weekend categories) a margin of 5% difference has been admitted.

The categories of hourly and daily criteria are defined based on previous studies of consumption characterization [22] and on the regulation of tariff periods in Spain [23]. The monthly criteria rankings are based on the Spanish Technical Code for Building Construction [24], also used by [25].

3 ALGORITHM APPLICATION AND RESULTS

3.1 Case Study: Individual Household

The algorithm has been tested in a residential building located in Codos (41.29° N, -1.37° W), north of Spain. First, the characterization of energy consumption was performed based on the hourly consumption data for the year 2022, obtained from the user area at the utility website [26]. Figures 3 and 4 present the daily and monthly consumption distribution, respectively. The reddish, yellowish, and greenish areas represent periods of higher, medium, and minimum consumption, respectively.

These Figures show that it is not a typical residential consumption, since the peak periods are at night (between 1 and 4 am), on Saturdays with a higher average consumption on weekends, and in the winter months, especially between November and January when there is a significant consumption of energy throughout the day. On the other hand, the periods of lower consumption are during the morning of weekdays and during the daytime of the summer months, which are precisely the periods of greater irradiance and consequently greater PV production. Therefore, the building's consumption is characterized as nighttime, on weekends and winter.

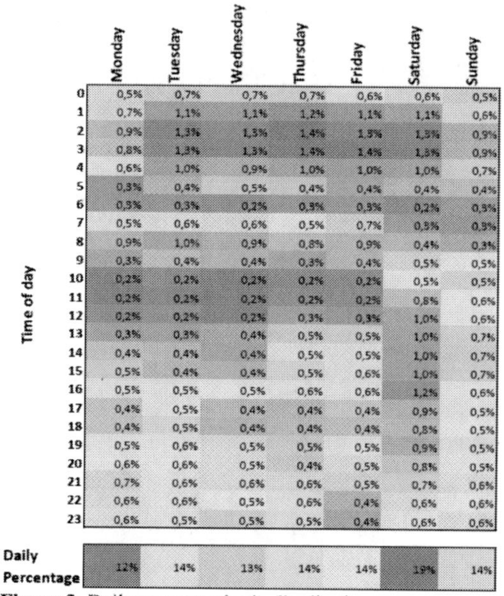

Figure 3: Daily consumption's distribution

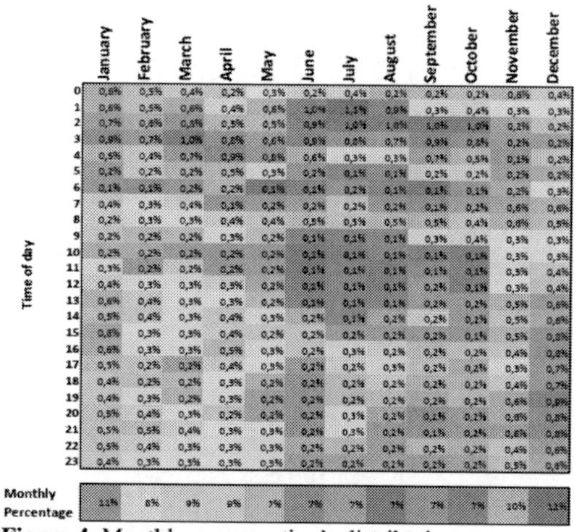

Figure 4: Monthly consumption's distribution

The Lorenz curves have been obtained for such consumption distribution and the effective irradiation over a south-oriented PV generator, tilted 30°. The daily Lorenz curve (Figure 5) shows a very low fit between both distributions, with $Il_d = 0.75$, while the annual curve (Figure 6) indicates a good fit, with $Il_a = 0.2$. This means that solar production and energy consumption coincide very little on an hourly basis, but their distributions are very similar if they are analyzed monthly. From these results it can be predicted that the SCR will be very low, but it can significantly increase adding a battery to the system.

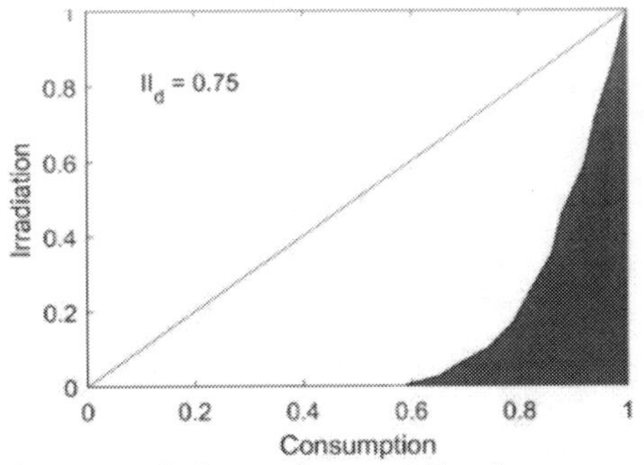

Figure 5: Daily Lorenz Curve and Il_d value

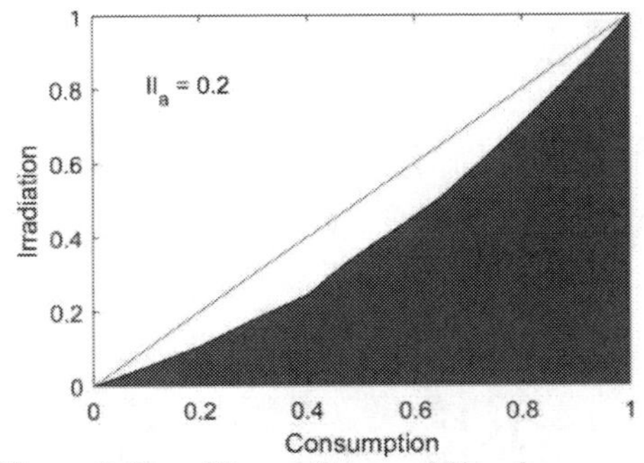

Figure 6: Annual Lorenz Curve and Il_a value

3.2 Energy Flow Analysis

For the case study reported on 3.1, the PV energy production has been obtained from the SISIFO simulation tool [27] for the geographic coordinates of the building, considering the typical meteorological year. The annual consumption of the household is 5,486 kWh and annual effective irradiance is 1,845 kWh/m². Other simulation parameters are shown in table 2.

Considering the PV generation equivalent to the annual average daily consumption ($C_{PV} = 1$), the obtained nominal PV power is 3.26 kWp. On the other hand, for $C_{ST} = 1$ it is necessary a storage capacity of 3.12 kWh.

Table 2: Simulation parameters

Parameter	Value
Azimuth	0°
Tilt angle	30°
NOCT	41°
Temperature coefficient	-0,0026 %/°C
Inverter performance	98%
System Losses	15%

3.2.1 Self-consumption without battery

Figure 7 shows how the energy flow behaves with the PV generator sizing variation, considering a system without storage ($C_{ST} = 0$). The graph shows that the SCR (dashed green line) declines almost exponentially when C_{PV} increases, while the SSR (dashed blue line) increases gradually. The grid sale (yellow line) increases proportionally to the increase in C_{PV}, approaching 100% for $C_{PV} = 1.3$. The grid purchase (red line) slowly declines, tending to stability. As expected, when$C_{PV} = 1$, SCR and SSR are equivalent since PV production equals consumption. As can be seen, the absence of storage in this case lead to both, very low SSR and SCR (0.28 for$C_{PV} =$ 1).

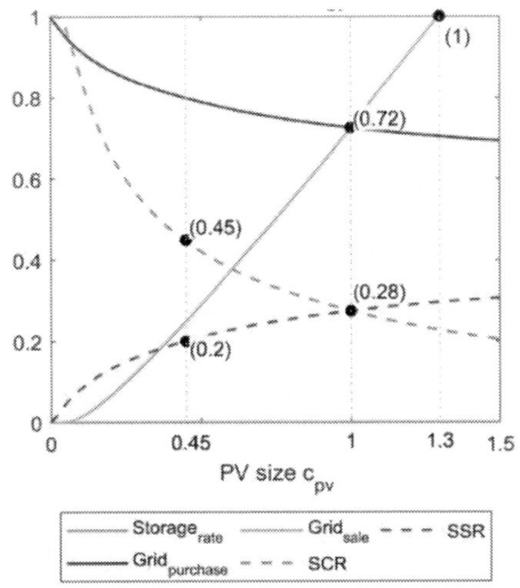

Figure 7: Energy flow for $C_{ST} = 0$

3.2.2 Self-consumption with battery

Considering $C_{ST} = 1$ (Figure 8), the SCR is close to 100% up to $C_{PV} = 0.65$, at which point the storage rate and SSR decrease the slope of its growth. As SCR decreases, the grid sale starts to increase.

The use of such storage system leads to higher SSR and SCR figures (82% for $C_{PV} = 1$), indicating that the energy consumption is supplied by 55% from the backup storage, 27% from the PV generator and 18% from the grid. In addition, there is an energy surplus (18%) that is sold to the grid.

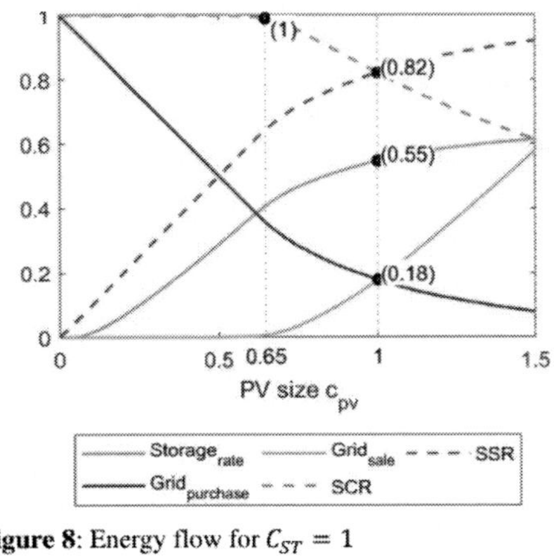

Figure 8: Energy flow for $C_{ST} = 1$

10.4229/EUPVSEC2023/4BV.4.24
020352-004

Finally, a flow analysis varying C_{ST} is achieved fixing a PV system size of $C_{PV} = 0.45$. Figure 9 shows that the increase of the backup storage capacity significantly improves energy use, and the system stabilizes from $C_{ST} = 0.4$, where SSR is 44% and the SCR approaches the maximum point with 98%.

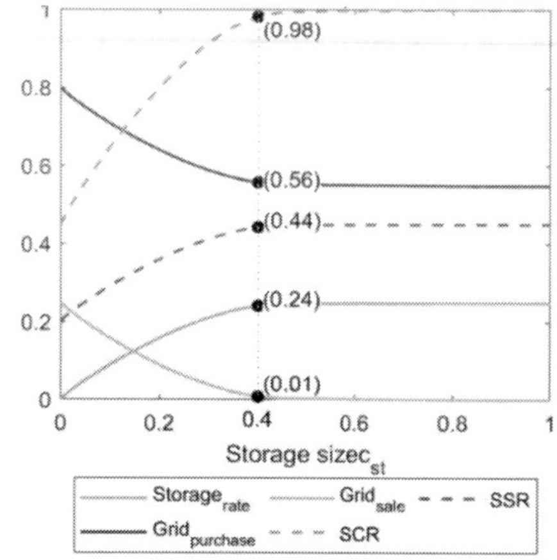

Figure 9: Energy flow for $C_{PV} = 0.45$

3.3 Comparative Analysis and Results

A comparative analysis of the energy use on three different buildings (residential, public administration and industry) is presented below using real energy consumption data. Table 3 presents the characterization of the energy consumption for each building.

Table 3: Characterization of the energy consumption of buildings

Building	Hourly	Daily	Monthly
Residencial	Daytime	All week	Winter
Industrial	Morning	Weekday	Summer
Public building	Daytime	All week	All year

An initial assessment has been carried out evaluating the variation of II depending on three different PV support structures: south-oriented 30° tilted; east-west oriented 30° tilted; and one horizontal axis tracker (see Figure 10). Note that the PV tracker is used only as comparative reference system, since trackers are not usually installed on rooftops. The SCR and SSR indicators have been evaluated in a scenario with PV generator $C_{PV} = 0.75$and storage capacity $C_{ST} = 0.2$ (see Figure 11).

The daily analysis shows that the east-west oriented structure obtains better results than the south-oriented one, approaching the optimal result of the solar tracker. However, in the annual analysis, the south-oriented structure improves the indicator of inequality in residential and public building consumption, with values close to the equality line (0.12 and 0.13, respectively).

Regarding PV energy use, with a storage system size of $C_{ST} = 0.2$, the three structures present similar SCR and SSR results. However, the east-west oriented structure obtained a slightly better result.

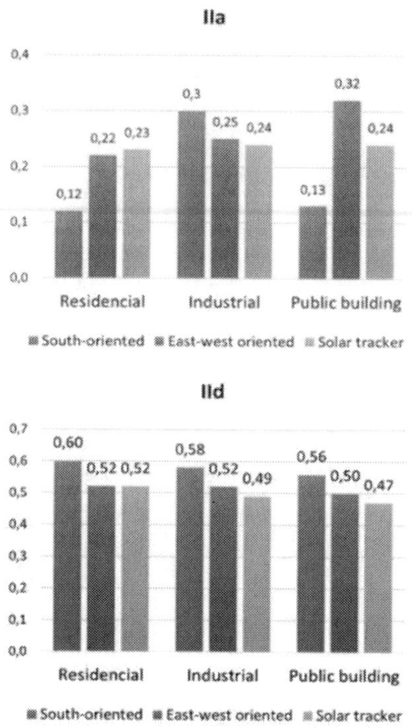

Figure 10: Annual and daily II values for three different support structures

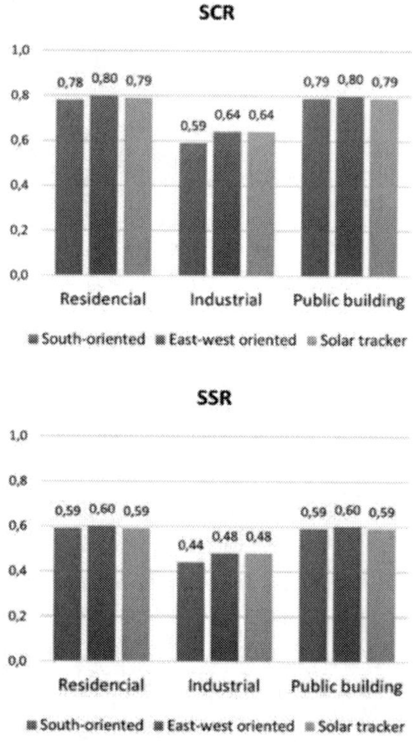

Figure 11: SCR and SSR indicators from PV system with PV generator $C_{PV} = 0.75$ and storage capacity $C_{ST} = 0.2$

10.4229/EUPVSEC2023/4BV.4.24
020352-005

Therefore, in the overall analysis, the east-west structure shows a slightly better performance considering the three selected cases. These results highlight the importance of studying the energy consumption pattern during the sizing phase. Such an examination aids in informed decision-making, not only for achieving optimal sizing but also for selecting the most compatible PV structure that aligns with the specific energy demands of the building.

4 CONCLUSIONS

This article presents an hourly energy flow assessment tool for PV self-consumption systems with or without backup storage. The core contribution of this paper lies on a novel approach to estimate energy flows considering hourly energy consumption and meteorological data, and the mathematical modeling of a PV self-consumption system.

Furthermore, three metric indicators of energy use have been used to assist in decision making, namely *II*, SCR and SSR. The introduction of the *II*, derived from the Gini Coefficient, is an innovative addition to the study of energy flows, allowing the assessment of the alignment between energy consumption and effective solar irradiance. The SCR and SSR parameters offer a comprehensive view of the system's performance.

Through a case study involving a residential building, the paper illustrates the practical application of the developed tool. Results show how varying PV system sizes, with or without backup storage, impacts energy flow and energy use indicators. Additionally, a comparative analysis across different building categories (residential, public building, and industrial) underscores the importance of tailoring PV system designs to specific consumption patterns.

This work sheds light on the methodologies to size self-consumption PV systems with a useful tool to help decision-making when designing and sizing self-consumption PV systems. In addition, it opens the door to the characterization of PV self-consumption systems and to the comparative analysis between different projects and / or different design solutions.

5 ACKNOWLEDGEMENTS

This work has been possible thanks to the Project IND2022/AMB-23718 funded by the Comunidad de Madrid, and the funding from the European Union's LIFE programme: Joining Actors for LOcal development of New large-scale regional energy communities (JALON), under grant agreement LIFE21-CET-ENERCOM-JALON/101076395.

REFERENCES

[1] J. Weniger, J. Bergner, T. Tjaden, and V. Quaschning, "Economics of Residential PV Battery Systems in the Self-Consumption Age," in *EU PVSEC*, 2014, pp. 3871–3878.

[2] "Photovoltaic self-consumption installed in Spain grew by 108% compared to 2021," *Global Solar Council*, Feb. 27, 2023. https://www.globalsolarcouncil.org/photovoltaic-self-consumption-installed-in-spain-grew-by-108-compared-to-2021/ (accessed Jul. 28, 2023).

[3] T. Beck, H. Kondziella, G. Huard, and T. Bruckner, "Assessing the influence of the temporal resolution of electrical load and PV generation profiles on self-consumption and sizing of PV-battery systems," *Appl Energy*, vol. 173, pp. 331–342, Jul. 2016, doi: 10.1016/J.APENERGY.2016.04.050.

[4] S. Silvestre, "Review of System Design and Sizing Tools," *Practical Handbook of Photovoltaics: Fundamentals and Applications*, pp. 543–561, Jan. 2003, doi: 10.1016/B978-185617390-2/50023-4.

[5] E. Zarezadeh, H. Fakharzadegan, A. Ghorbani, and H. Fathabadi, "A probabilistic approach to determine PV array size and battery capacity used in grid-connected PV systems," *ICEE 2015 - Proceedings of the 23rd Iranian Conference on Electrical Engineering*, vol. 10, pp. 1533–1538, Jul. 2015, doi: 10.1109/IRANIANCEE.2015.7146463.

[6] O. Erdinc and M. Uzunoglu, "Optimum design of hybrid renewable energy systems: Overview of different approaches," *Renewable and Sustainable Energy Reviews*, vol. 16, pp. 1412–1425, 2012, doi: 10.1016/j.rser.2011.11.011.

[7] T. Khatib, A. Mohamed, and K. Sopian, "A review of photovoltaic systems size optimization techniques," 2013, doi: 10.1016/j.rser.2013.02.023.

[8] C. Ammari, D. Belatrache, B. Touhami, and S. Makhloufi, "Sizing, optimization, control and energy management of hybrid renewable energy system-A review," *Energy and Built Environment*, vol. 3, pp. 399–411, 2022, doi: 10.1016/j.enbenv.2021.04.002.

[9] S. Alsadi and T. Khatib, "Photovoltaic Power Systems Optimization Research Status: A Review of Criteria, Constrains, Models, Techniques, and Software Tools," *Applied Sciences 2018, Vol. 8, Page 1761*, vol. 8, no. 10, p. 1761, Sep. 2018, doi: 10.3390/APP8101761.

[10] J. Linssen, P. Stenzel, and J. Fleer, "Techno-economic analysis of photovoltaic battery systems and the influence of different consumer load profiles q," 2015, doi: 10.1016/j.apenergy.2015.11.088.

[11] S. Quoilin, K. Kavvadias, A. Mercier, I. Pappone, and A. Zucker, "Quantifying self-consumption linked to solar home battery systems: Statistical analysis and economic assessment q," 2016, doi: 10.1016/j.apenergy.2016.08.077.

[12] L. Ciabattoni, M. Grisostomi, G. Ippoliti, and S. Longhi, "A Fuzzy Logic tool for household electrical consumption modeling," *IECON Proceedings (Industrial Electronics Conference)*, pp. 8022–8027, 2013, doi: 10.1109/IECON.2013.6700474.

[13] P. Escobar, E. Martínez, J. C. Saenz-Díez, E. Jiménez, and J. Blanco, "Modeling and analysis of the electricity consumption profile of the residential sector in Spain," *Energy Build*, vol. 207, p. 109629, Jan. 2020, doi: 10.1016/j.enbuild.2019.109629.

[14] J. Ortiz, F. Guarino, J. Salom, C. Corchero, and M. Cellura, "Stochastic model for electrical loads in Mediterranean residential buildings: Validation and applications," *Energy Build*, vol. 80, pp. 23–36, Sep. 2014, doi: 10.1016/J.ENBUILD.2014.04.053.

[15] M. Kavgic, A. Mavrogianni, D. Mumovic, A. Summerfield, Z. Stevanovic, and M. Djurovic-Petrovic, "A review of bottom-up building stock models for energy consumption in the residential sector," *Build Environ*, vol. 45, no. 7, pp. 1683–1697, Jul. 2010, doi: 10.1016/J.BUILDENV.2010.01.021.

[16] E. Nyholm, J. Goop, M. Odenberger, and F. Johnsson, "Solar photovoltaic-battery systems in Swedish households – Self-consumption and self-sufficiency," *Appl Energy*, vol. 183, pp. 148–159, Dec. 2016, doi: 10.1016/J.APENERGY.2016.08.172.

[17] The MathWorks Inc, "MATLAB." The MathWorks Inc., Natick, Massachusetts, United States}, 2022.

[18] A. L. Mata García, "Estudio de los flujos energéticos en una instalación fotovoltaica conectada a la red conectada a la red con baterías de respaldo," Trabajo Fin de Grado, Universidad Politécnica de Madrid, Madrid, 2022.

[19] R. Luthander, J. Widén, D. Nilsson, and J. Palm, "Photovoltaic self-consumption in buildings: A review," *Appl Energy*, vol. 142, pp. 80–94, Mar. 2015, doi: 10.1016/J.APENERGY.2014.12.028.

[20] C. Gini, "On the measurement of concentration and variability of characters," *Metron - International Journal of Statistics*, vol. LXIII, no. 1, pp. 1–38, 2005.

[21] J. Fellman, "Estimation of Gini coefficients using Lorenz curves," *Journal of Statistical and Econometric Methods*, vol. 1, no. 2, pp. 2241–0376, 2012, Accessed: Jul. 14, 2023. [Online]. Available: https://www.researchgate.net/publication/259870644

[22] B. A. Smith, "A Simple Way to Use Interval Data to Segment Residential Customers for Energy Efficiency and Demand Response Program Targeting Use of Segmentation Schemes within the Utility Industry: A Review", Accessed: Jul. 13, 2023. [Online]. Available: http://my.epri.com/portal/server.pt?Abstract_id=000000000001016386.CLASSIFYmanualsandreportsareavailableonlineatwww.epri.com.

[23] "BOE-A-2020-1066 Circular 3/2020, de 15 de enero, de la Comisión Nacional de los Mercados y la Competencia, por la que se establece la metodología para el cálculo de los peajes de transporte y distribución de electricidad." https://www.boe.es/buscar/act.php?id=BOE-A-2020-1066 (accessed Jul. 17, 2023).

[24] "BOE-A-2019-18528 Real Decreto 732/2019, de 20 de diciembre, por el que se modifica el Código Técnico de la Edificación, aprobado por el Real Decreto 314/2006, de 17 de marzo." https://www.boe.es/diario_boe/txt.php?id=BOE-A-2019-18528 (accessed Jul. 17, 2023).

[25] R. Barrella, A. Cosin, E. Arenas, J. I. Linares, J. C. Romero, and E. Centeno, "Modeling and analysis of electricity consumption in Spanish vulnerable households," *2021 IEEE Madrid PowerTech, PowerTech 2021 - Conference Proceedings*, Jun. 2021, doi: 10.1109/POWERTECH46648.2021.9494785.

[26] "E-distribuición Redes Digitales, S.L," 2023.

[27] Universidad Politécnica de Madrid, "SISIFO: An online simulator of PV systems." 2023. Accessed: Jul. 31, 2023. [Online]. Available: https://www.sisifo.info/es/default

10.4229/EUPVSEC2023/4BV.4.24
020352-007

[1]BIFACIAL BIPV BALUSTRADES – SHADOWING, PERFORMANCE AND BENEFITS

Fabio Parolini[a], Gabi Friesen[a], Teodosio del Caño[b], Elena Rico[b], María Jiménez[b], Ebrar Özkalay[a], Pierluigi Bonomo[a], Tian Shen Liang[a], Francesco Frontini[a]
[a] SUPSI - University of Applied Sciences and Arts of Southern Switzerland - Department of Environment Constructions and Design - Via Flora Ruchat-Roncati 15, CH 6850, Mendrisio, Switzerland
[b] ONYX SOLAR – R&D Department - C/ Río Cea 1, 46 - 05004 Ávila, Spain.

ABSTRACT: In the construction sector, more and more opportunities offer the possibility to produce energy from photovoltaic conversion, both in building skin and other parts such as balconies and terraces. Mainly, among the possible PV modules, glass-glass modules used as balustrades, positioned vertically, with bifacial solar cells, are examined in the present paper . However, these photovoltaic fences are not always installed in an optimal position for the best available solar radiation. Due to different orientations, building locations and architecture, they are affected by external and self-shading. Since previous researches have already addressed the topic of bifacial balconies (1), our work focuses on the effect of the shadow on the back of the solar cells to understand their operation. The study was conducted by creating a C-shape prototype on a full-scale mock-up with six modules (two for each orientation W-South-West). The geometry of the prototype interferes with the expected energy yield due to the shadow casting on the rear side. The shadows reduce power because of the intervention of the bypass diodes to protect hotspots. The effects of continuous shading (bypass diodes activations), non-uniformity temperature on the rear part of the module, and increased temperature in the junction box on the system lifetime have been analysed. The comparison between the energy production of the different bifacial modules on the full-scale prototype and temperature analysis provides information on these systems' actual behavior.

Keywords: BIPV performance assessment, shading and hotspot effects, bifacial PV

1 INTRODUCTION

According to the European Commission, "buildings are responsible for around 40% of EU energy consumption and 36% of energy-related greenhouse gas emissions" (2). Buildings are, therefore, the single largest consumer of energy in Europe. The European Energy Performance of Buildings Directive (EPBD) (3) requires EU countries to ensure that all new buildings are nearly zero energy and properly exploit solar energy. Building integrated photovoltaic (BIPV) systems, one of the sectors in strong expansion and development in recent years, supports achieving these objectives.

Glass-glass photovoltaic modules used as balustrades on balconies and terraces are possible BIPV options. Most of the modules used as balustrades consist of mono-facial crystalline silicon solar cells. Nevertheless, the adoption of bifacial crystalline solar cells has grown significantly in recent years. In fact, according to the latest market analysis, bifacial cells currently occupy about 25% of the photovoltaic market (4). As reported by Guerrero-Lemus et al., who conducted a study of nearly 400 papers published since 1979 (5), bifacial solar cells have been around for some time, and related research has begun since the 1960s.

The irradiance absorbed by both sides of the bifacial module allows for higher electricity production than a balustrade built with mono-facial modules. Still, specific attention should be given to limiting shading on both sides (front and rear sides).

Possible shading on the surface of the PV module causes power reduction on both sides. It is precisely the vertical positioning of the bifacial PV module which creates the phenomenon. In slope bifacial modules, the rear side is affected by scattered light and rarely by direct irradiation, as with vertical ones. During the morning, a vertically mounted and East exposure bifacial module will receive direct irradiation on the front side. Direct irradiation will be on the rear side in the afternoon, while diffuse radiation will be on the front.

An example regarding the topic is represented by the work of Valckenborg et al. (1). It deals specifically with bifacial glass-glass modules used as balconies and terrace balustrades where the effects of shading due to the projection of module shadows caused by the laying geometry are analysed through real data and simulations of different scenarios.

Based on a full-scale mock-up of balustrades with ONYX Solar glass-glass bifacial PV modules installed on the rooftop of the SUPSI Mendrisio campus building, the work presented in this paper is to investigate the behaviour of bifacial balustrades under real installation conditions both from the aspect of energy production as well as the stress it is subjected due to the continuous shadowing.

With more than one year of data collection, in addition to establishing the annual energy production and yield for the different orientations, this paper evaluates the impacts of shadows generated on the rear part of the solar cells on the performance and module temperature. Shading on the back of the bifacial cells causes periodic intervention of the bypass diodes to protect the modules, which could affect their durability and reliability.

The work forms part of the BIPVBOOST project, which the European Union funded under the Horizon2020 programme.

2 MOCK-UP AND RESEARCH DESIGNS

The C-shaped balustrade mock-up is a full-scale structure using six glass-glass bifacial PV modules (M1 – M6) as fall protection elements. Reproducing the typical terraces and balconies operating scenario, the balustrade shape causes interference on the glasses due to self-

1

10.4229/EUPVSEC2023/4BV.4.25
020354-001

shading and reflections. Two modules have been oriented towards the South, the other two towards the East and the West, respectively, as shown in Figure *1*.

Figure 1 Balustrade mock-up in full-scale created in SUPSI.

All modules have been made of laminated glass of 10 mm each with the same dimensions: 1190 mm x 850 mm. The junction box, where the bypass diode has been inserted, is positioned at the edge of the glass. All junction boxes and cable connections for power and sensors have been hidden inside the handrail. Below, Figure *2* shows the glass's features.

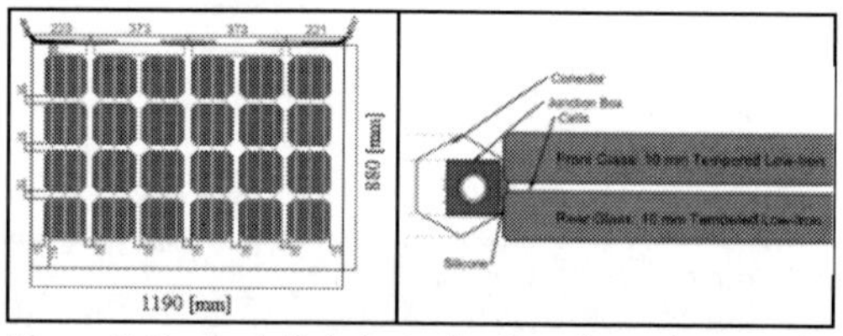

Figure 2 On the left: balustrade glass dimension. On the right: glass thickness and junction box.

Before installing the mock-up, all PV modules were measured in the laboratory with a Class A+A+A+ solar simulator for their performance under standard test conditions (STC) and bifaciality. A test procedure for capacitive modules was applied to avoid any measurement artefacts due to the high cell capacitance, which is typical for high-efficiency modules. The average maximum power, P_{max}, for six modules is 103 W; bifaciality (φ_{Pmax}) is 93.9%.

Table 1 shows the Initial indoor measured STC modules parameters (Pm, Isc, and Voc) and bifaciality factor φPmax with its respective measurement uncertainties.

Table 1 Indoor measured for all PV modules.

Module	P_m [W]	V_{oc} [V]	I_{sc} [A]	φ_{Pmax}
M1	103.37	16.95	8.073	
M2	102.91	17.02	8.018	
M3	102.57	17.03	8.036	
M4	103.09	17.02	8.100	
M5	102.95	17.09	8.012	
M6	102.65	17.12	7.988	
M7	103.45	17.14	8.044	0.939
mean	**103.00**	**17.05**	**8.039**	
standard deviation	**0.32%**	**0.39%**	**0.47%**	
meas. uncert. u[k=2]	*±2.7%*	*±2.7%*	*±2.6%*	*±2.0%*

Electroluminescence images were acquired for all modules, showing a slight cell mismatch and minor defects without impacting the performance. No pre-conditioning/stabilisation was performed.

Each module was connected to a maximum power point tracker (MPPT3000) with an integrated I-V tracer for the electrical performance analysis.

Compared to an inclined South-facing surface, which reaches the solar radiation's peak in the summer, vertically installed modules receive higher radiation in different seasons depending on the orientation and positioning.

The irradiance was therefore measured with four pyranometers towards all directions (South, North, East, and West) with calibrated Kipp & Zonen pyranometers. The pyranometers were positioned to be unaffected by shading, representing the reference irradiance under optimal conditions.

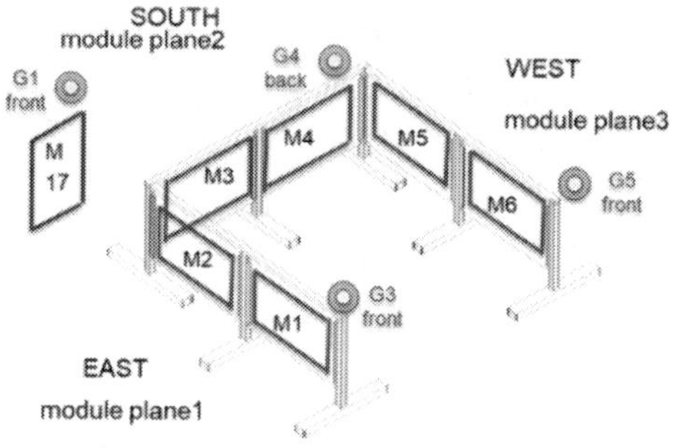

Figure 3 Schematics of the modules and irradiance sensors.

The temperatures of all the photovoltaic modules were acquired via PT100 contact sensors. During the research, the development of shadows was evaluated, which allowed the temperature sensors to be positioned in the right place on the modules subject to shading.

The correct positioning of the sensors allows the acquisition of information relating to any temperature non-uniformities both on the surface of the modules and in the junction boxes, as well as detecting any hotspots.

Figure 4 shows the positioning of the sensors on each glass, the MPPT associated with it and the type of sensor.

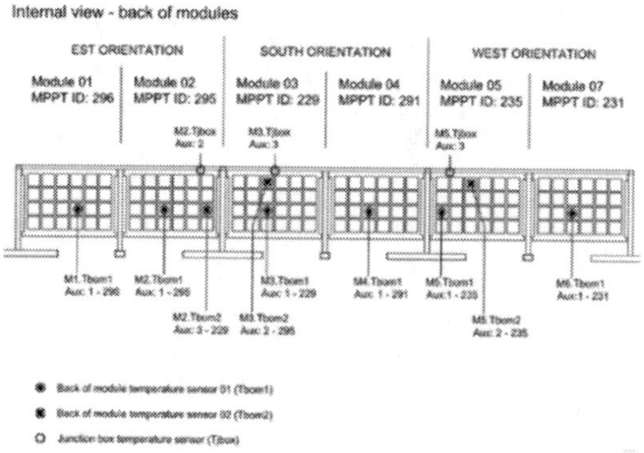

Figure 4 Positions of the PT100 temperature probes on the bifacial PV modules.

10.4229/EUPVSEC2023/4BV.4.25
020354-002

3 RESULTS AND DISCUSSIONS

3.1 Environmental parameters

3.1.1 Irradiance

The irradiance on PV balustrade modules has been measured in five-minute intervals.

Figure *5* shows the for a whole year measured plane of array irradiance G_{poa}, measured on the south G_{south} and north G_{north} of the South-facing modules. In winter, the modules received a maximum front side irradiance of up to 1000 W/m^2.

This is due to the lower solar height that produces perpendicular rays on modules. In the summertime, when the sun is higher in the sky, the radiation angle is low, and the minimum values are detected.

During the winter, the irradiance G_{north} on the rear side of the south-facing module is very small and is mainly the sum of the reflected light from the ground and the surroundings. This value increases in the summer when the sun's altitude is high, and the reflected light becomes more intensive.

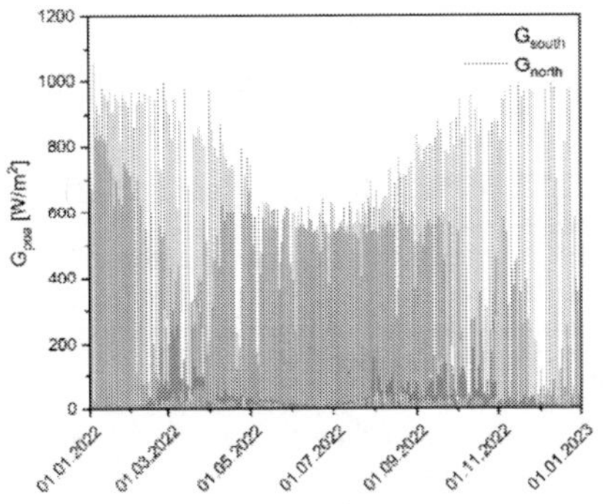

Figure 5 Five-minute time series of the measured irradiances for the South-facing modules.

In reverse, the East- and West-facing modules received maximum irradiance in the period from spring to autumn with an irradiance value of around 800 W/m^2. At this time of year, the sun is not yet at its peak, and the rays hit the modules with good perpendicularity.

The curve created by the irradiance appears flattened without a central peak. The annual irradiance for the East and West orientation modules for a whole year is reported in Figure *6*.

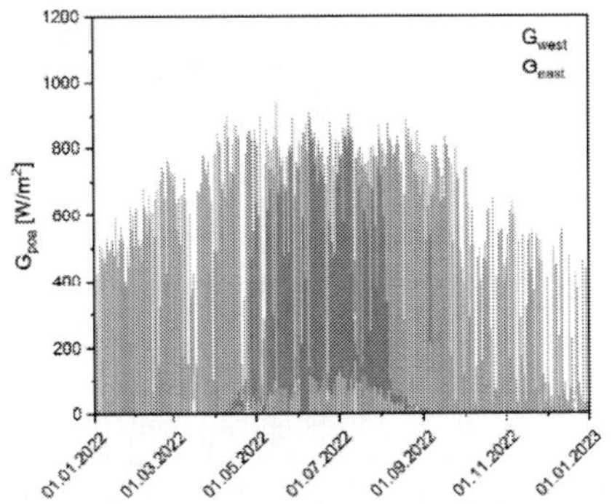

Figure 6 Five-minute time series of the measured irradiances for the East/West-facing modules.

A theoretical irradiance (G_{eff}), consisting of front (G_{front}) and rear irradiance (G_{back}) multiplied by the bifaciality coefficient of the module is calculated:

$$G_{eff} = G_{front} + G_{back} \cdot \varphi_{P_{max}}$$

The effective irradiance provides a measure of the additional solar resource available for the bifacial module respect to a monofacial module taking into account the bifaciality of the module. Figure *7* shows the measured and effective monthly irradiation values (H) obtained respectively with G_{south}, G_{north} and G_{eff}. Figure *8* shows the monthly irradiation values for the East- and West-facing modules.

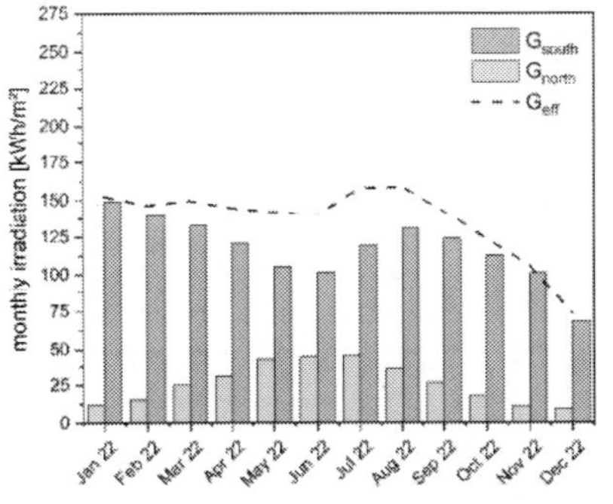

Figure 7 Measured (south, north) and calculated monthly irradiation values for the South-facing modules.

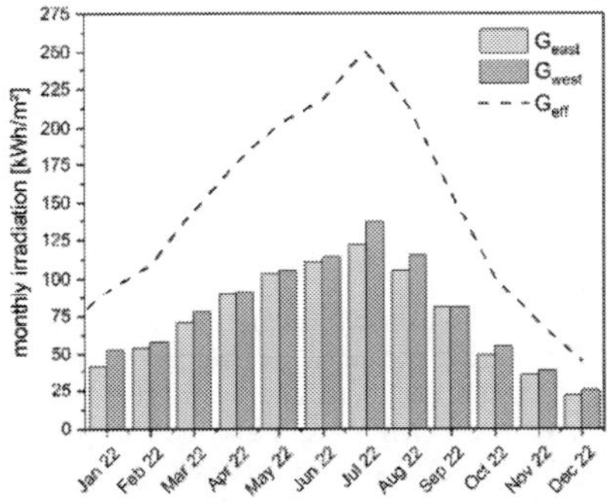

Figure 8 Measured (east, west) and calculated effective monthly irradiation values for the East/West-facing modules.

East/West-facing modules have the highest gain. It is possible to visualize the result by observing the graphs in Figure 9 and Figure *10*.

The first graph shows the annual cumulated irradiation [$H_{cum,south}$ and $H_{cum,eff}$] - (blue lines) and the irradiance distribution [$G_{freq,south}$ and $G_{freq,eff}$] (orange bars) for a south facing bifacial module.

The second one shows the annual cumulated irradiation [$H_{cum,east}$ and $H_{cum,eff}$] (blue lines) and the irradiance distribution [$G_{freq,east}$ and $G_{freq,eff}$] (orange bars) for an east facing bifacial module.

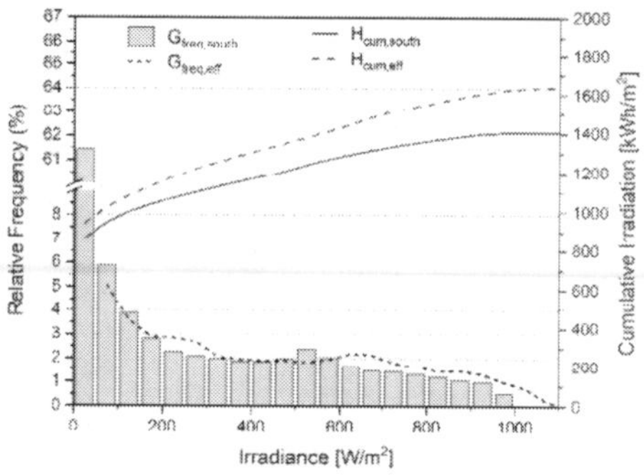

Figure 9 Distribution of the incoming front and effective irradiance G (data below 50W/m² have been filtered out) together with the cumulated irradiation H (blue lines) for South-facing modules.

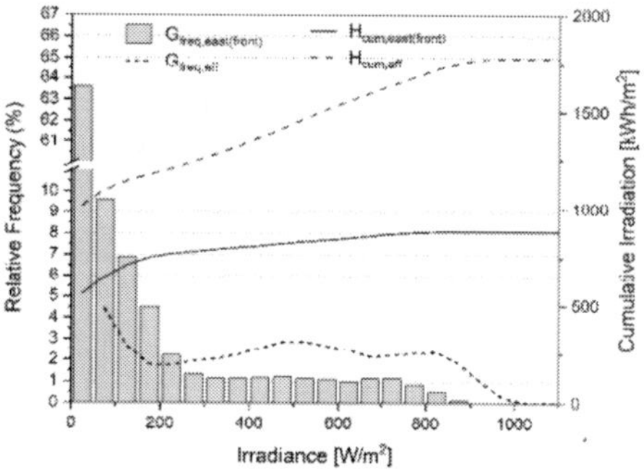

Figure 10 Distribution of the incoming front and effective irradiance G (data below 50W/m² have been filtered out) together with the cumulated irradiation H (blue lines) for East-facing modules.

In both graphs, the dotted lines represent the effective irradiation.

Due to the bifaciality, during the monitoring year, the East/West-facing modules theoretically received 4.6 % more irradiation than the South-facing ones.

Nevertheless, self-shading and, to a small extent, due to obstacles around the modules reduced the available irradiance.

3.1.2 Temperature

The module temperature is mainly the result of the total incoming radiation, the ambient temperature, and the back-side ventilation.

The modules most affected by shading were equipped with a second temperature sensor, as shown in Figure *4* in order to measure the shading-induced hotspot for each orientation.

Figure *11* shows the frequency plot of the all temperatures on the module. M2 and M5 have two probs for the temperature measurement. No critical overheating was observed throughout the operation of the modules.

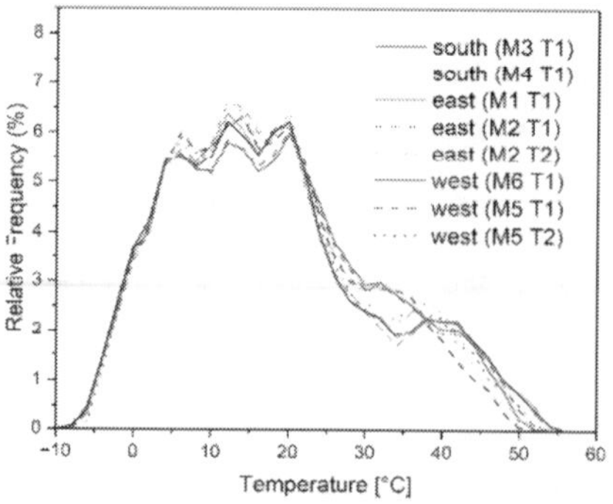

Figure 11 Frequency plots of measured temperatures on all modules. Due to shadowing, M2 and M5 have two temperature probes, T1 and T2.

The probes attached to the rear of the solar cell led to a good measurement when the sun is in front of the module, but they are affected by direct irradiance when the sun is behind the module.

It must also be considered that the contact temperature probes are applied on glass with a thickness of 10mm. That means that the temperature value measured is different from the actual one on the solar cell surface.

Therefore, the temperatures measured under these conditions do not represent the real module temperature. Alternative solutions for measuring vertically mounted bifacial modules will be developed to overcome this problem.

In the future, temperature probes with a smaller contact surface may be chosen to minimize the impact. Alternatively, infrared temperature detection techniques may be used and improved.

Figure *12* shows an example of temperature measurments on the west-facing modules (M5 and M6) together with ambient temperature.

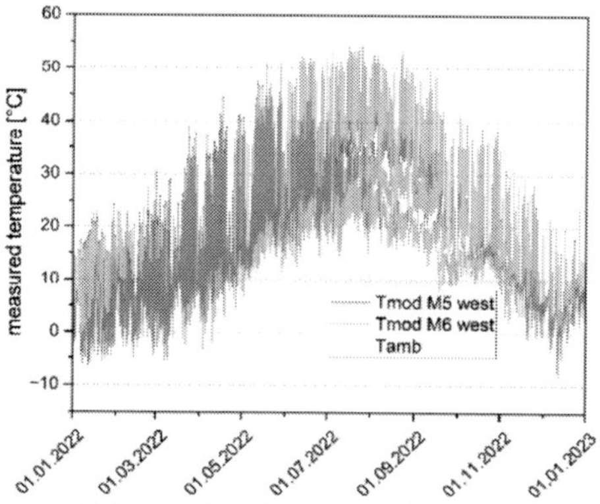

Figure 12 Time series of T_{module} and T_{amb} of the west-facing PV modules.

3.2 Electrical Performance

3.2.1 Performance Ratio (PR)

The module's electrical performances are evaluated based on the Performance Ratio (PR). Figure *13* shows the daily PR of the two South-facing modules, M3 and M4, calculated with the average P_{max} and the G_{eff} compared to an open-rack mounted mono-facial reference module M17 form another close-by test stand.

10.4229/EUPVSEC2023/4BV.4.25

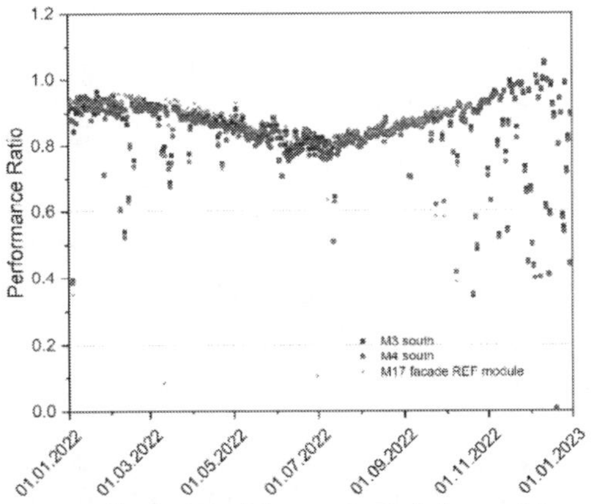

Figure 13 Effective PR of the two South-facing modules, M3 and M4 compared to reference module M17.

The PR of these modules are very close, with slightly lower values for the South-facing module M4. The clear-day PR values are in the range of 0.78 – 0.97. The graph shows the typical seasonal variation due to the higher module temperatures in summer.

The most outstanding performance occurs in winter with low temperatures and direct radiation more perpendicular to modules.

Figure *14* shows the average daily PR of the East-facing modules, M1 and M2, which are very similar to the West-facing ones.

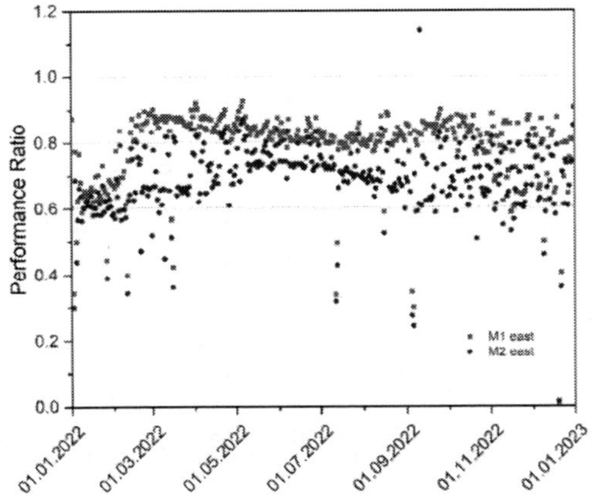

Figure 14 Effective PR of the East-facing module, M1 and M2.

The seasonal variation for East/West modules is less pronounced due to lower temperatures in the winter months than South-facing modules.

The difference in the PR of the two East modules, M1 and M2, is mainly due to different shading conditions. In particular, M2, the corner module, was strongly affected by the shading of the South-facing module M3, leading to a significant PR reduction for M2. M1 is affected to a minor extent by shading in the winter months by external obstacles and mounting structures.

3.2.2 Energy Yield

Figure 15 and Figure 16 show, respectively, the monthly and annual specific energy yield of the six balustrade modules (M1-M6) for the monitoring year.

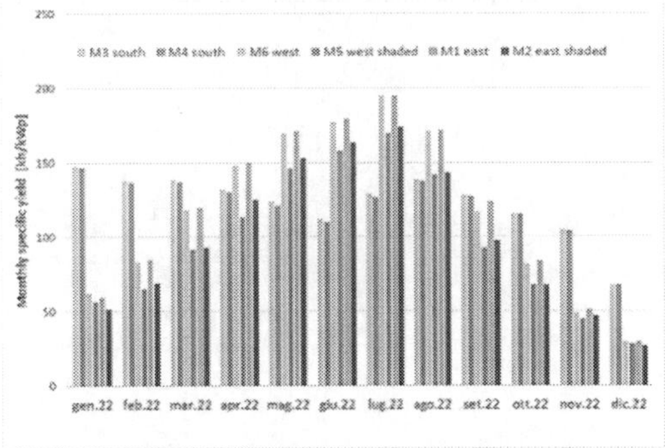

Figure 15. Monthly specific energy yield for the six ONYX balustrade modules.

With two modules for each orientation, it is possible to evaluate the effect of self-shading of the balustrade causes on the adjacent one by comparing the yield of the shaded module to the next one (unshaded).

The corner modules, M2 and M5, affected the most by shading, demonstrated an annual loss of -14.6 % and -16.1 %, respectively, as shown in Figure 16. The difference between the South and East/West-facing modules is -5.2% in favour of the South-facing modules, caused by the higher winter contribution and lower module temperatures in summer.

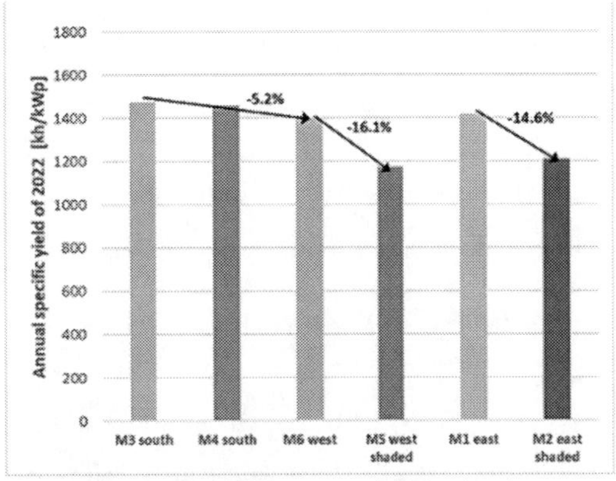

Figure 16. Annual specific energy yield for the six ONYX balustrade modules.

To better understand the shading phenomenon, Figure *17* shows an image of the projection of the shadow that the South-facing module M4 projects onto the East-facing module M2.

In the picture, it is possible to see the thermographic analysis carried out to establish the temperatures of the solar cells.

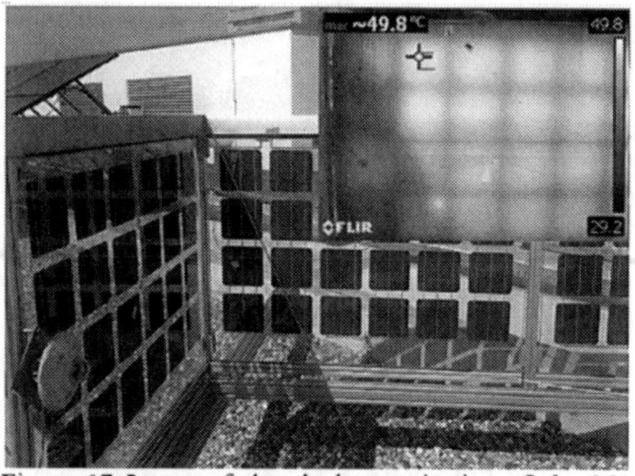

Figure 17 Image of the shadow projection of the M4 module (South-facing) on the M2 module (East-facing).

For a deeper understanding of the power loss due to shading of the modules' East-facing and the modules' West-facing, a Heat Map graph was created by hour and day of the year.

The graph represents the power difference between the unshaded module and the shaded corner module, together with a bar graph of the daily difference weighted by the annual average for the unshaded module.

The heat map was shown for both East and West-facing modules. As shown in Figure 18 and Figure 19, for the West exposure, the M5 module is subject to shading caused by the structure's shape, while the M6 module is almost unaffected.

The same applies to the East-facing modules M1 (unshaded) and M2 (shaded).The shading losses for M2, reaching up to almost 60 W, are observed in the afternoon after 14:00 and from February to October.

The daily losses are the highest in the middle seasons. In the summer, late afternoon, M1 was affected by shading from the mock-up. Figure 19 shows a mirrored picture for the West-facing modules, with similar losses for M5 occurring in the morning until noon and from February to October.

An influence from the surrounding building or obstacles on the West side is observed in the late afternoon, which explains the slight difference between East and West.

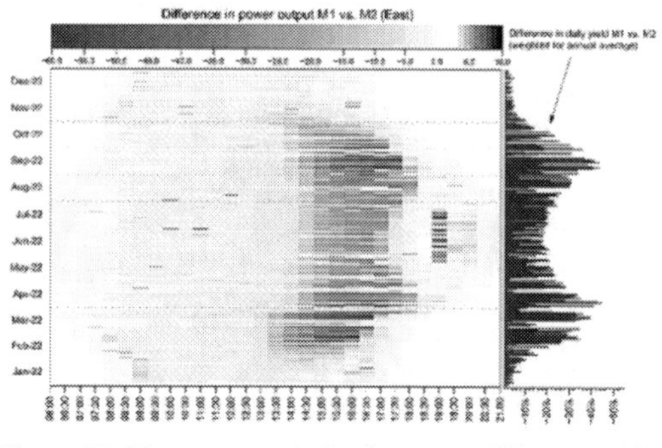

Figure 18. Heat map graph for the power difference with daily difference weighted by the annual average for the East-facing modules.

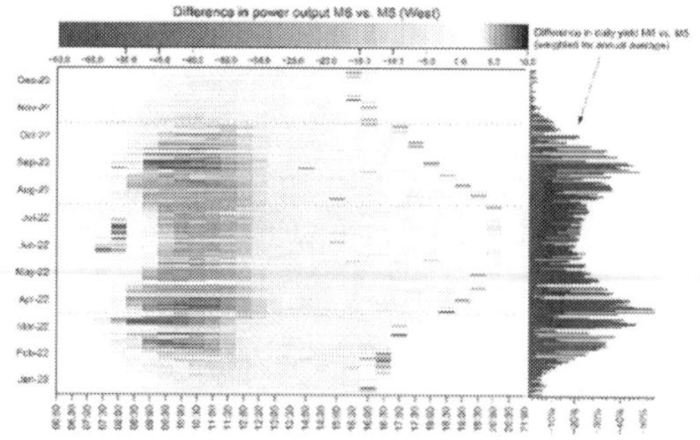

Figure 19. Heat map graph for the power difference with daily difference weighted by the annual average for the West-facing modules.

3.3 Shading and Bypass Diode Activation

Self-shading resulted in almost daily activation of the bypass diodes and the generation of hot spots detected with MPPT trackers and infrared thermography.

Figure *20* shows the measured current, voltage and power curves at maximum power and the temperature profile given by probes on the back of the module and on the JBs. All data have been collected by MPPT trackers on the 09th of June for the two East-facing modules in the moment of partial shading of module M2.

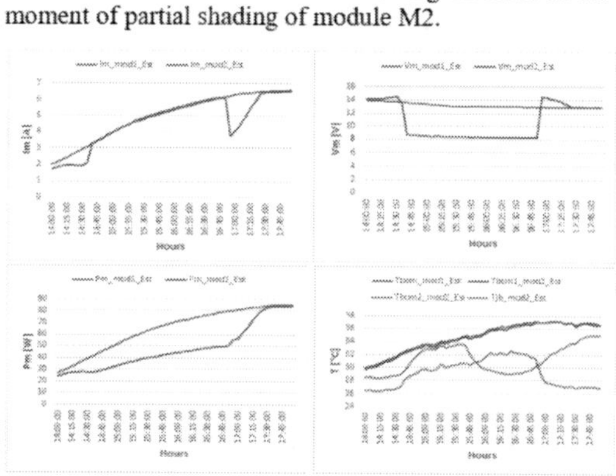
Figure 20 Curves of Im, Vm, Pm and temperature profiles for East-facing modules.

The graph of Im shows an influence on its value due to shading before and after the activation of the bypass diode.

During the activation of the diode, however, the value turns out to be the operating value of the module due to the bypass made by the diode.

On the contrary, the voltage value is reduced due to the diode, which excludes the electrical circuit subject to shading. At the same time, it rises before and after due to the shift of the curve due to the pursuit of maximum power.

Consequently, during the activation period of the bypass diode, there is a power loss of approximately one-third (three electrical circuits or three bypass diodes) of the shaded module.

The temperature curves show how the temperature profile rises due to shading and the relative operation of the bypass diode.

To better comprehension, the modules with East-facing were taken as reference. In the case of activation of the bypass diode in module M2 (subject to shading), an

10.4229/EUPVSEC2023/4BV.4.25
020354-006

increase of six degrees in temperature was observed (about 21 %).

However, as expressed previously, it must be considered that the glass's thickness can influence the actual measurement of the temperature value on the back of the cell. For the temperature measured on the JB, it must be specified that the value is that of contact with the casing.

The temperature value of the bypass diode, inserted in the plotting, could be much higher.

For future developments, it will be necessary to agree with the manufacturers to insert the probes in the appropriate positions during the modules' manufacture.

For more detail, Figure *21* shows that the maximum power point of the West-facing module is achieved at a lower voltage, and the bypass diode is activated due to daily self-shadowing on the rear side.

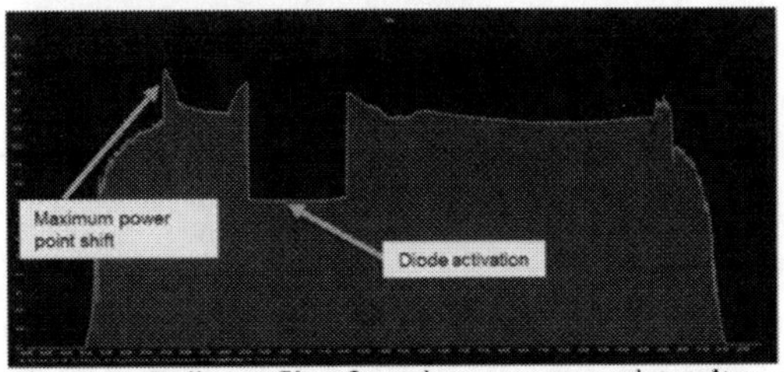

Figure 21 Daily profile of maximum power point voltage (Vm) of shaded module M5 comparison with unshaded M6 module.

As mentioned above, since this module has three diodes, the maximum power point voltage drops to about one-third. In this case, since the shaded string is bypassed, the power generated from the unshaded cells in this string is dissipated as heat (hot spot).

In Figure *17*, it can be seen that the hot spot temperature is within the expected operating temperature range of a PV module since there are only 12 cells per bypass diode and the amount of irradiation at the module surface is low (around 150 W/m^2 and 500 W/m^2 on the front and rear sides, respectively) concerning the optimised orientation.

These effects cause, over the day on the shaded module, a significant loss of power compared to the unshaded module, easily visible on the graph in Figure *22*.

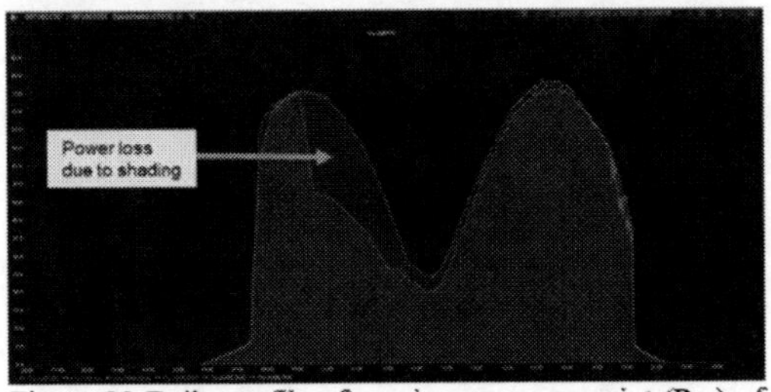

Figure 22 Daily profile of maximum power point (Pm) of shaded module M5 comparison with unshaded M6 module.

Moreover, since the current generated from the unshaded strings flows through the active diode, it causes the bypass diode to heat up.

One of the reasons affecting the amount of this heating is the amount of current flowing through. Since balustrades are mounted vertically, the irradiation on their surface is less than the optimum orientation. Therefore, the current proportional to the radiation is relatively low (around 3A to 5A).

Because of this, there is no significant temperature difference between the modules exposed to recurrent self-shading (East/West-facing) and those of South-facing ones. However, the frequent operation of the bypass diode is worthy of specific attention since it could increase the risk of its failure and associated risks.

4 CONCLUSIONS

The C-shape full-scale balustrade mock-up with six glass-glass bifacial PV modules was built with two modules facing each orientation (South, East, and West).

The electrical and thermal performance was monitored for over a year. The metrics for electrical performance evaluation include performance ratio and energy yield; thermal performance evaluation consists of measuring module and junction box temperatures and investigating the causal effect between shading, bypass diode activation and electrical output.

Pyranometers were also positioned on the balustrade, facing all the orientations to measure the theoretical optimal irradiance received by modules. As the measurement shows, South-facing modules received the highest irradiance during winter, up to 1000 W/m^2; for East- and West-facing modules, the highest receivable irradiance was during summer.

A potential irradiance gain of 4.6% compared to the South-facing bifacial module and 87% respect to an East or West-facing monofacial module highlighted the benefit of using bifacial modules for East and West orientation.

The measured irradiance was used to calculate the effective irradiance, i.e., the theoretical solar resource available by a bifacial module.

Nevertheless, the effective irradiation is not fully translated into the modules' electrical output due to the shading of adjacent modules and obstacles around the modules.

The shading has resulted in the hotspot and activation of bypass diodes, which were detected based on the electrical and temperature measurements of modules and junction boxes. Shading losses of -14.6 to -16.1% have been observed.

Corner modules, M2 and M5, suffered a significantly higher Performance Ratio drop and loss of annual energy yield by more than 10% due to shading.

The junction box temperature of shaded East-facing modules increased by around 20% when the current generated from unshaded string flows through the bypass diode and heats the diode.

Although the temperature rise was within the acceptable range, frequent diode activation has to be put about an adequate lifespan and component reliability to avoid any related safety risk.

Finally, it should be underlined that the analysis carried out only considered the shading due to the shape of the fence. This analysis did not consider shading due to possible anthropic activity that could be generated on the protective balustrades.

Furthermore, not being part of the project, a comparison analysis with single-sided modules in the same conditions was not carried out. The mono facial modules are not subject to power losses with rear shading and could (to be verified) offer better performance in cases with marked shading on the back of the modules. The authors leave this possibility for possible future developments on the topic.

ACKNOWLEDGMENT

This project "BIPVBOOST" has received funding from the European Union's Horizon 2020 research and innovation program under grant agreement No 817991. The sole responsibility for the content of this publication lies with the authors. It does not necessarily reflect the opinion of the European Commission. The European Commission is not responsible for any use that may be made of the information contained therein.

7 REFERENCES

[1] R. Valckenborg, S. Villa, B. B. Van Aken and T. Burgers. BIFACIAL PV BALCONY AND TERRACE FENCES, 8th World Conference on Photovoltaic Energy Conversion (WCPEC-8), Milano, 26 – 30 September 2022.

[2] European Commission. An official website of the European Union - https://energy.ec.europa.eu/topics/energy-efficiency/energy-efficient-buildings/energy-performance-buildings-directive_en

[3] Energy Performance of Buildings Directive. DIRECTIVE 2010/31/EU OF THE EUROPEAN PARLIAMENT AND OF THE COUNCIL of 19 May 2010 on the energy performance of buildings.

[4] Exawatt. Trends in bifacial cell and module manufacturing, Virtual BifiPV workshop, July 2020.

[5] R. Guerrero-Lemus, R. Vega, Taehyeon Kim, Amy Kimm, L.E. Shephard, Bifacial solar photovoltaics – A technology review, Renewable and Sustainable Energy Reviews, Volume 60, 2016, Pages 1533-1549, https://doi.org/10.1016/j.rser.2016.03.041.

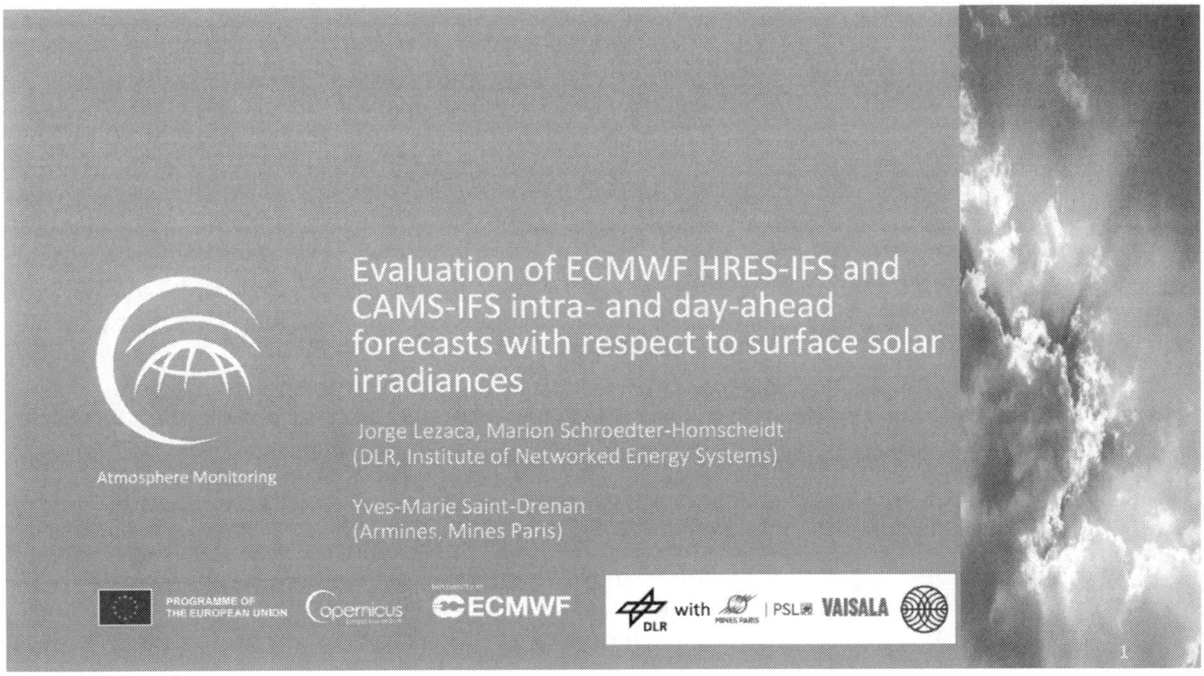

020356-001

40th European Photovoltaic Solar Energy Conference and Exhibition

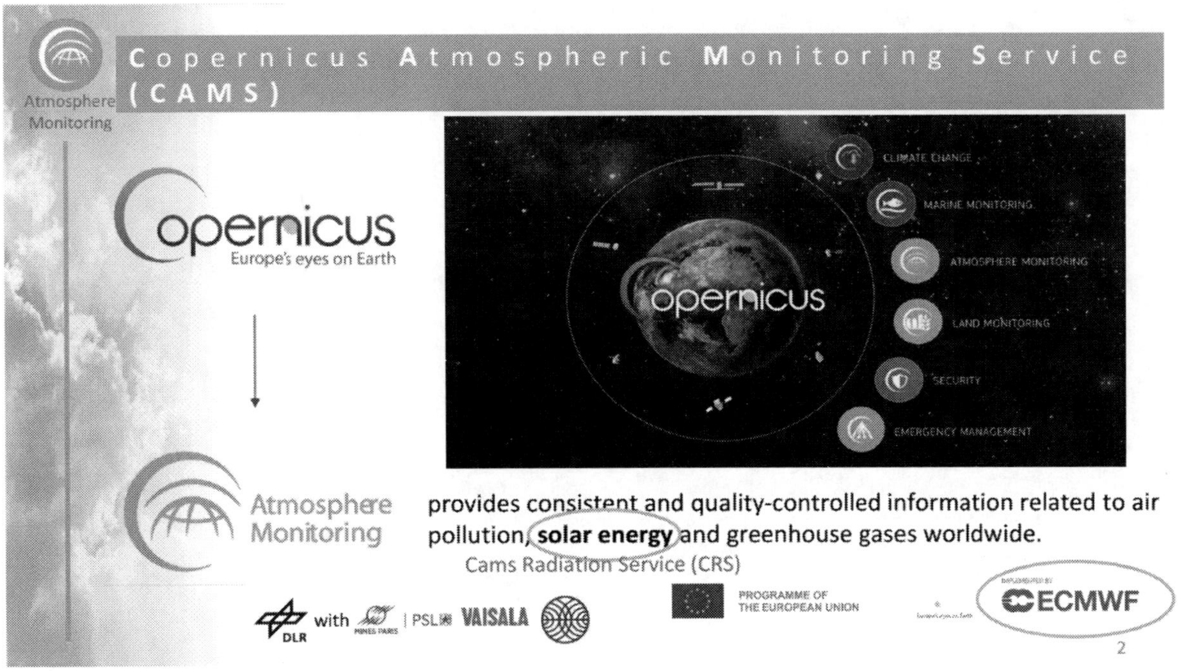

020356-002

Atmosphere Monitoring

CRS products

on-the-fly processing of time series

- global, diffuse, and direct normal irradiation
- time series 2004 until 2 days ago
  1 min, 15 min, 1 hour, 1 day, 1 month temporal resolution
- any point within MSG satellite field of view (soon HIMWARI and GOES)
- interactive access on CAMS ADS and user portal SoDa
- since 2019 via open source library pvlib
- access to all intermadiate quantities if 1 min is selected : Clear sky, cloud type, cloud coverage, AOD, ...

pre-calculated gridded data

- 15 min temporal resolution
- 2005-2022
- 0.2° grid step
- update foreseen 0.1° grid

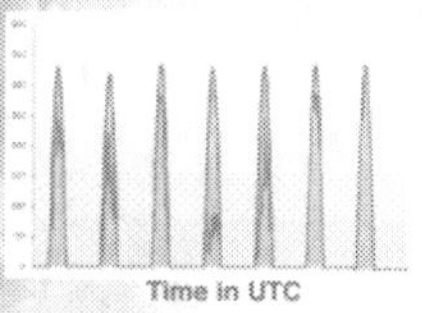

pvlib

pvlib.iotools.get_cams

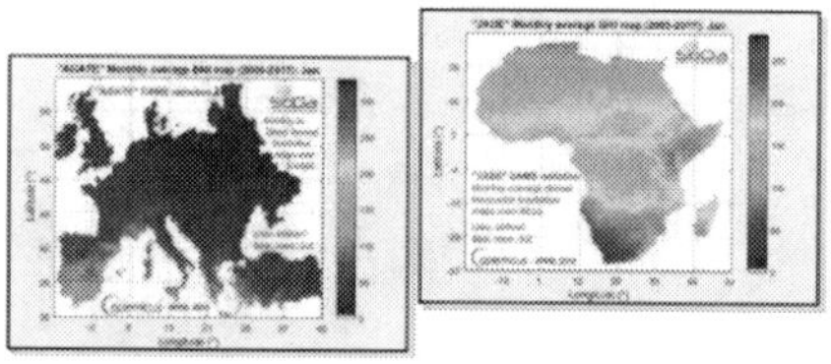

implemented by SoDa Solar radiation Data

PROGRAMME OF THE EUROPEAN UNION

IMPLEMENTED BY ECMWF

https://www.soda-pro.com/web-services/radiation/cams-mcclear

https://ads.atmosphere.copernicus.eu/cdsapp#!/dataset/cams-solar-radiation-timeseries

020356-003

40th European Photovoltaic Solar Energy Conference and Exhibition

Atmosphere Monitoring

The CAMS radiation service: Poster

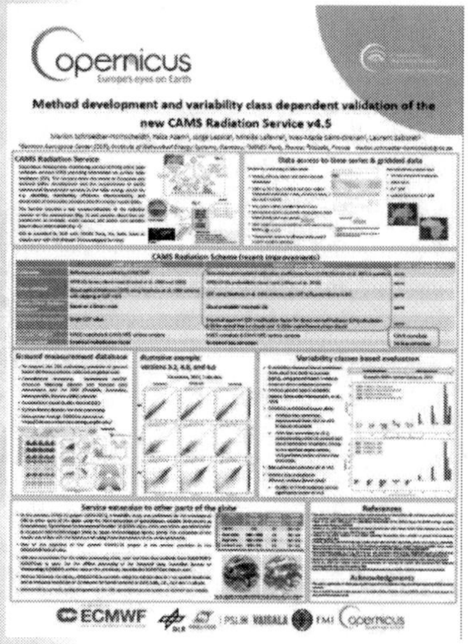

4DV.4.22
Thursday 21.09 @ 15:15 – 16:45

PROGRAMME OF THE EUROPEAN UNION

IMPLEMENTED BY ECMWF

020356-004

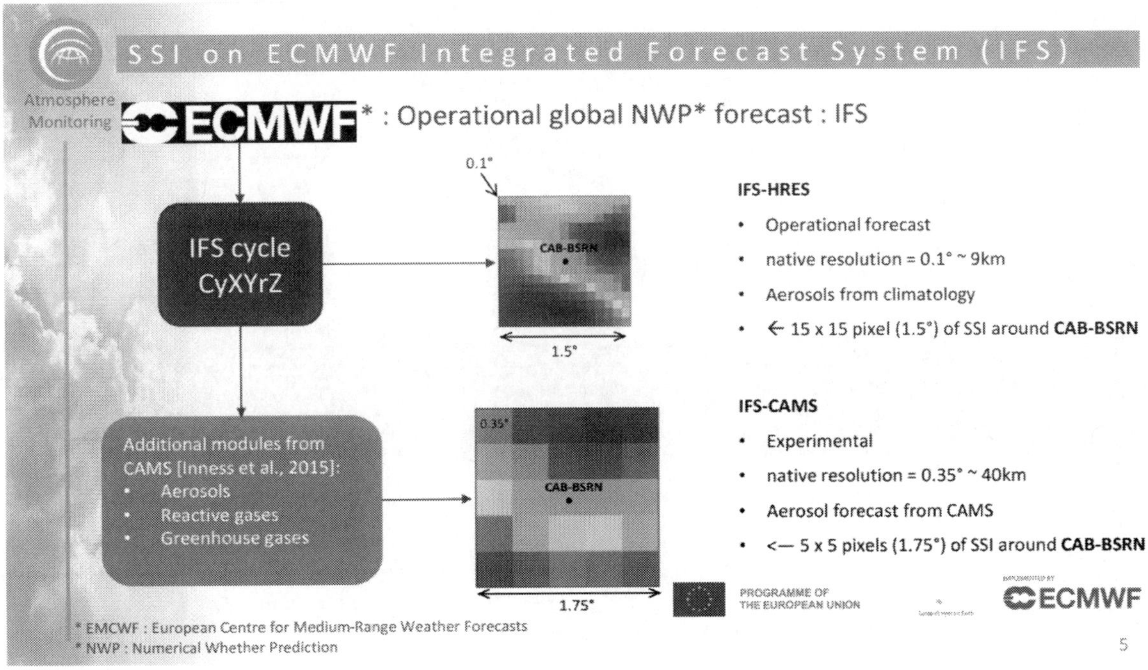

020356-005

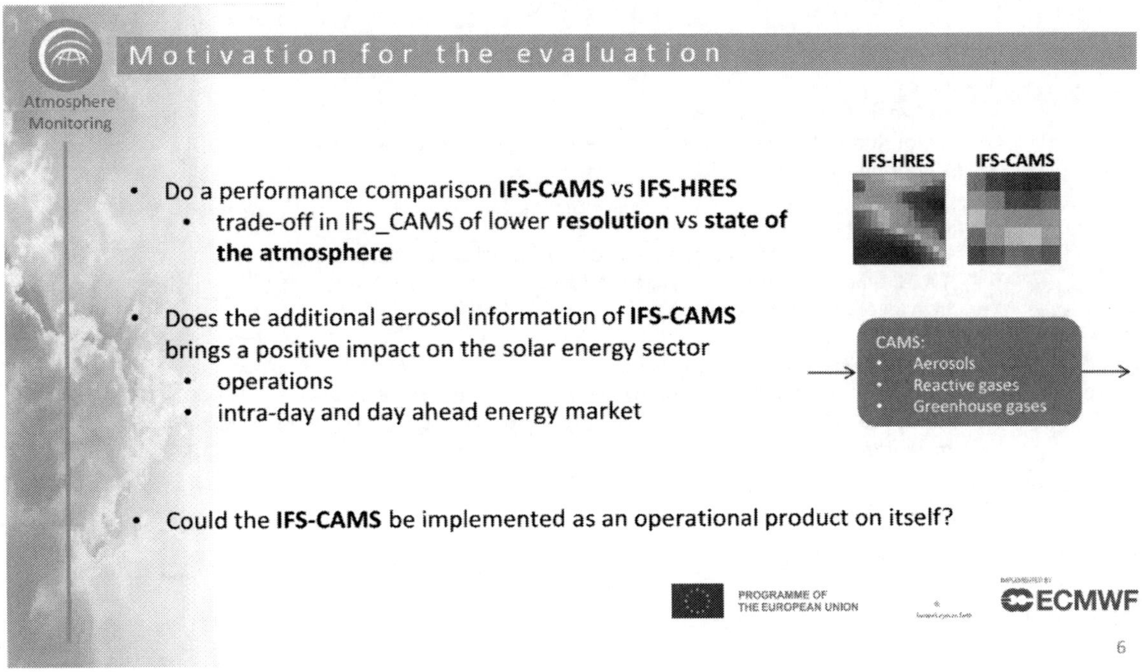

020356-006

Atmosphere Monitoring

Study design

PROGRAMME OF THE EUROPEAN UNION

ECMWF

7

020356-007

40th European Photovoltaic Solar Energy Conference and Exhibition

Atmosphere Monitoring

Evaluation study design

- NWP forecast:
  - **Full year 2021** on a global scale
  - Beam Normal Irradiance (**BNI**) and ~~Global Horizontal Irradiance (**GHI**)~~
  - **1 hour** time resolution
- Ground observations :
  - Research quality solar radiation networks (internal CAMS database)
  - **1 min.** time resolution → averaged **to 1 hour**
  - **high level QC** procedure
- Solar energy applications :
  - Concentrated on the intraday and day-ahead range (up to 48 hours)
  - **run at 0:00 UTC** which ensures that the forecasts availability for all intraday and day ahead market transactions

PROGRAMME OF THE EUROPEAN UNION

ECMWF

8

020356-008

Atmosphere Monitoring

Ground observations

46 sites that satisfied:

- level L1 (QC)
- > 8 months (2021)

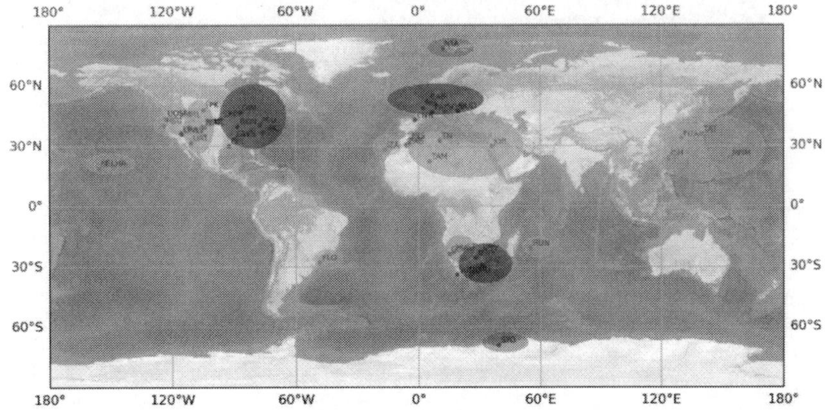

sites classification:

- desert
- polar
- subpixel : island/coasts (< 10km)
- continental
- mountainous (> 1500 m)

PROGRAMME OF THE EUROPEAN UNION

ECMWF

9

020356-009

40th European Photovoltaic Solar Energy Conference and Exhibition

Atmosphere Monitoring

Results all sky situations

PROGRAMME OF THE EUROPEAN UNION

ECMWF

10

020356-010

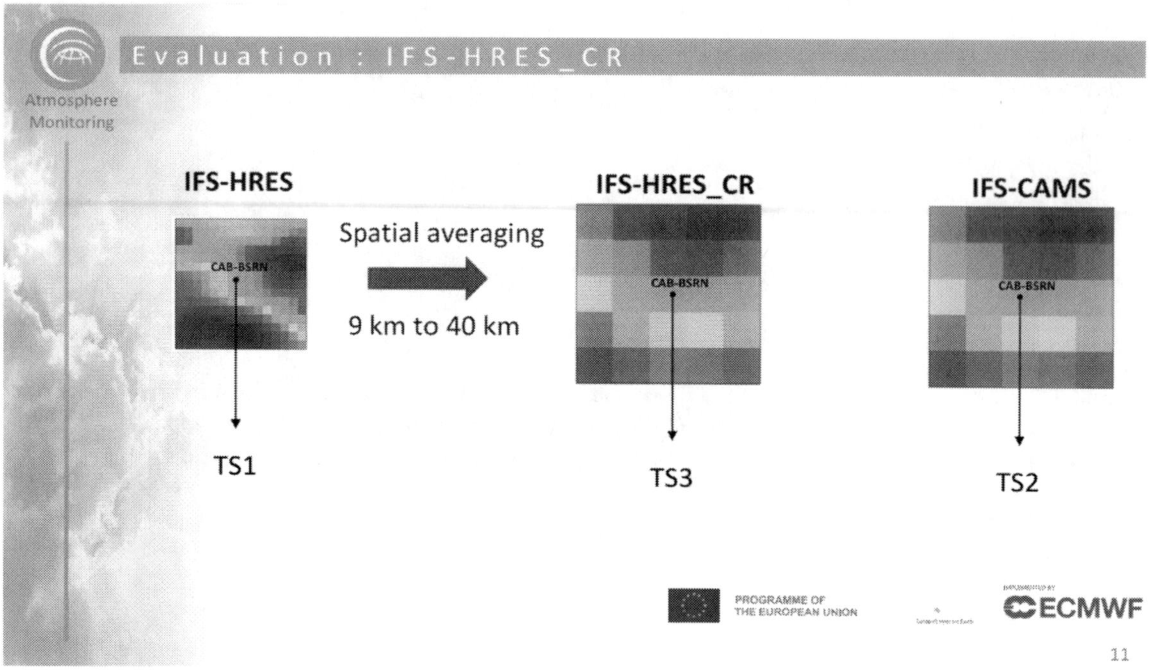

020356-011

40th European Photovoltaic Solar Energy Conference and Exhibition

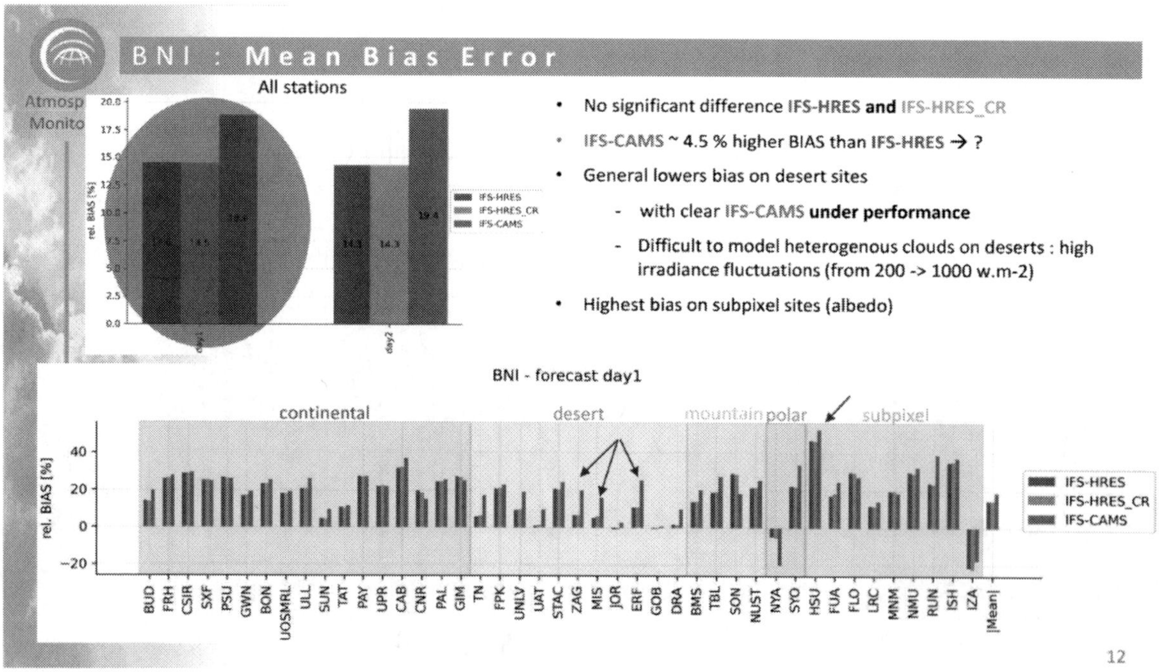

020356-012

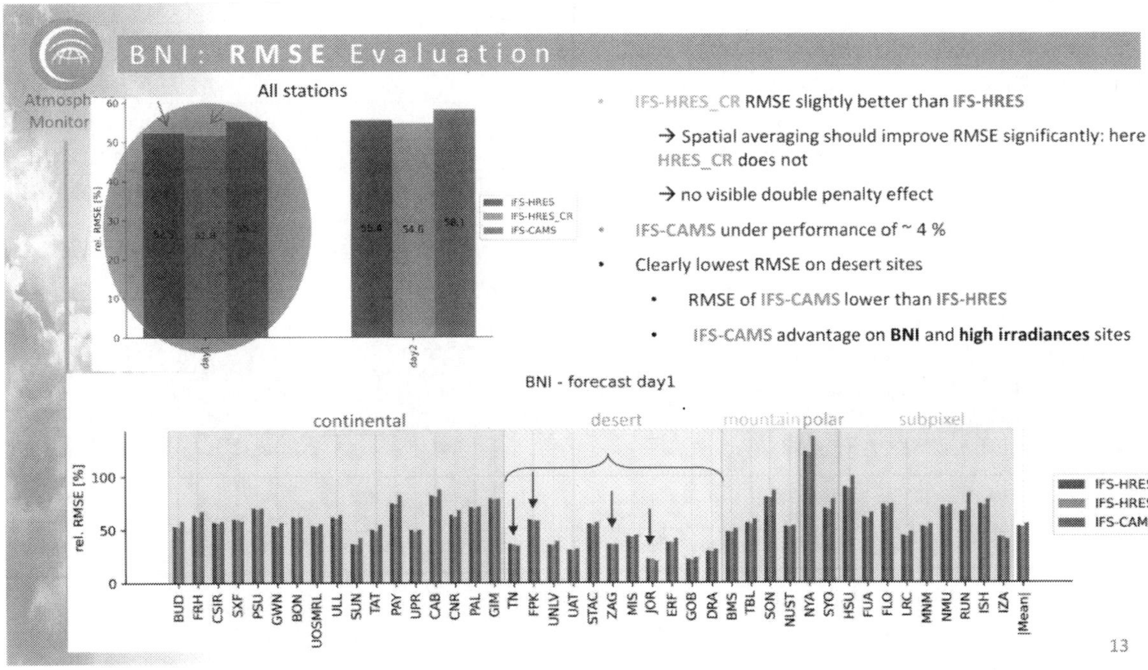

020356-013

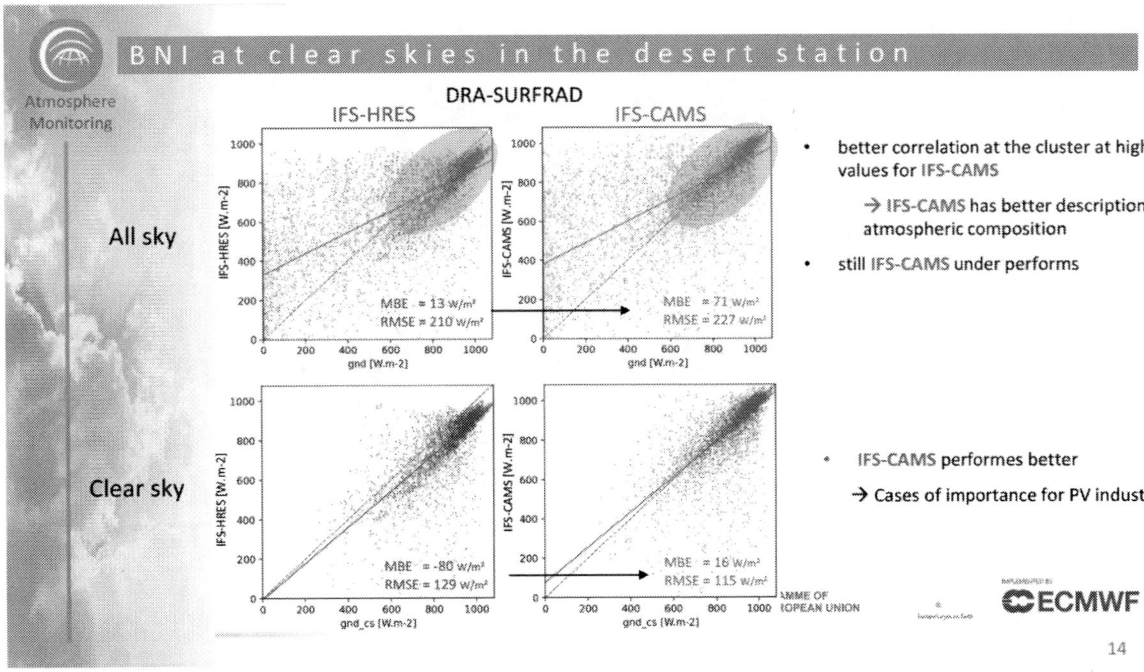

020356-014

Atmosphere Monitoring

Results Clear sky situations

PROGRAMME OF THE EUROPEAN UNION

ECMWF

15

BNI on Clear sky : MBE

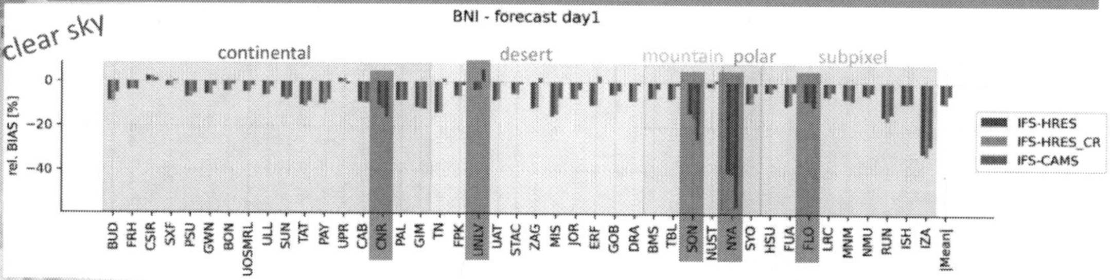

BNI In Clear sky situations

- The IFS-CAMS bias is lower than HRES IFS in **all sites except :** CNR, UNLV, FLO (subpixel), SON (2800 m) and NYA (polar)
- IFS-CAMS presents the lowest mean bias ~ 5% (lower by ~ 4 % with respect to IFS-HRES)
- Better aerosol description improves modelling of clear sky (high radiance) instances

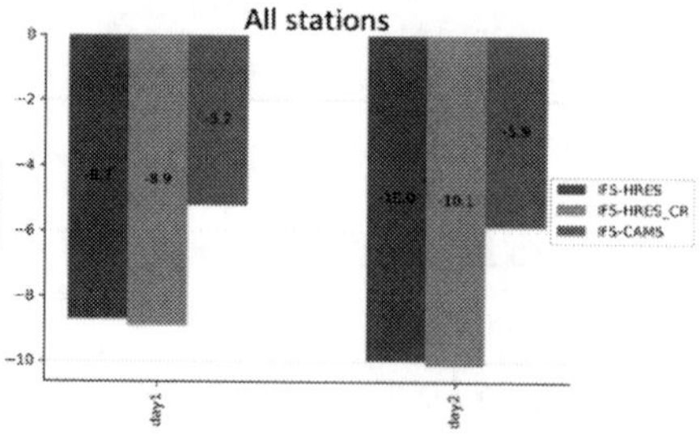

16

BNI on Clear sky: RMSE

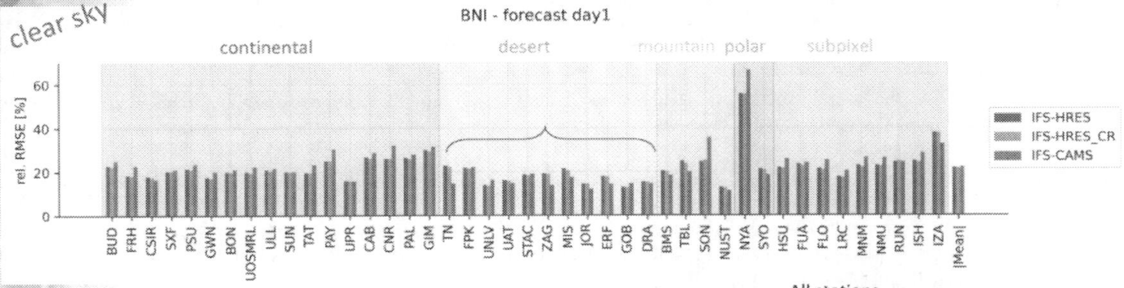

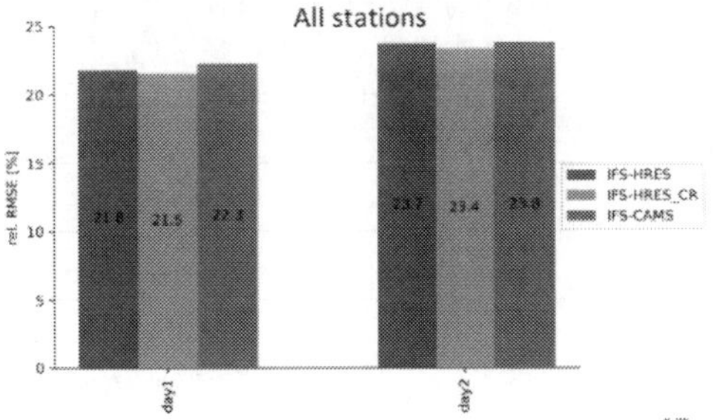

BNI in Clear sky situations

- The IFS-CAMS has in general a lower RMSE than IFS-HRES on the desert regions (high Irradiance)
- But is not the case in all other regions
- IFS-CAMS and IFS-HRES present a very similar general behaviour with a general under performance of IFS-CAMS of 0.5 % in RMSE

→ Due to non desert sites

17

020356-017

Conclusions

Atmosphere Monitoring

- all sky : **IFS-HRES** performs better than the **IFS-CAMS** for BNI (GHI)
- clear sky : **IFS-CAMS** presents some advantages with
  → reduced BNI biases
  → more accurate BNI forecasts in high irradiance (GHI)
  → instances of **special interest of solar community**
- Users focusing on regions with high irradiance may prefer the **IFS-CAMS** forecasts
- Solar stakeholders may use **IFS-HRES** together with the **IFS-CAMS** forecasts.
  - trust more the **IFS-HRES** to identify cloud-free conditions
  - take the **IFS-CAMS** forecasts into account in such situations
- Interesting to run IFS-CAMS in a native resolution of 0.1° → can it be integrated operationaly

PROGRAMME OF THE EUROPEAN UNION

ECMWF

18

020356-018

Atmosphere Monitoring

References

- Inness, A., Blechschmidt, A.-M., Bouarar, I., Chabrillat, S., Crepulja, M., Engelen, R. J., Eskes, H., Flemming, J., Gaudel, A., Hendrick, F., Huijnen, V., Jones, L., Kapsomenakis, J., Katragkou, E., Keppens, A., Langerock, B., de Mazière, M., Melas, D., Parrington, M., Peuch, V. H., Razinger, M., Richter, A., Schultz, M. G., Suttie, M., Thouret, V., Vrekoussis, M., Wagner, A., and Zerefos, C.: Data assimilation of satellite-retrieved ozone, carbon monoxide and nitrogen dioxide with ECMWF's Composition-IFS, Atmos. Chem. Phys., 15, 5275–5303, https://doi.org/10.5194/acp-15-5275-2015, 2015
- Heliosat-4 method
  - Qu et al., Fast radiative transfer parameterisation for assessing the surface solar irradiance: The Heliosat-4 method, Contr. Atm. Sci., 2017
  - Schroedter-Homscheidt et al., Surface solar irradiation retrieval from MSG/SEVIRI based on APOLLO Next Generation and HELIOSAT-4 methods, Contr. Atm. Sci., Vol. 31 No. 6 (2022), p. 455 – 476, DOI: 10.1127/metz/2022/1132
- McClear method
  - Lefèvre et al., McClear: a new model estimating downwelling solar radiation at ground level in clear-sky conditions, AMT, 2013
  - Gschwind et al., Improving the McClear model estimating the downwelling solar radiation at ground level in cloud-free conditions – McClear-v3, Contrib. Atm. Sci., 2019

PROGRAMME OF THE EUROPEAN UNION

ECMWF

19

020356-019

40th European Photovoltaic Solar Energy Conference and Exhibition

Atmosphere Monitoring

- general inquiries and user requests: copernicus-support@ecmwf.int
- specific for the Solar Radiation Service team: marion.schroedter-homscheidt@dlr.de
- User's Guide at http://atmosphere.copernicus.eu/documentation

Thank you for your attention

PROGRAMME OF THE EUROPEAN UNION

ECMWF

20

020356-020

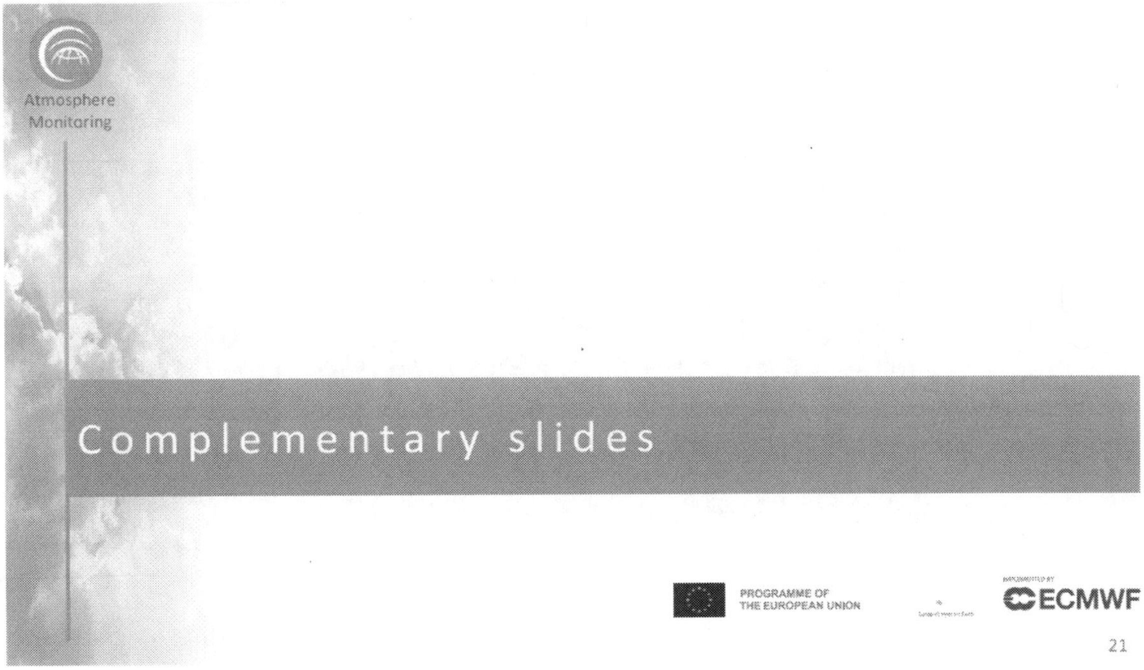

020356-021

40th European Photovoltaic Solar Energy Conference and Exhibition

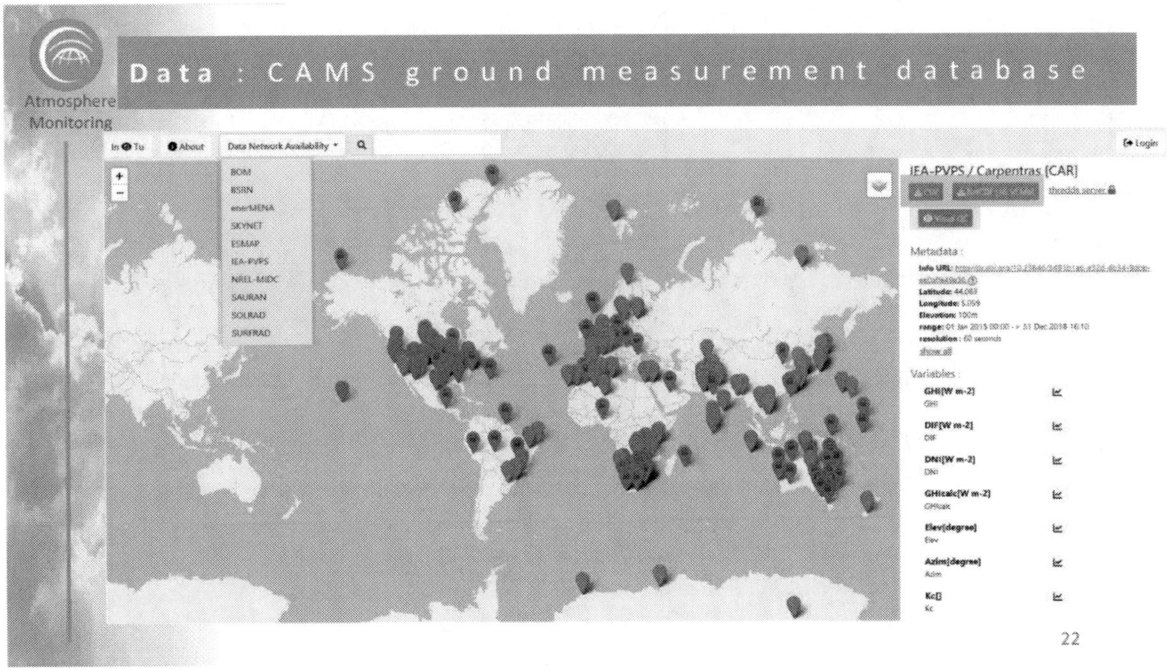

020356-022

Atmosphere Monitoring

DATA : Ground observations

- 79 stations found on the Thredds server for the year 2021
- Different climates / Latitudes / surroundings

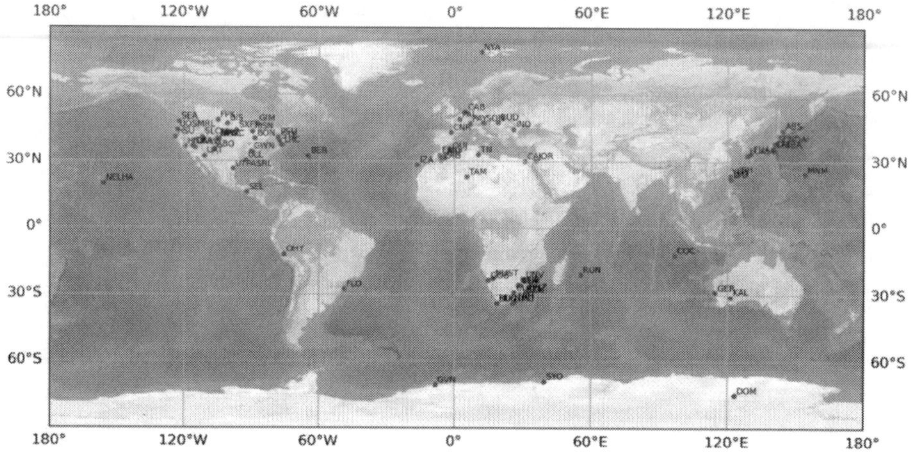

23

Atmosphere Monitoring

Data: Complete input

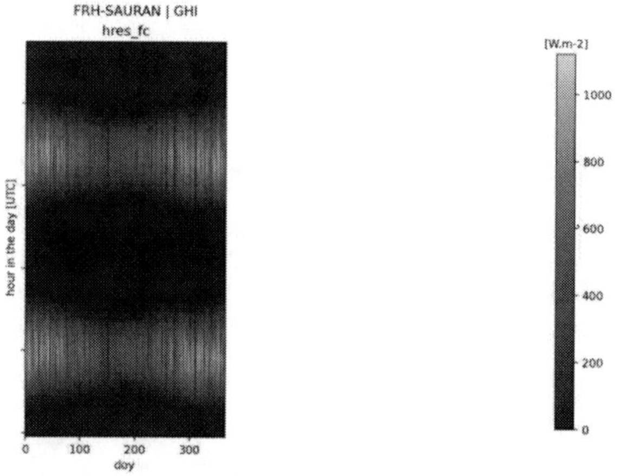

PROGRAMME OF THE EUROPEAN UNION

IMPLEMENTED BY ECMWF

24

Atmosphere Monitoring

Quality Checks on ground obs.

PROGRAMME OF THE EUROPEAN UNION

ECMWF

25

020356-025

40th European Photovoltaic Solar Energy Conference and Exhibition

QC dashboard

Atmosphere Monitoring

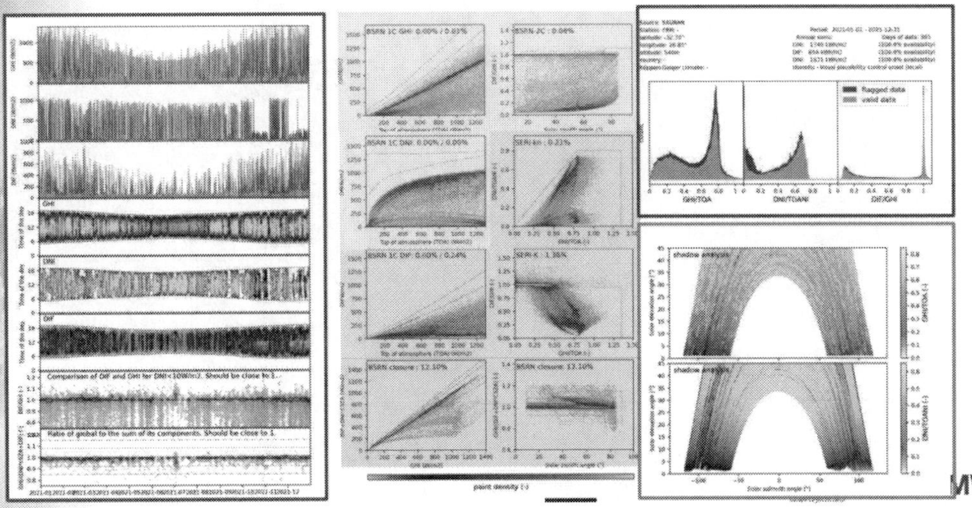

26

020356-026

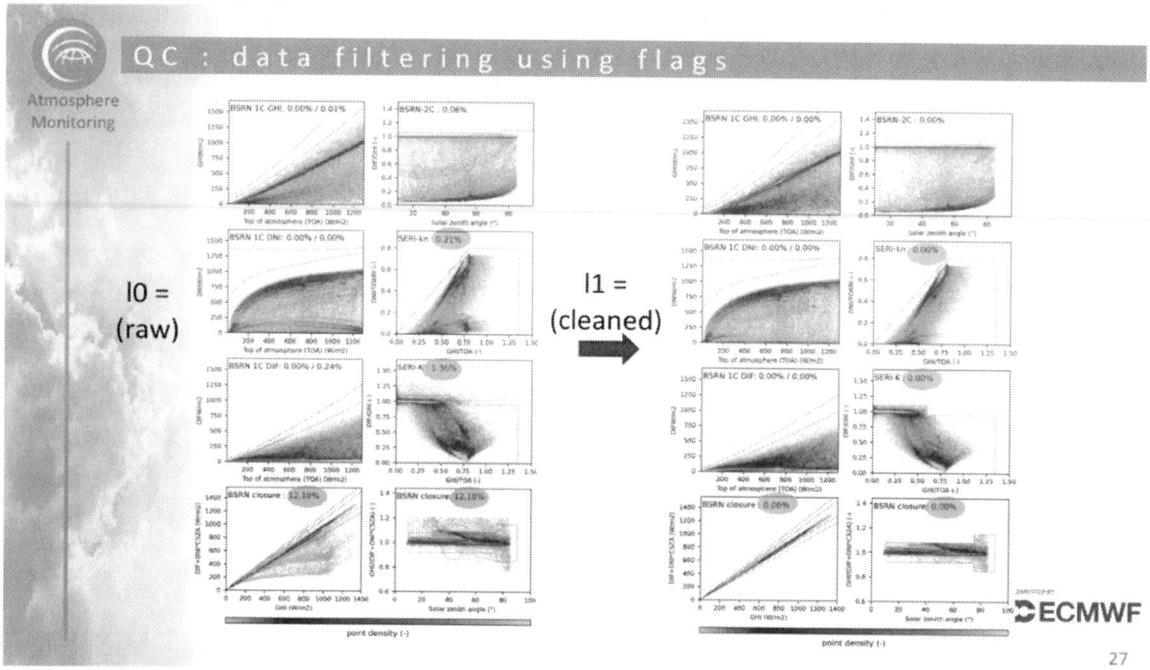

020356-027

40th European Photovoltaic Solar Energy Conference and Exhibition

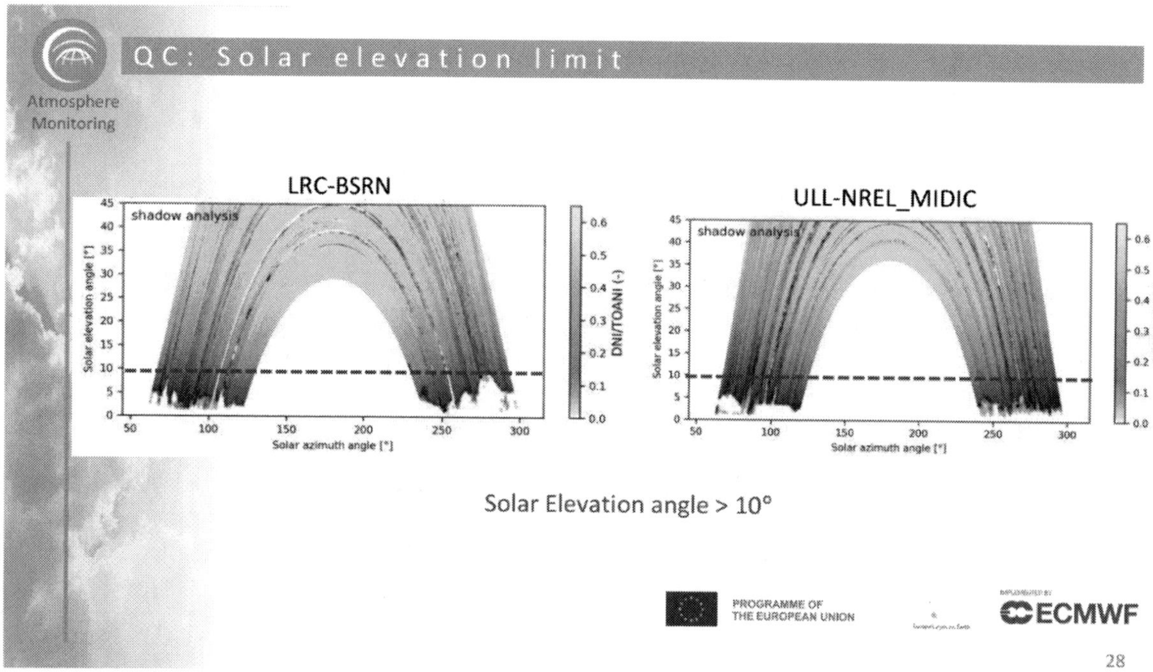

020356-028

Atmosphere Monitoring

Data : IFS through ECMWF MARS API

IFS-HRES

Parameter	Value	Comment
class	od	operational class
stream	oper	operational model
expver	0001	experiment version
type	fc	forecast
levtype	sfc	surface
param	169, 228021	ssrd, fdir
grid	0.1/0.1°	native grid (~ 9 km)
time	0	run at 0 UTC
step	0/to/48/by/1	hourly
date	20210101/to/20211231	full year 2021

IFS-CAMS

Parameter	Value	Comment
class	mc	CAMS (old MACC)
stream	oper	operational
expver	0001	experiment version
type	fc	forecast
levtype	sfc	surface
param	169, 228021	ssrd, fdir
grid	0.35°/0.35°	native grid (~ 40 km)
time	0	run at 0 UTC
step	0/to/48/by/1	hourly
date	20210101/to/20211231	full year 2021

29

020356-029

40th European Photovoltaic Solar Energy Conference and Exhibition

Atmosphere Monitoring

The CAMS radiation service (CRS)

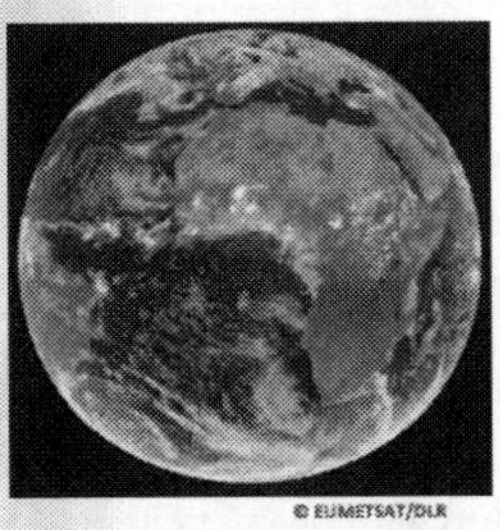

© EUMETSAT/DLR

clouds from satellite

SSI* retrieval scheme: physical approaches, fast radiative transfer (Heliosat-4 and McClear)

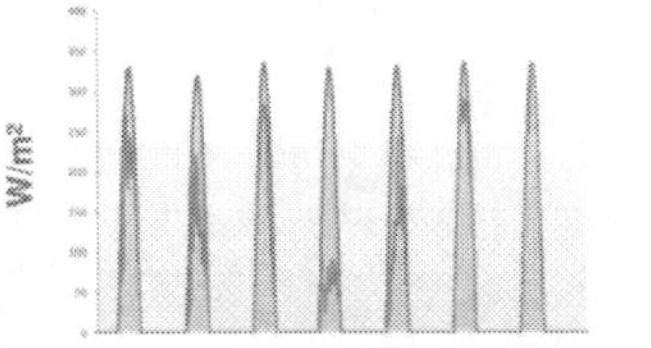

cloud free irradiance
all sky irradiance

© MEEO

aerosol H_2O, O_3 from model

method papers
Qu et al., Contrib. Atm. Sci., 2017
Schroedter-Homscheidt et al., Contrib. Atm. Sci., 2022
Lefèvre et al., Atm. Meas. Tech., 2013
Gschwind et al., Contrib. Atm. Sci., 2019

PROGRAMME OF THE EUROPEAN UNION

ECMWF

* SSI : Surface Solar Irradiance

020356-030

Atmosphere Monitoring

Error distribution of BNI on Desert Region

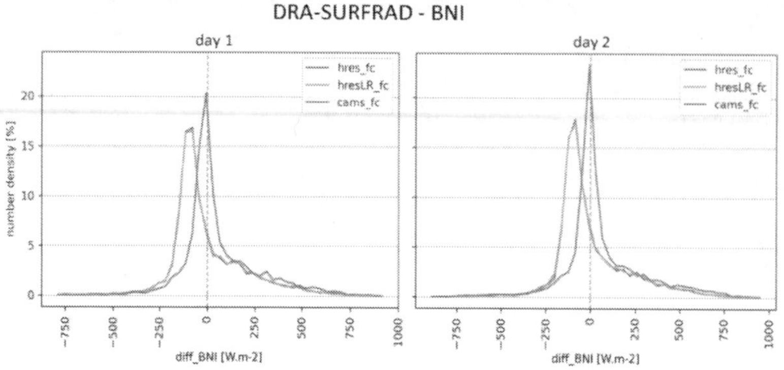

- BNI error distributions mean ~ 0
- The peak of the distribution is for both days 4% higher for the CAMS that for HRES
- BNI better modeled on these regions

PROGRAMME OF THE EUROPEAN UNION

ECMWF

31

020356-031

Atmosphere Monitoring

Error Metrics All Sky situations

- The IFS-CAMS has a similar behavior than the IFS-HRES
- IFS-HRES outperforms IFS-CAMS ~ 4 % the RMSE
- We did not found a noticeable double penalty effect when averaging the IFS-HRES forecast
- How to quantify the effect of the additional aerosols module on the performance?
    - Better modelling of BNI seen in desert regions → dominated with high irradiance values

PROGRAMME OF THE EUROPEAN UNION

ECMWF

32

020356-032

GHI: Mean Bias Error

Atmosphere Monitoring

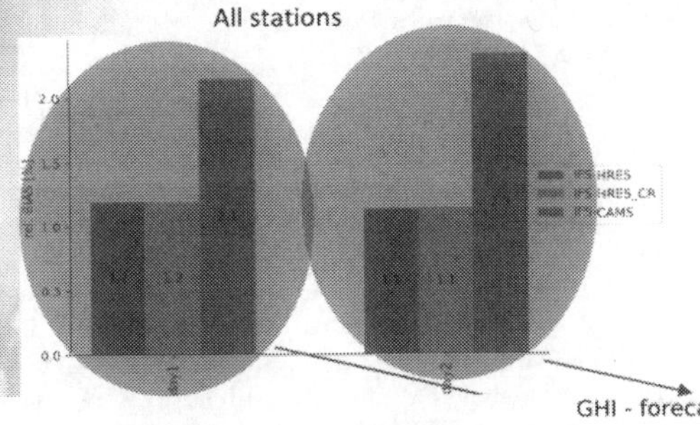

- No noticable difference between IFS-HRES and IFS-HRES_CR
- IFS-CAMS ~ 1% higher bias → ?
  - IFS-CAMS bias comparable on continental and desert
  - IFS-CAMS high bias on mountainous and subpixel
    - **SON** : 3108 m of altitude
    - **RUN** : 1km from the coast
    - Difficult to define **albedo** for this regions

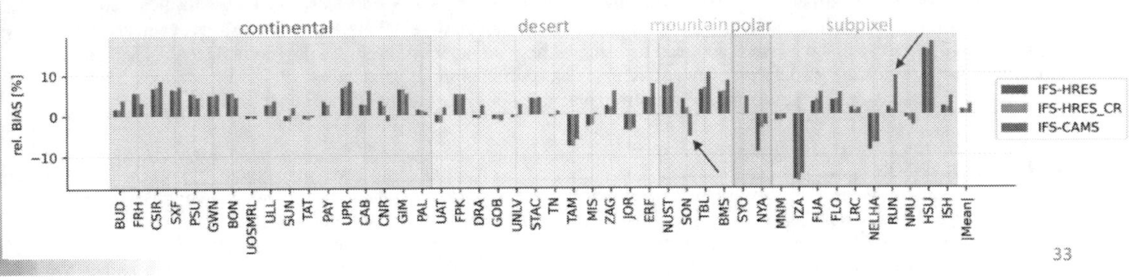

33

GHI: RMSE

Atmosphere Monitoring

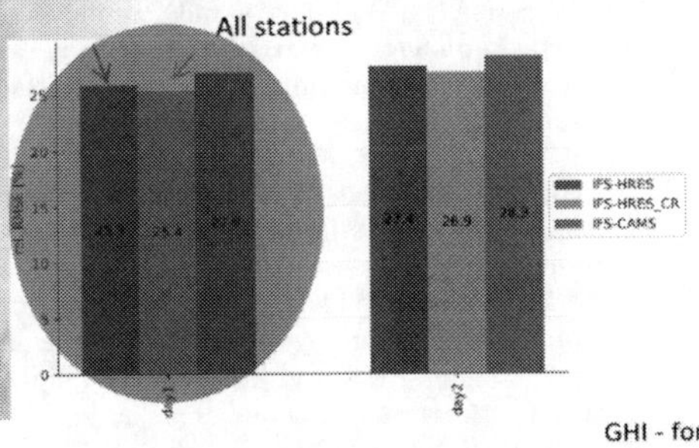

- IFS-HRES_CR RMSE slightly better than IFS-HRES
  - → Spatial averaging should improve RMSE significantly: here IFS-HRES_CR does not
  - → no visible double penalty effect
- IFS-CAMS shows a slight overall under performance of ~ 1.5 %
- Tendency for lower RMSE on desert sites
- Tendency for higher RMSE on polar / subpixel (albedo)

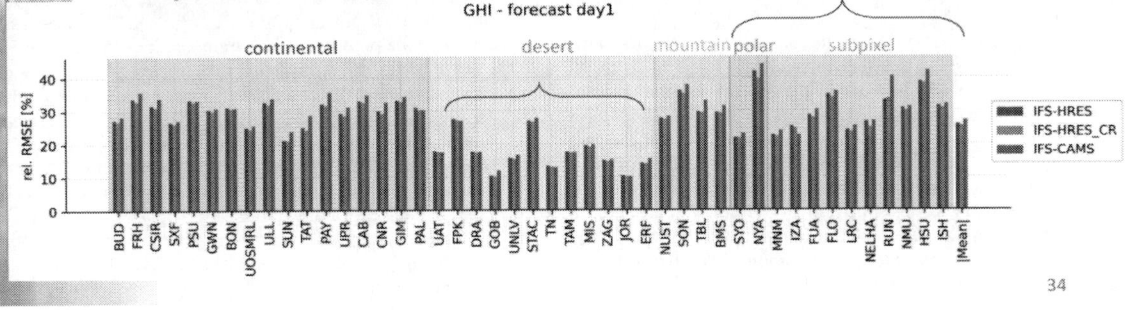

34

COMBINATION OF A NOVEL ALL SKY IMAGER BASED APPROACH FOR HIGH-RESOLUTION SOLAR IRRADIANCE NOWCASTING WITH PERSISTENCE AND SATELLITE NOWCASTS FOR INCREASED ACCURACY

[1]Nils Straub, [1]Wiebke Herzberg, [1]Anna Dittmann, [1]Elke Lorenz
[1]Fraunhofer Institute for Solar Energy Systems ISE,
Heidenhofstr. 2, 79110 Freiburg, Germany

ABSTRACT: Electricity grids with high PV-penetration benefit from the consideration of intra-minute and intra-hour variabilities via nowcasts (shortest-term forecasts). In this study we present approaches for blending all sky imager (ASI), persistence and satellite based nowcasts for increased nowcasting accuracy within the 15-minute interval. Our ASI method is a novel machine-learning (ML) based model, trained on spatially distributed irradiance measurements in Freiburg, Germany. These come from an irradiance measurement network of eight stations within a radius of ~10 km around the camera position, which we also use to evaluate our forecasting results.
Our ASI method exhibits a significantly lower root mean square error (RMSE) than the satellite-based method up to a lead time (LT) of 11 minutes ahead and a lower mean absolute error (MAE) throughout the entire interval. Using a LT dependent linear combination of the individual models RMSE improvement scores of $5-13\%$ and MAE improvement scores of up to 6% (for $LT \geq 5\ min$) could be achieved over the respective optimal individual method. Further improvements in RMSE (up to 2%) and MAE (up to 7.5%) were achieved by using a Lasso model with 3rd degree polynomial features and including cloudiness, sun position and variability of clear-sky index as additional input parameters.
Keywords: All Sky-imager, irradiance nowcasting, machine learning, satellite data, model blending

1 INTRODUCTION

Due to the fast extension of installed PV capacity worldwide, the generation characteristics within our electricity grids are in the process of undergoing a fundamental transformation: From a centralized electricity generation in relatively few major power plants towards a distributed generation on multiple plants of variable sizes. With the integration of large shares of distributed variable energy sources ([1] name of a share beyond 15% of annual energy production within the grid) significant changes in the operation of electrical grids become essential.

Synoptic and local weather patterns make solar irradiance a highly variable resource. Its forecasts hence make an important contribution to the integration of large shares of PV into the electricity grid [2].

On a local scale, solar irradiance can fluctuate with high amplitude on very short time scales due to passing clouds, also referred to as irradiance ramps. These ramps can result in sudden and heavy fluctuations in the power production of large PV-powerplants or distributed PV within quarters. This sets new challenges for all stakeholders including plant and grid operators.

Forecasts of solar irradiance based on numerical weather predictions (NWP) and the images of meteorological satellites have become established approaches to forecast solar irradiance days ahead and hours ahead respectively with a wide field of applications. However, neither of them can resolve single small-scale cloud structures and movements and therefore they are not suited for predicting irradiance ramps.

Nowcasts are high-resolution short-term irradiance forecasts for the immediate future [2], herein up to 15 minutes ahead. Based on these, fine-grained control applications can be performed to handle short-term fluctuations of PV-power in a beneficial way for the grid and other stakeholders involved. The control of heat pumps and cooling systems [3], as well as management of large battery storages, offgrid PV-Diesel systems and the short-term electricity market (up to 5 minutes ahead, [2]) are just a few examples of applications that can profit from accurate nowcasting.

In this study we investigate approaches of blending nowcasts of a novel all sky imager (ASI) model, persistence and satellite-based nowcasts.

Nowcasts are computed for a dataset spanning over one entire year and evaluated against measurements from our irradiance measuring network. This network includes eight measuring stations within a radius of ~10 km around the camera position.

Our first nowcasting approach is a persistence model, assuming persistence of clear-sky index at the respective position. The clear-sky index k^*_{clear} is defined as the ratio between current global horizontal irradiation (GHI) and GHI under clear-sky conditions [2]. Secondly, we use a satellite-based (SAT) approach [4] based on the well-established Heliosat method [5]. Using images of Meteosat Second Generation (MSG) forecasts with a temporal resolution of 15 minutes up to a few hours ahead can be computed. Within this study we only use the first LT step (15 minutes ahead), which is linearly interpolated to create nowcasts that match the 1-minute resolution of the ASI.

With its capability to predict short-term irradiance variability an ASI model is the centerpiece of our blending experiments. Here we use the novel ASI model (initially) reported in [6]. ASIs capture images of the sky in regular intervals to track clouds and their motion. Future cloud situations are predicted by extrapolating these movements into the future. Nowcasts are computed by modelling radiative effects of clouds either by applying statistical models to (short-term) historic irradiance measurements or, as done here, using machine learning (ML) methods, and applying the information to predicted cloud situations. To create areal forecasts clouds must be geolocated either using the cloud base height (CBH) from ceilometer measurements or multiple ASIs with overlapping fields of view to perform stereo photography. Subsequently, cloud shadows are projected onto their real-world positions on the ground exploiting cloud and sun position. Thereby spatial coverage and forecasting horizon can be extended by using multiple or even networks of ASIs [7]. Nevertheless many ASI methods reported of only generate

10.4229/EUPVSEC2023/4CO.8.3
020357-001

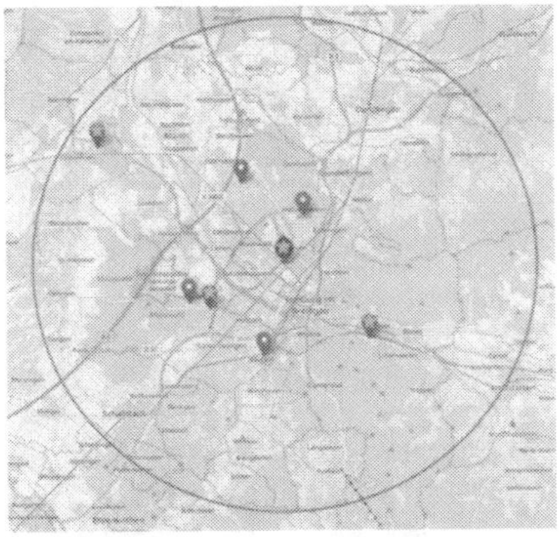

(a)

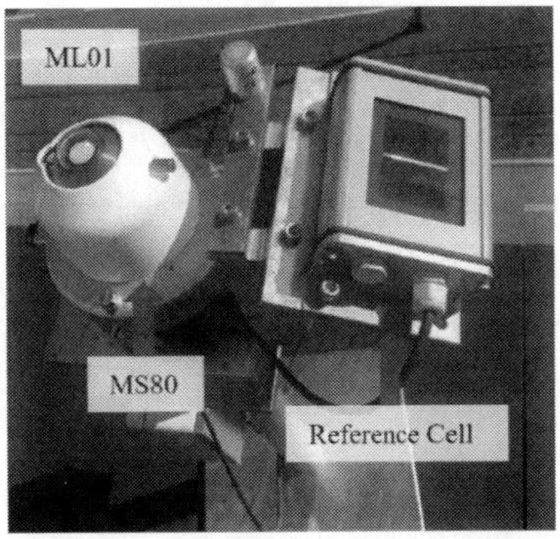

(b)

Figure 1: (a) Map of our irradiance measuring network. The blue circle has a radius of 10 km, the ASI is located in the center (marked red). Stations with a star are equipped with both MS80 and ML01 pyranometers, the remaining stations only with ML01 pyranometers; (b) Measurement station with EKO MS80, ML01 pyranometers and tilted reference cells. Within this work we only use EKO MS80 and ML01.

point nowcasts (usually for the camera position) not providing any spatial resolution [8–10].

Blending of different forecasting models is a common approach that reduces forecasting errors compared to the respective individual methods [11–13]. Model blending approaches within the short-term interval, however, have so far been hardly investigated.

As a baseline for our hybrid model, we use a lead time (LT) dependent linear combination of the three individual methods (see also [6]). The weights are determined via a linear regression on each LT using the three respective nowcasts as input and the corresponding measurement as target value. Here, in addition we investigate more sophisticated blending approaches beyond linear models, including additional input features such as cloudiness and variability of clear-sky index are investigated.

2 DATASET

We use GHI measurements from a network of eight measuring stations within Freiburg to train our ASI- and hybrid models and to evaluate forecasting results. These measuring stations are located within a radius of approximately 10 km around the camera position (Figure 1 (a)). An off-the-shelf surveillance camera (Vivotek FE9381) located in the center of the map is used as ASI, recording images every 10 s. Our measuring setup is shown in Figure 1 (b). Four of the stations (tagged with a star) are equipped with a ventilated thermal pyranometer (EKO MS-80) which has a particularly low response time ($< 0.5\ s$) and a silicon pyranometer (EKO ML-01). The remaining stations are equipped with silicon pyranometers only. All sensors record instantaneous GHI with second resolution. Wherever they are available, measurements from the thermal pyranometers are used, as they exhibit lower uncertainties [14, 15].

Moreover, our ASI-method uses ceilometer measurements of cloud base height (CBH) from a nearby ceilometer (distance to ASI $d_{ASI} \approx 1\ km$) operated by the German weather service Deutscher Wetterdienst (DWD).

All irradiance measurements used within this work undergo an extensive quality control (QC) procedure [16] to ensure data quality. The scope of our QC method is to identify flawed measurements and, if possible, correct or otherwise discard them.

The measuring station at the position of the sky imager is equipped with a sun tracker from EKO (STR-22G) measuring DNI with a pyrheliometer (MS-56) and diffuse horizontal irradiance DHI with a pyranometer shaded by a shadow ball (MS-802F). This sophisticated setup provides the highest measurement quality within the measuring network and serves as reference within our QC.

2.1 Training and test data

Our measuring campaign started in February 2018 and is running ever since. Within this study one entire year from (including) Mai 2021 to April 2022 was selected. Using one year of data aims at giving a balanced representation of the seasonal course of meteorological conditions on our testing site.

The dataset split for the ASI and the hybrid model is illustrated in Table 1. In total the dataset was split twice: For training of the ASI irradiance retrieval method (compare section 3.1) all odd days of the year have been used. ASI, SAT and persistence nowcasts were computed in the remaining even days of the year (in the following nowcasting-dataset). Subsequently the nowcasting-dataset was split to create a training and test dataset for our hybrid method. As four entire are days missing in our nowcasting-dataset we use every second day *available* to split it into evenly sized train and test datasets.

Both, measurements and nowcasts used in this work are instantaneous values. Nowcasts are computed throughout the entire previously described periods for sun zenith angles of $\leq 80°$ and with a temporal resolution of 1 minute. In principle, our ASI method can provide finer resolutions (down to 10 s) but as comparison to and blending with SAT nowcasts is the scope of this work minute resolution was chosen here. Nowcasts of all methods are computed with a base-time (starting time of each nowcast run) frequency of 15 minutes due to the update frequency of satellite images.

Nowcast horizon is set to 15 minutes ahead. The maximum possible nowcasting horizon of our ASI method depends on the prevailing cloud conditions. It can be longer or less than the intended 15 minutes ahead. Truncated nowcast

10.4229/EUPVSEC2023/4CO.8.3
020357-002

runs with a maximum LT of less than 15 minutes are considered in the evaluation. For these instances SAT and persistence nowcasts are filtered to the same maximum LT to ensure best possible comparability among our methods.

Table 1: Illustration of train test split of the ASI and hybrid model training. Splitting scheme continues after the depicted period throughout the entire dataset.

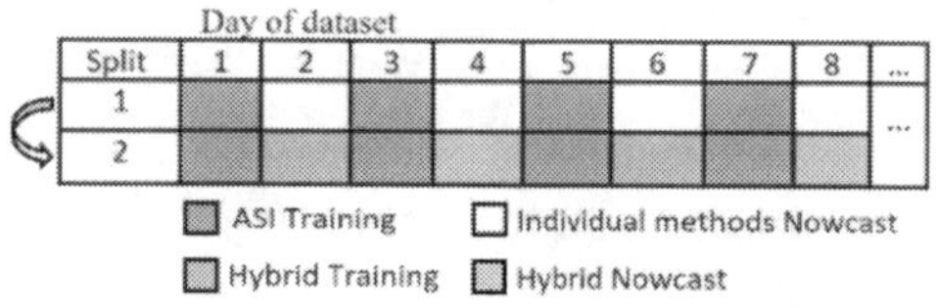

3 METHODOLOGY

3.1 ASI nowcasting

Our ASI nowcasting method uses camera images, local irradiance measurements at the position of the camera and CBH measurements to generate spatially and temporally highly resolved ($1\ minute$ and $\approx 2-10\ m$, depending on CBH) areal nowcasts of GHI. Our ASI nowcasting chain is illustrated in Figure 2. It builds on the ASI method described in [17] and includes the following steps:

a) Recording of all sky image
b) Rectification to compensate fisheye distortion
c) Computation of clear-sky index images using ML-method
d) Identifying cloud movements by computing optical flow from two consecutive rectified ASI images
e) Prediction of future clear-sky index images by applying the computed optical flow to (c)
f) Forecast of GHI maps by projecting clear-sky indices onto their real-world position and scaling them with computed clear-sky irradiation.

In [6] we extend the original approach from point to areal nowcasting by geolocating clouds and projecting their shadows onto real-world positions. Instead of the statistical irradiance retrieval from recent GHI histograms, an adapted version of the ML method from [16] is used to assign distinct GHI levels to each image pixel (hereinafter irradiance algorithm). The remaining steps (a, b, d) and algorithms are equivalent to the original approach. The single steps are described in [6] in more detail.

3.2 Satellite based nowcasts

Our satellite based nowcasting method (SAT-method) is based on [4] and uses data from the geostationary MSG satellite. We use the well-established Heliosat method, in a version based on [5] to infer irradiance. This method mainly exploits the broadband visible channel to determine the irradiance transfer through clouds. At that, the pixel intensity is a crucial parameter that serves as a measure for the backscattering and transmission within the clouds. The brighter a pixel appears, the stronger the backscattering ensuing lower irradiance levels on the ground. To detect and forecast cloud movement, we use block matching on two consecutive satellite images.

With this approach forecasts up to several hours ahead can be computed. The spatial resolution of the forecasts is 1 km x 1 km at sub-satellite point and approximately 1.2 km, x 2.2 km in the region of Freiburg. The temporal resolution is determined by the 15-minute time lapse between two consecutive satellite images. To reach the required 1-minute resolution, forecasted clear-sky indices are linearly interpolated (upsampling).

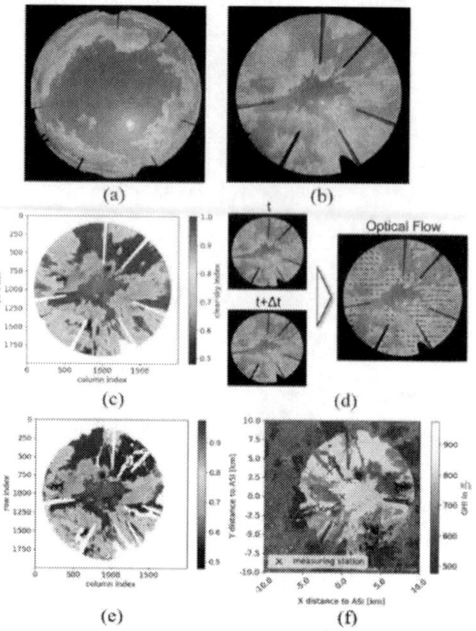

(a) (b) (c) (d) (e) (f)

Figure 2: Overview ASI nowcasting method. (a) recording of ASI image, (b) image undistortion, (c) computation of clear-sky index image, (d) computation of optical flow, (e) forecasting clear-sky index image, (f) forecasting irradiance map. Note, that from step (e) to (f) the clear-sky image is mirrored along the y-axis resulting from the projection.

3.3 Persistence

Persistence nowcasts assume persistence of prevailing conditions. Taking into account diurnal variations of irradiance, i.e. assuming persistence of clear-sky index, nowcasted GHI_{per} at a given time t is given by [2]:

$$GHI_{per}(t) = GHI_{clear}(t) \cdot k^{*}(t_0) \qquad (1)$$

Thereby $k^{*}(t_0)$ denotes the clear-sky index at the current moment. In this work we use persistence based on the clear-sky index derived from measurements for each of the stations of the network. This is consistent with the common persistence definition used in most publications. That way persistence nowcasts have the same quality for each measuring station.

In an operational scenario real-time irradiance measurements are not necessarily available at each position of interest. In that case persistence of clear-sky index of the closest irradiance measurement can be applied instead.

3.4 Model blending

Starting point of our model blending experiments is a LT-dependent linear combination of the individual nowcasts, in the following referred to as stepwise linear hybrid model. The hybrid nowcast from this basic model therefore is the weighted sum of the individual nowcasts, whereas the weights of each method vary with LT.

10.4229/EUPVSEC2023/4CO.8.3

To derive the coefficients, data from our hybrid training dataset is binned by LT-step. Separate regressions are performed on each LT-step, resulting in a separate set of coefficients for each instance.

3.4.1 Advanced model blending

After establishing a basic reference model, we investigate more elaborate ML-methods and additional input features to further enhance our results.

Besides LT, the performance of our individual models might depend on other parameters. These dependencies are not considered in our baseline approach. Here, cloudiness, variability of clear-sky index within the past 15 minutes and cosine of sun zenith angle are added as additional input features. Moreover, LT is used as an input feature instead of separate models for each LT. To reflect the presumably non-linear relations between model in- and output more sophisticated (non-linear) methods are investigated. We use the same set of input features in all our advanced blending approaches. These are listed in Table 2. Cloudiness is computed as the fraction of cloud-classified pixels in undistorted ASI images. Thereto, we use the cloud classification algorithm from [17]. Variability of clear-sky index is computed according to [18].

Six methods where preselected and hyperparameter optimized. Our six candidates include scikit-learn's [19] python implementations of gradient boosting regression (GBR), extremely randomized trees regression (Extra Trees) and a random forest regressor (Random Forest) with default parameters each. Moreover, k-nearest neighbors regression (kNN) with distance weights and 50 neighbors, and a Lasso (Lasso) model with 3rd degree polynomial features and $\alpha = 0.5$ (scikit-learn) and a neuronal network (NN) with two hidden dense layers of 200 nodes each (tensorflow) are deployed.

Table 2: List of input features for our advanced combi-approaches

Name	***Description***
Individual Nowcasts	ASI, SAT and Persistence nowcasts
Nowcasting LT	LT step in minutes
Cloudiness	Fraction of cloud classified pixels over number of valid pixels in undistorted image
$\cos(\theta_{sun})$	Cosine of sun zenith angle at ASI position
Variability k^*	Variability of clear-sky index [18] within the past 15 minutes

3.5 Error metrics

The error between an observation x_i and the corresponding prediction y_i is given by:

$$\epsilon_i = y_i - x_i \tag{2}$$

To compare model performances, we consider the root mean squared error (RMSE) and mean absolute error (MAE) and their normalized counterparts ($RMSE_{rel}$ and MAE_{rel}):

$$RMSE = \sqrt{\frac{1}{n}\sum_{i=0}^{n} \epsilon_i^2} \tag{3}$$

$$RMSE_{rel} = \frac{RMSE}{\bar{x}} \cdot 100\% \tag{4}$$

$$MAE = \frac{1}{n}\sum_{i=0}^{n} |\epsilon_i| \tag{5}$$

$$MAE_{rel} = \frac{MAE}{\bar{x}} \cdot 100\% \tag{6}$$

In the context of forecasting the skill evaluates the forecast performance compared to a trivial reference method with regard to a certain error metric (e.g. RMSE) [2]. It is calculated from the forecast score S_{fc} in the respective metric, the score of the reference forecast S_{ref} and the score of a prefect forecast S_{perf} (e.g. $S_{perf,RMSE} = 0$):

$$Skill = \frac{S_{ref} - S_{fc}}{S_{ref} - S_{perf}} \tag{7}$$

The skill is positive if the respective score shows an improvement compared to the reference method, zero if its equivalent and negative if it performs worse. It can reach a maximum value of 1 implying a perfect Score ($S_{fc} = S_{perf}$). When compared to a non-trivial method as reference this metric is referred to as improvement score. Here, we evaluate the RMSE skill of our methods in comparison to persistence and the improvement of the hybrid model compared to the single models with respect to RMSE and MAE.

4 RESULTS

4.1 Individual nowcasts and linear blending

Nowcasting results of our individual methods and hybrid model evaluated on all measuring stations are shown in Figure 2. Regarding the three individual methods the persistence nowcast performs best in terms of RMSE for lead times (LTs) up to 2 min. (Figure 3 (a)). For $LT = 0$ the persistence nowcast is the measured value at this exact instance. Starting at zero its RMSE quickly rises with LT. Subsequently the ASI takes over until $LT = 11\,min$ after which the satellite gains a slight edge compared to the ASI. Generally forecasting performances of all types of models decrease with increasing LTs. Opposed to that, a decline of RMSE and MAE with increasing LTs can be observed in the satellite curve (Figure 3 (a) and (b)). As outlined in section 2.1, data is filtered to situations where nowcasts of all methods are available. ASI nowcasts tend to drop out over LT in situations with high cloud velocities and thereby high variability. This causes a decline in mean variability over LT in the filtered dataset. On the remaining more stable situations lower error metrics are observed for all methods, causing a decline in RMSE and MAE for the SAT nowcasts.

When performing the same evaluation but considering complete nowcast runs only an increase of RMSE and MAE over LT could be observed for all methods (not shown). This way the dataset (variability-)composition remains unaltered over LT.

10.4229/EUPVSEC2023/4CO.8.3
020357-004

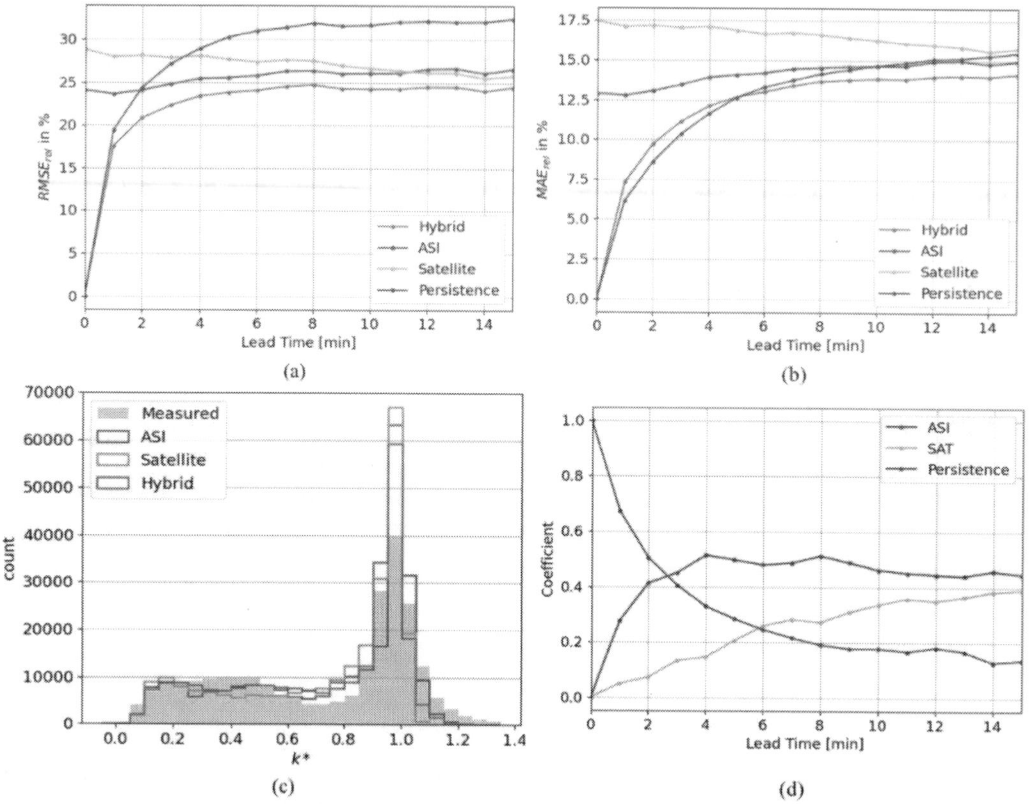

Figure 3: Nowcasting results of the four different methods on the test dataset (a) relative RMSE, (b) relative MAE, (c) Distribution of clear-sky indices of nowcast and measurement all LT combined and (d) contributions of each method to the stepwise linear hybrid model over LT.

All in all, the performances of the three methods match the respective temporal scale they are designed for and illustrate the sense of purpose of blending different methods with varying contributions over LT.

Regarding relative MAE (Figure 3 (b)), our persistence method shows the best individual performance until $LT = 9\ min$. After that, our ASI method takes over for the rest of the nowcasting interval. The SAT method shows the highest relative MAE for all LT. RMSE penalizes larger errors and outliers more than MAE, which tends to promote smoother nowcasts such as the SAT nowcast. Different error distributions of our models result in different courses of MAE and RMSE curves. Persistence reflects the full range of GHI values observed, resulting in a higher number of outliers which is penalized by RMSE.

Clear-sky index distributions of ASI, SAT and our stepwise linear hybrid model for all LTs combined are shown in Figure 3 (c). The distribution of our persistence nowcasts is inherently (almost perfectly) equivalent to the measured distribution and is therefore left out here. Measurements exhibit a notable spread around $k^* = 1$. This can be caused by variations in turbidity as well as reflections on cloud bottom sides that cause irradiance levels beyond clear-sky conditions. These aspects are insufficiently reflected within our nowcasting methods and clear-sky model, resulting in a sharp peak at $k^* = 1$. Values beyond that point are underrepresented. This underrepresentation is more significant for the SAT than for the ASI method. By design our SAT-method does not predict $k^* > 1.2$. Its initial 15-minute values are linearly interpolated to a 1-minute resolution, such that all instances except $LT = 0$ and $LT = 15\ Min$ are interpolated values (compare section 3.2). This creates a smooth nowcast with an over proportional share of values in the intermediate range ($\approx 0.6 - 0.9$) and an underrepresentation for $k^* > 1$ correspondingly. The k^*-distribution of our ASI-nowcasts is influenced by the RMSE optimized training procedure. To avoid large errors the ASI method favors intermediate values over extreme values likewise causing an overrepresentation in the intermediate range. This overrepresentation is however less pronounced than for the SAT method. Besides lower MAE and RMSE (within certain LT intervals) an advantage of the ASI over the SAT-method is the possibility to have higher update frequencies than the 15 minutes prescribed by the satellite.

Our stepwise linear hybrid model exhibits the lowest RMSE compared to any of the individual models over the entire interval. Regarding MAE it performs best for $LT \geq 5\ min$. The contribution of each method as a function of LT is depicted in Figure 3 (d). The stepwise linear method benefits from all three nowcasting methods, exploiting their varying contributions throughout the short-term interval. As expected, these contributions are correlated to the model's performances: The lower the RMSE (Figure 3 (a)) the higher the corresponding models' contribution. For very short LT ($\leq 3\ min$) the persistence nowcast makes by far the largest contribution, which quickly

10.4229/EUPVSEC2023/4CO.8.3
020357-005

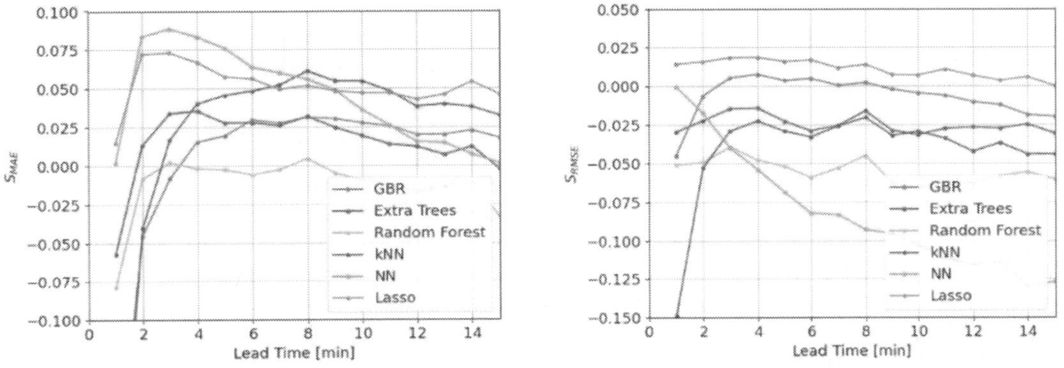

(a) (b)

Figure 4: MAE (a) and RMSE (b) improvement scores of our advanced compared to our basic stepwise linear hybrid model.

decreases throughout the 15-minute time window. The contribution of ASI quickly increases within the first minutes. From $LT \approx 3\ min$ onwards it is the dominant method within the hybrid nowcast. By LT dependent linear combination of our individual nowcasting results an RMSE improvement score of $5-13\%$ compared to the respective optimal individual method was achieved. A positive MAE improvement score of up to 6% can be achieved for $LT \geq 5\ min$. Our individual methods differ in their error distributions and therefore exhibit different courses of RMSE and MAE curves (Figure 3 (a) and (b)). Hence our RMSE optimized stepwise linear model does not minimize MAE, resulting in a lower MAE than RMSE improvement score. A much more detailed investigation of model performance including a variability analysis is given in [6].

4.2 Advanced model blending

RMSE and MAE improvement scores of our advanced blending approaches with our stepwise linear hybrid model as reference are depicted in Figure 4 (a) and (b).

Positive RMSE improvement score is shown by our Lasso method throughout the entire interval and by the GBR within $3\ \text{min} \leq LT \leq 6\ min$. Other than that, RMSE improvement scores are all negative. Our Lasso method exhibits the highest positive values with $\approx 0-2\%$.

Higher improvements can be observed regarding MAE. For $LT \leq 9\ min$ our NN shows the highest improvement scores of up to $\approx 8\ \%$ which quickly declines to 0 outside this interval. Our Lasso approach exhibits positive MAE improvement scores throughout the entire interval ranging from $\approx 4-7.5\%$. Within certain LT intervals positive MAE improvement score is shown by all remaining methods except Random Forest.

As mentioned, nowcasting performance was enhanced significantly less in terms of MAE with our stepwise linear blending approach compared to the individual methods (Figure 4 (b)) leaving higher potential for improvement regarding this metric.

With our advanced blending models, we could achieve a significant, yet somewhat limited performance improvements compared to our linear hybrid model. One cause for that limitation might be, that the selected additional parameters in the advanced approach are not separating the performances of the individual models well. As a result the hybrid model cannot learn virtues of the individual models under specific conditions resulting in only slight improvements. Nowcasting performance can however only be improved to a certain degree with this blending approach as our individual nowcasts themselves have limited accuracy.

We found that higher complexity models such as our neuronal network and others like transformers (in preliminary experiments) tend to perform worse than simpler models. Thereby adding or replacing further additional features likewise led to performance deterioration, probably due to overfitting All in all, we therefore conjecture that there is limited potential for further improvement with this approach.

5 SUMMARY AND CONCLUSION

In this work a novel ASI method for high resolution areal nowcasting was benchmarked against persistence and SAT nowcasts with LTs up to $15\ min$. To match the ASI resolution, linear interpolation was applied to the SAT nowcasts. Different approaches were investigated to blend nowcasts of these three individual methods. Areal nowcasts were computed every $15\ min$ with a temporal resolution of $1\ min$ for LTs up to $15\ min$ ahead on every fourth day throughout one entire year. The validation was performed with a network of eight measuring stations in the city region of Freiburg, within a radius of $\approx 10\ km$ around the ASI position.

Regarding our three individual methods persistence exhibits the lowest RMSE for $LT \leq 2\ min$ and lowest MAE for $LT \leq 9\ min$. The SAT method shows the lowest RMSE from all individual methods towards the end of the interval ($LT \geq 12\ min$) and highest MAE throughout all LT. It produces very smooth nowcasts on a minute timescale due to its linear interpolation, which results in a better performance regarding RMSE than MAE.

Within $2\ \text{min} < LT < 12\ min$ our novel ASI method exhibits the lowest RMSE and for $LT \geq 10\ min$ the lowest MAE of the individual methods. Clear-sky index distributions indicate that our ASI nowcasts exhibit some smoothing but significantly less pronounced as the SAT method. Compared to SAT nowcasts ASI nowcasts have the additional advantage of a higher updating frequency, which wasn't examined in detail in this work. One way to increase the forecasting horizon and spatial coverage of ASIs is the use of multiple ASIs with adjacent fields of view. Another approach is to use a hybrid model that fills up gaps in ASI nowcasts with nowcasts from other methods. This way individual methods, designed for

10.4229/EUPVSEC2023/4CO.8.3
020357-006

different spatio-temporal scales, can be used to complement each other.
In addition to the lead time dependent linear blending several more advanced approaches were tested and compared. Our stepwise linear hybrid model outperforms the three individual methods on any LT by an RMSE improvement score of $5\% - 13\%$ compared to the respective optimal method. A positive MAE improvement score of up to 6% can be achieved for $LT \geq 5\ min$ (see also [6]. By blending our nowcasts with a Lasso model with cloudiness, sun position and variability as additional inputs and 3rd degree polynomial features we could further improve our stepwise linear approach by an RMSE improvement score of $\approx 1\%$.More remarkable is the MAE improvement score of $\approx 4 - 7.5\%$ for $LT \geq 2\ min$. More sophisticated models like transformers with further input information performed worse in preliminary experiments. Based on these experiments, we presume that the potential of the model blending approach is mostly tapped with the proposed Lasso model and there is little room for further improvement.
However, we see great potential in investigating novel image-based ML-based methods for predicting future cloud situations. In the field of short-term precipitation nowcasting novel methods based on convolutional neuronal networks have led to great improvements and outperformed state-of-the-art methods within recent years [20, 21].

6 ACKNOWLEDGEMENTS

Parts of this research were funded by the European Union as part of the project SERENDI-PV. Special thanks go out to the German Federal Environmental Foundation (DBU) for supporting this work with a fellowship.

7 REFERENCES

[1] Denholm P, Margolis R. Energy Storage Requirements for Achieving 50% Solar Photovoltaic Energy Penetration in California; 2016.

[2] Sengupta M, Habte A, Wilbert S, Gueymard C, Remund J. Best Practices Handbook for the Collection and Use of Solar Resource Data for Solar Energy Applications: Third Edition. Golden, USA: National Renewable Energy Lab. (NREL); 2021 NREL/TP-5D00-77635.

[3] Dittmann A, Lorenz E, Kober P, *et al.* Optimierung der PV-Speisung von Wärmepumpen durch Kurzfristprognosen mit Wolkenkameras; 2019.

[4] Kühnert J, Lorenz E, Heinemann D. Satellite-Based Irradiance and Power Forecasting for the German Energy Market. In: Kleissl J, editor. Solar Energy Forecasting and Resource Assessment. Elsevier 2013; 267–97.

[5] Hammer A, Heinemann D, Hoyer C, *et al.* Solar energy assessment using remote sensing technologies. Remote Sensing of Environment 2003; 86(3): 423–32 [https://doi.org/10.1016/S0034-4257(03)00083-X]

[6] Straub N, Herzberg W, Dittmann A, Lorenz E. Blending of a Novel All Sky Imager Model with Persistence and a Satellite Based Model for High-Resolution Irradiance Nowcasting (Submitted Manuscript). Freiburg, Germany 2023.

[7] Schmidt T, Stührenberg J, Blum N, Lezaca J, Hammer A, Vogt T. A network of all sky imagers (ASI) enabling accurate and high-resolution very short-term forecasts of solar irradiance. IET Conference Proceedings 2022: 372-378(6).

[8] Liandrat O, Cros S, Braun A, Saint-Antonin L, Decriox J, Schmutz N. Cloud cover forecast from a ground-based all sky infrared thermal camera. In: Comerón A, Kassianov EI, Schäfer K, editors. Cloud cover forecast from a ground-based all sky infrared thermal camera; 2017. SPIE; 19–31.

[9] Pedro H, Coimbra C, David M, Lauret P. Assessment of machine learning techniques for deterministic and probabilistic intra-hour solar forecasts. Renewable Energy 2018; 123: 191–203 [https://doi.org/10.1016/j.renene.2018.02.006]

[10] Huang J, Kahn MM, Qin Y, West S. Hybrid Intra-hour Solar PV Power Forecasting using statistical and Skycam-based Methods. IEEE 46th Photovoltaic Specialists Conference (PVSC) 2019.

[11] Kühnert J. Development of a photovoltaic power prediction system for forecast horizons of several hours. PhDUniversity 2015.

[12] Wolff B. Support Vector Regression for Solar Power Prediction. PhDUniversität Oldenburg 2017.

[13] Nouri B, Blum N, Wilbert S, Zarzalejo LF. A Hybrid Solar Irradiance Nowcasting Approach: Combining All Sky Imager Systems and Persistence Irradiance Models for Increased Accuracy. Solar RRL 2021: 2100442 [https://doi.org/10.1002/solr.202100442]

[14] L., Batlles, F., Olmo F. Solar radiation resource assessment by means of silicon cells. Solar Energy 1995: 183–91.

[15] N. Errassi et al. An inexpensive and accurate solar irradiance sensor based in a small calibrated PV module. 33rd European Photovoltaic Solar Energy Conference and Exhibition 2017: (pp. 1588-1592).

[16] Straub N. Sky imager based modelling of solar irradiance using spatially distributed irradiance measurements and machine learning. Master thesis, Institut für Mikrosystemtechnik, Albert-Ludwigs-Universität, Freiburg 2020.

[17] Dittmann A, Holland N, Lorenz E. A new sky imager based global irradiance forecasting model with analyses of cirrus situations. metz 2020 [https://doi.org/10.1127/metz/2020/1024]

[18] Marquez R, Coimbra CFM. Proposed Metric for Evaluation of Solar Forecasting Models. J. Sol. Energy Eng. 2012; 135(1) [https://doi.org/10.1115/1.4007496]

[19] Pedregosa F, Varoquaux G, Gramfort A, *et al.* Scikit-learn: Machine learning in Python. the Journal of machine Learning research 2011; (12, 2825-2830.).

[20] Ayzel G, Scheffer T, Heistermann M. RainNet v1.0: a convolutional neural network for radar-based precipitation nowcasting. Geosci. Model Dev. 2020; 13(6): 2631–44 [https://doi.org/10.5194/gmd-13-2631-2020]

[21] Ravuri S, Lenc K, Willson M, *et al.* Skilful precipitation nowcasting using deep generative models of radar. Nature 2021; 597(7878): 672–7 [https://doi.org/10.1038/s41586-021-03854-z][PMID: 34588668]

COMBINING DEEP LEARNING AND PHYSICAL MODELS FOR SOLAR NOWCASTING

Yann Fabel, Bijan Nouri, Stefan Wilbert, Niklas Blum, Dominik Schnaus, Rudolph Triebel, Luis F. Zarzalejo, Enrique Ugedo, Julia Kowalski, Robert Pitz-Paal

EU PVSEC, 20th September 2023, Lisbon

1 Yann Fabel, Institute of Solar Research, EU PVSEC 20.09.2023

020359-001

40th European Photovoltaic Solar Energy Conference and Exhibition

Agenda

- Motivation & Introduction
- Hybrid Nowcasting Models
- Validation
- Conclusion

2 Yann Fabel, Institute of Solar Research, EU PVSEC 20.09.2023

020359-002

020359-003

40th European Photovoltaic Solar Energy Conference and Exhibition

Motivation

DLR

What is the problem?

- Short-term irradiance fluctuations induced by clouds pose a challenge for the power system

How to deal with this?

- Apply solar nowcasting to anticipate changes in solar irradiance and adapt plant operation

What are potential benefits?

- Optimized real-time trading
- Ramp rate control
- Frequency control

What are the requirements?

- Cloud information in spatially and temporally high resolutions

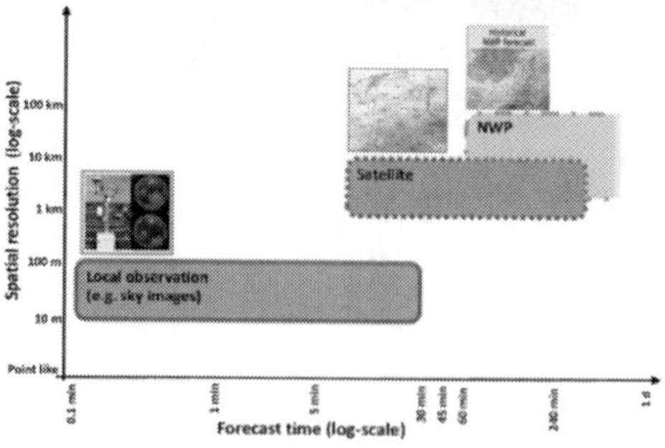

4

Yann Fabel, Institute of Solar Research, EU PVSEC 20.09.2023

020359-004

ASI Nowcasting

- All-Sky Imager
  - Ground-based camera observing complete hemisphere using fish-eye lens
- Image analysis
  - Physical approach
    - Explicit modelling of clouds, their motion and transmittance
  - Data-driven approach
    - Model learns correlation of clouds and irradiance directly from images
- Nowcast: intra-hour irradiance forecast

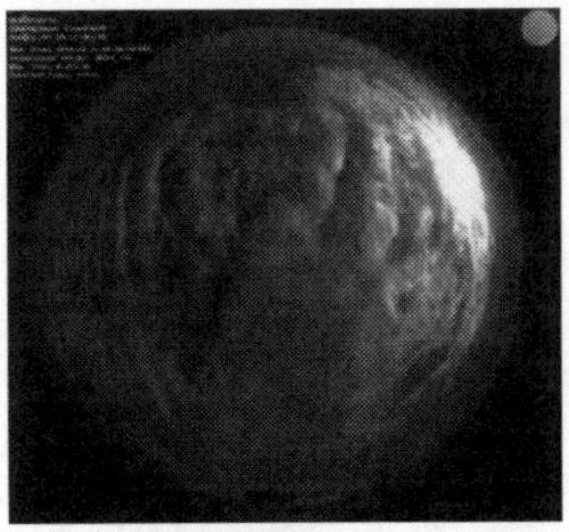

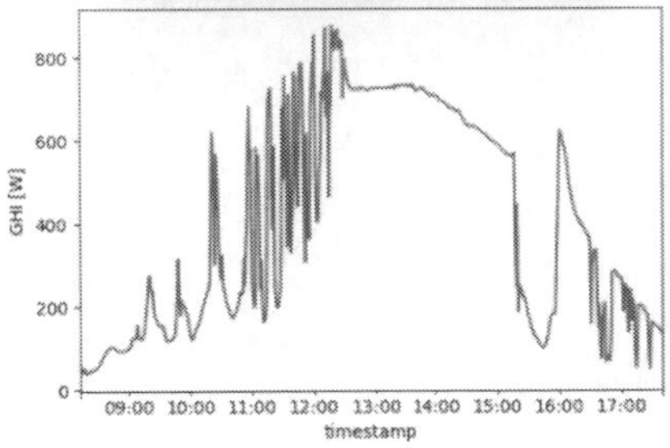

5 Yann Fabel, Institute of Solar Research, EU PVSEC 20.09.2023

020359-005

40th European Photovoltaic Solar Energy Conference and Exhibition

020359-006

Hybrid Model

DLR

Physical model

7

Yann Fabel, Institute of Solar Research, EU PVSEC 20.09.2023

Hybrid Model

DLR

Physical model

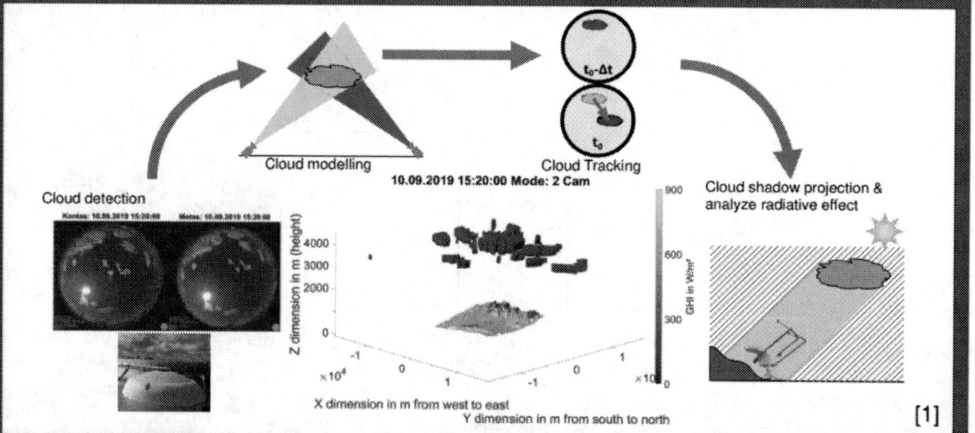

[1]

8

Yann Fabel, Institute of Solar Research, EU PVSEC 20.09.2023

Hybrid Model

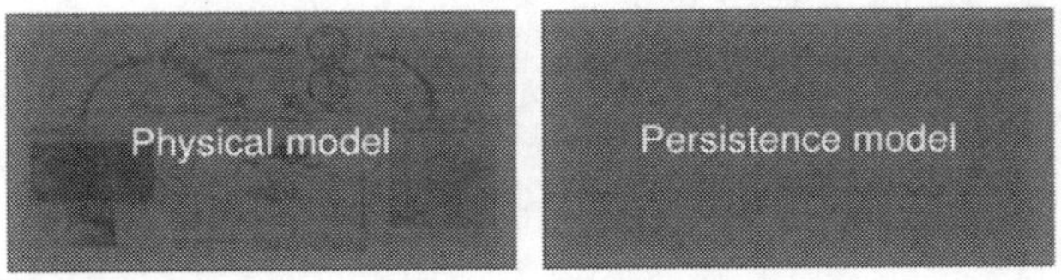

9

Yann Fabel, Institute of Solar Research, EU PVSEC 20.09.2023

020359-009

40th European Photovoltaic Solar Energy Conference and Exhibition

Hybrid Model

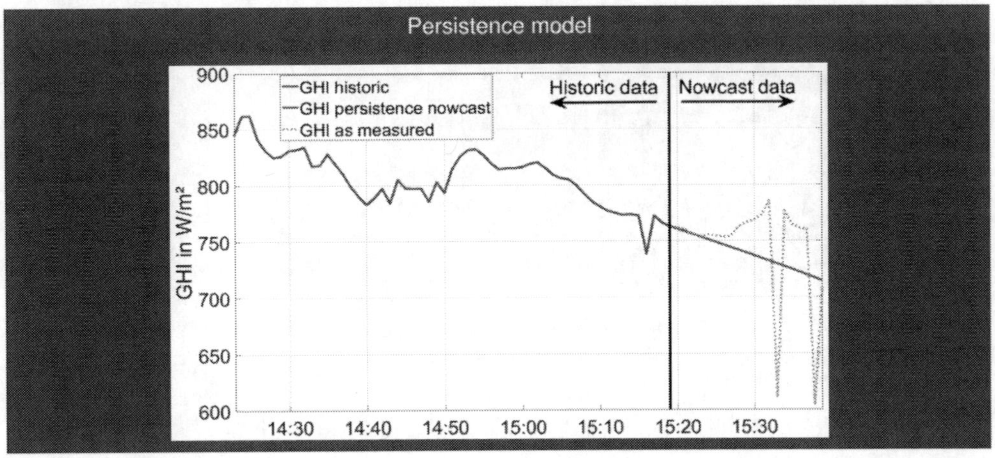

10

Yann Fabel, Institute of Solar Research, EU PVSEC 20.09.2023

020359-010

Hybrid Model

DLR

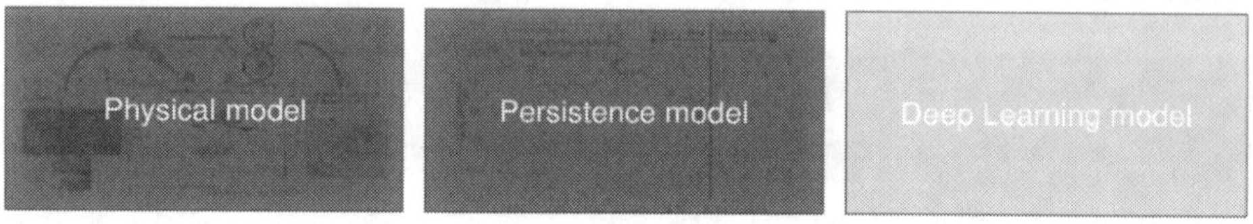

11

Yann Fabel, Institute of Solar Research, EU PVSEC 20.09.2023

020359-011

40th European Photovoltaic Solar Energy Conference and Exhibition

Hybrid Model

DLR

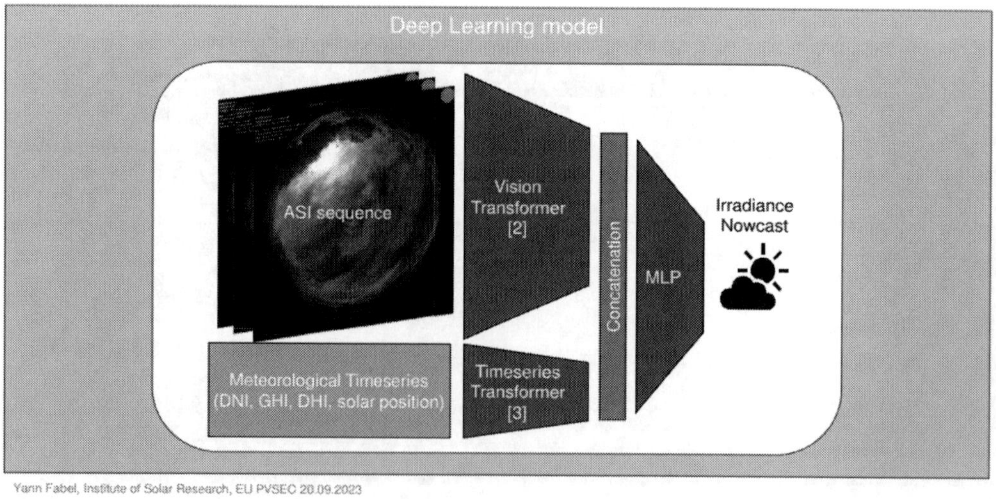

12

Yann Fabel, Institute of Solar Research, EU PVSEC 20.09.2023

020359-012

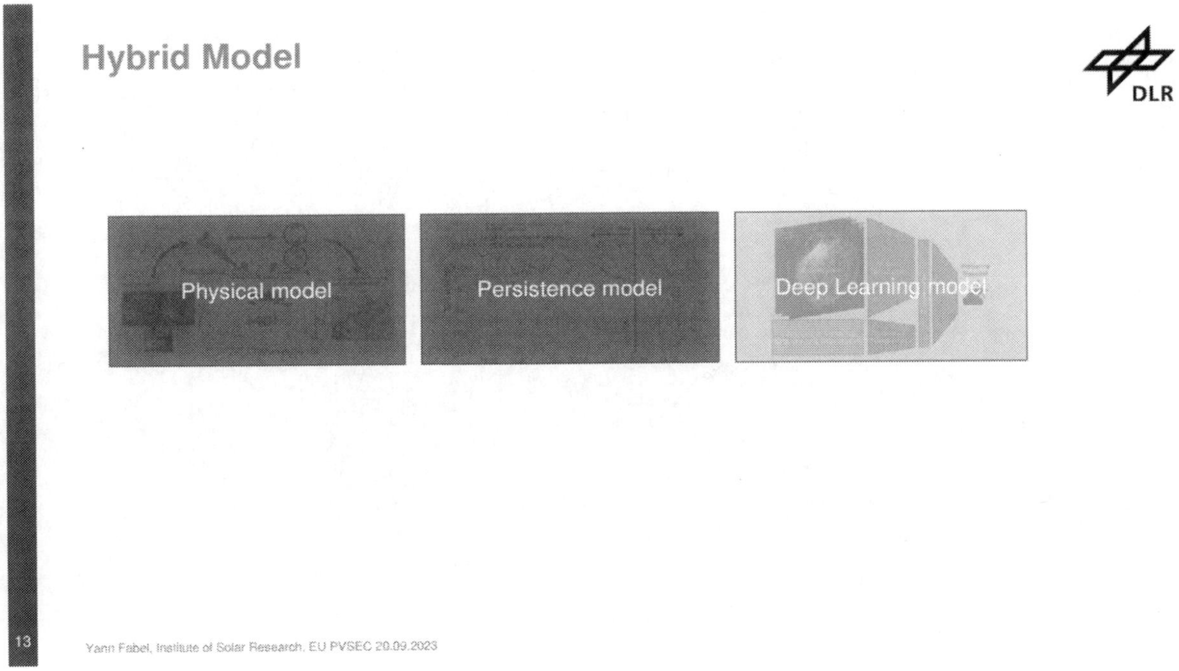

020359-013

40th European Photovoltaic Solar Energy Conference and Exhibition

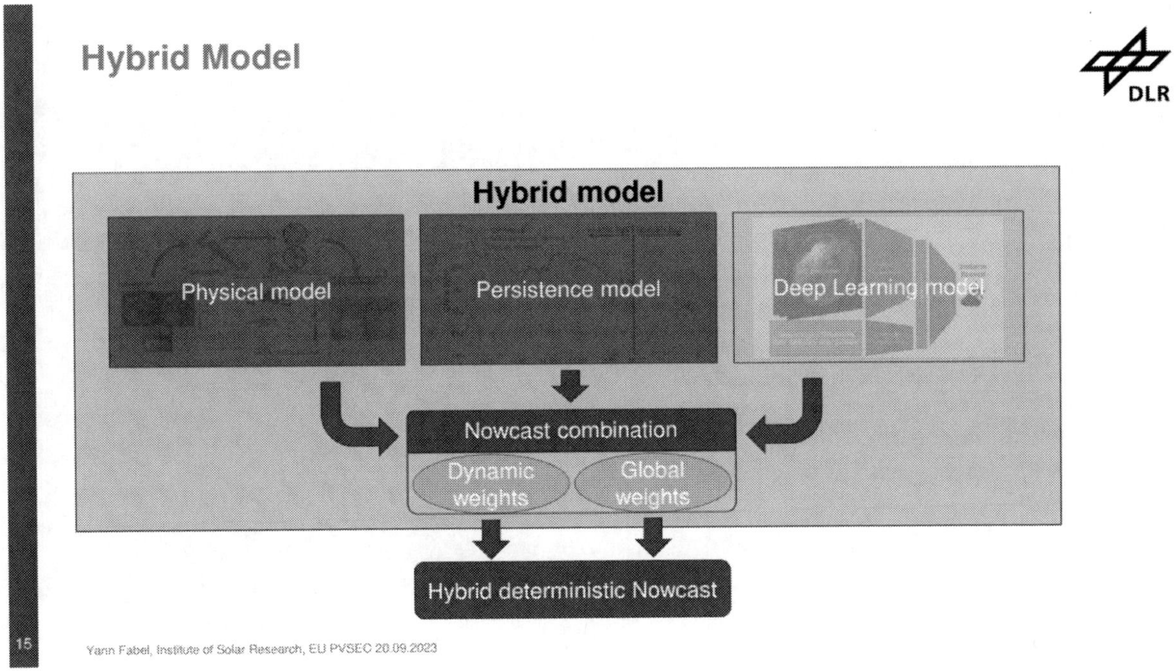

020359-014

020359-015

40th European Photovoltaic Solar Energy Conference and Exhibition

Validation

- Comparison of standalone deep learning model with hybrid models (global vs dynamic weights)
- Reference data
  - GHI field average from 9 measurement stations at CIEMAT's PSA (Spain)
  - Benchmark dataset [4]: 28 selected days from 2019 covering different atmospheric conditions
- Validation metric:
  - Forecast skill based on persistence

$$FS = 1 - \frac{error_{model}}{error_{persistence}}$$

17

Yann Fabel, Institute of Solar Research, EU PVSEC 20.09.2023

020359-016

Validation Results

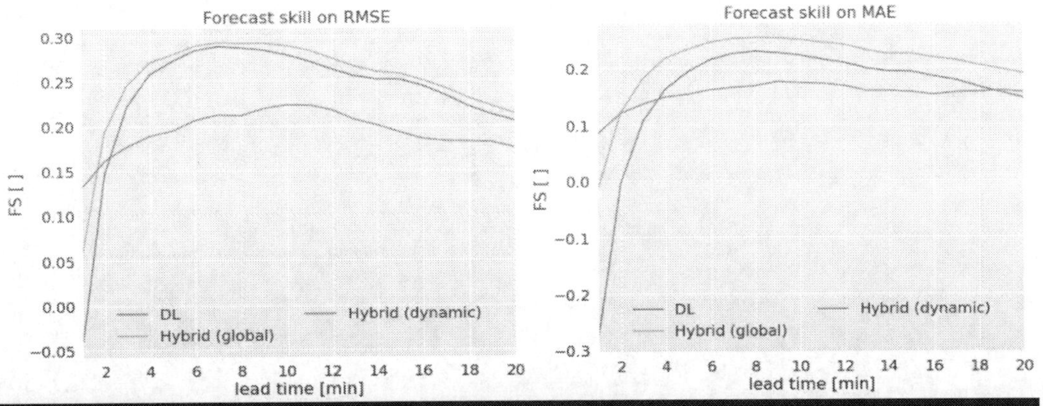

- DL point forecasts work well on field average (optimization on MSE leads to low RMSE)
- Forecast quality of hybrid models depends on combination method
- Generally persistence is already quite good in the first minutes → lower FS for all models

18 Yann Fabel, Institute of Solar Research, EU PVSEC 20.09.2023

020359-017

Validation Results

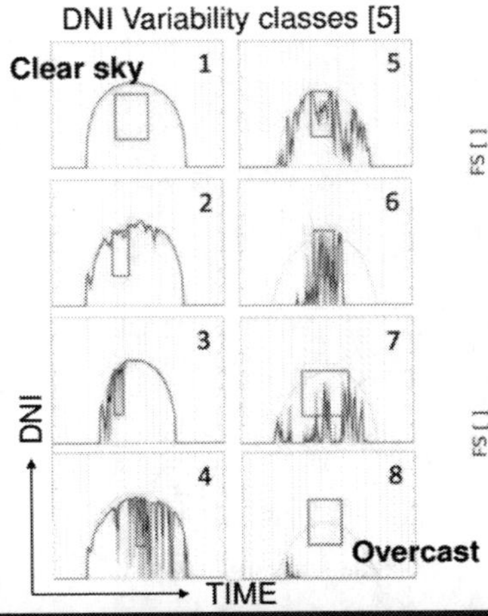

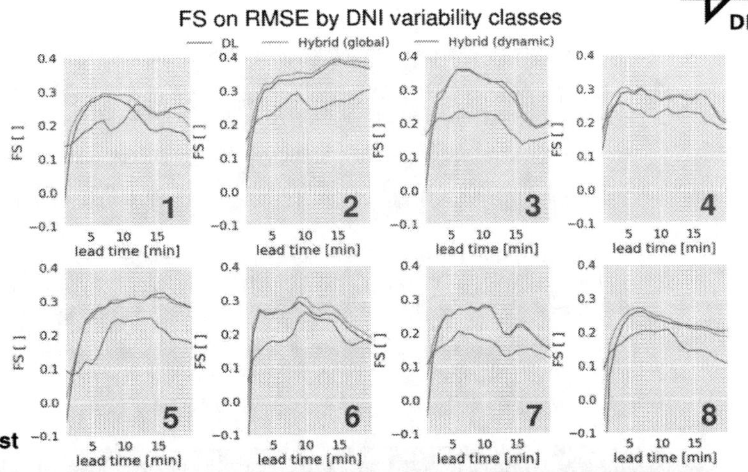

- DNI variability classification as representation of different cloud conditions
- Generally higher FS for more variable conditions
- Difference between Hybrid (dynamic) and Hybrid (global)/DL greater for variable conditions

19 Yann Fabel, Institute of Solar Research, EU PVSEC 20.09.2023

020359-018

020359-019

40th European Photovoltaic Solar Energy Conference and Exhibition

Conclusion

DLR

- Study presents a benchmark of purely data-driven and hybrid models confirming potential for data-driven models in solar nowcasting
- For limited regions (~1km^2) data-driven point forecasts can compete with spatial forecasts from physical models in terms of FS
- Advantage of hybrid models strongly depends on combination of nowcasts
- Forecast skills assess nowcast quality only partially
→ further metrics need to be established in solar nowcasting community
- Analysis of probabilistic nowcasts shows benefit from deep learning model [6]

21 Yann Fabel, Institute of Solar Research, EU PVSEC 20.09.2023

020359-020

020319-021

40th European Photovoltaic Solar Energy Conference and Exhibition

References

DLR

1. Nouri, B. / Blum, N. / Wilbert, S. / Zarzalejo, L. F., 2022
   A Hybrid Solar Irradiance Nowcasting Approach: Combining All Sky Imager Systems and Persistence Irradiance Models for Increased Accuracy
2. Bertasius, G. / Wang, H. / Torresani, L., 2021
   Is Space-Time Attention All You Need for Video Understanding?
3. Zerveas, G. / Jayaraman, S. / Patel, D. / Bhamidipaty, A. / Eickhoff, C., 2021
   A Transformer-based Framework for Multivariate Time Series Representation Learning
4. Logothetis, S. A. / Salamalikis, V. / Wilbert, S. / Remund, J. / Zarzalejo, L. F. / Xie, Y. / Nouri, B. / Ntavelis, E. / Nou, J. / Hendrikx, N. / Visser, L. / Sengupta, M. / Po, M. / Chauvin, R. / Grieu, S. / Blum, N. / van Sark, W. / Kazantzidis, A., 2022
   Benchmarking of solar irradiance nowcast performance derived from all-sky imagers
5. Schroedter-Homscheidt, M. / Kosmale, M. / Jung, S. / Kleissl, J., 2018
   Classifying ground-measured 1 minute temporal variability within hourly intervals for direct normal irradiances
6. Fabel, Yann / Nouri, Bijan / Wilbert, Stefan / Blum, Niklas / Schnaus, Dominik / Triebel, Rudolph / Zarzalejo, Luis F. / Ugedo, Enrique / Kowalski, Julia / Pitz-Paal, Robert, 2023, *(Under review)*
   Combining deep learning and physical models: a benchmark study on all-sky imagerbased solar nowcasting systems

23 Yann Fabel, Institute of Solar Research, EU PVSEC 20.09.2023

020319-022

DYNAMIC GRAPH MACHINE LEARNING FOR MULTI-SITE SOLAR FORECASTING

Rafael E. Carrillo, Baptiste Schubnel, Renaud Langou and Pierre-Jean Alet
CSEM, Sustainable Energy Center, Rue Jacquet-Droz 1, 2002 Neuchâtel, Switzerland
Emails : rafael.carrillo@csem.ch, baptiste.schubnel@csem.ch, renaud.langou@csem.ch, pierre-jean.alet@csem.ch

ABSTRACT: In this paper, we present DIGERATI, CSEM's solution for intraday forecasting (six-hour horizon). DIGERATI is based on graph neural networks (GNN). It extends previous algorithms developed by CSEM and enables their use in real operating conditions. These GNN algorithms outperformed the state of the art in forecasting photovoltaic production, while relying only on real, imperfect power measurements. In previous works the same nodes were present throughout training and evaluation. However, in a commercial application, nodes (e.g., photovoltaic systems) would be frequently added or removed as new customers sign up or physical assets change. To meet this constraint, DIGERATI exploits dynamic GNNs, where nodes and edges can be added or removed over time, thus providing a robust and scalable solution for real-life operations. Additionally, DIGERATI uses advanced GNNs to fuse information from heterogeneous data sources (PV power, irradiance, wind speed and temperature) measured at different locations to produce probabilistic forecasts of irradiance (or PV power) in desired locations. The live demonstrator can produce probabilistic forecasts at any location in Switzerland and the Netherlands every 15 minutes with a temporal resolution of 15 minutes.
Keywords: Solar radiation, forecasting, machine learning, photovoltaic

1 INTRODUCTION

Renewable energy sources (RES), such as wind power production or photovoltaic (PV) power production, play a key role in reaching a fully decarbonized electricity production. However, operating power systems with large amounts of RES raises many challenges to grid operators due to the variability of their power production. To ensure an adequate operation of the power grid, e.g., to anticipate and alleviate grid congestion, forecasting of power generation with fine temporal and spatial resolution is required [1].

State-of-the-art approaches for solar resources forecasting combine numerical weather predictions (NWP), satellite images and ground measurements with physical or machine learning models. The main limitation of current approaches is that they can only achieve precise spatial and temporal resolution at the expense of high computational and storage load. An alternative approach is to use past measurements from a network of ground sensors (e.g., PV or meteorological stations) distributed in some region of interest and exploit their spatiotemporal relations to improve the forecast accuracy and resolution [2]. Graph neural networks (GNN) have the capability to infer information from complex data defined in irregular domains (such as networks) by using graphs to model their interdependence. This capability makes them suitable to exploit the spatiotemporal correlations of ground sensors. GNNs have indeed demonstrated excellent performance in forecasting tasks on such data [3]–[6]. However, published forecasting approaches that use GNNs assume that their input data come from a fixed network of sensors (nodes). This assumption is a severe limitation for practical applications since nodes include renewable power plants, rooftop PV systems and weather stations, which can all be frequently added or removed as new customers sign up or physical assets change.

To overcome this drawback, CSEM has developed the DIGERATI solution to provide probabilistic forecasts of solar sources (irradiance or PV power) based on dynamic graph machine learning models that uses heterogeneous data sources as inputs, such as meteorological stations or rooftop PV systems, and can adapt to dynamic changes in the input sensor network. The solution can produce forecasts for new "unseen" nodes, e.g., recently added PV stations or irradiance forecasts in sites where input data is not available.

In this paper, we present the DIGERATI solution for solar forecasting and compare it to state-of-the-art methods for a forecasting horizon of six hours. Benchmarking showed that the proposed GNN models outperform state-of-the-art commercial forecasting services. The live demonstrator yields forecasts of solar irradiance on 128 sites across Switzerland every 15 minutes with a temporal resolution of 15 minutes. In addition, DIGERATI is capable of forecasting irradiance or PV power output in any location in Switzerland and we have demonstrated its scalability.

2 DIGERATI: GNN-BASED SOLUTION FOR INTRADAY FORECASTING

DIGERATI is CSEM's solution for intraday forecasting (six hours ahead horizon), based on dynamic GNNs and a ground-based network of sensors. DIGERATI extends our previous works [3]–[6] on multi-site PV forecasting by adding: (i) probabilistic forecasts capabilities by producing forecasts of the 5%, 50% and 95% quantiles for each forecast point in time, (ii) fusion of heterogeneous data sources (PV power and meteorological measurements), and, (iii) use of dynamic graph machine learning to address the problem of changing (sensor) network, where nodes and edges can be added or removed dynamically. In the following, we detail the main components of the system.

2.1 System description

The main problem addressed in DIGERATI is the forecast of irradiance on Q desired locations (points of interest) from past measurements of a network of N_k sensors (data sources, i.e., PV systems and weather stations). The points of interest might be different from the locations of the sensors in the network and, moreover, the set of nodes from the sensor network might be constantly changing in time. Let t denotes the current time, M the number of past discrete time samples used as input and H

10.4229/EUPVSEC2023/4CO.8.5
020360-001

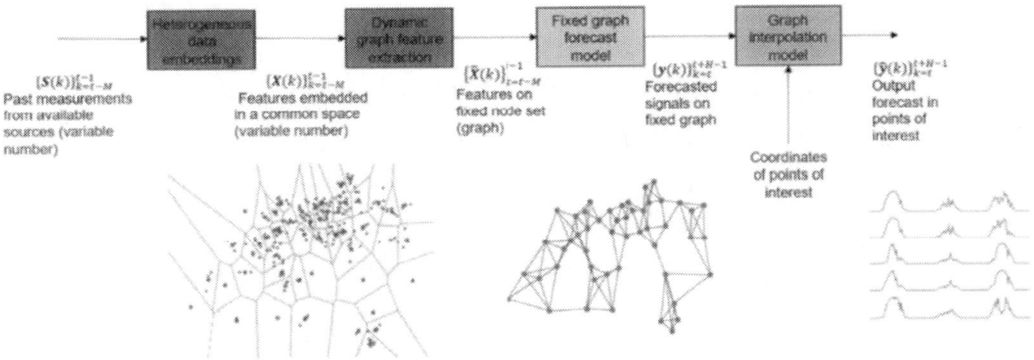

Figure 1. Block diagram of DIGERATI.

be the number of discrete time samples we want to forecast in the future. The problem is to map the sequence $\{\boldsymbol{S}(k)\}_{k=t-M}^{t-1}$ of past measurements (with varying number of nodes N_k) onto the sequence $\{\hat{\boldsymbol{y}}(k)\}_{k=t}^{t+H-1}$ of future irradiance (or power) for the Q points of interest.

A block diagram of the overall solution is depicted in Figure 1. The first block, the heterogeneous data embeddings, takes the raw data from all available sources (PV systems or meteorological stations) and computes an embedding into a common vector space. The input data is assumed to come from a network with a variable number of nodes at every time step. The second block, the dynamic graph feature extraction, computes features from the time varying network onto a fixed network that represents all the salient information for forecasting. The third block, the fixed graph forecast model, is a GNN forecast model that produces probabilistic forecasts of irradiance of the 5%, 50% and 95% quantiles for each forecast point in time for all nodes in the fixed network. In this block we use the graph convolutional LSTM (GCLSTM) model from [4] as the fixed forecast model with a modified loss function to extend the model for quantile learning [7]. The fourth block, the graph interpolation model, receives as inputs the forecasts on the fixed network and the locations (coordinates) of the desired (and possibly unseen) nodes to compute the forecasts of the desired nodes. We use in this block the graph interpolation algorithm proposed by CSEM in [3].

2.2 Heterogeneous data embeddings

The purpose of this block is to fuse the information from heterogeneous data sources. Indeed, we use power production data from the available PV systems as well as data from weather stations. Let us denote by $\boldsymbol{p}(k) \in \mathbb{R}^{N_p}$ the vector of PV power production of all PV systems at time step k and by $\boldsymbol{w}(k) \in \mathbb{R}^{w_f \times N_w}$ the matrix of weather data of all weather stations at time step t, where N_p denotes the number of PV systems, N_w denotes the number of weather stations and w_f the number of weather variables (e.g. irradiance, wind speed, etc.). The embedded vector $\tilde{\boldsymbol{x}}_v(k)$ is then computed as

$$\tilde{\boldsymbol{x}}_v(k) = \begin{cases} f^p(p_v(k)) & \text{if } v \text{ is a power node,} \\ f^w(w_v(k)) & \text{if } v \text{ is a weather node,} \end{cases}$$

where f^w and f^p are multilayer perceptron (MLP) networks that learn the mapping from the original space onto the common space. All vectors $\tilde{\boldsymbol{x}}_v(k)$ are stacked into the sequence of matrices $\{\boldsymbol{X}(k)\}_{k=t-M}^{t-1}$ that might have a varying number of nodes and are passed to the block.

2.3 Dynamic graph feature extraction

The dynamic feature extraction block takes the embedded features from the data fusion block and computes the sequence of features $\{\hat{\boldsymbol{X}}(k)\}_{k=t-M}^{t-1}$ on a fixed network with C nodes. The main steps of our approach to tackle the dynamic network problem are:

1. Cluster all nodes' locations (PV stations and weather stations) into C regions.
2. Construct an undirected fixed graph based on the C centroids locations.

The clustering process must be done just once before the training of the model to define the clusters. If new nodes are aggregated to the forecasting system, we don't need to retrain the whole model, we just need to assign a cluster to the new nodes and add the new systems as inputs to the heterogeneous data embedding block.

3 RESULTS

3.1 Use of heterogeneous data sources

We start by studying if (and by how much) the accuracy of the models is improved by using different data sources. The first dataset used consists of power data from 308 PV systems scattered irregularly across Switzerland. In addition to the power production values, we used data from 98 automatic weather stations from the Meteoswiss network. The measurements from the weather stations considered are: global horizontal irradiance, air temperature, relative humidity and wind speed and direction. The spatial distribution of the PV systems and the weather stations is shown in Figure 2. We used two years of historical data, from January 2016 to January 2018 and used the first year for training and the second year for evaluation. The models were trained to forecast the power on the 308 PV systems using as inputs only power or both power and weather measurements.

The results of the study are shown in Figure 3, which shows the daytime normalized root mean square error (NRMSE) over the forecasting horizon using only power data and combining power data and data from the weather stations. Weather data provided a clear improvement on the forecast accuracy. This improvement might be related to the Swiss case: PV systems are scarce in the mountain regions, which is compensated by the presence of the weather stations (see Figure 2). Moreover, by using weather data, the model yields better accuracy in the south and west parts of Switzerland.

10.4229/EUPVSEC2023/4CO.8.5
020360-002

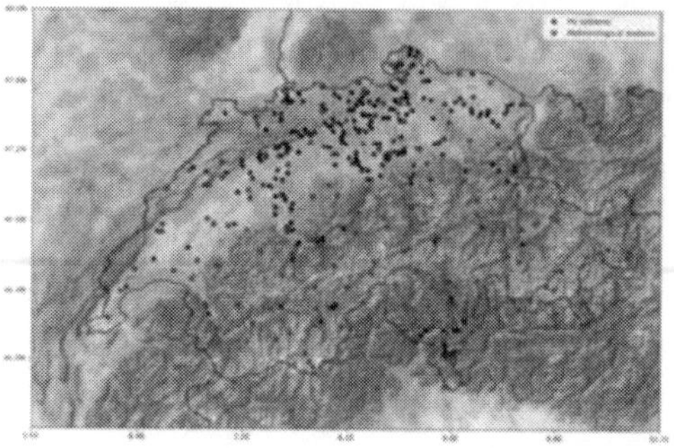

Figure 2. Spatial distribution of the PV systems (blue) and weather stations (red) used.

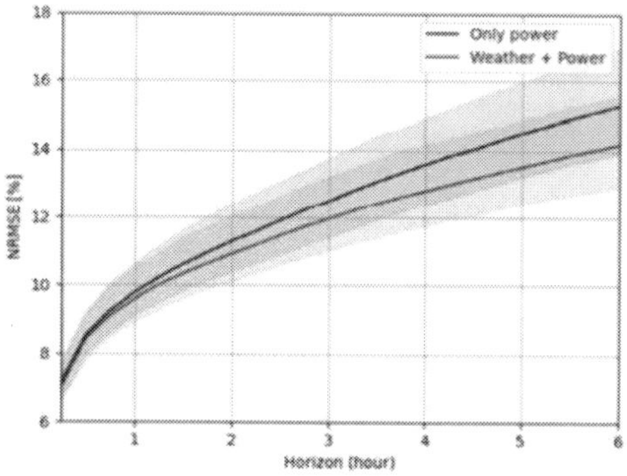

Figure 3. NRMSE results over the forecasting horizon using only power data (blue) and using both weather and power data (red).

3.2 Comparison with a commercial system

We also compare the accuracy of DIGERATI against a commercial forecast solution based on cloud motion tracking that uses satellite images and weather models. A representative set of 18 locations and 21 days were selected to cover the range of possible conditions in Switzerland. The study includes comparisons indifferent seasons with at least five representative weather days per season (see [5] for more details on the evaluation set-up). Figure 4 shows the results of the comparison. DIGERATI yielded lower NRMSE than the cloud tracking method for forecasting horizons above one hour.

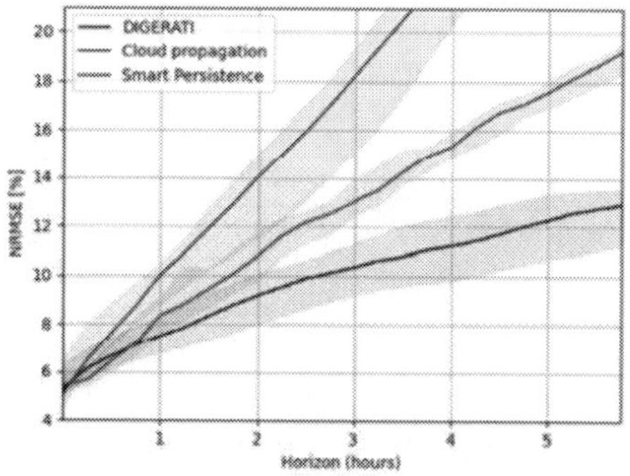

Figure 4. Accuracy comparison between DIGERATI and a commercial forecasting service.

3.3 Probabilistic forecasts

To evaluate the quality of the confidence intervals (5%-95%) produced by the probabilistic forecasts, we evaluated the reliability, resolution and sharpness by computing forecasts over one year of historical data. The metrics used for the evaluation are the prediction interval coverage probability (PICP) and the continuous rank probability score (CRPS).

Figure 5 shows the PICP for the desired forecasting horizon for all nodes. As expected, the PICP degrades as the forecasting horizon increases with median values ranging from 95% (15 minutes ahead) to 93% (six hours ahead). This means that the observations lie within the predicted intervals more than 92% of the time. These results indicate that our forecasting system based on the GCLSTM models and quantile regression yields reliable upper and lower bounds since we are targeting an interval that covers 90% of the probability mass.

As a global metric of the probabilistic models, we use the CRPS. Figure 6 reports the CRPS as a function of the forecasting horizon. The CRPS are reported as percentage of the peak value for each node. As expected, the CRPS increases as the horizon increases. Overall, the normalized CRPS shows that the forecasting system manages to reproduce the 5%, 50% and 95% quantiles of the empirical distribution of the data for the evaluation year.

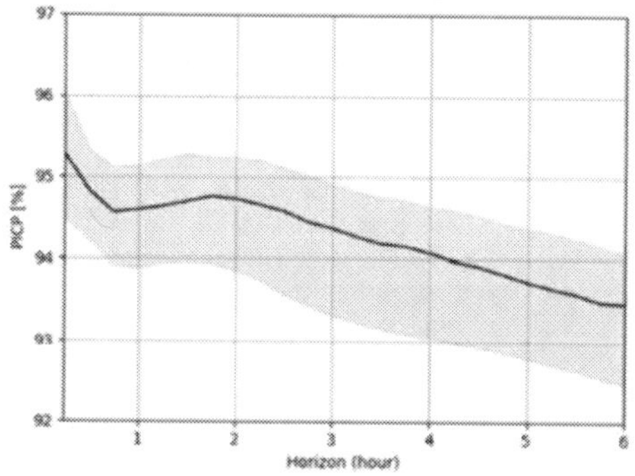

Figure 5. PICP vs forecasting horizon. Solid line is the median across all nodes and the shaded region indicates the interquartile distance across all nodes.

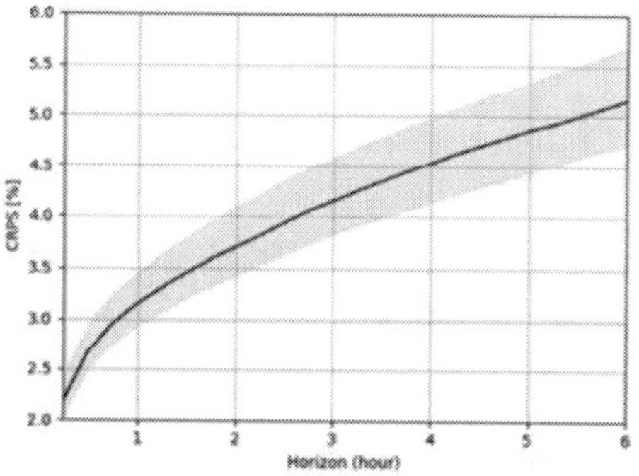

Figure 6. CRPS vs forecasting horizon. Solid line is the median across all nodes and the shaded region indicates the interquartile distance across all nodes.

Figure 7 shows an example of the forecasts yielded by DIGERATI for summer day (early morning). For the first hours in the morning the lower bound (5% quantile) seems quite conservative but as the day advance the bounds become more meaningful by predicting a drop in production due to passing clouds.

10.4229/EUPVSEC2023/4CO.8.5
020360-003

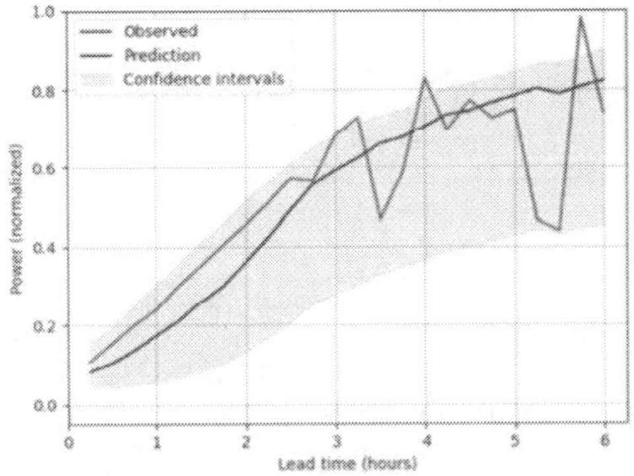

Figure 7. Example of a probabilistic forecasts over the prediction horizon. Solid lines denote the median (50% quantile) estimate. Shaded regions denote the confidence intervals between the 5% and the 95% quantiles.

4 LIVE DEMONSTRATOR

The live demonstrator[1] uses data from 192 nodes, 64 PV stations and 128 meteorological stations (data from Meteoswiss[2]), as inputs to the forecasting system. The system generates a forecast every 15 minutes for the six hours ahead horizon, with a resolution of 15 minutes, of the irradiance in the locations of the 128 meteorological stations. Based on these forecasts, the system is able to forecast the irradiance of any location in Switzerland. The spatial resolution of the system will be improved as more nodes (PV systems or meteorological stations) are added. We also developed a demonstrator for the Netherlands, which uses data from 608 nodes, 572 PV stations and 36 meteorological stations (data from KMNI[3]). As in the Swiss demonstrator, the system is trained to forecast irradiance in the 36 locations of the meteorological stations as a fixed network.

5 CONCLUSIONS

This paper presented the DIGERATI the solution for intraday solar forecasting. DIGERATI produces probabilistic forecasts of solar resources (irradiance or PV power) from heterogeneous data sources as inputs and can adapt to dynamic changes in the network of input sensors. The system yields a daytime normalized root-mean-square error (NRMSE), over all sites, between 7% and 14% for 15 minutes and 6 hours ahead prediction, respectively.

The unique aspects of the DIGERATI solution are:

- It takes a global approach to make localized forecasts. GNNs model the spatio-temporal correlations of the heterogeneous input data (PV power and meteorological measurements) to produce accurate and localized forecasts for any location in Switzerland.
- It is robust to real-life data. The developed solution can cope with incomplete and noisy data and is constantly monitoring faulty nodes with missing information. Additionally, due to the use of dynamic GNNs, it can adapt to the changing conditions of the network of stations when nodes are added or dropped during operation.

6 ACKNOWLEDGEMENTS

This research was co-financed by Innosuisse – the Swiss innovation agency – project 57765.1 IP-EE: DIGERATI.

7 REFERENCES

[1] P.-J. Alet *et al.*, 'Forecasting and Observability: Critical Technologies for System Operations with High PV Penetration', in *32nd European Photovoltaic Solar Energy Conference and Exhibition*, 2016.

[2] L. Benavides Cesar, R. Amaro e Silva, M. Á. Manso Callejo, and C.-I. Cira, 'Review on Spatio-Temporal Solar Forecasting Methods Driven by In Situ Measurements or Their Combination with Satellite and Numerical Weather Prediction (NWP) Estimates', *Energies*, vol. 15, no. 12, no. 12, Jan. 2022, doi: 10.3390/en15124341.

[3] R. E. Carrillo, M. Leblanc, B. Schubnel, R. Langou, C. Topfel, and P.-J. Alet, 'High-Resolution PV Forecasting from Imperfect Data: A Graph-Based Solution', *Energies*, vol. 13, no. 21, p. 5763, 2020.

[4] J. Simeunović, B. Schubnel, P.-J. Alet, and R. E. Carrillo, 'Spatio-Temporal Graph Neural Networks for Multi-Site PV Power Forecasting', *IEEE Transactions on Sustainable Energy*, vol. 13, no. 2, pp. 1210–1220, 2022, doi: 10.1109/TSTE.2021.3125200.

[5] R. E. Carrillo, P.-J. Alet, S. Müller, and J. Remund, 'A Computationally Light Data-driven Alternative to Cloud-Motion Prediction for PV Forecasting', in *8th World Conference on Photovoltaic Energy Conversion*, WIP, Dec. 2022, pp. 1268–1271. doi: 10.4229/WCPEC-82022-4BV.4.11.

[6] J. Simeunović, B. Schubnel, P.-J. Alet, R. E. Carrillo, and P. Frossard, 'Interpretable temporal-spatial graph attention network for multi-site PV power forecasting', *Applied Energy*, vol. 327, p. 120127, Dec. 2022, doi: 10.1016/j.apenergy.2022.120127.

[7] F. Rodrigues and F. C. Pereira, 'Beyond Expectation: Deep Joint Mean and Quantile Regression for Spatiotemporal Problems', *IEEE Transactions on Neural Networks and Learning Systems*, vol. 31, no. 12, pp. 5377–5389, Dec. 2020, doi: 10.1109/TNNLS.2020.2966745.

[1] https://digerati.portal.csem.ch/

[2] https://www.meteoswiss.admin.ch/

[3] https://www.knmi.nl/over-het-knmi/about

40th European Photovoltaic Solar Energy Conference and Exhibition

This presentation was selected by the Sc. Committee of the EU PVSEC 2023 for submission of a full paper to one of the EU PVSEC's collaborating peer-reviewed journals.

Evaluation of Irradiance Decomposition Models and Their Predictors

Jacob K. Thorning*, Thøger K. Hass, Sergiu V. Spataru
Email: J. K. Thorning jkrtho@dtu.dk
Technical University of Denmark (DTU), Department of Electrical and Photonics Engineering
Frederiksborgvej 399, 4000 Roskilde, Denmark

ABSTRACT: This study presents a comprehensive approach to developing models for decomposing global solar radiation in to its diffuse and direct components, a critical step in solar energy applications. Utilizing data from multiple geographical regions and Köppen-Geiger climate zones, the research ensures a broad understanding of the environmental factors affecting solar radiation. The study meticulously selects predictors through multiple feature selection techniques, focusing on capturing essential information while minimizing redundancy. Single-predictor models are developed to provide insights into the relationships between selected predictors and the diffuse fraction, while two multi-predictor models demonstrates the potential for more accurate estimations by leveraging the collective predictive power of multiple variables. One of the developed models, called Nested, performs better than other state-of-the-art universal models, while performing slightly worse than the best climate-specific model in terms of three performance metrics. The research lays a solid foundation for future studies and practical applications in the solar energy sector, emphasizing the need for further testing of techniques and procedures for decomposition model development.
Keywords: solar radiation, diffuse fraction, predictors, feature selection, model development

1 INTRODUCTION

Decomposition of Global Horizontal Irradiance (GHI) into Diffuse Horizontal Irradiance (DHI) and Direct Normal Irradiance (DNI) components is crucial for various applications, including solar energy resource assessment, solar collector performance evaluation, and climate studies. Accurate estimation of diffuse and direct solar irradiance improves the design and efficiency of photovoltaic (PV) and concentrating solar power (CSP) systems. Understanding the behavior of these components enables better modeling of the Earth's radiation budget and its effects on the climate. Improved predictions facilitate lower uncertainty in investment in PV projects, more efficient planning and allocation of energy resources, contributing to cost-effectiveness. Furthermore, enhanced grid integration through better synchronization and balancing of supply and demand is vital for grid stability and reliability. Since G_{poa} has the most significant influence on the expected power output, it is essential to estimate it reliably. However, due to the high costs associated with measuring DHI or DNI, it is more common to measure only GHI [1]. Consequently, estimating G_{poa} requires two modeling steps: (1) decomposing GHI into DHI and DNI with a decomposition model, which generally follows the form of Equation 1, where f is a decomposition model, t is date-time, lat is latitude and lon is longitude.

$$DHI = GHI \cdot f(GHI, t, lat, lon) \quad (1)$$

(2) transposing DHI, DNI, and ground-reflected irradiance into Global Tilted Irradiance (GTI) with a transposition model [2] and for more accuracy, shading simulations [1, 3]. These modeling steps often introduce the highest uncertainty in PV system performance simulations [4–6].

Over the past six decades, numerous decomposition models have been proposed to estimate diffuse and direct solar irradiance components from GHI measurements [7, 8]. These models have evolved to address various challenges, such as the need for higher accuracy, higher temporal resolution, improved performance under diverse climatic conditions, and incorporation of additional factors influencing solar irradiance. The earliest well recognized model for solar irradiance decomposition was proposed by Liu and Jordan [9]. This simple empirical model, based on the ratio of daily diffuse, k_d, to global solar radiation, laid the foundation for future developments in the field. In Skartveit, Olseth, and Tuft [10] a model for estimating the hourly diffuse fraction considering the variability in the atmospheric conditions and surface albedo was introduced. This model marked a significant improvement over previous approaches by being highly analytical and incorporating variability in the ratio GHI to extraterrestrial GHI, called clearness index k_t, and regional surface albedo, which affect k_d beyond k_t. A significant development came from Boland, Scott, and Luther [11], where the authors used a logistic function to map k_t values to k_d. The inspiration for using a logistic function was a visual impression through a k_d-k_t scatterplot. The "BRL" model, presented in Ridley, Boland, and Lauret [12], built upon previous research by incorporating multiple predictors in a logistic function to estimate k_d. This approach allowed for improved accuracy in modeling

k_d under various climatic conditions. The model from Engerer [13] was designed to estimate k_d with minute resolution for southeastern Australia. Being the first model developed with 1-minute average data, this model provided a higher temporal resolution, making it suitable for the accuracy and timescale of modern system performance simulations, and set the standard for decomposition model time resolution and accuracy going forward [7]. Starke et al. [14] proposed models for each category of Köppen-Geiger climate zones (KGCZs) based on a single framework, with focus on high performance during Cloud Enhancement (CE) events. In Yang and Boland [15] satellite-augmented models were proposed, leveraging DHI and GHI estimated from satellite imagery to improve the accuracy of the k_d estimation. In Yang [16], a temporal-resolution cascade model was introduced, where k_d is estimated with a time average of k_d as a predictor. Recently, two new models of different complexity were proposed in Paulescu and Paulescu [17], based on rational functions, named M1 and M2. M1 uses 8 predictors and M2 uses 3 predictors, included as a more accessible alternative to most recent models which use 6 or more predictors. In each work the authors found that at least one of the proposed models outperformed the established models.

In this study, we address the concerns raised by prior research regarding the potential for overfitting and the absence of a scientific foundation in solar irradiance decomposition models. In Gueymard and Ruiz-Arias [7] the authors observed that adding predictors generally improves model precision and repeatability; however, they also noted that there could be exceptions due to overfitting when incorporating numerous predictors without a clear basis. Similarly, Yang and Boland [15] emphasized the importance of understanding the atmospheric physics governing diffuse radiation, rather than merely modifying regression-based models or specifying alternative predictor sets.

The objective of this work is to establish a systematic procedure for predictor selection and model development. Through an in-depth analysis of the empirical relationships between predictors and the diffuse fraction, this study aims to establish a scientifically informed basis for the development of decomposition models. Employing feature selection techniques Variable Importance in Projection (VIP) of Partial Least Squares (PLS) regression, Maximum Relevance - Minimum Redundancy (mRMR), and Recursive Feature Elimination (RFE). This methodological approach is anticipated to contribute to the development of more accurate and reliable models, thereby enhancing the precision of solar irradiance predictions.

The rest of this paper is organized as follows: Section 2 describes the datasets used in this work, Section 3 presents the methods and results of predictor selection, Section 4 describes how single and multi-predictor models are developed, Section 5 presents a validation of two proposed models against models from the literature, Section 6 is a systematic discussion of the methods and results for each step of the process presented, and finally Section 7 presents the conclusions to this work.

2 DATA

The datasets used in this work are from 10 stations of the Baseline Solar Radiation Network (BSRN) network listed in Table I, where data from stations 1-5 are split randomly in to 80%/20% and used for model training and testing respectively, and data from stations 6-10 are used for model validation. Each station's data was sampled uniformly to match the station with the smallest sample size, ensuring equal representation across all stations. The training and testing datasets are used as a whole for the feature selection. Since the testing data is used in the iterative model development process, a fully independent dataset is required for validation. The sampling is performed after calculating the required predictor values. The training and testing data comprise 231662 samples from each of the 5 stations, amounting to a total of 1158310 samples. Similarly, for validation, each of the 5 stations contribute 232635 samples, yielding a total validation set of 1163175 samples. These post-Quality Control (QC) 1-minute datasets are from Gueymard et al. *BSRN data set for IEA-PVPS Task-16 Activity 1.4 Quality Control* [18]. The authors have made a significant contribution to the field by providing high-quality, reference datasets from various locations across the globe. Their efforts in collecting, curating, and sharing these valuable resources are gratefully acknowledged. The reader is referred to [19] for more information.

Using data from multiple locations with varying climates is crucial for developing a robust and generalizable model, as it captures a broad range of environmental and climatic conditions, enabling a model to identify underlying universal patterns and eliminate location-specific biases. This contributes to a more reliable predictor importance analysis and predictor selection process, as the methods will prioritize predictors that are significant across multiple climates and regions rather than those that are relevant only to a specific location.

3 PREDICTOR SELECTION

Table II compiles the symbols and expressions (or references) of predictors commonly employed in the literature [7, 8].

10.4229/EUPVSEC2023/4CO.9.2
020362-002

Table I: Selected BSRN stations representing multiple continents and Köppen-Geiger climate zones. Data from the first five stations are used for training and testing, while data from the last five stations are used for validation.

#	Code	Name	Country	Zone	Zone description
1	COC	Cocos Island	Australia	Aw	Equatorial savannah with dry winter
2	GOB	Gobabeb	Namibia	BWk	Desert climate, cold
3	GCR	Goodwin Creek	USA	Cfa	Warm temperate climate, fully humid, hot summer
4	TOR	Toravere	Estonia	Dfc	Snow climate, fully humid, cool summer and cold winter
5	NYA	Ny-Ålesund	Norway	ET	Tundra climate
6	BRB	Brasilia City	Brazil	Aw	Equatorial savannah with dry winter
7	FPE	Ford Peck	USA	BSk	Cold steppe climate
8	IZA	Tenerife	Spain	Csb	Warm temperate climate with dry summer
9	TIK	Tiksi	Russia	Dsd	Snow climate with dry summer, extremely continental
10	GVN	Dronning Maud Land	Antarctica	EF	Frost climate

Table II: Summary of predictors used in solar irradiance decomposition models. The table includes symbols and expressions (or references) of the predictors, which are derived from global horizontal irradiance, date, time of day, latitude, and longitude.

Symbol		Expression
Irradiance derivatives / k-indices		
k_t	=	E_{gh} / E_{0h}
k_{csi}	=	E_{gh} / E_{ghc}
k_{tc}	=	E_{ghc} / E_{0h}
Δk_{tc}	=	k_{tc} - k_t
k'_t	=	$\frac{k_t}{(1.031e^{-1.4/(0.9+9/AM)}+0.1)}$
k_{de}	=	max(0, 1 - E_{ghc}/E_{gh})
Derivatives of time and location		
θ_z		See [20, p 15]
Z	=	90 - θ_z
AM		See [21]
AST		See [20, p 11]
Time-series dependent		
K_t	=	$\frac{1}{60}\sum_{n=-30}^{30} k_{t,n}$
$\overline{\mathrm{K}_t}$	=	$\frac{1}{1440}\sum_{n=0}^{1440} k_{t,n}$
V^{SO}	=	$[(k_{csi} - k_{csi-1})^2 + (k_{csi} - k_{csi+1})]^{1/2}$
V^P	=	$(\lvert k'_t - k'_{t-1}\rvert + \lvert k'_t - k'_{t+1}\rvert)/2$
ψ	=	$(k_{t,-1} + k_{t,+1})/2$
Cascade modelling		
k_d^s	=	${E_{dh}}^{sat} / {E_{gh}}^{sat}$
k_d^m		k_d estimated with e.g. model from Engerer [13]

We have limited our work to predictors derived from any combination of measured E_{gh}, date-time (year, day, and time of day), and geographical location (latitude, longitude), with E_{gh} being the sole measured variable.

3.1 Methods for predictor evaluation and selection

Partial Least Squares (PLS) regression is a method that combines the ideas of Principal Component Analysis (PCA) and linear regression to model the relationship between the predictor variables (features) and k_d (the target variable). PLS projects both the input features and the output target onto a new set of latent variables (which are analogous to the components of PCA), called PLS components. These components are chosen to maximize the covariance between the features and the target. VIP scores are a measure of the importance of each feature in the projection of the data using PLS regression [22]. The highest VIP scores indicate the most significant predictors in the PLS model.

Maximum Relevance - Minimum Redundancy (mRMR) is a filter-based method that aims to select predictors that are highly correlated with the target variable (maximum relevance) while minimizing the inter-predictor correlation (minimum redundancy). It does so by ranking predictors based on their relevance to the target variable and iteratively selecting predictors that contribute the most new information while minimizing the redundancy between them. The relevance of each predictor to k_d is assessed using the importance scores provided by a random forest regression model (from [23]). For each pair of predictors, the redundancy score is computed using Spearman correlation. Recursive Feature Elimination (RFE) operates by iteratively fitting the model, ranking the predictors based on their importance, and removing the least important predictor in each iteration. This process continues until the desired number of predictors is reached. In this work, decision trees are used as the regression model.

An aggregated average score is calculated by first normalizing the scores from each feature selection method individually, such that the best predictor has a score of 1 and all other predictors has a score less than 1 but greater than or equal to 0. The scores are combined by taking the average score for each predictor. Finally, the predictors are ranked based on

10.4229/EUPVSEC2023/4CO.9.2
020362-003

their average score across the three metrics.

3.2 k_d predictor selection results

A heatmap of the normalized scores derived from various feature selection methods is shown in Figure 1.

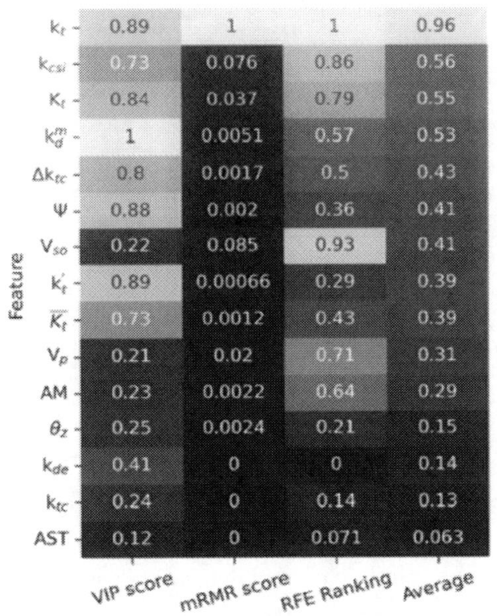

Figure 1: Heatmap of the normalized feature selection scores across different methods, sorted by their overall rank. The color intensity represents the relative importance of each predictor, with lighter colors indicating higher importance.

The top-ranked predictor, k_t, is identified as crucially important across all methods, with a high average score of 0.97. This is in line with our expectations [7, 8] k_t has the maximum mRMR score, indicating that it has the highest relevance to k_d. The low mRMR scores for the other predictors imply that these predictors either do not provide as much information about the outcome as k_t does, or, more likely, they may be somewhat redundant, sharing information that's already provided by other predictors in the predictor set. While other predictors might still be significant in their contribution to the model, their unique contributions are low. E.g. θ_z and Air Mass (AM) contain the same information, as they may be calculated as functions of each other. Due to the ranking being based on average scores, beyond k_t the only significant feature selection methods are VIP and RFE.

The second-ranked predictor, k_{csi}, also exhibits strong performance across all methods, when considering its mRMR score relative to the other predictors except k_t. The importance of K_t highlights the value of considering the temporal aspect of k_d estimation, reflecting the historical context and trends, capturing gradual changes in atmospheric conditions. ψ is ranked fourth, with an average score of 0.53. This predictor's importance also emphasizes the role of time-series information in predicting k_d, as ψ accounts for inertia in the atmosphere, similar to K_t. The fifth-ranked predictor, k_d^m, achieves the highest VIP score, but a relatively low RFE ranking, opposite the case of k_{csi}. This shows the importance of using multiple feature selection methods. The sixth-ranked predictor Δk_{tc} achieves an average score of 0.49, demonstrating a moderate level of importance across VIP and RFE. It is worth noting that some predictors, such as Apparent Solar Time (AST), exhibit consistently low scores across all methods, indicating that they may not contribute meaningfully to the prediction of k_d.

The analysis offers an understanding of the importance of different predictors in predicting k_d. By selecting the top six predictors, k_t, k_{csi}, K_t, k_d^m, Δk_{tc} and ψ, we have ensured a balance between capturing essential information and minimizing redundancy. This selection will provide a solid foundation for constructing models that accurately maps the relationships between these predictors and k_d.

4 DEVELOPMENT OF NEW MODELS

4.1 Single-predictor model development

In the pursuit to model k_d using the selected predictors individually, a modified logistic function is tailored. The basic logistic form in Equation 2 [12], where k_d is the target variable and x represents a predictor, is used as a starting point.

$$k_d = \frac{1}{1 + \mathrm{e}^{(\beta_0 + \beta_1 x)}} \tag{2}$$

Recognizing that a more sophisticated approach may be advantageous, we started an iterative process to fine-tune this function, experimenting with replacing constants and adjusting parameter placement. The end result is Equation 3 where C, β_0, and β_1 are parameters optimized for each predictor.

$$k_d = C + \left| \frac{1}{1 + \mathrm{e}^{(\beta_0 + \beta_1 x)}} - C \right| \tag{3}$$

In this formulation, C is a critical parameter positioned both within and outside of the absolute function. This placement provides the model with flexibility, allowing the function to pivot at a base level, C, and to start increasing with the predictor variables instead of the traditional decrease seen in basic logistic functions. Equation 3 is applied to k_t, Δk_{tc}, k_{csi}, K_t, and ψ independently, and each is fitted a set of optimized parameters. For the predictor k_d^m, we found a simple linear function with β_0 as intercept

10.4229/EUPVSEC2023/4CO.9.2
020362-004

and β_1 as slope to be the most fitting representation. The parameters are listed in Table III.

Table III: Optimized parameter values for the mathematical functions fitted for each predictor variable. The columns C, β_0, and β_1 represent the parameters of the logistic and linear functions as applicable. For the linear function applicable to k_d^m, C is not applicable (N/A).

Predictor	C	β_0	β_1
k_t	0.16130	-5.53387	9.42544
k_{csi}	0.19767	-7.48200	8.34717
K_t	0.13453	-5.62334	9.44307
ψ	0.16197	-5.47208	9.29277
k_d^m	N/A	-0.04510	1.00128
Δk_{tc}	0.20215	0.78571	-12.56507

Figure 2 displays the relationships between each predictor and k_d, visually justifying the functions chosen to map the predictors.

4.2 Multi-Predictor model development

In pursuit of achieving a more accurate estimation of the diffuse fraction, we incorporate multiple predictors simultaneously, leveraging their collective predictive capacity. Our approach has been two-fold.

In the first approach, we have selected the predictor of the single-predictor model with the highest R^2 value (coefficient of determination) as the principal predictor. Subsequently, additional predictors are incorporated into the logistic function iteratively, each being included in the exponential part of the denominator. For instance, the function to include one additional predictor with a logistic relationship is Equation 4.

$$k_d = C + \left| \frac{1}{1 + e^{(\beta_0 + \beta_1 x_1 + \beta_2 x_2)}} - C \right| \quad (4)$$

For the inclusion of k_d^m, which has a linear relationship with k_d, it is added as a separate term as illustrated in Equation 5

$$k_d = C + \left| \frac{1}{1 + e^{(\beta_0 + \beta_1 x_1)}} - C \right| + \beta_2 x_2 \quad (5)$$

In these functions, x_1 and x_2 represent the predictors, while β_0, β_1, β_2, and C are the model parameters. The β_1 values in Table III serve as the initial guesses for each new parameter in this iterative process. After calculating R^2 of incorporating each so-far unused predictor individually, the predictor which provides the largest improvement in R^2 is added to the model and the process is repeated until all predictors are included. This model is referred to as Nested, due to the incorporation of the predictors in the argument of the same logistic function. This iterative process provides information about the value of including each of the predictors considered post feature selection, and it ensures that the parameter solution is globally optimal at all times.

The second approach, in contrast, utilizes a weighted sum of all the individually fitted single-predictor models, as shown in Equation 6 where w_i denotes the weight of the i-th function in the order shown in Table III, f_i is the function, x_i is the corresponding predictor values, and $\mathbf{p}_i$ represents the set of parameters for the function.

$$k_d = C + \sum_i (w_i \cdot f_i(x_i, \mathbf{p}_i)) \quad (6)$$

Each model function is associated with a weight parameter, which is determined through an optimization process. This ensemble-based approach aims to harness the collective predictive power of all the predictors by adjusting their individual influence based on the associated weights. This model is called Summed.

4.3 Model development results

The Nested model demonstrated a strong performance when used on the test data, with an R^2 score of 0.87034. The order of predictors included and the R^2 at the end of each iteration is shown in Table IV.

Table IV: Improvement in coefficient of determination (R^2) with sequential addition of predictors in the developed model. Each row represents a new iteration where a predictor is added to the model.

Iteration	Added predictor	R^2	Improvement
1	k_t	0.81477	-
2	Δk_{tc}	0.85071	4.41%
3	K_t	0.86471	1.65%
5	k_d^m	0.86730	0.30%
4	ψ	0.87659	1.07%
6	k_{csi}	0.87713	0.06%

To further improve its predictive power, we capped the predicted values at 1 and introduced an additional parameter, C_2, in the numerator of the function. These modifications increased the R^2 score to 0.88452. Since multi-predictor fitting is significantly different from single-predictor fitting, we tested fitting the parameters with three modifications: 1) changing the 1 in the denominator to a parameter, 2) removing C, 3) removing the absolute function and the C with a negative sign.

There was nearly no difference on the four functions in terms of R^2, which leads us to nominate the simplest version, Equation 7, which had an R^2 of 0.88525 when applied to the test data. The param-

10.4229/EUPVSEC2023/4CO.9.2
020362-005

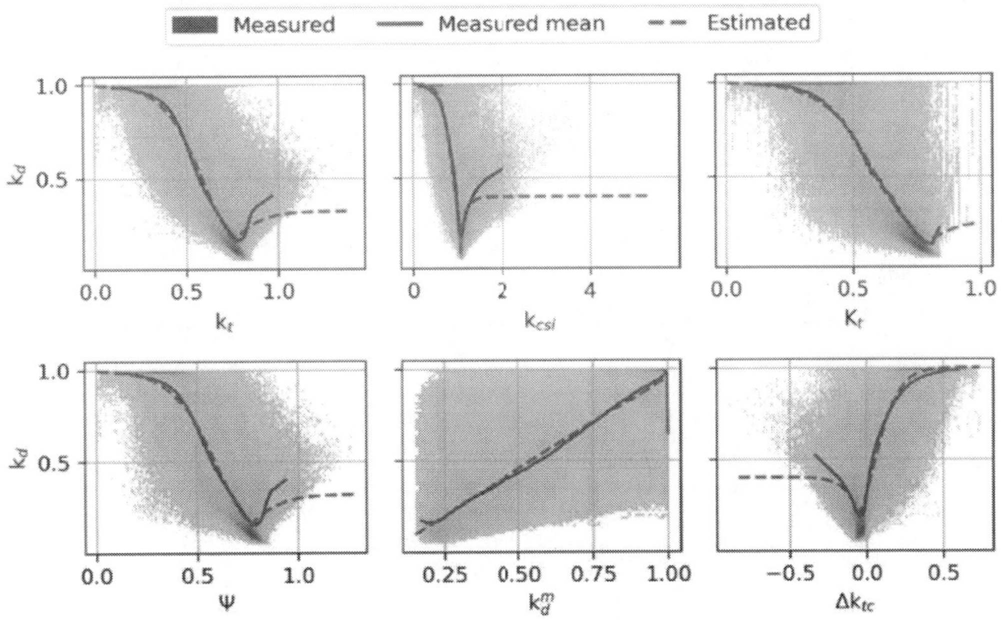

Figure 2: The relationships between the diffuse fraction (k_d) and each predictor variable. The blue histograms represent the measured data. The solid green line represents the mean trend, calculated within bins of the predictor variable. The dashed red line represents the estimated relationship derived from the respective functions for each predictor variable. The consistency between the green and red lines signifies the accuracy of our singe-predictor.

eters that produced the best fit are C: 0.45292, β_0: -13.49189, β_1: -6.48912, β_2: 1.48927, β_3: 12.03733, β_4: 4.07416, β_5: 8.08428, β_6: 0.53947.

The Summed model resulted in an R^2 score of 0.88167. The optimal weights are C: -0.09954, ω_0: 0.13129, ω_1: -0.00440, ω_2: 0.37380, ω_3: 0.16500, ω_4: 0.03460, ω_5: 0.51928.

5 MODEL VALIDATION

To assess the performance of the Nested and Summed models, five models from literature are selected. The models selected are Engerer2 [13], Engerer4 [24], Paulescu [25], Starke3 [26], Yang4 [16]. The selection is based on the best performing models in [8], from which the naming convention is also used. The recently proposed models from [17] are not included due to substantial discrepancies observed between the the first model's (M1) estimated k_d values and the corresponding measured values, perhaps caused by a typo in the parameters listed in the publication.

To understand the performance of each model, three metrics were employed: Normalized Mean Bias Deviation (nMBD), Normalized Root Mean Square Deviation (nRMSD), and coefficient of determination (R^2). The nMBD measures the average deviation of predicted values from the target values, normalized by the mean of the observations. It provides a measure of the overall bias in the predictions; a value of 0 indicates no bias. The nRMSD measures the average spread of the residuals, normalized by the mean of the observations, and thus provides a measure of the accuracy of the predictions; the lower the value, the better the prediction. The R^2 measures the proportion of the variance in the dependent variable that is predicted from the independent variables; a value close to 1 indicates a strong predictive power.

Tables V, VI, and VII present the R^2, nMBD, and nRMSD results respectively, for each model across each station, and the results are visualized in Figure 3. Performance was averaged across all stations to create a singular rating for each model per metric, and each model was accordingly assigned a rank. For nMBD the relative performance, averages and ranks are based on absolute values of the nMBD, i.e. mean absolute deviation. The relative performance of each model is indicated row wise in green.

The Nested model consistently presents excellent performance across the majority of stations and

$$k_d = \min\left(1, \frac{C}{1+e^{(\beta_0+\beta_1 k_t+\beta_2\Delta k_{tc}+\beta_3 K_t+\beta_4\psi+\beta_5 k_{csi})}} + \beta_6 k_d^m\right) \tag{7}$$

Table V: R^2 values for different models across various zones. Each cell represents the R^2 value of the corresponding model at a specific station. The highest value in each row, indicating the best performing model for that station, is highlighted.

Data	Zone	Engerer2	Engerer4	Paulescu	Starke3	Yang4	Nested	Summed
BRB	A	0.937	0.945	0.940	0.954	0.937	0.950	0.948
FPE	B	0.951	0.969	0.965	0.971	0.953	0.975	0.974
GVN	E	0.804	0.810	0.783	0.893	0.806	0.828	0.838
IZA	C	0.964	0.979	0.976	0.979	0.966	0.986	0.977
TIK	D	0.960	0.956	0.945	0.963	0.965	0.963	0.964
Average	All	0.923	0.932	0.922	0.952	0.925	0.940	0.940
Rank	All	6	4	7	1	5	2	3

Table VI: nMBD [%] values for different models across various zones. Each cell represents the nMBD value of the corresponding model at a specific station. The relative performance of each model is indicated row wise in green. The relative performance, averages and ranks are based on absolute values of the nMBD, i.e. mean absolute deviation.

Data	Zone	Engerer2	Engerer4	Paulescu	Starke3	Yang4	Nested	Summed
BRB	A	5.606	-1.748	-2.274	-2.259	8.161	3.182	4.803
FPE	B	8.472	-4.065	-5.200	-6.527	9.579	-0.655	0.422
GVN	E	-23.496	-46.549	-55.449	-31.123	-24.151	-42.025	-41.186
IZA	C	9.094	4.361	-12.332	-8.372	7.923	-0.075	12.359
TIK	D	-6.960	-12.659	-9.479	-7.346	-5.169	-9.511	-7.387
Average	All	10.726	13.876	16.947	11.125	10.996	11.090	13.231
Rank	All	1	6	7	4	2	3	5

Table VII: nRMSD [%] values for different models across various zones. Each cell represents the nRMSD value of the corresponding model at a specific station. The relative performance of each model is indicated row wise in green.

Data	Zone	Engerer2	Engerer4	Paulescu	Starke3	Yang4	Nested	Summed
BRB	A	40.591	38.014	39.645	34.652	40.769	36.311	36.794
FPE	B	35.709	28.595	30.158	27.378	35.107	25.393	25.997
GVN	E	71.696	70.715	75.462	52.891	71.444	67.286	65.308
IZA	C	30.896	23.390	25.253	23.610	29.876	19.111	24.733
TIK	D	32.533	34.147	38.113	31.255	30.463	31.001	30.571
Average	All	42.285	38.972	41.726	33.957	41.532	35.820	36.681
Rank	All	7	4	6	1	5	2	3

climatic zones. Compared to other universal models (i.e. not KGCZ-specific) it displays superior ability to generalize across different climate conditions in terms of all three metrics, confirming its robust nature. Overall the best performing model is Starke3, which utilizes parameters fitted for each KGCZ, which makes the Nested the best performing universal model based on the specific metrics and validation dataset. The poor performance of Yang4 in terms of R^2 and nRMSD is surprising, as it has recently been crowned the best performing model in a major study utilizing data from 126 different stations [8].

6 DISCUSSION

6.1 Predictor selection

The predictor selection process employs a combination of methods, including PLS regression, VIP scores, mRMR, and RFE. While the integration of multiple methods is great for its comprehensiveness, there are several aspects that warrant critical discussion. While PLS regression is effective in handling multi-collinearity, its reliance on linear relationships may not capture complex non-linear interactions between predictors and the target variable. Additionally, VIP scores, which are derived from PLS, may not always provide clear-cut distinctions between relevant and irrelevant features, especially in cases

10.4229/EUPVSEC2023/4CO.9.2
020362-007

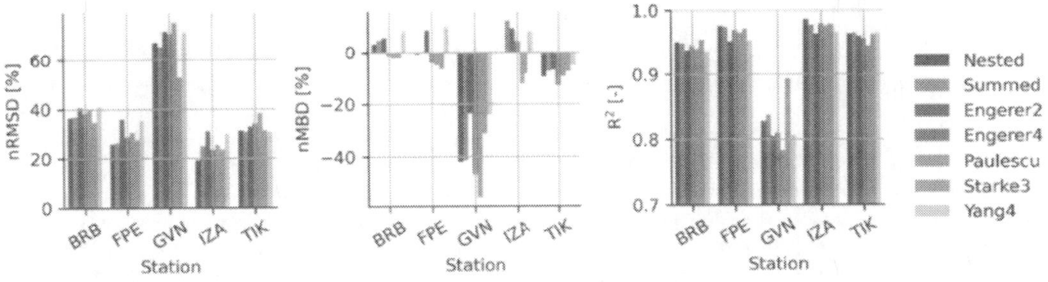

Figure 3: Station-specific model performance evaluated across different metrics. The bars represent model performance at each station in terms of Normalized Root Mean Square Deviation (nRMSD), Normalized Mean Bias Deviation (nMBD), and Coefficient of Determination (R^2).

where the number of features is large. mRMR does not take into account potential interactions between features, which can be critical in complex systems such as solar radiation prediction. While RFE is a powerful feature selection technique, it is computationally expensive, especially for large datasets. Additionally, by iteratively removing features, there is a risk of losing information that might have been captured by subsets of features. Using machine learning for feature selection may not account for the fundamental mathematical relationships in regression models. While machine learning focuses on optimizing performance through complex relationships, regression models often seek to capture simpler, interpretable mathematical patterns. This discrepancy necessitates careful feature selection in regression modeling.

The feature selection process is not validated against a separate dataset, so it is unknown if the results are specific to the combination of climates which the 5 stations belong to, i.e. the selected predictors may be overfitted to the training dataset.

6.2 Model development

While the logistic function is traditionally used for classification problems, for decomposition models it has been adapted for regression since its first use [11]. The logistic function inherently has an S-shape, which might not be the best representation for the relationships between predictors and k_d in all cases, particularly for k_t values above 0.7, where CE events occur. To circumvent this an absolute term was introduced, which appeared a good fit for single-predictor regression, but in the Nested model the absolute term was actually not utilized, which became apparent as fitted parameter values were identical with and without the absolute term. While introducing additional parameters outside of the exponential part of the denominator such as C or C_1 and C_2 may improve the fit, the introduction of additional parameters without a theoretical basis can lead to overfitting, where the model may fit the noise in the training data rather than capturing the underlying trend.

In terms of evaluating the models during development, reliance on R^2 as the sole metric might not be sufficient. R^2 can sometimes be a misleading indicator of model performance, especially in cases with noise, data clustering, or when the model is overfit. Other metrics such as nMBD or nRMSD could have been used in addition to R^2, however this was avoided for simplicity. The Summed model takes a more ensemble-based approach. It is opportune that this approach tries to harness the collective predictive power of all single-predictor models. However, it is essentially a linear combination of the outputs of single-predictor models. This linearity could limit the model's ability to capture complex, non-linear relationships between predictors and the target variable, and a different ensemble method could prove superior.

6.3 Validation

Since Yang [8], Yang4 has been anticipated to exhibit superior performance among the models. However, this expectation was not met in the current study. Interestingly, a similar discrepancy was observed in the work of Paulescu and Paulescu [17]. In Paulescu and Paulescu [17], Yang4 was found to be among the least effective models across the majority of the 36 stations included in their analysis, with Engerer2 performing the poorest. This observation aligns with the findings of the present study. In Yang [8], the stations utilized for validation are anonymized and denoted using codes such as A1, A2, B1, B2, etc., where the letter represents the major KGCZs and the number serves as a unique identifier within that zone. In contrast, [16, 17, 24, 26] provides explicit station codes, allowing for the identification of the specific stations from which the evaluation metrics are derived. The lack of transparency in station identification in Yang [8] precludes a direct comparison of results with those of Paulescu and Paulescu [17] or the present study. This is particularly regrettable given that both the current study and

Paulescu and Paulescu [17] observe a suboptimal performance of Yang4, contrary to expectations. The inability to directly compare results hampers the assessment of whether the observed discrepancies are attributable to the limited number of stations or other factors.

Since the validation dataset is chosen to be rather small relative to recent works on decomposition models, it is pertinent to discuss the contrasting methodologies employed for model validation, specifically focusing on the utilization of data from meteorological stations across various KGCZs. Traditionally [7, 8, 17, 26] model validation is conducted using data from numerous different stations spanning various KGCZs. The number of samples per station and the number of stations per climate zone are typically dictated by the data available, with authors leveraging all accessible data to ensure as extensive a validation as possible. Such an approach has advantages, as utilizing a large dataset from numerous stations provides a more comprehensive overview of the model's performance across diverse conditions. The extensive dataset can help in identifying patterns and trends that may not be evident with a smaller dataset, thereby contributing to the robustness of the validation. However, the varying number of samples per station can lead to inconsistencies in the validation results, and some climate zones may be over-represented due to the availability of more data, which can bias the validation results.

In contrast, in this work, a different approach is adopted where a uniform number of samples from five different stations, one station from each major KGCZ, is used for model validation. By using a uniform number of samples, the validation process is standardized, which can lead to more consistent and comparable results, and selecting one station from each major KGCZ ensures a balanced representation of different climate zones, preventing any single zone from dominating the validation. This approach is not perfect either, as using only five stations may limit the scope of the validation and may not capture the full range of variability within each climate zone, and with a smaller dataset, the validation results may be more sensitive to outliers or anomalous data points. A judicious combination of both approaches, ensuring both breadth and uniformity, and using a standard set of stations, could be considered for more comparable model validation in the future, and rule out cherry picking.

Employing a more appropriate climate classification scheme than the Köppen-Geiger climate classification could improve the consistency in model validations, as also concluded in Starke et al. [26]. Utilizing a classification system with greater relevance could facilitate more meteorologically adaptive modeling.

7 CONCLUSIONS

In the last decade, the literature on decomposition modeling has largely lacked data- or physics driven basics for the methods used for developing new models. The work presented in this paper addresses this gap by offering a systematic approach to the model development process.

The development of single-predictor models provided insights into the relationships between individual predictors and the diffuse fraction. However, it was the multi-predictor model, Nested, that showcased the potential for more accurate estimations by leveraging the collective predictive power of multiple variables. Notably, the Nested model outperformed other recent universal models during validation.

While this study represents an advancement in modeling the diffuse fraction of solar radiation, it is important to recognize its limitations. The models are based on the limited data used, and their performance may vary with different datasets, especially from locations where they have not been validated.

Furthermore, it is crucial to acknowledge that the methods employed in this work are not beyond scrutiny. Critical evaluation and transparent discussion of the choices made in model development are essential for the advancement of scientific knowledge and the development of models that are both robust and interpretable. This, in turn, contributes to the credibility and utility of models in practical applications and decision-making.

In conclusion, this study has laid a foundation for the modeling of the diffuse fraction and has opened avenues for further research. It is encouraged that others aid in the search for better methods of predictor selection, model development and standardization of model validation. We hope that this work will serve as a reference, ultimately benefiting solar energy applications.

DECLARATION OF COMPETING INTEREST

The authors declare that they have no known competing financial interests or personal relationships that could have appeared to influence the work reported in this paper.

APPENDIX, NESTED PYTHON CODE

```
def nested(kt,dktc,Kt,
    psi,kdm,kcsi):
        b0 = -13.49189475455595
        b1 = -6.489121933141542
        b2 = 1.489274316229536
        b3 = 12.037329137008419
        b4 = 4.074156126269681
        b5 = 8.08428258256891
        b6 = 0.5394717724137723
```

```
C = 0.45292421404358085

EXP = b0+b1*kt+b2*dktc+\
        b3*Kt+b4*psi+b5*kcsi
denum = 1 + np.exp(EXP)
kd = C/denum + b6*kdm
return np.minimum(kd,1)
```

References

[1] C.A. Gueymard. "Solar Radiation Resource: Measurement, Modeling, and Methods". In: *Reference Module in Earth Systems and Environmental Sciences*. Elsevier, 2021. ISBN: 978-0-12-409548-9. DOI: https://doi.org/10.1016/B978-0-12-819727-1.00101-1. URL: https://www.sciencedirect.com/science/article/pii/B9780128197271001011.

[2] Dazhi Yang. "Solar radiation on inclined surfaces: Corrections and benchmarks". In: *Solar Energy* 136 (2016), pp. 288–302. ISSN: 0038-092X. DOI: https://doi.org/10.1016/j.solener.2016.06.062. URL: https://www.sciencedirect.com/science/article/pii/S0038092X16302432.

[3] Marc A. Anoma et al. "View Factor Model and Validation for Bifacial PV and Diffuse Shade on Single-Axis Trackers". eng. In: *2017 Ieee 44th Photovoltaic Specialist Conference (pvsc)* (2017), pp. 1549–1554. ISSN: 01608371.

[4] Martin János Mayer and Gyula Gróf. "Extensive comparison of physical models for photovoltaic power forecasting". In: *Applied Energy* 283 (2021), p. 116239.

[5] Giorgio Belluardo et al. "Evaluation of uncertainty in PV project design: Definition of scenarios and impact on energy yield predictions". In: *2017 IEEE 44th Photovoltaic Specialist Conference (PVSC)*. IEEE. 2017, pp. 3360–3365.

[6] Clifford W Hansen and Curtis E Martin. "Photovoltaic system modeling: Uncertainty and sensitivity analyses". In: *Sandia Report (SAND2015-6700)* (2015), p. 82.

[7] Christian A. Gueymard and Jose A. Ruiz-Arias. "Extensive worldwide validation and climate sensitivity analysis of direct irradiance predictions from 1-min global irradiance". In: *Solar Energy* 128 (2016). Special issue: Progress in Solar Energy, pp. 1–30. ISSN: 0038-092X. DOI: https://doi.org/10.1016/j.solener.2015.10.010. URL: https://www.sciencedirect.com/science/article/pii/S0038092X15005435.

[8] Dazhi Yang. "Estimating 1-min beam and diffuse irradiance from the global irradiance: A review and an extensive worldwide comparison of latest separation models at 126 stations". In: *Renewable and Sustainable Energy Reviews* 159 (2022), p. 112195.

[9] Benjamin YH Liu and Richard C Jordan. "The interrelationship and characteristic distribution of direct, diffuse and total solar radiation". In: *Solar energy* 4.3 (1960), pp. 1–19.

[10] Arvid Skartveit, Jan Asle Olseth, and Marit Elisabet Tuft. "An hourly diffuse fraction model with correction for variability and surface albedo". In: *Solar Energy* 63.3 (1998), pp. 173–183. ISSN: 0038-092X. DOI: https://doi.org/10.1016/S0038-092X(98)00067-X. URL: https://www.sciencedirect.com/science/article/pii/S0038092X9800067X.

[11] John Boland, Lynne Scott, and Mark Luther. "Modelling the diffuse fraction of global solar radiation on a horizontal surface". In: *Environmetrics: The official journal of the International Environmetrics Society* 12.2 (2001), pp. 103–116.

[12] Barbara Ridley, John Boland, and Philippe Lauret. "Modelling of diffuse solar fraction with multiple predictors". In: *Renewable Energy* 35.2 (2010), pp. 478–483. ISSN: 0960-1481. DOI: https://doi.org/10.1016/j.renene.2009.07.018. URL: https://www.sciencedirect.com/science/article/pii/S0960148109003012.

[13] N.A. Engerer. "Minute resolution estimates of the diffuse fraction of global irradiance for south-eastern Australia". In: *Solar Energy* 116 (2015), pp. 215–237. ISSN: 0038-092X. DOI: https://doi.org/10.1016/j.solener.2015.04.012. URL: https://www.sciencedirect.com/science/article/pii/S0038092X15001905.

[14] Allan R. Starke et al. "Resolution of the cloud enhancement problem for one-minute diffuse radiation prediction". In: *Renewable Energy* 125 (2018), pp. 472–484. ISSN: 0960-1481. DOI: https://doi.org/10.1016/j.renene.2018.02.107. URL: https://www.sciencedirect.com/science/article/pii/S0960148118302593.

[15] Dazhi Yang and John Boland. "Satellite-augmented diffuse solar radiation separation models". In: *Journal of Renewable and Sustainable Energy* 11.2 (2019), p. 023705. DOI: 10.1063/1.5087463. eprint: https://doi.org/10.1063/1.5087463. URL: https://doi.org/10.1063/1.5087463.

[16] Dazhi Yang. "Temporal-resolution cascade model for separation of 1-min beam and diffuse irradiance". In: *Journal of Renewable and Sustainable Energy* 13.5 (2021), p. 056101.

[17] Eugenia Paulescu and Marius Paulescu. "Minute-Scale Models for the Diffuse Fraction of Global Solar Radiation Balanced between Accuracy and Accessibility". In: *Applied Sciences* 13.11 (2023), p. 6558.

[18] Christian A Gueymard et al. *BSRN data set for IEA-PVPS Task-16 Activity 1.4 Quality Control*. data set. 2022. DOI: 10.1594/PANGAEA.939988. URL: https://doi.org/10.1594/PANGAEA.939988.

[19] Anne Forstinger et al. "Expert quality control of solar radiation ground data sets". In: *ISES Solar World Congress*. 2021.

[20] John A. Duffie. *Solar Engineering of Thermal Processes, Photovoltaics and Wind, Solar Engineering of Thermal Processes Photovoltaics and Wind*. eng. John Wiley & Sons, 2020.

[21] Fritz Kasten and Andrew T Young. "Revised optical air mass tables and approximation formula". In: *Applied optics* 28.22 (1989), pp. 4735–4738.

[22] Tahir Mehmood et al. "A review of variable selection methods in partial least squares regression". In: *Chemometrics and intelligent laboratory systems* 118 (2012), pp. 62–69.

[23] F. Pedregosa et al. "Scikit-learn: Machine Learning in Python". In: *Journal of Machine Learning Research* 12 (2011), pp. 2825–2830.

[24] Jamie M. Bright and Nicholas A. Engerer. "Engerer2: Global re-parameterisation, update, and validation of an irradiance separation model at different temporal resolutions". In: *Journal of Renewable and Sustainable Energy* 11.3 (2019), p. 033701. DOI: 10.1063/1.5097014. eprint: https://doi.org/10.1063/1.5097014. URL: https://doi.org/10.1063/1.5097014.

[25] Eugenia Paulescu and Robert Blaga. "A simple and reliable empirical model with two predictors for estimating 1-minute diffuse fraction". In: *Solar Energy* 180 (2019), pp. 75–84. ISSN: 0038-092X. DOI: https://doi.org/10.1016/j.solener.2019.01.029. URL: https://www.sciencedirect.com/science/article/pii/S0038092X19300386.

[26] Allan R. Starke et al. "Assessing one-minute diffuse fraction models based on worldwide climate features". In: *Renewable Energy* 177 (2021), pp. 700–714. ISSN: 0960-1481. DOI: https://doi.org/10.1016/j.renene.2021.05.108. URL: https://www.sciencedirect.com/science/article/pii/S0960148121007916.

10.4229/EUPVSEC2023/4CO.9.2
020362-011

A BENCHMARK OF SIMPLE DIFFUSE AND DIRECT IRRADIANCE MEASUREMENT SYSTEMS

Niklas Blum[1*], Felix Maas[1], Jonas Stührenberg[2], Ra'al Broda[1], Paul Matteschk[1], Michael Meinel[3], Bijan Nouri[1], Laura Campos Guzmán[1], Andreas Kazantzidis[4], Mounir Abraim[5], Abdellatif Ghennioui[5], Martina Calais[6], Aron Habte[7], Mário Pó[8], Luis F. Zarzalejo[9], Stefan Wilbert[1]

EU PVSEC 2023, 20.09.2023, Lisbon

1 German Aerospace Center (DLR), Institute of Solar Research, niklas.blum@dlr.de
2 DLR, Institute of Networked Energy Systems
3 DLR, Institute for Software Technology
4 University of Patras, Department of Physics, Laboratory of Atmospheric Physics
5 Green Energy Park research platform (GEP, IRESEN/UM6P)
6 Murdoch University, School of Engineering and Energy, College of Science, Technology, Engineering and Mathematics
7 National Renewable Energy Laboratory (NREL), Power Systems Engineering Center
8 EKO Instruments Europe, Middle East, Africa, South America
9 CIEMAT Energy Department, Renewable Energy Division

1

020343-001

40th European Photovoltaic Solar Energy Conference and Exhibition

Outline

- Motivation: Overview on options to determine direct and diffuse irradiance
- Are some sensors more accurate than others?
- Which sensor is suited for a given application / site?

2

020343-002

Irradiance parameters for solar applications

DLR

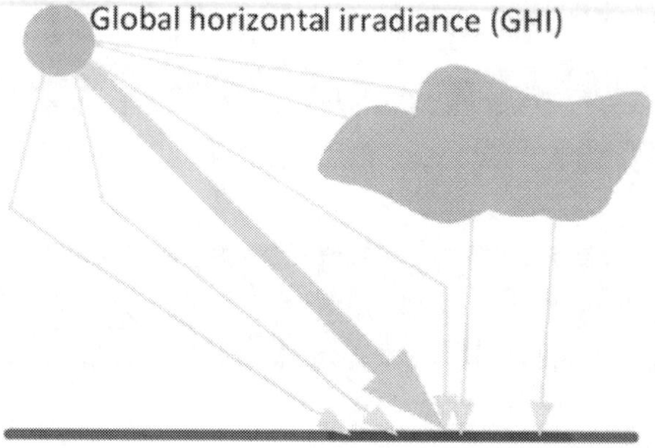

3

020363-003

40th European Photovoltaic Solar Energy Conference and Exhibition

Irradiance parameters for solar applications

DLR

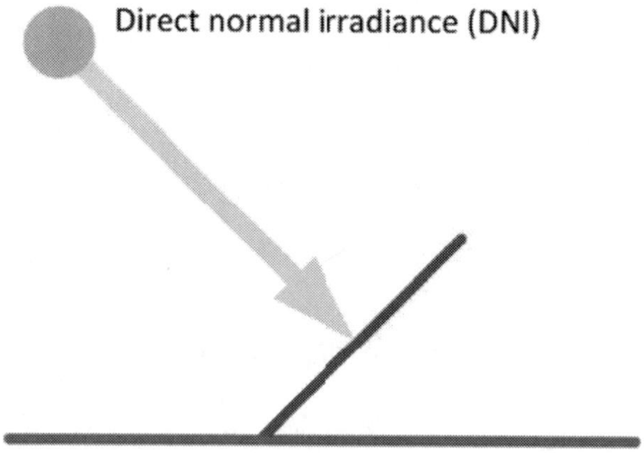

4

020363-004

Irradiance parameters for solar applications

DLR

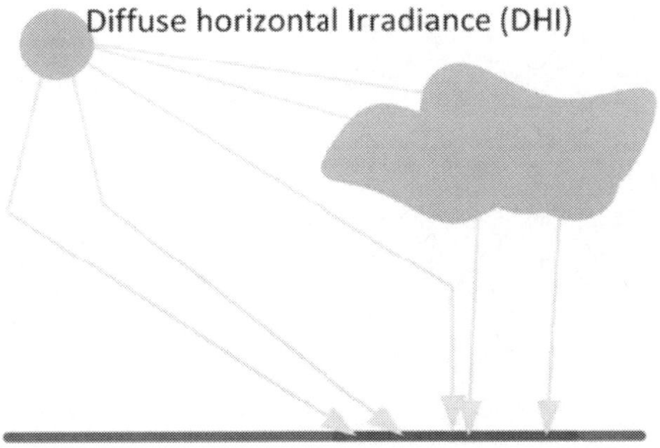

5

020343-005

40th European Photovoltaic Solar Energy Conference and Exhibition

Irradiance parameters for solar applications

DLR

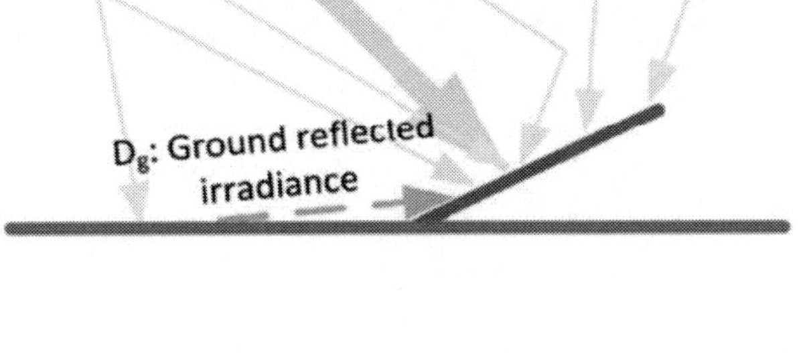

6

020343-006

What options do we have to determine DNI, DHI?

- Remotely-sensed

7

020343-007

What options do we have to determine DNI, DHI?

- Remotely-sensed
- Estimated from GHI by decomposition

8

020343-008

What options do we have to determine DNI, DHI?

DLR

- Remotely-sensed
- Estimated from GHI by decomposition

  Often too inaccurate,
  DHI on-site measurement
  recommended for bifacial PV plants [1]

- Measurement on-site

Image taken at PSA. The research site Plataforma Solar de Almería (PSA) is owned and operated by the Spanish CIEMAT

9

[1] International Electrotechnical Commission (2021). *Photovoltaic System Performance - Part 1: Monitoring* (IEC 61724-1:2021). IEC. iec.ch

020343-009

40th European Photovoltaic Solar Energy Conference and Exhibition

What options do we have to determine DNI, DHI?

DLR

- Remotely-sensed
- Estimated from global irradiance (decomposition)
- **Measurement on-site**
  - Sun tracker
    - Expensive
    - Prone to failures
    - Permanent data checks
    - High cleaning demand

    → Good reference system if operated properly – otherwise reduced accuracy

Image taken at PSA. The research site Plataforma Solar de Almería (PSA) is owned and operated by the Spanish CIEMAT

10

020343-010

What options do we have to determine DNI, DHI?

- Remotely-sensed
- Estimated from global irradiance (decomposition)
- **Measurement on-site**
  - Sun tracker
    - Expensive
    - Prone to failures
    - Permanent data checks
    - High cleaning demand
    - → Good reference system if operated properly – otherwise reduced accuracy
  - **Simpler sensors**
    - Intend to avoid the above shortcomings

11

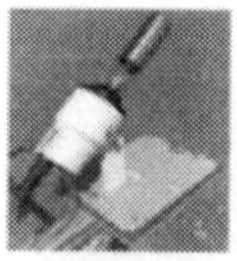

Images taken at PSA. The research site Plataforma Solar de Almería (PSA) is owned and operated by the Spanish CIEMAT

020343-011

40th European Photovoltaic Solar Energy Conference and Exhibition

Simple sensors for DNI, DHI

Delta-T SPN1

- 7 thermopile pyranometers
- Special shading mask
- Always ≥1 pyranometer shaded & ≥1 unshaded
- Measures GHI, DHI → calculates DNI

12

Image taken at PSA. The research site Plataforma Solar de Almería (PSA) is owned and operated by the Spanish CIEMAT

020343-012

Simple sensors for DNI, DHI

DLR

Rotating
Shadowband
Pyranometer (RSP)

- Fast-response photodiode pyranometer
- Shadow band
- when pyranometer unshaded: GHI; shaded: DHI
- DNI calculated

13

Image taken at PSA. The research site Plataforma Solar de Almería (PSA) is owned and operated by the Spanish CIEMAT

020343-013

Simple sensors for DNI, DHI

DLR

EKO MS-90

- Special rotating mirror
- Pyroelectric sensor
- measures DNI
- + thermopile pyranometer for GHI → DHI

14

Image taken at PSA. The research site Plataforma Solar de Almería (PSA) is owned and operated by the Spanish CIEMAT

020343-014

Simple sensors for DNI, DHI

DLR

EKO MS-90

- Special rotating mirror
- Pyroelectric sensor
- measures DNI
- + thermopile pyranometer for GHI → DHI

15

Image taken at PSA. The research site Plataforma Solar de Almería (PSA) is owned and operated by the Spanish CIEMAT

0203i3-015

40th European Photovoltaic Solar Energy Conference and Exhibition

Simple sensors for DNI, DHI

DLR

EKO MS-90

+

- Special rotating mirror
- Pyroelectric sensor
- measures DNI
- + thermopile pyranometer for GHI → DHI

16

Image taken at PSA. The research site Plataforma Solar de Almería (PSA) is owned and operated by the Spanish CIEMAT

0203i3-016

Simple sensors for DNI, DHI

DLR

Sunto CaptPro

- 11 R,G,B,IR photodiodes measure GHI and POA in 10 planes
- Proprietary algorithm → DNI, DHI

17

Image taken at PSA. The research site Plataforma Solar de Almería (PSA) is owned and operated by the Spanish CIEMAT

40th European Photovoltaic Solar Energy Conference and Exhibition

Simple sensors for DNI, DHI

DLR

PyranoCam (own development [2,3])

[2] Blum, N. B., et al. (2022). "Measurement of diffuse and plane of array irradiance by a combination of a pyranometer and an all-sky imager." Solar Energy 232: 232-247.

[3] Broda, R. (2022) Development of A Machine-Learning-Based Correction for Cloud-Camera-Based Solar Radiation Measurement. Master thesis, RWTH Aachen.

- Pyranometer → GHI
- All-Sky Imager sky radiance → raw DHI
- Corrections by combination → more accurate DHI
- DNI calculated

18

Benchmark setup

*The research site Plataforma Solar de Almería is owned and operated by the Spanish CIEMAT

DLR

- Tabernas*, southern Spain (37°N)
- Cold-desert climate
- 12 months measurements
- Mean irradiances over dataset:
  - 456 W/m² (GHI)
  - 541 W/m² (DNI)
  - 144 W/m² (DHI)

- Oldenburg, northern Germany (53°N)
- Temperate oceanic climate
- 12 months measurements
- Mean irradiances over dataset:
  - 279 W/m² (GHI)
  - 269 W/m² (DNI)
  - 135 W/m² (DHI)

In the images some sensors masked to comply with NDAs

19

020343-019

Are some sensors more accurate than others?
Sunto CaptPro

DLR

- Sensor broke after 4th month of the benchmark in Tabernas
- Repair attempt by manufacturer failed

Image taken at PSA. The research site Plataforma Solar de Almería (PSA) is owned and operated by the Spanish CIEMAT

20

020343-020

Are some sensors more accurate than others?
Sunto CaptPro

DLR

- Limited database of 4 months from Tabernas
- Overall accuracy unsatisfactory
- Maybe useful if POA should be assessed

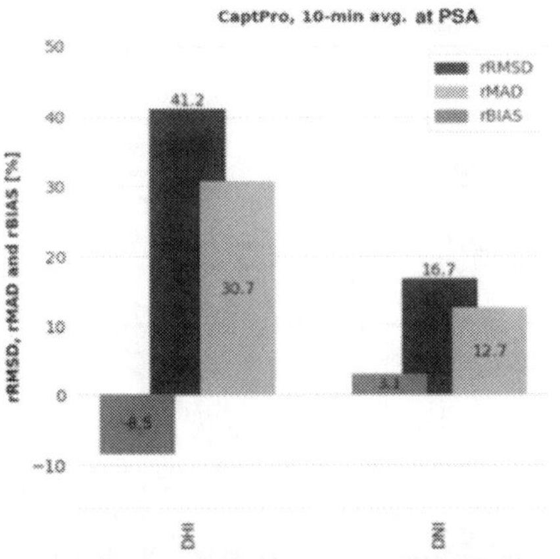

$$rRMSD = \frac{100\%}{\overline{ref}}\sqrt{\overline{(test-ref)^2}} \qquad rMAD = \frac{100\%}{\overline{ref}}\overline{|test-ref|} \qquad rBIAS = \frac{100\%}{\overline{ref}}\overline{test-ref}$$

21

40th European Photovoltaic Solar Energy Conference and Exhibition

Are some sensors more accurate than others?

DLR

- Evaluation in Tabernas

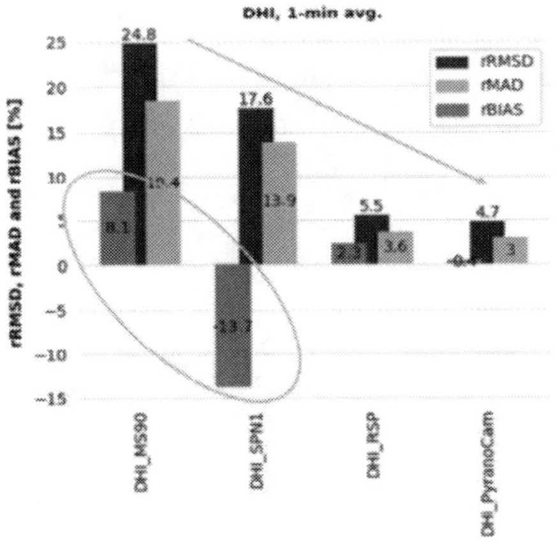

Evaluation in Oldenburg

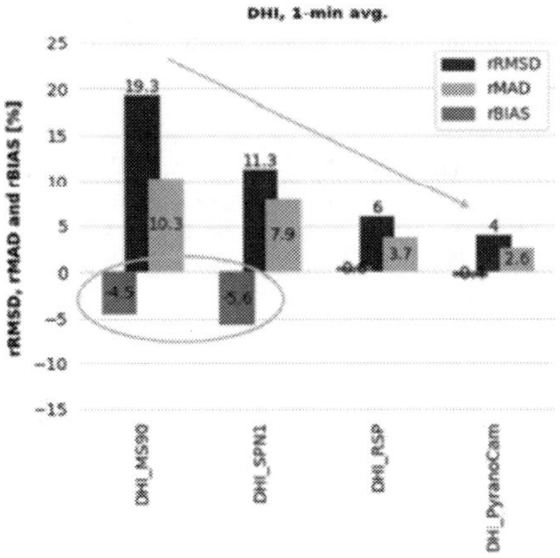

22

Are some sensors more accurate than others?

DLR

- Evaluation in Tabernas

Evaluation in Oldenburg

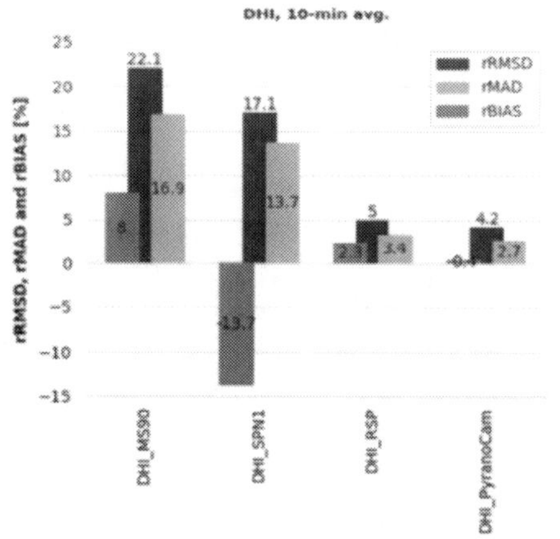

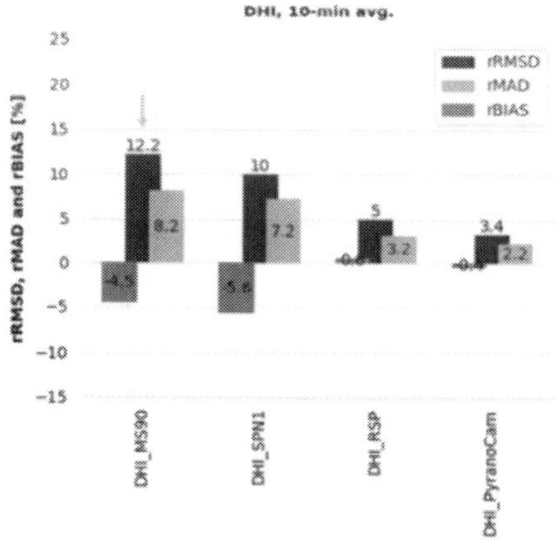

23

40th European Photovoltaic Solar Energy Conference and Exhibition

Are some sensors more accurate than others?

DLR

- Evaluation in Tabernas

Evaluation in Oldenburg

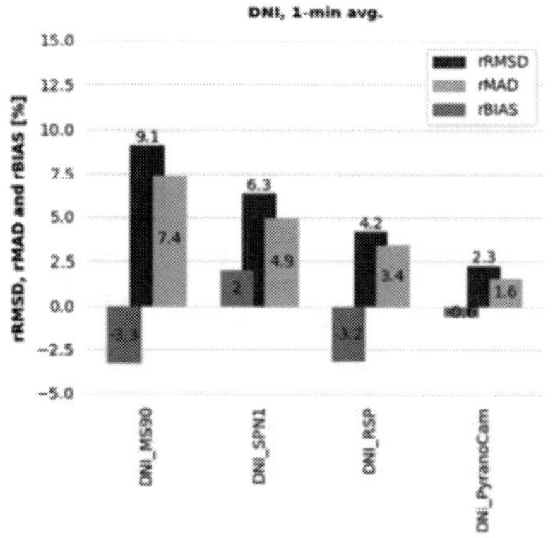

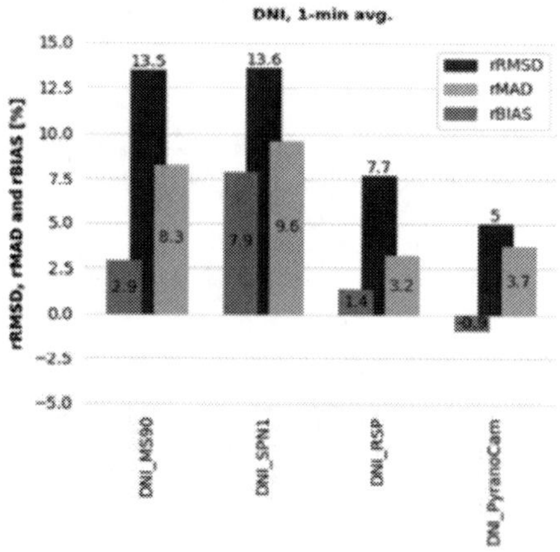

24

Are some sensors more accurate than others?

DLR

- Evaluation in Tabernas

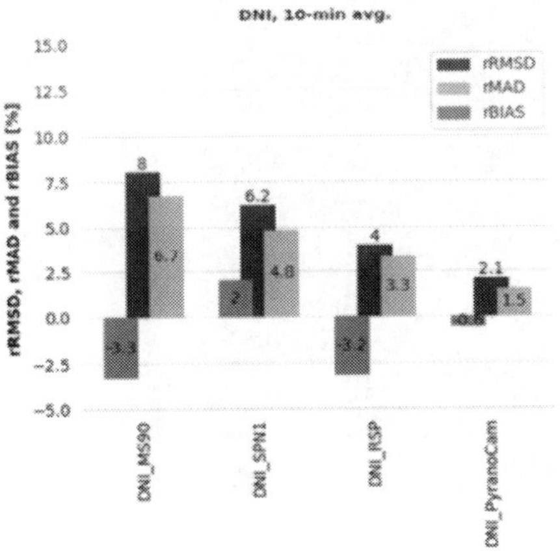

Evaluation in Oldenburg

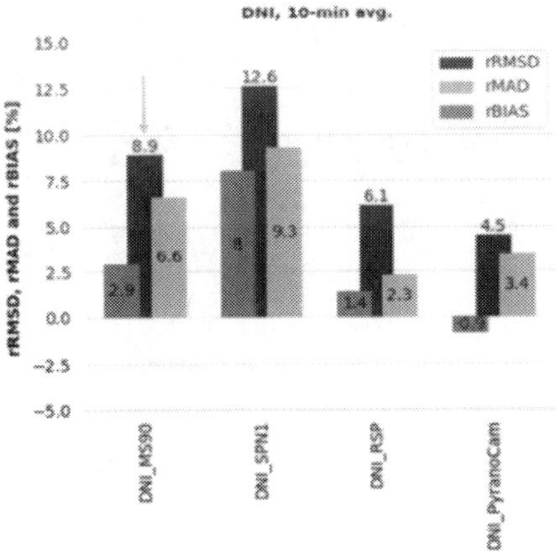

25

020343-025

Which sensor is suited for a given application / site?

DLR

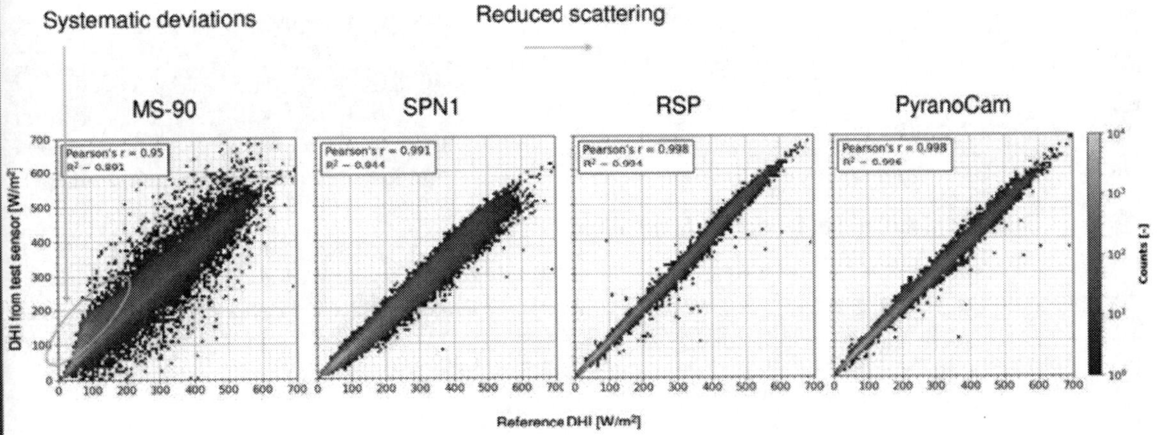

26

020343-026

Which sensor is suited for a given application / site?

DLR

- Linke turbidity
- Sun elevation
- **Cloud cover**

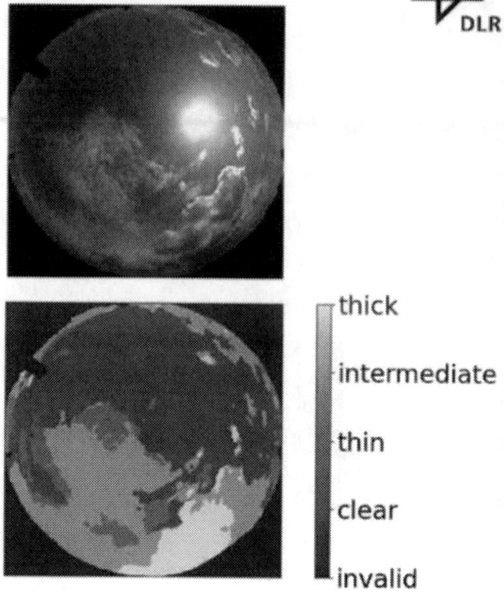

Image taken at PSA. The research site Plataforma Solar de Almeria (PSA) is owned and operated by the Spanish CIEMAT

27

020343-027

40th European Photovoltaic Solar Energy Conference and Exhibition

Which sensor is suited for a given application / site?

DLR

- Linke turbidity
- Sun elevation
- Cloud cover
- **Spectral composition of GHI, DHI, DNI**

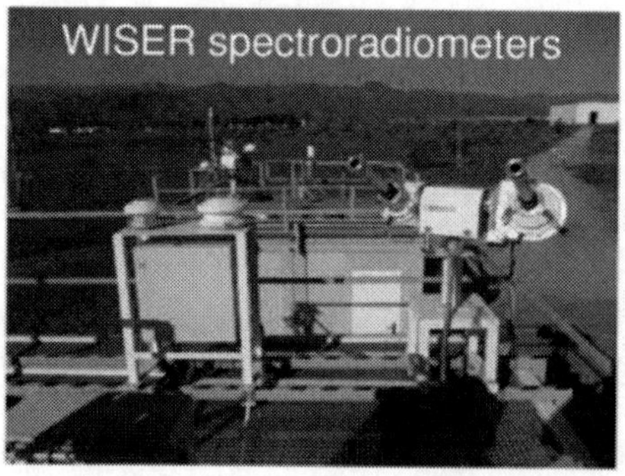

Image taken at PSA. The research site Plataforma Solar de Almeria (PSA) is owned and operated by the Spanish CIEMAT

28

020343-028

Influences on the sensors' accuracies

DLR

- Linke turbidity
- Sun elevation
- Cloud cover
- Spectral composition of GHI, DHI, DNI
- **Circumsolar irradiance**

Image taken at PSA. The research site Plataforma Solar de Almería (PSA) is owned and operated by the Spanish CIEMAT

29

020343-029

40th European Photovoltaic Solar Energy Conference and Exhibition

Which sensor is suited for a given application/ site?
Example: SPN-1

Largest influence SPN1, DHI – circumsolar irradiance (sun angular distance 2.5°…5°)

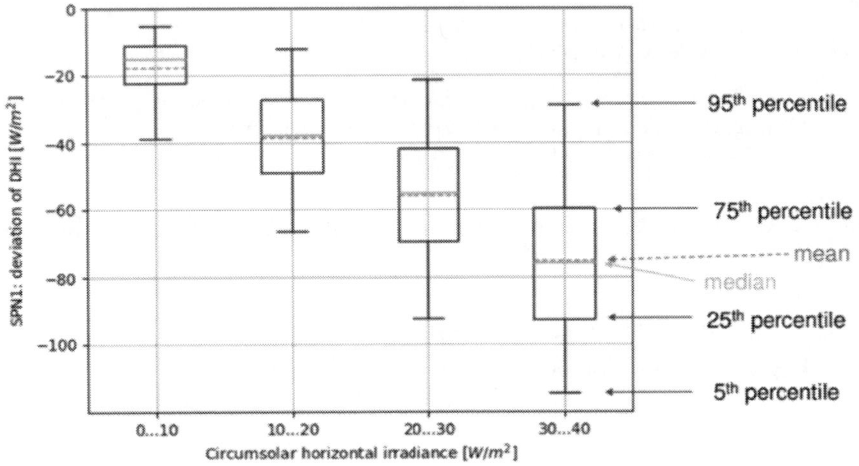

Image taken at PSA. The research site Plataforma Solar de Almería (PSA) is owned and operated by the Spanish CIEMAT

30

020343-030

Which sensor is suited for a given application/ site?
Findings

DLR

Specifics of a site and application need to be considered when choosing a sensor:

- SPN-1 considerable shortcomings if more circumsolar irradiance present (clouds, aerosols)
- MS-90 only recommended if DNI and exclusively longer clear periods of interest
- RSP high accuracy in our test,
  - → But deviations known when calibration transferred to different atmospheric conditions due to narrowband spectral response
- PyranoCam performs well under all conditions tested so far

31

020343-031

Conclusion

DLR

- 3 accuracy groups identified, e.g. by rRMSD (10-min avg.):
  - RSP, PyranoCam in high accuracy group:
    - RSP: 5% (DHI), 4-6% (DNI)
    - Pyranocam: 3-4% (DHI), 2-5% (DNI)
  - SPN-1, MS-90 larger deviations:
    - SPN-1: 10-17% (DHI), 6-13% (DNI)
    - MS-90: 12-22% (DHI), 8-9% (DNI)
  - CaptPro based on our experience problematic for DNI, DHI
    - rRMSD (10-min avg.): 41% (DHI), 17% (DNI)
- RSP, MS-90, SPN1 have shortcomings in certain atmospheric conditions
  - e.g. high circumsolar irradiance
- PyranoCam provides promising accuracy
  - → Should be verified at further validation sites

32

020343-032

Outlook

Validation of PyranoCam: 6 sites, 4 continents

DLR

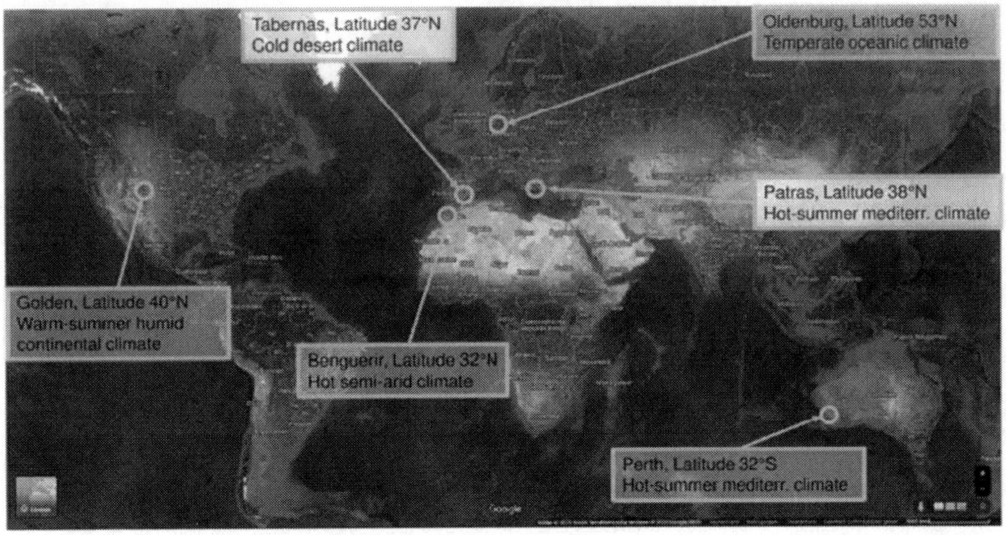

33

020343-033

40th European Photovoltaic Solar Energy Conference and Exhibition

Outlook

DLR

- Validation of PyranoCam is ongoing at 6 sites worldwide
  - Preliminary results suggest similar accuracy as in Tabernas, Oldenburg
  - Detailed analysis with 1-year datasets planned/ required
- Further evaluation of atmospheric influences for SPN1, RSP, MS-90, PyranoCam
  - will allow users to estimate a sensor's accuracy at their site
- Journal publication under preparation

34

020343-034

Outlook

- Validation of PyranoCam is ongoing at 6 sites worldwide
  - Preliminary results suggest similar accuracy as in Tabernas, Oldenburg
  - Detailed analysis with 1-year datasets planned/ required
- Further evaluation of atmospheric influences for SPN1, RSP, MS-90, PyranoCam
  - will allow users to estimate a sensor's accuracy at their site
- Journal publication under preparation

Thank you for your attention!
Further questions? niklas.blum@dlr.de

35

020343-035

This presentation was selected by the Sc. Committee of the EU PVSEC 2023 for submission of a full paper to one of the EU PVSEC's collaborating peer-reviewed journals.

AEROSOL ESTIMATION IN ARID REGION FOR SOLAR RESOURCE APPLICATIONS

Dunia A. Bachour[a], Daniel Perez-Astudillo[a], Christos Fountoukis[a], Antonio Sanfilippo[a]
[a] Qatar Environment & Energy Research Institute, P.O.Box 34110, Doha, Qatar.
dbachour@hbku.edu.qa, dastudillo@hbku.edu.qa, cfountoukis@hbku.edu.qa, asanfilippo@hbku.edu.qa

ABSTRACT: Atmospheric aerosols highly affect the energy balance and atmospheric radiation transfer. They induce a direct effect on the climate by absorbing or scattering the radiation, and also have an indirect effect by modifying the radiative properties and lifetime of clouds. Both effects influence the amount and quality of the incoming solar radiation incident on the earth surface. For solar energy applications, this influence is more pronounced in regions dominated by low cloudiness and where aerosols dominate the atmosphere. Due to their high uncertainty, accurate knowledge of aerosols properties and distribution is needed to understand and quantify their impact.

Aerosol Optical Depth (AOD) is commonly used to characterize solar radiation attenuation. Here, we describe a ground-based estimation of AOD using a CIMEL sun photometer across one year in Qatar, a hot desert climate region with high aerosol loads. In addition, AOD products available from two databases are used for comparison with ground-derived data: the Copernicus Atmosphere Monitoring Service (CAMS), and the Weather Research and Forecasting with Chemistry (WRF-Chem) model application to our region of interest. For validation, the AOD data is used as input to two clear-sky models to estimate direct solar radiation and compare with ground measurements of the same.

Keywords: AOD, sun photometer, aerosols, CAMS, clear sky, SMARTS, ESRA

1 INTRODUCTION

Atmospheric aerosols are tiny particles suspended in the atmosphere, and have a large impact on the energy balance and atmospheric radiation transfer. Depending on their properties, they induce a direct or indirect effect on the climate. More specifically, aerosols can absorb solar radiation causing heating and warmer atmospheric conditions, scatter the radiation causing a cooling effect, or act indirectly on modifying the radiative properties through interactions with clouds. Both direct and indirect interactions influence the amount and quality of the incoming solar radiation incident on the earth surface. For solar energy applications, this influence is even more pronounced in regions dominated by low cloudiness conditions and where aerosols are the main components of the atmosphere and induce high level of variability. Thus, accurate knowledge of the aerosol properties, as well as knowledge of their spatial and temporal distribution is needed to understand and quantify their impact on solar radiation estimation and forecasting.

The study presented here is conducted in Doha, Qatar, located in a hot desert climate region where dust storms happen frequently, holding a large amount of aerosols in the atmosphere and causing poor visibility and a significant reduction of incoming solar radiation; this is generally more intense towards the spring and summer seasons.

The aerosol optical depth (AOD) is one of the main optical parameters used to characterize aerosol properties. It describes the interaction of aerosols with the incoming solar radiation, and represents a measure of the total extinction by the aerosols in the vertical column. Other parameters, such as the single scattering albedo (SSA) and the Angstrom exponent (AE) are also used to describe aerosol properties. SSA is defined as the ratio between scattering and total extinction. AE provides a good indication of the size distribution of aerosols.

Aerosol properties can be retrieved using satellite instruments such as MODIS (MODerate resolution Imaging Spectroradiometer) [1], or MISR (Multi-angle Imaging SpectroRadiometer) [2], that measure the atmospheric radiation in different wavelengths and use aerosol retrieval algorithms to provide aerosol data. AOD can also be estimated using numerical weather prediction models and chemical transport models. The Weather Research and Forecasting with Chemistry (WRF-Chem) model instance is a 3-D Air Quality model simulating the physics in the atmosphere and land surface, accounting for the emissions of pollutants and their sources along with their transport and interactions to predict aerosol properties [3]. The Goddard Earth Observing System (GEOS)-Chem (GC) model is a global chemical transport model designed to study tropospheric ozone and related species [4]. More recently, reanalysis models are gaining more insights since they use a combination of data sources such as satellite observations, in-situ measurements, and numerical models. The Copernicus Atmosphere Monitoring Service (CAMS) database, developed under the European Commission in the framework of its FP7 program, is one example of reanalysis model and it is considered a well-suited database for the provision of aerosol data for solar radiation calculations, with an acceptable temporal resolution of ~3 hours [5].

In this contribution, we describe a ground-based estimation of AOD using a CIMEL sun photometer, deployed in Doha, Qatar. This instrument is used to estimate the optical properties of the atmosphere; through the calculation of the atmospheric physics equations, it provides the optical properties of the aerosols. We then use the AOD products derived from two different model-based sources, CAMS and WRF-Chem, to compare with AOD derived from the ground measurements. Finally, and for validation purposes, AOD data are used as inputs to clear-sky models to estimate solar radiation at ground level and compare with ground-measured data.

2 METHODOLOGY

A CIMEL CE318T sun photometer is installed on the rooftop of the Qatar Environment and Energy Research Institute (QEERI), in Doha, Qatar. This instrument is part

10.4229/EUPVSEC2023/4CO.9.4
020364-001

of a solar resource assessment facility and research in QEERI, where a high-quality solar radiation monitoring station is also operating and collects global, diffuse and direct solar radiation data [6].

The principle of measurements of the sun photometer for AOD derivation is based on a spectrally selective detector measuring the spectral extinction of the direct (beam) radiation. According to the Lambert-Beer equation:

$$I_\lambda = \left(\frac{I_{0\lambda}}{d^2}\right) \exp -(\sigma_\lambda * m) \qquad \text{(Eq.1)}$$

Where,

I_λ is the measured irradiance at wavelength λ, $I_{0\lambda}$ is the irradiance at the top of the atmosphere, σ_λ is the total optical thickness including molecular (Rayleigh) scattering, trace gas absorption (such as ozone), and aerosol extinction, m is the optical air mass, and d is the ratio of the actual to the average Earth–Sun distance.

For AOD values derived here, the sun photometer is programmed to operate continuously on a mode called "sun measurements", which is made in a field of view of 1.2 degrees, in 9 spectral bands at 340, 380, 440, 500, 675, 870, 937, 1020, and 1640 nm with two channels: silicon and InGaAs. These direct measurements are completed in a preprogrammed sequence of ~10s, starting at air mass 7 in the morning and ending at air mass 7 in the evening. A sequence of 3 sun measurements are taken 30 seconds apart to create a triplet observation per wavelength, used as a check for cloud contamination. For large air mass periods, direct sun measurements are made at 0.25 air mass intervals, while at smaller air mass periods the interval between measurements is typically 15 minutes.

In this study, the diurnal data of AOD derived at the wavelength 500 nm is used, which is the closer to the reference wavelength of 550 nm commonly used to be a good representation of the overall scattering properties of aerosols in solar radiation applications. For comparison purposes, we use two different AOD products at 550nm, one derived from the renowned online database CAMS, and another one derived from the use of the 3-D WRF-Chem model implemented in QEERI and applied over the Middle East region with an enhanced grid resolution over the State of Qatar [7]. To validate the results and set the ground truth for the AOD derived here, we analyze the correlation between direct normal irradiance (DNI) measured and estimated under clear-sky conditions, using the different AOD products as inputs to the clear sky model, DNI being the solar radiation component most affected by the atmospheric attenuation. The application would be to validate AOD data where only model-based AOD and ground solar data are available. This work complements previous studies on AOD estimation done by QEERI [8].

3 RESULTS

3.1 AOD derived from sun photometer

In this section, we study the ground AOD products derived by the sun photometer. Figures 1 and 2 show the daily profiles of hourly averaged AOD during the month of June for two consecutive years (2021-2022), for all the days of the month for which AOD values were derived, excluding periods of instrument downtime and cloudy days when AOD data are not obtained. The data of two years were shown to capture inter-annual variability. Since the range of AOD data in 2021 and 2022 was similar, the discussion hereafter is valid for both figures. Each colored line represents the variations of the AOD within one day indicated on the legend, the graph being shown here for daytime values only (5 am to 6 pm local time). It can be seen that in general there is a slight variation of AOD during the day, except some days where AOD variation is clear between morning and afternoon time. The observed range of AOD values during the studied month is between 0.1 and 0.9, which indicates a dynamic fluctuation of AOD during the one-month period, a variation that is translated to a varying amount of solar radiation received at the Earth's surface for the same period. Indeed, the maximum direct normal irradiance DNI can be reduced by an amount ~ 5 % for an increase of 0.1 in the AOD value, as can be seen in Figure 3 that shows the daily profiles of DNI simulated with the SMARTS2 model for different values of AOD (0.05 to 3). SMARTS stands for Simple Model of the Atmospheric Radiative Transfer of Sunshine, and predicts clear-sky spectral irradiances for defined atmospheric conditions. Some information about the solar radiation calculation shown here using SMARTS2 is described in the next sections, but for a more complete reference the reader can refer to [9].

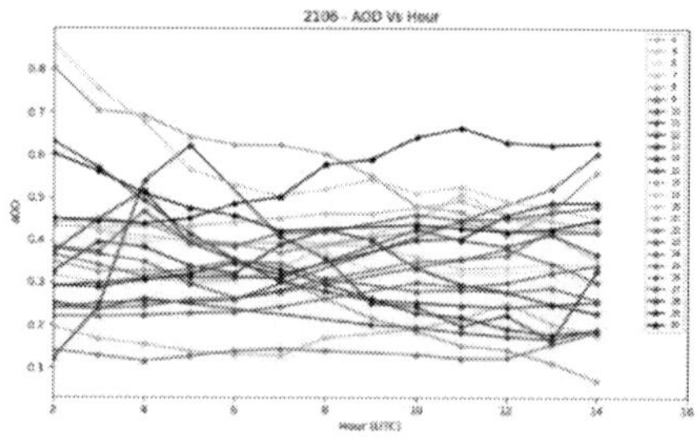

Figure 1: Daily profiles of hourly-averaged AOD calculated by a sun photometer in Doha, Qatar, during a one-month period (June 2021).

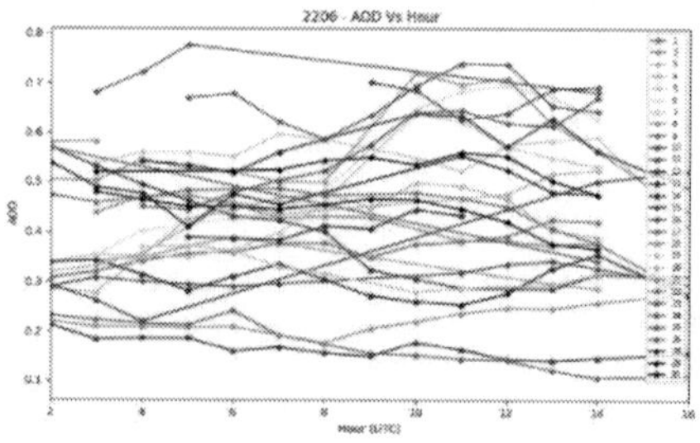

Figure 2: Daily profiles of hourly-averaged AOD calculated by a sun photometer in Doha, Qatar, during a one-month period (June 2022).

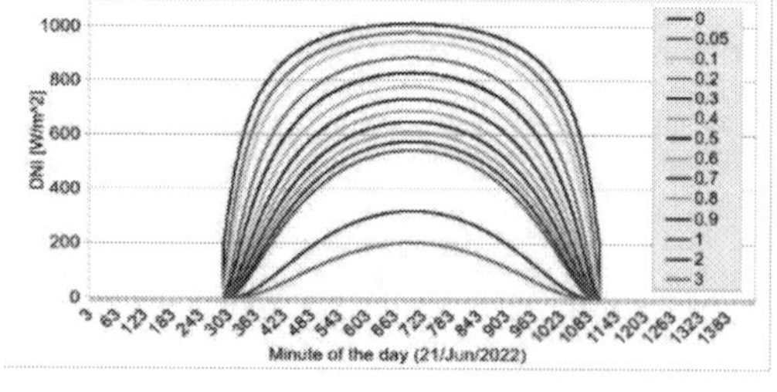

10.4229/EUPVSEC2023/4CO.9.4
020364-002

Figure 3: DNI profiles for 21/June/2022, i.e. around the summer solstice, for different AOD values.

In order to have a more complete assessment on the AOD values that can be observed along the year in the studied location, Figure 4 shows the hourly means of AOD values along with their standard deviations depicted in the Figure as the error bars shown on each data point. Two distinct periods are shown here, one in red for mornings before solar noon (AOD AM) and the other shown in green for the afternoon period (AOD PM), to discern AOD variations in one day. It is seen that AOD values are in general higher in the afternoon, with more or less a similar variation, except for the month of May, where the variations are more pronounced in the morning, indicating more pronounced dust and haze events in the morning time during this month. It is worth mentioning that in Doha dust storms occur more frequently between March and April. The AOD mean value during the whole period is 0.4941 with a standard deviation of 0.27.

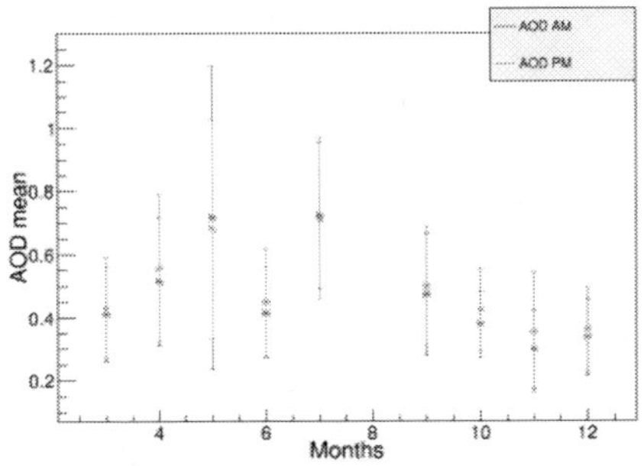

Figure 4: Hourly AOD averages per month, for morning (am) and afternoon (pm) periods.

3.2 AOD derived by CAMS

AOD data were obtained from the CAMS database using the CAMS Radiation Service v4.6 all-sky irradiation (derived from satellite data). 1-minute total AOD values at 550 nm were calculated as the sum of the AOD products of: dust, black carbon, sea salt, organic matter, sulphate, nitrate, ammonium, and secondary organic particles. Hourly AOD (AOD-CAMS) values were derived from the minute values in order to compare with AOD derived from the sun-photometer, denoted hereafter as AOD-Ground.

Figure 5 shows the frequency distribution of hourly AOD-CAMS (Green) compared to AOD-Ground (Red) for an almost one-year period (2022). It is seen that there are differences mainly in the tail part of the two histograms. High values of AOD between 0.7 and 1 seem to be more frequent in the CAMS data, while for AOD values higher than 1, the ground measurements are more prone to capture these typical values seen here in Doha during dust events. This is more obvious when studying the frequency distribution separately for each of the months of the year (not shown here).

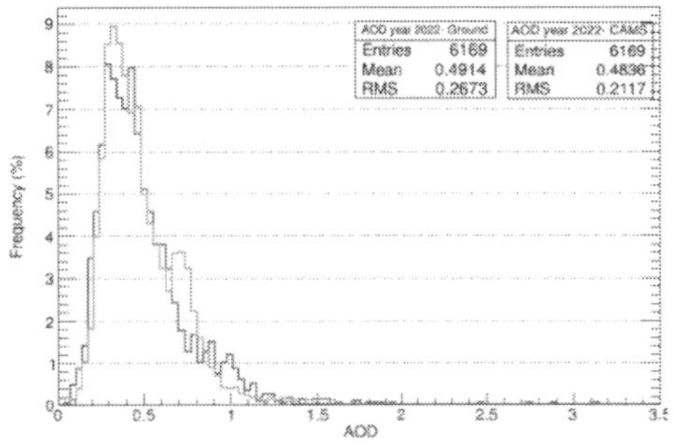

Figure 5: Frequency distribution of hourly AOD for ~ a one-year period (excluding August): ground AOD (top) and AOD CAMS (bottom).

3.3 AOD derived by WRF-Chem

In this study, we are also testing AOD derived by the WRF-CHEM model implemented in QEERI for air quality purposes. This 3-D Air Quality model is applied over the Middle East region with an enhanced grid resolution over the State of Qatar. A domain setup of three grids using a two-way nesting configuration is used. AOD is calculated based on aerosol concentrations and aerosol optical properties, taking into account the aerosol sizes and refractive index. A parameterized Mie theory [10] is adopted in WRF-Chem to calculate the aerosol optical properties.

The AOD is calculated by default in WRF-Chem at 4 different wavelengths (300, 400, 600 and 1000 nm), while through calculation of the Angstrom exponent, optical depths at other wavelengths in between the above can also be calculated. An optical depth for each of the model's vertical layers is calculated and by simply taking the sum of all layers, the total column AOD is finally calculated. The AOD at 550 nm is used here.

Figure 6 shows the frequency distribution of AOD derived from WRF-CHEM (top) as compared to AOD derived by the sun photometer (bottom) for the month of May 2022. It is evident that AOD distribution derived by WRF-CHEM model does not conform to the ground-derived AOD, with large discrepancies mainly in the part with high AOD values; Moreover, the mean value during this month is 57 % lower than the AOD derived by the sun photometer. This shows the challenges associated with predicting local dust events in a desert environment using WRF-Chem [11]. From the surface particulate matter measurements in the same location (not shown here) it is clear that May 2022 was a particularly dusty month with frequent local dust episodes. Further ongoing work includes testing WRF-Chem applications with multiple dust/aerosol scheme options in an effort to improve the AOD retrieval accuracy as well as comparisons during multiple months of 2022 and 2023.

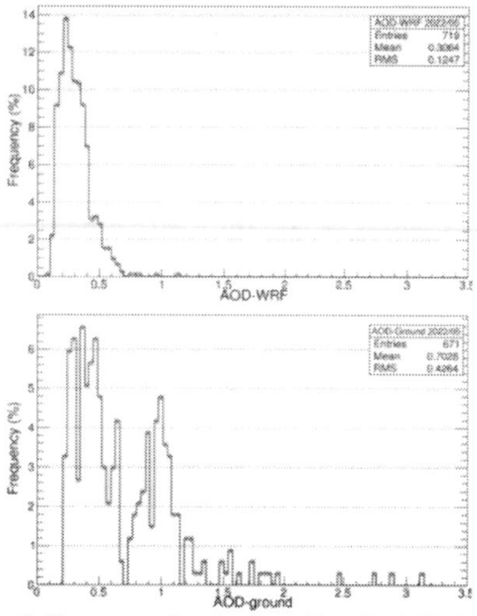

Figure 6: Frequency distribution of hourly AOD for May 2022: AOD-WRF (top) and AOD-Ground (bottom).

3.4 Comparison with ground AOD data

For the analysis hereafter, we compare AOD-CAMS and AOD-Ground. In order to quantify the difference between the AOD products, statistical parameters are evaluated in terms of the mean bias error (Bias) and the root mean square error (RMSE). The difference and its square between the two AOD values are calculated at the same time stamps. These differences and their squares are then added to determine the corresponding mean Bias and mean RMSE; the mean values are then used to calculate the relative values (rBias, rRMSE) with respect to the mean values calculated from the ground measurements. The error parameters of the comparison between AOD-Ground and AOD-CAMS have been calculated for all the months of year 2022. Table 1 shows the results of some of the months, reporting also the number of data points.

Table 1: Statistical parameters for the comparison of AOD_cams with AOD_ground shown for selected months.

Month	Bias	rBias(%)	RMSE	rRMSE(%)	Data points
Mar	0.013	3.28	0.10	25.54	248
May	-0.13	-20.09	0.33	51.12	409
Jun	0.041	10.00	0.12	29.13	333
Nov	0.084	27.15	0.18	57.03	296

When comparing the statistics separately for each of the months, it is noted that CAMS is sometimes overestimating AOD, and for some months it is under-estimating AOD. The relative bias is in general less than 10 % except for the two months of November and March, for which the relative RMSE is also very high (values shown in red in table 1). In order to have a closer evaluation on the comparison for these two months and eliminate the possibility of cloud contamination in the ground data, the plot of the comparison of AOD-CAMS and AOD-Ground at the same time stamp for the two months is shown in Figure 7. To note that the data of the x-axis, denoted as 'time' in the plots, while chronologically sorted, only include points when AOD data exists for both sources at the same timestamp. It is observed that the local variability seen in the ground data (in red) is not very well represented in the CAMS data (in blue). In addition, the higher AOD values (>1) in the ground data are not properly captured by CAMS. The days for which the rBias and rRMSE values were high were identified and further studied in the next section, where a validation of the retrieved AOD data is proposed.

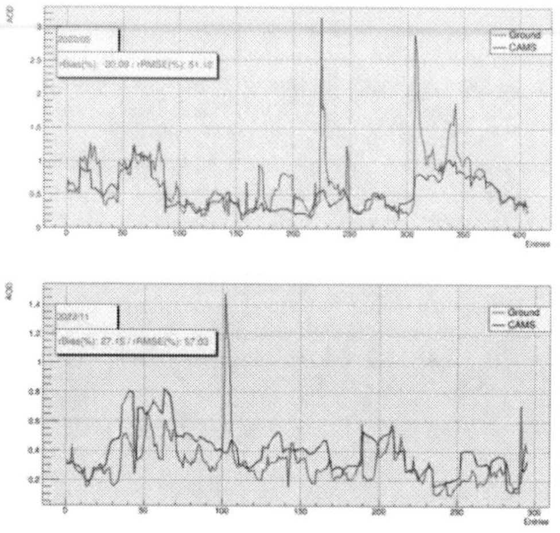

Figure 7: Hourly AOD variations during two one-month periods: May 2022 (top), Nov 2022 (bottom).

4 AOD VALIDATION THROUGH DNI

In order to validate the AOD derived from the sun photometer and from CAMS, we use ground-measured solar radiation data and more precisely the direct component DNI and compare with modeled solar radiation data derived from clear sky models using the AOD data derived here as input. The measured solar radiation is acquired by a high-precision solar radiation monitoring station at the same location of the sun photometer, measuring the three components of solar radiation.

The Modeled clear sky DNI is derived using two clear sky models. The SMARTS2 clear-sky model is used here with predefined input parameters and desert aerosol types corresponding to the atmospheric conditions seen in Doha. Table 2 summarises the input parameters used in SMARTS2 with 1-min AOD values (from ground and CAMS) to generate broadband DNI (280 to 4000 nm), which is then averaged hourly and denoted hereafter DNI-SM. The AOD values are the products described in section 3, namely at 500 nm for the sun photometer ground, and 550 nm for CAMS. The solar position values are changed for every simulation run, that is the sun-earth distance correction, sun zenith and sun azimuth angles.

Table 2: Values changed from defaults, constant for each SMARTS2 simulation run.

Atmospheric Pressure	1013.25 mb
US Standard Atmosphere	O3 abundance of 0.34
Altitude	20 m
Precipitable water	1.42 cm
CO2 amount	370 ppmv
Aerosol Model	DESERT_MIN

The other clear-sky model used here for the validation is the European Solar Radiation Atlas (ESRA) clear-sky model [12]. The ESRA model uses the Linke turbidity parameter to describe the attenuation of the direct component of radiation, combining the scattering by aerosols and the absorption by gas. In order to get the clear-sky maxima of the solar components, we use a Linke turbidity value of 1. The clear sky modeled DNI (DNI-ESRAc) is then modulated by the AOD values to calculate DNI (DNI-ESRA) and compare with measured DNI (DNI-Meas). The equation used for estimating the DNI with the ESRA model is as shown in Eq. 2, where m is the air mass and AOD is the value derived from the sun photometer and CAMS.

$$\text{DNI-ESRA} = \text{DNI-ESRAc} * \text{Exp}(-m*\text{AOD}) \qquad \text{(Eq.2)}$$

Figures 8 and 9 show the comparison of DNI for some of the days identified in section 3, for which the AOD values were high with high errors when comparing ground and CAMS data. Figure 8 shows the results using SMARTS2 as clear sky model, while figure 9 shows the results using ESRA as clear sky model.

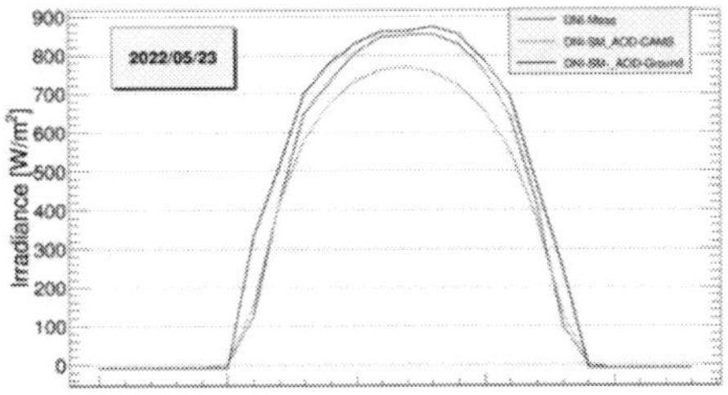

Figure 8: Comparison of measured and modeled DNI using AOD values as input to the SMARTS2 model.

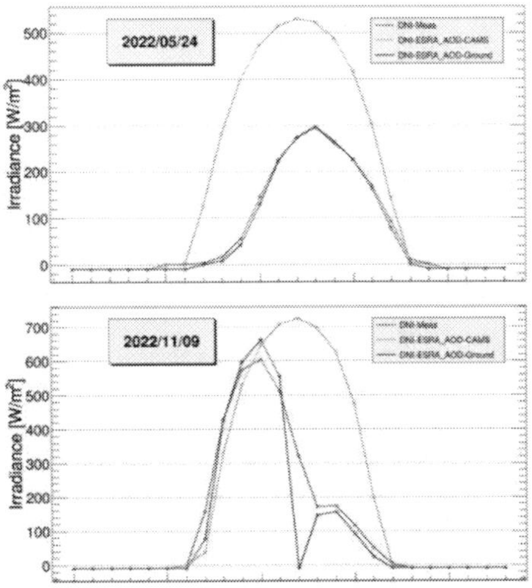

Figure 9: Comparison of measured and modeled DNI using AOD values as input to the ESRA model.

We can see in figures 8 and 9 that modeled DNI using AOD_Ground is in good agreement with measured DNI, in the case of the three days selected in March and November (night and missing values are shown with a value = -10 on the y-axis). DNI modeled using AOD_Ground is closer to measured DNI, yet with some discrepancies. In order to evaluate the comparison on a larger scale than the daily profiles shown here, we represent the whole month in a scatter plot in figure 10, where the measured DNI data are on the x-axis and the modeled DNI data are shown on the y-axis. The red line represents the one-to-one line between the measured and modeled DNI. As expected, the DNI modeled using the AOD_Ground shows good agreement with the measured DNI (plot on the top). An overestimation is observed for high DNI values; this can be explained by the fact that when modeling DNI in equation 1, only AOD is considered, whereas other constituents such as water vapor are present in the atmosphere and can attenuate further the solar radiation when taken into account in this calculation. For the modeled DNI using AOD-CAMS, DNI is clearly overestimated (bottom plot), with a large dispersion of the data during this month.

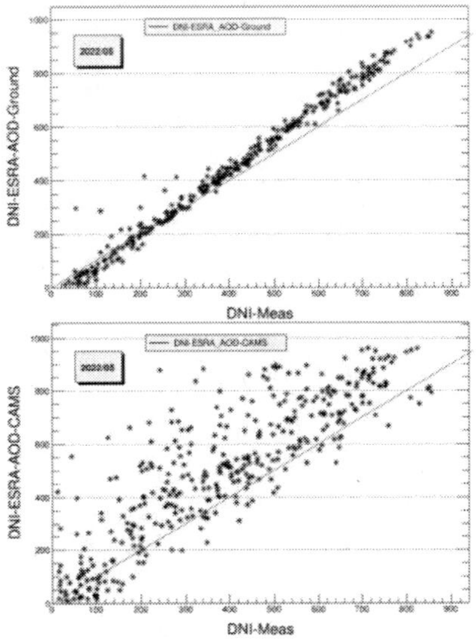

Figure 10: Correlation between measured and modeled DNI for the month of May 2022.

5 SUMMARY

Atmospheric aerosols are one of the main components that drive the variability of solar radiation, under cloud-free sky conditions. In this contribution, we describe the use of a ground-based estimation of AOD with a CIMEL sun photometer, deployed in Doha, Qatar. The sun photometer measures the spectral irradiances in 9 channels, and by using the atmospheric physics equation, it can provide an estimation of the optical properties of the aerosols including AOD. We also used the AOD products derived from the CAMS database, and the AOD derived from the WRF-Chem model to compare with AOD measured by the sun photometer. For validation, AOD data were used as inputs to clear-sky models to estimate the DNI component of the solar radiation and compare with ground measured DNI data.

The AOD values during a period of one year (excluding August due to ground data unavailability) as measured by the sun photometer have a peak around 0.5, though with high variability. AOD values higher than one are observed during months where dust events are more frequent, and even higher values are reached during days with dust storms. CAMS AOD data are not in full

10.4229/EUPVSEC2023/4CO.9.4
020364-005

agreement with the ground data and need to be calibrated before using them to model solar irradiance in regions with similar atmospheric conditions. Preliminary results of AOD derived by WRF-Chem are also presented showing high discrepancy with ground data. The WRF-Chem will be reconfigured with new parameters reflecting the atmospheric conditions seen in Doha, and calibrated with ground data with the ultimate goal to use the AOD forecast operationally for several applications including solar power plant solar or soiling forecasting.

6 REFERENCES

[1] Levy, R.C., Mattoo, S., Munchak, L.A., Remer, L.A., Sayer, A.M., Patadia, F. and Hsu, N.C., 2013. The Collection 6 MODIS aerosol products over land and ocean. Atmospheric Measurement Techniques, 6(11), 2989-3034. DOI:10.5194/amt-6-2989-2013

[2] Kahn, R.A., Gaitley, B.J., Martonchik, J.V., Diner, D.J., Crean, K.A. and Holben, B., 2005. Multiangle Imaging Spectroradiometer (MISR) global aerosol optical depth validation based on 2 years of coincident Aerosol Robotic Network (AERONET) observations. Journal of Geophysical Research: Atmospheres, 110, D10S04. DOI:10.1029/2004JD004706, 2005

[3] Grell G.A., Peckham, S.E., Schmitz, R., McKeen, S.A., Frost, G., Skamarock, W.C., Eder. B., 2005. Fully coupled 'online' chemistry in the WRF model. Atmos. Environ., 39, 6957-6976.

[4] Bey, I., Jacob, D., Yantosca, R., Logan, J., Field, B., Fiore, A., Li, Q.B., Liu, H., Mickley, L., Schultz, M., 2001. Global Modeling of Tropospheric Chemistry with Assimilated Meteorology: Model Description and Evaluation. Journal of Geophysical Research: Atmospheres, 106. DOI:10.1029/2001JD000807.

[5] Gschwind, B., Wald, L., Blanc, P., Lefèvre, M., Schroedter-Homscheidt, M., Arola, A., 2019. Improving the McClear model estimating the downwelling solar radiation at ground level in cloud-free conditions – McClear-v3. Meteorol Z, 28, 147-163. Doi: 10.1127/metz/2019/0946, 2019.

[6] Perez-Astudillo, D., Bachour, D., Sanfilippo, A., Al-Hajri H., 2022. Management and Operation of Qatar's Solar Radiation Monitoring Network. EuroSun 2022, ISES Conference Proceedings 2021. DOI:10.18086/eurosun.2022.15.06

[7] Fountoukis, C., Mohieldeen, Y., Pomares, L., Gladich I., Siddique, A., Skillern, A., Ayoub, M.A. 2022. Assessment of High-resolution Local Emissions and Land-use in Air Quality Forecasting at an Urban, Coastal, Desert Environment, Aerosol and Air Quality Research, 22 (6), doi: 10.4209/aaqr.220001.

[8] Bachour, D., Perez-Astudillo, D., Sanfilippo, A., 2022. Aerosol Optical Depth Estimation in Doha Using a Multifilter Rotating Shadowband Radiometer. In: Sustainable Energy-Water-Environment Nexus in Deserts. Advances in Science, Technology & Innovation. Springer, Cham. Doi: 10.1007/978-3-030-76081-6_61

[9] Gueymard, C., 1995. SMARTS2, A simple model of the atmospheric radiative transfer of sunshine: Algorithms and performance assessment. Technical Report FSEC-PF270-95, Florida Solar Energy Center.

[10] Ghan, S.J., Zaveri, R.A., 2007. Parameterization of optical properties for hydrated internally mixed aerosol. Journal of Geophysical Research 112, D10201. DOI: 10.1029/2006JD007927, 2007.

[11] Fountoukis, C., H. Harshvardhan, I. Gladich, L. Ackermann, M.A. Ayoub, Anatomy of a severe dust storm in the Middle East: Impacts on aerosol optical properties and radiation budget, Aerosol and Air Quality Research, 20, 155-165, 2020, doi: 10.4209/aaqr.2019.04.0165.

[12] Rigollier, C., Bauer, O., Wald, L., 2000. On the clear sky model of the ESRA - European Solar Radiation Atlas - with respect to the heliosat method. Solar Energy, 68, 1, 33-48. DOI: 10.1016/S0038-092X(99)00055-9.

10.4229/EUPVSEC2023/4CO.9.4
020364-006

EUPVSEC
European Photovoltaic Solar Energy Conference and Exhibition

Ciências ULisboa

INSTITUTO DOM LUIZ

Raising the limits to PV penetration:

Why we need battery swapping for the future energy and transport systems

António Vallêra

Universidade de Lisboa (IDL, Ciências)

References:

- *The transition: Why we need battery swapping for the future energy and transport systems*, A.M. Vallera, Universidade de Lisboa, Faculdade de Ciências, Instituto Dom Luiz, Lisboa, Portugal, 2023, ISBN 978-972-9348-24-2, https://doi.org/10.56526/10451/55274
- *Why we need battery swapping technology*, A.M. Vallera, P.M.Nunes, and M.C. Brito, Energy Policy 157, October 2021, 112481, https://doi.org/10.1016/j.enpol.2021.112481

1

020365-001

40th European Photovoltaic Solar Energy Conference and Exhibition

Abstract

The central problem of most future electric systems, with increasing penetration of wind and solar PV, is grid balance. Grid balance limits such non-dispatchable generation to relatively low levels of penetration, because higher levels require curtailment and such massive investments in grid-balancing technologies, such as storage and long-distance import/export, that the global cost becomes too high.

In this work, we start by stating that decarbonization of road mobility will have such an impact on the electric system that road mobility becomes crucial for its future design. We therefore define possible road transport models

1. ICE, or BAU - vehicles run on fossil fuels, the base model
2. Plug-in - all vehicles are battery electric, and refuel by connecting to a charger
3. Hydrogen - vehicles are powered by fuel cells running on electrolytic hydrogen
4. BSwap - all vehicles are battery electric, but refuel by swapping their low-charge batteries for charged ones at battery swapping stations

and test their impact on the electric system.

The result: the (overlooked, dismissed, understudied) BSwap model outperforms all others in ease of vehicle use, grid stability, wind and PV penetration, and cost. The main reason: centralized battery charging turns transport demand into a fully flexible load, with enough energy and power capacities to absorb all "excess" generation solar or wind peaks. Additionally, 2nd-life battery storage brings economic benefits (e.g., import/export away from peak demand/generation, at more favourable prices) and grid security. Large scale storage is much reduced, just as the need for high capacity cross-border interconnections. Most of the investment is paid for by transport, and the cost of electricity with BSwap is cheapest. And cheap electricity favours decrbonization in other sectors.

This is exemplified and discussed at the level of (i) a national system and of (ii) a small region where PV is the only viable energy resource. In this last case, we demonstrate that PV penetration, with a BSwap model, is ~3x that of a BAU model, and ~60% over that allowed for Plug-in, with a substantial gain in global cost.

2

020365-002

Outline:

A. A national system

1. The inspiration: carbon neutrality in Portugal 2050
2. The central problem of a system dominated by solar and wind generation
3. The impact of decarbonizing transport in a decarbonized electric system: the outcomes of different road transport models (ICE, Plug-in, H2 and BSwap)

B. A small region

- Served by the national transport grid, with PV the only local, viable, renewable resource
- How much PV may be installed?
- How is this affected by different road transport models?

3

020365-003

40th European Photovoltaic Solar Energy Conference and Exhibition

Portugal: present decarbonization policy

Redução de emissões por setor até 2050

> Trajetórias de neutralidade traduzem-se na descarbonização profunda da produção de eletricidade, da mobilidade e transportes e dos edifícios nas próximas duas décadas (2020-2040)

> Os setores da indústria e da agricultura apresentam um potencial de descarbonização mais reduzido, contribuindo ainda assim com reduções significativas no período 2020-2050, com especial ênfase no período 2040-2050

> A gestão agroflorestal eficaz é fator determinante para o objetivo da neutralidade carbónica em 2050

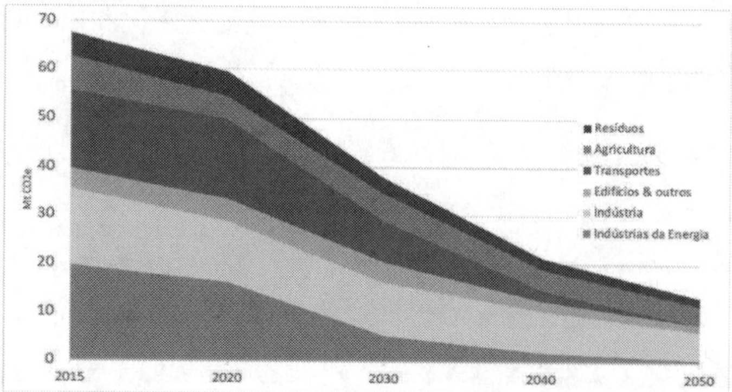

Gráfico da redução de emissões por setor até 2050

4

020365-004

Our focus:

- Power system
- Road Transport

- In a country aiming at the simultaneous decarbonization of both systems

- Portugal 2050 as an example of a future electric system dominated by solar PV (44%) and wind (39%) generation

5

020365-005

40th European Photovoltaic Solar Energy Conference and Exhibition

The energy matrix

		2015 TWh	2050 TWh
Consumption	**Total Consumption**	50	90
	General	50	70
	Mobility	0	20
Generation	Coal	14	-
	Gas	10	-
	Hydro	10	10
	Wind	12	37
	Solar PV	1	42
	WtE/Biomass	4	7
	Geothermal	0	0
	Wave	0	0
	Tidal	0	0
	Other	0	0

Main difference to RNC 2050: No gas
(for simplicity purposes only; retaining gas for some years makes sense)

6

020365-006

Generation is dominated by sun (~45%) and wind (~40%), which are plentiful and cheap:

3 to 7 cents/kWh

7

020365-007

40th European Photovoltaic Solar Energy Conference and Exhibition

- The central issue of the power system isn't anymore the (economic, environmental) cost of generation,
- The central issue is now

system balance

(a consequence of the dominance of solar and wind, variable and non-dispatchable)

Let us quantify this problem:

8

020365-008

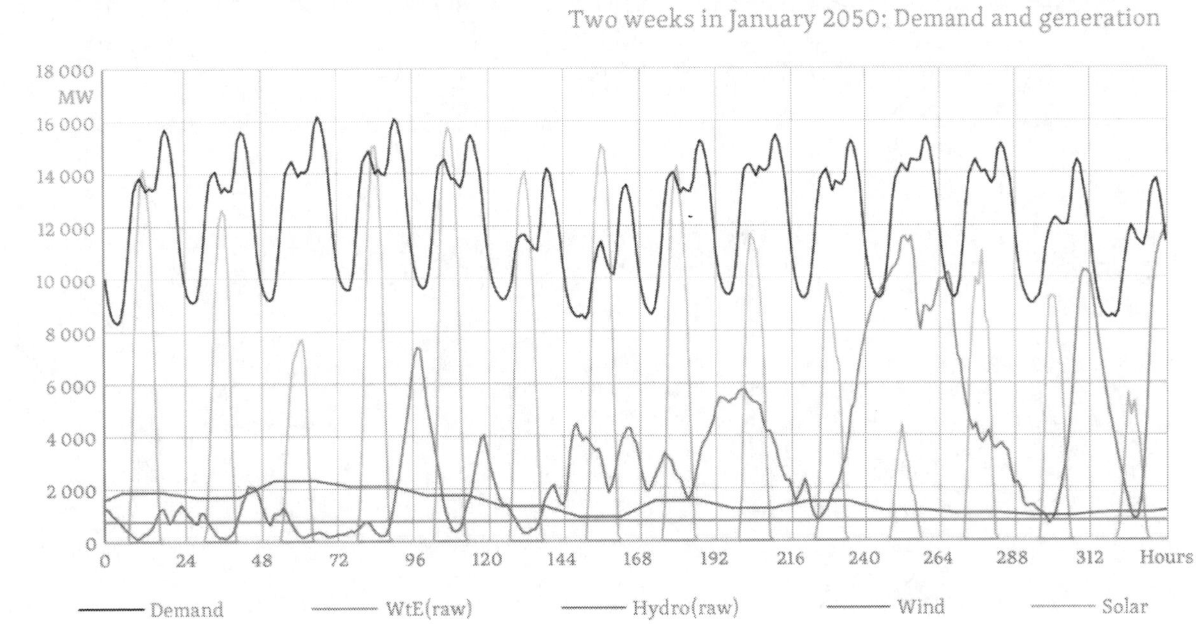

020365-009

40th European Photovoltaic Solar Energy Conference and Exhibition

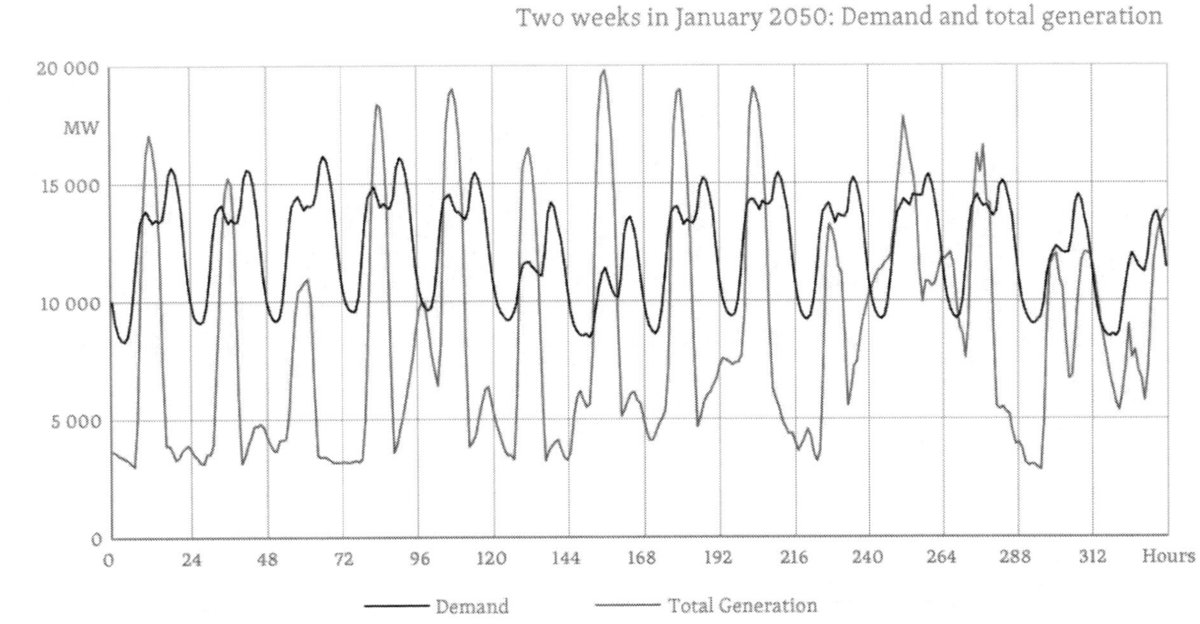

020365-010

We define an *imbalance* function as

Imbalance = Demand – Generation

which must be brought to zero at all times:

11

020365-011

40th European Photovoltaic Solar Energy Conference and Exhibition

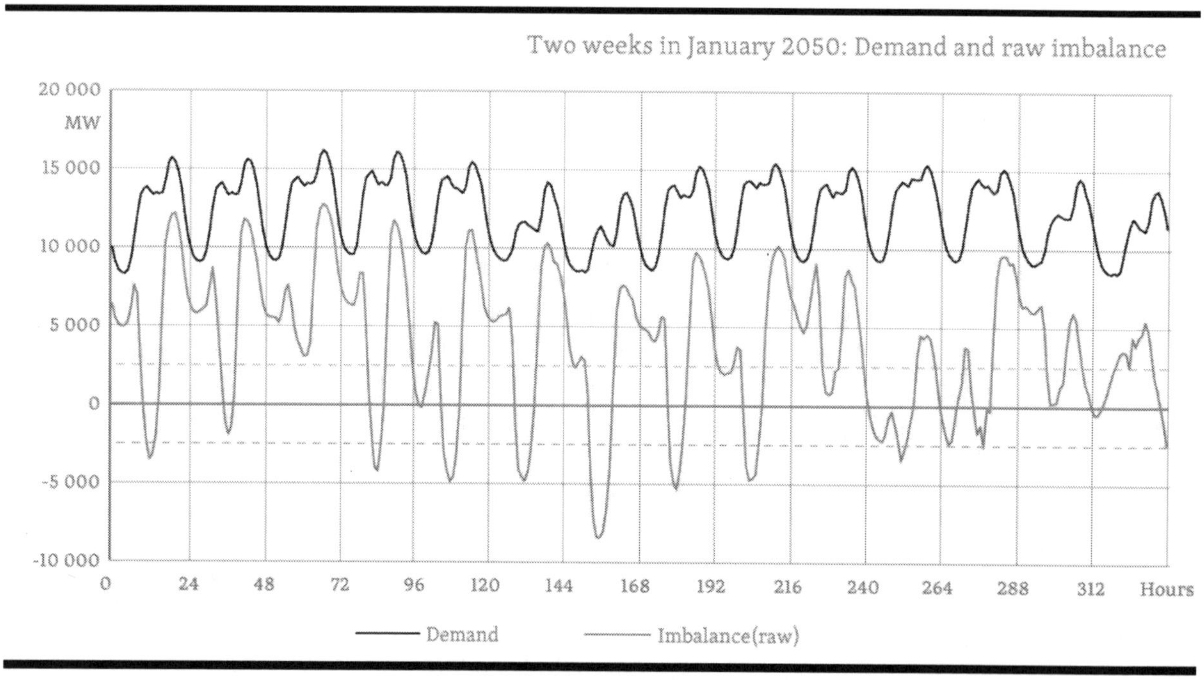

020365-012

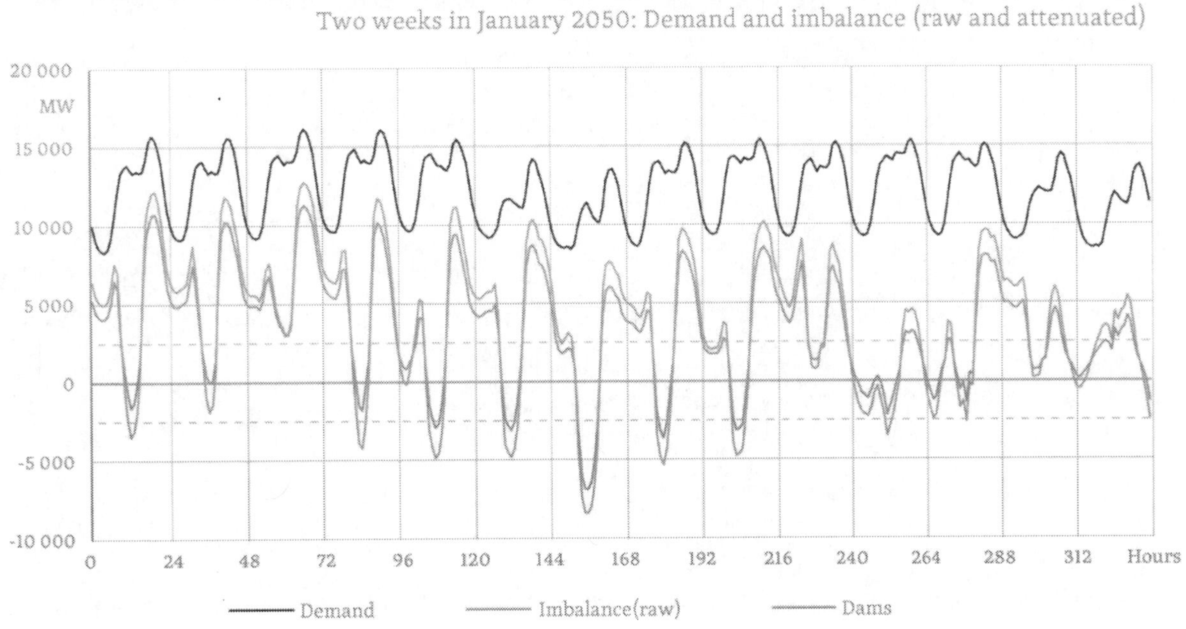

020365-013

40th European Photovoltaic Solar Energy Conference and Exhibition

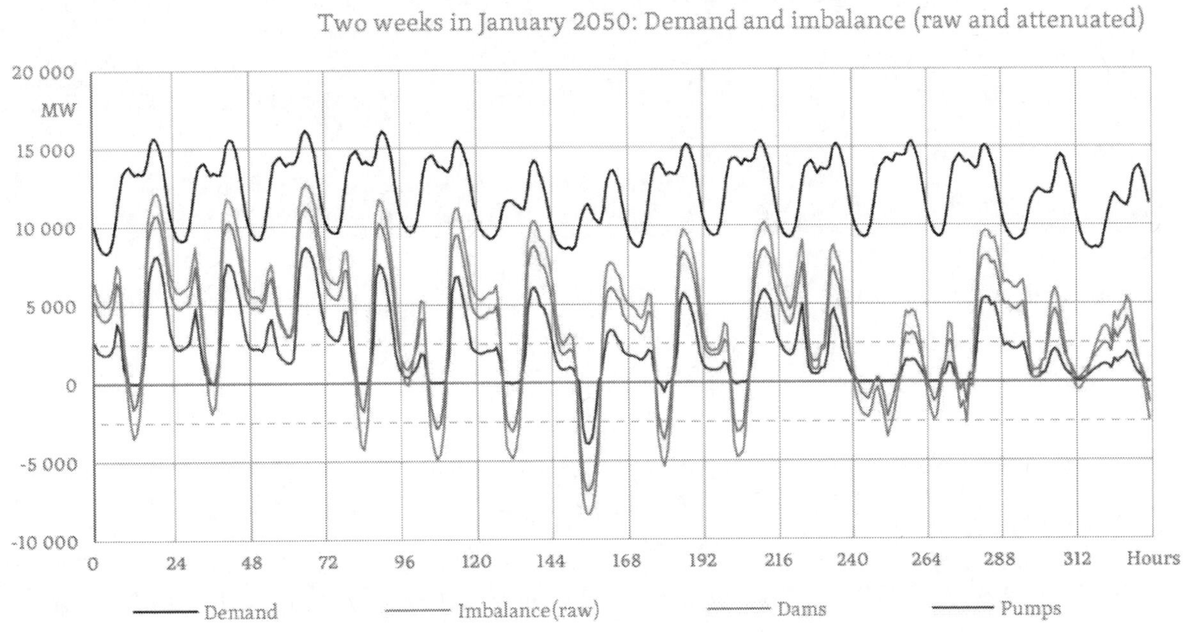

020365-014

Conclusion: After using all "classical" balancing means available, we still have a massive problem;

Since we do not solve the imbalance problem by trying to adapt generation to demand, why don't we look instead at demand flexibility, and try to adapt demand to generation?

Flexible demand possibilities:

- General demand management (*e.g.*, thermal systems): far too insufficient.
- Hydrogen: possible, but ... (How much H_2 would be produced? How much would it cost?)
- Road transport

15

40th European Photovoltaic Solar Energy Conference and Exhibition

The impact of decarbonization of road transport in the electric system:

1. Energy for road transport

	Energy spent on road transport TWh	Electric energy consumption TWh	(Energy consumed in road transport) / (Electric energy consumption) %
World	32 230	24 881	130 %
Portugal	65	48	137 %
EU	3 556	2 647	134 %
USA	5 981	3 961	136 %
China	4 467	6 753	66 %

Need to increase generation by ~60% to satisfy this additional demand - if we adopt efficient battery electric vehicles

16

EU: The impact of decarbonization of Eurupean trucks on the electric system

Ref: Transport and Environment, 2018

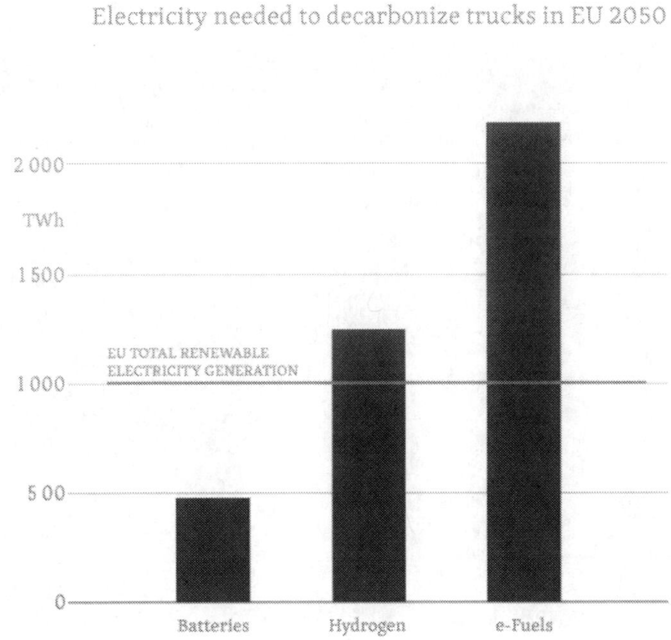

020365-017

40th European Photovoltaic Solar Energy Conference and Exhibition

The impact of decarbonization of road transport in the electric system:

2. Power

- Power of motors on wheels is ~100 times the average power of electric demand.
- With substitution of present vehicles by battery eletric vehicles, the power capacity of batteries would be about the same.
- If 10% of these batteries were grid connected, their nominal power capacity would still be ~10 times the average electric demand

18

020365-018

Road transport models studied:

1. **ICE** – Vehicles run on diesel or petrol
2. **Plug-in** – Vehicles are battery electric, and charge by connecting to a charger. Realistic charging flexibility and V2G are allowed.
3. **H2** – Vehicles are powered by fuel cells than run on electrolytic hydrogen.
4. **BSwap** – Vehicles are battery electric, and refuel by swapping their low-charge batteries by charged ones, at Battery Swapping Stations. Demand is fully flexible within battery residence time in BSS (*e.g.*, 24h). Storage and battery-to-grid are allowed with 2nd-life batteries.

19

020365-019

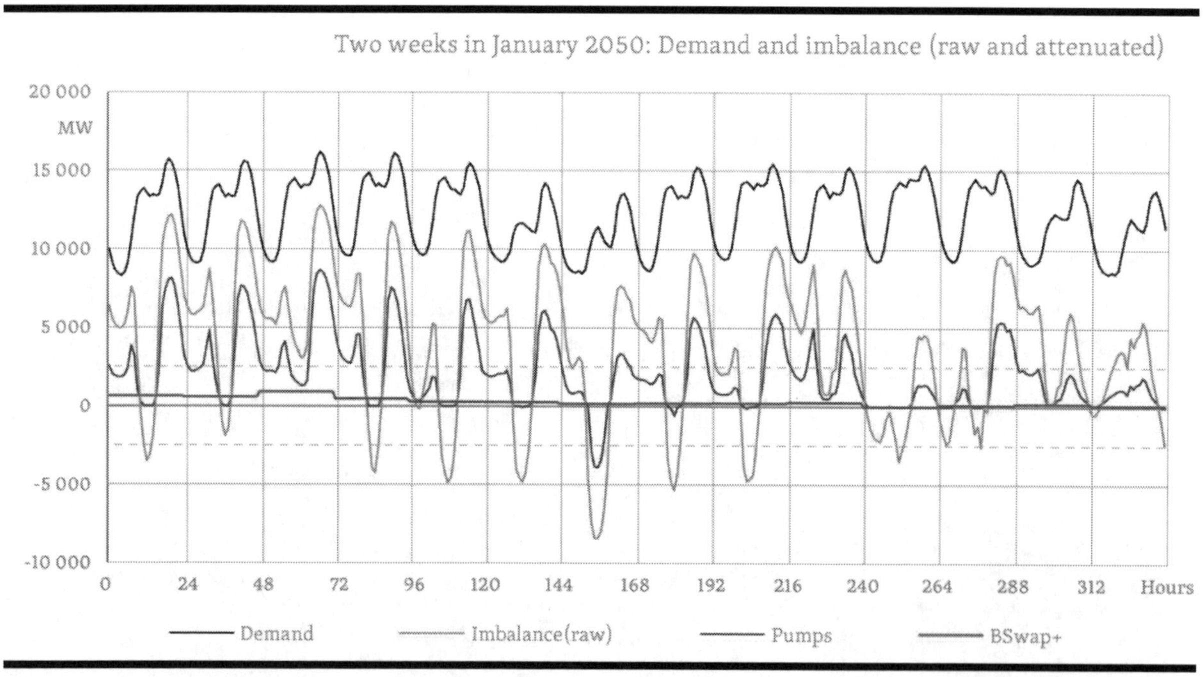

020365-020

Is this a miracle?!

No, simply the result of applying to road transport a Battery Swap model.

21

020365-021

40th European Photovoltaic Solar Energy Conference and Exhibition

The Battery Swap model:

- Vehicles refuel by swapping their low-charge batteries for charged ones at Battery Swapping Stations (BSS).
- Discharged batteries are inserted into chargers, and reside in the BSS for *e.g.* 24h, and so are charged when most convenient (for grid balance and lowest price).
- A large fraction of demand (20-30%) becomes flexible, and its power capacity is sufficient to absorb all "excess" wind or solar peaks. Demand flexibility is the main cause of the improvement in grid balance.
- 2nd-life battery storage is a second order effect, but important for economic and for security of supply reasons.
- Together with longer term hydro pumped storage, the result is a balanced grid. No need for further large-scale storage.
- The beauty of it: this benefit to grid stability is paid for by transport.
- The result: electricity will be cheap.

22

020365-022

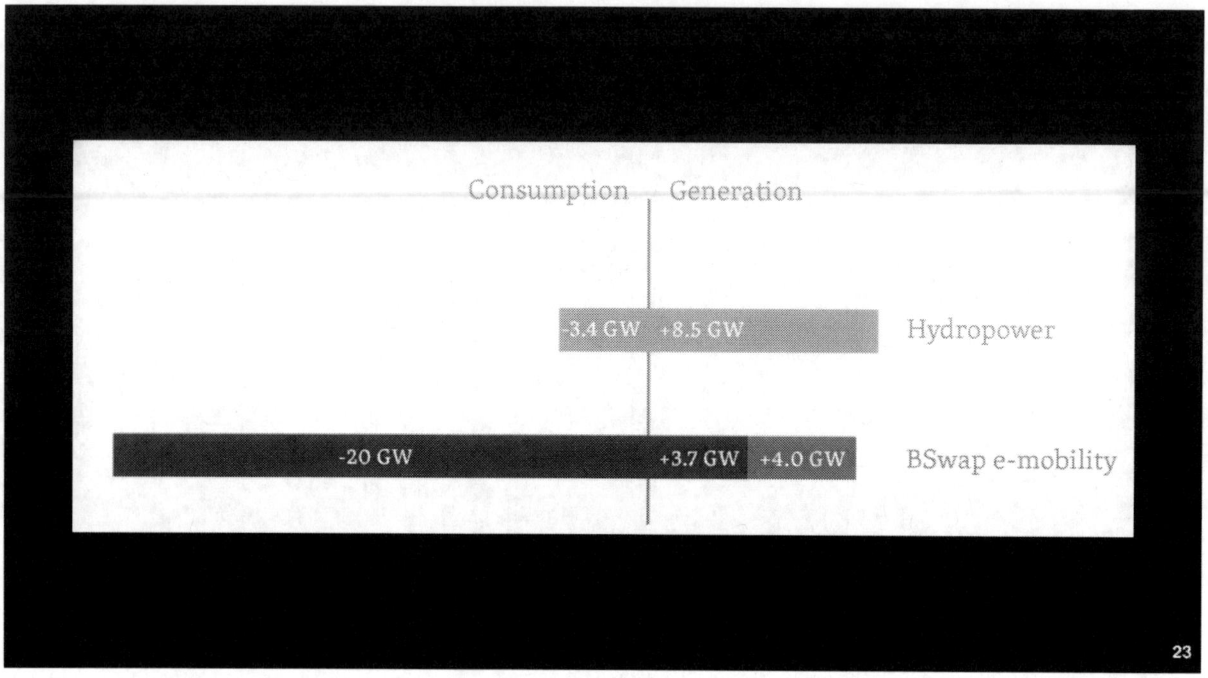

020365-023

40th European Photovoltaic Solar Energy Conference and Exhibition

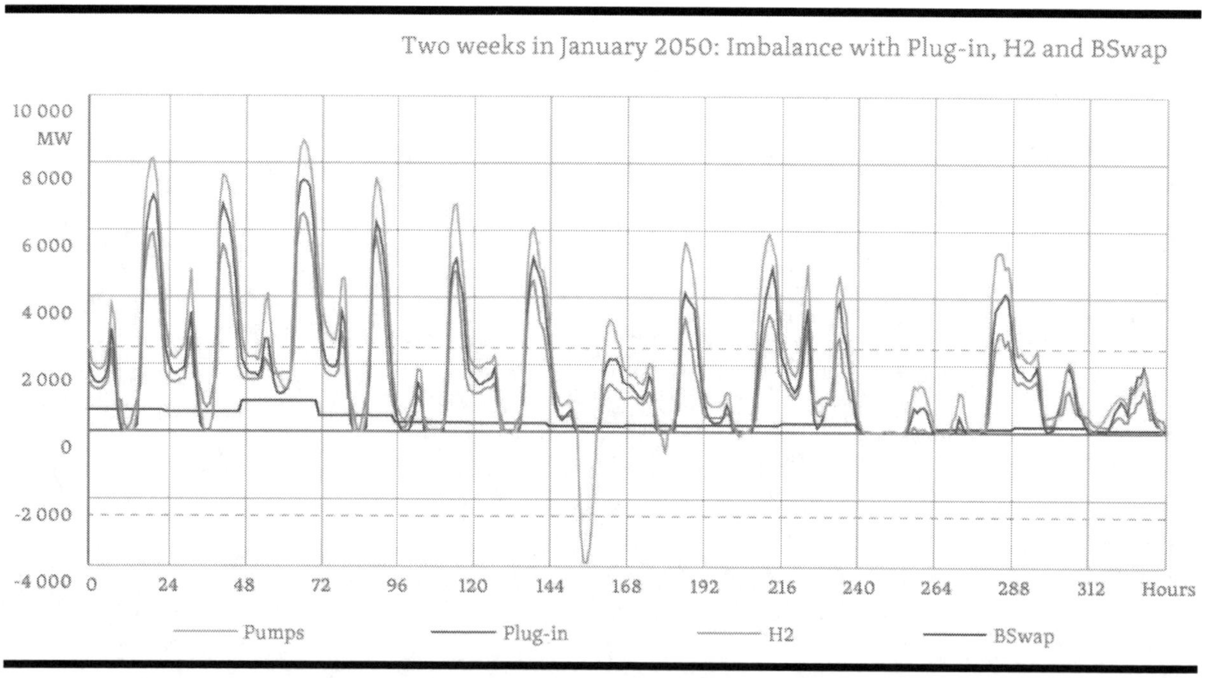

020365-024

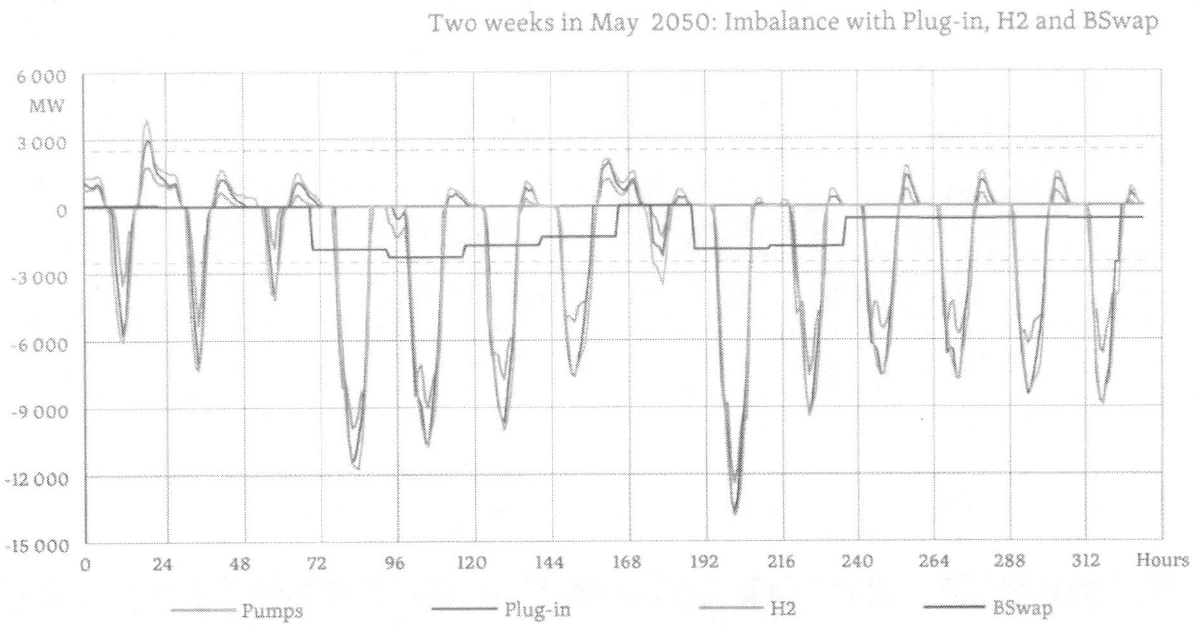

020365-025

40th European Photovoltaic Solar Energy Conference and Exhibition

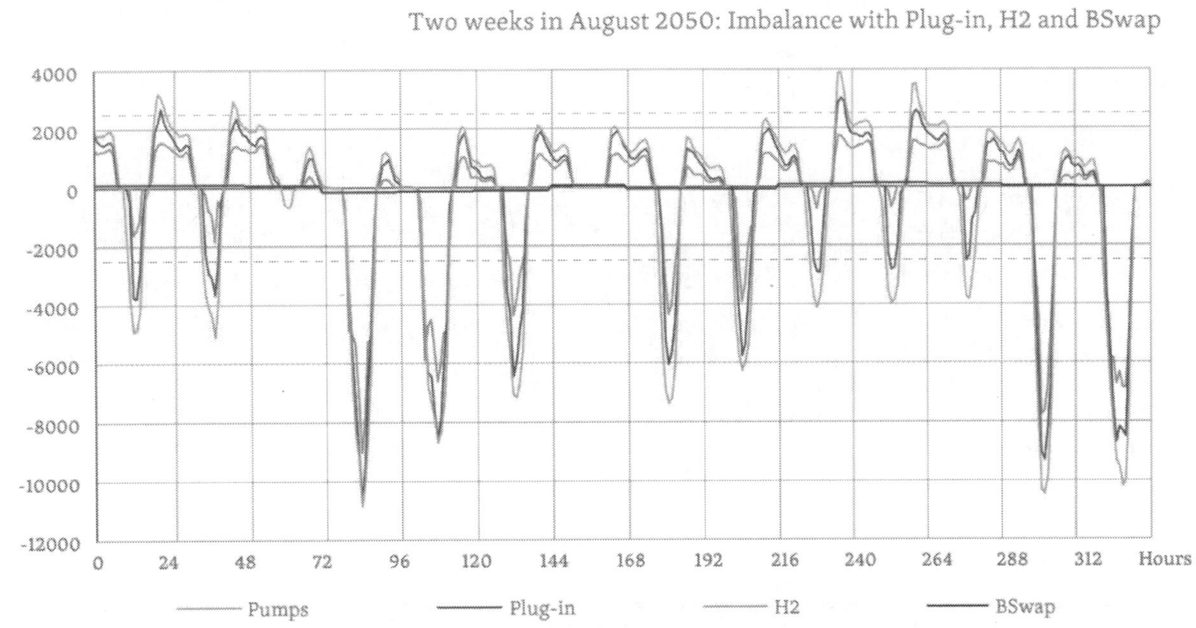

020365-026

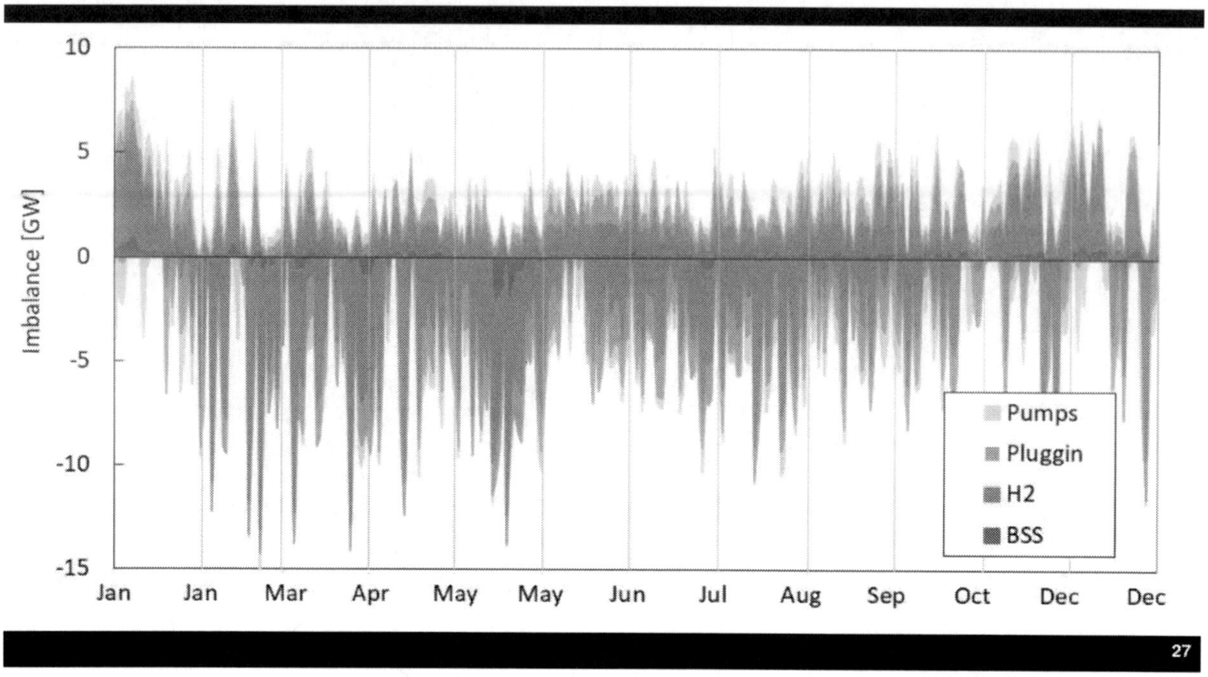

020365-027

40th European Photovoltaic Solar Energy Conference and Exhibition

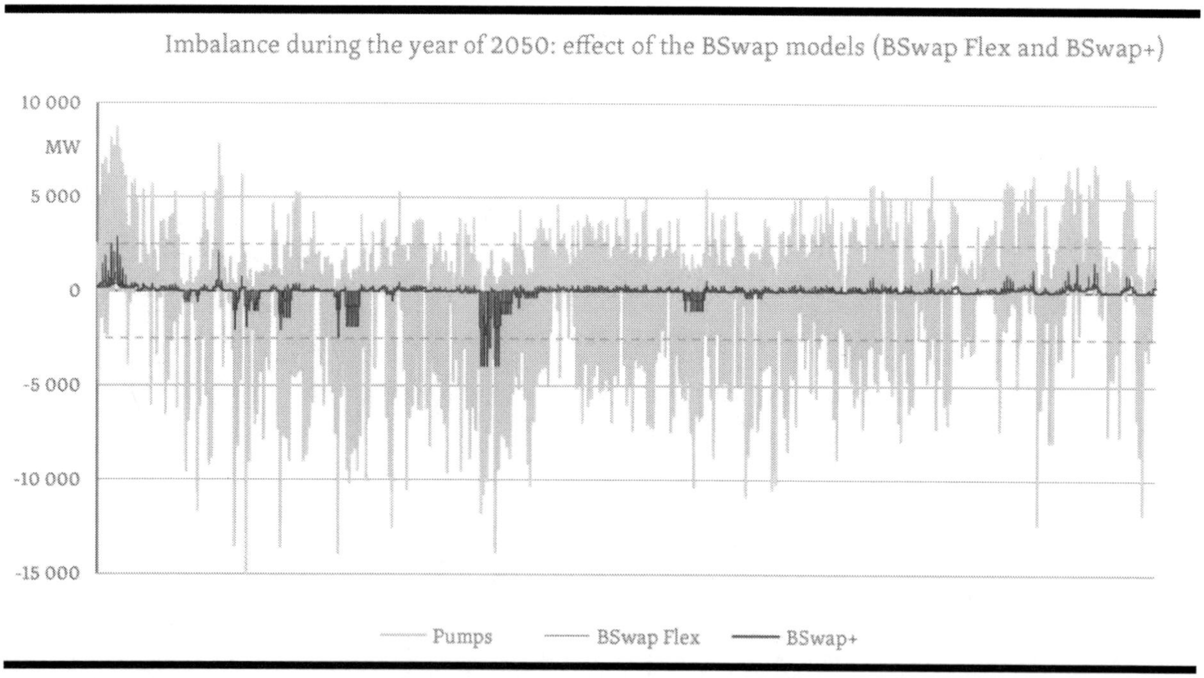

020365-028

Conclusion:

Intelligent satisfaction of road transport demand solves the imbalance problem.

29

020365-029

What about **costs**?

30

020365-030

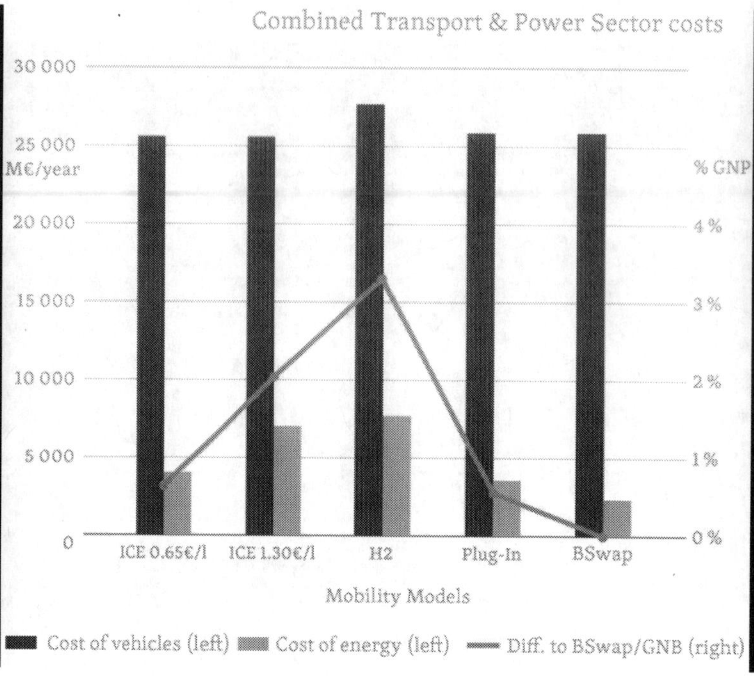

31

020365-031

With the Battery Swapping model

- The electric system becomes essentially balanced
  - No need to curtail, or sell at very low prices, a large fraction of solar and wind generation; nor to buy massively at peak demand times
  - No need for further large-scale storage
  - The need for high capacity cross-border connections is limited
- The grid becomes resilient
- Electrification of heavy vehicles is easily achieved
- Acceleration of transport electrification occurs
- Low cost electricity from a stable electric system encourages electrification of many other human activities

32

020365-032

Conclusion:

The 2 great structural changes in the future energy system by 2050,

- The dominance of non-dispatchable renewables,
  and
- Electric mobility

Taken together, are the solution to

- The total descarbonization of road transport,
  and
- The stability of the decarbonized power system

33

020365-033

40th European Photovoltaic Solar Energy Conference and Exhibition

B. A smaller region

With a population of, say, 200 000, served by the national electric grid, with

- General Demand of electric energy = 1TWh/year
- Fuel demand for transport = 0.72 TWh/year
- Transport decarbonization: electric energy demand for different transport models

Assume that PV is the only significant renewable electric energy resource.

Questions: How much PV may be economically installed?
How does this depend on road transport model?

34

020365-034

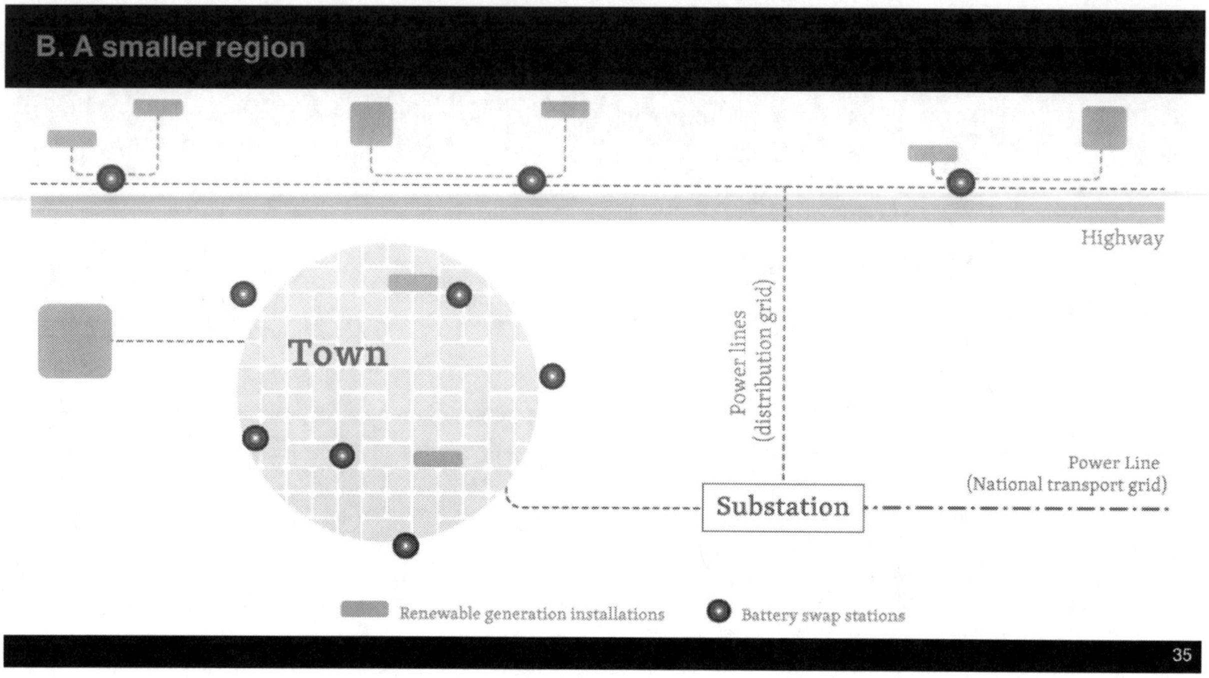

020365-035

40th European Photovoltaic Solar Energy Conference and Exhibition

Road Transport models considered:

1.	BAU (or ICE)	- Transport relies on fossil oil, mostly on diesel and petrol.
2.	Plug-in	- Vehicles are all battery electric, and charge by connecting a plug into the socket of a charger.
3.	Plug-in Flex + Storage	- " , with 10% demand flexibility and added 2nd-life battery storage.
4.	BSwap Flex	- Vehicles refuel by swapping their low-charge batteries by charged ones, at Battery Swapping Stations.
5.	BSwap Flex + Storage	- " , with added 2nd-life battery storage.
6.	BSwap Flex + double Storage	- " , with twice the previous storage capacity.

(A Hydrogen powered transport model was not considered because its cost was estimated as far too high.)

36

020365-036

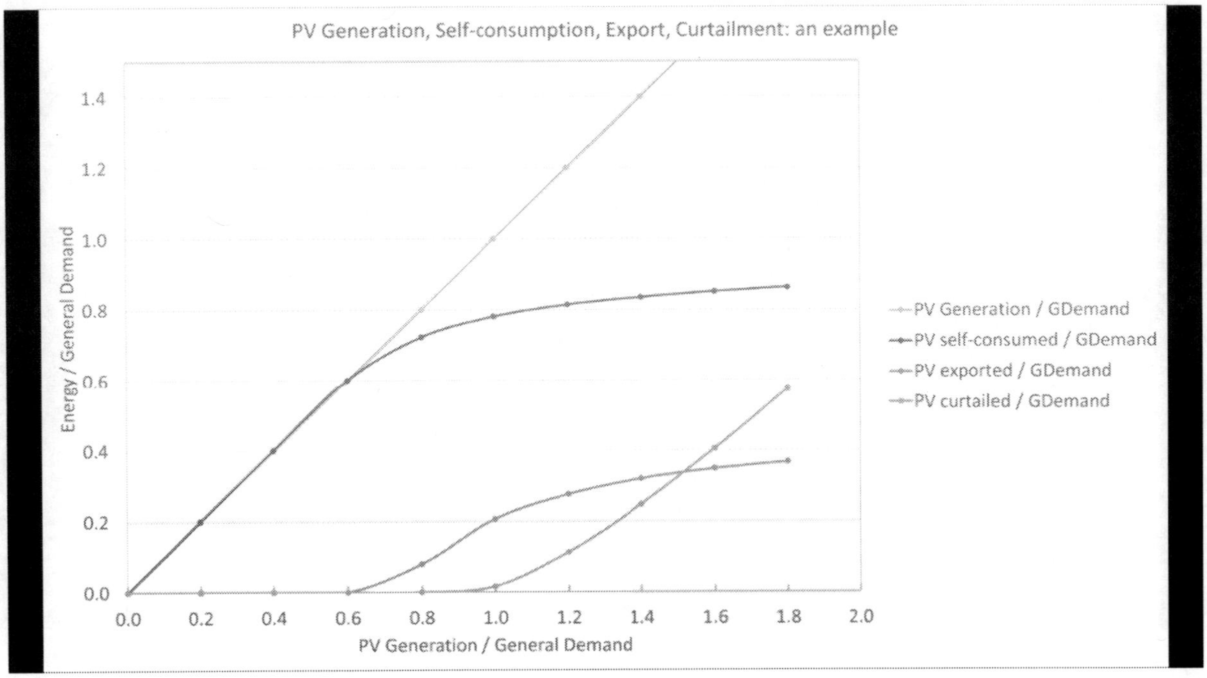

020365-037

40th European Photovoltaic Solar Energy Conference and Exhibition

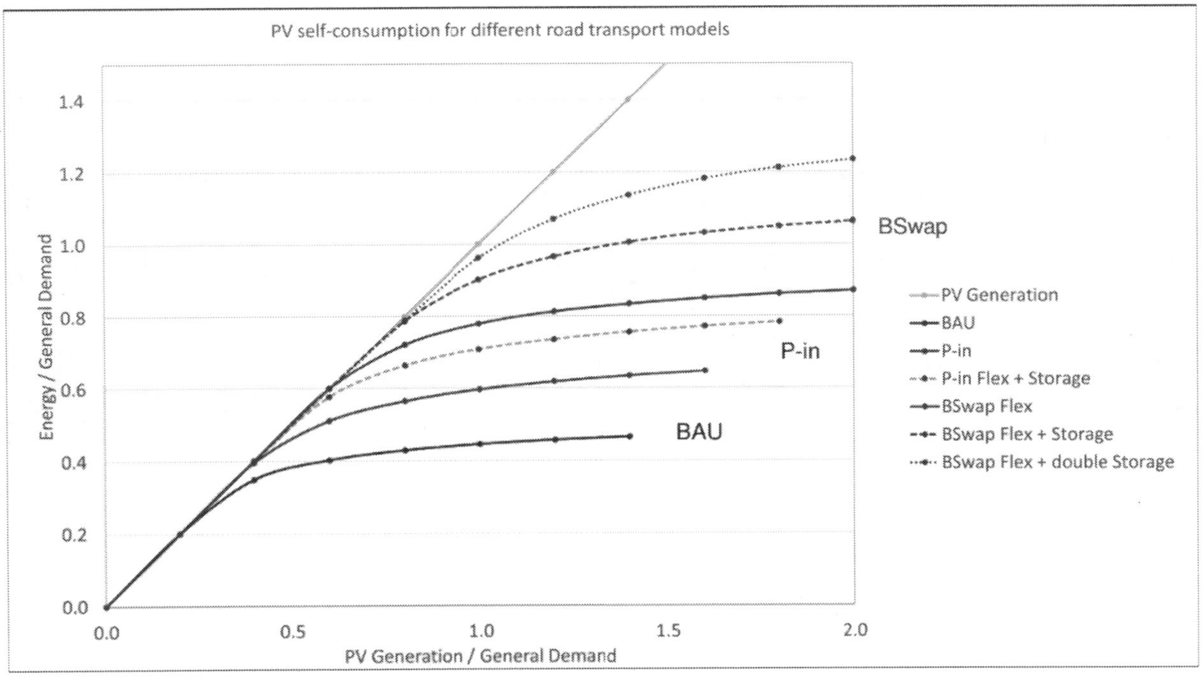

020365-038

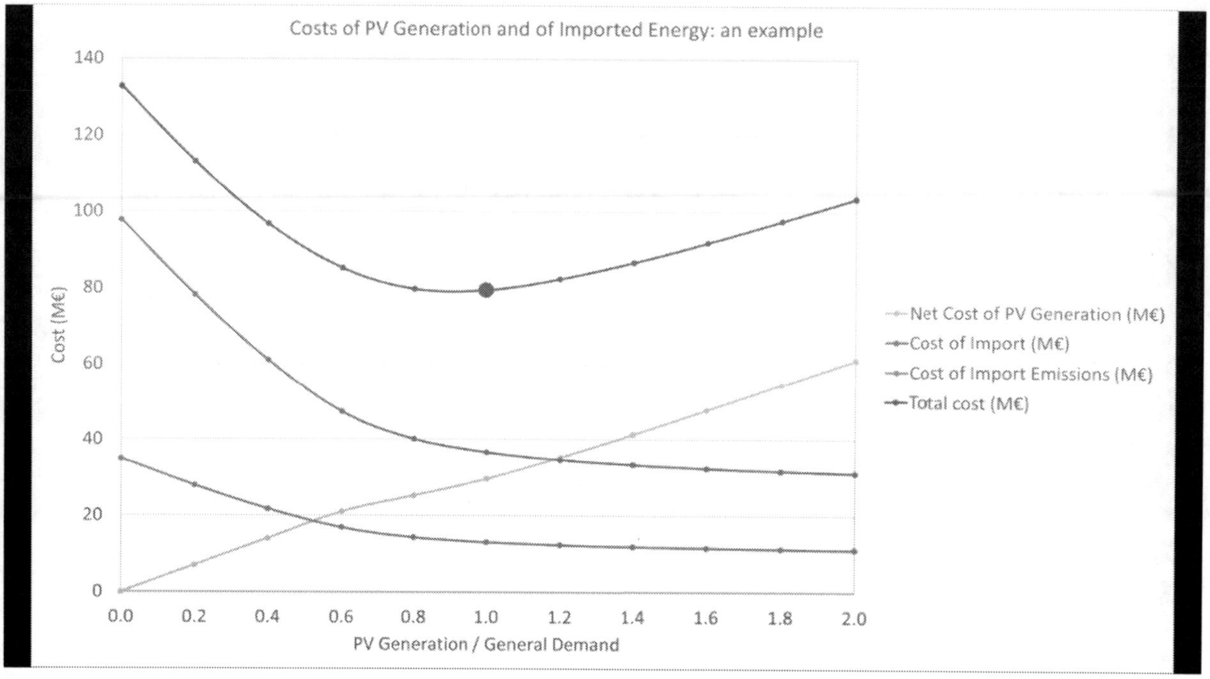

020365-039

40th European Photovoltaic Solar Energy Conference and Exhibition

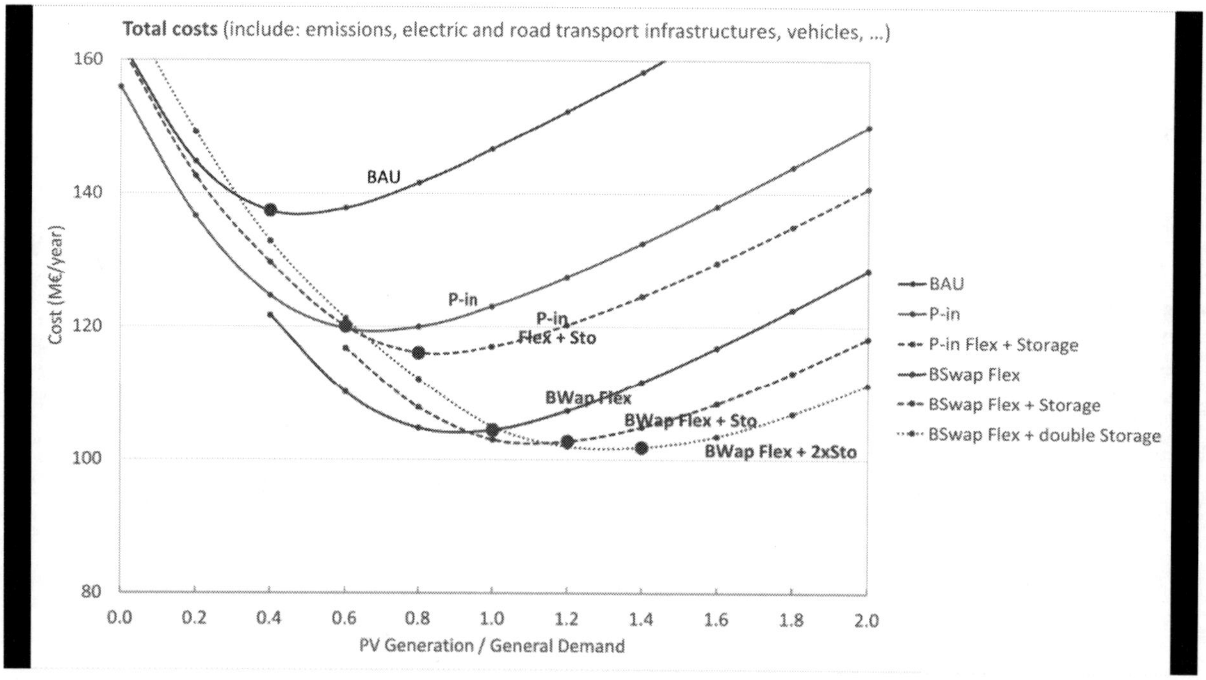

020365-040

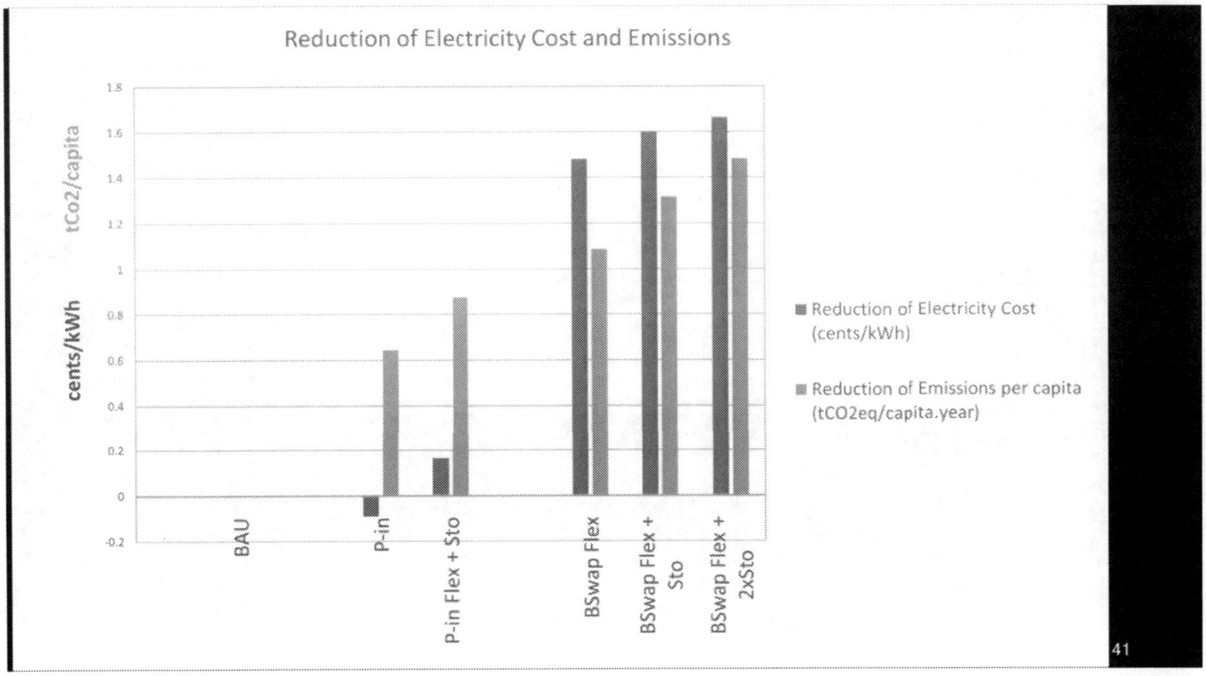

020365-041

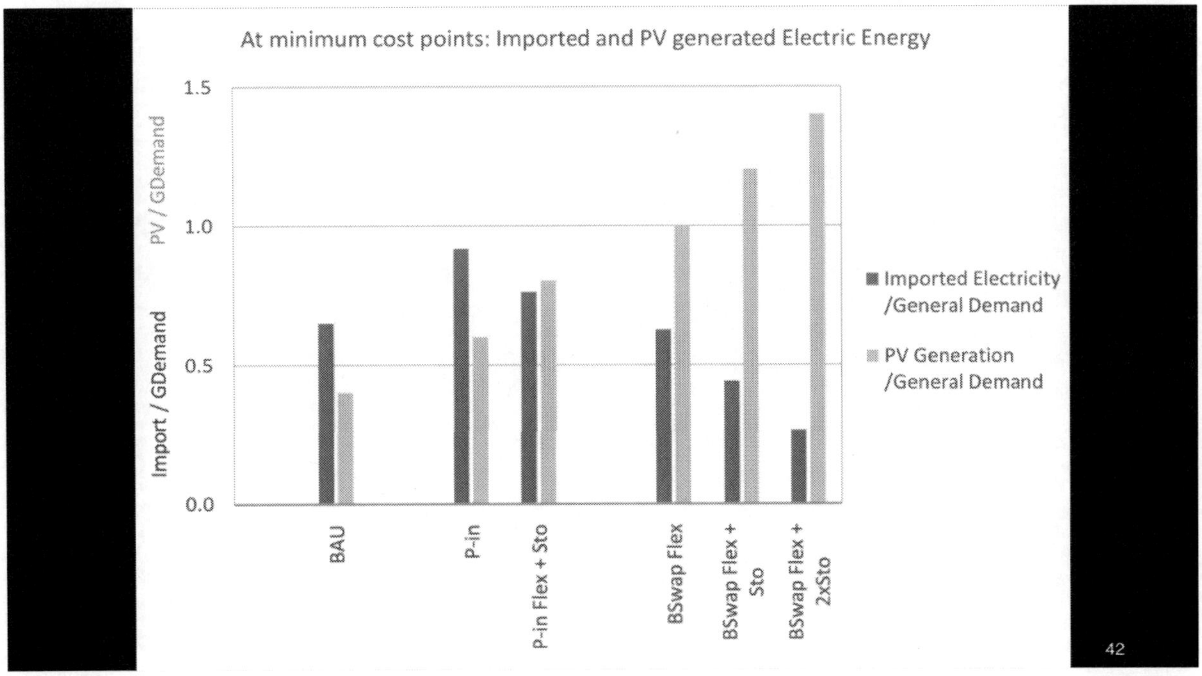

020365-042

Thank you

EUPVSEC
European Photovoltaic Solar Energy Conference and Exhibition

2023 4CP1_1 António Vallera, "*Raising the limits to PV penetration*"

43

020365-043

MAPPING THE RELEVANCE OF DIGITALISATION FOR PHOTOVOLTAICS

ETIP Photovoltaics

Pierre-Jean Alet, CSEM (chair)

Venizelos Efthymiou, University of Cyprus (vice-chair)

Giovanna Adinolfi, ENEA; Greg Arrowsmith, EUREC;
Grazia Barchi, EURAC; Gofran Chowdhury, 3E;
Saverio De Vito, ENEA; Giorgio Graditi, ENEA;
Jonathan Leloux, Lucisun; Ioannis Tsanakas, CEA

Thomas Garabetian, SolarPower Europe (secretariat)

020366-001

The European Technology and Innovation Platform for Photovoltaics

etip-pv.eu

WHAT IS "DIGITALISATION"?

ETIP Photovoltaics | csem

WHAT IS "DIGITALISATION"?

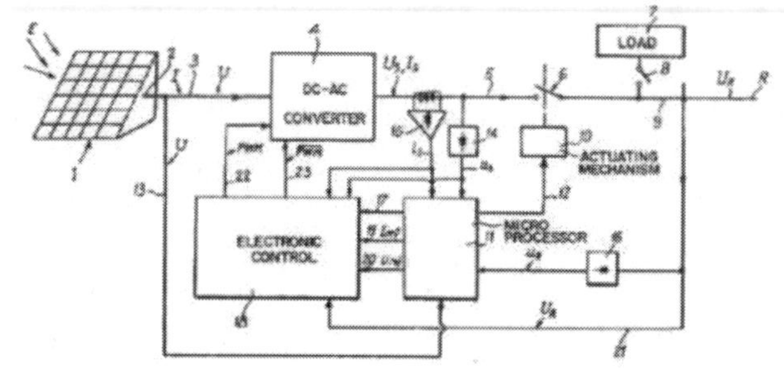

R. Corbefin and G. Vacelet, 'Process and system for producing photovoltaic power', US4390940A, Jun. 28, 1983

- **Digitisation** is the process of changing from analog to digital form, also known as digital enablement. Said another way, **digitization** takes an analog process and changes it to a digital form without any different-in-kind changes to the process itself.
- **Digitalisation** is the use of digital technologies to change a business model and provide new revenue and value-producing opportunities; it is the process of moving to a digital business.

Source: Gartner IT glossary

ETIP Photovoltaics | :: csem

40th European Photovoltaic Solar Energy Conference and Exhibition

WHAT IS "DIGITALISATION"?

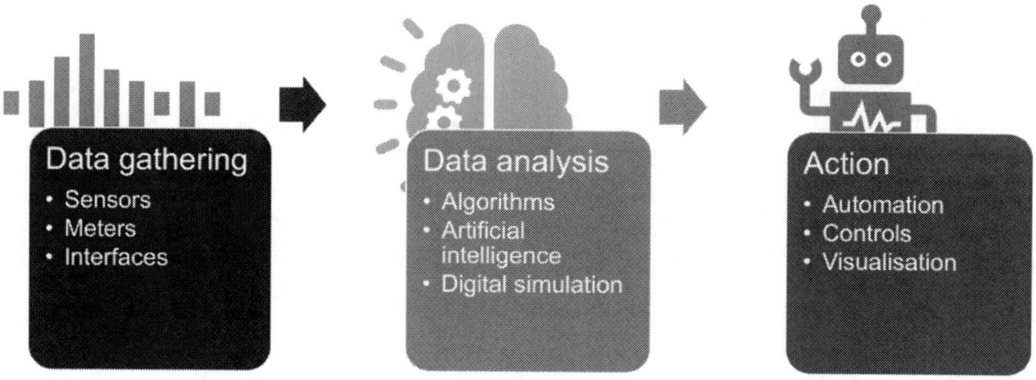

IEA (2019), Energy efficiency and digitalisation, IEA, Paris
https://www.iea.org/articles/energy-efficiency-and-digitalisation

ETIP Photovoltaics | :: csem

SCIENCE FICTION? MEET YOUR AI AGENTS

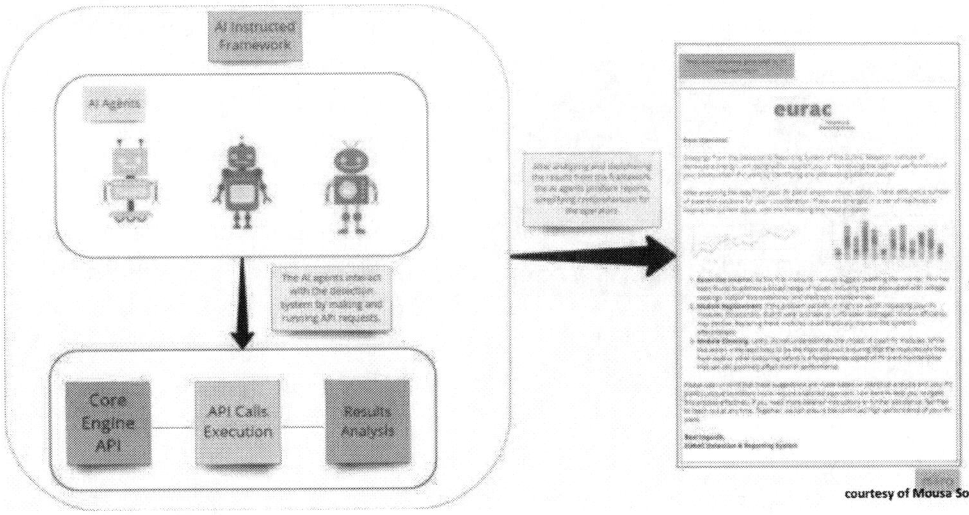

courtesy of Mousa Sondoqah, Eurac Research

eurac research | ETIP Photovoltaics | csem

DIGITALISATION: SHORT-TERM BENEFITS

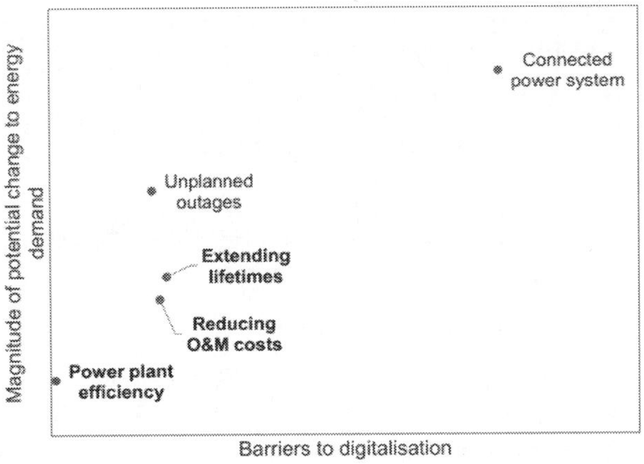

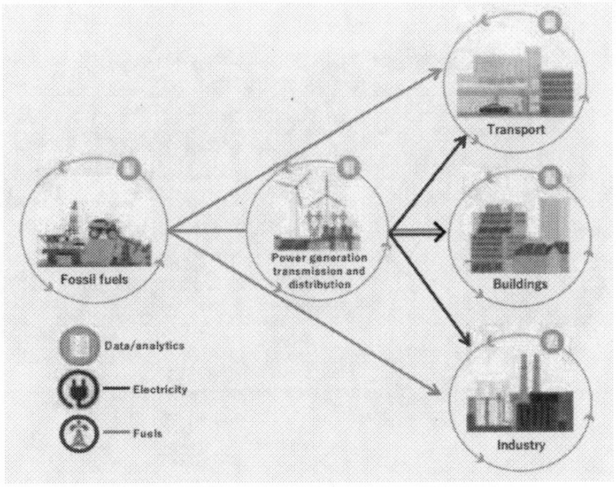

IEA (2017), Digitalisation and Energy, IEA, Paris https://www.iea.org/reports/digitalisation-and-energy

ETIP Photovoltaics | csem

MONITORING DATA LIMITATIONS

- Time series from inverters, instrumentation
- Insufficient spatial granularity, especially in utility-scale PV plants.
- At component / subsystem level:
  - Undetected losses or failures
  - Triggered false-negatives
  - Root-cause analysis practically impossible, time- and labour- intensive.
- Typically ~30% of PV plants still underperform, at global scale, according to 3E.

40th European Photovoltaic Solar Energy Conference and Exhibition

MONITORING DATA LIMITATIONS

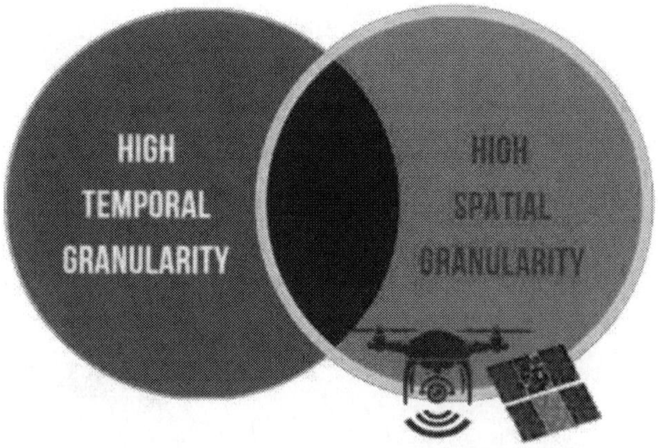

- Very high spatial granularity
- From system and string level, down to module and submodule/cell level :
  - Detection, diagnosis.
  - Root-cause analysis possible.
  - Time- and labour- efficient.
  - 1.5x up to 4x higher "diagnostic capacity"
- Limitations:
  - Snapshot assessment
  - Decoupled from PV monitoring → assessment rather than root-cause analysis
- Solution: **data fusion**

DATA FUSION AND AUGMENTATION

- Data augmentation with open-source data can increase accuracy of anomaly detection and classification algorithms
- Example:
  - OpenWeatherMap
  - Clear Sky Model
  - COPERNICUS:
    - infrared imaging
    - cloud cover
  - Passive reflectance at 550 nm → information on aerosol density

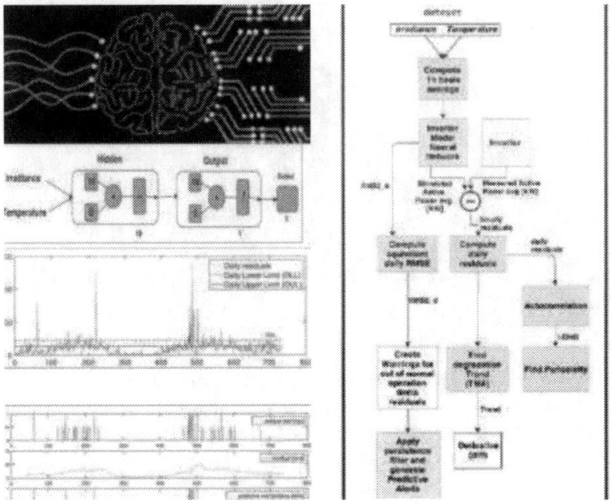

M. De Benedetti et a., Neurocomputing 310 (2018), 59-68

OVERCOMING MACHINE LEARNING BOTTLENECK DATA REPOSITORIES EXAMPLES

- Portugal: Observatório da Energia (observatoriodaenergia.pt). High-quality data on energy and PV
- IEA PVPS Task 13-ST2.5: PLR Determination Benchmark Study (https://osf.io/vtr2s/)
- France: BDPV (bdpv.fr) residential PV system database; 15k+ PV systems, 10+ years data
- Europe: Collaborative Platform for Simulation and Monitoring - COPLASIMON (coplasimon.eu)
- Europe: Pearl PV CKAN repository (ckan.pearlpv-cost.eu)

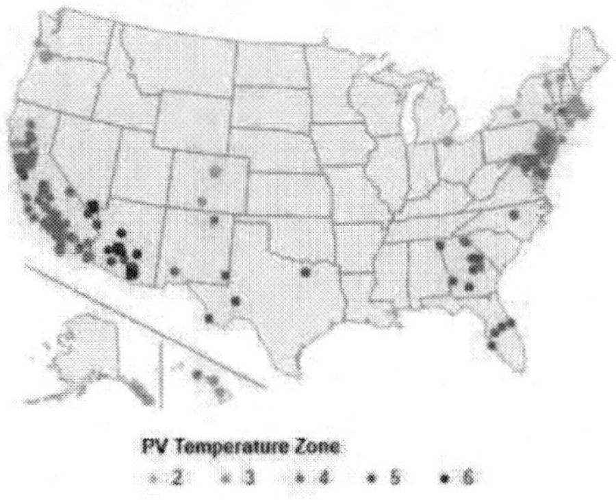

US PV Temperature Zone for each system, based on Karin et al. 2019 and pvcz 0.2.4

US: PV fleet performance data initiative (https://www.nrel.gov/pv/fleet-performance-data-initiative.html)

OVERCOMING GRID AND PLANNING BOTTLENECKS WITH OPEN DATA: SOURCES

- Power production units; Source: Swiss Federal Office of Energy, Swisstopo.
- Available: https://www.uvek-gis.admin.ch/BFE/storymaps/EE_Elektrizitaetsproduktionsanlagen

- Rooftop PV potential; Source: Swiss Federal Office of Energy, Swisstopo.
- Available: https://www.uvek-gis.admin.ch/BFE/sonnendach/index.html?featureId=4728178

40th European Photovoltaic Solar Energy Conference and Exhibition

OVERCOMING GRID AND PLANNING BOTTLENECKS WITH OPEN DATA: USE CASES

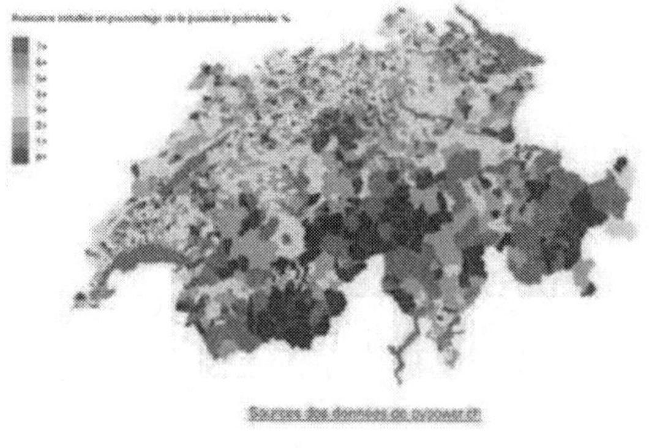

- Policy advocacy, benchmarking
- Comparison of installed power vs. potential per municipality

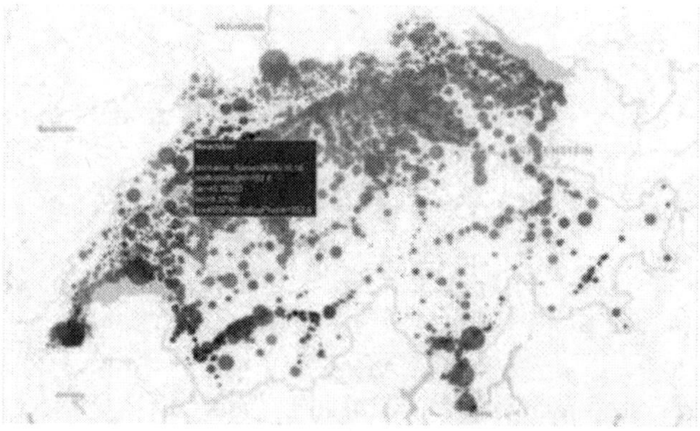

- Strategy, prospective studies
- Hour-by-hour simulation of energy balance by municipality in 2050 (source: CSEM)

ETIP Photovoltaics | :: csem

DATA SHARING FOR O&M: POWER IN NUMBERS

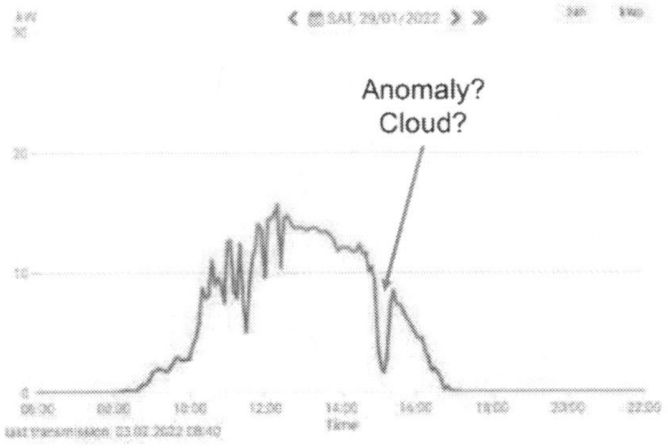

Data from single system: low value

40th European Photovoltaic Solar Energy Conference and Exhibition

DATA SHARING FOR O&M: POWER IN NUMBERS

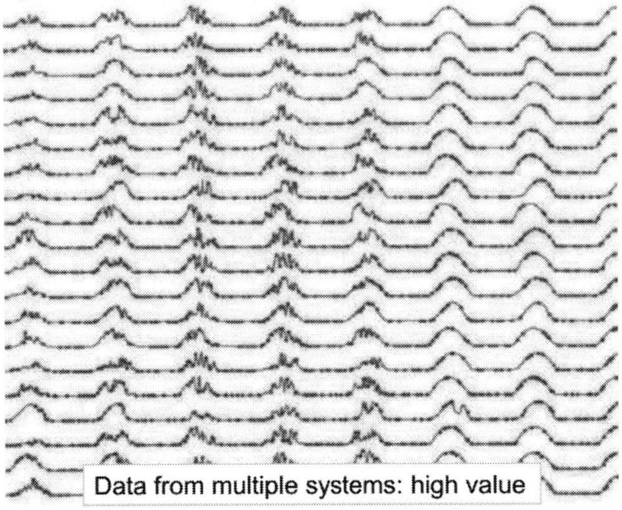

40th European Photovoltaic Solar Energy Conference and Exhibition

DATA SHARING FOR O&M: POWER IN NUMBERS

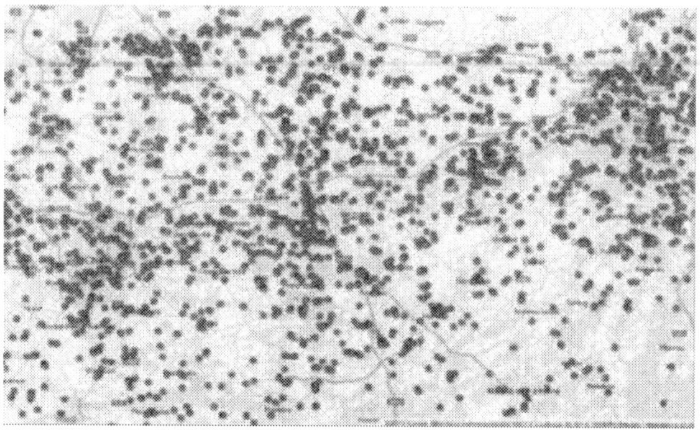

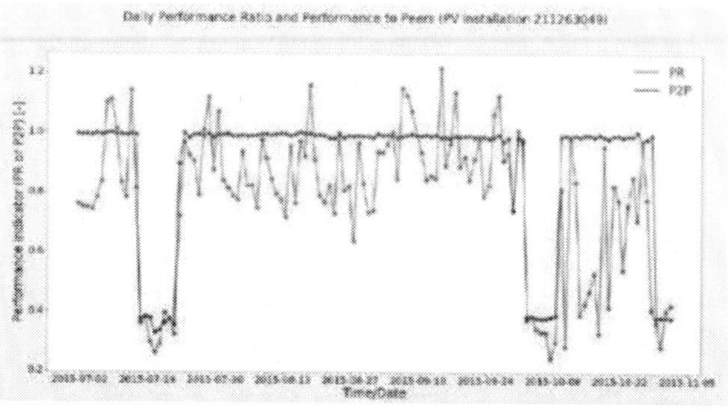

- Network of PV systems around Brussels
- Performance-to-peers indicator more stable and discriminating than performance ratio

J. Leloux, L. Narvarte, A. Desportes, and D. Trebosc, 'Performance to Peers (P2P): A benchmark approach to fault detections applied to photovoltaic system fleets', *Solar Energy*, vol. 202, pp. 522–539, May 2020, doi: 10.1016/j.solener.2020.03.015.

LUCISUN ETIP Photovoltaics csem

40th European Photovoltaic Solar Energy Conference and Exhibition

FROM DATA TO INSIGHTS: INVERTER PERFORMANCE

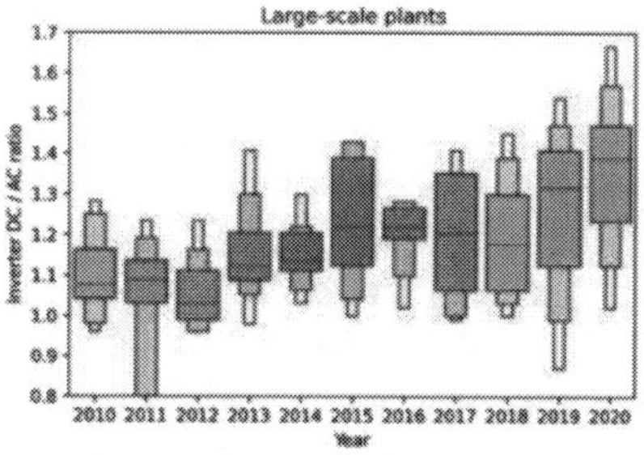

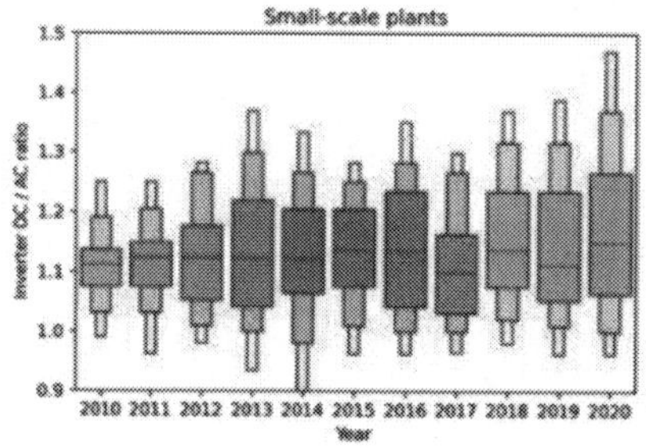

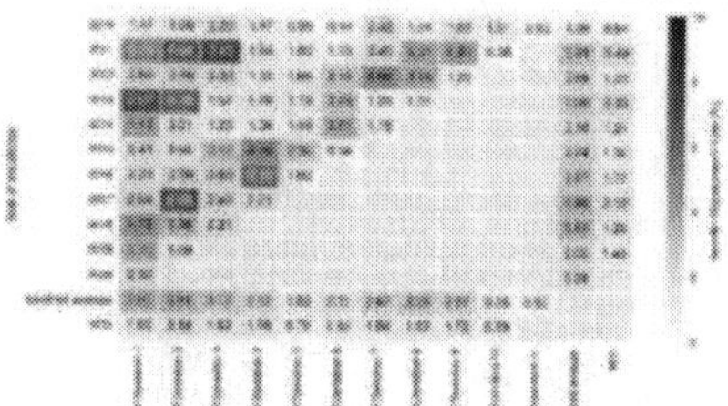

- DC/AC ratio: large-scale plants
- DC/AC ratio: small-scale plants
- Inverter replacement

Detailed report available on the TRUST-PV website: https://trust-pv.eu/reports/733/

TRUSTPV 3E ETIP Photovoltaics csem

MULTI-ACTOR TRUSTFUL DATA EXCHANGE

- Central approach in the EU's digital strategy: **data spaces**
- Core concepts:
  - Data sovereignty
  - Decentralisation
  - Connectors
- First deployment of data space for PV sector in the **PILATUS project**
  - Objective: total quality management and process improvement along entire supply chain
  - First four connectors deployed

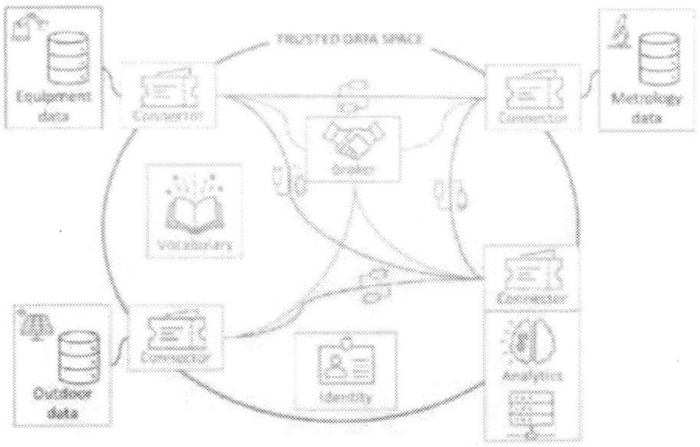

The PILATUS project is funded by the European Union under grant number 101084046. Views and opinions expressed are however those of the author(s) only and do not necessarily reflect those of the European Union. Neither the European Union nor the granting authority can be held responsible for them.
This work has received funding from the Swiss State Secretariat for Education, Research and Innovation (SERI)

40th European Photovoltaic Solar Energy Conference and Exhibition

ACCELERATING EUROPEAN FRAMEWORK

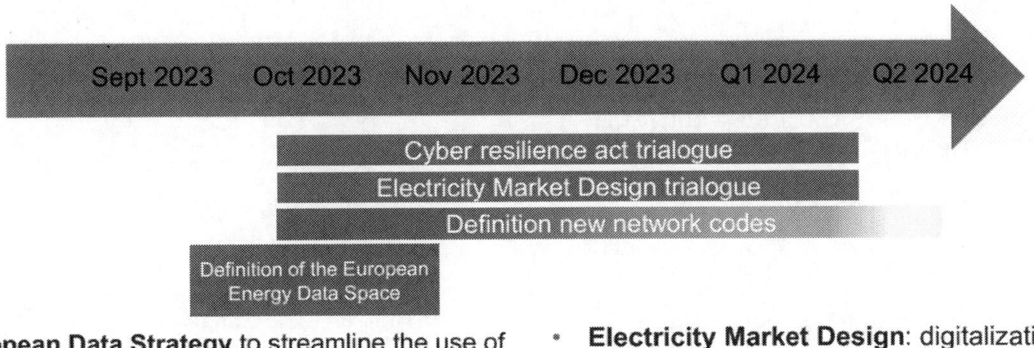

- **European Data Strategy** to streamline the use of data, define ownership and better allocate value from data production, enable new business, applications & processes
- **Cyber Resilience Act** defining key requirements for the PV industry, notably utility scale projects (e.g., localisation of data storage)
- **Electricity Market Design**: digitalization and consumers are at the heart of the revised EU laws on the electricity market
- **Energy Data Space**:
  - Facilitate energy data exchange, enable the energy transition "behind the meter", unlock flexibility
  - Expert group on energy data space by end of 2024 (focus likely on consumers/rooftop PV)

DIGITALISATION: LONG-TERM BENEFITS

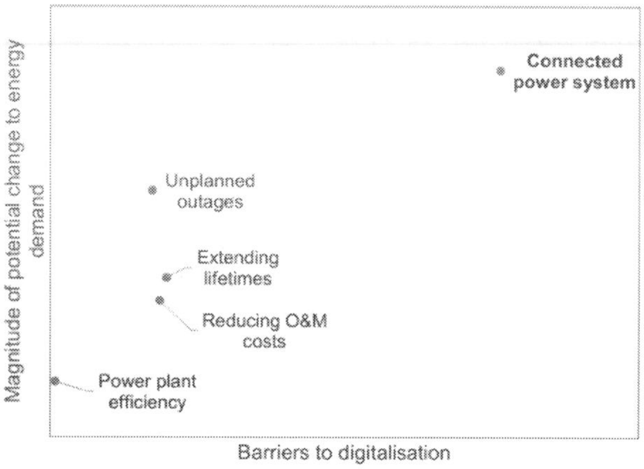

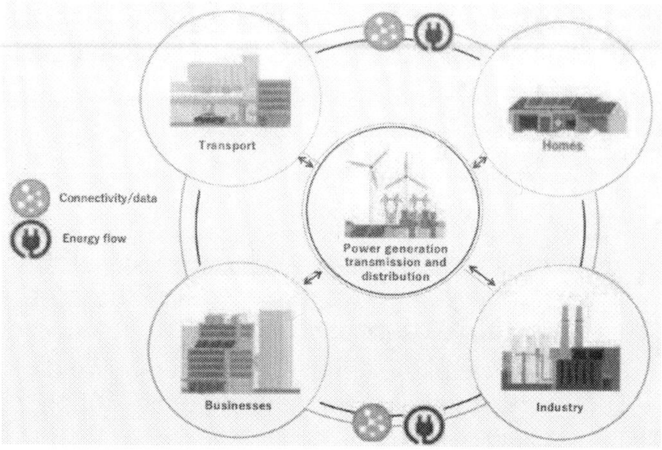

IEA (2017), Digitalisation and Energy, IEA, Paris https://www.iea.org/reports/digitalisation-and-energy

40th European Photovoltaic Solar Energy Conference and Exhibition

MANAGING ENERGY COMMUNITIES THROUGH SMART DATA SPACES

Energy Communities:

- Possible use cases facilitated through smart use of data
- Appropriate tools currently being deployed.
- Data providers:
  - Smart meter data from end users
  - Distributed generation data
  - Storage system data
  - EV charging data
  - DSO: Shareable system data
  - Publicly available weather data

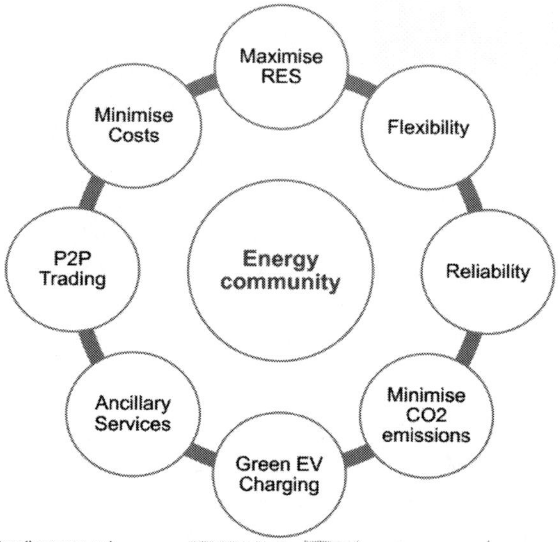

https://eneuron.eu/

ENRICHING BUSINESS POSSIBILITIES THROUGH ENHANCED DIGITALIZATION

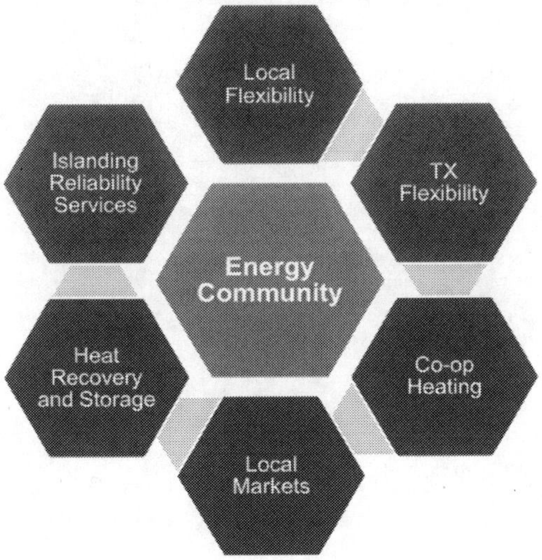

Estimated business benefits of digitalisation:

- +10% increase in value of PV electricity
- +50% increase in penetration levels in given grid
- 15% to 35% reduction in annual electricity costs through P2P trading and energy communities

https://eneuron.eu/

eneuron FOSS ETIP Photovoltaics :: csem

40th European Photovoltaic Solar Energy Conference and Exhibition

PILOT DEPLOYMENT: ENEA'S MISSION PROJECT

- **MISSION project: multi-vector integrated smart systems and intelligent microgrids for accelerating the energy transition** (MISSION) project.
- Objective: to study, design and implement technological solutions enabling the transition of networks towards integrated and **smart multi-energy distribution systems** based on different energy carriers.
- Digitalisation can contribute to guarantee an adequate level of:
  - energy security
  - adequacy
  - reliability
  - Resilience

  in grids and microgrids also facing anomaly and fault conditions.

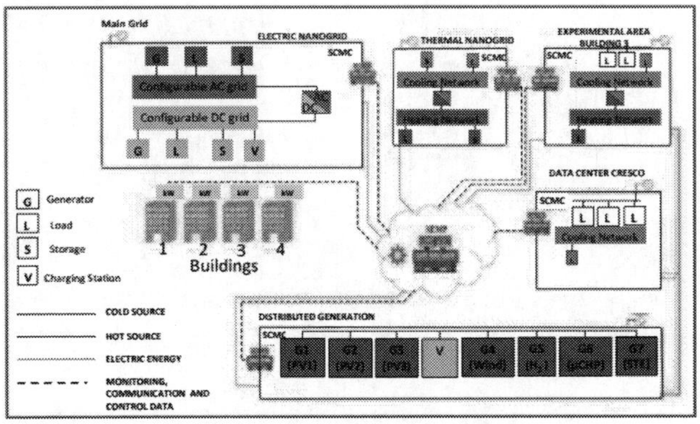

Schematic architecture MISSION demonstrator at Research Centre ENEA Portici (Naples)

ENEA ETIP Photovoltaics :: csem

40th European Photovoltaic Solar Energy Conference and Exhibition

You are invited!

Data, AI, IoT
Opportunities and Challenges for PV Innovation

21/09/2023, 17:00-18:00 – Room 5B

17:00-17:10 Introduction of ETIP PV and the Strategic Research and Innovation Agenda for Photovoltaics (SRIA PV) with focus on PV digitalization related roadmaps

17:10-17:50 Three rotating round table **discussions** about:

- **Digitalization of PV manufacturing**
- **Digitalization of PV systems**
- **Smart Energy Integration of PV**

17:50-18:00 Conclusions

020366-023

The European Technology and Innovation Platform for Photovoltaics

etip-pv.eu

40th European Photovoltaic Solar Energy Conference and Exhibition

Public consultation: where should PV R&D go?

2024 update of PV SRIA

SCAN ME to answer the survey by 31/10

Survey link

The European Technology and Innovation Platform for Photovoltaics

etip-pv.eu

40th European Photovoltaic Solar Energy Conference and Exhibition

This presentation was selected by the Sc. Committee of the EU PVSEC 2023 for submission of a full paper to one of the EU PVSEC's collaborating peer-reviewed journals.

COMPREHENSIVE ANALYSIS OF THE INTERACTIONS BETWEEN A CONCENTRATING PHOTOVOLTAICS (CPV) MODULE AND 5-JUNCTION SOLAR CELLS: INDOOR AND OUTDOOR SUB-CELL CURRENT MEASUREMENTS

Ruediger F. Loeckenhoff [1, a)], Philipp Schroth[1)]
[1]AZUR SPACE Solar Power GmbH
[a)] AZUR SPACE Solar Power, Theresienstr.2, 74072 Heilbronn Germany,
+497131672832vRuediger.Loeckenhoff@azurspace.com

ABSTRACT: We developed the 5C46 5-junction solar cell [1] to maximize the energy yield of concentrating photovoltaic modules. The advantage of the new cell was confirmed by an output power improvement of 18% demonstrated with AZUR C3PV Fresnel modules [2].
We achieved this remarkable gain by designing the cell for real operating conditions. Those include 90°C junction temperature, silicone as optical medium and the optical efficiency of the CPV module, which peaks in the visible spectral range. In contrast, this 5-junction solar cell adjusted for AM1.5d conditions at 25°C junction temperature in air would become current-mismatched under real operation under a changed spectrum within the module and at elevated junction temperatures.
We thoroughly elaborated this concept using proven and novel measurement techniques. The present publication focusses on a high throughput indoor measurement technique for the sub-cell currents, an outdoor measurement of sub-cell currents within a CPV module and a special analysis module that tracks IV parameters of 60 individual receivers, direct spectrum, wind, and temperature. These results allow us to confirm and fine tune our cell and the design of the C3PV module.

1 INTRODUCTION

1.1 The C3PV module

AZUR SPACE has developed the C3PV module technology to provide a CPV reference system in a franchising approach [1]. The module is designed for ease of assembly while achieving high precision with simple techniques. It uses "silicone on glass" Fresnel lenses glued to a trough-shaped module in which 60 cell receivers are arranged and series connected.

Figure 1 shows an improved version of the C3PV module with reduced inactive area, 3.3% extra active area, reduced aluminum weight and only 4 instead of 8 aluminum parts as compared to the original design.

Figure 1: Improved C3PV module

1.2 The 5C46 solar cell

The 5C46 cell is an upright metamorphic 5-junction solar cell grown on Germanium wafer (Fig. 2) that also forms the bottom sub-cell for absorbing infrared part of the spectrum.

It is designed considering realistic operating conditions within a CPV module. These include an elevated junction temperature of 90°C, silicone rather than air as optical medium surrounding the solar cell and the change in spectrum within the CPV module as compared to the sunlight spectrum incident on the Fresnel lens. The uppermost four sub-cells are semi-transparent and thus allow for a sub cell current adjustments by changing the sub-cell thicknesses. The Germanium cell always has enough excess current and will not become current limiting under realistic operating conditions. More details can be found in [2].

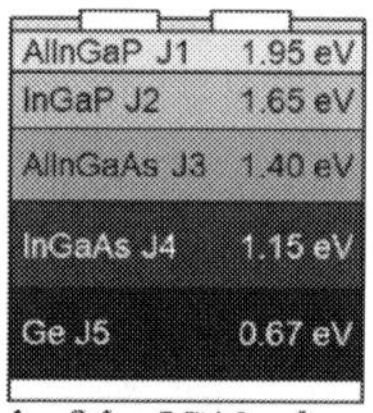

Figure 2: Layer stack of the 5C46 solar cell

With the 5C46 cell we achieved an 18% output power increase comparing to AZUR 3C44 triple-junction CPV solar cell [2] leading to increase of the module power from 335W to 392W under concentrator standard operating condition CSOC (outdoor-rated at Fraunhofer ISE [3]).

2 INDOOR SUB CELL CURRENT MEASUREMENT

2.1 Prior art: EQE and boost measurements

All methods for the measurement of sub-cell currents require a single sub-cell in the series connection to be current limiting. This is commonly achieved with narrow band boost lamps that provide extra current to the other sub cells or filters that suppress a single sub cell.

One of these methods is the external quantum efficiency measurement EQE [4] that uses chopped monochromatic light to determine the probability that a single photon of a certain wavelength will generate a single elementary charge flow through a certain sub-cell.

Other methods we evaluated are:

- constant light simulators tuned for AM1.5d conditions [5, 6]
- pulsed-light, high intensity (flash) simulator measurements with 4 filters to suppress the sub cells individually [7]
- similar measurements with 4 pulsed light sources to boost the sub cells individually [8].

10.4229/EUPVSEC2023/4CP.3.3
020367-001

All these methods work well and are important in their own way. Yet they are comparably slow, and the cells usually need to be contacted manually.

2.2 New method: Downsized sun simulator with fast LED-boost variation

This new method is based is a downsized version of the constant light simulator with LED boost lights as mentioned above (Fig. 3). The light of an incandescent lamp is collimated and passes through six reflective filters that reduce long wavelength light and thus generate a constant light spectrum close to AM1.5d (Fig. 4). The light is focused and passes through a prism into a homogenizing fused silica rod (10x12x150 mm^3). This constant light combines with the light of 4 fiber coupled LEDs 455 nm, 660 nm, 850 nm, 940 nm that were chosen to boost J1, J2, J3 and J4 respectively. J5 has excess current and doesn't need to be boosted. The rod passes the light on into a probe card just above a solar cell. The power density of the filtered incandescent light on the solar cell is close to 1000 W/m^2 and the LEDs can boost this to about 1600 W/m^2. This means that the power density is a bit above natural sunlight but nowhere close to concentrating PV conditions.

This setup is ready for the use with an automated prober that can measure about 1800 cells per hour. The prober may even be combined with an automated flash simulator that passes focused light into the input aperture of the homogenizing fused silica rod and performs its normal measurements. The charging time of the capacitor bank may be used for the low intensity measurement of the sub cell currents.

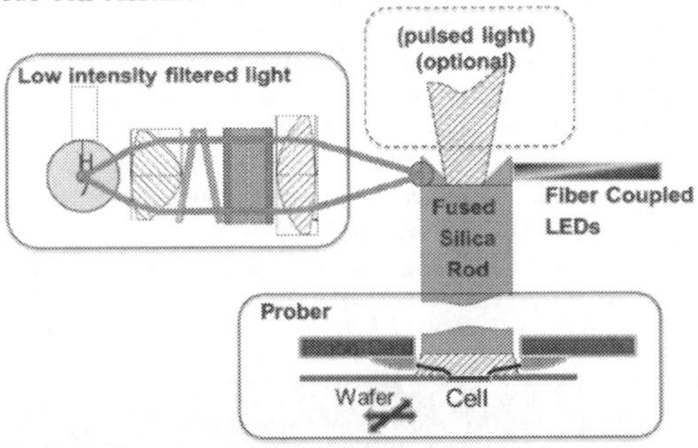

Figure 3: Schematic of the downsized constant light simulator. The integration in a pulsed light simulator is optional.

The example in Fig. 5 illustrates how the sub-cell currents of J1 to J4 may be determined with this setup. In a prior experiment the LED settings for J1 to J4 sub-cell current matching in a particular 5-junction-cell were determined. Now, all LEDs but one remain at this setting while the power of this one LED is reduced. Consequently, the corresponding cell will become current limiting (e.g., the 455 nm LED1-curve in Fig. 5 showing the current of J1 sub-cell).

We can now adjust four LED settings with J1, J2, J3 and J4 current limitations (black circles) for high volume measurements. The complete measurement for 4 sub-cells takes only 0.8 seconds. We then compare the resulting sub-cell currents with those of a reference cell.

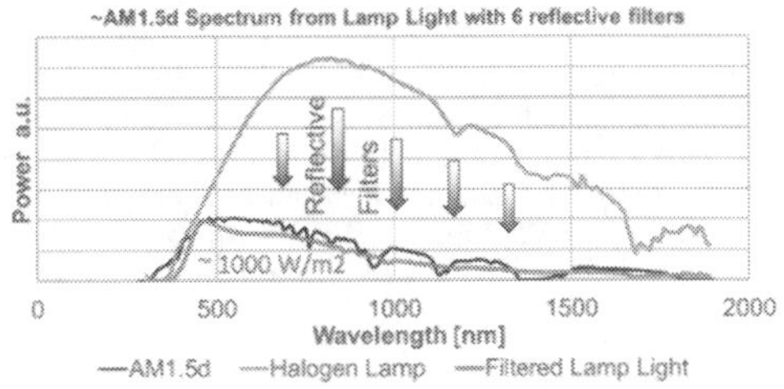

Figure 4: Approximating AM1.5d with incandescent lamp light.

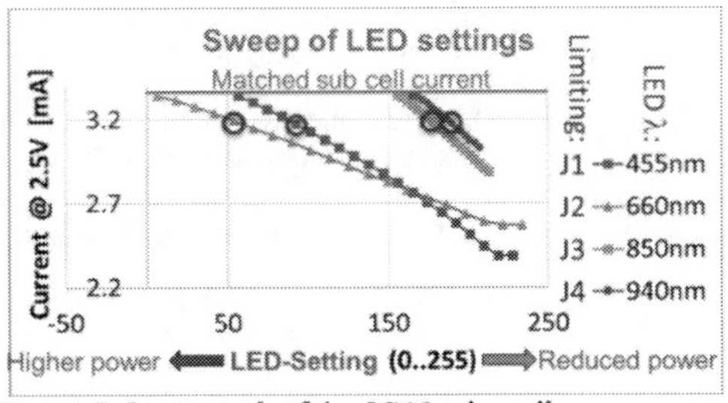

Figure 5: Layer stack of the 5C46 solar cell

Overall, this approach is not a rigorous AM1.5d sub-cell-current measurement but a pragmatic and highly valuable method for comparative measurements and process control in a production.

3 OUTDOOR SUB CELL CURRENT MEASUREMENT

3.1 Experimental setup

The method in section 2.2 can also be applied outdoors where natural sunlight replaces the incandescent lamp. In [9] this approach was used to build a pseudo isotype sensor that measures the direct light on one single 3-junction solar cell with varying LED boosts to gain information about the solar spectrum and intensity. We take this concept further by integrating the LED boosts in an actual CPV module (Figure 6).

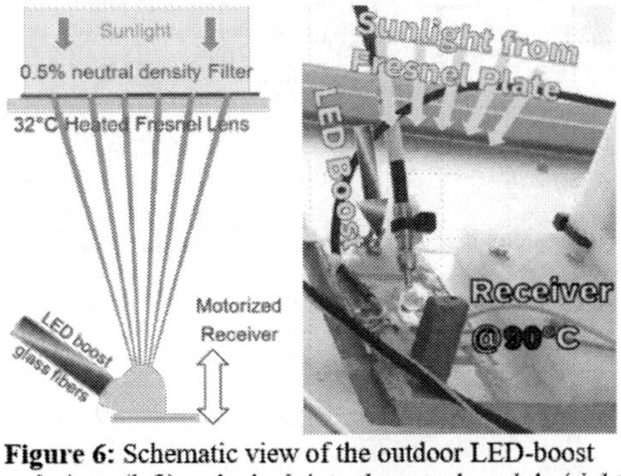

Figure 6: Schematic view of the outdoor LED-boost technique (left) and a look into the actual module (right)

Only one receiver is active and connected to the data acquisition. Again, we take one sub-cell into limitation by boosting the other three upper sub-cells. Again, J5 doesn't need to be boosted.

Our fiber coupled LEDs isn't capable to provide the 100 kW/m^2 intensity that would be required to effectively

10.4229/EUPVSEC2023/4CP.3.3
020367-002

boost sub-cells at actual CPV conditions. We therefore reduce the sunlight intensity with a 0.5% neutral density filter. A window is glued to the side of the secondary optical element to couple the LED-light into the receiver. The receiver and the Fresnel plate are heated to realistic operating temperatures of 90°C and 32°C respectively. Furthermore, the distance between the receiver and the Fresnel lens is adjustable to evaluate its influence and look for the best value depending on the Fresnel lens temperature.

3.2 X-Talk Correction

Figure 7 shows an example of the external quantum efficiencies characteristics (EQE) of the first four sub cells within the module.

Since the EQEs of the sub cells overlap (semi-transparent design), it is not possible to boost only one sub cell with an LED. For instance, LED2 will mostly boost J2 but it will also boost J1 and J3 to some extent. This unavoidable boost is referenced to as X-Talk. It needs to be measured and we use the result to correct the sub-cell currents we obtained with the LED-boost method.

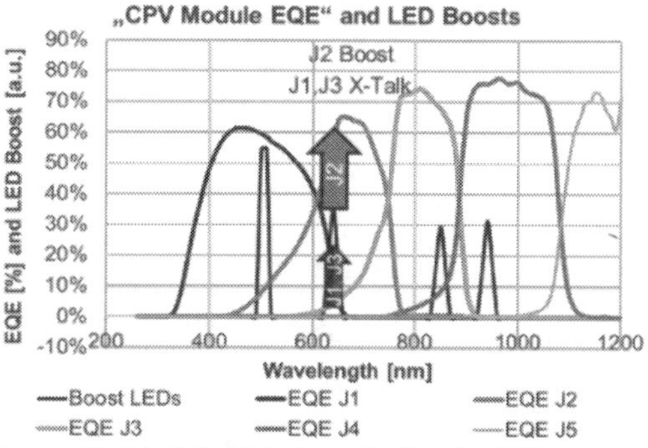

Figure 7: Sub-Cell EQEs, boost LEDs. The EQE overlap prevents single sub-cell boosts without X-Talk.

Our method for X-Talk measurement is illustrated in Figure 8. LEDs 1 and 4 are always on and LED3 is always off. When LED2 is ramped up, we see a steep increase in solar cell current which shows that J2 has been current limiting. At an LED2 setting of 95 (a.u.), J3 becomes current limiting and the slope above 95 represents the X-Talk of LED2 on J3. If we define the sensitivity (slope) of J2 on LED2 to be 100%, the sensitivity of J3 on LED2 is still 6.6%. We call this the X-Talk 2→3.

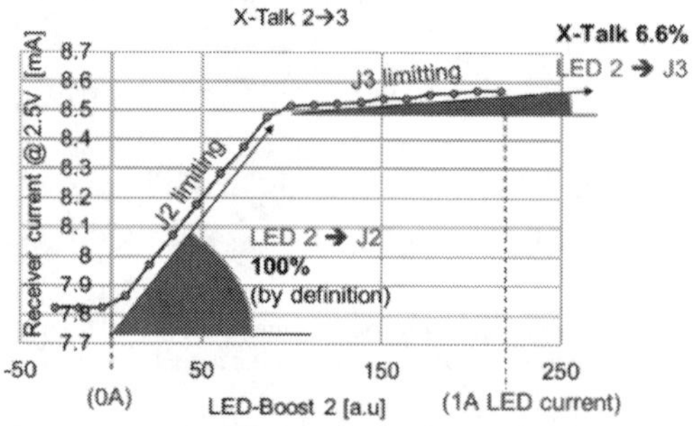

Figure 8: The X-Talk of LED2 on sub-cell J3 measured with a sweep of the LED2-boost.

Applying this principle to all combinations of dark LEDs and ramped LEDs gives us the X-Talk matrix shown in Figure 9. In this case it is the X-Talk-Matrix for outdoor measurements and the low signal X-Talks, such as 1→4, show a lot of noise due to atmospheric variations and may even appear to be negative. This is why they are best assumed to be zero. The incandescent light source in section 2, in contrast, is very stable and reproducible, even for small X-Talks.

	J1	J2	J3	J4
LED 1	100%	1.5%	0.4%	-0.3%
LED 2	17.6%	100%	6.6%	0.0%
LED 3	0.4%	0.2%	100%	5.9%
LED 4	0.4%	0	24.7%	100%

Figure 9: X-Talk matrix. The blue and the green value are extracted from Fig. 8.

3.3 Comparison to currents calculated from spectral data and EQE.

As a reference for our measured currents, we calculate currents from spectral data and EQE. The spectral data is provided by a commercial Spectrafy D2 sensor and combined with the EQE, to generate the sub cell currents.

To obtain correct results, the EQE in Figure 7 is corrected for a junction temperature of 90°C, the module optical efficiency obtained in ray-tracing simulations, the 0.5% neutral density filter and the measured change in reflectivity when the cell is operating in silicone rather than air.

Figure 10 compares the measured currents (boost method) to the calculated currents obtained from the Spectrafy software. For receiver 1 we achieve a good match while for receivers 2 and 3 the boost measurements seem to be too high (or the Spectrafy calculations are too low). Probably this difference is caused by a "relative EQE measurement". When the chopped monochromatic light spot doesn't hit the solar cell in the center, some light may be lost and the EQE may appear too low, even so its relativeshape is still correct.

	J1	J2	J3	J4
Receiver 1	101.0%	100.3%	101.5%	99.2%
Receiver 2	105.8%	105.6%	104.1%	99.2%
Receiver 3	106.2%	107.3%	109.2%	101.1%

Figure 10: Comparison of outdoor current measurement vs. sub-cell currents in the module calculated from the EQE.

3.4 Applications of the outdoor boost method

The outdoor boost method allows us to test whether the sub-cell current balance is correct for a certain application under a certain spectrum.

Secondly, we have kept twin receivers without secondary lens for each receiver in Figure 10 (same region of the same wafer). These twin receivers can be used as reference cells for indoor measurements such as in section 2.

Finally, there are methods to make relative EQEs absolute in lab measurements that should be applied to the receivers before the secondary lens is glued on. However, we might yet apply this correction to the twin receivers and assume the same EQEs for the twins.

10.4229/EUPVSEC2023/4CP.3.3
020367-003

The method is applicable to all kinds of CPV systems. The microcontroller unit that handles the LED current drivers and IV-tracing is easy to replicate with very low hardware costs based on the Arduino microcontroller ecosystem.

4. ANALYSIS MODULE WITH 60 INDIVIDUAL RECEIVERS

The boost method is a very powerful way to perform outdoor sub-cell current measurements. However, it is not applicable to long term measurements for power rating and comparison of different receivers and module adjustments. The intensity is low, and the test does not provide real IV curves under concentrated light. Furthermore, the temperatures of the receiver and the Fresnel lens are set with heating elements and temperature sensors rather than being the result of the outdoor conditions such as irradiation air temperature and wind.

Therefore, we have made special analysis modules as an alternative (Fig. 11). 60 receivers are connected to a multiplexer and can be addressed for single IV-curve acquisition. The multiplexer cycles through all receivers and takes IV-curves. One full cycle takes about 2 minutes. The receivers differ in the number of junctions (3 or 5), the 5-junction sub cell current balance, the distance to the Fresnel plate and intentional lateral displacements to evaluate a variety of adjustments and application conditions.

Again, we are using an Arduino microcontroller and a high precision AD-converter to measure a single receiver IV-curve while the receiver is charging a capacitor bank. The microcontroller calculates the IV parameters using an interpolation algorithm [10]. Furthermore, the microcontroller is also reading data from temperature sensors, an anemometer, a GPS unit and a Spectrafy sensor. Thus, a full set of environmental data is acquired and stored on an industrial grade SD card together with the IV-curve parameters.

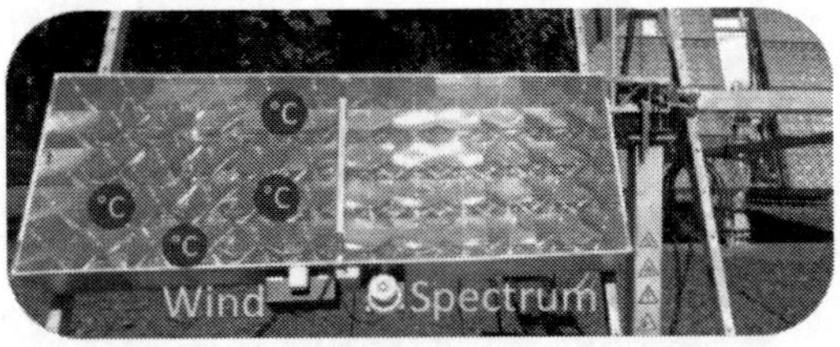

Figure 11: Analysis module that tracks 60 individual receiver currents, spectrum temperatures and wind.

Fig 12 shows the inside of the module with the heat shields removed which usually protect the wires against of axis beams.

We aim to acquire data from different locations and climates. This will allow us to further validate our cell and module design and identify further room for fine tuning. One module has been acquiring data at Freiburg and another one at Heilbronn in Germany for several months. These modules are presently being shipped to high DNI sites in Santa Fe, USA and Port Elizabeth, South Africa.

Figure 12: View into the analysis module. Multiplexed matrix of 60 individual receivers.

5. SUMMARY

The 5C46 5-junction solar cell boosts our C3PV module technology by 18% from 335 W to 392 W rated power (CSOC conditions). We investigated the interactions between the cell and the module that allows this remarkable gain of output power, and we identified room for fine tuning.

This paper particularly focusses on a low intensity, high speed method to measure sub-cell-currents in an industrial prober setup, an outdoor measurement of sub-cell currents in a CPV module and special analysis modules that measure individual IV curves on a variety of 60 individual receivers within the same module.

5. CREDITS

We acknowledge the contribution of our colleagues at AZUR SPACE and Fraunhofer ISE. This work was also supported by the Federal Ministry for economic affairs and climate action under contract No 03EE1088 "QuintuMod".

6 REFERENCES

[1] R. Löckenhoff et al., „C3PV: Mass Produced EFA Receivers for a Franchised Module Technology", in: .com/index.php/en/company/company-publications, 2021.

[2] P.Schroth et al., "AZUR's New 5C46 CPV Cell: Final Design for Optimized Outdoor Performance", Proceedings of the 17th International Conference on Concentrator Photovoltaics, Freiburg, Germany, 2021

[3] G. Siefer, T. Gandy, M. Schachtner, A. Wekkeli and A. W. Bett, "Improved grating monochromator set-up for EQE measurements of multi-junction solar cells," 2013 IEEE 39th Photovoltaic Specialists Conference (PVSC), Tampa, FL, USA, 2013, pp. 0086-0089, doi: 10.1109/PVSC.2013.6744105.

[4] M.Steiner et al., "Translating outdoor CPV I–V measurements to a CSTC power rating and the associated uncertainty", https://doi.org/10.1002/pip.2590, Progress in Photovoltaics, Vol. 23, Iss 11, p. 1557-1571, 2015.

[5] S.K.Reichmuth et al. "Measurement Uncertainties in I–V Calibration of Multi-junction Solar Cells for Different Solar Simulators and Reference Devices",

IEEE JOURNAL OF PHOTOVOLTAICS, VOL. 10, NO. 4, pp.1076-1083, JUNE 2020
[6] Schachtner Met al. "A New Tool to Measure Monolithic Multi Junction Solar Cells with up to Six Subcells". In: Proceedings of the 26th European Photovoltaic Solar Energy Conference and Exhibition; 2011, pp. 199–204.
[7] F.Gerstenberger et al. "Spectrally Accurate Calibration of Pulsed Solar Simulators for Multi-junction CPV Measurements", AIP Conf. Proc. 1679, 050012 (2015)
[8] M. Schachtner et al. "Analysis of a four lamp flash system for calibrating multi-junction solar cells under concentrated light", Proceedings of the 17th International Conference on Concentrator Photovoltaics, Freiburg, Germany, 2021.
[9] Norman Jost et al., "From component to multi-junction solar cells for spectral monitoring" AIP Conf. Proc. 2012, 100002 (2018)
[10] H.Akima, "A New Method of Interpolation and Smooth Curve Fitting Based on Local Procedures", Journal of the Association for Computing Machinery, Vol. 17, No. 4, October 1970.

This presentation was selected by the Sc. Committee of the EU PVSEC 2023 for submission of a full paper to one of the EU PVSEC's collaborating peer-reviewed journals.

MAKING EQUATION-BASED THERMAL MODELS DYNAMIC: THE FILTER-EWM-MBE (FEM) CORRECTION APPROACH

Bert Herteleer, Anastasios Kladas, Jan Cappelle
Research group ELECTA Ghent, Department of Electrical Engineering, KU Leuven - bert.herteleer@kuleuven.be

ABSTRACT

This paper illustrates the Filter-Exponential Weighted Mean (EWM) - Mean Bias Error (MBE), i.e. FEM approach, which can be applied to steady-state equation-based thermal models to make these dynamic. Based on the input and definition of the Exponential Weighted Mean which uses the (thermal) time constant τ, the EWM of irradiance and wind speed signals can be determined, even for different time resolutions; the calculation of the EWM is easily performed in python. It is shown how the different steps of the FEM approach contribute to the final result, with particular attention paid to the impact of the time resolution on modelling results. From the evaluated datasets, the coefficients for WM1, k_{dyn} and d_{dyn} are on average $16\,\%$ larger than their steady-state equivalents, with a Mean Bias Error of $-2.1\,\mathrm{K}$ in dynamic conditions; in the context of Faiman's model, $U_{0,dyn,avg} = U_{0,ss,avg} * 0.86$ and $U_{1,dyn,avg} = U_{1,ss,avg} * 0.9$. This has important implications for contractual evaluations, if steady-state models and coefficients are applied on measured data at shorter timescales.

1 INTRODUCTION

The world is moving into a new era, of high deployment of renewables (wind and solar PV), and improved computational power requiring increasing sophistication by investors, asset owners, and grid operators. Consequently, the importance of accurate temperature estimations of PV modules is also moving towards shorter time scales. With improved computational power and software and faster measurement time resolutions in use in PV systems, it is now feasible and desirable to *accurately* estimate the temperature of PV modules and systems at these faster time scales.

PV modules and systems are constantly subjected to changing weather circumstances. When the plane-of-the-array irradiance (G) and wind speed (WS) change slowly (or have datasets with slower time resolution), thermal models such as those by Ross [1], King et al [2] and Faiman [3] have seen high uptake in the literature, due to their ease-of-use and quite high accuracy. These equation-based models are "static" or suited for steady state conditions, in that they do not incorporate the thermal lag of PV modules. When significant irradiance or wind speed changes occur within short timescales ($1\,\mathrm{s}$ to $5\,\mathrm{min}$), these models then see temporarily large errors, particularly when evaluating these at second-to-minute time resolution. Incorporating the thermal lag in a model is then a useful avenue, resulting in improved error metrics (Root Mean Square Error RMSE, Mean Absolute Error MAE, and Mean Bias Error MBE), as seen in the work of Veldhuis et al [4], and Prilliman et al [5]. However, these methods can be overly complex to be implemented and understood widely.

Instead, we use a simpler approach described in [6], based on a simplified conceptual RC-equivalent thermal model, shown in Figure 1. As a PV module can be seen to be composed of RC networks, an equivalent thermal time constant τ can be found, by looking at the temperature response (change) from sustained irradiance step changes.

2 DATA AND METHODS

For this work, we use module temperature and weather data from NIST at $10\,\mathrm{s}$ [7] and KUL at $1\,\mathrm{s}$ [8] resolution; more details are given in Table 1.

By filtering according to wind speed in bins of $1\,\mathrm{m/s}$, and performing a linear regression of the measured change in over-temperature ΔT_o versus the step change in irradiance ΔG for 20-minute time intervals, the thermal time constant τ as a function of wind speed can be found [6], as shown in Figure 3. This thermal time constant τ can then be used to calculate the exponential weighted mean (EWM) signal of the irradiance and wind speed respectively, to make the equation-based

10.4229/EUPVSEC2023/4CV.1.1
020369-001

Table 1: Locations, data sources and key mounting characteristics of arrays used for thermal models. All sites used have plane-of-the-array irradiance data, T_a, T_m (typically backsheet), and wind speed data. More detailed system descriptions and metadata can be found in the cited references.

Organisation & site name	Location	Data used	Time step[a]	Mounting[b], tilt & azimuth	Module sensor(s)[c]	Ref
KUL TC Ghent Rooftop	Ghent, Belgium	2015/05-2016/03	1 s	Flat roof 18° S	PV052-5x4 C, BS	[9, 8]
NIST Ground	Maryland, USA	2016 Jan-Dec	10 s	Ground 20° S	BS RTD4	[7, 10]
NIST Ground	Maryland, USA	2016 Jan-Dec	10 s	Ground 20° S	BS RTD8	[7, 10]
NIST Canopy W	Maryland, USA	2016 Jan-Dec	10 s	Raised 5° W	BS RTD4	[7, 10]
NIST Canopy E	Maryland, USA	2016 Jan-Dec	10 s	Raised 5° E	BS RTD4	[7, 10]

[a] Recording resolution as stored in the dataset. NIST measured at 1 s resolution, but stored data at 10 s.
[b] Raised = parking canopy
[c] C = cell, BS = backsheet. RTD = Resistance temperature detector. Sensor name or number noted if multiple sensors available in dataset. 5x4 = Row 5, column 4 counting from top left, viewed from front (=centre cell of module).

thermal model of choice dynamic.

The EWM of a signal uses the smoothing parameter α which combines the time step or time resolution Δt and the thermal time constant τ (both expressed in seconds): α can thus be adapted to the time resolution used for modelling, whether this is 1 s or 5 min. To illustrate, the calculation of the EWM irradiance is as per Equation (2), for each time step i. This is easily applied in python, e.g. with the pandas package [11, 6], and other software,such as MATLAB.

$$\alpha = 1 - e^{-\Delta t/\tau} \text{ with } 0 \leq \alpha \leq 1 \quad (1)$$

$$G_{EWM} = \frac{\sum_{i=0}^{t=span} G_i \cdot (1-\alpha)^i}{\sum_{i=0}^{t=span} (1-\alpha)^i} \quad (2)$$

Recognising that most models incorporate the ambient temperature T_a, the over-temperature T_o defined by Kurnik et al [12] and given in Equation (3) will be used throughout this work. Equations (4) to (6), (8) and (9) are then written with that notation.

$$T_o = T_m - T_a \quad (3)$$

$$T_{o,Ross} = k \cdot G \quad (4)$$

$$T_{o,King} = G \cdot e^{a+b \cdot WS} \quad (5)$$

$$T_{o,Faiman} = \frac{G}{U_0 + U_1 \cdot WS} \quad (6)$$

Equations (4) to (6) can thus be made dynamic by applying the EWM to the irradiance and (if present in the data) the wind speed. To illustrate using Faiman's model:

$$T_{o,Faiman,EWM} = \frac{G_{EWM}}{U_0 + U_1 \cdot WS_{EWM}} \quad (7)$$

We further separate the heating and cooling components in these models, and use these to determine the coefficients from measured data. Based on this, two wind models (WM1 and WM2) are given in Equations (8) and (9) [6]:

$$T_{o,WM1} = k \cdot G \cdot \Big(\overbrace{1}^{heating} \underbrace{-(1 - e^{\frac{-WS}{d}})}_{convective\ cooling} \Big)$$

$$T_{o,WM1} = k \cdot G \cdot e^{\frac{-WS}{d}} \quad (8)$$

$$T_{o,WM2} = G \cdot (k - h \cdot WS|_{WS \leq 8}) \quad (9)$$

where k is the (modified) Ross heating coefficient, d is the wind speed divisor in WM1, and h is the convection coefficient in WM2. When using identical coefficients, WM1 is identical to the Sandia model [6]. While the approach advocated here requires two linear regression steps instead of one as presented by the original authors (e.g. Ross, Faiman, Sandia), it generally results in improved error Key Performance Indicator (KPI) values for the model used. To further improve these KPI metrics, the average radiative heat loss from the module must be corrected for, which is done using the MBE of the initial training part of the dataset used. The Filter-EWM-MBE correction (FEM) approach is thus:

- **Filter**: Determine model coefficients through the use of appropriate filters and regressions, in quasi-steady state (5 min averaged data). An example is shown in Figure 2.

- Irradiance heating coefficient k, $k = e^a = \frac{1}{U_0}$: $WS \approx 0$, i.e. $WS < 0.5$ m/s, and variable irradiance $G = 200\,\mathrm{W/m^2}$ to $1000\,\mathrm{W/m^2}$, linear regression of $T_{o,measured,filtered}$ versus $G_{filtered}$.
- Wind cooling coefficient $d = \frac{-1}{b}$ or U_1: $G = 1000 \pm 100$ $\mathrm{W/m^2}$ and variable wind speeds $WS = 0\,\mathrm{m/s}$ to $8\,\mathrm{m/s}$. $U_1 = \frac{G/T_o}{WS}$, $-\frac{1}{d} = b = \frac{ln(T_o/G)}{WS}$. WM2 requires two linear regressions and filtering steps: 0.5<WS<8 m/s, $20\,\mathrm{W/m^2}$ bins, G = 200-1000 $\mathrm{W/m^2}$, $m_W = \frac{T_o(G)}{WS}$, and G = 200-1000 $\mathrm{W/m^2}$ for $k_W = \frac{m_W}{G}$

- **EWM**: Make the irradiance and wind speed signals in thermal model equations dynamic, using the thermal time constant τ to calculate the EWM signal. From [6], $\tau = (6.3 \pm 1.0)$ min for fixed free-standing PV systems. Alternatively, τ can be found from measured data as shown in Figure 3, or calculated from material properties [6].
- **MBE correction**: Remove the residual bias in the training dataset, corresponding to the average radiative heat loss from the module, particularly in dynamic conditions ($\Delta t \leq \tau$).

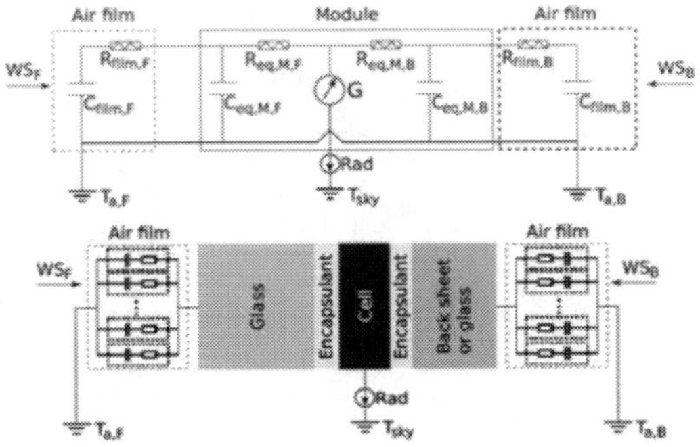

Figure 1: Conceptual RC model with simplified module cross section (not to scale). The ambient temperature and the wind speed on the front and back of the module may differ, depending on mounting conditions. The equivalent air films are in series to the glass or backsheet for convective heat transfer.

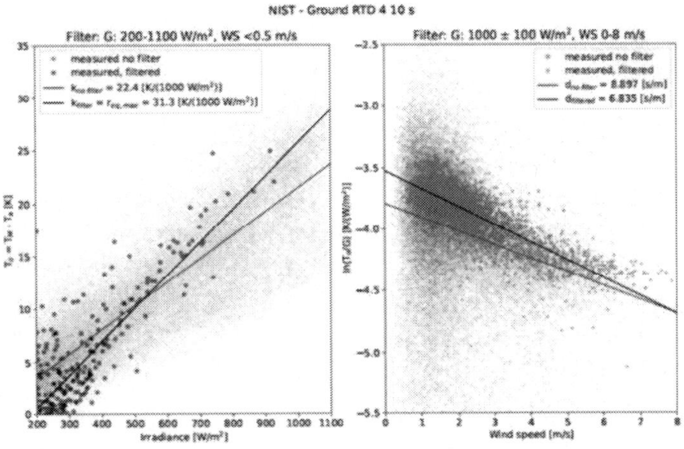

Figure 2: Determination of k and d for NIST ground mount array, RTD 4 [7].

10.4229/EUPVSEC2023/4CV.1.1
020369-003

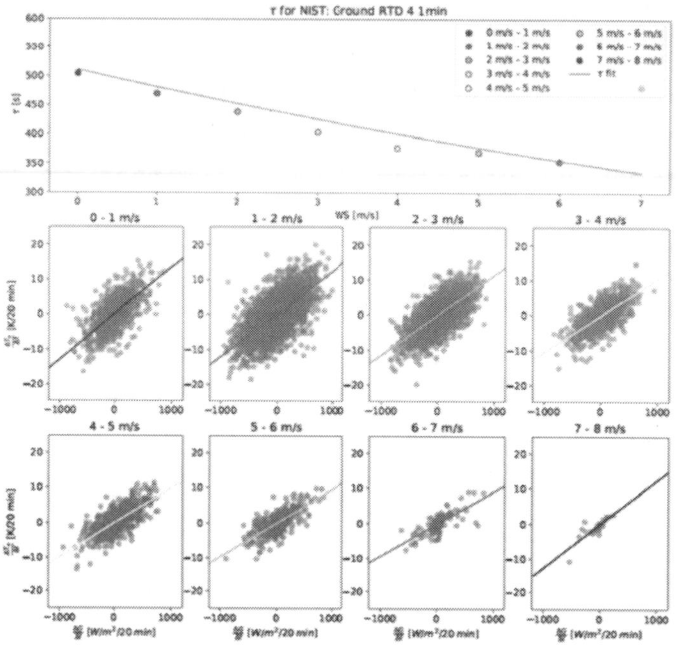

Figure 3: Determination of τ for NIST ground mount array, RTD 4 [7], $\tau_{used} = 334\,\mathrm{s}\ (5.5\,\mathrm{min})$. Each coloured dot in the top pane is the respective regression line coefficient found at that wind speed bin.

Splitting a dataset into a training dataset (weekdays) and testing dataset (weekend days), the FEM formulation for equation-based models is then as Equation (10), and illustrated for the Sandia model in Equation (11):

$$T_{o,FEM} = T_{o,EWM,test} + MBE_{train} \quad (10)$$

$$T_{o,FEM,Sandia} = G_{EWM} \cdot e^{a+b \cdot WS_{EWM}} + MBE_{training} \quad (11)$$

3 RESULTS AND DISCUSSION

Figure 4 illustrates the FEM approach applied to WM1 for 10 s data, where for most of the day, the error is smaller than $\pm 2.5\,\mathrm{K}$. By contrast, the non-EWM error can be up to 15 K (or higher for shorter timescales), as irradiance *and* wind speed changes are transmitted directly to the thermal model, essentially as noise, whereas the EWM smoothing step aligns the model with measured behaviour. Similarly, the MBE correction shifts the filter-EWM model output (up or down, depending on the $\mathrm{MBE}_{training}$ value), aiming to achieve $\mathrm{MBE}_{test} = 0$. Consequently, the MBE correction is most reliably applied when measured data is available.

Figure 5 shows how the five models (WM1, WM2, Ross, Sandia, and Faiman) benefit from the FEM approach, compared to using the filtering step, without the EWM and MBE corrections. As seen in Figure 4, the largest benefit comes from the EWM step, yet the combination of the Filter-EWM-MBE correction steps delivers the best and most consistent results. The model of Ross sees next to no change in the FEM RMSE and MAE for changing time resolution, as the model is "blind" to the wind speed. By contrast, the largest improvements can be seen for WM2, with the MAE reducing by more than 50 % at 10 s resolution from Filter-only to FEM, which likely is due to the direct linear relationship of the wind cooling coefficient. The lowest volatility change is seen for WM1, Sandia and Faiman, explaining their continued popularity in the PV industry: even if another model (e.g. WM2) *could be* better at short time resolutions, the likelihood of WM1/Sandia/Faiman delivering large errors is limited, and therefore more reliable.

10.4229/EUPVSEC2023/4CV.1.1
020369-004

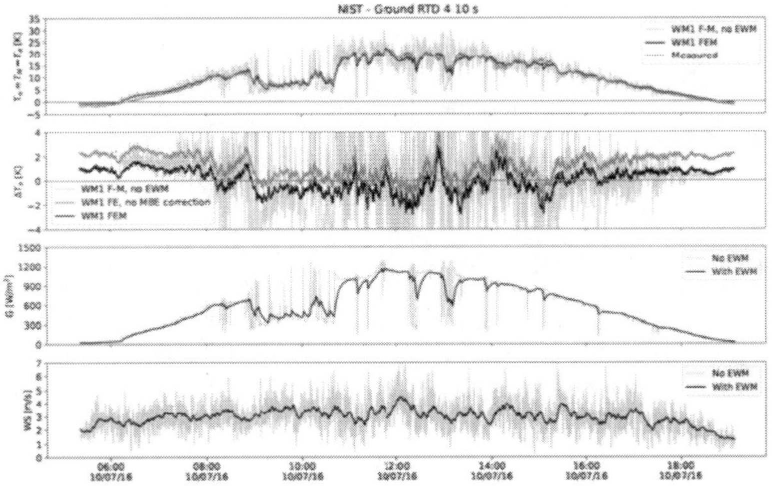

Figure 4: Illustration of WM1 for one day for NIST Ground RTD 4 at 10 s resolution. Note the impact of the EWM and MBE steps; the k and d coefficients were determined with the filtering step.

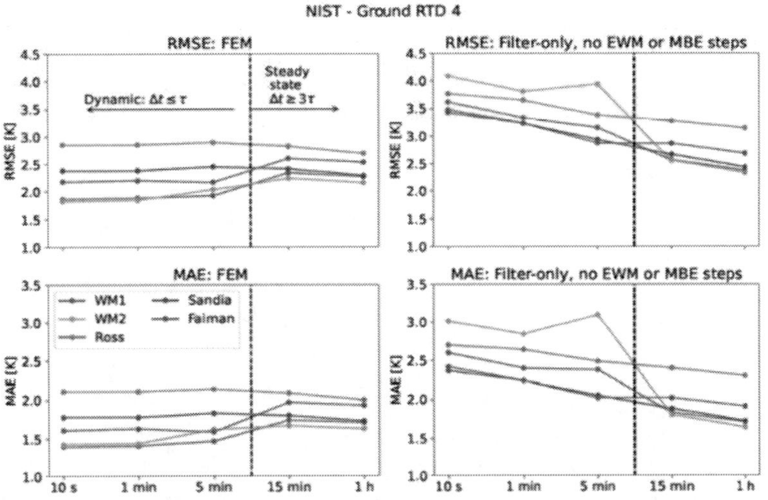

Figure 5: NIST Ground RTD 4 comparison of FEM, versus static (Filter-only) RMSE and MAE, for the five models. Note how the FEM RMSE and MAE values improve or stabilise as time scales reduce, whereas the steady state approach sees the RMSE and MAE generally worsen when $\Delta t \to 0$.

From Figure 2, the benefit of filtering and using separate regressions for the irradiance heating and wind speed cooling coefficients becomes apparent, as illustrated for WM1. For shorter time resolution data ($\leq$5 min), the EWM step provides the largest benefit, which grows in importance as the time steps shorten, as shown in Figure 6. The resulting MBE (MBE_{test}) for FEM remains steady over the changing time resolutions, whereas it shows a large step change when moving from dynamic modelling ($\Delta t \leq \tau$) to steady-state conditions ($\Delta t \geq 3 \cdot \tau$). From a methodological perspective, it does show that the filtering step correctly identifies the coefficients in steady-state conditions.

A further point to consider is how the modelling coefficients (in this case for WM1) k, d and the value of the MBE to use for the MBE correction vary as a function of the time resolution at which the thermal modelling is done.

Figure 7 shows how k, d and MBE_{train} are roughly constant for $\Delta t \leq 5\,\mathrm{min}$[1], and change strongly as the time resolution moves to $15\,\mathrm{min}$ and then $1\,\mathrm{h}$. This hints at why the MBE correction is needed in dynamic (i.e., non-steady-state) conditions, compared to steady state.

Moreover, this further elucidates the potential impact and risk of using a digital twin or contractual PV system model, developed and used in steady-state for the design and financing stages (e.g. at $1\,\mathrm{h}$ time resolution), at shorter time scales once the system is operational. In short: using thermal model coefficients that are appropriate, or have been extensively validated, at $1\,\mathrm{h}$ resolution, likely will lead to model errors when those coefficients are used with measured data at shorter timescales, e.g. $5\,\mathrm{min}$ data for Performance Ratio (PR) evaluations.

This difference is quantified in Table 2, where both k_{dyn} and d_{dyn} are on average $16\,\%$ larger than their steady state average values, and that the MBE in steady state is $-0.2\,\mathrm{K}$, yet this changes to $-2.1\,\mathrm{K}$ in dynamic conditions, i.e. an average change of $-1.9\,\mathrm{K}$. Applying the same approach for Faiman's model and the same sites, and using $U_0 = 1/k$ (while respecting the units of k and U_0), this indicates that $U_{0,dyn,avg} = \frac{U_{0,ss,avg}}{1.16} = U_{0,ss,avg} * 0.86$. For $U_{1,dyn,avg} = \frac{U_{0,ss,avg}}{1.11} = U_{1,ss,avg} * 0.9$.

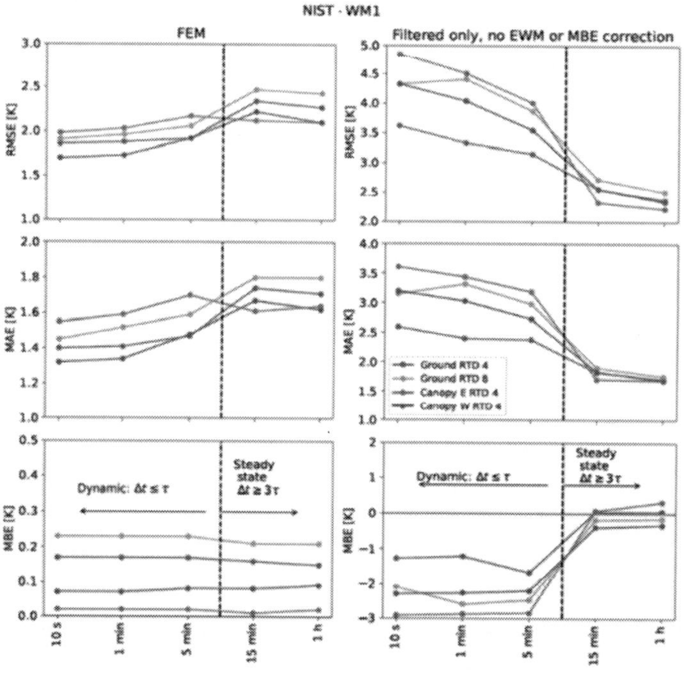

Figure 6: Comparison of the FEM method versus filtered only for NIST sites, RMSE, MAE and MBE. Note that the y-axes are different between the FEM and the Filter-only column, to better highlight changes.

[1]This is a logical result from the methodology, which uses $5\,\mathrm{min}$ averaged data, even for datasets with $\Delta t < 5\,\mathrm{min}$.

10.4229/EUPVSEC2023/4CV.1.1
020369-006

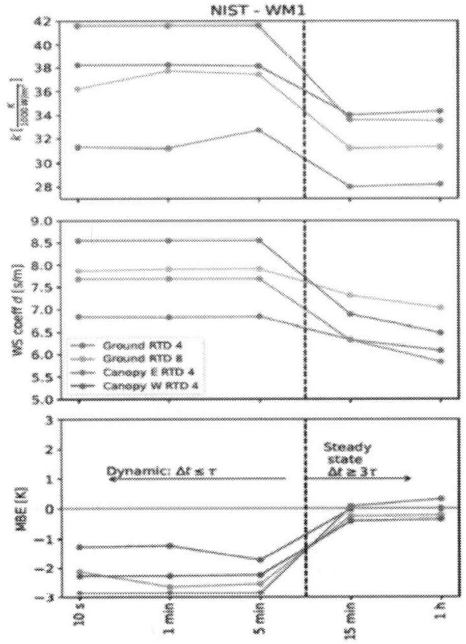

Figure 7: Impact of the time resolution on coefficients k, d, and MBE_{train} to use for the MBE correction for NIST sites, using WM1.

How much the impact can be when using steady state coefficients at faster time scales with the FEM approach, is illustrated in Table 3 using the KUL measurement set-up data, as an example of what happens when a 1 h model is compared with 5 min measured data. Overall, the impact on the model KPIs and the PR modelling error is relatively limited in this example, noting that $k_{dyn} = k_{ss} * 1.11$ and $d_{dyn} = d_{ss} * 1.07$ (i.e. the coefficients differ less than for the NIST sites). However, a change in the PR by 0.5 % points as in this example, purely from the "unforced error" of using steady-state coefficients in dynamic conditions (or vice versa) should be avoided where possible.

The impact of time resolution needs to be further investigated from an attribution perspective on where losses occur from photon to electron, such as in often-used Sankey diagrams for the PR.

Table 2: WM1 FEM coefficients for KUL Roof and NIST sites. dyn = dynamic: average of $\Delta t \leq 5$ min, ss = steady state: average of $\Delta t \geq 15$ min. $k\left[\frac{mK}{W/m^2}\right] = \left[\frac{K}{1000\,W/m^2}\right]$, $d\left[\frac{s}{m}\right]$, MBE_{train} [K]

Org	Site	k_{dyn}	k_{ss}	$\frac{k_{dyn}}{k_{ss}}$	d_{dyn}	d_{ss}	$\frac{d_{dyn}}{d_{ss}}$	MBE_{dyn}	MBE_{ss}	ΔMBE
KUL	Roof	35.12	31.3	1.12	7.02	6.55	1.07	-1.56	-0.56	-1.0
NIST	Ground RTD 4	31.73	28.1	1.13	6.83	6.21	1.1	-1.41	0.0	-1.41
NIST	Ground RTD 8	37.1	31.25	1.19	7.89	7.17	1.1	-2.42	-0.22	-2.2
NIST	Canopy E RTD 4	41.6	33.55	1.24	7.68	6.08	1.26	-2.85	0.2	-3.05
NIST	Canopy W RTD 4	38.17	34.15	1.12	8.55	6.68	1.28	-2.25	-0.37	-1.88
Avg		36.74	31.67	1.16	7.59	6.54	1.16	-2.1	-0.19	-1.91

Table 3: Evaluating the use of steady state coefficients for WM1 from Table 2, compared to dynamic coefficients, with the FEM approach. KUL data at 5 min resolution.

Method	RMSE [K]	MAE [K]	MBE [K]	PR[a] error
ss coeff	1.59	1.15	-0.02	−0.4 %
dyn coeff	1.42	1.12	-0.01	0.13 %

[a] Naïve error, i.e. where over- and under-estimations are deemed to cancel each other, instead of identifying each as an error from the underlying true behaviour. [6]

4 CONCLUSIONS AND FUTURE WORK

The FEM approach as described here shows how each step, from using appropriate filters for coefficient determination, to applying the Exponential Weighted Mean (EWM) to irradiance and wind speed signals, and finally applying the Mean Bias Error (MBE) correction, results in equation-based thermal models with significantly improved error metrics (RMSES, MAE, and MBE). The impact of the EWM step is large, particularly at shorter timescales (1 s to 5 min), reducing instantaneous errors from up to 20 K to generally less than 4 K. As such, the RMSE and MAE in dynamic conditions ($\Delta t \leq \tau$) is stable, even for shorter timescales, as opposed to the steady-state implementation. When using thermal coefficients, due care must be taken to ensure that the coefficients correspond to the time resolution for which these have been found, as $k_{dyn} = k_{ss} * 1.16$ and $d_{dyn} = d_{ss} * 1.16$, and the MBE changes from -0.2 K in steady state conditions, to -2.1 K in dynamic conditions. In the context of Faiman's model, $U_{0,dyn,avg} = U_{0,ss,avg} * 0.86$ and $U_{1,dyn,avg} = U_{1,ss,avg} * 0.9$. This has important implications for contractual evaluations, if steady-state models and coefficients are applied on measured data at shorter timescales.

Future work will evaluate the impact of the changing time resolution on more sites and datasets.

REFERENCES

[1] R. Ross Jr, "Interface design considerations for terrestrial solar cell modules," in *12th Photovoltaic Specialists Conference*, 1976, pp. 801–806.

[2] J. A. Kratochvil, W. E. Boyson, and D. L. King, "Photovoltaic array performance model." [Online]. Available: https://www.osti.gov/biblio/919131

[3] D. Faiman, "Assessing the outdoor operating temperature of photovoltaic modules," *Progress in Photovoltaics: Research and Applications*, vol. 16, no. 4, pp. 307–315, 2008.

[4] A. J. Veldhuis, A. M. Nobre, I. M. Peters, T. Reindl, R. Ruether, and A. H. Reinders, "An empirical model for rack-mounted PV module temperatures for Southeast Asian locations evaluated for minute time scales," *IEEE journal of photovoltaics*, vol. 5, no. 3, pp. 774–782, 2015.

[5] M. Prilliman, J. S. Stein, D. Riley, and G. Tamizhmani, "Transient weighted moving-average model of photovoltaic module back-surface temperature," *IEEE Journal of Photovoltaics*, vol. 10, no. 4, 5 2020. [Online]. Available: https://www.osti.gov/biblio/1634788

[6] B. Herteleer, A. Kladas, G. Chowdhury, F. Catthoor, and J. Cappelle, "Investigating methods to improve photovoltaic thermal models at second-to-minute timescales," *Solar Energy*, vol. 263, no. 111889, 2023.

[7] M. Boyd, T. Chen, and B. Dougherty, "NIST Campus Photovoltaic (PV) Arrays and Weather Station Data Sets," National Institute of Standards and Technology, 2017.

[8] B. Herteleer and J. Cappelle, "Replication Data for: Investigating methods to improve photovoltaic thermal models at second-to-minute timescales," 2022. [Online]. Available: https://doi.org/10.48804/RVTSD4

[9] B. Herteleer, B. Huyck, F. Catthoor, J. Driesen, and J. Cappelle, "Normalised efficiency of photovoltaic systems: Going beyond the performance ratio," *Solar Energy*, vol. 157, pp. 408–418, 2017. [Online]. Available: https://www.sciencedirect.com/science/article/pii/S0038092X1730717X

[10] M. Boyd, "High-speed monitoring of multiple grid-connected photovoltaic array configurations," *NIST Technical Note 1896*, 2015.

[11] W. McKinney, "Data structures for statistical computing in python," in *Proceedings of the 9th Python in Science Conference*, vol. 445. Austin, TX, 2010, pp. 51–56.

[12] J. Kurnik, M. Jankovec, K. Brecl, and M. Topic, "Outdoor testing of pv module temperature and performance under different mounting and operational conditions," *Solar Energy Materials and Solar Cells*, vol. 95, no. 1, pp. 373–376, 2011.

10.4229/EUPVSEC2023/4CV.1.1
020369-008

EVALUATING THE ACCURACY OF INVERTER BASED STRING IV MEASUREMENTS

Martin Bartholomäus, Luca Morino, Peter B. Poulsen and Sergiu V. Spataru
Technical University of Denmark, Department of Electrical and Photonics Engineering
Roskilde, 4000, Denmark

ABSTRACT: Inverter measurements of photovoltaic (PV) string current-voltage (IV) characteristic curves can support performance monitoring of PV plants without the need for additional equipment by using the intrinsic hardware capabilities of the inverter. This paper investigates the accuracy of such inverter based IV measurements of PV strings under various conditions, including shade, compared to a commercial PV array IV tracer. We find very good agreement for high, stable irradiance conditions with a mean relative P_{mp} error of 0.6% at a maximum of 4%. At low irradiance, the error increased to 2.7% on average at a maximum of 6.7%. Changing irradiance led to higher uncertainty around the irradiance conditions at each measurement and therefore to higher errors with I_{sc}, I_{mp}, and P_{mp}, while V_{oc} is usually well agreed. In limited shade conditions, the agreement is good, whereas the inverter started misinterpreting the shape of the IV curve when shade got more severe. At a G_{POA} of 220 W/m^2 or less, the inverter failed to take measurements or missed inflection points caused by the non-uniform shade. This work demonstrates that the inverter can produce IV curves comparable to a high-accuracy IV tracer under non-shaded and stable irradiance conditions.
Keywords: IV-curve, inverter, string-IV, IV-tracer, accuracy

1 INTRODUCTION

Commercial PV inverters frequently measure string IV curves to perform maximum power point tracking. Researchers and manufacturers have recently started showing interest in utilizing string IV data from inverters for performance monitoring and fault detection [1, 2]. The full IV characteristic curve of a PV device contains much more information on the state of a PV string/module than just the maximum power point (MPP), because most faults have a characteristic imprint on the IV curve outside of the MPP, for example they might also impact I_{sc}, V_{oc} or the slopes at these points. Therefore, it is possible to identify specific faults in PV modules by analyzing their IV curve [3-6]. Research has also demonstrated the use of string IV curves for condition monitoring and fault detection in field conditions [7, 8]. Other work showed the use of inverter IVs in combination with a wireless sensor network for fault diagnosis on module level [9]. The possibility of fully automating diagnostic systems based on IV data combined with machine learning algorithms has also been explored [10, 11]. Contrary to what was studied in [12] and [13], using inverters for condition monitoring bears the advantage that no further hardware is required. In addition, inverter IV data can also be utilized to feed a fully automated condition monitoring system that requires no manual labor. However, there is a lack of research on the accuracy and quality of inverter-based string IV measurements to this day. Therefore, we investigate such string IV compared to a high-accuracy commercial PV array IV tracer under stable irradiance, changing irradiance and artificial shade.

2 EXPERIMENTAL SETUP

2.1 Testing equipment

The experiments were conducted on a PV plant located at the Technical University of Denmark's (DTU's) Risø campus in Denmark (lat. 55.7, long. 12.1) on two south-facing fixed-tilt arrays with a tilt angle of 25°. The mounting structures accommodate five individual strings of different configurations connected to an inverter. Figure 1 gives an overview of the PV strings utilized for this study, which feature three different types of modules.

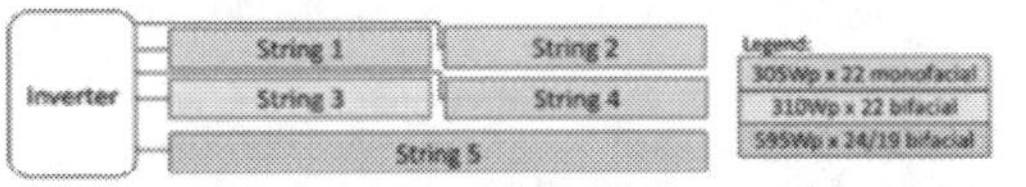

Figure 1: Array layout of tested strings at DTU Risø campus. The inverter can measure the IV-curve of each string. For measurements with the IV tracer, the strings were first disconnected from the inverter.

Strings 1, 2 and 4 each consist of 22 monofacial 295 Wp panels, string 3 of 22 bifacial 310 Wp panels, and string 5 was utilized with both 24 and 19 large bifacial 595 Wp panels. For some measurements, strings 1 and 2 were connected in parallel to form a larger string (string 1+2).

The inverter is a Huawei SUN2000-40KTL-M3, a commercial 40 kW inverter with IV scanning capabilities shown in Figure 2. IV curve measurements with the inverter consist of 64 data points per sweep. IV curve measurements from the inverter are compared to IV measurements taken with the Solmetric PVA-1500V4 portable IV tracer. The device has a voltage accuracy of ±0.5% or ±0.25 V and a current accuracy of ±0.5% or ±40 mA (both at 0 to 55°C) and measures a sweep in less than 250 ms, recording either 100 or 500 data points per sweep [14].

2.2 Description of the measurements

String IV measurements were taken by triggering the IV-scanning function of the inverter, which measures all attached strings. The procedure includes an initialization time, which creates a delay between when a sweep is triggered and the moment of the actual IV sweep, which is around 30 seconds. Therefore, the exact time when the string is measured is uncertain.

Figure 2: To the left: Huawei SUN2000-40KTL-M3 inverter with IV sweep capabilities. To the right: Solmetric PVA-1500V4 portable IV tracer. Image source: [14]

10.4229/EUPVSEC2023/4CV.1.4
020370-001

After an inverter sweep, the string under test was disconnected from the inverter and connected to the IV tracer. The total delay time between measurements caused by this procedure was approximately one to three minutes, with a maximum of five minutes. At each IV curve measurement, the irradiance in the plane of array, G_{POA}, and module temperature, T_{mod}, were measured, for which we used Solmetrics' Solsensor, a silicon photodiode mounted in the plane of array and one type K thermocouple temperature sensor per module type (one sensor is used to represent T_{mod} in string 1, 2 and 4).

Measurements taken under ambient conditions were corrected to common temperature and irradiance conditions following procedure 1 of IEC 60891, after which the new, corrected parameters (I_{sc}, V_{oc}, V_{mp}, I_{mp}, P_{mp} and FF) were calculated from the corrected curve. Here, P_{mp} is found as the observation that maximizes the product of current and voltage, V_{oc} is determined as the data point that is closest to a current value of 0 A, and I_{sc} analogous to that as the point where voltage is closest to 0 V. Then, the relative error is calculated with respect to the IV tracer. We captured measurements under different irradiance (from G_{POA} = 70 W/m^2 to G_{POA} = 1250 W/m^2) and temperature conditions (from T_{mod} = 21.6°C to T_{mod} = 41.9°C) as well as conditions that include artificial non-uniform shade and simulated row-to-row shading that were created using cardboard (0% light transmittance) or mesh (50% light transmittance), respectively.

The shade scenarios were performed on the monofacial strings 1 and 2, where each of the 22 panels mounted in portrait orientation has 60 full cells and three bypass diodes per module. The different shade scenarios include:

- Shading one cell on two modules with cardboard
- Shading two cells on two and on four modules with cardboard. The shaded cells are on different cell strings within each module.
- Shading a full row of cells with 50% transmittable mesh

Measurements were acquired on three different days: one day had intermittent clouds, one day had clear sky conditions, and the last day was overcast.

In addition, a second G_{POA} sensor (pyranometer with a photodiode detector, ISO 9060 fast response Class C [15]), which was mounted in the plane of array of string 5 and records G_{POA} every minute, was used to control the Solmetrics' sensor reading. To do that, we calculate a 'sensor disagreement' parameter, which is the absolute difference between the two irradiance sensors. This parameter is influenced by the response time of the sensors, the moment in which the measurement is read (the Solsenors' G_{POA} measurement was read from the software display, where the reading is only updated every few seconds) and the speed at which the conditions change. We also quantified the stability of the irradiance conditions, by computing a 'rate of change' parameter, RoC, as the absolute difference between consecutive measurement points.

3 RESULTS AND DISCUSSION

We present the results starting with measurements on unshaded strings at high and low irradiance conditions, then fast-changing irradiance conditions and finally the IV curve measurements on shaded strings.

3.1 Unshaded, high irradiance

Under stable irradiance conditions with G_{POA} > 650 W/m^2, RoC < 15 W/m^2 and sensor disagreement < 50 W/m^2 we performed 16 string IV measurements (n = 16). These include measurements on strings 1+2, 3, 4 and 5. Figure 3 displays all the IV curves, where each pair of corresponding measurements is corrected to common temperature and irradiance conditions. Figure 4 summarizes the same measurements by showing the relative differences between corresponding measurements at each relevant point and for the fill factor. The results show a mean P_{mp} error of -0.63% with a maximum of -3.95%. The currents have a higher error span of 4.91% at an average error of -0.44%, compared to the voltages' span of 0.81% with an average value of 0.18%. The higher span originates from the fact that the irradiance level almost linearly influences the currents, and the uncertainty of the irradiance measurements is greater. In general, there is a good agreement of the shape of the IV curves in clear sky conditions, apart from a noticeable difference between I_{sc} and P_{mp} on the measurements on string 1+2.

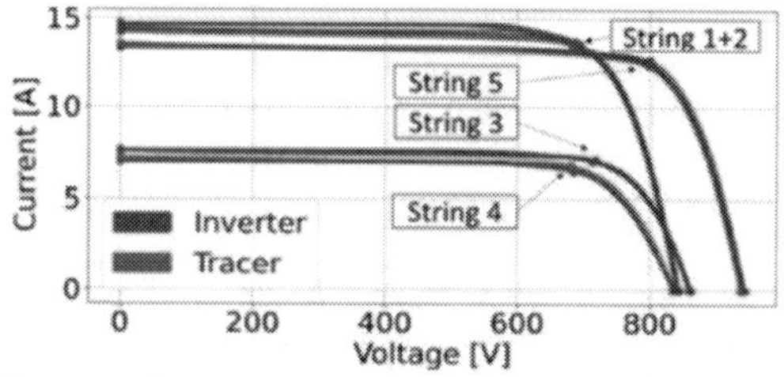

Figure 3: IV curves of non-shaded strings in high, stable irradiance conditions. 4 different strings are tested.

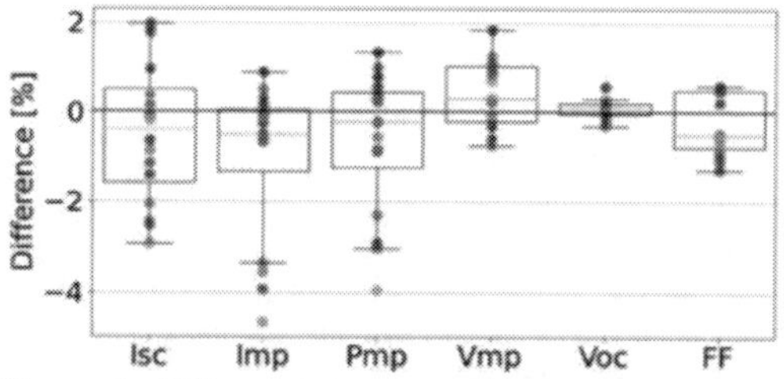

Figure 4: Differences between IV-tracer and inverter-measured IV curves across the IV parameters. Measurements on four different strings in high, stable irradiance without shading. Positive numbers mean the inverter overestimated. Unique colors indicate the same measurement.

3.2 Unshaded, low irradiance

Additional measurements (n = 12) were conducted in lower irradiance conditions at G_{POA} < 400 W/m^2, RoC < 15 W/m^2 and a sensor disagreement < 50 W/m^2, performed on string 1, 2, 3 and 5. The IV curve measurements and corresponding relative errors are displayed in Figures 5 and 6. Compared to the high irradiance scenario, the relative error increases. P_{mp} shows a mean deviation of -2.7% at a maximum of -6.73%. Again, the error is more prominent on the current side, whereas V_{oc} is very well agreed on. In some IV curves, there is an increasing slope in the inverter measurement between I_{sc} and P_{mp}, which can be caused by ramping irradiance during the measurement sweep.

10.4229/EUPVSEC2023/4CV.1.4
020370-002

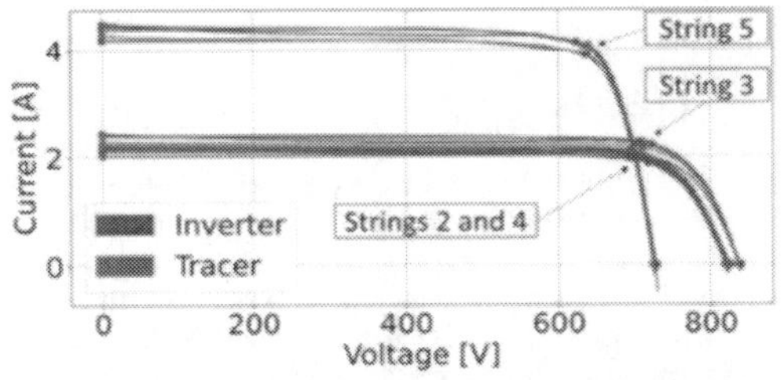

Figure 5: IV curves of non-shaded strings in low, stable irradiance conditions. 4 different strings are tested.

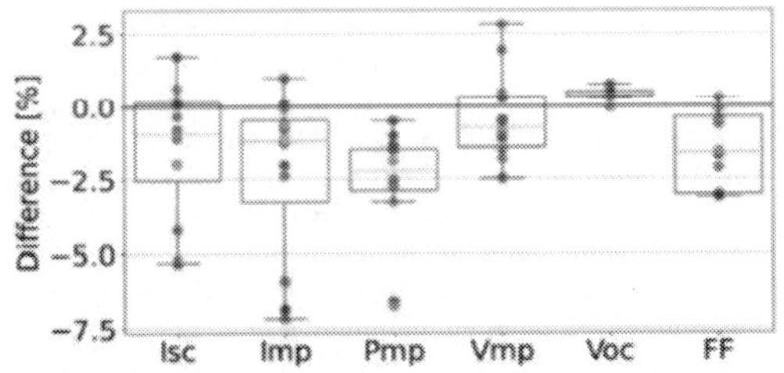

Figure 6: Differences between IV-tracer and inverter-measured IV curves across the IV parameters. Measurements on four different strings in low, stable irradiance without shading. Positive numbers mean the inverter overestimated. Unique colors indicate the same measurement.

3.3 Unshaded, fast-changing irradiance

On a day with intermittent clouds, measurements were selected with a change rate > 15 W/m^2 (n=4). The general level of sensor disagreement was around 150 W/m^2 for each measurement. The IV curves from string 2, 3, 4 and 5 and their summary statistics are depicted in Figures 7 and 8. The results show higher errors of around 8% on average for I_{sc}, I_{mp} and P_{mp}. This error is caused by uncertainty around the irradiance conditions at the inverter measurement, which originates from the fact that the exact moment of measurement was unknown. The large current level shift influences the position of the MPP on the curve and therefore causes an impact on P_{mp} as well.

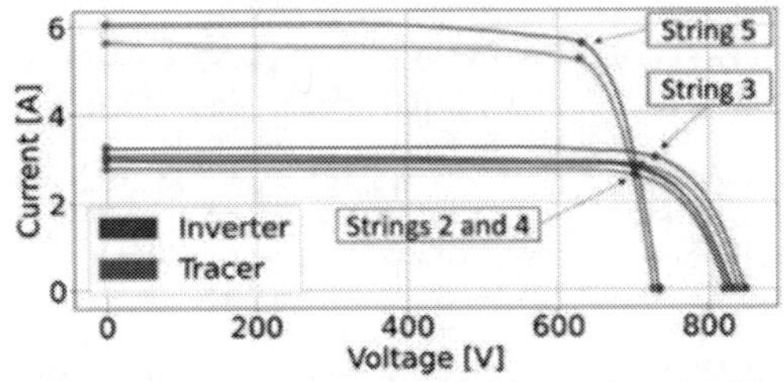

Figure 7: IV curves of non-shaded strings in fast-changing irradiance conditions. 4 different strings are tested.

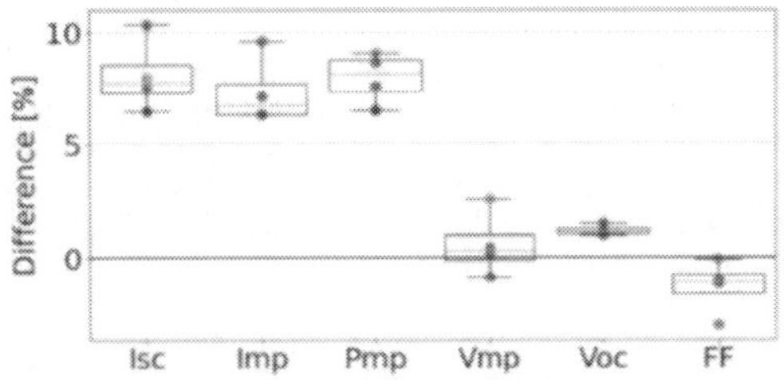

Figure 8: Differences between IV-tracer and inverter-measured IV curves across the IV parameters. Measurements on four different strings in fast-changing irradiance without shading. Positive numbers mean the inverter overestimated. Unique colors indicate the same measurement.

3.4 IV measurements in the presence of non-uniform shade

Figures 9 to 12 show IV curve measurements influenced by artificial non-uniform shade, with increasing severity from top to bottom. These measurements were done on strings 1 and 2, which feature monofacial panels in portrait mode with 60 full cells and three bypass diodes.

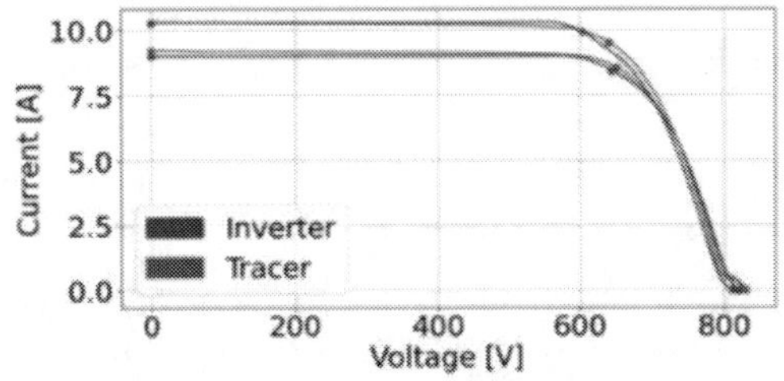

Figure 9: IV curves from measurements with shade: one cell was shaded on two out of 22 modules using cardboard.

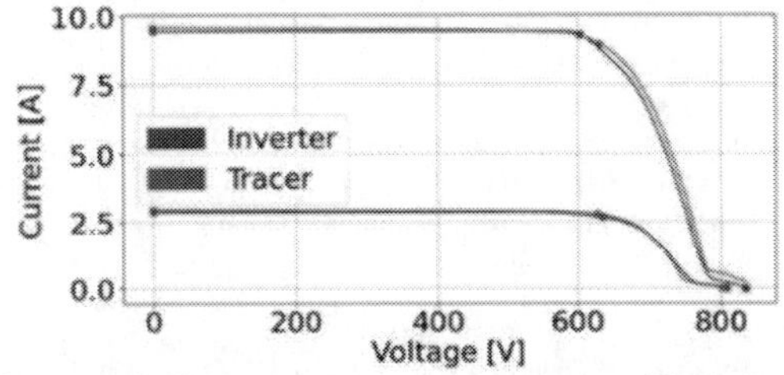

Figure 10: IV curves from measurements with shade: using cardboard, two cells were shaded on two out of 22 modules. The two shaded cells are on different cell strings within the modules.

The results show a good agreement between the IV curves under light shade, displayed in Figures 9 and 10. We see slight deviations around the maximum power point, where the inverter tends to display a lower, flatter curve. When one cell is shaded on two modules in a string of 22 modules, the IV tracer displays a clear inflection from the impact of the shade near V_{oc}, whereas the inverter does not show this inflection as clearly, instead, the curve is only slightly bent near V_{oc}. This implies that shade detection from the inverter IV curve based on inflection points could be negatively affected, leading to lower sensitivity than the IV tracer. If two modules are shaded on two cells each, the inflection becomes clearer and both methods display similar bending of the curves.

Under more severe shade, the inverter displays larger deviations, as seen in Figure 11, where in one of the measurements, the inverter-measured a flat curve between I_{sc} and P_{mp}, which appears to be a measurement artifact. In the second measurement, the full IV curve was recorded, but there are large deviations in V_{oc} as well as deviations in the shape and I_{sc} level.

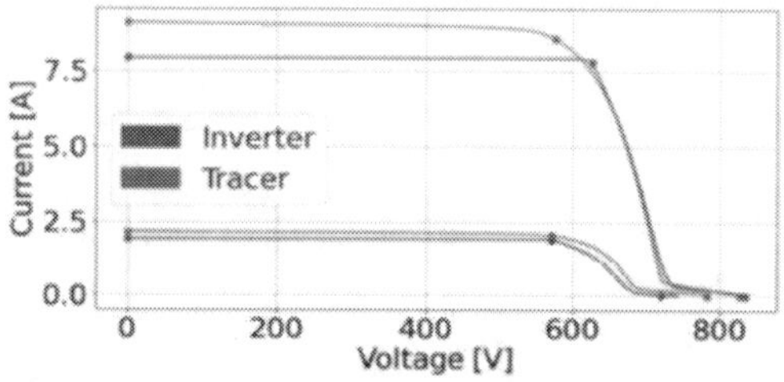

Figure 11: IV curves from measurements with shade: using cardboard, two cells were shaded on four out of 22 modules. The two shaded cells are on different cell strings within the modules.

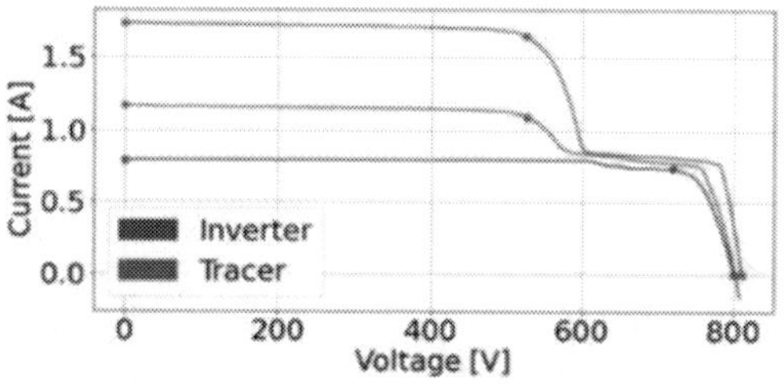

Figure 12: IV curves from measurements with shade: a full cell row on five out of 22 modules was shaded using a mesh with 50% light transmittance.

In the experiments shown in Figure 12, we used a mesh as a shading object (a material with 50% light transmittance) and the inverter failed to take a total of four attempted measurements, which were done at a very low irradiance level with a G_{POA} between 220 W/m^2 and 75 W/m^2. In the case in which the inverter completed the measurement, its IV curve shows a large deviation from the IV curve of the IV tracer, missing the inflections caused by the shade.

4 CONCLUSIONS

In high and stable irradiance conditions, we observed that inverter measured IV curves are very similar to the ones from the IV tracer. Relative P_{mp} differences were at 0.6% on average, with a maximum of 4%. In low but stable irradiance, an increase of the mean P_{mp} error to 2.7% and maximum error to 6.7% is observed.

In fast-changing irradiance conditions caused by intermittent clouds, we observed an error of around 8% on I_{sc}, I_{mp} and P_{mp}. Due to changing conditions during the measurement, uncertainty around measuring the irradiance conditions is the main contributor to the error.

In general, the disagreement was higher on the current side of the IV curves, caused by the higher uncertainty around G_{POA} measurements in our measurement method, where the exact time of measurement of the inverter is uncertain by about 30 seconds. In addition, the Solsenors' G_{POA} measurement was read from the software display, where the reading is only updated every few seconds. Therefore, V_{oc} is more agreed on across all measurements than I_{sc}.

Under non-uniform shade, the quality of the inverter IV curves was good when the irradiance conditions were stable and the shaded area small (no more than two cells on two modules fully shaded). When influenced by small shade the inverter did not display the inflection points as clearly. More severe shade led to the inverter showing faulty or no IV curves. The lower limit of the inverter was reached at irradiance conditions of $G_{POA} < 220$ W/m^2 in combination with shade on the panels, where the inverter struggled to produce an IV curve or missed the inflection points caused by the shade.

The real differences between the tracer and inverter are probably lower, because the error shown here includes uncertainty with the irradiance and temperature conditions of the string as well as from the correction procedure. Also, quick irradiance changes (e.g., from intermittent clouds) contributed to the uncertainty. The main limitation of the measurement method presented here is that measurements were not taken at the same time, which in turn caused the uncertainty around the conditions at each measurement and the need to correct to common conditions.

The measurement setup could be improved so that the strings can be measured with both methods at nearly the same time, for example with a switchboard or parallel measurements with both devices.

5 ACKNOWLEDGEMENTS

This research has been carried out within the EUDP research project "DigitalPV - Digitalisation of PV operations", case no. 64021-2082.

6 REFERENCES

[1] Kopacz, Csaba, et al. "Remote and Centralized Monitoring of PV Power Plants." 2014 International Conference on Optimization of Electrical and Electronic Equipment. Optim 2014. IEEE Computer Society. p. 6851005. 2014. DOI:10.1109/OPTIM.2014.6851005.

[2] A. Augusto, A. Killam, S. G. Bowden, and H. Wilterdink. "Measuring outdoor I–V characteristics of PV modules and systems." Progress in Energy. vol. 4. no. 4. p. 042006. 2022. DOI: 10.1088/2516-1083/ac851c.

[3] M. Köntges, S. Kurtz, C. Packard, U. Jahn, K. A. Berger, K. Kato, T. Friesen, H. Liu, and M. V. Iseghem. "Review of failures of photovoltaic modules." IEA Photovoltaic Power Systems Programme. 2014.

[4] Y. Zhu and W. Xiao. "A comprehensive review of topologies for photovoltaic I-V curve tracer." en. In: Solar Energy 196. pp. 346–357. Jan. 2020. ISSN: 0038092X. DOI: 10.1016/j.solener.2019.12.020.

[5] Ma, Mingyao, et al. "Fault Diagnosis of PID in Crystalline Silicon Photovoltaic Modules through I-V Curve." Microelectronics Reliability. vol. 126. Elsevier Ltd. p. 114236. 2021. DOI:10.1016/j.microrel.2021.114236.

[6] El Basri, Y. et al. "A Proposed Graphical Electrical Signatures Supervision Method to Study PV Module Failures." Solar Energy. vol. 116. Elsevier Ltd. 2015. pp.

247–56. DOI:10.1016/j.solener.2015.02.048.
[7] Sergiu Spataru et al. "Monitoring and Fault Detection in Photovoltaic Systems Based On Inverter Measured String I-V Curves." 31st European Photovoltaic Solar Energy Conference 2015 (Eupvsec). WIP - Renewable Energies. 2015. pp. 1667–74. DOI:10.4229/EUPVSEC20152015-5BO.12.2.
[8] Sergiu Spataru et al. "Diagnostic method for photovoltaic systems based on light I–V measurements." en. In: Solar Energy. vol. 119. Sept. 2015. pp. 29–44. ISSN: 0038092X. DOI: 10.1016/j.solener.2015.06.020.
[9] J. Zhang et al. "Module-Level Fault Diagnosis of Photovoltaic Array based on Wireless Sensor Networks and Inverter Activated I-V Scanning." 2022 IEEE 5th International Electrical and Energy Conference (CIEEC). Nangjing. China. 2022. pp. 2766-2771. DOI: 10.1109/CIEEC54735.2022.9846806.
[10] Haohui Liu et al. "Automatic IV Curve Diagnosis with Deep Learning." en. In: 2021 IEEE 48th Photovoltaic Specialists Conference (PVSC). Fort Lauderdale. FL. USA: IEEE. June 2021. pp. 2242–2246. ISBN: 978-1-66541-922-2. DOI: 10.1109/PVSC43889.2021.9519033.
[11] Michael W. Hopwood et al. "Neural Network- Based Classification of String-Level IV Curves From Physically-Induced Failures of Photovoltaic Modules." en. In: IEEE Access 8. 2020. pp. 161480–161487. ISSN: 2169-3536. DOI: 0.1109/ACCESS.2020.3021577.
[12] A. Marquis, M. Gostein and B. H. King. "Validation of In-Situ I-V Measurement Unit for PV System Monitoring Applications." 2022 IEEE 49th Photovoltaics Specialists Conference (PVSC). Philadelphia. PA. USA. 2022. pp. 0291-0294. DOI: 10.1109/PVSC48317.2022.9938898.
[13] Vervaart, Mark, et al. "Dynamic IV Analysis System for Diagnosis of PV-Module Strings in a Large Scale Pvpower Plant." 33rd European Photovoltaic Solar Energy Conference and Exhibition. Proceedings of the International Conference. WIP - Renewable Energies. 2017. pp. 2013–15.
[14] Solmetric Inc. "PVA-1500V4 PV Analyzer Kit". en. 2021. URL: https://www.solmetric.com/pva1500.html. (visited on 05/08/2023).
[15] OTT HydroMet B.V. "SP Lite2 Pyranometer". en. 2023. URL: https://www.kippzonen.com/Product/9/SP-Lite2-Pyranometer. (visited on 01/18/2023).

Health Monitoring of Solar Photovoltaic Plants

Gaetana Anamiati[1], Lars Landberg[2], Gerardo Guerra[1], Pau Mercade Ruiz[1]
[1]GreenPowerMonitor a DNV company, Gran Via de les Corts Catalanes, 130, Barcelona, Spain;
Email: gaetana.anamiati@dnv.com, pau.mercade@dnv.com, gerardo.guerra@dnv.com
[2]DNV Denmark, Tuborg Parkvej 8, Hellerup, Denmark; Email: lars.landberg@dnv.com

WHAT?

Generate a machine learning model able to monitor the performance of a photovoltaic (PV) plant. The model identify underperformances comparing a PV to its neighbors. The model used is a feed forward Neural Network (NN) based on [1] able to estimate the power output (P) of a specific PV plant from four input features coming from the neighbors of the plant. The procedure has been tested and validated at several plants, using historical data collected for a period of one and a half year and have a 5-minute sampling period.

WHY?

The idea is to identify PV's under/over performances by comparing it to its neighbors since it is very difficult to do so considering only its own data. In this way, we can provide industries with a valid tool on the correct functioning of PV.

HOW?

For this analysis, a NN consisting in 4 inputs is trained for each plant. The inputs, coming from 15 neighbors are: irradiance (I), ambient temperature (T), elevation and azimuth angle. The final output is the power (P) of the plant we are analyzing.

The following steps are crucial before to train the model:

1. **Deep data cleaning.** It consists in discarding missing data, out-of-range data, night time and discrepancies between P and the I-T area as in ref.[1] and finally finding the best irradiance measurements for each plant following the procedure in ref.[2].
2. **Finding the neighbors.** The concept of "neighborhood" is broadened beyond the mere geographical sense. It includes a conception of similarity that consider 10 dimensions. So, given a plant with its own specification the features used to find its neighbors are: longitude, latitude, elevation, global horizontal irradiance, temperature, turbidity, number of trackers, number of inverters, DC power, AC power.
3. **Dealing with the curtailment.** Curtailment is the deliberate reduction in output below what could have been produced in order to balance energy supply and demand or due to transmission constraints. So, the idea was to create a function to recognize which is the interval of time for which the power was curtailed, predict the power for those points using a quadratic regression: $P = I + I\cdot T + I^2$ [3].

RESULTS AND CONCLUSIONS

Performance ratio is, see [4] for definition, the useful energy (the power measured) with respect to the energy which would be produced if the system was continuously working at its nominal standard condition (STC), like temperature 25º, irradiance 1000 watt/m2, and air mass 1.5. Since we are interested in comparing the PV plant PR with that one of its neighbors, the PR calculated by us is the power of the PV plant (power measured) over the power predicted by the model (with the inputs coming from neighbors). The complete study consists in the analysis of three dataset of plants, considering two different regions:

- Plants with trackers in region 1.
- Plants without trackers in region 1.
- Plants with trackers in region 2.

Here we show the result regarding the first case. Tab.1 shows the numerical results for the PR of one plant, the mean of the PR of the plant and its neighbors, the mean of the PR of all plants belonging to the dataset. Fig. 1 shows the predicted power of a plant after training the model over its 15 neighbors. Fig. 2 shows a 2D histogram plot of predicted power vs power both for the plant and for each of the 15 neighbors. Fig. 3 shows the distribution of the PR of all plants tested.

REFERENCES

[1] G. Guerra, P. Mercade Ruiz, and L. Landberg, "A data-driven model for solar inverters", 37th EU PVSEC, 2020.

[2] G. Guerra, P. Mercade Ruiz, G. Anamiati, and L. Landberg, "A study of the impact of multiple irradiance measurements on the performance of data-driven models", 8th World conference on photovoltaic energy conversion, 2022.

[3] S. Moslehi, T. Agami Reddy, Srinivas Katipamula, "Evaluation of data-driven models for predicting solar photovoltaics power output", Energy, Volume 142, 2018.

[4] A. M. Khalid, I. Mitra, W. Warmuth, V. Schacht, "Performance ratio – Crucial parameter for grid connected PV plants", Renewable and Sustainable Energy Reviews, Volume 65, 2016.

	Plant
Performance ratio (days)	1.0111
Performance ratio (weeks)	1.0170
Performance ratio (months)	1.0246
	Mean (line of neighbors)
Performance ratio (days)	1.002
Performance ratio (weeks)	1.006
Performance ratio (months)	1.009
	Mean (57 plants)
Performance ratio (days)	1.0153
Performance ratio (weeks)	1.0192
Performance ratio (months)	1.0204

Tab. 1: Numerical results separated in three blocks. At the top, the PR of one single plants. In the middle, the mean of the PR of one plant and its neighbours. At the bottom, the mean of the PR of all plants. All blocks show results sampled in days, weeks and months.

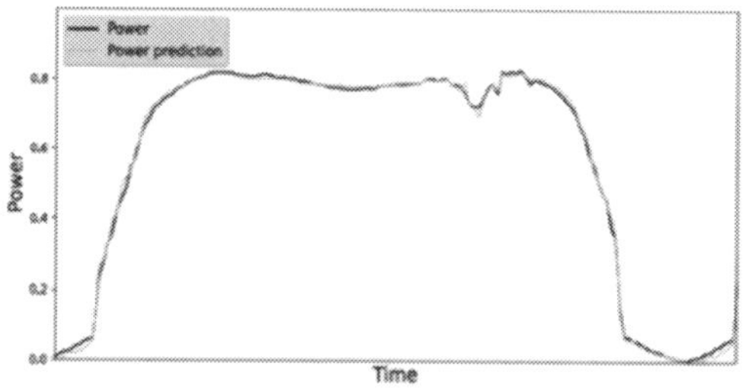

Fig. 1: Power output from the model. The blue line is the power of the plant, the light blue one is the power predicted by the model. The interval of time is one day.

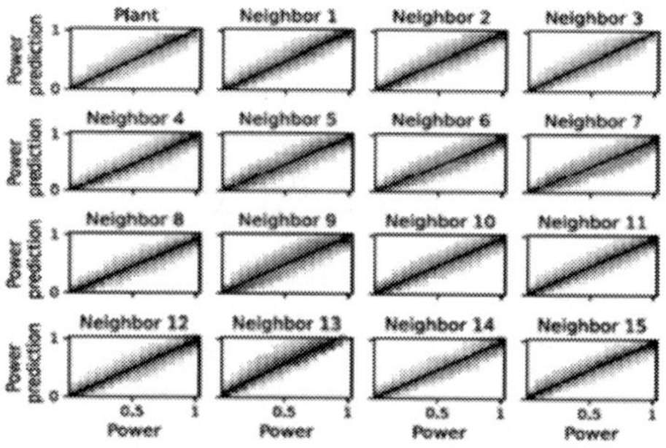

Fig. 2: 2D histogram of predicted power vs power for a plant and all its neighbours. The black line is the linear regression.

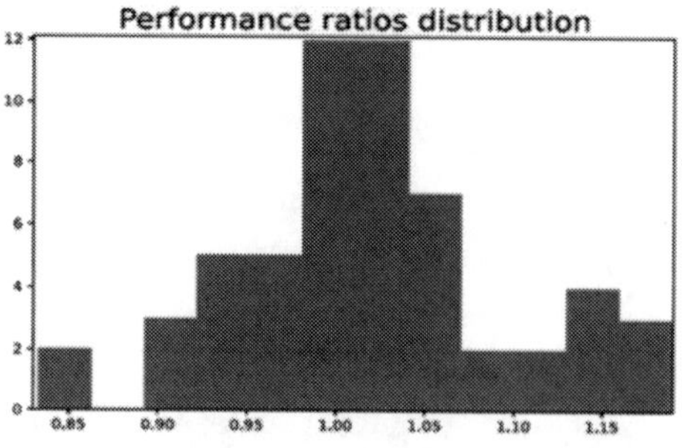

Fig. 3: Distribution of the PR of all the plants with trackers located in region 1.

CHARACTERIZING THE PERFORMANCE OF DAYLIGHT FILTERS FOR ELECTROLUMINESCENCE IMAGING OF CRYSTALLINE SILICON PV MODULES

Gisele A. dos Reis Benatto, Alejandra A. Mayordomo, Thøger Kari, Rodrigo Del Prado Santamaría, Peter B. Poulsen, Sergiu V. Spataru.
Department of Electrical and Photonics Engineering, Technical University of Denmark (DTU), Frederiksborgvej 399, 4000 Roskilde, Denmark.

ABSTRACT: To perform outdoor daylight electroluminescence (EL) and photoluminescence (PL) of PV modules for defect and fault detection, it is necessary to filter away any other light source apart from the solar cell's luminescence emission. Image processing is required in broad daylight to increase the signal-to-noise ratio digitally. However, it is reported in the literature that an optical (often bandpass) filter in front of the camera lens or sensor is also needed to overcome the 4-5 orders of magnitude more intense sunlight. The specs of such a "daylight filter" are often mentioned but are rarely the same between research groups and inspection companies. To our knowledge, a study has not addressed the main factors necessary to define an optimal filter for daylight luminescence imaging of PV modules. In this work, we evaluated the spectroscopic characteristics of four filter configurations. We performed daylight EL imaging to estimate their performance in signal-to-noise ratio and EL image quality.
Keywords: Electroluminescence, PV modules, Outdoor inspection.

1 INTRODUCTION

It is reported in the literature that outdoor electroluminescence (EL) and photoluminescence (PL) imaging under sunlight requires a "daylight filter" [1] to avoid most of the sunlight being captured by the camera at the same time. However, its required specs are not strict or optimized. Most of the reported ones are bandpass (BP) filters with center wavelength ranging from 1140 to 1160 nm, bandpass width of 25 to 30 nm, optical density ≥ 4, and bandwidth transmission $\geq 90\%$ [2]–[5]. Guada et al. report using a bandpass filter with a bandwidth of 150 nm [6], while Silverman et al. report using an Indium Gallium Arsenide (InGaAs) camera along with a 1000 nm long-pass (LP) filter only, thereby allowing a full detection range of 1000 to 1700 nm of the sunlight spectrum [7]. These reports indicate that a broader range of filters can be successfully used to acquire EL and PL images in broad daylight. However, which bandwidth range will be optimal to minimize the number of images required to reach a desirable signal-to-noise ratio (SNR) is still a question to be answered. The interest in investigating this issue occurs especially when additional noise is present due to camera motion (in the case of aerial drone inspections) and sun variations (environmental conditions). In this work, we compare the filter performance for daylight EL (dEL) image acquisition using two BP filters and two longpass and shortpass (SP) filter sets (which form a custom BP filter) and a No filter scenario as a reference. EL images are obtained with a range of exposure times to evaluate the correlation with possible contributing factors for the filter performance.

2 METHODOLOGY

2.1 Experimental details

The specifications of the studied filter configurations, provided by the suppliers, can be seen in Table I. For the BP filters, spectral transmission details beyond the 200 to 1500 nm interval was not available, which does not cover the entire InGaAs spectral sensitivity range.

For dEL imaging, three modules were put to the test: an aluminum back surface field (Al-BSF) module, a passivated emitter and rear cells (PERC) module, and an interdigitated back contact (IBC) module. These samples were chosen for low, medium and high EL emission scenarios, respectively, with Al-BSF modules having a cell V_{OC} of 0.62 V, PERC module with a cell V_{OC} of 0.66 V, and IBC module with a cell V_{OC} of 0.72 V.

Table I: Specs of the four filter configurations studied.

Filter configuration	Central wavelength	Bandwidth	Transmission	Optical density
BP25	1150 nm	25 nm	>90%	>4
BP50	1150 nm	50 nm	>90%	>4
LP1050+SP1250	1150 nm	200 nm	>97%	>5 + >4
LP950+SP1250	1100 nm	300 nm	>97%	>5 + >4

The spectral data was acquired using an InGaAs B&W Tek spectrometer Sol 1.7 (900-1700 nm range) with five averaged acquisitions. The measured spectra were dark-subtracted, and the absolute irradiance values were defined via calibration using an incandescent lamp.

The dEL image acquisitions were performed with an InGaAs 640x512 pixel model C-RED 3 from First Light Imaging. The forward current bias was applied using a Electro-Automatik PSI 9000 DC power supply unit with an embedded waveform generator, modulating the current from 0 A to Isc with a 20 Hz square waveform. The image acquisition framerate was set to 160 or 320 fps to accommodate the exposure time interval, which varied from 1 to 5 ms in steps of 1 ms. The sun irradiance at the plane-of-the-array during the image acquisition ranged from 530 to 750 W/m^2 and was verified before and after the acquisition to check for significant changes. For these irradiances, a lens equipped with an iris was a requirement to avoid sensor saturation. Measurements during which the irradiance variation was higher than 10 W/m^2 were discarded.

2.2 Image processing and SNR metrics

The dEL images were processed using Fast Fourier Transform (FFT) [8], and and to evaluate the amount of random solar noise in the signal, the SNR_{Kari} metric was used [9]. A satisfactory signal for image extraction was considered when SNR_{Kari} = 1 and upward. Values of $SNR_{Kari} > 4$ ensured a high-quality image, while values below one resulted in poor quality.

The filter performances were numerically evaluated using SNR_{Kari} metrics, plotted against the background solar variation range. Other factors, such as iris aperture and sun irradiance, were considered but no correlation was found with the filter performance evaluations.

10.4229/EUPVSEC2023/4CV.1.6
020372-001

3 RESULTS AND ANALYSIS

3.1 Filter spectroscopic evaluation

Fig. 1 shows the spectra with part of an Al-BSF cell EL emission transmitted by the studied filter configurations at the same spectral range detected by an InGaAs camera. The transmission ratio between the total EL emission of the PV cell in relation to the total emission (No-filter spectrum – gray line) is shown in Table II. It is important to note that the LP+SP filter sets transmit more than 90% of the EL emission and block all the remaining wavelengths of sunlight detected by an InGaAs sensor. The inset on the graph shows the EL image of the cell, indicating how much of the signal is dimmed by the filters.

Fig. 2 shows the portion of direct sunlight spectra blocked by the different filter configurations placed in front of the spectrometer's probe. This measurement revealed that BP25 and BP50 do not block the sunlight at 1520-1680 nm. This was unknown since the manufacturer only informs the transmission and blocking properties from 200-1500 nm. Table II shows the sunlight blockage ratio of the different filters configurations compared with the No-filter measurement of Fig. 2 regarding the spectral range detected by an InGaAs sensor.

Blocking the noise from the sun also means blocking a significant portion of the signal from the crystalline silicon (c-Si) EL emission. Filters tested transmit 34-94% of the EL emission spectrum and 88-49% of the sun within the InGaAs detector sensitivity range (900-1700 nm). It is important to notice that BP25, which was expected to be the filter that blocked most of the solar spectrum in comparison with the other filters, blocks 88%, while BP50 blocks 90% of the sunlight. This occurs due to the unexpected transmission of the BP25 filter in the 1520-1680 nm range, where BP50 has a narrower transmission.

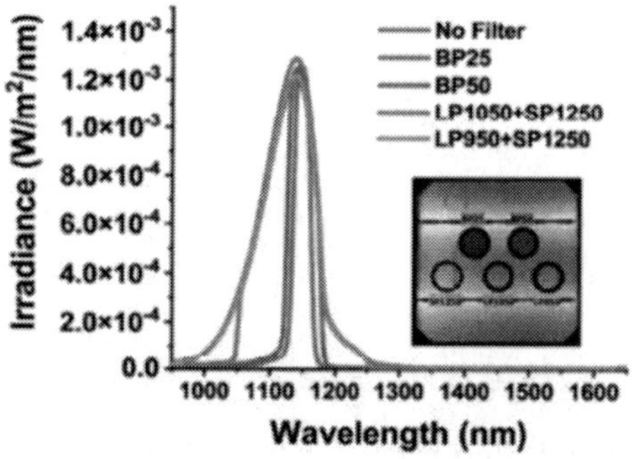

Figure 1:. Al-BSF solar cell emission spectra transmitted through each filter configuration. The inset shows the Al-BSF cell EL image with the filters on top of it, showing their effect in practice.

Table II: EL emission transmission and sunlight blockage ratio at the 900-1700 nm spectral range.

Calculated from the spectral data below	EL Emission Transmission Ratio	Sun Blockage Ratio
No Filter	100%	0%
BP25	34%	88%
BP50	51%	90%
LP1050+SP1250	90%	62%
LP950+SP1250	94%	49%

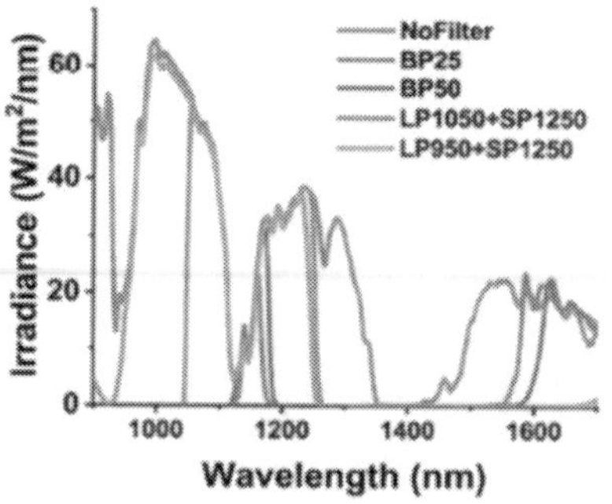

Figure 2:. Direct sunlight transmission spectra for each filter configuration.

3.2 Filter dEL imaging performance

Fig. 3 shows the dEL images of the Al-BSF, PERC, and IBC modules with exposure times from 1-5 ms performed with No filter and the four filter configurations. The obtained dEL images for the low EL emission Al-BSF module show low quality for the No and BP25 filters. For the BP25 at 3 and 5 ms most of the features, such as finger failures and cracks, can be well defined, but the fact that the images display much noise for 2 and 4 ms indicates that this filter is not efficient in providing robust dEL images for this type of module.

The images for BP50 and the LP+SP filter combinations are robust for ≥ 2ms exposure time, and the most stable throughout the exposures was BP50. Only the No filter option provided low and unstable quality images for the medium EL emission PERC module. As for the IBC, with its high EL emission, the dEL images are excellent and as good as indoor EL images in all instances of exposure time and filter configuration.

3.3 Filter SNR performance

Fig. 4 visualizes the filter configuration performance in terms of SNR_{Kari} marker for the different PV modules and exposure times. Sun variation range is also shown, to investigate possible correlations toward the SNR, because it has previously been observed, that it can be a key factor in adding noise to dEL image sequence acquisition [10]. For the Al-BSF module, BP50 and LP+SP filter configurations are the only ones that provide satisfactory SNR_{Kari} values for all exposure times, with the SNR in general doubled compared to BP25 and No filter cases. It is worth noting that BP25 and No filter show similar SNR values, which is unexpected due to their highly different transmission and blocking specs. An apparent inverse correlation between SNR value and sun variation range can be observed for this module for the images acquired with BP5o and LP+SP filters. However, more statistics would be required to confirm it since the opposite is the case for the PERC and IBC modules. The lack of a clear correlation with sun variation in these experiments could be due to the relatively low ranges of this variation compared to previous experiments.

A good correspondence between the dEL image quality and the SNR_{Kari} can be observed for the PERC module. High-quality images and $SNR_{Kari} \geq 4$ were acquired for all filter configurations, with only the No Filter case providing low SNR values. BP50 and LP+SP filter configurations follow the same trend as the Al-BSF module, showing SNR values two times higher than the

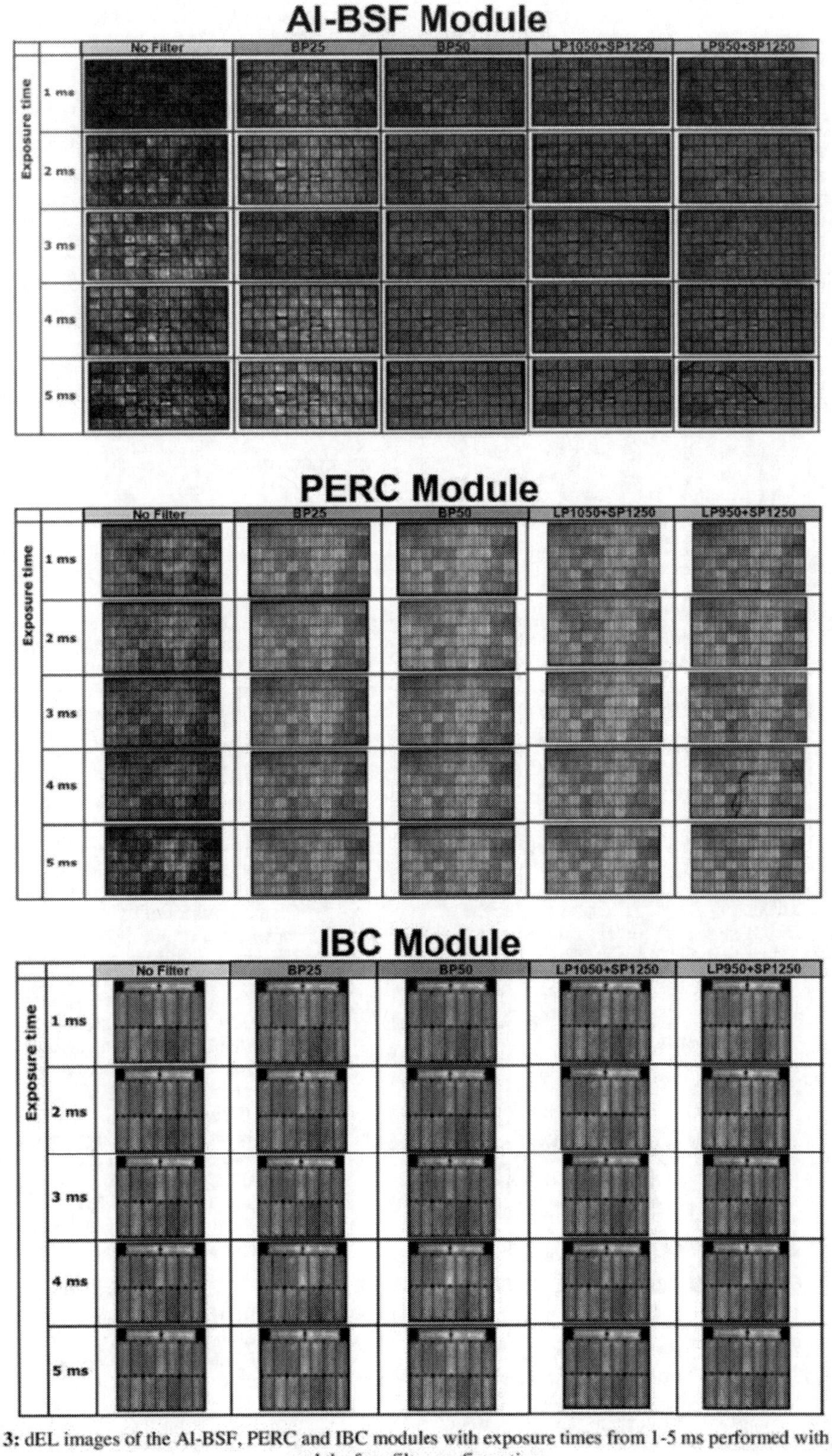

Figure 3: dEL images of the Al-BSF, PERC and IBC modules with exposure times from 1-5 ms performed with no filter and the four filter configurations.

10.4229/EUPVSEC2023/4CV.1.6
020372-003

image quality is confirmed with $SNR_{Kari} > 10$ for all cases. This time, LP+SP shows a better SNR value than BP25, and BP50 shows more clearly the highest SNR.

4 CONCLUSIONS

For Al-BSF and PERC modules (Cell V_{OC} = 0.62 V and 0.66 V, respectively), an optical filter was required to obtain a satisfactory EL signal ($SNR_{Kari} \geq 1$) – and good quality dEL image – from the sunlight noise at the studied conditions. A filter was unnecessary for the IBC module (Cell V_{OC} = 0.72 V).

Using BP50 and the LP+SP filter combinations doubled the SNR values for the three modules studied. BP50 showed the most stable signal and dEL image quality results for the Al-BSF, where the daylight filter had the highest performance. The LP+SP filter combinations had the subsequent best performance, indicating that transmitting most of the EL emission and blocking the remaining wavelengths detected by the InGaAs camera is the essential role of the optical filter.

Although some correlation between sun variation range and image quality has previously been observed, no significant interaction was found with the SNR metric in this study. The experiments' relatively small sun variation range could be a possible reason for this. In the future, more statistical data is needed for a comprehensive SNR analysis and filter optimization.

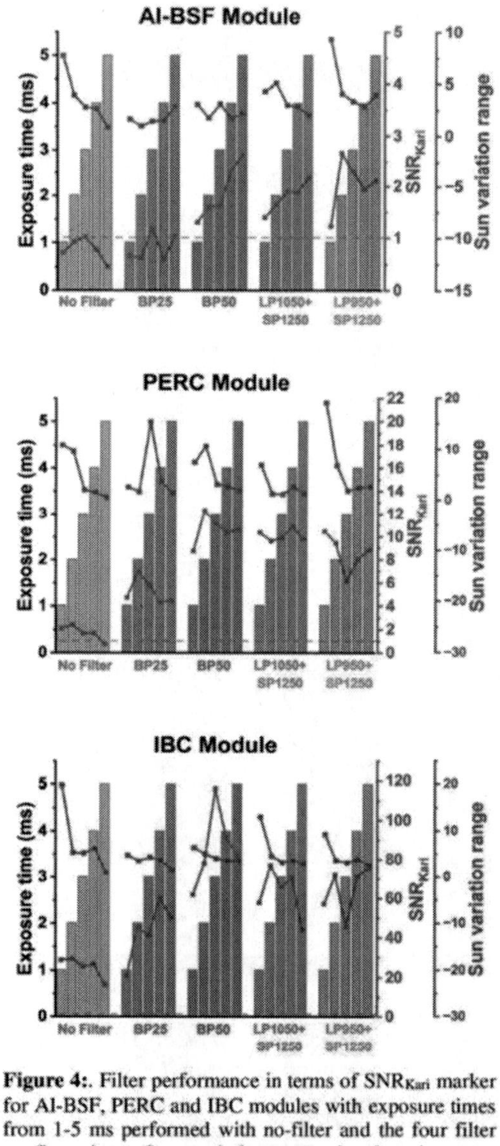

Figure 4:. Filter performance in terms of SNR_{Kari} marker for Al-BSF, PERC and IBC modules with exposure times from 1-5 ms performed with no-filter and the four filter configurations. Sun variation range is also shown to consider its influence on SNR.

5 ACKNOWLEDGEMENTS

This research has been carried in the Eurostars project no. 115687, Automated Daylight Electroluminescence Inspection of Large Photovoltaic Systems (ADELI).

6 REFERENCES

[1] L. Stoicescu, M. Reuter, and J. H. Werner, "DaySy: Luminescence Imaging of PV Modules in Daylight," in *29th European Photovoltaic Solar Energy Conference and Exhibition*, 2014, pp. 2553–2554. doi: 10.4229/EUPVSEC20142014-5DO.16.2.

[2] R. Bhoopathy, O. Kunz, M. Juhl, T. Trupke, and Z. Hameiri, "Outdoor photoluminescence imaging of photovoltaic modules with sunlight excitation," *Prog. Photovoltaics Res. Appl.*, vol. 26, no. 1, pp. 69–73, Jan. 2018, doi: 10.1002/pip.2946.

[3] G. A. dos Reis Benatto *et al.*, "Drone-Based Daylight Electroluminescence Imaging of PV Modules," *IEEE J. Photovoltaics*, vol. 10, no. 3, pp. 1–6, 2020, doi: 10.1109/jphotov.2020.2978068.

[4] Y. Augarten, A. Wrigley, A. Gerber, B. Pieters, and U. Rau, "Quantitative Outdoor Imaging: Analysis of Solar Modules in Daylight," in *32nd European Photovoltaic Solar Energy Conference and Exhibition*, 2016, pp. 1846–1848. doi: 10.4229/EUPVSEC20162016-5BV.1.17.

[5] M. Vuković, I. Eriksdatter Høiaas, M. Jakovljević, A. Svarstad Flø, E. Olsen, and I. Burud, "Photoluminescence imaging of silicon modules in a string," *Prog. Photovoltaics Res. Appl.*, vol. 30, no. 4, pp. 436–446, 2022, doi: 10.1002/pip.3525.

[6] M. Guada *et al.*, "Daylight luminescence system for silicon solar panels based on a bias switching method," *Energy Sci. Eng.*, no. June, pp. 1–15, 2020, doi: 10.1002/ese3.781.

[7] T. J. Silverman, M. G. Deceglie, K. Vansant, S. Johnston, and I. Repins, "Illuminated Outdoor Luminescence Imaging of Photovoltaic Modules," *IEEE Photovolt. Spec. Conf.*, 2017.

[8] Thøger Kari Hass, S. Spataru, A. A. Santamaria Lancia, H. Parikh, P. B. Poulsen, and G. A. dos Reis Benatto, "Computer Vision Method for Extracting an Induced Electroluminescence Signal from Photovoltaic Modules in Daylight Conditions Using Drone-Captured Images," in *37th European Photovoltaic Solar Energy*

Conference and Exhibition, 2020, pp. 1573–1579. doi: 10.4229/EUPVSEC20202020-5CV.3.44.

[9] G. A. dos Reis Benatto *et al.*, "Daylight Electroluminescence of PV Modules in Field Installations: When Electrical Signal Modulation is Required?," in *8th World Conference on Photovoltaic Energy Conversion*, 2022, pp. 735–739. doi: 10.4229/WCPEC-82022-3BV.3.44.

[10] T. Kari, R. D. P. Santamaria, G. A. dos R. Benatto, P. Koelblin, L. Stoicescu, and S. V. Spataru, "Evaluation of Motion-Induced Noise and Pixel Bleeding in Electroluminescense Field Inspection of PV-Modules," in *IEEE 50th Photovoltaics Specialists Conference (PVSC), San Juan, PR, USA*, 2023.'

DAYLIGHT ELECTROLUMINESCENCE IMAGING METHODOLOGY COMPARISON

Gisele A. dos Reis Benatto[1], Thøger Kari[1], Rodrigo Del Prado Santamaria[1], Sergiu V. Spataru[1], Cristian Terrados[2], Diego González-Francés[2], Julián Anaya[2], Kabir Sulca[2], Victor Gómez-Alonso[2], Miguel Ángel González[2], Oscar Martínez[2]
[1]Department of Electrical and Photonics Engineering, Technical University of Denmark (DTU), Frederiksborgvej 399, 4000 Roskilde, Denmark.
[2]GdS-Optronlab group, Dpto. Física de la Materia Condensada. Universidad de Valladolid. Edificio LUCIA. Paseo de Belén, 11. 47011 Valladolid, Spain.

ABSTRACT: Electroluminescence imaging is a powerful technique for PV fault diagnosis and for this reason there is a growing interest of the O&M industry to make it widely available for in site inspections instead of randomly choosing samples throughout the PV plant. Few groups in the world disclose methodologies for daylight EL (dEL) image acquisition of installed PV modules, using different equipment and image processing procedures. In this work, we aim to compare methodologies from two research groups that have been working on practical ways to perform robust dEL imaging in parallel. We present the results from acquisitions made with similar but not identical equipment, weather conditions and PV module technologies. Image processing on the same datasets from both laboratories is performed using their own methods, indicating if the image processing methods and approach provide consistent results or not. While each laboratory has developed its own signal-to-noise ratio markers, both present good consistency with signal at an exposure time analysis, but not always a direct correspondence to the image quality. This intercomparison will lead to improvements on each laboratory procedure, expand the discussion to other groups and initiate a discussion on the establishment of dEL image quality and SNR metric baselines.
Keywords: Daylight electroluminescence, PV modules, PV reliability.

1 INTRODUCTION

Electroluminescence (EL) is often used as ground truth for detection defects in PV modules and cells thanks to its characteristics of detecting a wide range of faults and high spatial resolution [1]. Nighttime EL doesn't provide major challenges, but there's still a big interest in performing it during daytime, as it assures accessibility and safety in dealing with high voltage equipment and cabling. Only a few groups worldwide have the capability for daylight EL (dEL) image acquisition of installed PV modules using different equipment and image processing procedures, and even less of them have described methodologies in publications [2]–[4].

Currently, there's no agreement or standard for dEL practices, making the development of the technique challenging in terms of robustness for different weather conditions and other (possibly unknown) factors. This work should pinpoint key considerations for robust dEL imaging and establish pathways for a future interlaboratory comparison of image processing and experimental methods aiming the best outcome for automatic power output assessment based on EL images [5], [6].

2 METHODOLOGY

2.1 Experimental details

To achieve and evaluate the methods for dEL imaging, here we compare the experimental data that were performed by two research laboratories, one located at the Technical University of Denmark (DTU) and another at the University of Valladolid (UVa). At DTU, the current bias was applied using an Electro-Automatik PSI 9000 DC power supply unit (PSU). At UVa, the module is powered by the PSU Electro-Automatik 91500-30 3U, which can generate DC voltages up to 1500 V and currents up to 30A, as long as 15000 W is not exceeded. Different types of waveforms can be programmed in both power supplies such as square, sinusoidal, or triangular waveforms, and even an arbitrary shape can be programmed by software. For this work square and sinusoidal current bias waveforms of different frequencies have been selected. The current amplitude was set to oscillate from 0 A to ~ Isc for both waveforms and at both laboratories. DTU and UVa used multi-crystalline silicon Aluminum Back Surface Field (Al-BSF) modules (60 cells and 72 cells respectively) as devices under test for a low EL signal scenario. Additionally, modules with higher open circuit voltage solar cells (Passivated Emitter and Rear Contact - PERC) were also used for a high EL signal scenario. In this case, DTU used a monocrystalline silicon PERC module and UVa a multi-crystalline PERC. The same type of cells in the compared modules should assure that the EL signal magnitude is equivalent when the same current bias is applied to them. Particularly, Al-BSF modules represent the most demanding scenario since the EL signal has the lowest luminescence emission from the large-scale commercially available PV module.

The dEL image sequences were acquired using SWIR cameras equipped with 640x512 InGaAs sensors and with 600 frames per second maximum frame rate. Both laboratories had available similar cameras from First Light Imaging: a C-RED 3 camera is used by DTU while UVa uses a C-RED 2 Lite camera. For all image data to be more efficiently used by the image processing method developed by DTU, the number of images per current bias modulation period at the PV module to be even, and particularly for the Fast Fourier Transform (FFT) method it was advantageous that the sequences had 8 images minimum per period, as it is shown in Fig. 1. Besides this ratio requirement expressed by equation (1) to be equal to 8, there's no type of active synchronization between the camera and the PSU.

$$\#\ of\ images\ per\ period = \frac{Frame\ rate}{Modulation\ frequency} \quad (1)$$

10.4229/EUPVSEC2023/4CV.1.7
020374-001

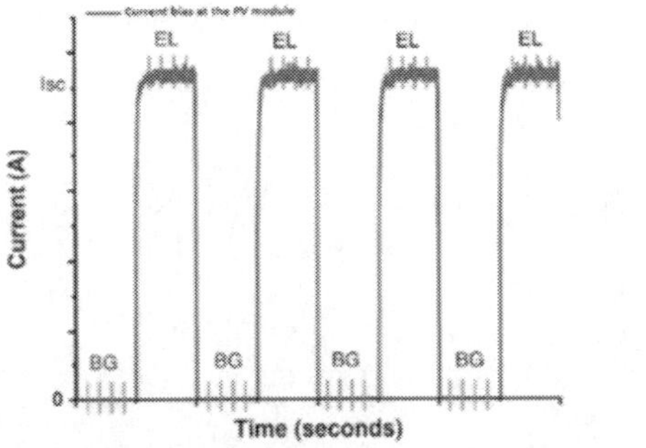

Figure 1: Current bias from a squared wave modulation (blue line) with 8 images acquired per period: 4 background images (BG) and 4 EL images.

As an exception, image sequences acquired with 2 ms exposure times had 16 images per period, to make better use of the high framerate available on the C-RED cameras. If this requirement is fulfilled, the DTU image processing method should provide its best results, even though the processing can also be performed when the requirement is not fulfilled. For the UVa processing method, this requirement also helps but is not a strict condition.

2.2 Image processing and signal-noise ratio (SNR) metric

For dEL, it is worth mentioning that the image processing is attached to the acquisition method itself, and this is also one of the important points of this laboratory intercomparison.

From the DTU side, the dEL images were processed with software developed in-house, which can use several different image processing methods. One of them used at this work was FFT, which generates a signal-to-noise ratio marker SNR_{Kari} [7]. A satisfactory signal for image extraction was extracted when $SNR_{Kari} = 1$ and upward. Values of $SNR_{Kari} > 4$ ensured a high-quality image, while values below one resulted in poor quality [7]. The software can process the images using a prevailing method found in the literature called Lock-in method which is described in detail by Adam et. Al. [8] which is an analysis made on time domain, opposing FFT which is a frequency domain analysis. For the Lock-in, a SNR marker SNR_{AVG} is reported in the literature [2], [3], [7], [9], which has been proven to present many inconsistencies in being an efficient signal and image quality metric [2], [7].

At UVa, the images are processed using two different methods: Frequency Domain (FD) and Time Domain (TD). In FD, image sequences acquired with both sinusoidal and square modulation waveforms can be processed, while for TD, only square waveform acquisitions can be considered. The SNR value for FD is calculated from the peak amplitude at modulation frequency and the variance of the noise level as:

$$SNR = \begin{cases} 10 log_{10} \left(\frac{(Peak\ amplitude)^2}{Var(Noise)}\right) & if > 10 \\ 0 & if \leq 10 \end{cases} \quad (2)$$

where the peak amplitude is defined as the difference between the value at excitation frequency and the average of the [f/2,2f] \{f} window, and the variance is calculated also in the [f/2,2f] \{f} window, both calculated from the absolute value of the FFT resultant frequency domain data. Here 10 dB is defined as the minimal SNR to recognize that the signal can be distinguished from the noise level, which roughly corresponds to a peak signal ~3 times higher than the standard deviation of the noise level. Finally, the quality metric is the average SNR of the pixels with SNR values in the lower quartile (SNR_{25}), i.e. those with the weakest SNR. Therefore, the higher that value the better we can distinguish the weakest part of the image from noise.

For TD, the SNR value is calculated from average difference between the *On* and *Off states* and the variance of the noise level as:

$$SNR = \begin{cases} 10\, log_{10} \left(\frac{(On_{avg} - Off_{avg})^2}{Var(Noise)}\right) & if > 3 \\ 0 & if \leq 3 \end{cases} \quad (3)$$

Here 3 dB is defined as the minimal SNR to recognize that the signal can be distinguished from the noise level, which corresponds to roughly ~92% certainty that the pixel signal has been modulated by the external waveform. For satisfactory image quality $SNR_{25} > 15$dB should be the indication for this marker when FD is used and a $SNR_{25} > 5$ dB when TD is used.

In principle, the FFT and FD can be considered equivalents, dealing with the EL signal extraction on the frequency domain, the same way that TD and Lock-in image processing method interpret the signal in time domain.

3 RESULTS COMPARISON AND DISCUSSION

At this section, the qualitative image quality is evaluated comparing examples of image sequence acquisitions from both groups with different image processing methods and modulation waveforms performed by DTU (section 3.1) and UVa (section 3.2). Quantitative comparison of the SNR markers is made for different exposure times and module technologies in section 3.3.

3.1 DTU image processing method

In Fig. 2, the comparison between similar experiments performed at DTU and UVa is shown using squared and sinusoidal waveform current bias modulation. In Fig. 2 and Fig. 3, the exposure time is given in milliseconds (ms), framerate in frames per second (fps), current bias modulation frequency in hertz (Hz) and approximate solar irradiance in the plane-of-the-array (POA) at the time of the measurement in watts per square meter (W/m^2). For the DTU experiment, FFT image processing for dEL method presents slightly better quality than Lock-in- for both waveforms. The dEL resulting images clearly show major cracks and electrically isolated areas, but only the FFT processed may present similar image quality as the indoor EL image which is the reference for the quality to work for automatic defect detection algorithms. For UVa data, Lock-in shows better results, but neither of the processing methods show high quality images similar to the indoor data. Regarding the SNR markers indication, SNR_{Kari} is higher than 1 for both measurements, but the image quality itself might not be enough for proper diagnosis evaluation even from an inspector. The marker does show higher values for the square waveform, and the image quality is also better in the same way.

3.2 UVa image processing method

Fig. 3 shows the dEL images from sequences acquired at UVa and DTU with current bias modulation using different waveforms, compared with their indoor EL images (same indoor images as in Fig. 2).

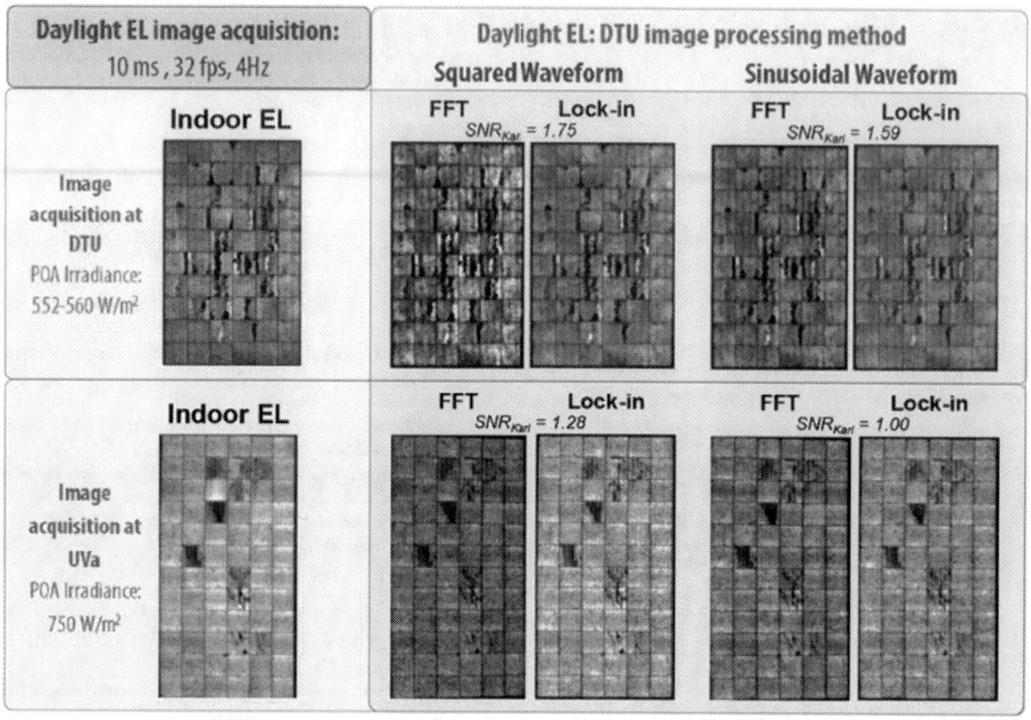

Figure 2: Indoor EL and dEL images acquired at both laboratories and processed by DTU's image processing method.

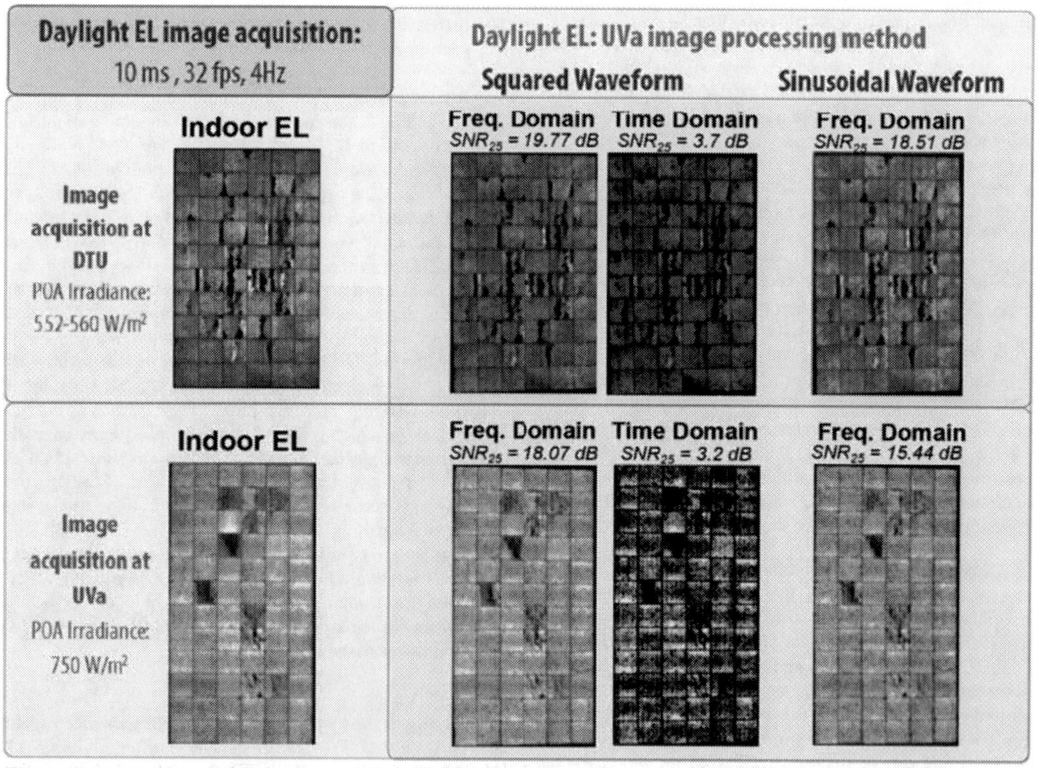

Figure 3: Indoor EL and dEL images acquired at both laboratories and processed by UVa's image processing method.

10.4229/EUPVSEC2023/4CV.1.7
020374-003

better contrast and similar quality as the indoors EL image. For the DTU data, both waveforms resulted in almost identical dEL images, however both with lower contrasts and sharpness than the indoors EL image. The SNR_{25} marker presents satisfactory values, although UVa dEL image with sinusoidal waveform may not be considered satisfactory qualitatively due to excessive granularity. For TD image processing, possible to be run for the square waveform, the dEL images quality was inferior to FD, with SNR_{25} equal to 3.2 and 3.7 dB for UVa and DTU data respectively, which were below the SNR_{25} good quality threshold.

3.3 SNR markers

The need to have an SNR marker that indicates quantitatively the quality of dEL images is paramount for large-scale automatic image quality evaluation, development of efficient image processing methods and understanding of the environmental factors that contribute in different proportions to generate noise during image sequence acquisition.

As observed in previous studies [2], [7] the first SNR marker for dEL mentioned in the literature SNR_{AVG}, presents high level of inconsistencies, having high values for low quality images at low exposure time (i.e. low signal), and low values for high quality dEL images. Fig. 4 and 5 show how the newer markers SNR_{Kari} and SNR_{25} perform with exposure time and different module technologies.

As mentioned previously, Al-BSF modules present a low EL signal due to its lower open circuit voltage compared to newer silicon cell technologies; therefore, it is the most challenging crystalline silicon PV module to perform dEL. For such modules, the dEL image processing method in general provided markers with good correspondence to exposure time and therefore with the amount of EL signal. It's worth noting that square waveforms tend to yield slightly better results compared to sinusoidal waveforms. In the case of the UVa processing, this seems to be related to the fact that the sinusoidal excitation is very weak in some frames of the cycle to contribute to the FFT, while the square excitation tends to be better resolved in its dominant frequency.

For strong EL emission PERC modules, the markers' correlations with exposure time seems to be consistent especially for the square waveforms, and for the UVa image processing, for the TD type. However, we observe a worsening of the SNR_{25} with the exposure time for the FD analysis that is related to the presence of light coming from reflections on the terrain surface. It should be noted that these panels emit EL intense enough to illuminate the scene, resulting in a weak modulated light coming from regions outside the panel. Since UVa is not cropping the images to the panel region only but keeping the whole 640 x 512 pixels from the camera, the increased exposure time leads to the inclusion of more weakly illuminated pixels

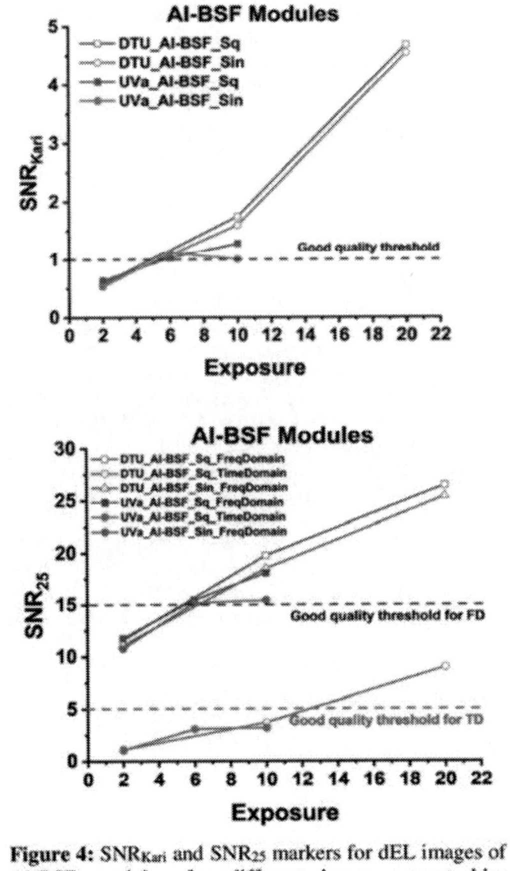

Figure 4: SNR_{Kari} and SNR_{25} markers for dEL images of Al-BSF modules for different image current bias waveforms and exposure times.

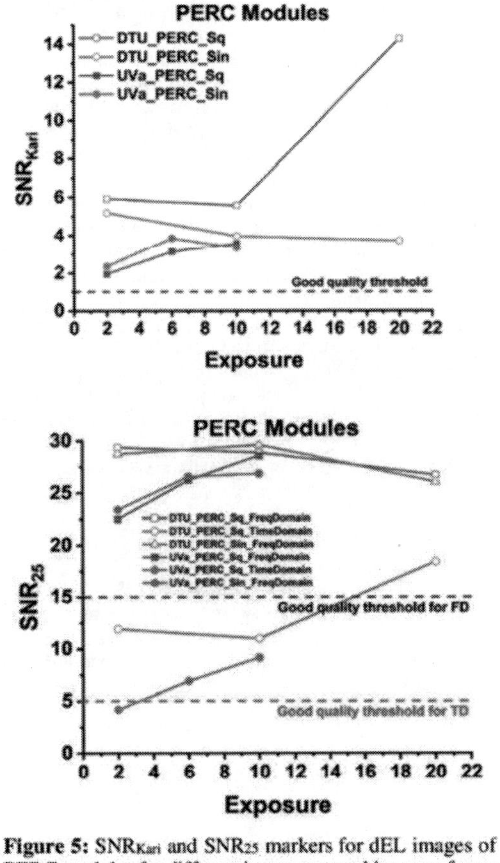

Figure 5: SNR_{Kari} and SNR_{25} markers for dEL images of PERC modules for different image current bias waveforms and exposure times.

from outside the panel area. Consequently, the SNR_{25} deteriorates in the 20 ms exposure case compared to the 10 ms case due to the increased presence of weak pixels outside the panel. The same effect is reasonable to be also occurring for SNR_{Kari} in the case of the sinusoidal waveform and for the square waveform at 10 ms exposure time.

This phenomenon is not observed in the case of the Al-BSF panels, as their emission is not intense enough to significantly illuminate the scene and be detected as a signal. Therefore, increasing the exposure time results in improved image quality markers for these panels.

In the context of the TD analysis, which is less sensitive to capturing modulated light from the panels, it cannot effectively distinguish the reflected light outside of the panels beyond the 3 dB cutoff limit. As a result, we do not observe the deterioration of SNR_{25} with increasing exposure time in the TD analysis.

4 CONCLUSIONS

In summary, both SNR marker approaches are more consistent for square waveforms than for sinusoidal. This same trend is also observed in the quality of the images, where frequency domain approaches (FFT and FD) is the more efficient to generate a dEL for qualitative evaluation of the PV modules health.

The SNR metrics appear to show consistent trends with exposure time, which hints at predictability. These deviations are internally consistent, and the experimental reasons can mostly be traced back and explained. Yet, when compared to each other, the image quality between laboratories cannot have a proper quantitative measure due to the different SNR marker definitions, bringing up again the problem of only relying on the observation of the dEL image by a human to define if a processing method is better or worse than the other.

For automated processing, the results of this intercomparison underline the paramount importance of establishing an SNR metric for dEL imaging that can be used universally and is robust for different approaches of image processing.

This intercomparison will lead to improvements on laboratory procedures, expand the discussion to other groups, and lay the groundwork for future round-robin discussions, intended for comparison of experimental practices, image processing approaches, and most importantly the establishment of dEL image quality and SNR metric baselines for daylight luminescence.

5 ACKNOWLEDGEMENTS

The UVa group has been financed by the Spanish Ministry of Science and Innovation, under project PID2020-113533RB-C33, and by the Regional Government of Castilla y León (Junta de Castilla y León) and by the Ministry of Science and Innovation and the European Union NextGenerationEU / PRTR under the project "*Programa Complementario de Materiales Avanzados*". The DTU group is supported by the Eurostars project Automated Daylight Electroluminescence Inspection of Large Photovoltaic Systems (ADELI).

6 REFERENCES

[1] L. Koester, S. Lindig, A. Louwen, A. Astigarraga, G. Manzolini, and D. Moser, "Review of photovoltaic module degradation, field inspection techniques and techno-economic assessment," *Renew. Sustain. Energy Rev.*, vol. 165, no. November 2021, p. 112616, 2022, doi: 10.1016/j.rser.2022.112616.

[2] G. Alves Dos Reis Benatto *et al.*, "Drone-Based Daylight Electroluminescence Imaging of PV Modules," *IEEE J. Photovoltaics*, vol. 10, no. 3, pp. 872–877, 2020, doi: 10.1109/JPHOTOV.2020.2978068.

[3] M. Guada *et al.*, "Daylight luminescence system for silicon solar panels based on a bias switching method," *Energy Sci. Eng.*, no. June, pp. 1–15, 2020, doi: 10.1002/ese3.781.

[4] T. J. Silverman, M. G. Deceglie, K. Vansant, S. Johnston, and I. Repins, "Illuminated Outdoor Luminescence Imaging of Photovoltaic Modules," *IEEE Photovolt. Spec. Conf.*, 2017.

[5] T. Kropp, M. Schubert, and J. H. Werner, "Quantitative prediction of power loss for damaged photovoltaic modules using electroluminescence," *Energies*, vol. 11, no. 5, 2018, doi: 10.3390/en11051172.

[6] R. del Prado Santamaría, T.K. Hass, G. A. dos R. Benatto, J. W. Ng, P. B. Poulsen, and S. V. Spataru, "Machine Aided Estimation of Solar Cell Crack Caused Power Loss from Electroluminescence Images," in *8th World Conference on Photovoltaic Energy Conversion*, 2022, pp. 480–487. doi: 10.4229/WCPEC-82022-3BO.11.6.

[7] G. A. dos Reis Benatto *et al.*, "Daylight Electroluminescence of PV Modules in Field Installations: When Electrical Signal Modulation is Required?," in *8th World Conference on Photovoltaic Energy Conversion*, 2022, pp. 735–739. doi: 10.4229/WCPEC-82022-3BV.3.44.

[8] J. Adams *et al.*, "Non-Stationary Outdoor EL-Measurements with a Fast and Highly Sensitive InGaAs Camera," *32nd Eur. Photovolt. Sol. Energy Conf. Exhib.*, 2016.

[9] C. Mantel *et al.*, "SNR Study of Outdoor Electroluminescence Images under High Sun Irradiation," in *2018 IEEE 7th World Conference on Photovoltaic Energy Conversion (WCPEC) (A Joint Conference of 45th IEEE PVSC, 28th PVSEC & 34th EU PVSEC)*, Jun. 2018, no. 1, pp. 3285–3289. doi: 10.1109/PVSC.2018.8548264.

POLYMER DEGRADATION IN DIFFERENT CLIMATE ZONES - IMPACT ON PERFORMANCE AND OPERATION IN SOUTH AFRICA AND GERMANY

C. Buerhop-Lutz[1], E.E. van Dyk[2,3], O. Stroyuk[1], M. Vumbugwa[2,3], E. Wittmann[1], F.J. Vorster[2,3], J.L Crozier McCleland[2,3], I.M. Peters[1]
[1] Forschungszentrum Jülich GmbH, HI ERN, Erlangen, Germany
[2] Nelson Mandela University, Port Elizabeth, South Africa
[3] PVinsight, Port Elizabeth, South Africa

ABSTRACT: A field investigation of a multi-MWp PV power station installed in the sun-rich, warm climate of South Africa is presented. For correlating polymer degradation with the electrical properties of PV modules, we identified the polymers on-site using near-infrared absorption spectroscopy, UV-fluorescence, and visual inspection, as well as measured insulation resistance and IV curves on a string level. Four PET-based multilayer backsheet (BS) types were identified, including two with a fluorinated coating (FC1 and FC2), polyvinyl fluoride (PVF), and polyvinylidene fluoride (PVDF) air sides. A visual inspection of 5,736 modules revealed severe corrosion and heavy cracking of the BS of all inspected modules with FC2 BSs. The impact on the electrical properties differs depending on the BS type. For strings with PV modules with PVF, PVDF, FC2-BSs the insulation resistance R_{iso} was found to be 30-40 MΩ increasing with higher temperature (lower humidity). For the PV modules with cracked and corroded FC1-BS, the insulation resistance was below 1 MΩ and did not show the same recovery trend as the other BS types. These observations confirm the previous findings in Germany. At that, degraded BSs impact the leakage current and therefore safety aspects and inverter availability but not the power outcome, once the inverter starts operation.

Keywords: backsheets; polymer degradation; cracks; mear-infrared spectroscopy; insulation resistance

1 INTRODUCTION

Cracks in backsheets (BSs) are clear signs of degradation. They are associated with insulation and safety issues. Cracks are typically reported for polyamide (PA) BSs and are rarely (or never) observed for fluorinated multi-layer BSs. In this study, we report crack structures in the inner layers of two polyester multi-layer BSs: (i) **FC-PET-PP** consisting of a thin fluorinated coating (FC) as air layer, polyethylene terephthalate (PET) as core layer and polypropylene (PP) as inner layer and (ii) **PVF-PET-PE:** consisting of a polyvinyl fluoride (PVF) air layer, a PET core layer and an inner layer made of polyethylene (PE). Both BSs suffer from inner cracks and corrosion. Both BSs are part of the bill-of-materials (BOM) of PV modules supplied by two TIER1 manufacturers.

2 METHODS

With this study, we point out the relevance of BS cracks for PET-based BSs as an indicator of the degradation of polymers and the loss of insulation resistance. We present field-collected evidence for the occurrence of inner cracks in a PET-based BS with a PVF-PET-PE structure, which has not been mentioned in the literature, to the best of our knowledge. To better classify the findings, we compare the measured data from two PET-based BBs with internal cracks, PVF-PET-PE and the well-reported FC-PET-PP.
We inspected 13% of a multi-MWp PV power station in the Central part of South Africa with an age of about eight years. Our inspection tools included imaging, spectroscopy, and electrical measurements of 5,736 modules in 239 strings. For the identification of the polymers and existing BOM versatility, we used near-infrared absorption (NIRA) spectroscopy [1], as well as UV-fluorescence imaging. The electrical data array includes string power and insulation resistance measured onsite using a Solmetric IV-tracer and a Benning PV2. The collected and evaluated data are set in relation to previous degradation studies in Germany. Furthermore, weather data was tracked using a mobile DAVIS weather station.
The cracked BSs are from two TIER1 module manufacturers, A and B. For PV power station A in Germany (G) (10 years of operation, modules A, BS: FC-PET-PP), modules with a capacity of 6 MWp were characterized, including an analysis of the plants' monitoring data. For PV power station B in Southern Europe (SE) (eight years of operation modules B, BS: PVF-PET-PE), 2 MWp of modules were characterized. In Southern Europe e.g. temperatures and irradiances are higher (I_{SE} = 3.2 – 9.7 kWh/m², T_{SE} = 0.6 – 29°C, RH_{SE} = 45.8 -92.9%) than in Germany (I_G = 0.3 – 8.4 kWh/m², T_G = -2.4 – 25.6°C, RH_G = 45.7 – 94.3%) evaluated for 2%- and 98%-percentiles for daily values of ambient temperature T, relative humidity RH, and irradiance I [2].

3 RESULTS AND DISCUSSION

In this study, we investigate the performance of PV modules installed in South Africa, where the estimated cumulative PV capacity reached 4.3 GWp at the end of 2021. Of this capacity, 2.3 GWp is found in utility-scale PV-power stations. The projected additions to the national utility grid in 2023 would bring the capacity of utility-scale PV to 3.4 GWp.
The rapid growth in the deployment of PV plants often results in decreasing the importance of an established quality control regime on PV modules, especially when it comes to checking the BOM. We aimed to gain insights into the BOM, especially concerning the polymers, ethylene vinyl acetate (EVA) co-polymer encapsulant and backsheets (BSs), their degradation modes, and the impact on the electrical parameters, power, and insulation resistance, which are important for inverter availability and power production.
We note that the present study provides a unique opportunity to compare the performance of PV modules from the same TIER1 manufacturer and similar age

operated in two climate zones – South Africa and Germany. In South Africa (ZA) the average temperatures and UV irradiance levels are higher than in Germany (G) [2]. The present study highlights the impact of these climatic differences on the degradation of the polymers in PV modules.

Visual inspection showed strong cracking of the polymers of 13% of PV modules, which could be mistaken as glass breakage because of broad and colored cracks (Figure 1 A-1D). The cracking was often accompanied by strong coloration in bright green and brown due to the corrosion of metal interconnects.

Compared to the observed destruction patterns in South Africa, the reported fine micro-crack net and local corrosion patterns seen in German /European PV modules are less distinct [3], see Figure1 E-1H.

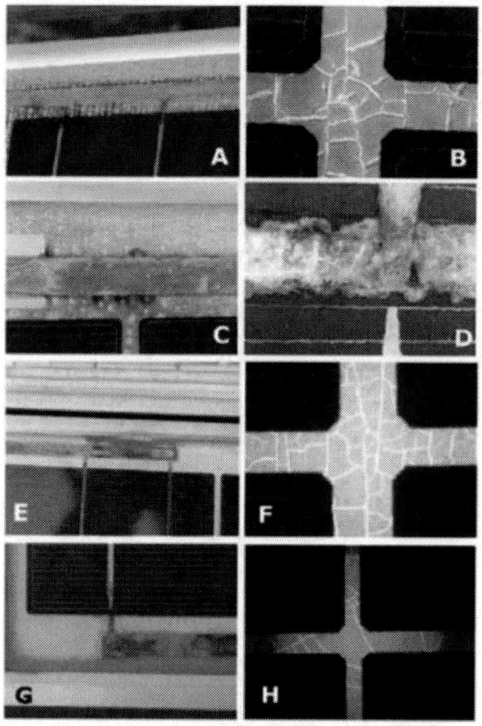

Figure 1. Visual images of degradation patterns for PV modules with FC1-based BS were found in South Africa (A, B, C, D) in year eight of operation and Germany (E, F, G, H) in year nine of operation. Modules have the same labels.

An inspection using NIRA spectroscopy revealed four multi-layered PET-based BS types, see Table 1, based on the existing information in our database. Three BS types showed no visual deficiencies. Only FC1-BSs showed cracking and corrosion. Findings of the visual inspection include also different browning patterns on the front side, see Table 1, which occurred exclusively for a certain BS-type.

We observed either browning across the cell area, brown rings close to the cell edges, or no browning at all. Furthermore, using NIRA a considerably higher water content was detected in South African BSs as compared to ones in Germany. The increased water content was detected explicitly in FC1 and PVF BS, but not in the corresponding EVA layers (Figure 2).

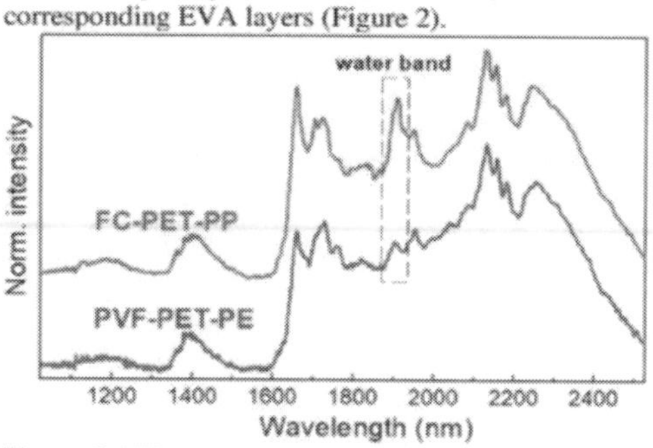

Figure 2. NIRA spectra illustrating the water content in FC1-PET-PP or PVF-PET-PE BSs.

The impact of different BS types on the electrical properties was studied by measuring the insulation resistance and IV curves of various strings with known BSs. We measured the insulation (isolation) resistance R_{iso} on two days for 103 strings on the first day and 136 strings on the second day with an overlap of 64 strings. The measurements continued for approx. 3 h at ambient temperature T(day1) = 21-34°C (T(day2)=22-32°C) and relative humidity RH(day1) = 48-24% (RH(day2) = 63-42%); during the night there was heavy rainfall.

As reported elsewhere [4, 5], R_{iso} changes with changing environmental conditions. It increases with increasing T and decreasing RH. Furthermore, the evolution of R_{iso} is strongly dependent on the BS material, see Figure 3.

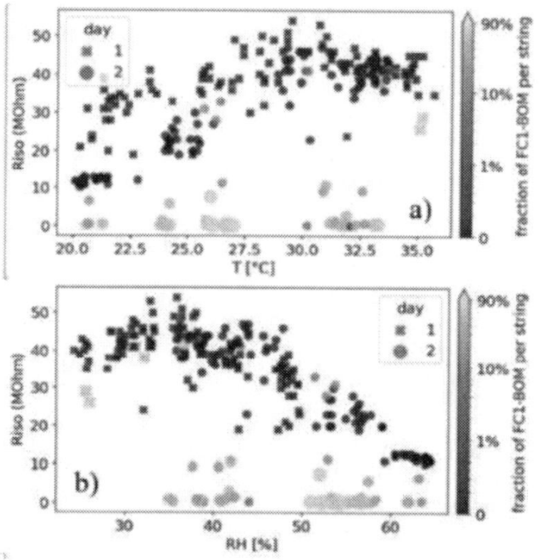

Figure 3. Insulation resistance R_{iso} as a function of ambient temperature a) and relative humidity b) for PV strings with BS-types quantified by the amount of FC1-BS within the strings.

Modules with PVF- or PVDF-BS started in the morning with R_{iso} around 20 MΩ and the values increased linearly with increasing temperature (linear regression fit: $R_{iso} = 2.3\,T - 35$) and decreasing humidity (linear regression fit: $R_{iso} = -1.1\,RH + 83$).

Thus, after repeating the measurements after two hours with ΔT= 20 K, R_{iso} increased from 20 MΩ to 40-50 MΩ.

This behavior was observed for both days of measurements.

Table I: Identified BS-architectures with layer thickness *d*, fraction in PV-system, and notes on visual anomalies

BS-name	*PVDF*	*PVF*	*FC1*	*FC2*
Air layer	PVDF+R d = 25-30 µm	PVF+R d = 25 µm	FC+R d = 5-10 µm	FC+R d = 8-9 µm
Core layer	PET d = 250-260 µm	PET d = 260-270 µm	PET d = 140 µm	PET d = 150 µm
Inner layer	PP+R d = 65-70 µm	PE d = 50-60 µm	PP+R d = 70 µm	PP+PE +R d ≈ 100 µm
fraction	62%	22%	13%	2%
Notes	Brown areas	Brown rings	Cracks, corrosion	Brown rings

Abbreviations: PET - polyethylene terephthalate, FC - fluorinated coating, PE - polyethylene, PVF - polyvinyl fluoride, PP - polypropylene, PVDF - polyvinylidene fluoride, R - rutile (TiO_2).

In contrast, the R_{iso} of strings with FC1-based BSs is low with R_{iso} < 5 MΩ for most strings and remains on that level even though the temperature increases. We assume that the R_{iso} recovery is much slower for these BSs. On day #1 before the rainfall, R_{iso} was much higher at comparable environmental conditions (see figure 2a). In the morning at low temperatures, on day #1: R_{iso} scatters about 30 MΩ, on day #2: R_{iso} = 10-20 MΩ for none FC1-strings. For strings with FC1-BSs at high temperatures on day #1: R_{iso} ≈ 30 MΩ, on day #2: R_{iso} < 5 MΩ. More detailed studies are needed for details of the R_{iso} recovery kinetics.

The data show a moderate linear correlation between R_{iso} and the fraction of modules with FC1-BOM within the strings, with a correlation coefficient of ρ = -0.52 and statistically significant linear regression on a 5%-level, which has been reported for strings in Germany, too [3].

Furthermore, we evaluated the string power P, measured at E = 950 - 1050 W/m² and T =30 - 35°C. All modules have the same nominal power of 240 W which results in a nominal power of the string of $P_{nominal,\ st}$ = 5,760 W. Strings with FC1-based BSs have a mean P = 5405±107 W while strings with PVF- or PVDF-based BSs have a mean P = 5415±83 W, see the power distribution in Figure 4.

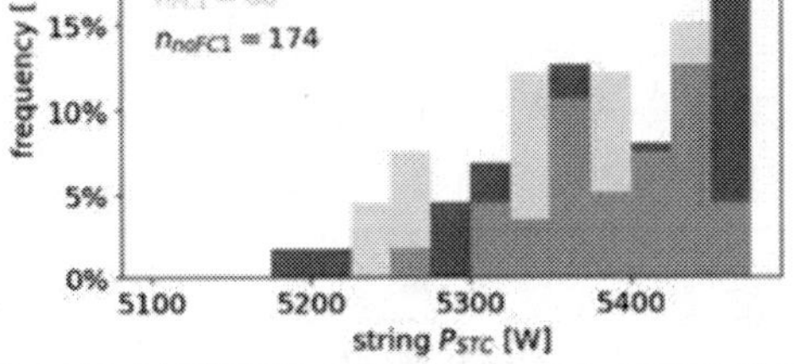

Figure 4. Distribution of measured power P for strings with BS-types classified by modules with and without FC1-based BSs within the string.

R_{iso} and P were measured at almost the same ambient conditions within 15 min. We expect that BS type and BS failures do not affect power directly and immediately. Indeed, under good operating conditions, there are no significant differences in string power. But we also expect that low R_{iso} will cause inverter tripping, reducing energy yield, and potentially inverter availability as well as posing safety issues for the personnel of PV plants.

4 SUMMARY

The BS variability of PV modules installed in South Africa was found to be comparable to that of earlier commissioning years in Germany. For the same module manufacturer, the same four BS-types PVDF, PVF, and FC1, FC2- PET-based BSs were found.

All BS types appear visually intact except for BSs with FC1 air layer. Modules with FC1-BS can be associated with strong corrosion, massive cracking as well and low insulation resistance.

All of these appear more distinct in South Africa than in Germany, indicating an overall faster backsheet degradation in ZA for this particular BS type. We correlate this with higher irradiance and ambient temperature in South Africa. At that, we expect low, critical insulation resistance, power reduction, and inverter tripping as well as the need for countermeasures in South Africa at a much earlier operation time (1-2 years). In general, PV operation in sun-rich and hot regions is more strongly and earlier affected by cracked BSs than in the moderate German climate. For evaluating climate-dependant degradation dynamics of the polymers more on-site data need to be collected.

5 REFERENCES

[1] O. Stroyuk, et al.; Solar Energy Materials and Solar Cells 216 (2020) 110702.
[2] Global Modeling and Assimilation Office, MERRA-2 tavg1_2d_slv_Nx: 2d,1-Hourly,Time-Averaged,Single-Level,Assimilation,Single-Level Diagnostics V5.12.4, Greenbelt, MD, USA, Goddard Earth Sciences Data and Information Services Center (GES DISC).
[3] C. Buerhop-Lutz, et al.; Solar Energy Materials and Solar Cells 246 (2022) 111913.
[4] C. Buerhop, et al.; SOLMAT 231 (2021).
[5] C. Buerhop, et al.; PIP (2022).

10.4229/EUPVSEC2023/4CV.1.10
020376-003

TOWARDS CLIMATE-SPECIFIC O&M FOR PV PLANTS: GUIDELINES AND BEST PRACTICES

John (Ioannis) A. Tsanakas[1]*, Bert Herteleer[2], Ulrike Jahn[3]
[1] CEA-INES, France; [2] Katholieke Universiteit Leuven, Belgium, [3] VDE Renewables, Germany

*Corresponding author: ioannis.tsanakas@cea.fr

ABSTRACT: Considering that nearly half (40%-50%) of the photovoltaic (PV) plants are less than just 5 years old, the cumulative knowledge and return-of-experience (REX) from PV operations and maintenance (O&M) are at their relative infancy; particularly where it relates to the impact of different climatic profiles on the budgeting, technical risks and customization of PV O&M. Besides, one of the main challenges in tailoring O&M for a specific PV application in a certain climate zone is the present scarcity of comprehensive guidelines and/or standards, which only clarify the minimum requirements of climate-specific O&M and their implementation. Recently, experts of the IEA PVPS Task 13 have gleaned best practices and guidelines for O&M in different climate zones/profiles, covering all key aspects of the typical PV O&M agenda, based on field experiences of the contributing PV and O&M experts from multiple countries representing the climate zones addressed. In this paper, we provide the key findings of this work: a comprehensive guide on setting up a customized O&M practice for PV plants in seven different climate zones, four of which are general (moderate, hot and dry, hot and humid, desert in high elevation), and three which are more specific to extreme conditions (flood-prone region, cyclonic region, snowy region). In terms of novelty, we particularly emphasize on (existing and new) knowledge and practices for PV O&M in the last three climatic profiles/regions, which present distinct challenges from a PV performance and reliability point of view. As such, the climate-specific O&M guidelines presented in this report aim to serve different functions to various stakeholders depending on their roles in the entire value chain of PV.

Keywords: *PV systems; Operations & Maintenance (O&M); PV field experience; PV systems reliability.*

1 INTRODUCTION – RATIONALE

In view of the (ongoing) rapid growth of photovoltaic (PV) installations worldwide, nearly 60% of the PV plants are less than just 5 years old [1], with more GW-scale PV markets in different climatic regions each year. With PV technical (service) lifetimes expected to reach 25 to 40 years or more, it becomes clear that the cumulative knowledge and return-of-experience (REX) from PV operations and maintenance (O&M) are at their relative infancy; particularly where it relates to different climatic regions, i.e. climate-specific O&M. For instance, multiple (individual or combined) stress factors such as temperature, humidity, exposure to UV light, rain, and wind would indeed contribute differently to the occurrence of PV degradation or failures in diverse climate zones and conditions worldwide; consequently, dictating different strategies for PV O&M.

Given that O&M costs constitute a large portion of total Operational Expenditures (OPEX) at 30%-70% [2] *and* that future O&M costs are inherently uncertain, much uncertainty in the real versus modelled Levelized Cost of Electricity (LCOE) remains. In a globalised world where investors and owners, O&M providers and installers (EPCs) operate in multiple countries with varying climates, it pays to understand and apply general best practices for O&M, supplemented or modified with climate- and site-specific practices.

Typically, PV plant operators adopt a rather standard scope of O&M services (Fig. 1) that could be replicated easily across to reach an optimal point between minimizing O&M efforts (and therefore OPEX), while maximizing PV plant uptime, performance and durability. However, a shift from the one-size-fits-all approach to a customized O&M approach could offer an advantage that the O&M activities are adapted for the needs of the PV plant, focusing on what maintenance activities are necessary for that specific plant and climate zone [1].

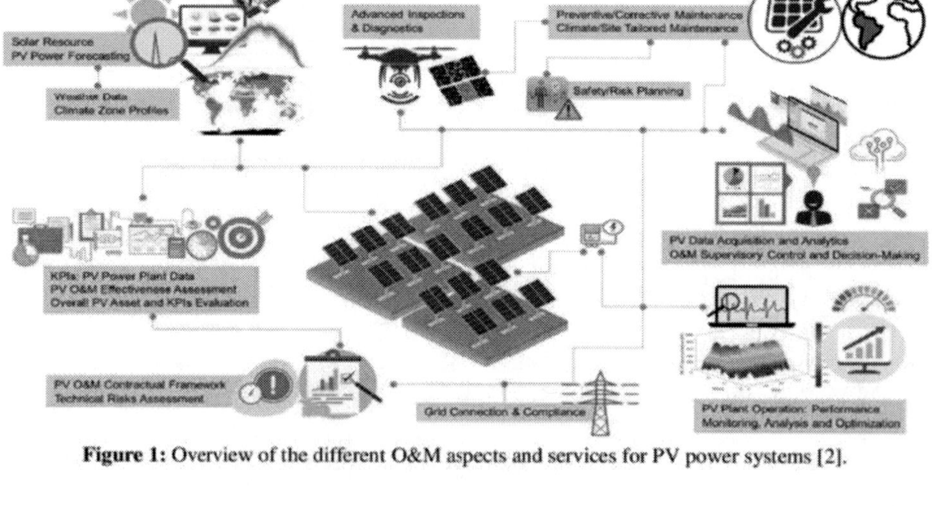

Figure 1: Overview of the different O&M aspects and services for PV power systems [2].

2 SCIENTIFIC RELEVANCE – APPROACH

The keystone towards PV O&M tailored to specific climate zone(s) lies upon comprehensive REX from PV developers and O&M service providers on worldwide scale. At present, insights into climate-specific PV O&M remain scarce, sporadically reported, and therefore relevant guidelines have yet to be developed. To this end, and with these "knowledge gaps" in mind, experts of the IEA PVPS Task 13 have gleaned best practices and guidelines for O&M in different climate zones/profiles, covering all key aspects of a typical PV O&M agenda (Fig. 1), ranging from the regulatory, risk/safety and asset management level, up to the operational level (notably the monitoring/inspections, data analytics, maintenance, and optimization).

This paper presents the key findings of this work, the basis of which is formed by O&M guidelines recommended for all PV plants and illustrated for moderate climates, with an additional six climate types or regions selected for their specific characteristics and PV installation (and penetration) potential. The selected climatic regions and their Köppen-Geiger classification are given in Fig.2. The first four refer to conditions which broadly prevail in large parts of the world (moderate, hot and dry, hot and humid, desert at high elevation), while the latter three refer to extreme conditions (flood-prone, cyclonic and snowy regions).

These guidelines and best practices can assist PV plant engineers and designers, financing parties, and investors in designing and maintaining PV plants, as well as in determining operational risk related to investment decisions. The O&M provider must aim to perform its contractual duties within the context that certain system design choices have been previously fixed at the design and EPC stages, subject to the knowledge at that time, which includes assumptions about climate change and the likelihood of extreme weather events (EWEs). O&M providers may have different levels of sophistication and provide different strata of quality, based on the agreed-upon contractual framework and fees.

Region and climate profile	Köppen-Geiger classification
Moderate (e.g. mainland Europe)	Csa-Csb-Cfa-Cfb-Cfc
Hot and dry (deserts, e.g. Israel, Australia)	BWh or BWk
Deserts in High Altitudes (e.g. Atacama)	BWn-BWk-BWk'-ETH
Hot and Humid (e.g. South-East Asia)	Af
Flood-prone (e.g. monsoons India)	Am
Cyclonic (e.g. Japan, Caribbean)	Cfa, Aw
Alpine/Snow-prone (e.g. Alaska)	Dfb-Dfc-Dfd-ET-EF

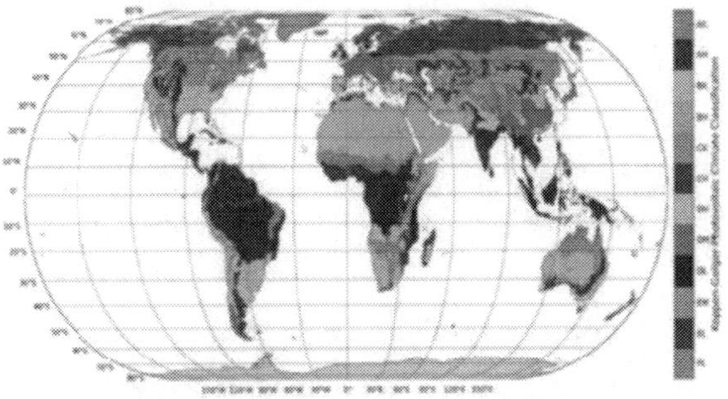

Figure 2: Left: Regions investigated for the climate-specific O&M guidelines. Right: Climate classification and implications to worldwide mapping for PV O&M [3].

3 KEY FINDINGS – DISCUSSION

3.1 Maintenance actions and cost considerations for PV

Preventive maintenance (PM) action plans that exclude redundant activities can bring costs down. The PM plan should seek to optimize the overall PV plant and O&M budgeting, depending on the plant´s size, design, complexity, and environment. In overall, the most important – commonly reported – maintenance actions include periodic sampling of individual electrical measurements at module level, soiling and snow mitigation, site and vegetation management as well as upkeeping of balance of systems (BOS) and SCADA (supervision control and data acquisition) monitoring systems.

During PV operational years, it is important to consider vegetation growing on adjacent PV assets, especially if these will cause shading on PV arrays, which can then result in hot spots (Fig. 3). Site and vegetation management, depending on the site/climatic conditions and the size of the PV plant, can be both challenging and labour/cost intensive, particularly in moderate and tropical (hot, humid) climatic profiles. On the other hand, in arid or alpine areas for instance, mowing may be unnecessary with proper soil stabilization. For areas where mowing is required, there is the added risk of projectiles damaging modules. In the latter case, vegetation management "strategies" that have gained increasing popularity in the last years include the collaboration with local farmers in using certain species of grazing animals.

Similarly, regular cleaning of modules from dust (soiling) or snow is an imperative PM action in PV plants installed in certain geographical locations and climates, e.g. hot arid, desert and alpine regions. As well known, soiling (Fig. 4) can result in considerable losses in the PV energy yield on both seasonal and annual scale. Also, as in the case of unwanted shading from vegetation, cases of uneven, localized soiling (dirt e.g. from bird droppings, Fig. 3, right) can lead to hot-spot effects on PV modules and risks of follow-up failures or even fire hazards. Upon cleaning of PV arrays, care should be taken, in compliance with manufacturer's recommendations, to avoid damaging the different PV components. For instance, cleaning solutions shall be based on plain demineralized water and mild detergents.

Figure 3 : Examples of shading (left, middle) and excessive soiling from bird droppings (right) caused by poorly managed vegetation and absence of cleaning respectively, in PV plants installed in Europe (moderate climate) [4].

Independently of the climatic zone of a PV plant is installed, O&M service providers tend to overestimate PV O&M budgets to offset both their own margins and uncertainties or technical assumptions. On the other hand, PV plant developers typically follow a more moderate or underestimating budgeting approach for PV O&M, to allow higher PV plant valuations, more "motivating" revenue prospects.

As aptly noted by Enbar et al. [5], such contrasting O&M budgeting viewpoints can ultimately undermine a PV plant's lifecycle performance economics. In Table 1, we provide a non-exhaustive list of representative costs and remarks associated with the main PM tasks, as publicly reported in the years 2021/2022, specific for utility-scale PV plants.

Table I: Representative costs of main preventive maintenance tasks in PV plants [2].

PM Task	Costs	Remarks
Base O&M scope	6 - 14 €/ kWp/ year	Includes: full preventive maintenance scope, regular module cleanings, security (remote or on-site); excl. IR and EL scans. Varying highly with the site characteristics, labor, and frequency of activity.
Cleaning/washing of PV modules [83]	0.5 - 2.5 €/ kWp/ year	Varying with the module technology, labor, cleaning solution and method, climatic conditions (affecting the frequency), etc.
IR scans	0.5 – 3.0 € / module	Includes drone inspections, analysis, and reporting
EL scans	3.0 – 10.0 € / module	

Figure 4: Example of (sources of) soiling in PV installations [2].

3.2 O&M best practices and recommendations in different climates/regions

O&M in flood prone regions

To date, few if any such plants have been designed with repeated flood endurance in mind. Yet, climate change and extreme weather events, combined with limited availability of land in certain parts of the world, mean that in some areas nowadays, PV plants may be subjected to flooding on an annual or even quarterly basis.

PV plants are yet to be flood-hardened through the design and orientation of the mounting system to likely water flow direction, selection of equipment (e.g. PV modules with IP68 rated junction boxes) and raising of inverters and transformers above likely flood levels. Floods can damage module mounting clips, PV modules, the lamination on panels, and even uproot foundations (Fig. 5). Soiling damage is particularly difficult to repair when water has been standing on the PV modules for a long time. Inverters connected to flooded systems may see repeated arcing and have increased fire risks, despite not being submerged; submerged inverters typically see short-circuits and require significant repairs. Fast flowing water can also cause debris impact because most PV mounting systems present a high resistance to flowing water.

10.4229/EUPVSEC2023/4CV.1.12

Consequently, O&M activities for flooding events may have a proactive nature (shutting a PV plant down before a flood arrives, if enough time is available), as well as a potentially large reactive set of actions. Once flood waters recede, the plant must be inspected. O&M staff must be given appropriate Personal Protective Equipment (PPE) and training for post-flood site visits, and repair and cleaning activities. Depending on the scale or nature of the flooding event, and the degree of insurance coverage, different approaches can be taken. For "light" flooding events and/or self-insured systems, the O&M provider will have to look to restore the system to its operational state. For this, spares management may be key, and the location of storage of spares a flow-on risk.

Figure 5: Damage to modules and structures (a-left) and structures' foundations (b-left) due to submersion with fast-flowing waters [2]. Another case of a PV plant after having been submerged under fast flowing water, showing limited damage on the upper part of its PV arrays (a-right), with a view of its all-concrete structure underneath (b-right).

O&M in snow prone regions

PV systems in snowy regions may be subjected to significant electrical mismatches (shading) and snow loads that exceed the module qualifications according to IEC 61215. A suggested limit for snow accumulation on panels is 0.7m. PV racking systems can also be damaged by extremes between winter and summer temperatures. In this case steel racking is preferred over aluminium racking. Even for systems where snow loads are below the module threshold, issues may appear.

To minimize such snow-load associated risks, appropriate PV array design strategies can be beneficial for "passive" snow mitigation, such as in the case of high tilted or vertical installations (Figure 6, left) [6]. If active cleaning measures such as mechanical cleaning (brushing or similar) are implemented, care must be taken not to scratch the glass as well. As an example, Fig. 6-right shows how inappropriate cleaning techniques can damage PV modules [7]. Similarly, (micro-) cracks can be initiated by snow loads, which can be then detected using electroluminescence (EL) inspections. This thus requires a different or additional set of skills and equipment by the O&M provider or their subcontractors.

Recent research and in-field experiments have been conducted to heat PV modules by applying a controlled forward voltage to melt the snow [8]. This would require careful load control and the use of weather forecasts. In this study, the forward biased PV modules' abilities to melt snow was assessed both in a climate chamber as well as in an operating rooftop PV system covered with snow. The main purpose of the PV system investigated was to mitigate snow from under-dimensioned roofs before, during, or after heavy snowfalls, and to keep the roof from collapsing from snow loads.

Figure 6: Left: Test site in Piteå, Northern Sweden (65°N) showing the impact of PV array tilt angle on the snow accumulation, for the studied case of frameless bifacial PV modules [6]; Middle and Right: PV module damage due to inappropriate snow cleaning of a PV array [7].

O&M in remote PV plants in hot arid regions

PV plants in such regions are routinely subjected to high temperatures and high irradiance (with high UV content); and, in some cases, high temperature variations between day and night hours. Typical risks and failures include increased soiling losses, cell cracks, optical degradation, damages to equipment such as cables (e.g.

due to termites). Adequate ventilation during hot months is crucial for good operating conditions of the inverters installed in such sites.

Inspections and site management are typically scheduled in the early morning hours and must consider numerous risks e.g. heat strokes, the presence of stray wild livestock, venomous insects, etc. Vegetation management is typically reactive and implemented together with fire breaks to facilitate access to the PV arrays.

When vegetation is present, soiling of PV modules is usually less of a problem in hot and dry climates; instead, yield losses due to rapid crop growth after rain events are possible. On another note, fire risks can also be posed by agricultural activities, e.g. field clearing, in the proximity of such PV plants. In view of these risks, in these areas, vegetation management is also implemented together with fire breaks (physical barriers that impede the spread of bushfires, such as cleared roads at the edges of PV plants) to facilitate access to the PV arrays. Besides, vegetation management is often reactive, i.e. following rapid vegetation growth after rainfall events, and the contracts often reflect these reactive activities.

With regard to soiling mitigation, due to the relative remoteness of such PV sites and their impact on the system maintainability, combined with relatively low pollution losses, the use of automatic soiling management systems (e.g. cleaning robots) has been limited to date. On the other hand, on hot and dry climates without vegetation, soiling losses are usually higher, and therefore the need for automated module cleaning solutions has increased, particularly with fully automated dry brushing (in view of high costs and low availability of appropriate low-mineral water). Cleaning schemes can decrease production losses of PV modules by as much as 6-8% during summer months in these PV sites.

Figure 7: Left: Vegetation management of PV plant in Central Australia [2]; Middle: PV module (from a PV array in Central Australia) damaged during vegetation management due to pebbles [2]; Right: Automated cleaning of soiled PV modules in the Negev Desert, Israel [2].

4 CONCLUDING REMARKS – OUTLOOK

Through this study, Task 13 experts shed light on PV O&M best practices and guidelines, which are indispensable for reaching (or even exceeding) the expected service lifetime of PV systems in different climatic profiles and site characteristics.

A combination of well-designed O&M specifications, proactive monitoring systems and a flexible and tailored O&M regime that considers both climactic impact on PV systems as well as possible changes to grid requirements are good practices to ensure that PV systems reach or even exceed the expected lifetime. Taking these recommendations into account in operating a PV power plant e.g. in flood-prone conditions, hot arid (and remote) regions or in sites subjected to significant snowfalls will help proactive asset owners and O&M providers in identifying problems before they become serious and help prevent losses from avoidable circumstances.

Overall, calculating the risk profile of PV power plants in different climates may prove beneficial. On that note, we reckon that PV asset owners need to re-evaluate their risk appetite and the willingness to pay for higher resilience of PV systems (e.g. physical elevation of inverters, installing flood barriers, installing hardware and software to proactively shut down PV plants in case of extreme weather events).

ACKNOWLEDGEMENTS

The authors would like to acknowledge the excellent and valuable contribution of all IEA PVPS Task 13 experts involved in the SubTask 3.3 and the drafting of the report IEA PVPS T13 25:2022. The report itself was supported by the German Federal Ministry for Economic Affairs and Energy (BMWi) under grant number 0324304A and 0324304B, by the Austrian Federal Government, represented by the Austrian Research Promotion Agency (FFG) under contract no. 876736, by the New Energy and Industrial Technology Development Organization (NEDO), Japan, under contract #15100576-0 and by the Swiss Federal Office of Energy (SFOE) under contract no.: SI/501788-01. Part of this paper was also supported by the French National Program "Programme d'Investissements d'Avenir - INES.2S" under Grant Agreement ANR ANR-10-IEED-0014-01.

REFERENCES

[1] Jahn, U., Herteleer, B. 2022. Climate-specific O&M for PV power plants. PVTech Power Magazine, Solar Media.

[2] Jahn, U., et al. (2022). Report IEA-PVPS T13-25:2022.

[3] Ascencio-Vásquez, J. et al. (2019). Solar Energy, vol.

10.4229/EUPVSEC2023/4CV.1.12
020377-005

191, pp. 672-685, 2019.
[4] W. Herrmann et al. (2022) Qualification of Photovoltaic (PV) Power Plants using Mobile Test Equipment
[5] Enbar, N. et al. (2016). Report SAND2016-0649R, Electric Power Research Institute (EPRI) & Sandia National Laboratories.
[6] A. Granlund, J. Narvesjö and A. M. Petersson, "The Influence of Module Tilt on Snow Shadowing of Frameless Bifacial Modules," in 36th European Photovoltaic Solar Energy Conference and Exhibition, pp.1646-1650, ISBN 3-936338-60-4, Marseille, France, 2019.
[7] E. Andenæs et al. (2018). Solar Energy, vol. 159, pp. 318-328, 2018.
[8] Aarseth, B.B. et al. (2018). In: Proceedings of the 35th EUPVSEC, pp. 1630-1635.

10.4229/EUPVSEC2023/4CV.1.12
020377-006

UNMANNED AERIAL VEHICLE (UAV) DECISION-MAKING FOR PHOTOVOLTAIC (PV) PLANT DIAGNOSTICS USING IMAGE AND ELECTRICAL DATA ANALYSIS

Andreas Livera[1], Anna Michail[1], Juan Luis Carús Candás[2], Diego Fernández Martínez[2], Angelos Antonopoulos[3], George Petrakis[3], Achilleas Tripolitsiotis[3], Panagiotis Partsinevelos[3], Eftichis Koutroulis[3], and George E. Georghiou[1]
[1] FOSS Research Centre for Sustainable Energy, Photovoltaic Technology Laboratory, Department of Electrical and Computer Engineering, University of Cyprus, Nicosia, 1678, Cyprus
[2] TSK Electrónica y Electricidad S.A., Gijón, Spain
[3] Technical University of Crete, Chania, Crete, Greece

ABSTRACT: This paper aims to develop an unmanned aerial vehicle (UAV) decision-making platform for accurate photovoltaic (PV) plant diagnosis and optimum operation and maintenance (O&M) activities. A modular approach was followed to support the full plant characterization and performance assessment of the PV system under study by combining image processing with electrical data analytics and advanced computing capabilities. The proposed UAV platform incorporates an electroluminescence camera, a Red-Green-Blue camera, and a thermographic sensor. The system is controlled by a user-friendly geo-visualization interface with advanced reporting and operational functionalities. Robot operating, processing and decision-making capabilities are also included, adding the possibility to perform aerial flight inspection and predefined actuations by the UAV platform. The platform can propose remedy field actions to minimize the response and resolution times (i.e., the timing since an anomaly is detected and certain actions are carried out to mitigate the incident). Therefore, it enables near real-time monitoring of the PV plant, early detection, accurate localization of faults and geolocation of defects, leading to time- and cost-efficient PV plant diagnosis. The proposed platform was validated in a real environment that manages the complete process of data and images acquisition, cleaning and storage of the received information, image processing and data analysis for early fault detection, generation of recommendations for maintenance and prioritization/scheduling of field O&M actions. The results showed the efficacy of the utilized equipment and incorporated functionalities for raw data management, near real-time fault detection and localization of faulty modules, and accurate geolocation of the defects within the PV system using UAV-captured images complemented by electrical data analysis.
Keywords: decision making, failure analysis, image analysis, online diagnosis, operation and maintenance, outdoor inspection, performance, photovoltaic, unmanned aerial vehicle.

1 INTRODUCTION

Effective diagnosis of any fault, remains a technical and economic challenge, especially when dealing with large-scale photovoltaic (PV) plants [1]. Currently, PV plant monitoring is carried out by either electrical performance measurements or image processing [2]. Electrical data analysis methods suffer from low fault classification/categorization accuracy. In addition, fast identification of the physical location of the fault (or identification of the faulty PV module within a string) is challenging due to the nature of monitored data (usually inverter/array level data are available) [3]. In the same token, Infrared Thermography (IRT) imaging has already been used for the characterization of PV module failures, but its setup and processing are rather complex, and an experienced human operator is usually required [2]. To ensure high defect identification rate and minimize false alarms, the IRT imaging can be complemented by a dedicated electroluminescence (EL) camera, and a Red-Green-Blue (RGB) colour camera. The full characterization of the PV plant and detailed diagnosis (inspection, detection, classification and localization) can be thus achieved by integrating EL, RGB, IRT imagery and electrical data analytic methods in a single platform.

In the operation and maintenance (O&M) segment, the yearly costs of a traditional PV inspection are estimated to be around 11.27 €/kW$_p$ [4], from which only 1.47 €/kW$_p$ are the costs associated to the planned inspection and monitoring. To inspect and monitor large areas with fielded PV systems, latest trends and innovations include the utilization of drones (e.g., Remotely Piloted Aircrafts, RPAs or Unmanned Aerial Vehicles [UAVs]), equipped with multi-spectral cameras, advanced sensors and control systems [5]. Monitoring of installed PV arrays by aerial techniques offers many advantages, including fast detection, low cost and large area coverage, high-resolution imagery, high flexibility, remote and relatively easy operations, and ability to operate in harsh environments and unreachable locations [6]. These performance characteristics lead to significant reduction of the inspection and defect detection time. Consequently, the performance of large-scale PV systems is increased as optimized O&M activities are carried out in short time after the defect characterization and localization. Bearing the above in mind, a modular UAV platform is proposed for improved and cost-efficient PV plant diagnosis, to address the main commercial challenge of creating a solution tailored to the user (PV plant operator) needs for inspection and diagnosis of large PV areas.

2 METHODOLOGY

2.1 Approach

The aim of this work is to develop an advanced UAV platform (see Fig. 1) with on-board processing capabilities that will support sensor fusion, real-time PV performance monitoring, accurate fault diagnosis and optimized maintenance. The proposed solution integrates EL and IRT image processing algorithms, complemented with RGB imagery. Machine learning algorithms were also integrated to combine the image processing outcomes with the electrical data analysis results (see Fig. 2) [7,8]. This combination enables the detection of almost all known PV fault types, maximizing the fault classification rate/accuracy and minimizing false alarms.

10.4229/EUPVSEC2023/4CV.1.13
020379-001

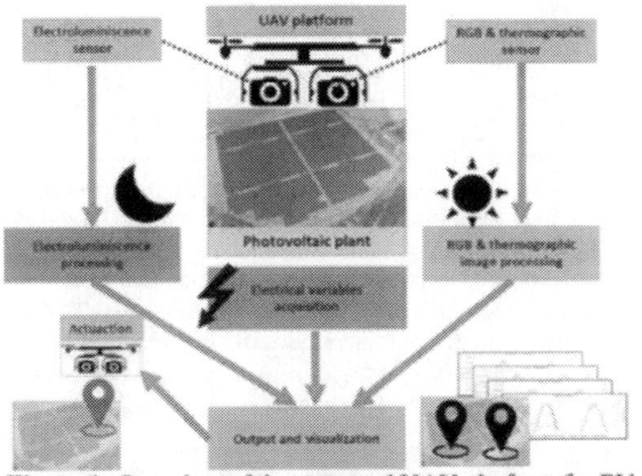

Figure 1: Overview of the proposed UAV platform for PV plant diagnosis and maintenance.

On-board processing capabilities were also included to support autonomous flight operation, precise geolocation of faulty PV modules and decision-making for fault diagnosis and optimum O&M actions.

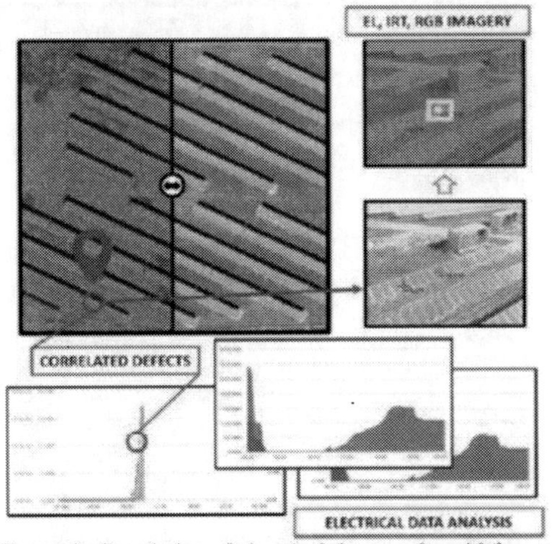

Figure 2: Correlation of electrical data results with image processing analysis to enable the diagnosis and geolocation of faults and defects with high accuracy.

2.2 Decision-making platform architecture and description

The proposed UAV platform advances current state-of-the-art in the following aspects: (1) integration of RGB, Infrared (IR) and Short Wavelength InfraRed (SWIR) camera payload, (2) accurate UAV geolocation implementing real-time kinematic (RTK) processing techniques, (3) defect geolocation based on the UAV position and the panel size and (4) design of dedicated maintenance actuators.

Currently, the market lacks a UAV platform with the ability to carry the aforementioned payload. To overcome this, an exhaustive analysis was carried out to determine the requirements in terms of flight time, payload weight, hovering accuracy, height, and orientation with respect to the solar panels, on-board processing capabilities, automated operations, etc. At first, the payload which meets these specifications was identified and purchased. More specifically, the DJI Zenmuse H20T and the Alvium 1800 U-030 cameras were selected. The first one carries a quad-sensor (thermal camera, range finder, RGB wide angle lens and zoom function), whereas the latter relies upon the Sony IMX991 sensor, which captures images in the visible and SWIR spectrum. Given the payload, two UAV system architecture designs were evaluated. The first incorporates manufacturing of a custom-made UAV platform, whereas the second consists of a modified commercial product. Due to cost and time implementation restrictions, the second design was implemented using the DJI Matrice M300 RTK platform (see Fig. 3).

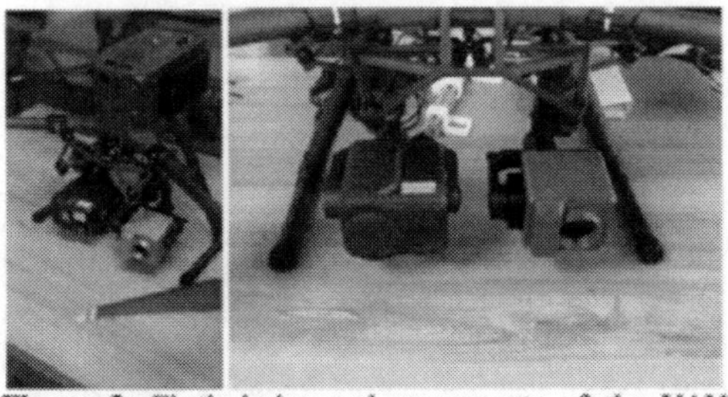

Figure 3: Final design and components of the UAV platform for PV plant diagnostics.

To accomplish the automated capture, archival and broadcasting of the motion imagery, an embedded processing unit with advanced computing capabilities was integrated into the commercial platform. The proposed platform enables the usage of third-party payloads by utilizing the DJI Payload SDK. The developed visible and SWIR along with the EL payload can operate and interface with the aerial platform with the usage of a Raspberry Pi 4 embedded computer. The final design of this module was designed and implemented, while the frame of the payload was fabricated with Acrylonitrile Styrene Acrylate (ASA) via additive manufacturing. This module takes advantage of the Payload SDK capabilities and allows interoperability with other onboard payloads such as the H20T. In such manner, all payloads can perform simultaneous coordinated actions.

The aerial platform has a maximum takeoff weight (MTOW) of 9 kg and can hover for about 30 minutes with both payloads (H20T, EL) installed. In such configuration, the aircraft can reach a maximum speed of 17 m/s while also providing an RTK augmented position accuracy of 1.5 cm in 3D space. The included optics achieve similar vertical field-of-view (vFOV) on both payloads.

Finally, a user-friendly graphical interface (GUI) was developed. The GUI of the flight planning application is depicted in Fig. 4.

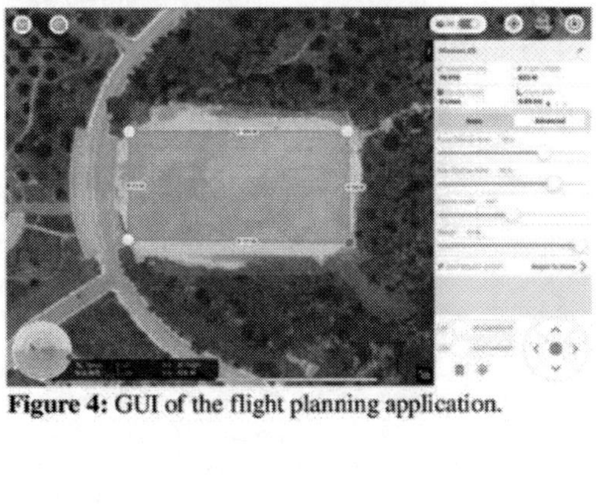

Figure 4: GUI of the flight planning application.

10.4229/EUPVSEC2023/4CV.1.13
020379-002

2.3 Description of geolocation, fault diagnosis and decision-making algorithms

For the geolocation of the defects, a two-step approach was followed. The first step was carried out during the UAV flight and focused on the accurate and precise determination of each image position and orientation. In the second step, coordinates of each image pixel (and subsequently the defect) were attributed in a global geodetic reference system (i.e., WGS'84). During the first stage, appropriate algorithms were employed, and associated software was developed to retrieve imagery from each camera sensor. Apart from the imagery, vital metadata (i.e., location, orientation, etc.) were also collected and archived. Horizontal and vertical accuracy of some centimeters was secured via the RTK global navigation satellite system (GNSS) receiver on-board the M300. Accurate orientation was carried out by processing and analyzing the measurements provided by the gimbal's rotary encoders and the UAV's Attitude and Heading Reference System (AHRS). At the second stage, dedicated and delicate image analysis was carried out to identify the defects in the acquired imagery. To geolocate these defects, a transformation of the image coordinate system to a global geodetic one must be carried out. Given the dimensions and shape of the solar panels, a Projection and Point (PnP) algorithm was employed to determine the defect's coordinates.

Apart from the geolocation algorithms, data-driven and image processing algorithms were also developed and integrated to the platform for PV fault diagnosis and maintenance. In this context, a deep learning algorithm (i.e., two-stage model based on a training phase built through a data augmentation process) was developed along with an image processing pipeline for detecting and classifying defects (e.g., bypass diode failures, hotspots, potential induced degradation [PID], cracks and microcracks, etc.) in PV systems. A preprocessing step was included in the pipeline to guarantee the quality of the results and to avoid existing failures in the acquisition process. Description of the data-driven diagnostic algorithms can be found in [1,8,9], while the image-driven algorithms are described in [10–12]. Available open-source image processing and diagnostic algorithms [12–14] were also integrated to the UAV platform.

To further enhance the accuracy of the UAV diagnostic system, the results of the electrical analysis were correlated with the outcomes from the image processing. All the processed information (captured images and electrical signals) and results (obtained from images and analysis of outdoor field measurements) are then transferred to the decision-making system. The hybrid PV power plant failure decision entity thus enables early defect detection, accurate categorization and localization of defects in PV systems as well as proposition of remedy field actions.

2.4 UAV platform benchmarking

The developed algorithms and capabilities of the UAV system were benchmarked in both laboratory and operational environment. More specifically, EL and thermal images obtained from crystalline silicon PV modules at the PV Technology Laboratory of the University of Cyprus were used for benchmarking the image processing algorithms. To evaluate the developed algorithms, the precision metric was used [15].

In parallel, the UAV platform was benchmarked in an operational environment (i.e., test PV power plants), using the available electrical and meteorological data along with the aerial data and images taken by the UAV.

The main experimentation was conducted in a test PV plant, located inside the campus of Technical University of Crete in Greece (see Fig. 5). This study area covers about 1.1 hectares and includes a total of 800 PV modules installed in 20 blocks (40 panels/block). Each PV module contains 120 monocrystalline solar cells, while being able to deliver a maximum power output of 375 W_p, at a voltage of 24 V.

Experiments were also conducted in a PV power plant of 772.8 kW_p capacity, located in the province of León and the municipality of Santa Marta in Spain. The plant consists of polycrystalline silicon panels and covers an area of 2.8 hectares. It was commissioned in the year 2010.

Figure 5: Test PV power plant in Technical University of Crete.

3 RESULTS

3.1 Benchmarking of image processing algorithms

Test crystalline silicon PV modules were disassembled and imaged indoors via EL to assess the performance of the image processing algorithms. The EL images were stored in an image failure database and revealed defects (e.g., PID, broken cells, cracks, etc.) not visible with the naked eye. An example of an EL image showing a PID affected PV module is shown in Fig. 6.

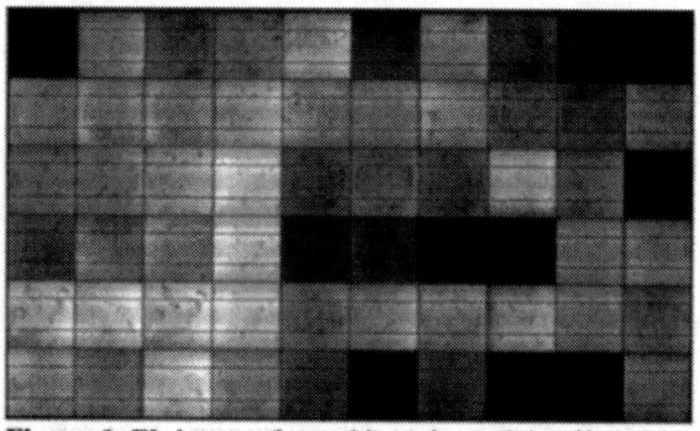

Figure 6: EL image of a problematic module affected by PID.

The taken EL images were then processed by the developed image diagnostic algorithms. The results showed the capability of the integrated algorithms for detecting PID, cracks, bypass diode failures and dead cells within the test PV modules. The algorithms were also capable of indicating the location of the problematic cells. Detected problematic cells by the algorithm (for the EL

10.4229/EUPVSEC2023/4CV.1.13
020379-003

image depicted in Fig. 6) are shown in Fig. 7. In total, 100 EL images (exhibiting different defects) were processed by the developed image processing algorithms, achieving an average precision of 89%. Further research is though required to improve the algorithm's precision.

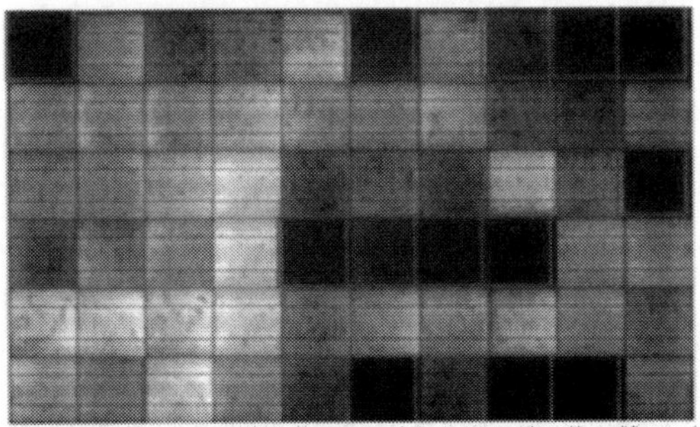

Figure 7: Processed EL image with detected cells affected by PID coloured in red.

3.2 Benchmarking of geolocation algorithms

Using appropriate photogrammetric algorithms and software tools, the 3D model of the test PV plant in Crete was initialy developed (see Fig. 8).

Figure 8: 3D model of the test PV power plant in Crete used for testing the geolocation algorithms.

A screenshot of the user-interface indicating the identification of PV array's corners and coordinates is shown in Fig. 9.

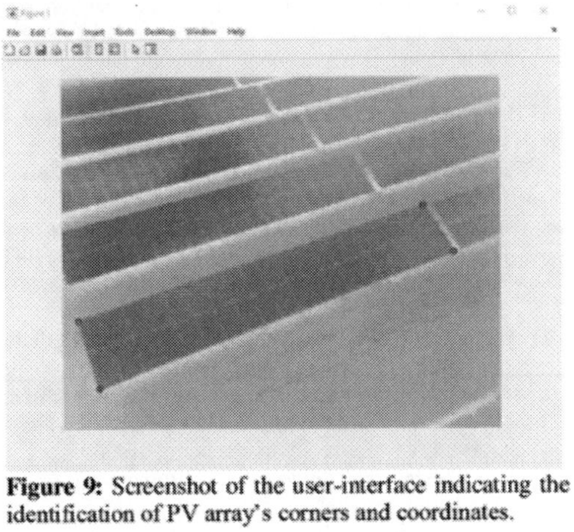

Figure 9: Screenshot of the user-interface indicating the identification of PV array's corners and coordinates.

To ease the procedure for geolocating the defects, the algorithm then applied a perspective correction (see Fig. 10).

Figure 10: Perspective correction applied to the selected area under study.

The results showed that the geolocation algorithms perform sufficiently with the proposed aerial platform, achieving a 30 cm error figure in 3D space when the procedure is conducted at a distance of 15m from the panels. This error is mainly produced by the angular error figure of the gimbal's yaw axis which is about ±2°.

3.3 Flight testing, aerial inspection, and diagnosis in PV systems

Experiments with autonomous flight plans were conducted. At first, a desired flight plan is designed while afterwards, the aerial platform is able to perform multiple flights autonomously (see Fig. 11 showing the execution of a flight route), following the pre-defined trajectory with satisfactory accuracy in a level of ±10 cm due to the RTK GNSS module.

Figure 11: GUI of the UAV platform during aerial inspection – execution of a flight route.

To obtain the geolocation of the diagnosis, an approach based on the metadata of the acquired images, RTK GPS data, the flight direction and other mathematical calculations was followed. The layout of the plant was subsequently used to determine the affected element of the plant. An example is shown in Fig. 12 for the test PV system plant in Spain, depicting the detected defects, their type/category and their location.

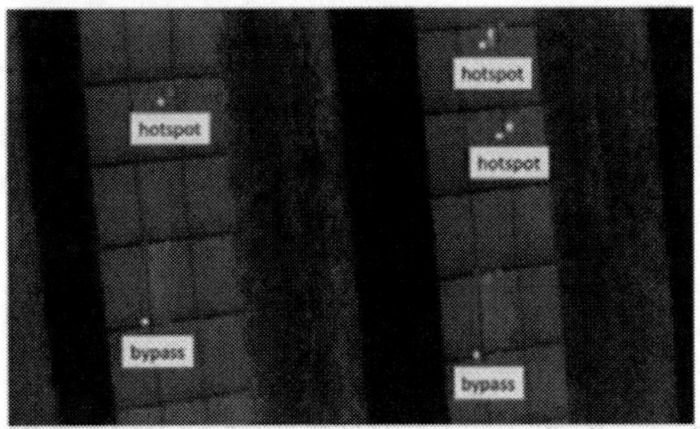

Figure 12: Defects detection, categorization and localization using obtained UAV images for a test PV plant in Spain.

Experiments with autonomous flight plans were also conducted to validate the performance of the UAV

10.4229/EUPVSEC2023/4CV.1.13
020379-004

platform for fault diagnostics. In total, 122 IRT images were taken from 35 m height to cover the surface of the PV plant in Spain. The results showed that 424 individual defects were detected and classified in two groups: 365 hot spot defects (not necessarily in different solar panels) and 59 bypass diode failure. The trained algorithm successfully detected and located more than 95% of the defective defects.

An online application dedicated to the management of the photographed flights has been built that allows the geopositioned visualization of the defects found. Defects can be visualized on a map or an orthophoto built with the flight images as shown in Fig. 13.

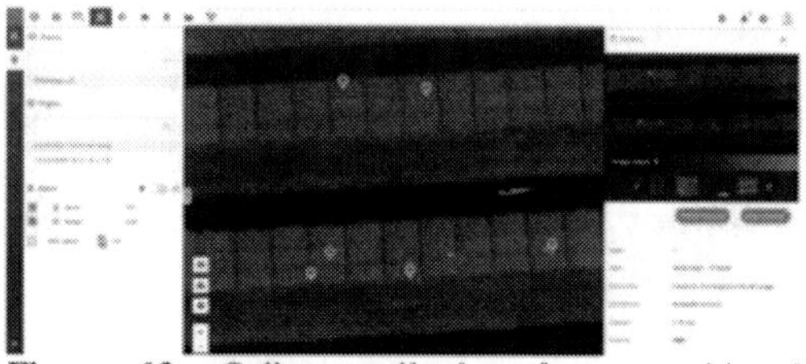

Figure 13: Online application for geopositioned visualization of the defects.

4 CONCLUSION

In this work, a decision-making system is proposed to support online PV plant diagnosis and time-efficient O&M activities. This study contributes towards the development of a UAV platform for reduced Levelized Cost of Electricity (LCOE), cost-effective O&M.

The developed UAV platform was validated in a real environment that manages the complete process of data and images acquisition, cleaning and storage of the received information, image and data analyses for early fault detection, generation of recommendations for maintenance and prioritization of field actions.

The results showed the efficacy of the utilized equipment and incorporated functionalities for raw data management, near real-time fault detection, localization of faulty modules, and accurate geolocation of the defects within the PV system using UAV-captured images.

Future publications will report additional benchmarking fault diagnostic results obtained by the UAV platform from the combination of image processing outcomes with the electrical data analysis results.

5 ACKNOWLEDGEMENTS

This work was funded by the AID4PV project. Project AID4PV is supported under the umbrella of SOLAR-ERA.NET Cofund 2 Additional Joint Call by the Centre for the Development of Industrial Technology (CDTI, IDI-20210170) in Spain, the General Secretariat for Research and Technology (GSRT, T12EPA5-00042) in Greece and the Research and Innovation Foundation (RIF, P2P/SOLAR/1019/0012) in Cyprus. The SOLAR-ERA.NET Cofund 2 Action is supported by funding from the European Union's HORIZON 2020 Research and Innovation Programme (Grant Agreement N° 786483).

6 REFERENCES

[1] A. Livera, M. Theristis, L. Micheli, E.F. Fernández, J.S. Stein, G.E. Georghiou, Operation and maintenance decision support system for photovoltaic systems, IEEE Access. 10 (2022) 42481–42496. https://doi.org/10.1109/ACCESS.2022.3168140.

[2] A. Livera, M. Theristis, G. Makrides, G.E. Georghiou, Recent advances in failure diagnosis techniques based on performance data analysis for grid-connected photovoltaic systems, Renew. Energy. 133 (2019) 126–143. https://doi.org/10.1016/j.renene.2018.09.101.

[3] A. Livera, M. Theristis, L. Micheli, J.S. Stein, G.E. Georghiou, Failure diagnosis and trend-based performance losses routines for the detection and classification of incidents in large-scale photovoltaic systems, Prog. Photovoltaics Res. Appl. 30 (2022) 921–937. https://doi.org/10.1002/pip.3578.

[4] G.M. Masters, Renewable and Efficient Electric Power Systems., 2004.

[5] SolarPower Europe [SPE], Operation & Maintenance Best Practice Guidelines / Version 5.0, 2022.

[6] H.P.C. Hwang, C.C.Y. Ku, J.C.C. Chan, Detection of malfunctioning photovoltaic modules based on machine learning algorithms, IEEE Access. 9 (2021) 37210–37219. https://doi.org/10.1109/ACCESS.2021.3063461.

[7] A. Livera, J.G. Franquelo, R.G. Bernal, G.E. Georghiou, Cloud-based decision support system for operation and maintenance in photovoltaic systems, in: 8th World Conf. Photovolt. Energy Convers., 2022: pp. 1339–1344. https://doi.org/10.4229/WCPEC-82022-4BV.5.3.

[8] A. Livera, G. Tziolis, F. Jose G., G. Ruben Gonzalez, Bernal George E., Cloud-based platform for photovoltaic assets diagnosis and maintenance, Energies. 15 (2022) 7760. https://doi.org/https://doi.org/10.3390/en15207760.

[9] A. Livera, Á. Fernández-solas, J.G. Bessa, J. Montes-romero, E.F. Fernández, V. Papaeconomou, G.E. Georghiou, Reducing the photovoltaic operation and maintenance costs through an autonomous control operation center, in: 2023 IEEE 50th Photovolt. Spec. Conf., 2023.

[10] H.M. Al-Otum, Automatic Defect Detection and Classification in Electroluminescence Images of PV Cells Using Convolutional Neural Networks, SSRN. (2023) 1–33. https://doi.org/10.2139/ssrn.4422413.

[11] U. Jahn, M. Herz, M. Köntges, D. Parlevliet, M. Paggi, I. Tsanakas, J.S. Stein, K.A. Berger, S. Ranta, R.H. French, M. Richter, T. Tanahashi, Review on infrared and electroluminescence imaging for PV field applications: International Energy Agency Photovoltaic Power Systems Programme: IEA PVPS Task 13, Subtask 3.3: report IEA-PVPS T13-12: 2018, 2018.

[12] A software utility for automatically finding cracks in UV Fluorescence images of solar panels (OpenUVF) by Southern Company R&D, GitHub. (2019). https://github.com/southern-company-r-d/OpenUVF (accessed September 6, 2023).

[13] Image analysis tool for solar modules, assisted by deep learning (pv-vision), GitHub. (2022). https://github.com/hackingmaterials/pv-vision (accessed September 6, 2023).

[14] Model Photovoltaic Fault Detector based in

model detector YOLOv.3 by simplemap.io (former Rentadrone.cl), GitHub. (2020). https://github.com/RentadroneCL/Photovoltaic_Fault_Detector (accessed September 6, 2023).
[15] A. Dong, Y. Zhao, X. Liu, Fault Diagnosis and Classification in Photovoltaic Systems Using SCADA Data, in: 2017 Int. Conf. Sensing, Diagnostics, Progn. Control Fault, 2017. https://doi.org/10.1109/SDPC.2017.31.

10.4229/EUPVSEC2023/4CV.1.13
020379-006

TOWARDS A ROBUST PERFORMANCE LOSS RATE ESTIMATE: MINIMISING THE UNCERTAINTY IN THE ANALYSIS OF PHOTOVOLTAIC SYSTEM DEGRADATION

Hugo Quest [1,2*], Christophe Ballif [1,3] Alessandro Virtuani [3]
[1] École Polytechnique Fédérale de Lausanne (EPFL), Institute of Electrical and Micro Engineering (IEM), Photovoltaics and Thin-Film Electronics Laboratory, Neuchâtel, Switzerland
[2] 3S Swiss Solar Solutions AG, Thun, Switzerland
[3] Swiss Centre for Electronics and Microtechnology (CSEM), Sustainable Energy Centre, Neuchâtel, Switzerland
[*] Corresponding author: hugo.quest@epfl.ch

ABSTRACT: The performance loss rate (PLR) is a key parameter in the assessment of photovoltaic (PV) systems' long-term performance and reliability. However, achieving robust, reproducible results with minimal uncertainty remains a challenge. Although there is no industry-wide consensus on how to compute the PLR and, thus, no standardised approach, the year-on-year (YoY) method is often deemed the most robust regression analysis. This work proposes a novel approach, the multi-annual YoY (multi-YoY), which reduces the statistical uncertainty of the metric through increased usage of available data. The concept is straightforward: the data points used in the standard YoY method are compared to all available subsequent years rather than just the year ahead, increasing the number of available comparisons. The methodology is validated using synthetic data and tested on high-quality datasets made available by IEA PVPS Task 13. Results show a 92% decrease in the size of the confidence interval (CI) comparing the standard YoY with the proposed method. Furthermore, comparisons with the IEA benchmark PLR values show good agreement with their ensemble method and minimised uncertainty. Further improvements and validations steps are envisaged, including comparisons of meteorological data sources, varying data lengths or quality, and accounting for performance ratio (PR) uncertainties.

Keywords: Performance Loss Rate (PLR), Degradation, Uncertainty, Year-on-Year (YoY)

1 INTRODUCTION

PV system lifetime and performance are described through the Performance Loss Rate (PLR), which represents the relative change in power output over time and is expressed as a percentage loss per year. The PLR is a vital metric for PV module manufacturers and installers, as it defines the performance and health status of systems in a single, comprehensive value. However, its estimation is notoriously complex, as there is no standardised methodology or definition of the metric. In this work, the PLR is defined as the change in annualized energy performance index relative to the first year [1], where the chosen energy performance index is the performance ratio (PR). On top of the challenge of defining the PLR, errors due to data quality or statistical methods generate large uncertainties, which are often not considered in current literature. Moreover, with the current exponential growth of installed PV capacity and the rapid product innovation and scale-up of the PV terawatt age, the majority of PV systems installed are and will be new, lacking the three years of performance data previously required to assess long-term performance and reliability [2]. It is therefore vital to establish robust and data-efficient solutions to estimate system performance over time.

The main steps for computing the PLR summarised by the IEA PVPS Task 13 experts consist of (i) input data cleaning, (ii) data filtering, (iii) performance metric selection, adjustments, and aggregation, and (iv) use of a statistical modelling method to calculate the PLR [3]. Importantly, data quality and data-related uncertainties play a significant role in the PLR values, adding a layer of complexity in the determination of a single, robust quantification of a system's health [4]–[6]. An early attempt was made at the IEC level to establish a standard or technical reference for calculating degradation using IEC 61724-4, but this first activity was abandoned [7]. Although there is no agreed-upon standard approach, the Year-on-Year (YoY) methodology is commonly considered the most reliable regression analysis method for long-term performance assessment of PV systems [8]. It relies on the comparison of normalised yields through subsequent years, and results in a distribution of PLR values. This allows for statistical analysis by evaluating the histograms of the PLRs. In this work, an enhanced version is proposed: the multi-annual YoY (hereby referred to as multi-YoY). The multi-YoY is a simple, efficient solution to make the most out of available performance data to estimate the PLR with minimised uncertainty. The new method relies on multi-annual year-on-year comparisons, effectively reducing the statistical confidence interval.

2 METHODOLOGY

Fig. 1 summarises the simplified analysis pipeline for both the standard and multi-annual YoY methods:

(1) Raw system data is gathered, requiring at least system yield (power output) and irradiance data. This stage already involves potential errors and uncertainties from the data sources, stemming from sensor accuracy, drift, or post-processing issues.

(2) Computing the performance ratio (PR), defined as the ratio between system output power and rated power with an irradiance correction term [3], see equation (1) where P_{meas} [W] is the measured power, P_{rated} [Wp] is the rated power, G_{meas} [W·m^{-2}] the measured irradiance, and G_{STC} = 1000 W·m^{-2} the irradiance at standard test conditions (STC). Filtering steps, such as removing low irradiances (< 50 W/m^2), can be included at this stage to reduce noise. The PR is then

10.4229/EUPVSEC2023/4CV.1.14
020381-001

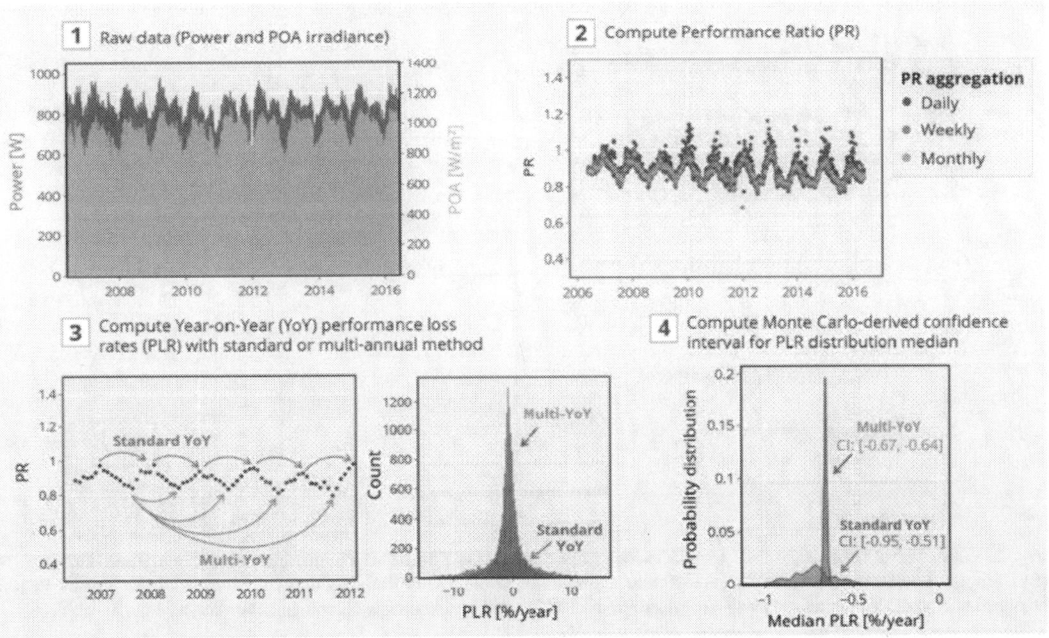

Fig. 1. Step-by-step decomposition of the proposed method: **(1)** Access PV system data (DC power and in-plane irradiance). **(2)** Compute the performance ratio (PR) at a given aggregation rate. **(3)** Compute the year-on-year PLR, with either the standard or newly proposed multi-YoY method, which compares all available data points instead of only looking at points separated by one year (here monthly PR values are shown). **(4)** Derive the Monte Carlo confidence intervals for the system median PLR.

aggregated to lower sampling rates (e.g., daily, weekly, monthly or yearly aggregations).

$$PR = \frac{P_{meas}/P_{nom}}{(G_{meas}/G_{STC})} \quad (1)$$

(3) The PLR can now be determined through various statistical methods. **Fig. 1.3** visually represents the difference between the standard and proposed multi-YoY methods, using a monthly PR time series as an example. For the standard YoY, each monthly PR point will be compared to the subsequent year (e.g., January 2007 vs. January 2008), and the relative difference represents one instance of PLR [%/year]. Applying this to all PR points, a histogram of PLRs is obtained, from which the final system value can be deduced through the median. For the multi-YoY, each PR point is compared to all subsequent years, normalising the relative difference to the number of years separating the points. This will also yield a distribution of PLRs, with an increased number of comparisons. The advantage is dual: since the PLR determination process is statistical in nature, increasing the number of counts will reduce the confidence interval. Moreover, the multi-annual approach increases the balance in terms of data usage since each PR point is compared to all subsequent years. For the standard YoY method, edge years (the first and final available years) are only used in one comparison, while other years will be used twice (previous and subsequent years).

(4) The final step in the analysis is the determination of the median PLR and its confidence interval (CI). The median of the distribution is chosen as it is more robust to outliers than the mean. However, given that the underlying distribution of the PLR values is unknown and that sample sizes can be relatively small (especially for lower aggregations), the CI of the median PLR cannot be analytically calculated. In this case, the Monte Carlo method is applied, which consists of generating random samples from the distribution of PLRs and using these samples to estimate the CI. Specifically, each random resampling generates a median PLR value, and the CI is deduced from the standard deviation of this new median PLR distribution [3], [7], [9].

3 RESULTS AND DISCUSSION

3.1 Validation

In order to validate the newly proposed multi-YoY method, a synthetic dataset with known PLR is generated. **Fig. 2** summarises the validation steps and results, following the defined methodology of **Fig. 1**. First, synthetic PV system power output is simulated using satellite-based irradiance data from the CAMS solar radiation service [10]–[12] and the Python pvlib library [13]. An artificial -1%/year PLR is added to the data, creating a known "true" value of relative performance loss trend. The second step consists of the computation of the performance ratio (PR), and multiple aggregation strategies are compared (daily, weekly, monthly and yearly mean values). From the PR trends, the standard YoY and multi-YoY methods are then applied to extract the PLR values, generating statistical distributions from which the median PLRs are extracted using Monte Carlo analysis. The distributions of median PLRs provide the final value of PLR, along with the confidence interval

10.4229/EUPVSEC2023/4CV.1.14

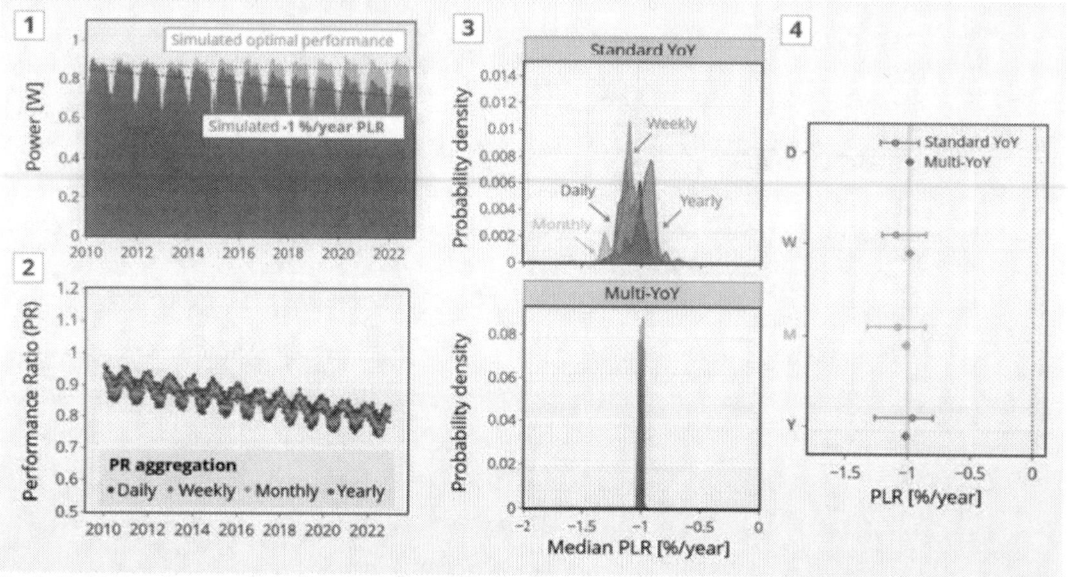

Fig. 2. Validation of new multi-YoY method with synthetic data. **(1)** Simulated PV production with artificial -1%/year PLR. **(2)** Computed performance ratio (PR) at different aggregation levels. **(3)** Probability density function of the median PLRs for standard and multi-YoY methods. **(4)** Estimated median PLRs and uncertainties comparing standard and multi-YoY.

(using 95% confidence level). **Fig. 2.3** shows the resulting median PLR distributions using the standard and multi-YoY methods. The standard YoY results in a wide distribution of PLR values, with significant variations depending on the level of aggregation, while the multi-YoY methods shows tighter distributions and better agreement with all sampling levels. Finally, **Fig. 2.4** summarises the PLR values derived from both methods, along with the uncertainties (CI). For each aggregation level, the top value represents the standard YoY results, while the bottom value represents the multi-YoY. The "true" PLR of -1%/year is highlighted in green. **Table I** presents the resulting PLR values comparing both methods at all aggregation levels. Confidence intervals are not reported directly but can be compared using **Fig. 2.4**. The main outcomes are as follows:

- The average relative error (RE) compared to the "true" PLR is 8% for the standard YoY, and only 1% for the multi-YoY.
- Comparing the 95% confidence intervals of both methods, the multi-YoY yields 92% decreased uncertainty compared to the standard YoY.
- Overall, the multi-YoY therefore increases both accuracy and precision.

The reduction in CI is mainly due to the statistical nature of the Monte Carlo method, and can be explained by the following factors: (i) a larger sample size generally results in a more accurate estimation of the measured statistic (in this case, the median), thus as the sample size increases with the multi-YoY, the sampling distribution of the median becomes narrower, which leads to a smaller CI; (ii) with a larger sample size, the variability in the estimates of the median decreases, and the estimates of the median become more stable and consistent; (iii) the Central Limit Theorem states that the distribution of sample medians approaches a normal distribution as the sample size increases, regardless of the underlying distribution of the data – increasing the sample size with the multi-YoY method makes the distribution properties tend towards a normal distribution, which has narrower tails compared to many other distributions, leading to a narrower confidence interval.

Table I. Resulting PLR values for the standard and multi-YoY methods applied to the synthetic dataset with known PLR of -1%/year, and relative errors (RE) compared to the true value.

Aggregation	Statistical method			
	Standard YoY		**Multi-YoY**	
	PLR [%/year]	**RE [%]**	**PLR [%/year]**	**RE [%]**
Daily	-1.11	11	-1.00	0
Weekly	-1.10	10	-0.99	1
Monthly	-1.08	8	-1.01	1
Yearly	-0.96	4	-1.02	2

3.2 IEA PVPS Task 13 datasets

In order to further validate and test the new multi-YoY method, a high-quality dataset from the IEA PVPS Task 13 report on PLR estimations is used as a first benchmark study [3]. Three of the datasets made available are used:

- EURAC: Ground-mounted polycrystalline silicon system with a nominal power of 4.2 kWp, fixed 30° tilt and orientation 8.5° west of South. Installed at the airport of Bolzano (Italy) with data from February 2011 to January 2019.
- FOSS: Ground-mounted monocrystalline silicon (mono-Si) PV system (rated at 1025 Wp) installed at the outdoor test facility of the University of Cyprus in Nicosia, Cyprus. The system is south-facing, at the optimum tilt angle of 27.5°, in open-field mounting. In total,

ten years of data are made available (from June 2006 to June 2016).

- RSE pc-Si: Ground-mounted polycrystalline silicon PV system with nominal power of 1.61 kWp, south orientation and tilt of 30°. Data available from June 2009 to January 2019.

For all datasets the electrical outputs and meteorological data are monitored following the IEC 61724-1 standards. **Table II** shows the results of the applied PLR analysis pipeline, where the standard and multi-YoY approaches are compared. For the two YoY methods, the average of all aggregation levels is shown. The mean PLR from the IEA PVPS Task 13 methodology is used as benchmark. Overall, the standard YoY PLRs align well with the IEA ensemble method proposed by Lindig et al., and falls within the defined CIs. The multi-YoY also shows a 90% decrease in CI compared to standard-YoY (not shown). However, the multi-YoY PLRs show slight deviations (especially for the FOSS dataset). Further analysis should focus on understanding the differences observed between the methodologies and confirming the robustness of the newly proposed multi-YoY.

Table II. PLR results for the three PV systems from the IEA PVPS Task 13 datasets, comparing the IEA mean from the study [3], standard YoY and multi-YoY.

System	PLR [%/year]		
	IEA Mean	Standard YoY	Multi-YoY
EURAC	-0.85	-0.82	-0.94
FOSS	-0.71	-0.80	-0.52
RSE pc-Si	-0.96	-0.94	-1.01

4 CONCLUSION

To conclude, this work proposes a simple solution to improve the statistical confidence interval of the YoY PLR, which we call the multi-YoY approach. Through the utilisation of all comparison points available instead of only those separated by one year, the median PLR confidence interval is decreased by 92% compared to the standard YoY, and results indicate that the sampling level does not impact the PLR estimation in the multi-YoY. In addition to the validation with a synthetic dataset, the novel method is tested with high-quality IEA PVPS Task 13 system data, showing 90% lowered uncertainty. Further datasets will be included in future work to quantify the robustness of this new method, and the impact of data quality, length, non-linearity or varying filters will be investigated.

REFERENCES

[1] M. G. Deceglie *et al.*, 'Perspective: Performance Loss Rate in Photovoltaic Systems', *Solar RRL*, vol. n/a, no. n/a, p. 2300196, Jun. 2023, doi: 10.1002/solr.202300196.

[2] D. C. Jordan, N. Haegel, and T. M. Barnes, 'Photovoltaics module reliability for the terawatt age', *Prog. Energy*, vol. 4, no. 2, p. 022002, Apr. 2022, doi: 10.1088/2516-1083/ac6111.

[3] S. Lindig *et al.*, 'International collaboration framework for the calculation of performance loss rates: Data quality, benchmarks, and trends (towards a uniform methodology)', *Progress in Photovoltaics Research and Applications*, vol. 29, pp. 573–602, Mar. 2021, doi: 10.1002/pip.3397.

[4] S. Lindig, M. Theristis, and D. Moser, 'Best practices for photovoltaic performance loss rate calculations', *Prog. Energy*, vol. 4, no. 2, p. 022003, Apr. 2022, doi: 10.1088/2516-1083/ac655f.

[5] S. Lindig, A. Louwen, D. Moser, and M. Topic, 'Outdoor PV System Monitoring—Input Data Quality, Data Imputation and Filtering Approaches', *Energies*, vol. 13, no. 19, Art. no. 19, Jan. 2020, doi: 10.3390/en13195099.

[6] S. Lindig, J. Ascencio-Vasquez, J. Leloux, D. Moser, and M. Topic, 'Climate related dependence of performance losses of over (3,500) PV systems', Sep. 2020.

[7] D. C. Jordan, M. G. Deceglie, and S. R. Kurtz, 'PV degradation methodology comparison — A basis for a standard', in *2016 IEEE 43rd Photovoltaic Specialists Conference (PVSC)*, Jun. 2016, pp. 0273–0278. doi: 10.1109/PVSC.2016.7749593.

[8] D. C. Jordan, C. Deline, S. R. Kurtz, G. M. Kimball, and M. Anderson, 'Robust PV Degradation Methodology and Application', *IEEE Journal of Photovoltaics*, vol. 8, no. 2, pp. 525–531, Mar. 2018, doi: 10.1109/JPHOTOV.2017.2779779.

[9] E. Hasselbrink *et al.*, 'Validation of the PVLife model using 3 million module-years of live site data', in *2013 IEEE 39th Photovoltaic Specialists Conference (PVSC)*, Jun. 2013, pp. 0007–0012. doi: 10.1109/PVSC.2013.6744087.

[10] Copernicus Atmosphere Monitoring Service (CAMS) Atmosphere Data Store (ADS), 'CAMS solar radiation time-series'. Jun. 01, 2023.

[11] Z. Qu *et al.*, 'Fast radiative transfer parameterisation for assessing the surface solar irradiance: The Heliosat-4 method', *Meteorologische Zeitschrift*, pp. 33–57, Feb. 2017, doi: 10.1127/metz/2016/0781.

[12] B. Gschwind, L. Wald, P. Blanc, M. Lefèvre, M. Schroedter-Homscheidt, and A. Arola, 'Improving the McClear model estimating the downwelling solar radiation at ground level in cloud-free conditions – McClear-v3', *Meteorologische Zeitschrift*, pp. 147–163, Jun. 2019, doi: 10.1127/metz/2019/0946.

[13] W. F. Holmgren, C. W. Hansen, and M. A. Mikofski, 'pvlib python: a python package for modeling solar energy systems', *Journal of Open Source Software*, vol. 3, no. 29, p. 884, Sep. 2018, doi: 10.21105/joss.00884.

IMAGE QUALITY EVALUATION OF CONTACTLESS OUTDOOR PHOTOLUMINESCENCE BASED ON STRING INVERTER'S IV CURVE SWEEP CAPABILITY

Gisele A. dos Reis Benatto[1], Rodrigo Del Prado Santamaría[1], Sergiu V. Spataru[1], Marija Vuković[2], Mathea Salmi Marjavara[2], Espen Olsen[2], Ingunn Burud[2].

[1]Department of Electrical and Photonics Engineering, Technical University of Denmark (DTU), Frederiksborgvej 399, 4000 Roskilde, Denmark.

[2]Faculty of Science and Technology, Norwegian University of Life Sciences (NMBU), Drøbakveien 31, 1430 Ås, Norway.

ABSTRACT: Luminescence imaging is widely used as ground truth for defect and fault detection in many instances of research and PV inspections. Cable handling and PV shading practices for field inspections can correspond to most of the inspection time and be responsible for the highest cost and safety concerns for the procedure. Performing photoluminescence (PL) during an IV sweep that has been already programed to happen in an inverter is an excellent strategy to reach high throughput for PV luminescence-based inspections without significant additional demands in terms of cost and inspection complexity. In this work, we present a detailed image quality analysis with image acquisition performed at different sun excitation intensities, show what is the level of image quality that can be achieved compared with daylight electroluminescence, and the available possibilities to acquire PL images during an inverter IV sweep procedure.

Keywords: Outdoor photoluminescence, PV module characterization, Inverter IV sweep.

1 INTRODUCTION

Outdoor luminescence-based imaging has the highest accuracy in the evaluation of photovoltaic (PV) modules health, providing data for preventive maintenance, fault severity and power loss estimation in solar energy production. However, the methods available in the operations and maintenance (O&M) industry provide very limited throughput of images [1]. When the process of luminescence imaging acquisition is optimized, the required inspection time can mainly be undertaken for plant electrical design location and cable handling. This is the case for both electroluminescence (EL) and photoluminescence (PL) imaging techniques that require electrical connections for powering and controlling the PV modules operational condition. The inverter market has recently implemented current-voltage (IV) curve sweep on a string level for preventive maintenance and better maximum power point tracking. During the IV sweep, the PV modules are in the different operational modes required for extraction and removal of reflected sunlight out for the daylight PL images [2], [3]. This work aims make use of the inverter IV sweep functionality by acquiring images with an Indium Gallium Arsenide (InGaAs) camera; exploring the possibilities to extract luminescence signal embedded in sunlight noise. This study extends the proposed PL imaging procedure during IV curve sweeps [3] to another type of inverter. With our inverter's scanning procedure, hereafter referred to as IV curve acquisition process, an investigation into PL images acquired during constant operating conditions outside an actual IV curve sweep is possible. We present an image quality analysis using the suitable signal-to-noise ratio (SNR) metric available in the literature [4], [5] on image acquisitions performed at different sun excitation intensities and sun variation scenarios. The level of image quality is also compared with daylight EL (dEL) images regarding detection of defects and faults for PV diagnosis.

2 METHODOLOGY

2.1 Experimental details

From the PV string side, the IV curve acquisition process is commanded to run by the inverter (Huawei SUN2000 -30KTL-M3), while the current at the string is monitored using an oscilloscope to verify when it has started. After having estimated the inverter response time, the IV curve acquisition process started in coordination with the camera acquisition. The camera was positioned with its field-of-view covering two passivated emitter and rear cells (PERC) bifacial PV modules. The exposure time was set to 2 ms to 2.5 ms to avoid saturation from the reflected sunlight. During the experiment, the irradiance was measured using a calibrated reference cell. An optical filter with 50 nm bandwidth centred at 1150 nm was used. The camera used was a C-RED 3 from First Light Imaging, which had the capability of acquiring 600 frames per second (fps). To make sure that the image sequence covered the whole IV curve acquisition process, the image acquisition was performed for about 6 s, regardless of the frame rate. This means, for example, that 3000 frames were acquired for the tests with image acquisition of 480 fps. Fig. 1 shows how the average pixel number acquired by the camera behaves when synchronized with the current produced by the PV modules connected to the inverter. This example corresponds to Experiment 1 and shows how the inverter is programed to acquire the IV curve data. It sets the PV modules at different operational points during this experiment, differently than what was observed previously for another inverter brand [3].

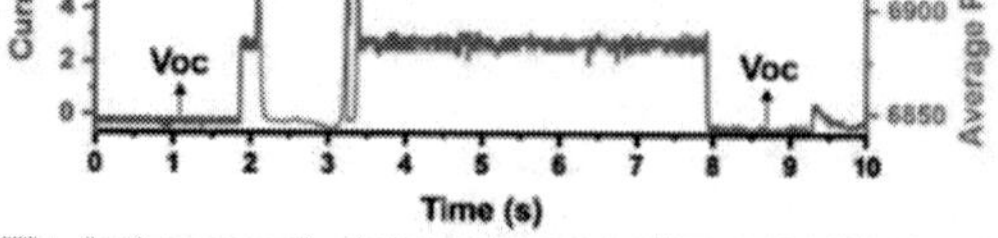

Fig. 1: Average pixel count corresponding to the PL signal levels extracted from the images acquired in the field during the inverter IV curve acquisition process

10.4229/EUPVSEC2023/4CV.1.19
020383-001

(Experiment 1).Experiment 1 was performed with 497 W/m^2 sun irradiance and the samples imaged were two PERC glass-glass bifacial 595 W modules constituted of half cells and with cell open circuit voltage (V_{OC}) equal to 0.69 V. One of these modules was not-degraded and the other had clear signs of degradation, with power loss measured in a solar simulator of 7.1%. Analysing the average pixel count of the images acquired during the IV curve acquisition process (Fig. 2), it was verified that the inverter functionality sets the PV string at three main operation points, generating at least three well distinguishable average pixel count levels in the image sequence. The highest is correspondent to V_{OC} and the lowest to Isc [2] and the difference of the images at these two levels were called here ΔPL_1. As the lowest level can be considered the minimum luminescence condition of the PV module, it is where the images are taken as background (BG). An intermediate level of current extraction was also detected in this experiment but could not be well defined. The images acquired in this level were taken into consideration for image quality evaluation, generating ΔPL_2 images.

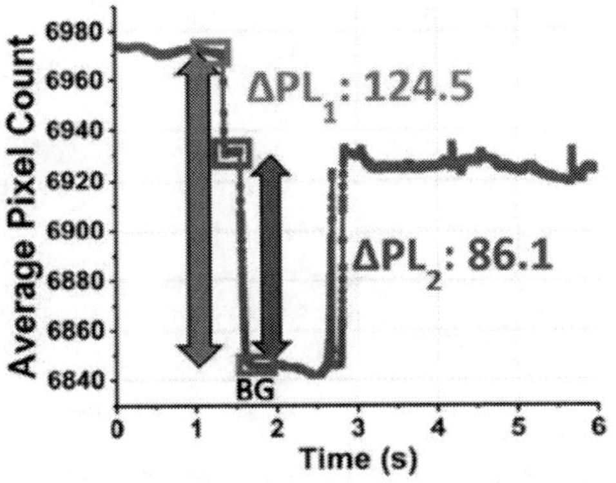

Fig. 2: Average pixel count from Experiment 1 and ΔPL_1 and ΔPL_2 levels studied.

For Experiment 2, the same type of modules as for Experiment 1 were used, a non-degraded and a degraded module (power loss of 8.3%). The sun irradiance in this case was 912 W/m^2 and for this scenario, no intermediate level is observed (Fig. 3 – top). Relatively stable sunlight conditions can be verified for these two experiments. At the bottom of Fig. 3 the part of the IV curve acquisition process which shows the IV curve sweep is zoomed in. Image pairs and BG images that are analysed in section 3.2 are indicated. In Experiment 3, two monofacial 305W PERC modules with cell V_{OC} of 0.66 V containing cracked cells (power loss of 12%) were imaged. The sun irradiance was measured to be 1080 W/m^2 just before image acquisition. But the average pixel count changes during the image sequence observed in Fig. 4, which is an indication that this was a scenario with significant and fast sun irradiance variation. In Fig. 2-4, the ΔPL in pixel counts is provided, but only as a reference of signal magnitude.

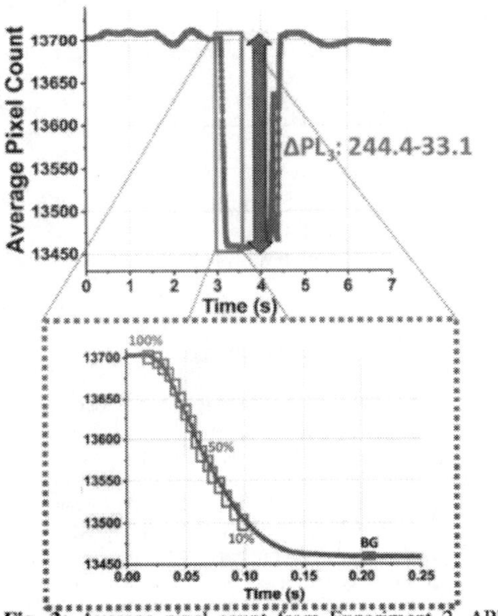

Fig. 3: Average pixel count from Experiment 2, ΔPL_3 interval studied (top) and zoom-in on the ΔPL_3 percentage levels studied throughout the IV sweep (bottom).

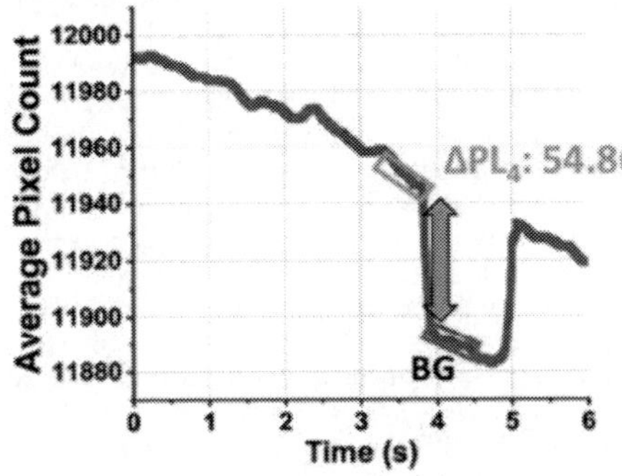

Fig. 4: Average pixel count from Experiment 3, ΔPL_4 interval studied.

2.2 Image processing and SNR metric

The image sequences acquired during the IV curve acquisition processes consist of batches of images with low (Isc) or high (V_{OC}) luminescence levels and the voltage sweep with a range of luminescence levels, i.e., current extraction conditions. This is not a modulation type of acquisition, which is the most common way to perform PL imaging using sunlight as excitation source [2], [6]–[8], as it provides an optimal scenario to eliminate sunlight noise. For PL imaging performed during an IV curve acquisition process from our particular inverter, batches of identical images with high PL and low PL (background – BG) can be averaged and subtracted. However, the quality of the final ΔPL image is more subjected to changes in the ambient conditions, mainly sunlight irradiance. For the IV sweep itself, only one image pair is taken per level, as even with high framerate, each image in the sweep shows different levels of PL intensity (in pixel counts – Fig. 3).

The available signal and image quality metrics for this image acquisition profile are the SNR_{50} for individual image pairs and SNR_{AVG} for image batches [4], [9]. SNR_{50}

10.4229/EUPVSEC2023/4CV.1.19
020383-002

was originally developed for EL acquired at low light conditions (nighttime or lab conditions) and SNR_{AVG} is its expansion when several image pairs are averaged and subtracted. For boths SNR markers, the minimal acceptable value for outdoor luminescence imaging is $SNR \geq 5$.

For the three experiments described, the image quality evaluation was performed respectively regarding:

1. Different number of image pairs at different levels of current extraction,
2. Different levels of current extraction along the IV curve sweep,
3. Unstable irradiance levels due to partial cloud cover.

3 RESULTS AND DISCUSSION

3.1 Experiment 1: Image quality per number of image pairs

It is expected that the SNR and the image quality will improve with the number of image pairs used to generate a ΔPL image. For ΔPL_1 and ΔPL_2, the SNR_{AVG} marker can be observed in Fig. 5. The SNR_{AVG} does increase with the number of image pairs for both ΔPL intervals, with the only exception being from 40 to 50 images pairs, where the ΔPL_1 has an indication of having worse signal than for 30 image pair. Yet, the SNR value is far beyond the minimal requirement and in practice providing good quality even when an individual image pair is subtracted. When observing ΔPL_1 images for 30 and 50 image pairs in Fig. 6, a "3D-like" anomaly can be observed in both; however, the SNR is so high in both cases that no diagnosis information is different for both instances.

In Fig. 7, the ΔPL_2 images generated with minimum and maximum number of image pairs is shown. This ΔPL interval does provide lower SNR values in general (Fig. 5 bottom), but they still indicate, high quality images, even if slight anomalies are visible. Even though ΔPL_2 has about 70% of the luminescence intensity in pixel counts of ΔPL_1, which can be translated linearly to current extraction levels, the diagnosis information of their ΔPL images is very similar. It is very clear to observe that in all images in Fig. 6 and 7 the degraded module is the one to the left and that no particular cells present lower emission relative to the brighter cells at the same module when we compare Fig. 6 to 7. This is an effect that can be detected for this kind of degraded module at lower than I_{SC} current injection for the case of indoor EL images.

3.2 Experiment 2: ΔPL images along the IV sweep

Fig. 8 shows the comparison of ΔPL_3 images of the degraded module obtained from single image pairs at different levels of current extraction along the IV sweep (indicated as percentages in Fig. 3). They are compared with equivalent current injection levels for dEL. PL is known to have a relatively lower signal than EL, for which current injection is highly controllable [11]. Additionally, dEL is performed with a modulated signal that, as mentioned previously, is more efficient to enhance the SNR.

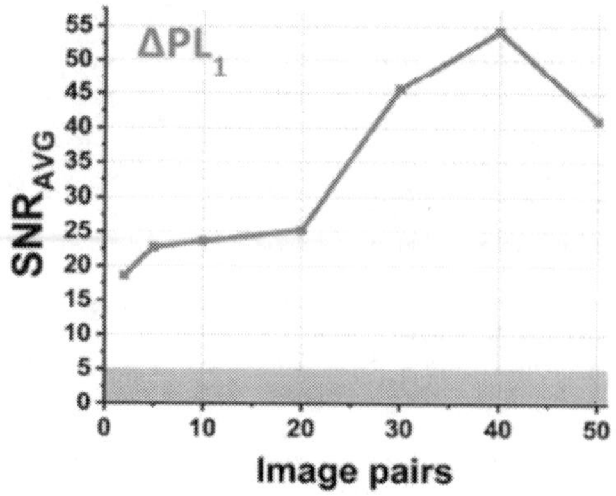

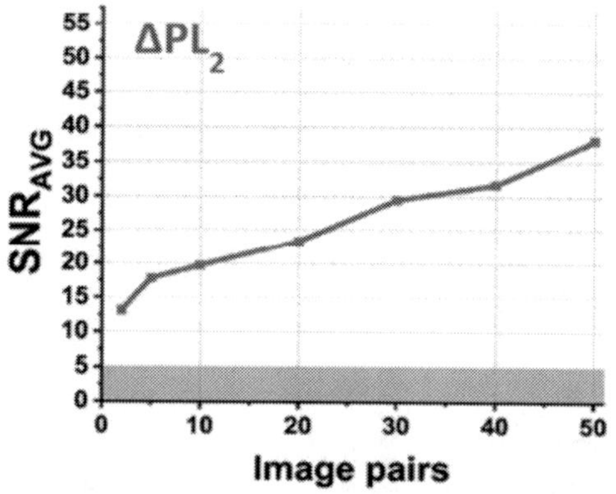

Fig. 5: SNR_{AVG} marker as a function of number of image pairs used for averaging and subtraction of ΔPL_1 (top) and ΔPL_2 (bottom).

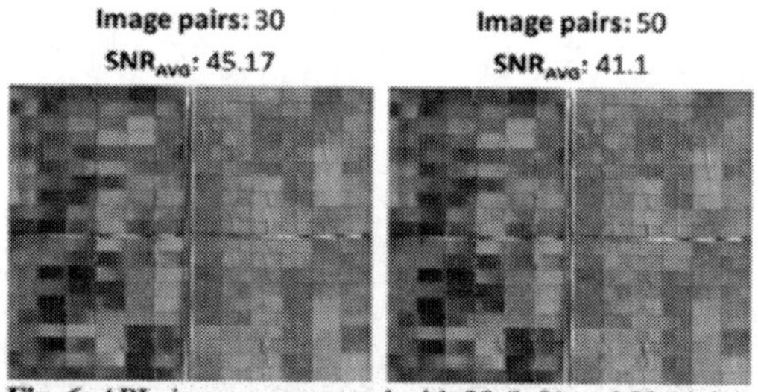

Fig. 6: ΔPL_1 images generated with 30 (left) and 50 (right) images pairs.

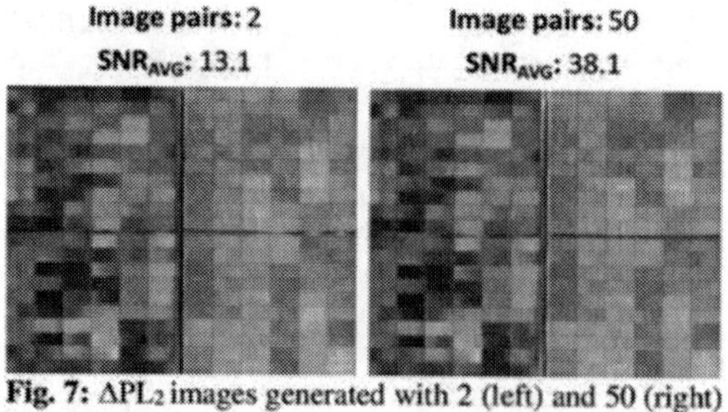

Fig. 7: ΔPL_2 images generated with 2 (left) and 50 (right) images pairs.

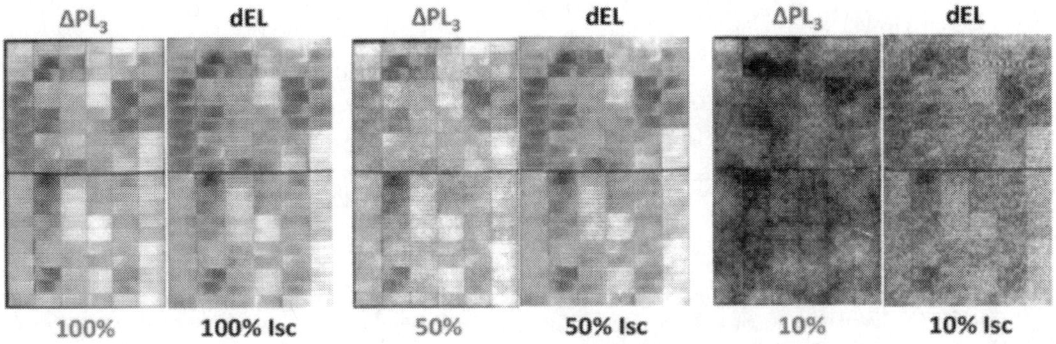

Fig. 8: ΔPL_3 images of the degraded module at different levels of current extraction along the IV sweep compared with dEL images with correspondent percentage of current injection.

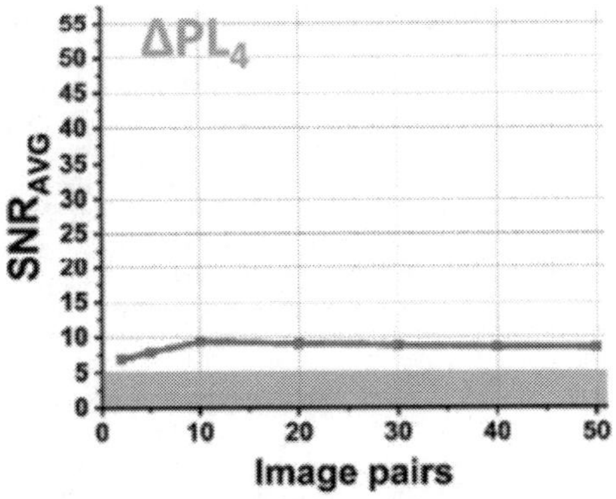

Fig. 9: SNR_{AVG} marker as a function of number of image pairs used for averaging and subtraction of ΔPL_4.

For 100% ΔPL_3 current extraction and 100% Isc current inject dEL, the diagnosis information is close to equivalent. dEL shows higher contrast for degraded cell at the left side of the top half of the degraded module, while the image quality is slightly better for ΔPL_3. For 50%, ΔPL_3 image starts to be significatly noisier than dEL and lose contrast essential for the diagnostic of the PV module. Here the SNR_{AVG} is 13.6 and it decreases with the current extraction. However, 50% ΔPL_3 was the last image in the sweep that alowed proper PV module diagnosis. At 10%, both images are mainly noise. For ΔPL_3, at 20% the SNR_{AVG} was 6, and in this case, this SNR marker presents an inconsistensy with the image quality.

3.3 Experiment 3: High sunlight variation scenario

The changing ambient conditions at the short-wave infrared range captured by the InGaAs camera (which includes the sunlight spectrum) is the most challenging scenario to calculate the SNR and enhance luminescence image quality. Fig. 9 shows the evolution of the SNR_{AVG} marker toward the number of image pairs for such a challenging scenario. Additionally, PV modules used in Experiment 3 had lower signal than the modules shown in Experiments 1 and 2 due to lower Voc. The SNR value range in the y-axis is the same as the ones shown for Fig. 5. The SNR values are very close to the minimum required, but all higher than 5, so it is expected all images had adequate quality.

Fig. 10 shows the ΔPL_4 images for 2, 10 and 50 image pairs and their correspondent SNR_{AVG} value. Here all the SNR_{AVG} values were inconsistent with the image quality. SNR value over 5 was obtained for an image with failures indistinguishable from the noise (2 image pairs). The ΔPL_4 image with the highest SNR_{AVG} value of 9 (10 image pairs) is clearly noisier than the one with SNR_{AVG} of 8.5 (50 image pairs). It is an intrinsic weakness of the SNR_{AVG} marker [6] that if the ambient conditions change consistently throughout the acquisition sequence, it will not affect the SNR calculation but will affect drastically the image quality. Even though SNR_{AVG} was developed for EL imaging, it can be useful for high signal and low sunlight variation scenarios. However, in the most challenging scenario where it will be the most necessary, it cannot be relied on.

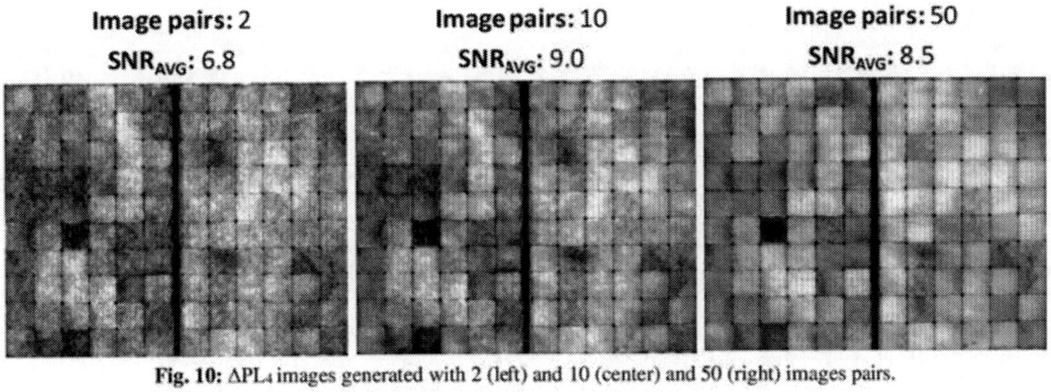

Fig. 10: ΔPL_4 images generated with 2 (left) and 10 (center) and 50 (right) images pairs.

10.4229/EUPVSEC2023/4CV.1.19
020383-004

4 CONCLUSIONS

Regardless of the challenges that the different experiment scenarios imposed on removing sunlight reflection, good quality ΔPL images were consistently obtained from all stages of inverter IV curve acquisition process in all experiments presented, i.e., all experiments provided ΔPL images with consistent $SNR_{AVG} > 5$.

When compared with the more established outdoor luminescence imaging technique (dEL), the ΔPL images presented corresponding diagnosis information of the PV modules.

For this ΔPL imaging technique to be more viable for inspections in practice, future development will have to tackle data analysis and engineering aspects of the technique, such as i) automatic noise removal of ΔPL image sequences; ii) development of a proper SNR marker for outdoor ΔPL; iii) inverter to camera synchronization; and iv. ΔPL image based automatic defect detection.

5 REFERENCES

[1] O. Kunz *et al.*, "Outdoor luminescence imaging of field-deployed PV modules," *Prog. Energy*, vol. 4, no. 4, 2022, doi: 10.1088/2516-1083/ac9a33.

[2] R. Bhoopathy, O. Kunz, M. Juhl, T. Trupke, and Z. Hameiri, "Outdoor photoluminescence imaging of photovoltaic modules with sunlight excitation," *Prog. Photovoltaics Res. Appl.*, vol. 26, no. 1, pp. 69–73, Jan. 2018, doi: 10.1002/pip.2946.

[3] M. Vuković, M. Jakovljević, A. S. Flø, E. Olsen, and I. Burud, "Noninvasive photoluminescence imaging of silicon PV modules in daylight," *Appl. Phys. Lett.*, vol. 120, no. 24, p. 244102, Jun. 2022, doi: 10.1063/5.0097576.

[4] C. Mantel *et al.*, "SNR Study of Outdoor Electroluminescence Images under High Sun Irradiation," in *2018 IEEE 7th World Conference on Photovoltaic Energy Conversion (WCPEC) (A Joint Conference of 45th IEEE PVSC, 28th PVSEC & 34th EU PVSEC)*, Jun. 2018, no. 1, pp. 3285–3289. doi: 10.1109/PVSC.2018.8548264.

[5] M. Guada *et al.*, "Daylight luminescence system for silicon solar panels based on a bias switching method," *Energy Sci. Eng.*, no. June, pp. 1–15, 2020, doi: 10.1002/ese3.781.

[6] L. Stoicescu, M. Reuter, and J. H. Werner, "DaySy: Luminescence Imaging of PV Modules in Daylight," *29th Eur. Photovolt. Sol. Energy Conf. Exhib.*, pp. 2553–2554, 2014, doi: 10.4229/EUPVSEC20142014-5DO.16.2.

[7] T. J. Silverman, M. G. Deceglie, K. Vansant, S. Johnston, and I. Repins, "Illuminated Outdoor Luminescence Imaging of Photovoltaic Modules," *IEEE Photovolt. Spec. Conf.*, 2017.

[8] M. Vuković, I. Eriksdatter Høiaas, M. Jakovljević, A. Svarstad Flø, E. Olsen, and I. Burud, "Photoluminescence imaging of silicon modules in a string," *Prog. Photovoltaics Res. Appl.*, vol. 30, no. 4, pp. 436–446, 2022, doi: 10.1002/pip.3525.

[9] IEC, "IEC TS 60904-13 Photovoltaic Devices - Part 13: Electroluminescence of Photovoltaic Modules," http://www.iec.ch/. [Online]. Available: http://www.iec.ch/

[10] T. Kropp, M. Berner, L. Stoicescu, and J. H. Werner, "Self-Sourced Daylight Electroluminescence from Photovoltaic Modules," *IEEE J. Photovoltaics*, vol. 7, no. 5, pp. 1184–1189, 2017, doi: 10.1109/JPHOTOV.2017.2714188.

10.4229/EUPVSEC2023/4CV.1.19
020383-005

SOILING LOSSES IN BIFACIAL TECHNOLOGY INSTALLATIONS IN THE SOLAR PLATFORM OF THE ATACAMA DESERT

Abel Taquichiri[1*], Pablo Ferrada[1, 2], Douglas Olivares[1], Jorge Rabanal-Arabach[1], Luis A. Conde[2], Erik Mella[3], Carlos Portillo[1, 2], Edward Fuentealba[1,2]
[1]Centro de Desarrollo Energético Antofagasta, Universidad de Antofagasta, Antofagasta, Angamos 601, 1270300 Antofagasta, Chile
[2]Solar Energy Research Center, Universidad de Chile, Tupper 2007, 8370451 Santiago, Chile.
[3]ATAMOSTEC, Pérez Valenzuela 1635, 7500028 Providencia, Chile.
*Corresponding author: abel.taquichiri@uantof.cl

ABSTRACT: This work shows the results concerning the soiling losses in bifacial modules operating in the Solar Platform of Atacama Desert (PSDA). The characteristic parameters such as maximum power (P_{mpp}), short circuit current (I_{sc}), and open circuit voltage (V_{oc}) were obtained from monofacial and bifacial photovoltaic modules installed in a fixed structure and tilted at an angle of 20° to the north and in a Horizontal Single Axis Tracker (HSAT). The results correspond to data collected between July and December 2022. They show that the soiling losses of the fixed installations are higher than those of the HSAT. The ratio between the power generated by the PV module (E_{dc}) and the irradiance on the module (H_{POA}) is between 2.3% and 6.67% difference between July and December for the soiled PV modules in the fixed installation. However, there is a percentage difference of less than 2% between July and December for the soiled PV modules in the HSAT installation.

Keywords: bifacial, soiling, desert, fixed, tracker

1 INTRODUCTION

Dust particles deposited on the glass surface of photovoltaic (PV) modules absorb and scatter the incident light on the solar cells, resulting in a reduction of the energy production [1]–[3]. The influence of soiling on PV performance has been extensively described and is known to be dependent on climatic conditions [4], so the amount of soiling is influenced by complex interactions of many factors that can vary significantly from site to site, including environmental parameters, PV module properties, site and installation characteristics, among others [1]. Previous studies conducted at the Solar Platform of Atacama Desert (PSDA) (24.09°S, 69.93°E) report a 19% loss of short-circuit current (I_{sc}) in multicrystalline modules during 6 months of exposure [5], and another bifacial technology study performed in a vertical and fixed installation shows a power loss of 0.05%/day and 0.20%/day, respectively. [6].

In this study, soiling loss is quantified by the soiling rate (SR) described in the international standard IEC 61724-1 [7]. which is defined as the ratio of the real output power/current of the module under certain conditions of soiling, to the power/current of the module totally clean and without soiling. Obtaining and monitoring soiling losses in PV installations contributes to establish operation and maintenance protocols. It allows adapting cleaning periods and protocols to optimize resources and guarantee an acceptable energy yield.

2 METHODOLOGY

This study presents the soiling losses produced in four different PV module technologies, as described in Table 1. The modules under study correspond to Passivated Emitter and Rear Contact (PERC), Bifacial Passivated Emitter and Rear Contact (PERC+), bifacial n-type Passivated Emitter and Rear Totally Diffused (nPERT) and bifacial Heterojunction (HJT). These modules are installed at the Solar Platform of Atacama Desert (PSDA), which has the facilities for testing PV technology in desert and high radiation conditions [8], [9] (see Fig. 1). The experimental installation at the PSDA is equipped with instruments for measuring meteorological parameters, measuring the current-voltage (I-V) curve of the PV modules, and a data acquisition system to record all measured variables.

In addition, two different types of structures were considered (Fig. 1); a fixed system with a tilt angle of 20° to the north and a Horizontal Single Axis Tracker (HSAT) system.

Table 1. Main characteristics of the modules studied.

Characteristics	PERC	PERC+	nPERT	HJT
P_{mmp} [W]	346	342	348	355
I_{sc} [A]	9.16	9.2	9.3	7.4
V_{oc} [V]	47.9	47.8	47.7	52.5
Bifaciality	None	>65%	>85%	>88%

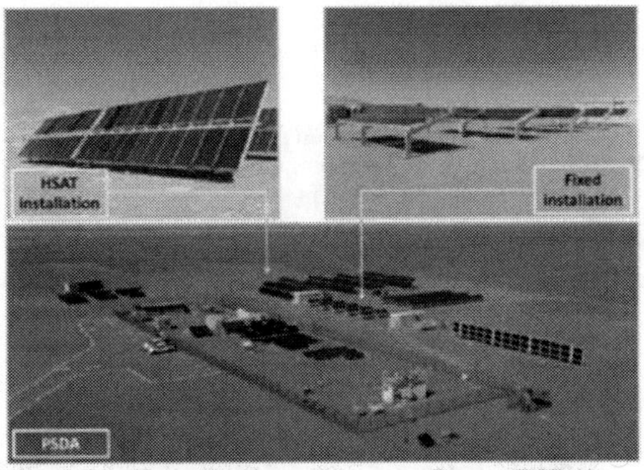

Fig. 1. The Solar Platform of Atacama Desert (PSDA).

A total of 16 PV modules were considered for this study, which have been distributed into 8 groups. Each group consists of two similar modules, and is divided by technology type (PERC, PERC+, nPERT, HJT) and installation (Fixed, HSAT) (see Table 2).

10.4229/EUPVSEC2023/4CV.1.20
020385-001

Table 2. Groups of modules studied.

Group	Technology	Bifacial	Installation
1	PERC	X	Fixed
2	PERC+	✓	Fixed
3	nPERT	✓	Fixed
4	HJT	✓	Fixed
5	PERC	X	HSAT
6	PERC+	✓	HSAT
7	nPERT	✓	HSAT
8	HJT	✓	HSAT

The results over a 6-month period between July - December 2022 are presented. At the beginning of the experimental period, all PV modules were cleaned. Then, a weekly cleaning protocol was implemented for only one of the PV modules in each group, which is used as a reference. The data collected for P_{mpp}, I_{sc}, and V_{oc} from each I-V curve were normalized for comparison according to the following equations.

$$P_{mSTC} = P_{mpp} \cdot (G_{STC}/G_e)/[1+\gamma \cdot (T_{pv}-T_{ref})] \quad (1)$$

$$I_{scSTC} = I_{sc} \cdot (G_{STC}/G_e)/[1+\alpha \cdot (T_{pv}-T_{ref})] \quad (2)$$

$$V_{ocSTC} = V_{oc} \cdot [1+\beta \cdot (T_{pv}-T_{ref})] \quad (3)$$

Where, α, β and γ are the temperature coefficients for I_{sc}, V_{oc} and P_{mpp} respectively, T_{pv} is the PV module temperature, T_{ref} is the reference temperature equal to 25 °C, G_{STC} is equal to 1000 W/m^2, and finally G_e represents the equivalent irradiance, obtained from the global irradiance in the array plane (G_{POA}), and Global Rear Irradiance (G_{RI}) and the module bifaciality coefficient (φ) [10], according to Eq. (4).

$$G_e = G_{POA} + \varphi \cdot G_{RI} \quad (4)$$

On the other hand, the SR is calculated from the ratio of the short-circuit current of the soiled PV module ($I_{sc,soil}$) to the short-circuit current of the clean PV module ($I_{sc,clean}$), as expressed in Eq. 5.

$$SR = I_{sc,soil} / I_{sc,clean} \quad (5)$$

3 RESULTS

The Fig. 2 shows the environmental parameters collected for a clear summer day. The global horizontal irradiance (GHI) is above 1000 W/m^2 for approximately 4 hours during the day, the ambient temperature reaches 32 °C after midday and high wind speed values are observed after noon compared to those recorded during the morning.

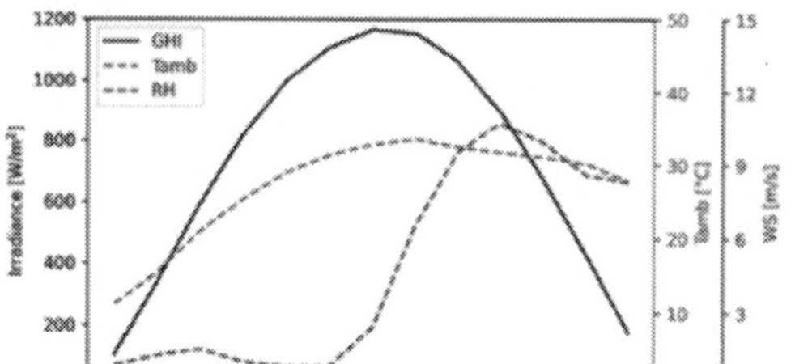

Fig. 2. Fig. 2. Environmental parameters for a clear summer day at PSDA.

In addition, the irradiance on the front and rear surface of the PV Fixed and HSAT installation were measured. In Fig. 3, it is evident that the HSAT installation has an advantage over the fixed installation by capturing more irradiance on the front surface; however, the irradiance received by the rear is similar in both installations.

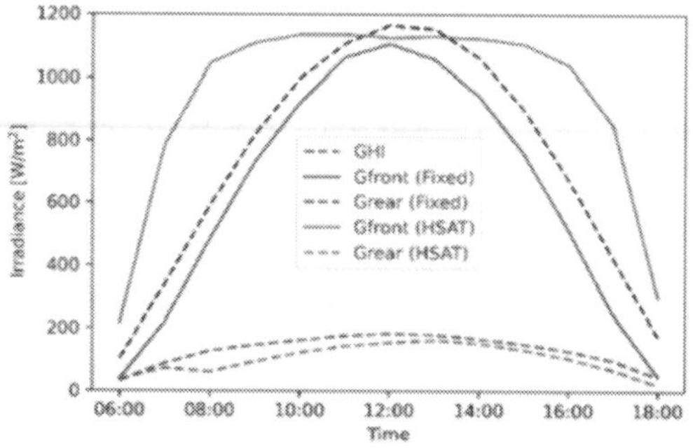

Fig. 3. Global Horizontal Irradiance (GHI) and irradiance on surface Front and Rear of the Fixed/HSAT installation.

The high irradiance values on the HSAT installation promote the electrical generation above the fixed installation; nevertheless, to demonstrate the influence of the soiling, the ratio of the energy generated by the PV module (E_{dc}) to the irradiance on the module (H_{POA}) during one day was calculated. These results are shown in Fig. 4 for the fixed installation and Fig. 5 for the HSAT installation. The percentage difference between July and December is shown in Table 3, where the maximum percentage variation for the fixed installation is given by the PERC+ technology, this absolute value is equal to 2.63% for the clean PV module and 6.67% for the soiled PV module.

In the HSAT installation (Fig. 5), the clean and soiled PV modules exhibit the same pattern, except for the low values registered in October and December for the PERC+ PV modules. The latter is attributed to the low values of P_{mpp} registered in the measurement, which influence the calculation of the energy generated during this day. However, except for these dates, there is a percentage difference of less than 2% between July and December for the clean and soiled modules, which indicates that they are less affected by losses due to Soiling.

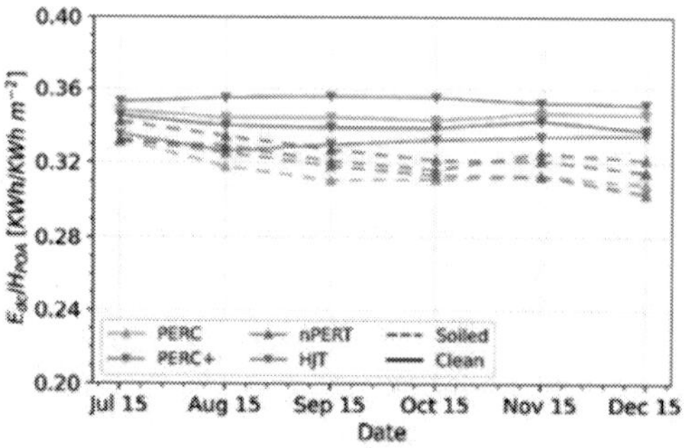

Fig. 4. Ratio E_{dc}/H_{POA}, obtained between July-December in the PV fixed installation.

10.4229/EUPVSEC2023/4CV.1.20
020385-002

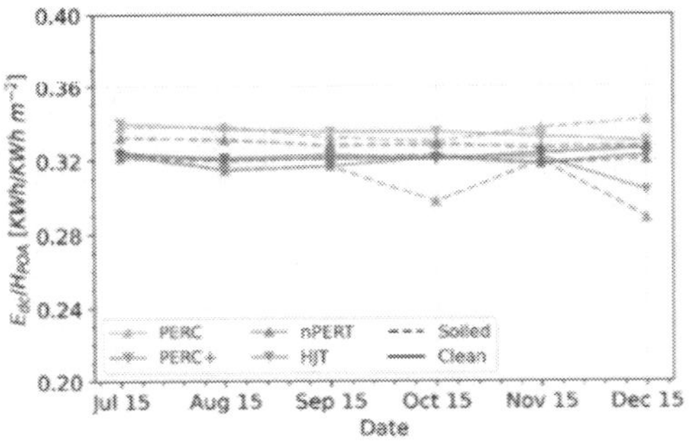

Fig. 5. Ratio E_{dc}/H_{POA}, obtained between July-December in the PV HSAT installation.

Table 3. Percent difference between July-December of (E_{dc} / H_{POA}).

Installation	PERC	PERC+	nPERT	HJT
Fixed(clean)	-0.42%	-2.63%	0.73%	1.04%
Fixed(soiled)	3.05%	6.67%	5.84%	2.3%
HSAT(clean)	1.89%	3.60%	-1.88%	-0.48%
HSAT(soiled)	-1.03%	8.09%	1.10%	-0.24%

The SR was calculated from the I_{sc} measurements. These values are shown for the half solar day during the experimental period. Fig. 6 shows that the SR decreases over time for the PV fixed installation, although there is evidence of cleaning events that appear to increase the SR. They occurred on two particular days, September 1 and November 11, when the SR increases most noticeably. These increases were attributed to cleaning events, which appear to be related to the high relative humidity at sunrise. At the HSAT facility, there was a greater existence of this type of increase in SR (Fig. 7), which shows a greater dispersion.

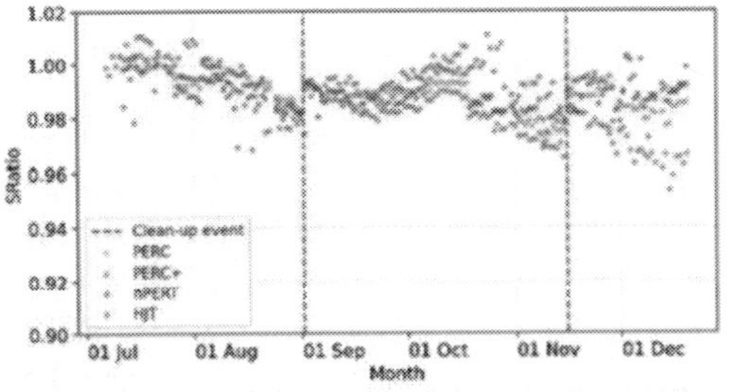

Fig. 6: Soiling coefficient (SR) obtained in the PV fixed installation (July - December).

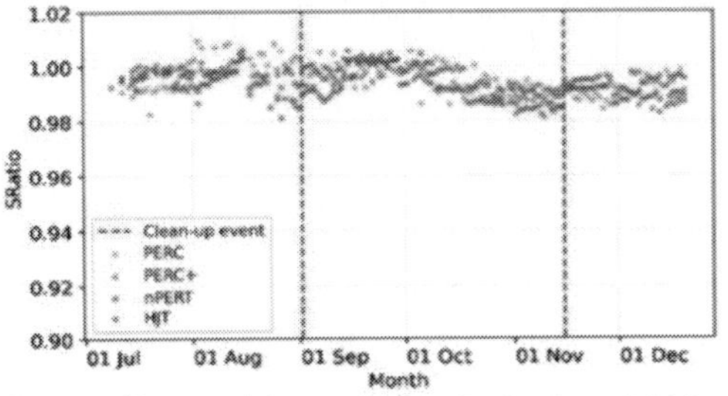

Fig. 7. Soiling coefficient (SR) obtained in the PV HSAT installation (July - December).

4 CONCLUSIONS

The effects of soiling through the SR on 4 different PV module technologies were evaluated. This study included a fixed PV installation and HSAT, during a period of 6 months, between July and December 2022. The daily ratio between generated energy and irradiance on the PV module shows a decreasing behavior in the soiled PV modules in the fixed installation. However, this does not occur in the HSAT installation, so it is concluded that they are less affected by losses due to fouling.

The SR parameter was calculated using the Isc (dirty/clean) ratio, which complements the trend of higher losses due to soiling in the PV fixed installations compared to HSAT system. In addition, there are particular cases in which SR increases, days in which high relative humidity values were recorded at sunrise.

5 ACKNOWLEDGEMENTS

This work was supported by the Chilean Economic Development Agency, CORFO, with the contract no. 17PTECES-75830 under the framework of the project ATAMOSTEC and the projects ANID/FONDAP/1522A0006 "Solar Energy Research Center"- SERC-Chile. The author thanks the doctoral program in Solar Energy of the Universidad de Antofagasta (UA), Project Engineering 2030 Code 16ENI2-71940 and the CDEA for the accessibility of the data and the use of their facilities.

6 REFERENCES

[1] K. K. Ilse, B. W. Figgis, V. Naumann, C. Hagendorf, and J. Bagdahn, "Fundamentals of soiling processes on photovoltaic modules," *Renew. Sustain. Energy Rev.*, vol. 98, no. September, pp. 239–254, 2018, doi: 10.1016/j.rser.2018.09.015.

[2] E. F. Cuddihy, "Soiling Mechanisms and Performance of Anti-Soiling Surface Coatings," *Part. Surfaces 1*, pp. 91–111, 1988, doi: 10.1007/978-1-4615-9531-1_8.

[3] B. Laarabi, Y. El Baqqal, N. Rajasekar, and A. Barhdadi, "Updated review on soiling of solar photovoltaic systems Morocco and India contributions," *J. Clean. Prod.*, vol. 311, no. September 2020, p. 127608, 2021, doi: 10.1016/j.jclepro.2021.127608.

[4] S. C. S. Costa, A. S. A. C. Diniz, and L. L. Kazmerski, "Dust and soiling issues and impacts relating to solar energy systems: Literature review update for 2012-2015," *Renew. Sustain. Energy Rev.*, vol. 63, pp. 33–61, 2016, doi: 10.1016/j.rser.2016.04.059.

[5] D. Olivares *et al.*, "Determination of the Soiling Impact on Photovoltaic Modules at the Coastal Area of the Atacama Desert," *Energies*, vol. 13, no. 15, pp. 1–17, 2020, doi: 10.3390/en13153819.

[6] E. Cabrera *et al.*, "AtaMoS TeC Project: Soiling Impact on Bifacial Modules with Different Mounting Geometry in the Atacama Desert in Chile," *2018 IEEE 7th World Conf. Photovolt.*, 2018, doi: 10.1109/PVSC.2018.8547697.

[7] IEC, "IEC 61724-1: Photovoltaic system performance - Part 1: Monitoring 2017." 2017.

[8] E. Cabrera *et al.*, "ATAMOSTEC Project : Atacama

Desert PV Platform," no. 17, p. 15110019, 2019.
[9] E. Urrejola, M. Riquelme, D. Muñoz, R. Kopecek, and E. Fuentealba, "Atamostec : Desert Label for Pv Technologies," *37th Eur. Photovolt. Sol. Energy Conf.*, vol. 15110019, no. October, p. 2021, 2020, doi: 10.4229/EUPVSEC20202020-4AV.1.36.
[10] International Electrotechnical Commission, "IEC TS 60904-1-2," *Webstore International Electrotechnical Commission*. 2019.

SMART MONITORING OF PHOTOVOLTAIC PLANTS WITH CLOUD COMPUTING AND IOT USING MACHINE LEARNING

M.Emamian[1], A. Eskandari[1], A. Nedaei[1], A. Moradi Sizkouhi[2], Killian Lobato[3], Jafar Milimonfared[1], and M. Aghaei[4,5*]

[1] Department of Electrical Engineering, Amirkabir University of Technology, Tehran 15119-43943, Iran
[2]Department of Electrical and Computer Engineering, Concordia University, Montreal, QC H3G 1M8, Canada
[3]Faculty of Sciences, Department of Geographic Engineering, University of Lisbon, 1649-004 Lisboa, Portugal
[4]Department of Ocean Operations and Civil Engineering, Norwegian University of Science and Technology (NTNU), 6009 Ålesund, Norway
[5]Department of Sustainable Systems Engineering (INATECH), University of Freiburg, 79110 Freiburg, Germany

*mohammadreza.aghaei@ntnu.no; aghaei@studysolar.uni-freiburg.de

ABSTRACT: The proposed Intelligent Monitoring System (IMS) for Photovoltaic (PV) systems is a cost-effective and easy-to-implement solution for monitoring large-scale PV power plants. It utilizes IoT for data handling and communication and includes a personal cloud server for data computation and storage. The IMS also has a web monitoring system through open-source software that provides information to multiple users. It employs deep ensemble models for fault detection and power prediction in PV systems, using an LSTM ensemble neural network to predict output power and machine learning-based models to detect faults. The fault diagnosis process involves feature analysis and an ensemble learning model that includes Naive Bayes, K-Nearest Neighbors, and Support Vector Machine with feature selection to enhance performance. The IMS has been tested and validated in a real PV system, and is a scalable and replicable solution for comprehensive monitoring of PV plants, including data acquisition, storage, pre- and post-processing, fault diagnosis, performance assessment, energy yield, and output power prediction.

Keywords: cloud computing; ensemble learning; intelligent monitoring system; internet of things; power prediction; fault detection; autonomous monitoring

1 INTRODUCTION

Photovoltaic (PV) monitoring systems have gained significant attention in recent years due to the need to diagnose operation and maintenance issues in PV systems, which have led to a shortened lifespan of these systems [1, 2]. An effective monitoring system is crucial to increase produced power and mitigate the challenges of monitoring PV systems, such as the high cost of commercial data acquisition and the need for a personal computer for real-time monitoring. Variations in output power can occur due to failures and defects in PV modules, making it necessary to have a smart monitoring system and accurate prediction model to identify and repair any detected issues promptly [3]. Accordingly, in recent years, many studies related to the PV monitoring systems have been carried out [4].

Recent years have seen a significant growth in cost-optimizing hardware for PV monitoring. A system designed by Pereira et al. [5] used the RaspberryPi platform and a cloud service for a decentralized PV plant, but with hardware complexity. Another approach uses an IoT platform with 3G connectivity for monitoring solar systems. The collected data is sent to a server and stored in a cloud, but its effectiveness is limited by proximity to a 3G station [6, 7]. Several machine learning-based models for fault detection and power output prediction in PV plants have been developed in recent years. These models use techniques such as artificial neural networks (ANNs), support vector machines (SVM), k-nearest neighbors (KNN), ensemble framework, multilayer perceptron (MLP), and fuzzy classification system. The models are important for evaluating the performance of PV systems. However, the models have limitations such as sensitivity to noisy data, high computational demand, and performance issues with high dimensional datasets. Some models also have limitations with time series data and are restricted to specific environmental conditions. There is still room for improvement in the accuracy and reliability of these models [8, 9]. However, these methods have limitations such as high computational demand, high cost, sensitivity to noise and weather conditions, and accuracy of models. There is a need for a more efficient, low-cost, and accurate solution for PV monitoring systems.

Here, we propose the Intelligent Monitoring System (IMS), consisting of an IoT platform, a cloud service, and a web monitoring software for end-users. Accordingly, we present a cloud service for data processing (cloud computing), cloud storage for recording data, and use an IoT platform for data transmission. Moreover, the proposed method employs a long short-term memory (LSTM) ensemble neural network to predict the output power PV systems for one day ahead. The LSTM model is appropriate for sequential and time series problems. The proposed PV monitoring system forecasts output power of PV systems, analyzes information, and creates detailed reports automatically for repair and maintenance in PV plants. In this study, an ensemble learning-based model is also used for the fault diagnosis of PV arrays to improve classification accuracy and diagnostic performance. The model uses an IoT platform and cloud computing to detect and classify open circuit faults, array degradation faults, and string degradation faults. The ensemble learning model consists of three algorithms: Naive Bayes (NB), K-Nearest Neighbors (KNN), and Support Vector Machine (SVM). These algorithms were chosen because they have been proven to be effective for detecting and classifying various faults in PV systems and are capable of handling both multi-class and binary classification problems. SVM was selected for complex classification problems using kernel tricks [10].

In summary, the proposed Intelligent Monitoring System (IMS) is designed to detect and classify three types of electrical faults (open circuit faults, array degradation faults, and string degradation faults) and predict the output

10.4229/EUPVSEC2023/4CV.1.21

power of PV arrays. It consists of an IoT platform for data transmission, a cloud service for data processing and storage, and a web monitoring software for end-users. The IMS employs a long short-term memory (LSTM) ensemble neural network for output power prediction and an ensemble learning model (Naive Bayes, K-Nearest Neighbors, and Support Vector Machine) for fault diagnosis. The system is cost-efficient, flexible, and suitable for standalone and remote PV plants.

2 PROPOSED INTELLIGENT MONITORING SYSTEM (IMS)

The aim of this study is to design an affordable and high precision monitoring sys-tem in order to perform the precise monitoring of PV plants. Consequently, an IMS has been proposed and designed to predict the output power of PV modules on the day ahead and diagnose fault events for improving the performance of the PV plant.
Figure 1 depicts the structure of the IMS proposed in this paper as well as the location of environmental and electrical sensors. The IMS collects meteorological and electrical data through sensors installed on PV modules, PV arrays, and the environment. As shown in Figure 1, the data logger consists of ESP8266 and Microcontroller boards. The electrical data is sent to the ESP8266 module for initial processing. This step is required for data preparation to be transferred to the server. The ESP8266 is a cost-efficient mini-Wi-Fi device based on ESP8266EX, which is popular for IoT applications. The ESP8266 module is able to be connected to the internet via the access point to post data to the cloud service for recording and ultimate processing (cloud computing). The device is set up in order to connect different types of internet access points.

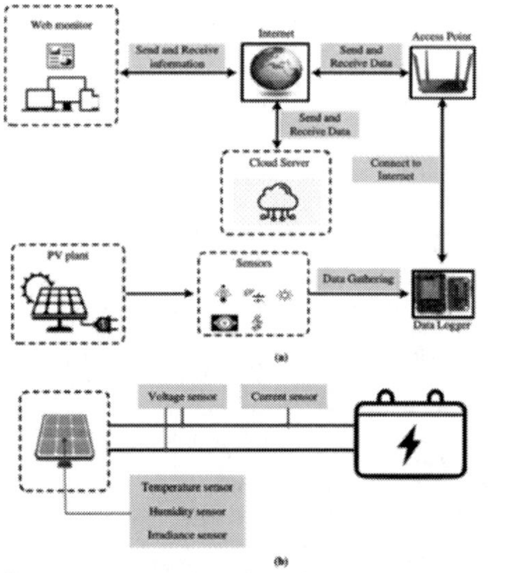

Figure 1: An overview of Intelligent Monitoring System (ISM) architecture; (b) sensors distribution.

2.1 IMS Architecture

As Figure 2 provides an in-depth overview of how the proposed monitoring system has been arranged. In this experimental research, a stand-alone PV system has been deployed to test and analyze the performance of the IMS. As shown in Figure 2, the designed IMS consists of PV modules, microcontrollers, a data logger, and sensors with wired connections. The equipment used in hardware section include 18 PV modules with 10 W total output power, a resistance load, an access point, mini-Wi-Fi device based on ESP8266EX, a microcontroller, and different sensors. By using the presented IMS structure, different PV monitoring tasks, e.g., data collection, can be done quickly and efficiently through the connected microcontroller to a computational system via the internet.

Figure 2: PV experimental setup.

2.2 Internet of Thing (IoT) Platform

The IoT is a new technology that is envisioned as a global network of interconnected machines, devices, objects, people, and even animals interacting with each other and transferring and sharing data over the network. IoT technology is one of the most important technologies in the future and has attracted the attention of a wide range of industries. A flowchart demonstrating the process of the IoT platform process in the pro-posed IMS is provided in Figure 3. In the first step, the IoT file is loaded by the microcontroller. The IoT file includes posts and gets commands for data transfer between the cloud and the microcontroller. After uploading the file to the microcontroller, using an Internet connection, the system then connects to the server and executes the script (.js file) on the server. In case the connection is lost, the platform starts transferring data from the microcontroller to the server again for recording and processing. If the data transfer fails, an URL error will be displayed to the user.

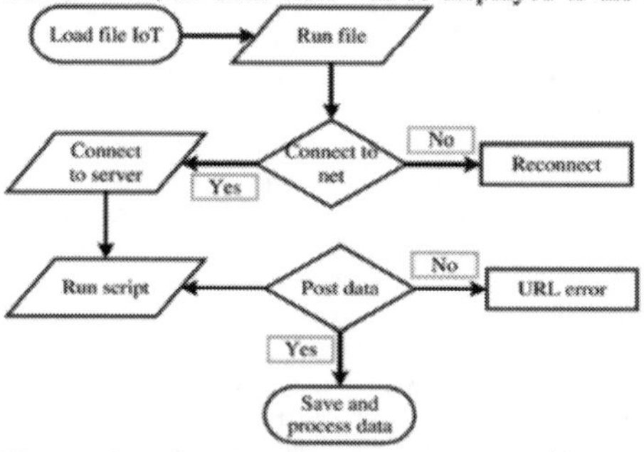

Figure 3: The flowchart of data transfer by the IoT platform.

10.4229/EUPVSEC2023/4CV.1.21

2.3 Cloud Server

This paper focuses on powerful cloud tools, which are presented by cloud servers such as coding packages, cybersecurity, admin page, process units, etc. Service models consist of three types, namely Software as a Service (SaaS), Platform as a Service (PaaS), and Infrastructure as a Service (IaaS). In this study, the SaaS model (Google Cloud) has been used for the cloud com-putation unit. The user has access to control the deployment of end applications. Howev-er, the user is not allowed to control the operating infrastructures. Cloud computing is a model for enabling ubiquitous, convenient, and on-demand network access to a shared pool of configurable computing resources (e.g., networks, servers, storage, applications, and services) that can be rapidly released with minimal management effort or service provider interaction.

2.4 Cloud Data Logger

Cloud data loggers communicate over the Internet connection with a remote cloud-based server, which can be deployed anywhere in the world. The cloud-based server uses the cloud data logger as a container for recording data. Users can take advantage of any devices supporting a web browser to access data on the cloud. This paper proposes a new structure for the data logger based on an open-source and lightweight software and a customized cost-efficient hardware.

2.5 Web Monitor

The term 'Web monitoring' signifies the process in which the information that end-users can interact with on a website or a web application via the internet is systematically analyzed and displayed (See Figure 4 for a general overview of a web monitoring process). Accordingly, the cloud server loads the scripts, which is followed by receiving a CSV-format version of the data file from the database and subsequently posting it to the backend for further processing. The backend is responsible for running programs in scripting languages like Python. The task includes execution of the ensemble learning algorithms code on the flask, users' identification, and data processing in the backend. Finally, the processed data is displayed in the frontend. The task of the frontend layer is to run web languages such as HTML, CSS, and JavaScript in web browsers. Figure 7 shows the overview of the frontend layer used for data display to users. If the error occurs in the process layers, the user receives the URL error through the browser.

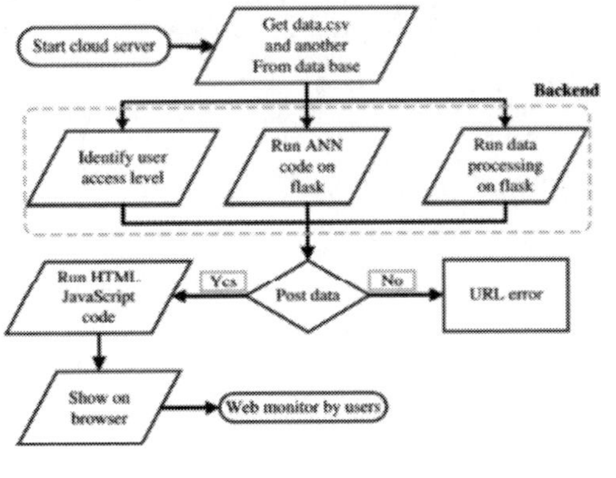

Figure 4: Performance intelligent web monitor flowchart.

3 IMS USING MACHINE LEARNING TECHNIQUES

To develop the IMS in order to obtain an accurate prediction of the output power and a precise detection of PV faults, we have applied novel machine learning techniques (MLTs) in a way that a long short-term memory (LSTM) model helps the IMS in the pro-cess of power prediction while, on the other side, an ensemble learning (EN) classifier which is based on the I-V curve measurement of the PV arrays is employed to provide a more accurate detection of the faults.

3.1 Power Forecasting Architecture

This study aims to achieve an efficient prediction model by composing ensemble neural networks and LSTM. The ensemble methods improve the dynamics and enhance the accuracy of the neural network models. Accordingly, this experimental research has developed an LSTM ensemble architecture for output power prediction of PV Plants. In the following, we describe three main components of the LSTM model, which are the structure of LSTM, LSTM ensemble architecture, and the evaluation methods, see Figure 5.

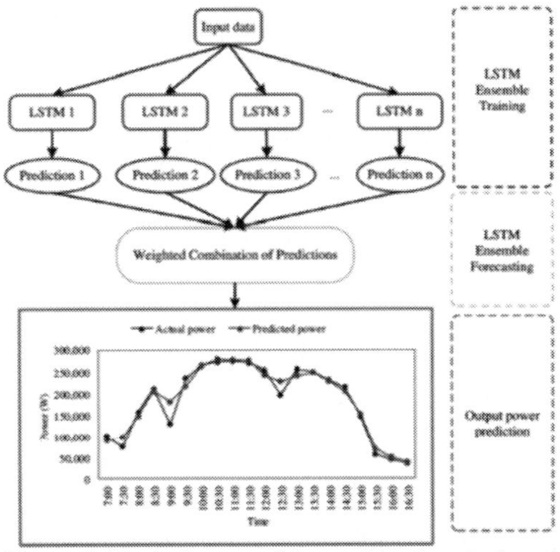

Figure 5: Block diagram of the quadrotor's attitude and positions dynamics.

3.2 Fault Detection and Classification

In this paper, an ensemble learning (EN) model for fault diagnosis of PV arrays is investigated in order to achieve superiority in classification accuracy. Figure 10a shows an overview of the required steps to build an EN-based fault classification model. As a supervised machine learning technique, the EN-based model relies on a certain number of fault features extracted from the raw data obtained. To achieve a high diagnostic performance with a low computational burden, a feature selection algorithm is utilized. To evaluate the performance of the model, the labeled samples are divided into training and validation sub-sets. The training set is used to train the model, while the validation set is used to measure the performance of the model. Then, we have tested the learning algorithm by an unseen dataset with dissimilar input parameters to the original dataset.

4 EXPERIMENTAL RESULTS

In this study, EN_1 and EN_2 models are compared

10.4229/EUPVSEC2023/4CV.1.21
020387-003

with other deep NN models such as LSTM, MLP, and Conv1D from different perspectives. The EN-1 model is an LSTM ensemble network with an updated weight based on RMSE error. Accordingly, the EN-2 model is an LSTM ensemble network with an updated weight based on MAE error. Figure 6 shows the performance of the trained models. According to Figure 6, by comparing the errors of the models in two metrics (RMSE and MAE), the ensemble model has a better performance than other models in both indices.

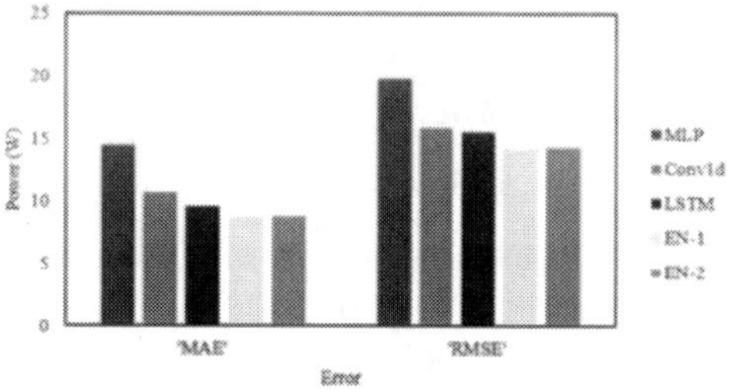

Figure 6: RMSE and MAE errors of models.

Figure 7 shows the actual power diagram of the PV plant and the predicted power by the predictor models for one day. Accordingly, the graph indicates that the output of the ensemble models is closer to the actual value compared with the rest of the models.

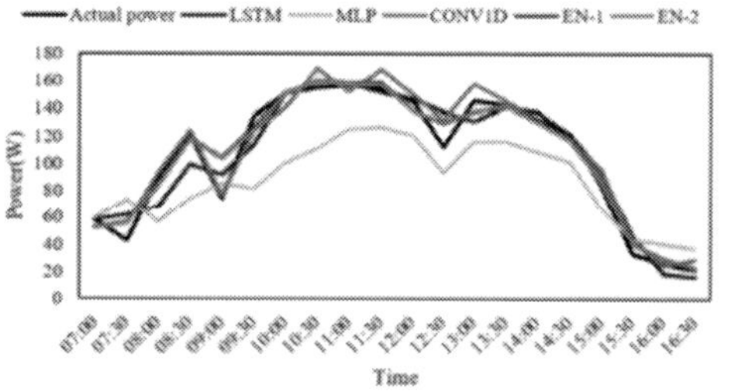

Figure 7: Forecasting and actual power diagram of deep learning models for one day.

The fault detection layer is perfectly trained in two previously mentioned scenarios. In addition, we deployed a 10-fold Cross-Validation (CV) algorithm to be reassured about the performance of both feature selection and learning processes.

Moreover, in order to evaluate the performance of the proposed method, we have devised two scenarios, one with and one without the feature selection algorithm for both layers. Figure 8 compares the performance of the algorithms in both scenarios and clearly shows that our proposed method is 100% accurate in the second scenario.

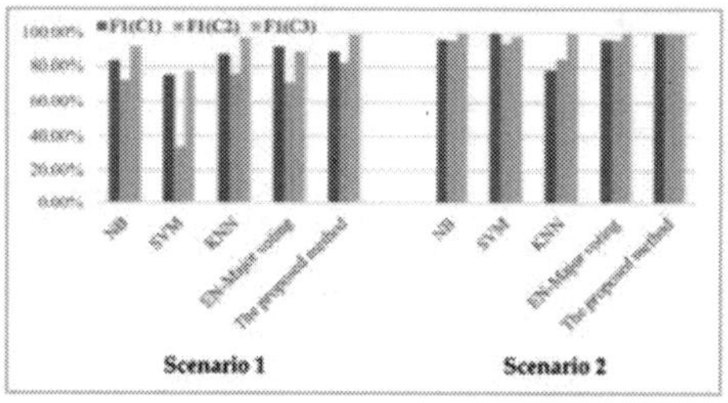

Figure 8: The results of testing the algorithms under two scenarios in the process of fault classification.

4 CONCLUSIONS

The Intelligent Monitoring System (IMS) is an IoT-based monitoring system developed for photovoltaic (PV) plants. The goal of IMS is to provide a smart and cost-efficient solution for monitoring PV systems. The system uses deep ensemble models for fault detection and power prediction. The faults in the PV system were detected and classified using a combination of Naive Bayes, Support Vector Machine, and K-Nearest Neighbors. The IMS is scalable, replicable, and interoperable and can be accessed using IoT applications. The results show that the LSTM ensemble network has higher accuracy in output prediction, failure identification, and decision-making compared to other models, with an average accuracy of 96.56% and 96.89%.

ACKNOWLEDGMENT

The authors gratefully acknowledge COLLECTiEF-Collective Intelligence for Energy Flexibility project that has received research funding from European Union's H2020 research and innovation programme under Grant Agreement No 101033683.

5 REFERENCES

[1] Eskandari, A et al. Autonomous Monitoring of Line-to-Line Faults in Photovoltaic Systems by Feature Selection and Parameter Optimization of Support Vector Machine Using Genetic Algorithm. *Appl. Sci.* **2020**, *10*, 5527.

[2] A. Eskandari,et al, "Line-Line Fault Detection and Classification for Photovoltaic Systems using Ensemble Learning Model Based on I-V Characteristics," Sol. Energy, vol. 211, pp. 354–365, 2020.

[3] Gonzalo, A.P.; Marugán, A.P.; Márquez, F.P.G. Survey of Maintenance Management for Photovoltaic Power Systems. *Renew. Sustain. Energy Rev.* **2020**, *134*, 110347.

[4] Ansari, S.; et al. A Review of Monitoring Technologies for Solar PV Systems Using Data Processing Modules and Transmission Protocols: Progress, Challenges and Prospects. *Sustainability* **2021**, *13*, 8120.

[5] Pereira, R.I.S.; et al. IoT Embedded Linux System Based on Raspberry Pi Applied to Real-Time Cloud Monitoring of a Decentralized Photovoltaic Plant. Measurement 2018, 114, 286–297.

[6] López-Vargas, A.; Fuentes, M.; García, M.V.; Muñoz-Rodríguez, F.J. Low-Cost Datalogger Intended for Remote Monitoring of Solar Photovoltaic Standalone Systems Based on ArduinoTM. IEEE Sens. J. 2019, 19, 4308–4320.

[7] López-Vargas, A.; Fuentes, M.; Vivar, M. IoT Application for Real-Time Monitoring of Solar Home Systems Based on ArduinoTM with 3G Connectivity. IEEE Sens. J. 2018, 19, 679–691.

[8] Das, U.K.; Tey, K.S.; Seyedmahmoudian, M.; Mekhilef, S.; Idris, M.Y.I.; van Deventer, W.; Horan, B.; Stojcevski, A. Forecasting of Photovoltaic Power Generation and Model Optimization: A Review. Renew. Sustain. Energy Rev. 2018, 81, 912–928.

[9] Samara, S.; Natsheh, E. Intelligent Real-Time Photovoltaic Panel Monitoring System Using Artificial Neural Networks. IEEE Access 2019, 7, 50287–50299.

[10] M. Emamian, et al. "Cloud Computing and IoT Based Intelligent Monitoring System for Photovoltaic Plants Using Machine Learning Techniques," Energies, vol. 15, no. 9, p. 3014, 2022.

SOILING AND IRRADIANCE MEASUREMENTS IN BIFACIAL PV SYSTEMS USING IN-SITU REFERENCE MODULES AND REAR-SIDE REFERENCE CELLS

Michael Gostein[1*], Marine Bila[2], Robert Campbell[2], Paul Wolffersdorff[2], Josh Horst[1]
[1]Atonometrics, Austin, USA
[2]EDF Renewables, San Diego, USA
*Corresponding author: michael.gostein@atonometrics.com

ABSTRACT: We report on soiling and irradiance measurement in a bifacial photovoltaic (PV) power plant using in-situ I-V measurements of module power together with front and rear-side reference cells. Soiling ratio is the ratio of actual module power output to expected power under clean conditions. However, precise determination of expected module power output for clean conditions is challenging for bifacial systems because of the need to account not only for front-side but also for rear-side irradiance contributions which have greater sources of variability. We initiated a field test to demonstrate two methods, denoted "module-module" and "module-cell", of determining soiling ratio using soiled module power measurements. In both methods we use in-situ module I-V measurement to directly measure power output of a soiled reference module within the plant, yielding the numerator of soiling ratio. In the module-module method, for the denominator of soiling ratio, the expected module power, we use measurements from another reference module that is routinely cleaned. In the module-cell method, for the denominator we use module power estimated from front and rear-side cleaned reference cells. Our aim is to determine the precision of soiling ratio measurements in a bifacial system given the challenge of normalizing for rear-side irradiance. We find that the module-module and module-cell methods produce nearly identical results, within ~0.5% during our 11-week study to date.
Keywords: photovoltaic (PV) systems, soiling, irradiance, bifacial PV

1 INTRODUCTION

The accumulation of dust and other contaminants on photovoltaic (PV) modules, known as soiling, is one of the principal energy loss factors in PV systems. Soiling is quantified by the soiling ratio (*SR*), the ratio of actual PV power output to expected output under clean conditions [1]. Early work on direct measurement of soiling ratio compared the output of soiled modules to regularly cleaned modules [2], using short-circuit current as a proxy for module power. This concept has also been extended to comparing short-circuit current of soiled and clean mini-modules or reference cells. In recent years soiling sensors employing all-optical detection of accumulated dust have been developed [3][4][5]. These approaches work well when soiling accumulations are assumed to be uniformly distributed on module surfaces. However, soiling accumulations are often non-uniform, with concentrated bands of soiling at module edges, especially bottom edges [6]. For non-uniform soiling, more accurate measurements of soiling ratio are achieved by measuring module power and calculating the ratio of measured soiled power to expected power [7][8][9] for clean conditions. Expected power, the denominator of *SR*, can be determined either by using clean modules identical to the soiled modules [8] ("module-module") or by using periodically cleaned reference cells [9] ("module-cell").

For bifacial PV systems, measurement of soiling ratio based on measured soiled module power becomes more challenging than for monofacial systems. To compute the denominator of *SR*, a module's expected power when clean must be determined considering the impact of total irradiance including contributions from both front and rear sides. In the module-module approach, a clean module

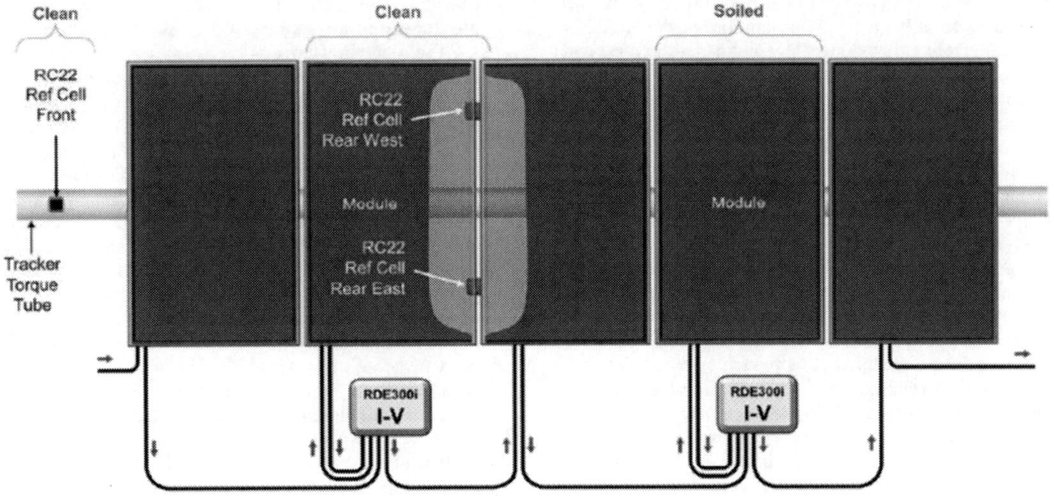

Figure 1: Setup in single-axis tracking bifacial PV array, with in-situ I-V of clean and soiled modules and front and rear reference cells for total effective irradiance. In actual installation, clean/soiled modules are farther from torque tube end.

10.4229/EUPVSEC2023/4CV.1.22
020389-001

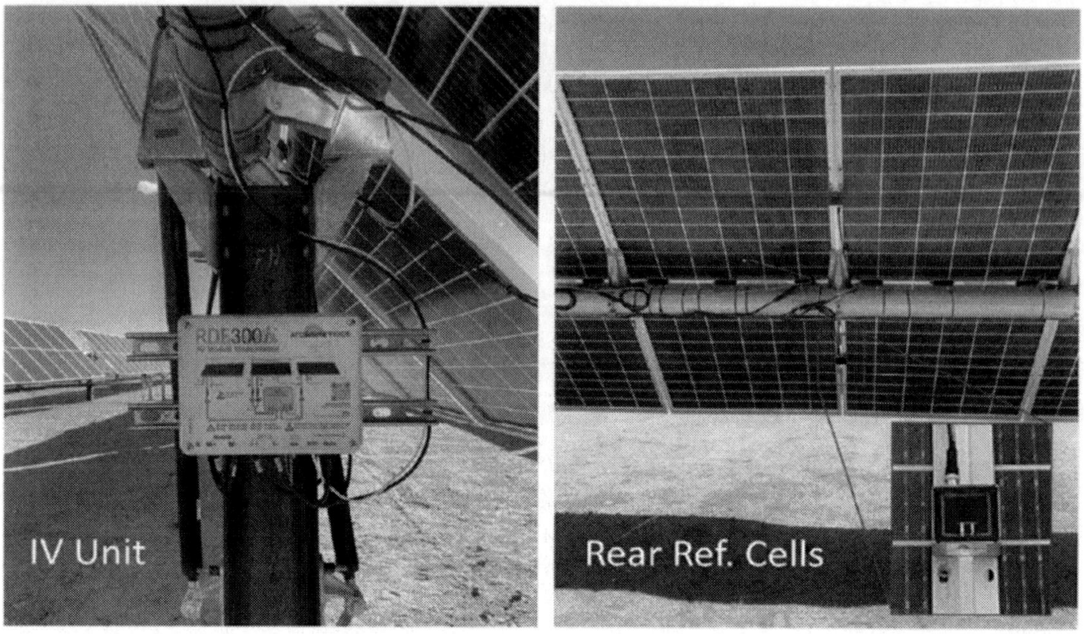

Figure 2: Left: in-situ I-V unit (RDE300i) mounted beneath module racking, wired to measure periodic I-V sweeps on one of the modules without disconnection from the array. Right: rear-side reference cells (RC22), which are designed to mount compactly within the module frame area in multiple positions without shading the rear side of the PV cells.

identical to the soiled module should be used. For bifacial systems care should be taken that factors affecting rear-side irradiance contributions are identical between soiled and clean modules. These factors include position within a module row, albedo variations, and variations in structural shading. In the module-cell approach, reference cells should be positioned on both the front and rear sides of the module, and rear-side reference cells should be placed in optimal locations for measuring average rear-side irradiance across the module area, as in [10].

Equipment for measuring soiled module power output is also evolving. In-situ module-level I-V measurement systems [11][12] now provide the ability to directly measure maximum power of modules while they are still connected to a PV array. This eliminates the need for standalone reference modules for soiling measurement and other applications and ensures that soiling reference modules are always identical to standard modules within an array.

The module-module and module-cell approaches for evaluating soiling losses each have potential advantages. Conceptually, the module-module approach is simple and easily understood, since it compares modules otherwise identical except for cleanliness state. However, routine cleaning of a full-size clean reference module is a burden for busy operations and maintenance (O&M) personnel, leading to irregular or infrequent cleaning schedules. In contrast, cleaning of small reference cells for the clean reference in the module-cell approach is relatively easy and similar to other common sensor cleaning procedures performed by O&M staff. Furthermore, cell cleaning can easily be automated using an automatic reference cell washing system [13] or even eliminated and replaced with compensation by an optical soiling sensor [14].

In this work we report on our recently initiated field study of soiling and irradiance measurement in a bifacial PV system using in-situ I-V measurement to determine soiled module power. We evaluate both module-module and module-cell methods for normalizing soiled module power to clean expectations based on total front plus rear-side irradiance contributions. Our primary goals are to demonstrate effective normalization methods for total irradiance contributions and to quantify expected precision levels for soiling ratio measurements in bifacial systems.

2 EXPERIMENT

We are performing measurements for this study at a utility-scale single-axis tracking bifacial PV power plant operated by EDF Renewables in North America. Measurements began in May 2023 and are ongoing.

Fig. 1 and Fig. 2 depict the instrumentation setup. Two modules in the array are selected for measurement. One of these is designated as "clean" and the other as "soiled". In-situ I-V measurement units (RDE300i) periodically measure I-V sweeps during short time slices and determine standard I-V parameters, including short-circuit current (I_{sc}) and maximum power (P_{max}). In between I-V measurements the modules remain connected to the array and power-producing. Each RDE300i unit includes a back-of-module Pt1000 RTD affixed to the module rear side for temperature measurement. Three bifacial-optimized PV reference cells (RC22) measure irradiance. A single cell designated "front" measures irradiance in the front plane of array (POA) while two cells, designated "rear east" and "rear west", measure rear POA irradiance.

To ensure that the clean and soiled modules see equal levels of ground-reflected irradiance, both modules are located at least three from the end of the torque tube, minimizing discrepancies in ground illumination.

The clean module and all three reference cells were cleaned approximately every 2 weeks by site O&M staff.

10.4229/EUPVSEC2023/4CV.1.22
020389-002

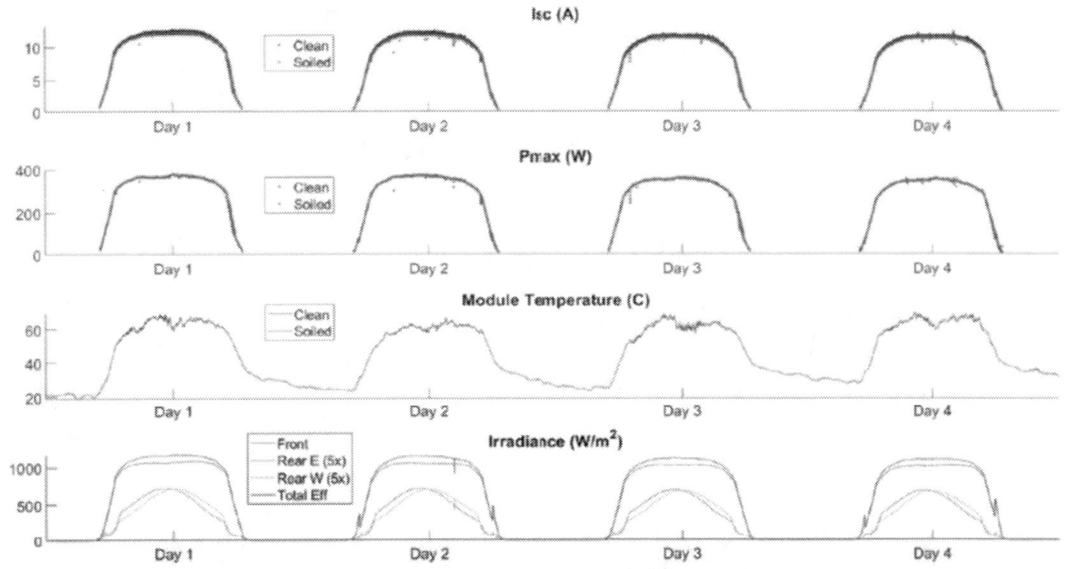

Figure 3: Time series data from a representative four-day period, showing I_{sc}, P_{max} and temperature of the designated soiled and clean modules and irradiance readings from the three reference cells. Note the correspondence between total effective irradiance and module I_{sc} and P_{max}. On Day 2 and Day 4 trackers followed a different program during early/late hours.

The cleaning schedule was adequate due to the very low rate of soiling at the site, which is also beneficial towards our goal of detecting any drift in the measurements.

3 EQUATIONS

We calculate soiling ratio as the ratio of the measured power of the soiled module to its expected power when clean, using the formula

$$SR = \frac{P_{\mathrm{max}}^{\mathrm{soiled}}}{P_{\mathrm{max,0}}^{\mathrm{soiled}} \cdot \left(1 + \gamma \cdot \left(T^{\mathrm{soiled}} - T_0\right)\right) \cdot \left(G_{\mathrm{tot}}/G_0\right)} \quad (1)$$

where $P_{\mathrm{max}}^{\mathrm{soiled}}$ is the maximum power of the soiled module determined from its I-V curve, $P_{\mathrm{max,0}}^{\mathrm{soiled}}$ is the calibrated maximum power of the soiled module when it is completely clean and at reference conditions of irradiance and temperature, γ is the module's maximum power temperature coefficient, T^{soiled} is the module's measured temperature, T_0 is the reference condition temperature, G_{tot} is the total effective front plus rear POA irradiance, and G_0 is the reference condition irradiance. Here "reference condition" refers to an irradiance and temperature for calibration and could mean standard test conditions (STC), bifacial standard test conditions (BSTC), or another condition such as a field calibration condition.

The module-module and module-cell methods differ in how G_{tot} for eq. (1) is determined.

In the module-module method, we use the clean module's I_{sc} to estimate the effective G_{tot} seen by the soiled module, by the formula

$$G_{\mathrm{tot}} = \frac{I_{\mathrm{sc}}^{\mathrm{clean}}}{I_{\mathrm{sc,0}}^{\mathrm{clean}} \cdot \left(1 + \alpha \cdot \left(T^{\mathrm{clean}} - T_0\right)\right)} \cdot G_0 \quad (2)$$

where $I_{\mathrm{sc}}^{\mathrm{clean}}$ is the clean module's measured short-circuit current, $I_{\mathrm{sc,0}}^{\mathrm{clean}}$ is the clean module's calibrated short-circuit current at the reference condition, α is the module's short-circuit current temperature coefficient, and T^{clean} is the clean module's temperature. Note that eq. (2) includes the contribution of both front and rear POA irradiance since both contribute to the module's Isc.

In the module-cell method, we use the three reference cells to estimate the average total effective irradiance illuminating the soiled module from front and rear, according to the formula

$$G_{\mathrm{tot}} = G_{\mathrm{front}} + \varphi \cdot \left(G_{\mathrm{rear,E}} + G_{\mathrm{rear,W}}\right)/2 \quad (3)$$

where G_{front} is the irradiance measured by the front reference cell, $G_{\mathrm{rear,E}}$ and $G_{\mathrm{rear,W}}$ are the irradiances measured by the rear east and west cells, and φ is the PV module bifaciality factor. Averaging the two rear-side cells allows for estimating the average irradiance across the module's rear plane [10]. The bifaciality factor φ appears in eq. (3) to account for the lower responsivity of the module to rear side light relative to front side light.

4 RESULTS

4.1 Time series data

Fig. 3 depicts example time series data recorded over a representative four-day period. Recorded parameters include the I_{sc}, P_{max}, and temperature of each of the modules and the front and rear reference cell irradiance readings. The reference cell readings are added per eq. (3) to compute total effective irradiance. Note that the rear reference cell readings are scaled by 5x for better visibility in the plot. The figure shows that the two modules have nearly identical electrical output, I_{sc} and P_{max}, and that their output profiles generally match the profile of the

10.4229/EUPVSEC2023/4CV.1.22

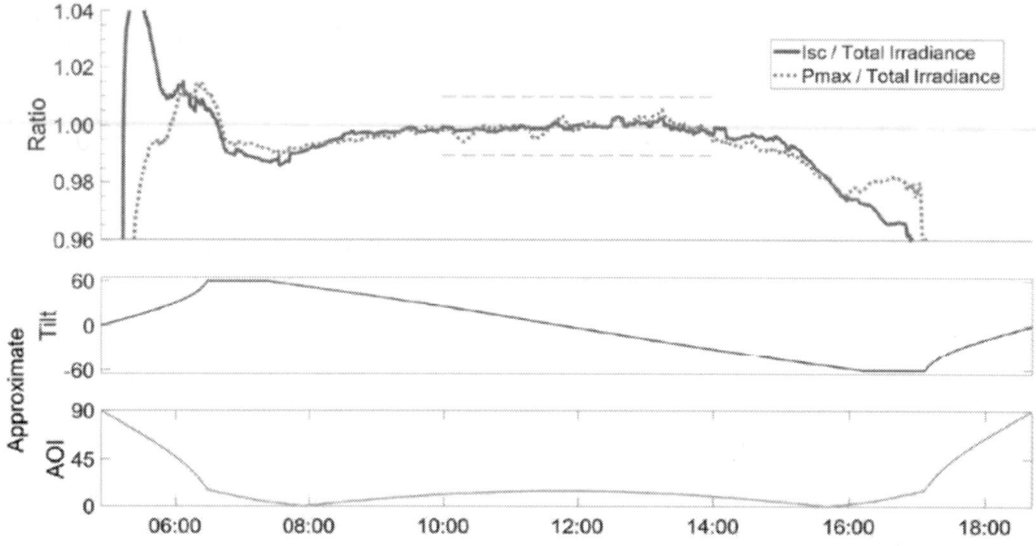

Figure 4: Proportionality between clean module temperature-corrected I_{sc} and P_{max} and total effective irradiance determined from the reference cells. Proportionality is nearly constant during mid-day when tilt and angle of incidence (AOI) are low.

reference cell total effective irradiance, with some exceptions.

On days 2 and 4 the total effective irradiance recorded by the reference cells jumps up towards the extremes of the day, while the module output does not show the same pattern. On these days the trackers were programmed to maximize power at low solar elevation by reducing angle of incidence at the expense of shading the lower halves of the half-cell modules and using only the top halves. Hence the reference cell readings diverge from the module output during these early and late minutes of the day, since the modules are only half illuminated at those times.

4.2 Proportionality of irradiance, current, and power

We wish to determine the extent to which module total effective irradiance can be represented by the reference cell readings, i.e. over what range do eqs. (2) and (3) yield equivalent results?

This question is addressed in Fig. 4. The top panel shows the normalized ratios of the temperature-corrected clean module I_{sc} and P_{max} to the reference cell total irradiance over a representative day. The selected day was one during which standard back-tracking was performed to eliminate row-to-row shading, as in day 1 and 3 in Fig. 1. During the middle ~6 hours of the day the ratios in the top panel of Fig. 4 are relatively constant (<0.5-1% variation), showing that the reference cell total irradiance profile well represents the module output in this time range, and that I_{sc} and P_{max} are proprtional to each other and to total irradiance. However, during the early morning and late afternoon the ratios diverge significantly, showing that the proportionalities no longer hold.

The loss of proportionality in the early morning and late afternoon can be understood by considering several geometric effects. The middle and lower panels in Fig. 4 depict the approximate tilt angle of the modules and resulting solar angle of incidence (AOI) throughout the day. The plots are only approximate because the tilt angle is not measured and is calculated from a generic backtracking model, rather than from the system design software. However, these representative values illustrate that the deviations from 1.0 of the ratios in the top panel occur when AOI is high and tilt is large. We believe that the front reference cell was mounted with a small angular offset (~1°) from the module plane, being tilted slightly more to the west. Therefore in the morning/afternoon the cell combination sees relatively less/more light than the module, explaining the asymmetry of the ratios in the top panel [2]. The effect of this small angular misalignment is negligible at low AOI in the middle of the day.

Additional effects that could contribute to the divergence of the ratios in Fig. 4 include non-uniformity of rear-side illumination at high tilt angle and the potential reduction of module efficiency (reduction in P_{max}/I_{sc} ratio) at low irradiance. We are exploring these topics but have not yet reached conclusions.

In addition, tracking algorithms that intentionally introduce shading can also result in a divergence between reference cell readings and module output, as discussed above.

Fig. 4 suggests that the module-cell and module-module methods will correlate best for data limited to the central portion of the day when AOI is low. Similar effects are observed in monofacial systems when two devices are being compared for soiling measurement [1][2].

4.3 Soiling ratios

Fig. 5 shows daily average soiling ratios calculated for the module-module and module-cell methods over the 11 weeks of the study to date. The data set has two gaps when equipment was offline.

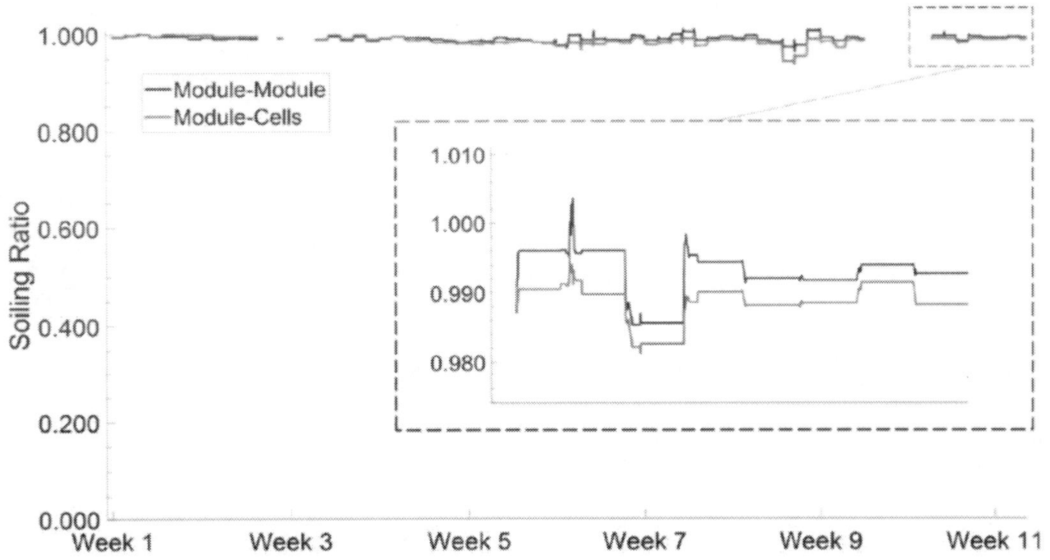

Figure 5: Trailing 24-hour average soiling ratios calculated using measured soiled module P_{max} compared to expected P_{max} determined from a clean module ("module-module") or from a combination of front and rear reference cells ("module-cell") showing overall agreement of the methods within ~0.5%.

To calculate the daily average soiling ratios, at each one-minute time point we use eq. (1), substituting either eq. (2) (module-module) or eq. (3) (module-cell) for G_{tot}, but we separately average the numerator and denominator of eq. (1) over all the time points in a trailing 24-hour period prior to calculating the ratio. This converts the ratio into a trailing 24-hour average. Only points within the central ~6 hours of the day are included. I.e., at noon, the ratio includes readings from the last three hours of the morning of the current day and the first three hours of the afternoon of the previous day. Rather than using a clock, the time window is determined approximately by filtering for times when the rear east and west reference cells agree within a target threshold.

Daily average soiling ratios were calibrated to 1.00 on a selected clear-sky day near the start of the study period for each of the methods.

Fig. 5 shows that the soiling ratio values calculated using either the module-module or module-cell methods agree well with each other over the entire period. Deviations on some days in weeks six through nine are associated with cloudy days where irradiance was unstable or there were few points passing the filter conditions, leading to noisier results. Overall, as seen in the inset for the final week, the soiling ratio values differ by less than ~0.5%.

Expanding the time window to include the entire day, while imposing an irradiance threshold of 100 W/m^2, produced similar results, with differences between the methods still on the order of ~0.5%. However, the limited time window is likely preferred due to the proportionality observations summarized in connection with Fig. 4.

Note that the soiling ratio values remained close to 1.00 during the study period to date, meaning that we did not observe much soiling at this site, which has a very low soiling rate. However, our aim here is not necessarily to detect soiling but to assess uncertainties in the soiling ratio measurements based on ability to account for the combined front and rear irradiance using different methods.

As the study progresses, we plan to periodically clean the soiled module to check how closely the soiling ratio returns to the baseline value of 1.00 for both methods. Any drift in the baseline readings will elucidate uncertainties in the measurements.

5 CONCLUSIONS

We have demonstrated the use of in-situ module I-V measurement to determine performance parameters of power-producing modules within a bifacial PV array for assessing soiling and irradiance. We consider two methods for accounting for total front plus rear irradiance illuminating a soiled module, using either the short-circuit current of a clean reference module (module-module) or a combination of readings from front and rear reference cells (module-cell). These methods produce average soiling ratio results agreeing within ~0.5%. The results to date therefore validate using either module-module or module-cell instrumentation setups for soiling measurement.

These findings are also relevant to the potential use of in-situ I-V measurements for long-term tracking of module degradation, which is similar to soiling ratio measurement in that a module's power output must be assessed over time against an expected reference value. Using a combination of reference cells, which can be easily cleaned and maintained and removed for recalibration, could provide an effective method of

10.4229/EUPVSEC2023/4CV.1.22
020389-005

normalizing module power for long-term degradation tracking.

6 ACKNOLWEGEMENTS

This material is partially based upon work supported by the U.S. Department of Energy's Solar Energy Technologies Office under Award Number DE-SC0020831.

We wish to thank Nick Diomedes of EDF for very valuable support at the site.

7 REFERENCES

[1] "IEC 61724-1 Ed. 1.0 en:2017 - Photovoltaic system performance - Part 1: Monitoring."

[2] M. Gostein, J. R. Caron, and B. Littmann, "Measuring soiling losses at utility-scale PV power plants," in *2014 IEEE 40th Photovoltaic Specialist Conference, PVSC 2014*, 2014, pp. 885–890, doi: 10.1109/PVSC.2014.6925056.

[3] M. Gostein, S. Faullin, K. Miller, J. Schneider, and B. Stueve, "Mars Soiling Sensor™," in 2018 IEEE 7th World Conference on Photovoltaic Energy Conversion, WCPEC 2018 - A Joint Conference of 45th IEEE PVSC, 28th PVSEC and 34th EU PVSEC, 2018, pp. 3417–3420, doi: 10.1109/PVSC.2018.8547767.

[4] M. Gostein, B. Bourne, F. Farina, and B. Stueve, "Field Testing of Mars™ Soiling Sensor," in *Conference Record of the IEEE Photovoltaic Specialists Conference*, Jun. 2020, vol. 2020-June, pp. 0524–0527, doi: 10.1109/PVSC45281.2020.9300975.

[5] M. Korevaar, J. Mes, P. Nepal, G. Snijders, and X. van Mechelen, "Novel soiling detection system for solar panels," in *33rd European Photovoltaic Solar Energy Conference*, 2017, pp. 2349–2351.

[6] Y. Cui, J. Xiao, J. Xiang, J. Sun, and E. M. Vitucci, "Characterization of Soiling Bands on the Bottom Edges of PV Modules," *Front. Energy Res.*, vol. 9, no. April, pp. 1–7, 2021, doi: 10.3389/fenrg.2021.665411.

[7] M. Gostein, B. Littmann, J. R. Caron, and L. Dunn, "Comparing PV power plant soiling measurements extracted from PV module irradiance and power measurements," *Conf. Rec. IEEE Photovolt. Spec. Conf.*, pp. 3004–3009, 2013, doi: 10.1109/PVSC.2013.6745094.

[8] M. Gostein, T. Duster, and C. Thuman, "Accurately measuring PV soiling losses with soiling station employing module power measurements," *2015 IEEE 42nd Photovolt. Spec. Conf. PVSC 2015*, Dec. 2015, doi: 10.1109/PVSC.2015.7355993.

[9] S. Kagan *et al.*, "Impact of Non-Uniform Soiling on PV System Performance and Soiling Measurement," *2018 IEEE 7th World Conf. Photovolt. Energy Conversion, WCPEC 2018 - A Jt. Conf. 45th IEEE PVSC, 28th PVSEC 34th EU PVSEC*, pp. 3432–3435, Nov. 2018, doi: 10.1109/PVSC.2018.8547728.

[10] N. Riedel-Lyngskar, M. Bartholomaus, J. Vedde, P. B. Poulsen, and S. Spataru, "Measuring Irradiance With Bifacial Reference Panels," *IEEE J. Photovoltaics*, Nov. 2022, doi: 10.1109/JPHOTOV.2022.3201468.

[11] J. E. Quiroz, J. S. Stein, C. K. Carmignani, and K. Gillispie, "In-situ module-level I-V tracers for novel PV monitoring," *2015 IEEE 42nd Photovolt. Spec. Conf. PVSC 2015*, Dec. 2015, doi: 10.1109/PVSC.2015.7355608.

[12] A. Marquis, M. Gostein, and B. H. King, "Validation of In-Situ I-V Measurement Unit for PV System Monitoring Applications," in *2022 IEEE 49th Photovoltaics Specialists Conference (PVSC)*, Nov. 2022, pp. 0291–0294, doi: 10.1109/PVSC48317.2022.9938898.

[13] M. Gostein *et al.*, "Soiling measurement station to evaluate anti-soiling properties of PV module coatings," *Conf. Rec. IEEE Photovolt. Spec. Conf.*, vol. 2016-November, pp. 3129–3131, Nov. 2016, doi: 10.1109/PVSC.2016.7750242.

[14] M. Gostein, W. Stueve, and S. Faullin, "Measuring soiling and non-uniform soiling of PV modules without a clean reference device," United States Patent 10,715,081, 2020.

10.4229/EUPVSEC2023/4CV.1.22
020389-006

PHOTOVOLTAIC ARRAY OPERATIONAL FAULT ANALYSIS WITH MACHINE LEARNING USING PRETRAINED DEEP LEARNING MODELS FOR FEATURE SELECTION

Tarikua Mekashaw Zenebe*[1], Ole-Morten Midtgård[1], Steve Völler[1], Berhane Darsene Dimd[1]
[1]Norwegian University of Science and Technology, Department of Electric Energy, Norway
*Corresponding author: tarikua.zenebe@ntnu.no, +4792972271

ABSTRACT: The growing global installed capacity of solar photovoltaic (PV) requires advancements in fault analysis methods for safe and reliable electricity supply. Machine learning are becoming more popular in this field, particularly for detecting and classifying faults using I-V curve data. However, manual feature selection can be challenging. Pretrained deep learning models simplify this process by automatically selecting features, making them valuable in the feature selection stage of machine learning-based fault detection and classification methods. This paper evaluates and compares less-explored models (EfficientNet, DenseNet, SqueezeNet, and ShuffleNet) to commonly used models (ResNet-50, AlexNet, GoogLeNet, VGG-19, and VGG-16) for detecting and classifying operational faults like line-to-line, and line-to-ground with different mismatch levels using I-V curve data. The paper also places special emphasis on modeling the PV modules in detail to consider the effect of the bypass diode, which is often overlooked in machine learning-based fault analysis using simulation data. The fault detection and classification method and the PV array model are simulated and evaluated in MATLAB/Simulink. All the models achieved a test accuracy of over 99.5%. The best model selection thus require other parameters such as processing time, power, and memory requirements. SqueezeNet and ShuffleNet are particularly promising, especially for small-scale PV system applications where cost and memory are important factors.
Keywords: Feature selection, I-V curve data, machine learning model, operational fault, pretrained deep learning model.

1 INTRODUCTION

The continued use of fossil fuels will result in environmental disasters and contribute to the climate change, posing a significant threat to humankind's very existence. Moreover, as we have witnessed due to the current political, economic and geographical situation, to fulfill the current electricity demand and the future, finding alternative and renewable energy sources is an indispensable solution. According to the IEA [1], the global solar photovoltaic (PV) installed capacity will increase by 370GW annually until 2030 if countries fulfill their climate pledges. This means that PV systems must be even more reliable and safe to sustain the electricity demand without interruption. However, faults can occur in PV arrays due to variable and harsh outdoor operating conditions. Some of these faults include short circuits (line-to-line and line-toground), open circuits, direct current arc, partial shading, aging degradation, hot spots, and micro-cracks, among others [2]. Since standard protection devices, such as circuit breakers, ground fault protection devices, and arc protection devices, are designed to clear faults at high irradiance levels, faults in PV arrays that occur at lower irradiance, lower mismatch, and high impedance conditions will be undetected. In addition, aging, partial shading, and other mechanical faults can only be identified through a visual and detailed performance analysis [3].

As the PV array and other system components are designed to withstand higher current, faults in PV arrays at lower irradiance, lower mismatch, and high impedance are often neglected as they do not cause immediate damage or safety issues, only power loss. However, if left undetected, these faults can lead to hotspots and permanent damage to the PV system and surrounding property, potentially causing a fire. To address this issue, researchers are exploring the use of data-driven methods such as machine learning and recently deep learning, which have proven effective in fields such as wind power generation and health systems.

Most faults in PV arrays can be detected using electrical parameters [4]. Recently, getting current-voltage (I-V) curve data has become easier as they can be obtained from common components such as power optimizers and inverters [5], making it easier and more promising to implement fault detection and diagnosis methods based on I-V curve data. This data is also highly informative and easily accessible, so I-V curves data have been widely used with machine learning techniques. However, the challenge is selecting the best influencing parameters when using machine learning methods. While increasing the number of independent features improves accuracy, having too many features decreases accuracy and increases processing time. To address this, researchers often use principal component analysis (PCA) for feature reduction and only select deterministic features. For example, [6] studied the impact of PCA on accuracy. Another solution is using deep learning methods, which have the advantage of automatic feature selection [7]. When using deep learning models on I-V curve data, researchers typically use a 1D (dimension) version of the model or convert the curve to 2D spectrograms or use wavelet transform [8]. In [9], raw I-V curve data was used together with irradiance and temperature data to construct a 40x4 matrix, which was then resampled to a 2D data point and used as input to a simplified ResNet deep learning model for fault detection and classification.

As training deep learning models require a lot of data usually in millions, several researchers have tried to use transfer learning to detect PV panel faults. In [10] two classes, normal and defect, are detected using transfer learning (fine-tuned AlexNet). The defects analyzed include mechanical faults like crystal breakage, micro-cracks, and scratches, faults that result in shading like dirt, and spotted past, and other faults line burned panels. Even if the applicability of the method using real data sets is in question as they used images from the internet with low quality, they have achieved a good performance with

10.4229/EUPVSEC2023/4CV.1.24
020390-001

validation accuracy of 93.3%. Furthermore, they have evaluated the performance of the method with more classes and its ability to classify the type of defect.

From the reviewed research papers, we conclude that while automatic feature selection is possible with deep learning methods, it is not widely used with I-V curve data despite its richness in information about the health of the PV array. As indicated in the comprehensive review paper [11], which investigated deep learning methods for fault detection and classification in papers published from 2000 to 2018, there remains limited investigation of pretrained deep learning models, such as EfficientNet, DenseNet, SqueezeNet, and ShuffleNet, for PV array fault diagnosis. In one hand this is due to the need for a lot of image data with fault which is not cheap and easy to get. On the other hand, most pretrained models which are suitable to be used with small data set are recently developed [11 - 15] as we see in Figure 1.

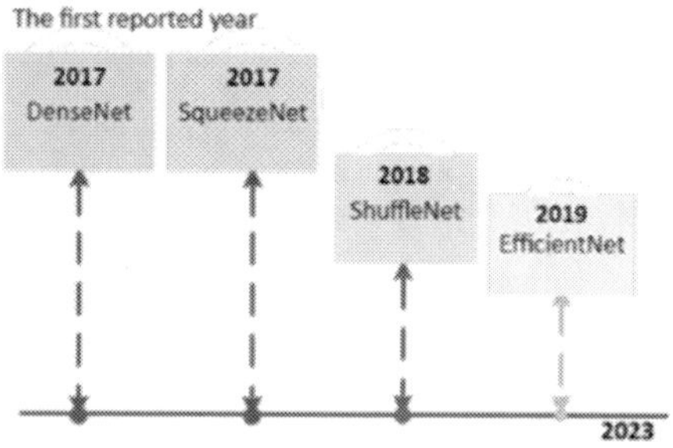

Figure 1: The first reported year of the less pretrained deep learning model: DenseNet, SqueezeNet, ShuffleNe and EfficientNet,.

In this work, we have investigated the performance of various pretrained deep learning models by replacing the manual feature engineering stage in the fault detection and classification process. The operational faults studied are line-to-line and line-to-ground faults with different mismatch levels, and the models are trained using I-V curve data, which is relatively easier to obtain without additional sensors or devices such as infrared cameras or drones. Additionally, as the main gap in previous studies on simulation-based machine learning and deep learning models for photovoltaic (PV) system fault detection and classification is the lack of detail on the importance of modeling PV modules while considering the integrated substring diode, in this paper the PV modules are modeled in detail to consider the effect of the bypass diode. Considering the integrated diode during line-to-line faults result in different I-V curve data. Our focus is on small-scale PV systems, where cost is a major concern, and where image-based deep learning models are generally expensive. After the feature selection stage, we use support vector machine (SVM) to reduce model complexity and speed up the fault classification process. We achieved test accuracy greater than 99.5% in all of the investigated pretrained deep learning models. The remaining sections of the paper are organized as follows: Section 2 describes the method and the tool used. Then the method applied for the fault detection and classification is elaborated in detail in section 3. The performance of the fault detection and classification method is discussed in section 4. Finally, the paper concludes by highlighting the main findings and future work in section 5.

2 MATERIAL AND METHODS

One of the aims of this paper is to investigate the performance of pretrained deep learning models such as EfficientNet, DenseNet, SqueezeNet, and ShuffleNet for automating the feature selection stage in detecting and classifying faults using the support vector machine (SVM) algorithm. By doing so, it is possible to extract all relevant information from the relatively easily available I-V curve data in small scale PV systems, without requiring expert knowledge or the use of expensive devices such as drones and cameras to collect image data. Second, based on our investigation, the PV module models used to collect simulated fault data for machine learning methods often lack detail. Therefore, in this paper, the simulation data is collected by first modeling the PV modules in MATLAB/Simulink as thoroughly as possible, as shown in Figure 2.

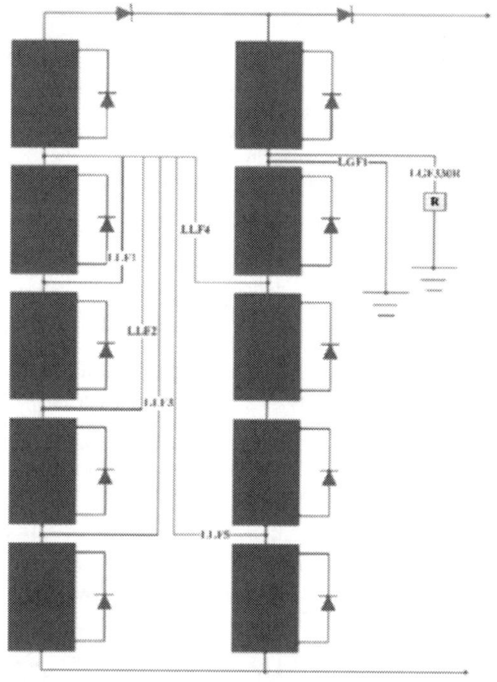

Figure 2: The modeled PV array to collect the I-V curve data at different operational faults condition.

The electrical data of the PV modules at standard test conditions (STC) is provided in Table 1. Each module is modeled using a single diode cell model, with four bypass diodes connected per 24 solar PV cells. The PV array consists of five PV modules connected in series and two strings connected in parallel, with a bypass diode for each module and a blocking diode for each string. The PV module therefore has a total of 96 solar PV cells. Next, I-V curve data is collected under different fault conditions. The workflow illustrated in Figure 3 is then used to extract features from the I-V curve data using the pretrained deep learning model, followed by using SVM to detect and classify the operational faults.

Table I: The PV module specification at standard test conditions (STC).

Parameters	Value
Open circuit voltage (V_{oc})	69.7V
Short circuit current (I_{sc})	6.07A
Voltage at maximum power point (V_{mpp})	58V
Current at maximum power point (I_{mpp})	5.7A
Temperature coefficient of V_{oc}	-0.25%/deg.C
Temperature coefficient of I_{sc}	0.3%/deg.C

The operational faults simulated include line-to-line fault with one module in one string (LLF1), line-to-line fault with two modules in one string (LLF2), line-to-line fault with three modules in one string (LLF3), line-to-line fault with one module across two strings (LLF4), line-to-line fault with three modules across two strings (LLF5), line-to-ground with four modules (LGF1), and line-to-ground with four modules (LGF330R) with resistance (see Figure 2).

To assess the performance of the models, we have utilized two key metrics: accuracy and prediction time. Accuracy is evaluated using Equation 1 [16], while prediction time is calculated by measuring the duration between the model receiving I-V data and returning the fault status.

$$Accuracy = \frac{TP + TN}{TP + TN + FP + FN} \quad \text{Equation (1)}$$

where, TP (True Positive) indicates instances correctly identified as positive. TN (True Negative) represents cases accurately predicted as negative. FN (False Negative) denotes instances mistakenly classified as negative. FP (False Positive) refers to instances mistakenly categorized as positive.

3 THE PROPOSED FAULT DETECTION AND CLASSIFICATION METHOD

To initiate model training, the I-V curve data must first undergo conversion into an image format. Following this, the image will undergo scaling, and its data format will be tailored to suit pre-trained deep learning algorithms. For instance in SqueezeNet, the image input size is 227-by-227 [14]. Then, the pre-trained deep learning algorithm will be loaded and fine-tuned for automatic feature selection. Subsequently, the machine learning algorithm, SVM, known for its speed and outstanding performance, will undergo training. This will be succeeded by a performance evaluation utilizing accuracy and time requirements as key metrics. Once satisfactory performance levels are attained, the model will be deployed for the tasks of fault detection and classification. To implement the proposed method for fault detection and classification, the I-V curve data must be converted into image format and subjected to thorough preprocessing. Following this, both the feature selection and the detection and classification models will be loaded and executed. During the model's execution, it will furnish an outcome indicating whether a fault has occurred. Furthermore, if a fault is detected, the model will provide specific information regarding the type of fault. The above procedure is shown in flowchart in Figure 3.

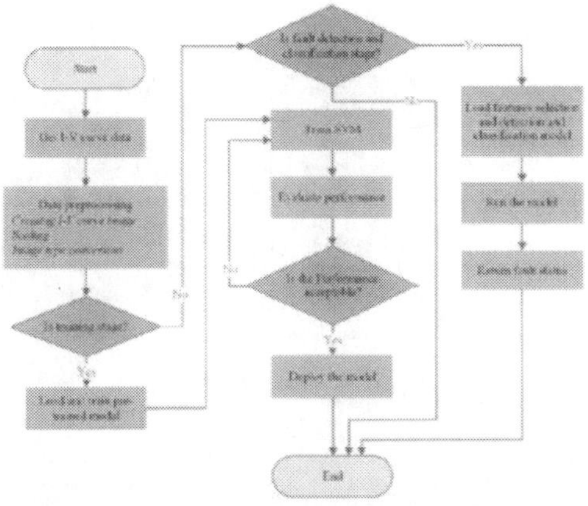

Figure 3: The simplified flowchart for the proposed fault detection and classification method using pretrained deep learning models for feature selection stage, and SVM for detection and classification stage.

4 RESULTS AND DISCUSSIONS

In this work, the performance of less-studied pretrained deep learning models (EfficientNet, DenseNet, SqueezeNet, and ShuffleNet) has been evaluated and compared to commonly used models (ResNet-50, AlexNet, GoogLeNet, VGG-19, and VGG-16) for use in the feature selection stage of fault detection and classification. From each class a total of 800 I-V curve images were used, with 70% for training and 30% for testing. And the performance evaluation incorporated both accuracy and time requirements. The accuracy of the proposed method is evaluated during both the training and testing phases. In the training phase, we have implemented cross-validation, calculating the average error across 10 folds to mitigate overfitting and gauge the models' ability to generalize.

Figure 4 illustrates the training, cross-validation, and testing accuracy of the models. For easy comparison, the accuracy and prediction time results are presented in the bar chart in Figure 4a and 4b. During training most of the models are able to predict the faults accurately. But while using cross validation, splitting the data in to 10 folds and taking the average of all the accuracy obtained, they perform slightly different. The cross validation and testing accuracy of SqueezeNet and ShuffleNet are slightly less than EfficientNet, DenseNet. However, as it shown in Figure 4b, they can predict the fault faster than the others. In general, the data indicates that all examined models achieved a test accuracy exceeding 99.5%, and most of them result in comparable accuracy. Furthermore, as depicted in Figure 4b, all the models can detect and classify the faults within 3 seconds.

10.4229/EUPVSEC2023/4CV.1.24
020390-003

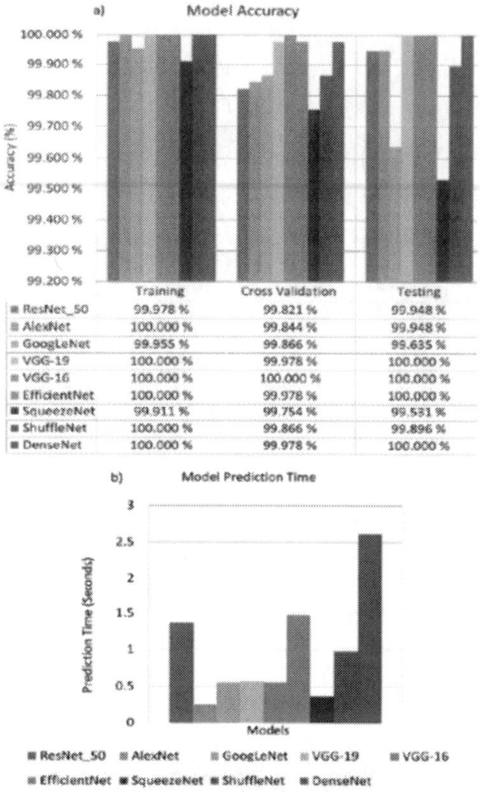

	Training	Cross Validation	Testing
ResNet_50	99.978 %	99.821 %	99.948 %
AlexNet	100.000 %	99.844 %	99.948 %
GoogLeNet	99.955 %	99.866 %	99.635 %
VGG-19	100.000 %	99.978 %	100.000 %
VGG-16	100.000 %	100.000 %	100.000 %
EfficientNet	100.000 %	99.978 %	100.000 %
SqueezeNet	99.911 %	99.754 %	99.531 %
ShuffleNet	100.000 %	99.866 %	99.896 %
DenseNet	100.000 %	99.978 %	100.000 %

Figure 4: The performance of the fault detection and classification model while using various pretrained deep learning model for feature selection stages, a) Model accuracy, b) model prediction time.

In order to look at the number of fault status that are correctly predicted and not, confusion matrix is the best way of showing this. For instance, we can look at Figure 5 which displays the confusion matrix from the testing stage while using ShuffleNet for feature selection and Support Vector Machines (SVM) for operational fault detection and classification. The fault detection and classification method is capable of correctly detecting all fault status, except for two instances where it incorrectly identified the fault as LLF3 instead of LLF5. These faults involve three modules within. The main distinction lies in the placement; LLF3 is within a single string, whereas LLF5 across two strings (see Figure 2). Due to this resemblance, the I-V curve data appears quite similar. Consequently, the model faces challenges in accurately classifying these faults, leading to occasional confusion. Figure 6a shows this similarity in I-V specially at lower irradiance level. Looking into more detail about the behavior of the PV array during this fault, the power-voltage (P-V) curve also show similar trend, having the same number of picks and valleys (see Figure 6b).

Looking at the accuracy results we can conclude that, selecting the best model necessitates consideration of other parameters, such recall, precision, F1-scores, processing power, memory, and time requirements. In general, SqueezeNet and ShuffleNet demand less computational power and memory, with ShuffleNet being specifically engineered for mobile applications.

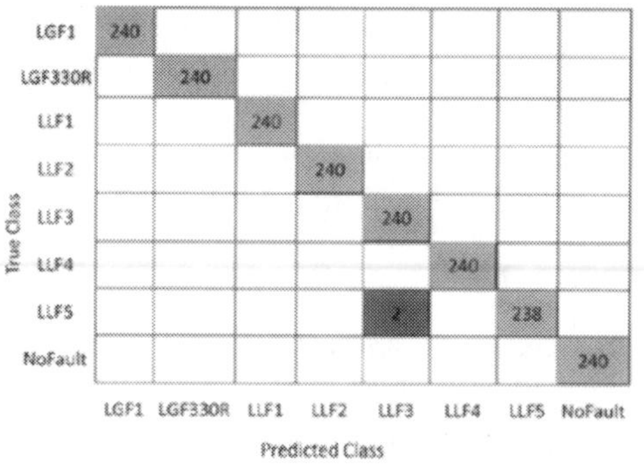

Figure 5: The confusion matrix during testing (using 30% (240 images) of the images) while using ShuffleNet for feature selection stage.

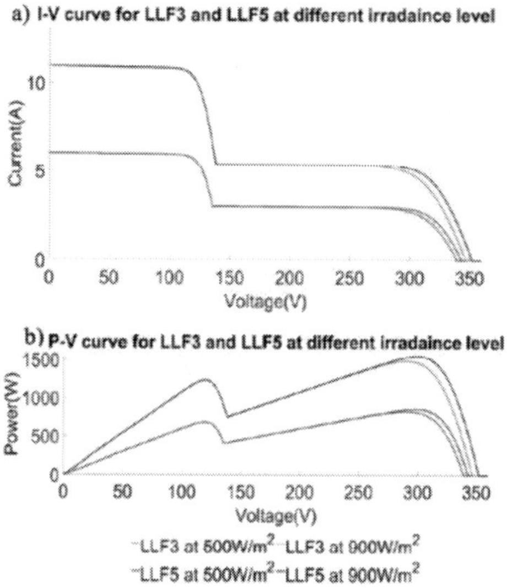

Figure 6: The current-voltage (I-V) curve (a) and the power-voltage (P-V) curve (b) of the PV array when LLF3 and LLF5 occurs at different irradiance level (500W/m^2 and 900W/m^2).

5 CONCLUSIONS AND FUTURE WORK

In this paper, the performance of pretrained deep learning models such as EfficientNet, DenseNet, SqueezeNet, and ShuffleNet were examined for fault detection and classification related to PV arrays, exploiting the rich information contained in I-V curves data. The deep learning models are used to replace the manual feature selection required when using machine learning algorithms. The results show higher test accuracy (over 99.5%) in detecting and classifying operational faults such as line-to-line and line-to-ground with different mismatch levels. The accuracy obtained is also comparable to that of commonly used models such as ResNet-50, AlexNet, GoogLeNet, VGG-19, and VGG-16. However, considering time requirements and memory size, among the less explored pretrained models, SqueezeNet, and ShuffleNet have superior performance, particularly ShuffleNet as it is being specifically engineered for mobile applications. Therefore, the results obtained are promising for operational fault analysis in

10.4229/EUPVSEC2023/4CV.1.24

small-scale solar PV systems. As the models have comparable accuracy the authors plan to conduct a comprehensive study of the models' performance, incorporating additional metrics such as recall, precision, and F1-scores, as well as assessing processing time, power consumption, and memory requirements. Furthermore, we intend to investigate and compare the performance of various machine learning models, including those with manual feature selection and those utilizing pre-trained deep learning models for the detection and classification stages.

6 REFERENCES

[1] IEA (2022), "World Energy Outlook 2022, " IEA(2022), Paris https://www.iea.org/reports/world energy-outlook-2022, License: CC BY 4.0 (report); CC BY NC SA 4.0 (Annex A)

[2] Chen, Zhicong, Yixiang Chen, Lijun Wu, Shuying Cheng, and Peijie Lin, "Deep residual network based fault detection and diagnosis of photovoltaic arrays using current-voltage curves and ambient conditions, " Energy Conversion and Management 198 (2019): 111793, https://doi..org/10.1016/j.enconman.2019.111793.

[3] IEA, "IEA PVPS Task 13: Performance, Operation and Reliability of Photovoltaic Systems: Guidelines for Operation and Maintenance of Photovoltaic Power Plants in Different Climates 2022, IEA (2022), Report IEA-PVPS T13-25:2022, ISBN 978-3-907281-13-0.

[4] Hong, Ying-Yi, and Rolando A. Pula, "Methods of photovoltaic fault detection and classification: A review, " Energy Reports 8 (2022): 5898- 5929, https://doi.org/10.1016/j.egyr.2022.04.043.

[5] Li, Baojie, Claude Delpha, Anne Migan-Dubois, and Demba Diallo, "Fault diagnosis of photovoltaic panels using full I–V characteristics and machine learning techniques," Energy Conversion and Management 248 (2021):114785, https://doi.org/10.1016/j.enconman.2021.114785.

[6] R. K. Mandal, N. Anand, N. Sahu and P. Kale, "PV System Fault Classification using SVM Accelerated by Dimension Reduction using PCA," 2020 IEEE 9th Power India International Conference (PIICON), 2020, pp. 1-6, doi: 10.1109/PIICON49524.2020.9112896.

[7] Hajji, Mansour, Mohamed-Faouzi Harkat, Abdelmalek Kouadri, Ka- maleldin Abodayeh, Majdi Mansouri, Hazem Nounou, and Mo- hamed Nounou, "Multivariate feature extraction based supervised machine learning for fault detection and diagnosis in photovoltaic systems, " European Journal of Control 59 (2021), pp. 313-321, https://doi.org/10.1016/j.ejcon.2020.03.004.

[8] M. A. Khan, A. Haque and V. S. B. Kurukuru, "Machine Learning Based Islanding Detection for Grid Connected Photovoltaic System, " 2019 International Conference on Power Electronics, Control and Automation (ICPECA), 2019, pp. 1-6, doi: 10.1109/ICPECA47973.2019.8975614.

[9] Chen, Zhicong, Yixiang Chen, Lijun Wu, Shuying Cheng, and Peijie Lin, "Deep residual network based fault detection and diagnosis of photovoltaic arrays using current-voltage curves and ambient conditions," Energy Conversion and Management 198 (2019): 111793, https://doi..org/10.1016/j.enconman.2019.111793.

[10] I. Zyout and A. Oatawneh, "Detection of PV Solar Panel Surface Defects using Transfer Learning of the Deep Convolutional Neural Networks, "2020 Advances in Science and Engineering Technology International Conferences (ASET), 2020, pp. 1-4, doi:10.1109/ASET48392.2020.9118384.

[11] Li, Baojie, Claude Delpha, Demba Diallo, and A. Migan-Dubois, "Application of Artificial Neural Networks to photovoltaic fault detection and diagnosis: A review, " Renewable and Sustainable Energy Reviews 138 (2021): 110512, https://doi.org/10.1016/j.rser.2020.110512.

[12] Tan, Mingxing, and Quoc Le, "Efficientnet: Rethinking model scaling for convolutional neural networks, " International conference on machine learning. PMLR, 2019.

[13] G. Huang, Z. Liu, L. Van Der Maaten and K. Q. Weinberger, "Densely Connected Convolutional Networks, " 2017 IEEE Conference on Computer Vision and Pattern Recognition (CVPR), Honolulu, HI, USA, 2017, pp. 2261-2269, doi: 10.1109/CVPR.2017.243.

[14] Iandola, Forrest N., Song Han, Matthew W. Moskewicz, Khalid Ashraf, William J. Dally, and Kurt Keutzer, "SqueezeNet: AlexNet-level accuracy with 50x fewer parameters and¡ 0.5 MB model size," " arXiv preprint arXiv:1602.07360 (2016).

[15] X. Zhang, X. Zhou, M. Lin and J. Sun, "ShuffleNet: An Extremely Efficient Convolutional Neural Network for Mobile Devices, " 2018 IEEE/CVF Conference on Computer Vision and Pattern Recognition, Salt Lake City UT, USA, 2018, pp. 6848-6856, doi:10.1109/CVPR.2018.00716.

[16] Kubat, M., & Kubat, J. A. (2017), "An introduction to machine learning" (Vol. 2, pp. 321-329), Cham, Switzerland: Springer International Publishing, dio: 10.1007/978-3-319-63913-0.

DEGRADATION ANALYSIS OF SMALL-SCALE PV SYSTEMS UNDER DIFFERENT CLIMATIC CONDITIONS

G. Papageorgiou[1], M. Schütze[1], D. Buß[1], M. B. Koentopp[1], M. Queck[1], R. Zimmermann[1],
G. Makrides[2], M. Norton[2], G. E. Georghiou[2]
[1]Hanwha Q Cells GmbH, Sonnenallee 17-21, 06766 Bitterfeld-Wolfen, Germany
[2]PV Technology Laboratory, FOSS Research Centre for Sustainable Energy,
Department of Electrical and Computer Engineering, University of Cyprus, 1678 Nicosia, Cyprus
G.Papageorgiou@q-cells.com

Hanwha Q-Cells (HQC) is performing extensive outdoor testing of its products to evaluate module aging and ensure superior performance, energy yield and durability. This work assesses the performance of two degradation rate (R_D) analysis techniques, RdTools [1], [2] and solar-data-tools [3] based on field data of small-scale photovoltaic (PV) systems, located in Germany (Central European climate) and Cyprus (Mediterranean climate). The median annual R_D is calculated by using both actual and modelled clear-sky data. The results show that data from the Cypriot test field yield stable results in close agreement with the expected degradation model for the PV modules under test. On the contrary, analysis of the data from the German test field shows less stable results which deviate significantly. This is attributed to the limited number of sunny days in central Europe, weather instability, differences between modelled and measured irradiance data. In this case, data sets with at least five years of field data can aid to minimise the uncertainties in the calculations. Finally, comparison to the indoor standard test condition (STC) values for the systems installed in Cyprus highlights the critical influence of measurement system differences, extra components in PV systems and variations in the pipeline techniques of each method.
Keywords: degradation, outdoor, clear-sky, RdTools, solar-data-tools, yoy

1 INTRODUCTION

The degradation rate is a crucial performance metric that offers valuable insights into the assessment of lifetime yield, reliability, and economic feasibility of photovoltaic systems. While numerous methods exist, there is limited knowledge regarding the precision, applicability, and stability of these methods under varying climatic conditions. By enhancing our understanding of the applicability and accuracy exhibited by each method, we can assist investors, operators, and manufacturers in effectively planning and evaluating the economic properties of PV systems.

To date, a significant amount of scientific research has been conducted to identify the causes of outdoor degradation in PV modules, gain insights into the fundamental physical mechanisms involved, and assess their influence on the evaluation of PV system performance [4, 5, 6, 7, 8, 9, 10, 11, 12, 13]. Yet, there is no universally accepted method for determining R_D of PV systems based on field performance data. The challenge in finding one stems from the fact that the resulting degradation value is often affected by data errors, seasonal variations, noisy data, and measurement uncertainties. Moreover, as the majority of assessments do not include comprehensive validation of the actual PV performance, the true degradation value remains unknown, making it challenging to assess the accuracy of each method. One of the methods that has been extensively used for analysing outdoor PV data is the ordinary least squares [7, 10, 14, 15]. However, although its simplicity makes its application very appealing, this method is highly susceptible to errors caused by seasonality, variable trends, outliers, and missing data. To address this issue, recent studies have focused on the elimination of seasonal variations from the observed values by employing the classical series decomposition or the seasonal and trend decomposition using Loess methods [7, 10, 16]. Alternatively, the robust principal component analysis has been employed in many studies [4, 11, 17] to calculate R_D by reducing and extracting the underlying structure of PV data.

Although many open-source libraries have already been developed using the Python programming language [18] employing some of the abovementioned methods, the impact of data originating from significantly different climatic conditions on the performance of various pipeline techniques and models has not been thoroughly explored yet. To this end, this study focuses on evaluating the performance of two open-source Python libraries, RdTools and solar-data-tools. Both libraries are equipped with tools to effectively address common challenges that arise when working with real-world performance data. These challenges include issues such as sensor drifts, variations in shading or soiling throughout different seasons, truncated power curves, and changes in data caused by maintenance events. For the purpose of this work, field data from small-scale PV systems with capacities ranging from 1-3 kWp located in Germany (Central European climate) and Cyprus (Mediterranean climate) are used. The degradation rate is calculated based on either sensor-based or modelled data. Finally, the robustness of each method and the accuracy of the degradation rate calculation is evaluated.

2 METHODOLOGY

2.1 Data set

The German test fields are monitored and maintained by HQC with more than 50 systems of various technologies, where both experimental and standard products, are being tested. The Cypriot outdoor facility is monitored and maintained by FOSS Research Centre for Sustainable Energy of the University of Cyprus (UCY) comprised of 16 PV systems of different technologies. In both test field locations, the PV modules are mounted in portrait arrangement and due south. Meteorological data (global irradiance, ambient and module temperature, relative humidity, wind speed and direction) are continuously collected via sensors installed accordingly. The systems are grid connected and electrical data are measured as high-resolution time series. To further facilitate sufficient normalisation of outdoor data, there are several reference cells and pyranometers, module temperature sensors and weather stations installed on-site.

For the purpose of this study, systems of 1-3 kWp of capacity were used. Data from four systems (UCY1-08 and -10, UCY2-01 and -03) installed in Cyprus and twelve systems (THA1-27, -28, -29, -52, -54, -55, -61, -62, -64, -73, -82, -83) installed in Germany are analysed. Every system consists of four to twelve modules connected in series in a string topology, with module technologies ranging from conventional HQC market products of different product generations to experimental modules.

2.2 Calculation of annual degradation rate using Python

The open-source Python libraries RdTools and solar-data-tools are primarily used for working with time series data of PV systems. They are equipped with tools and functions for analysing PV system performance and calculating various metrics, such as degradation rate, performance ratio and energy yield. Specifically, for the calculation of the median annual R_D over a specific timeframe, the DC maximum power point power (P_{MPP}) data points are used. Initially, the data are loaded, cleaned and pre-processed to handle missing values, outliers and other data quality issues. Subsequently, the data are normalised to indoor STC rated values. In the case of RdTools, the measured P_{MPP} is extrapolated to STC conditions using the temperature coefficient. Additionally, measured or modelled clear-sky irradiance data are used to get the datasheet expected P_{MPP} at STC conditions and to extract the normalised value. The modelled clear-sky irradiance and temperature are calculated using the pvlib [19] Python module and PV system longitude, latitude and azimuth values as inputs.

On the contrary, solar-data-tools having P_{MPP} values from the measured datasets as the only input, employs an alternative normalisation approach. In this case, the modelled clear-sky P_{MPP} values are generated from the measured datasets using a machine learning approach. This is done using signal analysis techniques to identify clear-sky days. Specifically, the daily signals are evaluated using two criteria, the smoothness of the daily P_{MPP} curve and the power output. In this way, the calculation of R_D does not require any actual or modelled estimates of irradiance or weather data. The theoretical advantages of clear-sky approaches are their ability to address sensor drifts or calibration problems and reduce the reliance on a multitude of input parameters. Finally, the data are aggregated to daily intervals and the median annual R_D is calculated using the Year-on-Year (YoY) method [2]. According to the YoY decomposition method, the slope is calculated for two points at the same calendar day between two subsequent years and repeated for the whole time series. The median value of the distribution of all slopes is, then, the most representative value of the true annual degradation over the analysed period. This method reduces the impact of seasonal factors such as soiling, shading, and temperature bias. In this study, the median annual R_D of each PV system are calculated through three different approaches: a) the sensor-based RdTools (R_DT-S) method, which utilises actual irradiance and temperature data; b) the clear-sky-based RdTools (R_DT-CS) method, which relies on modelled clear-sky irradiance and temperature data; c) and the solar-data-tools (SDT-CS) method, which utilises modelled clear-sky P_{MPP} data.

2.3 Evaluation of results

To investigate the robustness of each method, the median annual R_D of each system is calculated for an increasing number of years in the field. Subsequently, the relative performance to the rated power is determined by multiplying R_D by the number of years for each timeframe and plotted for an increasing number of years. The degree of perturbation for each method is assessed visually and, additionally, the data points are fitted using equation Eq. (1):

$$100 + R_D * x = 100 + \left(R_{D0} + R_{D1} * (x - 1)\right)$$

where R_{D0} is the initial degradation rate occurring in the first year, R_{D1} the long-term degradation rate for the additional years in the field and x the number of years in the field. As R_D calculations require a minimum of two years' worth of field data, Eq. (1) is only valid for values of x that are greater than two. This formula is describing the performance warranty function accounting for various degradation mechanisms (Figure 1). For the remainder of this text, R_D*x will be referred as "relative performance".

Solar modules once installed outside and exposed to sunlight, temperature cycles, humidity and mechanical stress, inevitably experience a gradual decline in their power generation capacity. An initial decline in the first months is primarily attributed to light induced degradation effects [20]. During the initial year of operation, a substantial power loss of up to 2-3% can be observed depending on module type and technology. In the subsequent years, the annual reduction in power output is considerably less pronounced with several other factors contributing to the degradation, including the deterioration of solder joints [21], discolouration of encapsulating materials [22], hotspots [23], increase in series resistance due to exposure to UV light [24], and various other factors [14]. For the modules under test, the values for R_{D0} and R_{D1} can range from zero to -3% and -0.6% per year, respectively. These values serve as general, indicative measures of annual power output loss.

In the case of the Cypriot test field the calculated median annual R_D of the outdoor PV systems are compared to results obtained from indoor calibrated precision module power measurements. Specifically, two modules from each PV system are demounted and cleaned and indoor measurements are conducted at STC. The PV parameters are then referenced to measurements obtained from identical modules stored in dark. In this way, possible deviations in calibration of the solar simulator can be partly compensated.

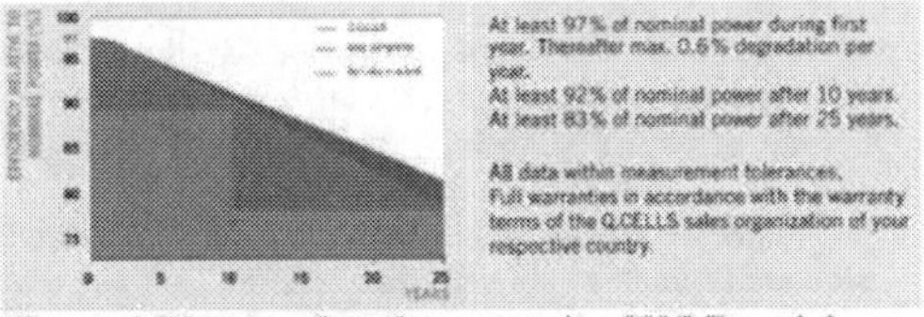

Figure 1 Warranty function example of HQC module Q.PRO-G2 260-265 included in some PV systems under examination. Although the production of this module type is discontinued, HQC is still performing long-term assessment of its performance to continuously evaluate module ageing and ensure superior performance, energy yield and durability of its products

10.4229/EUPVSEC2023/4CV.1.25

3 RESULTS AND DISCUSSIONS

3.1 Mediterranean climate: Degradation analysis of PV systems in Cyprus

The relative performance is calculated for an increasing number of years using the using RDT-S, RDT-CS and SDT-CS approaches. The results from systems UCY1-08 and -10, UCY2-01 and -03 are shown in Figure 2. When comparing the relative performance values of each system, significant variations become evident depending on the methodology employed. As the number of years increases, however, the differences gradually decrease. This observation is more evident in the findings from the central European climate (following section). This trend suggests that there is a higher tolerance for uncertainties as more years of data are included. The decreasing differences can be attributed to factors such as the higher dataset quality and the influence of long-term environmental factors and how each method handles them. With a larger dataset spanning multiple years, the quality of the data improves which helps to mitigate some of the uncertainties and variations that are present in shorter-term observations. Additionally, solar system performance is influenced by various environmental factors that can vary from year to year, such as weather patterns and irradiance. Over a longer duration, the impact of these factors tends to average out, resulting in a more stable estimation of R_D. Moreover, the statistical significance of the analysis increases with more years of data. A larger dataset allows for more robust statistical analyses, providing more accurate estimates and confidence intervals. However, it is important to note that the decreasing differences between the methods do not indicate a decrease in uncertainties themselves. While the convergence of the methods suggests increased consistency, uncertainties related to factors like measurement errors, data collection methods, and environmental variability still exist and should be considered when interpreting the results.

The individual curve fittings demonstrate a good fit to the expected degradation model, yet disparities in the initial and long-term degradation rates are evident. Notably, the RdTools sensor-based approach consistently computes a relatively low R_{D0} value. In contrast, the RdTools clear-sky-based approach yields variable results, whereas the solar-data-tools clear-sky-based approach consistently exhibits higher initial degradation rates leading to shallower curves. It is important to highlight the impact of the normalisation step on the computed R_D and relative performance values. Specifically, the outdoor measured output is normalised to indoor measured STC values. Consequently, since disparities are anticipated between the two measurement systems, variations in the calculated degradation values should be expected. This would result in a vertical offset of the data points, showing a different initial degradation R_{D0}. In conclusion, the fitted R_{D0} values should be treated cautiously, while R_{D1} values are less susceptible to such discrepancies.

Notably, comparison between the indoor STC and the calculated values shows that there's a noticeable consistency between the slopes of SDT-CS and STC values. This suggests that even if SDT-CS doesn't precisely determine the initial performance decrease due to some bias in its computations, it successfully gauges the long-term performance decline in line with the STC measurements. On the other hand, a consistent agreement can only be attained with the RdTools sensor-based approach for systems UCY2-01 and UCY2-03. Here, it's important to emphasise that the STC measurements were performed on two out of the five modules comprising each system. Furthermore, variations in the PV performance among the two modules within the same system of up to 2.2% are observed. This suggests that the actual degradation rate of each system might diverge from that of its individual modules. Furthermore, PV systems include additional components, such as cables, connections or inverters which can contribute to the calculated P_{MPP} values and result in different relative performance trends.

An alternative approach to evaluate the performance of each method is by comparing the Year-on-Year distributions. Figure 3 presents the distributions of RDT-S and RDT-CS of system UCY1-08. Regrettably, a similar plot for the solar-data-tools analysis cannot be generated due to the current difficulty in accessing YoY data stored by this Python module. It is noteworthy that RDT-S results exhibit a considerably narrower distribution and this was evident in all systems under investigation. These findings indicate a reduced level of uncertainty in the R_D calculation using this method.

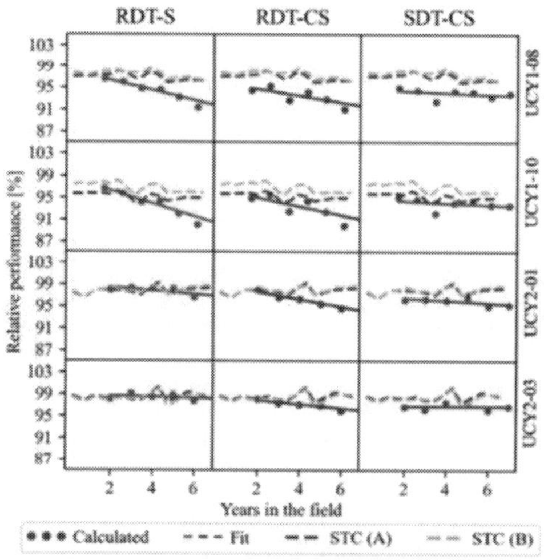

Figure 2 Mediterranean climate: plots and curve fittings with the expected degradation model of the relative performance for an increasing number of years in the field. STC relative performance data are also shown for comparison.

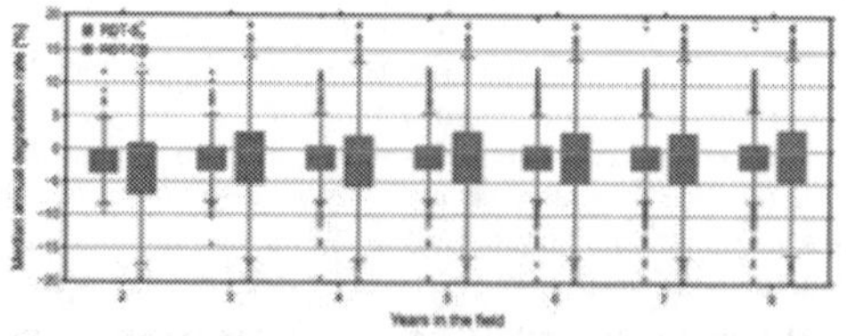

Figure 3: Mediterranean climate: Boxplots of YoY R_D using sensor- and clear-sky-based RdTools approaches

3.2 Central European climate: Degradation analysis of PV systems in Germany

Figure 4 and Figure 5 show representative plots of the relative performance of PV systems installed in the German test field. It is evident that the degradation trend can deviate significantly depending on the system under examination and on the employed method. Apart from the high degree of instability in the calculated values especially in datasets with less than five years in the field, in many cases, (unexpected) positive degradation rates are observed. However, similarly to the Mediterranean climate analysis, the differences between the calculated R_D values for the three methods gradually decrease with increasing number of years in the field. In general, these disparities tend to minimise in datasets containing at least five years of field data.

One factor contributing to these discrepancies arises from the unpredictable weather fluctuations throughout the day, which result in unstable sensor readings and inaccurate performance estimations. Additionally, the scarcity of sunny days in the central European climate reduces the statistical significance of clear-sky modelled approaches and results in increased uncertainty in YoY R_D calculations. This issue might affect the solar-data-tools clear-sky approach at a higher degree, as the accuracy of the modelled PV performance under clear-sky conditions is strongly tied to the frequency of sunny days in the dataset. Lastly, in the case of RdTools clear-sky approach, a discrepancy of up to 5% between modelled and measured irradiance is observed, for reasons that remain unclear. This disparity can significantly impact the reliability of the normalization process and, consequently, the calculation of R_D. As a result, in such climatic conditions, sensor-based approaches might be advantageous over clear-sky-based due to the comparatively lower uncertainty in the calculation of R_D. To mitigate these effects, it is advisable to utilise larger datasets collected from systems with a minimum of five years of field data. However, it should be noted that bigger datasets do not eliminate falsely estimated YoY values, but rather reduce the effect that they have on the extracted median value of the distribution.

In contrast to the results from the Mediterranean test field, further curve fittings using the expected degradation model are not meaningful and are omitted from this section due to the aforementioned reasons. Thus, an estimation of the initial and the long-term degradation rates cannot be carried out.

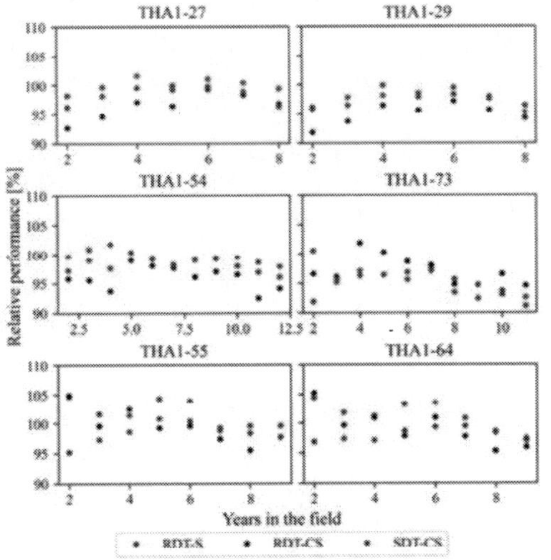

Figure 4 Central European climate: plots of the relative performance calculated using all three methods under test for an increasing number of years

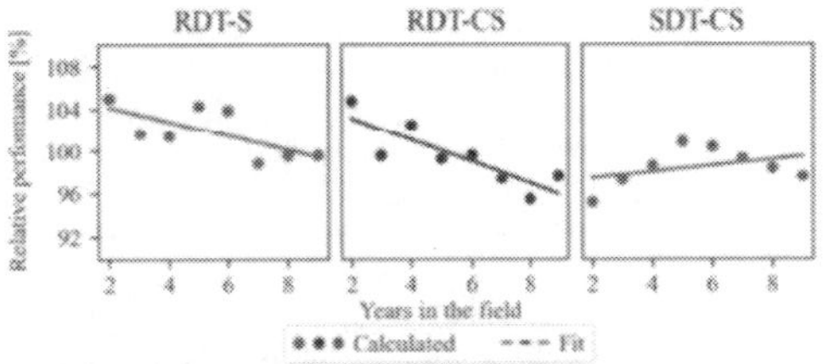

Figure 5 Central European climate: plots of R_D calculated using all three methods under test for an increasing number of years (system THA1-55)

4 CONCLUSIONS

The assessment of the performance of RdTools and solar-data-tools Python libraries, utilising both actual data and clear-sky model data from systems installed in Germany and Cyprus, yields a mixed set of outcomes. The findings reveal that the data collected from the Cypriot test site produce relatively stable R_D values. Additionally, the respective PV performance ratios exhibit a good fit with the expected degradation model for the PV modules under examination. In contrast, the analysis of data from the German test site yields less consistent results. This discrepancy is attributed to the lower number of sunny days, weather instability and lower statistical significance of the filtered data sets leading to increased uncertainty in the R_D calculations. Furthermore, observed differences between modelled and measured irradiance can contribute to a biased normalization of outdoor data. As a result, sensor-based approaches could be advantageous over clear-sky approaches due to the higher uncertainty associated with clear-sky methods caused by limited sunny days in central European climates.

In both cases, it is beneficial to have access to larger datasets from systems with a minimum of five years of field data. This aids to reduce the uncertainties contributing to the calculated R_D values, ultimately enhancing the reliability of the computation. Finally, it should be stressed that the comparison to indoor STC measurements for the systems installed in the Cypriot test field shows that a) although solar-data-tools does not determine the initial performance loss precisely, it successfully computes the long-term performance decline in line with the STC measurements and b) RdTools sensor-based approach exhibits good agreement for half of the total systems installed. Reasons that could result in misalignments between indoor and outdoor data could be the observed variations in the STC PV performance reaching up to 2.2% between the two modules within the same system, additional degradation contributions in the case of PV systems and differences in the pipeline techniques and the normalization steps.

5 REFERENCES

[1] M. G. Deceglie, A. Nag, A. Shinn, G. Kimball, D. Ruth, D. Jordan, J. Yan, K. Anderson, K. Perry, M. Mikofski, M. Muller, W. Vining and C. Deline, "RdTools 2.2.0-beta.1," [Online].

[2] D. Jordan, C. Deline, S. Kurtz, G. Kimball and M. Anderson, "Robust PV Degradation Methodology

10.4229/EUPVSEC2023/4CV.1.25
020391-004

and Application," in IEEE Journal of Photovoltaics, 2018.
[3] B. E. Meyers, E. Apostolaki-Iosifidou and L. T. Schelhas, "Solar Data Tools: Automatic Solar Data Processing Pipeline," in IEEE Photovoltaic Specialists Conference (PVSC), 2020.
[4] D. Jordan, T. Silverman, B. Sekulic and S. Kurtz, "PV degradation curves: non-linearities and failure modes," Progress in Photovoltaics: Research and Applications, vol. 25, no. 7, pp. 583-591, 2016.
[5] G. Belluardo, P. Ingenhoven, W. Sparber, J. Wagner, P. Weihs and D. Moser,, "Novel method for the improvement in the evaluation of outdoor performance loss rate in different PV technologies and comparison with two other methods.," Solar Energy, vol. 117, pp. 139-152, 2015.
[6] J. Lopez-Garcia, A. Pozza and T. Sample, "Analysis of crystalline silicon PV modules after 30 years of outdoor exposure," in 31rd EUPVSEC 2015, Hamburg, Germany, 2015.
[7] G. Makrides, B. Zinsser, M. Schubert and G. E. Georghiou, "Performance loss rate of twelve photovoltaic technologies under field conditions using statistical techniques," Solar Energy, vol. 103, pp. 28-42, 2014.
[8] G. Makrides, M. Theristis, J. Bratcher, J. Pratt and G. E. Georghiou, "Five-year performance and reliability analysis of monocrystalline photovoltaic modules with different backsheet materials," Solar Energy, vol. 171, pp. 491-499, 2018.
[9] T. Ozden, B. Akinoglu and R. Turan, "Long term outdoor performances of three different on-grid PV arrays in central Anatolia–An extended analysis," Renewable energy, vol. 101, pp. 182-195, 2017.
[10] A. Phinikarides, G. Makrides, B. Zinsser, M. Schubert and G. Georghiou, "Analysis of photovoltaic system performance time series: Seasonality and performance loss.," Renewable Energy, vol. 77, pp. 51-63, 2015.
[11] E. Pieri, A. Kyprianou, A. Phinikarides, G. Makrides and G. Georghiou, "Forecasting degradation rates of different photovoltaic systems using robust principal component analysis and ARIMA," IET Renewable Power Generation 11, vol. 11, no. 10, pp. 1245-1252, 2017.
[12] A. Pozza and T. Sample, "Crystalline silicon PV module degradation after 20years of field exposure studied by electrical tests, electroluminescence, and LBIC," Progress in Photovoltaics: Research and Applications, vol. 24, no. 3, pp. 368-378, 2016.
[13] A. Frick, G. Makrides, M. Schubert, M. Schlecht and G. Georghiou, "Degradation rate location dependency of photovoltaic systems," Energies, vol. 13, no. 24, p. 6751, 2020.
[14] D. C. Jordan and S. R. Kurtz, "Photovoltaic degradation rates—an analytical review," Progress in photovoltaics: Research and Applications, vol. 21, no. 1, pp. 12-29, 2013.
[15] I. Romero-Fiances, E. Muñoz-Cerón, R. Espinoza-Paredes, G. Nofuentes and J. de la Casa, "Analysis of the Performance of Various PV Module Technologies in Peru," Energies, vol. 12, p. 186, 2019.
[16] E. Solís-alemán, J. de la Casa, I. Romero-Fiances, J. Silva and G. Nofuentes, "A study on the degradation rates and the linearity of the performance decline of various thin film PV technologies," Sol. Energy, vol. 188, pp. 813-824, 2019.
[17] A. Kyprianou, A. Phinikarides, G. Makrides and G. Georghiou, "Definition and Computation of the Degradation Rates of Photovoltaic Systems of Different Technologies with Robust Principal Component Analysis," IEEE J. Photovoltaics, vol. 5, pp. 1698-1705, 2015.
[18] G. Van Rossum and F. L. Drake Jr, Python reference manual, Centrum voor Wiskunde en Informatica Amsterdam, 1995.
[19] W. F. Holmgren, C. W. Hansen and M. A. Mikofski, "pvlib python: a python package for modeling solar energy systems," Journal of Open Source Software, 2018.
[20] J. Lindroos and H. Savin, "Review of light-induced degradation in crystalline silicon solar cells," Solar Energy Materials and Solar Cells, pp. 115-126, 2016.
[21] M. A. Quintana, D. L. King, T. J. McMahon and C. R. Osterwald, "Commonly observed degradation in field-aged photovoltaic modules," in IEEE Photovoltaic Specialists Conference, 2002.
[22] F. J. Pern, "Ethylene-vinyl acetate (EVA) encapsulants for photovoltaic modules: Degradation and discoloration mechanisms and formulation modifications for improved photostability," Die Angewandte Makromolekulare Chemie: Applied Macromolecular Chemistry and Physics, vol. 252, no. 1, pp. 195-216, 1997.
[23] D. C. Jordan, T. J. Silverman, J. H. Wohlgemuth, S. R. Kurtz and K. T. VanSant, "Photovoltaic failure and degradation modes," Progress in Photovoltaics: Research and Applications, vol. 25, no. 4, pp. 318-326, 2017.
[24] K. Morita, T. Inoue, H. Kato, I. Tsuda and Y. Hishikawa, "Degradation factor analysis of crystalline-Si PV modules through long-term field exposure test," in 3rd World Conference onPhotovoltaic Energy Conversion, 2003.
[25] M. Theristis, A. Livera, C. Jones, G. Makrides, G. Georghiou and J. Stein, "Nonlinear photovoltaic degradation rates: Modeling and comparison against conventional methods," IEEE Journal of Photovoltaics, vol. 10, no. 4, pp. 1112-1118, 2020.
[26] W. F. Holmgren, C. W. Hansen and M. A. Mikofski, "pvlib Python: A python package for modeling solar energy systems," Journal of Open Source Software 3(29): 884., 2018.
[27] A. Dobos, "PVWatts version 5 manual," National Renewable Energy Lab.(NREL), Golden, CO (United States).

10.4229/EUPVSEC2023/4CV.1.25
020391-005

FAULT DIAGNOSIS OF GRID-CONNECTED PHOTOVOLTAIC SYSTEMS BASED ON UNSUPERVISED ENSEMBLE CLUSTERING AND MULTI LAYER PERCEPTRON MODEL

Mohsen Zargarani[3,1,*], Anne Migan-Dubois[1],, Demba Diallo[1], Claude Delpha[2], Sara Zermani[3], Chabakata Mahamat[3] and Laurent Linguet[3]

Université Paris-Saclay, CentraleSupélec, CNRS, Sorbonne Université, GeePs, Gif Sur Yvette, 91192, France
2. Université Paris-Saclay, CNRS, CentraleSupélec, L2S, Gif Sur Yvette, 91192, France
3. University of French Guiana, 97323 Cayenne cedex, French Guiana,
*Corresponding Author: mohsen.zargarani@centralesupelec.fr

ABSTRACT: Fault diagnosis in PV systems is studied broadly from various aspects. There are different techniques for FFD in PV systems. Among them unsupervised methods are appreciated for not needing labeled datasets or huge amounts of historical data. Clustering the data to study the similarities and differences is one of reliable unsupervised methods. To obtain proper accuracy in this paper, we introduce new integrated techniques for fault diagnosis and prediction. By generating different PV fault scenarios, different clustering techniques yield labels separately and all these clustering labels are gathered in an ensemble area to obtain best labels by voting. Voting majority empowers our ensemble clustering techniques. By interpreting the centroids of consensus clusters based on electrical characteristics of faults, we can generate fault labels. The final classification is outlined by neural networks (MLP). The results and experiments in a real test PV system show higher accuracy than other state of art unsupervised techniques.

Keywords: Photovoltaic (PV), fault diagnosis, consensus clustering, multilayer perceptron (MLP).

1 INTRODUCTION

Fault diagnosis in PV systems is studied broadly from various aspects.. Among them unsupervised methods are appreciated since they dispense with labeled datasets or huge amounts of historical data. Honglu Zho et al. proposed a fault diagnosis approach which was designed based on fuzzy clustering and probabilistic neural networks [2]. Shengyang Liu et al introduced an integrated method based on Gaussian kernel fuzzy C means clustering and back propagation neural networks [8]. Abbas M et al. presented an ANFIS system trained by means of subtractive clustering and grid partition technique [9]. In general, integration of clustering and neural networks played a novel role in data analysis and in different areas of science. Tanmoy Chakraborty applied both classification and clustering together and evaluated all by an optimization function [7]. It must be clarified that most of above mentioned methods applied clustering as conventional sole clustering inside their analysis and did not study the ensemble clustering with their integrated diagnosis methods.In fact, to the best of our knowledge, the proposed method in this paper is the first time that ensemble clustering and neural networks integrate in an overall diagnosis architecture. To better understand the state of art in the ensemble clustering part, the vast majority of research and methods in ensemble clustering are verified. Strehl and Ghosh developed the problem of ensemble clustering by utilizing three new techniques for solving cluster ensemble target problem. They introduced an objective function based on the ANMI and also algorithms based on meta clustering and the hyper graph partitioning [4].Khedairia and Khadir used an iterative voting process presented in a multiple clustering combination method [6]. Compared to our proposed method, the ensemble technique in their method is somehow divergent since they used only jaccard index while in our method, all Jaccard, overlapping, and similarity coincidence indices are calculated in an iterative algorithm to distinguish the sub-clusters and similarity sets in a smarter routine. Ayad and Kamel introduced a general formulation for the voting problem as a multi-response regression problem. Using probabilistic mapping in their voting process makes their method dissimilar to the other methods and different from the method in this paper[5].

In this paper, we introduce new integrated techniques for ensemble clustering and fault diagnosis to obtain superior accuracy. Generating different clustering yields inputs to the voting process. The algorithm interprets the centroids of final consensus clusters. The MLP accomplishes classification and fault prediction. The results show competitive accuracy than other state of art unsupervised routines.

Table I: Nomenclature

Abbreviation	description
MLP	Multi-Layer Perceptron
ANMI	Average Normalized Mutual Information
ANFIS	Adaptive Neuro-Fuzzy Inference System
D&EC	Diagnosis based on Ensemble Clustering
D&SC	Diagnosis based on Spectral Clustering
D&AC	Diagnosis based on Agglomerative Clustering
D&KC	Diagnosis based on K-means Clustering
RF	Random Forest
DT	Decision Tree
GNB	Gaussian Naïve Bayes
LR	Logistic Regression

2 STUDY AREA AND DATA

In this paper, we propose an approach for fault diagnosis in a context of use in French Guyana, where the climate is tropical and even "in some regions, the mean daily potential exceeds 5000 Wh/m² " [1]. To study the effect of internal and external (environmental) factors on PV faults, maintenance and performance, a model based on a real PV plant in French Guiana is simulated and designed. Studies on real plant and empirical results from different methods encouraged us to select unsupervised and clustering techniques. In addition, for global challenging fault scenarios, we designed different fault

10.4229/EUPVSEC2023/4CV.1.27
020392-001

category tests in a PV test module installed in SIRTA atmospheric observatory center located in Paris.

Figure 1: PV test module, SIRTA, Paris

3 METHODOLOGY

Overall unsupervised diagnosis comprises steps that are listed in the following algorithm table.

Algorithm I: Overall unsupervised diagnosis

---

description

Input: *data points, features extracted from the I-V curves (Isc, Voc, Imp, Vmp)*
Output: *Predicted Fault from classification*

Steps:

I. *Feature Correction and Normalization) by using Equations (1) and (2)*
II. *Various clustering generations*
III. *Consensus clustering based on new coincidence voting majority*
IV. *Centroid interpretation*
V. *Fault declaration & Relabeling (Fault label generation)*
VI. *Fault Classification : Neural Networks Classification*

3.1 Normalization

I-V curve features are corrected according to the standard IEC 60891 at STC condition of environment for temperature and irradiance [3]. All corrected values are normalized based on related nominal values. The collected I-V curves are corrected based on the following procedure [3]:

$$I_2 = I_1(1 + \alpha_{rel}(T_{m2} - T_{m1})\frac{G_2}{G_1} \quad (1)$$

$$V_2 = V_1 + V_{oc1}\left(\beta_{rel}(T_{m2} - T_{m1}) + a.\ln\left(\frac{G_2}{G_1}\right)\right) - R_s(I_2 - I_1) - k.I_2(T_{m2} - T_{m1}) \quad (2)$$

Where V_1 and V_2, I_1 and I_2 represent voltage and current before correction and after correction. The same are for T_{m1} and T_{m2}, G_1 and G_2 which mean temperature and irradiance before and after correction. β_{rel} and α_{rel} are relative temperature coefficients for V_{oc} and I_{sc} [3].

3.2 Base clustering generation

Among numerous clustering methods, three dissimilar techniques are preferred. K-means, spectral clustering and agglomerative clustering are selected because of their better performance and variety in their techniques and calculation philosophy.

3.3 Consensus clustering and Coincidence voting

Given two clusters from different partitions, their similarity measure as Jaccard index can be represented as:

$$Jx^{i,j}_{n,m} = \frac{|Z^i_n \cap Z^j_m|}{|Z^i_n \cup Z^j_m|} \quad (1)$$

Where Z^i_n in eq. (1) represents the cardinality of elements with cluster label n in partition i and Z^j_m means cardinality of elements with cluster label m in partition j. Obviously, $Jx^{i,j}_{n,m}$ is between zero and one. Here, $i = 1$ and $j = 1, \dots P$. Values of n and m vary between one and K. Another similarity measure is interiority or overlapping index. The interiority index is measured by eq. (2) [11].

$$Ix^{i,j}_{n,m} = \frac{|Z^i_n \cap Z^j_m|}{min\{|Z^i_n|,|Z^j_m|\}} \quad (2)$$

Although the Jaccard index seems to be a compelling tool in quantifying similarity and intersection, it is limited in considering the interiority between two clusters of separate partitions. For instance, two sample cluster sets may have the same Jaccard index but different interiority. The influence of interiority in I-V curve features can be better figured out when two clusters have the maximum interiority. In this case, it means one cluster from partition i is inside another cluster that belongs to partition j. In this instance, from a fault diagnosis point of view, they naturally exist in the same fault category. But in intersection clusters, they may either belong to the same fault category or different fault categories. The intersection clusters are more likely to be misleading in PV systems since some different faults have partly common behaviors. For example, open circuit fault and aging fault have both similar declines in voltage that may outline intersection areas of different faults while they obviously belong to different categories. For instance, two clusters of different data partitions that have similar Jaccard index but different interiority index are considered. In this case, higher interiority index value is preferred because it can demonstrate the relevant similarity of I-V curve more precisely. The coincidence similarity index covers both intersection and interiority concerns, which represents both by multiplying the two values as eq. (3) [10].

$$Cx^{i,j}_{n,m} = Jx^{i,j}_{n,m}.Ix^{i,j}_{n,m} \quad (3)$$

After running diverse clustering algorithms, all cluster labels are listed separately in the new matrix. From this cluster matrix, three different performance indices make the proper stand for confusion matrix. The confusion matrices are of Jaccard, Interiority and Coincidence index

$$A^{i,j} = \begin{bmatrix} Cx^{i,j}_{1,1} & \cdots & Cx^{i,j}_{1,k} \\ \vdots & \ddots & \vdots \\ Cx^{i,j}_{k,j} & \cdots & Cx^{i,j}_{k,k} \end{bmatrix} \quad (4)$$

$$A = [A^{1,2} \quad \dots \quad A^{1,j} \quad \dots \quad A^{1,P}] \quad (5)$$

10.4229/EUPVSEC2023/4CV.1.27
020392-002

To secure final consensus clustering, first, the algorithm groups similar labeled clusters of the first clustering method Z^i in rows and the clusters of other methods as Z^j in columns. Then, it calculates the cardinality of intersection for common members of grouped clusters to make intersection sets as **S**. In the next step, it calculates Jaccard index J_x, Interiority Index I_x and Coincidence similarity index C_x for each member of intersect clusters. It will be followed inside intersection sets **S** until to come by similarity matrix **A**. The process attempts to realize the maximum value of coincidence similarity index, C_x in each row of similarity matrix **A**, then, it generates new labels for voting. By voting process, a voted list and a non-voted list will be provided separately. Non-voted list builds the new dataset and this loop frequently iterates until all remaining rows of non-voted clusters compare their intersect Coincidence similarity index and become relabeled. This method outperforms all previous methods that rely only on a single Jaccard index. However, coincidence index can reach a more sensitive and more accurate comparison base than Jaccard index or interiority index while logically covering both of them with product operation.

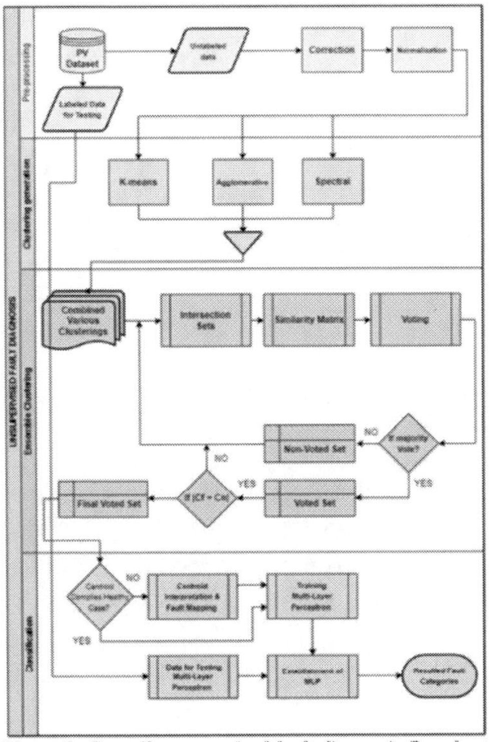

Figure 3: Overall unsupervised fault diagnosis flowchart

3.4 Mapping

After labeling the elements by ensemble clustering, the process interprets each cluster centroid for electrical characteristics to verify they belong to which fault category.

3.5 Classification

Since the mapping of the dataset is done, the process generates a labeled dataset. Proposed MLP undertakes classification and fault prediction. Based on the empirical results, for the activation functions, RELU is used for input and hidden layers, Softmax for output layer and Adam is selected as optimizer.

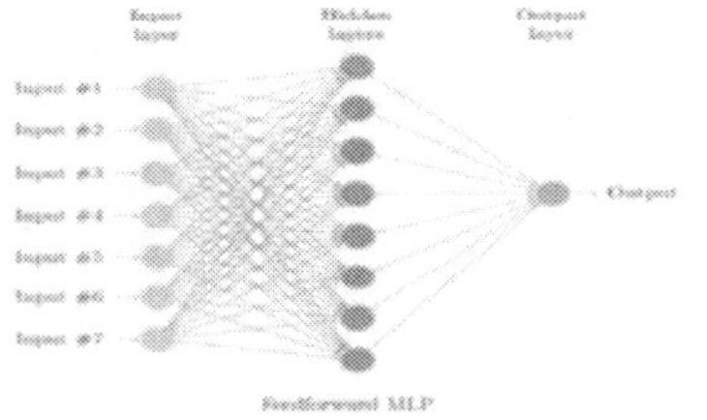

Figure 2: Feedforward MLP

4. EXPERIMENTAL SETUP FOR TEST

An experimental setup is built to verify the efficiencies of the introduced diagnosis technique. All test scenarios of PV panels are carried out for different faults and numerous weather conditions. The test platform comprises 8 PV modules which are arranged four by four in parallel. Each four panels are connected in series.

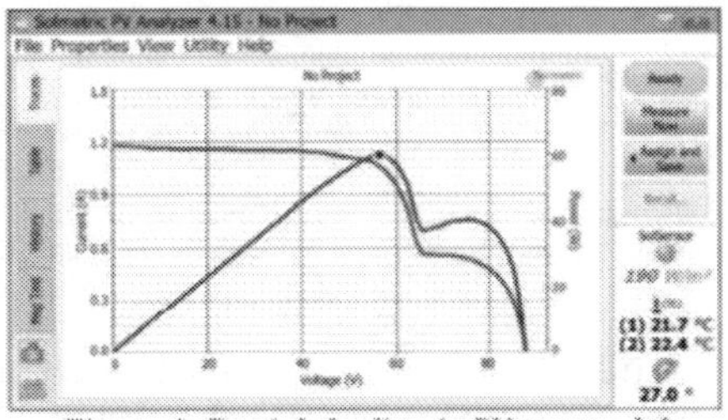

Figure 4: Partial shading in PV test module.

5. RESULT ANALYSIS

This study mainly intended to develop a novel fault diagnosis for PV systems using "ensemble-clustering then labeling algorithm" which is integrated with MLP for classification. The proposed algorithm accepts four main I-V features (I_{sc}, V_{oc}, I_{mp} and V_{mp}) plus T (temperature) and G (irradiance), as inputs.

Table II: Accuracy comparison of proposed unsupervised ensemble clustering with other unsupervised clustering methods

	D&EC	D&SC	D&AC	D&KC
Open circuit	90.83	77.59	79.0	78.5
Average aging	89.31	68.12	69.71	64.08
Partial shading	92.22	75.03	76.50	66.67

Table III: Accuracy comparison of D&EC and supervised methods

	D&EC	RF	DT	GNB	LR
Open circuit	90.83	97.83	97.29	95.08	90.22
Average aging	89.31	96.42	97.11	91.67	90.45
Partial shading	92.22	100	100	97.97	93.91

10.4229/EUPVSEC2023/4CV.1.27
020392-003

To illustrate our main contribution, compared to the state of art methods [2],[4],[5],[6] and [8], we must point out that, on the one hand, we integrated neural networks and ensemble clustering together, in a novel architecture, and on the other hand, we designed smarter philosophy for similarity analysis embedded inside iterative voting process. The performance of overall unsupervised diagnosis can be observed in tables. II and III. The accuracy of ensemble clustering diagnosis outperforms other clustering based diagnosis methods. Furthermore, the accuracy of the proposed unsupervised method is competitive to conventional supervised methods.

6. CONCLUSION

Integrated ensemble-clustering and neural networks method shapes the simultaneous advantages of both neural networks and novel ensemble-clustering notion, that improves the capability in carrying out diagnosis of the PV plant without historical knowledge about their past faults.

REFERENCES

[1] Linguet, Laurent & Atif, Jamal. A Bayesian approach for solar resource potential assessment using satelite images. 2014, IOP Conference Series, DOI: 10.1088/1755-1315/17/1/012171.

[2] Honglu Zhu et al. 2018. Fault diagnosis approach for photovoltaic arrays based on unsupervised sample clustering and probabilistic neural network model. doi.org/10.1016/j.solener.2018.10.054

[3] Baojie Li, Anne Migan-Dubois, Claude Delpha, Demba Diallo. Analysis of the Performance of the I-V Curve Correction Methods in the Presence of Defects. 37th European Photovoltaic Solar Energy Conference and Exhibition (EU PVSEC 2020), Sep 2020, Lisbon, Portugal. ffhal-02903352

[4] Strehl, A., Ghosh, J. 2002. Cluster ensembles—a knowledge reuse framework for combining multiple partitions. J. Mach. Learn. Res. 3, 583–617.

[5] Ayad, H.G., Kamel, M.S., 2010. On voting-based consensus of cluster ensembles. Pattern Recogn. 43, 1943–1953

[6] Khedairia Soufiane et al. 2022. A multiple clustering combination approach based on iterative voting Process. Computer and Information Sciences 34 (2022) 1370–1380.

[7] T Chakraborty, F Pierazzi, VS Subrahmanian. 2017. Ensemble clustering and classification for predicting android malware families - IEEE Transactions on Dependable and Secure.

[8] Shengyang Liu et al. 2019. Photovoltaic array fault diagnosis based on Gaussian Kernel Fuzzy C-Means clustering algorithm. Sensors 2019, 19, 1520; doi:10.3390/s19071520

[9] Muhammad Abbas and Duanjin Zhang. 2021. A smart fault detection approach for PV modules using adaptive Neuro-Fuzzy inference framework. Energy reports 7 (2021) 2962-2975

[10] Luciano da Fontoura Costa. 2022. A brief Guide to the coincidence similarity and its applications. DOI: 10.13140/RG.2.2.16322.25285/7

[11] M.K.Vijaymeena and K.Kavitha. 2016. A survey on similarity measures in text mining. Machine learning and applications. MLAIJ Vol.3, No.1, March2016

10.4229/EUPVSEC2023/4CV.1.27
020392-004

AUTHOR INDEX

40th European Photovoltaic Solar Energy Conference and Exhibition (EU PVSEC 2023)

Lisbon, Portugal
18-22 September 2023

Volume 3 of 3

ISBN: 979-8-3313-1553-5

Printed from e-media with permission by:

Curran Associates, Inc.
57 Morehouse Lane
Red Hook, NY 12571

Some format issues inherent in the e-media version may also appear in this print version.

WIP – Renewable Energies
Sylvensteinstr. 2
81369 Munchen
Germany

Phone: +49 89 72012735
Fax: +49 89 72012791

wip@wip-munich.de

Additional copies of this publication are available from:

Curran Associates, Inc.
57 Morehouse Lane
Red Hook, NY 12571 USA
Phone: 845-758-0400
Fax: 845-758-2633
Email: curran@proceedings.com
Web: www.proceedings.com

TABLE OF CONTENTS

VOLUME 1

VOLUME 2

VOLUME 3

ANALYSIS OF THE THERMAL BEHAVIOUR OF A MISMATCHED MODULE IN A STRING

M. Vumbugwa*, E.E. van Dyk, F.J. Vorster and J.L. Crozier McCleland,
Nelson Mandela University, PO Box 77000, Summerstrand, 6031, Port Elizabeth, South Africa, +27 41 504 4989,
monphias.vumbugwa@mandela.ac.za

ABSTRACT: Thermal and electrical measurements of photovoltaic (PV) modules and strings offer a guide to the state of a PV plant and its ability to produce electrical power. One of the most important issues that is revealed in these measurements is the level of mismatch between series-connected cells and modules that are operating in a PV plant. PV modules operate under changing operating conditions which can influence the results and interpretation of thermal and electrical characterisation measurements. This study analyses the dynamics of abnormal thermal signatures observed on thermal infra-red (TIR) images of a polycrystalline PV module that were captured at different operational current-voltage (I-V) points. Significant cell mismatch, which causes abnormal thermal signatures, occurs when the module's operational point is on the horizontal section of the I-V curve, close to short circuit current (I_{SC}). When the operational point of the module is on the vertical section of the I-V curve, close to open circuit voltage (V_{OC}) the effect of the mismatch is minimal, causing the bad cells to appear as good cells on TIR images. When a PV module underperforms, it is beneficial for the operational point voltage to be higher than maximum power voltage (V_{MP}), this will prevent the bad cells from becoming abnormally hot to cause detrimental effects such as module back sheet damage, cracks, risk of fire and increased module degradation rate. Not all series-connected cells or modules will operate at their maximum power points due to cell or module mismatch. A mismatched module can operate at a voltage lower than its V_{MP} and deliver a large current causing a significant current mismatch between good and underperforming cells. This results in most cell anomalies emerging easily and presenting themselves as abnormal thermal signatures, thus enhancing the observation of mismatched cells in TIR imaging. This study describes the dynamics of abnormal thermal signatures when the module's operational point changes and can improve TIR imaging inspection methods as well as the interpretation of results. In this way, the reliability and performance of crystalline PV modules operating in PV plants can be improved.

Keywords: Thermal infra-red imaging, operational current-voltage point, current mismatch, abnormal thermal signature

1 INTRODUCTION

Photovoltaic (PV) modules installed in the field generate electrical power under dynamic climatic conditions [1][2], which can stress the modules and cause cell anomalies that can impact on performance and longevity of the modules. Series connected cells in substrings, or modules in a string, are meant to generate and deliver equal power, however dynamic environmental conditions can cause a mismatch in the operation of the cells and modules. The dynamic conditions such as changes in soiling and shading levels among other things, influence the operational electrical characteristics of PV cells and modules in strings, and result in external cell mismatch. Inherent cell mismatch can occur when the series connected cells have non-uniform composition of the material. One of the main consequences of mismatch is the inability of cells to produce the same current with the result that underperforming cells are forced to operate at lower voltages, or even negative (reverse bias) voltages, resulting in performance degradation of the PV module [3]-[5]. PV module inspections such as current-voltage (I-V) measurements, electroluminescence (EL) and Thermal Infrared (TIR) imaging can reveal the quality and performance of the modules and hence can identify modules having mismatched cells [4][5]. Bad or underperforming cells can result in I-V curves having bumps, abnormal hot cells on TIR images and dark cells on EL images [4][5]. EL occurs in a PV cell when it is forward biased, receiving an electrical input and outputting an electromagnetic emission spectrum whose intensity is related to the cell defects. Non-defective cell will have bright light distribution while the defective cell or cell part appears dark [5][6]. Bumps on I-V curves reflect the activation of bypass diodes due to mismatched substrings [7]. Bypass diodes are incorporated in crystalline PV modules to minimise the mismatch effects [8], by diverting the module's current around the mismatched substring, thus protecting the poor-performing or bad cell from reaching reverse breakdown voltage and avoiding further degradation of the module. However, when the bypass diodes are activated, they cause the affected PV module to present multiple local I-V maxima points, but only one of them relates to the global maximum of the module string or array [9]-[12]. Any factor that can change the operational point of PV modules will influence the level of current mismatch between good and bad cells and cause a change in the I-V curve and the TIR image of a module, thus complicating the interpretation of these measurements. The factors can include, as previously stated, partial shading and change in load conditions. Shading due to soiling is dynamic and is unavoidable since PV modules are installed outdoors at an inclination for optimum capture of solar radiation and conversion efficiency. When the load conditions change, it causes the change in power (current and voltage) delivery from, as well as the operational current and voltage point of, the PV modules.

A bad cell emerges as a hot cell on TIR images when it gets reverse biased and dissipates the electrical power of good cells as heat, resulting in the cell operating at an elevated temperature. The use of TIR imaging is popular, especially in large PV plants since the inspection method is non-destructive and conducted in-situ [13]-[15]. High-quality TIR images and I-V measurements are captured under steady environmental conditions; clear sky, calm air flow, stable ambient temperature with irradiance above 600 W.m^{-2} [5][16]-[18]. TIR imaging still has the potential to become an essential inspection technique in operational PV modules, however, the TIR measurements can be influenced by a shift in the operating point on a cell, module or string. This paper focuses on the study of the behaviour of thermal characteristics of PV cells under partial shading and at different module's operational points along the I-V curve.

2 EXPERIMENTAL PROCEDURE

The dynamics of thermal signatures on a SolarWorld SW 240 poly module were studied by TIR imaging the polycrystalline PV module with and without partial shading on one cell, when the module was generating power at different operational voltages, 0V through V_{MP} to V_{OC}. The operational voltages were changed by varying the load resistor to observe the effect of changing the module's operational current-voltage points on the abnormal thermal signatures when cell 30 was unshaded and when the cell was under a shading level of 25% with the module inclined and receiving irradiance of the order 700-1000 W.m^{-2}. Figure 1 shows the inclined module on a supporting structure with opaque shading material of 25% on the front side of cell 30.

Figure 1: A polycrystalline silicon PV module, with partial shading of 25% on one cell, inclined for TIR imaging.

The TIR camera (FLIR T620) was positioned at about 4 meters (m) with a capturing angle of almost 15° to the normal of the targeted module's front surface, as recommended in the FLIR Thermal Imaging Guidebook [19]. TIR images were recorded under calm environmental conditions, and a thermal equilibrium time of 10 minutes was allowed after adjusting the operational voltage of the module. The I-V curves of the module were taken with and without the 25% partial shading on cell 30.

3 RESULTS AND DISCUSSION

The effects of shifting the module's operational point on the dynamics of the abnormal thermal signatures are discussed. EL images of the module were taken to check for any pre-existing cell anomalies which can cause inherent mismatch. Figure 2 shows the EL images of the module recorded at (a) high current (110% I_{SC} - High EL image) and (b) low current (10% I_{SC} - Low EL image). There are minor cracks on cells 1, 20, 47 and 52 which run diagonally across the cells. Cells 1, 20, 47 and 52 are enlarged, in Figure 2(c), (d), (e) and (f) respectively, to easily identify the cracks. All fingers of the cracked cells appear to be connected to the busbars and there are no dark or isolated cell parts, thus the cracks did not impact the performance of the module. There are no dark cells on the low EL image, hence no shunt paths or PID are present on the cells.

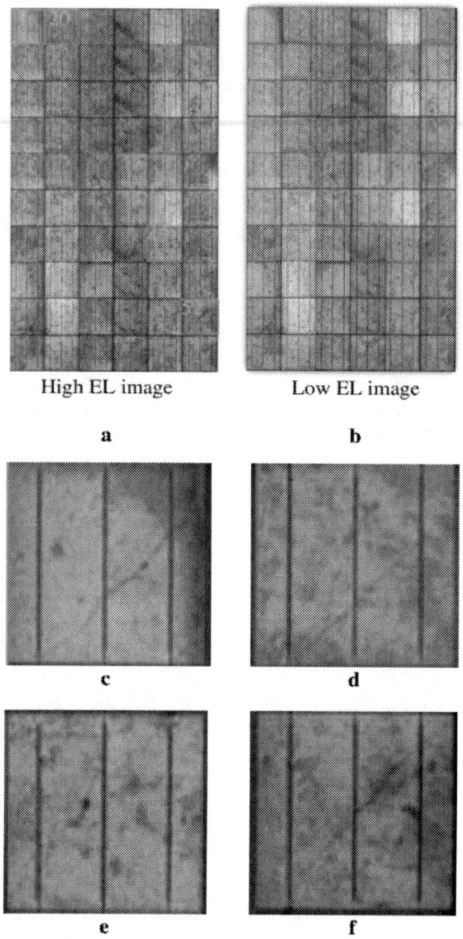

Figure 2: EL images of the polycrystalline PV module taken at (a) 110% and (b) 10% of the I_{SC}. Cells 1, 20, 47 and 52 are enlarged in (c), (d), (e) and (f) respectively.

Figure 3 shows TIR images that were captured when the module was generating power under different operational currents and voltage points without partial shading on cell 30. The TIR images are linked to the I-V curves of the module which were taken under no shading conditions. TIR image in Figure 3(a) shows several abnormal hot cells, in a patchwork pattern, on the TIR image taken when the module was unshaded and at operational point 0.2 V, 7.1 A, close to I_{SC}. The number of abnormally hot cells decreases when the module's operational point is changed, along the I-V curve, towards V_{OC}, as shown on TIR images (b), (c), (d), (e) and (f). As the operational point of the module shifts from I_{SC} through MPP towards V_{OC}, the abnormally hot cells disappear until all cells in the module appear to operate normally with uniform temperature at points closer to V_{OC} (26.0 V;6.7 A to 33.5 V; 0.5 A), as shown on TIR images (e) and (f). Cell 34 is highlighted in TIR images (a), (b), (c) and (d) and shows the reduction in its operational temperature as the module's voltage increases from 0.2 V to 25 V. Thus, the

10.4229/EUPVSEC2023/4CV.1.30
020393-002

operational temperature of the abnormally hot cells decreases with increase in the operational voltage of the module. The effects follow the decrease in the amount of current generated but are contrary to the increase in voltage supplied by the module and are indicated by the operating point or position of each TIR image along the I-V curve of the module. The abnormally hot cells on TIR image (a) can be a result of underperforming (bad) cells, due to cell material deformities, which are operating at MPP different from the normal cells. When the module's operational point is closer to I_{SC}, a significant current mismatch occurs between the normal and underperforming cells resulting in the bad cells appearing abnormally hot on the TIR image, Figure 3(a).

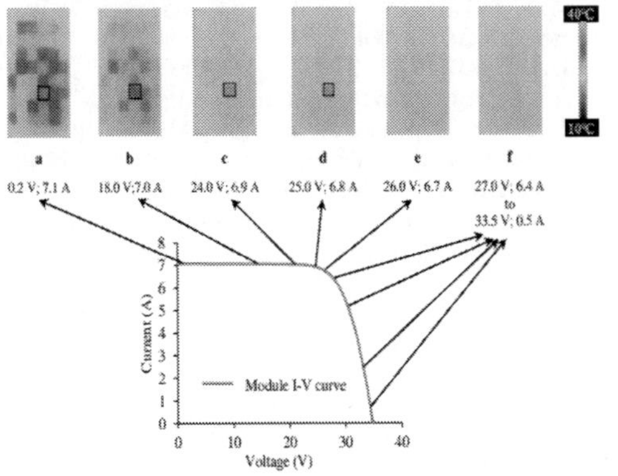

Figure 3: Thermal images of the polycrystalline silicon PV module captured at different operational current and voltage points of the module under no shading conditions.

Similar results, as shown in Figure 3, were discussed by Vumbugwa et al. 2022 [20], when module current and voltages of individual cells were monitored at different operational points of the module. To clarify these dynamics of hot cells, Vumbugwa et al. simulated the I-V curves of three cells that operated differently. Figure 4 shows the simulated I-V curves of the three cells, A, B and C. In the paper [20], cell A seemed normal, cell B emerged to be warm and cell C appeared abnormally hot on the TIR image taken when the module was operating under short circuit conditions. Figure 4 shows that a change in the module's current influences the operational points of individual cells. The change in the module's current results in a marginal shift in operational points of a good cell A, and a significant change in operational points of cells B and C. The operational voltage of cell B changes between positive and negative, while the voltage of cell C shifts significantly and stays negative when the operational points shift.

Figure 5 shows TIR images and the I-V curve of the polycrystalline PV module taken at different operational points when a shading level of 25% was on cell 30. The I-V curve shows a bump due to the partial shading, signifying the expected activation of a bypass diode on substring (SS2) having the shaded cell 30.

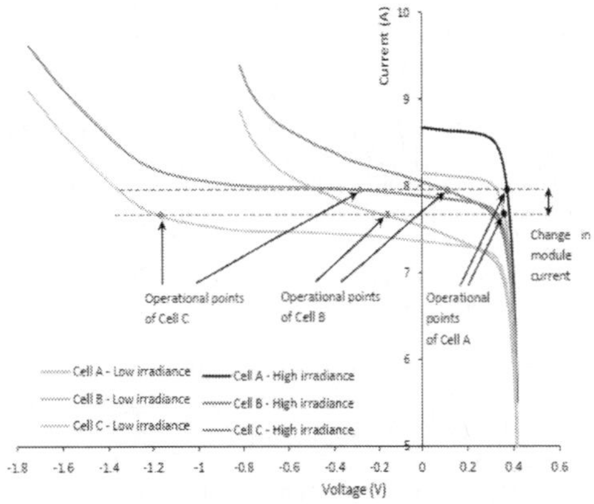

Figure 4: I-V curves of three representative mismatched cells A, B and C, indicating a change in operational points when module current changes [20].

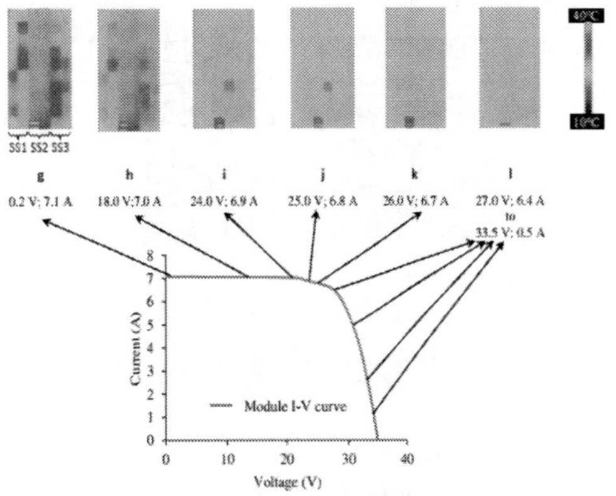

Figure 5: Thermal images of the polycrystalline silicon PV module captured at different operational current and voltage points of the module when cell 30 was under partial shading.

TIR image (l) in Figure 5 shows that the operational temperature of cell 30 is not influenced by the 25% shading, but a hot strip is seen on cell 30, on the respective TIR image, due to the difference in emissivity of the module glass (0.92) and the opaque shading material (paper: cardboard 0.81) [21]. The shading material radiates higher thermal energy than the module glass and hence appears hotter on the TIR image. The thermal image (k) was taken at operational point 26.0 V; 6.7 A and shows the whole cell 30 operating at an elevated temperature due to the 25% partial shading. At this point, the partial shading caused the cell to operate in reverse bias and significant current mismatch occurs resulting in the cell dissipating the module's generated electrical energy and the whole cell being abnormally hot. As the module's operational point shifts towards I_{SC} (25.0 V; 6.8 A to 0.2 V; 7.1 A), the temperature of cell 30 increases as shown on TIR images (j), (i), (h) and (g). Additionally, the bad cells appear abnormally hot and their number increases, as previously discussed. However, TIR images (g) and (h), at points 0.2 V; 7.1 A and 18.0 V; 7.0 A, show that the abnormally hot cells in substring SS2 which emerged under no shading conditions on TIR images (a) and (b) in

10.4229/EUPVSEC2023/4CV.1.30
020393-003

Figure 3, appear as normal operating cells. The partial shading applied on cell 30 influenced current mismatch between the good and bad cells in substring SS2 and the results on TIR images (g) and (h). The bypass diode on substring SS2 diverts the module current around the shaded substring SS2 to circumvent the heating effect on the partially shaded cell 30. When the bypass diode conducts the module's current it becomes abnormally hot and is seen on the TIR images (g) and (h). When the bypass diode is activated, a minimal current mismatch occurs between bad and good cells in substring SS2, resulting in the bad cells appearing as normal operating cells on TIR images (g) and (h). Thus, as previously discussed, a current mismatch between good and bad cells can easily occur when the module's operational voltage is less than V_{MP} and closer to 0 V than when the voltage is greater than V_{MP} and closer to V_{OC}.

Other researchers, Vumbugwa et al. 2023, [22] showed similar results where bad cells would not show their abnormal thermal signatures under partial shading conditions. Figure 6 shows the thermal images of a module with (a) no soiled cells and (b) one soiled cell in each of the three substrings.

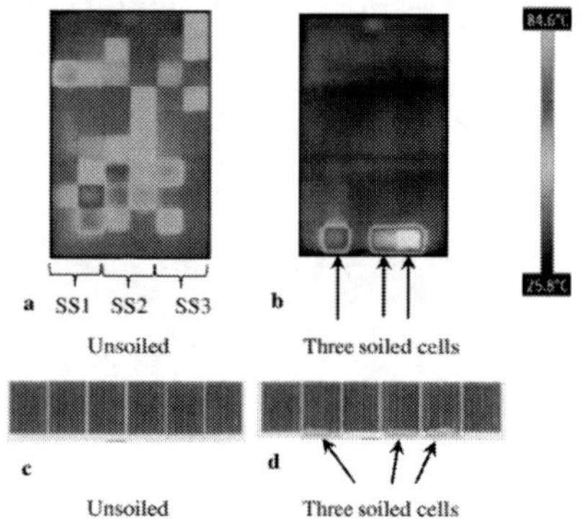

Figure 6: Thermal images of a polycrystalline module recorded under short circuit conditions with (a) no soiled cells and (b) three soiled cells in each of the three substrings. The enlarged visual images, (c) and (d), show the soiling conditions of the cells [22].

The level of soiling, < 10% [22], on the three cells is shown in visual images in Figure 6(c) and (d). The soiled cells in substrings SS1, SS2 and SS3, Figure 6(d), become abnormally hot on the TIR image shown in Figure 6(b) and the rest of the cells in the module appear normal on the TIR image regardless of the anomalies highlighted on the TIR image in Figure 6(a). Partial soiling can therefore mislead the interpretation of TIR images by hiding inherent cell anomalies. These results were obtained under short circuit conditions; however, this scenario is unlikely to occur under real field conditions. Nevertheless, as discussed before, the operational point of the module will change due to a change in field conditions, thus the module will not always operate at its MPP.

PV modules operating in a string under real field conditions will behave similarly to cells in a substring. The modules in a string will not contribute equal power since they will operate at different MPPs. If the operational point of a mismatched module is less than its MPP, then the bad cells in the module will easily emerge on TIR images. The bad cells will not present their abnormal thermal signatures but will appear as good cells when the mismatched module's operational point is higher than its MPP. Therefore, when the module is TIR imaged while the module's operational point is < MPP, then the mismatched cells will easily emerge as hot cells on TIR images unlike when imaged while at operational point > MPP.

4 CONCLUSIONS

The operational conditions of the modules, such as changes in load conditions and partial shading, can influence the operational current and voltage points of the module and individual cells, and can result in the dynamic behaviour of the abnormal thermal signatures. The shift in the operational point of a mismatched module operating in a string will affect the level of current mismatch between good and bad cells and when a bad cell emerges as a hot cell on TIR images. Partial soiling influences the current mismatch and dynamics of hot cells in the substring containing shaded cells and can mislead decisions during TIR imaging. When the operational point of a mismatched module is higher than the module's MPP, closer to V_{OC}, then mismatch will be minimal for bad cells to appear abnormally hot on TIR images. Therefore, taking TIR images on clean PV modules operating closer to short circuit conditions is beneficial as bad cells can certainly be identified. This paper can contribute to the operations and maintenance of PV modules in the field when TIR imaging inspections are carried out for long-term reliability and an improved return on investment of the PV modules.

REFERENCES

[1] A. Singla, K. Singh, and V. K. Yadav, "Environmental effects on performance of solar photovoltaic module," in *2016 Biennial International Conference on Power and Energy Systems: Towards Sustainable Energy (PESTSE)*, Bengaluru, 21-23 January 2016, IEEE 2016, pp. 1–6, doi: 10.1109/PESTSE.2016.7516480.

[2] C. Ferrara and D. Philipp, "Why do PV modules fail?," *Ener. Proc.*, vol. 15, pp. 379–387, 2012, doi: 10.1016/j.egypro.2012.02.046.

[3] S. Bowden and C. Honsberg, "PVCDROM | PVEducation." [Online]. Available: http://pveducation.org/pvcdrom, (accessed 20 August 2023).

[4] M. Köntges, S. Kurtz, C.E. Packard, U. Jahn, K. Berger, K. Kato, T. Friesen, H. Liu and M. Van Iseghem, (2014). Review of failures of photovoltaic modules, IEA-Photovoltaic Power Systems Programme, IEA-PVPS, 2014. https://iea pvps.org/wp content/uploads/2020/01/IEAPVPS_T13 01_2014_R eview_of_Failures_of_Photovoltaic _Modules_Final.pdf.

[5] U. Jahn, M. Herz, M. Köntges, D. Parlevliet, M. Paggi, I. Tsanakas, J. S. Stein, K. A. Berger, S. Ranta, R. H. French, M. Richter and T. Tanahashi, (2018). Review on Infrared and Electroluminescence Imaging for PV Field Applications, IEA-Photovoltaic Power Systems Programme, IEA-PVPS. https://iea-pvps.org/wp content/uploads/2020/01/Review_on_IR _and_EL_Imaging_for_PV_Field_Applications_by_ Task_13.pdf.

[6] J. L. Crozier, E. E. van Dyk, and F. J. Vorster, "Evaluating the Relationship Between Electroluminescence (EL) Imaging and the Power Output of Photovoltaic Modules," in *3rd South.*

African Sol. Energy Conf. Proc., Kruger National Park, 11-13 May 2015, SASEC, 2015, pp. 24–28.

[7] Solmetric Corporation, "Guide To Interpreting I-V Curve Measurements of PV Arrays," 2011. http://resources.solmetric.com/get/Guide to Interpreting I-V Curves.pdf (accessed 17 August 2023).

[8] S. Silvestre, A. Boronat, and A. Chouder, "Study of bypass diodes configuration on PV modules," *Appl. Energy*, vol. 86, no. 9, pp. 1632–1640, 2009, doi: 10.1016/j.apenergy.2009.01.020.

[9] S. Bowden and C. Honsberg, PVCDROM | Bypass Diodes | PVEducation, [Online]. Available: https://www.pveducation.org/pvcdrom/modules-and-arrays/bypass-diodes, (Accessed 09 April 2022).

[10] R. G. Vieira, F. M. U. de Araújo, M. Dhimish and M. I. S. Guerra, "A Comprehensive Review on Bypass Diode Application on Photovoltaic Modules," *Ener.*, 13(10), 2472, 2020, doi: 10.3390/en13102472.

[11] B. B. Pannebakker, A. C. de Waal and W. G. J. H. M. van Sark, "Photovoltaics in the shade: one bypass diode per solar cell revisited," *Prog. in Phot.*, 25(10), 836-849, 2017, doi: 10.1002/pip.2898.

[12] Solaredge, Technical Note Bypass Diode Effects in Shaded Conditions, [Online]. Available: https://www.solaredge.com/sites/default/files/se_technical_bypass_diode_effect_in_shading.pdf (accessed 09 April 2022).

[13] Tyutyundzhiev, N., Lovchinov, K., Moreno, F. M., Leloux, J., Narvarte, L., 2015. Advanced PV modules inspection using multirotor UAV. In: 31st EUPVSEC Proceedings, Hamburg 14-18 September 2015.

[14] Aghaei, M., Quater, P. B., Grimaccia, F., Leva, S., Mussetta, M., 2014. Unmanned Aerial Vehicles in Photovoltaic Systems Monitoring Applications. In: 29th European Photovoltaic Solar Energy Conference (EUPVSEC) Proceedings, Amsterdam, 22-26 September, 2014. https://doi.org/10.4229/EUPVSEC20142014-5BV.1.45.

[15] Tsanakas, J. A., Ha, L., Buerhop, C., 2016. Faults and infrared thermographic diagnosis in operating c-Si photovoltaic modules: A review of research and future challenges. Renew. Sust. Ener. Rev. 62, 695–709. https://doi.org/10.1016/j.rser.2016.04.079.

[16] Buerhop, C., Pickel, T., Dalsass, M., Scheuerpflug, H., Camus, C., Brabec, C.J., 2016. aIR-PV-check: A quality inspection of PV-power plants without operation interruption. In: 43rd Photovoltaic Specialist Conference (PVSC), Proceedings, Portland, 5-10 June 2016. https://doi.org/10.1109/PVSC.2016.7749909.

[17] Vollmer, M., Möllmann, K.-P., 2018. Infrared Thermal Imaging, Fundamentals, Research and Applications, 2nd ed. WILEY-VCH 1708.

[18] IEC TS 62446-3 Ed.1, 2017. Photovoltaic (PV) systems - Requirements for testing, documentation and maintenance - Part 3: Photovoltaic modules and plants - Outdoor infrared thermography.

[19] FLIR Systems, "Thermal Imaging Guidebook for building and renewable energy applications," 2011. [Online]. Available: http://www.flirmedia.com/MMC/THG/Brochures/T820325/T820325_EN.pdf. (Accessed: 15 April 2022).

[20] Vumbugwa, M., Vorster, F. J., Crozier McCleland, J. L., van Dyk, E. E., 2022. Effects of changing partial cell shading on the electrical and thermal characteristics of crystalline silicon photovoltaic module. Sol. Energy 240, 147-156. https://doi.org/10.1016/j.solener.2022.05.031.

[21] FLUKE, "Emissivity values of common materials," 2007. [Online]. Available: https://www.bergeng.com/mm5/downloads/fluke/Emissivity-Values-of-Common-Materials-Chart.pdf. (Accessed 12 October 2021).

[22] Vumbugwa, M., Vorster, F. J., Crozier McCleland, J. L., van Dyk, E. E., "Common misinterpretations of thermal signatures on polycrystalline PV modules under different operational conditions. Sol. Energy 263 (2023) 111957. https://doi.org/10.1016/j.solener.2023.111957.

HOME ASSISTANT: A VALUABLE TOOL FOR ENERGY MANAGEMENT IN SELF-CONSUMPTION PV SYSTEMS

M. Alonso Abella
CIEMAT
Avda. Complutense 40, 28040 Madrid (Spain). E-mail:miguel.alonso@ciemat.es

ABSTRACT: Home Assistant, HA, is a free open source software platform oriented to home automation. It can be easily installed and configured. A wide range of Internet of Things, IoT, devices can be directly incorporated without no need of code programming. Home Assistant can manage most of commercial inverters among other house power management and control devices, mainly having wifi control, as switches, air conditioning, aerothermal heating-cooling units. The operation all these home consumptions can be optimized through Home Assistant automations.
Keywords: Home Assistant, energy management, self-consumption.

1 INTRODUCTION

Home Assistant [1], also known as HA, HAS or Hassio, is a free and open-source [2] python software designed to be a central control system for smart home devices, with a focus on local control and privacy. It can be accessed through a web-based user interface using companion apps for Android and iOS, or by voice commands via supported virtual assistants such as Google Assistant or Amazon Alexa.

Home Assistant can be installed in a wide range of platforms, Mac, Intel, Rasberry Pi, and operating systems, windows, linux. Devices using different IoT technologies can be integrated in Hassio. Bluetooth, Zigbee, Z-Wave, Serial, I2C, SMBus, UART or USB ports can be used to communicate. Also wifi communication between IoT (Internet of Things) devices and Hassio is used in different ways. Directly to local devices or indirect through hubs/gateways/bridges, or cloud services.

The web interface, or front-end dashboard, see Figure 6, offers different cards to display information and control devices. The user has to build its own interface. All work is almost visual or with a simple Hassio code, called YALM code [3]. Additional resources created by community members are available [4].

Home Assistant is the central controller combining in a single software all home IoT devices named as entities. Custom script routines based on trigger events, conditions and actions allows the home automations.
Most PV inverters can be integrated with Hassio via local wifi, manufacturer cloud, or direct serial RS485 communication [5]. PV installation data, such as voltage, current, power, and energy values, can be easily managed using specific add-ons developed by the community (some examples can be found in [6][7][8][9][10]). Home energy management, including energy measurement and device on/off control, is officially supported by Hassio[11]. A wide range of power monitoring IoT devices [12], such as Sonoff [13], Tasmota [14], e-welink or Tuya Smartlife [15] devices, can be used. An increasing number of appliances have integrated wifi management. Air conditioning with heat pumps or aerothermal units (air/water), and water heater tanks for domestic hot water can be integrated in Hassio. Other home appliances can also be controlled for optimal energy management. The hourly energy price can usually be accessed online.

Hassio includes the ability to program automations in response to events, performing energy management. Air conditioning or car charging can be switched on/off based on available solar energy. Optimal energy and economic management can also be achieved. An online daily, weekly, monthly, and yearly economic balance between energy purchased/sold from/to the grid is available, and decisions can be programmed. For example, the air conditioning can be turned on during the night hours until the economic balance is zero, in order to take advantage of excess daytime energy.

2 HOME ASSISTANT AND PV SYSTEMS

This work is based on a 8.19 kWp PV self-consumption system installed in a private home. It composed by 18 half-cells PV modules model JAM72S20-455/MR and a Huawei SUN200-6KTL-L1 hybrid PV inverter. No batteries are installed. Results here presented to this specific inverter manufacturer, taken as example, can be extended to any others. Home Assistant is installed in an Oracle free cloud server [16] running Ubuntu. Also it could be installed almost in any machine, i.e. locally in a PC, mini-PC, Raspberry Pi or VirtualBox running in a local PC or notebook.
Figure 1 shows the general configuration of the PV system and different IoT sensors and energy meters. The inverter has its own wattmeter measuring grid energy in order to control inverter grid injection. Two additional wattmeters have been installed. One at the inverter output, with wifi and *Tuya* [15] controlled. The other installed to measure the grid interchange, a bidirectional SDM230 model. A ESP8266 controller with a RS485 TTL board connected to SDM230 RS4845 port is used to read MODBUS data (some libraries can be found in *github*, i.e.[17] and being adapted to a developed specific software running in the ESP8266) and send it to an *emoncms* [18] site running in a Oracle cloud server.

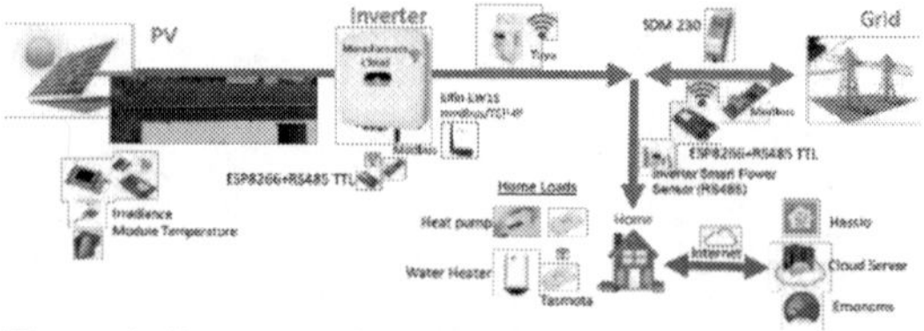

Figure 1. General configuration, PV system and IoT wifi sensors.

10.4229/EUPVSEC2023/4CV.1.33
020396-001

MODBUS TCP can be used to read inverter data in Home Assistant. Other options are available, but in this case it was used an ESP8266+RS485 TTL board directly connected to the RS485 inverter port. Data is also uploaded to the emoncms cloud server.

ESP8266 controllers can be easily used to develop own cheap IoT sensors. Irradiance and PV modules and ambient temperature is monitored using an ESP8266 microcontroller [19]. The ADS115 [20, 21] offers 4 analogue inputs of 16 bits resolution to read the mV output a PV cell with a shunt, in order to monitor irradiance incident on the PV generator plane. DS18B20 [22] with one-Wire sensor is used to monitor temperatures. In the same way that the inverter and SDM230 data is sent to an *emoncms* [23] cloud server. The server can be local (i.e. raspberry pi or Linux Ubuntu micro-pc) or even be in a cloud (Oracle, Amazon AWS,..). In this case simple oracle cloud free tier [24] Ubuntu virtual machines of 1 GB memory, 50 GB disk, are enough for running both, the *emoncms* server and the Home Assistant instance.
Home Assistant has an *emoncms* integration [25] to read directly the data.

To connect the inverter to Home Assistant other options are available, for example, using an Elfin-EW11 modbus/TCP-IP converter [26], with direct TCP-IP communication with the inverter or dataloguer [27,28] from Home Assistant server.

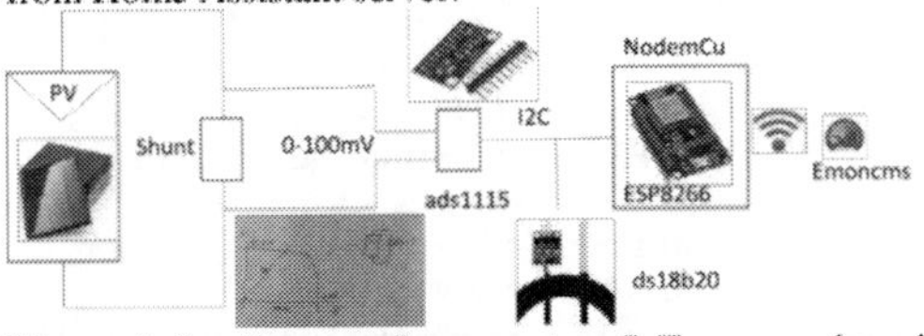

Figure 2. Irradiance and temperature IoT sensors, based on the ESP8266 controller, the 16bits ADS1115 board and the DS18B20 1-Wire temperature sensor.

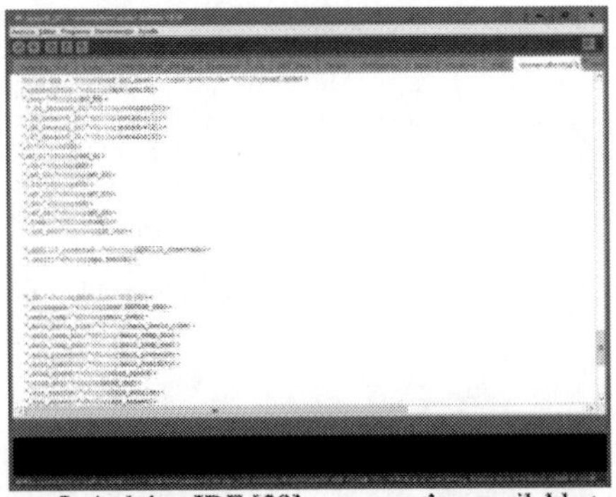

Figure 3. Arduino IDE [29], among other available tools, can be used to upload the control program code to the ESP8266 NodeMCU board to monitor the irradiance and temperature data.

Once all the sensors and devices data and control is available in Home Assistant, different needed automations can be easily programmed in function of the user energy needs and economic optimization. For example, home loads as the water heater, air conditioner heat pumps,... can be turn on/off in function of the levels of available PV power.

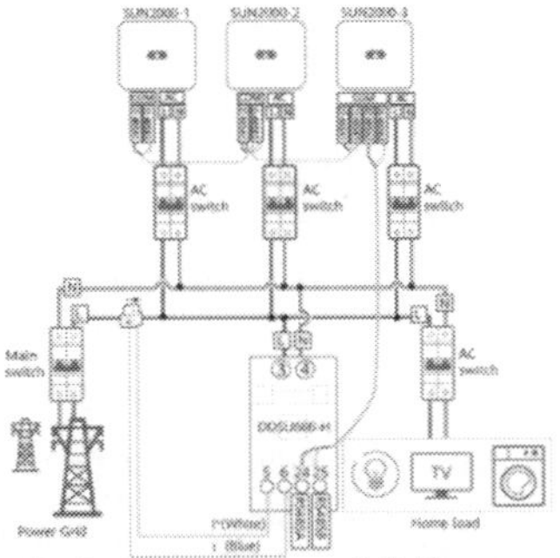

Figure 4. Typical installation of Self-consumption PV inverters and grid power-meter for control. Source: Huawei SUN2000 [30].

If the home has a regulated PVPC [31] electricity tariff contract with an electric company and hourly data can be online [32] accessed through an API. Each home case can be configured and adapted to the own specific economic contract with the electrical marketer company. There are companies that offers to its clients virtual wallets, also known as *virtual batteries*, where the money is stored along a time (year or more), when excess production, to be used in low energy periods. Monetary production exceeds, wallets, can be stored in summer to be used in wintertime. Figure 5 show the energy bill corresponding to a summer month. A total of 147.17€ are generated, corresponding to 1337.88 kWh paid at 0.11€/kWh. Because of most of the energy consumed in the house is generated (and self-consumed) by the PV system, the monthly energy consumed from the grid was only 122.26 kWh (8.44 kWh in peak hours, 14.44 in flat hours and 57.82 in valley hours. Peak, flat and valley hors are regulated in legislation [33]). The monthly energy bill was 13.34€, that are discounted from the solar wallet. Finally zero € are paid and 136,16€ are added to the general wallet.

Detalle de la Factura

	Unidades	Precio Unitario	Importe
Potencia (2023-05-31 a 2023-06-30)			**10,71 €**
Punta	3,450 kW * 30 días	0,085981 €/kW/día	8,90 €
[illegible]	[illegible]	[illegible]	[illegible]
[illegible]	[illegible]	[illegible]	[illegible]
[illegible]	[illegible]	[illegible]	[illegible]
Valle	3,450 kW * 30 días	0,017401 €/kW/día	1,81 €
[illegible]	[illegible]	[illegible]	[illegible]
[illegible]	[illegible]	[illegible]	[illegible]
[illegible]	[illegible]	[illegible]	[illegible]
Energía Activa (2023-05-31 a 2023-06-30)			**0,00 €**
Punta	8,44 kWh	0,211236 €/kWh	1,79 €
[illegible]	[illegible]	[illegible]	[illegible]
[illegible]	[illegible]	[illegible]	[illegible]
[illegible]	[illegible]	[illegible]	[illegible]
Llano	12,44 kWh	0,159999 €/kWh	1,99 €
[illegible]	[illegible]	[illegible]	[illegible]
[illegible]	[illegible]	[illegible]	[illegible]
[illegible]	[illegible]	[illegible]	[illegible]
Valle	57,82 kWh	0,124903 €/kWh	7,23 €
[illegible]	[illegible]	[illegible]	[illegible]
[illegible]	[illegible]	[illegible]	[illegible]
[illegible]	[illegible]	[illegible]	[illegible]
Mecanismo de ajuste			0,00 €
[illegible]	[illegible]	[illegible]	[illegible]
[illegible]	[illegible]	[illegible]	[illegible]
[illegible]	[illegible]	[illegible]	[illegible]
Excedente de energía			
[illegible]	[illegible]	[illegible]	[illegible]
[illegible]			[illegible]
Otros conceptos			**1,95 €**
Bono Social	30 días	0,038455 €/día	1,15 €
Alquiler de Equipos	30 días	0,026667 €/día	0,80 €
Impuestos			**0,68 €**
Impuesto Eléctrico	10,71 €	0,50 %	0,05 €
IVA (REDUCIDO)	12,71 €	5,00 %	0,63 €
Total factura			**13,34 €**
Uso Solar Wallet			**-13,34 €**
Total a pagar			**0,00 €**

Figure 5. Example of summer month energy bill.

10.4229/EUPVSEC2023/4CV.1.33

Figure 6. Example of Home Assistant web interface.

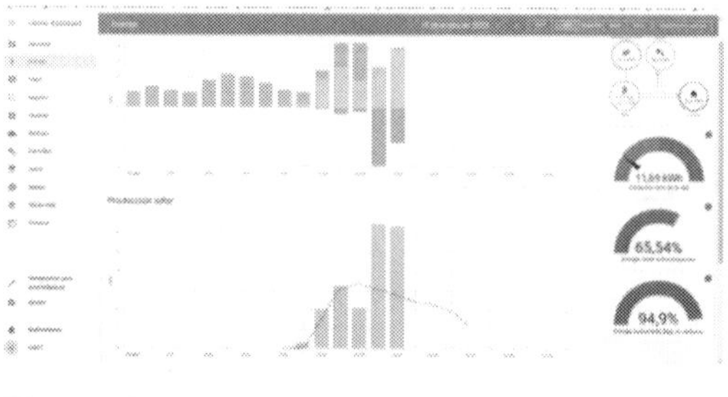

Figure 7. Energy balances.

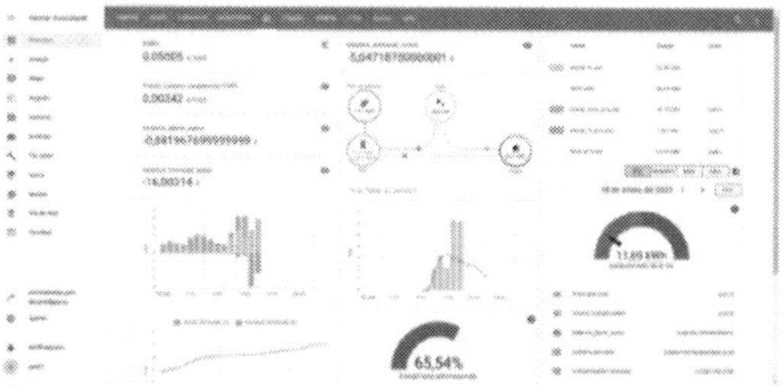

Figure 8. Energy and economic data.

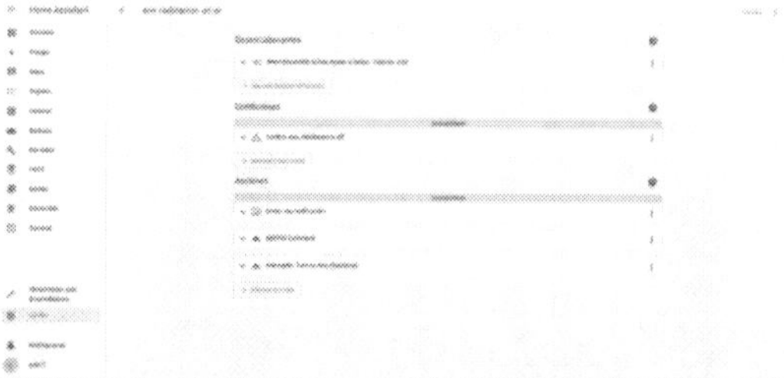

Figure 9. Example of Home Assistant automations programming.

Figure 10, data directly obtained from the inverter monitoring software [34], shows the results of Home Assistant automations acting as energy management system or Active load control in two winter days, cloudy an sunny. When enough PV power is available, two heat pumps are sequentially switched-on in order to heat in wintertime, or cold in summer time, the home.

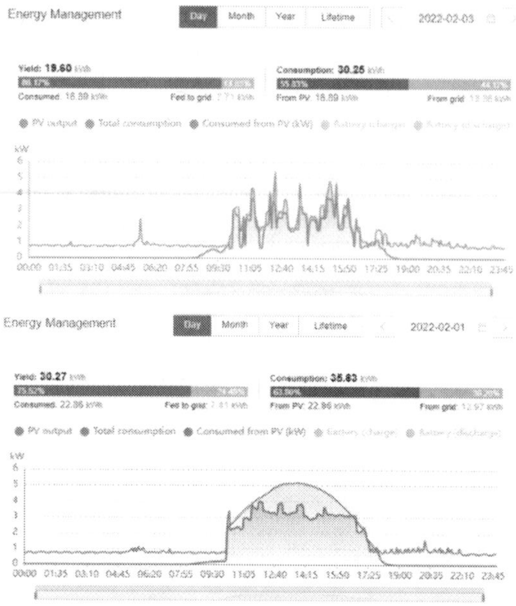

Figure 10. Example system loads management in a winter cloudy and sunny days, heat pumps are switched on when available PV power.

4. PV ENERGY PRODUCTION MODELLING

It is possible to use theoretical modelled that allow the estimation of the voltage, current and power values at MPP I-V curve of a PV generator from the measured values of the Irradiance, *G*, and cell temperature, *Tc*. *Isc*, Voc, and Pm values of the PV generator at STC conditions (*G**, *Tc**) should also be known.

In the operation of an inverter connected to a real PV array, the input DC voltage is not constant, it varies along the day in function of the irradiance and temperature conditions. During the test, PV module irrradiance and temperature are acquired simultaneously with the rest of the operation electrical parameters in order to estimate the theoretical values of MPP voltage, current and power using basic equations (1) to (11) [35], where * refers to STC conditions and α, β and δ art the temperature coefficients of *Isc*, *Voc* and *Pm*, respectively. It should be mentioned that *Vm* has a stronger decreasing dependence with the cell temperature increase than *Voc*, as predicted by equation (4) and experimentally confirmed [36].

$$I_{sc} = \frac{G}{G^*} I_{sc}^* + \alpha\left(T_c - T_c^*\right) \quad (1)$$

$$V_{oc} = V_{oc}^* + m v_t ln\left(\frac{G}{G^*}\right) - \beta\left(T_c - T_c^*\right) \quad (2)$$

$$V_{oc} = V_{oc}^* + \left(\frac{E_g}{q} - V_{oc}^*\right)\left(1 - \frac{Tc}{T_c^*}\right) + m v_t \left[\ln\left(\frac{G}{G^*}\right) - \xi \ln\left(\frac{T_c}{T_c^*}\right)\right] \quad (3)$$

$$V_m \cong V_{oc} - v_t \gamma Ln\left(\frac{\gamma}{1-\gamma}\right) - I_{sc} R_s\left[1 - (1-\gamma)^{\gamma}\right] \quad (4)$$

$$I_m \cong Isc\left[1 - (1-\gamma)^{\gamma}\right] \quad (5)$$

$$\gamma = \left(\frac{V_{oc} - 2R_s I_{sc} + v_t}{V_{oc} - 2R_s I_{sc} + 2v_t}\right) \quad (6)$$

$$v_t = \frac{KT_c}{q} \quad (7)$$

$$P_m = P_m^* \frac{G}{G^*}\left[1-\delta\left(T_c - T_c^*\right)\right] \quad (8)$$

Pm can be calculated using the simplified model of equation (8), only valid if constant PV modules efficiency was assumed, higher error at low irradiance values, of as the product $Pm(\gamma)=ImVm$. Other model, FF model, can be obtained using equations (9) to(11). FF model is valid for irradiances higher than 1000 W/m², but also shows errors when estimating MPP at low irradiances.

$$P_m = I_{sc}V_{oc}FF \quad (9)$$

$$FF = \left(1-\frac{mv_t}{V_{oc}}\right)\left(1+\frac{mv_t}{V_{oc}}\ln\left(\frac{mv_t}{V_{oc}}\right)-\frac{I_{sc}R_s}{V_{oc}}\right) \quad (10)$$

$$R_s = -\left(\frac{\partial I}{\partial V}\right)^{-1}_{V=V_{oc}} - \frac{mv_t}{I_{sc}} \quad (11)$$

Power estimation from measured irradiance an module temperature can be used as an scada tool to advice and program an alarm if difference is out a defined range. It can be also combined with forecast services [37] available for HA,

Figure 11 shows an example of the PV power measured and the estimated value using the simplified model.

Figure 11: Example of PV power and simplified theoretical model, Eq.(15).

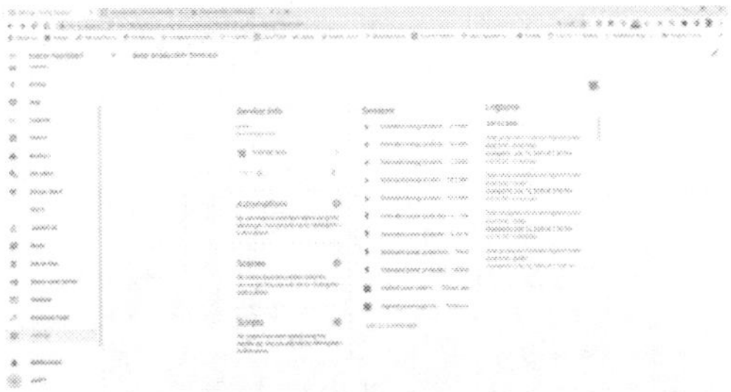

Figure 12: Forecast.Solar service can be configured in HA. Hourly and daily (today and tomorrow) energy production estimations are provided.

4. CONCLUSIONS

Energy management in self-consumption PV systems can improve its energetic and economic efficiency. Home Assistant is a valuable software to house control and monitor. Home Assistant allows straight integrations of PV self-consumption systems for monitoring, but also for energy management through its automations, leading to an optimal economic performance.

5. REFERENCES

[1] https://en.wikipedia.org/wiki/Home_Assistant
[2] https://github.com/home-assistant
[3] https://www.home-assistant.io/docs/configuration/yaml/
[4] https://hacs.xyz/
[5] https://domotica.solar/inversor-solar-home-assistant
[6] https://github.com/wlcrs/huawei_solar
[7] https://www.home-assistant.io/integrations/fronius/
[8] https://github.com/MatterVN/ModbusTCP2MQTT
[9] https://www.home-assistant.io/integrations/growatt_server/
[10] https://www.home-assistant.io/integrations/sma/
[11] https://www.home-assistant.io/blog/2021/08/04/home-energy-management/
[12] https://www.home-assistant.io/integrations/#all
[13] https://sonoff.tech/products/
[14] https://tasmota.github.io/docs/
[15] https://www.tuya.com/solution/hardware/electrical
[16] https://www.oracle.com/cloud/free/
[17] https://github.com/reaper7/SDM_Energy_Meter
[18] https://emoncms.org
[19] https://es.wikipedia.org/wiki/ESP8266
[20] https://es.aliexpress.com/w/wholesale-ads1115.html
[21] https://github.com/addicore/ADS1115
[22] https://randomnerdtutorials.com/esp8266-nodemcu-mqtt-publish-ds18b20-arduino/
[23] https://docs.openenergymonitor.org/
[24] https://www.oracle.com/es/cloud/free/
[25] https://www.home-assistant.io/integrations/emoncms/
[26] https://www.amazon.com/-/es/Elfin-EW11-Elfin-EW11-0-Dispositivos-inal%C3%A1mbricos-servidor/dp/B0833WD8H8
[27] https://github.com/StephanJoubert/home_assistant_solarman/
[28] https://github.com/wlcrs/huawei_solar
[29] https://www.arduino.cc/en/software
[30] https://support.huawei.com/enterprise/es/doc/EDOC1100129799
[31] https://www.ree.es/es/actividades/operacion-del-sistema-electrico/precio-voluntario-pequeno-consumidor-pvpc
[32] https://api.esios.ree.es/indicators/1739. https://www.esios.ree.es/es/analisis
[33] https://www.boe.es/diario_boe/txt.php?id=BOE-A-2020-1066
[34] https://eu5.fusionsolar.huawei.com/
[35] Luque A. Solar cells and optics for photovoltaic concentration. CRC Press, 1989.
[36] M. Alonso-Abella, F. Chenlo. Testing microinverters according to EN 50530. Proc. 29th European Photovoltaic Solar Energy Conference and Exhibition, Amsterdam 2014; 3104 – 3109.
[37] https://forecast.solar/. https://www.home-assistant.io/integrations/forecast_solar.

10.4229/EUPVSEC2023/4CV.1.33
020396-004

A GIS Application for Managing a Photovoltaic Plant: First advances for the EU PVSEC 2023

Andressa de Sousa Cardoso, María Isabel Ramos Galán, Juan Manuel Jurado Rodríguez and Francisco Feito Higueruela.
University of Jaén
Campus Las Lagunillas s/n, 23071 Jaén
(+34) 632993717

ABSTRACT: Currently, photovoltaic (PV) energy is considered a promising alternative as a clean energy source. Spain has a strong commitment to renewable energy, particularly photovoltaics (PV). As of 2023, the country has a significant amount of installed PV capacity, and many new projects are being developed and implemented. The Spanish government offers incentives and subsidies for solar energy development, making it an attractive market for investment. Overall, PV is expected to play an important role in Spain's energy mix in the coming years. However, most PV plants are hundreds or even thousands of hectares in size and are costly and time-consuming to maintain. Thermography with drones in photovoltaic installations allows effective monitoring, saving time and costs, detecting possible anomalies in the photovoltaic modules. As a result of the flights, various geospatial data are generated. On the other hand, there are Geographic Information Systems (GIS) that allow the integration, visualization and analysis of spatial and non-spatial data to support decision-making processes. Thus, the objective of this article is to present a useful methodology for the inspection and maintenance of a photovoltaic plant, using geomatics tools. It deals with the first advances in the application of this methodology to a specific case study: a photovoltaic plant in Jaén. The drone flight used a camera with two sensors: thermographic and RGB. These data were processed, and cartographic products were generated, these were geoprocessed through a GIS, then loaded into a database, and explored through a web map viewer.
Keywords: Solar Energy, Geographic Information System, UAV, Web Map.

1 INTRODUCTION

In 2018, the Intergovernmental Panel on Climate Change (IPCC) emphasized that global warming had already exceeded 1.5 °C above pre-industrial levels. The report underscored the urgent need for a worldwide effort to curb greenhouse gas emissions, mitigate the risks of climate change and poverty, and promote sustainable development [1].

The transition to renewable energy sources is a critical step in mitigating these problems and ensuring a sustainable energy future. Clean or renewable energies offer a compelling opportunity due to their limitless availability and their ability to mitigate the release of widely recognized greenhouse gasses into the atmosphere, significantly contributing to the fight against global warming [2].

Photovoltaic (PV) energy has emerged as a prominent contender in this shift due to its eco-friendliness and growing economic viability. Following a period of stagnation that extended until around 2018, Spain has embarked on an ambitious investment initiative in photovoltaic energy [3]. This commitment arises from the goal of transitioning to a carbon-free energy system by 2050, the consistent reduction in technology costs, and the favorable solar irradiance levels across its land [4].

Spain, in particular, has shown a strong commitment to renewable energy, with a focus on PV installations. Currently, the Spanish Government has enacted Royal Decree Law 23/2020, of June 23 [5], with the purpose of promoting the transition of the Spanish energy system towards climate neutrality. The main objective of this decree is to promote renewable energies as a fundamental element in achieving this objective, in addition to contemplating measures in other areas for economic reactivation [6]. As of 2023, the country boasts a substantial installed PV capacity, driven by government incentives and subsidies, making it an attractive destination for PV investment. Spain is one of the countries with the highest radiation index, which makes it an ideal territory for the construction of photovoltaic plants [2]. This trend is expected to further shape Spain's energy landscape in the coming years.

Despite the promising prospects of PV energy, the maintenance of PV plants, which often span hundreds or thousands of hectares, presents significant challenges in terms of efficiency and cost-effectiveness.

Once a solar plant is in operation, damaged and overloaded cells can induce irregular heat generation. Later, the whole module may short-circuit, which will reduce the overall production. This is due to the fact that all PV modules in a solar plant receive the same amount of radiation, and those modules or cells that cannot efficiently convert photons into electricity end up dissipating this energy in the form of heat. Despite this, defects and losses can occur in solar panels caused by various factors [7].

PV systems that have been in operation for several years often exhibit critical behavior, with failures that can negatively affect the performance and energy production of power plants. Therefore, these systems can derive great benefits from new technologies and tools used to monitor and control solar plants [8]. To address these challenges, innovative approaches are required, and this paper presents a case study focusing on the use of Geographic Information Systems (GIS) and drone-based thermography as a solution.

Drones have the capability to fly over large areas quickly and accurately, capturing high-resolution images of solar panels [7]. The RGB cameras on drones capture conventional visual images, enabling the identification and visual analysis of solar panels. Through image processing algorithms and computer vision techniques, automatic detection of solar panels in images is possible, even in large solar fields where scale and complexity make human analysis challenging [8]. Many studies have shown the potential of automating the inspection process,

10.4229/EUPVSEC2023/4CV.1.37
020398-001

as the speed at which inspections are conducted generates a large number of images to be processed and analyzed [9], [10].

By combining the visual and thermal information collected by drones, a comprehensive and accurate assessment of photovoltaic installations can be performed. This data can be processed, enabling automatic fault detection, report generation, and data-driven decision-making [11]. This streamlines maintenance and repair processes, reduces downtime, and maximizes the overall performance of photovoltaic systems.

Thermal cameras on board drones are capable of detecting temperature differences in solar panels. This thermal information is valuable for identifying potential faults or failures, such as defective solar cells, loose connections, hotspots or cold spots, and obstructions caused by dirt or shadows, among other issues. Automated thermal analysis allows for the early detection of faults and the implementation of corrective measures before they cause significant efficiency losses. The use of unmanned aerial vehicles (UAVs) for obtaining infrared (IR) images of photovoltaic systems to monitor the health status of photovoltaic modules has been identified as a cost-effective approach that offers inspection times 10 to 15 times shorter than conventional techniques [12].

RGB and thermographic sensors carried by UAV are used to check the normal operability and the existence of failures in solar modules [7]. From here, the orthophotos were generated together with temperature information. The analysis of temperature fluctuations showed that abnormal cells exhibit greater values, thus proving the benefits of using thermal infrared sensors onboard UAVs for the inspection and safety control of rapidly growing solar power installations.

Elidrissi et al. [13] studied the automated field detection and localization of defective solar PV modules using orthorectified RGB images generated from UAV datasets. First, orthophotos were generated from images collected using photogrammetry, and then individual modules were identified to obtain a single image per module. A Deep Learning (DL) model was trained for defect detection, achieving an Average Mean Accuracy of 73%.

The integration of geospatial data, collected through drone flights equipped with thermal and RGB sensors, into a GIS provides a powerful platform for effective monitoring and maintenance of PV plants. This approach not only saves time and costs but also enables the early detection of anomalies in the photovoltaic modules. The data generated through these flights are processed, mapped, and analyzed, facilitating informed decision-making processes.

This paper presents a comprehensive methodology developed for the inspection and maintenance of a PV plant, with a specific case study conducted in Jaén, Spain. The workflow includes the selection of the study area, data acquisition using a DJI Matrice 210 drone equipped with a Zenmuse XT2 thermal imaging sensor, and subsequent data processing using Pix4D software. The project emphasizes the creation of Digital Terrain Models (DSM) and the identification of thermal anomalies, vital for effective plant management.

Furthermore, the study integrates various geospatial data into a GIS, leveraging the capabilities of QGIS for visualization, analysis, and data management. This spatial data is structured in a spatial database using PostgreSQL with the PostGIS extension, ensuring secure and efficient data access. The ultimate goal is to empower decision-makers with a user-friendly web map viewer to aid in the inspection and maintenance of the PV plant.

This paper underscores the significance of geomatics tools in the renewable energy sector, showcasing how the integration of various technologies and methodologies can streamline the monitoring and management of large-scale PV installations. Additionally, it hints at the potential for future advancements in thermal imaging and database expansion to further enhance PV plant operations. Through this case study, the paper contributes to the broader discussion of sustainable energy and the pivotal role of geospatial technologies in its realization.

2 MATERIALS AND METHODS

2.1 Workflow

The workflow of this project was developed in several key stages to achieve its objectives. First, the study area was carefully selected in Úbeda, Andalusia, Spain, considering its accessibility and the necessary authorizations for drone flights. Then, data acquisition was carried out using a DJI Matrice 210 drone equipped with a Zenmuse XT2 [14] thermal imaging sensor. This sensor provided valuable data for the project. The images, both RGB and thermal images, were then processed using Pix4D software [15]. This process involved automatic aerial triangulation, the creation of densified point clouds and the generation of Digital Terrain Models (DSM). In addition, specific adjustments were applied to the thermal images.

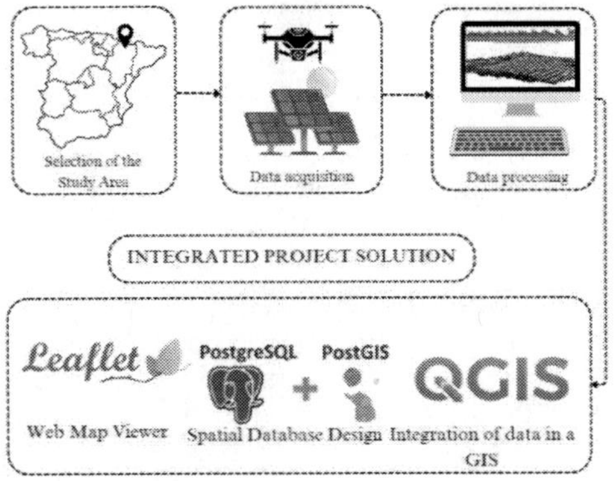

Figure 1: Location map of the study area

Once the imagery had been processed, the project moved on to the integration phase of the project, where the aim was to provide a comprehensive solution. This included the integration and visualization of geospatial data in a geographic information system (GIS) using QGIS. Vector data were generated, such as the identification of possible thermal anomalies, which were essential for the management of the photovoltaic plant. These data were stored in a structured manner in a spatial database using PostgreSQL with the PostGIS extension to ensure security and efficient access to the information.

2.2 Study area

The study area (Figure 2) is located in the city of Úbeda, in the province of Jaén, in the Autonomous

10.4229/EUPVSEC2023/4CV.1.37
020398-002

Community of Andalusia (ES). The approximate coordinates of the area are: (X,Y) (479744.74 m, 4195614.57 m) - (Reference System: ETRS89 - Zone 30N).

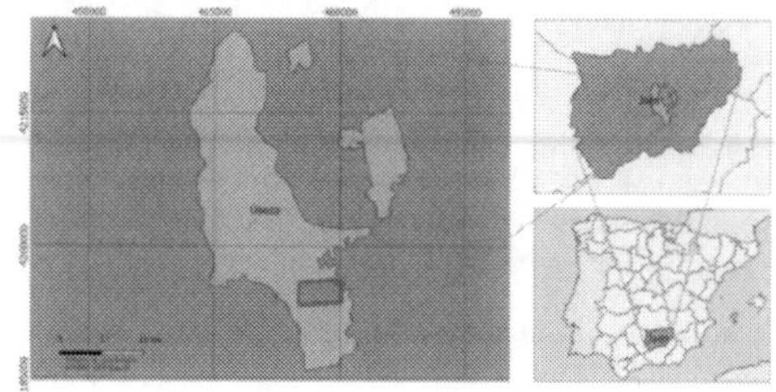

Figure 2: Location map of the study area

2.3 Data acquisition

In order to achieve the objectives defined in this project, a series of data are required that will need to be collected in the field.

The drone model used in this project was the DJI Matrice 210 (Figure 3), a radio-controlled quadcopter that according to its designers has high-performance motors, coupled with 17-inch propellers, which ensure stable flight in high wind speeds of up to 35 km/h.

Figure 3: DJI 210 Drone and example of images from both thermal and RGB sensors

The sensor used was the Zenmuse XT2 thermal imaging sensor, which combines a FLIR Tau 2 thermal sensor and a 4K visual camera.

The Zenmuse XT2 has complete radiometry, that is, the temperature measurement can be made in any pixel of the image. It is considered one of the best thermal cameras to be used in drones.

Table I: Zenmuse XT2 Sensor Features

	RGB	Thermal
Focal Distance (mm)	9	19
Image size (px)	4000x3000	640x512

2.3. Images Processing

In this section, all stages will be described for the processing of RGB and thermal images using the Pix4D software. This software presents a workflow to process the images as shown in Figure 4.

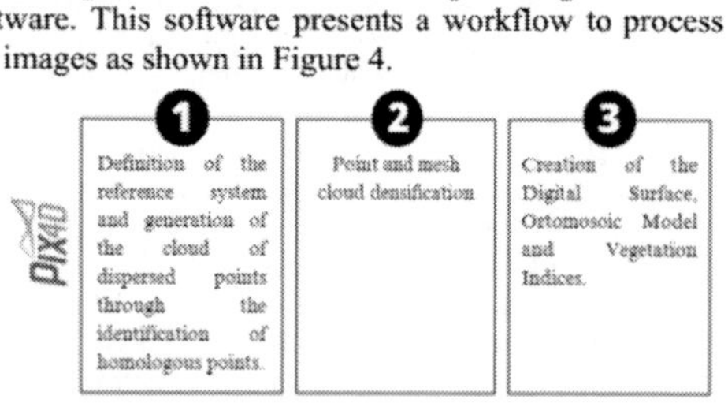

Figure 4: workflow for Pix4D image processing

In the first part, the calculation of the location and parameters of the images are made by means of a processing methodology: Automatic Aerotriangulation (ATA) and a Simultaneous Block Adjustment (ABS). What the software does is the extraction and correlation of common points between two or more images. In addition to this, an optimization of the camera model (calibration of focal distance and distortion parameters) is made. As a result, what you get is a Low-dense point cloud (Figure5).

Figure 5: Low-dense Point Cloud.

In step 2: Point cloud and mesh, the densified point cloud is created through the low-density point cloud created in the previous step (Figure 6). According to Torrado et al. [16] in this second stage: "The objective is to determine the coordinates of the land of Each of the common points is extracted from the images and then determines the coordinates of the land for each of the pixels in the images through what is called cloud densification of points."

Figure 6: Densified Point Cloud.

Digital Surface Models (DSM) are created in Step 3: DSM, Orthomosaic and Indexes. This is done from the rasterization of the point cloud.

Figure 7 shows the created DSM and the orthoimage.

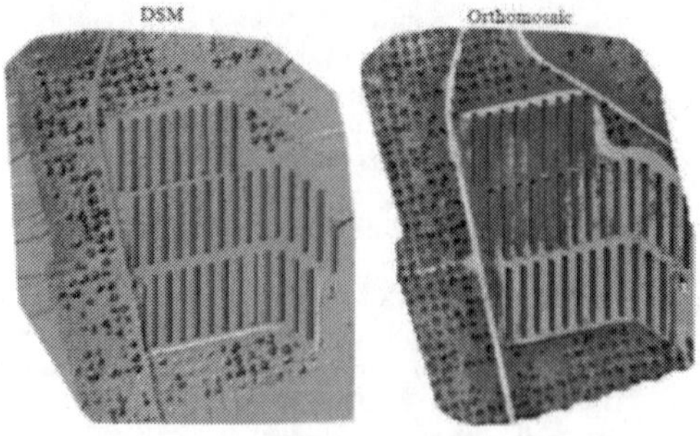

Figure 7: DSM and Orthomosaic

A mosaic corresponds to a set of images taken from

one or several cameras, which have areas of overlap between them, and which are joined and combined into a single image to expand the range of view of the scene (Cheng, Xue and Li, 2007). When the mosaic is corrected for distortions caused by terrain relief and the objects in it, it is called an orthomosaic.

In short, Pix4D finds the homologous points in the images, then correlates them as there are the overlapping areas. It then performs triangulation, by means of which the 3D coordinates of each point are obtained.

For thermal imaging, the process was the same, but with some changes to the processing options. The thermal images captured by the Zenmuse XT2 camera are in RJPG format, which according to the support of the processing software (Pix4D) "is a JPG radiometric software used for processing (Pix4D) "is a radiometric JPG image, i.e. a .jpg image with radiometric data embedded in the image metadata. of the image. This is the recommended image format for thermal imaging."

In Figure 8, you can see the sparse and dense point clouds created in the processing.

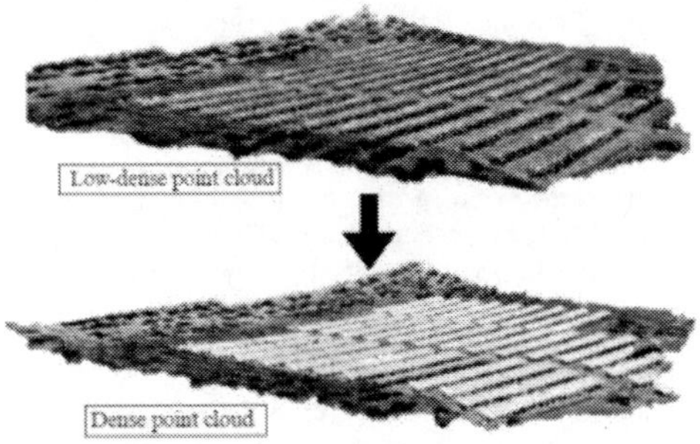

Figure 8: Low-dense and dense point clouds

3 INTEGRATED PROJECT SOLUTION

The objective of this section is the generation of an integral solution for the project, that is, a solution that offers a methodology that involves from the generation of the data to its final visualization in a web map viewer.

As can be seen in Figure 9, the methodology followed was composed of the following steps: 1) Integration and visualization of the data in a GIS (QGis)[17]; 2) Storage of the well-structured information in a Database (PostgreSQL/PostGis); 3) Import of the data to a Web Map Server (Geoserver); 4) Generation of a Web map viewer (Leaflet) and connection to the Server.

3.1 Integration and visualization of data in a GIS (QGis)

Every cartographic project is based on the implementation of a GIS project. GIS project, which enables the storage, management, analysis and visualization of spatial data in an integrated way in a GIS desktop software. In addition, it is by means of which information that is not needed is eliminated, and new data is created. information that is not necessary and also creates new data.

In this project, the GIS software used was QGis (which is a free and Open Source desktop GIS software). GIS software). In addition, its geoprocessing tools were used for geoprocessing tools to create shapefiles important in this project, edit the tables of the project, editing of attribute tables, spatial analysis and other processing.

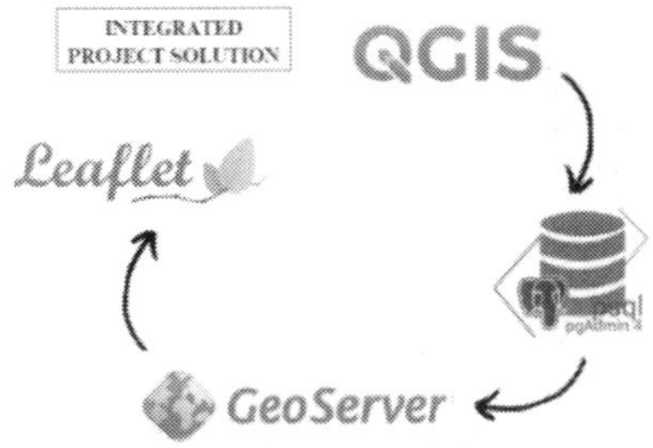

Figure 9: Integrated Solution Project Methodology

3.2 Vector Data Generation

The geoprocessing tools of the QGis software were used to generate vector data. The main objective of this section is to generate useful information for the analysis and management of the photovoltaic plant in question.

After that, a threshold for temperature was defined, i.e. all pixels with temperature higher than 30ºC were considered as "With possible anomaly". Using the QGis tool called raster calculator the following equation was applied (Figure 10):

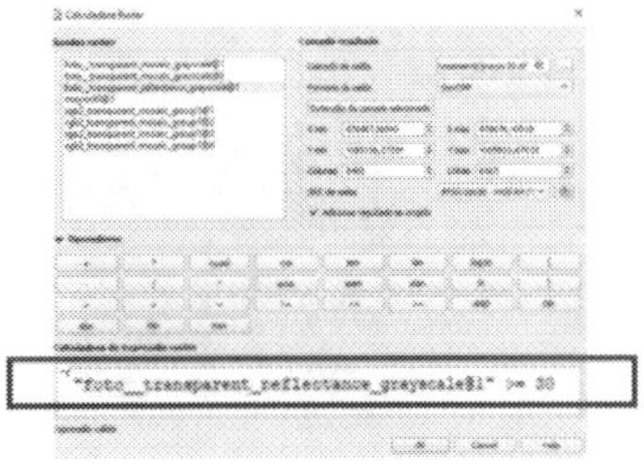

Figure 10: Raster calculator (QGis)

The result is shown in Figure 47, where a new raster was obtained, where pixels above 30ºC are represented by 1 (white) and pixels below 30ºC are represented by 0 (black). less than 30ºC are represented by 0 (black). In other words, a binary image was generated. By approaching one of the photovoltaic panels, it is possible to see in more detail how this sorting was done (Figure 11).

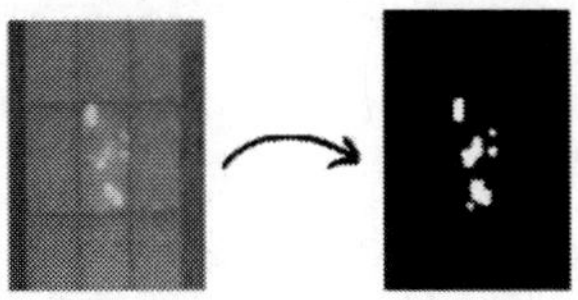

Figure 11: Classified temperature raster

The next step was to transform the raster into a vector, using the QGis "Polygonize" tool, shown in Figure 12 below.

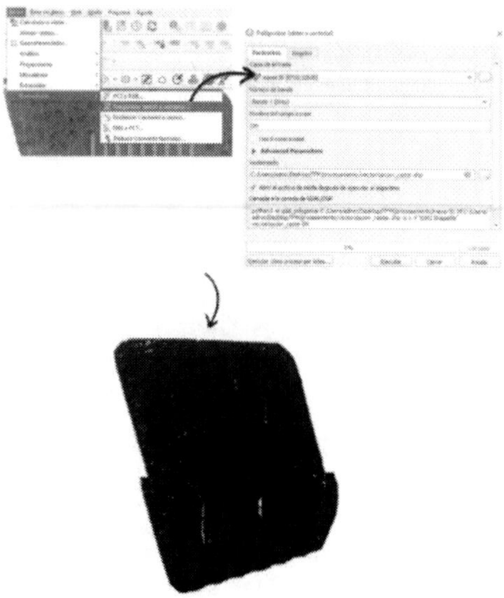

Figure 12: QGis "Polygonize" tool

Then, the vectors corresponding to DN=0 were deleted, leaving only those of DN=1 corresponding to the "possible anomalies". The result was as follows (Figure 13):

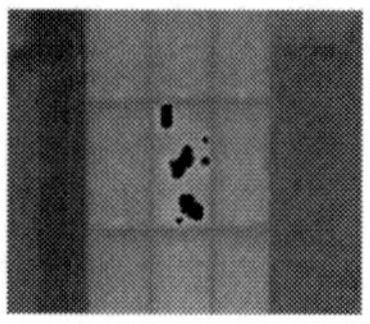

Figure 13: Vector created

As this data will serve as the basis for the management of the photovoltaic plant, it is necessary to geolocate the anomalies so that they can be easily found in an on-site inspection, i.e. when the managers go to the field to find out what is happening with the panels with high temperatures.

Therefore, before generating the geographic coordinates of these anomalies, and as the automatically generated layer is in polygon format, what was done was to generate the centroid of each one and then generate its geographic coordinates. For the generation of the centroid, the QGis tool is called "Centroid". The result was as follows (Figure 14). After generating the center of each hot spot, the ID and the X and Y coordinates of all hot spots were calculated by means of the "Field Calculator" tool.

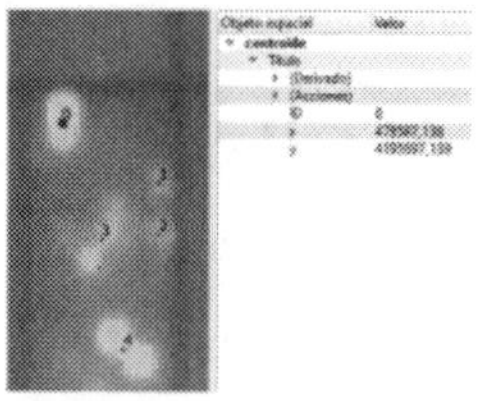

Figure 14: Centroids with their respective geographic coordinates

Another process that was carried out in order to generate more information useful for the monitoring of the photovoltaic plant was the digitization of the photovoltaic cells (figure 15).

Figure 15: Digitized plates

3.3 Storage of information in a database

Before insertion into the database the data needs to be analyzed, structured and sorted. That is, all unnecessary information must be deleted and the important information that has been generated must be stored in a safe and correct way. In this sense, there is the option of storing this data in a geo-referenced database. In this article, Postgre-SQL was used, together with PostGis which is a spatial extension for PostgreSQL that allows the management of geographic objects. The administration platform used was PGAdmin.

3.3.1 Spatial Database Design

Designing a database means developing a data model that represents a reality in detail. It is useful and important in storing a set of spatial data, so that the relationships between them are established correctly. The importance of storing data in a database lies mainly in ensuring the security of the data, as well as making it possible to share it with other systems.

According to GEOPAISA (2018), in the design of a spatial database, three phases are generally distinguished for modeling, where each phase is a more advanced abstraction of the previous one. Before the design of the database it is necessary to determine all the data that will be stored. This is followed by Conceptual Design (1st phase), Logical Design (2nd phase) and Physical Design (3rd phase).

When working with geographic data in a GIS or in a database, they have two types of structure: vector and raster. A geographic object presents its coordinates, its reference system, its shape, size, i.e. its spatial or geometric component. But it also presents the non-spatial information, which describes its attributes. The good thing is that GIS and databases nowadays allow data of different types to be stored in the same place, or in the same database.

There is also the addition of a new component: Temporal. Data changes over time and it is important to update it, but at the same time to keep the old data for comparisons. In this project, as only one flight was made, the temporal component will not be taken into account.

There are some components of the data that must be described before the generation of the database, in order to allow a correct correlation between them. In summary, the data to be integrated in the database were classified according to their discrete or continuous nature. The data of a continuous nature for this project presents data with a raster structure and those of a discrete nature, vectorial. The complete classification of the data for this project is presented in Table 2.

Table 2: Standing waves ratio

	Structure		Nature	
Variable	Raster	Vector	Space	Temporary
Thermal Ortho	grid		Cont.	x
RGB Ortho	grid		Cont.	x
Anomalies Points		Points	Disc.	x
Boundary Farm		Polygon	Disc.	x
Polygons Anomalies		Polygon	Disc.	x
Cell Polygon		Polygon	Disc.	x

The description of the entities are as follows:

- ANOMALY POINTS: this entity stores the geographical coordinates of each point and its unique identifier.
- ANOMALY POLYGONS: this entity stores the area of the anomalies found in the plates and their geographical coordinates.
- CELLS: this entity stores the percentage of anomalies in each cell and the number of pixels with anomalies and total.
- FARM BOUNDARY: this entity delimits the study farm where the photovoltaic plant is located.
- ORTHOMOSAICS: this entity stores the data of the orthomosaics processed in the project: RGB and Thermal.

In the conceptual design, the relationships between the data are determined, so prior to defining the model it is necessary to identify and adequately define each of the entities that will make up the database. In this section, the outline of the entity-relationship model of the database (DB) of the project will be presented.
The existing relationships between these entities are shown in the diagram (Figure 16). The model used was the Entity-Relationship model, where the entities are illustrated with a rectangle. The relationships, on the other hand, are usually a verb, are represented by a rhombus and link the entities, while the attributes are represented in ovals and those that are primary keys are underlined.

Figure 16: Entity-Relationship Model

The second phase is the definition of the Logical Model, which "refines" the conceptual model. According to Plana et al (2009), "The logical model is the complete and detailed development of the conceptual model in which there is a detailed description of each of the entities, the design of the tables with their attributes, identifiers, relationships, data type, which constitute the database".
The logical design or Relational Graph can be seen in Figure 17. It is important to say that this is an initial and basic database, since the objective is to analyse the application of the Integral Solution methodology of the project, later on it is ideal to add also "non-cartographic" information, such as, for example, the Date of technical revision, Quantity of KW produced, solar radiation data and others.

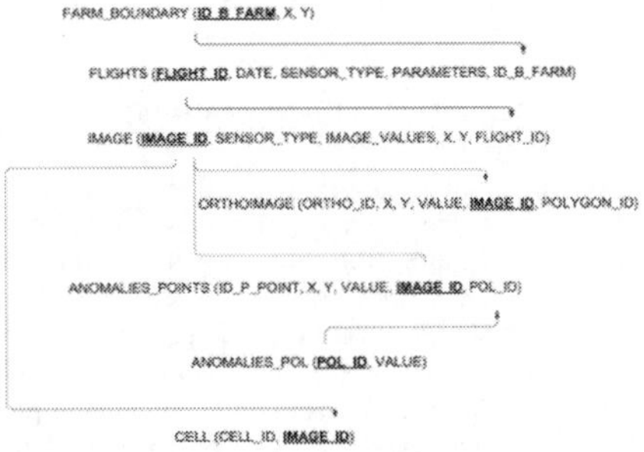

Figure 17: Entity-Relationship Model

According to Plana et al (2009), "The physical model corresponds to the implementation of the database in a specific programme or software. database in a specific programme or software. The physical model converts the logical model relationships into tables, and the foreign foreign relationships following the relational model".
So, differently from the two previous phases, the physical schema depends on the SGBD to be worked with. In this project, Postgres SQL (with the POSTGIS extension) and PGAdmin as manager. The implementation of The implementation of the data will be described in the following sections.

10.4229/EUPVSEC2023/4CV.1.37
020398-006

The first step before inserting the data into the database is, obviously, to create the database. For this purpose, PostgreSQL was installed, together with its PostGis extension (which allows the management of geographic data). In addition, the PGAdmin software, which is used to manage the database, is automatically installed. The next step was to create the server and the database.
As mentioned in the previous sections, all the geographic data created were organised and structured in QGis. In fact, this GIS software also includes a tool that makes the direct connection with PostgresSQL/PostGIS and allows the import of shapefile and raster files into the database in a simple and fast way.
All data were imported into the database.

Figure 18: Screenshot of the Database in PgAdmin

3.4 Import of data into Geoserver
For the import of the data into Geoserver a connection to the database where the data is stored will be made.
But first there are some steps that must be carried out, which include the creation of a workspace with the WMS service, which allows the graphical representation of information and metadata stored on the server. Subsequently, a warehouse is created that supplies the data to Geoserver, using PostGis as the source of the data. Once the database is connected, all the data layers are published individually in Geoserver.

4 CREATION OF THE WEB MAP VIEWER

The objective of making a web map viewer is to facilitate the inspection in the field by the managers. This is because when generating the anomaly points and passing them on to the client, the client will need to go to the field to find out what is wrong with the plates. Therefore, it is interesting to have this data on a Tablet with geolocated data, because as you walk around the plant, you can update and view the data of each module or photovoltaic panel.
In addition to that, in the visualiser all the information generated can be viewed overlaid, organized, with their styles set in a simple and useful way, and it does not require the user to have any knowledge of databases or GIS software. For the creation of the web viewer the open source JavaScript library Leaflet was used.
For editing the web page, HTML (for inserting elements such as the map and images) and CSS (for defining the styles, such as colors, sizes and position of these elements) were used.
For the definition of the tools to be implemented, the general idea was to have a visualiser as simple and useful as possible for a PV plant manager.

So, the aspects required for the design were:

- Cartographic elements: Coordinates, Scale and Reference System.
- Possibility to control the layers to be displayed.
- Possibility to draw on the map.
- Possibility to measure areas and distances.
- View the map in full screen.
- Possibility of changing the zoom of visualization.
- Legend
- Printing

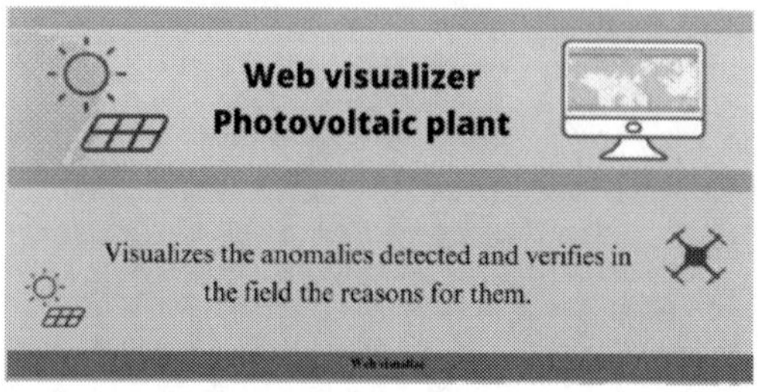

Figure 19: Web Map Viewer (Part 1)

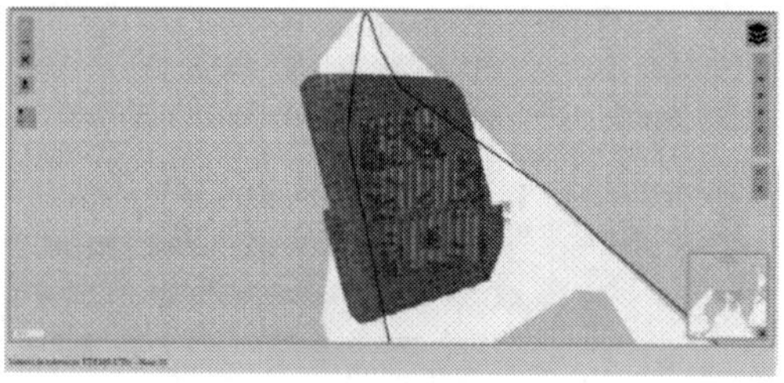

Figure 19: Web Map Viewer (Part 2)

5 CONCLUSION

This paper presented an useful result for the monitoring or inspection of a photovoltaic plant. As seen, the decrease of all prices in the world, brings as a consequence that the installations of photovoltaic plants are increasingly larger. In this sense, the use of geomatic tools that facilitate and accelerate the time of analysis of the situation of a photovoltaic plant with hectares of extension is shown as a feasible, cheap and fast solution.
In this project, it was possible to see to what extent content integration can be useful in various areas, including inspections of photovoltaic plants.
The steps involved range from the flight plan to the processing of the drone images to obtain the orthoimages, which requires knowledge of photogrammetry applied to drones. The data was then imported into QGIS and the generation of other vector data, which includes knowledge of Geographic Information Systems.
Then the database was generated, connected to QGis and Geoserver and the web map viewer was created, which requires knowledge in Geospatial Web Services, data modeling, and database creation. In short, the use of a fairly broad set of geomatics tools is noticeable.
The QGis software was a great ally in this work. In addition to allowing the generation of the secondary data, it allowed the direct connection with PostGis, which greatly facilitated the import of the data into the database. Even the raster data, by means of the add-on mentioned in the work.

It is important to say that future work will seek to further analyze thermal images and the potential of these types of sensors. In addition, the current system has a general structure but is ready to host more information and grow by increasing its functionality. More interactive web viewers and the extension of the database to store several data captures can be done.

5 REFERENCES

[1] Ipcc, *Global Warming of 1.5°C: IPCC Special Report on Impacts of Global Warming of 1.5°C above Pre-industrial Levels in Context of Strengthening Response to Climate Change, Sustainable Development, and Efforts to Eradicate Poverty*, 1.ª ed. Cambridge University Press, 2022. doi: 10.1017/9781009157940.

[2] N. S. Pérez y J. Alonso-Montesinos, «Economic and environmental solutions for the PV solar energy potential in Spain», *J. Clean. Prod.*, vol. 413, p. 137489, ago. 2023, doi: 10.1016/j.jclepro.2023.137489.

[3] I. Mauleón y M. J. García, «Friends or foes? Exploring the performance of incumbent energy providers and the expansion of renewable energy in five European countries», *Energy Res. Soc. Sci.*, vol. 76, p. 102058, jun. 2021, doi: 10.1016/j.erss.2021.102058.

[4] I. Mauleón, «Optimising the spatial allocation of photovoltaic investments: Application to the Spanish case», *Energy Convers. Manag.*, vol. 291, p. 117292, sep. 2023, doi: 10.1016/j.enconman.2023.117292.

[5] C. Fernández-Espinar Muñoz, «Comentario al Real decreto ley 23/2020 de 23 de junio por el que se aprueban medidas en materia de energía y en otros ámbitos para la reactivación económica», *Actual. Jurid. Ambient.*, n.° 103 (Julio), pp. 99-127, 2020.

[6] J. Gutiérrez *et al.*, «Multi-Criteria Methodology for the Location of Photovoltaic Solar Energy Production Facilities in Tenerife (Spain)», *Infrastructures*, vol. 7, n.° 3, Art. n.° 3, mar. 2022, doi: 10.3390/infrastructures7030028.

[7] D. H. Lee y J. H. Park, «Developing inspection methodology of solar energy plants by thermal infrared sensor on board unmanned aerial vehicles», *Energies*, vol. 12, n.° 15, 2019, doi: 10.3390/en12152928.

[8] F. Grimaccia, S. Leva, y A. Niccolai, «PV plant digital mapping for modules' defects detection by unmanned aerial vehicles», *IET Renew. Power Gener.*, vol. 11, n.° 10, pp. 1221-1228, 2017, doi: 10.1049/iet-rpg.2016.1041.

[9] E. Molenbroek, D. W. Waddington, y K. Emery, *Hot spot susceptibility and testing of PV modules*, vol. 1. 1991, p. 552 vol.1. doi: 10.1109/PVSC.1991.169273.

[10] P. B. Quater, F. Grimaccia, S. Leva, M. Mussetta, y M. Aghaei, «Light Unmanned Aerial Vehicles (UAVs) for cooperative inspection of PV plants», *IEEE J. Photovolt.*, vol. 4, n.° 4, pp. 1107-1113, 2014, doi: 10.1109/JPHOTOV.2014.2323714.

[11] L. Morando, C. T. Recchiuto, J. Calla, P. Scuteri, y A. Sgorbissa, «Thermal and Visual Tracking of Photovoltaic Plants for Autonomous UAV Inspection», *Drones*, vol. 6, n.° 11, 2022, doi: 10.3390/drones6110347.

[12] Y. Zefri, I. Sebari, H. Hajji, y G. Aniba, «In-depth investigation of applied digital photogrammetry to imagery-based RGB and thermal infrared aerial inspection of large-scale photovoltaic installations», *Remote Sens. Appl. Soc. Environ.*, vol. 23, p. 100576, ago. 2021, doi: 10.1016/j.rsase.2021.100576.

[13] H. Elidrissi, H. Achakir, Y. Zefri, I. Sebari, G. Aniba, y H. Hajji, «Automatic on Field Detection and Localization of Defective Solar Photovoltaic Modules from Orthorectified RGB UAV Imagery», en *2022 6th International Conference on Green Energy and Applications (ICGEA)*, mar. 2022, pp. 46-50. doi: 10.1109/ICGEA54406.2022.9791946.

[14] «Zenmuse XT2 - DJI», *DJI Official*. https://www.dji.com/es/zenmuse-xt2 (accedido 11 de septiembre de 2023).

[15] «Pix4D Manual – Support». https://support.pix4d.com/hc/en-us/sections/360003718992-Manual (accedido 11 de septiembre de 2023).

[16] «(2) (PDF) Orthomosaics and digital elevation models generated from images taken with UAV systems». https://www.researchgate.net/publication/320501113_Orthomosaics_and_digital_elevation_models_generated_from_images_taken_with_UAV_systems (accedido 14 de septiembre de 2023).

[17] Q. Project, «QGIS Desktop 3.28 User Guide», *Guía Usuario QGIS*, Accedido: 21 de agosto de 2023. [En línea]. Disponible en: https://docs.qgis.org/3.28/es/docs/user_manual/index.html

10.4229/EUPVSEC2023/4CV.1.37
020398-008

PERFORMANCE DIFFERENCES OF INDIVIDUAL MODULES IN A PV ARRAY

Li Feng[1,2,*], Nowshad Amin[2], Jingwei Zhang[3], Kun Ding[3], Frank U. Hamelmann[1,*]
[1]Solar Computing Lab, Bielefeld University of Applied Sciences and Arts, Artilleriestraße 9, 32427 Minden, Germany
[2]Institute of Sustainable Energy, Universiti Tenaga Nasional (The Energy University), 43000 Kajang, Malaysia
[3]College of Mechanical and Electrical Engineering, Hohai University, 213200 Changzhou, China
[*]Corresponding author: li.feng@hsbi.de; frank.hamelmann@hsbi.de

ABSTRACT: As photovoltaic technologies continue to be improved, the manufacturers and researchers are paying more attention to the performance of PV modules in commercial PV arrays/systems. There are still many differences in the performance of individual PV modules in the laboratory and under outdoor conditions. And the current monitoring and evaluation of the PV system performance is mainly focused on the string level and system level, which lack performance analysis of PV system at module level. Thus, it becomes important to analyze the performance differences between the individual modules.This study presents an approach to analyze the performance differences in the voltage and temperature of junction box for individual modules in the same PV array. The data from individual modules are collected and further analyzed for performance differences between individual modules based on the performance differences calculation method. The results show that the temperature difference is smaller for each module with a smaller mounting angle compared to the substring with a larger mounting angle. The final performance of a PV module depends to a large extent on the location of the PV module as well as the surrounding obstacles. However, the difference in voltage difference was not significant. Performance evaluation at the module level helps in analyzing the natural degradation of PV modules and re-cycling them once they have reached their expected lifetime.
Keywords: Module performance, voltage difference, temperature difference, PV array

1 INTRODUCTION

Performance evaluation of photovoltaic (PV) module is of paramount importance in the realm of solar energy generation, and playing a pivotal role in ensuring the efficiency, reliability, and sustainability of solar power systems [1]. As the world increasingly pivots towards renewable energy sources to combat climate change and reduce our dependence on fossil fuels, the significance of rigorously assessing PV module performance cannot be overstated [2]. Evaluating the performance of PV modules is a crucial aspect of ensuring the efficiency, reliability, and longevity of solar power systems. However, this process is not without its challenges, as the complexity of PV technology, environmental factors, and evolving standards present unique obstacles that must be addressed [3],[4].

Solar panels are exposed to a wide range of climates, from extreme heat to freezing cold, and from high humidity to arid desert conditions [5]. These fluctuations can impact module efficiency, making it challenging to establish a consistent baseline for performance evaluation. Furthermore, the amount of sunlight a PV module receives depends on weather patterns and seasonal changes. Cloud cover, shading [6], and other meteorological factors can cause significant fluctuations in energy output. Evaluating performance accurately requires accounting for these external variables, which can be complex and time-consuming. Besides, over time, PV modules experience degradation due to exposure to UV radiation, temperature fluctuations, and other environmental factors. This aging process can result in decreased efficiency. Evaluating how different modules degrade at varying rates and identifying the causes of degradation pose considerable challenges [7].

Despite these issues, the ongoing development of PV module performance evaluation methods and standards is crucial for the growth of the solar industry. Researchers, manufacturers, and certification organizations continue to work together to address these issues, improve measurement accuracy, and develop predictive models that enhance the reliability and longevity of solar power systems. As the world increasingly relies on solar energy to combat climate change and transition to sustainable power sources, overcoming these issues will remain a top priority for the renewable energy sector.

In this research, a methodology for assessing variances in voltage and temperature among individual modules within a single PV array is introduced. Subsequently, the data from individual modules is collected and subjected to in-depth analysis to identify performance disparities utilizing a specific calculation method.

2 METHOD

2.1 Monitoring Data

A smart PV system is employed to autonomously gauge the current and voltage of individual modules at their maximum power point (MPP), record the temperature within the junction box, and store this information in a local database, which serves as a key resource for this study. The primary objective is to facilitate module-level monitoring, enabling the collection and transmission of telemetry data.

The measured voltage and temperature in the junction box in every 15 minutes of two PV strings from the monitoring system are shown in Fig.1 and Fig.2. In Fig.1, module number 3-7-12 has lower voltage between 7:00 am and 10:30 am, which is caused by the shadow from the front row module. At sunset between 5:00 pm and 7:00 pm, the voltages of modules 3-7-5 and 3-7-6 are reduced due to the shadows from the higher roof. The backsheet of PV modules 3-7-15 and 3-7-18 is close to a wall, which may lead to air flow being affected by the narrow space, so the temperature in junction boxes of these two modules is lower than the others. In Fig.2, modules 3-6-6 and 3-6-7 are near two walls and the shading from the walls reduces their voltages in the morning and evening. However, the temperatures of these

10.4229/EUPVSEC2023/4CV.1.38
020399-001

21 modules have no fast changes during the whole day.

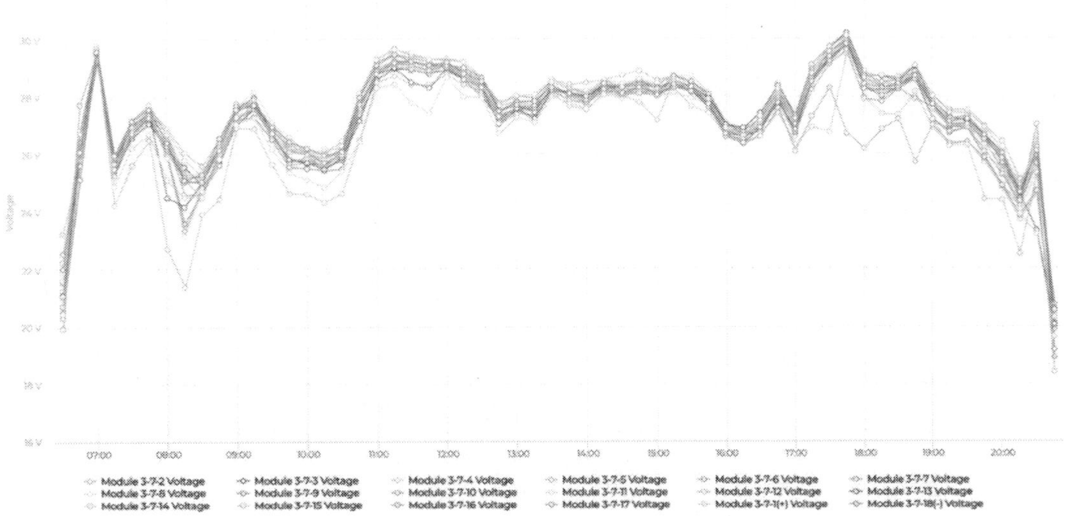

(a) Voltage of the different module in a PV string

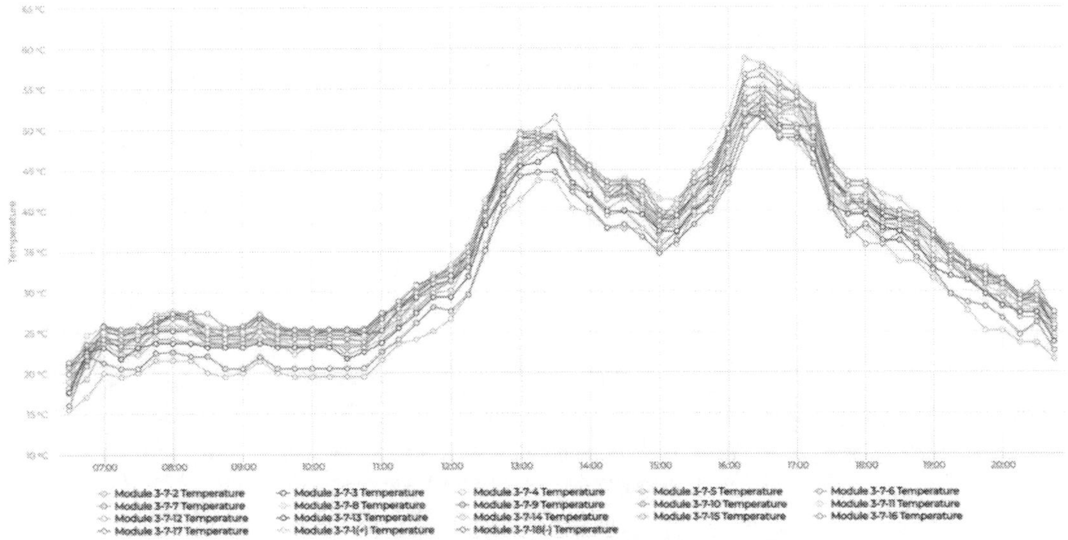

(b) Temperature of the different module in a PV string

Figure 1: Voltage and temperature of the different module in PV string 7

10.4229/EUPVSEC2023/4CV.1.38
020399-002

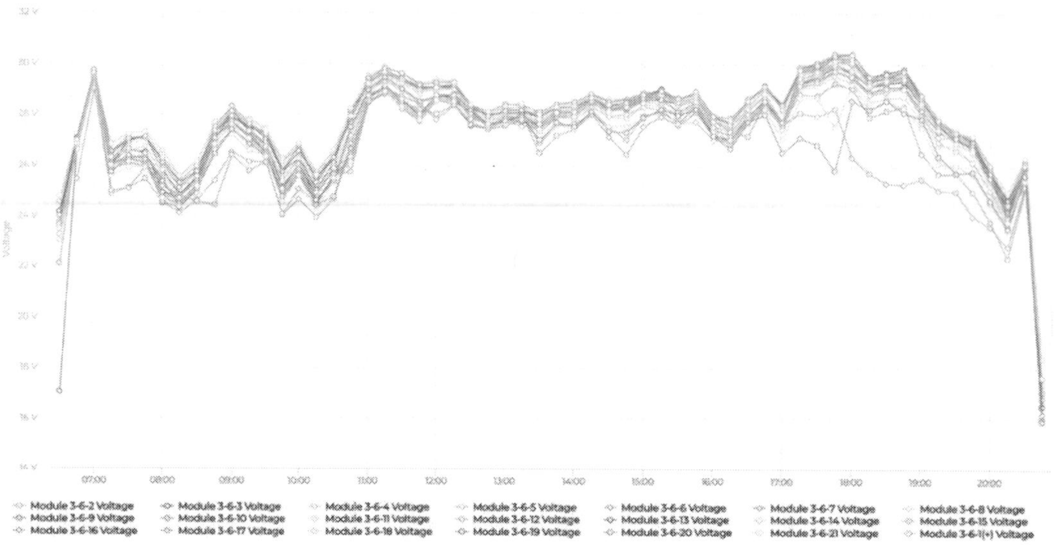

(a) Voltage of the different module in a PV string

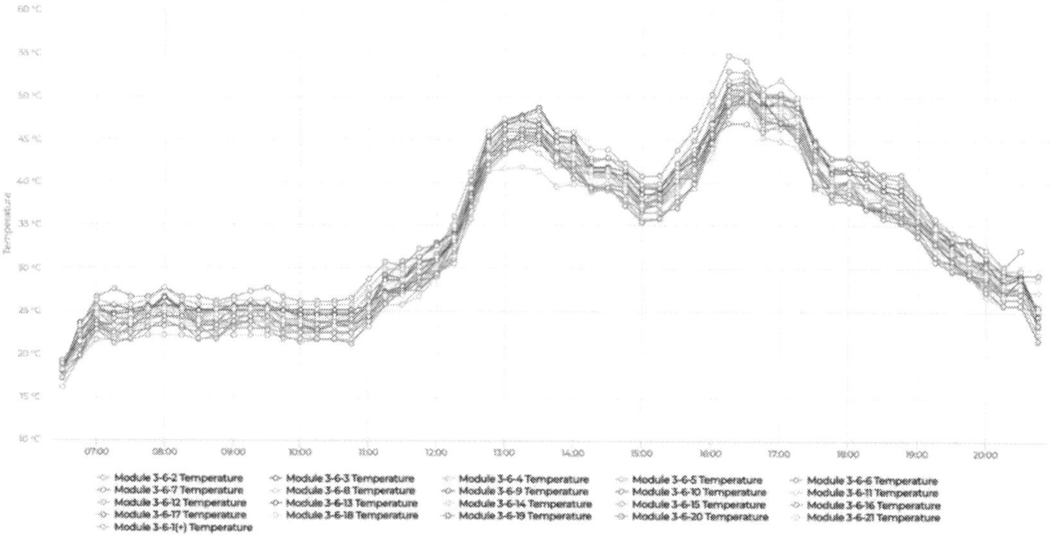

(b) Temperature of the different module in a PV string

Figure 2: Voltage and temperature of the different module in PV string 6

2.2 Performance Analysis Method

In order to further analyse the performance differences between individual modules, the voltage and temperature differences in Fig.1 and Fig.2 are calculated by the following equations:

$$\varepsilon_V(i) = \left| V(i) - \frac{V_{\max} + V_{\min}}{2} \right| \tag{1}$$

$$\varepsilon_T(i) = \left| T(i) - \frac{T_{\max} + T_{\min}}{2} \right| \tag{2}$$

where, i is the time, $\varepsilon_V(i)$ is the voltage differences at i^{th} moment of single PV module, $\varepsilon_T(i)$ is the temperature differences in junction box at i^{th} moment of single PV module, $V(i)$ is the voltage at i^{th} moment of the module on whole day, V_{max} is the maximum voltage of the module at MPP on whole day, V_{min} is the minimum voltage of the module at MPP on whole day, $T(i)$ is the temperature at i^{th} moment in the junction box of the module on whole day, T_{max} is the maximum temperature in the junction box of the module on whole day, T_{min} is the minimum temperature in the junction box of the module on whole day.

The average voltage and temperature differences are defined as below:

$$\Delta V = \sum_{i=1}^{M} \varepsilon_V(i) / M \tag{3}$$

$$\Delta T = \sum_{i=1}^{M} \varepsilon_T(i) / M \tag{4}$$

Where, ΔV is the average voltage differences of single PV module, ΔT is the average temperature differences

10.4229/EUPVSEC2023/4CV.1.38
020399-003

single PV module, M is the total number of data points.

3 RESULTS

The data in every 15 minutes from the above smart PV system is used to explore the differences of each PV module in the same array. The results are shown in Fig.3 and Fig.4. The Fig.3(a)-(b) and Fig.4(a)-(b) reflect the difference of each module in the same PV strings at the same time based on the measured data and the equations of ε_V and ε_T. The differences in ε_V and ε_T for string 7 (mounting angle in 30°) in every 15 minutes are varying from 0.12V to 2.33V, 0.14℃-to 5℃, and they change in string 6 (mounting angle in 10°) from 0.12V to 2.3 V and 0.04 to 3.9 ℃. In substring 7, the voltage difference between morning and evening is large because the small roof on the left side of the substring and the railing on the right side and front-row modules will create shadows on the surface of the PV modules in the morning and evening when the sun angle is low. In substring 6, a large voltage differences was generated only in the evening because of the shadow from the small rooftop on the left side. Then the temperature differences for both substrings throughout the day are more balanced.

The average values of the ε_V and ε_T for all modules in every 15 minutes are 0.1V-1.4V, 0.13 ℃ -2.8 ℃ in string 7 and 0.2V-1.6V, 0.8 ℃ -1.6 ℃ in string 6. In substrings 6 with smaller mounting angles, the temperature difference between each module is smaller than that in substring 7 with larger mounting angles. But the differences in voltage are the opposite. The final performance of the PV module are highly dependent on the manufacturing differences, location of the PV module and surrounding obstacles. The faults detection algorithms and localization rules over the next few months are planned to be developed based on the performance differences of each module in the PV array.

At the same time the performance evaluation will be of great help for the future analysis of the natural degradation of PV modules and their reuse after end-of-life.

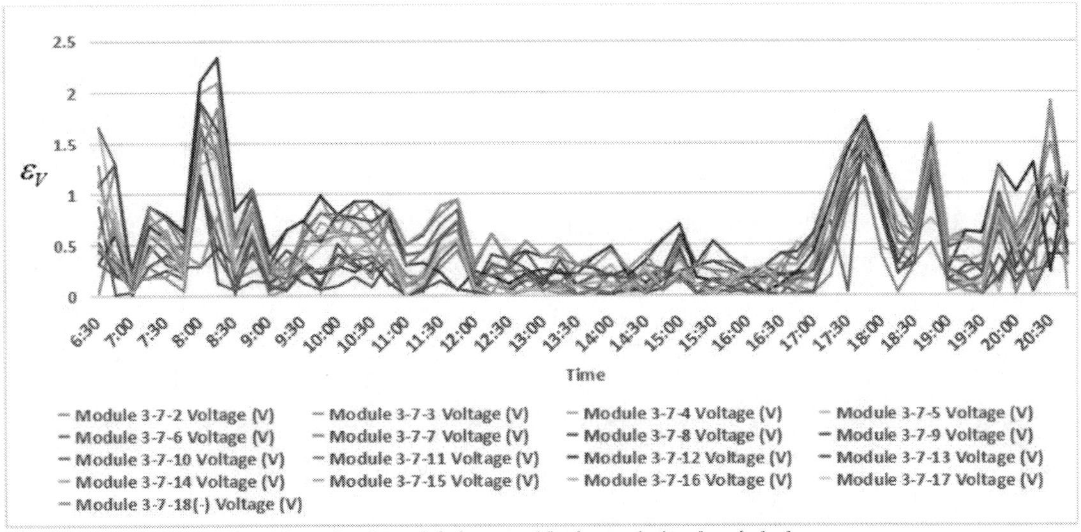

(a) ε_V of each module in every 15 minutes during the whole day

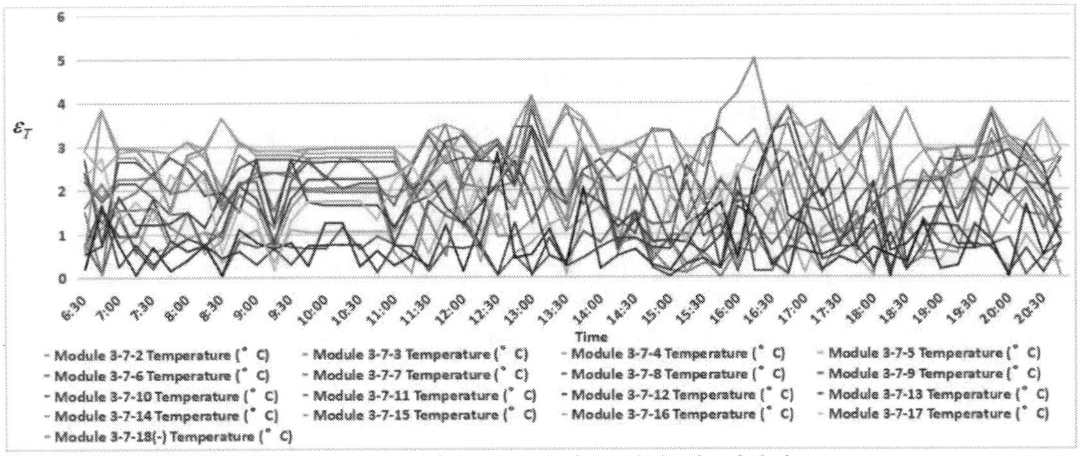

(b) ε_T of each module in every 15 minutes during the whole day

10.4229/EUPVSEC2023/4CV.1.38
020399-004

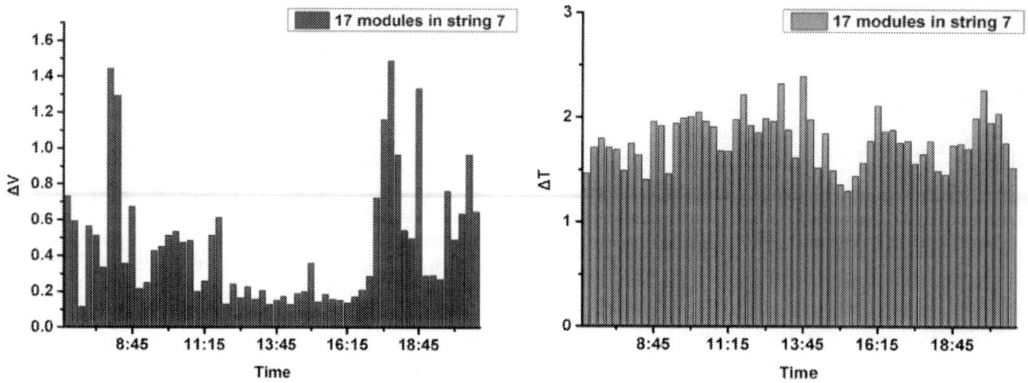

(c) ΔV of 17 modules in every 15 minutes (d) ΔT of 17 modules in every 15 minutes

Figure 3: Differences in the calculation of individual modules in the same string 7

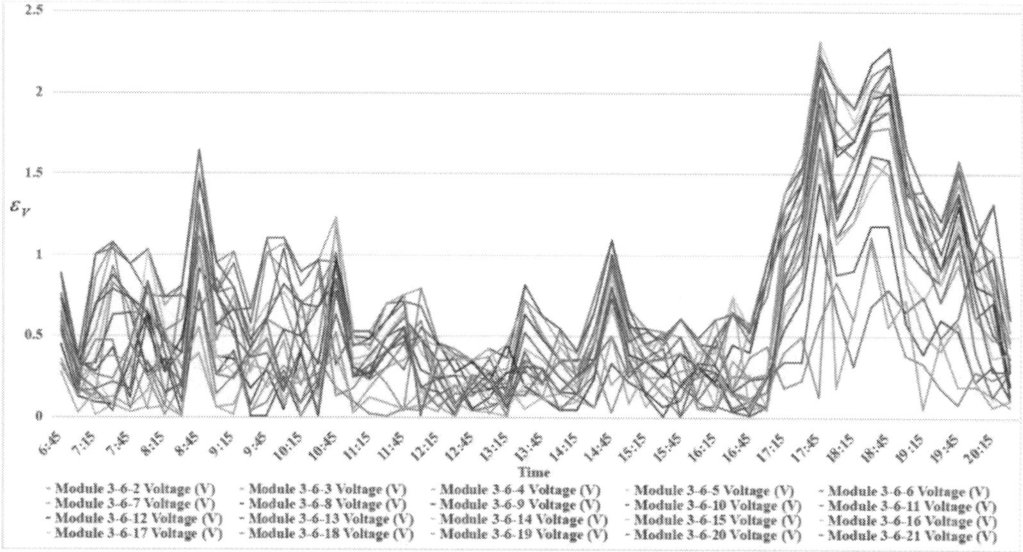

(a) ε_V of each module in every 15 minutes during the whole day

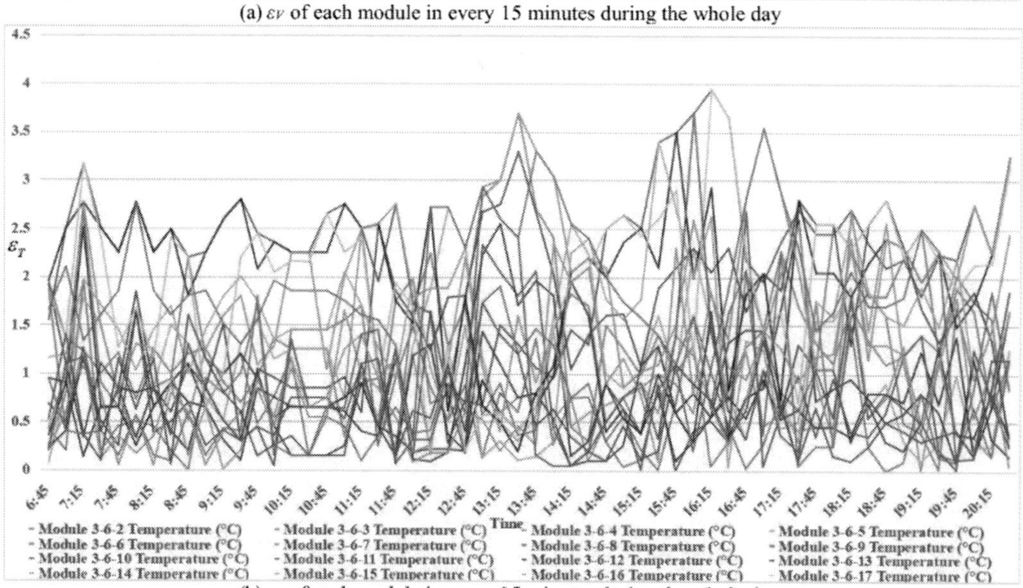

(b) ε_T of each module in every 15 minutes during the whole day

10.4229/EUPVSEC2023/4CV.1.38
020399-005

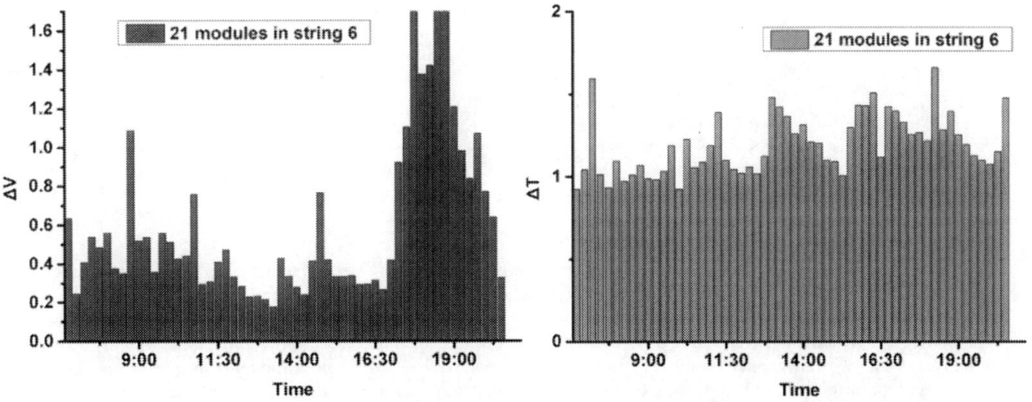

(c) ΔV of 21 modules in every 15 minutes (d) ΔT of 21 modules in every 15 minutes

Figure 4: Differences in the calculation of individual modules in the same string 6

4 COLUSION

This research introduces a method for assessing variations in voltage and temperature within a junction box of individual modules in a PV array. The data from individual modules is gathered and subjected to an in-depth analysis using a performance differential calculation technique. The findings indicate that modules with a shallower mounting angle exhibit reduced temperature disparities compared to those with a steeper mounting angle. However, the variations in voltage differences were found to be statistically insignificant.

The final performance of the PV modules is highly related to the manufacturing differences, the location of the PV modules and the surrounding obstacles. Faults detection algorithms and localization rules over the next few months are planned to be developed based on the performance differences of each module in the PV array. In addition, the performance evaluation is very important for the future study of the degradation in PV modules and their recycling after end-of-life.

5 REFERENCES

[1] O.A. Alimi, E.L. Meyer, O.I. Olayiwola. Solar Photovoltaic Modules' Performance Reliability and Degradation Analysis-A Review, Energies 15 (2022) 5964.

[2] A.L.M. Abdali, B.A. Yakimovich, V.V. Kuvshinov. Hybrid power generation by using solar and wind energy. Energy 2 (2018) 26-31.

[3] M.G. Gebreslassie. Development and manufacturing of solar and wind energy technologies in Ethiopia: Challenges and policy implications. Renew. Energy 168 (2020), 107-118.

[4] R. Srivastava, A.N. Tiwari, V.K. Giri. An Overview on Performance of PV Plants Commissioned at Different Places in the World. Energy Sustain. Dev. 54(2020) 51-59.

[5] D.H. Daher, L. Gaillard, M. Amara, C. Ménézo. Impact of Tropical Desert Maritime Climate on the Performance of a PV Grid-connected Power Plant. Renew. Energy 15 (2018), 729-737.

[6] D. Pera, J. A. Silva, S. Costa, M. Joao, M. Serra. Investigating the impact of solar cells partial shading on photovoltaic modules by thermography,2017 IEEE, 1979-1983.

[7] M. Malvoni, N.M. Kumar, S.S. Chopra, N. Hatziargyriou. Performance and Degradation Assessment of Large-scale Gridconnected Solar Photovoltaic Power Plant in Tropical Semi-arid Environment of India. Sol. Energy 203 (2020) 101-113.

10.4229/EUPVSEC2023/4CV.1.38
020399-006

IN OPERATION ENERGY AUDIT TO SUPPORT BIPV SUSTAINABILITY AND RELIABILITY

Simon BODDAERT[1,*], Simon THEBAULT[1], Alexandre MATHIEU[1], Laurent PRIEUR[2], David TREBOSC[3]

[1] CSTB, Energy & Environment Direction, France
[2] CERTISOLIS, Testing laboratory and certification body, France
[3] BDPV, Association BDPV, France

* E-mail to: simon.boddaert@cstb.fr; phone: +33 (0)680 58 1001

ABSTRACT: The knowledge of the performance over time of energy systems and mainly BIPV is a major concern for public policy orientations. The development of solutions to achieve decarbonation and energy objectives must meet ever higher requirements for increasingly ambitious objectives in a context of development at an unprecedented pace. The answers provided focus on claimed performances supported by numerical approaches and controls at delivery stage. However, the performance of an energy system is affected over time by non-homogeneous and erratic phenomena that can only be apprehended by on-site measurement to assess the operating state of photovoltaic systems. In addition to characterizing the level of actual performance compared to expected, these operations can highlight the causes of the observed differences and determine solutions for correction. The purpose of this work is to identify ways to improve the overall quality of the BIPV sector through the energy performance indicator and to position BIPV solutions as a crucial contributor to achieve the objectives of the energy transition. Eventually, this energy indicator must interact with economic and environmental indicators. With self-consumption solutions, monitoring solution, often linked to the feed-in tariff are less and less accessible, which makes the means of performance monitoring and control "blind". This article presents the work done to develop the method of quality control in operation of new and existing BIPV installations whose objective is to determine their real energy contribution and their role in the energy mix and expected production over the time.
Keywords : BIPV, Operating condition, Performance, Guarantee

1 INITIAL WORK

Since many years, CSTB is involved in IEA PVPS T15 program with the support of ADEME (French Energy Agency) to support development of BIPV solutions. Even more in current energy share and cost, the knowledge of global quality generation is a key point to give the correct drivers from decision markers to privateers to adopt PV and BIPV energy and disseminate over the countries sustainable and reliable solutions
[*1*],[2].

Despite massive works on certifications (TC 82 – JWG11) dealing with specific BIPV concern with ongoing revision of EN 50583-1[3], regulations and construction codes, BIPV solutions meet barriers for a massive adoption, not exclusively based on cost. To overcome these barriers and have a representative inventory of the current quality and durability of PV and BIPV, a national survey addressed to French BIPV users is initiated, with the support of the BDPV® association. More than 14,000 installations are requested, and more than 4000 answers are collected and evaluated to specifically identify four categories of malfunctions. The data collected is cross-referenced with the data and classifications of insurers to ensure that the data collected can meet the challenges of the market. The first analysis of the data allows us to point out the representativeness of each category and to emphasis their impact on the degradation of performance and this over time.

Four main categories are identified even if large progress have been done, many messes are revealed and remain relevant and highlight progress to be made.

Important notes highlight the extreme low rate of remaining defaults (close to 10% of installation). Fists results underline that main issues are not found on expected categories as water leakages or fire risks but mainly on energy production and electrical defaults. Two well managed considered categories. Less remarkable these two categories significantly affect installation numbers and results in negative contribution on LCOE, ROI and global warming reduction rate.

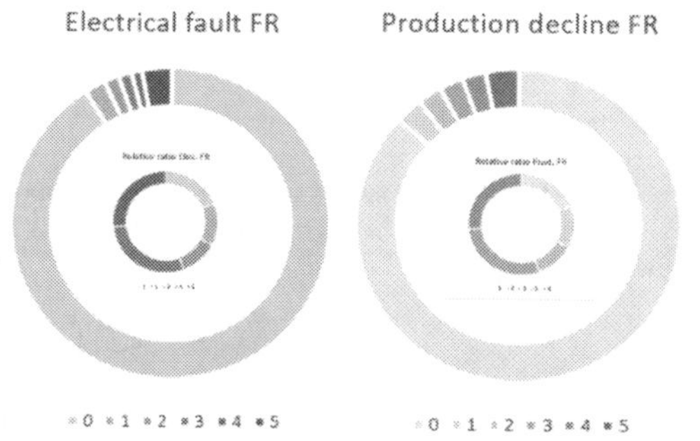

Figure 1 : Electric and energy production defaults detected. Outer ring expresses global rate of defauts, inner ring the relative rate

The two main representative categories affected by defaults are showed on Figure 1 with two different rings expressing the global rate of risk from 0 (null) to 5 (extreme). In the inner ring, are displayed the relative default rate (from 1 to 5) to underline the weight of default gravity. At any stages defaults are well identified.

For these two fists categories, the rate of appearance is 11% and 16% respectively for Electric and Production defaults. The rate reaches 8% and 4,8% on respectively Water leakages and Fire defaults, as showed on Figure 2. Considered as mastered defaults are unfortunately the most present with long terms effects and could contribute largely to discriminate the use of BIPV. These results highlight the large work performed to develop specific

10.4229/EUPVSEC2023/4CV.1.40
020401-001

assessment to reduce risks of water or mechanical damages and fire risks. Extended work performed by experts of T15 provided relevant inputs to give correct drivers for BIPV solutions. The last report on fire risks and labs involved to assess correctly BIPV components[4], is a part of this global improvement to manage and avoid fire hazards.

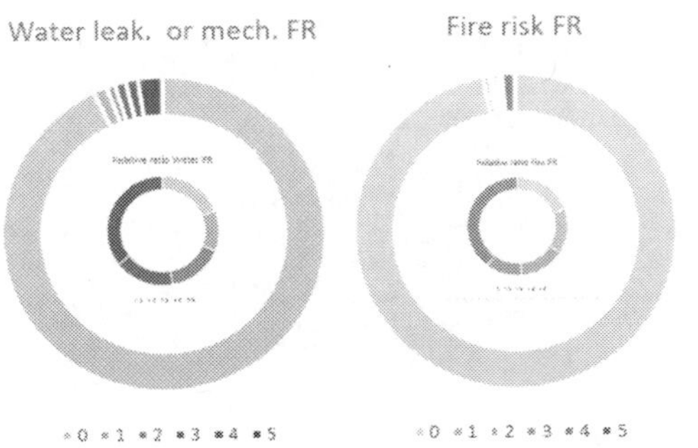

Figure 2 : Water leakages and Fire risk defaults detected. Outer ring expresses global rate of defauts, inner ring the relative rate.

This first investigation highlights the lack of knowledges or the lack of technical supervision on BIPV installations, where defaults are estimated as unsignificant. Even if default rate is about 1% on global amount of PV installation, we could assume that 1.61 GWc over the EU countries are affected by and cut energy production by 1,58 TWh in 2022 [5], and the current growth of BIPV market could expect to much more impact. To find a way to reduce these issues we conducted research on default identification methodology to prevent default and support maintenance work for an optimal energy production of BIPV installation.

2 PRELIMINARY METHODOLOGY

The development and implementation of this method is based on a preliminary work based on literature review [6] with two main objectives, a) characterize failure modes and detection methods and b) Rank and identify the most critical failure modes. This State of the art on photovoltaic failure modes and detection methods allows to count 26 failure modes and up to 13 detection methods. Interactions between failures and detection methods are showed on Figure 3. A rank failure methodology is expressed via FMECA analysis using three scores parameters: Severity (S), Occurrence (O) and Detection (D).

A Risk Priority Number is expressed (RPN) on the base of these three parameters, with D is an addition of market maturity parameter (M) and the intervention cost parameter (C). RPN is expressed as the following equations and can be computed.

$$RPN = S.O.D$$
Eq. 1

$$RPN_{RMS,K} = \sqrt{\frac{1}{N_S}\sum_{n=1}^{N_S} {RPN_{k,n}}^2}$$
Eq. 2

This method aims at mapping BIPV failures and ranking them based on their risk and identify the severity of defaults, the occurrence rate, and the detection ability. RPN is tie linked with LCOE and cost impact of defaults. To combine results of this literature review and main BIPV defaults collected from field, on site validation is necessary to fine tuning the methodology to specific BIPV sets and expectations. Validation campaigns are conducted on CSTB experimental platforms and real installation sites to validate approach.

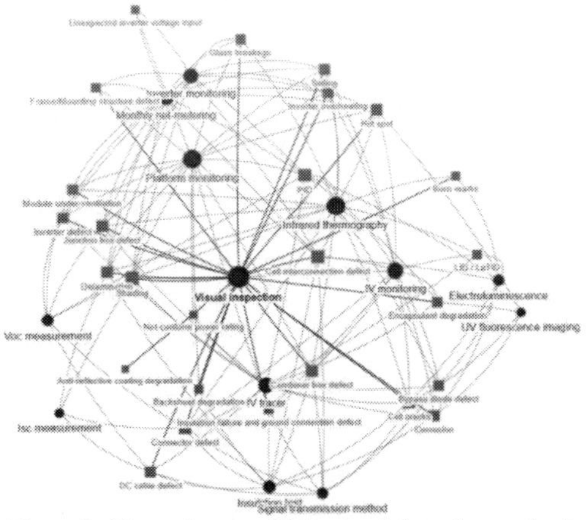

Figure 3 : Network map of failure and detection method.

3 VALIDATION PROTOCOL

As expressed in the preliminary work, on site check and validation is necessary to bring confidence in RPN score and control the failure mode. The ranking must be improved by on site audits and visual controls with appropriate equipment. The Figure 4 shows general RPN score for any PV installation, a dedicated figure will be generated after on site measurement to consolidate feedbacks and method accuracy.

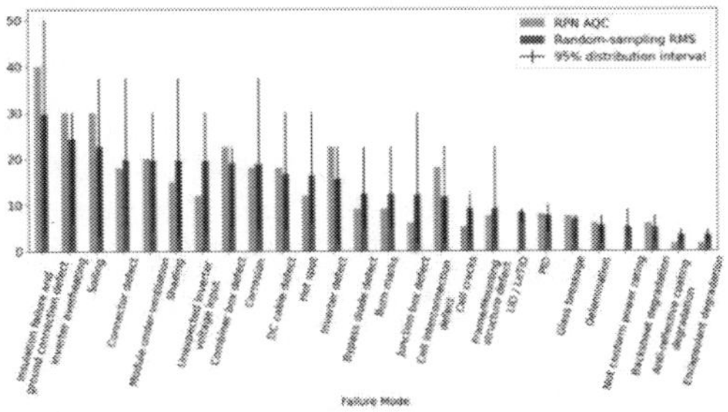

Figure 4 : RPN score according to statistic failure mode.

For each BIPV installation the RPN score will be to validate in front of the complexity of each BIPV category as described in [3] and according to the extended list of each failure case *k*, conditions of the power systems and to check their technical sets. Indicators, ENERGY and ELECTRICAL are the two main elements that are audited so far. To do so, an adapted equipment and an intervention procedure is currently being finalized and validated on test platforms to determine the duration, accuracy of sensors, measurement solutions, components and equipment. Duration and costs of such operations are still monitored.

The verification of the effective quality of an installation should have the lower possible impact on the

10.4229/EUPVSEC2023/4CV.1.40
020401-002

LCOE and should allow improving the reliability and durability of BIPV installations.

The large-scale deployment of this method, on new as well as existing installations, must above all identify chronic malfunctions and provide corrective solutions that will allow the improvement of the entire sector based on dynamic and statistically representative feedback.

These corrections are as much physical as numerical, since the measurements carried out on sites will be compared with the estimates of productions carried out with the numerical tools and will make it possible to bring corrective actions based on real data and thus to improve the base of knowledge of the operation of the BIPV.

3.1 Validation in controlled conditions

First stage of validation is performed on experimental platforms and PV lab of CSTB to validate in high end accuracy conditions measurements and visual control. Additional investigations on damaged module, measured and provided by CERTISOLIS, French certification body, as reference measurements are also carried out to control effect of default typologies and well-known damages (cracks, soiling, yellowing) and validate impact on expected energy generation. CSTB experimental setups, allow to validate methodology on many different BIPV solutions, as roofs, cladding façades, insulated glazing solution and canopy.

Figure 5 : Canopy mockup on CSTB plateformes – Onyx Solar® modules in BIPVBOOST project – 0,9 kWc.

Controlled conditions in laboratory allow to validate methodology and identify the weight contribution of any identified issue including damage or climatic and surrounding conditions. The Figure 6 shows the gap between STC conditions and in real operating conditions including all production hazards and aging effects.

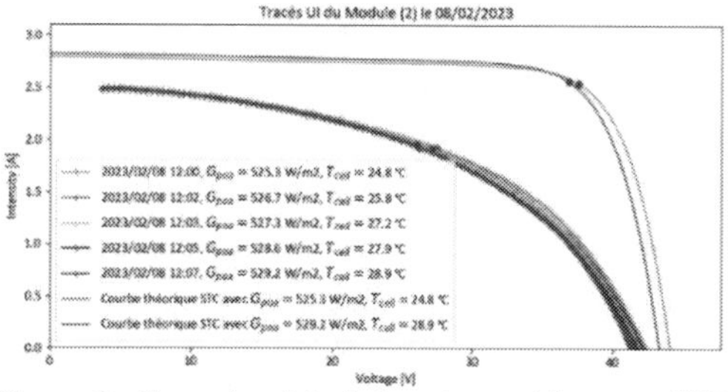

Figure 6 : Exemple of deviation observed between STC condition and in operation conditions.

Correction are applyed on calculation prediction taking into account natural deviation of PV parts and external contribution from surrounding conditions.

3.2 Validation in real operating conditions

Second stage aims to face to face controlled conditions in laboratory to real operating conditions to correct methodology with constraints from real field, as operating time, access constraints and unknow data. Main part is to take real operating conditions as temperature balance, local climates conditions, and shadowing effects. Several sites are selected to have relevant representativity of French BIPV installations. Key point is to have as close as possible similarities between BIPV categories to define or detect underperformance by category for a higher results accuracy.

Figure 7 : Canopy on ADEME building – 2ES company – 26 kWc

Same investigation process is applied on operating installations as showed on Figure 7, a large canopy in operation from almost two years. Measurements on strings, DC parts, bring us comparison results and estimation of possible degradation of self-aging, soiling and temperature balance tied to integration mode.

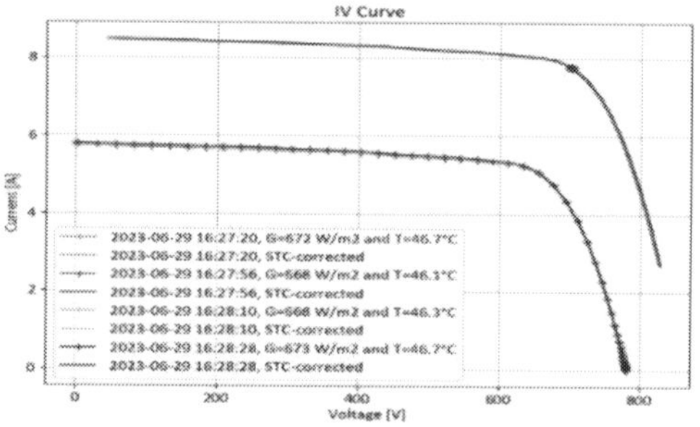

Figure 8 : STC correction of on-site measurement. Gap integrated in numerical expected production.

In lab or experimental platforms results and in real operating condition results are then compared in order to consolidate methodology validation of defaults detection or deviation to identify possible solution to resolve underperformances and maintain the highest possible level of energy production.

This approach aims to distinguish issues coming from BIPV system themselves and those coming from surrounding environment. Another item is to point out undercalculation or mistakes or inappropriate data in the use digital tools.

4 RESULTS REPRESENTATION AND OVER THE TIME DEPENDENCE

A generic graphic representation is attempted to make the investigations carried out more intelligible. The technical dependence of the category will be an important point for the final representation of the current graphs.

4.1 Graphic presentation results

A color code is used to express every stage of methodology with the location of energy performance index level by a colored circle. Projection on expected energy is also represented to give trend of optimal normal evolution and lower expected estimation (colored parallelogram) over the time. Deviations from initial calculation, commissioning and different audit carried out on site are also displayed to underline gaps between any stages and impact on final energy production. Following figures express the step-by-step progress of the methodology, starting with the Figure 9, and the representation of expected energy produced by the installation according to installer's calculation and reference simulation using reference tools.

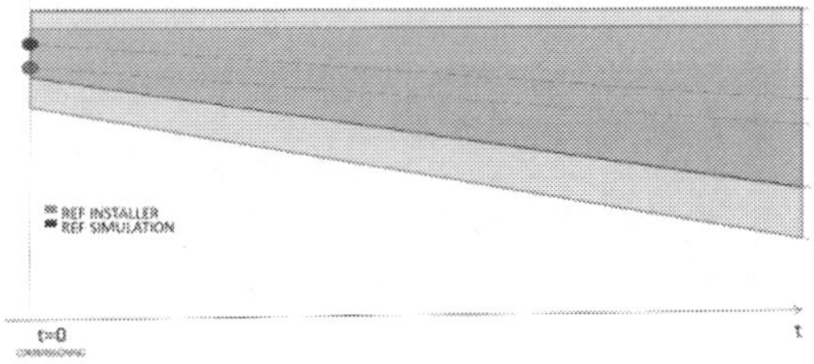

Figure 9 : Reference values (green) of expected energy from installer and by simulation (red), time projection.

On-site audit provide crucial data that allow corrections to be made on simulation results, linked with real operating conditions. Th shift induced by on site data correction is showed on Figure 10 and could be significant.

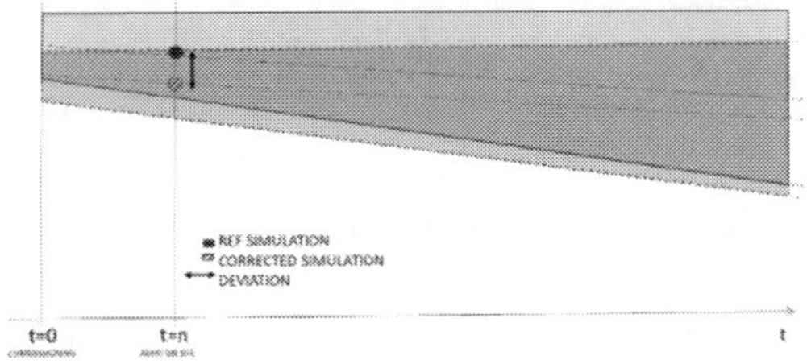

Figure 10 : Corrected simulation (striped red) including collected data and measurements during on-site audit. Deviation displayed (black).

4.2 On site measure implementation

Then, the next main stage is to perform on-site measurement in order to display results in operating conditions. As showed on Figure 11, the difference between ideal energy forecast (green) and on-site measurements are quite different, and expected prevision over the time could be far each other.

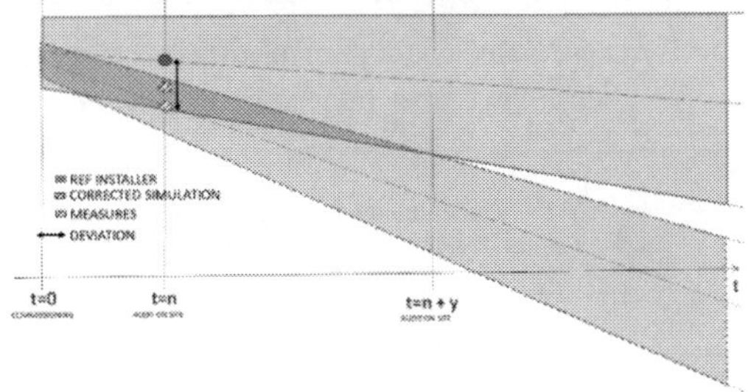

Figure 11 : On site measurement (blue) energy status compared with reference (green) and corected simulation (striped red) - Final deviation dysplayed (black).

An on-site carried out measurement appear to be an essential step of this methodology, allowing to integer in the PR calculation or on the Energy Performance Index all underperformances that could affect a BIPV installation not reachable without an on-site investigation. If long terms degradation as soiling, dusting, or aging could be simple taken into, more unpredictable degradations could significantly affect BIPV production without an accurate identification of causes. Interest of regular audits is to provide higher level validation of the methodology presented, including BIPV category, and bringing more confidence in the identification of causes of energy degradation.

4.3 Time dependance

Stages of this methodology aims to identify BIPV power underperformances and effects on quantifiable indicators as PR, Energy Performance Indicator or Energy Yield. A significant step forward is to consider not only commissioning time but to integer significant influence of time, often considering as aging but not only.

If standard deviations are due to component or system themselves, identified as photovoltaic losses (temperature losses, natural aging, inverter losses…), another set of factors comes from environment stresses or from initial faults (design, poor quality commissioning…).

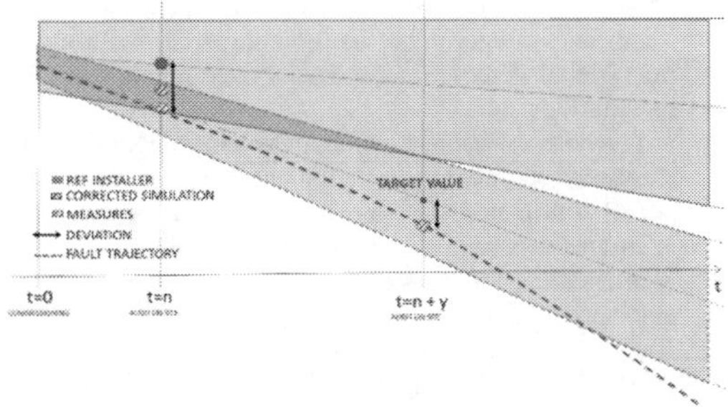

Figure 12 : Over the time audits, expression of faults trend and deviation on expected trajectory. Target value (blue dot) and on site measurement (blue circle).

As an exemple, t=n+y time audit could give a state of large deviation far from expected deviation (target value) and to bring un inappropriate deviation (blue dashed line) and affect significantly ordinate value as showed on Figure 12. Faced with this type of deviation, an investigation on site is carried out to identify the causes of deviation and Figure 3 network is also updated to make future defect identification estimates or prediction more accurate.
This data processing.

4.3 Lessons learned.

Validation of this methodology to support in operation performance measurement of BIPV solution, provide us many lessons learned on necessity to control and supervise quality and evolution over the time on BIPV. First underline the role of audits that are essential not only to validate methodology but also to point out exact issues. Then audits are strengthening even more the accuracy of the self-detection brings by this method and the statistical issue identification, linked with BIPV category.

Secondly, this incremental loop is essential in this

10.4229/EUPVSEC2023/4CV.1.40
020401-004

process and make more precise methodology results to support in operation performance of BIPV installation over the time.

Even if performance guarantee for BIPV solution is crucial for supporting large dissemination, issues or abnormal deviations are still regular and a best know of possible defaults is essential to bring confidence in the use of renewable energy and achievement of green energy contribution in the energy mix.

7 RESULTS AND CONCLUSIONS

Stages of this methodology aims to identify BIPV power underperformances and effects on underproduction and collateral effects on energy payback time and contribution on climatic change.

This method is currently being applied on the experimental platforms of the CSTB where different BIPV solution are under supervision and selected real installations for a final validation. Equipment and the intervention procedures are crucial points to perform a state-of-the-art audit according to relevant standards [8], [9] and provide valuable data for the quality state in operation but also extrapolated expected energy generation.

The identification of possible defects and the modalities of corrections to be made will also be organized and classified to feed a database that can be exploited and extended beyond France. To support these investigations, dysfunctional modules have been provided by CERTISOLIS (case studies) and are used to identify pre-identified categories of pathologies that will make it possible to determine the level of dysfunction of BIPV installations.

On site validation appears to be essential at the current stage of methodology development, but expected results is a full automated data processing allowing to provide accurate estimation of energy deviation with an identification of possible defaults occurring to define appropriate action.

8 NEXT STEPS FORWARDS

Firsts results are convincing on the detection of many deviations from initial study to operating condition. Defaults from the beginning, starting from commissioning, to issues over the time are affecting significantly BIPV installation and must be detected with a high accuracy to mandate maintenance and preserve expected energy generation over the time. Computation of all collected data will allow to predict possible deviation or by identification of deviation trend, and cause identification will provide more confidence in the use of BIPV by increasing sustainability and reliability.

to make this method completely operational and functional, and easily adopted by stakeholder, capability to learn over the time and integer new BIPV solutions or new influencing parameters not already identified is important. Next step forward will be to introduce BIPV sub-family categories.

Final expectations of this methodology are not only dealing with energy performance even if it's the first concern. Results obtained are expected to provide recommendation for all BIPV value chain, from planners to installers, and give the good drivers for investors and insurers. As a direct output, identification of weak points and ranking of default sensibility of BIPV could be applied. Maintenance recommendation as well as specific training could also be identified to streamline on-site intervention and reduce downtime of energy generators.

Continuous implementation of issues will make possible to determine the weight or contribution of each defect on performance indicator and to propose targeted progress operations dedicated to each BIPV categories.

Validation on further sites is under progress for a general application, including weather or climate sensitivity, aging, BIPV category, maintenance operation. Investigation with other countries with different concerns, regulations or codes for a universal solution is in the scope of the next works.

REFERENCES

[1] IEA PVPS T15 : Enabling Framework for the Development of BIPV; https://iea-pvps.org/research-tasks/enabling-framework-for-the-development-of-bipv/

[2] https://iea-pvps.org/wp-content/uploads/2021/09/IEA-PVPS-T15-12_2021_BIPV-categorization_report.pdf

[3] NF EN 50583-1:2016, Photovoltaics in buildings - Part 1 : BIPV modules

[4]Fire safety of BIPV: International mapping of accredited and R&D facilities in the context of codes and standards. Report IEA-PVPS T15-15:2023. ISBN 978-3-907281-39-0

[5] 21th EurObserv'er report. The state of renew able energies in Europe, Ed. 2022. Online report available on www.energies-renouvelables.org

[6] A. Mathieu, S. Boddaert and Al. Failure risk analysis of photovoltaic systems based on literature review. EUROSUN 2022.

[7] Categorization of BIPV applications Breakdown and classification of main individual parts of building skin including BIPV elements. Report IEA-PVPS T15-12:2021. ISBN 978-3-907281-21-5

[8] NF EN 61853-1:2011. Photovoltaic (PV) module performance testing and energy rating - Part 1 : irradiance and temperature performance measurements and power rating

[9] NF EN 61724-1:2021. Photovoltaic system performance - Part 1 : monitoring

ACKNOWLEDGMENTS

Authors would like to thank IEA PVPS T15 members for supporting this work with valuable inputs, disseminating survey and contributions.

This project has received funding from ADEME, French National Agency for the Environment and Energy.

PHOTOVOLTAIC INSPECTION TOOL FOR SOLAR PANELS DEFECTS ON THE FIELD

Daniel Jason Castillo Patton*, Alejandro Barrera del Pozo, Fernando García, Lucas Viani, Ignacio Javier Fernández López, Sofía Rodríguez-Conde
Enertis Applus+, Parque Empresarial Las Mercedes, C/ de Campezo, 1, 28022 Madrid, Spain
Universidad Carlos III de Madrid, Av. de la Universidad, 30, 28911 Leganés, Madrid, Spain
*e-mail: daniel.castillo.p@enertisapplus.com

ABSTRACT: This article presents an innovative approach to automated defect detection in solar photovoltaic (PV) plants using deep learning and infrared imaging technology. The proposed tool is a deep learning neural network that utilizes infrared images to accurately locate and classify defects in installed PV panels. The temperature information extracted from the thermal image metadata enriches the model and enhances the detection accuracy by highlighting temperature differences between defects and other elements in the plant. Unlike traditional computer vision techniques, this deep learning model allows for accurate inference without the need for additional costly phases. The tool presented in this work represents a strategic contribution to the energy industry, enabling plant owners to quickly and efficiently identify and mitigate losses, thus reducing the need for manual labor. Additionally, this research lays the groundwork for future advancements in defect detection methodologies for solar PV plants, thereby driving industry efficiency and sustainability.

1 INTRODUCTION

In recent years, the advancement of renewable energies has undergone significant expansion, witnessing an exponential surge in energy production, particularly in solar photovoltaic technologies, with certain regions of the world, such as China, reaching an annual output of 400 TWh over the last decade [1]. This momentum can be attributed to novel technologies that enable the production of robust and cost-effective solar panels, boasting a projected operational lifespan of 25 to 30 years [2]. Throughout the operational life of these panels, a myriad of potential factors (e.g., weathering, climatic influences, and handling procedures) come into play, leading to the manifestation of defects that can adversely affect their energy generation performance [3]. Such deviations in energy production encompass a broad spectrum, ranging from diminished overall performance to notable losses in power conversion efficiency.

Today, with the proliferation of new installations, as well as the maintenance of old ones, the sector requires the right tools to overhaul thousands of PV modules to keep production levels at their highest. As production increases, it becomes crucial to minimize the margin of error. For instance, in 2022, the United States achieved a production of 204.29 TWh [1]. A mere 1% error in production could result in a significant loss of 2 TWh annually, which is equivalent to powering an entire small city. Therefore, precise and accurate production processes are essential to avoid such substantial losses.

Since the manpower to perform such tasks is very difficult and very costly, the development of new artificial intelligence models is necessary, as well as new analytical techniques to perform these inspections in an efficient and automated workflow. In this work we propose the use of a Deep Neural Network that is capable of handling different types of solar panels in different setups thanks to the large dataset created for this purpose with samples of images in all types of scenarios, including different types of ground, panel inclinations, solar reflections, gradient palette, table structure, etc. It ensures that the model presented in this work is able to process images taken in solar plants located in any region of the globe, while drastically reducing the manpower needed to perform an inspection. The tool presented is a strategic contribution to the energy industry, allowing plant owners to detect and mitigate their losses in a short period of time.

Figure 1. Thermal images captured from a drone from various inspections in an Enertis Applus+ project.

Once the panels are installed in the field, there is a need for methodologies capable of inspecting the PV modules at a rapid pace. In this sector, one of the most widely used and extended technologies is infrared image analysis with the use of drone imaging., because the ease of detecting defects from afar is a key element when performing inspections, taking into account that these involve the analysis of thousands of panels. With this, a deep learning neural network is presented as a detection model with infrared (IR) images of installed panels since the visualization of defects in solar PV plants by using drones (**Figure 1**) is one the most efficient ways to identify defects in the field [5]. A vast majority of the state of the art techniques in the PV industry focus on the detection of defects using electroluminescence (EL) images, due to the quality and efficiency in the search for patterns and defects [4], especially in the fabrication phase.

These artificial intelligence models are increasingly pervasive in the industry, aiming to overcome the most common challenges encountered, such as module shadows, vegetation, ground interference, solar

10.4229/EUPVSEC2023/4CV.1.42
020402-001

reflections, and more [5, 6]. The dataset plays a pivotal role in generating relevant machine learning or deep learning models, as the selection of images is paramount to enable the model to distinguish defects from other effects affecting each panel, as previously mentioned.

In this study, we primarily focus on developing a deep learning model for defect detection in photovoltaic (PV) solar panels, utilizing temperature metadata acquired through infrared (IR) camera imaging during image capture. By incorporating temperature data, our aim is to mitigate some of the most detrimental effects that can affect defect detection, as it provides a more accurate representation of the panel, devoid of optical distortions caused by other factors.

2 RELATED WORKS

2.1 Detection of PV panels

The search for defects in solar panels has been a field of continuous research since the inception of solar energy generation. In this domain, computer vision algorithms, both simple and complex, have been widely utilized. One of the most prevalent methods in computer vision [6] relies on thresholding to efficiently identify contour candidates resembling rectangles in images to detect each of the cells of the panel. Subsequently, texture features are extracted from these contours, enhancing the precision and discrimination in the detection process. The contours are then classified using a Support Vector Machine (SVM), accurately discerning those corresponding to photovoltaic modules.

Nevertheless, a significant limitation of these methods is their reliance on conventional image processing techniques, which involve subjective judgments, rule-based strategies, and hyperparameter tuning. As a result, their performance on unseen images is compromised. This underscores the pressing need for more advanced and adaptable methodologies to effectively detect photovoltaic modules in diverse scenarios. This highlights the need for more sophisticated and adaptive approaches to detect photovoltaic modules in various scenarios.

2.2 Machine learning for panel detection

Deep learning has emerged as a powerful tool in recent years, significantly addressing many of the challenges in this field. Neural networks are commonly employed as detectors [14], providing object localization and classification in images. This methodology is well-suited for defect detection in solar panels due to the wide range of models and architectures available. One widely adopted architecture is Mask R-CNN [6], which effectively addresses detection and localization issues by employing segmentation masks, ensuring accurate and precise detection of panels while avoiding false positives. Other works attempt to tackle panel detection using a ResNet-34-based network [7] combined with semantic segmentation. However, a challenge with semantic segmentation lies in the 'panel' class, where all panel pixels are grouped together without distinguishing individual panels. Additional development is required to address this limitation and achieve finer segmentation. In conclusion, deep learning techniques, particularly Mask R-CNN, have significantly advanced defect detection in solar panels, providing more accurate and efficient results compared to traditional image processing methods. The ongoing development of sophisticated architectures and innovative approaches continues to drive progress in this important field.

2.3 Defects detection in PV panels

As previously mentioned [6], defect detection in PV solar panels has been extensively investigated over time. Thermal images have proven to be a highly valuable tool for capturing anomalies based on temperature variations. Some studies have focused on specific defects, such as the identification of hotspots using a color ramp [8][9]. These hotspots are areas on the solar panel with significantly higher temperatures, which can indicate potential issues or defects in the panel's operation. Other works have explored methods involving binary classification-based filters to distinguish between defective and non-defective regions in thermal images [10]. However, these methods assume that only solar panels are present in the images, which may not always hold true in real-world scenarios where other objects or structures could be present in the background.

2.4 Machine learning for defects detection in PV panels

Machine learning techniques have also been applied to defect detection. For instance, one approach involves segmenting the thermal image to identify characteristic points using a k-means clustering algorithm [6][11]. However, this method is classified as unsupervised learning, meaning it does not require labeled data during training. While this approach can be effective in some cases, it may not provide robust solutions for complex backgrounds or crowded regions with multiple defects. As for the utilization of neural networks, YOLO has been employed in defect detection for solar panels [13]. YOLO is a popular one-stage object detection algorithm that can process images rapidly and detect multiple objects simultaneously. However, one of its limitations lies in the accurate detection of small defects in solar panels, as the use of anchor boxes for object localization may hinder the precise detection of smaller anomalies [14]. Another neural network-based model, utilizing a detection network with reNet-50, has been successfully employed to identify various thermal defects in solar panels [5]. The model achieved an impressive 90% precision in classifying ten different types of defects, showcasing the potential of deep learning methods in the domain of solar panel defect detection. It's important to acknowledge that while YOLO and neural network-based models offer promising results, they also present certain limitations, such as the need for large annotated datasets, challenges in generalizing to new defect types, and potential computational demands during inference.

3 METHODOLOGY

In this section, we will describe in detail the process of obtaining the data used in this study, covering both its acquisition and its inherent characteristics. Subsequently, we will carry out a concise review of the selected model, highlighting its advantages for the execution of the proposed detection task and explaining its mode of operation.

10.4229/EUPVSEC2023/4CV.1.42
020402-002

3.1 Data acquisition and structure

The dataset data is composed of a series of images collected from several inspections owned by Enertis Applus+. These plants belong to several different regions of the world, including Europe and South America, so the context of the images (soil, type of panel, vegetation, etc.) is different, ensuring heterogeneity in the dataset.

An inspection of a solar power plant consists of unmanned aerial vehicles (UAVs) flying over a solar photovoltaic plant along the strings of panels. The speed and frequency of the camera triggering depends on each inspection and each technician, adjusting the parameters according to altitude and time. The acquisition of the images is done with thermographic cameras that capture infrared radiation, usually from DJI or FLIR manufacturers. This is important, since each manufacturer has its own algorithms to decode the temperature information of the metadata, which as we will see later is crucial in this work. A predominant feature in this imaging genre is the uniform resolution of 640x512 pixels across all images. This choice of resolution is due to the delicate balance between thermal imaging capture resolution and practical considerations of sensor dimensions and data processing capabilities.

After the data acquisition stage, images exhibiting defects are collected and subjected to a thorough evaluation by solar panel experts. These technicians conduct a thorough visual review of the images, accompanied by detailed annotations covering all defects identified in each snapshot. Once this phase is completed, a robustly enriched set of images is available, designed to feed the artificial intelligence model with relevant and diverse information. The data and classes can be seen in the (see Table I).

Table I: Dataset.

Data	Train	Validation	Total
Images	3948	986	4934
Dft: Total	10808	2779	13587
Dft: Spot	5021	1349	6370
Dft: Diode	2435	624	3059
Dft: Multiple spot	2817	673	3490
Dft: Open circuit	535	133	668

*Dft = Defect

3.2 Deep Learning Model - FASTER R-CNN

The selection of an appropriate model architecture for the detection of solar panel defects in thermal images is a critical aspect of the successful accomplishment of this task. In this context, we have chosen to use the Faster R-CNN (Region-based Convolutional Neural Network) architecture, implemented in the Detectron2 library [17], based on the following justifications:

1. High performance in object detection:
The Faster R-CNN architecture has demonstrated outstanding performance in the object detection domain, emerging proficiently compared to other alternatives. Its two-stage approach, involving region of interest (RoI) generation followed by classification and regression, has proven to be effective in accurately and robustly capturing objects of different dimensions and geometries that are inherent in thermal images of solar panels [15].

2. Versatility for detecting different size areas of interest:
Solar panels exhibit a range of flaws that encompass various sizes and shapes, including larger areas of concern for searching defects (a defect can appear in a small corner of a panel or it can be the whole panel). The two-step approach of Faster R-CNN, which involves identifying potential problem regions and then enhancing detections using classification and regression, inherently proves to be well-suited for handling this array of defect dimensions. This characteristic establishes it as a relevant choice for effectively tackling the task of identifying defects in solar panels through thermal imaging. [15][16].

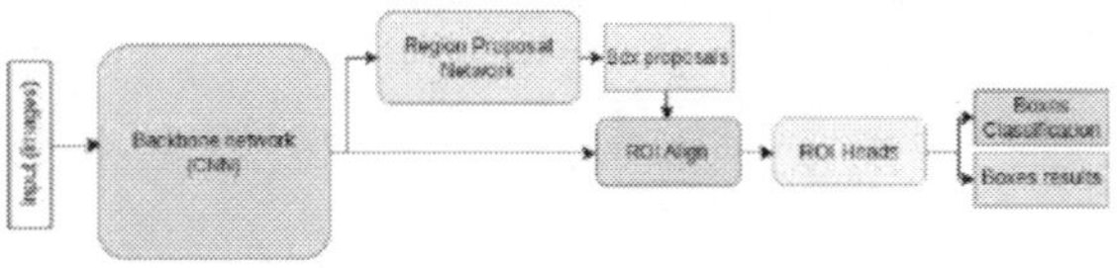

Figure 2: Basic architecture for a Faster RCNN model, showing the different steps performed to realize a detection.

3. Adaptability and customization capability:
Detectron2 provides a modular and flexible platform that empowers efficient adaptation of important model components and hyperparameters, such as the backbone used, feature extraction techniques, anchor sizes, etc.[17]. Such versatility makes it possible to precisely optimize model performance for the task at hand.

4. Availability and reach with the scientific community:
Detectron2, an open source library developed by Meta's Artificial Intelligence research group, ensures access to comprehensive documentation, tutorials, and implementation examples. The presence of an active community of scientists and researchers further consolidates its worth by providing a constant flow of improvements and solutions to specific challenges. This supportive ecosystem facilitates effective problem-solving and iterative model optimization for solar panel defect detection in thermal imaging.

In summary, the choice to employ the Faster R-CNN architecture implemented in Detectron2 for the task of solar panel defect detection in thermal imaging is supported by its empirical performance [15], its suitability for the detection of objects of various dimensions [18], its adaptability, and its foundation in a robust documentation base and a dedicated scientific community. These premises corroborate that the choice of Faster R-CNN in Detectron2 is a valid and justified option to successfully address solar panel defect detection in thermal imaging.

4 TEMPERATURE DATA INTEGRATED WITH THE FASTER R-CNN

The use of a color scale in images captured by thermal cameras presents immediate visual advantages by highlighting temperature differences in a scene in an intuitive manner. However, this chromatic representation does not accurately reflect precise numerical temperature values and can lead to misunderstandings in interpretation. Since different palette systems may assign colors differently to the same temperature range, the consistency and correlation between colors and temperatures may vary [19].

In addition, the limited color resolution in thermal maps can lead to confusion when representing small temperature differences with similar colors, which could compromise accuracy in applications that require accurate detection and diagnosis. This visual coding can also be misleading by suggesting continuous gradations in temperature that do not exist in reality, potentially leading to erroneous decisions in any type of application. Therefore, it is essential to carefully consider the disadvantages of color-based representations and weigh them against the need for accuracy and understanding in each application context.

In this paper, our approach is aimed at enhancing the precision of the model by integrating temperature data, thus allowing an accurate estimation of the defects present in the panels. To achieve this purpose, we have enriched the Faster R-CNN architecture with the addition of an additional channel specially dedicated to thermal data. This design strategy **(Figure 3)** ensures that during the process of assigning weights for feature extraction, the actual temperature values associated are considered with each specific region of the image, resulting in a much more effective system as a whole.

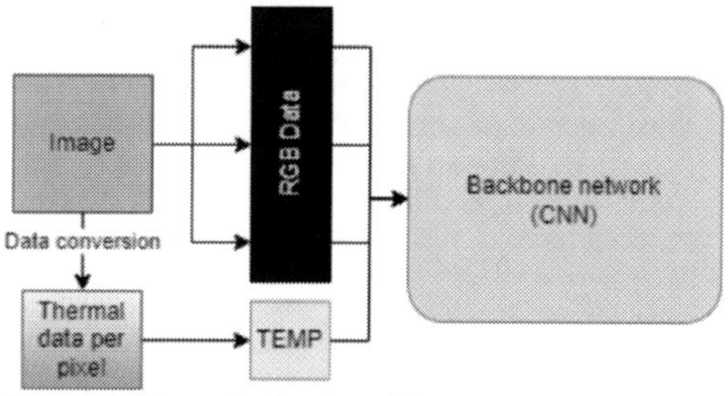

Figure 3: Example for the additional channel included into the architecture of the model, after being processed to extract the temperature data.

This extension of the input channels in the Faster R-CNN architecture plays an essential role in improving both the accuracy and robustness of the model in detection tasks. The inclusion of additional channels, such as temperature records in the context of solar panel inspection, enriches the captured information and is crucial for the detection of potential defects. Recent research [20] in the field of computer vision on Faster R-CNN, has conclusively confirmed the effectiveness of incorporating extra channels to boost feature representation and achieve remarkable improvements in object detection. Furthermore, studies following similar structures as the Faster R-CNN [21], explore specific optimizations for the same framework and strongly support the notion that the inclusion of specialized input channels can boost overall model performance and adaptability to various application contexts. Also, this technique of treating the data minimizes the visual impact of some of the effects that undermine the capability of defect detection, such as the sun reflections as we can see in the **Figure 4**.

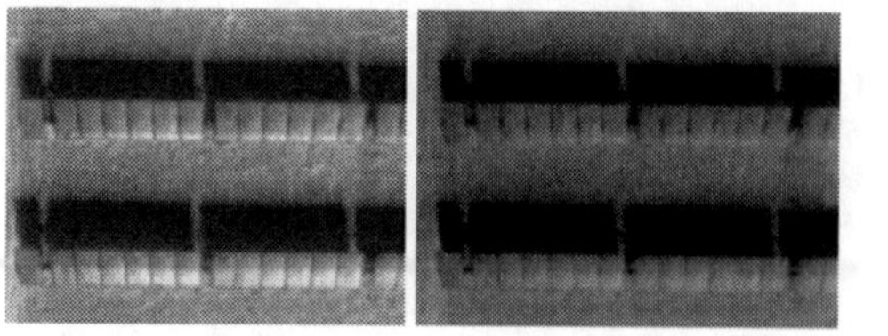

Figure 4: Left: Image in the standard visualization of a thermal image. Right: Image processed with the temperature data.

In the context of thermography and computer vision, radiometric calibration plays a fundamental role. Its objective is to establish a quantitative correlation between the radiant intensity captured by the camera and the actual radiation emitted by the surface under analysis [22]. The focus of some works explored methodologies aimed at achieving a meticulous calibration, taking into account environmental factors and implementing emissivity corrections. These measures are essential to ensure high fidelity thermographic measurements as the valor for emissivity, as an incorrect value that is not adjusted for the task to perform can provide incorrect data (1) as Tcorr is the corrected value taking into consideration the object with an emissivity E (each material has its own predefined emissivity). The process of validation through experimentation is of fundamental importance, as it supports the reliability and consistency of thermographic measurements [6]. This validation not only reinforces the reliability of the measurements, but also strengthens the ability of thermal cameras to provide accurate temperature estimates in a variety of industry-relevant applications.

$$(1) \quad Tcorr = \frac{Tmeasured}{\sqrt[4]{Emissivity}}$$

To acquire temperature data, we retrieve metadata extracted from either a DJI or FLIR camera (depending on the inspection conducted at that moment), and employing the manufacturer's library, we extract the encoded information from the metadata, capable of converting pixel data into actual temperature values. Through the utilization of these transformed images, we obtain our fourth channel, with each channel characterized as follows:

- R: intensity of the red color in each pixel
- G: intensity of the green color in each pixel
- B: intensity of the blue color in each pixel
- T: real temperature value in each pixel

5 MODEL TRAINING AND EVALUATION

In this section, we will explore the data used in the model training process, with the purpose of providing a concise explanation of the most salient parameters that have been selected for a more effective understanding.

5.1 Training

For the training of the Faster-RCNN model, the adjacent table shows the selected characteristics (see Table 2).

Table II: Training selection.

Parameters settings	
Learning Rate	0.001
Iterations	100000
Batch Size	2
Loss Function	SmoothL1Loss + CrossEntropyLoss
Optimizer	SGD
Data Augmentation	Horizontal Flipping
Backbone	ResNet101
Regularization	L2 (0.0001)
Learning Schedule	(70000,99999)
Activation Function	ReLU

The loss function is the one that allows us to measure the discrepancy of the model with the real data we are using to train it. In this case, the functions we use are SmoothL1Loss and CrossEntropyLoss. The former is used to quantify the discrepancy between the anticipated coordinates of the bounding boxes and their real counterparts. This function mitigates the adverse consequences of outliers, thus ensuring accurate estimation of locations. On the other hand, the CrossEntropyLoss is used to calibrate the discrepancy between the probability distributions projected by the model and the true distributions of the target categories. In a detection context, this function concerns the categorizing predictions of each generated anchor, contributing fundamentally to the appropriateness of label assignments to regions of interest.

L2 regularization is used in model detection to penalize larger weights during training, thus controlling overfitting and improving model generalization.

The selected hyperparameters have been chosen to optimize the model output in order to obtain the best results.

The search for a more agile initial parameter setting and the need to maintain stability during the training process emerge as primary considerations. Opting for a moderate learning rate allows a steady progression towards the global minimum of the loss function, mitigating the inherent risks of divergence that may materialize when approaching higher learning rates.

Then, after having detailed the context of the parameters chosen for the training process, we proceed to expose the approach adopted to evaluate the model. This approach stands out for its ability to corroborate the veracity and consistency of the actions taken, thus supporting the integrity of the information presented.

5.2 Evaluation metrics

For the evaluation of the model, an algorithm has been created based on the definitions of Precision (2), Recall(3) and F1-Score(4).

$$(2) \quad Precision = \frac{TP}{TP+FP}$$

Although these metrics are widely known, the accuracy allows us to see how much we are wrong in our prediction of each class, and whether we are confusing panel defects with each other.

$$(3) \quad Recall = \frac{TP}{TP+FN}$$

In the field of defect detection, recall in our case is the metric we consider most important of all, as it determines the total number of defects found in the image as a whole. For our case, it is paramount to find all defects during an inspection, and although it is important not to make a mistake with the defect class, it is something that is easily correctable a posteriori, while failure to find a defect and miss it is a major problem in the long run.

$$(4) \quad F1\,Score = 2 \cdot \frac{Precision \cdot Recall}{Precision + Recall}$$

The F1-score, is a key metric in the evaluation of classification models, provides a balanced measure of accuracy and recall. Its contribution lies in considering both false positives and false negatives, providing a complete picture of the model's performance in class identification, especially in unbalanced scenarios. In summary, the F1-score helps to understand how the model balances accuracy and completeness in its classification capability.

5.3 Results

For our evaluation results, we selected a dataset containing available metadata and generated two distinct datasets: one comprising basic RGB images captured by the thermal camera (standard), and another including the same images augmented with temperature information as detailed in point 4.

Presented below are the performance metrics for each class across both models. Firstly, the outcomes for the standard model (see Table III):

Table III: Standard model results.

Class	Precision	Recall	F1-Score
spot	0.836	0.885	0.856
diode	0.917	0.951	0.934
multiple_spots	0.823	0.861	0.841
open_circuit	0.75	0.789	0.769
Average total	0.832	0.871	

Moving on to the results for the temperature-augmented model (see Table IV):

Table IV: Temperature model results.

Class	Precision	Recall	F1-Score
spot	0.885	0.902	0.893
diode	0.906	0.959	0.932
multiple_spots	0.823	0.889	0.855
open_circuit	0.855	0.791	0.822
Average total	0.867	0.885	

Considering this comparative analysis, we synthesize the results to facilitate direct metric comparison (see Table V):

10.4229/EUPVSEC2023/4CV.1.42

Table V: Comparative results between the models.

Metric	Standard Model	Temperature Model
Iteration	100K	100K
AP	0.832	**0.867**
Recall	0.871	**0.885**
F1-Score	0.851	**0.876**

It's important to note that the data volume and training process remain consistent between both cases, with the only variance being the additional channel.

Let's review some comparative cases to see how the model works in each case (**Figures 5-7**):

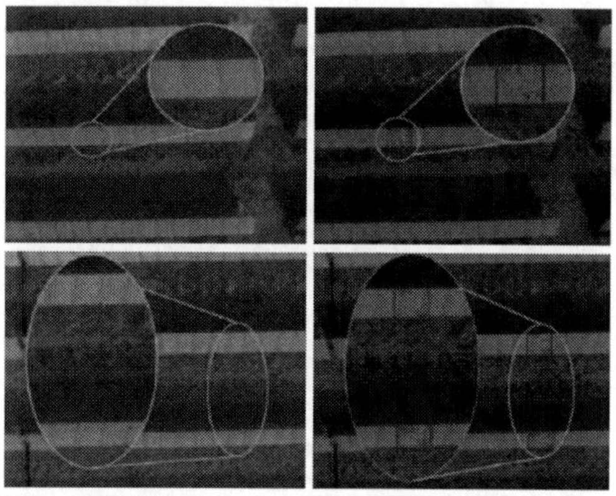

Figure 5: Differences between models: false negatives. Top and bottom left: image treated with the trained standard model. Top and bottom right: image treated with the trained temperature model. Defect found in temperature model versus the standard model.

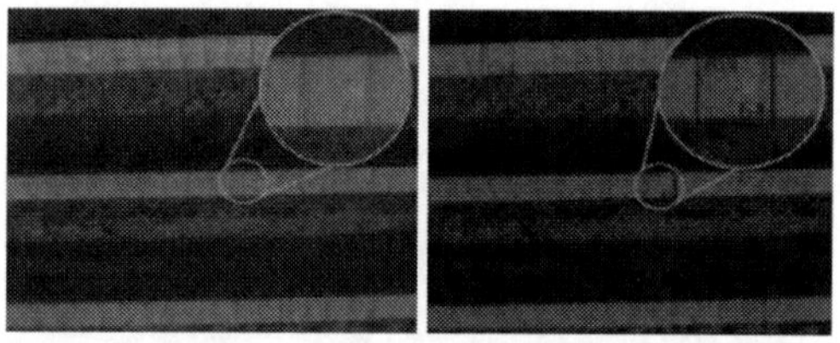

Figure 6: Differences between models: false negatives. Top and bottom left: image treated with the trained standard model. Top and bottom right: image treated with the trained temperature model. Additional detections found in the temperature model.

As we can see in this case, the temperature model has been able to detect a defect that is not seen in the standard model.

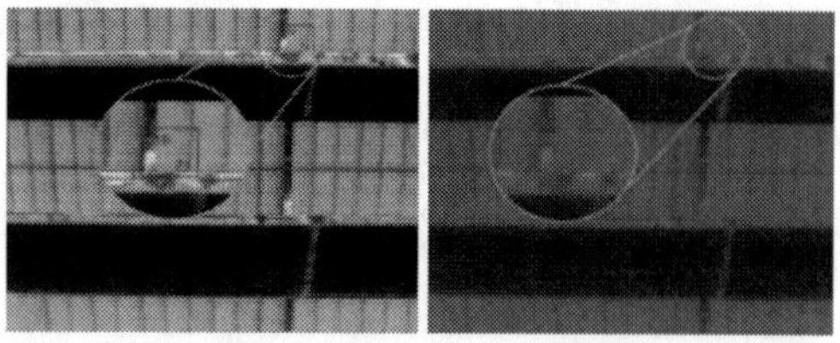

Figure 7: Differences between models: false positives. Left: image treated with the trained standard model. Right: image treated with the trained temperature model.

In this second case (**Figure 7**), we can see that in the standard model a detection appears, which is actually a false positive. This happens because the pattern of the cut between the series of panels (strings) appears on the floor, and the temperature of the floor becomes higher than that of the panels. The color palette representation accentuates this difference more, making the area more noticeable, and causing the model to make an incorrect prediction. With the temperature model, when we have the images with the temperature data we see that this accentuation does not occur ergo this erroneous detection does not appear.

6 CONCLUSIONS AND FUTURE WORKS

In this study, we undertook an evaluation and comparison of two object detection models: the standard model and the temperature-augmented model. Our inquiry centered on pivotal metrics including Average Precision (AP), Recall, and F1-Score to gauge the efficacy and quality of both approaches. Our findings indicate that the temperature-augmented model exhibited superior performance across all evaluated metrics compared to the standard model. This outcome suggests that integrating temperature information into the training methodology led to notable enhancements in both accuracy and robustness of the object detection system. These findings underscore the significance of considering advanced strategies, such as temperature augmentation, to optimize detection model performance. Importantly, these outcomes hold practical implications for applications requiring precise object detection under diverse conditions and environments.

As a future line of research, the adaptability of the temperature approach to other model architectures and datasets could be explored, as well as investigating its application in other types of scenarios to further validate its usefulness and effectiveness.

7 REFERENCES

[1] Hannah Ritchie, Max Roser and Pablo Rosado (2022) - "Renewable energy - Breakdown of renewables in the energy mix". Published online at OurWorldInData.org. Retrieved from: 'https://ourworldindata.org/energy' [Online Resource]

[2] Gerarden, T. D. (2023). Demanding innovation: The impact of consumer subsidies on solar panel production costs. *Management Science.* https://doi.org/10.1287/mnsc.2022.4662

[3] Subarnan, Gayathri M.; Damodaran, Manimegalai; Madhu, Karthikeyan, Recent Advances in Electrical & Electronic Engineering (Formerly Recent Patents on Electrical & Electronic Engineering), Volume 15, Number 8, 2022, pp. 607-620(14), Bentham Science Publishers. https://doi.org/10.2174/2352096515666220620093933

[4] Deitsch, S., Buerhop-Lutz, C., Sovetkin, E., Steland, A., Maier, A., Gallwitz, F., & Riess, C. (2021). Segmentation of photovoltaic module cells in uncalibrated electroluminescence images. Machine Vision and Applications, 32(4). https://doi.org/10.1007/s00138-021-01191-9

[5] Bommes, L., Pickel, T., Buerhop-Lutz, C., Hauch, J., Brabec, C.; Peters, I. M. (2021). Computer vision tool for detection, mapping, and fault classification of photovoltaics modules in aerial IR videos. Progress in Photovoltaics: Research and Applications, 29(12), 1236–1251. https://doi.org/10.1002/pip.3448

[6] Vega Díaz, J. J., Vlaminck, M., Lefkaditis, D., Orjuela Vargas, S. A.; Luong, H. (2020). Solar panel detection within complex backgrounds using thermal images acquired by uavs. Sensors, 20(21), 6219. https://doi.org/10.3390/s20216219

[7] Zhang, H., Hong, X., Zhou, S., & Wang, Q. (2019). Infrared Image segmentation for photovoltaic panels based on Res-UNet. *Pattern Recognition and Computer Vision*, 611–622. https://doi.org/10.1007/978-3-030-31654-9_52.

[8] Dhimish, M.; Alrashidi, A. Photovoltaic degradation rate affected by different weather conditions: A case study based on pv systems in the uk and australia. Electronics 2020, 9, 650. [CrossRef]

[9] Libra, M.; Daneˇcek, M.; Lešetický, J.; Poulek, V.; Sedláˇcek, J.; Beránek, V. Monitoring of defects of a photovoltaic power plant using a drone. Energies 2019, 12, 795.

[10] Liao, K.C.; Lu, J.H. Using Matlab real-time image analysis for solar panel fault detection with UAV. In Journal of Physics: Conference Series; IOP Publishing Ltd.: Bristol, UK, 2020; Volume 1509.

[11] Uma, J.; Muniraj, C.; Sathya, N. Diagnosis of photovoltaic (PV) panel defects based on testing and evaluation of thermal image. J. Test. Eval. 2019, 47.

[13] Greco, A.; Pironti, C.; Saggese, A.; Vento, M.; Vigilante, V. A deep learning based approach for detecting panels in photovoltaic plants. In Proceedings of the ACM International Conference Proceeding Series, Las Palmas de Gran Canaria, Spain, 7–9 January 2020.

[14] Diwan, T., Anirudh, G. & Tembhurne, J.V. Object detection using YOLO: challenges, architectural successors, datasets and applications. *Multimed Tools Appl* 82, 9243–9275 (2023). https://doi.org/10.1007/s11042-022-13644-y

[15] Ren, S., He, K., Girshick, R., & Sun, J. (2017). Faster R-CNN: Towards real-time object detection with region proposal networks. *IEEE Transactions on Pattern Analysis and Machine Intelligence*, *39*(6), 1137–1149. https://doi.org/10.1109/tpami.2016.2577031

[16] Girshick, R. (2015). Fast R-CNN. *2015 IEEE International Conference on Computer Vision (ICCV)*. https://doi.org/10.1109/iccv.2015.169

[17] Wu, Y., & Kirillov, A. (2019). Detectron2. GitHub Repository.

[18] Zhu, J., Deng, Z., Ke, G., & He, X. (2020). End-to-End Object Detection with Transformers. arXiv preprint arXiv:2005.12872.

[19] Mantiuk, R., Myszkowski, K., & Seidel, H.-P. (2006). A perceptual framework for contrast processing of high dynamic range images. *ACM Transactions on Applied Perception*, *3*(3), 286–308. https://doi.org/10.1145/1166087.1166095

[20] Lin, T.-Y., Goyal, P., Girshick, R., He, K., & Dollar, P. (2017). Focal loss for dense object detection. *2017 IEEE International Conference on Computer Vision (ICCV)*. https://doi.org/10.1109/iccv.2017.324

[21] He, K., Gkioxari, G., Dollar, P., & Girshick, R. (2017). Mask R-CNN. *2017 IEEE International Conference on Computer Vision (ICCV)*. https://doi.org/10.1109/iccv.2017.322

[22] James CA, Richardson AJ, Watt PW, Maxwell NS. Reliability and validity of skin temperature measurement by telemetry thermistors and a thermal camera during exercise in the heat. J Therm Biol. 2014 Oct;45:141-9. doi: 10.1016/j.jtherbio.2014.08.010. Epub 2014 Sep 3. PMID: 25436963.

[23] Fendri, E., Boukhriss, R. R., & Hammami, M. (2017). Fusion of thermal infrared and visible spectra for robust moving object detection. *Pattern Analysis and Applications*, *20*(4), 907–926. https://doi.org/10.1007/s10044-017-0621-z

CHALLENGES OF AERIAL DRONE ELECTROLUMINESCENCE IN SOLAR PHOTOVOLTAIC FIELD INSPECTIONS

Rodrigo del Prado Santamaría, Gisele A. dos Reis Benatto, Thøger Kari, Luca Morino, Peter B. Poulsen and Sergiu V. Spataru
Department of Electrical and Photonics Engineering, DTU Electro, Technical University of Denmark
Frederiksborgvej 399 4000 Roskilde, Denmark

ABSTRACT: This work aims to give insight into the challenges of performing aerial daylight drone Electroluminescence (EL) inspections in PV farms. In this context, we have compared an EL inspection process using a tripod to record the images to an aerial drone EL inspection system. We focus on the challenges of the "Stop and Go" procedure for recording EL images with the drone, which consists of stopping the drone module by module during the inspection flight. The drone inspection tests resulted in an average module inspection time of 24 seconds per module and around 8 minutes per PV string. Regarding image processing challenges, drone inspection requires an added module tracking algorithm due to the drone motion which introduces some artifacts in the image but does not lower the image quality and diagnostic information of the images compared to stationary daylight EL. Images acquired with both the drone and tripod fail in identifying microcracks in the images but do well in detecting larger B and C cracks. For obtaining the best possible image quality, it is essential to have a calibrated camera, avoid shadows from the drone when recording, and avoid taking images with fast-changing irradiance.
Keywords: Electroluminescence, drone, cell cracks, motion, irradiance

1 INTRODUCTION

Photovoltaic (PV) energy installed capacity is growing year after year, and a big part of this growth comes from the commission of large utility-scale PV plants. With the increase in the size of PV installations come new challenges in reliability. Faster and automated ways of inspecting PV panels are crucial for maintaining the performance of the plants over the years. Electroluminescence imaging (EL) is a widely used diagnostic method for PV panels that, since its introduction [1] has been used to identify material defects in PV modules and cells. EL imaging is typically performed in dark conditions; however, the need for a better diagnostic on-site has popularized this imaging application in broad daylight.

High image throughput is important for optimizing plant maintenance operations and reducing costs when performing daylight EL inspections. Conventional outdoor EL imaging inspections in the field are done by mounting the camera on a tripod [2]. However, that is a limiting factor of image throughput because of the time and manpower it takes to move the camera from module to module and from string to string. For that reason, improvements in the camera carrier device are being made, the most promising being quad and hexacopter drones [3]–[6]. These have been successfully integrated with other PV diagnostic methods, such as infrared thermography (IRT) [7].

Using stationary camera carriers for EL inspections, a specialized PV plant inspection company with an optimized setup can have a high image throughput. It has been reported that these inspections can obtain EL images of as much as 400 kWh per 8 hours of the working day, which approximately corresponds to 1 min per module [8].

Drone carriers allow a faster inspection time; however, their development introduces other concerns in the inspection process. A differentiation can be made between inspection procedures using drones:

• *Stop and Go*: This procedure consists of flying over the modules with the drone and stopping above each individual PV module for image acquisition.

• *Drone scan*: This procedure consists of flying across a whole PV string and recording EL images without stopping at any time.

In this work, a drone EL inspection system has been developed, integrating the EL camera into a rotating gimbal and investigating the "*Stop and Go*" methodology. The tests have been carried out at DTU's PV field in Risø, Denmark, using the drone and a tripod as camera carriers to compare the methods and output images. Furthermore, indoor EL images of some of the inspected modules have been recorded as a baseline for comparison. With the introduction of motion in the measurement system, extra challenges should be addressed for obtaining EL images of good quality that can be used to diagnose the PV module's health. The following sections will discuss these inspection challenges, recommendations, and diagnostic potential of the resulting EL images.

2 METHODOLOGY

2.1 EL daylight inspection

The presence of irradiance noise in the image sequences captured in daylight EL inspections typically requires injecting a modulated current signal into the modules and extracting the EL signal in postprocessing. In these inspections, the PV string current was modulated between I_{sc} and $-I_{sc}$, as shown in Fig. 1. To reach $-I_{sc}$, a bidirectional power supply, Delta Elektronik 15 kW, has been used.

For the daylight EL inspection using the tripod as a camera carrier, a modulation frequency of 30 Hz was used, while a modulation frequency of 40 Hz was applied for the drone EL inspection. For all experiments, indoor EL, daylight EL with the tripod, and drone EL, a C-RED 3 InGaAs camera from First Light Imaging was used. On top of the camera, a sapphire lens of 25 mm focal length and an optical bandpass filter of 1150 nm center wavelength and 25 nm bandwidth were used. For the daylight EL inspection using the tripod, the camera was running at 480 fps. Thus, this configuration acquires 16 image frames per modulation period, recording eight images at I_{sc} and eight at $-I_{sc}$. In the drone inspection, the camera was running at 320 fps. We record eight image frames per period with a

10.4229/EUPVSEC2023/4CV.1.43
020403-001

modulation frequency of 40 Hz, acquiring four images at I_{sc} and four at $-I_{sc}$.

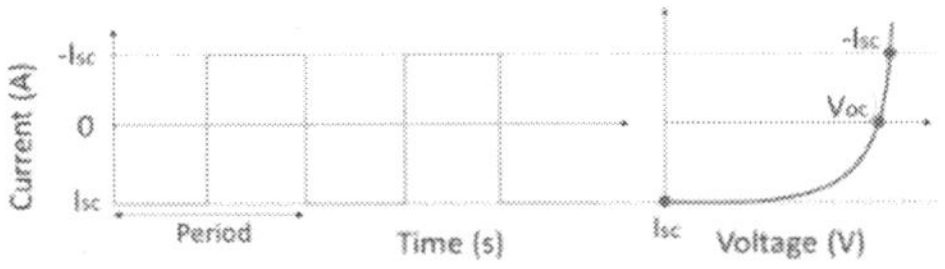

Figure 1: Electrical modulation applied to the PV strings. Current modulation from I_{sc} to $-I_{sc}$.

Fig. 2 shows the inspection procedure using the drone platform as the camera carrier. The drone was flying approximately 630 cm from the PV string and at a height of 270 cm while the modules were tilted 27°. Fig. 3 shows the field of view of the camera in the drone during the inspection, where the module being inspected has been highlighted in yellow. The measured plane of array (POA) irradiance was around 1000 W/m^2 for the case of the stationary inspection and between 600 W/m^2 and 1000 W/m^2 when the drone inspection was carried out.

Figure 2: Drone daylight electroluminescence inspection in process. Drone position and height relative to the PV string shown in white. PV string stands at 27° tilt.

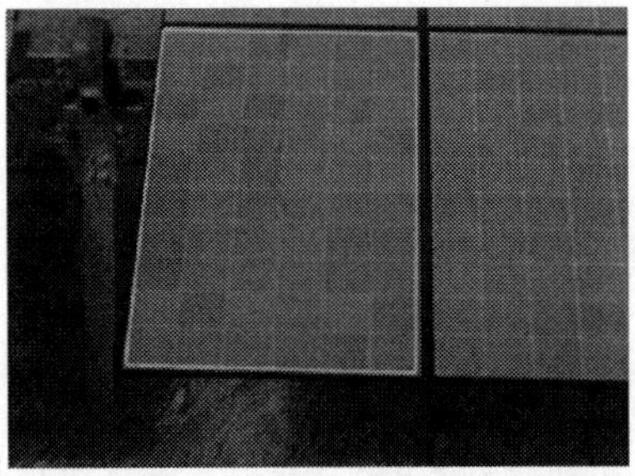

Figure 3: Drone camera field of view during inspection. One module being imaged. Module selection marked in yellow overlay.

2.2 Drone specifications

The drone setup consists of a DJI Matrice 600 pro with a payload attached to it. The payload contains the same InGaAs camera, lens, and filter described previously.

In the field, the whole drone setup consists of a ground station for remote control, where one operator supervises the drone flight, and another controls the gimbal and image acquisition. The ground station is connected via radio to the payload, where an embedded PC controls the gimbal and the camera GUI and records and saves the images locally on the drone. Fig. 4 shows a diagram of the drone payload components and their connections.

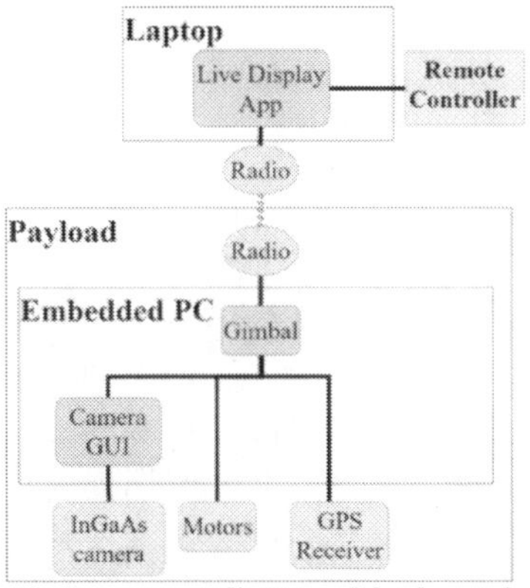

Figure 4: Components diagram of the drone gimbal.

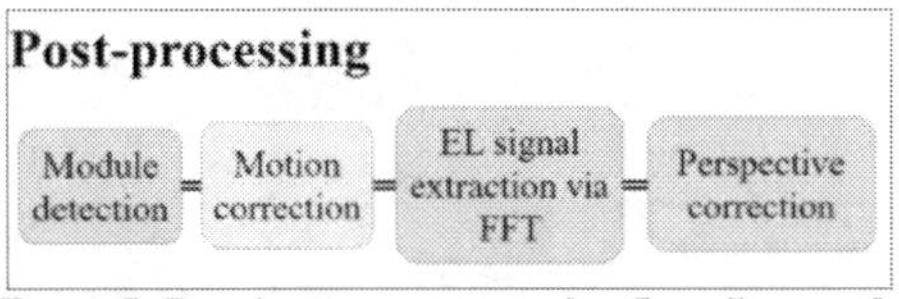

Figure 5: Raw image postprocessing flow diagram for both stationary and drone daylight EL inspection.

2.3 Image processing

The image processing steps differ in the drone EL inspection from the stationary. Fig. 5 shows the postprocessing steps of the raw image sequences recorded in the field.

When inspecting with the drone, the whole process consists of module detection and motion correction to correct the movement from frame to frame, both from the drone and gimbal displacement. Despite this being the *Stop and Go* methodology, slight motion due to drone and gimbal drift cannot be avoided and needs to be dealt with in postprocessing.

For the extraction of the EL signal, different algorithms are available in the literature, such as lock-in and Fast-Fourier Transform (FFT) [9]–[11]. The FFT-based algorithm introduced by T. K. Hass et al. [9] has been used for these inspections. Finally, the modules' perspective is corrected to obtain the final EL image.

For the stationary inspection, the same methodology applies except for the motion correction algorithm, which is not necessary in this case, marked in yellow in Fig. 5.

Throughout this work, a metric defined as "*motion*" will be presented together with the output EL image from the drone. This metric is the motion caused by the drone and gimbal movement captured in the raw images, expressed as the average pixel displacement of a module per image frame (px/frame).

Another metric that will be shown throughout the analysis, both for drone and tripod inspections, is the so-called "*irradiance variation*". This metric represents how much the irradiance changed over the time it took to complete the recording of the raw sequence. This metric is a calculation based on the mean pixel value of each frame in the recording as expressed in (1), where $\bar{\Phi}_f$ is the mean pixel value of each frame *f* of the raw sequence, with the EL signal excluded through a suitable interpolation.

$$Irr.Variation = \max(\bar{\Phi}_f) - \min(\bar{\Phi}_f) \quad (1)$$

3 RESULTS AND DISCUSSION

3.1 EL imaging inspection comparison

Fig. 6 compares the EL images acquired indoors, in daylight stationary, and with the drone of three PV modules from the inspected string. One cell with B and C cracks zoomed in for the same module in the different EL images is shown in Fig. 7.

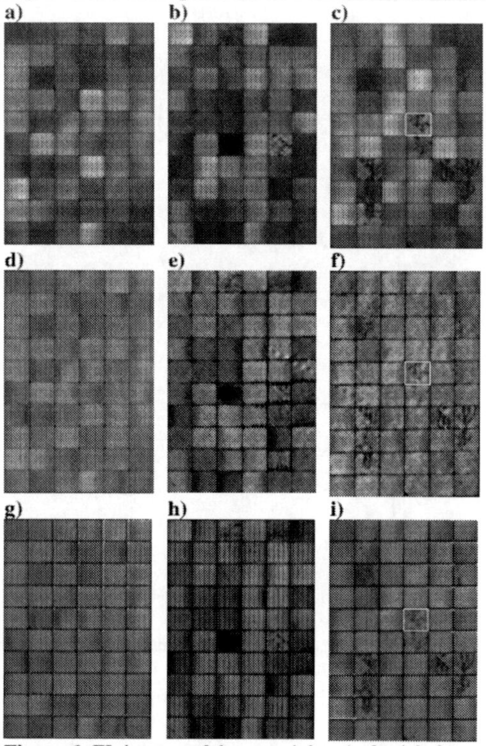

Figure 6: EL images of three modules, **(a, b, c)** indoor EL, **(d, e, f)** daylight stationary EL, **(g, h, i)** daylight drone EL. **g)** motion 0.23 px/frame. **h)** motion 0.35 px/frame. **i)** motion 0.28 px/frame.

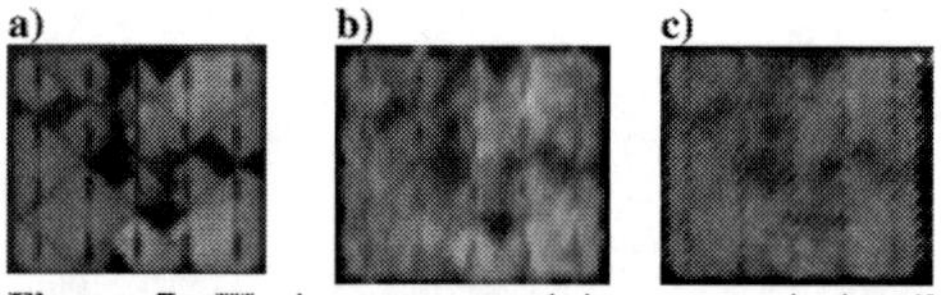

Figure 7: EL image zoomed into a cracked cell (marked with a white square) from Fig. 6. **a)** Indoor EL. **b)** Daylight stationary. **c)** Daylight drone.

The first important takeaway is that microcracks are hard to detect in any of the daylight EL images. In the case of the daylight tripod inspection, it has been previously reported that microcracks can be detected [3], [8]. On the other hand, drone and tripod can detect larger defects such as B and C cracks.

The drone images tend to present a pattern with diagonal lines, evident in the spacing between cells. This effect comes from the perspective correction and module tracking algorithms. It can be minimized by having a camera field of view perpendicular to the module and by minimizing the motion to keep the drone and the gimbal stable.

Furthermore, overexposure in the image recording must be avoided, as it can occur in high irradiance conditions if the camera exposure time is not chosen accurately. In Fig. 6e the effect of overexposure in the image can be seen as entirely dark regions in the cells. Such image artifacts could be mistaken as B or C cracks.

3.2 Challenges of Drone EL inspection

Several extra challenges were identified during the PV plant inspection and should be addressed for an optimal inspection process.

As previously reported by T. Kari et al. [12] camera calibration is crucial for obtaining a usable EL image with the current image processing methodology.

The camera needs to be calibrated appropriately with bias frame and flat-field correction. Pixel errors will not appear after processing the stationary EL images. However, in images with motion, such as the ones acquired with the drone, the pixel offset errors will introduce image artifacts that will move around and appear in the final image. Fig. 8 shows the same module in the PV string, inspected with the drone with and without camera calibration. After processing, a clear line crossing the module can be seen in the uncalibrated image.

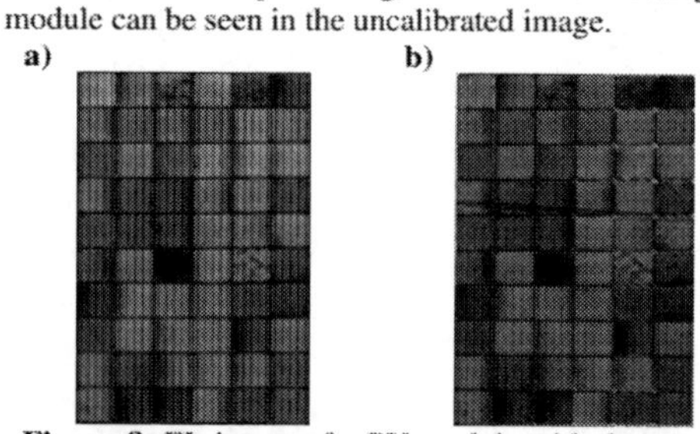

Figure 8: EL image of a PV module with the camera calibrated with bias and flat-field correction **(a)** and without calibration **(b)**.

An extra challenge when flying with the drone is casting its shadow over the PV modules. When irregular lighting patterns appear in the camera field of view, the EL signal extraction and the module tracking algorithms can fail, resulting in a distorted EL image with dark regions. Fig. 9 shows such a situation where the drone shadow appears over the imaged module. Despite being a small shadow in the corner, it is enough to degrade the final EL image.

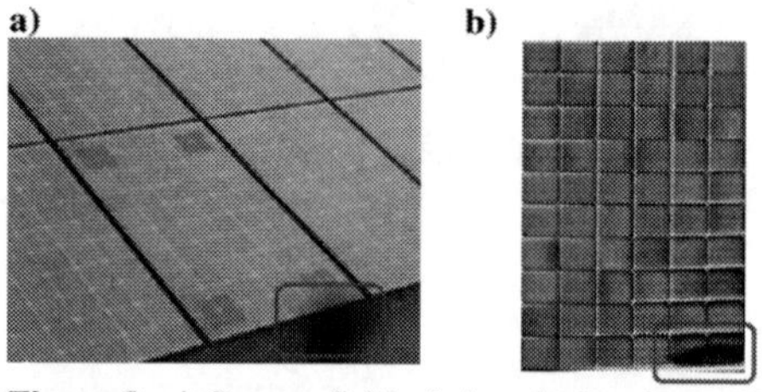

Figure 9: a) Camera field of view during drone flight where the drone is casting a shadow over the PV module under inspection. **b)** Resulting EL image after FFT and module tracking post processing. The image appears distorted and dark in the bottom right corner due to the shadow, marked in red.

Despite these challenges, obtaining very good quality EL images from the drone is still possible. These images

10.4229/EUPVSEC2023/4CV.1.43

have a comparable quality in terms of sharpness and fault detection possibilities as indoor lab EL images.

Fig. 10 shows an example of EL images from two modules recorded during the drone inspection, and Fig. 11 shows the zoom-in of two of their cells. These images look much sharper and clear, with fewer diagonal lines. Cell defects and busbars are clearly defined, and they present a perfect example of images that could potentially be used for further automated analysis.

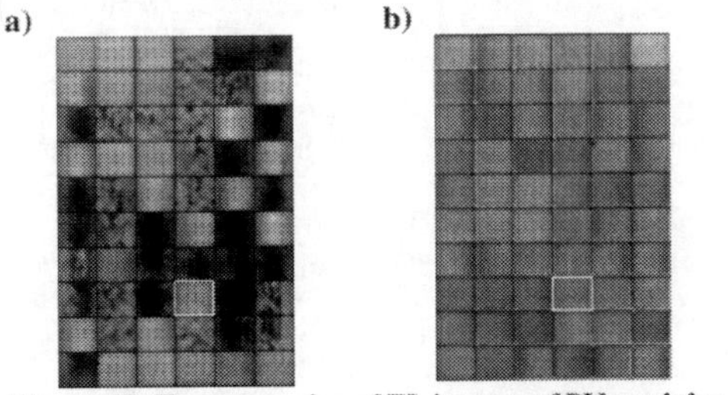

Figure 10: Two examples of EL images of PV modules recorded with low motion in the image (< 0.2 px/frame) and low irradiance variation (< 60 mean pixel-value difference during the recording).

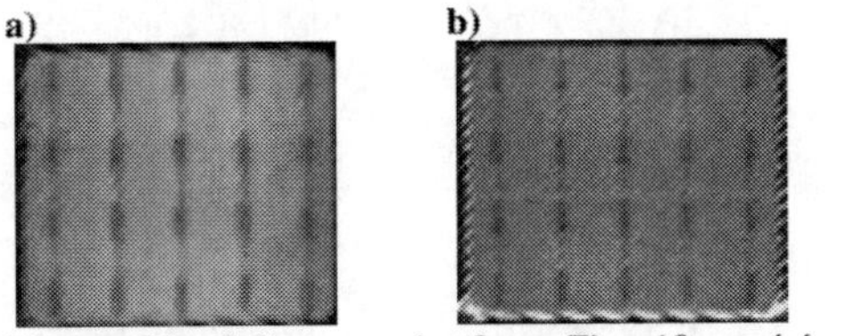

Figure 11: Cells zoom-in from Fig. 10 modules marked in white.

The key to obtaining these images is low motion during the image acquisition. The measured motion in the raw sequence of these images is < 0.2 pixels per frame. The irradiance variation during the acquisition is the factor that most largely influences the image quality of the EL images postprocessing. Fig. 12 shows an example of another module with low motion but where the image recording happened during a period of high irradiance variation (> 100 mean pixel-value difference during the recording). This case results in blurrier images with less sharp busbars and defects. Irradiance variations during image acquisition will affect the resulting images of stationary inspections the same way, and it is not a drone-only factor to avoid. For these specific series of images, it has been observed that the output image quality starts being questionable from a variation of more than 100 in mean pixel value.

Regarding the full inspection time, the drone as camera carrier inspection resulted in an average module inspection time of 24 seconds per module and around 8 minutes per PV string. Note that at the current state of development, the drone inspection process is not yet optimized and these numbers can be lowered in future inspections with more practical experience. On the other hand, the tripod average inspection time per module ranges between 35 seconds and 1 minute. However, the primary advantage of the drone is saving human resources to move around the camera stationary equipment between rows and the setup time of the camera when imaging a new string, where camera position and focus need to be adjusted. In contrast, the drone can move to the following string and continue the inspection once the power supply is ready. Furthermore, with the drone, the modules can be more easily imaged normal to the plane of array; therefore, perspective correction issues can be minimized.

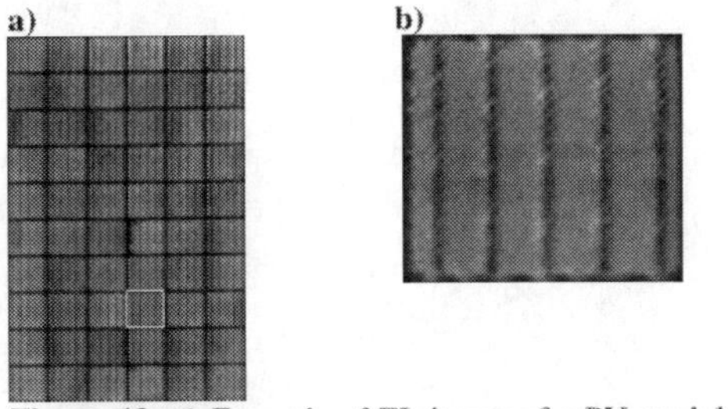

Figure 12: a) Example of EL image of a PV module with low motion (< 0.2 px/frame) but with high irradiance variation (> 100 mean pixel-value difference during the recording). **b)** Zoom-in of cell marked in white.

4 CONCLUSIONS AND OUTLOOK

This work investigated the challenges, limitations, advantages and disadvantages of drone EL inspections. In order to do so, a drone EL inspection has been carried out at DTU's PV field. Furthermore, a stationary inspection using a tripod has also been performed in the same PV strings. Finally, for benchmarking, some of the modules inspected in daylight were also imaged in indoor dark conditions.

Whereas the time required to image a PV module is similar using both a tripod and a drone, the drone presents advantages in inspection time savings regarding manpower needed and readjustment of the camera between PV strings. However, due to the motion from the camera carrier, drone EL inspections require an extra module tracking algorithm on top of the EL signal extraction algorithm, which increases the complexity of the postprocessing.

The final image quality of the drone and tripod stationary EL inspection is similar. Microcracks are hard to detect in the generated EL images in daylight, no matter the camera carrier, but large defects such as B and C solar cell cracks can be identified. Extra care is needed with the drone in order not to shade the modules, which ruins the EL signal extraction. Recording in fast-changing irradiance conditions should be avoided in any daylight EL inspection. Similar to indoor EL, good quality images can be obtained by minimizing the drone motion (< 0.2 px/frame) and irradiance variations.

Drone EL inspections, accounting for all challenges, can produce good quality images with higher throughput and less manpower than stationary inspections. Future works should address the possibility of using these EL images for automated fault detection and quantitative analysis of defects, such as power loss. Furthermore, recent progress in contactless photoluminescence (PL) is imaging using the I-V sweep functionality of specific string inverters and by shading control cells in the PV modules to obtain PL signal [13], [14]. If these imaging techniques were to be applied in a drone carrier, the considerations presented in this work would also apply.

5 ACKNOWLEDGEMENTS

This work is supported by the Eurostars project no. 115687, Automated Daylight Electroluminescence Inspection of Large Photovoltaic Systems (ADELI).

6 REFERENCES

[1] T. Fuyuki, H. Kondo, T. Yamazaki, Y. Takahashi, and Y. Uraoka, "Photographic surveying of minority carrier diffusion length in polycrystalline silicon solar cells by electroluminescence," *Appl Phys Lett*, vol. 86, no. 26, p. 262108, Jun. 2005, doi: 10.1063/1.1978979.

[2] J. Adams *et al.*, "Non-Stationary Outdoor EL-Measurements with a Fast and Highly Sensitive InGaAs Camera," in *32nd European Photovoltaic Solar Energy Conference and Exhibition*, 2016, pp. 1837–1841.

[3] G. Alves dos Reis Benatto *et al.*, "Drone-Based Daylight Electroluminescence Imaging of PV Modules," *IEEE J Photovolt*, vol. 10, no. 3, pp. 872–877, May 2020, doi: 10.1109/JPHOTOV.2020.2978068.

[4] G. A. dos R. Benatto *et al.*, "Luminescence Imaging Strategies for Drone-Based PV Array Inspection," in *Proceedings of 33rd European Photovoltaic Solar Energy Conference and Exhibition*, 2017. [Online]. Available: http://www.photovoltaic-conference.com

[5] G. A. dos Reis Benatto *et al.*, "Development of outdoor luminescence imaging for drone-based PV array inspection," in *2017 IEEE 44th Photovoltaic Specialist Conference (PVSC)*, IEEE, Jun. 2017, pp. 2682–2687. doi: 10.1109/PVSC.2017.8366602.

[6] S. Koch, T. Weber, C. Sobottka, A. Fladung, P. Clemens, and J. Berghold, "Outdoor electroluminescence imaging of crystalline photovoltaic modules: Comparative study between manual ground-level inspections and drone-based aerial surveys," in *Proc. 32nd Eur. Photovolt. Solar Energy Conf. Exhib.*, 2016, pp. 1736–1740.

[7] U. Jahn *et al.*, "Review on Infrared and Electroluminescence Imaging for PV Field Applications," 2018.

[8] L. Stoicescu, M. Reuter, and J. H. Werner, "DaySy: Luminescence Imaging of PV Modules in Daylight," in *Proceedings of the 29th European Solar Energy Conference* , 2014, pp. 2553–2554.

[9] T. K. Hass, S. Spataru, A. A. Santamaria Lancia, H. Parikh, P. B. Poulsen, and G. A. dos R. Benatto, "Computer Vision Method for Extracting an Induced Electroluminescence Signal from Photovoltaic Modules in Daylight Conditions Using Drone-Captured Images," in *Proceedings of 37th European Photovoltaic Solar Energy Conference and Exhibition*, 2020, pp. 1573–1579. doi: 10.4229/EUPVSEC20202020-5CV.3.44.

[10] M. Guada *et al.*, "Daylight luminescence system for silicon solar panels based on a bias switching method," *Energy Sci Eng*, vol. 8, no. 11, pp. 3839–3853, Nov. 2020, doi: 10.1002/ese3.781.

[11] H. R. Parikh *et al.*, "A Photovoltaic Module Diagnostic Setup for Lock-in Electroluminescence Imaging," in *2019 IEEE 46th Photovoltaic Specialists Conference (PVSC)*, IEEE, Jun. 2019, pp. 0538–0543. doi: 10.1109/PVSC40753.2019.8981255.

[12] T. Kari Hass, R. del Prado Santamaria, G. A. dos Reis Benatto, L. Stoicescu, S. V. Spataru, and P. Koelblin, "Evaluation of Motion-Induced Noise and Pixel Bleeding in Electroluminescense Field Inspection of PV-Modules," in *Unpublished*, 2023.

[13] R. Bhoopathy, O. Kunz, M. Juhl, T. Trupke, and Z. Hameiri, "Outdoor photoluminescence imaging of photovoltaic modules with sunlight excitation," *Progress in Photovoltaics: Research and Applications*, vol. 26, no. 1, pp. 69–73, Jan. 2018, doi: 10.1002/pip.2946.

[14] M. Vuković, I. Eriksdatter Høiaas, M. Jakovljević, A. Svarstad Flø, E. Olsen, and I. Burud, "Photoluminescence imaging of silicon modules in a string," *Progress in Photovoltaics: Research and Applications*, vol. 30, no. 4, pp. 436–446, Apr. 2022, doi: 10.1002/pip.3525.

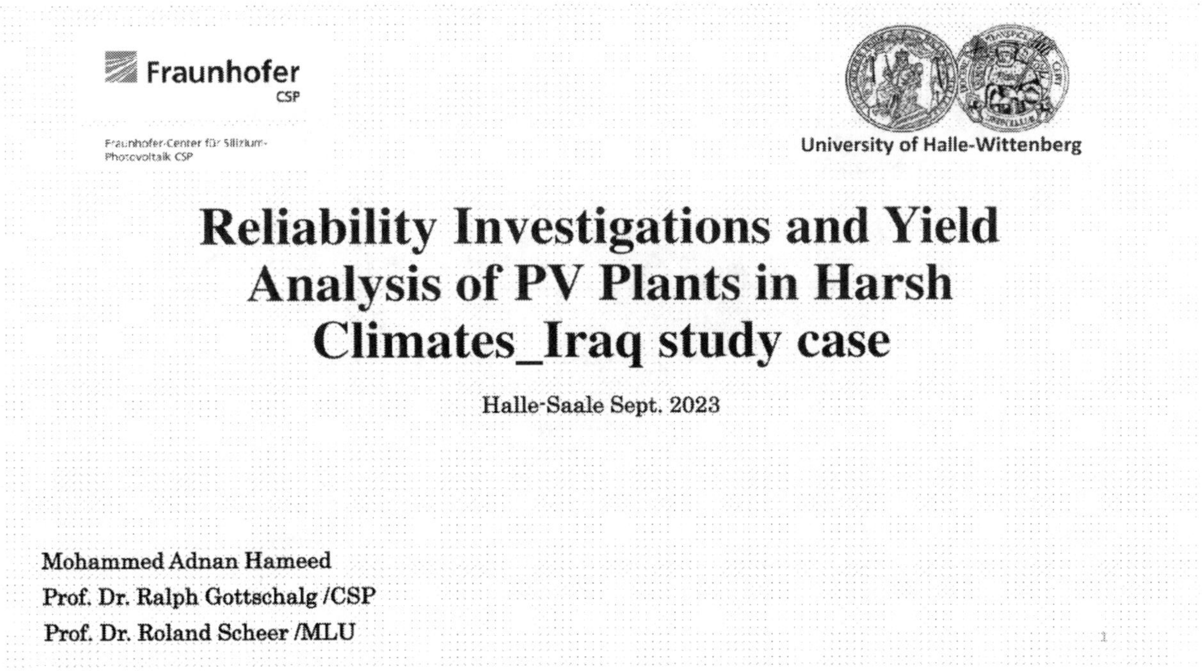

020404-001

40th European Photovoltaic Solar Energy Conference and Exhibition

020404-002

1. Introduction

01 Solar irradiance

Long sunny hours 02

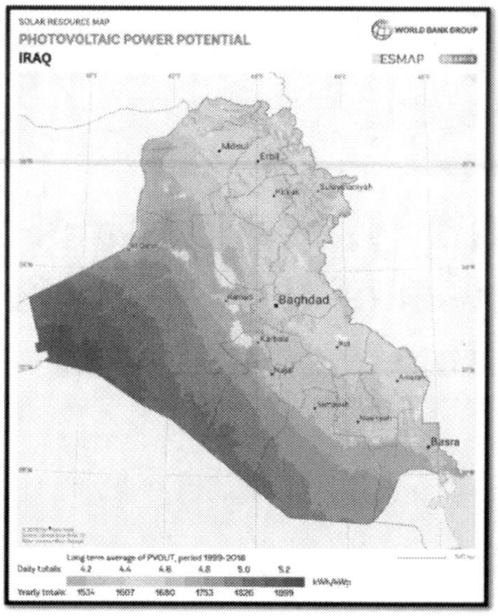

04 Power demand

Other issues 03

020404-003

1.1 Energy Transition

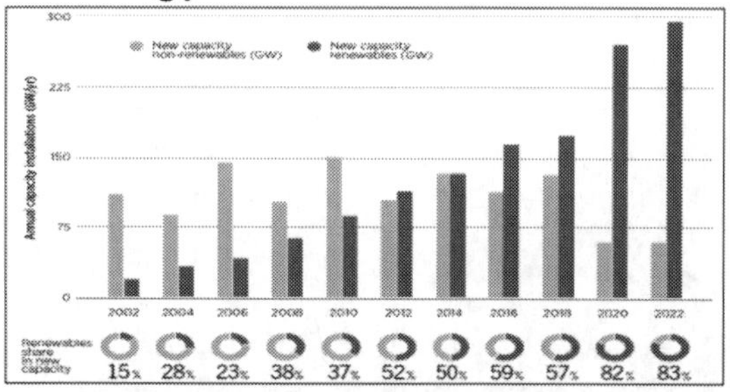

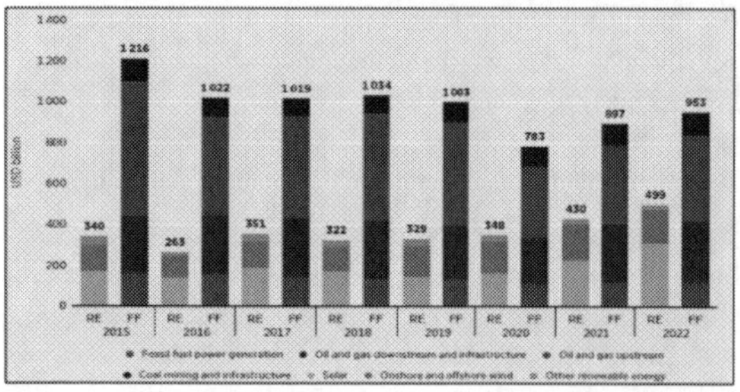

1.2 History of Theory

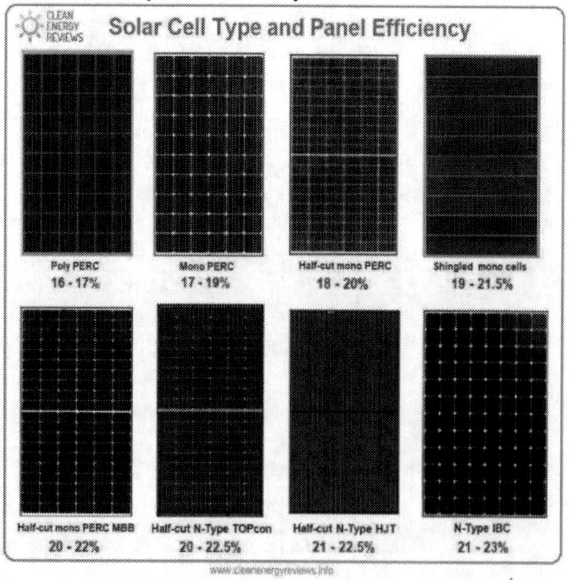

4

020404-004

2. Reliability ?

How to improve the reliability?

How it started...

how it's going.....

Focus on IEC Type Approval

Focus on
Efficiency Measurement

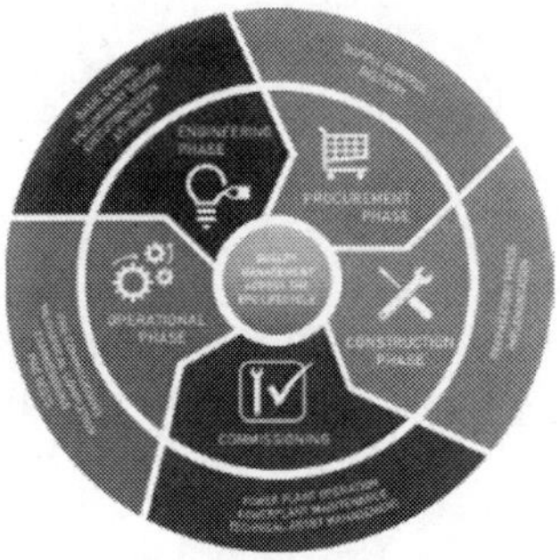

Holistic Quality
Management

Seite 5

020404-005

40th European Photovoltaic Solar Energy Conference and Exhibition

3. Timeline

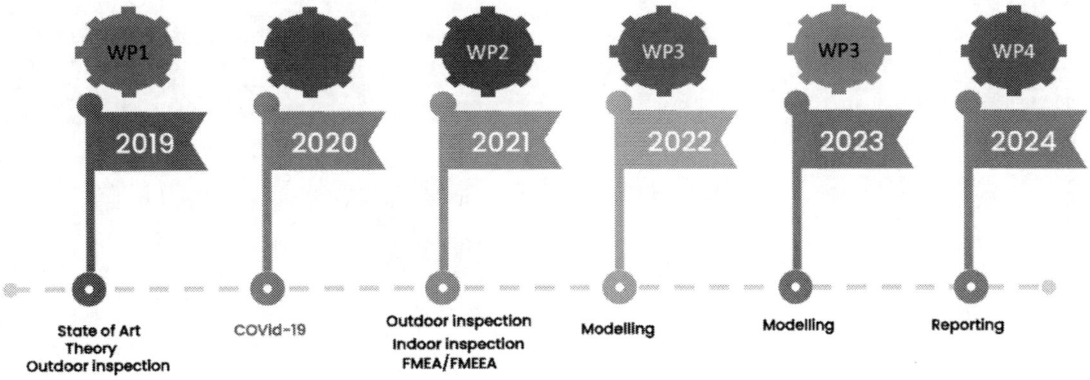

6

020404-006

Process/outcomes

Modelling the failure modes
Analysis and prediction of the performance and reliability of PV modules installed in harsh climates

Reporting the results

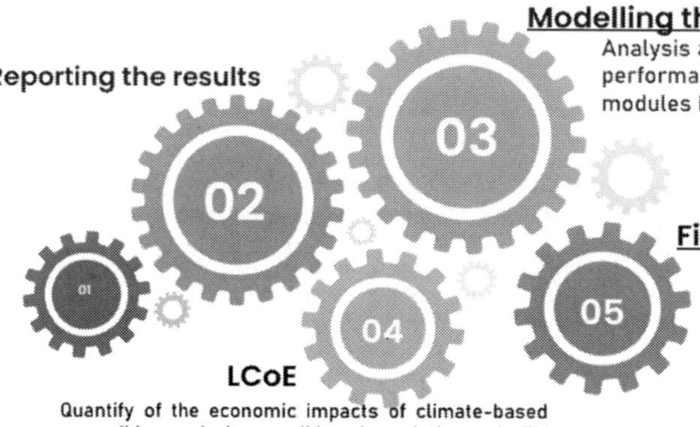

Figure the requirements

Survey
Define the most common failure modes and its economic impacts

LCoE
Quantify of the economic impacts of climate-based reversible and irreversible degradation of PV systems in harsh environmental conditions of Iraq

7

4. Site activity

4.1 Outdoor inspection/testing

- Visual inspection
- IV curve
- Continuity test

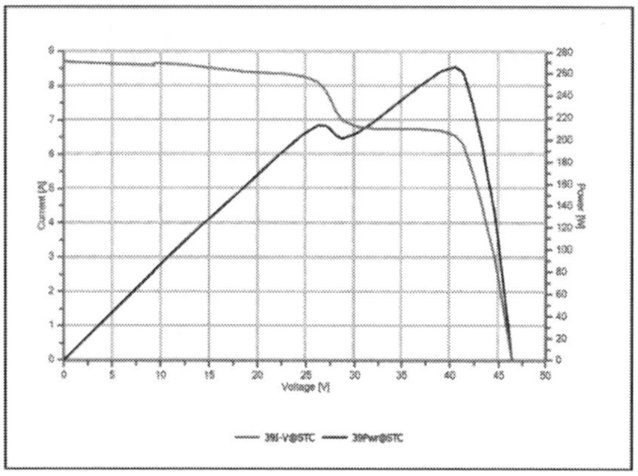

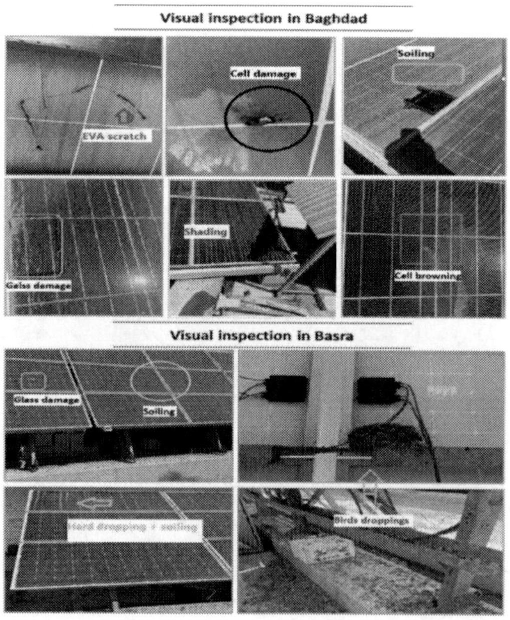

8

4.2 Indoor Inspection

- EL test in CSP
- Iraqi fielded modules in CSP

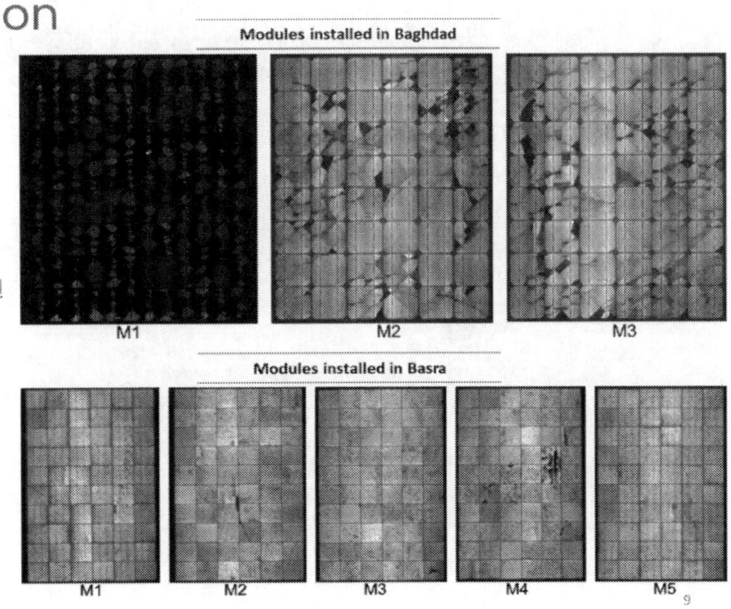

The most common failure modes revealed by the study:

- Delamination.
- Discoloration in EVA.
- Backsheet defects
- Cell browning.
- Solder bond fatigue.
- Micro-cracks.
- Frequent soiling.

9

Lab characterization: Electrical

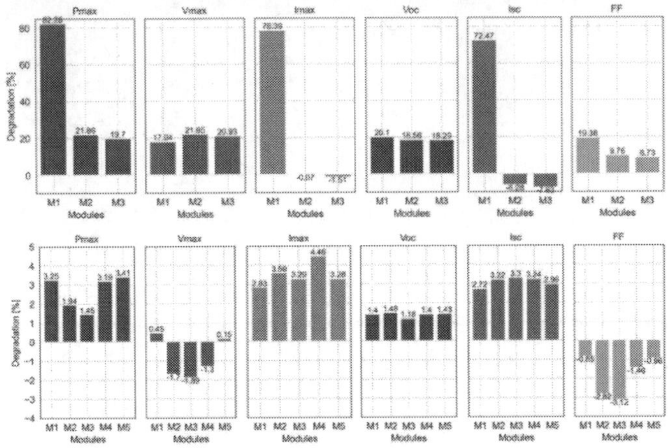

❑ Electrical characterization

- Increase in Fill factor and Vmax indicates a decrease in shunt resistance which can be due to cell cracks.
- Failures due to Imax and Isc indicate increase in series resistance due to cell cracks or solder bond failure and transmitance loss due to mechanical abrasion due to soiling, cleaning and can also be linked discoloration/browning of the encapsulant.

10

Lab characterization: Electrical

❑ Electrical characterization: Baghdad Vs Basrah

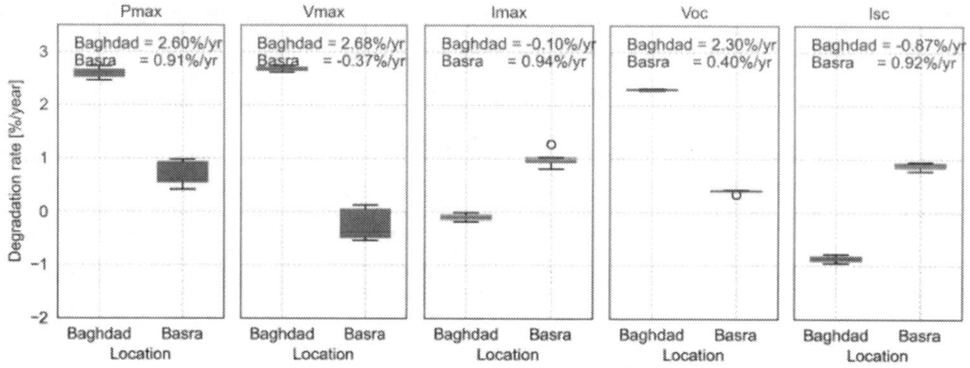

5. Physical Based Model [1]

❑ Used a physics based model the evaluated the effects of combined stress factors:

- Module operating temperature (T).
- Maximum module temperature (Tmax).
- Temperature cycling (ΔT).
- Relative humidity (RH).
- UV dose (UV).
- Soiling (S).

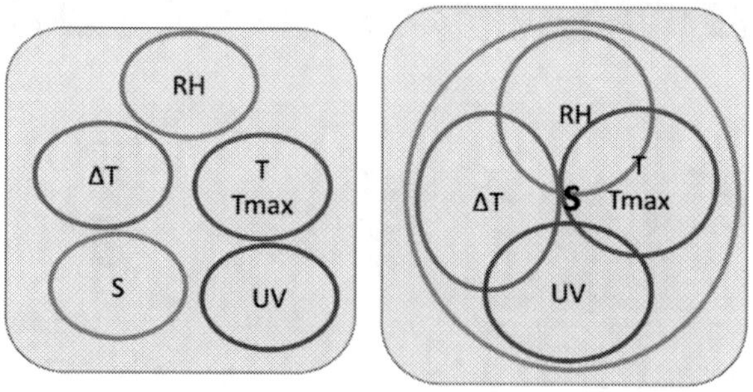

$$DR_H(T, RH) = A_H \cdot \exp\left(\frac{-E_{aH}}{k_B \cdot T}\right) \cdot RH^n$$

$$DR_P(U, V, TRH) = A_p \cdot UV^y \cdot (1 + RH^{n_1}) \cdot \exp\left(\frac{-E_{aP}}{k_B \cdot T}\right)$$

$$DR_{Tm}(\Delta, TT_{max}) = A_T \cdot (\Delta T + 273)^x \cdot C_r \cdot \exp\left(\frac{-E_{aT}}{k_B \cdot T_{max}}\right)$$

$$\mathbf{DR\,(T, RH, \Delta T, T_{max}) = A_N \cdot (1 + DR_H(T, RH)) \cdot (1 + DR_P(UV, T, RH)) \cdot (1 + DR_{TC}(\Delta T, T_{max})) - 1}$$

[1]. I. Kaaya et al. "Modeling Outdoor Service Lifetime Prediction of PV Modules: Effects of Combined Climatic Stressors on PV Module Power Degradation," in *IEEE Journal of Photovoltaics*, vol. 9, no. 4, pp. 1105-1112, July 2019, doi: 10.1109/JPHOTOV.2019.2916197.

- **5.1 Climate stress modelling [2,3]**

Maximum temperature | Relative humidity | Annual GHI | Module temperature

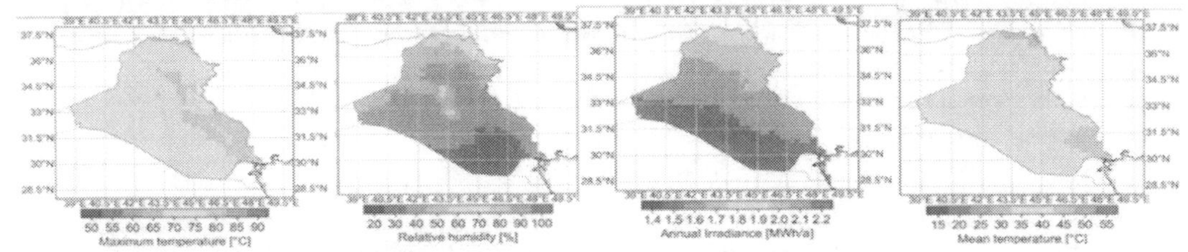

- **5.2 Degradation Modelling**

Hydrolysis degradation | Photodegradation rate | Thermomechanical rate | Total degradation rate

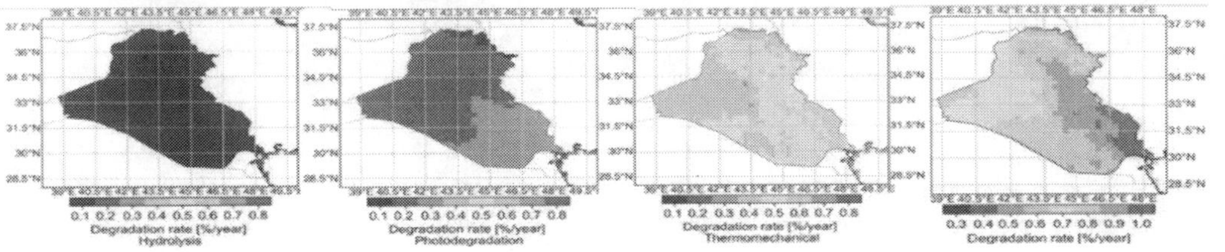

[2] J. Ascencio-Vásquez, I. Kaaya, K. Brecl, K.-A. Weiss, and M. Topič, "Global Climate Data Processing and Mapping of Degradation Mechanisms and Degradation Rates of PV Modules," *Energies*, vol. 12, no. 24, p. 4749, Jan. 2019, doi: 10.3390/en12244749.
[3] "Climate reanalysis | Copernicus." https://climate.copernicus.eu/climate-reanalysis (accessed Jan. 30, 2022).

13

Degradation rate modelling

❑ Using the simulation: For degradation rate zones are proposed as shown in the map

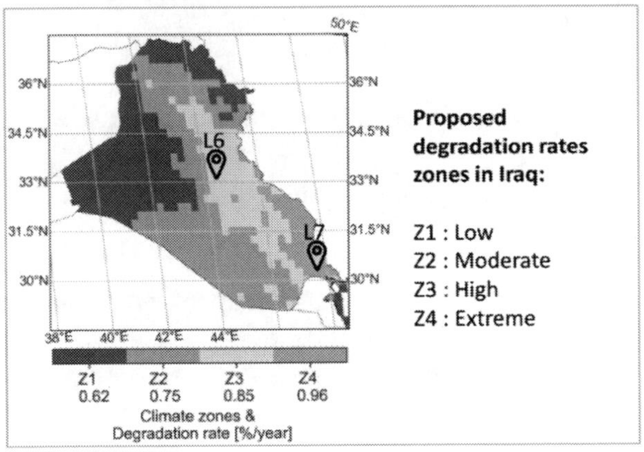

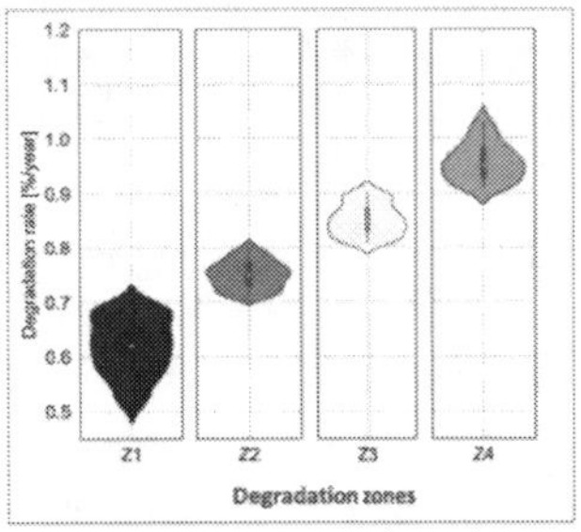

This show how the degradation zones are determined: They are medium values of the simulated ranges (please look at the paper for details)

14

5.3 Energy yield modelling[4]

❑ How the degradation zones impact decision making?

- Eight locations are selected based on total annual irradiance degradation rate zones
- Lifetime energy yield is calculated to compare the long-term energy benefits
- Best PV performance locations are selected best on the lifetime energy yield

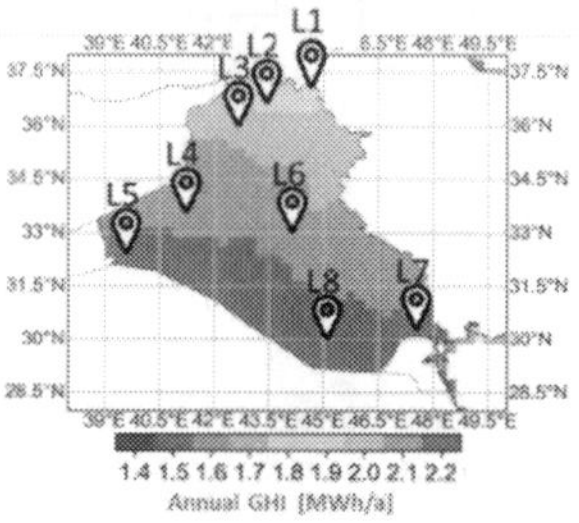

Location	L1	L2	L3	L4	L5	L6	L7	L8
DR [%/year]	0.62	0.62	0.75	0.62	0.62	0.85	0.96	0.75
FT [years]	32	32	26	32	32	23	20	26
Gpoa [kWh/m²/a]	1939.78	2215.97	2318.15	2416.15	2444.49	2341.5	2393.65	2428.3
Yield_FT [kWh/kWp]	49674.47	55433	47921.2	62389.3	63376.3	42186.7	37679.7	50184.5
Yield_30 [kWh/kWp]	45246.01	50547.8	53524	56806	57698.2	51652.9	56033	54644.6
Tmod [°C]	31.79	39.41	37.5	34.9	34.01	40.94	41	38.61
Tmax [°C]	60.61	67.57	63.85	58.3	58.85	64.9	65.31	62.11

- The study shows that under normal circumstances, module installed in locations L4 and L5 will have long lifetime and hence in general produce more energy during the systems lifetime compared to modules installed in other locations.

[4] "PVGIS data sources & calculation methods." https://joint-research-centre.ec.europa.eu/photovoltaic-geographical-information-system-pvgis/getting-started-pvgis/pvgis-data-sources-calculation-methods_en (accessed Jul. 11, 2023).

15

5.4 Effect of soiling [5]

❑ Modeled soiling impact including the effect of rain and manual cleaning

- Analysed the soiling data from 2004 to 2022 from ERA-5
- Assumed different soiling rates and cleaning cycles

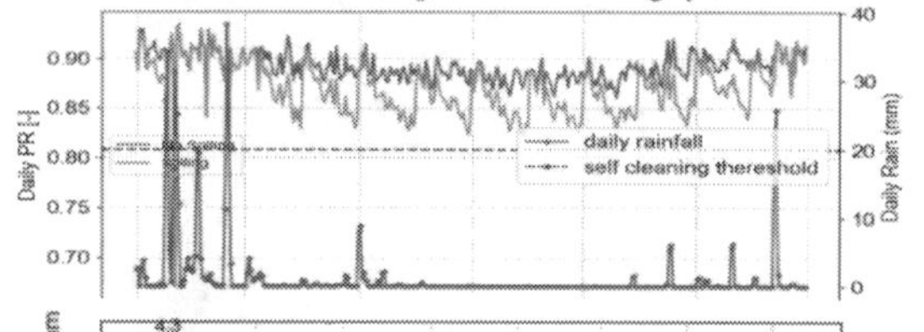

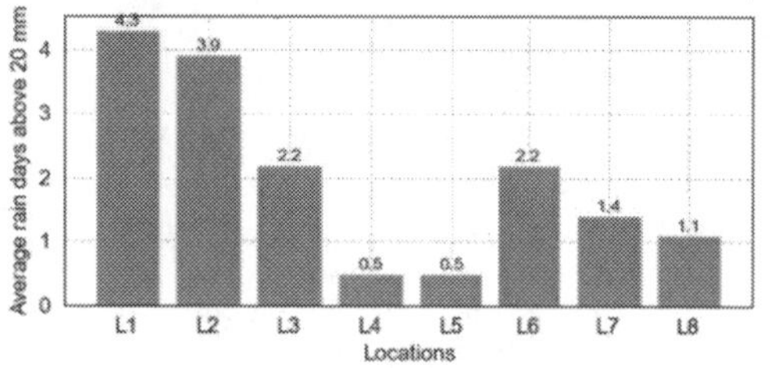

Locations with more rainfall have reduced soiling losses due to self-cleaning by rainfall

Soiling effect: location L1

Cleaning time [Months]	1	2	3	4	5	6
0.2 [%/day]	2.74%	5.34%	7.72%	9.95%	12.14%	13.02%
0.3 [%/day]	4.17%	8.23%	12.04%	15.71%	19.39%	20.88%
0.4 [%/day]	5.64%	11.29%	16.73%	22.10%	27.64%	29.93%
0.5 [%/day]	7.14%	14.15%	21.82%	29.24%	37.12%	40.04%

Soiling effect: location L4

Cleaning time [Months]	1	2	3	4	5	6
0.2 [%/day]	3.04%	6.22%	9.52%	12.98%	16.40%	19.98%
0.3 [%/day]	4.63%	9.62%	15.00%	20.83%	26.80%	33.30%
0.4 [%/day]	6.27%	13.26%	21.05%	29.84%	39.24%	49.94%
0.5 [%/day]	7.96%	17.14%	27.77%	40.31%	54.38%	69.80%

16

6. Deliverables

Ph.D Title ´´ Reliability investigation and yield analysis of PV plants in Iraq´´

A- Accomplished

- Risks Identification and its Economic Impact for PV Plant in Harsh Climate: Baghdad_Site Case.
- Analysis and prediction of the performance and reliability of PV modules installed in harsh climates: Iraq_site case.

B- ONGOING

- Assessing the influence of the PV systems installation design on the energy yield and PV lifetime in harsh climates: Iraq_site case.
- Assessing (quantification) of the economic impacts of climate-based reversible and irreversible degradation of PV systems in harsh environmental conditions of Iraq.

17

020404-017

40th European Photovoltaic Solar Energy Conference and Exhibition

7. Conclusions & recommendations

- One PV module design and unified qualification test procedure are not applicable for all climates region.
- Update the qualification test procedure according to **IEC 61215** specially for harsh climates.
- Categorize the production basing on GIS site.
- Over engineering kills SUSTAINBILITY.
- Create staff with more "Obsensiv" of PV.
- The degradation can be higher if we implement the BoS.
- The harsh climate will not limit the deployment of PV plants if the right O&M implemented.
- Avoid applying the manufacturers recommended degradation rates in financial models as this might lead to poor decisions but rather use adjusted degradation rates based on locations .

020404-018

Thanks for Your Attention!
Contact:
Mohammed A. Hameed
Email: m.adnan.scop@gmail.com
Mobile: +964 7901 423 007

020404-019

A TWO-STEP PROCEDURE TO ASSESS PHOTOVOLTAIC MODULE PERFORMANCE

Andreea Sabadus[1,*], Marius Paulescu[2]
[1]Institute for Advanced Environmental Research, West University of Timisoara, Timisoara, Romania
[2]Faculty of Physics, West University of Timisoara, Timisoara, Romania
*Corresponding author: andreea.sabadus@e-uvt.ro

ABSTRACT: Accurate modeling of photovoltaic (PV) power generation is of great importance when it comes to planning and operating an electric grid system. Usually, a PV module is delivered with a datasheet provided by manufacturers. Using the scarce information provided by a PV module datasheet, the mathematical description of its I-V characteristic and thus the modeling of PV power, become a challenge. This study is focused on the assessment of a PV module performance under real weather conditions using a simple two-step procedure. The PV module is considered a black box, not presuming any knowledge of the internal characteristics and processes. The two steps are as follows: (1) estimating the model parameters of the PV module based on an explicit version of Shockley equation and (2) applying a simple model that estimates the output power of the PV module from in-plane measured solar irradiance and air temperature. Data measured on the Solar Platform of the West University of Timisoara are used to conduct this study.
Keywords: solar cell, one-diode model, explicit Shockley equation, PV power

1 INTRODUCTION

In 2022, there was an increase in solar power capacity, with a 45% growth compared to the previous year, as 239 GW were installed [1]. As solar power capacity is predicted to experience further growth, PV modules efficiency plays a pivotal role in maximizing energy generation. The efficiency and performance of a PV module are primarily affected by two key factors: the amount of incoming solar energy and the temperature of its constituent solar cells. In order to accurately estimate the electrical output of a PV module when exposed to the complexities of real weather conditions, gaining a comprehensive understanding of its I-V characteristic becomes paramount. Typically, a PV module is delivered with a datasheet provided by manufacturers. It commonly contains information about the module's electrical behavior, particularly focusing on its I-V characteristic. The three notable parameters featured on the datasheet are the short-circuit current, open circuit voltage, and the maximum power point, all measured at standard test conditions (STC). STC imply a normal incoming solar radiation of 1000 W/m^2 under AM1.5G spectrum [2] and a cell temperature of 25°C. The limited imformation often proves insuffiecient for complex modeling, requiring further analysis and experimentation in order to obtain a better understanding of the PV module's performance.

To describe an operating PV module, almost all mathematical models are rooted in Shockley theory of the illuminated p-n junction [3]. The one-diode model (Fig. 1) is the most common and mainly used due to its apparent simplicity. Based on the one-diode equivalent circuit from Fig. 1, probably the most popular equation for the I-V characteristic of a solar cell is:

$$I = I_L - I_S\left[\exp\left(\frac{q\left(V + IR_S\right)}{mk_BT}\right) - 1\right] - \frac{V + IR_S}{R_P} \quad (1)$$

where q is the electron charge, k_B is the Boltzmann constant, and T is the cell temperature. The photocurrent I_L, diode saturation current I_S, diode ideality factor m, serial resistance R_S, and shunt resistance R_P form a set of five parameters (specific for a cell), which give the name of the model, respectively the five-parameter model.

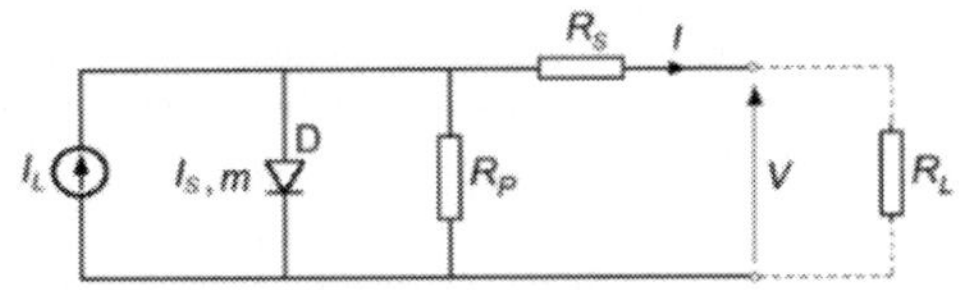

Figure 1: One-diode equivalent circuit for a solar cell.

There are several methods dealing with the extraction of parameters of the one-diode model [4-5]. For solar engineering applications, accuracy and computational accessibility are the two main aspects taken into account when selecting a method for obtaining the parameters of the one-diode model.

The performance of a PV module is greatly influenced by environmental factors, specifically temperature and solar irradiance. Since weather conditions can vary significantly from STC, having accurate knowledge of these parameters is of paramount importance in order to ensure quality control and performance assessment of PV modules.

The aim of this study is to assess a PV module's performance under real weather conditions using a simple two-step procedure: (1) estimating the model parameters of the PV module based on an explicit version of Shockley equation [6] and (2) applying a simple model [7] that estimates the output power of the PV module from in-plane measured solar irradiance and temperature.

The paper is organized as follows. In Section 2, we introduce the explicit equation for the five-parameter model. The parameters sets obtained at STC are presented in Section 3. Section 4 focuses on estimating parameters under real weather conditions. The performance assessment of a PV module is discussed in Section 5. Finally, Section 6 gathers the main conclusions of this study.

2 EXPLICIT ONE-DIODE MODEL EQUATION

This study introduces a noteworthy form of the equation employed for parameter extraction [6]. Specifically, it presents an explicit five-parameter equation, which stands out as a distinctive aspect of the research.

10.4229/EUPVSEC2023/4CV.1.45
020405-001

Table I. Two different sets of one-diode model parameters which accurately replicate the experimental characteristics.

Parameters	Set-1	Set 2
I_L [A]	7.67	7.67
I_S [A]	10^{-7}	10^{-10}
m	1.061	0.770
R_S [Ω]	0.0010	0.0028
R_P [Ω]	13	15
R^2	0.997	0.997

$$I = \frac{I_L - I_S\left[\exp\left(\frac{qV}{mk_BT}\right)-1\right]-\frac{V}{R_P}}{1+\frac{R_S}{R_P}+I_S\frac{qR_S}{mk_BT}\exp\left(\frac{qV}{mk_BT}\right)} \qquad (2)$$

Eq. (2) explicit nature offers a straightforward approach for parameter extraction, making algorithms that deal with this problem more accessible. The results presented in [6] also underline the feasibility of the proposed equation.

3 PARAMETER ESTIMATION AT STC

In this paper, data collected from the experimental setup located at the Solar Platform at West University of Timisoara was used [8]. The FORPV setup (Fig. 2) was designed to test and evaluate models used for estimating and forecasting PV power. It comprises a PV system with a total capacity of 540 Wp, consisting of four Cleversolar SPR-130 PV modules positioned at a 30-degree angle. These modules are connected in a series-parallel configuration and directly supply power to a resistive load. To measure the in-plane solar irradiance, a high quality Kipp & Zonen SMP-10 pyranometer (class A, ISO9060:2018) is used. Furthermore, meteorological data is recorded using a Delta-T Devices WS-GP2 weather station. All these instruments undergo a 24/7 monitoring, with data recorded at regular intervals of 4 seconds. Also, an EKO Instruments ASI-16 all-sky imager is used for monitoring the state-of-the-sky at 1-min sampling.

To replicate the measured I-V characteristic of one of the Cleversolar SPR-130 PV modules installed at the Solar Platform at the West University of Timisoara at STC a successive discretization algorithm (SDA) [9] and Eq. (2) were used. Among many solutions, two sets of parameters were retained.

Both sets replicate the experimental I-V characteristic with high accuracy (Fig. 3), with a coefficient of determination R^2 of 0.997. Examining Table 1, we observe notable variations in the saturation current I_S and the diode ideality factor m. We considered sub-unitary values for m to highlight that even nonphysical parameters set in Eq. (2) can accurately replicate the I-V characteristic.

The question is: How does parameters estimation at STC affect PV module performance assessment?

Figure 2. FORPV experimental setup located at the Solar Platform at West University of Timisoara.

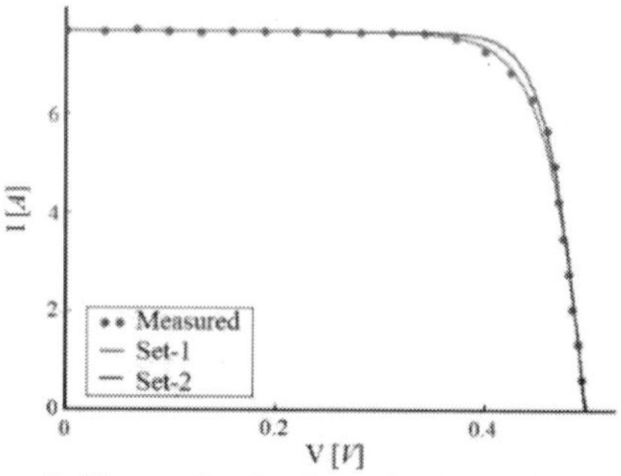

Figure 3: Measured and estimated I-V characteristic.

Table II. The cleanliness coefficient.

Cleanliness degree	τ_c
Perfect	1.00
Proper	0.98
Medium	0.96
Low	0.92

4 PARAMETER ESTIMATION UNDER REAL WEATHER CONDITIONS

The procedure [7] used for the calculation of the I-V curve under real conditions relies on the five-parameter model of the solar cell and requires a minimal set of input data obtained under STC, typically available from manufacturers. Additionally, the cleanliness degree of the module's surface is considered a key parameter with the model. This approach offers notable advantages, characterized by its simplicity and accuracy.

The procedure implies the parameters computed at STC and the following assumptions:

1. The effective solar irradiance G_{ef} is determined taking into account the cleanliness degree of the PV module:

$$G_{ef} = \tau_c G \qquad (3)$$

where τ_C is the cleanliness coefficient.

2. The operating temperature of the solar cell is calculated as:

$$T = T_a + \frac{NOCT - 20°C}{800} G_{ef} \qquad (4)$$

3. The photocurrent is linearly linked with the incident solar irradiance and can be obtained by considering the thermal coefficient of the short-circuit current α_I :

10.4229/EUPVSEC2023/4CV.1.45
020405-002

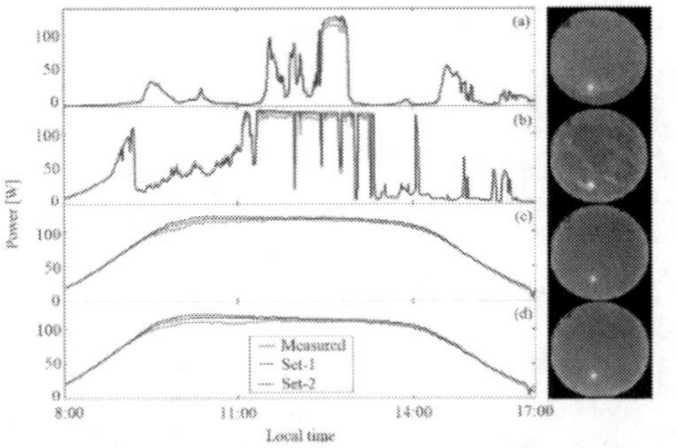

Figure 4: Measured vs estimated output power delivered by the PV system for four days in March 2022: (a) 04, (b) 08, (c) 22, (d) 24. The output power is estimated with the two sets of parameters obtained from the five-parameter model (Table 1).

Table III. Normalized root mean square error (nRMSE) of the sets of parameters applied for the four testing days from March 2022 (Day-1 4th of March, Day-2 8th of March, Day-3 22nd of March, Day-4 24th of March).

	nRMSE	
Parameter set	Set-1	Set-2
Day-1	0.108	0.152
Day-2	0.066	0.087
Day-3	0.030	0.045
Day-4	0.041	0.061

$$I_L(G,T) = I_{L,STC}\left[1+\alpha_I(T-T_{STC})\right]\frac{G_{ef}}{G_{STC}} \qquad (5)$$

4. The saturation current is determined as:

$$I_S(G,T) = I_L(G,T)\exp\left[-\frac{qV_{OC,STC}\left[1+\alpha_V(T-T_{STC})\right]}{mk_BT}\right] \qquad (6)$$

where α_V is the thermal coefficient.

5. The diode ideality factor, the series resistance and shunt resistance, respectively, remain the same as at STC:

$$m = m_{STC}, R_S = R_{S,STC}, R_P = R_{P,STC} \qquad (7)$$

For any set of parameters, this method facilitates the computation of the I-V characteristic of the PV module. By employing this procedure, we can obtain a better understanding of how the PV module performs under varying conditions.

5 PERFORMANCE ASSESSMENT

The two-step method was employed to assess the Cleversolar SPR-130 PV module's output power for four days in March 2022 (specifically, on the 4th, 8th, 22nd and 24th). These days exhibit notable differences in terms of the solar radiative regime, with the 4th and 8th being characterized as unstable while 22nd and 24th as stable.

Experimental data were measured on the Solar Platform of the West University of Timisoara. The in-plane solar irradiance and the air temperature values were retrieved from measurements and used to compute the output power. The output power calculation was conducted with the assumption that the cleanliness degree of the PV module was considered "proper".

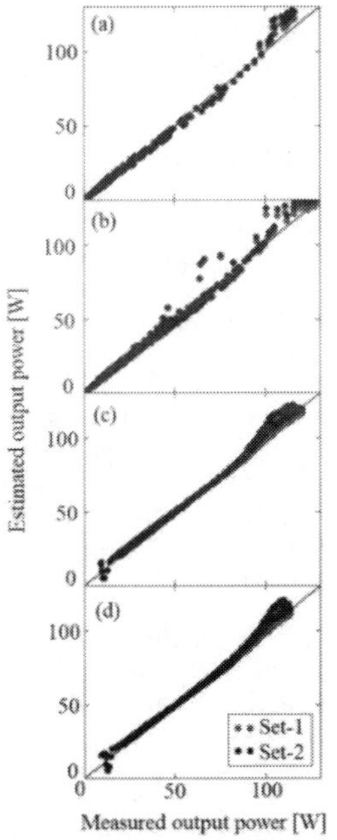

Figure 5: Estimated vs measured output power delivered by the PV module in four days of March 2022: (a) 04, (b) 08, (c) 22, (d) 24.

Further, the parameters estimated under real weather conditions, as determined by Eq. (5)-(7), were applied to estimate the output power of the PV module.

Fig. 4 presents a comparison between the estimated and the measured output power delivered by the PV module. It can be seen that regardless of the set of parameters used, the estimated power replicates the measured one with high accuracy. There can be seen a very good overlap between the values. However, slight differences can be observed for high values.

The results are further detailed in Table III, which provides a comprehensive overview of the estimated power under both unstable and stable weather conditions. Table III presents the normalized root mean square error (nRMSE), Eq. (8), of the sets of parameters applied for the four testing days.

$$nRMSE = \frac{\left(N\sum_{i=1}^{N}\left(P_i - M_i\right)^2\right)^{1/2}}{\sum_{i=1}^{N} M_i} \qquad (8)$$

where M_i and P_i are the i-th recorded and modeled amounts, respectively. N represents the number of records taken into account. The indicator is normalized to the measurements average value.

The results show that the estimated output power exhibits slight fluctuation when compared to the measured values. Specifically, on the meteorological unstable days (Day-1 and Day-2) the largest biases are observed. The largest bias (15.2%) is experienced during Day-1 by estimates evaluated with the STC parameters Set-2, while the smallest bias is experienced by estimates evaluated with Set-1. Similarly, in the case of the stable days (Day-3 and Day-4) the hierarchy remains consistent (Fig. 5). Therefore, the choice of various parameters sets evaluated

10.4229/EUPVSEC2023/4CV.1.45
020405-003

under STC can result in varying biases in estimation, with strong dependence on the stability in the state-of-the-sky. As expected, the model performance is better in stable meteorological conditions.

6 CONCLUSIONS

The results underline the practical utility of the suggested two-step procedure in the field of solar engineering, offering a reliable mean to accurately evaluate a PV module's performance under real weather conditions.

The proposed procedure successfully estimates the output power of a commercial PV module, regardless of weather conditions.

However, it is important to emphasize the importance of careful one-diode model parameters set selection. Choosing the appropriate parameters set is crucial to ensure the accuracy of the PV module's output power estimation.

Further studies are needed to determine the necessary corrections that can enhance the accuracy of output power estimation, particularly when dealing with high power output values.

7 ACKNOWLEDGMENTS

This work was supported by grants from FSS "Conferinţa Smart Diaspora în Învăţământul Superior, Ştiinţă, Inovare şi Antreprenoriat", West University of Timisoara Grant Develop and FSI 0105 Grant.

8 REFERENCES

[1] Global Market Outlook for Solar Power 2023-2027 Report, SolarPower Europe 2023.
[2] Reference Air Mass 1.5 Spectra https://www.nrel.gov/grid/solar-resource/spectra-am1.5.html, 2023 (accessed September 2023).
[3] Nelson J. Imperial College Press: London, UK, 2013.
[4] Rashed M, Mohammed OY, Shibab S, Al-Adili A. J. Phys.: Conf. Ser. 2021, 1795, 012043.
[5] Sabadus A, Paulescu M. Energies 2021, 14, 3974.
[6] Sabadus A, Paulescu M. 10.4229/WCPEC-82022-2BV.2.48.
[7] Paulescu M, Badescu V, Dughir C. Energy 2014, 70, 49-57.
[8] Solar Platform of the West University of Timisoara, Romania http://solar.physics.uvt.ro/srms/ (accessed September 2023).
[9] Cotfas DT, Deaconu AM, Cotfas PA. Energy Convers. Manage. 2019, 196, 545-556.

Assessing the Influence of Tilt and Azimuth Angle Variations on Plane of Array Irradiance in Brazilian Cities

Eslam Mahmoudi [a*], Isabelle Nunes [a], M.R.M. Neves [a], Rodrigo Garcia [b], Francisco C. Marques [c], Marcelo Gradella Villalva[a], Tarcio Andre dos Santos Barros [a]

[a]Department of Systems and Energy, Faculty of Electrical and Computer Engineering, University of Campinas, Brazil

[b] BYD energy, Campinas, Brazil

[c] Gleb Wataghin Institute of Physics, University of Campinas, Brazil

ABSTRACT: Photovoltaic (PV) generation is fundamentally contingent upon the amount of incident solar radiation that PV modules receive, a factor heavily influenced by the site's geographical location and the PV array's position. This paper examines how changes in the position of the PV modules, as determined by tilt and azimuth angles, impact the received irradiance levels by PV arrays across different regions of Brazil. To this end, a simulation tool has been developed that uses hourly Global Horizontal Irradiance (GHI) data of Brazilian cities to estimate the hourly, daily, monthly, and annual received solar irradiation under possible variations in tilt and azimuth angles. The plane of array (POA) irradiances for each pair of tilt and azimuth angles are estimated by combinations of decomposition models (i.e., Erbs, DISC, and DIRINT models) and transposition models (i.e., Liu and Jordan, Klucher, Hay and Davies, Reindl, and Perez models). Accordingly, the developed simulation tool generates graphs of the estimated POA that illustrate the relative POA received at any pair of azimuth and tilt angles compared to the optimal angles. Simulation results indicate that, for slope angles of up to approximately 20°, variations in azimuths or tilt angles did not generally lead to significant losses in annual POA. Thus, installing the PV modules under this tilt angle limitation and any azimuth angle can maintain an acceptable annual POA level. However, the results also show that for this tilt angle limitation, significant POA losses are observed in certain months of the year, while in other months, there are no POA losses, akin to the overall annual POA. The developed simulation tool can be used by PV installers and designers to quickly evaluate the effect of specific tilt and azimuth angles on the monthly and yearly solar irradiation incident for a particular location.

1 INTRODUCTION

Energy plays a pivotal role in every nation's economic and social advancement. Its availability significantly enhances the overall quality of life, whether sourced from traditional fossil fuels like oil and gasoline, nuclear power, or any form of renewable energy [1]. Out of all the renewable energy resources, solar energy is quickly establishing itself as one of the most promising green energy technologies on a global scale [2]. Solar energy is environmentally friendly and does not emit harmful gases that contribute to the greenhouse effect or disturb the delicate ecological equilibrium of our planet [3]. The utilization of PV solar energy conversion has experienced substantial growth in recent years, especially in its applications when integrated into public electricity grids [4].

Nevertheless, architects and building planners often overlook incorporating PV panels into their design plans due to the challenge of identifying an ideal orientation for PV installation on buildings [5]. However, it remains feasible to attain acceptable performance even with suboptimal PV orientations, ensuring that the advantages of PV technology can still be harnessed [6]. A commonly used rule of thumb for the optimal orientation of PV modules suggests setting the tilt angle equal to the location's latitude.

Additionally, it advises orienting the azimuth angle towards the southern direction in the northern hemisphere and the northern direction in the southern hemisphere [7]. Nevertheless, adopting this PV positioning approach may result in increased costs stemming from system oversizing if implemented without a thorough analysis [8-9]. Moreover, due to architectural factors involved in integrating PV modules into building structures, orientations other than the optimal one may also be worthy of consideration. For instance, aligning PV modules with the angle of a sloping roof, despite resulting in reduced solar radiation intake, could be a viable option [10-11]. Considering that a solar tracking system may not be the preferred choice due to its associated capital and maintenance costs, various studies suggest that it is feasible to determine the optimal tilt and azimuth angles for each location on Earth, factoring in its distinct radiation characteristics, to maximize solar energy absorption to the greatest extent possible [12-16]. Given that non-optimal tilt and azimuth angles can still yield satisfactory energy generation levels, it becomes essential to evaluate location-specific losses to ascertain the constraints of these angles [17-19].

Numerous studies have been undertaken to explore the architectural considerations of integrating PV systems into buildings [20-23]. These studies aim to assess the annual and seasonal energy generation from PV installations across various architectural configurations. The paramount factor influencing the efficacy of integrated PV systems is the optimization of monthly and annual solar irradiance capture. Given that seasonal and annual solar irradiance levels are significantly contingent upon the geographical location of the PV system, the ideal orientation and tilt angles for PV modules are inherently location-specific [24]. Therefore, architects and PV system designers should consider site-specific data and irradiance calculation techniques when evaluating the effectiveness of integrated PV systems. In light of the available GHI data for desired locations, it's important to consider decomposition and transposition methods to estimate the POA irradiance at these locations, accounting for the azimuth and tilt angles of the PV systems [25].

10.4229/EUPVSEC2023/4CV.1.46
020406-001

Consequently, this approach allows for the estimation of the POA and PV generation losses in various orientations specific to each location, facilitating the assessment of PV installations at non-ideal orientations and tilt angles. Hence, PV system designers and installers require monthly and annual POA calculations tailored to specific locations, accounting for potential variations in PV module orientation and slope angles. Simulation tools offer a valuable means to assess how a particular surface position in a given geographical location affects the reduction in monthly and annual incident solar radiations for non-optimal surface orientations. This paper proposes a Python-based simulation tool to explore different plane positions' influence on received POA in this context. The tool provides estimates of monthly and annual irradiance for surfaces with varying tilt and azimuth angles across all cities of Brazil. This simulation tool generates graphical representations to illustrate potential POA losses resulting from changes in inclination and azimuth angles for PV module positioning. The created simulation tool can be valuable for PV system designers, installers, and architects in making informed decisions regarding module placement.

2 METHODOLOGIES

In the course of conducting simulations to evaluate the performance of a PV system, historical data on GHI, DNI, and DHI irradiance components plays a pivotal role in accurately estimating the POA irradiance on the inclined surface. However, given the complexity and cost associated with measuring DNI and DHI, the GHI data is more readily accessible. Mathematical models have been devised to estimate the absent components based on measured GHI data. Consequently, these three irradiance components are employed by the transposition model to estimate the POA irradiance on a tilted plane.

2.1 Decomposition models

This study employs Erbs, DISC, and DIRINT decomposition models for decomposing GHI into DNI and DHI.

2.1.1 Erbs model

The model Erbs [26] introduces a factor denoted as f_d, which serves to estimate the DHI. This factor exhibits a direct correlation with the clear index, k_t. This correlation is divided into three distinct zones:

$$f_d = \begin{cases} 1 - 0.09k_t & 0 \leq k_t \leq 0.22 \\ 0.9511 - 0.1604k_t + 0.4388k_t^2 & \\ -6.638k_t^3 + 12.338k_t^4 & 0.22 \leq k_t \leq 0.8 \\ 0.165 & 0.8 < k_t \end{cases} \quad (1)$$

According to this correlation factor, the DHI is calculated by

$$DHI = f_d GHI \quad (2)$$

The DNI can be calculated based on GHI and DHI by using the equation (x)

$$DNI = \frac{GHI - DHI}{cos(\theta_z)} \quad (3)$$

Where the θ_z is the zenith angle.

2.1.2 DISC model

The Direct Insolation Simulation Code (DISC) model proposed by Maxwell [27] is termed" quasi-physical" as it combines a clear sky model with experimental fits for other sky conditions. The model has the form of a clear sky irradiance attenuated by a function of the clearness index, k_t. The direct normal irradiance is given by:

$$DNI = I_0(k_n - (a + be^{c.am})) \quad (4)$$

Where I_0 is the normal incident of extraterrestrial irradiance and k_n is given by

$$k_n = 0.866 - 0.122am + 0.0121am^2 - 0.000653am^3 + 0.000014am^4 \quad (5)$$

Where am is the airmass and defined as

$$am = \frac{1}{sin\,\alpha + 0.15(\alpha + 3.885)^{-1.253}} \quad (6)$$

Where α is the solar elevation. Finally, the coefficients a, b, and c that are functions of the clear index k_t, are calculated for two following conditions:

If $k_t \leq 0.6$:

$$\begin{cases} a = 0.512 - 1.56k_t + 2.286k_t^2 - 2.222k_t^3 \\ b = 0.37 + 0.962k_t^3 \\ c = -0.28 + 0.932k_t - 2.048k_t^2 \end{cases} \quad (7)$$

If $k_t > 0.6$:

$$\begin{cases} a = -5.743k_t + 21.77k_t^2 + 11.56k_t^3 \\ b41.4 - 118.5k_t + 66.05k_t^2 + 31.9k_t^3 \\ c = -47.01 + 184.2k_t - 222k_t^2 + 73.81k_t^3 \end{cases} \quad (8)$$

2.1.3 DIRINT model

Perez et al. developed a decomposition model as an improvement to the DISC model, namely the DIRINT model [28]. This model used the parameter k_t' that is expressed in equations (9).

$$k_t' = \frac{k_t}{1.031\,\exp\left(\frac{-1.4}{0.9 + \frac{9.9}{am}}\right) + 0.1} \quad (9)$$

The DIRINT model introduces a new coefficient $\delta k_t'$ that is named the stability index and varies according to time, as shown in the equation (10), in which $k_{t.next}'$ is the future value of the series k_t' and $k_{t.last}'$ and o value of the series k_t'.

$$\delta k_t' = 0.5(|k_t' - k_{t.next}'| + |k_t' - k_{t.last}'|) \quad (10)$$

10.4229/EUPVSEC2023/4CV.1.46
020406-002

If any of the parameters of the series of $\delta k_t'$ is not available, the equation (10) is modified by the equation (11).

$$\delta k_t' = \left|k_t' - k_{t.next/last}'\right| \tag{11}$$

Another variable that is added to the DIRINT model is the dew point temperature (T_d). This factor estimates atmospheric precipitation, which influences the absorption of irradiance. In this way, the atmospheric precipitation index can be obtained through the equation (12).

$$W = exp(0.07T_d - 0.075) \tag{12}$$

Therefore, the DNI is obtained through the equation x.

$$DNI = DNI_{DISC}X(k_t', Z, W, \delta k_t') \tag{13}$$

Where $X(k_t', Z, W, \delta k_t')$ is a function of coefficients that represents the four parameters of the insolation condition. This function is obtained through Table 1.

Table 1: Bins of the DIRINT model.

Bin	k_t'		z		W		$\delta k_t'$	
	min	max	min	max	min	max	min	max
1	0	0.24	0	25	0	1	0	0.015
2	0.24	0.4	25	40	1	2	0.15	0.035
3	0.4	0.56	40	55	2	3	0.035	0.07
4	0.56	0.7	55	70	3	∞	0.07	0.15
5	0.7	0.8	70	80	0	∞	0.15	0.3
6	0.8	1	80	90			0.3	1
7							0	1

The DHI can be calculated based on GHI and DNI by using the equation (14).

$$DHI = GHI - DNIcos(\theta_z) \tag{14}$$

2.2 Transposition models

The global incident POA on a tilted surface, I_t, with a tilt angle of β degree can be calculated by

$$I_t = I_b + I_g + I_d \tag{15}$$

Where I_b represents the direct irradiance on the tilted plane that can be obtained through equation (16) according to the relation between the direct irradiance in the horizontal plane and the angle of incidence of the sun rays on the tilted plane, θ.

$$I_b = DNI.\cos\theta \tag{16}$$

I_g is ground-reflected irradiance that is calculated based on GHI, ground reflectance index, *albedo*, and β by using the following equation:

$$I_g = GHI.albedo.\frac{1-\cos\beta}{2} \tag{17}$$

I_d is the diffuse radiation from the sky, which refers to the diffuse irradiance on the inclined plane, that is calculated as:

$$I_d = DHI.R_d \tag{18}$$

Where, R_d is the diffuse transposition factor that can be calculated by solar transposition models. This paper utilizes the five dominant transposition models of Liu & Jordan, Klucher, Hay & Davies, Reindl, and Perez, which are described as follows.

2.2.1 Liu & Jordan model

The Liu & Jordan transposition model [29] assumes that the diffuse irradiance propagates uniformly, with the same intensity from the sky dome in all directions. Hence, the circumsolar or horizon brightening factors are not taken into account. In this model, the R_d is calculated as below:

$$R_d = \frac{1+\cos\beta}{2} \tag{19}$$

2.2.2 Klucher model

Klucher [30] developed a model that considers two factors: the horizon brightening and the circumsolar component. This model estimates the R_d by equation (20).

$$R_d = (1 + f'cos^2\theta sin^3\theta_z)\left(1 + f'sin^3\left(\frac{\beta}{2}\right)\right)\left(\frac{1+\cos\beta}{2}\right) \tag{20}$$

Where $f' = 1 - (DHI/GHI)^2$.

2.2.3 Hay & Davies model

In the Hay & Davis model [31], diffuse radiation from the sky comprises an isotropic component and a circumsolar one. Horizon brightening is not taken into account. An anisotropy index, A_i, is considered to quantify a portion of the diffuse radiation treated as circumsolar, with the remaining portion of diffuse radiation assumed to be isotropic. The Hay & Davis model comprises two components of diffuse radiation from the sky: isotropic and circumsolar. Horizon brightening is not considered in this model. To quantify the portion of diffuse radiation treated as circumsolar, an anisotropy index, A_i is defined as below.

$$A_i = \frac{DNI}{I_0} \tag{21}$$

The remaining portion of diffuse radiation is assumed to be isotropic. Hay & Davis model estimates R_d using equation (22).

$$R_d = A_i\frac{\cos\theta}{\cos\theta_z} + (1 - A_i)\frac{1+\cos\beta}{2} \tag{22}$$

10.4229/EUPVSEC2023/4CV.1.46
020406-003

2.2.4 Reindl model

Reindel [32] expanded the Hay model by incorporating the horizon brightening as shown in the following equation:

$$R_d = A_i \frac{\cos\theta}{\cos\theta_z} + (1 - A_i)\frac{1+\cos\beta}{2} + \left(1 + \sqrt{\frac{DNI\cos\theta_z}{GHI}}\, sin^3\left(\frac{\beta}{2}\right)\right) \quad (23)$$

2.2.5 Perez model

In contrast to the previously described models, the Perez model [33] offers a more comprehensive analysis of isotropic diffuse radiation, circumsolar radiation, and horizon brightening. This is achieved by incorporating empirically derived coefficients, enabling a more detailed and accurate representation of these components. This model describes the R_d by equation (24).

$$R_d = F_1\frac{a}{b} + (1 - F_1)\frac{1+\cos\beta}{2} + F_2\sin\beta \quad (24)$$

Where F_1 and F_2 represent the sky brightness for the circumsolar area and the region above the horizon, respectively. These coefficients are:

$$F_1 = F_{11}(\varepsilon) + F_{12}(\varepsilon)\Delta + F_{13}(\varepsilon)\theta \quad (25)$$

$$F_2 = F_{21}(\varepsilon) + F_{22}(\varepsilon)\Delta + F_{23}(\varepsilon)\theta \quad (26)$$

Where the sky clearness parameter ε and the sky brightness parameter Δ are described by equations (27) and (28), respectively.

$$\varepsilon = \frac{\frac{DHI+DNI}{DHI} + 1.041\theta^3}{1 + 1.041\theta^3} \quad (27)$$

$$\Delta = \frac{DHI.am}{I_0} \quad (28)$$

The Perez coefficients F_{ij} are the function of ε and Δ. Moreover, the coefficients a and b that consider the angle of incidence of the sun onto the inclined surface are calculated by the following equations:

$$a = max(0, \cos\theta) \quad (29)$$

$$b = max(\cos 85^{\circ}, \cos\theta_z) \quad (30)$$

2.3 CALCULATING THE POA ON INCLINED PLANE

The decomposition and transposition models are combined to calculate the POA irradiance on an inclined plane under specific tilt and azimuth angles, using the ground-measured GHI for estimating the. This process involves two key steps. First, a decomposition model estimates the horizontal DNI and DHI from the measured GHI. Second, a transposition model calculates the POA utilizing the horizontal GHI, DNI, and DHI. Therefore, 15 combinations exist for the aforementioned decomposition and transposition models to estimate the hourly POA for each pair of tilt and azimuth angles. Subsequently, the daily POA and monthly average daily POA can be calculated for each month of the year. Fig. 1 shows the processes of calculating the monthly average daily POA using a combination of decomposition and transposition models for each pair of tilt and azimuth angles.

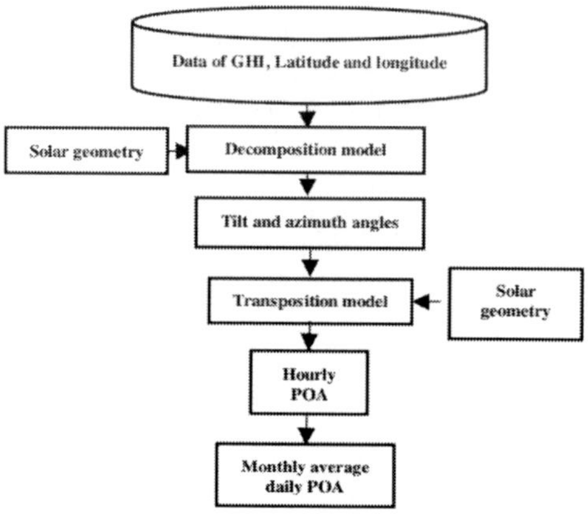

Figure 1: Flowchart of calculating POA irradiance.

2.4 GENERATING POA GRAPHS

To generate the POA graphs, the tilt and azimuth angles are varied within their respective permissible ranges at specific intervals. Consequently, the tilt angle spans from the 0°to 90°, while the azimuth angle ranges from -180° to +180°. Using the hourly GHI data for a full year, in accordance with the process outlined in Fig. 1, the hourly POA for the year is calculated for each unique combination of tilt and azimuth angles for the selected combination of decomposition and transposition models. Accordingly, the monthly average daily POA for each month can be calculated based on the estimated hourly POA data.

3. SIMULATION RESULTS

The developed simulation tool is employed to generate the graphs of the monthly average daily POA for the Brazilian cities. As a case study, the ground-measured hourly GHI data of the year 2020 at the city of Curitiba is used to estimate the hourly POA over the year. Considering the interval variation of 1° in tilt and azimuth angles, the hourly POA and monthly average daily POA for a combination of 91× 361 are calculated for any pair selection of a combination of decomposition and transposition models. In this section, the monthly average daily POA computed by the Dirint decomposition and Perez transposition models are used to generate the POA graphs for the months of 2020. These graphs that are plotted from the 91× 361 matrices of monthly average daily POAs are shown in Fig. 2 to Fig. 13.

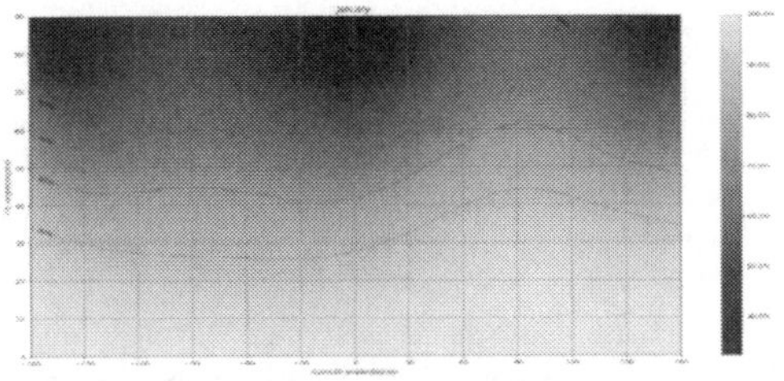

Figure 2: Monthly average daily POA graph-January

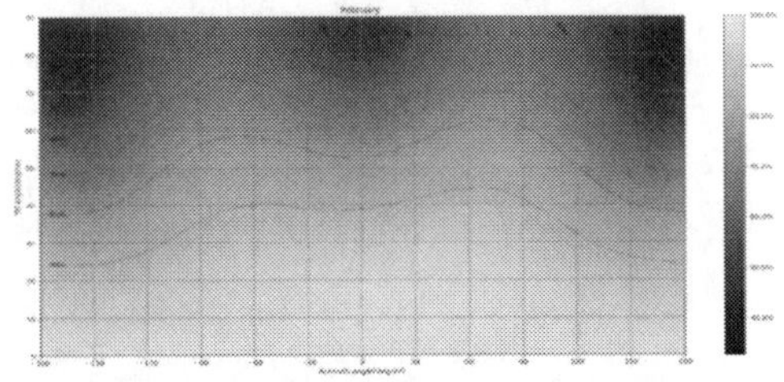

Figure 3: Monthly average daily POA graph-February

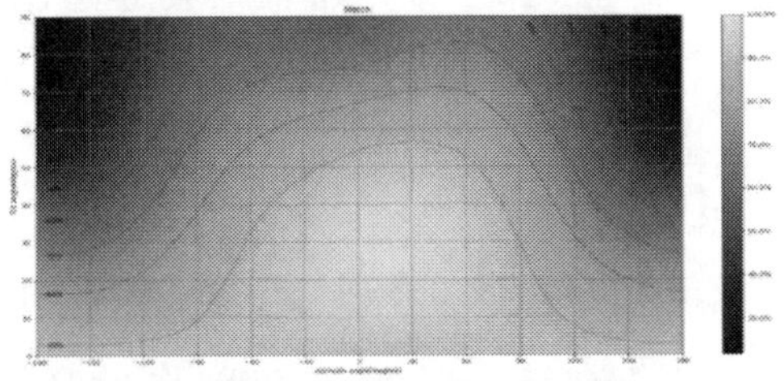

Figure 4: Monthly average daily POA graph-March

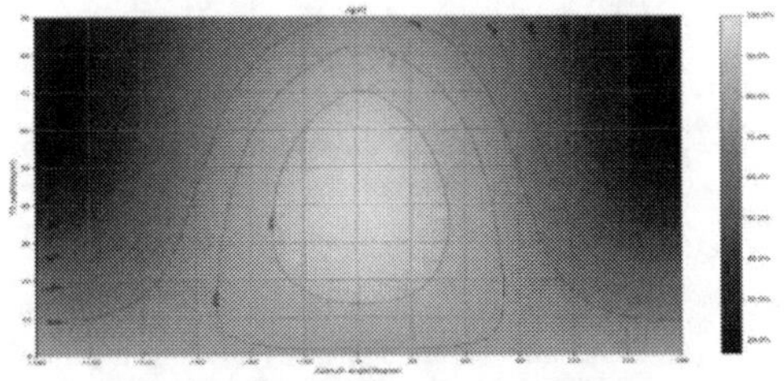

Figure 5: Monthly average daily POA graph-April

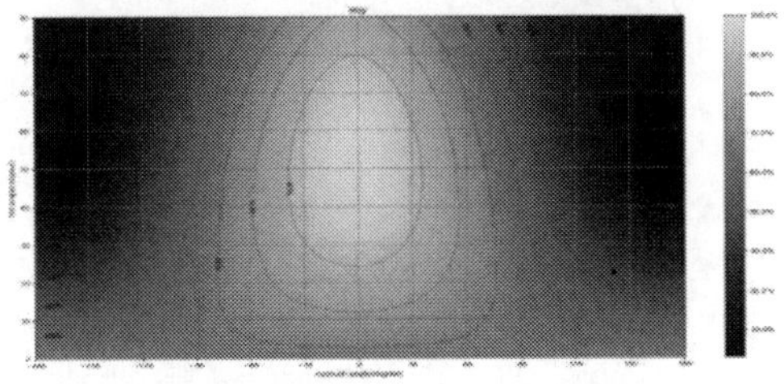

Figure 6: Monthly average daily POA graph-May

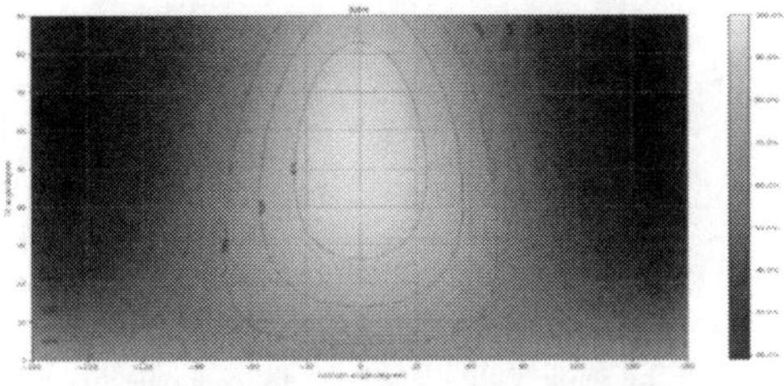

Figure 7: Monthly average daily POA graph-June

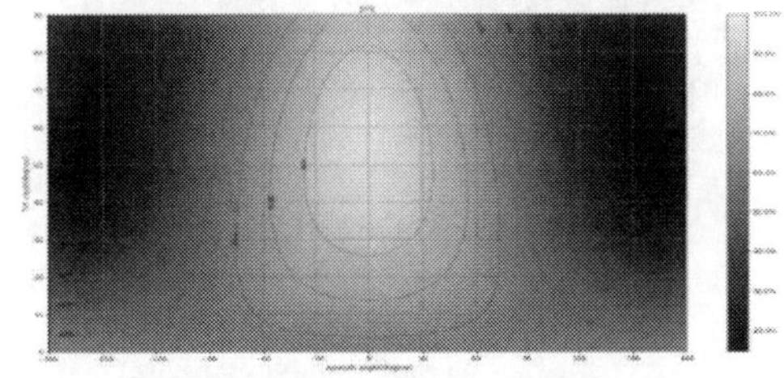

Figure 8: Monthly average daily POA graph-July

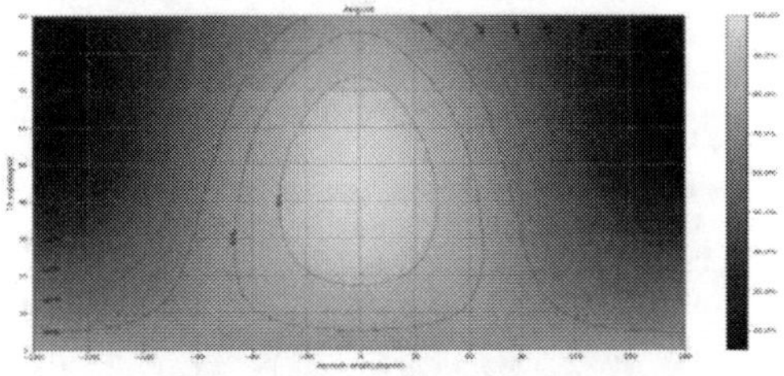

Figure 9: Monthly average daily POA graph-August

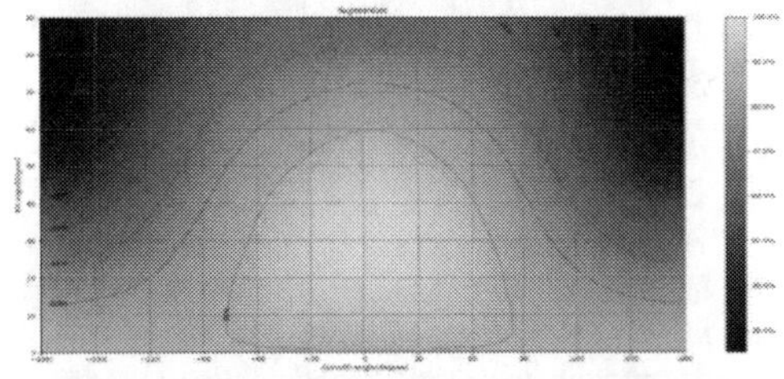

Figure 10: Monthly average daily POA graph-September

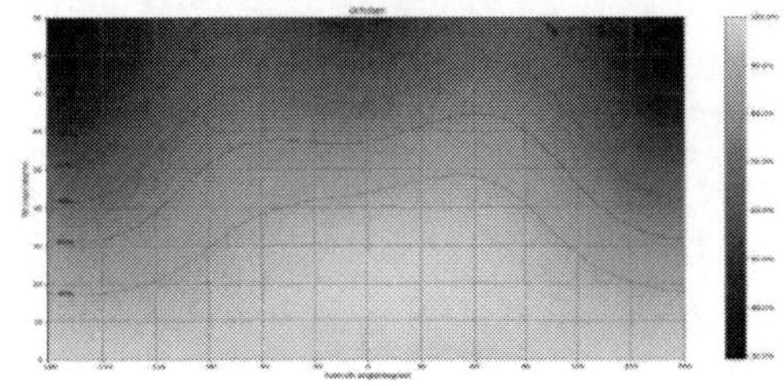

Figure 11: Monthly average daily POA graph-October

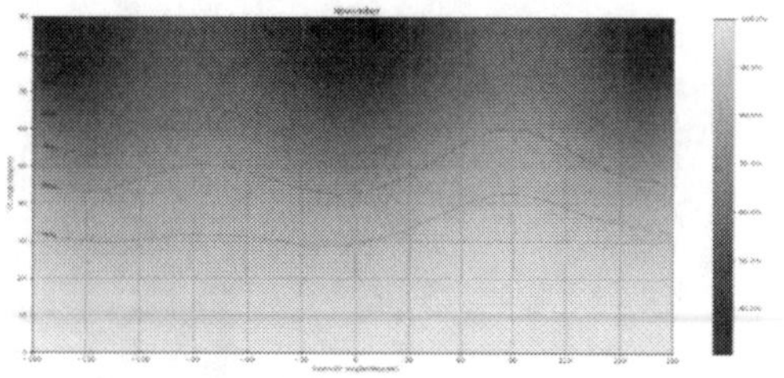

Figure 12: Monthly average daily POA graph-November

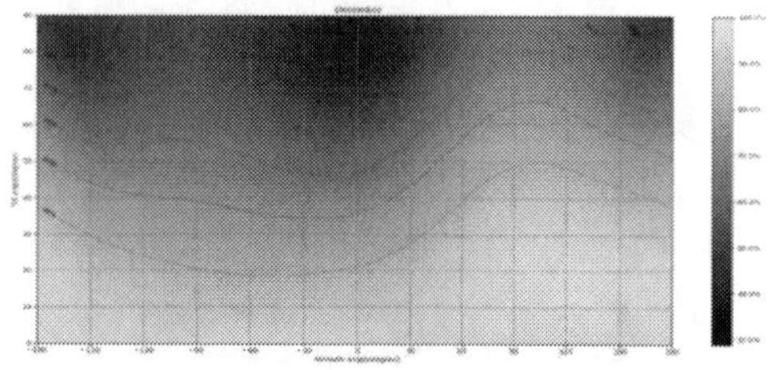

Figure 13: Monthly average daily POA graph-Dezember

Fig. 14 shows the graph of the annual average daily POA computed by the Dirint decomposition model and the Perez transposition model.

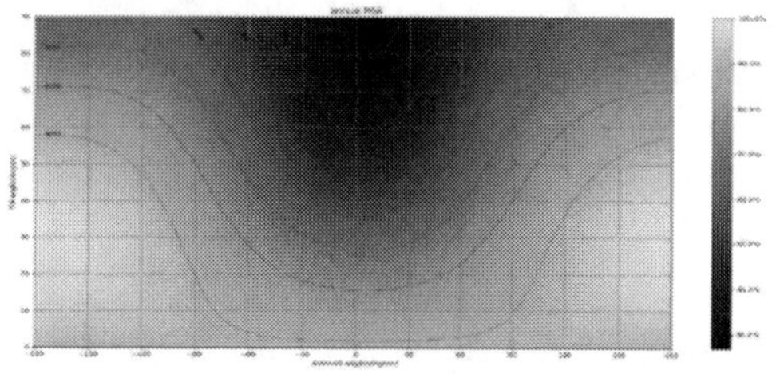

Figure 14: Annual average daily POA graph

4. CONCLUSION

This study developed a simulation tool to generate graphs that allow a quick and simplified evaluation of installing PV systems in Brazilian cities. The results demonstrate that variations in azimuth for annual and some months of the year do not lead to significant POA losses, even up to tilt angles of approximately 20 degrees. However, this tilt angle limitation for specific months of the year makes significant losses monthly average daily POA. The results illustrate that the best orientation for installing the PV systems in the case study city generally involves an inclination close to the local latitude and an azimuth angle close to zero degrees, oriented towards the north. The developed simulation tool can efficiently analyze POA levels for fixed PV systems, especially for PV retrofitting in building-integrated PV systems.

6. REFERENCES

[1] Yaqoob Majeed, Muhammad Usman Khan, Muhammad Waseem, Umair Zahid, Faisal Mahmood, Faizan Majeed, Muhammad Sultan, Ali Raza, Renewable energy as an alternative source for energy management in agriculture, Energy Reports, Volume 10, 2023, Pages 344-359.

[2] Shijing Nan, Yuchen Huo, Chien-Chiang Lee, Assessing the role of globalization on renewable energy consumption: New evidence from a spatial econometric analysis, Renewable Energy, Volume 215, 2023, 118974.

[3] Sevda Kuşkaya, Faik Bilgili, Erhan Muğaloğlu, Kamran Khan, Mohammad Enamul Hoque, Nurhan Toguç, The role of solar energy usage in environmental sustainability: Fresh evidence through time-frequency analyses, Renewable Energy, Volume 206, 2023, Pages 858-871.

[4] Md Masud Rana, Moslem Uddin, Md Rasel Sarkar, Sheikh Tanzim Meraj, G.M. Shafiullah, S.M. Muyeen, Md. Ariful Islam, Taskin Jamal, Applications of energy storage systems in power grids with and without renewable energy integration — A comprehensive review, Journal of Energy Storage, Volume 68, 2023, 107811.

[5] Hongying Zhao, Rebecca Jing Yang, Chengyang Liu, Chayn Sun, Solar building envelope potential in urban environments: A state-of-the-art review of assessment methods and framework, Building and Environment, Volume 244, 2023, 110831.

[6] Rebecca Jing Yang, Samarasinghalage Tharushi Imalka, W.M. Pabasara Wijeratne, Gayashan Amarasinghe, Nilmini Weerasinghe, Sujan Dev Sureshkumar Jayakumari, Hongying Zhao, Ziheng Wang, Chathuri Gunarathna, Justin Perrie, Chengyang Liu, Ron Wakefield, Digitalizing building integrated photovoltaic (BIPV) conceptual design: A framework and an example platform, Building and Environment, Volume 243, 2023, 110675.

[7] H.Z. Al Garni, A. Awasthi, M.A.M. Ramli, Optimal design and analysis of grid connected photovoltaic under different tracking systems using HOMER, Energy Convers. Manag. 155C, 42-57.2018.

[8] Guo, M.; Zang, H.; Gao, S.; Chen, T.; Xiao, J.; Cheng, L.; Wei, Z.; Sun, G. Optimal Tilt Angle and Orientation of Photovoltaic Modules Using HS Algorithm in Different Climates of China. Appl. Sci., 7, 1028. 2017.

[9] Aybüke Taşer, Başak Kundakcı Koyunbaba, Tuğçe Kazanasmaz, Thermal, daylight, and energy potential of building-integrated photovoltaic (BIPV) systems: A comprehensive review of effects and developments, Solar Energy, Volume 251, 2023, Pages 171-196.

[10] Ceylin Şirin, Jamie Goggins, Magdalena Hajdukiewicz, A review on building-integrated photovoltaic/thermal systems for green

5 ACKNOWLEDGMENT

This work was supported by BYD Energy Brazil (through the PADIS/MCTI program, Funcamp project N.5779) and Total Energies (through ANP - Brazilian Oil & Gas Agency, Funcamp project N.6002).

10.4229/EUPVSEC2023/4CV.1.46
020406-006

buildings, Applied Thermal Engineering, Volume 229, 2023, 120607.

[11] Linda G. Teka, Mehdi Zadshir, Huiming Yin, Mechanical analysis and design of large building integrated photovoltaic panels for a seamless roof, Solar Energy, Volume 251, 2023, Pages 1-12.

[12] Neelam Baghel, K. Manjunath, Anil Kumar, Performance evaluation and optimization of albedo and tilt angle for solar photovoltaic system, Computers and Electrical Engineering, Volume 110, 2023, 108849.

[13] M. A. Z. Abidin, M. N. Mahyuddin and M. A. A. M. Zainuri, "Optimal Efficient Energy Production by PV Module Tilt-Orientation Prediction Without Compromising Crop-Light Demands in Agrivoltaic Systems," in IEEE Access, vol. 11, pp. 71557-71572, 2023.

[14] G. Veljanovski, M. Atanasovski, A. K. Mateska, S. Sabotkovska, K. Bilbiloska and H. Petreska, "Analysis on the Optimum Tilt and Azimuth Angles for Fixed-Tilt PV Systems," 2023 58th International Scientific Conference on Information, Communication and Energy Systems and Technologies (ICEST), Nis, Serbia, 2023, pp. 179-182.

[15] N. Mignoni, R. Carli and M. Dotoli, "An Optimization Tool for Displacing Photovoltaic Arrays in Polygonal Areas," IEEE EUROCON 2023 - 20th International Conference on Smart Technologies, Torino, Italy, 2023, pp. 573-578.

[16] Yanis Prunier, David Chuet, Sylvain Nicolay, Gwenaëlle Hamon, Maxime Darnon, Optimization of photovoltaic panel tilt angle for short periods of time or multiple reorientations, Energy Conversion and Management: X, Volume 20, 2023, 100417.

[17] Salman Habib, Muhammad Tamoor, Muhammad Ans Zaka, Youwei Jia, Assessment and optimization of carport structures for photovoltaic systems: A path to sustainable energy development, Energy Conversion and Management, Volume 295, 2023, 117617.

[18] A. Barbón, C. Bayón-Cueli, L. Bayón, C. Rodríguez-Suanzes, Analysis of the tilt and azimuth angles of photovoltaic systems in non-ideal positions for urban applications, Applied Energy, Volume 305, 2022, 117802.

[19] Tamoor, M.; Habib, S.; Bhatti, A.R.; Butt, A.D.; Awan, A.B.; Ahmed, E.M. Designing and Energy Estimation of Photovoltaic Energy Generation System and Prediction of Plant Performance with the Variation of Tilt Angle and Interrow Spacing. Sustainability 2022, 14, 627.

[20] Rafaela A. Agathokleous, Soteris A. Kalogirou, Status, barriers and perspectives of building integrated photovoltaic systems, Energy, Volume 191, 2020, 116471.

[21] Maghrabie, H.M.; Abdelkareem, M.A.; Al-Alami, A.H.; Ramadan, M.; Mushtaha, E.; Wilberforce, T.; Olabi, A.G. State-of-the-Art Technologies for Building-Integrated Photovoltaic Systems. *Buildings* 2021, *11*, 383.

[22] De Medici, S. Italian Architectural Heritage and Photovoltaic Systems. Matching Style with Sustainability. Sustainability 2021, 13, 2108.

[23] WanTing Wang, Hongxing Yang, ChangYing Xiang, Green roofs and facades with integrated photovoltaic system for zero energy eco-friendly building – A review, Sustainable Energy Technologies and Assessments, Volume 60, 2023, 103426.

[24] Sheikh Suhail Mohammad, Sheikh Javed Iqbal, Rayes Ahmad Lone, Improved Utilization of Solar Energy Using Estimated Optimal Tilt Factor and Trackers, Energy Reports, Volume 8, Supplement 14, 2022, Pages 175-183.

[25] Herman Böök, Anders V. Lindfors, Site-specific adjustment of a NWP-based photovoltaic production forecast, Solar Energy, Volume 211, 2020, Pages 779-788.

[26] ERBS, D. G.; KLEIN, S. A.; DUFFIE, J. A. Estimation of the diffuse radiation fraction for hourly, daily and monthly-average global radiation. Solar Energy, v. 28, n. 4, p. 293-302, 1982.

[27] MAXWELL, E. L. A Quasi-Physical Model for Converting Hourly Global Horizontal to Direct Normal Insolation. Golden, Colorado, 1987.

[28] R. Perez, "Dynamic global to direct conversion models," ASHRAE Trans. Res. Series, vol. 98, no. 1, pp. 354–369, 1992.

[29] B.Y. Liu, R.C. Jordan, The long-term average performance of flat-plate solar-energy collectors Solar Energy, 7 (1963), pp. 53-74.

[30] T. Klucher. Evaluation of models to predict insolation on tilted surfaces Solar Energy, 23 (1979), pp. 111-114.

[31] J.E. Hay. Calculating solar radiation for inclined surfaces: practical approaches Renew. Energy, 3 (1993), pp. 373-380.

[32] D. Reindl, W. Beckman, J. Duffie. Evaluation of hourly tilted surface radiation models Solar Energy, 45 (1990), pp. 9-17.

[33] R. Perez, P. Ineichen, R. Seals, J. Michalsky, R. Stewart Modeling daylight availability and irradiance components from direct and global irradiance Solar Energy, 44 (1990), pp. 271-289.

REAR-SIDE IRRADIANCE MEASUREMENT FOR BIFACIAL PV SYSTEMS: A CASE STUDY OF DIFFERENT SENSOR INSTALLATION SETUPS

Imke Meyer, Mott MacDonald Ltd, Imke.Meyer@mottmac.com
Jose Muñoz, Mott MacDonald Spain jose.muñoz@mottmac.com
Ricardo Velasco, Mott MacDonald Colombia, ricardo.velascomunar@mottmac.com

Pertinent industry standards for assessing bifacial module performance remain continuously under review. This research investigates the placement of rear-side Plane of Array (PoA) pyranometers to establish a representative positioning as called for under the IEC 61724-1:2021, [1]. Data from three case studies located in Japan, Namibia and Mexico have been evaluated. The analysis highlights dispersion in rear-side measurements for the case studies in Namibia and Mexico although the pyranometers in each plant for each case study are installed across the plant with the same local environmental factors (vegetation, soil, location within array and distance from internal or external roads). The case study in Japan includes three rear-side PoA pyranometers at different heights and thus used to highlight the irradiation dispersion across the rear-side of bifacial modules and aligning with the IEC 61724-1:2021 recommendation stating the need of three rear-side pyranometers for at each meteorological station. From the case studies located in Namibia (three meteorological stations) and Mexico (15 meteorological stations), the results were unable to identify specific drivers for dispersion. Highest dispersion was noted during low irradiance periods. Lower dispersion from the mean value was observed in the Mexico plant where 15 meteorological stations are present, where seven sensors were positioned within one standard deviation from the mean. These findings highlight the importance of multiple meteorological stations to capture the rear side irradiance across the site for an accurate PR calculation as limited measurements may not capture the fluctuating albedo and irradiation reflectance across a site and thus not accurately reflect the plant performance in the performance ratio calculation.

Keywords: Bifacial solar PV, Pyranometers, rear-side irradiance measurements, IEC 61724-1:2021

1 INTRODUCTION

The utility-scale solar PV market has observed an uptake in the utilization of bifacial solar photovoltaic (PV) modules over the past 5 years, currently holding a 30% market share. This market share for bifacial modules is projected to escalate to 70% by 2033 [2]. While industry bodies constantly review the pertinent standards and guidelines for performance measurement in projects involving bifacial modules, there persists some uncertainty regarding the appropriate positioning of rear-side irradiance sensors in the IEC 61724-1:2021 standard titled "Photovoltaic system performance – Part 1: Monitoring" [1].

This paper investigates the placement of rear-side pyranometers, seeking to increase understanding of what constitutes the representative positioning for these sensors. Specifically, the study attempts to contribute towards the knowledge base of what is deemed representative rear-side pyranometers placement as called upon within the IEC 61724-1:2021 for the measurement of bifacial Performance Ratio (PR).

2 AIM

The objective of this study is to assess how the positioning and configuration of rear-side Plane of Array (PoA) irradiance sensors impact the irradiance measurement captured by these sensors. This parameter holds importance in determining the PR of bifacial PV systems. This research attempts to offer insights into discrepancies detected in rear-side irradiance measurements and evaluates the influence of various system design factors on this measurement .

The paper focuses on factors related to design, installation, and maintenance. These encompass the placement of rear-side irradiance sensors across PV plants, considering proximity to internal roads and local surrounding variables such as vegetation and soil. These aspects can potentially affect rear-side irradiance measurements as each of these aspects impact the amount of irradiation that is reflected to the PoA rear side pyranometer. The research strives to identify the most representative sensor locations and setups for rear-side irradiance pyranometers, conforming to the guidelines set by the IEC 61724-1:2021 [1].

These parameters are investigated through the analysis of three case studies from operational PV plants situated in different locations. Noteworthy variations between the measurements of the rear-side pyranometers within each plant are presented and where possible linked to environmental influences, sensor placement or configuration, or maintenance practices.

3 CASE STUDIES

Data from three case studies located in Japan, Namibia and Mexico is evaluated. The three case studies present the following meteorological station setups.

Case study 1: Japan
Fixed tilt PV plant. A single meteorological station is installed on site encompassing the following irradiance measurement devices:

- 1 Albedometer (1 up-facing pyranometer and 1 down-facing pyranometer)
- 3 rear-side PoA pyranometers at 1.88m, 2.2m and 3m height from the ground level (marked as no.2 on Figure 1 below)
- 1 front-side PoA pyranometer at 3.7m height from the ground level (marked as no.3 on Figure 1 below)
- 1 Diffuse Horizontal Irradiance (DHI) pyranometer
- 1 horizontal downfacing pyranometer

All pyranometers are Class A according to ISO 9060:2018 standard [3]. The configuration, using three rear-facing pyranometers distributed at different heights along the rear side mounting structure, follows the recommendation of Table 2 in the IEC 61724-1:2021 standard [1]. The available dataset spans from February 2020 to September 2020 with data samples acquired at a 5-minute time resolution.

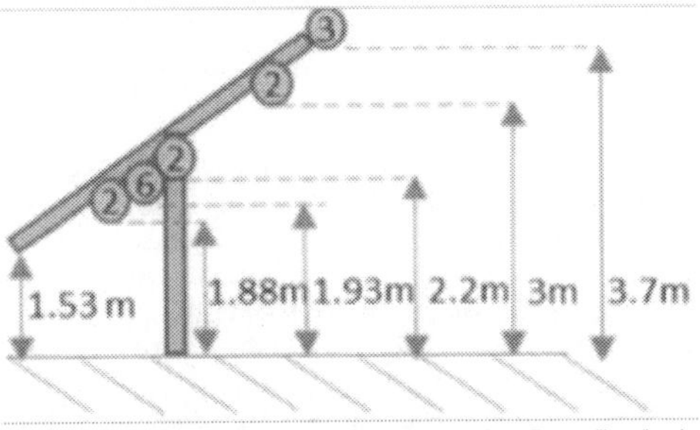

Figure 1: Lateral view configuration for Case Study 1

Case study 2: Namibia

Single axis tracking PV plant. Three meteorological stations located throughout the plant. Each meteorological station consists of the following irradiance measurement devices:

- 1 Front-side PoA pyranometer
- 1 Rear-side PoA pyranometer
- 1 Global Horizontal Irradiance (GHI) pyranometer

All pyranometers are Class A according to ISO 9060:2018 standard [3]. The pyranometers are mounted on N-S single axis trackers, on inner table of the tracker and close to the centre (E-W) of the tracker table. Data available from April 2022 to June 2023, at a 5-minute time resolution. However, due to operational issues one of the meteorological stations went offline on January 2023.

Figure 2: Configuration of the in-plane front and rear-side pyranometers for Case Study 2

Case study 3: Mexico

Single axis tracking PV plant equipped with 15 meteorological stations distributed across the PV plant. Each of the 15 different meteorological stations have the following irradiance measurement devices:

- 1 Front-side PoA pyranometer
- 2 Rear-side PoA pyranometers

All pyranometers are Class A according to ISO 9060:2018 standard [3]. The pyranometers are mounted at the end of the tracker (N-S) with the single front-side PoA pyranometer installed in the centre (E-W) of the tracker table and the two rear-side facing pyranometers installed at the top and bottom edges of the tracking table (one towards east and another one towards west). The analysed samples cover a time period of nine months, from January 2022 to September 2022. The samples were gathered at a 5-minute time resolution and all sensors are reported to have worked properly over that period.

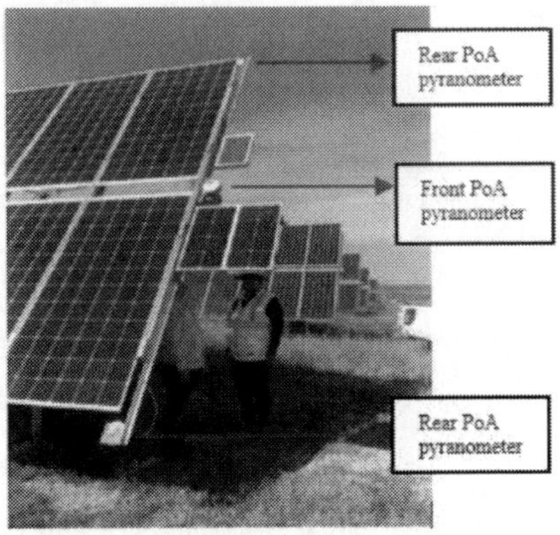

Figure 3: Configuration for Case Study 3

For all three case studies it has been confirmed that all installed sensors hold a valid calibration certificate, that maintenance activities are performed identically on each of the meteorological stations within each plant (i.e., weekly cleaning of rear-side irradiance sensors) and the location and setup of the meteorological stations and its sensors is consistent across each PV plant.

3 METHODOLOGIES

3.1 Data analysis

Data-driven and a systematic approach is used to assess the impact of different factors (i.e., location, setup, etc.) on the rear-side irradiance measurements.

Initial raw data processing (data cleaning, removal of data under no irradiance conditions, inspection of gaps and physical limits) is carried out. All data is collected at a 5-minute temporal resolution.

The data analysis strategy encompasses:

- Relative comparison of pyranometers at top, middle, and bottom altitudes from Case Study 1.
- Relative compassion among different locations within the same PV plant from Case Studies 2 and 3.
- Intraday trends for clear sky and overcast sky conditions.
- Seasonal variations over the sampled period for each PV plant.

Statistic parameters namely mean, standard deviation, and coefficient of variation are implemented during the data analysis. Similarly, absolute and relative differences were estimated to establish trends in between pyranometers' measurements for a given location. These metrics are used to help inform what may be deemed the most representative location for the placement of rear-facing pyranometers.

4 RESULTS

4.1 Case Study 1: Japan Testbed

Table I summarizes the primary dataset characteristics

10.4229/EUPVSEC2023/4CV.1.52

of the analysed measurements.

Table I: Statistics of the different rear-side irradiance measurement sensors over the complete sampled period

Sensor location	Mean [W/m²]	STD [W/m²]	Variation[1] [%]	Dev. from avg. [%]
Met. Station 1 (Bottom)	8.33	14.34	172.08	-31.89
Met. Station 2 (Middle)	12.69	19.75	155.68	3.76
Met. Station 3 (Top)	15.67	25.71	164.01	28.13

During the sampled period, data gaps were found for each measuring device due to reported temporal malfunctions of on-site data collection PCs. Nonetheless, the results presented in Table I correspond to time intervals where data readings were available for every sensor. During the data cleaning it was observed that the data gaps are distributed across all months, thus avoiding any significant gap concentration that could introduce significant bias to the measurements and analysis.

The mean values of the rear-facing pyranometers increase as the ground clearance height becomes larger, as anticipated and physically proving the consistency of the measurement. A similar rationale applies to the standard deviation, owing to the higher irradiance values measured by the sensor installed at higher height from the ground.

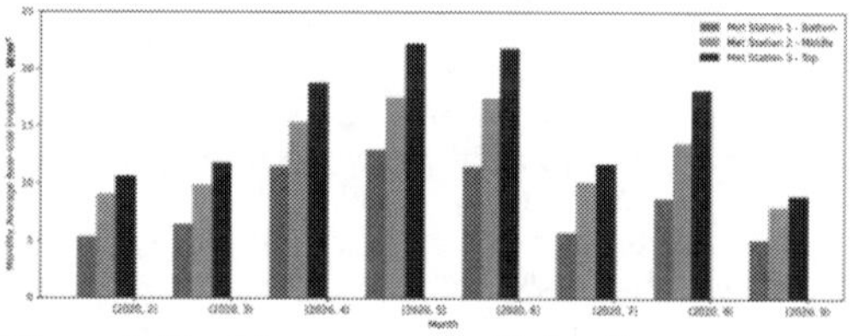

Figure 4: Monthly average rear-side irradiance measurements

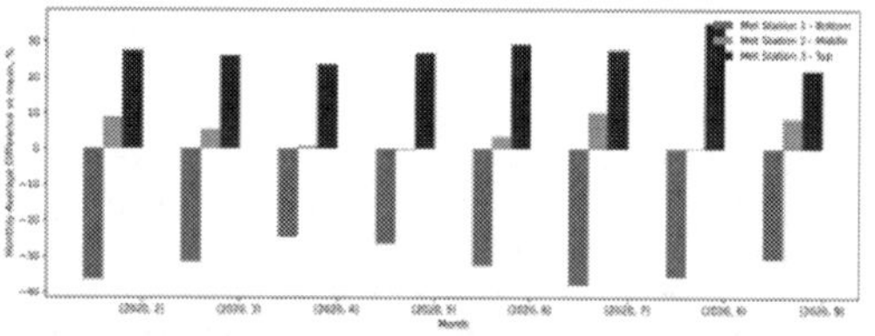

Figure 5: Monthly average rear-side irradiance relative difference from the mean value

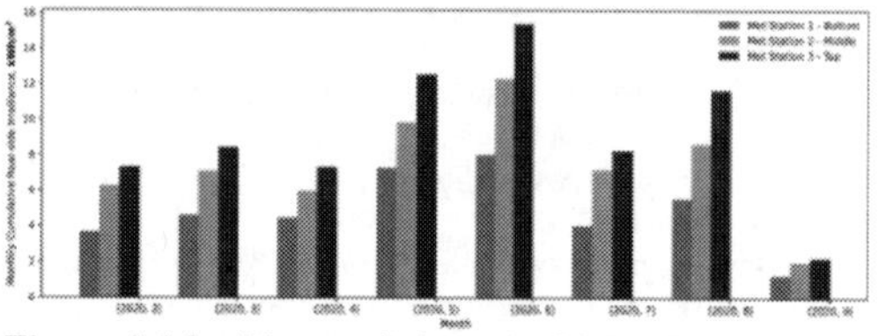

Figure 6: Monthly cumulative rear-side irradiance measurements

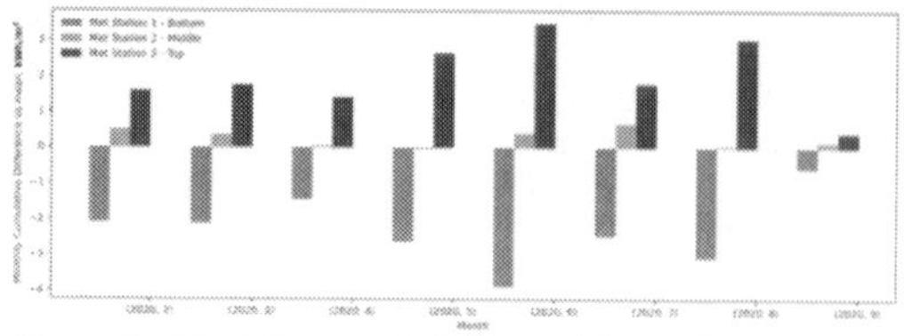

Figure 7: Monthly cumulative rear-side irradiance deviation from the mean value

From a statistical perspective, the centrally positioned rear-facing pyranometer closely aligns with the mean of the three pyranometers and also exhibits the lowest coefficient of variation throughout the entire sampled period (refer to Table I). This attribute strongly correlates with a narrower spread of data across all three pyranometers under analysis.

When scrutinized on a monthly basis, the mean 5-minute irradiance and monthly cumulative irradiation, as depicted from Figure 4 to Figure 7, further emphasize that the centrally placed pyranometer consistently hews closest to the mean value for each month. Meanwhile, the others display monthly absolute deviations ranging from 2 kWh/m² to 4 kWh/m², or up to 30% in relative terms.

The deviation is positive for the uppermost rear-facing pyranometer and negative for the lowermost rear-facing one. This trend persists across varying levels of front-side PoA irradiance, with daily aggregation yielding analogous results. The centrally placed pyranometer remains closest to the mean, although for low front-side PoA irradiance levels, variations increase across all three pyranometers, as illustrated in Figure 8.

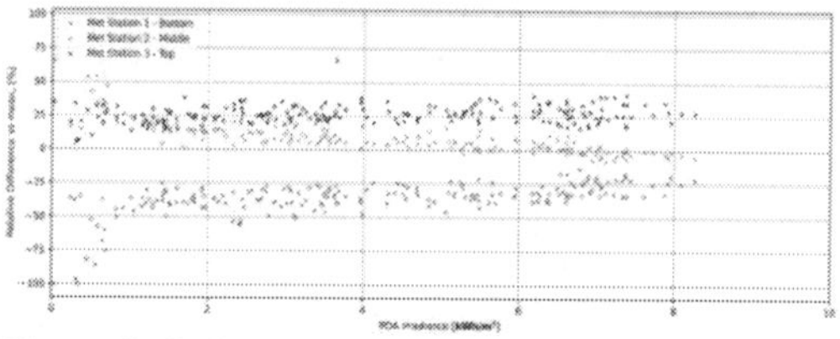

Figure 8: Daily sum rear-side irradiance measurements relative difference from the mean value vs front-side PoA

The meteorological station configuration of this case study is aligned with the guidelines provided in the IEC 61724-1:2021 standard [1], where a minimum of three rear-side irradiance measurement points per meteorological station is recommended to capture the irradiance dispersion on the rear side of the PV modules. The results illustrate the importance of having three rear-side in-plane pyranometers to capture the variance of irradiance falling on the rear side of a bifacial module.

Although the central rear-facing pyranometer algins with the mean, it cannot be assumed that this is the most representative position across the rear side of the module array. We also highlight that there is no second meteorological station for comparison or pyranometers installed part way between the middle and upper/lower pyranometers which may provide further insights into the understanding if a single location can capture representative irradiance on the rear side of the modules.

[1] The coefficient of variation represents the ratio of the standard deviation to the mean of any data series and is useful for comparing the degree of variation from one data series to another, even if the means are drastically different from one another.

10.4229/EUPVSEC2023/4CV.1.52
020408-003

4.2 Case Study 2: Namibia PV Plant

One of the rear-facing pyranometers went offline on January 27, 2023, thus the comparative data analysis is carried out with the three online sensors before this event and then with the two remaining sensors after this event.

The malfunctioning of the second rear-facing pyranometer inadvertently sheds light on the influence of a larger or smaller quantity of measuring devices on the assessment of rear-side PoA irradiance, and emphasises the importance of following at least the minimum requirements in terms of number of on-site sensors as specified in the IEC 61724-1:2021 standard [1], which for a system of 40MW or less – as is the case for this case study – recommends two sets of three rear-facing pyranometers.

The rear-facing pyranometers are placed at what is visually observed to be in uniform surrounding conditions and at the same location on the array (same distance from the ground, same tilt). The following tables show the data captured for the two periods and data deviations measured.

Table II: Statistical parameters of the different irradiance measurement sensors up to January 27, 2023.

Sensor location	Mean [W/m²]	STD [W/m²]	Variation [%]	Dev. from avg. [%]
Met. Station 1	61.77	41.00	66.37	-6.82
Met. Station 2	70.12	43.51	62.05	5.77
Met. Station 3	66.98	42.01	62.72	1.05

Table III: Statistical parameters of the different irradiance measurement sensors from January 28, 2023, onwards.

Sensor location	Mean [W/m²]	STD [W/m²]	Variation [%]	Dev. from avg. [%]
Met. Station 1	55.12	35.10	63.67	-7.72
Met. Station 2	N/A	N/A	N/A	N/A
Met. Station 3	64.32	40.08	62.30	7.72

The data from this PV plant exhibits no reported issues with data gaps. This dataset encompasses 43,250 5-minute data points for the period culminating on January 27, 2023, and 21,153 5-minute data points for the subsequent sampled period after the malfunction of the second rear-facing pyranometer.

Upon inspecting the results presented in Table II and Table III, it becomes evident that the deviation from the mean increases when dealing with only two rear-facing pyranometers, and what would be the representative value cannot be established. During the period when all three rear-facing pyranometers were operational, meteorological station 3 pyranometer stood as the closest to the mean, displaying a 1.05% deviation from the mean over that timeframe. From visual inspection of the surrounding areas of the rear-facing pyranometers there is no obvious reason for the dispersion of the measurements such as significant change in soil colour or vegetation coverage or proximity to an internal road or similar or why the meteo station 3 pyranometer should provide a more representative reading. Surroundings of meteorological station 1 and 2 can be seen in Figure 9 and Figure 10.

Figure 9: Surrounding around meteo station 1

Figure 10: Surrounding around meteo station 2

Examining the data on a monthly granularity of rear-side irradiance measurements reveals that relative differences decrease in summer months while increasing in winter ones, as represented in Figure 12 and Figure 14. Which is expected as the absolute deviation on rear side measurement in kWh/m² translates into a lower relative difference (in %) from the average value due to higher accumulated rear side PoA irradiance during summer months. In fact, a closer look at the below figures illustrates how the difference seen during the winter months is around a third of the difference seen during the summer months.

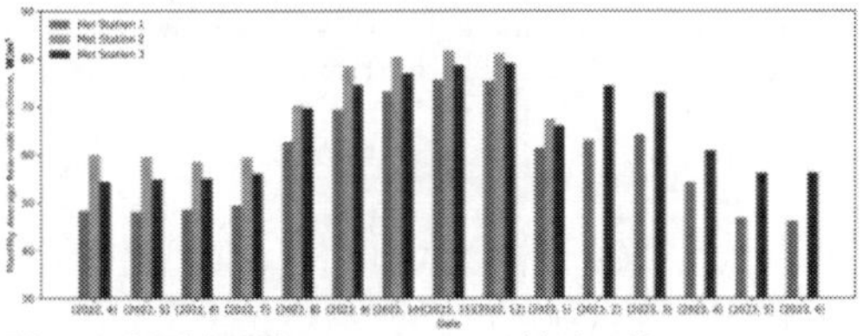

Figure 11: Monthly average rear-side irradiance measurements

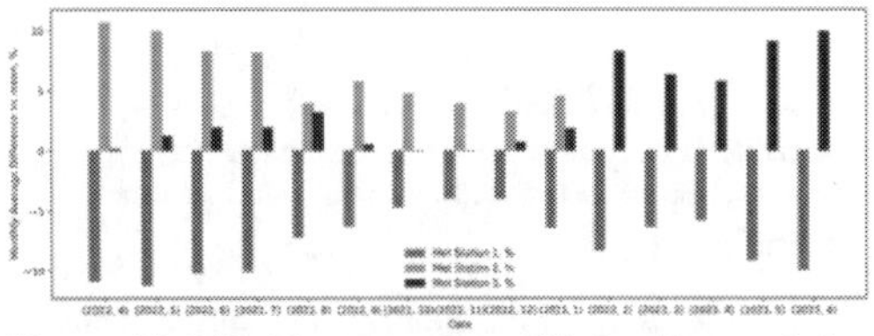

Figure 12: Monthly average rear-side irradiance relative difference from the mean value

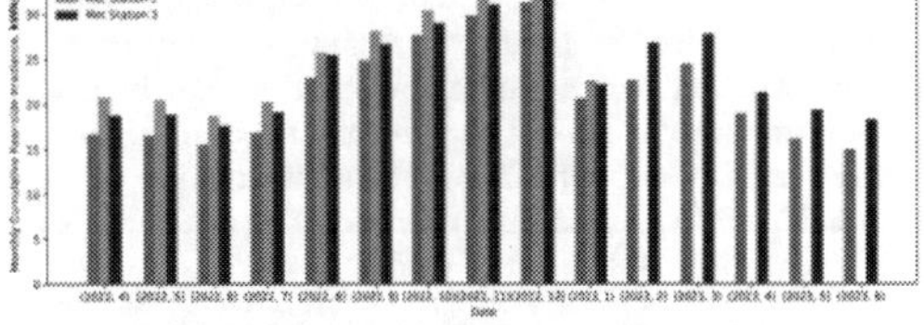

Figure 13: Monthly cumulative rear-side irradiance measurements

10.4229/EUPVSEC2023/4CV.1.52
020408-004

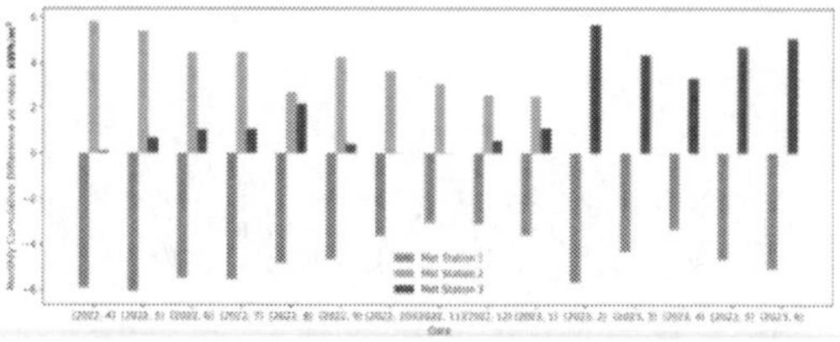

Figure 14: Monthly cumulative rear-side irradiance deviation from the mean value

At a higher time resolution – daily – the previously outlined patterns persist, where meteo station 3 rear-side sensor remained closest to the mean when all three rear-side irradiance sensors were operational, as depicted in Figure 15. Once meteo station 2 rear-side irradiance sensors malfunctioned, daily deviations increased, particularly for meteo station 3 sensor, inducing a mirror effect in relative deviations due to the presence of only two sensors, as illustrated in Figure 16.

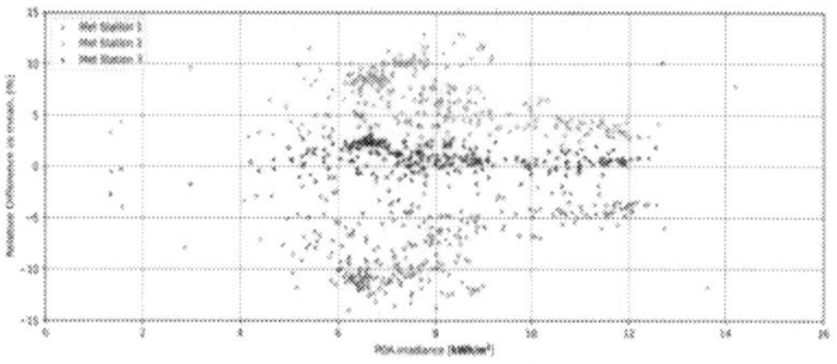

Figure 15: Daily cumulative rear-side irradiance measurements difference from the mean up to January 27, 2023

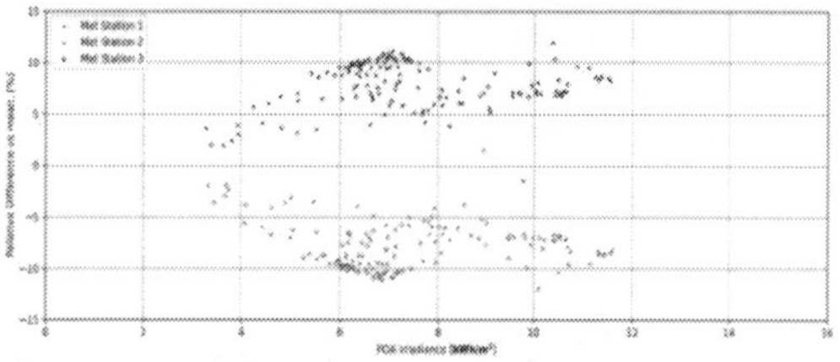

Figure 16: Daily cumulative rear-side irradiance measurements difference from the mean after malfunction of meteo station 2 rear-side irradiance sensors

Single day views plotting the 5-min resolution measurements of the rear-facing irradiance deviations under clear sky and overcast day conditions are represented in Figure 17 to Figure 19, respectively. The patterns in absolute deviations (W/m^2) from the mean still holds during most of the day for clear-sky conditions (Figure 18), except for low irradiance conditions such as at sunrise and sunset, where deviations trends shift. These low irradiance conditions can be extended to the overcast day scenario in Figure 20 with same fluctuating deviation trends and no clear identification of a device with lowest deviation from the mean of readings.

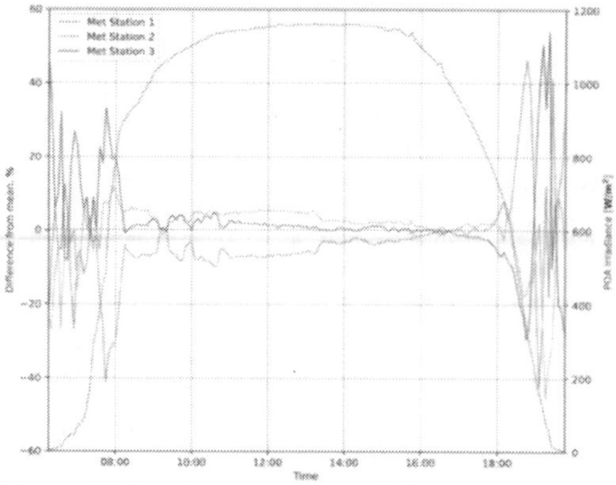

Figure 17: Intraday relative rear-side irradiance deviations from the mean under clear sky conditions on January 5, 2023. (Dashed red line: front PoA irradiance)

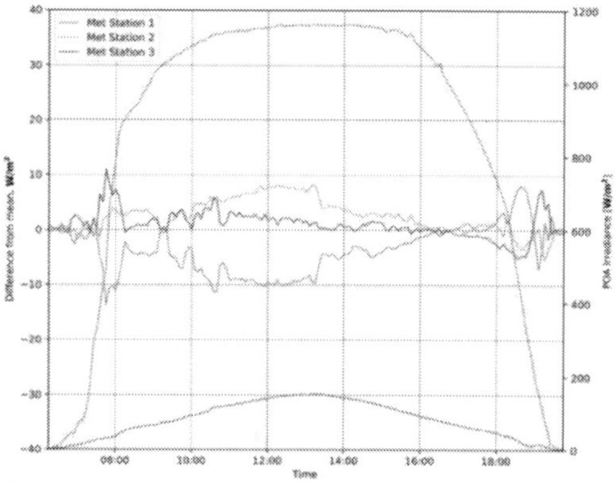

Figure 18: Intraday absolute rear-side irradiance deviations from the mean under clear sky conditions on January 5, 2023. (Dashed red line: front PoA irradiance, dashed blue line: mean rear PoA irradiance)

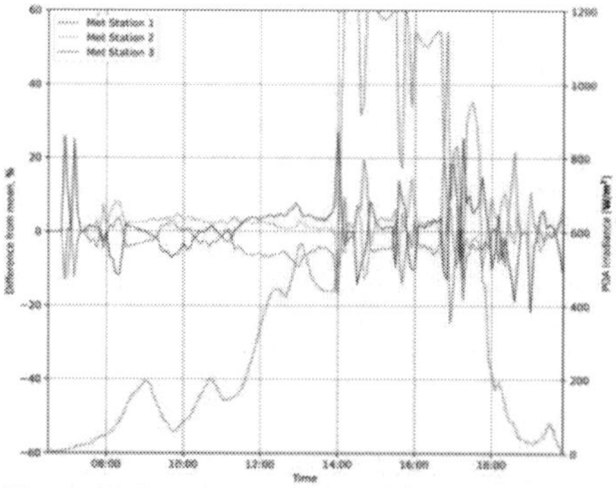

Figure 19: Intraday relative rear-side irradiance deviations from the mean under overcast conditions on January 23, 2023. (Dashed red line: front PoA irradiance)

10.4229/EUPVSEC2023/4CV.1.52
020408-005

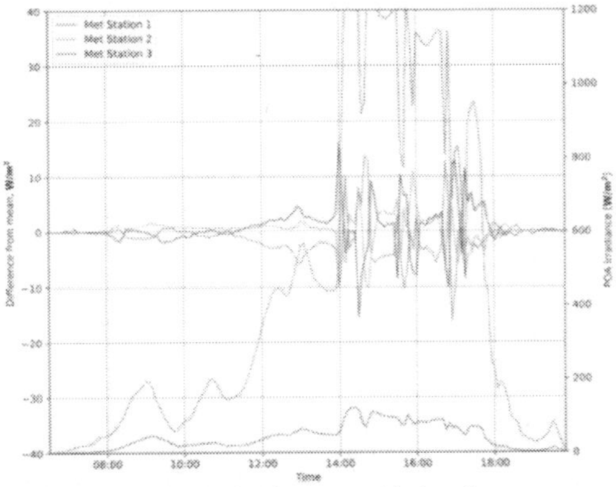

Figure 20: Intraday absolute rear-side irradiance deviations from the mean under overcast conditions on January 23, 2023. (Dashed red line: front PoA irradiance, dashed blue line: mean rear PoA irradiance)

The aforementioned representations of monthly, daily and 5-min deviations highlight that deviation between the rear-facing pyranometers is present at all temporal resolutions, whether employing three or two functional rear-side irradiance pyranometers. Given that all sensors are situated at uniform ground clearance with similar surroundings (as seen in Figure 9 and Figure 10), the disparities at different granularities remain lower than those observed in the Japan case study, thereby exemplifying the impact stemming from alterations in the sensor configuration.

Since each rear-facing pyranometer is exposed to what appears to be uniform environmental conditions, the discrepancies described do not appear to be attributed to obvious environmental factors such as proximity to an internal road which would cause station one to have higher measurements or differing soil conditions which may result in different localised albedo values. It is thus challenging to identify a specific source of variation or driving factors that impact the measurements and should be avoided on site.

4.3 Case Study 3: Mexico PV Plant

The PV plant in Mexico is equipped with 15 different meteorological stations, each with two rear-side facing pyranometers installed at the top and bottom edges of the tracking table (one towards east and another one towards west).

The primary focus was dedicated to evaluating and observing differences concerning:

- Spatial variances across the 15 locations of meteorological stations.
- Orientation differences given that each location presented one rear-facing pyranometer at the east edge and another one at the west edge of the table tracker.
- Temporal variances spanning the nine months of data collection.

To characterize each location with a singular rear-facing irradiance value across the nine months, the mean rear-side PoA irradiance was calculated from the east and west irradiance measurements, depicted in Figure 21. This approach enhances the clarity and quantification of spatial differences, facilitating the identification of any anomalies for a specific location, if any.

Table IV: Statistical summary of rear-facing pyranometers at the different sensor locations for Case Study 3.

Met Station	East-West mean [W/m²]	Deviation from avg [W/m²]	Deviation from avg [%]
PS02	97.49	0.36	-0.37
PS07	100.64	3.51	3.61
PS09	92.11	-5.02	-5.17
PS12	92.41	-4.73	-4.87
PS15	102.88	5.75	5.92
PS18	101.90	4.77	4.91
PS23	89.65	-7.48	-7.70
PS25	99.07	1.93	1.99
PS27	90.61	-6.52	-6.72
PS31	101.77	4.63	4.77
PS34	97.61	0.48	0.49
PS36	102.81	5.68	5.84
PS40	94.80	-2.33	-2.40
PS43	95.89	1.24	-1.28
PS47	97.35	0.21	0.22
Average	**97.13**		

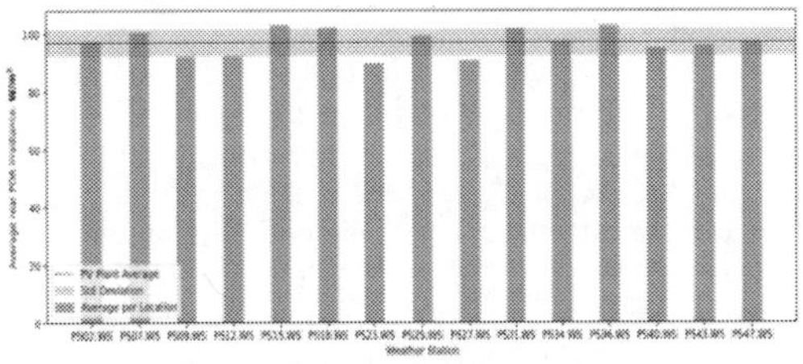

Figure 21: Mean rear-side irradiance measurements per location for the complete time frame.

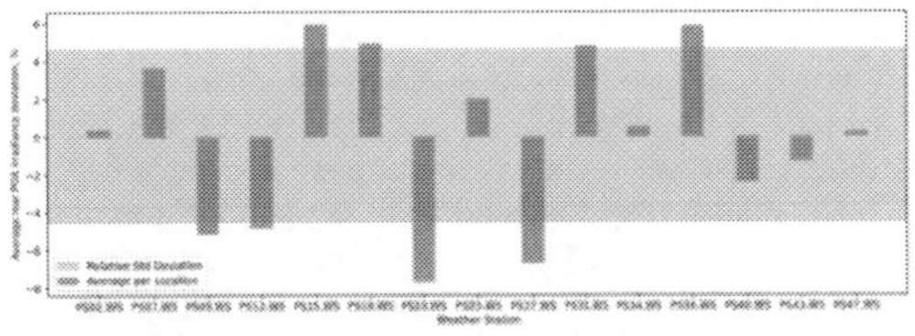

Figure 22: East-West mean rear-side irradiance measurements relative difference from the average for the complete time frame.

Over the nine-month dataset, the average rear-side PoA irradiance value for the entire PV plant is estimated at 97.13W/m², while the standard deviation across the 15 locations is calculated at 4.46W/m². Among the 15 locations, 7 meteorological stations were positioned within one standard deviation. Four stations overestimated rear-side PoA irradiance, while the remaining four stations underestimated it. This underlines the spatial variability evident in this case study in a distributed manner, albeit reduced due to uniform environmental conditions prevalent across the PV plant area.

Figure 23 depicts the mean rear-side PoA irradiance for each meteorological station, categorized by east and west sensors, over the nine-month duration, while Figure 24 highlights the relative difference of each location edge against the mean value calculated per edge. Meteorological station PS43 exhibited the smallest difference for the east-placed rear-side pyranometer (0.17% vs the *east* mean), while station PS34 displayed the lowest difference for the west-placed rear-side

10.4229/EUPVSEC2023/4CV.1.52
020408-006

pyranometer, measuring 0.26% from the *west* mean. The largest differences per edge were seen at PS23 (-9.38%) and PS36 (8.58%), for the east and west edges, respectively. These differences remain relatively minor, given that the mean difference among all meteorological stations stood at 3.33% for the east edge and 4.62% for the west edge. Figure 24 also shows that 12 out of the 15 locations present similar patterns for both locations, that is both rear-facing pyranometer presented readings either above or below the edge mean, showing the consistency per location across the PV plant.

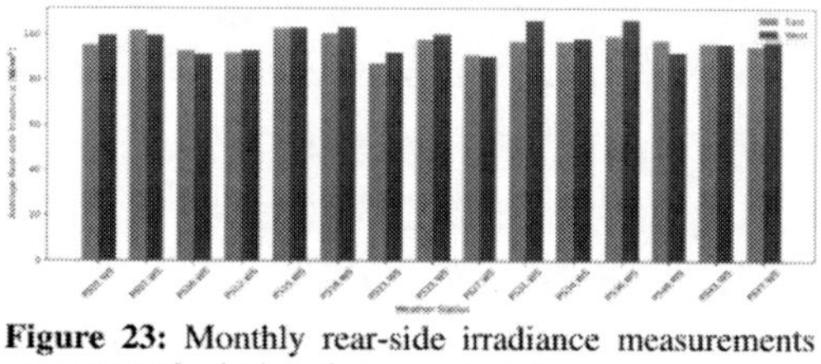

Figure 23: Monthly rear-side irradiance measurements per meteorological station per orientation.

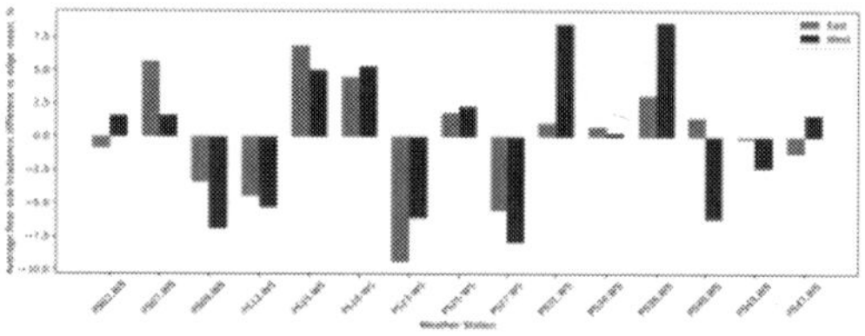

Figure 24: Monthly relative difference in rear-side irradiance measurements per meteorological station per edge.

Evaluating the east west orientation sensors on a monthly basis, a seasonal trend is observed in Figure 25, with rear-side irradiance increasing during the initial five months of the year before taking a descending trajectory. Except for September, when east west sensors differences peaked at 23.69W/m² between the sensors, the absolute differences over the year remain steady. The most substantial difference, other than September, was seen in June 2022, at 9.87W/m². Despite the anomaly in September 2022, differences remain relatively stable, with no distinct trend for either east or west pyranometers across the study period.

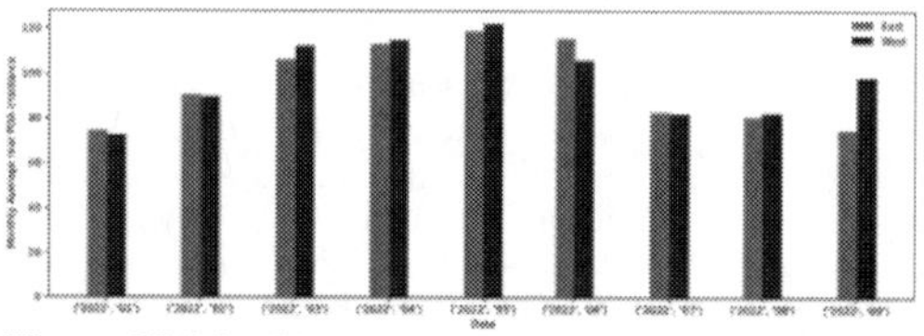

Figure 25: Monthly rear-side irradiance measurements per edge.

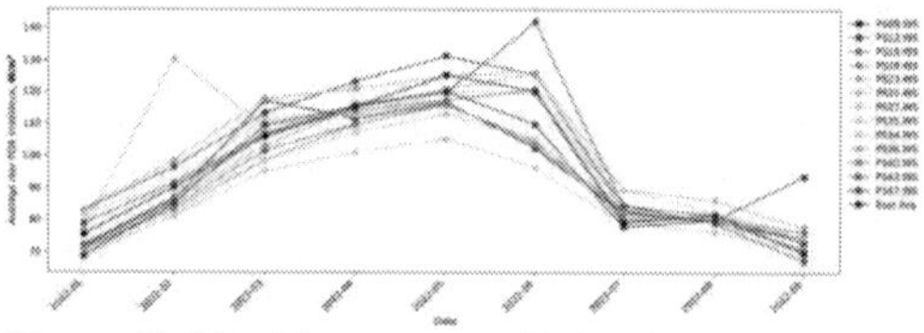

Figure 26: Monthly mean rear-side irradiance measurements at East edge.

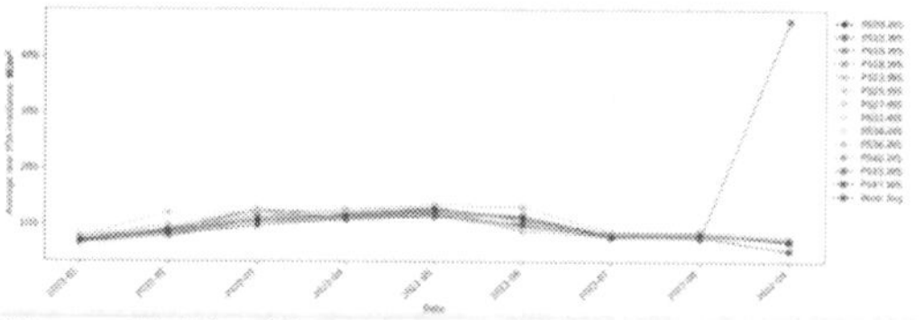

Figure 27: Monthly mean rear-side irradiance measurements at West edge.

Figure 26 represents the monthly average rear PoA irradiance per location for all east-placed rear-facing pyranometer, while Figure 27 similarly does the same for the west-placed sensors. The monthly behaviour seen for each sensor at each edge in the above Figures can be classified as consistent with no significant fluctuations among the sensors' readings and trends that are regular across the different months. Some outlying readings were collected at the east edge for sensors PS25 in February 2022, PS43 in June 2022 and PS46 in September 2022, whereas at the west edge the clearest anomaly was found for PS43 in September. Despite these variations, sensors readings' exhibit constant patterns over the studied period.

In relative terms, the standard deviation of the east-west average values for each location stands at 4.60%, which is above the associated uncertainty for a Class A pyranometer (2% for daily sums and 3% for hourly sums [3]).

Based on Figure 23 and Figure 25, it becomes apparent that neither a specific location nor an east-west preference can be identified.

All the meteorological stations are located at the end of the tracker rows and subject to irradiation reflectance from the internal roads. There are no meteorological stations within the arrays thus it cannot be assessed if the measurements captured are higher than what would actually be representative for this plant.

5 CONCLUSIONS

Analysis of the rear-facing plane of array irradiation was carried out for three case studies. One fixed tilt PV system and two single axis tracking PV systems. Each case study exhibited short comings in respect to either number of rear-facing pyranometers per meteorological station or the number of meteorological stations present on site.

Case study 1, located in Japan and fixed tilt plant configuration, with three rear-side pyranometers in one location illustrates the importance of having three rear-side pyranometers to capture the variation of irradiance falling on the rear-side of the modules. Although the centrally located pyranometer trends to the mean value of all three pyranometers, it cannot be concluded that this is the most representative location as there is no second meteorological station installed for comparison. Similarly, there are no pyranometers installed part way between the middle and upper/lower pyranometer which may provide further insights into the understanding if a single location can capture representative irradiance on the rear side of the modules.

Case study 2, located in Namibia, single axis tracking configuration, has three meteorological stations with one rear-facing pyranometer at each station. For this case, no convergence of rear-side measurements was observed, nor could any specific environmental factors be attributed to the deviations observed. Nonetheless analysed data

10.4229/EUPVSEC2023/4CV.1.52
020408-007

showed that the relative deviation (%) between rear side irradiance measurements is lower during summer months due to higher accumulated rear side PoA irradiance.

Case study 3, located in Mexico, single axis tracking configuration has 15 meteorological stations with two rear-facing pyranometer at each station.

Overall deviations from the average across each of the 15 meteorological stations ranges from -7.7% to 5.9%, with 11 meteorological stations with deviations lower than 5% from the mean, which is considered reasonable dispersion as per WMO Guide to Meteorological Instruments and Methods of Observation [4].

Although no conclusion can be made on what would constitutes a representative rear-facing pyranometer position, the study findings highlight the importance of installing multiple rear side irradiance sensors, at least the minimum number as specified in the IEC 61724-1:2021 standard, in order to properly evaluate the Performance Ratio of a bifacial PV system.

All three case studies show that the dispersion of the rear-side irradiation measurements increase during periods of low irradiation and periods of high irradiance fluctuation.

The analysis highlights that dispersion in rear-side measurements is to be anticipated and further detailed research needs to be carried out in controlled conditions to better understand local environmental influences on the measurements.

6 REFERENCES

[1] International Electrotechnical Commission, IEC 61724-1:2021 Photovoltaic system performance - Part 1: Monitoring, International Electrotechnical Commission, 2021.

[2] D. M. Fischer, de *ITRPV CellTech*, Berlin, 2023.

[3] International Organization for Standardization (ISO), ISO ISO 9060:2018. Solar energy — Specification and classification of instruments for measuring hemispherical solar and direct solar radiation, 2018.

[4] World Meteorologcial Organization. WMO Guide to Meteorological Instruments and Methods of Observation, 2021 edition. Available: https://library.wmo.int/idurl/4/41650

10.4229/EUPVSEC2023/4CV.1.52
020408-008

IMPROVING IV CURVE CLASSIFICATION BY MACHINE LEARNING METHODS USING DEEP AUTOENCODERS

Maximilian Schönau[1,2], Darwin Daume[2], Bernd Hüttl[2], Dieter Landes[2]
[1]Smartblue AG, Kistlerhofstraße 75, Munich, Germany
[2]Coburg University of Applied Sciences, Dept. of Electrical Engineering and Computer Sciences, Friedrich-Streib-Straße 2, Coburg, Germany
Maximilian.Schoenau@smartblue.de, Darwin.Daume@hs-coburg.de

ABSTRACT: On-site current-voltage (IV) measurements will play an essential role in the online monitoring of PV systems. However, challenging measurement conditions like inconsistent irradiance levels on PV arrays (e.g., due to local shading) can distort IV curves, leading to inaccurate characterizations. By accurately detecting deformed IV curves, the reliability of outdoor IV measurements is significantly enhanced. For this purpose, several classifiers were evaluated using 4104 manually labeled IV measurements on a mc-Si-PV array. Machine learning techniques perform much better than a traditional rule-based filter, with accuracy above 99 %. A deep Autoencoder was employed to reduce IV measurements into a set of 7 features, which encoded the shape of the curves into a low dimensionality. The IV-Autoencoder improved the classification of IV curves, yielding better results than a feature reduction with Principal Component Analysis. The proposed classifier is able to sort out on-site IV measurements under unsatisfactory environmental conditions, benefiting the online monitoring of PV systems. It may also be used as an indicator for faulty PV strings.
Keywords: IV Curves, Machine Learning, Classification, Autoencoder

1 INTRODUCTION

Current-voltage (IV) measurements are state-of-the-art for evaluating photovoltaic systems [1]. Typically, these measurements are done at standard test conditions (STC) [2]. They may also be used on-site as a diagnostic tool for operating PV arrays. With inverter manufacturers increasingly launching smart inverters that can remotely measure IV curves of PV arrays [3]–[5], on-site IV measurements will become an essential online monitoring method for PV systems, promising cheap and accurate diagnostic data for the asset manager.

However, on-site IV measurements are inadvertently performed under unpredictable test conditions, such as varying and inhomogeneous irradiances. These lead to deformed IV curves. Figure 1 displays examples of measurements at inadequate test conditions.

An adequate IV curve interpretation requires test conditions that should be as homogeneous as possible. In particular, irradiance variations within the spatially extended PV array and during the measurement should be avoided or as small as possible. This can be enforced to a certain degree by using spatially distributed pyranometers to detect and measure at adequate conditions.

However, typical PV site sensor configurations use only a few selected sensors for many PV arrays, which cannot detect inhomogeneities such as partial shading or temperature differences within the PV string. Many PV sites do not have well-functioning sensors at all, forcing the asset manager to rely on unreliable weather predictions.

Consequently, it is desirable to filter IV curves into adequate and deformed measurements, employing only the data of the smart inverter as a criterion. If an online monitoring system can reliably and automatically filter out deformed IV measurements of problematic test conditions, multiple measurements can be performed such that only adequate IV curves are used for diagnostic purposes without depending on additional sensors.

When measuring IV remotely, deformed IV curves are mainly a result of inhomogeneous irradiances. However, such measurements may also result from different types of degradation. IV curve classifiers must, therefore, distinguish permanently deformed curves due to degradation from temporarily deformed IV curves due to inhomogeneous irradiation [6]. By employing this distinction, separating deformed from adequate measurements not only enables the identification of suitable IV curves for characterizing PV systems but also serves as a valuable diagnostic tool.

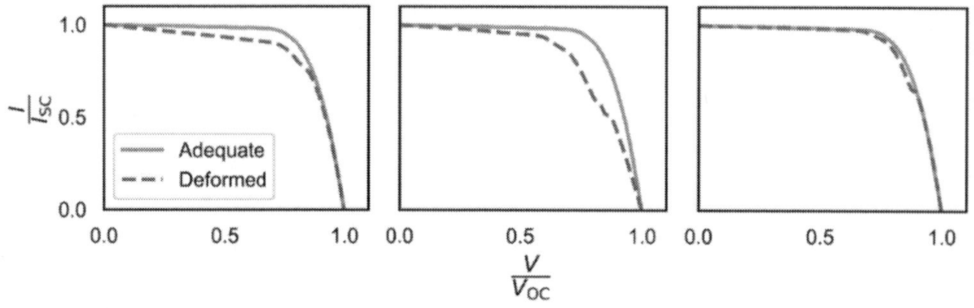

Figure 1: Examples of IV curve deformations resulting from inhomogeneous irradiances.

2 METHODOLOGY

This work implements different classification methods for determining deformed IV curves using various binary classifiers on multiple representations of the IV measurements.

2.1 Data

4104 IV measurements were employed for classification on a string of 12 mono-crystalline south-facing silicon modules with 20° elevation at various operating conditions. These were gathered at multiple outdoor measurement campaigns conducted over two years. The measurements were manually labeled into deformed and adequate curves.

80 % of the IV curves are deformed because of inadequate outdoor conditions as well as artificially induced fault scenarios. 20 % of the measurements are adequate IV measurements at homogeneous conditions at various irradiation levels. All measurements were performed at short circuit currents above 100 mA. To enable high applicability in the field, we intentionally did not employ additional sensor data to correct or classify the IV curves.

2.2 Preprocessing

Classification techniques involving the entire IV curve have high computational costs due to high input dimensionality. This dimensionality increases the intricacy of classification algorithms since they must account for the interdependencies among IV data points. This problem can be addressed, among others, by Convolutional Neural Networks (CNNs), which are suited to capture spatial patterns within the IV curves [7].

The classification can be simplified by extracting features of IV curves. We used the short-circuit current (I_{SC}), open-circuit voltage (V_{OC}), voltage and current at MPP (V_{MPP} and I_{MPP}) and fill factor (FF) for our classification of IV curves as a benchmark. These parameters can reduce the dimensionality of an IV curve and are relatively easy to interpret for humans. However, only using these parameters loses information about measurement, such as relevant slopes and knees within the IV curve. Thus, when considered in isolation, the electrical parameters have been inadequate for a reliable rule-based classification of IV curves [6]. Similarly, as pointed out by Hopwood et al. [7], when using an IV curve for diagnostic purposes, utilizing only a small set of parameters of an IV measurement could negatively influence the performance of the employed methodologies [8].

Principal Component Analysis (PCA) can drastically reduce the size of IV curve data, which improves the classification of IV measurements [9]. However, deep Autoencoders were demonstrated to be superior to PCA in reducing data dimensionality [10]. Autoencoders use neural networks to learn a low-dimensional representation of high-dimensional input vectors by identifying patterns in the training data. With IV curves being very similar and even faulty curves often having comparable deformations, Autoencoders are well-suited for the creation of dense representations of IV measurements.

Using an Autoencoder, we were able to compress our IV curves. These were normalized to short-circuit current and open-circuit voltage to specialize the IV-Autoencoder into encoding the shape of the curves. Our IV measurements were reduced into 7 features using a deep Autoencoder with 2021 neurons.

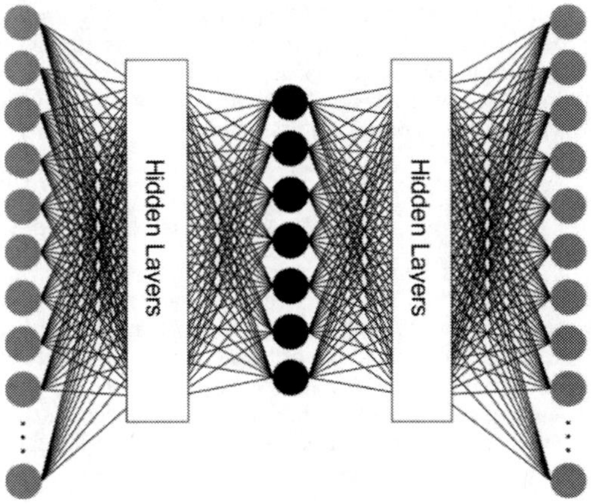

Figure 3: Schematic of the IV-Autoencoder

Figure 3 schemes the architecture of the IV-Autoencoder: Blue nodes represent the input layer, capturing the IV curve data. The hidden dense layers transform the input, resulting in a dense representation of the IV curve marked by black neurons. The right-hand side of the displayed IV-Autoencoder is the decoder, which is able to reconstruct the seven features back into an IV curve. The IV-Autoencoder did not utilize the adequate / deformed label. Only the IV measurements were used for training. The resulting 7-dimensional, non-humanly readable, but very dense encoding improved the classifiers' performance.

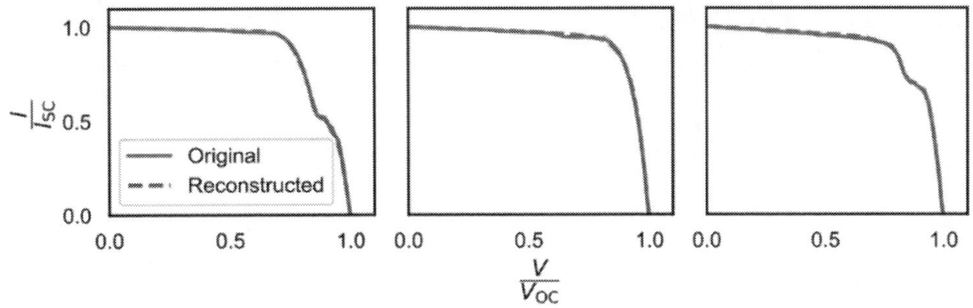

Figure 2: Examples of IV curves and their representation of the Autoencoder

10.4229/EUPVSEC2023/4CV.1.53

2.3 Classifiers

The IV curve classifications were performed with three feature sets (see Table 1) trimmed using univariate feature selection.

Table 1, Feature sets for the binary classifiers

Electrical Parameters	I_{SC}, V_{OC}, V_{MPP}, I_{MPP}, and FF
PCA	Electrical parameters and 3 Principal Components of the IV curves
Autoencoder	Electrical parameters and 7-dimensional IV encoding

Out of various binary classifiers, the Random Forest [11], Extra Trees [12], and Boosting [13] performed the best and were finetuned using grid search for optimal results for each feature set. Using the raw IV curve (in addition to the electrical parameters) as a feature set had no positive impact on the accuracy of the employed classifiers, which were too simplistic to identify interrelationships within the data points of the IV curves.

3 RESULTS AND DISCUSSION

Figure 2 shows examples of IV measurements and their reconstruction, produced using the 7 features incorporated in our IV-Autoencoder model. The encoding represents the general shape of the IV curves very well, ensuring a high level of accuracy. There are only slight differences between the original data and the reconstruction, with the reconstruction of the 20 % test split having an RMSE of 0.83 % compared to the original curves.

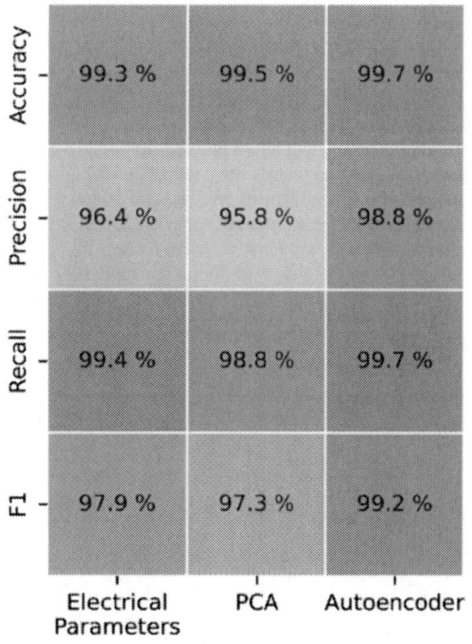

Figure 4: Performance comparison of the three feature sets (see Table 1)

Figure 4 shows the resulting performance of the Extra Trees Classifier for the different feature sets. The parameters were calculated as the mean value of classifying the 20 % test split out of 20 runs. The electrical parameters as a feature set resulted in an astonishing accuracy of 99.3 %, outperforming the previously published rule-based method with an accuracy of 96.4 % [6]. Using Principal Components to compress the IV curves yielded a slight increase in accuracy compared to the electrical parameters, with declines in precision and recall.

The best overall performance was achieved when using the 7 features of the IV-Autoencoder in combination with the electrical parameters. This improved the accuracy to 99.7 %, in addition to a substantial increase in precision and recall.

4 CONCLUSION

The classification of IV curves into deformed and adequate curves can be done very reliably by training binary classifiers instead of implementing a rule-based approach. The classification was improved by compressing the measurements into a small parameter space using an Autoencoder. Compared to PCA, the IV-Autoencoder showed superior capability for the feature extraction of IV curves.

Our proposed methodology enhances the utility of onsite IV measurements and holds promise for classifying various other error types.

ACKNOWLEDGEMENTS

The authors gratitude the Bavarian Research Foundation for financial support of the project Kick-PV: "AI-Based Characterization and Classification of PV-Plants for Predictive Maintenance."

LITERATURE

[1] *IEC 61829:2015, Photovoltaic (PV) array: on-site measurement of current-voltage characteristics*, Edition 2.0. Geneva, Switzerland: IEC, 2015.

[2] *IEC 60904-1:2020, Photovoltaic devices. Part 1, Measurement of photovoltaic current-voltage characteristics*, Edition 3.0. Geneva, Switzerland: International Electrotechnical Commission, 2020.

[3] Huawei, "Smart I-V Curve Diagnosis." Jun. 28, 2023. Accessed: Jun. 28, 2023. [Online]. Available: https://solar.huawei.com/~/media/Solar/attachment/pdf/eu/datasheet/IV-Curve.pdf

[4] Growatt, "Growatt I/V Curve Diagnosis." Jun. 28, 2023. Accessed: Jun. 28, 2023. [Online]. Available: http://www.growatt.pe/upload/file/contents/2019/06/5cf8adfa1d465.pdf

[5] SMA, "SUNNY BOY 3.0-6.0 / SUNNY TRIPOWER CORE1 / SUNNY TRIPOWER 3.0-6.0 / SUNNY TRIPOWER 8.0-10.0, I-V and P-V diagnostics function: Using the inverter to determine the generator's characteristic curve to detect faults in PV modules." Jul. 05, 2023. Accessed: Jul. 05, 2023. [Online]. Available: https://files.sma.de/downloads/SB_STP-Diagnosefunktion-TI-en-12.pdf

[6] M. Scheler *et al.*, "Precise On-Site Power Analysis of Photovoltaic Arrays by Self-Reference Algorithm," *8th World Conference on Photovoltaic Energy Conversion; 1070-1073*, p. 4 pages, 28190 kb, 2022, doi: 10.4229/WCPEC-82022-4DO.1.4.

[7] M. W. Hopwood, T. Gunda, H. Seigneur, and J. Walters, "Neural Network-Based Classification of String-Level IV Curves From Physically-Induced Failures of Photovoltaic Modules," *IEEE Access*, vol. 8, pp. 161480–161487, 2020, doi: 10.1109/ACCESS.2020.3021577.

[8] G. A. Susto, A. Cenedese, and M. Terzi, "Time-Series Classification Methods: Review and Applications to Power Systems Data," in *Big Data Application in Power Systems*, Elsevier, 2018, pp. 179–220. doi: 10.1016/B978-0-12-811968-6.00009-7.

[9] M. Hopwood, T. Gunda, H. Seigneur, and J. Walters, "An assessment of the value of principal component analysis for photovoltaic IV trace classification of physically-induced failures," in *2020 47th IEEE Photovoltaic Specialists Conference (PVSC)*, Calgary, AB, Canada: IEEE, Jun. 2020, pp. 0798–0802. doi: 10.1109/PVSC45281.2020.9300601.

[10] G. E. Hinton and R. R. Salakhutdinov, "Reducing the Dimensionality of Data with Neural Networks," *Science*, vol. 313, no. 5786, Art. no. 5786, Jul. 2006, doi: 10.1126/science.1127647.

[11] L. Breiman, "Random Forests," *Machine Learning*, vol. 45, no. 1, Art. no. 1, Oct. 2001, doi: 10.1023/A:1010933404324.

[12] P. Geurts, D. Ernst, and L. Wehenkel, "Extremely randomized trees," *Machine Learning*, vol. 63, no. 1, Art. no. 1, Apr. 2006, doi: 10.1007/s10994-006-6226-1.

[13] Y. Freund and R. E. Schapire, "A Decision-Theoretic Generalization of On-Line Learning and an Application to Boosting," *Journal of Computer and System Sciences*, vol. 55, no. 1, Art. no. 1, Aug. 1997, doi: 10.1006/jcss.1997.1504.

LAB MEASUREMENTS OF POWER OPTIMIZER EFFICIENCY AND PERFORMANCE SIMULATION OF PARTIALLY OPTIMIZED SYSTEMS AFFECTED BY SHADING

Cyril Allenspach, Fabian Carigiet, Franz Baumgartner*

ZHAW, Zurich University of Applied Sciences, School of Engineering, IEFE
www.zhaw.ch/~bauf, Technikumstr. 9, CH-8401 Winterthur, Switzerland
*phone: +41 58 934 72 32; email: bauf@zhaw.ch

ABSTRACT: Photovoltaic (PV) systems encounter operational challenges when subjected to partial shading conditions, resulting in non-uniform conditions within the module array. Traditional PV systems employing string inverters experience power loss, as they lack the capability to individually monitor and optimize each photovoltaic module. Conversely, module-level power electronics address this issue but introduce increased overall system losses. In order to assess the efficiency of power optimizers and string inverters, comprehensive measurements were conducted in a controlled environment at the ZHAW indoor laboratory. The findings revealed that the Huawei Optimizer achieved an efficiency of 99.5% ± 0.4% under specific operating conditions (kI = 1.00), while the Tigo product reached 99.8% ± 0.3% (k = 1). Importantly, these efficiencies are only realized when the module-level power electronics operate without switching. In scenarios involving switching operations, the efficiencies exhibited a notable reduction of more than 1%. These empirical results were then integrated into the ZHAW PV simulation tool, enabling the tool to simulate PV systems equipped with independent power optimizers (indMLPE). This research demonstrates that a PV system with partial optimizer deployment can exhibit superior annual performance, particularly in medium shading scenarios such as those encountered with dormer window shading.

Keywords: Module-Level Power Electronics, MLPE, Power Optimizer, Efficiency, Performance, Testing, Shading

1 INTRODUCTION

To tackle the challenges of global warming, the installation of photovoltaic (PV) generators is the most viable solution, due to the reason that its flexibility for installation is unparalleled. However, residential buildings come in nearly innumerable varieties and in many cases, the PV systems will be encountering shading. which can lead to inhomogeneous operating conditions within the PV array. Traditionally, a centralised control scheme is realised by the use of conventional string inverters (SINV), which includes a single maximum power point tracking (MPPT) for the entirety of the connected array. Consequently, in situations with partial shading, it is a frequent occurrence that not every module is operated in its individual maximum power point (MPP). Accordingly, in these cases an MPPT at the module-level would increase the power of the partially shaded modules and thus, improve the overall PV system performance, which is where module-level power electronics (MLPE) come into play.

In this paper, three different types of PV systems are analysed and compared. The baseline is set by the conventional string inverter system (SINV), whereas in addition two types of PV systems with power optimizers are examined. First, a system where every PV module is equipped by a buck/boost-type power optimizer (allMLPE), and secondly, a system that is controlled by a string inverter, while a few PV modules that are affected by shading are equipped with independent power optimizers of the buck-converter type. The respective configurations are visualised in Figure 1, 2 and 3.

To accurately calculate the differences, power optimizer and string inverter efficiency measurements were conducted in the ZHAW indoor laboratory [1]. With these results, simulations are performed by the use of the ZHAW MLPEShade tool for a Module-Level Power Electronics (MLPE) system with power optimizers at every module. Additionally for the first time, MLPE systems with independent optimizers at a few affected modules (indMLPE) were simulated by the use of MLPEShade for two cases with weak and heavy shading and compared to standard string inverter rooftop photovoltaic system (SINV). For the analysis, a MLPE system with P370 power optimizer and a 3500H inverter was chosen and compared to a SINV system with a SUN2000-3.68KTL-L1 string inverter, without the use of module-level power electronics. The results of the annual simulations for the weak shading case showed a gain of the module-level power electronics system between - 0.9% and 1.4% (14 modules) or -0.2% and 0.8% (13 modules), depending on the position of the chimney. In other works of the ZHAW PV research group, the shading adaption efficiency, a method of quantifying the annual performance for shading situations by applying weightings to a few indoor measured performance values of power electronic components, was developed, and published [2].

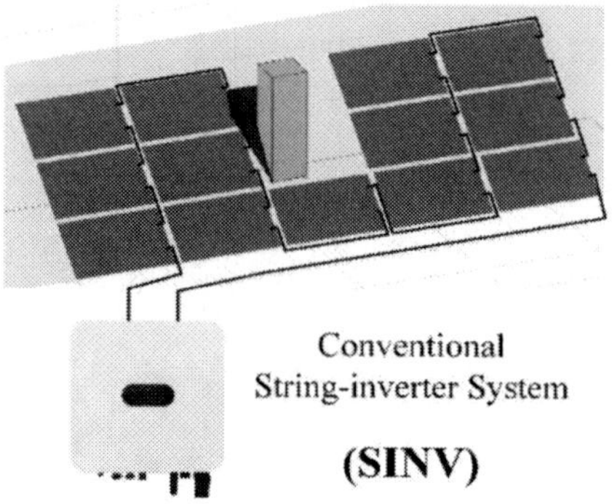

Figure 1: Electrical setup of the conventional string inverter system without any power optimizers.

10.4229/EUPVSEC2023/4DO.1.1
020412-001

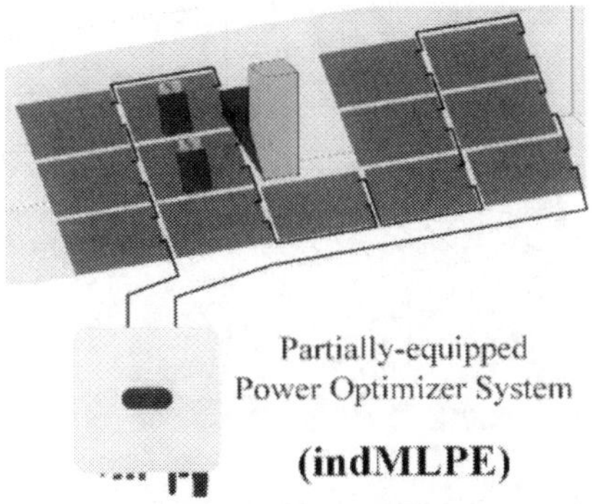

Figure 2: Electrical setup of the MLPE system, which is partially equipped with two power optimizers at the PV modules most affected by shading.

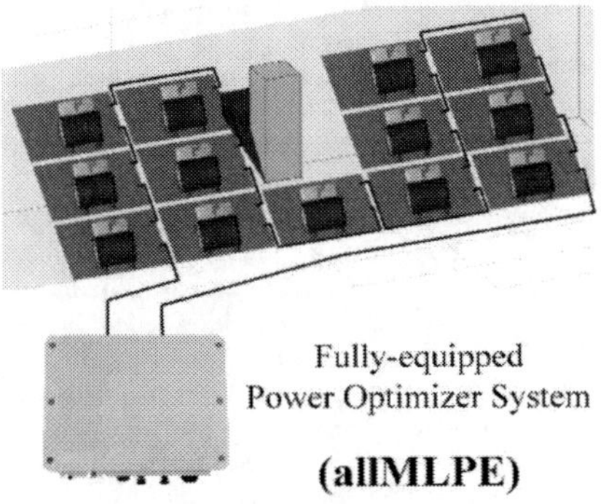

Figure 3: Electrical setup of the fully equipped MLPE system with 10 power optimizers.

2 INDOOR LABORATORY TESTING

In the IEFE laboratory at the ZHAW, a measurement and testing setup for MLPE was developed, which can be operated as a state-of-the-art residential photovoltaic plant consisting of 10 PV-Modules (shown in Figure 4).

The measurement system can mimic unshaded and shaded PV modules with the use of 10 solar array simulators (SAS) by providing different current-voltage (IV) curve settings, according to the irradiance and temperature conditions on the PV module level. Every workstation has several types of MLPE by different manufacturers installed, which are powered by a Keysight E4360A SAS and measured with a Power Analyzer PPA1530 at the electrical input and output terminals. During testing, the desired operating points of the PV modules are entered in the SAS devices, which calculate respective IV-curves for the given inputs to each MLPE. Accordingly, the SAS generate DC-currents and voltages for the device-under-test based on I-V curves of PV-modules. Further descriptions of the measurement setup are given in "Performance analysis of shaded PV module power electronic systems" and "Performance of New Photovoltaic System Designs" [3, 4].

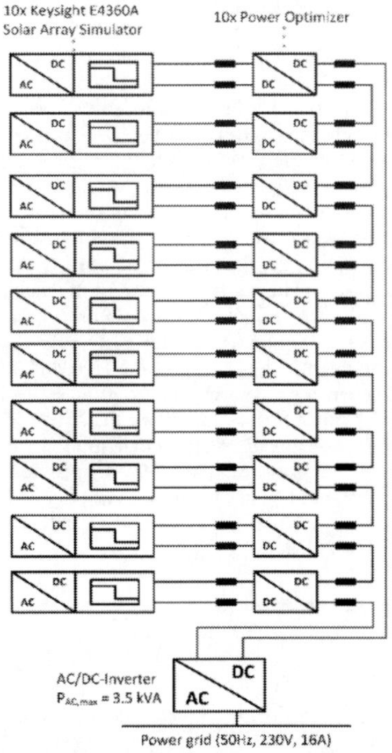

Figure 4: Electrical setup of the analysed MLPE system with 10 power optimizers.

The losses in the power electronic converters occur in the individual components such as ohmic losses in the inductors and capacitors, ohmic and switching losses in the semiconductor switches and the socket losses due to the power supplies of all active components, e.g., drives, control and communication circuits. Equation (1), (2), (3) and (4) describe the power loss of the buck-boost type power optimizer, which is used in the simulation as underlying model.

$$D_{cycle} = \frac{I_{IN}}{I_{OUT}} \quad (1)$$

$$P_{loss,buck} = a + b \cdot U_{IN} \cdot I_{IN} \cdot D_{cycle} + c \cdot I_{OUT}^2 + d \cdot U_{IN}^2 \quad (2)$$

$$P_{loss,boost} = a + b \cdot U_{IN} \cdot I_{OUT} \cdot |1 - D_{cycle}| + c \cdot I_{IN}^2 + d \cdot U_{IN}^2 \quad (3)$$

$$P_{loss,passthrough} = a + c \cdot I_{IN}^2 \quad (4)$$

With the use of high-precision power analyzers, the DC/DC-efficiency of the MLPE are determined with an accuracy in the range of 0.2 - 0.5 % (k=1) in the ZHAW IEFE indoor laboratory. Accordingly, the whole operating range of buck-boost converter-type Power Optimizers were measured. In Figure 5, the DC/DC-efficiencies of the type S500B for current ratios: $k_I = I_{OUT} = I_{IN}$ ranging from 0.5 to 2, and input power: P_{IN} ranging from 0 W to 500 W, are visualized. In detail, the resulting efficiencies $\eta_{DC/DC}$ are between 95 %, at low input power $P_{IN} \approx 18$ W, and 99.6 %, at $P_{IN} = 200$ W and $k_I = 1$. However, distinct differences in efficiency can be identified between, the region where $k_I = 1$, and all other operating points. In detail, in the narrow range around $k_I = 1$ (see red dots in Figure 5), the efficiencies are more than 0.8 % higher than everywhere else, due to the reason that the converter is not operating either in buck or boost mode, in the very few operation points. Therefore, the MOSFETs are inactive, and no switching losses occur. In other words, the input is directly connected to the output and only ohmic losses of the lines, inductor and the MOSFET $R_{DS(on)}$ will occur.

The efficiency mapping in Figure 5 shows that values above 98 % only occur in a limited field of the operating range, while the values mentioned by the manufacturers are either, the maximum efficiency with 99.5 % or the weighted efficiency of 98.9 % [5]. Furthermore, the latter mentioned value could not be reached in the buck- or boost-mode, for which the component was designed.

The measurement results of the 450W-P2 buck-type power optimizer, illustrated in Figure 6 show a similar picture at $k_I = 1$ compared to the S500B results in Figure 5, especially with efficiency values below 97 % at about one tenth of full load. A similar picture is shown by the measurement results $\eta_{DC/DC}$ of the Tigo TS4-R-O in Figure 7, whereas a greater variation in efficiencies is identified in buck-mode than is shown for the 450W-P2. Finally, it is important to realize that for determining the total MLPE system performance, the DC/AC efficiencies of the inverter would be needed to be multiplied with the values presented in this section, which would result in lower overall efficiencies.

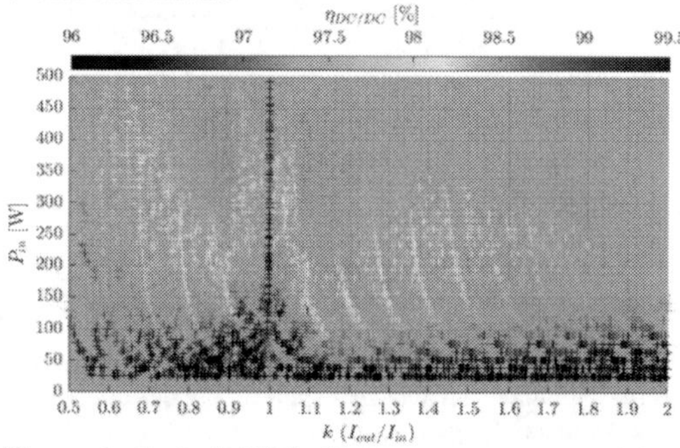

Figure 5: Static DC/DC-efficiency measurement of the buck/boost-type power optimizer S500B at static input voltage U_{IN} = 35 V relative to input/output current ratio.

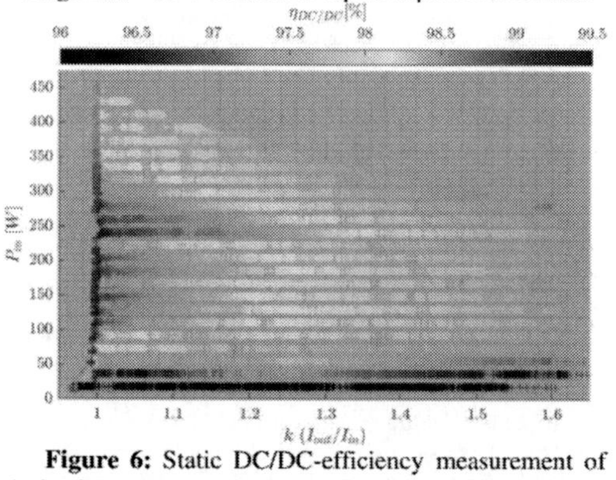

Figure 6: Static DC/DC-efficiency measurement of the buck-type power optimizer of model 450W-P2 at static input voltage U_{IN} = 35 V relative to the input/output current ratio.

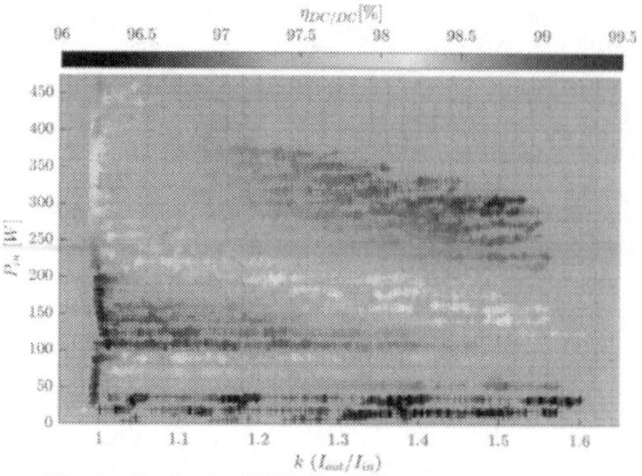

Figure 7: Static DC/DC-efficiency measurement of the buck-type power optimizer of model TS4-R-O at static input voltage U_{IN} = 35 V relative to the input/output current ratio.

3 ANNUAL PV SHADING SIMULATION

3.1 Power Loss Model and Shading Adaption Efficiency

The results were obtained by a combination of laboratory testing and measurement, as well as using an in-house simulation tool called "MLPEShade" at the ZHAW School of Engineering [1]. The tool uses real meteorological data and calculates the shading of a PV plant based on time, date, longitude, latitude, and irradiance models. The performance of the PV modules is determined by physical models that consider changing weather and shading conditions. The resulting quantities are congregated by three different methods to show the power and yield of a photovoltaic plant, firstly for a SINV system, secondly for a fully equipped MLPE system (allMLPE) and finally for a MLPE system with independent optimizers (indMLPE). Moreover, the PV shading simulations are based on laboratory measurements and provide an accurate representation of the performance of power electronic components.

The results of the MLPEShade simulation tool are presented as a MLPE gain, which represents the superior performance of the optimizer system compared to the SINV for a given year. In addition, the shading adaptation efficiency (SAE) is used (as described in Equation 5) to compare the performance of SINV and MLPE systems based on the same reference, which is the theoretical performance of a lossless system with module-level tracking for a specific shading case.

$$\eta_{shad,a}(SAE) = \sum_{t=0}^{T=1yr.} \frac{P_{AC}}{\sum_{i=1}^{k} P_{mod,i}} \tag{5}$$

$\eta_{shad,a}$ = Shading Adaption Efficiency (SAE) for a shading case "a"

P_{AC} = Effective AC-Power (after losses) of a PV system with shading

$P_{mdo,i}$ = Max. Power of module "i" (without losses / with shading)

In addition to the SAE, a measure for the degree of shading of a PV plant is defined by shading index $SI_{DC,Max}$, described in Equation 6. In detail, the equation is based on the shading index, SI, developed by Hanson et al. [6]. In contrast to the original formulation, the $SI_{DC,Max}$ is calculated by the sum of the maximum power at each PV module for every timestep of a year under the influence of shading, which is divided by the sum of the DC power and

10.4229/EUPVSEC2023/4DO.1.1
020412-003

duration, for the same PV system but without shading. Finally, the quotient is subtracted from 1 and shown in percentages.

$$SI_{DC,Max}[\%] = \left(1 - \frac{\sum_{t=0}^{T=1yr.} \sum_{i=1}^{nMod} P_{Mod,i,shaded,t}}{\sum_{t=0}^{T=1yr.} \sum_{i=1}^{nMod} P_{Mod,i,unshaded,t}}\right) \cdot 100 \qquad (6)$$

$P_{Mod,i,shaded,t}$= Power at time "t" of module "i" at MPP with shading

$P_{Mod,i,unshaded,t}$ = Power at time "t" of module "i" at MPP without shading

Notably, the PV system shown in Figure 8 without shading object was modelled in MLPEShade and PVSyst, to set a baseline of how similar the results of the tools are in general. In detail, a conventional string inverter system with a Huawei SUN2000-3.68KTLL1 and thirteen Trina Solar Honey M Plus TSM-DD5A.08 (II) 300Wp modules was simulated with the same meteorological inputs (with an annual solar plane-of-array irradiation of 1486.5 kWh/m2) provided by MeteoSchweiz for Kloten, Switzerland. In PVSyst, the annual DC-yield of the 30 ° tilted south-facing PV plant (for semi-integration) resulted in 5,171 kWh with a Performance ratio (DC) of PR_{DC} = 89.2%, while in MLPEShade, 5,260 kWh with PR_{DC} = 91.1% (for open-rack) and 4,919 kWh with PR_{DC} = 85.2% (for close roofmount) was calculated. Meanwhile, the AC-values by PVSyst resulted in 5,028 kWh with PR_{AC} = 86.7%, whereas for MLPEShade, 5,142 kWh with PR_{AC} = 89.1% (open-rack) and 4,809 kWh and PR_{AC} = 83.3% (close roof-mount) was calculated. Consequently, the difference of the annual energy yield (DC) in percentages are in the range of +1.7 to -4.9 %, while the annual energy yield (AC) differed by +2.3 to -4.4% relative to the PVSyst results.

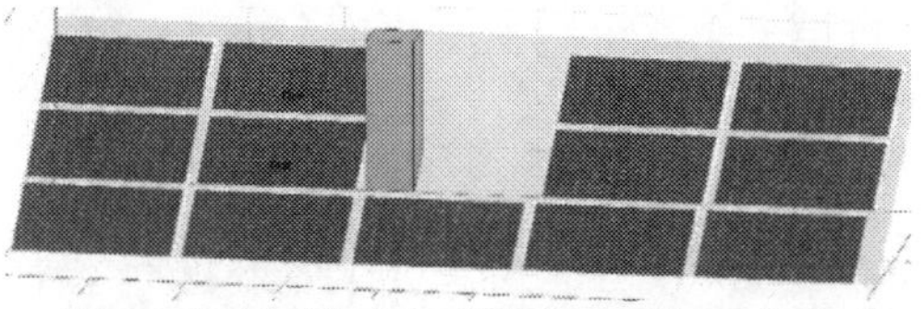

Figure 8: Model of the PV plant with weak shading by a chimney. The two red boxes and the annotation: "Opt", indicate the PV modules which are equipped with power optimizers for the indMLPE system.

With a focus on the annual simulation of the case with weak shading by a chimney (as visualised in Figure 8), the results show a gain of +0.6% for the allMLPE system, and 1% for the MLPE system with independent optimizers (indMLPE), when compared to the string inverter system. Although the total DC/DC losses of the system with optimizers at each PV module are higher than the indMLPE system, the DC/AC losses are lower, which reduces the yield gain gap between the two. The reason for the lower inverter conversion losses is based on the fact, that the analysed SolarEdge SE3500H does not have a built-in boost converter, which results in fewer switching losses. In the comparison of the three software tools, the focus is on the annual AC yield differences by comparing the same PV system and shading situation for the different system configurations.

4 DISCUSSION

Indoor efficiency laboratory measurements of independent power optimizers, Huawei SUN2000-450W-P and Tigo TS4-R-O, for a wide range of different working points were conducted. The maximum measured efficiency at very limited operation points at a current ratio (Iout/Iin) of k_I = 1.00 of the MLPE optimizer Huawei resulted in 99.5% ± 0.4% (k = 1), whereas the Tigo product reached 99.8% ± 0.3% (k = 1). Thus, when the MLPE device has no switching operation and is only acting as a low ohmic resistance. Based on these activities, annual simulations for a weak shading case were performed. In detail, the yield variations of a PV system with shading by a chimney (weak shading) were analysed for three different PV system configurations: a Solar Edge P370 power optimizer system (allMLPE), a MLPE system with a few Huawei SUN2000-450W-P independent optimizer (indMLPE) and a string inverter system (SINV). Additionally, the simulated energy yields from MLPEShade are compared to the results of the commercial simulation software, PVsyst and PVSOL. Notably, these tools highly overestimate MLPE efficiency by using data sheet values which are only valid for a few operation points, which do not represent the performance in real operation conditions. In the case of weak shading, the allMLPE system with optimizers installed at every module, shows an increase of annual yield by 0.6% and a yield gain by the independent optimizer system by -1%, when compared to the conventional string inverter system in MLPEShade. In contrast, the PVsyst results show an MLPE yield gain of +3.3% (allMLPE) and +1.6% (indMLPE) respectively, while the PVSol simulation results in +4.3% (allMLPE) and +2.1% (indMLPE). Consequently, the overstimation in solar yield forecasts in a several percentage range of commercial tools should not be neglected to protect the clients from paying for underperforming PV systems.

ACKNOWLEDGEMENT

Within the IEC TC 82/WG 6, the shading adaption efficiency was presented [7]. To support the development of a future technical specification, the ZHAW is involved in the analysis of shading scenarios, to identify characteristic, benchmark shading situations, as a part of the collaboration by the International Energy Agency (IEA) PVPS Task 13 ST2.5 [8]. The MLPE research of the ZHAW is thereby funded by the Swiss Federal Office of Energy, with Project Number: SI/502247-01 [9].

REFERENCES

[1] C. Allenspach, «Module Level Power Electronics Dynamic and Static Performance in Partial Shaded Photovoltaic Systems» (Master Thesis), ZHAW School of Engineering, Winterthur, Jan. 2023.

[2] C. Allenspach, F. Carigiet, A. Bänziger, A. Schneider and F. Baumgartner, « Power Conditioner Efficiencies and Annual Performance Analyses with Partially Shaded Photovoltaic Generators Using Indoor Measurements and Shading Simulations», Wiley Solar RRL 2200596, [Online] DOI: doi.org/10.1002/solr.202200596 (2022).

[3] F. Baumgartner, R. Vogt, C. Allenspach, and F. Carigiet. Performance analysis of shaded pv module power electronic systems. In Proceedings of the 38th European Photovoltaic Solar Energy Conference and Exhibition (EUPVSEC), Online, pages 650 – 654, Munich, Germany, 2021. WIP Renewable Energies.

[4] M. Littwin, U. Jahn, F. Baumgartner, et al., Performance of New Photovoltaic System Designs. Technical Report Report IEA-PVPS T13-15:2021 (ISBN 978-3-907281-04-8), International Energy Agency (IEA) PVPS Task 13, 2021.

[5] SolarEdge Technologies Inc. P400-P505 Power Optimizer Module Add-On. https://knowledge-center.solaredge.com/sites/kc/files/se-P5-series-add-on-power-optimizer-datasheet-na.pdf. [Online: accessed 03.09.2023].

[6] J. Hanson, et al., «Partial-shading assessment of photovoltaic installations via module-level monitoring», IEEE Journal of Photovoltaics, 4(6):1618–1624, 2014.

[7] International Electrotechnical Commission (IEC), TC82: Solar photovoltaic energy systems - WG6: Balance-of-system components, https://www.iec.ch/dyn/www/f?p=103:14:605685340133361::::FSP_ORG_ID,FSP_LANG_ID:2749,25. [Online; accessed 08:09:2023].

[8] International Energy Agency (IEA) PVPS, Task 13: Performance, Operation and Reliability of Photovoltaic Systems. www.iea-pvps.org/research-tasks/performance-operation-and-reliability-of-photovoltaic-systems/. [Online; 08.09.2023].

[9] Swiss Federal Office of Energy (SFOE). EFFPVSHADE - Effizienzanalyse von dezentraler Photovoltaik Leistungselektronik bei Teilbeschattung [EFFPVSHADE - Analysis of efficiency of decentralized phototvoltaic power electronics under partial shading]. SI/502247-01, 2021-2023.

10.4229/EUPVSEC2023/4DO.1.1
020412-005

40th European Photovoltaic Solar Energy Conference and Exhibition

This presentation was selected by the Sc. Committee of the EU PVSEC 2023 for submission of a full paper to one of the EU PVSEC's collaborating peer-reviewed journals.

LIGHT AND SHADE – A HEMICUBE APPROACH FOR EFFICIENT SHADING CALCULATIONS IN UTILITY-SCALE PV PLANTS

Anja Neubert[1], Mike Hamer[2], Javier Lopez-Lorente[3]
[1]DNV, Emsstraße 18 c/o Glaswerk, 26135 Oldenburg, Germany
[2]DNV, Avon Street, Temple Quay One Linear Park, BS2 0PS Bristol, UK
[3]DNV, Utrechtseweg 310 B50, 6812 AR Arnhem, Netherlands
Corresponding author: anja.neubert@dnv.com

ABSTRACT: Accurate estimations of the energy production in the planning and design stage of photovoltaic (PV) plants require advanced models to simulate shading effects. This is particularly important for plants located in complex and irregular terrain, or plants that are affected by nearby shading obstacles. This paper presents a hemicube approach for calculating shading effects and assesses the model accuracy and performance for utility-scale PV plants. The validation of the model includes a sensitivity analysis of the model's computational efficiency and modelling accuracy for several fictitious power plants of diverse nominal capacities. The advanced shading model presented in this work can accurately estimate energy yield in PV plants in the pre-construction design and engineering stages to reduce financial risk for projects in locations with complex shading characteristics.
Keywords: modelling, shading, PV plant design

1 MOTIVATION

Utility-scale PV plants are being installed at locations where shading effects are becoming a significant driver for the layout design [1]. An accurate energy production assessment is utterly important to make financial decisions, where near shading effects considerations are made to evaluate the cost for civil terrain grading to reduce shading effects versus the energy production of the plant in more complex terrain without grading.

Simple models designed to estimate row-to-row shading in simple terrain are very fast, but not designed for modelling edge effects in PV plants and cannot consider complex shading effects for plants installed on complex terrain or shading from arbitrarily complex shading obstacles. In this work, we discuss the ability of a hemicube shading model to estimate shading effects and look at computational efficiency versus accuracy of a hemicube model when simulating utility-scale PV plants of diverse nominal capacities.

2 HEMICUBE MODEL

A hemicube can be used to represent a 360 ° view from a surface or point, here some specific target points on a PV module. For each target point we place an imaginary hemicube (half a cube) centred on the target point. For each face of the cube, we start with a white face and then render any object that can be broken down into triangles, like the terrain, complex 3D shading objects and racks with PV modules, into plain black silhouettes where the objects project on to the hemicube face. The hemicube reference system is defined by the orientation of the PV module so this can vary across a PV plant or even be time-dependent when modelling tracker systems. Each azimuth and zenith combination of the sun's position can be mapped to a pixel on one of the sides of the hemicube once the sun position is converted into the hemicube reference system. If the pixel is white, we can say the target point is not shaded, or if black it is shaded by an obstacle for a given sun position.

2.1 Rendering the hemicube sides

Four renderings (one for each of the side faces of the hemicube) are performed. Therefore, four camera positions are required – the camera's up vector for each face is pointing directly up the z-axis from the x-y plane. The camera's direction vectors point North, East, South and West from the target point for each of the four side faces.

Figure 1 shows the virtual hemicube (shown using dotted lines). The camera's viewport is shown with blue lines. The render surface is shown in red.

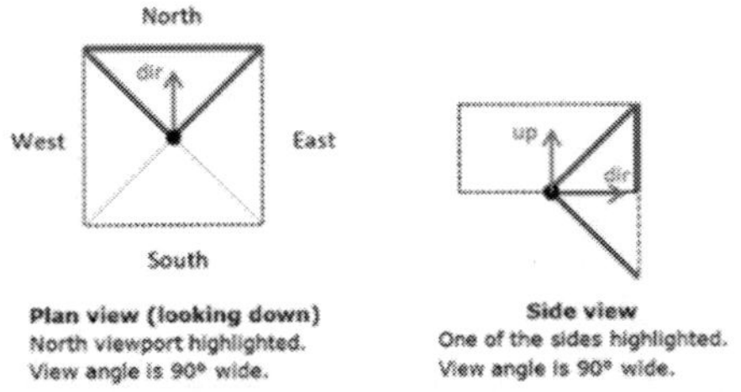

Figure 1: Sketch of hemicube for rendering the side face pointing North.

The view angle is 90 ° wide (45 ° to the left and right, 45 ° up and down). The render canvas is square. Only the top half (above the horizon) is used. This results in 4 black and white images, one for each side, as shown in Figure 2. These can be combined into a single side image.

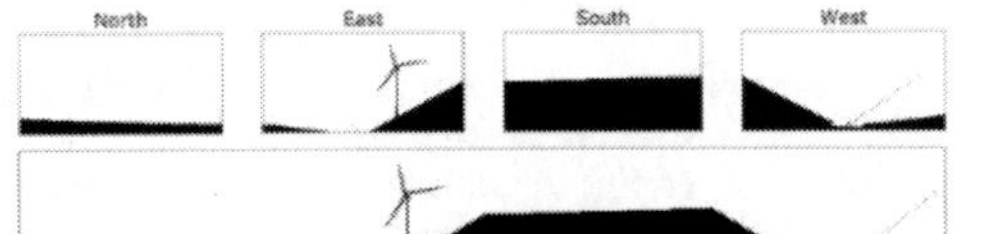

Figure 2: Images of rendered shading scene for hemicube sides and resulting side image.

For a given sun azimuth γ_z and zenith angle θ_z defined in the hemicube reference system, we can calculate an index into the side image for a fixed cube resolution *CubeRes*. The x_{side} of the image for azimuth values between -45 and +45 is calculated as:

10.4229/EUPVSEC2023/4DO.1.2
020414-001

$$x_{side} = \frac{CubeRes}{2} + \left(\tan \gamma_s * \frac{CubeRes}{2}\right)$$

Then, depending on the quadrant's azimuth, the appropriate multiple of *CubeRes* is added to offset into the correct quadrant of the image.

The y_{side} of the image depends on both the θ_z and γ_z values and is calculated as:

$$y_{side} = \frac{CubeRes}{2} * \frac{\cot \theta_z}{\cos \gamma_s}$$

For certain combinations y_{side} will be greater than the height of the image, especially when θ_z gets below a certain angle (when the sun is higher in the sky). In this case we must index into the top face of the cube.

2.2 Rendering the hemicube top face

This time the camera is pointed directly upwards (perpendicular to the x-y plane) with a view angle of 90 ° as illustrated in Figure 3.

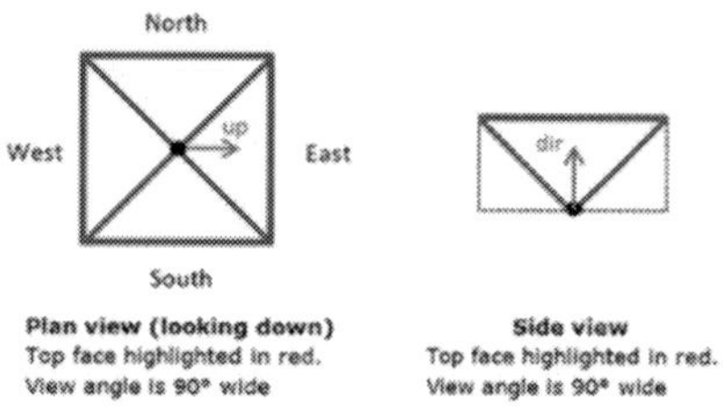

Figure 3: Sketch of hemicube for rendering the top face for cases when solar zenith angle is higher in the sky.

To index into the top image (of resolution *CubeRes* * *CubeRes*):

$$x_{top} = (\tan \theta_z \cos \gamma_s + 1) \times \frac{CubeRes}{2}$$

$$y_{top} = (\tan \theta_z \sin \gamma_s + 1) \times \frac{CubeRes}{2}$$

For all sun azimuth γ_z and zenith angle θ_z combinations, it is possible to index into the appropriate image (side or top) and extract the pixel value (black for shaded or white for unshaded) to estimate whether a target point is shaded.

2.3 Diffuse shading

The hemicube rendering used in the direct shading calculation above can also be used to determine the contribution by diffuse light from the sky for a target point. However, this needs to compensate for the hemicube's shape (the effect that perspective has on the areas of the pixels) and also take the angle that the light is falling on the module's surface into account. A weighting map is constructed that contains the appropriate weighting for each pixel in the hemicube, so they we can more accurately sum their contribution to the diffuse light shining on the target point, see Figure 4.

Figure 4: Weighting map for the diffuse light contribution for each pixel of the hemicube

For each target point we use the hemicube rendering used previously for the direct shading calculation. For each lit pixel (showing the white sky) in the hemicube we sum the corresponding weighting value from the weighting map. This results in a value between 0.0 and 1.0. This is the diffuse contribution for the target point.

2.4 Storing the results in a lookup table

The beam shading and diffuse values for each hemicube are compressed and stored in an efficient lookup table. Then, for the solar position of each timestamp of the energy calculation, a lookup is done for each target point for each timestamp to determine if it is shaded, rather than executing a complex shading calculation. This allows to easily handle multi-year and sub-hourly calculations without any additional shading calculations.

3 MODEL ACCURACY FOR BEAM SHADING

The most shaded cell in a PV submodule determines how much current can pass through the whole submodule as all the cells are connected in series. Therefore, the irradiance of the most shaded cell is needed to calculate I-V curves for all individual strings in a PV plant to estimate electrical mismatch and inverter losses; and so multiple hemicubes need to be placed on each submodule. To investigate the accuracy of the model to detect row-to-row shading cases, the proportion of a short row of modules directly shaded from a much longer row of modules in front has been compared to an infinite sheds model for a fixed-tilt site with monofacial modules in horizontal terrain for different resolutions of shading target points. The site details are summarised in Table 1.

Table 1: PV design characteristics of the test site.

Module	Number of cells	60
	Number of bypass diodes	3
	Orientation	Portrait
	Technology	Si-mono, monofacial
Rack system	Tilt	40 °
	Azimuth	180 °
	GCR	0.5

We have implemented the hemicube method in DNV's SolarFarmer software [2] and use this package to generate the results presented below. We compare the detection of row-to-row shading cases to an infinite sheds model [3], [4] where this is derived geometrically. The simulation period is 1 year and the time resolution is 1 hour. Additionally, the irradiance of the most shaded cell in the sub-modules is used in the comparison as this defines the

I-V curves going into the power calculation.

The number of hemicube shading target points used in the comparison is varied between 1 per module and 9 per cell. The location of the target points on a single module is shown in Figure 5.

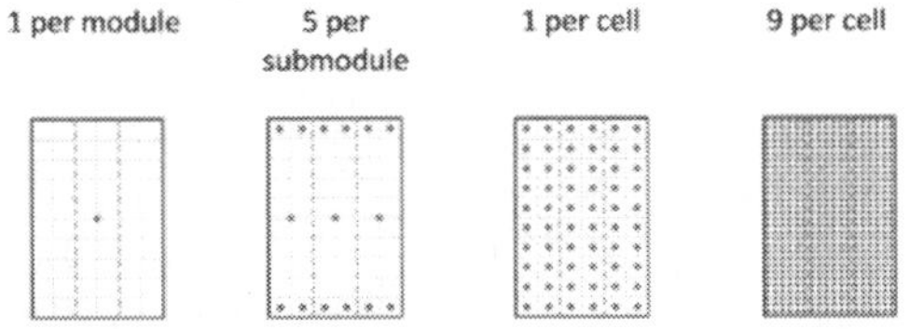

Figure 5: Location of shading target points on a single module.

The number of time stamps by time of day where shading of the direct irradiance is detected is shown in Figure 6 for the hemicube model with different shading target point resolutions and an infinite sheds model. Given the location of the target point in the module centre, the "1 per module" resolution can only detect shading cases where at least 50 % of the module is shaded. This results in no shading cases being detected in the morning where the average shaded proportion of the module is small according to the infinite sheds model. Adding target points in the corner cells of the submodule as in the "5 per submodule" or "1 per cell" resolutions improves the detection of the shading cases. However, given the location of the lowest target point at 5 % of the module height for a module with 60 cells, only cases where the shade line exceeds this proportion can be detected. This can be seen at 09:00 where the average shading proportion is only 4 % so both methods can only detect some of the shading cases. Using a "9 per cell" resolution allows to detect cases where less than 5 % of the module are shaded so this resolution gets close to the infinite sheds model.

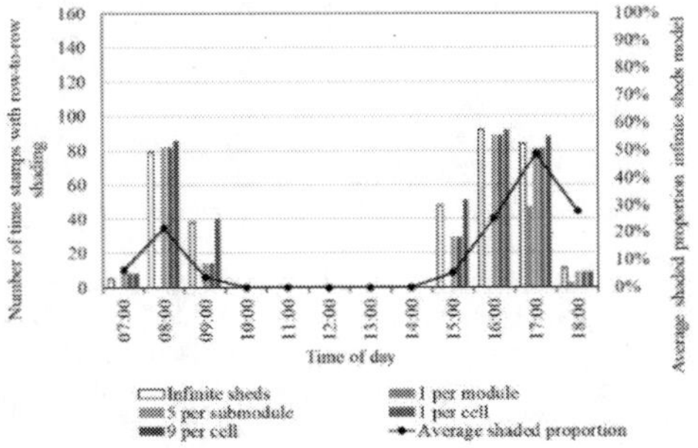

Figure 6: Number of row-to-row shading cases (bars) detected by hemicube model for different target point resolutions and infinite sheds model and average shaded proportion from infinite sheds model (line) by time of day.

The irradiance for the most shaded cell in a submodule is needed for correctly modelling the power output so the methods are compared for the irradiance of the most shaded cell in Figure 7. The most shaded cell irradiance considers shading of the direct and diffuse components so the hemicube model cannot be directly compared to the infinite sheds model as the shading for the diffuse irradiance is calculated differently. With the "1 per module" method not accurately detecting row-to-row shading cases, it tends to overpredict the irradiance while all other resolutions give almost the same results.

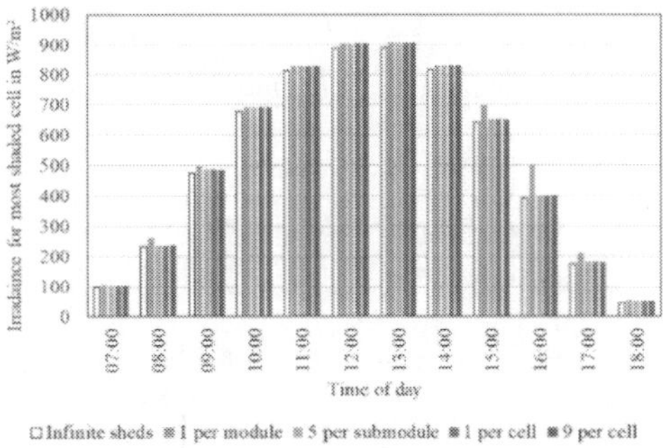

Figure 7: Irradiance for most shaded cell in back-row modules for infinite sheds model and hemicube model with different target point resolutions by time of day.

The detected number of row-to-row shading cases of the direct irradiance and the irradiance of the most shaded cell are summarised in Table 2 for all models and resolutions. The "1 per module" resolution detects only 14 % of the row-to-row shading cases for the direct irradiance compared to the infinite sheds model and overpredicts the irradiance for the most shaded cell by 3.5 % compared to the highest resolution. The additional use of the submodule corner points improves the detection of row-to-row shading cases and brings the modelled irradiance of the most shaded cell very close to the results of the "9 per cell" resolution. The hemicube model with "9 per cell" resolution detects slightly more shading cases than the infinite sheds model. This is caused by the chosen hemicube resolution.

Table 2: Number of detected direct shading cases and irradiance for most shaded cell for infinite sheds and hemicube model with different target point resolutions. The annual average global unshaded irradiance is 544.3 W/m².

Shading model	Number of detected shading cases	Average irradiance for most shaded cell in W/m²
Infinite sheds	357	511.9
Hemicube		
1 per module	50	540.4
5 per submodule	312	522.0
1 per cell	312	522.0
9 per cell	374	522.0

4 MODEL PERFORMANCE

We have run simulations for utility-scale test sites with SolarFarmer and measured the effect on the results for an annual energy yield estimation and the time to run the shading part of the simulation for different target point resolutions. The test sites are all theoretical sites in flat terrain without any shading obstacles. The site statistics are summarised in Table 3. The total number of shading target points ranges from 16,800 for the smallest site modelled with the "1 per module" resolution to

196,992,000 for the largest site modelled with the "9 per cell" resolution.

Table 3: Characteristics of tests sites used for performance testing

AC capacity in MW	5	20	100
Number of modules	16,800	73,440	364,800

The increase in the shading calculation time relative to fastest method with only 1 per module is shown in Figure 8. As expected, the simulation time scales directly with the number of shading target points and the number of modules in the PV plant.

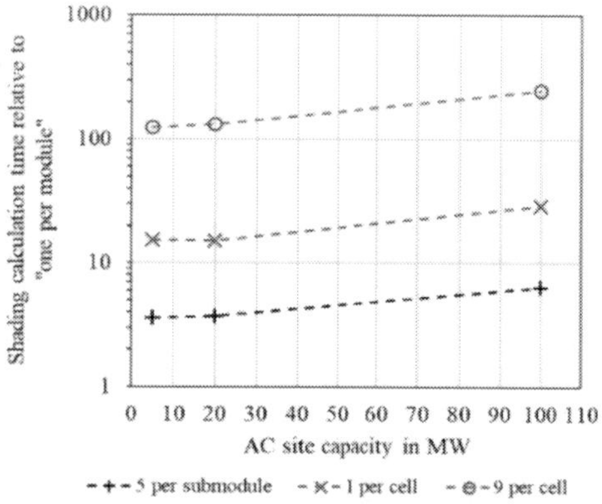

Figure 8: Increase in calculation time for different target point resolutions relative to the fastest "1 per module" method.

To assess the sensitivity of the modelled annual energy production depending on the shading target point resolution, the energy yield in kWh/kWp has been compared to the results using the highest resolution of 9 shading target points per cell as this is expected the most accurate. As shown in Figure 9, the "1 per module" resolution has the largest difference of 0.1 % to 0.2 % whereas the higher resolution methods show almost no difference compared to simulations with the "9 per cell" resolution.

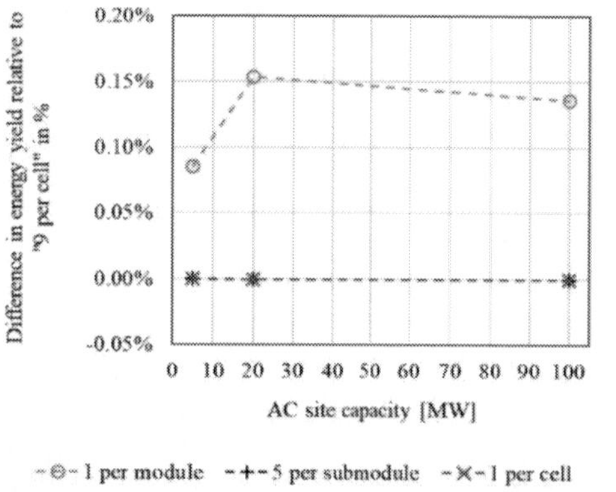

Figure 9: Difference in energy yield relative to "9 per cell" shading target point resolution.

5 CONCLUSIONS

The aim of this paper was to present a hemicube-based rendering model to account for near shading in fixed-tilt PV plants. The sensitivity of relevant parameters (e.g., target point resolution) of the proposed model was evaluated in terms of accuracy and computational performance as a function of PV project capacity. The results demonstrated that hemicube-based rendering models are suitable to estimate shading effects for utility-scale PV sites that are affected by row-to-row shading or shading caused by nearby terrain features or shading obstacles. Although such models can be computationally costly, a smart choice of shading target points helps to reduce simulation time while keeping an appropriate model accuracy to enable PV plant designers making the layout choice and reduce financial risk. It was found that using a shading target point resolution of "5 per submodule" represented a good compromise between accuracy and simulation time for energy assessments of utility-scale PV plants.

More information about the hemicube model, extended simulation results, and validation against operational data are available in [5].

REFERENCES

[1] T. E. K. Zidane *et al.*, "Identifiability Evaluation of Crucial Parameters for Grid Connected Photovoltaic Power Plants Design Optimization," *IEEE Access*, vol. 9, pp. 108754–108771, 2021, doi: 10.1109/ACCESS.2021.3102159.

[2] P. Neubert, Anja; Mikofski, Mark A.; Hamer, Mike; Rainey, "Validation of the SolarFarmer software with operational data," in *37th European Photovoltaic Solar Energy Conference and Exhibition*, EU PVSEC, 2020, pp. 1621–1625. doi: 10.4229/EUPVSEC20202020-5CV.4.10.

[3] M. A. Mikofski, R. Darawali, M. Hamer, A. Neubert, and J. Newmiller, "Bifacial Performance Modeling in Large Arrays," in *2019 IEEE 46th Photovoltaic Specialist Conference (PVSC)*, IEEE, Ed., Chicago, IL: IEEE, 2019.

[4] B. Marion *et al.*, "A Practical Irradiance Model for Bifacial PV Modules," in *2017 IEEE 44th Photovoltaic Specialist Conference (PVSC)*, IEEE, Jun. 2017, pp. 1537–1542. doi: 10.1109/PVSC.2017.8366263.

[5] A. Neubert, M. Hamer, and J. Lopez-Lorente, "Shading in Utility-Scale Photovoltaic Plants: A Hemicube Approach for Efficient Shading Calculations," *Solar RRL*, Jul. 2023, doi: 10.1002/solr.202300473.

10.4229/EUPVSEC2023/4DO.1.2
020414-004

KU LEUVEN

imec

Investigating the temperature distribution of a PV module under various environmental factors and exploring novel module geometries for the reduction of thermal losses

Gauthier du Boullay, Gofran Chowdhury, Jef Poortmans

EU PVSEC 2023
21/09/2023

Temperature impact of PV – the problem exposed

Problem

Energy losses
power coefficient =0.25 – 0.5 %/°C

Thermal aging
ΔT_avg = 5°C → Lifetime×2

Modelling error
Up to 4.3% energy yield overestimations

Solutions ?

Active cooling
- × Too high complexity and cost
- × Require parasitic energy consumption
- × Require O&M

Passive cooling
- × Ineffective and unreliable
- × Too high complexity
- × Too costly for low positive impact

No research analyzing the spatial temperature variations (cell level) found → simulation needed

imec

KU LEUVEN

Experimental setup

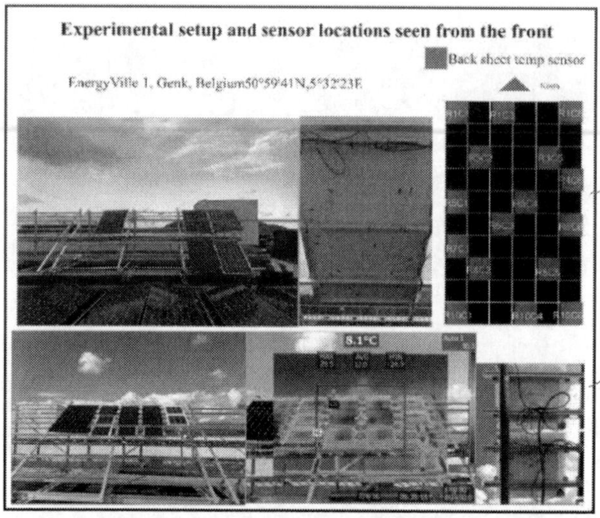

16 thermal sensors attached to the back of a single full-scale PV module.

Multiple mini modules with 5 thermal sensors

➔ Investigate T distribution = Investigate the T mismatch

imec KU LEUVEN

020415-003

40th European Photovoltaic Solar Energy Conference and Exhibition

Experimental results

Time series data

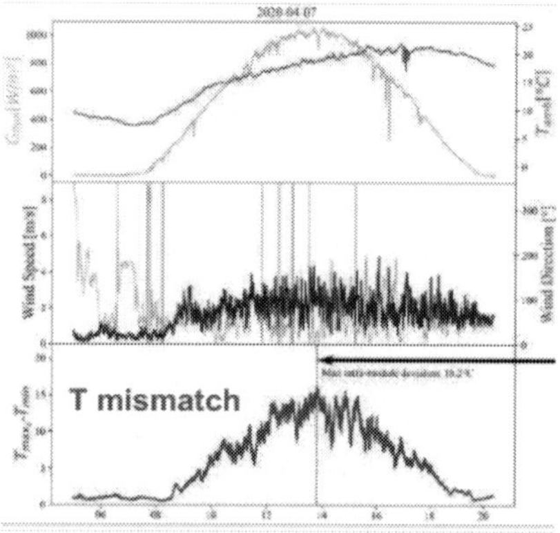

- Cell temperature (mismatch) depends on Irradiance, wind speed, and direction

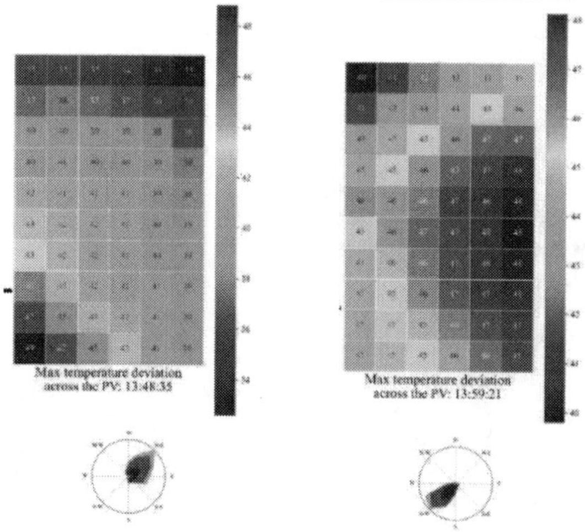

T mismatch = 16°C   T mismatch = 8°C

- Wind direction has a major impact on temperature mismatch

imec KU LEUVEN

020415-004

Simulations and proposed geometrical modifications

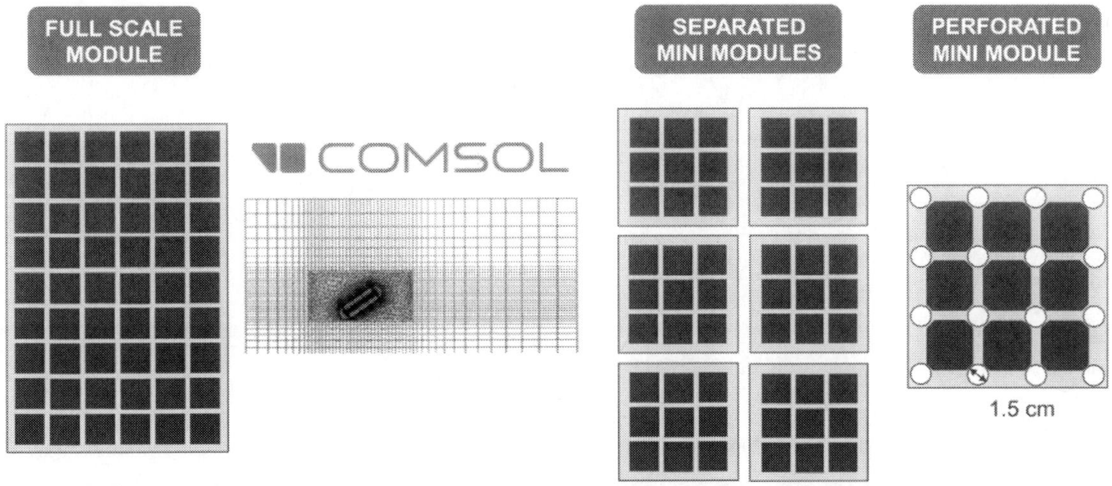

imec KU LEUVEN

020415-005

40th European Photovoltaic Solar Energy Conference and Exhibition

Results – Full scale module

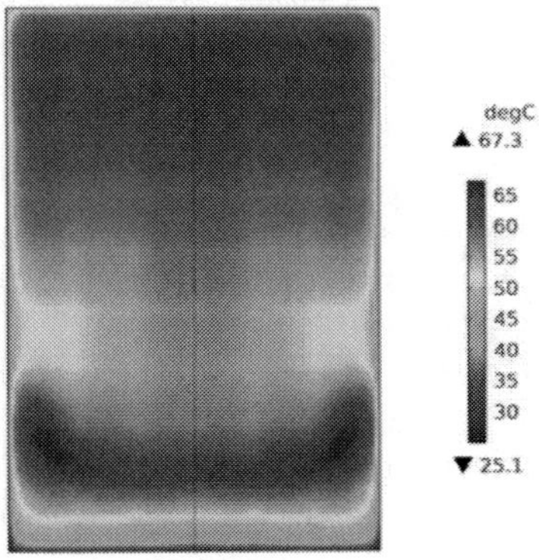

- Highly non uniform temperature pattern
  - 1st part: Flow separation bubble
  - 2nd part: boundary layer development

imec KU LEUVEN

020415-006

Results – Full scale module

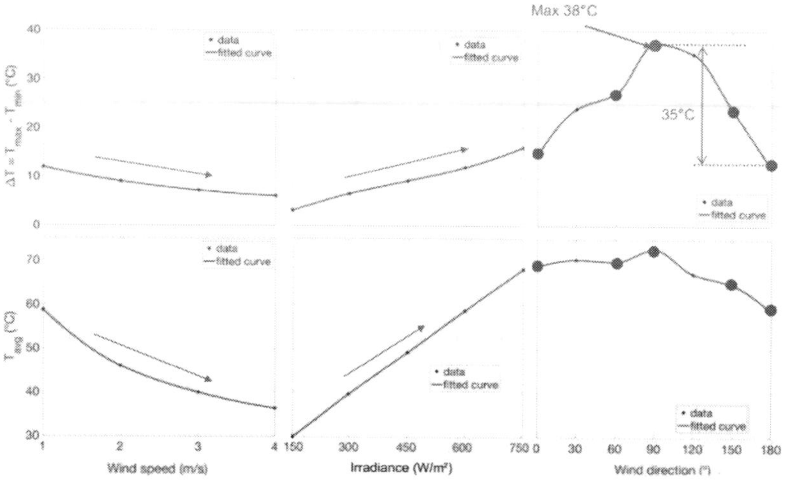

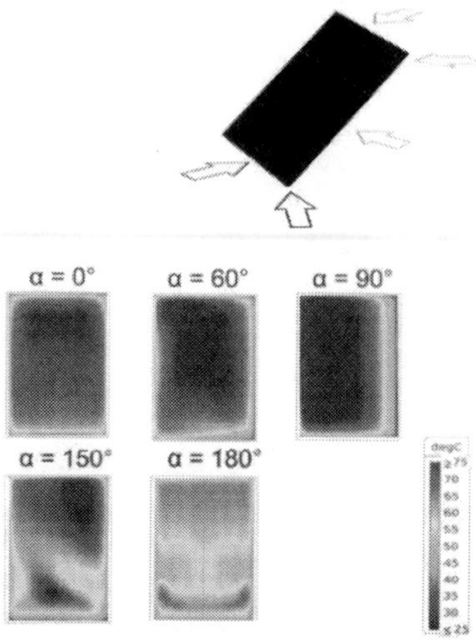

➢Most influential factor is wind direction (+ tilt angle)

imec KU LEUVEN

40th European Photovoltaic Solar Energy Conference and Exhibition

Results – Separated mini modules

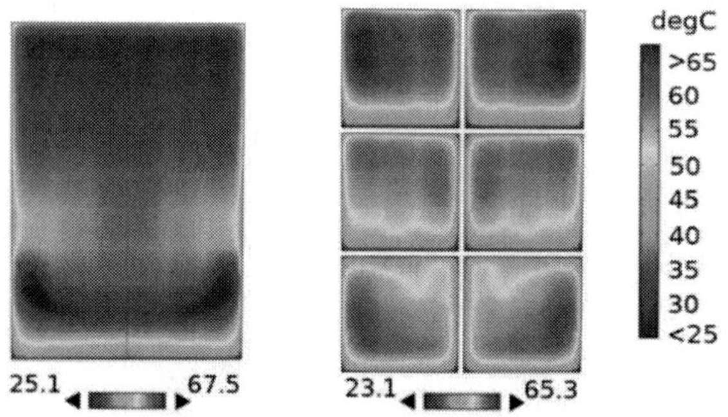

- Optimal spacing = 2.5cm
  - Area increase of 14%
- Cooling mechanisms
  - High-velocity jets
  - Heat evacuation path
  - Breaking boundary layer and separation bubble

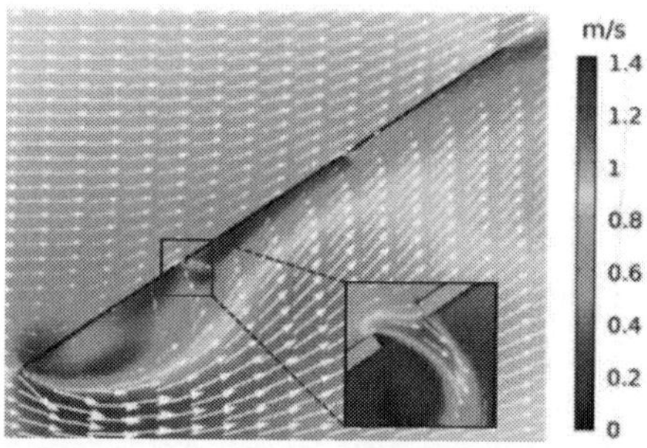

imec KU LEUVEN

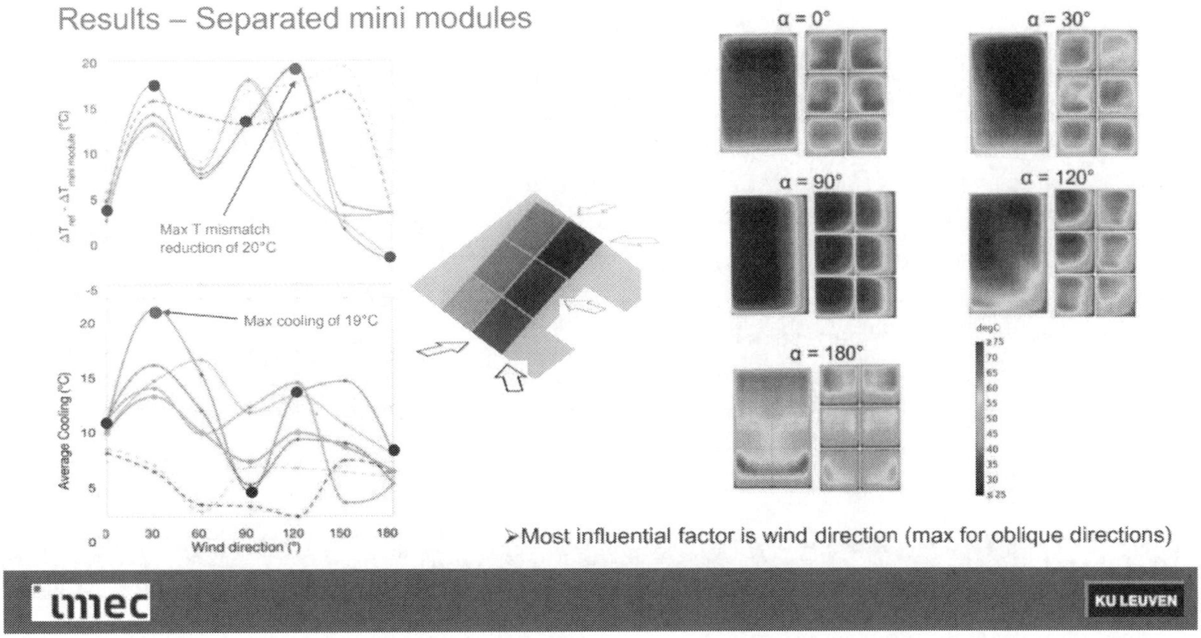

020415-009

40th European Photovoltaic Solar Energy Conference and Exhibition

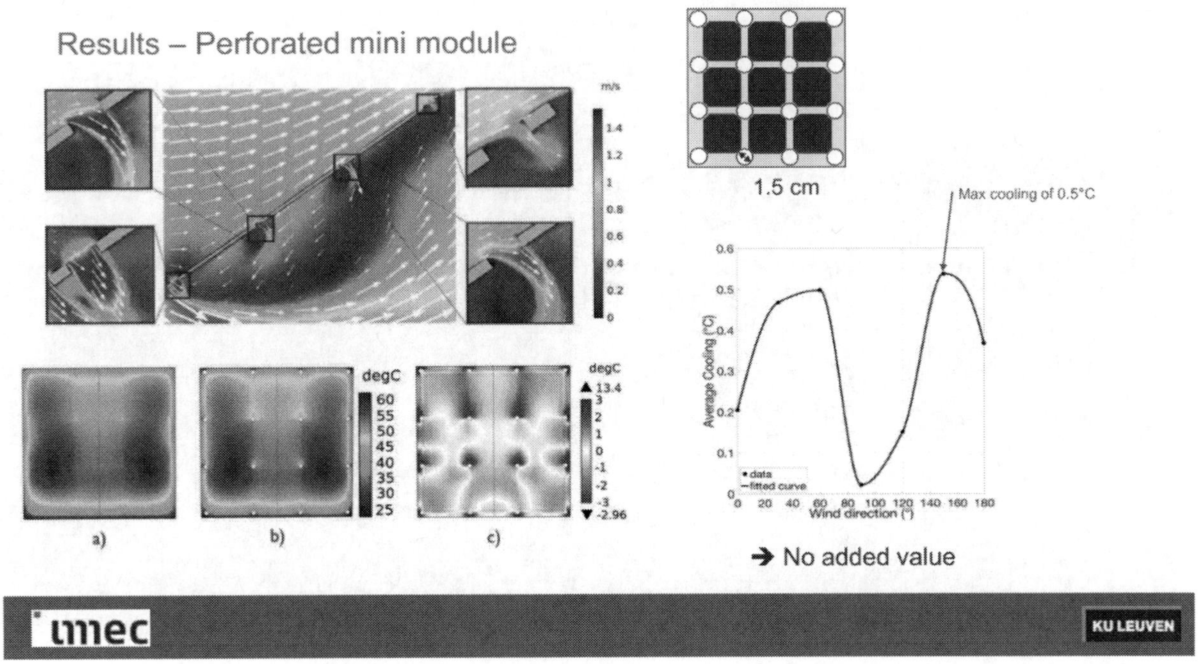

020415-010

Key messages

- The temperature of a PV module is highly non uniform. Temperature mismatch can reach values up to 38°C.
- The most influential parameter on temperature pattern is wind direction.
- The perforations do not show any added value
- Separating a full scale module by spaced mini modules has a clear beneficial impact
  - Cooling enhancing (max 19°C)
  - Temperature mismatch reduction (max 20°C)

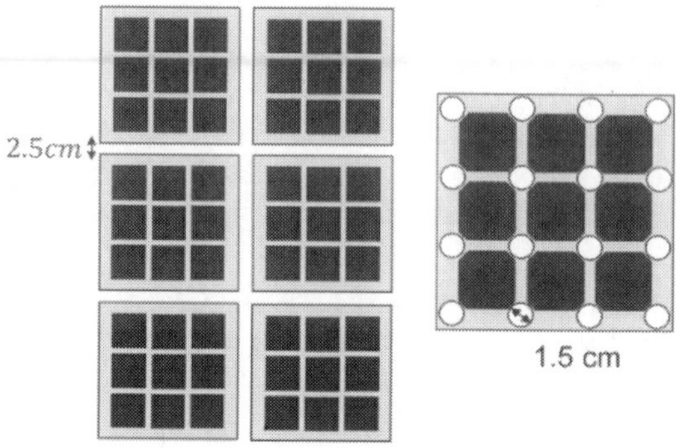

020415-011

40th European Photovoltaic Solar Energy Conference and Exhibition

020415-012

Module Type Impacting Electromagnetic Emissions from Photovoltaic Systems

EU PVSEC 2023 - Lisbon

Sept. 2023

Désirée Kroner[1,2] and Urban Lundgren[3]

desiree@du.se urban.Lundgren@ri.se

1 DALARNA UNIVERSITY 2 LULEÅ UNIVERSITY OF TECHNOLOGY 3 RISE

Co-financed by Swedish Energy Agency

40th European Photovoltaic Solar Energy Conference and Exhibition

Introduction

PV EME

Radiated

Radio

Amateurs

C2000

Research

In-situ measurements

My contribution

Systematic in-situ measurements

On realistic, flexible and comparable Systems

C2000

- Antenna Emergency Service
- Mobile Emergency Service
  - Ambulance
  - Fire brigade
  - Police
- Photovoltaic system with high emissions

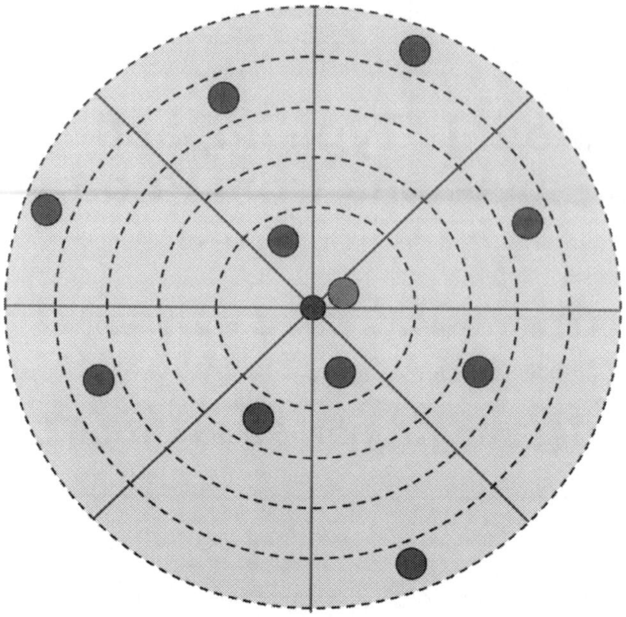

Co-financed by Swedish Energy Agency 3 Désirée Kroner desiree@du.se DALARNA UNIVERSITY

C2000 reduction by 80%

- Antenna Emergency Service
- Mobile Emergency Service
  - Ambulance
  - Fire brigade
  - Police
- Photovoltaic system with high emissions

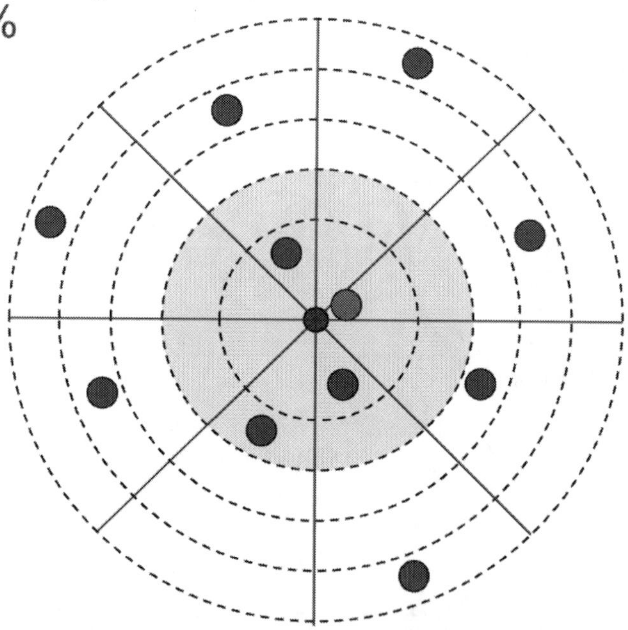

Co-financed by Swedish Energy Agency 4 Désirée Kroner desiree@du.se DALARNA UNIVERSITY

Introduction

PV EME	Radio	Research	My contribution
Radiated	Amateurs	In-situ measurements	Systematic in-situ measurements
	C2000		On realistic, flexible and comparable Systems

40th European Photovoltaic Solar Energy Conference and Exhibition

Photovoltaic system

PV module *Inverter*

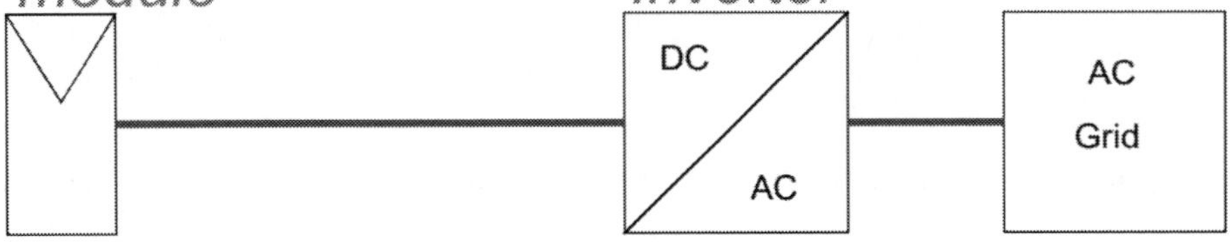

Mono or Bifacial String or Optimizer

Components	*Inverter* *PV module*	***Installation details*** *Small loop area, with func. equ. earthing and zero distance between string cables*

Solar Inverter concepts

String inverter

Inverter with optimizers

String inverter with one optimizer

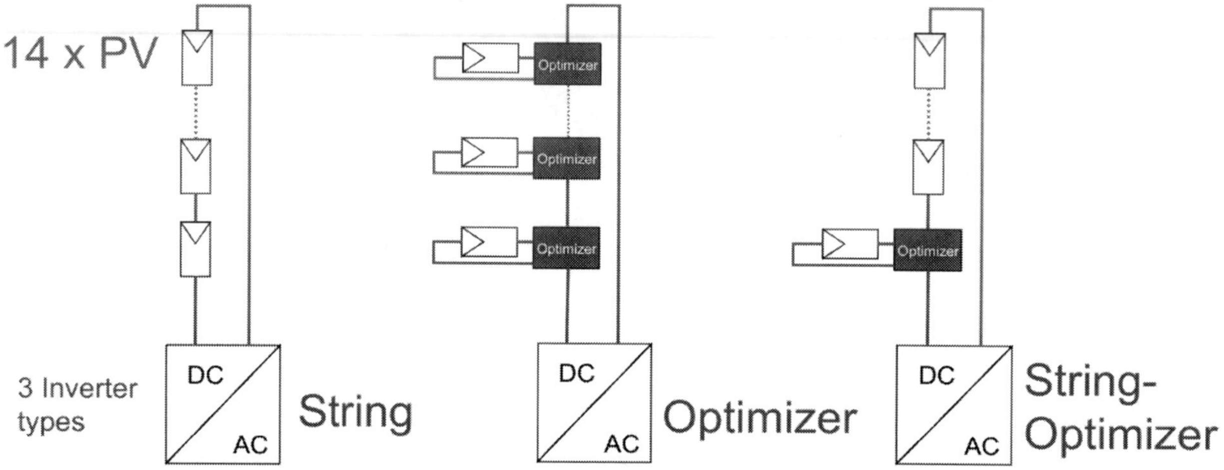

Module Type

Mono

Half-cut module

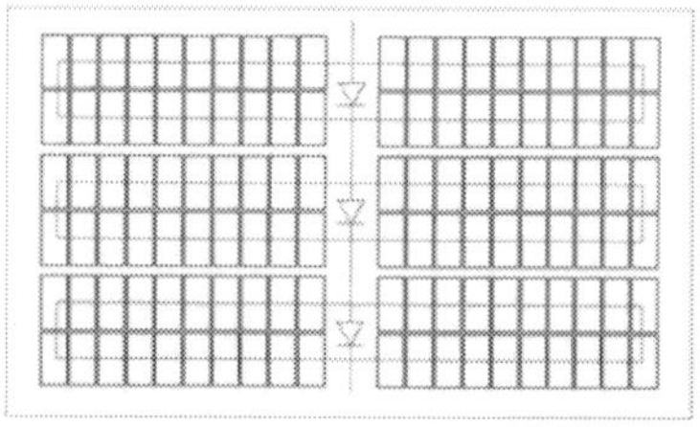

Major Difference

Wiring

Bifacial

Full-cell module

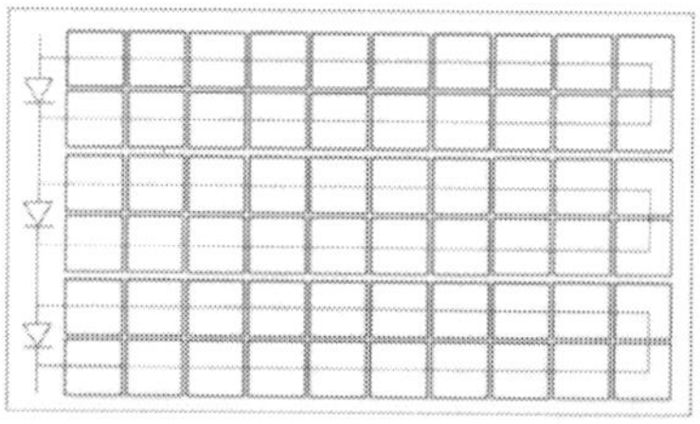

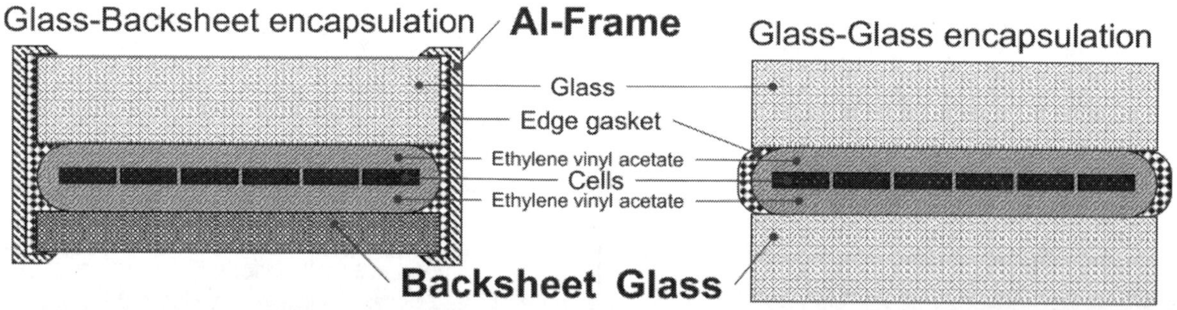

Measurement setup
Location Borlänge

- Loop antenna:

x-direction

y-direction

z-direction

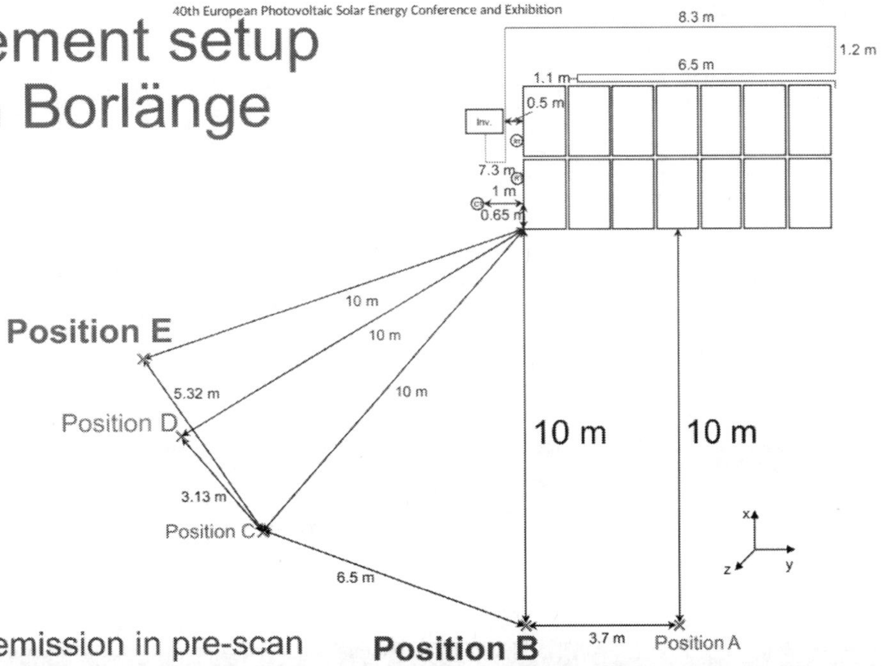

- Point of highest emission in pre-scan

Measurement Method
Full Time domain 150 kHz – 30 MHz

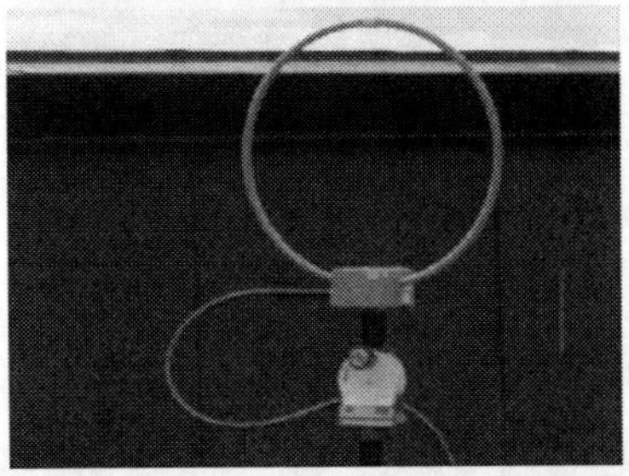

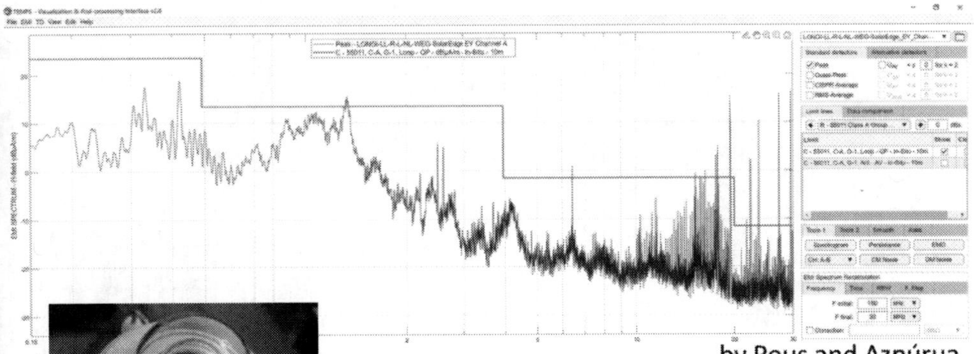

by Pous and Azpúrua

Data analysis method

Position E - Z

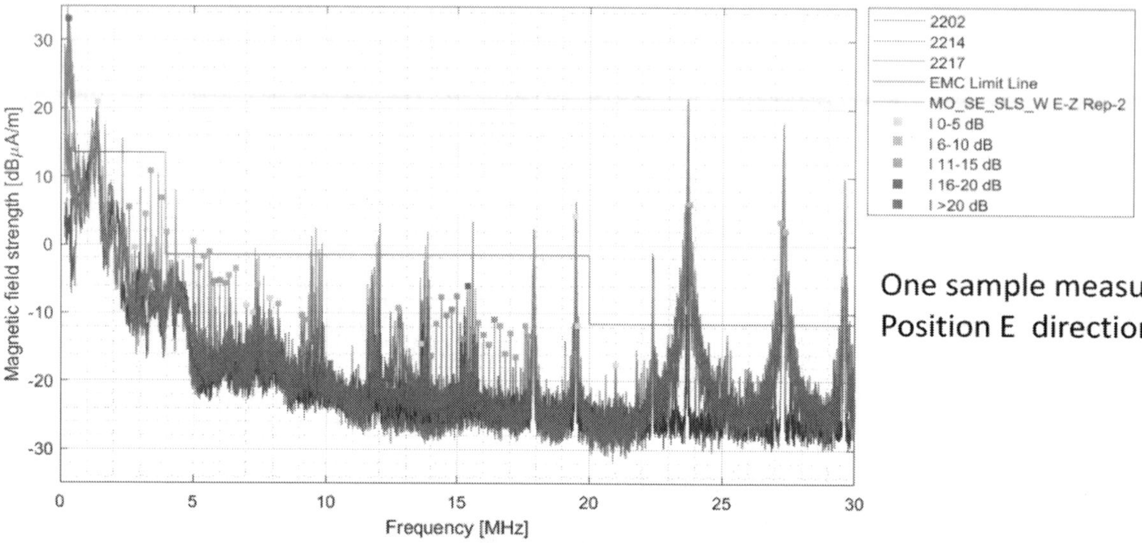

One sample measurement
Position E direction Z

Data analysis method

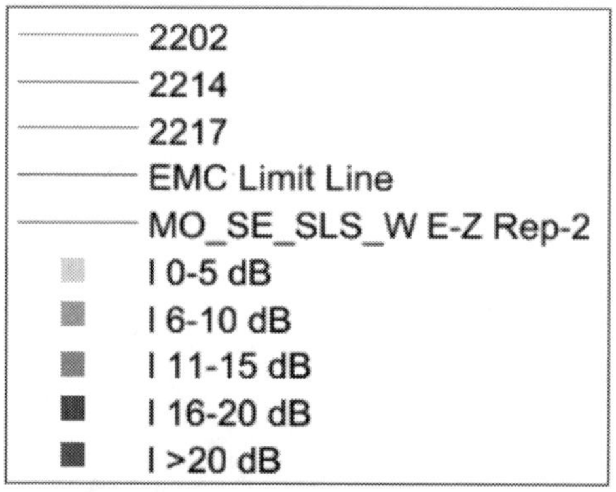

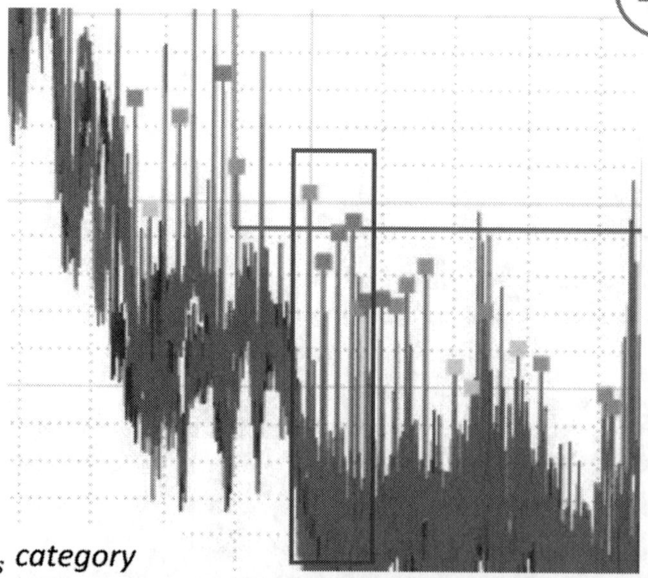

peak >10 dB

margin between peaks M_{sys} and background measurements $BG2_i$

peak to background *margin* P_{sys} *category*

40th European Photovoltaic Solar Energy Conference and Exhibition

Data analysis method

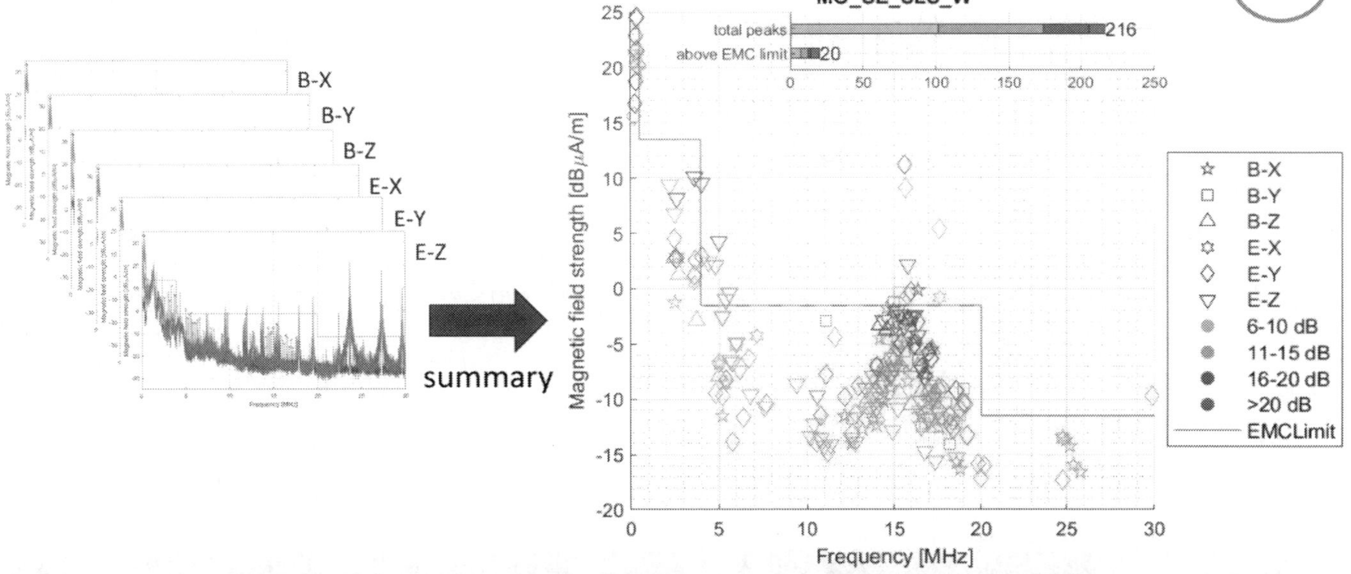

Co-financed by Swedish Energy Agency 13 Désirée Kroner desiree@du.se DALARNA UNIVERSITY

40th European Photovoltaic Solar Energy Conference and Exhibition

Result inverter concept

Optimizer

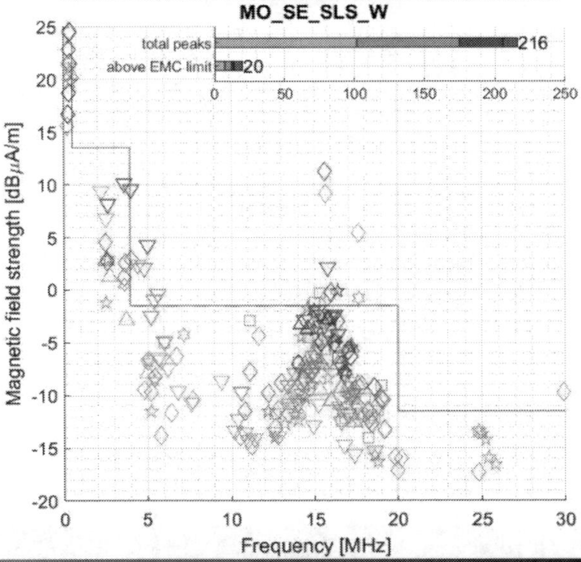

String

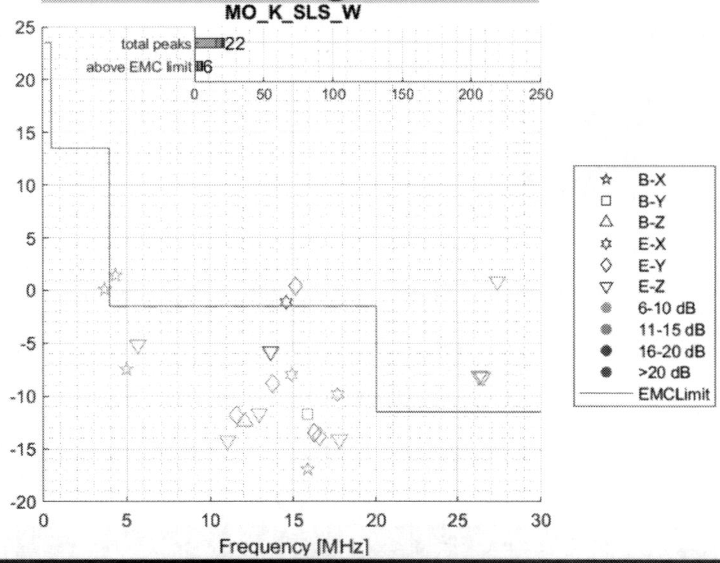

Co-financed by Swedish Energy Agency 14 Désirée Kroner desiree@du.se DALARNA UNIVERSITY

Module Type

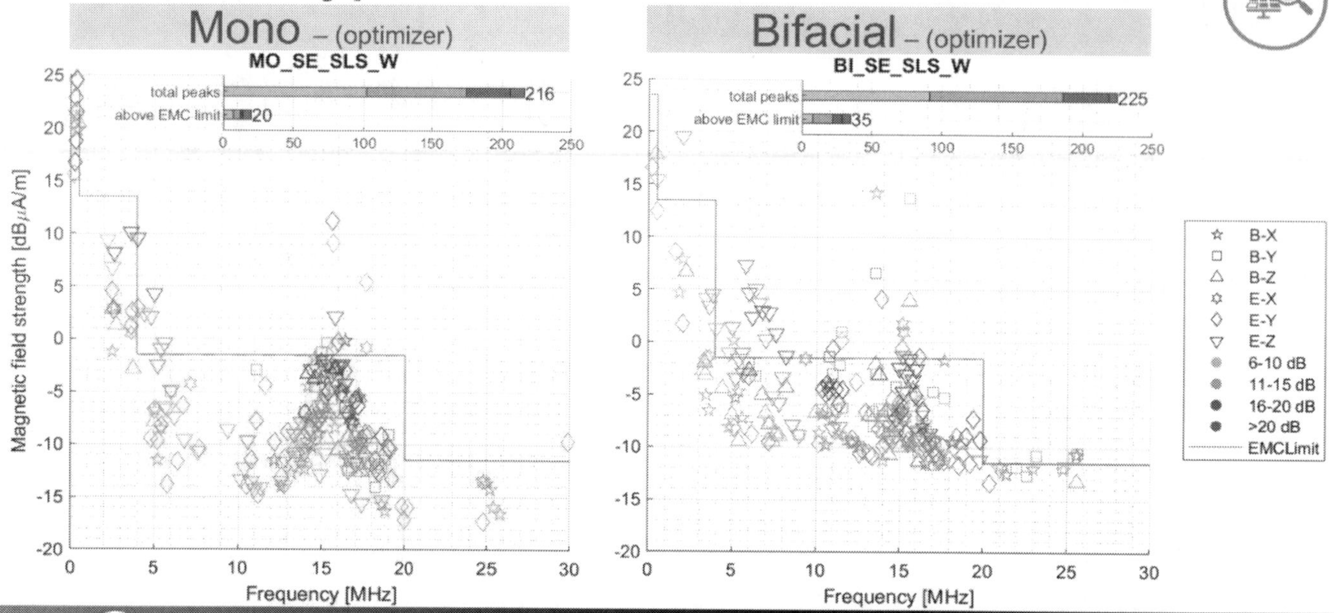

Co-financed by Swedish Energy Agency 15 Désirée Kroner desiree@du.se DALARNA UNIVERSITY

40th European Photovoltaic Solar Energy Conference and Exhibition

Summary

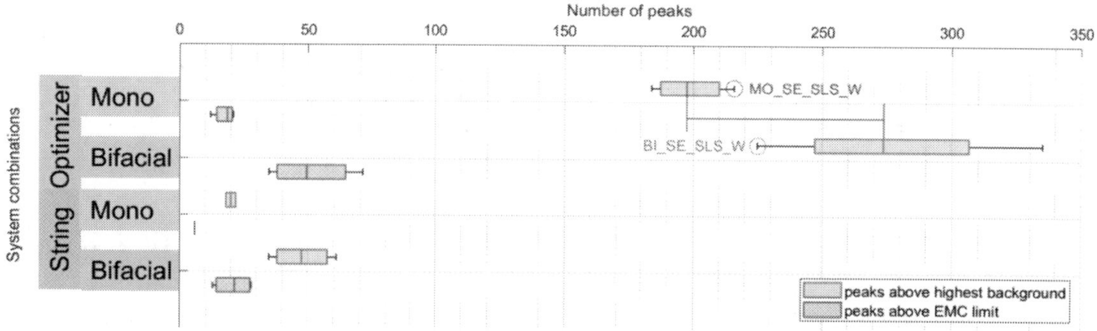

- String inverter lower emissions than optimizer systems
- Module type has a significant impact on emissions
- String-optimizer has similar low emissions as String
- Installations details have lesser impact on EME than expected.

Co-financed by Swedish Energy Agency 16 Désirée Kroner desiree@du.se DALARNA UNIVERSITY

Take away

Low electromagnetic emissions matters when scaling up PV

String inverter

Glass-backsheet with frame

Module optimizer

Glass-glass no frame

Désirée Kroner
desiree@du.se

Pre-print article

Acknowledgements

Funding

Swedish Energy Agency — Methods to detect and to prevent EMI from PV installations (solEMC) (project number P2020-90239)

Solar Electricity Research Center, Sweden (SOLVE). SOLVE is co-financed by the Swedish Energy Agency (project number P2021-90275) as a national center of excellence.

Cooperation

Swedish National Electrical Safety Board

Assistance

Measurement campaign

Internship students Nadine Eisinger and Hugo Serris
Joar Nyman from Solfamiljen
and diverse solar energy engineering master students from Dalarna University .

Supervisors

Thanks to Math Bollen, Luleå University of Technology and André Augusto, Dalarna University for their useful advice regarding the article.

Questions

Low electromagnetic emissions matters when scaling up PV

String inverter

Glass-backsheet with frame

Module optimizer

Glass-glass no frame

Désirée Kroner
desiree@du.se

Pre-print article

PVRADAR by Virtuous Re

Cleaning optimization for photovoltaic powerplants: A novel approach combining techno-economic modelling with historic rain and soiling.

Virtuous-Re GmbH, Munich, Bayern, info@virtuous-re.com/ 85560, Germany

Franco Clandestino – fc@virtuous-re.com
Thore Müller – tm@virtuous-re.com
Konstantin Pogorelov – kp@virtuous-re.com

EUPVSEC

020417-001

40th European Photovoltaic Solar Energy Conference and Exhibition

AGENDA

1. **PVRADAR**
   Software solutions for decision making
2. **An optimal cleaning strategy is key**
   How to achieve it?
3. **Methodology**
   Soiling assessment and cleaning optimization based on historic data
4. **Example case**
   Cleaning optimization for a 100MW plant in Portugal
5. **Key takeaways**

9/21/2023

020417-002

VIRTUOUS-RE HAS DEVELOPED PVRADAR. THE SOFTWARE TOOL THAT FINDS THE BEST STRATEGIES FOR YOUR PORTFOLIO

PVRADAR

- Optimal cleaning strategy in just a few clicks.
- Integrated techno-economic models for highest lifetime value.
- Intuitive interface and interactive graphics to enhance collaboration.

020417-003

40th European Photovoltaic Solar Energy Conference and Exhibition

020417-004

THE DECISION ON HOW TO CLEAN YOUR PV PLANT IS CRITICAL

Standard approach

- Cheaper cleaning technology
- Previous experiences
- Visual inspections

Optimal strategy

- Soiling conditions
- Meteorological conditions
- PV plant parameters: Technical and financial
- Cleaning technologies parameters: Technical and financial

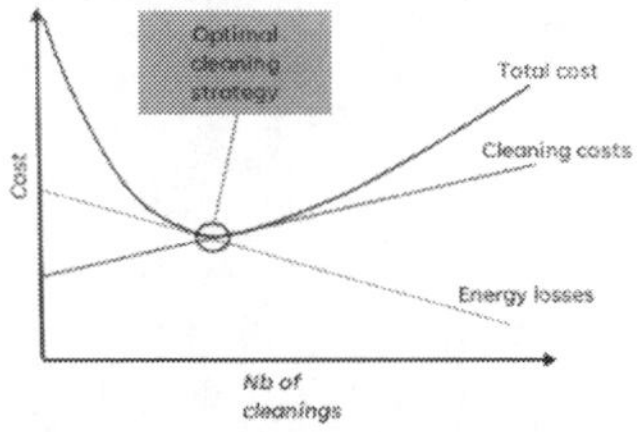

Interactive demonstration:

Cleaning optimization process for a PV plant in Portugal

Which technology? Which model? Which frequency? Which dates?

40th European Photovoltaic Solar Energy Conference and Exhibition

CLEANING OPTIMIZATION METHODOLOGY

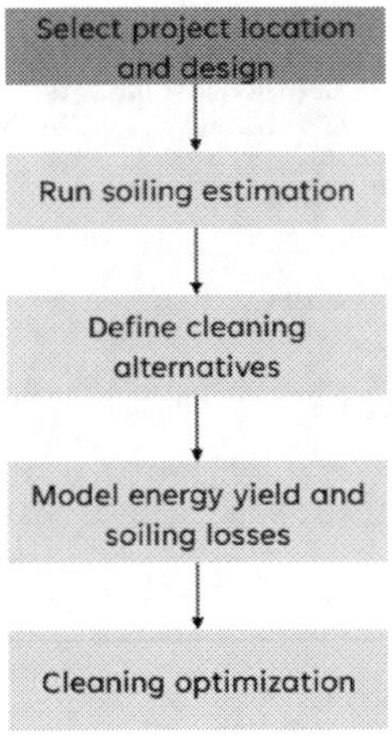

Soiling depends on both site conditions and mounting configuration of the plant.

CLEANING OPTIMIZATION METHODOLOGY

- Select project location and design
- **Run soiling estimation**
- Define cleaning alternatives
- Model energy yield and soiling losses
- Cleaning optimization

Satellite re-analysis source MERRA-2

21 years of historical data

Meteorological data
- T amb
- W speed and direction
- RH

Particulate matter
- Concentrations
- Components

Rainfall data
- Intensity
- Dates

Soiling estimation models

Historic soiling profile

9/21/2023 PV RADAR 7

020417-007

40th European Photovoltaic Solar Energy Conference and Exhibition

CLEANING OPTIMIZATION METHODOLOGY

- Select project location and design
- Run soiling estimation
- **Define cleaning alternatives**
- Model energy yield and soiling losses
- Cleaning optimization

Components lifetime

Operational parameters

Costs

Cleaning performance

Different characteristics for different technologies, manufacturers and models

9/21/2023 PV RADAR 8

020417-008

CLEANING OPTIMIZATION METHODOLOGY

Select project location and design
↓
Run soiling estimation
↓
Define cleaning alternatives
↓
Model long term yield and soiling losses
↓
Cleaning optimization

2010
2011
2012
2013
2014
2015
2016
2017
2018
2019
2020
2021
2021

→ Long-term conditions forecast →

Production profile for each year

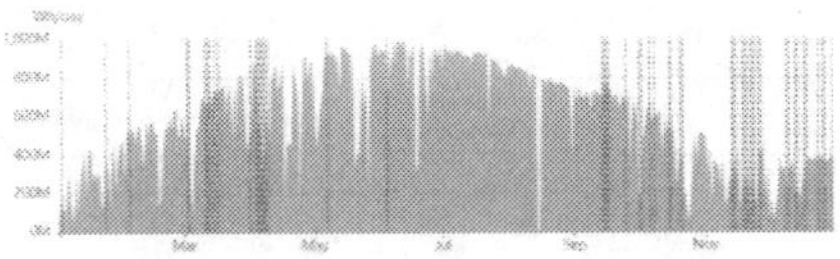

- Soiling losses
- Clipping losses

40th European Photovoltaic Solar Energy Conference and Exhibition

CLEANING OPTIMIZATION METHODOLOGY

Select project location and design
↓
Run soiling estimation
↓
Define cleaning alternatives
↓
Model long term yield and soiling losses
↓
Cleaning optimization

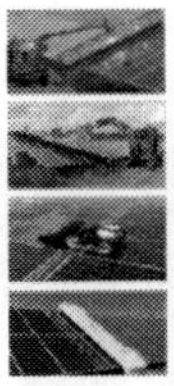

Nb of systems
Nb of cleanings
Cleaning dates

$$\min C_{TOTAL} = C_{Energy} + C_{Cleaning}$$

Total cost — Cost of energy losses — Cost of performed cleanings

Lifetime total cost

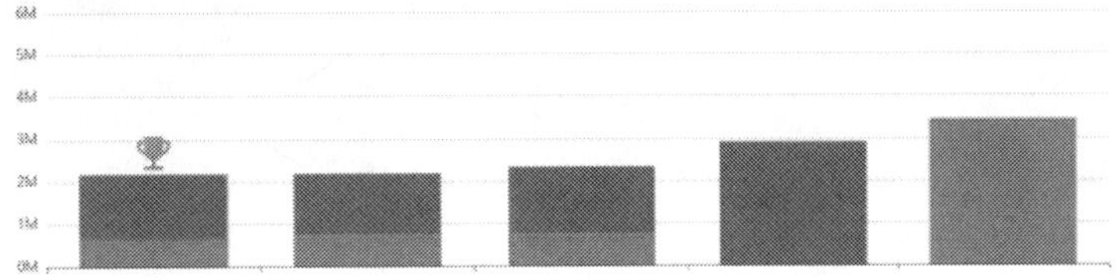

020417-011

40th European Photovoltaic Solar Energy Conference and Exhibition

EXAMPLE CASE: 100MW PV PLANT IN PORTUGAL

TECHNICAL

System DC capacity:	100.01 MW
System AC capacity:	111 MW
DC to AC ratio:	0.901
Module nominal power:	400 W
Mounting configuration:	Fixed tilt 10°

FINANCIAL

Energy Sales Price:	40 USD/MWh
Project lifetime:	30 years
Discount rate:	4%
Labour cost:	10000 USD/FTE
Fuel:	1.5 USD/Liter
Water:	3 USD/m³

Project location

Latitude * 37.3538

Longitude * -8.0818

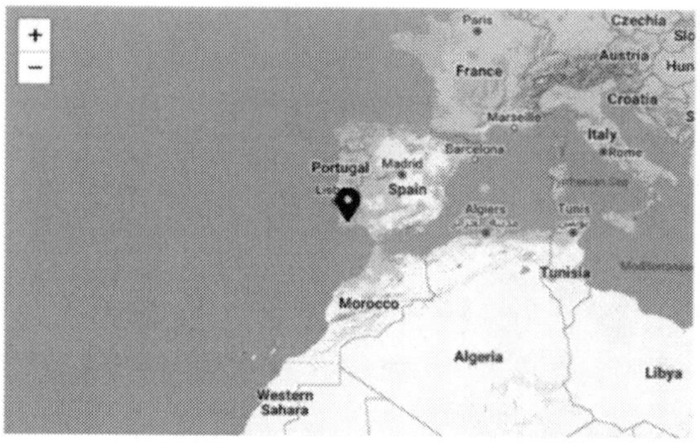

9/21/2023 PV RADAR 12

020417-012

HISTORIC RAINFALL DATA

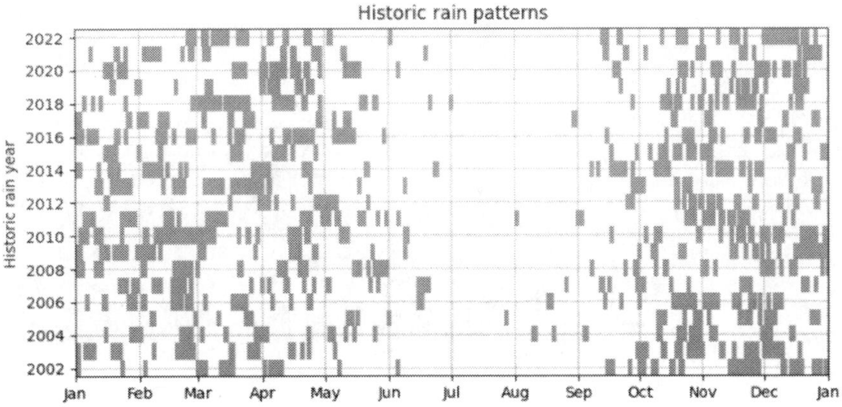

Rainfall registered for the last 21 historical years. Data sourced from satellite-based reanalysis MERRA-2.

HISTORIC MONTHLY SOILING LEVEL

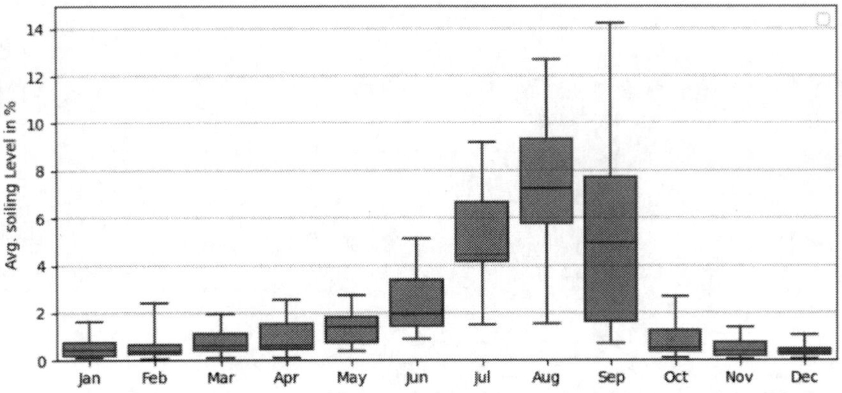

Soiling levels derived from soiling estimation models based on particulate matter and meteorological satellite data registered for the last 21 years.

HISTORIC AVG. YEARLY SOILING LEVEL

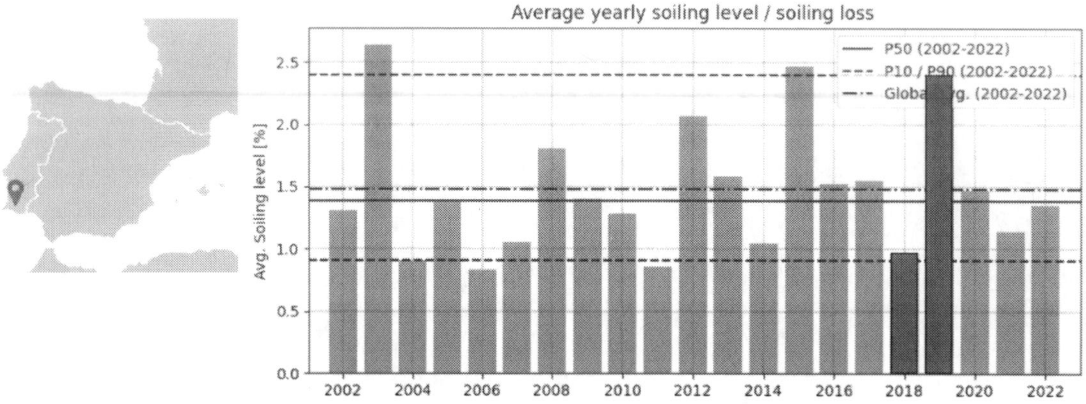

020417-015

020417-016

CLEANING TECHNOLOGIES AND BUSINESS MODELS

PV RADAR

	Speed	Cleaning efficiency	Water consumption	Fuel consumption	Operators per system	Components
Wet-cleaning tractor	6.0 km/day	95%	1 L/module	250 L/100km	2	• Tractor • Brush • Water treatment system
Dry-cleaning tractor	6.2 km/day	90%	0	250 L/100km	2	• Tractor • Brush
Semi-autonomous robot	2.0 km/day	90%	0	0 L/100km	2	• Robot • Brush • Batteries
Fully-autonomous robot*	-	90%	0	0	0	• Robot • Brush • Batteries • Docking station • Table connectors

9/21/2023 PV RADAR 17

020417-017

40th European Photovoltaic Solar Energy Conference and Exhibition

OPTIMAL CLENING STRATEGY AND LIFETIME BENEFIT

TECHNICAL ASSUMPTIONS

- Nominal DC Power: 100.0 MWp
- Nominal AC Power: 87.0 MWac
- DC/AC ratio: 1.15
- Monofacial
- Fixed tilt 2V
- Tilt: 10°

FINANCIAL ASSUMPTIONS

- Energy Sales Price: 40 USD/MWh
- Project Lifetime: 30 years
- Discount rate: 4%
- Labour cost: 10000 USD/FTE
- Fuel: 1.5 USD/Liter
- Water: 3 USD/m³

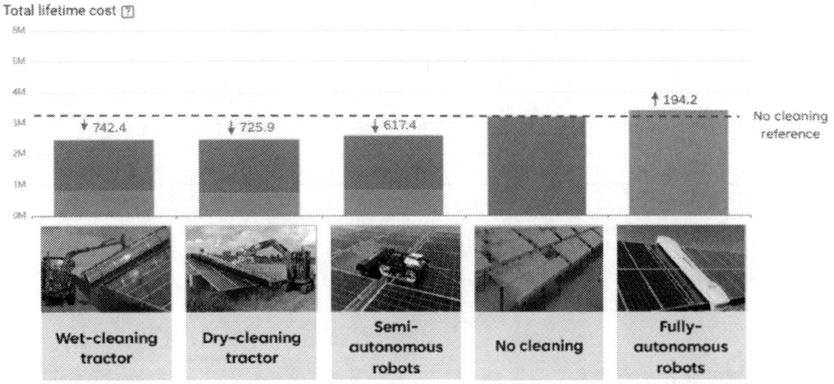

	Wet-cleaning tractor	Dry-cleaning tractor	Semi-autonomous robots	No cleaning	Fully-autonomous robots
ΔNPV (kEUR)	742.4	725.9	617.4		-194.2
Nb of systems	3	3	10		1 every 4 tables
Nb of cleaning/year	2	2	2		92 (every 4 days)

9/21/2023 PV RADAR 18

020417-018

OPTIMAL CLENING STRATEGY AND LIFETIME BENEFIT

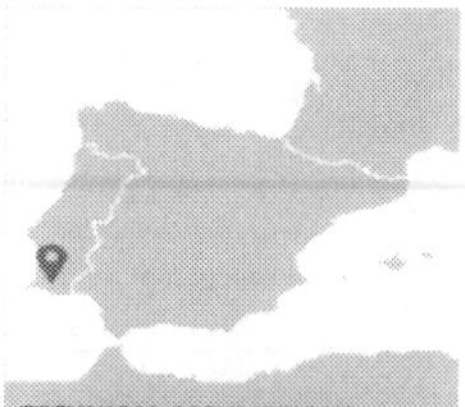

TECHNICAL ASSUMPTIONS

- Nominal DC Power: 100.0 MWp
- Nominal AC Power: 87.0 MWac
- DC/AC ratio: 1.15
- Monofacial
- Fixed tilt 2V
- Tilt: 30°

FINANCIAL ASSUMPTIONS

- Energy Sales Price: 40 USD/MWh
- Project Lifetime: 30 years
- Discount rate: 4%
- Labour cost: 10000 USD/FTE
- Fuel: 1.5 USD/Liter
- Water: 3 USD/m³

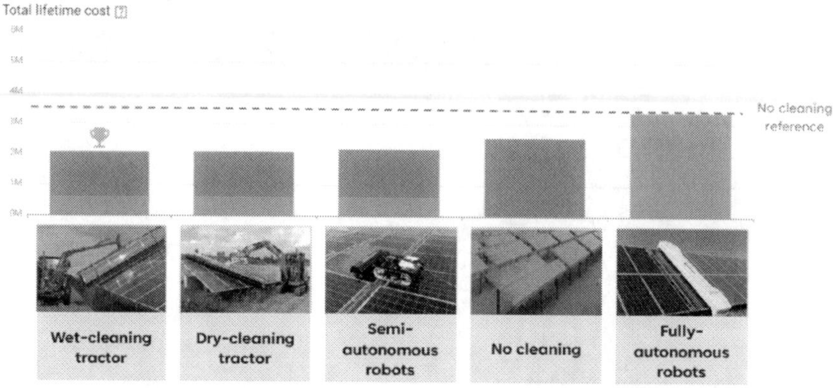

	Wet-cleaning tractor	Dry-cleaning tractor	Semi-autonomous robots	No cleaning	Fully-autonomous robots
ΔNPV (kEUR)	480.8 ↓	466.2 ↓	384.5 ↓		-833.7 ↓
Nb of systems	3	3	9		1 every 4 tables
Nb of cleaning/year	1	1	1		92 (every 4 days)

9/21/2023 PVRADAR 19

TILT MODIFICATION AS A MITIGATION STRATEGY AGAINST SOILING

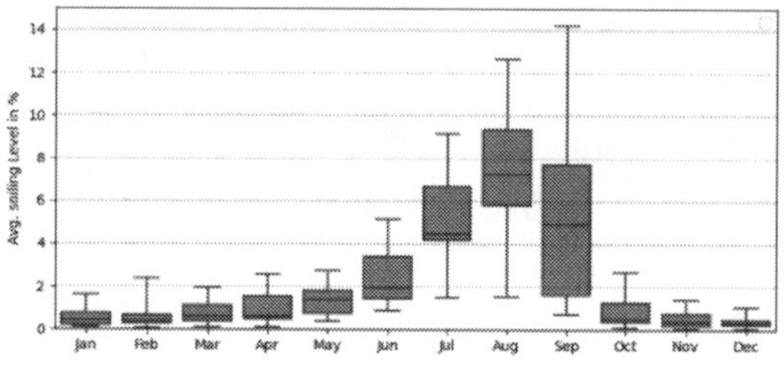

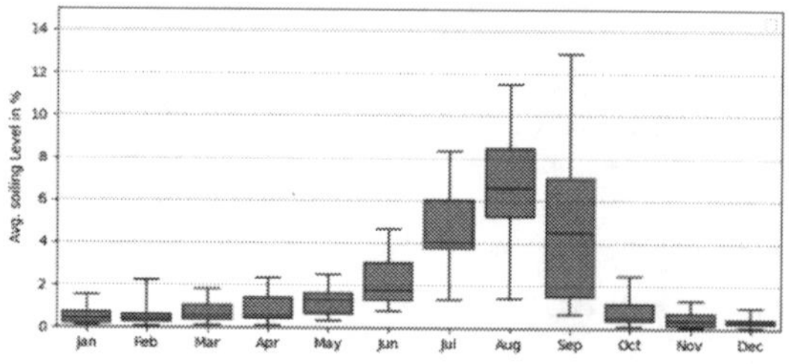

9/21/2023 PVRADAR 20

OPTIMAL CLENING STRATEGY AND LIFETIME BENEFIT

TECHNICAL ASSUMPTIONS
- Nominal DC Power: 100.0 MWp
- Nominal AC Power: 87.0 MWac
- DC/AC ratio: 1.15
- Bifaciality 80%, albedo 30%
- Fixed tilt 2V
- Tilt: 30°

FINANCIAL ASSUMPTIONS
- Energy Sales Price: 40 USD/MWh
- Project Lifetime: 30 years
- Discount rate: 4%
- Labour cost: 10000 USD/FTE
- Fuel: 1.5 USD/Liter
- Water: 3 USD/m³

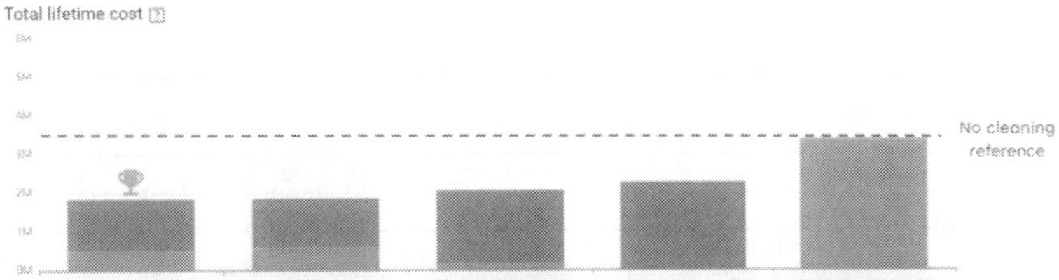

	Dry-cleaning tractor	Wet-cleaning tractor	Semi-autonomous robots	No cleaning	Fully-autonomous robots
ΔNPV (kEUR)	413.1	407.2	204.7		-1120.7
Nb of systems	2	2	9		1 every 4 tables
Nb of cleaning/year	1	1	1		52 (every 7 days)

OPTIMAL CLENING STRATEGY AND LIFETIME BENEFIT

TECHNICAL ASSUMPTIONS
- Nominal DC Power: 100.0 MWp
- Nominal AC Power: 87.0 MWac
- DC/AC ratio: 1.15
- Bifaciality 80%, albedo 30%
- Fixed tilt 2V
- Tilt: 30°

FINANCIAL ASSUMPTIONS
- Energy Sales Price: 40 USD/MWh
- Project Lifetime: 30 years
- Discount rate: 4%
- Labour cost: 10000 USD/FTE
- Fuel: 1.5 USD/Liter
- Water: 3 USD/m³

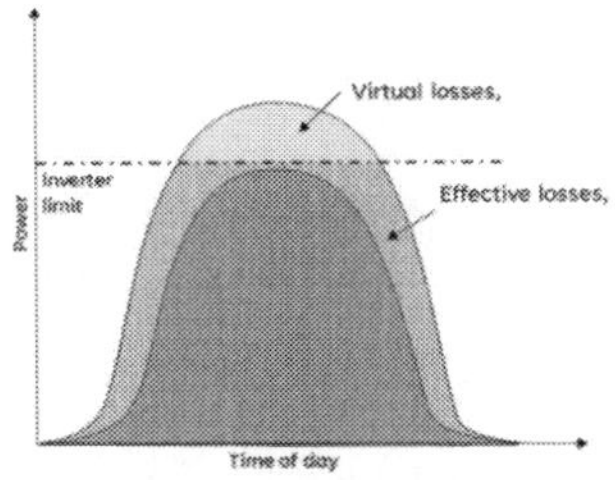

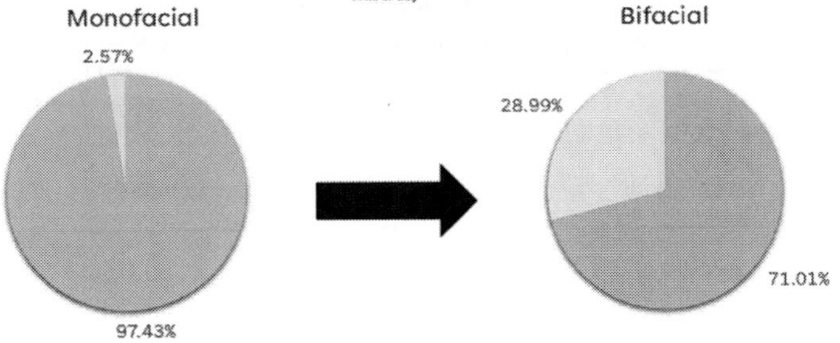

OPTIMAL CLENING STRATEGY AND LIFETIME BENEFIT

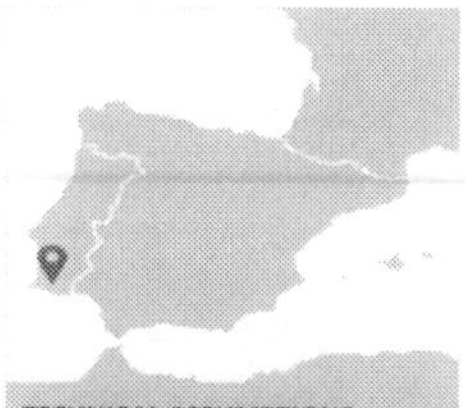

TECHNICAL ASSUMPTIONS

- Nominal DC Power: 100.0 MWp
- Nominal AC Power: 110.0 MWac
- DC/AC ratio: 0.9
- Bifaciality 80%, albedo 30%
- Fixed tilt 2V
- Tilt: 30°

FINANCIAL ASSUMPTIONS

- Energy Sales Price: 60 USD/MWh
- Project Lifetime: 30 years
- Discount rate: 4%
- Labour cost: 20000 USD/FTE
- Fuel: 1.5 USD/Liter
- Water: 3 USD/m³

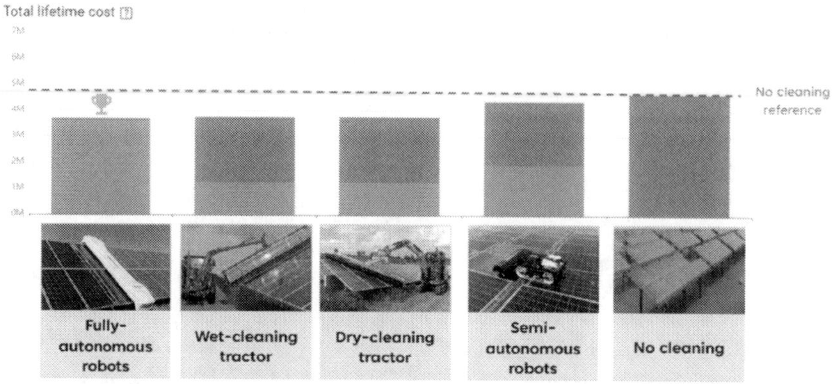

ΔNPV (kEUR)	**950.1** ↑	**900.8** ↑	**885.8** ↑	**281.2** ↑	
Nb of systems	1 every 4 tables	**3**	3	10	
Nb of cleaning/year	**121** (every 3 days)	3	2	2	

KEY TAKEAWAYS

Soiling assessment

- Soiling depends on the meteorological conditions but also in the characteristics of the plant. Design parameters such as tilt or DCAC ratio can have a significant impact on the losses.
- It is recommended to always perform an historical analysis using soiling estimation models. Measurements catch local events improving the accuracy but can't identify the year-by-year variation.

Cleaning optimization

- Different projects will have different optimal cleaning strategies. Dependency on many factors: the site conditions and techno-economic characteristics of the PV plant and the technologies .
- All the characteristics of the cleaning technologies can make a difference when it comes to define the best one. In the future, also effects such as the impact of cleaning over the modules (ARC degradation or mechanical stress shall be taken into account).
- Different strategies can generate various economic effects. The optimal strategy can be significantly better than non-optimal while some of them can even be detrimental for the economic evaluation of the project.

Thank you very much!

Thore Müller
Co-Founder and CEO
tm@virtuous-re.com

Konstantin Pogorelov
Co-Founder und CTO
kp@virtuous-re.com

Franco Clandestino
Co-Founder and Head of Product
fc@virtuous-re.com

Virtuous-Re GmbH
pvradar.com
pvradar in

PVRADAR by Virtuous-Re

Fraunhofer CSP

Mitigation of soiling by understanding and control of dust particle cementation

Guido Willers[1*], Muhammad Z. Khan[1], Volker Naumann[1], Ralph Gottschalg[1,2], Klemens Ilse[1]

[1]Fraunhofer-Center for Silicon-Photovoltaics CSP, 06120 Halle (Saale), Germany
[2]Anhalt University of Applied Sciences, Faculty EMW, Koethen (Anhalt), Germany

020418-001

40th European Photovoltaic Solar Energy Conference and Exhibition

Agenda

1. **Introduction**
2. **Experimental setup**
2. **Results**
   - Soiling studies
   - Microstructural Characterization
   - Smart heating
3. **Conclusions**

Seite 2 18.09.2023 Fraunhofer CSP

020418-002

Introduction: Soiling of PV modules

Soiling – a truly complex problem

- Dust deposits can lead to a total loss of energy yield.
- The growth of installed PV is particularly observed in areas with high dust loads, such as the USA, China, India, North Africa and the Middle East.

➡ Avoiding soiling is important!

Simulation in laboratory

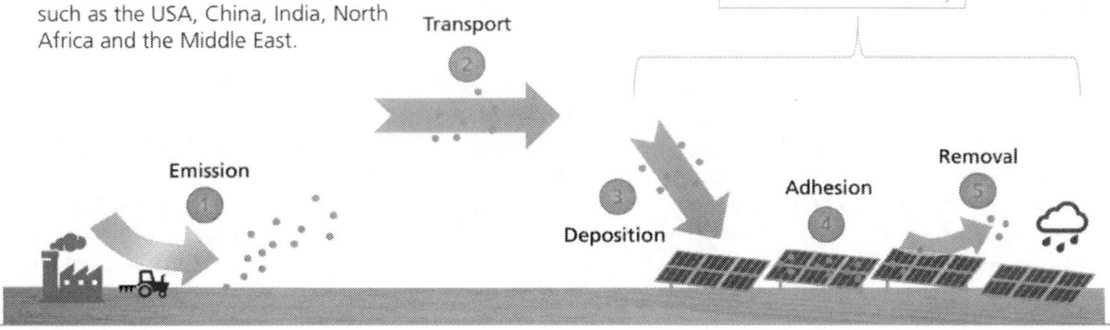

3

Fraunhofer CSP

020418-003

40th European Photovoltaic Solar Energy Conference and Exhibition

Experimental setup

Soiling procedure

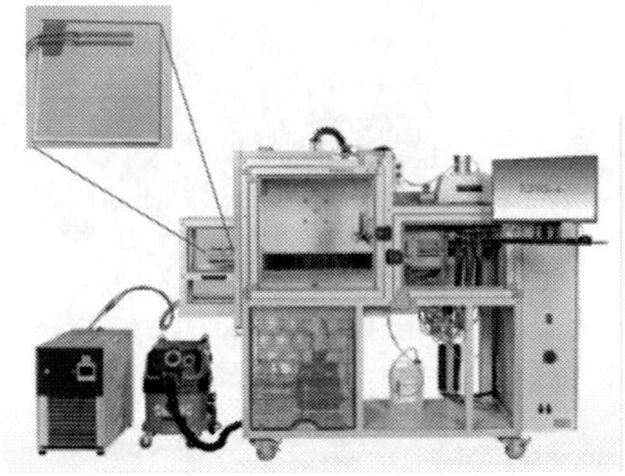

Solar glass sample with sensing electrodes and soiling test device

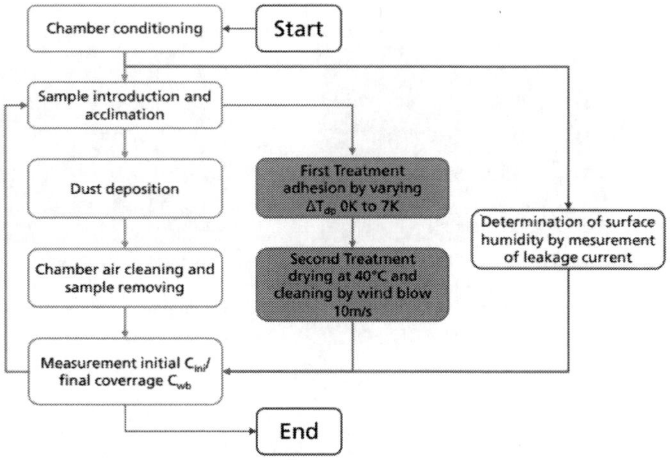

*Flow diagram of general test procedure in accordance with the VDI3956**

4

* https://www.vdi.de/richtlinien/details/vdi-3956-blatt-1-bewertung-der-verschmutzungseigenschaften-von-oberflaechen-pruefverfahren-fuer-das-staubbedingte-verschmutzungsverhalten-solarer-energiesysteme

TOPAS Fraunhofer CSP

020418-004

Results: Soiling study

Dust Coverage Ratio

The critical dew point difference

- cleaning efficiency is 82% for ME Test dust and 73% for Qatar, after a dry treatment cycle and a windspeed of 10m/s
- Dust adhesion increase sharply du to cementation at a critical dew point difference ($\Delta T_{dp/critical}$)
- $\Delta T_{dp/critical}$ is for ME Test Dust 1.9 K and for Qatar dust 2.4 K.
- In some cases, higher surface coverage was observed than was initially detected. Is the sample contaminated with extra dust?

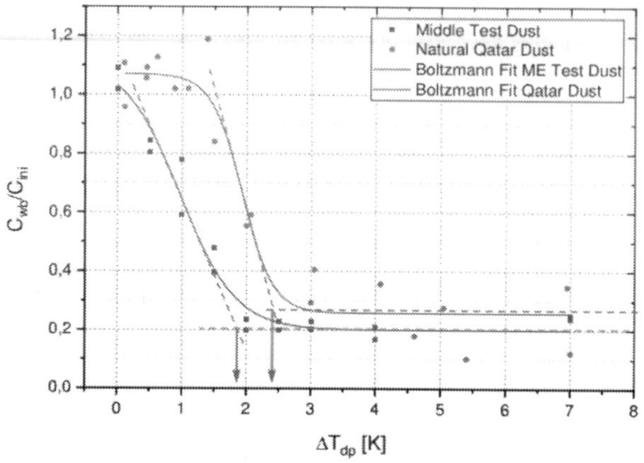

Fraunhofer CSP

020418-005

40th European Photovoltaic Solar Energy Conference and Exhibition

Results: Soiling Study

Microstructural Characterization

Cemented ME Test Dust (Lab)

- Sample created with soiling test device
- First treatment: 20 min at ΔT_{dp} =0K
- Appearance: 5µm to 30µm sized rods and plates

Cemented Qatar Dust (Lab)

- Sample created with soiling test device
- First treatment: 20 min at ΔT_{dp} =0K
- Appearance: sub-micron needles, agglomerated particles

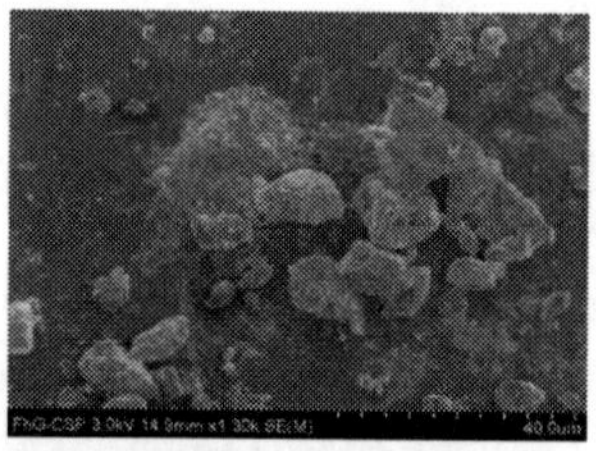

Cemented Qatar Dust (Outdoor)

- One day outdoor exposure in Qatar/Doha
- Sample temperature difference to dew point ΔTdp ≈ -1K
- Approximately 6 hours dew cycle
- Appearance: sub-micron needles, agglomerated particles

Fraunhofer CSP

020418-006

Results: Soiling study
Surface Humidity Sensor Signal

Correlation of surface humidity sensor signal to ΔTdp

- In logarithmic scale the sensor signal has a linear trend
- The graphs slope is slightly different between the dust types
- The level of the sensor signal at 0K differs by one decade.
- critical sensor current for ME Test Dust is 0,3e-9A and for Qatar 8,0e-9A

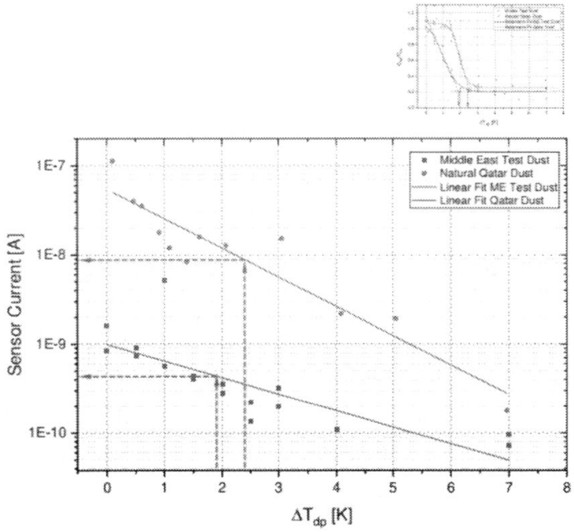

Fraunhofer CSP

Approach for mitigation of soiling
Laboratory set-up of a controlled heating

Direct surface humidity controlled smart heating PV-Modul

- Simple and inexpensive sensing system
- No additional heating elements needed
- Low annual energy consumption
  - by smaller process window
  - lower heating power
  - self-optimizing
- Additional positive aspects
  - less expensive cleaning
  - Less or zero water consumption

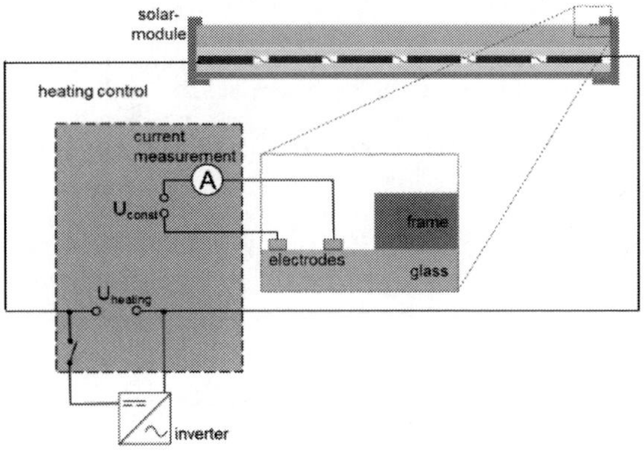

General design and electrical circuit layout

intern

Fraunhofer CSP

Approach for mitigation of soiling

First results: controlled smart heating

Smart Heating in a laboratory test

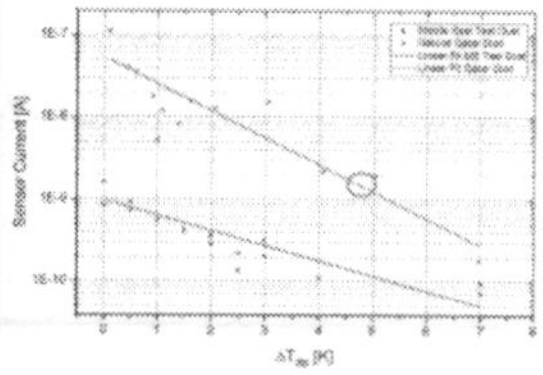

- Set up four tests with identical conditions, the only difference being that the heating control is switched on.
  - Used Dust: Natural Qatar Dust
  - The unheated surface temperature is one kelvin above dew point $\Delta T_{dp} = 1K$ (T_{sample}=14,9°C)
  - Heat sink and sample are semi-coupled thermally to enable heating
  - Heated sample temperature is lower than ambient temperature (T_{sample}=18,4°C< T_{amb}=22°C at 60%RH)
- Reduction of coverage by 60 % due to heating
- The energetic benefit is the focus of the next planned investigations.

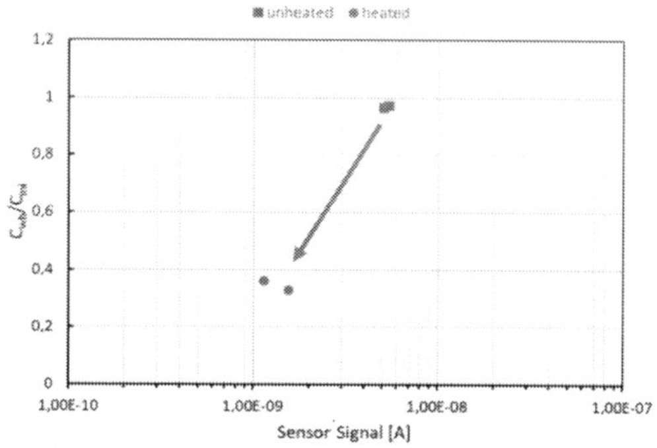

Fraunhofer CSP

Summary

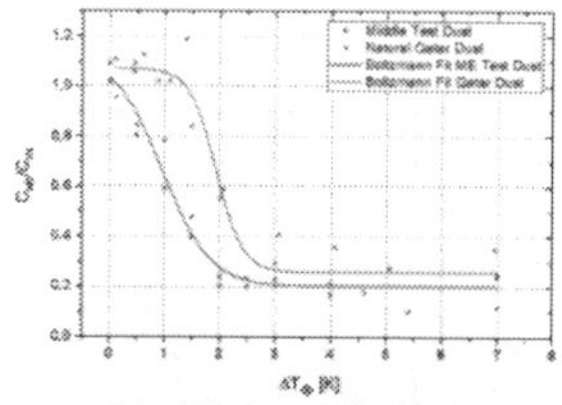

- The study showed that cementation occurred already 2.4K for Qatar Dust and at 1.9K for ME Test Dust above the dew point temperature.
- Cementation showed very different appearance and prevented natural particle resuspension by an air stream of 10 m/s.
- A novel surface humidity sensor was developed that directly indicates the critical surface humidity.
- A prototype of a heated PV module provided a 60% reduction in dust coverage due to dew avoidance.
- The efficiency of dust mitigation by smart heating depends strongly on the ratio of energy input to energy output. This will be investigated in future outdoor experiments.

This work was partially funded by the German Federal Ministry of Education and Research (BMBF) within the VIP+ validation project ETA-Tau (Contract no. 03VP08430).

Fraunhofer CSP

020418-011

40th European Photovoltaic Solar Energy Conference and Exhibition

020418-012

OPTIMAL SIMULATION-BASED DESIGN OF A PHOTOVOLTAIC SYSTEM FOR INTEGRATION INTO AN AGRIVOLTAIC FIELD

Julien Ancousture, Université de Lorraine, INRAE, Laboratoire Sols et Environnement, 54000 Nancy, France.
Univ. Grenoble Alpes, CEA, LITEN, DTS, LAM, F-38000 Grenoble.
julien.ancousture@univ-lorraine.fr Tél +33 (0)3 72 74 41 19
Pierre Leglize, Université de Lorraine, INRAE, Laboratoire Sols et Environnement, 54000 Nancy, France,
pierre.leglize@univ-lorraine.fr
Emile Benizri, Université de Lorraine, INRAE, Laboratoire Sols et Environnement, 54000 Nancy, France,
emile.benizri@univ-lorraine.fr
Ioannis Tsanakas , Univ. Grenoble Alpes, CEA, LITEN, DTS, LAM, F-38000 Grenoble, ya-brigitte.assoa@cea.fr
Ya Brigitte Assoa, Univ. Grenoble Alpes, CEA, LITEN, DTS, LAM, F-38000 Grenoble, ya-brigitte.assoa@cea.fr

ABSTRACT: Industrial brownfield sites are an opportunity to install photovoltaic systems. These sites are often polluted but can be rehabilitated through agromining. Coupling these two processes, Voltagroming would optimize the ecosystem services provided by these environments. To this end, simulations and design tests need to be carried out to establish the balance between the two processes. In this work, a parametric study carried out with the PVsyst software predicted the monthly and annual electrical energy production of two photovoltaic systems at two sites in France with different meteorological conditions, to define optimal configurations. Several parameters were considered such as the inclination of the photovoltaic modules, their technology, the season, and the geographical location of the sites. These preliminary simulations showed maximum energy production for a tilted opaque system compared to a semi-transparent bifacial system, with optimum energy production for a module slope of 30°. Regarding ongoing work, the first results on the development of the plant cover are presented. Photovoltaic modules seem to have a protective effect on the development and survival of the plants against thermal stress. In terms of prospects, the photovoltaic system configurations tested outdoors in combination with agromining will validate the Voltagroming system on industrial brownfield sites.
Keywords: Modelling, Ecovoltaics, Hyperaccumulating plant, *Noccaea caerulescens*

1 INTRODUCTION

In the context of global climate change, the European Union has committed itself to achieving carbon neutrality by 2050 [1]. To this end, there must be a reduction in the use of fossil fuels in favour of renewable energy sources. To achieve this energy transition, the development of photovoltaics (PV) is therefore promising. In France, photovoltaics represents a small part of the annual energy mix estimated at around 3%, with a production capacity of around 15 GW [2, 3]. Consequently, the French Ministry of Ecological Transition has set a total production capacity target of 100 GW for 2050 [4]. However, the optimal production of electrical energy by photovoltaic systems requires most of the time, a large area.

Various ways of integrating PV exist (*e.g.,* buildings, cars) but only ground-mounted installations would be able to fully meet this requirement. Nevertheless, there is a limited amount of land available per country. In recent years, agrivoltaics, which aims to define a coupling and compromise between agriculture and PV energy production using different PV systems, has progressed and represents an opportunity to increase PV systems electricity production capacity [5]. This application optimizes the space underneath the PV modules, which is usually not used, although, the implementation of PV fields artificializes the sites and could affects the yields of the local agricultural activities [6, 7]. Moreover, the French government has set an objective of "Zero Net Artificialisation" of Natural, Agricultural and Forest areas (NAF) under the Biodiversity Plan [8].

Thus, another opportunity to develop PV is to realize energy production on abandoned soils, such as post-industrial brownfield sites. In France, the surface area of these sites is over 150,000 ha [9]. These areas therefore represent a high potential of available and valuable surfaces. However, industrial brownfield lands are abandoned surfaces marked by ancient anthropic (*i.e.,* human) activities, particularly characterised by metal contaminations, which can provide potentially few ecosystem services to humans (*e.g.,* erosion control, carbon storage, additional habitats for greater biodiversity) [10, 11, 12]. For this reason, these sites are usually rehabilitated using physical-chemical techniques or by excavating the contaminated soils [13]. However, these methods are expensive and not very environmentally friendly, unlike phytoremediation techniques such as agromining, which combines the phytoextraction of metals by hyperaccumulating plants with a specific metallurgical process adapted to the bio-ore harvested [14, 15]. Agromining responds to the challenges of managing the pollution of industrial brownfields, but this process is slow and not financially profitable in the short term. In the case of agromining of an industrial brownfield site, an ecovoltaics approach, involving the development of currently unused areas under the PV modules of PV parks as part of an applied scientific ecology programme, can be implemented [16].

The global system coupling PV electricity production and agromining, called Voltagromining, would make the site profitable through the production of electrical energy during the soil phytoremediation process. In the long term, industrial brownfield sites will be rehabilitated to provide other ecosystem services. The global approach of Voltagroming is therefore to improve the ecosystem services provided by industrial brownfields and their long-term use for other activities. In this respect, it is necessary to determine the best compromise between the PV systems electrical energy production, the development of hyperaccumulating plant and the soil phytoremediation. For this purpose, the development of a hyperaccumulating plant, *Noccaea caerulescens*, under PV modules will be

10.4229/EUPVSEC2023/4DO.2.1

characterized on a pilot site. On this site, the impact of plant cover on PV energy production will be evaluated numerically and experimentally to define optimal configurations. Further to the experiments on the pilot site, the Voltagroming global system will be implemented on a large-scale demonstration site on an industrial brownfield to assess the feasibility of this global system, while validating numerical approaches developed in pilot site, evaluating the impacts on the ecosystem and the ecosystem services provided by the industrial brownfield site.

In this work, the preliminary simplified design of the PV systems to be tested on both sites is presented. The first results on the effect of PV modules on the development of hyperaccumulating plant cover at the pilot site are also presented.

2 METHODS

2.1. Experimental sites and pilot site experimental set-up

In the Voltagroming project, the two experimental sites are located in France, over 400 km apart. The pilot site is located at the Institut National de l'Energie Solaire (INES) in Le Bourget-du-Lac (73000, Savoie), and the demonstration site will be located on an industrial brownfield site in Homécourt (54000, Meurthe-et-Moselle).

Experimentation on the pilot site began on July 13, 2023, and opaque tilted PV modules from Panasonic and Solarworld, and semi-transparent bifacial PV modules from Jolywood were installed facing south. On the pilot site, 21 culture trays of 1 m² each with 49 *Noccaea caerulescens* plants were positioned. The culture trays PAN and SOL were positioned under opaque tilted PV modules, and LG trays under semi-transparent bifacial PV modules. Moreover, 3 culture trays without the influence of PV modules shade (*i.e.,* S tray) were positioned as shown in Figure 1. Among the culture trays, the PAN.O, SOL.O and LG.O trays are in the shade of the PV modules, while the PAN.S, SOL.S and LG.S trays are exposed to sun and shade of PV modules during the day.

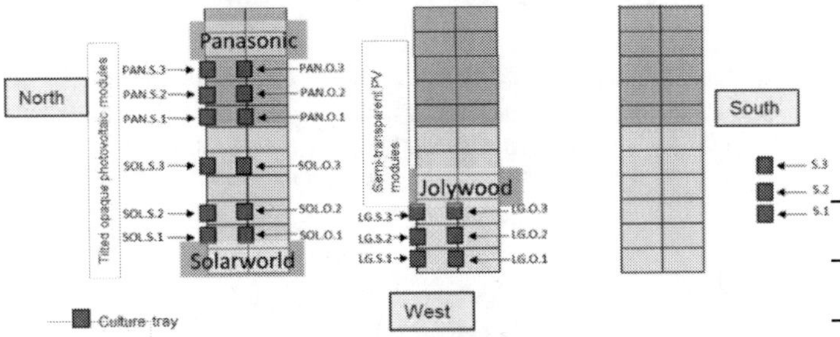

Figure 1: Experimental pilot site set-up at the INES site in Le Bourget-du-Lac (France). The PV modules from Panasonic, Solarworld and Jolywood face south and the PAN, SOL and LG cultivation trays are positioned under the PV modules. The S culture trays are positioned to avoid shading.

2.2. Simulation of electrical energy production

The preliminary simplified design of the electrical energy productions of 50 m² ground-mounted opaque and semi-transparent bifacial monocrystalline PV modules were simulated and compared on a monthly and an annual bases by using the PVsyst software (version 7.3) with weather data associated to the software. The slope of the PV modules, the season (*i.e.,* warm one from April to September and cold one from October to March), the PV modules technology and the site were considered with the data simulation provided by PVsyst software.

2.3 Measuring the area covered by *Noccaea caerulescens*

The area of *Noccaea caerulescens* plant coverage was measured using pictures taken in the visible wavelength spectrum. Using the PHENOPSIS macro (LEPSE-INRAE, Montpellier, France) of the ImageJ software, the pictures were processed semi-quantitatively to measure only the area of plant cover and number of plants per tray.

The data obtained were analysed on the RStudio environment, using R v.4.0.3 (RStudio Team, 2020). The normality of the data was checked with Shapiro's test, and the homoscedasticity of variances with Bartlett's test. Once these conditions had been met, an Analysis of Variances (ANOVA) was performed using Tukey's post hoc test.

3 RESULTS

3.1 Simulation of electrical energy production

From the simulations performed on PVsyst, a difference in energy production was obtained between tilted opaque PV modules and semi-transparent PV modules. Indeed, tilted modules are almost twice as efficient as semi-transparent modules regardless of the production site, the season, or the slope of the PV modules, since their transparency ratio is of 50 %.

Also, the slope has been shown to have an impact on the energy production throughout the year and seasonally. Thus, for the Homécourt site, in warm season, the best yield will be obtained with a slope of 30°, while in cold season, the optimum slope will be 45°. For the Bourget-du-Lac site, in warm season, the best yield will be obtained with a slope of 15°, while in cold season, the optimal slope will be 45°. For both sites, if the modules are not destined to have different slopes during the year, then the best compromise will be to maintain a slope of 30° all year round, regardless of the PV system studied (Table 1). This can be explained by the sun height and the total solar radiation variations along the year.

Table 1: Simulation of electrical energy production by the two PV systems on each site according to the seasons and the slope.

		Homécourt – PV system output (kWh/year)			**The Bourget-du-lac** – PV system output (kWh/year)		
	Tilt	Warm season	Cold season	Total year	Warm season	Cold season	Total year
Tilted PV modules	45°	6018	2831	8849	6843	4165	11008
	30°	6364	2501	8865	7287	3921	11208
	15°	6316	2371	8687	7341	3576	10917
Tilted semi-transparent PV modules	45°	3011	1416	4427	3421	2082	4549
	30°	3174	1356	4530	3643	1959	5602
	15°	3158	1186	4344	3651	1684	5335

Furthermore, the geographical location of the production sites also has an impact on the yield. On average, Le Bourget-du-Lac site provides 25.5 % more electricity with titled PV modules and 23.6 % more with semi-transparent PV modules than the Homécourt site. This variation is explained by a higher yearly incident total

10.4229/EUPVSEC2023/4DO.2.1
020427-002

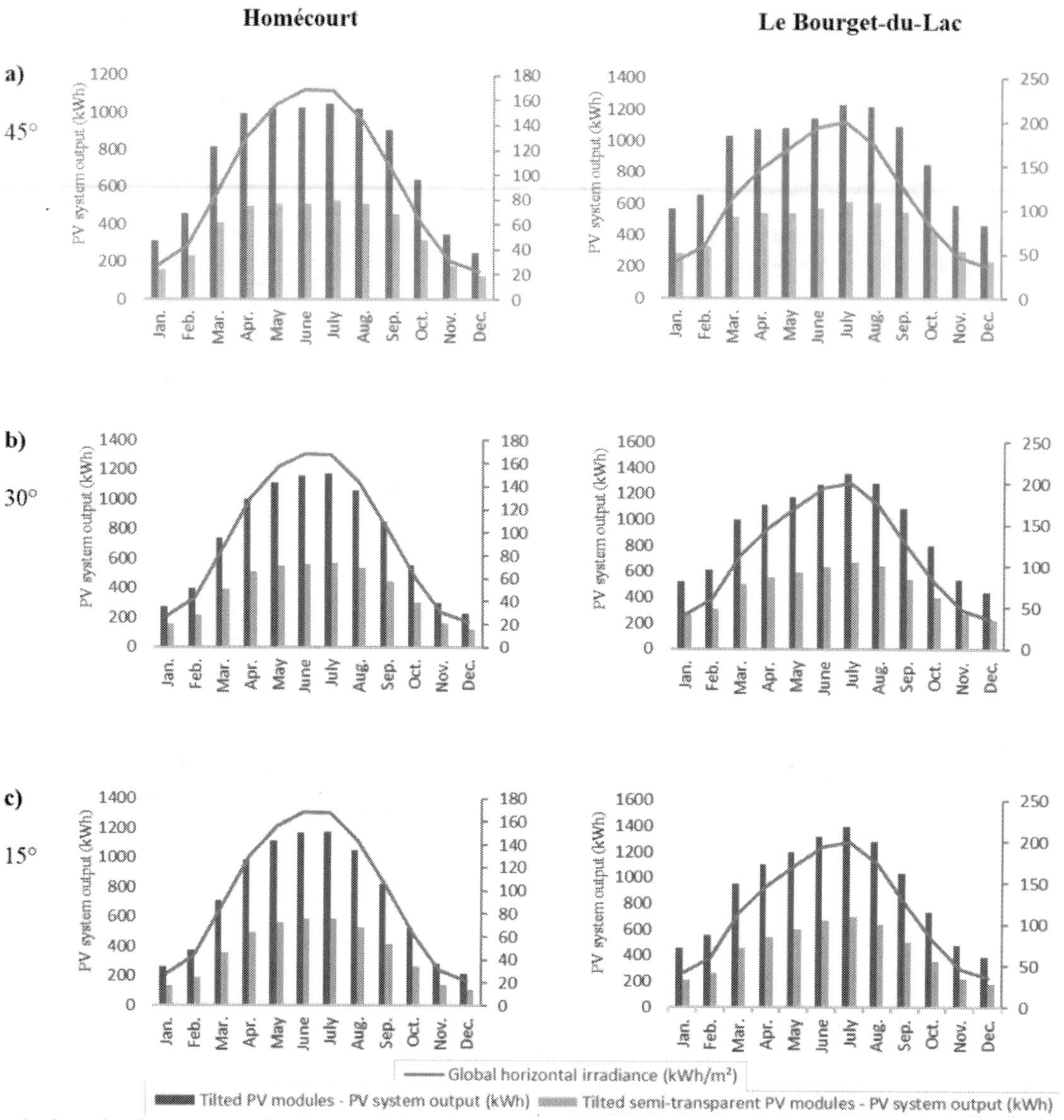

Figure 2: Simulation of the electrical energy produced by the two PV systems and total incident horizontal solar energy over a year according to the slope and the site.

horizontal solar energy at Le Bourget-du-Lac than at Homécourt (1390.41 kWh/m^2 and 1141.23 kWh/m^2 respectively) (Figure 2).

3.2 Measuring the area covered by *Noccaea caerulescens*

After 46 days of cultivation, the preliminary results on the development of plant cover have been analysed. On average, the area covered by *Noccaea caerulescens* was greater under the opaque PV modules than under the semi-transparent PV modules, followed by the control condition. For opaque PV modules, the coverage areas are on average greater in the shade of the modules PV (SOL.O and PAN.O) than in the sun and shade condition (SOL.S and PAN.S). In addition, there was a significant difference between the plant coverage area in the shade of the opaque Panasonic PV modules (PAN.O) and the control condition (S) (Figure 3).

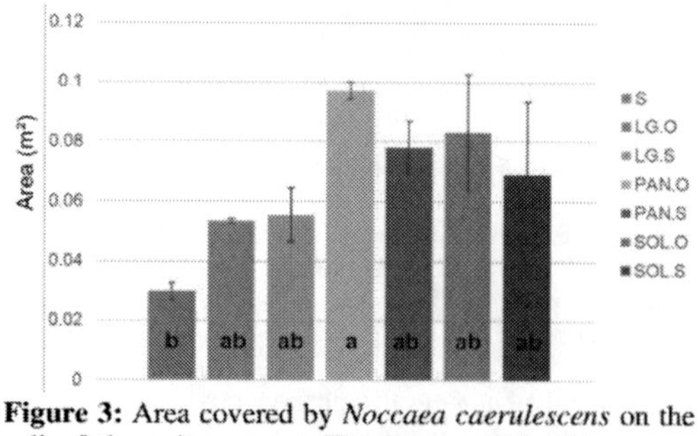

Figure 3: Area covered by *Noccaea caerulescens* on the soil of the culture trays. The letters on the histograms correspond to the significant difference between the averages for an α at 5% (Tukey test).

10.4229/EUPVSEC2023/4DO.2.1
020427-003

However, losses in the number of *Noccaea caerulescens* plants between the start of the experiment and the measurement of the area covered at the end of August 2023 were found. On average, the sunny (S) control trays were the most affected with 29.25% of *Noccaeae caerluscens* plants losses, followed by the trays under tilted opaque PV modules (SOL and PAN) and the semi-transparent bifacial PV modules (LG). In addition, there was a significant difference in the number of dead *Noccaeae caerulescens* plants between S trays and LG.S, PAN.S trays. Moreover, culture trays in the shade of the modules (LG.O, PAN.O and SOL.O) appear on average to have higher plant mortality than their associated sun and shade condition (LG.S, PAN.S, SOL.S). (Figure 4).

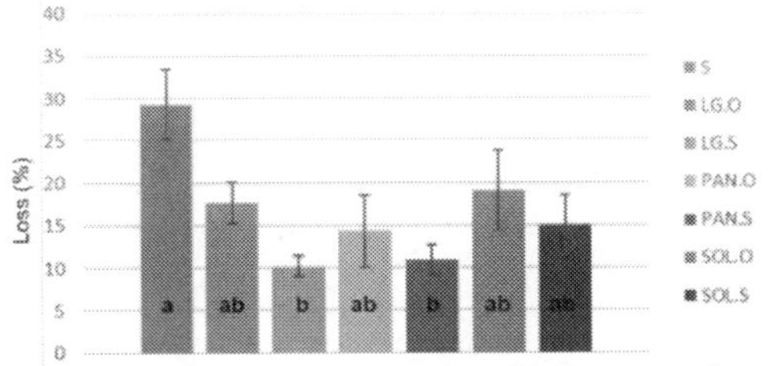

Figure 4: Loss of *Noccaea caerulescens* plants per culture trays. The letters on the histograms correspond to the significant difference between the averages for an α at 5% (Tukey test).

The area covered by each *Noccaea caerulescens* plant must be considered due to the mortality (Figure 5). Indeed, it is possible for one tray to have more *Noccaea caerulescens* plants than the other tray given the same area of cover. In this case, these results are similar to the results of area covered by *Noccaea caerulescens* on the soil of the culture trays (Figure 3). In fact, the area per *Noccaea caerulescens* plant is greater under the opaque PV modules compared to the semi-transparent PV modules, followed by the control condition. In addition, there was also a significant difference between the plant coverage area in the control condition (S) and the shade of the opaque Panasonic PV modules (PAN.O).

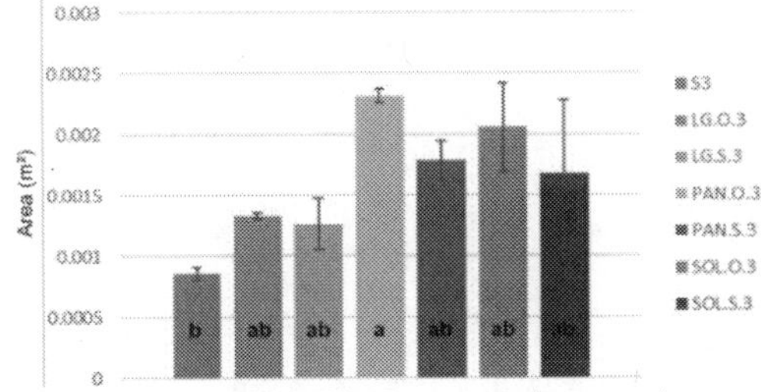

Figure 5: Area of a *Noccaea caerulescens* plant per culture trays. The letters on the histograms correspond to the significant difference between the averages for an α at 5% (Tukey test).

4 CONCLUSION AND PERSPECTIVES

In fine, the simulations regarding energy production have shown that the PV system, the slope of the PV modules, the seasons and the sites all have an impact on PV systems electrical energy production.

As outlooks on pilot site, the measurement of physical parameters (*e.g.* relative humidity in the plant cover, temperature of the plant cover, soil surface and PV modules, radiation reflected by the plant cover) and weather parameters (*e.g.* wind speed, solar radiation, ambient temperature) will be used to develop a dynamical thermal and electrical model of the PV systems to define the effects of plant cover on PV modules energy production. The results of this modelling will be compared with experimental data on PV energy production at pilot site scale in Le Bourget-du-Lac and at larger scale on industrial brownfield site in Homécourt to determine the best configuration for the Voltagroming solution.

Concerning the development of plant cover area by *Noccaea caerulescens*, the preliminary measurements suggest that PV modules have a protective effect against thermal stress on the survival and development of *Noccaea caerulescens*. These first results must be compared with the results of other plant parameters (*e.g.*, plant photosynthetic pigment content, fresh and dry biomass, concentrations of minor and major elements, carbon, and nitrogen content) and also with physicochemical and biological soil parameters. These data will inform our study of the phytoextraction capacity of *Noccaea caerulescens* depending on the conditions under which the plants are cultivated with PV systems. The data will enable evaluation of the effects of the PV modules on plant development at the pilot site before expanding to a larger scale.

In the end, all these approaches combined will lead to find the best compromise in the Voltagroming project between agromining and PV energy production to obtain a set of ecosystem services on industrial brownfields (*e.g.*, supply services with bio-based metals and electrical energy production).

5 REFERENCES

[1] Kuhn, T.E., Erban, C., Heinrich, M., Eisenlohr, J., Ensslen, F., Neuhaus, D.H., 2021. Review of technological design options for building integrated photovoltaics (BIPV). Energy Build. 231, 110381. https://doi.org/10.1016/j.enbuild.2020.110381

[2] Gestionnaire du reseau de transport. (2021). Production totale. https://bilan-electrique-2021.rte-france.com/production_totale/

[3] International Renewable Energy Agency. (2022). Renewable Capacity Statistics 2022. https://www.irena.org/publications/2022/Apr/Renewable-Capacity-Statistics-2022

[4] Vie publique. (2022). Déclaration de M. Emmanuel Macron, président de la République, sur la politique de l'énergie, à Belfort le 10 février 2022. https://www.vie-publique.fr/discours/283773-emmanuel-macron-10022022-politique-de-lenergie

[5] Mamun, M.A.A., Dargusch, P., Wadley, D., Zulkarnain, N.A., Aziz, A.A., 2022. A review of research on agrivoltaic systems. Renew. Sustain. Energy Rev. 161, 112351. https://doi.org/10.1016/j.rser.2022.112351

[6] Weselek, A., Bauerle, A., Hartung, J., Zikeli, S., Lewandowski, I., Högy, P., 2021. Agrivoltaic system impacts on microclimate and yield of different crops within an organic crop rotation in a temperate climate. Agron. Sustain. Dev. 41. https://doi.org/10.1007/s13593-021-00714-y

[7] Fu, J., Li, C.-H., Zhao, J.-R., Ma, L., Liu, T.-X., 2009. Shade-tolerance indices of maize: Selection and evaluation. Chin. J. Appl. Ecol. 20, 2705–2709.
[8] Portail de l'artificialisation des sols. (2023). Parution des données de consommation d'espaces 2009-2022. https://artificialisation.developpement-durable.gouv.fr/parution-des-donnees-consommation-despaces-2009-2022
[9] Cerema. (2023). Cartofriches : plus de 8300 sites en friches répertoriés et caractérisés, l'API données foncières du Cerema en accès libre. https://www.cerema.fr/fr/actualites/cartofriches-plus-8300-sites-friches-repertories
[10] Morel, J.L., Chenu, C., Lorenz, K., 2014. Ecosystem services provided by soils of urban, industrial, traffic, mining, and military areas (SUITMAs). J. Soils Sediments 15, 1659–1666. https://doi.org/10.1007/s11368-014-0926-0
[11] Salemaa, M., Vanha-Majamaa, I., Derome, J., 2001. Understorey vegetation along a heavy-metal pollution gradient in SW Finland. Environ. Pollut. 112, 339–350. https://doi.org/10.1016/S0269-7491(00)00150-0
[12] Salisbury, A., Gallagher, F.J., Parag, H.A., Meneses-Florián, L., Holzapfel, C., 2021. Plant diversity increases in an urban wildland after four decades of unaided vegetation development in a post-industrial site. Urban Ecosyst. 24, 95–111. https://doi.org/10.1007/s11252-020-01018-x
[13] Liu, L., Li, W., Song, W., Guo, M., 2018. Remediation techniques for heavy metal-contaminated soils: Principles and applicability. Sci. Total Environ. 633, 206–219. https://doi.org/10.1016/j.scitotenv.2018.03.161
[14] Ali, H., Khan, E., Sajad, M.A., 2013. Phytoremediation of heavy metals—Concepts and applications. Chemosphere 91, 869–881. https://doi.org/10.1016/j.chemosphere.2013.01.075
[15] Morel, J.-L., 2013. Using plants to "micro-mine" metals. In: http://www.inra.fr/en/Scientists-Students/Biomass/All-the-news/Using-plants-to-micro-mine-metals
[16] Grison, C., 2021. Photovoltaïsme, agriculture et écologie: de l'agrivoltaïsme à l'écovoltaïsme, Collection écologie-environnement. ISTE editions, London.

6 ACKNOWLEDGEMENTS

This work was partially funded by the call for projects 2022 "Compétences recherche de la Région Grand-Est". This work realized within INES.2S was partially supported by the French state under the Investments for the Future programme (ANR-10-IEED-0014-01).

COMPARISON OF DIFFERENT AGRIPV LAYOUTS IN TERMS OF PHOTOVOLTAIC ENERGY YIELD OUTPUT

Hugo Sánchez Ortiz[1], Sebastian Dittmann[1,2], Carlos Meza[1,2], Ralph Gottschalg[1,2]
[1]Hochschule Anhalt University of Applied Sciences, Bernburger Str. 55, 06366, Köthen, Germany
Email: hugo.sanchez@hs-anhalt.de
[2]Fraunhofer Center for Crystalline Silicon Photovoltaics CSP

ABSTRACT: Agrivoltaics (AV) presents an economical solution to address both food security and the decarbonization of our energy systems. This synergy between agricultural crops and photovoltaic (PV) systems holds the potential to increase land productivity significantly, drawing considerable interest from academia and the industrial sector. However, despite the promise of this dual-use approach, numerous questions remain unanswered regarding the optimization of light harvesting systems. The multitude of possible configurations necessitates thorough analysis to identify the most productive system layouts.

This study presents a case study involving computational simulations that utilize measured meteorological data to optimize energy yield for various system topologies based on the DIN SPC 91434 regulation. These topologies include elevated fixed systems (CAT.1 and CAT.2), vertical bifacial systems with east-west orientation (14m and 24m spacing), and single-axis tracking systems, all assessed in the context of a field located in Bernburg, Germany. The study establishes a comparative framework for evaluating different system integration layouts and assesses the remaining ground irradiance for each configuration.

Our design optimization approach prioritizes energy yield for these systems while considering its impact on the light profile for crop cultivation. This analysis is pivotal for comprehending the real-world productivity of innovative AV schemes and providing guidance for future AV plant installations. We specifically focus on comparing energy yield estimates, employing high-quality measured data, and considering field constraints related to land utilization, such as the operability of machinery for crop farming. The ensuing discussion aims to offer insights into current and future regulations and inform stakeholders involved in the decision-making process for AV system design. Ultimately, our goal is to provide precise tools that promote the adoption of agrivoltaics across various stakeholders, contributing valuable solutions to the challenges of energy transition and food security.

KEYWORDS: agrivoltaics, bifacial, design optimization, energy yield.

1 INTRODUCTION

Agrivoltaics (AV) is a solution to address food security challenges and the decarbonization of energy generation. It has yielded multiple benefits since the term first appeared in the early 1980s. [1]. The recent interest in research and industrial groups [2, 3] required studying the light-sharing tradeoff between the crops and the PV modules, as well as the land's productivity factor. These analyses will enhance the accuracy of planning tools for various layouts.

While significant advances have been made in the PV sector in terms of improving overall power density, agrivoltaics necessitates further research on the comparison of various agrivoltaic layouts [4] . Recent studies [5–8] have proposed a design optimization analysis. However, most studies rely on satellite data as a reference and assume a constant value for parameters such as albedo. This approach introduces inaccuracies in energy yield predictions due to the variability and impact of albedo. [9, 10]. Works such as [6] and [8] concur on the necessity of conducting a variety of analyses to enhance planning tools for agrivoltaics.

At present, most authors are focused on developing demonstrators for various layouts, while questions regarding the optimization design for the plants remain open. [11]. The optimization design process for the AV design corresponds to a multivariable function in terms which factor wants to be optimized. Authors in [12] offers an interesting review as a starting point for optimization analysis, considering factors such as PV panel density, height spacing, checkboard arrangement, and the available ground irradiance for agricultural processes. Additionally, some optimization efforts have been directed toward bifacial systems. [13] and vertical mounted AV [14]. All of the discussions take into consideration the Land Equivalent Ratio (LER) as a measure of optimization. However, it's worth noting that some practical constraints, such as the impact of using machinery on the lands, are still not thoroughly studied.

On the other hand, the industry is increasingly demanding the implementation of these systems. In the case of Germany, the country rapidly transitioned from lacking a defined legal framework for AV system [15] to define an support scheme for these systems DIN SPEC 91434 [16]. Under this regulation, systems deployed on land up to 2.5 hectares, ideally with a nearby grid connection point, receive preferential treatment. This means that a building application can be approved without the need for prior preparation of a development plan. To meet these regulations, the system must satisfy the following criteria:

- At least 85% of the area must be allocated for agricultural purposes.
- The agricultural yield must be proven to reach at least 66% of the normal yield.
- For elevated systems, the lower edge of the modules must be positioned at a minimum height of 2.1 meters above the ground.

In this paper, we present a computational analysis of various layouts to compare them in terms of PV energy yield output and the trade-off between light-sharing and the remaining irradiance on the ground. The computational analysis covers the requirements outlined in DIN SPEC 91434, as well as special considerations for machinery use on the land.

10.4229/EUPVSEC2023/4DO.2.2
020428-001

2 APPROACH

This work aims to introduce a computational investigation setup for a location in Bernburg, Germany, situated near the site where our demonstrator will be constructed. [17]. Additionally, this location serves as our PV Verification Plant, where meteorological data have been continuously measured. It's important to note that Germany is currently deliberating a new regulation regarding the definition of Agrivoltaic Systems. In this context, our present work will analyze the following scenarios, taking into account an available area of 2.5 hectares, which corresponds to the maximum allowable area for an AV system according to the discussed specification DIN SPEC 91434.

1. Utilizing the entire area solely for the installation of an optimized PV plant (PV Opt) with a clearance height of 0.5 meters. In this case, the optimization parameter corresponds to a distance of 4 meters between PV tables to prevent self-shadowing effects between rows.
2. Tilt Elevated Angle AV CAT.1 (TEA Cat.1) System with a clearance height of 2.1 meters, utilizing 85% of the field for agricultural activities. To comply with DIN SPEC 91434, a row-to-row distance of 8 meters is taken into account.
3. Tilt Elevated Angle AV CAT.2 (TEA CAT 2) system with the same clearance height as system 1 but adhering to the requirements specified for system 2. Due to machinery limitations, a distance of 14 meters between the tables is considered in this scenario.
4. Vertical Bifacial East West oriented systems with a distance between rows of 14m that allow also the use of machinery (VBPV 14m)
5. Vertical Bifacial East-West oriented systems with a distance between rows of 24m (VBPV 24m). In this case, it considers more space between rows for the utilization of bigger machinery.
6. Single Axis Tracker: In this case a seasonal one single axis is implemented in case of the TEA CAT 1.
7. Not PV System and let the field just for agriculture.

Fig 1. shows an schematic view for the systems analyzed.

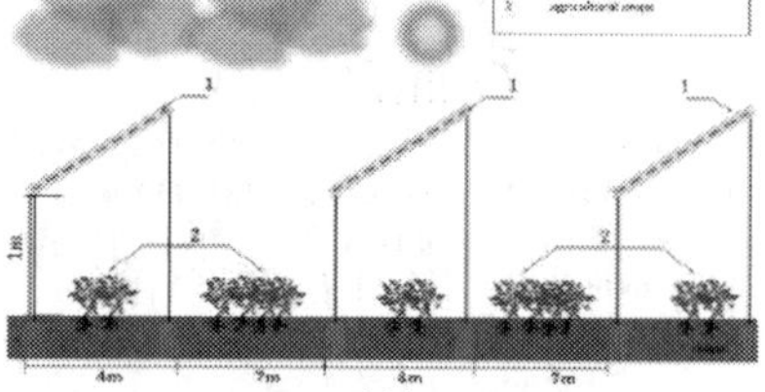

(A)

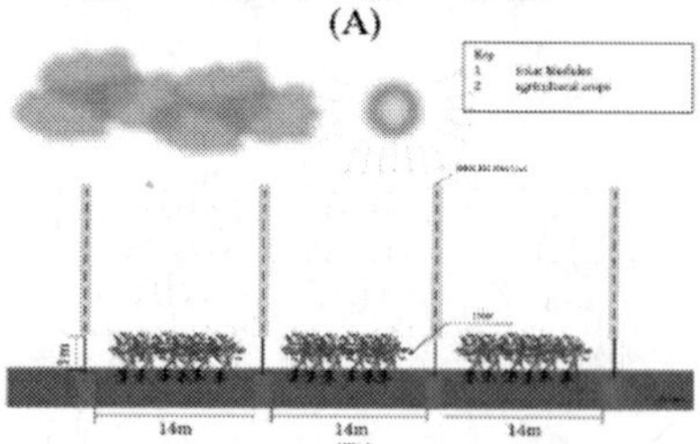

(B)

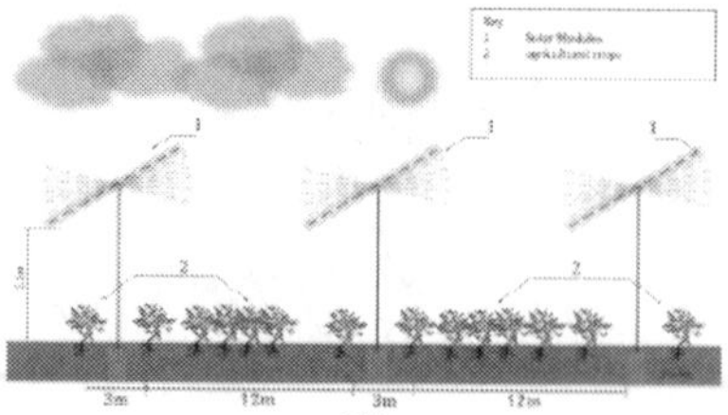

(C)

Figure 1: Different configurations for the array in Agrivoltaics. (A) Elevated tilt angle, (B) Vertical East West, (C) Solar trackers.

The simulations were conducted using meteorological data collected from May 2021 to April 2022 at the PV Verification Plant in Bernburg, Germany. This dataset includes measurements of temperature, relative humidity, wind speed, Global Horizontal Irradiance, Rear Global Horizontal Irradiance, and albedo. The details of the measurement setup can be found in reference [9]. Following a thorough verification analysis, the experimental data has been transformed into a Typical Meteorological Year (TMY) dataset. Figure 2 illustrates the monthly distribution of each measured variable.

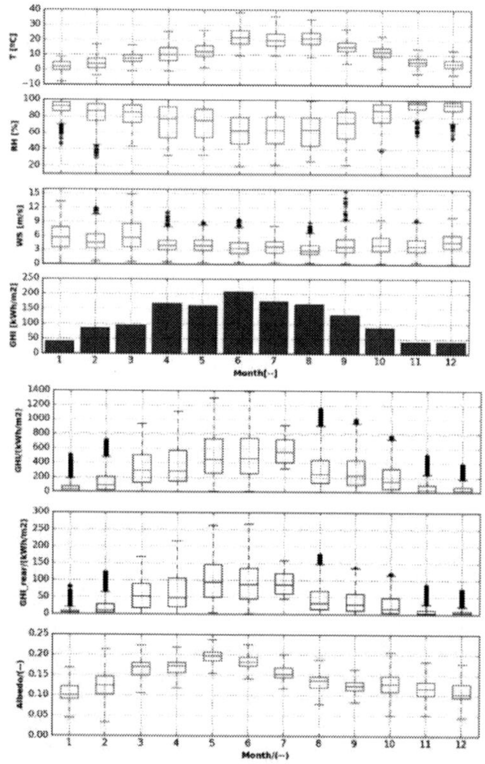

Figure 2: Input Meteodata measured at the PV verification Plant in Bernburg Germany.

The data is inputted into our modeling tool to calculate energy yield production. For energy yield prediction, we employ a Python library with the main frameworks being PVLib and Factors. The input data includes Global Horizontal Irradiance, Average Temperature, Wind Speed, and albedo. Several limitations and assumptions are taken into account for the simulation:

10.4229/EUPVSEC2023/4DO.2.2
020428-002

- Tilt angles of 30° are assumed for systems 1, 2, and 3.
- All systems are integrated into a 175 kWp string inverter with a CEC efficiency of 97.5%.
- A seasonal variation in albedo is considered based on Figure 2.
- Spectral mismatch at different heights is not considered.
- Clipping and thermal losses are not accounted for in the inverter.
- The bifaciality factor, representing the ratio of nominal efficiency at the rear side to the front side, is set at 90%.
- Losses in the rear side contact are not factored in.

This computational investigation will primarily focus on energy yield predictions and a comparison in terms of available irradiance on the ground, taking into consideration the footprint of the systems.

3. RESULTS

The summary of results is presented in Table I, including the balance between the energy production as well as the irradiance on the ground

Table I: Yearly Energy Yield for the simulations

System	Installed capacity (kWp)	Yearly Energy Production (kWh/m2)	Specific Energy Yield (kWh/kWp)	GHI in the ground (kWh/m2)
PV Opt	3019	15.45	1280	400.79
TEA Cat.1	1499	7.51	1230	750.7
TEA Cat.2	816	4.20	1279	896.8
VBEW 14m	816	3.72	1140	996.2
VBEW 24m	459	2.16	1151	1036.9
SA Tracker	1499	8.10	1323	790.7
Not PV	0	0	0	1137

A visualization of this information is presented in Figure 3, where the specific energy yield and the total installed capacity are compared. As expected, the system without PV has no value in these results. Additionally, the system with a seasonal tracker generates more electricity compared to a fixed elevated installation (CAT 1). In the case of PV Opt and TEA CAT 2, their performance is quite similar, with the primary difference being the distance between the PV tables. Finally, the vertical systems exhibit the lowest values in terms of energy yield, but as we will discuss below, they offer interesting opportunities for the integration of AV systems.

As the productivity of AV systems is not solely determined by energy yield, Figure 4 presents a sensitivity comparison between energy generation and the remaining global irradiance estimated at the ground, which can be translated as the available resource for crop cultivation. Notably, a higher module density (PV Opt) results in lower remaining irradiance on the field, as expected. However, the SA Tracker system stands as an exception due to its reduced shading impact on the ground compared to a fixed installation.

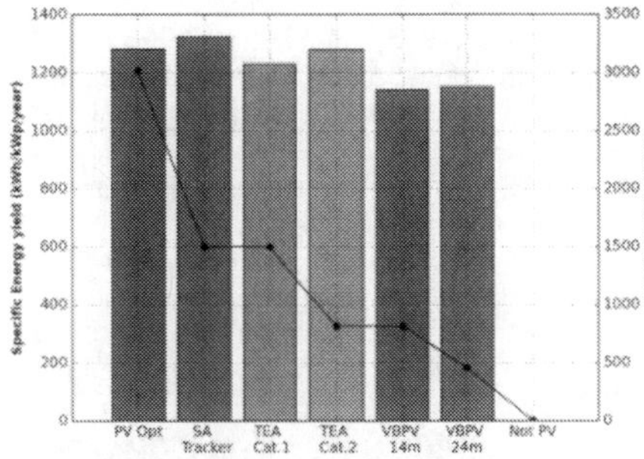

Figure 3. Yearly energy yield and installed capacity for all the systems considered.

A particularly case emerges in the transition from TEA CAT 2 to VBPV 14m. These two systems yield nearly identical energy production, but there is a substantial difference in the remaining ground irradiance. This finding hints at the potential of vertical systems for AV integration schemes, allowing for up to 52% of the irradiance to be available on the ground compared to less than 10% of the energy yield. This disparity presents compelling opportunities for integration.

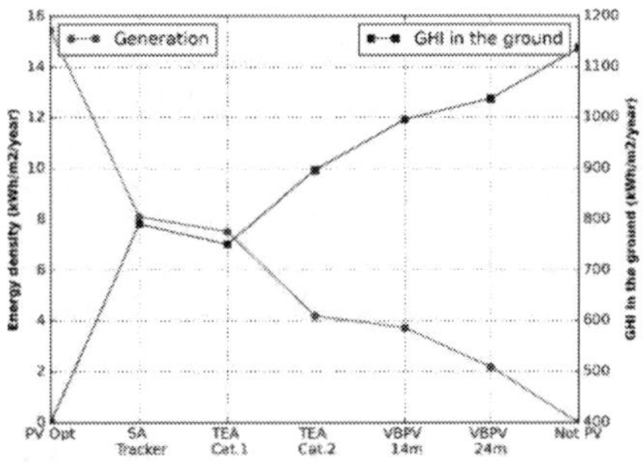

Figure 4. AC energy yield density and yearly ground irradiation for all the PV topology.

Achieving a balance between energy yield and crop production is a crucial factor in the integration of AV systems. In this context, it's apparent that the performance of the PV component is not significantly affected by the availability of light on the ground for agricultural activities. As demonstrated, the performance of systems ranging from PV Opt to TEA Cat.1 and TEA Cat.2 remains quite similar, even with an increase in available irradiance on the ground. However, it's worth noting that these systems may pose challenges for the use of machinery due to their design.

In this regard, a VBPV system emerges as an interesting solution that can accommodate the dynamics of machinery without causing a considerable loss in energy production. This suggests that vertical systems like VBPV hold promise for achieving a harmonious integration of agricultural and energy production activities. Such as the reduction of the evapotranspiration due to the windbreak effect [18].

4 OUTLOOK

In this paper, we have investigated the design of various layouts for Agrivoltaic (AV) systems in

10.4229/EUPVSEC2023/4DO.2.2
020428-003

accordance with the DIN SPEC 91434 Standard. This computational analysis was conducted on a 2.5-hectare land in Bernburg, Germany, and it provides insights into seven potential AV system configurations based on their energy yield and the remaining irradiance on the ground. Our aim is to contribute to the ongoing discussion regarding the optimization of AV system layout design, with a specific focus on achieving a balance between energy yield and crop production.

We explored two extreme scenarios: a field optimized for PV generation and a field without PV. Additionally, we defined five intermediate scenarios following the DIN SPEC 91434 standard. It's evident from our findings that modifications to the system topology do not have a significant impact on energy yield performance (with a 10% difference between the PV-optimized case and VBPV). However, these modifications do play a substantial role in terms of the remaining irradiance on the ground, with potential gains of up to 52% for the same configuration.

Furthermore, we are actively developing demonstrators to experiment with the trade-off between crop cultivation and electricity generation, aiming to validate the models proposed in this work. AV systems offer innovative solutions to address current global challenges, and it is imperative to conduct accurate and comprehensive studies to understand both the constraints and benefits associated with their implementation.

5 ACKNOWLEDGMENTS

This work is supported by the German Federal Ministry of Education and Research under the project "Biodiversity in solar parks - Innovative concepts and construction of demonstrators for a better compatibility of photovoltaic systems, nature conservation and agriculture (BIODIV-SOLAR)". Funding code: 13FH133KX0

6 REFERENCES

[1] A. Goetzberger and A. ZASTROW, "On the Coexistence of Solar-Energy Conversion and Plant Cultivation," *International Journal of Solar Energy*, vol. 1, no. 1, pp. 55–69, 1982, doi: 10.1080/01425918208909875.

[2] H. Dinesh and J. M. Pearce, "The potential of agrivoltaic systems," *Renewable and Sustainable Energy Reviews*, vol. 54, pp. 299–308, 2016, doi: 10.1016/j.rser.2015.10.024.

[3] C. Toledo and A. Scognamiglio, "Agrivoltaic Systems Design and Assessment: A Critical Review, and a Descriptive Model towards a Sustainable Landscape Vision (Three-Dimensional Agrivoltaic Patterns)," *Sustainability*, vol. 13, no. 12, p. 6871, 2021, doi: 10.3390/su13126871.

[4] M. Reasoner and A. Ghosh, "Agrivoltaic Engineering and Layout Optimization Approaches in the Transition to Renewable Energy Technologies: A Review," *Challenges*, vol. 13, no. 2, p. 43, 2022, doi: 10.3390/challe13020043.

[5] M. H. Riaz, H. Imran, and N. Z. Butt, "Optimization of PV Array Density for Fixed Tilt Bifacial Solar Panels for Efficient Agrivoltaic Systems," in *2020 47th IEEE Photovoltaic Specialists Conference (PVSC)*, 2020.

[6] M. H. Riaz, H. Imran, R. Younas, M. A. Alam, and N. Z. Butt, "Module Technology for Agrivoltaics: Vertical Bifacial Versus Tilted Monofacial Farms," *IEEE J. Photovoltaics*, vol. 11, no. 2, pp. 469–477, 2021, doi: 10.1109/jphotov.2020.3048225.

[7] M. H. Riaz, H. Imran, H. Alam, M. A. Alam, and N. Z. Butt, "Crop-Specific Optimization of Bifacial PV Arrays for Agrivoltaic Food-Energy Production: The Light-Productivity-Factor Approach," *IEEE J. Photovoltaics*, vol. 12, no. 2, pp. 572–580, 2022, doi: 10.1109/jphotov.2021.3136158.

[8] K. A. K. Niazi and M. Victoria, *Comparative analysis of PV configurations for agrivoltaic systems in Europe*: arXiv, 2022.

[9] S. Dittmann and H. Sanchez, *Comparative Analysis of Albedo Measurements (Plane-of-Array, Horizontal) at Multiple Sites Worldwide*, 2019. [Online]. Available: https://www.researchgate.net/profile/sebastian-dittmann/publication/337242908_comparative_analysis_of_albedo_measurements_plane-of-array_and_horizontal_at_multiple_sites_worldwide/links/5dccf5e892851c382f3af755/comparative-analysis-of-albedo-measurements-plane-of-array-and-horizontal-at-multiple-sites-worldwide.pdf

[10] H. Sanchez, S. Dittmann, C. Meza, and R. Gottschalg, "An experimental comparison between view factor and ray tracing models for energy estimation of bifacial modules," in *2022 IEEE 49th Photovoltaics Specialists Conference (PVSC)*, 2022.

[11] *Combining PV and food crops to Agrophotovoltaic–optimization of orientation and harvest*, 2012. [Online]. Available: https://publica.fraunhofer.de/bitstreams/d853a27b-d675-4639-9f4f-07c5744c7557/download

[12] A. Perna, E. K. Grubbs, R. Agrawal, and P. Bermel, "Design Considerations for Agrophotovoltaic Systems: Maintaining PV Area with Increased Crop Yield," in *2019 IEEE 46th Photovoltaic Specialists Conference (PVSC)*, 2019.

[13] O. A. Katsikogiannis, H. Ziar, and O. Isabella, "Integration of bifacial photovoltaics in agrivoltaic systems: A synergistic design approach," *Applied Energy*, vol. 309, p. 118475, 2022, doi: 10.1016/j.apenergy.2021.118475.

[14] P. E. Campana, B. Stridh, S. Amaducci, and M. Colauzzi, "Optimisation of vertically mounted agrivoltaic systems," *Journal of Cleaner Production*, vol. 325, p. 129091, 2021, doi: 10.1016/j.jclepro.2021.129091.

[15] J. Vollprecht, M. Trommsdorff, and C. Hermann, "Legal framework of agrivoltaics in Germany," *AIP Conference Proceedings*, vol. 2361, no. 1, p. 20002, 2021, doi: 10.1063/5.0055133.

[16] Deutsches Institut für Normung, *Agri-Photovoltaik-Anlagen – DIN SPEC 91434:2021-05. Anforderungen an die landwirtschaftliche Hauptnutzung*, 2021.

[17] Sánchez H., Dittmann S., Tosello S., Meza C., Dullau S., Meyer M.H., Scholz P., Tischew S., Gottschalg R., Ed., *Novel measurement concept for AgriPVPlus Systems- A triple use approach*, 2022.

[18] R. Bruhwyler *et al.*, "Vertical agrivoltaics and its potential for electricity production and agricultural water demand: A case study in the area of Chanco, Chile," *Sustainable Energy Technologies and Assessments*, vol. 60, p. 103425, 2023, doi: 10.1016/j.seta.2023.103425.

3D MODELLING OF LIGHT-SHARING AGRIVOLTAIC SYSTEMS FOR ORCHARDS, VINEYARDS AND BERRIES

Ibrahim El Boujdaini[1], Roxane Bruhwyler[2], Prithivi Rajan[1], Jesus Robledo Bueno[1], Babacar Sarr[1], Jonathan Leloux[1,*], Frédéric Lebeau[2], Christian A. Gueymard[1,3]
[1]*LuciSun, Villers-la-Ville, Belgium,*
[2]*University of Liège, Gembloux Agro-Bio Tech, Belgium,*
[3]*Solar Consulting Services, Colebrook, United States*
[*]*jonathan.leloux@lucisun.com*

ABSTRACT: This study focuses on the modelling and simulating of agrivoltaic (Agri-PV) systems, where solar photovoltaic (PV) and agriculture coexist. Systems aimed at orchards demonstrate land-efficient use and weather protection, whereas those aimed at vineyards improve wine quality through reduced solar exposure. LuciSun's LuSim, a specialized and detailed simulation and visualization tool, is used to analyze real Agri-PV projects in southern France, including orchards, vineyards, and berry fields. LuSim's high-resolution irradiance profiling reveals design parameter trade-offs, emphasizing the need for advanced modelling. Complexities surrounding PV module selection and 3D modelling are explored, along with initial investigations into the potential benefits of semi-transparent PV panels and light diffusers. This research underscores the significance of relying on specialized modelling tools when performing the critical task of optimizing projected Agri-PV systems.

Keywords: Agrivoltaics, light sharing, semi-transparent PV, GPU-based modelling, vineyards, orchards, berries

1 INTRODUCTION

Agrivoltaics (Agri-PV) is a novel application of solar photovoltaic (PV) power generation where the PV modules are installed in the same field as where crops are also cultivated. This is advantageous because this combines land use while keeping attractive productions from both agricultural crops and power generation.

Several configurations currently exist for these Agri-PV installations, including ground-mounted PV plants and PV installations on greenhouses. In recent years, the bifacial PV technology has experienced a swift increase in its share of the PV market, which opens the door to performance improvements, and even to new kinds of solutions.

For orchards, Agri-PV projects offer several important benefits, including increased land-use efficiency, reduced water stress, combined PV energy generation and crops harvest, crops protection from extreme weather conditions (such as hail or excessive sunlight), attenuated temperature extremes, extended growing seasons (by providing a more controlled microclimate), or income diversification for the farmer.

In the case of vineyards, Agri-PV projects have the same advantages, but additionally contribute to reducing the amount of solar irradiation received by the vines. This helps reduce the sugar content of the grapes, which in turn decreases the alcohol content in wine. This is increasingly important for winemakers in southern Europe because the increasing trend in the overall solar resource over the last few decades (known as "brightening") has led to the production of wine with excessive alcohol levels, which must be mitigated by other means. Reducing the amount of sunlight by controllable measures using Agri-PV technologies is therefore very attractive to that industry.

Regarding the production of berries, Agri-PV constitutes another very attractive possibility because they generally require a relatively low amount of solar radiation, and can thus tolerate high amounts of shading.

Finally, in most projects, the use of Agri-PV technologies also makes it possible to recoup the cost of the support structures for the nets that are used for protection against insects and birds.

All these reasons explain the recent surge in agrivoltaic applications and the increasing interest and popularity for this novel concept. However, in the current state-of-the-art PV simulation tools, the added complexity introduced by Agri-PV and bifacial technologies is still not fully covered. The associated solar projects include many complex structural elements that need to be accurately modeled so that the amount of light that reaches the crops can be accurately evaluated despite various sources of shading or light reflections from the plants to the backside of bifacial modules. Sunlight and skylight are partially transmitted through the glass of PV modules, semi-transparent modules, or diffusing materials, whose transmittance needs to be modeled accurately. Moreover, the shape and size of the crops can vary, which induces further complexity in the overall modelling.

This work presents the application of a novel simulation tool (LuciSun's LuSim) to the case of Agri-PV projects with bifacial PV modules installed above orchards (pear trees), vineyards, or berries. In particular, the role of several key variables that drive the irradiance received by the crops and reflected onto the backside of PV modules is discussed in detail. Overall, this work presents an overview of the direct experience gained from recent modelling exercises applied to several real-world case studies of light-sharing Agri-PV projects. In particular, several cases of Agri-PV applications in various types of farms (orchards, vineyards, and berries) located in southern France are explored, and the key findings from the modelling exercises are discussed.

2 METHODOLOGY OVERVIEW

The assessment of shading profiles affecting both vegetation and PV modules, along with the calculation of bifacial energy gain (BEG), is conducted using the LuSim simulation tool. This tool leverages cutting-edge 3D evaluation libraries integrated into the Graphic Processor Units (GPUs) found in modern computers. Although initially developed for the video game industry, these libraries offer several compelling advantages in the context of bifacial PV applications. The achievable spatial resolution rivals that of backward ray-tracing techniques but demands only a fraction of the latter's simulation time. The methodology followed to employ GPUs in solar energy applications has been detailed in previous contributions, such as for the assessment of intricate shading issues [1], bifacial irradiance [2], or the energy simulation of vertical bifacial PV systems in agrivoltaics [3]. The irradiance distribution profiles are assessed at high spatial resolution, either at the leaf scale or the PV cell level, and with a relatively high temporal resolution of 10 minutes. This study builds upon prior research involving light-sharing 3D modelling applied to Agri-PV applications within greenhouse environments [4]. In this context, a distinct 3D model is created for each configuration under investigation. The 3D view-field method is employed for the comprehensive evaluation of the irradiance field, both incident and reflected, that involves the ground and PV modules on a component-by-component basis. The incident irradiance profile for each PV cell comprising the PV system is obtained at 10-minute intervals throughout the year. Then, this irradiance data time series is transformed into electrical power using a PV simulation model that accounts for conversion losses within the entire system. In most cases, conventional simulation routines, such as those contained in pvlib [5], are sufficient to model these energy losses.

To assess the impact of the actual solar resource on crop yields, each configuration is compared to a reference model. The latter is intentionally designed to closely resemble the base case, where trees are planted in rows, and vertical structures support protective nets. However, in this base case, neither PV panels nor their supporting structures are present.

3 3D MODELLING

Creating a 3D model for an agrivoltaic system can be a challenging task. It involves modelling several elements, including the PV modules, their support structures, racking elements, crops, the ground surface, and any other potential surrounding objects. Each object must be assigned specific optical properties such as albedo, transmittance, or optical porosity. As illustrated in Figure 1, the latter quantity is a crucial parameter when modelling agrivoltaic systems because it determines both the amount of sunlight that passes through a particular plant before potentially reaching another, as well as its impact on the irradiance that plant leaves can effectively collect and use for photosynthesis. For each type of crop, the optical porosity typically undergoes seasonal changes in direct relation with the growth and shedding of leaves. Using LuSim, this temporal evolution is modeled based on monthly values. This approach strikes a balance between simplicity and accuracy, while offering the advantage of using readily available monthly data inputs.

Figure 1: Illustration depicting the concept of optical porosity, which represents the fraction of sunlight that is transmitted through a canopy [6].

In LuSim, the 3D objects that constitute the scene can be imported from specialized CAD tools, such as SketchUp or AutoCAD, or from tools commonly used in the solar energy industry, such as PVcase. Subsequently, each object is assigned a texture containing its optical properties. A mesh associated with these textures defines the spatial resolution used in irradiance modelling.

Figure 2 illustrates a 3D model of an Agri-PV pear-tree orchard. Two types of trellises are included in the model: T-mode and V-mode, resulting in distinct shapes for pear trees. The modelling process accounts for light interception by structures and PV modules at every point of the trees, considering each trellis mode and PV module installation geometry.

Figure 2: Example of 3D modelling of an agrivoltaic pear-tree orchard with T-mode and V-mode trellises.

When assessing plant growth, the incident irradiation must be integrated separately for specific plant zones. In the realm of 3D modelling, several key questions arise regarding how to best represent plants and define these zones of interest. For plant shapes, it is possible to select either simple shapes, which approximate the outer boundaries of the crops, or more intricate shapes, which attempt to faithfully replicate the geometry of plant organs and leaves in detail. Basic geometric shapes, such as parallelepipeds, cylinders, spheres, or cones, can be employed to represent the outer envelopes, whereas shapes of varying complexity between the simplest and most detailed forms are also viable options. Each approach comes with its own set of advantages and disadvantages.

Complex geometries attempt to realistically represent the shape of crops. They facilitate the utilization of more intricate models used to evaluate crop photosynthesis and good estimates of the 3D optical porosity (see Section 5).

10.4229/EUPVSEC2023/4DO.2.3
020430-002

However, this approach demands significantly higher computational resources because of the concomitant substantial increase in required spatial resolution and of the number of points where irradiance must be assessed. It also restricts the use of simpler agronomic models that have been developed based on a preliminary evaluation of the irradiance incident on the external canopy envelope.

In contrast, the use of basic shapes that depict the external envelope of crops reduces the computational complexity significantly by reducing the number of points where irradiance calculations are necessary. This approach also facilitates the direct utilization of parametric models that assess photosynthesis in the canopy based on the solar radiation reaching its outer envelope. When employing these straightforward models, optical properties—including optical porosity—cannot be directly modeled, but must be incorporated through a parametric model attached to the object's texture.

In most agrivoltaic applications modeled using LuSim, experience has favored the use of basic geometric shapes alongside parameterized optical properties. If necessary, the optical porosity can be initially modeled using a high-resolution 3D representation of the plant under scrutiny, and the results can then be applied to all simple shapes employed in modelling the entire agrivoltaic system.

Figure 3 shows a simple geometric shape used to represent the external envelope of a pear tree in an orchard with T-mode trellises. The subdivisions used for the modelling of the incident light are also indicated.

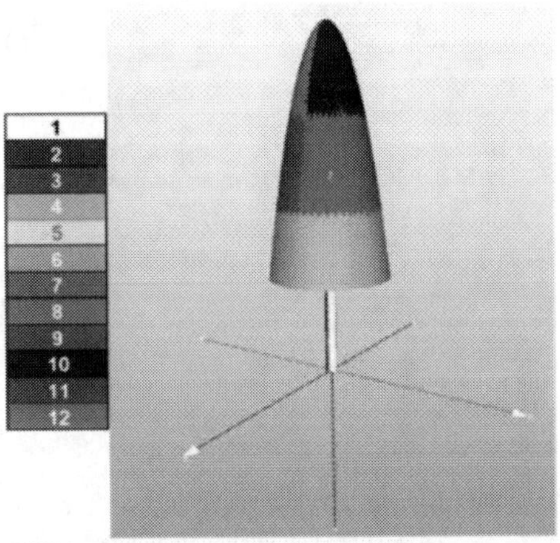

Figure 3: External envelope used for pear-tree orchards with T-mode trellises. The twelve subdivisions used for the modelling of the incident light are indicated.

Figure 4 illustrates the use of a parallelepiped to model a vineyard. The height of the plant is separated into three equal zones per side, for which the irradiance incident on the plants is calculated. Additionally, the horizontal upper face constitutes a seventh zone.

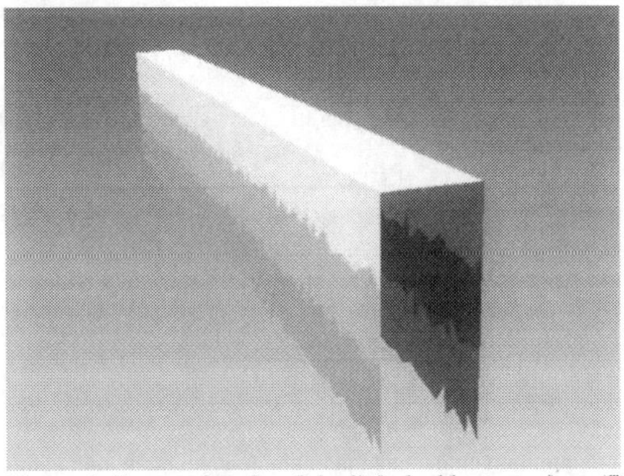

Figure 4: Zones of study of the light incident on vines (7 zones).

Figure 5 shows the shadows cast by the PV system and its structure at one specific moment and on one specific part of a pear-tree agrivoltaic system with T-trellises. The mesh adopted here for the trees is made visible on one of them.

Figure 5: Shadows cast on a pear-tree agrivoltaic system, showing the mesh on one of the trees.

4 IRRADIANCE MODELLING

Once the 3D mesh model of the agrivoltaic system has been completed, the irradiance simulations are carried out at each instant and for each radiation component (direct, isotropic sky diffuse, circumsolar sky diffuse, and ground reflected). For instance, Figure 6 illustrates the evaluation of the direct irradiance component reaching the agrivoltaic system at 3 PM during a clear summer day. The irradiance values are then aggregated into areas of interest within the 3D scenario, as well as over periods of time that are relevant to the crop yield to be evaluated. Those periods depend on the type of crop and the corresponding growth and harvest seasons. These integrations are typically done on a daily, monthly, or yearly basis, depending on what needs to be evaluated. As an example, Figure 7 shows the results of the temporal integration of the direct irradiance over one whole year. Similarly, Figure 8 shows the yearly-integrated isotropic diffuse irradiation, and Figure 9 shows the yearly-integrated global irradiation.

10.4229/EUPVSEC2023/4DO.2.3

Figure 6: Evaluation of the direct irradiance reaching the agrivoltaic system during a clear summer day.

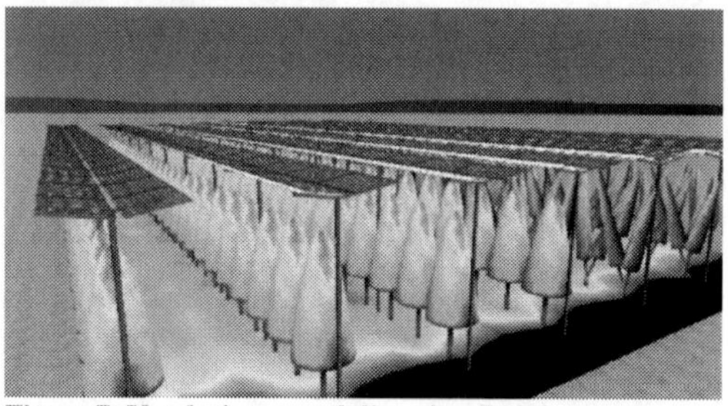

Figure 7: Yearly-integrated direct irradiation.

Figure 8: Yearly-integrated isotropic diffuse irradiation.

Figure 9: Yearly-integrated global irradiation.

5 REFERENCE SCENARIOS FOR CROP YIELD

The assessment of the solar resource's impact on photosynthesis can be approached in various ways. The scientific literature has introduced agronomic models tailored to specific plants, varying in complexity from basic agronomic models to intricate Functional Structural Plant Models (FSPMs). In agrivoltaic system modelling, simple agronomic models often suffice, offering the advantage of wider applicability across plant types.

The modeled irradiance incident on trees is not directly usable for agronomic interpretation because it strongly depends on the 3D model used. In particular, the surface area of the tree's outer envelope increases when the complexity of the 3D shape model also increases. Hence, complex shapes lead to larger areas, whereas simple shapes lead to smaller areas. In reality, however, the total solar irradiation (in Wh) that truly reaches a tree is relatively independent from the complexity of its actual shape. Consequently, using a complex 3D tree geometry induces a smaller irradiation per unit area (in Wh/m^2), since the total irradiation is intercepted by a larger surface. Carrying out the agronomic interpretation relatively to a base-case scenario is thus more reliable. In the LuSim framework, this is done by comparing a reference case (without PV panels) to each of the configurations under scrutiny, always using the same 3D tree models. This normalizes most of the model-induced irradiation variations noted above, thus providing more robust results.

The base-case scenario needs critical attention because the PV yield directly depends on it. Hence, it has to accurately represent the canopy and other objects in the absence of any specific PV structure, including support bars and mounting racks. For instance, Figure 10 depicts the reference scenario for the pear-tree orchards discussed earlier. This scenario consists of the canopy and supporting structures for the nets that are typically used to protect the crops from insects and birds. For standard nets used in agriculture, a typical constant transmittance (90%) is applied to estimate their impact on the entire sky view.

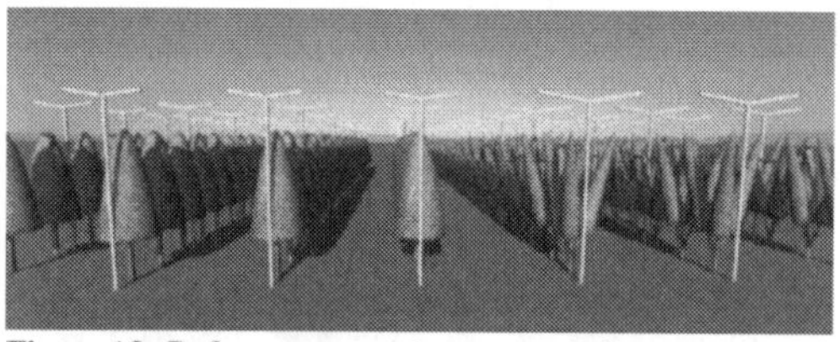

Figure 10: Reference scenario corresponding to the pear-tree orchards shown in Figure 5.

6 PV MODULE SEMI-TRANSPARENCY

The use of semi-transparent PV modules is becoming increasingly popular in agrivoltaic projects because of several potentially important advantages. These include (i) the possibility to install PV modules above the plants without much change with respect to the common installation practices in standard PV projects, and (ii) the desirable outcome of a more homogeneous distribution of light around the crop. PV modules offering various semi-transparency attributes are available on the market, thus offering more flexibility to adapt to the needs of each specific project. However, the use of semi-transparent PV modules also conveys some important disadvantages, including a higher price per nameplate power. This is because the manufacturing costs of such PV modules are not proportional to the amount of PV cells but also depend on the balance of materials, which is roughly dependent on the total surface of the PV modules, and can be relatively more expensive than PV cells.

This section explores some key observations gathered while using LuSim in the context of simulating the pear-tree Agri-PV systems introduced earlier, for which semi-transparent modules are considered as an option. This experience might provide guidance on the main parameters to consider, and help decide whether this kind of module is suitable or not for similar agrivoltaic projects.

Different configurations are tested here, with a combination of design parameters including PV modules

10.4229/EUPVSEC2023/4DO.2.3
020430-004

with semi-transparency of either 27% or 51%, the installation of PV modules in portrait or landscape mode, and the installation of either one or two PV modules per row. Figure 11 shows the 3D model from the top side of the PV installation, where the semi-transparency of the PV modules is visualized.

Figure 11: Semi-transparent PV installation above a pear-tree Agri-PV system.

Figure 12 shows the pear trees with the panel support structures in portrait mode. Clearly, the structures intercept a significant fraction of the trees' field of view. This shading is caused by the width of the bars, particularly for the bars that are just above the chosen viewpoint, but, even more so, by their depth. This is observed for the bars that are above the neighboring tree rows because these bars block a significant part of the sky, including the light rays that arrive obliquely.

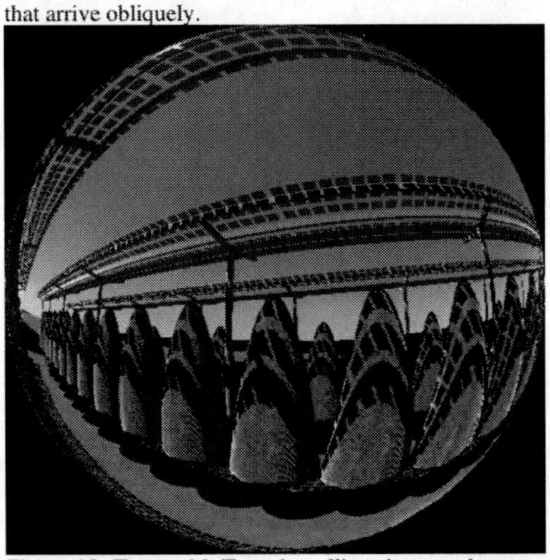

Figure 12: Trees with T-mode trellis using panel support structures in portrait mode.

Figure 13 shows an example of incident light profile for a specific Agri-PV configuration over the course of a clear summer day. The blue curve represents the simplest configuration where only *one* tree is present and only the vertical support structures are included. This case obviously corresponds to the highest solar resource that can be captured by that tree. The second configuration (orange curve) corresponds to the presence of *all* trees and vertical structures, but still with no PV panels or support bars. In summer, this configuration results in a resource almost as high as that for a single tree. The third configuration (green curve) corresponds to the case of panels installed in portrait mode, with a semi-transparency factor of 27%. Finally, the last configuration (red curve) corresponds to panels installed in landscape mode, with a semi-transparency factor of 51%. Interestingly, a more dramatic reduction in transmitted light is noticeable in landscape mode compared to portrait mode, mainly caused by the presence of 4 support bars rather than 2. The curves also show a trend with two peaks separated by a trough around local noon, which corresponds to the passage of the sun over the PV panels. The green curve shows a slight asymmetry, which is caused by the panels being tilted towards the west. Conversely, the installation in landscape mode (red curve) shows relative symmetry, because half of the panels are tilted towards the east and the other half are facing west.

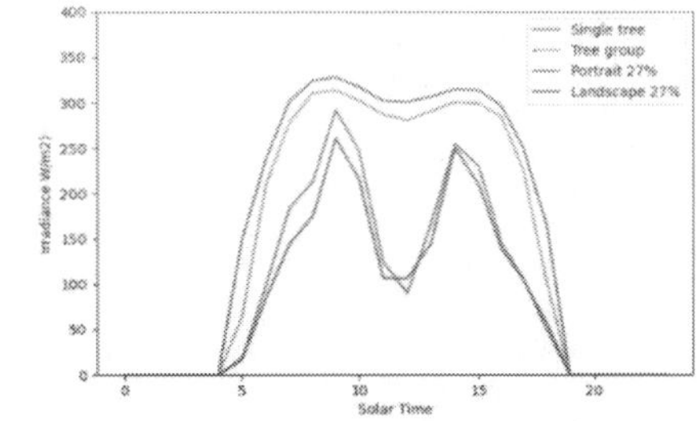

Figure 13: Solar irradiance on trees for different Agri-PV configurations, during a clear summer day.

Figure 14 presents the comparison of the solar resource evaluated for different scenarios: the base-case reference (no PV, blue curve), PV with overall panel transparencies of either 51% or 27%, and PV modules installed in either portrait or landscape mode. The solar resource decreases as expected depending on the transparency factor of the modules. Remarkably, it is found that the landscape configuration with 51% transparency leads to a slightly lower resource than the portrait configuration with 27% transparency. This is caused by the high impact of the modules' support bars, which are twice as many in landscape than in portrait configuration.

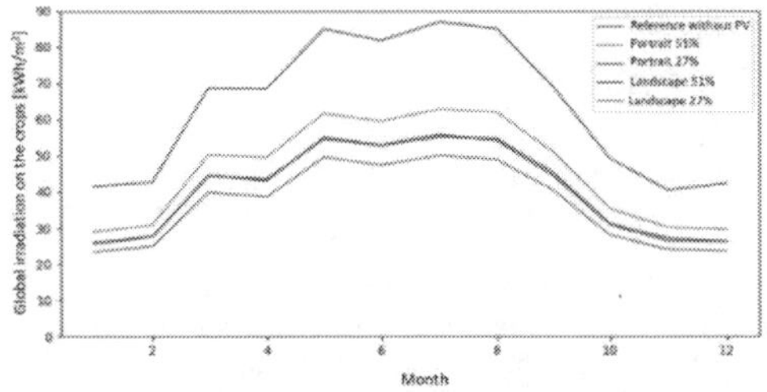

Figure 14: Comparison of the resource evaluated for the reference scenario, panel transparencies of 51% or 27%, and PV modules installed in portrait or landscape mode.

These modelling exercises have shown that the shading caused by panel support bars plays a major role in reducing the resource. The relative impact of semi-transparency is therefore reduced compared to what it would be without the presence of the support bars. On the other hand, these simulations assume that a reduction of ≈6% in the transmittance of glass-glass PV modules is caused by soiling. This might be a relatively optimistic scenario if the practice of spraying calcined kaolinite clay as a protective measure against insects is carried out in the standard way.

10.4229/EUPVSEC2023/4DO.2.3
020430-005

Overall, the present results suggest that the use of semi-transparent modules might not be optimal, at least for some agrivoltaic projects, and that rather relying on conventional PV modules could lead to a better technical-economic optimum. If semi-transparent modules are used, then a detailed design optimization of the supporting structures, and in particular the module fixing bars, should be carried out to reduce their impact on the incident light.

7 EVALUATION OF THE CROP YIELD

Photosynthetic efficiency can be described by the well-known irradiance-photosynthesis function, which expresses the ability of plants to respond linearly to low irradiance levels, and to rather saturate at high irradiance. This property can be advantageously used to drastically improve the productivity of some appropriate crops under PV shades because the irradiation magnitude drops drastically when affected by the shadows created by the panels and structure.

Obtaining a good homogeneity in the photosynthetic irradiance field (i.e., highly scattered light) around the plants is important to maximize the crop yields because this helps the plant limit the need to translocate metabolites towards the fruits located in the shaded zones. For the pear-tree agrivoltaic system under scrutiny, the analysis of this homogeneity is done at two tree levels, and its results appear in Table 1 for the month of April. This period is key for the tree development, and ultimately for the crop yield. In general, the presence of projected shadows from the modules and structure improves the light distribution homogeneity by reducing the resource more in the upper part of the trees than in their lower part.

Table 1: Irradiation received per unit area at two tree levels, for the T-shaped trellises.

Monthly effective solar irradiation on pear trees with T-shaped trellises			
	TMY April [kWh/m²]		
	Bottom of pear trees	Top of pear trees	Difference vs average
Control scenario	21.8	27.6	23%
Portrait 51	21.0	26.5	23%
Landscape 51	18.4	23.1	23%
Portrait 27	18.6	23.2	22%
Landscape 27	16.6	20.6	21%

In the case of pear trees, the simulations show that irradiance reductions caused by the presence of PV-induced shadows (i.e., by the modules and their supporting structure) do not present a major risk for the agricultural production in the planned implantation area. In fact, measurements made with an Agri-PV prototype for apple trees in France have shown that an average irradiation reduction of 50–55% does not significantly impact the quality of the harvest [7]. In parallel, simulations made for raspberry shrubs show very favorable results under most possible configurations. Raspberry is an understorey plant found in temperate and cold regions that is particularly well adapted to low light levels. In its natural environment, it is actually able to maintain itself under very low light levels. For instance, the results in [8] attest to the presence of raspberry plants at residual irradiance levels of only 7% of the photosynthetic flux density above the canopy. Raspberry bushes remain abundant above the level of 25% residual irradiance and even increase their vigour by up to 40%. This illustrates the raspberry's great adaptability to a wide range of solar illumination levels. Such analyses leave no doubt as to the feasibility of growing raspberries with a satisfactory yield and quality when their cultivation is combined with an Agri-PV system. Various geometric configurations and panel types offer multiple combinations. The remarkable adaptation of the raspberry plant to the available solar resource means that it is probably possible to take advantage of most, if not all, proposed configurations.

For vineyards, the modelling exercises carried out with LuSim for various PV configurations show that, for the typical climate of southern France, it is important to use relatively modest Ground Cover Ratios (GCRs), typically under 30%, to maintain an optimum yield for the crops. Changing the panel size, or using semi-transparency PV modules, only alters the frequency of the irradiance modulation. In parallel, introducing diffusing components between the PV cells or modules can greatly contribute to homogenize the amount of sunlight received by each part of the plant at any time, leading to the beneficial phenomenon called "even lighting". From that standpoint, the potential benefits offered by light diffusers are further discussed in the next section.

8 POTENTIAL ADVANTAGES OF LIGHT DIFFUSERS

The crops' need for light is a major constraint in many agrivoltaic projects. The design of such a system is tedious because finding the optimal balance between PV array density and crop yield is difficult to achieve. This stems from the fact that the shadows cast by the solar panels reduce the light available to the plants for several hours a day on sunny days, which are precisely the days that determine growth, most usually. Conversely, during hours of high irradiance, the light received by plants might be far beyond their photosynthetic saturation threshold (particularly in summer). This excess light does not contribute to plant growth. To the contrary, it reduces growth through the phenomena known as "photoinhibition" and "stomatal closure", possibly combined with episodes of water stress. In this context, agrivoltaic systems could greatly benefit from the introduction of light diffusing elements between the PV cells or modules, by redistributing the light more evenly over all the crops. Several approaches are possible, including quasi-isotropic diffusing surfaces or partially diffusing materials. More elaborate solutions also exist, such as lenticular diffuser sheets that are specially designed to maintain photosynthesis to its virtually unchanged potential—a concept known as "even-lighting agrivoltaics". These novel solutions remain largely unexplored, but at least this work presents some first results modeled for one specific agrivoltaic system.

9 BORDER EFFECTS

For Agri-PV projects, LuSim is an appropriate tool to simulate the irradiance's field inhomogeneity around plants as a function of their location relative to the PV system. This leads to the precise quantification of edge effects in the spatial light resource. As above, these edge effects are easier to assess and visualise in relative terms, i.e., by comparison with a no-PV base-case scenario.

10.4229/EUPVSEC2023/4DO.2.3
020430-006

Figure 16 illustrates the results of one modelling exercise that is representative of typical agrivoltaic projects involving orchards, vineyards, or berries (e.g., an agrivoltaic system similar to the ones shown in Figures 2 and 17). Figure 16 displays the normalised annual irradiance received by each plant, in relation with the minimum overall irradiance. The x-axis represents the distance in an east-west (E-W) direction, and the y-axis represents the distance in a north-south (N-S) direction. The LuSim results show that the overall resource available to the plants located in the centre of the installation is about a quarter of what would be available in the absence of the agrivoltaic system.

From the standpoint of pure agricultural yield, it is pragmatically considered that an edge effect occurs when the solar resource available at one point of an Agri_PV project is more than 10% greater than the resource available at its centre. According to this definition and the typical system under scrutiny, an edge effect occurs when plants are located less than 5–6 m from the east or west edges, or less than 3–4 m from the north or south edges. In simplified terms, significant edge effects occur wherever the distance from the edges is less than ≈5 m. Such results provide a critical estimate of the radiation field's overall homogeneity, as well as some guidance on the minimum size that an agrivoltaic demonstration plant would need to have so that its results can be statistically meaningful.

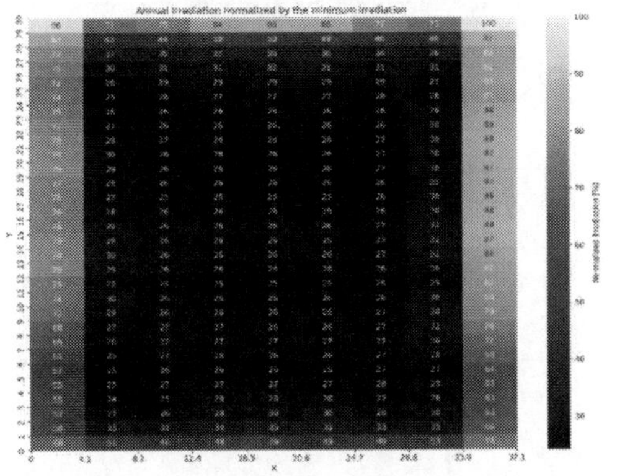

Figure 16: Example of border effects in an Agri-PV system, represented through the annual received irradiation normalized by the minimum irradiation. The red color indicates strong border effects.

10 DESIGN OPTIMIZATION: A CASE STUDY

Considering the lack of appropriate design tools for Agri-PV systems, it is desirable to investigate LuSim's capabilities in terms of design optimization. To that effect, a modelling exercise is conducted here to evaluate the respective merits of several possible configurations for an agrivoltaic system to be installed in a vineyard:

0) Reference (E-W and N-S): Trees alone without structures or panels.
1) E-W panels: Standard (full) PV panels
2) E-W panels: PV panels with 50% semi-transparency.
3) E-W panels: Standard PV panels on the west side alternating with a row of diffusing translucent polycarbonate panels on the east side.
4) E-W panels: Alternating standard panels and diffusing polycarbonate panels (1 standard PV panel, 1 polycarbonate panel) of the same panel size (alternating on the N-S axis). The west modules are staggered in relation to the east modules.
5) N-S panels: Standard PV panels on the south alternating with a row of polycarbonate panels on the north.

Figure 17 shows the scenario corresponding to the PV installation case with semi-transparent panels.

Figure 17: PV installation using semi-transparent panels.

Similarly, Figure 18 shows the scenario corresponding to a PV installation with alternating standard PV panels and light-diffusing panels.

Figure 18: PV installation with alternating standard panels and light-diffusing panels.

Figure 19 shows views from above and below in the case of Scenario 2 with semi-transparent panels. They provide the plants with a better view to the sky than Scenario 1.

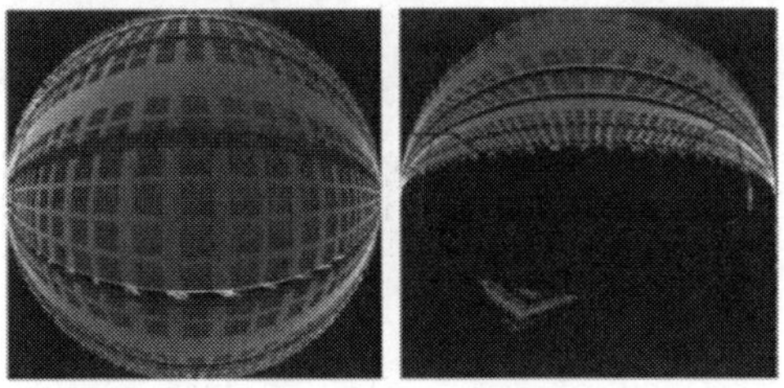

Figure 19: Hemispherical view from above (left) and from the side of the plant (right) for Scenario 2 (semi-transparent panels).

Scenarios 3 and 5 are simulated for two different possible positions of the trees: either beneath a row of PV panels or under a diffusing polycarbonate sheet. The upper face of the plant receives significantly more diffuse irradiance in the latter case than in the former case, since a significant portion of the light is intercepted by the PV

10.4229/EUPVSEC2023/4DO.2.3
020430-007

panels. Conversely, on vertical crop surfaces, more light is received when neighboring rows of plants are covered by a diffuser rather than a row of PV panels. However, the vertical faces of the plants receive less irradiance when they are positioned just below a diffuser, because then the adjacent rows are under the panels, resulting in a reduction in the diagonal light reaching the plants. Additionally, to study the effect of structures, these scenarios are also simulated both with and without diffuser support structures.

As mentioned just above, Scenarios 3 and 5 are simulated in different ways, firstly to evaluate how the position of the tree impacts the results, and secondly to study the effects of the diffuser's supporting structure. Figure 20 shows the fraction of the sky seen from the top of the plant as a function of the position of the tree.

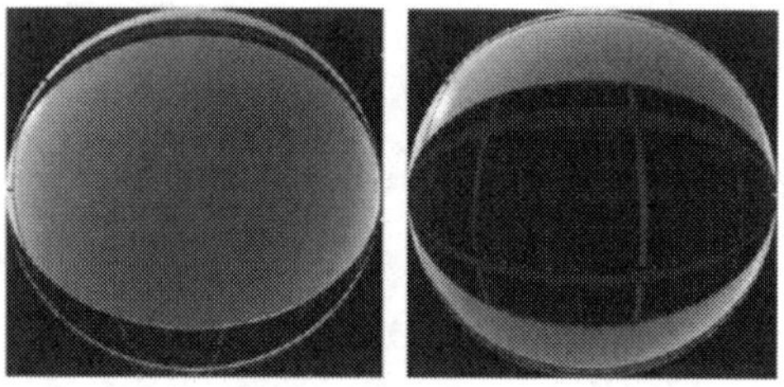

Figure 20: Spherical views of the top of the plant, for Scenario 3, with the plant under either the diffuser (left) or the row of modules (right).

To complement this information, Figure 21 shows the spherical views from a vertical face of the plant. The general finding is that the irradiance received by the different zones of the plants (see Figure 4) is significantly different depending on the row in which they are located. More specifically, the vertical plant faces that see the largest fraction of the sky are those located under the row of PV panels.

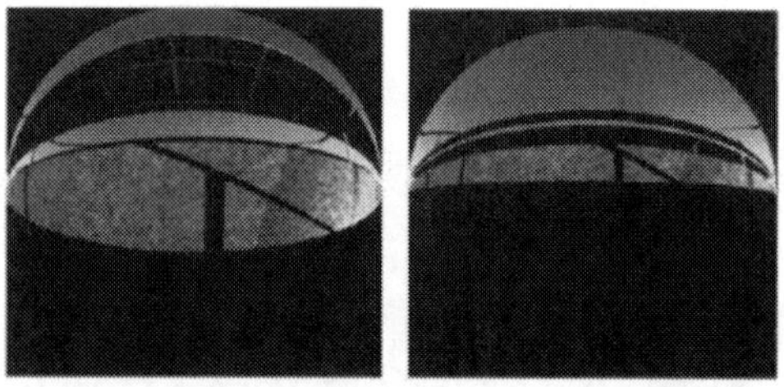

Figure 21: Spherical views of the plant, for Scenario 3 with the plant under either the diffuser (left) or the row of modules (right).

The presence of supporting structures also affects the results because of their shadowing effect. Figure 22 shows the spherical views of the side of the plant either with or without support structure. The structures intercept a non-negligible fraction of the field of view perceived from any point on a plant. This interception is partly caused by the width of the bars and, even more so, by their depth. Overall, these bars block a significant proportion of the light rays that arrive obliquely onto the plant.

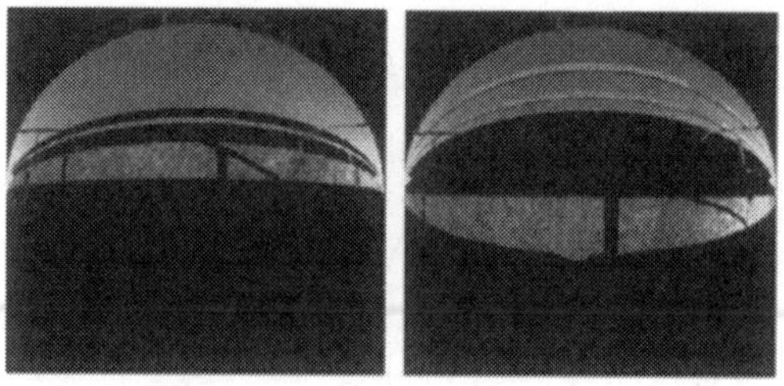

Figure 22: Spherical views of the plant, for Scenario 3 with the plant under a diffuser either without structure (left), or with structure (right) to support it.

Simulations made with LuSim evaluate the direct and diffuse components of the irradiance reaching each of the plant's seven zones (from Figure 4) at hourly temporal resolution or better. These results are then aggregated on a monthly basis, but also at the level of the complete plant by integrating the results from the 7 zones.

Figure 23 provides a general overview of the irradiance received by the plant (averaged across zones 1–6, with contributions from zone 7 neglected) on a clear spring day for all 5 scenarios. The global horizontal irradiance (GHI) that is incident above the crops is significantly reduced in all scenarios, primarily because of its transposition onto the tree's vertical sides, and also because of the nearby sources of shading from neighboring trees, structures, and rows of PV panels. Additionally, this reduction accounts for the optical porosity of the trees (the fraction of light passing through vegetation or not available for photosynthesis) and the effective fraction of irradiance that can be utilized by the plants for photosynthesis.

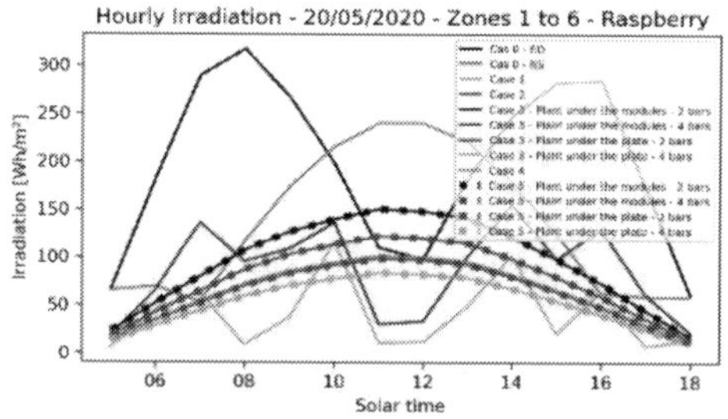

Figure 23: Irradiance received and useable by plants on a clear spring day for all scenarios.

The irradiance received by the tree sides is higher in the cases of the reference scenarios with trees alone without structures or solar panels, as could be expected. The scenario with the lowest irradiance values is Scenario 1, i.e., with standard PV modules. Scenario 2, with semi-transparent modules, provides more light to the plants. It is relatively difficult to distinguish between Scenarios 3 and 5 for the same position of the tree and the same number of bars, because the only difference between these two scenarios is the orientation of the agrivoltaic system. Moreover, the polycarbonate diffuser is assumed to transform all the direct component into isotropic diffuse, which therefore makes the results not very sensitive to orientation. Scenarios 3, 4, and 5 also show the smoothest

10.4229/EUPVSEC2023/4DO.2.3
020430-008

irradiance curves, thanks to the polycarbonate sheet acting as a pure diffuser.

In contrast, Scenario 2, with semi-transparent modules, has the roughest irradiance curve because of the absence of a diffuser, unlike the other scenarios. In addition to being associated with irradiance variations that are not smooth over the day, Scenario 2 provides the least homogeneous light to the plants, again compared with Scenarios 3, 4, and 5 with diffusers.

For still the same clear day in spring, the most favorable cases for the plants correspond to Scenarios 3 and 5, which are somewhat ideal because there are no supporting structures for the diffusers. The scenarios in which the trees are located under the row of panels offer more interesting results than when they are located under the diffuser. That is because the vertical surfaces of the crops capture most of the solar irradiation, and the light reaches them mainly obliquely.

The monthly analyses provide a more general and meaningful comparison of the results in terms of solar potential and crop yield. Figure 24 represents the monthly irradiation received and useable by the plant (after integration of the incident irradiance on zones 1 to 6 only) for each of the scenarios.

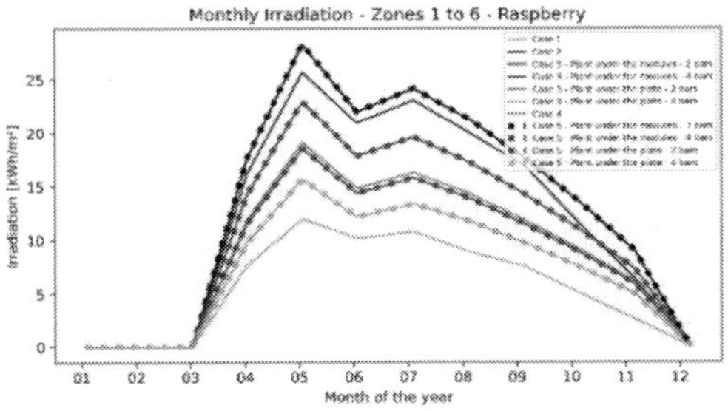

Figure 24: Monthly irradiation received and useable by plants for all scenarios.

Whereas the incident GHI above the canopy is always non-zero, the monthly irradiance received and useable by plants is zero for different scenarios in January, February, March, and December. This is because the trees cannot make use of the solar irradiation to grow or generate fruits in the absence of foliage. Moreover, the reference cases labelled E-W and N-S are the scenarios that offer plants the most irradiation, thanks to the absence of structures and PV panels.

Scenario 1 with standard PV modules is the least favourable case for plant growth. This is simply because the panels do not let any light through, and the installation is almost entirely covered with PV.

The simulation results also show the significant reduction in irradiance received by the plants and through the diffusing panels for Scenarios 3 and 5, when a support structure is considered for the diffusers (indicated by "4 bars" in the legend) compared with the scenarios without a support structure (compare the blue curve and the green curve, or the red curve and the orange curve). As already observed in the analysis of hourly irradiances, Scenarios 3 and 5 without supporting structures and with the trees located under the row of panels (blue curve) provide the plant with the largest monthly irradiance.

Scenario 2 with semi-transparent modules (violet curve) provides only slightly less monthly irradiance to the plant than the most optimal case. However, its irradiance is more uneven, which is less favourable for the crop.

In scenarios where plants are placed alternately between rows of panels and diffuser plates, one row out of two receives a larger solar irradiance, whereas the other row receives less. This difference, depending on the strategy adopted in terms of crop layout and harvesting method, could result in either an advantage or a disadvantage; hence, a more in-depth study would be necessary to provide a more definitive answer.

11 CONCLUSION

This investigation has consisted of a succession of modelling exercises carried out in the context of several agrivoltaic projects related to applications for orchards, vineyards, and berries in southern France. Various configurations of the PV systems were considered, leading to various comparisons of both their electric output and their effects on the crops growth and yield. The results have shown that the modelling of such Agri-PV systems leads to several important challenges that are difficult to tackle with conventional modelling tools. The use of LuSim has made it possible to model the irradiance profiles on the crops with a high spatial resolution, and subsequently to investigate the trade-offs between the main design parameters of Agri-PV systems and their implications on the solar radiation field's homogeneity and the yield of the crops.

The study's results have also demonstrated that many important questions deserve further investigation, such as whether the design of semi-transparent PV modules can be optimized depending on each crop's requirements, or whether standard PV modules with sufficiently low GCR can be used to maintain the solar resource at a high-enough level to preserve or even improve the yield of the crops underneath.

Details of the 3D modelling of the agri-PV system have also been discussed, including the application of the optical porosity approach, the specific crop modelling, and the mutual interaction between the different model parts. A major finding is that simple crop models possess many important advantages over more elaborate ones. Finally, the potential beneficial role of light diffusers has been explored through a preliminary study, showing that they might not be advantageous in all situations, contrary to expectations. Hence, further analyses should be undertaken to expand the present results and obtain more specific recommendations.

REFERENCES

[1] J. Robledo et al., From video games to solar energy: 3D shading simulation for PV using GPU, Solar Energy, https://doi.org/10.1016/j.solener.2019.09.041, 2019

[2] J. Robledo et al., Dynamic and visual simulation of bifacial energy gain for photovoltaic plants, European Photovoltaic Solar Energy Conference and Exhibition (EU PVSEC), 2021.

[3] J. Robledo et al., Lessons learned from simulating the energy yield of an agrivoltaic project with vertical bifacial photovoltaic modules in France, European Photovoltaic Solar Energy Conference and Exhibition (EU PVSEC), 2021.

[4] J. Robledo et al., Key parameters for the simulation of agrivoltaics in greenhouses with bifacial PV modules, WCPEC-8 Conf., Milan, Italy, 2022.

[5] W.F. Holmgren et al., pvlib python: a python package for modeling solar energy systems, Journal of Open Source Software, 3(29), 884, 2018.

[6] J.G. Velarde et al., Estimation of optical porosity or canopy structure of two species of tree with hemispherical and vertical images, WSEAS Transactions on Environment and Development, 14, 112-124, 2018.

[7] P. Juillion et al., Impact of shading with photovoltaic panels on fruit quality. Case study in apple orchard, Agrivoltaics Conference and Exhibition, Piacenza, Italy, 2022.

[8] J.P. Ricard et al., Abundance, growth and allometry of red raspberry (Rubus idaeus L.) along a natural light gradient in a northern hardwood forest, Forest Ecology and Management, 81, 153-160, 1996.

10.4229/EUPVSEC2023/4DO.2.3
020430-010

From the design to the implementation, a symbiosis where PV and agriculture can have a mutually beneficial relationship

Date
21.09.2023
EU PVSEC
4DO.2
Agrivoltaics: Advances in Design, Modelling and Performance Analysis

Prepared by:
David Moser

Symbiosyst

eurac research

Funded by the European Union

Funded by the European Union. Views and opinions expressed are however those of the author(s) only and do not necessarily reflect those of the European Union or CINEA. Neither the European Union nor the granting authority can be held responsible for them.

020432-001

40th European Photovoltaic Solar Energy Conference and Exhibition

SYMBIOSYST

Project Information

SYMBIOSYST
Grant agreement ID: 101096352

DOI
10.3030/101096352

Start date 1 January 2023 — **End date** 31 December 2026

Funded under
Climate, Energy and Mobility

Total cost
€ 5 775 143,42

EU contribution
€ 4 827 668,00

Coordinated by
ACCADEMIA EUROPEA DI BOLZANO
Italy

eurac research

3E, aleo, ET Solare Italia, ENEA, above, etaflorence renewable energies, FOTONIQ, KUBO, KU LEUVEN, imec, Laborelec, Laimburg, LUCISUN, Südtiroler Bauernbund, TU Delft, valmont

2

020432-002

Why what we do is timely and important

020432-003

40th European Photovoltaic Solar Energy Conference and Exhibition

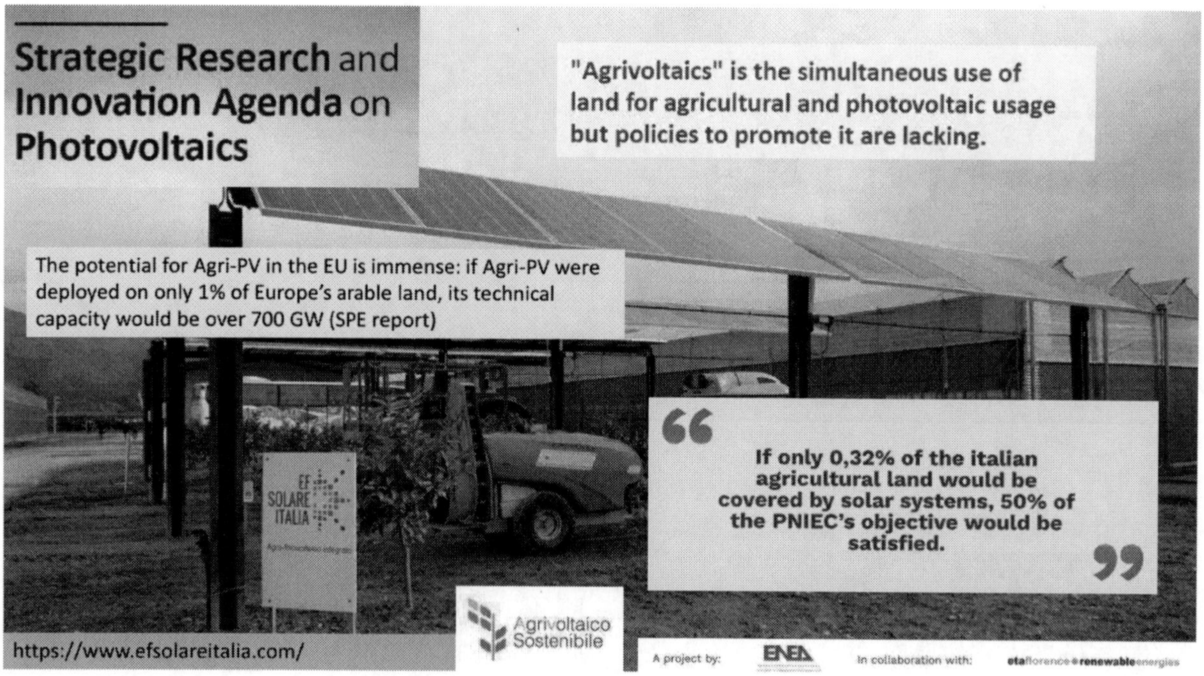

020432-004

Agri-PV as a local strategic choice for regions with limited available areas

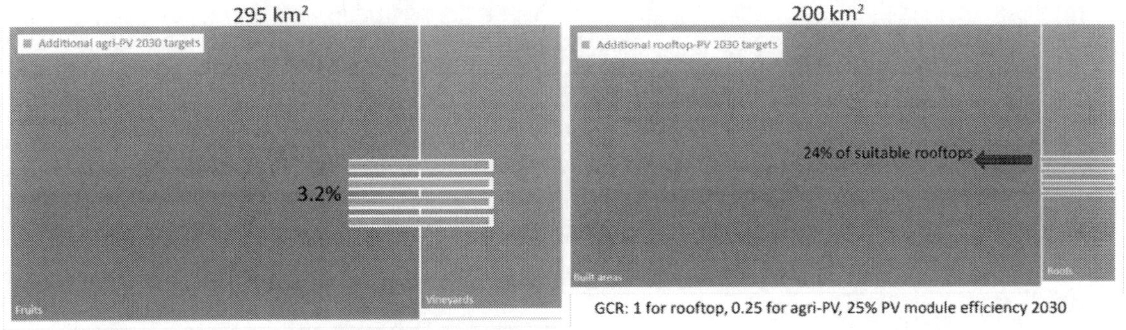

GCR: 1 for rooftop, 0.25 for agri-PV, 25% PV module efficiency 2030

Case study: South Tyrol, north of Italy

020432-005

40th European Photovoltaic Solar Energy Conference and Exhibition

SYMBIOSYST learns from other sectors

020432-006

020432-007

40th European Photovoltaic Solar Energy Conference and Exhibition

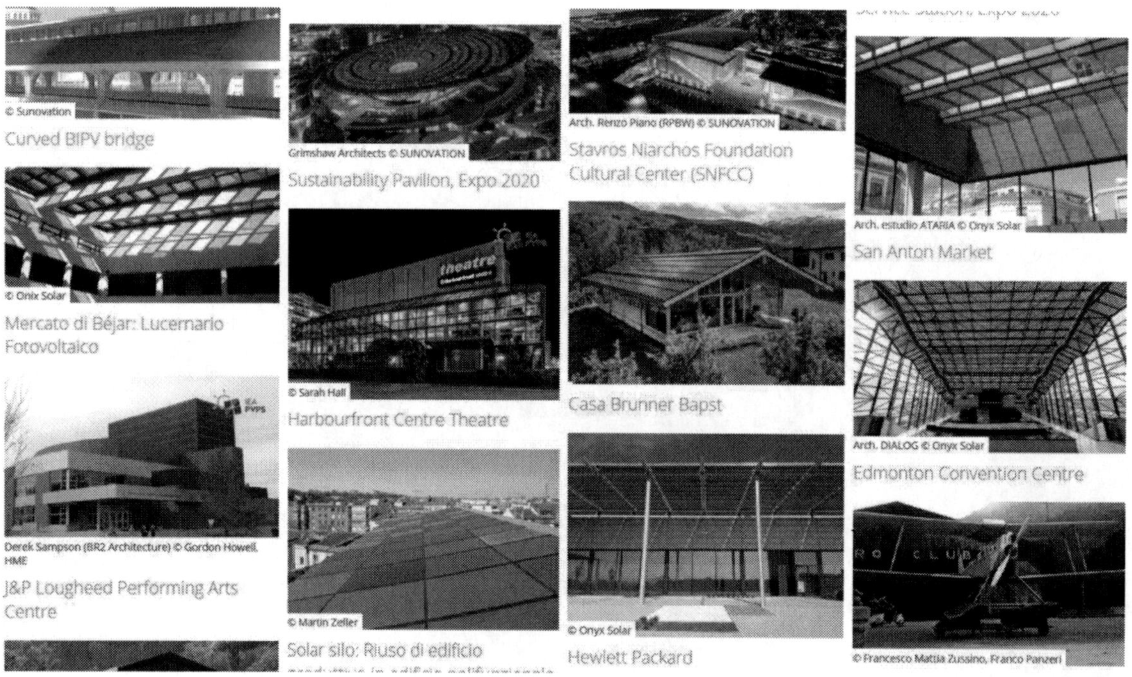

020432-008

40th European Photovoltaic Solar Energy Conference and Exhibition

020432-009

40th European Photovoltaic Solar Energy Conference and Exhibition

https://integratedpv.eurac.edu/it

020432-010

Premises

The rate of installation of PV rooftop systems will not be fast enough to support a rapid transition

Large utility-scale PV installations in open fields are not an option in many regions

Several decades of efforts developing Building Integrated PV (BIPV) products and systems for rooftops and facades has taught us is that there are multiple levels of integration and various stakeholders across the value chain which can all generate significant barriers to widespread adoption

Agri-PV is in fact in the unique position to learn from the past mistakes made with BIPV which delayed the development of cost-effective BIPV solutions and market acceptance.

High risk that agri-PV is incorrectly employed to justify "fake" systems that mainly generate electricity and not agricultural products. This eventuality would cause a rapid transformation of the landscape which might be nor sustainable, neither acceptable by society.

11

020432-011

SYMBIOSYST: Concepts, ambition

12

020432-012

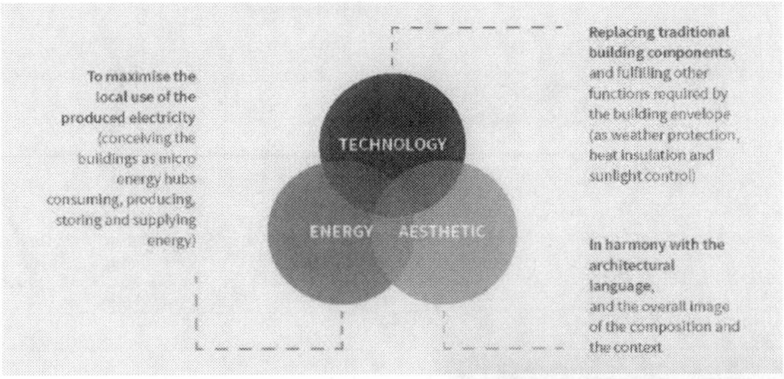

Multiobjective optimisation

Discussion with stakeholders

Impact of customisation on costs

Definitions

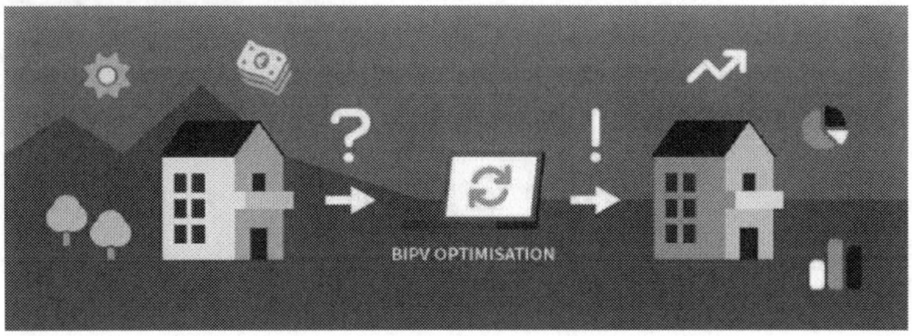

40th European Photovoltaic Solar Energy Conference and Exhibition

Objectives

Obj.	Description	Driver	Outputs
1	**To develop (system) technologies adaptable** to plant variety, geography, and landscape to maximize crop-electricity synergy.	Technology	**Multifunctional mounting system** based on 1-axis tracker **Innovative PV modules** with optimised spacing and semitransparency. Innovative PV modules with functional coatings. **Sensing solutions** to gather combined data on energy and crop yield and improve tracking algorithms.
2	**To increase the design accuracy along the whole value chain** of novel agro-photovoltaics systems to optimize both electricity generation and crop yield.	Reliability	**Modelling tools along the whole value chain** tailored for various stakeholders
3	**To increase the performance and reliability** and the ease of maintenance in novel agro-photovoltaic systems.	Reliability	**A standardised monitoring protocol** for orienting and assessing the quality of the agrivoltaic system
4	**To map, involve, and train the European PV value chain** by leveraging on the experience from the BIPV sector and develop innovative and sustainable business models.	Business	**Bespoke business cases for agri-PV** to study the end-use of electricity and the combined economic value of crop + PV. **Training of professionals.**
5	**To develop sustainable synergic use and system integration** for novel agro-photovoltaics to minimize the Life Cycle Assessment	Sustainability	**LCA methodology for agri-PV** plants including the impact on the biodiversity
6	**To increase societal acceptance and energy justice** through a constant technical – social dialogue between stakeholders.	Social	A complete set of tools to **properly engage a large variety of stakeholders.** **Development of co-governance of Agrivoltaics**

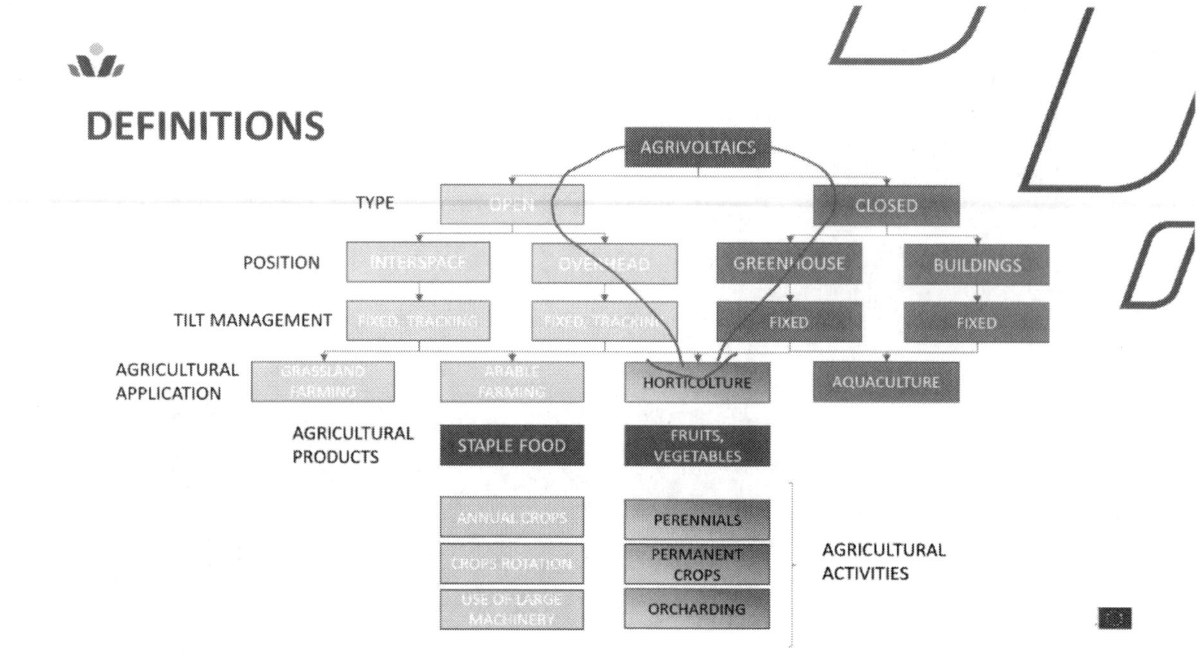

020432-015

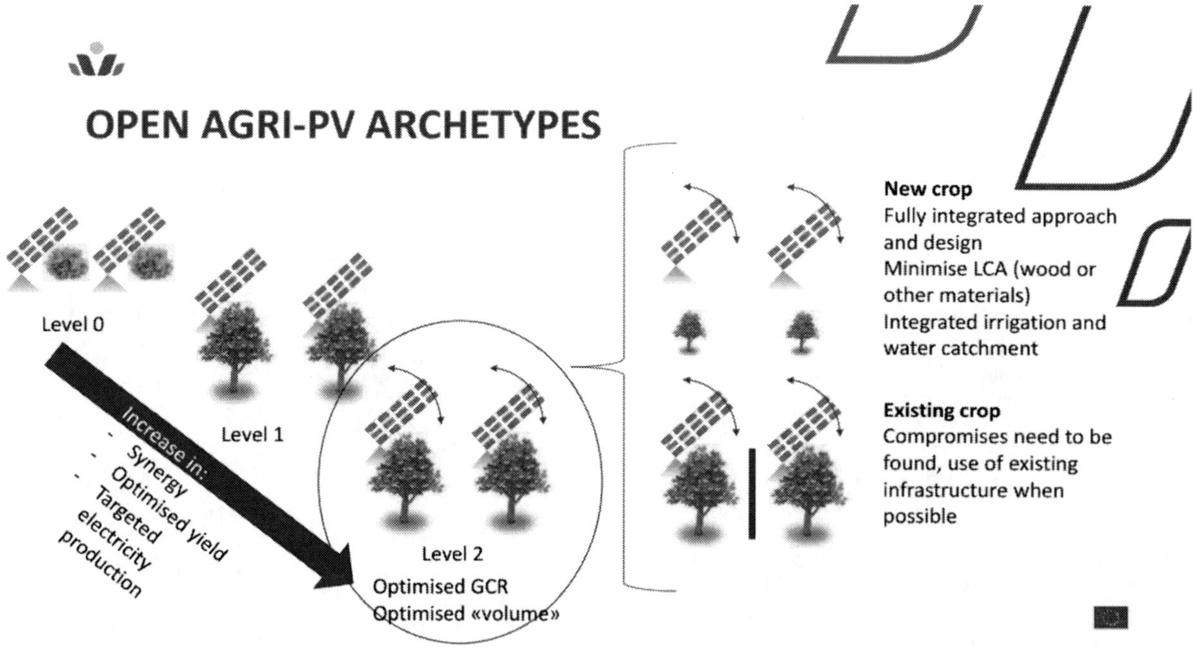

020432-016

OPEN AGRI-PV ARCHETYPES

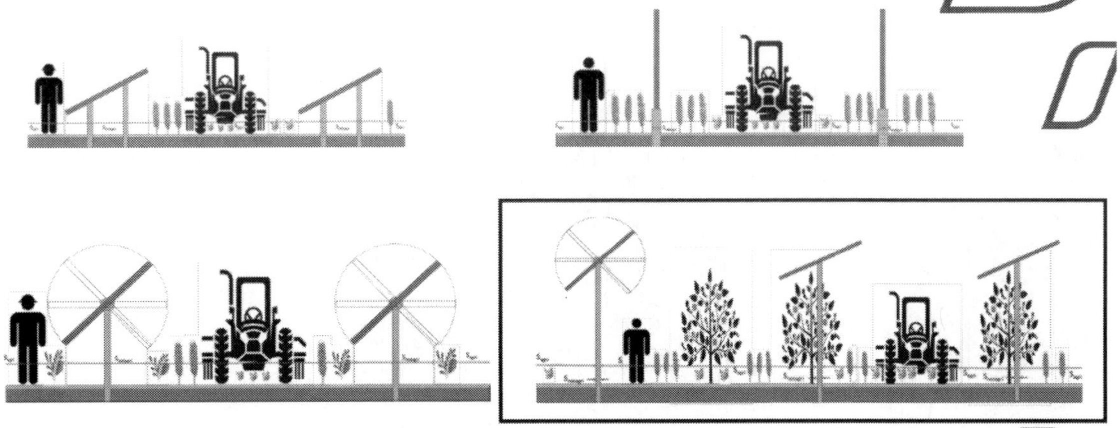

17

Figures: courtesy of EFSOLARE

020432-017

40th European Photovoltaic Solar Energy Conference and Exhibition

Concept and ambition

In SYMBIOSYST we will not develop solutions that are highly customized as this would drive the costs up

In SYMBIOSYST we will innovate by adapting standardized cost-effective solutions in terms of PV modules, mounting structures and Operation and Maintenance (O&M) practices to the specific needs of various crops in different climates and landscapes.

We will find aesthetically pleasing solutions that can be mass manufactured and integrated with the agricultural land in a harmonious way while maintaining the primary goal of farming

We will aim at obtaining a 2-way optimization where the agri-PV systems can be adapted to climate and crops, but also the culture underneath and the culture method can be adapted to fully benefit from the presence of the PV system. To this extent we will use agri-PV to develop (i) the orchard of the future and (ii) nearly zero energy greenhouses.

Our ambition is to create a fully integrated solution, from the design to the implementation, a symbiosis where PV and agriculture can have a mutually beneficial relationship.

18

020432-018

SYMBIOSYST

LEARN

INNOVATE

TRANSFER

020432-019

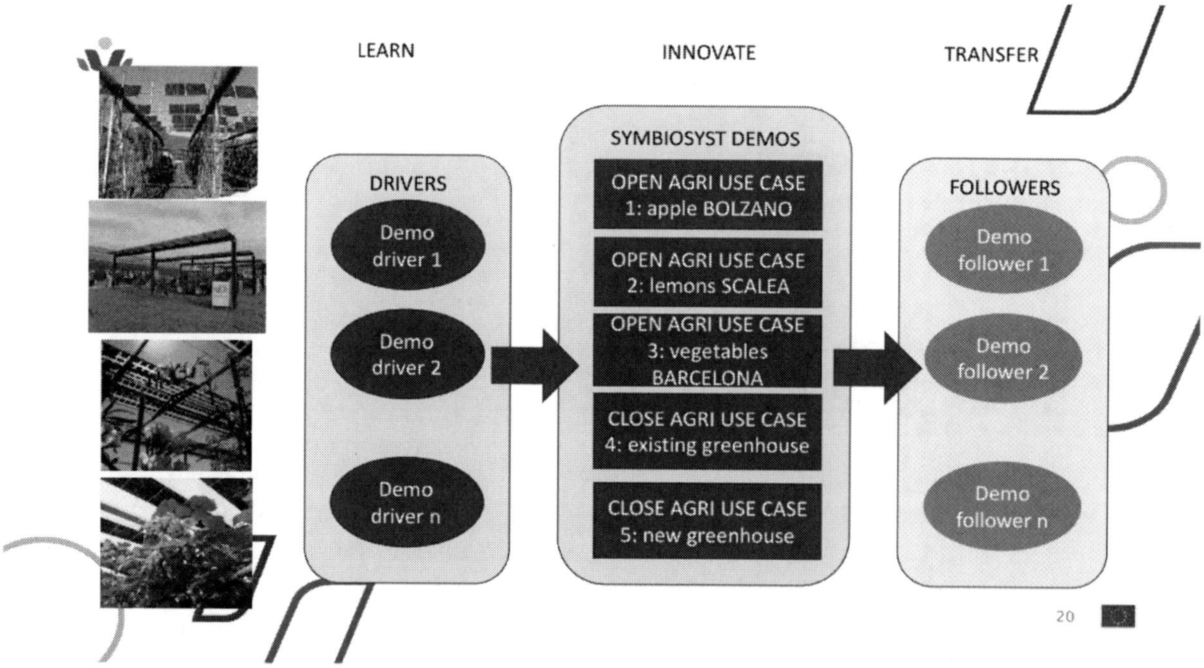

020432-020

SYMBIOSYST LEARNS

020432-021

40th European Photovoltaic Solar Energy Conference and Exhibition

020432-022

Ora, Italia

- **Crop**: apples
- **Type**: Overhead / fixed
- **PV modules**: Standard
- **Features**: In operation since 2008. No impact on crop

020432-023

40th European Photovoltaic Solar Energy Conference and Exhibition

020432-024

Bierbeek, Belgium

Crop: pears

Type: Overhead / fixed

PV modules: Semitransparent

Structure: integrated with hail protection

Features: Hail protection. Night temperatures 1-2 degrees higher. During summer, shading with 1-2 degress cooler. Harvesting can be spread over more weeks

40th European Photovoltaic Solar Energy Conference and Exhibition

WE WANT TO LEARN FROM YOU!

If you have an interesting demo please let us know

26

SYMBIOSYST INNOVATES

27

020432-027

MODELLING

- Light sharing in open agrivoltaics: Modelling of semi-transparency and light diffusion
- Agrivoltaics on greenhouses: several challenges for light modelling
- Structural elements blocking the sunlight
- Modelling of the Albedo / Reflectance
- 3D modelling of the crops: high spatial resolution approach

28

020432-028

020432-029

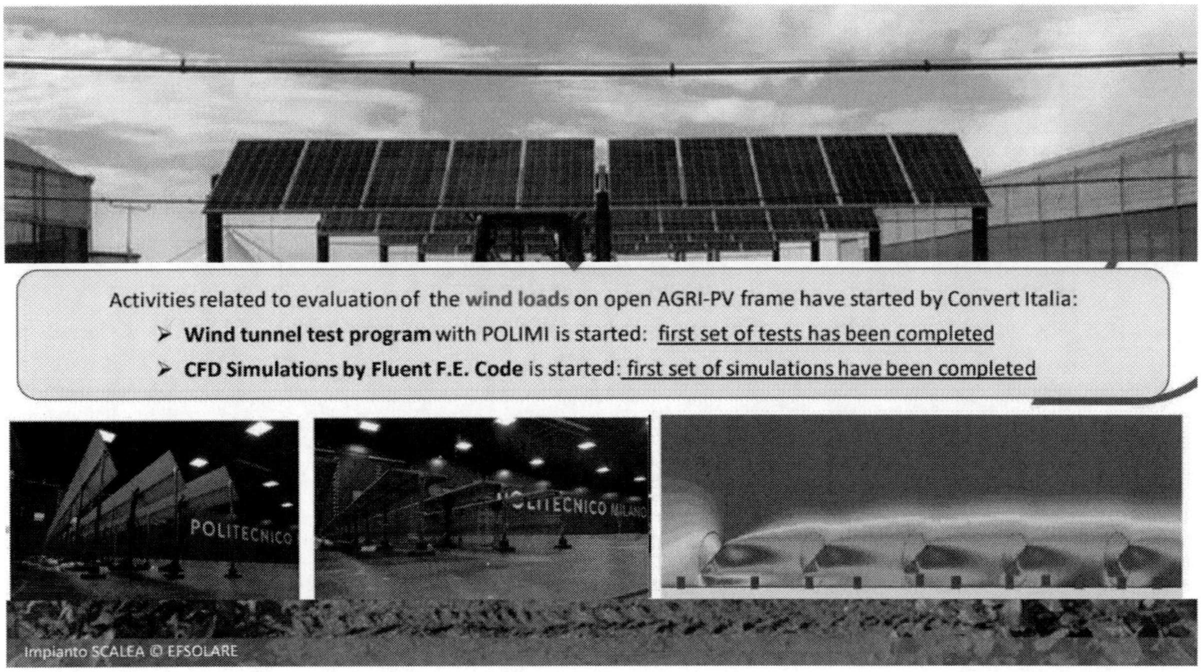

020432-030

ENVIRONMENTAL AND SOCIAL JUSTICE IN AGRI-PV

- Methodology: semi-structured interview
- Case study: Italy (followed by Spain and the Netherlands)

Symbol of the clean energy transition	Industrialisation and damage of natural landscape
Biodiversity alongside agricultural activities	Risk of agricultural activity becoming marginal

Productive diversification rather than intensification of monocultures

Energy and economic independence for farmers	Large investor benefits economically at the expense of small farmers
Anticipate policies involving stakeholders	Risk of higher land prices

31

020432-031

40th European Photovoltaic Solar Energy Conference and Exhibition

Cross European Cooperation for Agri-PV

We search for relevant Agrivoltaic experiences and pilot plants in all EU Member States as for example other Agrivoltaic projects funded by Horizon EU (or other programs), stakeholder networks, pilot developers, etc.

We aim at knowledge sharing activities, good practices exchanges and presentations in webinar and the physical events.

The final aim is to start building a truly European community for Agri-PV development

Please when exiting the room reserve collect the card with the bar code through which provide your contacts and feedbacks on critical topics/issues in Agri-PV development for future research agenda

32

020432-032

Symbiosyst Horizon EU Project

European Cross Cooperation for Agri-PV development

Please through the QR code provide your contacts and feedbacks on critical topics/issues in Agri-PV development and be part of the network

etaflorence renewableenergies 33

020432-033

40th European Photovoltaic Solar Energy Conference and Exhibition

Media exposure!

Produrre contemporaneamente energia e mele, la scommessa in Alto Adige

Symbiosyst. Il fotovoltaico a misura di campo

Con il progetto europeo Symbiosyst al via l'agrivoltaico "su misura"

Enea: «Soluzioni tecnologicamente avanzate in grado di contribuire al raggiungimento degli obiettivi di decarbonizzazione ... etitiva e sostenibile»

DINERO

¿Es viable combinar agricultura y paneles?

• La agrovoltaica plantea aplicar ambas actividades en un mismo terreno para aprovechar sinergias

:o Symbiosyst

ltaico EF Solare, che si
ovative solu...

SOLARE B2B

SENEC

HOME CHI SIAMO SOLARE B2B SOLARE B2B WEEKLY DOCUMENTI VIDEO E PODCAS

Eurac Research: al via il progetto Symbiosyst per lo sviluppo dell'agrivoltaico in Europa

written by Editore 24/01/2023

020432-034

Thank you!

symbiosyst.eu

Funded by the European Union

Funded by the European Union. Views and opinions expressed are however those of the author(s) only and do not necessarily reflect those of the European Union or CINEA. Neither the European Union nor the granting authority can be held responsible for them.

020432-035

CHALLENGES AND POTENTIAL FOR OFFSHORE FLOATING SOLAR IN THE DUTCH NORTH SEA

Wim Soppe, Minne de Jong, Marco Huisman, Wiep Folkerts, Jan Kroon*,
TNO Energy and Materials Transition – Solar Energy, Westerduinweg 3, 1755 LE, Petten, the Netherlands
*Corresponding author: jan.kroon@tno.nl

ABSTRACT: The past few years there is an increasing attention for the development of Offshore Floating Solar as it is a technology with substantial large power generation potential for emission-free production of electricity for the Netherlands. Especially when this can be applied in the Dutch Exclusive Economic Zone (EEZ), in particular in the areas used by offshore wind farms. However, it is still in a in a relatively early stage of development. In this contribution an overview will be given of the main research challenges which can be classified into four main categories: 1) Design and construction of offshore solar systems 2) Operation and Maintenance 3) Power production and LCOE and 4) Societal acceptance. Several offshore floating solar system concepts that are currently under development by commercial enterprises will be introduced together with near future plans of pilots and demonstrator projects in the North Sea in the period of 2023-2027
Keywords: floating solar, offshore, reliability, LCOE, operation and maintenance

1 INTRODUCTION

Floating Photo-Voltaics (FPV) is a rapid developing technology and worldwide there are dozens of companies active in this field, offering commercial solutions with a high maturity level for inland based systems. This has led to a global installed Floating PV capacity of around 6 GWp in 2023, mainly concentrated in South East Asia countries like China, Japan, Singapore and India. Europe is catching up quite rapidly with the Netherlands as frontrunner with an installed capacity of 230 MWp in Q1 2023 [1]. The installations are mainly built on static fresh water bodies like sand extraction lakes, drinking water reservoirs and ponds, where the circumstances are relatively mild in terms of wind and wave-induced loads.

In order to fully exploit the potential of FPV as an important contributor to the large scale deployment of PV, solutions for larger area waterbodies are needed. These waterbodies are characterized by significantly higher wave heights which require more robust system designs while keeping the capital expenditure (CAPEX) costs low. Therefore pilot studies with small scale systems are ideal to investigate the most important technical challenges with respect to performance, reliability and lifetime of the components while keeping the costs to an affordable level. We have carried out pilot studies on several Floating PV concepts since 2017 on our Field Labs at near-coast inland water lakes: the Slufter and Oostvoornse Lake in the Netherlands. Despite the fact that fully mature FPV systems have not been commercially demonstrated yet on these type of more exposed inland waterbodies, there is an increasing attention for the development of offshore floating solar.

Relevant parties in the Netherlands recognize that in view of land scarcity Offshore Floating PV (OFPV) could have an enormous potential for emission-free production of renewable electricity. Especially when this can be applied in the Dutch Exclusive Economic Zone (EEZ), in combination with offshore wind farms, by sharing electrical infrastructure, profiting from the synergies in construction and maintenance and increasing the capacity factor of the grid (cable pooling) as a result of the complementary wind and solar generation patterns [2]. The integration of OFPV is now also actively being promoted in the recent tender conditions for future offshore wind farms [3].

However, the technology is still in a relatively early stage of development and the challenges that need to be resolved to accelerate commercialization to a level that makes it competitive in terms of costs, energy yields and reliability to other technologies for renewable electricity production are enormous.

The major challenges that need to be overcome to make OFPV commercially viable are:

- Reduction of Levelized Cost of Energy (LCOE) to a competitive level (increase power yield, decrease CAPEX and operational expenditure (OPEX), increase economic lifetime)
- Reduction of risks: (Reliability, lifetime of system and components, societal acceptance)
- Minimize environmental impact (Low Carbon Footprint, improvement of circularity of materials and components)
- Stakeholder acceptance. Make evident that OFPV has no or minimal negative ecological impact; minimize negative effects for other users of the sea area.

Within this context, We have carried out a desktop study commissioned by the Dutch government (RVO) with the aim to create the guidelines for funding instruments to stimulate the development of OFPV on the North Sea. This has resulted in a report describing the main challenges and potential of OFPV and is based on publicly available knowledge and on own insights [4].

In this contribution a summary will be given of these main challenges which are classified into four main categories: 1) Design and construction of OFPV systems 2) Operation and Maintenance (O&M) 3) Power production and LCOE and 4) Societal acceptance. Several OFPV system concepts that are currently under development by commercial enterprises will be introduced together with near future plans of pilots and demonstrator projects and long term ambitions for OFPV in the North Sea.

2 DESCRIPTION OF SYSTEMS CONCEPTS

OFPV is not a fully established technology yet, but there are a number of conceptual development directions at the moment. For each concept/typology, one or more companies in the Netherlands and in other European countries are developing prototypes and demonstrators

10.4229/EUPVSEC2023/4DO.3.1
020433-001

that are of will be tested on several offshore test sites amongst others on the North Sea.

2.1 Elevated Truss typology

The elevated (truss) platform is equipped with floaters and mooring. A platform with solar panels is placed on the truss structure, supported by floating elements (Figure 1). The platform is located on a certain height above the waterline. Waves can move underneath the truss structure. The concept aims at avoiding direct contact between (slamming) waves and the solar panels. Examples of companies working on this concept in a construction or development phase are SolarDuck (Netherlands), Moss Maritime (Norway), Seavolt (Belgium), Swimsol (Austria), SINN Power (Germany), Solarinblue (France).

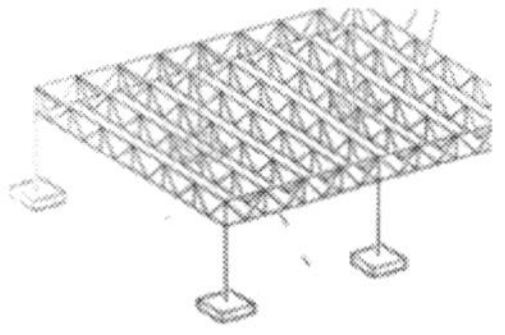

Figure 1: Schematics of an elevated truss typology

2.2 Pontoon typology

The *rigid pontoon* concept is based on rigid floating substructures (Figure 2) that are mechanically coupled into a massively modular structure. The complete floating structure is moored as one unit. Solar panels are mounted on the pontoons, and are positioned relatively close to the water level reducing the wind load on the PV panels and the substructure. Part of the structure may be overflown with high waves. The companies Oceans of Energy (The Netherlands), Heliorec (France) and Fred Olsen (Norway) are constructing and developing variations of this typology. Oceans of Energy already has demonstrated an off-grid pilot system up to 1 MWp installed on the North Sea, 12 km offshore from Scheveningen (The Netherlands)

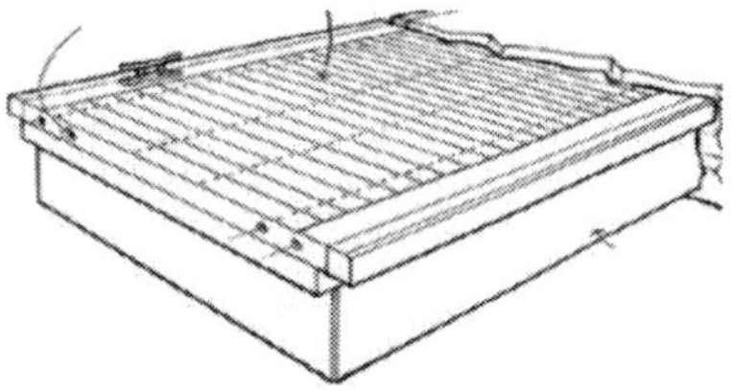

Figure 2: Schematics of a floating pontoon

The *flexible pontoon* concept: the idea of this concept is to let the floating structure move with the waves as much as possible, minimizing the transfer of mechanical energy from the waves to the structure thereby reducing mooring forces (Figure 3). The concept is based on the usage of inflatable mattresses as floaters, waterbags as stabilizers and light weight (semi-)flexible PV modules. These must be able to follow the deformation of the floater. Similar to the rigid pontoon concept, part of the structure may be overflown by high waves. Being near the water surface, it is less sensitive to wind loads. Bluewater and GENAP (The Netherlands) are developing this explorative concept.

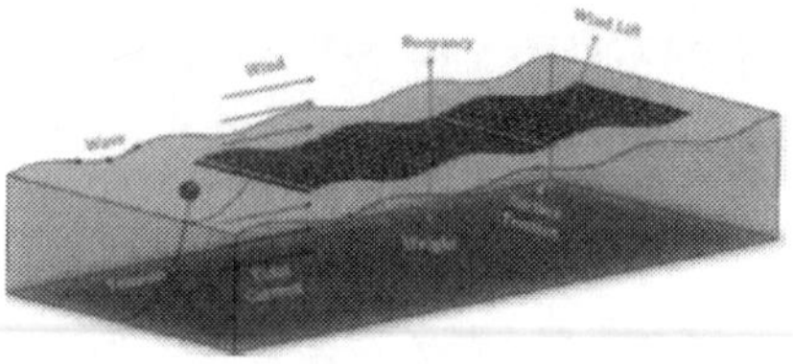

Figure 3: Schematics of the flexible pontoon concept

2.3 Floating Foil typology

The *membrane or fish farm* concept (Figure 4) is an example of a Floating foil concept: this concept builds on experience from near-shore fish farming. A floating ring equipped with a membrane is moored. Rigid solar panels are mounted on the membrane. Ocean Sun (Norway) is developing this concept and already demonstrated their system on various inland and offshore locations around the globe.

Figure 4: Schematics of the floating foil membrane concept

3 CHALLENGES FOR OFFSHORE FLOATING PV

3.1 Main identified challenges

The main challenges that we have identified can be divided into four categories and are listed in Table I:

Table I: Main identified challenges for OFPV

1. Design and construction
• Interconnects between substructures (floaters, mooring): corrosion, fatigue • Dynamic power cable connections (learn from floating wind) • Marine growth (buoyancy and drag) • Minimize non-accessible area • Materials choice: balance between costs, longevity and ecology
2. Operation and Maintenance
• Preventive and corrective maintenance • Performance losses, galvanic corrosion of electrical contacts • Insulation faults (cable burn out) • Potential Induced Degradation (PID) caused by negative system voltage • Mechanical degradation of solar cells by frequent deformations • Fouling – may cause hot spots • Salt residues on glass (less transmission, permanent degradation)

10.4229/EUPVSEC2023/4DO.3.1
020433-002

3. Power performance and LCOE
Validated power yield and LCOE models • System design challenges • Optimization of the combined production of solar, wind and H2 production • Power transfer and electrical storage at an offshore site (cable pooling) • Mutual interaction effects of Solar and Wind on production • Minimize/avoid Wave Induced Mismatch Losses • Fouling and resulting yield reduction • Factors that determine costs (PR, Plant availability ratio, CAPEX, OPEX
4. Societal acceptance
• Circularity of systems and components • Ecological effects • Permitting aspects • Inclusion of stakeholders and communities depending on maritime economic activities

From Table I it becomes clear that the identified challenges have both a technical and non-technical nature and requires a multi-disciplinary R&D effort from amongst others PV and material scientists, electrical and structural mechanical engineers, economists, ecologists and legal experts. In this paper, where the targeted readers are from the PV community, the aspects related to the PV part will be discussed in the sections 3.2-3.4: the reliability of PV components in a new PV application domain like OFPV, PV system performance and LCOE.

3.2 Reliability of PV components

Current IEC qualification standards were designed to identify early failures related to module design but not to predict long term durability and these standards do not address synergistic effects of multiple stresses in the field. Improved durability test programs including sequential testing of different stressors or longer duration single stress testing are required to prevent large scale field failures and to provide the information to investors, bankers, project developers for a more reliable risk assessment. PVEL and TüV provide these extended test programs [5,6].

The design of meaningful accelerated lifetime testing strategies for new applications is not straightforward and requires extensive long term field experience for a good correlation. For emerging application fields like (offshore) Floating PV, reliability datasets are not available yet so that direct laboratory-field testing correlations cannot be made.

It is anticipated that environmental stresses as moisture (in combination with salt), strong mechanical loads induced by wind and wave actions and partial shading due to bird droppings or other fouling mechanisms could have a stronger impact on module and component degradation in a floating environment compared to various other land-based operating environments.

Despite the lack of established testing standards, early recommendations for qualification testing as well as material/component choices are provided as a minimum requirement for application in an OFPV application to reduce the risk of early failures: PV modules should follow as a base case the minimum requirements of certification for module design and safety like IEC 61215 (module design), IEC 61730 (module safety), IEC 62970 (junction box safety) and IEC 62852 (connector safety) and IEC 61701 (salt mist corrosion). PV Modules that perform well in extended product qualification programs(PQP) and/or quality assurance programs from PVEL and TüV beyond the current standards offer a higher confidence level.

Much more scientific research and field studies via pilots and demonstrations is required to build up the information and knowledge base on the potential failure modes and subsequent definition of mitigation strategies as a basis towards recommended practices, international standards and bankable products.

Despite the fact that there are no clear design guidelines yet for the manufacturing of OFPV specific modules that are qualified for long term offshore use, some recommendations can already be given to reduce the risk of early failures [7]. At this stage, most of the recommendations given below are still qualitative, as the research towards seaworthy panels and components is yet in an explorative phase, but give potential directions for future material R&D:

- Glass-glass modules have desirable (mechanical) features for FPV applications but also potential downsides like weight. It is important to note that the mechanical robustness of a PV module is strongly dependent on how the panels are mounted on the floating structure. Mounting structures that would allow substantial deflections in modules, caused by continuous wind and wave actions in extreme offshore conditions, pose an increased risk of cell cracking and module breakage due to fatigue of the frame. To limit these risks it is recommended that for the various FPV concepts under development (pontoons, elevated truss structures) the (glass-glass) modules should preferably be mounted on a sufficiently rigid support to withstand the impact of strong wind and wave actions (like wave slamming), and avoid deflections beyond a certain threshold that could lead to undesired module failures in the early phases of a project

- In order to minimize moisture ingress via the edges of the PV modules, hydrophobic encapsulants and suitable protective edge sealants for this applications are recommended.

- PV modules in a floating environment could suffer from shading effects due to bird droppings, or other fouling phenomena (algae growth, limescale deposition). This will not only lead to yield losses but can also lead to potential irreversible degradation (hot cells and hot spots!) and safety (fire!) issues. With the development of new generations of high power PV modules with increasing cell and module sizes, more cells per bypass diode (24 instead of 20 per substring), hot cell temperatures may increase and thereby the risk of accelerated degradation of the module materials and decrease of performance. In order to mitigate these potential risks, advanced PV module designs that will limit the heat generation, materials with a higher relative temperature index and anti-soiling coatings should be considered in the selection of modules for a project.

10.4229/EUPVSEC2023/4DO.3.1
020433-003

- Junction boxes and connectors should have IP(X)68 rating to minimize the risk of corrosion
- Equipment such as string inverters (but also transformers and other units) shall be designed in accordance with applicable standards (IEC61683, IEC62109-1, IEC612109-2), as a base case. Equipment should be positioned in a suitable position on the OFPV system, sheltered as much as possible from the environment and preferably be certified to NEN-EN-IEC 60068-2-52. Enclosures for equipment should have an IP(X)68 rating, unless it can be demonstrated by a system provider that the equipment is suitably protected by other means.

3.3 PV system performance

The electricity production of a FPV system is determined by a number of factors like irradiation, temperature, orientation of the PV modules in a plant, movements caused by waves, soiling, fouling and degradation effects. This can be expressed in the so-called Performance Ratio (PR) which is defined as the annual yield divided by the theoretical maximum yield.

For an OFPV system on the Dutch North Sea the theoretical PR can be somewhat higher than for land since the average yearly irradiation level is 4-8% higher compared to Dutch land area [8], while ambient temperatures on sea are somewhat lower. The latter could lead to a positive performance or "cooling" effect although this is strongly dependent on the specific design and construction of the Floating PV system and the wind conditions.

Floater movements induced by waves will under specific circumstances result in strings of PV panels with different orientations. This could potentially result in mismatch losses with negative effects on the PR, depending on the configuration. Theoretical calculations have shown that these losses can amount up to 7 % for North Sea conditions for OFPV systems using one floater per module [9].

A maximum energy output can be achieved when PV modules are oriented towards the south with a certain tilt angle depending on the latitude of the system location, which is in the order of 40° in the Netherlands. However, OFPV systems are designed with smaller tilt angles ranging from 0 to 15° making them more robust against harsh wind conditions and mostly oriented in an east-west configuration distributing energy production over a full day. This will lead to power output compared to a solar park with an ideal tilt angle.

For the pontoon and foil concepts the panels are oriented horizontally. For the elevated truss concept usually a tilted East-West configuration is chosen. Simulations show that for the North Sea the PR is virtually the same for both configurations.

Soiling and fouling (bird droppings) of the PV panels can also lead to temporary PR losses or even permanent damage through hot spots if modules are not timely cleaned.

All these effects described above, system reliability and the Plant Availability ratio will determine the overall PR of an OFPV plant. This is unknown yet and be subject of upcoming pilot studies as will be described in section 4.

3.4 Levelized Cost of Electricity (LCOE)

At this stage of the development it is not possible to make accurate estimations of the LCOE of OFPV as there are no commercial large scale systems yet. The LCOE is determined by factors as the initial annual utilisation in year 0 without degradation (in kWh/kWp), CAPEX, OPEX, degradation rate and weighted average cost of capital (WACC). The most important challenge is to reach a competitive LCOE for a reliable and robust OFPV system with a lifetime of >25 years. A number of technical challenges are listed below:

- The CAPEX and installation costs are estimated to be significantly higher than for land as more expensive materials and components should be selected that are resistant against the harsh and corrosive circumstances on sea, while installation (and decommissioning) of OFPV is more complex than on land.
- OPEX is expected to be a significant cost factor as it is difficult and expensive to perform predictive and corrective maintenance on an offshore park that is not easy accessible, so a smart O&M strategy need to be devised to minimize these costs. Redundancy of critical components will also be taken into consideration.
- PV modules and Balance of System (BOS) components (inverters, cables) need to be qualified for offshore use to keep degradation rates low. Meaningful accelerated testing protocols need to be developed based on field experiences towards standardized procedures (see section 3.2).

An important strategy that will be followed for large scale OFPV deployment in the North Sea is to integrate it with offshore wind energy parks. The Dutch Government has already formulated the ambition to have 21 GW wind power installed on the North Sea by 2030 up to around 72 GWp in 2050 [10]. Solar and Wind Energy have largely complementary generation profiles and will increase the capacity factor of the electricity cable towards shore ("cable pooling"). This will lead to cost benefits but will introduce new system integration challenges like amongst others optimization of combined production of solar and wind (and hydrogen!) and minimize curtailing, power transfer and electrical storage, mutual interaction effects of solar and wind on production.

The long term perspective is that when all the aforementioned challenges are addressed and tackled in cooperative and extensive R&D between energy companies, concept developers, knowledge institutes and other important stakeholders, this will lead to the expected development of the sector that is required for the scale-up and reduction of the LCOE. In the next section a number of examples of joint R&D efforts in the North Sea region are summarized.

4 OFFSHORE R&D IN THE NORTH SEA

Several pilot and/or demonstration OFPV projects ranging from 0.5 to 5 MWp are planned on the North Sea from 2023 to 2027 (see Figure 5) to address the challenges as outlined in Table I and described in the former sections. On the Offshore Test Site of the North Sea Farmers, the three Dutch system developers Oceans of Energy, SolarDuck and Bluewater are building small scale pilot systems up to 1 MW in 2023 and 2024 and collaborate in joint development projects with research institutes, universities and companies to build up the knowledge base and accelerate the learnings that will lead to improvement of their designs towards realization of fully mature, scalable and low cost offshore solar park solutions.

In addition, Oceans of Energy and the Belgian company Seavolt are planning to build a 3 MWp and a <0.5 MWP pilot near the North Sea coast of Oostende.

In 2025 and 2026, two OFPV pilots of 0.5 MWp and 5 MWp are planned within the offshore wind parks of Hollandse Kust Noord (HKN) and Hollandse Kust West (HKW) operated by Crosswind [11] and RWE [12] respectively. Oceans of Energy and SolarDuck from the Netherlands are the selected OFPV system builders for these pilots. These will be first small scale demonstrations of offshore solar integrated in wind farms in the Dutch North Sea.

All these pilot and demonstration activities will pave the way for large scale deployment of OFPV after 2030. The Dutch Ministry of Economic affairs (EZK) has set an ambitious target of 3 GWp offshore solar deployment in 2030. This has already resulted in the inclusion of OFPV in the tender conditions and proposal rating approach for the next upcoming Dutch offshore wind farm tender, for IJmuiden Ver, which will close early 2024 [3].

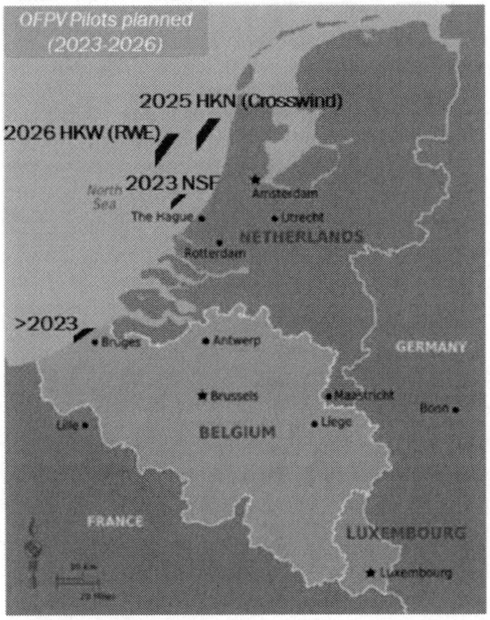

Figure 5: OFPV pilot and demonstrator projects planned in the period of 2023-2027

5 CONCLUDING REMARKS

OFPV is a promising development and can be one of the solutions for large scale solar deployment as an important contribution to the energy transition. However, many challenges need to be overcome to fulfil its potential. Various system concepts are currently under development and are in the beginning of their learning curves. Pilots and demonstrators are needed to accelerate the learnings with (pre-competitive) research programmes to address the main challenges. Pilot and demonstration OFPV projects up to 5 MWp have been planned in the period 2023-2027 in the Dutch and Belgian North Sea. These studies will generate a strong knowledge base and learning experiences that should finally lead to OFPV solutions (1) robust enough against the harsh North Sea conditions, (2) cost effective (3) easy to integrate within offshore wind parks and ready for large scale deployment after 2030.

6 ACKNOWLEDGEMENTS

The content of this proceeding paper is the result of:
(1) a study that TNO carried out as an assignment from the Dutch organization TKI Wind op Zee with financial support from RVO[4];
(2) a collaborative project on offshore solar between TNO and Crosswind C.V, with funding provided by CrossWind C.V

7 REFERENCES

[1] Market Trends of Floating Solar (rvo.nl)
[2] S.Z.M. Golroodbari et al., *Pooling the cable: A techno-economic feasibility study of integrating offshore floating photovoltaic solar technology within an offshore wind park*, Solar Energy 219 (2021) 65–74.
[3]https://www.rvo.nl/onderwerpen/windenergie-op-zee/ijmuiden-ver#tenderinformatie
[4] https://topsectorenergie.nl/nl/kennisbank/challenges-and-potential-for-offshore-solar/
[5] https://scorecard.pvel.com/
[6] https://www.tuv.com/world/en/quality-controlled-pv.html
[7] World Bank Group, ESMAP and SERIS. 2019. *Where Sun Meets Water: Floating Solar Handbook for Practitioners.* Washington, DC: World Bank
[8] J.F. Meirink, W. Koetse, C. Oudshoorn, 2018 KNMI-activiteiten Zon Op Water (zoninlandschap.nl)
[9] M. Dörenkämper et al., *Influence of Wave Induced Movements on the Performance of Floating PV Systems,* Proceedings 36th European Photovoltaic Solar Energy Conference and Exhibition, Marseille, France 2019.
[10]https://www.rvo.nl/onderwerpen/windenergie-op-zee/plannen-windenergie-op-zee
[11] https://www.crosswindhkn.nl/
[12] https://www.rwe.com/en/press/rwe-renewables/2022-07-19-hollandse-kust-west/

Analysis of climate conditions upon the driving distance of photovoltaics-powered vehicles

Masafumi Yamaguchi[1*], Taizo Masuda[2], Tsutomu Tanimoto[3], Yosuke Tomita[3], Yasuyuki Ota[4], Kenji Araki[4], Kensuke Nishioka[4], Christian Thiel[5], Anastasios Tsakalidis[5], Arnulf Jaeger-Waldau[5], Takashi Nakado[6], Kazumi Yamada[6], Tatsuya Takamoto[7], Takashi Mabuchi[2], Kenichi Okumura[2], Akinori Satou[2], Yusuke Zushi[4], Kyotaro Nakamura[1], Ryo Ozaki[1], Nobuaki Kojima[1] and Yoshio Ohshita[1]

[1] Toyota Technological Institute, Nagoya, Japan
[2]Toyota Motor Corporation, Susono, Japan
[3] Nissan Motor Corporation, Yokosuka, Japan
[4] Miyazaki University, Miyazaki, Japan
[5] European Commission - Joint Research Centre, Ispra, Italy
[6] Toyota Motor Corporation, Toyota, Japan
[7] Sharp Corporation, Nara, Japan

Abstract

The development of vehicles powered by vehicle integrated photovoltaic (VIPV) is desirable and very important for reducing CO_2 emissions from the transport sector to realize a decarbonized society. Although long-distance driving of PV-powered vehicles without electricity charging is expected in a sunny region, the driving distance of PV-powered vehicles is affected by climate conditions such as solar irradiation and air-conditioner usage. In this paper, analytical results for system efficiency of PV-powered vehicles and effects of climate conditions such as solar irradiation and usage of air-conditioners are presented by using our test data for Toyota Prius and Nissan Van demonstration cars. The potential of PV-powered vehicles to be deployed in major cities in Japan and the world is also analyzed. Mild weather cities such as Miyazaki in Japan and Sydney in Australia are thought to have long driving ranges even under the usage of air-conditioners.

1. Analysis for System Efficiency and Toyota Prius and Nissan Van Demonstration Cars

In this study, the effects of climate conditions upon the driving distance of VIPV-powered vehicles are presented by using analytical results for Toyota Prius and Nissan Van demonstration cars installed with high-efficiency InGaP/GaAs/InGaAs 3-junction solar cell modules with an efficiency of more than 30%. Table 1 compares the Toyota Prius and Nissan Van demonstration cars. The following equation calculated the daily driving distance (DD) of VIPV-EV [1]:

$$DD\ [km/day] = SI[kwh/m^2/day] x SE x \eta [\%] x A[m^2] x EM[km/kWh], \quad (1)$$

where SI is the global horizontal irradiation, SE is the efficiency of the PV system, η is the solar cell module's efficiency, A is the solar cell module's area, and EM is the electric mileage of the PV-EV. At first, PV system efficiencies SE were analyzed using power yield [kWh] and global horizontal irradiation SI [kWh/m²/day] in the Toyota Prius and Nissan Van demonstration cars.

Table 1. Comparison of Toyota Prius and Nissan Van demonstration cars

Model	Toyota Prius demonstration car	Nissan Van demonstration car
Photo		
Vehicle	PHEV	BEV
Solae Cell	InGaP/GaAs/InGaAs 3-junction solar cell	InGaP/GaAs/InGaAs 3-junctio solar cell
Module efficiency	More than 30%	More than 30
Installed area	Roof, hood, back door	Roof, hood, rear gate
Output power	860 W	1150 V
Battery capacity	8.8 kWh	40 kw
Electric mileage	~9.35 km/kWh	~6.6 km/kW

2. Effects of Usage of Air-conditioner upon Driving Distance of Toyota and Nissan Demonstration Cars

In general, passenger cars operate under air-conditioning, cooling, and heating; as a result, the electric millage is decreased, and thus, driving distances are shortened. The power consumption of the air-conditioner is changed dramatically depending on the outside temperature. Although the Toyota Prius demonstration car is under test driving, an average daily driving distance of 17.0 km/day at average solar irradiation of 4.0 kWh/m²/day has been demonstrated [2]. The average daily driving distance of 17.0 km/day is lower than the estimated daily driving distance of 26.6 km/day at average solar irradiation of 4.0 kWh/m²/day, and the effect of air-conditioner usage is thought as the cause of such a difference.

Power loss due to air-conditioning was determined from changes in electrical mileage EM using eq. (1), 0.739 as SE, 860 W as output power, and actual driving data. As calculated results, power loss was calculated by using eq. (2) in the case of average outside temperature T_{ave} and eq. (3) in the case of maximum outside temperature T_{max}, respectively.

$$EM = EM_0(16 - T_{ave}) x (\pm 0.4)\ [km/kWh/°C], \quad (2)$$

$$EM = EM_0(19 - T_{max}) x\ (\pm 0.4)\ [km/kWh/°C], \quad (3)$$

where EM_0 is the original EM of PV-EV in the case of no operation of air-conditioners.

Such values are fitted with experimental values, although the reference value is about -0.25 km/kWh/°C [3]. Especially the Toyota Prius demonstration car showed a large power loss of 6.4 km/kWh in August in Nagoya. EM decreased from 9.35 km/kWh originally to 6.6 km/kWh average as a result of air-conditioner usage in Nagoya in the Toyota Prius demonstration car. However, the decrease rate of EM from 6.62 km/kWh originally to 5.97 km/kWh average is lower than the Nissan Van demonstration car in Yokohama. Less EM reduction in the case of the Nissan Van demonstration car is thought to be due to mild climate conditions in

10.4229/EUPVSEC2023/4DO.4.1
020434-001

Yokohama and the driver's intention for power saving [4].

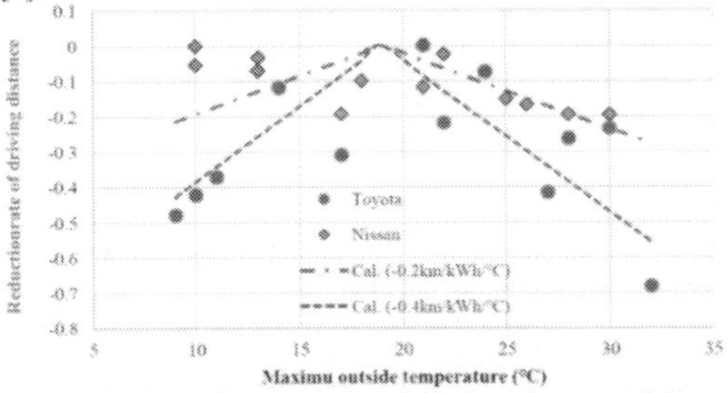

Figure 2. The reduction ratio of driving distance of the Toyota Prius demonstration car and the Nissan Van demonstration car during one year of driving due to using an air-conditioner versus maximum outside temperature during one-month driving and calculated reduction ratio of driving distance.

Figure 2 shows the reduction ratio of driving distance of the Toyota Prius demonstration car and the Nissan Van demonstration car during one year driving due to using an air-conditioner versus maximum outside temperature during one month of driving and the calculated reduction ratio of driving distance. As shown in Fig. 2, the reduction ratio of driving distance of the Toyota Prius demonstration car is dependent on -0.4 km/kWh/°C with the maximum outside temperature. In the case of the Nissan van demonstration car, the reduction ratio of driving distance seems to be dependent on -0.2 km/kWh/°C with maximum outside temperature and shows less reduction in driving range even in low outside temperature seasons. Such a situation is thought to depend on climate conditions, the performance of the air-conditioner unit, and the driver's intention.

3. Estimating Driving Distance of PV-EV under Usage of Air-conditioners in the Major Cities in Japan and the World.

Here, the driving distance of PV-EV in the major cities in Japan as a result of the usage of air-conditioners was analyzed by using eq. (4) and (5).

$DD[km/day]=SI[kwh/m^2/day]xSEx\eta[\%]xA[m^2]xEM_0(19-T_{max})x(\pm0.2)$ [km/kWh/°C], (4)

$DD[km/day]=SI[kwh/m^2/day]xSEx\eta[\%]xA[m^2]xEM_0(19-T_{max})x(\pm0.4)$ [km/kWh/°C], (5)

where 1kW was used as the output power of the VIVP module, 0.739 was assumed as SE, and 10 km/kWh was assumed as EM_0. Data for ambient temperature [°C] and daily average global solar irradiation [kWh/m²/day] were taken from the reference [5] in the major cities in Japan and the world.

Figure 3 shows annual driving distances calculated by considering power loss due to the usage of air-conditioners for PV-EV installed with 1kW VIVP module and with EM of 10 km/kWh in the case of driving in major cities in the world. Although sunshine cities such as Phoenix, Dubai, and Cairo show longer annual driving ranges of more than 15,000 km, the high reduction ratio of driving ranges of 48 - 57% and an annual driving range of 7,000- 8,000 km under usage of air-conditioners are estimated. On the other hand, mild weather cities such as Sydney and Sao Paulo are thought to keep longer annual driving ranges with 11,500- 12,00 km by less reduction ratio of driving range with 13 - 17% even under usage of air-conditioners. Populous cities such as Tokyo, New York, and Beijing show an annual driving range of 8,000- 9,000 km by reduction ratio of driving range with 24 - 34% even under usage of air-conditioners.

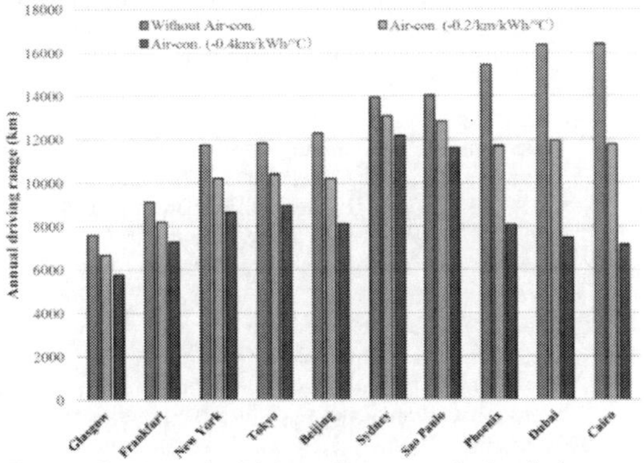

Figure 3. Annual driving distance calculated by considering power loss due to air-conditioner usage for PV-EV installed with 1kW VIVP module and driven in major cities worldwide.

5. Summary

This paper presented the effects of air-conditioner usage analyzed by using our test data for Toyota Prius and Nissan Van demonstration cars installed with high-efficiency 3-junction solar cell modules with a module efficiency of more than 30%.

Reduction ratios of electric mileage of the Toyota Prius demonstration car and Nissan Van demonstration car under usage of air-conditioners were estimated to be dependent on -0.4 km/kWh/°C and -0.2 km/kWh/°C with the maximum outside temperature. Such a situation is thought to depend on climate conditions, the performance of the air-conditioner unit, and the driver's intention.

Using power loss of Toyota Prius and Nissan Van demonstration cars, driving ranges of VIPV-EV under usage of air-conditioners in significant cities in Japan and the world were estimated. Sunshine cities such as Phoenix showed a high reduction ratio of driving range with 48 - 57% under usage of air-conditioner. On the other hand, mild weather cities such as Sydney showed less reduction ration of driving range with 13 - 17% even under usage of air-conditioners.

The other losses such as temperature rise of VIPV modules and partial shading was sugggested.

Acknowledgments

The authors sincerely thank the NEDO for supporting R&D.

References

[11 M. Yamaguchi et al., Prog. Photovolt. **29**, 684 (2021).

[2] T. Masuda et al., presented at the 49th IEEE Photovoltaic Specialists Conference, Philadelphia, June 5-10, 2022.

[3] T. Kobayashi et al.. Transactions of the Society of Automotive Engineers of Japan, **51**, 906 (2020). (in Japanese).

[4] Weatherspark, https://ja.weatherspark.com/

[5] T. Tanimoto, presented at the PVSEC-33, Nagoya, Japan, November 14-17, 2022.

10.4229/EUPVSEC2023/4DO.4.1
020434-002

This presentation was selected by the Sc. Committee of the EU PVSEC 2023 for submission of a full paper to one of the EU PVSEC's collaborating peer-reviewed journals.

SOLAR ENERGY HARVESTING FOR A CUSTOM SAILING YACHT, A CASE STUDY FOR PROJECT ZERO

Alwin Krul[1,2], Hans Wieggers[1], Elham Shirazi [2,*]

1) Vitters Shipyard BV, 8064 PD Zwartsluis, The Netherland
2) University of Twente, 7522 NB Enschede, The Netherland
akrul@vitters.com, hwieggers@vitters.com, e.shirazi@utwente.nl

ABSTRACT: The yachting industry aims to become more sustainable by incorporating sustainable energy resources such as solar energy on yachts. This research proposes a model to calculate yield the photovoltaic (PV) system integrated into the selected concept areas of a custom sailing yacht during different scenarios including location, time, and type of movement namely in harbour, at anchor or sailing. Shading is of large influence on the PV yield due to decrease in plane of array irradiance. As yacht is a non-stationary object and it will be on different locations from Mediterranean to Caribbean, during different times of the year, modelling shading is challenging. In this study therefore a model will be presented which provides insight in the potential of PV integration into the selected concept areas on the case study yacht considering the effect of shading caused by masts, booms and rigging of the yacht. The results show a large potential for PV integration into the yacht; however, the integration should be done in an aesthetically pleasing way, so as not to disturb the luxury exterior look of the yacht.
Keywords: Photovoltaic yield, Integrated PV, sailing yacht, shading.

1 INTRODUCTION

Efficient use of energy and alternative sources of energy are new trends in the yachting industry as it tries to become more sustainable [1]. Vitters Shipyard [2] works on a custom sailing yacht which does not rely on fossil fuels to provide electricity. It will use a large battery pack to provide electricity to the systems onboard. To be less dependent on charging infrastructure or other means of charging, integration of Photovoltaic (PV) technology to the yacht has been studied for the case study yacht, project Zero. Photovoltaics are already being used on yachts, as a source to charge the batteries, so to have less dependency on a generator or connection to shore power. However, it was found that the PV is often installed on roofs of the yachts as these are often not used for other purposes, have relatively large surface area and are not obviously visible from outside of the yacht. Due to the limited space for integration of PV, it is important to make best use of available surface to achieve the highest yield possible.

As yacht is a moving object, the shading pattern and consequently its effect on PV generation is more complicated. The yield of integrated PV technology as well as the shading for the yacht, a moving object, will be modelled in this study.

2 PV INTEGRATION OPTION

There is a set of requirements to be fulfilled while choosing areas for the integration of PV, mainly focused on not disturbing the selected exterior design and interference with the crew and guests. The concept areas should work for the most important scenarios which are 'at harbour' and 'at anchor'. PV could be integrated into many areas such as hull, mast, boom, deck, coach houses, shades and biminis. Based on the expected impact on comfort, design, or workload some ideas have been discarded. The sections below discuss the options that have passed the test against the design requirements although some of the concepts are a more radical design when compared to the requirements. The foredeck, spreaders, coach roofs, cap rail and bimini's. Some of these options have been looked at by the project but are further researched to be able to compare to the additional surfaces. The cap rail and spreaders are the areas that would seem too small, but further research will show the potential of these areas.

The chosen areas for PV integration are shown in **Figure 1** and include owners bimini, owners coach roof, main bimini, main coach roof, spreaders, foredeck, and cap rail. The selected areas are all horizontally placed to reduce visibility from outside the yacht. The roofs and sunshades provide the most surface area but are expected to be largely influenced by the shading of the masts, booms and rigging of the yacht.

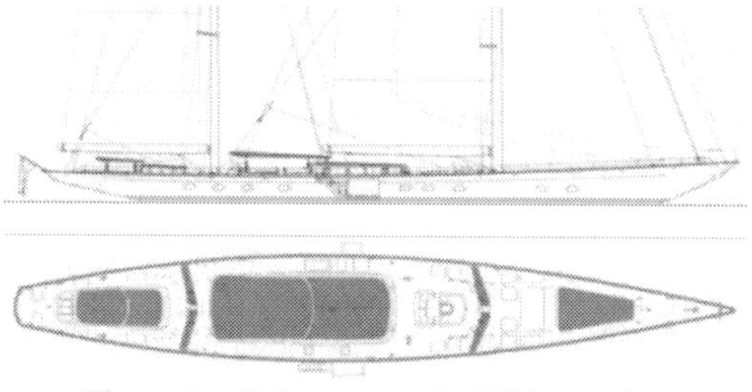

Figure 1. All chosen areas for PV integration

Total areas of chosen concepts for our case study is shown in **Table I**.

Table I total area of chosen concepts

Concept area	surface areas (m2)
Main coach roof	41.21
Owners coach roof	12.98
Foredeck	37.08
Main bimini	56.60
Owners bimini	14.33
Cap rail	25
Spreaders	37.12

* Corresponding Author

10.4229/EUPVSEC2023/4DO.4.3
020435-001

3. PV YIELD MODELING

3.1. Shading

Shading is of large influence on the yield of photovoltaic cells specially in the yachts due to decrease in plane of array (POA) irradiance. Even partial shading can reduce the yield of a PV panel drastically as solar cell with the least irradiance limits the other cells placed in series. Bypass diodes can help to reduce the effect of partial shading, although optimum number and architecture of bypass diode is not in the context of this study.

Different parts of the sailing yacht are the main causes of shading. The masts, booms, sails will shade large areas of the yacht and thus possibly parts of the installed PV panels. The layout of the yacht together with the design preference of the client will determine the possible locations of the PV panels and thus the shading as a result. The chosen masts and booms can determine the sizing of shade in different situations. When a roof or surface is closer to the boom for example, the shade will move less if the yacht is changing direction. Also, the size of the shading is determined by the size of the surrounding objects which are in line with the sun, and how fare they are situated from the concept area. When a yacht has 2 masts, the amount of shading can be more than if the yacht has only 1 mast. The number of small details on a yacht could reduce the possibility to place PV panels. If the PV panels are chosen, the yacht could be designed in such a way, the yield is expected to be highest. For the project of Vitters, the PV panels would be added to an existing design, which results in less options for PV integration then when the yacht would be designed with PV in mind or even with PV as highest priority.

3.2. Location

The yacht can be located in a lot of different locations, in each location the weather or even climate can be different. This can influence the use of the yacht drastically, as well as the PV yield. For the project yacht at Vitters the expected travel locations are mainly the Mediterranean and the Caribbean. The selected locations for this study are shown in **Figure 2**.

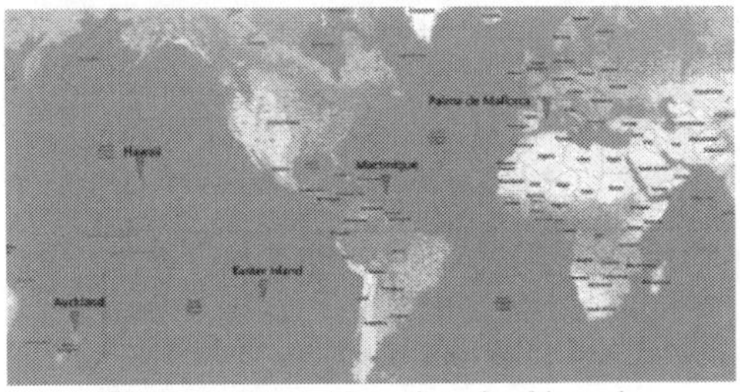

Figure 2. Selected Locations for this study

3.3. PV Technology and Modelling

Besides the abovementioned factors, the PV technology itself plays an important role in yield modelling. There are various PV technologies, with different specifications. The most important parameters which should be included in the modelling are Nominal Power (P_{mpp}) [W], Open-Circuit Voltage (Voc) [V], Short-Circuit Current (Isc) [A], Power Temperature Coefficient (Tp) [mW/°C], Voltage Temperature Coefficient (Toc) [mV/°C], Current Temperature Coefficient (Tsc) [mA/°C]. Depending on the requirements of the Shockley single diode model, some extra parameters of the PV technology are required.

In this study, different PV panels have been chosen to include in the PV yield model. The PV panels are chosen on different grounds. The Sharp NUJC-365 has been chosen because of the even black colour, standard sizing and use of half cut cells which should reduce internal losses, hotspot risks and have improved shading performance. The SunPower Maxeon 5 com 440W was later added because it was used to compare results, this panel also represents a standard PV panel and has a high efficiency of up to 21.7%. The SunPower SPR-E-Flex 110 has been chosen as a flexible option which would be used on smaller recreational yachts. The SunPower Maxeon Gen 3 solar cells are high efficiency cells which can be used to create a custom design for a PV installation. These types of cells are likely to be used for a custom solution for project Zero. A thin film option has been researched but due to its design it is hard to compare to other types of PV and thus hard to integrate in the PV yield model. Future development can include this type of PV in the model.

Bypass diodes, MPPT technology or the use of optimizers for the panels will influence the PV yield drastically, especially the Max Power Point Trackers are expected to increase yield during partial shading compared to panels without MPPT's.

3.4. Yacht Model

A 3D CAD model is used to study the influence of the shading/irradiance on the selected concept areas. A flexible setup is used in the CAD software to simulate many scenarios. By running the irradiance simulation for different orientations of the yacht and at different elevation angles of the sun, these results can be referred to for a large selection of scenarios when the photovoltaic yield is calculated.

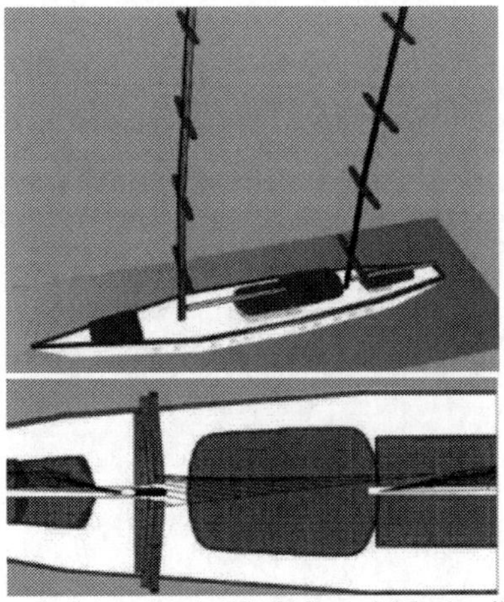

Figure 3. Grid points in 3D CAD model for irradiance simulation.

The irradiation that reaches each concept area is imported into MATLAB for further calculations. The expected PV output for each irradiance category per concept area is summed to get the total expected PV yield per concept area. The calculations are performed based on inputs regarding the to be simulated scenario. The inputs required from the user of the model are the timeframe of

10.4229/EUPVSEC2023/4DO.4.3
020435-002

the simulation, the chosen location, the scenario of the yacht (in harbour, at anchor or sailing), the orientation of the yacht in relation to the North and the type of PV type, of which several have been included in the model. Weather data with hourly resolution was chosen as daily weather data would not be sufficient for a moving yacht at anchor. The month August in Palma the Mallorca is mainly used for the simulation as this location and period is used by the project team of Vitters Shipyard.

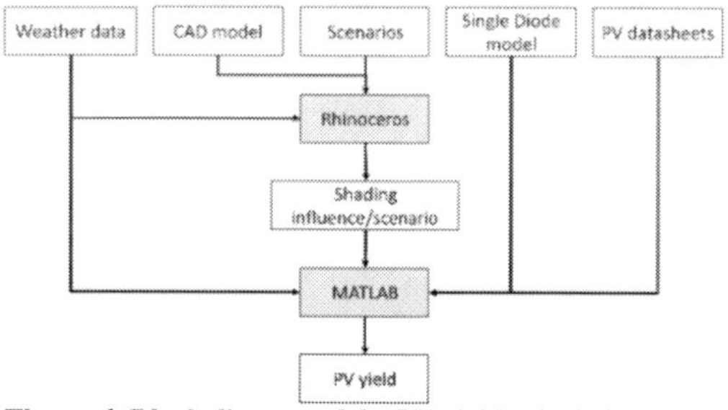

Figure 4. Block diagram of the PV yield calculation

The block diagram is showed in **Figure 4**, which will be used to calculate the expected PV yield based on the selected parameters. The effects of shading and irradiance can be taken from the earlier stages of the model.

4 RESULTS AND DISCUSSION

Two locations (Fort-de-France, Martinique and Palma de Mallorca, Spain) have been chosen in the areas that the yacht is expected to visit the most (Mediterranean and Caribbean). Three other locations are chosen to include to offer more comparing options (Hawaii, Easter Island, Auckland). The effect of shading is linked to the reduction in irradiance and thus lower values for PV yield. SunPower Maxeon Gen 3 cells [3] are the main focus for the PV type used, as these cells could be arranged in a custom pattern to fill the selected areas with larger efficiency. The shading pattern can be seen in the simulations as well, based on which, the future placement of panels can be based. The changing shading pattern due to changing orientation will determine the partial shading effect on the yacht. The shading pattern will change based on the orientation, but the shaded area does not change drastically, as it is self-shading by masts, booms, spreaders and the bimini. The shading due to the booms are parallel with the length of the yacht and the shading due to the spreaders are perpendicular to the length of the yacht. The shading of the masts will rotate as the orientation of the yacht or sun change. The shading pattern of yacht under to scenarios harbour and sailing has been shown in **Figure 5**.

(a) (b)

Figure 5. Shading patterns in (a) harbour (b) sailing

When smaller sized PV panels are used the effect of shading was found to be reduced. As the PV yield model will set the amount of PV panels included based on the size of the PV panel and the available surface area, a larger panel will fill the available surface area less efficient. For a single PV panel, the PV yield reduces with 5.22% while for a scenario with two PV panels in series the yield reduces with 10.85%. The scenario simulated is a single hour on the first of August in Palma de Mallorca for the foredeck concept at 0 degrees orientation. The yield has also been simulated for each of the selected locations. As expected, the location with the highest GHI in a year has the highest calculated yield for each of the concept areas. Auckland does have a higher yield than Easter Island, due to the lower temperature combined with a similar GHI value.

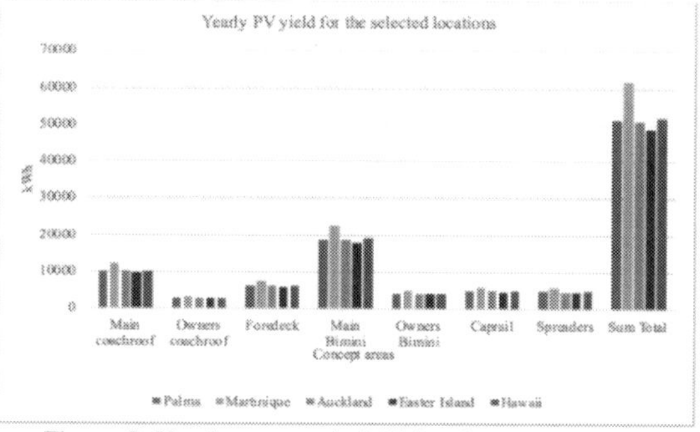

Figure 5. Yearly expected yield for the concept areas in different locations.

Depending on the chosen reference value (862 or 511 kWh/24h) the modelled concepts could provide 25.4% or 42.9% of the hotel load when at anchor. This was calculated for the first week of August in Palma de Mallorca. The reduction in PV yield when comparing the yield for the best and worst week of August in Palma de Mallorca is 24.14% for the owners coach roof. This is the most shaded surface. The reduction in yield is less than the reduction in irradiance, as expected, due to the reduction in average temperature. The main bimini, the largest surface will have the highest PV yield, however, the least shaded area will have the highest yield per m^2. For the first week of August, the foredeck has the highest yield per square meter, the cap rail and spreaders are outperforming the smaller bimini and could be further researched for PV integration. The standard test conditions (STC) have been validated for the included PV types. This resulted in a maximum deviation of 4.5% compared to the datasheet of the manufacturer. When the yield of a non-shaded scenario is compared to the result of Helioscope the created PV yield model outputs a 2% higher yield. More research regarding validation options is recommended.

5 CONCLUSION

The integration of Photovoltaic panels has shown a great potential for yachts. Although the shading has large influence in reduction of irradiance reaching the potential areas to install PV on the yacht. It was found that for custom sailing yachts the design of the yacht is an important aspect. Integration of PV technology on a custom sailing yacht is therefore largely influenced by the chosen design of the yacht. Using the created PV yield

10.4229/EUPVSEC2023/4DO.4.3
020435-003

model, the concepts that have been selected can cover up to 42.9% of the electric energy load at anchor for the case study yacht, which is a significant part of the hotel load. In addition to the expected output for many different scenarios, the model also provides insight in the potential of each of the selected concept areas, for a moving object. Smaller sections of PV arrays will provide more energy under partial shading with having the right bypass diode configuration. A custom PV cell solution is expected as the custom exterior design will not match conventional PV panels. The decision can be made based on the trade-off between energy generation and the aesthetics which matches to exterior design of the yacht.

6 REFERENCES

[1] "Sustainable Shipping - EMSA - European Maritime Safety Agency." [Online]. Available: https://www.emsa.europa.eu/we-do/sustainability/environment/sustainable-shipping.html. [Accessed: 02-Feb-2023].
[2] "Our story - Vitters." [Online]. Available: https://vitters.com/. [Accessed: 01-Feb-2023].
[3] "SunPower Maxeon Gen III Solar Cells | SunPower." [Online]. Available: https://us.sunpower.com/solar-resources/sunpower-maxeon-gen-iii-solar-cells. [Accessed: 02-Feb-2023].

10.4229/EUPVSEC2023/4DO.4.3
020435-004

In with the Electrons, Out with the Heat: Evaluation of a Spectrally – Selective Foil in Sunroof – Integrated PV

Apostolos Bakovasilis[1,2,3], Jonathan Govaerts[1,2,3], Arttu Tuomiranta[1,2,3], Arvid van der Heide[1,2,3], Koen Strijckmans[5], Georgi Yordanov[2,4] and Michaël Daenen[1,2,3]

(1) Imec, imo-imomec, Energy systems, Genk, Belgium, (2) EnergyVille, imo-imomec, Genk, Belgium, (3) Hasselt University, imo-imomec, Hasselt, Belgium
(4) KU Leuven, Leuven, Belgium
(5) GroupMAM, Zele, Belgium

020436-001

40th European Photovoltaic Solar Energy Conference and Exhibition

Storm Daniel 4-9-2023 – 12-9-2023

020436-002

Problem Statement

How to optimise the design and material stack for VIPV

Different variations **of encapsulants, interconnections and spectrally selective foils** have an impact on the **PV performance and the temperature** within the vehicle.
The evaluation of **different designs** for the proper selection of the final design, in realistic conditions, would be **timely and costly**

In this work we want to

- **validate the modeling for curved modules**
- **extend the model with coated foils for spectral reflection to assess their impact on VIPV performance and temperatures**

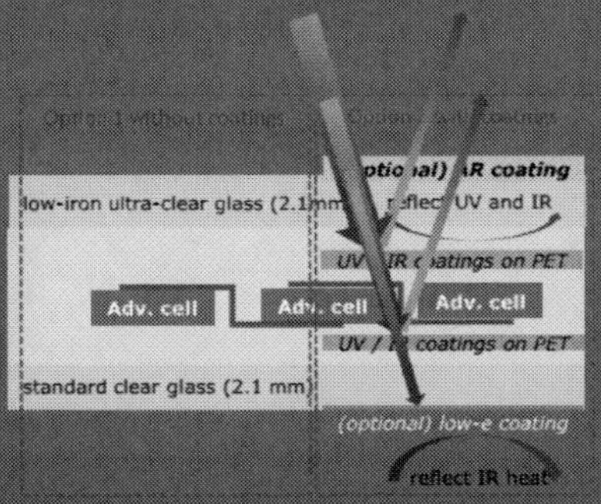

IMO-IMOMEC UHASSELT imec Energy Ville

Curved Model Validation

The PV simulation framework is validated with outdoor measurements for a curved sunroof

IMO-IMOMEC UHASSELT imec Energy Ville

Energy yield simulation framework

- Irradiance model – ray tracing
- Optical properties of module layers
- Different module interconnection topologies/ different array and configuration etc

- Coupled electro-thermal model

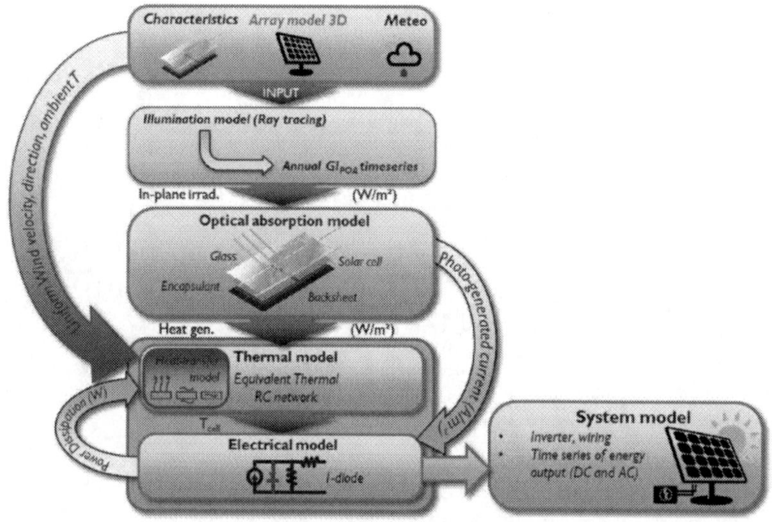

IMO-IMOMEC UHASSELT imec Energy Ville

6

Module Cell Configuration without spectral filtering

- **6 x 12** bifacial PERC cells of 158.75 x 79.375 x 0.16 mm
- Transparent cell spacing of 2 mm
- **STC cell voltage**
  V_{oc}: 0.68 V, V_{mp}: 0.59 V, β_{Voc}: -0.3 %/K
- **STC cell current** (I_{sc}, I_{mp}, γ_{Isc}) derived from the reference measurements
- **3 by-pass diodes**

Thermal parametrization

Datasheets + Literature (thermal resistivity)

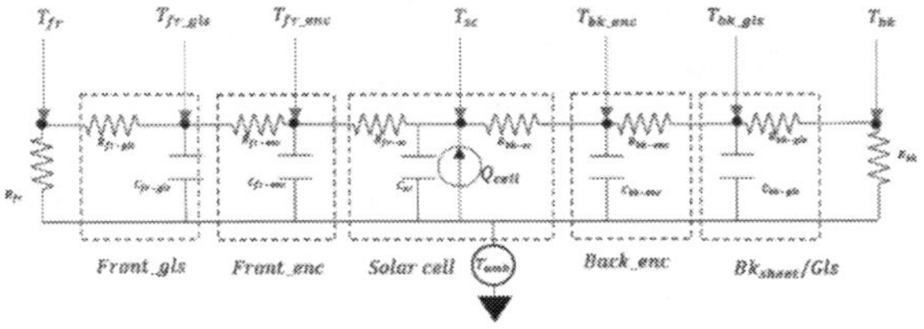

		thickness [mm]	density [kg dm⁻³]	resistivity [K m W⁻¹]	specific capacitance [kJ kg⁻¹K⁻¹]
	ultra clear glass	2.10	2.5	1.1	0.7
	clear glass	2.10	2.5	1.1	0.7
Thermoplastic foils	PET950	0.08	1.4	3.6	2.0
	PET1150	0.08	1.4	3.6	2.0
Encapsulants	4015	0.4-0.8	0.9	4.4	2.1
	R331C	0.80	0.9	4.4	2.1
	MW182	0.10	0.9	4.4	2.1
	cell	0.16	2.3	0.0	0.7

Module Calibration

IV and IAM measurements on outdoor tracker setup

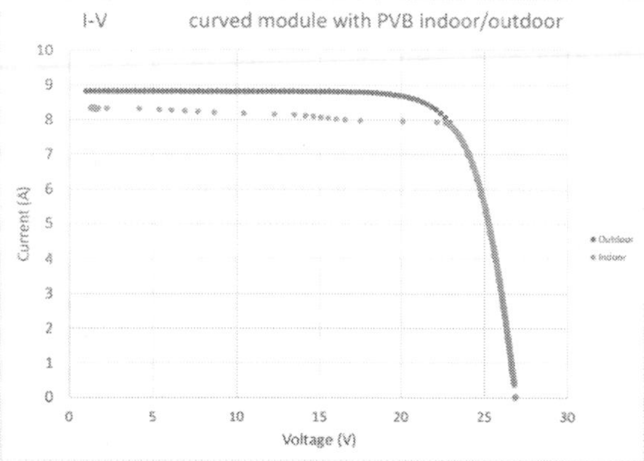

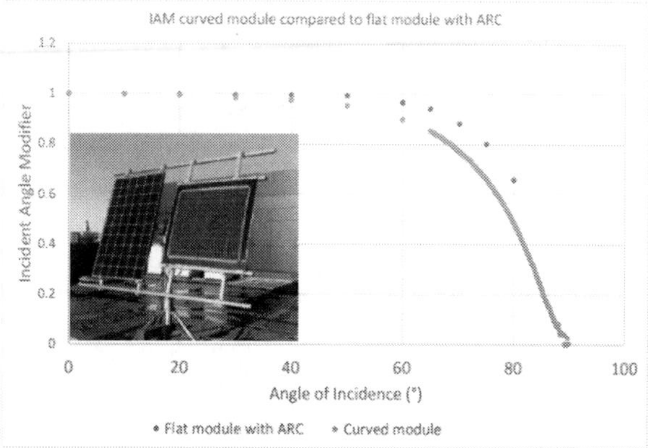

Instantaneous angle of incidence
power measurements on tracker

IMO-IMOMEC UHASSELT imec EnergyVille

A. Van der Heide et al., Accurate measurement of the angular dependence of PV module performance using a dual axis solar tracker, WCPEC-8, Milan, 2022

9

Curved model validation

- 3D drawing of the curved sunroof and the outdoors measurement setup
- Mean bias error of 2.7% between measurement and simulation
- Noteworthy that simulations of flat modules result in 13% increase of energy yield
  - Current mismatches due to intrinsic curvature

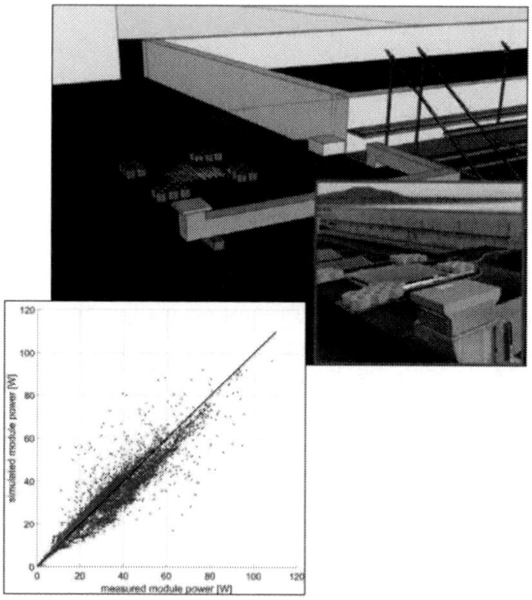

IMO-IMOMEC UHASSELT imec EnergyVille

10

Extending the model with coated foils for spectral reflection

IMO-IMOMEC
UHASSELT imec
Energy Ville

020436-011

40th European Photovoltaic Solar Energy Conference and Exhibition

Module design variations

PET foil front-side (1)

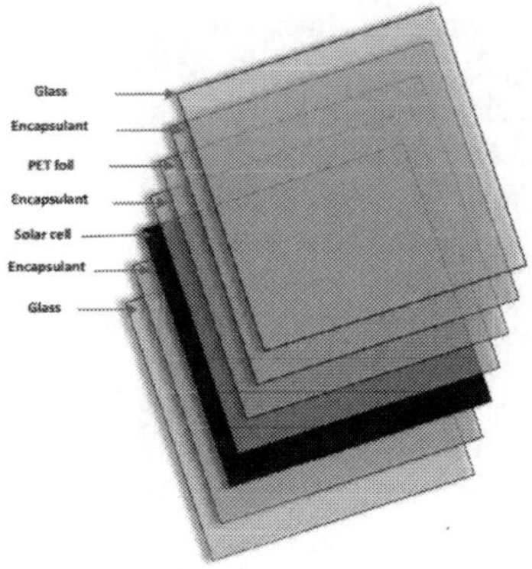

Reference (2) – no foil

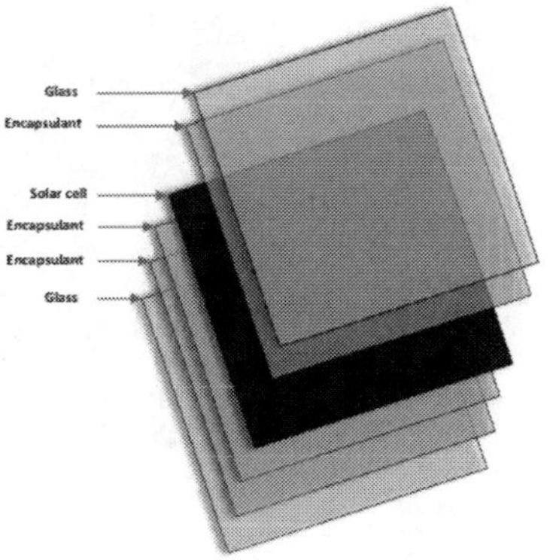

PET foil back-side (3)

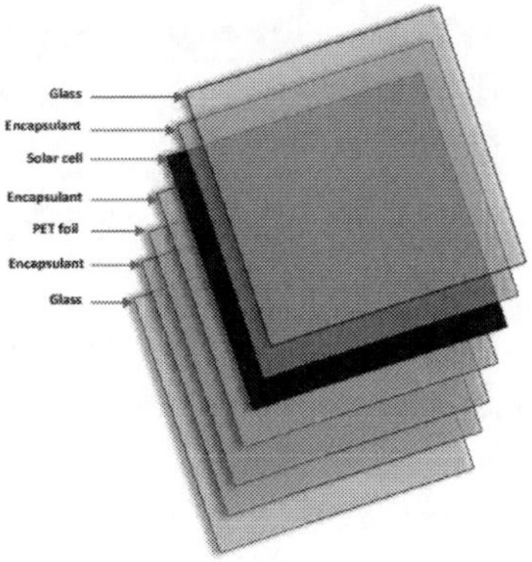

IMO-IMOMEC
UHASSELT imec
Energy Ville

12

020436-012

Spectral reflectance of PET foils

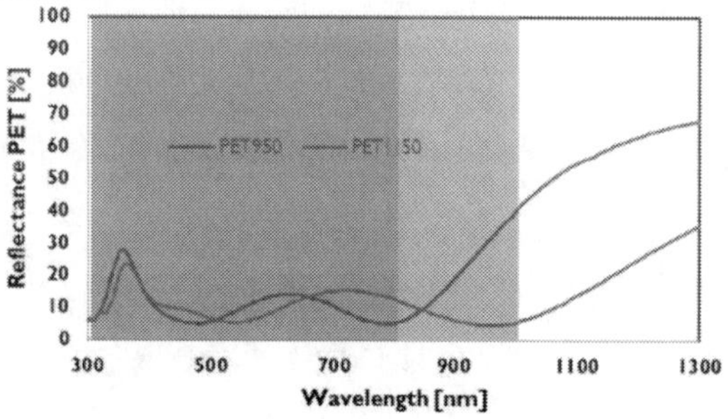

- PET1150: front side of SCs
  - It should transmit as many photons as possible in SC range
  - Infrared (IR) (>1200nm) should be reflected as much as possible
- PET950 back side of SCs
  - It should reflect as much as possible photons that are not absorbed by the cell

IMO-IMOMEC UHASSELT imec EnergyVille

13

40th European Photovoltaic Solar Energy Conference and Exhibition

Design variations matrix – corresponding module

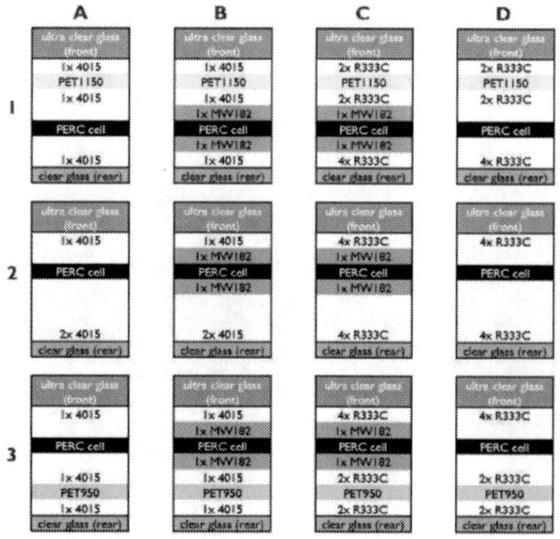

Two types of polyolefin encapsulants with different haziness level

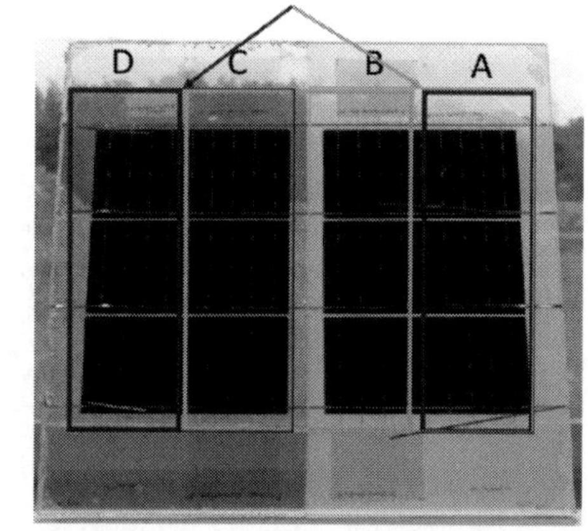

- Variations A, D: standard tabbing stringing
- Variations B, C: Multiwire foil interconnect

IMO-IMOMEC UHASSELT imec EnergyVille

14

Optical characterization of PET + Encapsulant

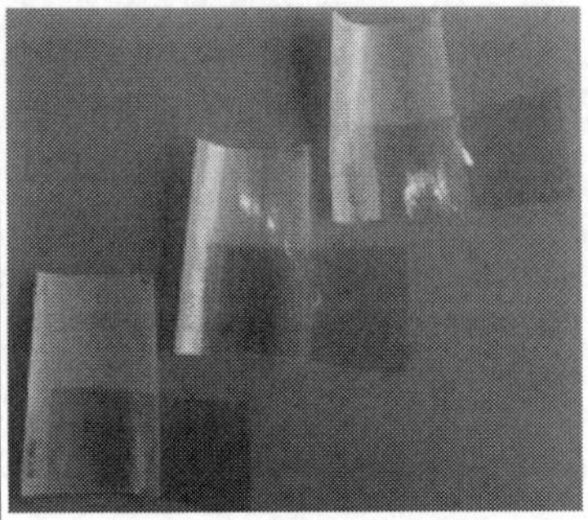

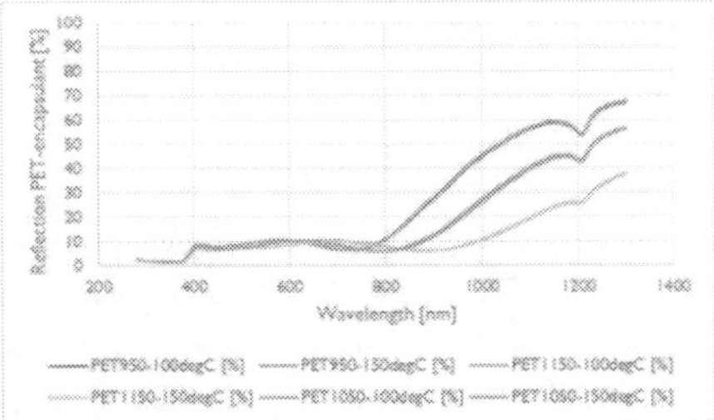

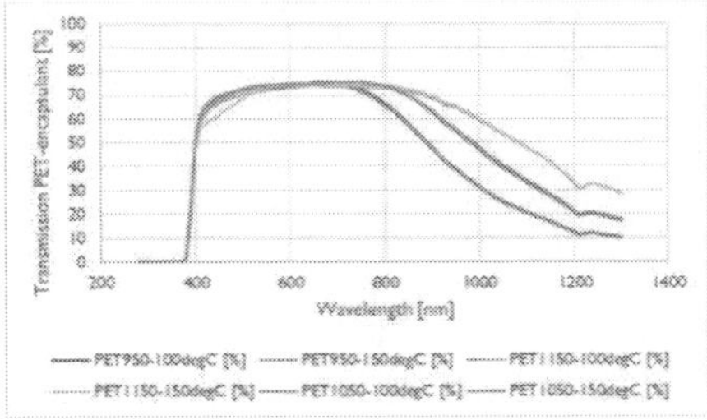

Reflectance and transmission measurements are used to calibrate the refractive indices of the front and rear stack in the ray tracing software

		4015 (UV clear)		R333C (UV cut)	
		988	MW	MW	988
		A	B	C	D
PET1150 (front)	1	12.2	13.0	11.1	12.1
Reference (no PET)	2	8.8	8.7	8.3	8.3
PET950 (rear)	3	9.1	9.2	8.7	8.7

IMO-IMOMEC UHASSELT imec EnergyVille

15

40th European Photovoltaic Solar Energy Conference and Exhibition

Vehicle scenario

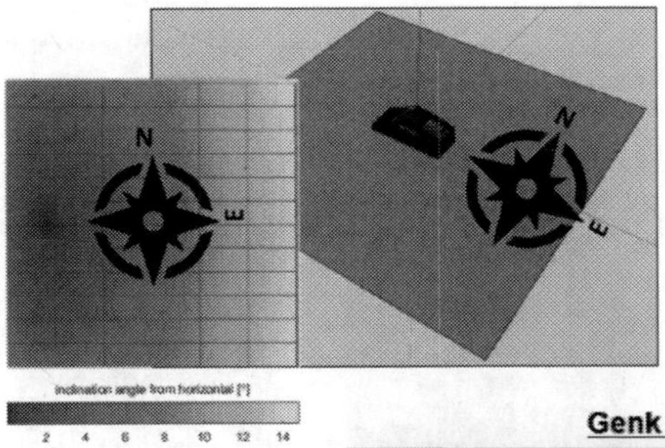

6x12 PERC cells

- Module facing east at a mean tilt angle of 6.6°
- Ground surface reflectance: 0.2
- Vehicle surface reflectance: 0.05

Genk (EV)

Phoenix

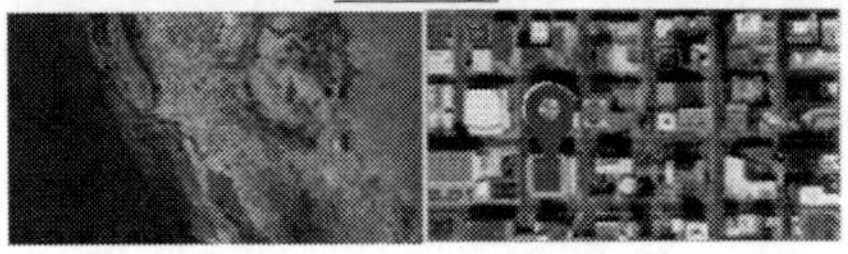

- **2 considered sites**
  - EnergyVille, Genk, Belgium (temperate oceanic, GHI: 1.1 MWh/m²/yr)
  - Phoenix, Arizona, USA (hot desert, GHI: 2.1 MWh/m²/yr)
- **1 shading scenarios**
  - Not shaded

IMO-IMOMEC UHASSELT imec EnergyVille

16

Simulation workflow

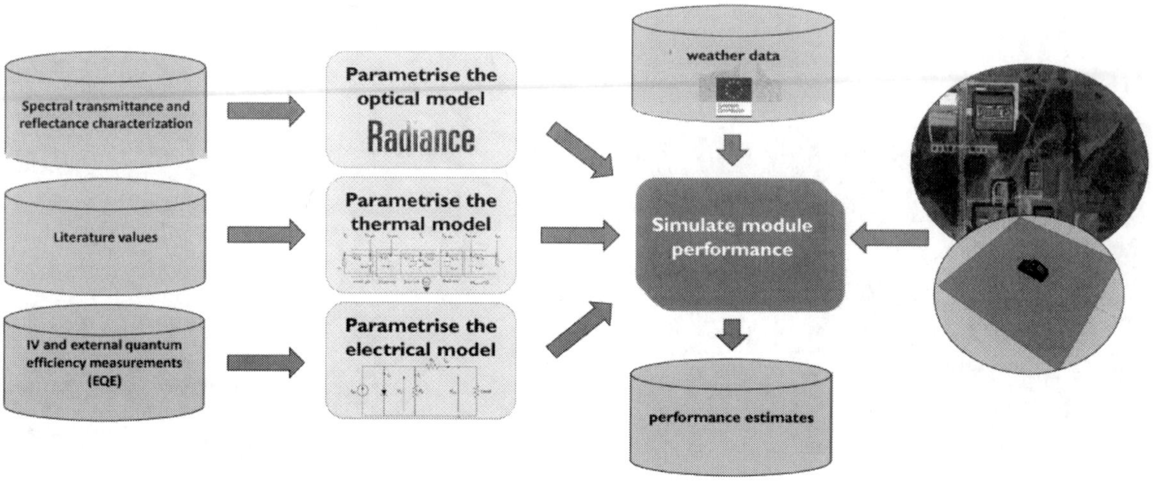

IMO-IMOMEC UHASSELT imec EnergyVille

17

020436-017

40th European Photovoltaic Solar Energy Conference and Exhibition

Results
Unshaded Annual Energy Yield

Test matrix:

	4015 (UV clear)		R333C (UV cut)	
	9BB	MW	MW	9BB
	A	B	C	D
PET1150 (front) 1	12.2	13.0	11.1	12.1
Reference (no PET) 2	8.8	8.7	8.3	8.3
PET950 (rear) 3	9.1	9.2	8.7	8.7

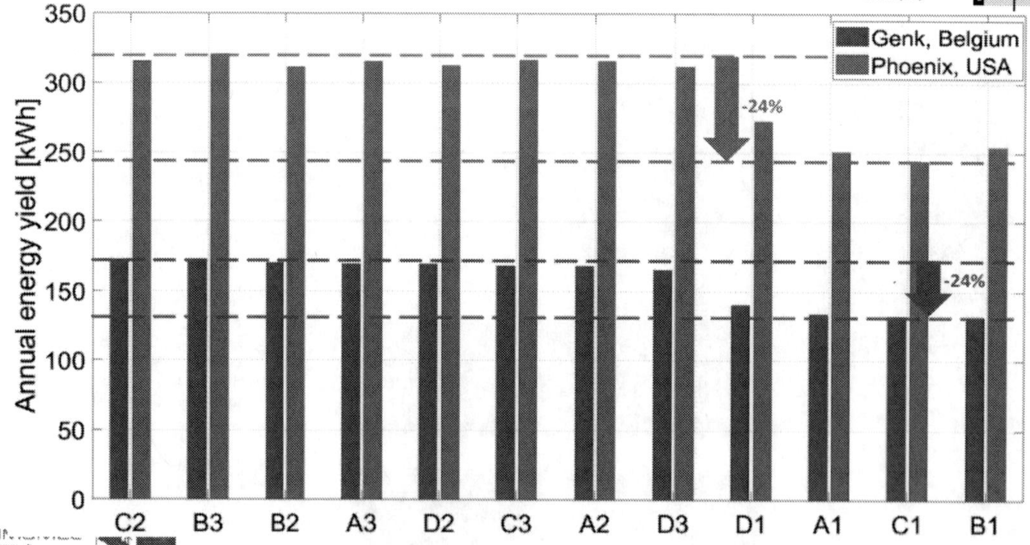

IMO-IMOMEC UHASSELT imec EnergyVille

18

020436-018

Potential estimation: Black Reference vs Al mirror

IMO-IMOMEC
UHASSELT imec
Energy Ville

Metal coating vs black PET background foil

Sample buildup

- Front glass ultraclear
- Encapsulant 2x RC999A
- SHJ cells + MW182 vs PERC5BB
- Encapsulant 1x RC999A
- PET Al-coated black vs black
- Encapsulant 1x RC999A
- Rear glass clear

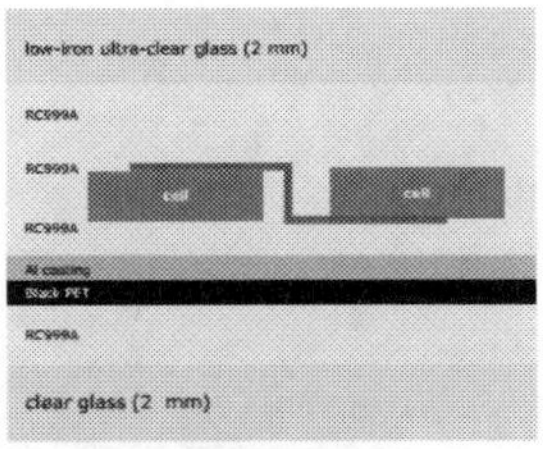

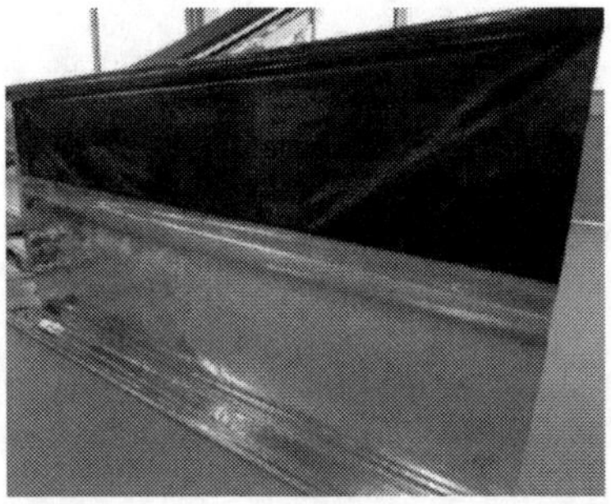

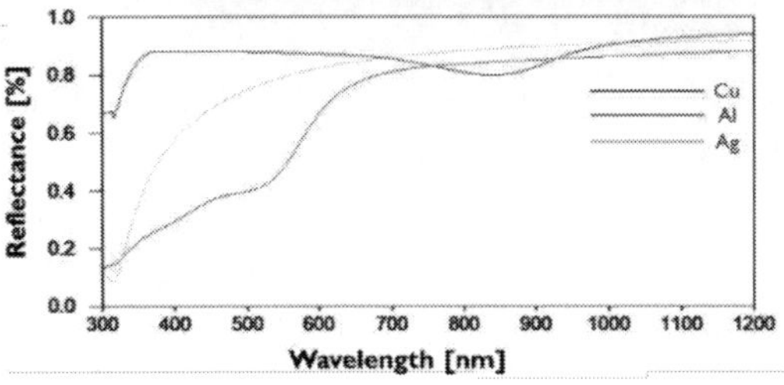

IMO-IMOMEC
UHASSELT imec
Energy Ville

20

Impact of cell type and reflector foil on Isc (full vs shielded vs EQE)

"shielded" ➔ no impact from edges (reflections from outside cell area)

"EQE" ➔ 2x2 cm² spot size in centre

	SHJMW-RC999A-black	SHJMW-RC999A-Al	increase [%]	
Isc_full [mA]	8804	9151	+3.9	Combined impact of rear and edge reflections
Isc_shielded [mA]	8778	9030	+2.9	Impact of only rear reflections (NB: semisquare corners)
Isc_EQE [mA/cm2]	36.57	37.14	+1.6	

	PERC5BB-RC999A-black	PERC5BB-RC999A-Al	increase [%]	
Isc_full [mA]	9627	9814	+1.9	Impact of reflections at edge
Isc_shielded [mA]	9559	9571	+0.1	No impact of reflector (PERC full rear metallization)
Isc_EQE [mA/cm2]	40.01	40.08	+0.2	

➔ ~ 2% increase due to rear reflector (for SHJMW)

IMO-IMOMEC UHASSELT imec EnergyVille

21

020436-021

40th European Photovoltaic Solar Energy Conference and Exhibition

Conclusions

- The physics based E-yield framework was validated on curved modules
  - MBE -2.7%
- A comparison was made between 12 options of module designs with reflective coated foils
  - Front-coated module types >20% lower yield than the other types
    - ~15-20% absorbance across range of 400-110nm
  - cell temperature is not influenced directly by rear coating
- A 2% Gain in Isc can be obtained in case of a full metal Al rear coating
- The influence on the vehicle temperature can further be studied and coupled to the model

IMO-IMOMEC UHASSELT imec EnergyVille

020436-022

Thank you for your attention

Author contact: Apostolos.Sakovasilis@imec.be

Supported by VLAIO: SNRoof

VLAIO

GROUP MAM

KU LEUVEN

IMO-IMOMEC

UHASSELT imec

Energy Ville

Thanks!

IMO-IMOMEC

UHASSELT imec

Energy Ville

Thermal model input

	thickness [mm]	density [kg dm^{-3}]	resistivity [K m W^{-1}]	specific capacitance [kJ kg^{-1}K^{-1}]
Ultra clear glass	2.1	2.5	1.1	0.7
Clear glass	2.1	2.5	1.1	0.7
PET950	0.08	1.4	3.6	2.0
PET1150	0.08	1.4	3.6	2.0
Enlight 4015	0.4-0.8	0.9	4.4	2.1
Apolhya R33C	0.8	0.9	4.4	2.1
MW182	0.1	0.9	4.4	2.1
Solar cell	0.16	2.3	0.0	0.7

IMO-IMOMEC UHASSELT imec EnergyVille

Electrical – Optical - Thermal model parametrization

IV and external quantum efficiency measurements (EQE)

Parametrization based on literature values

Spectral transmittance and reflectance characterization

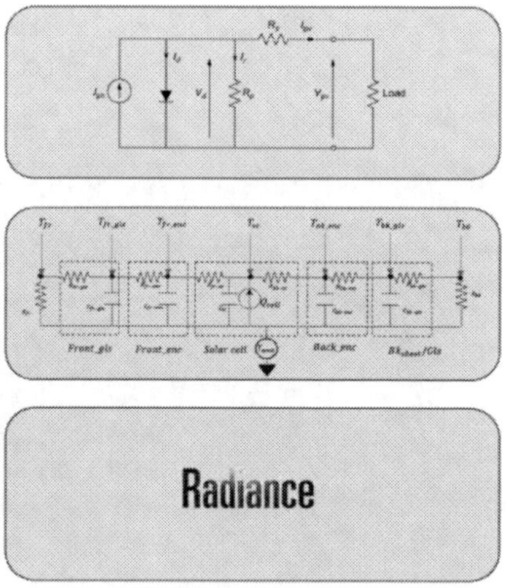

IMO-IMOMEC UHASSELT imec EnergyVille

Results
Cell Temperature

Test matrix:

		4015 (UV clear)		R333C (UV cut)	
		9BB	MW	MW	9BB
		A	B	C	D
PET1150 (front)	1	12.2	13.0	11.1	12.1
Reference (no PET)	2	8.8	8.7	8.3	8.3
PET950 (rear)	3	9.1	9.2	8.7	8.7

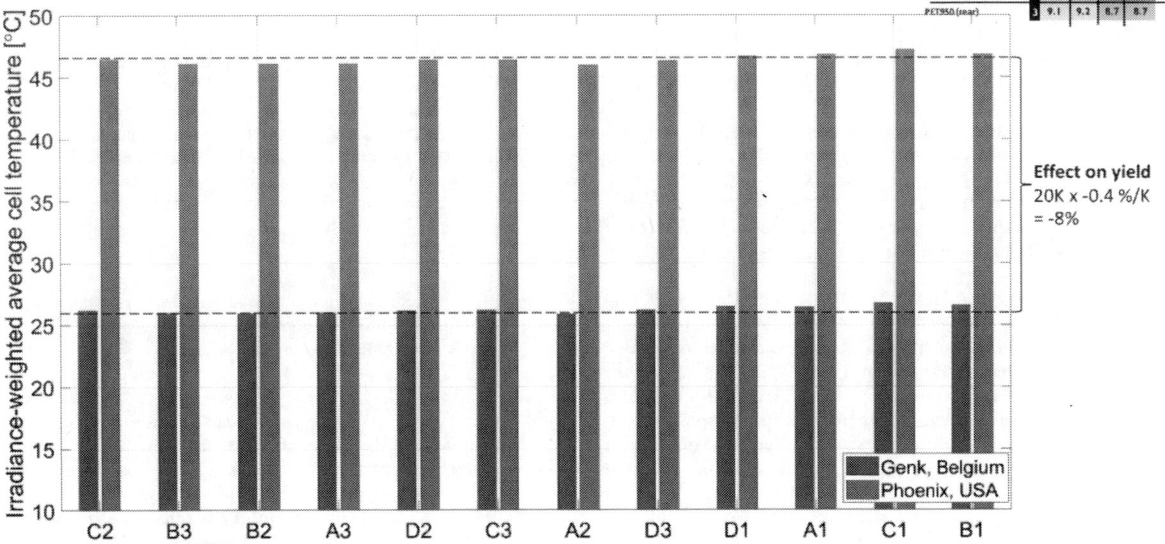

IMO-IMOMEC UHASSELT imec EnergyVille

27

020436-027

40th European Photovoltaic Solar Energy Conference and Exhibition

This presentation was selected by the Sc. Committee of the EU PVSEC 2023 for submission of a full paper to one of the EU PVSEC's collaborating peer-reviewed journals.

SOLAR FOR RAILWAYS - INVESTIGATION OF THE PV POTENTIAL ON THE GERMAN RAIL INFRASTRUCTURE

Magnus Herz*[1], Annett Sepanski[1], Uwe Hupach[1], Bernd Schönauer[1], Svend Ulrich[1], Jürgen van der Weem[1]
[1]TÜV Rheinland, Cologne, Germany, Magnus.Herz@de.tuv.com

ABSTRACT: The aim of this project is to determine the technical potential of photovoltaic applications on and in the rail infrastructure in Germany that can be considered as renewable generation systems for direct feed-in. Examples are the horizontal PV use in the track bed (track integrated) as well as the possibility to integrate PV applications in noise protection systems. For this purpose, the German railway network and the infrastructure facilities along the track will generally be investigated with regard to the applicability of PV systems. Taking into account the special requirements of the railroad system in operation, inspection and maintenance, the effects of PV systems on the superstructure or affected regulations will be identified and recommendations for action derived. Overall, the results show that PV application in railroad infrastructure offers various field-tested possibilities to enable energy supply to railroads by means of renewable energies on a larger scale.

1 INTRODUCTION

The national climate protection targets are formulated in the Climate Protection Plan 2050 adopted by the German Cabinet in November 2016 and contain a substantive orientation for the entire transport sector. According to the estimates of the Federal Ministry for Environment, greenhouse gas emissions for the transport sector in 2019 were 163.5 million metric tons of CO2 equivalent, almost the same level as in 1990 (164 million metric tons). This corresponds to a share of around 20.3% of total greenhouse gas emissions in Germany, The energy sector, for comparison, accounted for 254 million t in 2019, or 31.5% of total greenhouse gas emissions. By 2030, emissions in the transport sector are expected to fall to 95-98 million t. CO2-equivalent, a 42-40% reduction from 1990 levels. These targets are also anchored in the German government's current Climate Protection Program 2030. It specifies concrete measures to reduce emissions as well as further fields of action, e.g. decarbonization through the electrification of further rail lines or the use of alternative drive systems.

In terms of climate protection targets, the rail mode of transport can make a substantial contribution to greenhouse gas reduction through the high proportion of existing electrification in general. The efforts to provide or operate the transport service as well as the rail infrastructure itself in a climate-neutral manner require a steady expansion of the areas of application for the use of renewable energies. The current share of renewable energies in the rail electricity mix of 60.1% is achieved, among other things, through the purchase of electricity volumes with a certificate of origin from the public grid or, for example, from the railroad's own hydroelectric power plants.

As a further component for increasing the share of renewable energies, the direct feed-in of renewable energy plants into the 16.7 Hz traction current grid is currently being tested. For the extensive public rail network with a line length of about 33,500 km and an electrification level of about 61 %, direct feeding into the 15 kV overhead line results in additional application potential for RE energy generation directly on or in the rail infrastructure, in particular through higher overall efficiency as well as elimination of electricity transport costs.

2 METHODOLOGY

The study is divided into three work packages:

1. Market research of railroad relevant PV systems,
2. Quantification of technical potential, and
3. Assessment of requirements

In the first work package, we systematically investigate the PV systems relevant for use on or in rail infrastructure, including all necessary components. The investigation covers PV systems available on the global market and systems that are being or have been tested worldwide on or in rail infrastructure. The research revealed a large number of international research activities and test projects related to the increased use of photovoltaics in railroad operations.

Based on these results, in the second work package we quantify the PV potential and determine the existing areas on and in Germany's rail infrastructure that can be used for PV applications. Based on the available information and the defined criteria, a GIS-based determination and evaluation of the theoretical and technical usable PV potentials of the rail infrastructure was carried out. As a result, the feed-in potential of suitable PV systems on and in the rail infrastructure was presented in both spatial and temporal context.

The third work package deals with the requirements for PV systems, in particular from a railroad engineering perspective. The relevant railroad regulations are named and, if necessary, concrete recommendations are given for necessary adaptations due to the integration of PV systems on and in the rail infrastructure with consideration of direct feed.

3. RESULTS

3.1 MARKET RESEARCH OF RAILROAD RELEVANT PV SYSTEMS

The literature research revealed a large number of international research activities and test projects related to the increased use of photovoltaics in rail operations. The goal of the published activities is generally to supply electrified lines with 100% renewable energy. Where no economically viable electrification is possible, alternative propulsion systems (with traction battery) are being investigated, often with energy storage to be charged with renewables. The focus of the researched PV applications in operation or under construction for supplying energy to railroads is on larger free-field systems and rooftop PV systems. These mainly feed into the general 50 Hz grid. In countries with their own 15 kV / 16.7 Hz railroad grid, the topic of direct feed is also being pushed (Austria,

Switzerland) and tested in previously smaller applications. There are also some test projects with PV systems integrated in the track bed. Especially in India, the supplementary energy supply of trains via PV systems on the roofs of the carriages is applied. The search for analyses and research work yielded several completed works and some still in progress on the use of renewable energies with a focus on photovoltaics for traction power supply.

3.2 QUANTIFICATION OF TECHNICAL POTENTIAL

For the installation or integration of PV systems on and in the rail infrastructure, five application classes were defined according to the installation location (Figure 1). Each application class was described and the relevant characteristics were shown. Implemented examples were pointed out.

Based on the results of work package 1 in terms of requirement profiles and rough system designs, the design requirements for suitable PV systems were specified. Various criteria were defined, which enabled the exclusion of areas that obviously cannot be used economically for photovoltaics from the determined gross area. The structural conditions and special requirements in the railroad area were also taken into account in the selection criteria.

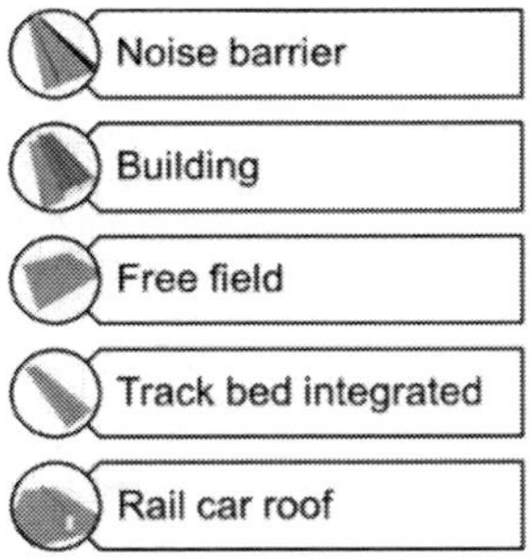

Figure 1: Defined application classes.

Following the published average LCOE for standard PV systems in Germany that feed into the public three-phase national grid, the additional costs required for integration into the rail system were estimated for the application classes described, based on average cost components and expert opinion. In each case, a range of expected power generation costs is given. The lowest additional costs and also the lowest LCOE are expected for the application classes of free-field systems and rooftop systems with the assumed usability of standard components for the PV generator. However, other criteria such as land availability and special boundary conditions on site also play a role for the best solution in each case in the specific project.

With the available information and the defined criteria, a GIS-based computation and evaluation of the theoretical and technical PV potentials of the rail infrastructure was carried out. As a result, the feed-in potential of suitable PV systems on and in the rail infrastructure was presented in both spatial and temporal context. A total of 37000 km of rail tracks and close to 14 km^2 were identified as theoretically suitable areas. Figure 2 shows an example of the identification of a potential roof area based on GIS data.

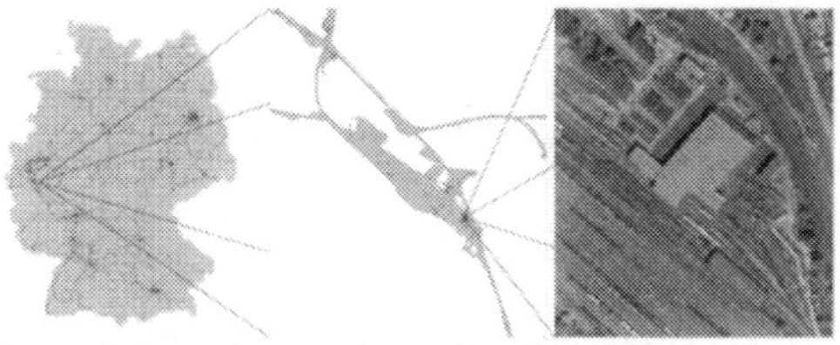

Figure 2: Exemplary roof areas based on GIS data

After the theoretically suitable and the technically and economically preferable areas and routes were determined, the feed-in potential was calculated for three different scenarios. The calculation was based on validated models used in the energy yield calculation of PV systems. A total installable capacity of 420 MWp - 3200 MWp can be expected, leading to an annual total yield between 380 GWh and 2920 GWh (Table 1). The number of systems to be realized ranges between 13000 and 26000.

Table 1: Determined installable capacity and total yield

Application Class	Installable PV Capacity	Total Yield
AC1: Noise barrier	70-470 MWp	50-350 GWh/a
AC2: Building	250-660 MWp	240-620 GWh/a
AC3: Free field	100-460 MWp	90-440 GWh/a
AC4: Track bed integrated	0-1600 MWp	0-1500 GWh/a
AC5: Rail car roof	0-10 MWp	0-10 GWh/a
Total	**420-3200 MWp**	**380-2920 GWh/a**

3.3 ASSESSMENT OF REQUIREMENTS

The main requirements for the PV systems to be integrated into the railroad infrastructure are PV systems to be integrated into the rail infrastructure are the regulations contained in the Deutsche Bahn guidelines (DB RiL) regulations. The relevant guidelines were identified and the contents briefly named. The guidelines are named that are to be continued due to the possible introduction of PV systems with direct feed into the overhead line should be continued and supplemented.

For the integration of PV systems in the railroad infrastructure the regulations of the railroads as well as the regulations that are independent of the railroads be taken into account. These are, in particular, the generally applicable technical regulations, legal regulations regarding the PV grid connection and energy and energy marketing, regulations on nature conservation and environmental protection, and building requirements. The essential requirements with regard to PV application are presented..

In order to identify further requirements resulting from regulations that have not yet been taken into account or from many years of professional and operational experience, railroad experts were questioned in interviews. The resulting effects, on the one hand on the existing railroad infrastructure and on the other hand on the PV systems, the resulting restrictions and obstacles as well as possible synergies were addressed in the interviews. In this way, it was possible to compile a comprehensive body of expert knowledge. Thus, for the

10.4229/EUPVSEC2023/4DO.5.1
020437-002

planning, construction and operation of PV systems in the railroad infrastructure, many essential aspects have been pointed out, which have to be considered for a PV integration into the railroad network in order to enable a promising construction and operation of the systems.

For the application class "installation in the track bed", a high energetic potential is generally seen due to the large available area. This assessment is confirmed by the analyses carried out in work package 2. However, there are also very high requirements, particularly with regard to the robustness of the PV modules and with regard to the maintenance and inspection of the track systems. There are many hurdles to overcome, and there is a high risk that can hardly be assessed so far.

The requirements for the application class "On or in a noise barrier" are also very high. In addition to the required robustness of the PV systems, the inspection capability of the noise barrier is also an important aspect to be considered. In addition, this application class was already identified in work package 1 as the one with the highest financial expenditure.

For the integration of PV systems on a noise barrier and next to the railroad track, comparatively low railway-specific requirements have to be fulfilled. In many cases, these installations can certainly be considered as largely independent of the railroad system. The same applies to installations on or adjacent to buildings.

Feeding the PV energy into the overhead line still poses a challenge. Here, some difficulties that have not yet been solved are still seen, especially concerning the topics of overhead line protection, power control, grid stability and EMC. Uniform regulations must be created here for stable operation of the railroad network. In addition, the overhead contact line with its comparatively low voltage is not suitable for transporting electrical energy over long distances.

4 CONCLUSIONS

Overall, the results to date show that PV application in rail infrastructure offers several field-proven opportunities to enable larger-scale renewable energy supply to railroads.

However, before PV application in rail infrastructure can be introduced on a larger scale, installations must first be set up as pilot projects. For rapid, considerable progress in the expansion of renewable energy, it is advisable to build pilot plants preferably at locations where the potential negative impacts, both on rail operations and on PV installations, can be expected to be low. With the experience gained and the resulting improvements introduced, PV applications can continue to expand until the appropriate level of expansion is reached in terms of economy and carbon footprint.

5 ACKNOWLEDGEMENTS

This work is supported by the German Center for Rail Transport Research (DZSF) at the Federal Railway Authority (EBA).

6 REFERENCES

[1] Biesenack, George, Hofmann, Schmieder u.a. (2006): Energieversorgung elektrischer Bahnen. B. G. Teubner Verlag / GWV Fachverlage GmbH, Wiesbaden

[2] Deutschen Zentrums für Schienenverkehrsforschung (2022): Berichte des Deutschen Zentrums für Schienenverkehrsforschung - Potenzialstudie Photovoltaik-Anwendungen an der Schieneninfrastruktur - Zwischenbericht zu Arbeitspaket 1.

[3] Steimel, A. (2017): Elektrische Triebfahrzeuge und ihre Energieversorgung. 4. Auflage, ITM Innotech Medien GmbH, Augsburg

[4] Deutsche Bahn AG (2021): Deutsche Bahn - Daten Fakten 2021

[5] Deutsche Bahn AG: Infrastrukturregister der DB [Online], Zugriff am: 2022-11-04. Verfügbar unter: https://geovdbn.deutschebahn.com/isr

[6] Statista GmbH: Anzahl der Güterwagen im Bestand der Deutsche Bahn AG in den Jahren 2007 bis 2021 [Online], Zugriff am: 2022-10-19. Verfügbar unter: https://de.statista.com/statistik/daten/studie/13346/umfrage/anzahl-der-gueterwagen-im-bestand-der-db-ag/

[7] DB Fernverkehr AG (2021): Die Fahrzeugstrategie der DB Fernverkehr - Für eine Starke Schiene [Online], Zugriff am 2022-10-19. Verfügbar unter: https://www.ews.tu-berlin.de/fileadmin/fg98/aushaenge/2021-sose/20210426_EWS_Nagl_DB_Die_Fahrzeugstrategie_der_DB_Fernverkehr.pdf

[8] Deutsche Bahn AG (2022): Milliardeninvestition: Deutsche Bahn bestellt 43 neue ICE [Online], Zugriff am 2022-10-19. Verfügbar über: https://www.deutschebahn.com/de/presse/pressestart_zentrales_uebersicht/Milliardeninvestition-Deutsche-Bahn-bestellt-43-neue-ICE-7252994

[9] DB Vertrieb GmbH: Die wichtigsten Fakten und Daten zum neuen ICE L [Online], Zugriff am 2022-10-19. Verfügbar über https://www.bahn.de/service/ueber-uns/inside-bahn/news/ICE-L

[10] DB Vertrieb GmbH: Züge im Fernverkehr und Bahnen im Ausland [Online], Zugriff am 2022-10-19. Verfügbar über: https://www.bahn.de/service/ueber-uns/zugtypen

[11] Reise, Christian & Müller, Björn & Armbruster, Alfons & Reich, Nils & Kiefer, Klaus. (2012): Ist eine Performance Ratio von mehr als 90 Prozent machbar?.

[12] Geofabrik GmbH Karlsruhe: OpenStreetMap [Online], Zugriff am: 2022-11-08. Verfügbar über: https://www.geofabrik.de/de/geofabrik/openstreetmap.html

[13] Geographisches Institut der Universität Heidelberg: OpenStreetMap Datenqualität

10.4229/EUPVSEC2023/4DO.5.1
020437-003

40th European Photovoltaic Solar Energy Conference and Exhibition

This presentation was selected by the Sc. Committee of the EU PVSEC 2023 for submission of a full paper to one of the EU PVSEC's collaborating peer-reviewed journals.

PHOTOVOLTAIC NOISE BARRIERS AS ENERGY GENERATING INFRASTRUCTURE: FUNCTIONAL OVERVIEW ABOUT FIVE SOLUTIONS

Jacob Forster[1], Giorgi Tsutskiridze[1], Cornelius Herr[1], Jonas Huyeng[1], Felix Basler[1],
Reinhard Kohlhauer[2],
Dirk Holger Neuhaus[1], Martin Heinrich[1], Li Carlos Rendler[1]
[1]Fraunhofer Institute for Solar Energy Systems ISE, Heidenhofstr. 2, 79110 Freiburg, Germany
[2]R. Kohlhauer GmbH, Draisstraße 2, 76571 Gaggenau, Germany
Corresponding Author: Jacob Forster | jacob.forster@ise.fraunhofer.de

ABSTRACT: Photovoltaic Noise Barriers (PVNB) serve as suitable example for Integrated Photovoltaics, as they make use of existing area of infrastructure to create a secondary function in form of energy generation. Since this secondary function must cohere with the requirements of the primary function of noise protection and additional fulfill mechanical requirements, PV integration into noise barriers is a field of development for special solutions depending on different use cases. Within the project "PVwins", we developed in collaboration with industry partners five different solutions to be build up for testing and monitoring at the end of the project. This paper provides a comprehensive overview of their use case, working principle, impact on acoustical properties, estimated power generation per meter length, and achievable energy yield per meter length. In addition to presenting measurements and simulations, we showcase prototypes. The solutions range from simple street-averted vertical integration of standard modules, to a construction method involving the mounting of standard modules on top of noise barriers. We also developed customized modules with an unique layout incorporating acoustic absorbers between PV strings. For optimal functionality in terms of acoustics and photovoltaics, small-width custom modules are mounted on triangular acoustic absorbers in the cassette solution. Lastly, we introduce an innovative R&D solution featuring a transparent microperforated absorber that offers sound absorption capabilities and enables the full area integration of bifacial modules.
Keywords: Integrated Photovoltaics, Road Infrastructure PV (RIPV), Photovoltaic Noise Barriers, Module Design

1 INTRODUCTION

To tackle the climate crisis and generate affordable energy, photovoltaic will become one major renewable energy source in terms of installed capacity [1]. Achieving this aim requires large areas of land. In countries with high population densities, integrating photovoltaic systems into areas with other primary functions can mitigate land-use conflicts with natural habitat, agriculture, and urbanization [2]. Among integrated PV applications, Photovoltaic Noise Barriers (PVNB) are a good example for Road Infrastructure PV (RIPV) [3]. The current German Federal Government announced plans to utilize the area next to highways stronger for renewable energies and make use of the wind and solar potential for new highway constructions if possible [4]. Also, rising interest from local municipalities, institutions such as BASt (Germany) [5], ASFINAG (Austria) [6], ASTRA (Switzerland) [7], or the European Union with the EU solar energy strategy [8] bring PVNB into the perspective as part of future infrastructure. Photovoltaic noise barriers have been demonstrated in European pilot projects since around 30 years [9] but did not become the state-of-the-art solution for new noise barrier infrastructure yet. On the one hand, organizational obstacles such as the tender design, planning approval process, insufficient political pathways and restrictive legal, as well as technical regulations prevented large scale implementation in Germany [10]. On the other hand, maintaining the main technical function of sound absorption together with all additional requirements still pose challenges to design solutions fulfilling these in an economically attractive way for different use cases.

Within the project "PVwins" [11] (public funded by BMWK) technical solutions for PVNB were explored and developed in cooperation between Fraunhofer ISE and the industry partners R. Kohlhauer GmbH, IGRA Power GmbH and Megasol Energie AG. On the advisory side helped the associated partners Bundesanstalt für Straßenwesen EGIS eG and the Eisenbahn Bundesamt EBA. From this collaborate work we showcase five technical solutions for PVNB.

2 TECHNICAL REQUIREMENTS

2.1 Primary function of noise reduction

As depicted in Figure 1, a noise barrier reduces the sound between the place of emission and the place of immission through reflection and absorption. What remains is a transmitted (not reflected or absorbed) and diffracted part of the emitted sound. Thereby the required noise reduction at the place of immission needs to be larger than 25 dB according to ZTV Lsw 22 [12].

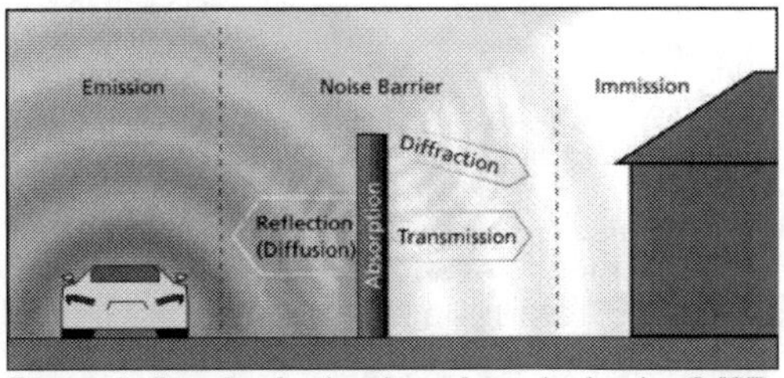

Figure 1: Sound reduction through a noise barrier © ISE

With a hard surface such as concrete or glass, sound can be reflected at the noise barrier. However, this is not feasible at all locations, as it increases the sound on the other side of the street, where potentially also noise protection is required. Therefore, as seen in Table 1, noise barriers often must be sound absorbing. Depending on the local situation, absorption requirements are defined in levels of requirements groups which vary from <4 dB (not absorbing = reflecting), 4 to 7 dB (absorbing) and > 8 dB (highly absorbing) [12]. Since PV modules act with their glass surface highly reflective and are considered as secondary function of a noise barrier, solutions to incorporate PV need to vary depending on the local requirements. For example, distinguishing between street facing- and averted installations is important in terms of noise functions.

Table I: Absorption groups according to ZTV Lsw 22 [12]

Group	Noise Absorption DLa	Explanation
A1	< 4 dB	Not absorbing
A2	4 to 7 dB	Absorbing
A3	8 to 11 dB	Highly absorbing
A4	> 11 dB	

2.2 Additional mechanical requirements

Apart from the noise-related requirements, there are numerous other norms that need to be fulfilled. Since photovoltaic integration is not yet a standard practice, it is not mentioned in the existing norms, and therefore must adhere to regulations which are already in place for long time without being adjusted for the needs of photovoltaics.

One of the main challenges for photovoltaic integration are stone- and snow impact resistance [12]. As safety by design measure, we avoid deploying photovoltaic modules lower than 1.5 m to 2 m above ground. This also is an issue for the susceptibility of vandalism such as graffiti and prevents shading from ground vegetation.

Other concerns of disturbing optical sun reflection on the glass surface of PV modules need to be analyzed site specific and can be solved with anti-reflective glass surfaces. On street averted sides PV integration is more relaxed regarding those acoustic and mechanical requirements. Besides that, different solutions are required for new installations in contrast to retrofit installations of PV modules on existing noise barriers.

For the variety of use cases, solutions for different szenarios are required, of which our five solutions are a demonstration of possibilities. The start of such a design process is an evaluation of the local sound absorption requirements, the orientation of the street and possible designs that fulfill mechanical, aesthetical, and economical requirements.

3 FUNCTIONAL OVERVIEW ABOUT FIVE PHOTOVOLTAIC NOISE BARRIER SOLUTIONS

We present five technical solutions for PVNB in different settings of usage and with different requirements. Furthermore, preliminary estimations regarding acoustical behavior and energy yield are shown, based on tests and simulations. For all five noise barrier concepts it is assumed, that the noise barrier is 4.5 m high, and the bottom 1.5 m above ground are kept clear from PV integration.

3.1 Rear side integrated solution

As an attractive option for noise barriers located on the south side of the street, we propose a road-averted integration of low-cost standard photovoltaic modules. In this particular use case, simplified requirements for optical anti-reflection and stone or snow impact resistance can be advantageous, as custom modules with thicker glass may not be necessary. Also, noise protection is achieved through full-surface absorbers facing the road. The rear side integrated solution is made from two conventional modules, which are integrated into a special designed noise absorber housing, filled with stone wool on the rear of the modules. Three of these elements are then stacked onto one another.

Figure 2: Rendering of rear side integrated solution © ISE

3.2 Top-mount solution

A very flexible solution for existing noise barriers and new ones was developed by our project partner R. Kohlhauer GmbH with the approach of top-mounted PV modules [13]. This solution ensures that the noise absorption capabilities remain unchanged compared to conventional noise barriers. Depending on the specific requirements of the location, PV modules can be mounted with flexible orientations, including inclinations towards the south, east-west, or horizontally. Horizontal integration is particularly beneficial for noise barriers with curved trajectories, as it helps prevent current mismatch.

Figure 3: Top-mount solution: Rendering, principle illustration (© ISE) and photo of demonstrator at company R. Kohlhauer GmbH

To understand the noise diffraction behavior with variation of the module inclination from -35° to 35°, we simulated the single value of diffraction. The results in Figure 4 show, that the module decreases the noise behind the noise barrier particularly at higher inclination and are therefore beneficial for sound protection.

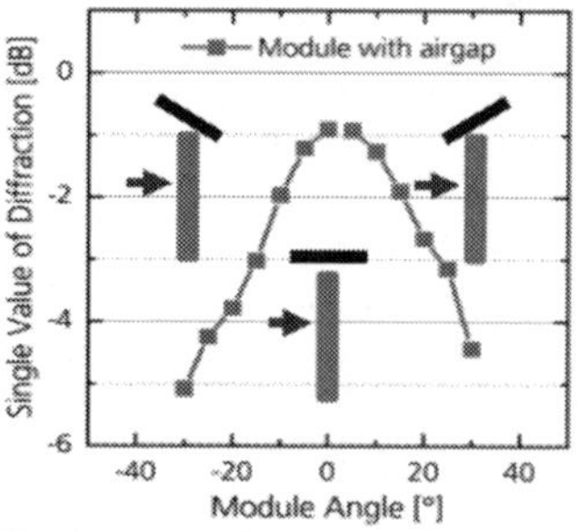

Figure 4: Diffraction variation in relation to module tilt angle

10.4229/EUPVSEC2023/4DO.5.2
020439-002

3.3 Combination solution

The combination solution offers a balance between the area dedicated to sound absorption and the area allocated for photovoltaic generation with the use-case of noise barriers with northern position towards the street. Both functional areas are placed vertically towards the street in an alternating parallel configuration. Therefore, we developed a custom module in oversize of 3.9 m length, with two individual substrings, seen in Figure 5 below. The substring between the absorbers is bifacial, the other one is monofacial on the rear of the absorbers, facing to the north.

Figure 5: Rendering of combination solution with a sketch of the substring interconnections (view from rear) © ISE

To investigate the impact of parallel area combinations on acoustic performance, we conducted hall chamber measurements on various configurations with different ratios of PV area to absorber area. After evaluation, we selected a ratio that features alternating absorber stripes corresponding to approximately 40% PV area, shown in Figure 6.

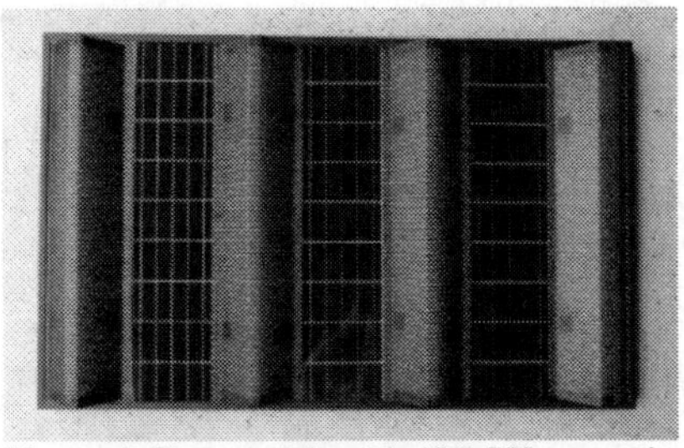

Figure 6: Test module with 40% PV area ratio

This choice was made among others because lower PV ratios were found to be economically unattractive. The absorption coefficient of this configuration was measured and is shown in Figure 7. A single value of absorption of 4.8 dB was calculated. It is important to note that the hall chamber measurements could not capture low frequencies, resulting in a lower single value of absorption as these frequencies were considered as zero. Additionally, it is not straightforward to compare measurements conducted in a hall environment with in-situ measurements, such as the Adrienne method, which is the new measurement standard in DIN EN 1793-5 and ZTV-Lsw 22. Therefore, in-situ measurements still need to be performed for validation. However, we project that this solution is suitable for absorption group A2, which requires an absorption of >4 dB.

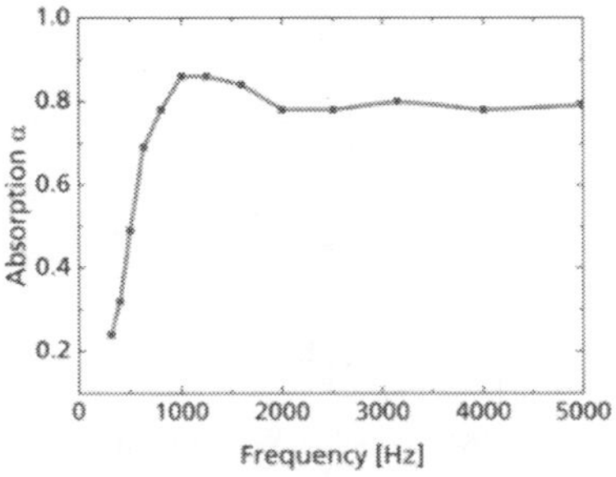

Figure 7: Absorption measurement 40% PV

3.3 Cassette solution

The cassette solution is an effective way to align the functions of noise protection and energy generation in a way that optimizes their individual performance.

Figure 8: Rendering cassette solution © ISE

With the triangular cross-section, the absorbers are facing towards the road as origin of noise and the PV modules are mounted on the side facing towards the sun in an angle of 35°. Acoustically, the PV modules also improve noise protection by reflecting a part of the sound towards the sky. Since the basic concept is already commercially available [14], we focused on a solution that is not slided into the H-beams but can be mounted in front of existing noise barriers. Thereby it is a retrofit option for noise barriers which should be improved regarding their acoustic performance with additional PV integration.

Figure 9: Prototype of cassette solution with 1m length

We simulated the sound absorption using COMSOL and calculated single values of absorption between 9 and 10 dB. The simulated sound pressure is visualized for a frequency of 1000 Hz in Figure 10.

10.4229/EUPVSEC2023/4DO.5.2
020439-003

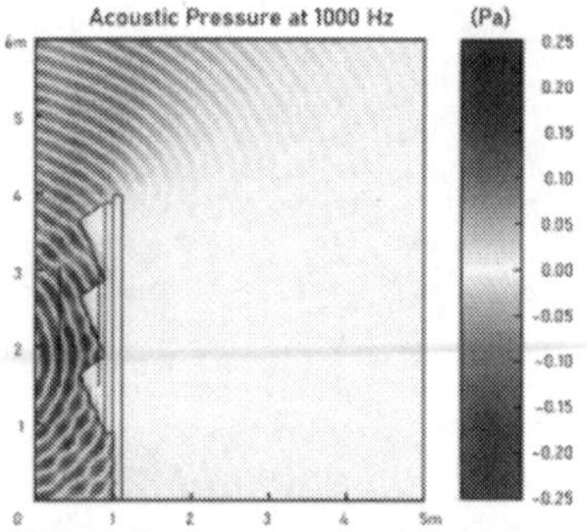

Figure 10: Test setup and absorption measurement

3.4 Microperforated absorber solution

A more innovative attempt is utilizing a sound absorption principle known from room acoustics in interior architecture known as microperforated absorber [15]. It consists of a transparent acrylic plate with micro-perforation of 40.000 holes per square meter, which is kept by distance holders with an airgap towards the sound reflective glass surface of the PV modules. The tested hole diameters ranged between 0.6 and 0.8 mm. Thereby, the functions sound absorption and energy generation are integrated in a serial way and are not parallelized as in the combination solution.

Figure 11: Rendering MPA solution © ISE

The sound absorption works according to the principle of a Helmholtz resonator. The sound waves propagate through the holes, get reflected at the glass surface of the PV module and a small resonating mass-spring-system evolves. The holes do not only let the sound pressure into this resonating system, but also cause friction to the moving air mass, which essentially dissipates a part of the sound energy into heat. The microperforated absorber (MPA) has four main parameters which influence the system behavior strongly and can be used to adjust the absorption with respect to frequency. As shown in Figure 10, these are the hole diameter, the distance between the holes, the thickness of the plate and the air gap distance.

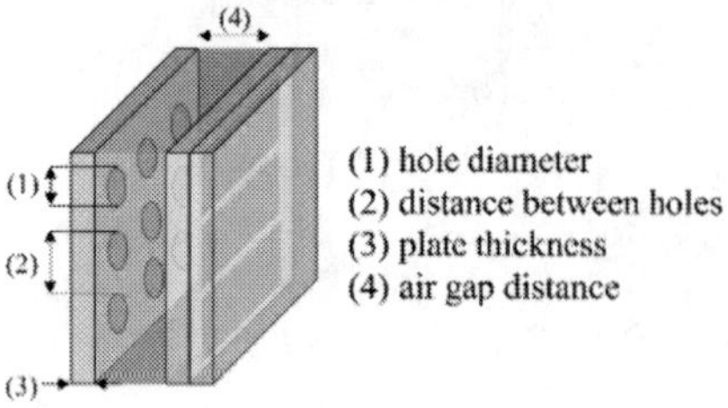

Figure 12: Parameters to adjust MPA system behavior

With these possibilities to vary the bandwidth of the absorption, the frequency of the absorption peak and the absorption maximum, we concentrate in this paper only on the effect of the air gap distance. In Figure 13 is shown, how the peak of the absorption is changing over the frequency range by variation of the airgap distance between 2 mm and 20 mm. The measurement is conducted in an impedance tube, in which a speaker is sweeping the frequency between 50 Hz and 5000 Hz and a microphone array measures the back reflected sound an calculates impedance and absorption values. It can be observed that increasing the air gap distance decreases the frequency of the absorption peak.

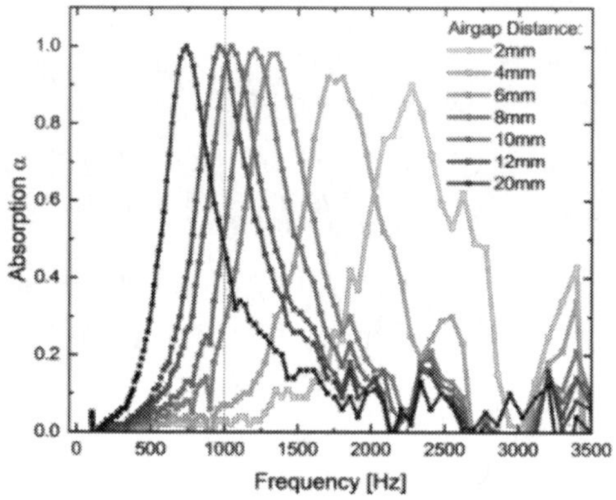

Figure 13: Absorption peaks for a MPA with parameters: (1) = 0.8mm, (2) = 5mm, (3) = 4mm, (4): Variation

Our experimental optimization objective was to achieve a high and broad absorption peak at 1000 Hz, as this frequency range has the most significant impact on the single value of absorption [16]. In the impedance tube experiments, the absorption peaks show small bandwidths, resulting in a single value of absorption of approximately 1.68 dB for the finally chosen plate parameters, still falling within the absorption group A1. However, through further research and development, we anticipate that by optimizing the microperforated absorber or combining it with absorbing frames, a higher absorption value of 4 dB will be achieved. The potential feasibility demonstrated Leistner et. al. [17] with an absorption value of 6 dB for a single layer MPA construction and even 9 dB with a second MPA layer.

Figure 14: Prototype of MPA solution with 1 m length

10.4229/EUPVSEC2023/4DO.5.2

3.5 Comparison of reachable absorption groups

In Table 2, a short summary of the reachable absorption groups of the discussed five solution is given.

Table 2: Comparison of reachable absorption groups

Nr.	Solution	Absorption Group
1	Rear-side integrated	A3
2	Top-mount	A3
3	Combination	A2
4	Cassette	A3
5	MPA	A1 → A2 (aim)

4 PHOTOVOLTAIC PERFORMANCE

To compare the energy generated by each solution we calculated installable rated power per meter, energy yield per installed kWp and energy yield per meter length shown in Figure 15. The yield estimations are an approximation calculated under the assumptions in Table 3 with the tool *PVGIS [18]* (https://re.jrc.ec.europa.eu/pvg_tools/en/).

Table 3: Assumptions for energy yield calculations

Location	48.004, 7.778 (Freiburg)
Shading of environment	none
Orientation	South
Module Efficiency	21 %
Performance Ratio PR	85 %
Bifaciality Factor BF	80 %

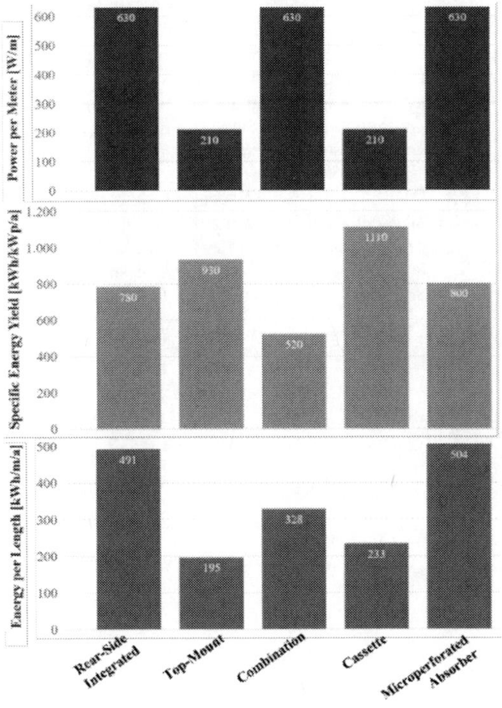

Figure 15: Calculated PV potential. Blue bars indicate the installable power per meter. Orange bars show the specific energy yield per installed kWp per year. The green bars depict the energy yield per meter per year.

The energy calculations of combination solution and MPA solution required extra assumptions. For the MPA solution additional optical absorption losses due the MPA plate were considered with a transparency factor τ_{MPA} of 80 % on the southward oriented front side. The yield of the rear side of the bifacial modules was calculated with a bifaciality factor BF of 80 % with respect to an otherwise monofacial yield on the northern vertical side of 222 kWh/kWp/a.

For the combination solution the energy yield of the two substrings was calculated with the geometrical area ratios of the strings to the whole module. For the bifacial substring, besides the bifaciality factor a shading factor SF of 90 % was also considered for self-shading due to the absorber slides.

Summing up, the MPA- and rear-side integrated solutions generate the highest energy yield per meter, since they accommodate PV modules over the whole available vertical area from 1.5 m above ground. However, also the top-mount and cassette solution could both together be integrated into a noise barrier with some modifications and could reach high yield and high noise absorption.

5 DEMONSTRATOR

A demonstrator with the discussed five PVNB solutions will be build up at the Fraunhofer ISE outdoor performance lab in Merdingen [19], as seen in the rendering in Figure 16. It will be placed with a H-pole distance of 4 m over a length of 16 m. The size of this demonstrator is the minimum required for conducting acoustical in-situ measurements [20] and understand how the PV integration solutions compare against a conventional aluminum absorber filled with stone wool for reference. Furthermore, energy monitoring will be conducted of each individual module using power optimizers.

Figure 16: Rendering of demonstrator © ISE

6 ACKNOWLEDGEMENTS

This work was supported by the Federal Ministry for Economics and Climate Action (BMWK) under contract number 03EE1062A, with the acronym PVwins. We would like to express our gratitude to our project partners at R. Kohlhauer GmbH for their assistance in designing and constructing the noise barrier and to IGRA Power GmbH for their contributions to the electrical system configuration and safety considerations. Lastly, we would like to thank Megasol Energie AG for building custom modules for our demonstrator.

10.4229/EUPVSEC2023/4DO.5.2
020439-005

7 REFERENCES

[1] IEA, "World Energy Outook 2015 Factsheet - Global energy trends to 2040," 2015.

[2] Dr. Harry Wirth, Fraunhofer ISE, "Aktuelle Fakten zur Photovoltaik in Deutschland," 2022.

[3] Jonas D. Huyeng, Jacob Forster, Felix Basler, Pascal Romer, Andreas J. Beinert, Christian Schill, Martin Heinrich, "Technical Aspects for Road Integrated PV Towards a More Sustainable Mobility Sector," 2022.

[4] Entwurf eines Gesetzes zur Beschleunigung von Genehmigungsverfahren im Verkehrsbereich und zur Umsetzung der Richtlinie (EU) 2021/1187 über die Straffung von Maßnahmen zur rascheren Verwirklichung des transeuropäischen Verkehrsnetzes: Drucksache 20/6879, 2023.

[5] Bär, F., Kaspar, F., Streek, P., Gersdorf, F., Auerbach, M., Rieck, D., "Abschätzung des potenziellen jährlichen Energieertrags von PV-Anlagen an Lärmschutzeinrichtungen der Verkehrswege," Jul. 2022. [Online] Available: https://www.bmdv-expertennetzwerk.bund.de/DE/Publikationen/Kurzberichte/TF5/Baer_et_al_07-2022.html.

[6] ASFINAG, Die Autobahn als Sonnenkraftwerk: Mit der Lärmschutzwand Strom produzieren. [Online] Available: https://www.ioeb-innovationsplattform.at/challenges/detail/die-autobahn-als-sonnenkraftwerk-mit-der-laermschutzwand-strom-produzieren/.

[7] ASTRA, "Studie über das Potenzial der Lärmschutzwände entlang von Autobahnen und Bahnstrecken für die Produktion von Solarenergie," 2021.

[8] European Commission, EU Solar Energy Strategy: Communication from the commission to the european parliament, the council, the european economic and social committee and the committee of the regions: Document 52022DC0221. [Online] Available: https://eur-lex.europa.eu/legal-content/EN/TXT/?uri=COM%3A2022%3A221%3AFIN&qid=1653034500503.

[9] Thomas Nordmann and TNC Consulting AG, "Drei integrierte PV-Schallschutz Versuchsfelder," 2002.

[10] Elena Chvanova, Dr. Birgit Haller, Dr. Ole Langniß (Dr. Langniß Energie & Analyse), Prof. Dr. Uwe Leprich (E&E Consult GbR), Ursula Mayr, Christian Mayr, Christian Oßwald (MaxSolar GmbH), Dr. Martin Altrock, Dr. Christian Gemmer, Dr. Sascha Michaels, Julia Wagner (Becker Büttner Held PartGmbB), "Verkehrsträgerübergreifender Austausch von Erneuerbarer Energie Forschungsvorhaben FE 69.0009/2020: Schlussbericht," Bundesanstalt für Straßenwesen (BASt), Stuttgart, Oct. 2021.

[11] Dr. Martin Heinrich, Dr. Jonas Huyeng, Jacob Forster, PVwins – Entwicklung von wandintegrierten PV-Elementen für den Lärmschutz. [Online] Available: https://www.ise.fraunhofer.de/de/forschungsprojekte/pvwins.html.

[12] Forschungsgesellschaft für Straßen- und Verkehrswesen, Zusätzliche Technische Vertragsbedingungen und Richtlinien für die Ausführung von Lärmschutzwänden an Straßen: ZTV Lsw 22, 2022nd ed. Köln: FGSV Der Verlag, 2022.

[13] Reinhard Kohlhauer, KOHLHAUER VOLTA ROOF, die effektive Nachrüstung durch Standard-Module. [Online] Available: https://www.ioeb-innovationsplattform.at/challenges/detail/die-autobahn-als-sonnenkraftwerk-mit-der-laermschutzwand-strom-produzieren/detailIdea/kohlhauer-volta-roof-die-effektive-nachruestung-durch-standard-module/.

[14] Etway, Sound solar panel. [Online] Available: https://soundsolarpanel.com/pdf/SSP_Barriera-Antirumore-Fotovoltaira_ENG.pdf.

[15] Helmut V. Fuchs, Xueqin Zha, Xiaoru Zhou, "Schallabsorber und Schalldämpfer - Innovatorium für Maßnahmen zur Lärmbekämpfung und Raumakustik, Teil 5: Mikroperforierte Bauteile," Ernst & Sohn Verlag für Architektur und technische Wissenschaften GmbH & Co. KG., 2003.

[16] DIN EN 1793-3 Lärmschutzeinrichtungen an Straßen - Prüfverfahren zur Bestimmung der akustischen Eigenschaften - Teil 3: Standardisiertes Verkehrslärmspektrum.

[17] X. Z. Philip Leistner, "Transparente, schallabsorbierende Lärmschutzwände," Fraunhofer Institur für Bauphysik, 2009.

[18] EU science hub PVGIS, PVGIS data sources & calculation methods. [Online] Available: https://joint-research-centre.ec.europa.eu/photovoltaic-geographical-information-system-pvgis/getting-started-pvgis/pvgis-data-sources-calculation-methods_en.

[19] Dr. Christian Reise, Solartestfeld in Merdingen bei Freiburg. [Online] Available: https://www.ise.fraunhofer.de/de/geschaeftsfelder/photovoltaik/photovoltaische-module-und-kraftwerke/photovoltaische-kraftwerke/solartestfeld.html.

[20] DIN EN 1793-5 Lärmschutzvorrichtungen an Straßen - Prüfverfahren zur Bestimmung der akustischen Eigenschaften - Teil 5: Produktspezifische Merkmale - In-situ-Werte der Schallreflexion in gerichteten Schallfeldern.

New precision pressure point test method to detect Si cell breakage in PV modules

Application to the Solar Road module

M. Vite, P. Messaoudi, R. De Bettignies, D.R. Heslinga, E. Voroshazi, E. Coquelle
Marion.Vite@cea.fr

wattway COLAS ines INSTITUT NATIONAL DE L'ENERGIE SOLAIRE

40th European Photovoltaic Solar Energy Conference and Exhibition

WHY A NEW MECHANICAL TEST?

Growing class of **new lightweight PV panels** 'Integrated PV'
Without glass → **polymers and/or composites**
as front- and back-plate

Quid of the mechanical robustness of such panels?

In certification protocols IEC61215 and IEC 71730,
no criteria or check of Si cell damage: only visual inspection

Need for a mechanical characterization method:

- ✓ Specifically targets **Si cell breakage**
- ✓ Easy to perform
- ✓ Versatile
- ✓ Quantified

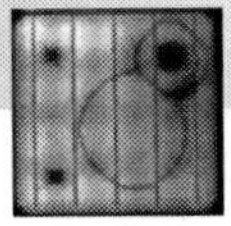

New Pressure point test useful in different ways:

- ✓ Tool for process development: **selection of the BOM**
- ✓ **Qualification** of lightweight modules
- ✓ **Production testing** of Si PV modules: monitoring
- ✓ Future: include into **certification protocols**?

2

HOW DOES THE PRESSURE POINT TEST WORK?

- Using a classic tensile stress testing bench
  → monitoring of the **Load/Displacement curve**

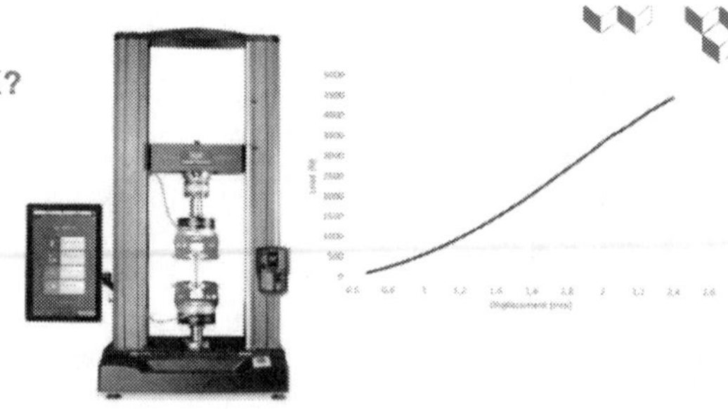

'Push instead of pull'

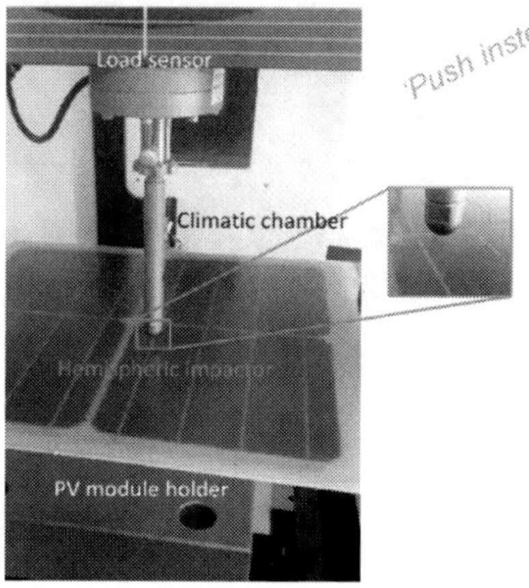

Test protocol parameters:

- Impactor shape:
  - Hemispheric
  - Sphero-conical (Rockwell)
  - Pyramidal (Vickers)

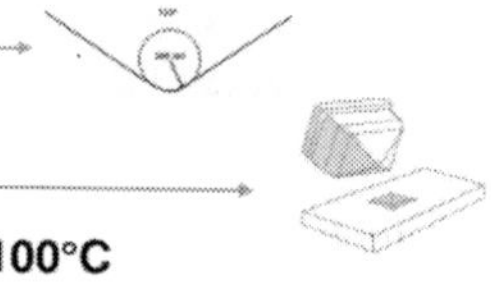

- Temperature variation: **-40°C to 100°C**
- Position: on busbar / between busbar (precision 0.2 mm)
- Load ramp speed: static/dynamic (1 to 1000 mm/min)
- Support material below the module (representing the final application)
- Dimensions of samples: 20x20 to 40x40 cm² mini-modules

3

020440-003

WHICH ARE THE EXTRACTED PARAMETERS?

- Ramp-up / ramp-down load cycle → **4 key parameters** from the Load/Displacement curve :
  - ✓ Instantaneous identification of the Si cell breakage event
  - ✓ Maximum penetration depth: combination of deformation, flexion and compression of the different polymers of the PV module
  - ✓ Hysteresis: refers to the visco-elasto-plastic behavior of the combined polymers in the PV module
  - ✓ Residual deformation: plastic deformation of encapsulants and/or breakage of composites

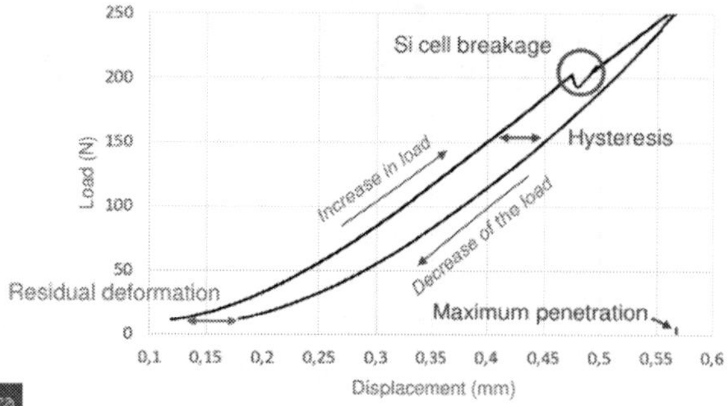

Load → stress concentration on the module:
1000 N ≈ 100 MPa

Stilettos effect

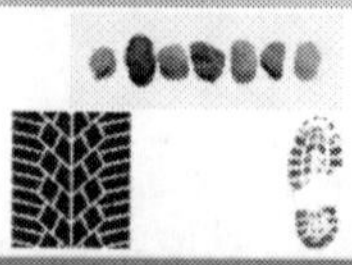

Asperity – contact area	Truck	Pedestrian
Without asperity – natural area	1 MPa	0.01 MPa
Asperity (1 x 1 mm²)	400 MPa	100 MPa

4

020440-004

APPLICATION: MECHANICAL ROBUSTNESS OF THE SOLAR ROAD PV MODULE

Optimization of the Bill Of Materials: choice of the frontsheet-backsheet composite

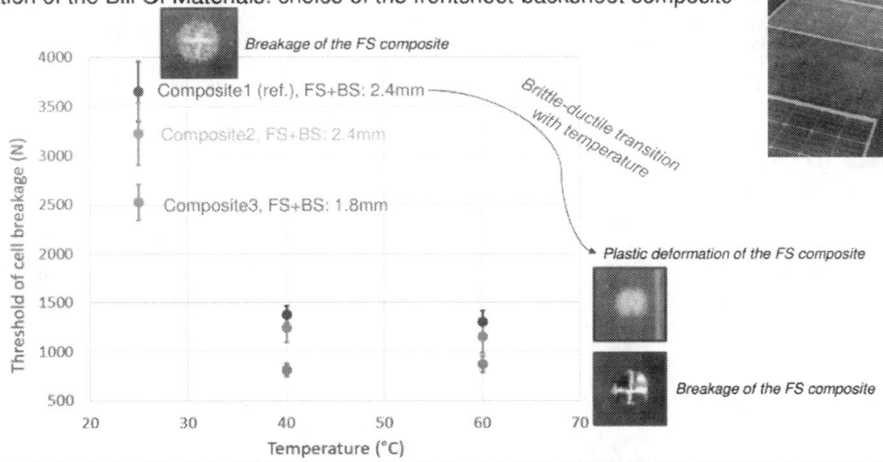

Precise differentiation of mechanical performance as a function of type/thickness of composite
Strong sensitivity to temperature

EXTENSION TO -20°C

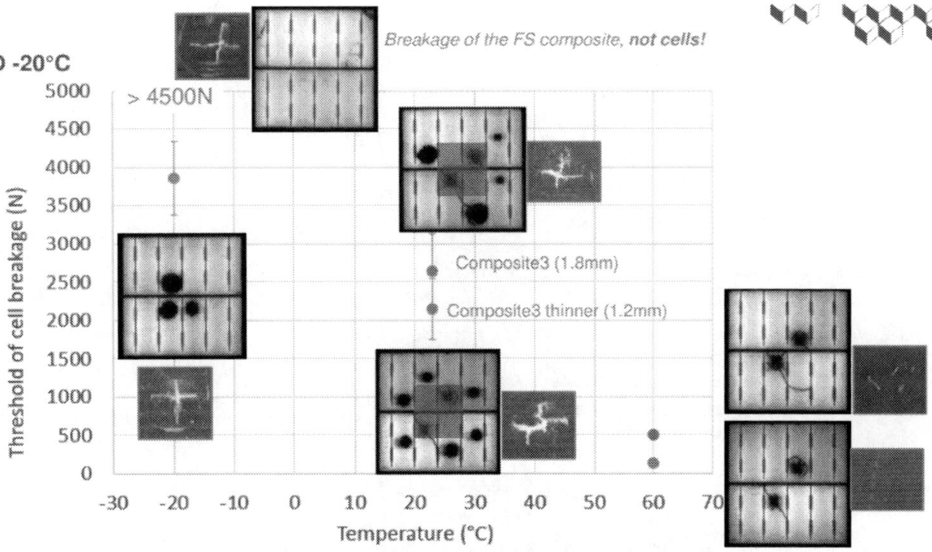

Huge robustness at -20°C, largely better than room temperature
→ Open at other areas in the world where medium ambient temperature is low

HOW TO REPRESENT REAL MOUNTING CONDITIONS

• *Simulated conditions*

✖ *Real conditions*

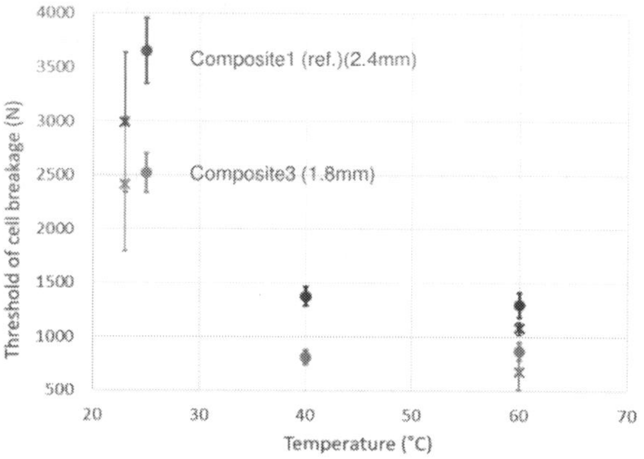

Real mounting conditions:
Slightly lower cell breakage threshold
But it confirms the high level of robustness

SUMMARY / PERSPECTIVES

New test method to characterize the mechanical robustness of Si PV modules:

- A **precise and versatile** method
- **Easily adapted** to a commonly available tensile stress test equipment
- Large range of temperature: **-40°C to +100°C**
- From the Load/Displacement curve, 4 quantifiable key parameters:
  - ✓ **Si cell breakage threshold**
  - ✓ Penetration depth
  - ✓ Hysteresis
  - ✓ Residual deformation

  } Future investigation topics (Penetration depth, Hysteresis, Residual deformation)

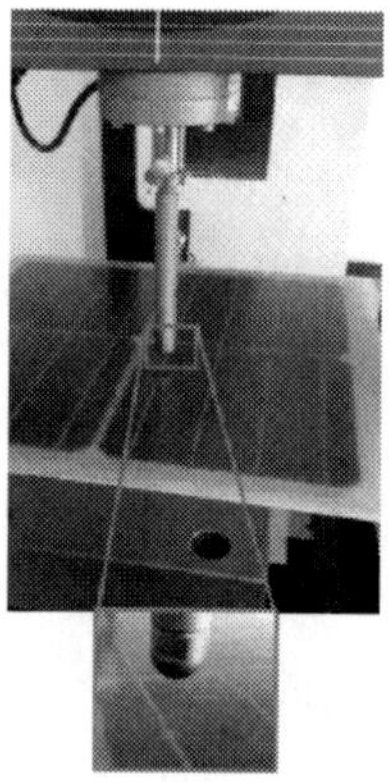

Results of this method demonstrated on the Solar Road PV module:

- **Optimization** of the module's robustness to the harsh use conditions
- **High level of cell breakage threshold** at room temperature and significantly reduced but **remains satisfactory upto 60°C**

Perspectives:

- Great **process development** tool
- **Benchmark** of other lightweight Si PV modules on the market
- Comparison of **frontside and backside impact**
- Incorporation into **future PV module certification protocols?**

Follow us : ines-solaire.org

THANK YOU

Areas of expertise

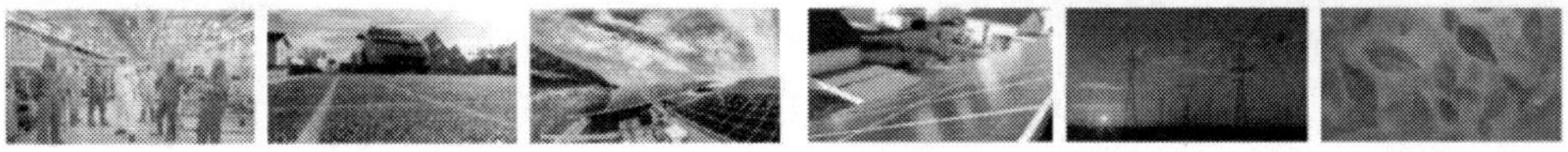

You can contact me:

Marion.Vite@cea.fr

cea

ines
INSTITUT NATIONAL DE L'ENERGIE SOLAIRE

40th European Photovoltaic Solar Energy Conference and Exhibition

Call for Abstracts: Submit your latest results for SiliconPV and nPV by **November 4**!

SiliconPV 2024
14th International Conference on Crystalline Silicon Photovoltaics

npworkshop Chambéry 2024

bifiPV2024 workshop

Advanced technologies, materials and concepts for crystalline Si solar cells and modules

April 15 – 19, 2024 | Chambéry, France

Three PV Events in One Week – Save the Date!

SiliconPV 2024: April 15 – 17
www.siliconpv.com

nPV Workshop 2024: April 17 – 18
www.2024.npv-workshop.com

bifiPV Workshop 2024: April 18 – 19
www.bifipv-workshop.com

Venue: Chambéry, France

Registration will be open online in January.

© Grand Chambéry Alpes Tourisme © Centre de Congrès Le Manège

Conference Host: cea, ines INSTITUT NATIONAL DE L'ENERGIE SOLAIRE

Conference Organizer: conexio pse

URBANBOX THE LIGHTWEIGHT RETRACTABLE PV PLANT

Arthur Büchel[1], Franz Baumgartner[2], Roland Fehr[2], Simon Metzger[2], Matthias Loup[2], Fabian Carigiet[2], Robert Eberlein[2]
[1] iWorks AG, Oberweilerstrasse 36, 9491 Ruggell, Liechtenstein
Email: arthur.buechel@iworks.li, Website: www.urbanbox.li
[2] ZHAW, Zurich University of Applied Sciences, School of Engineering, IEFE and IMES
www.zhaw.ch/~bauf, Technikumstr. 9, CH-8401 Winterthur, Switzerland
Email: bauf@zhaw.ch, Website: zhaw.ch/=bauf

ABSTRACT: Recently a study of Swiss utilities showed a Photovoltaic (PV) economic potential of 3 GW and a three times higher technical potential on classical infrastructure areas in Switzerland. Therefore, there exists a large opportunity for suitable dual-use PV systems. However, the requirements, such as long span between foundations, investment cost and operating cost for cost-effective power production have been challenging for existing fixed roof type solutions. The retractable URBANBOX PV Roof is a new approach, offering light weight design for the structure and foundation, integrated automated PV module cleaning and a modular factory-prefabricated box concept for the storage of the PV carriers, drives and controls. In 2022 the URBANBOX demonstration system was installed in Liechtenstein and test results and result of a research project for further optimisation of the lightweight structure are presented in this paper. They include the optimization of steel for the mounting structure (<70 kg/kWp) for the URBANBOX prototype to be built in 2024. This project also investigated and developed a built-in automatic panel cleaning system and weather-based control algorithm for the retraction of the PV module carriers.
Keywords: PV Module Mounting, retractable PV Plant Systems, automated PV panel cleaning, BOS, Innovation, Soiling, Testing, long span PV structure

1 INTRODUCTION

Globally, the demand for PV power plants is unbroken and the potential for infrastructure areas will also have to be raised for the renewable electricity hunger, as the buildings are not sufficient. Recently, the Swiss Association of Electricity Companies estimated the PV potential in Switzerland at 3 GW on infrastructure sites, which is slightly less than one tenth of the total PV expansion potential [2]. The same authors have also recently calculated the potential for PV carports at 6 to 10GW, i.e. double or triple the above potential [3]. With a nominal PV power of about 2kWp per parking lot in a typical Swiss medium-sized city of 23 000 inhabitants PV carport on medium and large parking areas will account for 5.3MW PV power [4].

Successful PV innovations overcome the bottleneck of the rising cost shares in state-of-the-art photovoltaic power plants. This rising cost share is not found in the photovoltaic module itself but in the mounting, cabling system and the installation process, summaries as Balance of System (BOS). In the last decade global photovoltaic module prices declined by a factor of ten due to economy of scale in the fabs. But the BOS costs do not change so much. Therefore, in the period 2013 to 2019 the cost for PV modules and inverter decreased by one third from 55% to 37% of share of total plant costs in Switzerland for larger rooftop PV plants between 100 to 300kW in power. Reducing BOS costs is a global trend, and it calls for substantial innovation in and new system approaches. If the fixed PV module mounting structures are reaching the marginal material cost limits of metal and concrete, new BOS systems must be developed. Today's PV carports with fixed support structure, typically require up to 200kg of metal and about half a ton of concrete per kW PV power, resulting in double to triple the construction costs compared to industrial roofs [4]. This calls for new systems using a minimum amount of metal used in the mounting structure, and not only innovate one part.

In recent years, projects have been implemented in Switzerland to investigate the potential of mounting modules on steel cables in order to save the amount of metal. This was combined with single and dual axis mechanical trackers and the successful power supply of a ski lift was demonstrated (Table I).

Table I: Experience with dual use PV mounting projects initiated by the co-authors [5,6,7].

Year		PV Power
2009		648 kWp Lonza Solarpark, Waldshut, Germany
2011		70kW, Solar Ski Lift, Tenna, Switzerland
2013		20kW Light Energy Plant, Balzers, Liechtenstein
2022		50 kW URBANBOX Bendern, Liechtenstein

In 2013, with the development of a retractable PV module system, as is common in a similar form for shading in and on buildings, the static PV mounting was deviated from for the first time, even for higher PV power. This reduces the maximum mechanical loads caused by wind, hail and snow, as the modules are

10.4229/EUPVSEC2023/4DO.5.5
020441-001

retracted in a protected area during these weather conditions. The current paper is the first to present a different retractable PV module system, where the modules are stacked on top of each other in a box rather than folded as they were a decade ago, with potential economic benefits.

2 URBANBOX CONCEPT

In the current URBANBOX prototype, 12 PV module carriers, each with 10 PV modules can be stepwise horizontally extracted onto the solar roof and retracted back into the buffer box, where they are vertically stacked [8]. The carriers move on a steel guide rail, and they are designed to be fully compatible to standard mainstream PV modules, a big advantage, when compared to other retractable PV system solutions (Fig. 1).

Figure 1: Top view of the URBANBOX demonstration system installed in 2022 in Bendern Liechtenstein.

Table III: Technical features URBANBOX demo system and future prototype/commercial URBANBOX systems

	Demo2022	Prototype
PV Power		
Number of PV carriers	12	up to 20
Power per carrier (kWp)	3.7	4.3
Power on roof of box (kWp)	6	5
Power of plant (kWp)	50	80 to 91
Dimensions		
Length (m)	28	up to 38
Width (m)	11	11.7
Ground clearance (m)	4.5	4.5
Automatic PV module cleaning	YES	YES
Structure	Wood	Wood/Steel
Foundation	Screw-type	Screw-type, other

Based on the experience of the 2022 demo system in Bendern the prototype solution 2024 in the way to the commercial system will lead to a lightweight construction of 70kg steel used per kW PV power for 20 carriers each equipped with 10 PV high efficient commercial modules at a plant power of 85W for each URBANBOX (Table II).

3 LIGHTWEIGT CONSTRUCTION

The technical requirements for the structural mechanical design result from the reduced wind and snow loads due to the principle of retraction and the long-term reliability of the moving parts (Fig. 3). The consideration of corrosion and soling for the longitudinal movement of the module carriers led to sectional steel/sheet metal profiles that are placed on wooden longitudinal beams as structural elements, while complying with the economic framework conditions. Thus, wood as a renewable building material is primarily used in the housing and base construction (Fig. 2). The classical consideration of bending stiffness, mechanical bending and tensile stresses have been applied and maintenance-free and low-wear drives are used. This results in a steel input of 41kg/kW for the module support and 29kg/kW for the remaining metal support structure, which will be integrated into the wooden construction of the URBANBOX prototype to be realised in 2024.

Figure 2: Wood is used as relevant construction material in the URBANBOX base construction.

Figure 3: URBANBOX Concept of horizontal and vertical PV module movement.

10.4229/EUPVSEC2023/4DO.5.5
020441-002

It is important to point out that such systems can be realized in a similar light weight design also in hurricane regions with extreme wind conditions, because only the small area of the structure itself is exposed to the elements, but not the large surface area of the solar panels.

4 AUTOMATIC PANEL CLEANING SYSTEM

Soling may strongly depend on the location and may reduce the PV output by several percent. The natural weather cleaning effect also depends on factors like inclination of the PV modules and density of rain or snow for example. One of the most comprehensive studies of state of the art in soiling in PV is given by the resent IEA Task 13 Report 2022. They estimated on the studies that soling losses will leading to subsequent financial losses in the range of four to seven billion euros by 2023 on the global scale based on 4-5% energy production losses [9].

4.1 Rotating brush cleaning system

A key element of the URBANBOX concept is a simple, in-expensive implementation of fully automated high quality PV module cleaning [8]. The development of the appropriate hardware and process technology of cleaning was studied within the ongoing Innosuisse research project 100.944 IP-EE and led to the brush-based solution without artificial water supply. With indoor and outdoor tests, the final solution was found with a brush diameter of 180mm (Fig. 4). Good results were achieved with rotating speeds between 100 and 500rpm at low horizontal speed (3 to 6 cm/s) of passing carriers. The frequency of cleaning, whether it is on a weekly basis and depends on rainfall, is still in the development phase and is primarily based on tests carried out with the first URBANBOX demo system in Bendern. Moreover, cleaning tests are planned with the rain sensor installed on the system, allowing also the use of rainwater on the panels for optimized cleaning.

Figure 4: -The rotating brush is placed on the top of the PV module carrier entrance of the box.

The irradiation losses during the cleaning process can be neglected, as the algorithm selects low irradiation periods below 50 W/m^2. Thus, due to the low average degree of soiling achieved by the automatic cleaning, even stubborn layers of dirt repeatedly dried by dew and dust can be significantly avoided.

Together with the Soiling sensor and in particular its sensitivity and reproducibility, these cleaning operations can be reduced even further, eliminating the need to schedule service activities for an operating life of 25 years for typical applications in Central Europe. In this region, the use of automatic URBANBOX cleaning is therefore expected to result in a PV yield gain of 1 to 5% compared to conventional uncleaned systems.

It is important to note, that the new fully integrated cleaning system has particular relevance for arid regions such as Middle East countries, Southwest USA, parts of India and many other regions, where dust, sometimes in combination with due, represents a big problem for PV and mandate frequent costly PV panel cleaning especially for dual-use PV system with high ground clearance. This serious problem is greatly alleviated with URBANBOX because the panels are be wiped off in the evening, protected to a large extent from dust and due during the night in the box and cleaned again in the morning when the carriers are extracted onto the structure.

4.2 Development of integrated soling sensor

The performance of the most proven commercial dust /soling sensors have been analysed recently in [9] but with their typical costs of 3000$ and more, will not fit into economics of URBANBOX. They mostly use the optical principle of transmission change and are partly equipped with complex image recognition and self-calibration techniques and achieve a measurement uncertainty of no less than 1%.

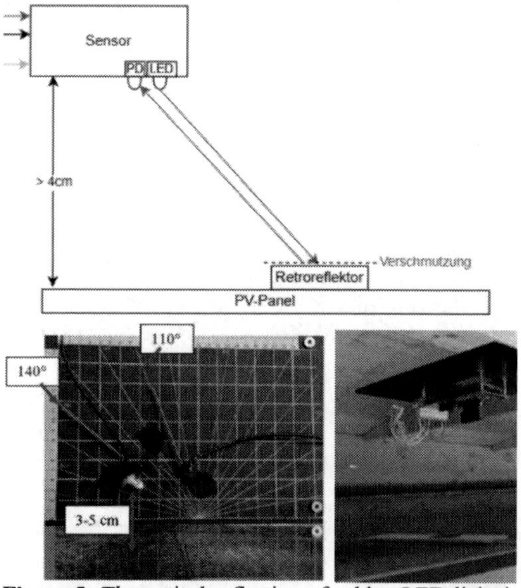

Figure 5: The optical reflection of a blue LED light is measured as the functional principle of the soling sensor and the indoor experimental test setup and the position within the BOX on top of the PV module is shown in the photographs below.

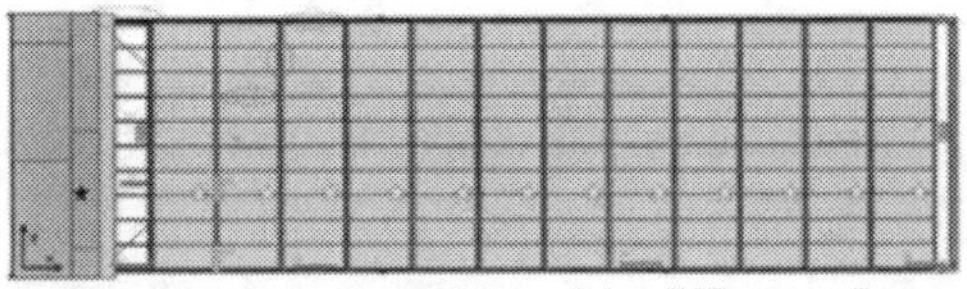

Figure 6: In yellow the position of the different reflectors for the blue light is shown mounted between tow PV modules on each carrier.

Within the research project ZHAW developed an optical dust sensor and tested a first version in the URBANBOX demo system. The optical sensor head contains both the LEDs and the photodiodes, the structure of which concentrates the beam path from the transmitter and receiver (Fig. 5). The optical reflector, whose soiling is measured, is placed in the area between two PV modules and passes the optical sensor head when the carriers are automatically moved in and out (Fig. 5).

To reduce stray light from outside the box, the drive current of the blue LED was modulated with a carrier frequency of 1280 Hz and the signal of the receiver photodiode was filtered with a bandpass filter after Opamp amplification. The self-calibration, processing of the output variables and communication with the URBANBOX's PLC control unit are carried out by an integrated microcontroller.

The sensor concept has proven itself in the first tests and was able to achieve deviations in reproducibility of typically one percent.

5 WEATHER BASED CONTROL

The local weather sensors, such as wind, rain and snow sensors, mounted on the URBANBOX currently provide the input variables to control the entry and exit of the PV carriers using a robust algorithm. To develop the algorithm, various historical weather profiles, wind and solar radiation were analyzed, with the results from three locations in the Switzerland being presented here. The locations are in the Swiss plateau near cities which does not include a mountainous region and include weather data that is available every second over three years.

The results of a simple, robust algorithm will be presented here, which switches to the retraction state at a wind speed threshold of 12m/s and automatically extends again when the wind speed falls below 10m/s with a hysteresis time of 15 minutes if the irradiation is above 50W/m^2.

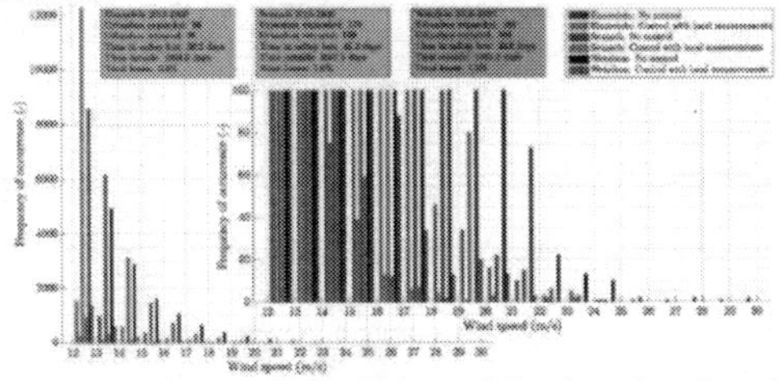

Figure 7: Histogram of wind speed for three locations in Switzerland.

The solar yield losses are less than one percent. Due to this algorithm, 33, 55 and 56 additional trips (Fig. 7) can be expected for the various location Einsiedeln, Wetzikon and Seuzach, every year. During these retraction processes, gusts of wind can affect the partially exposed PV module carrier with max wind speeds of 25, 22 and 19m/s.

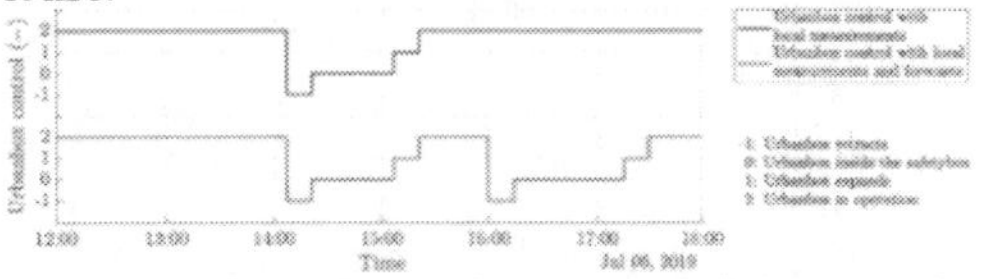

Figure 8: Time course of the states of the retraction process of the module carriers with and without weather forecast.

Tests using cost-effective weather forecasts available today have not resulted in a significant improvement in wind exposure, but have significantly increased the number of trips, which is not conducive to the longest possible operating time of over 25 (Fig. 8).

The URBANBOX system is already configured with a commercial monitoring system, with cloud data processing, monitoring 24/7 the sensors on the system states variables and the drive hardware and PV performance. Moreover, alarms by email are automatically generated, if the actual parameter differ from the pre-set parameters range. For the URBANBOX 2024 prototype a new weather control algorithm combining local weather data with area specific weather forecast data is planned to control the system from the cloud storage solution and further enhance the performance. Moreover, future URBANBOX systems installed nearby my exchange data via the cloud solution for instant monitoring and control.

6 RESULTS AND OUTLOOK

The automatic cleaning system of the URBANBOX brings economic advantages over locations with high pollution, such as dry agricultural areas with up to 11% solar power losses per month, as well as desert areas. [9].

The fact that the URBANBOX can be set up and fully tested in a fully functional way indoors at the production site with specialists, and that no specialists are needed for the installation apart from the suppliers, opens relevant advantages of binding delivery at the most diverse locations in times of great shortage of skilled workers.

The limitation of the use of metal to, for example, 70kg per kW for the URBANBOX undercuts the requirement for fixed assembly solutions with a comparable height and span by a factor of 2. This enables raw material costs to be minimized, which become more important with green steel, and helps produce solar electricity with a small amount CO_2 rucksack to reach.

With increasing climate change, the risks of local storms are rising, which very often lead to damage to fixed PV parks that is usually not covered by insurance.

Investors who require a reliable power supply and who are concerned about preserving the value of their investment could take advantage of the protected solar modules in the URBANBOX when it is retracted in safety mode during a storm or hailstorm.

Further development of the URBANBOX in combination with integrated battery storage allows the construction of an off-grid power supply, which can contribute to disaster relief operations, charging stations for electric vehicles in remote areas or the supply of settlements, for example.

10.4229/EUPVSEC2023/4DO.5.5

ACKNOWLEDGEMENT

The realization of URBANBOX was made possible through financial support from the government of Liechtenstein and the Swiss Climate Foundation and the authors thank for these contributions.

The URBANBOX development project is a joint research project of the two ZHAW institutes IEFE and IMES in Winterthur together with the startup company iWorks, supported by the funded by Innosuisse, Switzerland under project number 100.944 IP-EE. Thanks to the students Jonas Eisenring and Jonas Kaufmann at ZHAW for the valuable results they achieved in their bachelor thesis 2023 for the optimisation of the PV soiling sensor and the supervision of colleague Hanspeter Hochreutener at ZHAW SoE ISC.

REFERENCES

[1] L. Grüter, C. Nikiforos, S. Probst, L. Konersmann, Solarstrom auf Infrastrukturanlagen und Konversionsflächen, Energie Zukunft Schweiz, 23.07.2021; https://www.energiezukunftschweiz.ch [Online: accessed 12.09.2023].

[2] M. Hochreutener, L. Grüter, C. Nikiforos, L. Konersmann, Solarstrom auf Parkplatzüberdach-ungen, Energie Zukunft Schweiz, 04.05.2022; https://pubdb.bfe.admin.ch/ [Online: accessed 12.09.2023].

[3] H.-M. Neumann, D. Schär and F. Baumgartner, The potential of PV Carport, Prog. Photovolt: Res. Appl. (2011) DOI: 10.1002/pip.1199

[4] M. Loup, C. Allenspach, H. Hofmann, R. Vogt, F. Carigiet, F. Baumgartner, KEY PERFORMANCE INDICES OF PHOTOVOLTAIC CARPORTS, In Proceedings of the 38th European Photovoltaic Solar Energy Conference and Exhibition (EUPVSEC). WIP Renewable Energies, online, 2021, poster; 6CV.4.31

[5] F. Baumgartner, Arthur Büchel, F. Carigiet, T. Baumann, R. Epp, A. Wirtz, A. Huegli, U. Graf, In Proceedings of the 28th European Photovoltaic Solar Energy Conference and Exhibition (EUPVSEC). WIP Renewable Energies, Paris, France, 2013, talk 4CO.12.2.

[6] F. Baumgartner, A. Büchel, R. Bartholet, In Proceedings of the 27th European Photovoltaic Solar Energy Conference and Exhibition (EUPVSEC). WIP Renewable Energies, Frankfurt, Germany, 2015, poster; 4BV.1.60

[7] F. Baumgartner et al., PV Carport, WePo.9.33LN, 6th World Conference on Photovoltaic Energy Conversion (WCPEC-6); 26th Nov 2014, Kyoto, Japan.

[8] Arthur Büchel, world patent WO 2019 144248A1.

[9] C. Schill, U. Jahn et al., Soiling Losses, Report IEA-PVPS T13-21:2022; ISBN 978-3-907281-09-3, www.iea-pvps.org

10.4229/EUPVSEC2023/4DO.5.5
020441-005

PERFORMANCE OF PV ARRAYS WITH BIFACIAL PV MODULES INSTALLED IN EAST-WEST STRUCTURES FOR RELIABLE IRRIGATION APPLICATIONS

Luis Miguel Carrasco, Pablo Merodio, Francisco Martínez-Moreno, Rita Hogan, Luis Narvarte
Instituto de Energía Solar, Universidad Politécnica de Madrid (IES-UPM)
Campus Sur UPM, Crta. Valencia km7, Madrid, 28031, España.
luismiguel.carrasco@upm.es, francisco.martinezm@upm.es, rita.hogan@upm.es, luis.narvarte@upm.es

ABSTRACT: Bifacial PV technology contributes to increasing energy production in PV facilities compared with mono-facial PV modules, using the same occupied surface. Applications such as PV pumping for irrigation can also take advantage of using bifacial PV modules if the design of the PV system is adapted. Professional irrigation in agriculture requires water production during the dry season and constant operation during the day to maintain a regular flow and pressure, mainly in direct irrigation systems. Static support structures oriented to west and east (EW) with tilt angle of 60° reach energy profiles adapted to the irrigation needs and constitute a reliable solution compared to other systems, such as trackers. Bifacial PV modules installed in such EW structures can improve the performance and increase energy production for PV irrigation, meeting the technical requirements and maintaining a high degree of reliability of the system. This paper reports on the characterization of an EW static structure using bifacial PV modules, the so-called EWB(60). Measurements from a prototype installed in the facilities of the IES-UPM in Madrid have been taken over a full year of sun exposure using two bifacial PV modules, previously calibrated, with the aim of recording irradiance and cell temperature. The results provide some key conclusions in terms of energy production, constancy of the daily energy profile and the evaluation of the energy losses.

1 INTRODUCTION

PV technology applied to irrigation must guarantee quality in terms of water supply and be capable of maintaining constant flow and pressure regimes, as close as possible to the service provided by the electrical grid [1]. However, the variation of solar radiation throughout the day and the solar intermittency caused by passing clouds [2], are a limitation to meet optimal irrigation requirements compared to the paradigm of the electrical grid. . One of the improvements introduced for water pumping stability in PV irrigation has been the employment of horizontal north - south axis trackers (1xh) [3] which, during the irrigation periods, mainly in summertime, are able to generate horizontal daily energy profiles in clear sky days [4] instead of the bell-shaped energy profiles generated by equator oriented static structures (Figure 1).

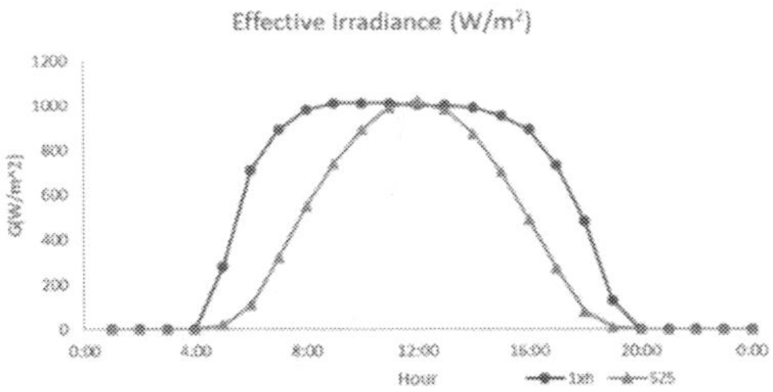

Figure 1: Effective irradiance of two different supporting structures: one horizontal north - south axis tracker (1xh) and one equator (south) oriented static structure with tilt angle of 25° (S25). Both curves have been obtained with the simulation tool SISIFO (www.sisifo.info) in Madrid (40.39N, -3.63W) on June 17 using as meteorological data the typical meteorological year (TMY).

Trackers allow a constant daily energy profile in summertime sunny days and extend the available hours for irrigation [1]. However, trackers present maintenance needs [5] that static structures do not have, and this is an issue for stand-alone applications in remote and decentralized areas with lack of operation and maintenance structures. To solve that, a previous work presented a solution to reach horizontal daily energy profiles employing static structures [6], which do not present maintenance needs. The so-called delta structure, ΔS(β), is a static structure made up of two halves, one oriented to the east and the other one to the west, achieving a constant irradiance daily profile similar to that obtained by a 1xh tracker when the tilted angle of PV modules is $\beta = 60°$. To evaluate the constancy of the irradiance daily profile, that work proposed a "constancy index", k_c, based on the coefficient of variation (CV) of such irradiance:

$$k_c = 1 - CV = 1 - \frac{\sigma}{\mu} = 1 - \frac{\sqrt{\frac{1}{N}\sum_{i=1}^{N}(x_i - \mu)^2}}{\frac{1}{N}\sum_{i=1}^{N} x_i} \quad (1)$$

for a given time series of N values, and x_i describing a general variable profile, where μ and σ are the mean and the standard deviation respectively. k_c values close to 1 indicates daily plane irradiance profiles.

The work published in [6] reported a $k_c = 0.98$, measured at the latitude of Madrid on a clear sky day close to the summer solstice. It must be noted that ΔS(60) applied to irrigation requires a peak power higher than that needed in a 1xh tracker to achieve the same water volume that an 1xh during the irrigation period, but lower than that needed in a S25 : 1.75 for ΔS(60) and 2 for S25 (Figure 2).

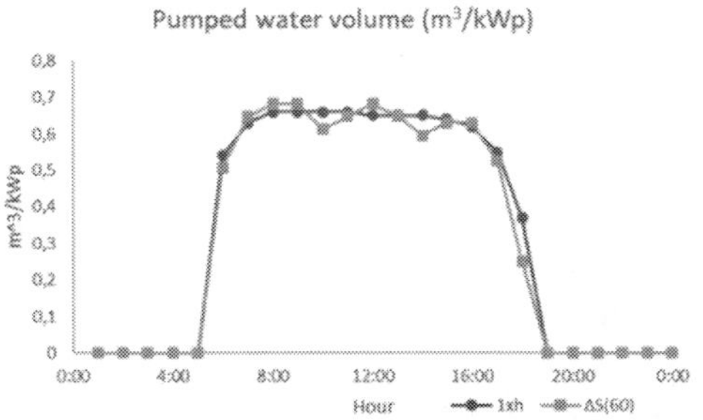

Figure 2: Hourly water pumped volume for 1xh and ΔS(60) structures in Madrid (40.39N, -3.63W) on June 17,

10.4229/EUPVSEC2023/4DO.12.1
020419-001

simulated with SISIFO and using TMY.

Performance of such east-west PV structures applied in water pumping systems for irrigation can be improved using bifacial PV modules. This technology installed in such structures can increase energy production, while meeting PV irrigation technical requirements and maintaining a high degree of system reliability.

This paper reports on the performance of a new concept of east-west structure using bifacial PV modules, the so-called EWB(β). A prototype has been installed on the rooftop of the Solar Energy Institute of the Technical University of Madrid (IES-UPM) facilities in Madrid and measurements have been taken using two bifacial PV modules, previously calibrated and transformed into reference modules, with the aim of recording the front and rear irradiances and cell temperatures of both halves over a full year of sun exposure.

The experiment has been useful to characterize this new design concept, obtaining results such as quantifying the energy gain compared to the delta structure and evaluating both the constancy index and the energy losses when using just a single maximum power point tracker (MPPT) due to the different operating temperatures of the PV modules of each half.

2 EAST-WEST STRUCTURE WITH BIFACIAL PV MODULES CONCEPT

The concept is similar to the ΔS(β) basis. It is a ground-mounted structure with half of the PV array oriented to the east and the other half to the east, both parts with the same tilt angle ($\beta = 60$). However, the two halves must be separated, as shown in Figure 3, to allow that the PV modules' rear sides obtain a good irradiation uptake, capturing both the albedo and global solar irradiation.

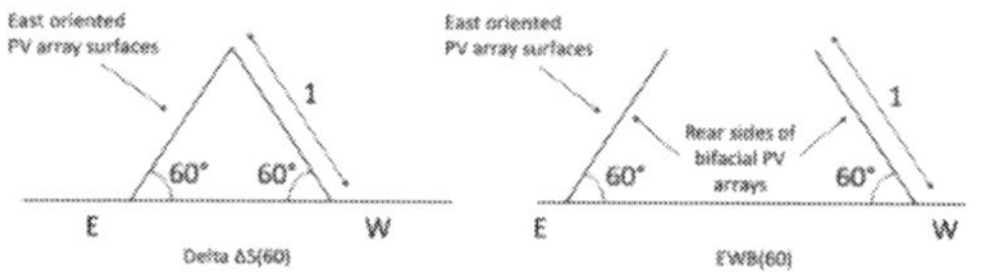

Figure 3: ΔS(60) and EWB(60) structure concepts.

The tilt angle $\beta = 60$ allows that the in-plane direct irradiance (B) on the front sides is close to constant during the period comprising 4 hours before and 4 hours after solar noon, which leads to in-plane global irradiance (G) reasonably approaching constancy for 8 hours per day. To obtain the total irradiance measured by both bifacial PV modules, the part captured by the rear sides must be added to such direct irradiance.

The experiment carried-out on the IES-UPM rooftop has been conceived to measure the front and rear sides in-plane irradiances and the cell temperatures in both east and west oriented PV modules throughout a full year with measuring slots of 5 minutes. Two bifacial PV modules (manufacturer JASOLAR, STC power $P^* = 370\ Wp$) have been modified to convert them into bifacial reference modules [7] working as sensors to measure front irradiance (G_{FRONT}), rear irradiance (G_{REAR}) and bifacial cell temperature (T_c) just dividing the module electrically into three parts [8]. The two end parts are short-circuited through calibrated resistances. A black vinyl sheet is directly glued onto the front (rear) side and then, protected with a white vinyl sheet (see Figure 4). This way, the corresponding short-circuit current, $I_{SC,REAR}$ ($I_{SC,FRONT}$), becomes sensitive to only G_{REAR} (G_{FRONT}). The central part is left as open-circuit [9] running as a sensor of cell temperature.

Figure 4: Bifacial 60° tilt angle east-west prototype EWB(60). IES-UPM facilities (Madrid).

3 RESULTS

3.1 Energy gain

Energy increase achieved using bifacial PV modules can be analyzed just comparing solar irradiance measured on both front and rear sides with that measured only on the front sides of both PV modules of the EWB(60) prototype.

Figure 5 shows the set of irradiance profiles measured by the prototype on July 20, 2022, in a clear sky day. It can be appreciated the front and rear irradiances ($G_{E,FRONT}$, $G_{W,FRONT}$, $G_{E,REAR}$ and $G_{W,REAR}$) of both PV modules and the sum of all of them (G_{TOTAL} in W/m^2).

The EWB(60) gain in terms of solar irradiation (Wh/m^2) compared with only the front sides in this particular day is 20.7% (note that a bifaciality factor equal to 1 has been considered).

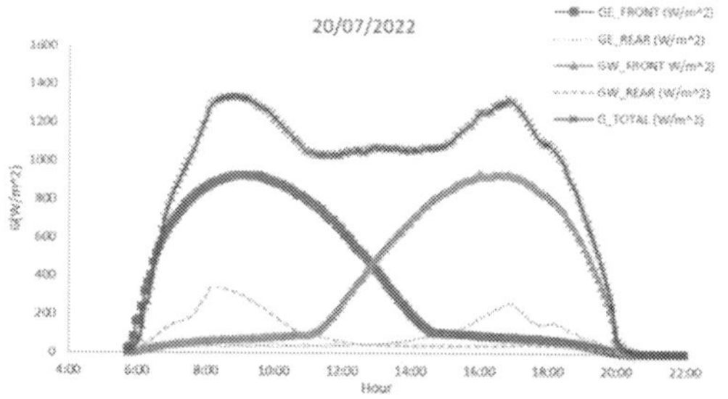

Figure 5: Solar irradiance profiles corresponding to the front and rear sides of each bifacial PV module in EWB(60), and total irradiance integrating all the system, measured on July 20, 2022.

Taking into account the measurements corresponding to the hypothetical irrigation period (here considered as the period between March and September equinoxes in the Northern Hemisphere), the total energy gain of the EWB(60) structure has been 21.5% .

These results must be underlined since energy gains in PV systems with bifacial technology but without changes in the design of the support structure, such as equator-oriented static structures, can simply achieve 5-7% more energy [10].

10.4229/EUPVSEC2023/4DO.12.1
020419-002

3.2 Constancy index.

The constancy index k_c has been obtained for each day of the year. It has been measured over the total irradiance (G_{TOTAL}) for the central 8 hours of each day (4 hours before and after noon).

Taking the measurements of July 20, 2022, shown in Figure 5, the corresponding constancy index has been $k_c = 0.91$. However, this figure is lower than that achieved if considering only the front sides, whose constancy index in this particular day would be $k_c = 0.96$. The reason for such a difference, as can be observed in Figure 5, is the effect of adding the rear side irradiances to G_{TOTAL}, producing two peaks, one in the morning and other one in the afternoon.

k_c values have been extracted from the irradiation profiles for each day of the year, as shown in Figure 6. Points below 0.8 indicates cloudy days or seasons with low solar declination (winter, and part of spring and autumn), while points greater than 0.8 correspond to days of clearer skies, which are concentrated in the summer season. The mean value of the latter has been calculated, $\mu(k_c) = 0.89$, and maximum k_c value was reached on July 14, 2022, with $k_{c,max} = 0.95$.

In the same way, looking at the results obtained considering only the front faces, the mean value would be $\mu(k_c) = 0.91$ and the maximum $k_{c,max} = 0.97$.

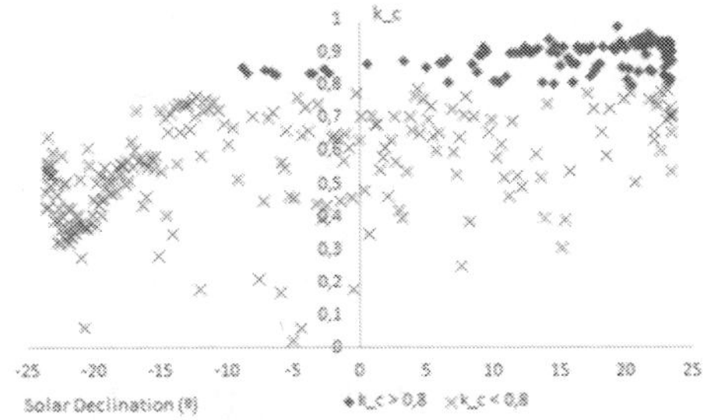

Figure 6: Distribution of the measured k_c values for each day of the year. Points under 0.8 indicate cloudy days and low solar declination values, corresponding, mainly, to the winter, autumn and spring seasons. Highest k_c values are concentrated during the summer season.

Thus, it can be concluded that EWB(60) dramatically improves the energy gain but at the expense of a reduction of the k_c value. However, it can be accepted that despite this lower k_c, these results are consistent with the requirement of constant irradiance profile for PV irrigation, suggesting that the EWB(60) structure is appropriate for such applications.

3.3 Energy losses

Operating conditions of the EWB(60) two halves are different since they are working at different incident irradiance ($G_{E,FRONT}$, $G_{W,FRONT}$, $G_{E,REAR}$ and $G_{W,REAR}$) and cell temperature ($T_{c,E}$ and $T_{c,W}$).

It must be considered that mismatching losses (MML) depend also on the non-uniformity of the rear side of the bifacial PV modules caused by both support structure shading (see Figure 4) and variations of the view factor from the ground to the different points of the PV module. In this regard, the experimental observations reported on [11] indicate that such MML represent about $0.1 - 0.2$ % of the energy produced by a PV module.

Considering a hypothetical EWB(60) PV generator and assuming that all the bifacial PV modules of the same string are installed on the same half, in addition that each one of the two halves of the PV generator are operated at the same maximum power point (MPP), the power delivered by each half, P_i, is given by:

$$P_i = P_i^* \cdot \frac{G}{G^*} \cdot [1 + \gamma(T_c - T_c^*)] \quad (2)$$

Where, subscript i designates the east or west side of the EWB(60) structure, the superscript "*" means Standard Test Conditions, T_c indicates cell temperature, γ is the power temperature coefficient of the PV module and G corresponds to the incident irradiance considering the module's bifaciality (with bifaciality factor φ_P):

$$G = G_{FRONT} + \varphi_P \cdot G_{REAR} \quad (3)$$

$$\varphi_P = \frac{P_{REAR}^*}{P_{FRONT}^*} \quad (4)$$

The total power delivered by the EWB(60), P_{EWB}, is defined by the sum of the two halves:

$$P_{EWB} = P_W + P_E \quad (5)$$

In irrigation applications, PV generators are generally connected to a single frequency converter which integrates just one maximum power point tracker (MPPT). Since the MPP voltage varies with both irradiance and cell temperature, the MPP voltage of the two PV generator halves is different. Then, electrical mismatching losses appear, and Eq. (2) must be corrected. In accordance with [6], the electrical mismatching losses (F_{ML}) are obtained as the difference between the power obtained with two and one MPPT:

$$F_{ML} = \frac{P_{EWB} - P_{MPP}^{EWB}}{P_{EWB}} \quad (6)$$

where P_{MPP}^{EWB} represents the PV power considering just one MPPT for both EWB(60) halves and is obtained as the sum of both halves, as done in Eq. (5). In P_{MPP}^{EWB} each half operates at a voltage given by the weighted average of the corresponding values of the two halves, while P_{EWM} is the sum of the two halves operating at the corresponding voltage given by their both irradiance and cell temperature. Figure 7 illustrates the daily energy losses derived from the recorded data by the EWB(60) prototype. Each single point represents the total losses of a particular day. It can be observed that energy losses rise in the summer period (high declination values), where the mean value between the two equinoxes with positive declination (March and September) is $\mu(F_{ML}) = 1.66\%$.

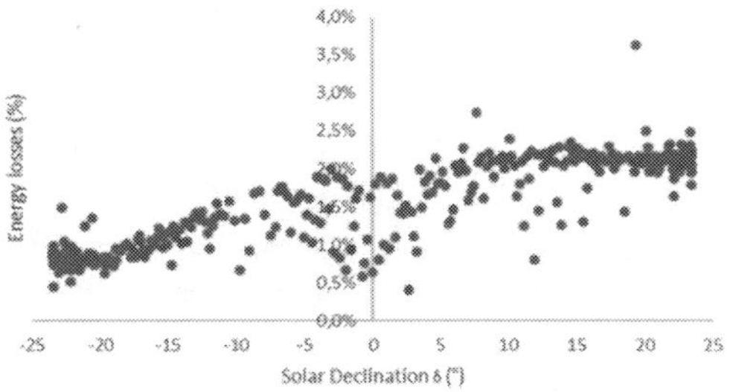

Figure 7: Energy losses caused by mismatching due to the different operating conditions between the two EWB(60)

10.4229/EUPVSEC2023/4DO.12.1
020419-003

halves working both with a single MPPT. Each single point represents the losses for a particular day, measured over the entire year of recordings.

4 CONCLUSIONS

A prototype of a new PV generator concept, integrating bifacial PV modules, has been conceived as a ground-mounted structure with half of the PV array oriented to the east and the other half to the west, both parts with the same tilt angle ($\beta = 60$) and separated to allow that the PV modules' rear sides obtain a good irradiation uptake.

The so-called East-West Bifacial EWB(60) prototype has been tested at coordinates 40.39N, -3.63W in Madrid for a full year using two bifacial PV modules modified as reference modules to measure front and rear irradiances, and cell temperature. The experiment has allowed to characterize the performance of the EWB(60) structure in terms of energy gain, constancy index and energy losses caused by mismatching between the two halves of such PV generator.

The results show that EWB(60) increases energy production in terms of solar irradiation by 21.5% during the irrigation period compared to the same structure concept but using monofacial PV modules (the so-called Delta ΔS(60) structure).

The constancy index obtained by the experiment for clear sky days during the summer period (namely irrigation period) presents a mean value of 0.88, achieving a maximum figure of 0.94. The results are lower than those obtained by the ΔS(60) structure, but are consistent with the requirement for a constant irradiance profile for PV irrigation, suggesting that the EWB(60) structure is appropriate for such applications.

Finally, the energy losses due to the mismatching between both east and west halves of the PV generator have been characterized, yielding a mean value during the irrigation period of 1.66%.

These preliminary results open the door to a widespread use of EWB(60) structures for PV irrigation and future lines of work will be explored for the development of energy production simulation tools and their validation with full-scale prototypes.

5 ACKNOWLEDGMENTS

This work has been possible thanks to the funding from the European Union's Horizon 2020 research and innovation programme in the project Accessible, reliable and affordable solar irrigation for Europe and beyond (SolAqua), under grant agreement nº 952879.
This work has been supported by the Spanish MCIN/AEI/10.13039/501100011033 through Project GREASE (Proyectos de I + D + i Retos Investigación 2020), with grant number PID2020-113533RB-C31.

5 REFERENCES

[1] L. Narvarte, J. Fernández-Ramos, F. Martínez-Moreno, L.M. Carrasco, R.H. Almeida, I.B. Carrêlo. Solutions for adapting photovoltaics to large power irrigation systems for agriculture, *Sustainable Energy Technologies and Assessments*, Vol. 29 (2018) 119-130.

[2] J. Fernández-Ramos, L. Narvarte, R. López-Soria, R.H. Almeida, I.B. Carrêlo. An assessment of the proportional-integral control tuning rules applied to Photovoltaic Irrigation Systems based on Standard Frequency Converters, *Solar Energy*, Vol. 191 (2019) 468-480.

[3] Barata Carrelo, I. A., Hogan Teves de Almeida, R., Carrasco Moreno, L. M., Martínez Moreno, F., Narvarte Fernández, L. A 360 kWp PV irrigation system to a water pool in Spain, Proceedings 33rd *European Photovoltaic Solar Energy Conference*, (2017).

[4] N.A. Engerer, F.P. Mills, KPV. A clear-sky index for photovoltaics, *Solar Energy*, Vol. 105 (2014) 679-693.

[5] Muñoz-Cerón, E., Lomas, J. C., Aguilera, J., De la Casa, J. Influence of Operation and Maintenance expenditures in the feasibility of photovoltaic projects: The case of a tracking pv plant in Spain, *Energy policy*, Vol. 121 (2018) 506-518.

[6] Almeida, R. H., Narvarte, L., & Lorenzo, E. PV arrays with delta structures for constant irradiance daily profiles, *Solar Energy*, Vol. 171 (2018) 23-30.

[7] Martínez-Moreno, F., Lorenzo, E., Muñoz, J., Moretón, R. On the testing of large PV arrays, *Progress in Photovoltaics: Research and applications*, Vol. 20(1) (2012) 100-105

[8] Martínez-Moreno, F., Rossa, C. H., Ledesma, J. R., & Merodio, P. Aspectos de la no uniformidad de la irradiancia trasera en generadores fotovoltaicos bifaciales, *CIES2020-XVII Congresso Ibérico e XIII Congresso Ibero-americano de Energia Solar*, (2020) 271-279.

[9] Ledesma, J. R., Almeida, R. H., Martinez-Moreno, F., Rossa, C., Martín-Rueda, J., Narvarte, L., & Lorenzo, E. A simulation model of the irradiation and energy yield of large bifacial photovoltaic plants, *Solar Energy*, Vol. 206 (2020) 522-538.

[10] Podlowski, L., Bifacial PV Technology - Ready for Mass. Development, *PI Berlin* (2019).

[11] Rossa, C., Martínez-Moreno, F., & Lorenzo, E. Experimental observations in mismatch losses in monofacial and bifacial PV generators, *Progress in Photovoltaics: Research and Applications*, Vol. 29(12) (2021) 1223-1235.

Task 2.3: Bifacial PV Trackers

Sandia National Laboratories is a multimission laboratory managed and operated by National Technology & Engineering Solutions of Sandia, LLC, a wholly owned subsidiary of Honeywell International Inc., for the U.S. Department of Energy's National Nuclear Security Administration under contract DE-NA0003525.

This work was authored [in part] by the National Renewable Energy Laboratory, operated by Alliance for Sustainable Energy, LLC, for the U.S. Department of Energy (DOE) under Contract No. DE-AC36-08GO28308. Funding provided by the U.S. Department of Energy's Office of Energy Efficiency and Renewable Energy (EERE) under Solar Energy Technologies Office (SETO) Agreement Number 34910. The views expressed in the article do not necessarily represent the views of the DOE or the U.S. Government.

IEA PVPS

IEA PVPS Task 13 techno-economic study of bifacial photovoltaic systems on single axis trackers

Joshua Stein (Sandia), Giosuè Maugeri (RSE), Silvana Ovaitt (NREL),
Nicholas Riedel-Lyngskær (European Energy), Matthew, Berwind (ISE) Daniel Riley (Sandia)

Technology Collaboration Programme by iea | EU PVSEC, Lisbon 21 Sept 2023 | SAND2023-09017C

IEA PVPS Task 13 Activity 2.3 Bifacial Tracking

- IEA PVPS Task 13 is focused on reliability and performance of PV systems.
  - Subtask 1: Reliability of Novel PV Materials, Components, and Modules
  - **Subtask 2: Performance and Durability of PV Applications**
  - Subtask 3: Techno-Economic Key Performance Indicators
- Bifacial PV tracking systems have the lowest LCOE for >90% of the world.

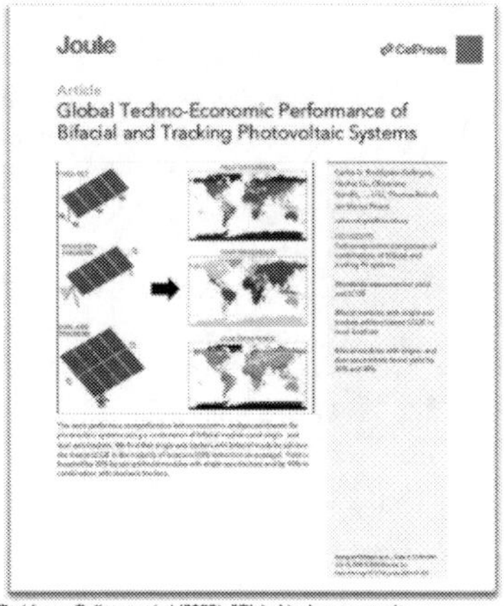

Rodriguez-Gallegos, et al (2020). "Global techno-economic performance of bifacial and tracking PV systems." Joule.

PVPS

2

Activity Relevance: Technology Trends

Bifacial PV and tracker is growing in market share over time

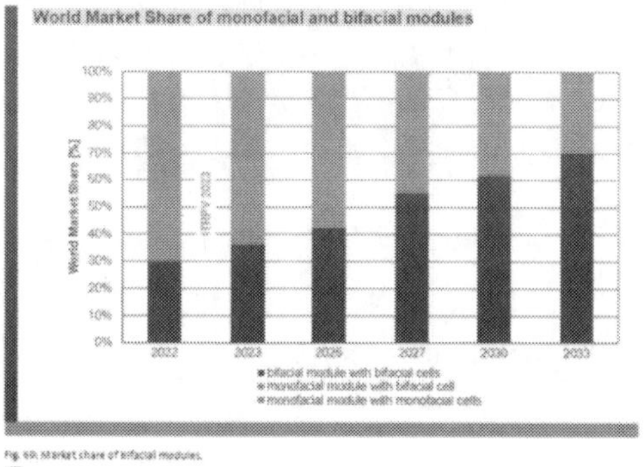

Fig. 69: Market share of bifacial modules.

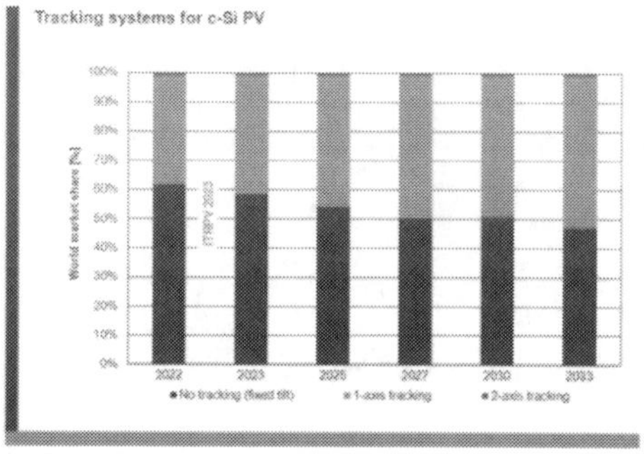

Fig. 73: Market share of tracking systems for PV power plant installations.

PVP

Fischer et al., "International Technology Roadmap for Photovoltaics: 2022 Results", 2023 Edition

3

Activity 2.3 Report Outline

18 contributing authors, 6 chapters

1. **Current technologies and emerging trends**
   - Industry survey
2. **System designs for optimal yield and value**
   - Tracking algorithms & control
   - System layouts
   - Albedo optimization
   - Dual-use Applications
3. Performance monitoring and evaluation
   - Instrumentation best practices
   - Challenges with capacity/acceptance testing
4. **Performance modeling and yield assessment**
   - Modeling intercomparison
5. Reliability considerations overview
6. Technical and Financial Optimization

PVPS

4

Activity 2.3 Report Outline

18 contributing authors, 6 chapters

1. **Current technologies and emerging trends**
   - Industry survey
2. **System designs for optimal yield and value**
   - Tracking algorithms & control
   - System layouts
   - Albedo optimization
   - Dual-use Applications
3. Performance monitoring and evaluation
   - Instrumentation best practices
   - Challenges with capacity/acceptance testing
4. **Performance modeling and yield assessment**
   - Modeling intercomparison
5. Reliability considerations overview
6. Technical and Financial Optimization

PVPS

5

Tracker Overview

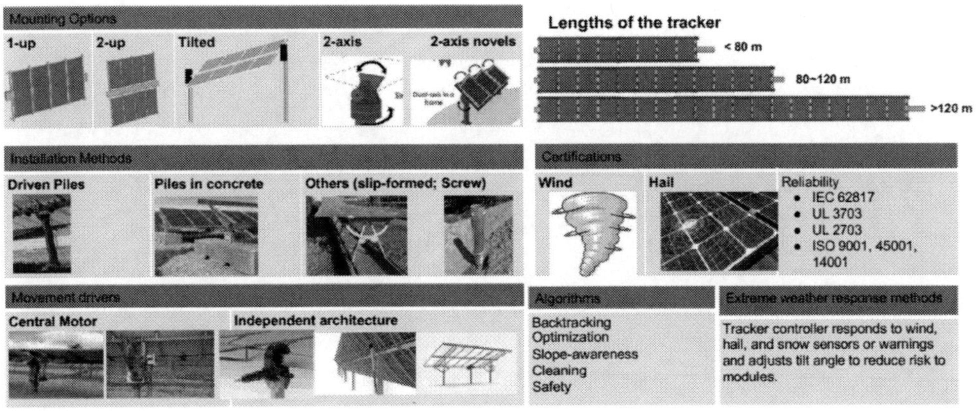

6

Current Tracker Technology and Industry Trends

Data was obtained from interviews with 17 tracker companies (>87% of global market share from 2012-2021) and review of the 2022 Wood Mackenzie Global Solar PV Tracker report.

45 questions covered topics that include:

- Company overview (region, scale, history)
- Product specifications and features (including bifacial-related features)
- Tracking algorithms (backtracking, terrain-aware, diffuse, etc.)
- Sustainability

This presentation covers the highlights from the survey. Our final report will cover more of the details.

PVPS

7

020421-007

40th European Photovoltaic Solar Energy Conference and Exhibition

Tracker Companies are International

- 70% of companies have been in business for at least 10 years.
- ~50% of companies sell trackers in more than 20 countries.
- >80% of companies sell in more than 10 countries.

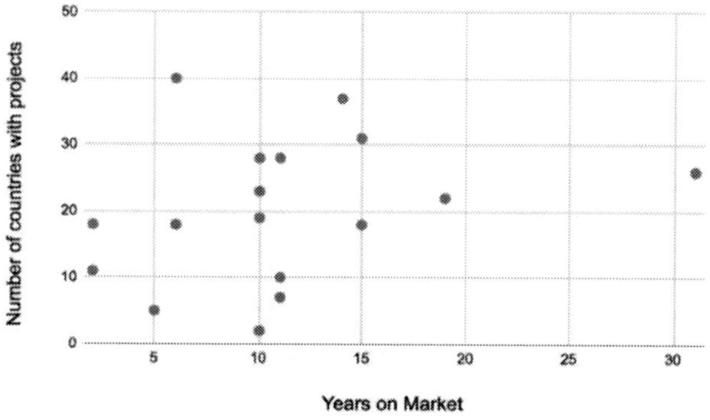

PVPS

8

020421-008

Companies Overview

Number of companies that took part in the survey: 17

Number of projects finished: >210 GW

Number of projects in developments: >47 GW

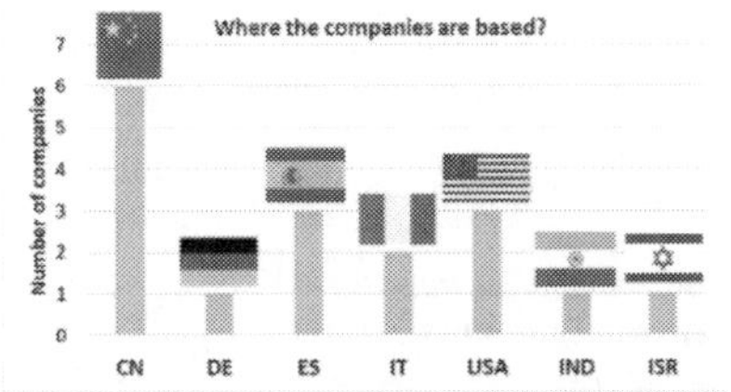

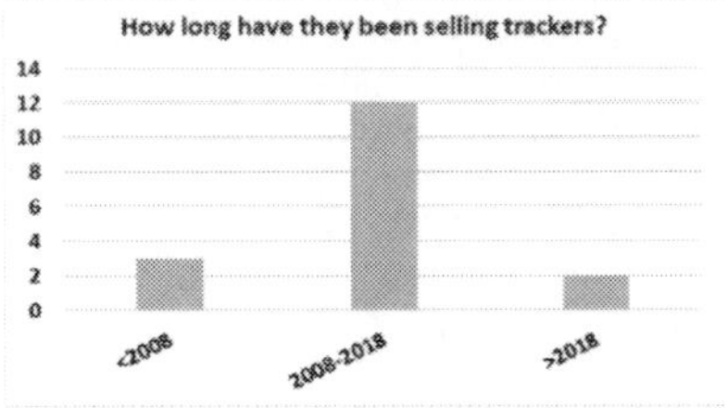

PVPS

9

Companies Overview

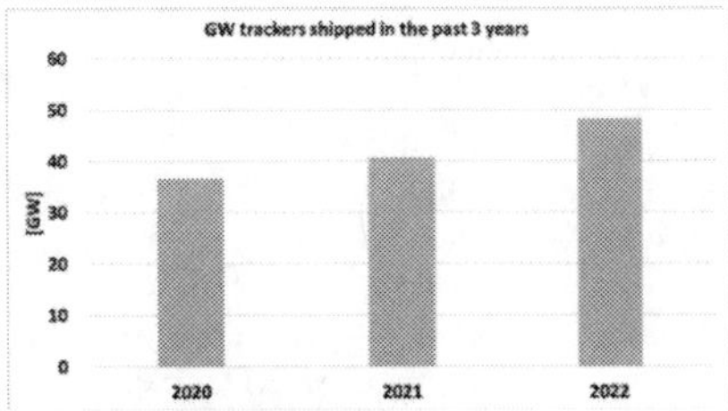

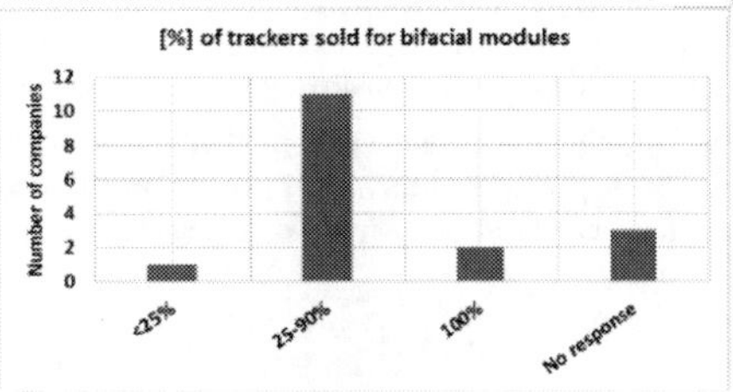

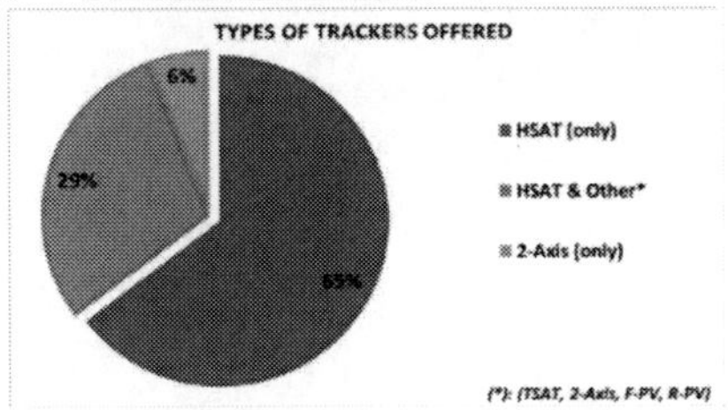

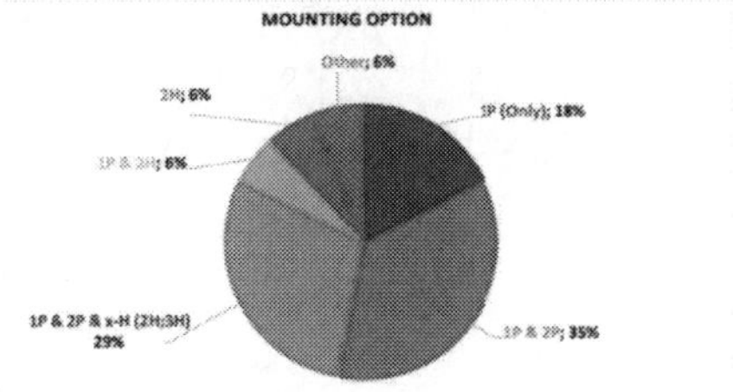

PVPS

10

ESG and diversity   Environmental, Social, Governance

- ESG is being requested by clients
- Nearly all companies have ESG strategies and a few have comprehensive reports.
- Strategies include:
  - ➢ Increasing efficiency of electric motors
  - ➢ Carbon footprint certificates
  - ➢ Reducing materials usage (e.g., concrete, steel)
  - ➢ Eliminating child labor via supply chain requirements
  - ➢ Promotion of local manufacturing.
  - ➢ Partnering with local universities for workforce development.

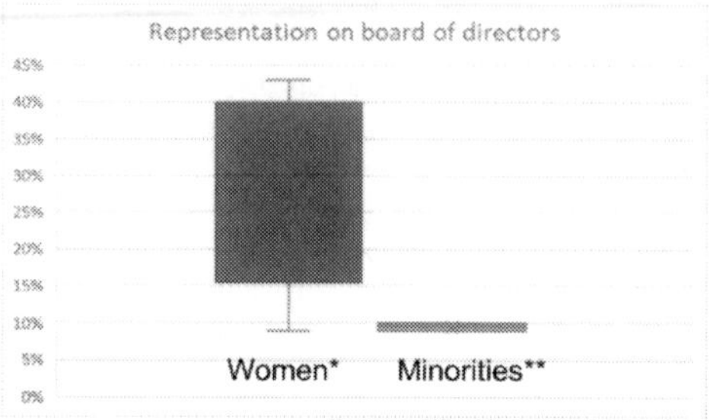

*based on 50% of the surveyed companies
**only two self-declared values

PVPS

11

40th European Photovoltaic Solar Energy Conference and Exhibition

Market trends and drivers

- **Reported Cost Range** of 0.06 – 0.14 $/W for 1-axis, and 0.2 – 0.4 $/W for 2-axis trackers.
  - Wind speed risk, cost of concrete, design factors affect cost (2P is more expensive)
  - In general, SAT systems increase yields by ~20% over fixed-tilt systems (location and site-specific)
- **Supply chain issues** and market prices are important (i.e. cement and steel).
- Some companies use **local providers** of these two materials to offset cost and carbon emissions.
- Developers value **reliable delivery schedule** and **availability of equipment** and are willing to pay more.
- Companies are focusing on certain market sectors (e.g., dual-uses for AgriPV, deployment on non-agricultural or usable land, highly sloped terrains). **Divergent perspectives on land-use and value.**
- Companies cite many testing certifications and wind tunnel tests.
  - UL 2703 "Mounting Systems, Mounting Devices, Clamping/Retention Devices, and Ground Lugs for Use with Flat-Plate Photovoltaic Modules and Panels",
  - UL 3703 "Solar Trackers",
  - IEC 62817 "Photovoltaic Systems - Design Qualification Of Solar Trackers",
  - ISO 9001, ISO 45001, ISO 14001

PVPS

12

Activity 2.3 Report Outline

18 contributing authors, 50 pages, 6 chapters

1. **Current technologies and emerging trends**
   - Industry survey
2. **System designs for optimal yield and value**
   - Tracking algorithms & control
   - System layouts
   - Albedo optimization
   - Dual-use Applications
3. Performance monitoring and evaluation
   - Instrumentation best practices
   - Challenges with capacity/acceptance testing
4. **Performance modeling and yield assessment**
   - Modeling intercomparison
5. Reliability considerations overview
6. Technical and Financial Optimization

PVPS

13

40th European Photovoltaic Solar Energy Conference and Exhibition

Backtracking and Sloped Land

- All tracker companies surveyed offer backtracking.
- Complex terrain presents challenges for certain tracker designs.
  - Slope changes in the direction normal to the rows requires adjustment to each rows tilt angle.
  - Slope changes parallel to rows requires flexible couplings on the torque tubes.
- Some designs can easily accommodate complex terrain. Some companies offer multiple designs optimized to different environments.

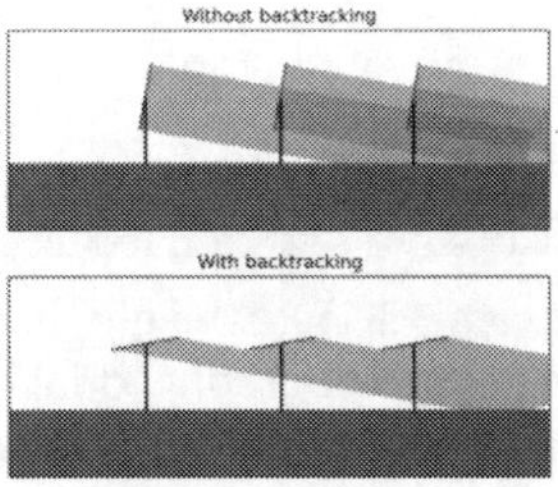

Example 1: Each row tracked separately

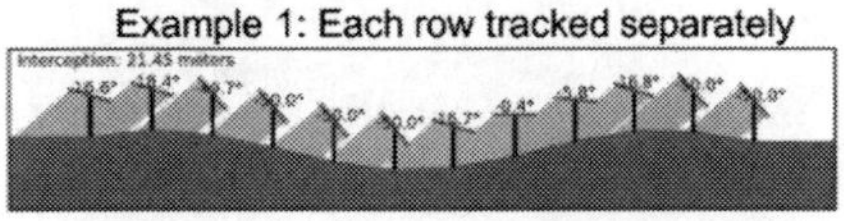

Example 2: Four rows connected

Figures from Kevin Anderson, Sandia

PVPS

14

Weather Resilience Responses

- Tracker controllers receive signal from wind (or sometimes hail) sensors dispersed in the field.
  - Short duration events - controllers automatically place trackers into or out of a defensive stow position
  - long-duration events (like hurricanes) - Stow planning features, system come out of stow only by user intervention
- Tilt adjustments to protect systems and modules.
  - Rapidly move to maximum tilt – e.g., hail or snow
  - Horizontal position - in the case of wind gusts to reduce the sail effect.
    - Passive vs. active stow

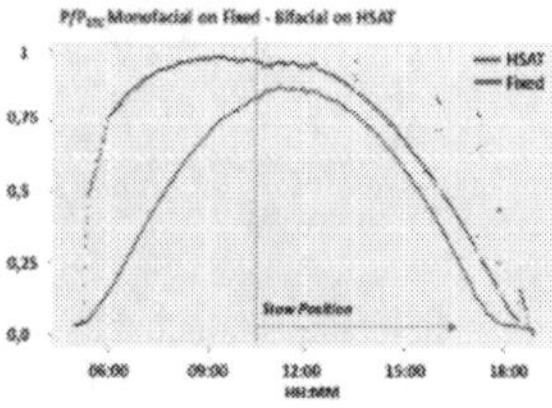

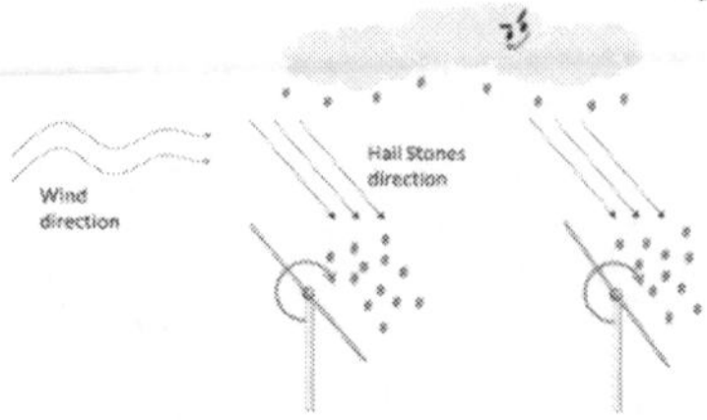

TRACKER FACE AWAY FROM WIND DIRECTION

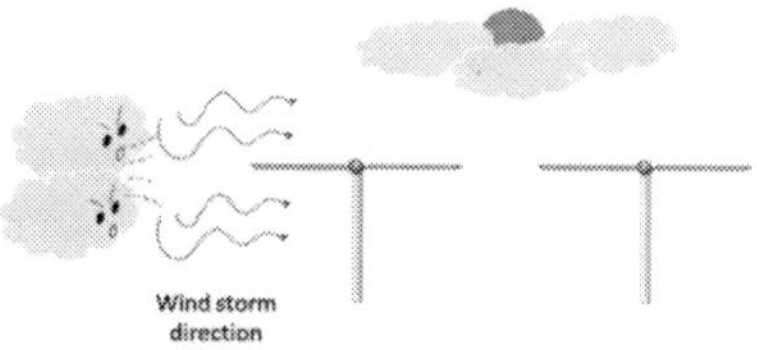

TRACKER WIND STORM POSITION

PVPS

15

020421-015

40th European Photovoltaic Solar Energy Conference and Exhibition

Conclusions / Next Steps

- Bifacial + Tracking systems will continue to grow in market share.
- Trackers are selected for both technical and non-technical reasons.
- Trackers will increasingly have to adapt to more complex and constrained sites (e.g., topography, dual use, extreme weather, etc.).
- Ease of installation, availability of materials, and supply chain issues will be more prominent as PV markets continue to grow.

- IEA PVPS Task 13 will summarize this survey in our report (2024)
- PV performance modeling tool comparison is underway and open for participation. (Details on next slides).

PVPS

16

020421-016

Activity 2.3 Report Outline

18 contributing authors, 6 chapters

1. **Current technologies and emerging trends**
   - Industry survey
2. **System designs for optimal yield and value**
   - Tracking algorithms & control
   - System layouts
   - Albedo optimization
   - Dual-use Applications
3. Performance monitoring and evaluation
   - Instrumentation best practices
   - Challenges with capacity/acceptance testing
4. **Performance modeling and yield assessment**
   - Modeling intercomparison
5. Reliability considerations overview
6. Technical and Financial Optimization

PVPS

17

40th European Photovoltaic Solar Energy Conference and Exhibition

Modeling Intercomparison - Due October 1

All are invited to submit PV bifacial tracking simulation results

- Seven scenarios that explore system design variations and modeling approaches.
- Hourly meteo data and module performance parameters are provided
- Participants have several options for submitting their model results (full year or selected days). Templates provided.
- **Results due by October 1, 2023**

https://tinyurl.com/TrackingModeling2023

Example tracker system layout for simulation

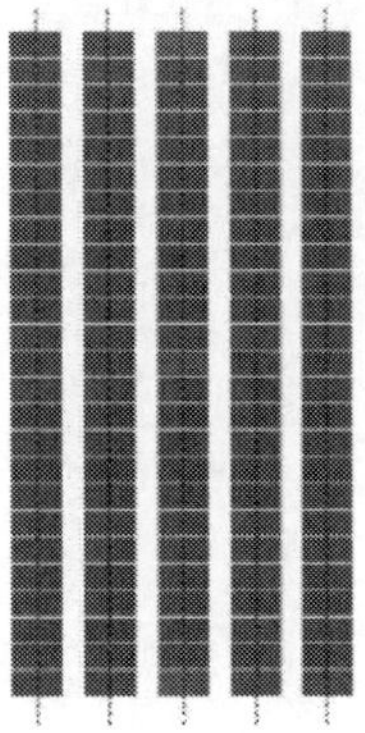

PVPS

18

Modeling Scenarios

https://tinyurl.com/TrackingModeling2023

- Scenarios are hypothetical and cover variations in:
  - GCR
  - Albedo
  - Hub height
  - Configuration
  - Ground slope

1P

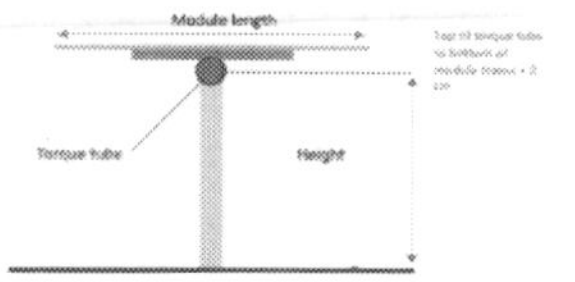

2P

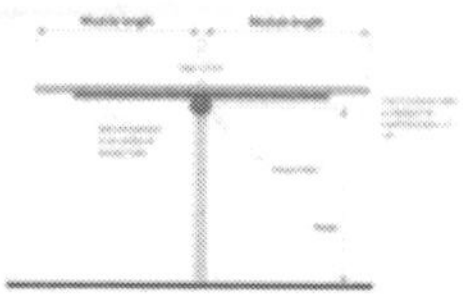

Scenario number	Scenario name	GCR*	Albedo	Hub Height	Configuration	Ground surface
1	Ref-A	0.4	0.2	1.5 m	1-Up portrait	Horizontal
2	A1	0.25	0.2	1.5 m	1-Up portrait	Horizontal
3	A2	0.4	0.5	1.5 m	1-Up portrait	Horizontal
4	A3	0.4	0.2	3.5 m	1-Up portrait	Horizontal
5	A4	0.4	0.2	1.5 m	1-Up portrait	10% grade* down to the East
6	A5	0.4	0.2	1.5 m	1-Up portrait	10% grade* down to the SW
7	Ref-B	0.4	0.2	3.5 m	2-Up portrait	Horizontal

PVPS

19

020421-019

40th European Photovoltaic Solar Energy Conference and Exhibition

www.iea-pvps.org

Joshua S. Stein
jsstein@sandia.gov

Giosuè Maugeri
giosue.maugeri@rse-web.it

Dan Riely
driley@sandia.gov

Silvana Ovaitt
silvana.ovaitt@nrel.gov

Nicholas Riedel-Lyngskær
nri@europeanenergy.com

Matthew Berwind
matthew.berwind@ise.fraunhofer.de

Thank you

IEA PVPS

Technology Collaboration Programme by iea

020421-020

www.iea-pvps.org

Come to the European PVPMC Workshop
November 8-9, 2023

SUPSI, Mendrisio, Switzerland

https://pvpmc.sandia.gov

IEA
PVPS

Technology Collaboration Programme
by iea

40th European Photovoltaic Solar Energy Conference and Exhibition

Content

Motivation

40th European Photovoltaic Solar Energy Conference and Exhibition

Motivation
Bifacial Gain (BG) Improvement to reduce the LOCE

Higher bifacial gains have the potential to reduce the LCOE

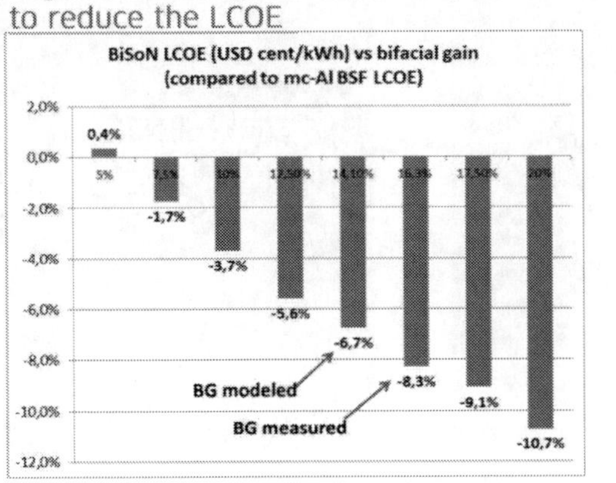

D.Berrian et al., MoBiDiC: simulations and LCOE, ISC Konstanz e.V., Bifacial Workshop, 11.10.2017

Trackers are expected to have more market share but fix till still dominates

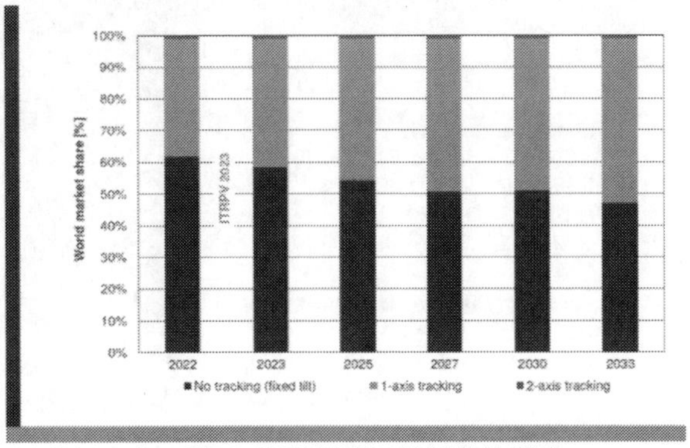

Verband Deutscher Maschinen- und Anlagenbau, International technology roadmap for photovoltaic, 2023, URL https://www.vdma.org/international-technology-roadmap-photovoltaic.

Motivation
Ways to Enhance the Bifacial Gain

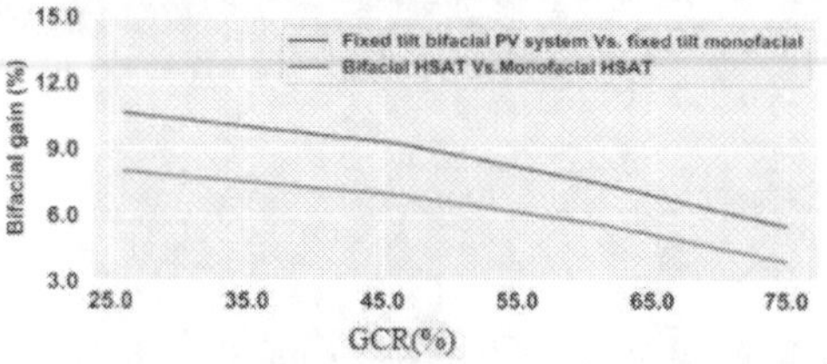

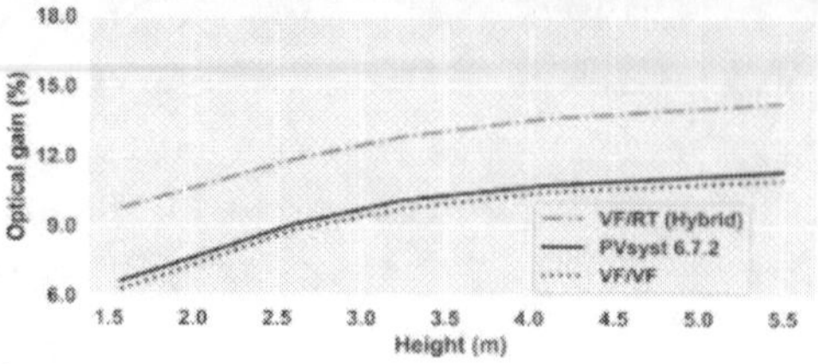

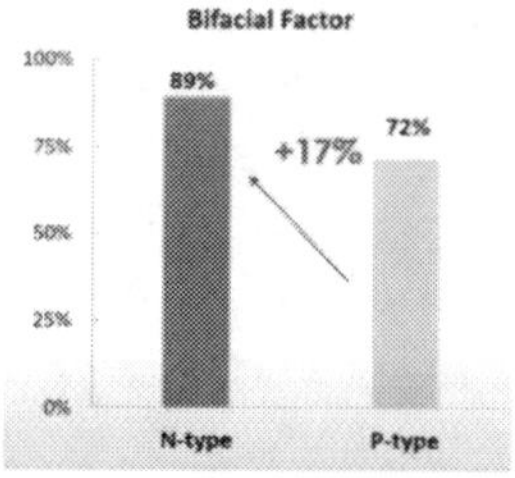

[1] Berrian, Diaber, et al. "Performance of bifacial PV arrays with fixed tilt and horizontal single-axis tracking: Comparison of simulated and measured data." IEEE Journal of Photovoltaics 9.6 (2019): 1583-1589.

[2] Kopecek, R., 2018. Photos from bifacial PV power plant "La Hormiga". Personal Communication

[3] Image. Solmax Geosynthetics GmbH

[4] Linko. Innovation in Bifacial P and N-type Modules to Boost Efficiency and Power Density

40th European Photovoltaic Solar Energy Conference and Exhibition

Motivation
Bifacial Gain Improvement Through Mounting Systems Design

NOMS Characteristics

- High number of beams per PV module
- Beams are close to the active area
- Purlins shade the active area

OMS Characteristics

- Reduced number of beams
- Spaced beams from the PV module active area
- Purlins do not shade the active area

1. How much is the improvement in BG due to the OMS design?
2. What factors contribute to BG improvement?

→ D. Berrian et al. "Enhanced Bifacial Gain with Optimized Mountings in Photovoltaic Systems." Solar RRL - Wiley Online Library (2023)

Non-optimized mountings (NOMS)

Optimized Mountings (OMS)

Methodology

Methodology
Site Description

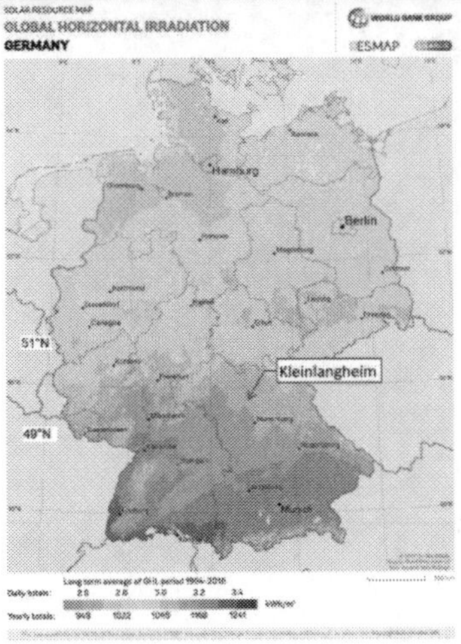

Top view of the bifacial PV plant in Kleinlangheim (5 MWp)

3P, Fix South, 20° tilt
OMS substructures are highlighted in red, (42 m long)
GCR approx. 55% (12.4 m pitch), average albedo 20%

Methodology
Modeling & Simulation

bifacial_radiance (ray tracing) → Construct the 3D scene of the bifacial PV plant with OMS and NOMS designs → Structure shading factors (S) of OMS and NOMS (to calibrate bifacial VF) → (view factor) bifacialVF

→ Incorporation of the empirical electrical mismatch model into bifacial_VF → Incorporation of electrical model + thermal models into bifacial VF → Hourly simulations of OMS and NOMS energy yield based 8760 timestamps (bifacial gain)

Two important changes made to bifacial VF

1. calibration of bifacial VF

$$G_{\text{eff}} = G_{\text{front}} + \beta \times (1 - \frac{S}{100}) \times G_{\text{rear}} \longrightarrow S(\%) = (1 - \frac{G_{\text{rear}}(\text{with structure shading})}{G_{\text{rear}}(\text{without structure shmading})}) \times 100$$

2. Incorporation of the empirical electrical mismatch

$$MM(\%) = \exp(1.57 \times \ln(\sigma[\%]) - 2) \times \frac{FF}{FF_0} \longrightarrow \sigma[\%] = \frac{1}{\bar{G}_{\text{eff}}} \times \sqrt{\frac{\sum(G_{\text{eff, i}} - \bar{G}_{\text{eff}})^2}{n}} \times 100$$

[1] G. J. Ward. The radiance lighting simulation and rendering system. In Proceedings of the 21st annual conference on Computer graphics and interactive techniques, pages 459–472, 1994
[2] A. P. Silvana, & C. Deline. "Bifacial_radiance: a python package for modeling bifacial solar photovoltaic systems." Journal of Open Source Software 5.NREL/JA-5K00-75222 (2020).
[3] Berrian, D., & Libal, J. (2020). A comparison of ray tracing and view factor simulations of locally resolved rear irradiance with the experimental values. Progress in Photovoltaics: Research and Applications, 28(6), 609-620.
[4] Marion, Bill, et al. "A practical irradiance model for bifacial PV modules." 2017 IEEE 44th Photovoltaic Specialist Conference (PVSC). IEEE, 2017.
[5] Deline, Chris, et al. "Estimating and parameterizing mismatch power loss in bifacial photovoltaic systems." Progress in Photovoltaics: Research and Applications 28.7 (2020): 691-703.
[6] D. Faiman. Assessing the outdoor operating temperature of photovoltaic modules. Progress in Photovoltaics: Research and Applications, 16(4):307–315, 2008.
[7] W. De Soto, S.A. Klein, and W.A Beckman. Improvement and validation of a model for photovoltaic array performance. Solar energy, 80(1):78–88, 2006.

40th European Photovoltaic Solar Energy Conference and Exhibition

Results & Discussion

Results & Discussion
The impact of the structure shading on the backside irradiance

Structure shading factor breakdown (19.2%, 13.1%)

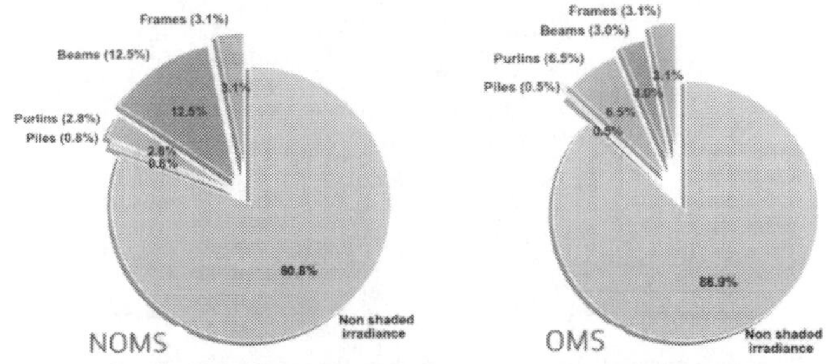

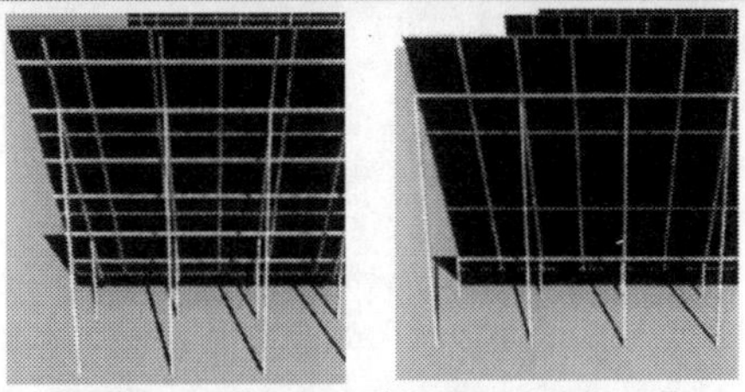

3D Scenes from Radiance 21st June, 12:00 (Top figures)

- Beams are the most source of shading in the NOMS design.
- To improve the substructure design, one should consider reducing the number of beams and increasing the number of purlins.
- There is potential of the rear irradiance boost for frameless PV modules up to 3.1% (This means -> 3.1% relative boost in BG)

Results & Discussion
Electrical mismatch (MM) loss of NOMS and OMS designs

Yearly electrical mismatch loss NOMS vs. OMS

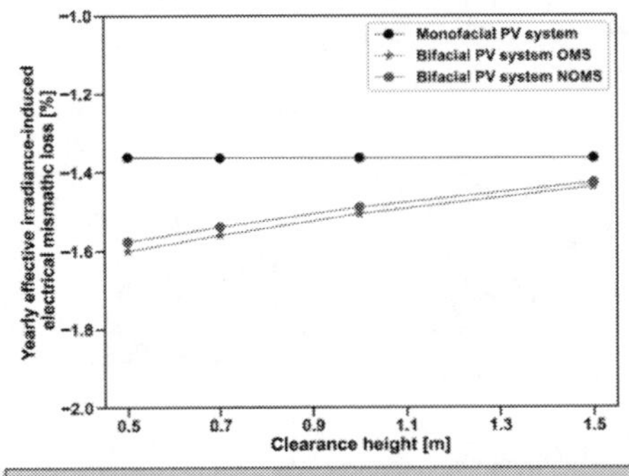

Additional electrical MM loss due to the structural shading elements

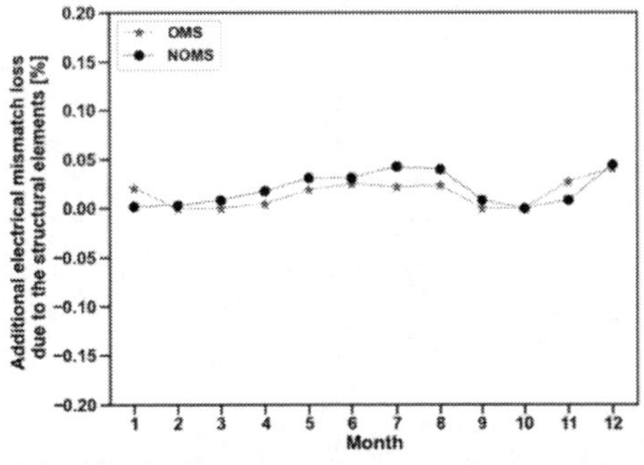

- The electrical mismatch in monofacial PV systems is independent of the clearance height.
- Bifacial PV systems can add more electrical mismatch losses (0.05 – 0.24% abs.).
- The additional electrical MM loss due to the structural shading elements in NOMS and OMS design is negligible (0.02% vs 0.24%).

Results & Discussion
Yearly Bifacial Gain Improvement

Yearly bifacial gain improvement at different GCR and albedo values (0.7m)

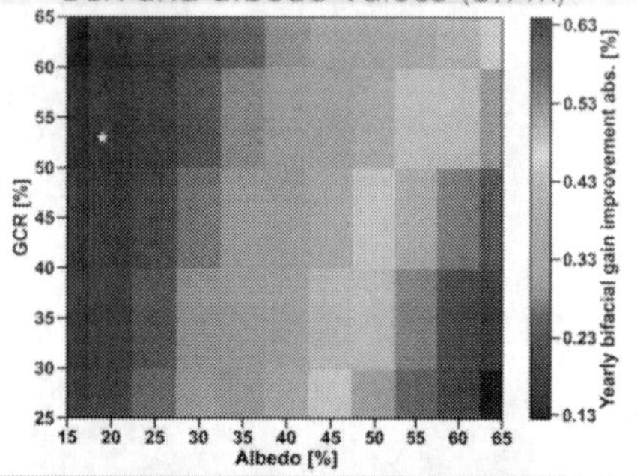

Yearly bifacial gain improvement at different GCR and albedo values (>3m)

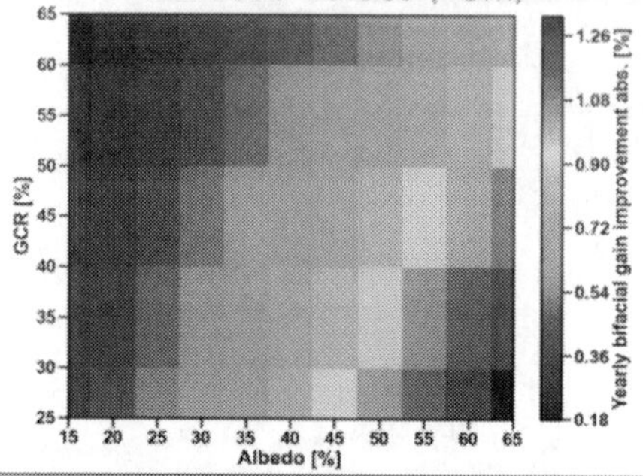

- The improvement in bifacial gain due to **OMS design** depends on the **clearance height**, **albedo** and **GCR**.
- For a given clearance height, **low GCR** and **high albedo** values allow for high improvement in bifacial gain (compared to NOMS).
- For our setup, the **OMS design** can achieve a maximum of **0.63%** or a minimum of **0.13%** (absolute) **additional BG**.
- **Elevated PV application**, such Agri-PV can achieve **high bifacial gain improvement due to OMS exceeding 1%** (absolute).

Results & Discussion
Measured Bifacial Gain Improvement (GCR 55, Alb. 18, H. 0.7m)

Monthly measured and modeled BG improvement

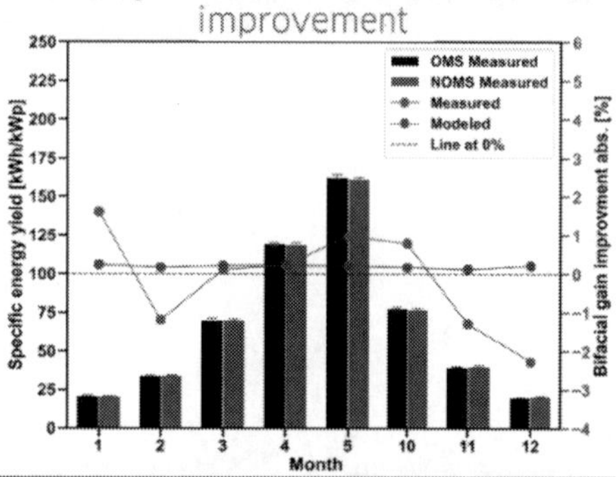

Cumulative measured and modeled BG improvement

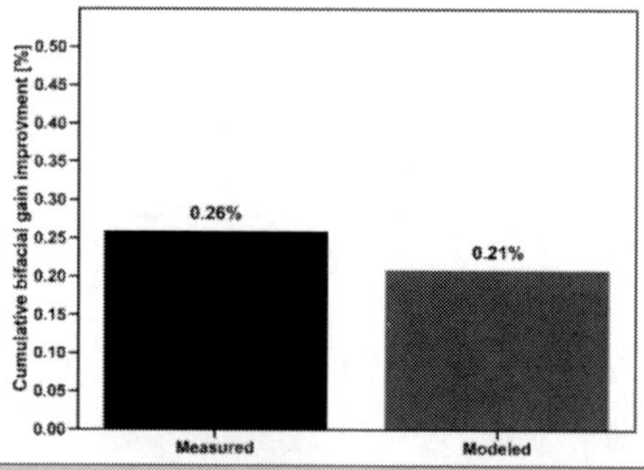

- Stronger **seasonal dependency** of the **measured bifacial gain improvement** than the simulated values.
- The highest bifacial gain improvement of **approximately 1%** (+/- 0.02%) was recorded in May.
- Relatively **high match between measured and modeled bifacial gain improvement** is observed from **spring toward summer**.
- The measured and simulated **bifacial gain improvements** over the entire period are **0.26%** (+/- 0.02%) and **0.21%**, respectively.

Conclusions

40th European Photovoltaic Solar Energy Conference and Exhibition

Conclusions
Summary of key takeaways

- **Redesigning standard mounting systems** to reduce backside shading in bifacial PV systems increases the annual bifacial gain by **at least 0.13% absolute**, as proven by simulations and experiments.
- Depending on **the geometrical configurations** (GCR, albedo, and clearance height), the yearly bifacial gain improvement can be **as high as 1% absolute** (i.e., 1% increase in energy yield).
- Reducing **the number of horizontal beams** and increasing **the number of purlins** leads to less backside shading, and hence, a **high bifacial gain**.
- The **main driving factor** for **high bifacial gain improvement** in the optimized mounting systems is **the structural shading factor** (not the electrical loss mismatch reduction).
- Frameless PV modules have **the potential for a further increase of the bifacial gain up to** approx. **3.1% relative.**

Thank you for your attention

Dr. Djaber Berrian
Innovation & Development Projects Leader

For any questions
djaber.berrian@belectric.com

BELECTRIC

Optimizing Vertically Oriented Solar PV in Northern Climates

EU PVSEC 2023, September 18-22

Christopher Pike[1], Henry Toal[1], Daniel Riley[2], Laurie Burnham[2]
[1]University of Alaska Fairbanks, Fairbanks, United States of America
[2]Sandia National Laboratories, Albuquerque, United States of America

This project is part of the Arctic Regional Collaboration for Technology Innovation and Commercialization (ARCTIC) Program, an initiative supported by the Office of Naval Research (ONR) Award # N00014-19-1-2235

020423-001

Vertical East/West Bifacial PV

- Vertical bifacial arrays provide more desirable production curves.
- Nearly identical cumulative yearly energy.

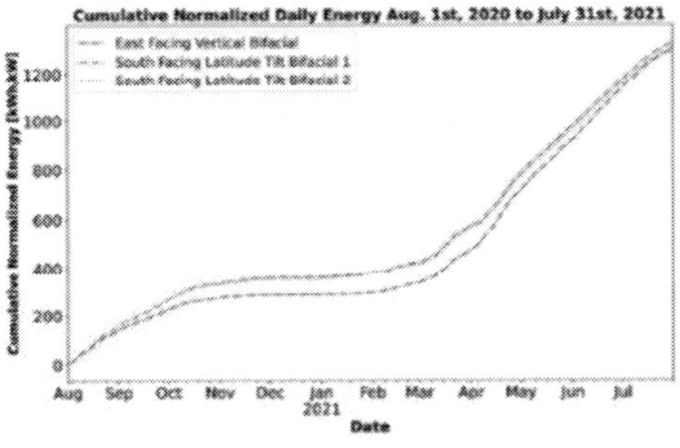

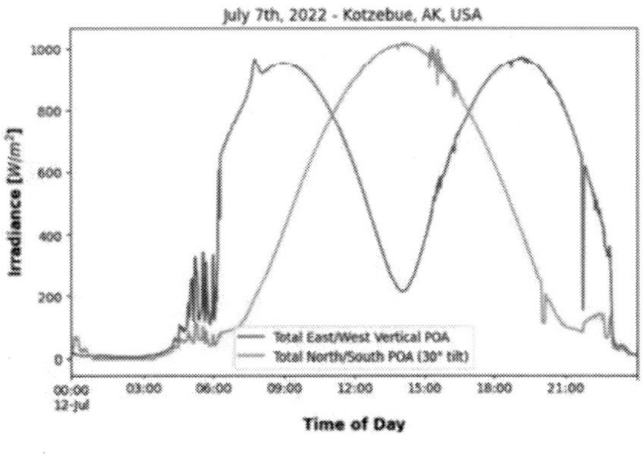

020423-002

The Goal of this Research

A problem: flat environments with significant wind and snow cause snowdrifts, especially around objects perpendicular to the wind.

- How does this influence optimum east/west vertical bifacial PV design?
  - How much shading loss due to snowdrifts?
  - Can snowdrift shading be eliminated by increasing the height of the array?
  - Does it make economic sense to do so?
- Does increasing array elevation lead to more energy generation regardless of shading?
- How much energy is lost to shading, especially in a high-albedo environment?

Large snowdrift near one section of the Kotzebue Alaska community solar array.

020423-003

40th European Photovoltaic Solar Energy Conference and Exhibition

Data

- Kotzebue, Alaska, USA
  - 67° North, 162° West
  - ~3100 people
  - 1.2 MW of bifacial PV
- 11 months of 1-minute averaged data from instrumentation co-located with PV array from 04-2022 to 02-2023:
  - East and west plane-of-array (POA) irradiance at **1, 2, and 3 meters** and front/back POA for local 30° tilt PV array
  - GHI, wind speed/direction, ambient temperature, albedo
  - Daily images of PV array at solar noon

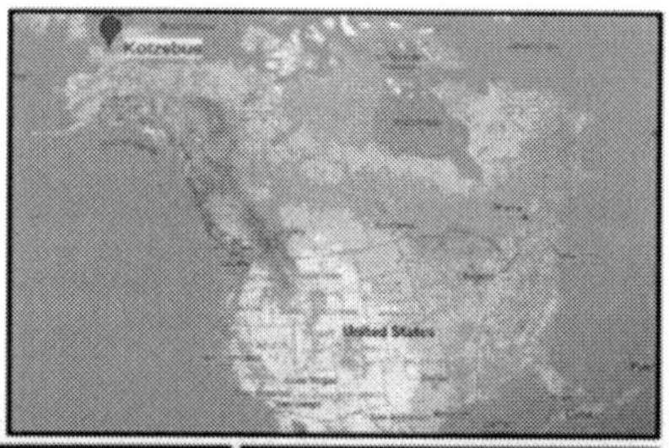

Location of Kotzebue, Alaska. ©2023 Google, INEGI, TMap Mobility

Chris Pike (ACEP) with the east/west reference cell installation

MET station at PV array site.

020423-004

Model Structure and Inputs

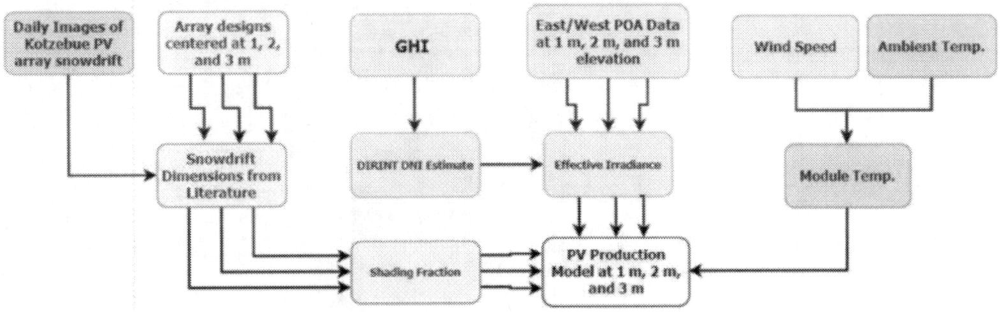

- 100 kWp array modelled using pvlib
  - Single module high, landscape orientation
  - single diode model
- 1x2m, 400-Wp LG400n2t-j5 bifacial module (same as the Kotzebue array)

020423-005

Estimating Shading from Snowdrifts

- A snowdrift has two dimensions which effect its shading potential: height ***(h)*** and distance from the top of the drift to the array ***(d)***.
- These dimensions are impacted by the total array height ***(H)*** and the gap between the modules and the ground ***(g)***.

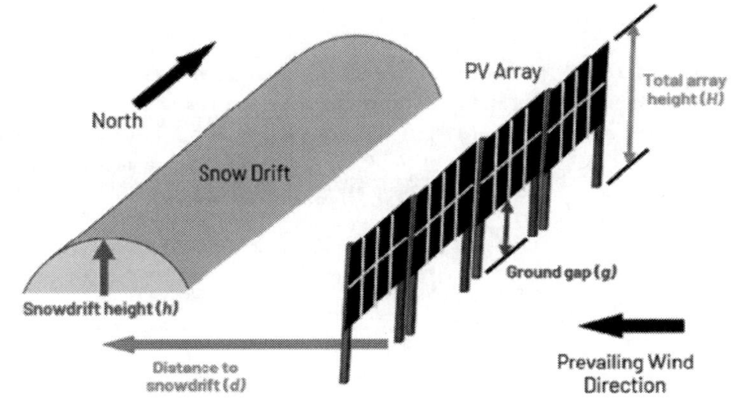

020423-006

Calculating Snowdrift Dimensions

- Multiple studies on drifts around snow fences (structures designed to create snowdrifts) support an equilibrium snowdrift height of **1.2 times** total fence height [1].
    - Snowdrifts around PV arrays follow a similar maximum drift height profile [2].
- We can estimate relative drift height throughout the year from daily images.
- Representative heights for each month were calculated

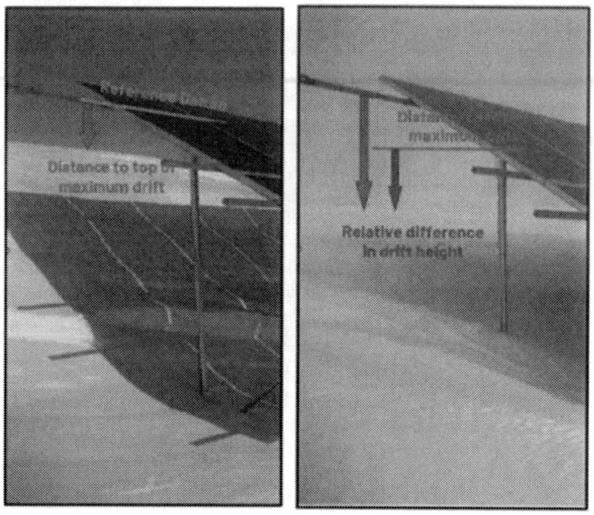

020423-007

40th European Photovoltaic Solar Energy Conference and Exhibition

Snowdrift Dimensions with Varying Gap Size

- A max drift size of 1.2 times fence/array height assumes minimal gap under fence or dense support structures.
- Max drift height and distance *(d)* for drifts around fences shaped similarly to vertical bifacial arrays vary with gap *(g)*.

- (Martinelli, 1964) collected imperial measurements of similar structures [3].
- We extrapolate their relationships between *H* and *d* with *g* to find values for our models at 1, 2, and 3 meters *(g = 0.5 m, 1.5 m, 2.5 m, H = 1.5 m, 2.5 m, 3.5 m)*

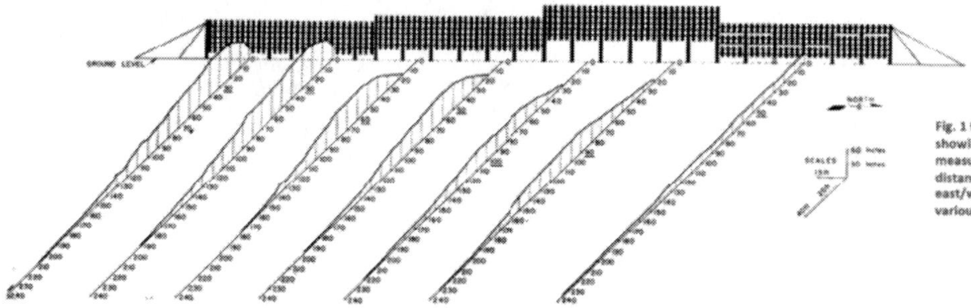

Fig. 1 from Martinelli, 1964 [3] showing downdrift measurements at various distances from vertical east/west snow fences with various gap sizes.

020423-008

Drift Dimensions used in Models

Elevation	Distance to top of drift (*d*)	Relative maximum drift height (*x/H*)
1 Meter	15 m	1.12
2 Meters	28 m	0.64
3 Meters	41 m	0.35

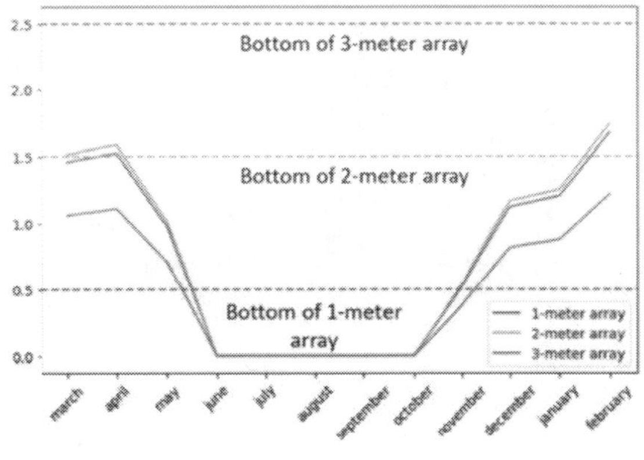

Modelling System Production

- Snowdrift shadow was calculated geometrically to obtain shading fraction.
- Diffuse fraction was calculated using modelled DNI from pvlib's DIRINT implementation.
- Bifaciality of 70%.
- Individual module simulation done in pvlib to simulate losses from partial shading.
- Extrapolated for full 100 kWp array.
    - Total system efficiency of 96% used as estimate.

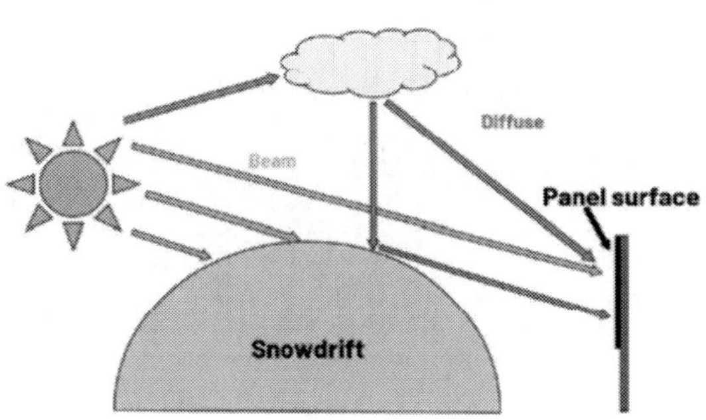

Model Results

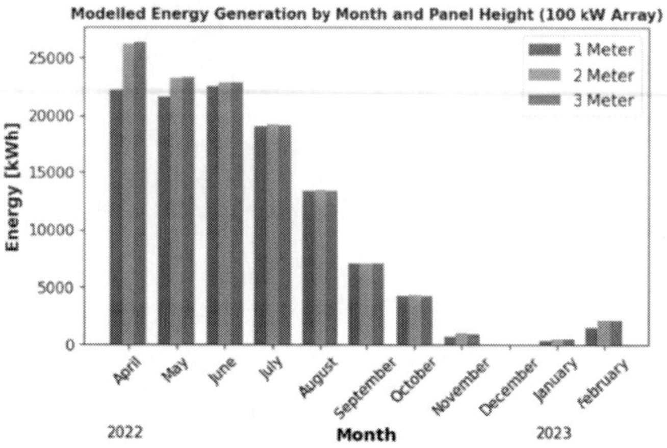

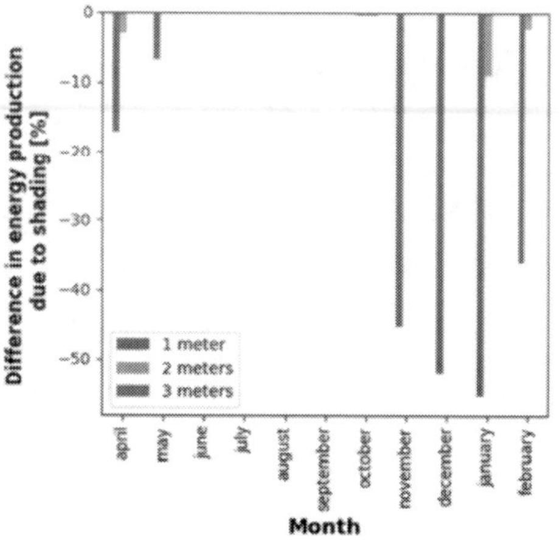

More Model Results

- **5.6%** total shading energy loss for 1-meter array
- 0.3% for 2-meter array
- None for 3-meter

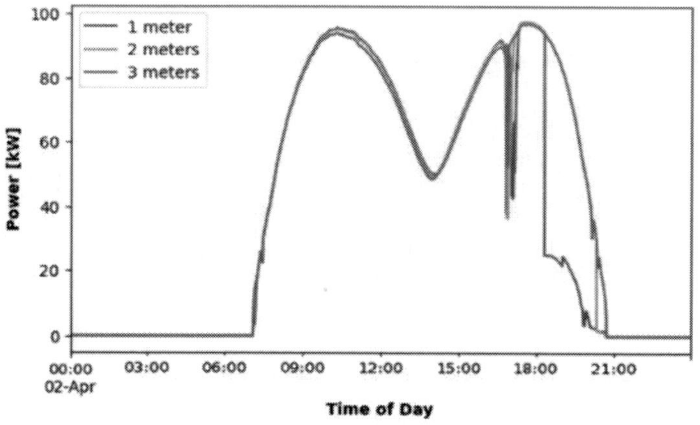

System Cost Estimates

- A base value of **$3.25 USD / Watt** (based on similar installations in Kotzebue)
- **10%** additional cost for vertical bifacial racking (based on limited inquiries with local installers)
- Raising the height of the array was assumed to increase balance-of-systems (BOS) and labor costs proportionally to the increase in required material. BOS/labor assumed to be **36%** of total cost [4].

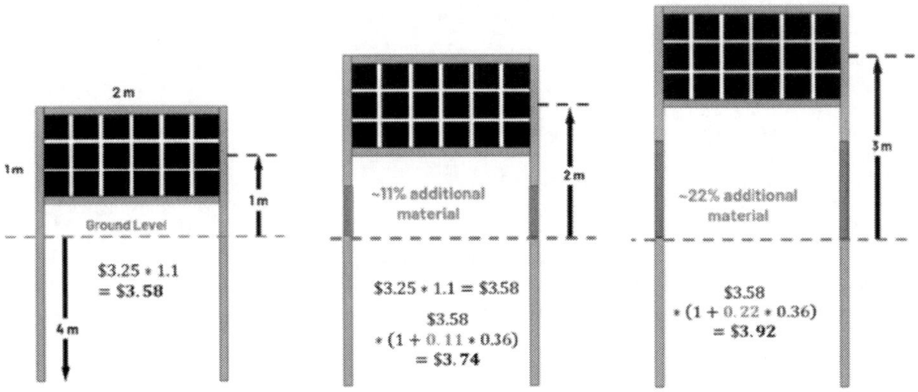

LCOE

- A simple levelized cost of energy (LCOE) value was calculated for each array elevation using estimated scaled system costs and capacity factors.
  - **25-year** lifetime
  - **0.5%** degradation rate
  - **$100 USD/kWp/year** OEM cost (based on similar installations in Alaska)

Elevation	Capacity Factor	System Cost ($USD/Watt)	LCOE ($USD/MWh)	% Improvement from 1-Meter Configuration
1 Meter	11.52%	3.58	301	N/A
2 Meters	12.40%	3.74	286	**+ 5.2%**
3 Meters	12.39%	3.92	298	+1.0%

Future Research and Considerations

Limitations of this Study

- Small snowdrifts on the upwind side not considered
- Major assumptions about increased costs at various heights
- Very rough estimates of racking cost data in general
    - **If you have any, get in touch!**

Further Considerations

- What about sites with north/south prevailing wind?
- Are extrapolations of existing
- Sites with lots of diffuse irradiance
    - Bad for vertical E/W PV
    - Better for low-tilt PV
    - Optimum ratio based on yearly energy diffuse fraction?
- More detailed shading study
- Computational Fluid Dynamics (CFD) study for arrays at various heights/angles to the wind?

40th European Photovoltaic Solar Energy Conference and Exhibition

References and Contact Info

Henry Toal: ehtoal@alaska.edu
Chris Pike: **cpike6@alaska.edu**

[1] R. Tabler Tabler, "Snow Fence Guide," Strategic Highway Research Program , Oct. 1991. Available: https://onlinepubs.trb.org/onlinepubs/shrp/shrp-h-320.pdf

[2] I. Frimannslund, T. Thiis, A. Aalberg, and B. Thorud, "Polar solar power plants – Investigating the potential and the design challenges," *Solar Energy*, vol. 224, pp. 35–42, Aug. 2021, doi: https://doi.org/10.1016/j.solener.2021.05.069.

[3] M. Martinelli, "INFLUENCE OF GAP WIDTH BELOW A VERTICAL SLAT SNOW FENCE ON SIZE AND LOCATION OF LEE DRIFT," *International Association Of Scientific Hydrology. Bulletin*, Dec. 1964, doi: https://doi.org/10.1080/02626666409493687.

[4] V. Ramasamy *et al.*, "U.S. Solar photovoltaic system and energy storage cost benchmarks, with minimum sustainable price analysis: Q1 2022," 2022. Available: https://www.nrel.gov/docs/fy22osti/83586.pdf

UA is an affirmative action/equal opportunity employer, educational institution and provider and prohibits illegal discrimination against any individual: www.alaska.edu/nondiscrimination.

PHOTOVOLATIC AND GREEN ROOF –ENERY YIELD PER AREA OF VERTICALLY INSTALLED BIFACIAL MODULS IN DENSE ARRANGEMENTS

Hartmut Nussbaumer, Roger Hiltebrand, Selina Pfyffer, Markus Klenk
Zurich University of Applied Science, SoE, Institute of Energy Systems and Fluid Engineering
Technikumstrasse 9, 8401 Winterthur, Switzerland)

ABSTRACT: Green roofs have several advantages compared to other flat roof solutions such as water retention, additional insulation, cooling of the building and therefore reduction of cooling needs, air humidification and promotion of the biodiversity. Building regulations supporting green roofs are being implemented around the world and in some regions green roofs are mandatory. Flat roofs are as well valuable areas for the installation of photovoltaic systems. Therefore, the combination of photovoltaics and green roofs is of great interest. In addition to the typical installation of PV modules at low tilt angles, vertically installed bifacial modules in combination with green roofs turned out to be an attractive photovoltaic concept, as the entire roof area can be used for greening, and the maintenance effort can be kept relatively low. For the first time the energy yield of vertically installed photovoltaic systems in dense arrangements with ground cover ratios of 100 % and more are presented and compared to east/west and south oriented systems with monofacial modules at low tilt angles as reference systems. Measurement data of real systems on roof tops collected over a duration of over one and a half year, is compared to measurements of different test systems and to yield simulations. We show that even an increase of the ground cover ratio GCR from 0.5 to 1 can be reasonable for vertical bifacial PV, depended on the cost for modules and the sub-construction.

Keywords: flat roof, green roof, bifacial, vertical, LCOE, simulation, GCR

1 INTRODUCTION

As the share of photovoltaic energy on the total energy generation has to be increased significantly in order to meet the targets for renewable energy generation, the valuable area of flat roofs should be used as efficient as possible. In urban areas however also other aspects, such as the insulation of buildings, energy demand for cooling, air purification, water retention and biodiversity play an important role in a holistic approach to optimize the area utilization.

Green roofs promise a wide range of benefits compared to other flat roof solutions such as the reduction of peak water runoff (water retention), protection of the roof seal, additional insulation, cooling, and air humidification of the ambient atmosphere, enhancing biodiversity, filtering air pollutants and other effects [1]. Currently, building regulations that support the green roof concept are being implemented around the world [2, 3]. These regulations aim to reduce rainwater runoff, to improve air quality and to reduce the need for cooling in urban areas [3]. Sealed floor surfaces in cities, waste heat from houses and engines enhance the warming of the air during the day and reduce the cooling of the air during night. Thus, summer nights in cities are four to five degrees warmer compared to rural regions. The maximum cooling at street level due to green roofs ranges from 0.03 to 3°C according to 17 studies that provide data on urban heat island reduction [3]. Besides the use of the roof as green oases, flat roofs are well suited for the installation of photovoltaic systems. The combination of photovoltaics and roof greening therefore is an interesting fusion, because the advantages of a green roof can be combined with the local electrical energy production at the place of consumption.

To avoid a conflict-of-interest the photovoltaic systems should be combined with roof greening in an optimized way to achieve the maximum benefit of a photovoltaic system. Relevant factors are matching power generation and consumption profiles, intact vegetation on the entire area and low maintenance cost for the green roof [4]. In this work we give an overview of photovoltaic system designs in combination with green roofing. We analyze the energy yields of different system designs by measuring the time resolved power generation of installed modules on large area systems, by using miniaturized module arrays and by energy yield simulations using different software tools. The analyzed systems vary in orientation, tilt angles, GCRs, weather conditions and ground reflectance. The distinctiveness of our research study is the profound data base of almost two years operation and a focus on highest ground cover ratios, maximum power generation and optimized generation profile. For the comparison of the various systems, we also make use of an own-developed miniaturized test rig consisting of various arrangements of miniaturized photovoltaic modules, enabling one-to-one comparisons of systems under identical environmental conditions [5].

2 SYSTEMS UNDER INVESTIGATION AND MEASUREMENT RESULTS

2.1 Different solutions for PV systems in combination with green roofs

Using a conventional photovoltaic system in combination with a green roof with tilted modules in south or east-west direction and low installation height can be problematic due to partial shading the collector surface or limited access for maintenance.

One approach to address these issues is to separate the area of the green roof from the area of the PV plant, see Fig. 1 a), However, this either limits the green roof area or the installed PV capacity. In order to retain the green roof character other system designs have been used as shown in Fig. 1 b) and c). A straightforward approach to combine a photovoltaic system with a green roof is to mount the modules on a higher position under a low tilt angle with the front side facing towards south (northern hemisphere). The increased module height enables regular maintenance to prevent undesired plant growth and module shading.

10.4229/EUPVSEC2023/4DO.12.5
020424-001

a) Modules with low tilt angles

b) Modules with low tilt angles at elevated position

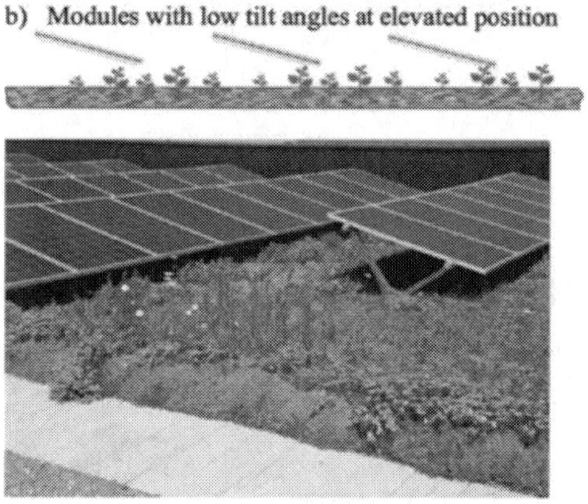

c) Vertically installed bifacial modules

Figure 1: PV system designs combined with green roofing. a) System with low tilt angles, low installation height. b) Modules south oriented with sub-construction for moderate tilt angles and elevation. c) Vertically installed bifacial modules.

A more recent approach uses vertically installed photovoltaic modules, see figure 1 c). This solution has the advantage that the entire roof area can be considered as an almost uncovered green roof, regardless of the ground cover ratio (GCR). Obvious disadvantages are higher wind loads and increased self-shading, depending on the GCR.

The first PV system under our evaluation that combined green roofing and vertically installed bifacial modules had a peak power of 9.09 kWp (see Fig. 2) and was installed in Winterthur, Switzerland. The bifacial modules used had only 20 monocrystalline cells in two rows to achieve a narrow layout with reduced wind load and visibility. The system had an azimuth angle of −65° for the front side, a GCR of 36.7 % and an albedo of about 0.2. The GCR was chosen to avoid mutual shading of adjacent rows of PV modules. The photovoltaic system reached a specific yield of 942 kWh/kWp over one year, close to typical values of 1000 kWh/kWp for east west orientated modules under low tilt angels installed in this area [4]. However, with a GCR of only 36.7 % the advantage of vertically installed modules compared to tilted ones is limited, because "tilted systems" with the same GCR can still be considered green roofs with a large proportion of uncovered area. Moreover, their energy yield is superior compared to the system with the vertically installed modules.

Figure. 2: PV system in Winterthur using vertically installed bifacial modules with a GCR of 36.7 % [4].

If the GCR is further increased to maximize the total energy yield for a given flat roof, the different designs show specific characteristics in relation to the remaining green roof area, as shown in Figure 3.

For the case as described by figure 3 a) it is obvious that the actual green roof area is increasingly limited for higher GCR values. In the second design (figure 1 b)) the GRC is limited, even though the conditions for the plants are better compared to the situation in figure 1 a), because a certain spacing is required to suppress mutual shading and to ensure sufficient access for maintenance of the greenery. Using vertically installed modules enables the installation of modules in dense arrangements with GCRs in the range of 0.5 up to 1.5 without significant reduction of green roof area, which is a unique property compared to the other approaches. The dense arrangement reduces the specific yield of the modules in kWh/kWp. However, with increasing GCR also the energy yield per area can be enhanced, which motivates the investigation of specific and total energy yield.

Mutual shading is very pronounced for vertically installed modules and is also particularly dependent on the spacing of the module rows. Systems of "taller" vertically

10.4229/EUPVSEC2023/4DO.12.5
020424-002

mounted modules must have greater row spacing to obtain similar shading losses compared to narrower designs.

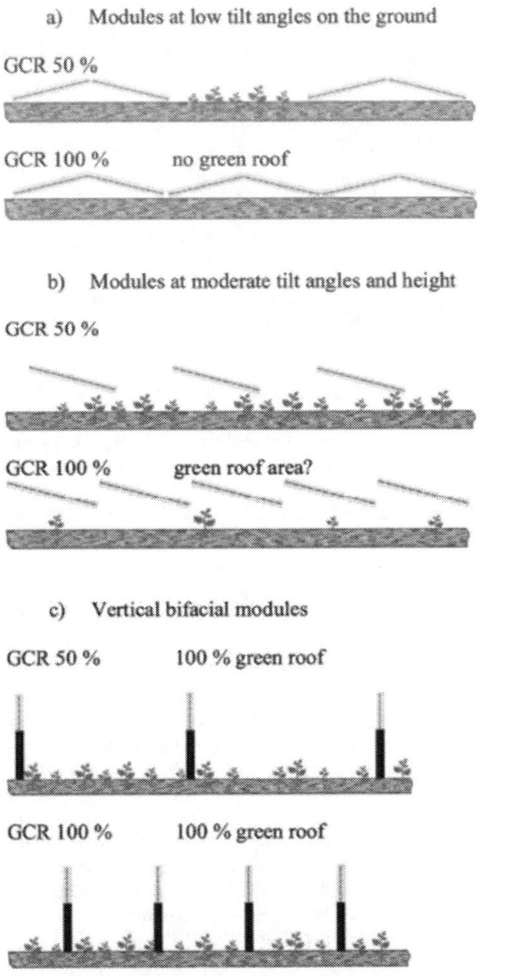

Figure. 3: Combinations of PV systems and green roofs with GCRs of 50 % and 100 % respectively.

In a newer test system in Mattenbach, Winterthur, bifacial modules with 72 (6 x 12 strings) and 48 (4 x 12 strings) cells were installed vertically in "landscape" orientation. Together with the vertical systems also more typical installation types with south-orientation and butterfly configuration (mono- and bifacial) were implemented to enable a direct comparison.

A schematic plan of the layout is shown in Fig. 4. with the vertical modules in the dominating central area. The spacing between the large vertical modules was just under 2 m, corresponding to a GCR of 0.5, with a distance to the ground of 40 cm. For the narrower modules, a row spacing of 1.25 m was chosen for design reasons, which corresponds to a GCR of 0.55. The distance between the lower edges of the modules and the ground is 30 cm.

In contrast to the described first project with comparatively expensive custom-made solar modules and substructures, larger bifacial modules with 72 and 48 cells were now used for the vertical subsystems to implement typical industrial module dimensions and an intermediate format (48-cells). Larger formats are advantageous in terms of system price, as the number of substructures required is reduced. The use of industry-typical formats, such as the 72-cell bifacial modules installed here, is preferable for cost reasons. The installed total peakpower of the system in Mattenbach is 93 kWp. Three module types from the manufacturer "Hevel" were used in different configurations:

- 34 monofacial HJT-modules with 60 cells and 330 Wp in portrait orientation on butterfly sub-construction, 15° tilt each. L x W: 1.67 m x 1 m.
- 113 bifacial modules with 72 cells and 390 Wp in landscape orientation. Vertical and butterfly (15° tilt) arrangement. L x W: 2 m x 1 m.
- 148 bifacial modules with 48 cells and 255 Wp, vertical, landscape. L x W: 2 m x 0.69 m.

The bifacial modules (48 and 72 cells) contain the same solar cells which are arranged in strings of 12 cells. The monofacial modules (60 cells with three strings of 20 cells) are equipped with heterojunction cells that are connected by "SmartWire" technology.

Two modules at a time are connected via a readable optimizer. In addition to the different substructures (vertical, portrait and butterfly), half of the roof was covered with two different substrate types, which also differ in their brightness and albedo. The northern half of the roof was covered with the darker substrate type. At the northern end of the roof, monocrystalline 60-cell modules were installed in portrait and butterfly configuration. At the other end of the roof, 72-cell bifacial modules (as in vertical) were installed on butterfly substructures.

For visual reasons, the PV system should be oriented to the roof geometry. The architect found the vertical modules, which protrude over the edge of the roof, to be an excellent match for the structure of the old commercial building with vertical struts. Even if, in the special case of the old printing plant, visibility is even desired for the reasons mentioned, it is generally to be assessed as rather problematic in the context with local regulations. The orientation along the building geometry leads to the fact that the test facility is rotated by 28° in relation to the south orientation. Thus, there is neither a typical south orientation of the monofacial modules, nor an east/west orientation of the vertical installation and the butterfly system.

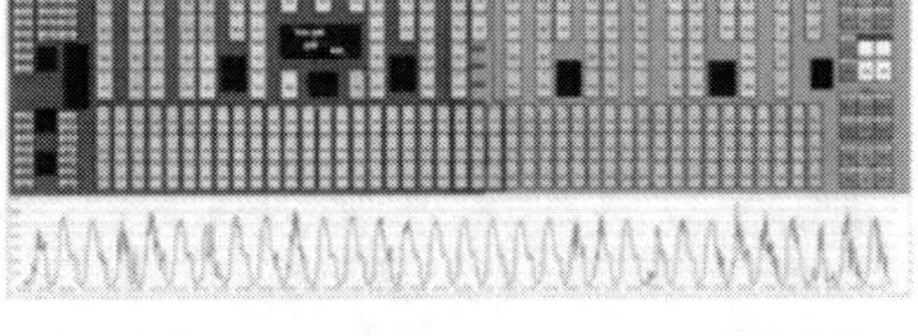

Figure 4: Schematic plan of green roof with different PV subsystems. Brown and yellow color represent darker and lighter ground. The locally achieved specific yield relative to the other system variants can be recognized by the colored representation of the modules (darker → higher specific yield in kWh/kWp). The specific yield for July 2022 is shown as an example.

The effect of the PV plant alignment to the building orientation on the power output is explementary shown in Fig. 5., in which the typical camel hump peak of an E/W-

10.4229/EUPVSEC2023/4DO.12.5
020424-003

aligned vertical system is shifted towards an almost single peak. For the modules at the narrow ends of the roof (monofacial south and butterfly, as well as bifacial butterfly), the determination of a GCR is more difficult due to specific circumstances (few rows, edge areas, numerous superstructures). However, especially for the monofacial systems with a flat tilt angle of 15°, the GCR plays a neglectable role in terms of yield anyway. Nevertheless, it is included in the consideration.

In the lower area of figure 4, the system output (total, selected subarrays, or single modules) for each time stamp, day or month can be monitored. The course and changes of the local yield can be continuously observed by the colored representation via slide controller.

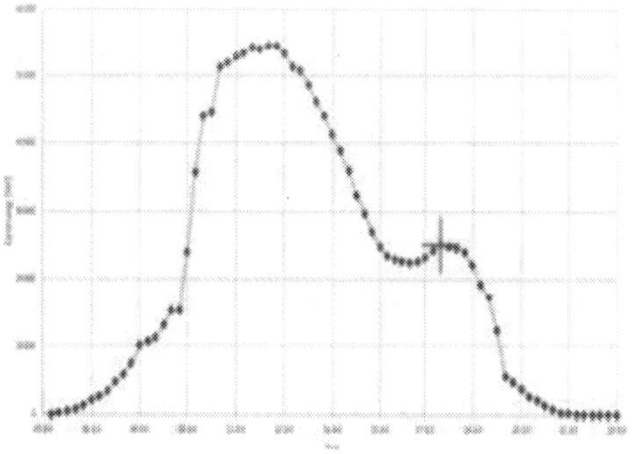

Figure 5: Power of vertical modules at a clear sky day (July 2nd). The rotated system alignment causes a transformation of the typical camel hump peak towards an almost single peak.

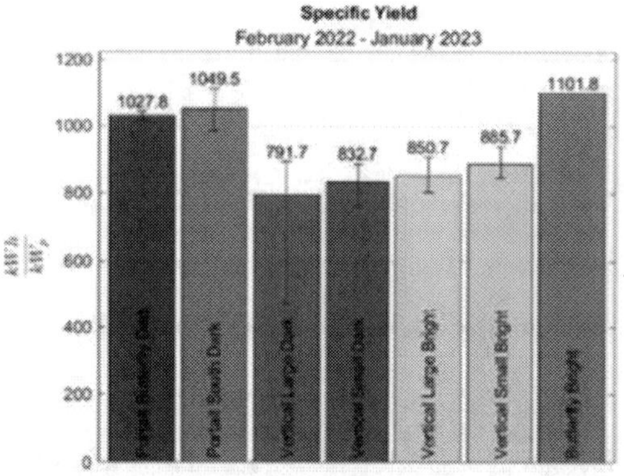

Figure 6: Specific energy yield in kWh/kWp for the systems analyzed at the roof in Mattenbach for one complete year (February 2022 to January 2023). The CGR of the vertical systems are 0.5 and 0.55.

The specific yield in kWh/kWp for the different systems during one year of operation beginning February 2022 until February 2023 is shown in Fig. 6. Typical values of around 1000 kWh/kWp for the Winterthur area have been measured for the modules with a butterfly like sub-construction and tilt angles of 15°. As expected, the specific yields of the vertical installations are reduced by about 10% compared to the "tilted systems". No significant difference of the yield was measurable for the systems with different grounds (dark, bright) as the ground was covered by plants after spring resulting in similar albedos. The differences between all the systems are lower in the winter month as can be seen from the monthly specific energy yields shown in Fig 7.

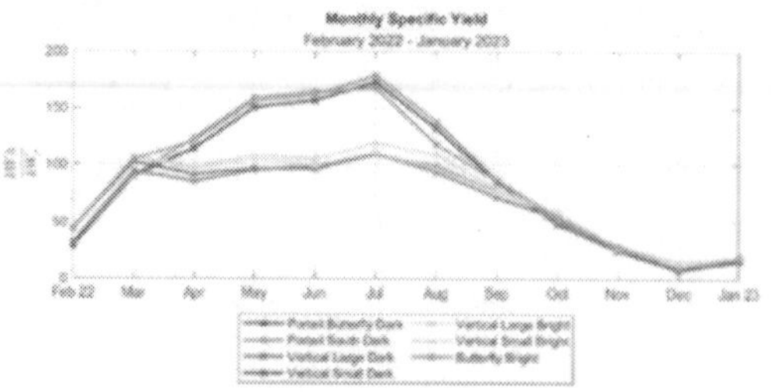

Figure 7: Monthly values of the specific energy yield in kWh/kWp for the different systems.

The specific energy yield in kWh/kWp of a PV system directly relates the yield to the installed peak capacity. This property is of particular interest if the module prices (peak capacity) are a dominating cost factor. (Remark: A discussion of economic aspects will follow in section 3 of this work.)

With regard to flat roofs and the inherently limited area it is also reasonable to calculate the total energy yield per roof area (kWh/m^2_{roof}). This can be done in several ways by means of the already introduced or known relations.

$$\frac{kWh}{kWp} * \frac{kWp}{m^2_{module}} * GCR = \frac{kWh}{kWp} * \frac{kWp}{m^2_{module}} * \frac{m^2_{module}}{m^2_{roof}} = \frac{kWh}{m^2_{roof}}$$

The first expression in the respective multiplications is the specific energy yield for a given system with given modules, as for example depicted in Fig. 6. The second expression reflects the module efficiency and the last multiplicator the GCR or module coverage per roof area.

In order to enable a comparability, the same modules or adapted values (kWp/m^2_{module}) have to be used.

Example: For a system with modules lying flat on the ground with a GCR of 100 % an installed peak power of $200Wp/m^2_{module}$, and a specific yield of 1000 kWh/m^2 would result in a yield of 200 kWh/m^2_{roof}.
If we apply this calculation (assuming the same module with $200Wp/m^2_{module}$) for the specific yield of 850.7 kWh/kWp, as measured for the system with vertical modules (large bright) and 55% GCR the yield per area would be 93.6 kWh/m^2_{roof}.

2.2 Dense arrangements of vertical bifacial modules

In order to investigate dense arrays (large GCR) of vertically installed bifacial systems we operate several PV test systems simultaneously on the roof of the school of engineering in Winterthur (Fig. 8).

One of the systems which are relevant for this work consists of rotating industrial bifacial modules orientated in east/west direction with a ground cover ratio of close to 100 %. The modules permanently rotate around the middle axis to measure the energy yield for various tilt angels. This system (Biforot) is shown in Fig 8. For this work only the measurements at the vertical position are relevant, which are performed every two minutes.

In parallel we measure several miniaturized systems with different configurations. The miniaturized test systems are described elsewhere and use mini modules with a size of 10 x 16 cm^2 [6, 7, 8]. In former investigations we have shown that the specific energy yield measurements using the miniaturized test rig correspond quite well with similar systems using industrial sized modules [5].

Figure 8: Test rigs for the measurement and comparison of different PV systems. Systems which are relevant for this work are marked red.

To further evaluate how PV systems using vertical bifacial modules perform under dense arrangements we have simulated the power output of the measured systems with the software PVCase (Yield and Ground Mount)

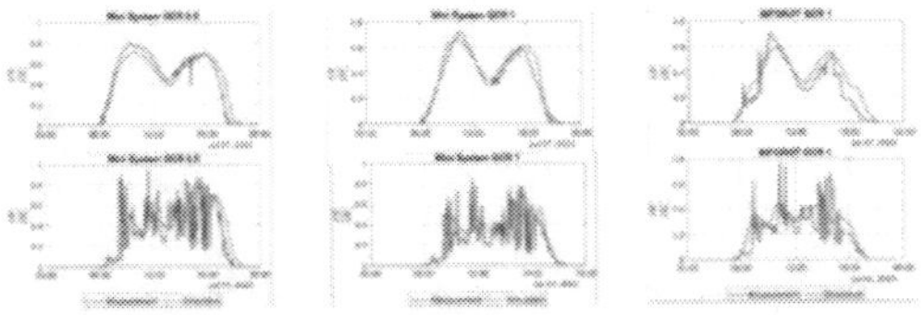

Figure 9: Measured power for single days compared to simulations by PVCase (Yield and Ground Mount). Due to the increasing self-shading of adjacent modules the simulation of dense arrays (vertical bifacial) is increasingly challenging. Here GCRs of 0.5 and 1. Even for these unfavorable conditions there is still reasonable accordance between simulation and measured data.

Using PV case (Yield and Ground Mount) the energy yield per area of vertically installed bifacial modules oriented in east/west in dense arrangements was simulated [9]. In this simulation we used a typical flat roof area element of 10 x 10 m^2. In Fig. 11 the dependence of the yearly specific energy yield as a function of the GCR is shown in blue crosses. As expected, the specific energy yield decreases with increasing GCR as the self-shading of adjacent modules increases. In addition, the energy yield per area by red crosses is shown (module nominal power 200 W/m^2_{module}).

Interestingly the energy yield per area still slightly increases even for GCR values of one. One reason for this behavior might be the irradiance on the modules from sunlight from south directions shining in between the rows. As we have simulated a limited area of 10 x 10 m^2 the array of modules has an unshaded south face enabling the penetration of light in between the rows. If we compare the energy yield per area for a GCR of 0.5 with the value related to GCR 1 we observe an increase of the energy yield per roof area of about 35 % (~95 kWh/m^2_{roof} → ~130 kWh/m^2_{roof}).

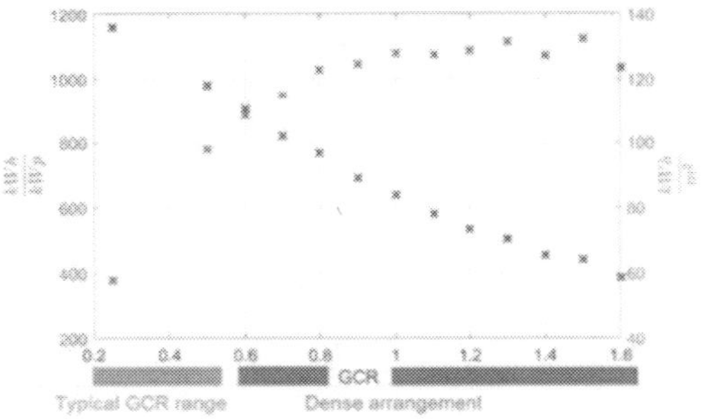

Figure 11: Simulation of an east/west-oriented array of vertical bifacial modules as a function of the GCR for the location of Winterthur (albedo 0.2). The annual specific energy yield is plotted in blue and the total energy yield per roof area in red.

The simulated results concerning the relative increase of the energy yield per roof area were compared to the measurement results of the miniaturized test rigs. Interestingly, the measurements indicate that the simulated increase in kWh/m^2_{roof} may be underestimated. Because the measured gain is much more pronounced than expected (48 % at clear sky conditions), it was decided to carry out a more detailed analysis of the measurement data prior to a publishing.

3 ECONOMIC EVALUATIONS

The specific energy yield (kWh/kWp) is of particular interest if the module prices are a dominating cost factor. Due to the permanently decreasing module prices during the past decades systems with lower specific energy yields became increasingly interesting. So, system designs like the ones in Fig. 3 a) with almost 100 % GCR are a common installation type on large flat roofs. Other examples are all systems with alignments deviating from the formerly dominating south orientation, or also most BIPV systems. In all these examples, the optimally achievable specific energy yield is sacrificed in order to achieve other goals, here optimized total energy yields or façade utilization.

Similar conditions apply to vertical modules on green roofs. It is known that the specific energy yield decreases due to mutual shading and that the albedo on green roofs is comparatively low. On the other hand, this approach enables 100 % green roofs and is advantageous for maintenance. In addition, a dense arrangement of vertical modules can be considered as an option to maximise the total energy yield for a given roof area while maintaining the character of a green roof. The relevant factor is then the energy yield per roof area (kWh/m^2_{roof}).

The output per module, the specific energy yield (kWh/kWp) will always be lower for vertically installed modules in dense arrangements compared to typical systems with low or moderate tilt angle. (Remark: Free standing vertical bifacial modules can exceed the specific energy yields of modules in classic "tilted" systems).

It is obvious that the question if it is reasonable to install a vertical bifacial system on a green roof cannot be reduced to an optimized specific energy yield. If this would be the case, it would be the best to install very few modules with optimized south orientation or virtually free-standing bifacial ones in vertical position. However, the total output would be extremely limited then.

10.4229/EUPVSEC2023/4DO.12.5
020424-005

For PV systems with only a partial share of the total roof area, as shown in Figure 3, it makes sense to realize limited areas with more classic, non-vertical systems, preferably as shown in Figure 1 b) and 3 b). But then the energy yield is also limited and with increasing GCR the actual green roof area is increasingly lost.

As another example vertical systems with a GCR of 0.5 and 1 are compared. It has to be pointed out, that already a GCR of 0.5 is very dense for a vertical bifacial system, a GCR of 1 is quite extreme, particularly if compared to for example vertical Agri-PV installations. It is well known that in a vertical installation, the chosen row spacing and thus the GCR is a crucial factor. Such a dense arrangement would probably be generally considered too loss-prone, which motivates an analysis.

In Fig. 12 the relative cost of a PV system in combination with a green roof for the project in Mattenbach, Winterthur (GCR 0.5) is shown. The used modules are still not standard type, and the sub-construction is a custom-made product.

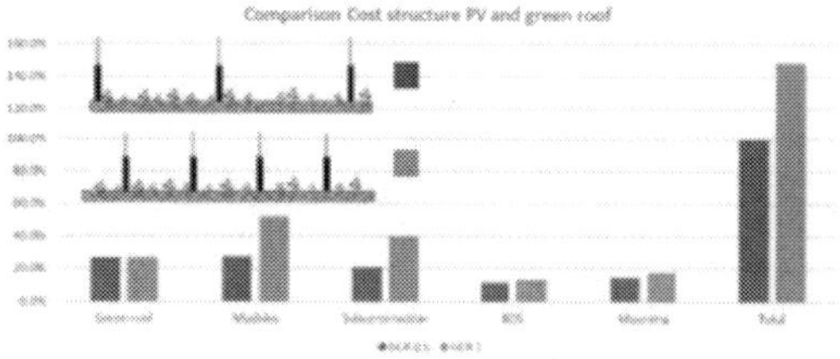

Figure 12: Cost distribution for a real PV system on a green roof with vertical bifacial modules and a GCR of 0.5 (blue) in comparison to an estimated system with GCR of 1 (orange). Data normalized to the total cost of the system with a CGR of 0.5.

The total cost of the system with GCR 1 is about 48% higher compared to the system with GCR 0.5. Thus, if the energy yield of the system with GCR 1 is at least 48% higher, the levelized cost of energy of this system would therefore be favorable. According to our simulated data the increase of the energy yield is about 35 %. The measured increase for a GCR of 1 is higher and coincidentally corresponds exactly to the postulated 48 % under clear skies. Further investigations will be necessary for a more detailed analysis.

Again, this is an important, but not necessarily the most relevant aspect in this regard. One must keep in mind that the described approach enables a considerably higher total energy output while maintaining an almost 100 % green roof area. If this is the goal the approach to plan a vertical system with a GCR of 1 instead of 0.5 may be a reasonable option. Taking lower module prices and a more standardized sub-construction in the future into account the additional cost may even be overcompensated by the higher energy yield per area.

4 CONCLUSION AND SUMMARY

In this paper, different designs of PV systems in combination with green roofs were presented and discussed. This combination is not new and corresponding systems have been installed for a long time. The typical more classic approaches with "tilted" modules have several advantages, as they rely on standard modules, sub-constructions, and other proven standard components. On the other hand, there are some disadvantages with regard to maintenance and shading by the plants. These effects can be countered, for example, by choosing substructures that offer a certain height. Another possibility is to realize a sufficient distance between the module rows. However, the green roof character of systems with inclined modules will be increasingly lost as the degree of ground cover increases.

One possibility is to limit the roof area covered by PV modules. The modules can for example be concentrated in PV islands. This of course limits the peak PV power installed, but if this aspect is not a priority and the remaining green space is considered sufficient, this is a reasonable approach.

The type of PV installation that is in the focus of the work is the vertical installation of bifacial modules in dense arrangements. The vertical installation is also not new and often seen in Agri-PV systems. In such systems however the ground cover ratios are low, which suppresses negative effects due to mutual shading of module rows. As for the before mentioned PV islands a corresponding installation with low ground cover ratio (few modules) would result in a too low total energy yield for a given roof area. The well-known detrimental effect of mutual shading for vertical PV systems motivated a closer analysis based on measurements and simulations.

Two corresponding and already realized systems on green roofs with moderate GCRs of ~0.4 and ~0.5 were briefly introduced. For one of the systems a direct comparison of different installation types is available based on 1.5 years of measurement data.

The interest in this project from many sides also triggered the investigation of even more densely arranged vertical PV systems in order to increase the total output. The results and measurements indicate that increasing the GCR from 0.5 to 1 does not make sense in terms of LCOE at this time, but the difference is not as pronounced as one might think at first glance for such an extremely dense array. On the other hand, with this approach a comparatively high total energy yield can be combined with a 100 % green roof area.

It is a non-technical trade-off to what extent the benefits outweigh the additional costs due to the larger number of modules in this design option. However, given the limited roof area, it makes sense to optimize the energy yield per area, even if this is not the most cost-effective option. Other aspects, such as energy distribution and avoided snow cover, can be considered additional advantages.

Table I: Comparison of the major key figures of different PV systems in combination with green roof under investigation for location Winterthur, Switzerland. The range of energy yields could even be higher for the vertically installed systems if the orientation differs from east/west (see system Mattenbach and values in Fig. 6)

System type	GCR	kWh/kWp	kWh/ $m^2_{\_}$
Vertical bifacial*	0.5	~878–979	~90
Vertical bifacial*	1	~558–640	~120
Monofacial, tilt 10°*	0.5	~890–990	~95
Monofacial south 15°	0.5	~950–1050	~100

*east/west orientated

Note: A vertical bifacial system with a GCR of 50 % and 100 % is currently being installed in Landquart, Switzerland. The results will be reported elsewhere.

10.4229/EUPVSEC2023/4DO.12.5
020424-006

5 ACKNOWLEDMENT

This research is supported by the Swiss Federal Office of Energy in the following projects:

- PV systems on flat roofs with highest energy yields per area, SI/502309-01
- Development and comparative test of a complete solution for bifacial PV systems on green roofs, SI/502213-01

6 REFERENCES

[1] Francis, L.F.M., Jensen, M.B., 2017. Benefits of green roofs: a systematic review of the evidence for three ecosystem services. Urban For. Urban Green. 28, 167–176. https://doi.org/10.1016/j.ufug.2017.10.015

[2] Snow, J., 2016. Green Roofs Take Root Around the World [WWW Document]. National Geographic News. URL https://news.nationalgeographic.com/2016/10/sanfrancisco-green-roof-law/

[3] Azeñas, V.et. al. 2018 Thermal regulation capacity of a green roof system in the mediterranean region: the effects of vegetation and irrigation level. Energy Build. 164, 226–238. https://doi. org/10.1016/j.enbuild.2018.01.010.

[4] Thomas Baumann et. al. "Photovoltaic systems with vertically mounted bifacial PV modules in combination with green roofs", Solar energy 190, 2019, pp. 139-146

[5] H. Nussbaumer, M. Klenk, M. Morf, and N. Keller, "Energy yield prediction of a bifacial PV system with a miniaturized test array," Sol. Energy, vol. 179, 2019 pp. 316

[6] H. Nussbaumer et al., "Accuracy of simulated data for bifacial systems with varying tilt angles and share of diffuse radiation," Sol. Energy, vol. 197, pp. 6–21, Feb. 2020, doi: 10.1016/j.solener.2019.12.071.

[7] H. Nussbaumer et al., "Miniaturized Test Array as a Means to Determine the Energy Harvest of Bifacial Installations," in IEEE Conference Publications, 2016, pp. 2698–2703, doi: DOI: 10.1109/PVSC.2016.7750140.

[8] Miniaturized test rig: https://sc9d9463193920cbd.jimcontent.com/download/version/1652095568/module/14669904332/name/PVPO_Re07.pdf

[9] PVcase simulation software tools: https://pvcase.com/yield/ https://pvcase.com/ground-mount/

This presentation was selected by the Sc. Committee of the EU PVSEC 2023 for submission of a full paper to one of the EU PVSEC's collaborating peer-reviewed journals.

QUANTIFYING INHERENT SOURCES OF ELECTRICAL MISMATCH IN UTILITY-SCALE BIFACIAL SYSTEMS

Keith R. McIntosh, Malcolm D. Abbott and Ben A. Sudbury
PV Lighthouse
Coledale, NSW, Australia

ABSTRACT: We use ray tracing and circuit simulation to examine inherent mismatch losses in bifacial PV systems arising from nonuniform irradiance. We calculate both cell-to-cell (CTC) and module-to-module (MTM) mismatch for three example system types: (i) a 1P single-axis tracker (SAT), (ii) a 2P fixed system, and (iii) a 1P east–west oriented wave. When situated in Portugal, we find that the annual CTC mismatch loss of these example systems is 0.3–0.5%. The loss is mostly due to row-to-row shading of diffuse light and of direct light when the sun is very low in the sky, and, to a lesser extent, to nonuniformity on the rear irradiance. The mismatch loss varies greatly from hour to hour, reaching a maximum of 3% for SATs and ~30% for the other systems. We also find that MTM mismatch is significant (~0.4%) in the fixed system when the upper and lower modules are strung in series. The results of this paper show how the complicated behavior of electrical mismatch can be understood and quantified by combining ray tracing and circuit simulation.
Keywords: mismatch, ray tracing, bifacial, yield, systems

1 INTRODUCTION

Electrical mismatch reduces the yield of a PV system. It arises from a variability in the performance of the cells within a module (cell-to-cell mismatch), or the modules within a system (module-to-module mismatch). Such variability prevents some cells or modules from operating at their maximum power point, and thus the output power of the system is reduced.

Utility-scale power plants—the focus of this study—mitigate many sources of mismatch. For example, the installed cells and modules are sorted into sufficiently tight bins [1–3]; modules are regularly cleaned, which limits nonuniform soiling; and modules are installed far from trees and buildings to prevent localized shading. Nevertheless, some sources of mismatch are unavoidable. In particular, shading from neighboring modules and support structures (e.g., posts, rafters, torque tubes, clamps, frames) introduce nonuniform irradiance on modules, and hence a spatial variability in cell and module performance. The resulting mismatch is markedly different depending on whether the sunlight is direct or diffuse, and depending on the time of day and year.

This inherent mismatch loss is significant in most systems and therefore view-factor yield programs, such as PVSyst, SAM, PlantPredict and SolarFarmer, require the user to enter mismatch loss factors for the front and rear irradiance. How can those factors be determined?

It is very difficult to measure the mismatch losses in a PV system [4–6] but they can be calculated via simulation. In recent years, this has been achieved with a combination of ray tracing and SPICE solving for 1P and 2P single-axis trackers (SATs) [7–13]. In this study, we build on [7], accounting for all structural components, spectral effects, the internal optics of the module, and thermal effects (though not variability of temperature within modules) to compute both the cell-to-cell (CTC) and module-to-module (MTM) mismatch within a bay of modules within an infinite system. Three system types are explored: (a) a 1P east–west SAT, (b) a 2P conventional south-facing fixed system with purlins and rafters with a 25° tilt, and (c) a 1P east–west oriented wave system with a 10° tilt based on 5B's Maverick [14]. The unit systems are illustrated in Fig. 1.

While these results are specific to the design and location of the systems, the examples provide an opportunity to understand the differences between system types, and the variability in mismatch induced by its dependence on time of day and year, solar zenith and diffuse fraction.

2 SIMULATION DETAILS

We locate the systems in Santiago do Cacém, Portugal (38° N, 8.7° E), where plans are afoot to build Europe's largest PV power plant (1.2 GW) [15]. We generate typical mean year (TMY) weather from Solcast [16] in one-hourly intervals. This weather set gives an annual resource of 1.76 MWh/m² with a diffuse fraction of 30.4%. We also use a Solcast time-series weather dataset for an example day, 18-Sep-2022, in five-minute intervals.

Fig. 1 illustrates the types of unit systems and Table I lists their dimensions. These unit systems are repeated *ad infinitum* to represent a utility-scale power plant. In this case, the unit system can be considered a bay of modules, as illustrated by Fig. 2 for an SAT. The SAT includes ideal backtracking to prevent row-to-row shading of direct light.

We use SunSolve-Yield [17] to perform the simulations. This software program harnesses 600 parallel cores in the cloud to solve the annual yield. The immense

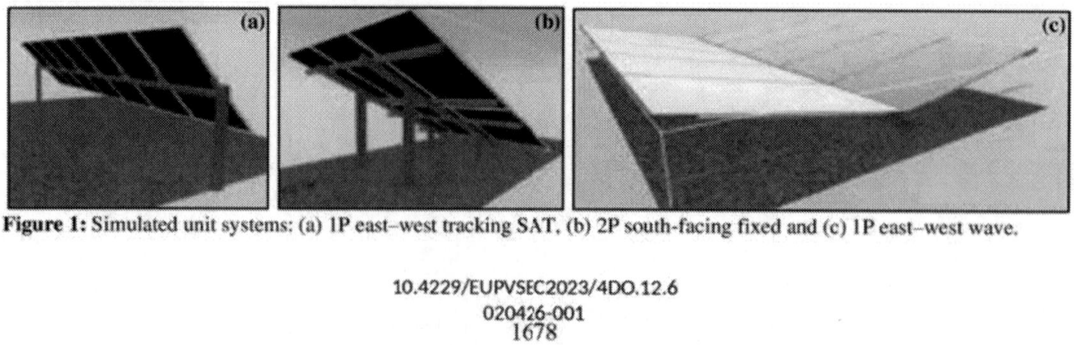

Figure 1: Simulated unit systems: (a) 1P east–west tracking SAT, (b) 2P south-facing fixed and (c) 1P east–west wave.

Figure 2: Example results for the bay of seven modules (the unit system) within an infinite field.

computing power helps reduce computation times, thus enabling greater complexity than past studies [7–13]. The current solution times for an annual yield simulation in hourly intervals is either ~10 or ~1 minute, depending on whether the internal optics of the cells and modules are included.

The ray tracing accounts for spectral effects, where the spectrum at each time of day is computed by the SPECTRL2 model [18] from atmospheric inputs. In this work, we take the precipitable water vapor, atmospheric pressure and far-field albedo from Solcast; we set the ozone and turbidity to values used to generate the AM1.5g spectrum; and we calculate the air mass from the solar position. The spectral reflectance of the ground and structural supports is plotted in Fig. 3, where the data is from the ECOSTRESS library [19] for green rye grass and galvanized steel. The ground and supports are 100% and 50% Lambertian scattering, respectively. We apply the Hay–Davies diffuse-sky model [20].

By ray tracing a unit system we account for local edge effects like the spacing between bays within an SAT or wave, but not the global edge effects at the boundaries of the system.

The ray tracing extends into the module, accounting

Table I: System dimensions.

	SAT	Fixed	Wave
Modules in unit system	1 × 7	2 × 3	2 × 5
Module facing	E–W	S	E–W
Row pitch (m)	6.139	10	4.577
Tilt (°)	–60 to 60	25	10
Module sep, lateral (mm)	15	0	33
Module sep, longitudinal (mm)	n.a.	0	64
Bay spacing (mm)	188	0	616
Min height above ground (mm)	446	230	258
Ground slope (°)	0	0	0

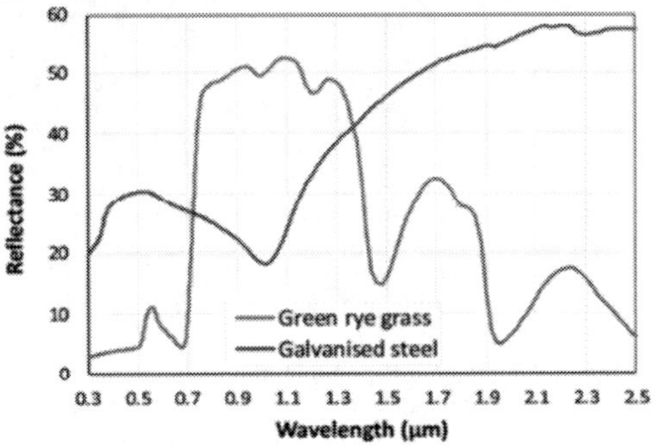

Figure 3: Spectral reflectance of the ground and structural supports taken from the ECOSTRESS library [17].

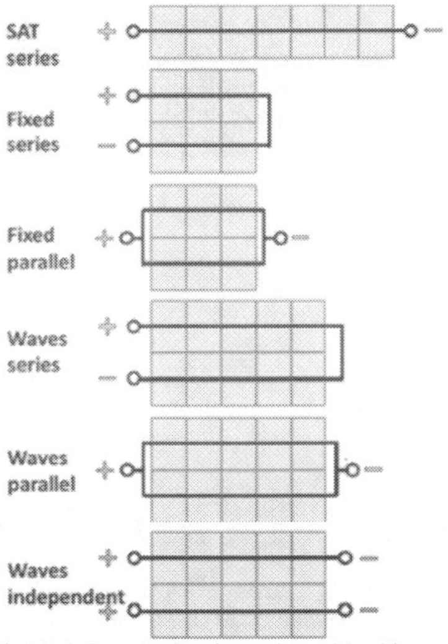

Figure 4: Module stringing investigated in this work.

for the complicated optics of a modern Longi LR5 530 W bifacial module with 144 half-cut cells (anti-reflection coatings, surface texture, etc.). The optical inputs of the modules are very similar to those listed in [7] except the cells are half-cut. (We note that the irradiance plots in Fig. 10 of [7] are incorrect, except the plot at midday when the module is horizontal.)

After the ray tracing has computed the generation current in every cell of the unit system, the electrical circuit is solved with a SPICE model. This circuit replicates the cell stringing and bypass diodes within a 144-cell module, as well as the stringing of the modules within the unit system. We examine the module circuits shown in Fig. 4, setting the inter-module resistance to be negligible (10^{-4} Ω).

At the completion of the SPICE solving, CTC mismatch loss is determined as the relative difference between the module power when (i) the cells are electrically connected by the stringing and bypass diodes, and (ii) the cells are electrically independent. MTM mismatch loss is determined as the relative difference between the power of (i) the entire string and (ii) the sum of the power of each individual module if held at maximum power. We also quantify the mismatch due to front irradiance by blocking the light from entering the rear of the module.

Finally, we make all cells and modules identical and neglect soiling, so no mismatch arises from variance in device quality or shading other than the inherent shading from structural supports, frames and neighboring modules.

3 RESULTS

3.1 Number of rays

Ray tracing simulates the optical behavior of a PV system by tracing a large number of discrete rays. If insufficient rays are traced, a nonuniform irradiance results due to the stochastic nature of ray tracing (e.g., rays are initiated in a random location), thereby introducing additional electrical mismatch.

Fig. 5 evaluates this stochastic source of mismatch by plotting the electrical mismatch on our example day against the number of rays traced at each 5-minute

10.4229/EUPVSEC2023/4DO.12.6
020426-002

interval. It shows that the mismatch saturates at about 10^8 rays, and that the additional 'stochastic mismatch' is about 0.1% for 10^7 rays. We note that the stochastic mismatch is roughly proportional to the number of rays because for low-to-moderate current variability, mismatch loss is roughly proportional to the square of the standard deviation within a string of cells [21–23], which is itself proportional to the square-root of the number of rays.

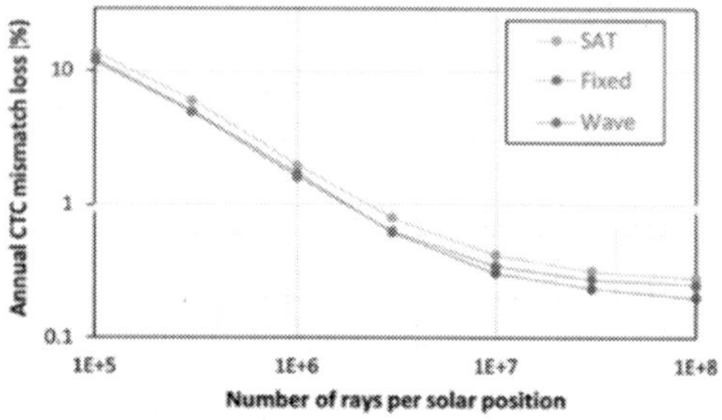

Figure 5: CTC mismatch vs number of rays on the example day (18-Sep-2022) for the unit-systems of Fig 1.

3.2 Number of solar positions

In the above section, the day was simulated in five-minute intervals, requiring ~150 solutions. When ray tracing 10^7 rays into the cells of the system, SunSolve currently completes that simulation in less than 10 minutes with 600 parallel cores. To simulate a year in one-hourly intervals, however, requires ~4500 solutions, which would take several hours to solve with the same computing power. We can reduce this time by a factor of ten by neglecting the ray tracing into the module, or by using ten times the number of cores. Another way to reduce the time is to run sufficient solutions to represent the solar positions throughout the year, and then to interpolate the results at each desired time. Ernst *et al.* described a similar approach in [24]. This interpolation introduces some nonuniformity (especially when one or more of the interpolated points have row-to-row shading of direct light), which itself introduces mismatch.

Fig. 6 plots how the annual mismatch depends on the number of solar positions traced. Consistent with Ernst *et al.* (who use a similar binning approach), we find that the mismatch saturates beyond 400 solar positions.

We conclude that for the unit-systems of this study, the optimal trade off between accuracy and speed is to run 200, 300 and 200 solar positions for SAT, fixed and wave systems, with 10^7 rays per position. This approach introduces stochastic annual CTC mismatch of $\leq 0.1\%$.

3.3 Example irradiance maps

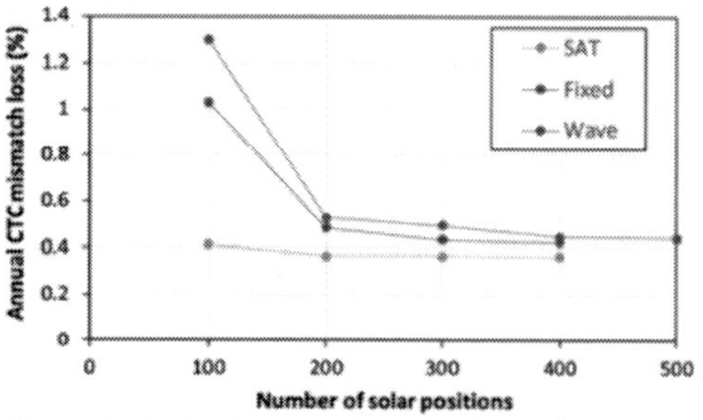

Figure 6: CTC mismatch vs number of solar positions with 10^7 rays per solar position.

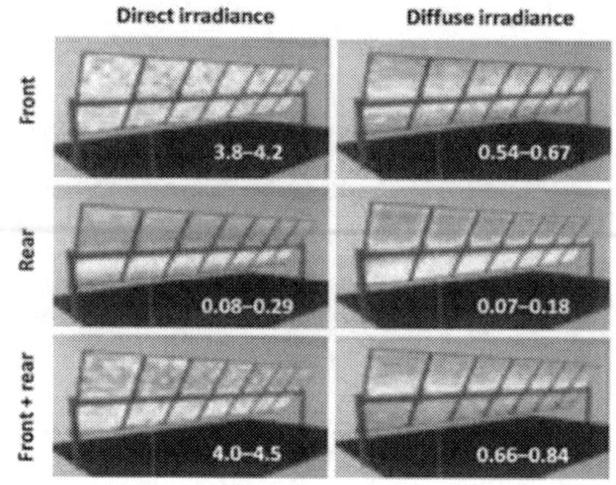

Figure 7: Current generated by each cell in the SAT unit-system for front and rear under direct and diffuse light at 3 pm on 18-Sep-2022. Values show min and max I_L in A.

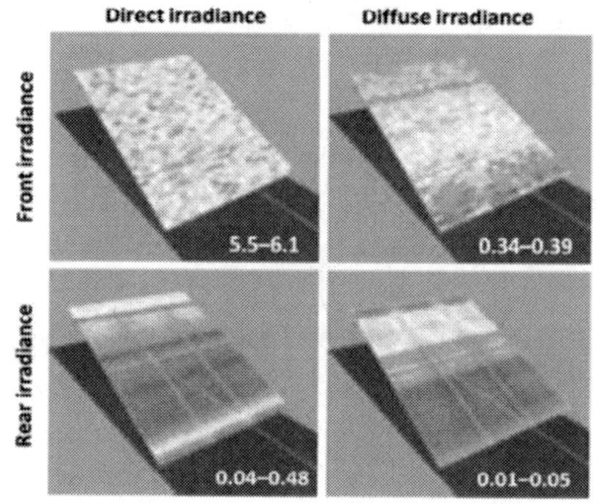

Figure 8: Current generated by each cell in the fixed unit-system for front and rear under direct and diffuse light at 1 pm on 18-Sep-2022. Values show min and max I_L in A.

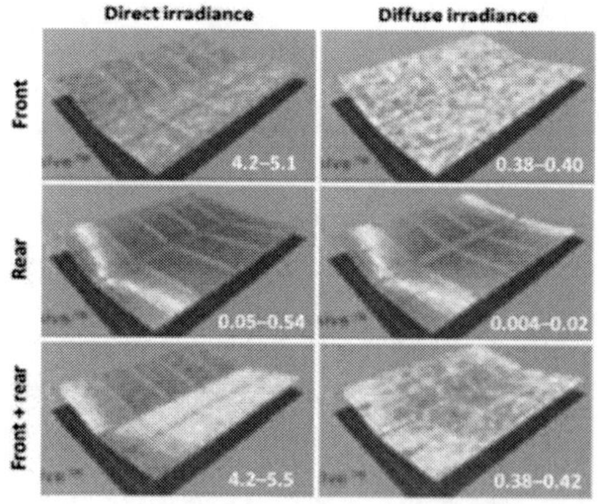

Figure 9: Current generated by each cell in the wave unit-system for front and rear under direct and diffuse light at 1 pm on 18-Sep-2022. Values show min and max I_L in A.

10.4229/EUPVSEC2023/4DO.12.6
020426-003

Figs. 7–9 plot irradiances map for our three unit-systems on the example day at one time with 10^7 rays. The results are presented for both direct and diffuse sources of irradiance, as well as for front and rear irradiance. Each pixel represents a solar cell in the module, and the redder the pixel, the higher the light-generated current I_L within the cell (and hence the higher the incident irradiance). The values in Fig 7–9 give the maximum and minimum I_L in each map. Here are some important trends:

(i) Direct irradiance on the front of the module is almost uniform, where variations in I_L of up to ±5% arise from stochastic error. As found in Section 3.1, this evenly distributed random error introduces CTC mismatch of ~0.1%. There is also a small systematic variation of light whereby the cells at the module edges receive more light from backsheet reflection [25, 26]. Not shown here are the rare occasions for fixed and wave systems when a large variation in I_L results from row-to-row shading of direct light when the sun is low in the sky.

(ii) The nonuniformity of the diffuse irradiance on the front of the modules is predominantly due to row-to-row shading; the steeper the tilt (and the smaller the pitch), the greater the nonuniformity [27–29]. Thus, the cloudier the day, the greater the nonuniformity of the front irradiance.

(iii) The rear irradiance—whether from direct or diffuse light—is always nonuniform for many reasons. These include shading and reflection from the structural supports, ground-reflected light being nonuniform due to shadows of the modules, and row-to-row shading of the diffuse light emanating from the sky. The images show that the rear irradiance is highly nonuniform on the fixed system due to the purlins and rafters, and on the wave system because the edge modules receive ground-reflected light from the sunlight falling between bays of modules.

3.4 Bifacial gain

Fig. 10 plots the annual gain before and after mismatch for our example scenario. Before mismatch, the annual gains are 6.3%, 3.3% and 2.0%, and after mismatch they are 6.2%, 3.0% and 1.8%.

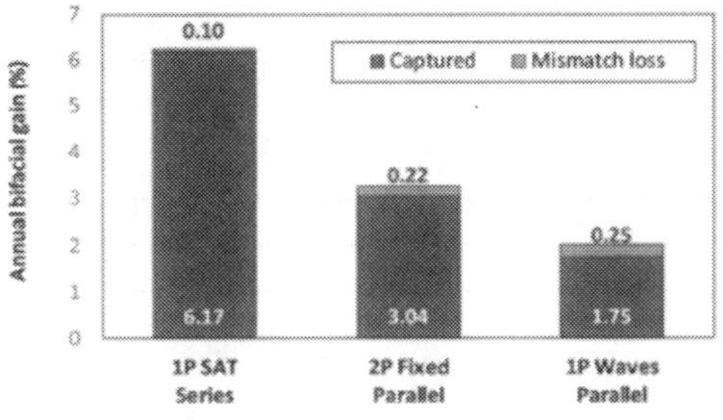

Figure 10: Annual bifacial gain for each unit system.

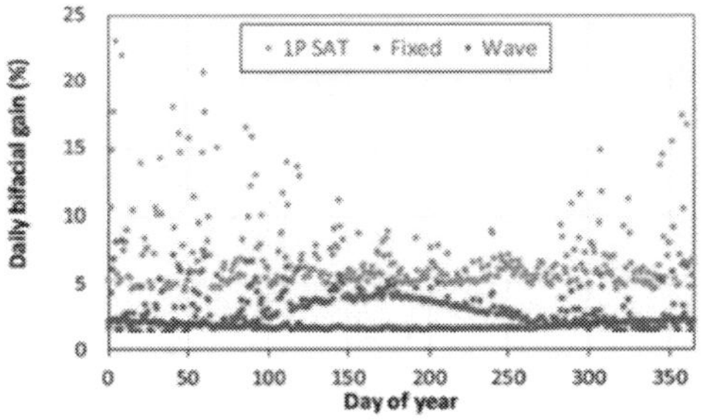

Figure 11: Daily bifacial gain for each unit system.

Fig. 11 plots the daily bifacial gain after mismatch, and Fig. 12 plots the hourly gain after mismatch against solar zenith and diffuse fraction. We find that the bifacial gain of the SAT varies greatly from hour to hour (1–40%), primarily due to the variation of its tilt from horizontal to 60°. The gain for fixed systems varies substantially less (1–8%), except when the sun is very low in the east or west and hence a large fraction of direct light impinges on the ground; and sometimes in summer, the gain is very high (up to 110%) because the direct light is incident to the rear of the modules. The gain for the waves is much less variable (1.5–7%) because the modules are static and close to the ground.

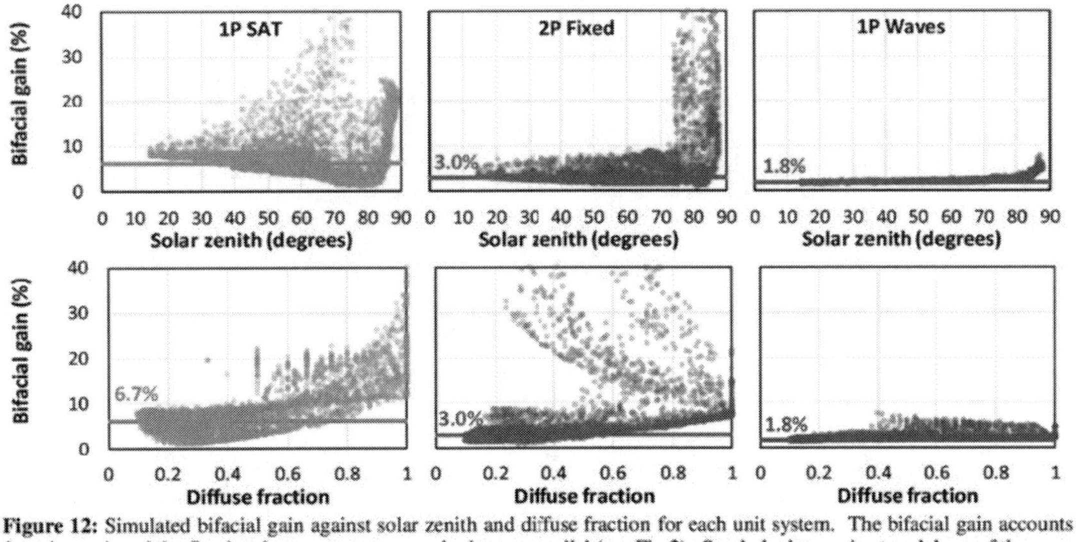

Figure 12: Simulated bifacial gain against solar zenith and diffuse fraction for each unit system. The bifacial gain accounts for mismatch and the fixed and wave systems are wired up as parallel (see Fig 2). Symbols show gain at each hour of the year, line and value show annual gain.

10.4229/EUPVSEC2023/4DO.12.6
020426-004

The bifacial gains are significant for all systems, even for the wave system. Moreover, despite the nonuniformity in the rear irradiance, only a minor fraction of this additional energy is lost. That is, contrary to the oft-cited 'rule of thumb' that a string is limited to its poorest-performing cell or module, in this example, about 90% of that energy is gained. (A closer approximation is to use the standard deviation 'rule of thumb' [21–23].) The reasons for the high collection rate of the bifacial illumination relates to the complicated nature of the irradiance map, the stringing of the modules into substrings with bypass diodes, the orientation of the modules, and the electrical interaction between cells.

These results highlight the value of using both the ray tracing and SPICE solver to evaluate the system.

3.5 Mismatch results — annual

Fig. 13 plots the annual CTC and MTM mismatch for the various cases. We emphasize three points: (i) these results are specific to the example location, module type and the system designs; (ii) the mismatch is calculated from the inherent nonuniform irradiance of the system (i.e., the unavoidable nonuniformity) but neglects any contribution from the variation in cell and module quality (whether due to installed variation or degradation), of nonuniform soiling, temperature variation within the module, and system-level edge effects; and (iii) up to 0.1% (abs) of the CTC mismatch is introduced by stochastic effects.

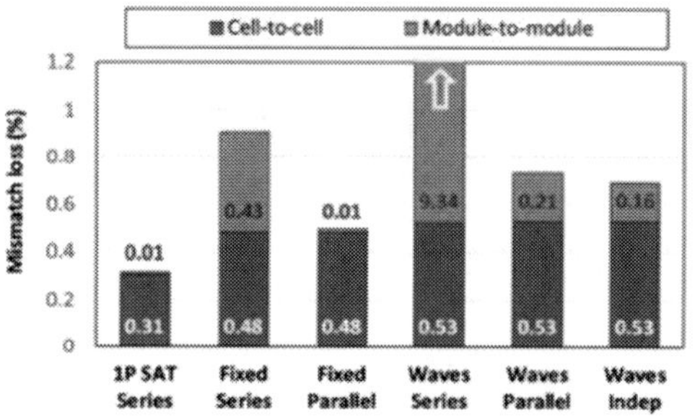

Figure 13: Annual mismatch loss in the example unit-systems at Santiago do Cacém. Cells and modules are assumed identical, and mismatch arises solely from the inherent irradiance nonuniformity on the unit system.

We find that inherent CTC mismatch is not insignificant for any design, being about 0.3% for SATs and 0.5% for the fixed and wave systems. In all cases, the CTC mismatch is due more to nonuniformity in the front irradiance rather than the rear irradiance. (When solving with front-only irradiance, the CTC mismatch is 72%$_{rel}$ for SATs, 56%$_{rel}$ for fixed, and 83%$_{rel}$ of the value calculated for bifacial irradiance.)

By contrast, MTM mismatch changes greatly depending on the system type and, more particularly, on the way the modules are strung together:

For the 1P SAT, the irradiance on all modules is very similar (with the outer modules receiving a very small amount of additional rear irradiance), and thus MTM mismatch is negligible. This would not be true of 2P SATs.

For the fixed system, the upper modules receive more irradiance than the lower modules, particularly on diffuse days, but there is no lateral difference in irradiance. Hence, the MTM mismatch is significant (0.43%) when strung in series (combining upper and lower modules) but not in parallel.

For the waves system, modules in the east and west rows receive significantly different direct-light irradiance, except at times near the middle of the day. This leads to very high MTM mismatch (9.3%) when the modules are strung in series. But even in parallel, there is a small degree of mismatch because the edge modules receive more illumination (0.2%). This compares to a study by Staudacher, who also connected the strings of wave modules at a 15° tilt to an inverter, calculating 0.3–0.5% mismatch for monofacial east–west connected in parallel rather than independent [30]; this mismatch loss reduced below 0.1% once inverter behavior was considered.

3.6 Mismatch results — daily

Fig. 14 plots the daily mismatch throughout the year, illustrating how mismatch tends to be higher in the winter, when the sun is lower, but also in the summer for fixed system. The daily loss is up to 1%, 1.5% and 2.1%,

3.7 Mismatch results — hourly

Figs. 15 and 16 plot the mismatch against solar zenith and the diffuse fraction. The symbols represent each hour of the year, and the line represents the annual loss (equivalent to a weighted average). The figure makes clear that both CTC and MTM mismatch vary greatly depending on the solar zenith (or to be more precise, solar position), and to a lesser extent on the diffuse fraction. It exceeds 3% for SATs and >30% for fixed and wave systems, particularly when the sun is low in the sky and row-to-row shading of direct light ensues.

These results highlight the value of determining mismatch losses at each hour of the year, rather than assuming a constant loss throughout the year.

3.8 Mismatch results — variability

Table II lists the range of inherent mismatch loss for each example system.

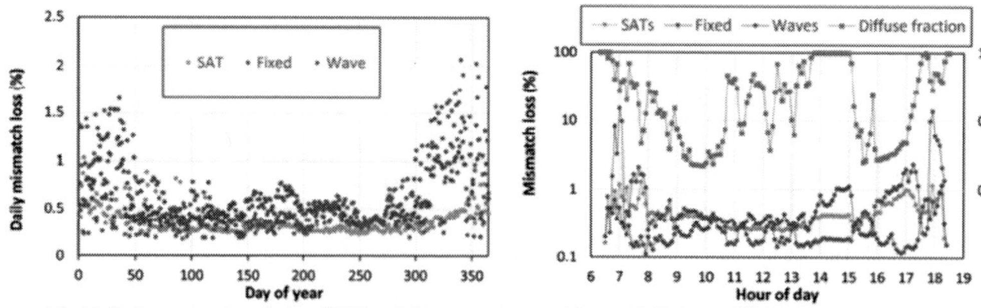

Figure 14: (a) Daily energy loss during TMY and (b) power loss on 18-sep-2022 due to CTC mismatch.

10.4229/EUPVSEC2023/4DO.12.6
020426-005

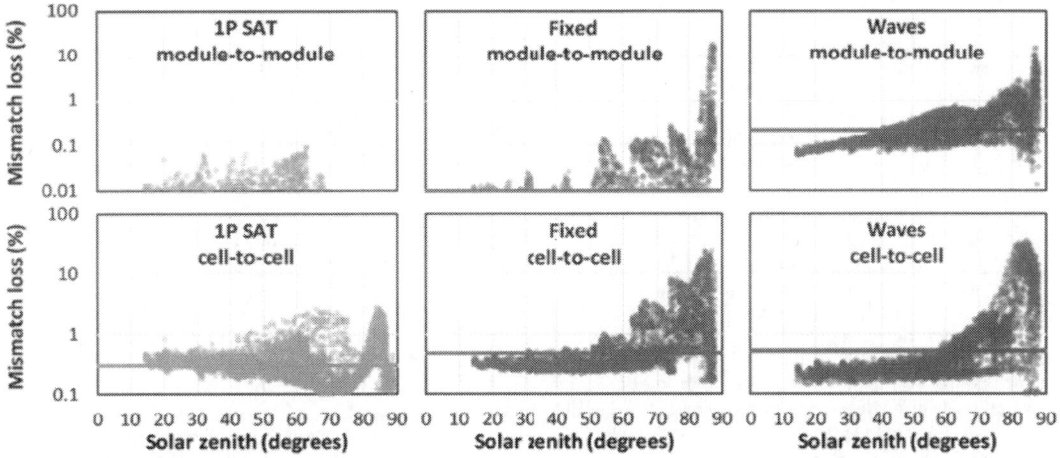

Figure 15: Inherent mismatch loss against solar zenith for each example unit system, where fixed and wave systems are wired in parallel (see Fig 2). Symbols plot mismatch at each hour of the year, lines plot the annual mismatch loss.

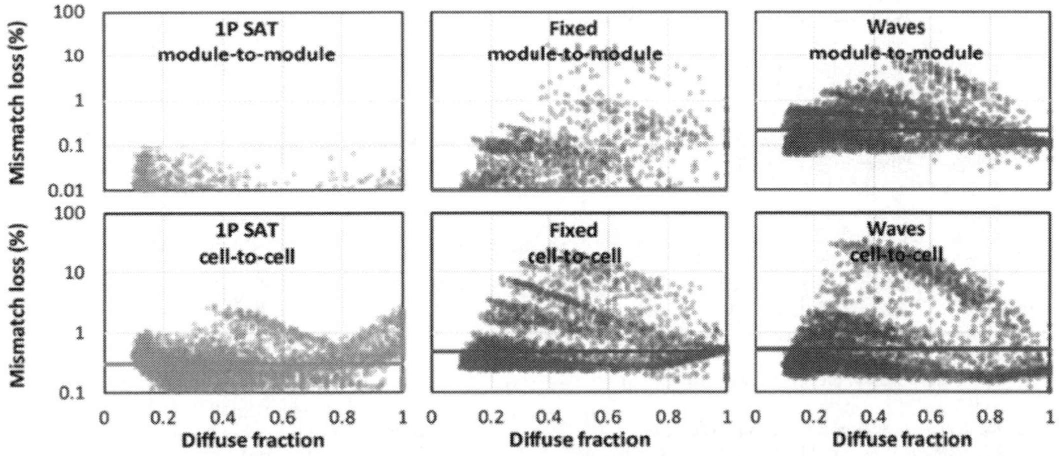

Figure 16: Inherent mismatch loss against diffuse fraction for each example unit system, where fixed and wave systems are wired in parallel (see Fig 2). Symbols plot mismatch at each hour of the year, lines plot the annual mismatch loss.

Table II: Range of inherent mismatch loss (CTC + MTM) in the three example unit systems.

	Hourly	Daily	Annual
1P SAT	0–2.7%	0.2–1.1%	0.32%
2P Fixed, series	0–40%	0.4–5.1%	0.91%
2P Fixed, parallel	0–36%	0.3–1.5%	0.49%
1P Waves, series	0–49%	0.3–15.5%	9.81%
1P Waves, parallel	0–36%	0.3–2.6%	0.74%

4 CONCLUSION

With ray tracing and SPICE solving, we calculated the inherent CTC and MTM mismatch in example SAT, fixed and waves systems due to unavoidable nonuniform illumination. When located at Santiago do Cacém, the annual CTM was 0.3% for the SAT and 0.5% for the other two system types, but it varied markedly from hour to hour, depending on the position of the sun in the sky and the diffuse fraction. The MTM mismatch was insignificant for the 1P and parallel-wired fixed system because, in this example, there is little variation in irradiance between modules. Series-connected strings had significant mismatch when the strings received differing amounts of irradiance, like the upper and lower modules of the 2P fixed system, or the east and west modules of the 1P wave system. In these example systems, the mismatch arose primarily due to variability on the front of the modules, and only a small fraction (<12%) of the bifacial gain was lost due to mismatch.

These results underscore the complexity of electric mismatch, and how it can be understood and quantified by combining ray tracing and SPICE solving.

5 REFERENCES

[1] F. Spertino, J.S. Akilimali, "Are manufacturing I–V mismatch and reverse currents key factors in large photovoltaic arrays?" IEEE Transactions on Industrial Electronics, 56 (2009) 4520.

[2] R. Evans, M. Boreland, M.A. Green. "A holistic

10.4229/EUPVSEC2023/4DO.12.6
020426-006

review of mismatch loss: From manufacturing decision making to losses in fielded arrays." Solar Energy Materials and Solar Cells 174 (2018) 214.

[3] Niazi, Kamran Ali Khan, Yongheng Yang, and Dezso Sera. "Review of mismatch mitigation techniques for PV modules." IET Renewable Power Generation 13 (2019) 2035.

[4] Wilson, K., De Ceuster, D. and Sinton, R.A., 2006, May. Measuring the effect of cell mismatch on module output. In 2006 IEEE 4th World Conference on Photovoltaic Energy Conference (Vol. 1, pp. 916-919). IEEE.

[5] C. Sainsbury, R.A. Sinton, D. Jordan, "On the ambiguity of using Isc for analyzing Suns-Voc data on modules," Proceedings of 47th IEEE Photovoltaic Specialists Conference, (2020) 1441.

[6] M.K. Forsyth, M. Mahaffey, A.L. Blum, W.A. Dobson, R.A. Sinton, 2014, Use of the suns-Voc for diagnosing outdoor arrays & modules. In 2014 IEEE 40th Photovoltaic Specialist Conference (PVSC) (pp. 1928-1931). IEEE.

[7] K.R. McIntosh, M.D. Abbott, B.A. Sudbury, J. Meydbray, "Mismatch loss in bifacial modules due to nonuniform illumination in 1-D tracking systems," IEEE Journal of Photovoltaics 9 (2019) 1504.

[8] S.A. Pelaez, C. Deline, J.S. Stein, B. Marion, K. Anderson, M. Muller, "Effect of torque-tube parameters on rear-irradiance and rear-shading loss for bifacial PV performance on single-axis tracking systems," Proceedings 46th IEEE Photovoltaic Specialists Conference, Vol. 2 (2019) 1.

[9] C. Deline, S. Ayala Pelaez, S. MacAlpine, C. Olalla, "Estimating and parameterizing mismatch power loss in bifacial photovoltaic systems." Progress in Photovoltaics, 28 (2020) 691.

[10] C. Zhao, J. Xiao, Y. Yu, J.N. Jaubert, "Accurate shading factor and mismatch loss analysis of bifacial HSAT systems through ray-tracing modeling," Solar Energy Advances 1 (2021) 100004.

[11] G. Raina, S. Sunanda, "A comprehensive assessment of electrical performance and mismatch losses in bifacial PV module under different front and rear side shading scenarios," Energy Conversion and Management 261 (2022) 115668.

[12] A.C. Russell, C.E. Valdivia, C. Bohémier, J.E. Haysom, K. Hinzer, "DUET: A Novel Energy Yield Model With 3-D Shading for Bifacial Photovoltaic Systems," IEEE Journal of Photovoltaics 12 (2022) 1576.

[13] T.J. Coathup, M.R. Lewis, A.C. Russell, J.E. Haysom, C.E. Valdivia, K. Hinzer, "Impact of torque tube reflection on bifacial photovoltaic single axis tracked system performance," Optics Express 31 (2023) 6143.

[14] www.5b.co.

[15] https://taiyangnews.info/markets/europes-largest-solar-power-farm-planned-in-portugal/, Taiyang News, 1-Feb-2023.

[16] www.solcast.com.

[17] https://pvlighthouse.com.au/sunsolve.

[18] R.E. Bird and C. Riordon, "Simple solar spectral model for direct and diffuse irradiance on horizontal and tilted planes at the Earth's surface for cloudless atmospheres," Journal of Climate and Applied Metrology 25 (1986) 87.

[19] S.K. Meerdink, S.J. Hook, D.A. Roberts and E.A. Abbott, "The ECOSTRESS spectral library version 1.0", Remote Sensing of Environment 230 (2019) 111196.

[20] J.E. Hay and J.A. Davies, "Calculations of the solar radiation incident on an inclined surface," Proceedings 1st Canadian Solar Radiation Data Workshop, 1980.

[21] L. L. Bucciarelli, "Power loss in photovoltaic arrays due to mismatch in cell characteristics," Solar Energy, vol. 23, pp. 277-288, 1979.

[22] Evans, Rhett, Kyung H. Kim, Xiaoyi Wang, Adeline Sugianto, Xiaochun Chen, Rulong Chen, and Martin A. Green. "Simplified technique for calculating mismatch loss in mass production." Solar Energy Materials and Solar Cells 134 (2015): 236-243.

[23] R.A. Sinton, FMT-500 Flash Module Tester manual, 2020 p86.

[24] M. Ernst, G.E. Conechado, C.A. Asselineau, "Accelerating the simulation of annual bifacial illumination of real photovoltaic systems with ray tracing," iScience 25 (2022).

[25] M.R. Vogt "Development of physical models for the simulation of optical properties of solar cell modules," Doctoral dissertation, ISFH, Chapter 4.

[26] K.R. McIntosh, J. Jung, M.D. Abbott, B.A. Sudbury, "Determination and evaluation of a backsheet's intrinsic reflectance" AIP Conference Proceedings 1999 (2018).

[27] K.R. McIntosh, M.D. Abbott, G. Loomis, B.A. Sudbury, A. Mayer, C. Zak, J. Meydbrey, "Irradiance on the upper and lower modules of a two-high bifacial tracking system," Proceedings 47th IEEE Photovoltaic Specialists Conference (2020) 1916.

[28] K.R. McIntosh, M.D. Abbott, B.A. Sudbury, "The optimal tilt angle of monofacial and bifacial modules on single-axis trackers," Journal of Photovoltaics 12 (2021) 397.

[29] J. Appelbaum, "The role of view factors in solar photovoltaic fields," Renewable Sustain. Energy Review 81 (2018) 161.

[30] D. Staudacher, "Efficient east-west orientated PV systems with one MPP tracker," Proceedings of 25th EU PVSEC, 2010.

MEASURING DIFFUSE HORIZONTAL IRRADIANCE USING A STATIC SENSOR ARRAY

Michael Gostein[1]*, Bruce H. King[2], Josh Horst[1]
[1]Atonometrics, Austin, USA
[2]Sandia National Laboratories, Albuquerque, USA
*Corresponding author: michael.gostein@atonometrics.com

ABSTRACT: We report on a field trial in progress to test measurement of diffuse horizontal irradiance (DHI) using a low-cost combination of static sensors with no moving parts. The system pairs two tilted reference cells, one of which measures global irradiance and the other of which is modified with an isolator tube that admits only diffuse light contributions from a limited region of the sky. Using reference data from a research-grade tracking diffusometer, we trained a neural network model to use the readings of the two static sensors to estimate DHI. Testing over a 3-month period shows the static sensors coupled with the neural network model yielded DHI values with significantly lower root-mean-square deviation (RMSD) versus the reference diffusometer as compared to results obtained from standard estimations of DHI from decomposition of global horizontal irradiance (GHI) using the DIRINT model.
Keywords: photovoltaic (PV) systems, irradiance, diffuse

1 INTRODUCTION

Predicting and assessing the performance of a photovoltaic (PV) system requires modeling its response to measured irradiance. As PV systems become more sophisticated, irradiance modeling is becoming more complex. For example, in single-axis tracking systems, predicting and optimizing performance during row-to-row shading or cloudy sky conditions requires a detailed assessment of diffuse light contributions. And in bifacial systems, modeling must include both front-side and rear-side irradiance, which may have components of direct beam, sky diffuse, and ground-reflected light. Often irradiance quantities must be transposed from a measurement plane to a plane of array (POA), also requiring modeling of direct and diffuse irradiance components.

The most complete assessments of solar irradiance include determination of both direct and diffuse irradiance, typically measured as direct normal irradiance (DNI) and diffuse horizontal irradiance (DHI). Together with measurements of ground-reflected irradiance, this allows projection of irradiance onto any plane.

High-accuracy systems for measuring DNI and DHI typically employ sensors on a two-axis sun-tracking motorized stage or other, simpler, motorized systems. However, due to the cost and maintenance requirements of such equipment, DNI and DHI measurements are typically omitted from resource assessment and PV power plant meteorological stations. Instead, DNI and DHI are usually estimated at these sites from global measurements, such as GHI or POA irradiance, using one of a number of decomposition models [1]. Estimation via decomposition models is often incorporated directly into PV performance modeling software as an intermediate step in calculations. For example, in both pre-construction resource assessment and operating plant meteorological stations, global horizontal irradiance (GHI) may be measured and then used to predict POA irradiance. If measured DNI and DHI are not available, modeling software may estimate the DNI and DHI components of GHI as a step in transposing GHI to POA. However, the accuracy of decomposition models is limited [1], so this approach may add to uncertainties in

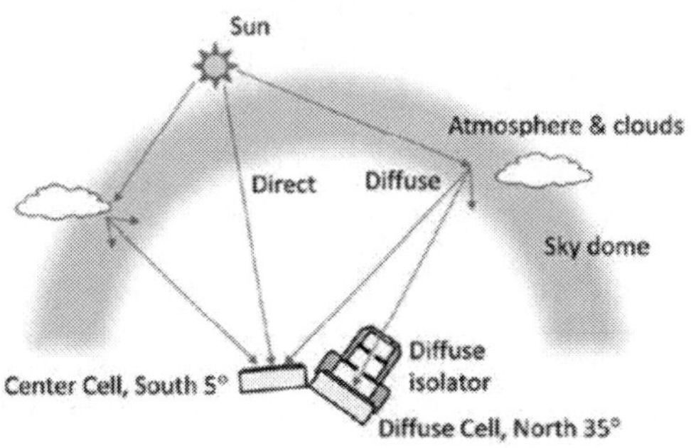

Figure 1: Measurement concept comprising two sensors. The north-tilted cell with diffuse isolator senses diffuse contributions from a limited region of the sky which always excludes the sun. The center tilted cell responds to whole sky diffuse and direct light.

Figure 2: Static sensor array deployed at Sandia National Laboratories in Albuquerque, New Mexico, USA. The array comprises five sky-facing PV reference cells, four tilted north, south, east, and west at 35° and one center cell tilted south at 5°. The north-facing cell is fitted with a diffuse isolator tube blocking all direct and ground-reflected light.

10.4229/EUPVSEC2023/4DV.4.2
020443-001

overall PV system performance modeling. Therefore, the availability of low-cost instrumentation with moderate accuracy for DNI or DHI measurement could help lower PV performance modeling uncertainties in commercial sites. Only one of DNI and DHI, plus GHI, is required, since these three quantities are related by $\mathrm{GHI} = \mathrm{DNI} \cdot \cos(Z) + \mathrm{DHI}$, where Z is the zenith angle.

A DHI measurement device with static (non-moving) sensors has been developed using multiple photodiodes which are alternately shaded during the day by a cutout pattern in a shade screen. This device has been widely used with reported good performance [2], but has some cost and technical limitations motivating interest in alternatives.

Our aim in the present work is to test a low-cost concept for DHI measurement using an array of static sensors based on PV reference cells. We previously reported results from related work using an array of reference cells in different orientations [3]. In this work we report on an updated configuration with better sensitivity to low DHI.

2 EXPERIMENT

Fig. 1 depicts the concept for the current work. A PV reference cell tilted 5° south measures global tilted irradiance. A second reference cell tilted 35° north is covered with a diffuse isolator tube comprising a series of apertures which only admit diffuse light contributions from a limited region of the northern sky. We denote the reading of this cell "diffuse polar". Together these two readings may be used to estimate DHI, as explained below. Note that south/north tilts assume northern hemisphere placement and would be reversed in the southern hemisphere.

Fig. 2 shows a reference cell array prototype we have installed at Sandia National Laboratories in Albuquerque, New Mexico, USA for testing the new concept. This prototype is an updated version of that described in our previous work [3]. It comprises five sky-facing reference cells (RC22), four facing nominally north, east, south, and west on a 35° tilt, and one center cell on a 5° tilt to the south. An additional ground-facing cell is not shown. The north cell additionally has a diffuse isolator tube as discussed above and shown in Fig. 1. The shunt resistance of this cell was increased to enable sensitivity to the very small irradiance levels reaching the cell through the reduced angular acceptance of the isolator tube. The central cell is intended to be nominally horizontal, with a 5° tilt for water roll-off. In the present work, only the north and center cells of the reference cell array are used.

Figure 3: Tracking pyrheliometer and diffusometer at Sandia National Laboratories. The tracked ball blocks direct light from reaching the pyranometer on the right.

Co-located with the reference cell array is a secondary-standard thermopile pyranometer (SR30) measuring GHI.

The site also includes a research-grade two-axis sun-tracking system (SOLYS2) with a pyrheliometer for DNI measurement and a shaded pyranometer for DHI measurement, which we denote a tracking diffusometer, shown in Fig. 3. We used the tracking diffusometer as a reference instrument for training and testing the prototype sensor array measurements. The tracking diffusometer is located approximately 150 meters from the sensor array.

Data acquisition began in May 2023 and is ongoing. Data from all sensors is recorded once per minute.

As a precursor to analyzing the RC22 reference cell data from the sensor array, we experimentally measured the RC22 device's angular responsivity. This is quantified by the incidence angle modifier (IAM), the ratio of the sensor's reported irradiance to the product of direct beam irradiance and cosine of angle of incidence, normalized at normal incidence. IAM measurements were performed on four separate RC22 units mounted in each of four clock orientations on a dual-axis tracker system over three clear days. Procedures were similar to those described in [4], [5]. Fig. 4 shows the average of the measured IAM. Only minor deviations were observed between the different sensors and measurement days. Results shown in Fig. 4 were used in the subsequent analysis of the sensor array data as discussed below.

3 ANALYSIS MODEL

We trained a neural network model to process the sensor array data to estimate DHI. The model contains the following five inputs selected as predictors:

- Diffuse polar irradiance signal. This is the output signal of the north-facing reference cell modified with the diffuse isolator tube.
- Global tilted irradiance reading of the center RC22. This signal measures the irradiance received from the whole sky on the 5° tilted plane.
- Expected clear sky reading of the center RC22. Using the known latitude, longitude, altitude, and orientation of the center cell, and its incidence angle response (Fig. 4), we calculate the expected irradiance reading of the cell under clear sky

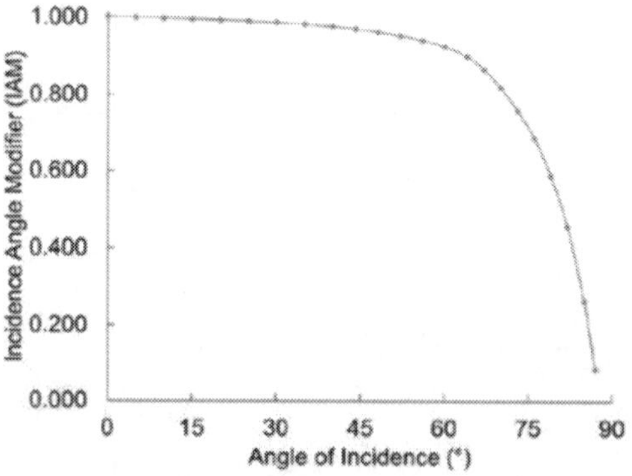

Figure 4: Measured average incidence angle modifier (IAM) of the RC22 reference cell.

conditions using the Ineichen clear sky model in the open source MATLAB PV_LIB toolbox [6][7]. With the model we calculate both beam and diffuse irradiance, apply the IAM to the beam component, and then project the beam and diffuse components onto the tilted cell.

- Trailing 10-minute standard deviation of the diffuse polar signal.
- Trailing 10-minute standard deviation of the center cell irradiance.

The selection of these five signals can be understood as follows. Under clear sky conditions, the diffuse polar irradiance signal is a strong predictor of DHI. Under partly cloudy conditions, the diffuse polar signal may over- or under-estimate DHI at any given time point, depending on whether clouds are present within the field of view of the diffuse isolator tube. In these conditions, comparing the global tilted irradiance reading of the center cell with its expected clear-sky reading allows an estimate of overall cloudiness, which is a whole-sky predictor of DHI. In addition, the two trailing 10-minute standard deviations indicate the degree of partly cloudiness and therefore the degree to which instantaneous readings of the diffuse polar signal may over- or under-estimate DHI.

The neural network regression model was implemented in MATLAB and contains the five inputs listed above, two intermediate layers containing ten nodes each, and a single output for DHI. The rectified linear unit (ReLU) activation function is used for each node.

We trained the neural network model using four weeks of data beginning in early May 2023. The training data comprises the five input signals and reference values for DHI obtained from the tracking diffusometer. Training was performed in the MATLAB regression learner toolbox by optimizing the model to minimize the root mean square deviation (RMSD) between the predicted and reference values of DHI.

4 RESULTS

Following the training period, we process the input data to estimate DHI using the neural network model and compare the estimated values against the reference readings of the tracking diffusometer.

Fig. 5 shows a representative period of hourly-averaged results spanning approximately four weeks, which does not include the training data period. The figure indicates general agreement between the sensor array results and the reference instrument. Note the good agreement during stable, clear-sky conditions when DHI is low.

For most commercial PV projects, the alternative to low-cost DHI measurements is simply to estimate DHI from decomposition models, as discussed above. Various decomposition models are in use. An evaluation of decomposition model accuracy [1] found that the DIRINT model was among the best performing.

In Fig. 6 we compare the accuracy of the sensor array results for DHI with those of the DIRINT decomposition model calculated using the MATLAB version of the open source PV_LIB toolbox [6][7]. The figure plots hourly average predicted DHI for both DIRINT and the sensor array plus neural network model against the hourly-average DHI measured by the tracking diffusometer. The plot contains all data recorded over a three-month period beginning in early June 2023 and excludes the four-week neural network training data period beginning in early May 2023. The DIRINT DHI values were computed from GHI readings of the SR30 pyranometer co-located with the sensor array. Note that the DIRINT model tends to over-

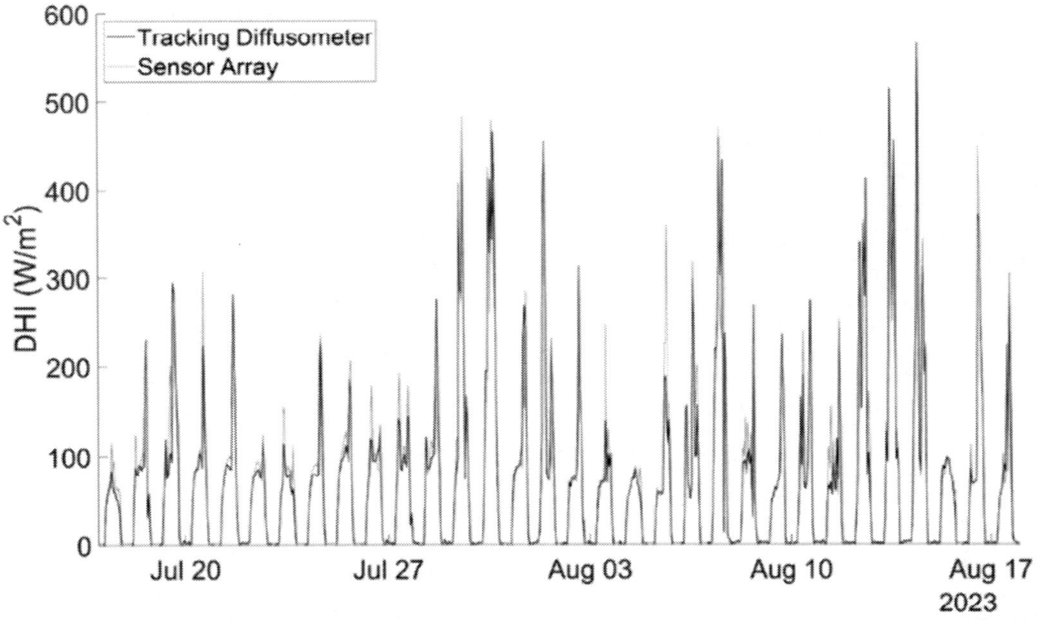

Figure 5: Hourly averaged time series data from a representative ~four-week period, comparing DHI determined from the sensor array to DHI measured by the tracking diffusometer.

10.4229/EUPVSEC2023/4DV.4.2
020443-003

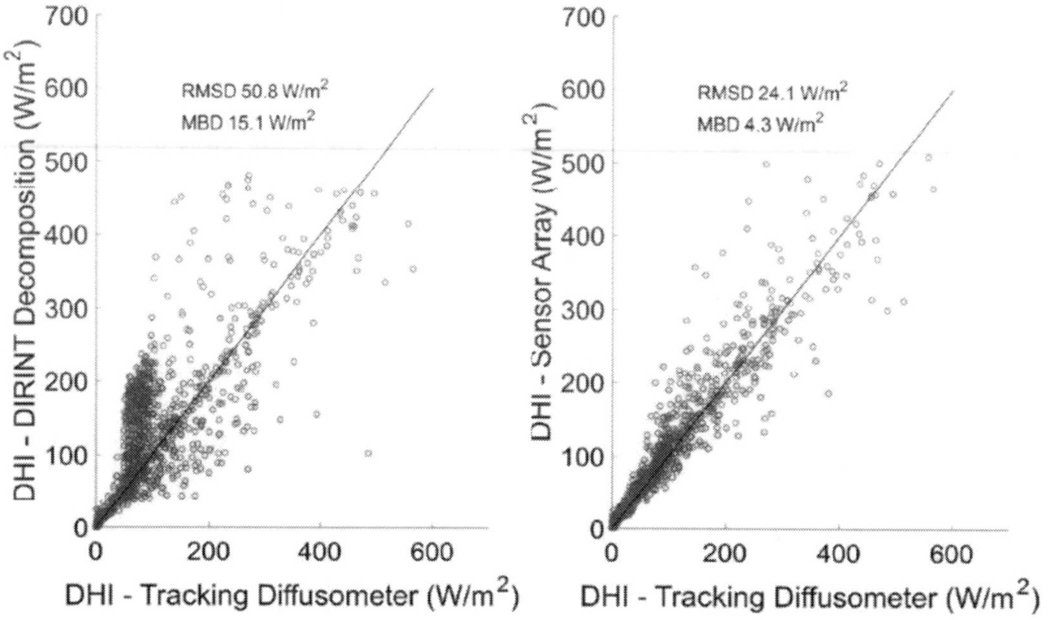

Figure 6: Comparison of DHI prediction accuracy for the DIRINT decomposition model (left) versus the results of the sensor array and neural network model (right). Each plot shows the predicted DHI versus the reference value of DHI from the tracking diffusometer, for hourly averaged results. Results are from a three-month period beginning in June 2023 and exclude the neural network training period from May 2023.

predict DHI during clear-sky (low-DHI) conditions. Mean bias deviation (MBD) and root-mean square deviation (RMSD) correlation statistics are listed on the plots.

Expressed as a percentage of mean GHI, the DIRINT MBD and RMSD are 4.8% and 16.1%, respectively, while the sensor array MBD and RMSD are 1.4% and 7.6%, significantly better. The DIRINT RMSD result is consistent with observations in [1] over a collection of sites, although it is at the upper end of the range. Note that the RMSD values for both the DIRINT and sensor array results are likely elevated at least partially by the fact that the GHI pyranometer and sensor array are both co-located approximately 150m from the tracking diffusometer, and therefore may indeed register different amounts of diffuse light during any time interval during partly cloudy conditions.

5 DISCUSSION AND ONGOING WORK

Results presented in Fig. 5 and Fig. 6 are promising and show the potential ability of the sensor array approach to produce DHI results with significantly improved accuracy versus using decomposition models alone. Compared with our previous approach in [3], fewer sensors are required and the results have better performance particularly for clear-sky conditions with low DHI. Therefore, the new concept could provide a low-cost DHI measurement device suitable for pre-construction resource assessment and PV power plant meteorological stations, helping reduce PV performance modeling uncertainties.

However, additional work is needed to fully test this concept.

We envision that the two sensors used in the present work would be deployed together with a processing unit that would automatically calculate DHI results from the trained neural network model. It will be important to demonstrate that the neural network model trained at one site can be deployed at other sites and remain accurate without retraining.

Therefore, we plan to deploy the sensor array experiment to additional sites for further testing. Additional sites will be chosen to have different latitude than the Albuquerque site, since the diffuse polar irradiance signal may depend on angular separation between the sun and the observed sky region, which depends on latitude, time of day, and time of year. Testing therefore also will be conducted over a longer period that includes more seasonal variation.

6 ACKNOLWEGEMENTS

This material is partially based upon work supported by the U.S. Department of Energy's Solar Energy Technologies Office under Award Number DE-SC0020831 and 38268.

Sandia National Laboratories is a multi-mission laboratory managed and operated by National Technology and Engineering Solutions of Sandia, LLC, a wholly owned subsidiary of Honeywell International Inc., for the U.S. Department of Energy's National Nuclear Security Administration under contract DE-NA0003525.

7 REFERENCES

[1] M. Lave, W. Hayes, A. Pohl, and C. W. Hansen,

"Evaluation of global horizontal irradiance to plane-of-array irradiance models at locations across the United States," *IEEE J. Photovoltaics*, vol. 5, no. 2, pp. 597–606, Mar. 2015, doi: 10.1109/JPHOTOV.2015.2392938.
[2] V. Estelles, D. Serrano, S. Segura, J. Wood, N. Webb, and M. Utrillas, "Evaluation of the Delta-T SPN1 radiometer for the measurement of solar irradiance components," *Proc. EGU Gen. Assem. 2016, Vienna, Austria*, 2016.
[3] M. Gostein, A. Hoffman, B. H. King, and A. Marquis, "Measuring Global, Direct, Diffuse, and Ground-Reflected Irradiance Using a Reference Cell Array," in *2022 IEEE 49th Photovoltaics Specialists Conference (PVSC)*, Nov. 2022, pp. 0285–0290, doi: 10.1109/PVSC48317.2022.9938489.
[4] B. H. King, D. Riley, C. D. Robinson, and L. Pratt, "Recent advancements in outdoor measurement techniques for angle of incidence effects," *2015 IEEE 42nd Photovolt. Spec. Conf. PVSC 2015*, Dec. 2015, doi: 10.1109/PVSC.2015.7355849.
[5] B. H. King and C. D. Robinson, "Differential Analysis of the Angle of Incidence Response of Utility-Grade PV Modules," in *2019 IEEE 46th Photovoltaic Specialists Conference (PVSC)*, Jun. 2019, pp. 77–81, doi: 10.1109/PVSC40753.2019.8981355.
[6] "PV Performance Modeling Collaborative | PV_LIB Toolbox." https://pvpmc.sandia.gov/applications/pv_lib-toolbox/ (accessed May 19, 2022).
[7] W. F. Holmgren, C. W. Hansen, and M. A. Mikofski., "pvlib python: A python package for modeling solar energy systems," *J. Open Source Softw.*, vol. 3, no. 29, p. 884, 2018, doi: 10.21105/joss.00884.

NRSDB AND PVGIS VALIDATION USING WEATHER STATIONS INSTALLED IN NORTHEAST BRAZIL

Luiz Reis[1, a)], *Lúcio Paiva*[1, b)], *Luis Castro*[1, c)] *and Rodrigo Santos*[1, d)]
[1]*Casa dos Ventos Energias Renováveis, Fortaleza, Brazil*
[a)]*luiz.reis@casadosventos.com.br*
[b)]*lucio.paiva@casadosventos.com.br*
[c)]*guilherme.castro@casadosventos.com.br*
[d)]*rodrigo.raimundo@casadosventos.com.br*

ABSTRACT: The availability of high-quality satellite data plays a pivotal role in expediting decision-making processes and propelling global efforts towards energy transition and decarbonization. Such data serves to streamline assessments of solar resources, effectively reducing both costs and time investments. However, evaluating the precision of satellite data in specific regions can prove to be a formidable task. Previous research has predominantly focused on Europe and North America. This study aims to break new ground by validating global horizontal irradiance (GHI) and temperature data for Northeast Brazil on an unprecedented scale, utilizing information gathered from twenty-three meteorological stations. To assess data performance and gauge data uncertainty, this study employs evaluation metrics such as mean bias error (MBE) and root mean squared error (RMSE). Furthermore, the study explores the fitting of hourly satellite data by comparing it to annual averages sourced from the Global Solar Atlas 2.0. After adjustments, the results reveal an estimated uncertainty level below 5% for GHI values and under 9% for temperature values, both on a monthly aggregate. These findings suggest that these data sets are indeed viable for conducting monthly and seasonal analyses in the evaluated region.
Keywords: Validation; Satellite; Solar.

1 INTRODUCTION

Use of satellite data is extremely important for expanding temporal and spatial coverage during prospecting stages of new photovoltaic plant projects. The increase in the amount of data mitigates the uncertainties related to resource's spatial and temporal variation. According to [1], the time series should ideally cover the history of the last 10 to 20 years when it intends to be used for solar resource assessment.

Some sources, including PVGIS and NSRDB V3.2.0, provide users with access to hourly global horizontal irradiance (GHI) and long-term temperature data via an API, providing broad temporal and spatial coverage. According to studies carried out in [2], satellite-derived irradiance data have an advantage over terrestrial measurements due to their temporal homogeneity over large regions and several years. However, evaluating the performance of these sources in specific regions of the globe can be a challenge, since scientific validation work has focused mainly on Europe and North America.

The primary aim of this study is to enhance the analysis of satellite data sources in Northeast Brazil. This enhancement involves a deeper investigation into the accuracy of the GHI and temperature data supplied by the NSRDB V3.2.0 and PVGIS V5.2.

Additionally, we present a novel methodology for refining hourly satellite data by utilizing long-term averages of global horizontal irradiance and air temperature at a 2 meter above the ground, as sourced from the Global Solar Atlas 2.0. This methodology employs the ratio method, described in the works of [1], to adjust the data, with the overarching goal of assessing potential enhancements in the precision of uncertainty estimates for each data source.

2 DATASETS

2.1 NSRDB V3.2.0

The National Renewable Energy Laboratory (NREL) is responsible for developing the NSRDB V3.2.0. The irradiance database covers most of the Americas and provides data from 1998-2020. The Solar Physical Model (PSM) developed by NREL in collaboration with the University of Wisconsin and the National Oceanic and Atmospheric Administration (NOAA) calculates global horizontal irradiance (GHI) using the visible and infrared channel [3] . According to [4] PSM v3 is considered one of the most advanced models for converting satellite data into irradiance data.

2.2 PVGIS - SARAH-2

The Photovoltaic Geographical Information System (PVGIS) is a tool developed by the Joint Research Center of the European Commission. PVGIS, which is now at version 5.2, is under constant development. According to According [5.] in the current version, data and irradiance for South America come from Surface Solar Radiation Data Set Heliosat Edition 2 (SARAH-2) while temperature values are derived from ERA5-Land. The dataset covers the period from 2005 to 2020. Unlike other data sources used in this work, the data available in PVGIS have hourly values estimated from the analysis of a single image per hour, thus being an instantaneous value.In order to compare this data with the ground measurements, the PVGIS values have been considered to be averages.

2.3 Global Solar Atlas 2.0

The Global Solar Atlas v2.0 is a series of GIS data layers with irradiance and temperature information. The atlas, which aims to provide simplified information on

10.4229/EUPVSEC2023/4DV.4.3
020444-001

solar resource, was funded by the World Bank and the International Finance Corporation and developed by SOLARGIS.

2.4 Measuring stations

The stations used for validation are part of the photovoltaic resource measurement campaign of Casa dos Ventos Energias Renováveis. Each of the stations has two Kipp&Zonen superior category pyranometers, classified according to ISO 9060 (2018) as a secondary standard and with calibration tracked by the World Radiometric Network, a thermohygrometer Galltec KPC 1.S/6-ME, a rain gauge and a pre-configured datalogger to read the sensors every second and store the average, maximum, minimum, and standard deviation values every one or ten minutes.

During the entire measurement period, the station's pyranometers were cleaned with a minimum frequency of 15 days. Table 3, in appendix, provides detailed information on the stations used for validating the satellite data. The Köppen and Geiger classification for the climate of cities was completed according to [6.]. In Figure 1, the geographic distribution of measuring stations can be seen.

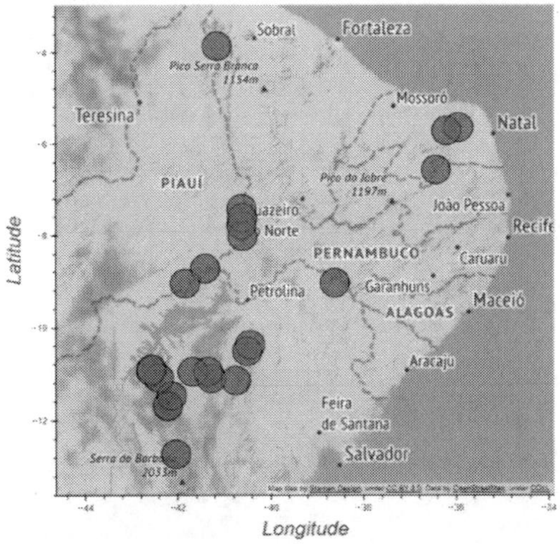

Figure 1: Geographical distribution of measuring stations

3 ACQUISITION AND PROCESSING OF DATA

3.1 Satellite data

The satellite data used in this study were acquired using the functions 'get_pvgis_hourly' and 'get_psm3' from the Python pvlib library developed in [7.]. After data collection, a pre-processing step was performed, involving the renaming of columns and treatment of time labels, facilitating the comparison between all data sources.

3.2 Satellite data adjustment methodology

In studies carried out in [1] it was found that nearly half of the hourly Root Mean Square Deviation (RMSD) for Global Horizontal Irradiance (GHI) and Direct Normal Irradiance (DNI) could be attributed to spatial scale mismatch. To address this issue, the present work proposes a solution involving the utilization of annual averages obtained from the Global Solar Atlas 2.0 geoTIFF files to correct data from the Photovoltaic Geographical Information System (PVGIS) and the National Solar Radiation Database (NSRDB V3.2.0) using the ratio method. This correction process is made feasible by the consistent behavior of normalized Mean Bias Error (nMBE) values over the years, as observed in the study by [4].

According to [8.] the spatial resolution of the data is improved to 250 meters after downscaling, as opposed to data from other satellite sources used in this work, which maintain a spatial resolution of the order of 4 to 5 kilometers. Consequently, two new data series were generated, denoted by the "WB" suffix to indicate that they have been adjusted using annual averages from the atlas.

In the pursuit of evaluating potential enhancements to satellite data, the ratio method was employed to scale the annual averages of GHI and temperature to align with the values from the Global Solar Atlas 2.0. This approach facilitated the adjustment of the original data, resulting in the creation of two new datasets known as NRSDB_WB and PVGIS_V5.2_WB.

The adjustment process involved a comparison of long-term averages of global horizontal irradiance and air temperature at 2 meters above the ground with corresponding values from the satellite array. A scaling factor was derived by comparing the World Bank data with each of the satellite data sources, which was subsequently applied to the entire long-term hourly dataset. This yielded corrected values for these parameters.

It is essential to recognize that this methodology has certain limitations and primarily focuses on rectifying discrepancies in the long-term mean values of the datasets rather than addressing all possible data discrepancies.

3.3 Acquisition of satellite data

For the terrestrial measurement data, we sourced information from 23 high-quality weather stations. These stations employ data loggers to remotely transmit their measurements to central servers. As part of our rigorous quality control process, we subject the data collected by these solarimetric stations to thorough scrutiny, with a particular focus on identifying and addressing potential instrument failures.

Our quality control procedure comprises two key stages. In the first stage, we apply data flags based on well-established methodologies and criteria as outlined in references [9.], [10.], and [11.]. The purpose of this step is to flag data points that may be affected by anomalies or errors. In the second stage, we conduct a visual inspection of the data, considering the flags applied in the first stage. The data specialist reviews these flags and can either approve or reject them based on their assessment.

Following this meticulous validation process, we calculate hourly averages for Global Horizontal Irradiance (GHI) and temperature values obtained from meteorological stations. To ensure the accuracy and reliability of our data, we exclusively utilize GHI values corresponding to solar zenith angles greater than 75°. This approach aligns with a methodology similar to that described in [12.] and helps eliminate nocturnal periods and address any inconsistencies that may occur at the

10.4229/EUPVSEC2023/4DV.4.3
020444-002

beginning and end of each day.

To ensure the robustness of our analyzes and mitigate possible biases arising from incomplete years with limited sample sizes, we adhered to the methodology established in [4], exclusively using complete calendar years for our investigations.

4 RESULTS AND DISCUSSION

4.1 Statistical assessment

The statistical metrics utilized for data analysis included MBE and RMSE, and the overall uncertainty of the dataset was calculated following the same approach as that adopted by [12.]. The expression for calculating each of the metrics can be seen Table 4 in appendix.

4.2 Results

After collecting and processing the data, it was finally possible to calculate the performance metrics for the series. According to the data presented in Table 1, Table 2 and Table 5, in appendix, it is possible to state that the adjustments made with the Atlas Global data resulted in a subtle improvement in all parameters, being more relevant in the GHI values of the NSRDB V3.2.0 source and in the values PVGIS temperature.

Table 1: Validation metrics of GHI values.

Metrics	NSRDB V3.2.0	NSRDB V3.2.0 WB	PVGIS V5.2	PVGIS V5.2 WB
MBE	4.69	-0.45	0.41	0.35
RMSE	18.27	17.55	15.91	15.78

Table 2: Validation metrics of temperature values.

Metrics	NSRDB V3.2.0	NSRDB V3.2.0 WB	PVGIS V5.2	PVGIS V5.2 WB
MBE	3.42	3.32	3.27	2.85
RMSE	9.40	9.12	8.29	6.92

For a better analysis, this metric was also evaluated from the perspective of frequency distribution. Evaluating Figure 2, 3, 4 and 5, it is possible to verify a greater concentration in the distribution of MSE and RMSE values when we compare the adjusted data with the unadjusted data. The statement is valid for both GHI values and temperature values.

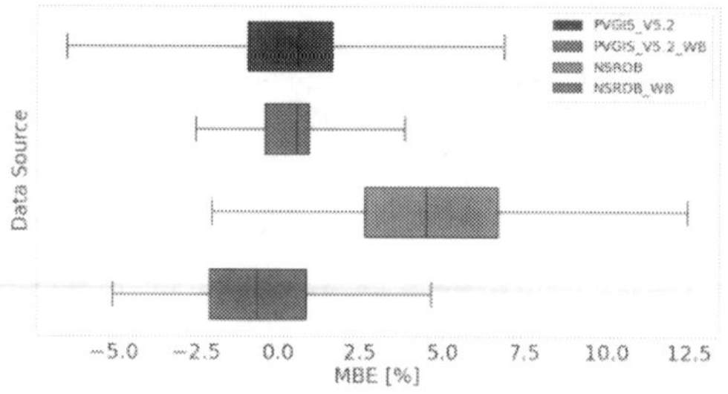

Figure 2: Distribution of MBE values for the GHI data

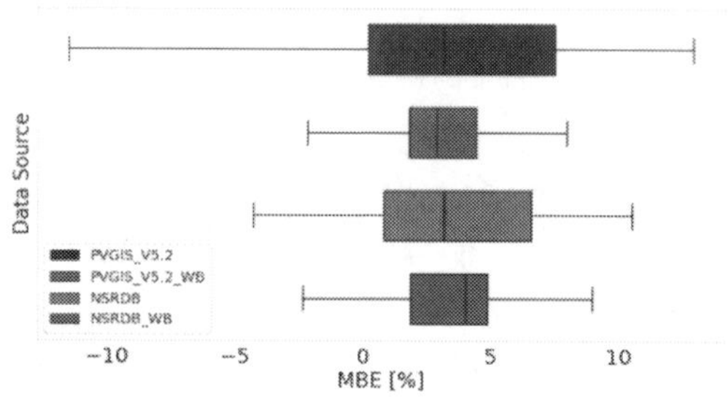

Figure 3: Distribution of MBE values for the temperature data.

Upon evaluating the distribution of MBE values, we can draw some noteworthy conclusions regarding the original and adjusted series. It becomes evident that the original series tend to exhibit an overestimation of both GHI and temperature values for northeastern Brazil. Conversely, the adjusted series display errors that are notably more predictable in nature compared to the original series. These errors are characterized by distributions that are more tightly concentrated and closer to zero.

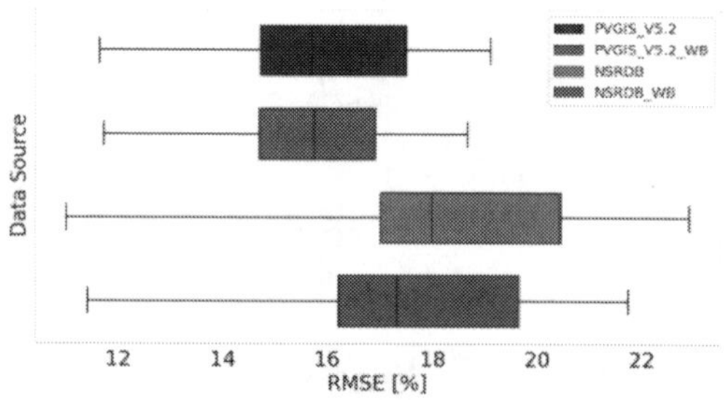

Figure 4: Distribution of RMSE values for the GHI data

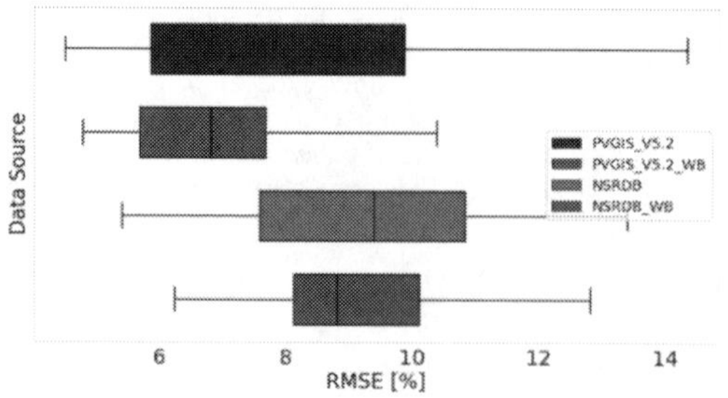

10.4229/EUPVSEC2023/4DV.4.3
020444-003

Figure 5: Distribution of RMSE values for the temperature data.

When examining the RMSE values, another interesting observation emerges. There is a reduced occurrence of outliers in the adjusted series, and these series tend to exhibit slightly lower RMSE values when contrasted with the original series. This suggests that the adjustments have led to improved accuracy and precision in our data, resulting in a more robust and reliable dataset.

To assess the estimated uncertainty of the U_{meas} solar station, we consider several key elements:

Pyranometers: We accounted for a standard uncertainty of 1.25%, as indicated on the calibration certificate. Furthermore, an extra standard uncertainty of 0.5% was considered to cover the calibration process itself. Furthermore, it is important to point out that all stations were equipped with the Eol Zenith model datalogger manufactured by Kintech, which has a 12-bit analog-to-digital converter. An additional standard uncertainty of 0.04% was assigned to the radiation data acquisition process.

Station Thermometers: Similarly, the station thermometers were assigned a standard uncertainty of 5%.

These uncertainties were meticulously taken into account to provide a comprehensive assessment of the measurement uncertainty at the Umeas station. This meticulous approach ensures the most accurate and reliable characterization of the data collected from this source.

Uncertainty was further evaluated for different periods of data aggregation. The results of the analyzes can be seen in Figures 6, 7, 8 and 9.

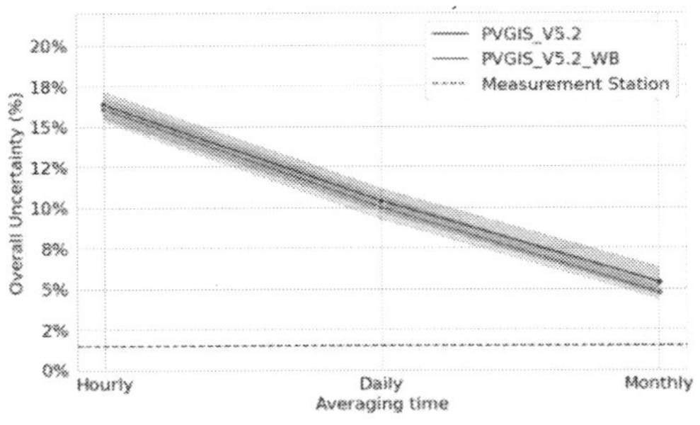

Figure 6: The 95% confidence interval of the expanded uncertainties estimated for the PVGIS GHI values in the 23 locations evaluated.

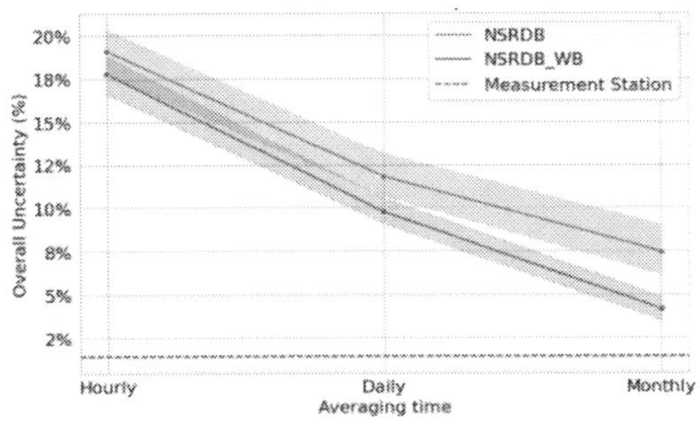

Figure 7: The 95% confidence interval of the expanded uncertainties estimated for the NRSDB GHI values in the 23 locations evaluated.

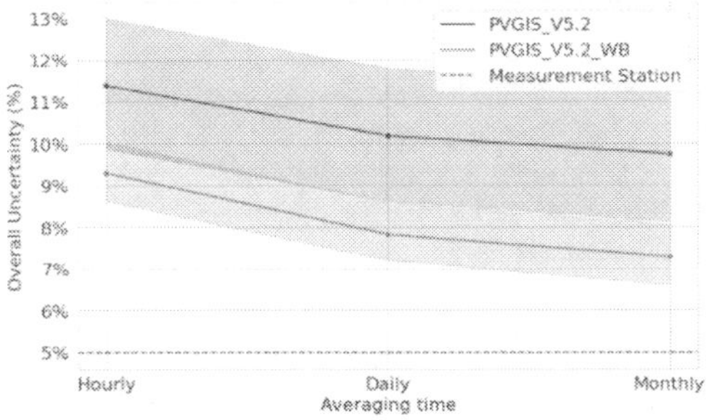

Figure 8: The 95% confidence interval of the expanded uncertainties estimated for the PVGIS temperature values in the 23 locations evaluated.

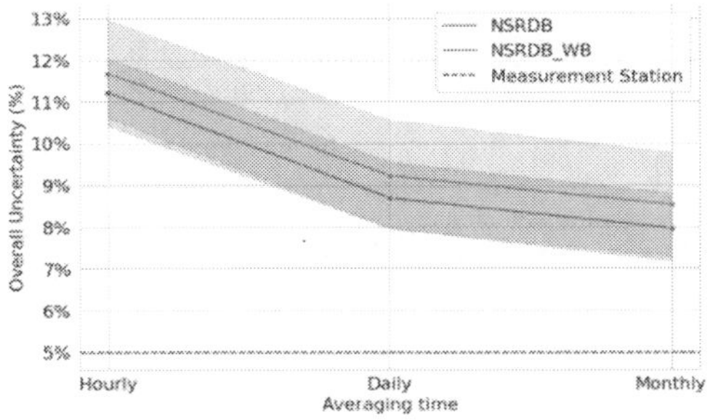

Figure 9: The 95% confidence interval of the expanded uncertainties estimated for the NRSDB temperature values in the 23 locations evaluated.

As previously demonstrated, the adjustments made to the Altas Solar data led to a substantial improvement in the accuracy of the NSRDB V3.2.0 GHI values and the PVGIS temperature values. Another noteworthy conclusion is that the estimated uncertainty values for monthly aggregated data are closely aligned with the estimated uncertainty associated with the measurement station itself.

5 CONCLUSIONS

In general, it can be concluded that the original data from the satellite sources analyzed tend to overestimate temperature and radiation values at the specified points in the northeastern region of Brazil when compared to ground measurements. Furthermore, a simple adjustment of the data using the averages derived from Atlas Solar 2.0 as a reference point may offer greater predictability in the observed errors and a subtle reduction in these discrepancies.

The study has also revealed that the estimated uncertainties for hourly aggregated data tend to be relatively high. However, when considering daily or monthly aggregations, these uncertainty levels exhibit a significant reduction. This suggests that analyses based on annual or seasonal averages using this data are likely to be highly accurate, with estimated uncertainties falling below 5% for GHI values and below 9% for temperature

10.4229/EUPVSEC2023/4DV.4.3
020444-004

values, particularly when aggregated on a monthly basis.

6 ADDITIONAL COMPONENTS

6.1 References

[1] Cebecauer, Tomas & Suri, Marcel. (2016). Site-adaptation of satellite-based DNI and GHI time series: Overview and SolarGIS approach. 1734. 150002. 10.1063/1.4949234.

[2] Darragh Kenny, Stephanie Fiedler, Which gridded irradiance data is best for modelling photovoltaic power production in Germany?, Solar Energy, Volume 232, 2022, Pages 444-458, ISSN 0038-092X, https://doi.org/10.1016/j.solener.2021.12.044.

[3] Sengupta, M., Y. Xie, A. Lopez, A. Habte, G. Maclaurin, and J. Shelby. 2018. "The National Solar Radiation Data Base (NSRDB V3.2.0)." Renewable and Sustainable Energy Reviews 89 (June): 51-60.

[4] Dazhi Yang, Jamie M. Bright, Worldwide validation of 8 satellite-derived and reanalysis solar radiation products: A preliminary evaluation and overall metrics for hourly data over 27 years, Solar Energy, Volume 210, 2020, Pages 3-19, ISSN 0038-092X, https://doi.org/10.1016/j.solener.2020.04.016.

[5.] Gracia Amillo, A.M.; Taylor, N; Martinez A.M.; Dunlop E.D.; Mavrogiorgios P.; Fahl F.; Arcaro G.; Pinedo I. Adapting PVGIS to Trends in Climate, Technology and User Needs. 38th European Photovoltaic Solar Energy Conference and Exhibition (PVSEC), 2021, 907 - 911.

[6.] Alvares, C.A., Stape, J.L., Sentelhas, P.C., Gonçalves, J.L.M.; Sparovek, G. Köppen's climate classification map for Brazil. Meteorologische Zeitschrift, v. 22, n. 6, p. 711-728, 2013.

[7.] Dreisse, A.; Jain, P.; Harrison, S.; Beyond the Curves: Modeling the Electrical Efficiency of Photovoltaic Inverters. Proc. IEEE Photovoltaic Specialists Conf., 2008.

[8.] Global Solar Atlas 2.0 : Validation Report (English). Energy Sector Management Assistance Program Washington, D.C. : World Bank Group. http://documents.worldbank.org/curated/en/507341592893487792/Global-Solar-Atlas-2-0-Validation-Report

[9.] A. Forstinger, S. Wilbert, A. Jensen, B. Kraas, C. M. Fernandez-Peruchena, C. Gueymard, ´ D. Ronzio, D. Yang, E. Collino, J. Polo, J. Ruiz-Arias, N. Hanrieder, P. Blanc, and Y.-M. SaintDrenan, "Expert quality control of solar radiation ground data sets," 10 2021.

[10.] M. Journee and C. Bertrand, "Quality control of solar radiation data within the rmib solar mea- ´ surements network," Solar Energy, vol. 85, no. 1, pp. 72–86, 2011.

[11.] WMO, Guide to Meteorological Instruments and Methods of Observation. World Meteorological Organization.

[12.] Habte, Aron, Sengupta, Manajit, and Lopez, Anthony. Evaluation of the National Solar Radiation Database (NSRDB V3.2.0): 1998-2015. United States: N. p., 2017. Web. doi:10.2172/1351859.

[13] Huld, T., Müller, R. and Gambardella, A., 2012. "A new solar radiation database for estimating PV performance in Europe and Africa". Solar Energy, 86, 1803-1815.

[14] Brazil GISdata LTAym Yearly Totals GEOTIFF obtained from the "Global Solar Atlas 2.0, a free, web-based application is developed and operated by the company Solargis s.r.o. on behalf of the World Bank Group, utilizing Solargis data, with funding provided by the Energy Sector Management Assistance Program (ESMAP). For additional information: https://globalsolaratlas.info

6.2 Appendix

Table 3: General information about the weather stations used in the study

S/No	Station	Elev [m]	Clim	Temporal resolution	Time period	Invalid Hourly data
1	Tianguá / CE	745	As	1 min	2017	0.92%
2	Angicos / RN	145	BSh	10 min	2020	0.00%
3	Lajes / RN	209	BSh	10 min	2020	1.12%
4	Picuí / PB	619	BSh	10 min	2017	0.00%
5	Caldeirão Grande do Piauí / PI	774	BSh	1 min	2016-2017	0.00%
6	Araripina / PE	793	BSh	1 min	2016-2017	1.43%
7	Araripina / PE	799	BSh	10 min	2017	0.16%
8	Curral Novo do Piauí / PI	830	BSh	1 min	2016-2017	2.94%
9	Lagoa do Barro do Piauí / PI	683	BSh	1 min	2016-2019	1.48%
10	Rodelas / BA	478	BSh	1 min	2019-2020	0.01%
11	Dom Inocêncio / PI	710	BSh	1 min	2016-2017	0.02%
12	Campo Formoso / BA	934	BSh	10 min	2017	0.05%
13	Campo Formoso / BA	831	BSh	1 min	2020	1.13%
14	Xique Xique / BA	504	BSh	1 min	2019	2.01%
15	Xique Xique / BA	499	BSh	1 min	2016-2019	0.39%
16	Morro do Chapéu / BA	756	BSh	1 min	2016-2017	0.00%
17	São Gabriel / BA	612	BSh	1 min	2016-2017	0.85%
18	Itaguaçu da Bahia / BA	494	BSh	1 min	2016-2017	0.06%
19	Várzea Nova / BA	1014	BSh	1 min	2019-2020	0.33%
20	Jacobina / BA	815	BSh	1 min	2015-2017	3.14%
21	Uibaí / BA	1072	BSh	10 min	2017	1.23%
22	Ibipeba / BA	1066	BSh	10 min	2017	0.80%
23	Novo Horizonte / BA	1295	As	1 min	2019-2020	0.29%

Table 4: Metrics for satellite data analysis.

Metric	Expression	
MBE [%]	$\frac{\left(\frac{1}{n}\times\sum_{i=1}^{n}(x_i-y_i)\right)}{\left(\frac{1}{n}\times\sum_{i=1}^{n}(y_i)\right)} * 100$	(1)
RMSE [%]	$\sqrt{\frac{\left(\frac{1}{n}\times\sum_{i=1}^{n}(x_i-y_i)^2\right)}{\left(\frac{1}{n}\times\sum_{i=1}^{n}(y_i)^2\right)}} * 100$	(2)

U_{95} [%] $$\pm 2 * \sqrt{(U_{meas})^2 + \left(\frac{MBE}{2}\right)^2 + \left(\frac{RMSE}{2}\right)^2} \quad (3)$$

Where x_i represents the hourly means of data derived from satellite sources, y_i corresponds to high-accuracy hourly measurements obtained from weather stations, n stands for the number of verification samples and U_{meas} represents the estimated uncertainty for the surface measurement.

Table 5: General results about the weather stations used in the study by datasource

	PVGIS 5.2		PVGIS 5.2 WB		NSRDB V3.2.0		NSRDB V3.2.0 WB	
S/No	MBE [%]	RMSE [%]	MBE [%]	RMSE [%]	MBE [%]	RMSE [%]	MBE [%]	RMSE [%]
1	1.43	17.60	-0.28	17.62	1.65	19.78	-1.85	19.62
2	6.85	18.31	0.65	16.79	6.79	18.07	-0.37	16.09
3	4.25	15.14	-0.95	14.58	4.02	14.97	-3.60	14.22
4	3.14	16.45	-1.44	16.42	3.98	22.20	-5.03	21.67
5	-0.60	17.36	0.62	17.43	3.07	19.52	-1.42	19.09
6	-1.27	16.42	0.84	16.48	6.61	20.26	0.87	19.09
7	1.33	17.80	1.75	16.04	6.76	20.63	1.69	19.65
8	-3.84	17.92	0.54	17.95	4.49	18.51	-1.30	17.92
9	-0.28	14.16	-0.51	14.17	1.45	15.36	-2.61	15.38
10	1.86	14.61	0.73	14.52	6.82	16.59	1.87	15.40
11	2.34	15.54	-0.17	15.34	2.21	17.60	-0.84	17.32
12	1.07	14.70	0.46	14.67	6.59	21.23	-0.90	19.97
13	0.45	17.80	3.84	18.18	12.44	22.89	4.65	20.73
14	-1.26	11.64	-1.91	11.72	-2.03	11.01	-3.89	11.42
15	1.25	12.93	0.68	12.89	1.61	13.27	-0.14	13.13
16	1.21	15.70	1.58	15.73	5.84	17.98	0.80	16.92
17	0.32	14.97	1.13	15.03	5.79	17.44	2.05	16.44
18	-0.32	15.35	0.43	15.39	3.16	17.76	0.84	17.37
19	-1.36	15.88	1.01	15.84	7.38	17.66	0.52	16.76
20	-6.41	19.09	-2.51	18.65	1.56	20.83	-3.02	20.63
21	0.58	14.71	0.37	15.72	5.40	22.58	-0.63	21.73
22	1.96	14.61	2.97	14.71	8.78	17.99	4.35	16.80
23	-3.37	17.19	-1.77	16.99	3.42	16.32	-2.32	16.33

10.4229/EUPVSEC2023/4DV.4.3
020444-007

QUALITY CONTROL METHODS & GAP FILLING RULES FOR REAR-SIDE IRRADIANCE MEASUREMENTS ON LARGE-SCALE PV POWER PLANTS

Giuliano L. Martins*, Sarah Southern, Elena Koumpli, Anastasia Panoui and Jan Muller
Statkraft UK Ltd, 22 Bishopsgate, London, UK
*Corresponding Author: giuliano.luchettamartins@statkraft.com

ABSTRACT: Rear-side irradiance is a core component of bifacial photovoltaic (PV) power plants. The IEC 61724-1 has defined the bifacial Performance Ratio (PR_{bifi}) formula, which considers the rear-side irradiance contribution. However, quality control (QC) and gap filling rules for rear-side irradiance are scarce in the literature. This work presents and validates a novel QC method, where anomalous rear-side irradiance was identified efficiently. Four different gap filling methods were assessed, where alpha interpolation was found to be the most accurate one. In addition, this study highlights the high variability of albedo values within a single PV power plant, at least in its early operational stages. The impact of rear-side irradiance QC in the PR_{bifi} was assessed, showing that deviations around 5% can occur, which can be especially compromising if caused by wrongly installed sensors. The long-term effect of specific site issues (e.g. tracker stows) is left for future work, but initial results indicate that the impact on long-term PR_{bifi} tends to be minor.

Keywords: data quality, performance ratio, irradiance, solar resource

1 INTRODUCTION

Bifacial Photovoltaic (PV) systems are becoming the industry standard. By now, bifacial solar cells may already have reached more than 60% of the global market share [1]. Literature on bifacial PV has substantially increased in the last few years. Modelling rear-side irradiance is still one of the core challenges in the broader acceptance of bifacial PV.

Literature has shown, until now, that the single use of satellite data does not provide accurate outputs for bifacial PV systems simulations, however, combining multiple satellite sources assists in mitigating uncertainty [2]. In addition, ground measurement campaigns, when of at least 1-year period, improves uncertainty [3].

Modelling rear-side irradiance from front-side irradiance is still a challenge; different models have been presented and recent work on validation is showing promising results [4]. Although uncertainties on bifacial projects are still relatively high, such technology already is a significant part of the current and to-be-installed PV power plants.

Even though bifacial PV is an industry reality, rear-side irradiance has not been proposed a set of quality control (QC) metrics. BSRN QC standards [5] for front-side irradiance are broadly accepted, however, no official adaptations for rear-side irradiance are available. Data quality control is vital, both for research and industry, as rear-side irradiance contributes to the energy yield, especially at higher latitudes [6].

The IEC 61724-1 [7] defines the bifacial Performance Ratio (PR_{bifi}). Thus, such key performance index (KPI) may be used by the industry as a contractual guarantee; however, the lack of literature on quality control routines and gap filling rules, specifically on the rear-side irradiance, creates uncertainties on applying this KPI as a guarantee.

In this work, the authors adapted the physical limit to rear-side irradiance conditions, using the infinite sheds model [8]. Four different gap filling methods are tested and the potential impact of QC in the PR_{bifi} is evaluated.

2 METHODOLOGY

This section comprises detailed information about the dataset evaluated, the quality control method proposed, the assessed gap filling rules, and relevant formulas.

The dataset has a 15-minute time resolution and data points are filtered for daylight values, where zenith angles < 75°.

A set of 50 sensors was evaluated, which are installed throughout different European countries. The sensors were installed both for albedo measurement campaigns and at large-scale (> 30MWp) PV Power plants. The sensors were cleaned weekly. The data timespan varies depending on sensor and location, it ranges between 3 to 12 months.

2.1 Quality control method

Based on the physical limits for Global Horizontal Irradiance (GHI), Diffuse Horizontal Irradiance (DHI) and Direct Normal Irradiance (DNI), the novel method is proposed. These values are historically used to filter anomalous front-side irradiance data [9].

The authors propose an adaptation for the rear-side case, using the physical limits and the yearly albedo to create a maximum threshold.

For instance, it is recommended that the yearly albedo is multiplied with its expanded uncertainty, to mitigate effects of seasonal deviations, or satellite inaccuracy. Several studies have already been done to assess albedo expanded uncertainty in relation to satellite data, and can be used as a reference [10,11].

The values shall be inputted in the infinite sheds model, available in PVLib [8].

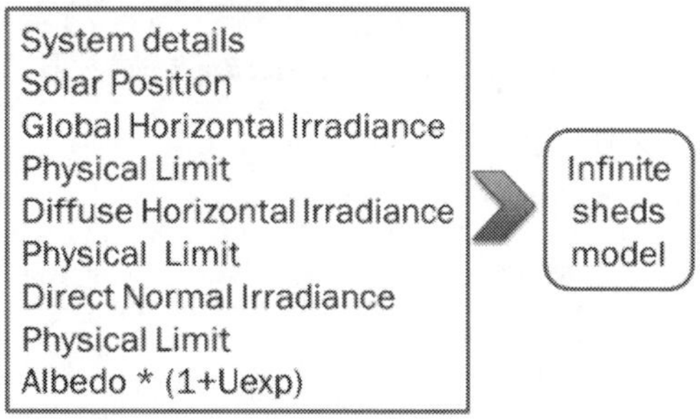

Figure 1: Relation of inputs required for modelling rear-side irradiance using infinite sheds model.

10.4229/EUPVSEC2023/4DV.4.5

The objective of the proposed method is not to identify minor deviations in data, for example, due to ground shading. However, to identify major outliers in relation to expected results, which may be caused by direct sunlight, as an effect of badly installed sensors.

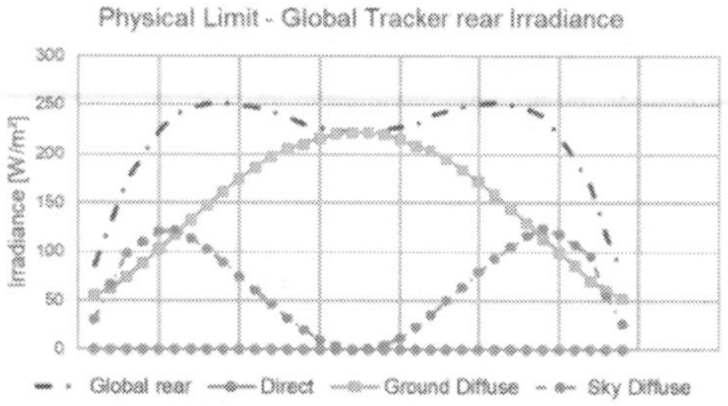

Figure 2: Modelled rear-side irradiance physical limit, showing each irradiance component stratified.

It is important to highlight that the physical rear-side irradiance limit may present a different shape from real measurements, especially from tracker rear-side data. It occurs because the DHI physical limit is much higher than the ground measured.

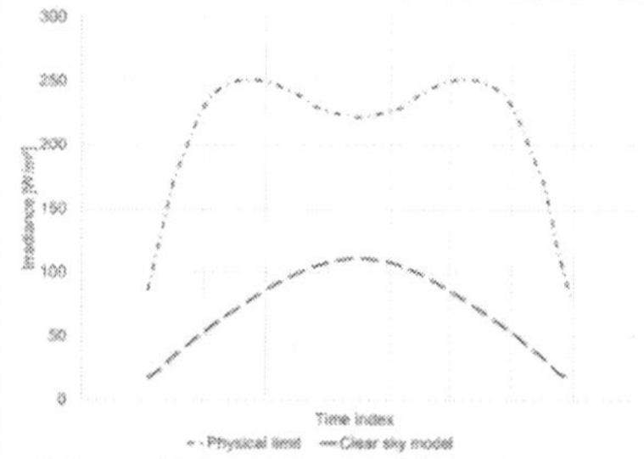

Figure 3: Rear-side irradiance modelled using physical limit and clear-sky inputs, for a tracker site.

2.2 Gap filling method

Two gap intervals will be assessed:

- 15min < x ≤ 60min
- 1h < x ≤ 8h

A Python script was written to generate random gaps within the specific time intervals.

After the data is gap filled, it is compared with the real measurement to assess the normalized Mean Absolute Error (nMAE) and the normalized Root Mean Square Error (nRMSE).

Four different gap filling methods will be evaluated:

- Linear interpolation of rear-side irradiance
- Average of remaining available rear-side irradiance sensors on site
- Alpha linear interpolation (using on-site sensor data)
- Alpha linear interpolation (using satellite data)

Where alpha (α) is defined as:

$$alpha\ (\propto) = {G_{rear}}/{G_{front}} \qquad \text{Eq.1}$$

In this case, G_{front} refers to GHI or GPOAI measured on-site or from satellite. Thus, the fill value will be calculated from the interpolated alpha multiplied by the device's measured irradiance.

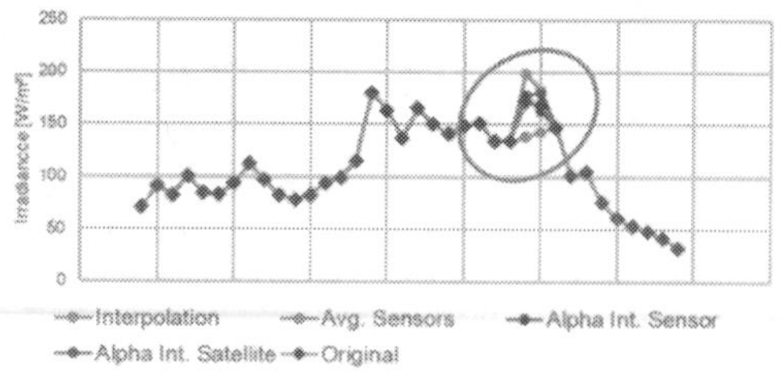

Figure 4: Example of each gap filling rule applied.

2.3 Performance ratio evaluation

The bifacial performance ratio (PR_{bifi}) defined in the IEC 61724-1 [7] will be evaluated when calculated with anomalous data (rawPR), and after treated and gap filled data (qcPR). The difference between the former and the latter is defined in this paper as varPR.

3 RESULTS & DISCUSSION

This section comprises the main results and findings from this work. It has a detailed discussion of the core reasons for the outcomes observed from the analyzed dataset.

3.1 Quality control method

The rear-side physical methodology showed itself efficient in identifying large deviations in rear-side irradiance data. Around 1% of inclined rear-side irradiance data was flagged as anomalous.

However, when analyzing horizontal rear-side irradiance most of the data was not flagged. The reason is that the method provides a very conservative threshold, therefore anomalous data caused by ground shading and small magnitude nearby reflections will not be flagged. Furthermore, due to the position of the sensors, direct sunlight will not be cast directly at the sensor.

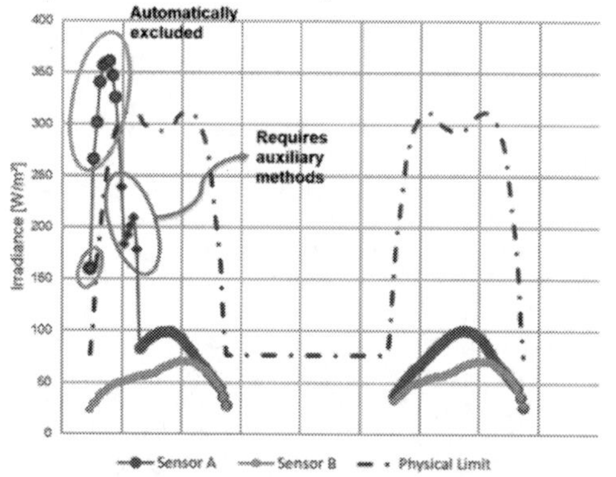

Figure 5: Illustration of quality control threshold identifying anomalous rear-side irradiance data.

The method proved itself efficient for highlighting days with anomalous data, and especially identifying sensors badly installed, thus affected by edge effect.

On the other hand, identifying anomalous data for rear-side irradiance is still a challenge, as it could be observed that further auxiliary methods are needed, to identify other anomalous conditions.

Rear-side irradiance measurement uncertainty and low magnitude of values are current challenges. However, rear-side irradiance modeling and nearby sensor comparison can be useful approaches. However, such an assessment is left for future work.

3.2 Gap filling results

The gap filling methods were evaluated, for the two proposed time intervals, and are presented in the table below.

Table I: Gap filling method errors for two gap intervals.

Error per gapfilling rule & Interval				
	15min < X ≤ 60min		1h < X ≤ 8h	
	nMAE	nRMSE	nMAE	nRMSE
Interpolation	8%	19%	20%	38%
Available Sensor average	21%	36%	21%	35%
Alpha Int. Sensor	7%	27%	17%	61%
Alpha Int. Satellite	9%	22%	18%	35%

From **Table I**, it is possible to observe that the alpha interpolation provided the most accurate results. However, it is noticeable that the nRMSE is high for all methods. This occurs due to the magnitude of the data, as small absolute deviations will generate high percentual deviations.

As expected, for linear interpolation related methods, the error increases for higher gap intervals. On the other hand, when using the average of remaining sensors the error is stable, independently of the gap interval.

The average of the remaining sensors presented high errors, which is explained by the high variation of albedo within a large-scale PV power plant.

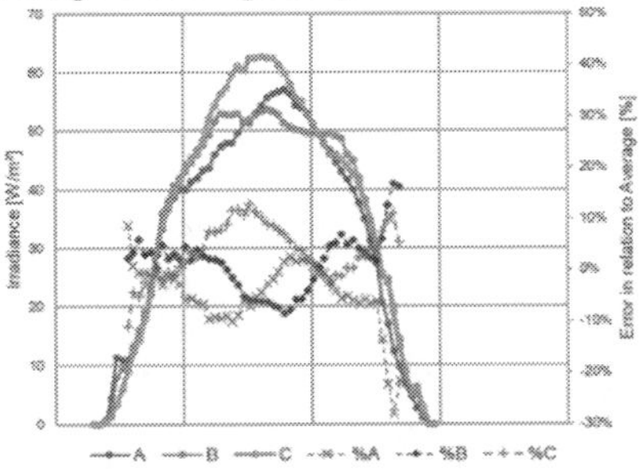

Figure 6: On-site measure rear-side irradiance data, and error relative to average. Sensors within one photovoltaic power plant.

The image above illustrates a day where three different rear-side sensors, installed in the same power plant, present significant deviations.

It is important to note that these are recently built power plants, and larger datasets are needed to ensure that such relations will be maintained over time. Thus, if the albedo of the plant will become more similar after years of operation.

3.3 Impact on Performance Ratio

As mentioned previously, the available dataset from large-scale PV power plants is recent, and thus it was not possible to perform an extensive assessment of the impact of quality control and gap filling of rear-side sensors in the PR_{bifi}.

However, it was possible to observe that systematical errors in sensor installation can have a long-term negative effect. The table and figure below illustrate the variation for one specific day.

Table II: Relation of daily bifacial performance ratio when rear-side irradiance data is treated or not.

rawPR bifi	68.87%
qcPR bifi	73.59%
varPR	4.72%

Therefore, if sensors are not correctly installed, the guaranteed PR can be negatively impacted, as data will not be treated. Which can be problematic if front-side irradiance is not treated either [12].

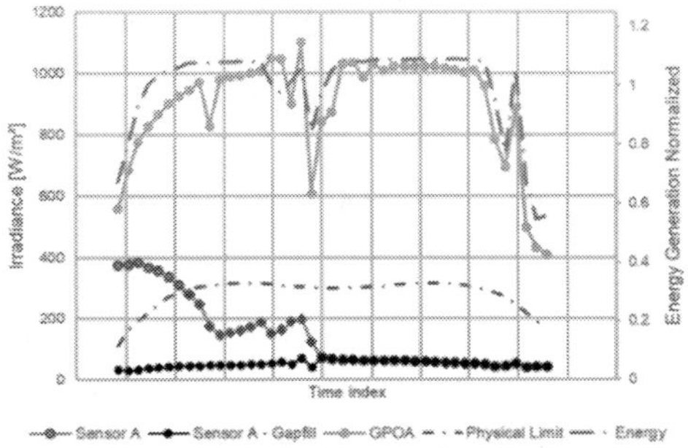

Figure 7: Plot showing raw and quality-controlled rear-side irradiance data, alongside the physical limit threshold and other relevant variables for performance ratio calculations.

On the other hand, due to the relatively small contribution of the rear-side irradiance data, when compared to the front-side, specific fault events, such as tracker stows, may have a negligible effect in the long term. However, a more extensive assessment, with a large dataset is needed to be more conclusive.

4 CONCLUSIONS

To conclude, this paper presents a novel rear-side irradiance quality control method, based on irradiance physical limits and albedo. The method proved itself efficient for flagging big deviations in rear-side irradiance data, however, new methods are still needed for identifying problems such as ground shading and nearby reflections.

Four different gap filling methods were evaluated, being the use of alpha, based on on-site sensor data, the most accurate of them. In addition, it could be observed that albedo varies significantly within a single large-scale PV power plant, at least in the first months of operation.

In addition, an assessment of PR_{bifi} was done. It can be already observed that wrongly installed rear-side sensors can have a systematic negative impact on PR calculations. Furthermore, a larger dataset is needed to assess the long-term effects of specific events on site, such as tracker stows.

5 REFERENCES

[1] IEA PVPS task 1, Masson G, Kaizuka I, Bosch E, Plaza C, Scognamiglio A, et al. Trends in Photovoltaic Applications — 2022. 2022.

10.4229/EUPVSEC2023/4DV.4.5
020445-003

[2] Blakesley JC, Koutsourakis G, Douglas S, Holder JKL, Torry J, Mukadam FA, et al. Effective Spectral Albedo from Satellite Data for Bifacial Gain Calculations of PV Systems. 37th Eur Photovolt Sol Energy Conf Exhib 2020:1292–7.
[3] Lara-Fanego V, Gueymard CA, Cebecauer T, Suri M. EFFECTIVENESS OF SHORT GROUND-BASED MEASUREMENT CAMPAIGNS OF SURFACE ALBEDO, n.d., p. 999–1003. https://doi.org/10.4229/WCPEC-82022-4BO.16.4.
[4] Riedel-Lyngskær N, Berrian D, Alvarez Mira D, Aguilar Protti A, Poulsen PB, Libal J, et al. Validation of Bifacial Photovoltaic Simulation Software against Monitoring Data from Large-Scale Single-Axis Trackers and Fixed Tilt Systems in Denmark. Appl Sci 2020;10:8487. https://doi.org/10.3390/app10238487.
[5] Ohmura A, Gilgen H, Hegner H, Müller G, Wild M, Dutton EG, et al. Baseline Surface Radiation Network (BSRN/WCRP): New Precision Radiometry for Climate Research. Bull Am Meteorol Soc 1998;79:2115–36. https://doi.org/10.1175/1520-0477(1998)079<2115:BSRNBW>2.0.CO;2.
[6] Martins GL, Dittmann S, Rodríguez-Gallegos CD, Park H, Oh S-Y, Kim WK, et al. Energy Contribution of Rear-Side Irradiance for Bifacial Photovoltaic Modules, 38th European Photovoltaic Solar Energy Conference and Exhibition; 2021, p. 556–9. https://doi.org/10.4229/EUPVSEC20212021-4BO.2.4.
[7] International Electrotechnical Commission. IEC 61724-1:2021 Photovoltaic system performance - Part 1: Monitoring n.d.
[8] Mikofski MA, Darawali R, Hamer M, Neubert A, Newmiller J. Bifacial Performance Modeling in Large Arrays. 2019 IEEE 46th Photovolt. Spec. Conf., IEEE; 2019, p. 1282–7. https://doi.org/10.1109/PVSC40753.2019.8980572.
[9] Long CN, Dutton EG. BSRN Global Network recommended QC tests, V2. n.d.
[10] Marion B. Measured and satellite-derived albedo data for estimating bifacial photovoltaic system performance. Sol Energy 2021;215:321–7. https://doi.org/10.1016/j.solener.2020.12.050.
[11] Lara-Fanego V, Gueymard CA, Ruiz-Arias JA, Cebecauer T, Betak J. Extensive Evaluation and Uncertainty Estimation of Albedo Data Sources. Conf Rec IEEE Photovolt Spec Conf 2022;2022-June:945–7. https://doi.org/10.1109/PVSC48317.2022.9938504.
[12] Martins GL, Koumpli E, Southern S, Panoui A, Muller J. IMPACT OF DATA QUALITY ON THE PERFORMANCE EVALUATION OF LARGE-SCALE PV PLANTS, Milan: 2022. https://doi.org/10.4229/WCPEC-82022-4BV.4.24.

6 ACKNOWLEDGEMENTS

The authors acknowledge funding from the Horizon 2020 programme of the European Union under GA. No. 952957-Project TRUST PV. Would also like to acknowledge Statkraft for the site data.

SHORT-TERM GROUND-BASED ALBEDO MEASUREMENTS FOR LOW-UNCERTAINTY ENERGY YIELD ESTMATIONS IN UTILITY-SCALE BIFACIAL PHOTOVOLTAIC SYSTEMS

P. Merodio[1,*], F. Martínez-Moreno[1], R. Moretón[2], E. Lorenzo[1]

[1]Instituto de Energía Solar, Universidad Politécnica de Madrid (IES-UPM)
[2]Qualifying Photovoltaics (QPV)
*Corresponding author: pablo.merodio@upm.es

ABSTRACT: This research analyses the uncertainty issues associated to different short-term ground-based albedo measurement procedures, looking for deriving recommendations for albedo measurements in the frame of commercial photovoltaic plants. The relation between albedo measurement uncertainty and subsequent energy yield estimations uncertainty is established via the albedo sensitivity coefficient. The latter is analyzed through a representative yield simulation exercise performed for three different sites (Chile, Spain and Sweden) and two types of photovoltaic arrays: static and single-axis tracker. A formula for considering the impact of the Ground Covering Ratio and Inverter Loading Ratio is also proposed.

An experimental campaign allowed to observe the relation between the albedo measurement and the associated uncertainty in the corresponding measurements. The results of this campaign led to 6 measurement recommendations concerning important variables such as measurement time and maximum solar zenith angle.

Furthermore, a representative case of albedo measurement practises at the current photovoltaic market scene is also presented in the form of a commercial albedo measurement project. Additional uncertainty sources relevant for utility-scale projects are identified.
Keywords: albedo, bifacial, uncertainty, PV performance modelling

1 INTRODUCTION

The deployment of bifacial photovoltaic (PV) modules for utility-scale PV plants has quickly proliferated in recent years. As they can harness the irradiance received at the rear side of bifacial generators -which depends on the albedo of the reflector ground-, bifacial modules provide higher productivity than monofacial modules, in exchange for a low increase in manufacturing costs. However, the reduction of PV performance estimations uncertainty still represents a main concern for both PV and financial actors, which seek to reduce financial risks of bifacial PV systems.

Despite this need, best practices are not yet completely defined, and proper procedures for albedo measurement are still under debate [1]. For instance, the ASTM E1918-21 norm [2] establishes a standard test method for measuring solar reflectance in the field, but it is not especially designed for the PV energy community and imposes certain measurement restrictions that are questionable.

The irradiance received at the rear side of bifacial generators, G_{REAR}, depends on the albedo of the reflector ground, ρ, whose value is a key input parameter for energy yield estimations. ρ is usually measured by two paired horizontal pyranometers: one facing up towards the sky and one facing down towards the ground. The resulting albedo is calculated dividing the ground-facing pyranometer output measuring the reflected irradiance, $G^{\uparrow}(0)$, by the sky-facing pyranometer output measuring the incident irradiance, $G^{\downarrow}(0)$.

$$\rho = \frac{G^{\uparrow}(0)}{G^{\downarrow}(0)} \quad (1)$$

where the zero in parentheses denote the tilt angle, that is, horizontal surfaces.

Because PV modules are angular and spectrally sensitive, G_{REAR} needs to be adjusted for angle of incidence losses and spectral mismatch. Current PV software usually includes some model for the Incidence Angle Modifier, IAM [3], but not for the spectral correction. Hence, an appealing alternative is to measure $G^{\uparrow}(0)$ not with a pyranometer but with a reference solar cell. This way, resulting ρ corresponds only to the relevant part of the spectrum [4] and accounts for the field of view of the PV modules.

The aim of this work is to provide a better understanding of the impact that albedo measurements have on PV yield estimations uncertainty in order to significantly reduce it. For that, we have proceeded along two complementary lines. On the one hand, we have simultaneously measured albedos with pyranometers and with reference solar cells, which allows to obtain three useful different ρ values, depending on the sensors chosen to measure $G^{\uparrow}(0)$ and $G^{\downarrow}(0)$. We have carried out two different experimental testing campaigns: i) In-lab tests on the IES-UPM rooftop, performed using samples of surfaces covering a wide range of albedo values (artificial grass -polypropylene fibres on a latex matrix-, conventional gravel, white gravel, a high-reflective plastic layer -PVC-P- and snow). Here, we have thoroughly analysed the influence of the testing conditions. ii) In-field tests on terrains where real PV plants are planned to be constructed. Hence, the measurements have been performed with natural soils, assessing additionally the effects of surface moisture, the non-uniformity of the ground along the full PV plant site, and the annual evolution of the surfaces' properties. On the other hand, we have performed a thorough simulation exercise, addressing static and single axis tracking PV arrays at three different locations (Chile, Spain and Sweden), to analyse the sensibility of annual energy yield estimations to albedo values. The observed albedo variations during the experimental campaigns are understood as the main uncertainty source in the albedo results from short-term ground-based measurement campaigns (typically, few hours). Finally, the uncertainty in albedo is translated into uncertainty in annual energy yield estimations, which

10.4229/EUPVSEC2023/4DV.4.6
020446-001

provides the basis for proposing albedo measurement requirements and recommendations.

2 ALTERNATIVE DEFINITIONS FOR ALBEDO

Figure 1 shows the experimental arrangement for measuring the albedo at the rooftop of the IES-UPM building. It includes a pair of pyranometers and a pair of reference solar cells for measuring both $G^{\uparrow}(0)$ and $G^{\downarrow}(0)$.

Figure 1: (a) Albedometer setup and dimensions for a white gravel ground surface. (b) Close-up of the irradiance sensors.

That allows for calculating four different albedo values. The ratio between the measure of two pyranometers is the standard and today most referred ρ value. The ratio between the measure of a reference cell for $G_{RC}^{\uparrow}(0)$ and the measure of a pyranometer for $G_{PYR}^{\downarrow}(0)$ gives a ρ value adjusted for spectral mismatch and angle of incidence sensor losses (soiling is not included because clean sensors is a condition to validate ρ measurements). By analogy with nomenclature commonly used for irradiance, we denote these two ρ values as 'global' and 'effective', ρ^{G} and ρ^{E}, respectively. It is worth noting that using ρ^{E} as input for simulation exercises excludes the need of considering angular and spectral corrections in the calculation of the effective rear irradiance. Moreover, to the best of our knowledge, today PV simulation software typically considers angular but not spectral correction in G_{REAR}. Apart from that, a solar cell is much low-priced than a pyranometer. For the same reason, the ratio between the measure of two solar cells, that we denote ρ 'photovoltaic', ρ^{PV}, is the most cost-effective value to measure. Finally, the ratio between the measure of a reference cell for $G_{RC}^{\downarrow}(0)$ and the measure of a pyranometer for $G_{PYR}^{\uparrow}(0)$ does not represent any advantage either in terms of meaning or in terms of cost. Hence, we have disregarded this alternative.

3 ALBEDO UNCERTAINTY PROPAGATION ON ENERGY YIELD ESTIMATIONS

Looking for understanding the impact of albedo uncertainty, $u(\rho)$, on the uncertainty of annual energy yield estimations, $u(E_{AC})$, we have performed a comprehensive simulation exercise to derive the corresponding sensitivity coefficient, c_ρ. According to the Guide to the Expression of Uncertainty in Measurements [5], c_ρ is calculated as the change in the output variable, $\delta(E_{AC})$, when changing the value of the input variable, $\delta(\rho)$, while keeping the remaining variables constant, and is described [6] by the partial derivative of E_{AC} with respect to ρ:

$$c_\rho = \frac{\delta(E_{AC})}{\delta(\rho)} = \frac{\partial E_{AC}}{\partial \rho} \quad (2)$$

Given the linear dependence of G_{REAR} -and, therefore, $G_{TOTAL} = G_{REAR} + G_{FRONT}$- on ρ, we have (see **¡Error! No se encuentra el origen de la referencia.**):

$$E_{AC}(\rho) = E_{AC}(0)[1 + c'_\rho \cdot \rho] \quad (3)$$

where -as it is convenient to deal with values relative to $E_{AC}(0)$, the energy yield at the inverter output for the bifacial PV system with zero albedo- we define the relative sensitivity coefficient, c'_ρ, as

$$c'_\rho = \frac{c_\rho}{E_{AC}(0)} \quad (4)$$

Strictly speaking, the relative sensitivity coefficient should be defined with respect to the actual total yield $E_{AC}(\rho)$ instead of $E_{AC}(0)$. Nevertheless, our definition benefits from two important advantages: 1) It does not depend on albedo, which makes it much easier to handle and 2) It provides a conservative value for the sensitivity coefficient that overestimates the uncertainty in energy yield estimations, which ensures that actual errors are lower than those estimated in this work.

Note that, since ρ is dimensionless, c_ρ has the same units as E_{AC}, while c'_ρ is also dimensionless. Hereafter, c'_ρ is expressed in % per 0.01 of ρ in a scale of 0.00 to 1.00.

Seeking representativity, we have analysed c'_ρ through energy yield simulations comprising three sites of different latitudes (see Table I, which summarizes the relevant characteristics of each site's solar climate) and two types of PV array: static and single axis tracking -with activated backtracking-, both in 2P (portrait) configuration, placed over horizontal terrain. The height of both bifacial arrays (at the midpoint and at the axis for the static and tracker structures respectively) has been set to 2.3 m.

Table I: Solar climate characteristics of the three analysed sites. The quotient $D_y(0)/G_y(0)$ indicates the yearly diffuse to global irradiance ratio on a horizontal surface.

	Lat.(°)	Gy(0)(kWh·m^{-2})	Dy(0)/Gy(0)
Chile	-24.0	2.680	0.15
Spain	39.9	1.786	0.32
Sweden	57.8	1.017	0.48

The Typical Meteorological Year for each site has been obtained either from PVGIS or Solargis. For each site

10.4229/EUPVSEC2023/4DV.4.6
020446-002

and type of PV array, we have first simulated a baseline scenario, defined by selected values for main features of the PV system, namely: DC/AC or Inverter Loading Ratio, ILR_0, Ground Coverage Ratio, GCR_0 and, for the case of static PV arrays, the tilt angles β_0. Then, the simulations have been extended for each configuration to the ranges $0.10 \leq \rho \leq 0.90$ (keeping constant the corresponding baseline ILR_0 and GCR_0), $0.84 \leq ILR \leq 1.50$ (keeping constant ρ_0 and GCR_0) and $0.30 \leq GCR \leq 0.80$ (keeping constant ρ_0 and ILR_0). The relevant characteristics of the PV module are: temperature coefficient of power, $\gamma = 0.35$, and bifacial power coefficient, $\varphi_P = 0.70$, which are typical of current prevailing technology (Si-p, PERC). Table II summarizes the baseline features for each scenario, denoted with the subscript "0".

Table II: Baseline simulation parameters for the PV plants considered in this paper.

		Static		Tracker
	ILR_0	β_0(°)	GCR_0	GCR_0
Chile	0.8	20	0.6	0.4
Spain	0.8	25	0.5	0.4
Sweden	0.8	30	0.4	0.4

The module tilt and GCR_0 for the fixed structures have been optimized using a good compromise for the North-South pitch, accounting for economic considerations. All the scenarios imply little losses due to both mutual shading and inverter clipping. In the future, assuming that PV prices decline will continue, GCR and ILR are expected to rise[1], which will decrease the influence of ρ in energy production, thus lowering c'_ρ. A formula is proposed below to consider this effect. Moreover, technological changes from current PERC to TopCon or HJT technologies will increase φ_P and consequently, c'_ρ. Since the relationship between E_{AC} and φ_P is essentially linear, a reasonable approximation for corresponding c'_ρ variations is given by

$$c'_\rho(\varphi_{P,2}) = c'_\rho(\varphi_{P,1})\frac{\varphi_{P,2}}{\varphi_{P,1}} \tag{5}$$

Simulations have been performed with SISIFO, an open-source PV simulation software (basic version available at https://sisifo.info) developed by the IES-UPM.

For the sake of illustration, Figure 2 plots E_{AC} and c'_ρ versus (a) ρ, (b) GCR and (c) ILR for the case of the 2P tracker located at Chile without applying the angular correction.

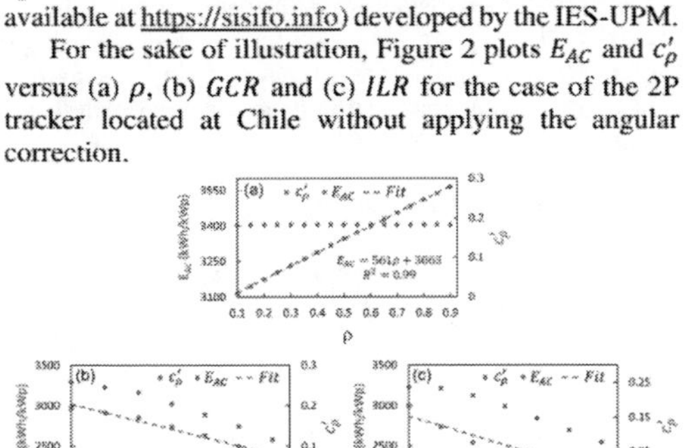

Figure 2: (a) Energy yield E_{AC} and sensitivity coefficient c'_ρ as a function of ρ. A linear fit for E_{AC} is provided. (b) E_{AC}, c'_ρ and a linear fit for the latter as a function of GCR. (c)E_{AC}, c'_ρ and a linear fit for the latter as a function of ILR.

The fits of c'_ρ in Figure 2 (b)-(c), with $R^2 = 0.99$ and $R^2 = 0.93$ respectively, show also that the dependence of c'_ρ on GCR and ILR can be considered linear to a very good approximation. Therefore, it is easy to expand c'_ρ in a first-order Taylor series about its value for each particular baseline case (characterised by GCR_0 and ILR_0):

$$c'_\rho = c'_{\rho,0} + a(GCR - GCR_0) + b(ILR - ILR_0) \tag{6}$$

where $c'_{\rho,0} = \frac{1}{E_{AC(0)}} \cdot c_\rho(GCR_0, ILR_0)$, $a = \frac{1}{E_{AC(0)}} \cdot \frac{\partial c_\rho}{\partial GCR}(ILR_0)$ and $b = \frac{1}{E_{AC(0)}} \cdot \frac{\partial c_\rho}{\partial ILR}(GCR_0)$ are easily estimated from the slopes of the linear fits shown in Fig. 2 (b)-(c) and the y-intercept of Figure 2 (a). Table III provides $E_{AC}(0)$ and c'_ρ defining parameters of Eq. (6) for the here considered cases, where $c'_{\rho,0}$ corresponds to % per 0.01 ρ and $E_{AC}(0)$ is measured in kWh/kWp.

Table III: Energy yield for a bifacial PV plant for $\rho = 0$, $E_{AC}(0)$, and the Taylor coefficients of Eq.(6) for the six PV systems simulated in this paper.

		$E_{AC}(0)$	$c'_{\rho,0}$	a	b
Chile	Static	2390	0.20	-0.26	-0.21
	1×h	3063	0.18	-0.24	-0.27
Spain	Static	1739	0.24	-0.31	-0.16
	1×h	2063	0.22	-0.23	-0.23
Sweden	Static	1056	0.30	-0.34	-0.11
	1×h	1228	0.26	-0.19	-0.15

In addition to the output energy yield for any design, $E_{AC}(\rho, GCR, ILR)$, these results and Eq. (6) allow to determine the associated uncertainty, $u_{E_{AC}}(GCR, ILR)$, which is a main objective of this paper. Indeed, considering that $E_{AC}(0)$ and ρ are independent, their corresponding uncertainties $u_{E_{AC}(0)}$ and u_ρ are independent too -that is, we assume that the uncertainties have different uncorrelated sources [7]-, which enables us to apply the rule of squares to compute the resulting combined uncertainty:

$$u_{E_{AC}} = [u^2_{E_{AC}(0)} + [c'_\rho u_\rho]^2]^{1/2} \tag{7}$$

where the uncertainty component of E_{AC} due to the albedo is:

$$u^\rho_{E_{AC}} = c'_\rho u_\rho \tag{8}$$

Continuing with the example of the tracker located at Chile for the baseline scenario, u_ρ= 0.01 results in $u^\rho_{E_{AC}}$ = 0.18%. That is, the uncertainty in ρ ($100 \cdot \frac{0.01}{0.3} = 3.3\%$) is divided by 18 (3.3/0.18 ≈18) when translated to the albedo contribution to the uncertainty in E_{AC}. It is anticipated that such significant uncertainty reduction allows for a wide margin of freedom regarding albedo measurements conditions into the frame of PV engineering practices.

[1]For instance, ILR for commercial PV plants is usually in the range [1.1, 1.5].

4 CHARACTERIZATION OF THE ALBEDO

An outdoor testing campaign was performed between 19 February 2019 and 10 May 2021 at the IES-UPM building's rooftop, whose coordinates are: 40.4° N, 3.6° W, 660 m above sea level, to characterize the surface albedo of different materials: artificial grass, conventional gravel, white gravel, a high-reflective plastic layer and snow. The experimental arrangement is shown in Figure 1.

As a representative example, Figure 3 shows the evolution of ρ^G, ρ^E and ρ^{PV} for a test performed a clear day the 5 May 2021 with a white gravel ground, along with these ρ values versus θ_{ZS} for the range $\theta_{ZS} \leq 70^0$, which is a generally applied filter for solar radiation measurements [8]. Time evolution of $G^{\downarrow}_{PYR}(0)$ measured by the pyranometer is also shown in (a), being around 970 W/m^2 at noon.

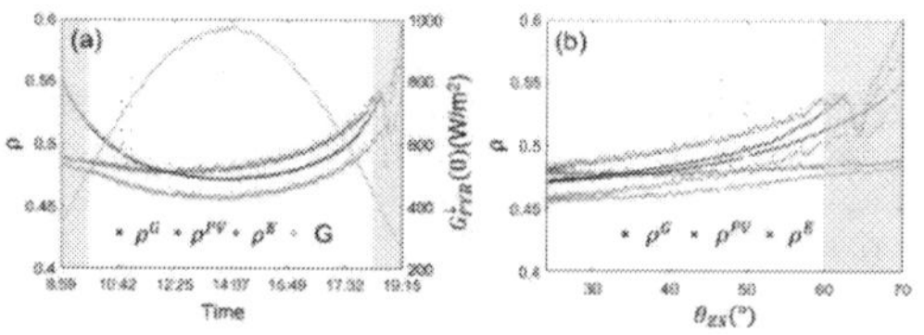

Figure 3: (a) Time evolution for ρ^G, ρ^E, ρ^{PV} and ρ^{PV} *and* $G^{\downarrow}(0)$measured the 05/05/2021 for a white gravel ground and h = 100 cm. (b) ρ^G, $\rho^E$and ρ^{PV}as a function of θ_{ZS}. Intervals for θ_{ZS}≥60 are grey shaded in both graphs. The spurious dip at $\theta_{ZS} \approx 65°$ is caus by the shade casted by a nearby PV generator.

This kind of experimental results allows to answer some questions of practical relevance. The crux consists of relating the u_ρ resulting from albedo variations, $\Delta\rho$, with the corresponding $u^{\rho}_{E_{AC}}$, by means of the c'_ρ values calculated in the previous section. Each application might require its own limit for this uncertainty component, depending on the precision needed for subsequent energy yield estimates. Hereafter, we establish the limit $u^{\rho}_{E_{AC}} \leq$ 1.5%, which implies an increase in $u_{E_{AC}}$ from 5% to 5.2%. Practical recommendations for ρ measurements can then be derived, accounting for the corresponding limit allowed for u_ρ. It must be understood that the corresponding error in E_{AC} comes from the error between the measured ρ value used as input for the E_{AC} simulation exercise, and the "true" *daily* ρ. Since the final purpose of our work is the calculation of the total energy produced by a PV plant, the "true" ρ value must be the ratio between reflected and incident horizontal *irradiations* received during the considered time interval. Since current PV software rely on the assumption that ρ remains constant along the day, irrespective of the solar radiation peculiarities, the "true" ρ is given by $[G^{\uparrow}_d(0)/G^{\downarrow}_d(0)]$, where the subscript d indicates the irradiance integration period of one day. However, measuring this value requires a full day, which is a too long period for practical purposes; it is necessary to select a ρ value corresponding to a time interval shorter than a day. Hence, an interesting question arises concerning the measuring time: when and how long to measure. The observations throughout our testing campaign, described below, help answer this question.

4.1 Measurement time

A first observation is that, for the same θ_{ZS}, forenoon values of the albedo are slightly lower than the values after noon. This asymmetry in the diurnal variation of surface albedo has already been reported [9] and was attributed to the effect of multiple factors, namely dew, frost, precipitation, evaporation, and wind. However, the impact of such asymmetry in E_{AC} estimations is frankly low. For example, $\rho^E(\theta_{ZS} = 45^o) = 0.47$ before noon and 0.48 after noon. That is, a difference of $\Delta\rho \equiv u(\rho) = 0.01$, which leads to $u^{\rho}_{E_{AC}} \leq$ 0.3% for all the simulated cases. For the rest of materials this asymmetry is even smaller. These results suggest that it does not matter measuring before or after noon.

Figure 4 (a) plots the resulting ρ versus the length of the period Δt_ρ, always centred at noon, along which irradiances are integrated to determine the albedo. The resulting 10-minute centered moving average values, CMA10, are provided at noon as a reference, for the three albedos: $\rho^G(\Delta t_\rho = 0) = 0.48$, $\rho^{PV}(\Delta t_\rho = 0) = 0.47$ and ρ^E $(\Delta t_\rho = 0) = 0.46$.

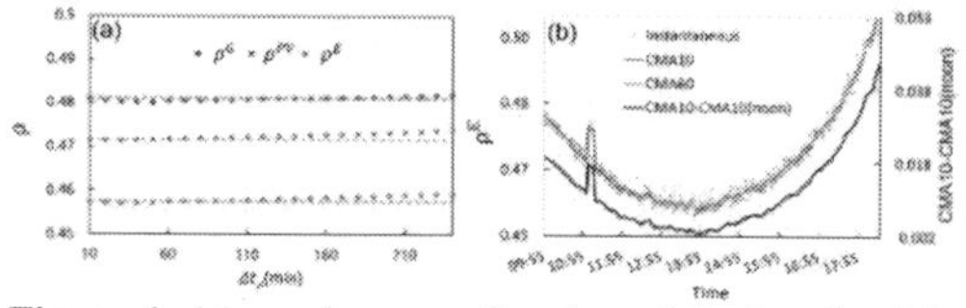

Figure 4: (a) ρ values as a function of the length of the period Δt_ρ, along which irradiances are integrated (dots). Corresponding CMA10 noon albedos are provided as a reference (dashed lines). (b) Instantaneous (green dots), CMA10 (green solid line), CMA60 (yellow solid lines) and the difference CMA10-CMA10(noon) (black solid line) ρ^E values VS time.

The good news is that the deviations from these values for $\Delta t_\rho \leq 4h$ are $\Delta\rho(\Delta t_\rho) \leq 0.002$ making $u^{\rho}_{E_{AC}}$ negligible. For the rest of materials, $\Delta\rho(\Delta t_\rho)$ when Δt_ρ goes from 10 min to 4h ranges from 0.00 to 0.01, making $u^{\rho}_{E_{AC}} \leq 0.3\%$ for the six simulated configurations. These results suggest that short measuring times suffice. Figure 4 (b) shows CMA10 and CMA60 ρ^E values. CMA20 and CMA30 have been analysed too, but as in the case of CMA60, no significant differences are apparent. For this reason, and seeking for practicality, we propose using the CMA10 as a recommended low-pass filter for avoiding unwanted noise, so that the minimum measuring time is just 10 minutes.

4.2 Maximum solar zenith angle

On the same lines, the maximum deviation between the $CMA10(t)$ at a time t and the $CMA10(noon)$ at noon, $\Delta\rho = \Delta CMA10(t) = max|CMA10(t) - MA10(noon)|$ depends on the restriction imposed to the solar zenith angle θ_{ZS}, in the form of its maximum value allowed for measurements at time t. Table IV shows these *measured* deviations for ρ^{PV}, which exhibited the maximum differences with respect to its values at noon, for three θ_{ZS} values, for the white gravel, which is the surface that showed the highest differences between noon albedo and the rest of albedos. Corresponding $u^{\rho}_{E_{AC}}(\%)$ are also provided for the three locations analyzed in this paper, and for the worst-case scenarios, that is, static structures with the highest c'_ρ.

Table IV: Albedo variations -$\Delta\rho^{PV} = \Delta CMA10_t$- from noon up to θ_{ZS}= 45°, 60° and 70°, and the associated yield uncertainty components for static PV arrays located at the three sites assessed.

$\max(\theta_{ZS})$ (°)	$\Delta\rho^{PV}$	$u^{\rho}_{E_{AC}}$(%)		
		Chile	Spain	Sweden
45	0.02	0.4	0.5	0.6
60	0.06	1.2	1.5	1.8
70	0.12	2.4	2.9	3.6

These results led us to propose $\theta_{ZS} = 60^0$ as the maximum angle allowed for albedo measurements. It is less restrictive than that of the ASTM norm, enlarging the yearly time availability for albedo measurements, yet it implies low $u^{\rho}_{E_{AC}}$. The increase from 45° (ASTM standard) to 60° may seem small, but at mid-latitudes like Madrid (41° N), this means doubling the daylight time available for measurement. We stressed that these uncertainties have been obtained for the most conservative cases, namely white gravel, c'_ρ with respect to $E_{AC}(0)$, ρ^{PV} and static structures. $\theta_{ZS} = 60^0$ can therefore be considered as a safe and meaningful limit entailing measurement flexibility and low impact on the yield uncertainty. Note that $u^{\rho}_{E_{AC}}$ for Sweden slightly exceeds the established limit of 1.5. However, although multi-megawatt-scale bifacial PV projects are starting to be deployed in latitudes as far north as Denmark (56° N) [3], this is still an extreme case that does not represent the majority of PV plants installed.

4.3 Instrumentation

$CMA10(noon)$ measured values of the three types of albedo for the different materials were further used to compute their maximum differences and the corresponding maximum $u^{\rho}_{E_{AC}}$for each simulated site. Except for snow, the derived maximum uncertainties $\max(u^{\rho}_{E_{AC}})$ are lower than 1% for the three simulated sites. This indicates that it does not matter to measure with reference solar cells or with pyranometers. Nevertheless, it is opportune remembering that, because current PV software does not include spectral corrections, ρ^E should be preferably used without further angular correction as input data for yield simulation exercises.

4.4 Sensor's height above the ground

We have also analysed the influence of the albedometer sensor's height above the ground on the albedo measurement, and we have found that, for the range of heights analysed, it is negligible, especially concerning subsequent yield uncertainty. For instance, modifying the sensors height from 100 to 58.7 cm led to $\Delta\rho^G$=0.005 for the white gravel, which would result in $u^{\rho}_{E_{AC}} \leq 0.2\%$.

As a final note, the generalization of the results to other locations is based on the relation between observed ρ and θ_{ZS}, which can reasonably be assumed as latitude independent. Nevertheless, a warning must be made, since, as all the authors' experimental experience has been gained in latitudes below 45°, the application of the recommendations of this article to higher latitudes should be taken with caution. Indeed, as albedo is a complex surface characteristic that is also function of actual illuminating and atmospheric conditions, although unlikely its specific time dependence could significantly differ from the behaviour described in this paper.

5 FIELD MESUREMENTS FOR UTILITY-SCALE BIAFACIAL PV PROJECTS

In the framework of several commercial agreements, the PV engineering firm QPV has carried out a number of albedo measurement campaigns, implementing the four-sensor procedure described in this paper. Overall, from May 2019 to October 2020, the albedo has been characterised in-field for 52 PV projects, totalling 137 locations and 1.91 GW of future bifacial PV plants.

A first consideration is that ρ is usually demanded to be measured *before* the construction of the PV plant; thus, the resulting values can only be approximations of the actual albedo that will characterise the future PV system, as civil work and land use change can strongly modify the features of the ground where the PV plant is going to operate. Not to mention other natural changes that might alter the reflective properties of the ground through the year with season, due to vegetation growth[2], etc.

The procedure followed by QPV for characterizing the albedo of a power plant site begins by selecting specific spots on the ground that are representative of the diversity of conditions that can be found throughout its extension. This meets the ASTM standard, which specifies that to adequately represent the solar reflectance of an inhomogeneous site, measurements must be conducted at a minimum of three widely spaced test surfaces. It is also in agreement with the IEC 61724-1:2021 norm, which requires to choose the location of the sensors in such a way that it is representative of the albedo of the site.

As a general requirement of project owners, each measurement is performed, at a 1.5 m height above the ground, because it is representative of the height at which the torque tube axes are usually installed in the case of a 1P tracker, or the lower module in the case of a 2P tracker (which additionally meets the IEC 61724-1:2021 standard). At one of the spots, considered as "primary", additional measurements are carried out, varying the sensors' height to 0.5 m, in agreement with the ASTM norm. Furthermore, the effect of ground humidity is also checked at the primary spot, by measuring the albedo after abundantly water-spraying an area of the ground that covers an angle of vision from the sensors of at least 60°.

In general, albedo is measured when $G^{\downarrow}(0) \geq 600\frac{W}{m^2}$, for at least one hour at the primary location and 30 min at the rest of the spots, using a 5 s sampling. Albedo is then calculated as the ratio of the upward to the downward *irradiations*. Note that these conditions are somewhat more restrictive than those recommended above.

5.1 Spatial variations

As a representative example, Figure 5 shows the location of the spots selected to measure the albedo for a

[2]In general, PV plants tend to favour vegetation growth at arid zones, for two reasons: on the one hand, PV modules are efficient dew condensers, providing water, and, on the other hand, their shade reduces evapotranspiration.

10.4229/EUPVSEC2023/4DV.4.6
020446-005

future 50 MW bifacial PV plant that is being developed at Monforte del Cid (Alicante, Spain, Latitude = 38.5°N). The one identified as "Spot 1" has been declared the primary spot, while the other two have been considered secondary spots.

Figure 5: Picture with a Google Earth satellite image, showing the location of the points selected for the albedo measurements. The green contour line shows the boundary of the land of the future plant. Each spot comes with an inset showing an image of the specific ground and the experimental setup, taken in July 2022.

This is a particularly interesting example, since measurements for this project were performed once a month for a period of 6 months, from 28 May 2021 to 17 November 2021, which allowed to assess the effect of the seasonal evolution of the vegetation cover on albedo.

The outstanding result is that, by far, the main uncertainty source is the spatial variation, which gives $u^{\rho}_{E_{AC}}$ values comparable to $u_{E_{AC}(0)}$. It should be noted that this result is site-specific peculiarity that can vary a lot from one site to another. In principle, that can be counterbalanced by using the ρ spatial-weighted average, being the surface of each type of ground the weighting factor. However, we expect that land use change will entail an uncertainty of the same order. A sensible alternative would be to measure the albedo in nearby PV plants -even monofacial- that are already operational and lie over stabilized ground, even though we recognize that it is likely very difficult to fit into current commercial practises.

The albedo variation when the height of the sensors is reduced to 0.5 m, to meet the ASTM standard is $\Delta\rho = 0.01$, which has a negligible effect on the yield uncertainty $u^{\rho}_{E_{AC}} \leq 0.3\%$. The height of the sensors is therefore unimportant for measurements, at least for the range $0.5m \leq h \leq 1.5m$. On the other hand, wetness of the ground lowers the albedo, which will affect rear irradiance in the periods with greater frequency of rains. The measured albedo variation after wetting the ground is also $\Delta\rho$= 0.01. These are insignificant differences, although according to QPV's experience, albedo reductions of up to 0.04 and 0.05 can be found in other similar terrains, as long as there is no vegetation cover that cushions the effect.

5.2 Seasonal variations

Seasonal albedo variations associated to the time evolution of land cover due to rainfalls, snowfalls, changes in the vegetation, etc., cannot be assessed by means of a single short ground-based measurement [10]. Instead, monthly ρ values are available from specialized solar consulting services, based on satellite observations.

Table V summarizes the albedo values measured in six months at spot 1 (mostly for sunny days), which is considered the most representative of the plant as a whole. It also includes the albedo values from satellite data provided by the Solargis database Prospect, for every month, which shows the whole annual evolution of albedo.

The ASTM condition $\theta_{ZS} \leq 45°$ could be fulfilled for the first five measurements, but not for the last one (17th November), since such low zenith angles cannot be reached from October the fifth for the latitude where the experimental campaign was carried out.

Table V: Monthly global albedo values for spot 1, obtained both from in-field measurements and satellite data retrieved from Solargis. Measurement days were 28 May, 07 July, 06 August, 09 September, 01 October and 17 November in 2019.

Month	May	Jul	Ago	Sep	Oct	Nov
ρ^G	0.21	0.23	0.23	0.23	0.24	0.23
ρ^E	0.19	0.21	0.21	0.21	0.21	0.21
ρ^{PV}	0.20	0.21	0.21	0.21	0.22	0.22
Solargis	0.23	0.24	0.24	0.24	0.23	0.24

On average, the maximum difference between the satellite values and those measured on the ground has been $\Delta\rho$ =-0.02 ($u^{\rho}_{E_{AC}} \leq 0.6\%$), which agrees with the uncertainty estimated by other authors for this albedo data source [11]. Spot 1 presents low differences with the satellite values, in contrast to spots 2 and 3, whose measured albedos strongly differ from Solargis. These differences may be due to multiple factors such as the different surface area analysed by satellite, typically 1 km × 1 km squares, which comprises different vegetation covers and soil types. Actually, long-term time series of satellite estimates can be empirically adjusted according to the recorded observational data -assumed to be the most accurate source of radiation data-, which is a correction method commonly accepted by the solar industry to provide a reliable characterization of the albedo [10]. In any case, it could be said that spot 1 is effectively the most representative of the whole terrain, whose average albedo might be assimilated to the satellite value.

The *measured* maximum time variation throughout the period is $\Delta\rho^G = 0.03$, resulting in $u^{\rho}_{E_{AC}} \leq 0.9\%$, which is low, especially compared to the spatial variation. Again, this is a very local result, due to the little vegetation covering spot 1, which is relatively barren and calcareous (normal in dryland farmland according to QPV's experience is around 0.06). In contrast, this maximum variation has reached 0.07 ($u^{\rho}_{E_{AC}} \leq 2.1$) for spot 3 (not shown for brevity), which is consistent with what QPV observed in similar lands. Other sites, especially at high latitudes, exhibit much larger seasonal albedo variations.

6 CONCLUSIONS

This paper analyses the uncertainty issues associated to different albedo measurement procedures, looking for deriving recommendations for testing methods in the frame of commercial PV plants. The relation between albedo measurement uncertainty and subsequent energy yield estimations uncertainty is established via the albedo sensitivity coefficient, c'_{ρ}. We provide an analytic formula for the latter, relating albedo measurements and

10.4229/EUPVSEC2023/4DV.4.6
020446-006

subsequent energy yield estimations uncertainties, considering the impact of Ground Coverage Ratio and Inverter Loading Ratio. A key point is that albedo uncertainty is significantly reduced when translating to uncertainty in final energy yield. That allows for a wide margin of freedom regarding the procedure used to perform albedo measurements into the frame of PV engineering practices.

We define three types of albedo values as input data for yield simulation exercises, using pyranometers and reference cells, which enables to account for spectral mismatch and angle of incidence sensor losses and reduce costs.

Establishing a limit of 1.5% for $u^{\rho}_{E_{AC}}$ leads to the following measurement recommendations: **1)** Use the CMA10 as a low-pass filter for avoiding unwanted noise, so that the minimum measuring time is just 10 minutes. **2)** Limit the measurements to $\theta_{ZS} \leq 60°$, which is less restrictive than prescribed in ASTM E1918-21 and allows for enlarging the yearly time available for albedo measurements. **3)** Reference solar cells can be used instead of pyranometers, which reduces the instrumentation price. Furthermore, the preferred alternative consists of combining a pyranometer for measuring $G^{\downarrow}(0)$ and a reference cell for measuring $G^{\uparrow}(0)$. This way, spectral effects, which are disregarded in standard PV software, are included in the calculation of PV yield. **4)** Sensors' height above the ground can be chosen in the range [0.5m-1.5m]. **5)** Measurements before or after noon are equally suitable. **6)** Measurements performed on cloudy and unstable days are still valid providing that the data is filtered by the condition $G^{\downarrow}(0) \geq 400\ \mathrm{W/m^2}$and the majority of the instantaneous $G^{\downarrow}_{PYR}(0)$ values is greater than 300 W/m^2.

Concerning energy yield simulations, we suggest using ρ^{G} and ρ^{PV} as input data for PV simulation software which do not include spectral corrections and angular corrections are applied by default and SISIFO, and ρ^{E} only for SISIFO, with angular corrections disabled. We consider that this latter alternative offers the greatest proximity between modelling and experimentation.

A representative case of ρ measurement practises at the current PV market scene is also presented in the form of a commercial albedo measurement project. Overall, the most relevant uncertainty source found is associated to the variety of ground types encountered on all the surface of the future PV plant. However, this fact cannot be generalized to other sites, as it is a site-specific characteristic. An important consideration is that ρ is usually demanded to be measured *before* the construction of the PV plant. Because of that, in grounds whose reflectance features are expected to change with the change in the use of land, the derived uncertainty predominates over all the uncertainty sources associated to the albedo measurement conditions. Similarly, long-term seasonal variations can become the dominant uncertainty source for high latitude locations, and the combination of ground-based albedo measurements with satellite data is recommended to reduce the corresponding energy yield uncertainty.

ACKNOWLEDGEMENTS

The authors gratefully acknowledge Javier Ramírez Ledesma for his SISIFO software development, supporting the simulations of the bifacial systems analysed here. They deeply thank Vicente Lara-Fanego (Solargis s. r. o., Bratislava, Slovakia) for the high-quality data and subsequent analysis concerning seasonal variations. They also wish to thank Miguel León for helping in the preparation of the reflective surfaces and Marta Martí for her albedo data processing contribution.

This work was supported by the Spanish MCIN/AEI/ 10.13039/501100011033 through Project GREASE (Proyectos de I + D + i Retos Investigación 2020), with grant number PID2020-113533RB-C31.

REFERENCES

[1] M. Gostein, B. Marion and B. Stueve, "Spectral Effects in Albedo and Rearside Irradiance Measurement for Bifacial Performance Estimation," in *IEEE 47th Photovoltaic Specialists Conference (PVSC)*, d. d.

[2] ASTM, *E1918-21. Standard Test Method for Measuring Solar Reflectance of Horizontal and Low-Sloped Surfaces in the Field,* 2021.

[3] N. Riedel-Lyngskær, D. Berrian, D. Alvarez Mira, A. Aguilar Protti, P. Behrensdorff Poulsen, J. Libal and J. Vedde, "Validation of Bifacial Photovoltaic Simulation Software against Monitoring Data from Large-Scale Single-Axis Trackers and Fixed Tilt Systems in Denmark," *Applied Sciences,* vol. 10, no. 23, 2020.

[4] A. Luque, E. Lorenzo and G. Sala, "Diffusing Reflectors for Bifacial Photovoltaic Panels," *Solar Cells,* vol. 13, pp. 277-292, 1984.

[5] ISO/IEC, *Guide98-3, 2008. Evaluation of measurement data - Part 3: Guide to the expression of uncertainty in measurement,* 2008.

[6] JCGM, *100:2008. Evaluation of measurement data-Guide to the expression of uncertainty in measurement.,* 2008.

[7] V. Lara-Fanego , J. A. Ruiz-Arias , A. Skoczek, C. A. Gueymard , T. Cebecauer and M. Suri, "Annual Energy Production Uncertainty of Bifacial PV Plants Caused by Inaccuracies in Albedo Data: Case Studies using SAM," in *49th IEEE PVSC,* Philadelphia, 2022.

[8] M. Sengupta, A. Habte, S. Wilbert, C. Gueymard and J. Remund, "Best Practices Handbook for the Collection and Use of Solar Resource Data for Solar Energy Applications: Third Edition," 2021.

[9] P. Minnis, S. Mayor, W. L. Smith, Jr and D. F. Young, "Asymmetry in the Diurnal Variation of Surface Albedo," *IEEE TRANSACTIONS ON GEOSCIENCE AND REMOTE SENSING,* 1997.

[10] V. Lara-Fanego, C. Gueymard, T. Cebecauer and M. Suri, "Effectiveness of short ground-based measurement campaigns of variable duration for the characterization of local surface albedo," in *WCPEC-8*, Milan, 2022.

[11] V. Lara-Fanego, C. A. Gueymard , J. A. Ruiz-Aias, T. Cebecauer and J. Betak, "Extensive Evaluation and Uncertainty Estimation of Albedo Data Source," in *49th IEEE PVSC*, Philadelphia, 2022.

10.4229/EUPVSEC2023/4DV.4.6
020446-007

EVALUATION OF THE DURATION OF ALBEDO MEASUREMENT CAMPAIGNS

Stefan Mau, Adam Sharpe, Christian Gertig, Christophe Campistron, Julia Garcia, Nicolas Chouleur
Everoze SL, Calle Edgar Neville 6, 28020 Madrid, Spain
Tel: +34 679 319 876
stefan.mau@everoze.com, adam.sharpe@everoze.com, christian.gertig@everoze.com,
christophe.campistron@everoze.com, julia.garcia@everoze.com, nicolas.chouleur@everoze.com

ABSTRACT: Bifacial PV modules are increasingly used in commercial solar PV projects. With the use of the energy available at the module´s rear surface, the correct estimation of the albedo during the pre-construction phase of commercial projects becomes more important. The intention of the study is to determine the duration of an albedo measurement campaign to reduce the uncertainty with respect to satellite albedo, leading to a 'beneficial campaign'. Therefore, albedo measurements from seven sites in the USA have been used to estimate the impact of 1-day and 7-day measurements (short-time measurements) on the simulated annual energy production. Simulations based on monthly albedo data obtained from a recognized satellite provider have been compared with simulations where the albedo has been corrected based on short term measurements. It has been found that measurement campaigns in winter generally do not reduce the error of the annual energy production simulation. A 1-day campaign is often not beneficial, causing an increase in uncertainty with respect to satellite albedo. Instead, a 7-day campaign undertaken in summer appears to often reduce the uncertainty.

Keywords: albedo, measurement campaign, on-site

1 INTRODUCTION

The following different procedures are possible ways to estimate the annual irradiation available on the rear side of solar modules:

1) Long-term on-site measurement of the module rear-side solar irradiance.
2) Long-term on-site measurement of the albedo and use of an optical model to obtain the module rear-side solar irradiance.
3) Use of long-term satellite derived albedo together with an optical model.
4) Adaptation of long-term satellite derived albedo with short-term on-site measurement on site together with an optical model.

Based on the author's experience, the latter two procedures are mostly used in commercial projects with bifacial modules. Albedo stations are often installed in Europe for bifacial sites for a short period of time before construction or during site commissioning, while in some countries, more and more albedo stations prior to development can be witnessed. There is various literature dealing with best practice of measurements or their impact on simulated energy [1] [2] [3]. Short-time albedo measurement campaigns often comprise only one or a couple of days. With the information obtained during this measurement period, a long-term satellite-derived albedo dataset is then adapted, with a consistent offset applied to the satellite data based on the offset obtained from the onsite measurement during the measurement period. While [4] asserts that such measurements are suitable to accurately determine monthly albedo data, [5] reports that short-term measurements may introduce an increased uncertainty. This paper assesses the error in the simulated annual energy yield involved with such short-term measurement campaigns. The analysis is based on publicly available long-term albedo measurement data at seven sites accessible at Surface Radiation Budget Network (SURFRAD).

2 APPROACH

In this work, albedo measurements from the following seven sites in the USA accessible at SURFRAD have been used. The stations are generally considered of good quality with the ground-reflected irradiation measured at 10 m height. Albedo measurements in commercial projects are instead often undertaken at 1.5 m to 2 m height. It can be seen that the annual GHI is generally high for these sites, with an exceptionally high value at Desert Rock, which is comparable to central and southern Europe. Five of the sites show considerable snowfall in winter. Only Desert Rock and Goodwin Creek show snow events comparable to central Europe. The climate conditions, together with the ground vegetation and albedo measurement at a height of 10m, may mean the results of this analysis cannot be applied to all sites in Europe.

Site	Coordinates	Terrain type	Long-term satellite derived annual GHI / kWh/m²	Long-term satellite derived albedo
Bondville - 'Bv'	40.052N, 88.373W	Native grasses	1,504	0.247
Boulder - 'Bd'	40.125N, 105.237W	Sandy with exposed rocks, spare grass	1,695	0.199
Desert Rock - 'D R'	36.624N, 116.019W	Fine rock and scattered creosote bush	2,088	0.211
Fort Peck - 'F P'	48.308N, 105.102W	Native grasses	1,446	0.247
Goodwin Creek - 'G C'	34.255N, 89.873W	Pasture grass and sparse deciduous trees	1,674	0.200
Penn State - 'P S'	40.720N, 77.931W	¾ grass and ¼ crops	1,390	0.252
Sioux Falls - 'S F'	43.734N, 96.623W	Native grasses	1,479	0.238

Table 1: Site overview

Measurement data for horizontal GHI and ground-reflected irradiation in 1-min resolution for the full year 2020 have been downloaded. Daily albedo has been calculated by dividing the sum of daily ground-reflected irradiation by the sum of daily GHI and applying the following criteria:

10.4229/EUPVSEC2023/4DV.4.11
020448-001

1) GHI irradiance >= 50 W/m2
2) Daily albedo < 0.5
3) Data availability > 5h/d

Considering that a PV plant starts operation at GHI values of around 50 W/m2, lower irradiance values have been excluded.
During winter months, a large scattering in the daily albedo data has been observed. Everoze considers that this was due to snowfall. Most of the time, however, scattering was low and albedo values were in the range of 0.2 to 0.25. Assuming that albedo assessments do not provide representative values if the soil or the modules are covered by snow and that measurement campaigns would not be undertaken under such conditions, days with albedo equal to or higher than 0.5 have been excluded from the analysis.
Even though data availability has generally been observed to be high, days with a data availability of less than 5 hours have been excluded. This is due to the fact that the albedo depends on the zenith angle and missing data i.e., during the morning period will therefore cause a bias in the daily albedo.
Everoze has observed that albedo assessments often correct monthly long-term albedo values based on a one-day measurement. If this measurement is e.g., 10% higher than the value of the corresponding month, values for all 12 months are increased by 10%. Therefore, the objective in this work was to compare the simulated annual energy of a bi-facial one-axis horizontal tracking system, undertaken with a state-of-the-art simulation tool, for the following cases:

1) Measured albedo (six years, 'reference albedo').
2) Monthly albedo data obtained from a satellite provider 'satellite albedo').
3) Monthly albedo data corrected by 1-day measurements ('corrected albedo').
4) Monthly albedo data corrected by 7-day measurements ('corrected albedo').

Measurement periods of 1 and 7 days have been analysed knowing that measurements are often undertaken in commercial projects during only one day. For the 7-day period, in a first step, day one to seven have been assumed as measurement periods, then iteratively days two to eight, and so on for the whole year (see figure below). In this way, a January dataset allows for 25 measurement campaigns if no days were excluded. For each campaign sample period, the deviation to the monthly 'satellite albedo' has been calculated.

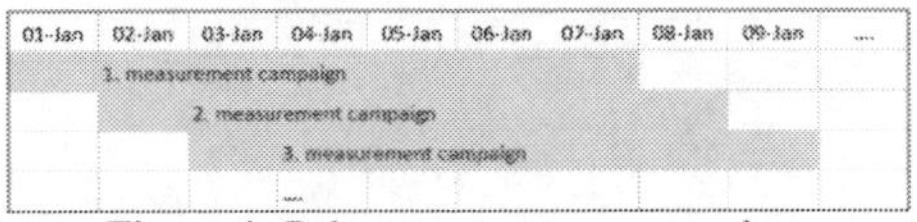

Figure 1: 7-day measurement campaigns

3 RESULTS

Figure 2 and 3 show the monthly distribution of albedo for Bondville and Sioux Falls. The grey lines indicate the albedo measured during the individual six years, whereas the red line represents the averaged albedo over the period, the 'reference albedo'. The black line indicates the 'satellite albedo'. It is observed that snow leads to a high scattering of the albedo during winter months. For six of the seven sites, the 'satellite albedo' was lower than the 'reference albedo'. For Sioux Falls, the 'satellite albedo' fitted best with the 'reference albedo'.

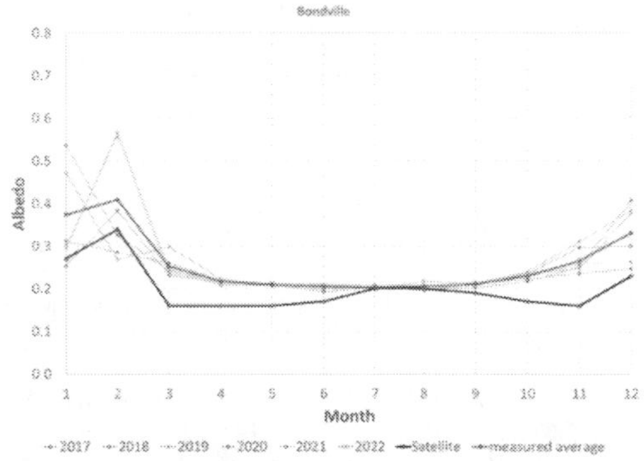

Figure 2: Monthly albedo rates, Bondville

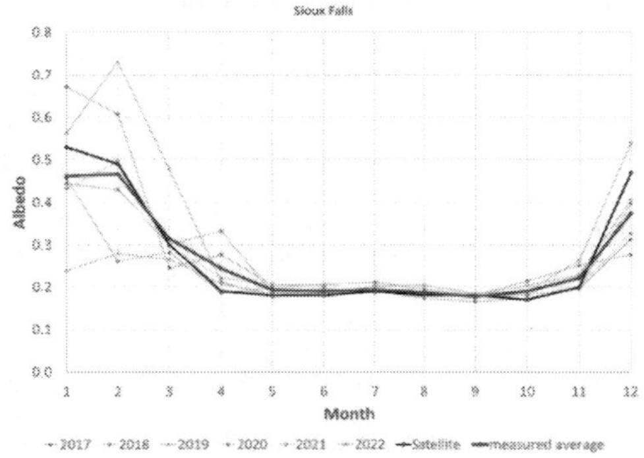

Figure 3: Monthly albedo rates, Sioux Falls

The 'satellite albedo' has subsequently been corrected based on the measurement campaigns. As an example, for this, the correction factors for Bondville for June for a 1-day measurement campaign are shown in the figure below. A total number of 178 days from the six years were available after applying the exclusion criteria. The remaining two days have been rejected due to data availability of less than 5 hours. One can see that the correction factor is always > 1, indicating that the 'satellite albedo' for June is always lower than the values obtained from the measurement campaign. Therefore, the 'satellite albedo' is always corrected upwards. In order to quantify the uncertainty related to the measurement campaigns, the red, blue and green points, that indicate the P10, P50 and P90 correction factors, have been used to correct the 'satellite albedo'. E.g. in order to determine the P50 'corrected albedo', each of the 12 monthly values of the 'satellite albedo' have been multiplied with 1.21. In this way, three corrected albedo datasets have been obtained for the annual energy simulation, only based on the measurements undertaken in June.

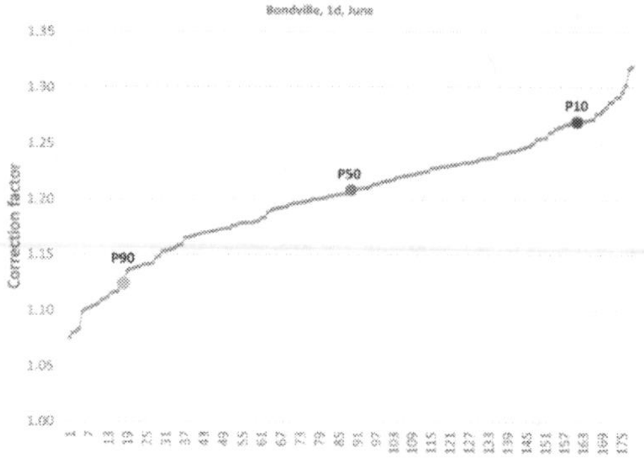

Figure 4: Correction factor in June for a 1-day campaign in Bondville

The following energy simulations have been undertaken for a bi-facial horizontal one-axis tracking system. From these, it has been observed that a relative 10% change in the albedo assessment leads to an approximately 0.2% change in the annual energy assessment. The annual energies calculated with the 'reference albedo', 'satellite albedo' and 'corrected albedo' are called 'reference energy', 'satellite energy' and 'corrected energy', respectively, in the following figures. The difference between 'reference energy' and 'satellite energy' has been 0.4% on average, being the latter in all cases lower.

How the albedo impacts the energy simulation is presented in Figure 5 for 1-day and Figure 6 for 7-day measurement campaigns, exemplary for Bondville. The x-axis represents the months and the y-axis the error with respect to the 'reference energy'. The horizontal continuous black line indicates the error of the 'satellite energy' (-0.6%) and the dashed line mirrors this. The red, blue and green lines, respectively, represent the error of the 'corrected energies' P10, P50 and P90. As soon as a line is between the two horizontal lines, the 'corrected energy' is closer to the 'reference energy', thus the correction improved the results with respect to the 'satellite albedo'. One can see this e.g. for measurements campaigns undertaken from April to June, meanwhile all measurement campaigns undertaken in February would lead to a worse result. Increasing the measurement period to 7 days brings the three lines closer together as the impact of individual outlier days is reduced. If the P10, P50 and P90 lines are within the two horizontal lines, we consider this a 'beneficial campaign', meaning that the likelihood of an increased error due to the campaign is theoretically less than 20%, but often much lower as for example for measurements from April to June. Beneficial are 7-day campaigns in April to June and August to October. In May for example a 7-day campaign reduces the 0.6% error of the 'satellite albedo' with an 80% likelihood to less than 0.15%. It should be noted that, in July and August, the campaigns barely reduce the error for this site. As can be seen from Figure 2 above, the 'satellite albedo' for Bondville for the months June and August is already so close to the 'reference albedo' that the campaigns mostly increase the uncertainty. On the other hand, from April to June, the 'satellite albedo' is considerably below the 'reference albedo' and even a 1-day measurement reduces the error.

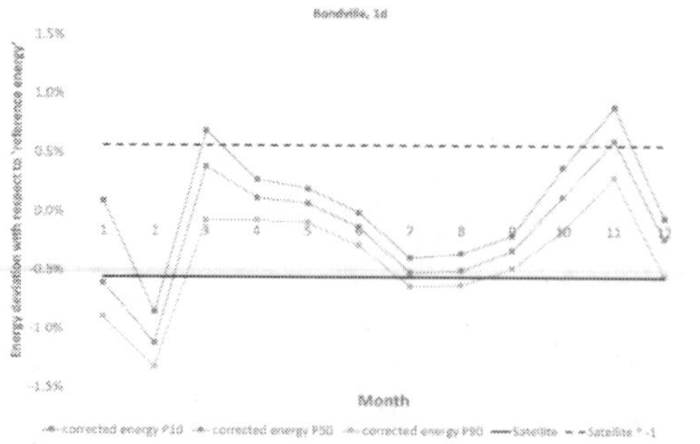

Figure 5: Error with respect to 'reference energy', Bondville 1-day

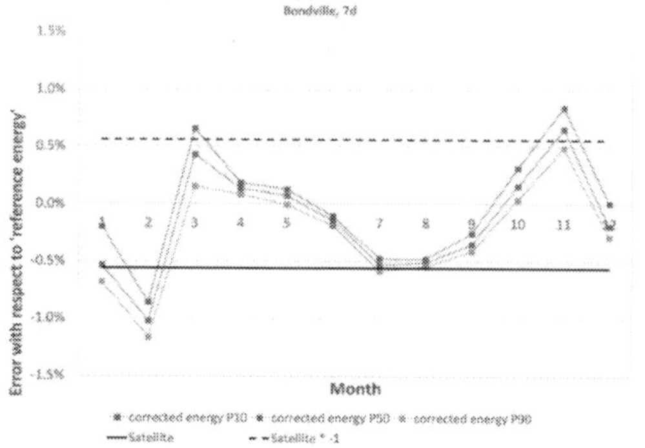

Figure 6: Error with respect to 'reference energy', Bondville 7-day

Whether a campaign is beneficial or not is presented in Figure 7 for a 1-day and Figure 8 for 7-day campaign for all 7 sites. The color code is as follows:

- beneficial campaign, low snow month
- no beneficial campaign, low snow month
- beneficial campaign, high snow month
- no beneficial campaign, high snow month

It can be seen that for Sioux Falls, a campaign is never beneficial. The reason is that the 'satellite energy' was already so close to the 'reference energy' that any correction caused an increased error. In general, campaigns undertaken in winter are not beneficial. A 1-day campaign undertaken in summer appears to be beneficial for some of the sites. However, there is still a considerable risk that it is not beneficial. If the campaign is extended to 7 days, it is often beneficial.

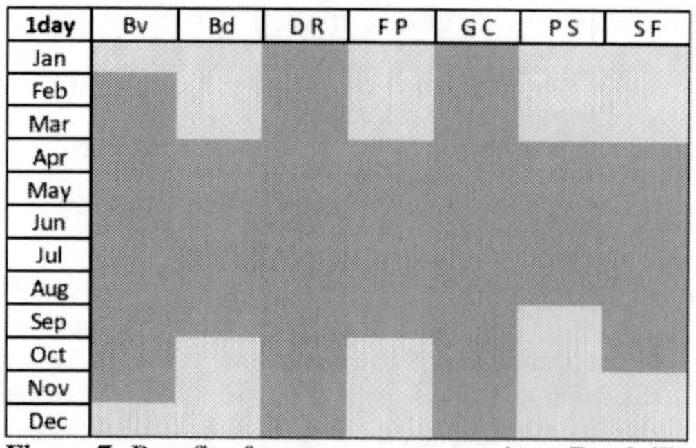

Figure 7: Benefit of measurement campaigns, **Bondville** 1-day

10.4229/EUPVSEC2023/4DV.4.11
020448-003

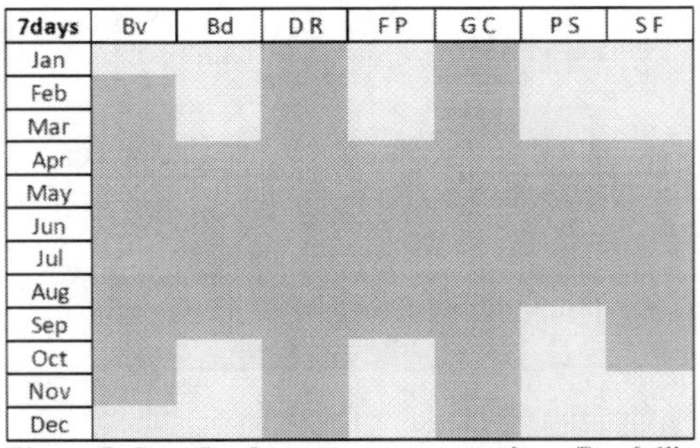

Figure 7: Benefit of measurement campaigns, Bondville 7-day

4 CONCLUSION

Based on the measured albedo values from seven sites in the USA accessible at SURFRAD, the impact of correcting pure satellite albedo data with results from a 1-day and 7-day measurement campaign has been investigated. It has been found that:

- Measurement campaigns in winter are generally not beneficial for more accurate energy yield predictions.
- A 1-day campaign is often not beneficial, causing an increase in uncertainty with respect to satellite albedo.
- Everoze recommends avoiding 1-day measurements.
- A 7-day campaign undertaken in summer is often beneficial.

It should be noted that the climate conditions together with the ground vegetation and measurement setup of the SURFRAD sites located in the USA may not be representative for other sites in the rest of the world.
Everoze recommends careful adherence to best practice when undertaking albedo measurement campaigns.

5 REFERENCES

[1] Manajit Sengupta et al., "NREL, Best practices handbook for the collection and use of solar resource data", NREL, 2021

[2] Eva-Maria Grommes et al., "The impact of albedo measurements on power density calculations of bifacial modules", 37th EUPVSEC, 2020

[3] Ben Bourne, "Ground Albedo Field Measurements", NREL Bifacial PV Workshop, 2018

[4] Alvaro Payán et al. "ATA-Renewables-Albedo-Methodology-Whitepaper", 2020

[5] Vicente Lara-Fanego et al. "Effectiveness of short ground-based measurement campaigns of surface albedo", WCPEC-8, 2022

10.4229/EUPVSEC2023/4DV.4.11
020448-004

Cloud Cover Forecasting for Prediction of Short Term PV Power Generation using Hybrid Models

Ruchita Korgaonkar, Varun Thakore, Ashay Awate, Sharat Chandran, Narendra Shiradkar
Indian Institute of Technology, Bombay
Mumbai, India
ruchita@iitb.ac.in, varunt@iitb.ac.in, ashay88@gmail.com, sharat@cse.iitb.ac.in, naren@ee.iitb.ac.in

***Abstract*—Increased penetration of photovoltaics in the grid comes with challenges due to its significant dependence on weather parameters. The intermittent output of solar power plants can cause unpredictable grid instabilities. In order to do optimal management of energy storage, spinning reserves and demand response, precise photovoltaic power forecasting is important. Reliable photovoltaic power forecasting is essential for smart grid applications like optimal dispatch, grid regulation, etc. PV power generation can be predicted accurately on clear-sky days but to forecast the PV power generation accurately on cloudy days is challenging due to different reasons like uncertain cloud formation, local terrain complexity, etc. In this paper, the accuracy achieved by training different statistical models for the prediction of cloud cover as well as GHI for 2-hour ahead forecast is shown.**

***Index Terms*—Short-term Forecasting, Hybrid Model, Cloud Cover, GHI (Global Horizontal Irradiance)**

I. INTRODUCTION

Changing climate, increase in energy demand, national financial incentives and continuous reduction in the cost has led to rapid growth of photovoltaics. But increased penetration of photovoltaics in the grid comes with challenges due to its significant dependence on weather parameters. The intermittent output of solar power plants can cause unpredictable grid instabilities. In order to do optimal management of energy storage, spinning reserves and demand response, precise photovoltaic power forecasting is important. Reliable photovoltaic power forecasting is essential for smart grid applications like optimal dispatch, grid regulation, etc.

PV power generation can be predicted accurately on clear-sky days but to forecast the PV power generation accurately using a physical model on cloudy days it is essential to have an accurate forecast of the irradiance. This can be achieved by the prediction of cloud cover as measured using sky cameras. Prediction of cloud cover is complex due to different reasons like local terrain complexity, uncertain cloud formation, etc. Available NWP models have resolution of the order of kms whereas predicting cloud formation and deformation would require much better resolution for PV power generation.

In order to do reliable PV power generation forecasting using a physical model, it is important to have forecasts of explanatory variables like irradiance (hence cloud cover), temperature, etc. In this paper, accuracies obtained using different statistical models by using different combinations of inputs from weather station for short term (30-minute ahead) cloud cover forecasting are explored and presented. The data used for this study was obtained from Shell.ai Hackathon for Sustainable and Affordable Energy 2021.

II. FORECASTING ASPECTS

A major aspect of PV generation forecasting is characterized by prediction horizon, it can be divided in 4 categories as follows:

- Very short term forecasting (for next few seconds to minutes)
- Short term forecasting (for next few hours)
- Medium term forecasting (for weeks to month)
- Long term forecasting (for month to years)

Each category has different applications. Among these, short term forecasting i.e. for hours to day-ahead forecasting is very crucial for scheduling of conventional generations and many national grid regulatory authorities require precise power forecasting.

Another major aspect of forecasting is the type of the model used for forecasting, it is classified into 3 categories as shown below:

- Physical Model
- Statistical Model
- Hybrid Model

Physical Models use physics based models for forecasting, statistical models predicts on the basis of the statistics of historical data and hybrid models use a combination of these two models for forecasting of PV power generation.

On clear-sky days, irradiance and PV power forecasting can be done very accurately, but on cloudy days, even 1-hour to 2- hour ahead forecasts are challenging.

In this paper, different four models has been trained for different ranges of total cloud cover and A hybrid model consisting of a clear-sky model and a statistical model is used for the 2-hour ahead forecast of cloud cover and Global Horizontal Irradiance (GHI).

III. DATASET

A. Training Dataset

A total of 1-year weather data was used for cloud cover forecasting with parameters: GHI (Global Horizontal Irradiance), DNI (Direct Normal Irradiance), Temperature, RH (Relative Humidity), wind speed, wind direction, azimuth

10.4229/EUPVSEC2023/4DV.4.13
020449-001

Model	Inputs	Output	Accuracy(%)
Persistence Model	1. cloud cover	30-minute ahead cloud cover	88.67
Linear Regression	1. cloud cover	30-minute ahead cloud cover	87.95
Linear Regression	1. cloud cover 2. DNI	30-minute ahead cloud cover	87.94
Linear Regression	1. cloud cover 2. wind speed 3. wind direction 4. azimuth angle	30-minute ahead cloud cover	87.79
Linear Regression	1. Change in cloud cover (from last 30 to 60 mins)	Change in cloud cover after 30 min	88.05
Neural Network	1. cloud Cover 2. 30-mins before cloud Cover 3. 60-mins before cloud Cover 4. 40-mins before cloud Cover 5. Pressure 6. Temperature 7. Moisture	30-minute ahead cloud cover	89.44

TABLE I: Models along with inputs and the achieved accuracy

angle, moisture, precipitation, pressure, total cloud cover (in %). This total percentage cloud cover is taken from embedded software in total sky imager for pre-processing raw images. Also, corresponding total sky images of daytime with 10-minute time intervals are available.

B. *Testing Dataset*

Data was tested using a total 300 different testing scenarios where for every scenario 6 hours of weather data with 1-minute sampling interval and 2 hours of total sky images with 10-minute sampling interval was available.

(The data used for this study was obtained from Shell.ai Hackathon for Sustainable and Affordable Energy 2021)

IV. METHODOLOGY

Statistical models using linear regression and neural networks with different combinations of inputs from weather parameters were trained and evaluated using mean absolute error (MAE).

$$MeanAbsoluteError(MAE) = \frac{1}{N}\Sigma_{i=1}^{N}|\hat{x_{i+1}} - x_i| \quad (1)$$

$$Accuracy = 100 - 100 \times \frac{1}{N}\Sigma_{i=1}^{N}\frac{|\hat{x_{i+1}} - x_i|}{x_i} \quad (2)$$

A. *Cloud Cover Forecasting*

The correlation of different weather parameters with 30-minute ahead cloud cover was calculated, it was found that historical cloud cover, change in cloud cover and DNI have significant correlation with 30-minute ahead cloud cover as shown in fig.1.

To forecast cloud cover different weather parameters are explored using linear regression model and neural network. Simple linear regression models were fitted using python library- sklearn. 3-layer sequential neural networks were trained with different combinations of inputs using python library- keras. It was observed that linear regression models were not able to achieve the accuracy which is achieved by persistence model (simplest model). Neural networks trained with historical cloud cover, pressure, moisture and temperature achieve better accuracy than persistence model.

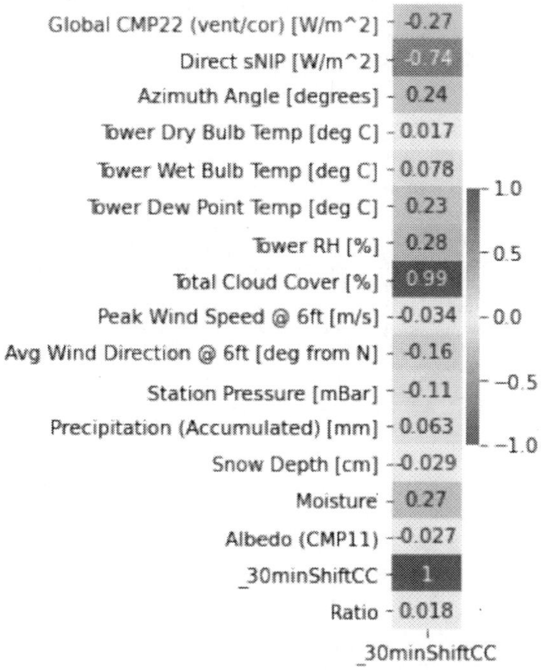

Fig. 1: Correlation Coefficients of different weather parameters with 30-minute ahead cloud cover

Table I summarizes the details and accuracies of different models. It shows inputs and the output for which the statistical model is trained. Persistence model is the simplest model where the forecasted value is the same as that of the present/ current value of the parameters as shown in equation:

$$\hat{x_{i+1}} = x_i \quad (3)$$

where, $\hat{x_{i+1}}$ - is the forecasted value at (i+1)th time instance and x_i is the actual value at (i)th time instance.

10.4229/EUPVSEC2023/4DV.4.13
020449-002

B. GHI Forecasting

It was observed that training only one model for all the data hampers the accuracy of the overall model. Hence, to forecast Global Horizontal Irradiance (GHI) accurately, four different neural networks were trained based on the total cloud cover percentage as shown in table II.

Model	Range of Total Cloud Cover
Model-1	Total cloud cover< 11%
Model-2	10%¡=Total cloud cover<50%
Model-3	50%¡=Total cloud cover<75%
Model-4	75%¡=Total cloud cover

TABLE II: Models for different cloud cover ranges

As it was mentioned in the earlier section that, using weather parameters- historical cloud cover, pressure, moisture and temperature neural network gave better accuracy. These parameters are used to train the four models using a neural networks along with the clear sky GHI and clear sky DNI as additional inputs. With these 4 models 90.16% accuracy was achieved on the dataset.

V. SUMMARY

The method presented in this paper uses the cloud cover obtained using total sky imager. Trained neural network using the 1-hour historical cloud cover and the other weather parameters-moisture, temperature, pressure and clear sky model could predict cloud cover with accuracy 89.44% for 2-hour ahead cloud cover forecast. Trained neural network using the 1-hour historical cloud cover and the other weather parameters- GHI, DNI, moisture, temperature, pressure and clear sky model could predict GHI with accuracy 90.16% for 2-hour ahead GHI forecast. It was observed that for the above dataset, including clear sky model estimated Global Horizontal Irradiance (GHI), Direct Normal Irradiance (DNI) (from 81.3% to 90.16%) and the percentage cloud cover from total sky imager (TSI), improves the accuracy of the model. This model gives 20% nRMSE error whereas the available commercial tool gives 22% nRMSE.

ACKNOWLEDGEMENT

The Data that used for training and testing was given for the Shell Hackathon 2021.

This work is supported by National Centre for Photovoltaic Research and Education (NCPRE) funded by Ministry of New and Renewable Energy (MNRE), Government of India through the Project No. 31/09/2015-16/PVSE-R&D dated 15th June 2016.

REFERENCES

[1] Data Source: Shell Hackathon, 2021

[2] Mosaico, G., and Saviozzi, M., "A hybrid methodology for the day-ahead PV forecasting exploiting a Clear Sky Model or Artificial Neural Networks," International Conference on Smart Technology, 2019.

[3] Wang, J., Zhonga, H., Lai, X., Xia, Q., Wang, Y., and Kang, C., "Exploring Key Weather Factors from Analytical Modeling Toward Improved Solar Power Forecasting," IEEE Transactions on Smart Grid, Vol. 10, No. 2, 2019.

[4] Xiyun, Y., and Song, C., "A Combination Method in Photovoltaic Power Forecasting Based on the Correlation Coefficient," IEEE International Conference of Software Engineering and Service Science, 2014.

[5] Yuehui Huang, Jing Lu, Chun Liu, Xiaoyan Xu, Weisheng Wang and Xiaoxin Zhou, "Comparative Study of Power Forecasting Methods for PV Stations," International Conference on Power System Technology, 2010.

[6] https://solcast.com/forecast-accuracy

ANALYSIS OF THE CORRELATION BETWEEN CLOUD PROPERTIES AND SOLAR IRRADIANCE IN SOUTHERN GERMANY BASED ON METEOROLOGICAL DATA OF THE DEUTSCHER WETTERDIENST (DWD)

Fabian Pauls [a,*], Mike Zehner [a], Fabian Flade [b]
[a] Rosenheim Technical University of Applied Sciences, Technology Centre for Energy & Buildings, Working Group Energiy Meteorology, PV Systems und Grids, Hochschulstraße 1, 83024 Rosenheim, Germany
[b] Bavarian Association for the Promotion of Solar Energy, Friedrich-List-Str. 88, 81377 Munich, Germany

ABSTRACT: This paper analyses the correlation between different cloud properties and solar irradiance. The aim is to gather further insight on how solar irradiance is influenced by different cloud types and which cloud types are responsible for which phenomena in given irradiance data. Additionally, this research focusses on the nature of irradiance enhancement events and their relation to different cloud properties. The analysis is based on measurements which have been conducted by the Deutscher Wetterdienst (DWD) in six separate locations in southern Germany. These datasets contain multiyear 1-minute-resolution measurements of solar irradiance and 60-minute-resolution synoptic observations for various cloud properties. The data is prepared, analyzed, visualized, and statistically evaluated via MATLAB. As a clear sky reference an open-source MATLAB implementation of both the Haurwitz and the Ineichen/Perez model is used. The study suggests that the main cause of irradiance enhancement events are cumulus, cirrus and cirrostratus clouds and that a medium to high total cloud cover is responsible for these events. Lastly a new method for classifying clouds is proposed via a scatter plot visualization of both irradiance gradients and irradiance enhancements. Here, distinct patterns and clusters can be observed in different cloud types. The results of this study can lead to improved solar irradiance forecasts with respect to different cloud properties and to a better understanding of their underlying principles.
Keywords: Solar Irradiance, Irradiance Enhancements, Cloud Types, Irradiance Volatility, Clouds

1 INTRODUCTION

1.1 Problem description

As the demand for independent and renewable energy sources increases, so does the development of solar energy plants, particularly photovoltaics (PV). Therefore, comes the need to find solutions to the problems and challenges associated with this increasingly prevalent technology. A large factor in this is the volatility of solar irradiance, which is caused to a large extent by shading and cloud cover (Caldas & Alonso-Suárez, 2019), (Paulescu, et al., 2014), (Paulescu & Paulescu, 2019). A related phenomenon is irradiance enhancement, which has been receiving more attention from researchers in recent years. Irradiance enhancements (IE) are events where the global horizontal irradiance exceeds a theoretically expected irradiance value (for example, a corresponding clear-sky reference curve), according to definition (Zehner, et al., 2011). While the volatility of solar irradiance and corresponding irradiance events like IEs may not be of major importance for solar thermal systems due to their inert nature, photovoltaic networks and photovoltaic systems are particularly sensitive to the volatility of solar irradiance and resulting spikes in irradiance. This makes the unpredictable nature of clouds a significant factor for these systems. Although these effects may be negligible in many cases for European distribution networks, they can still pose a threat to energy plants in worst-case scenarios. Zehner et al. (2011) determined for example, that in some scenarios the output power of PV-modules could be up to 30% higher than under STC conditions. This is further aggravated through the occurrence of even extremer events, considering that power peaks of up to 2000 W/m^2 have been reportedly observed (Gueymard, 2017a).

Thus, the need arises to better understand, predict, and model these events and solar irradiance in general.

1.2 Related studies

Clouds influence solar irradiance and while this conclusion might be trivial in the case of shading, there is also the opposite effect, namely that clouds, under certain circumstances, can also generate higher irradiance values than the expected clear sky reference. In addition, the volatility of solar irradiance associated with clouds is a major factor in solar irradiance profiles. The cause of these effects and their influence on surface solar irradiance has already been investigated by numerous researchers.

There is still no uniform terminology in the literature when it comes to irradiance enhancements. Various authors have named this effect, among other things, as "silver lining" (Norris, 1968), "super-irradiance", "cumulus solar-irradiance reflection", "cloud gap effect" (Gu, et al., 2001), "cloud edge" (Wirth, et al., 2015) and "irradiance spikes" (Chicco, et al., 2016). Gueymard (2017a) and Järvelä, et al. (2020) refer to this phenomenon as "cloud enhancements". The term "irradiance enhancements" was used by Pecenak et al. (2016) and Zehner et al. (2011).

The irradiance enhancement effect can be observed anywhere in the world. Among others, extreme values of up to 1832 W/m2 have been observed in Ecuador at 3400 meters above sea level (Emck & Richter, 2008), 2000 W/m2 at 40 degrees latitude north at an altitude of 1829 meters (Gueymard, 2017a), 1565 Wm^{-2} in the western Ghat mountains on the Indian pen-insula (Padmakumari, et al., 2022) as well as 1600 Wm^{-2} (Yordanov, et al., 2015) at 60° latitude north at sea level.

The cause of these events is still disputed in the literature, but they are generally ascribed to certain types of clouds (e.g., cumulus or cirrus) or to certain cloud reflection or scattering effects. Iqbal noted in "An Introduction to Solar Radiation", that reflections of solar irradiance on clouds could potentially cause a higher solar irradiance than theoretically expected regarding the extraterrestrial solar irradiance (Iqbal, 1983).

10.4229/EUPVSEC2023/4DV.4.14
020450-001

Investigations by Quiroz & Cameron (2012) and Balfour & Shaw (2011) found that reflection and refraction of solar irradiance by cloud edges and clouds in general contribute a large part to irradiance enhancement events. Yordanov et al. (2013) attributes irradiance enhancement phenomena to be caused by strong Mie forward scattering within the cloud and finds that the strongest IE events occur when a narrow gap of thin clouds is surrounded by a 5° radius around the sun disk.

Another term found in literature is solar variability, often seen in relation with ramp rates. According to Perez et al. (2016), the cause for this variability is firstly to be found in the predictable and traceable apparent motion of the sun in the sky and earth's distance from the sun and secondly by the motion of the clouds and weather systems. Especially in relation to the second cause, Inman et al. (2016) and Lave et al. (2012) point out that the fast passage of obscuring clouds in front of or past the sun generates strong power ramp rates that are often accompanied by irradiance enhancement events (Almeida, et al., 2014), (Ranaweera, et al., 2014), (Roy, 2015), (Zehner, et al., 2011).

Several researchers have already attempted to determine the primary cloud types responsible for the aforementioned irradiance phenomena. Norris (1968) attributes them to the reflection of solar irradiance by "broken cumulus clouds". Zehner et al. (2010), (2011) and Tzoumanikas et al. (2016) also suggests a connection between cumulus clouds and irradiance enhancements. Thuillier et al. (2013) finds that under certain conditions cirrus clouds can also be responsible for the mentioned effects, whereas Pecenak et al. (2016) and Schade et al. (2007) suggest that cirrus clouds might not be effective enough to be responsible.

A study by Stut et al. (2020) tried to investigate the relationships of solar irradiance volatility, irradiance enhancements, cloud types and other meteorological parameters based on high-resolution DWD and MIM data sets.

1.3 Purpose and outline

This study is about making the general relationships between the behavior of solar radiation and selected cloud properties (genera, cloud layer, height, and total cloud coverage) more visible and understandable.

The focus of this work is to evaluate solar irradiance measurements and synoptic cloud observation data from DWD stations in the Alpine foreland of southern Germany. An attempt is made to visualize the irradiance data for the six mentioned locations and then to relate them to the corresponding cloud properties occurring at the same time. For this purpose, the raw data is processed in MATLAB and then suitable visualizations are sought to make the relationships visible. The focus here is mainly on the jump height of the global horizontal irradiance $\Delta(GHI_{mes},GHI_{mod})$ in Wm^{-2}, the gradient of the irradiance ∇GHI in $Wm^{-2}min^{-1}$, the cloud type, as well as the total cloud cover (okta). Other quantities such as sun height, temperature and humidity have already been dealt with in other studies (e.g., Stut et al. (2020)). Factors that influence radiation in addition to clouds should be minimized or eliminated. For example, site-dependent and temporal factors are largely excluded by statistical analysis. Furthermore, this work also investigates the irradiance behavior of clouds in general. It is examined whether certain clouds lead to certain recurring patterns in the irradiance data and how they are composed.

This work offers a completely new type of representation of cloud and irradiance data. It differs from previous analyses of the relationships between cloud types and solar radiation in that it firstly specifically attempts to gain insights into solar meteorology for locations in the South German Alps, secondly seeks to find new insights into the behavior of clouds, and thirdly provides a tool for future possible analyses (of other locations) by innovative visualization methods.

2 DATA

The analyzed datasets are part of the observation network of Deutscher Wetterdienst (DWD). The basis for this work is the irradiance and synoptic cloud data of this measuring network for six distinct locations in the German pre-Alpine region, whereas the individual stations are listed in Table I and illustrated in Figure. 1.

The stations were selected due to their location in the pre-Alpine region and their data availability. While the measured data of the solar irradiance are available in high-resolution 1-minute intervals, synoptic cloud data only has an hourly measuring interval.

To be able to carry out the analysis anyway, the synoptic data is assumed to be constant over the course of an hour and therefore, have been interpolated with the value of the previous measurement until a new value is available. Since the rate of change of clouds and their properties is assumed to be relatively inert and these changes only occur slowly (e.g., compared to the volatile nature of solar irradiance), the hourly data interval and the resulting interpolation should nevertheless be sufficient for the purpose of this analysis.

It should be noted that the synoptical cloud data has also been shifted by -10 minutes to match the corresponding UTC time, as indicated in the DWD data description.

The solar irradiance data contains the global irradiance, the direct and diffuse irradiance share and the sun height. For the irradiance measurements the following sensors have been used: a scanning pyrheliometer/pyranometer (SCAPP) in Freiburg and Kempten and a CG4 Pyrgeometer by Kipp&Zonen in Fürstenzell, Hohenpeißenberg, Konstanz and Weihenstephan.

Table I: Overview of the DWD measuring stations including location and climate zone according to Köppen-Geiger

Location	abbr.	Longitude	Latitude	Elevation a.s.l.	Climate
Freiburg	FB	7.8343°E	48.0232°N	237 m	Dfb
Fürstenzell	FZ	13.3532°E	48.5451°N	476 m	Cfb
Hohenpeißenberg	HP	11.0108°E	47.8009°N	977 m	Cfb
Kempten	KP	10.3348°E	47.7233°N	705 m	Cfb
Konstanz	KS	9.1307°E	47.6952°N	428 m	Cfb
Weihenstephan	WS	11.6946°E	48.4024°N	477 m	Cfb

10.4229/EUPVSEC2023/4DV.4.14
020450-002

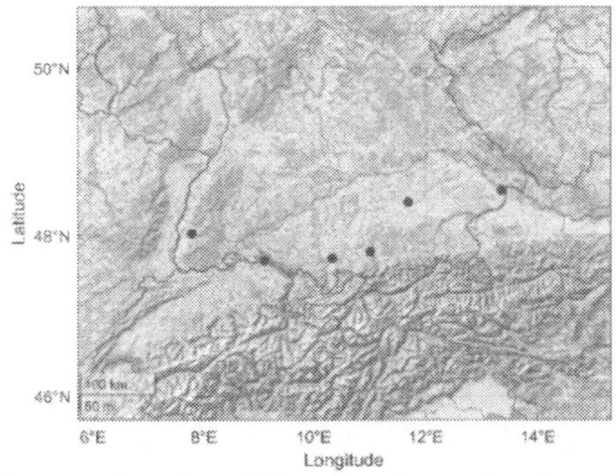

Figure 1: Locations of the measuring stations in southern Germany (L-R: Freiburg, Konstanz, Kempten, Hohenpeißenberg, Weihenstephan, Fürstenzell)

The synoptic cloud data contains the total cloud coverage of all layers (in okta), as well as the cloud type (cloud code), the cloud height (in meters) and the cloud coverage of the individual layers (from cloud layer 1 to cloud layer 4, also in okta). These cloud layers do not cover a fixed altitude range but refer to the first layer in which clouds are present. This means that, for example, Cirrus clouds can also be present in the lowest cloud layer 1 despite being typically high clouds. The cloud types are divided into the ten cloud types following the classification into cloud genera according to the WMO cloud atlas, which are cirrus, cirrocumulus, cirrostratus, altocumulus, altostratus, nimbostratus, stratocumulus, stratus, cumulus, and cumulonimbus.

The synoptic cloud data has only been recorded by the DWD for a specific time period. Since many stations are no longer manned by personnel and now carry out automatic instrument measurements, there is no more synoptic cloud observation data from a certain point in time, at least as far as the recording of the individual cloud types is concerned. Furthermore, the highly resolved irradiance data is only available from a certain year on. This results in an overlap of the two datasets in a certain period. These periods for the individual stations can be found in table II. Outside of these, either no highly resolved irradiance data or recorded cloud types are available, which is why these times cannot be considered in the analysis. In addition, when analyzing two of the six locations (Freiburg and Weihenstephan), it was found that there is no overlap of the required data at all, so these stations could not be included in the analysis of the cloud types as well.

Table II: Overview of the data availability of the six DWD measuring stations

Station	Data availability
Freiburg	no overlap in data
Fürstenzell	2007-2014
Hohenpeißenberg	2010-2018
Kempten	2007-2009
Konstanz	2007-2017
Weihenstephan	no overlap in data

Additional details on the data composition can be found in the corresponding data description of the DWD. To be able to set the measured solar irradiance in relation to a comparison value, a clear sky reference irradiance must be modeled. In previous related research (e.g., by Stut et al. (2020)) this irradiance was modeled using the software libRadtran. In this work, however, the PV_LIB Toolbox MATLAB library by Sandia National Laboratories is used to model the clear sky reference irradiance for reasons of integrability and simplicity. This also avoids the need for a data interface between two distinct programs.

The corresponding functions use, on the one hand, the Haurwitz model (Haurwitz, 1945), (Haurwitz, 1946) and, on the other hand, the Ineichen/Perez model (Ineichen & Perez, 2002). The former was evaluated by Reno et al. (2012) and has a very good performance according to this evaluation, but does not give any diffuse irradiance share, which is why the latter model was used as necessary.

Furthermore, in the analysis other functions of the PV Lib Toolbox are used – in particular, functions that are necessary for the execution of the two clear sky models and, in addition, a function for the determination of leap years. All other scripts were designed and implemented for this study in MATLAB.

3 ANALYSIS OF SOLAR IRRADIANCE ENHANCEMENTS AND REDUCTIONS IN RELATION TO CLOUD PROPERTIES BY THE MEANS OF IRRADIANCE ENHANCEMENT PLOTS

3.1 General analysis of irradiance enhancements and reductions

For a general analysis, the measured solar irradiance is used. To minimize the influences of daily and yearly change of solar irradiance and to extract only the irradiance enhancements, measured values and modelled, theoretical values are compared. Representatively for all the available data the year 2012 data of the DWD station in Fürstenzell is considered and used for all visualizations in this study. Figure 2 shows the measured solar irradiance compared to the corresponding envelope of modelled global horizontal irradiance. The abscissa and the ordinate show the hour of day (UTC) and the day of the year, so that a 3D representation of the yearly irradiance course is generated.

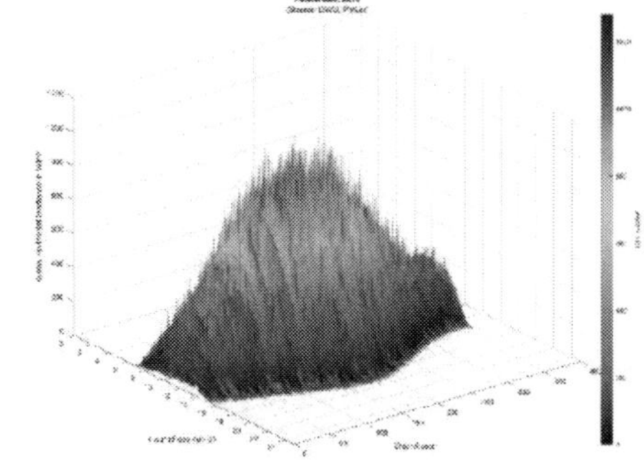

Figure 2: Yearly course of the measured GHI vs. the modelled GHI

On the applicate the solar irradiance in Wm^{-2} is applied. Measured values crossing the envelope and thereby exceeding the theoretically possible solar irradiance relate to irradiance enhancements. Measured

10.4229/EUPVSEC2023/4DV.4.14
020450-003

values within the envelope are irradiance reductions. Measured values on or a little under the envelope correspond to near clear sky conditions.

From this absolute depiction of the solar irradiance another visualization is derived by calculating the difference between modelled and measured irradiance for given points in time, thus highlighting the irradiance enhancements and reductions as shown in Figure 3.

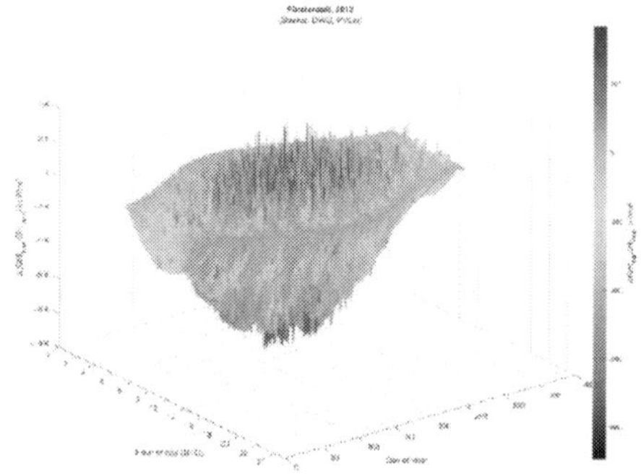

Figure 3: Yearly course of the irradiance enhancements and reductions

While the time axes are unchanged, the applicate is replaced by the difference of the measured and the modelled GHI. Thus, values above 0 Wm^{-2} correspond to irradiance enhancements, while values below 0 Wm^{-2} correspond to irradiance reductions.

At this point these plots already indicate that irradiance enhancements mainly appear in the summertime and furthermore during the middle of the daytime.

Because the required information, i.e., conclusions regarding the irradiance enhancements, are conveniently depicted in this type of illustration, it is further utilized for the analysis of the correlation between solar irradiance and synoptic cloud data in chapter 3.2 and chapter 3.3.

3.2 Analysis of irradiance enhancements in relation to the total cloud cover (TCC)

It is to be assumed that the total cloud cover is a large factor for the irradiance behavior in general and therefore for the appearance of irradiance enhancements. For this reason, the irradiance shall be put in relation to the total cloud cover (TCC). Furthermore, the total cloud cover shall be statistically analyzed for the available locations and several years.

For the first step, the total cloud cover is used as the color grading in the irradiance difference plots yielding Figure 4.

3.3 Yearly course of irradiance enhancement or reduction effects in relation to the total cloud cover

Figure 4 shows the yearly course of the total cloud cover in relation to the irradiance enhancements. Several phenomena can be observed here, i.e., that the total cloud cover is tendentially higher in the wintertime (as one might expect) and lower in the summertime, especially in the evening hours. Moreover, examining the irradiance enhancements in relation to the total cloud cover, the plot suggests, that not the completely clear days and hours, but the scattered to broken cloud conditions are most likely responsible for the highest irradiance enhancement events. The data also suggests that the clear sky intervals, i.e., time frames of minimal cloud cover, heavily correlate with the least difference between modelled and measured irradiance, as one would expect.

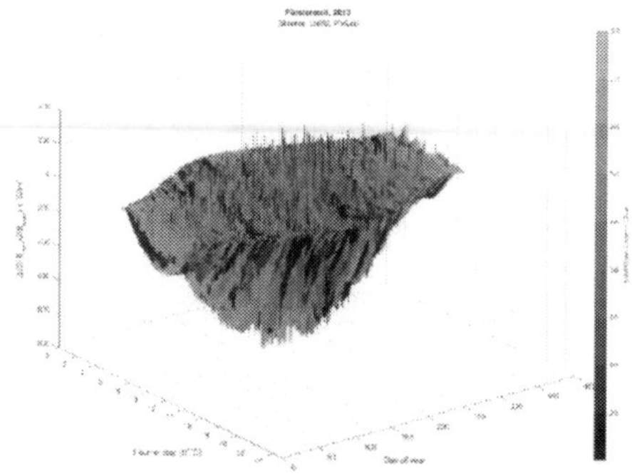

Figure 4: Yearly course of the irradiance enhancements and reductions in relation to the total cloud cover

3.4 Statistical evaluation of total cloud cover frequencies regarding irradiance enhancements

At the beginning and end of individual days, unexpected behavior within the data was observed. In these areas of the plot, the inaccuracies of the difference between modelled and measured values tend to increase. This leads to the apparent existence of irradiance enhancements, where there should not be any or in much smaller extent. For this reason, the statistical evaluation of the total cloud cover frequencies is done for all irradiance enhancement events larger than a threshold of 15 Wm-2. This gets rid of most of these early-day and late-evening artifacts but leaves the irradiance enhancements during the day mostly intact.

Table III: Irradiance enhancement (IE) / reduction (IR) frequencies in relation to the total cloud cover

Total Cloud Cover in okta	total TCC freq.	IE TCC freq.	IR TCC freq.
0/8	7.98 %	3.63 %	4.13 %
1/8	9.74 %	18.16 %	8.35 %
2/8	4.95 %	11.12 %	3.50 %
3/8	5.58 %	12.01 %	4.65 %
4/8	4.29 %	9.71 %	3.19 %
5/8	4.56 %	10.09 %	4.01 %
6/8	6.89 %	14.46 %	8.42 %
7/8	21.96 %	18.84 %	27.88 %
8/8	34.05 %	1.97 %	35.86 %

The statistical analysis of the total cloud cover levels results in a frequency distribution as shown in Table III. It shows that, together, around 50 percent of the observed time a total cloud cover of 7 or 8 okta is present. The remaining frequencies are distributed relatively evenly over all other cloud cover levels. If the frequency distribution is filtered by irradiance enhancements, it shows that the frequencies of the coverage levels 0/8 and 8/8 decreases by large amounts. This means that a completely covered or a completely clear sky has a very reduced probability for irradiance enhancements. However, there is a very strong increase in frequencies for the medium coverage levels between 1/8 and 7/8, which indicates that it is mostly mixed cloud coverage conditions that are responsible for irradiance enhancements. If the frequencies are filtered by the irradiance reductions, the frequency is greatly increased for high coverage levels. Low coverage levels only make up a small part of the shading here, as expected.

3.5 Analysis of irradiance enhancements in relation to cloud types

Beside the total cloud cover, a large influence of the current cloud type on the irradiance is suspected as already pointed out by other studies. Thus, to examine this correlation, the color grading of the plot is firstly replaced by the distinct cloud types and secondly statistically evaluated.

3.6 Yearly course of irradiance enhancements in relation to cloud types

To examine the correlation of cloud types to solar radiation, Figure 3 is equipped with a color scheme that makes the cloud types of the first cloud layer visible.

The further cloud layers (layer 2 to layer 4) are not shown but were also analyzed. The cloud type color grading is divided into three different color groups, which correspond to the typical cloud height of these individual clouds, with red tones corresponding to high, green tones corresponding to medium, and blue tones corresponding to low clouds. Furthermore, there is a gray tone for missing values and a black tone for instrument measurements without cloud type determination. The result is shown in Figure 5.

Firstly, it can be observed that the most common cloud type over the year is of the genus cumulus. It can further be seen that there is a strong correlation between the occurrence of irradiance enhancements and the presence of cumulus, altocumulus, and cirrus clouds. This is confirmed by filtering out the irradiance reductions to highlight the enhancements. Similar results could be found for all available measurement stations and all analyzed years.

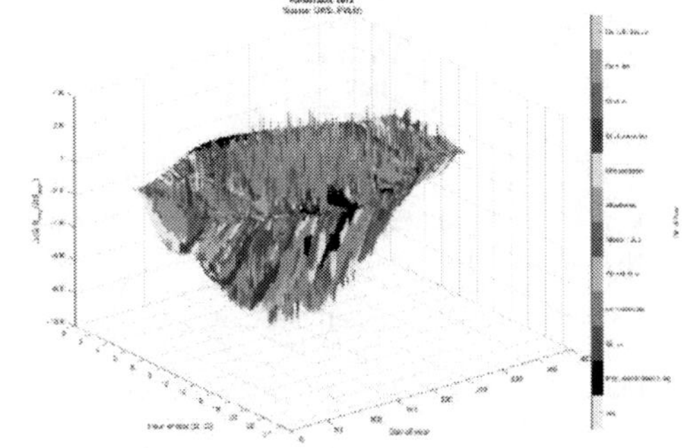

Figure 5: Yearly course of the irradiance enhancements and reductions in relation to the cloud types

3.7 Statistical evaluation of cloud type frequencies regarding irradiance enhancements

Table IV shows the frequency distribution of clouds. In the overall distribution, cumulus is the most common, with over 30%, followed by stratus, stratocumulus, and altocumulus clouds, each at about the same percentage.

When the frequencies of the individual cloud types are filtered according to the irradiance enhancements, a different frequency distribution shows up. The significant increase in the proportion of cirrus, cumulus, and cirrostratus clouds compared to the base amount indicates that these clouds contribute the most to irradiance enhancement effects. In contrast, there is a significant decrease in the frequency of altostratus, nimbostratus, stratocumulus, and stratus clouds with respect to irradiance enhancements. When the filter is changed to irradiance reductions, only a slight change in distribution is seen. Cirrus is considerably less common here, and in addition, cumulus clouds are slightly increased.

Table IV: Irradiance enhancement (IE) / reduction (IR) frequencies in relation to the cloud types of cloud layer 1 (CT1)

Cloud type of cloud layer 1	Total CT1 frequency	IE CT1 freq.	IR CT1 freq.
Cirrus	8.34 %	14.04 %	5.35 %
Cirrocumulus	0.24 %	0.37 %	0.27 %
Cirrostratus	5.30 %	8.01 %	4.35 %
Altocumulus	17.13 %	13.61 %	16.51 %
Altostratus	1.18 %	0.13 %	0.95 %
Nimbostratus	0.84 %	0.00 %	0.84 %
Stratocumulus	17.66 %	4.95 %	15.71 %
Stratus	18.06 %	2.76 %	19.72 %
Cumulus	31.25 %	56.13 %	36.31 %

The cloud type data is separated into four different cloud layers. To consider this in the analysis, a heatmap (Figure 6) is created. This heatmap includes the absolute amount of cloud type datapoints when irradiance enhancements that exceed 20 Wm^{-2} were measured, for the first two cloud layers. Cloud layers 3 and 4 are not considered in this analysis, because they only include a small amount of total data points. It should be noted though that cloud layer 4 includes the cumulonimbus data. The heatmap shows the interaction of the first two cloud layers and their relationship to the irradiance enhancement phenomenon. The biggest influence on IEs, according to this heatmap, can be found for the case of cumulus clouds in layer 1 with no clouds present in layer 2, cirrus clouds in layer 1 with no clouds present in layer 2, cumulus clouds in layer 1 with altocumulus clouds in layer 2, cumulus clouds in layer 1 with stratocumulus clouds in layer 2 and cirrostratus clouds in layer 1 with no clouds in layer 2.

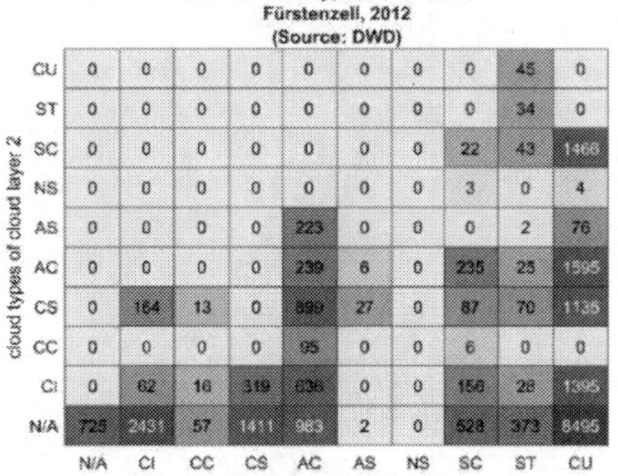

Figure 6: Heatmap containing the absolute count of cloud types when irradiance enhancements larger than 20 Wm^{-2} were detected

10.4229/EUPVSEC2023/4DV.4.14
020450-005

4 ANALYSIS OF SOLAR VARIABILITY IN RELATION TO CLOUD PROPERTIES BY THE MEANS OF IRRADIANCE GRADIENT PLOTS

4.1 General analysis of irradiance gradients

To better analyze the volatility of solar irradiance, the difference of the measured and the modelled GHI is replaced with the gradient of measured solar irradiance.

Figure 7 shows that the highest irradiance gradients and thus the highest volatility of solar irradiance can be observed in times where the solar irradiance is generally high, which is especially the case during the day and in the summer. This suggests that the influence of clouds on solar volatility is particularly high during these times. This also confirms the similar results found by Stut et al. (2020). Like with the plots in the last chapter, the color grading of the gradient plot is also replaced by the total cloud cover and the cloud types in the following subchapters.

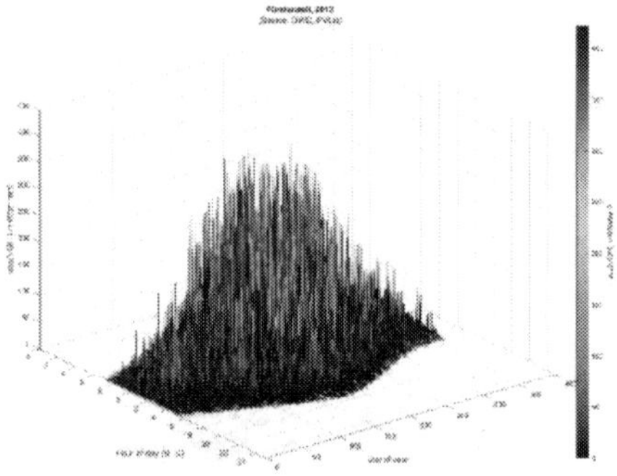

Figure 7: Yearly course of the absolute values of solar irradiance gradients

4.2 Irradiance gradients in relation to the total cloud cover (TCC)

The yearly gradient plot (Figure 8) shows no clear correlation between the total cloud cover and the solar irradiance gradients, except for days that are either completely clear or completely covered. These two cases seem to lead to overall lower gradients. The graphic also indicates that a medium total cloud cover leads to the highest gradients, but this cannot be conclusively determined from the graphic alone.

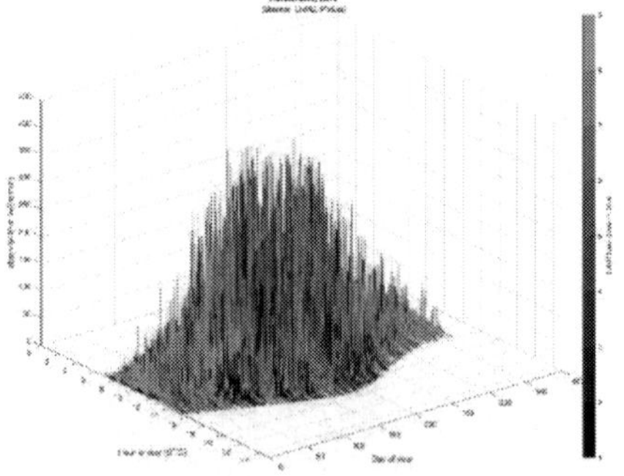

Figure 8: Yearly course of the absolute values of solar irradiance gradients in relation to the total cloud cover

4.3 Irradiance gradients in relation to cloud types

An analysis of the yearly irradiance gradients course (Figure 9) indicates that most irradiance gradients are caused by, or associated with, cumulus clouds. However, it should be noted that cumulus clouds predominantly occur during the day when gradients tend to be high anyway. In addition, an occasional influence from stratocumulus, altocumulus, and cirrus clouds can be observed. A heatmap analysis of the cloud types of cloud layers 1 and 2 with gradients larger than 35 $Wm^{-2}min^{-1}$ has been conducted and the findings suggest a similar pattern to the heatmap analysis of irradiance enhancements.

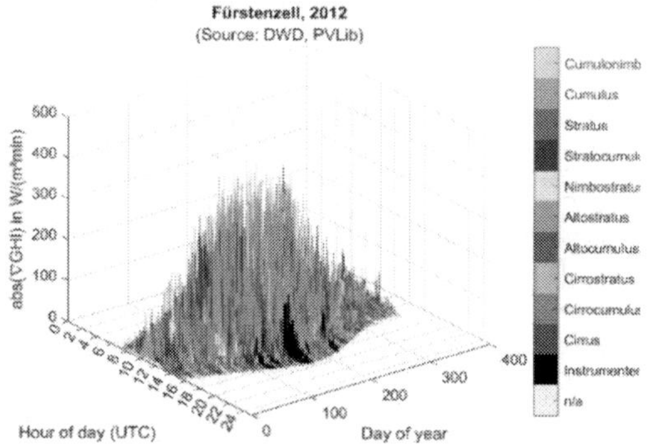

Figure 9: Yearly course of the absolute values of solar irradiance gradients in relation to the cloud type

The most influence on gradients could be found with cumulus clouds in layer 1 with no clouds in layer 2, cumulus clouds in layer 1 with altocumulus clouds in layer 2, cumulus clouds in layer 1 with stratocumulus clouds in layer 2 and both cumulus clouds in layer 1 with cirrostratus clouds in layer 2, as well as altocumulus clouds in layer 1 with no clouds in layer 2.

5 CLOUD TYPE CLASSIFICATION AND IRRADIANCE VOLATILITY ANALYSIS BY IRRADIANCE ENHANCEMENT/REDUCTION AND GRADIENT SCATTER CHARTS

Following a data preparation procedure suitable for future machine learning approaches, two features have been selected to give detailed insight into the irradiance behavior of different cloud types. To visualize these features the gradient of the irradiance and the irradiance enhancements are plotted onto the abscissa and the ordinate of a scatter plot. Filtering the data for the different cloud types then yields the depiction like shown in fig. 10. The plot shows the irradiance behavior of all available cloud types of cloud layer 1.

The patterns that emerge in this type of depiction can be observed in all analyzed measuring stations and years, thus suggesting that these patterns are cloud-specific, at least as far as locations within the pre-Alpine regions in southern Germany are concerned.

5.1 Cirrus

Cirrus clouds appear to have a medium variance when it comes to the gradient. Besides a few outliers, most points accumulate inside the ±100 $Wm^{-2}min^{-1}$ range. The irradiance enhancement behavior is characterized by points that lie mostly within the range of

10.4229/EUPVSEC2023/4DV.4.14
020450-006

150 Wm^{-2} and -200 Wm^{-2}, which means that cirrus clouds contribute to a substantial amount of medium irradiance enhancements and also account for a good portion of irradiance reductions that can – in extreme cases – go as low as -800 Wm^{-2} and more, which then seem to mainly gather in the gradient area around 0 $Wm^{-2}min^{-1}$ with very small variance. While the most extreme values here could indeed be attributed to analysis inaccuracies due to the prior assumption of constant cloud types within an hour and the inaccurate nature of measurements at the beginning and end of a day, the general trend can be observed in all stations and years, which suggests at least a certain correlation.

5.2 Cirrocumulus

Cirrocumulus besides from accounting only to a small absolute occurrence, generally seem to have little variance in both the gradient and the irradiance enhancement axes. The data points can be mostly found in the range of -300 Wm^{-2} and 100 Wm^{-2} in the case of irradiance enhancements and show a similar variance of the irradiance gradient to the cirrus clouds.

5.3 Cirrostratus

Cirrostratus clouds behave like cirrus clouds, but without the extreme outliers regarding the irradiance reductions. The main cluster is located within the ±100 $Wm^{-2}min^{-1}$ gradient range. Irradiance enhancements up to around 100 Wm^{-2} can be observed while reductions reach to around -450 Wm^{-2}. The total amount of data points for this cloud type is relatively small.

5.4 Altocumulus

Altocumuli show a higher variance in both the gradient and the irradiance enhancements. While most points are still concentrated around a gradient of 0 $Wm^{-2}min^{-1}$, they also seem to be increasingly prevalent in higher gradients compared to the previous three cloud types, reaching up to ±400 $Wm^{-2}min^{-1}$. Irradiance enhancements are also substantially higher and can reach values of up to 300 Wm^{-2} in comparison to clear sky conditions. The irradiance reductions reach as far as -700 to -800 Wm^{-2} also with a decreasing variance of the gradient the higher the irradiance reductions become.

5.5 Altostratus

Altostratus clouds have a very narrow pattern where the main cluster of points seem to be located within a gradient of 50-100 $Wm^{-2}min^{-1}$. Apart from a few outliers, there are mainly irradiance reductions reaching as far as -700 Wm^{-2}. This suggests that altostratus has little influence on irradiance enhancement events or solar irradiance volatility.

5.6 Nimbostratus

Like in altostratus clouds, nimbostratus also has a very narrow irradiance pattern however more extreme in its extent. There are very few to no irradiance enhancement events and the cluster of points is almost entirely located within a thin band with very small gradients and a large irradiance reduction effect resulting in reductions as low as -800 Wm^{-2}. This pattern is consistent with the expected behavior of nimbostratus clouds, which mostly appear as contour less greyish-blue sky-covering cloud structures, thus blocking a lot of sunlight for extended time periods.

5.7 Stratocumulus

Stratocumulus clouds show a similar pattern to altocumulus, but thinner in its gradient wise extent and reaching a little lower in terms of irradiance reductions. As in altocumulus, they also show a wide spread of outliers and a substantial amount of irradiance enhancement events reaching around 200 Wm^{-2}. The effect of irradiance reduction can reach as low as -900 Wm^{-2}, suggesting a high sunlight blocking effect at certain times.

5.8 Stratus

Stratus is very similar to stratocumulus and a differentiation between their scatter plots is difficult. They share a similar main cluster, outliers, and ranges both regarding their gradients and irradiance enhancements and reductions, although the main cluster of stratus seems to be a little narrower than in stratocumulus.

5.9 Cumulus

Cumulus clouds show extreme behavior in several aspects. Firstly, they account by far for the most data points, leading to the largest point cloud of all cloud types. The gradient lies within the range of up to ±400 $Wm^{-2}min^{-1}$, which is also the highest in all cloud types and suggests a very high volatility. Irradiance reductions can reach as low as -900 Wm^{-2} and irradiance enhancements can reach values of 450 Wm^{-2}.

The scatter plot is widely spread in both dimensions, combining points of both high gradient values as well as high reduction or enhancement effects. This suggests a very high influence of cumulus clouds on both irradiance volatility as well as irradiance enhancement effects.

5.10 Cumulonimbus

Cumulonimbus clouds are only recorded within the fourth cloud layer and are therefore not yet considered in this analysis.

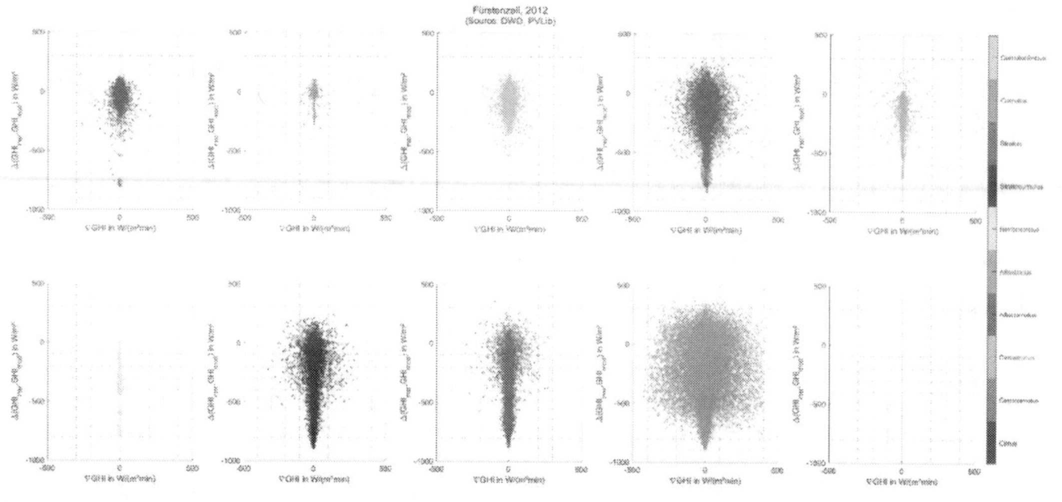

Figure 10: Irradiance enhancements, reductions and gradients scatter chart for the clouds in the first cloud layer

6 CONCLUSION AND OUTLOOK

The volatility of solar irradiance is a key factor in photovoltaics systems. In addition to previous studies, this analysis serves as a finalizing step to better understand irradiance phenomena regarding clouds and cloud cover, especially in the German pre-Alpine region.

The data confirmed the suggestions of some of the earlier research, that certain cloud types indeed seem to be mainly responsible for solar volatility and irradiance enhancement events, namely cumulus and cirrus clouds. Additionally, all clouds with cumulus-like properties and furthermore, cirrostratus clouds show this behavior to some extent. It is also shown that nimbostratus and altostratus have very little to no impact on irradiance enhancements and volatility, but a high impact on irradiance reductions.

The analysis also showed that most clouds form distinct recurring patterns when plotted in a gradient and irradiance enhancement/reduction scatter chart. This novel type of visualization method can form the basis for further analyses, especially for future machine learning classifications of cloud types and the improved forecasting of irradiance from past data.

However, it should be noted that the data basis was reduced because of data availability in the six observed measurement stations. Consequently, future work might focus on analyzing new measurement locations and more data with the methods proposed in this study, to validate the results and to examine the behavior of clouds and irradiance in other climate zones.

As some irradiance peaks can be on timeframes of seconds or even lower, higher resolution data (both synoptic cloud data as well as sub-second irradiances) could greatly improve the analysis. At this point, the study is still limited by the resolution of the available data. It should be also noted that the synoptic data has been manually recorded, so fluctuations in the accuracy might be possible for different stations and years.

Because of the way the DWD classifies clouds into different cloud layers and how cumulonimbus clouds are handled, these clouds have been neglected in this study and might be a topic of future research.

Furthermore, the analyses might be taken a step further by including new features like the clear sky index or the fractal dimension of irradiance curves to find possibly hidden patterns within the cloud properties.

REFERENCES

Almeida, M., Zilles, R. & Lorenzo, E., 2014. Extreme overirradiance events in São Paulo, Brazil. *Solar Energy*, Issue 110, pp. 168-173.

Balfour, J. & Shaw, M., 2011. *Introduction to photovoltaic*. s.l.:Jones Bartlett Learn.

Caldas, M. & Alonso-Suárez, R., 2019. Very short-term solar irradiance forecast using all-sky imaging and real-time irradiance measurements. *Renewable Energy*, Issue 143, pp. 1643-1658.

Chicco, G. et al., 2016. Error assessment of solar irradiance forecasts and AC power from energy conversion model in gridconnected photovoltaic systems. *Energies*, Issue 9.

Emck, P. & Richter, M., 2008. An Upper Threshold of Enhanced Global Shortwave Irradiance in the Troposphere Derived from Field Measurements in Tropical Mountains. *Journal of applied meteorology and climatology*, Issue 47, pp. 2828-2845.

Gueymard, C., 2017a. Cloud and albedo enhancement impacts on solar irradiance using high-frequency measurements from thermopile and photodiode radiometers. Part 1: Impacts on global horizontal irradiance. *Solar Energy*, Issue 153, pp. 755-765.

Gu, L. et al., 2001. Cloud modulation of surface solar irradiance at a pasture site in southern Brazil. *Agric. For. Meteorol.*, Issue 106, pp. 117-129.

10.4229/EUPVSEC2023/4DV.4.14
020450-008

Haurwitz, B., 1945. Insolation in Relation to Cloudiness and Cloud Density. *Journal of Meteorology*, Volume 2, pp. 154-166.

Haurwitz, B., 1946. Insolation in Relation to Cloud Type. *Journal of Meteorology*, Volume 3, pp. 123-124.

Ineichen, P. & Perez, R., 2002. A New airmass independent formulation for the Linke turbidity coefficient. *Solar Energy*, Volume 73, pp. 151-157.

Inman, R., Chu, Y. & Coimbra, C., 2016. Cloud enhancement of global horizontal irradiance in California and Hawaii. *ScienceDirect*, Issue 130, pp. 128-138.

Iqbal, M., 1983. *An introduction to solar radiation*. Canada: Academic Press.

Järvelä, M., Lappalainen, K. & Valkealahti, S., 2020. Characteristics of the cloud enhancement phenomenon and PV power plants. *Solar Energy*, Issue 196, pp. 137-145.

Lave, M., Kleissl, J. & Arias-Castro, E., 2012. High-frequency irradiance fluctuations and geographic smoothing. *Solar Energy*, Issue 86, pp. 2190-2199.

Norris, D., 1968. Correlation of solar radiation with clouds. *Sol. Energy*, Issue 12, pp. 107-112.

Padmakumari, B. et al., 2022. High temporal variability of surface solar irradiance due to cloud enhancement effect over the Western Ghat mountains in peninsular India. *Journal of Atmospheric and Solar-Terrestrial Physics*, Issue 232.

Paulescu, M. et al., 2014. Nowcasting solar irradiance using the sunshine number. *Energy Conversion and Management*, Issue 79, pp. 690-697.

Paulescu, M. & Paulescu, E., 2019. Short-term forecasting of solar irradiance. *Renewable Energy*, Issue 143, pp. 985-994.

Pecenak, Z. et al., 2016. Simulating irradiance enhancement dependence on cloud optical depth and solar zenith angle. *Sol. Energy*, Issue 136, pp. 675-681.

Perez, R. et al., 2016. Spatial and Temporal Variability of Solar Energy. *Foundations and Trends ® in Renewable Energy*, 1(1), pp. 1-44.

Quiroz, J. & Cameron, C., 2012. *Technical analysis of prospective photovoltaic systems in Utah*, United States: N.p..

Ranaweera, I., Midtgård, O. & Yordanov, G., 2014. *Short-term intermittency of solar irradiance in southern Norway*. Amsterdam, The Netherlands, 29th European Photovoltaic Solar Energy Conference and Exhibition.

Reno, M., Hansen, C. & Stein, J., 2012. *Global Horizontal Irradiance Clear Sky Models: Implementation and Analysis*, s.l.: Sandia National Laboratories SAND2012-2389.

Roy, S., 2015. Statistical estimates of short duration power generated by a photovoltaic unit in environment of scattered cloud cover. *Energy*, Issue 89, pp. 14-23.

Schade, N., Macke, A., Sandmann, H. & Stick, C., 2007. Enhanced solar global irradiance during cloudy sky conditions. *Meteorol. Zeit.*, Issue 16, pp. 295-303.

Stut, N. et al., 2020. *Analyse von Einstrahlungsvolatilitäten und -überhöhungen in hochaufgelösten Datensätzen des DWD und MIM zur Untersuchung von Korrelationen zu meteorologischen Messdaten*. 35. Symposium Photovoltaische Solarenergie, s.n.

Thuillier, G., Perrin, J.-M., Keckhut, P. & Huppert, F., 2013. Local enhanced solar irradiance on the ground generated by cirrus: measurements and interpretation. *APPRES*, Issue 7, pp. 073543-073543.

Tzoumanikas, P., Nikitidou, E., Bais, A. & Kazantzidis, A., 2016. The effect of clouds on surface solar irradiance, based on data from an all-sky imaging system. *Renew. Energy*, Issue 95, pp. 314-322.

Wirth, G. et al., 2015. Modeling the maximum power output of a distributed PV fleet. *Progress Photovolt. Res. Appl.*, Issue 23, pp. 1164-1181.

Yordanov, G. et al., 2013. Overirradiance (cloud enhancement) events at high latitudes. *IEEE J. Photovoltaics*, Issue 3, pp. 271-277.

Yordanov, G., Saetre, T. & Midtgård, O., 2015. Extreme overirradiance events in Norway: 1.6 suns measured close to 60°N. *Solar Energy*, Issue 115, pp. 68-73.

Zehner, M. et al., 2010. *Systematic analysis of meteorological irradiation effects*. Proc. 25th EU PVSEC Conf., Valencia, Spain, s.n.

Zehner, M. et al., 2011. *Energy Loss Due to Irradiance Enhancement*. Proceedings of the 26th Photovoltaic Solar Energy Conference, Hamburg, s.n.

10.4229/EUPVSEC2023/4DV.4.14
020450-009

IMPACT OF DATA GRANULARITY ON NOWCASTING THE SOLAR RESOURCE

Sergiu-Mihai Hategan[1,2,*], Marius Paulescu[1]
[1]Faculty of Physics, West University of Timisoara, Romania
[2]Institute for Advanced Environmental Research, West University of Timisoara, Romania

[*]Corresponding author: sergiu.hategan98@e-uvt.ro

ABSTRACT: The smart grid concept is becoming increasingly relevant as renewable energy sources like PV become more prevalent in power grids worldwide. Accurate forecasting of PV power is essential for effective grid management and balancing with traditional sources. To maintain cost-effectiveness, computation and data recording costs must be factored into the forecast. This study aims to examine the impact of data granularity on machine learning models for solar resource forecasting. For a forecast horizon of 15 minutes, four different granularities and four machine learning models: gradient boosted decision trees (GBT), random forest (RF), multilayer perceptron (MLP), and long short-term memory (LSTM) were tested. The results show that models such as GBT and LSTM are less affected by changes in granularity, while RF and MLP models show some sensitivity. This study provides valuable insights for developing more accurate solar power forecasting models and optimizing the efficiency of the power grid.
Keywords: machine learning, nowcasting, solar resource, data granularity

1 INTRODUCTION

The innate variability of photovoltaic (PV) power, an outcome of the unpredictable and inherent variability of solar irradiance due to weather conditions, may inhibit the PV electricity's contribution to the overall energy mix [1]. To manage electricity grids effectively, which integrate a significant proportion of PV plants, the need for precise short-term predictions of PV power output becomes critical [2]. Because the output power of a PV plant is strongly correlated to the in-plane solar irradiance, nowcasting this resource is highly dependent on accurate forecasts of horizontal solar irradiance [3].

For intra-hour forecasts of solar irradiance there are several methods currently in use both in practice and in literature. All these methods have shown various degrees of success and reliability [4]. First there is the class of statistical methods, such as autoregressive integrated moving average (ARIMA), exponential smoothing or random walk. Another class of models are sky imager-based methods and finally machine learning models [4].

Recently, machine learning methods have been applied to time series applications. A subset of machine learning models are decision trees ensembles: random forest (RF) and gradient boosted trees (GBT) being two such examples. GBT was used for intra- hour global horizontal irradiance forecast with significant success [5]. Artificial neural network models (ANN) also represent a current topic in solar radiation forecasting, having on average similar or better performances than classical statistical models for the intra-hour time horizon [6]. The current study considers both ensembles of decision trees and ANN models.

This paper departs from the results and ideas obtained in previous research [3]. In this study, statistical models like moving average and random walk were found to not exhibit a clear influence from data granularity. The two-state model, however, benefited from a finer sampling rate for the same lead time. In the present paper, the studied machine learning models were developed and tested with data recorded on the Solar Platform of the West University of Timisoara, Romania [7].

As to our knowledge, there are few to none studies on the impact of data granularity on solar resource forecasting. The current study links in an innovative manner the performance in solar resource forecasting to a dataset granularity. The results represent a tool for solar engineers to choose the appropriate forecast method depending on the available data granularity.

The paper is organized as follows: Section 2 presents the overall methodology, describing each model employed. Section 3 presents the data used for the construction and testing of the models. Section 4 presents the results and findings.

2 METHODOLOGY

Previous research was preoccupied with studying the influence of granularity on classical statistical models such as ARIMA, random walk or exponential smoothing. For the present study, we seek to identify correlations between the performance of four machine learning models and the data granularity available, for a set forecast horizon. The forecast horizon of 15 minutes was chosen, being relevant for energy providers across the European grid [8]. For this horizon and a lead time of 15 minutes, we train and test the models at four different granularities. Thus, the possible time steps are: 1 minute, 3 minutes, 5 minutes and 15 minutes.

Since we operate with machine learning models, we can utilize all the available variables when constructing the models. These variables depend on the capabilities of each measurement station, and in Section 3 we enumerate the measured variables in the data. The models and their predictions operate on the clearness index, defined as the ratio between GHI and the extraterrestrial irradiance. The reason is to eliminate the cyclical behavior of solar irradiance, and to facilitate generalization of the models. Since the extraterrestrial irradiance is a deterministic variable, the conversion from predicted clearness index to predicted GHI is trivial.

10.4229/EUPVSEC2023/4DV.4.15
020452-001

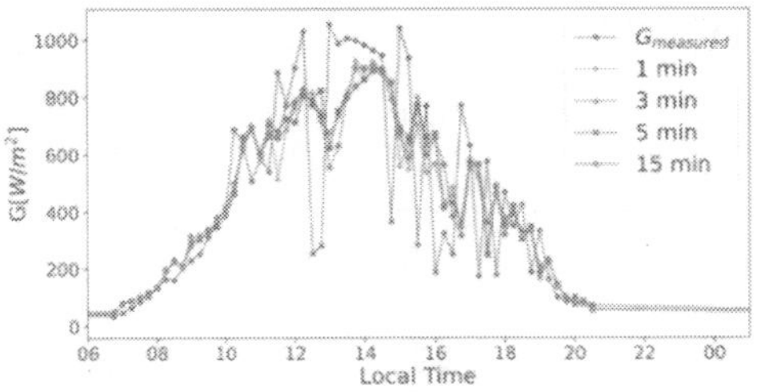

Figure 1. Performance of the GBT model at the four given granularities.

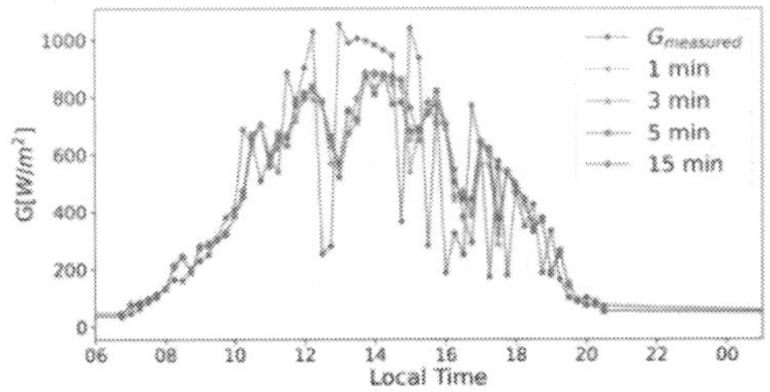

Figure 2. Performance of the RF model at the four given granularities.

2.1. Gradient Boosted Trees (GBT)

GBT, short for Gradient Boosting Trees, is a model that is easy to train and requires minimal preprocessing. It has a quick training time and achieves improved performance compared to individual decision tree regression models. GBT is an ensemble model consisting of multiple decision tree regression models, where each subsequent model aims to optimize the objective function where previous models may have failed. It has been successfully applied in solar resource and PV power forecasting, both for deterministic and probabilistic forecasts [5].

In the specific study mentioned, the GBT model used 200 estimators, a maximum tree depth of 3, and a gamma value of 0.7 for all time horizons. We implemented it through the open source Python library xgboost [9].

2.2 Random Forest (RF)

Random Forest is a machine learning algorithm that can be sufficiently accurate in short-term solar resource forecasting. It operates by constructing an ensemble of decision tree models and aggregates their predictions to make accurate forecasts. As all machine learning models, it can capture complex nonlinear relationships between various meteorological and environmental variables, for a precise solar resource forecast.

When building the model, we employed the following hyperparameters for all granularities: 200 estimators and a maximum depth of 5, as we found the other hyperparameters to have less of an impact on performance.

2.3 Multilayer Perceptron (MLP)

The Multilayer Perceptron (MLP) neural network, a popular class of artificial neural networks, holds great promise for short-term solar forecasting. It is the

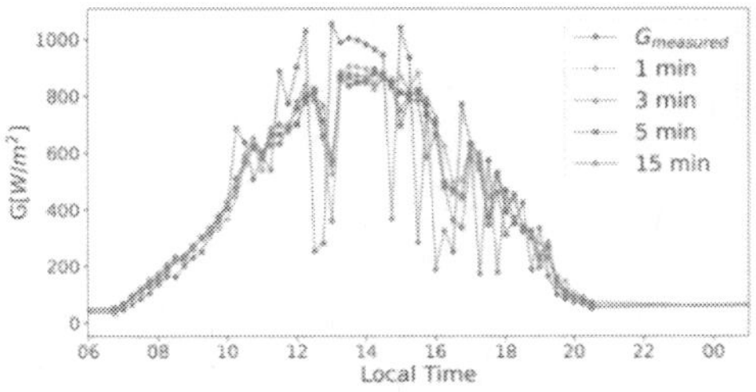

Figure 3. Performance of the LSTM model at the four given granularities

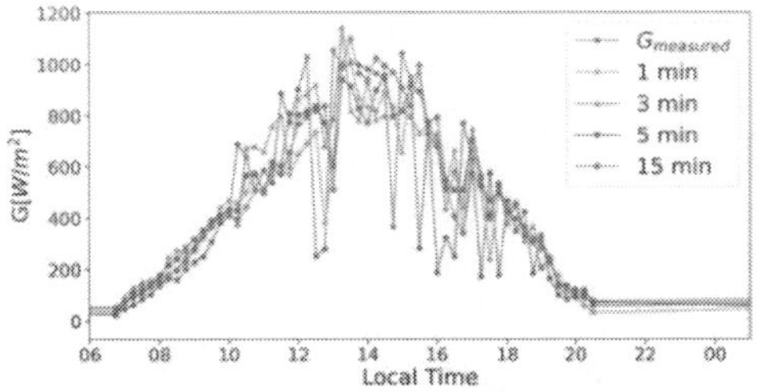

Figure 4. Performance of the MLP model at the four given granularities

simplest neural network architecture, consisting of interconnected artificial neurons [10]. In short-term solar forecasting, the MLP neural network can be employed to model and predict the intricate patterns and dependencies present in solar irradiance, temperature, and other relevant meteorological variables. Its ability to learn and generalize from historical data allows it to generate accurate forecasts.

For this study, we found that deeper networks did not produce satisfactory results, so we remained at an architecture of two hidden layers of 16 neurons each. The training was realized over 100 epochs with batch size of 32.

2.4. Long Short-Term Memory (LSTM)

LSTM networks, a subset of deep recurrent neural networks, are highly regarded for their ability to address predictive models in various domains like image recognition, automatic speech recognition, and natural language processing [11]. They excel at overcoming limitations of standard data-driven models by capturing short-term and long-term dependencies between desired outcomes and historical variables. This makes them well-suited for assimilating and learning data across different time scales, particularly in solar resource forecasts. However, they can have long training times and require more computational power.

For simplicity, our model's architecture included only one LSTM layer and a dropout layer with a dropout value of 0.2, and a final layer with a single neuron using linear activation. The mean squared error was optimized using the ADAM optimizer, with 100 epochs and a batch size of 32 performing the best.

3 DATA

This study was conducted with data of high quality recorded on the Solar Platform of the West University of Timisoara, Romania [7]. Radiometric (global and

10.4229/EUPVSEC2023/4DV.4.15
020452-002

Table I: Performance of the models when forecasting GHI, grouped by granularity.

Gran.	1min	3 min	5 min	15 min
GBT				
nRMSE[%]	32.88	32.86	32.89	32.56
SS[%]	17.31	17.33	17.26	18.10
nMBE	-0.030	-0.017	-0.013	-0.001
RF				
nRMSE[%]	32.99	33.83	33.65	34.07
SS[%]	17.03	14.91	15.30	14.31
nMBE	0.005	0.007	0.008	0.011
LSTM				
nRMSE[%]	32.78	33.03	32.91	33.17
SS[%]	17.54	16.92	17.21	16.55
nMBE	0.004	0.005	0.005	0.004
MLP				
nRMSE[%]	34.73	33.92	34.56	34.08
SS[%]	12.64	14.69	13.07	14.29
nMBE	-0.006	0.027	-0.007	0.016

Table II: Performance of the models when forecasting clearness index, grouped by granularity.

Gran.	1min	3 min	5 min	15 min
GBT				
nRMSE[%]	30.93	30.89	31.04	30.78
SS[%]	16.68	16.78	16.39	17.08
nMBE	-0.015	-0.004	-0.002	0.014
RF				
nRMSE[%]	31.60	32.14	32.25	32.57
SS[%]	14.87	13.42	13.13	12.24
nMBE	0.019	0.020	0.021	0.024
LSTM				
nRMSE[%]	31.18	31.27	31.25	31.82
SS[%]	16.02	15.78	15.82	14.28
nMBE	0.019	0.018	0.026	0.029
MLP				
nRMSE[%]	33.32	33.23	33.24	32.97
SS[%]	10.23	10.46	10.47	11.19
nMBE	0.007	0.055	0.016	0.031

diffuse solar irradiances) and meteorological (air temperature and wind speed) data were recorded with a Delta-T Devices WS-GP2 weather station equipped with a sunshine pyranometer SPN1. Other variables recorded were the clearness index and sunshine number (SSN), defined as a binary variable taking the value of 1 when direct normal irradiance exceeds 120 W/m^2, and 0 otherwise.

As customary, we separate the data into a training and a test set. The training set consists of data measured in the period 5-11 May 2023, while the test set was recorded during six days of May 2023, 21-23 and 25-27. The test days were properly selected to experience representative variability in the state-of-the-sky.

4 RESULTS AND DISCUSSIONS

The metrics used to quantify the performance of the models are normalized root mean squared error (nRMSE), normalized mean bias error (nMBE) and skill score (SS). Skill score is defined in comparison with persistence forecast.

The general behavior of the models, as influenced by granularity, can be inferred from Figs 1-4. The GBT models closely follow the same curve, while the RF and LSTM models slowly diverge as the time step increases (from 1 minute to 15 minutes). The MLP models show significant variation between predictions.

The numerical results, presented in terms of three statistical indicators are shown in Tables I and II. Table I showcases the statistical metrics of the models for the GHI forecast. We observe that the GBT maintains a consistent performance in terms of nRMSE and skill score, with skill score values between 17 and 18%. The bias however, decreases, as the data points become sparser, at the 15 minute granularity. The RF models, perform worse at this granularity, with a decrease in skill score of 2.72% from the best performance at the granularity of 1 minute. The bias follows an inverse trend, with a slow increase as the time step interval increases.

LSTM models also show a similar behavior to RF, however, their degradation is not as significant, with a difference of 1%. The bias remains constant at about 0.005 regardless of data granularity, suggesting good reliability. The MLP models show the most chaotic behavior, performing the worst overall regardless of data granularity.

Table II presents the same metrics for the raw, clearness index forecasts. The behavior is largely similar to the GHI forecasts. We identify that small skill score differences between the clearness index forecasts translate into larger performance differences for the GHI forecasts.

Overall, the highest performances were obtained by the GBT and LSTM models, with the GBT model's bias being more affected by granularity. The LSTM model provided slightly lower performance, with consistent low bias across granularities.

5 CONCLUSIONS

This study was focused on the different impact of data granularity on several machine learning methods. Building on previous work, which was preoccupied with classical statistical models, we turned our attention to the emerging class of machine learning methods. We tested two decision tree regressors ensembles, GBT and RF, and two ANN architectures: LSTM and MLP.

A fixed forecast horizon of 15 minutes was set for all models. For this horizon of 15 minutes, GBT and LSTM models were not significantly affected by changes in granularity, with differences of maximum 1% in terms of skill score. RF and MLP were more dependent on granularity, with MLP providing lower performances and higher randomness than the other models. In terms of bias, GBT performed better at the 15 minutes granularity, while the others were mostly unaffected, considering the forecast horizon of 15 minutes.

Studying the performance metrics of both the clearness index forecasts and the GHI forecasts, we observed that small differences in the clearness index forecasts translate to greater effects in the final GHI forecasts. Future studies could cover different forecast horizons, or different classes, such as sky-imagery.

6 ACKNOWLEDGMENTS

This work was supported by IOSUD—West University of Timisoara. The work was developed within a grant of the Romanian Ministry of Education, UEFISCDI, project number PN-III-P2-2.1-PED-2021-0544

7 REFERENCES

[1] Blaga R, Paulescu M. Sol. Energy 2018, 174, 606.
[2] Rodríguez F, Galarza A, Vasquez JC, Guerrero JM. Energy 2022, 239 B, 122116.
[3] Paulescu M, Paulescu E. Renew. Energy 2019 143, 985.
[4] Chu, Y, Li M, Coimbra, CFM, Feng D, Wang, H. iScience 2021, 24, 10.
[5] Pedro HTC, Coimbra CFM, David M, Lauret P. Renew. Energy 2018, 123, 191.
[6] Blaga R, Sabadus A, Stefu N, Dughir C, Paulescu M, Badescu V. Prog. Energy Combust. Sci. 2019, 70, 119-144.
[7] Solar Platform of the West University of Timisoara, Romania. http://solar.physics.uvt.ro/srms/ (Accessed August 2023).
[8] ENTSO-E: Single Intraday Coupling (SIDC) https://www.entsoe.eu/network_codes/cacm/implementation/sidc/ (Accessed August 2023).
[9] Chen, T, Guestrin, C. In Proceedings of the 22nd ACM SIGKDD International Conference on Knowledge Discovery and Data Mining, San Francisco, CA, USA, 13—17 August 2016, 785–794.
[10] Haykin, S. Neural Networks: A Comprehensive Foundation; Prentice Hall PTR: Hoboken, NJ, USA, 1994.
[11] Hochreiter S, Schmidhuber, J. Neural Comput. 1997, 9, 1735–1780.

A MULTI-MODEL APPROACH TO NOWCASTING CLOUD SHADOW

Marius Paulescu[1,*], Sergiu-Mihai Hategan[1,2], Eugenia Paulescu[1], Andreea Sabadus[2], Ciprian Dughir[3]
[1]Faculty of Physics, West University of Timisoara, Romania
[2]Institute for Advanced Environmental Research, West University of Timisoara, Romania
[3]Faculty of Electronics, Telecommunications and Information Technologies, Politehnica University Timisoara, Romania

[*]Corresponding author: marius.paulescu@e-uvt.ro

ABSTRACT: Sunshine number (SSN) is defined as a binary quantifier for the relative position of the Sun and clouds: SSN =1 when the Sun shines and SSN = 0 otherwise. The state SSN = 0 directly quantifies the presence of shadow on the ground, specifically even on the place where a PV system could be operational. Consequently, accurate nowcasting of SSN could contribute decisively to intra-hour nowcasting of solar irradiance and PV power. Being a binary quantity, nowcasting SSN is a great challenge. This study proposes an ensemble forecast model with several innovative features: diversity in nature of the component models, decreasing the persistence by bringing into the equations data resulting from a sky imager and the parameters adaptation to the actual atmospheric conditions and a periodic adjustment of the parameters. The tests indicate that the multi-model approach demonstrates a visibly better performance than the individual models. The study was conducted with high-quality radiometric data and high-resolution all-sky images recorded on the Solar Platform of the West University of Timisoara, Romania
Keywords: nowcasting; clouds; sunshine number; solar resource

1 INTRODUCTION

The uncontrolled fluctuation of the photovoltaic (PV) power limits the share of PV electricity in the energy mix. This is due to the intrinsic variability of solar irradiance, caused by weather, which is essentially erratic in nature [1]. Smart management of the power grids, comprising a substantial segment of PV plants, requires accurate short-term forecasts of PV power production (see e.g. [2]). Because the output power of a PV plant is strongly correlated to the in-plane solar irradiance, nowcasting the relative position of the Sun and clouds is an essential task for accurate nowcasting of solar irradiance [3] and PV power [4].

In this context, the present work deals with intra-hour forecasting of sunshine number (SSN), a binary quantifier showing whether the Sun shines (SSN = 1) or not (SSN = 0) [5]. The state SSN = 0 represents a basic indicator for the existence of cloud shadow over the modules of a PV system. Being a binary quantity, nowcasting SSN always constitutes a challenge. A large research effort was committed for accurately nowcasting SSN (e.g. [6]). However, there still is plenty of room for improving the quality of forecasts.

This paper reports the results obtained in continuation of the research presented at WCPEC-8 [6]. An innovative ensemble model for intra-hour SSN forecasting is presented. Forecasts issued by three models are combined into a new multi-model forecast. The three individual models are very different in their nature: a classic statistical model (Autoregressive Integrated Moving Average ARIMA [7]), a machine learning model (Gradient Boosted Trees GBT [8]) and an adapted model based on sky imagery [9]. The ensemble was developed and tested with data recorded on the Solar Platform of the West University of Timisoara, Romania [10] The results (see Sec. 3) show that the proposed multi-model approach demonstrates a visibly better performance than the individual models.

The paper is organized as follows. Section 2 introduces the ensemble model. Section 3 summarizes the relevant data. Section 4 presents the results. The main conclusions are gathered in Sec. 5.

2 ENSEMBLE FORECAST MODEL

The three models included in the ensemble were chosen based on our previous results in forecasting SNN, solar irradiance and PV power (e.g. [4], [6]). The ensemble forecast model was built in two steps. At the first step each individual model was calibrated to issue accurate SSN forecasts for the test location. At the second step, the ensemble itself was devised. Three model equations were established, with the objective of maximizing the accuracy of SSN forecasts. The coefficients of the three independent models in the final forecast were evaluated based on historical data, frozen and always used in the future. The difference between the three models is given by the way in which the coefficients of each model are computed. The construction methods considered in this study are: arithmetic mean, linear regression and Ridge linear regression. The three individual models are briefly next.

2.1 ARIMA

The acronym ARIMA defines a class of time-series models, very often used in practice. An ARIMA(p,d,q) model includes the following terms: (1) the autoregressive part of order p; (2) the moving-average part of order q, and (3) the non-seasonal differencing of order d. Even if the general equation of ARIMA models seems to be complicated, in fact, a practical model is a simple linear equation. Based on the parsimony principle, the ARIMA(1,1,1) model was selected in this study:

$$\hat{u}_t = u_{t-1} + \phi_1(u_{t-1} - u_{t-2}) + \theta_1(\hat{u}_{t-1} - u_{t-1}) + \varepsilon_t \quad (1)$$

The coefficients were obtained by using the maximum likelihood method [8]. ARIMA forecasts are not binary numbers. The forecasts are projected into the binary set considering the switch from 0 to 1 at 0.5.

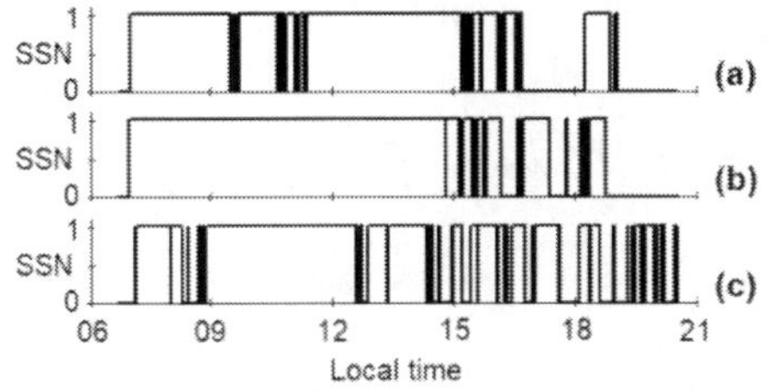

Figure 1: Variation of the sunshine number (SSN) in three days from May 2023: (a) 21, (b) 22 and (c) 23.

Table 1: Definition of the misclassification table

		Model-based class	
		SSN = 1	SSN = 0
Measurement-based class (truth)	SSN = 0	*p*00	*p*01
	SSN = 1	*p*10	*p*11

2.2. GBT

GBT is a model that is easy to train and requires minimal preprocessing steps. It has a short training time compared to other models. GBT is an ensemble of decision tree regression models. Each individual model in the GBT ensemble aims to optimize the objective function where the previous models might have failed. This allows GBT to achieve better performance than using just a single decision tree regression model. Previous research has demonstrated that GBT performs well in solar resource and PV power forecasting [8].

2.3 SI

The proposed sky imagery method is constructed from two algorithms, principal component analysis (PCA) and extremely randomized trees (ET) [9]. For issuing a forecast at a time instant, the method needs to input two consecutive sky images: the current image and the image recorded at the previous time step. The two images are then compressed into a vector of size 128x128x3x2. The PCA algorithm is used to reduce the vector dimension to 50. A decision tree classifier ensemble, in this case ET, is used to forecast the SSN, based on the time series of consecutive images.

3 DATA

This study was conducted with high-quality data recorded on the Solar Platform of the West University of Timisoara, Romania [10]. Radiometric (global and diffuse solar irradiances) and meteorological (air temperature, wind speed) data were recorded with a Delta-T Devices WS-GP2 weather station equipped with a sunshine pyranometer SPN1. Series of SSN values were derived from radiometric measurements on the basis of the WMO sunshine criterion [11]: the Sun is shining at time t if the direct-normal solar irradiance $G_{dn,t}$ exceeds 120 W/m^2, i.e.:

$$SSN_t = \begin{cases} 1 & \text{if} \quad G_{dn,t} > 120 \text{ W/m}^2 \\ 0 & \text{otherwise} \end{cases} \quad (2)$$

The all-sky imager ASI-16 from EKO Instruments monitors the state-of-the-sky at 1-minute resolution.

Table 2: Implemented models

Model	Definition	Aggregation
ARIMA	Sec. 2.1	-
GBT	Sec. 2.2	-
SI	Sec. 2.3	-
ENa	Ensemble	Arithmetic mean
ENw	Ensemble	Linear regression
ENR	Ensemble	Ridge linear regression

Table 3: Ensemble coefficients (Eq.3) for different forecast horizon *FH* [min]

Ensemble	*FH*	*a*	*b*	*c*	*d*
ENw	5	0.004	0.003	0.990	0.003
	15	-0.216	0.269	0.785	0.086
	30	-0.164	0.366	0.698	0.046
ENR	5	0.157	0.164	0.361	0.182
	15	0.090	0.114	0.157	0.374
	30	0.052	0.091	0.102	0.438

The raw fisheye images were processed (defishing, cloud identification, binary conversion) with FindClouds Trinity software provided by the ASI-16 manufacturer.

The training set of data was recorded between 5 and 11 May 2023, while the test set was recorded during six days of May 2023, 21-23 and 25-27. The test days were selected to experience different variabilities in the state-of-the-sky (Fig. 1).

4 RESULTS AND DISCUSSIONS

This section summarizes the results. Firstly, the metric used to evaluate the models' performance is introduced. Then the three proposed ensembles are presented and assessed.

4.1 Misclassification probability

The accuracy of forecasting a binary variable, like SSN, can be appropriately examined as a classification problem. The results are displayed as in Table 1, where *p*00, *p*01, *p*10 and *p*11 represent the probabilities of occurrence of the corresponding events. For instance, *p*01 represents the probability as the measured value SSN = 0 (the Sun is covered by clouds) to be falsely forecasted as SSN = 1 (the Sun shines on the sky). The total misclassification $TM = p01 + p10$, is a relevant quantifier for measuring the quality of a model.

4.2 Individual models and proposed ensembles

Table 2 lists the individual models and the proposed ensembles. The ensembles are defined by a simple linear equation:

$$\text{EN} = a\cdot\text{ARIMA} + b\cdot\text{GBT} + c\cdot\text{SI} + d \quad [3]$$

For ENa $a = b = c = 1/3$ and $d = 0$. For ENw and ENR, the coefficients are listed in Table 3.

4.3 Performance assessment

The individual models and the proposed ensembles were evaluated at forecast horizons *FH* between 5 and 30 minutes. Figures 2 and 3 display the results for the marginals forecast horizons, i.e. *FH* = 5 min and 30 min.

10.4229/EUPVSEC2023/4DV.4.16
020453-002

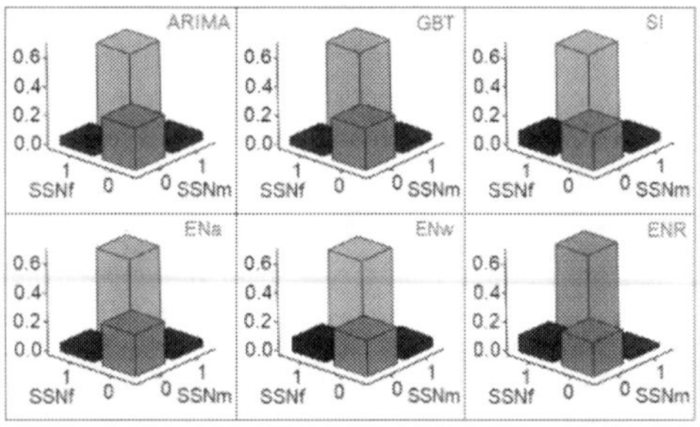

Figure 2: SSN misclassification at a forecast horizon of 5 minutes.

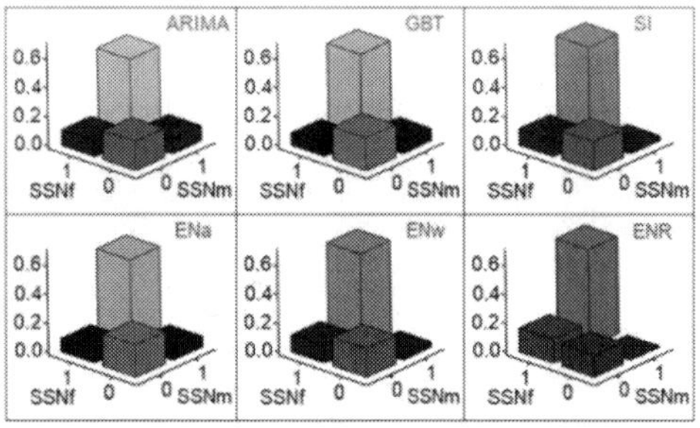

Figure 3: SSN misclassification at a forecast horizon of 30 minutes.

At first glance, irrespective of the model, a small percentage of forecasts is misclassified. Generally, the ensembles show a better performance than the individual models.

Table 4 presents the total misclassification of SSN forecasts at three forecast horizons: *FH* = 5 min, *FH* = 15 min and *FH* = 30 min. At *FH* = 5 min the minimum total misclassification is achieved by GBT with *TM* = 0.103. In other words, 10.3% of the issued forecasts are wrong, i.e. GBT forecasts sunny when the Sun is covered by clouds or vice versa. At *FH* = 15 min, ENa achieves the best performance, with *TM* = 0.156. At *FH* = 30 min, ENw achieves the best performance, with *TM* = 0.157. The most sensitive model to the forecast horizon is ARIMA. When *FH* increases from 5 to 30 min, *TM* increases twice from 0.115 to 0.224. GBT is less sensitive to *FH*, while SI is almost invariant to *FH*. This behavior of SI is transferred to ENw and ENR, which ascribe a large weight to SI in the aggregation of individual models (Table 2). Increasing the total misclassification with the forecast horizon does not necessarily represent a rule.

Differently, the variability in the state-of-the-sky controls the accuracy of SSN forecasting. Table 5 presents the total misclassification of SSN forecasts in two test days, 05/21 and 05/23, experienced a very different variability in the-state-of-the-sky. According to Fig. 1, 05/21 is a mostly clear sky day with passing clouds, the day 05/23 faced a high variability in the state-of-the-sky. The state-of-the-sky in the two days is illustrated by the fisheye photos taken at15:00 local time (Figure 4). Visual inspection of Table 5 shows that irrespective the model and time horizon the models' performance decrease considerably with the increasing in the variability of the state-of-the-sky.

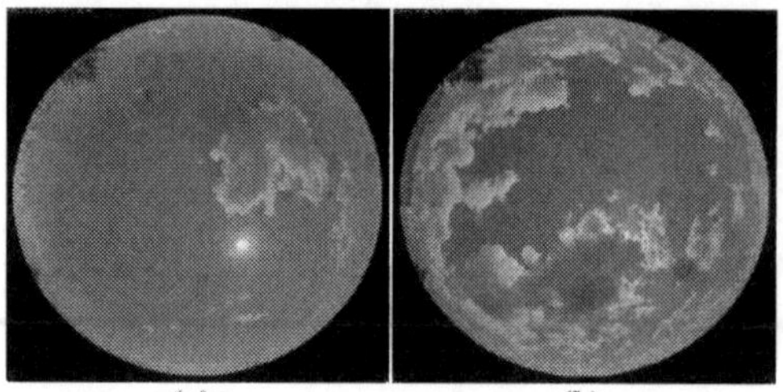

Figure 4: Fisheye photos of the sky taken on two test days at 15:00 local time: (a) 21/05/2023 (b) 23/05/2023.

Table 4: Total misclassification of SSN forecasts for three forecast horizons *FH* [min]

FH [min]	ARIMA	GBT	SI	ENa	ENw	ENR
5	0.115	0.103	0.157	0.107	0.157	0.133
15	0.171	0.168	0.162	0.156	0.162	0.174
30	0.224	0.174	0.157	0.173	0.157	0.175

Table 5: Total misclassification of SSN forecasts in two test days for three forecast horizons *FH* [min]

FH [min]	5		15		30	
Day	05/21	05/23	05/21	05/23	05/21	05/23
ARIMA	0.09	0.17	0.06	0.26	0.12	0.39
GBT	0.08	0.16	0.05	0.28	0.12	0.14
SI	0.08	0.21	0.00	0.25	0.04	0.21
ENa	0.09	0.16	0.02	0.26	0.04	0.17
ENw	0.09	0.28	0.00	0.25	0.04	0.21
ENR	0.07	0.28	0.00	0.25	0.08	0.21

5 CONCLUSIONS

Sunshine number (SSN) is defined as a binary quantifier for the relative position of the Sun and clouds. The state SSN = 0 is a basic indicator for the occurrence of cloud shadow at ground. This paper reported three ensemble forecast models for SSN forecasting. Each ensemble linearly aggregates three individual models very different in their nature: a classic statistical model (ARIMA), a machine learning model (GBT) and a model based on sky imagery. The results show that the ensembles seem to be less sensitive to the forecast horizon than the individual models. None of the SSN forecasting models can be ranked the best. At forecast horizons of minutes, the statistical model and machine learning model perform with reasonable accuracy. At larger forecast horizons the all-sky imager-based model becomes highly competitive, transferring its performance to the ensembles. The models' performance decreases with the increase in the variability of the state-of-the-sky. With all the effort devoted to decrease the persistence, there is still room enough for mitigation its effect.

6 ACKNOWLEDGMENTS

This work was supported by a grant of the Romanian National Authority for Scientific Research and Innovation, UEFISCDI project number PN-III-P2-2.1-PED-2021-0544.

10.4229/EUPVSEC2023/4DV.4.16
020453-003

7 REFERENCES

[1] R. Blaga, M. Paulescu. Sol. Energy 174 (2018) 606.
[2] F. Rodríguez, A. Galarza, J. C. Vasquez, J. M. Guerrero. Energy 239 B (2022) 122116.
[3] M. Paulescu, E. Paulescu. Renew. Energy 143 (2019) 985.
[4] M. Paulescu, N. Stefu, C. Dughir, A. Sabadus, D. Calinoiu, V. Badescu. Renew. Energy 195 (2022) 322.
[5] V. Badescu, M. Paulescu. Atmos. Res. 101 (2011) 194.
[6] M. Paulescu, N. Stefu, C. Dughir, A. Sabadus. In Proc. WCPEC-8, 1264-1267. DOI: 10.4229/WCPEC-82022-4BV.4.10
[7] G.E.P. Box, G.M. Jenkins, Time series analysis. Forecasting and Control. Holden-Day (1970).
[8] [H.T.C. Pedro, C.F.M. Coimbra, M. David, P. Lauret. Renewable Energy 123 (2018) 191-203
[9] A. Al-lahham, O. Theeb, K. Elalem, T.A. Alshawi, S.A. Alshebeili. Electronics 9 (2020) 700. https://doi.org/10.3390/electronics9101700.
[10] Solar Platform of the West University of Timisoara, Romania. http://solar.physics.uvt.ro/srms/ (Accessed August 2023).
[11] World Meteorological Organization WMO-No.8/2008 https://community.wmo.int/activity-areas/imop/wmo-no_8, (Accessed August 2023).

QATAR SOLAR ATLAS WEB APPLICATION

Nassma Mohandes, Daniel Perez-Astudillo, Sachin Jain, Dunia Bachour, Antonio Sanfilippo
Hamad Bin Khalifa University/Qatar Environment and Energy Research Institute
nsalim@hbku.edu.qa, dastudillo@hbku.edu.qa, sajain@hbku.edu.qa, dbachour@hbku.edu.qa,
asanfilippo@hbku.edu.qa

ABSTRACT: The Qatar Environment and Energy Research Institute (QEERI) launched a solar resource assessment research and development program in 2012. The solar data resources produced at QEERI have already been used by commercial companies in the development of plans for the 800 MW PV plant in Al Kharsaah. Through a collaboration agreement with Qatar Meteorological Department (QMD), QEERI has deployed a network of 13 high-quality solar radiation monitoring stations across Qatar. The network provides reliable measurements of the three components of solar radiation: direct normal irradiation (DNI), global horizontal irradiation (GHI), and diffuse horizontal irradiation (DHI). The solar radiation measurements from the 13 stations offer a comprehensive coverage of solar resources across Qatar that forms a primary source of knowledge for the development and management of solar energy applications in the country. The QATAR SOLAR ATLAS WEB APPLICATION is a comprehensive web portal developed to provide real-time solar data for the state of Qatar. The portal is built using the latest web technologies, including JavaScript, and utilizes a Postgres database as the source of data. This makes the portal a valuable tool for researchers and practitioners studying solar energy in Qatar.

Keywords: Forecasting, Solar Resources, System Engineering, Solar Maps, Web application.

1. PROBLEM STATEMENT

Solar resources characterization and assessment is a prerequisite when investing in solar energy. For any solar energy application, solar radiation is the main input parameter. Given that solar radiation varies in space and in time, it is fundamental to characterize it accurately. This is particularly important for solar energy production systems, both at a small-scale rooftop system and for a larger solar energy power plant.

By nature, solar radiation varies in time, from day to day throughout the year and from year to year, and in space, from one location to another. Continuous measurements and modeling of the solar resources is needed to capture solar radiation variability and continuously record the current solar radiation data. When ground measurements of solar radiation are not available, satellite-based solar radiation databases are an alternative. However, these sources are produced by models and reanalysis, and they have a greater degree of uncertainty when compared to ground measurements specially in complex atmospheric conditions such as in Qatar, negatively affecting the results of the application that relies on satellite-based data. This is particularly critical for solar plant developers where precise estimates of solar radiation are key inputs to the design and implementation of their projects. Significant errors in the input data may greatly affect the overall financial costing of the project and may result in its mismanagement when in operation. Furthermore, due to the high temporal variability characteristic of solar radiation, there are important challenges to meeting the dynamic power production and demand curves. For this reason, a reliable solar resource forecasting solution is a very important asset for the operation and management of solar power plants. For the integration of solar energy resources into power grids and energy markets, reliable forecasting of the solar radiation at different time horizons, from near real-time to several days ahead is required.

2. AIM AND APPROACH

The aim of the QATAR SOLAR ATLAS WEB APPLICATION is to provide real-time and historical solar data for the state of Qatar collected from a network of 13 solar radiation monitoring stations and to provide solar forecasts obtained with these data. The data collected and the forecasts derived from them can be used by researchers, practitioners, and energy companies to study solar energy patterns and make data-driven decisions. The approach taken to develop the portal consists of using the latest web technologies, including JavaScript, and Postgres database hosted on Azure cloud server. The portal is built with a focus on data visualization and a user-friendly interface. The web app is designed and prepared to use real-time data when available. In the current configuration of the data acquisition from the monitoring network, the database receives new data every 2 hours, and the web portal is updated accordingly.

The portal includes authentication and authorization features that allow for user management, including the ability to add new users and assign roles such as admin and user. Users can login to the portal with a secure cookie-based authentication system implemented using Spring Security. The portal also includes an authorization feature that adjusts existing pages so that they only show information that the logged-in user is authorized to see.

3. WEBSITE PAGES

The QATAR SOLAR ATLAS WEB APPLICATION includes several pages that are designed to provide a user-friendly interface and an easy access to the data.
Figure 1 shows the website map.

10.4229/EUPVSEC2023/4DV.4.21
020454-001

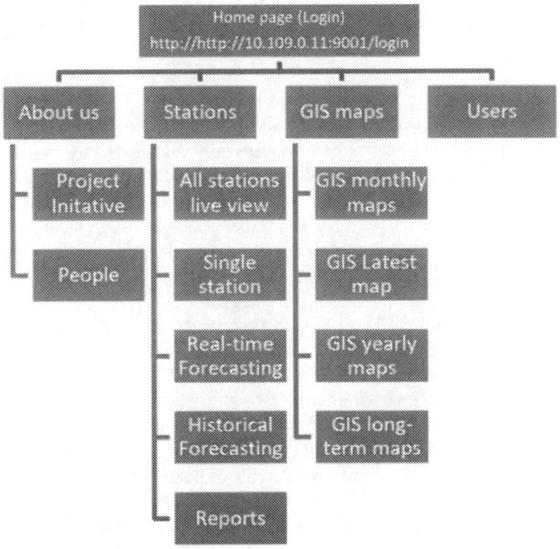

Figure 1. Site Map

3.1 All Stations Live View

This page, shown in Figure 2, displays current data on the map for the distinct locations of the network in Qatar. It shows the latest values read from the real-time database for GHI, DNI, and DHI values. The map is styled using google static map for an improved user experience. Users can view the data for different stations on the map in real-time. If there is no data available to show from today or for the last two hours, the script will look for measurements from the previous day and add the symbol "*" next to the measurements, showing the data of yesterday at today's time. If there is no data available in that location from yesterday at the same time, it shows ("N/A").

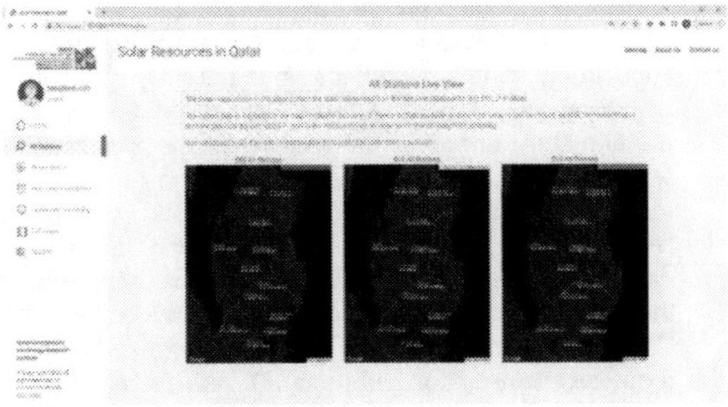

Figure 2. Stations live view.

3.2 Single Station

This page shows one day of data for a selectable location and component, with line charts using chart (java chart) and selectable date filter, allowing to plot a daily profile. Users can select any location from the 13 sites available in the dropdown menu (Figure 3). The user can specify to visualize current day or historical data. This feature is particularly useful for researchers studying a specific location or station.

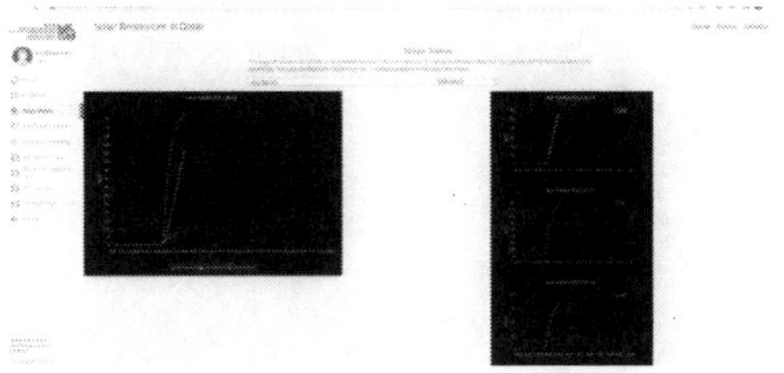

Figure 3. Single Station Page.

3.3 Real-time Forecasting

This page contains the 12-hour ahead GHI forecasting for all the stations. For more details on the GHI forecasting model developed here, see [1, 2]. GHI forecasts are visualized as a chart containing an hourly average of real-time GHI measurements for the 13 stations, with the 12-hour ahead forecasting, that are visualized separately for each of the 13 stations (Figure 4).

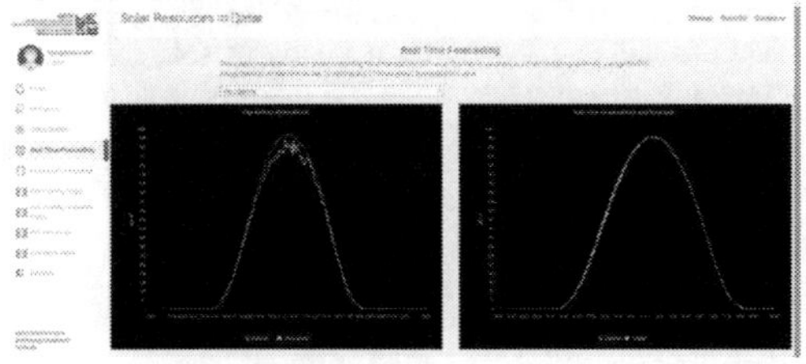

Figure 4. Real-time Forecasting.

3.4 Historical Forecasting

This page as shown in Figure 5, shows 12-hours ahead GHI forecasting for historical data. It is visualized as two daily charts, one for the hourly averages of GHI measurements and one for the forecasted data, that can be visualized for any of the 13 stations and historical date selectable in the menu. This feature is also useful for energy companies and researchers studying solar energy and evaluating the performance of the forecasting algorism over a period.

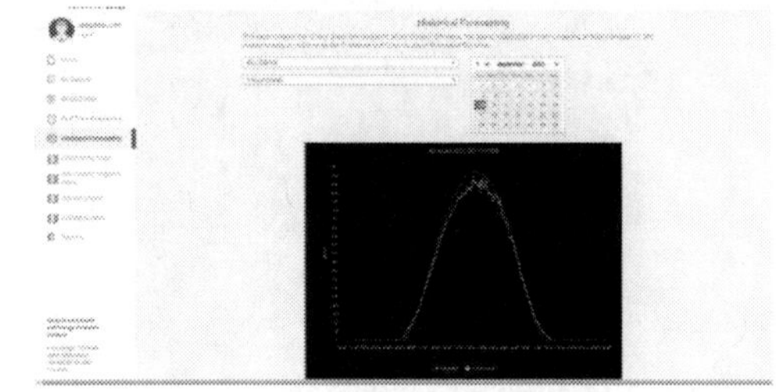

Figure 5. Historical Forecasting.

3.5 Data Reports

This page allows the user to query the stations data and download it as raw files (.csv file) selecting a specific time and station name from the 13 stations. This feature is useful for the users to analyze and study the solar radiation data minute by minute (Figure 6).

10.4229/EUPVSEC2023/4DV.4.21

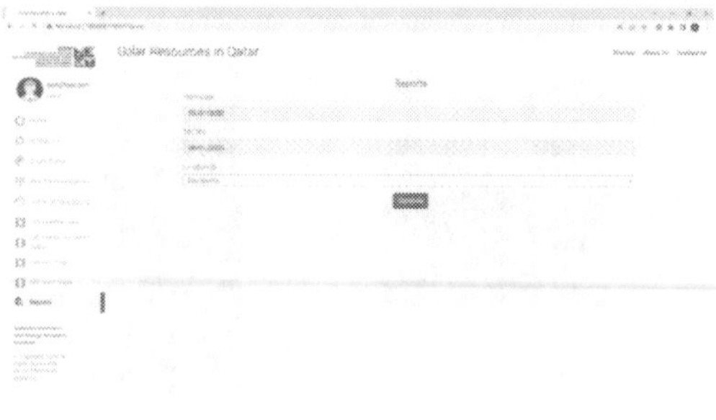

Figure 6. Reports Page.

3.6 GIS maps

This page will display all GIS-based maps produced from the data of the 13 monitoring stations including yearly monthly, and hourly real-time maps (Figure 7 and Figure 8).

This feature is useful for researchers studying long-term solar energy patterns and providing a broader perspective on solar energy data. It allows researchers to analyze and understand the solar energy patterns over the course of a month and the seasonal variation. This can be useful in understanding the long-term trends in solar energy and how it varies over time. The GIS maps page also provides a spatial context to the data, allowing researchers to analyze how solar energy varies across different regions and terrains.

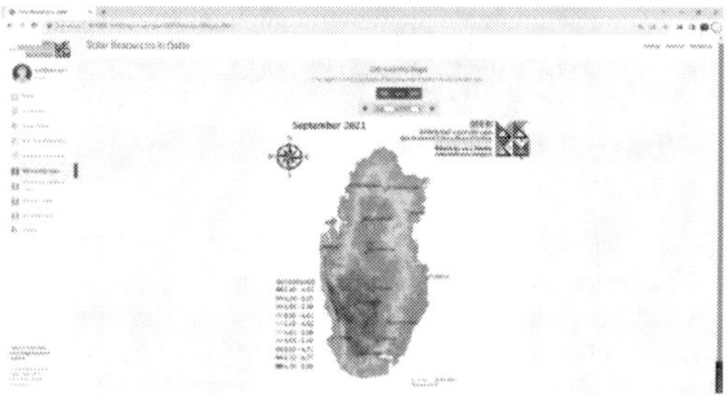

Figure 7. GIS Monthly Map.

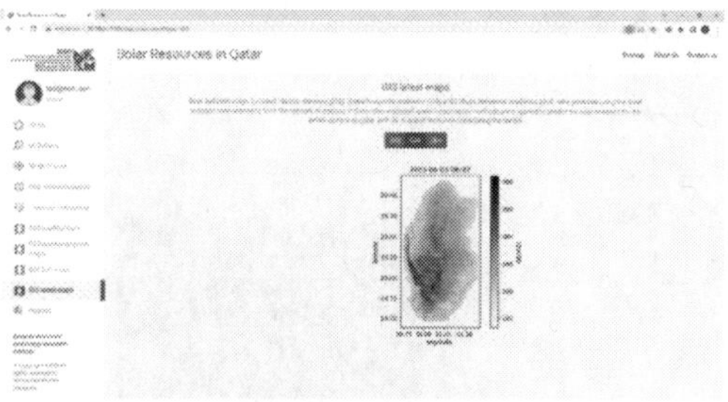

Figure 8. GIS Latest Real-time Map.

4. SCIENTIFIC INNOVATION AND RELEVANCE

Accurate measurements of the incoming solar irradiance are essential for solar energy projects design and implementation. This is achieved through ground measurements which are relatively complex when compared to other meteorological measurements, and require specific technical skills for the instrument selection, installation, site design, and operations and maintenance (O&M), including calibration of sensors.

In this project, a dense network of high-quality radiometric stations has been designed and deployed across Qatar. Figure 9 shows samples of solar radiation monitoring stations deployed in Qatar by QEERI. QEERI's network of solar monitoring stations provides the only source of local high quality solar radiation data for Qatar, offering high temporal resolution measurements of the three components of solar radiation: direct, global, and diffuse. Furthermore, the project includes the development and improvement of novel methods and procedures [3] for the whole operation of solar monitoring stations, from the data collection to the final processing and analysis with the objective of obtaining the most accurate measurements in this climate.

The QATAR SOLAR ATLAS WEB APPLICATION is a web portal service built with the latest web technologies and is deployed on cloud infrastructure, which provides several benefits for its users. The use of virtual machines allows for better performance and load balancing, as well as providing redundancy and failover capabilities, ensuring that the application is always available to the users. The use of cloud infrastructure for hosting the QATAR SOLAR ATLAS WEB APPLICATION provides several benefits. One of the key benefits is scalability. Cloud infrastructure allows for easy scaling of resources, such as CPU and memory, to meet the changing needs of the users. As the number of users increases, resources can be quickly and easily added to ensure that the application can handle the increased load. This eliminates the need to invest in expensive hardware and reduces costs associated with maintaining it. Another benefit of cloud infrastructure is reliability. Cloud providers typically have multiple data centers located around the world, which allows for better redundancy and failover capabilities. In the event of a failure in one data center, the application can automatically failover to another data center, ensuring that the application is always available to the users.

In addition, deploying the application on cloud infrastructure improves the security of the application by separating the application and its data. In this case, two virtual machines were deployed on cloud, one for the frontend and another for the backend hosting the databases. This approach enhances security by minimizing the risk of unauthorized access to sensitive information. Cloud infrastructure also provides cost benefits by reducing the need for additional IT staff for maintenance and upgrades, reducing costs for the organization and increasing efficiency. This makes it more environmentally friendly as well, as it reduces the carbon footprint by eliminating the need for physical servers and reducing energy consumption.

Current approaches to solar forecasting include physics-based and stochastic models. Numerical Weather Prediction (NWP) models, and local sensing models are physics-based techniques, more appropriate for several days ahead forecasts where information about atmospheric information is needed. Machine learning and stochastic models use instead carefully designed regression methods able to produce accurate forecasts up to 32 hours ahead [4-6]. In this project, a patent pending algorithm has been developed [7], using an ensemble classification approach

10.4229/EUPVSEC2023/4DV.4.21
020454-003

to select the best set of machine learning models for each choice of time series data input. This system of models can decide which specific approach among the ones that it includes will produce the most accurate prediction. Doing so further improves the overall precision of the forecasting process compared to other more classical approaches.

Overall, the use of virtual machines on cloud infrastructure for the QATAR SOLAR ATLAS WEB APPLICATION is a scientifically innovative and relevant approach that provides several benefits for its users. The scalability, reliability, security, and cost-effectiveness of cloud infrastructure make it an ideal choice for hosting this application, providing a high-performance user experience, and ensuring that the application is always available to the users. The use of cloud infrastructure has a lower impact on the environment than other strategies as well.

Figure 9 Solar radiation monitoring stations deployed in Qatar by QEERI [8]

5. CONCLUSION AND FUTURE WORK

The Qatar solar atlas web application is a comprehensive web portal created to supply real-time solar information for the state of Qatar. The portal is built utilizing the most recent web advances, counting JavaScript, and utilizes a Postgres database as the source of information. The database collects information in real-time from a network of 13 stations deliberately set all through Qatar, covering diverse regions and landscapes. This portal can be used by researchers and companies studying solar energy in Qatar and the region.

The web portal application will be maintained and constantly updated with real-time measurement and model data providing a streaming data service to inform real-world solar energy applications and solar energy research.

References

[1] Scabbia, Giovanni, et al. "Exploring the limits of machine learning in the prediction of solar radiation." Sustainable Energy-Water-Environment Nexus in Deserts: Proceeding of the First International Conference on Sustainable Energy-Water-Environment Nexus in Desert Climates. Cham: Springer International Publishing, 2022.

[2] Scabbia, Giovanni, et al. "Exogenous Parameters in Solar Forecasting." 2020 47th IEEE Photovoltaic Specialists Conference (PVSC). IEEE, 2020.

[3] Astudillo, D., Bachour, D., Martin-Pomares, L. (2018). Improved quality control protocols on solar radiation measurements. Solar Energy 169, 425-433. DOI:10.1016/j.solener.2018.05.028

[4] Sanfilippo, A., Martin-Pomares, L., Mohandes, N., Perez-Astudillo, D., Bachour, D. (2016). An Adaptive Multi-Modeling Approach to Solar Nowcasting. Solar Energy, 125, 77-85.

[5] Scabbia, G., Sanfilippo, A., Perez-Astudillo, D., Bachour, D. and Fountoukis, C., 2022. Exploring the Limits of Machine Learning in the Prediction of Solar Radiation. In Sustainable Energy-Water-Environment Nexus in Deserts (pp. 381-384). Springer, Cham.

[6] Scabbia, G., Sanfilippo, A., Bachour, D. and Perez-Astudillo, D., 2020, June. Exogenous Parameters in Solar Forecasting. In 2020 47th IEEE Photovoltaic Specialists Conference (PVSC) (pp. 0894-0896). IEEE.

[7] Sanfilippo, A., Martin-Pomares, L., Mohandes, N., Perez-Astudillo, D., Bachour, D. (2016). Method of Forecasting for Solar Power Systems. International Patent Application No.: PCT/US2016/03897; 2016.

[8] D. Perez-Astudillo, D. Bachour, A. Sanfilippo, H. Al-Hajri. Management and Operation of Qatar's Solar Radiation Monitoring Network. EuroSun 2022, 25-29 Sep/2022

10.4229/EUPVSEC2023/4DV.4.21
020454-004

VIABILITY OF VARIOUS TEMPERATURE MODELS FOR FAÇADE PV IN MODERATE CLIMATES

Torsten Ernst[1*], Bianca Lim[1], Carsten Schinke, Robert Puknat[1], Timo Gewohn, Raphael Niepelt[1]
[1]Institute for Solar Energy Research, Am Ohrberg 1, 31860 Emmerthal, Germany
ernst@isfh.de

ABSTRACT: This work presents the results of an investigation into the viability of several models that are commonly used to simulate the PV module temperature. In the past, these models, namely, Faiman-, King- and NOCT-model, were validated mostly for rooftop PV and in hotter climates. We test these models for façade PV in moderate climates and show that the annual average deviation between the PV module temperature simulated with those models and the measured PV module temperature is below 3.1 K between the hours of 09:00 – 16:00 for all models used. With exception of the NOCT-model all models include fitting coefficients. Fitting those coefficients to measurement data obtained in moderate climate and for façade PV does not yield an improvement (i.e. no smaller annual average deviation between simulated and measured module temperature within measurement precision). Applying a simple 10 min. rolling average to the simulated temperature data yields similar or better results compared to the fitting of the model coefficients to the dataset. Even though the NOCT-model does not include the effect of wind speed, in contrast to the other models mentioned, it showed a very competitive performance.

1 Introduction

In addition to ground-mounted and rooftop PV, façade PV offers great potential to increase PV installations and consequently electricity generated by PV (238 TWh/a for Germany compared to 504 TWh/a for rooftop PV [1]). Accurate yield predictions are required to identify suitable façade areas for PV installations and quantify economical aspects. Module temperature influences the electrical output of a PV module and thus must be considered during yield prediction calculations.

Many temperature models for PV modules can be found in literature. The vast majority is empirical and several of the more widely used temperature models were developed and tested mostly in hotter climates and for ground-mounted PV or rooftop PV (Faiman [2], King [3]) and their use for façade PV in moderate climates has not been rigorously tested yet.

This work adds to the body of research that investigates the viability of three different temperature models for simulating the PV module temperature of façade PV in moderate climates for building-attached and building-integrated PV.

2 Equations and test sites

2.1 Temperature model equations

We use three different temperature models to simulate the PV module temperature.

Faiman [2]:

$$T_{\mathrm{Mod}} = T_{\mathrm{Amb}} + \frac{GVI}{U_0 + U_1 \cdot v} \tag{1}$$

With module temperature T_{Mod}, ambient air temperature T_{Amb}, Global, vertical irradiance (irradiance in the module plane) *GVI*, wind velocity v, and fitting coefficients U_0 and U_1.

King [3]:

$$T_{\mathrm{Mod}} = T_{\mathrm{Amb}} + GVI \cdot e^{a+b\cdot v} \tag{2}$$

With fitting coefficients a and b.

NOCT [4]:

$$T_{\mathrm{Mod}} = T_{\mathrm{Amb}} + \frac{GVI}{GVI_{\mathrm{NOCT}}}(T_{\mathrm{Mod,NOCT}} - T_{\mathrm{Amb,NOCT}}) \tag{3}$$

With the Nominal Operating Cell Temperature (NOCT) $T_{\mathrm{Mod,NOCT}}$ and the ambient air temperature at which the NOCT can be measured $T_{\mathrm{Amb,NOCT}}$.

Fitting coefficients were determined by rearranging the respective temperature model equation and solving it with linear regression. As an example, for the case of the Faiman model:

$$h = \frac{GVI}{T_{\mathrm{Mod}} - T_{\mathrm{Amb}}} = U_0 + U_1 \cdot v \tag{4}$$

The temperature models investigated in this study are steady-state models and thus neglect the effect of the module thermal mass. Additionally, the effect of infrared back radiation (radiation cooling) and natural convection (buoyancy) is neglected. For this reason, in order to determine the fitting coefficients, it is necessary to pre-filter the data (*GVI*, v) following the IEC standard for module operating temperature measurements (IEC 61853-2, 2016) [5]. The filter criteria used to filter the data sets, which in turn were used to determine the fitting coefficients of the temperature models, are as follows [5]:

- Neglect irradiance values below 400 W/m².
- Neglect irradiance values in 10-min. interval after the irradiance varies by more than 10% from the maximum value to the minimum value during the preceding 10 min.
- Neglect wind speed values in a 10-min. interval after and including a deviation of the instantaneous wind speed to below 0.25 m/s or gusts larger than 200% from a 5-min. running average.
- Low and high wind speed filter: Neglect wind speed data when the 5-min. running average is less than 1 m/s or greater than 8 m/s.

The annual (or monthly) average deviation between simulated and measured module temperature was calculated in the following way:

10.4229/EUPVSEC2023/4DV.4.24
020456-001

$$\Delta T_{Aver} = \sum \frac{|T_{\text{Sim}} - T_{\text{Meas}}|}{N} \quad (5)$$

In this paper the value of ΔT_{Aver} was determined between the hours of 09:00 – 16:00, because at the location of data recording (Germany) 9:00 hours is always after sunrise and 16:00 hours is always before sunset. Please note that we did analyze ΔT_{Aver} also for data sets that included times before sunrise and after sunset and observed that the value for ΔT dropped significantly. However, times without solar irradiance are irrelevant for electricity generation via PV. Thus, it is much more important that a temperature model describes the PV module temperature correctly during daytime than during the full 24 hours of a day. Deviations from this procedure will be explicitly mentioned. A comparison of ΔT for all hours and hours between 09:00 – 16:00 is given in Table 1.

2.2 Test sites

For this study, we use data from two different test sites that are located at the Institute for Solar Energy Research (ISFH), 31860 Emmerthal, Germany. Data collection at a third test site, located in Hanover, started mid-July 2023 and we will begin to analyze the data once four months of data are available.

The first test site at ISFH is called "A: Versuchshaus" and the other one "B: SolarTec". Test site "A: Versuchshaus" is composed of 2 commercial Solitek monocrystalline PV modules (glass-cell-glass). This type of PV module installation is regarded as "building-attached". One of the PV modules has a white textile laminated with a weather-resistant layer as a front cover onto the front glass, thus giving that PV module a grey color. This method of post-coloring PV modules is called CoTex [6] and was developed by Gewohn et al. at the Institute for Solar Energy Research.

Figure 1: Test site "A: Versuchshaus" with 2 PV modules (PV module on the left is not part of this study) and pyranometers (blue), anemometer (green), hygro-thermometer (red) and PT100 thermo-elements (red)

The other PV module acts as a reference. The PV module temperature (PT100) as well as weather data (pyranometers measure global and diffuse irradiance in plane of the façade), wind speed (anemometer) and ambient temperature (PT100) are measured at site once every minute. The module temperature is measured at the center and the corner of the PV modules and the average of both values is used as the 'measured' temperature value to compare the simulated value against. We use an MPP tracker to determine the current-voltage characteristic every minute and from that extract the electrical current and voltage at the MPP.

Test site "B: SolarTec" is composed of 6 HeliArtec Pixasolar BIPV Colored AEP-L PV modules (dyed glass-cell-glass, white) with aluminum heat exchange plates mounted on the back that make them photovoltaic-thermal (PVT) modules. The PVT module installation type is "building-integrated". Data points, that correspond to times when the PVT modules were actively cooled or heated by the heat exchange plates, were excluded, in order to make the results comparable with those from test site "A: Versuchshaus". The test site features pyranometers for global and diffuse irradiance and infrared back radiation, Si irradiance sensors, anemometers, hygrometer and thermometer (PT100) for ambient temperature as well as PT100 at the back center of every PVT module and at two PVT modules at the corners, too. Electrical current and voltage is measured at the MPP via an MPP tracker. Measurement points are recorded once every second.

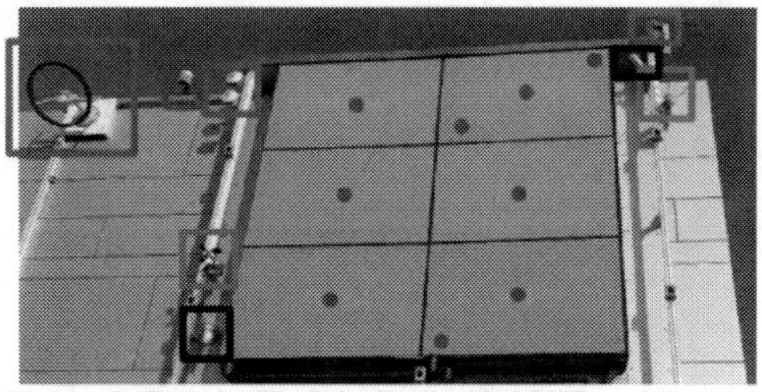

Figure 2: Test site "B: SolarTec" with 6 PVT modules and pyranometers (blue), anemometers (green), hygro-thermometers (red), thermometers (black) and PT100-thermo-elements (red)

The third test site is called "C: Dänischer Pavillon" and is located in 30539 Hanover, Germany. There are 33 HeliArtec Pixasolar BIPV Colored AEP-L PV modules (dyed glass-cell-glass, white) installed at this test site. 4 of these 33 modules have aluminum heat exchange plates mounted on the back that make them photovoltaic-thermal (PVT) modules. However, these 4 PVT modules are not connected to a fluid circuit and thus are not actively cooled or heated. A weather station installed at this test site includes a pyranometer for global horizontal irradiance, an anemometer for wind speed and a PT100 to measure the ambient air temperature. Additional PT100 elements were placed on the back side of certain PV modules as depicted in Figure 3.

Figure 3: Test site "C: Dänischer Pavillon" with 33 PV modules and PT100 thermo-elements (red)

The recording of the PV module temperature at this site started on 2023.07.17. The data recorded at this test site will be used at the end of the year 2023 to test the temperature models with self-fitted coefficients. The data acquisition rate is 1 minute.

10.4229/EUPVSEC2023/4DV.4.24
020456-002

3 Results

3.1 Test site "A: Versuchshaus" – building-attached PV

The data set of test site "A: Versuchshaus" of the year 2021 was pre-filtered using the NOCT criteria as listed above and then used to determine the fitting coefficients of the Faiman- and King-model applying equation (4) (Faiman) or the equivalent of it for the King model. After that the PV module temperature was simulated using the NOCT-model and the Faiman- and King-model with literature as well as self-fitted values for fitting coefficients.

The annual average difference ΔT_{Aver} between the hours of 09:00 – 16:00 for $\text{Faiman}_{\text{Literature}}$, $\text{King}_{\text{Literature}}$ and NOCT is 2.76 K, 2.01 K and 1.96 K (see Table 1) respectively. Assuming a PV module temperature coefficient of Γ= -0.36 %/°C, this would translate into an uncertainty of the simulated module power of ΔP_{Aver} 1 %, 0.72 % and 0.71 %.

When we perform the simulations using the self-fitted coefficients, the values of ΔT_{Aver} for $\text{Faiman}_{\text{Self-fitted}}$, $\text{King}_{\text{Self-fitted}}$ reduce to 1.86 K and 1.88 K (see Table 1) or in terms of ΔP_{Aver} 0.67 % in both cases.

Full 24 hours	$\Delta T_{\text{Literature}}$	$\Delta T_{\text{Self-fitted}}$
Faiman	1.61 K	1.19 K
King	1.27 K	1.19 K
NOCT	1.23 K	
09:00 – 16:00		
Faiman	2.76 K	1.86 K
King	2.01 K	1.88 K
NOCT	1.96 K	

Table 1: Annual average temperature difference ΔT_{Aver} between simulated and measured PV module temperature for different temperature models and fit coefficients values. The NOCT-model has no fitting coefficients. ΔT_{Aver} for all hours and for hours between 09:00 – 16:00. Data was recorded at test site "A: Versuchshaus".

Please note that ΔT_{Aver} gives means of absolute values, where fluctuations in the deviation between measured and simulated temperatures do not cancel each other out.

In addition, we plot the difference between measured PV module temperature of test site "A: Versuchshaus" of the year 2020 and the simulated values for that year in Figure 4. The data points plotted in Figure 4 are the monthly averages between the hours of 09:00 – 16:00.

From Figure 4 (and 5) it can thus be seen whether the simulations over- or understate the PV module temperature.

The average monthly difference ΔT_{Aver} between simulated and measured PV module temperature is below |2.5 K| for all temperature models. Assuming a PV module temperature coefficient of Γ= -0.36 %/°C, this would translate into an uncertainty of the simulated module power ΔP_{Aver} < 0.72 %. The NOCT-model and the Faiman- and King-model with literature values (short $\text{Faiman}_{\text{Literature}}$, $\text{King}_{\text{Literature}}$) for the fitting coefficients tend to overestimate the PV module temperature.

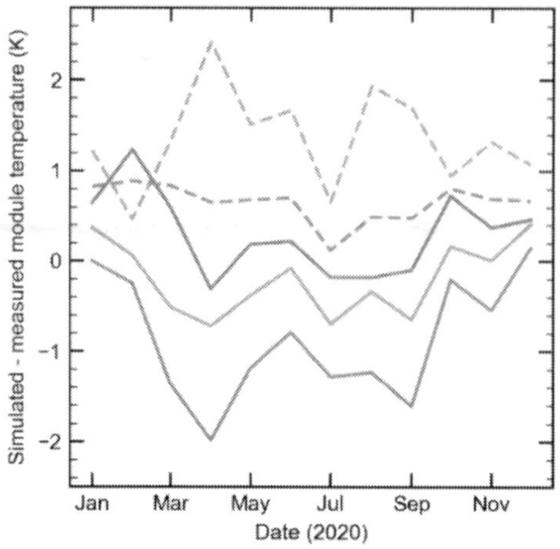

Figure 4: Difference between the simulated and measured PV module temperature ΔT for each month of the year 2020 for data recorded at test site "A: Versuchshaus". Monthly data points are mean values for that respective month. The PV module temperature was simulated using the Faiman- (orange), NOCT- (green) and King-model (purple). The temperature models were applied using literature values (dashed lines) or self-fitted values (continuous lines) for the fitting coefficients.

3.2 Test site "B: SolarTec" – building-integrated PV

At the time this study was conducted, only six months of data had been recorded at the test site (see Figure 5).

The average monthly difference ΔT_{Aver} between simulated and measured PV module temperature is below |4.5 K| for all temperature models. The data points plotted in Figure 5 are the monthly averages between the hours of 09:00 – 16:00. ΔT_{Aver} for $\text{Faiman}_{\text{Literature}}$, $\text{King}_{\text{Literature}}$ and NOCT is 2.88 K, 2.95 K, 3.08 K (see Table 2) or in terms of ΔP_{Aver} 1.03%, 1.06% and 1.1%. The values for ΔT_{Aver} for $\text{Faiman}_{\text{Self-fitted}}$, $\text{King}_{\text{Self-fitted}}$ are 3.49 K and 3.38 K or in terms of ΔP_{Aver} 1.26% and 1.21% and are thus higher than the values for simulations run with the literature values for the model coefficients (see Table 2).

10.4229/EUPVSEC2023/4DV.4.24
020456-003

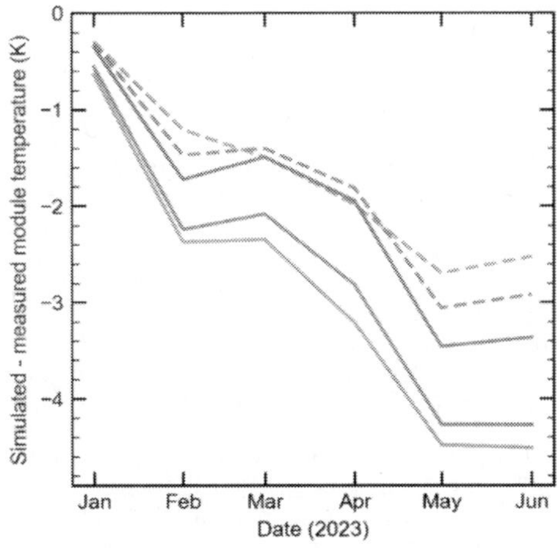

Figure 5: Difference between the simulated and measured PV module temperature ΔT for each month of the year 2023 for data recorded at test site "B: SolarTec". Monthly data points are mean values for that respective month. The PV module temperature was simulated using the Faiman- (orange), NOCT- (green) and King-model (purple). The temperature models were applied using literature values (dashed lines) or self-fitted values (continuous lines) for the fitting coefficients.

09:00 – 16:00	$\Delta T_{Literature}$	$\Delta T_{Self\text{-}fitted}$
Faiman	2.88 K	3.49 K
King	2.95 K	3.38 K
NOCT	3.08 K	

Table 2: Annual average temperature difference ΔT_{Aver} between simulated and measured PV module temperature for different temperature models and fit coefficients values. The NOCT-model has no fitting coefficients. ΔT_{Aver} for hours between 09:00 – 16:00. Data was recorded at test site "B: SolarTec".

3.3 Evaluation of 10 min. rolling average

Since the investigated temperature models neglect the thermal mass and thus the thermal inertia of the PV modules, we tested the impact of applying a 10-min. rolling average to the simulated data. For this we use data recorded at test site "A: Versuchshaus", which has an original sampling rate of 1 min.

Figure 6 shows the difference between simulated and measured module temperature. The cyan line shows the temperature difference ΔT between averaged data from the Faiman-model using literature values for the model coefficients, referred to as $\text{Faiman}_{Literature\text{-}Aver}$, and measured data. The orange line in Figure 6 shows ΔT for non-averaged simulated data using the Faiman-model and self-fitted model coefficients, denoted $\text{Faiman}_{Self\text{-}fitted}$.

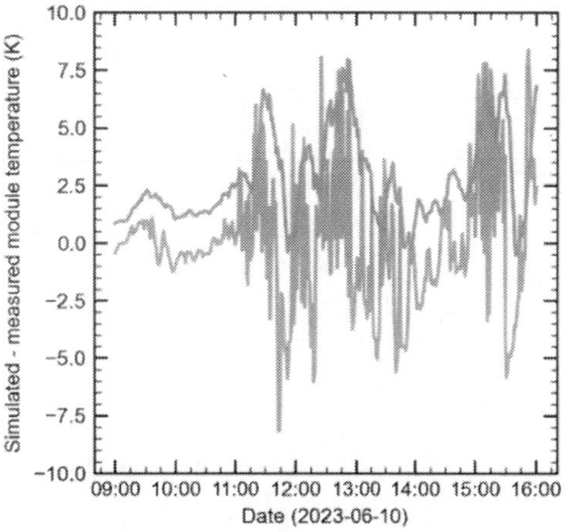

Figure 6: Difference between the simulated and measured PV module temperature on 2023.06.10. Data recorded at test site "A: Versuchshaus". The PV module temperature was simulated using the Faiman-model with self-fitted model coefficients (orange) and literature values with a 10 min. rolling average applied (cyan).

ΔT_{Aver} for $\text{Faiman}_{Self\text{-}fitted}$ and $\text{Faiman}_{Literature\text{-}Aver}$ is 1.86 K, and 1.93 K (see Table 2) or in terms of ΔP_{Aver} 0.67% and 0.69%.

Figure 7 shows the difference between (averaged) simulated and measured data for the King-model. The cyan line shows the temperature difference ΔT between averaged data from the King-model using literature values for the model coefficients, referred to as $\text{King}_{Literature\text{-}Aver}$, and measured data. The orange line in Figure 7 shows ΔT for non-averaged simulated data using the King-model and self-fitted model coefficients, denoted $\text{King}_{Self\text{-}fitted}$.

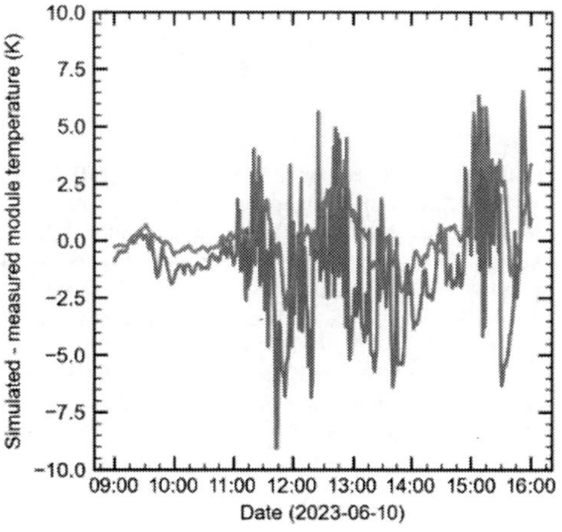

Figure 7: Difference between the simulated and measured PV module temperature on 2023.06.10. Data recorded at test site "A: Versuchshaus". The PV module temperature was simulated using the King-model with self-fitted model coefficients (purple) and literature values with a 10 min. rolling average applied (cyan).

ΔT_{Aver} for $\text{King}_{Self\text{-}fitted}$ and $\text{King}_{Literature\text{-}Aver}$ is 1.88 K, and 1.38 K (see Table 2) or in terms of ΔP_{Aver} 0.67% and 0.5%.

09:00 – 16:00	$\Delta T_{Literature}$	$\Delta T_{Self\text{-}fitted}$
Faiman	2.76 K	1.86 K
King	2.01 K	1.88 K
09:00 – 16:00	$\Delta T_{Literature\text{-}Aver}$	$\Delta T_{Self\text{-}fitted\text{-}Aver}$
Faiman	1.93 K	1.19 K
King	1.38 K	1.22 K

Table 3: Annual average temperature difference ΔT_{Aver} between simulated and measured PV module temperature for different temperature models ($\Delta T_{Literature}$, $\Delta T_{Self\text{-}fitted}$) and 10-min. rolling average ($\Delta T_{Literature\text{-}Aver}$, $\Delta T_{Self\text{-}fitted\text{-}Aver}$). ΔT_{Aver} for hours between 09:00 – 16:00. Data was recorded at test site "A: Versuchshaus".

4 Discussion

Overall the annual average discrepancy between simulated and measured PV module temperatures ΔT_{Aver} is rather small with values below 3 K for data of test site "A: Versuchshaus" and values below 3.1 K for data of test site "B: SolarTec". In terms of power ΔP_{Aver} this means less than 1.1% .

The value of ΔT_{Aver} for $Faiman_{Self\text{-}fitted}$ and $King_{Self\text{-}fitted}$ for the data of test site "A: Versuchshaus" is 0.9 K and 0.13 K lower than the value for $Faiman_{Literature}$ and $King_{Literature}$, giving a smaller reduction for the King-model by using self-fitted model coefficients. This phenomenon is known in the literature, Kohl et. al. [7] reported that the Faiman-model is more sensitive to climate zones when it comes to model coefficient fitting and especially to the influence of the wind velocity. The literature values were fitted to data recorded in Israel (Faiman-model [2]) and New Mexico, USA (King-model [3]), which differ in their climatic conditions from Germany, where the data for this study was recorded.

In order to evaluate the effect of self-fitted model coefficients, the measurement uncertainty has to be taken into account. The measurement uncertainty was calculated as follows:

$$\Delta T_{Mod} = \sqrt{(\frac{\partial T_{Mod}}{\partial T_{Amb}}\Delta T_{Amb} + \frac{\partial T_{Mod}}{\partial GVI}\Delta GVI + \frac{\partial T_{Mod}}{\partial v}\Delta v)} \quad (6)$$

The measurement uncertainty of the PT100 element used to measure the ambient air temperature is 0.2 K, for the anemometer 0.2 m/s, for the pyranometer 2.8%. The average GVI values between the hours of 09:00 – 16:00 applying the data filter criteria mentioned before is 664 W/m², the average wind speed is 1 m/s. The uncertainty ΔT_{Mod} is 1.1 K with those values in case of the Faiman-model and 0.6 K in case of the King-model. That means within measurement precision there is no reduction of ΔT_{Aver} by application of self-fitted model coefficients instead of literature values neither for the King- nor the Faiman-model.

Different data sets were used to determine the fitting coefficients (data set "A: Versuchshaus, 2021") and testing the fitting coefficients (data set "A: Versuchshaus, 2020", "B: SolarTec, 2023") to reduce the error of overfitting. For that reason, the validity of the self-fitted coefficients will additionally be tested using a data set recorded at the test site "B: SolarTec" and at the end of the year 2023 using a data set recorded at the test site "C: Dänischer Pavillon".

Using the temperature models with self-fitted model coefficients to simulate the PV module temperature of test site "B: SolarTec" yields larger ΔT_{Aver} compared to using literature values for the model coefficients. As soon as enough data is recorded at test site "C: Dänischer Pavillon" (at the end of 2023) the accuracy of the temperature models with self-fitted model coefficients will be tested using that data set. So far, this study gives no indication that using self-fitted model coefficients yields better results in simulating PV module temperatures in moderate climate regions compared to using literature values.

A 10-min. rolling average was applied to the PV module temperature values that were calculated using the Faiman- and King-model (see Table 3). Within measurement precision the value ΔT_{Aver} for the 10-min. rolling average with literature values for model coefficients ($Faiman_{Literature\text{-}Aver}$ = 1.93 K) was identical with the value of ΔT_{Aver} calculated with the Faiman-model with self-fitted model coefficients ($Faiman_{Self\text{-}fitted}$=1.86 K). In case of the King-model the 10-min. average value of ΔT_{Aver} is even lower than the one for self-fitted coefficients ($King_{Literature\text{-}Aver}$ = 1.38 K, $King_{Self\text{-}fitted}$ = 1.88 K). A simple procedure such as applying a 10-min. rolling average to the simulated data yielded as good or better results as the more time-consuming procedure of self-fitting the model coefficients. The best results though were obtained by combining self-fitted model coefficients with 10-min. rolling averages (see Table 3).

5 Conclusion

Simulating the PV module temperature with the Faiman-, King- and NOCT-model yielded ΔT_{Aver} below 3.1 K or ΔP_{Aver} below 1.1%. All three models showed overall a similar accuracy. It is noteworthy that the simplest model (NOCT) did so without taking into account the effect of wind speed as do the Faiman- and King-model.

Within measurement precision running the temperature simulations with self-fitted model coefficients gave no better results compared to using literature values.

Lastly, applying a simple 10-min. rolling average to the simulated PV module temperature data yielded comparable or better results compared to the use of self-fitted coefficients for test site "A: Versuchshaus". The best results however were obtained by combining both approaches.

6 Acknowledgement

The collaborative project "Retrospective BIM approach for life cycle optimized integration of BIPV systems in the building envelope" (BIMPV, FKZ 03EN1010B), on which this publication is based, is funded by the German Federal Ministry of Economic Affairs and Climate Action. In addition, the authors gratefully acknowledge the support of the State of Lower Saxony. The responsibility for the content of this publication lies solely with the authors.

References

[1] J.-B. Eggers, "PV-Ausbauerfordernisse versus Gebäudepotenzial: Ergebnis einer gebäudescharfen

Analyse für ganz Deutschland," in *35. PV-Symposium*, Bad Staffelstein, 2020.

[2] D. Faiman, "Assessing the Outdoor Operating Temperature of Photovoltaic Modules," *Progress in Photovoltaics: Research and Applications,* vol. 16, pp. 307-315, 2008.

[3] D. King, "Photovoltaic Array Performance Model," Sandia National Laboratories, Albuquerque, 2004.

[4] M. Mattei, "Calculation of the polycrystalline PV module temperature using a simple method of energy balance," *Renewable Energy,* vol. 31, p. 553–567, 2006.

[5] "IEC 61853-2, Photovoltaic (PV) module performance testing and energy rating – part 2: Spectral responsivity, incidence angle and module operating," IEC, 2016.

[6] T. Gewohn, "Postproduction Coloring of Photovoltaic Modules With Imprinted Textiles," *IEEE JOURNAL OF PHOTOVOLTAICS,* pp. 2156-3381, 2020.

[7] M. Koehl, "Modeling of the nominal operating cell temperature based on outdoor weathering," *Sol. Energy Mater. and Sol. Cells,* vol. 95, pp. 1638-1646, 2011.

10.4229/EUPVSEC2023/4DV.4.24
020456-006

LEVERAGING GRADIENT-BOOSTED TREES TO COMBINE FFT, CORRELATION AND FRÉCHET-DISTANCE FEATURES FOR CLEAR-SKY DETECTION

Tobias Veihelmann[1], Masoud Seifaei[1], Maximilian Lübke[1], Norman Franchi[1]
[1] Chair of Electrical Smart City Systems, Friedrich-Alexander-Universität Erlangen-Nürnberg
Cauerstr. 7, 91058 Erlangen

ABSTRACT: Clear-sky detection has become a widely researched sub-domain of clear-sky modeling and solar forecasting. Current approaches differ in ground truth, criteria, and threshold values. With clear-sky detection being a classification problem, this work introduces a new machine learning-based approach with automatic and accurate feature selection and threshold values. The features include correlation, discrete Fréchet distance, fast Fourier transform, and their Haurwitz clear-sky global horizontal irradiance (GHI) time series counterparts. For classification, we employ a gradient-boosted trees method. An irradiance dataset from Folsom, California, in the US, with a one-minute resolution, serves as the training (2014) and evaluation (2015-2016) dataset. The results are compared to a prior clear-sky detection approach based on machine learning to show the positive effect of features based on fast Fourier transform and correlation. With an F1-score of 90.1 percent, the method exceeds the 86.7 percent of the benchmark approach. A qualitative analysis confirms the positive impression from the quantitative results, suggesting that GHI measurements are sufficient for accurate clear-sky detection.
Keywords: Clear-sky detection, machine learning, fast Fourier-transform (FFT), Solar resource assessment

1 INTRODUCTION

Clear-sky detection (CSD) has gained much attention over the past years and is a preceding step for different evaluation studies where irradiance under clear-sky conditions is necessary [1]. A prominent example is evaluation studies on clear-sky radiation models (e.g., [2]). However, there are still some challenges in CSD:

First, in an extensive evaluation study, many clear-sky models did not perform as well as expected, indicating that CSD still needs to be improved [1]. Second, there is the problem of obtaining a reliable ground truth: While CSD algorithms are often evaluated on sky images (e.g., [1], [3]), ground truth from sky images can be erroneous [1]. Therefore, also the visual inspection-based manual labeling of irradiance measurements remains a frequently applied approach (e.g., [4], [5], [6]). Third, the different ways to create a ground truth lead to difficulties in the quantitative evaluation of CSD approaches. Thus, Bright et al. [4] and Gueymard et al. [1] present and propose additional qualitative evaluation – i.e., comparing CSD approaches based on plotted measurement curves with their respective clear-sky classification. Our paper addresses the objective of developing a more accurate CSD method: We provide an innovative machine learning (ML)-based approach to CSD with a novel feature combination and evaluate it with ground truth data based on visual inspection.

2 SCIENTIFIC CONTRIBUTION AND RELEVANCE

2.1 State of research

To improve the methodology of performing CSD, recalling the basic working principle of current approaches is essential: Reported methods have employed one (e.g., [7]) or many (e.g., [6]) criteria with corresponding threshold values. There has been a wide range of criteria, including correlation [7], [8], comparisons of global horizontal irradiance (GHI) and direct normal (DNI) or diffuse horizontal irradiance (DHI) [3], [9]. Besides, multi-criteria approaches have been introduced using GHI's mean, maximum, and standard deviation in specific periods (e.g., 10 min), line length, and maximum difference between changes. The threshold values then apply to the difference between those calculations for measured GHI and its counterpart for clear-sky conditions (GHI_{cs}) [6], [10]. A more recent criterion is Fourier transform [11]. In a nutshell, there is a wide range of criteria without a consensus on the most appropriate.

Aside from the criteria, the data required classifies a CSD approach. While some methods only require GHI data, other techniques also include criteria based on DHI or DNI data. The more substantial impact of cloud events on DHI curves compared to GHI curves (e.g., [4]) leads to conclusions that the inclusion of irradiance sub-components implies superior CSD [2]. Indeed, recent methods with promising results involve DHI (e.g., [3], [4]). However, DNI or DHI data need additional ground measurements and instrumentation, which are often unavailable.

Thus, research still focuses on more appropriately classifying irradiance time series with only GHI measurements. Lewis et al. [11] introduced the Fourier transform to clear-sky detection to target the shape of irradiance curves. However, they classified 15-minute periods into five classes such that their results differ from (binary) CSD methods. While Liu et al. [3] present a version of their method that works with GHI data only, it performs slightly inferior to the model with an additional DHI-based feature. Their main contribution is bringing ML to CSD for a more efficient combination of features and determination of threshold values.

2.2 New approach and contributions

This work proposes an ML-based approach: Classification (which is by principle applicable to "classifying" into clear-sky and cloudy periods) is a fundamental ML task (see, e.g., [12]). Since the criteria in CSD are equivalent to features in ML, ML can perform two crucial CSD tasks simultaneously: Feature selection provides insights into the most appropriate criteria. The subsequent training does not only return the best threshold values but also a more sophisticated criteria interaction than traditional approaches. While the threshold values and model structure may depend on the labeling of the dataset, the features used to represent the characteristics of irradiance curves might be much more consistent. Hence, the paper delivers the following contributions:

First, we introduce new features based on fast Fourier transform (FFT) and discrete Fréchet distance and combine them with a gradient-boosted trees classifier

(GBC), resulting in the FFT-GBC approach presented in Fig. 1. Second, we evaluate our resulting ML model against the – to the best of our knowledge – only other ML model for CSD, published by Liu et al. [3]. They trained a random forest and compared it to conventional approaches [3]. Third, we qualitatively compare our method with the traditional CSD approaches from Reno and Hansen [6], Ellis et al. [10], and Alia-Martinez [8]. We make our code publicly available in [13] to facilitate upcoming evaluations and applications of our technique.

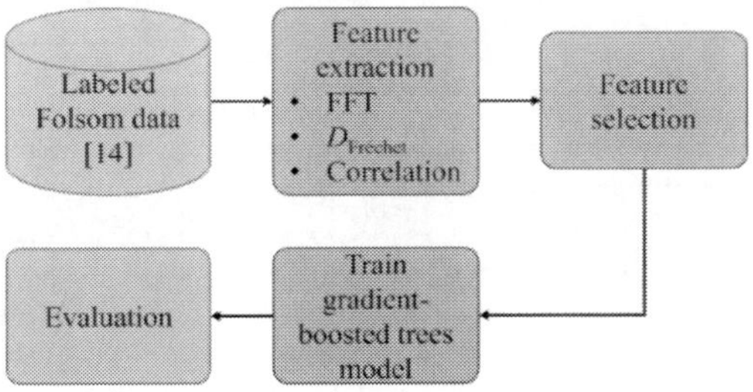

Figure 1: Procedure of the FFT-GBC approach

3 METHODS

3.1 Dataset and reference approach

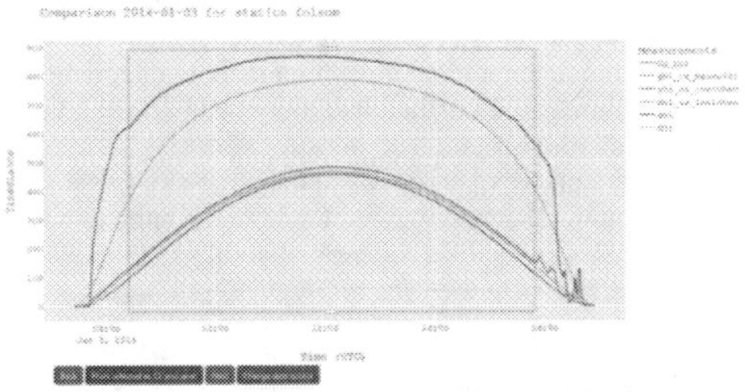

Figure 2: User interface of the labeling tool [13]

This work builds on the Folsom irradiance dataset [14], which covers GHI, DNI, and DHI with a one-minute resolution from 2014 to 2016. The labels result from the visual inspection of GHI, DNI, and DHI measurements next to Ineichen (GHI_{cs}, DNI_{cs}) [15] and Haurwitz (GHI_{cs}) [16] clear-sky estimates. Fig. 2 displays the labeling tool's user interface with the decision-relevant plots for one day. 2014 is the training and test dataset, and 2015 and 2016 are the additional evaluation datasets. The evaluation data includes 187,007 positive (clear-sky) and 316,026 negative (cloudy) samples; the 2014 data consist of 100,593 positive and 142,896 negative observations. Under-sampling the training and test (2014) dataset for balanced data results in 201,186 samples with half positive and half negative samples. To replicate the approach by [3], we strictly follow the procedure described in their paper: The features are the clearness index (K_t), the ratio of DHI and GHI (K_d), the difference of adjacent GHI measurements (ΔGHI), the standard deviation of a 10-minute GHI time series (Std10) and the cosine of the solar zenith angle (μ). In contrast with [3], we do not apply a grid search to determine optimal hyperparameter values. Instead, the grid search serves as a starting point for choosing parameters that lead to a leaner model with slightly better results on the evaluation dataset. Table I shows the final parameters.

Table I: Hyperparameters; abbreviations: N_{Trees} denotes the number of trees in the ML model, $Depth_{max}$ is the maximum depth, Split is the minimum number of samples per split, and $N_{Features}$ is the number of input features.

	N_{Trees}	$Depth_{max}$	Split	$N_{Features}$
Liu et al. [3]	190	30	20	5
This work	100	8	20	20

3.2 Feature extraction

For the new approach, 20-minute time series of measured and Haurwitz [16] GHI with a one-minute resolution are prepared for feature extraction. The features include fast Fourier transform (FFT) based calculations, the correlation, and the discrete Fréchet distance between each 20-minute GHI and GHI_{cs} time series.

The intuitive definition of the Fréchet distance is a person walking a dog on a leash. Both are not allowed to backtrack. The Fréchet distance is the shortest possible length of the leash, allowing the master and dog to complete their path [17]. In CSD, dog and master match measured and clear-sky irradiance. To transfer irradiance values and time points into a metric space for calculating Fréchet distance, we let a one-minute spacing correspond with one W/m² difference.

The FFT features result from the following procedure: First, the FFT is applied to each 20-minute time series of GHI and GHI_{cs} time series separately, producing values for 17 frequency bins, respectively. Fig. 3 and 4 show exemplary clear-sky and measured time series and their FFT representations.

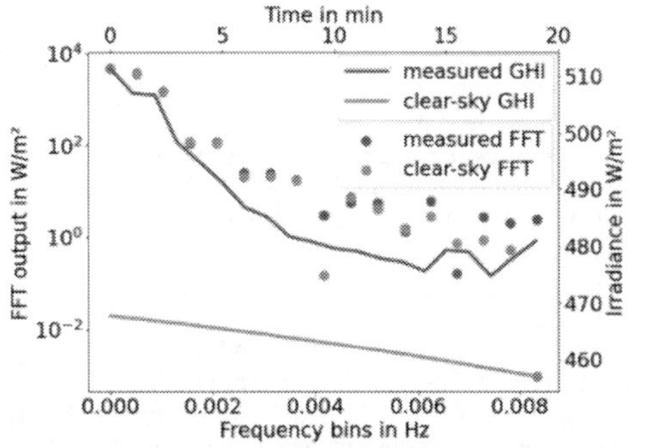

Figure 3: FFT representation of 20-minute measured and clear-sky time series for cloudy period

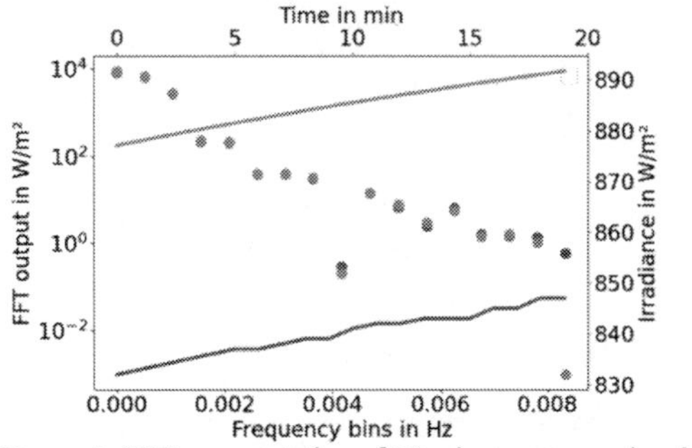

Figure 4: FFT representation of 20-minute measured and clear-sky time series for clear-sky period

10.4229/EUPVSEC2023/4DV.4.27
020458-002

Second, equation (1) returns the FFT feature X_{FFT} for the ith frequency bin fi for a 20-minute GHI time series GHI and is corresponding GHI_{cs}.

$$X_{\mathrm{FFT},fi} = \frac{\left| |\mathrm{FFT}(GHI)|_{fi} - |\mathrm{FFT}(GHI_{cs})|_{fi} \right|}{|\mathrm{FFT}(GHI_{cs})|_{fi}} \quad (1)$$

That way, each frequency bin has a feature representing the difference between the clear-sky and measured irradiance at its frequency domain. To increase the FFT's resolution without significantly extending the time series, the absolute difference between each feature value and its equivalent from the preceding time series forms another feature. This comparison applies to the two preceding and subsequent time series, resulting in five features for each frequency bin. Table II represents the structure of the training dataset.

Table II: Training dataset; abbreviations: $D_{Fréchet}$ is the Fréchet distance, r the correlation coefficient, $X_{FFT,fi}$ the FFT-output for the frequency bin with index i (i=0: 0 Hz, i=16: 0.0083 Hz), and is_cs indicates the label (clear-sky: 1, cloudy: 0).

Index	$D_{Fréchet}$	r	$X_{FFT,f0}$	…	$X_{FFT,f16}$	…	is_cs
1	0.16	0.18	0.25		0.87		0
2	0.17	0.18	0.23		0.90		0
…							
201 186	0.80	0.68	0.34		0.99		1

3.3 ML model and parameterization

An *ExtraTreesClassifier* from the Python library sklearn [18] with 50 estimators selects 20 out of 87 initial features. Since the class distribution in CSD may change a lot from location to location, a classifier robust to class imbalances is required. We thus choose the gradient-boosted trees technique (see [19]) and utilize sklearn's *HistGradientBoostingClassifier* implementation. After exploring different parameterizations with five-fold cross-validation, we opt for a maximum of 100 iterations, a maximum depth of 8, and a minimum of 20 samples per split. Both models are fitted to the 2014 dataset and predict the 2015 and 2016 classes for evaluation. Samples with a solar zenith angle below 85° are excluded.

3.4 Evaluation metrics

To provide an evaluation that does not heavily depend on the class balance in the evaluation dataset, we employ precision, recall, and F_1-score, which follow from the values and variables in Table II with the equations (2), (3), and (4).

$$\mathrm{Precision} = \frac{\mathrm{TP}}{\mathrm{TP}+\mathrm{FP}} \quad (2)$$

$$\mathrm{Recall} = \frac{\mathrm{TP}}{\mathrm{TP}+\mathrm{FN}} \quad (3)$$

$$F_1 = 2\frac{\mathrm{precision}\cdot\mathrm{recall}}{\mathrm{precision}+\mathrm{recall}} \quad (4)$$

4 RESULTS AND DISCUSSION

4.1 Quantitative evaluation

The benchmark model scores a precision of 81.9 percent and a recall of 92.0 percent on the evaluation dataset, resulting in an F1-score of 86.7 percent. The FFT-GBC model, in comparison, yields a precision of 85.1 percent, a recall of 95.6 percent, and an F1-score of 90.1 percent.

Table III shows the corresponding confusion matrix scores. While both models are similar regarding true negatives and true positives, the FFT-GBC model returns significantly fewer false negatives at 1.7 to 3.1 percent and slightly fewer false positives at 6.4 to 7.7 percent.

Table III: Confusion matrix scores. True (T) negatives (N), false (F) positives (P), etc., for the benchmark [3] and this work's model, shown as absolute numbers and percentages (in brackets).

	TN	FP	FN	TP
Liu et al. [3]	262 714 (54.2%)	37 398 (7.7%)	14 917 (3.1%)	169 430 (35.0%)
FFT-GBC (This work)	269 245 (55.5%)	30 876 (6.4%)	8 086 (1.7%)	176 271 (36.4%)

4.2 Visual evaluation

For qualitative analysis, the FFT-GBC model faces conventional models relying on GHI-based features alone. The only comparison model with a DHI-based variable is the ML model from Liu et al. [3]. Fig. 5 reveals that the FFT-GBC model labels fewer points as "clear-sky" than the benchmark models. The Alia-Martinez model [8] looks most relaxed, followed by Ellis et al. [10], Reno and Hansen [6], and Liu et al. [3]. The labeling from the Liu model appears very closely aligned to the clear-sky curve. While it also labels local minima, this is only when the local minimum is on the clear-sky curve. The FFT-GBC model only labels points in periods where the shape of the measurement curve is similar to the clear-sky curve. There is no visually fluctuating period where the approach labels some points as clear. In this regard, it differs from all other methods.

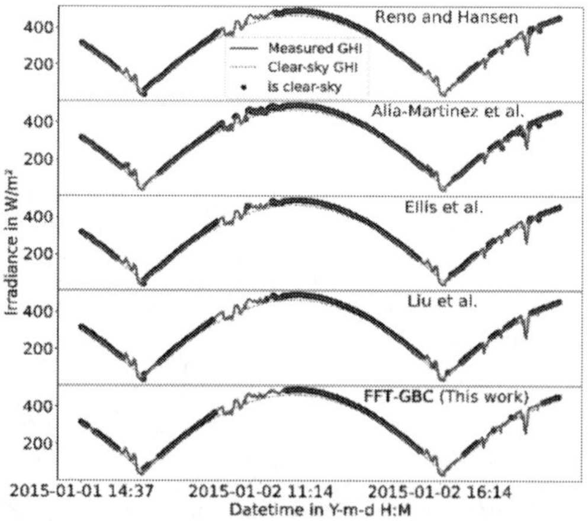

Figure 5: Qualitative evaluation; comparison for three days of the Reno [6], Alia-Martinez [8], Ellis [10], Liu [3], and the FFT-GBC model. Labels of the first three methods come from Bright's implementations in his GitHub repository [20] with Haurwitz [16] GHI_{cs} estimates.

4.3 Discussion

The increased sensitivity to minor changes in

10.4229/EUPVSEC2023/4DV.4.27
020458-003

irradiance curves may be subject to including FFT features that target the frequencies in the irradiance series. Frequencies are neither addressed in the multi-criteria approaches by Reno and Hansen [6] or Ellis et al. [10] nor in the correlation-based approach by Alia-Martinez. Seeing the differences in labeling irradiance measurements that fluctuate near the clear-sky estimate, we reason that including FFT features is highly desirable.

Comparing the FFT-GBC approach to prior approaches, the assumption that improvements in CSD are only possible with DHI or DNI-based features seems arguable: While the benchmark ML-based approach includes the ratio of DHI and GHI as features, the FFT-GBC approach solely contains features that can be derived from GHI. Still, the benchmark ML approach remarkably only labels points close to the clear-sky curve while being the only model without clear-sky estimates, but only the zenith angle, as an input feature. However, it was outperformed by the FFT-GBC approach in quantitative evaluation, which proves that adequate classification is possible without atmospheric variables except for GHI and a simple (e.g., Haurwitz) clear-sky model.

Finally, this work extends the literature on ML-based clear-sky detection. While Liu et al. [3] already showed similar accuracy to the Bright-Sun [4] model, our model fits better with our data and presents visually comprehensive results. With our and their features being quite different, the valid conclusion is that ML will work well with a wide range of features. Nevertheless, it requires labeled data not only for evaluation but also for training. Labeled data might be the critical challenge for the success of ML in CSD since there is the problem of getting reliable evaluation data. Liu et al. [3] completely re-trained their model to evaluate it on another dataset. Established conventional approaches, in contrast, have been validated with different datasets in different evaluation studies without any adjustment of threshold values. To provide some starting point for investigations on the necessity for re-training for different measurement devices, weather stations, or climate zones, we provide our model trained on the Folsom dataset for further evaluation. In addition, we provide the code for re-training the model with other or additional data [13].

5 CONCLUSIONS

In summary, we have brought new insights to the field of CSD by building on existing multi-criteria approaches by combining correlation with discrete Fréchet distance and FFT features. We applied an ML feature selection procedure to choose appropriate features and added more extensive rules, threshold values, and weightings by using those as input for a GBC method. Quantitative and qualitative results indicate that this approach delivers state-of-the-art performance while requiring only GHI and GHI_{cs} input.

In the future, we plan to address the question of reliable ground truth data and, subsequently, to extensively compare our model to existing models with various input features. Furthermore, we will investigate the model's generalizability to other sensor types, weather stations, and climate zones. Once ML models can generalize to new stations and climate zones, such that no re-training and newly labeled datasets are required, the potential of ML in the field of CSD may finally unfold.

6 REFERENCES

[1] C. A. Gueymard, J. M. Bright, D. Lingfors, A. Habte, and M. Sengupta, "A posteriori clear-sky identification methods in solar irradiance time series: Review and preliminary validation using sky imagers," *Renew. Sustain. Energy Rev.*, vol. 109, pp. 412–427, 2019, doi: 10.1016/j.rser.2019.04.027.

[2] X. Sun, J. M. Bright, C. A. Gueymard, B. Acord, P. Wang, and N. A. Engerer, "Worldwide performance assessment of 75 global clear-sky irradiance models using Principal Component Analysis," *Renew. Sustain. Energy Rev.*, vol. 111, pp. 550–570, 2019, doi: 10.1016/j.rser.2019.04.006.

[3] M. Liu, X. Xia, D. Fu, and J. Zhang, "Development and validation of machine-learning clear-sky detection method using 1-min irradiance data and sky imagers at a polluted suburban site, xianghe," *Remote Sens.*, vol. 13, no. 18, 3763, 2021, doi: 10.3390/rs13183763.

[4] J. M. Bright, X. Sun, C. A. Gueymard, B. Acord, P. Wang, and N. A. Engerer, "BRIGHT-SUN: A globally applicable 1-min irradiance clear-sky detection model," *Renew. Sustain. Energy Rev.*, vol. 121, 109706, 2020, doi: 10.1016/j.rser.2020.109706.

[5] F. Antonanzas-Torres, R. Urraca, J. Polo, O. Perpiñán-Lamigueiro, and R. Escobar, "Clear sky solar irradiance models: A review of seventy models," *Renew. Sustain. Energy Rev.*, vol. 107, pp. 374–387, 2019, doi: 10.1016/j.rser.2019.02.032.

[6] M. J. Reno and C. W. Hansen, "Identification of periods of clear sky irradiance in time series of GHI measurements," *Renew. Energy*, vol. 90, pp. 520–531, 2016, doi: 10.1016/j.renene.2015.12.031.

[7] J. Polo, L. F. Zarzalejo, L. Martín, A. A. Navarro, and R. Marchante, "Estimation of daily Linke turbidity factor by using global irradiance measurements at solar noon," *Sol. Energy*, vol. 83, no. 8, pp. 1177–1185, 2009, doi: 10.1016/j.solener.2009.01.018.

[8] M. Alia-Martinez, J. Antonanzas, R. Urraca, F. J. Martinez-de-Pison, and F. Antonanzas-Torres, "Benchmark of algorithms for solar clear-sky detection," *J. Renew. Sustain. Energy*, vol. 8, no. 3, 033703, 2016, doi: 10.1063/1.4950948.

[9] M. Lefèvre *et al.*, "McClear: A new model estimating downwelling solar radiation at ground level in clear-sky conditions," *Atmos. Meas. Tech.*, vol. 6, no. 9, pp. 2403–2418, 2013, doi: 10.5194/amt-6-2403-2013.

[10] B. H. Ellis, M. Deceglie, and A. Jain, "Automatic Detection of Clear-sky Periods Using Ground and Satellite Based Solar Resource Data," *IEEE 7th World Conf. Photovolt. Energy Conversion, WCPEC*, pp. 2293–2298, 2018, doi: 10.1109/PVSC.2018.8547877.

[11] C. Lewis, J. M. Strauss, and A. J. Rix, "A solar resource classification algorithm for global horizontal irradiance time series based on frequency domain analysis," *J. Renew. Sustain. Energy*, vol. 13, no. 3, 036102, 2021, doi: 10.1063/5.0045032.

[12] T. M. Mitchell, *Machine Learning*, 1st ed. New York (United States): McGraw-Hill, 1997.

[13] T. Veihelmann, "Clear-sky detection repository." Gitlab repository, 2023. [Online]. Available: https://gitlab.rrze.fau.de/ty90rize/cs_detection

[14] H. T. C. Pedro, D. P. Larson, and C. F. M. Coimbra, "A comprehensive dataset for the accelerated development and benchmarking of solar forecasting methods," *J. Renew. Sustain. Energy*, vol. 11, no. 3, 036102, 2019, doi: 10.1063/1.5094494.

[15] P. Ineichen and R. Perez, "A new airmass independent formulation for the linke turbidity coefficient," *Sol. Energy*, vol. 73, no. 3, pp. 151–157, 2002, doi: 10.1016/S0038-092X(02)00045-2.

[16] B. Haurwitz, "Insolation in Relation to Cloud Type," *J. Meteorol.*, vol. 3, pp. 123–124, 1946.

[17] T. Eiter and H. Mannila, "Computing discrete Fréchet distance," Wien, Austria, 1994.

[18] F. Pedregosa *et al.*, "Scikit-learn: Machine Learning in Python," *J. Mach. Learn. Res.*, vol. 12, pp. 2825–2830, 2011.

[19] J. H. Friedman, "Greedy function approximation: A gradient boosting machine," *Ann. Stat.*, vol. 29, no. 5, pp. 1189–1232, 2001, doi: 10.1214/aos/1013203451.

[20] J. M. Bright, "Clear-Sky Detection Library." Github repository, 2018. Accessed: Jul. 26, 2023. [Online]. Available: https://github.com/JamieMBright/csd-library

10.4229/EUPVSEC2023/4DV.4.27
020458-005

EVALUATION OF METHODS FOR OPTIMIZING A RAY TRACING SIMULATION WITH A BIFACIAL VERTICAL FIXED-TILT AGRI-PV SYSTEM

Oliver Pfeiffer[1], Eva-Maria Grommes[1], Ulf Blieske[1]
[1]University of Applied Sciences Cologne
Cologne Institute for Renewable Energy - CIRE
Cologne, Germany
oliver.pfeiffer@th-koeln.de, eva-maria.grommes@th-koeln.de, ulf.blieske@th-koeln.de

ABSTRACT: This work deals with the application of methods from the Global Sensitivity Analysis framework on a PV simulation use case. Four methods (EET, FAST, RSA, PAWN) are used to analyze the influence of nine input factors on the output of a ray tracing simulation done with bifacial_radiance in the example of a bifacial vertical fixed-tilt agri-PV system. Albedo and row spacing are identified to have the highest influence. Furthermore, it is found that a system size of 10 rows and 7 modules per row meets a semi-finite assumption. These results are used to derive optimization measures for the ray tracing simulation. The results obtained are very consistent throughout all methods. Since RSA and PAWN allow for factor mapping it is recommended to use one of them in combination with a method based on a different sampling strategy for cross-validation. The general use of GSA methods in the context of PV modelling is emphasized since the methods apply to all kinds of simulation models.

1 INTRODUCTION

The integration of bifacial photovoltaic technology in agricultural areas has gained significant attention as a sustainable approach to address the dual challenges of energy production and food security [1]. Vertical PV setups have emerged as an innovative solution, maximizing land use efficiency, and having advantages in load-shifting in comparison to titled monofacial setups [2]. To optimize the performance of bifacial Agri-PV systems and assess their energy generation potential, accurate modelling, and simulation of solar irradiance on the PV modules are essential. Since modelling is a costly activity, profound knowledge of the model's input-output relation is key, to save time and energy.

BifacialSimu [3] is a python-based simulation tool for bifacial PV yield prediction. It offers the opportunity to use a RADIACNE [4] ray tracing algorithm through the bifacial_radiacne [5] library, to determine irradiance on PV modules. Since computation time and output variance depend on the system size and the level of modelling detail, influential input factors need to be identified. Therefore, the input-output relations of nine parameters are investigated using four methods from the Global Sensitivity Analysis (GSA) framework [6] to find out which input parameters are most important for simulation and to gain insight into parameter combinations that lead to extreme model behaviour. Hence the proposed approach is a novelty and applicable to all input-output models, it provides added value for all PV modellers.

The objective of this work is to provide a comprehensive overview and a best-practice approach for deriving optimisation measures for PV simulations using GSA methods. This is done in the example of a ray tracing simulation with vertical fixed-tilt agri-PV systems.

A general overview and workflow of GSA is provided first. Then the four methods are described, followed by the input factors. After that the results of the GSA are presented. Subsequently, optimisation measures are derived. Then, the result plausibility is discussed, the GSA methods are compared and their general use for PV modelling is evaluated. Finally, a conclusion is drawn.

2 METHODOLOGY

2.1 Global Sensitivity Analysis

Global Sensitivity Analysis (GSA) describes a set of methods that can be used to analyse the relationship between multiple inputs and the output variance of a (numerical) model [6].
GSA is used to answer three questions, which are called settings in the terminology of Saltelli et al. [6]:

- Factor Ranking (FR): determining the most influential input factors,
- Factor Fixing (FF): determining input factors that are neglectable.
- Factor Mapping (FM): determine input factor combinations that promote extreme model behaviour.

Moreover, a GSA consists of three steps, as Pianosi et al. describe. The first step is input sampling. Ranges and distribution functions must be defined for each input factor that is analysed. Representative sets of these inputs are then generated according to the methods used. In a second step, simulations are done with each sample and finally, the results are analysed. The analysis can be done qualitatively by analysing the results visually, or quantitatively by calculating sensitivity measures. [7]

In this work, the input factor sampling and the analysis are done with a python library called *safepython-toolbox* [8].

It should be noted that GSA has known applications in environmental modelling [7] and production logistics simulation [9], but has not been used on PV simulations before.

2.2 Methods

This section describes basic principles of the GSA methods used.

2.2.1 Elementary Effects Test (EET)

EET is a method that can be used to rank input factors based on their relative importance on the output (factor ranking).
An elementary effect (EE) is a finite difference between two outputs that can be attributed to a variation of a specific input factor. To generate an EE, all the inputs must be sampled with a One-At-a-Time strategy (OAT). In

10.4229/EUPVSEC2023/4DV.4.28
020459-001

Input factors	Pseudonym	Range	Allocation	Unit
Number of rows	*rows*	*[1,50]*	*Discrete*	*[-]*
Modules per row	*modules*	*[1,50]*	*Discrete*	*[-]*
Albedo	*albedo*	*[0.15,0.55]*	*Continuous*	*[-]*
Row distance	*row_gap*	*[5,20]*	*Continuous*	*[m]*
Clearance height	*height*	*[0,1.5]*	*Continuous*	*[m]*
Module distance (horizontal)	*X_gap*	*[0, 0.2]*	*Continuous*	*[m]*
Module distance (vertical)	*Y_gap*	*[0, 0.2]*	*Continuous*	*[m]*
Cell distance (horizontal)	*XCell_gap*	*[0, 0.02]*	*Continuous*	*[m]*
Cell distance (vertical)	*YCell_gap*	*[0, 0.02]*	*Continuous*	*[m]*

Table I: Input factors that are analysed in with the GSA methods.

terms of sample size, Pianosi et al. recommend 10 to 100 samples per input factor. [7]

Therefore 500 OAT input factor combinations were generated. Then simulations were made with every input combination that was sampled, and the results were analysed.

Two sensitivity measures can be derived by analysing all the EEs that correspond to an input. On one hand, the mean elementary effect indicates the overall impact of an input factor on the output. On the other hand, the standard deviation, reveals, if the input factor can be attributed to a great range of output results. [6]

2.2.2 Fourier Amplitude Sensitivity Test (FAST)

FAST is a variance-based sensitivity analysis that was first introduced by Cukier et al. [10].
Like EET, it can be used in the FR and FF settings. The basic idea behind the FAST method is to periodically sample each input factor at a different frequency. The output variance is then analysed by Fourier transformation to see which frequencies are detectable. The contribution of an input factor to the output variance is called the main effect and is a measure of sensitivity. A main effect is calculated with the following equation,

$$\alpha_{xi} = \frac{V_{x_i}}{V^{x_i,\ldots,x_n}}$$

where V_{xi} is the partial variance caused by one input factor and $V^{xi,\ldots,xn}$ is the total output variance. V_{xi} is determined by summing up the Fourier amplitudes that correspond to the input factor frequency up to its fourth harmonic. [11]

2.2.3 Regional Sensitivity Analysis (RSA)

RSA can be used for FR, FF and for the analysis of how a region in the input range affects the output (factor mapping FM).
Here, the input factors are sampled with an All-At-a-Time strategy (AAT) [6]. In this work, 9000 samples were generated, even though 100 samples per input factor are sufficient [7]. Simulations are then made with all sampled input combinations. For analysis, the output range is divided into 10 equally spaced intervals, which is called the Freer approach [12].

Then the distribution of every input factor for a specific interval is calculated. By analysing the shape of the cumulated distribution function (cdf) over the input factor range, a ranking can be made, and other valuable insights can be derived. The median distance of the cdfs for extreme values is a quantitative sensitivity measure. [6]

2.2.4 PAWN

The PAWN method can like RSA be used in all three settings. It was developed by Pianosi et al. and falls under the category of density-based methods [13].
The idea is to generate outputs where all input factors are varied equally (unconditional outputs) and compare them to outputs where one input factor is fixed to a certain value and all other inputs are varied (conditional output). The comparison can be made qualitatively by analysing the corresponding cumulative distribution functions (cdf), or by calculating the Kolmogorov-Smirnov statistics between conditional and unconditional outputs. [13]

To apply this method, Pianosi et al. propose a minimum of 1000 input samples per input [7]. Therefore 9000 samples were created and analysed.

2.3 Simulating vertical bifacial east/west systems

The previously described GSA methods are used to optimise a ray tracing simulation at the example of a vertical fixed tilt east/west system. The simulations are done using the python library *bifacial_radiance* which is integrated into the simulation tool *BifacialSimu*.

The general layout of the simulated system is oriented to a real system in south Germany [14]. The basic unit consists of two vertically stacked bifacial modules in landscape format. The clearance height is 0.8 metres and there is a 0.1 metre vertical gap and a 0.2 metre horizontal gap between the modules. Without the mounting structure and module gaps, this setup is two metres high and 2 metres wide. The modules are east/west oriented, with the module's front side facing east. Nine input factors are then chosen to simulate different versions of that system type (explained in the next section).

Each simulation uses the same weather data for the location of the real PV system in southern Germany, provided by the German Weather Service (DWD) [15]. The simulations are done with the cumulative sky procedure that is described in [16] and utilized by the bifacial_radiance library.

2.4 Input factors

Table 1 gives an overview of the input factors that were chosen. When sampled with the described methods, they are varied within the listed ranges. The ranges for number of rows and module per row are chosen to mimic experimental to utility-size PV systems. The allocation in their interval must be discrete since module rows must be whole numbers. The distribution for all factors is chosen to be uniform. In terms of the albedo, the range corresponds to typical values found on cropland [17] and concrete [18], since vertical PV systems are used in the agri-PV context or as fences in urban areas [14]. The range for row distance is based on calculations for the minimum row distance of the basic unit, described in section 2.3. and on common widths of big agricultural land machines. The clearance height was varied to simulate plant growth. Its range is inspired by the real PV system mentioned in section 2.3. Its maximum value was exaggerated to see if the input could have a significant influence. The last four input factors are chosen to simulate a high level of modelling detail. Their values are also based on the real system in South Germany.

10.4229/EUPVSEC2023/4DV.4.28
020459-002

3 RESULTS

3.1 Interpretation of Global Sensitivity Analysis results

In this section, the results obtained with the GSA methods are interpreted. It should be noted, that all results correspond to the module front side since the assessment for the rear side was qualitatively similar.

3.1.1 Results of Elementary Effects Test

Fig. 1 shows the results of the EET. On the x-axis, there is the mean elementary effect (mean EE). When an EE has a comparatively high value, it means that a relatively small variation of an input factor causes a relatively large output variation. Since EEs are calculated over the input factors whole range, the mean EE is a meaningful measure for its relative importance. On the y-axis, there is the corresponding standard deviation, which reveals, if the input factor can be attributed to a great range of output results. When the standard deviation is comparatively high, a relatively small change in input causes either great or small changes in the output. This indicates that the output was rather influenced by the choice of the other input factors. Accordingly, it is a measure of how much an elementary effect depends on the choice of other input factors.

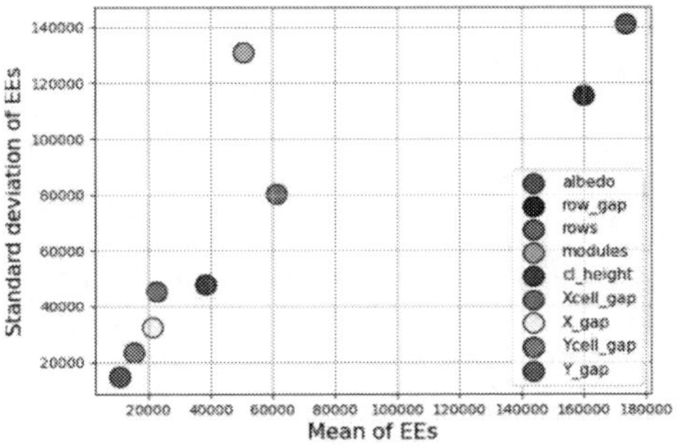

Figure 1: Results of the Elementary Effects Test regarding the module front side

In the first setting, the results are used for factor ranking, to determine influential input factors. Albedo exceeds the other factors in both the mean EE and the standard deviation, closely followed by row distance, making them the most influential factors in this consideration. Thus, the score for standard deviation is the highest too, the EEs strongly depend on the choice of other factors. This indicates, that in some configurations, albedo and row distance have a very strong influence on the output and in others not. The lowest overall scores for both EE and standard deviation have the parameters that define module and cell gaps, making them the least influential.

3.1.2 Results of Fourier Amplitude Sensitivity Test

Fig. 2 shows the main effects of the input factors regarding the module's front side. A main effect is the ratio of the partial output variance (caused by an input) to the total output variance (caused by all inputs), as explained in section 2.2.2

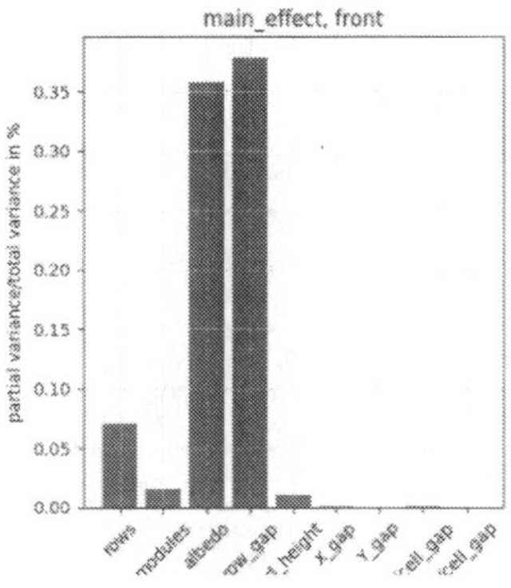

Figure 2: Results of the Fourier Amplitude Sensitivity Test regarding the module front side

Albedo and row distance exceed the other factors, each causing 35 % to 40 % of total output variance. Next, but with a large distance comes the number of rows, scoring 5 %. The number of rows and clearance height score even lower to about 1 % and 2 % and the gap-related input factors score 0. Regarding the second scenario, the back side was analysed showing similar results. The only difference is that albedo scores higher than row distance. The absolute results however are in the same region as before.

3.1.3 Results of Regional Sensitivity Analysis

Fig. 3 shows the result of the RSA. Each plot corresponds to one input factor, showing its cumulative distribution function (cdf) over its range for various output intervals. While the x-axis displays the input factor range, the y-axis shows the cdf. Each plot has ten coloured lines corresponding to an interval in the output Y, with Y being the averaged cumulated solar irradiance per module area in W/m^2, which is calculated by bifacial_radiance. In the first setting, the results are used for factor mapping to analyse how the input factors are distributed when the output is in a certain range. This is particularly useful when analysing how extreme values are generated. First look at the ranges for the coloured output intervals on the right-hand side of the figure. The ranges for the lowest interval (deep blue) and the highest (deep red) are much larger than the others. This indicates that certain combinations of input factors produce highly variable extreme values. With this information in mind, an analysis of the corresponding cfds of the input factors gives an insight into how these values are generated.

The first graph gives an important insight into the relationship between the number of rows and the highest output values (red graph). Here, $\Delta cdf = 0.5$ for a range of 1 to 5 rows. This means that 50 percent of the outputs in the deep red interval were generated with 1 to 5 module rows. The other 50 per cent were generated within 6 to 50 rows. This indicates that a low number of rows promotes extremely high output values. A similar result can be obtained by analysing the number of modules per row. As the graph starts at 0.2, 20 percent of all extreme values were generated with only 1 module per row. However, the rest of the range is evenly distributed, indicating that more modules do not affect the result. Next, the albedo is considered. The deep red graph starts at 0.35, which means that all the extremely high values were generated with an albedo greater than 0.35. The last

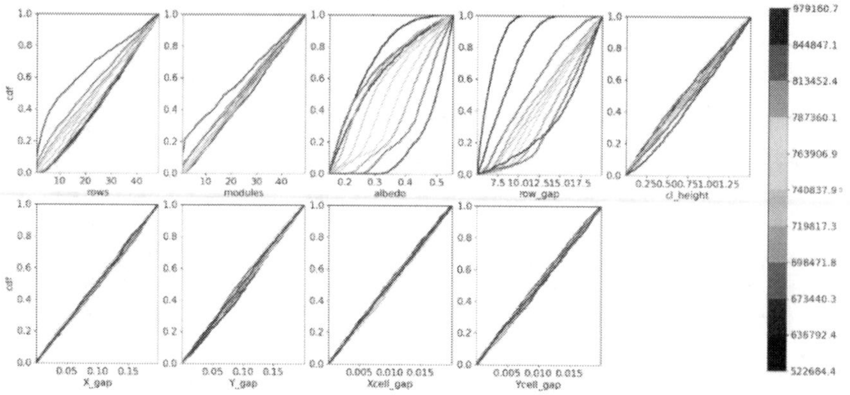

Figure 3: Regional Sensitivity Analysis result: cumulative distribution function (cdf) for each input parameter regarding the module front side

influencing value is r_dis. Here, we can see that 80% of the extreme values occurred in a range of 15 to 20 metres. To summarise: Extremely high and fluctuating values are favoured when the number of rows is less than 5, there is only one module in a row, the albedo is greater than 0.35 and the row distance is greater than 15 metres. A similar analysis can be done concerning extremely low output values. Here, 80 percent of the values were generated with albedos below 0.3 and all values were in a range of 10 metres with regards to the row distance.

As an interim conclusion, it is important to avoid the input combinations described above when aiming for a stable and moderate simulation result. In the second setting, the results are used for factor ranking to determine influential input factors. The maximum vertical distance "mvd" of the two cdfs for a point on the x-axis indicates the importance of an input factor. In plot three, mvd is large, because low albedo values cause low output values and high albedo high cause high outputs. Following that logic, the albedo and the row distance are regarded as influential factors, driving the outputs behaviour. In contrast, mvd is very small in plots five to nine. The lines for all intervals are on the same level, which means that any point in the input range can cause both large and low outputs. Accordingly, these input factors are regarded as non-influential since they have no (or very little) effect on the output.

3.1.4 Results of PAWN

Fig. 4 shows a selection of results obtained from PAWN. As described in section 2.2.4 unconditional outputs are compared to conditional outputs. In the analysis, the input range is divided into equally sized intervals, which are displayed on the x-axis. For each interval, the maximum distance between the cumulative distribution function of the conditional and unconditional outputs is plotted. The abbreviation KS (displayed on the y-axis) refers to the "Kolgomorov-Smirnov statistic", which is the name of this kind of analysis. A high KS value indicates a high impact of the input factor in the specified range. A value of zero implies that the input factor has no influence.

The PAWN results reinforce the findings of previous GSA methods in terms of factor ranking. Albedo and row spacing are the most important, followed by the number of rows and modules per row. This is followed, with some distance, by clearance height and then, in last place, by the gap-related input factors. Overall, the factor ranking shows plausible results due to the cross-validation with other methods. On the other hand, the analysis differs from the previous results in one respect. Row distance was considered the most important factor overall, while the other methods ranked albedo highest.

In terms of factor mapping, the results are comparable to RSA. Both methods find that low values for number of rows and modules per row favour extreme values in the output. However, it should be noted that the same sampling data set was used in both cases. The results for the factor fixing were comparable to the FAST method. Both show that the factors related to clearance height and gap are the least influential. While FAST identified these factors as having no influence, PAWN identified them as having very little influence. Another point to emphasise is that the PAWN method is relatively new and not well-established. However, the results are transparent and easy to understand because of the factor mapping. This underlines their plausibility.

3.2 Derivation of optimisation measures

GSA revealed, that the impact of the number of rows and modules per row is system size dependent. They have an impact when their value is lower than 10 (rows) or 7 (modules) (cf. Fig. 4) and only in combination with high values for albedo and row distance (cf. Fig. 3). When analysing utility-scale systems, the system size can therefore be set to 10 rows and 7 modules. Furthermore, it is sufficient to analyse the centre module of the centre row only, since all the other rows do not meet the semi-finite assumption and distort the output. This saves computation time and will not influence the output accuracy. In experimental scale systems, however, the actual system size should be simulated.

GSA further showed that the albedo and the row distance have the largest overall impact. As described in section BifacialSimu's ray tracing mode however does not offer the possibility to consider a variable albedo. Moreover, the row distance can only indirectly be set via the ground cover ratio. It is therefore clear that a variable albedo must be included and that the program should be enabled to set the row distance directly.

The analysis showed that the gap-related input factors have little to no influence on the simulation output.

10.4229/EUPVSEC2023/4DV.4.28
020459-004

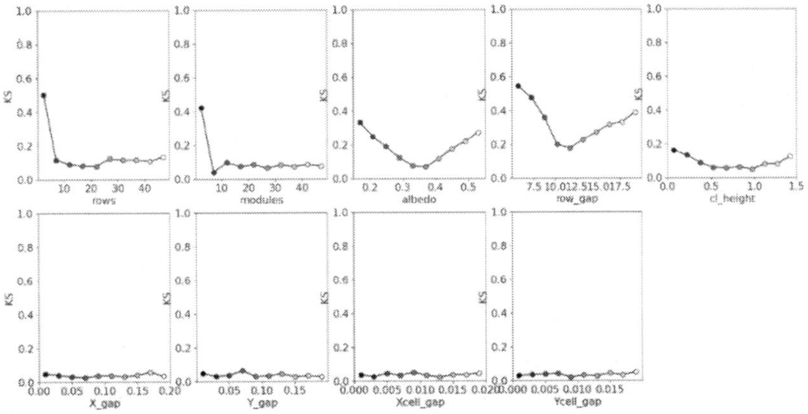

Figure 4: PAWN result: Kolgomorov-Smirnov statistic for each input factor regarding the module front side

Therefore, they can be neglected. Furthermore, the clearance height can be neglected when simulating vertical east/west systems.

4 DISCUSSION

4.1 Result consistency

In terms of cross-validation, the results are very consistent. Four GSA methods were used to investigate the importance of the input factors, based on 500 (EET), 1289 (FAST) and 9000 (RSA, PAWN) model evaluations. Through all the methods the results for factor raking show high homogeneity. The input factors albedo and row distance were ranked as the most influential parameters. Except for the FAST method, the albedo was always identified as the most influential factor regarding the module's front side, closely followed by row distance. With great distance, the next most influential factor is the number of rows followed number of modules per row. This finding is consistent through all methods. RSA and PAWN further showed that these input factors are only influential in their lower range and combination with either very high or very low values for albedo and row distance. Next comes the clearance height which evaluation showed little to very low influence in every method. All gap-related input factors were ranked equally low in all methods.

4.2 Assessment of result plausibility

On a physical level, the GSA results seem plausible. As the albedo is the ratio of irradiance reflected by the ground, it is logical that it has a large influence on the amount of irradiance hitting the module surface. The same applies to row spacing, as it determines how much shading occurs from module to module. The lower impact in their mid-range (cf. Fig. 4) can be explained when taking into consideration that the results are not absolute but relative to all other input factors.

The relatively low influence of number of rows and modules per row can be explained by the selection of their range. Their impact is greatest when number of rows is below 10 and number of modules is below 7. When simulating with higher values, the influence of edge-brightening effects decreases. Since they varied from 1 to 50, their impact on methods like EET and FAST is considered low.

Clearance height was considered an influential input factor for tilted setups [19] but its influence on vertical setups was considered very low in this work. When modules are tilted, the down-facing side is mostly illuminated by ground-reflected irradiance. In vertical setups, both module sides are affected by direct, diffuse and ground-reflected irradiance. Since this is independent of the clearance height, its low impact is plausible.

The very low influence of all gap-related parameters is also plausible since gaps slightly increase irradiance passing through a module. Also, the selected ranges are very narrow which decreases their impact.

The high consistency of the results in all methods emphasises their plausibility. It should be noted, however, that they are only valid for the investigated use case, which is vertical bifacial east/west systems, and when varying the input factors in the defined ranges. If the input factor ranges were different, the results could be different as well. Especially, when differing the ranges for albedo and row distance their relative importance would vary, which can be seen in Fig. 4. Furthermore, it is not clear if the results transfer to other PV system types, including tilted modules or single axis tracking and different module azimuths.

4.3 Method comparison

The four methods used, namely EET, FAST, RSA and PAWN have advantages and disadvantages. All require a different minimum number of model runs, which makes some of them very time-consuming. On the other hand, the level of information and insight is also different. The EET method is discussed first. It requires the least number of model evaluations, and the results can be used for factor ranking. In this paper, 500 simulations were run to obtain the results, but fewer model evaluations would have been possible (see section 2.2.1). Factors are ranked based on their mean Elementary Effect (EE) score and corresponding variance. As an EE is a finite difference between two model outputs produced by varying an input factor, its interpretation requires little statistical knowledge, and its validity can be easily assessed. However, the information obtained is limited compared to the other methods, since factor mapping is not possible. EET can be used to identify input factors with low

10.4229/EUPVSEC2023/4DV.4.28
020459-005

influence, but it is not possible to characterise a factor as having no influence.

Secondly, the FAST method is evaluated. It provides more information than EET at a moderate level of model evaluation. Its sensitivity coefficient is more meaningful since it shows the contribution of an input factor to the total output variance. It differs from the other methods in that the minimum number of model evaluations can be calculated. Therefore, there is no uncertainty as to whether the number of evaluations is sufficient. The disadvantage of this method is that it requires a deep knowledge of statistics and mathematics to understand how the results are generated. It also provides less information than the remaining two methods.

The third method to be discussed is RSA. It can be used for factor ranking, factor fixing and factor mapping and is therefore more versatile than the previously described methods. Although the minimum number of model evaluations required is small, it provides much more information than EET and RSA. In this work, 9000 model evaluations were made since the PAWN method required his number and the simulation results could be used for both methods. It should be noted however, that 1000 evaluations would meet the minimum criteria (cf. sec. 2.2.3). The main advantage of this method is, that the results can be used to identify regions in the input space that promote extreme model behaviour. This is very useful information as these regions can then be avoided. In addition, the method is relatively easy to understand, and the results are easy to interpret.

The final method is PAWN. It is as versatile as RSA and has the advantage that it can be used for factor fixing since a KS value of zero is a sufficient criteria for an input factor to be non-influential [13]. The main problem is the extremely high number of necessary model evaluations. In this work, 9000 evaluations were performed, which is the minimum recommended number, as described in section 2.2.4. As the number of model evaluations scales with the number of input factors, only a relatively small number of factors can be effectively analysed.

4.4 Evaluation of use for PV simulations

Lastly, this section evaluates, if the GSA results were useful for optimizing the ray tracing simulation and if it could be useful for other PV simulations.

The identification of the system size that meets a semi-finite assumption has a great impact. When simulating utility-scale systems, the system size can be safely set to this value. This leads to a dramatic decrease in the simulation time since it scales with the number of rows over which the irradiance is to be analysed.

The findings that the gap-related parameters are non-influential lead to the conclusion that they should not be included in the simulation procedure. The same is true for the clearance height. However, these findings only apply to vertical bifacial east/west systems, as already explained in the previous section.

The finding that albedo is the most influential input factor emphasises the need for albedo measurements to produce accurate simulations for a specific location. The row distance was also found to be important. This is essential information when designing a system using bifacial_radiance.

The example shows that the application of GSA methods in the PV simulation use case led to valuable optimisation measures. As the methods apply to any PV simulation model that produces an output from multiple inputs, they are of general use to PV modellers. The results can be used to identify influential input factors and to identify specific combinations or sub-ranges that promote extreme model behaviour. With this knowledge, such behaviour can be effectively avoided, which makes the simulation more reliable. In this respect, the use of factor mapping becomes clear. Accordingly, RSA or PAWN methods should be applied. To cross-validate the results a methods using a different sampling strategy like EET or FAST should be used too.

5 CONCLUSION

In this work, the use of GSA methods for PV modellers has been exemplarily demonstrated. The EET, FAST, RSA and PAWN methods were used to derive optimisation measures for a ray tracing simulation with bifacial_radiance using a vertical bifacial east/west PV system as an example. Three optimisation measures were found to reduce the simulation time, simplify the simulation model and increase the reliability of the results. The use of GSA methods in PV modelling was discussed. It was found that these methods are of general use, since they apply to any model, help to identify important and negligible input factors and detect regions in the input space that promote extreme model behaviour. The use of at least two methods can be recommended for cross-validation purposes. To exploit the full potential of GSA, a method that allows factor mapping should be used in combination with at least one other method that uses a different sampling strategy.

However, in terms of factor ranking, the results obtained are only valid for the use case considered. Different ranges of input factors or different choices of input factors will change the results.

Future research may show whether comparable results are obtained for inclined systems and other combinations of input factors.

6 References

[1] M. Trommsdorff, S. Gruber, T. Keinath, M. Hopf, C. Hermann, and F. Schönberger, "Agri-Photovoltaik: Chance für Landwirtschaft und Energiewende: Ein Leitfaden für Deutschland | Stand April 2022," Apr. 2022. [Online]. Available: https://www.ise.fraunhofer.de/content/dam/ise/de/documents/publications/studies/APV-Leitfaden.pdf

[2] M. H. Riaz, H. Imran, R. Younas, M. A. Alam, and N. Z. Butt, "Module Technology for Agrivoltaics: Vertical Bifacial Versus Tilted Monofacial Farms," *IEEE J. Photovoltaics*, vol. 11, no. 2, pp. 469–477, 2021, doi: 10.1109/jphotov.2020.3048225.

[3] E.-M. Grommes and U. Blieske, "BifacialSimu: Holistic Simulation of large-scale Bifacial Photovoltaic Systems," *JOSS*, vol. 7, no. 78, p. 4443, 2022, doi: 10.21105/joss.04443.

[4] G. J. Ward, "The RADIANCE lighting simulation and rendering system," in *Proceedings of the 21st annual conference on Computer graphics and interactive techniques - SIGGRAPH '94*, Not Known, 1994, pp. 459–472.

[5] S. Ayala Pelaez and C. Deline, "bifacial_radiance: a python package for modeling bifacial solar photovoltaic systems," *JOSS*, vol. 5, no. 50, p. 1865, 2020, doi: 10.21105/joss.01865.

10.4229/EUPVSEC2023/4DV.4.28
020459-006

[6] A. Saltelli *et al., Global Sensitivity Analysis. The Primer*: Wiley, 2007.
[7] F. Pianosi *et al.,* "Sensitivity analysis of environmental models: A systematic review with practical workflow," *Environmental Modelling & Software*, vol. 79, pp. 214–232, 2016, doi: 10.1016/j.envsoft.2016.02.008.
[8] F. Pianosi, F. Sarrazin, and T. Wagener, "A Matlab toolbox for Global Sensitivity Analysis," *Environmental Modelling & Software*, vol. 70, pp. 80–85, 2015, doi: 10.1016/j.envsoft.2015.04.009.
[9] A. Ghannouchi, T. Stittgen, and J. H. Schleifenbaum, "Sensitivity Analysis of a Simulation Model for the Determination of the Utilization of a Production Environment with the LPBF-Process," 2021.
[10] R. I. Cukier, C. M. Fortuin, K. E. Shuler, A. G. Petschek, and J. H. Schaibly, "Study of the sensitivity of coupled reaction systems to uncertainties in rate coefficients. I Theory," *The Journal of Chemical Physics*, vol. 59, no. 8, pp. 3873–3878, 1973, doi: 10.1063/1.1680571.
[11] C. Xu and G. Gertner, "Understanding and comparisons of different sampling approaches for the Fourier Amplitudes Sensitivity Test (FAST)," *Computational statistics & data analysis*, vol. 55, no. 1, pp. 184–198, 2011, doi: 10.1016/j.csda.2010.06.028.
[12] J. Freer, K. Beven, and B. Ambroise, "Bayesian Estimation of Uncertainty in Runoff Prediction and the Value of Data: An Application of the GLUE Approach," *Water Resources Research*, vol. 32, no. 7, pp. 2161–2173, 1996, doi: 10.1029/95WR03723.
[13] F. Pianosi and T. Wagener, "A simple and efficient method for global sensitivity analysis based on cumulative distribution functions," *Environmental Modelling & Software*, vol. 67, pp. 1–11, 2015, doi: 10.1016/j.envsoft.2015.01.004.
[14] Next2sun, *Projects.* [Online]. Available: https://next2sun.com/projekte/
[15] Deutscher Wetterdienst, *CDC - Climate Data Center.* [Online]. Available: https://cdc.dwd.de/portal/
[16] D. Robinson and A. Stone, "Irradiation modelling made simple: the cumulative sky approach and its applications," in *PLEA 2004: The 21st Conference*, pp. 19–22.
[17] P. Sieber, S. Böhme, N. Ericsson, and P.-A. Hansson, "Albedo on cropland: Field-scale effects of current agricultural practices in Northern Europe," *Agricultural and Forest Meteorology*, vol. 321, p. 108978, 2022, doi: 10.1016/j.agrformet.2022.108978.
[18] H. Li, J. Harvey, and A. Kendall, "Field measurement of albedo for different land cover materials and effects on thermal performance," *Building and Environment*, vol. 59, pp. 536–546, 2013, doi: 10.1016/j.buildenv.2012.10.014.
[19] S. A. Pelaez, C. Deline, S. M. MacAlpine, B. Marion, J. S. Stein, and R. K. Kostuk, "Comparison of Bifacial Solar Irradiance Model Predictions With Field Validation," *IEEE J. Photovoltaics*, vol. 9, no. 1, pp. 82–88, 2019, doi: 10.1109/JPHOTOV.2018.2877000.

10.4229/EUPVSEC2023/4DV.4.28
020459-007

SENSOR DEVELOPMENT AND APPLICATION OF HIGHLY PRECISE MOBILE IRRADIANCE MEASUREMENTS

C. Braun, A. Kleinhans, F. Basler, M. Kaiser
Heidenhofstr. 2, 79110 Freiburg, Germany
Fraunhofer Institute for Solar Energy Systems ISE
christian.braun@ise.fraunhofer.d

ABSTRACT: In 2019 it was estimated that 24% of CO2 emissions came from the transport sector. Out of those 24% almost three quarters are due to the road sector. Electric vehicles have the potential to significantly reduce CO2 emissions from transport. More and more governments are setting policies and incentives to help move the sector to zero emission (f.e. typical tasks to reach for new vehicles until 2025). To judge the overall impact on the electrical grid the electrical efficiency of the vehicles is a major key factor especially also looking at heavier transport vehicles like buses, coaches and trucks.

To boost the effective electrical efficiency, it has been suggested to add solar on-board electricity generation to those vehicles. Some studies have suggested a significant potential for this type of electricity generation to offer cost-effective solutions. A comprehensive compilation of studies has been published by IEA Task17 on PV in Transportation in 2021, *State of the Art and Expected Benefits of PV-Powered Vehicles* [1].

As a basis for this type of studies irradiance measurements for mobile applications are required and it is suggested here to combine very precise mobile solar irradiance measurements taking into account rapidly changing shadowing situations with advanced satellite derived irradiance data for solar potential modelling.

This paper deals with the conceptualization, research & development as well as preliminary results of such highly precise solar irradiance measurement devices.

Keywords: mobile irradiance measurements, solar potential for mobile applications, mobile sensor development

1 AIM AND APPROACH

Vehicle integrated photovoltaics (VIPV) has the potential to make the dream of travelling without use of additives come true. To this aim many studies have been conducted and in experimental setups like in the "World Solar Challenge (crossing the desert in Australia with vehicles driven by solar energy)" ranges of 100s of kms have been shown to be possible. Now many studies have been conducted based on modeling the incoming power source – solar energy. Whereas for static applications through many years of research, solar potential maps and atlases have been established with reliable data, the reliability and data uncertainty being dependent on the area and the research that has taken place for this area.

For the case of a "moving target", a moving photovoltaic plant, this poses a range of hardly studied questions namely the degree to which shading influences the available radiation. Furthermore the range of measurement campaigns for this special case of a moving PV installation are very limited [2]. This clearly shows the requirement for more detailed analyses in which the shadowing situation needs to be taken into consideration.

To meet this challenge in the framework of a "citizen scientist" project for the evaluation of the solar potential for Germany road network [3] a highly precise mobile irradiance sensor has been developed. This device is self powered by a Fraunhofer ISE inhouse laminated solar module and reference cell. The module feeds into a battery pack of 3 x 5200mAh to be able to maximize measurement time when the vehicle is shaded.

Energy consumption has been optimized by activating several levels of sleeping states for the processor and minimizing time on air through compressing data. The device features a 3-axis acceleration sensor which allows changing measurement resolution depending on the driving state. In the mentioned project we used a 1/s resolution for driving state and one minute averages for static state. Data transfer is done via the mobile LTE network and sending status is only activated in static state and transmitting intervals and quantities can be accustomed to the needs of the project.

Figure 1: Picture and schematics of ISE developed mobile irradiance sensor

2 PROJECT OVERVIEW

As mentioned in the well compiled IEA Task 17 report, not enough and precise enough mobile solar irradiance data has been collected to make generalizations on the characteristics of solar irradiance for mobile applications. This is why FHG-ISE has developed a self-powered sensor with reference cells that are calibrated in Fraunhofer ISE's certified solar module measurement lab. The sensor has proven to be usable by non-scientists and the fleet of sensors has collected almost 40 Mio data points from a road network of about 400Tkm until end of 2022.

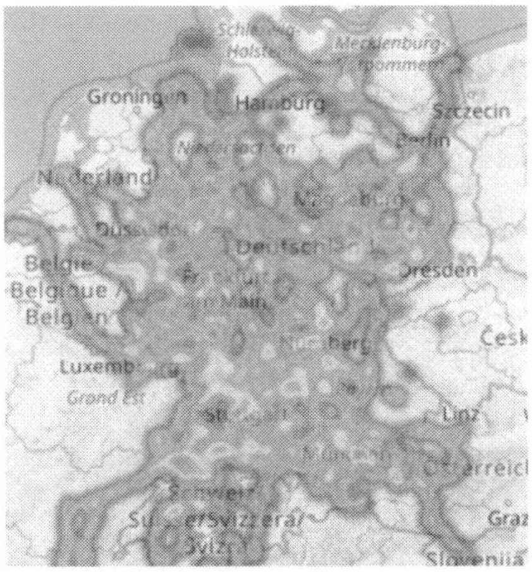

Figure 2: Heatmap of area of german road network covered by the measurements

3 RESULTS AND CONCLUSIONS

Example results for a clear sky day with partial shadowing:

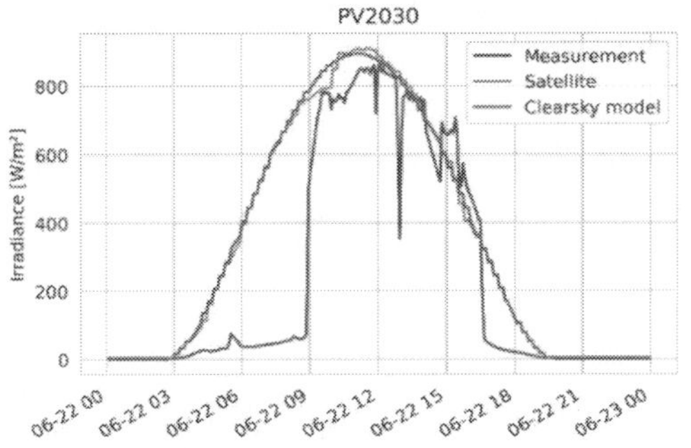

Figure 3: Clear Sky (cls) day comparing measurement data, satellite and clear sky model

In Fig 3 the measured (meas) global horizontal irradiance is compared with the satellite (sat) data derived irradiance and a clear sky model calculation in partially shaded example. To be able to deal with these different shading situations, a method has been developed for automatized shadow detection after data resampling to 15 min and application of this filter:

$$\frac{G_{sat}}{G_{cls}} - \frac{G_{meas}}{G_{cls}} > 0.3$$

Comparing the total energy of all available data for one sensor shows about 28% less energy than in the satellite data, about 8% unshaded data and shading influence on about 70% of all data. This would amount to an energy loss of about 20%.

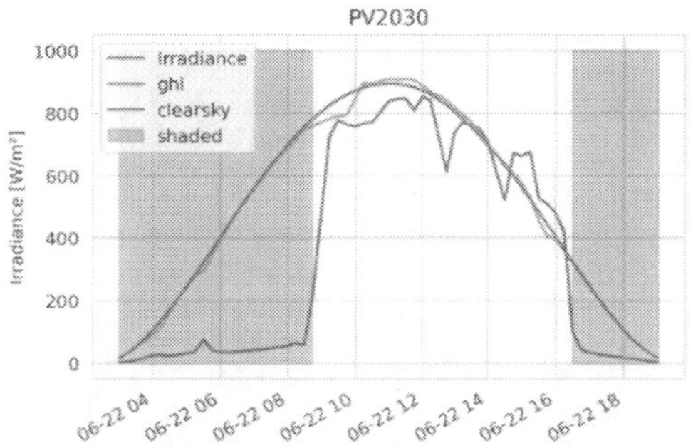

Figure 4: Clear Sky (cls) day comparing measurement data, satellite and clear sky model

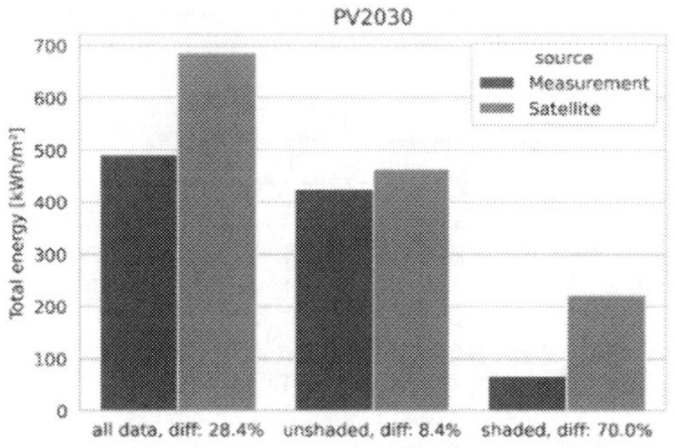

Figure 5: Comparing energy calculation of measurement & satellite data

These results are very much in alignment with studies the cited studies and measurements campaigns. Now we are looking very much forward the extended results in the EU Horizon project “Solar Moves” where the fleet of sensors will – after recalibration in the lab- find their next application cases on cars, coaches, buses and trucks in at least 2 european countries in different climate zones and shading situations.

References

[1] https://iea-pvps.org/key-topics/state-of-the-art-and-expected-benefits-of-pv-powered-vehicles/

[2] IEA Tracking Transport Report 2020, Jacob Teter, May 2020, Tracking Transport 2020 – Analysis – IEA

[3] https://pv2go.org/

10.4229/EUPVSEC2023/4DV.4.29
020460-002

EVALUATING 2D VIEW-FACTOR SOLUTIONS TO ESTIMATE GROUND-REFLECTED IRRADIANCE IN LARGE BIFACIAL PV PLANTS

Juan Santamaría-Sancho, Javier R. Ledesma
Instituto de Energía Solar, Universidad Politécnica de Madrid
juan.santamariasan@upm.es, javier.ledesma@upm.es

ABSTRACT. In large photovoltaic (PV) bifacial plants, the relationship between energy-yield modelling and subsequent on-site measurement campaigns remains an ongoing challenge. One particular area of difficulty concerns the energy exchange between the ground and the PV module surfaces. This process is effectively captured through simplified two-dimensional view factor models, being the segment-to-segment method an effective approach to compute them. Through this approach it is provided a direct outcome, however, concerns arise regarding the loss of the non-uniformity of the rear irradiance received on the module surfaces. This issue presents challenges for on-site measurements, prompting the exploration of an alternative approach, which preserves non-uniformity information: the segment-to-point method. In this work it is performed a comparative analysis between the two view factor methods, both implemented in SISIFO (an open simulation tool), and it is conducted using a specific case study, a real solar plant in Egypt.
Keywords: two-dimensional view factor model, segment-to-segment method, non-homogeneous rear irradiance, bifacial PV modules, open free PV simulation tool.

1 INTRODUCTION

In 1984, a groundbreaking moment in solar energy occurred when Isofoton Inc. introduced the world to the first-ever industrial production of bifacial modules in Spain [1]. Since then, the solar industry has witnessed a significantly surge in bifacial PV plants, with market shares projected to soar up to a 40% by 2028 [2]. However, this remarkable growth of bifacial solar technology also involves a challenging activity for both the scientific and the engineering communities, who must address the differences between monofacial and bifacial modules and develop accurate mathematical tools to model energy production, and to establish a reliable framework of confidence for enterprises and customers. This challenge begins with a non-perfunctory understanding of the distinctive feature of bifacial modules: the energy exchange between the ground, which acts like an emitting surface of irradiation, and the rear surface of the PV modules. It is worth noticing that there is also an energy exchange between the ground and the frontal surface of the modules, that impacts both monofacial and bifacial PV generators and requires a comprehensive analysis too.

Prior to any explanation, it is required to underscore two preliminary considerations which will foster the analytical study of the underlying physical phenomena, and, in turn, it will lead to a simplified version of the model employed for estimating the energy yield of the PV plants. Firstly, it will be considered that the exchange of radiative energy between surfaces is unaffected by the medium that separates them. Secondly, the ground and surroundings will be regarded as ideal diffuse reflector surfaces, which means that the reflected radiation is uniformly distributed irrespective of the nature of the incident radiation. Within the framework of these two postulates, the geometric relationships governing the energy transfer between the ground and the PV module surface can be explained through the concept of *view factors*, defined as the ratio of diffuse energy leaving the emitting surface A_i intercepted by the receiving surface A_j, and the entire diffuse energy leaving A_i. The view factor between two infinitesimal surface elements dA_i and dA_j will be referred to as $dF_{dA_i-dA_j}$, and it is defined as

$$dF_{dA_i-dA_j} \equiv \frac{(\boldsymbol{n}_i \cdot \boldsymbol{s}_{ij})(\boldsymbol{n}_j \cdot \boldsymbol{s}_{ji})}{\pi S^4}, \qquad (1)$$

where $\boldsymbol{n}_i$ (or $\boldsymbol{n}_j$) is the surface normal to dA_i (or dA_j), $S = |\boldsymbol{s}_{ji}|$, $\boldsymbol{s}_{ji} = \boldsymbol{r}_j - \boldsymbol{r}_i$, and $\boldsymbol{r}_i$ (or $\boldsymbol{r}_j$) is the position vector of dA_i (or dA_j), as it is shown in **Figure 1**. Nevertheless, it is worth remembering that the angle θ_i (or θ_j) between the surface normal $\boldsymbol{n}_i$ (or $\boldsymbol{n}_j$) and the line connecting dA_i and dA_j obeys the expression $\cos\theta_i = \frac{\boldsymbol{n}_i \cdot \boldsymbol{s}_{ij}}{|\boldsymbol{s}_{ij}|}$, so that it can easily be proved that

$$dF_{dA_i-dA_j} \equiv \frac{\cos\theta_i \cos\theta_j}{\pi S^2} dA_j. \qquad (2)$$

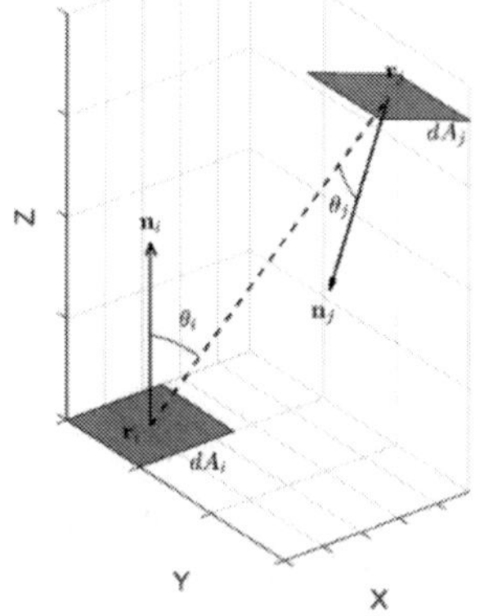

Figure 1: Radiative exchange between two infinitesimal surface elements.

The aim of this work is to use view factors over a solar plant model consisting of multiple arrays, each composed by a large number of relatively long rows of equally spaced bifacial photovoltaic modules positioned on the ground. Consequently, the later equation (2) may not be the most suitable choice for calculating the view factors in this context. This is primarily due to its high computational demands, even though it finds use in three-dimensional procedures like the ray-tracing simulation, which delves

10.4229/EUPVSEC2023/4DV.4.30
020462-001

into much detail features [3]. A more pragmatic and feasible approach involves considering two-dimensional view factors. Therefore, in the subsequent discussions, the solar plant model will be a projection onto a vertical plane Π_{x-z} orthogonal to dA_i and containing $\boldsymbol{n}_i$, that is, orthogonal to the rows that constitute a photovoltaic array with bifacial modules (i.e. forming an angle with the meridian plane equal to the azimuth of the rows), as it is shown in **Figure 2** and **Figure 3**. It should be noticed that this projection plane typically coincides with the North-South direction in the case of static arrays but with the East-West direction in the case of the one-axis tracked arrays. Aside from that, every row of solar PV modules is regarded as an undivided planar surface with its centre elevated at a distance H above the ground, which is supposed to be flat and parallel to the edge of the rows (but not necessarily horizontal), and the impacts of the mounting structures are not factored into the analysis. In addition, the model incorporates two more hypothesis: an infinite number of rows, each with an also infinite length. Nonetheless, these last two assumptions force to disregard an effect that occurs at the row edges due to the increase of rear irradiance on the modules located close to the boundaries of the photovoltaic plants. Regardless of this discrepancy between reality and the model constructed, some studies suggest that this edge effect is scarcely relevant in the case of large photovoltaic arrays [4], and thus, it is adopted as it helps to create a new and more manageable framework, both computationally and analytically. Under these considerations, there are still different paths to get an accurate computation of the view factors.

On one hand, a more straightforward approach to the calculus of the view factors involves the use of the segment-to-segment method, which will be discussed in Section 2.2. However, this method contrast with on-site measuring campaigns due to its implicit loss of the information related to the non-uniformity of the irradiance received on the rear PV module surface, in favour of a simplified integration. This happens since the rear irradiance is not evenly distributed and it is measured through solar cells, which can be treated as point entities in comparison to the dimension of the modules. On the other hand, a second method surges to overcome this issue. It proposes a way of calculating the view factors through a segment-to-point approach (discussed in Section 2.1), so that it is possible to make a comparison between measured and theoretical results. However, it also presents some drawbacks, such as the need to choose a suitable number of points on the rear surface of the PV module to ensure a proper integral estimation that accurately represents the entire module. In the subsequent sections, both methods will be subjected to an analytical comparison, and studied for a specific case of interest.

2 METHODOLOGIES

In this section, it will be exclusively explained the computation of view factors from the ground to the rear surface of the solar module, albeit the procedure to obtain the view factors between the ground and the frontal surface module is entirely analogous. Before starting the calculus, it should be noticed that every length is normalized with respect to the row width, and that the array geometry of the photovoltaic plants is defined by the tilt angle, the row-to-row space L, and the row center height H. Additionally, the abscissa axis (referred to as x-axis) is positioned over the line defining the ground, while the ordinate axis (designated as z-axis) is set to pass through the centre of an arbitrary chosen row of modules, which will be assigned the number zero. Likewise, the rows located on the right (or left) of this zero-row will be assigned the number $N \in \mathbb{Z}$ just prior to (or following) the number of the row immediately to their left (or right). Furthermore, each N-th row of modules is bounded by the points referred to as $E_{p_1}^N \equiv (E_{p_1,x}^N, E_{p_1,y}^N) \in \mathbb{R}^2$ and $E_{p_2}^N \equiv (E_{p_2,x}^N, E_{p_2,y}^N) \in \mathbb{R}^2$, such that $\overline{E_{p_1}^N E_{p_2}^N} = 1$ and $E_{p_1,x}^N < E_{p_2,x}^N$, considering that values to the right (left) and upward (downward) from the origin are positive (negative). Concerning angles, any line forms a positive angle with the x-axis (or with the normal to the x-axis) when, in bringing the line to the axis (or the normal) through the

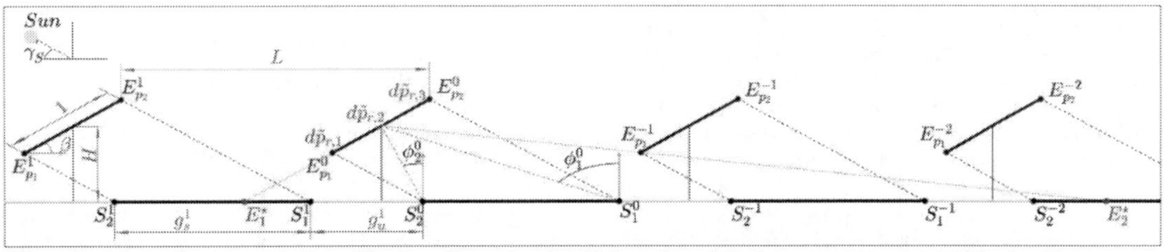

Figure 2: Schematic representation of the two-dimensional model employed for the segment-to-point method, with accompanying notation. The colour code employed for the notation corresponds to the colour of the respective elements it denotes, with the array of PV modules being depicted in a deep blue.

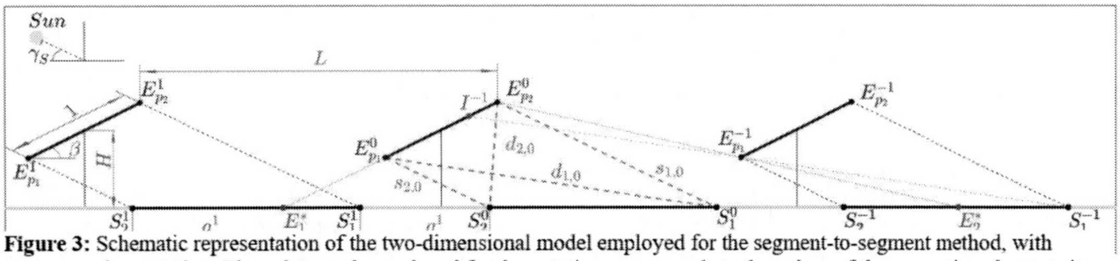

Figure 3: Schematic representation of the two-dimensional model employed for the segment-to-segment method, with accompanying notation. The colour code employed for the notation corresponds to the colour of the respective elements it denotes, with the array of PV modules being depicted in a deep blue.

10.4229/EUPVSEC2023/4DV.4.30

shortest path, the rotation is performed counter clock. In an analogous way, the angle will be negative if the rotation is performed in a clockwise direction.

The regular disposition of the rows translates into a basic pattern of shaded and unshaded segments over the ground, which repeats with a period equal to L and that, obviously, depends on the solar position. Therefore, there are two types of irradiation emitting segments: the illuminated segments and the shaded ones. It is important to note that, due to the projection of the relevant surfaces onto the Π_{x-z} plane forming a segment, it is employed a slight abuse of notation by referring to the irradiating surfaces as irradiating segments and by referring to the module surface as module segment.

The shadow segment produced by the N-th module during the hour h of the day d, is limited by the points $S_1^N \equiv (S_{1,x}^N, S_{1,z}^N)$ and $S_2^N \equiv (S_{2,x}^N, S_{2,z}^N)$. To define these points it is considered that γ_S' is the projection onto the plane Π_{x-z} of the angle formed between the abscissa axis and the line connecting the centres of the Sun and the Earth, and that β denotes the angle formed between the x-axis and the line connecting the points E_{p1}^N and E_{p2}^N during the hour h of the day d. Then, if $\gamma_S' \neq \pm\frac{\pi}{2}$, the latest points S_1^N and S_2^N can be defined as [1]

$$S_1^N \equiv \left(\frac{1}{2}cos\beta \pm \frac{H - \frac{1}{2}sin\beta}{|tan\gamma_S'|} + NL, 0\right), \quad (3)$$

and

$$S_2^N \equiv \left(-\frac{1}{2}cos\beta \pm \frac{H + \frac{1}{2}sin\beta}{|tan\gamma_S'|} + NL, 0\right), \quad (4)$$

where the positive sign is used when the Sun's position falls on the x-axis's negative side (prior to noon, for flat terrains), while the negative sign is used otherwise. If $\gamma_S' = \pm\frac{\pi}{2}$, then $S_1^N = -S_2^N = \frac{1}{2} + NL$.

It is important to note that ground global irradiance is non-uniform due to the variability of its diffuse component, however, this issue can be addressed by subdividing the irradiative segments into portions of smaller length, an adjustment which does not affect the main purpose of the explanation. Therefore, the irradiance over the shaded (or illuminated) ground is assumed to be uniform. This enables the definition of the sum of the non-normalized view factors between each shaded (or illuminated) irradiative segment of ground and the rear segment module, and it will continue to be useful in calculating the rear irradiance attributed to the shaded (or unshaded) ground. It must be highlighted that the number of these contributing segments is not infinite, and that only those who fall within a defined field of view, delineated by the extreme angles ϕ_{E_1} and ϕ_{E_2} [1], contribute to this process. This range depends on the specific technique employed for the view factors computation, and it will be explained later in Section 2.1 and Section 2.2. In the upcoming subsections, it is temporarily revisited the three-dimensional model for calculating the view factor between the ground and the module surfaces.

2.1 Segment-to-point method

Let $d\tilde{g}_s^N \not\subset \Pi_{x-z}$ be an infinitesimal surface element of the shaded ground surface $\tilde{g}_s^N$, whose projection onto Π_{x-z} is contained in the segment $g_s^N \equiv \overline{S_2^N S_1^N}$. Taking into account that this surface element is a parallelogram, its orientation is chosen such that two of its sides align with the axis orthogonal to the Π_{x-z} plane, (referred to as y-axis). Similarly, let $d\tilde{p}_r \not\subset \Pi_{x-z}$ denote the analogous infinitesimal surface element positioned on the rear surface $\tilde{p}_r$ of the zero-assigned module (which means that the projection of $d\tilde{p}_r$ onto the Π_{x-z} plane is contained in the segment $p_r \equiv \overline{E_{p_1}^0 E_{p_2}^0}$), following the same orientation procedure. For this purpose, it is introduced the notation of $strip_{g_s}^N$ for the resultant infinitesimally narrow strip obtained by extending the infinitesimal surface element $d\tilde{g}_s^N$ over the entire y-axis, and $strip_{p_r}$ referring to the corresponding strip formed by extending $d\tilde{p}_r$ in the same manner. It is worth noting that the projection onto the Π_{x-z} plane of both $strip_{g_s}^N$ (or $strip_{p_r}$) and $d\tilde{g}_s^N$ (or $d\tilde{p}_r$) is the same, and will be denoted as dg_s^N (or dp_r). In addition, it is known that [5]

$$dF_{d\tilde{g}_s^N - strip_{p_r}} = \frac{1}{2}cos\phi d\phi, \quad (5)$$

where ϕ is the angle between the normal to dg_s^N and the line connecting dp_r and dg_s^N. However, since $dF_{d\tilde{g}_s^N - strip_{p_r}}$ does not vary along $strip_{g_s}^N$, it follows immediately that

$$dF_{d\tilde{g}_s^N - strip_{p_r}} = dF_{strip_{g_s}^N - strip_{p_r}}. \quad (6)$$

Consequently, based on the last equality, the analysis of view factors between the ground and the module surfaces can be conducted exclusively using elements within the Π_{x-z} plane, hence they are denoted as two-dimensional view factors, even though their influence extends between surfaces. Thus, in the subsequent study, the computations are conducted exclusively on the Π_{x-z} plane, with reference to the surfaces being made solely when naming the respective view factors.

In the remainder of this section, it is explained an approach to compute the rear irradiance $G_{REAR}^{g_s}$ on the solar rear module originating from the shaded ground surfaces at a specific time h on a particular day d (these parameters will define the angle β governing the calculations, as it has already been explained). Following the ordinary procedure to obtain the view factors,

$$F_{\tilde{g}_s^N - d\tilde{p}_r} = \frac{1}{g_s^N}\int_{\phi_2^N}^{\phi_1^N} dF_{d\tilde{g}_s^N - d\tilde{p}_r} =$$
$$= \frac{1}{2g_s^N}\int_{\phi_2^N}^{\phi_1^N} cos\phi d\phi = \frac{1}{2g_s^N}(sin\phi_1^N - sin\phi_2^N), \quad (7)$$

where ϕ_1^N (or ϕ_2^N) is the angle between the normal to the x-axis positioned in S_1^N (or S_2^N) and the line connecting S_1^N (or S_2^N) with the position vector of dp_r (see **Figure 2**).

At this point, it is necessary to make an observation regarding the computation of the final exchanged energy. To expound on this matter, it must be defined the set $M^* = \{N \in \mathbb{Z} : S_{2,x}^N \geq E_1^* \text{ and } S_{1,x}^N \leq E_2^*\}$, where E_1^* and E_2^* are, respectively, the minimum and the maximum points of the x-axis within the field of view of dp_r. It must be noticed that in the case in which $S_{2,x}^N \leq E_1^*$ (or $S_{1,x}^N \geq E_2^*$), then $g_s^N \neq \overline{S_2^N S_1^N}$ and instead is defined as $g_s^N = \overline{E_1^* S_1^N}$ (or $g_s^N = \overline{S_2^N E_2^*}$). Considering that x_F is the position of dp_r on the u-axis constructed along the segment $\overline{E_{p_1}^0 E_{p_2}^0}$ which defines the zero-row of modules, and that $\beta \neq 0$, E_1^* and

$E_2^\star$ adhere to the equations [1]

$$E_1^\star = \pm(H + x_F|sin\beta|)\,tan\left(\frac{\pi}{2} - |\beta|\right) + x_F cos\beta, \quad (8)$$

and

$$E_2^\star = \pm\frac{(H + x_F|sin\beta|)\left(L - \left(\frac{1}{2} + x_F\right)cos\beta\right)}{\left(\frac{1}{2} + x_F\right)|sin\beta|} + x_F cos\beta, \quad (9)$$

where the negative sign is employed when the Sun's position is on the negative side of the x-axis (before noon, in the case of non-sloping terrains), and the positive sign otherwise. It is important to note that when $\beta > 0$, the definitions of $E_1^\star$ and $E_2^\star$ must be interchanged. Also, when the modules become parallel to the ground, the number of viewed segments becomes infinite. To avoid this problem, the view field is restricted from $-89°$ to $89°$, since the error associated to this constrain is deemed negligible [1]. The accurate definition of $E_1^\star$ and $E_2^\star$ entails referencing the limit angle of the glass of the PV module, a detail that will not be elaborated on here (for further details, refer to [1]).

At this juncture, sufficient information has now been acquired to proceed with the energy-related observation. It is important to notice that the rear irradiance on the shaded ground G_{SHADED} [1] is conventionally expressed in power per unit area, and that the view factor is normalized respect to the emitting surface. Therefore, the irradiance $G_{d\tilde{p}_r}^{\tilde{g}_s}$ received over $d\tilde{p}_r$ from the shaded ground $\tilde{g}_s$, can be computed as

$$G_{d\tilde{p}_r}^{\tilde{g}_s} = \sum_{i=1}^{M^*} \frac{\mathcal{F}^\star_{\tilde{g}_s^i - d\tilde{p}_r}}{g_s^i}\left(g_s^i G_{SHADED}\right), \quad (10)$$

where it is introduced the factor $\mathcal{F}^\star_{\tilde{g}_s^N - d\tilde{p}_r}$ defined as

$$\mathcal{F}^\star_{\tilde{g}_s^N - d\tilde{p}_r} = g_s^N F_{\tilde{g}_s^N - d\tilde{p}_r}, \quad (11)$$

being $F_{\tilde{g}_s^N - d\tilde{p}_r}$ the view factor per unit length relative to the N-th ground segment, defined in (7). In addition, it is also defined $\mathcal{F}^\star_{\tilde{g}_s - d\tilde{p}_r}$ as the sum of the factors defined in (11) associated to all contributing segments, that is

$$\mathcal{F}^\star_{\tilde{g}_s - d\tilde{p}_r} = \sum_{i=1}^{M^*} \mathcal{F}^\star_{\tilde{g}_s^i - d\tilde{p}_r}. \quad (12)$$

Furthermore, following a similar process to the one that will be employed for the regular view factors, the factor $\mathcal{F}^\star_{\tilde{g}_s - \tilde{p}_r}$ between the shaded ground and the rear module surface can be computed as

$$\mathcal{F}^\star_{\tilde{g}_s - \tilde{p}_r} = \int_{-\frac{1}{2}}^{\frac{1}{2}} \mathcal{F}^\star_{\tilde{g}_s - d\tilde{p}_r}\, du. \quad (13)$$

Hence, the irradiance $G_{REAR}^{\tilde{g}_s}$ over the rear module surface due to the shaded segments of ground can be calculated as follows,

$$G_{REAR}^{\tilde{g}_s} = \mathcal{F}^\star_{\tilde{g}_s - \tilde{p}_r} G_{SHADED}. \quad (14)$$

This last equation epitomizes the analogous role between view factors and $\mathcal{F}^\star_{\tilde{g}_s - \tilde{p}_r}$ in calculating the rear irradiance. In fact, the factor $\mathcal{F}^\star_{\tilde{g}_s - \tilde{p}_r}$ is referred to as the view factor in [1].

Before concluding this subsection, it is worth mentioning that the integral (13) is approximated using a straightforward discretization approach. To achieve this, it is chosen a set of K points (that is, a set of K differential elements $d\tilde{p}_{r,i}$ over the module, with $i \in \{1, \ldots, K\}$), leading to the approximated factor $\tilde{\mathcal{F}}^\star_{\tilde{g}_s - \tilde{p}_r}$ defined as

$$\tilde{\mathcal{F}}^\star_{\tilde{g}_s - \tilde{p}_r} = \sum_{i=1}^{K} \alpha_i \cdot \mathcal{F}^\star_{\tilde{g}_s - d\tilde{p}_{r,i}}, \quad (15)$$

where $\sum_{i=1}^{K} \alpha_i = 1$. It is worth highlighting that through this method, the considered irradiating surface might vary for each $d\tilde{p}_{r,i}$ with $i \in \{1, \ldots, K\}$, owing to the dependency of $E_1^\star$ and $E_2^\star$ on x_F.

The process for analysing the energy exchange between the illuminated ground and the PV module is quite similar, hence, it is not explained further.

2.2 Segment-to-segment method

The segment-to-segment method, formulated by Hoyte Clark Hottel in the first half of the 20-th century, is a technique designed to compute view factors of systems characterized by long enclosures with constant cross-section. It is indeed the case of the studied model, so that it can easily be applied that for the hour h of the day d [5],

$$F_{\tilde{g}_s^N - \tilde{p}_r} = \frac{d_{1,N} + d_{2,N} - s_{1,N} - s_{2,N}}{2g_s^N}, \quad (16)$$

where $d_{1,N} = \overline{S_1^N E_{p_1}^0}$, $d_{2,N} = \overline{S_2^N E_{p_2}^0}$, $s_{1,N} = \overline{S_1^N E_{p_2}^0}$ and $s_{2,N} = \overline{S_2^N E_{p_1}^0}$ (see **Figure 3**). Following an analogous procedure as in the previous subsection, it will be defined the factor $\mathcal{F}^\star_{\tilde{g}_s^N - \tilde{p}_r}$ as

$$\mathcal{F}^\star_{\tilde{g}_s^N - \tilde{p}_r} = g_s^N F_{\tilde{g}_s^N - \tilde{p}_r}, \quad (17)$$

and

$$\mathcal{F}^\star_{\tilde{g}_s - \tilde{p}_r} = \sum_{i=1}^{M^*} \mathcal{F}^\star_{\tilde{g}_s^i - \tilde{p}_r}. \quad (18)$$

In this case, the set M^* is defined as $M^* = \{N \in \mathbb{Z} : S_{2,x}^N \geq E_1^\star \text{ and } S_{1,x}^N \leq E_2^\star\}$ and, if $\beta \neq 0$

$$E_1^\star = \pm\left(H + \frac{1}{2}|sin\beta|\right)tan\left(\frac{\pi}{2} - |\beta|\right) + \frac{1}{2}cos\beta, \quad (19)$$

and

$$E_2^\star = \pm\frac{\left(H + \frac{1}{2}|sin\beta|\right)(L - cos\beta)}{|sin\beta|} + \frac{1}{2}cos\beta, \quad (20)$$

where, just like in the point-to segment case, the negative sign is employed when the Sun's position is on the negative side of the x-axis (before noon, in the case of non-sloping terrains), and the positive sign otherwise. Again, when $\beta > 0$, the definitions of $E_1^\star$ and $E_2^\star$ must be interchanged, and if $S_{2,x}^N \leq E_1^\star$ (or $S_{1,x}^N \geq E_2^\star$), the segment g_s^N is not represented by $\overline{S_2^N S_1^N}$ and instead, it is defined as $\overline{E_1^\star S_1^N}$ (or $\overline{S_2^N E_2^\star}$). Moreover, when the modules become parallel to the ground, the view field is also restricted from $-89°$ to

10.4229/EUPVSEC2023/4DV.4.30

89°.

An essential point to highlight is that the segments identified by the indexes included in M^* are those that fall within the field of view of the entire module. Nonetheless, the remaining shaded segments (or parts of shaded segments) within the field of view from $-89°$ to $89°$ will still have visibility of a portion of the module. In order to compute this, it is defined the intersection point $I^N \equiv (I_x^N, I_y^N)$ as the point where the line passing through the points S_1^N (or S_2^N if the Sun's position is on the positive side of the x-axis) and $E_{p_1}^1$ (or $E_{p_2}^{-1}$ if the Sun's position is on the positive side of the x-axis) intersects with the segment defining the zero-row of modules $\overline{E_{p_1}^0 E_{p_2}^0}$. Then, it is applied the equation (16), defining $d_{1,N} = \overline{S_1^N E_{p_1}^0}$ (or $d_{1,N} = \overline{S_1^N I^N}$ if the Sun's position is on the positive side of the x-axis), $d_{2,N} = \overline{S_2^N I^N}$ (or $d_{2,N} = \overline{S_2^N E_{p_2}^0}$ if the Sun's position is on the positive side of the x-axis), $s_{1,N} = \overline{S_1^N E_{p_2}^0}$ (or $s_{1,N} = \overline{S_1^N I^N}$ if the Sun's position is on the positive side of the x-axis) and $s_{2,N} = \overline{S_2^N I^N}$ (or $s_{2,N} = \overline{S_2^N E_{p_1}^0}$ if the Sun's position is on the positive side of the x-axis). This computation allows for a partial assessment of the non-uniformity of the rear irradiance.

Regardless the slight differences compared to the point-to-segment method, it is important to emphasize that still holds this relationship between $G_{REAR}^{\tilde{g}_s}$ and G_{SHADED},

$$G_{REAR}^{\tilde{g}_s} = \mathcal{F}_{\tilde{g}_s-\tilde{p}_r}^{*} G_{SHADED}. \qquad (21)$$

Once again, it is important to emphasize that the calculations explained are specific to a particular time h on a given day d. The process of analyzing the energy exchange between the illuminated ground and the PV module follows a similar procedure and is not further elaborated here.

2.3 Non-uniformity assessment

This work originates from the concern about the non-uniformity of the rear irradiance measured on PV modules. To assess this, it can be employed a proposed method from [6], that uses the equation,

$$GNU = \frac{G_{MAX} - G_{MIN}}{G_{MAX} - G_{MIN}}, \qquad (22)$$

where, G_{MAX} and G_{MIN} are, respectively, the maximum and minimum values of irradiance along the rear side of the row.

3 RESULTS

The results explained above have been integrated into SISIFO, an open-source software for general photovoltaic system performance simulations freely available at www.sisifo.info. In particular, a comparative assessment between the segment-to-point method and the segment-to-segment method has been conducted for a specific case study involving a solar facility located in Egypt.

The solar photovoltaic technology of this PV plant is characterized for having horizontal tracking structures, designed to do backtracking [7]. The installation is set over a flat terrain and its geometry is characterized by the values $H = 1$ and $L = \frac{221}{99}$.

3.1 View factors

The segment-to-point method has been incorporated into SISIFO. To accomplish this, three points ($K = 3$) were selected at positions $x_F = -0.49$, $x_F = 0$, and $x_F = 0.49$ along the u-axis, and an arithmetic average of their contributions ($\alpha_i = \frac{1}{3}$ for $i \in \{1,2,3\}$) has been computed to determine $\tilde{\mathcal{F}}_{\tilde{g}_s-\tilde{p}_r}^{*}$. Following this, the segment-to-segment method has been implemented to calculate the factor $\mathcal{F}_{\tilde{g}_s-\tilde{p}_r}^{*}$. **Figure 5** displays the results of these factors. Similarly, referring to the unshaded ground as $\tilde{g}_u$ and designating the frontal surface of the zero-assigned module as $\tilde{p}_f$, **Figure 6**, **Figure 7** and **Figure 8** present the results obtained for the factors $\tilde{\mathcal{F}}_{\tilde{g}_u-\tilde{p}_r}^{*}$ and $\mathcal{F}_{\tilde{g}_u-\tilde{p}_r}^{*}$, $\tilde{\mathcal{F}}_{\tilde{g}_s-\tilde{p}_f}^{*}$ and $\mathcal{F}_{\tilde{g}_s-\tilde{p}_f}^{*}$, and $\tilde{\mathcal{F}}_{\tilde{g}_u-\tilde{p}_f}^{*}$ and $\mathcal{F}_{\tilde{g}_u-\tilde{p}_f}^{*}$, respectively. The input data for these computations were sourced from PVGis [8], reflecting a typical meteorological year (TMY), and the results are presented for two representative days, the spring equinox (20/03/TMY) and the winter solstice (22/12/TMY). The legend in the graphics uses SS to denote the segment-to-segment method, and SP for the results obtained by averaging the outcomes achieved through the application of the segment-to-point method over the three selected points.

Interestingly, there is almost no distinguishable difference between the results obtained from the segment-to-segment method and the segment-to-point method. In fact, it has been compared the factors between the shaded ground and the rear surface module ($\tilde{g}_s - \tilde{p}_r$) computed with the SP method against those obtained through the SS approach. To do so, it has been computed the respective Root Mean Square Error (RMSE) across the entire Typical Meteorological Year (TMY) dataset. Remarkably, the RMSE value consistently showed an order of magnitude smaller than the magnitude of the factors analysed. Additionally, detailed RMSE results are summarized in **Table I** for three more specific scenarios, denoted as $\tilde{g}_u - \tilde{p}_r$, $\tilde{g}_s - \tilde{p}_f$, and $\tilde{g}_u - \tilde{p}_f$, respecting the notation introduced above.

Table I: RMSE between the two pair of the methods used (SS-SP) for the factors computed between the shaded or unshaded ground and the frontal or rear module surface.

	RMSE /[]
$\tilde{g}_s - \tilde{p}_r$	$1.2 \cdot 10^{-2}$
$\tilde{g}_u - \tilde{p}_r$	$9.5 \cdot 10^{-3}$
$\tilde{g}_s - \tilde{p}_f$	$3.4 \cdot 10^{-3}$
$\tilde{g}_u - \tilde{p}_f$	$3.4 \cdot 10^{-4}$

For a more comprehensive understanding of **Figure 5**, **Figure 6**, **Figure 7** and **Figure 8**, it is advised to refer to **Figure 4**, where the evolution of the tracker angles for the two representative days is depicted. Notably, owing to the symmetrical movement, the evolution of the view factors also displays symmetry. Therefore, it is only explained the view factor behaviour until noon, which coincides with the beginning of the symmetric character since the terrain is horizontal.

10.4229/EUPVSEC2023/4DV.4.30
020462-005

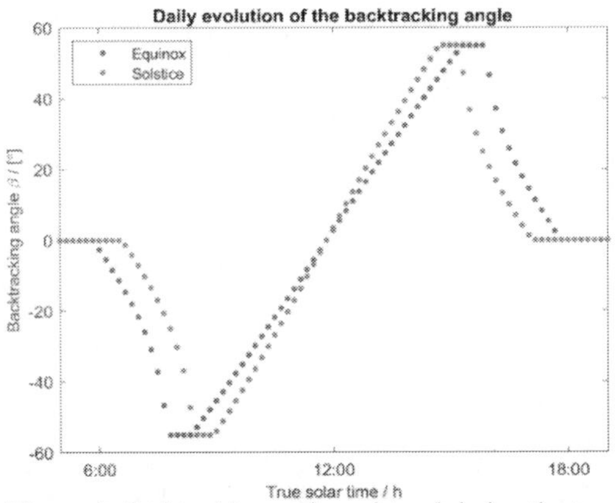

Figure 4: Backtracking angle registered during the two representative days.

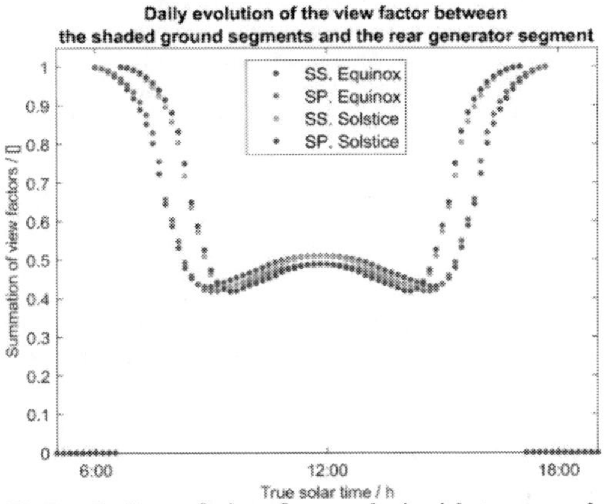

Figure 5: Sum of view factors obtained between each shaded segment of ground and the rear PV generator segment computed through the two different methods, applied for the two representative days.

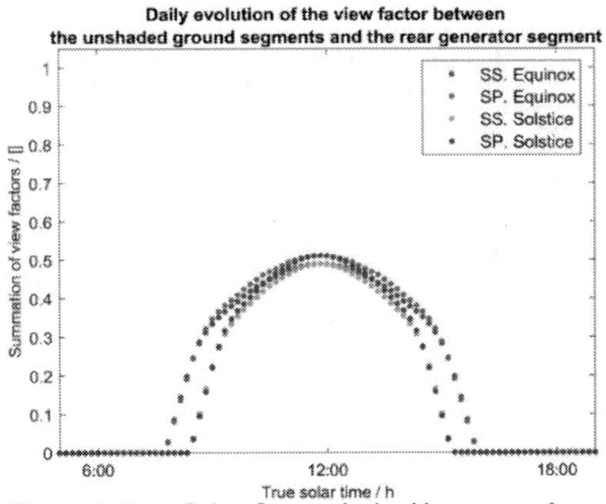

Figure 6: Sum of view factors obtained between each unshaded segment of ground and the rear PV generator segment computed through the two different methods, applied for the two representative days.

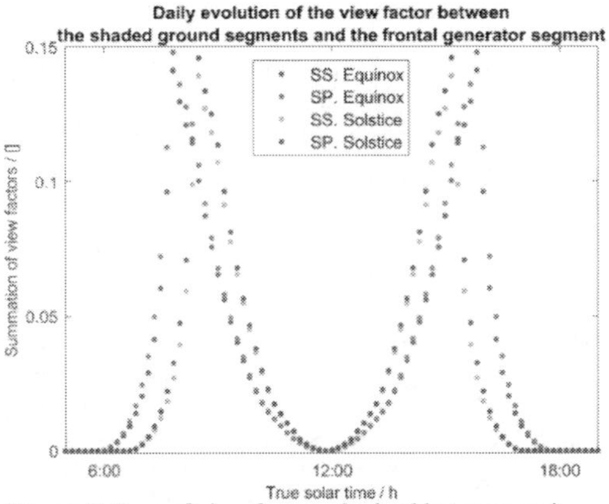

Figure 7: Sum of view factors obtained between each shaded segment of ground and the frontal PV generator segment computed through the two different methods, applied for the two representative days.

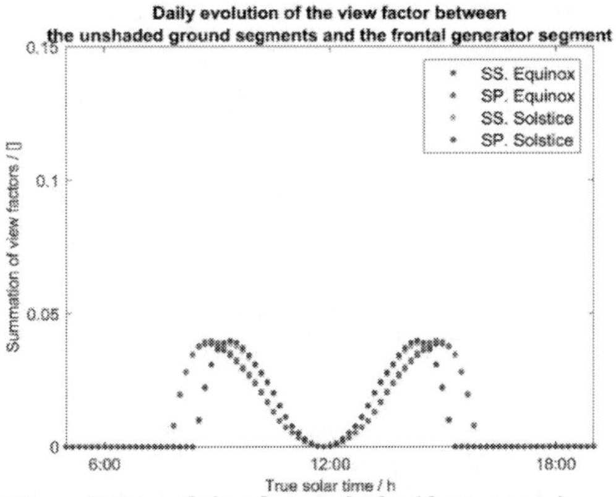

Figure 8: Sum of view factors obtained between each unshaded segment of ground and the frontal PV generator segment computed through the two different methods, applied for the two representative days.

As depicted in **Figure 5**, it is evident that the factors $\mathcal{F}^{\star}_{\tilde{g}_s-\tilde{p}_r}$ and $\mathcal{F}^{*}_{\tilde{g}_s-\tilde{p}_r}$ are nearly equal to unity during dawn and dusk, as the entire ground is shaded and the module is in a horizontal position. It is worth noting that during the backtracking process, these factors decrease due to the increase of $|\beta|$, which results in greater distance from the PV module to the ground. After the backtracking, as $|\beta|$ decreases again, these factors increase until they reach a local maximum at noon.

Figure 6 illustrates that the factors $\tilde{\mathcal{F}}^{\star}_{\tilde{g}_u-\tilde{p}_r}$ and $\mathcal{F}^{*}_{\tilde{g}_u-\tilde{p}_r}$ between the unshaded regions and the rear module surface are zero during the backtracking phase, due to the absence of illuminated areas. Subsequently, as the absolute value of β increases, these factors also increase until noon, similar to the pattern observed in **Figure 5**.

Regarding the frontal case, **Figure 7** illustrates that at dawn, the computed factors $\tilde{\mathcal{F}}^{\star}_{\tilde{g}_s-\tilde{p}_f}$ and $\mathcal{F}^{*}_{\tilde{g}_s-\tilde{p}_f}$ are zero due to the horizontal orientation of the module. Subsequently, as $|\beta|$ increases during the backtracking phase, both $\tilde{\mathcal{F}}^{\star}_{\tilde{g}_s-\tilde{p}_f}$ and $\mathcal{F}^{*}_{\tilde{g}_s-\tilde{p}_f}$ also increase (the field of view between the frontal module surface and the ground increases). Once the backtracking finishes, as $|\beta|$ decreases, these factors also decrease (the frontal surface observes less portion of ground). This trend continues until noon when the module's frontal face becomes horizontal, rendering it unable to observe the ground.

The analysis of **Figure 8** indicates once again that the factors $\tilde{\mathcal{F}}^{\star}_{\tilde{g}_u-\tilde{p}_f}$ and $\mathcal{F}^{*}_{\tilde{g}_u-\tilde{p}_f}$ between the illuminated ground and the module are zero during the backtracking period, due to the absence of unshaded regions. Once the backtracking concludes, the tracker remains static for a few minutes and, thus, these factors gradually increase. This is due to the gradual expansion of the illuminated ground as a result of the daily solar motion, regardless of the tracker's static state. Subsequently, as the tracker resumes its movement, the angle $|\beta|$ decreases, causing $\tilde{\mathcal{F}}^{\star}_{\tilde{g}_u-\tilde{p}_f}$ and $\mathcal{F}^{*}_{\tilde{g}_u-\tilde{p}_f}$ to decrease, as expected, since the field of view between the PV module and the ground becomes worse. These factors reach their zero value at noon, corresponding to the module's horizontal orientation.

3.2 Non-uniformity analysis

In Section 2.1, it was explained that certain segments beyond the field of view of the entire module also contribute to the rear irradiance received by the module. It was expected that these segments' contribution could help in evaluating the non-uniformity of the rear irradiance. Nevertheless, it has been demonstrated that this approach is not an efficient method to quantify such non-uniformity. According to equation (25), it has been searched the maximum and the minimum of the rear irradiance received by the PV module among each contributing segment. Nevertheless, the contribution from the more distant ones is almost negligible and, consequently, the GNU factor yields values close to the unity, as depicted in **Figure 9**. In addition, when there is only one contributing segment, as in the case of the backtracking period, the GNU is inherently equal to zero.

In contrast, the segment-to-point method allows the assessment of the maximum and minimum rear irradiance calculated at each of the selected points $d\tilde{p}_{r,i}$, with $i \in K$. As it is shown in **Figure 9**, this approach yields significantly different and more precise results compared to those obtained through the segment-to-segment method.

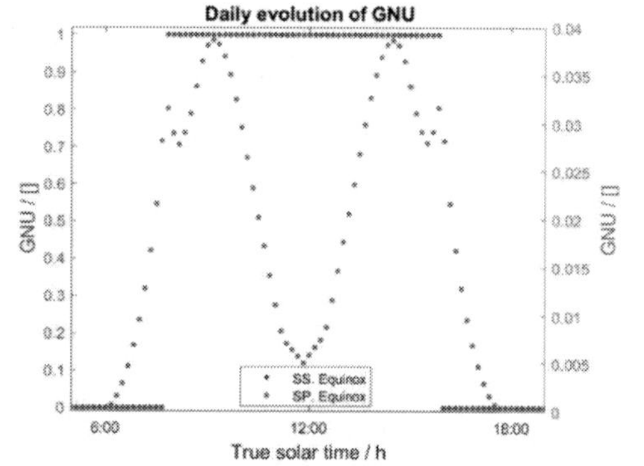

Figure 9: Assessment of the non-uniformity in rear irradiance for both employed methods, considering one representative day.

4 CONCLUSIONS

The conclusions drawn from this study are based on the observations made using a real PV plant as the scenario for the simulations conducted on SISIFO.

While the segment-to-segment method simplifies integration, it results in a significant loss of the information related to the non-uniformity of the rear irradiance received on the PV module, that cannot be adequately assessed. On the other hand, the segment-to-point method retains this information. Furthermore, the process of averaging the outcomes derived from applying the segment-to-point method over three strategically positioned points at the centre and the edges of the PV module result in a not significant discrepancy with the outcome of the segment-to-segment method. Additionally, the segment-to-point approach facilitates experimental measurements at these specific positions, providing feedback to the model results.

5 ACKNOWLEDGEMENTS

The authors express their sincere appreciation for the financial support provided by the Comunidad de Madrid through Project IND2022/AMB-23718 and funding received from the European Union under grant agreement No 101076395. They are also grateful to EU PVSEC for the opportunity to present their work at the 2023 edition held in Lisbon. Special thanks go to the research staff at the Instituto de Energía Solar of the Universidad Politécnica de Madrid for their valuable contributions.

6 REFERENCES

[1] J. Ledesma, E. Lorenzo, R. Almeida, F. Martinez-Moreno, C. Martín-Rueda, R. J. and L. Narvarte, "A simulation model of the irradiation and energy yield of large bifacial photovoltaic plants," *Solar Energy,* vol. 206, pp. 522-538, 2020.

[2] "VDMA - German Engineering Federation. International Technology Roadmap for Photovoltaic," [Online]. Available: https://www.vdma.org/international-technology-roadmap-photovoltaic.

[3] A. Asgharzadeh, T. Lubenow, J. Sink, B. Marion, C. Deline, C. Hansen, J. Stein and F. Toor, 2017 IEEE 44th Photovoltaic Specialist Conference (PVSC), 2017, pp. 3333-3338.

[4] C. Hansen, C. Deline, S. MacAlpine, B. Marion, A. Asgharzadeh Shishavan, F. Toor and J. Stein, Analysis of Irradiance Models for Bifacial PV Modules, 2016.

[5] M. F. Modest, Radiative Heat Transfer, Third ed., San Diego, CA: Academic Press, 2013.

[6] C. Deline, S. MacAlpine, B. Marion, F. Toor, A. Asgharzadeh Shishavan and J. Stein, "Evaluation and field assessment of bifacial photovoltaic module power rating methodologies," pp. 1-6, 06 2017.

[7] E. Lorenzo and M. H. Macagnan, "Considerations in the design of a one-axis tracking photovoltaic system," *Progress in Photovoltaics,* vol. 2, pp. 45-55, 1994.

[8] "Photovoltaic Geographical Information System (PVGIS)," 01 06 2023. [Online]. Available: https://joint-research-centre.ec.europa.eu/photovoltaic-geographical-information-system-pvgis_en.

10.4229/EUPVSEC2023/4DV.4.30
020462-007

BATTERY STORAGE ANALYSIS DEMANDS BASED ON NOVEL HIGH TEMPORAL RESOLUTION WEATHER DATA PLATFORM.

Kasper Mayntz Paasch
Department of Mechanical and Electrical Engineering
University of Southern Denmark
Alsion 2, DK-6400 Sønderborg, Denmark
Email: paasch@sdu.dk

Henrik Andersen*
Department of Mechanical and Electrical Engineering
University of Southern Denmark
han@sdu.dk

Søren Bækhøj Kjær
Danfoss Power Electronics and Drives A/S
sbk@danfoss.com

Morten Nymand
Department of Mechanical and Electrical Engineering
University of Southern Denmark
mny@sdu.dk

ABSTRACT:
This paper has two purposes.

A) It describes a novel database for standardized inverter and storage system analysis based on two yearlong high speed data recording campaign and compiled from two sets of parallel recording of more than 400 parameters with 1-second interval. The data included are among others, 19 parameters form a weather station, PV-plant irradiance, irradiance change, PV plant ambient and cell temperatures, DC/AC voltage and current levels etc.

B) It is demonstrated how to apply the use of said database by applying it to a battery storage system with special focus on rapid charging and discharging requirements to compensate for fast irradiation changes due to clouds. The simulation involves modeling the solar power generation based on data from the database such as solar irradiance and temperature, while considering the dynamic behavior of the battery storage system including charge and discharge rates, state of charge, and efficiency characteristics. The main question to answer is if it is possible to smoothen the negative contributions from clouds moving over the plant by adding a battery storage system and how stable the DC link can become.

The overall purpose of the developed database is to be a standard database covering one year of recordings with a 1-second temporal solution, based on real-life recordings of weather data. This temporal resolution allows the use of the database for simulation and laboratory tests.
An investigation of the use of Markov Transition Field together with the database will also be done. Here the question is if the difference in irradiation over days can be distinguish from each other.

1 Database

The aim was to establish a representative one-year database based on high-speed real-life weather and inverter data for simulation and testing of power electronics for PV-applications, such as inverters and battery-based energy storage systems and demonstrate it on a battery storage system. Normally irradiation and other weather data are only available in intervals of several minutes, not covering fast changes due to moving clouds (a normal encounter in Northern Europe). The developed data set is based on 1-second interval logging recordings of weather/inverter parameters from a weather station and a total of 16 3-phase 15kW string inverters in Denmark, containing more than 400 data series with rapidly changing data due to moving clouds. The data were recorded with a custom designed recording system including communication with the Solar inverters monitored. All data were stored on a server on an hourly basis and then post processed in MATLAB. The recorded data sets from two monitored PV-plants (a 2MW and a 60kW plant) in the southern part of Denmark are merged into one data structure with a 1-second temporal resolution. The data set contains irradiation, windspeed, air, panel temperature, DC and AC inverter data per string/phase (for each of the inverters) etc.
An example showing recordings from two different days, one without clouds and one with many moving clouds is shown below.

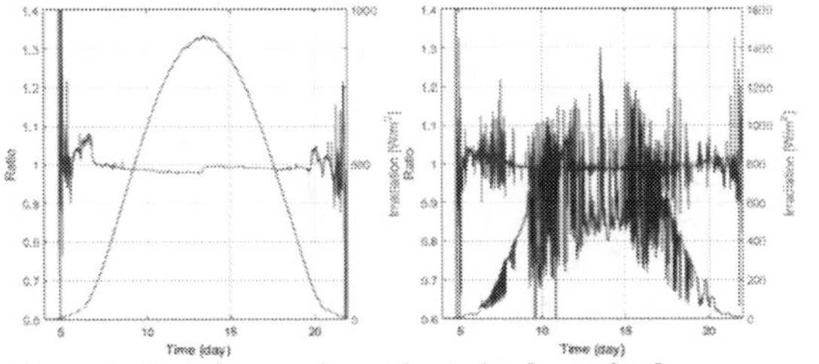

Figure 1 - Left shows a day without clouds - right shows a day with many clouds. [1]

10.4229/EUPVSEC2023/4DV.4.38
020464-001

An example of the irradiation slope proportional with PV current generation from a PV-system, determining the required battery charge/discharge dynamics is shown in figure 2. It illustrates the distribution of a full year data set with 1-second resolution.

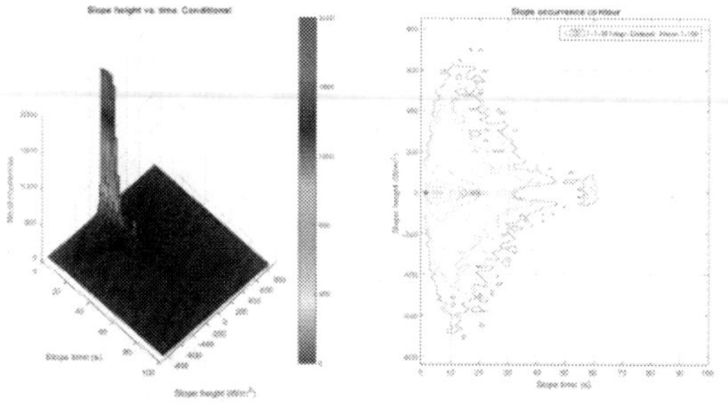

Figure 2 - Irradiation slope proportional with PV current. [1]

The DC voltage and current data can for example be applied to simulation and accelerated testing of stress/reliability/lifetime of the input stage components in PV inverters. The fast-changing DC currents cause nonperiodic thermal stress cycles on the switching components.

It is shown that the weather data sets currently available do not give a sufficiently clear picture of an actual PV-site performance, as all fast events are not present in those data set.

1.1 Data collection

The data collection system was designed around an industry field-PC with proprietary software, communicating directly with the inverters and a Davies Vantage Pro weather station for weather data acquisition. Raw data was recorded continuously and stored in files containing recordings for one hour. AC-output data were in parallel monitored with a Class 0.2 instrument (WatsOn) on selected inverters. The raw data files were streamed to a server for storage. Post-processing and building of the data structure was subsequently done in MATLAB.

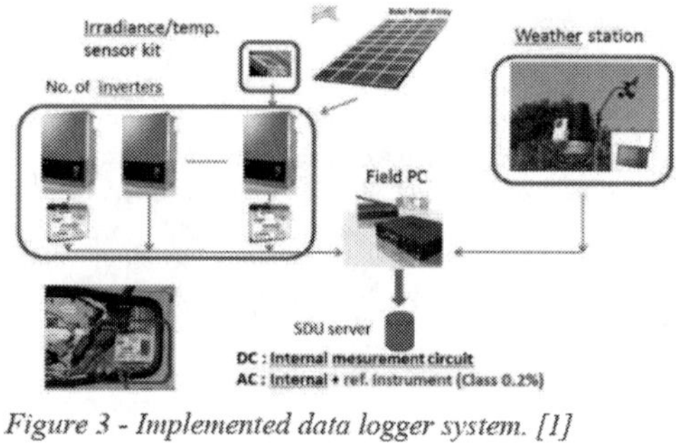

Figure 3 - Implemented data logger system. [1]

2 Irradiation data.

PV technology has emerged as a pivotal player in sustainable energy systems, offering a renewable and clean source of electricity. The efficiency and performance of a PV plant are intricately linked to several factors, with solar irradiation being a primary determinant. Solar irradiation, defined as the electromagnetic energy incident on a surface per unit area, profoundly influences the output and efficiency of PV systems. Understanding how varying irradiation levels impact the behavior and performance of a PV plant is crucial for optimizing its design, operation, and integration within the larger energy infrastructure.

In this section, we describe the simulation of a PV plant and explore the dynamic relationship between solar irradiation and the plant's response using MATLAB Simulink. MATLAB Simulink is a powerful tool for modeling and simulating complex systems, providing a platform to represent the behavior of the PV plant under different irradiation conditions. Through this simulation, we aim to investigate how changes in solar irradiation affect the electrical output and overall behavior of the PV plant, thus contributing to a deeper understanding of its operational dynamics and potential for harnessing solar energy effectively.

Other factors that influence irradiation levels are geographical location, weather patterns, time of day, and atmospheric conditions. Geographical positioning significantly impacts irradiation, with equatorial regions experiencing higher solar irradiance due to the direct angle of sunlight. Seasonal and diurnal variations further emphasize the need for meticulous design and planning to harness solar energy optimally.
It is common to assume a uniform irradiation function for a given place. For example:

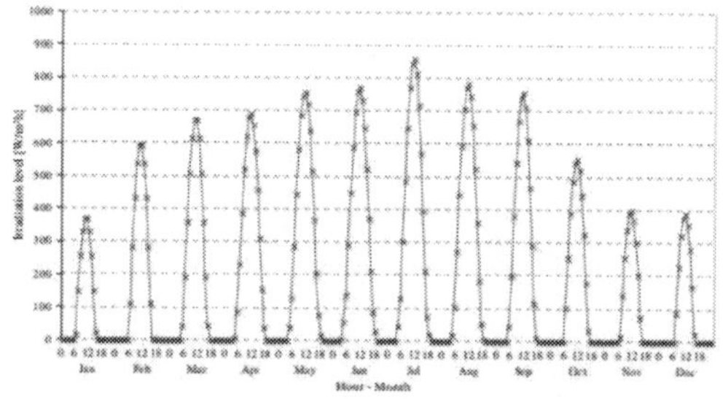

Figure 4 – General irradiation data [7]

But this approach is lacking the potential high variance in the irradiation data. Using data from existing PV plants is indirectly incorporating all the important data for the geographical location of the plant. From the data collected over 2 years a section of days has been taken going from the best day to the worst day seen from the PV plants point of view.

10.4229/EUPVSEC2023/4DV.4.38
020464-002

In the figures below different days of irradiation data can be seen.

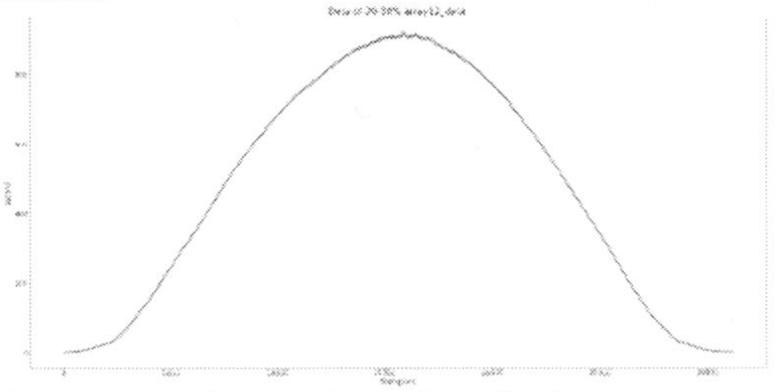

Figure 5 - Irradiation - day without clouds.

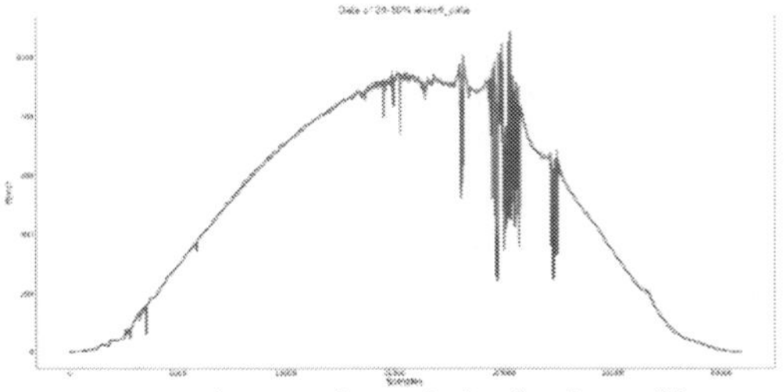

Figure 6 - Irradiation - day with clouds after midday.

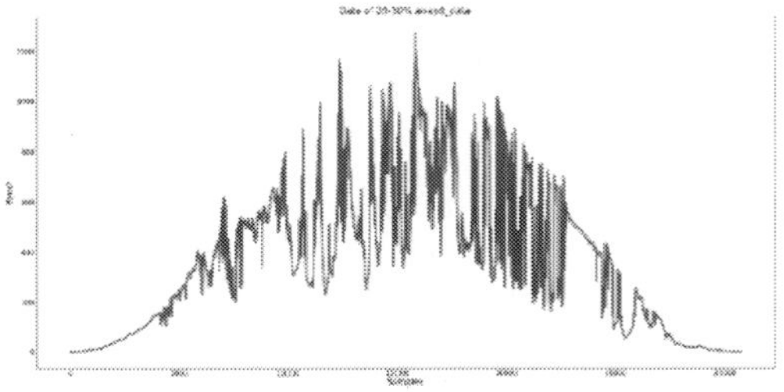

Figure 7 - Irradiation - day with a mix of sun and clouds.

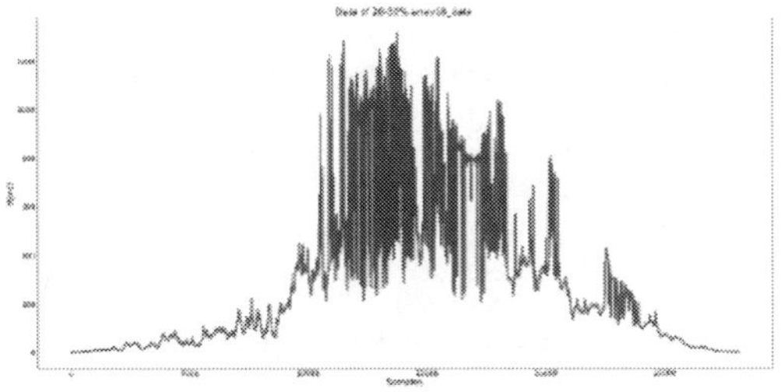

Figure 8 - Irradiation - day with mainly clouds.

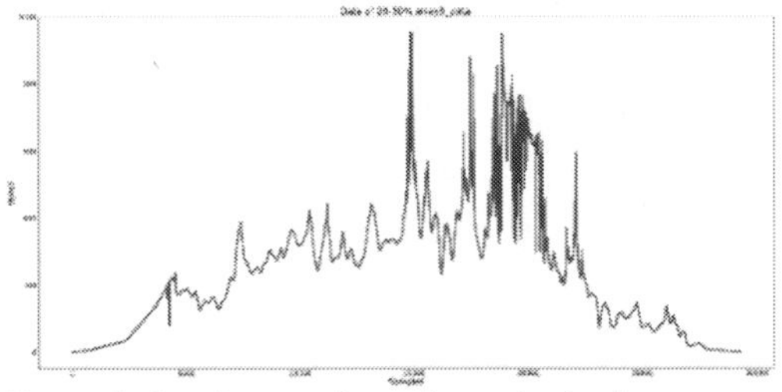

Figure 9 - Irradiation - day with mainly clouds.

As can be seen in the figures above the irradiation is fluctuating a lot depending on the weather conditions. This will have a negative effect on a PV systems DC link.

To evaluate this, a series of simulations has been made in MATLAB Simulink and Simscape Electrical. [4]

3 Simulation

The simulation is using the irradiation data with the length of one day.

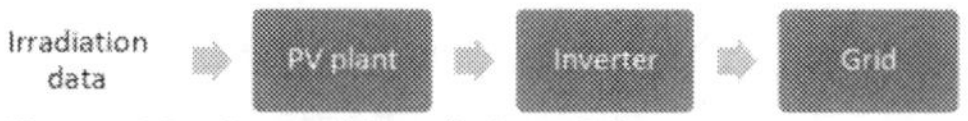

Figure 10 - General Simulink model

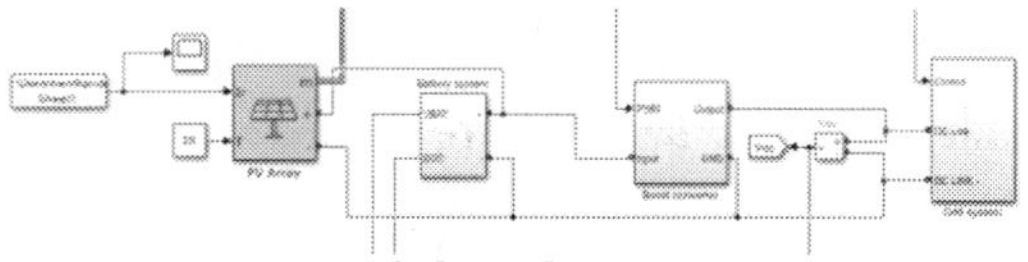

Figure 11 - Main Simulink simulation.

3.1 PV plant

The PV plant is set up with the PV module SunPower 305. 6 in series per string and 80 strings in parallel
The SunPower 305 has the following data:

- Maximum power = 305 W
- Cells per module = 96
- Open circuit voltage = 64.2 V
- Short circuit current = 5.96 A
- Voltage at MPP = 54.7 V
- Current at MPP = 5.58 A

3.2 Boost converter

A boost converter is used to change the DC level from the PV plant and battery system to a 600 V DC level that is fed to the grid system. This converter is trying to stabilize the fluctuations but without the battery it doesn't stand a chance.

Figure 12 - DC Link voltage and duty cycle.

3.3 Simulation results

The results from the simulation with irradiation and the PV plant specifications can be seen in the figure below.

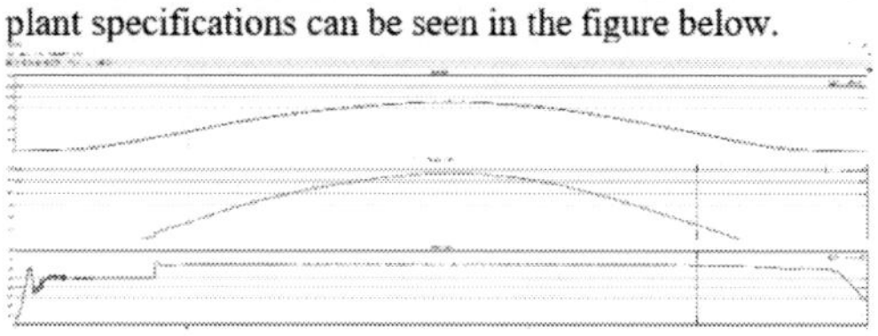

Figure 13 - Simulation with irradiation data from figure 5

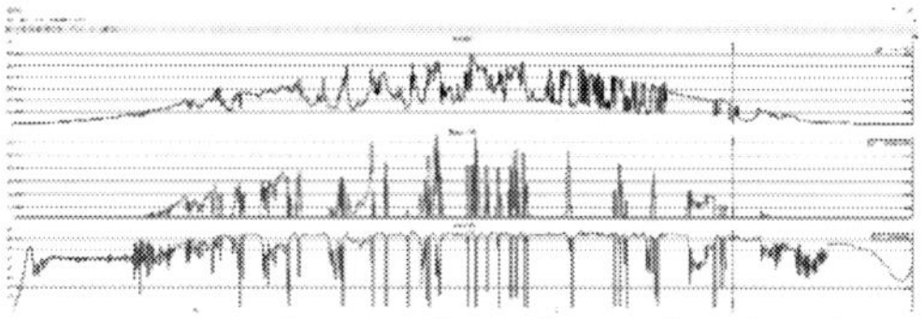

Figure 14 - Simulation with irradiation data from figure 7.

10.4229/EUPVSEC2023/4DV.4.38
020464-003

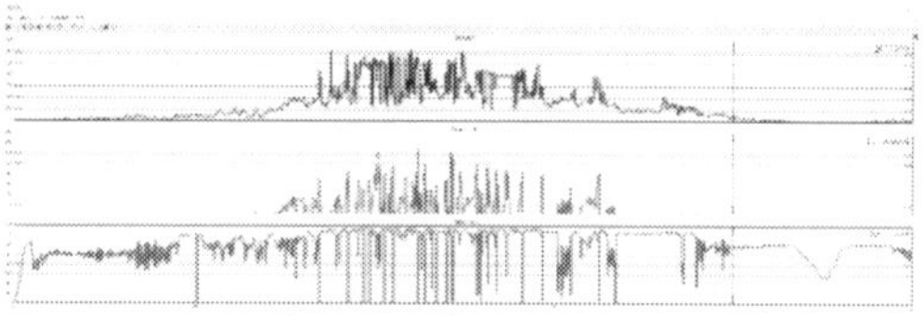

Figure 15 - Simulation with irradiation data from figure 8.

The top plot shows the irradiation data taken from the database, the middle plot shows the power production from the PV plant and the bottom plot shows the DC Link voltage. As can be seen the DC link and the power production from the PV plant are fluctuating due to the fluctuation from the irradiation data. Only the good day, without clouds, behaves as expected from a PV plant.

3.4 Battery storage

The symbiosis of PV systems and energy storage technologies holds promise in revolutionizing the landscape of sustainable energy generation and consumption. The inclusion of battery storage presents a transformative approach that mitigates the challenges posed by solar intermittency and enhances the overall effectiveness and resilience of PV systems. [2] [3]

The battery pack design details are:

- Battery type = Lithium-Ion
- Nominal Voltage = 300 V
- Rated capacity = 800 Ah
- Initial state of charge (SOC) = 80 %

3.5 Simulation result with battery storage

Adding battery storage to the system has the potential to stabilize the fluctuation on the DC link and having all the other benefits from a storage device. [5] [6]

Figure 16 - Simulink model with battery storage

The results from the simulation with battery storage added can be seen below. Note that the battery is keeping the voltage level in the start and end of the simulation.

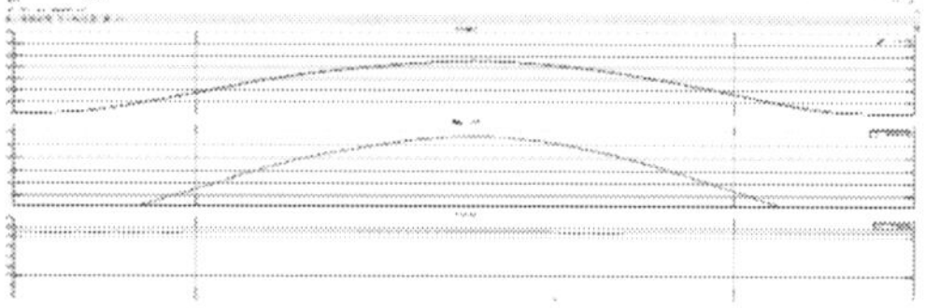

Figure 17 - Simulation with battery storage and irradiation data from figure 5.

As can be seen in the simulation result in figure 18 and 19, the battery storage is stabilizing the DC link, and this will help the switching components.

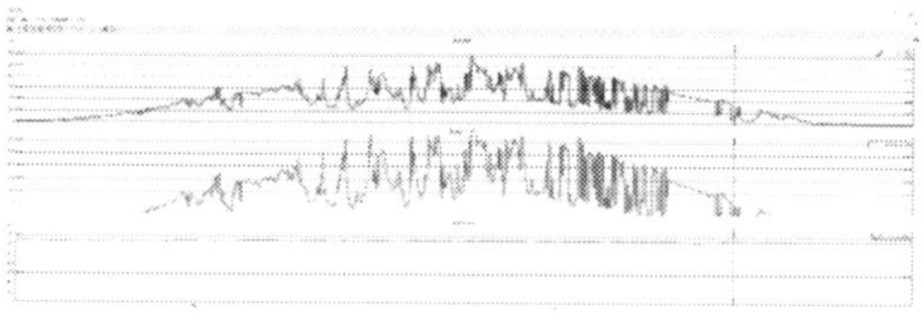

Figure 18 - Simulation with battery storage and irradiation data from figure 7.

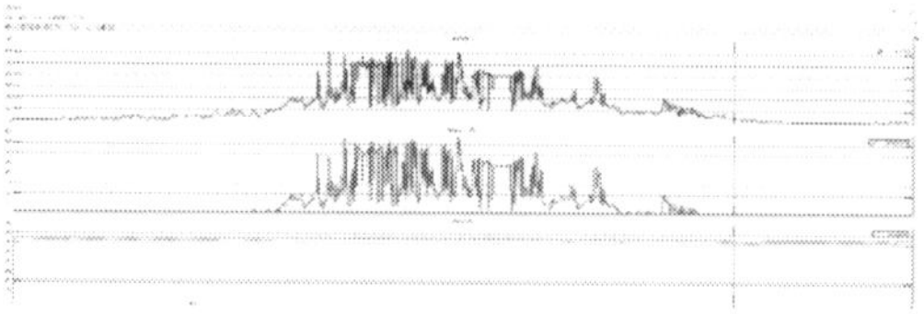

Figure 19 - Simulation with battery storage and irradiation data from figure 8.

With the size of battery storage, the energy cost of stabilizing is limited. The SOC plot of the worst-case scenario from (figure 19) can be seen below.

Figure 20 - SOC after stabilizing figure 19.

The initial state of charge is set to 80% and after stabilizing it is at 79.9%.

4 Classification of days using Markov Transition Field

Markov Transition Fields (MTFs) constitute a powerful framework for analyzing time-series data, offering valuable insights into sequential patterns, transition behaviors, and underlying dynamics. Originally rooted in the field of image processing, MTFs have been successfully adapted and applied to various domains, including signal processing, finance, healthcare, and environmental sciences. This section provides an overview of the Markov Transition Field and how it can be used to classify the data from a PV plant.

Markov Transition Fields represent a structured representation of a time-series by quantifying the transitions between different states or levels within the data. The core idea involves discretizing the continuous or categorical values of the time-series into a set of states or symbols. Subsequently, transition probabilities between these states are computed, resulting in a transition matrix. By visualizing this matrix in the form of an image, the Markov Transition Field provides a comprehensive and intuitive representation of the time-series dynamics. [8]

The process of constructing a Markov Transition Field involves the following steps:

10.4229/EUPVSEC2023/4DV.4.38
020464-004

Discretization: The time-series data is discretized into distinct states or levels (bins), depending on the nature of the data. For the time-series X each x(i) is assigned to the corresponding bin q(j).
An example of this can be seen in figure 21.

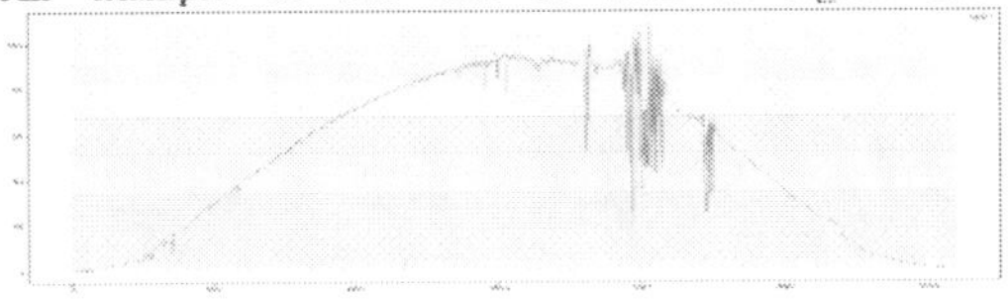

Figure 21 - Example of MTF bins done on data from figure 6.

Transition Probability Calculation: Transition probabilities between the discretized states are computed, typically by analyzing the frequency and patterns of transitions in the time-series.

Markov Transition Field Generation: The transition probabilities are organized into a matrix, call Markov Transition Matrix (MTM) and denoted W (eq.1), which can be visualized as an image known as the Markov Transition Field (matrix M) [9] [10]. Each element of the matrix represents the transition probability from one state to another, as shown below

$$M = \begin{bmatrix} W_{ij}|x_1 \in q_i, x_1 \in q_j & \cdots & W_{ij}|x_1 \in q_i, x_n \in q_j \\ W_{ij}|x_2 \in q_i, x_1 \in q_j & \cdots & W_{ij}|x_2 \in q_i, x_n \in q_j \\ \vdots & \ddots & \vdots \\ W_{ij}|x_n \in q_i, x_1 \in q_j & \cdots & W_{ij}|x_n \in q_i, x_n \in q_j \end{bmatrix} \quad (1)$$

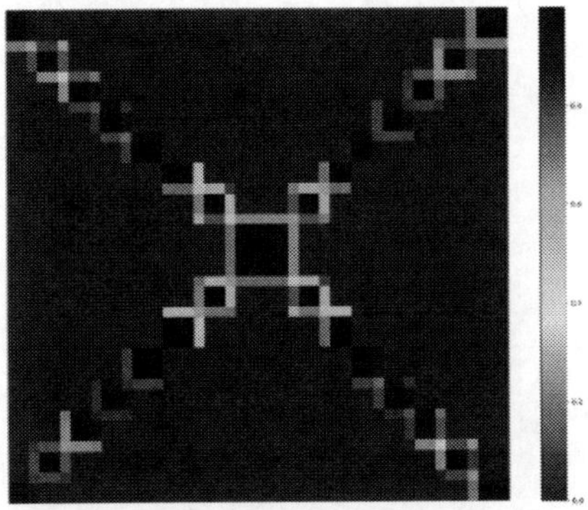

Figure 22 - MTF of figure 5.

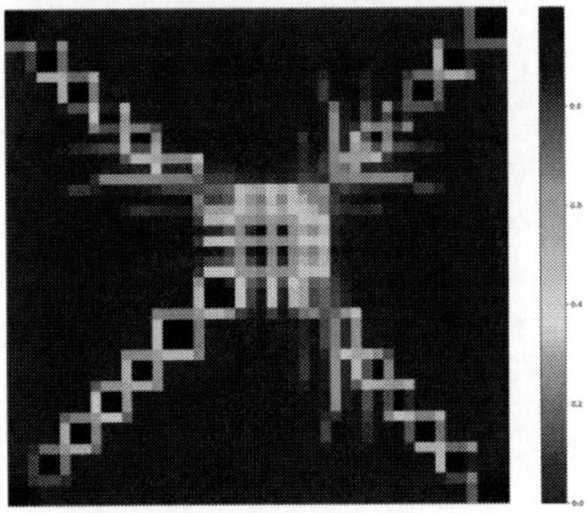

Figure 23 - MFT of figure 6.

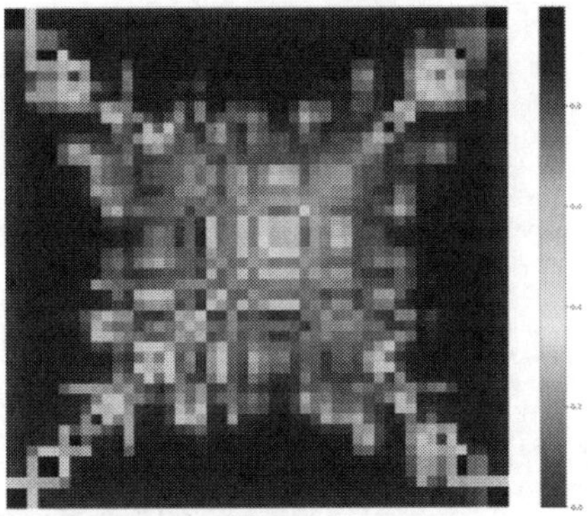

Figure 24 - MTF of figure 7.

It can be noted that the cloudy day where the probabilities of switching between bins are easily identified using this method. For example, the MTF of figure 7 seen in figure 24. This will be subject to an upcoming paper where an investigation into AI classification using Convolutional Neural Network (CNN) and if this can be used to make a charge/discharge strategy.

5 Summary and conclusion

The paper underscores the critical importance of incorporating irradiation fluctuations into the design and operation of photovoltaic (PV) plants. Solar irradiation, a fundamental environmental factor, greatly influences the performance and efficiency of PV systems. This research demonstrates that ignoring irradiation fluctuations during the design phase can lead to suboptimal system performance and inefficient energy production. Understanding and integrating the variability of solar irradiation into the design process is imperative for maximizing the potential of PV plants.

Through meticulous simulation and analysis using MATLAB Simulink, the paper elucidates the dynamic relationship between solar irradiation and PV plant behavior. The findings emphasize that accounting for irradiation fluctuations enables better system design and highlights the benefits of having battery storage included in the PV design. The result from this simulation is that the impact from fluctuations in the irradiation, due to clouds moving over the plant, can't be neglected. It has been showed that the DC link is experiencing fluctuations which can be critical for the boost converters inside the inverters.

Adding a battery storage system will smoothen the DC link and the energy cost for the battery is very small. Furthermore, getting the benefit of having storage and the added benefits this can bring to both the owner and the grid.

The MTF process is expected to be a suitable tool to classify the difference in irradiation over days and is suitable for further research into that topic.

10.4229/EUPVSEC2023/4DV.4.38
020464-005

5 References

[1] Paasch, Kasper. Power Electronic System for Multi-MW PV Sites. Ph.D-thesis. University of Southern Denmark 2016.

[2] A. Chaurey, S. Deambi, Battery storage for PV power systems: An overview, Renewable Energy, Vol. 2, Issue 3, (1992), 227-235.

[3] Iromi Ranaweera, Ole-Morten Midtgård, Optimization of operational cost for a grid-supporting PV system with battery storage, Renewable Energy, Vol. 88 (2016), 262-272.

[4] T. Salmi , M. Bouzguenda , A. Gastli and A. Masmoudi , "MATLAB/Simulink Based Modeling of Photovoltaic Cell", International Journal Of Renewable Energy Research, vol. 2, no. 2, pp. 213-218, Jun. 2012

[5] X. Li, D. Hui and X. Lai, "Battery Energy Storage Station (BESS)-Based Smoothing Control of Photovoltaic (PV) and Wind Power Generation Fluctuations," in IEEE Transactions on Sustainable Energy, vol. 4, no. 2, pp. 464-473, April 2013, doi: 10.1109/TSTE.2013.2247428.

[6] Iqbal, Mirza Mursalin, and Kafiul Islam. "Design and simulation of a PV System with battery storage using bidirectional DC-DC converter using Matlab Simulink." International Journal of scientific & Technology research 6.07 (2017): 403-410.

[7] Bortolini, M., Gamberi, M., Graziani, A. (2014). Technical and economic design of photovoltaic and battery energy storage system. Energy Conversion and Management, Vol. 86, 81–92.

[8] Zhiguang Wang and Tim Oates. (2015). Imaging Time-Series to Improve Classification and Imputation. Twenty-Fourth International Joint Conference on Artificial Intelligence.

[9] John Cristian Borges Gamboa. (2017). Deep Learning for Time-Series Analysis. CoRR. http://arxiv.org/abs/1701.01887

[10] Campanharo, A., Sirer, M.I., Malmgren, R.D., Ramos, F.M., Amaral, L.A.: Duality between time series and networks. PloS one 6(8), e23378 (2011)

Acknowledgements.

This work was supported by Interreg Deutschland-Danmark with funds from the European Regional Development Fund via the PE-Region Platform project (ref. 098-1.1-18) and the SmartPowerConversion project (ref. 16-2.1-22 1). Find further information on Interreg Deutschland-Danmark on www.interreg5a.eu.

10.4229/EUPVSEC2023/4DV.4.38
020464-006

DATABASE AND VERSION COMPARISON OF PVSYST SOFTWARE IN DIFFERENT INSTALLATIONS IN BRAZIL

João Frederico S. de Paula[1], João Lucas de Souza Silva[1], Guilherme C. S. Prym[1], Geyciane Pinheiro de Lima[1], Mendelsson Rainer Macedo Neves[1] , Hugo Alvarez[2], Rodrigo Garcia[2], Francisco C. Marques[2], Tárcio André dos Santos Barros[1], and Marcelo G. Villalva[1]

[1]University of Campinas - FEEC, Campinas, São Paulo, Brazil, joaofredericosp@gmail.com

[2]Build Your Dreams (BYD), Campinas, São Paulo, Brazil, rodrigo.garcia@byd.com

ABSTRACT: With the growing demand for the insertion of photovoltaic (PV) systems and different existing technologies, it is necessary to use simulations. The simulations contribute to defining an energy generation value close to the real one, but it is influenced by factors such as software modeling and the database used, making its configuration a challenge. PVsyst is one of the best-known software in the literature and is used in Brazil, which allows you to carry out studies, design, and analysis of PV systems. In this way, the objective of this paper is to compare different versions of PVsyst and meteorological data. In this study, ten PV systems were simulated in version 6.8 and version 7.2, varying the version of the meteorological base with data from the National Institute for Space Research (INPE). As a result, PVsyst 7.2 proved to be more conservative than 6.8. As a contribution, the present study helped PV integrators choose the database and how the PVsyst software behaves in different versions.

Keywords: pvsyst, database, photovoltaic.

1 INTRODUCTION

PVsyst [1] is a software that allows you to carry out studies, dimensioning and analysis of photovoltaic (PV) systems. The results of the simulations are shown in a report, containing information such as the analysis of system losses, financial analysis, performance, and the total power generated (kWh/year) by the projected PV system [2]. As it is a software that is continually updated, new advances in PV modeling can be added, and it is interesting to verify the differences in the results obtained in different versions.

However, one of the factors that affect the simulations is the database used, this is a factor studied in the literature [3]. Thus, the impact of different meteorological databases on simulation results must be taken into account. Simulations of PV systems are very sensitive to the quality of the input data. For example, it was noted that in Brazil, PV integrators, have differences in the versions of the National Institute for Space Research (INPE) database, a well-used source in Brazil due to the confidence in the data.

With this challenge of understanding the behavior of the different versions of PVsyst, as well as the INPE database and helping PV integrators, this study was proposed. The objective of this paper is to simulate several PV installations in different versions of the PVsyst and INPE base. For this purpose, ten PV systems were configured and simulations were performed in versions 6.8 and 7.2 of PVsyst, varying the version of INPE's meteorological base. As a technical/scientific contribution there is: Show PV integrators the differences between INPE's base in Brazil in the PVsyst software; Verify the behavior of PV modeling in different versions of PVsyst; Evaluate the impacts of different database versions within the software.

2 METHODOLOGIES

For the simulations, it was proposed to use the software PVsyst version 6.8 and 7.2, chosen by the evolution in the different versions and its modeling. As for the database, the one from old INPE, and the year 2017, was used.

Among the main differences between these two versions of the PVsyst software, the version 7 proposes a new completely overhauled user interface and complete localization, support of 64-bit architectures that allows the software to handle large projects and shading scenes, an unlimited number of sub-arrays is now allowed to consider in the development of the project. Some other upgrades in the latest version are the new conversion of fixed tables to trackers, in the case of output AC circuit brings the definition of MV and HV transformers, the software simulation shows the result values and graphs in real time while the simulation is running and the errors and warnings are enable in the results, economic evaluation is done to add with the results and brings a new complete report of the simulation, a new transposition model that applies to both Hay and Perez model, and module layout was reworked to handle half-cell models with updated electrical shading calculations [4].

Ten real PV systems were simulated in the state of Goiás, Minas Gerais and the Federal District in Brazil. Table I summarizes the power of each simulated installation. The same parameters were adopted for the simulations with both versions of PVsyst employed.

For comparison between simulations, the percentage differences between the results were also calculated, varying the values according to the simulated software, and graphs were generated from a dashboard for comparison.

10.4229/EUPVSEC2023/4DV.4.40
020466-001

Table I: Rated Powers and Locations of the Projects Studied

Installations	State	Total Rated Power
Installation 1	GO	36 kWp
Installation 2	GO	50,9 kWp
Installation 3	MG	57,2 kWp
Installation 4	MG	127 kWp
Installation 5	MG	76,3 kWp
Installation 6	GO	29,68 kWp
Installation 7	GO	36 kWp
Installation 8	GO	36.6 kWp
Installation 9	MG	111 kWp
Installation 10	DF	35 kWp

3 RESULTS AND DISCUSSION

3.1 Comparison of INPE databases

The Table II shows the monthly average of irradiation for each location, along with the percentage difference between the outdated INPE database and the 2017 database showed smaller values in relation to the outdated database. This proves the hypotheses raised by photovoltaic integrators (in Brazil) about the suspicions that the irradiation of the current INPE database showed lower values of solar irradiation. The percentage difference margin analyzed between the databases is between a range of -1.33% to -7.28%.

Table II: Monthly Average of Global Horizontal Irradiation (kWh/m²)

Installations	INPE Outdated	INPE 2017	Percentage Difference
Installation 1	170,0	161,0	-5,28%
Installation 2	164,2	159,8	-2,69%
Installation 3	161,2	159,1	-1,33%
Installation 4	166,7	159,2	-4,49%
Installation 5	157,1	145,7	-7,28%
Installation 6	168,1	161,9	-3,69%
Installation 7	169,4	160,2	-5,39%
Installation 8	169,4	159,9	-5,60%
Installation 9	164,4	156,5	-4,82%
Installation 10	168,3	160,5	-4,65%

3.2 PVsyst Version Comparison

Figure 1 presents the summary of the results regarding the annual energy generated by the PV installations in each project, with the 6.8 and 7.2 versions of PVSYST and the INPE databases. It is noticed that version 7.2 presents, in general, more conservative results than the simulations of version 6.8.

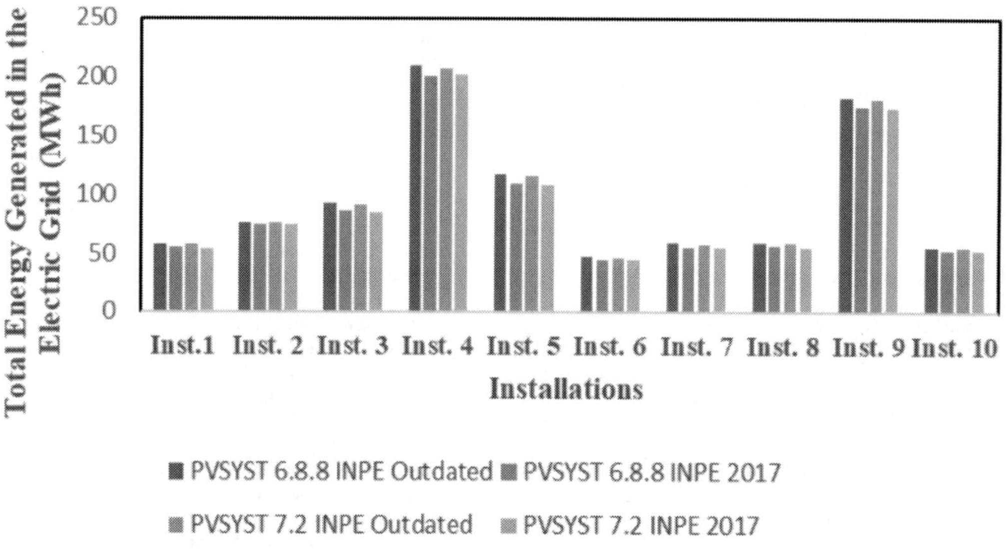

Figure 1: A comparison of the annual total energy generated in the electric grid (MWh) of each software version and different INPE database.

10.4229/EUPVSEC2023/4DV.4.40
020466-002

4 FINAL CONSIDERATIONS

This paper simulated ten PV systems with PVsyst versions 6.8 and 7.2, varying the INPE database. It was noticed that the horizontal global solar irradiation averages of the INPE 2017 database are smaller (with a variation from -1.33% to -7.28%) in relation to those of the outdated INPE database, directly influencing the energy generation results in the projects. The greatest percentage errors of irradiation in relation to its monthly average found in the two databases were detected in the month of June, with the exception of installation 1 with the old INPE database.

The level of dispersion of INPE 2017 global horizontal solar irradiation data is lower in all locations studied in relation to the old database, showing greater stability in relation to irradiation variation throughout the year. As for the results of energy generation, the software in version 7.2 was more conservative in general, with lower generation. This may be an indication of differences in modeling.

5 REFERENCES

[1] Features of PVSyst, https://www.pvsyst.com/ (current jan. 14, 2023)

[2] J. L. de Souza Silva, T. S. Costa, K. B. de Melo, E. Y. Sakô, H. S. Moreira and M. G. Villalva, "A Comparative Performance of PV Power Simulation Software with an Installed PV Plant," 2020 IEEE International Conference on Industrial Technology (ICIT), Buenos Aires, Argentina, 2020, pp. 531-535, doi: 10.1109/ICIT45562.2020.9067138.

[3] J. Ernesto and R. Rüther. The influence of the solar radiation database and the photovoltaic simulator on the sizing and economics of photovoltaic-diesel generators. Energy Conversion and Management, 210, 2020, 112737, 10.1016/j.enconman.2020.112737.

[4] Differences between version 7 and version 6, https://www.pvsyst.com/help/version7_6_differences.htm (current jan. 15, 2023)

ACKNOWLEDGMENT

The authors gratefully acknowledge support from BYD Energy Brazil through the PADIS/MCTI program, Project No. 5779/FUNCAMP; CAPES; ALSOL; in part by TotalEnergies/CEPETRO under FUNCAMP/UNICAMP agreement 6002.4; and FEEC.

10.4229/EUPVSEC2023/4DV.4.40
020466-003

GLOBAL SENSITIVITY ANALYSIS: RATING THE INFLUENCE OF PARAMETERS IN A PV PERFORMANCE TOOL

Katherine Alvino[1,2], Nicolas Zalachas[2], Rodrigo Amaro e Silva[3], Philippe Blanc[3], Maroun Nemer[1]
[1] Mines Paris, Centre Efficacité Énergétique des Systèmes, 5 rue Léon Blum, 91120 Palaiseau, France
[2] BBS Slama, 23 Rue du Mont Mouchet, 63000 Clermont-Ferrand, France
[3] Mines Paris, Centre Observation, Impacts, Énergie, 1 Rue Claude Daunesse, 06560 Valbonne, France
katherine.alvino@minesparis.psl.eu

ABSTRACT: We applied a global sensitivity analysis (GSA) to a PV performance tool to describe the influence of four parameters (orientation, tilt, distance between rows and albedo, along with the interaction effects between them) on two intermediary results of the estimation: yearly effective irradiance and yearly energy generation.
In the case of yearly effective irradiance, the most influential factors where orientation, distance between rows and tilt, in order of influence, followed by two interaction effects between them. For the yearly energy output, the most influential factors where distance between rows, orientation and tilt, with the first parameter noticeably exceeding the rest of influence sources. We underlined the importance of using this information to obtain relevant visual representations of the model's behaviour.

Keywords: PV performance tools, global sensitivity analysis (GSA), Sobol', optimisation

1 INTRODUCTION

Photovoltaic (PV) project developers increasingly use optimisation tools for planning, sometimes without knowing neither the most influential parameters nor the interaction effects between them. But, do these tools point out the most influential parameters or the interaction effects that can occur between them? And, why would this information be useful?

Global sensitivity analysis (GSA) is a tool to enhance the knowledge of not only individual parameters but also of the interactions between them. When optimising a configuration, GSA detects the most influential parameters and describes their effect, allowing to prioritise the most important factors.

A few articles entailing sensitivity analysis applied to photovoltaics are found in the literature. Andrei et al. [1] studied in 2012 the first-order derivative sensitivity of six output variables: short circuit current (I_{SC}), open circuit voltage (V_{OC}) and maximum power point (P_{MPP}, V_{MPP}, I_{MPP}), to the temperature and irradiance factors, using them to produce and validate a set of analytical relations. In 2013, Li-nan et al. [2] applied static (the research object is time-independent) and trajectory (dynamic) sensitivity methods for parameter identification of grid-connected photovoltaic inverters. In a third study published in 2015, Hansen et al [4] performed a sensitivity analysis by means of scatterplots to evaluate the uncertainty of the output to the model choice. More recently, Zhang et al.[3] applied a Sobol algorithm in order to study the influence of three input factors, yielding to similar first-order and global sensitivity values for each parameter and obtaining a ranking of sensitivities (S): $S_{irradiance} >> S_{module\ temperatur} > S_{air\ rel.\ humidity}$.

Generally, most sensitivity analyses within the field of earth sciences apply a local approach [5]. Global methods evaluate more than one factor simultaneously, thus providing the added capacity to evaluate whether there are interaction effects between two or more input parameters, besides of the perturbation of each parameter individually. That is a particularly useful feature for PV applications since the optimal positioning of solar modules cannot be made considering each parameter individually. With this in mind, the focus of this research work will be to evaluate the influence of parameters related to the physical configuration on two steps of the energy estimation. Which parameters among orientation, inclination, distance between rows and albedo influence the most the energy output? What are the interaction effects between them? How can we apply this knowledge in the development of tools?

2 SCIENTIFIC OBJECTIVES

To give an answer to the previously exposed questions, we explored the application of global sensitivity analysis to analyse PV performance methods and propose a methodological framework to identify parameter influences and their interaction effects.

3 EXPERIMENTAL SETUP AND METHODOLOGY

We simulated a PV system that explores a surface of 378 m² (limited by the dark grey contour enclosing the panels in Figure 1), located in Marseille, France (43.3°, 5.37°). The weather database (year 2005) was obtained from Meteonorm 7. Two buildings are placed nearby, from which the blue one is 5 m tall and placed to the west, while the grey building is 3 m tall, due south.

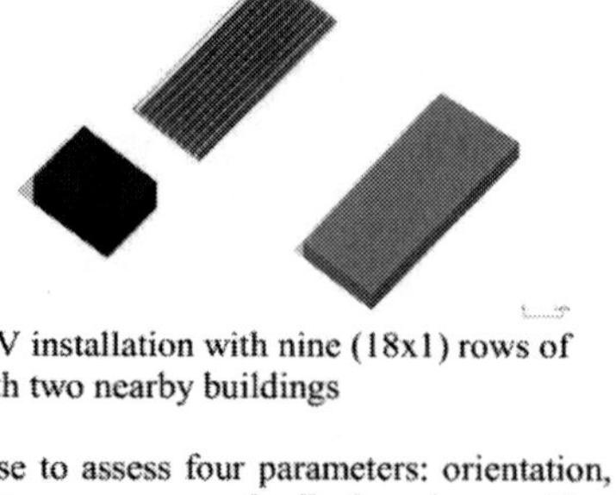

Figure 1: PV installation with nine (18x1) rows of modules with two nearby buildings

We chose to assess four parameters: orientation, tilt, distance between rows and albedo, along with their intervals of variability (see Table I). Then, the sensitivity analysis was performed in three steps (see Figure 2): first, we applied the quasi-random Sobol' method to generate

10.4229/EUPVSEC2023/4DV.4.41
020468-001

over 40000 combinations of the parameters, using the library SaLib in Python [7], [8]. Then, we evaluated all the scenarios with the PV Prod tool. Finally, we performed a Sobol' GSA by using two tools developed by the *Centre OIE (Observation, Impacts, Énergie)*. The first algorithm is based in MATLAB and the second one in Python.

Table I: Parameter variabilities for optimisation case

Parameter	Distribution	Boundaries
Orientation of modules	Uniform	-90° to 90°
Tilt of the modules	Uniform	0° to 45°
Distance between rows	Uniform	0 to 1 m
Ground albedo	Uniform	0.1– 0.3

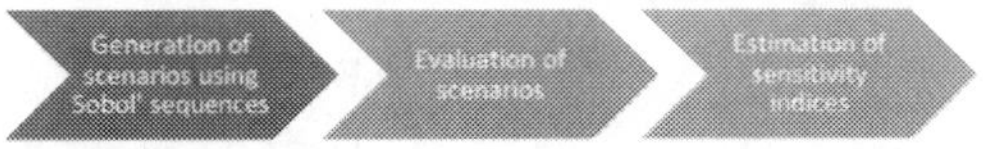

Figure 2: Steps of the estimation of the Sobol' indices

We also confirmed the sufficiency of data for the sensitivity results.

4 RESULTS

The GSA has been applied for two steps of the method (outputs of interest): yearly irradiation first, followed by yearly energy output.

We used another python-based tool from the OIE group that confirmed the results of the sensitivity analysis, and additionally we obtained the RMSD values to explain the part of RMSE that corresponds to using a function that only considers the indicated parameters.

4.1 Yearly irradiation

The modules orientation is by far the most influential factor (See Figure 3a), followed by the distance between rows and module tilt. The most important interaction effects come after them and albedo comes only at the seventh place of importance, almost with no addition to the cumulated effects. Note that the sum of the first-order indices, those corresponding to the four parameters, is less than one. Thus, we can consider that the method is a non-additive model.

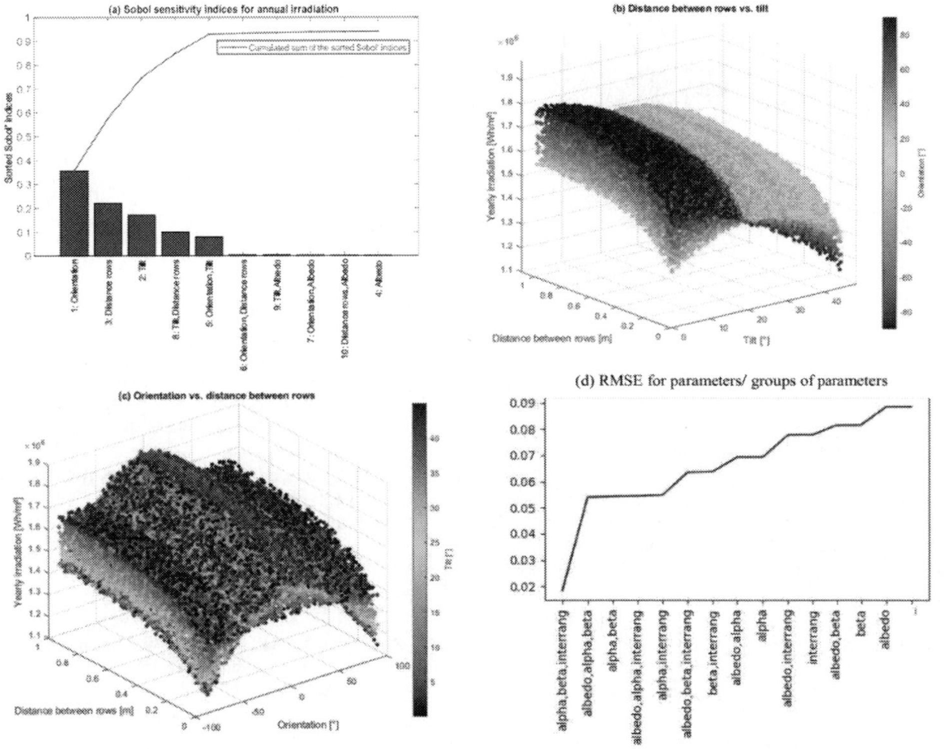

Figure 3: a) Sobol' indices, b) and c) dependency on the four parameters of the effective yearly irradiation; d) part of the RMSE caused by each parameter/group of parameters.

Figure 3b presents a green layer of points corresponding to a south orientation, showing that , the further the panels are, generally the irradiation per square meter increases, which is not so true when the tilt is zero or when the modules are oriented away from the equator, as is shown by the red and blue points. An intersection of points occurs when the tilt is equivalent to around 20°. Fig. 3d follows similar characteristics, only differing in the fact that albedo practically doesn't produce any change in the values.

Figure 3c presents that the yearly irradiation increases as the distance between rows decreases, while it reaches a maximum when the modules are facing a point around the south. As the inclination reduces (blue points), the orientation loses influence.

The highest irradiation point (1.8 Gh/m²/year)

corresponds to an orientation of -4.0°, a distance between rows of 0.99 m, and a tilt equivalent to around 35°. The more straightforward way to find it is by observing Fig. 2c, which it describes graphically the relationship between the two most influential factors.

4.1 Yearly energy output [Wh/year]

In this step, the distance between rows is the most influential parameter (see Fig. 2a), since it strongly determines the number of modules that fit in a limited area.

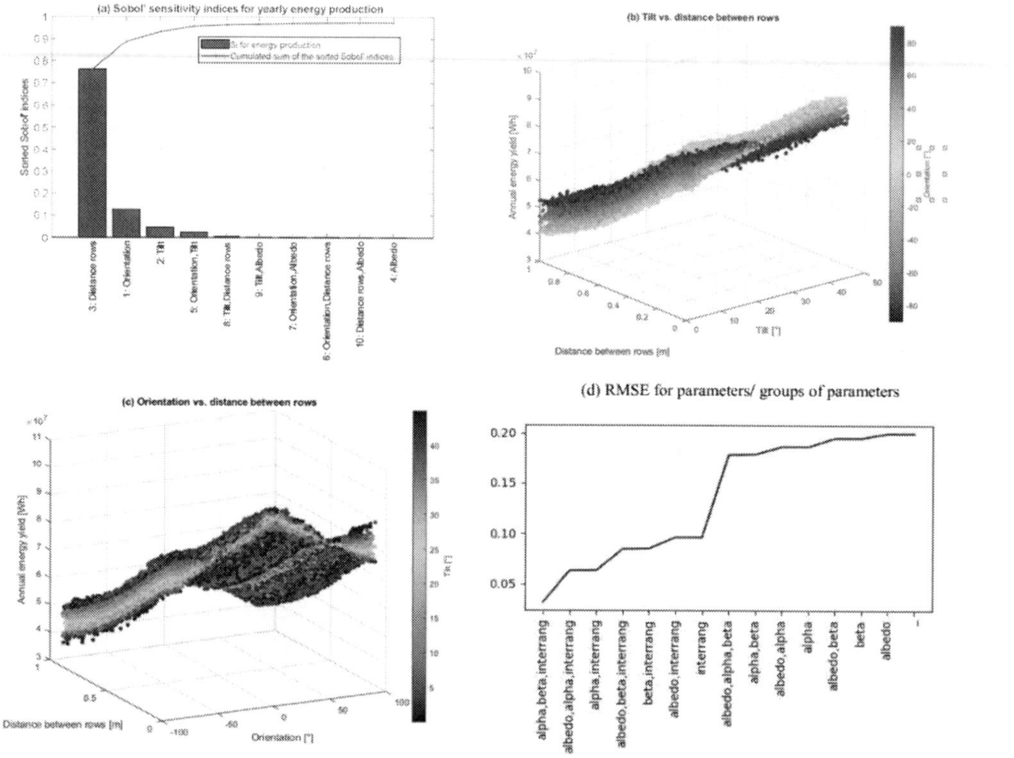

Figure 4: a) Sobol' indices; b) and c), dependency on the four parameters of the effective yearly irradiation; d) part of the RMSE caused by each parameter/groups of parameters.

We are no longer considering energy per square meter but the total energy provided, so the energy yield increases as the modules get closer, in opposition to the previous step.

The highest point, 0.1 GWh/year, corresponds to 4° of azimuth and 0 m of distance between rows, with a tilt around 30°, suggesting that we might need to take a look at the cost function instead of only maximising energy yield.

5 DISCUSSION

It is clear that Figs. 2c and 3c in both steps provided the clearest behaviours, making easy to visually identify the highest point since they consider the most influential factors. In contrast, figs. 2d and 3d do not provide the same facility.

We can observe that there is an overall symmetry between east and west orientations, since the obstacle in the scenario is 8m far and only 5m tall. Other applications with closer shading factors should be able to find a more visible asymmetry.

Soiling losses were not considered in this method, which should be taken into account afterwards. There exist issues to fix in our evaluation model and we aim to obtain more outputs to make further analyses (energy output, self-consumption, self-production rates or even performance ratio indicators).

Many other questions originate from these results. For example, how does this ranking vary for different locations and weather databases?

6 CONCLUSIONS

The Sobol GSA method was applied to a system with fixed-tilt, mono-facial modules confined in a rectangular area, in order to make a step-by-step optimisation.

The possibility to change the output of interest allows to study the dependencies of the final energy output, self-consumption, self-production rates or even performance ratio indicators. We underline the importance of analysing the relevant graphs and of being aware of the configuration compromises that can be taken, either for economic reasons or other constraints. It only takes to set the output of interest to rate the parameters according to their influence on it.

The number of parameters increases the difficulty to plot and to visually interpret them. There lies the benefit of using Sobol' to identify the parameters with the higher

10.4229/EUPVSEC2023/4DV.4.41
020468-003

sensitivity indices that will be useful in order to find an optimal point in a parsimonious way.

7 ACKNOWLEDGEMENTS

We thank Julien Liochon for the development of the evaluation tool and we thank Benoit Gschwind and Nathalie Saint-Geours for helpful discussions.

This research was supported by a CIFRE convention and BBS Slama (CIFRE N° 2021/0487).

REFERENCES

[1] H. Andrei, T. Ivanovici, E. Diaconu, M. R. Ghita, O. Marin, and P. C. Andrei, 'Analysis and experimental verification of the sensitivity of PV cell model parameters', in *2012 International Conference on Synthesis, Modeling, Analysis and Simulation Methods and Applications to Circuit Design (SMACD)*, Seville, Spain: IEEE, Sep. 2012, pp. 129–132. doi: 10.1109/SMACD.2012.6339434.

[2] Q. Li-nan, Z. Ling-zhi, S. Tao, and L. Jing, 'Identification of photovoltaic power system based on sensitivity analysis', in *2013 IEEE PES Asia-Pacific Power and Energy Engineering Conference (APPEEC)*, Kowloon, Hong Kong: IEEE, Dec. 2013, pp. 1–5. doi: 10.1109/APPEEC.2013.6837296.

[3] C. Zhang, Y. Zhang, J. Zhao, and J. Luo, 'Sensitivity Analysis of Weather Factors Affecting PV Module Output Power Based on Artificial Neural Network and Sobol Algorithm', in *2020 IEEE/IAS Industrial and Commercial Power System Asia (I&CPS Asia)*, Weihai, China: IEEE, Jul. 2020, pp. 246–250. doi: 10.1109/ICPSAsia48933.2020.9208612.

[4] C. Hansen and C. Martin, 'Photovoltaic System Modeling. Uncertainty and Sensitivity Analyses', SAND2015--6700, 1211576, 598837, Aug. 2015. doi: 10.2172/1211576.

[5] T. Wagener and F. Pianosi, 'What has Global Sensitivity Analysis ever done for us? A systematic review to support scientific advancement and to inform policy-making in earth system modelling', *Earth-Science Reviews*, vol. 194, pp. 1–18, Jul. 2019, doi: 10.1016/j.earscirev.2019.04.006.

[6] R. Amaro E Silva and M. C. Brito, 'Spatio-temporal PV forecasting sensitivity to modules' tilt and orientation', *Applied Energy*, vol. 255, p. 113807, Dec. 2019, doi: 10.1016/j.apenergy.2019.113807.

[7] T. Iwanaga, W. Usher, and J. Herman, 'Toward SALib 2.0: Advancing the accessibility and interpretability of global sensitivity analyses', *SESMO*, vol. 4, p. 18155, May 2022, doi: 10.18174/sesmo.18155.

[8] J. Herman and W. Usher, 'SALib: An open-source Python library for Sensitivity Analysis', *JOSS*, vol. 2, no. 9, p. 97, Jan. 2017, doi: 10.21105/joss.00097.

10.4229/EUPVSEC2023/4DV.4.41
020468-004

MODELING OF INVERTER DERATE TEMPERATURE AND CLIPPING USING THE PVLIB LIBRARY FUNCTIONS

Luiz Reis[1, a)], Lúcio Paiva[1, b)], Luis Castro[1, c)] and Rodrigo Santos[1, d)]
[1]Casa dos Ventos Energias Renováveis, Fortaleza, Brazil
[a)]luiz.reis@casadosventos.com.br
[b)]lucio.paiva@casadosventos.com.br
[c)]guilherme.castro@casadosventos.com.br
[d)]rodrigo.raimundo@casadosventos.com.br

ABSTRACT: To accurately model the energy performance of a connected PV system and predict its losses, it is crucial to incorporate the behavior of inverters under saturation conditions that occur during periods of high irradiance and elevated temperatures. Today, industry standard software such as PVsyst effectively models inverter behavior under such conditions. However, this remains a challenge for certain open source tools such as pvlib. The main objective of this study is to perform energy modeling of a commercial inverter using the functionalities available in pvlib. This workflow is based on data extracted from the ".OND" file provided by the inverter manufacturer, enabling the simulation of temperature derating and curtailment effects. To validate this approach, a set of more than 200 simulations was performed in PVsyst, varying the DCAC ratio from 0.75 to 1.50 for different string lengths. The results were later compared with those obtained using pvlib. The proposed workflow, which combines two pvlib functions, demonstrated greater consistency with the results obtained by PVsyst. Root mean squared error (RMSE) and mean bias error (MBE) average values were 0.029% and 0.0046%, respectively.
Keywords: Inverter; Clipping; Derating.

1 INTRODUCTION

The inverter is responsible for converting the DC energy generated by the photovoltaic panels into AC energy, respecting a set of energy quality criteria. To accurately determine the output power of this device, the mathematical model must be able to evaluate the inverter performance under limiting conditions, reproducing the cuts that occur during peak irradiance and high ambient temperature conditions.

Inverter clipping is a critical aspect of modeling and is directly linked to the relationship between the output power of the DC array and the maximum AC output power indicated in the inverter's technical data sheet. This relationship is commonly called the DC/AC ratio.

Notably, there is a growing trend in the design of PV systems, as indicated by [1], to increase the DC/AC ratio due to the continuous decline in module prices over the years. In a comprehensive study detailed in [2], it was observed that among 156 large-scale projects that began operations in the US in 2021, the median DC/AC ratio was 1.34.

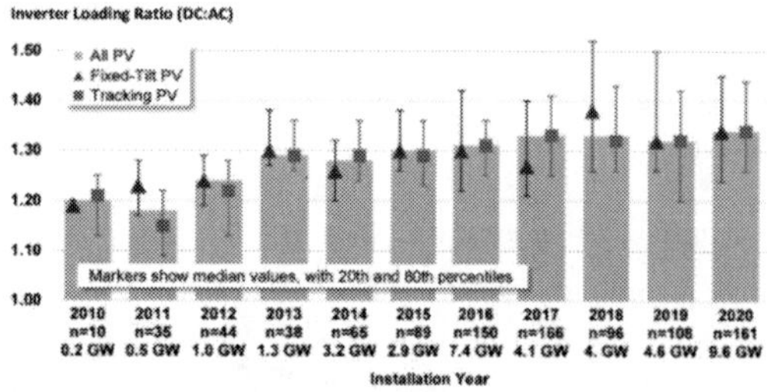

Figure 1: DCAC ratio per installation year for projects in the United States. Source: [2].

Although industry-leading software like PVsyst now models inverter properly under such conditions, it remains a challenge for some open source tools like pvlib.

The objective of this work is to carry out the energy modeling of a commercial inverter using the functions available in pvlib. This modeling will be based on data obtained from the ".OND" file provided by the inverter manufacturer, aiming to simulate the impacts of temperature derating and clipping.

2 TOOLS USED

The workflow proposed in this study for inverter modeling is based on the pvlib library, an open source Python library developed in [3]. The library aims to model solar photovoltaic systems. Within pvlib, there are two main options for inverter modeling: the Sandia model and the PVWatts model.

The inverter's efficiency curve is critical in determining its ability to convert energy efficiently. The research carried out in [4] and [5] focuses on modeling these efficiency curves, which are then integrated into the "inverter.pvwatts" and "inverter.sandia" functions of the pvlib library, respectively.

These models play an important role in simulation and return accurate estimates for inverter output under DCAC ratio values less than or close to 1. However, for limit conditions, where the overload ratio is greater than 1, derating and temperature clipping events become more frequent. As the models currently implemented in pvlib cannot reproduce these conditions accurately, the greater the DCAC ratio and the clipping of the simulated scenario, the greater the errors presented when comparing the results with the PVsyst outputs.

2.1 PV Watts model

The work detailed in reference [4] played a key role in the implementation of the "invert.pvwatts" function.

10.4229/EUPVSEC2023/4DV.4.42
020470-001

This model is based on efficiency data from the California Energy Commission (CEC) database.

Using inverter data available in the CEC database, an efficiency curve for a typical inverter was established. This representative inverter simplifies the modeling process for users by allowing them to calculate the inverter's output power by requiring only the DC power for each timestamp and two constant values: the inverter's maximum input DC power and the efficiency equipment rating. This simplified approach makes calculations faster for users.

The value of the two constants, the inverter nominal efficiency "eta_inv_nom" and the inverter DC input limit "pdc0", necessary to model a commercial inverter using pvlib functions, can be easily obtained in files with ".OND" extension, provided by inverter manufacturers. These values can be calculated using the following equations in terms of the maximum inverter efficiency "EfficMax" and the nominal inverter power "PNomConv" or maximum inverter power "PMaxOUT", all found in the ".OND" files:

$$eta\_inv\_nom = \frac{EfficMax}{100} \quad \text{(i)}$$

$$pdco = (PMaxOUT \times 1000) \div eta\_inv\_nom \quad \text{(ii)}$$

There are some controversies related to the use of "PMaxOUT" or "PNomConv" in equation (ii) that have led to a comprehensive evaluation of this parameter. To evaluate its impact, more than 200 simulations were carried out using PVsyst, where we varied the number of modules per string and the number of strings connected to the inverter. This variation made it possible to cover DC/AC ratio values ranging from 0.75 to 1.5.

The simulation results were then used for comparative analysis with two pvlib functions. Within the function "inverter.pvwatts" the effective energy at the output of the array "EArrMPP" was used as the DC power "pdc" and the constant parameter "pdc0" was obtained as a function of "PMaxOUT" and "PNomConv". The results were compared with the energy available at the inverter output "EOutInv" obtained from the PVsyst and the MBE between the values can be seen in Figure 2.

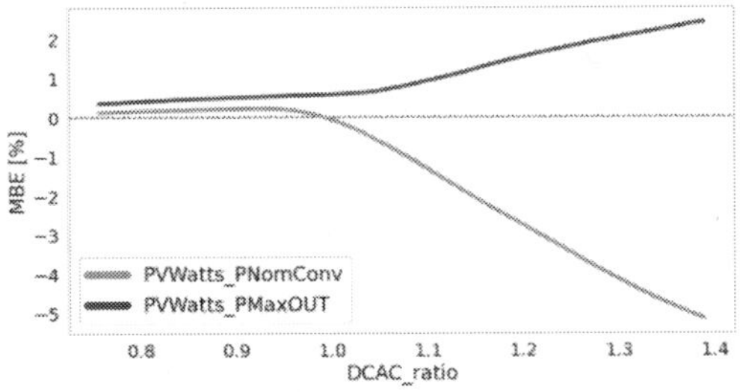

Figure 2: MBE value of comparing the outputs of the function inverter.pvwatts with the output "EOutInv" of PVSyst as a function of the DC/AC ratio.

As mentioned previously, the models implemented in pvlib perform extremely accurate calculations for DC/AC ratio values less than or close to 1. For systems without inverter overloading, both strategies tend to overestimate the inverter output with respect to the reference model. For these conditions, the results presented indicated an average bias of 0.16% using PNomConv to calculate the maximum DC power and 0.46% using PMaxOUT.

However, as expected, by not incorporating corrections related to temperature, the analysis of overloaded conditions indicated higher errors when compared with the "EOutInv" values obtained with PVsyst. When "PNomConv" is used in these conditions, the points where the inverter output power could be higher than the nominal power are clipped, resulting in a pessimistic production estimate.

With "PMaxOUT" it was possible to obtain results with slightly smaller biases, however, as the effects of reduction in inverter production caused by temperature derating are ignored, production estimates become optimistic, generating a more pronounced positive bias when compared with the reference values.

2.2 Sandia model

The implementation of the "inverter.sandia" function was heavily influenced by the studies developed in [5]. This function employs a quadratic model to derive the efficiency curve for the inverter. To calculate the model coefficients, the "inverter.fit_sandia" function is applied, using input values such as DC voltage, the effective energy at the output of the array, and AC output. Once these coefficients are obtained, they can be used to determine the inverter's output power. This is achieved by supplying the main function with the efficiency curve coefficients, along with the DC input voltage and power values.

To model a commercial inverter using the Sandia model, the initial step involves acquiring the DC voltage, DC power and AC output power points that characterize the inverter efficiency curve. The ".OND" file has three curves with these parameters obtained for three reference DC voltage levels. The "ProfilPIOV1", "ProfilPIOV2" and "ProfilPIOV3" curves can be used in the "inverter.fit_sandia" function to obtain the necessary parameters for inverter simulation using the main "inverter.sandia" function. The reference DC voltage level associated with each of these curves can be obtained from the variable "VnomEff." Additionally, the user needs to provide the model with the values of the inverter's nominal AC power "p_ac_0" and nighttime power consumed while the inverter is not supplying AC power "p_nt", which can be calculated in terms of the maximum inverter power "PMaxOUT" and night consumption "Night_Loss", both obtained from the ".OND" file:

$$p\_ac\_0 = (PMaxOUT \times 1000) \quad \text{(iii)}$$

$$p\_nt = Night\_Loss \quad \text{(iv)}$$

In a similar approach to that used for the PVWatts function, the use of the "PMaxOUT" or "PNomConv" parameters in equation (iii) was evaluated. The results can be seen in Figure 3.

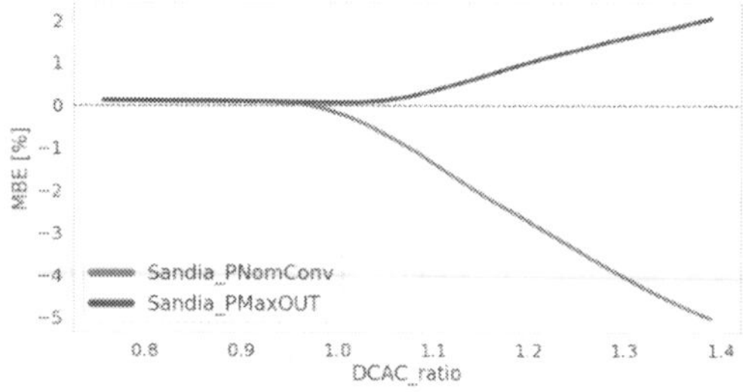

Figure 3: MBE value of comparing the outputs of the function inverter.sandia with the output EOutInv of PVSyst as a function of the DC/AC ratio.

As it is a more detailed model that adheres to the inverter curve, the results for the no-overload condition were even better than those obtained with the previous function. For both strategies in the calculation of equation (iii) the average bias remained below 0.1%.

A behavior similar to PVWatts was observed for conditions in which the inverter is overloaded. With more pronounced biases and overestimated estimates using "PMaxOUT" in the calculation of (iii) and underestimated using "PNomConv".

2.3 Preliminary assessment of the models

In both scenarios, the negative biases associated with "PNomConv" arise because the models tend to truncate the data points when the AC power of the inverter exceeds its nominal power capacity. Errors associated with the use of "PMaxOUT" arise from the models' inability to adjust to variations in ambient temperature when determining maximum power.

Operating at high temperatures can induce changes in the inverter's efficiency curve due to the increase in the internal temperature of its components. This phenomenon is commonly referred to as "temperature reduction". Power reduction is performed because prolonged operation under high temperature conditions can exacerbate the risk of equipment failure and lead to a shorter lifespan of the components involved.

The objective of this work is twofold. Firstly, it aims to model a commercial inverter using the pvlib library, using the data available in a ".OND" file. Secondly, it seeks to establish a calculation workflow to adjust the values at which the inverter suffers both temperature clipping and derating.

3 PROPOSED WORKFLOW

So far it has been possible to verify that the Sandia model is more accurate than the PVWatts model, however there are some difficulties associated with implementing temperature reduction in the Sandia model. Since, it would also be necessary to move the operating points on the I-V curve of the arrangement at the cutting moments, obtaining DC voltage values different from those at the MPPT point.

Given the complexity of the procedure described above, an alternative option was to replace the constant parameters of "eta_inv_nom" and "pdc0" of the PVWatts model with parameters calculated for each timestamp of the simulation.

Thus, to calculate the hourly values of "eta_inv_nom", an interpolation model was obtained using power and DC efficiency from the "ProfilPIOV1," "ProfilPIOV2," and "ProfilPIOV3" curves using the interpolation function interp1d from the numpy python library. With the model in hand, it was possible to interpolate the efficiency values for each two DC input power values.

The model certainly presents difficulties in calculating efficiency for low power values, however, for output powers greater than PNomConv, the interpolation presents an RMSE of 0.06% when compared to inverter efficiency "EffInvR" obtained from the PVsyst outputs.

To calculate the hourly values of the variable "pdc0", we started with the "PMaxOUT" parameter and then corrections were applied using the hourly ambient temperature values of the location, as well as the inverter's thermal derating curve. The power x temperature curve of the inverter was obtained with the multiple linearization of the following parameters from the ".OND" file ("PmaxOUT", "TPMax"), ("PNomConv", "TPNom"), ("PLim1", "TPLim1") and (0, "TPLimAbs").

$$if\ t_{amb,i} < TPMax$$
$$PMaxOUT_i = PMaxOUT \quad (v)$$

$$if\ TPMax \leq T_{amb,i} < TPNom$$
$$PMaxOUT_i = t_1 \times T_{amb,i} + t_2 \quad (vi)$$

$$if\ TPNom \leq T_{amb,i} < TPLim1$$
$$PMaxOUT_i = t_3 \times T_{amb,i} + t_4 \quad (vii)$$

$$if\ TPLim1 \leq T_{amb,i} < TPLimAbs$$
$$PMaxOUT_i = t_5 \times T_{amb,i} + t_6 \quad (viii)$$

$$if\ t_{amb,i} \geq TPLimAbs$$
$$PMaxOUT_i = 0 \quad (ix)$$

With t_1, t_2, t_3, t_4, t_5 and t_6 given by:

$$t_1 = \frac{(PMaxOUT - PNomConv)}{(TPMax - TPNom)} \quad (x)$$

$$t_2 = PMaxOUT - (TPMax \times t_1) \quad (xi)$$

$$t_3 = \frac{(PNomConv - PLim1)}{(TPNom - TPLim1)} \quad (xii)$$

$$t_4 = PNomConv - (TPNom \times t_3) \quad (xiii)$$

$$t_5 = \frac{(PLim1)}{(TPLim1 - TPLimAbs)} \quad (xiv)$$

$$t_6 = PLim1 - (TPLim1 \times t_5) \quad (xv)$$

Finally, we can recalculate the "pdc0" parameter by applying each of the hourly values in equation (ii).

10.4229/EUPVSEC2023/4DV.4.42
020470-003

3 RESULTS AND DISCUSSION

To evaluate the results, scatter plots were made for each of the inverter models, comparing them with the PVSyst EOutInv output. Since the objective of this work is to analyze limit conditions, in which the inverter suffers from clipping, a scenario with a DC/AC ratio equal to 1.389 was selected to carry out the analyses.

Given that the original scenarios PVWatts_PNomConv and Sandia_PNomConv do not perform well under these conditions, they were not included in this analysis. The PVWatts_PMaxOUT model will be called PVWatts with constant parameters, while the model proposed for PVWatts will be called PVWatts with variable parameters.

During the evaluations it was noted that the proposed PVWatts model with variable parameters presented difficulties in predicting the output power in the lower range. The investigations indicated that this problem is a result of the aforementioned difficulties that the interpolation model has in estimating the efficiency for these operating points. Thus, a fourth model was evaluated, merging the results obtained with the Sandia model with the results of the new model. Thus, for values lower than PNomConv, the results of the Sandia model were adopted and for higher values, the values of the new model were used. The combined model will be called Sandia - PVWatts in later analyses.

Figure 7 in appendix shows the difficulty that the sandia and pvwatts functions, with constant parameters, have in calculating the power points at which the temperature derating acts, as well as the problem of the efficiency curve obtained through interpolation in the calculations using low values of DC power. It is also clear that the combination of the two calculation methods is quite advantageous compared to the separate methods.

Figure 4, 5 and 6 also shows the biases obtained for the new model under varied overload conditions. The results indicate constancy in the MBE and RMSE for different DCAC ratio values, indicating reproducibility of the results.

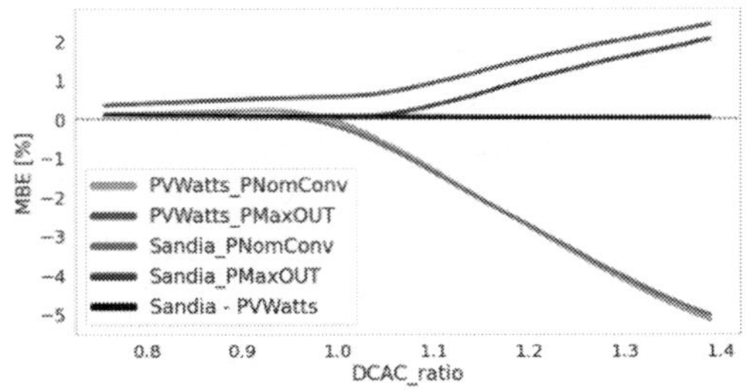

Figure 4: MBE value for comparing the outputs of all models with the PVSyst's EOutInv output as a function of the DC/AC ratio.

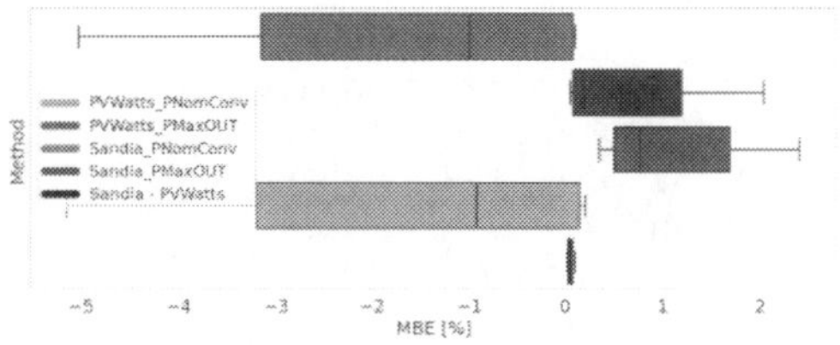

Figure 5: Distribution of MBE values comparing the outputs of all models with the PVSyst EOutInv output as a function of the DC/AC ratio variation from 0.75 to 1.5.

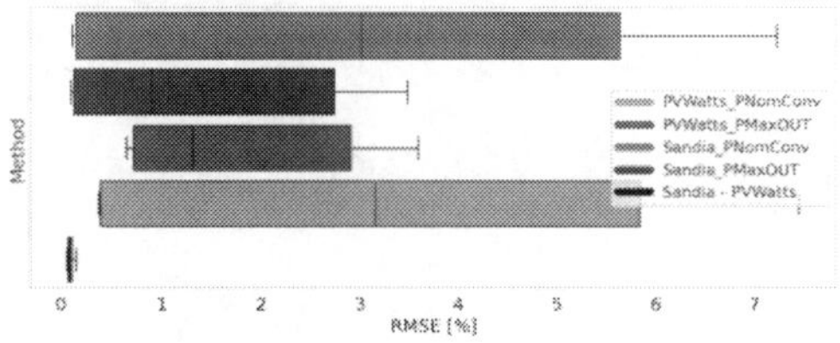

Figure 6: Distribution of RMSE values comparing the outputs of all models with the PVSyst EOutInv output as a function of the DC/AC ratio variation from 0.75 to 1.5.

4 CONCLUSIONS

The proposed methodology, which combines two pvlib functions, demonstrated greater consistency with the results obtained by PVsyst for different project configurations. The root mean square error (RMSE) and mean bias error (MBE) values were 0.029% and 0.0046%, respectively. Thus, it can be used as an alternative for production estimates under high overload conditions.

5 ADDITIONAL COMPONENTS

5.1 References

[1] A. Villoz, B. Wittmer, A. Mermoud, M. Oliosi, A. Bridel-Bertomeu, A Model Correcting the Effect of Sub-Hourly Irradiance Fluctuations on Overload Clipping Losses in Hourly Simulations, 8th WCPEC, Session 4EO.2.2, Pages 1151 - 1156, 2022, ISBN 3-936338-86-8, 10.4229/WCPEC-82022-4EO.2.2

[2] Bolinger, Mark, Seel, Joachim, Warner, Cody, and Robson, Dana. Utility-Scale Solar, 2021 Edition: Empirical Trends in Deployment, Technology, Cost, Performance, PPA Pricing, and Value in the United States [Slides]. United States: N. p., 2021. Web. doi:10.2172/1823604.

[3] William F. Holmgren, Clifford W. Hansen, and Mark A. Mikofski. "pvlib python: a python package for modeling solar energy systems." Journal of Open Source Software, 3(29), 884, (2018). https://doi.org/10.21105/joss.00884

[4] A. P. Dobos, "PVWatts Version 5 Manual," http://pvwatts.nrel.gov/downloads/pvwattsv5.pdf (2014).

[5] Dreisse, A.; Jain, P.; Harrison, S.; Beyond the Curves: Modeling the Electrical Efficiency of Photovoltaic

Inverters. Proc. IEEE Photovoltaic Specialists Conf., 2008.

[6] PVsyst SA. PVsyst Photovoltaic Software Version 7.3. http://www.pvsyst.com. 2023.

5.2 Appendix

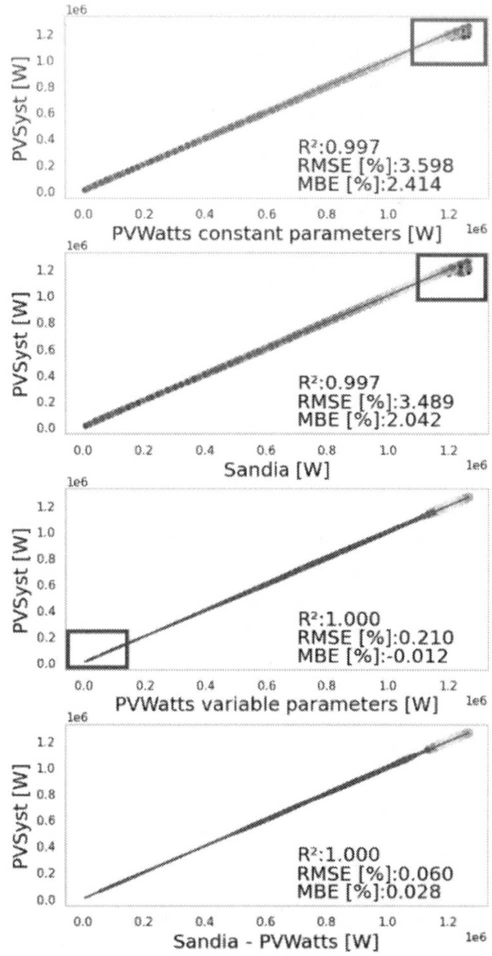

Figure 7: The scatter plot of each of the models in comparison with the variable EOutInv of the PVSyst for the condition of DC/AC ratio equal to 1.389.

ACCOUNTING FOR SUB-HOURLY IRRADIANCE FLUCTUATIONS IN HOURLY PERFORMANCE SIMULATIONS

Michele Oliosi, Bruno Wittmer, André Mermoud, Agnes Bridel-Bertomeu
PVsyst SA
Route de la Maison-Carrée 30 - 1242 Satigny – Switzerland, support@pvsyst.com

ABSTRACT: Hourly performance simulations tend to report higher yields than sub-hourly performance simulations. One reason is that the clipping losses due to sub-hourly irradiance fluctuations will be underestimated in hourly simulations. In a previous work [1], we developed a model to estimate these extra clipping losses; it can be used to remove most of the yearly discrepancy between minute-level and hourly results.
In this work we show how part of the remaining yearly discrepancies are an artefact from applying transposition models at the sub-hourly level. Taking as example the widely used Perez model, we propose a way to correct the diffuse decomposition coefficients in the minute simulation, which further reconciliates hourly- and minute-level simulations.
Keywords: Simulation, System Performance, Subhourly, Irradiance, Clipping

1 MOTIVATION

Hourly simulations are widely used to estimate the performance of PV systems at different stages of their development and operation. While simulations using shorter time scales are also possible and used in some instances, hourly simulations have the advantage of having a shorter simulation time, and a wide availability of data, since hourly weather data is easier to collect.

However, weather, and in particular irradiance, may also exhibit important sub-hourly fluctuations. The effects of these fluctuations are therefore not necessarily well captured by hourly simulations. As several studies (see references collected in [1]) have shown, using hourly data leads on average to overestimating yearly yields by up to a few percent. This mostly depends on the DC:AC ratio of the system, in addition to the details of the sub-hourly weather data.

Indeed, for higher DC:AC ratios, sub-hourly fluctuations of the weather may drive momentarily the DC output across the threshold for inverter clipping. When simulating in hourly time steps, these sudden drops or peaks in the DC production may be missed. This results in missing clipping losses in the hourly simulation. After accumulating the missed clipping losses over a full simulation, the yearly performance may then be overestimated by a few percent. This makes it one, if not the main cause for the discrepancy between hourly and sub-hourly simulations.

In a previous study [1], we have presented a model that estimates the missing clipping losses at each hour, based on 3 hourly parameters fitted from the input horizontal-plane minute global irradiance data. The advantage of this method is that the simulation may be then run in hourly steps, thus keeping a lower overall simulation time, while reasonably modeling the missed clipping losses.

After correcting the clipping losses, however, there remained a bias towards higher yields in hourly simulations. The remaining discrepancies are shown in Figure 1 for different DC:AC ratios, but usually remain below the percent level.

In this paper, after first explaining our methodology in Section 2, we show in Section 3 that the remaining discrepancy is partially attributable to applying certain transposition models to the minute data. This implies that by applying these same transposition models to minute data, the minute-level simulations, previously used as a reference, may be biased towards underestimating the yield of the system. We then investigate how the discrepancy between minute-level and hourly transposition is generated. We identify the non-linearity of certain transposition steps as the main cause for the averaging error.

Based on this observation we then propose in Section 4 a simple method to calculate new coefficients to be used within the widely used Perez model when applied to minute data. The method itself may be straightforwardly applied to wider datasets than the one used in this paper and may be also adapted to other coefficient-based models. We then discuss the results and conclude in Section 5.

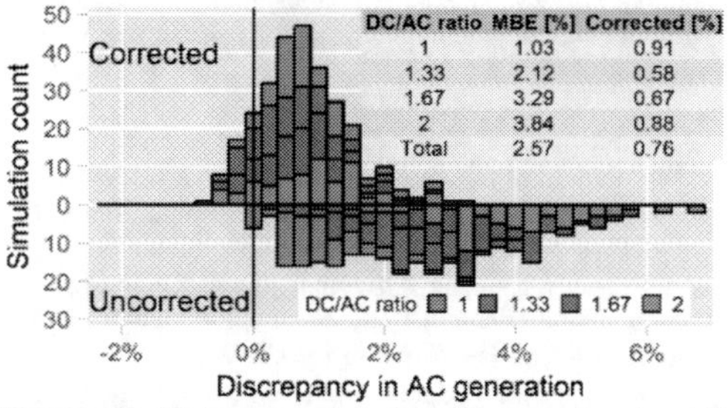

Figure 1: Histograms for the discrepancies in AC generation, across all simulated cases (4 DC:AC ratios, 18 orientations, 4 sites). Different DC:AC cases are highlighted in different colors. The transposition model is the Perez 1990 model [2]. Positive counts have had their clipping losses corrected with the model [1], negative counts are uncorrected.

2 METHODOLOGY

With the goal of achieving a better understanding of how the discrepancies are generated throughout the simulation process, and especially at the transposition step we reanalyze the same dataset that was studied in [1], and complement it with new results.

2.1 Input and system data

We use as input data the minute-level measurements of global horizontal (GHI) and diffuse horizontal (DHI) irradiances, from four stations in the U.S. listed by the NREL Measurement and Instrumentation Data Center (MIDC). These are the NREL SRRL Baseline Measurement System [3], in Golden, Colorado

10.4229/EUPVSEC2023/4DV.4.43
020471-001

(abbreviated NRELSRRL); the University of Oregon [4], in Eugene, Oregon (abbreviated UniOregon); the Natural Energy Laboratory of Hawaii Authority [5], in Kailua-Kona, Hawaii (abbreviated Hawaii); and the University of Louisiana at Lafayette [6], in Lafayette, Louisiana (abbreviated Lafayette). These four sites belong to four different Köppen climate classes: respectively, semi-arid continental (BSk), warm-summer Mediterranean (Csb), tropical semi-arid (BSh), and humid subtropical (Cfa). The year considered is 2020 for all results, except the determination of new Perez coefficients in Section 4, for which we have used 2018, 2019, 2020 for NRELSRRL and UniOregon and 2019 and 2020 for Hawaii and Lafayette.

As PV systems, we consider four different DC:AC ratios (1, 1.33, 1.67. 2) and the seventeen fixed orientations of [1], and an additional single-axis tracker orientation. The inverter nominal power is fixed at 9kVA, and the DC:AC ratio is varied by considering different number of PV module strings. When averaging over multiple scenarios or climates, we have given the same weight to all cases.

2.2 Simulation procedure

For both minute-level and hourly simulations, we use PVsyst version 7.4.0. The hourly simulation is handled natively by the software, but the minute-level simulation is performed using the procedure described in [1]. This procedure works on the premise of combining 60 hourly simulations to mimic a single minute-level simulation. For the minute simulations, the array temperature is assumed to be stable during each hour. To obtain the value for the array temperature for the minute simulations, we use the hourly simulation results. Finally, most, albeit not all, output variables are available in the release version of the software.

2.3 Transposition models

The default transposition model used in PVsyst is the Perez 1990 model [2], which we abbreviate here "Perez model" for simplicity. The base for this model is the decomposition of the horizontal irradiance into different components, denoting different sky sectors: direct, circumsolar, and isotropic diffuse. In addition to these, the horizon band component, which is not measurable in the horizontal plane, is estimated from the horizontal irradiances.

The decomposition and estimation of the components on the horizontal (or vertical) plane follows simple equations,

$$\begin{aligned} \text{direct} &= \text{GHI} - \text{DHI}\,, \\ \text{isotropic diffuse} &= (1 - F_1)\,\text{DHI}\,, \\ \text{circumsolar} &= F_1\,\text{DHI}\,, \\ \text{horizon band} &= F_2\,\text{DHI}\,, \end{aligned}$$

where F_1 and F_2 are determined from

$$\begin{aligned} F_1 &= F_{11}(\epsilon) + F_{12}(\epsilon)\,\Delta + F_{13}(\epsilon)\,Z\,, \\ F_2 &= F_{21}(\epsilon) + F_{22}(\epsilon)\,\Delta + F_{23}(\epsilon)\,Z\,, \end{aligned}$$

where Δ denotes the sky brightness, Z the sun zenith angle, ϵ the sky clearness, and where the 6 factors $F_{ij}(\epsilon)$ have values over 8 clearness bins, in total 48 numeric coefficients. In the 1990 publication [2], these coefficients have been fitted with hourly and 15-minute measurements. The definitions for sky brightness and sky clearness are to be found in the same article.

The Hay-Davies model [7] (which we abbreviate "Hay model" in certain parts of the text) is simpler, in that it does not consider a horizon brightening band. Starting from the direct normal irradiance DNI and the extraterrestrial irradiance G_{ext} it defines an anisotropy factor $A_i = \text{DNI}/G_{ext}$, which is used to separate the horizontal diffuse irradiance into the isotropic and circumsolar parts,

$$\begin{aligned} \text{direct} &= \text{GHI} - \text{DHI}\,, \\ \text{isotropic diffuse} &= (1 - A_i)\,\text{DHI}\,, \\ \text{circumsolar} &= A_i\,\text{DHI}\,. \end{aligned}$$

We use the Hay-Davies model in section 3.3 to illustrate the impact of the transposition model on the discrepancy between minute-level and hourly data.

In both the Perez and Hay-Davies, the multiplicative transposition factors used to go from horizontal and vertical-plane irradiances to the plane of array are the same, i.e.,

$$\begin{aligned} \text{direct}: &\ \frac{\max(0, \cos\alpha)}{\cos Z} \\ \text{circumsolar}: &\ \frac{\max(0, \cos\alpha)}{\max(\cos 85°, \cos Z)} \\ \text{isotropic diffuse}: &\ \frac{1 + \cos\theta}{2} \\ \text{horizon band}: &\ \sin\theta \end{aligned}$$

where α is the incidence angle between the normal to the plane of array and the sun direction, Z the zenith angle, and θ the tilt of the plane of array.

Both models are implemented in PVsyst making it possible to run PV performance simulations with either of these.

3 TRANSPOSITION DISCREPANCIES

3.1 Propagation of the discrepancy

The discrepancy between hourly and minute-level simulations can be evaluated at different stages of the simulation process. The good correlation between the discrepancies at the transposition, DC energy, and AC energy evaluation stages for all simulation cases, can be considered as strong evidence for the importance of the transposition stage in the generation of the discrepancies. We summarize the results at the different stages in Figure 2, where we illustrate by a linear fit the correlation of the discrepancies. A better fit is achieved between the transposition stage and the DC energy evaluation stage. Denoting as $b_{E_{AC}}$, $b_{E_{DC}}$, b_{GlobInc} the yearly relative discrepancy in AC energy, DC energy, and transposed irradiance respectively, we find the following fits:

$$\begin{aligned} b_{E_{DC}} &= 1.01\, b_{\text{GlobInc}}, \quad \sigma = 0.07\%, \quad R^2 = 0.99, \\ b_{E_{AC}} &= 0.77\, b_{\text{GlobInc}}, \quad \sigma = 0.47\%, \quad R^2 = 0.57, \end{aligned}$$

The latter fit, i.e. between the discrepancy at the AC stage and at the transposition stage, shows that despite the clipping correction model, a certain discrepancy is still generated by the DC to AC conversion model. Note that these fits are specific to the set of simulations chosen, and may be different for a different set of simulations. We therefore interpret them qualitatively, without investigating the detailed values.

The average discrepancy across the simulations at the 3 stages is reported in Table 1 As we will show in Sections 3.2 and 4, it is possible to reduce the AC power discrepancy to about 0.5% on average, either by choosing another transposition scheme for both minute-level and hourly simulation, or by adapting the transposition model when it is applied to the minute-level data.

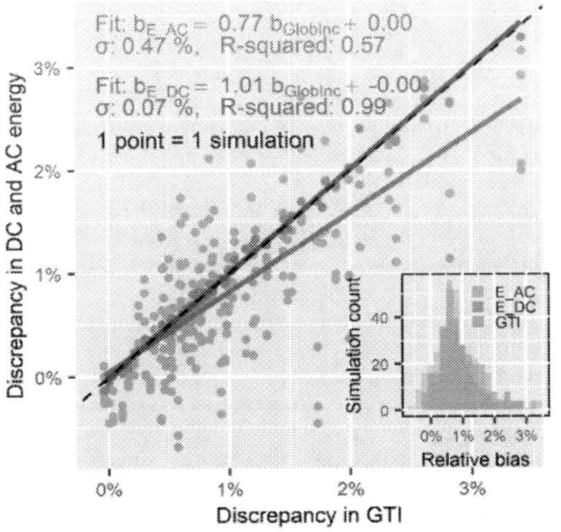

Figure 2: Correlation between the discrepancy in the global transposed irradiance (GTI), in blue in the histogram, and the discrepancies in the DC energy (green) and AC energy (red). Each point represents one simulation. Linear fits have been added to each series of points, they reflect the propagation of the discrepancy throughout the simulation.

Table 1: Average discrepancy across simulations for three different simulation stages.

Stage	Mean discrepancy	Std dev.
GTI	0.96 %	0.69 %
DC energy	0.95 %	0.71 %
AC energy	0.78 %	0.71 %

3.2 Transposition model comparison

Since the transposition of the irradiance components already leads to differences between the minute-level and hourly simulation, it is natural that the choice of the transposition model will have an impact on the discrepancy. To illustrate this, we run the same set of simulations but using the Hay transposition model instead of the Perez transposition model. The Hay transposition model is chosen for this comparison because it is simpler than the Perez model, leading to a smaller chance of accumulating averaging errors.

The comparison of the yearly results from the two transposition models is shown in Figure 3. The mean values and overall behavior of the histogram show that the Hay model in general will have less discrepancies.

In Figure 4, we show the discrepancies by site, and discriminating between irradiance components, in addition to showing the GTI values only. Indeed, the Hay transposition handles the direct component in the same way as the Perez transposition. Differences therefore show up rather in the handling of the diffuse.

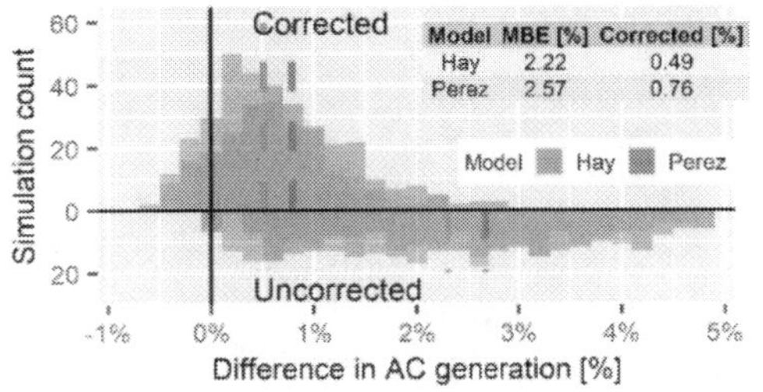

Figure 3: Histograms for the discrepancies in AC generation, across all simulated cases. We use two transposition models, the Perez 1990 model [2] in blue, and the Hay-Davies mode [7] in red. Positive counts have had their clipping losses corrected with the model [1], negative counts are uncorrected. See Figure 1 for the Perez data with a breakdown of the DC:AC cases.

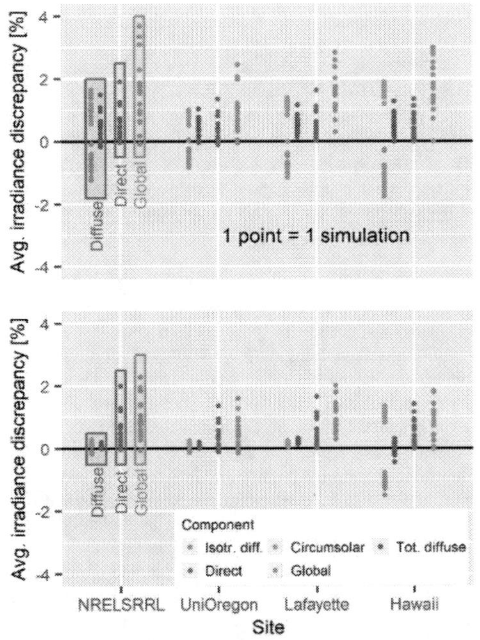

Figure 4: Site by site breakdown of the average transposed irradiance discrepancies, by simulation. We separate the different irradiance components: the global in grey, is composed by the sum of direct (orange) and diffuse irradiances (green). The diffuse can be further reduced to the sum of circumsolar (yellow) and isotropic diffuse (blue), which includes here the horizon band component.

It is however interesting to note that for both the Hay and Perez model, the transposition of the direct component generates a sizeable part of the discrepancy, however less so at common orientations such as 30°-South. This is illustrated further in Figure 5, in which we show the discrepancies in the transposed direct component as a function of the azimuth and tilt. We interpret the direct component transposition procedure as an instantaneous procedure, because it entirely depends on geometric variables such as the sun position. It is therefore preferentially applicable to shorter time scales, for which these variables vary less within a time step. Therefore, the transposition of the direct component should be corrected in the hourly simulations.

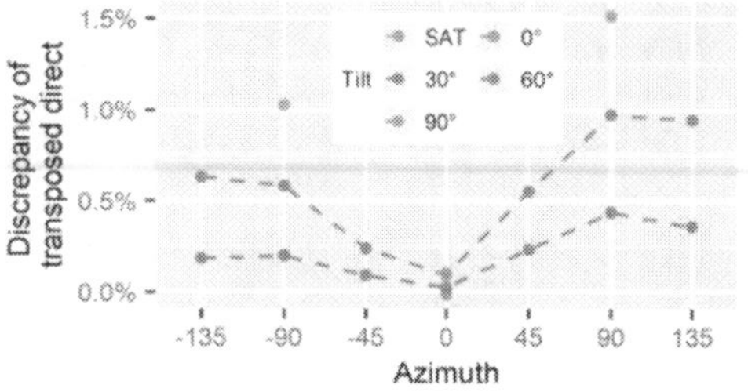

Figure 5: Discrepancies generated at the transposition of the direct irradiance component, as a function of the azimuth, and colored by tilt. The case of a single North-South-axis tracker (SAT) is shown in red. Typical south-facing orientations are the least affected by this type of transposition bias. In this case, we interpret the minute values as the reference, i.e. the hourly values need to be corrected.

Regarding the diffuse components, the sum of isotropic diffuse and circumsolar components produces a negligible discrepancy in the Hay model. These two components themselves have negligible discrepancies for all sites except Hawaii. Contrary to the Hay model, the Perez transposition, has a sizeable discrepancy in the diffuse components roughly of the same magnitude as the direct component.

It is instructive to visualize the differences between components when averaging discrepancies separately over each hour of the day. We show this in Figure 7, by choosing a specific site, year, and selecting the 30°-South orientation. By doing so, one can visually identify the difference between direct and diffuse irradiances. The direct irradiance generates a pronounced discrepancy around sunrise and sunset exclusively. The reason for this is still unclear to us, but this may be related either to the handling of the sun crossing the horizon line, or to the fact that at these hours the incidence angle and transposition factor change more drastically than in the middle of the day, for a typical equator-facing orientation.

In the Perez model, the discrepancies for the horizontal decomposed diffuse components are already quite marked. However, due to how the decomposition of circumsolar and isotropic diffuse is defined, the respective discrepancies compensate each other. However, because the geometrical transposition factor given in Section 2.3 is different for each component, once these have been transposed the isotropic diffuse and circumsolar discrepancies do not compensate each other any more on the plane of array. The horizon band generates a discrepancy already present in the vertical plane. It is not compensated by other factors therefore it remains present once transposed to the plane of array. The Hay model, with example profiles shown in Figure 8, generates negligible discrepancies for the diffuse components in the horizontal plane. Once the transposition factors are applied, the discrepancies remain small.

To summarize, the discrepancy in the diffuse components appears, already at the level of the decomposition into horizontal (and vertical) contributions. This is dependent on the specific transposition model and is more prevalent in the Perez model with respect to the Hay model. The transposition to the plane of array simply exacerbates their impact. The discrepancy of the direct component appears when transposing in the plane of array in both models. This suggest that it depends mainly on the orientation, and does not depend on the transposition model. This further suggests that the transposed direct irradiances are more accurate when considering minute-level data.

4 PEREZ-MINUTE COEFFICIENTS

4.1 Correction philosophy

As seen in the previous section, the transposed diffuse irradiance averaging errors originate mainly from the diffuse decomposition into horizontal and vertical components. The latter non-linear step is at the core of the Perez model. While the detailed mechanism of the generation of a bias via the individual averaging errors is still elusive, it seems inevitable that differences will arise between the minute-level and hourly data transpositions, because of the non-linearities involved.

The model [2] has originally been built using hourly and 15-minute irradiance measurements, with the aim of reproducing the observed plane of array irradiance values. Therefore, when comparing the two sets (minute-level and hourly transposed data) it seems justified to use the hourly transposed values as a reference, since this is the experimentally verified case. This raises the question of whether a correction of the transposition process for minutes could be achieved, with the goal of bringing the hourly and minute results closer to each other, and especially to remove the bias component of the discrepancy.

In what follows we suggest a way to correct the minute transposition via a new set of coefficients $F_{ij}(\epsilon)$, computed using the hourly transposed values as a reference. Schematically, the procedure is depicted in Figure 6. This is of course one among many possible correction methodologies, but it has the advantage of not modifying the model fundamentally, but rather to adjust a few numeric coefficients while retaining the qualitative aspects of the model. We further note that the fitting of coefficients in the Perez model is a common practice suggested in [2], for example to determine coefficients specific to a given climate or region (see e.g. [8] or [9]).

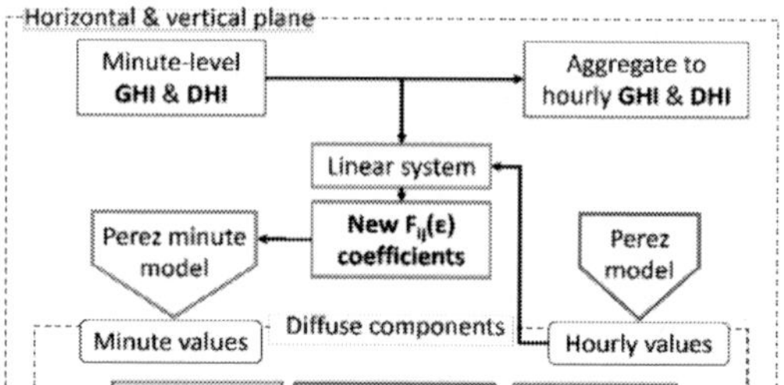

Figure 6: Scheme of the generation method for new Perez coefficients, to be used with minute data and fitted to minimize the mean square differences with the hourly diffuse decomposition determined with the original coefficients.

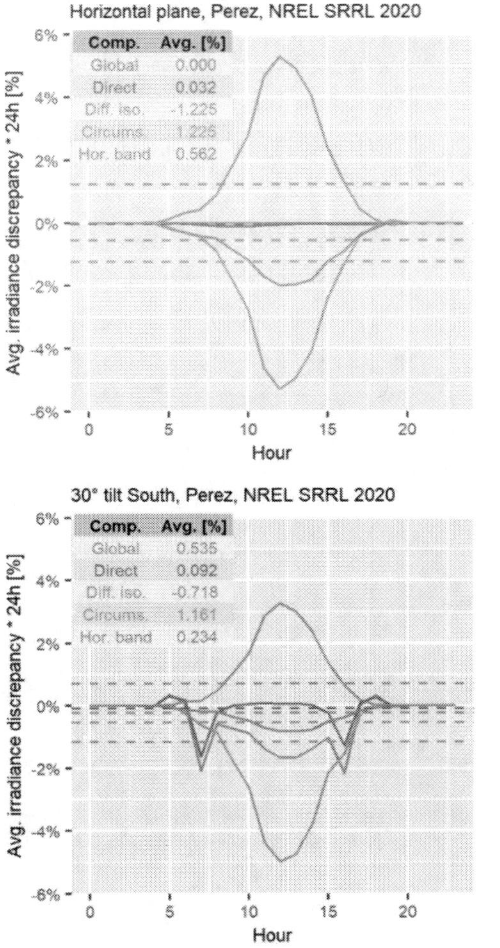

Figure 7: Daily profile of the discrepancies for different irradiance components with the Perez transposition. An average for the whole simulation by hour of the day has been taken. Once transposed, components that compensated each other in the horizontal plane now generate a bias.

4.2 Perez coefficient fitting

For each hour, the difference between the minute level and the hourly estimate of the horizontal circumsolar irradiance is given by $\langle F_1\mathrm{DHI}\rangle_H - F_{1H}\,\langle\mathrm{DHI}\rangle_H$, where H denotes a given hour in the dataset, and $\langle\ \rangle_H$ is an average over that hour. An equivalent form can be written for F_2 for the vertical-plane horizon band irradiance component. We choose as best fit the one that will minimize the mean-square of these differences. It is possible to add weight filters to the individual hours. We choose to fit over data for which the $\mathrm{DHI} > 10\ \mathrm{W/m^2}$, to eliminate most night data, and choose an equal weight for all remaining hours. We accumulate the mean-square of the differences using data described in Section 2, including several years.

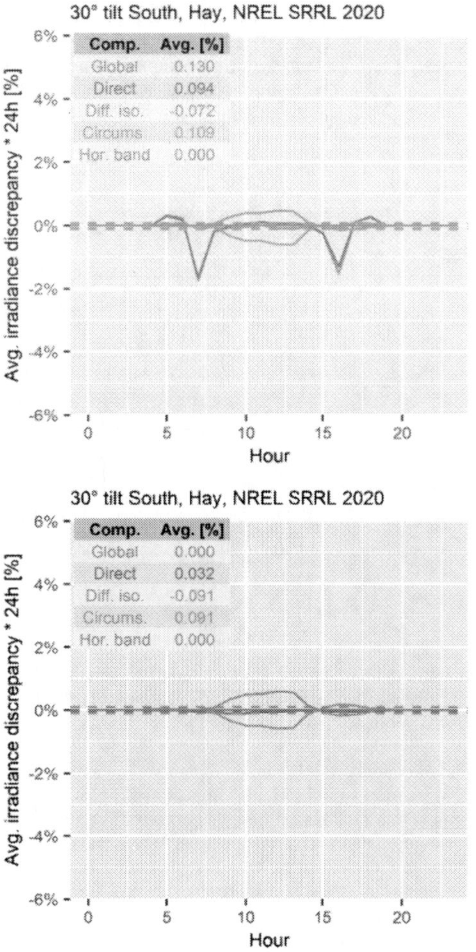

Figure 8: The same as Figure 7, but with the Hay model. Discrepancies are much less prevalent at the decomposition stage than with the Perez model. After transpositions, the discrepancies stay small.

The fitting procedure amounts to solving a pair of linear systems of equations,

$$\frac{\partial}{\partial F_{1jk}} \sum_{\text{sites, H}} \left(\langle F_1\mathrm{DHI}\rangle_H - F_{1H}\,\langle\mathrm{DHI}\rangle_H\right)^2 = 0\ ,$$
$$\frac{\partial}{\partial F_{2jk}} \sum_{\text{sites, H}} \left(\langle F_2\mathrm{DHI}\rangle_H - F_{2H}\,\langle\mathrm{DHI}\rangle_H\right)^2 = 0\ ,$$

for all $j = 1{,}2{,}3$ and $k = 1, \dots, 8$ the index denoting the clearness ϵ bin, between values $\epsilon = 1$, 1.065, 1.23, 1.5, 1.95, 2.8, 4.5, 6.2, and ∞. Indeed, formally it is possible to write

$$F_i = \sum_k \Pi_k(\epsilon)(F_{i1k} + F_{i2k}\Delta + F_{i3k}Z)\ ,$$

where Π_k is the Heaviside Pi function for bin k. Each of the 48 F_{ijk} will be subject to the fitting procedure, leading to a new set of 48 F_{ijk} deviating from the original set in [2].

The result of the fitting procedure is summarized in Figure 9, and numerical coefficients are reported in Appendix A. Of course, applying this procedure to a wider minute-level dataset can provide more statistical significance to the fits. It is notable that the fitting procedure does not modify extensively the parameters, excepting central clearness bins for F_{12} and F_{22}.

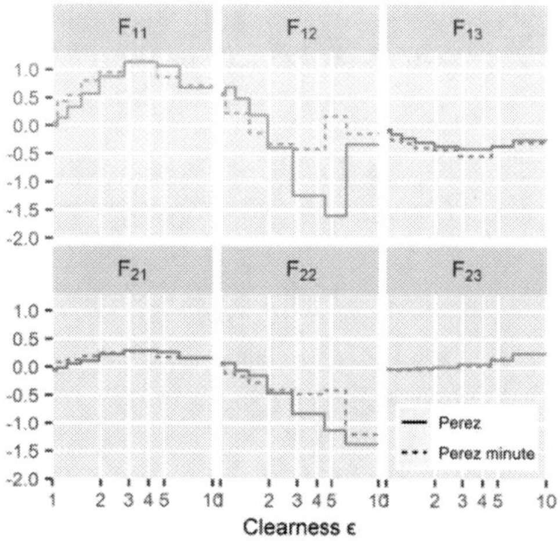

Figure 9: New Perez-minute coefficients, as obtained from the procedure in Figure 6, compared to the original 1990 publication ones [2], used for the hourly simulation.

4.3 Procedure validation

In order to illustrate the impact of the new coefficient on the minute transposition, we reproduce Figures 7 and 8, i.e. aggregate the results for a specific site and one specific year, for each hour of the day, this time however after using the new Perez-minute coefficients for the minute level calculations. The result is depicted in Figure 10. The new transposition coefficients help achieving a much better agreement with the hourly Perez transposition results. Since the discrepancy is reduced already for horizontal irradiances and for the vertical-plane horizontal band irradiance, after applying the usual transposition factors, the discrepancy remains small of the order of 0.1% for the global irradiance.

Note that because of the mildly tilted and south-facing orientation, the impact of the direct irradiance transposition factor is small. For different orientations, the transposition factors may have a bigger impact. Overall, the discrepancies are expected to be comparable with the more linear Hay-Davies model, as in Figure 8.

6 DISCUSSION AND OUTLOOK

6.1 Summary

In this work, we have first complemented and reanalyzed the same data as in [1], i.e. minute-level and hourly simulation results based on four sites with different climates, a wide range of orientations, and four DC:AC ratios. Despite having corrected the hourly simulations for the missing clipping losses due to sub-hourly fluctuations, the minute-level and hourly simulated AC energies still differ by on average 0.8% more for the hourly results.

In order to find the cause for this remaining bias, we have studied the propagation of the discrepancy at different stages of the simulation. We have found that the discrepancy is already largely present after the transposition of the irradiances, and that ulterior simulation steps do not add much to the average discrepancies.

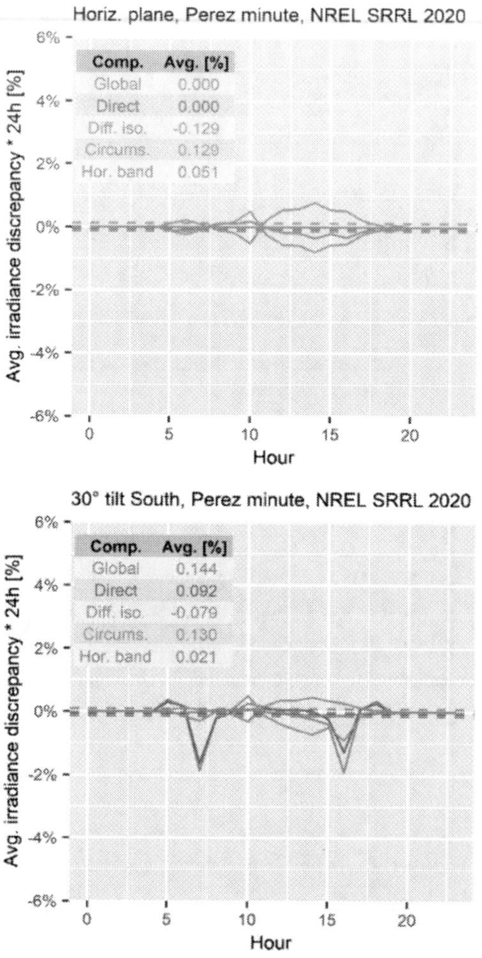

Figure 10: Daily profile of the discrepancies for different irradiance components with the new Perez minute coefficients used for the minute-level transposition. See Figure 7 for comparison with the Perez original coefficients.

By comparing the Perez and Hay transpositions, as well as decomposing the discrepancies into different irradiance components, we have recognized that two mechanisms cause significant discrepancies at the transposition stage: the geometric transposition of the direct component, and the horizontal and vertical-plane decomposition of the diffuse components, the latter effect being only prevalent in the Perez transposition model, due to the non-linear nature of the decomposition.

In order to address the issue of the discrepancy generated by the separation of the diffuse components in the Perez model, we have proposed a method that allows to obtain a new set of 48 coefficients, based on minute-level horizontal global and diffuse irradiance data. These new coefficients allow to use the Perez model on minute-level data, and to recover transposed diffuse irradiances close to the respective hourly values. We applied the method on a limited set of data, with the intent of providing a proof of concept. The method should however be applied over larger datasets (including several climates, sites, and years) to gain in statistical significance.

6.2 Outlook

Based on this study, we recognize two directions in which this analysis can be improved, which will impact both minute-level and hourly simulations. First, it is important to address the discrepancies generated by the transposition of the direct irradiance. While for more typical south-facing orientations this effect is not so prevalent, this is critical for other common orientations, e.g. for vertical East-West bifacial arrangements. We believe that this should be addressed in the hourly simulations, since the direct transposition is by nature best applied over shorter time scales. Second, after correcting as much as possible for transposition effects, it should be possible to reanalyze the discrepancies related to clipping losses. As we have suggested in [1], there exists a wide range of different correction models for hourly simulations based on coefficients extracted form the minute data. The one studied in [1] is simple and efficient, but it could be improved.

In addition to these two directions, the exact mechanism behind the generation of a bias component from the Perez diffuse decomposition remains to be elucidated. Indeed, this could come among others from the non-linearity of the clearness determination, which depends for example on the zenith angle to the third power, from the binning in clearness, which could lead to jumps, or simply in the overall shape of the decomposition coefficients F_{ij} as functions of the clearness values.

7 REFERENCES

[1] A. Villoz, B. Wittmer, A. Mermoud, M. Oliosi and A. Bridel-Bertomeu, "A Model Correcting the Effect of Sub-Hourly Irradiance Fluctuations on Overload Clipping Losses in Hourly Simulations," *8th World Conference on Photovoltaic Energy Conversion; 1151-1156,* 2022.

[2] R. Perez, P. Ineichen, R. Seals, J. Michalsky and R. Stewart, "Modeling daylight availability and irradiance components from direct and global irradiance," *Solar Energy,* vol. 44, p. 271–289, 1990.

[3] T. Stoffel and A. Andreas, *NREL Solar Radiation Research Laboratory (SRRL): Baseline Measurement System (BMS); Golden, Colorado (Data),* Not Available, 1981.

[4] F. Vignola and A. Andreas, *University of Oregon: GPS-based Precipitable Water Vapor (PWV),* NREL-DATA (National Renewable Energy Laboratory - Data (NREL-DATA), Golden, CO (United States)), 2013.

[5] K. Olson and A. Andreas, *Natural Energy Laboratory of Hawaii Authority (NELHA): Hawaii Ocean Science & Technology Park; Kailua-Kona, Hawaii,* NREL-DATA (National Renewable Energy Laboratory - Data (NREL-DATA), Golden, CO (United States)), 2012.

[6] NREL, *University of Louisiana at Lafayette,* NREL-DATA (National Renewable Energy Laboratory - Data (NREL-DATA), Golden, CO (United States)).

[7] J. E. Hay and J. A. Davies, "Calculations of the solar radiation incident on an inclined surface," *Proc. of First Canadian Solar Radiation Data Workshop, 59. Ministry of Supply and Services, Canada.,* 1980.

[8] D. Yang, Z. Ye, A. M. Nobre, H. Du, W. M. Walsh, L. I. Lim and T. Reindl, "Bidirectional irradiance transposition based on the Perez model," *Solar Energy,* vol. 110, p. 768–780, December 2014.

[9] M. Meshkinkiya, R. C. G. M. Loonen, R. Paolini and J. L. M. Hensen, "Calibrating Perez Model Coefficients Using Subset Simulation," *IOP Conference Series: Materials Science and Engineering,* vol. 556, p. 012017, July 2019.

[10] C. W. Hansen, J. S. Stein and D. Riley, "Effect of time scale on analysis of PV system performance," *SANDIA Report,* 2012.

A NUMERICAL PEREZ-MINUTE COEFFICIENTS

We report in Table 2 the coefficients obtained by minimizing the mean-squared error between hourly horizontal and vertical-plane diffuse components (reference) and the respective minute-level values. The dataset used is described in Section 2.

Table 2: Legend

Clearness bin	F_{11}	F_{12}	F_{13}	F_{21}	F_{22}	F_{23}
[1-1.065)	0.0489	0.5429	-0.1035	-0.0356	0.0466	-0.0353
[1.065-1.23)	0.4339	0.2185	-0.2529	0.0814	-0.1142	-0.0462
[1.23-1.5)	0.5423	0.2124	-0.3100	0.1236	-0.1676	-0.0424
[1.5-1.95)	0.8067	-0.1334	-0.3941	0.1894	-0.2816	-0.0332
[1.95-2.8)	0.9534	-0.3256	-0.4268	0.2405	-0.4068	-0.0095
[2.8-4.5)	1.1437	-0.4193	-0.5341	0.2747	-0.4772	0.0262
[4.5-6.2)	0.8618	0.1698	-0.3524	0.1706	-0.4145	0.1544
[6.2-∞)	0.7136	-0.1367	-0.2966	0.1579	-1.1983	0.2392

10.4229/EUPVSEC2023/4DV.4.43
020471-007

EXPERIENCE FEEDBACK ON PRECONSTRUCTION ENERGY YIELD ASSESSMENTS: CASE STUDY ON 125 PV PLANTS (1.3 GWP)

Marion Jude
Eoltech
8 rue du colonel Driant 31400 Toulouse, France, Tel. +33 5 61 15 44 13, marion.jude@eoltech.fr

ABSTRACT: The present project is a feedback study on energy yield assessments carried out at the preconstruction stage of photovoltaic plants. It consists in comparing the P50 value issued from these assessments (theoretical production capacity), to the actual P50 value of the plants estimated from their production data. Production data recorded at the utility meter of each plant were analyzed over the entire historical period available and a long-term prediction was carried out based on a specific solar energy index. This process enabled to estimate the actual production capacities of the plants, which have been compared to the theoretical yield assessments. This feedback study also provided additional results regarding statistics on actual availability rates as well as performance degradation factors, as numerous plants offer significant historical data. The study sample is comprised of 125 operating PV plants with a capacity of over 1MWp each, representing a total power >1.3 GWp.
Key words: energy yield assessment, feedback study, actual P50, availability rates, performance degradation

1 INTRODUCTION

The objective of the project is to get a feedback study on energy yield assessments performed during the development phase of PV projects. To this end, theoretical P50 values issued from the pre-construction yield assessments were compared to the actual long-term production capacities of the plants estimated from the analysis of their production data (actual P50 values resulting from a post construction yield assessment).

The deviation between both values (ΔP50) allows to quantify the gap between the expected production capacity (generally considered for the project financing) and the actual production capacity of the plant (effective value of the asset).

This deviation can be linked to the following causes:

1-The uncertainty on the preconstruction yield assessment (irradiation sources, losses assumptions, simulation settings, etc.)

2-The actual performance of the plant

3-Actual availability rates of the plant versus theoretical assumptions

4-The variation of the long-term solar resource.

The gap between the actual and the theoretical availability rates was accounted for in the frame of the study, and the variation of the long-term resource is rather low. Hence, the remaining deviation is linked to any or both of the two first causes.

The objective of the study is to assess the deviations on a significant sample of plants in order to improve the process of pre-construction energy yield assessments and the estimation of the associated uncertainty level.

2 METHODOLOGY

The theoretical P50 values are issued from the energy yield assessments performed at the development stage of the project by an independent consultant or by the developer itself. The corresponding reports were reviewed in the frame of this study.

The actual P50 values were estimated by Eoltech from the analysis of 10 min power data recorded at the utility meter of each plant (post-construction yield assessment). To do so, a solar energy index was set up specifically for each plant, taking into account the characteristics of the plant, the irradiance, the temperature and the seasonality of main losses. This index enabled to normalize the production measured at the meter of each plant and thus estimate a long-term production capacity from operational data at different ages of the plant. Availability issues were identified through different consistency analyses (coherence production vs resource mainly), and associated production losses were quantified, leading to an estimation of actual availability rates experienced by the plants.

The main purpose of the analysis was to estimate the deviation between the P50 values corresponding to the beginning of operation (i.e., equivalent to year 1), the performance degradation over time being analyzed separately. The comparison was made on the one hand taking into account the net P50 (i.e., including the actual average availability rates experienced by the plant over the historical period considered), and on the other hand under similar availability assumptions (i.e., P50 values adjusted to 100% availability),

For each plant, the deviations ΔP50 (net values) and $\Delta P50_{100\%\ avail.}$ (adjusted to 100% availability) were established as follows:

$$\Delta P50 = \frac{Theoretical\ P50}{Actual\ P50} - 1$$

$$\Delta P50_{100\%\ avail.} = \frac{Theoretical\ P50_{100\%\ avail.}}{Actual\ P50_{100\%\ avail.}} - 1$$

Thus, a positive ΔP50 corresponds to an overestimated yield at the preconstruction stage (Theoretical P50 > Actual P50), whereas a negative value indicates an underestimated yield (Theoretical P50 < Actual P50).

Deviations were analyzed considering the entire sample (144 cases, 125 plants), but also according to various criteria (age of the assessment, type of plant, solar irradiation source taken into account, etc.), leading to conclusions aiming at improving preconstruction yield assessment processes.

This analysis also enabled to get actual availability rates observed on a large sample of operating plants as well as actual performance degradation factors, as numerous plants offer significant historical data.

10.4229/EUPVSEC2023/4DV.4.47
020473-001

3 SAMPLE

3.1 Global sample

PV plants operating for more than 18 months and corresponding to an installed power >1MWp each were considered for the survey. The main required elements were the following:

- Characteristics of the plants
- Production data recorded at the utility meter (10 min time step)
- Pre-construction energy yield assessments (full report and/or main results).

15 owners/operators have participated in this study, providing preconstruction assessments and operational data of 125 PV plants representing a total installed power >1.3GWp. Most participants are mentioned in Table 1:

Table 1: Participants in the survey (agreement to release the company name provided)

Owners/Operators participating in the survey		
Engie Green	Total Energies	Neoen
Allianz Capital Partners	Valeco	CVE
BayWa r.e	Sonnedix	Valorem
VSB energies nouvelles	Mirova	CNR
GEG	Solvent	

For several plants, two distinct pre-construction assessments were provided, leading to 144 assessments available for the comparison pre-construction versus post construction. These assessments were performed either by independent consultancies (70% of the cases, 20 distinct consultancies involved) or by the developer itself (30% of the cases).

Table 2: Global sample

	Global sample
Owners/Operators	15
PV Plants	125
Total installed power	> 1.3 GWp
Pre-construction yield assessments	144

3.2 Geographical distribution

Except for 4 plants located in Germany, Spain and Portugal, all plants are located in France with a regional distribution representative of the total power capacity installed in the country:

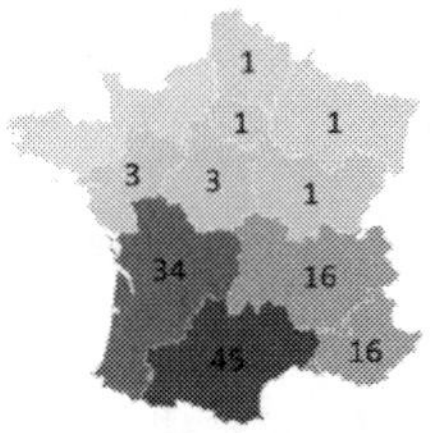

Figure 1: Regional distribution of the plants (France)

3.2 Characteristics of the plants

The following figure presents the plants distribution depending on their characteristics: installed power, age, type of installation (ground-mounted plants with a fixed-tilt, shelters, rooftops or trackers), type of cells, DC/AC ratio, number of inverters/MWp):

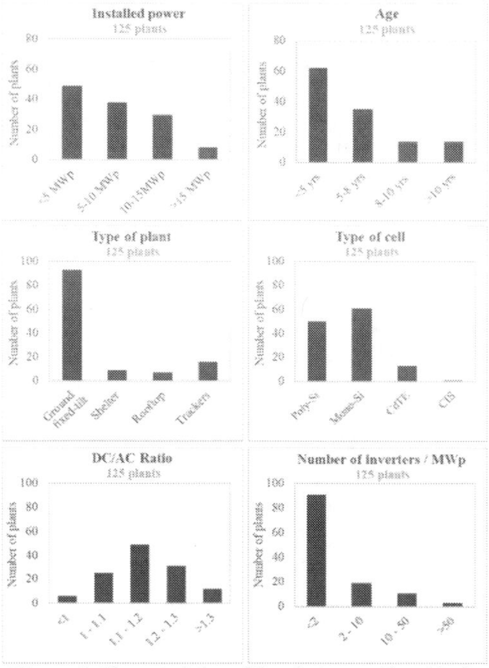

Figure 2: Characteristics of the plants considered in the survey

Table 3 indicates that the plants concerned by two distinct pre-construction assessments corresponds to ground-based plants with a fixed-tilt.

Table 3: Sample by type of plant

Number of	Plants	Assessments
All	125	144
Ground based fixed-tilt	93	112
Shelters	9	9
Rooftops	7	7
Trackers	16	16

4 MAIN RESULTS

4.1 Global deviation on the P50

The distribution of the deviations between the Theoretical P50 and the Actual P50 for all cases including availability losses (i.e., net P50 values) is displayed in Figure 3.

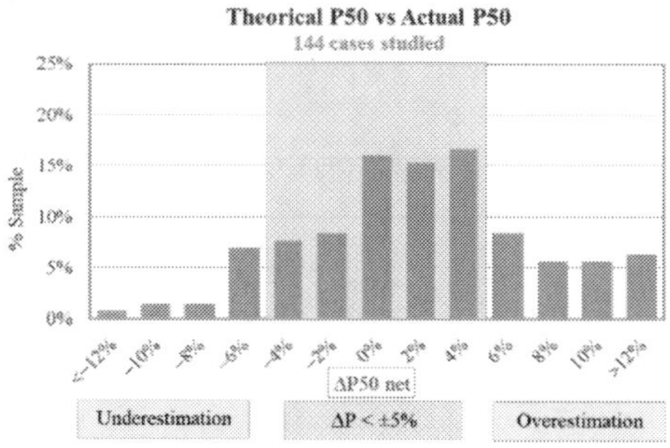

Figure 3: Global distribution of the deviations ΔP50 net

10.4229/EUPVSEC2023/4DV.4.47
020473-002

Figure 4 presents the distribution of the deviations considering the P50 estimates adjusted to 100% availability (i.e., under similar availability assumptions):

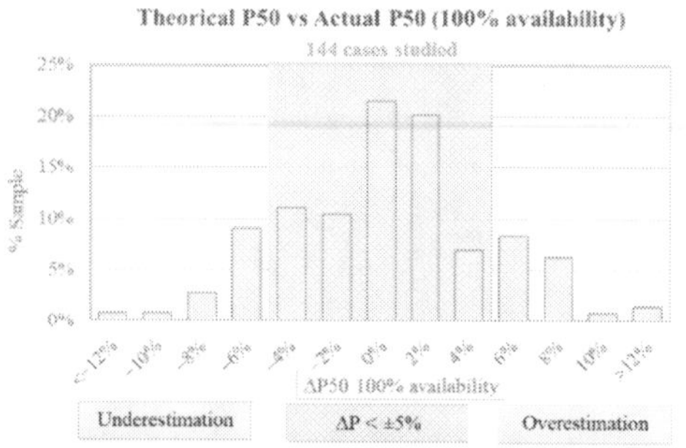

Figure 4: Global distribution of the deviations ΔP50 adjusted to 100% availability

The median values of the deviations obtained in both cases as well as the first quartile (Q1), third quartile (Q3), first decile (D1) and ninth decile (D9) are displayed in Table 4 and in Figure 5

Table 4: Distribution of the deviations

ΔP50	Net	100% avail.
Min	-17.3%	-17.5%
D1	-5.0%	-5.5%
Q1	-1.4%	-2.8%
Median	**+2.1%**	**+0.3%**
Q3	+5.0%	+2.8%
D9	+9.5%	+6.4%
Max	+31.9%	+20.0%

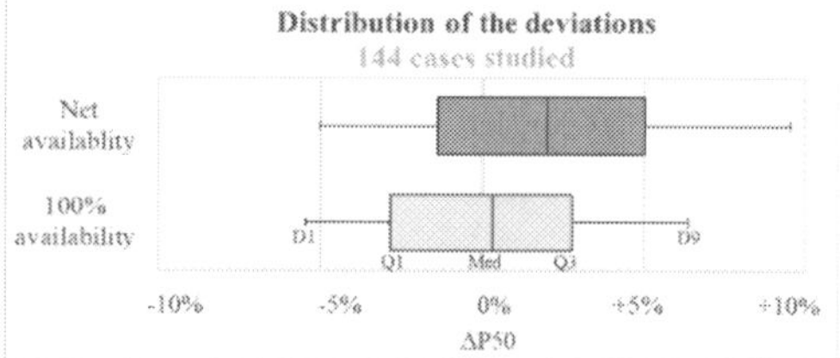

Figure 5: Global distribution of the deviations ΔP50 (Boxplot displaying D1, Q1, median, Q3, D9)

The following points should be highlighted regarding the deviations of the net P50 values:

- The median deviation is +2.1%, reflecting a trend towards overestimation of the theoretical assessments.
- The deviation is within the range ±5% in 64% of the cases, within the range ±7% and ±10% respectively in 80% and 90% of the cases.

Considering the deviation between the P50 values adjusted to 100% availability, the main outcomes are the following:

- The median reaches +0.3%, meaning than on average, the actual P50 matches the theoretical P50 when availability issues are not accounted for. This seems to indicate that theoretical studies generally overestimate availability rates.
- The deviation is within the range ±5% in 70% of the cases, within the range ±7% and ±10% respectively in 90% and 98% of the cases.
- Extreme deviations are exceptional (deviations exceeding ±10% in only 2% of the cases), but can reach very high levels (see Min and Max values in table 4).

Moreover, it should be noted that the consideration of net P50 values not only leads to a trend of an overestimated theoretical yield, but is also associated with a higher spread of the deviations than the P50 values adjusted to 100% availability.

In the following sections, in order to exclude the part of deviation due to availability issues (actual losses vs theoretical assumptions), the comparisons are made for the P50 values adjusted to 100% availability, the base case being the whole sample (144 cases). A focus is made specifically on availability issues in section 5.1.

4.2 Breakdown of the deviations (ΔP50 adjusted to 100% availability)

4.2.1. Age of the preconstruction assessment

A focus was made on the distribution of the deviations (ΔP50 adjusted to 100% availability) for most recent plants (theoretical preconstruction assessments carried out after 2017, i.e., almost half of the cases) in order to analyze the evolution of the distribution of the deviations over time.

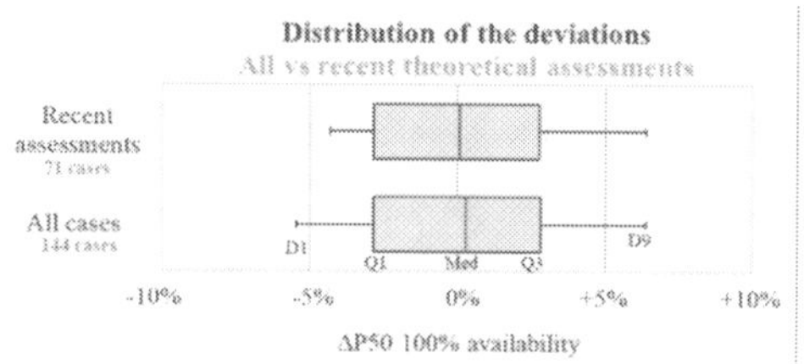

Figure 6: Distribution of the P50 deviations for most recent theoretical assessments vs base case

Most recent yield assessments show a slight improvement on the median deviation (+0.1% versus +0.3%), linked to less cases of underestimation. However, the part of cases overestimated remains stable (no significant improvement observed).

4.2.2. Type of plant

The deviations between the Theoretical P50 and the Actual P50, both adjusted to 100% availability, were analyzed depending on the type of plant. Figure 7 shows the results obtained for fixed tilt (either ground-mounted plants, shelters or rooftops) and trackers systems.

10.4229/EUPVSEC2023/4DV.4.47
020473-003

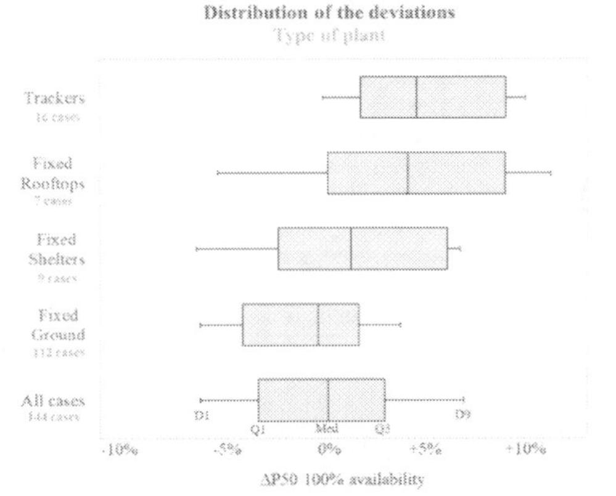

Figure 7: Distribution of the deviations depending on the type of plant

The following points can be highlighted from this analysis:

- The ground-mounted plants with a fixed tilt (most common type in the sample) are associated with the lowest deviations. The median value is -0.1%, and less overestimations of the yield are observed compared to other types of installation. In almost 80% of the cases for this type of installation the deviation remains within the range ±5%.
- Shelters and rooftops (respectively 9 and 7 cases), show a significant spread of the deviations and a trend to an overestimated yield, especially for rooftops (median deviation of +3.9%).
- The theoretical P50 is higher than the actual P50 in most cases for tracker systems (D1=+0.1%, meaning that ΔP50 is >0 for 14 out of the 16 cases involving tracker systems). The median deviation is +4.3%, and the deviation remains within the range ±5% for half on the cases only.

Table 5 below summarizes the median value and the part of cases for which the deviation remains within the range ±5% for each type of installation:

Table 5: Distribution of the deviations (type of plant)

ΔP50. (100% avail.)	Sample size	Median deviation	Part of sample within ±5%
All cases	144	+0.3 %	70%
Fixed Ground	112	-0.1%	80%
Fixed Shelters	9	+1.3%	33%
Fixed Rooftops	7	+3.9%	43%
Trackers	16	+4.3%	50%

4.2.3 Other tests

The deviations were analyzed for the plants depending on their DC/AC ratio (plants with a high DC/AC ratio versus plants with a low DC/AC ratio), but no significant trend was highlighted.

The deviations were analyzed for plants located on flat terrain versus those located on an undulating or sloping terrain. No specific trend due to terrain slope was pointed out on the sample studied.

5 ADDITIONAL RESULTS

5.1 Actual availability

Actual availability rates were estimated from the analysis of operational data over the entire period provided for each plant. These rates were compared to the assumptions taken into account in the theoretical energy yield assessments carried out at the preconstruction stage.

It appears from this comparison that the energy yield assessments carried out before the commissioning overestimate significantly the availability rates of the plants, as shown in the following table and figure:

Table 6: Availability rates theoretical vs actual

Energy availability	Theoretical	Actual
Min	96.5%	82.0%
D1	98.0%	94.0%
Q1	98.0%	96.4%
Median	**98.5%**	**97.8%**
Q3	100.0%	98.7%
D9	100.0%	99.4%
Max	100.0%	99.9%

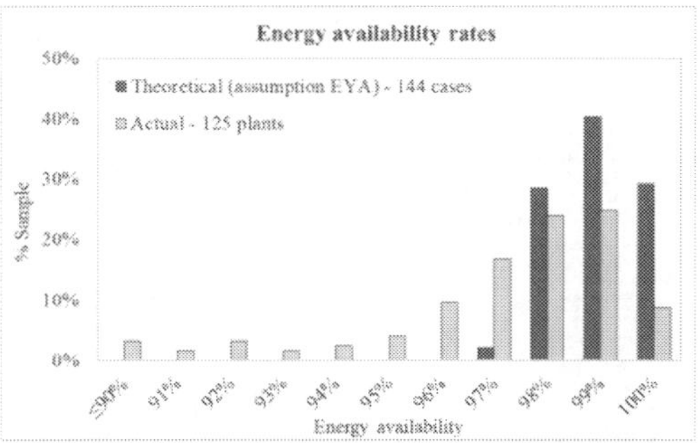

Figure 8: Theoretical versus actual availability rates

Figure 8 shows a wide spread of actual availability rates compared to theoretical assumptions. Actual availability rates are below 94% for 10% of the plants (most of them being fixed ground-mounted plants).

The analysis carried out by type of plant shows that actual availability rates are quite high for trackers and rooftops with a median value close to 98.5%.

Table 7: Actual availability rates by type of plant

Actual availability rates	Sample	Median
All plants	125	97.8%
Fixed Ground-mounted	93	97.6%
Fixed Shelters	9	97.3%
Fixed Rooftops	7	98.5%
Trackers	16	98.5%

Actual availability rates were analyzed depending on the type of inverter (number of inverters per MWp).

Table 8: Actual availability rates by type of inverter

Actual availability rates	Sample	Median
All plants	125	97.8%
<10 inverters per MWp	110	97.1%
10-20 inverters per MWp	7	97.5%
>20 inverters per MWp	8	99.3%

10.4229/EUPVSEC2023/4DV.4.47
020473-004

Table 8 seems to highlight that actual availability rates are significantly higher for plants equipped with smaller inverter and/or high density of inverter per MWp (>20 inverters per MWp). However, this observation is linked to a rather low sample (conclusion to be considered with caution).

5.2 Actual performance degradation

The actual performance degradation of the plants was assessed by analyzing the evolution of the actual production capacity over time (previous normalization to 100% availability and 100% solar energy index). The degradation rate was assessed only for plants with more than 4 years of operational data. It corresponds to the average degradation rate over the entire period analyzed. Table 9 and Figure 9 displays the actual performance degradation rates determined in the frame of this analysis. It should be noted that these rates correspond to the global performance degradation of the plant which should be mainly due to the degradation of the modules but could also include performance drifts of other components of the plant. As a reminder, availability losses have been corrected upstream this analysis, meaning that the observed drift should not be linked to availability issues.

Table 9: Actual performance degradation (>4 years of historical data available)

Actual performance degradation	Sample size	Median value	Standard deviation
> 4 years of data	59	0.5 %/year	0.5 %
4 to 8 years of data	48	0.4 %/year	0.5 %
>8 years of data	11	0.7 %/year	0.4 %

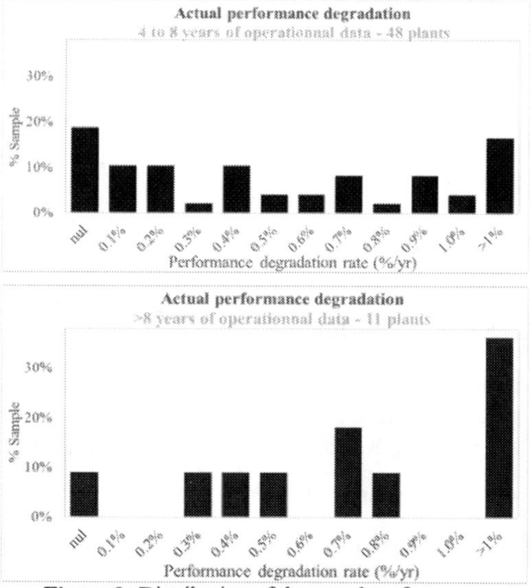

Figure 9: Distribution of the actual performance degradation observed for the plants

The main results are the following:

- The median value observed for all plants is 0.5%/year, which is in line with the theoretical assumptions (median value assumed in the theoretical yield assessment 0.5%/year).
- The spread in the observed degradation rates is significant, with some plants showing almost no degradation over time and other a degradation rate >1%/year.
- Older plants show a higher degradation rate than most recent ones (0.7%/year vs 0.4%/year). While focusing on this sample (11 plants), we observe that one plant shows very stable performance over time (no degradation highlighted), others show a degradation rate >0.3%/year, and 4 show a degradation rate >1%/year. For 10 of the 11 plants concerned by a performance degradation over time, 8 of them seem to be subject to an acceleration of the degradation over time (Degradation rate higher between years 5 to 8 than between years 1 to 4).

The actual performance degradation rate depending on the type of cells is presented in table 10 (plants with >4 years of historical data available):

Table 10: Actual performance degradation by type of cell (>4 years of historical data available)

Actual performance degradation	Sample size	Median value	Standard deviation
All	59	0.5 %/year	0.5 %
Poly-Si	39	0.4 %/year	0.5 %
Mono-Si	17	0.9 %/year	0.6 %
Thin film (CdTe, CIS)	3	0.7 %/year	0.1 %

Only 3 plants with over 4 years of data are equipped with thin film cells, however the degradation rate observed for these 3 plants is quite similar (0.7%, 0.7% and 0.8%/year).

On the sample analyzed, the degradation rates observed are significantly higher for plants equipped with mono-crystalline cells compared to those with poly-crystalline cells (0.9%/year versus 0.4%/year).

4 CONCLUSIONS

The main outcomes resulting from this feedback study are the following:

- Putting aside availability issues (i.e., considering the P50 estimates normalized to 100% availability), the median deviation observed between the actual P50 value of the plant estimated from operational data and the theoretical P50 value issued from the preconstruction yield assessment is close to zero (+0.3%). For 70% of the projects, the observed deviations range within ±5%, which is approximately the uncertainty range considered for preconstruction assessments. Accordingly, 30% of projects present a deviation range exceeding ±5% and 10% of them above ±7%.
- Regarding the net P50 values (i.e., including availability losses), theoretical energy yield assessments tend to overestimate the actual net production capacities of the PV plants. Assumptions of availability rates considered in the preconstruction yield assessments are globally too optimistic, leading to a trend to overestimate the net P50 (median deviation +2.1%). For 10% of the plants considered in the sample, the actual availability rate on the analyzed period is below 94% and the median

10.4229/EUPVSEC2023/4DV.4.47
020473-005

availability rate observed is 97.8%, whereas theoretical assessments assume availability rates globally ranging between 98% and 100%.

- Actual performance degradation rates were estimated from production data when over 4 years of data were available. The median degradation rate observed is close to 0.5%/year, which is in line with the theoretical assumptions. However, the analysis shows a wide spread in results, with several plants subject to a degradation rate >1%/year. Oldest plants (>8 years) generally show a higher degradation rate and an increase of the degradation rate over time. Plants equipped with mono-crystalline cells seem to show higher degradation rates than the ones equipped with poly-crystalline cells.

While the present study is based on a significant sample (125 plants, 144 assessments), complementary work would be interesting to consolidate the observations made when specific types of installation are considered.

10.4229/EUPVSEC2023/4DV.4.47
020473-006

PV MODELING IN IDA ICE (5.0 BETA) PERFORMANCE ANALYSIS OF BASED ON MEASURED DATA IN HIGH LATITUDES

Marieke Rynoson[1], Chris Bales[1], Joakim Munkhammar[2]
[1]University of Dalarna, School of Information and Engineering, Falun SE-79188
[2]Uppsala University, Department of Engineering Sciences, SE-751 21 Uppsala
* Corresponding author: E-mail: mary@du.se, Tel.: +46 72 444 9050

ABSTRACT:
In recent years, the widespread adoption of photovoltaic (PV) installations across various sectors has created a growing demand for accurate PV design tools. In this pursuit, the latest version of IDA Indoor Climate and Energy (IDA ICE 5 beta) emerges as a contender, offering advanced PV modeling capabilities, together with the advanced building simulation capabilities it already has. This study evaluates the accuracy of PV modeling within IDA ICE by comparing predicted power outputs to real-world data from three existing PV systems located at the Research Institute of Sweden (RISE) in Borås. To achieve this, weather files are created using historical weather and radiation data. As the measured radiation data is only available as total irradiation on a tilted plane, it was deconstructed into direct and diffuse components on the horizontal plane using a modified version of the model by Erbs et al.. Additional known parameters are the geometry of the PV array, and the characteristics of PV panel and inverter, based on their product data sheets. The accuracy of the PV design tool in IDA ICE is evaluated by comparing the power output of unshaded arrays against the measured data from RISE. The calculated power output is compared to the measured power output and analyzed through ASHRAE 14-2014 guidelines for performance evaluation. It was found that the software gives an accurate prediction of both panel temperature and PV power production. A study on shading effects is an open problem to improve the generality of the results in this study.
Keywords: PV Array, PV Modeling, Renewable Energy Planning, IDA ICE, Building Energy Simulation

1 BACKGROUND

PV installations have increased rapidly in the past years for utility-scale, commercial and industrial, as well as residential use [1]. As awareness of the technology also increases as a result of it, prices both for modules and turnkey installations decrease [2], and as inverter and optimizer technology improves and becomes available, previously undesirable building surfaces are being considered to be used for PV installations, including facades for building integrated solutions. It is therefore important for planners to have tools for PV design at hand, that can accurately calculate the predicted output of a PV system under more complex conditions.

The newest version of IDA Indoor Climate and Energy (IDA ICE) by EQUA Simulation AB [3], which is a popular building simulation tool in Northern Europe, utilizes a different PV calculation model than previous versions and now claims to be on the same level of accuracy as established PV design tools, such as PVsyst [4], Polysun [5], or PV*SOL [6].

The aim of this paper is to investigate the accuracy of the PV modelling in IDA ICE 5 (*beta*) by comparing the predicted power output to the measured power output of three existing PV systems, located at RISE Research Institute of Sweden in Borås in southern Sweden [7].

As IDA ICE is traditionally used to determine a building's energy demand and the indoor climate, this software could give planners good insights on the relationship between energy demand and production over the course of a year. IDA ICE's model also includes calculations of the thermal connection between the building envelope and the panels, which is especially interesting for building integrated PV. The indoor climate model and energy demand simulation is not relevant to this paper, as this part of the software is well established and validated since the middle 2000s [8, 9]. Experiments are ongoing with typical shading and validation of performance with shading will be included in a future paper.

2 METHOD

To model existing systems in IDA ICE, historic irradiance data on the tilted plane was decomposed and transposed to obtain diffuse horizontal (DHI) and beam normal irradiance (DNI), which is the relevant input for the software. Additional necessary climate data was taken from the Swedish Meteorological and Hydrological Institute (SMHI). Parameters of PV system components were gathered from product data sheets. The results generated by IDA ICE were compared to measured and cleaned power output data.

2.1 Case study setup

Research Institutes of Sweden (RISE) have conducted a long-term study of nine different PV systems at their facility in Borås, Sweden from June 1st, 2020, to May 31st, 2021. In the final report the test rig is described in detail [7]. As a part of this test, the systems' energy production on the AC side has been logged during the whole year of operation in 10-minute intervals, amounting to a total of 52560 data points. Additional measurements were made and logged in the same time intervals for panel temperature, ambient temperature, and two sets of total irradiance, measured with pyranometers, in the 45-degree tilted plane (GTI), facing south.

The data of five out of the nine systems was provided to the researcher for the purpose of this study, with permission of the Swedish Energy Agency. Out of these, three were chosen to be modeled, two applied systems with Mono-Si panels and one building integrated PV system with CIGS panels in landscape orientation ("C"). Out of the Mono-Si systems, one system has full cells, is in landscape orientation, and is equipped with optimizers ("B"), and the other system has half-cut cells and is installed in portrait orientation ("A").

As per standard output settings, and since weather data is commonly hourly, IDA ICE gives the predicted power output in hourly intervals. Therefore, the measured power output was averaged to hourly values, after being cleaned of outliers by using the moving median method [10]. It is important to note that the hourly average was determined by averaging the values between, for instance, 9.30 to 10.30, as IDA ICE defines the timestamp value for the hour around 10 o'clock, in this example.

The two sets of irradiance measurements, one next to system A, and the other next to systems B and C, were also cleaned with the moving median method and then averaged to hourly intervals as described above, so that they could be used for the creation of a weather file with hourly historical data. The ambient temperature showed 228 data gaps in the 10-minute interval data, which were interpolated based on SMHI measurements of the closest weather station in Borås.

As the irradiance was measured as global irradiance on a 45-degree tilted plane, which is the same tilt as the panels, it had to be deconstructed into beam normal (DNI) and diffuse horizontal irradiation (DHI), which is the type of input that is required by IDA ICE. The deconstruction and transposition of measured GTI was done based on the insights by Yang et al. [11], who described that more complex decomposition models do not outperform simpler models and that conversion results from tilt to horizontal are sufficiently accurate. Therefore, the simple model by Erbs et al. [12] was chosen for the decomposition of total irradiance from the south-facing panels GTI. The clearness index k_T is defined in **Equation 1** as the ratio of DHI to global horizontal irradiance GHI:

Equation 1

$$\frac{DHI}{GHI} = \begin{cases} 1.0 - 0.09k_T \\ 0.9511 - 0.1604k_T + 4.388k_T^2 \\ -16.638k_T^3 + 12.336k_T^4 \\ 0.165 \end{cases} if \begin{cases} k_T \leq 0.22 \\ 0.22 < k_T \leq 0.80 \\ k_T > 0.80 \end{cases}$$

In order to obtain a "tilted clearness index" k_S that can be used in the decomposition with Erbs et al, the measured GTI was divided by the calculated extraterrestrial irradiance G0T on the equally tilted surface [13]. With K_S, GTI is separated into diffuse (DTI) and beam (BTI) components on the tilted surface. Exceptions to this rule were made if K_S is smaller than 0.3, which indicates an overcast sky, and if the $cos\theta$, where θ is the incidence angle and for that timestep is smaller than 0.1, which indicates a very low sun angle. In these cases, the GTI was assumed to be fully diffuse. The beam irradiance on the horizontal plane (BHI) was calculated as in **Equation 2** with the geometrical factor of the beam irradiance R_b [14]:

Equation 2

$$BHI = \frac{BTI}{R_b}$$

From BHI, DNI was derived, and the anisotropy index A_i can be calculated as in **Equation 3** with the extraterrestrial irradiance on the horizontal surface G0:

Equation 3

$$A_i = \frac{BHI}{G0}$$

The diffuse component DHI is calculated in **Equation 4** with both A_i, and the geometric factors R_b and R_d [14]:

Equation 4

$$DHI = \frac{DTI}{(1 - A_i) * R_d + A_i * R_b}$$

The resulting values of the re-calculated GTI with the Hay and Davis model and Liu and Jordan model were then compared to the original measured GTI. These models were chosen as common methods to calculate irradiance on the tilted surface and allow for a good estimation for both isotropic (Liu and Jordan) and anisotropic sky models (Hay and Davis). They were chosen to showcase the general potential of the decomposition and transposition method. The total differences between them were analyzed based on their Mean Bias Error (MBE) and coefficient of variation of the Root Mean Square Error (CV RMSE) analysis for hourly data as described in appendix C2 in ASHRAE 14-2014 [15].

Other weather data that was needed as input for an IDA ICE compatible climate file were the relative humidity, wind speed from x- and y-direction, and the sky cover in percent. The closest SMHI station with available historical weather data for these parameters was Rångedala, ca. 16 km east of Borås. Data on wind speed v [m/s] and direction θ_{WD} [rad] was decomposed in **Equation 5** into the required wind speed from x- and y-direction:

Equation 5

$$v(x) = v * \cos(\theta_{WD})$$
$$v(y) = v * \sin(\theta_{WD})$$

2.2 Implementing in IDA ICE

The prepared climate file was inserted in IDA ICE 5 (beta 30) and a model with one zone as part of a building body with a 45-degree tilted roof, facing south, was built. IV-curve information and other data, such as panel geometry and number of panels in the string, was added to the PV model in the software, based on information available in the product data sheets. The same was true for inverter and optimizer parameters.

The simulated building was set to have no heating energy load. Electrical loads, such as facility electricity or lighting, were also not considered. The energy simulation was run for the full year, although January 1st through May 31st of 2021 were set before June 1st of 2020, as IDA ICE only runs full year simulations from January to December. The irradiance calculation was set to be done with the ASHRAE model.

2.3 Data Analysis

The calculated results of the panel temperature and power output were then compared to the measured panel temperature and AC power output that was provided by RISE. Both the total yearly and hourly power output was compared, the latter by using MBE and CV RMSE according to the ASHRAE guidelines for whole building performance simulation [15].

In the case of power output data, MBE and CV RMSE were calculated from the absolute errors between measured and predicted output for daytime-measurements and neglecting 3.5 – 3.7 % of daytime data points, which are labelled outliers based on the

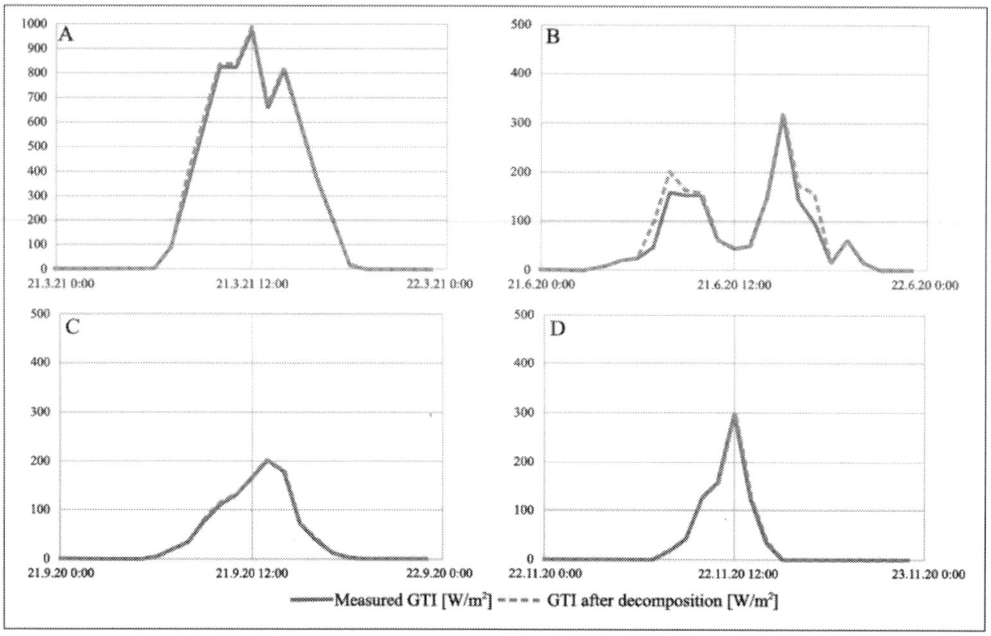

Figure 1 Example plots of measured GTI and GTI after decomposition; The subplots show irradiance profiles for A: March 21, 2021, B: June 21, 2020, C: September 21, 2020, November 22, 2020; Location: Borås, Sweden

Median Method [16]. This method was chosen to reduce large errors that may have occurred due to unreliable input data, human error, or faulty data logging of the power output.

The Median Method was chosen over Tukey's method [17], as Carling [16] found that it performs better in non-Gaussian distribution, which was the case in this study. As Hampel described in [18], "1-10% gross errors in routine data" are to be expected in a dataset, with up to 20 % for datasets under 3000 data points. As the logged data was smoothed out with the moving median, and was averaged to 8760 data points, with 4668 daytime data points, a percentage on the lower end of the expected number of outliers was chosen.

In the case of panel temperature data, MBE and CV RMSE were calculated with 4.9 - 5.0 % of data points labelled as outliers.

3 RESULTS

3.1 Decompositions and Transposition

It was found that the approximated values for horizontal irradiance parameters derived from the decomposition were reasonable, although some outliers had to be corrected. The results of the comparison between measured GTI and re-calculated GTI are presented in Table I.

Table I: MBE and CV RMSE for re-calculated GTI after decomposition

	GTI calculation	MBE [%]	CV RSME [%]
Measurement A	anisotropic	3.4	5.9
	isotropic	-1.4	9.4
Measurement B/C	anisotropic	3.4	6.1
	isotropic	-1.2	9.4
Requirement (ASHRAE 14-2014)	-	*±10*	*30*

Figure 1 shows the irradiance curves for the measured GTI (blue), and the re-calculated GTI with the anisotropic sky model (orange) during the equinoxes and solstices. As the month of December had both low irradiance and mainly overcast days, November 22nd, 2020, was chosen instead.

3.2 Full-year simulation of systems in IDA ICE

The full-year simulation was made for three existing PV systems A, B, and C, that were described in section 2.1. The resulting predictions by IDA ICE for panel temperature and power output were compared to measured data. Both the yearly total and the hourly data were analyzed for the power output, while the panel temperature was only analyzed with hourly data. The hourly values were investigated through the means of MBE and CV RMSE and visual perception of the resulting curves. **Table II** gives the calculated accuracy and precision for the panel temperature simulation,

Table III shows the comparison of the yearly total power output for the three modeled systems when outliers are removed, and **Table IV** gives an overview of the calculated errors from the hourly data comparison.

Table II: MBE and CV RMSE values for IDA ICE panel temperature simulation

System	MBE [%]	CV RMSE [%]
A	0.4	16.6
B	-0.9	26.3
C	-0.9	25.5
Requirement (ASHRAE 14-2014)	*± 10*	*30*

10.4229/EUPVSEC2023/4DV.4.48
020475-003

Table III: Comparison of yearly total power output between measured and IDA ICE-predicted values

System	Measured [kWh]	IDA ICE [kWh]	% of Measured
A	3133.5	3079.5	98.3
B	3515.6	3587.4	102.0
C	2346.9	2393.8	102.0

Table IV: MBE and CV RMSE values for IDA ICE AC power production

System	MBE [%]	CV RMSE [%]
A	-6.0	15.6
B	9.8	19.1
C	8.2	16.6
Requirement (ASHRAE 14-2014)	*± 10*	*30*

The panel temperature has a high degree of accuracy, with MBEs lower than +/- 1 %, and high, but acceptable values for precision, with CV RMSE between 16.6 and 26.3 %.

From the yearly power output comparison, IDA ICE was overestimating the annual power production by 2 %, for systems B and C. For system A, the total AC power output is underestimated by 1.7 %. The correlation coefficients for correlation between the differences in measured and predicted power output and panel temperature was found to be -0.14 to 0.21 and are therefore negligible in all three studied systems.

When analyzing the accuracy and precision through MBE and unbiased RMSE, system A shows a negative bias, with B and C having a positive bias, when the outliers are limited to the in chapter 2.3 described 3.5 - 3.7 %. This confirms the impression given by the total yearly estimate. When increasing the number of outliers to ca. 10 %, which Hampel [18] described as also to be expected, all MBEs become lower and the precision improves.

When visually investigating the power output curves to look deeper into times with high deviations, it can be seen that the most extreme deviations occur in the form of overestimations during the colder months, with an example of this being found on February 2nd, 2021, on systems B and C (see **Figure 2**).

Underestimations were found on clear sky days in all three systems during spring, fall, and winter months (see **Figure 3**). During the summer, the curves lined up remarkably close or were slightly overestimated (**Figure 4**). On days where the irradiance indicated that the sky was partially cloudy to varying degrees, both over- and underestimations could be found (**Figure 5**).

4 DISCUSSION

4.1 Decomposition and transposition

Total irradiance measured on the tilted surface decomposed using a modified version of the Erbs model [12] and transposed unto the horizontal plane. Based on the derived statistical values according to the ASHRAE standard for whole building performance simulation, the accuracy and precision are acceptable. As some of the generated decomposed values resulted in extreme values, further investigation will be made for a correction and validation of this approach.

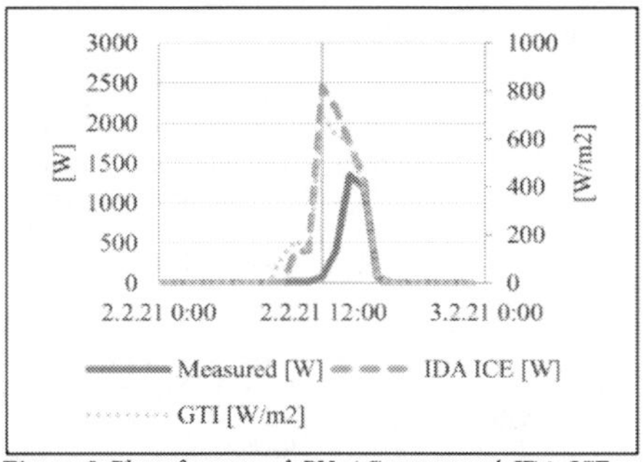

Figure 2 Plot of measured PV AC power and IDA ICE prediction on February 2nd, 2021, for system B

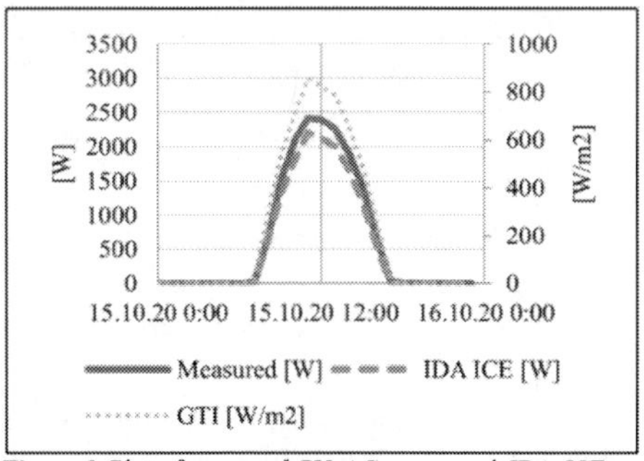

Figure 3 Plot of measured PV AC power and IDA ICE prediction on October 15th, 2020 for system C

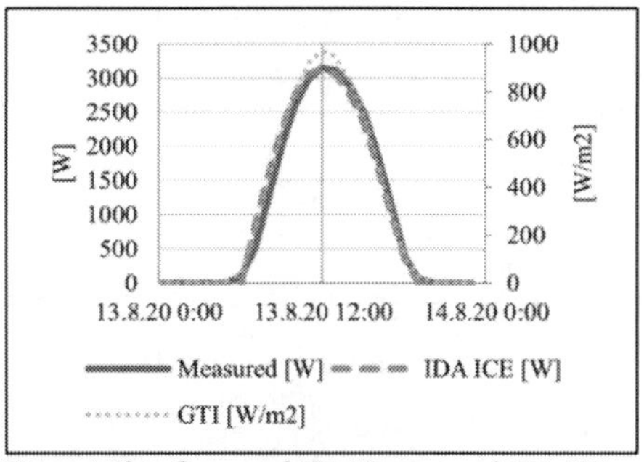

Figure 4 Plot of measured PV AC power and IDA ICE prediction on August 13th, 2020, for system A

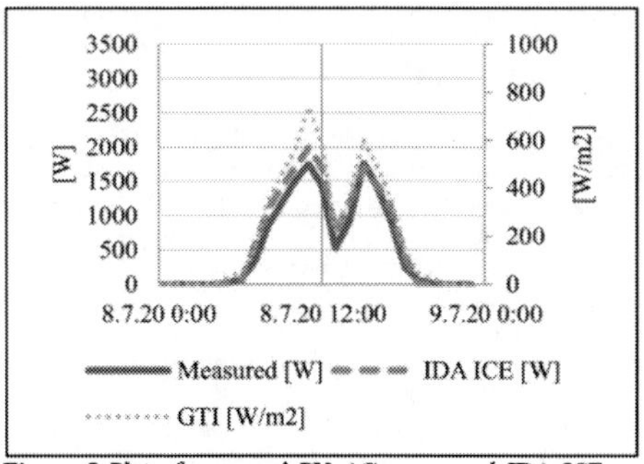

Figure 5 Plot of measured PV AC power and IDA ICE prediction on July 8th, 2020 for system C

10.4229/EUPVSEC2023/4DV.4.48
020475-004

4.2 Simulation of existing PV systems in IDA ICE

Three existing PV systems were modeled in IDA ICE 5 (beta 30), based on historic climate data and parameters gathered from the system components' data sheets. The panel temperature predicted by IDA ICE is highly accurate, and the precision of the temperature was sufficient. Imprecision in panel temperature prediction was not found to correlate to deviations in the predicted power output. Deviations may have been caused by a microclimate of the weather station which differs from that on the site of the systems.

Based on the provided input data, the total annual power output was overestimated in two out of three systems, with the highest overestimation in terms of MBE being found in system B, which had optimizers. A possible hypothesis may be that the software is currently overstating the influence of optimizers, as sudden drops in irradiance on partially cloudy days seem to affect the existing system more than IDA ICE predicted. But as this tendency is also noticeable in system C, albeit not as pronounced, part of this overestimation may well be due to overestimated irradiance input. The possible impact of overestimated irradiance input was not investigated.

The higher accuracy of system A may be due to its and the corresponding pyranometer's location. Unlike system B and C, which are mounted on a roof structure that is ground-mounted, system A's substructure is placed on the roof of the testing facility, where surrounding buildings or tree patches would have no influence on the system or the irradiance measurements. While the facilities and trees are located far enough away that they should not cast a shadow directly on the systems, they may still have an influence on pyranometer measurements, as the diffuse irradiance aspect is lower when the sky view is limited [19].

Extreme positive deviations that were found mostly in systems B and C are mainly found during the colder month. Based on precipitation and temperature data, these deviations likely come from snow that is covering the panels, but which either melts or is removed during the course of the day.

Most differences are found on cloudy days, which is likely due to the fact that the IDA ICE software internal ASHRAE irradiance model does not take cloud cover into account and treats the input irradiance as though it was measured on a clear day. In addition, the software does not take cloud enhancement or changing albedo into account, which can have an effect on the "overall GTI uncertainty at any specific location" [13]. Depending on how the logged 10-minute irradiance data was cleaned and averaged, to which the authors had no insight, irradiance peaks or drops may have been taken into account or discarded, causing further uncertainty and potentially adding to the bias.

5 CONCLUSIONS

It was determined that the derived values for diffuse horizontal and beam normal irradiance from the decomposition and transposition approach could be used for the creation of a historic hourly weather file that sufficiently accurately represents the weather conditions at the time of measurement. Investigation is ongoing for the validation of the method.

Based on measured data, a study was made to investigate the accuracy of IDA ICE 5 (beta 30). It was found that the software gives an accurate prediction of both panel temperature and PV power production. A study of suboptimal and shaded condition is an open problem to improve the generality of the results in this study.

6 ACKNOWLEDGEMENTS

We acknowledge funding from the Solar Electricity Research Center, Sweden (SOLVE). SOLVE is co-financed by the Swedish Energy Agency (project number 52693-1) as a national center of excellence. The authors would like to thank EQUA Simulation AB for providing the software to the researchers and RISE for supplying the data which was the foundation of this study.

7 REFERENCES

[1] IEA, "Solar PV," vol. 2022, ed. Paris: IEA, 2022.

[2] A. O. W. Johan Lindahl, "National Survey Report of PV Power Applications in Sweden " in "IEA PVPS Task 1," IEA, 2021. Accessed: 2022-10-31.

[3] "EQUA." EQUA Simualtion AB. https://www.equa.se/en/ (accessed.

[4] "PVsyst." PVsyst SA. https://www.pvsyst.com/ (accessed.

[5] "POLYSUN." Vela Solaris. https://www.velasolaris.com/?lang=en (accessed.

[6] "PV*SOL." Valentin Software GnbH. https://pvsol.software/en/ (accessed.

[7] RISE, "Provning av solelsystem för villor - Slutrapport 2021," RISE Research Intitute of Sweden AB, Eskilstuna, 2P03024-03/ P109947, 2021-10-05 2021. [Online]. Available: https://www.energimyndigheten.se/4acd02/globalassets/tester/slutrapport-systemtester-solel-sign-pk_jp_cm.pdf

[8] G. Z. Sven Kropf, "Validation of the Building Simulation Program IDA-ICE: According to CEN 13791 „Thermal Performance of Buildings - Calculation of Internal Temperatures of a Room in Summer Without Mechanical Cooling - General Criteria and Validation Procedures“," HTA Luzern, Luzern, 2007.

[9] S. Moosberger, "IDA ICE CIBSE-Validation: Test of IDA Indoor Climate and Energy version 4.0 according to CIBSE TM33, issue 3," HTA Luzern
Zentrum für Integrale Gebäudetechnik, Luzern, SAP-Nr.: 1120213, 2007-12-17 2007. Accessed: 2023-09-05.

[10] C. L. do Lago, V. F. Juliano, and C. Kascheres, "Applying moving median digital filter to mass spectrometry and potentiometric titration," *Analytica Chimica Acta,* vol. 310, no. 2, pp. 281-288, 1995/06/30/ 1995, doi: https://doi.org/10.1016/0003-2670(95)00130-R.

[11] D. Z. Yang, Z. B. Dong, A. Nobre, Y. S. Khoo, P. Jirutitijaroen, and W. M. Walsh, "Evaluation of transposition and decomposition models for converting global solar irradiance from tilted surface to horizontal in tropical regions," (in English), *Solar Energy,* vol. 97, pp. 369-387, Nov 2013, doi: 10.1016/j.solener.2013.08.033.

[12] D. G. Erbs, S. A. Klein, and J. A. Duffie, "Estimation of the diffuse radiation fraction for hourly, daily and monthly-average

global radiation," *Solar Energy*, vol. 28, no. 4, pp. 293-302, 1982/01/01/ 1982, doi: https://doi.org/10.1016/0038-092X(82)90302-4.

[13] C. A. Gueymard, "Cloud and albedo enhancement impacts on solar irradiance using high-frequency measurements from thermopile and photodiode radiometers. Part 2: Performance of separation and transposition models for global tilted irradiance," *Solar Energy*, vol. 153, pp. 766-779, 2017/09/01/ 2017, doi: https://doi.org/10.1016/j.solener.2017.04.068.

[14] W. A. B. J. A. Duffie, *Solar Engineering of Thermal Processes, 2nd edition*. New York: John Wiley & Sons, Inc., 1980.

[15] *ASHRAE Guideline 14-2014: Measurement of Energy, Demand, and Water Savings*, ASHRAE, 2014.

[16] K. Carling, "Resistant outlier rules and the non-Gaussian case," *Computational Statistics & Data Analysis*, vol. 33, no. 3, pp. 249-258, 2000/05/28/ 2000, doi: https://doi.org/10.1016/S0167-9473(99)00057-2.

[17] J. W. Tukey, *Exploratory data analysis*. Reading, MA: Addison-Wesley, 1977.

[18] E. M. R. Frank R. Hampel, Peter J. Rousseeuw, Werner A. Stahel, *Robust Statistics: The Approach Based on Influence Functions* (Wiley series in probability and mathematical statistics. Probability and mathematical statistics, ISSN 0271-6356). New York: John Wiley & Sons, Inc, 1986.

[19] I. García, M. de Blas, B. Hernández, C. Sáenz, and J. L. Torres, "Diffuse irradiance on tilted planes in urban environments: Evaluation of models modified with sky and circumsolar view factors," *Renewable Energy*, vol. 180, pp. 1194-1209, 2021/12/01/ 2021, doi: https://doi.org/10.1016/j.renene.2021.08.042.

ALL-SKY IMAGES FOR SOLAR RESOURCE ASSESSMENT IN QATAR

Daniel Perez-Astudillo, Dunia Bachour
Qatar Environment and Energy Research Institute/Hamad bin Khalifa University/Qatar Foundation
PO Box 34110, Doha, Qatar. dastudillo@hbku.edu.qa, dbachour@hbku.edu.qa

ABSTRACT: All-sky cameras take images of the complete sky dome at their installation site, allowing to visually assess the local sky conditions; through the analysis of these images, different kinds of information can be obtained, for instance cloud cover, which is one of the main factors affecting solar radiation and thus the available input for solar power technologies. This work presents a study of all-sky images taken in the visible spectrum in Doha, Qatar, across three months, and covering different sky conditions, from sunny to overcast and rain. A software workflow was developed in-house to allow automated processing of the high number of images, with some adjustable input parameters. With the code, each image can be analysed in either of the colour channels (RGB, i.e. red, green, blue), so that the channel providing the best contrast can be found and used for further analysis. The final processing provides a "cloudiness index" for each image, which is then compared with different solar resource and meteorological parameters to study possible correlations, a basis to develop solar nowcasting algorithms.
Keywords: all-sky imager, solar resources, cloudiness

1 INTRODUCTION

Clouds have the largest effect on the amount and quality of solar radiation at ground level, which is the fuel of solar power production technologies. Therefore, a determination of cloudiness in the local sky is of great relevance for several fields and in particular for solar power applications. It is then of high importance to develop an automatic way to derive a cloudiness parameter for use and application in solar energy, especially at high temporal and spatial resolutions. The study presented here describes an automated processing scheme to analyse a large number of sky images. The approach is applied on sky images acquired by an all-sky camera installed in Doha, Qatar, in order to derive a cloudiness index that is then compared with different solar resource parameters for validation, and correlation is also studied with other meteorological parameters. The methodology shown here will be further developed for integration with solar nowcasting algorithms using sky images.

2 DATA

2.1 All-sky imager

All-sky imagers, also called sky cameras or all-sky cameras, capture photos (usually visible or infrared) of the whole sky dome at their location, and are useful tools for visually assessing local sky conditions at any given time; by post-processing the images, different kinds of information may be obtained; the most direct result is cloud cover (which is the main factor affecting solar radiation at ground level, thus impacting the available input for solar power technologies), but other common applications include modelling [1] and forecasting [2,3] irradiance.

In this work we study visible-spectrum all-sky images taken in Doha, Qatar (25° N, 51° E), throughout the period from 24/Oct/2022 to 31/May/2023, covering different sky conditions going from sunny to dusty to overcast with and without rain.

The images were captured with an SRF-02 all-sky imager from EKO Instruments (note that this sky imager is no longer commercialised by EKO), taking a 2272x1702 pixel image every minute during daytime and part of the night, resulting in 880 images per day, around 190,000 possible images in the studied period. For this study, as no geometric corrections are applied to the images, only the images taken at local solar noon were used, so that the sun disk size and shape are not greatly different between images. This resulted in 219 images used for the study (there are 220 days in the above-mentioned period, but the image of 13/Nov/2022 at noon was not available).

2.2 Additional data

At the same location, solar measurements were taken also with 1-minute resolution: a thermoelectric solar radiation monitoring station with sun tracker provides direct normal irradiance (DNI), global horizontal irradiance (GHI), and diffuse horizontal irradiance (DHI) [4,5], while other sensors, also at the same site, provide air temperature, barometric pressure and relative humidity. In addition, the top-of-atmosphere (also called "extra-terrestrial") beam and horizontal solar radiations (ETb and ETh, respectively) were calculated for each minute, in order to obtain the clearness index

$$Kt = \frac{GHI}{ETh} \quad (1)$$

and the beam clearness index

$$Kn = \frac{DNI}{ETb} \quad (2)$$

Finally, the diffuse fraction

$$Kd = \frac{DHI}{GHI} \quad (3)$$

was also calculated. ETh and ETb are analogous to GHI and DNI but before the sunlight crosses the earth's atmosphere [6]. ETb only depends on the distance between the sun and the earth, and ETh is simply the projection of ETb on a horizontal plane. Kt and Kn provide an indication of how much solar radiation was lost through the atmosphere due to all atmospheric components, and the largest losses are in most cases due to clouds, more so on DNI than on GHI (an opaque cloud covering the sun reduces DNI to zero, while GHI is less affected, and DHI is increased).

10.4229/EUPVSEC2023/4DV.4.49
020477-001

Figure 1: All-sky imager and solar radiation monitoring station at QEERI's facilities in Doha.

3 ANALYSIS METHODOLOGY

3.1 'Original' method

A software workflow based on CERN's ROOT data analysis framework (https://root.cern) was developed in-house to allow an automated processing of the high number of images, with some adjustable input parameters such as the colour channel to use, included image area, or normalisation. With this code, each image can be analysed in either of the colour channels (RGB, i.e. red, green, blue), so that the best, namely the one providing the best contrast, can be found and used.

To process the images, only the pixels inside the circular field of view of the camera are included. The brightness levels of these pixels (in a given colour channel) are first normalised from the original range (0 to 255) to the range 0 to 1, and then these values are averaged for the image, so that each image is assigned a value between zero and one, hereby called its "cloudiness index". This index is then compared against different solar resource and meteorological parameters to study possible correlations, a basis to develop solar nowcasting algorithms.

3.2 Modified method test

Trying to reduce the influence of the "clearest sky" and any static structures in the images, a modified approach was tested, in which a lower threshold was introduced on the pixel brightness included in the calculation of cloudiness of each of the images, so that the contributions of the "base" sun and structures might be removed. Although the location of the sky imager used here is almost ideal, some distant structures can be seen in the captured images, obscuring the horizon although only at very low elevations on the left (eastern) half side of the view.

In the red channel (see Results section), the lowest cloudiness with the original approach was 0.0292, on 30/Mar/2022 (left side of Figure 2). However, with the modified approach even using low threshold, namely by removing pixel brightness below 0.1, the lowest derived cloudiness when using this method was 0.774, obtained for 6/Jan/202, which the captured image (right side of Figure 6) shows that in fact was an overcast day with rain. Therefore, no lower thresholds were introduced in the rest of the study.

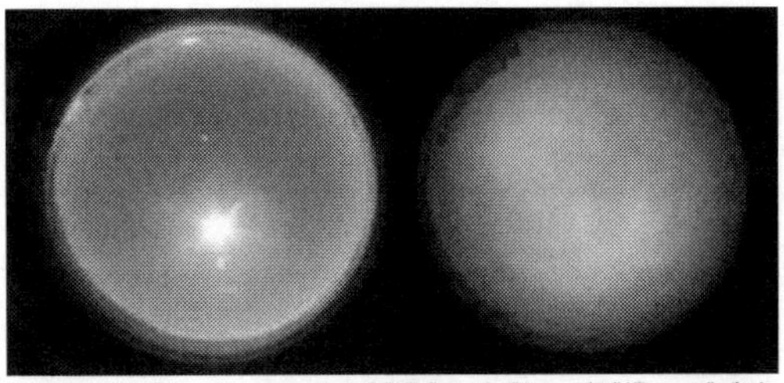

Figure 2: Sky images for 30/Mar (left) and 6/Jan (right), corresponding to the minimum derived cloudiness indices from, respectively, the "original" and "modified" approaches.

4 RESULTS

4.1 Colour channels and cloudiness determination

Figure 3 shows a sample sky image, from 28/Oct/2022 at solar noon, i.e. 11:18 am local time (note that the bottom of the view is not perfectly aligned with south, but this has no effect on the method used in this work), and the corresponding RGB channels. The brightest pixels correspond to the sun and clouds; it is noted that the red channel provides better contrast to identify cloud features, so this channel was chosen for the rest of the analysis shown here. Figure 4 shows the resulting normalised values (in the colour scales shown on the right sides) for the same images of Figure 3.

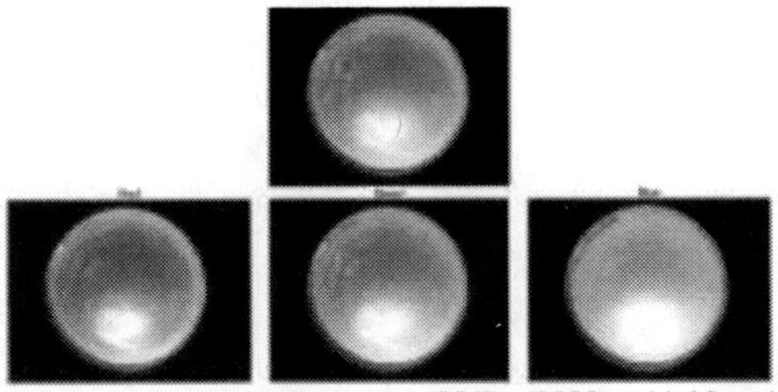

Figure 3: A sample sky image (28/Oct/2022 at 11:18 am) and its red, green and blue channels.

Figure 4: Normalised brightness values (colour scale) of the pixels within the FOV for the same images of Figure 2 (RGB channels for 28/Oct/2022 at 11:18 am).

Figure 5 shows the frequency distribution of the cloudiness index derived from the red channel for the 219 images (days) included. Even though Qatar has generally a high frequency of cloudless conditions, clouds are frequent between November and March, with some (but not very common) completely overcast and rainy days.

10.4229/EUPVSEC2023/4DV.4.49
020477-002

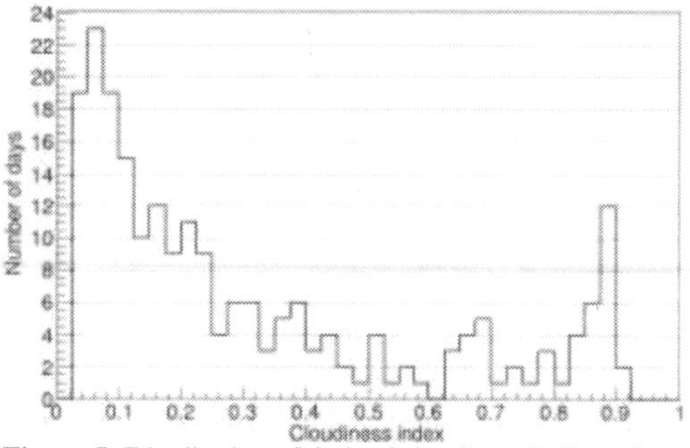

Figure 5: Distribution of derived cloudiness indices, from the red channel, for the studied images.

4.2 Cloudiness vs solar and meteorological data

The plots in Figure 6 show the cloudiness index (x-axis) obtained from the red channel for the image at solar noon of all the included days (219 days, or data points) applying the 'original' approach, against solar irradiances and their derived indices Kn (beam clearness), Kt (clearness) and Kd (diffuse fraction). As is expected, the cloudiness index shows inverse relations with DNI and GHI, and a positive relation with DHI; indeed, more clouds block more effectively the direct solar radiation, reducing slightly the global radiation and increasing the diffused light.

Similarly, Figure 7 shows the cloudiness index against the meteorological parameters: temperature (in degrees Celsius), relative humidity (in %), and barometric pressure (in mbar). For these variables, no clear correlations are noticeable.

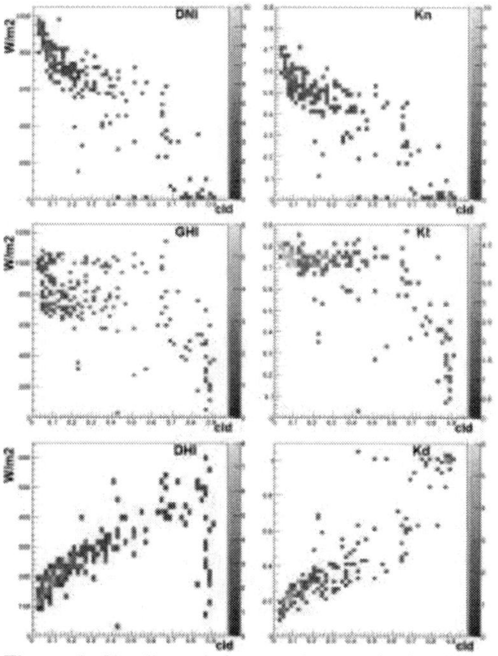

Figure 6: Cloudiness index (x-axis) vs solar irradiance components and derived indices, at solar noon. The colour scales represent the number of entries (days).

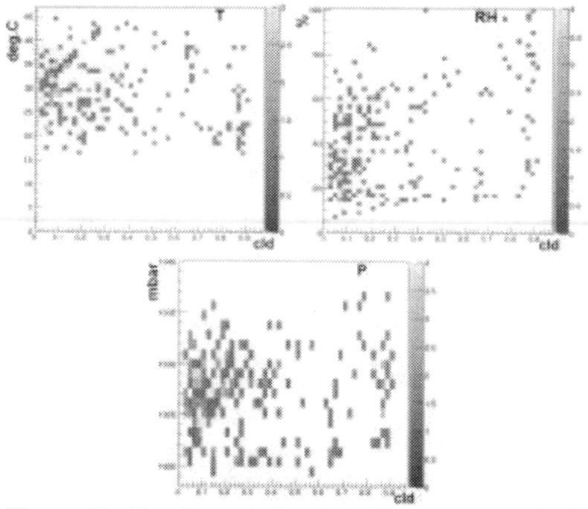

Figure 7: Cloudiness index (x-axis) vs measured air temperature (T), relative humidity (RH) and air pressure (P), at solar noon. The colour scales represent the number of entries (days).

5 CONCLUSIONS AND OUTLOOK

Although commercial sky imagers are usually provided with some analysis software, to batch process large numbers of images a home-made solution has to be developed, as has been done in the work presented here. As the developed workflow was done in-house, all aspects can be controlled and adjusted at will, providing added flexibility and features. By selecting the colour channel with best contrast, a cloudiness index is derived for each image in a set taken at solar noon spanning over 7 months in Qatar. By comparing this index with various concurrent solar and meteorological variables, good correlations are found with the solar irradiances and clearness indices, with no clear correlations with the studied meteo variables. A simple model based on local measurements may be derived to allow for modelling solar radiation from sky image observations, and further refinements may allow for nowcasting, by first predicting cloud coverage and then deriving solar radiation. The simple method shown here can be further improved by using identification techniques to exclude the sun and other undesired (for this application) features, such as static structures in the FOV, and by studying zenith angle dependencies.

6 REFERENCES

[1] R. Chauvin, J. Nou, S. Thil, S. Grieu. Modelling the clear-sky intensity distribution using a sky imager. Solar Energy 119 (2015) 1-17. DOI: 10.1016/j.solener.2015.06.026

[2] Chia-Lin Fu, Hsu-Yung Cheng. Predicting solar irradiance with all-sky image features via regression. Solar Energy 97 (2013) 537-550. DOI: 10.1016/j.solener.2013.09.016

[3] Chi Wai Chow, Bryan Urquhart, Matthew Lave, Anthony Dominguez, Jan Kleissl, Janet Shields, Byron Washom. Intra-hour forecasting with a total sky imager at the UC San Diego solar energy testbed. Solar Energy 85 (2011) 2881-2893. DOI: 10.1016/j.solener.2011.08.025

[4] D. Perez-Astudillo, D. Bachour. DNI, GHI and DHI Ground Measurements in Doha, Qatar. Energy Procedia 49 (2014), 2398-2404. DOI: 10.1016/j.egypro.2014.03.254
[5] Perez-Astudillo, D., Bachour, D., Sanfilippo, A., Al-Hajri H., 2022. Management and Operation of Qatar's Solar Radiation Monitoring Network. EuroSun 2022. ISES Conference Proceedings. DOI: 10.18086/eurosun.2022.15.06
[6] Duffie, J. A., Beckman, W. A., 2013. Solar engineering of thermal processes, 4th ed. Wiley & Sons, New Jersey.

10.4229/EUPVSEC2023/4DV.4.49
020477-004

SMOOTHING OF POWER OUTPUT FLUCTUATIONS IN THREE UTILITY SCALE PV PLANTS LOCATED IN DIFFERENT CONTINENTS

Jonathan Lehmann[1], Quentin Van Nieuwenhoven[1]
1: ENGIE Laborelec, Solar Lab, Rue de Rhode 125, 1630 Linkebeek, Belgium
e-mail: jonathan.lehmann@engie.com

ABSTRACT:
The intermittent nature of the power generated by photovoltaic plants introduces new challenges for the stability of electricity grids. This article presents a study of the variability of the power output fluctuations of three utility scale PV plants. By using the Fourier transform, the data was decomposed from the time domain into frequency domain so that the power spectral density (PSD) of the different PV plants can be analysed and compared. The study is performed using one year of 1-min power and irradiance data from plants located in France, Brazil and India with respective power of 25.6 MWp, 32.1 MWp and 75 MWp. For each plant, by combining the power from several inverters, several sizes of PV plant have been considered. Our work confirms that regarding power output, PV plants act as a low-pass filter for the irradiance. However, our analysis shows that this behaviour is highly site dependent. Indeed, we have demonstrated that the smoothing of high frequency variations is significantly stronger for the PV plant located in France when compared to the PV plants located in India or Brazil. The underlying mechanism of the site dependence is still to be understood.
Keywords: variability, grid, photovoltaic, power.

1 INTRODUCTION

Solar PV plants have an inherent variability in the power production caused by clouds, diurnal intermittency, and seasonal variability. At large scale, the intermittent character of power generated from solar photovoltaic (PV) systems can affect power quality and reliability[1][2]. One of the way to study power fluctuations relies on using the Fourier transform to decompose the time domain data into frequency domain data so that the power spectral density (PSD) can be calculated [4][5][6][7][8]. The PSD of the output of large-scale PV plants provides information on both cyclic (daily and seasonal) and non-cyclic (weather-related) power fluctuations.

This approach was first applied to PV plants by Curtight et al [4] to study the intermittency of two PV arrays located in Arizona. By studying the power spectra, their work made the hypothesis that PV plant attenuate high frequency fluctuations of the irradiance due to the time required for a cloud shadow to cross the full array. Marcos et al [5][6] analyzed in the frequency domain one year of 1 second irradiance and output power data coming from 6 PV plants located in the region of Navarra, Spain, with power ranging from 1MW to 9.5MW. Their analysis confirmed that considering power output, PV plants smooth high frequency variations of the irradiance, thus acting as low pass filter in the frequency domain. Moreover, it was shown that this filtering is a function of the PV plant area, with larger plant inducing greater smoothing. Subsequently, by fitting empirical data, they presented a model to determine the power output of a PV plant of any size using only irradiance measurements. Finally, Klima et al used the PSD to study geographic smoothing by examining the power output of 20 utility scale PV plants located in the Indian State of Gujarat [5] and 15 utility-scale and rooftop PV plant located in the US states of California, Nevada, and Arizona [6]. By comparing power spectrums, they studied how and to what extend the variations of the power output of PV plants can be reduced by geographical averaging.

In this paper, the power output fluctuations of three utility scale PV plants will be studied in the frequency domain. The study is performed using one year of 1-min power and irradiance data from plants located in France, Brazil and India with respective power of 25.6 MWp, 32.1 MWp and 75 MWp. For each PV plant, the authors will analyse their low-pass filter nature, i.e., how high frequency variations of the irradiances are smooth when examining power output. By combining the power from several inverters, several sizes of PV plant will be considered for each studied location. To the author's best knowledge, this is the first examination of the low pass filter nature of PV plants in different climates.

2 METHODOLOGY

The analysis is built around three utility scale PV plant that we named: Bordeaux, for the one located in France, Natal for the one located in Brazil and Jodhpur for the one located in India. Their power ranges from 25 MWp to 75 MWp (Table 1). For each plant, plane of array irradiance is measured every minute as well as DC power from each inverters. By combining the power from several inverters, several sizes of PV plant will be considered for each studied location. For example, for Bordeaux, combining the output of the 800 kWp inverters will allow us to consider any PV plant size from 0.8 to 25.6 MWp. Finally, regarding the configuration of the modules, Bordeaux and Jodhpur are fix tilt south oriented PV plants while Natal is equipped with single axis trackers with east-west motion.

Table I: Characteristics of the studied PV plants

Name	Power [MWp]	Inverter power [kWp]	Type	Size [ha]	Latitude
Natal	32,1	1070	East-west one axis tracking	70	-5.81
Bordeaux	25,6	800	Fix tilt	40	44.83
Jodhpur	75	1080	Fix tilt	140	26.26

The method used to estimate the PSD has been described previously[3][5][7]. Fig. 1 shows for Natal the spectrum of the irradiance as well as the power spectrums for: all inverters and for only 5 inverters. Before calculating the PSD, the datasets have been normalized to their maximum value. A first peak representing the daily variability of solar resource is observed at a frequency of $1{,}6\text{x}10^{-5}$ Hz, corresponding to 24h. Harmonics at 12h, 8h

10.4229/EUPVSEC2023/4EO.2.2
020478-001

and 6h are also visible. For frequencies lower than 1×10^{-3} Hz, it can be seen that the three curves share the same dynamics, corresponding to a linear region that can be fitter with a function in the form $f^{-0.64}$. Looking at the spectrums of power output, an attenuation of the variations in the high frequency region is clearly visible, evidencing the low pass filter effect of the PV plants. It can be noted that this behavior is dependent from the size of the PV plant, with a stronger smoothing of the variations obtained when considering 30 inverters as compared to the power obtained considering only 5 inverters. In the high frequencies range, the power spectrum can be fitted with functions of the $f^{-1.4}$ or $f^{-1.14}$ depending on the number of inverters considered. The cut-off frequency of the low-pass filter is defined as the cross point between the trendline at high frequencies and the trendline at low frequencies.

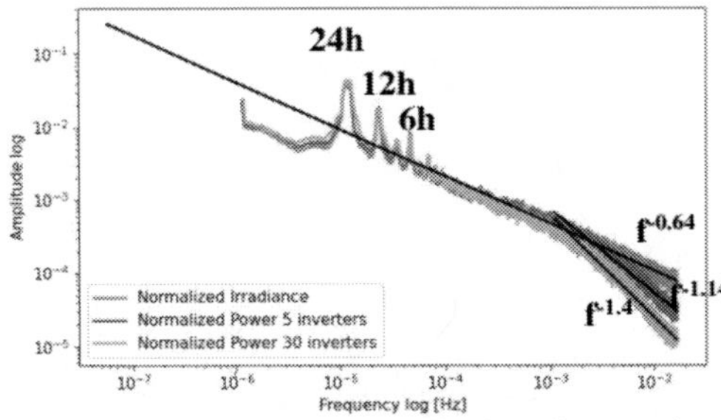

Figure 1: Spectrum of the normalized irradiance and power of 5 and 30 inverters for Natal

Eq 1 represents the function transfer associated to the filter:

$$P = \frac{K}{\left(\frac{1}{2\pi f_c}\right)s+1} \times Irr \qquad (1)$$

Where P is the power in kW, Irr the irradiance in W/m², K is the gain and f_c is the cut-off frequency in Hz.

Marcos et al showed that the cut-off frequency can be calculated through the PV plant surface:

$$f_c = a \cdot S^b \qquad (2)$$

Where S is the total surface of the plant in hectares (ha), and a and b are coefficients obtained from fitting in [5]. Their values, deduced from the observation of 6 PV plants in the region of Navarra, Spain, are a=0,0204 and b= -0,4997. It can be seen that the cut-off frequency is directly related to the size of the PV plant: increasing the size of the plant increases f_c, thus accentuating the low pass filter effect.

3 RESULTS AND DISCUSSIONS

Figure 2 shows for each studied PV plant the cut off frequency calculated as a function of the size of the PV plant. As stated previously, different sizes of PV plants have been considered by combining the power output from one inverter to the power output from all the inverters. As all inverters have the same power rating, it was considered that the size of the plant varies linearly with the number of inverters connected. Data from Marcos et al [5] was added for comparison.

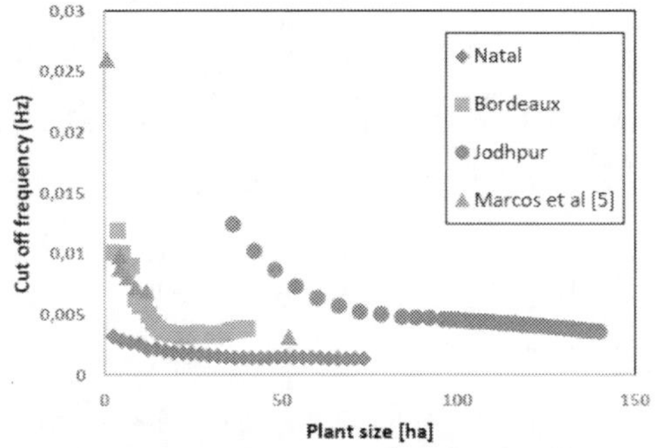

Figure 2: Cut-off frequency fc, as a function of the PV plant area for the three studied location. Marcos et al [5] indicated for comparison

A common behaviour is visible for the three studied sites: cut-off frequencies and thus attenuation increases when increasing the size of the PV plant. In addition, for each curve, a similar phenomenon is observed: when increasing the size of the PV plant, a rapid decrease of the cut-off frequency is observed, followed by a relative stagnation of the cut-off frequency. However, it can be clearly noted that the dynamics of the curves are site dependent. Indeed, it can be seen that Bordeaux shares a dynamic similar to the PV plants studied by [5]. It is interesting to note that all those plants have a similar climate, being located at 350 km from each other. On the other hand, for Jodhpur, the smoothing of the variations is observed only for PV plant sizes larger than 40 ha. Finally, for Natal, the smoothing effect seems to be relatively small. It is worth noting that the latitude of the PV plants is inversely proportional to the smoothing effect, with Bordeaux being the PV plant located the furthest from the equator.

Table 2 shows, for each studied site, the coefficients a and b of equation (2) obtained from fitting. Data from Marcos et al [5] was added for comparison purpose.

As observed in Figure 2, it can be seen that Bordeaux and Marcos et al [5] share similar dynamics, as evidenced by the similar values of a and b obtained. On the other hand, the low values obtained for Natal illustrate the small smoothing of the variations of the irradiance observed.

Table II: For each studied site, coefficients a and b of equation (2) obtained from fitting . Marcos et al [5] indicated for comparison.

	a	b
Bordeaux	0.0184	-0.496
Jodhpur	0.1768	-0.791
Natal	0.0045	-0.27
Marcos et al [5]	0.0204	-0.4997

4 CONCLUSION

In this work, three utility scale PV plant located in France, Brazil and India with respective power of 25.6 MWp, 32.1 MWp and 75 MWp have been analysed in the frequency domain. By combining the power from several inverters, PV plants of different size were considered. In agreement with what was demonstrated by [5] we have observed that regarding power output, PV plants act as a low-pass filter for the irradiance. However, our analysis shows that this

10.4229/EUPVSEC2023/4EO.2.2
020478-002

behaviour is highly site dependent. Indeed, we have demonstrated that the smoothing of high frequency variations is significantly stronger for the PV plant located in France while for the PV plant located in India, the smoothing started only for important sizes of the PV plant and for the PV plant located in Brazil the attenuation of variations was very limited. The similarity of dynamics observed between the PV plant located in France and the PV plant studied by [5] as well as the decrease of smoothing when moving away from the equator could point towards a mechanism such as weather or latitude.

5 REFERENCES

[1] R. Shah, N. Mithulananthan, R.C. Bansal, V.K. Ramachandaramurthy,
"A review of key power system stability challenges for large-scale PV integration" ,Renewable and Sustainable Energy Reviews, Volume 41,2015

[2] A. Fernández-Guillamón, E. Gómez-Lázaro, E. Muljadi, A. Molina-García,*"Power systems with high renewable energy sources: A review of inertia and frequency control strategies over time"*,Renewable and Sustainable Energy Reviews,Volume 115,2019, 109369

[3] Apt J. The spectrum of power from wind turbines. Journal of Power Sources 2007; 169(2): 369–374.

[4] Curtright E, Apt. J. *"The character of power output from utility-scale photovoltaic systems."* Progress in Photovoltaics: Research and Applications 2008; 16: 241–247

[5] J. Marcos, L. Marroyo, E. Lorenzo, D. Alvira and E. Izco, *"From irradiance to output power fluctuations: the pv plant as a low pass filter"* Progress in Photovoltaics: Research and Applications, vol. 19, p. 505–510, January 2011

[6] J. Marcos, L. Marroyo,E. Lorenzo, M. Garcia, *"Power output fluctuations in large PV plants."* International Conference on Renewable Energies and Power Quality, 2012 (ICREPQ'12)

[7] K. Klima and J. Apt, *"Geographic smoothing of solar PV: Results from Gujarat"* Environ. Res. Lett. 10, 104001, 2015

[8] K. Klima, J. Apt, M. Bandi, P; Happy, C. Loutan, R. Young, *"Geographic smoothing of solar photovoltaic electric power production in the Western USA"* Journal of renewable and sustainable energy 10, 053504, 2018

10.4229/EUPVSEC2023/4EO.2.2
020478-003

CONTROLLING POOLS OF PV AND WIND POWER PLANTS CONNECTED TO THE DISTRIBUTION GRID TO PROVIDE LOCAL ANCILLARY SERVICES

Achour OUDJEDI, Vincent KRAKOWSKI
Hespul association
Quentin Denoeud, Engie Green
Daniel Zahn, Kaspar Knorr, Sven Liebehentze, Fraunhofer Institute for Energy Economics and Energie System Technology
achour.oudjedi@hespul.org, vincent.krakowski@hespul.org

ABSTRACT: The REgions project, financed through European ERA-Net Smart Energy System program between 2019 and 2022, aimed at demonstrating the participation of fluctuating RES (fRES), such as PV and wind power plants, aggregated through a virtual power plant (VPP), in congestion management and voltage support.
In this paper we describe the French demonstration which focused on the distant activation of PV and wind power plants connected to the medium voltage distribution grid HTA (20 kV) to solve congestion and voltage issues on the high voltage transmission grid HTB1 (63 kV). These demonstrations took place in five different "areas" (each area being compounded of one or two HV/MV substations and the MV grid downstream), with the participation of 12 wind farms and 5 PV power plants for more than 150 MW of overall capacity. 20 tests were realized on those areas by sending activation orders to the power plants through the Fraunhofer's VPP. The Fraunhofer Institute developed a smart dispatch that enables to send orders to a pool of power plants, gathering both PV and wind plants, and to optimize the activation of each plant considering its Available Active Power (AAP) at each time.
Keywords: VPP, Ancillary services, fRES.

1 INTRODUCTION

The French electricity mix is expected to face important changes in the coming years regarding renewable energy sources (RES) development set in the law. The recently adopted Energy-Climate law (November 8th 2019) [1] sets the following objectives concerning the electricity sector by modifying the legislative part of the energy code [2]: by 2030, RES should represent 40% of the electricity production (this objective was already inscribed in the Energy transition law of 2015 [3]) compared to 22,7% in 2018, of which 12.4% is coming today from hydro [4].

To reach the 2030 objectives, new power plants capacity will mainly come from wind and solar as no new major development is expected in hydro (potential is practically exhausted, production is falling every year due to climate change) and biomass or biogas power plant capacities are increasing slowly. Growth of wind installed capacity should slightly to about 2GW/year [5] (currently around 1,5GW/year [4]), while growth of solar power capacity should dramatically increase from around 2.5GW/year today [4] to 4GW/year [5].

The French TSO RTE, covering the continental part of France, established scenarios of grid development up to 2035 based on these objectives of RES development [6]. RTE will have to face in the coming years the first large-scale renewal of its transmission lines, the reinforcement of high voltage (HV) lines and the development of new substations in order to host RES capacities. Up to 50GW of renewable capacities (wind and solar) are expected to be reached in 2025, RTE is optimistic that grid optimization can avoid large-scale modifications of the grid [6], through Dynamic Line Rating, Real Time Redispatching (via in-line opening and closing switches on HV lines to redistribute production on the most loaded lines), and RES production curtailment.

In 2019, RTE identified that 7 billion euros could be saved by 2035 through RES curtailment for 0.3% of annual production loss. In 2017, the main French DSO, Enedis, carried out a cost-benefit analysis on various smart solutions that could be deployed by 2030. It is indicated in this report, named "Smartgrids economic valuation", that the use of reactive power local adjustment of power plants connected to the MV grid would induce 500 k€ of costs saving for only 60 k€ of investment costs. On the contrary, centralized solutions for the voltage regulation would induce only 125 k€ of costs saving (for an investment of 215 k€). Thus, voltage management through MV RES plants is less expensive for a significant economic saving.

In this paper, we will focus on how a Virtual Power Plant (VPP) can enable the participation of fRES in such ancillary services in particular Congestion management and Voltage Support.

This paper is structured in three sections. This section presents the topic and motivation of the study. The second section illustrates the methodology used during the project. The last section introduces the results and highlights of the study.

2 METHODOLOGY

2.1 Definition of electrical regions and corresponding ancillary services

The REgions project analyses regional aspects of ancillary services and corresponding markets. For this purpose, a definition of a region from an electrical point of view is necessary. The REgions project defines its terminology (sub-regional, regional and inter-regional) based on the voltage level rather than on geographical or administrative boundaries. Electrical regions are defined as areas powered by the same extra-high voltage to high voltage

10.4229/EUPVSEC2023/4EO.2.3
020480-001

transformer (EHV/HV). From a geographical point of view, boundaries of these high-voltage grid areas are difficult to clearly identify due to their meshing: only the TSO has knowledge of the "normal" dispatch scheme. In contrast, medium-voltage grid areas are usually clearly geographically defined due to their mostly radial topology (see Figure 1 below).

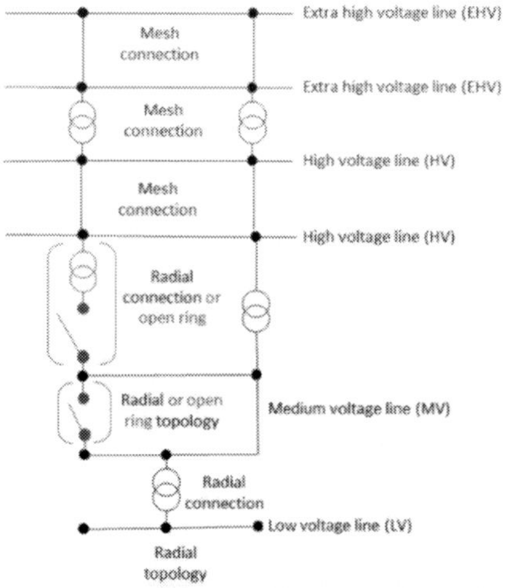

Figure 1: Typical grid topology / grid structure. Source: Fraunhofer IEE deliverable 3.1.1 of REgions project

In order to distinguish medium-voltage grid areas from the electrical regions in which they are located, the REgions project uses the term "sub-regional" or "local" for medium-voltage grid areas. Based on this definition, the terminological definition of the associated ancillary services and markets can be defined as shown in figure 2 below. The figure also points out that the real voltage values of the different voltage levels as well as the denominations of transmission and distribution grids can vary from country to country and are therefore not suitable for a general definition of electrical regions and their ancillary services and markets.

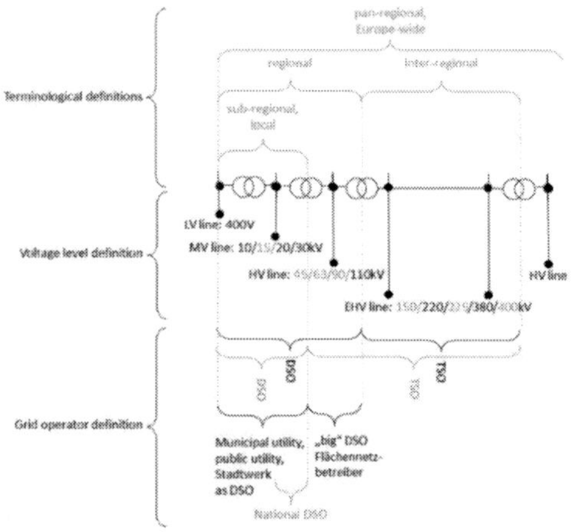

Figure 2: Definition of regions terminology (Blue), voltage levels and grid operators in Germany (Black) and France (Orange). Source: Fraunhofer IEE, Hespul.

Based on the above considerations, ancillary services and markets can be classified according to the following table.

Table 1: Regional classification of ancillary services analyzed in REgions (FR – France, DE – Germany)

Regionality	Beneficiary	Ancillary service
Sub-regional/ local	DSO (voltage level up to 40kV)	N/A (no use case on this level in REgions)
Regional	TSO (FR)/ DSO (DE, AT) (voltage level 40 to 110 kV)	Congestion management (FR) Voltage control (FR) (Constrained) Balancing reserve (FR/DE)
Inter-regional	TSO	Congestion management (DE) Constrained) balancing reserve (DE/FR)
Pan-regional/ Europe wide	TSO	Balancing reserve (AT)

2.2 Plants connection to the VPP

A large number of different power plants have been selected for their integration into the VPP. The integration of power plants can be very work-intensive due to the lack of standardized interfaces and the establishment of new secure communication links via a virtual private network to the different plant operators; some interfaces had to be individually customized and extended according to the use-case (UC) requirements. The interfaces used are shown in Figure 3. Consequently, not all of the numerous power plants selected could in the end be integrated into the VPP. Figure 3 shows an extract of the finally integrated power plants and the various interfaces developed for this purpose.

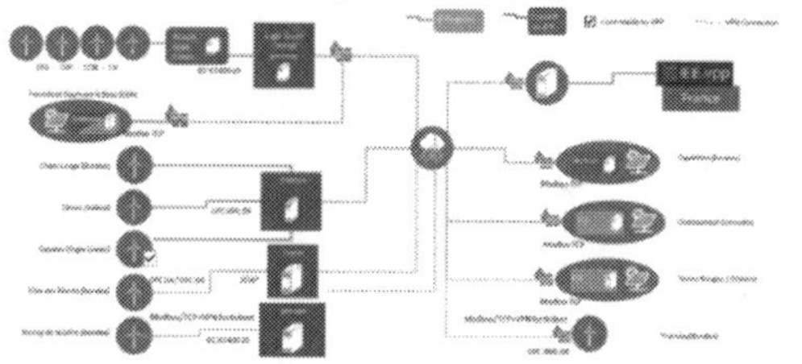

Figure 3: Plant integration Status Quo in France - Wind & PV Parks integrated in IEE.vpp

The VPP was already existing but had to be upgraded in various respects: a system architecture of the VPP for the different demonstration regions in France and Germany was developed, the plants and other components (such as power forecasts) were integrated into the VPP, and new functionalities were developed and added to the VPP. Among these new functionalities, Smart Dispatch was the most important. The project's use cases make it necessary to develop a procedure that, in the event of failure or poor operation of a plant, assigns its task of set point fulfillment to another suitable plant in the pool. This process is referred to as Smart Dispatch and reduces the gaps between the set point and the actual value of the VPP's feed-in.

10.4229/EUPVSEC2023/4EO.2.3
020480-002

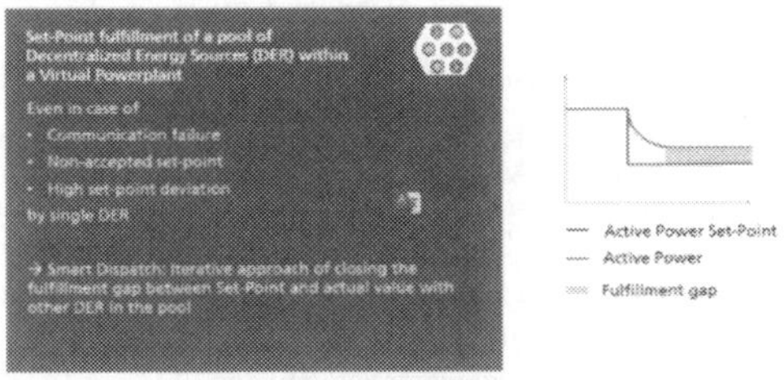

Figure 4: Principe of Smart Dispatch

Smart Dispatch starts with the preparation of a merit order list based on the marginal costs of the plants. After that, for each plant, the SCADA communications status and controllability is checked, the past and present acceptance of set points is considered (latest set point success), and the integral of the past deviations from the set point (performance indicator) is determined. If a plant does not meet the requirements, it is excluded from the smart dispatch process and the set point is distributed among the remaining plants according to the prioritization list (priority sequence), which is calculated from the marginal costs and the performance indicators. Ramp operation can also be specified as an option in smart dispatch. The drop or deselected plants from the priority sequence can be either receive a reset signal (back to normal operation) or a turn-off signal (set-point to zero MW).

2.3 Execution of the tests

In France, 20 field tests in 4 different regions were realized on 17 real power plants for a total of more than 150 MW of installed capacity. To conduct these tests in good conditions, we adopted a planning strategy for each of them.

Maximal energy loss definition: The power plants involved in the demonstrator are operated by several producers: Engie Green as funded project partner within REgions and Boralex, Valeco and Sonnedix as associated non-funded partners. A threshold for the maximum non-injected energy to the grid - due to downward activation of the power plants, mainly for active power control (UC1) and secondly for reactive power control (UC2) - was defined for each producer. To quantify the energy curtailed during the tests, we implemented for each plant an algorithm to estimate its expected production (especially for PV sites since the Wind plants had such estimation from design) and consequently create a new real time parameter: Available Active Power (AAP).

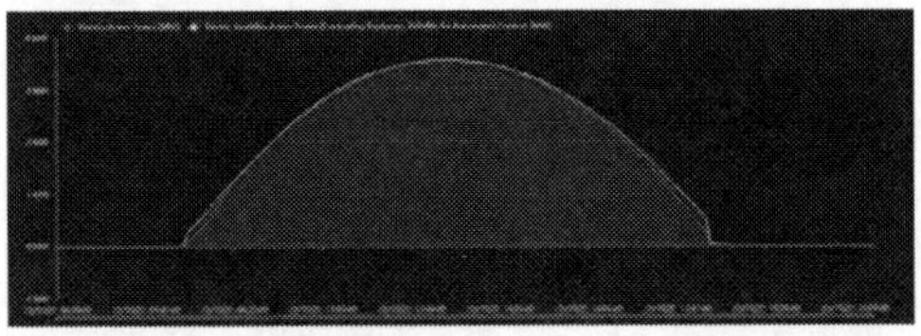

Figure 5: Example of AAP estimation and power production of one PV plant (in blue measured production and in magenta AAP estimation)

Check Wind and PV Forecasts: All performed tests required a minimal power produced by the power plants to be successful. It is then necessary to integrate Wind and PV forecasts within the test for planning purpose. To do so, the producers and Fraunhofer IEE use two main tools:

- Windy, a wind and weather add-on within Fraunhofer Virtual Powerplant application EnergyConnect.

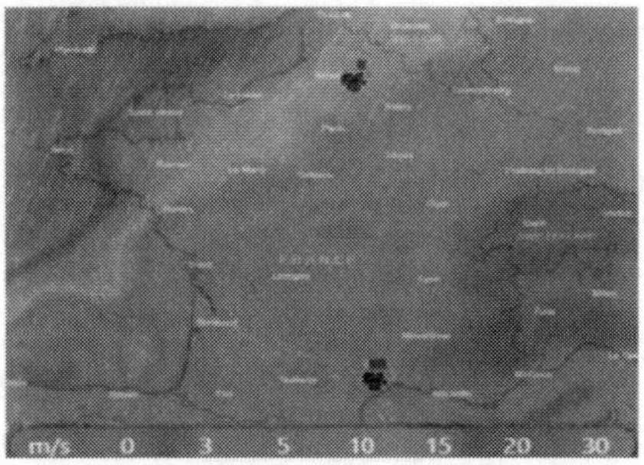

Figure 6: Example of forecast from windy.com embedded on the Fraunhofer VPP web-ui.

- Forecasts by Armines integrated in the Fraunhofer VPP on a plant level.

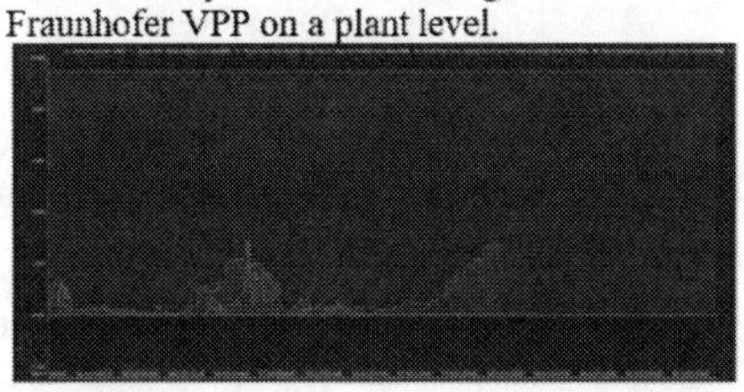

Figure 7: Example of forecast integration for one wind power plant. Forecasts are shown in red and realized production in bleu.

Internal and external communication: A validation workflow has been defined with the producers to examine the prerequisites for successful test execution. This allows each producer to validate plants availability, to check potential maintenance works, birds/noise management, grid curtailment or communication failure in order to ensure the feasibility of the test.

Figure 8: Example of communication failure to solve before a test.

Once the availability of the plants is checked, two main methods were performed depending on the use-case: The tests were either performed by script handled by Fraunhofer operators of the VPP or by using the web user interface (Web-UI) directly. For voltage support, the control of reactive power was exclusively done by the script because the Web-UI did not allow this possibility. For congestion management, either the script or the Web-UI were used to realize the control of the active power. The script was used to schedule a series of set-points and their duration on the VPP before the tests (preventive curtailment mode). This method allows only the operator of the VPP to control the pool of plants. The Web-UI allows a VPP operator as well as other stakeholders to send and update set-points in real time (curative curtailment and Delta control modes).

2.4 Evaluation of the tests (KPI)

To evaluate and make comparable the quantity and quality of reactive power and redispatch delivery, key performance indicators (KPIs) were developed. a Streamlit application (https://streamlit.io/) was written which, on the basis of the VPP behavior in the individual tests, provides a statement as to how well the VPP itself and the respective participating plants performed in the test. These KPI have also been used for active power control tests as well for reactive power control tests. In the following sections, the KPI parameterization for the active power tests is described as an example.

Two main aspects were considered in the evaluation:

2.4.1 KPI Control Deviation

This KPI evaluates how well the specified active power setpoint was reached. For this purpose, the corresponding integral is calculated for overshoot and undershoot in the times of active control. An active control time is defined by the fact that a setpoint specification is set unequal to the MPP (Maximum Power Point) specification. This can be transformed for the test period into an averaged power value for the time period, which, normalized to the installed power, results in a factor that enables a higher-level comparability of the individual field tests.

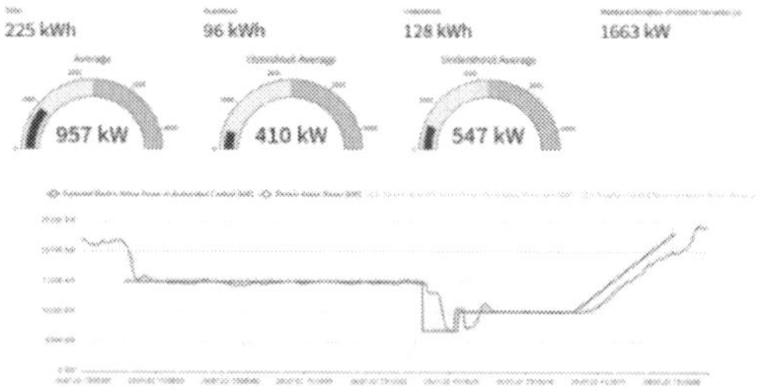

Figure 9: Example KPI Control deviation with overshoots and undershoots of the setpoint value (in red the setpoint and in blue the measured production).

. The following ratings for performance were defined in this KPI:

- A deviation less than or equal to 5% of the installed active power (Pnom) is rated as good and is based on the requirements for feed-in management from the grid connection guidelines for high voltage in Germany (VDE-AR-N 4120 Application Rule:2018-11).
- A deviation greater than 5%/Pnom and less than or equal to 15%/Pnom is classified as moderate. This additional evaluation criterion is intended to create a transition corridor to absolutely undesirable high target deviations.
- A deviation greater than 15%/Pnom is assessed as insufficient due to the significant deviation from the respective target value.

2.4.2 KPI Control Activation times

This KPI is used to evaluate how quickly the active power specifications could be realized. An activation time is defined for this purpose. The activation time per plant is defined by the time required for a plant to reach an active power corridor around a given setpoint for the first time after a set point command has been sent.

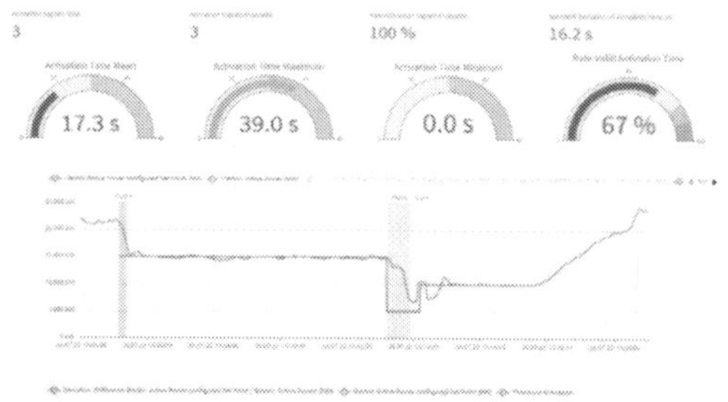

Figure 10: Example of Control Activation time KPI (in red the setpoint and in blue the measured production).

In this KPI, the following ratings for the activation times were defined:

- An activation time less than or equal to 40 seconds is good from an evaluation point of view and fulfills the requirements from the grid connection guidelines for generation units in the feed-in management case.
- An activation time greater than 40 seconds and less than or equal to 80 seconds is poor and does not meet the requirements from the grid connection guidelines point of view, but offers a better insight into possible average target value deviations due to the differentiation analogous to the control deviation KPI.
- An activation time greater than 80 seconds is rated as insufficient due to the significant deviation from the desired target value.

3 RESULTS AND DISCUSSION

In this section we will discuss the results of the tests performed on the plants through the VPP developed by the Fraunhofer IEE. As previously mentioned, a total of 20 field tests were conducted on 17 power plants (150 MW). We can't present all of them in this paper, so we will give an example of analysis for each use case studied in the French demonstrator: Congestion Management (UC1) and Voltage Support (UC2). We will also, share the main important findings over the 20 tested realized for each use case.

3.1 UC1: Congestion Management

Three kind of control mechanisms have been tested to demonstrate the potential of VPP's cluster control to address congestion management. Each control mechanism addresses a grid operator need and are associated to sub use-cases. Since none of the grid operators were involved in the demonstrator, these controls have been tested on emulated congestion and corresponding grid operator's needs.

10.4229/EUPVSEC2023/4EO.2.3
020480-004

Preventive curtailment: Active power control of a local cluster (pool) through a predefined schedule. Setpoints are likely to be defined by grid operators thanks to simulations to avoid forecast congestion during period of maintenance or work on the grid that limit the flow on a specific node of the network. Such schedule is defined in advance and can't be update in real time.

Curative Curtailment: Active power control of a local cluster (pool) in real time, when grid flows have to be controlled and limit to resolve a current congestion. With the use of real time measurements and close surveillance by grid dispatchers, the setpoints are updated in real time to optimize the system and resolve the congestion.

Delta control: Active power control of a local cluster (pool) defining a fixed and stable difference between the feed-in power and the available power the cluster would have produced without control. Such "delta control" are more likely to be used to match reserve products that can be defined on markets.

3.1.1 Active power control on Lodève Zone

This test was performed on one cluster of PV and Wind plants connected to the same electrical region as defined previously. It pretends to solve congestions which would be anticipated by DSO (preventive curtailment - see above). Obviously, in our case, the congestions were emulated as no DSO was involved in the project.

The Cluster was composed by 3 wind and 2 PV plants. It represents respectively 30.81 MW and 14.2 MW installed capacity. In total, we had 45 MW of installed capacity that was controlled remotely.

As the curtailment was preventive, the test was done by applying a single setpoint to the whole pool during the test. The setpoint is represented in yellow line, the AAP in green one, the measured production in blue and the availability of the wind turbines in blank (see Figure 11).

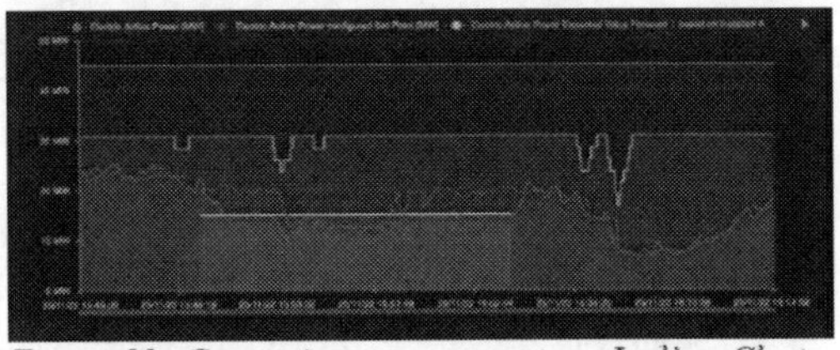

Figure 11: Congestion management on Lodève Cluster through a VPP.

Weather conditions: The lack of production observed in PV plants (which only reached 5% of its installed capacity) was compensated by favorable conditions of Wind production (which reached 72% of its installed capacity). Note that the test was performed on November, that' why the PV plants were at their minimum level.

Shut down events: As we can observe on the chart above, two shut down of wind turbines occurred during the test (blank line). It was due to the bird management mechanism implemented in the turbines. In the first one, 2 wind turbines were impacted at 15h53. In the second one, only one was stopped. These events allowed us to challenge our "Smart Dispatch" optimization algorithm.

"Smart Dispatch" behavior: This algorithm, as it is described above, is configured in order to minimize the losses (no injected energy due to the curtailment) and compensate the lack of reaction of one or more plants in the cluster by curtailing deeply the others. The "Smart Dispatch" considers the production capability of each plant (AAP) and their availability to define the optimized setpoint for each one and maintain the global setpoint fixed for the pool. The setpoint sent to the plants is updated every 3 seconds.

Thanks to the "Smart Dispatch" a production which would have otherwise been curtailed if the setpoint was the same all the time, as it is usually practiced today, was saved (see the fixed setpoint in red line on the chart). Once some wind turbines shut down, the algorithm adapts the setpoint sent to the plants which could produce more in order to increase their production.

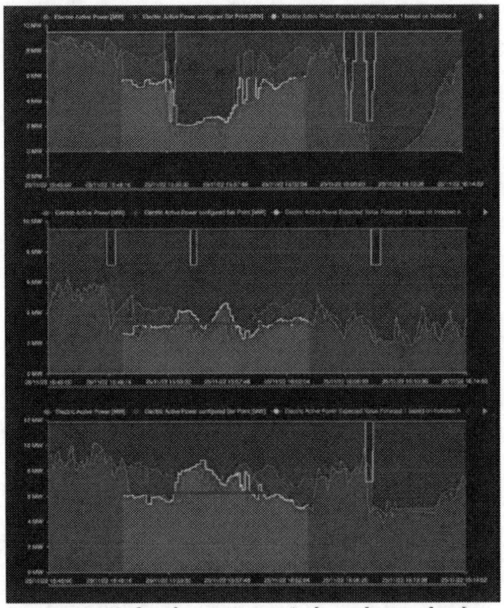

Figure 12: Updated setpoints on 3 plants during shutdown events.

Lack of production: A test was performed as the production of solar plants was very low. Among the 2 PV plants, one of them was not controllable. The "Smart Dispatch" curtailed the controllable one by 10 % in average proportionally to its production and compensate the lack of control of the other by curtailing deeply the wind plants involved in the test.

Figure 13: No controllable PV plant in the cluster.

10.4229/EUPVSEC2023/4EO.2.3
020480-005

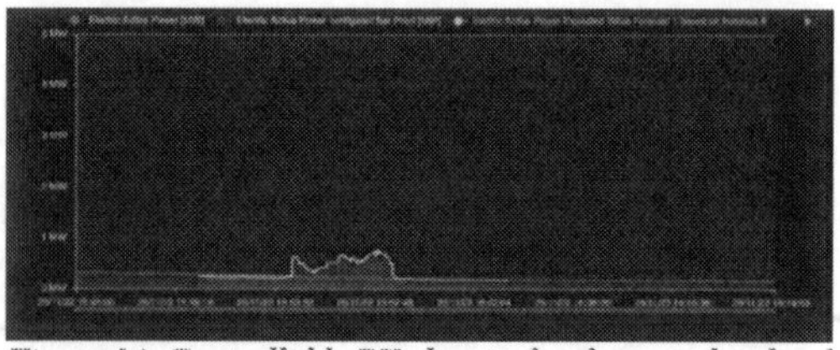

Figure 14: Controllable PV plant in the cluster at low level of production.

Control of the pool: The control of the pool was successful even if shut down occurred and one PV plant was not controllable thanks to the “Smart Dispatch”.
The time response of the pool to the setpoint was 47s and the deviation was 2% compared to setpoint. Since we deal with congestion, the deviation indicated here take into account only the deviation which exceeds the setpoint.
The AAP was revealed to be very important to the “Smart Dispatch”. Thus, a good precision is required to get finally a reliable and optimized control of the pool that solves congestion constraints and minimizes the loss of production.

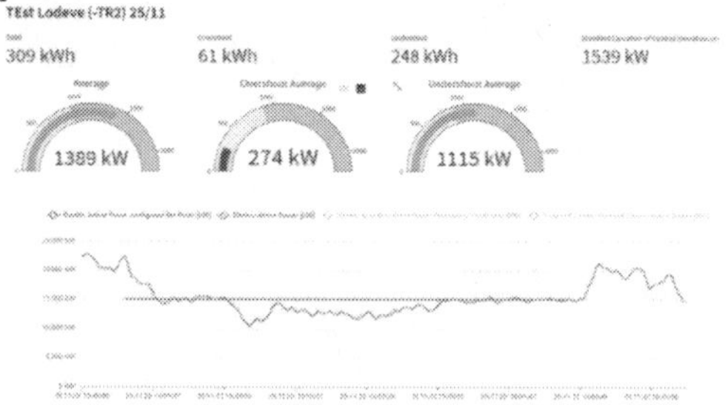

Figure 15: KPI control deviation for Lodève Cluste test.

Figure 16: KPI activation time for Lodève Cluster test.

Congestion Management highlights: The most important findings over the 13 tests performed on active power are summed-up below:

- The time response of cluster control varies between 8 and 43 seconds depending on the plants within the cluster and if ramps were configured or not.
- The VPP, through its “Smart Dispatch”, ensures set-point delivery with better resilience in case of control failures.
- The VPP cluster control enables to minimize the economic losses. Based on real production capability of each plant and their availability, it reduces the energy curtailed compared to the usual curtailment which uses fixed setpoint for each plant.
- The VPP can handle and harmonize specific control features such as ramping (up/down) and reserve control.

3.2 UC2: Voltage Support

Three kinds of control are usually used to control reactive power generation:

- Q setpoint control: The grid operators ask a power plant to inject or absorb a fixed Reactive Power Setpoint (in kVAr).
- Power factor (tan(phi)) Setpoint Control: The grid operators ask a power plant to control the Reactive power based on the Active Power feed in (setpoints as a tan(phi)=Q/P or cos(phi)=P/S).
- Q=f(U) control: The grid operators ask a power plant to control the reactive power based on the voltage at the point of coupling.

Such control at power plant level depends on the capabilities of internal SCADA to accept the different setpoints. It has been decided that the control of the reactive power of a whole cluster will use only Q setpoint control (in kVAr).

3.2.1 Reactive Power control on Lodève Zone

As for Active Power control, this test implied PV plant (one plant) as well as wind plant (3 plants) for a total of 35.81 MW of installed capacity.

The test was done by applying many different Q setpoints to the whole cluster. As the Reactive Power control has been demonstrated in REgions project for the first time, the Q setpoint definition is not handle by the VPP Web-UI but had to be defined through specific script defined by Fraunhofer IEE and sent to the energy assets involved via the VPP.

Figure 17: Test of Reactive Power control on Lodève Cluster.

Weather conditions: The test was performed in average conditions of production. The whole production of the cluster reached 34% of its installed capacity.
Shut down events: No shut down events occurred during this test.
"Smart Dispatch" behavior: The algorithm was not performed on reactive Power control contrary to what was done for Active Power control. Indeed, a model of estimated reactive power capability for each plant at each time is needed to perform such an optimized control as done for active power.
As a result, the control was not able to fully compensate the deviation to the setpoint of one plant by the other plants included in the cluster.

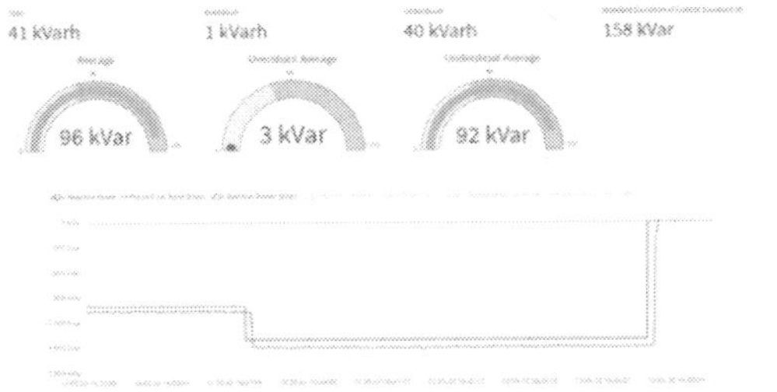

Figure 18: Reactive Power control deviation for PV plant with continuous offset between set-point and measured reactive power.

Control of the pool: The control of the pool was successful even if its response was fluctuating around the setpoint due to the deviation observed in some plants.
The deviation to setpoint was very low, 2% at its maximum (see Figure 19).

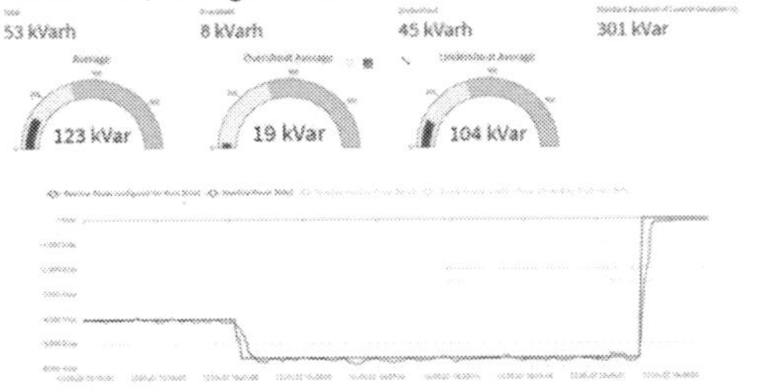

Figure 19: KPI Reactive Control deviation of Lodève zone test.

The response time was also quick. It varies between 6 and 25 seconds depending on the plants (see Figure 20). The time activation is discussed with further details in the conclusion including an analysis of communication's chain time reaction.

Figure 20: Overview of plants individual activation time for the test.

Voltage support highlights: The main findings over the 5 tests realized on reactive Power are summed-up below:

- The activation time for the reactive power control is less than 60s in all the tests performed as for Active Power control.
- The provision of Reactive Power is ensured even in case of low production condition (< 20%Pmax).
- The provision of Reactive Power is ensured under control of Active Power. Both powers could be controlled at the same time with good performances.
- The provision of Reactive Power is not ensured while the Wind farms plants are stopped. One test highlighted the impossible provision (or absorption) of reactive power when a setpoint P=0kW was sent to the plant. This behavior could not be representative of all Wind Farms since manufacturers could develop some optional "stand-in reactive capabilities" at customer requests.
- The Smart Dispatch can't be developed for Reactive Power control because of the unavailability of estimated Reactive Power on the plants. The main challenge in future is to find a model that could estimate the Reactive capability of a plant regarding the components of the plants, its internal capability and also the voltage measured at the point of common coupling.

4 CONCLUSION

This paper proposed to evaluate the contribution of the VPP in enabling the participation of fRES to local ancillary services such as Congestion Management and Voltage Support.

The French Demonstrator, that we presented here, has revealed the key role the VPP could play in the energy transition as more fRES will compose the electrical mix. Thanks to the VPP, the control could be more optimized and reliable with a reduction of costs for the grid operators and producers. Different features (ramp control, reserve mode control, etc.) could be harmonized and handled at the VPP level instead of individual plants. This allows plants with old controller to participate in such ancillary services and save money for those who would have to develop them to participate.
The VPP could be operated by a GRD or an entity market like aggregator which can make the link between the GRDs and the users. It reduces the IT interfaces between grid operators and users to concentrate them in one interface which's the VPP.

The time response and the deviation observed in the French demonstrator are considered sufficient to resolve congestion or voltage constraints. Nevertheless, the times' reaction could appear quite long but the time indicated in the analysis contains the total communication chain (between the VPP and the controller and between this one and wind turbines or inverters). For example, one controller of wind plants communicates every 10 seconds with the wind turbines. Thus, if a setpoint is sent at the beginning of this timeout (10s), hte setpoint will only be considered after 10s. In this case, 10s delays will be added to the time's reaction of the plant. In furthers works, it would be interesting to measure precisely each time in the communication chain.

Even if these results could not be generalized at all the existing plants, it demonstrates that fRES are able to participate in such ancillary services and that the VPP facilitates and optimizes this participation.

5. REFERENCES

[1] Energy-Climate Law n°2019-1147, November 8th 2019, Article I ([Online] https://www.legifrance.gouv.fr/affichTexte.do?cidTexte=JORFTEXT000039355955&fastPos=1&fastReqId=377625168&categorieLien=cid&oldAction=rechTexte#LEGISCTA000039358574)
[2] Legislative Energy code, Article 100-4 ([Online] https://www.legifrance.gouv.fr/affichCode.do?cidTexte=LEGITEXT000023983208&dateTexte=20110606)
[3] Energy Transition for Green Growth law n° 2015-992, August 17th, Article 1 III
[4] Bilan Electrique 2018, RTE [Online] https://www.rte-france.com/sites/default/files/be_pdf_2018v3.pdf / Similar data can be found in the "National

Survey Report of PV Power Applications in France 2018", IEA PVPS
[5] Le baromètre 2018 des énergies renouvelables électriques en France, Observ'ER [Online]
[6] Schéma décennal 2019, RTE. [Online] https://www.rte-france.com/fr/article/evolution-dureseau-electrique-francais-l-horizon-2035

10.4229/EUPVSEC2023/4EO.2.3
020480-008

HEIBRIDS · DLR · ZIB · Freie Universität Berlin · HZB Helmholtz Zentrum Berlin

APPLICATION OF NOWCASTING TO REDUCE THE IMPACT OF IRRADIANCE RAMPS ON PV POWER PLANTS

J. Schaible, B. Nouri, L. Höpken, T. Kotzab, M. Loevenich, N. Blum , A. Hammer, J. Stührenberg, K. Jäger, C. Becker, S. Wilbert

EU PVSEC 2023
4EO.2.4
22/Sep/2023

40th European Photovoltaic Solar Energy Conference and Exhibition

Introduction – Nowcasts – Dataset – Model performance – Control strategy – Results – Conclusion

WHY POWER SMOOTHING?

- **Solar energy challenges grid stability [1]**
  - e.g. **grid frequency balance [2] [3]**
- **Irradiance fluctuations cause ramps**
- **Proposed solution: Power smoothing [4] [5]**
  - (e.g. 10% power / min)

Example: Power Fluctuations

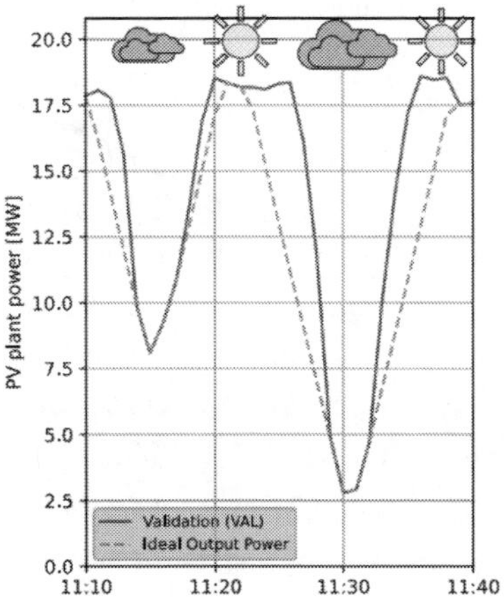

➢[1] Ottmar Edenhofer et al. IPCC, 2011: Summary for Policymakers.
➢[2] Esteban A. Soto et al. "Analysis of Grid Disturbances Caused by Massive Integration of Utility Level Solar Power Systems".
➢[3] N. Mithulananthan, R. Bansal, and V. Ramachandaramurthy, "A review of key power system stability challenges for large-scale PV integration".
➢[4] Remember Samu et al. "Applications for solar irradiance nowcasting in the control of microgrids: A review".
➢[5] Qianwei Zheng et al. "Overivew of grid codes for photovoltaic integration".

WHICH SOLUTIONS EXIST?

- **Current solution: energy storage**

  (e.g. batteries, …)

- **Nowcasting: short-term solar irradiance forecasting**
- **Ideal nowcasts can substitute battery storage completely [6]**

  But: Nowcasts with high resolution required

  & uncertainties are decisive

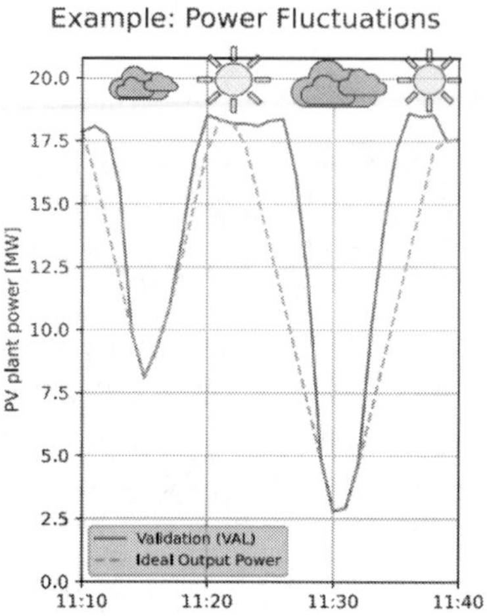

➢[6] Mojtaba Saleh et al. "Battery-less short-term smoothing of photovoltaic generation using sky camera".

PROBABILISTIC NOWCASTS

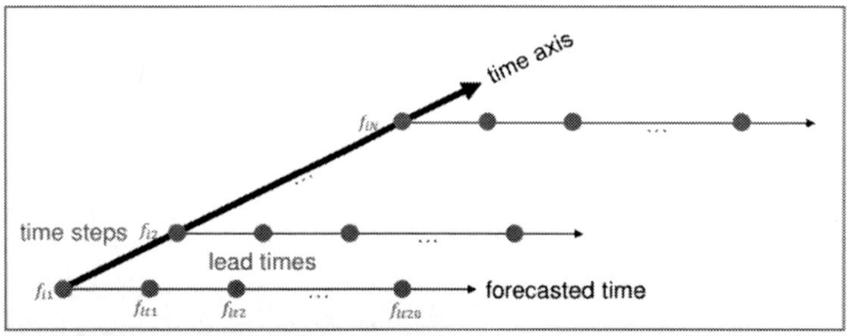

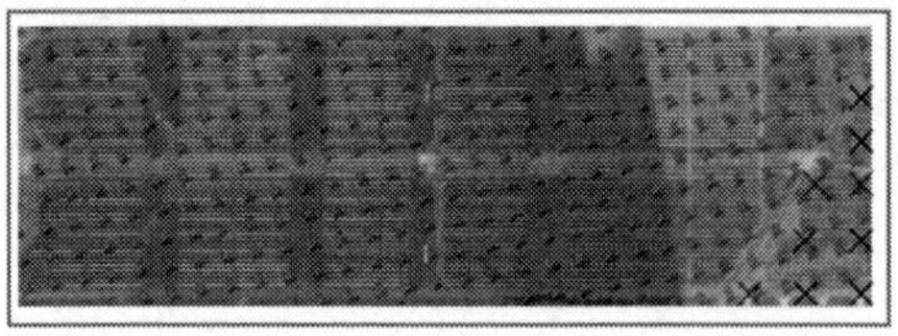

Parameter	ASI network
Spatial resolution	50 m
Extent	up to 156 km²
Forecasts update	30 s
Forecast step	1 min
Forecast horizon	20 min

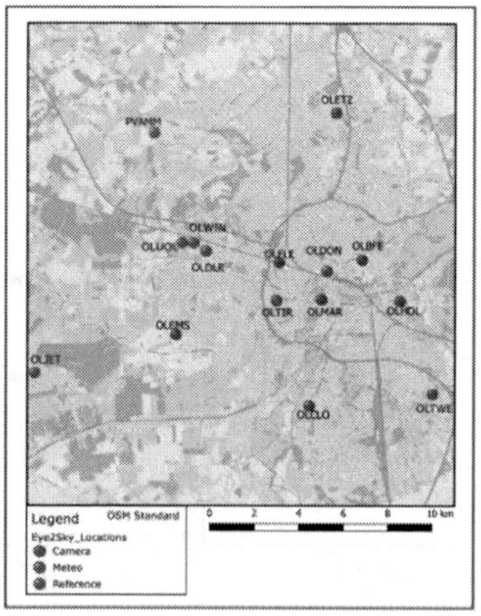

[13]

CLASSIFICATION

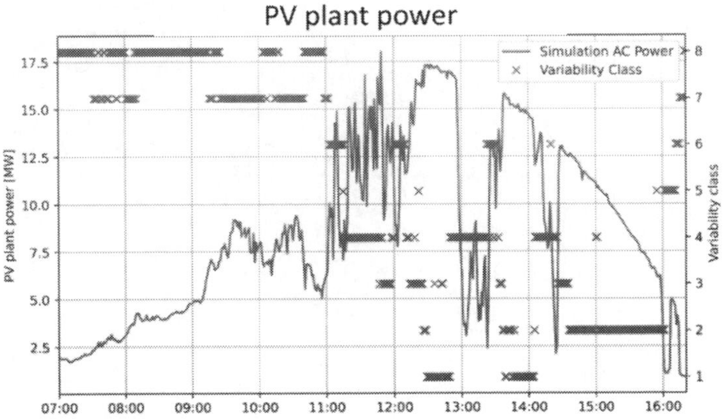

Class	Sky conditions	Variability
1	Mostly clear sky	Low variability
2	Almost clear sky	Low variability
3	Almost clear sky	Medium variability
4	Partly cloudy	High variability
5	Partly cloudy	Medium variability
6	Partly cloudy	High variability
7	Almost overcast	Medium variability
8	Mostly overcast	Low variability

DATASET

- **nowcasts for 18 test days with high variability conditions**

  Days have large range of classes and vary from each other

- **PV power plant production data (Validation)**

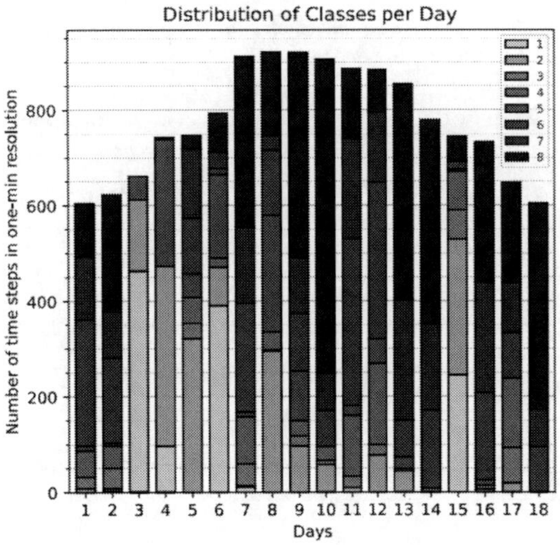

MODEL PERFORMANCE

- **Simulated power vs. validation data**
- **18 day test dataset**

  with various weather conditions
- **nRMSE between 2.7 and 21.7% (mean 12.0%)**

  (normalized to max power of day)
- **nBIAS of +0.5%**

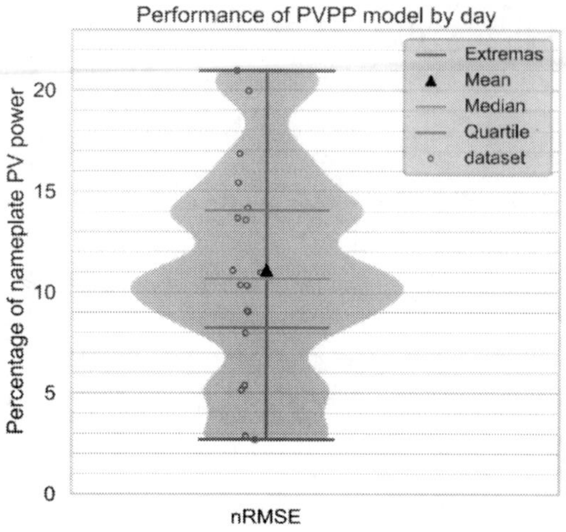

3 STRATEGIES: VAL, IDV & MIX

- **Battery works as a fail-save: only interferes in case of control strategy failure**

  System power = output power + battery

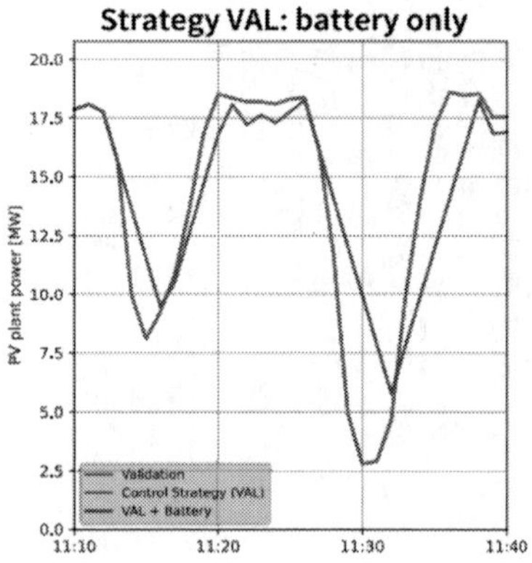

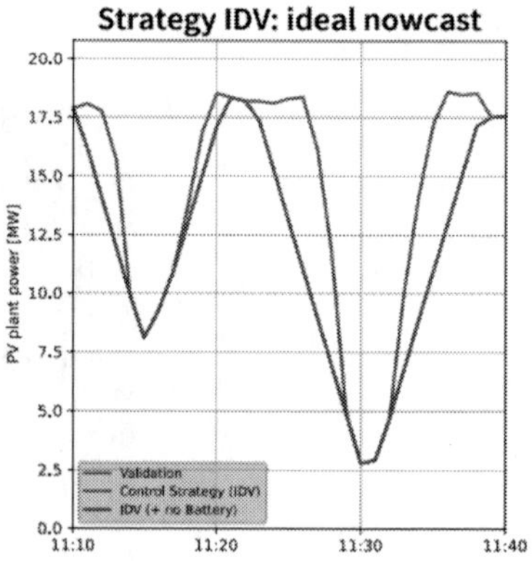

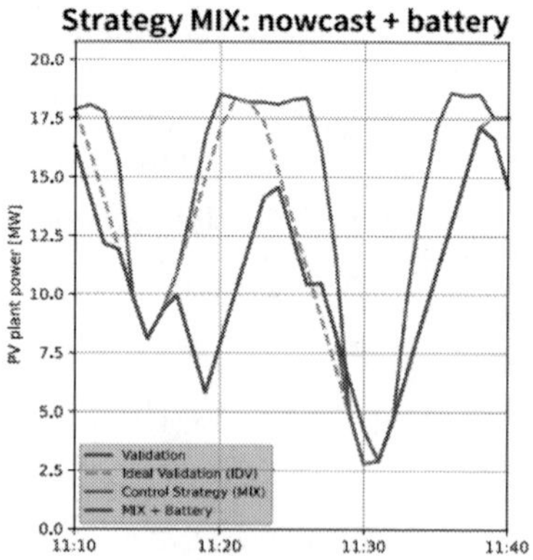

Introduction – Nowcasts – Dataset – Model performance – **Control strategy** – Results – Conclusion

3 STRATEGIES: VAL, IDV & MIX

- **Battery works as a fail-save: only interferes in case of con**

System power = output power + battery

Strategy VAL: battery only

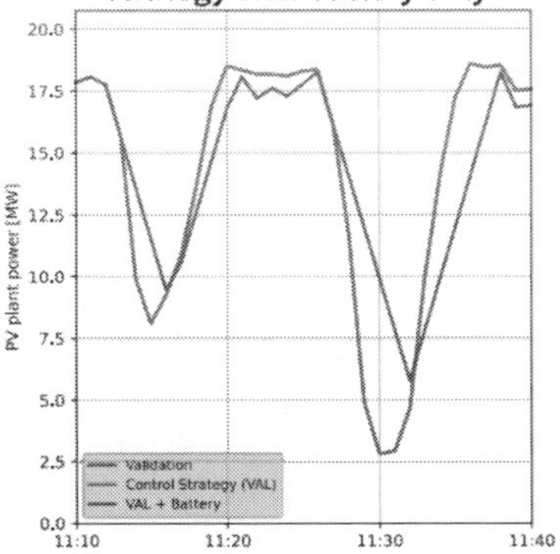

Strategy IDV:

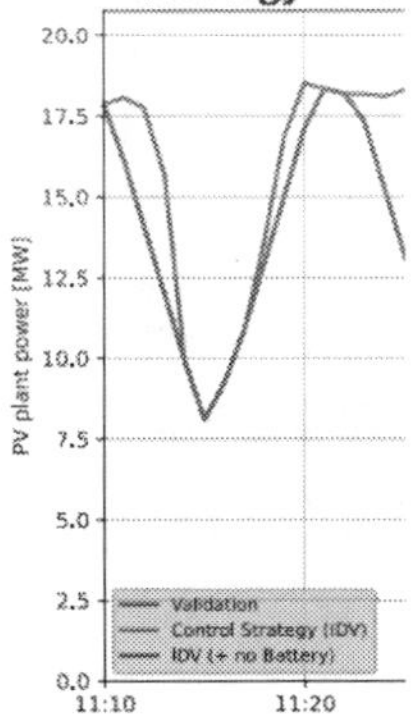

Strategy MIX: nowcast + battery

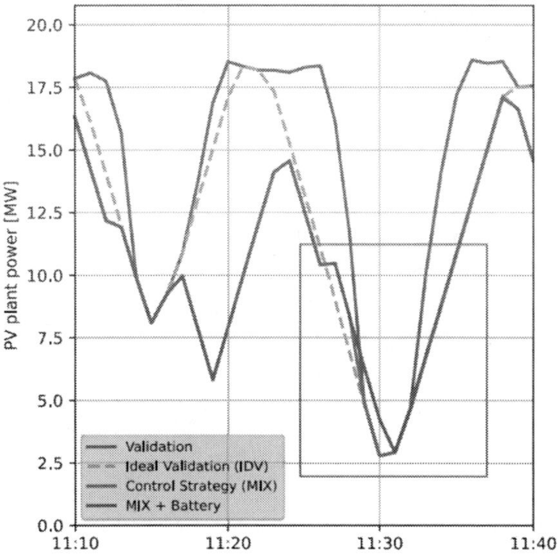

9

Introduction – Nowcasts – Dataset – Model performance – Control strategy – **Results** – Conclusion

18 TEST DAYS – ACCUMULATED

Three cases for comparison:

- **Strategy VAL: battery only**
- **Strategy IDV: ideal nowcast**
- **Strategy MIX: nowcast + battery**

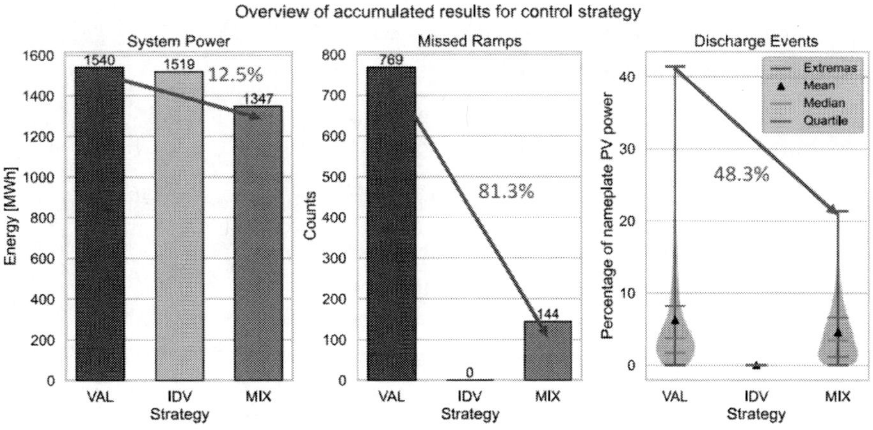

10

HZB Helmholtz Zentrum Berlin

UPSCALING TO ONE YEAR

Based on:

- 18 test days
- Class frequency

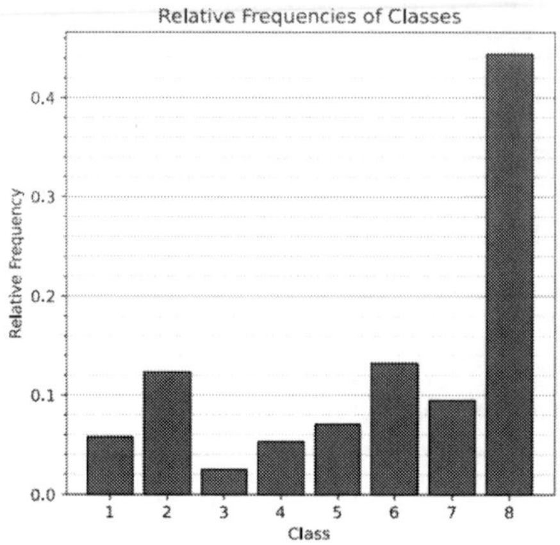

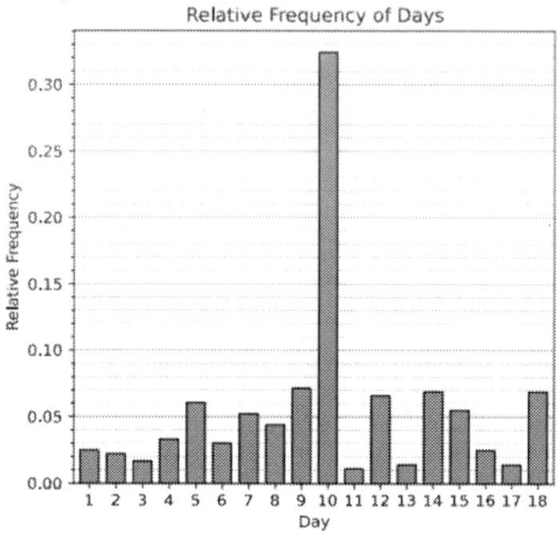

HZB Helmholtz Zentrum Berlin

40th European Photovoltaic Solar Energy Conference and Exhibition

Introduction – Nowcasts – Dataset – Model performance – Control strategy – **Results** – Conclusion

ECONOMIC EVALUATION

Key findings:

- **Energy loss reduced to 8.7% for one year**
- **Levelized-cost-of-energy (LCOE) reduced by 12.7% (MIX)**
- **Reduction potential of 34.8% (IDV)**

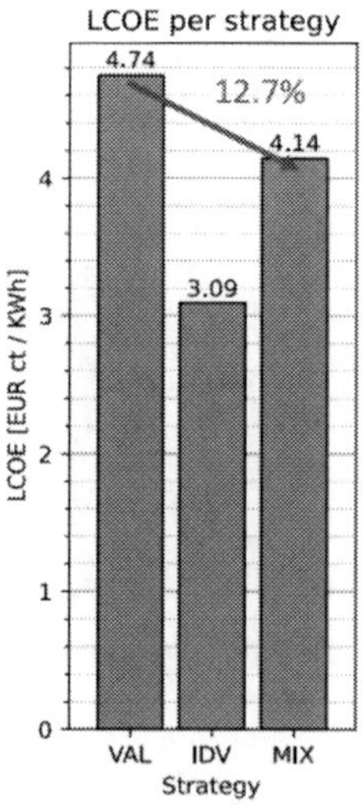

HZB Helmholtz Zentrum Berlin

CONCLUSION

- **Simulation study of nowcasts in power smoothing for Germany**
- **18 test days with variety of irradiance conditions**

  Mean model performance of 12.0% nRMSE (max)

  4 out of 5 ramps avoided by control strategy alone

  Fail-save battery with smaller capacity (-71.1%) and power (-48.3%)
- **Full year (scaled-up)**

  8.7% curtailment losses

  LCOE reduction of 12.7% with potential to 34.8%

40th European Photovoltaic Solar Energy Conference and Exhibition

THANK YOU FOR YOUR ATTENTION!

- **Jonas Schaible**

  Helmholtz Center Berlin

  jonas.schaible@helmholtz-berlin.de

- **Bijan Nouri**

  German Aerospace Center (DLR)

  bijan.nouri@dlr.de

HZB Helmholtz Zentrum Berlin HEIBRIDS DLR ZIB Freie Universität Berlin

UNIVERSITY OF CAMBRIDGE

Trap-assisted tunneling in ultra-thin GaAs solar cells for space applications

Armin Barthel[1], Larkin Sayre[1,2], Shin-Ichiro Sato[3], Takeshi Ohshima[3], Tetsuya Nakamura[4], Mitsuru Imaizumi[4,5], Louise C. Hirst[1,6]

[1] Department of Materials Science and Metallurgy, University of Cambridge, Cambridge, United Kingdom
[2] ORISE Fellow at U.S. Department of Energy, Washington D.C., United States
[3] National Institutes for Quantum Science and Technology (QST), Takasaki, Gunma, Japan
[4] Japan Aerospace Exploration Agency (JAXA), Tsukuba, Ibaraki, Japan
[5] Sanjo City University, Sanjo City, Niigata, Japan
[6] Department of Physics, University of Cambridge, Cambridge, United Kingdom

EU-PVSEC 2023

The Ultra-Thin Solar Cell

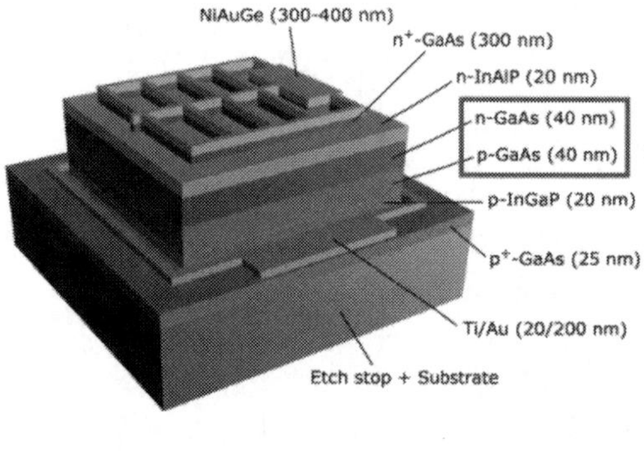

80 nm thick absorber layer → **1e17 cm^{-3}** / **1e18 cm^{-3}**

→ **High radiation tolerance for space applications**

high energy protons & electrons

UNIVERSITY OF CAMBRIDGE 2

The Ultra-Thin Solar Cell

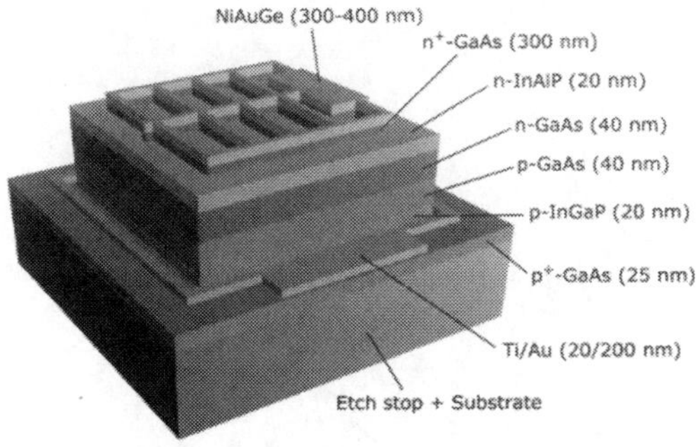

Radiation degradation of device performance

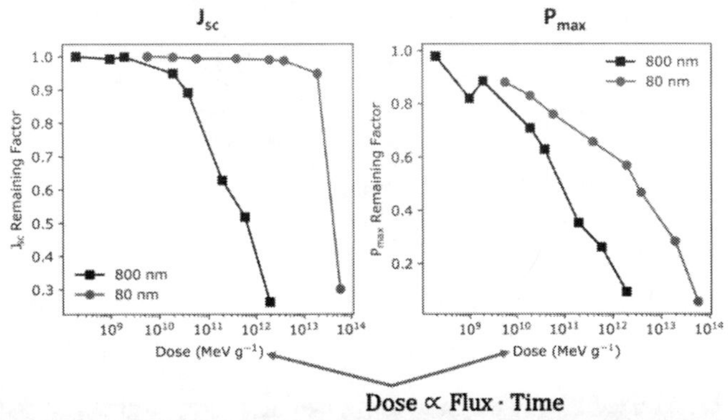

40th European Photovoltaic Solar Energy Conference and Exhibition

Diode Equation and Dark IV

2 Diode Model

$$J = \underbrace{J_{01}\exp\left(\frac{qV_j}{kT}\right)}_{\text{Diffusion}} + \underbrace{J_{02}\exp\left(\frac{qV_j}{2kT}\right)}_{\text{Recombination}} \qquad \underbrace{V_j = V - JR_s}_{\text{Series Resistance}}$$

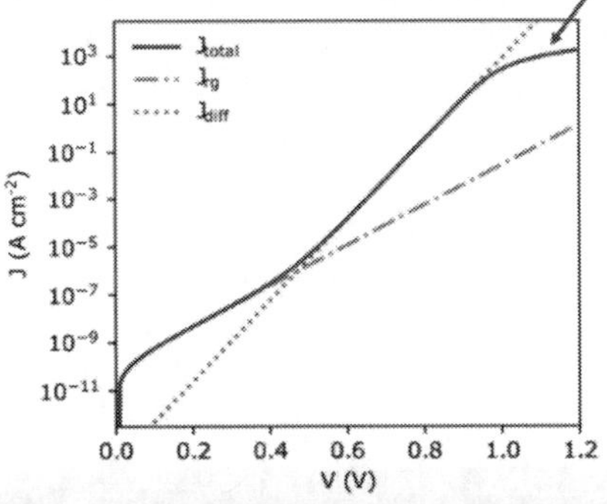

Local Ideality Factor

$$n = \frac{q}{kT}\left[\frac{d\ln(J)}{dV}\right]$$

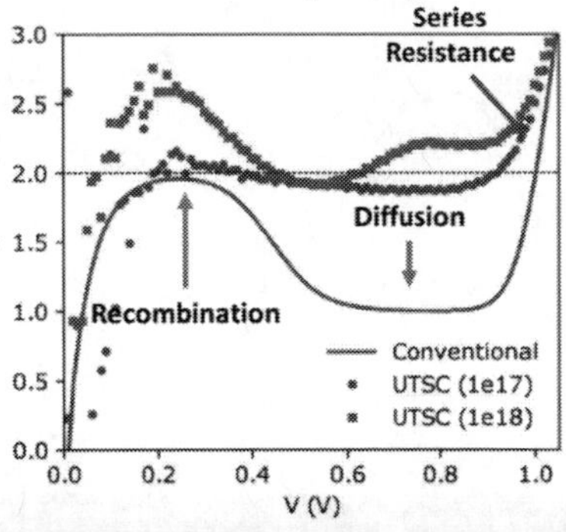

Radiation Damage and Shunting

Effect of Radiation (1e18 devices)

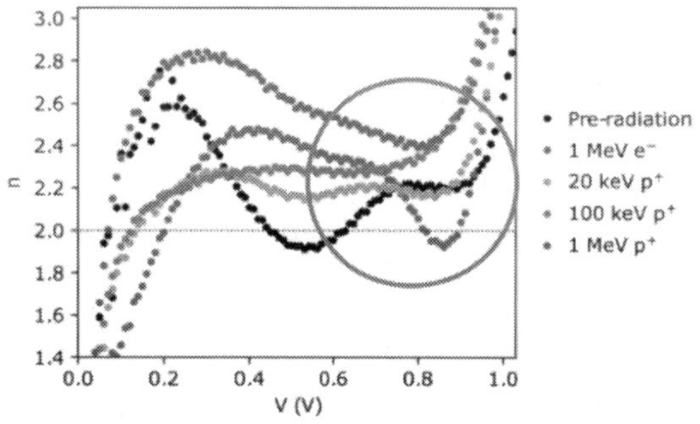

Effect of Shunting

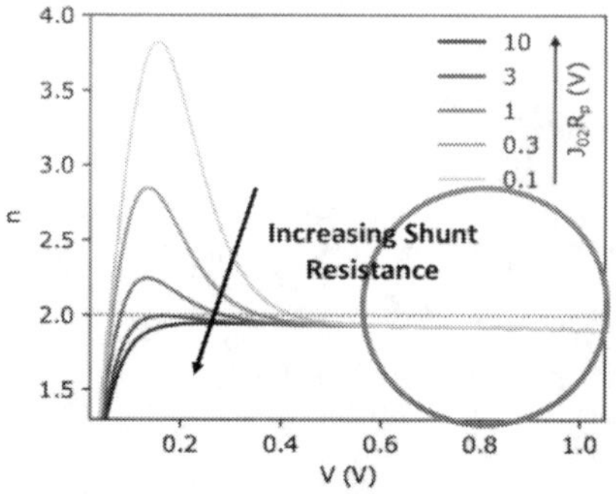

40th European Photovoltaic Solar Energy Conference and Exhibition

Cause of High Ideality Factor

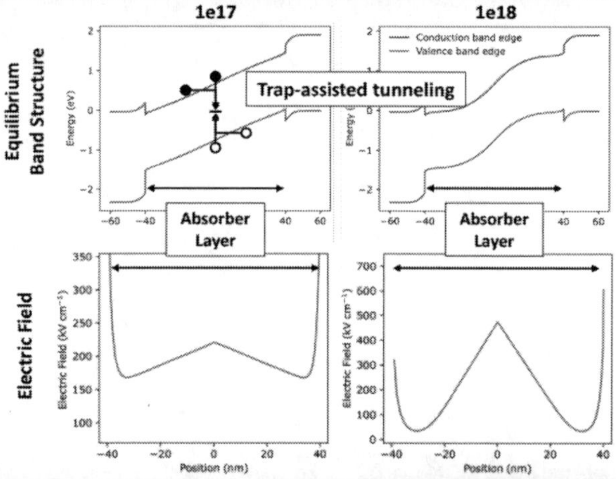

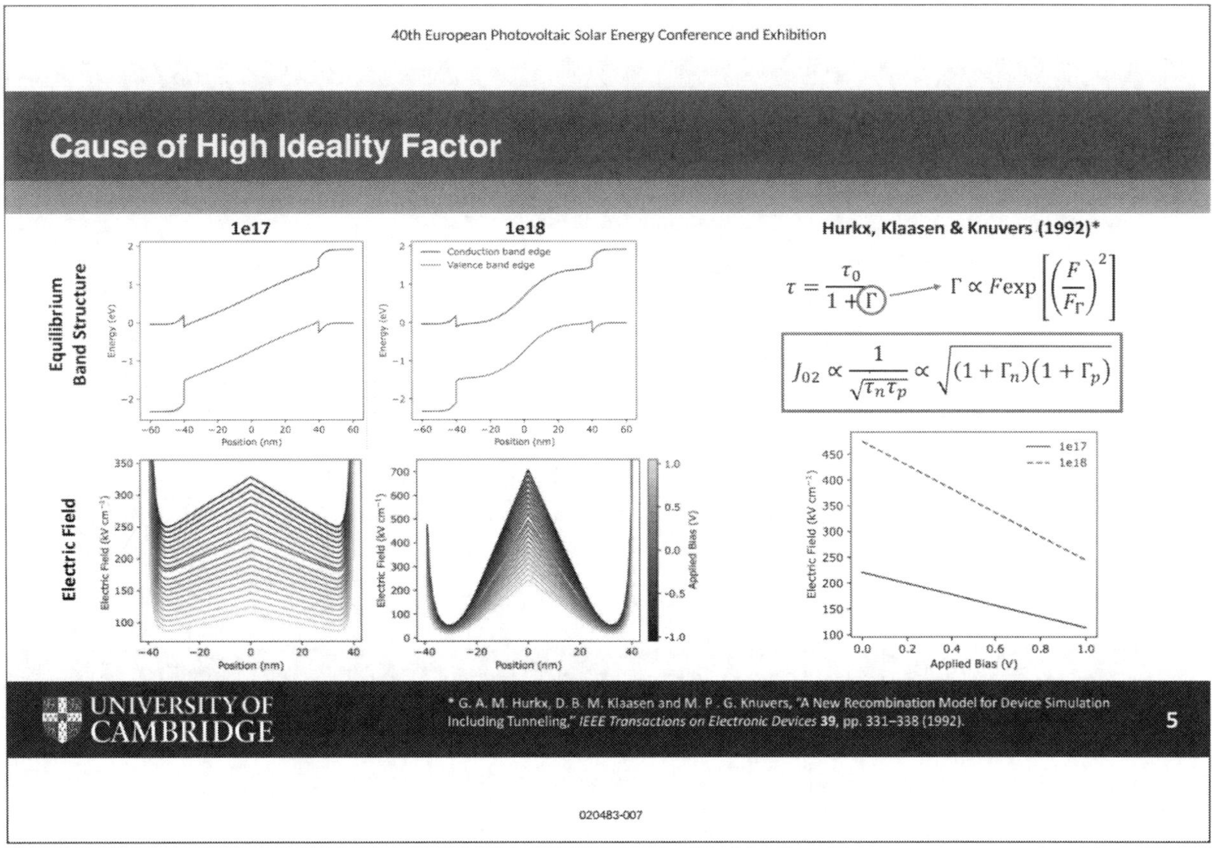
40th European Photovoltaic Solar Energy Conference and Exhibition
Cause of High Ideality Factor
1e17
1e18
Equilibrium Band Structure
Electric Field
Conduction band edge
Valence band edge
Energy (eV)
Position (nm)
Electric Field (kV cm⁻¹)
Applied Bias (V)
Hurkx, Klaasen & Knuvers (1992)*
$\tau = \frac{\tau_0}{1+\Gamma} \longrightarrow \Gamma \propto F\exp\left[\left(\frac{F}{F_\Gamma}\right)^2\right]$
$J_{02} \propto \frac{1}{\sqrt{\tau_n \tau_p}} \propto \sqrt{(1+\Gamma_n)(1+\Gamma_p)}$
1e17
1e18
UNIVERSITY OF CAMBRIDGE
* G. A. M. Hurkx, D. B. M. Klaasen and M. P. G. Knuvers, "A New Recombination Model for Device Simulation Including Tunneling," IEEE Transactions on Electronic Devices 39, pp. 331–338 (1992).
5
020483-007

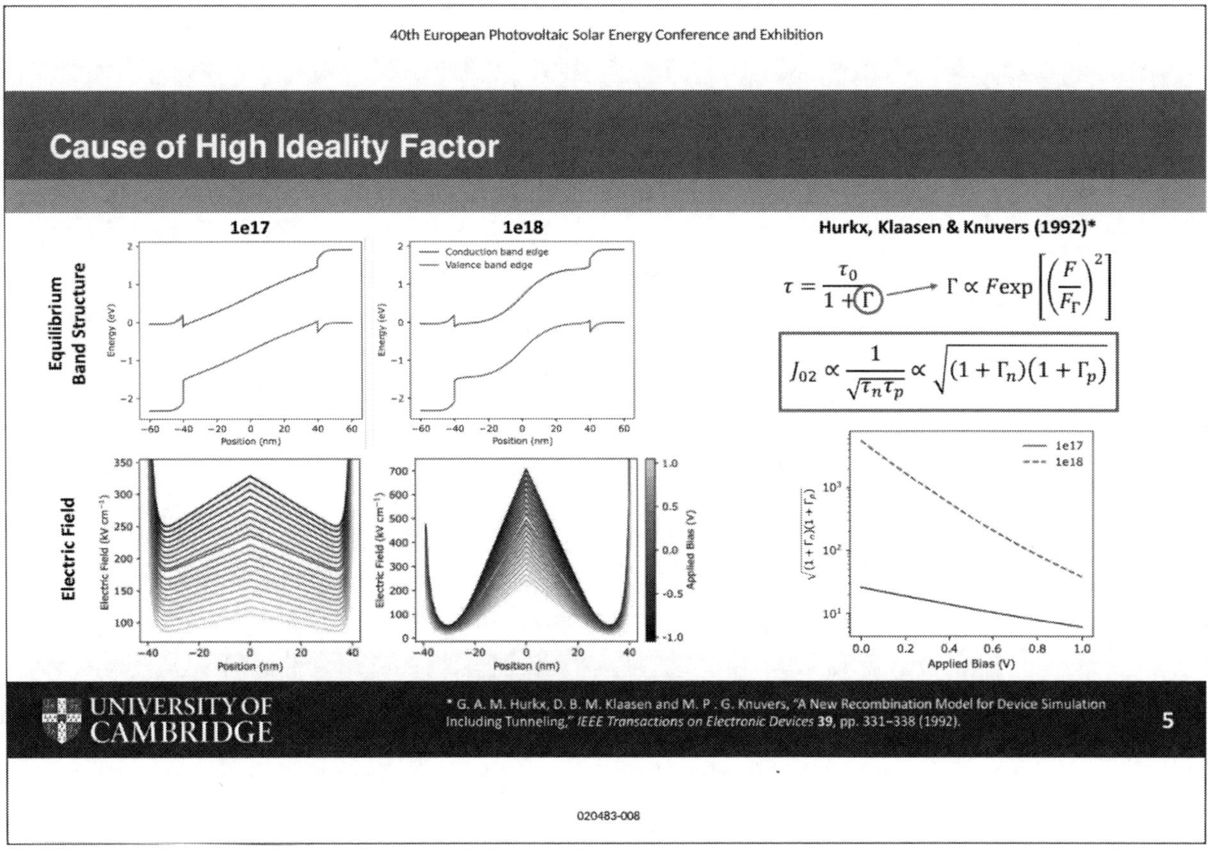
40th European Photovoltaic Solar Energy Conference and Exhibition
Cause of High Ideality Factor
1e17
1e18
Equilibrium Band Structure
Electric Field
Conduction band edge
Valence band edge
Energy (eV)
Position (nm)
Electric Field (kV cm⁻¹)
Applied Bias (V)
Hurkx, Klaasen & Knuvers (1992)*
$\tau = \frac{\tau_0}{1+\Gamma} \longrightarrow \Gamma \propto F\exp\left[\left(\frac{F}{F_\Gamma}\right)^2\right]$
$J_{02} \propto \frac{1}{\sqrt{\tau_n \tau_p}} \propto \sqrt{(1+\Gamma_n)(1+\Gamma_p)}$
$\sqrt{(1+\Gamma_n)(1+\Gamma_p)}$
1e17
1e18
UNIVERSITY OF CAMBRIDGE
* G. A. M. Hurkx, D. B. M. Klaasen and M. P. G. Knuvers, "A New Recombination Model for Device Simulation Including Tunneling," IEEE Transactions on Electronic Devices 39, pp. 331–338 (1992).
5
020483-008

Effect of Trap-Assisted Tunneling (TAT) on DIV

Same Carrier Lifetime for 1e17 & 1e18

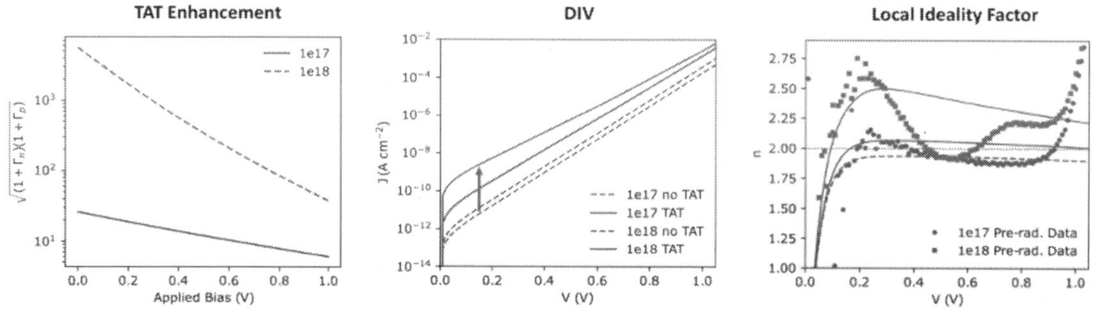

UNIVERSITY OF CAMBRIDGE

40th European Photovoltaic Solar Energy Conference and Exhibition

Effect of Defects on DIV

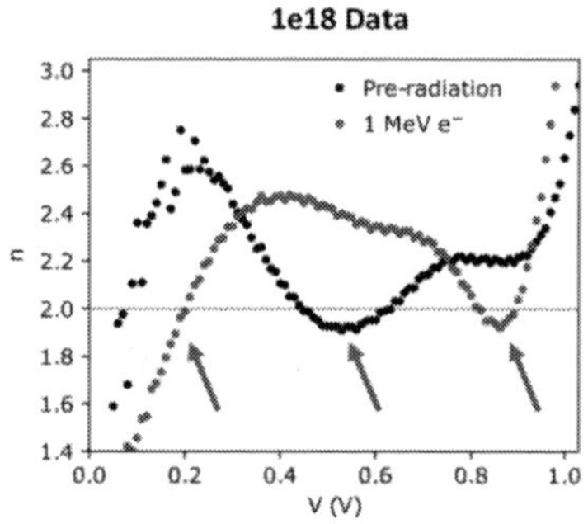

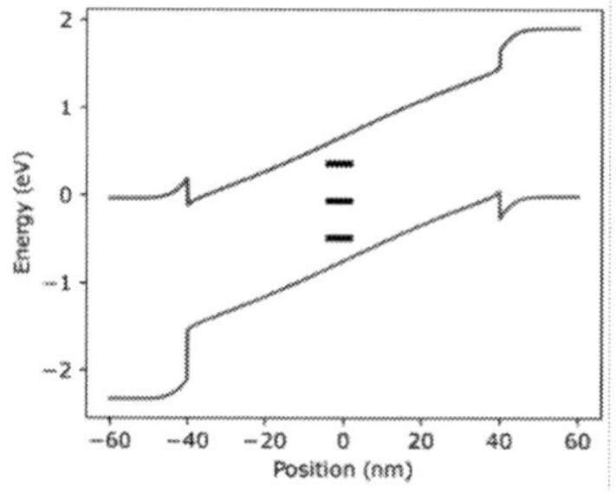

UNIVERSITY OF CAMBRIDGE

Further reading: C.-T. Sah, R. Noyce and W. Shockley, "Carrier Generation and Recombination in P-N Junctions and P-N Junction Characteristics," *Proc. IRE* **45**, pp. 1228–1243 (1957).

40th European Photovoltaic Solar Energy Conference and Exhibition

Effect of Defects on DIV

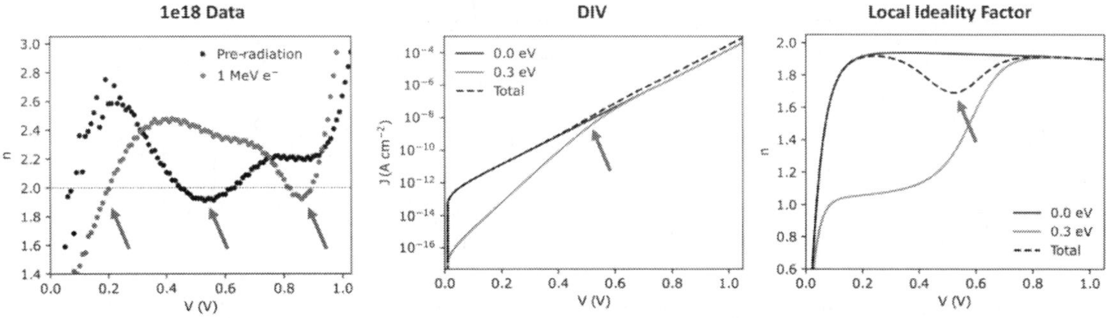

UNIVERSITY OF CAMBRIDGE

Further reading: C.-T. Sah, R. Noyce and W. Shockley, "Carrier Generation and Recombination in P-N Junctions and P-N Junction Characteristics," *Proc. IRE* **45**, pp. 1228–1243 (1957).

7

020483-011

40th European Photovoltaic Solar Energy Conference and Exhibition

Fitting DIV Data

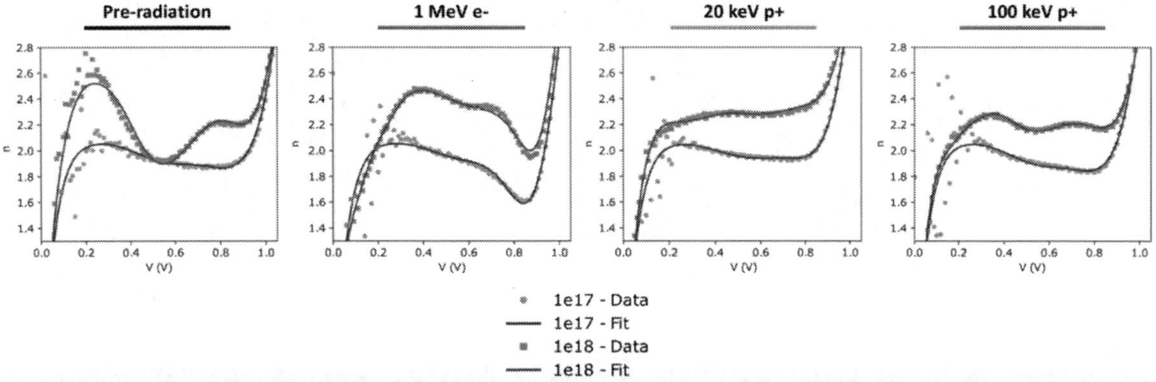

UNIVERSITY OF CAMBRIDGE

8

020483-012

TAT Effect on Radiation Tolerance

TAT

$$\tau = \frac{\tau_0}{1+\Gamma}$$

Lifetime Degradation

$$\frac{1}{\tau_0} \sim K_\tau \cdot \text{Dose}$$

TAT-Enhanced Lifetime Degradation

$$\frac{1}{\tau} \sim K_\tau(1+\Gamma) \cdot \text{Dose}$$

TAT-Enhanced J_{02} Increase → **TAT-Enhanced V_{oc} Degradation**

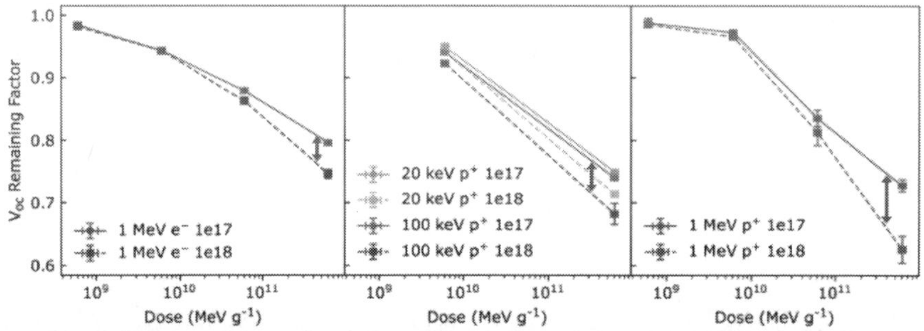

UNIVERSITY OF CAMBRIDGE

High Energy Proton Damage

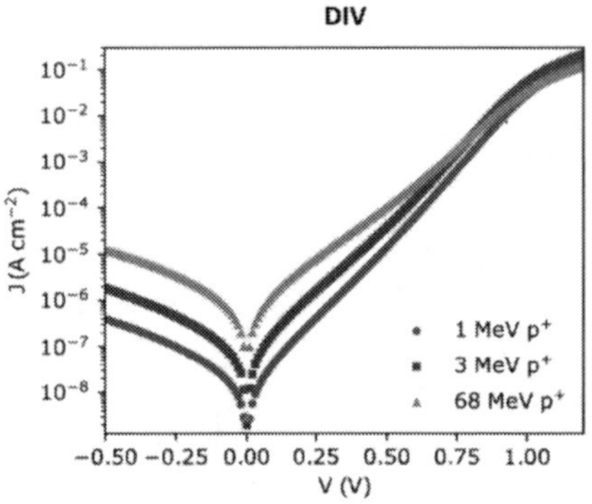

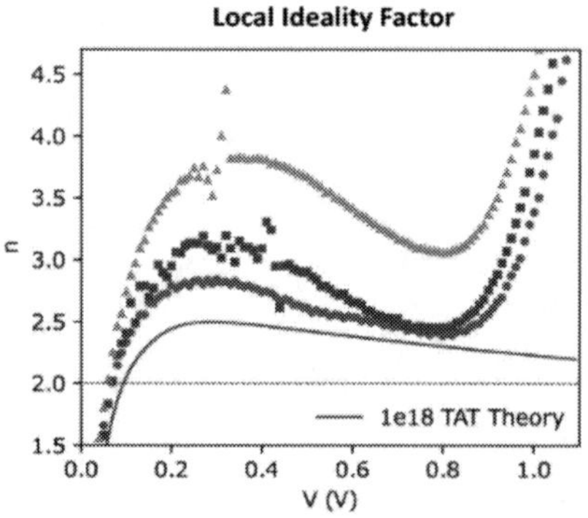

UNIVERSITY OF CAMBRIDGE

Summary

- Ultra-thin solar cells are of interest for space applications due to their high radiation tolerance.
- Ultra-thin solar cells display unconventional DIV behaviour, with n > 2.
- n > 2 can be accounted for by trap-assisted tunneling enhancement of the recombination rate.
- Trap-assisted tunneling accounts for the greater V_{oc} degradation rate in 1e18 over 1e17 devices.

Acknowledgements

Thank You for listening!
Any Questions?

Special thanks to:

- Carl Andrews for assisting with the 3 MeV proton irradiations, which were carried out at The University of Manchester's Dalton Cumbrian Facility (DCF), a partner in the National Nuclear User Facility, the EPSRC UK National Ion Beam Centre and the Henry Royce Institute.
- Felix Lang, Jürgen Bundesmann and Andrea Denker for carrying out the 68 MeV proton irradiations at the Department Protons for Therapy, Helmholtz-Zentrum Berlin.

MANCHESTER 1824 The University of Manchester Dalton Nuclear Institute

Funding Sources:

THE ROYAL SOCIETY

UKRI Engineering and Physical Sciences Research Council

erc European Research Council Established by the European Commission

UK SPACE AGENCY

UKRI EPSRC PhD Studentship: EP/R513180/1.
H2020 European Research Council: 853365.
The Royal Society: IEC\R3\193005.
UK Space Agency: PF2-012.

EVALUATION OF THE INFLUENCE OF DIFFERENT COMPONENT LAYOUTS ON THE SUITABILITY, LIFETIME, AND COST OF A HYBRID, HYDROGEN-BASED PV ENERGY SYSTEM

Marius C. Möller, Stefan Krauter
Faculty of Computer Science, Electrical Engineering and Mathematics, Warburger Str. 100, D-33098 Paderborn
Electrical Energy Technology - Sustainable Energy Concepts (EET-NEK), Paderborn University,
marius.claus.moeller@uni-paderborn.de, stefan.krauter@uni-paderborn.de

ABSTRACT: The achievement of a carbon-neutral society, set as a goal for the next few decades, goes directly hand in hand with the need for an energy system strongly based on renewable energies [1]. However, renewable energies are of intermittent nature and underlie seasonal fluctuations. Specifically, hydrogen is considered essential for achieving high CO_2 reduction targets [1]. Sufficient seasonal energy storage capacity is needed for large parts of the world, where only hydrogen is a comparatively cost-efficient form of storage compared to other energy storage systems [2]. Such an energy system, designed exclusively for renewable energies and based on hydrogen, is the subject of this research work and is investigated using a household energy system as an example. The energy system was designed for fully covering the year-round energy demand of a private household on the basis of electricity generated by a photovoltaic (PV) system, using a hybrid energy storage system consisting of a hydrogen unit and a lithium-ion battery. Hydrogen is produced with a Proton Exchange Membrane (PEM) electrolyser by PV surplus during the summer months and then stored in a hydrogen tank. Mainly during winter, in terms of lack of PV energy, the hydrogen is converted back into electricity and heat by a fuel cell (FC).
As part of an extensive parameter study the influence of different component layouts and weather data will be presented. A household energy system developed in Matlab/Simulink is used for the investigation, which is based on a whole year simulation and uses real measured household load profiles and weather profiles as input.
In addition to the energy balance, the model can also be used to predict the lifetime of the three main components: electrolyser, fuel cell and lithium-ion storage. The lifetime of the components depends on the layout and the control and has a mutually influencing effect, which will be proved by the presented results. A cost estimation is also integrated.
Keywords: hydrogen, hybrid energy system, PV system, modeling

1 AIM AND APPROACH

The whole energy system will have to be transformed to an environmentally friendly energy system with high shares of renewable energies in the foreseeable future. However, there are some hurdles in the transformation, such as the need for high seasonal energy storage capacity and a burden of the grids due to high load on the power grids at some times (e.g. during sunny days in midsummer, due to high PV output). In order to nevertheless achieve a rapid transformation, the load of renewable energy on the power grid should be minimized, which aims to a significant level of residential self-consumption. In addition, for example, Germany's low sun irradiance in winter leads to a high seasonal storage requirement.
Hydrogen, when produced by renewable energy sources (RES), is seen as one of the most promising means of transforming the energy system in a climate-friendly way [3]. However, a conversion of the energy system to renewable energies using hydrogen still requires clarification of some important issues. With regard to safety in the use of hydrogen, controllability has already been demonstrated [3]. However, increasing the service life, reducing the use of resources and increasing efficiency are current research priorities for hydrogen components. Furthermore, however, the investment costs and the use of resources will always play a role in the installation of such a system. With regard to the consideration of the two factors, capital investment and resource usage, the use of the components should be dimensioned as precisely as possible for the designated use case. For this purpose, it is crucial to work out recommendations concerning the capital investment, the resource usage and the service life. Such findings can be obtained in a targeted manner on the basis of a simulation. A coordinated design of the components prior to practical implementation is crucial in order to comply with the framework conditions of the intended use case. For this purpose, knowledge of the use case and its influence on the suitability of the energy system is of fundamental importance. Several factors are to be emphasized here: First, the energy use behavior of the considered household, second, the intended installation site with the prevailing climatic conditions, and third, the influence of the PV system orientation. In addition, the coordination of the dimensioning of the short-term storage in connection to the long-term storage also plays a role. This in turn has an influence on the overall service life of the system. The aim is to gain knowledge and conclusions about all these factors, which in turn can be used for optimal system design and future installations of such energy systems.
The energy system construction consists of a PV system as main energy source. It is oversized to ensure high excess energy production during summer. Excess energy is first used for charging the Lithium-ion battery (LIB) and when the LIB is fully charged the energy is used for hydrogen production by an electrolyser (ELY). The hydrogen is compressed and stored for usage in a fuel cell (FC), where the hydrogen is reconverted to electricity on demand. The hydrogen system is used for long-term storage, while the LIB is used for short-term energy demand, because the LIB can better deal with rapid load peaks compared to FCs. Therefore, a high level of self-sufficiency can be guaranteed. The heating system was also considered, as the focus was on ensuring the overall energy demand by PV. The waste heat of the FC is used inside the space heating system, while the waste heat of the ELY is used for the hot water heating demand. The remaining energy demands were met from electrical heat pumps.
The energy system model was created in Matlab/Simulink [4] and was designed for the analysis of a real data series over an entire year with a time resolution of 15 min. The Simulink model used is shown in figure 1. The model uses different input data sets like weather data and load profiles. Irradiance and temperature profiles with a time resolution

10.4229/EUPVSEC2023/5CO.4.2
020484-001

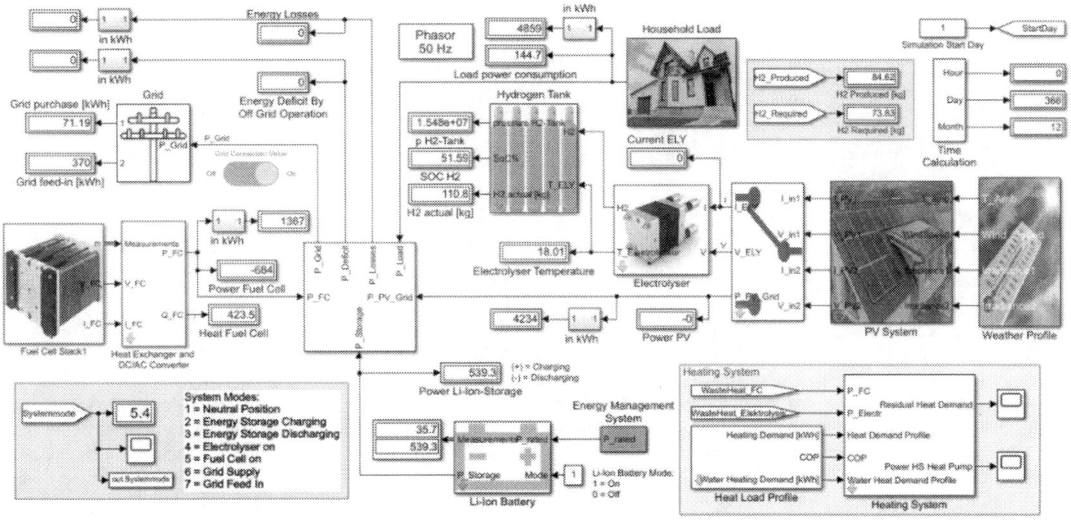

Fig. 1. Simulink model of the overall household energy system (image sources: [11, 12]).

of one hour have been granted by German Weather Service (Deutscher Wetterdienst—DWD) [5]. The data sets were recorded during 2015 in Wuerzburg (Germany). For the analysis of the lifetime behavior and the energy balance by performing an extensive parameter study, a representative real dataset recorded in Switzerland in 2012 was used as household load profile [6].
The energy system model has been presented in detail in our previous paper [8] and the integrated lifetime prediction analysis with description of the methodology is shown in another paper in detail [9]. In a next paper an investigation of different load characteristics, component dimensioning, and system scaling on the energy balance and the cost of the energy system have been presented [10].

2 SCIENTIFIC INNOVATION AND RELEVANCE

The topic "hydrogen in energy systems" is presently attaining increasing interest in the research community. Such systems are analyzed at various levels and in several component constellations. Many modelling studies deal with dynamic real-time operation and control, examining relatively short time scales of a few hours, a day, or up to two weeks, such as [13-16]. Certain component constellations have been examined quite often, such as hybrid energy systems designed for wind power as the only energy source [17]; others assume hydrogen as the only kind of energy storage [18].
This modelling approach offers some advantages compared to other modeling tools such as Homer Energy [19], ReMOD-D [20] or TRNSYS [21]. One advantage is the high resolution in combination with a simulation of a whole year, which allows a meaningful analysis. The system boundary can be arbitrary and targeted, and the model approach also allows an upscaling of the system boundary. Also, arbitrary additional objects of investigation (like a lifetime prediction model) and a perfectly tailored energy management system can be built with the usage of Matlab/Simulink. As the lifetime of the components highly depend on the energy management system, this was of vital importance.

No other model represents the lifetime of all three components in one modeling approach and enables cross-sector (both electricity and heat) energy balance analysis in the time frame of a whole year considering a sophisticated energy management system, which makes this energy system model unique. A lifetime prediction enables optimization of component sizing and can thus result in lifetime extension of the entire energy system. This investigation gains insights into mutual interdependency and effect.

3 RESULTS AND CONCLUSIONS

Within the scope of the research, the influence of different component dimensions on the energy system was investigated. The influence of different PV system orientations was also investigated. An extensive parameter study was carried out, the results of which are shown in figure 2.
The parameter study shows the change of lifetime expectations when changing the component layouts. It was shown that the variation of a component not only touches its intrinsic component lifetime, but also has an effect on the other components. For example, the lifetime of the FC could be significantly increased by increasing the LIB storage capacity, which is mainly caused by the less frequent switch-on and switch-off processes of the FC when recharging the LIB. But on the other hand, the utilization rate of the LIB (expressed in the full storage cycles) until the End of Life (EoL) decreased with an increase in storage capacity. In this case, the use-independent calendric aging process of the LIB became the decisive criterion for aging. In terms of the cost performance, a smaller LIB performs better. Therefore, an economic analysis for better assess the usefulness of increasing the capacity of the LIB to increase the lifetime of the FC should be performed. An integration of an economic analysis model into the energy system model is also planned for the next months.
Figure 2 shows that the size of the LIB also has a

Study Type	Ressource usage	ELY	FC	PV system	LIB	Self-Sufficiency	Net Self-Sufficiency	Net hydrogen utilization	ELY Degradation		FC Degradation		LIB Degradation	
									Years	Operating hours	Years	Operating hours	Years	Max. full cycles
ELY 1	100.00%	5 kW	700 W	6.82 kW (45°, South)	20 kWh	98.51%	97.22%	-4.42%	11.83	17290.03	15.66	44213.02	9.10	971.48
ELY 2	64.00%	3 kW	700 W	6.82 kW (45°, South)	20 kWh	98.45%	96.75%	-5.89%	12.24	21130.03	15.86	44780.37	9.11	960.78
ELY 3	38.00%	2 kW	700 W	6.82 kW (45°, South)	20 kWh	98.41%	92.03%	-25.93%	6.62	12806.39	15.90	44270.68	9.12	953.40
FC 1	175.00%	5 kW	1200 W	6.82 kW (45°, South)	20 kWh	99.78%	97.63%	-6.86%	12.01	17802.44	15.31	40938.73	9.11	1001.32
FC 2	100.00%	5 kW	700 W	6.82 kW (45°, South)	20 kWh	98.51%	97.22%	-4.42%	11.83	17290.03	15.66	44213.02	9.10	971.48
FC 3	45.00%	5 kW	300 W	6.82 kW (45°, South)	20 kWh	85.30%	85.30%	40.70%	11.94	16876.29	15.44	46735.21	9.28	839.66
FC 4	100.00%	5 kW	700 W No P. Adj.	6.82 kW (45°, South)	20 kWh	99.44%	95.80%	-13.01%	11.82	17714.28	14.07	31370.21	9.09	1015.60
PVS 1	100.00%	5 kW	700 W	6.82 kW (45°, South)	20 kWh	98.51%	97.22%	-4.42%	11.83	17290.03	15.66	44213.02	9.10	971.48
PVS 2	150.00%	5 kW	700 W	10.32 kW (45°, South)	20 kWh	98.77%	98.77%	54.92%	10.80	20744.99	17.98	41511.19	9.16	955.30
PVS 3	200.00%	5 kW	700 W	13.64 kW (45°, South)	20 kWh	98.95%	98.95%	65.15%	7.47	13048.96	18.33	37884.38	9.19	895.88
PVS-O 1	100.00%	5 kW	700 W	6.82 kW (45°, South)	20 kWh	98.51%	97.22%	-4.42%	11.83	17290.03	15.66	44213.02	9.10	971.48
PVS-O 2	100.00%	5 kW	700 W	6.82 kW (45°, East/West)	20 kWh	98.22%	84.88%	-57.80%	11.95	16616.08	14.97	51593.26	9.22	894.75
PVS-O 3	100.00%	5 kW	700 W	6.82 kW (60°, South)	20 kWh	98.46%	92.47%	-23.41%	12.01	16181.55	15.08	43445.71	9.08	978.61
PVS-O 4	100.00%	5 kW	700 W	6.82 kW (30°, South)	20 kWh	98.53%	98.53%	1.68%	11.79	17801.17	15.95	45980.68	9.11	960.53
LIB 1	25.00%	5 kW	700 W	6.82 kW (45°, South)	5 kWh	93.68%	88.85%	-12.97%	11.84	18580.47	6.25	25994.55	6.99	1860.67
LIB 2	50.00%	5 kW	700 W	6.82 kW (45°, South)	10 kWh	96.59%	93.65%	-8.97%	12.02	17787.80	9.73	32429.60	8.02	1446.84
LIB 3	75.00%	5 kW	700 W	6.82 kW (45°, South)	15 kWh	97.93%	95.90%	-6.66%	12.06	17705.55	12.60	37810.55	8.68	1168.47
LIB 4	100.00%	5 kW	700 W	6.82 kW (45°, South)	20 kWh	98.50%	97.22%	-4.42%	11.94	17440.36	15.63	44253.94	9.11	967.30
LIB 5	200.00%	5 kW	700 W	6.82 kW (45°, South)	40 kWh	99.27%	99.27%	0.14%	11.43	16468.04	21.30	57272.01	9.82	554.23
DOD 1	100.00%	5 kW	700 W	6.82 kW (45°, South)	20 kWh Max DOD = 90%	99.36%	97.72%	-5.62%	11.88	17147.77	15.74	45839.66	8.95	963.79
DOD 2	100.00%	5 kW	700 W	6.82 kW (45°, South)	20 kWh Max DOD = 80%	99.23%	97.58%	-5.68%	11.90	17195.11	15.64	44669.43	9.00	970.92
DOD 3	100.00%	5 kW	700 W	6.82 kW (45°, South)	20 kWh Max DOD = 70%	98.51%	97.22%	-4.42%	11.83	17290.03	15.66	44213.02	9.10	971.48

Fig. 2. Lifetime Comparison between all parameter studies (configurations with less self-sufficiency under 97% are marked red; resource usage is based on the number of cells (for ELY and FC), number of modules (PV) or capacity (LIB)). The base configuration is marked orange; the net self-sufficiency describes the self-sufficiency when an H_2 deficit (negative net hydrogen utilization) is compensated by the energy grid (calculation: $HHV \cdot \Delta H_2[kg] \cdot \eta_{FC,avg}$ (with HHV = higher heating value)); The net hydrogen utilization describes how much H_2 is required compared to the H_2 produced (minus: higher H_2 demand than production).

significant impact on hydrogen utilization. By increasing the total capacity from 5 kWh to 40 kWh, the net hydrogen utilization was brought from a deficit of −12.97% to + 0.14%. This is primarily due to the increased efficiency from the increased short-term storage by the LIB, which decreases the necessity of hydrogen based long-term storage.

The system model also allows an adjustment of the PV system orientation. This had a high impact on the net hydrogen utilization. For the considered use case, a better orientation for the summer month for reaching higher overall energy yield was more meaningful (see figure 2). An orientation towards south was also a way better than an east/west orientation.

The parameter study also showed that an increase in PV system size also led to an increase in FC lifetime and that a power-adjusted control of the FC also makes sense with regard to the FC lifetime and the degree of self-sufficiency.

The results demonstrate the benefits of a holistic modelling approach and enable a design optimization regarding the use of resources, lifetime and self-sufficiency of the system.

In another parameter study, the influence of varying weather data was investigated with Würzburg in Germany as location (see figure 3). For this purpose, weather data from four different years (2014 - 2017) were used. A slightly different hydrogen demand was found with a standard deviation of 4.48 kg. Hydrogen production, on the other hand, did not vary as much (standard deviation

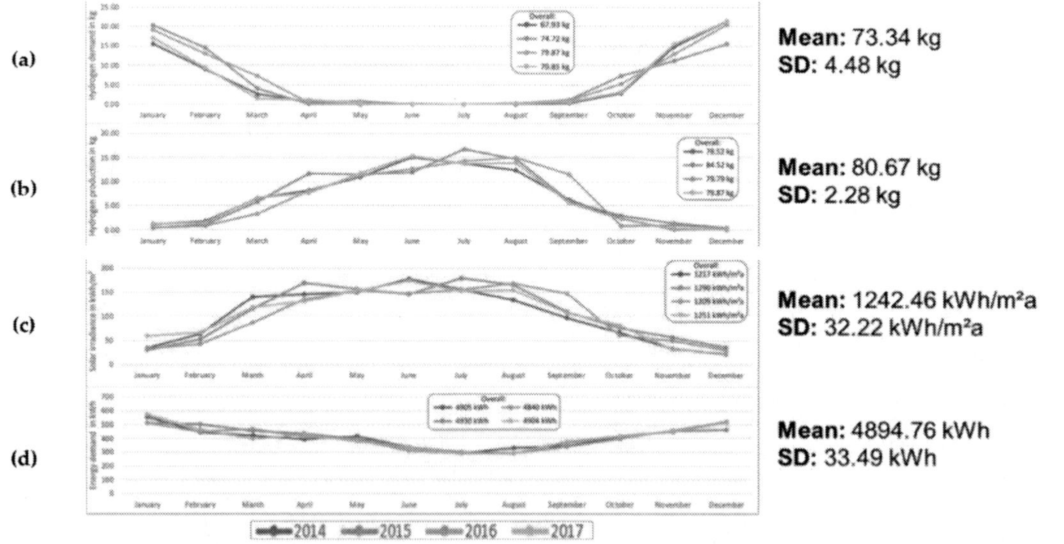

Fig. 3. Investigation of different years of interest (2014–2017 for Würzburg (Germany)): (a) Monthly hydrogen demand; (b) Monthly hydrogen production; (c) Monthly solar irradiation; (d) Monthly electricity demand.

10.4229/EUPVSEC2023/5CO.4.2
020484-003

2.28 kg). The average irradiation in the four years considered was 1242.46 kWh/m². The total energy demand was 4894.76 kWh on average, with slightly higher energy demand in winter due to the additional heat demand of 4000 kWh. The electrical energy demand for heat was 1407 kWh due to the use of heat pumps and a fuel cell heat recovery of 730 kWh.

Based on the data of figure 3, the monthly hydrogen demand (a) and hydrogen production (b) as a function of the monthly solar irradiation have been established in figure 4. With a coefficient of determination of 0.9219 for (a) and 0.9539 for (b), the curves show a very good approximation to the measurement points. Figure 4 (c) and (d) show the hydrogen demand and production curves under different heat demands and PV system scales, respectively. The layout according to the heat demand with aim of self-sufficiency is shown in figure 5. For the hydrogen demand (c), there is a largely uniform upward shift in the curve as the heat demand increases, while the hydrogen production curve (d) shifts largely uniformly as the PV system size increases.

Based on a cost estimation integrated into the model, the net present values for a purchase of the system in 2015 to 2031 for various heat demands is shown in figure 5. A period under consideration of 20 years was assumed. It was considered that LIB, ELY and FC would have to be replaced in this period due to the end of life, but that the replacement component would then already have a longer life expectancy due to advanced research. However, only the pro rata costs of the replacement component that fall within the period under consideration were applied (linear depreciation approach). All other components were given a life expectancy of 20 years. In the case of the ELY and the FC, the assumption was made that only the stack and not the entire system would have to be replaced at the end of life, which reduced the costs for this component somewhat. The hydrogen tank accounted for 58% - 65% of the overall costs. With rising heat demand, the price per kWh is increasing due to the fact that more energy has to be covered by the hydrogen part because of higher energy demand during winter. Therefore, a proportionally larger hydrogen tank is required. The costs include all components except of the heat pumps, since a reference energy system with a public grid connection would also require these.

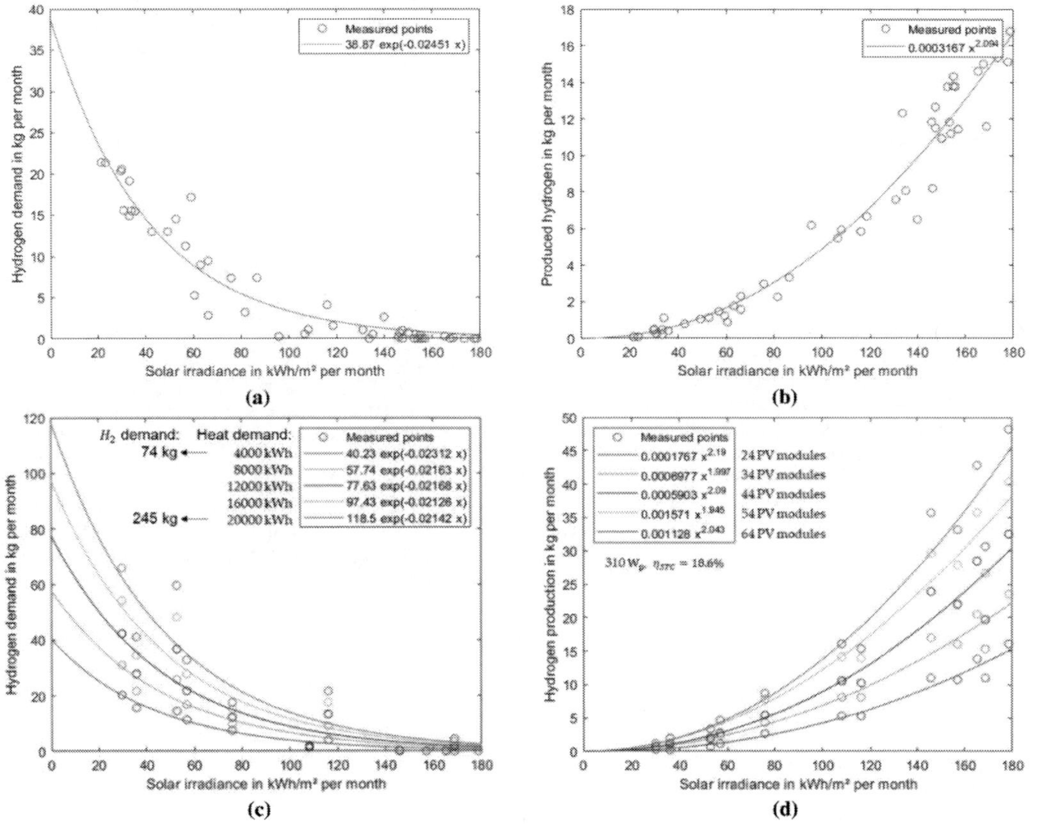

Fig. 4. Dependency between solar irradiation and hydrogen demand **(a)** respectively to hydrogen production **(b)** for a heat demand of 4000 kWh and dependency between solar irradiation and hydrogen demand **(c)** respectively to hydrogen production **(d)** for different heat demands

		Heat demand				
		4,000 kWh	8,000 kWh	12,000 kWh	16,000 kWh	20,000 kWh
Electrolyser nominal power		3.6 kW	5 kW	7 kW	9 kW	10 kW
Fuel cell nominal power		700 W	1200 W	2000 W	3000 W	3500 W
PV system size (south; elev. angle 45°)		24 modules	34 modules	44 modules	54 modules	64 modules
LIB capacity		20 kWh	20 kWh	25 kWh	30 kWh	35 kWh
Net present value if bought in:	2015	156,983 €	220,967 €	297,416 €	352,967 €	449,568 €
	2016	150,538 €	212,282 €	285,541 €	338,443 €	430,481 €
	2017	141,963 €	203,552 €	273,540 €	320,206 €	412,926 €
	2018	136,097 €	192,626 €	258,903 €	306,724 €	390,963 €
	2019	130,430 €	184,921 €	248,322 €	293,700 €	374,225 €
	2020	125,217 €	177,176 €	237,715 €	281,591 €	358,541 €
	2021	119,934 €	169,672 €	227,489 €	269,456 €	342,945 €
	2022	115,091 €	162,681 €	217,944 €	257,849 €	328,365 €
	2023	110,220 €	155,724 €	208,477 €	247,019 €	313,970 €
	2024	105,564 €	149,266 €	199,687 €	236,291 €	300,514 €
	2025	101,300 €	142,886 €	191,023 €	226,347 €	287,352 €
	2026	97,078 €	136,984 €	182,986 €	216,592 €	275,125 €
	2027	93,250 €	131,227 €	175,179 €	207,653 €	263,261 €
	2028	89,650 €	125,963 €	168,031 €	199,253 €	252,368 €
	2029	86,140 €	121,046 €	161,347 €	191,162 €	241,946 €
	2030	83,034 €	116,348 €	154,995 €	183,935 €	232,567 €
	2031	80,184 €	112,182 €	149,355 €	177,091 €	223,992 €

Price per kWh	Overall energy demand within 20 years (in kWh)				
(€/kWh)	**95,089**	**127,881**	**162,306**	**198,190**	**230,945**
in 2023	1.16	1.22	1.28	1.25	1.36
in 2031	0.84	0.88	0.92	0.89	0.97

Fig. 5. Net present value of the energy system for different heat demands, when bought in 2015 up to 2031.

4 REFERENCES

[1] Wietschel, M.; Zheng, L.; Arens, M.; Hebling, C.; Ranzmeyer, O.; Schaadt, A.; Hank, C.; Sternberg, A.; Herkel, S.; Kost, C.; Ragwitz, M.; Herrmann, U.; Pfluger, B. (2021): Metastudie Wasserstoff – Auswertung von Energiesystemstudien. Studie im Auftrag des Nationalen Wasserstoffrats. Karlsruhe, Freiburg, Cottbus: Fraunhofer ISI, Fraunhofer ISE, Fraunhofer IEG (Hrsg.).

[2] Wurster, R.; Schmidtchen, U. DWV Wasserstoff-Sicherheits-Kompendium, 2011.

[3] Rosen, M.; Koohi-Fayegh, S. The prospects for hydrogen as an energy carrier: an overview of hydrogen energy and hydrogen energy systems. Energy, Ecology and Environment, 1, 10-29, 2016.

[4] MATLAB. Version 9.11 (R2021b); The MathWorks Inc.: Natick, MA, USA, 2021.

[5] Deutscher Wetterdienst (DWD) – Climate Data Center. Available online: https://opendata.dwd.de/climate_environment/CDC/ (accessed on 8 November 2022).

[6] Kleiminger, W.; Beckel, C. ECO data set (Electricity Consumption & Occupancy) – A Research Project of the Distributed Systems Group. 2016. ETH Zürich. Available online: https://rossa-prod-ap21.ethz.ch:8443/delivery/DeliveryManagerServlet?dps_pid=IE594964 (accessed on 8 October 2023).

[7] E-Control. Der Tarifkalkulator der E-Control. Available online: https://www.e-control.at/marktteilnehmer/infos/tarifkalkulator#p_p_id_com_liferay_journal_content_web_portlet_JournalContentPortlet_INSTANCE_k53Sz6jcdjfx_ (accessed on 8 October 2023).

[8] Möller, M. C.; Krauter, S. Hybrid Energy System Model in Matlab/Simulink Based on Solar Energy, Lithium-Ion Battery and Hydrogen. Energies 2022, 15, 2201. https://doi.org/10.3390/en15062201.

[9] Möller, M. C.; Krauter, S. Dimensioning and Lifetime Prediction Model for a Hybrid, Hydrogen-Based Household PV Energy System Using Matlab/Simulink. In Solar (Vol. 3, No. 1, pp. 25-48). Multidisciplinary Digital Publishing Institute, 2023.

[10] Möller, M.C.; Krauter, S. Investigation of Different Load Characteristics, Component Dimensioning, and System Scaling for the Optimized Design of a Hybrid Hydrogen-Based PV Energy System. Hydrogen 2023, 4, 408–433.

[11] ElringKlinger Kommt Mit Neuen Produkten Zur IAA. Available online: https://www.electrive.net/2018/09/14/elringklinger-kommt-mit-neuen-produkten-zur-iaa/ (accessed on 8 October 2023).

[12] Elektrolyse-Stacks SERIES S30. Available online: https://www.h-tec.com/produkte/ (accessed on 8 October 2023).

[13] Ram, V.; Salkuti, S.R. Modelling and Simulation of a Hydrogen-Based Hybrid Energy Storage System with a Switching Algorithm. World Electr. Veh. J. 2022, 13, 188.

[14] Trifkovic, M.; Sheikhzadeh, M.; Nigim, K. Modeling and control of a renewable hybrid energy system with hydrogen storage. IEEE Trans. Control Syst. Technol. 2013, 22, 169–179.

[15] Javadpoor, S.; Nazarpour, D. Modeling a PV-FC-Hydrogen Hybrid Power Generation System. Eng. Technol. Appl. Sci. Res. 2017, 7, 1455–1459.

[16] Utomo, O.; Abeysekera, M.; Ugalde-Loo, C.E. Optimal operation of a hydrogen storage and fuel cell coupled integrated energy system. Sustainability 2021, 13, 3525.

[17] Acakpovi, A.; Adjei, P.; Nwulu, N.; Asabere, N.Y. Optimal hybrid renewable energy system: A comparative study of wind/hydrogen/fuel-cell and wind/battery storage. J. Electr. Comput. Eng. 2020, 2020, 1756503.

[18] Villa Londono, J.E.; Mazza, A.; Pons, E.; Lok, H.; Bompard, E. Modelling and Control of a Grid-Connected RES-Hydrogen Hybrid Microgrid. Energies 2021, 14, 1540.

[19] HOMER Energy LLC. HOMER (Hybrid Optimization of Multiple Energy Resources) Software. Version 3.14.5. Available online: https://www.homerenergy.com/products/pro/index.html (accessed on 8 October 2023).

[20] Erlach, B.; Henning, H.M.; Kost, C.; Palzer, A.; Stephanos, C. Optimierungsmodell REMod-D. Materialien zur Analyse » Sektorkopplung « — Untersuchungen und Überlegungen zur Entwicklung Eines Integrierten Energiesystems; Schriftenreihe Energiesysteme der Zukunft: München, Germany, 2018.

[21] Beckman, W.A.; Broman, L.; Fiksel, A.; Klein, S.A.; Lindberg, E.; Schuler, M.; Thornton, J. TRNSYS The most complete solar energy system modeling and simulation software. Renew. Energy 1994, 5, 486–488.

10.4229/EUPVSEC2023/5CO.4.2
020484-005

eurac
research

The True Cost of Green Hydrogen: A Dynamic Simulation

Giovanni Borz, Grazia Barchi, Christian Breyer, Eero Vartiainen, David Moser

Lisbon, 20 September 2023

40th European Photovoltaic Solar Energy Conference and Exhibition

Outline

1. Introduction and the uses of green hydrogen
2. Purpose of this work
3. System
4. Method
5. Modelling
6. Simulations
7. Levelized Cost of Hydrogen (LCOH)
8. Costs
9. Results and Conclusion

Introduction (1)

- Need to decarbonize
- Green Hydrogen is a possible energy vector

- PV System as an Energy Source
- Electrolysis

40th European Photovoltaic Solar Energy Conference and Exhibition

Introduction (2): the Uses of Green Hydrogen

Green Hydrogen is used for:

- Hard to Abate Sectors (steel, cement, petrochemicals,..)
- Ammonia Production
- Heavy Transports
- And Others...

Purpose of this Work

The aim of this work is to create a model that can simulate the behaviour of an **Electrolyser** connected to a **Utility Scale PV System**

to

1) Calculate how much hydrogen can be generated from a MWp PV System
2) Calculate the Levelized Cost of Hydrogen **(LCOH)**

40th European Photovoltaic Solar Energy Conference and Exhibition

System (1)

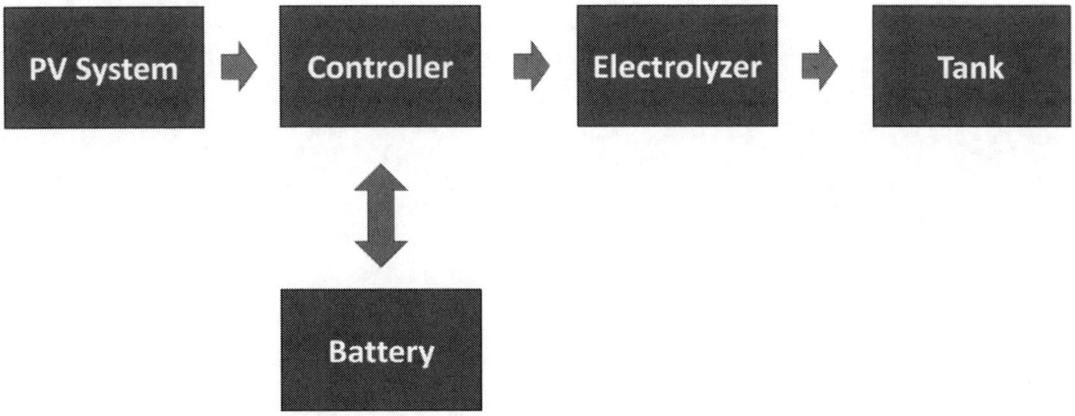

System (2): Electrolyzers - Alkaline and Solid Oxide

ALKALINE ELECTROLYSER

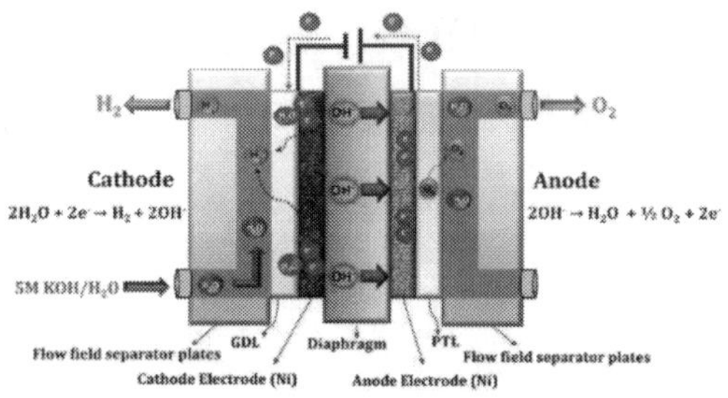

S. Shiva Kumar and H. Lim

SOLID OXIDE ELECTROLYSER

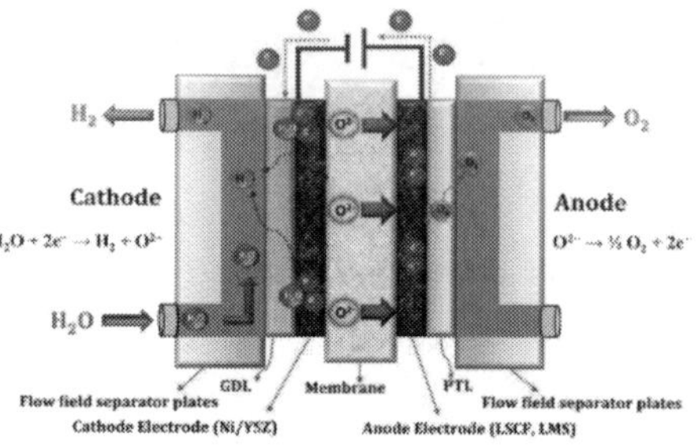

S. Shiva Kumar and H. Lim

eurac research 7

020486-007

40th European Photovoltaic Solar Energy Conference and Exhibition

System (2): Electrolyzers - Alkaline and Solid Oxide

	Alkaline Electrolyzer	Solid Oxide Electrolyzer
Pros	• Established Technology • Lowest Investment Cost • Does Not Requires Critical Raw Material • Longer Lifetime	• Higher Efficiency • Integration with waste Heat • Not Noble Catalists
Cons	• Lower efficiency • Quite Long Cold Start Time	• Still Under Development • High Cold Start Time • Shorter Lifetime • Highest Investment Cost

eurac research 8

020486-008

Method

1. Modelling: creating a two-dimensional model for each technology (Electrolyzer and PV)
2. Generate a 30-year simulation taking into account the degradation of each component and its replacements
3. Calculate the LCOH considering different growth scenarios of cumulative market volumes of the various technologies

40th European Photovoltaic Solar Energy Conference and Exhibition

Modelling (1)

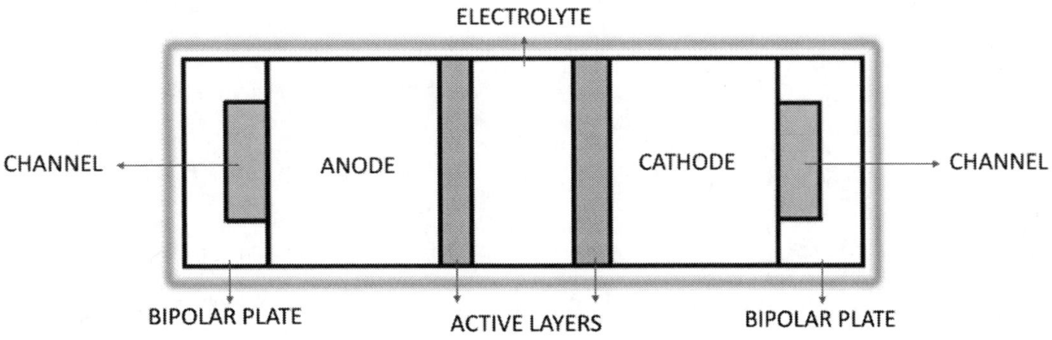

Approximations

a) Two-dimensional
b) Constant Pressures at the anode and at the cathode
c) Constant Temperatures at the anode and the cathode

Modelling (2): Polarization Curve

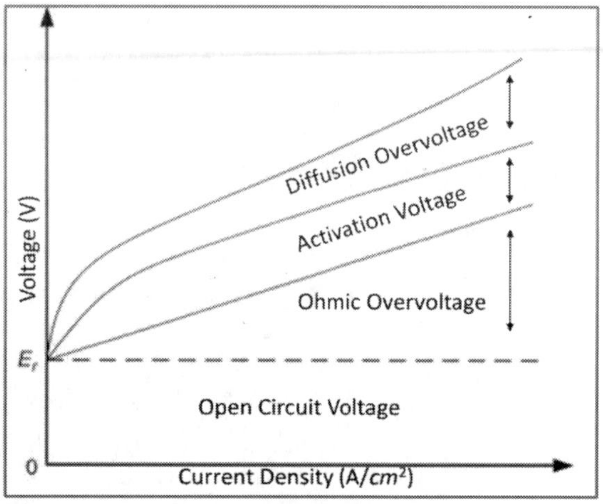

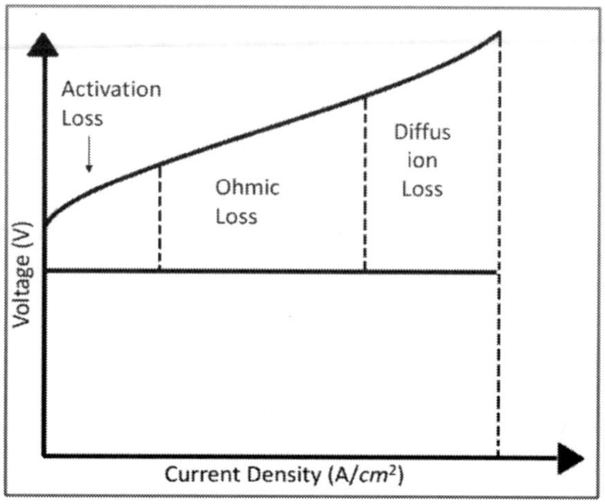

eurac research 11

40th European Photovoltaic Solar Energy Conference and Exhibition

Modelling (2): Polarization Curve

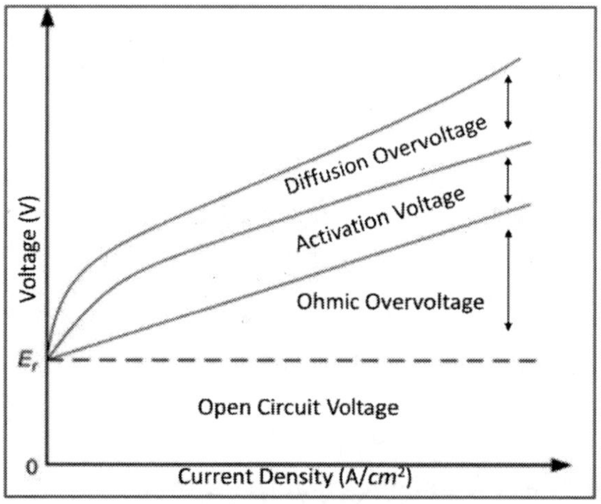

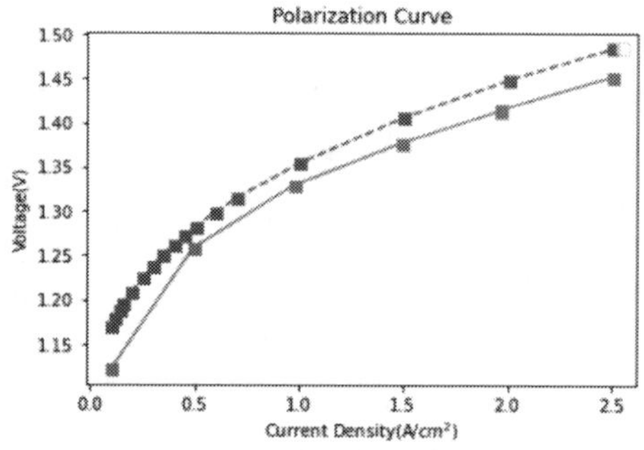

T_{cell} = 1023 K
P_{an} = 1 atm
P_{cat} = 1 atm

eurac research 12

Simulations (1)

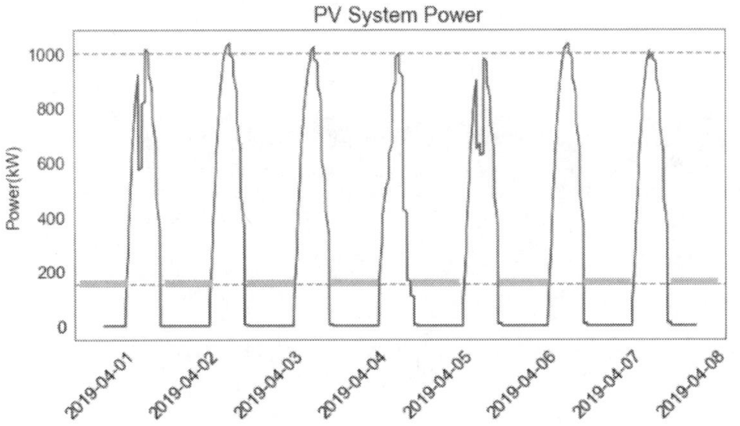

Electrolyzer in stand-by

---- Electrolyzer Working Range

Timestep = 1 min

STAND-BY CONSUMPTION = 2%

PRODUCTION EFFICIENCY:

- **SOE – 43.5 kWh/kg$_{H2}$**
- **ALK – 48.2 kWh/kg$_{H2}$**

40th European Photovoltaic Solar Energy Conference and Exhibition

Simulations (2): Locations and Timing

https://worldmapblank.com/it/mappa-muta-europa-cartina/

Locations:
- Catania
- Nürnberg

Timing:
- 2024 - 2054
- 2030 - 2060

Levelized Cost of Hydrogen (LCOH)

$$LCOH = \frac{CAPEX_{PV} + \sum_{t=1}^{30} \frac{OPEX_{PV}(t)}{(1+WACC)^t} + CAPEX_{EL} + \sum_{t=1}^{30} \frac{OPEX_{EL}(t)}{(1+WACC)^t} + CAPEX_{BESS} + \sum_{t=1}^{30} \frac{REPLACEMENTS}{(1+WACC)^t}}{\sum_{t=1}^{30} \frac{kg_{H_2}(t)}{(1+WACC)^t}}$$

Technology	Lifetime	Degradation (per year)	CAPEX (€/kWp/a)	OPEX (€/kWp/a)
PV	30 years	0.5 %	350 (2024)	12 €/kWp
Inverter	15 years		25	
BESS	10 years	2 %	200	
Alkaline	90000 h	0.3 %	480 (2024)	20 €/(kg/d)/y
SOE	25000 h	1.23 %	1250 (2024)	130 €/(kg/d)/y

Costs: Learning Rate and Cumulative Market Volume

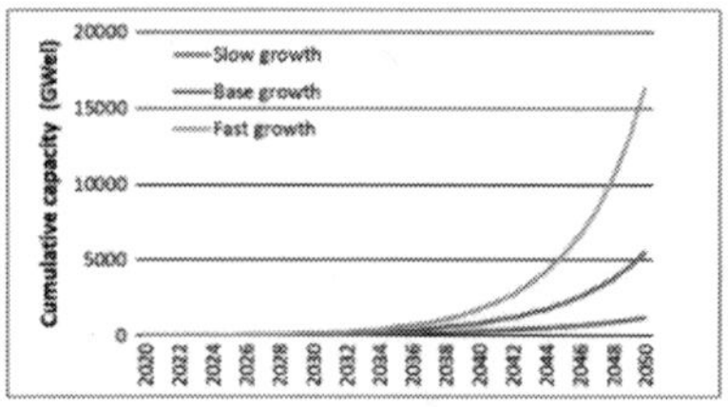

Global cumulative **Electrolyzer** capacity growth in three different scenarios.

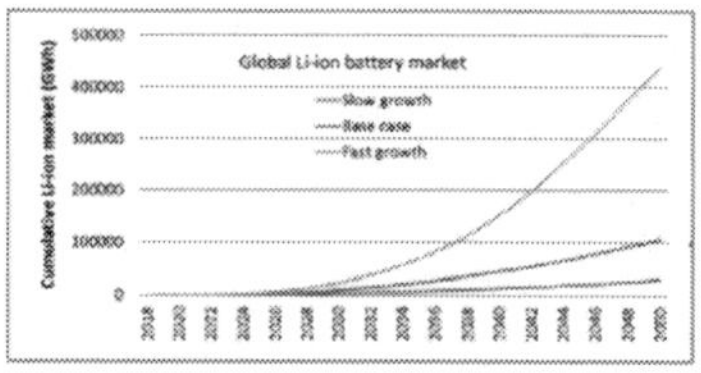

Global cumulative **Battery** volume with three growth scenarios.

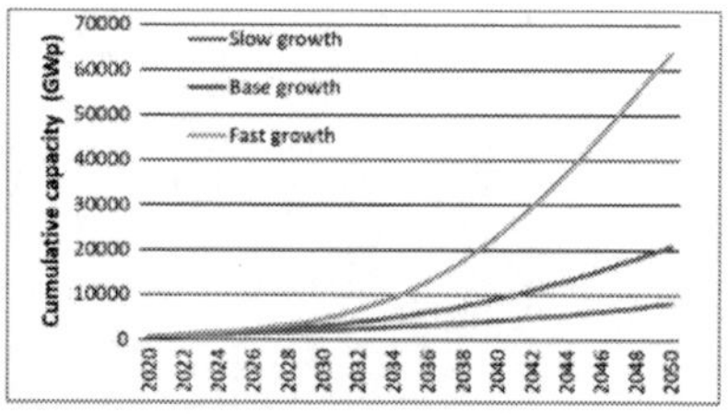

Global cumulative **PV** volume with three growth scenarios.

CAPEX Learning Rate = 28 %
OPEX Learning Rate = 10 %

CAPEX Learning Rate = 17.5 %
OPEX Learning Rate = 10 %

CAPEX Learning Rate = 25 %
OPEX Learning Rate = 10%

Results (1)

2024 - 2054

CASE	PRICE
SOE Catania Base	€ 5.01
SOE Catania Fast	€ 4.82
SOE Nuremberg Base	€ 7.55
SOE Nuremberg Fast	€ 7.27
ALK Catania Base	€ 2.80
ALK Catania Fast	€ 2.78
ALK Nuremberg Base	€ 3.33
ALK Nuremberg Fast	€ 3.31

2030 - 2060

CASE	PRICE
SOE Catania Base	€ 2.67
SOE Catania Fast	€ 2.14
SOE Nuremberg Base	€ 3.76
SOE Nuremberg Fast	€ 3.17
ALK Catania Base	€ 2.24
ALK Catania Fast	€ 1.98
ALK Nuremberg Base	€ 3.08
ALK Nuremberg Fast	€ 2.69

LCOH calculated considering a WACC of 7 %

Results (2)

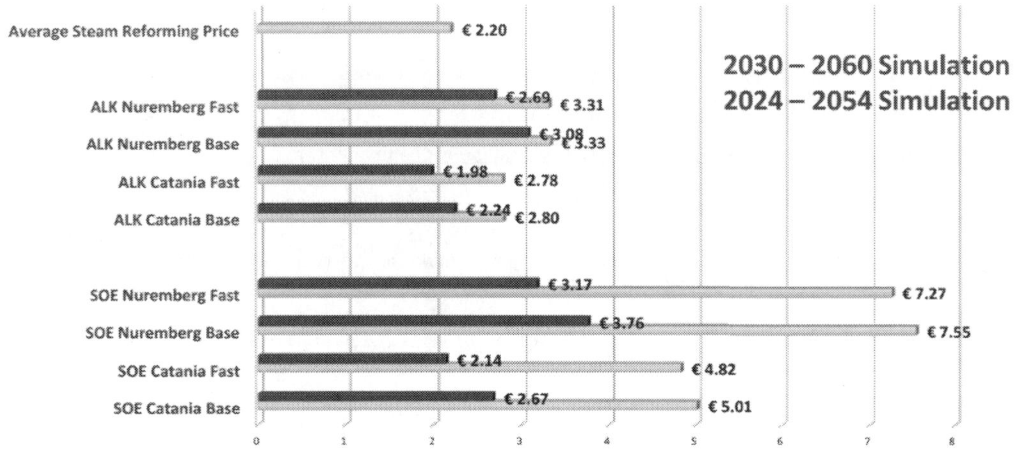

Thank you!

40th European Photovoltaic Solar Energy Conference and Exhibition

Bibliography (1)

- Böhm, H., Goers, S., Zauner, A. (2019). *Estimating future costs of power-to-gas – a component-based approach for technological learning*. International Journal of Hydrogen Energy, 44 (59), Pages 30789-30805
- M. Carmo, D. L. Fritz, J. Mergel, and D. Stolten. *A comprehensive review on PEM water electrolysis*. International Journal of Hydrogen Energy, 38(12):4901-4934, 2013.
- https://joint-research-centre.ec.europa.eu/pvgis-online-tool_en
- https://pvlib-python.readthedocs.io/
- A. Patonia and R. Poudineh, *Cost-competitive green hydrogen: how to lower the cost of electrolysers?* Oxford: The Oxford Institute for Energy Studies, 2022.
- J. P. Stempien, Q. Sun, and S. Hwa Chan. *Solid oxide electrolyzer cell modeling: A review. Journal of Power Technologies*

Bibliography (2)

- Strategic Research and Innovation Agenda 2021-2027, *Clean Hydrogen Joint Undertaking*
- *SRIA-Final Draft*, October 2020
- Ǿ. Ulleberg, *Stand-Alone power system for the future: optimal design, operation and control of solar-hydrogen energy systems*, 1998
- E. Vartiainen et al., "*True Cost of Solar Hydrogen*," Sol. RRL, vol. 6, no. 5, p. 2100487, May 2022, doi: 10.1002/solr.202100487.
- E. Vartiainen, G. Masson, C. Breyer, D. Moser, and E. Román Medina, "*Impact of weighted average cost of capital, capital expenditure, and other parameters on future utility-scale PV levelised cost of electricity*," Prog. Photovolt. Res. Appl., vol. 28, no. 6, pp. 439–453, Jun. 2020, doi: 10.1002/pip.3189.
- G.Matute, J.M. Yusta, L.C. Correas, Techno-Economic modelling of water electrolysers in the range of several MW to provide grid services while generating hydrogen for different applications: A case study in Spain applied to mobility with FCEVs, International journal of Hydrogen Energy, 5 July 2019, doi:https://doi.org/10.1016/j.ijhydene.2019.05.092

40th European Photovoltaic Solar Energy Conference and Exhibition

The True Cost of Green Hydrogen: A Dynamic Simulation

Giovanni Borz
giovanni.borz@eurac.edu

www.eurac.edu

Follow us!

Credits: NOI Spa - NOI TechPark Bolzano

OPTIMIZING RENEWABLE ENERGY INTEGRATION IN RHINELAND-PALATINATE: ASSESSING THE ROLE OF PUMPED HYDROELECTRIC ENERGY STORAGE AND DEMAND SIDE MANAGEMENT

Alexander Blinn, Kevin Kreitzer, Henrik te Heesen
Trier University of Applied Sciences, Environmental Campus Birkenfeld
Institute for Operations and Technology Management, Campusallee, 55768 Hoppstädten-Weiersbach, Germany
h.teheesen@umwelt-campus.de

ABSTRACT: Germany is undergoing a significant shift in its electrical supply system, transitioning from conventional fossil fuel-based power plants to decentralized renewable energy technologies, notably photovoltaic and wind energy plants. This shift brings challenges due to the intermittent nature of renewable sources, necessitating advanced grid management strategies, especially as electricity becomes more integral to heating and mobility sectors. Our study examines energy dynamics in Rhineland-Palatinate using a digital twin model of the region's energy system. We specifically evaluate the potential of integrating a pumped hydroelectric energy storage near the Moselle River and the impact of demand-side management. Our findings reveal that while the pumped hydroelectric storage effectively counteracts short-term power grid fluctuations, its utility is limited by capacity. On the other hand, demand-side management, which involves smart coordination of consumers and producers, offers a robust approach to balancing energy consumption and generation, ensuring flexibility and responsiveness. Our study underscores the importance of combining energy storage with demand-side management in propelling a sustainable energy transition, reaffirming the move towards a decentralized energy model. These insights are valuable not only for Rhineland-Palatinate but also have implications for policymakers, energy providers, and stakeholders in other regions.
Keywords: Demand Side Management, Pumped Hydroelectric Energy Storage, Renewable Energy, Energy Modeling

1 INTRODUCTION

As the global focus sharpens on green energy solutions, integrating renewable energy sources into existing power grids is one of the most critical and pressing challenges. As renewable energy sources gain traction, power grid operators face the challenge of integrating these variable and sometimes unpredictable sources into established systems. Research indicates that effective solutions lie in utilizing flexibility options, such as energy storage systems and demand-side management (DSM) strategies [1]. There is compelling evidence suggesting that DSM strategies, in tandem with energy storage systems, play a crucial role in enhancing the penetration rate of energy supply from renewable sources by mitigating short-term fluctuations due to their volatile nature [2]. Notably, the concept of employing such strategies is not novel. Research in the energy sector, as cited by [3, 4, 5], has long emphasized the importance of using methods like peak clipping, load shifting, and enhancing load flexibility to maintain the constant operation of inflexible base-load power plants.

By balancing generation and consumption patterns using DSM techniques, significant optimizations can be achieved before energy storage measures are considered, underscoring the strategic move toward a decentralized energy system. While several studies have explored the role of DSM in integrating renewable energies, its contribution to bolstering the power system remains a focal point. For instance, O'Connell et al. argue that DSM can reduce the need for electricity imports and exports and investments in conventional power plants and storage systems [6]. Similarly, Tveten et al. have demonstrated the potential of load shifting in specific regions, reducing the curtailment of renewable energy supply [7]. Moreover, Gils' research in the German context supports these findings, indicating that DSM not only minimizes renewable energy curtailment but also brings about a decline in overall system costs [8]. Diverging from these broader analyses, Dallinger et al. [9] and Heinrichs [10] specifically delve into the impact of electric vehicle charging behaviors on the power system. Their studies highlight that uncontrolled charging can introduce additional peak loads. However, with smart charging management, these peaks can be mitigated, and the integration of surplus renewable energy can be enhanced. Ladwig's research explores the potential of selected DSM applications across various categories [11]. Her research delves into the role of DSM in power grids with a high proportion of renewable energies, utilizing the linear optimization model ELTRAMOD to represent and further develop the German and European electricity markets.

Consumers, both novel ones like heat pumps and electric vehicles and conventional ones open to innovative approaches, hold the potential for partial flexibility in their utilization [12]. Load shifting finds applicability in diverse settings, encompassing industrial processes, residential needs, and service sectors. Industrial processes include ventilation and air conditioning systems, paper machines, cooling cycles in the food sector, and cement industry mills. Traditional applications in residential and service sectors, such as electric heating systems (night storage heaters, direct heaters, circulation pumps, boilers), air conditioning systems, household appliances (washing machines, dryers, dishwashers), and refrigeration and freezing devices, can also be adapted for load shifting. Electric vehicles and heat pumps represent the most impactful new electricity applications conducive to DSM [12]. In the case of electric vehicles, effective load shifting is achieved through optimized charging strategies, with access to charging infrastructure in public areas and workplaces playing a pivotal role. Moreover, the composition of the electric vehicle fleet, especially the ratio of pure electric vehicles to plug-in hybrid vehicles, has a significant influence on load shifting capabilities. Similarly, for heat pumps, load shifting often focuses on procuring energy for heat supply at the lowest feasible cost, and the integration of a heat storage system becomes imperative to temporally decouple electricity consumption from heat generation [13].

10.4229/EUPVSEC2023/5CO.5.1
020487-001

Despite the extensive research on DSM and its potential benefits, there remains a gap in understanding its holistic impact on the entire energy system, especially when considering integrating various DSM categories. To address this, our study integrates the PHES-Rio, a pumped hydroelectric energy storage (PHES) conceived in 2010. Although it faced implementation challenges due to economic and regulatory constraints, the PHES-Rio exemplifies a pivotal strategy to reduce energy and resource consumption compared to current standards [14]. Serving as a case study, it helps to further develop and refine a technological concept aligned with sustainable objectives.

This study, structured into five chapters, employs the simulation model UCB-SEnMod to depict a potential energy system for Rhineland-Palatinate based on renewable energies. Enhanced to simulate flexibility measures, the model uses time series data spanning a specific time frame, broken down into 15-minute increments, showing the average electric power during each interval from photovoltaics, wind power, hydropower, and combined heat and power plants. This is juxtaposed with the electrical energy demand from sectors such as industry, households, commerce, electromobility, and heat. The results, presented in subsequent chapters, emphasize the significant benefits of flexibility measures and energy storage in balancing the energy profiles, with discussions comparing these findings to other publications and offering critical reflections on the study's methodology.

Through a series of model calculations based on predefined assumptions, our research evaluates the potential of suitable DSM options (including load shifting, flexible production, and consumption) in balancing future power systems, contrasting their efficacy with pure energy storage solutions. Central to this research is the revelation that while initiatives like the PHES-Rio project hold promise, the true linchpin for future power system equilibrium lies in the judicious harnessing of flexible consumption and generation units. When synergized with adept strategies for load shifting, these units outpace the potential of standalone energy storage solutions and chart a transformative path for sustainable energy systems.

2 METHODOLOGIES

For our simulations, we used the computational methods described in our previous publications that have already proven successful in modeling energy systems [15, 16, 17, 18]. To model the energy cluster of Rhineland-Palatinate, we have created a digital twin of the region using our energy system model UCB-SEnMod. This digital replica provides the basis for our analytical framework to investigate the impact of the PHES-Rio and DSM in 2030. To establish a robust basis for our study, we sourced load profiles spanning the industrial, commercial, residential, electromobility, and Power-to-Heat sectors. These were informed by historical data provided by the network operator Stadtwerke Trier and supplemented with insights from a pilot project on the decentralization of the German energy system [19].

Using this model, we carry out energy simulations of clusters, considering energy producers and consumers and the influence of storage and DSM. The model architecture is depicted in Figure 1 and encompasses photovoltaic, wind power, run-of-river hydroelectric, and combined heat

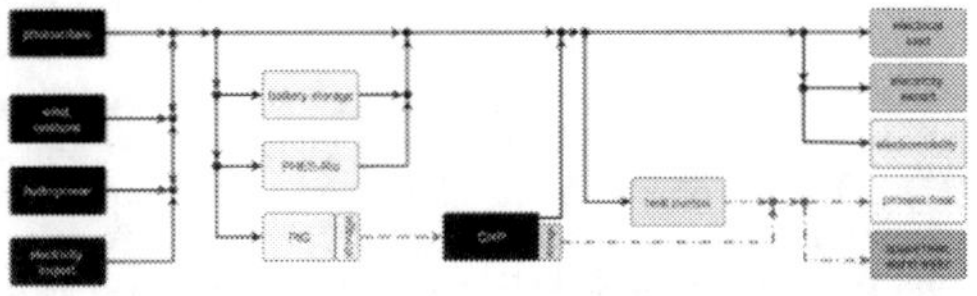

Figure 1: Design of the energy model UCB-SEnMod for mapping the energy system of Rhineland-Palatinate.

and power (CHP) plants. To counterbalance energy shortfalls, we considered electricity imports. Energy storage devices, such as batteries and the PHES-Rio, are integrated, with surplus electricity being transformed into hydrogen through electrolyzes for storage and later re-electrification. The heating aspect incorporates CHP units and heat pumps, paired with heat storage, catering to applications ranging from low-temperature industrial processes to space heating and warm water heating. The electricity produced addresses the needs of diverse sectors, with any surplus energy earmarked for export.

The outcomes from our model components, both for generation and consumption units, are presented as profiles spanning one year, broken down into 15-minute increments, showing the average electric power in that interval. The procedure for creating the profile pattern is described in the methodology paper from Blinn and te Heesen [15]. The time series are scaled using the total consumption and production values in 2030 shown in Table 1 and Table 2.

Table 1: Comparison of electrical energy supplied by various power plant types in 2021 and 2030, measured in GWh. Data for 2021 is sourced from [20], while 2030 data comes from [19].

Power plants	2021	2030
Photovoltaic systems	2 554	6 179
Onshore wind turbines	6 342	19 032
Run-of-river power plants	593	1 862
Gas power plants/CHP plants	883	1 427

Table 2: Comparison of electrical energy demanded by various consumer types in 2021 and 2030, measured in GWh. The data for 2021 is sourced from [20], while 2030 data comes from [19].

Consumers	2021	2030
Private households	6 623	6 300
Trade, commerce, and services	5 333	5 700
Industry	11 100	11 900
Electromobility	66	2 200
Power-to-Heat	246	2 000

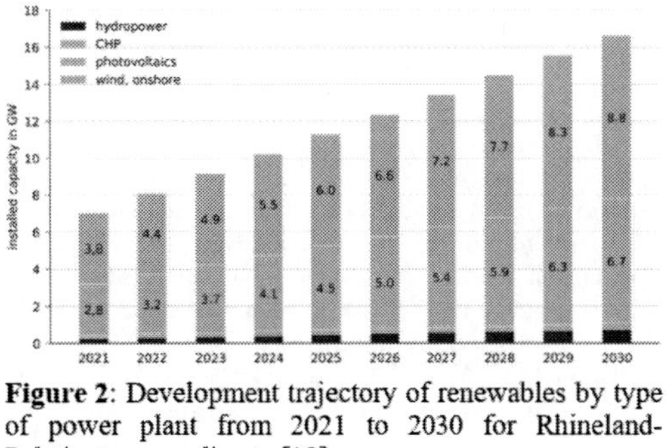

Figure 2: Development trajectory of renewables by type of power plant from 2021 to 2030 for Rhineland-Palatinate, according to [19].

10.4229/EUPVSEC2023/5CO.5.1
020487-002

The development trajectory for renewable energy in Rhineland-Palatinate is designed such that by 2030, energy production and consumption achieve a balanced state. Figure 2 illustrates this trajectory. The proportional growth of individual electric power plants is derived from [19]. The trajectory and the generated energy underscore the dominance of wind power in the renewable energy supply, accounting for 67 % with 2,153 full-load hours. This is followed by photovoltaic systems at 22 % and 918 full-load hours, hydroelectric power at 7 % with 2,513 full-load hours, and CHP units at 4 % with 4,536 full-load hours.

Table 3: Potential of flexibility measures in MW in 2030. The numbers serve as an upper limit for the optimization. The data sources for private households, commerce, industry, power-to-heat, and electromobility are taken from [13]; the potential for bidirectional battery storage is taken from [21]; the potential of hydrogen electrolyzes is taken from [22], and the potential of plannable CHP units fired with biomethane or hydrogen is taken from [19].

Flexibility measures	2030
Private households and Trade, commerce, and services	1750
Industry	225
Electromobility	481
Power-to-Heat	484
Bidirectional battery storage	500
Hydrogen electrolysis	145
Biomethane/Hydrogen CHP unit	100

The flexibility measures employed in the simulation serve the dual purpose of enhancing generation and load while facilitating load shifting. The operating times of these flexibilities were fine-tuned using the least squares method, chosen for its simplicity in reducing peaks over a daily time horizon spanning 2030. Details of their available capacities for Rhineland-Palatinate can be found in Table 3.

Our simulation framework relied on established practices in energy system modeling, utilizing 15-minute intervals aligning with energy market trading intervals. To ensure the fidelity of our model, systematic validation was conducted using historical regional data while also adhering to stringent data verification standards.

Our analysis spans three distinct scenarios:

1. Reference Scenario: Absent of DSM and PHES-Rio, this serves as our comparison point.
2. DSM-Only Scenario: This scenario introduces DSM, offering insights into the potential for optimized energy consumption.
3. Combined DSM and PHES Scenario: An advanced setup integrating both DSM strategies and PHES. This scenario explores the intricate balance and outcomes of demand adaptation paired with energy storage mechanisms.

It's imperative to recognize that the evolution of the energy system hinges on multifaceted variables, including energy demand, the selection of plants and technologies, and overarching societal developments. The assumptions within our model inherently carry uncertainties, and the results should not be construed as forecasts. A central theme emerging from our analysis is the significant potential of DSM. Our findings suggest more efficient and sustainable strategies than relying solely on storage solutions. The following chapter presents these results and the possible alternatives to a storage-only concept in more detail.

3 RESULTS

Using the simulation framework previously outlined, we examined three scenarios to elucidate the effects of DSM and PHES-Rio in the Rhineland-Palatinate region. In line with the works of O'Connell et al., which explain the potential of DSM to reduce electricity imports and exports [6], and Tveten et al., focusing on load shifting to decrease curtailment of renewable energy supply [7], our reference scenario depicts the energy system's characteristics without optimization.

In the "DSM-Only" scenario, while integrating system flexibilities, we excluded the PHES-Rio. Here, a discernible change in energy consumption patterns was observed, particularly in peak demand reduction, reinforcing Tveten et al.'s findings on the potential for load shifting [7].

Lastly, in our third scenario, where DSM strategies are paired with the PHES-Rio, our results revealed a synergistic effect on energy management. This aligns with Gils' observation in the German context, emphasizing that DSM contributes to minimizing renewable energy curtailment [8]. Our results highlight the potential of combined DSM and storage mechanisms in meeting immediate and delayed energy needs.

To visually represent these scenarios, refer to Figure 3, which clearly delineates the power profile dynamics across each. The red line represents the sorted annual duration lines of the electric power without any DSM and storage usage. In contrast to the red line, the green line considers the effects of flexibility measures, resulting in a noticeably smoother power profile. The differences between the red and green lines highlight the substantial potential of flexibility in refining power distribution and optimizing consumption patterns. In this context, observations on the right side of the graph are particularly noteworthy. Here, the negative deviations in the profile are significantly mitigated through the flexible operation of CHP units.

The distinctions between the green and blue lines are subtle. Even though the PHES-Rio boasts an electrical output of 300 MW and an upper basin storage capacity of 3.4 GWh, placing it among the larger pumped hydroelectric energy storages in Germany, its impact on the regional electricity equilibrium remains limited. However, compared to the reference scenario and combined with DSM, overproduction is reduced by

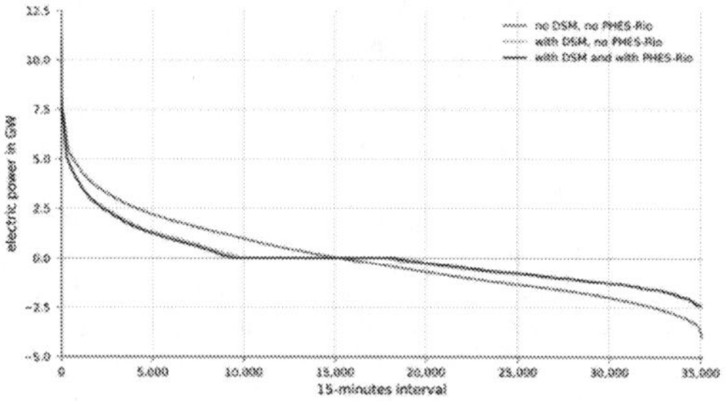

Figure 3: Generated power balances displayed as sorted annual duration lines. Red represents the scenario without DSM, green with DSM only, and blue with both DSM and PHES-Rio.

2,925 GWh (41.3 %) and underproduction by 2,937 GWh (41.5 %).

Furthermore, our observations on the left side of the graph indicate that overproduction peaks remain relatively unaffected by the flexibility and storage measures. This can be attributed to the limited capacity of the electrolyzes and the rapid saturation of the upper basin of the PHES-Rio during peak times, preventing further energy intake. However, this limitation is not critical, as in cases of overproduction, curtailment can be applied to maintain grid safety. In the center of the graph, the effects of flexibility measures in mitigating power imbalances become apparent. These measures can compensate for about a fifth of the 15-minute intervals over a year, emphasizing their central role in a balanced energy system.

Among the flexibility measures, those of households and businesses have a significantly higher impact on the electricity balance than the PHES-Rio. While they can provide up to 1,750 MW for load shifting, the actual shifted capacity is below 850 MW in about 70 % of cases. In contrast, the industrial sector consistently contributes almost its maximum potential power for load shifting. The load shifting patterns of electric mobility and power-to-heat units mirror those of households and businesses, exhibiting a pronounced decline when tapping into higher power levels. The CHP units and electrolyzes operate at nearly 100 % full supply. The CHP units achieved 4,483 full-load hours out of 4,695 operational hours. In contrast, the electrolyzes reached 2,540 full-load hours out of 3,143 operating hours.

Figure 4 illustrates the change in the import and export of electric energy in Rhineland-Palatinate. The upper graph depicts the prevailing electricity overproduction or underproduction in the scenarios without DSM and storage usage. In contrast, the lower chart compares electricity overproduction and underproduction when integrating DSM and the PHES-Rio. The color scale is chosen so that a transition toward red represents export, while a shift toward dark blue indicates import.

The upper graph clearly shows a pronounced overproduction in early March. The midday region stands

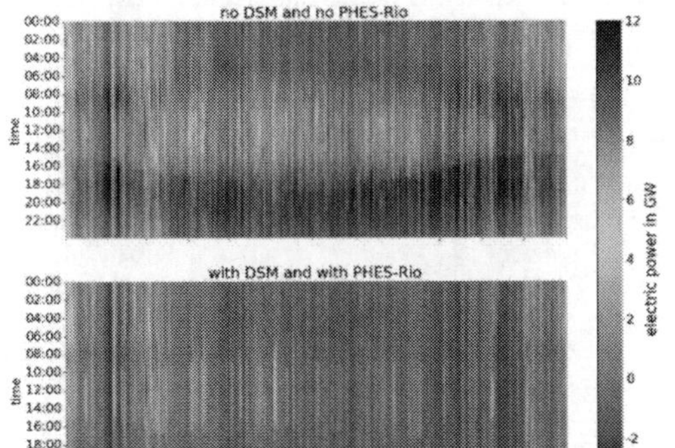

Figure 4: The upper section presents a heat map illustrating electricity imports and exports for Rhineland-Palatinate without DSM or storage intervention. The lower section depicts a heat map of electricity imports and exports, factoring in utilizing flexibilities and storage systems. The color scale on the right margin demonstrates the gradient of imported and exported energy. Negative values signify electricity imports, while positive values represent exports.

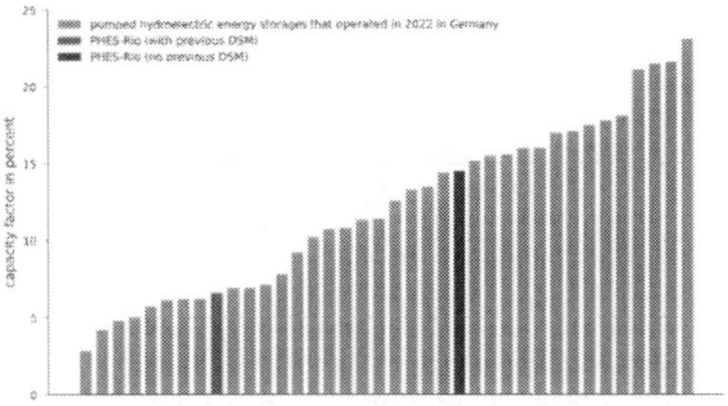

Figure 5: Comparison of capacity factors for pumped hydroelectric energy storages in Germany that operated in 2022. The capacity factor represents the ratio of full-load hours in generator operation to the total number of hours per year. The red bars show the capacity factor of the PHES-Rio in two scenarios: with the use of DSM (shorter bar) and without (taller bar).

out, colored prominently in green, indicating high levels of solar energy feed-in. This surge is evident from the morning and persists into the evening, especially during the summer months. Notably, in the evening hours, when photovoltaic generation drops and household consumption spikes, there's a marked underproduction. As illustrated in the lower graph, the chosen DSM strategy substantially reduces the midday overproduction and evening underproduction imbalance.

Furthermore, DSM has a significant impact on the charging and discharging processes of the PHES-Rio. Due to the restricted capacity of its upper basin, the PHES-Rio frequently reaches maximum capacity, primarily during January, February, March, April, and December. This behavior is aligned with the increased electricity supply from wind turbines during winter. Flexibility measures effectively manage disparities between energy production and consumption, resulting in decreased dependency on PHES-Rio. However, the summer months see the PHES-Rio operating at reduced reservoir levels due to diminished wind electricity output, which isn't sufficiently offset by rising photovoltaic contributions. This pattern necessitates a heightened operational demand on the PHES-Rio.

Considering the broader German energy infrastructure, this operational context becomes even more significant. Data from Fraunhofer ISE indicates that the full-load hours for the existing pumped hydroelectric energy storages in Germany average about 1,000 hours in 2022, translating to a capacity factor of approximately 11 % [23]. The capacity factors for these plants are depicted as gray bars in Figure 5, indicating the ratio of full-load hours to the total hours in a year. Specifically, the PHES-Rio's full-load hours presents 1,271 full-load hours (or 15 % capacity factor) without DSM and 577 full-load hours (or 7 % capacity factor) with DSM. These values are highlighted in red, illustrating the variance brought about by the integration of DSM, resulting in a significant reduction in the number of full-load hours.

In the reference scenario for 2030, both the energy surplus and deficit were 7,078 GWh. The DSM and PHES-Rio changed these values significantly. The energy surplus decreased to 4,154 GWh, while the deficit fell to 4,141 GWh. This represents a reduction in the surplus and deficit of approximately 40 %, underscoring the measures' effectiveness. These results serve as the basis for the following chapter and help to contextualize and interpret the results.

10.4229/EUPVSEC2023/5CO.5.1
020487-004

4 DISCUSSION

This research was conceived to probe the dynamics of Rhineland-Palatinate's energy system, primarily upon integrating the PHES-Rio and introducing various flexibilities, combining the electricity, heat, and transport sectors. Methodologies and findings, as detailed in the previous chapters, concentrate on minimizing the gap between electricity consumption and production in the region.

Our study builds on the data from publications, such as those by Koch et al. [19], Schröer et al. [22], and the reports of the four German transmission system operators [13] and [21]. In this chapter, we intend to contextualize our findings within existing literature and knowledge. By doing so, we offer a comparative view that delineates the expected shifts in the 2030 energy system of the analyzed region. This encompasses a review of the PHES-Rio's integration, and the deployment of flexibility measures like CHP, hydrogen electrolyzes, and sectoral load shifting in an energy system with a high penetration of renewable energy. Our analysis further refines the understanding of Rhineland-Palatinate's proposed renewable energy pathway, enriching the prevailing body of knowledge.

The growing emphasis on renewable energy sources, like solar and wind, for electricity supply underlines the need for efficient and adaptable energy storage solutions. Pumped hydroelectric energy storages, given their structural design, stand well-poised to meet these requirements. Positioned near renewable energy plants in particular, these storages can quickly compensate for grid fluctuations due to their large capacities and fast response times. Consequently, pinpointing suitable locations for such facilities has gained prominence, as highlighted by Görtz et al. [24].

The planned PHES-Rio near the Moselle River aims to moderate peak loads in its vicinity. This facility is projected to possess an electric power of 300 MW and a storage capacity of 3.4 GWh. This design ensures that surplus energy from renewables is stored efficiently and utilized when needed. When renewable energy is abundant, the system pumps water from the lower basin to the upper one. Conversely, when electricity is demanded, water flows downhill, spinning turbines connected to generators, producing electricity as described by Menke, Brechtken, and Gebauer [14].

Kruger et al. have set an annual benchmark of 1,000 full-load hours for the operation of pumped hydroelectric energy storage to be considered profitable. These plants are designed to meet various power requirements within their control spectrum. In addition to their primary functions, they often offer auxiliary services to network operators, such as adjusting for forecast errors. These additional roles, which aren't included in our present simulations, significantly extend operational hours [25]. With only 577 full-load hours, the PHES-Rio's operation, according to Kruger et al., might not be economically justifiable. This threshold is surpassed only in scenarios devoid of DSM, clocking 1,271 full-load hours.

Earlier assessments by Menke et al. from 2012 explored the integration of the PHES-Rio into the Trier region of Rhineland-Palatinate [14]. In their projections for 2030, the storage, in its 300 MW variation, could compensate for an energy deficit by delivering 478 GWh while storing a surplus of up to 713 GWh. It's pertinent to mention that these results don't factor in efficiency losses. After accounting for a 19 % inefficiency in pumping and turbine operations, the adjusted results are 570 GWh for charging and 382 GWh for discharging. Given the disparities in size and availability of flexibility measures between the Trier region and the more expansive Rhineland-Palatinate, drawing a direct parallel with earlier outcomes presents challenges. Nevertheless, during our model's validation stage, we fine-tuned it specifically for the region Trier, leading to projected figures of 283 GWh in discharging and 347 GWh in charging in 2030. This discrepancy of approximately 100 GWh is due to the integration of battery capacities in our study, covering an additional 77 GWh.

Our data indicates that Rhineland-Palatinate appears to be the most promising for establishing a pumped hydroelectric energy storage of this scale among the analyzed regions. Its performance, especially in non-DSM scenarios, places it in the upper third of Germany's pumped hydroelectric energy storages regarding capacity factors. Kruger's benchmark of 1,000 full-load hours, equivalent to an 11.4 % capacity factor, was met only in the Trier and Rhineland-Palatinate regions without flexibility deployments. Many of Germany's operational pumped hydroelectric energy storages, as shown in Figure 5, operate below this threshold. Here, the flexible operational capabilities of the power plants come into play. Pumped hydroelectric energy storage can be lucrative, even with lower operating duration, when one considers not just the pure net balance of a region. Factors like converting off-peak energy to peak energy, providing instantly deployable power reserves, rapidly absorbing surplus power during sudden major consumer failures, and applications for frequency stabilization, phase balancing, and phase-shifting operations play significant roles [26]. The comparison of the PHES-Rio explored in this study to currently operated pumped hydroelectric energy storages remains inconclusive since the simulations did not consider the aforementioned operational capabilities. These would likely positively impact both operational and full-load hours, directly influencing economic viability.

Conventional storage solutions complement promising approaches like demand-response and broader DSM strategies to balance electrical power supply and demand. Demand-response exploits short-term consumer behavior, leveraging consumer demand elasticity from factors like thermal inertia, demand flexibility, and physical storage [1]. The main goals of DSM are 1) to integrate high-efficiency devices and building structures, thus reducing overall energy consumption, and 2) to shift energy consumption patterns in end-users, especially during high-demand periods [27].

Research by O'Connell et al. and Strbac addresses how DSM impacts the integration of renewable energies into systems [6, 28]. Their analysis highlights its advantages in harmonizing inconsistent renewable energy inputs and curtailing peak loads. One salient benefit is the reduced reliance on electricity interregional transfers. Instead of just balancing systems, these connections become economically optimized. Regions with a high dependence on wind energy frequently require interconnected networks for stability. Meteorological trends dictate that heightened wind activities in one area often resonate in adjacent regions, leading to wind energy exports at reduced prices.

Significant costs in the energy sector are attributed to the procurement and maintenance of generation capacities. Leveraging DSM for system capacity needs can present marked cost reductions. By adeptly managing renewable energy variances and using load shifting to tackle peak loads, there's a diminished urgency for high-cost, often inefficient peak power stations and flexible units like open-cycle gas turbines [6].

In their research, Salpakari et al. evaluated the feasibility of using heat pumps and electrode boilers as flexible load mechanisms for integrating widespread photovoltaic energy in urban settings [29]. Their findings highlight that electrode boilers offer the most potential for assimilating photovoltaic surpluses. Additionally, electrolyzes can absorb excess renewable energy, classifying them under the category of load increase. Notably, many studies suggest these are only essential when the renewable share in the electricity system surpasses 60 %, primarily due to their high investment costs relative to their efficiency. In the context of this study, the renewable energy share significantly exceeds this 60 % threshold, justifying the use of electrolyzes.

Ladwig's research [11] underscores that current studies focus on individual consumers and the diverse DSM strategies, such as load reduction, shifting, and increasing. However, a comprehensive analysis of these strategies' interactions and their collective impact on the power system must be included. For instance, a surplus of renewable energy can be balanced through load shifting or load increase for hydrogen production, meaning these DSM categories directly compete. Literature reviews confirm Ladwig's observation about this gap in research. In his work, Ladwig examines the potential of specific DSM applications across these categories, projecting their medium to long-term evolution. This involves enhancing the linear optimization model ELTRAMOD, which depicts the German and European electricity markets. This research aims to accurately represent the technical characteristics of Demand Side Management, which dictate its application and flexibility within the ELTRAMOD model.

A cornerstone of this research is assessing the smoothing effect of DSM options on the residual load, especially with varying levels of renewable energy in the electricity system. The residual load represents the electric power that conventional power plants must supply to meet the demand. The provided data in this study emphasizes that implementing DSM measures can significantly stabilize fluctuations in the residual load.

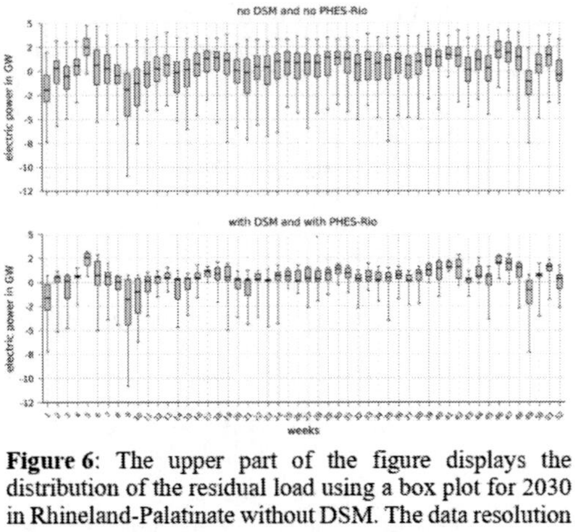

Figure 6: The upper part of the figure displays the distribution of the residual load using a box plot for 2030 in Rhineland-Palatinate without DSM. The data resolution spans one week. The lower graphic illustrates the data from the scenario with DSM similarly.

Figure 6 illustrates the changing distribution of the residual load in Rhineland-Palatinate due to the implementation of DSM options. The upper part of the figure displays the distribution as a box plot for each week in 2030. The box plot in the lower image represents the distribution using DMS and the PHES-Rio. When comparing both subplots, it's noticeable that the interquartile ranges in the scenario with DSM have been significantly reduced. Likewise, the upper and lower whiskers are less pronounced.

Despite the distinct choices of DSM options, energy storage variants, and the size of the systems considered by Ladwig and this study, their results have striking similarities. Ladwig's findings indicate that as the integration of renewable energy sources increases, and DSM potential is harnessed more, there's a consistent decline in the variability of the residual load. The scenarios examined in this research similarly demonstrate a significant decrease in the variability when DSM options are in play by approximately 20 %. However, diminishing returns are observed with the increasing use of further flexibility measures, which might be attributed to the variability inherent in available renewable production.

Similar results were found by Koch et al., who analyzed a renewable electricity-heat system in Rhineland-Palatinate, comparing the efficiency of decentralized and centralized electricity supply [19]. Part of their research focused on optimizing the Rhineland-Palatinate electricity system using flexibilities and storage to understand the advantages and disadvantages of a decentralized versus a centralized power supply. Koch et al. noted that efficient energy management can significantly reduce storage needs. Our results underline this statement. The literature review conducted for this chapter revealed that very few studies exist in energy system modeling examining the impact of various DSM options on an energy system. This highlights the innovative nature of this research and may serve as an impetus for further studies.

5 CONCLUSION

To achieve the German federal government's targeted 65 % reduction in CO_2 emissions by 2030, there is an urgent need to expand the deployment of volatile energy carriers. Addressing short-term fluctuations due to the volatile nature of renewables necessitates integrating appropriate storage systems and demand-side management into the energy system.

To comprehend the dynamic interaction between electrical energy supply and demand in Rhineland-Palatinate, we established a digital twin of the region's energy system for 2030. Our investigation explored the impact of strategically controlling electricity demand during peak or high influx from photovoltaic and wind energy systems. We also evaluated the influence of pumped hydroelectric energy storage in this region. For this purpose, we employed the UCB-SEnMod energy system model [15], a tool designed to analyze potentials for reducing energy demand. Additionally, it facilitates the integration of renewable energy sources and storage systems to enhance energy efficiency. We expanded the model during this study to incorporate various demand control options. The quarter-hourly comparison of supplied and demanded electrical energy, combined with integrating the pumped hydroelectric energy storage Rio and demand-side management, enabled a comprehensive assessment of the power system expansion. Our findings highlight that demand-side management can lead to a more balanced energy profile, significantly reducing the need for electrical storage.

A pivotal insight from this work relates to comparing the benefits between the pumped hydroelectric energy storage Rio and demand-side management. Especially in the context of decentralization and the evolution of smart grids, significant differences emerge concerning the performance and scalability of these two approaches. The pumped hydroelectric energy storage Rio serves a vital role as an energy storage solution, capable of offsetting short-term fluctuations within minutes in the electricity grid. However, upon closer examination, its limited capacity presents certain adaptability constraints. On the other hand, demand-side management offers a promising approach to address the challenges associated with a high penetration of renewables in an energy system. They encompass various measures, including intelligent control of consumers and producers. Integrating these flexibility measures allows for directing energy flows within certain bounds, adjusting to specific requirements. This enables more efficient utilization of renewable energy sources and reduces grid bottlenecks. A key advantage lies in the scalability and adaptability of this approach, allowing for flexibility in response to changing conditions. Demand-side management and the implementation of the pumped hydroelectric energy storage Rio significantly impacted the region's energy balance within the model's parameters. Set initially at 7,078 GWh for both surplus and deficit, the interventions reduced the energy surplus to 4,154 GWh and the deficit to 4,141 GWh. This translates to an approximate 40 % reduction, emphasizing the measures' effectiveness.

Our findings emphasize the increasing relevance of decentralized energy systems, with demand-side management and energy storage as integral components. Demand-side management emerges as a crucial strategy, bridging the gap between energy supply and demand to reduce storage capacities.

REFERENCE

[1] L. Tronchin, M. Manfren and B. Nastasi, "Energy efficiency, demand side management and energy storage technologies – A critical analysis of possible paths of integration in the built environment," *Renewable and Sustainable Energy Reviews,* vol. 95, pp. 341-353, 2018.

[2] A. Pina, C. Silva and P. Ferrão, "The impact of demand side management strategies in the penetration of renewable electricity," *Energy,* vol. 41.1, pp. 128-137, 2012.

[3] W. A. Qureshi, N.-K. C. Nair and M. M. Farid, "Impact of energy storage in buildings on electricity demand side management," *Energy Conversion and Management,* vol. 52, no. 5, pp. 2110-2120, May 2011.

[4] V. Hamidi, F. Li, L. Yao and M. Bazargan, "Domestic demand side management for increasing the value of wind," *China International Conference on Electricity Distribution,* 2009.

[5] C. Gellings, "The concept of demand-side management for electric utilities," *Proceedings of the IEEE,* vol. 73, no. 10, pp. 1468-1470, October 1985.

[6] N. O'Connell, P. Pinson, H. Madsen and M. O'Malley, "Benefits and challenges of electrical demand response: A critical review," *Renewable and Sustainable Energy Reviews,* vol. 39, pp. 686-699, November 2014.

[7] Å. G. Tveten, T. F. Bolkesjø and I. Ilieva, "Increased demand-side flexibility: market effects and impacts on variable renewable energy integration," *International Journal of Sustainable Energy Planning and Management,* vol. 11, Oct 2016.

[8] H. C. Gils, "Economic potential for future demand response in Germany – Modeling approach and case study," *Applied Energy,* vol. 162, pp. 401-415, January 2016.

[9] D. Dallinger, S. Gerda and M. Wietschel, "Integration of intermittent renewable power supply using grid-connected vehicles – A 2030 case study for California and Germany," *Applied Energy,* vol. 104, pp. 666-682, April 2013.

[10] H. U. Heinrichs, "Analyse der langfristigen Auswirkungen von Elektromobilität auf das deutsche Energiesystem im europäischen Energieverbund," KIT Scientific Publishing, 2013.

[11] T. Ladwig, "Demand Side Management in Deutschland zur Systemintegration erneuerbarer Energien," Series of the chair of energy economics, TU Dresden, Dresden, 2018.

[12] B. Özalay, H. Schuster, J. Kellermann, J. Priebe and A. Moser, "Wirtschaftlicher Vorteil der netzdienlichen Nutzung von Flexibilität in Verteilnetzen," Stadtwerke München Infrastruktur GmbH, 2019.

[13] 50Hertz Transmission GmbH; Amprion GmbH; TenneT TSO GmbH; TransnetBW GmbH, "Szenariorahmen für die Netzentwicklungspläne Strom 2030, Entwurf der Übertragungsnetzbetreiber," 2016.

[14] C. Menke, D. Brechtken and M. Gebauer, "Pumpspeicherkraftwerk Rio: Energiewirtschaftliche Betrachtung," ECOSCOP, Trier, 2012.

[15] A. Blinn and H. te Heesen, "UCB-SEnMod: A Model for Analyzing Future Energy Systems with 100 % Renewable Energy Technologies—Methodology," *Energies,* vol. 15, no. 12, p. 4383, 2022.

[16] A. Blinn and H. te Heesen, "Modellierung und Simulation von Energieclustern am Beispiel des europäischen Strom-, Wärme- und Verkehrssektors - Methodik," in *5. Regenerative Energietechnik Konferenz in Nordhausen*, Deutschland, 2022.

[17] A. Blinn and H. te Heesen, "Sector Coupled Energy Model for the European Electricity,

Heat and Transport Sectors - Methodology," *38th European Photovoltaic Solar Energy Conference and Exhibition,* pp. 1356-1363, 2021.

[18] A. Blinn and H. te Heesen, "Sector Coupled Energy Model for the European Electricity, Heat, and Transport Sectors - Results," *8th World Conference on Photovoltaic Energy Conversion,* pp. 1568-1573, 2022.

[19] M. Koch, M. Vogel, C. Heinemann, T. Hesse and D. Bauknecht, "Pilotprojekt Dezentralisierung - Stärkere Dezentralisierung des bundesdeutschen Strom-Wärme-Systems: Rechtliche und organisatorische Rahmenbedingungen sowie infrastrukturelle Folgen," Ministerium für Umwelt, Energie, Ernährung und Forsten, 2021.

[20] Energieaentur Rheinland-Pfalz, "Energieatlas Rheinland-Pfalz," [Online]. Available: https://www.energieatlas.rlp.de/earp/unser-portal/die-energieagentur. [Accessed 14 September 2023].

[21] 50Hertz Transmission GmbH; Amprion GmbH; TenneT TSO GmbH; TransnetBW GmbH, "Netzentwicklungsplan Strom 2030, erster Entwurf der Übertragungsnetzbetreiber," 2019.

[22] R. Schröer, E. Kötter and L. Schneider, "Die Rolle von Power-to-Gas in der zukünftigen Stromversorgung - Das optimierte Stromversorgungsystem bei hohen Anteilen Erneuerbarer Energien am Beispiel der Modellregion Trier-Amprion 5," 5. Bundesministerium für Wirtschaft und Energie, 2015.

[23] Fraunhofer Institut für Solare Energiesysteme, "Energy Charts," [Online]. Available: https://www.energy-charts.info. [Accessed 15 09 2023].

[24] J. Görtz, M. Aouad, S. Wieprecht and K. Terheiden, "Assessment of pumped hydropower energy storage potential along rivers and shorelines," *Renewable and Sustainable Energy Reviews,* vol. 165, September 2022.

[25] K. Kruger, A. Maaz, A. Moser and N. Rotering, "Forecasting for Pumped Storage in Germany," *HRW,* vol. 24, no. 6, 2016.

[26] J. Giesecke and E. Mosonyi, Wasserkraftanlagen. Planung, Bau und Betrieb, Vols. 5., aktualisierte und erweiterte Auflage, Springer Berlin, 2009.

[27] F. Boshell and O. Veloza, "Review of developed demand side management programs including different concepts and their results," *2008 IEEE/PES Transmission and Distribution Conference and Exposition: Latin America,* 2008.

[28] G. Strbac, "Demand side management: Benefits and challenges," *Energy Policy,* vol. 36, no. 12, pp. 4419-4426, 2008.

[29] J. Salpakari, J. Mikkola and P. Lund, "Importance of Demand Side Flexibility and Management for Large-Scale Variable PV Integration in Urban Environment," in *Proceedings of the 5th International Workshop on Integration of Solar Power into Power Systems,* Brussels, 2015.

10.4229/EUPVSEC2023/5CO.5.1
020487-008

ON THE EFFECTS OF NODAL PV POWER FORECASTING ERRORS ON THE OPERATION OF POWER SYSTEMS

Joao Gari da Silva Fonseca Junior[1], Tomoya Takeuchi[1], Tomoyo Saito[2], Segawa Shuhei[1], Kazuhiko Ogimoto[1]
1) The University of Tokyo, 113-8505, 4-6-1 Komaba, Meguro-ku, Tokyo, Japan
2) Kozo Keikaku Engineering, 164-001, 4-5-3, Chuo, Nakano-city, Tokyo, Japan

ABSTRACT: In this study, we propose a methodology to evaluate the forecasts from the point of view of power systems operators. Using this methodology, and considering a realistic scenario of high penetration of PV power, we analyzed the effects of day-ahead PV power forecasting errors, when used at the level of power transformer substation, to operate a power system. To do that we simulated one year of operation of the power systems of Kanto-Tohoku area in Japan, using a day-ahead security constrained unit commitment scheduling, SCUC, followed by an economical dispatch, SCED, simulating intraday operations. The power system configuration was simulated using an adapted version of the 10-machine and 18-node EAST10 model, published by the Institute of Electrical Engineers of Japan. The results of the simulations indicate that the PV power forecast based security constrained unit commitment scheduling often overestimated the occurrence of congestion in the nodes. For example, annually, over all nodes, day-ahead SCUC scheduling overestimated congestion more than in the case where there are no forecasting errors by 1.5%. Regarding the occurrence of last resort power supply, the frequency was low, happening only three times in the year and in quantities below 150 MWh per day. With respect to cost, day-ahead forecasting errors added deviations between the SCUC and SCED costs of up to 35% compared with a no error case. The results and the analysis presented indicate that the methodology proposed can be a useful tool for operators to characterize the effects of different PV forecasting techniques on their power systems.

Keywords: Energy System Integration; Photovoltaic Power Forecasting,

1 INTRODUCTION

The share of photovoltaic, PV, power generation in the energy matrices of many countries around the world continues to increase [1]. To support such growth, it is indispensable to investigate the impacts of PV power generation, when integrated in large scale, to regional power systems. In this regard, PV power forecasting techniques have an important role to fulfill as they are expected to help to mitigate eventual weather related instabilities in PV power generation. In related literature, there are studies on the use of forecasting of renewables' based power generation in the operation of regional power systems such Udagawa et al. [2], Ishizaki et al.[3], and Lowery et al. [4].

Nevertheless, the impact of using such forecasts in the operation of power systems is still insufficiently explored, which also hinders its efficient application by power system operators.

With this gap in knowledge in mind, in this study, we propose a methodology to evaluate the forecasts from the point of view of power systems operators. Using this methodology, and considering a realistic scenario of high penetration of PV power, we present an initial analysis of the effects of day-ahead PV power forecasting, at the level of power transformer substation, to operate a power system. To do that, we simulated one year of operation of the interconnected power systems of Kanto-Tohoku area in Japan. This system has the highest power demand in the country.

Its operation was simulated using a day-ahead security constrained unit commitment scheduling, SCUC, followed by an economical dispatch, SCED, simulating intraday operations. The configuration of the power system was based on a adapted version of the 10-machine and 18-node model, the EAST10 model, published by the Institute of Electrical Engineers of Japan [5].

In the following sections we describe the models used, section 2, the target area and data, section 3, the proposed methodology in section 4, and the results in section 5. The study is then concluded in section 6.

2 MODELS AND DATA

2.1 Forecasting

Day-ahead forecasts of PV power were done for each node of the power system with PV installed capacity. The forecast were done using numerical weather prediction data from the Japan Meteorological Agency, and support vector regression. The model is trained with 60 days of data preceding each target day, and it is used to forecast solar irradiance, which is then converted to PV power generation reaching the node, using a simple conversion procedure presented in [6]; or directly to forecast PV power depending on the availability of data. This method presents forecast accuracy at the same level or better than direct numerical weather prediction of solar irradiance as shown in [7]. Thus, we believe it provides typical accuracy of day-ahead forecasts of solar irradiance and PV power in Japan.

Wind and Power Demand Forecasts

To focus the study on the effects of PV power forecasts' errors, it was assumed that forecasts of wind power and power demand had no errors. Observed values of each quantity was simulated using power demand data published by Kanto and Tohoku power utilities. Wind power generation data was simulated using wind speed information for the locations assumed to have wind power. Details about the simulation are described in Azuma et al. [8].

10.4229/EUPVSEC2023/5CO.5.4
020489-001

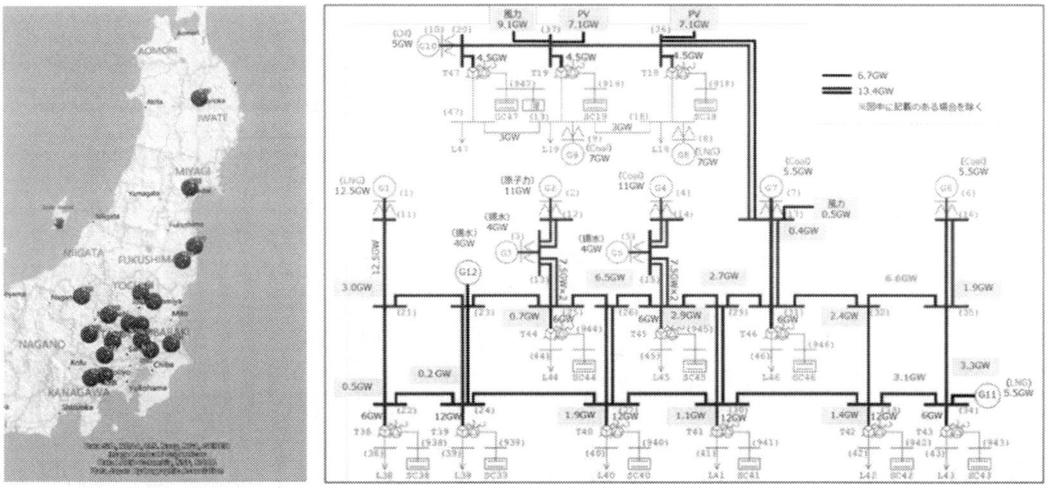

(a) (b)

Figure 1: Target area (a) and configuration of the modified EAST10 model (b), used to simulate a scenario of large penetration of PV and congestion.

2.2 Power System Modeling

A modified version of the EAST10 model, published by IEEJ [9] was used to simulate the power grid of the Kanto-Tohoku area. To simulate a scenario of high penetration of PV, 43.1 GW of installed capacity of PV was simulated. For wind it was assumed an installed capacity of 9.6 GW. The maximum load of the system was assumed to be 63.1 GW.

The simulation of the power system of the target area was done using security constrained unit commitment scheduling, SCUC and security constrained economic dispatch, SCED. These are optimization problems, aiming to schedule the operation of the system's generators at minimum operation and start costs, while meeting power demand and several security, frequency and regulation constraints. Both problems are solved as mixed-integer linear ones. Once the operation of the generators is scheduled, we assumed that they cannot be turned on and off, but only redispatched in intraday operations. The second problem, the SCED is solved, using as initial conditions the solution of the SCUC, and the observed values of renewable power generation and power demand.

To consider in both problems the power flow in the lines and congestion, their formulations employed DC optimal power flow, DC-OPF modelling, of active power only. In the following lines we present just a brief description of the model. More information about the SCUC and SCED implementations has been published in Udagawa et al. [10][11].

In Eq. 1 we show the objective function. The variable $pSt_{n,t}$ is binary, and it indicates the on/off status of generators at time t. The fuel cost $CF_{n,t}$ is calculated with Eq. 2, in which $a_{n,k}$ and $b_{n,k}$ are coefficients of the fuel cost function in the piece-wise linear interval k, and $p_{m,t}$ represents the power output of a thermal generator n, in MW, at time step t. The variable $pWk_{n,t}$ indicates if generator n is connected or not to the power grid at time step t.

Minimize

$$\sum_{t=1}^{T}\sum_{n=1}^{N}\left(CF_{n,t}+CS_{n,t}\cdot pSt_{n,t}\right) \quad (1)$$

where we have,

$$CF_{n,t}\geq a_{n,k}\cdot p_{n,t}+b_{n,k}\cdot pWk_{n,t},\ \forall n,\ \forall t,\ \forall k \quad (2)$$

Eq. 3, Eq. 4, and Eq. 5 describe the constraints affecting power flow in the branches,

$$\sum_{j=1}^{M}P_{i,j,t}+P_{d,i,t}=P_{g,i,t}\ \forall i,\ \forall j \quad (3)$$

$$P_{i,j,t}=B_{i,j}\left(\theta_{i,t}-\theta_{j,t}\right)\ \forall i,\ \forall j,\ \forall t \quad (4)$$

$$-P_{i,j,max}\leq P_{i,j,t}\leq P_{i,j,max}\ \forall i,\ \forall j,\ \forall t \quad (5)$$

In Eq. 3, Eq. 4, and Eq. 5, M is the total number of buses, P is the power output of generator g or of the necessary demand d on bus i. The power flow from bus i to bus j at time t is indicated by $P_{i,j,t}$. The variables $B_{i,j}$ and $\theta_{i,t}$ are the susceptance and the phase angle (in rad). The variable $P_{i,j,max}$ is the maximum transmission capacity from bus i to bus j.

3 TARGET AREA AND DATA

In Fig. 1, we present the location of the power system simulated and the structure/configuration of the modified EAST10 model.

As the EAST10 simulates the power grid of Kanto and Tohoku, these were the target areas of the study. For the meteorological, solar and power demand data, the meteorological year of 2019 was used. The SCUC and SCED simulated the operation of the power system in

30-minute steps considering data of the period, scaled to create a scenario of high penetration of PV power. Regarding data about the configuration of the power system and installed capacity of PV and wind power in different buses, they were set based in consultation with power utilities and expected scenarios for PV growth in the target areas.

4 IMPACT EVALUATION METHODOLOGY

Theoretically, one approach that could be used to evaluate the impact of the forecasts' errors would be to compare directly the day-ahead scheduling provided by the SCUC with the intraday operation simulated by the SCED. Nevertheless, this approach has some limitations. One problem is that both SCUC and SCED are different problems. For example, regulation and contingency reserves are part of the solution of the former, where in the SCED it is fixed, and entered in it as part of its initial conditions. Another important matter is the mip gap, or the gap between the upper and lower bound of the solution found by the mixed integer programming. Even with a tight value such as 1%, depending on how the power system is modeled, meaningful variations can occur between SCUC and SCED. This happens even when there is no forecasting errors. Due to these limitations, effects that are not caused by forecasting errors, might be mistaken as caused by it.

To mitigate this problem, we propose a simple methodology. Instead comparing directly SCUC solutions with SCED ones, we treat the SCUC/SCED solutions as a set, and evaluate different cases regarding the use of forecasts. The evaluation methodology is presented in Fig. 2.

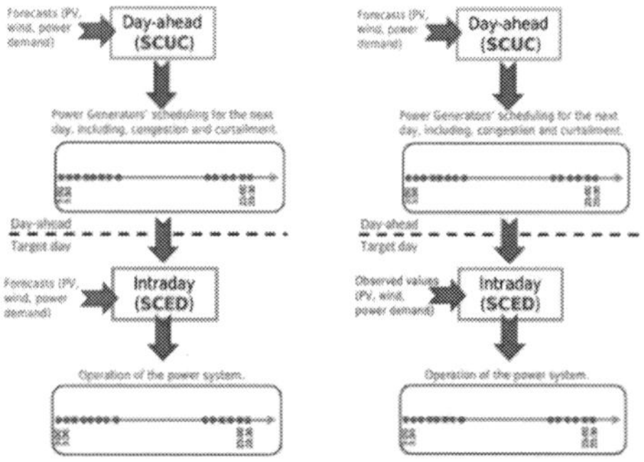

Figure 2: Comparison of sets of SCUC/SCED to evaluate the impact of forecasting errors in the operation of power systems. Base case with no forecasting errors (a). Test case with forecasting errors (b).

To account for the variability of the SCUC/SCED solutions we propose first to simulate the operation of the power system assuming that the forecast of PV power will be realized in the intraday operation. This is regarded as a base case, Fig. 2(a). Once this problem is solved, we solve a different problem where now the forecast is used only in the SCUC problem, and the SCED uses the observed value of PV power generation. This is the test case Fig. 2(b).

By comparing the solutions of both sets (base and test cases), we hypothesize that the effects of the forecasting errors can be separated from the natural variability of both solutions.

5 RESULTS

5.1 Impact on Curtailment Planning

The impacts of the forecasting errors on curtailment of PV can be seen in Fig. 3a and Fig. 3b, where the number of hours with curtailment, True +, without curtailment, True -, false non-curtailment, False -, and false curtailment, False +, of the base case are compared with those of the test case.

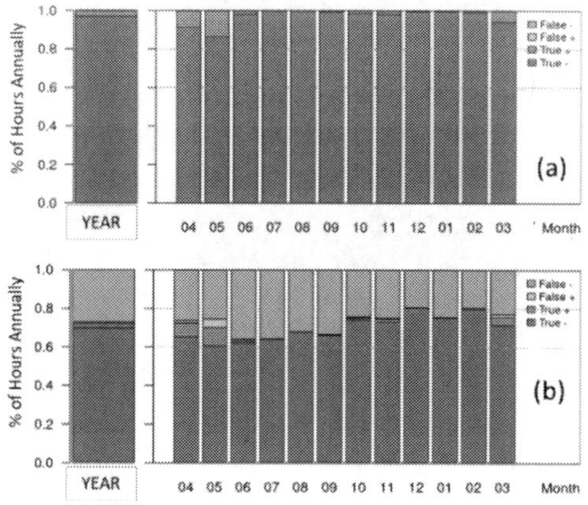

Figure 3: Curtailment Scheduling with (base case) (a) and without (test case) (b) forecasting error.

When there is no forecast error, Fig. 3A indicates that the day-ahead scheduling of PV curtailment is executed flawlessly. In this case, both SCUC and SCED solutions indicate that the system can be operated with a minimum of PV curtailment, 3%. Comparing now these results with the case where there is PV forecasting errors, the difference is clear. When there are forecasting errors, curtailment scheduling was poorly executed in 30% of the time. Not only that, in most of this time, there were false non-curtailment type errors. In other words, in almost 30% of the time, curtailment was not expected to happen, but did happen.

5.2 Impact on Economics

To evaluate the impact on economics, we calculated the difference between the daily running cost of the system of the day-ahead schedule with that of the intraday operation, for one year. The difference is plotted for the base case in Fig. 4 and for the test case in Fig. 5.

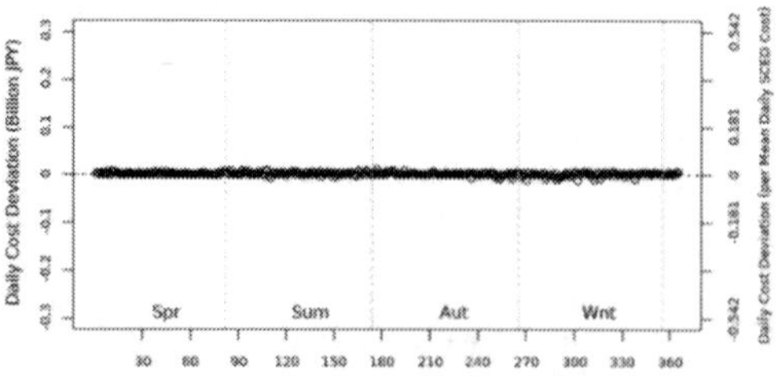

Figure 4: Difference of the running cost of the system of the SCUC versus the SCED (base case).

10.4229/EUPVSEC2023/5CO.5.4

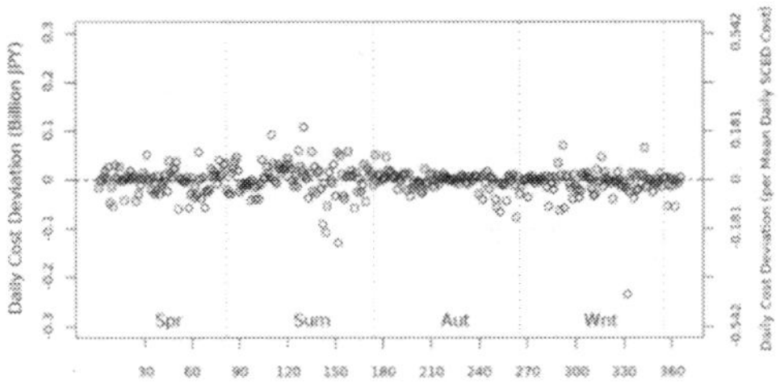

Figure 5: Difference of the running cost of the system of the SCUC versus the SCED (test case).

The cost results in Fig. 4 show that when there is no error, the SCUC and SCED solutions are very close to each other. A maximum variation of 2% happened in the year studied. The root mean square deviation, RMSD, between both costs was 0.4%. This means that, regarding cost, the SCUC models well what will happen in the SCED, when there is no forecast error.

However, a different picture appears when there is forecasting errors. In this case, Fig. 5 indicates that the SCUC cost is not always a good predictor of the intraday cost with SCED. The maximum variation was 38% and the RMSD was 3%, almost 10 times higher than the one of the base case. One can see here clearly the effects of the forecasting error.

5.3 Impact on Congestion

The impact on congestion, is evaluated using two metrics. The first one, is the annual and monthly total congestion frequencies (in 30-minute steps) with SCUC and SCED. The results related with this metric is in Fig. 6. The second one is the ability of the day-ahead scheduling to proper identify the time and location in the system where congestion will happen. This is presented in Fig. 7.

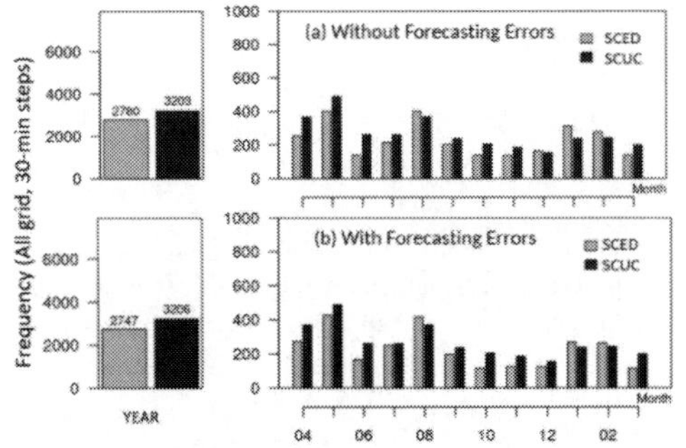

Figure 6: Annual and monthly frequency of congestion in the base case (a) and test case (b).

Starting with the results in Fig. 6, one can see two clear trends. Firstly, annually, the SCUC congestion occurrence overestimated the intraday occurrence considerably. The effect of the forecasting errors in this case was to increase the annual overestimation of SCED congestion with SCUC. Secondly, comparing both cases monthly, one can see a different trend only in December. In other months, the SCUC versus SCED trend was similar, with differences occurring only in the monthly totals. From these results the importance of comparing cases becomes clearer, as even without forecasting errors a lot of variation between SCUC and SCED occurs. Still, we can also see that with forecasting errors, the general trend regarding congestion was well modeled.

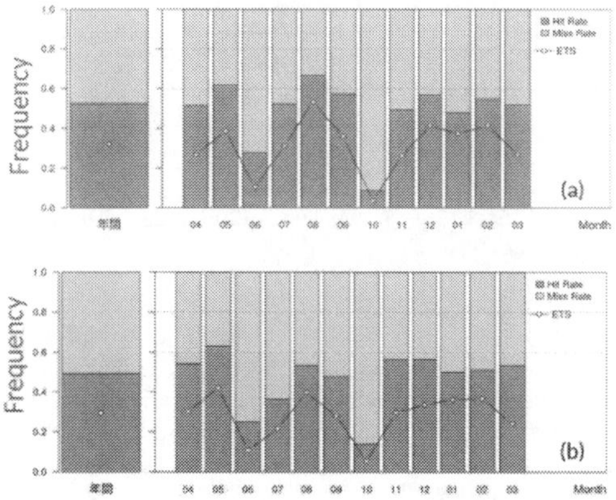

Figure 7: Detection of congestion and annual ETS in the base case (a) and test case (b).

Regarding congestion detection the natural variability of the modeling used, and between SCUC and SCED is even clearer. The results of the base case in Fig. 7a show that even without forecasting errors, the SCUC results are a poor predictor for the SCED congestion occurrence in time and location. Annually, 47% of congestion occurrence is missed. The equitable threat score, ETS [12], was used to evaluate how well the occurrence of congestion was forecasted with SCUC. Its value for the base case was 0.32. When there were forecasting errors, missed congestion increased to 50% and the ETS dropped to 0.29.

Taking the SCUC and SCED results by themselves without comparing with a base, would misleadingly make one to account on forecasting errors all the 50% of missed congestion. This when the effect in practice is a decrease in performance by just 3%.

5.4 Operational Stability (Last Resort Power Generation)

The last power system related metric used to evaluate the forecasting errors, was the occurrence of emergency, last resort, power generation to avoid black-out or loss of loads that might occur due to the mismatch between the day-ahead scheduling of power generators with SCUC and intraday operation with SCED. Last resort power generation is modeled as the amount of power generation the system cannot provide to maintain the balance between power supply and demand.

The frequency and amount of power requires with base and test cases are presented in Fig. 8. As the results show in Fig. 8a, when there is no forecasting errors, the power system operates within stability and security constraints. Thus, there is no need for last resort power generation. Now, the power system operates with forecasting errors, three days in the year, in three different months, the need for last resort power generation arises.

10.4229/EUPVSEC2023/5CO.5.4
020489-004

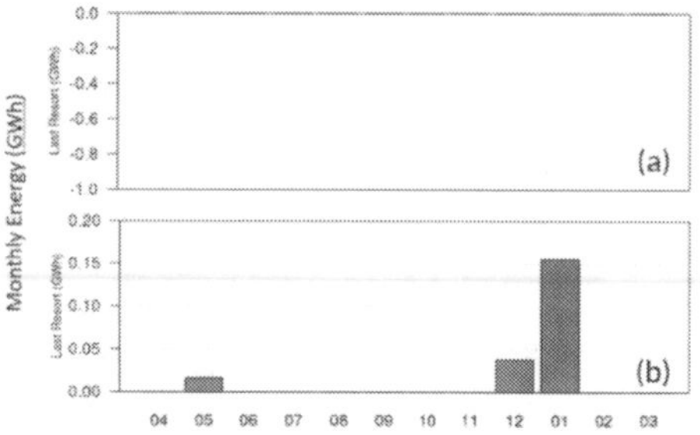

Figure 8: Last resort power generation with the base case (a) and test case (b).

This happened because the PV power forecasting errors in these days, associated with with power demand and the day-ahead scheduling of conventional power generators, created situations where it was not possible to meet the demand in some hours with what was available in the power system. This shows another clear effect of the PV forecasting errors. It should be noted though that the highest amount of emergency power generated was small, approximately 150 MWh in one day. As the power system simulated had a maximum power demand of 63 GW, we believe that this amount of emergency generation could be easily managed.

6. CONCLUSIONS

In this study we presented a simple methodology to evaluate the effects of PV power forecasting errors in the operation power systems. The novelty of the study was to focus not on metrics directly related with the forecasts themselves, such as MAE, etc, but on quantities related with the operation of power systems.

The results showed clearly that, at least, regarding congestion and cost, a direct comparison between the day-ahead scheduling with SCUC and the intraday operation with SCED is not advisable, as it will result in accounting to the forecasting errors more impact that it actually has. A comparison with a base case, allowed the correct evaluation of their effects.

Further exploration venues regarding this topic, include the consideration of wind and power demand forecasting errors, individually and together, as well as a better simulation of the target power system, using a configuration closer to the real power system. Regarding the SCUC and SCED modeling, next steps will include modeling the problem as an AC-OPF one where besides active power, reactive power is also regarded. Another improvement margin is with respect the modeling of curtailment scheduling, which currently is does not follow any specific policy when choosing the PV systems and branches subjected to it. We will be exploring several of these issues in following investigations.

ACKNOWLEDGMENT

This study is based on results obtained from a project, JPNP19002, commissioned by the New Energy and Industrial Technology Development Organization (NEDO)

REFERENCES

[1] "Snapshot 2023," IEA-PVPS. https://iea-pvps.org/snapshot-reports/snapshot-2023/ (accessed Sep. 20, 2023).

[2] Y. Udagawa, K. Ogimoto, J. G. da Silva, F. Junior, H. Ohtake, and S. Fukutome, "Economic impact of photovoltaic power forecast error on power system operation in Japan," in 2017 IEEE Manchester PowerTech, Jun. 2017, pp. 1–6. doi: 10.1109/PTC.2017.7981080.

[3] T. Ishizaki et al., "Interval quadratic programming for day-ahead dispatch of uncertain predicted demand," Automatica, vol. 64, pp. 163–173, Feb. 2016, doi: 10.1016/j.automatica.2015.11.002.

[4] C. Lowery and M. O'Malley, "Impact of Wind Forecast Error Statistics Upon Unit Commitment," IEEE Trans. Sustain. Energy, vol. 3, no. 4, pp. 760–768, Oct. 2012, doi: 10.1109/TSTE.2012.2210150.

[5] IEEJ, "Bulk Power System Models." http://www2.iee.or.jp/~pes/model/english/kikan/East10/east10.html (accessed Sep. 13, 2023).

[6] J. G. da S. Fonseca Junior, T. Oozeki, H. Ohtake, T. Takashima, and K. Ogimoto, "Regional forecasts of photovoltaic power generation according to different data availability scenarios: a study of four methods," Prog. Photovolt. Res. Appl., vol. 23, no. 10, pp. 1203–1218, Oct. 2015, doi: 10.1002/pip.2528.

[7] J. Gari da Silva Fonseca, F. Uno, H. Ohtake, T. Oozeki, and K. Ogimoto, "Enhancements in Day-Ahead Forecasts of Solar Irradiation with Machine Learning: A Novel Analysis with the Japanese Mesoscale Model," J. Appl. Meteorol. Climatol., vol. 59, no. 5, pp. 1011–1028, Apr. 2020, doi: 10.1175/JAMC-D-19-0240.1.

[8] Azuma Hitoshi, Isonaga Akira, Fukudome Suguru, Ogimoto Kazuhiko, Iwafune Yumiko, Urabe Chiyori, Udagawa Yusuke, Segawa Shuhei, "Building an Analysis Model power supply and demand for 2030 based on long-term energy supply and demand forecasts," presented at the 38th Energy System, Economics and Environment Conference of the Japan Society of Energy and Resources, 2022, pp. 758–763. In Japanese.

[9] IEEJ, "IEEJ EAST 10-machine System Model (50Hz),"http://www2.iee.or.jp/~pes/model/english/kikan/East10/east10.html (accessed Sep. 20, 2023).

[10] Yusuke Udagawa, Kazuhiko Ogimoto, Joao Gari da Silva Fonseca Junior, Shuhei Segawa, Isoshi Mukai, Tomoyo Saito, Suguru Fukutome: "A Preliminary Study of Security Constrained Unit Commitment Considering Grid Congestion", IEEJ2022, 6-091 (2022) (in Japanese).

10.4229/EUPVSEC2023/5CO.5.4
020489-005

[11] Yusuke Udagawa, Kazuhiko Ogimoto, Joao Gari da Silva Fonseca Junior, Shuhei Segawa, Isoshi Mukai, Tomoyo Saito, Suguru Fukutome: "A Preliminary Study of Impacts by Uncertainties on PV Power Generation Forecasts", Annual Conference of Power and Energy Society, IEEJ, (2022). (in Japanese).

[12] "Forecast Verification - Methods and FAQ." https://www.cawcr.gov.au/projects/verification/verif_web_page.html#Methods_for_foreasts_of_continuous_variables (accessed Sep. 21, 2023).

DEVELOPMENT OF ENERGY MANAGEMENT SYSTEM FOR PV-BESS SYSTEM FOR NON-INTERCONNECTED AREAS

Araya Francisco, Ha Duy-Long, Melliti Dhia, Riou Mael, Al-Shakarchi Franck
Entech Smart Energies
11 Allée Jean-François de la Pérouse, ZA de Menez Prat, 29000 Quimper
francisco.araya@entech-se.com

ABSTRACT: The primary function of an energy management system (EMS) is to enhance and to elevate the consistency of energy distribution to the power grid based on specific limitations (like grid protocols). Factors like meteorological conditions and daily cycles influence solar energy yields. Hence, exact PV predictions empower the EMS to estimate the available solar energy at any moment, leading to superior storage system management. Incorporating probabilistic photovoltaic (PV) forecasting into the control mechanisms lead by the EMS might enhance the efficiency and financial returns of a photovoltaic storage system (PV-ESS). This paper delves into the efficacy of a PV storage system when merged with probabilistic forecasting. Our study zeroes in on the influence of forecasting methods on energy equilibrium and maximizing profits. We contrast the results derived from probabilistic, deterministic and persistence-oriented over a 12-month simulation period. The data underscores the system's capacity to accentuate power distribution during high-demand intervals, leading to elevated power supply and minimized penalties. In essence, precise PV forecasting is pivotal for refining a PV storage system's operation and boosting financial gains.

Keywords: Energy management system, software in the loop, Energy Storage System, Probabilistic forecast.

1. INTRODUCTION

The global community's emphasis on renewable energy as the keystone for energy transition has been escalating. Consequently, countries have been proactive in adapting regulations and strategies to enhance the inclusion of renewable energy sources into the broader energy matrix. Nevertheless, the fluctuating characteristics of renewables, especially solar and wind energy, present notable grid integration challenges. A synergized approach involving battery energy storage (BES) and large-scale Photovoltaic (PV) systems (designated as PV-ESS) appears to offer a viable answer.

In this approach, producers craft a production strategy for subsequent day energy production and retrievals, taking solar radiation predictions into account. To fine-tune the system's output within defined boundaries, an energy management system (EMS) becomes indispensable. Accurate PV forecasting, as evidenced in [1], is paramount, and its importance is gaining traction as indicated in references [2]–[3]. The spotlight of this research is a grid-integrated PV system equipped with an energy storage system (PV-ESS) situated in a French overseas territory in the southern hemisphere, stemming from a call for tenders for non-connected regions in France. The system in question boasts a 5 MW terrestrial PV installation and an associated 10MWh storage component. This document introduces an EMS rooted in probabilistic predictions and devises an optimization challenge rooted in anticipated PV outputs to elevate PV-ESS functionality.

2. PV-ESS SYSTEM

2.1 Description of the installation

At the core of this investigation is a grid-integrated photovoltaic (PV) arrangement paired with an energy storage system (ESS). Situated in a southern hemisphere's French overseas territory, this installation came to fruition as a response to a call for tenders aimed at non-connected regions within France. The infrastructure encompasses a 5 MW PV facility along with a 10MWh ESS operate with feed-in tariff contract. Its primary function is to counterbalance the inaccuracies in forecasting and periods when PV is not operational, all the while adhering to stipulated grid regulations. The ESS configuration is segmented across two distinct battery enclosures, each linked to an inverter.

3. POWER INJECTION PROFILE AND GRID CODE

For the efficient operation and management of our the system, the power production and injection are subjected to the stipulations laid out by the grid code. These directives are encapsulated as mathematical constraints, ensuring:

a) Power Injection Constraints:

- During peak periods: Fluctuations in power injection should not exceed +/-10% of the installed capacity for each time step.
- During non-peak periods: Limitations are set to +/-5% of the installed power for every time step.

b) Adherence to Announced Production Plans:

There's an allowable deviation or error margin of +/-5% from the initially communicated production blueprint.

c) Announcement Timings and Adjustments:

Four pivotal time slots for announcements are defined: 4h, 10h, 14h, and 16h (R1, R2, R3 and D-1 respectively). This scheduling grants the flexibility to recalibrate plans in tandem with real data coming from the operation of the

10.4229/EUPVSEC2023/5CO.5.6
020490-001

system.

d) Regulations on Value Modifications Between Re-declarations:

- From D-1 to R1: Stable values from 6h to 6h10.
- From R1 to R2: Consistent figures between 12:00 to 12:10.
- From R2 to R3: Fixed values in the time frame of 4 pm to 4:10 pm.
- Between declarations R1, R2, and R3: Invariant data from 7 pm to midnight.

D-1 as the day prior to the actual day of power production or delivery.

e) Withdrawal Constraints:

The withdrawal process is limited at 5% of the installed PV power, accompanied by a tolerance span of 5%.

4. ESESOFT SOFTWARE

Entech's R&D department developed an innovative platform, ESEsoft, that boasts a myriad of functionalities and modules. These are meticulously designed to cater to the nuances of control, diagnosis, monitoring, and management of energy production. A distinctive feature of ESEsoft's capabilities is its inherent adaptability leveraging its various modules.

In the context of our PV-ESS application, the system's management adopts a structured, two-tiered control architecture. These levels comprise the EMS and the Power Management System (PMS). At its core, the EMS undertakes intricate computations to devise an optimize production plan based on the current supplied forecasts and principally the state of charge (SOC) of the storage system.

The system's operational framework can be broken down as follows:

- Planning Stage: Here, the production blueprint is sculpted over a pre-determined horizon, anchored by the projected Photovoltaic power and the battery's state of charge.

- Operational Stage: This phase is characterized by rigorous control measures to align operations with the pre-established plan.

In instances where adhering to the plan becomes unfeasible, adaptive control measures are implemented, tailoring responses to the current scenario, such as curtailment or disconnection.

The overarching objective of the EMS is dual-faceted: it aims to maximize revenue through energy injections into the grid, while either strictly adhering to grid code stipulations, as indicated by equation (1), or by channeling efforts to maximize energy transfers during peak demand periods, as outlined by the objective function in the following equation:

$$Revenue_{net} = \max \sum_{t=1}^{N} P\_\{syst\}\,(t)\, *\, Prix(t) - Penalties(t)$$

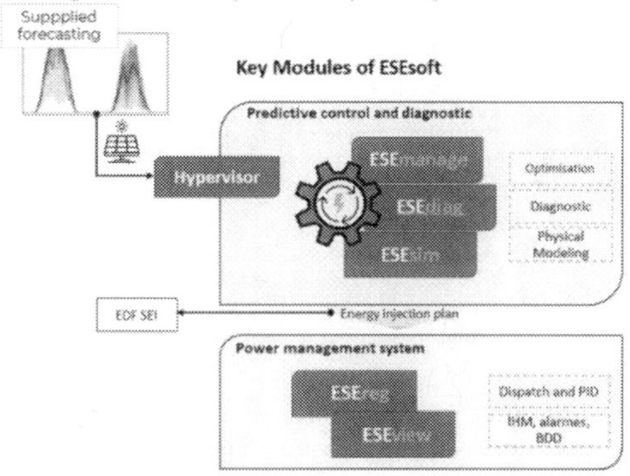

Figure 1 ESEsoft two control level architecture for case study representation.

5. PV POWER FORECASTING

Given set guidelines, energy plants are obligated to pre-announce their next day's energy injection to the grid manager. Therefore, an accurate forecast is crucial.

In our study, we gauge PV production using varied scenarios. These include:

Probabilistic forecasts ranging from quantile P0 to P100. Deterministic forecast, which provides a singular prediction without accounting for initial uncertainties. Persistence forecast, where the PV production profile for a day mirrors that of its predecessor. For reference, we also consider an "ideal" scenario where PV production is pre-known.

The forecasting data for this study is derived from satellite imagery via a third-party provider. Our day-ahead predictions utilize various numerical weather models. These datasets are merged to produce probabilistic local weather forecasts, subsequently translated into power production through a physical solar model.

6. RESULTS AND DISCUSSION

In order to evaluate the performance of our EMS, we run an annual simulation by input the current forecast and the ideal forecast. This allows us to calculate the performance by calculating the ratio of the obtained revenues as given by equation the following equation:

$$PR = \frac{Current\_income}{Ideal\_income}$$

From our year-long study, the PR using deterministic forecast approximated 91.1%. Introducing varied quantile forecasts facilitated an assessment of PR across both optimistic and pessimistic scenarios, defined by higher and lower quantiles respectively as shown in **Figure 2**. Key observations from our annual simulations included:

- PV Curtailment: There's a direct increase in PV curtailment with more pessimistic forecasts, with the persistence-based forecast witnessing the highest curtailment.

- Power Injection: Optimistic forecasts led to a surge in the total power injected into the grid.

Indeed, there was a marked difference of 27% between the highest quantile and a P30 quantile.

- Penalty Insights: As forecast quantiles veered towards a more conservative approach, penalties experienced a decline. Notably, forecasts within the P10 to P50 quantile spectrum induced the least penalties. A stark difference of 147% was observed between the most optimistic forecasts and the P30.

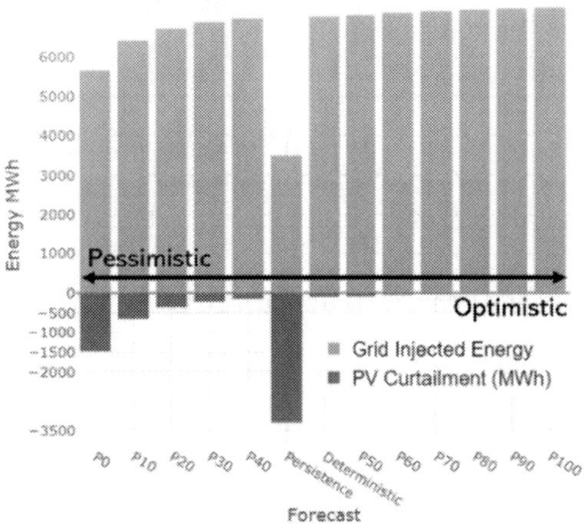

Figure 2 Annually total grid injected energy and curtailment as result from probabilistic forecasting.

This analysis underscores the balancing act between maximizing power injection, minimizing curtailment, and reducing penalties, all pivotal for optimizing the efficiency and economic viability of the system. These results imply yearly PR as shown in **Figure 3**:

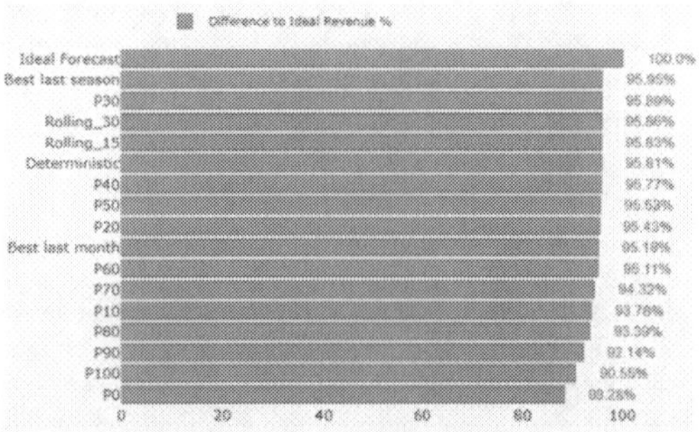

Figure 3 EMS annualy Performance Ratio obtained from the evatluation of different strategies of expliatation of the probabilistic forecast.

7. CONCLUSIONS

The results of our simulations underline the effectiveness of our control strategy, which guarantees optimum energy production and consumption management. Here are a few highlights from the analysis:

- The introduced EMS and optimization approach adeptly integrates renewable energy with storage mechanisms, reinforcing sustainable energy production.
- The system's performance, both in energy balance and revenue facets, hinges substantially on the forecasting strategy selected.
- When veering towards optimistic forecasting, we observe an uptick in power injections. However, this comes at the expense of elevated penalties and possibly diminished revenues.
- In contrast, a conservative forecast strategy paves the way for reduced penalties and peak returns.
- Among the forecasting techniques, the persistence-based approach falls short, resulting in amplified penalties, especially during peak hours. This highlights the cardinal role of precise forecasting in the EMS's efficient functioning.
- Prioritizing electricity injection during peak demand hours is central to the system, ensuring elevated electricity outputs while curtailing penalties.
- To sum up, our EMS with the optimization problem's formulation, presents a viable blueprint for seamlessly merging renewable energy sources with energy storage systems. This holistic approach facilitates sustainable and proficient energy consumption management.
- A future work will look for stochastic optimization pivoted on a handful of scenarios. An empirical validation against real-world data will be embarked upon post a year of operation to further cement our findings.

8 REFERENCES

[1] P.Besson, W. Mahmoud, D.Raynaud, E. Escudero-Ramos, G. Tremoy, T. Phuong Do, F. Bourry "Control optimization and sizing of energy storage for PV systems using probabilistic forecasts"5th International Hybrid Power Systems Workshop, 2021.

[2] R. B. Venkateswaran, A. Tomar, P. Nguyen, and A. Suma, 'An optimal energy management system using day-ahead prediction algorithm', in 2021 IEEE PES Innovative Smart Grid Technologies Europe (ISGT Europe), Espoo, Finland, 2021.

[3] Antonanzas, J., Osorio, N., Escobar, R., Urraca, R., (2016).

10.4229/EUPVSEC2023/5CO.5.6
020490-003

FLEXIBLE ELECTRICITY DEMAND FORECASTING BASED ON MRMR FEATURE SELECTION AND WT-LSTM

Koki Iwabuchi, Daichi Watari, Dafang Zhao, Ittetsu Taniguchi, Takao Onoye
Graduate School of Information Science and Technology, Osaka University, Japan
1-5 Yamadaoka, Suita, Osaka 565-0871, Japan

ABSTRACT: Solar power generation is difficult to match electricity demand. Therefore, energy management systems are attracting attention for their effective use of solar energy. Energy management systems automatically track and manage building electricity consumption. Highly accurate electricity demand forecasting is a very important factor for the system. This paper proposes an electricity demand prediction method based on systematic feature selection and WT-LSTM that switch MW. A systematic feature selection method, mRMR (minimal Redundancy Maximal Relevance), is adopted in this paper. The prediction model is a combination of WT and LSTM (Long Short-Term Memory). The model uses WT and inverse transform to decompose the time series by frequency, which is then trained and predicted by LSTM. In this model, MW were switched based on past forecast results and thus addresses cross-daily and cross-seasonal demand fluctuations. The proposed forecasting model of objective feature selection and switching MW is used for the first time not only to forecast electricity demand, but also among other things to be forecasted. In feature selection by mRMR and prediction by WT-LSTM, the proposed method of switching the MW improved the accuracy by up to 48.4% compared with the case of fixing the MW. The maximum error was smallest when the features are selected by mRMR, as compared with the case of forecasting only from past demand and the case of forecasting from all feature candidates. Thus, the proposed method, which performs feature selection by mRMR and flexible prediction by WT-LSTM, is useful for stable electricity demand prediction.
Keywords: Electricity demand forecasting, Energy management system, Feature selection, Wavelet transform, Long Short-Term Memory

1 INTRODUCTION

PV (Photovoltaic) power generation, which has become increasingly popular in recent years, has the advantage of being able to generate electricity without depending on depletable resources. On the other hand, it has the problem that the amount of electricity generation is greatly affected by weather and other natural conditions, making it difficult to generate electricity in accordance with demand. Therefore, energy management systems are attracting attention for their effective use of solar energy. Energy management systems are systems that install PV panels and storage batteries and enable centralized management of facilities within an electric power network with information and communication technology. Since energy management systems manage facilities based not only on future PV generation but also on future electricity demand, highly accurate electricity demand forecasting is a very important factor.

Figure 1 shows the change in electricity demand for one year in one of the buildings in the Graduate School of Information Science and Technology, Osaka University. As can be seen, the degree of change in electricity demand varies from period to period. For example, the summer months of July and August are intensely variable, while the winter months of January and February are mildly variable. In other words, the electricity demand time series is non-stationary.

Various studies have been conducted on electricity demand forecasting. In particular, the combination of machine learning and WT (Wavelet Transform) is being actively studied because it is very effective in predicting non-stationary time series [1-3]. Ofori-Ntow Jnr et al. proposed a three-level hybrid ensemble short-term demand forecasting method consisting of DWT (Discrete Wavelet Transform), PSO (Particle Swarm Optimization), and RBFNN (Radial Basis Function Neural Network) [1]. Compared with BPNN (BackPropagation Neural Network) and SaDE-ELM (Self adaptive Differential Evolution - Extreme Learning Machine), they highlighted the superiority of the proposed method. Li et al. proposed a model that combines WT and fuzzy inference system for predicting building electrical demand [2]. In this method, WT was used to obtain diurnal patterns and residual data for forecasting in order to better reflect the actual characteristics of electricity demand. Saoud et al. proposed a method to predict household electricity demand at different resolutions using a machine learning model based on SWT (Stationary Wavelet Transform) and transformers [3]. SWT and its inverse characteristics were used to decompose and reconstruct the actual and predicted household electricity demand data, respectively, while the deep transformer was used to predict the subbands of the SWT. WT has a kind of parameter called the mother wavelet (MW). The above studies used the same MW throughout all experimental periods, although different MW were used in each. However, electricity demand series can exhibit considerable variation over time. Therefore, a MW that is optimal for one period may not be optimal for another period.

Thus, we previously proposed a method for electricity price forecasting in which periodically switching the MW provides stable accuracy [4]. The model is a combination of WT and LSTM (Long Short-Term Memory) and uses WT and inverse transform to decompose the time series by frequency, which is then trained and predicted by LSTM. In this model, MW were switched based on past forecast results and thus addresses cross-daily and cross-seasonal price fluctuations. However, the historical electricity prices and demand used in the inputs to this model are selected subjectively, and no objective or systematic method is used. The selection of useful and non-redundant features is essential for machine learning models that are established by the relationship between inputs and outputs.

This paper proposes an electricity demand prediction method based on systematic feature selection and WT-LSTM that switch MW. Among systematic feature selection methods, mRMR (minimal Redundancy

10.4229/EUPVSEC2023/5CO.6.1
020491-001

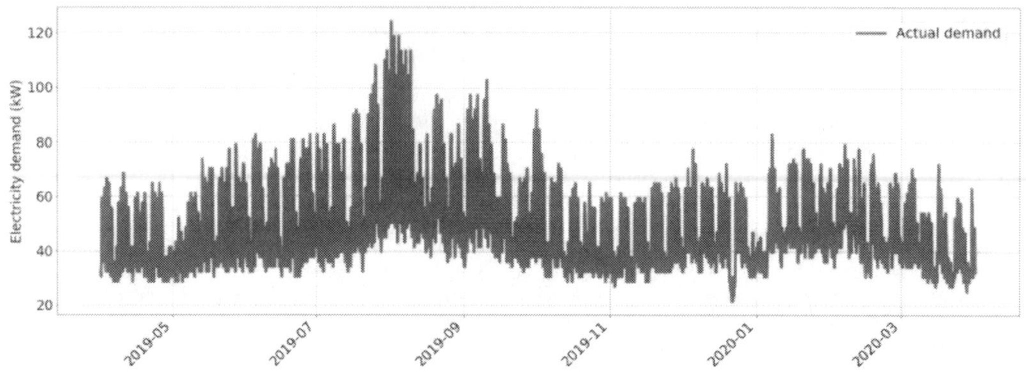

Figure 1. Electricity demand change in one building at Osaka University's Graduate School of Information Science and Technology for one year in FY2019.

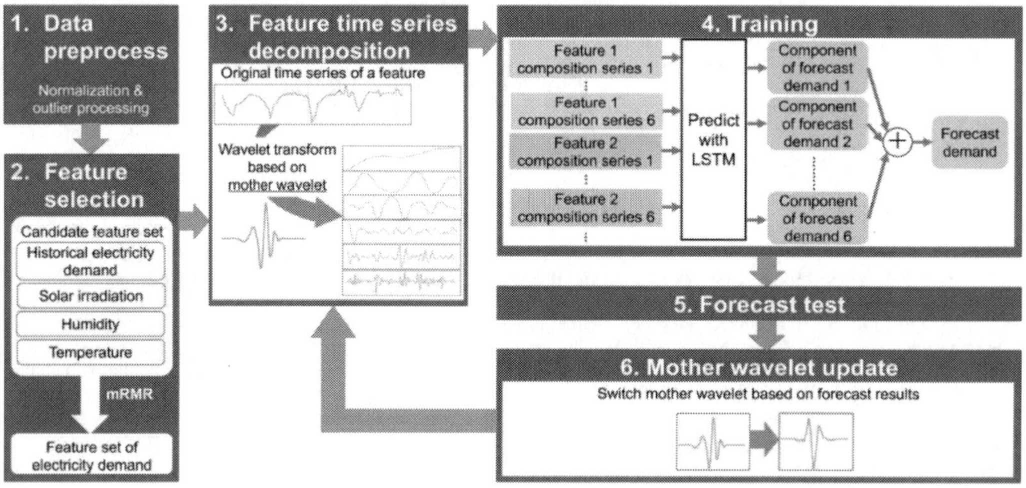

Figure 2. Overview of proposed method

Maximal Relevance), which not only ensures the correlation between features and target variables but also avoids excessive redundancy, is adopted in this paper. This feature selection is intended to help the method of switching MW, which has achieved stable accuracy in electricity price forecasting, to perform well in electricity demand forecasting as well. The proposed forecasting model of objective feature selection and switching MW is used for the first time not only to forecast electricity demand, but also among other things to be forecasted.

2 PROPOSED METHOD

Figure 2 shows an overview of the proposed method. Initially, normalization and outlier processing of all data is performed. Next, based on mRMR, we obtain a set of features that are not only highly correlated with electricity demand, but also non-redundant. Then, each feature series in the feature set is frequency decomposed by WT. Each decomposed time series is trained by LSTM to forecast future electricity demand, and the model is evaluated. Then the mother wavelet is switched based on the evaluation results. It is repeated again from the time series decomposition. Subsequent sections will discuss some of these important steps in more detail.

2.1 Feature selection by mRMR

Feature selection by mRMR selects the features that maximize the following equation. The first term calculates the mutual information between a feature and the target variable, and the more relevant they are, the more this term contributes to maximizing the overall equation. The second term computes the mutual information between the features, and the smaller their redundancy, the more this term contributes to maximizing the overall equation.

$$\max_{x_j \in X - S_{m-1}} \left[I(x_j; c) - \frac{1}{m-1} \sum_{x_i \in S_{m-1}} I(x_j; x_i) \right], \quad (1)$$

where x_j is j -th feature candidate, c is objective variable (electricity demand), X is set of feature candidates, $I(\cdot)$ is mutual information, and S_{m-1} is set of selected features of size $m-1$.

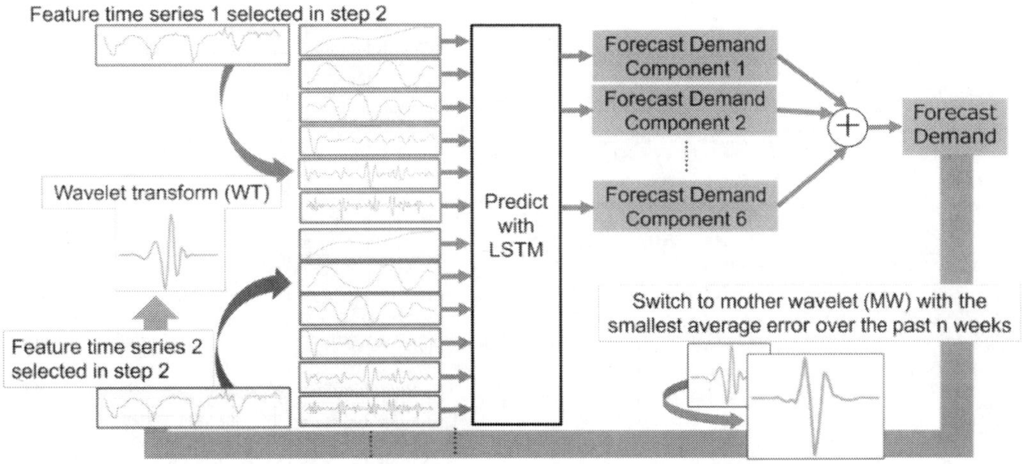

Figure 3. From time series decomposition to switching mother wavelets in the proposed method.

2.2 From time series decomposition to forecasting and switching mother wavelets

The detailed flow of steps 3 through 6 in Figure 2 is shown in Figure 3. First, each feature time series selected in one previous process is decomposed by frequency. Two time series are shown in Figure 3, but the same applies to the others. All component series are then input into LSTM to forecast future electricity demand for each component series. They are then added together to obtain the final forecasted electricity demand. This is done for each mother wavelet, and each week the switch is made to the mother wavelet with the best predictive test results for the previous n weeks.

3 SIMULATION

3.1 Setup

To verify the accuracy of the model proposed in this paper, we use electricity demand data from a building at the Graduate School of Information Science and Technology, Osaka University for fiscal year 2019. Training with four weeks of data and one-step-ahead prediction test with one week of data are repeated for a year. This experiment is performed with each of the following 19 mother wavelets: 9 ways to fix the MW with Daubechies (db) 1 to 6 and Symlet (sym) 3 to 5 and 10 ways when n, the number of weeks to be considered in switching the MW, is changed from 1 to 10 in the proposed method. In addition, in order to evaluate the effect of combining feature selection by mRMR with flexible prediction by WT-LSTM, we compare it with the case of predicting only from past demand and the case of predicting from all feature candidates. In this paper, the available data, which included electricity demand, solar irradiation, temperature, humidity, and their respective lag features, are used as candidate features. MAPE is introduced as performance metrics.

$$MAPE = \frac{100}{k}\sum_{i=1}^{k}\left|\frac{\hat{y}_i - y_i}{y_i}\right|, \tag{2}$$

where k is the number of data, y_i is the actual value, and $\hat{y}_i$ is the predicted value.

3.2 Results

Using the combined mRMR and WT-LSTM model, MAPE of the proposed method for switching the MW and the method for fixing to a specific MW are shown in Table 1. The best accuracy of $n = 8$ is shown as a representative value of the proposed method. Table 1 shows that the proposed method of switching the MW improves the error by 1.17% to 48.8% compared with the method of fixing it to a specific MW. Figure 4 shows the weekly accuracy ranking of each mother wavelet, with the higher position indicating higher accuracy. The proposed method and the mother wavelets that were highly accurate when fixed are shown as solid lines. Figure 4 shows that the proposed method always maintains the 8th place or higher, while all other methods fall below the 9th place depending on the week. This result indicates that switching the MW periodically is useful for reducing the deterioration of prediction accuracy in electricity demand forecasting.

Table 1. MAPE for each mother wavelet

Mother wavelet	MAPE (%)
Switching MW (n =8) [Proposed method]	1.569602
Symlet 5	1.588242
Daubechies 6	1.652728
Daubechies 4	1.660685
Daubechies 5	1.667753
Symlet 4	2.340044
Symlet 3	2.481672
Daubechies 3	2.519659
Daubechies 2	2.924541
Daubechies 1	3.063592

Figure 5 shows a comparison between the proposed method of selecting features with mRMR for flexible WT-LSTM prediction and the case of selecting only historical power demand data as features or all feature candidates. The vertical axis is the maximum weekly MAPE. For the proposed method with mRMR, electricity demand without time shift, 2 days, 6 days, and 4 weeks ago, and solar irradiation without time shift were selected. From Figure

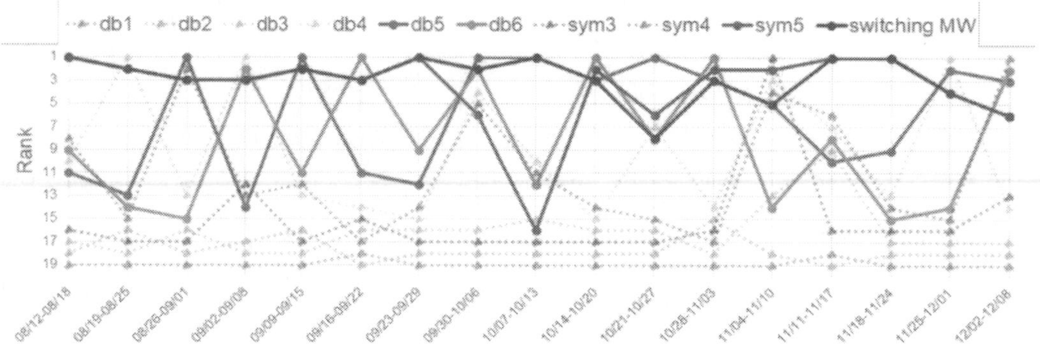

Figure 4. Accuracy ranking of each mother wavelet by week

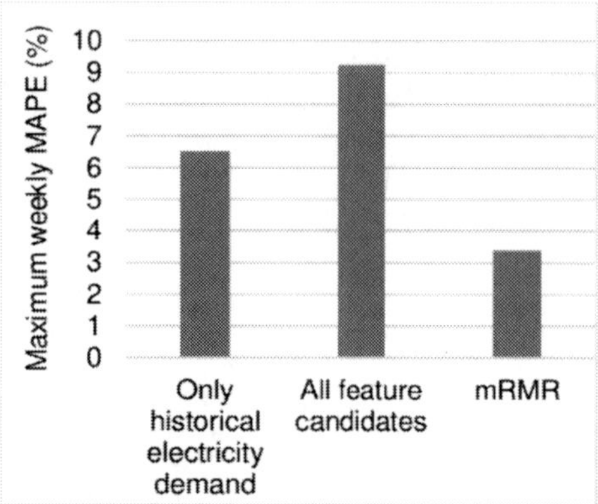

Figure 5. Maximum weekly MAPE of each feature selection

5, the maximum MAPE of mRMR is the smallest, and this means that appropriate feature selection by mRMR is effective for flexible prediction by WT-LSTM.

4 CONCLUSION

This paper proposes a new power demand forecasting method that combines feature selection by mRMR with flexible forecasting by WT and LSTM. A flexible forecasting method based on WT and LSTM, which achieves consistently accurate forecasting in electricity price forecasting, is applied to electricity demand forecasting. Furthermore, in order to select features suitable for electricity demand forecasting, mRMR is adopted. Experimental results show the effectiveness of the proposed method. Future work includes applications to other time-series forecasts needed for energy management systems, such as PV generation. It is also necessary to integrate this approach into the energy management system and evaluate its effectiveness with current accuracy.

REFERENCES

[1] Jnr, E. O. N., Ziggah, Y. Y., & Relvas, S. "Hybrid ensemble intelligent model based on wavelet transform, swarm intelligence and artificial neural network for electricity demand forecasting." Sustainable Cities and Society 66 (2021) 102679.

[2] Li, C., Tang, M., Zhang, G., Wang, R., & Tian, C. "A hybrid short-term building electrical load forecasting model combining the periodic pattern, fuzzy system, and wavelet transform." International journal of fuzzy systems 22(1) (2020) 156-171.

[3] Saoud, L. S., Al-Marzouqi, H., & Hussein, R. "Household Energy Consumption Prediction Using the Stationary Wavelet Transform and Transformers." IEEE Access 10 (2022) 5171-5183.

[4] Iwabuchi, K., Kato, K., Watari, D., Taniguchi, I., Catthoor, F., Shirazi, E., & Onoye, T. "Flexible electricity price forecasting by switching mother wavelets based on wavelet transform and Long Short-Term Memory." Energy and AI 10 (2022) 100192.

10.4229/EUPVSEC2023/5CO.6.1
020491-004

IMPACT OF TIME RESOLUTION OF SOLAR AND METEOROLOGICAL DATA ON CLIPPING LOSSES AND ENERGY YIELD SIMULATION

Jozef Rusnak[1], Branislav Schnierer[1], Marcel Suri[1], Martin Opatovsky[1], Giridaran Srinivasan[2]
[1] Solargis s.r.o., Bratislava, Slovakia
[2] Solargis Americas Inc., Toronto, Canada

ABSTRACT: The design of a photovoltaic (PV) system should minimize the investment and operating costs, while complying with the requirements of the technical regulations and standards related to the safe and environmentally friendly operation of the system, and the reliable supply of high-quality electricity. From a designer's point of view it is a multicriteria optimization problem with functional constraints. A particular objective is to minimize operational losses within the entire PV system, often with the help of simulation tools. Critical inputs to the PV simulation models are solar and meteorological data. This article shows how time resolution of input data impacts the simulated clipping losses and energy yield considering PV systems with various DC to AC ratios located at different climate zones (according to the Köppen-Geiger classification), and demonstrates the importance of using high-frequency data for optimum design of many aspects of a PV system (inverter design, battery sizing, grid control applications, etc.).
Keywords: solar data and meteorological data, time resolution, PV system simulation, clipping losses, energy yield

1 INTRODUCTION

PV system simulation based on mathematical modelling of the conversion process from solar irradiance input to electrical energy output is a key part of the PV system design process. Solar and meteorological time series data are crucial inputs, and for the purpose of PV design optimization they are typically derived from satellite-based solar radiation models and from global meteorological models. Note that a good practice for large projects is to enhance the accuracy of these solar and meteorological model time series data by adapting them to the site conditions using ground measurements acquired over a period of minimum 12 months.

A comprehensive overview of the solar model used to derive the solar radiation time series used in this work is available in [1]. The methodology is also described in [2], [3].

Considering meteorological data, the best source are continuous measurements from high-accuracy sensors installed on a meteorological station. However this option is usually available only for a short period during the development phase of the project. The only alternative is to derive historical meteorological data from meteorological models. Several models are available; a good option is to use the ERA5 European atmospheric reanalysis (source ECMWF) covering long period of time with continuous data [4]. Ground albedo is derived from the MODerate-resolution Imaging Spectroradiometer (MODIS) albedo data product, version 6 (MCD43A3) [5], [6], [7] post-processed by the Solargis model.

Solar irradiance time series calculated using geostationary satellite data can only be computed at the time slots for which satellite scans are available. Hence, solar irradiance is typically computed only at 10-, 15-, or (for older satellites) 30-minute time steps, depending on the satellite mission. However, fast moving clouds lead to solar irradiance changes at time scales shorter than 10 minutes. These changes are highly determinant to proper design of integral parts of PV systems such as battery sizing and inverter tuning.

Solar irradiance time series can be alternatively represented by their frequency distribution or, being conveniently normalized, by the probability density function (PDF). For applications that require accurate estimates of peak solar irradiance in the long run, there exist methods that focus on the improvement of the PDFs of the Global Horizontal Irradiation (GHI) and Direct Normal Irradiation (DNI) components. The most widely used are stochastic methods based on auto-regressive, gaussian, or *Markov* processes. The Solargis synthetic generator is based on the latter approach (i.e., Markov processes) inspired by Hoffman et al. (2014) [8].

PV simulation used for the purpose of this article is based on hourly, 15-minute, and 1-minute input data. This data was used as an input to the PV simulation chain consisting of several consecutive calculation steps: optical losses (shading, pollution, angular, and spectral losses), conversion losses and electrical part losses (DC, inverter DC/AC conversion, inverter power limitation, AC cable, and transformer losses). Optical simulation is based on ray-tracing method [9], [10].

Optimal design of a PV system requires choosing a suitable combination of PV modules and PV inverters to maximize the utilization of solar radiation falling on the PV modules. Combination of PV modules and PV inverters is characterized by DC to AC ratio, which can be calculated as the ratio between the PV modules' maximum power according to IEC 60904-1 and reported at Standard Test Conditions (STC), to the maximum AC power rating of the inverter(s) at reference conditions. The choice of DC to AC ratio directly affects inverter saturation losses that represent losses of electrical energy due to the inverter clipping which can be (based on [11]) defined as a state when the inverter output is limited by its capability rather than by the input power from the PV array.

Among the above-mentioned losses, different time resolution of input solar and meteorological data affects the clipping losses estimated by the PV yield simulation. The different time resolutions of the yield simulation input data can thus affect PV system design decisions. This effect is examined in the following chapters.

2 CONTEXT, OBJECTIVE AND APPROACH

Several articles dealing with the impact of the time resolution of solar data on clipping losses and energy yield simulation can be found below. They are based on:

Datasets from meteorological stations measurements:

In [12] datasets from two sites (Freiburg in Germany and Florianopolis in Brazil) were used. Site in Germany: 10-sec measured data, and 1-min and hourly calculated

10.4229/EUPVSEC2023/5CO.6.2
020493-001

averages were used. Site in Brazil: measured 1-min averaged data and calculated hourly averages were used.

Conclusions: With increased time resolution of solar irradiation data there are higher calculated losses due to inverter under-sizing. Hourly averages hide important irradiation peaks that need to be considered.

In [13] three sites (Lanai, Las Vegas, and Albuquerque), measured 3-sec or 1-min data, and calculated 5-min, 15-min, and hourly averages were used.

Conclusions: Energy calculated from time-averaged weather is subject to error that increases as the averaging interval lengthens. At hourly intervals, error in energy ranges from -0.3% to 2%, with values increasing as the frequency of variable irradiance conditions (i.e. partly cloudy) increases, and with greater overestimates possible for modules with stronger nonlinear response.

In [14] one site (San Diego, California), 1-sec measured data, and 15-min and hourly averaged data were used.

Conclusions: From 1-sec to 1-hr, the percentage of total yearly energy loss to total yearly energy production decreases from 2.65% and 2.2% for 7 kW and 5 kW inverters respectively, to less than 0.94% and 0.66%. This indicates that smoothing of small timescale fluctuations in plane-of-array global irradiance leads to underestimations of the energy losses due to inverter saturation.

In [15] modelling of 10 sites across the United States to cover a range of climatic regions, the predicted production error was examined for hypothetical PV plants with DC:AC ratios ranging from 1.0 to 2.0. 1-min measured data and hourly averaged weather data were used.

Conclusions: The hourly averaged prediction error is proportional to the DC:AC ratio. The error response curve has a natural-log-shaped dependence on ratios between 1 and 2.

Datasets from DC power measurements:

In [16] one site (Birmingham, Alabama, US) with four real PV arrays with different orientation, 1-sec measured data averaged to 1-, 5-, 10-, 15-min data were used.

Conclusions: PV models using hourly inputs would have overestimated annual power production by approximately 2% for a range of PV installations with 1.4 DC:AC ratios.

Our objective is to demonstrate the clipping effects using satellite-based solar model data at 20 different geographical sites. While natural time resolution of this type of data is 15 minutes, we also make use of synthetically generated 1-minute data and aggregated 60-minute time series data. To show how the time resolution of meteorological and solar data impacts the predicted clipping losses and energy yield, PV simulations were performed for:

1. Different geographical sites (see Table I) which cover four (A, B, C, D) of the five main groups of the Köppen-Geiger climate classification [17]. Group E, polar climate, was excluded due to low practical significance.
2. Single axis tracker system, with north-south axis orientation and backtracking with rotating limits of ±60°.
3. Different DC:AC ratios (1.1, 1.3 and 1.5). The main details of the PV systems are listed in Table II.

Table I: List of simulated sites and climate classifications

Site name (ID)	Country	Climate group: Subgroup
Aswan (S17)	Egypt	B (Dry): Bwh
Bengaluru (S11)	India	B (Dry): Bsh
Bratislava (S7)	Slovakia	D (Continental): Dfa
Calama (S1)	Chile	B (Dry): Bwk
Cambridge (S16)	United Kingdom	C (Temperate): Cfb
Da Nang (S8)	Vietnam	A (Tropical): Am
Doha (S4)	Qatar	B (Dry): Bwh
Douala (S5)	Cameroon	A (Tropical): Am
Edmonton (S6)	Canada	D (Continental): Dfb
Kamishihoro (S10)	Japan	D (Continental): Dfa
Kuching (S18)	Malaysia	A (Tropical): Af
Lusaka (S3)	Zambia	C (Temperate): Cwa
Melbourne (S12)	Australia	C (Temperate): Cfb
Orlando (S2)	United states	C (Temperate): Cfa
Petrolina (S14)	Brazil	B (Dry): Bsh
Pofadder (S20)	South Africa	B (Dry): Bwh
Rosso (S13)	Mauritania	B (Dry): Bwh
Trapani (S15)	Italy	C (Temperate): Csa
Uppsala (S19)	Sweden	D (Continental): Dfc
Yumen (S9)	China	B (Dry): Bwk

Table II: Information about the simulated PV systems

PV modules type	Crystalline silicon, half-cut cells
String size	27 PV modules
Number of strings (DC:AC ratio of 1.1/1.3/1.5)	306 / 362 / 416
PV inverter	Centralized 3MW
DC cable losses	1% at STC
AC cable losses	1 % at STC
Inverter transformer	3.5 MVA, 655 V/33 kV, 5 kW no-load loss, 30 kW full-load loss

The following effects are investigated:

1. Effect of different input data time resolution and DC to AC ratio on clipping losses (CL).
2. Effect of different time resolution and DC to AC ratio on electrical power at the grid connection (PVOUT).

To perform the investigation, the following indicators were calculated, statistically processed, and visualized using boxplots:

- annual sum of Global Tilted Irradiation (GTI), annual CL,
- annual diffuse to global irradiation (D2G), annual difference in CL,
- annual difference in PVOUT.

Principally, three main links are presented in the following chapters: the impact of GTI on CL (Chapter 3.1), the impact of D2G on the difference in CL (Chapter 3.2) and the impact of CL and the difference in CL on the difference in PVOUT (Chapter 3.3). The impact of other indicators (such as std GTI, DIF, std DIF, precipitation level, average change between adjacent samples of GTI) was also examined and the analysis showed that the abovementioned factors are the most significant.

In each boxplot, the central line represents the median, and the box represents the interquartile range – from the 25th to the 75th percentile. The upper and the lower whiskers represents the most extreme data points, not considered outliers. The outliers are single data points which are lower than 1.5x value of the lower quartile or higher than 1.5x value of the upper quartile, They are plotted using the 'o' marker symbol.

The calculated medians were used to order the examined sites and Climate groups (groupings of sites) based on the indicators. This categorization enables the relative comparison of the Climate groups based on the probability of obtaining certain values of the indicators. For example it enables identification of the Climate group, where the (comparatively) largest overestimation of PVOUT can be expected when input data with low time resolution is used. This information is directly useful to PV designers, as it enables them to better understand and quantify the limitations of the PV yield simulations they are performing, hence avoiding underperformance in the operational stages of the projects.

3 RESULTS

3.1 Annual GTI and annual CL

The purpose of this chapter is to show the link between GTI at the sites from different Climate groups and CL calculated for different time resolutions and DC to AC ratios.

3.1.1 Annual GTI

Figure 1 shows the annual sums of GTI for the examined sites (the first five sites with the highest levels are highlighted). All highlighted sites are from Climate group B (Dry).

Figure 2 shows the annual sums of GTI using statistical representation in the form of boxplots. One boxplot represents GTI of all sites belonging to the corresponding Climate group. Statistically, sites with the highest GTI are in Climate group B (Dry). It can be expected that the highest level of CL would be also found at sites from Climate group B because of the highest saturation of inverters.

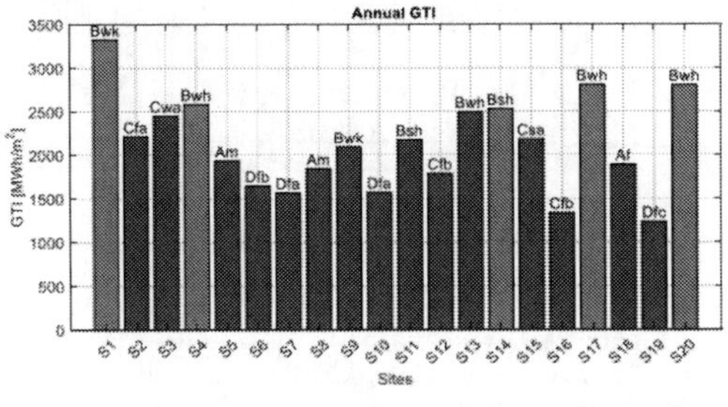

Figure 1: Annual sum of GTI for all examined sites

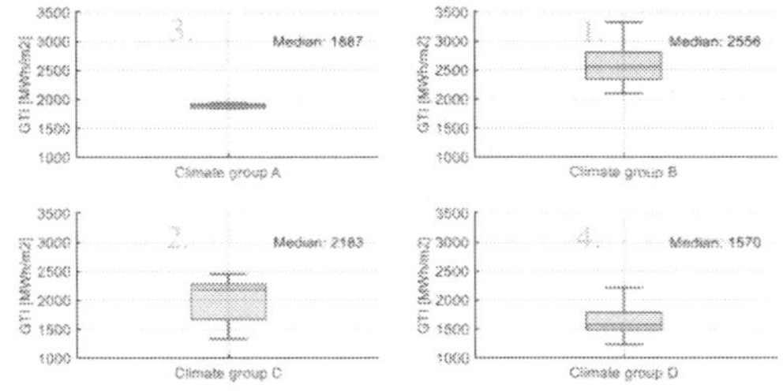

Figure 2: Boxplot of GTI for Climate groups, numbers indicate ranking of median values

3.1.2 Annual CL

Figure 3 shows the annual CL as a percentage of PVOUT for the examined sites, different time resolutions, and a DC to AC ratio of 1.3. It is obvious that there is a relation between: GTI and CL (statistically, the higher the GTI the higher the CL), DC to AC ratio and CL (the higher the DC to AC ratio the higher the CL), and time resolution and CL (the lower the time resolution the higher the CL).

In addition, it can be seen that the annual sum of GTI is not the only parameter determining the annual CL at a certain DC to AC ratio, e.g. annual GTI of S17 and S20 are almost the same but CL at S20 is significantly higher than at S17.

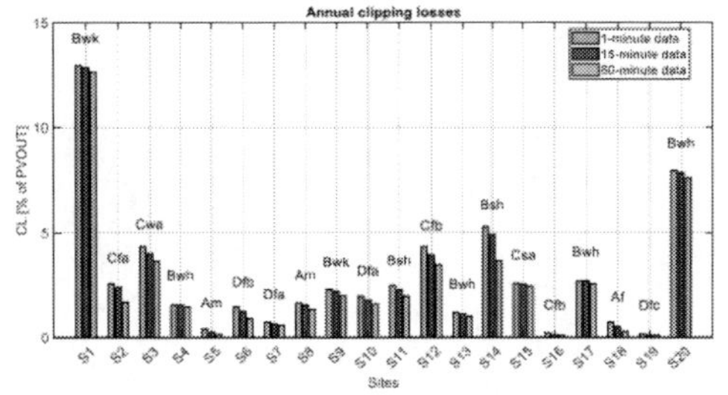

Figure 3: Annual CL for all examined sites (DC to AC ratio of 1.3)

Figure 4 (for DC to AC ratio of 1.3) and Figure 5 (for DC to AC ratio of 1.5) show the differences in CL for different time aggregations at each site. Results are based on CL determined using 1- (upper whiskers in boxplots), 15- (the median) and 60-minute data (lower whiskers in boxplot). The wider the boxplot the higher the difference between CL based on 1-minute and 60-minute data.

Note: Median ranking of CL for DC to AC ratio of 1.1 has the same order as for DC to AC ratio of 1.3.

The higher the DC to AC ratio the higher the impact of GTI on CL. In the case of DC to AC ratio of 1.1 and 1.3, the highest median is for Climate group C and after increasing the DC to AC ratio to 1.5 the highest median of CL is in Climate group B, the group with the highest GTI.

Figure 6 shows CL for examined sites at all time resolutions and DC to AC ratios. Upper (lower) whiskers in boxplots represent CL based on 1-minute data and DC to AC ratio of 1.5 (60-minute data and DC to AC ratio of 1.1). Medians of CL in percentage are shown in plots. The values were later used to sort sites by CL value.

The sites with the lowest annual GTI (S16 and S19) have the lowest medians. The sites from Climate group B have the widest boxplots, meaning that increasing DC to AC ratio causes the highest increase in CL, for some of them from ~0 % to ~10 % and higher.

10.4229/EUPVSEC2023/5CO.6.2
020493-003

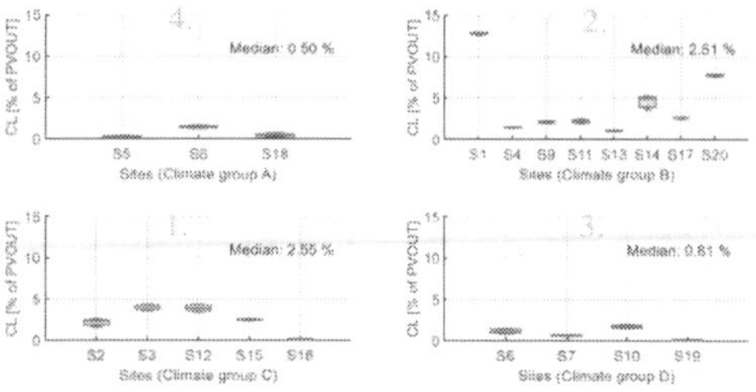

Figure 4:CL: Boxplots for examined sites and medians for each Climate group (DC to AC ratio of 1.3)

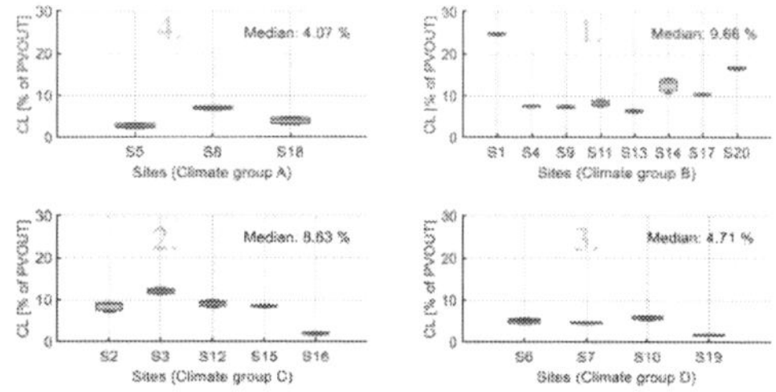

Figure 5: CL: Boxplots for examined sites and medians for each Climate group (DC to AC ratio of 1.5)

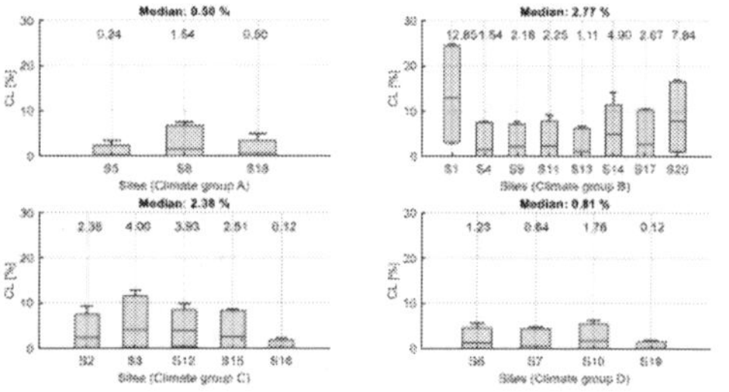

Figure 6: Boxplots of CL (all examined time resolutions and DC to AC ratios)

Figure 7 shows CL at all examined time resolutions and DC to AC ratios (same as Figure 6) aggregated for Climate groups. Upper whiskers and outliers in boxplots represent 1-minute data and DC to AC ratio of 1.5. Lower whiskers or bottom part of the boxplots represent 60-minute data and DC to AC ratio of 1.1.

The order of medians corresponds to the order in the case of DC to AC ratio of 1.5 (Figure 5). The first two medians (B: Dry and C: Temperate) are also the same as medians of the GTI (Figure 2). **The lower the annual sum of GTI level the lower impact of GTI on the CL level.**

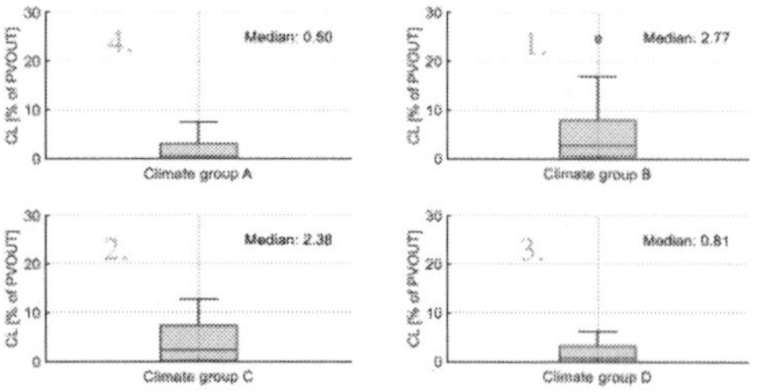

Figure 7: CL: Boxplots for climate groups and medians for each Climate group (all examined time resolutions and DC to AC ratios)

The proportion of time samples (as a percentage of all samples) when CL occurs was examined as an indicator of the pattern of GTI during the year.

Figure 8 shows the proportion of time samples when CL occurs for all examined DC to AC ratios and time resolutions. This is another parameter (in addition to the annual sum of GTI) determining the annual CL at a certain DC to AC ratio. The order of the medians is exactly the same as the order of the medians in the case of the CL based on all examined time resolutions and DC to AC ratios (Figure 6 and 8).

Main indicators having impact on the CL percentage for a certain DC to AC ratio are:

- ***annual sum of GTI (i.e. more irradiance means more clipping),***
- ***the proportion of time samples when CL occurs in all time samples.***

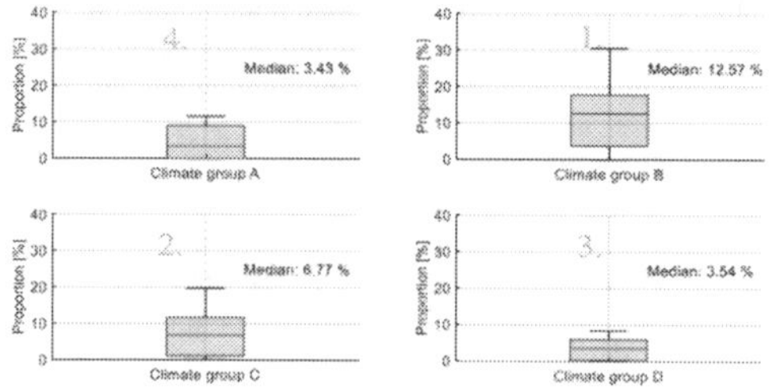

Figure 8: Boxplots of the proportion of time samples when CL occurs (all examined time resolutions and DC to AC ratios)

3.2 D2G ratio and difference in CL

The purpose of this chapter is to show the link between D2G at the sites from different Climate groups and the difference in CL calculated for different time resolutions and DC to AC ratios. D2G level is a measure of the content of molecules and particles in the atmosphere. These substances can lead to quick changes in irradiance and PVOUT when passing over the PV modules. It is expected that D2G level is an important factor affecting the difference in CL.

3.2.1 D2G ratio

Figure 9 shows the annual D2G for the examined sites (the first five sites with the highest levels are highlighted). The highlighted sites are from different Climate groups (A, B, C).

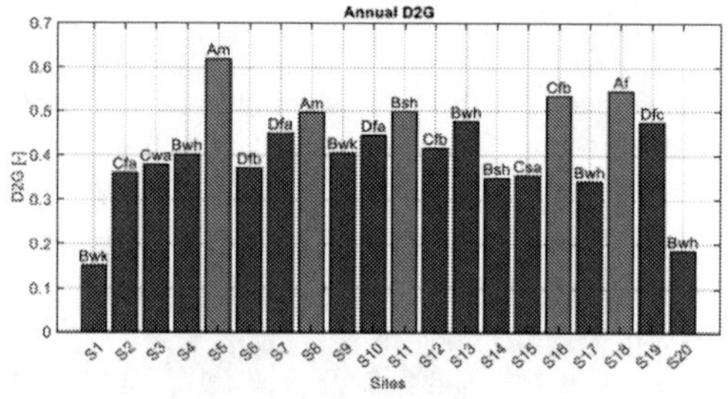

Figure 9: Annual D2G

Figure 10 shows D2G for all sites, grouped into Climate groups. The upper (lower) whisker in a boxplot represents the site with the highest (the lowest) D2G level.

10.4229/EUPVSEC2023/5CO.6.2
020493-004

The highest median as well as the site with the highest D2G level is in Climate group A (Tropical). Examined sites are sites with tropical monsoon (Am) and rainforest (Af) climate with high precipitation rates.

The lowest median as well as the site with the lowest D2G level is in Climate group B (Dry). These are sites with low precipitation rates.

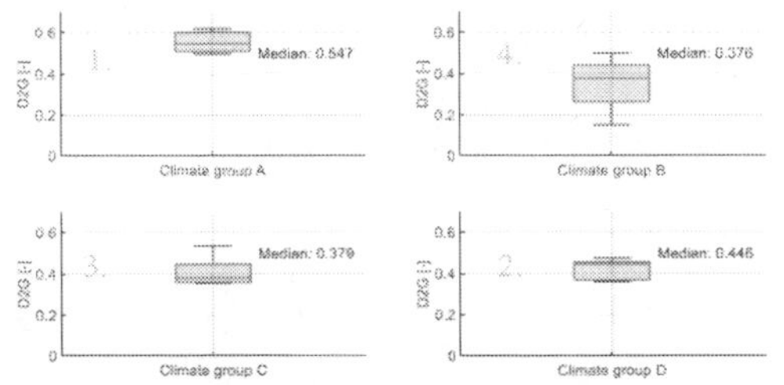

Figure 10: Boxplot of D2G for Climate groups

3.2.2 Annual difference in CL

To study the difference in CL resulting from using different time resolutions of input data, the annual percentage difference in CL was determined. It was calculated as the absolute difference between CL based on 15-minute (or 60-minute) data and 1-minute data divided by CL based on 1-minute data multiplied by 100.

Figure 11 shows the difference in CL for the examined sites. at time resolutions and DC to AC ratios. Medians based on percentage differences in CL are shown in plots. The values were later used to sort sites by difference in CL. The upper whiskers, outliers or quartiles represent difference in CL between 60-minute (in a few cases 15-minute) and 1-minute data at DC to AC ratio of 1.1 (e.g. in case of S2 there is 81 % difference). The lower whisker represents the difference in CL between 15-minute and 1-minute data at DC to AC ratio of 1.5 (e.g. in case of S2 there is 4 % difference).

Underestimation of CL was observed in the case of using input data with lower time resolution (15-minute and 60-minute in comparison with 1-minute input data)**.** With a lower DC to AC ratio, in some cases, only 1-minute data shows saturation of the inverter. **The higher the DC to AC ratio the lower the percentage differences in CL.**

The same order of medians is obtained for all examined DC to AC ratios. This means that the climatic differences in underestimation of CL are mostly caused by lower time resolution and do not depend significantly on the DC to AC ratio.

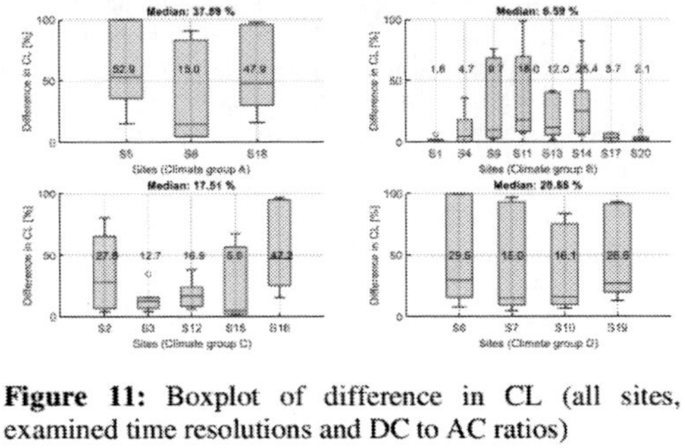

Figure 11: Boxplot of difference in CL (all sites, examined time resolutions and DC to AC ratios)

Figure 12 shows the difference in CL for all examined DC to AC ratios and time resolutions (same as Figure 11), aggregated for Climate groups.

The order of medians of differences in CL corresponds with the order of D2G values (Figure 10 and Figure 12) and it is the opposite to the order of medians of CL (Figure 6).

Diffuse irradiation is the factor contributing to a number of short-term exceedances of inverter capability and hence CL. Statistically, short-term exceedance of inverter capability is more often occurring in sites from Climate groups with lower levels of CL.

The ratio of annual diffuse irradiation is an important factor determining the difference in CL. Statistically, the higher the D2G the higher the difference in CL, i.e. with higher proportion of diffuse light the importance of high temporal resolution of data increases. The highest percentage differences in CL are observed in sites from Climate group A: Tropical.

D2G provides an indication of differences in CL. The order of medians of difference in CL corresponds with the order of D2G values and is the opposite to the order of medians of annual sum CL as a percentage of PVOUT. Underestimation of CL was observed in case of using input data with insufficient time resolution.

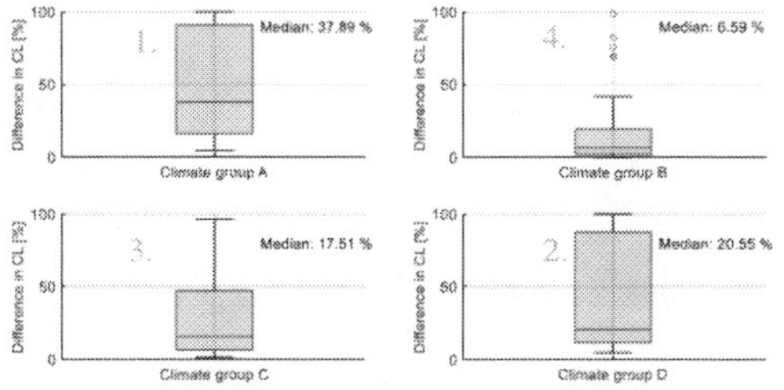

Figure 12: Boxplot of difference in CL (all examined time resolutions and DC to AC ratios)

3.3 Annual difference in PVOUT

The purpose of this chapter is to show the link between the abovementioned indicative factors (CL and difference in CL) and the difference in PVOUT calculated for different time resolutions and DC to AC ratios.

To study the difference in PVOUT due to different time resolutions of input data, the annual percentage difference in PVOUT was determined. It was calculated as the absolute difference between PVOUT based on 15-minute (60-minute) data and 1-minute data divided by PVOUT based on 1-minute data multiplied by 100.

Figure 13 (in three parts: a, b, c) shows the annual difference in PVOUT for the examined sites, different time resolutions, and DC to AC ratios (1.1: top figure, 1.3: middle figure, 1.5: bottom figure). The difference in PVOUT is proportional to the DC to AC ratio and the time resolution.

The percentage levels represent overestimation of PVOUT calculated using 15-minute (60-minute) data in comparison with 1-minute data.

10.4229/EUPVSEC2023/5CO.6.2
020493-005

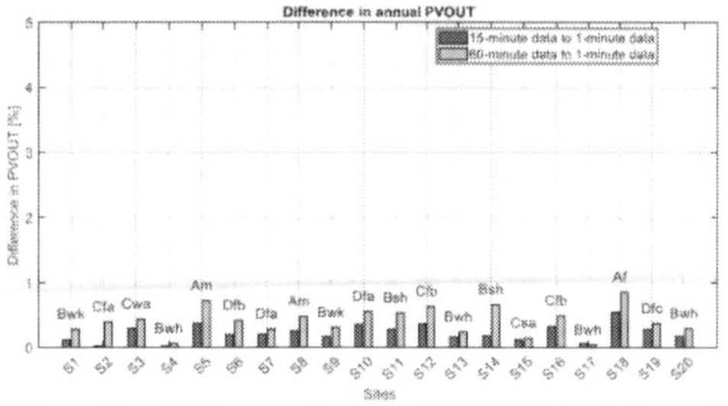

Figure 13a: Difference in annual PVOUT for DC to AC ratio of 1.1

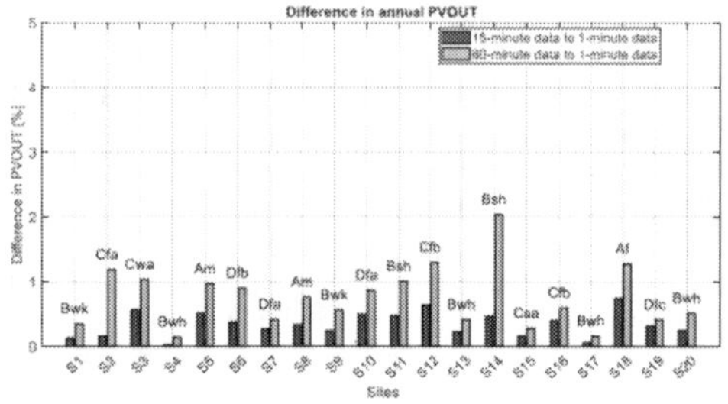

Figure 13b: Difference in annual PVOUT for DC to AC ratio of 1.3

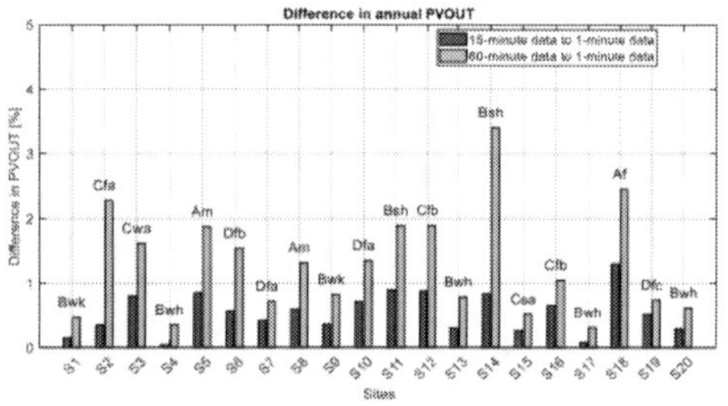

Figure 13c: Difference in annual PVOUT for DC to AC ratio of 1.5

Figures 14 and 15 show the percentage difference between PVOUT based on 15-minute data (and 60-minute data) and 1-minute data for the examined DC to AC ratios.

Nonlinear increase in the difference in PVOUT can be seen with increasing DC to AC ratios - the higher the DC to AC ratio, the higher the increment in difference. It is because of the higher the DC to AC ratio the higher the absolute difference (between CL based on 15-minute data (60-minute data) and 1-minute data) in CL. Abovementioned effect is statistically more significant in Climate groups A and C, where a higher difference in CL is observed. In practical terms this means that a larger DC to AC ratio leads to greater overestimation of PVOUT when using lower temporal resolution of input data, and this phenomenon is more pronounced in Climate groups A and C.

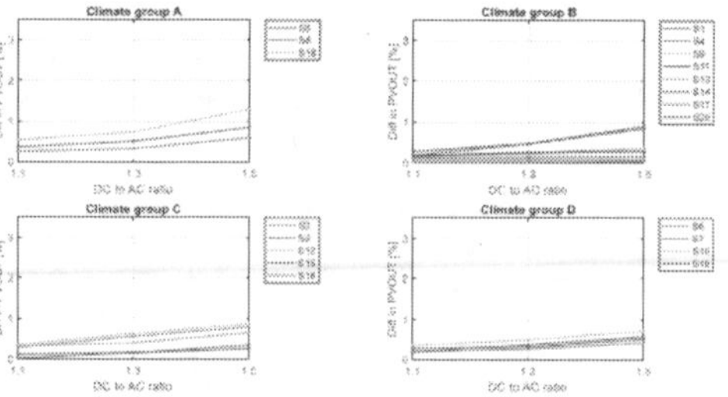

Figure 14: Difference in PVOUT for different DC to AC ratios (15-minute data to 1-minute data)

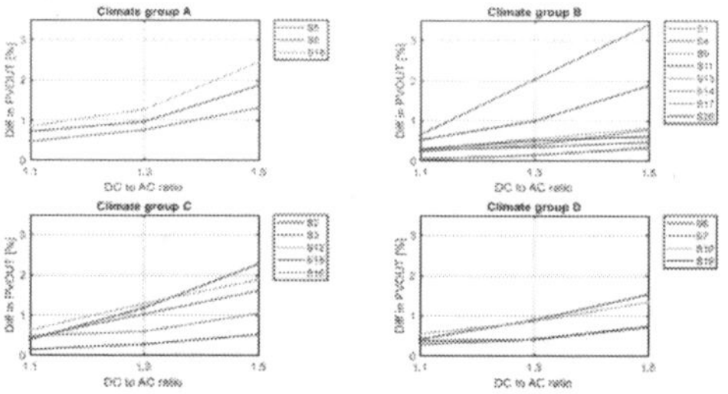

Figure 15: Difference in PVOUT for different DC to AC ratios (60-minute data to 1-minute data)

Figures 16, 17, and 18 show difference in PVOUT for the examined sites using 1-, 15- and 60-min data and DC to AC ratio of 1.1 (Figure 16), 1.3 (Figure 17) and 1.5 (Figure 18). The upper whisker represents the difference in PVOUT between 60-minute and 1-minute data. The lower whisker represents the difference in CL between 15-minute and 1-minute data.

The order of medians of difference in PVOUT is the same for all examined DC to AC ratios. The highest percentage differences in PVOUT (the highest PVOUT overestimation in comparison to 1-minute data) are observed in sites from Climate group A: Tropical (approx. 0.8 % PVOUT overestimation in case of DC to AC ratio of 1.3).

The higher the DC to AC ratio, the higher the difference in PVOUT based on the 60-minute and 15-minute data (a wider boxplot). Significant difference in the case of DC to AC ratio of 1.5 is observed at S14 where PVOUT based on 15-minute data is overestimated by 0.83 % and based on 60-minute data it is overestimated by 3.40 %.

Statistically, using 15-minute data instead of 60-minute data we can reduce PVOUT overestimation by approx. 50 % of the PVOUT value (a higher reduction occurs at sites with a wider boxplot and closer to x-axis). **This means that using 15-minute data is more accurate than 60-minute data, but there is still significant overestimation of PVOUT in comparison with 1-minute data.**

10.4229/EUPVSEC2023/5CO.6.2
020493-006

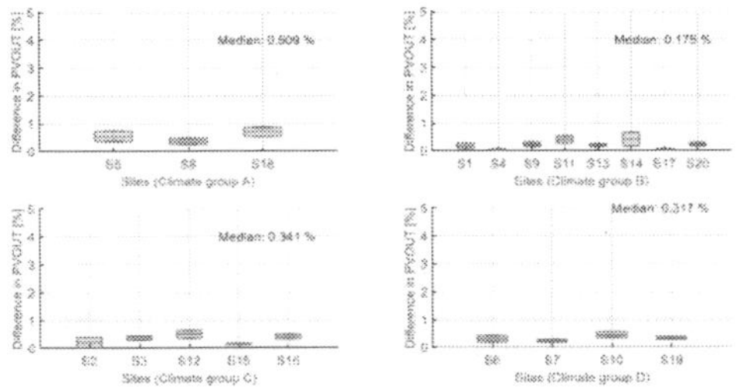

Figure 16: Boxplots of difference in PVOUT (DC to AC ratio of 1.1)

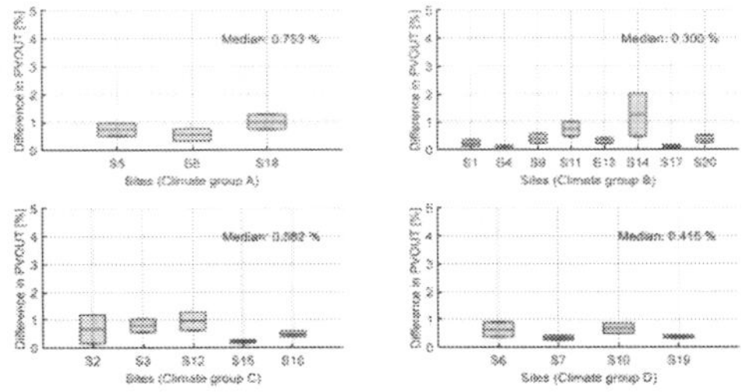

Figure 17: Boxplots of difference in PVOUT (DC to AC ratio of 1.3)

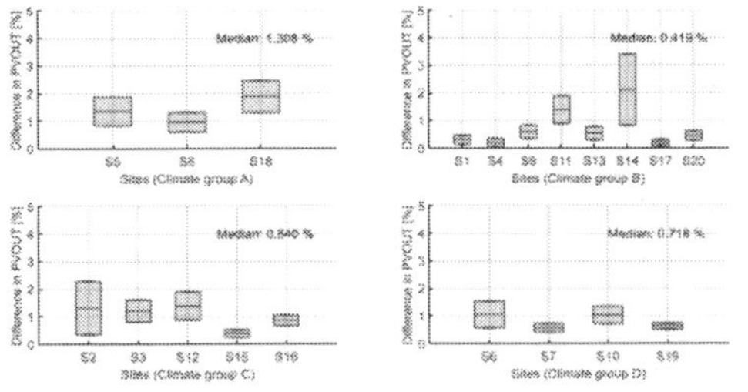

Figure 18: Boxplots of difference in PVOUT (DC to AC ratio of 1.5)

Figure 19 shows difference in PVOUT for the examined sites using all examined time resolutions and DC to AC ratios. Medians based on percentage differences in PVOUT are shown in plots. The values were later used to sort the sites by difference in PVOUT. The site with the highest median is site S18 where the median difference in PVOUT reaches 1.06 %.

The upper whisker and outliers represent the difference in PVOUT between 60-minute and 1-minute data and DC to AC ratio of 1.5. For example, at S14 it reaches 3.40 %. The lower whisker represents the difference in PVOUT between 15-minute and 1-minute data and DC to AC ratio of 1.1.

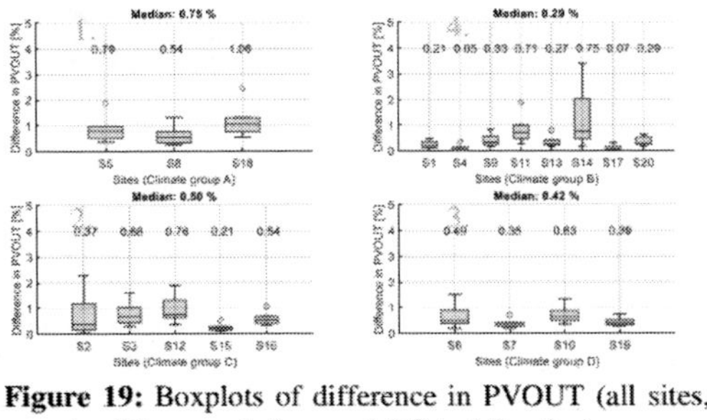

Figure 19: Boxplots of difference in PVOUT (all sites, examined time resolutions and DC to AC ratios)

Indicative factors of differences in PVOUT are:

- ***CL*** *(mainly depends on annual sum of GTI and the time duration when CL occur)*
- ***difference in CL*** *(depends on D2G level).*

Summarization of the examined factors ordered by their median values for the different Climate groups is listed in Table III. The results are based on statistical analysis, meaning that e.g. a site from Climate group A has comparatively the highest chance of obtaining the largest difference in PVOUT (at a certain DC to AC ratio) calculated using different time resolutions.

Table III: Examined indicative factors

Indicator		Order from the highest to the lowest value			
Annual sum of GTI [MWh/m²]		B (Dry)	C (Temperate)	A (Tropical)	D (Continental)
Annual CL [% of PVOUT]	DC:AC = 1.1, 1.3	C (Temperate)	B (Dry)	D (Continental)	A (Tropical)
	DC:AC = 1.5	B (Dry)	C (Temperate)	D (Continental)	A (Tropical)
Annual D2G [-]		A (Tropical)	D (Continental)	C (Temperate)	B (Dry)
Annual difference in CL [% diff to 1-min CL]		A (Tropical)	D (Continental)	C (Temperate)	B (Dry)
Annual difference in PVOUT [% diff to 1-min PVOUT]		A (Tropical)	C (Temperate)	D (Continental)	B (Dry)

To further investigate the dependence of the PVOUT differences on the other factors, a linear regression analysis was performed, and a trend line was laid over the datapoints. In an attempt to ensure better correlation, the data was split by the Climate groups, which have so far proven to be a useful grouping.

Figure 20 shows the dependence of difference in PVOUT on D2G and DC to AC ratio based on 15-minute data and 1-minute data. Similar calculations were performed for 60-minute and 1-minute data, however, the resultant trends were weaker. It can be seen that the order of the trend line gradients corresponds to the order of annual difference in PVOUT listed in Table 3: Climate group A, C, D and B (the higher the trend line gradient the higher the annual difference in PVOUT).

The results show that the linear fit is not a very good one, even within one Climate group. This is mainly obvious at sites from Climate group with the lowest difference in PVOUT (Climate group B). **Trend lines are poor indicators, the difference in PVOUT cannot be easily estimated based on a trend, and it is important to analyze each site individually in practice.**

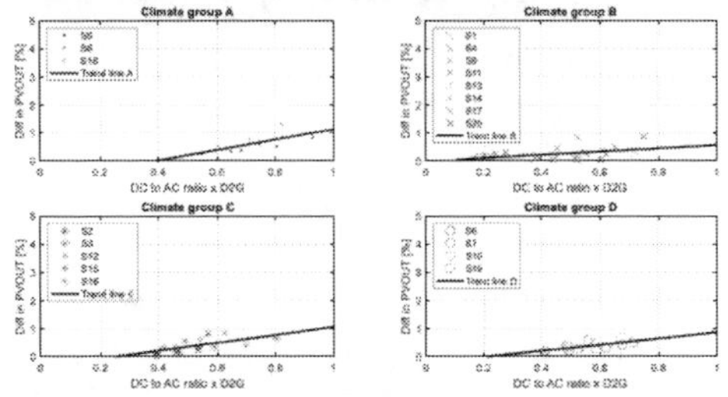

Figure 20: Trend lines for PVOUT differences (15-min. to 1-min. data)

4 DETAILED RESULTS FOR SELECTED SITES

To demonstrate the examined indicative factors, the sites with the highest and the lowest values of the factors (based on Figures 6, 11, and 19 respectively) were selected. The sites are listed in Table IV where lowest and the highest values are highlighted.

Table IV: Selected sites

Indicator/ Selected site	S19: Uppsala (Sweden), Dfc	S1: Calama (Chile), Bwk	S5: Douala (Cameroon), Am	S4: Doha (Qatar), Bwh	S18: Kuching (Malaysia), Af
Annual sum of GTI [MWh/m²]	Lowest	Highest	9. lowest of 20	17. lowest of 20	8. lowest of 20
Annual CL [% of PVOUT]	Lowest	Highest	3. lowest of 20	8. lowest of 20	4. lowest of 20
Annual D2G [-]	14. lowest of 20	Lowest	Highest	9. lowest of 20	19. lowest of 20
Annual difference in CL [% diff to 1-min CL]	14. lowest of 20	Lowest	Highest	4. lowest of 20	19. lowest of 20
Annual difference in PVOUT [% diff to 1-min PVOUT]	10. lowest of 20	4. lowest of 20	19. lowest of 20	Lowest	Highest

4.1 Annual GTI and D2G

Figure 21 shows GTI at S1, the site with the lowest level of D2G. A significant level of GTI variability is observed at site S5 with the highest annual level of D2G (Figure 22) and also at site S18 with the highest level of difference in PVOUT (Figure 23).

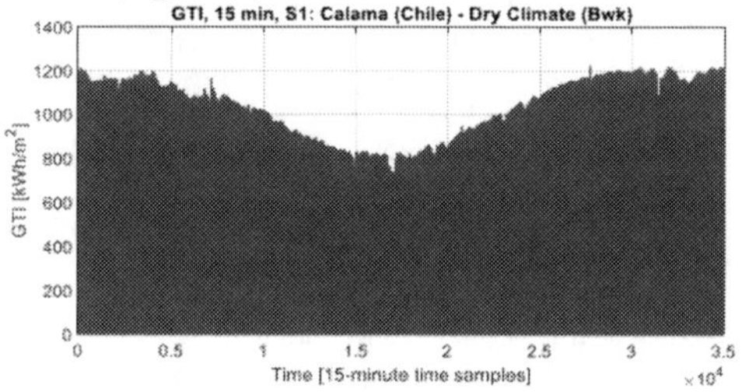

Figure 21: GTI (15-minute data): S1, site with the lowest D2G

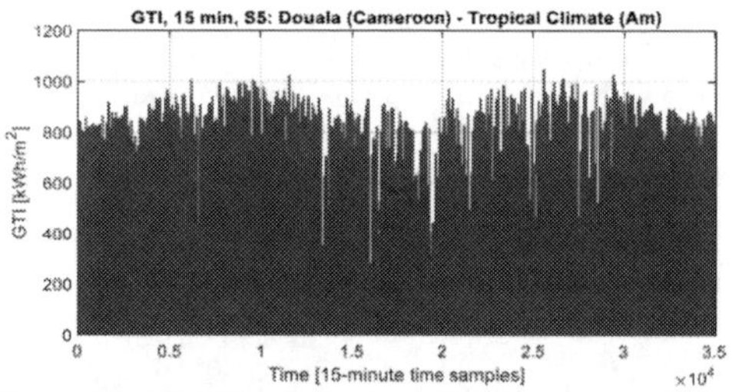

Figure 22: GTI (15-minute data): S5, site with the highest D2G

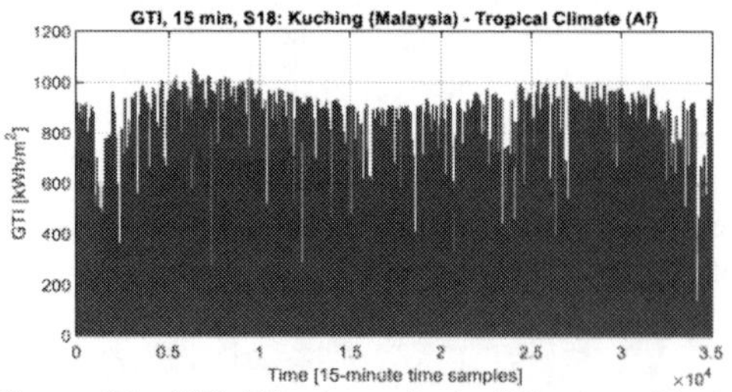

Figure 23: GTI (15-minute data): S18, site with the highest difference in PVOUT

4.2 Annual CL and difference in CL

The effect of different time resolutions and different DC to AC ratios on annual CL is shown in Figure 24.

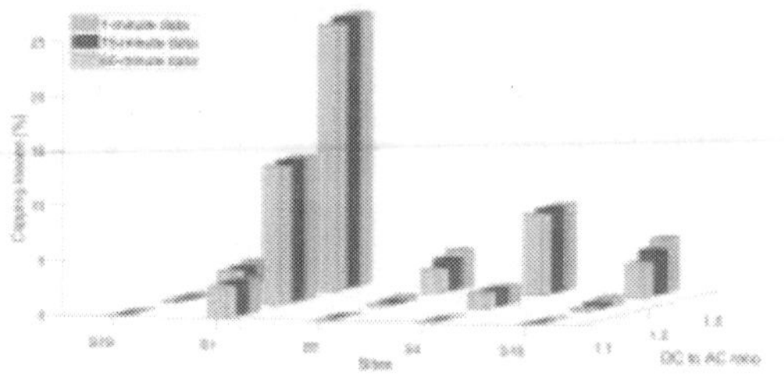

Figure 24: Effect of different time resolutions and different DC to AC ratios on annual CL

Two extreme sites are described in detail. Site S1 with the highest annual CL and site S5 with the highest annual difference in CL.

Site S1 – Calama, Chile in Dry climate is the site with the highest annual CL but also the lowest percentage difference between CL based on 60-minute or 15-minute in comparison with 1-minute data. It is the site with the lowest D2G. Despite the fact that the annual difference in CL is the lowest, the difference in PVOUT is not the lowest because of the high level of CL.

Site S5 – Douala, Cameroon in Tropical climate is the site with the third lowest annual CL (the first is S19) but the highest difference in CL. It is the site with the highest D2G. Despite the fact that CL is almost the lowest among the sites, the highest difference in CL is observed, leading to the second highest difference in PVOUT.

4.3 Monthly CL and difference in CL

The effect of different time resolutions on monthly CL is shown in Figure 25 for selected sites (the top figure for 5 selected sites and the bottom figure for S5 and S18 to show the results for these sites in more detail).

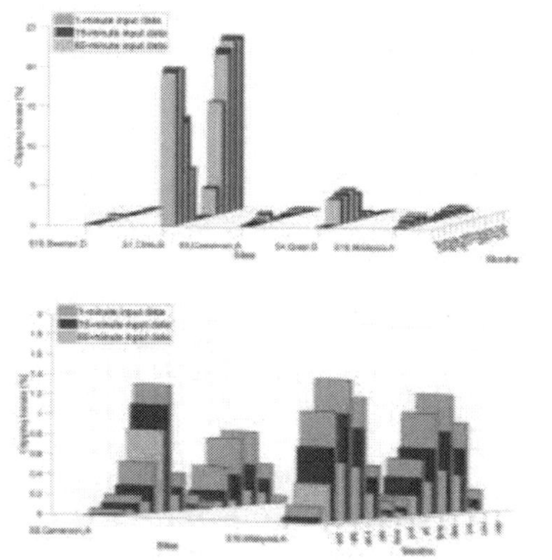

Figure 25: Effect of different time resolutions on monthly clipping losses (DC to AC ratio of 1.3)

Significant percentage difference between CL based on 60-minute (green bar) or 15-minute (blue bar) data in comparison with 1-minute (orange bar) data is shown at site **S5 (Cameroon) and S18 (Malaysia)**. This effect leads to the highest level of PVOUT difference at these sites (see Figure 27) even if the level of CL at these sites is relatively low.

The difference in CL has a significant impact on the difference in PVOUT also in the case when CL is relatively low. Therefore, low CL cannot be taken as an assurance of accurate PVOUT calculation, if data with low time resolution is used.

4.4 Difference in PVOUT

PVOUT at **site S18** based on 1-minute, 15-minute, and 60-minute data for one day is shown in Figure 26 as an example. The figure demonstrates inverter clipping in the upper part of the plot by limiting PVOUT to a specific level (because of inverter capability). It is obvious that PVOUT based on 60-minute data (green line) is not clipped. This leads to overestimation of PVOUT in the case of 60-minute data, in comparison with 1-minute or 15-minute data.

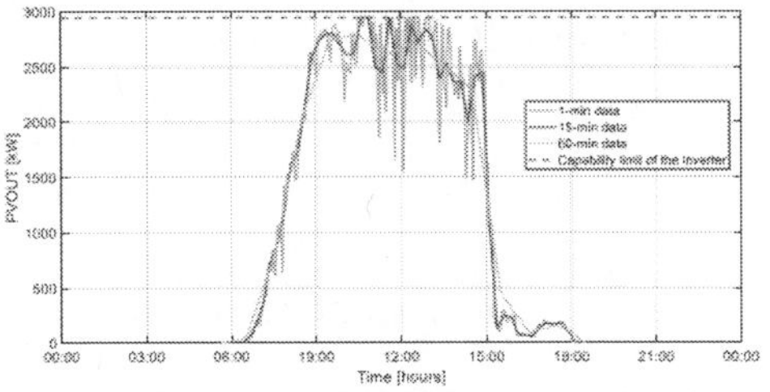

Figure 26: PVOUT at S18, DC to AC ratio 1.3: a day with inverter clipping occurring with 1-minute and 15-minute data

Overestimation of annual PVOUT in the case of 60-minute or 15-minute data in comparison with 1-minute data is shown in Figure 27. Relatively low level of CL at **sites S5** (third lowest CL level) and **S18** (fourth lowest CL level) but significant difference between CL based on 60-minute or 15-minute data and 1-minute data (see Figure 25) leads to significant overestimation of PVOUT based on 15-minute or 60-minute data in comparison with 1-minute data. PVOUT overestimation reaches more than 2 % at DC to AC of 1.5 and 60-minute data.

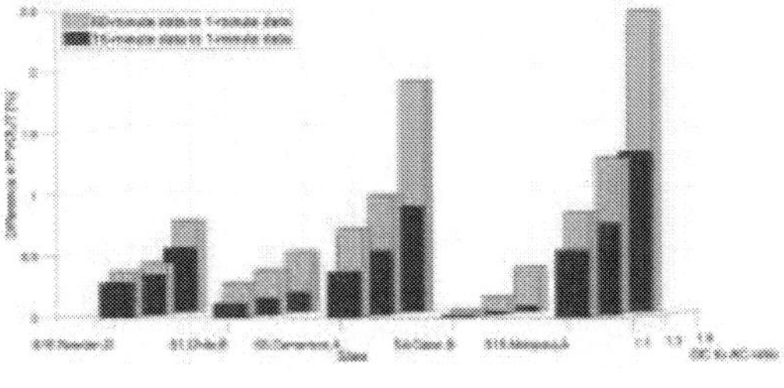

Figure 27: Overestimation of annual PVOUT for different time resolution and DC to AC ratio

4 CONCLUSUIONS

Results are presented for 20 selected sites located over the world in different Climate groups (A, B, C, and D). PVOUT simulations are based on satellite-based solar and meteorological model data and ray-tracing simulations. The effect of different time resolution of input data on clipping losses and PVOUT of the PV system model with DC to AC ratio of 1.1, 1.3, and 1.5 was examined. Percentage differences in CL (PVOUT) were based on the difference between CL (PVOUT) calculated using 15-minute (60-minute) data and 1-minute data.

The value of CL (mainly depending on GTI) together with the difference in CL for different time resolutions (mainly depending on D2G level) are indicative of the difference in PVOUT.

The higher the CL and difference in CL, the higher the difference in PVOUT, and the difference in CL has a much stronger influence than the CL. It means that relatively low level of CL but significant difference in CL leads to a significant percentage difference in PVOUT.

Patterns in values of indicative factors in connection with Climate groups can be observed. Statistically, the highest level of annual D2G/percentage difference in CL and in PVOUT are observed in sites from Climate group A: Tropical and the lowest level of annual D2G/percentage difference in CL and in PVOUT are observed in sites from Climate group B: Dry (see Table 3).

PVOUT overestimation in sites from Climate group A is from 0.3 % (0.5 %) in the case of DC to AC ratio of 1.1. and 15-minute (60-minute) data to 1.3 % (2.5 %) in the case of DC to AC ratio of 1.5 and 15-minute (60-minute) data. The site S18 (Kuching, Malaysia) is the site with the highest annual difference in PVOUT. It is the site with the second highest D2G level and the fourth lowest CL level.

Statistically, using 15-minute data we can reduce PVOUT overestimation to about half of the PVOUT overestimation if 60-minute data is used.

Numerical results presented in this article are close to results presented in articles based on datasets from meteorological stations readings and datasets from DC power measurements listed in Chapter 2 (for the sites with similar climatic conditions). For example: In [13] *at site of Lanai and hourly intervals, error in energy ranges to 2%* as it was mentioned in Chapter 2. The closest climatic conditions (based on GTI and D2G are in S12 (Melbourne, Australia) where the overestimation of annual power production was determined to be approx. 1.9 % for 1.5 DC:AC ratio. In [16] one site (Birmingham, Alabama, US) was studied *(at hourly inputs the overestimation of annual power production by approximately 2% for PV installations with 1.4 DC:AC ratio was achieved)* as mentioned in Chapter 2. The site with similar conditions is S2 (Orlando, USA), where with 60-minute input data the overestimation of annual power production was determined to be approx. 2.3 % for 1.5 DC:AC ratio (app. 2 % for 1.4 DC:AC ratio, Figure 15).

Each particular site is characterized by its own conditions which have an impact on the pattern of GTI and finally on the overestimation of annual power production using input data with insufficient time resolution. Statistical approach provides indicative view of PVOUT overestimation. In practice, it is important to analyze each site for a PV system individually.

Based on these results it is recommended to perform the PV yield simulations based on data with 1-minute resolution, to take into account the effect of clipping losses caused by short-term exceeding of inverter capability. Short-term variability of the solar resource is a natural feature occurring around the globe and PV yield will depend on this feature. It has to be considered during the design process to avoid underperformance in the operational stages of projects.

10.4229/EUPVSEC2023/5CO.6.2
020493-009

The future research will be focused on:

1. Expansion of number of sites, DC:AC ratios (to cover the range 0.9 - 2.0), and time resolutions (from 1-minute to 60-minute).
2. Comparison of the results at selected sites with meteorological stations readings and DC power measurements.
3. Examination of the impact of sky type on calculation accuracy of annual power production using different time resolutions.

REFERENCES

[1] Perez R., Cebecauer T., Šúri M., 2013. Semi-Empirical Satellite Models. In Kleissl J. (ed.) Solar Energy Forecasting and Resource Assessment. Academic press. http://www.sciencedirect.com/science/article/pii/B9780123971777000024

[2] Cebecauer T., Šúri M., Perez R., High performance MSG satellite model for operational solar energy applications. ASES National Solar Conference, Phoenix, USA, 2010. http://solargis.com/assets/publication/2010/Cebecauer-Suri-Perez--ASES2010--High-performance-MSG-satellite-model-for-operational-solar-energy-applications.pdf

[3] Šúri M., Cebecauer T., Perez P., 2010. Quality Procedures of Solargis for Provision Site-Specific Solar Resource Information. Conference SolarPACES 2010, September 2010, Perpignan, France. http://solargis.com/assets/publication/2010//Suri-Cebecaue--Perez-SolarPACES2010--Quality-procedures-of-solargis-for-provision-site-specific-solar-resource-information.pdf

[4] European Centre for Medium-Range Weather Forecasts. 2017, updated monthly. ERA5 Reanalysis. Research Data Archive at the National Center for Atmospheric Research, Computational and Information Systems Laboratory. https://doi.org/10.5065/D6X34W69. Accessed 30-04-2019.

[5] NASA MODIS website https://modis.gsfc.nasa.gov/data/

[6] AppEEARS Team. (2019). Application for Extracting and Exploring Analysis Ready Samples (AppEEARS). Ver. 2.24. NASA EOSDIS Land Processes Distributed Active Archive Center (LP DAAC), USGS/Earth Resources Observation and Science (EROS) Center, Sioux Falls, South Dakota, USA. https://lpdaacsvc.cr.usgs.gov/appeears/

[7] Schaaf, C., Wang, Z. (2015). MCD43A3 MODIS/Terra+Aqua BRDF/Albedo Daily L3 Global - 500m V006. NASA EOSDIS Land Processes DAAC. doi: 10.5067/MODIS/MCD43A3.006

[8] Hoffmann M., Reichelmann S., Crisosto C., 2014. Improved Synthesis of Global Irradiance with One-Minute Resolution for PV System Simulations, International Journal of Photoenergy, Volume 2015, Article ID 808509.

[9] Dvonc, L., Orosi, P., Cebecauer, T., Schnierer, B.: Implementation of ray tracing rendering technique for improved solar radiation modeling of bifacial PV modules, 8th World Conference on Photovoltaic Energy Conversion, Milan, 2022

[10] Dvonc, L., Orosi, P., Cebecauer, T., Schnierer, B.: Comparison of ray tracing rendering technique with ground measurements for improved solar radiation modeling, 2022, 15th PV Performance Modeling Workshop, Salt Lake City, Utah USA

[11] IEC TS 61724-3:2016: Photovoltaic system performance – Part 3: Energy evaluation method

[12] B. Burger, R. Ruther: Inverter sizing of grid-connected photovoltaic systems in light of local solar resource distribution characteristics and temperature, Sol. Energy. 80 (2006). pp. 32-45. doi: 10.1016/j.solener.2005.08.012.

[13] C. W. Hansen, J. S. Stein, and D. Riley, "Effect of time scale on analysis of PV system performance", Sandia National Lab, Albuquerque, NM, Tech. Rep. SAND2012-1099, 2012.

[14] J. Luoma, J. Kleissl, and K. Murray, "Optimal inverter sizing considering cloud enhancement", Solar Energy, vol. 86, no. 1, pp. 421-429, 2012.

[15] Performance Modeling. EPRI, Palo Alto, CA: 2018. 3002014718

[16] J. O. Allen and W. B. Hobbs: The effect of short-term inverter saturation on modeled hourly PV output using minute DC power measurements. Journal of Renewable and Sustainable Energy 14, 063503, 2022.

[17] Beck, H.E., Zimmermann, N. E., McVicar, T. R., Vergopolan, N., Berg, A., & Wood, E. F. - "Present and future Köppen-Geiger climate classification maps at 1-km resolution". Nature Scientific Data. DOI:10.1038/sdata.2018.214.

10.4229/EUPVSEC2023/5CO.6.2
020493-010

Fraunhofer ISE

Sensitivity of Global Warming Potential of PERC Solar System to Production Electricity Mix, Installation Location and System Lifetime

Abeer Ali Khan, Christian Reichel, Lorenz Friedrich, Dilara Maria Subasi, Pamela Molina, Holger Neuhaus, Sebastian Nold
40th EU PVSEC, 20th September 2023
www.ise.fraunhofer.de

020495-001

40th European Photovoltaic Solar Energy Conference and Exhibition

Motivation and Research Questions

Analysing *carbon footprint* of PV

EU Solar Energy Strategy: Photovoltaic (PV) modules expected in the EU market → **320 GW by 2025** and ~**600 GW by 2030** [1]

- Support **REPowerEU** plan [2]
- Reduce EU's dependence on fossil fuels

The carbon footprint currently used in commonly used databases are outdated.

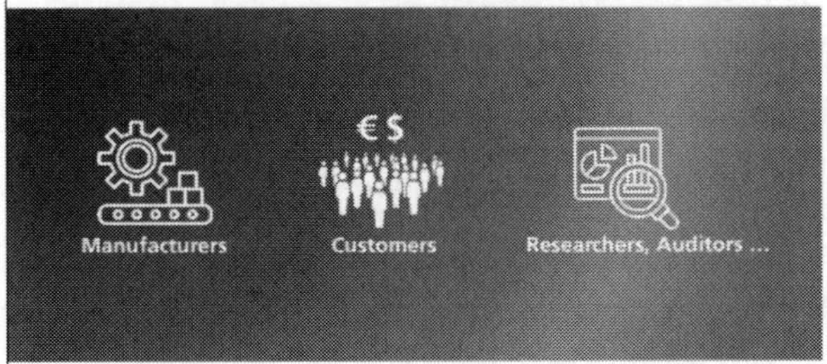

How is the *carbon footprint* of PV systems influenced by

- **Manufacturing location**
- **Installation location**
- **Lifetime**

Slide 2
17.09.2023

1. EU Solar Energy Strategy (2022). Communication from the commission to the European parliament, the council, the European economic and social committee and the committee of the regions.
2. REPowerEU (2022). Communication from the commission to the European parliament, the European council, the council, the European economic and social committee and the committee of the regions.

Fraunhofer ISE

020495-002

LCA of PERC PV System

Goal and Scope

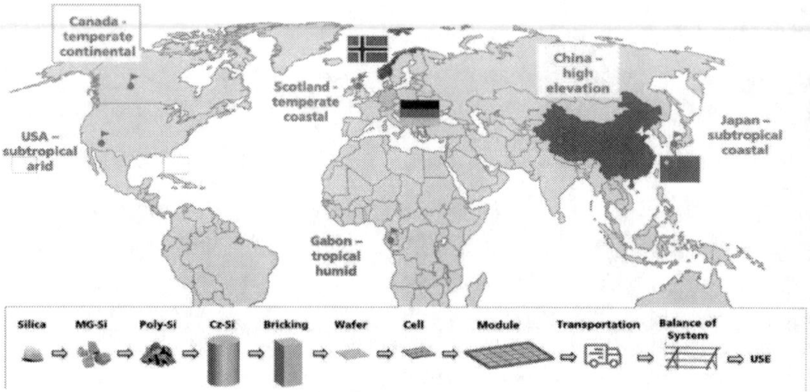

Cradle-to-use phase LCA

Sensitivity of GWP of a PERC 15 kWp slanted roof-top system (without battery) to

- **3** production locations
- **6** installation locations
- **3** PV system lifetimes (25, 30 and 40 years)

Methodology:

- *ISO 14040-4* [1,2]
- *IEA PVPS 12* "Methodology Guidelines for LCA on PV" [3]
- *Product Environmental Footprint Category Rules (PEFCR)* for Photovoltaics of the European Union [4]

Database: Ecoinvent 3.8 [5]

1. ISO 14040 (2009) Environmental management – Life cycle assessment – Principles and framework., 2009-11.
2. ISO 14044 (2006) Environmental management – Life cycle assessment – Requirements and guidelines., 2006-10.
3. Frischknecht, R.P. Stolz, G. Heath, M. Raugei, P. Sinha, and M. de Wild-Scholten, Methodology Guidelines on Life Cycle Assessment of Photovoltaic 2020: Task 12-18 2020.
4. European Union Commission, Product Environmental Footprint Category Rules (PEFCR): Photovoltaic Modules used in Photovoltaic Power Systems for Electricity Generation, 2019.
5. G. Wernet, C. Bauer, B. Steubing, J. Reinhard, E. Moreno-Ruiz, B. Weidema, The ecoinvent database version 3 (part I): overview and methodology, Int J Life Cycle Assess 21, 2016.

Fraunhofer ISE

LCA of PERC PV System

System Boundary

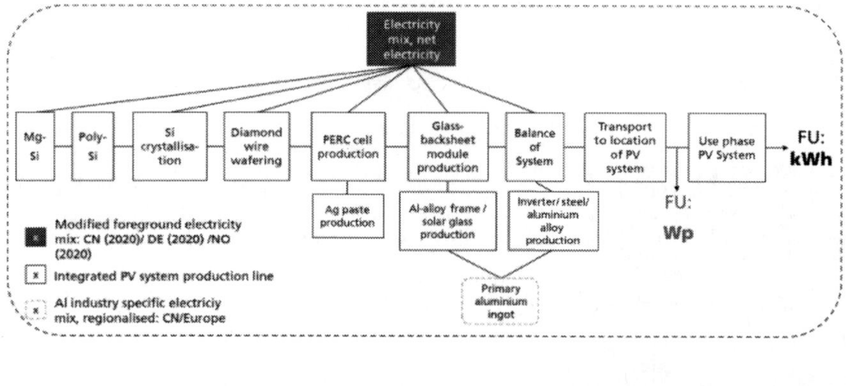

Modified production electricity mix

- o electricity used by different PV components directly in the factory.

GWP impact from electricity in production
63% → Chinese glass-backsheet module
52% → European glass-glass module. [1]

Unmodified electricity mix

- o electricity used to produce background materials such as auxiliaries, aluminium ingot production etc. that the manufacturers do not have a direct control over, is left unchanged.

1. A. Müller, L. Friedrich, C. Reichel, S. Herceg, M. Mittag, D.H. Neuhaus, A comparative life cycle assessment of silicon PV modules: Impact of module design, manufacturing location and inventory, Solar Energy Materials and Solar Cells 230 (2021) 111277.

Fraunhofer ISE

Production Electricity Mix and its GWP

Overview

The electricity mix is the **share of different energy generating technologies** used to produce electricity for specific ***location*** and ***year***.

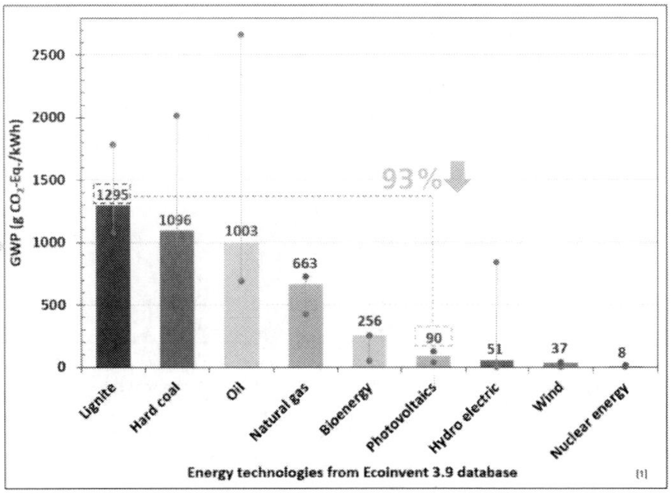

Slide 5
17.09.2023

1. GWP of different electricity producing technologies sourced from LCIA documentation of Ecoinvent 3.9 available in Ecoquery. Red and green points signify the maximum and minimum GWP values respectively for each technology.

Fraunhofer ISE

40th European Photovoltaic Solar Energy Conference and Exhibition

Sensitivity of the Grid Electricity Mix

Spatial and Temporal Changes

2014

2018

2018

- Biogas
- Biomass
- Geothermal
- Hard Coal
- Hydro
- Lignite
- Natural Gas
- Nuclear
- Photovoltaic
- Waste
- Wind

Net electricity mixes are modelled based on the electricity activities of Ecoinvent 3.8 and literature shares for the year 2020.

Slide 6
17.09.2023

1. China: CEC Statistics and Data Center, List of basic data of electricity statistics in 2020, [July 14, 2022]
2. Germany: Bruno Burger, Public Net Electricity Generation in Germany 2020: Share from Renewables Exceeds 50 percent, 2021.
3. Norway: Statistics Norway, SSB, Generation and consumption of electricity. GWh, [July 14, 2022].

Fraunhofer ISE

Sensitivity of the GWP of Grid Electricity Mix

Temporal and Spatial Changes

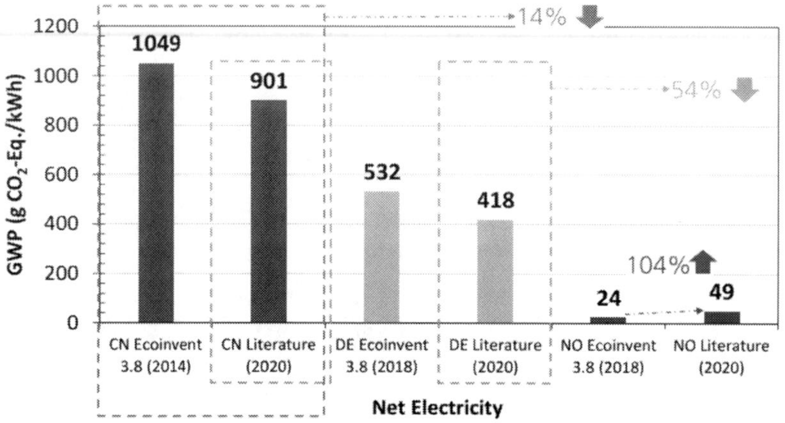

Modelled using Ecoinvent 3.8 database, System model: Cut-off, LCIA method: IPCC 2013, Software: Umberto 11.

Fraunhofer ISE

40th European Photovoltaic Solar Energy Conference and Exhibition

Consequent Impact on the GWP of PERC PV System LCA

Functional Units: Wp and kWh

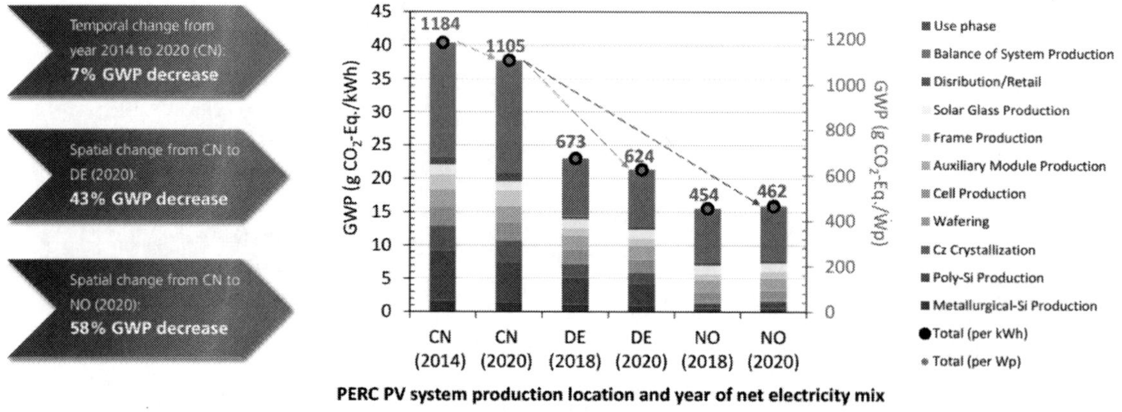

To reach kWh, this LCA follows the guidelines of Product Environmental Footprint Category Rules (PEFCR): Photovoltaic Modules used in Photovoltaic Power Systems for Electricity Generation, 2019. Installation is assumed at an average European location where the average total electricity generation during 30 years of system operation is 29.25 MWh/kWp and the average global tilted irradiation (GTI) per module square meter is 1,331 kWh/m². Impact assessment method: IPCC 2013 GWP 100a, with long term effects.

Fraunhofer ISE

Sensitivity of PERC CO_2 Footprint

To production electricity mix

- GWP of PV ranges from 6.6 to 54.8 g CO_2-Eq./kWh → **factor 8.**
- Installed at the same climatic zone, GWP differs by **~140%** based on module production location.

To installation location and system lifetime

For a PERC system produced in same location, different GWP is calculated:

- ~130% depending on the installation location
- **~53%** depending on system lifetime.

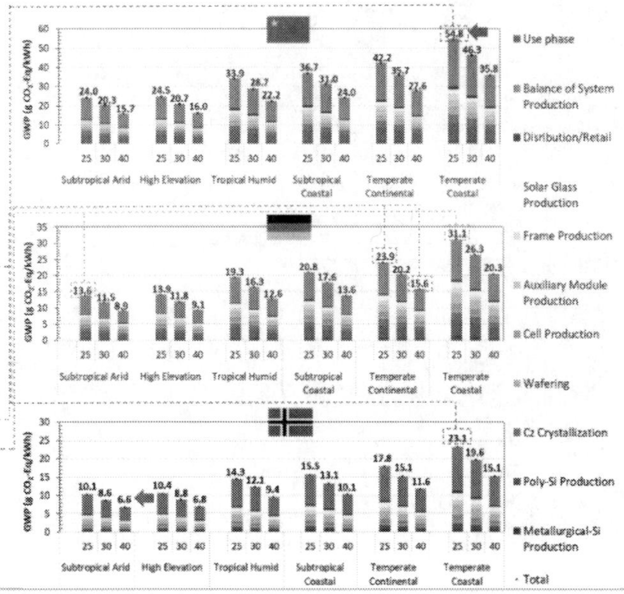

Climatic datasets:

Location	Global in-plane irradiation (kWh/m²a)
Subtropical Arid	2295
High Elevation (above 3000m)	2139
Tropical Humid	1678
Subtropical Coastal	1497
Temperate Continental	1266
Temperate Coastal	973

The energy yield calculations are in agreement with the IEC 61853 and based on- Alexander Aguilar Protti, Jibran Shahid, Max Mittag, Holger Neuhaus, Ulli Kräling, Martin Kaiser (Ed.), Virtual energy rating: a method for optimizing module performance through cell-to-module analysis, 2022.

Fraunhofer ISE

40th European Photovoltaic Solar Energy Conference and Exhibition

Sensitivity of PERC GWP to Production Electricity Mix, Installation Location and System Lifetime: Conclusion

- The production location, PV system lifetime have high influence on GWP reduction.
  - Ranges between **54.8** g CO_2-eq./kWh to **6.6** g CO_2-eq./kWh → **factor 8 difference**.
  - Current calculations show significantly lower GWP for PV than existing LCA databases (**90** g CO_2-Eq./kWh)
  - Module produced in CN, installed in EU with 30 years lifetime is **36** g CO_2-Eq./kWh.
- At locations with very good solar irradiation, the PERC system can have even lower GWP than *wind* energy.
- The GWP can be lowered even beyond what is shown, through optimized sourcing of the background materials (such as aluminium ingot production).

Fraunhofer ISE

Thank you!

Abeer Ali Khan
Photovoltaics, Techno-economic and ecological analyses
Tel. +49 17 4150-1608
abeer.ali.khan@ise.fraunhofer.de

Goal and Scope

Overview of the LCA

A Life Cycle Assessment (LCA) is the systematic analysis [1,2] of the potential environmental impacts of products or services during their entire life cycle (production, distribution, use and end-of-life phases).

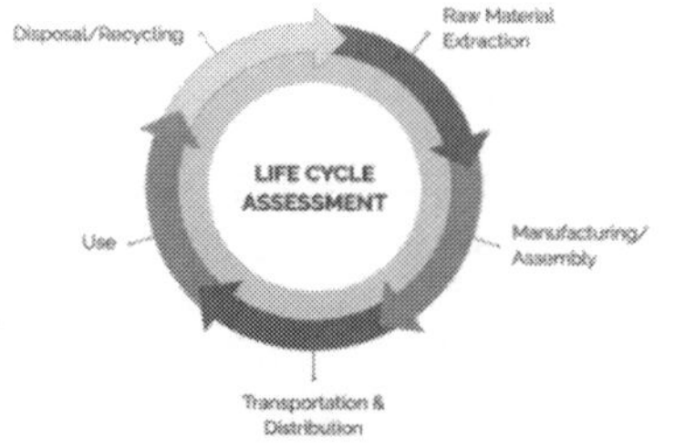

This LCA is aimed to analyse the sensitivity of **GWP of a PERC 15 kWp slanted roof-top system (without battery)** to • *production electricity mix,* • *installation location* and • *PV system lifetime.*

The GWP is analysed for a functional unit (FU) of

❖ 1 watt peak (Wp) of a PERC PV system and 1 kilowatt hour (kWh) of alternating current (AC) electricity generated by the PERC PV system.

Parameter	PERC Module		
Module area, (m^2)		1.83	
Module efficiency (%)		20.11%	
Lifetime (a)	25	30	40
Guaranteed power after 1st year (%)		98%	
Degradation following years (%)		0.54%	
Guaranteed power EoL (%)	85.00%	82.30%	76.90%
Remaining module power at EoL (%)	91.76%	90.41%	87.71%

Methodology: *ISO standards 14040-4* [1,2], *IEA PVPS 12* 'Methodology Guidelines for LCA on PV' [3] as well as the *Product Environmental Footprint Category Rules (PEFCR)* for Photovoltaics of the European Union [4].

[1] ISO 14040, Environmental management – Life cycle assessment – Principles and framework, 2009. [2] ISO 14044, Environmental management – Life cycle assessment – Requirements and guidelines, 2006. [3] Frischknecht, R.P. Stolz, G. Heath, M. Raugei, P. Sinha, and M. de Wild-Scholten, Methodology Guidelines on Life Cycle Assessment of Photovoltaic 2020. Task 12-18 (2020). [4] European Union Commission, Product Environmental Footprint Category Rules (PEFCR): Photovoltaic Modules used in Photovoltaic Power Systems for Electricity Generation, 2019.

Fraunhofer ISE

References

[1] ISO 14040, Environmental management – Life cycle assessment – Principles and framework , 2009.

[2] ISO 14044, Environmental management – Life cycle assessment – Requirements and guidelines., 2006.

[3] Frischknecht, R.P. Stolz, G. Heath, M. Raugei, P. Sinha, and M. de Wild-Scholten, Methodology Guidelines on Life Cycle Assessment of Photovoltaic 2020: Task 12-18 (2020).

[4] European Union Commission, Product Environmental Footprint Category Rules (PEFCR): Photovoltaic Modules used in Photovoltaic Power Systems for Electricity Generation, 2019.

[5] G. Wernet, C. Bauer, B. Steubing, J. Reinhard, E. Moreno-Ruiz, B. Weidema, The ecoinvent database version 3 (part I): overview and methodology, Int J Life Cycle Assess 21 (2016) 1218–1230. Activities:
- electricity, high voltage (RoW) [electricity production, nuclear, boiling water reactor]
- electricity, high voltage (RoW) [electricity production, wind, 1-3MW turbine, onshore]

[6] A. Müller, L. Friedrich, C. Reichel, S. Herceg, M. Mittag, D.H. Neuhaus, A comparative life cycle assessment of silicon PV modules: Impact of module design, manufacturing location and inventory, Solar Energy Materials and Solar Cells 230 (2021) 111277.

[7] Alexander Aguilar Protti, Jibran Shahid, Max Mittag, Holger Neuhaus, Ulli Kräling, Martin Kaiser (Ed.), Virtual energy rating: a method for optimizing module performance through cell-to-module analysis, 2022.

[8] IEC, IEC61853: PV module performance testing and energy rating. Part 1: Irradiance and temperature performance measurements and power rating., 2011.

[9] IEC, IEC61853: PV module performance testing and energy rating. Part 2: Spectral responsivity, incidence angle and module operating temperature measurements., 2016.

[10] IEC, IEC61853: PV module performance testing and energy rating. Part 3: Energy rating of PV modules, 2018.

[11] IEC, IEC61853: PV module performance testing and energy rating. Part 4: Standard reference climatic profiles., 2018.

[12] CEC Statistics and Data Center., List of basic data of electricity statistics in 2020, [July 14, 2022], https://www.cec.org.cn/detail/index.html?3-305140.

[13] Bruno Burger, Public Net Electricity Generation in Germany 2020: Share from Renewables Exceeds 50 percent, 2021.

[14] Statistics Norway, SSB, Generation and consumption of electricity. GWh, [July 14, 2022], https://www.ssb.no/en/energi-og-industri/energi/statistikk/elektrisitet.

Fraunhofer ISE

40th European Photovoltaic Solar Energy Conference and Exhibition

This presentation was selected by the Sc. Committee of the EU PVSEC 2023 for submission of a full paper to one of the EU PVSEC's collaborating peer-reviewed journals.

GROUND-MOUNTED OR ROOFTOP PHOTOVOLTAIC PLANT - EUROPEAN PRODUCTION OR CHINESE PRODUCTION: WHICH IS THE MOST ENVIRONMENTALLY SUSTAINABLE SYSTEM?

Elisabetta Brivio*, Andrea Danelli*, Pierpaolo Girardi*
*RSE SpA - Research on Energy System, Sustainable Development and Energy Sources, Milano, Italy
e-mail: elisabetta.brivio@rse-web.it
andrea.danelli@rse-web.it
pierpaolo.girardi@web-rse.it

ABSTRACT:
The study focuses on the potential environmental impact generated by PV energy production through two types of plants (utility-scale and rooftop) and considers two different energy mixes (European and Chinese) for the manufacturing of PV components (mono-Si ingot, wafer, cell and module), in order to analyze different scenarios and support the decision makers. The impacts are evaluated by considering the entire life cycle of the plant (from raw materials acquisition to PV plant decommissioning and disposal of materials) and several impact categories, including greenhouse gas emissions.
According to the results, the utility-scale PV plant appears beneficial with respect to the rooftop PV plant, allowing a reduction higher than 20% in most of the categories analyzed. Concerning the energy mix used in the PV component manufacturing, the outcomes highlight its relevant influence on the results; moving the entire silicon value chain to the European continent, whose electricity mix is characterized by a higher percentage of renewable energy instead of Chinese electricity mix, provides a remarkable increase in the environmental performances (516 $gCO_2eq./W_{DC}$ vs 714 $gCO_2eq./W_{DC}$).
Keywords: PERC module, PV Plant, Life Cycle Assessment

1 INTRODUCTION

In recent years, decarbonization policies are pushing for a significant increase in renewable energies for power generation applications. In 2022, as a part of REPowerEU plan, the European commission defined an EU solar energy strategy, which aims to identify remaining barriers and challenges in the solar energy sector in order to accelerate the deployment of solar technologies. The goal is to reach a total capacity of solar photovoltaic equal to over 320 GW by 2025 and almost 600 GW by 2030.
Thanks to easy installation and versatility (as it exploits global radiation and not only direct radiation as in the case of concentrator photovoltaics) with respect to other technologies, photovoltaic systems will be crucial to replace fossil fuels at the required speed.
Regarding currently installed technologies in Italy, polycrystalline silicon modules are the most widespread technology (approximately 65% in 2022), followed by monocrystalline silicon (30% in 2022). The amount of amorphous silicon (2%) and thin-film technologies are negligible [1]. Monocrystalline silicon technology includes a wide range of cell types, which differ on the position of contacts or specific treatments that increase the cell's efficiency and make it more suitable for particular installations. Currently, due to the higher efficiency and without an increase of production costs with respect to Al-BSF the most commercialized technology is the Passivated Emitter Rear Cell (PERC) technology which accounts for 78% of the world market in 2022 (**Figure 1**). [2]

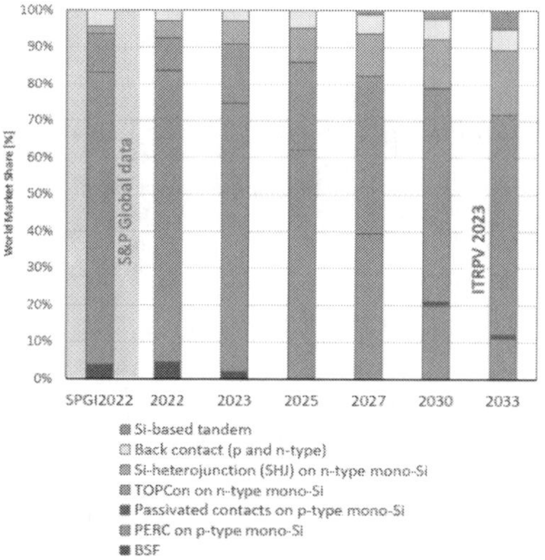

Figure 1 World market share for different cell technology. [2]

In this context, considering the important role of the PV sector to face climate change issues, it is not only important to figure out the main hotspots of PV system but also understand which is the best configuration from the environmental point of view.

The analysis aims to understand the environmental benefit introduced by the production of 1 kWh through two different plants configuration: a ground-mounted PV plant (84 MW) and a rooftop PV plant (3 kW). In this paper, the PERC (Passivated Emitter and Rear Cell) technology, among the most efficient on the market, is investigated; its use is assumed in both the case studies, utility scale and rooftop plant. The study is carried out by applying the Life Cycle Assessment methodology in accordance with ISO [3] and ISO 14044 [4] standards, and IEA-PVPS guidelines [5]. Both case studies are located in Italy and for each of them the entire life cycle, from the raw material extraction to the plant decommissioning and disposal, is considered. In addition, to assess the influence of the energy mix used in the manufacture of the PV components (mono-Si ingot, wafer, cell and module) on the LCA results, two different production sites, i.e. in China (current production site) and in Europe (hypothetical future production site), are explored.

2 MATERIALS AND METHODS

The study was performed according to the Life Cycle Assessment methodology which includes four different steps: (i) Goal and scope definition; (ii) Life cycle inventory; (iii) Life cycle impact assessment (LCIA); (iv) Interpretation of the results.

2.1 Goal and scope definition

The study aims to investigate the potential environmental impact generated by PV energy production through two types of plants (utility scale and rooftop) and considers two different energy mixes (European and Chinese) for the manufacturing of PV components (mono-Si ingot, wafer, cell and module), in order to analyze different scenarios and support the decision makers.

The study was performed in accordance with ISO 14040 e ISO 14044 standards and IEA-PVPS LCA guidelines [5], and considers as functional unit, that is a measure of the function of the system under study and it represents the unit to which all inputs and outputs of the studied system have to be referred to, 1 kWh produced. The F.U adopted allows to compare the impacts obtained in this study with the ones generated by other technologies. Regarding the systems boundary, the analysis considers all the life cycle of the two plants from the raw materials extraction to the end of life (EOL) (**Figure 2**). The last phase (EOL), due to the lack of specific information, includes only the modules' treatment.

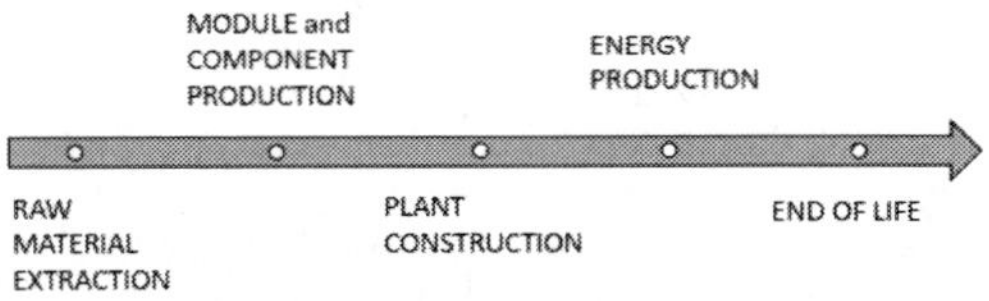

Figure 2 Life Cycle phases include in the analysis

The potential environmental impacts have been evaluated by resorting to the midpoint indicators of the Product Environmental Footprint Category Rules (PEFCR). The only exception is the ionizing radiation category which is omitted as it is not relevant for the system investigated.

The attributional approach [6], which is the most prevalent type used in the published LCAs [7], has been adopted.

2.2 Life Cycle Inventory (LCI)

Concerning the LCA model, the data involved in the analysis were acquired thanks to the collaboration with some Italian companies that realize and market PERC modules and, also to the participation in the Europe project GOPV, whose consortium includes photovoltaic operators and inverter and tracker manufacturers.

The LCA study considered two hypothetical photovoltaic plants (i.e. a Rooftop PV plant and a Utility-scale PV plant), equipped with PERC modules.

The two plants have a peak power of 3 kW and 84.73 MW respectively and a life time equal to 30 years defined according with the IEA-PVPS guidelines [5]. Both systems include the following components:

1. PERC modules;
2. Modules support structure (in the utility-scale plant, modules are mounted on fixed-tilt structure manufactured with weathering steel while in the rooftop plant the support structure is realized in aluminum);
3. High efficiency inverter

The main technical data about the PERC modules, inverter and modules support are reported from **Table 1** to **Table 4**.

Table 1 Technical data about the module

	PERC module
Cells	144 PERC half-cut cells
Dimension (mm)	2,094 x 1,038 x 35 mm
Area (m^2)	2.17
Lifetime (years)	30
Weight (kg)	23.8
Nominal power (W)	460
Efficiency	21.16%
Number of modules per plant	Utility-scale plant: 184.196 Rooftop plant: 7

Table 2 Main data regarding module support structure adopted in the utility-scale PV plant

	Modules support structure
Typology	Support for 26 modules in portrait configuration supported by 5 piles. Made of Corten steel
Number of support structure per plant	7085
Dimension	14 x 4.2 m
Life (years)	35
Fixed-tilt support weight (kg)	953.8
Mono-axial tracker weight (kg)	1082
Producer	Convert Italia SPA

10.4229/EUPVSEC2023/5CO.6.4
020496-002

Table 3 Main data regarding the Inverter adopted in the utility-scale PV plant

	Inverter
Typology	166 kVA Inverter (two stage)
Number per plant	426
Dimension	
Life (years)	17
Weight (kg)	100
Producer	REFU Elektronik

Table 4 Main data regarding structure and Inverter adopted in the rooftop PV plant

	Modules support structure*	Inverter
Typology	Aluminium support	3 kVA Inverter
Number	7	1
Life (years)	30	17
Weight (kg)	-	20.8
Producer	-	Elettronica Santerno S.p.A.

**The structure was modelled according to the ecoinvent v 3.8 database.*

The lifetime of the system, which is relevant parameter in PV LCAs [8] is set to 30 years, according to the IEA-PVPS guidelines [5].

The electricity produced during the operation time has been estimated using a model which considers, the data derived from PVGIS database [9], modules degradation, soling losses, thermal losses etc. Modules degradation is 3% the first year, 0.4% from the second to the twentieth year, and 0.5% in the following years. In the inventory phase, both primary data (provided by some manufacturers) and secondary data from Ecoinvent database v3.8, cut-off system model have been used [10].

3 RESULTS

The results of the study, referred to the production of 1 kWh by a PV plant located in Catania (South Italy), are here presented. To really understand which is the most sustainable PV plant, the rooftop plant is compared with two hypothetical utility-scale PV plants which are distinguished by the module structure (mono-axis solar tracker and fixed-tilt structure). The three plants are installed in the same location and are characterized by the same lifespan (30 years). The outcomes reveal that the utility-scale plant performs better than the rooftop plant in 10 out of 11 environmental indicators; the only exception is the Land use category. In fact, the installation of module on the building surface, like, roof or façade, do not require dedicated areas. On the other side, except the Land use category, the utility scale plant, thanks to the more favorable conditions, generates more energy that contributes to an improvement of the environmental performance with respect to a rooftop PV plant.

Table 5 Environmental impacts generated by the PV analyzed: Rooftop PV plant, Utility Scale PV plant equipped with fixed-tilt structure (US – FIXED TILT) and Utility Scale PV plant equipped by tracker (US TRACKER).

Unit	ROOFTOP	US - FIXED TILT	US - TRACKER
Climate change (kg CO2 eq)	2.50E-02	1.97E-02	1.65E-02
Ozone depletion (kg CFC11 eq)	1.80E-09	1.71E-09	1.51E-09
Photochemical ozone formation (kBq U-235 eq)	9.19E-05	7.23E-05	6.11E-05
Particulate matter (kg NMVOC eq)	1.68E-09	1.28E-09	1.03E-09
Acidification (disease inc.)	1.66E-04	1.19E-04	1.02E-04
Eutrophication, freshwater (CTUh)	1.01E-05	6.74E-06	6.40E-06
Eutrophication, marine (CTUh)	3.66E-05	3.06E-05	2.64E-05
Eutrophication, terrestrial (mol H+ eq)	3.18E-04	2.54E-04	2.13E-04
Land use (kg P eq)	9.96E-02	4.47E-01	3.84E-01
Resource use, fossils (kg N eq)	2.78E-01	2.31E-01	2.01E-01
Resource use, minerals and metals (mol N eq)	1.19E-06	8.79E-07	7.99E-07
CED (MJ)	3.25E-01	2.68E-01	2.33E-01

3.1 Contribution analysis

Focusing on the Climate change category (**Figure 3**), the contribution analysis reveals that the module is the most impactful component More specifically, the environmental burdens associated with the wafer production (from silicon extraction and treatment to wafer manufacturing) represent the main criticality of the entire PV plant life cycle. In the Resource use, minerals and metals impact category (**Figure 4**), the module and the inverter, due to the precious metals used in the electronic manufacturing, are the main contributors in the generation of impacts.

10.4229/EUPVSEC2023/5CO.6.4
020496-003

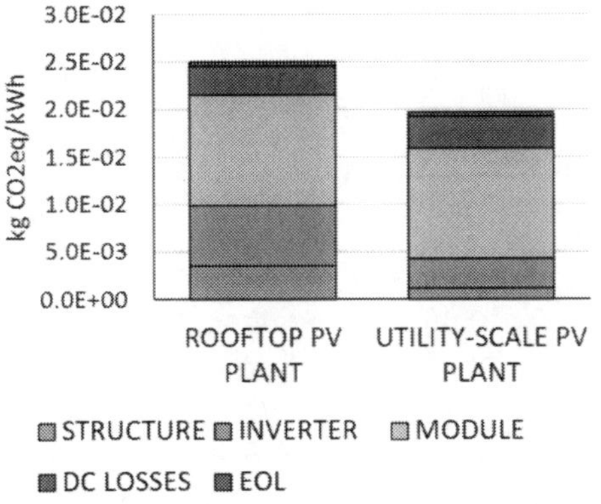

Figure 3 Contribution analysis related to the Climate Change category. The values are expressed per FU (1 kWh produced)

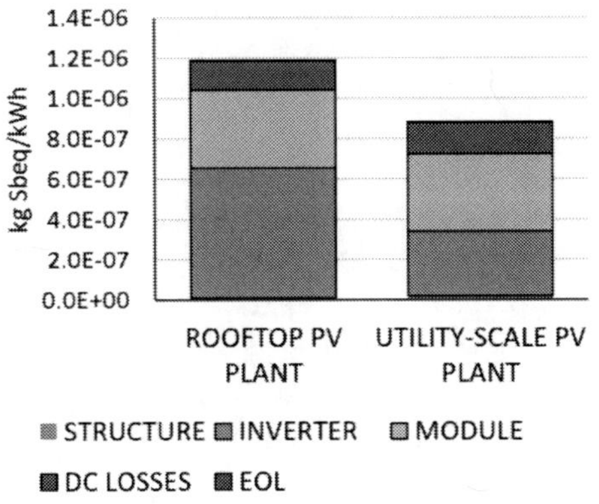

Figure 4 Contribution analysis related to the Resource Use, minerals and metals category. The values are expressed per FU (1 kWh produced)

3.2 Module scenarios

In order to analyze the impact variation achievable thanks to the relocation in Europe of the entire supply-chain of the module, three different module production scenario have been analyzed:

1. Scenario A: in which modules, cells and wafer are produced in China
2. Scenario B: modules and cells are produced in Europe while wafer is produced in China
3. Scenario C: modules, cells, and wafer are produced in Europe.

Table 6 shows the percentage variation of the environmental burden associated with the module production among the current situation (Scenario A) and the other two scenarios (Scenario B and Scenario C). The impacts variation was calculated according to eq. 1, where X is related to Scenario B or Scenario C

$$\Delta_{AX} = \left(sc_A - SC_X\right) / SC_A \qquad \text{eq. 1}$$

According to the outcomes, it can be noticed that the use of more clean electricity mix, used in the modules, cells and wafer production (Scenario C), brings a significant increase in the environmental performances (reduction ≥ -20% of the impacts in 6 out of 12 categories). This highlights the relevance of the composition of the energy mix used in the silicon supply chain on the results. Focusing on Scenario B, the variation of the impact in function of the energy mix considered is negligible (less than 10%): this is because the energy consumption in assembling the module is significantly less than that required in manufacturing the wafer and cell.

Table 6 Analysis of the production module locations. The results are expressed as a variation of the impacts generated by the production of 1 module with respect to the Scenario A.

IMPACT CATEGORIES	ΔAB	ΔAC
Climate change (kg CO2 eq)	-3.40%	-27.70%
Ozone depletion (kg CFC11 eq)	+0.60%	+6.10%
Photochemical ozone formation (kBq U-235 eq)	-3.80%	-32.10%
Particulate matter (kg NMVOC eq)	-5.90%	-44.00%
Acidification (disease inc.)	-3.20%	-29.30%
Eutrophication, freshwater (CTUh)	+3.40%	+30.40%
Eutrophication, marine (CTUh)	-2.70%	-22.80%
Eutrophication, terrestrial (mol H+ eq)	-4.10%	-34.10%
Land use (kg P eq)	-1.00%	-11.80%
Resource use, fossils (kg N eq)	-0.50%	-8.40%
Resource use, minerals and metals (mol N eq)	0.00%	-0.10%
CED (MJ)	-0.20%	-4.80%

4 CONCLUSIONS

To assess the reduction of GHG emissions and other environmental impacts, achievable by the increase of the contribution of PV electricity production in the energy sector, a Life Cycle Assessment was performed. The two plants analyzed, i.e. PV utility-scale plant and rooftop PV plant were modelled by using primary data collected from the PV modules, tracker and inverter manufacturers.

Focusing on the results, the utility-scale PV plant appears beneficial compared to rooftop plant, thanks to the higher electricity production, however due to land consumption during the PV utility-plant installation the latter brings an increase of the environmental burden in the Land use category (+55%). According to the comparison of the three plants configuration a further impacts reduction is achievable if the modules are mounted on mono-axial tracker instead of fixed-tilt structure.

The contribution analysis reveals the higher impacts associated with the module supply chain, especially due to the intensive energy consumption Czochralski process, widely used to produce the wafer. In this context from sensitivity analysis on the location of the production site emerges that the module and in particular wafer production in a country with a greener energy mix can lead to an improvement of the environmental benefits (-20% of the impacts in the Climate change category). Besides modules, electronics in the inverter contributes to the environmental burden in the Resource use, mineral and metals category.

4.3 References

[1] «RAPPORTO STATISTICO - SOLARE FOTOVOLTAICO». Gestore dei Servizi Energetici S.p.A. (GSE), aprile 2023.

[2] VDMA - Photovoltaic Equipment, «International Technology Roadmap for Photovoltaic (ITRPV) 2022 Results. 14. Edition». VDMA - Photovoltaic Equipment, aprile 2023. [Online]. Disponibile su: https://www.vdma.org/international-technology-roadmap-photovoltaic

[3] ISO, «The International Organization for Standardization, ISO 14040:2006: Environmental management - Life Cycle Assessment - Principles and framework.» 2006.

[4] ISO, «The International Organization for Standardization, ISO 14044:2006: Environmental management - Life Cycle Assessment - Requirements and Guidelines.» 2006.

[5] R. Frischknecht *et al.*, «Methodology Guidelines on Life Cycle Assessment of Photovoltaic 2020». IEA-PVPS T12-18: 2020, aprile 2020.

[6] T. Ekvall, A. Azapagic, G. Finnveden, T. Rydberg, B. P. Weidema, e A. Zamagni, «Attributional and consequential LCA in the ILCD handbook», *Int. J. Life Cycle Assess.*, vol. 21, fasc. 3, pp. 293–296, mar. 2016, doi: 10.1007/s11367-015-1026-0.

[7] D. Hsu *et al.*, «Life Cycle Greenhouse Gas Emissions of Crystalline Silicon Photovoltaic Electricity Generation», *J. Ind. Ecol.*, vol. 16, pp. S122–S135, mar. 2012, doi: 10.1111/j.1530-9290.2011.00439.x.

[8] V. Muteri *et al.*, «Review on life cycle assessment of solar photovoltaic panels», *Energies*, vol. 13, fasc. 1, 2020, doi: 10.3390/en13010252.

[9] T. Huld, R. Müller, e A. Gambardella, «A new solar radiation database for estimating PV performance in Europe and Africa», *Sol. Energy*, vol. 86, fasc. 6, pp. 1803–1815, giu. 2012, doi: 10.1016/j.solener.2012.03.006.

[10] G. Wernet, C. Bauer, B. Steubing, J. Reinhard, E. Moreno Ruiz, e B. Weidema, «The ecoinvent database version 3 (Part I): Overview and methodology», *Int. J. Life Cycle Assess.*, vol. 21, pp. 1–13, set. 2016, doi: 10.1007/s11367-016-1087-8.

10.4229/EUPVSEC2023/5CO.6.4
020496-005

PROCESSES OF CO_2 CAPTURE AND REDUCTION TO SOLAR FUELS COMPATIBLE WITH PHOTOVOLTAICS

A. González del Valle, P. García-Linares and A. Martí
Instituto de Energía Solar – Universidad Politécnica de Madrid
ETSI Telecomunicación, Ciudad Universitaria, 28040, Madrid, Spain.

ABSTRACT: Photovoltaics can be used to provide the energy needed to produce different chemicals from CO_2. This process is often known as solar fuel production and encompasses three subprocesses: (a) the photovoltaic conversion itself, (b) CO_2 capture and (c) the electrochemical reduction of CO_2 (CO2RR) to fuels. In this framework we study: two technologies for sub-process (b), one electrochemical and the other based on a phase-separation method, which we have replicated experimentally and four technologies for sub-process (c). We then propose the optimal method for sub-process (b) on the basis of the CO_2 capture ratio and the simplicity of the set-up, and the optimal technology for sub-process (c) based on the economic profitability of the produced solar fuel.

Keywords: direct air capture, carbon dioxide reduction reaction, isophorone diamine, carbamic acid, electrolyzer

1 INTRODUCTION

The photovoltaic production of solar fuels faces several challenges. The first of these challenges lies in the obtention of CO_2. In this regard, CO_2 can be captured from industrial point sources, where it is highly concentrated, or directly from air (DAC). Capturing CO_2 from industrial sources offers the advantage of high CO_2 concentration; however, it comes with the drawback of potential contaminants that must be eliminated before supplying the CO_2 to the electrochemical solar fuel production systems. These contaminants are specific to each industrial process, demanding a purification process adapted to the specific industrial process. On the other hand, supplying CO_2 from DAC offers the advantage of providing CO_2 that is nearly free of these contaminants. Moreover, DAC has the potential to enable the development of a universal process that could be deployed worldwide with only minor modifications, making it suitable for use across diverse geographic locations on Earth, and reducing the global atmospheric concentration of CO_2 [1]. However, DAC technologies are limited by the low concentration of CO_2 in the atmosphere (approx. 400 ppm [2]) which demands the consumption of energy to separate CO_2 from other air components. In section 2 of this work, we shall review the operation of a DAC system based on the use of K_2SO_4 and capable of capturing CO_2 from air using electrochemistry [3] and another DAC system based on the phase separation of liquid isophorone diamine (IPDA) and the corresponding solid carbamic acid (CA1) [4].

Another challenge of solar fuels production is associated with the development of appropriate electrolyzers, a concern that becomes significant when considering the reduction reaction of CO_2 (CO2RR). These reactions generally produce various organic species (CO, CH4, HCOOH, C2H2, CH3OH, etc.) [5]. Far from being advantageous, this lack of selectivity towards a specific product, is CO2RR biggest problem. The selectivity can be enhanced and it heavily depends on factors such as the electrolyzer configuration, materials of the electrodes and membranes, the reactants and solvents used on the reaction. In section 3 of this work, we shall review the operation of four CO2RR methods, which respectively produce carbon monoxide (CO), methane (CH_4), formic acid (HCOOH) and solid carbon (C).

Finally, we will outline the fundamentals of a thermodynamic analysis aimed at determining the cost competitiveness of organic compounds based on the energy required for their production.

2 DIRECT AIR CAPTURE TECHNOLOGIES

2.1 Electrochemical DAC method

One example of DAC based on electrochemistry has been provided by Wang et al [3]. The process occurs within a three-chamber electrolyzer, with the anodic chamber including the anode, a middle chamber, and the cathodic chamber accommodating the cathode. Since this approach uses directly electricity, it can be powered using photovoltaics (PV).

The process begins (Fig.1a) with an aqueous solution of K_2SO_4 in the middle chamber, which is separated from the other chambers by selective membranes. One of them is selective to cations (CEM), and allows K^+ to pass through towards the cathodic chamber. The other membrane (AEM) is selective to anions, thus permitting the SO_4^{2-} anions to cross towards the anodic chamber. The process follows with simultaneous reactions occurring at the anode and cathode. At the anode, two water molecules undergo oxidation reaction, releasing four protons into the aqueous solution, one oxygen molecule into the air, and four electrons into the electric circuit:

$$2H_2O \rightarrow 4H^+ + O_2 + 4e^- \quad (1)$$

The H^+ and the SO_4^{2-} form sulfuric acid. Simultaneously, the solar generator rises the electrochemical potential of the electrons liberated at the anode and transports them to the cathode, where these electrons reduce H_2O to produce H_2 and OH^- anions:

$$4H_2O + 4e^- \rightarrow 4\,OH^- + 2H_2 \quad (2)$$

A flow of air is fed into the cathodic chamber, where the CO_2 contained in the air reacts with the KOH formed in the cathodic reaction (2). The reaction between these species produces both K_2CO_3 (3) and $KHCO_3$ (4):

$$2CO_2 + 2K^+ + 2\,OH^- \rightarrow 2KHCO_3 \quad (3)$$
$$CO_2 + 2K^+ + 2\,OH^- \rightarrow K_2CO_3 + H_2O \quad (4)$$

K_2CO_3 and $KHCO_3$ diffuse towards an additional chamber

10.4229/EUPVSEC2023/5CO.6.5
020498-001

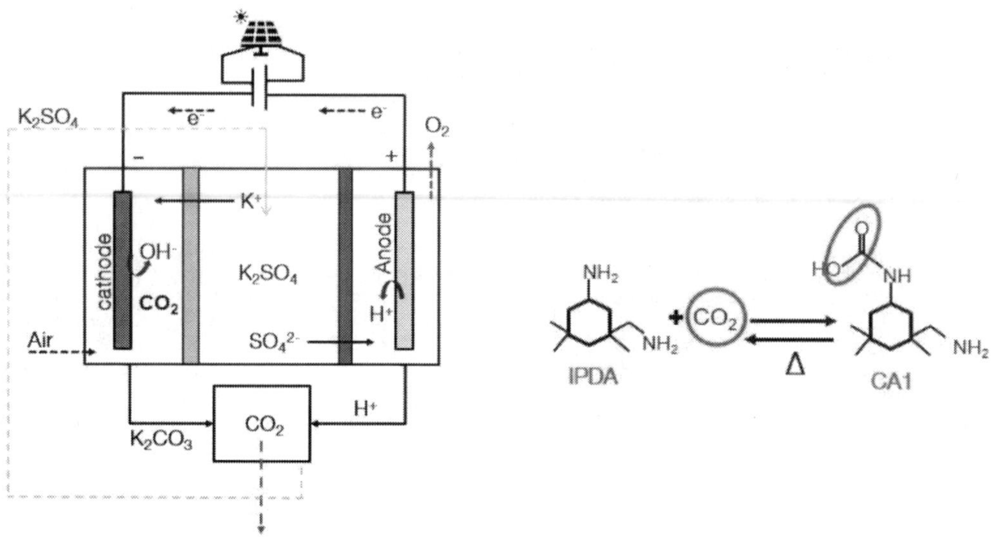

(a) (b)

Fig.1(a) *Scheme of electrochemical matrix for CO_2 enrichment using electrochemically generated OH^-/H^+ (adapted from Wang et al* [3]). (b) *Formation of the CA1 and dissociation into IPDA and CO_2 after heating.* (b) *Formation of the CA1 and dissociation into IPDA and CO_2 after heating. Carbon atom from CO_2 bonds with more accessible nitrogen atom from IPDA, capturing CO_2.*

where they encounter H^+ and SO_4^{2-} ions that diffused from the anodic chamber and the following reactions, releasing CO_2, takes place:

$$K_2CO_3 + 2H^+ + SO_4^{2-} \rightarrow CO_2 + H_2O + K_2SO_4 \quad (5)$$

$$2KHCO_3 + 2H^+ + SO_4^{2-} \rightarrow 2CO_2 + 2H_2O + K_2SO_4 \quad (6)$$

The resulting K_2SO_4 can be then recycled into the middle chamber.

This process requires at least a voltage difference of 2.1 V between anode and cathode for reactions (1) and (2) to occur. In this process 1.58×10^{-4} $molh^{-1}$ of CO_2 are captured, with an energy consumption of 1.68×10^{-5} kWh, that is 9.40 mol/kWh [3]. In other words, and for illustrative reasons, this technique would require 2,420 kWh to separate 1 CO_2 ton from air, i.e., more than 17 times the theoretical minimum energy of 145 kWh, according to the ideal reference values we have previously calculated [6].

2.2 Phase-separation DAC method

The second DAC technology is based on the phase separation between liquid amine and the solid carbamic acid formed through the absorption of CO_2 by the amine. Kikkawa et al. reported that the best performing amine for this purpose is liquid isophorone diamine (IPDA) [4] . The process starts with the bubbling of air into a solution of IPDA on dimethyl sulfoxide (DMSO) (Fig.2a) which is transparent at room-temperature. This drives CO_2 to react with IPDA, forming the corresponding carbamic acid (CA1) (Fig.1b). The carbon atom in CO_2 chemically bonds with the nitrogen atom in IPDA, effectively trapping the airborne CO_2. The CA1 product is insoluble in the IPDA and DMSO solution, resulting in its precipitation as a white solid (Fig. 2b). We have replicated this experiment in our laboratory and observed that after 2 hours, the initially transparent solution becomes cloudy, eventually leading to the formation and precipitation of white solid CA1. After the CO_2 absorption, the phases are immiscible, of which the CO_2-rich phase is suitable to release concentrated CO_2 by heating. In fact, the said chemical bond between carbon and nitrogen atoms breaks when CA1 is heated above 303 K. This reaction, in which CA1 transforms into IPDA and concentrated CO_2 is the reverse reaction to CO_2 capture and is known as CO_2 recovery. The captured CO_2 was completely desorbed at 333 K, as Kikkawa et al. report . It is a cyclical process, as IPDA is fully reusable for more than 5 heating/cooling cycles. The removal efficiency is superior to 99%, and the capture rate is 2.01×10^{-1} $molh^{-1}$ [4].

Comparing the two DAC methods, the phase separation method captures 1,270 times more CO_2 per hour than the electrochemical method. In addition, the phase separation method uses simpler equipment compared to the three-chamber electrolyzer used on the electrochemical DAC method. Additionally, the operation of the phase-separation method is more straightforward, as it only involves cyclically heating and cooling a solution. These various reasons make the phase separation method more appealing as a DAC technology.

3 CARBON DIOXIDE REDUCTION REACTION

The CO2RR involves reducing CO_2 (previously concentrated) to species with fewer oxygen atoms and more hydrogen atoms, or even just carbon. Although CO2RR can be achieved using technologies of different natures, it typically takes place in electrolyzers. This sub-process often generates different products within the same electrolyzer, making it a low-selective process. In fact, the lack of selectivity towards the production of a specific solar fuel is the primary challenge this process faces.

10.4229/EUPVSEC2023/5CO.6.5

Table I: Electricity cost per metric ton (t) of product assuming an electricity price of 35 €MWh^{-1} and that the minimum amount of energy allowed by thermodynamics is used to synthetize the product from CO_2 via the methods above stated. For the calculations the following data have been assumed: $\Delta_f G^0 = -394.359$ kJmol^{-1} for CO_2(g); $\Delta_f G^0 = -237.19$ kJmol^{-1} for H_2O(l); $\Delta_f G^0 = 0$ kJmol^{-1} for O_2 and H_2;[7].

Product	Reaction	Market price (€t^{-1})	$\Delta_f G^0$ (kJmol^{-1}) [8]	ΔW (kJmol^{-1})	Product molar mass, m (g) [8]	Electricity cost (€t^{-1})
Carbon monoxide (CO)	$CO_2(g)+H_2O(l) \rightarrow CO(g)+O_2(g)+H_2(g)$	184 [9]	-137.3	542.9	28.01	189
Methane (CH_4)	$CO_2(g)+4H_2(g) \rightarrow CH_4(g)+2H_2O(l)$	496 [10]	-50.72	808.02	16.04	490
Formic acid (HCOOH)	$2H_2O(l)+CO_2(g) \rightarrow 2HCOOH(l)+O_2(g)$	838[11]	-385.8	294.4	46.03	62
Carbon (C)	$CO_2(g) \rightarrow 2C(s)+O_2(g)$	32[12]	0	394.4	12.01	320

Nevertheless, selectivity can be improved, and it highly depends on the configuration of the electrolyzer, the materials used, especially electrode metals and selective membranes, as well as the reactants and solvents employed in the reaction. The election of the most suitable CO2RR technology will be based on the thermodynamic and the cost analysis of the solar fuel production.

3.1 Solar fuel production methods

The first CO2RR technology considered was proposed by Rosen et al. for CO production (1), and it takes place in an electrolyzer that uses an ionic liquid electrolyte (Emim-BF4) to decrease the formation energy of a reaction intermediate (CO2)$^-$, reducing the energy barrier of the reaction. The reaction intermediate is reduced to CO at the silver cathode and H_2O is also formed [13].

(a) (b)

Fig.2 *Photograph of* (a) *a DMSO solution of DMSA under air flow* (b) *of the corresponding CA1 after IPDA captures* CO_2*.*

For methane (CH_4) production from CO_2 (Eq. 2), there is a selectivity problem, which is lower due to C-H bond strength and the requirement of an 8-electron transfer step. Sun et al. proposed an electrochemical system with n-doped graphene-like materials (NGMs) electrodes, since they are chemically stable and low-cost in large-scale applications, also using ionic liquid (Bmim-BF4) as electrolyte. They also reported CO and H_2 as byproducts of the reaction when not optimal voltage is applied [14].

Formic acid (HCOOH) production by CO_2 reduction has previously been unsuccessful, but Zhu et al. proposed an electrochemical method based on Pb/Sn cathodes and ionic liquids (Bmim-PF$_6$) as electrolytes (Eq.3), with which they manage to produce HCOOH at 2.2 V. It was also reported that the presence of water traces is beneficial because it decreases electrode overpotentials and increases electrolyte conductivity [15]. All the above mentioned CO2RR technologies are electrochemical and can be directly powered via PV generators.

However, CO2RR can occur also through non-electrochemical technology. An example of such an approach was proposed by Zuraiqi et al. who use a method based on a liquid metal alloy (LM) with low melting point, such as eutectic GaIn (EGaIn). A concentrated stream of CO_2 is bubbled into the LM while being heated to 200°C. As a result of this reaction (4), the CO_2 is reduced to solid carbon, which can be easily separated from the liquid medium [16]. This is highly interesting from an environmental point of view.

4 THERMODYNAMIC AND COST ANALYSIS

Thermodynamics enables the calculation of the minimum energy required to produce a specific solar fuel from concentrated CO_2 and other reactants.

Let us consider the production of CO using the method proposed by Rosen et al. Initially, we have a system comprising 1 mole of CO_2 and 1 mole of H_2O in thermodynamic equilibrium with a pressure and thermal reservoir at 1 atm and 298 K (representing the atmosphere). Subsequently, the system is connected to a work source, an ideal PV generator, and allowed to undergo a reversible and quasi-static evolution to reach a state in which 1 mole each of CO, O_2, and H_2 are present, as illustrated in Eq.1. The reversible evolution of the system and the reservoir implies that, when considered as a whole, their total entropy does not increase, although entropy can be exchanged between them. It's important to note that work sources, by definition, do not introduce entropy into the system they are connected to.

In this context, according to thermodynamics, the minimum amount of work, ΔW, that the work source must provide to carry out the described process per mol of CO is determined by the change in Gibb's free energy:

$$\Delta W = \Delta_f G^0[\mathrm{CO}(g)]+\Delta_f G^0[\mathrm{O_2(g)}]+\Delta_f G^0[H_2(\mathrm{g})] - \Delta_f G^0[\mathrm{CO_2(g)}] - \Delta_f G^0[\mathrm{H_2O(l)}] \quad (7)$$

where $\Delta_f G^0$ is the Gibbs energy of formation of the compound indicated in square brackets. The resulting figure, taking data from [7] for CO is ΔW=542.9 kJmol^{-1}.

10.4229/EUPVSEC2023/5CO.6.5
020498-003

Obtaining 1 ton of CO, rather than 1 mole (28.01 g), would require 5.4 MWh of energy. Considering that photovoltaic modules in the market typically have an efficiency of around 21%, this translates to a cost of approximately 35 €MWh^{-1} of electricity when considering levelized cost of electricity (LCOE) for utility-scale crystalline silicon solar systems [17]. Therefore, producing 5.4 MWh of PV electricity would cost approximately 189 €. Comparing this cost to the current market price of CO, which is 184 €t^{-1},[9] it becomes evident that, based on thermodynamics, the production of CO from CO_2 and H_2O is not economically viable at present. This straightforward analysis leads to the conclusion that, given the current costs, it would not be feasible to produce CO as a solar fuel at competitive prices.
Table I illustrates the results of the calculation above for the case of CH_4 (resulting in a "thermodynamic" minimal cost of 490 €t^{-1} vs 496 €t^{-1} market cost), formic acid (HCOOH) (resulting in a "thermodynamic" minimal cost of 62 €t^{-1} vs 838 €t^{-1} market cost). In these cases, thermodynamics permits the production below current market prices with a margin of 6 and 776 €t^{-1}, respectively, to be spent in non-idealities of the system as well as CAPEX and OPEX costs. The production of solid carbon results in negative balance of -287.5 €t^{-1} since its production would lead to a cost of 319.5 €t^{-1} and its market fluctuates around 32 €t^{-1}.

Considering the above, the production of HCOOH via the electrochemical CO2RR technology seems the most suitable, as the product is the most profitable of all the solar fuel considered.

5 SUMMARY

In this work we have studied two technologies for DAC capture, one being electrochemical and the other based on phase-separation method. The latter is most suitable to scale-up to industry level due to better CO_2 capture rate and simpler set-up and functioning compared to electrochemical method. We have replicated the phase-separation method on laboratory and future work will let us determine the energy consumption of the technology. We have also described four CO2RR technologies for solar fuel production, specifically, CO, CH_4, HCOOH and solid carbon. We have also described the fundamental thermodynamic analysis that allows to calculate the minimum amount of energy needed to carry out the processes of the said CO2RR technologies and translate this energy into electricity cost. Assuming photovoltaic LCOE of 35 €MWh^{-1}, the most profitable solar fuel production method is the one proposed by Zhu et al.

5 REFERENCES

[1] E. S. Sanz-Pérez, C. R. Murdock, S. A. Didas, y C. W. Jones, «Direct Capture of CO 2 from Ambient Air», *Chem. Rev.*, vol. 116, n.º 19, pp. 11840-11876, oct. 2016, doi: 10.1021/acs.chemrev.6b00173.

[2] K. Z. House, A. C. Baclig, M. Ranjan, E. A. van Nierop, J. Wilcox, y H. J. Herzog, «Economic and energetic analysis of capturing CO 2 from ambient air», *Proc. Natl. Acad. Sci.*, vol. 108, n.º 51, pp. 20428-20433, dic. 2011, doi: 10.1073/pnas.1012253108.

[3] C. Wang, H. Liu, X. Li, y L. Zheng, «Importance of Ambient O 2 for Electrochemical Enrichment of Atmospheric CO 2», *Ind. Eng. Chem. Res.*, vol. 52, n.º 7, pp. 2470-2476, feb. 2013, doi: 10.1021/ie302991y.

[4] S. Kikkawa *et al.*, «Direct Air Capture of CO 2 Using a Liquid Amine–Solid Carbamic Acid Phase-Separation System Using Diamines Bearing an Aminocyclohexyl Group», *ACS Environ. Au*, vol. 2, n.º 4, pp. 354-362, jul. 2022, doi: 10.1021/acsenvironau.1c00065.

[5] G. Wang *et al.*, «Electrocatalysis for CO 2 conversion: from fundamentals to value-added products», *Chem. Soc. Rev.*, vol. 50, n.º 8, pp. 4993-5061, 2021, doi: 10.1039/D0CS00071J.

[6] A. Martí *et al.*, «CEOTRES: A PROJECT FOR THE PHOTOVOLTAIC GENERATION OF SOLAR FUELS».

[7] H. B. Callen, «Introducción a las teorías físicas de la termostática del equilibrio y de la termodinámica irreversible».

[8] D. D. Wagman, W. H. Evans, V. B. Parker, R. H. Schumm, y I. Halow, «The NBS Tables of Chemical Thermodynamic Properties. Selected Values for Inorganic and C1 and C2 Organic Substances in SI Units», NATIONAL STANDARD REFERENCE DATA SYSTEM, ene. 1982. Accedido: 1 de julio de 2022. [En línea]. Disponible en: https://apps.dtic.mil/sti/citations/ADD095481

[9] G. A. O. Tiago, I. A. S. Matias, A. P. C. Ribeiro, y L. M. D. R. S. Martins, «Application of Ionic Liquids in Electrochemistry—Recent Advances», *Molecules*, vol. 25, n.º 24, p. 5812, dic. 2020, doi: 10.3390/molecules25245812.

[10] «Methane market price». https://www.methanex.com/our-business/pricing (accedido 18 de enero de 2023).

[11] «Formic Acid Price Trend and Forecast». https://www.chemanalyst.com/Pricing-data/formic-acid-1242

[12] «Value of Carbon Market Update 2021». https://carboncreditcapital.com/value-of-carbon-market-update-2021-2/

[13] B. A. Rosen *et al.*, «Ionic Liquid–Mediated Selective Conversion of CO 2 to CO at Low Overpotentials», *Science*, vol. 334, n.º 6056, pp. 643-644, nov. 2011, doi: 10.1126/science.1209786.

[14] X. Sun *et al.*, «Very highly efficient reduction of CO 2 to CH 4 using metal-free N-doped carbon electrodes», *Chem. Sci.*, vol. 7, n.º 4, pp. 2883-2887, 2016, doi: 10.1039/C5SC04158A.

[15] Q. Zhu *et al.*, «Efficient Reduction of CO 2 into Formic Acid on a Lead or Tin Electrode using an Ionic Liquid Catholyte Mixture», *Angew. Chem. Int. Ed.*, vol. 55, n.º 31, pp. 9012-9016, jul. 2016, doi: 10.1002/anie.201601974.

[16] K. Zuraiqi *et al.*, «Direct conversion of CO 2 to solid carbon by Ga-based liquid metals», *Energy Environ. Sci.*, vol. 15, n.º 2, pp. 595-600, 2022, doi: 10.1039/D1EE03283F.

[17] D. Ray, «Lazard's Levelized Cost of Energy Analysis—Version 15.0», 2021.

This presentation was selected by the Sc. Committee of the EU PVSEC 2023 for submission of a full paper to one of the EU PVSEC's collaborating peer-reviewed journals.

MORE THAN RECYCLING: HOW SHOULD WE DEFINE CIRCULARITY GOALS FOR PV IN A GLOBAL ENERGY TRANSITION?

[1,2]Heather Mirletz, [2]Silvana Ovaitt, [2]Teresa Barnes, [3]Seetharaman Sridhar
[1]National Renewable Energy Laboratory, Colorado, US, [2]Advanced Energy Systems Graduate Program, Colorado School of Mines, Colorado, US; [3]Ira A. Fulton Schools of Engineering; Arizona State University, Arizona, US

ABSTRACT: Energy transition to carbon-free electricity is a crucial pillar of the Circular Economy [1]. Renewable energy reduces environmental impacts and decarbonizes the production of other goods. But, manufacturing renewable energy sources, such as photovoltaic (PV) modules, require energy inputs that are currently carbon intensive. So, how do we decarbonize and circularize these critical technologies to achieve a sustainable energy transition? This work proposes that effective capacity—the installed capacity accounting for degradation rates and failures—is a critical metric to evaluate renewable energy technologies on the path toward circular economy and energy transitions. Our analyses also emphasize the importance of examining a suite of metrics incorporating mass and energy flows to identify potential tradeoffs and inform design or lifecycle management decisions holistically.
Keywords: circular economy, energy transition, reliability

1 INTRODUCTION

Proposals for sustainable PV range from high-yield, high-efficiency paradigms, to short-lived but fully recyclable module designs, to long-lasting, reliable, indestructible modules [2]. Our work quantifies the benefits and tradeoffs of various PV design strategies in the energy transition context.

In this study, we analyze the mass and energy flows of several PV technologies on their ability to address energy transition and circular economy challenges, including:

- a high-efficiency module that is open-loop recycled at low rates (e.g., high-efficiency bifacial silicon),
- a short-lived module where components can be remanufactured or closed-loop recycled into a new module (e.g., perovskite),
- a long-lived, re-usable module with low recyclability (e.g. 50-year silicon module),
- a 100% recycled silicon wafer with a lower module efficiency [3]

We compare the ability of these module design scenarios to achieve energy transition deployment schedules and meet capacity targets. Then, we examine the quantities of virgin material demands, lifecycle wastes, energy demands, and energy generation. The module designs are compared on a module product basis and at an energy system scale. The analysis leverages the open-source, peer reviewed PV in Circular Economy (PV ICE) tool [4], [5]. PV ICE can quantify each module's embodied materials and energy and its energy generation capability. Additionally, the PV ICE tool can examine the effect of deploying each module design at the energy system scale, accounting for replacements, and quantifying progress toward circular economy and energy transitions.

2 RESULTS & CONCLUSIONS

Please see the full manuscript of this journal submitted to EUPVSEC EPJ for the expanded set of scenarios, methods, and results for the various mass and energy metrics. The complete paper should not exceed **1 MB including illustrations**. This is a very good proven capacity for final papers. Note that there is no limit in the number of pages. An adequate number is 3 to 6 pages.

3 RESULTS & CONCLUSIONS

Ideally, a renewable energy technology would be long-lived and fully circular—however, such a PV module does not yet exist, and aspects of one design priority may interfere with another (e.g., remanufacturing vs. indestructible). Therefore, this analysis compares PV design choices' material and energy impacts to meet and

Table I: Key parameters of 2022 module designs compared for sustainability in achieving Energy Transition. Module efficiency improves from 2022 to 205

Module Type	Lifetime (years)	Degradation Rate	Circular Features	Module efficiency	Bifacial Factor	Package
PV ICE Baseline [4]	35	0.5%	Downcycling	21%-25%	NA	Marketshare Weighted
SHJ	30	0.5%	Downcycling	23.5%-25.3%	0.9	Glass-glass
Idealized Perovskite Si-Tandem	15	1.47%	98% Remanufacture glass 98% Closed-loop Recycling Perovskite, Si, and Al frame	17.9%-32.5%	NA	Glass-glass
50-year PERC	50	0.445%	None, slowly increase to downcycling	21%-24.5%	0.7	Glass-glass
Recycled Si PERC [3]	25	0.5%	Glass downcycling 100% closed-loop recycled Si 98% closed-loop recycled Al frame	19% - 22%	0.6	Glass-glass

maintain deployment targets, prioritizing PV module improvements and lifecycle management. PV module designs under consideration are described in Table 1. This enables us to explore the extreme boundary cases of potential futures. PV ICE Baseline [4] is used as the comparison point throughout, as it captures historical and current trends of PV modules and represents a conservative projection of future PV module improvements.

This study analyzes these PV technologies' required mass and energy quantities to address energy transition and CE challenges. Furthermore, we compare the ability of these module designs to achieve energy transition deployment schedules and meet capacity targets. We compare the entailed quantities of virgin material demands, lifecycle wastes, energy demands, and energy generation for the Energy Transition scale deployments.

3 SCIENTIFIC AND INNOVATION RELEVANCE

Traditional CE metrics, such as the MCI [1], prioritize increased circularity over extension of the use phase and do not always align with a reduction in environmental impacts as measured by LCA [6]. However, the use phase for renewable energy technology is critical for both CE and Energy Transition goals; thus, a suite of mass and energy metrics should be used. Additionally, renewable energy technologies are infrastructure, so they should be circularized more like buildings than consumer products.

Five component materials of c-Si PV modules are tracked dynamically in the PV ICE tool and baselines: glass, silicon (Si), silver, aluminum (Al) frames, and encapsulated copper (excludes junction box). Historical shifts in material intensity are accounted for (e.g., decreasing silicon wafer thickness). The PV ICE tool accounts for manufacturing inefficiencies, quality or material purity levels, and several circular pathways for PV modules, including reuse, repair, remanufacture, and recycle in both open and closed loops. The energy flows parallel the mass flows, accounting for the energy demands throughout the lifecycle of the PV module.

ACKNOWLEDGEMENTS

This work was authored [in part] by the National Renewable Energy Laboratory, operated by Alliance for Sustainable Energy, LLC, for the U.S. Department of Energy (DOE) under Contract No. DE-AC36-08GO28308. Funding provided by the U.S. Department of Energy's Office of Energy Efficiency and Renewable Energy (EERE) under Solar Energy Technologies Office (SETO) Agreement 38269 and 38699. The views expressed in the article do not necessarily represent the views of the DOE or the U.S. Government. The U.S. Government retains and the publisher, by accepting the article for publication, acknowledges that the U.S. Government retains a nonexclusive, paid-up, irrevocable, worldwide license to publish or reproduce the published form of this work, or allow others to do so, for U.S. Government purposes. The funders had no role in study design, data collection and analysis, decision to publish, or preparation of the manuscript.

Inclusion and Diversity Statement: One or more of the authors of this paper self-identifies as an underrepresented ethnic minority in science. While citing references scientifically relevant for this work, we also actively worked to promote gender balance in our reference list.

Data and Code Availability Statement: Analyses process and source data for this publication support open-science and are available in the GitHub repository: https://github.com/NREL/PV_ICE/.

REFERENCES

[1] D. Smith and S. Jones, "Ellen Macarthur Circularity Indicators: Measuring Methodology," p. 64.

[2] F. Figge, A. S. Thorpe, P. Givry, L. Canning, and E. Franklin-Johnson, "Longevity and Circularity as Indicators of Eco-Efficient Resource Use in the Circular Economy," *Ecol. Econ.*, vol. 150, pp. 297–306, Aug. 2018, doi: 10.1016/j.ecolecon.2018.04.030.

[3] S. Bächle and P. Dold, "PERC Solar Cells from 100 Percent Recycled Silicon - Fraunhofer ISE," *Fraunhofer Institute for Solar Energy Systems ISE*, Feb. 07, 2022. https://www.ise.fraunhofer.de/en/press-media/press-releases/2022/solar-cells-from-recycled-silicon.html (accessed Feb. 09, 2022).

[4] S. Ovaitt, H. Mirletz, S. Seetharaman, and T. Barnes, "PV in the circular economy, a dynamic framework analyzing technology evolution and reliability impacts," *iScience*, vol. 25, no. 1, p. 103488, Jan. 2022, doi: 10.1016/j.isci.2021.103488.

[5] S. Ovaitt, H. Mirletz, D. Jordan, K. Anderson, M. Deceglie, and A. Hegedus, "NREL/PV_ICE." NREL, 2021. [Online]. Available: 10.5281/zenodo.4324010

[6] J. Brändström and M. Saidani, "Comparison between circularity metrics and LCA: A case study on circular economy strategies," *J. Clean. Prod.*, vol. 371, p. 133537, Oct. 2022, doi: 10.1016/j.jclepro.2022.133537.

[7] H. Mirletz, S. Ovaitt, S. Sridhar, and T. M. Barnes, "Circular economy priorities for photovoltaics in the energy transition," *PLOS ONE*, vol. 17, no. 9, Sep. 2022, doi: 10.1371/journal.pone.0274351.

[8] D. Liu, "2022 Global Solar PV O&M economics," Wood Mackenzie Power & Renewables, Dec. 2022.

[9] N. M. Haegel *et al.*, "Terawatt-scale photovoltaics: Transform global energy," *Science*, vol. 364, no. 6443, pp. 836–838, May 2019, doi: 10.1126/science.aaw1845.

40th European Photovoltaic Solar Energy Conference and Exhibition

This presentation was selected by the Sc. Committee of the EU PVSEC 2023 for submission of a full paper to one of the EU PVSEC's collaborating peer-reviewed journals.

UNLOCKING THE POTENTIAL OF PHOTOVOLTAIC ENERGY COMMUNITIES IN THE PUBLIC SECTOR: ACTION FOR THE PV COMMUNITY

Ana Belén Cristóbal[1*], Cristina Sanz-Cuadrado[1], Luis Miguel Carrasco[1], Marta Victoria[2], Zhe Zhang[2], Marko Topic[3], Matevž Bokalič[3], Afonso Cavaco[4], Pedro Horta[4] & Luis Narvarte[1]

[1] Instituto de Energía Solar,
Universidad Politécnica de Madrid
C/Alan Turing s/n, 28031, Madrid, Spain
*Email: anabelen.cristobal@upm.es

[2] Department of Mechanical and Production Engineering and
iCLIMATE Interdisciplinary Centre for Climate Change
Katrinebjergvej 89 G-F, 8200 Aarhus N, Denmark

[3]University of Ljubljana,
Faculty of Electrical Engineering,
Tržaška cesta 25, SI-1000 Ljubljana, Slovenia

[4]Universidade de Évora, Renewable Energies Chair,
Largo dos Colegiais, N° 2, 7004-516 Évora, Portugal

ABSTRACT: Decarbonizing cities in Europe is crucial for meeting its climate objectives. While technology and investments are essential, changing citizens' behavior is equally important to achieve Europe's energy goals. In pursuit of this mission, the public sector should serve as a "living lab," allowing people to directly experience and learn about energy challenges.

In this context, the AURORA-H2020 project is establishing these living labs in different countries by upgrading large pre-established local social hubs including university communities, school communities, and more. The goal is for these communities to become citizen science energy hubs.

This paper presents our journey in implementing such living labs, focusing specifically on public universities.

One of the key actions within these living labs is the installation of a photovoltaic (PV) facility at each location, funded through crowd-financing by hundreds of individuals connected to these sites via micro-investments starting from €20. Each installation helps the local authorities accelerate their decarbonization plans, reduce energy costs, and provide people with a first-hand experience of PV technology through various activities.

However, due to the innovative nature of this approach, the need for a legal framework for such living labs, as well as the necessary public-private partnerships, has posed several challenges. Our goal is to highlight and discuss the hurdles encountered, raise awareness, and initiate a discussion within the PV community about potential solutions and collaborations to make this concept a widespread reality in Europe.

Keywords: energy communities, universities, citizen science, implementation models, living labs, public-private partnerships.

1 INTRODUCTION

The quest for clean, affordable, and reliable energy has emerged as the defining mission of the 21st century, driven by a confluence of compelling factors. At its heart, this mission is a testament to our growing understanding of the planet's fragility. In this era, we have witnessed a profound awakening to the myriad environmental perils that threaten the Earth. The ominous specter of climate change, the insidious grasp of air and water pollution, and the relentless march of habitat destruction have all cast a shadow over our collective consciousness.

In response to these existential challenges, humanity has turned its gaze towards the promise of clean energy sources—majestic wind turbines harnessing the sky's power, radiant solar panels capturing the sun's inexhaustible energy, and the awe-inspiring force of hydroelectric power tapping into nature's abundance. These technological marvels offer not just a glimmer of hope but a beacon of possibility, illuminating a path toward curbing the ominous greenhouse gas emissions that threaten our climate and alleviating the toll exacted by traditional energy production on our fragile ecosystems.

The involvement of citizens in this mission over many decades, including participation in communities focused on energy, grassroots movements, and environmental advocacy groups, among others, serves as a clear demonstration of the strength of collective action. Moreover, we now witness citizens displaying a growing interest in energy transition issues, particularly when they understand how factors such as energy pollution, security, and prices impact their daily lives. This underscores the importance of recognizing that achieving a clean energy future requires more than just leadership from the top; it relies on the active participation and engagement of the general population.

Unfortunately, the role of citizens does not shine as brightly as that of technology. And there are inconsistencies between the policy-maker's speeches, the directives they promote, the documents they generate, and actions they make. For instance, the in-depth analysis [1] in support of the Commission Communication "A Clean Planet for all" in 2018, stated and highlighted the impact that consumers' choices would have in the achievement of the global emissions target presenting a scenario influenced by such attitudes. But it is noticeable, and even surprising, that the final challenging goals stated by the "2030 Climate Target Plan" [2] were designed with scenarios where people's behaviour is not considered. And even at some point is literally written in the European communication that "*Consumer choice was not explored in detail in the scenarios of this impact assessment due to the shorter time focus*"[3] (Shorter in relation to [1]).

Thus, while citizen engagement could be powerful in a quicker achievement of the climate objectives, the lion's share of change often stems from determined energy policies that "dictate" the path, creating at some point, reluctance, and non-desirable attitudes in certain sectors of the population, becoming climate change in another policy-position.

Consequently, within this evolving narrative, another theme emerges—one that underscores the critical role of education. Education stands as the linchpin in this grand endeavour a pivotal force for raising awareness, nurturing a profound understanding of environmental challenges, and endowing individuals with the knowledge and tools needed to make informed decisions. It empowers citizens to become champions of clean energy, enabling them to engage in sustainable practices and become agents of change at both personal and societal levels.

Therefore, we should be able to offer lifelong learning experiences to citizens, opportunities where they have the chance to experience clean energy first-hand. These living labs will be a way to change the vision of many citizens who can later make better-informed decisions. This is nothing new in marketing; many companies offer testing samples to potential customers or even allow them to test some goods for a period. Even customers who will never acquire the product due to a lack of resources can learn about the product and become ambassadors.

AURORA is a European project granted under the Green Deal call (101036418) [4]. Its spirit is to shape these living labs to engage thousands of people in clean energy experiences. Aurora leverages the energy community concept to operationalize such living labs. Our idea relies on using social communities already established around public services in Europe, such as universities, schools, hospitals, libraries, etc., and upgrading them to energy communities.

Many living labs and various types of educational initiatives are already operational in Europe based on memorandums of understanding among private and public bodies. However, AURORA takes a step further by advocating for these living labs to be genuine public-private partnerships with a defined legal structure. These partnerships truly commit the public administration to its public service, while allowing citizens to engage in activities within a living lab that would not otherwise be possible, such as collectively financing local renewable installations and participating more actively in the energy transition.

This paper introduces the different models adopted in different European locations to create such living labs and describes the hurdles we have encountered when our idea becomes a reality, focusing on universities as public bodies.

We also introduce the work that needs to be done to expand this model throughout Europe and the collaboration we would need from PV academics and researchers throughout Europe to smooth barriers and allow people to shine at the same level that technology does right now.

2 METHODOLOGY

2.1 Concept and design

When talking about energy, AURORA believes that the existing top-down system has room for improvement from the educational point of view to avoid certain attitudes shown by citizens, which sometimes are directly caused by the own system. Europe is bringing about behavioural changes because the public sector funds and subsidizes electric cars, private photovoltaic installations, and the replacement of heating boilers, among other things. The public sector should finance a technology when it is not yet profitable to enter the market, but it represents a genuine green alternative, and some pioneers are willing to take the risk and test the technology. Otherwise, we are promoting the Matthews effect, the phenomenon more commonly known as 'the rich get richer".

AURORA model would imply public administrations offering their buildings to allow every single person to invest in a local PV facility with a small return on investment (ROI). Obviously limiting the value of shares and number of shares should be a must so anybody cannot see this action as a way of doing business. Later on, the operational model would depend on each community, some of them can decide to devote themselves to selling electricity a at lower cost to fight against energy poverty in their local area, and others can decide this electricity is for self-consumption the public administration where the installation is settled, or even they can operate as a real energy community where all the investors can have access to buy the electricity. AURORA living labs design would enable public administrations to stop subsidizing private renewable installations, thus preventing an unfair energy transition in which only those who already possess resources, such as roofs and money, can participate and benefit from public funding, giving the false impression that they are more environmentally conscious than others.

2.2 Demo sites

Universities were chosen for several reasons. Firstly their educational mission for society and the high number of people associated with each of the more than 3,000 universities Europe has. Secondly, because they are centers where innovation, creativity, and thinking outside the box are part of their daily work. And, additionally, they bring together many young people, those who don't possess the resources to experiment with PV.

The four demos are: Aarhus University (AU), Univerza v Ljubljani (UL), Universidade de Évora (UEVORA), and Universidad Politécnica de Madrid (UPM).

3 RESULTS

3.1 Framework for the implementation

3.1.2 Energy Communities Directives

Each of the pilots collaborated with local legal advisors to create customized models for each location. AURORA encountered several challenges, with the most notable being the uncertainty regarding the status of public universities within the European Directives on Energy Communities. Specifically, EU Directive (EU) 2018/2001 on the promotion of the use of energy from renewable sources, known as REDII [5], states that energy communities are formed by the aggregation of any of these

10.4229/EUPVSEC2023/5DO.10.1

entities:

Natural persons, local authorities, including municipalities, or small enterprises and microenterprises, provided that for private undertakings their participation does not constitute their primary commercial or professional activity.

Can public administrations, such as those dependent on the central government or other public bodies like universities, be included? All our legal advisors aimed to justify their eligibility under the label of 'local authority' and sought clarification from national law, which proved to be even more complicated. In the end, the uncertainty led universities to adopt a conservative stance, and unfortunately, they abandoned the idea of an energy community as defined in the EU directives.

The legislation surrounding energy communities has opened avenues for citizens to engage in various energy-related activities beyond self-consumption, such as aggregation, storage, and commercialization. However, many existing energy communities were, and still are, built upon older national laws. These laws had already enabled citizens to form cooperatives, finance renewable projects, and sell excess energy to the grid. Therefore, as for our purpose, national self-consumption regulations, including shared self-consumption, would allow the operation of these 'crowdfunded facilities' on a smaller scale and would meet the objectives of our 'living labs' for educating citizens about photovoltaic energy.

In essence, while recognizing the need for better regulation under the energy communities' directives, the AURORA project found that national self-consumption laws provided a practical framework for implementing its local citizen hubs.

3.1.2 (Shared) Self-consumption national laws

In all our pilot areas, except Slovenia, the legal framework for shared and individual self-consumption from an installation that can be crowdfunded community structures is permissive. However, in Slovenia, it's limited to cooperative structures.

Consequently, initially, we explored the cooperative model involving the university and its members. However, we encountered legal barriers. In Slovenia, Article 10 of the Higher Education Act prevents universities from establishing or co-establishing new legal entities without explicit national parliament approval.

The cooperative model also raised concerns for Universidade Évora and Universidad Politécnica de Madrid. Cooperatives are designed for commercial purposes, while universities' objectives don't include generating profits. This restriction includes offering financial products with interest or payments derived from capital income to participants or outgoing cashflows. In Évora, AURORA considered offering participants discounts on university services, but the university management board rejected this approach, deeming it outside the scope of university actions.

We then explored the creation of communities through foundations or associations. Foundations and associations offer more flexibility than cooperatives, as they aren't required to distribute profits. Citizens' participation can be seen as a loan that the foundation must repay. Universities can be part of associations, but decisions must be made collectively, with no privileges granted to universities as public entities. In a living lab like the one proposed by AURORA, where hundreds of participants can join for a small fee, this framework caused some reluctance among the universities and was discarded directly. Apart from the fact that associations can't have a public character.

Foundations, however, provide more freedom in organizing votes and management. They can also have a public character, subject to rules governing public universities. This makes foundations more appealing to universities, offering greater legal guarantees.

3.2 Implementation in specific demo sites

3.2.1 Aarhus University

The model found in Denmark for establishing such a living lab with real engagement from hundreds of university members in crowdfunding a distributed PV facility arises from a confluence of factors that are not found in many other European universities, especially for buying renewable electricity.

AURORA has organized Aarhus University community members using the legal framework of a cooperative. Aarhus University is not a member of the cooperative due to the issues highlighted previously for public universities. The key point here is that the university does not own some of the buildings where its daily activity occurs. Consequently, while negotiations have not been easy, AURORA has managed to establish an agreement with the building owner. Under this agreement, the cooperative installs photovoltaic systems on privately owned buildings that the university rents. Due to the proximity of consumption and production, Danish regulations acknowledge the concept of a 'prosumer via third party,' which is what has ultimately been adopted in Aarhus. Thanks to this legal clause, the university can purchase electricity without the need for public energy tendering processes.

3.2.2 University of Ljubljana

On the contrary, in Slovenia, AURORA encountered a blockade when attempting to establish the living lab through such a public-private partnership. The fact that universities cannot engage in any new legal entity also prevented them from legally constituting the community with a legal figure. At that point, the University committed to shaping the living lab in the form of a Student Energy Club - ŠEK (Študentski energetski klub) and virtual crowd funding.

As for the PV installation, the University finances a dedicated facility with its resources – the public tendering was already launched - and club members will receive personalized information about the facility, considering potential investments they could have made. Regarding the rest of the activities performed by the living lab, there are no differences from the rest of the demos.

3.2.3 Universidade de Évora

Once the cooperative model was discarded to engage universities in a public-private partnership, foundations seemed to be the best choice. This way, the University of Évora and the Universidad Politécnica de Madrid started negotiations to legally organize their living labs under such a structure.

In the case of foundations, in both Spain and Portugal, AURORA universities ruled out creating an ad-hoc foundation for the project's objectives. There is no legal impediment to doing so, but in both countries, the reorganization of the public sector and the improvement of its efficiency have been regulated by law, partly supervised by Europe, for the receipt of European aid funds that these countries have received. In all sectors, without exception, including UEVORA and UPM, foundations have been merged or eliminated, and the creation of ad-hoc foundations for the energy community has been rejected. However, while the Universidad Politécnica de Madrid still owns a General Foundation, allowing us to explore this option, Universidade de Évora does not.

Consequently, AURORA joined the Évora University community in a living lab in the form of a cooperative, excluding the university as a member of this structure. They reached an agreement with Red Cross Evora to install a crowdfunded PV facility in the buildings owned by this organization in the city.

3.2.4 Universidad Politécnica de Madrid

As mentioned earlier, UPM still owns a Foundation, which, among its foundational purposes, includes the promotion and conservation of the environment. The proposed scheme involves individuals who are part of the university community lending money to the Foundation, which uses that money to provide UPM with a photovoltaic installation. While we are not technically speaking of an energy community, the idea is to organize an 'energy community' as a commission within the foundation. Two different financial approaches were explored: loans and participatory accounts. The Foundation would not profit from the business, as it serves as a mere intermediary between UPM, which entrusts its Foundation with the mission of establishing the facilities to save on energy costs, contribute to the energy transition, and promote education within its community.

Due to the Foundation's public nature, the proposal was presented to the regional government department in Madrid responsible for overseeing such organizations. Regrettably, they rejected the proposal for two primary reasons.

Firstly, their initial objection is grounded in the assumption that a foundation with 100% public participation should exclusively rely on assets owned or contributed by public entities. This stance poses challenges for including the proposed private funding sources. The second objection, as detailed in their report, is rooted in Spanish regulations. According to these regulations, the University is obligated to ensure that the foundation has adequate material and personnel resources to fulfill the mandate's purpose. The requirement for private resources signifies their inability to meet this obligation.

This report led the University to conclude that it had no interest in continuing the process of seeking a legal form for an energy community through its Foundation. However, external legal advisors believe that the regional government report's approach is more opinion-based than based on legal criteria.

As a result, UPM explored another alternative that could involve its university community in a legal living lab and offered the AURORA team the use of some buildings managed by third parties that currently serve for UPM activities. In this arrangement, the university community forms a cooperative, and people join through participatory accounts. This way, UPM will have the PV facility for its living lab on its own Campus but in a private builiding. Yet it still lacks a genuine public-private partnership.

3.3. Other challenges

Although there is a good willingness, Universities have not been very active in pushing the initiatives forward, though multiple colleagues have expressed how exciting the initiative is. When it comes to decision-making, the fact that the university is a large institution with responsibilities diluted among different employees makes it extremely difficult to have an efficient decision process when discussing the installation of the PV panels and the purchase of electricity. Technical difficulties also appear when some of the buildings at the campus are "heritage buildings", whose aspect cannot be altered in a way that has visual impact, which has been manifested in Évora and Aarhus; or when buildings have different energy meters so previous works with the distribution company must be done. The emergence of many other doubts and concerns is a reflection of the fear of the unknown that comes when a pioneer being, i.e. the return on investments for the public administration in such a long period, the responsibility for the maintenance of the roofs where the installations are set up, the status of the installations after the end of the contract for later exploitation, etc.

4. DISCUSSION

For the enormous change of mind on public administration vision we had relied on the energy communities' directives. However, AURORA has proven that both European Directives on energy communities [5,6] only refer to the participation from the public sector point of view to local authorities and municipalities. This is seen by organizations such as universities or research centers – and probably so many other public sector administrations- as a hurdle to deciding their participation in an energy community. They feel they are excluded- probably they are, but a clarification from the European Union should be recommendable, indicating the reasons of such a decision.

Fortunately, using national laws on shared self-consumption we could create legal structures such as cooperatives or any other type of legal figure, and form the public-private partnership. It seems that from the figures explored by AURORA – there could be some others- Public or Private Foundations (depending on the participation of each party) could drive such public-private formal cooperation for universities. But once the legal formula is found, we realized that other problems appear, i.e., new foundation creations are limited by the national governments, national laws prevent universities from engaging in any other legal structure, decision boards pass the documents to being assessed by public servants that prefer to be on the "safest side" and basing their reports on mere opinions, etc.

All our demos have been able to create their living labs – three of them with legal structure and one in a more informal way- but there is not a truly public-private partnership in any of them. Universities have paved the AURORA demos by supporting the project, i.e., by looking at buildings owned by third parties in the campuses to set up the facility – this way the PV facility is part of the local environment- or accepting purchase energy prices that are not a major saving on costs for the university.

Paradoxically, while universities are rejecting private money from participants in their own living labs (e.g., at

10.4229/EUPVSEC2023/5DO.10.1
020501-004

UPM where expressions of interest raised 140,000€ in the first communication campaign), what is happening is that universities are using European and national public funds to install PV facilities. However, beyond the financial savings and associated environmental benefits, these installations are not impacting the behaviour or education of their community any differently than a private house would.

During the AURORA study, we found a case [7] in which an European University has offered a public tender to install on their own rooftops 1.4 MWp by more than 2.5M€. However, the tender was solved by 141€. The key is that one of the big national providers offered to the university does not make any initial investment and buys the electricity from such PV installations for 10 years at a fixed price. After that period the institutions will become property of the university.

This makes AURORA think of another way to enable, encourage, and allow local people to act via public tenders. This solution can be the most comfortable for public bodies and the quickest for replication in any member state for any organization. However, we need to think on several critical points to really evaluate and ensure local people's participation. We also know that this solution yet still not a public-private partnership, and people become mere suppliers for the administration.

5. FUTURE WORK

There are different working pathways we are interested in exploring:

- Engaging with policymakers in a dialogue on how to promote the use of any public buildings for the establishment of these living labs.
- Engaging with policymakers in a dialogue on how to operationalize return on investments for those participants of the living lab investing their own money to have a greener local area and learn about PV technology. Limited shares in number and amount should be a must. Return on investment is a reward for being a committed citizen. Probably public governments are spending more when subsidizing private PV installations, electric cars, heat pumps, and other mature technologies while excluding many people from the energy transition.
- **Engaging with researchers and academics working on photovoltaic or renewable energy and aligned with the AURORA's vision** [8]. We want to explain to them in detail the models we have already implemented. Their mission would be to become an AURORA ambassador to take the idea to their organization leadership teams, receive feedback on its potential within such organizations, and share such feedback with us. Being aware of the challenges found in many different universities is essential. We need a clear roadmap to work on how to face them collectively.
- Later, those researchers/academics willing to step up and bring this idea to life and twinning any AURORA demo will be supported by the AURORA team. While recognizing that national legislation may differ, we're committed to offering you a clear roadmap and standing by your side, giving unwavering guidance and support at every turn.

6. CONCLUSIONS

In conclusion, it is important to acknowledge and understand the barriers that the energy transition is facing. The directives of energy communities have not been clear enough to inspire Member States to participate in the energy transition within any type of public administration. Despite this, public-private partnerships to establish local energy living labs have been explored in AURORA. These living labs, born from existing social local communities, evolve into energy hubs with the potential to increase access to renewable energy for everyone and promote lifelong learning among citizens.

Providing them with a legal structure allows for more active participation from citizens, enabling them, for instance, to crowdfund community PV installations using public buildings. Citizens would receive a moderate return on investment from the sale of the electricity and an educational experience that would not be possible for those lacking the physical or financial resources to experience PV technology. However, we have learned that the lack of clear regulations, the complex bureaucracy of public administration, and the fear of pursuing innovative legal approaches hinder the potential of this idea. This also prevents a fair transition, leaving many people behind, and restricts the use of public resources allocated to climate change in a more inclusive manner.

Could resolving this be more challenging than overcoming technological limitations?

Despite all the difficulties, AURORA has successfully created such living labs in four European universities with "alternative models" compared to a real public-private partnership as desired in the energy communities' directives. We have engaged approximately 3000 people across our five demos so far. We conclude that AURORA needs more academics or researchers interested in propelling these living labs into their working environments to facilitate the replication process throughout Europe and drive a bottom-up, tangible change. We encourage all readers of this paper to contact us just to explore how they could be a seed for driving change in their local areas. Becoming an ambassador of AURORA simply involves discussing the potential implementation of the idea in your own university or research center and sharing the feedback received with us to create a roadmap on the challenges that need to be addressed. Twinning agreements could hopefully follow later.

7. ACKNOWLEDGMENTS

This work has received funds from the European Union's Horizon 2020 research and innovation program under grant agreement No: 101036418.

C.Sanz-Cuadrado acknowledges the funding for her contract received by the Community of Madrid through the Program MADRID-PV2 (S2018/EMT-4308).

The authors want to acknowledge the work made by the legal teams hired, Piet Holtrop, Erik Christiansen, all staff of the universities, and our own team members involved in the process for the information shared and the support given to explore the approach of this work.

8. REFERENCES

[1] In-depth analysis in support of the commission communication COM(2018) 773. 28/11/2018.

[2] Communication From The Commission To The European Parliament, The Council, The European Economic And Social Committee And The Committee Of The Regions "Stepping up Europe's 2030 climate ambition. Investing in a climate-neutral future for the benefit of our people" {{SEC(2020) 301 final} - {SWD(2020) 176 final} - {SWD(2020) 177 final} - {SWD(2020) 178 final} }. COM(2020) 562 final. 17/09/2020.

[3] SWD(2020) 176 final. IMPACT ASSESSMENT Part 2/2. Accompanying the document Stepping up Europe's 2030 climate ambition 17/09/2020.

[4] https://www.aurora-h2020.eu

[5] Directive (EU) 2018/2001 of the European Parliament and of the Council of 11 December 2018 on the promotion of the use of energy from renewable sources.

[6] Directive (EU) 2019/944 of the European Parliament and of the Council of 5 june 2019 on common rules for the internal market for electricity and amending directive 2012/27/eu (recast), official journal of the European Union, 2019.

[7] Public Record PNSP 54/2022. Universidad Complutense de Madrid. https://short.upm.es/p606n

[8] AURORA Ambassadors form: https://short.upm.es/wm79o

SHINING LIGHT ON EUROPEAN BIPV

A SURVEY OF DEPENDENCE AND FRAGMENTATION IN THE EMERGING EUROPEAN VALUE CHAIN FOR BUILDING INTEGRATED PHOTOVOLTAICS

Gerardo López Pinto[1,3], Fredrik Özaras[1,3], Johnn Andersson[2], Björn Sandén[1], Amelia Oller Westerberg[3], Johan Lindahl[3*]

[1] Chalmers University of Technology, Sweden
[2] Uppsala University, Sweden
[3] The European Solar Manufacturing Council
*lindahl@esmc.solar

ABSTRACT: Building integrated photovoltaics (BIPV) is a technology that could support European climate neutrality and energy independence by making buildings sources of renewable energy. BIPV could further support these ambitions if sourced and produced locally. This study provides an overview of the status and import dependency of the European BIPV value chain. The analysis of the upstream value chain showed that the European BIPV industry is highly dependent on non-European countries for the supply of key components, especially the silicon-based cell technology, with 80% of interviewed companies sourcing these cells from China. Other main components are also sourced from manufacturers in China, such as junction boxes (67%) and encapsulants (63%). This dependency is further aggravated by a mismatch between producer requirements and domestic supply of main components. In the downstream value chain, BIPV producers are required to fulfill regulations for both PV products and buildings, which results in time-consuming and complicated processes. It was further identified that a lack of knowledge and specific expertise of BIPV exists among important actors, hindering its development and diffusion.

Keywords: Building integrated photovoltaics (BIPV), Value chain analysis, Technological Innovation Systems, Multi-level Perspective, Energy independence

1 INTRODUCTION

As a result of the Russian full-scale invasion of Ukraine, and the war that followed, the European Union (EU) has decided to accelerate the aim for energy independence and climate neutrality. This has resulted in EU launching the REPowerEU plan, which aims at "rapidly reducing our dependence on Russian fossil fuels by fast forwarding the green transition and joining forces to achieve a more resilient energy system and a true Energy Union" [1, p.1]. The REPowerEU plan contains various strategies and initiatives to further strengthen specific industries and energy sources, such as the EU Solar Energy Strategy. This initiative aims at doubling the European photovoltaic (PV) capacity by 2025, installing 600_{AC} GW by 2030, increasing the supply chain resilience and the domestic production capacity, as well as making rooftop solar mandatory for certain buildings, see Section 2.3 [2]. In addition, the Green Deal Industrial Plan for the Net-Zero Age was launched in 2023 with the goal of making the EU's net-zero industry[1] more competitive as well as supporting the transition towards climate neutrality [3]. The goal of increased value chain resilience is further enhanced by other initiatives that have been developed due to multiple supply chain disruptions in recent years, such as those that resulted from COVID-19 and the Suez Canal obstruction [4].

The European Union (EU) aims for climate neutrality by 2050 and a goal to increase the share of renewable energy has been set [5]. It is estimated that approximately 40% of the EU's energy consumption and 36% of the energy-related greenhouse gases come from buildings [6]. Therefore, it is essential to reduce the emissions and energy consumption related to this sector, for example through adoption of PV. Besides conventional PV modules, alternative PV technologies, such as building integrated photovoltaics (BIPV) are part of the urban capacity additions. The European Commission [7, p.13] has stated that:

> *"The potential of this sector remains to be unlocked through uptake by the construction sector and the related economies of scale. EU-wide deployment would require homogeneous certification for the affected products [...]"*

Hence, BIPV introduces an opportunity to increase the PV capacity in the built environment and society at large. It is therefore highly relevant to assess the current status of the European BIPV industry to understand what support is necessary to develop a resilient European BIPV industrial value chain and diffuse the technology into society [8].

Therefore, this study analyses the status of the European BIPV industry with a specific focus on the upstream value chain. Through stakeholder interviews and an examination of relationships and policies, the study seeks to uncover prevailing issues, barriers, and gaps in this industry. The research aims to describe the current structure of the European BIPV value chain, and asses its reliance on non-European imports.

1.1 Research Context

BIPV is a type of technology that has two major functions: (i) generate electricity, and (ii) act as a building material for the envelope of buildings. There are multiple definitions of how to classify BIPV. The International Energy Agency Photovoltaic Systems Programme (IEA-PVPS), via Task 15: Enabling Framework for the Development of BIPV, has reviewed several definitions and versions of how to define it, and compiled it into one [9, p.16]. This definition of BIPV is also used in this study:

[1] The Net-Zero Industry Act includes the following technologies: solar photovoltaic and solar thermal technologies, onshore and offshore renewable technologies, battery and storage technologies, heat pumps and geothermal energy technologies, electrolysers and fuel cells, sustainable biogas and methane technologies, carbon capture and stor age (CSS) technologies, and grid technologies [3].

10.4229/EUPVSEC2023/5DO.10.2
020502-001

"A BIPV module is a PV module and a construction product together, designed to be a component of the building. A BIPV product is the smallest (electrically and mechanically) non-divisible photovoltaic unit in a BIPV system which retains building-related functionality. If the BIPV product is dismounted, it would have to be replaced by an appropriate construction product.

A BIPV system is a photovoltaic system in which the PV modules satisfy the definition above for BIPV products. It includes the electrical components needed to connect the PV modules to external AC or DC circuits and the mechanical mounting systems needed to integrate the BIPV products into the building."

IEA Task 15 further lists the main elements that compose a BIPV module, which are the PV cells, encapsulates, front and back covers, and junction boxes [10]. The cells create the core of the module where the energy is generated, while being insulated by the encapsulants, which in turn are protected by the front and back covers; usually made of glass or polymers.

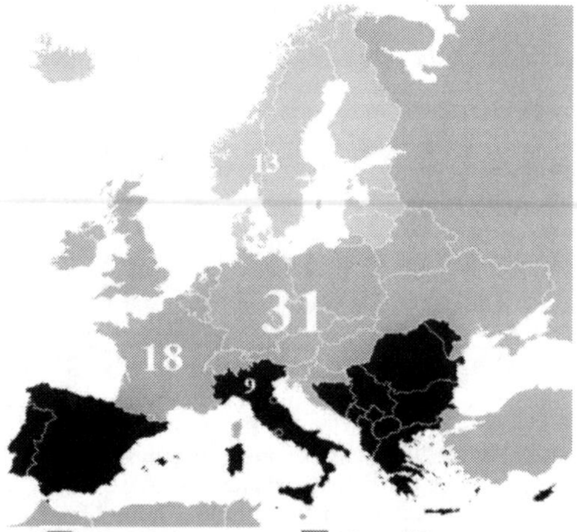

Figure 1: Overview of the location of European BIPV companies by region.

2 METHOD AND THEORY

2.1. Data collection

As suggested by Hellin and Meijer [11], a qualitative approach is preferred when analyzing value chains, such as performing interviews and questionnaires. The companies selected for this study are producers and manufacturers of BIPV modules, research facilities, and policymakers in Europe. During the period between January and May of 2023, 71 companies were contacted and asked to participate in an interview, with the goal of obtaining a description of the current European BIPV industry and value chain. Overall, 2 companies rejected participation in the interview, 46 did not respond to the request, and 23 participated.

By conducting semi-structured interviews with representatives of the BIPV companies, typically Chief Executive Officers, Production Managers, Chief Product Officers and Chief Technology Officers, information about the upstream and downstream value chain was obtained. The duration of the interviews varied between 30 and 60 minutes and was divided into four different sections: (i) general information and background of the company, (ii) value chain and production information, (iii) networks and relations of the company with other companies, institutions, and actors, and (iv) perceived challenges of the company and European BIPV industry. Furthermore, to keep the responses for the interviews anonymous, Europe was divided into different regions: Northern Europe, Central Europe, Western Europe, and Southern Europe, as seen in Figure 1. The system categorization for BIPV was also utilized to classify companies according to the function in the building envelope that their BIPV product performs.

To acquire background information about the European BIPV value chain and EU policies, various databases and websites were the main sources of information. Literature and information regarding how BIPV is currently understood and perceived in Europe was mainly gathered from the BIPVBoost research project and Research Task 15 from the IEA PVPS. A list of European companies producing BIPV was also provided by the ESMC, which was complemented by actors listed in the study of Corti et al. [12].

2.2. Multi-Level Perspective theory

Transitions can be understood as changes from one socio-technical system to another [13]. A socio-technical system is a system that provides a specific function for society [14]. Since transitions influence patterns, behaviors, regulation and even infrastructure, they involve multiple elements and engage with both technologies and society simultaneously. This co-evolution of technology and society can be seen as system innovations [15]. To understand what influences the transition to a new technological system, attention needs to be placed on both the internal mechanisms and interactions within the system, as well as on the externalities and the context in which the system develops. An approach to study these system transitions and technology evolutions, and their diffusion and adoption within society, is to utilize the multi-level perspective (MLP). The MLP provides a framework to study the transition of innovations through the interaction of technology and society [15].

The MLP presents a nested hierarchy composed of three levels: (i) micro-level, (ii) meso-level, and (iii) macro-level [13]. The micro-level is represented by niches. Technologies at this level do not compete with already established technologies as they are not mature nor efficient enough, which results in high uncertainty. Technologies in niches are formed through trial and error in a protected space [15]. The next level in the hierarchy is the meso-level, represented by the regime [13]. The regime can be understood as a set of established rules, patterns, and behaviors agreed upon by actors in society around a technology [16]. Technologies in this level are mature and stable, and the innovation that occurs at this level is of a slow and incremental nature, as opposed to the radical innovation from the micro-level niches. The last level of the hierarchy is the macro-level, which is represented by the landscape. The landscape is the broad context of society, including spaces and materials, such as infrastructure [15, p.684]. As the highest level of the MLP, the landscape is hard to change and cannot be directly influenced by actors.

The nested hierarchy of these levels originates from

10.4229/EUPVSEC2023/5DO.10.2

the constant interaction and embedment between them, which results in the levels influencing and affecting each other. More specifically, changes in the regime come from pressure applied by the landscape. This pressure can be of a material-, political-, perception-, or behavioral nature. When an opportunity opens in the landscape level, and pressure is put in the regime, technologies in the niche level can seize this pressure and gain more diffusion and acceptance [13].

The MLP framework was used to analyze the broader context in which the BIPV system is developing by identifying what other systems that interact with the industry and how this occurs.

3 RESULTS

3.1. European BIPV industry analysis

According to the system categorization of BIPV, it was identified that 54 of the analyzed companies produce roof BIPV, 39 façades and 35 external integrated devices (EID). However, some of the companies also produce multiple categories, or all of them.

The data gathered to describe the size of the industry (number of employees, turnover, and production capacity) does not consider all 71 companies, only those for which the data was available. The companies producing BIPV in Europe differ greatly in size, as can be seen in Figures 2–4. The largest company has 810 employees, while the smallest producer only has three. The average number of employees is 84 employees, while the median is 30 employees. However, it should be noted that some of the larger companies also produce regular modules, and that not all reported employees necessary work within the BIPV branch of these companies. The size difference is also reflected in the turnover, as seen in Figure 3. The lowest turnover of any company in 2021 was €4 250, the company with the highest turnover reached €46 800 000. The average and median turnover for the year amounts to €8 287 732 and €1 572 438, respectively. However, note that some of the companies exclusively produce BIPV while others offer it as part of a broader product catalog.

The production capacity of the companies varied greatly, as can be seen in Figure 4. The producers measure their capacity using different units; some companies measure it in megawatts per annum (MW/a), while others measure it in square meters per annum (m^2/a). This stems from the dual nature of BIPV, as some companies view their product as a PV system while others view it as a construction material. To get a uniform compilation, the manufacturing capacities reported in m^2/a were converted by the authors to MW/a by the average W/m^2 ratio of the company's BIPV products disclosed in their datasheets. In terms of MW, the largest capacity was found to be 200 MW/a, while the company with the lowest capacity produces 1.4 MW/a. The total combined yearly capacity of the 27 reporting companies was 1 313 MW/a, which on average gives a capacity of 49 MW/a and a median capacity of 30 MW/a per company. However, these figures do not include all the 71 identified BIPV companies, just the 27 companies that disclosed manufacturing capacities in the survey. In addition, it is important to stress that due to fluctuations in demand, some of the companies mentioned that they often produce only at a share of their maximum production capacity.

3.2. Value chain of silicon based BIPV

Silicon based BIPV is the conventional cell technology. Out of 71 contacted companies, 51 used

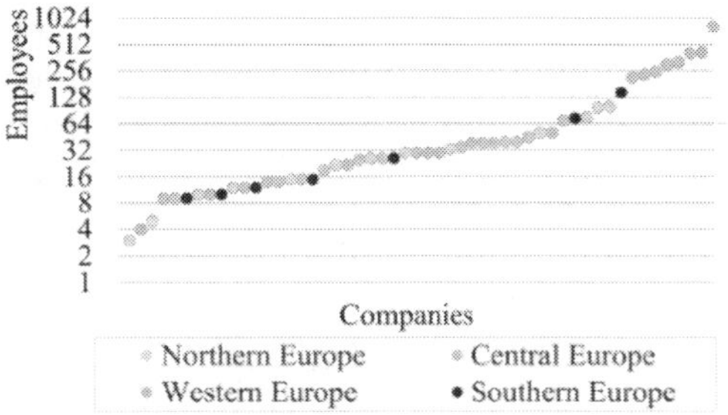

Figure 2: Number of employees per company of the 52 companies' data was acquired for.

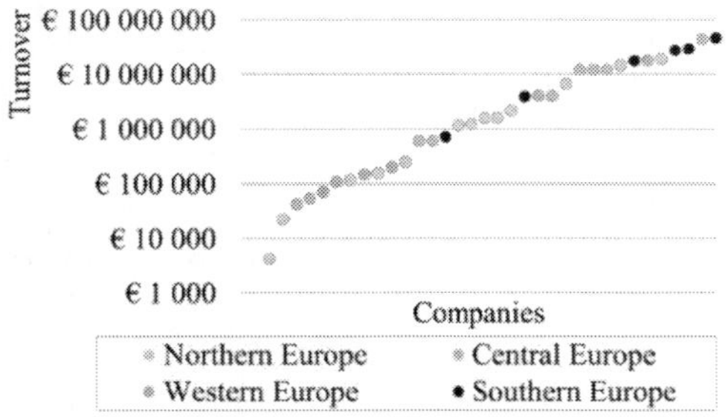

Figure 3: Turnover in 2021 per company of the 34 companies' data was acquired for.

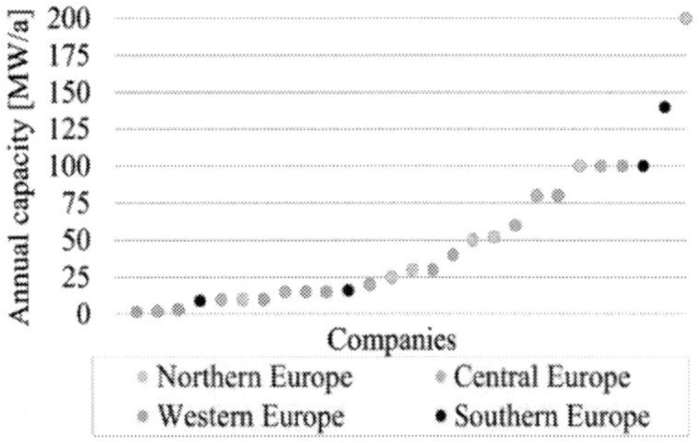

Figure 4: Yearly BIPV manufacturing capacity of the 27 companies' that were willing to share data.

silicon as their cell technology, whereof 16 were interviewed. It was revealed that 80% of interviewed companies source their silicon cells from China. Responding companies indicated that 67% and 63% of the sourcing of junction boxes and encapsulants, respectively, also were from China. Glass (for the front and back covers) dependence on China is lower, with only 27% mentioning it. Taiwan is also a significant silicon cell source; 38% of producers utilize it, while 44% use both China and Taiwan.

Notably, producers in Northern Europe also source silicon cells from the United States and the Philippines.

Regarding the other components, European silicon based BIPV producers acquire glass primarily from European countries (36%), such as Belgium, Germany, Italy, Spain and the Netherlands, as well as Türkiye, Vietnam, and India. junction boxes are primarily sourced from China (67%), while the secondary source for this component is European countries (unspecified). In addition, some interviewees also mentioned that they used to source junction boxes from Germany, but since the

10.4229/EUPVSEC2023/5DO.10.2

supplier closed its operations, this was no longer an option.

Encapsulants mostly come from non-European countries, namely China (63%) and Taiwan (13%), with 25% from Europe. Other components such as back sheets, wires, and foil vary in origin. The most mentioned suppliers were China and unspecified European countries. In addition, one also mentioned India as a supplier.

3.3. Value chain of CIGS- & perovskite-based BIPV

CIGS is the second most common cell technology used amongst the contacted companies, as nine out of 73 use this technology, whereof 3 were interviewed. Two of the contacted companies use perovskite cells, whereof one agreed to participate in this study.

The materials and components used in CIGS based BIPV differ from the ones used in silicon based BIPV, and the companies that participated did not share the origin in much detail. One company manufactures CIGS cells in-house, sourcing materials from Europe and Asia. The other two import the cells from the United States, however, one plans to change to a French supplier within 2–5 years. Regarding the other components, one producer mentioned challenges sourcing solar graded glass from Europe, the remaining sourced 50% from Europe and 50% from China. All firms source junction boxes and encapsulants from China. In addition, one evaluates a shift to a European supplier due to supply chain issues and expiration concerns regarding the encapsulants.

Lastly, the company that uses perovskite disclosed that their suppliers are mainly located in Europe, whereas the metal used is sourced from the UK but could be mined somewhere else. Although they did not specify which parts, the producer did mention that some specific parts were obtained from Japanese suppliers as well.

In summary, it can be concluded that the European BIPV value chain is strongly dependent on China, and Figure 5 illustrates the situation.

Figure 5: Visual representation of the countries that contribute to the upstream European BIPV value chain. Darker color indicates a stronger concentration of supply.

3.4. Upstream Challenges of European BIPV

A challenge that was mentioned throughout the interviews was the inability to source materials from Europe. This is not only due to a lack of suppliers in Europe, but also because few European suppliers can provide the quantity required by producers. There is, however, a very strong interest from European BIPV producers to source all materials and components from Europe, since it would decrease the risk of value chain disruptions, bring shorter lead times, and improve the environmental footprint of their products. Moreover, a clear mismatch was identified between European BIPV producers and possible European suppliers of materials and components. The two main drivers of this mismatch are: (i) that some of the materials and components are of a specific nature and cannot be found in Europe; and, (ii) that the prices offered by European suppliers are often not competitive with the international market, especially since the required volumes are too small to allow for economies of scale.

Another challenge mentioned by multiple interviewees was the limited control that they have when procuring materials, specifically silicon cells. It was stated that they perceive that the quality of the cells received does not match what was ordered, which could be due to more relaxed laboratory standards in China. At the same time, options are limited since the silicon cell supply is currently rather concentrated to a few countries.

3.5. Downstream Challenges of European BIPV

One identified downstream challenge is the general lack of awareness of BIPV in Europe. This was mentioned in connection with various actors, including installers, architects and policymakers. During all interviews, independently of the size of the producer, it was mentioned that policymakers and governmental entities lack awareness and understanding of BIPV, and therefore fail to regulate it properly. Firstly, BIPV must comply with both the PV and building regulations which differ throughout Europe. Secondly, there is no standardized regulation for BIPV in Europe and it is unclear for some actors if it legally should be treated as a PV system, a building material, or both. This misconception is further confirmed by the fact that some producers perceive BIPV mainly as either a PV system or a building material.

Multiple companies further expressed a general lack of knowledge of the architectural sector, especially those who do not have a relationship with architectural firms. This is important because architects are a primary actor, with a high influence for the diffusion of BIPV in Europe. The lack of awareness can further be seen in other aspects. For example, throughout the interviews, it has been mentioned that it is rather difficult to secure funding to further expand the production capacity. The interviewees further express that this obstacle may be due to a lack of understanding about the technology from investors, and that it may be considered too risky to invest in. Because of this, two companies have been pushed to close their production in Europe and move to China or the United States, where better support mechanisms exist for the technology, such as the Inflation Reduction Act in the United States.

Lastly, numerous producers noted the challenge of sourcing skilled installers. Some manage this by performing installations in-house or through close connections with the construction industry. In specific instances, BIPV project costs surge significantly due to installers and subcontractors, rendering the end price overly costly for customers. An interviewee even highlighted a potential tenfold price increase.

3.6 The niche of European BIPV

To summarize, the function that the socio-technical system of BIPV aims to provide is that of transforming energy-passive construction infrastructure, such as buildings, into active energy sources. Because of this, it can be interpreted that the two regimes that the BIPV socio-technical system interacts with are the electricity generation regime, including conventional PV, and the building regime. During the interviews, European BIPV was recurrently characterized as a niche, partly due to its specific function. In addition, BIPV is neither mature enough to compete directly with common construction

10.4229/EUPVSEC2023/5DO.10.2

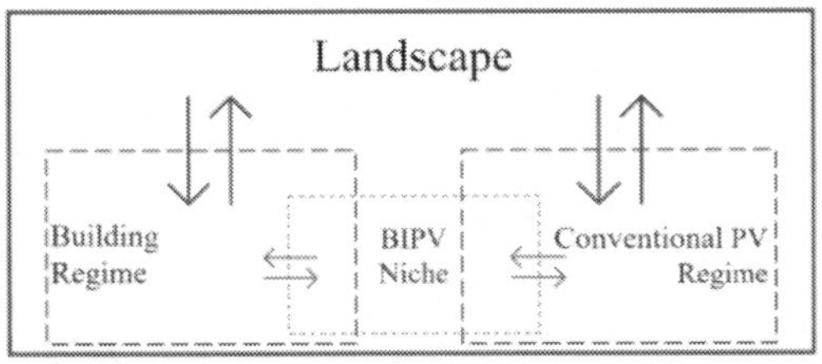

Figure 6: MLP characterization of the European BIPV industry.

materials, nor with more widespread PV systems, or any other energy source in general. It can therefore be argued that European BIPV found itself in the niche level of the MLP, as depicted in Figure 6. However, many of the interviewees responded that their goal with BIPV is not to compete against more diffused energy sources, but instead to serve a customer with specific needs, for which energy generation and aesthetics play an equally important role.

From the interviews, it can further be interpreted that the European BIPV industry is still going through a learning phase since there is still knowledge being developed that is essential to end the formative phase. This knowledge includes certification requirements and processes, as well as ways of working among actors, such as the sourcing and supply of materials and the approach taken when designing, developing and installing BIPV projects. Furthermore, from the perspective of BIPV producers, a lack of awareness from some important actors has been mentioned on multiple occasions, which suggests that more knowledge still needs to be developed.

From the interviews, it also can be inferred that European BIPV lacks a protected space to develop knowledge. One reason for this is that BIPV is usually categorized together with more conventional PV solutions, such as roof mounted modules and large-scale solar parks, which results in tension between the PV regime and the BIPV niche. This lack of differentiation and protected spaces hinders BIPV from developing its own standards and ways of working, and instead, the standards of the PV regime are applied to BIPV. Another reason mentioned in multiple interviews for the lack of protected spaces in the BIPV niche is the limited financing and incentives to invest in the industry.

Finally, the landscape in which the BIPV niche and the mentioned regimes interact involves the current energy supply infrastructure and the building infrastructure in Europe, as well as the current environmental, political and societal context. All these factors that shape the landscape put pressure on the regimes to transition into new ways of generating energy and constructing buildings. These factors also add pressure on Europe to become more energy-independent from foreign fossil fuels. An evident way in which this pressure is manifesting in the regimes is through the policies and regulations proposed by the EU, mentioned in the introduction. The interaction between the levels in the nested hierarchy presented is mainly driven by actors with different roles, responsibilities, influence, and interests, as seen in Figure 7, adapted from Corti et al. [12]. Furthermore, the actors are categorized according to their relationship with European BIPV producers, and are listed in Table I.

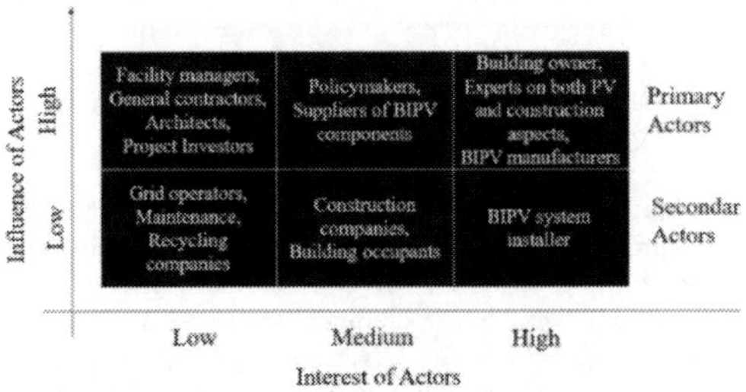

Figure 7: Classification of actors regarding their influence and interest of BIPV, based on Corti et al. [12].

4 DISCUSSON

The results from the value chain analysis indicate that the European BIPV industry relies heavily on foreign suppliers. The dominant country of origin, however, varies depending on cell technology. Silicon-based BIPV, which dominates the market, gets most key components from

Table I: Classification of actors within the European BIPV Industry.

Actors	Category & Definition
Research institutes, universities, R&D, PV and BIPV industry associations, and consortiums	**Knowledge Development and Sharing:** These actors are of importance because they support with research and development of BIPV. These developments could result in BIPV becoming more attractive due to higher efficiencies, new materials, or new forms to adapt the technology into the building.
Laboratories, governments, certification institutes	**Testing and Certification:** The actors in this group provide guidance and frameworks for certification of BIPV products so that the modules can be installed into buildings across Europe.
Governments, private investors, banks, and other companies	**Funding and Financing:** This group of actors are important to the producers of BIPV from an economical and financial perspective, as they provide financial support for companies to operate and expand.
Architects, installers, façade and roof companies, builders, distributors, construction companies, and developers	**Planning and Installation:** The actors in this group are of utmost importance for a successful implementation of a BIPV project. Each of them has a specific role throughout the process and collaborate closely among each other, and with the producers of BIPV.
Policymakers and entrepreneurs	**Industry development and support:** These actors have the ability to influence the development and diffusion of BIPV among all actors. They support with the legitimation of the technology, which can present itself in the form of legislation favoring BIPV, special demonstration of the technology, and sharing of BIPV success stories.
Companies, owners of buildings, and government	**Final customer:** These are the actors that purchase the BIPV products and for whom the BIPV project is performed.
Suppliers, international partnerships, sales and technical consultancy	**Other:** The actors in this category have different ways to support and collaborate with producers of BIPV. For example, they could include suppliers, or they may provide market knowledge, special partnerships for collaboration, or access to a wider range of contacts in other industries

10.4229/EUPVSEC2023/5DO.10.2
020502-005

manufacturers in China. CIGS cells, on the other hand, are sourced from suppliers in the United States or Europe. The strong dependency on foreign suppliers, especially Chinese, is due to the economies of scale and low cost achieved by these producers, which creates barriers for other producers to enter the market. Interviews do, however, indicate a preference for European suppliers if large enough quantities were available. Hence, there is an unmet demand for European upstream BIPV components and materials.

The European BIPV niche exists at the intersection of the conventional PV and building regimes. The expertise of these two regimes needs to be combined when implementing BIPV. This fusion could yield a distinct BIPV expertise and more well-defined roles and responsibilities for involved actors. Due to the diverging perceptions of BIPV as either a PV system or a building material, an industry-wide agreement on a definition of BIPV and EU-wide BIPV standards would likely facilitate implementation and certification processes. However, this would also make it easier for foreign producers to enter the European market, which could result in more competition and a weaker domestic value chain. Furthermore, a market niche, protected by policy, would enable the accumulation of knowledge and practical experience required for future competitiveness with conventional roof mounted PV modules as well as conventional building materials.

One limitation of this study is its relatively small sample size of participating manufacturers. Out of 71 contacted companies, only 23 (31%) engaged in the study, leading to a somewhat narrow scope. Among the 48 remaining companies, two declined, and the rest did not respond. Furthermore, of the 71 companies included, 51 utilized the silicon cell technology, nine employed CIGS cells, and two utilized perovskite cells. The study is therefore somewhat angled towards producers of silicon-based BIPV and their perception of the European BIPV market. However, silicon cells are the most used cell technology in the PV market, which makes the sample of producers in this study representative of the market.

Despite this bias, the geographically varied participants provide a comprehensive perspective, offering well-representative findings of European BIPV perceptions. Future studies regarding the European BIPV industry would benefit from a broader range of cell technologies and more European producers. Including the processes further up in the value chain, especially module components and raw material production, would add insight. Further investigations could potentially also delve into actor network dynamics within the industry.

There are, however, some limitations to the study. Firstly, due to practical constraints, not all European BIPV module manufacturers are included; the selection is based primarily on data from the European Solar Manufacturing Council (ESMC). Secondly, the study examines the upstream value chain in terms of component sourcing — restricted to the specific components of PV cells, junction boxes, front and back covers, and encapsulants — disregarding the origin of raw materials. Lastly, import volumes are not factored into the value chain evaluation.

5 CONCLUSIONS

The silicon cell technology is the most dominant technology used in European BIPV modules. Overall, most key components used in the module, such as silicon cells, junction boxes and encapsulants originate mainly from China, whereas glass is the only key component that is mainly sourced from European countries. Other cell technologies used in European BIPV are, for example, CIGS and perovskite cells. Although most of the producers of these technologies rely on imports, there are companies who either produce the cells themselves, source them from European producers, or will, in the near future, source them from European suppliers. Nevertheless, since the CIGS and perovskite BIPV segment is significantly smaller than the silicon one, there is an overall large import dependency. While many producers currently import components, interviewees demonstrated a preference for domestically sourced components. However, due to the mismatch between producer requirements and domestic supply, it is usually not possible to meet this demand. Policy packages, like the Green Deal Industrial Plan for the Net Zero Age, could potentially stimulate growth of domestic production capacity and expertise, via faster access to funding and a simplified regulatory environment.

The downstream value chain involves diverse actors, including investors, policymakers, architects, and installers. Interviews reveal a lack of BIPV awareness, which increases the perceived investment risks, pushing producers to expand in the United States and China. Measures to enhance general knowledge about BIPV is therefore key in order for the European BIPV industry to develop and expand. Certification poses another challenge. BIPV producers need to comply with both PV regulations and varying regional building regulations. Standardized EU-wide BIPV regulations could simplify certification. Nevertheless, this could also facilitate the entry of foreign producers into the European market, potentially leading to increased competition and a less resilient domestic value chain.

6 REFERENCES

[1] European Commission. (2022a). Questions and Answers on the REPowerEU. European Union, (August). https://ec.europa.eu/commission/presscorner/detail/en/qanda_22_3132

[2] European Commission. (2022b). REPowerEU: A plan to rapidly reduce dependence on Russian fossil fuels and fast forward the green transition.

[3] European Commission. (2023). The Net-Zero Industry Act: Accelerating the transition to climate neutrality. Retrieved May 16, 2023, from: https://single-market-economy.ec.europa.eu/industry/sustainability/net-zero-industry-act_en

[4] Özkanlisoy, Ö., & Akkartal, E. (2021). The Effect of Suez Canal Blockage on Supply Chains. Dokuz Eylül Üniversitesi Denizcilik Fakültesi Dergisi. https://doi.org/10.18613/deudfd. 933816

[5] European Commission. (2018). Directive (eu) 2018/2001 of the European parliament and of the council of 11 December 2018 on the promotion of the use of energy from renewable sources (recast).

[6] European Commission. (2021). Energy performance of buildings directive. Retrieved on March 13 2023, from: https://energy.ec.europa.eu/topics/energy-efficiency/energy-efficient-buildings/energy-performance-buildings-directive_en

[7] European Commission. (2022c). EU Solar Energy Strategy. Retrieved April 6, 2023 From: https://eur-lex.europa.eu/legalcontent/EN/TXT/HTML/?uri=CELEX:52022DC0221&from=EN
[8] European Solar PV Industry Alliance. (2022). Launch of the European Solar PV Industry Alliance - European Solar PV Industry Alliance. Retrieved March 13, 2023, from https: //solaralliance.eu/news/launch-of-the-european-solar-alliance/
[9] Berger, K., Cueli, A. B., Boddaert, S., Del Buono, S., Delisle, V., Fedorova, A., Frontini, F., Hendrick, P., Inoue, S., Ishii, H., Kapsis, C., Kim, J.-T., Kovacs, P., Chivelet, N. M., Laura, M., Machado, M., Schneider, A., & Wilson, H. R. (2018). International definitions of BIPV. Report IEA-PVPS T9-18: 2018, 32. https://iea-pvps.org/wpcontent / uploads / 2020 / 02 / IEA - PVPS _ Task _ 15 _ Report _ C0 _ International _ definitions_of_BIPV_hrw_180823.pdf
[10] Bonomo, P., Eder, G., Chivelet, N. M., Eisenlohr, J., Frontini, F., Kapsis, C., Scognamiglio, A., Wilson, H. R., & Yang, R. (2021). Categorization of BIPV Applications. IEA-PVPS T15-12:2021. https://iea-pvps.org/wp- content/uploads/2021/09/IEA-PVPS-T15-12_2021_BIPV categorization_report.pdf
[11] Hellin, J., & Meijer, M. (2006). Guidelines for value chain analysis, food and agriculture organization (FAO). UN Agricultural Development Economics.
[12] Corti, P., Bonomo, P., Frontini, F., Macé, P., & Bosch, E. (2020). Building Integrated Photovoltaics: A practical handbook for solar buildings' stakeholders Status Report, 107. www.solarchitecture.ch
[13] Geels, F. W. (2002). Technological transitions as evolutionary reconfiguration processes: A multi-level perspective and a case-study. Research Policy, 31 (8-9), 1257–1274. https: //doi.org/10.1016/S0048-7333(02)00062-8
[14] Bergek, A., Jacobsson, S., & Sandén, B. A. (2008). 'Legitimation' and 'development of positive externalities': Two key processes in the formation phase of technological innovation systems. Technology Analysis and Strategic Management, 20 (5), 575–592. https://doi. org/10.1080/09537320802292768
[15] Geels, F. W. (2005). Processes and patterns in transitions and system innovations: Refining the co-evolutionary multi-level perspective. Technological Forecasting and Social Change, 72 (6 SPEC. ISS.), 681–696. https://doi.org/10.1016/j.techfore.2004.08.014
[16] Rip, A., & Kemp, R. (1998). Technological change. In S. Rayner & E. Malone (Eds.), Human choice and climate change (pp. 327–399). Battelle Press.

IEA PVPS

Comparison of BIPV Innovation System Structures in Multiple Countries – Task 15.2 Subtask A

Michiel van Noord, Nuria Martín-Chivelet, **Momir Tabakovic**, Rebecca Yang, Nilmini Weerasing Otto Bernsen, Wilfried G.J.H.M. van Sark, Francesca Tilli, Angelo Baggini

Technology Collaboration Programme by iea

EU PVSEC, 21 Sep. 2023

020503-001

40th European Photovoltaic Solar Energy Conference and Exhibition

BIPV systems in a real building case (Example)

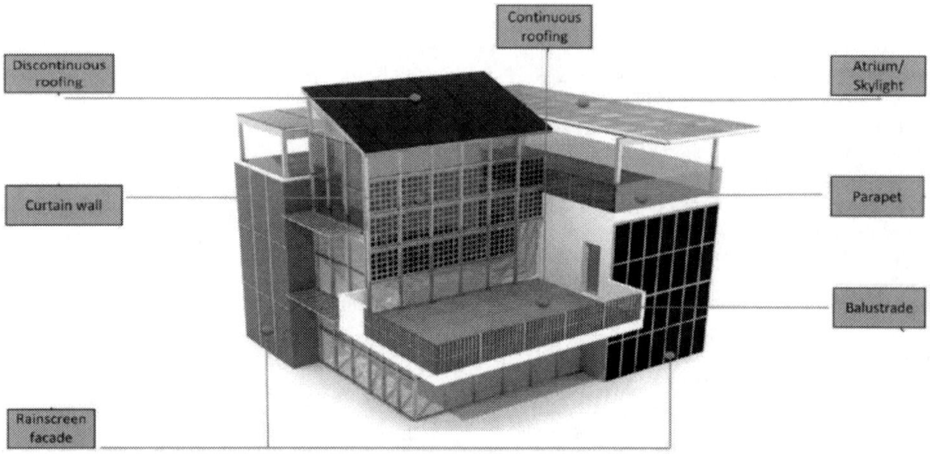

PVPS

Source: SUPSI

2

020503-002

Analyses for 7 countries

- Rebecca Yang, Nilmini Weerasinghe, Chathuri Gunarathna, Pabasara Wijeratne (RMIT)
- Momir Tabakovic, Stefan Savic, Andreas Türk, Thomas Schostal (FH Technikum Wien)
- Francesca Tilli (GSE), Angelo Baggini (Uni. Bergamo)
- Otto Bernsen (RVO), Tjebbe Vroon (UU/Stedin), Wilfried van Sark (Utrecht Uni.)
- Nuria Martín Chivelet (CIEMAT), Lucas García García, Estefanía Caamaño Martín, David Racero Patino (UPM)
- Michiel van Noord, Peter Kovacs, Malin Unger, Martin Warneryd (RISE), Bengt Stridh (MdU)
- Janne Halme, Roni Simola, Iiro Mikola (Aalto University)

PVPS

All reports will be published on the Task 15 webpage.

TIS report for Spain published

3

Analysis of the Technological Innovation System (TIS)

- Process

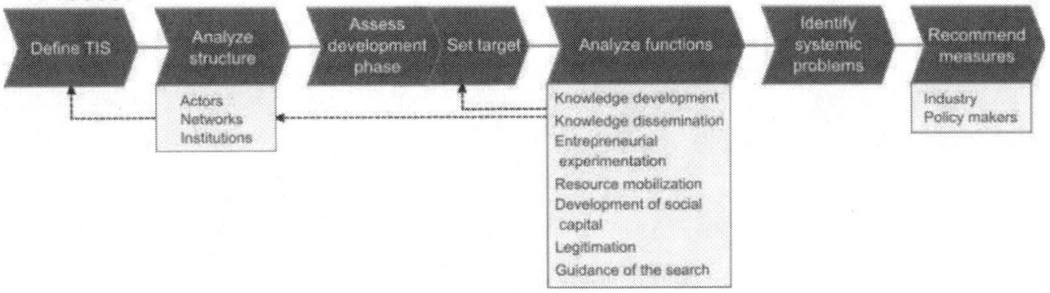

- The Guide for Technological Innovation System Analysis for BIPV is published on the IEA-PVPS Task 15 website. https://iea-pvps.org/key-topics/guide-for-technological-innovation-system-analysis-for-building-integrated-photovoltaics/
- The general TIS framework is well-described and anchored in academic literature (major contributions by M.P. Hekkert, A. Bergek)

Image: own adaptation of Bergek et al. (2008)

PVPS

4

Analysis of the Technological Innovation System (TIS) for BIPV

- Process and questions to answer:

What is BIPV?

Who is active? What boundary conditions?

What is the current market status? What do we want to reach?

What is done to create the conditions needed? Are these the right things to do? By the right actors? Is it enough?

What/whom is lacking? Who is not doing its job?

What can industry and policy makers do to advance BIPV?

- Data sources:

Interviews – Questionnaires – Projects - Patent applications – Publications – etc.

PVPS

Image: own adaptation of Bergek et al. (2008)

5

020503-005

40th European Photovoltaic Solar Energy Conference and Exhibition

Market development

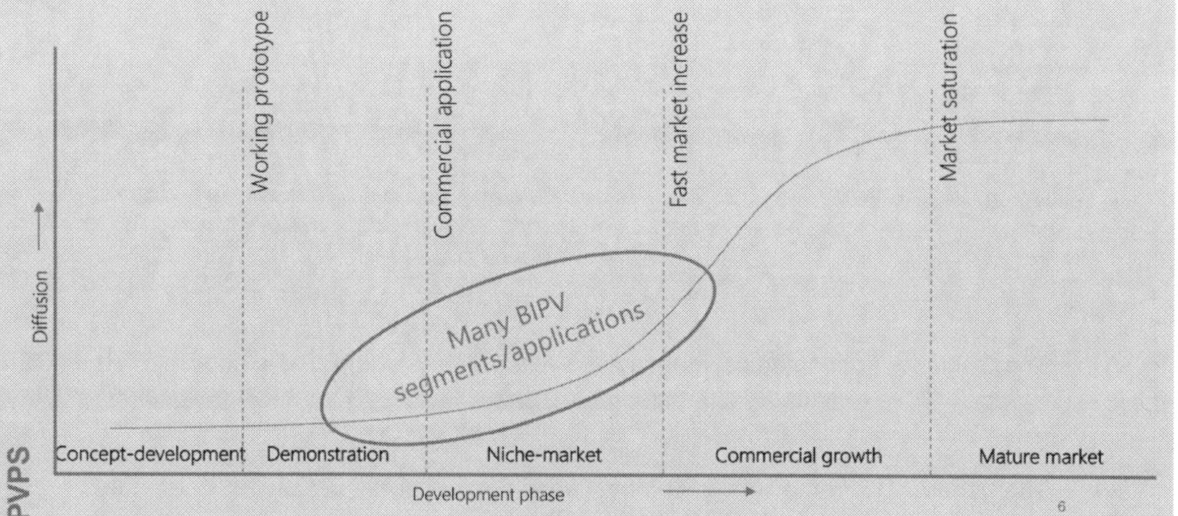

PVPS

6

020503-006

iea-pvs.org

Markets and boundary conditions

Structural analysis results

IEA PVPS

Technology Collaboration Programme by iea

40th European Photovoltaic Solar Energy Conference and Exhibition

Dominating BIPV applications

Nr.	Categories[1]	Austria	Australia	Italy	the Netherlands	Spain	Sweden
A1.r	Discontinuous roofing, regular modules						
A1.s	Discontinuous roofing, tiles						
A2	Continuous roofing						
B1	Atrium/Skylight						
C1	Rainscreen façade						
C2	Masonry wall						
C/D3	Double skin façade						
D1	Curtain wall						
D2	Window						
E1	Parapet						
E2	Balustrade						
E3	Canopy						
E4	Solar shading						

Colour map

- Concept dev.
- Demonstratio n
- Niche market
- Commercial growth

PVPS

[1] Source: https://iea-pvps.org/key-topics/categorization-of-bipv-applications/

8

Financial incentives

- **Specific BIPV support**
  - Bonus in capital subsidy for BIPV pergola in Spain
  - Bonus in capital subsidies (+30 % for innovative PV Systems) in Austria
- **PV in buildings support**
  - Tax deduction/Fiscal incentives: BAPV, NOT for BIPV in the Netherlands
  - Support for replacing asbestos roofs with PV or BIPV, especially in farms in the Netherlands and Spain
- **Support, not specific for BIPV nor PV in buildings**
  - Self-consumption scheme: Net- metering or net-billing scheme in Australia
  - Tax deduction on renovation labour for private houses (incl. PV) in Sweden

PVPS

9

020503-009

40th European Photovoltaic Solar Energy Conference and Exhibition

Patent and Utility Model (UM) applications

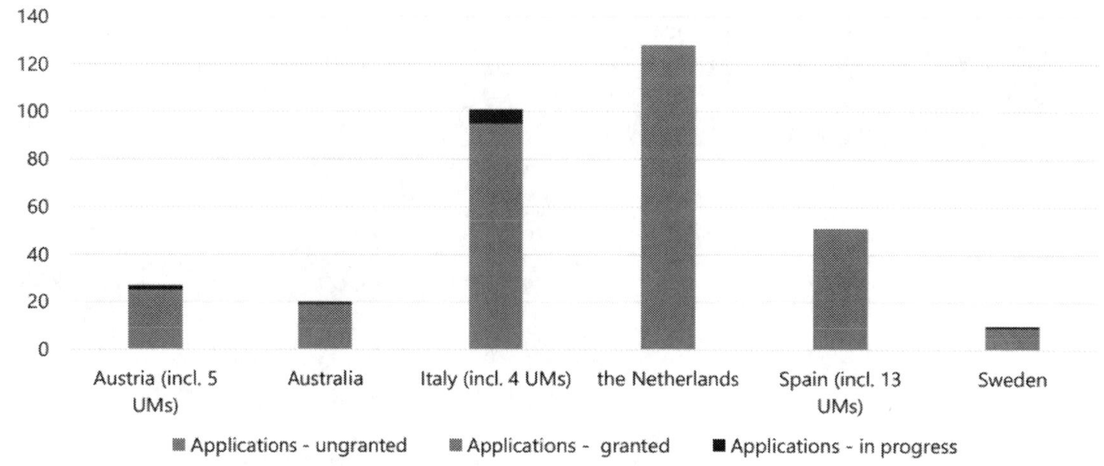

PVPS

10

020503-010

BIPV regulations

- **Specific BIPV support**
  - BIPV benefits from low-cost loans for PV products improving energy efficiency in dwellings in Australia
  - BIPV benefits from the government support for innovation and solar technologies R&D in the Netherlands
- **PV in buildings support**
  - In the Viennese building code an obligation to install renewable energy sources on all new buildings
  - Building permit to install PV modules in buildings is suppressed throughout Spain since July 2023 in Spain
  - The Energy Label is greatly improved by installing solar panels, not specifically BIPV in the Netherlands
- **Barriers**
  - All of them agreed that BIPV is not addressed at all in politics, only PV in all Countries
  - No specific funding for BIPV in all Countries
  - BIPV products are not considered construction materials in all Countries
  - No specific attention to BIPV in public procurement regulations: BIPV competes with BAPV, which economically does not usually favour BIPV.

PVPS

11

020503-011

40th European Photovoltaic Solar Energy Conference and Exhibition

Interested in methodology and more results?

Earlier publications:

- TIS-analys BIPV in Spain (2022)
- Guide for TIS-analysis (2023)

...and more to come:

- TIS-analyses for BIPV in Australia, Austria, Sweden, Netherlands, Italy (2023)
- Synthesis report (2023/2024)

https://iea-pvps.org/research-tasks/enabling-framework-for-the-development-of-bipv/

PVPS

12

020503-012

www.iea-pvps.org

Thank you!

FH-Prof. Ing. Momir Tabakovic PhD. MSc - Task 15
momir.tabakovic@technikum-wien.at

IEA PVPS

Technology Collaboration Programme
by iea

020503-013

DEVELOPMENT AND PILOTING OF A NOVEL DECENTRALIZED SOLAR PV CHARGING SYSTEM FOR SUSTAINABLE E-MOBILITY IN RURAL AFRICA

Korbinian Götz[1], Jakob Steger[1], Clemens Pizzinini[1], John (Ioannis) A. Tsanakas[2]*, Johan Giliomee[3], Tafesse Abera[4], Menelaos Meli[3], Arnold Rix[3], Thinus Booysen[3], Markus Lienkamp[1]

[1] Technical University of Munich, Institute of Automotive Technology, Germany
[2] Commissariat à l'énergie atomique et aux énergies alternatives (CEA) - Institut National de l'Énergie Solaire (INES), France
[3] Stellenbosch University, Stellenbosch, South Africa
[4] Adama Science And Technology University (ASTU), Adama ,Ethiopia

*Corresponding author: ioannis.tsanakas@cea.fr

ABSTRACT: This paper presents the rationale, objectives and preliminary research results of the EU-funded project SolChargE. The project addresses the two major barriers for rural communities in sub-Saharan Africa (SSA): (1) no access to electricity and (2) no access to mobility. Therefore, the project proposes a decentralized transportation hub (DTH) based on a stand-alone solar charging station to harness solar energy and replace transportation which depends on pollutant internal combustion engines. The project targets 33 million smallholder farmers in SSA who rely on transportation to improve the economic success of the agricultural sector in SSA. SSA has abundant access to sunlight, making it a perfect location for solutions based on photovoltaic (PV) technologies. First, the project will develop a stand-alone solar charging station to be able to use solar energy for electric mobility in a decentralized manner in rural areas. Smallholder farmers will benefit from the use of electric vehicles as part of a transportation hub to address their mobility challenges and access to economic success. The vehicles will be utilized to transport agricultural goods, using the electricity generated in the process. To make the service available to the users and run it profitable, the DTH is combined with a service model that provides on-demand transportation. The impact of the system will be tested in Ethiopia and South Africa and the whole DTH will be evaluated with a life cycle analysis (LCA). The LCA addresses the potential to reduce the environmental impact of electric vehicles and the solar charging station with the means of circular economy.

Keywords: *e-mobility; solar charging station; off-grid PV system; decentralized transportation hub; life cycle analysis.*

1 INTRODUCTION – CONTEXT

1.1 Sustainable transportation and Stand-alone Solar Charging Stations

Sustainable, affordable and reliable transportation is indispensable in the rural African context. Current transportation solutions in rural Africa are scarce and highly unsustainable where found. Where a transport solution is present, the majority of vehicles are imported near the end of their lifetime. As a result, these vehicles are fuel inefficient, highly pollutant and present high costs of spare parts and maintenance. Besides such means of rural transport are based on imported fuel, which leads to issues as increased greenhouse gas emissions, fuel cost currency dependency and sporadic unavailability during shortages.

Off-grid solar PV charging stations comprise a very topical and emerging solution for remote electrification, particularly in regions such as sub-Saharan Africa (SSA). Recent developments focus mostly on improving the energy management, sustainability, and bankability of such systems. Additionally, decreasing the installation and operations and maintenance (O&M) cost is also large focus point, as the cost of PV modules themselves has significantly decreased in recent years [1-5]. Stand-alone solar charging stations (SASCS) raise a number of technical issues, mostly related to the lifetime durability (and sustainability) of the batteries. It was shown that, since the end-user practically pays for the number of charges rather than for the energy use, batteries are often discharged below the appropriate discharge level. This, in turn, significantly shortens their lifetime [1]. Even when SASCSs are community-owned, the lack of users' technical training leads to deep discharges and thus lower service lifetimes in the long-term. Thus, another relevant challenge whose importance has been highlighted by several earlier studies is that of the need for local technical training of SASCS users and asset managers. In particular, in order to develop a solid business model, a solar charging station has to be monitored and serviced. This requires technical skills, which is not usually available locally, but that would otherwise be too costly if left unsupervised. From a market and end-user perspective, SASCS typically address the electricity demand of low income-low demand end users in the proximity, which cannot afford or access their own reliable electricity generation system. Thus, one can classify the market segment of solar charging stations as the first step towards electrification. The following step would be to own and operate such system(s), which would provide better flexibility and availability of the energy service (more power and at a lower cost per unit), yet require a higher income (either for the initial investment or, in the long-term if it is to be acquired e.g. by a microfinancing approach). Subsequently, solar PV-powered mini-grids would be the third step, providing another level of energy services, both in quantitative and qualitative terms (e.g. self-consumption, flexibility, etc), at an even lower cost [1].

On a broader view, there are also many socioeconomic challenges towards a greater market uptake of SASCS, as there is not yet one definite, sustainable and successful business model that can be replicated ubiquitously [6,7]. In fact, limited but existing research on SASCS shows that scalability is possible with such technology [8]. As true as this conclusion seems, questions remain regarding the distribution, exploitation, business modelling, and financing of SASCS – questions that have so far gone unanswered in the research. This means that the scalability of SASCS stands at a low TRL, i.e. around 2 to 3.

1.2 LCA Analysis of renewable energy systems in Africa

The current developments in transport emissions show the relevance of leapfrogging fossil fuels and easing the barriers of electric mobility in Africa: According to Ayetor et al. (2021), Africa's transport emissions are growing every year at a rate of about 7 %, compared to a growth of 0.8 % in the United States of America [9]. The authors state that with the current rate, the emissions in Africa will double by 2030 compared to the ones of 2015 if no measures are established. In addition, Africa is the second most populated continent with the lowest level of motorization in the world [10]. To overcome the connected burdens of a lack of mobility and rising emissions due to motorization, SolChargE proposes a local emission-free solution for rural transport with solar energy. To show the potential in mitigation of emissions the project proposes a Life Cycle Analysis (LCA) of the system regarding the decentralized transportation hub (DTH). Besides the investigation of the status quo of road vehicle emissions, [9] gives advice for future actions to lower transportation emissions. This implies the assessment of future transport systems and their impact on how they decrease emissions in a standard methodology as proposed by LCA. Farquharson [11] analysed the impact of the electrification of the motorcycle fleet in Kigali, Rwanda. Therefore, the author analysed trip data from a commercial ride sharing platform with 118 drivers in a 30-day period. The fleet is modelled with a comparable commercially available electric motorbike with a swappable battery concept. Different scenarios are analysed for battery charging, such as from the current electrical-grid generation mix, a peat-based and an increased solar-based electricity generation. The author assesses the resulting emissions and costs and concludes that electrification may not only offer emission and health benefits but also an economically feasible solution. Yet, the electricity mix for charging has the major influence on the emissions and shows the potential for solar charging solutions [11].

Schünemann et al. [12] proposed electrified cargo bikes as an eco-friendly freight transportation in cities which can be powered by commonly used mini grids in African countries. The authors raise the question if the cargo bikes are environmentally friendly, if they are assessed for their impact on the environment during the whole life cycle, and how they perform compared to other urban transport modes. The question is answered by conducting an LCA on the use case of an electric cargo bike used in Ghana. The LCA show savings of 4-8 % in greenhouse gas emissions with the inclusion of solar-powered e-cargo bikes in the Accra region. Their results show the relevance of LCA as well as solar powered charging [12]. The same results are shown by [13], who assessed the impact of replacing an ICE bus fleet with electric busses in Africa over their whole lifetime. The authors do not just see a reduction of lifetime costs up to a maximum of 47 % in South Africa due to solar charging, but also significantly lower CO_2-emissions. Charging the electric bus from the grid exemplary in South Africa causes 489 % more emission than with solar energy.

2 OBJECTIVES – SCOPE/METHODOLOGY

2.1 The project's motivation and objective

Communities in SSA face a multitude of interconnected challenges rooted in their lack of access to electricity (Fig.1). Rural areas are increasingly left behind; for instance in Ethiopia, only 1 % of the rural population has access to electricity, even though they make up for 85 % of the total population [14]. The most solutions for providing fast access to electrification are based on decentralized solutions, which are able to function without direct connection to major energy sources. In the context of SSA, many (if not most) of said decentralized solutions have come to rely on solar energy–and with reason. With its overabundance of sunlight (some regions average 6.5kWh/m2 of solar radiation per day), sub-Sahara is seemingly the perfect location for solutions reliant on photovoltaic (PV) technologies. However, it remains unclear whether decentralized solar-powered solutions will be able to tackle the challenges facing SSA today (and in the future).

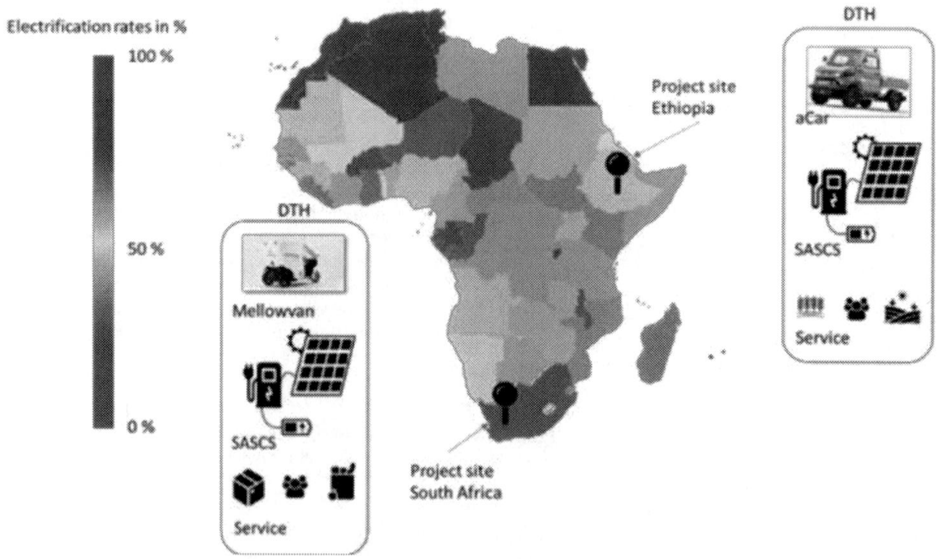

Figure 1: Electrification rates of the African continent; presentation of the project sites with the respective decentralized transport hubs (DTH) and their components.

10.4229/EUPVSEC2023/5DO.10.4
020504-002

We reckon that, while decentralized energy solutions may be one piece of the may be one piece of the puzzle, it alone cannot adequately confront 'wicked' problems currently impacting SSA's rural communities, to achieve sustainable development. This is especially true of the difficulties facing the continent's 33 million smallholder farmers who, even when given access to solar energy, continue to encounter barriers to economic success. The project proposes electric mobility in combination with decentralized solar energy, to improve the live and economic success of the rural population. The research question arises, how can the region's inhabitants be addressed by transport modes, which offer them the access to points of interest, in a manner which permits rural SSA to leapfrog over noxious modes of transportation reliant on polluting internal combustion engines?

The SolChargE project's answer is the investigation of the impact of a decentralized transportation hub (DTH) which includes a standalone solar charging station (SASCS), a battery electric vehicle (BEV) and an operational model. On this aspect, the project will seek to define and assess parameters specific to SSA (Ethiopia and South Africa in particular) and design a DTH (Fig. 1). The methodology for the system design of the three components (SASCS, BEV, operational model) must ensure scalability and parametrization to different locations and circumstances.

To assess the impact of the system, a Life Cycle Assessment (LCA) and the Life Cycle Costing (LCC) will be performed to ensure every step is able to include positives outweighing possible economic, social, and environmental negatives. The specific details associated with such analyses will be defined as the project moves forward and is further described in the next section. It will be our team's endeavour to bring the scalability of SASCS, especially in Africa, to a TRL of at least 5 thanks to our analyses' application within relevant environments.

2.2 Methodology of the project

The methodology for this research project is broken up into five axes/milestones: (1) understand, (2) develop, (3) production, (4) operation and (5) recycle.

- For the first milestone of the project, we start with the understanding phase. Requirement analysis is done across diverse work packages such as the SASCS development, business model design, capacity building and DTH definition. Literature review is conducted to understand contextual parameters which will influence these work packages. This includes technical parameters for the SASCS, a database on the best practices for the business model and the identification of local stakeholders. Concluding with interviews with experts in the field, parameters for these work packages are finalized and incorporated in a database.
- Succeeding requirement analysis, the development phase commences for the second milestone. Layout and prototyping for the SASCS is done by developing a technically feasible concept and building a minimum viable model to be tested in a lab environment. Optimal configurations are identified and adopted for the next phase. The business modelling is based on the full product life cycle and based on stakeholder needs, value chains and information flows. The business model contains a parametrizable cost model, which is adjustable to different locations and circumstances in order to investigate the necessary pricing strategy for the DTH operation and its payback period. Capacity building for each stakeholder comprises workshops and lectures offered on concept development. The development phase concludes with component and parametric system modelling for the DTH.
- Production starts with local sourcing and procurement assessment for the SASCS. Required materials and components are determined with a focus on regional or national sourcing of second-life components. An in-lab production facility to test innovative solutions is set up in South Africa and Ethiopia where a general quality measure for all components and materials is developed. Along with a production strategy developed for the business model, capacity training focusses on production. Finally, a fully functioning prototype is built by students and technicians involved in capacity building. Meanwhile, operational training for the following phase also commences. Simultaneously, a simulation model based on analysis of scalability, environmental impact and socio-economic impact is built to determine the optimal DTH configuration across a defined area.
- Operation commences with the implementation and evaluation of the DTH. It will prove successful by working seamlessly in the field and generating valuable data that can be analyzed with a focus on optimizing second-life potential. Feedback from all DTH stakeholders regarding the developed business model is utilized to develop a proof of concept thereof. Focus is put on the user-facing pricing strategy and transport services offered by the DTH. To prepare for the final recycling phase, capacity building regarding efficient recycling commences during the operation phase. As part of the DTH simulation assessment, a life cycle impact assessment with a special focus on business model transferability is conducted.
- The recycling of the DTH aims to ask an important question: How does the operation influence the quality of DTH components and their usability for a circular usage? Each component is categorized according to the different circular principles; whilst CEA performs further diagnostics on PV module reuse and battery degeneration. An optimal recycling strategy is developed accordingly and documented in an online handbook. Finally, all project-related findings are to be made publicly available and technical manuals made open source as part of knowledge dissimilation. The project concludes with results validation of the DTH simulation and impact assessment.

3 PRELIMINARY RESULTS

As shown in the methodology, the first steps in the project include the collection of requirements for the DTH in order to design a standalone solar charging station, which is the enabler for the BEV to run the operation. Since the requirement of the project is to offer a cost-efficient and scalable solution, it is evident that the cost of the standalone solar charging station plays an important role in the design process. Additionally, the design shall not be limited to one specific location.

10.4229/EUPVSEC2023/5DO.10.4
020504-003

3.1 Background

Standalone charging stations are necessary for the success of BEV adoption worldwide, especially in rural areas with no grid connection [15, 16]. Studies show that standalone PV systems emit less CO_2 over their lifetime compared to conventional systems run by fossil fuels [16, 17, 18]. However, the challenges of these stations are the economic viability and system reliability due to high capital cost, limited PV and battery capacity and the intermittency of the PV technology [19]. Therefore, the authors of [16, 20, 21] conclude that more accurate techniques for better energy management are necessary to make standalone PV stations competitive, with the potential being in more accurate energy scheduling for optimal energy utilization.

In the context of this research, the objective is to identify and analyse the electricity cost that standalone PV charging stations can provide for user-selected locations and different mobility scenarios in rural areas of SSA. Moreover, the research must show how reliability affects costs, with PV systems not being considered reliable compared to conventional fossil fuel methods [19].

3.2 Sizing of the standalone charging station

The stationary battery of the standalone charging station is one of the most significant cost contributors for PV systems with the purpose of enabling reliability during solar intermittency [18]. Therefore, the applied common sizing approach of [22], which uses autonomy factors based on the nocturnal energy demand, and the maximum days in a row without enough solar irradiation, tends to oversize the battery and is therefore not suitable for cost-sensitive applications. The main potential lies in oversizing the provided PV power to enable the stationary battery to be charged optimally in challenging solar periods [23].

The aim of the presented methodology is to fundamentally analyse how much PV power is constantly available to charge the BEV battery for the demanded mobility and the stationary battery for storage. This goal requires the correlation of the PV power with the capacity of the stationary battery. Hereby, a reasonable design of the whole charging station is achieved instead of only focusing on single components as the stationary battery.

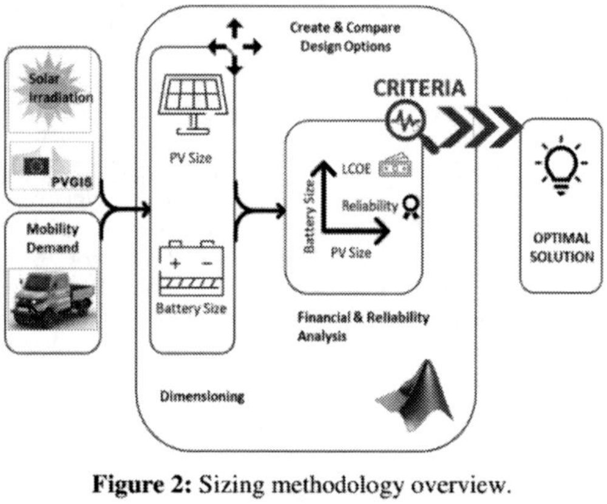

Figure 2: Sizing methodology overview.

Figure 2 gives an overview of the methodology leading to the optimal solution. The input consists of the mobility scenario to determine the energy demand and the solar irradiation data of the location where the station is operating. Those are used to size the PV power and the stationary battery capacity by generating multiple design options. The optimal design is then chosen by the reliability criteria and the Levelized Cost of Electricity (LCOE [€/kWh]).

The daily energy demand is determined from the chosen mobility scenario and is used as a basis to determine the standard size of a PV panel area according to [22]. For this project, the mobility scenario allows the vehicle to be charged over night from 23 p.m. to 8 a.m. and during noon from 12 a.m. to 2 p.m. (Fig. 3). During the rest of the day, the vehicle must be available for driving.

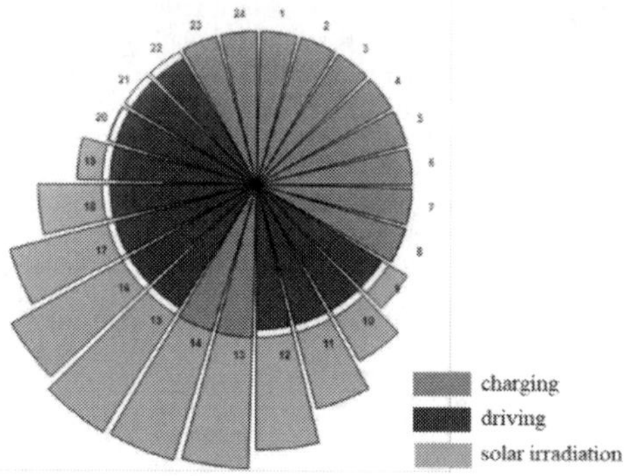

Figure 3: Mobility scenario over a day with 2 h of charging during the day.

In order to find the optimal design, multiple design options are generated: Those include the necessary PV power, being oversized by a factor ranging up to twice the initial value which is derived from the method of [22], and a range of fitting stationary battery capacities. The range of PV power and battery capacity is displayed on the horizontal and vertical axis in Fig. 4. All design options are then analysed for energy availability over the period of the year 2020 with hourly PVGIS solar irradiation data and hourly mobility demand data. The simulation analyses hourly if the vehicle must be charged and if either the PV panels provide the energy directly or the stationary battery. If no car is available for charging, the produced PV energy charges the stationary battery. Otherwise, if solar energy is available, but neither the vehicle nor the stationary battery can be charged, the excess energy is discarded.

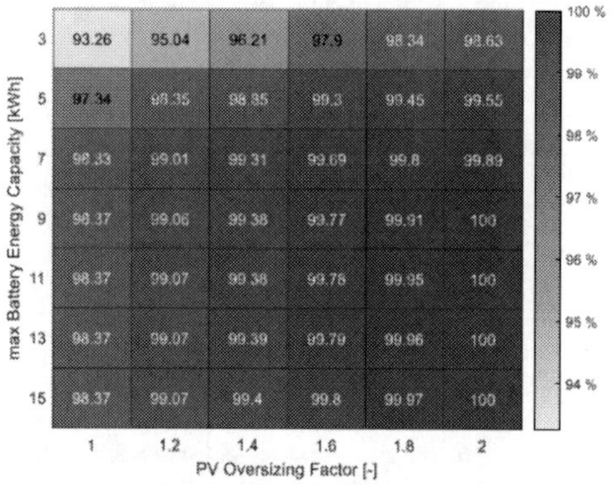

Figure 4: Reliability of the design options in providing the daily requested energy demand in percent over a year.

The simulation enables us to determine the times when not enough energy is available to charge the BEVs. Furthermore, the LCOE of each station design is

10.4229/EUPVSEC2023/5DO.10.4

calculated to enable a cost comparison of the different designs. The two criteria of (1) the reliability to provide energy over the course of a year and (2) the LCOE associated with the station designs determine which design option is optimal. For the purpose of this project, a reliability of a minimum 99% is chosen. Of all station designs with a minimum 99 % availability, the one with the least cost is the optimal design. Figure 4 displays an example of the reliability enabled by different station design options for the mobility concept of Fig. 3.

3.3 Sizing / Simulation Results

To demonstrate the results of the methodology, the authors simulated one battery electric vehicle of the type aCar with a 33 kWh battery, which operates with the mobility concept shown in Fig. 3. The aCar is assumed to drive 85 km every day with a consumption of 0.318 kWh/km and charged with 2.5 kW in its depot in Bishoftu, Ethiopia.

Figure 4 shows the simulated design options according to the methodology presented in Section 3.2. The optimal design with minimum 99 % reliability and least LCOE is given for the calculated scenario with an oversizing factor of 1.2 (4.4 kWpeak) and a7 kWh stationary battery.

In Fig. 5 the significance of the reliabilities of the charging station is shown on an annual basis: The figure displays the number of days in one year for each station design, where less than 80 % of the demanded energy is provided. The impact of oversizing the PV panel area significantly improves the reliability of the station design, for instance, oversizing with a factor of 2 decreases the number of days to one-third for the same battery capacity of 3 kWh. The increase of the stationary battery capacity has a great impact on smaller batteries; however, the improvement rate fades after the 7 kWh battery, which is in our case the optimal solution.

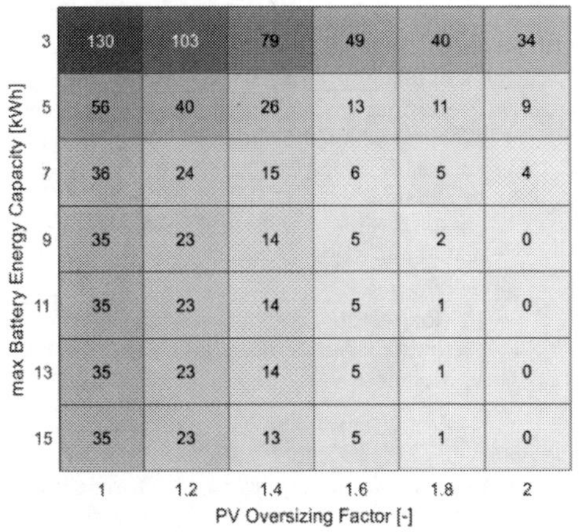

Fig. 5: Number of days per year where less than 80 % of the required energy can be provided.

The pie chart in Fig. 6 displays the percentage of the component's contribution to the Total Cost of Ownership (TCO) of the station. The PV Modules are the most significant cost contributor with 44 % of the station's costs, followed by the stationary battery costs with 27 %.

Figure 7a illustrates the solar energy of the PV system for the optimal standalone charging station over one year in more detail: The figure visualizes the excess energy, which is produced but not needed for the demand of the car in blue bars, whereas the missing energy, which is not available for the demand of the mobility scenario in black bars. Figure 7b shows the compensation of the daily energy lack by the stationary battery. The black in a) and b) are equivalent, showing the amount of energy lacking on the critical days of the year. The chosen capacity of 7 kWh of the stationary battery can compensate for some of the daily energy deficiency. However, there is still a deficit which is displayed with the red bars, due to the chosen reliability of 99 %. The bottom chart of Fig. 7c allows an evaluation of the number of days per year when not enough energy can be provided with a reliability of 99 %. It displays how much of the demanded daily energy is provided on the deficient days of the year. It is visible that even on the worst days of the year, still 75 % of the demanded energy can be provided.

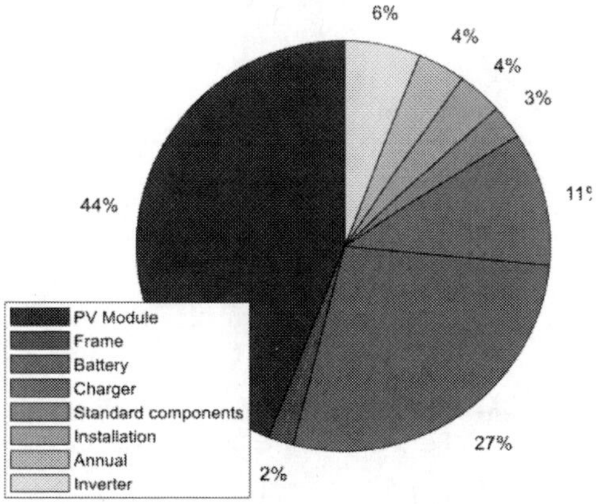

Fig. 6: Percentage of the cost contributing elements for the proposed charging station.

The methodology presented in this paper proposes a charging station design based on the variation of the most influential costs, which are the PV panels and the battery. The optimal charging station is selected based on reliability and energy costs. The reliability of a station design is assessed by following the energy utilization over the course of one year of operation, while costs are calculated based on the specific elements of the station. The methodology selects the optimal station design by the lowest LCOE, with a 99 % reliability. As a result, the methodology enables the determination of the charging station size, as well as the LCOE that can be offered by simply specifying the location and mobility demand.

The methodology was implemented in MATLAB with the input of PVGIS solar irradiation data and the mobility demand based on the scenario described by two mobility project partners operating in rural SSA. The optimal design on the chosen inputs is a PV power of 4.4 kWpeak and a stationary battery capacity of 7 kWh. The methodology can be adapted to different mobility scenarios and locations, in order to assure future research with more complex mobility scenarios. The presented methodology has the potential to contribute to future research efforts by determining the LCOE for BEVs in rural SSA and elsewhere. The insights can assist researchers in their investigations and support local decision-makers in making informed choices regarding investments in electric mobility.

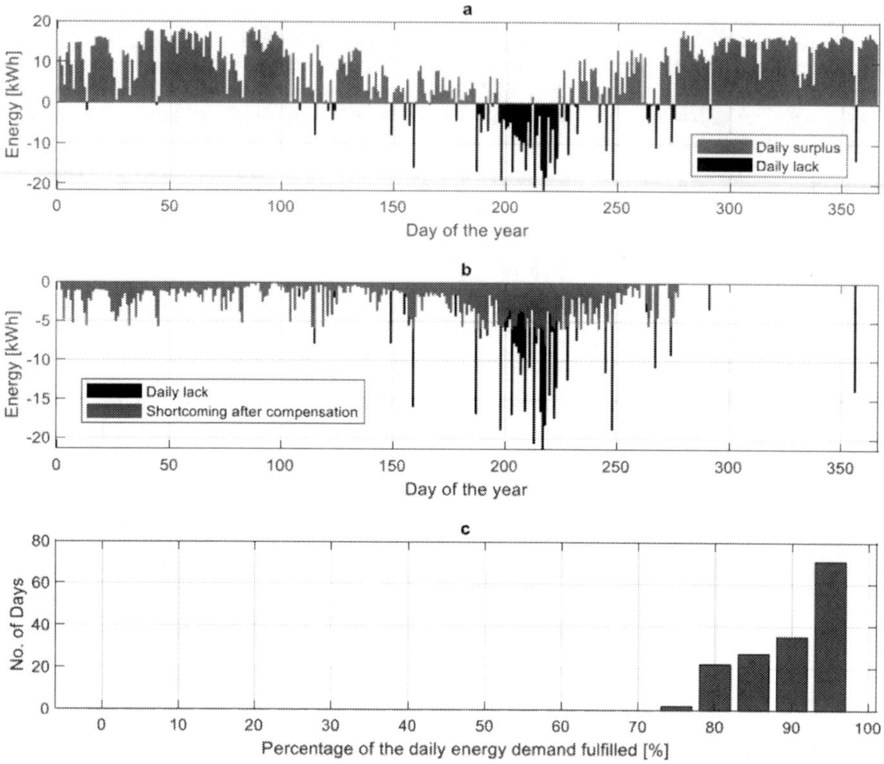

Figure 7: a) Visualization of the daily surplus energy and lack of energy; b) Compensation of the lack of energy with the stationary battery; c) Number of days where the energy demand is not met and % demanded energy provided on those days.

4 CONCLUDING REMARKS – FUTURE WORK

In the SolChargE project, we aim at developing a decentralized transportation hub (DTH), based on a stand-alone solar charging station (SASCS), to harness solar energy for sustainable e-mobility in rural sub-Saharan Africa. Towards this goal, the project will be implemented through five main research activities/steps: i) Development (sizing and design) of the SASCS, ii) Production (local sourcing based) of the SASCS system, iii) Installation and monitoring/assessment of the DTH in the field, iv) Recycling study for the DTH including life cycle analysis and v) business model development.

The sizing methodology represents the groundwork and current (ongoing) research of the SolChargE project, particularly on the SASCS development. The method enables a parametric design of various mobility scenarios and locations all over sub-Saharan Africa, to enable a scale-up of standalone charging of battery electric vehicles in areas with no or low electricity supply. However, it becomes evident that a more detailed mobility scenario, which is based on actual demand profiling and usage data is necessary to make a reliable design of the standalone charging station. Additionally, the component costs in the cost model for the LCOE differ enormously in different locations over sub-Saharan Africa or are not available. Therefore, the next steps of the SolChargE project, with regard to this activity, include the gathering of actual real-world mobility data of a BEV in an operational model and the component prices in the project locations of South Africa and Ethiopia, to enhance the sizing model. As soon as the standalone charging station is built, it can be evaluated in real-world scenarios.

Following this activity, in the next 12 months our research will focus on:

- In short-term: finalizing the drafting the technical layout (and specifications) as well as the open-source development guidelines for the SASCS production.
- In mid-term: Local SASCS production workshop and functional (installation-ready) SASCS demonstrator.
- In long-term: Installation and evaluation of preliminary monitoring/testing results of the DTH in operation.

As the SolChargE project progresses on these activities, in next year's EUPVSEC we aspire to present a comprehensive overview of key results, lessons learnt and return-of-experience from the design optimization, implementation and field operation of the SASCS and DTH in the two demo sites.

ACKNOWLEDGEMENTS

The authors acknowledge the valuable contribution of all partners of the SolChargE project. SolChargE is a project funded in the context of Pillar 1 call of the LEAP-RE project/program. The LEAP-RE project has received funding from the European Union's Horizon 2020

Research and Innovation Program, under Grant Agreement 963530.

For CEA team, part of this work was also supported by the French National Program "Programme d'Investissements d'Avenir - INES.2S" under Grant Agreement ANR ANR-10-IEED-0014-01.

REFERENCES

[1] Almeida, R.H., Brito, M.C. (2015). A review of technical options for solar charging stations in Asia and Africa. AIMS Energy 3(3):428-449.

[2] Erickson, L.E, et al. (2015). Social, economic, technological, and environmental impacts of the development and implementation of solar-powered charge stations. Environmental Progress & Sustainable Energy 34(6):1808-1813.

[3] Savari, G.F., Krishnasamy, V., Sugavanam, V., Vakesan, K. (2019). Optimal Charging Scheduling of Electric Vehicles in Micro Grids Using Priority Algorithms and Particle Swarm Optimization. Mobile networks and applications 24(6):1835-1847.

[4] Apostolaki-Iosifidou, E., et al. (2017). Measurement of power loss during electric vehicle charging and discharging. Energy 127: 730-742.

[5] Ashkrof, P., Homem de Almeida Correia, G., van Arem, B. (2020). Analysis of the effect of charging needs on battery electric vehicle drivers' route choice behaviour: A case study in the Netherlands. Transportation research part D: transport and environment 78, p. 102206.

[6] Knobloch, C., Hartl, J. (2014). The Energy Kiosk Model - Current Challenges and Future Strategies. Endeva Business Model Library.

[7] Hartl, J.M. (2014). The Energy Kiosk Model for Electrification - Status Quo and Future Strategies. Instituto Superior Técnico, Lisbon.

[8] Badea et al. (2019). Design and simulation of Romanian solar energy charging station for electric vehicles. Energies 12(1), 74.

[9] Ayetor, G. K., Mbonigaba, I., Ampofo, J., & Sunnu, A. (2021). Investigating the state of road vehicle emissions in Africa: A case study of Ghana and Rwanda. Transportation Research Interdisciplinary Perspectives, 11, 100409. doi:10.1016/j.trip.2021.100409

[10]Deloitte Africa Automotive Insights, Navigating the Afrian Automotive Sector: Ethiopia, Kenya and Nigeria. In: SCHILLER, M. D. A. D. T. (ed.).

[11]Farquharson, D. (2019), Sustainable Energy Transitions in sub-Saharan Africa: Impacts on Air Quality, Economics, and Fuel Consumption, Carnegie Mellon University, 10.1184/R1/9250325.V1

[12]Schünemann, J, et al. (2022). Life Cycle Assessment on Electric Cargo Bikes for the Use-Case of Urban Freight Transportation in Ghana, Procedia CIRP 105, 721–726.

[13]Ayetor, G. K., Mbonigaba, I., Sunnu, A., Nyantekyi-Kwakye, B., (2021), Impact of replacing ICE bus fleet with electric bus fleet in Africa: A lifetime assessment. Energy. doi: 10.1016/j.energy.2021.119852

[14]Bayu, E.S.; Khan, B.; Hagos, I.G.; Mahela, O.P.; Guerrero, J.M. Feasibility Analysis and Development of Stand-Alone Hybrid Power Generation System for Remote Areas: A Case Study of Ethiopian Rural Area. Wind 2022, 2, 68–86. https://doi.org/10.3390/wind2010005

[15]Dr. Simon Philipps, Fraunhofer ISE and Werner Warmuth, PSE Projects GmbH, "Photovoltaics Report," (accessed: 20.05.2023), 2023. [Online]. Available: https://www.ise.fraunhofer.de/content/dam/ise/de/documents/publications/studies/Photovoltaics-Report.pdf

[16]A. Bugaje, M. Ehrenwirth, C. Trinkl, and W. Zörner, "Electric Two-Wheeler Vehicle Integration into Rural Off-Grid Photovoltaic System in Kenya," *Energies*, vol. 14, no. 23, p. 7956, 2021, doi: 10.3390/en14237956

[17]C. Nsengimana, L. Kai, C. Yuhao, and L. Li, "Standalone photovoltaic and battery microgrid design for rural areas," *Energy Exploration & Exploitation*, vol. 40, no. 6, pp. 1617–1633, 2022, doi: 10.1177/01445987221102196.

[18]G. Badea *et al.*, "Design and Simulation of Romanian Solar Energy Charging Station for Electric Vehicles," *Energies*, vol. 12, no. 1, 2019, doi: 10.3390/en12010074.

[19]A. Mohammed, J. Pasupuleti, T. Khatib, and W. Elmenreich, "A review of process and operational system control of hybrid photovoltaic/diesel generator systems," *Renewable and Sustainable Energy Reviews*, vol. 44, pp. 436–446, 2015, doi: 10.1016/j.rser.2014.12.035.

[20]A. R. Bhatti, Z. Salam, M. J. B. A. Aziz, K. P. Yee, and R. H. Ashique, "Electric vehicles charging using photovoltaic: Status and technological review," *Renewable and Sustainable Energy Reviews*, vol. 54, pp. 34–47, 2016, doi: 10.1016/j.rser.2015.09.091.

[21]A. Bugaje, M. Ehrenwirth, C. Trinkl, and W. Zoerner, "Investigating the Performance of Rural Off-Grid Photovoltaic System with Electric-Mobility Solutions: A Case Study Based on Kenya," *J. sustain. dev. energy water environ. syst.*, vol. 10, no. 1, 2022, Art. no. 1090391, doi: 10.13044/j.sdewes.d9.0391.

[22]A. Wagner, Photovoltaik Engineering. Berlin, Heidelberg, 2015.

[23]A. F. Crossland, O. H. Anuta, and N. S. Wade, "A socio-technical approach to increasing the battery lifetime of off-grid photovoltaic systems applied to a case study in Rwanda," *Renewable Energy*, vol. 83, pp. 30–40, 2015, doi: 10.1016/j.renene.2015.04.020.

[24]P. Mohanty, T. Muneer, and M. Kolhe, Solar Photovoltaic System Applications. Cham: Springer International Publishing, 2016.

[25]International Renewable Energy Agency (IRENA), "Solar PV in Africa: Costs and Mar-kets," 2016.

[26]International Renewable Energy Agency (IRENA), "Renewable power generation costs in 2017," (ISBN: 978-92-9260-040-2), 2018.

[27]A. Korkovelos, B. Khavari, A. Sahlberg, M. Howells, and C. Arderne, "The Role of Open Access Data in Geospatial Electrification Planning and the Achievement of SDG7. An OnSSET-Based Case Study for Malawi," Energies, vol. 12, no. 7, p. 1395, 2019, doi: 10.3390/en12071395.

Assessing Energy Access and Energy Poverty in Cities: the Role of PV as Driver for Action

Palermo V., Pittalis M.

European Commission, Joint Research Centre, Ispra, Italy

EU PVSEC 2023, Lisbon, 21st September 2023

European Commission

020505-001

The Global Covenant of Mayors for Climate and Energy (GCoM)

- GCoM is the largest global alliance for city climate leadership, built upon the commitment of over 12,500 cities and local governments all over the world.
- GCoM cities and partners share a long-term vision of supporting voluntary action to combat climate change and act on mitigation, adaptation and energy access and energy poverty.

2

European Commission

020505-002

Sustainable Energy and Climate Action Plan (SECAP)

- The SECAP is the key instrument used by GCoM Signatories to translate their climate strategies into concrete actions
- The SECAP guides local authorities to: assess, set a target and develop actions in the domains of:
  - Mitigation
  - Adaptation
  - Energy Access and Energy Poverty

3

European Commission

020505-003

40th European Photovoltaic Solar Energy Conference and Exhibition

Developing the Energy Access and Energy Poverty Pillar (EAPP)

2018-2019

Worldwide mapping exercise of energy access and poverty

Production of the energy access concept note in the European context

2020 / **2021**

11 GCoM Regional and National Covenants

47 active GCoM alliance partners

120 total proposed EAPP indicators

2020

1st round of global consultation
Engagement with all GCoM partners and stakeholders

Development of **Consultation Note**

2021

2nd round of global consultation
Further engagement with GCoM partners and stakeholders

Identification of 14 Shortlisted global Indicators

2022

Finalization and launch of the EAPP

2023

Activation of the EAPP

4

European Commission

020505-004

Features of the EAPP

- The EAPP is the third pillar of the GCoM Common Reporting Framework
  - Requirements, indicators, and guidelines
- Helps GCoM signatories to:
  - Collect data and assess their status
  - Take action
  - Track progress

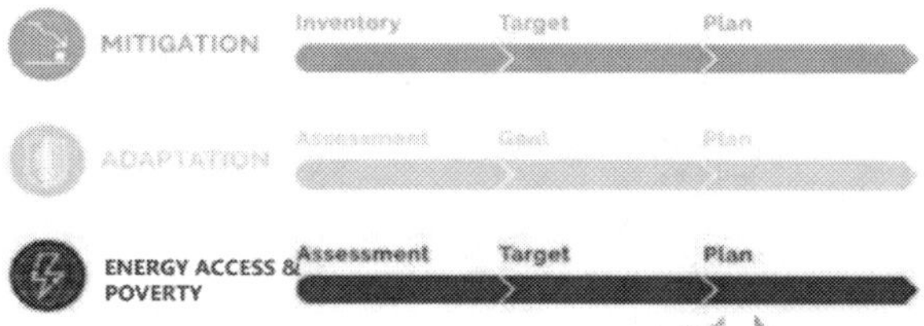

European Commission

5

Municipalities and the Energy Sector

Opportunities	Barriers	Power dimensions	Levers
• Proximity to local context • Narratives • Lived experiences	• Finance • Data • Institutional capacity and institutional structure • Policy landscape • Political leadership	• Set vision and policy • Own and operate • Set and enforce regulation • Dedicate and control budget	• Policies and regulations • Stakeholders and collaborations • Capacity building and data collection • Investment and secure financing • Programmes led by local governments

European Commission

6

Energy Access and Energy Poverty Examples of Municipalities Taking Action

Kenya
PV rooftop and energy-efficient technologies in high-consuming facilities

Italy
PV rooftop on public infrastructure

Hungary
Facilitated access to solar maps and PV system information to support installation in residential buildings

7

European Commission

020505-007

40th European Photovoltaic Solar Energy Conference and Exhibition

Methodology and Results

European Commission

020505-008

Workflow of Analysis

- Focus on CoM SSA
- Sample of 9 signatories from six countries
- Analysis of the SEACAPs
- Review of electricity access conditions
- Focus on electricity access actions
- Analysis of actions centred on PV technology

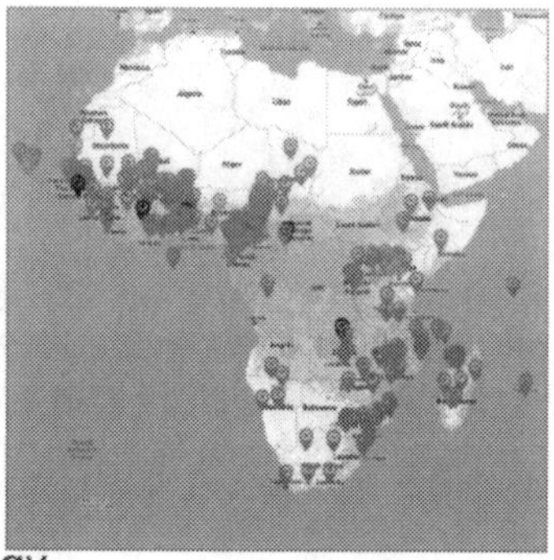

European Commission

40th European Photovoltaic Solar Energy Conference and Exhibition

Workflow of Analysis

- Focus on CoM SSA
- Sample of 9 signatories from six countries
- Analysis of the SEACAPs
- Review of electricity access conditions
- Focus on electricity access actions
- Analysis of actions centred on PV technology

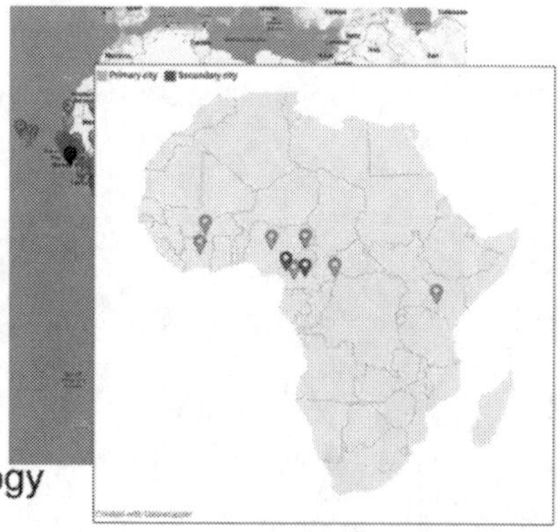

European Commission

Workflow of Analysis

- Focus on CoM SSA
- Sample of 9 signatories from six countries
- Analysis of the SEACAPs
- Review of electricity access conditions
- Focus on electricity access actions
- Analysis of actions centred on PV technology

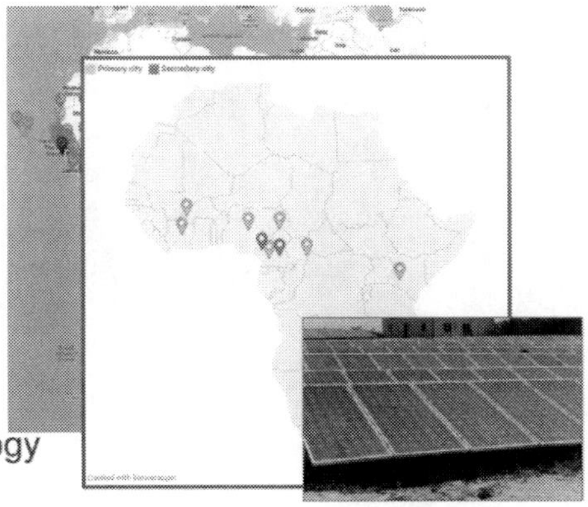

11

European Commission

40th European Photovoltaic Solar Energy Conference and Exhibition

Electricity Access: Baseline

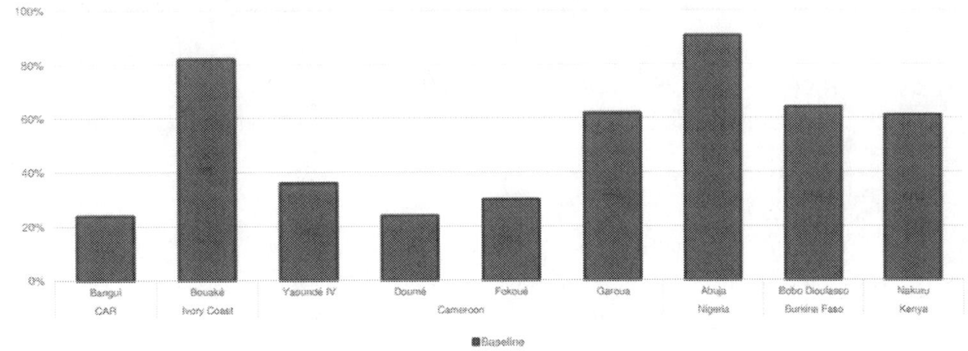

- Current levels of electricity access vary substantially among the sample

12

European Commission

Electricity Access: Ambition, Gap, Actions

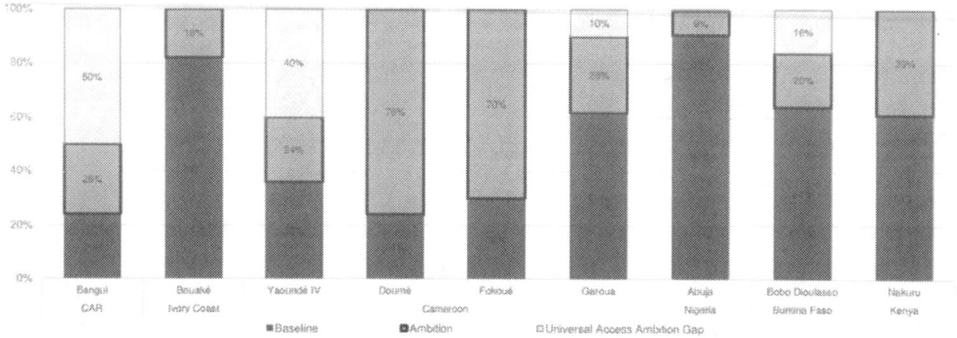

- Electricity access targets for 2030 reflect municipalities ambitions
- Not all signatories aim to reach universal access by 2030
- Municipalities aim to increase electricity access through various actions.

13 European Commission

40th European Photovoltaic Solar Energy Conference and Exhibition

Type of PV Actions

COUNTRY	MUNICIPALITY	PV TRAINING AND CAPACITY BUILDING	SOLAR LANTERNS	SOLAR HOME SYSTEMS AND STAND ALONE PV FOR HOUSEHOLDS	STAND ALONE PV FOR SOCIAL INFRASTRUCTURE	MINI GRIDS	SOLAR STREET LIGHTING
							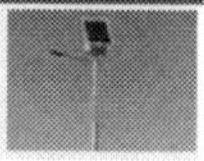
CAR	Bangui						
IVORY COAST	Bouaké						
CAMEROON	Yaoundé IV						
	Doumé						
	Fokoué						
	Garoua						
NIGERIA	Abuja						
BURKINA FASO	Bobo Dioulasso						
KENYA	Nakuru						

- Actions using PV technology constitute the majority of interventions planned by municipalities.
- Actions include "soft" and "hard" measures

14 European Commission

Type of PV Actions: “Soft” Measures

COUNTRY	MUNICIPALITY	PV TRAINING AND CAPACITY BUILDING	SOLAR LANTERNS	SOLAR HOME SYSTEMS AND STAND ALONE PV FOR HOUSEHOLDS	STAND ALONE PV FOR SOCIAL INFRASTRUCTURE	MINI GRIDS	SOLAR STREET LIGHTING
							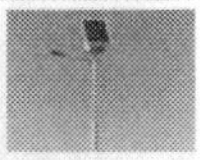
CAR	Bangui						
IVORY COAST	Bouaké						
CAMEROON	Yaoundé IV						
	Doumé						
	Fokoué						
	Garoua						
NIGERIA	Abuja						
BURKINA FASO	Bobo Dioulasso						
KENYA	Nakuru						

- “Soft” measures typically include the delivery of training and capacity building on the use, installation and maintenance of PV systems.

15

European Commission

40th European Photovoltaic Solar Energy Conference and Exhibition

Type of PV Actions: “Hard” Measures

COUNTRY	MUNICIPALITY	PV TRAINING AND CAPACITY BUILDING	SOLAR LANTERNS	SOLAR HOME SYSTEMS AND STAND ALONE PV FOR HOUSEHOLDS	STAND ALONE PV FOR SOCIAL INFRASTRUCTURE	MINI GRIDS	SOLAR STREET LIGHTING

CAR	Bangui						
IVORY COAST	Bouaké						
CAMEROON	Yaoundé IV						
	Doumé						
	Fokoué						
	Garoua						
NIGERIA	Abuja						
BURKINA FASO	Bobo Dioulasso						
KENYA	Nakuru						

- “Hard” measures typically include the construction of energy infrastructure or the distribution of PV equipment.

16

European Commission

PV Actions: From Plan to Project

- Nouakchott, Mauritania – PV system, water pump and drip irrigation system

17

European Commission

40th European Photovoltaic Solar Energy Conference and Exhibition

Conclusions and Next Steps

- A clear framework for action helps municipalities to structure their energy actions and have a strategy to move them into the operational phases.
- PV technology is central to electricity access actions, with nearly 90% of signatories examined in SSA adopting PV-based solutions.
- The activation of the GCoM energy access pillar empowers more municipalities to act. PV is expected to be a central element of their actions.
- There's a need for future research to critically assess energy actions globally, leveraging data from GCoM EAPP to guide such studies.

18

European Commission

Thank you

- JRC is open to cooperation and receiving PhD students and other researchers (as unpaid visiting scientists)
- JRC is constantly hiring new staff: https://recruitment.jrc.ec.europa.eu/
- https://www.globalcovenantofmayors.org/

Slide 9: i) electric board and switches, source: CoMSSA; ii) Rooftop PV system, source: Legambiente - https://www.comunirinnovabili.it/comune-di-cisternino/ ; iii) PV installation in private house, source: https://uj.budapest.hu/zold-budapest/klima-es-kornyezetvedelem/nappal-hajtva
Slide 17: PV minigrid and drip irrigation system, source: CoMSSA – https://comssa.org/en/news/the-feeding-and-greening-nouakchott-experience-urban-vegetable-gardens-in-a-desert-climate

19

European Commission

PHOTOVOLTAICS AND CITIZENS' PARTICIPATION AS KEY ENABLERS OF THE GLOBAL WARMING MITIGATION

Silvia Caneva, Duygu Celik[1], Iban Lizarralde, Mikhail Hamwi, Basma Samir[2], Volker Kromrey, Dimitri Vedel, Linda Lentzen[3], Andreas Schneller, Jan Fjornes, Kathrin Anger[4], Alice De Ferrari, Claudia Crippa[5], Eleanor Denny, Ivan Petrov[6], Sofía Mulero, Alejandro Hernández Serrano[7], Iban Lacoste[8], Maria Regidor, Rodrigo Ruiz[9], Rita Oliveira, Nuno Brito[10], Karl Harder[11], Velimir Segon, Tijana Šimek[12], Alexandru Costeniuc [13]

[1]WIP – RENEWABLE ENERGIES, [2]ESTIA INSTITUTE OF TECHNOLOGY, [3]BODENSEE-STIFTUNG /LAKE CONSTANCE FOUNDATION, [4]ADELPHI, [5]FONDAZIONE ICONS, [6]TRINITY COLLEGE DUBLIN, [7]CARTIF, [8]I-ENER, [9]ENERGETICA, [10]POWER PARITY, [11]ABUNDANCE, [12]REGEA, [13]TRACTEBEL

[1] WIP – Renewable Energies
Sylvenstein Strasse 2 Munich,
Germany
silvia.caneva@wip-munich.de

ABSTRACT

The UN Scientific Assessment Panel to the Montreal Protocol on Ozone Depleting Substances quadrennial assessment report confirms the phase out of nearly 99% of banned ozone-depleting substances. The Montreal Protocol was adopted in Montreal on 16 September 1987 with the aim of phasing out several groups of halogenated hydrocarbons to safeguard the ozone layer, leading to the recovery of the ozone layer in the upper stratosphere and decreasing human exposure to harmful ultraviolet (UV) rays from the sun. If current policies remain in place, the ozone layer is expected to recover to 1980 values by around 2066 over the Antarctic, by 2045 over the Arctic and by 2040 for the rest of the world. In addition to this extraordinary result, the Scientific Assessment Panel estimates avoiding 0.3–0.5°C of global warming by 2100 as consequence of the phase down of production and consumption of many hydrofluorocarbons (HFCs), which as halogenated hydrocarbons have been included in Montreal Protocol and are powerful climate change gases[1]. "*Over the last 35 years, the Montreal Protocol has become a true champion for the environment*" said MegSeki, Executive Secretary of the United Nations Environment Programme's Ozone Secretariat. Another destiny has been seen for the Kyoto Protocol, which was adopted on 11 December 1997 to implement the objective of the UNFCCC to reduce the onset of global warming by reducing greenhouse gas concentrations in the atmosphere to a level that would prevent dangerous anthropogenic interference with the climate system[2]. It is now clear that this vital target is not being achieved and further political actions such the Paris Agreement to limit the temperature increase to 1.5 °C above pre-industrial levels, recognizing that this would significantly reduce the risks and impacts of climate change, cannot be achieved without a strong citizens' participation. A strong citizens' participation is required to make sure that global warming will not lead to an irreversible status of the climatic crisis characterized by climate events, which will make the life of human being no longer sustainable on this planet. Photovoltaics combined with citizens' participation are the key to overcome the on-going climate emergency and provide a sustainable solution to the mitigation of global warming towards the achievement of 1.5 degree-goal. The citizens' participation in photovoltaic projects find its realisation in the business models of cooperatives, communities, aggregators, and crowdfunding platforms. As common denominator they have the concept of "power of many" to highlight that, through the union of small contributions, it is possible to build projects having positive environmental, economic and social impacts leading to the mitigation of global warming.

[1] Scientific Assessment Panel to the Montreal Protocol on Ozone Depleting Substances: Quadrennialassessment report (January 2023).
[2] Kyoto Protocol to the United Nations Framework Convention on Climate Change.

10.4229/EUPVSEC2023/5DO.10.6
020506-001

1 AIM AND APPROACH USED

This paper aims to provide the owners and managers of crowdfunding platforms, cooperatives, communities, and aggregators with new tools to further support their businesses to produce renewable energy through Photovoltaics and citizens' involvement. The tools presented with this paper have been developed in the framework of the H2020 European project SocialRES, which started in May 2019 and endedin December 2022, after 44 months. The aim of SocialRES was to close the non-technological research gaps that impede the widespread uptake of business models for citizens' participation in the European renewable energy sector.

The SocialRES consortium brought together thirteen partners who have been chosen specifically for their expertise and experience to ensure the successful implementation of the novel research proposed in the project and to guarantee a widespread geographical coverage.

The SocialRES project partners are briefly described in the following paragraph.

1. WIP – Renewable Energies (WIP), based in Germany, the project coordinator and social energy expert.
2. ESTIA Institute of Technology (ESTIA), based in France, with extensive experience in research aspect and innovative business models.
3. CARTIF, energy expert based in Spain.
4. Bodensee-Stiftung - Lake Constance Foundation (LCF), based in Germany, as networking and innovation expert and providers of case studies for RES cooperatives.
5. Adelphi, policy think tank based in Germany.
6. Fondazione iCons (ICONS), communication, dissemination and exploitation expert based in Italy.
7. Trinity College Dublin (TCD), University based in Ireland, with extensive experience in behavioural and energy economics, and consumer engagement.
8. I-ENER, RES cooperative based in France
9. EnergEtica, RES cooperative based in Spain.
10. Power Parity, RES cooperative and crowdfunding platform based in Portugal.
11. Abundance, RES crowdfunding platform based in United Kingdom.
12. REGEA, RES crowdfunding platform based in Croatia.
13. Tractebel, RES aggregator based in Romania.

2 SCIENTIFIC INNOVATION AND RELEVANCE

Business models for citizens' participation in the renewable energy sector will become more and more important in the upcoming years to mitigate global warming and provide a sustainable solution for the climate emergency. The business models investigated in this paper are cooperatives, aggregators, and crowdfunding platforms, which have been established in the market since several years (aggregators) or even decades as crowdfunding platforms and cooperatives. Nevertheless, further support is required to make sure that these forms of civic participation can further be optimized to involve more citizens in the future. The tools described in this paper provide relevant insights from the business models developed during the SocialRES project such as the ***Council Green Bond*** *and* ***the Peer to Peer (P2P) energy trading platform.*** Moreover, the paper provides an overview of the methodology developed within SocialRES to support the establishment of new business models of energy communities (***SOGUI framework), to adopt them*** (***Report on potential to adopt the developed business models***) and to ***assess*** them (***Toolkit for policy assessment of the proposed business models***).

3 RESULTS AND CONCLUSIONS

Council Green Bond

During the SocialRES project, the partner Abundance, a crowdfunding platform based in UK, has developed the world's first crowdfunding council bond, which is also the European first council green bond. The green bond supported the West Berkshire Council to raise up to £1,000,000 to fund the installation of photovoltaic panels on council owned buildings around West Berkshire, to help delivering the ambitious plans to become carbon neutral by 2030, 20 years ahead of the UK Government's target. West Berkshire has an installed capacity of 34.2 MW of renewable energy, almost all of it from solar power. With the green bond the aim was to deliver an additional 21.5 MW of renewable energy through council projects, and to encourage residents to generate another 20 MW.

Green bonds represent a relevant tool for the citizens' engagement in the clean energy transition supporting local projects having a direct impact on their life, in terms of a more sustainable life from the social, environmental and also financial point of view, with the return of investment from the bond. All details on the climate bond are available in the following document on the SocialRES website: Abundance Climate Bond.

Peer to Peer (P2P) energy trading platform

The partner Tractebel, an international engineering company based in Bucharest, Romania, has developed a ***Peer to Peer (P2P) energy trading platform*** to connect consumers and photovoltaic prosumers of the Amber Gardens Residence in Bucharest and to allow them to tradeenergy instead of purchasing energy from the local energy supplier Tractebel.

Simulations carried out for 1100 users (Fig.1) clearly show how P2P platforms can generate savings for the consumers, photovoltaic prosumers, small RES providers and for the utility managing the platform.

The earning of the utility has been quantified as TEAP revenues, where TEAP stands for Tractebel Energy Aggregation Platform. The saving for the users is quantified in RON, the Romanian Leu. P2P platforms provide citizens with the freedom to choose their energy supplier. On the other hand, P2P platforms represent a new opportunity for energy utilities to become service providers managing the P2P platform and leave the role of energy supplier. Detailed information on the TEAP, the Tractebel Energy Aggregation Platform, is available on the SocialRES website in the report Roadmap for virtual energy transactions platform TEAP (socialres.eu).

10.4229/EUPVSEC2023/5DO.10.6
020506-002

Variable	Unit	Consumers	Prosumers	SmallRES
Users	n	800	300	10
Months	n	12	12	12
Users Consumption	kWh	1,488,000	-568,800	0
Users Production	kWh	0	2,952,000	2,270,980
Users Energy Cost / Revenue Supplier	RON	-874,944	244,138	445,112
Users Energy Cost / Revenue TEAP	RON	-583,680	658,800	679,768
Users Gross Savings	RON	291,264	414,662	234,656
	%	33%	170%	53%
TEAP Fee	%	20%	20%	20%
	RON	58,253	82,932	46,931
Users Net Savings	RON	233,011	331,730	187,725
	%	27%	136%	42%
TEAP Revenues	EUR	11,650.6	16,586.5	9,386.3
TEAP Revenues total	EUR	37,623		

Fig. 1 Simulation for 1100 users of the TEAP - Tractebel Energy Aggregation Platform

The SOGUI framework

The SOUGI framework has the aim to support stakeholders in the creation of business models, such as renewable energy communities, cooperatives, crowdfunding platforms, or business models for aggregation, in the renewable energy sector. To facilitate and support the implementation of these energy-related projects, the SOGUI framework (Fig. 2) allows to define the pathway of the energy project to be developed. This will help to anticipate the organization of the different activities to be undertaken and the definition of different stakeholders depending on the needed skills and used technologies.

The SOUGI framework is divided in four parts:

- The technologies that can be used for each activity in the energy field are proposed: solar PV, wind, wood, solar thermal, biogas, hydro, sea energy, biofuel, electric vehicle, batteries, hydrogen and thermal.
- Different activities in the energy field are listed, to choose the ones that will be used to develop the defined project: electricity generation, electricity retailers, energy savings, heat generation, self-consumption, storage, mobility, demand response, electricity distribution, heat distribution, energy poverty, community building, citizens meet local authorities, aggregation and funding.
- The needed competences/skills to accomplish these activities are also proposed: finance, planning and governance, energy efficiency, renewable technics, energy flexibility, digital, logistics and transport, smart grid, energy markets, communication, legal skills, teamwork and conflict management, commercial and information analysis.
- Involved stakeholders for the different activities are defined: end user, energy cooperative, crowdfunding platform, aggregator, local authority, DSO, ESCO, SME, digital service provider, social institutions.

Those four parts are cross-referenced according to four criteria:

- Technical synergies (T) defines the specific technologies used for each activity of the project.
- Internal competences/skills (C) defines which are the needed competences that should be kept internally for each activity.
- Involved stakeholders (S) defines the involved stakeholder for each activity.
- Delegating competences (D) defines which are the needed competences that are planning to be subcontracted.

The methodology to be implemented to create new business models for the citizens' participation in the renewable energy sector based on the SOGUI framework is explained in details in the report "Guidelines for successful social innovations in the energy sector" available on the SocialRES website.

Methodology to adopt the developed business models

Another fundamental step is how to adopt the newly developed business model. The first step is about defining the context and discovering opportunities or challenges associated with the proposed business model and power system transition. The Sustainable Power System Triangle (SPST) is used for this scope. The triangle consists of three pillars indicated by the European Commission as guiding dimensions for the future power system design. The three dimensions are energy affordability, environmental sustainability, and security.

The triangle represents the trade-off between the three dimensions. Users map the influence of their project idea on the three pillars and determine any harmony and conflict. Once defined the position of the project in the Sustainable Power System Triangle (SPST), the next step for the adoption is to define the energy business model by supporting the visualization of different stakeholders and their relationships (Fig.3).

Example of adoption have been explained in details by the partner ESTIA in the "Report on potential to adopt the developed business models" available on the SocialRES website.

10.4229/EUPVSEC2023/5DO.10.6
020506-003

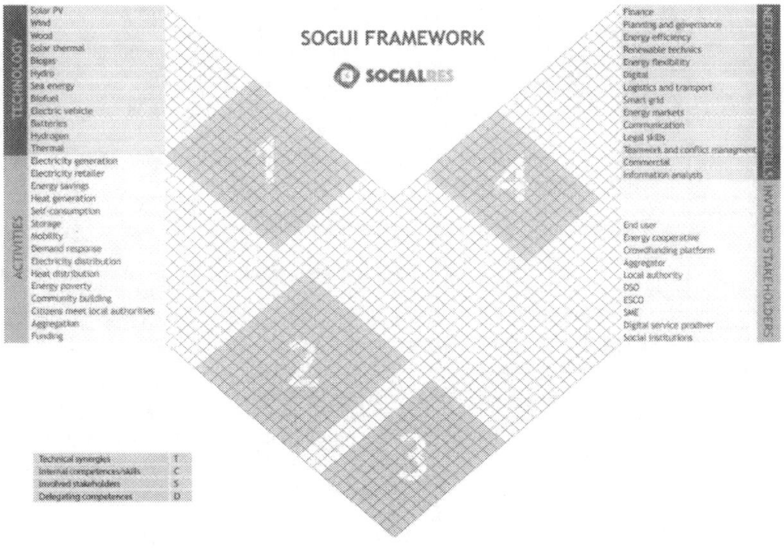

Fig.2 The SOGUI Framework

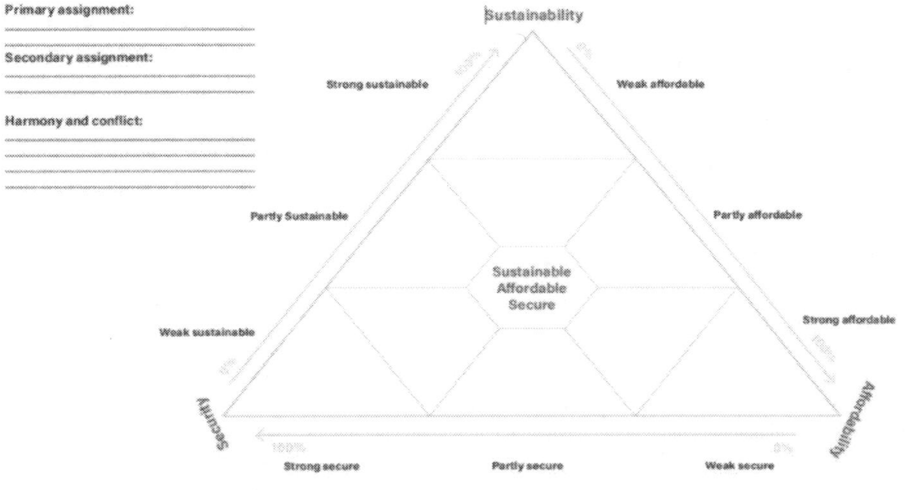

Fig. 3 The Sustainable Power System Triangle (SPST)

Toolkit for policy assessment

The toolkit for policy assessment developed by the SocialRES partner adelphi outlines a ***taxonomy to be used by stakeholders to quantify and assess the multiple impacts of the business models for citizens' participation also known as social innovations***. The taxonomy has been developed based on the literature on social innovations in the energy sector. The proposed taxonomy identifies a range of indicators in the broad categories of economic, social, environmental, and health impacts. The following table gives an overview of the measured impacts, which social innovation they are measured for and thereasoning behind their inclusion.

Further detailed information is available in the report available on the SoialRES website "Toolkit for Policy Assessment", which provides a detailed description on applications for use of the taxonomy, for policymakers, who may want to assess their social innovation support policies by looking beyond the outcomes at the impacts of the social innovations. Applications for use are also suggested for practitioners, as owners or managers of the social innovations in question. The toolkit also includes the evaluation of the impacts for each SocialRES case study representing cooperative, crowdfunding and aggregators (Table 1).

10.4229/EUPVSEC2023/5DO.10.6
020506-004

Table 1 Social Innovation Impact Assessment

INDICATOR	SOCIAL INNOVATION INDICATOR IS MEASURED FOR	REASONING
CONSUMER SAVINGS	RECs, Aggregators	RECs can lead to financial savings for members due to reduced energy prices compared to traditional utilities. Aggregators can improve energy efficiency at home and reduce expenditures that way.
JOB CREATION	RECs, Aggregators, Crowdfunding	Jobs created both during the implementation and construction phase of the projects, as well as during the operation phase which included management and maintenance.
AVOIDED GHG EMISSIONS	RECs, Crowdfunding, Aggregators	RECs and Crowdfunding provide a RE supply which may previously have stemmed from fossil fuels. Aggregators increase energy efficiency at the household level and can contribute to peak shaving across the entiregrid.
CLEAN ENERGY GENERATED	RECs, Crowdfunding, Aggregators	RECs and Crowdfunding build generating units and thus it is important to see how much RE is produced by these units. Aggregators do not directly generate RE but the amount produced by members can be measured.

4 ACKNOWLEDGEMENTS AND DISCLAIMER

The SocialRES project received funding from the European Union's Horizon 2020 research and innovation programme under grant agreement No. 837758. The sole responsibility for the content of this publication lies with the authors. It does not necessarily reflect the opinion of the European Commission. The European Commission is not responsible for any use that may be made of the information contained therein.

ADVANCED SOLAR PV MARKET FORECASTING WITH HIGH GRANULARITY FOR 33 EUROPEAN COUNTRIES

Adrien Van Rechem, Philippe Macé, Elina Bosch, Gaëtan Masson
Becquerel Institute
a.vanrechem@becquerelinstitute.eu

ABSTRACT: The Becquerel Institute developed an advanced forecasting method of European solar PV markets, at country level, with details available per segment. These forecasts by country and market segments are heavily substantiated and documented by an in-depth analysis of individual markets' trends and conditions. This allows to derive a consensus on the highest probability scenario. This comprehensive approach makes Becquerel Institute's advanced forecasting of European solar PV markets an indispensable planning reference for PV stakeholders. The results of the analysis for the 33 European countries show that around 13.5 GWp is expected to be installed in the next quarter (Q3 2023) in Europe, a 32 % year on year growth compared to Q3 2022, driven by recent prices dwindle, despite the pressure put by a lack of skilled labour, delays in supply of components, as well as permitting issues. Established markets (Germany, Spain, and Poland) are expected to be key drivers to this growth although we expect market development in almost all European countries. In total, around 52 GWp is expected to be installed in 2023 in Europe (47 GWp in the EU). By 2027, Europe's annual PV market is expected to exceed 80 GWp. This growth is expected to come evenly from the centralized and distributed segments.
Keywords: market analysis, market forecast, quarterly, market segments

1 INTRODUCTION

1.1 Objectives

The reason behind the creation of this forecasting method by the Becquerel Institute lies in the current difficulties of collecting and projecting PV market data. To cite a few of these difficulties:

- Is data for X country available?
- What is the format of the data (capacity, number of installations…)
- Is the market data reliable?
- Is it all grid-connected or only subsidized installations?
- Is the data in AC or DC?
- How are market segments defined?
- Are there multiple sources?

In this context, an advanced forecasting of European solar PV markets has been developed by the Becquerel Institute to support this data extraction, data verification and cleaning, context analysis and normalization of the market data. Its 3 main objectives are:

1. Consolidate and standardize quantitative data.
2. Review and cross-check with qualitative data.
3. Use insights for proper forecasting.

These quantitative forecasts by country and market segments are heavily substantiated and documented by an in-depth analysis of individual markets leading to defining a consensus on the highest probability scenario.

1.2 Methodology

The methodological approach of this forecasting of European solar PV markets relies on three main steps, as explained on the Figure 1.

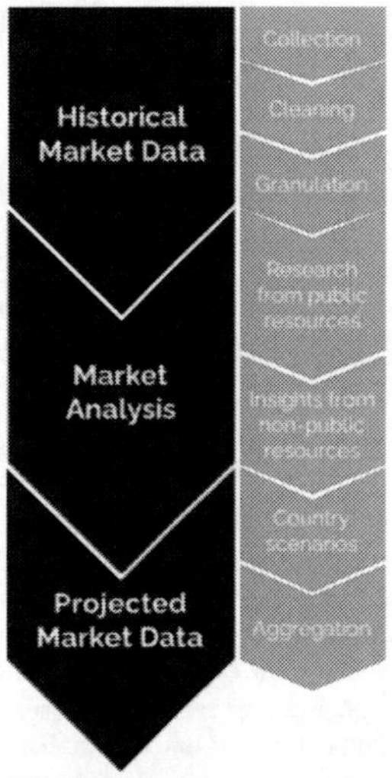

Figure 1: Methodological approach scheme

The seven steps followed are:

1. Collecting of historical capacity installations as stated on official sources, at country-level.
2. Cleaning, verifying and correcting historical capacity installations via non-official sources, (experts, organizations, etc).
3. Granulation of historical information:
   - Periods: annual & quarterly
   - Segments: distributed & centralized
4. Collecting information on main market influencing factors:
   - Regulation & policies
   - Announced solar PV or RE targets
   - Cost trends
   - Announced projects
   - Supply constraints
   - And more
5. Completing the information from publicly available resources with insights from non-

10.4229/EUPVSEC2023/5DO.11.2
020509-001

public resources, by discussing with local experts, installers, developers or trade organizations, e.g., on local projects' pipelines.

6. Defining country-level forecasts until 2027 for 33 European countries, considering segment distribution on a quarterly-based timeline.
7. Figures are aggregated to obtain values at the scale of the European continent as well as for the European Union.

The studied countries are the following: Austria, Belgium, Bulgaria, Croatia, Cyprus, Czech Republic, Denmark, Estonia, Finland, France, Germany, Greece, Hungary, Ireland, Italy, Latvia, Lithuania, Luxembourg, Malta, Netherlands, Norway, Poland, Portugal, Romania, Russia, Slovakia, Slovenia, Spain, Sweden, Switzerland, Turkey, Ukraine, and United Kingdom.

It should be noted that part of our methodology includes our regularly revising projections against available market data.

This comprehensive approach makes this advanced forecasting of European solar PV market an indispensable planning reference for PV stakeholders willing to drive their business with the highest level of precision possible.

2 RESULTS

2.1 Historical market data

Examining the historical development of the European PV market, we can observe a remarkable surge in 2022, surpassing the numbers seen during the prosperous early 2010s. To be precise, Europe witnessed the installation of **43 GWp of PV capacity last year**, reflecting a substantial 33% year-over-year increase compared to 2021. Main contributors to this market were Spain (8.5 GWp), Germany (7.2 GWp), Poland (4.9 GWp), the Netherlands (3.9 GWp) and France (3.0 GWp), with market growth experienced in almost all European countries. Significant increases were observed in Spain (+3.6 GWp absolute increase), Italy (+162% relative increase), Denmark (+119% relative increase), Finland (+174% relative increase) and Slovenia (+98% relative increase). On the other hand, the French market slightly contracted and underperformed compared to expectations, with just under 3 GWp installed (compared to 3.4 GWp in 2021). Overall, this growth came evenly from the centralized and distributed segments.

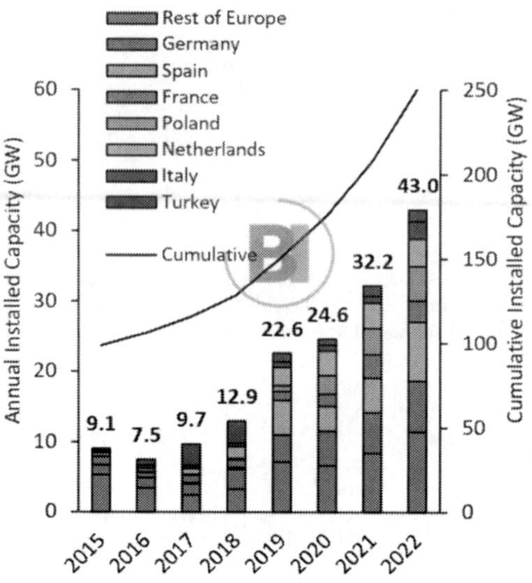

Figure 2: Historical evolution of the European PV market (in GW) per country

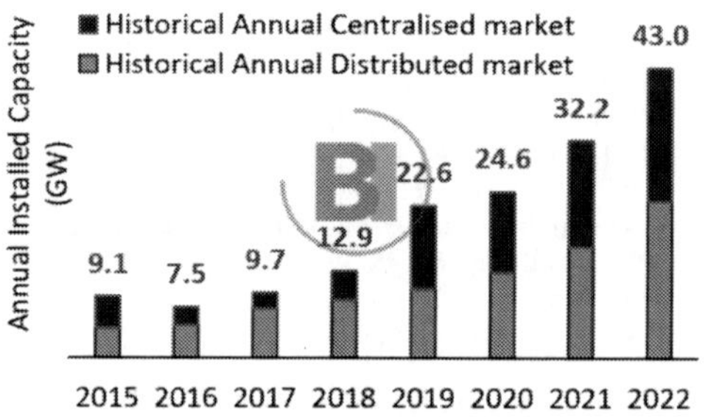

Figure 3: Historical evolution of the European PV market (in GW) per segment

2.2 Forecast market data

The results of the projection analysis for the 33 European countries show that around 13.5 GWp is expected to be installed in the next quarter (Q3 2023) in Europe, a 32 % year on year growth compared to Q3 2022, driven by recent prices dwindle, despite the pressure put by a lack of skilled labour, delays in supply of components, as well as permitting issues. Established markets (Germany, Spain, and Poland) are expected to be key drivers to this growth although we expect market development in almost all European countries. In total, around 52 GWp is expected to be installed next year in Europe (while 47 GWp for the European Union). By 2027, Europe is expected to have above 80 GWp of installed solar PV capacity annually. Overall, this growth is expected to come evenly from the centralized and distributed segments.

10.4229/EUPVSEC2023/5DO.11.2
020509-002

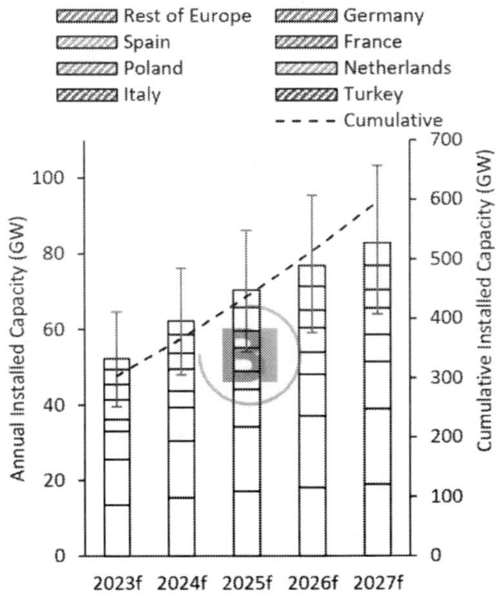

Figure 4: Projected evolution of the European PV market (in GW) per country according to our most probable scenario

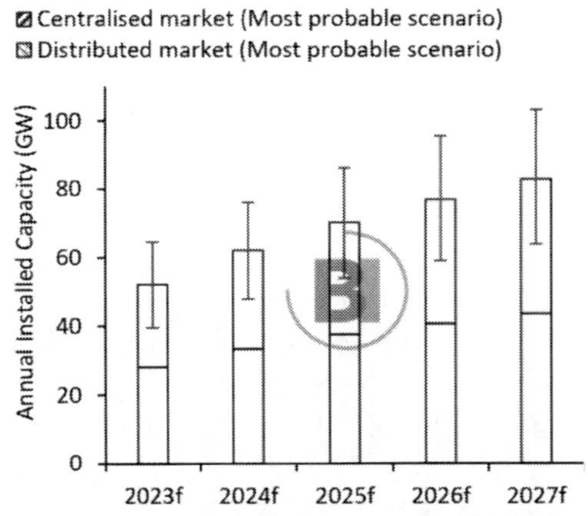

Figure 5: Projected evolution of the European PV market (in GW) per segment according to our most probable scenario

2.3 Comparison of the European situation in 2022 & 2025

The number of GW-scale markets in Europe was **12 countries in 2022 and should evolve to 16 countries in 2025.**

Top 10 Annual PV Markets

Country	GWp
Spain	8.5
Germany	7.2
Poland	4.9
Netherlands	3.9
France	3.0
Italy	2.5
Turkey	1.6
Denmark	1.6
Greece	1.4
Switzerland	1.1

Top 10 Cumulative PV Markets

Country	GWp
Germany	66.9
Spain	27.0
Italy	25.1
France	19.8
Netherlands	18.2
UK	15.2
Poland	12.5
Turkey	12.5
Belgium	8.1
Greece	6.0

Figure 6: Situation 2022

Top 10 Annual PV Markets

Country	GWp	Change
Germany	17.0	+1
Spain	10.0	-1
Italy	6.2	+3
Poland	6.2	-1
France	4.8	=
Turkey	4.6	+1
Netherlands	4.4	-2
Portugal	2.2	+6
Austria	1.7	+3
UK	1.7	+6

Top 10 Cumulative PV Markets

Country	GWp	Change
Germany	110.9	=
Spain	53.5	=
Italy	40.1	=
France	32.1	=
Netherlands	30.8	=
Poland	29.7	+1
Turkey	23.6	+1
UK	19.2	-2
Belgium	11.5	=
Greece	10.5	=

Figure 7: Situation 2025 (number given here in GW are according to our most probable scenario)

2.4 Example of country-level market data

As an illustration of a country-level outcome, we present the historical and projected progression of the Portuguese PV market below (**Figure *8***). Portugal's government aims to install 20.4 GW of solar by 2030 and is on track to achieve this target, mainly thanks to large scale ground mounted PV systems. Despite political willingness, various obstacles continued to impede the progress of distributed PV in Portugal. With centralized systems holding around 80% of the market share, it underscores the imperative need to move past a market that is highly dominated by centralized systems.

10.4229/EUPVSEC2023/5DO.11.2
020509-003

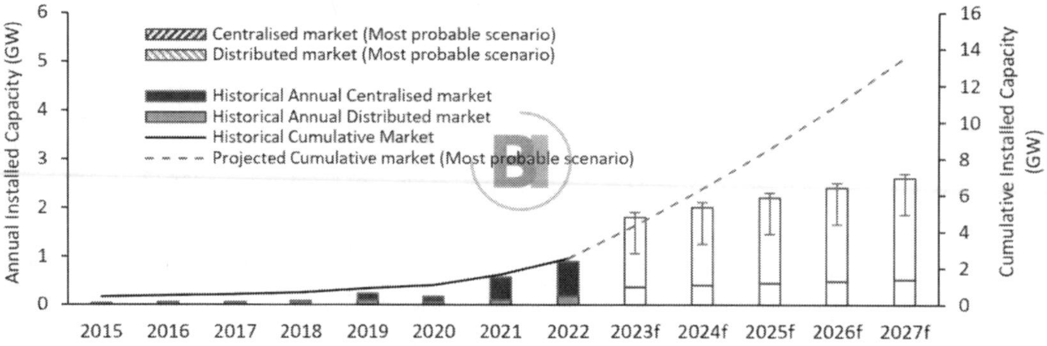

Figure 8: Historical and projected evolution of the Portuguese PV market (in GW) per segment according to our most probable scenario

2.5 Comparison previous projection vs actual 2022 market

As shown on Figure 9, past projections according to our "most probable scenario" and actual market performance differ. On the other hand, the actual market performance for 2022 remains in line with our "high scenario".

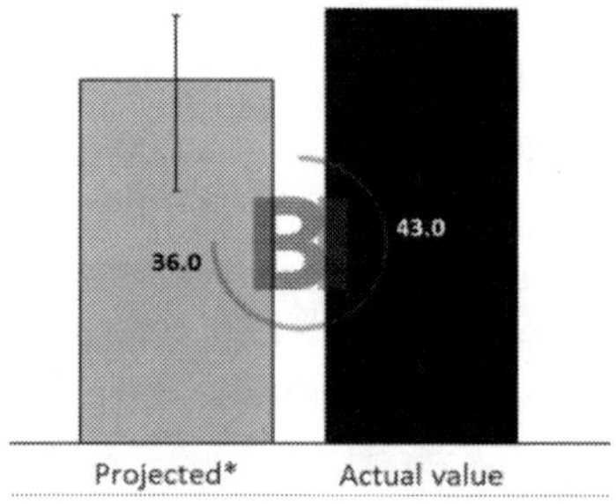

Figure 9: Comparison of the projected (according to our most probble scenario") and actual European market (in GW) in 2022

To explain this gap, one can mention the fact that in many countries, project developers had to postpone installations from 2021 to 2022. In terms of per country-gap, one can mention **Spain,** which achieved a record performance in 2022, partially attributed to the addition of approximately 2.5 GW of PV systems under its self-consumption regulations. Poland can also be highlighted. This country is currently experiencing an exponential growth curve, having begun from a nearly non-existent market until 2018. Poland demonstrated remarkable progress, setting it apart from the average growth trends, thanks to policy changes. Denmark is another outperformer that can be mentioned, where an unexpected market surge was witnessed, transforming to a GW-scale market for the first time in its history. On the side of underperformers, France can be pointed. Despite reaching approximately 3.4 GW in 2021, the **French** market was anticipated to exhibit modest growth. But in 2022, it did worse than expected and decreased, on an annual basis. Other examples of countries explaining the total gap are shown on Figure 11.

On the other hand, on **Table *1***, are shown a few countries whose PV markets were predicted with high level of accuracy.

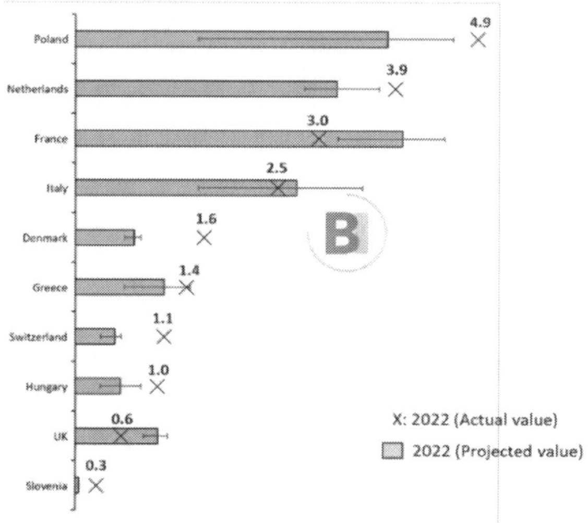

Figure 10: Unforeseen market trends (in GW) in 2022

Table 1: Anticipated key markets in 2022 (*according to our most probable scenario)

	Projected (GWp)*	Actual (GWp)
Germany	7.0	7.2
Portugal	1.0	0.9
Austria	1.0	1.0
Belgium	0.8	1.0

3 CONCLUSIONS

Our analysis shows that the European solar PV market is resilient, as demonstrated by the massive growth all over Europe. The future is also positive, although there will be some differences in pace between countries and across segments in function of the regulatory frameworks in place. This development is still hindered by (i) labour shortage, (ii) delays in supply of key components or (iii) grid congestion. Established markets (Germany, Spain, France and Poland) are expected to be key drivers to this foreseen growth but we expect market development in almost all European countries.

More details on these PV market historical performances and forecasts are available on Becquerel Institute's European Solar PV Data Hub (at https://datahub.becquerelinstitute.eu/)

10.4229/EUPVSEC2023/5DO.11.2
020509-004

TOWARDS A RECYCLABILITY INDEX FOR PHOTOVOLTAIC MODULES: METHODOLOGY, CHALLENGES AND POLICY IMPLICATIONS

Davide Polverini[1], Felice Alfieri[2], Christoforos Spiliotopoulos[2] and Aleksandra Arcipowska[2]
[1] European Commission, DG Internal Market, Industry, Entrepreneurship and SMEs, Brussels, Belgium
[2] European Commission, Joint Research Centre, Seville, Spain

ABSTRACT: Given the ongoing – and expected - rise of new photovoltaic (PV) installations, it is of primary importance to ensure that newly installed PV products in the European Union are environmentally-friendly and do not create future burdens on Europe's environment. In order to improve the sustainability of PV, not only voluntary goals have to be taken along, but compulsory regulations could potentially set limits or targets for the next PV modules generations to come to the market. The European Commission is currently analyzing the feasibility of Ecodesign and energy labelling requirements for these products. The measures would in particular concern the efficiency, durability, reparability and recyclability of PV modules. Specifically concerning recyclability, information on the design, the material content of PV modules as well as the dismantlability of precious, environmentally relevant and critical raw materials could facilitate the end-of-live recovery of these materials. This could be effected via a PV module 'recyclability index'. Once available, this index would represent a ready-made comparison tool to inform users and recyclers about the recyclability of the product. This paper elaborates on the methodology for the development of this index, and on the potential policy implications.
Keywords: Ecodesign, Circular Economy, Photovoltaics, Recyclability

1 INTRODUCTION

Photovoltaic (PV) is a key technological sector. The European Union (EU) cumulative installed capacity of 160 GW in 2021 is projected to grow significantly in the short and medium term. The REPowerEU Plan [1], presented by the European Commission in May 2022, increases the ambition of the 'Fit for 55 Package' and sets the target of achieving a 45% share of renewable energy by 2030. As part of the REPowerEU Plan, the EU Solar Energy Strategy sets the ambitious goal to install 600 GW of the solar PV capacity by 2030.

With particular regard to the minimization of the PV environmental impacts at the end of life (EoL), some quantitative considerations are needed. A PV module is constituted of several materials. In terms of weight, it is mostly made of glass and aluminum; valuable metals such as copper, silicon and silver only constitute a small amount of its mass [3]. Some of the materials used in PV modules are considered strategic to the EU's' economy and have high supply risk, thus they have been included in the list of the EU critical raw materials [9].

A number of studies and initiatives offer useful design guidelines and insights into relevant parameters for recyclability at generic level, such as those described in PolyCE [5], as well as specifically at PV level the Sustainability Leadership Standard for Photovoltaic Modules and Photovoltaic Inverters [11] and the PV Module Design for Recycling Guidelines from IEA [6]. Moreover, the European standard EN 45555 offers a method to assess recyclability of energy-related products, and proceeds with proposing a Recyclability Rate formula [7]. The approach followed in EN45555 is semi-quantitative and much based on mass, which might introduce some challenges in its application and use in regulatory context and at design stage. The initiative RecyClass [8] proposes a recyclability classification system (A-F) for plastic packaging. An index based on a recyclability assessment specifically for PV has not been identified.

In order to access the most valuable components and materials in the solar cells, other layers must be removed, such as the glass and the ethylene vinyl acetate (EVA) encapsulant layer. Optimized thermal delamination enables the intact recuperation of the silicon wafer, which provides the highest value in recycling. However, this approach is expensive in low volumes and furthermore the incineration of fluorinated backsheet materials requires adequate safety measures [2].

The design of current PV modules on the market is not optimized toward circularity [3]. They are not usually designed to be "re-opened" and the only way for recycling to take place is through destructive processes such as shredding. Such irreversible design severely limits not only the potential for repair/refurbishment, but also for the recovery of valuable materials. There are only currently limited examples of PV module design to support ease of disassembly or dismantling for recycling [3].

For smaller PV systems, such as the residential ones, two relevant design solutions to consider are those related to halogen free cables and backsheets. Separation and selective treatment of these parts can be beneficial to avoid harmful halogen smoke during incineration at the EoL. Moreover, plastic containing brominated flame retardants and external electric cables are subject to selective treatment for materials and components of waste electrical and electronic equipment according to in Article 8(2) of the WEEE Directive [10].

Currently the majority of module designs, in particular those related to silicon technologies, present various challenges at the moment of seeking to dismantle them to recovery materials for recycling [3]. Once the junction box and the aluminum frame (if present) have been removed, the main difficulty is to separate the encapsulated components as well as the soldered connections and the tabbing of the cells. This requires destructive thermal and mechanical processes to be used, which result in low grade, cross contaminated material recovery. Alternative module designs have been developed until the pre-commercial stage, that have eliminated the polymer encapsulants and laminates as well as the metal soldering that hinder dismantling. Pilot production modules have already been certified to have passed IEC 61215 design approval [3].

According to the Ecodesign preparatory study [3], the main feature of the design facilitating recycling is that the

10.4229/EUPVSEC2023/5DO.14.2
020510-001

components and their means of assembly allows for an easier separation of the components and materials, including ease of access to the cells without breakage. Materials need to carefully selected in order to allow for recycling. In the case of current module designs, which are more problematic to achieve a clean separation of materials, materials choices may hinder recovery processes.
It is therefore considered relevant to have information on the design, the material content of PV modules as well as the dismantlability of precious, environmentally relevant and critical raw materials. The purpose would be to facilitate future EoL recovery of valuable raw materials and to identify appropriate recovery routes, for example in the case of encapsulant and backsheet materials where the presence of fluorinated materials could create a processing hazard. A way to deliver the information on the PV module recyclability could consist in the definition of a 'recyclability index', to be calculated for each PV module model placed on the (EU) market. Once available, this index would represent a ready-made comparison tool to inform users and recyclers about the recyclability of the product. As such, it could be used in the context of environmental policies, such as the Ecodesign and energy labelling.

2 METHODOLOGY

First, some general considerations related to the usefulness of a recyclability index are presented, including having a means of assessing product aspects which are multilayered and communicating in an aggregated manner to various audiences. Policy making and facilitating sustainable purchase decisions could benefit from the availability of this index, as argued in the concluding section of this paper. PV constitutes a relevant product for application of such index, firstly due to its relevance from a circularity perspective, and also for the purpose of providing useful information to PV owners and recyclers in an aggregated manner. Methodologically, the approach proposed is similar to that followed for other circularity indices, notably that of the Reparability Scoring System developed by the Joint Research Centre (JRC) of the European Commission [4], however adapted to the particularities of recyclability as a concept. The method (Figure 1), as a first step, foresees the identification of key product components and materials relevant for recycling and continues with the identification of relevant recyclability parameters. The criticality of those parts/materials, as well as the identified parameters, are subjected to weighting before a system of scoring levels results in an aggregated recyclability value for a given product.
The approach described above has conceptual similarities with the previously cited work of the authors: the JRC Reparability Scoring System method presents a blueprint for the preparation of product-specific indexes (in the field of energy-related products), on parameters related to circular economy aspects. Published in 2019, the method consists of distinct steps, resulting in an aggregated reparability score. The first step entails an identification of priority parts for a product category, in order to limit the reparability assessment to a reasonable amount of effort, based on concrete criteria notably the failure rate and the functional relevance of the parts (higher rate and relevance meaning higher priority). Secondly, a set of parameters relevant to reparability are considered, and, again, the most important are selected for the next steps of the assessment. Two distinct categories of parameters are considered, (a) design-related referring to parameters which characterize the technical design of the product and affect its ability to be disassembled, and (b) service-related, which refer to provisions by the product manufacturer that deem the product reparable or not (such as disassembly information, or spare part availability). What follows is a scoring system for each parameter and, in the case of design-related parameters, also for each part. The reparability index for smartphones has been 'incorporated' in the energy label for these products.

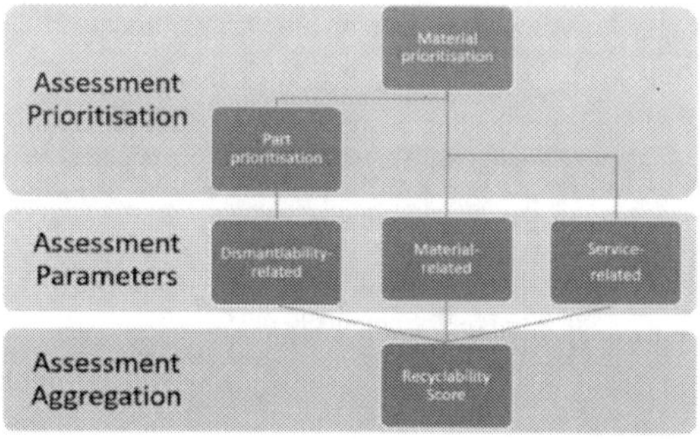

Figure 1: Generic approach proposed for the recyclability score

3 POLICY IMPLICATIONS AND CONCLUSIONS

At EU level, in light of the ambitious commitments to dramatically increase the pace of new PV installations in the short-medium term, attention is being brought on the fact that newly installed PV products should be environmentally-friendly. With the view to facilitate future EoL recovery of valuable raw materials and to identify appropriate recovery routes, it is therefore considered relevant to have information on the design, the material content of PV modules as well as the dismantlability of precious, environmentally relevant and critical raw materials. A way to deliver the information on the PV module recyclability with a 'condensed' figure could consist in the definition of a 'recyclability index', to be calculated for each PV module model placed on the market. Once available, this index would represent a ready-made comparison tool to inform users and recyclers about the recyclability of the product.

Several policy applications for a recyclability index for PV modules can – at least in preliminary terms – be envisaged, such as:

- information to be compulsorily displayed/made available by manufacturers for each PV module model placed on the EU market, as an effect of Ecodesign and/or energy labelling measures;
- public procurement: public procurers could use the index as awarding criteria to select best environmental performing products;
- a recyclability index could also be used to modulate fees under collective collection schemes according to the WEEE Directive, with lower fees applicable to higher recyclable products.

In any case, these policy approaches would likely influence design and purchase decisions; it can be envisaged that, with the right incentives, bifacial glass to glass module designs, which are predicted to account for 50-60% of the market by 2030, could evolve to adopt some features of the 'best-not-yet-available technology' module

10.4229/EUPVSEC2023/5DO.14.2
020510-002

designed for recycling [3]. As a further consideration, it has to be born in mind that the direct environmental benefits of a recyclability index would materialize mainly in the long term. In case a recyclability index would be in place in 2024, and considering a lifespan of 25 years, the decommissioning of models with recyclability index would start in 2058 (earlier benefits could be attained only in the case of PV modules with earlier mortality due to e.g. production defects or failure in operations).

4 DISCLAIMER

The views expressed in the article are personal and do not necessarily reflect an official position of the European Commission. Neither the European Union institutions and bodies nor any person acting on their behalf may be held responsible for the use which may be made of the information contained therein.

5 REFERENCES

[1] European Commission. REPowerEU Plan. COM/2022/230 final
[2] Bobba, Silvia & Carrara, S & Huisman, Jaco & Mathieux, Fabrice & Pavel, Claudiu. (2020). Critical Raw Materials for Strategic Technologies and Sectors in the EU - A Foresight Study. 10.2873/58081.
[3] Dodd N, Espinosa N et al. Preparatory study for solar photovoltaic modules, inverters and systems. https://publications.jrc.ec.europa.eu/repository/handle/JRC122431
[4] Cordella, M., Alfieri, F. and Sanfelix Forner, J., Analysis and development of a scoring system for repair and upgrade of products, EUR 29711 EN, Publications Office of the European Union, Luxembourg, 2019, ISBN 978-92-76-01602-1 (online), doi:10.2760/725068 (online), JRC114337.
[5] PolyCE project. Design for Recycling: Guidance for designers: https://www.polyce-project.eu/wp-content/uploads/2021/04/PolyCE-E-book-Circular-Design-Guidelines-2.pdf
[6] Jose I. Bilbao, Garvin Heath, Alex Norgren, Marina M. Lunardi, Alberta Carpenter, Richard Corkish, 2021, PV Module Design for Recycling, International Energy Agency (IEA) PVPS Task 12, Report T12-23:2021. ISBN 978-3-907281-27-7.
[7] EN 45555:2019 General methods for assessing the recyclability and recoverability of energy-related products
[8] RecyClass https://recyclass.eu/recyclability/methodology/
[9] European Commission, Critical Raw Materials Resilience: Charting a Path towards greater Security and Sustainability, COM/2020/474 final.
[10] European Union. Directive 2012/19/EU of the European Parliament and of the Council of 4 July 2012 on waste electrical and electronic equipment (WEEE). OJ L 197, 24.7.2012, p. 38–7.
[11] NSF International (2019). Sustainability Leadership Standard for Photovoltaic Modules and Photovoltaic Inverters. Available at https://globalelectronicscouncil.org/wp-content/uploads/NSF-457-2019-1.pdf

10.4229/EUPVSEC2023/5DO.14.2
020510-003

This presentation was selected by the Sc. Committee of the EU PVSEC 2023 for submission of a full paper to one of the EU PVSEC's collaborating peer-reviewed journals.

LIFE CYCLE ASSESSMENT OF PV MODULE BACKSHEETS

Paul de Wild, SmartGreenScans, Wagenmakersweg 22, 1873 GH Groet, The Netherlands,
paul@smartgreenscans.com, +31 6 10520901 (corresponding author),
Mariska de Wild–Scholten, SmartGreenScans, The Netherlands, mariska@smartgreenscans.nl,
Imco Goudswaard, Endurans Solar™, The Netherlands, Imco.Goudswaard@endurans.com.

Backsheets are very important regarding lifetime energy yield of the PV module, but the environmental impacts of their production, use and end-of-life (EoL) processing are largely neglected. In comparison with polyethylene terephthalate (PET) - based backsheets and fluoropolymer containing backsheets, poly-olefin (PO) - based backsheets perform best in terms of energy yield, reliability and environmental impacts. The production of fluoropolymer- and PET-based backsheets cause substantial environmental impacts, especially regarding climate change and ozone depletion. It was shown that pyrolysis of the spent backsheets potentially leads to much lower GWP when compared to incineration, especially for the PO-based backsheets. Incineration of the shredded and solid backsheet material causes direct emissions of CO2 with a limited heat recovery potential only. Deploying pyrolysis for spent PO-based backsheets significantly improves their life-time GWP per kWh produced. Pyrolysis offers the possibility to recover a large part of the polyolefin as an usable pyrolysis oil that might serve as feedstock for chemicals or as transportable liquid fuel for the generation of process heat in recovery boilers, thereby avoiding the use of new fossil resources. EoL pyrolysis or incineration of fluoropolymer-based backsheets is problematic due to presence of fluorinated hydrocarbons, leading to corrosive and/or toxic products.

Backsheets, poly-olefin, LCA

1 INTRODUCTION

Increased deployment of solar PV enables the transition to decarbonized energy systems, capable of tempering the dire consequences of global warming. While PV modules and materials are mostly selected on their €/Wp cost and minimum performance requirements, reliability and sustainability improvements are still underexploited. Even though backsheets are very important regarding lifetime energy yield of the PV module, the environmental impacts of their production, use and end-of-life (EoL) processing are largely neglected. Most polymeric backsheets are laminated multi-layer films based on an electrical insulating PET core film. Fluorine containing outer layers protect the backsheets against harsh outdoor conditions. Unfortunately, the use of fluorine, adhesives and multi-material stacks adversely affect the environmental impact and cost-efficient recyclability of backsheets. For instance, to date, at the end of their life, incineration of crushed spent PV modules, including the backsheet, to generate (process) heat is common practice. Regarding the (polymeric) backsheets, other technologies such as pyrolysis or solvolysis might be applied to recover value-added products such as valuable chemicals (e.g. monomers from which plastics may be produced) or pyrolysis oil (as fuel or as chemical feedstock) [1,2]. As part of a recently finalized Dutch national project EXTENSIBLE (Energy yield assessment of neXT gENeration and SustaInaBLE backsheets) the environmental impacts for 7 different polymeric backsheets have been evaluated via a life cycle assessment (LCA). The selected backsheets include 3 traditional polyethylene terephthalate (PET) - based backsheets with a fluorine containing outer layer (two white pigmented and one fully transparent). The other 4 backsheets are novel high-performance polyolefin (PO) - based backsheets, manufactured by Endurans Solar™, also including one transparent version. To date, this selection is representative for the wealth of backsheet types that is around. Figure 1 presents a non-exhaustive schematic overview of currently used materials and processes for polymeric backsheets [3].

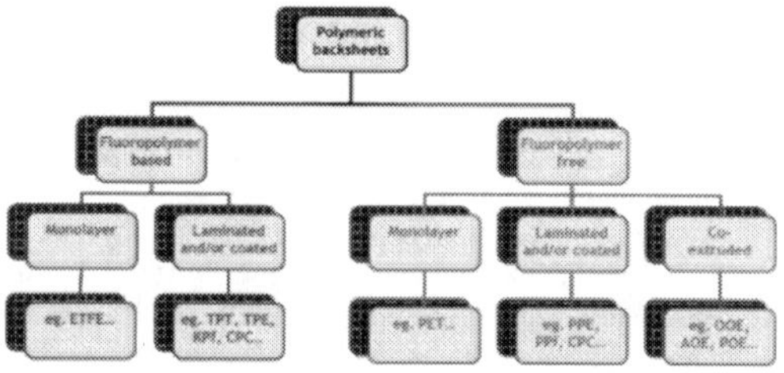

Figure 1: Polymeric PV module backsheet material classification (adapted from [3]). ETFE: ethylene tetra fluorethylene, TPT: Tedlar®-PET-Tedlar® , TPE: Tedlar®-PET-PE, KPf: Kynar®-PET-fluorine skin, CPC: coating*-PET-coating* PPE: PET-PET-PE, PPf: PET-PET-fluorine skin: OOE: PO-PO-PE, POE: PBT-PO-PE;, AOE: PA-PO-PE, Tedlar®: PVF (polyvinylfluoride (Dupont)), PET: polyethylene terephthalate, PBT: polybutylene terephthalate, PE: polyethylene, PO: ethylene/propylene copolymer or blend, PA: polyamide, Kynar®: PVDF (polyvinyldifluoride (Arkema)), coating*: proprietary coating (Feron)

2 LCA GOAL AND SCOPE

2.1 Goal

The aim of this work was a life cycle assessment (LCA) of 7 different backsheets, involving a critical evaluation of environmental impacts, benchmarking, sensitivity analysis and linking of material properties with module performance, lifetime and environmental impacts to identify guidelines for future material developments.

2.2 Scope

The work encompassed a full LCA (production – use – end-of-life) of the backsheets. The focus was to establish the environmental impacts including incineration and pyrolysis as end-of-life (EoL) phases. Pyrolysis offers the possibility to recover a large part of the polyolefin as an usable pyrolysis oil that might serve as feedstock for chemicals or as transportable liquid fuel for the generation of process heat in recovery boilers, thereby avoiding the use of new fossil resources [1].

EoL pyrolysis or incineration of fluoropolymer-based backsheets is problematic due to presence of fluorinated hydrocarbons, leading to corrosive and/or toxic products [4,5]. The LCA was conducted using the software package SimaPro (version 9.4.0.1) including the Ecoinvent database (version 3.8) and the (adapted) Environmental Footprint impact assessment method (EN15804+A2, version 1.03). Figure 2 briefly explains what LCA entails.

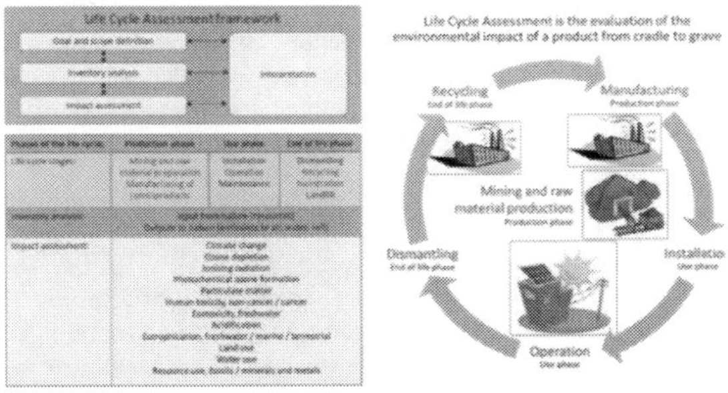

Figure 2: The LCA framework (SmartGreenScans©)

3 LIFE CYCLE INVENTORY

Data such as detailed backsheet material composition and the energy yield of the test modules in 5 climate zones (The Netherlands, Belgium, Singapore, India and Kuwait) including the measured degradation rate were provided by the project partners. The group of polymeric backsheets included 3 traditional polyethylene terephthalate (PET) - based backsheets with a fluorine containing outer layer. Two of them are white pigmented and 1 is fully transparent. The other 4 backsheets belong to a new group of co-extruded polyolefin (PO) backsheets. They all consist of a robust polyolefin core layer and differ by the outer layer polymer, i.e. polyamide (PA), polyester or polyolefin. Also, in this group of backsheets there is one transparent version.

4 LIFE CYCLE IMPACT ASSESSMENT

4.1 Overall environmental impact assessment

The contribution of all environmental impacts during the production phase of the backsheet was estimated and compared to the overall environmental impact from the production of a full module.

4.2 Global warming potential (GWP)

For the estimation of the GWP's for commercial backsheet types a compilation of typical characteristics of (especially non-fluoropolymer containing) commercial backsheets [5] was made to enable the calculation of their GWP's according to IPCC2013GWP100a (kg CO2-eq/kg) using the Ecoinvent database for comparison with EXTENSIBLE backsheets and typical solar glass. Because not all used materials are present in the Ecoinvent database, some approximations were made. Fluoropolymers are modelled as PVF. Urethane methacrylate resin is modelled as polymethylmethacrylate (PMMA). For the poly-olefins, N-olefins were used and polyamide was modelled as Nylon 6,6. The GWP for the production phase of the 7 backsheets was benchmarked against GWP's of various commercial backsheets from literature [6].

4.3 Assessment of GWP per kWh electricity produced

In a sensitivity analysis the effect of backsheet material, degradation rate, lifetime, EoL technology and location on the CO2-eq/kWh was assessed and material selection guidelines were identified. For the EoL phase via incineration and pyrolysis, products such as gases, liquids and solids are estimated using literature data for the processing of typical polymeric materials, representative for the materials that are used in the EXTENSIBLE backsheets. During one year, project partner IMEC has measured energy yield in kWh including the corresponding rate of degradation. Results were extrapolated to a lifetime of 30 years for mono-crystalline silicon based PV modules containing the EXTENSIBLE backsheet and used to calculate the overall greenhouse gas emission (from production and end-of-life phases) per kWh produced by the modules during their use-phase for 5 climate zones.

5 RESULTS

5.1 Impact contribution of backsheets

The contribution of the backsheet to the carbon footprint of a full module production is small (~1%). This is also valid for all other environmental impact categories next to climate change. The only exception is the ozone depletion in case of PET based backsheets due to the emission of methylbromide during PET production [7]. The relevance of this result with regard to commercial use should be checked with PET producers.

5.2 Global warming potential (GWP)

In general, the PO-based backsheet production shows significant lower GWP when compared to the two fluoropolymer containing-backsheets. Leaving out the TiO2 pigmentation further reduces the GWP for the transparent PO-based backsheet PO-T, manufactured by Endurans Solar™. It should be noted that both 2 and 3.2 mm solar glass have a GWP far beyond all other backsheets.

5.3 Comparison PO-based backsheets with traditional PET- and fluoropolymer- based ones

Due to its PO base and the absence of TiO2 the Endurans Solar™ PO-based transparent backsheet PO-T shows best environmental performance for all impact categories. For CO2 emissions assessment during the end-of-life phase, incineration (waste-to-energy) and pyrolysis were considered. When backsheets are incinerated, the amount of process heat generated (avoided CO2 emissions), is more or less the same for all backsheets. However, fluoropolymers release highly corrosive hydrogen fluoride (HF) [4,5], that should be handled in specialty incineration plants at significantly higher cost. Treatment of the PO transparent backsheet in a pyrolysis process results for 90% in oils and gases that can be used as feedstock again for production of chemicals. For the fluoropolymer-based transparent backsheet this is less than 45%.

5.4 Assessment of the GWP per kWh electricity produced

For assessing carbon emissions (GWP) during production and EoL phases and environmental impact per kWh, degradation rates and initial yields are translated into lifelong (30 yrs) energy yields for the 7 backsheets including incineration and pyrolysis as EoL technologies and for the 5 locations (Figure 3).

10.4229/EUPVSEC2023/5DO.14.5
020511-002

The lowest GWP per kWh produced is achieved with modules that contain the PO-based backsheets, irrespective of the EoL technology (incineration or pyrolysis), climate zone, backsheet pigmentation (transparent or white). Using pyrolysis instead of incineration as EoL technology, the lower impact of PO backsheets becomes more pronounced and the transparent PO does even reach negative emissions in high irradiation regions. From the environmental impact assessment over the cradle-to-grave lifecycle of the backsheets it is concluded that the PO backsheets show lowest impact per kWh produced for all categories. The production of fluoropolymer- and PET-based backsheets cause substantial environmental impacts regarding climate change and ozone depletion. The large ozone depletion potential of all backsheets containing PET film is due to the emission of methylbromide during the production of terephthalic acid [7]. Further validation of this high ozone depletion potential is recommended.

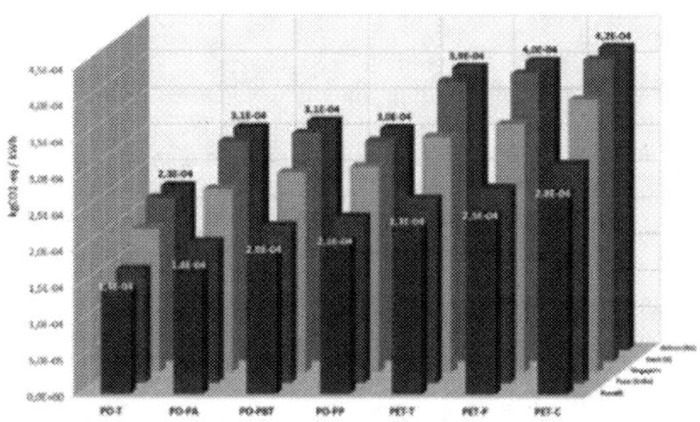

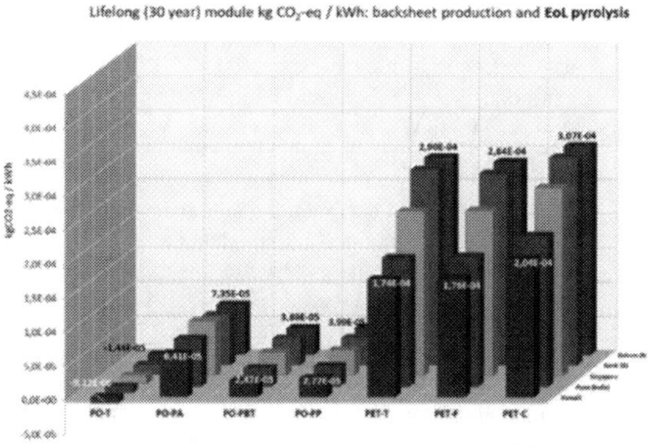

Figure 3: GWP of production and EoL phases per kWh produced as function of backsheet, location and EoL technology (PO-T: transparent polyolefin backsheet, PO-PA: white polyolefin backsheet with PA outer layer, PA: polyamide, PO-PBT: white polyolefin backsheet with PBT outer layer, PBT: polybutylene terephthalate, PO-PP: white polyolefin backsheet with PP outer layer, PP: polypropylene, PET-T: transparent PET backsheet with PVF outer layer, PVF: polyvinylfluoride, PET-F: white PET backsheet with PVDF outer layer, PVDF: polyvinyldifluoride, PET-C: white PET backsheet with coated fluorine outer layer)

6 CONCLUSIONS

Results from a cradle – to – grave life cycle assessment involving EN15804 environmental impact assessment indicate that in comparison with polyethylene terephthalate (PET), polyamide (PA), and various fluoropolymers containing backsheets, poly-olefin (PO)-based backsheets perform best in terms of energy yield, reliability and environmental impacts. This conclusion is corroborated by recent literature data [3,8]. The production and end of life processing of fluoro- and PET-based backsheets cause substantial environmental impacts regarding climate change and ozone depletion. In general, the PO-based backsheet production shows a significantly lower GWP when compared to the fluoropolymer containing-backsheets. Leaving out the TiO2 pigmentation further reduces the GWP for the transparent PO-based backsheet PO-T (manufactured by Endurans Solar™). It should be noted that both 2 and 3.2 mm solar glass have a GWP far beyond all other backsheets. The contribution of the backsheet to the carbon footprint of a full module production is small (~1%). This is also valid for all other environmental impact categories next to climate change. The only exception is the ozone depletion in case of PET based backsheets due to the emission of methylbromide during PET production. The relevance of this result with regard to commercial use should be checked with PET producers. Regarding the end of life phase of the backsheets, incineration for generation of process heat yields an amount of process heat that is more or less the same for all backsheet types. However, fluoropolymers release highly corrosive hydrogen fluoride (HF), that should be handled in specialty incineration plants at high cost. In contrast, treatment of PO-based backsheets in a pyrolysis process results for 90% in oils and gases that can be used as a usable fuel oil or as a feedstock for production of chemicals. For the fluoropolymer-based backsheet these options are problematic due to the presence of fluor and/or fluorinated compounds in all pyrolysis products. From a sensitivity analysis it was found that the lowest GWP per kWh produced, is achieved with modules that contain the PO backsheets, irrespective of the EoL technology (incineration or pyrolysis), climate zone and backsheet pigmentation (transparent or white). Using pyrolysis instead of incineration as EoL technology, the lower impact of PO backsheets becomes more pronounced and the transparent PO does even reach negative emissions in high irradiation regions.

7 REFERENCES

[1] Ina Vollmer, Michael J. F. Jenks, Mark C. P. Roelands, Robin J. White, Toon van Harmelen, Paul de Wild, Gerard P. van der Laan, Florian Meirer, Jos T. F. Keurentjes, and Bert Marc Weckhuysen, Beyond Mechanical Recycling: Giving New Life to Plastic Waste, Angew. Chem. Int. Ed. 10.1002/anie.201915651bb.

[2] Simon Hann and Toby Connock, Chemical Recycling: State of Play, Report for CHEM Trust, December 2020, Eunomia Research & Consulting Ltd 37 Queen Square Bristol BS1 4QS, UK, www.eunomia.co.uk.

[3] Gernot Oreski et al., IEA PVPS Task 13: Performance, Operation and Reliability of Photovoltaic Systems, Designing New Materials for Photovoltaics: Opportunities for Lowering Cost and Increasing

10.4229/EUPVSEC2023/5DO.14.5
020511-003

Performance through Advanced Material Innovations, Report IEA-PVPS T13-13:2021, April 2021.
[4] John T. Stapler, William J. Barnes, William E. C. Yelland, Thermal degradation of polyvinylidene fluoride and polyvinyl fluoride by oven pyrolysis, Technical Report 69 – 7- CM, United States Army, Natick laboratories, Clothing and Organics Laboratory C & OM 50, Project Reference Series: IT024401A329, C&W4-50, July 1968.
[5] Venkat Aryan, Mercè Font-Brucart, Daniel Maga, A comparative life cycle assessment of end-of-life treatment pathways for photovoltaic backsheets, Prog Photovolt Res Appl. 2018;26:443–459.
[6] Shravan K. Chunduri and Michael Schmela, TaiyangNews Market Survey on backsheets and encapsulation 2020, https://taiyangnews.info/reports/market-survey-backsheets-encapsulation-2020.
[7] Rogério A. F., Tomás, João C. M. Bordado and João F. P. Gomes, p-Xylene Oxidation to Terephthalic Acid: A Literature Review Oriented toward Process Optimization and Development, dx.doi.org/10.1021/cr300298j | Chem. Rev. 2013, 113, 7421−7469.
[8] Antonia Omazic et al., Increased reliability of modified polyolefin backsheet over commonly used polyester backsheets for crystalline PV modules, J. Appl. Polym. Sci. 2020, 137, 48899.

8 ACKNOWLEDGEMENTS

This work has been conducted as part of a recently finalized Dutch national project EXTENSIBLE (Energy yield assessment of neXT gENeration and SustaInaBLE backsheets). EXTENSIBLE was conducted from 01-03-2020 – 31-08-2022 by a consortium led by Endurans Solar™ (NL, project co-ordinator), IMEC / EnergyVille (B), DSM ACC (NL) and SmartGreenScans (NL) with subsidy from the Ministry of Economic Affairs and Climate Policy and the Ministry of Agriculture, Nature and Food Quality, National EZK- and LNV-subsidies Top Sector Energy, carried out by the Netherlands Enterprise Agency (RVO). SmartGreenScans gratefully acknowledges the subsidies of the Dutch government and the fruitful collaboration with the consortium partners.

COMPARATIVE AND PROSPECTIVE LCA OF PEROVSKITE-SILICON TANDEM PV MODULES

Cristina Polacchi[1,2], Atse Louwen[1], David Moser[1], Cristina Cornaro[2,3]
[1]Eurac Research, Institute for Renewable Energy, Bolzano, Italy,
[2]University of Rome Tor Vergata, Department of Enterprise Engineering, Rome, Italy,
[3]CHOSE, University of Rome Tor Vergata, Rome, Italy
cristina.polcacchi@eurac.edu; atse.louwen@eurac.edu; david.moser@eurac.edu; cornaro@uniroma2.it

ABSTRACT: The Perovskite-Silicon Tandem (PST) is currently one of the most promising PV technologies, especially among the silicon-based tandem cells. It has reached a module efficiency of 26%, which is expected to grow above 30% by 2030, and it can be produced with relatively simple additional processing steps compared to mono-crystalline silicon technologies, while offering much higher efficiency. The major weakness that has limited the Perovskite-based PV cells from entering the market for many years was their low stability in operating conditions. One proposed way to overcome this issue is the paring of a perovskite layer with a C-Si cell in a tandem configuration, in order to take maximum advantages from the high stability of the C-Si cells on one hand, and the high efficiency of the perovskite top absorber on the other hand. . The goal of the present study is the evaluation of the environmental impact of the emerging Perovskite Silicon Tandem PV modules, through the Life Cycle Assessment (LCA) methodology, focusing on the Global Warming Potential (GWP) impact. Both a Comparative LCA and a prospective dynamic LCA were performed, in order to evaluate how the impact on climate change of the PV modules analysed might change in the future, considering future trends that are relevant for the PV market until 2030.

Keywords: LCA, GHG, PV, Perovskite, Tandem

1 INTRODUCTION

It is well known that replacement of fossil energy sources with renewable ones will allow to reduce the impact of electricity generation on climate change. In fact, the GHG emissions per kWh produced using photovoltaic (PV) systems are almost 80% lower than the ones from coal powered plants, on a global average [1]. PV systems have about zero emissions during their operation. Although, during their manufacturing, GHG emissions are released as a result of direct and indirect energy use from a mix of fossil and renewable sources.
In order to limit the increase of global temperatures below 2 °C, not only a reduction in the emissions, but a complete cut in the GHG emissions, needs to be achieved by 2050. [2]. In a scenario where PV will be massively integrated in the energy systems worldwide, it is crucial to evaluate the impact on climate change by measuring the GHG emissions, but also other relevant effects on humans and the environment. A lifecycle approach should be used to evaluate the impact of the PV technologies that are expected to hit the market in the future, to guide industry and research towards the most environmental-friendly solution and ensure a sustainable supply chain of the solar energy, from the extraction of the required raw materials to the end-of-life operations.

Literature offers a good variety of LCAs focused on PV applications but, for what concerns the new perovskite-based technologies, most of the results available are based on laboratory-scale processes, only a few on industrial-scale, and even fewer studying prospective technology improvements of perovskite-based PV. Aim of our research is in fact to fill this gap, enriching the literature with a study of the industrial-scale manufacturing of perovskite technology, considering several future technological changes, and including a comparison with mainstream technologies.

Perovskite is a semiconductor material with a higher energy bandgap than silicon. It can convert high-energy photons into electrical energy, while the silicon cell converts the low-energy ones. In this way, a larger fraction of the energy in the solar light spectrum can be absorbed and converted into electricity, allowing higher power conversion efficiency than silicon cells alone.
Currently, modules based on crystalline silicon account for about 95% of the market share [3]. Among them, monocrystalline silicon-based PV modules have become the standard technology. Within this category, the Passivated Emitter and Rear Contact (PERC) cell design is the most common, which offers commercial module efficiencies approaching 21%. Technologies that offer higher module efficiencies, like silicon heterojunction (SHJ) technology, have also seen long-term development but have failed to gain substantial market shares due to their relatively high cost. Major advances towards module efficiencies of >25% are not expected for any monocrystalline silicon technology without the transition towards tandem cells. In tandem cells, several layers with different band gaps ensure that the energy available in the solar spectrum is used more efficiently.
Perovskite-based PV modules do not have a share in the market jet, but forecasts are showing that silicon-based tandem cells are expected to increase their share until about 5% by 2030, and growing more in the next years [4], the Perovskite-Silicon Tandem (PST) technology being one of the more promising among them. This technology has reached a module efficiency of 25%, which is expected to grow above 30% by 2030 [5].
PST PV cells consists of a high efficiency silicon bottom cell with a smaller bandgap, on which a perovskite layer with larger bandgap is deposited. The major weakness that has for years limited perovskite-based PV cells from entering the market was their low stability in real operating conditions. One proposed way to overcome this issue is the paring of a perovskite layer with a C-Si cell in a tandem configuration. This should allow manufacturers to guarantee a good stability under real operating conditions – for up to 25 years – which is almost comparable with the 30 year-lifetime of the crystalline silicon based PV modules [6]. PST cells can be produced with relatively low

additional environmental impact compared to monocrystalline silicon technologies, while offering much higher efficiency.

For this reason, manufacturing companies have started to invest in PST technology worldwide. E.g., manufacturer Oxford PV plans to build a facility in Germany with an annual expected capacity of up to 1 GW, while the French joint venture between the companies Voltec and IPVF has the target to build a 5 GW capacity facility, similarly to the Chinese company Microquanta Semiconductor.

Perovskite top cells can be realized with the same techniques used to produce thin-film cells. Those can be solvent-based and solvent-free. At first, solvent-based technologies were used to apply the perovskite absorber layer, such as inkjet-printing or spin coating. Those techniques allow to reach high performances and a good control in the injection of multiple materials at the same time, as required by the perovskite absorber deposition.
However, the most used solvents are toxic for humans and the environment, and a 2021 update of the EU REACH regulation is restricting its use in industry [7]. For this reason, the Horizon Europe project NEXUS aims at the development of a more sustainable solvent-free process for manufacturing PST cells, without jeopardizing cell efficiency.
Once the tandem cell is produced, the traditional standard methods are used to apply the contacts, encapsulate the module with polymer-based encapsulants, cover it with glass sheets, and apply the aluminium frame.

2 METHODOLOGY

The aim of the present research study is to evaluate the environmental impact of the emerging technology of Perovskite-Silicon Tandem (PST) PV modules, considering its whole lifecycle, and compare it to the current mainstream PV technologies based on Mono and Multi Crystalline Silicon (Mono and Multi C-Si) PV cells.
The performance indicators for the modules analyzed are summarized in Table I. The module efficiency of the perovskite is assumed to be the highest, according to what suggested by updated literature sources [8]. While the module efficiency values for the C-Si PV modules are evaluated as suggested in the IEA PVPS Report [9].
The lifetime average performance ratio includes the degradation ratio with time, conversion losses from DC to AC, and the other losses. The value of 0.85 performance ratio for the C-Si was chosen according to Müller et al., 2021[10], while the 0.736 value for perovskite-silicon tandem module was selected from M. Roffeis et al., 2022 [8], which is lower because of the higher degradation of the perovskite top cell in operating conditions.
The lifetime of the PST module is assumed to be 30 years in a first place, same as the lifetime assumed for C-Si. Then, two sensitivity analysis were performed: one considering a lifetime of 25 years, as suggested from literature sources [8], and another one considering the same lifetime of 30 years for the PV modules, and the same degradation ratio of 0.85 for the three technologies.
To better represent the current market conditions, China was selected as manufacturing country, but the impact of moving manufacturing to Europe was investigated as a sensitivity analysis as well.
Europe was considered as installation location, with an average annual horizontal irradiation on an optimal oriented module of 1391 W/m^2, as suggested in the Guidelines provided by IEA [11].

Table I: Performance indicators of the analyzed PV modules.

	LIFETIME	MODULE EFFICIENCY	LIFETIME AVERAGE PERFORMANCE RATIO
Mono C-Si PV	30 yr	20.5%	0.85
Multi C-Si PV	30 yr	18.0%	0.85
Perovskite -Si Tandem PV	25 yr (30 for sensitivity analysis)	26.0%	0.736

The study is focused on the Global Warming Potential (GWP) impact, representing the impact on climate change, and expressed in kilograms or grams of CO_2 equivalent greenhouse gas (GHG) emissions. As suggested in the IEA Guidelines, the most updated IPCC impact method was used for this analysis, the IPCC 2021 - GWP100 indicator, which is part of the Environmental Footprint (EF) 3.1 Method indicators available for the OpenLCA software.

Two types of LCA were performed. First, a comparative LCA where the PST technology is compared to Mono and Multi C-Si. Here, the contribution to the GWP impact of the most relevant lifecycle phases was analyzed.
Different functional units and system boundaries were considered. In fact, the GWP impact was expressed per unit area of module in a cradle-to-gate analysis, per kWp of installed power capacity, considering the different efficiencies of the modules, and per kWh of AC electricity, considering the different lifetimes of the modules, in a cradle-to-gate analysis which includes the manufacturing of the balance of system (BOS) components, and the end-of-life operations of module dismantling and partial recycling. The operating phase was not considered, since its impact is negligible compared to the other life stages.

In a second step of the study, a dynamic prospective LCA was performed, to evaluate how the impact of the PV manufacturing will change by 2030, with 1-year timesteps. This analysis is performed considering the technological advancements foreseen in the PV market. Specifically, three trends were modelled:

1. Decarbonization of the electricity mix
2. Increase in cell and module efficiency
3. Manufacturing technology advancements.

For the first point, the decarbonization of the electricity mix of China was modelled according to the ENTSO-E medium-decarbonization rate scenario [12], and this new electricity mix was used in the foreground processes to manufacture the modules.
On the other hand, an increase in the efficiency is expected for Mono C-Si and PST PV technologies, and was modelled according to the ITRPV 2023 Roadmap [13] for C-Si PV. For Multi C-Si PV, an increase of efficiency is not expected in this roadmap, due to technology saturation. For PST PV, the increase of efficiency was modelled considering the trends expected by the last Perovskite White Paper [5].
Lastly, the advancement of manufacturing technology was modelled considering one of the most important expected trends, according to the PV 2023 Roadmap [13], which is

10.4229/EUPVSEC2023/5DO.15.1
020512-002

the decrease in the polysilicon used to produce a C-Si module. This reduction is due to the wafer thickness reduction trend and to a reduction in the silicon kerf losses incurred while cutting the wafers and the cells.

OpenLCA was the software used for the analysis. The parametrization function of the software was used to model the selected trends foreseen by 2030 for the PV sector, and perform the dynamic prospective LCA.
As inventory for the foreground processes, meaning the ones from the metallurgical silicon production to the end-of-life operations, the most updated IEA PVPS Task 12 inventories [14,15] were taken as reference. For the background processes, the most updated Ecoinvent 3.9 database [16] was used.

3 RESULTS

3.1 Comparative LCA GWP Impact results

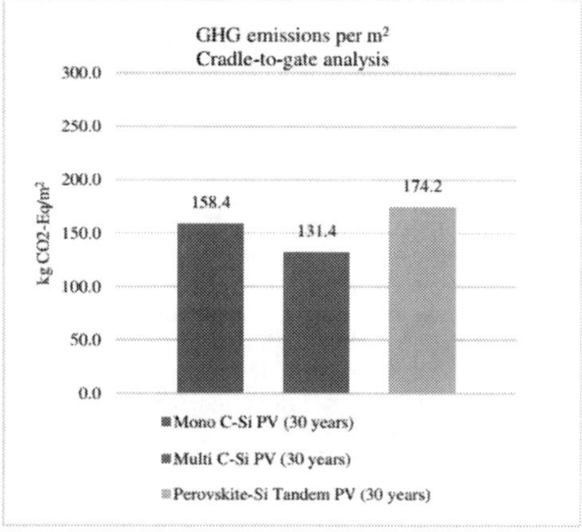

Figure 1

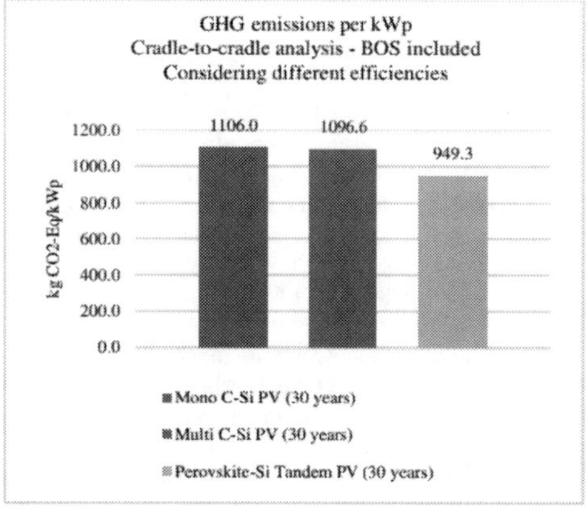

Figure 2

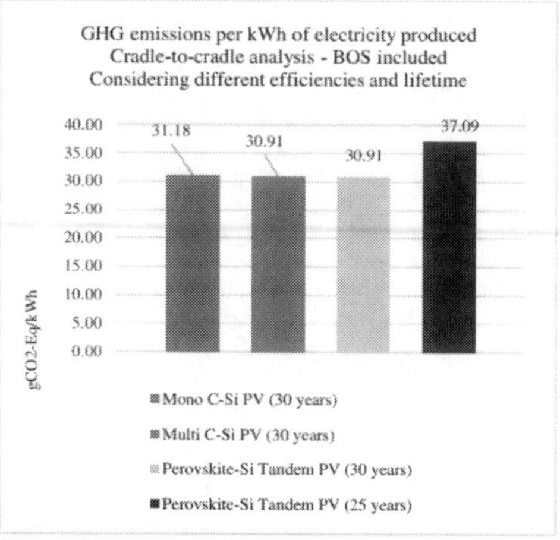

Figure 3

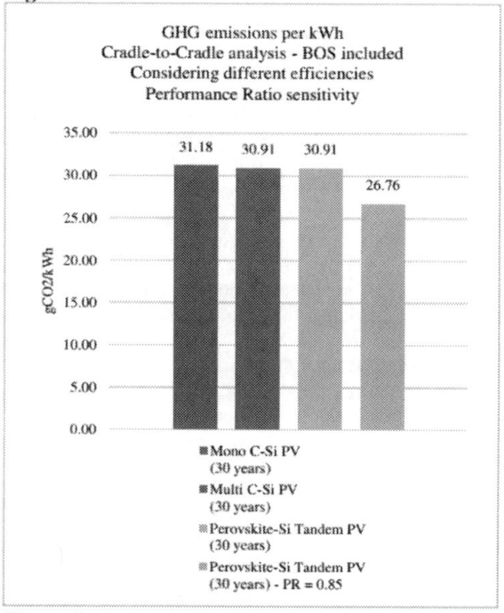

Figure 4

Figures 1-2-3-4 GWP impact per different functional units: 1 m² [Figure 1], 1 kWp [Figure 2], 1 kWh [Figure 3], and sensitivity analysis considering different module lifetime [Figure 3] and performance ratio (PR) [Figure 4]
In **Figure 1Figure 2Figure 3**, **Figure *4*** the results are expressed per different functional units, as explained in Section 2. It can be observed how the GWP impact of the PST PV is higher per square meter, in a cradle-to-gate analysis (**Figure 1**). But, when the impact is expressed per kWp of installed capacity (**Figure 2**) where the effect of the efficiency is considered, in a cradle-to-cradle analysis, the impact of the PST becomes the lowest, due to its higher efficiency compared to the Mono and Multi C-Si PV technologies.
On the other hand, by observing the cradle-to-cradle GWP impact per kWh of AC electricity generated (**Figure 3**), and considering different lifetimes for the PST PV modules, of 30 and 25 years, one can observe how the impact of the perovskite lasting 25 years becomes the highest. This is due to the fact that, in this case, the higher PST module efficiency does not repay in terms of GHG

10.4229/EUPVSEC2023/5DO.15.1
020512-003

emissions, the additional manufacturing process necessary to produce it.

Finally, in **Figure 4**, it is assumed a "best-case scenario" in which the Perovskite – Silicon Tandem PV module would reach the same performance ratio of the only-silicon ones (0.85). As a result, it is possible to observe how, also considering the same lifetime of the silicon module, the perovskite-based technology would have a further reduction in the GHG emissions per kWh of electricity produced.

In the previous results presented, China was chosen as a manufacturing location. In a sensitivity analysis, the manufacturing location was shifted to Europe. This shift has been modelled by using the updated Ecoinvent ENTSO-E electricity production mix in the foreground manufacturing processes, instead of the Chinese electricity mix. As a result, the lifecycle GHG emissions of the PV systems are reduced by about 16%, due to the higher share in renewable in the European electricity mix. This value has been obtained by modifying the electricity mix that is directly used in the main foreground processes, from the silicon production to the module assembly, buy not in the production of all the other components. This value would be higher if all the PV components would be produced in Europe as well.

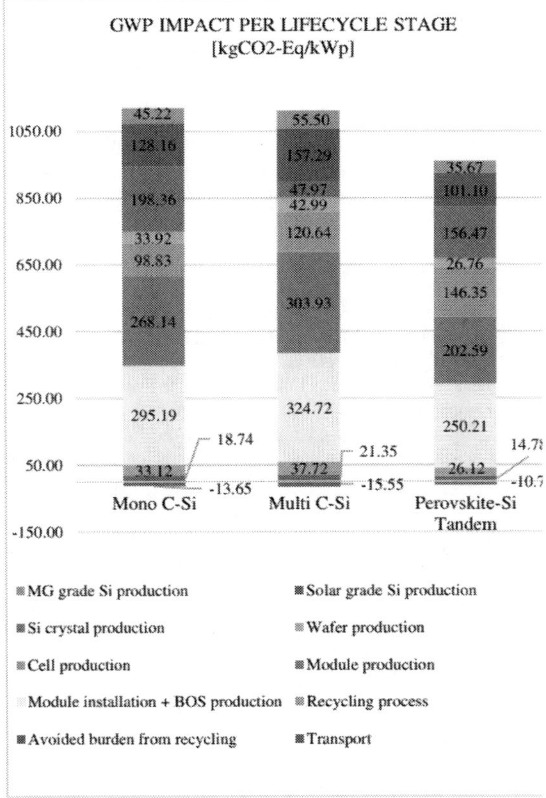

Figure 5 GWP impact per lifecycle stages per kWp

In **Figure 5** the GWP impact results of 1 kWp of PV installed capacity is represented for the three types of modules analyzed, and the contribution of each lifecycle stage is enlightened. We can notice how the impact contribution share is similar between the Mono C-Si and the PST modules, despite the PST is based on a Mono C-Si module, with one additional manufacturing step, meaning that the perovskite deposition process is not responsible for a high increase in the GHG emissions.

The Mono Crystal silicon production phase, so called Czochralski process, has a higher GWP impact compared to the casting process to obtain the Multi Crystal silicon rod.

The production of the silicon cell, from the metallurgical silicon, is responsible for about half of the lifecycle GHG emissions, while the transportation from China to Europe and the end-of-life stages, are responsible for only about 2% of the impact share each. Furthermore, the electricity used in the processes is responsible alone for about 40% of the GWP impact share, demonstrating the importance of the use of low-carbon sources to produce electricity, in order to reach the decarbonisation targets.

3.2 Prospective LCA results

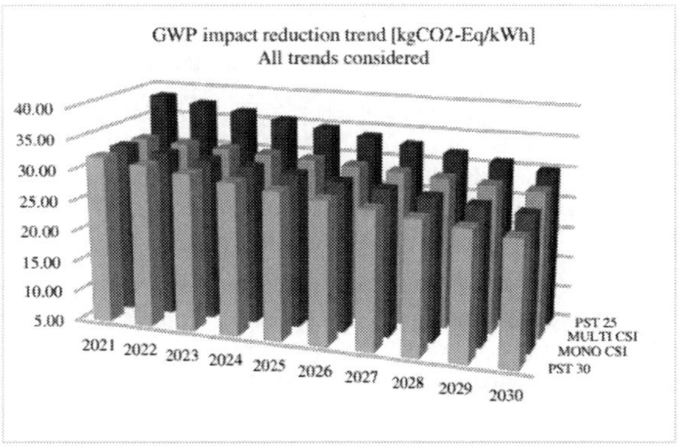

Figure 6 GWP impact reduction trend for different PV module technologies by 2030

Figure 6 represents a visualization on how the GHG emissions are expected to decrease by 2030, year by year, as a result of the foreseen trends for the PV market sector, as previously described in Section 2. One might observe that the highest decrease in the emissions is expected for the PST PV type, with an expected impact reduction of 22% from 2021 to 2030, compared to a 17% and 12% of GWP impact reduction, respectively for Mono and Multi C-Si PV types. Although, the PST PV technology would be competitive against then C-Si one, in terms of GHG emissions, only if a lifetime of 30 years will be guaranteed, otherwise, its impact per kWh will stay higher.

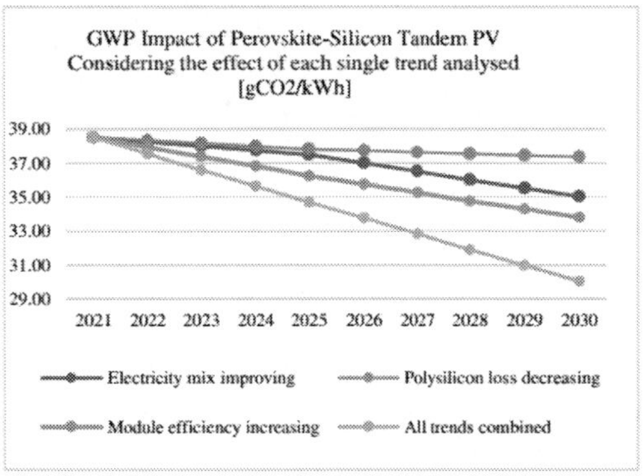

Figure 7 GWP impact reduction trend by 2030, for the Perovskite-Silicon Tandem PV technology.

In **Figure 7**, the contribution to the GWP impact of each single future trend analyzed, is plotted for the PST PV technology. It is possible to observe how all the three trends combined are responsible for a 22% of GHG emission reduction but, among the trends considered, the increase in module efficiency is responsible for the highest reduction of the emissions. In fact, it might allow to achieve a 12% of GHG emissions reduction from 2021 to 2030. The second major effect is the production of electricity sources, due to an increase in the renewables share and a decrease in the fossil fuel share. This effect would allow a GWP impact reduction of about 9%. Lastly, the improvement in the silicon cell manufacturing line technologies, modelled as a reduction in the silicon used to produce the wafer, would allow an emission reduction of about 3%.

4 CONCLUSION

To conclude, the Perovskite-Silicon tandem (PST) PV modules are produced by adding a manufacturing step to the Mono-Crystalline Silicon (Mono C-Si) PV modules.
For this reason, for the PST PV module there is a slight increase in GHG emissions compared to the production of crystalline-silicon-based modules, when the GWP impact is analysed per unit of module area produced, passing from 158 to 174 kgCO_2-eq/m^2, respectively for the Mono C-Si and the PST PV.
However, due to its higher efficiency, the lifecycle GWP impact per unit of electrical energy produced for the PST PV module is comparable to the Crystalline PV modules one (31 gCO_2-Eq/kWh), but only if the same lifetime of 30 years would be guaranteed for both module technologies. In fact, if a 25-year lifetime is considered for the PST technology, the lifecycle GHG emissions would be higher than the C-Si one (37 gCO_2-Eq/kWh).
Finally, assuming the "best-case scenario" for PST PV, in which the same lifetime and performance ratio of the C-Si modules is assumed, respectively 30 years and 0.85, the GHG emissions of PST PV would be very competitive with those of C-Si PV, reaching a value of 26.75 gCO_2-eq/kWh.

A reduction of about 22% in the GWP impact of the production of photovoltaic modules is expected by 2030, thanks to the PV market trends considered. The main cause of GWP impact reduction is represented by the foreseen increase in the module efficiency, up to 30% by 2030.

The results obtained from the dynamic prospective LCA analysis should not be taken as an exact prevision of what will happen in the future, since they are based on expected future trends and not precise forecasts, which is impossible to foresee. Although, they can provide some guidelines on the speed rate of the decarbonization process for the PV sector, and enlighten the most critical points in the PV supply chain from the GHG emissions point of view.

The lifecycle inventory based on the 2020 IEA Report [14] represents one of the most reliable and complete sources available in literature, since it is based on information coming from a large numbers of industrial level manufacturers worldwide. However, some inventory data are quite outdated – the oldest data being collected in 2011, and newest ones in 2019. Those data are based on the old back surface field technology (BSF), that is being surpassed by the passivated emitter and rear contact (PERC) one. While for the bottom silicon cell in tandem with the perovskite one, the silicon heterojunction (SHJ) configuration is the most promising one, in a 2 terminal or 4 terminal configurations [16], and this will be taken into account in further studies.
In fact, thanks to the NEXUS HORIZON EUROPE Project [17], a lifecycle inventory update with industrially relevant collected data is ongoing, and an updated LCA will be available by the end of the project, in October 2025. Aim of the project is to test innovative and more sustainable manufacturing processes employing solvent-free perovskite deposition, indium free and low-silver silicon bottom cell. Part of the scope of NEXUS project is to demonstrate their low environmental impact from a lifecycle point of view.
Furthermore, the analysis will be enriched with a discussion of other relevant impact for the PV sector, as suggested in the IEA Guidelines [11], together with an economic, social and circularity indicators.

5 AUTHOR CONTRIBUTION

C.P. performed the formal analysis, visualization and wrote the paper. A.L. contributed to the formal analysis and conceptualization of ideas, reviewed and supervised the work. D.M and C.C. contributed to the conceptualization of ideas, reviewed, and supervised the work.

6 DECLARATION OF COMPETING INTEREST

The authors declare no competing interests.

7 ACKNOWLEDGEMENTS

Funded by the European Union. Views and opinions expressed are however those of the author(s) only and do not necessarily reflect those of the European Union or RIA. Neither the European Union nor the granting authority can be held responsible for them. NEXUS project has received funding from the European Union's Horizon Europe research and innovation program under grant agreement No. 101075330.

8 REFERENCES

[1] 'Carbon Neutrality in the UNECE Region: Integrated Life-cycle Assessment of Electricity Sources', UNECE, Mar. 2022.
[2] 'Net Zero by 2050', IEA, Paris, 2021. [Online]. Available: https://www.iea.org/reports/net-zero-by-2050
[3] 'Solar PV Global Supply Chains', IEA, Paris, 2022. [Online]. Available: https://www.iea.org/reports/solar-pv-global-supply-chains
[4] 'International Technology Roadmap for Photovoltaic (ITRPV) – 2021 Results – 13th Edition March 2022'.
[5] 'Perovskite-based photovoltaics: A unique chance for European PV-industry – A White Paper from the European Perovskite Initiative (EPKI) – Sep. 2019'.
[6] 'Sustainable Energy Fuels, 2022,6, 2924-2940',

[7] 'EUROPEAN COMMISSION, «Commission Regulation (EU) 2021/2030 of 19 November 2021 amending Annex XVII to Regulation (EC) No 1907/2006 of the European Parliament and of the Council concerning the Registration, Evaluation, Authorisation and Restriction of Chemicals (REACH) as regards N,N-dimethylformamide (Text with EEA relevance)». 2021. https://eur-lex.europa.eu/legal-content/EN/TXT/?uri=CELEX:32021R2030'.
[8] 'M. Roffeis et al., «New insights into the environmental performance of perovskite-on-silicon tandem solar cells – a life cycle assessment of industrially manufactured modules», Sustain. Energy Fuels, vol. 6, fasc. 12, pp. 2924–2940, 2022, doi: 10.1039/D2SE00096B.'.
[9] 'R. Frischknecht, L. Krebs - IEA PVPS, «Environmental life cycle assessment of electricity from PV systems FACT SHEET 2021 PV POWER SYSTEM T12», 2021.'
[10] A. Müller, L. Friedrich, C. Reichel, S. Herceg, M. Mittag, and D. H. Neuhaus, 'A comparative life cycle assessment of silicon PV modules: Impact of module design, manufacturing location and inventory', *Sol. Energy Mater. Sol. Cells*, vol. 230, p. 111277, 2021, doi: https://doi.org/10.1016/j.solmat.2021.111277.
[11] 'R. Frischknecht, P. Stolz, G. Heath, M. Raugei, P. Sinha, M. de Wild-Scholten, 2020, Methodology Guidelines on Life Cycle Assessment of Photovoltaic Electricity, 4th edition, IEA PVPS Task 12, International Energy Agency Photovoltaic Power Systems Program'.
[12] 'EU-CHINA energy cooperation platform, «ENTSO-E Grid Planning Modelling Showcase for China». novembre 2021. - https://energy.ec.europa.eu/system/files/2022-03/02%20-%20ENTSO-E%20Grid%20Planning%20Modelling%20-%20Showcase%20for%20China.pdf'.
[13] 'International Technology Roadmap for Photovoltaic (ITRPV) – 2022 Results – 14th Edition April 2023'.
[14] 'R. Frischknecht, et al. «Life Cycle Inventories and Life Cycle Assessment of Photovoltaic Systems, Report T12», 2020'.
[15] 'Wambach K, Heath G, Libby C. 2018. Life Cycle Inventory of Current Photovoltaic Module Recycling Processes in Europe. IEA-PVPS Task 12 Report T12-12:2017. ISBN 978-3-906042-67-1'.
[16] 'Ecoinvent database v.3.9.1 www.ecoinvent.org, dec. 2022'.
[17] D. Yang *et al.*, '28.3%-efficiency perovskite/silicon tandem solar cell by optimal transparent electrode for high efficient semitransparent top cell', *Nano Energy*, vol. 84, p. 105934, 2021, doi: https://doi.org/10.1016/j.nanoen.2021.105934.
[18] 'https://nexus-pv.eu/'.

Modelling recycling in life cycle assessment of perovskite/silicon tandem modules

Lu WANG[1,2,3], Paula PEREZ-LOPEZ[1], Mathilde MARCHAND LASSERRE[1], Lars OBERBECK[2,3]

[1]Mines Paris, PSL University, Centre Observation Impacts Energie (O.I.E.), France
[2]TotalEnergies OneTech, France
[3]Institut Photovoltaïque d'Ile-de-France (IPVF), France

September 21, 2023

MINES PARIS | PSL TotalEnergies IPVF

Context

- **Photovoltaics (PV)** plays a key role in the global energy transition
- **Conventional Si cell** is still improving, but the power conversion efficiency has almost reached the limit
- **Perovskite on Si tandem** with a current record efficiency of 33.7% (KAUST) is among the most promising technologies to overcome the limit of power conversion efficiency

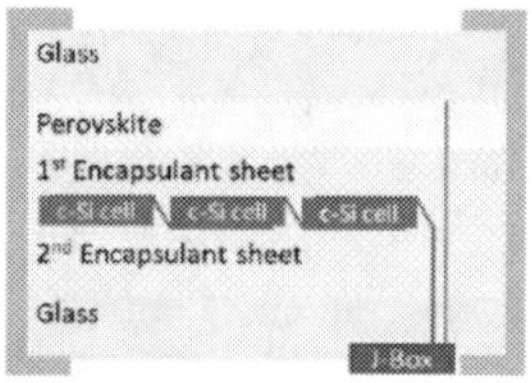

An example of 4T perovskite on c-Si tandem

Context

- **Photovoltaics (PV)** plays a key role in the global energy transition
- **Conventional Si cell** is still improving, but the power conversion efficiency has almost reached the limit
- **Perovskite on Si tandem** with a current record efficiency of 33.7% (KAUST) is among the most promising technologies to overcome the limit of power conversion efficiency

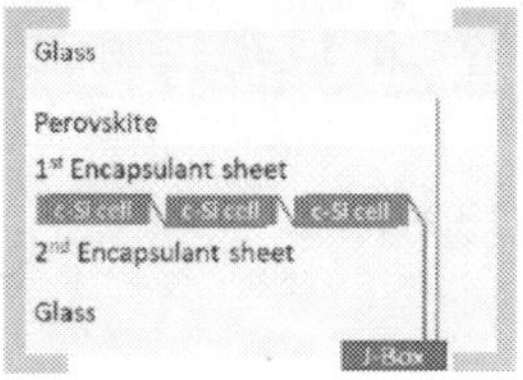

An example of 4T perovskite on c-Si tandem

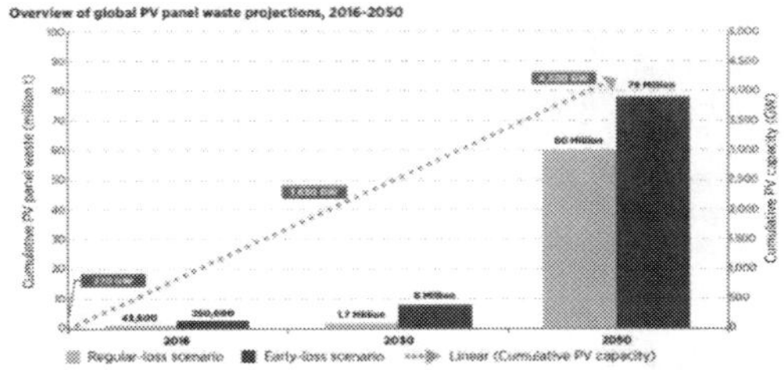

- Global **PV waste** will **increase** dramatically in the future, raising both challenges and opportunities
- An appropriate **end-of-life (EoL)** management such as **recycling** is critical to ensure the environmental soundness and avoid the shortage of mineral resources

Context

Using LCA to evaluate environmental impacts of perovskite/silicon tandem modules

- **Life cycle assessment (LCA)**: most common methodology to assess the potential environmental impacts of a technology over its entire life cycle (ISO 14040/14044)

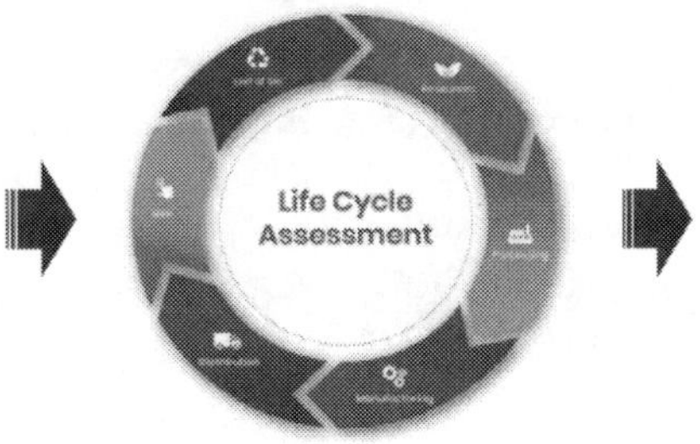

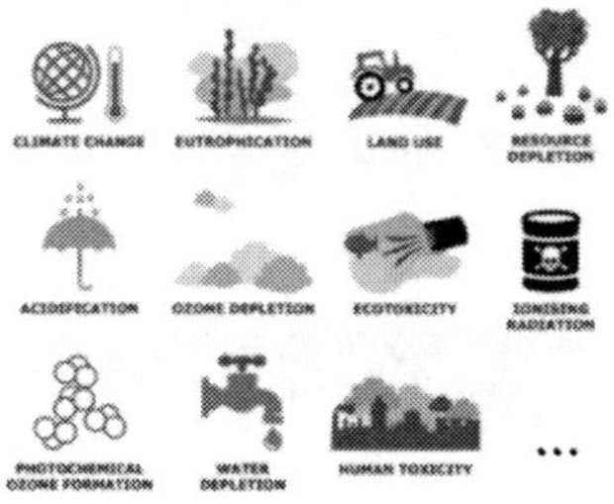

→ **Avoid shifting burdens between impact categories and between life cycle stages**

Challenge & Objective - EoL modelling in LCA

- **Challenge** Lack of consensus in selecting the EoL modelling approaches
- **Objective** Selecting the **adequate EoL modelling approach** for PV technologies

40th European Photovoltaic Solar Energy Conference and Exhibition

Challenge & Objective - EoL modelling in LCA

- **Challenge** Lack of consensus in selecting the EoL modelling approaches
- **Objective** Selecting the **adequate EoL modelling approach** for PV technologies

EoL modelling: Allocation between the product recycling at the EoL and the recycled material

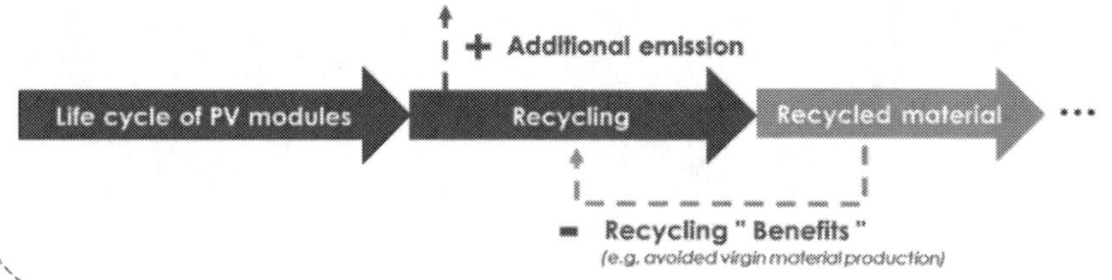

Other elements
Quality degradation
Market price
Market supply and demand ...

EoL modelling

- The **impacts** of EoL modelling: incentives to **"More recycled input?"** **<u>Or</u>** **"More recycling output?"**

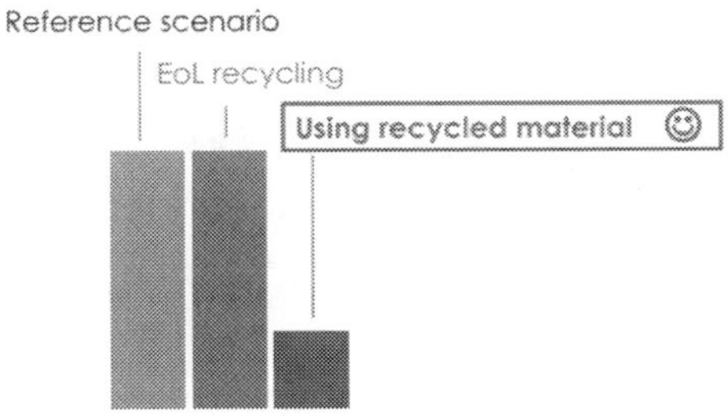

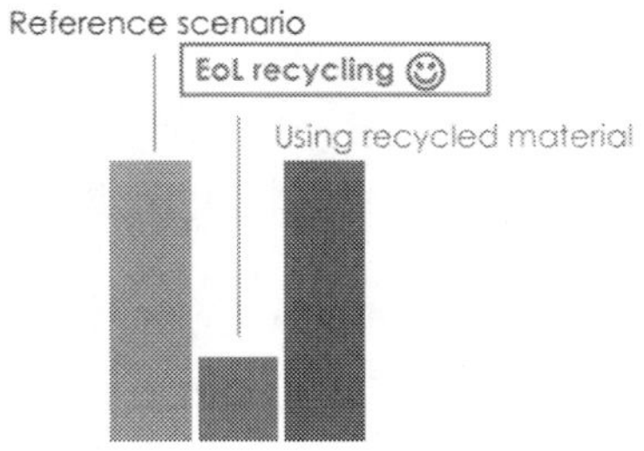

Methodology

- LCA is used to assess the environmental impacts of **1 m² of perovskite/Si tandem modules**

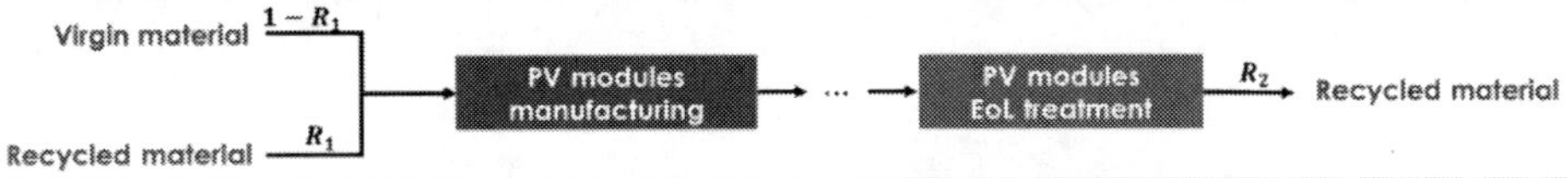

Recycled content ratio (R_1)		Recovered rate (R_2)		
Aluminium	74%	**Aluminium**	97%	
Glass	0%	**Glass**	93%	
Copper	20%	**Copper**	100% (Cable)	76% (Interconnection)
Silver	8%	**Silver**	100%	
Silicon	0%	**Silicon**	98%	
Lead	55%	**Lead**	99%	

Methodology

- LCA is used to assess the environmental impacts of **1 m² of perovskite/Si tandem modules**

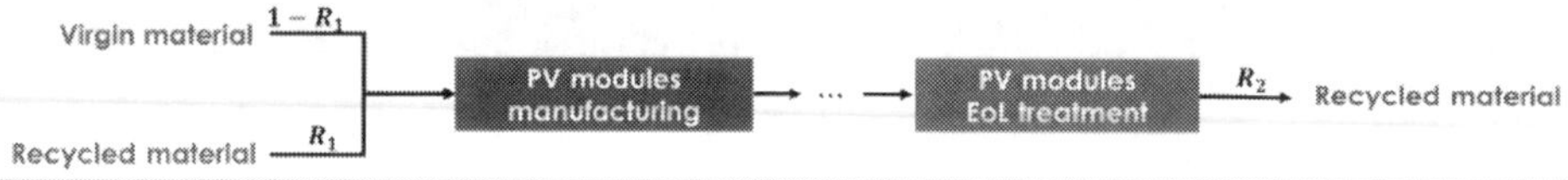

Recycled content ratio (R_1)		Recovered rate (R_2)		
Aluminium	74%	Aluminium	97%	
Glass	0%	Glass	93%	
Copper	20%	Copper	100% (Cable)	76% (Interconnection)
Silver	8%	Silver	100%	
Silicon	0%	Silicon	98%	
Lead	55%	Lead	99%	

- **Six common** approaches of EoL modelling are applied to model LCA

EoL modelling approach	Recommended by
Simple cut-off	International EPD system, PAS 2050, GHG Protocol
Cut-off with economic allocation	Dutch Handbook on LCA, IEA-PVPS T12
Closed-loop allocation	ISO 14067:2018, PAS 2050, GHG Protocol, IEA-PVPS T12
50/50 method	Nordic Guidelines on LCA
Open-loop allocation	ISO 14067:2018
Circular Footprint Formula (CFF)	Product Environmental Footprint Guide

IMPORTANT GUIDELINES IN LCA

| PSL IPVF

40th European Photovoltaic Solar Energy Conference and Exhibition

Result & Discussion – Impact on climate change

- **Result**

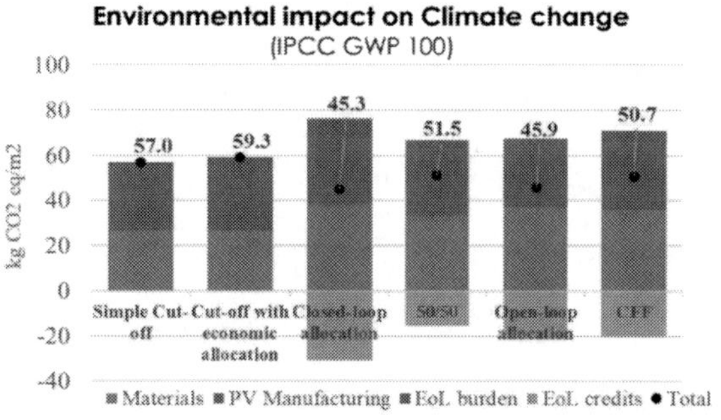

- **Discussion**
  - The result from the **cut-off with economic allocation** is **31%** higher than that of the **closed-loop allocation** for the climate change category
  - Simple cut-off and cut-off with economic allocation: allocation **only for the burden** of EoL recycling
  - Closed-loop allocation: **full credits** due to the output recycled material into the product
  - 50/50, open-loop allocation and CFF: **partial credits** into the product

| PSL IPVF

Result & Discussion – Impact on climate change

- **Result**

Environmental impact on Climate change
(IPCC GWP 100)

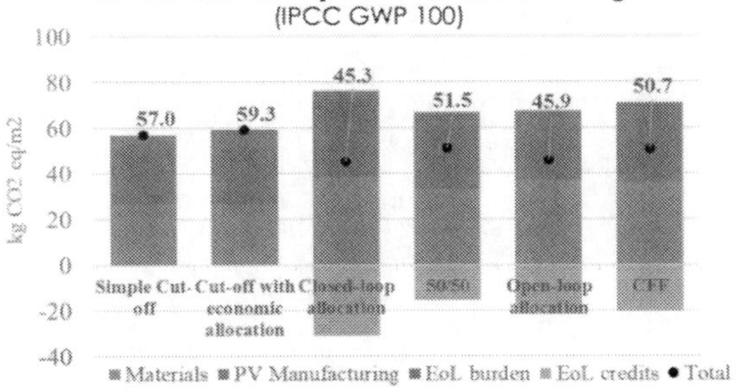

- **Discussion**
  - The result from the **cut-off with economic allocation** is **31%** higher than that of the **closed-loop allocation** for the climate change category
  - Simple cut-off and cut-off with economic allocation: allocation **only for the burden** of EoL recycling
  - Closed-loop allocation: **full credits** due to the output recycled material into the product
  - 50/50, open-loop allocation and CFF: **partial credits** into the product

- Usually, **cut-off approaches** give high incentives to **using recycled material**
- **Closed-loop allocation** gives inventives to **EoL recycling**
- **50/50, open-loop allocation & CFF** provide both incentives **to using recycled material** and **EoL recycling**

PSL IPVF

Result & Discussion – Impact on climate change

- **Result**

Environmental impact on Climate change
(IPCC GWP 100)

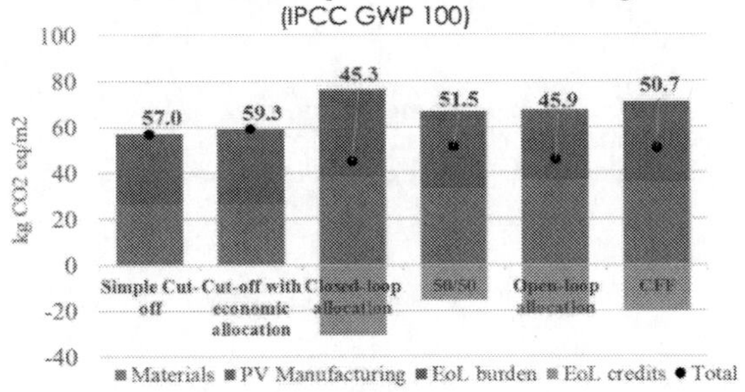

- **Discussion**
  - The result from the **cut-off with economic allocation** is **31%** higher than that of the **closed-loop allocation** for the climate change category
  - Simple cut-off and cut-off with economic allocation: allocation **only for the burden** of EoL recycling
  - Closed-loop allocation: **full credits** due to the output recycled material into the product
  - 50/50, open-loop allocation and CFF: **partial credits** into the product

- Usually, **cut-off approaches** give high incentives to **using recycled material**
- **Closed-loop allocation** gives inventives to **EoL recycling**

→ Sensitivity analysis

- **50/50, open-loop allocation & CFF** provide both incentives **to using recycled material** and **EoL recycling**

PSL IPVF

Result & Discussion – Impact on climate change

- **Result**

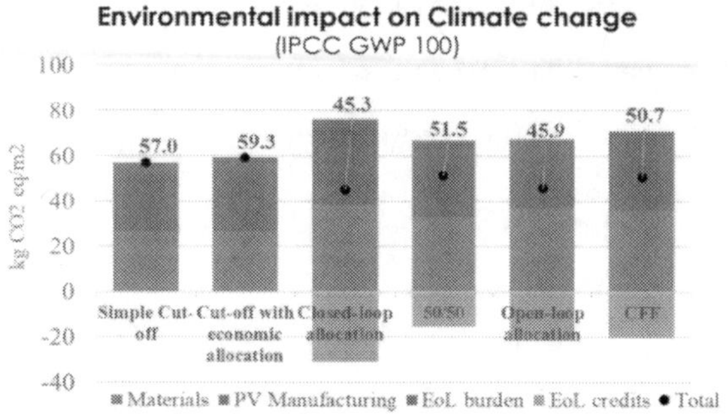

- **Discussion**
  - The result from the **cut-off with economic allocation** is **31%** higher than that of the **closed-loop allocation** for the climate change category
  - Simple cut-off and cut-off with economic allocation: allocation **only for the burden** of EoL recycling
  - Closed-loop allocation: **full credits** due to the output recycled material into the product
  - 50/50, open-loop allocation and CFF: **partial credits** into the product

If $E_V >> E_R$

- Usually, **cut-off approaches** give high incentives to **using recycled material**
- **Closed-loop allocation** gives inventives to **EoL recycling**

→ Sensitivity analysis

- **50/50, open-loop allocation & CFF** provide both incentives **to using recycled material** and **EoL recycling**

E_V: Environmental burden of virgin material production E_R: Environmental burden of recycling

Result & Discussion – Impact on ecotoxicity (fresh water)

- **Result**

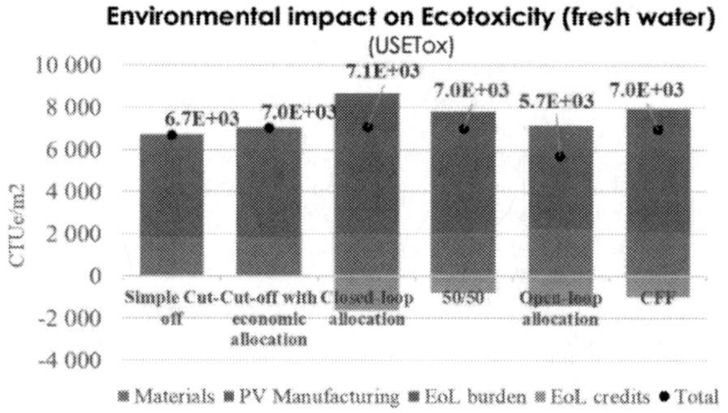

- **Discussion**
  - The results from the **closed-loop allocation** are **24%** higher than those of the **open-loop allocation** regarding ecotoxicity (fresh water) category
  - $E_R > E_V$, **various elements** (the allocation of E_R and E_V, the allocation factor and the quality of recycled material) contribute to the final results

Result & Discussion – Impact on ecotoxicity (fresh water)

- **Result**

Environmental impact on Ecotoxicity (fresh water)
(USETox)

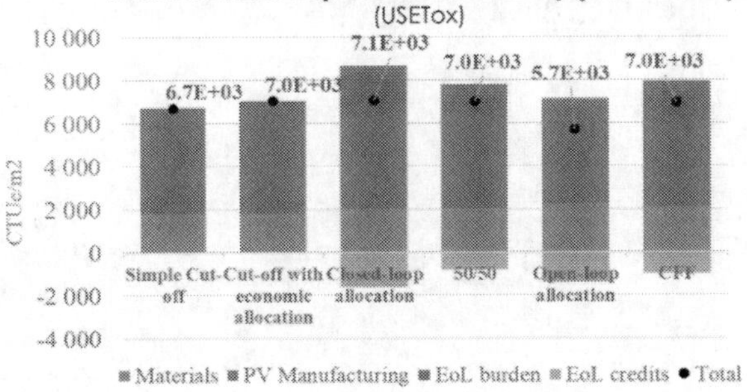

- **Discussion**
  - The results from the **closed-loop allocation** are 24% higher than those of the **open-loop allocation** regarding ecotoxicity (fresh water) category
  - $E_R > E_V$, **various elements** (the allocation of E_R and E_V, the allocation factor and the quality of recycled material) contribute to the final results

- Pay attention to cases where E_R is close to E_V or even greater than E_V
  A more prudent way would ask for employing all six EoL modelling approaches for the sensitivity analysis

E_V: Environmental burden of virgin material production E_R: Environmental burden of recycling

Conclusions

- **Six** common EoL modelling approaches were applied to perovskite/silicon tandem modules
- Recommendations for the choice of EoL modelling approach:
  - Choose **an intermediate solution,** namely CFF method, as the **default** EoL modelling approach;
    - CFF: Intermediate result, comprehensive elements, reflecting material characteristics
  - Apply sensitivity analysis to check the robustness of the results;
  - Use **at least** one **cut-off approach** and **closed-loop allocation** in parallel for sensitivity analysis; additional approaches can be introduced for sensitivity analysis if the burden of recycling is particularly high.

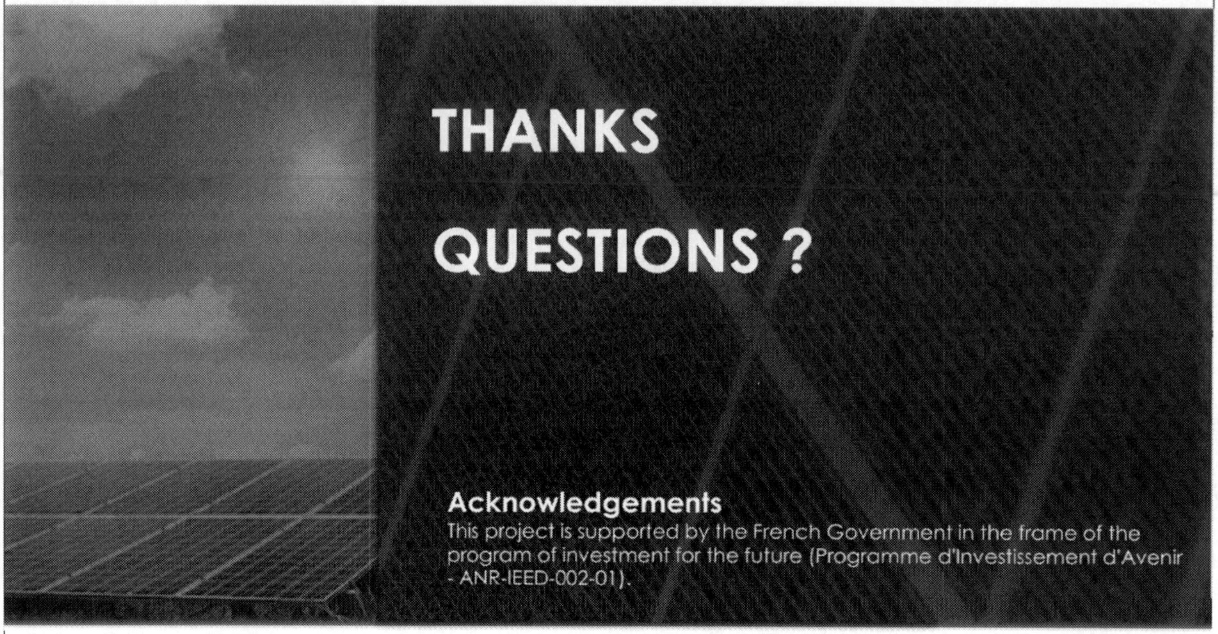

17

020514-017

RE-USE OF PV MODULES: PROGRESS IN STANDARDISATION AND LEARNINGS FROM A REAL CASE STUDY

Arvid van der Heide[1,2,3], Daniela Maria Godinho Ariolli[4], Guillermo Oviedo Hernandez[4], Serge Noels[5], Jan Clyncke[5]
[1]imec, imo-imomec, Thor Park 8320, 3600 Genk, Belgium
[2]Hasselt University, imo-imomec, Martelarenlaan 42, 3500 Hasselt, Belgium
[3]EnergyVille, imo-imomec, Thor Park 8320, 3600 Genk, Belgium
[4]BayWa r.e. Operation Services S.r.l., R&D department, via di Settebagni 390, Rome, Lazio 00139, Italy
[5]PV CYCLE, Brand Whitlocklaan 114/5, 1200 Brussels, Belgium
e-mail: arvid.vanderheide@imec.be

ABSTRACT: With the projection of a growing amount of waste from photovoltaic (PV) modules in coming years, it is crucial to address properly the challenges of their end-of-life management. Initiatives on re-use and recycling must lead the way. However, the re-use of decommissioned PV modules is quite a complex subject since there are concerns about performance, safety, financial viability and exports to non-regulated markets. To create clarity in the market, it is crucial that universal guidelines and standards for re-use are developed. First suggestions for requirements have been made by the H2020 project CIRCUSOL, which led a team of experts of the IEC Technical Committee 82 WG2 to start preparing a Technical Report entitled *´PV module re-use and circular economy´*. Such report will contain clear recommendations for PV module re-use and may serve as a basis to introduce future IEC Technical Specifications and eventually, Standards. In this paper, the latest progress on that report is presented and complemented with real field data from a selected case study where the *preparation for re-use* procedures were tested and validated. As a result, adapted and optimised on-site functionality test procedures are discussed and practical insights are given from the operation and maintenance (O&M) perspective.
Keywords: utility-scale PV, end-of-life, re-use, second life, O&M

1 INTRODUCTION

In Europe, many large PV fleets are approaching the end of their designed technical life (20-25 years) or being revamped/repowered with newer technology. According to IRENA, 78 million metric tons of PV module waste will be generated worldwide by 2050 [1]. With that, the discussion about strategies to manage this large volume of decommissioned PV components, such as recycling and re-use, is gaining more and more attention.

Currently, decommissioned PV components enter the waste stream directly, and are usually either disposed of or sent to recycling (downcycling) [4, 22, 23]. The European Waste from Electrical and Electronic Equipment (WEEE) Directive [2] influences this practice since it poses precise targets for the collection and recycling of decommissioned PV components. However, considering extending the lifetime of the components that are still functioning (via re-use and repair strategies) instead of sending them directly to recycling, has proven to reduce the environmental impact and contributes to higher levels of circularity [3-4].

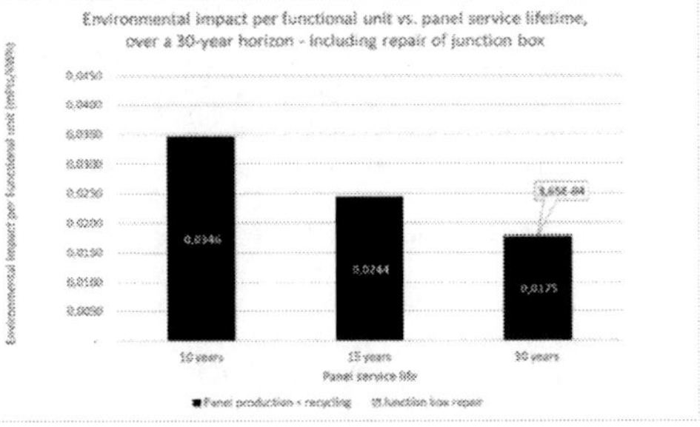

Figure 1. Environmental benefits of second life [3]

Figure 1 shows the environmental impact of repair activities needed to extend PV modules lifetime up to 30 years (e.g. junction box replacement every 10 years); Furthermore, it illustrates that the environmental impact generated by a PV module operated until the end of its 30 years lifetime is much lower than situations in which modules are replaced with more efficient ones every 10/15 years, even if a repair is required twice within this period.

1.1 The PV-waste challenge

During the operation phase of large-scale PV plants, there are basically three main activities that contribute to the generation of decommissioned PV modules and inverters [5]:

i. Preventive and corrective maintenance
ii. Revamping and repowering interventions
iii. End of life (EoL)

Regarding PV modules, a common practice in the solar industry is to consider a module as waste when it goes below the guaranteed power (see Figure 2*)* or as part of an insurance claim after damaged by severe weather events (e.g. hail).

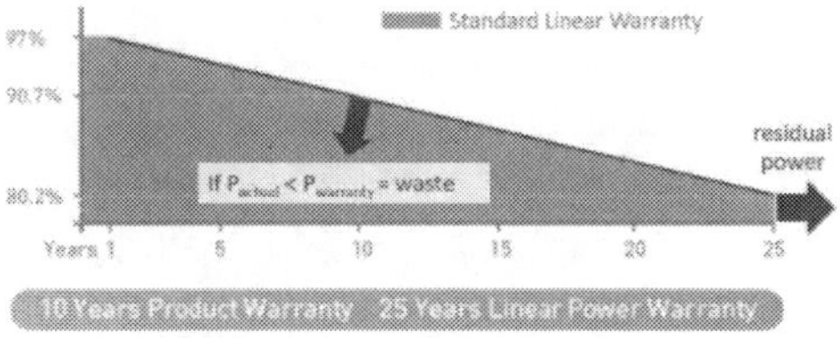

Figure 2. Standard power warranty

However, considering that when a PV module naturally reaches the end of its designed technical life (after at least 25 years of operation in the field), it is estimated that it still possesses approximately 80% of its initial power generation capacity and therefore it has remaining value to be exploited in a second life [6].

By the time this paper is being written (Q2 2023), the very first European PV fleets installed during the boom of

10.4229/EUPVSEC2023/5DO.15.6
020515-001

feed-in tariff schemes had passed midlife (10+ years old) and we are observing an unprecedented increase of revamping and repowering activities aimed at improving the performance of existing assets through the replacement of old PV modules and inverters by new ones.

On the one hand, revamping and repowering projects open up new opportunities for optimizing the performance of existing PV assets without compromising new land. On the other hand, this premature replacement of PV components contributes to increasing the number of decommissioned/used components before reaching the end of the natural lifetime of the PV plants. With that, it carries with it increasing challenges and poses questions such as: are O&M service providers, local e-waste consortia and recyclers prepared to handle properly the incoming volumes of decommissioned PV components? Is recycling the only option or could second life be taken seriously?

1.2 The re-use market: opportunities and challenges

The main reasons for reusing functional PV components are connected to the opportunities and advantages that this practice can bring [3-12; 16], such as:

- Prevent them from prematurely entering the e-waste stream
- Reduce the amount of e-waste generated.
- Reduce the unnecessary extraction of valuable raw materials
- Reduce the amount of energy and water needed for the production of new modules from zero
- Decrease the EU dependence on imports of new components manufactured elsewhere
- Decrease the environmental impact of the overall PV sector and enable wide-spread access to electricity, especially to poorly connected areas

Furthermore, according to the Waste Framework Directive 2008/98/EU [13], re-use has a higher priority than recycling and disposal (see Figure 3). Thus, ideally, recycling should be considered only when re-use is not possible, in situations where the components are damaged and cannot be repaired, for example. For that reason and to comply with the principles of circularity and sustainability, second life should be placed as a standalone phase separate from EoL. In other words, decommissioned PV components should only reach recycling if either unsafe or non-functional.

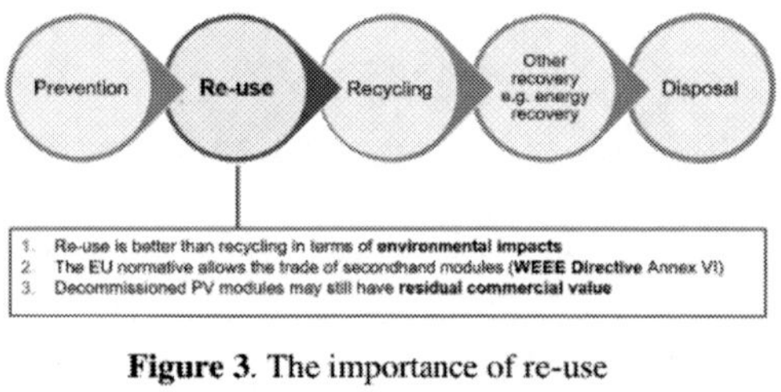

Figure 3. The importance of re-use

One of the main challenges of the re-use market is connected to the legislation, more specifically, the WEEE Directive. Even though the re-use option is also considered in it, there is no specific target for PV modules and inverters, which can be one of the reasons why the PV second-hand market is not widely developed yet [7,10].

Presently, the market for re-used PV modules is focused on low-income countries and regions, such as Afghanistan, Pakistan and several African countries [7, 10, 12]. In Europe, this market is small and usually concentrated in Western Europe, where used modules serve as spare parts in existing installations when damaged modules need to be replaced by very similar ones [7,12].

Besides the unmature legislation, the main barriers to the development of the second-hand market are mainly the low price of new PV modules and the absence of (extended) warranties of performance, functionality, safety and quality of used modules, due to the lack of worldwide-adopted standards on labelling and testing [3-12; 16]. Because of that, many companies are currently developing independently their own methods to test for re-use, all differing widely in safety and quality.

To start tackling some of these challenges, a guideline for preparing modules for re-use was published in 2022 by the H2020 European-funded project CIRCUSOL [5,12, 25]. This guideline contains a first attempt to define the steps needed for the application of several testing methods with the aim of defining what a functional module is and ensuring sufficient quality and safety while still being cost-effective. This document is now being used as a baseline by the IEC Technical Committee 82 to write a technical report (TR) as a first standardization effort.

In this paper, the current process of developing international technical specifications by the creation of an IEC Technical Report on re-use will be described.

Moreover, in the framework of the TRUST-PV H2020 research project [17], an actual revamping project of a utility-scale PV plant was used as case study for testing the re-use procedures that are under development. It was possible then to test the selection and classification procedures of the modules on-site, as well as re-thinking the packaging procedures, taking advantage of the expertise of a large O&M service provider.

2 STANDARDISATION PROGRESS IN IEC TC82

In this section, the progress on the standardisation concerning the re-use of PV modules will be outlined. Since the current status is the writing of an IEC Technical Report (TR), no pass/fail criteria are allowed, so the focus will be on a high-level description of the module qualification methods for re-use. In section 3, test methods will be described in more detail and suggestions for criteria will be given. These criteria were applied in practice for a real case study for module re-use, performed in Italy. The learnings are very useful for the writing of the TR.

At the end of 2021, a project team was created within the IEC Technical Committee 82 WG 2, that has the task to write an IEC Technical Report on the re-use of PV modules. The project team involves around 30 IEC experts from different continents. Since a Technical Report is not allowed to be normative, no pass/fail criteria can be included in it. However, recommendations for test procedures and suggestions for future limit selection can be provided. In this way, it can be used as a basis to create an IEC Technical Specification or Standard in the future, to guarantee the safety and performance of decommissioned PV modules intended for re-use.

2.1 General concepts in Technical Report on re-use of PV

Before discussing the concepts being explored in the Technical Report in more detail in the following sections, some of the most important terms and definitions in the

10.4229/EUPVSEC2023/5DO.15.6
020515-002

world of end-of-life, waste and re-use are given for further clarification in Table I. Most of the original formal definitions have been revised and adapted to the PV sector common nomenclature for a better understanding, using mainly as inspiration the definitions in [13] and [18].

Table I. Important terms and definitions

Decommi-ssioning	Process of removing a PV system or component from an active status. For a complete PV system, it means that it is deconstructed and the land is made ready for redevelopment or returned to its original use
End-of-life (EoL)	Natural or planned end of a PV system or component service lifetime
Performance Loss Rate (PLR)	Parameter indicating the decline of a PV module power output over time (in % per year)
Preparing for re-use	Checking, cleaning and/or repairing operations by which PV components are prepared so that they can be re-used
Recycling	Any recovery operation by which PV waste materials are reprocessed into products, materials or substances whether for the original or other purposes
Repowering	Increasing the nominal power of the PV system by replacement of old components (mainly modules and inverters) by new ones, to enhance its overall performance.
Revamping	Replacement of old components of a PV system (mainly modules and inverters) by new ones, to enhance its overall performance without substantially changing its nominal power and without compromising new land
Re-use	Any operation by which PV products or components that are not waste are used again for the same purpose for which they were conceived.
Re-used PV module	A PV module that has been removed from a first PV system and prepared for re-use in another PV system.
Waste	Any substance or object which the holder discards or intends or is required to discard.
E-waste	Electrical or electronic equipment which is waste, more formally '*waste electrical and electronic equipment (WEEE)*'
Functional PV module	A PV module that performs its primary function without posing a risk for the operators, the environment and to the system they form part of and its performance follows a natural degradation trend according to the manufacturer specifications

2.2 The re-use strategy being discussed

The general PV module testing strategy targeted is to minimise the testing as much as possible while still ensuring module safety and quality. A key question that needs to be answered upfront is whether *every single* module of a given PV plant must undergo all quality assurance tests, or whether a sampling approach can be used instead. Indeed, all modules from a utility-scale PV plant can be assumed to be similar in their remaining quality and it might be sufficient to use a sample. Such an approach would be limited to large plants that are still operational and all modules have degraded in a similar way, excluding of course any damage by extreme weather events before decommissioning. It is then clear that suitable criteria need to be formulated case-by-case, to apply a sampling approach that accounts for the history and actual status of the plant. Furthermore, to justify sampling, it is necessary to first make a general assessment of a PV plant based on historical monitoring and maintenance data (e.g. tickets, IR thermography reports, insulation errors reported by the inverters, etc.). For PV modules from a (weather-)damaged plant or that are already stacked in a warehouse, individual module testing is considered inevitable.

2.3 Module qualification based on sampling approach

The sampling approach is intended for PV plants in which the modules can be assumed to be all in similar condition. The clear benefit would be that the number of PV modules to be tested would be much smaller, reducing time and costs significantly. On the other hand, this is the most difficult approach, since it requires a procedure that will give sufficient guarantee about module safety, quality and performance to a customer, without measuring every single module. It is also more difficult to check if the sampling approach has been used in the correct way, and for a case where it was justified. It is different when sampling is used internally for qualification of re-use modules from a PV plant, so when the modules are to be used in another location/application from the same owner. In that case, the owner only needs to be convinced himself about the correctness of the approach and will also not fool himself.

The procedure should (as always) start with a visual inspection in the PV plant and an analysis of the monitoring data to check for the approximate PLR, possible outlier strings and possible insulation failures detected by the inverters. Based on this, and possible I-V and insulation checks for a few module strings (or modules), the decision about re-using the PV modules can be taken, before decommissioning.

The current recommendation for the module sampling is to base it partly on the checking of new plants in the "Engineering, & Construction Procurement Best Practice Guidelines" [27]. In Table II, the different tests to be applied for new modules are listed with their sample rates. These are based on the Acceptance Quality Limit (AQL) system, with sample rates defined in ISO2859-1. From this list, the maximum power determination at STC, electroluminescence inspection, (detailed) visual inspection and insulation test under wetting should be done at least. These tests can in principle all be done on-site, but for EL and module power testing during cloudy weather, a mobile indoor test set-up is required. The deeper technical checks can be done when desired, most of them will typically not be possible on-site. The sampling rates with a G (general) are higher than the S (special) rates. As an example, G1 means 200 for a 125.000 modules plant, while S3 would mean only 32 modules. These rates could of course be adapted/increased for modules to be re-used, but in the TR no limits can be set. The modules should of course be sampled across the area of the plant to obtain a representative sample. For the maximum power determination at STC, measurements can be done on module strings or modules with a handheld tool, or otherwise in the mobile PV test setup mentioned before.

Concerning the module powers, it is still open which variation should be the upper limit for sampling to be still justified. Concerning the visual test, in this case it is meant to be a detailed one, while a visual inspection on major issues should still be performed for each module during the decommissioning of the plant. Finally, the insulation test under wetting can be done with a hand-held insulation

10.4229/EUPVSEC2023/5DO.15.6
020515-003

tester and something to spray water onto the rear side of the module

Table II. Sampling rates for checking PV plants [adapted from 27]

Type of testing	Sampling rate (ISO 2859-1)		
	For new PV modules	Modules for re-use	
		On-site testing	Deeper technical check
Performance characterisation testing			
Maximum power determination at Standard Test Conditions (STC)	G 1	G 1	G 1
Efficiency loss at low irradiance	S 1	--	(S 1)
Electroluminescence inspection	G I	G 1	G 1
Qualification testing			
Visual Inspection	S 3	S 3	S 3
Insulation test under wetting (wet leakage test)	S 3	S 3	S 3
Degree of ethylene-vinyl acetate (EVA) cross linking	S 1	--	(S 1)
Adhesion strength EVA/backsheet	S 1	--	(S 1)
Power loss due to light induced degradation (LID)*	S 1	--	--
Power loss due to power induced degradation (PID)**	2 modules per BOM and test	--	--
Power loss due to light and elevated temperature induced degradation (LeTID)	2 modules per BOM and test	--	--
Reliability testing			
Design suitability (extended stress testing i. e. damp heat, thermal cycling, humidity freeze, UV exposure, mechanical load), relevant for all BOM used	2 modules per BOM and test	--	(2 modules per BOM and test)

*Can be less considered for n-type technology.
**Can be less considered for systems that have anti-PID solutions.

2.4 Module qualification based on individual testing

When a PV plant has been (partly) damaged, typically by a weather event like hail or strong winds, modules will not be in similar condition (also monitoring date from the past are useless due to the damage) so that measuring every single module is the only option. This is also the case when modules have already been stacked on piles in a warehouse. In this case, the approach is to perform first a visual check on major issues, followed by performing the following indoor tests: I-V curve tracing, dry insulation test, bypass diodes test and EL (with automatic evaluation). It should be mentioned that the dry insulation test should be performed in a faster way than in the IEC61215 standard (that mentions a voltage ramp rate of maximum 500 V/s) to increase the throughput. Since a faster testing would only increase the severity of the insulation test, this is an acceptable adaptation.

A factory test line performing these measurements like this can reach a throughput of around 150 modules/h (or 500 MW/y), while a mobile test setup can reach only 200 modules/d. For this reason, testing each module on-site is no viable option.

Recently some companies have started using such a testing line. They apply this individual testing also for modules from large PV plants that could have been suitable for a sampling approach. However, it is still unclear if this testing of every single module for all cases will be economically viable.

2.5 Final comments on overall TR contents

Both testing approaches will be included in the TR. For the future, it will probably be the best to create different standards for the different approaches, since it is to be expected that the sampling approach will require (much) more time and discussions than the every-module approach. To allow for the re-use of modules that do not meet the (dry) insulation requirement for the original system voltage (typically 1000 or 1500 V), or maybe also for certain types of visible defects, it is suggested to introduce also a "lowered system voltage" class. This voltage could be somewhere between 60 V and 150 V, the exact value to be determined in a future standard.

The labelling of the modules for re-use should be done in addition to the original one, so not covering or removing the original label. The requirements for the information on the label are still under discussion. There has been a discussion to put information about PID sensitivity on the label, but it has been concluded that this would be too difficult in practice due to the complexity of the subject (for example, modules can be PID-prone but not degraded when they have been mounted near the "right" polarity of the module string. Concerning repair, different options are mentioned and discussed (replacing bypass diodes, cables, connectors, junction box, etc.). It is most likely that only replacement of connectors could be allowable in a future standard since other repairs would require way too much testing.

3 PREPARING FOR RE-USE: METHODOLOGY SUGGESTION AND APPLICATION IN CASE STUDY

As mentioned in section 2.1, *Preparing for re-use* is defined in the European Waste Framework Directive as "checking, cleaning or repairing operations (…)" [13]. In this section, a methodology for the "checking" operations of PV modules is described and suggested based on the CIRCUSOL guidelines and on the most updated work on the IEC technical report. Test criteria are also suggested, including recommendations about pass/fail criteria. With this, the objective was to go a little beyond the TR, since this document does not allow this type of pass/fail criteria. Repair, refurbishment and relabelling of PV modules were left out of the scope due to a lack of industry consensus.

In order to be able to re-use PV modules and prevent them from entering the waste stream prematurely, they should be submitted to functionality testing procedures that are technically feasible, cost-effective and contextualized to the second-hand market, giving priority to safety over performance. Therefore, replicating the same qualification procedures for new modules (IEC 61215) should be avoided because it would be technically very challenging and simply not financially viable. The suggested tests and steps proposed for the preparation of PV modules for re-use are illustrated in Figure 4 and explained in detail in the following subsections.

10.4229/EUPVSEC2023/5DO.15.6
020515-004

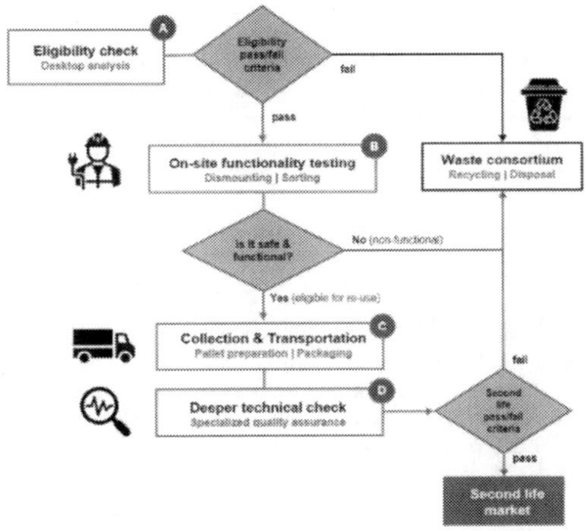

Figure 4. Preparing for re-use procedure

3.1 Eligibility check (Step A)

Consist of a desktop analysis carried out before going on-site, which based on electrical and weather variables and maintenance information (O&M historical data) is used to calculate and/or estimate the performance loss of the entire plant or sub-sections of it (at transformer, inverter or string level). This check is meant to be the first filter to decide whether it makes sense to perform additional checks on-site (and avoid therefore wasting valuable resources unnecessarily). Additionally, known failures should be assessed based on annual/monthly reports and punch lists made available by the O&M service provider and/or Asset Manager, such as IR thermographic and I-V curve measurement campaigns commonly done contractually at least once a year. Following this logic, O&M service providers that follow high-quality monitoring and maintenance practices [18] and with quality assurance evidence in their hands, should be able to avoid successive costly diagnostic techniques.

Please note that this check should make the most out of all the already existing documentation that defines the history of the plant, minimizing therefore any additional costs.

<u>Suggested eligible pass/fail criteria:</u>

- *Performance loss*: should be acceptable and in line with expected natural degradation. It is suggested to apply the Performance Loss Rate (PLR) described in [14] as main metric.

  Example: the PLR calculated with the last 5 years of monitoring data is estimated to be 10%/year, which deviates greatly from the 0.8 %/year defined by the manufacturer's power warranty. It is then decided to fail the re-use case and send the modules directly to recycling.
- *Residual value*: based on the remaining power, check if there is a use/business case that makes sense. The residual value is not an absolute fail/pass criterion and will depend on how value is defined.

  Example 1: During a revamping project of a 10-year-old plant, it is verified that the PLR is in line with the expected annual degradation rate provided by the manufacturer. Furthermore, the PV module type is retained of high value for a specific asset owner because such a module is not available anymore in the market and can be used as spare part. It is then decided to go ahead and continue with the re-use preparation steps.

  Example 2: after a PV plant has been damaged by hail, based on the significant decrease in performance (calculated by the O&M service provider using monitoring data) and based on third-party investigations done on behalf of the insurance company (e.g. Electroluminescence imaging of a representative sample of the plant), it is estimated that a very high percentage of modules (ca. 80%) is affected by cell breakage caused by excessive mechanical impact. Therefore, it is concluded that the decommissioned PV modules are not eligible for re-use and no further resources should be spent on additional on-site investigations.

3.2 On-site functionality testing (Step B)

This activity is meant to be carried out by site technicians shortly before and/or while decommissioning PV modules (e.g. during revamping or repowering activities). It includes but is not limited to functionality testing, dismounting, sorting and on-site temporary storage. *Functionality testing* is suggested to be done on a representative and cost-effective sample of PV modules and should cover at least the four tasks described in the following sub-sections.

3.2.1 Visual inspection

Dedicated visual inspection supported by a checklist (see Table III) that covers known failures that can be detected with the naked eye (for a comprehensive list of failures, please see [15]). This inspection does not require that the PV plant is under operation.

Table III. Typical failures detected by visual inspection. Adapted from [15]

Module component	PV module failures
Front of module	Bubbles, delamination, yellowing, browning, glass breakage
PV cells	Broken/cracked cell (snail tracks), discoloured anti reflection
Cell metallization	Burned, oxidized, corrosion
Frame	Bend, broken, scratched, misaligned
Back of module	Delaminated/peeled, bubbles, yellowing, scratches, burnt, chalking
Junction box	Loose, oxidation, corrosion
Cables-connectors	Broken, cut, detachment, brittle, exposed electrical parts, corrosion

3.2.2 Standard Test Conditions (STC) measurements

I-V curve tracing of representative and cost-effective sample of modules, according to IEC 61829:2015 *On-site measurement of current-voltage characteristics* [19], which requires minimum in-plane irradiance of 700 W/m^2 for measurements that are meant to be extrapolated to STC (using at least 400 W/m^2 as allowed by IEC 62446-1:2016 [20] is not advised since it is meant to use only for performance measurements, not for STC extrapolation). For a comprehensive list of PV module failures detectable by the shape of the I-V curve, please see [15, 20].

Commercially available I-V curve tracers are usually equipped with a reference cell and a module temperature sensor to be able to translate the measured I-V curve into the I-V curve at STC conditions, using IEC 60891:2021 [26]. It is advised to measure, additionally, the temperature in several other points of the module and then calculate the average temperature, to be used then for a more accurate

STC translation. Interpreting I-V curve shapes could be tricky, so it is advised to follow the guidelines in [15] and in Annex D of IEC 62446-1:2016 [20].

3.2.3Infrared (IR) thermography

Performed with a hand-held (low-cost) thermal camera or smartphone with integrated thermal sensor, according to IEC-TS 62446-3-2017 *Outdoor infrared thermography* [21], which requires a minimum in-plane irradiance of 600 W/m^2 for proper detection of anomalies and provides as well as a comprehensive list of thermal abnormalities.

3.2.4 Insulation resistance test

Carried out at string box level or module level if feasible, according to IEC 62446-1 *Requirements for testing, documentation and maintenance* [20]. Performing this test presents a potential electric shock hazard; therefore, it is important to fully understand the procedure before starting any work. The result is satisfactory where the insulation resistance is not less than the appropriate value given in Table IV.

Table IV. Minimum values of insulation resistance for PV arrays up to 10 kW$_P$. [20]

System voltage [V] (V_{oc} (stc) × 1.25)	Test voltage [V]	Minimum insulation resistance [MΩ]
<120	250	0.5
120 to 500	500	1
>500	1000	1

3.2.5 Suggested second life functionality pass/fail criteria for decommissioned PV modules

Despite the lack of international standards and industrial consensus concerning the re-use of PV modules, in this section a further attempt to provide clear and applicable guidelines for selecting PV modules good for re-use is presented.

Table V. Suggested pass/fail criteria for re-use of decommissioned PV modules

Field test	PV module defects and anomalies	Pass	Fail	Investigate further
Visual inspection	Any type of non-permanent (washable) soiling	X		
	Front of the module presents yellowing or browning			X
	Front of the module presents any of the following: bubbles, delamination or glass breakage		X	
	PV cells present discolored anti reflection coating	X		
	PV cells are broken/cracked (snail tracks)	X		
	Cell metallization is scratched or misaligned			X
	Cell metallization is burned, oxidized, corroded, bent or broken		X	
	Back of module (polymer) presents chalking or yellowing			X
	Back of module (polymer) presents any of the following failures: delamination/peeling, bubbles, scratches or burnt marks		X	
	Frame is bent or broken		X	
	Frame is scratched or just misaligned	X		
	Junction box is loose, oxidized or corroded		X	
	Cables and/or connectors are broken, cut, have exposed electrical parts or are corroded		X	
STC measurements	I-V curve follows the shape of the nominal curve	X		
	I-V curve shape presents slight deviations from the nominal curve			X
	I-V curve shape presents severe deviations from the nominal curve, such as steps or deep slope drops		X	
	Power loss at STC (after applying soiling correction) deviates more than -20% from the expected power loss according to the datasheet		X	
	Power loss at STC (after applying soiling correction) deviates more than -10% from the average power loss of the group of samples		X	
	Power loss at STC (after applying soiling correction) deviates less than -20% from the expected power loss according to the datasheet AND less than -10% from the average power loss of the sample	X		
Insulation resistance test	Dry insulation resistance is higher than 1 [MΩ] (for PV arrays up to 10 kWp)			X
	Wet insulation resistance is higher than 1 [MΩ] (for PV arrays up to 10 kWp)	X		
	Dry or wet insulation resistance is lower than 1 [MΩ] (for PV arrays up to 10 kWp)		X	
IR thermography	Hotspot(s) with $\Delta T < 10°C$	X		
	Hotspot(s) with $10°C < \Delta T < 40°C$			X
	Hotspot(s) with $\Delta T > 40°C$		X	
	One or more substrings of cells in open circuit (e.g. bypass diode failure)		X	

10.4229/EUPVSEC2023/5DO.15.6
020515-006

Please note that these criteria have been tailored for O&M providers in order to optimise field work and should take advantage of the fact that field technicians are already on-site executing module substitutions (e.g. due to corrective maintenance or revamping/repowering activities). In this way, the time spent on-site is optimized as little extra effort is needed. The goal was to create a cost-effective, practical and easy-to-apply checklist, focused on the functionality and safety of PV modules. In Table *V* the list of suggested pass/fail criteria is presented. In cases where there is unconclusive pass criteria (last column of the right), it is advised to further investigate by cross-checking with the other field test results.

3.3 Collection and transportation (Step C)

Once the PV modules have been piled up into two different categories (non-functional and eligible for re-use), it is time for proper pallet preparation and packaging. Logistics (from dismounting to proper packaging and pallet shipping) is not only fundamental to prevent handling and transportation damage but also to avoid the risk that PV modules eligible for re-use are 'confused' with e-waste by customs authorities or highway police. Special attention should be paid to the minimum requirements for shipments of used products laid down by the WEEE Directive- Annex VI (Minimum Requirements for Shipment) [2], designed to avoid unwanted transportation of e-waste to countries where recycling schemes are weak or non-existent.

For the reason mentioned above, it is suggested that a differentiated packaging procedure is done for functional and non-functional modules (see Figure 5). In addition, to minimize extra packing costs, it is advised to re-use the pallets and cardboard of the new batches of modules being installed during revamping activities, for example.

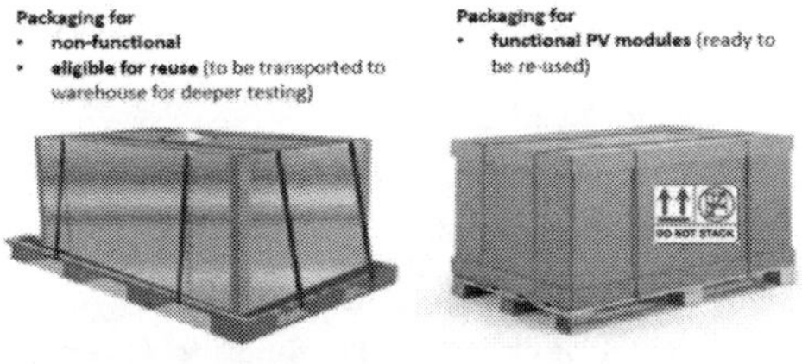

Figure 5. Packaging suggestions

3.4 Deeper technical check (Step D)

This last step will provide definitive and very accurate evidence on the health status of the modules that were categorized as eligible for re-use in previous steps. Complementary specialized and costly quality assurance done in the lab is meant to be applied only to a representative sample. Extensive testing shall be avoided as much as possible for cost reasons. Based on this check, the final value and selling price a of PV module can be set and agreed with the final client.

Lab tests might include but are not limited to: flash tests, IR thermography, electroluminescence (EL) imaging, photoluminescence (PL) imaging, diode tests, ultra violet fluorescence (UV-F) imaging, wet insulation testing, ageing tests in climatic chambers, etc.

<u>Suggested second life pass/fail criteria:</u>

- Is the PV module affected by failures such as severe PID (Potential Induced Degradation) and microcracks?
- Does the PV module have 'enough' residual value to be given a second life according to the final use to be given to it?
- Is the actual power acceptable for the price tag?

Example: After a deeper technical check, it is established that the actual power of the sample tested is 60% of the original nameplate and that modules can be placed on the second-hand market at a price 50% lower than newer modules. The used PV modules are then sold to a local installer that specializes in small rooftop and carport applications.

3.5 Real case study

The procedure described in *section 3.2. On-site functionality testing* was tested and validated at a utility-scale PV plant located in Tuscany, Italy. The PV plant, under operation since 2011, was undergoing a major revamping intervention at the time the tests were performed (see Table VI and Table VII).

Table VI. PV plant metadata

Installation type	Ground-mounted fixed tilt
Installed capacity	4,995.48 kW_p
Country	Italy
Commissioning year	2011
Years under operation	12
PV module nominal power	280 W_p
Number of modules	17,841

Table VII. PV module datasheet (nameplate)

Nominal Power (Pmax)	280 W
Model	LDK-280P-24
Voltage at Pmax (Vmp)	36.6 V
Current at Pmax (Imp)	7.72 A
Open Circuit Voltage (Voc)	44.3 V
Short Circuit Current (Isc)	8.26 A
Tolerance on Nominal Power	-0/+5 W
Maximum System Voltage	IEC EN:1000V / UL:600V
Cell efficiency	15.98%
Module efficiency	14.39%
Cell technology	Multicrystalline silicon
Number of cells	72 (6x12) 156x156 mm
Temp. coefficient for Pmax	-0.45 %/°C
Temp. coefficient for Voc	-0.33 %/°C
Temp. coefficient for Isc	0.06 %/°C

The on-site functionality test procedures were conducted in two days of July 2023. The tests and the equipment used, as well as the reference standards used are described in Table VIII.

A total of 16 measurements (196 PV modules) was used as sample size considered representative for this re-use case study, corresponding to 1.1% (54.88 kW) of the entire PV plant. It is worth noting that the sample size was decided based only on the available time of field technicians and on the weather conditions. It is advised that sampling for non-research purposes consider as well commercial variables such as the end selling price and the market segment targeted.

10.4229/EUPVSEC2023/5DO.15.6
020515-007

Table VIII. Tests and equipment

Test	Normative and references	Equipment used
Visual inspection	IEA PVPS Task 13 report *Review of Failures of Photovoltaic Modules, IEC 61215, IEC 61646*	None (naked eye)
STC measurement (I-V curve tracing)	IEC 61829:2015 *On-site measurement of current-voltage characteristics*	HT IV 500w
Insulation resistance test	IEC 62446-1 *Requirements for testing, documentation and maintenance*	HT PV Isotest
Infrared thermography	IEC-TS 62446-3-2017 *Outdoor infrared thermography*	CAT S62 PRO smartphone with FLIR Thermal camera

4 RESULTS AND DISCUSSION

4.1 Visual inspection

The visual inspection was carried out on the first day of the tests with the aim of identifying as many defect types as possible. This check is meant to be done to decide whether it makes sense to perform additional checks on-site, avoiding successive costly diagnostic techniques.

The following are the representative visual defects detected and documented during the test: peeled backsheet, snail trails, burned cells, broken glass, broken and/or corroded cables/connectors and damaged frames (see Figure 6).

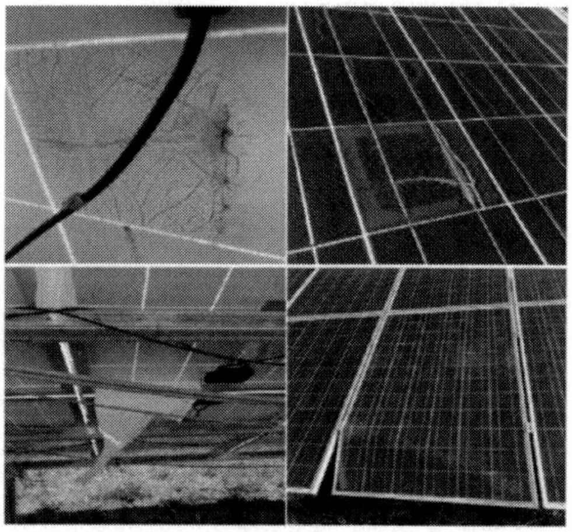

Figure 6. Common failures found (cracked and peeled backsheets, burnt cells, broken glass and bent frame). Source: BayWa r.e.

4.2 STC measurements and power loss calculation

As explained in section 3.5, a sample of 16 measurements was selected as being considered representative for the re-use case study. Measurements of PV modules with evident severe damage (broken glass) were excluded from the sample.

A summary of the STC electrical parameters is presented in table V, including power loss calculation before and after the soiling correction. Power loss was calculated as follows:

$$P_{loss} = \frac{(P_{STC} - P_{nom})}{P_{nom}}$$

In average, the module power at STC was found to be 216.45 W and the power loss calculated after the soiling correcting was **-19.61 %**. Figure 7 shows two representative I-V curves. As already stated in Table VIII, the measurements were performed with a portable I-V tracer commonly used in the industry, the HT IV 500w (calibration date 14/02/2023, STC power accuracy ±5%).

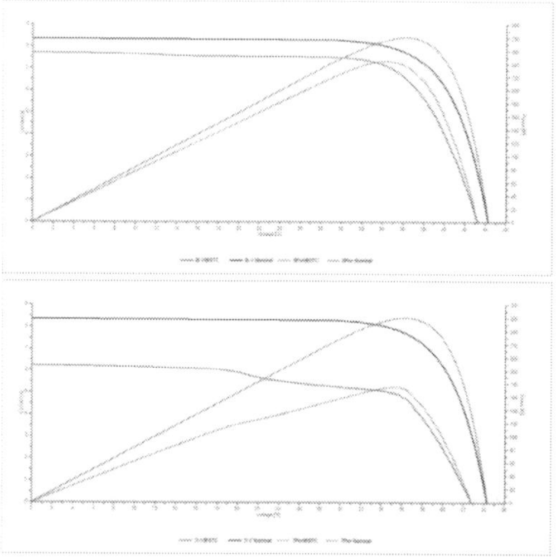

Figure 7. Best (top) and worst (bottom) performing PV modules, Meas. 7 and 6, respectively. Grey: STC datasheet curves. Green: STC measured curves

4.3 Soiling loss measurement

One factor hindering the performance of PV modules is soiling, defined as the accumulation of dust, sand, snow or any other particles on the surface of the modules that reduce irradiation collection. The soiling loss is calculated using the ratio of the actual power output of the PV array under given soiling conditions to the power that would be expected if the PV array were clean and free of soiling. For the purpose of this work, it was decided to follow the preferred method to calculate the soiling ratio (SR) from the power change after cleaning, as stated by the IEC 61724-1:2017 *Photovoltaic system performance - Part 1: Monitoring* [24], because it best represents the actual power loss due to soiling, and in particular it produces more accurate results when soiling may be nonuniform across the modules, especially for typical crystalline silicon modules.

Table IX. Summary of STC measurements and Power Loss calculations

	Pmax [W]	Vmpp [V]	Impp [A]	FF [%]	In-plane irradiance [W/m2]	Power Loss [%]			
Nominal	280.00	36.30	7.72	76		Measured	After soiling correction	Pass/Fail	
STC (Irrad. 1000 [W/m^2], Module Temp. 25[°C])								Actual vs. Expected	Deviation from average
Meas.1*	183.48	29.73	6.17	58	687	-34.47	-31.38	-20.67	-11.78
Meas.2*	182.72	30.19	6.05	57	672	-34.74	-31.65	-20.94	-12.05
Meas.3	209.60	33.79	6.20	64	794	-25.14	-22.05	-11.34	-2.45
Meas.4	221.79	33.59	6.60	70	799	-20.79	-17.70	-6.99	1.91
Meas.5	213.86	34.00	6.29	68	812	-23.62	-20.53	-9.82	-0.93
Meas.6	175.14	35.62	4.92	66	819	-37.45	-34.36	-23.65	-14.75
Meas.7	244.28	34.65	7.05	73	837	-12.76	-9.67	1.04	9.94
Meas.8*	227.86	34.55	6.60	66	890	-18.62	-15.53	-4.82	4.07
Meas.9*	218.66	35.16	6.22	63	893	-21.91	-18.82	-8.11	0.79
Meas.10*	227.73	34.24	6.65	66	903	-18.67	-15.58	-4.87	4.03
Meas.11*	244.64	34.70	7.05	70	888	-12.63	-9.54	1.17	10.07
Meas.12*	228.08	33.32	6.85	67	898	-18.54	-15.45	-4.74	4.15
Meas.13*	227.39	34.01	6.69	65	911	-18.79	-15.70	-4.99	3.91
Meas.14*	208.94	31.72	6.59	63	913	-25.38	-22.29	-11.58	-2.68
Meas.15*	215.31	35.09	6.14	61	947	-23.10	-20.01	-9.30	-0.41
Meas.16	233.73	34.75	6.73	65	941	-16.53	-13.44	-2.73	6.17
Average	216.45	33.69	6.43	65	850.25	-22.70	**-19.61**		

*String measurements (corresponding to the averaged values of 19 modules in series)
Measurements 1, 2 and 6 would not be eligible for re-use after applying the pass/fail criteria described in section 3.2.5.

The measurement followed was the following:

a) Choose a soiled PV module representative of the general soiling condition of the PV plant.
b) Measure the I-V curve of the soiled PV module following the minimum requirements according to IEC 61829:2015 On-site measurement of current-voltage characteristics [19].
c) Clean the module, following the manufacturer's recommendations.
d) Repeat b) for the cleaned PV module.
e) Calculate Soiling Ratio (SR) and Soiling Losses (SL) as follows:

$$SR = \frac{P_{soiled}}{P_{clean}} \times 100 \qquad SL = 100 - SR$$

P_{soiled} *is the STC power in W of soiled PV module*
P_{clean} *is the STC power in W of clean PV module*

4.4 Power degradation analysis

Unfortunately, for this case study, the PLR calculation (part of the eligibility check discussed in section 3.1) was not possible due to the lack of access to the historical monitoring data. Therefore, the power warranty values were used as baseline for this analysis, taken directly from the manufacturer's data sheet, reported in Table X. Furthermore, two different degradation scenarios were considered (see Figure 8).

Scenario A assumes a constant annual power degradation of 1.63%, calculated as the total power loss measured (see section 4.2) and spread evenly thought the operating life of the plant until the present (12 years). This same value of annual degradation was used to estimate the power degradation until year 25, which corresponds to the PV plant's theoretical end-of-life (see Table XI). In other words, it could be considered an *optimistic* scenario, because it assumes that the modules will not continue to degrade faster than they have done so far in the past 12 years. Scenario B is identical to Scenario A from year 1 to year 12 (present). But from year 13, it depicts instead a *pessimistic* or *conservative* scenario, where a constant value of 2% of annual degradation was arbitrary chosen by rule of thumb, thought to be realistic none the less.

Table X. 4-step manufacturer power warranty

Years	Power warranty [%]	Annual degradation [%]
5	94	1.2
11	90	0.67
18	85	0.71
25	80	0.71

10.4229/EUPVSEC2023/5DO.15.6
020515-009

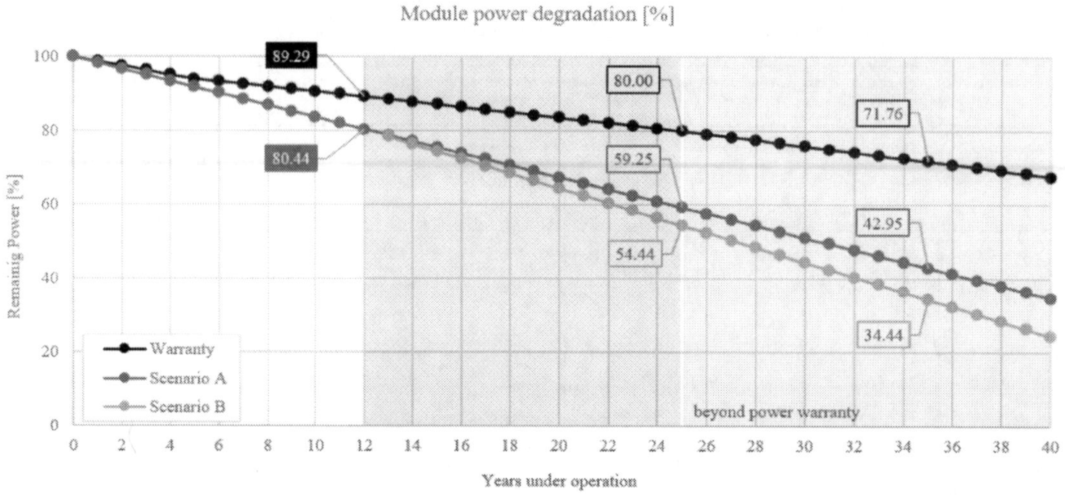

Figure 8. Power degradation analysis. Warranty baseline (black): average annual degradation of 0.83% according to datasheet. Scenario A (green): annual degradation of 1.63%, as per STC measurements. Scenario B (grey): from year 1 to year 12 annual degradation of 1.63% and 2% from year 13.

Table XI. Overview of power gradation analysis

		Warranty baseline Manufacturer's datasheet				**Scenario A** Based on field measurements			
		Power degradation		**Remaining power**		**Power degradation**		**Remaining power**	
Year		**Annual (%)**	**Cumulative (%)**	**%**	**Wp**	**Annual (%)**	**Cumulative (%)**	**%**	**Wp**
2011	0	0.00	0.00	100.00	280.00	0.00	0.00	100.00	280.00
2012	1	1.20	1.20	98.80	276.64	1.63	1.63	98.37	275.44
2013	2	1.20	2.40	97.60	273.28	1.63	3.26	96.74	270.87
2014	3	1.20	3.60	96.40	269.92	1.63	4.89	95.11	266.31
2015	4	1.20	4.80	95.20	266.56	1.63	6.52	93.48	261.74
2016	5	1.20	6.00	94.00	263.20	1.63	8.15	91.85	257.18
2017	6	0.67	6.67	93.33	261.33	1.63	9.78	90.22	252.62
2018	7	0.67	7.33	92.67	259.47	1.63	11.41	88.59	248.05
2019	8	0.67	8.00	92.00	257.60	1.63	13.04	86.96	243.49
2020	9	0.67	8.67	91.33	255.73	1.63	14.67	85.33	238.92
2021	10	0.67	9.33	90.67	253.87	1.63	16.30	83.70	234.36
2022	11	0.67	10.00	90.00	252.00	1.63	17.93	82.07	229.80
2023	12	0.71	10.71	**89.29**	250.00	1.63	19.56	**80.44**	225.23
2024	13	0.71	11.43	88.57	248.00	1.63	21.19	78.81	220.67
2025	14	0.71	12.14	87.86	246.00	1.63	22.82	77.18	216.10
2026	15	0.71	12.86	87.14	244.00	1.63	24.45	75.55	211.54
2027	16	0.71	13.57	86.43	242.00	1.63	26.08	73.92	206.98
2028	17	0.71	14.29	85.71	240.00	1.63	27.71	72.29	202.41
2029	18	0.71	15.00	85.00	238.00	1.63	29.34	70.66	197.85
2030	19	0.71	15.71	84.29	236.00	1.63	30.97	69.03	193.28
2031	20	0.71	16.43	83.57	234.00	1.63	32.60	67.40	188.72
2032	21	0.71	17.14	82.86	232.00	1.63	34.23	65.77	184.16
2033	22	0.71	17.86	82.14	230.00	1.63	35.86	64.14	179.59
2034	23	0.71	18.57	81.43	228.00	1.63	37.49	62.51	175.03
2035	24	0.71	19.29	80.71	226.00	1.63	39.12	60.88	170.46
2036	25	0.71	20.00	80.00	224.00	1.63	40.75	59.25	165.90

It can be seen that the measured remaining power in year 12 (the moment this plant was being revamped entirely by replacing all the PV modules by new ones), is 80.44%, considerably lower than the expected warranty value of 89.29% (main reason for the revamp action).

Furthermore, stretching a bit the case study as if the revamp wouldn't have taken place at all, the estimated remaining power in year 25 (when the PV plant will reach its designed lifetime), would still be of 59.25% and 54.44% for scenarios A and B, respectively.

4.5 Insulation resistance testing

Insulation testing was not done randomly by picking a sample of modules but was performed by targeting the most interesting cases of evident visual damage (as described in section 4.1). The idea was to test in the worst case scenario because testing 'healthy' modules wouldn't have added value to this case study.

Of all the measurements done, the most extreme one was selected for this discussion: a module with a hole and the front glass entirely broken (see Figure 9). After a first standard (dry) insulation test was performed that surprisingly resulted as being OK, a second (wet) test was done by spraying water onto the rear surface of the module, including the junction box, the frame and the connectors. This time the result was not OK (see Table XII).

Figure 9. Broken PV module used for insulation testing. Source: BayWa r.e.

Table XII. Insulation resistant test results (broken module)

Parameter	Dry test	Wet test
Test result	OK	Not OK
Vmis (+) [V]	1039	256
Vmis (-) [V]	1041	394
Ri + [MΩ]	>100	0.74
Ri - [MΩ]	>100	0.56
RP [MΩ]	>100	0.32
VPE [V]	18	21
VNE [V]	-18	-15
VPN [V]	36	36
R Lim [MΩ]	1	1
V0 nom [V]	1000	1000

As already stated in Table VIII, the measurements were performed with a portable PV insulation tester commonly used in the industry, the HT PV-ISOTEST (calibration date 16/03/2023, insulation resistance accuracy ±5%).

4.6 IR thermography

Satisfactory results were achieved even with a low-cost device and severe hot spots were easily detected through the field activities (see Figure 10 and Figure 11).

As stated in Table VIII, the measurements were performed with a CAT S62 PRO smartphone with a FLIR thermal camera (resolution 160 x 120).

Figure 10. Low-cost IR thermography with smartphone. Source: BayWa r.e

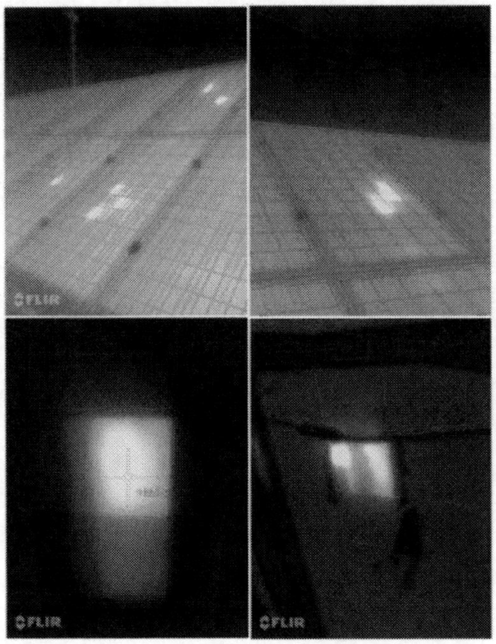

Figure 11. Example of hot spots detected during the test activities. Source: BayWa r.e.

10.4229/EUPVSEC2023/5DO.15.6
020515-011

5 CONCLUSIONS

The aim of this article was to report on the progress on the standardisation for the re-use of PV modules within IEC TC82 and to test these re-use procedures under development in a real case study, adapting them into the O&M reality. The final goal was to create a cost-effective, clear and easy-to-perform set of recommendations for the tests focused on the functionality and safety of decommissioned PV modules eligible for re-use.

On the one hand, for most PV plants under normal operation, this can be done in the simplest way by evaluating monitoring data, in combination with thermographic drone inspections. When monitoring data is not available or not accurate enough, actual I-V measurements can be performed when the weather conditions allow. The actual I-V measuring and insulation testing of modules can be limited to representative sampling in such a case, but all modules should undergo a visual inspection on major visual defects after removal from the racks. On the other hand, for modules that originate from a (partly) damaged plant, individual inspection of modules by means of visual inspection, I-V, insulation test, bypass diode test and (possibly) EL is inevitable. The same holds for modules that have already been removed and are stacked on piles in a warehouse. In the current draft of the IEC technical report, both the sampling approach and the inspection of every module are included. The sampling approach is proposed to be a slightly adapted version of the module sampling in a new plant, as described in a document published by Solar Power Europe [27]. It seems most logical that the two approaches will get their own international standard or technical specification in the future standardisation process.

To reduce the requirements for modules with some insulation resistance decrease and to avoid modules known to be PID-prone being used at too high voltage, the recommendation is to introduce also a reduced voltage class. Almost all repair of modules has been excluded as a viable option for the re-use of modules since the retesting requirements would involve excessive costs and effort. Only the replacement of damaged connectors could be an option, when the range of cable cross-section is respected.

Learnings from the real case study

The fact that PV modules just do not stop working (power = 0 W) at any time within a lifetime of 25-30 years, and that there is still not neglectable remaining power at the moment of decommissioning (year 12 for our case study), new opportunities for the second-hand market open up. The case study presented in this work suggests that even with a 50% of remaining power, a second life scenario could be envisioned, since the *used old modules* could still produce energy safely for many years. A study about the market segments was out of the scope of this paper, but possible options are applications where performance is not the main driver: off-grid systems with battery storage for basic lighting and refrigeration, household roof-tops and carports for e-mobility, public lighting and pumping solutions for agricultural irrigation, just to name a few.

It has been confirmed that in order to be able to re-use PV modules and prevent them from entering the waste stream prematurely, they should be submitted to functionality testing procedures that are technically feasible, cost-effective and contextualized to the second-hand market, giving priority to safety over performance. It is important to note that these criteria have been tailored for O&M providers in order to optimise field work and should take advantage of the fact that field technicians are already on-site executing module substitutions (e.g. due to corrective maintenance or revamping/repowering activities). In this way, the time spent on-site is optimized as little extra effort is needed, besides the fact that very short training (or none at all) is needed since field technicians (should) possess already the relevant know-how needed to carry out all the activities described in this paper. Furthermore, concerning the soiling analysis described in section 4.3, it has been shown that measuring only one reference module is not optimal. For this case study, one PV module was selected randomly only based on a visual inspection aiming at finding a 'healthy' module, but it turned out to be a module with severe failures (confirmed by the I-V curve). It is then suggested to measure soiling on at least 2 or 3 modules, possibly in different locations of the plant.

As demonstrated in section 4.5, on-site dry insulation testing might not be conclusive for re-use purposes and it is advised to perform wet insulation tests only when it doubt of safety (because it is time consuming). It has been proven that even broken modules with evident compromised electric insulation can yield positive insulation results without the presence of water (see Table XII). Of course, this is important input for the technical report.

Concerning the applicability of the sampling approach, it has been found in this plant that the power loss varied considerably between strings (range of 12% to 35%) due to the presence of many cell cracks, so that the assumption of more or less similar modules is quite questionable. So, it is clear that something like a criterion for the maximum spread in module power will be required to avoid an incorrect application of the sampling approach.

Finally, of the total of 196 modules tested for re-use, only 39 (19.9% of the total) didn't pass the particular pass/fail criteria proposed in this paper (see Table V) and could be categorized as *not eligible* for re-use. Therefore, it can be generalized that approximately 80% of the decommissioned PV modules could be re-used and given a second life. These numbers bring fresh evidence to back up the re-use concept and that take us a step further towards a more sustainable PV sector.

It should be clear that the main goal of the case study was to test the practicability of the functionality tests for decommissioned PV modules and not their extensive and systematic large-scale applicability. There is still work to do until an industry consensus is reached and an IEC standard is written and implemented.

6 ACKNOWLEDGEMENTS

The work of the authors was supported by funding from the European Union's Horizon 2020 research and Innovation Programme under Grant Agreement No. 952957, Project TRUST-PV (https://trust-pv.eu/).

7 REFERENCES

[1] IRENA, IEA-PVPS, 2016. End-of-life management: Solar Photovoltaic Panels. p.98.

[2] Directive 2012/19/EU. (2012) Waste electrical and electronic equipment -WEEE. (https://eur-lex.europa.eu/eli/dir/2012/19/oj/eng)

[3] PVPS, I.E.A., 2021. Preliminary Environmental & Financial Viability Analysis of Circular Economy Scenarios for Satisfying PV System Service Lifetime. p.69 Report IEA-PVPST12-21:2021.

[4] Wim Van Opstal, Anse Smeets, 2022, Circular economy strategies as enablers for solar PV adoption in organizational market segments, Sustainable Production and Consumption, Volume 35, 2023, Pages 40-54, ISSN 2352-5509.

[5] G Oviedo Hernandez et al 2022. Trends and innovations in photovoltaic operations and maintenance. Prog. Energy 4 042002

[6] Majewski P et al 2021 Recycling of solar PV panels-product stewardship and regulatory approaches Energy Policy 149 112062

[7] Tsanakas J A et al 2019 Towards a circular supply chain for PV modules: review of today's challenges in PV recycling, refurbishment and re-certification Prog. Photovolt. Res. Appl. 28 454–64

[8] Lempkowicz B et al 2021 RE-USE of PV modules, challenges and opportunities of the circular economy PV Cycle & IMEC (available at: https://pvcycle.be/wp-content/uploads/Press-Release-Reuse-08032021.pdf)

[9] Dodd N, Espinosa N, Van Tichelen P, Peeters K and Soares A 2020 Preparatory study for solar photovoltaic modules, inverters and systems: final report European Commission Publications Office (available at: https://data.europa.eu/doi/10.2760/852637)

[10] Godinho Ariolli D M 2021 Moving towards a circular photovoltaic economy in Europe: a system approach of the status, drivers, barriers, key policies and opportunities Master Thesis in the framework of the Erasmus Mundus Masters course in Environmental Sciences, Policy and Management (MESPOM)

[11] Hengky K S et al 2019 Drivers, barriers and enablers to end-of-life management of solar photovoltaic and battery energy storage systems: a systematic literature review J. Clean. Prod. 211 537–54

[12] van der Heide A et al 2021 Towards a successful re-use of decommissioned photovoltaic modules Prog. Photovolt. Res. Appl. 2021 1–11

[13] Directive 2008/98/EC. (2008) waste and repealing certain Directives. (https://eur-lex.europa.eu/legal-content/EN/TXT/?uri=celex%3A32008L0098)

[14] International Energy Agency, Photovoltaic Power Systems Programme (IEA-PVPS) - Assessment of Performance Loss Rate of PV Power Systems (https://iea-pvps.org/key-topics/assessment-of-performance-loss-rate-of-pv-power-systems/)

[15] Report IEA-PVPS T13-01:2014, International Energy Agency, Photovoltaic Power Systems Programme (IEA-PVPS) - Review of Failures of Photovoltaic Modules (https://iea-pvps.org/key-topics/review-of-failures-of-photovoltaic-modules-final/)

[16] Strategic Research and Innovation Agenda Photovoltaics (SRIA), 2022.

[17] TRUST-PV Solar PV, Performance & Reliability (https://trust-pv.eu/).

[18] SolarPower Europe (2021)– *Operation & Maintenance Best practices guidelines version 5.0* (https://www.solarpowereurope.org/insights/thematic-reports/o-and-m-best-practice-guidelines-version-5-0)

[19] IEC 61829:2016 *Photovoltaic (PV) array – On-site measurement of current-voltage characteristics.*

[20] IEC 62446-1:2016 *Photovoltaic (PV) systems – Requirement for testing, documentation and maintenance. Part 1: Grid connected systems- Documentation, commissioning testes and inspection.*

[21] IEC 62446-3:2017 *Photovoltaic (PV) systems – Requirement for testing, documentation and maintenance. Part 3: Photovoltaic modules and plants – Outdoor infrared thermography.*

[22] Farrell C C et al (2020). *Technical challenges and opportunities in realizing a circular economy for waste photovoltaic modules.* Renew. Sustain. Energy Rev. 128 109911

[23] Deng, R., Chang, N., Lunardi, M. M., Dias, P., Bilbao, J., Ji, J., & Chong, C. M. (2020). *Remanufacturing end-of-life silicon photovoltaics: Feasibility and viability analysis.* Progress in Photovoltaics: Research and Applications, pip.3376. https://doi.org/10.1002/pip.3376

[24] IEC 61724-1:2017 *Photovoltaic system performance - Part 1: Monitoring*

[25] CIRCUSOL Circular Business Models for the Solar Power Industry – Deliverable D3.2.: Labelling and certification protocols for second life PV modules (https://www.circusol.eu/files/Deliverables/D3-2_Labelling_and_certification_protocols_for_second_life_PV_modules.pdf).

[26] IEC 60891:2021 *Photovoltaic devices Procedures for temperature and irradiance corrections to measured I-V characteristics*

[27] SolarPower Europe (2021)– *Engineering, Procurement & Construction Best practices guidelines version 2.0* (https://www.solarpowereurope.org/insights/thematic-reports/epc-best-practice-guidelines-version-2-0)

10.4229/EUPVSEC2023/5DO.15.6
020515-013

COMPARISON BETWEEN HOME BATTERIES AND DISTRICT BATTERIES IN THE BELGIAN ELECTRICITY GRID

Cas Lavaert, Lies Debruyne, Bert Herteleer, Jan Cappelle
Research group ELECTA Gent, Faculty of Technology Engineering, KU Leuven
Gebroeders De Smetstraat 1, 9000 Gent
cas.lavaert@kuleuven.be

ABSTRACT:
This study examines the technical and economic feasibility of home and district batteries in the changing Flemish energy landscape, marked by the adoption of digital meters and energy sharing options. Consumption and production profiles from 12 Flemish households were measured at 1-second time resolution. A simple battery model assessed the impact on consumption profiles, self-consumption, self-sufficiency, and total electricity costs. Key findings reveal that home and district batteries can enhance self-consumption and self-sufficiency, with district batteries showing promise for local energy use. However, implementing home batteries led to significant electricity losses during charging and discharging processes. Economically, considering additional energy sharing costs charged by energy providers and battery payback periods, none of the scenarios (energy sharing, home battery, or district battery) proved attractive for the studied households. However, when disregarding these extra costs, both the energy sharing and the district battery scenarios show potential for electricity cost reductions. Nevertheless, results are sensitive to assumptions such as consumption profiles, electricity prices, and energy supplier choices, emphasizing the need for tailored evaluations.

Keywords: Energy sharing, Home battery, District battery

1 INTRODUCTION

Recent changes in the Flemish energy landscape caused a large increase in the number of installed home batteries. In 2019, Fluvius, the Flemish distribution system operator (DSO), started the rollout of the digital meter. The two-way registration of the energy exchange with the grid, and the high difference between consumption and feed-in tariffs, created awareness amongst the residential users about the importance of the self-consumption of the self-produced solar energy. The European energy crisis resulting from Russia's invasion of Ukraine resulted in a huge increase in energy prices and motivated people to use as little electricity as possible from the grid. Additionally, the Flemish government provided a subsidy, from the summer of 2019 to March 2023, for home batteries. The combination of the above resulted in a strong increase in the number of installed home batteries in Flanders. However, it is unclear to what extent these home batteries are an interesting/good investment in technical and economic terms. Furthermore, as of the beginning of 2023, households with digital meters have the ability to share excess electricity with others (Figure 1), with agreed-upon electricity prices. This could result in lower electricity costs, making the investment in a battery less interesting. This study aims to investigate, based on measured consumption and production profiles, to what extent a home battery (or district battery) could be technically or economically interesting in the new energy landscape with the digital meter and energy sharing possibilities.

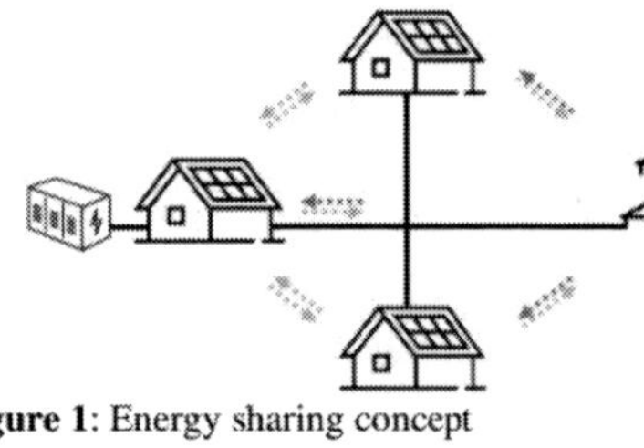

Figure 1: Energy sharing concept

2 METHODOLOGY

2.1 Data acquisition

In literature, several load profile generators that generate electrical load profiles based on a number of input parameters are available [1][2]. However, it is unclear to what extent these simulated profiles match reality and whether they match the electrical consumption of typical Flemish households. On top of that, these load profile generators often limit themselves to a time step of one minute or larger, which means that rapid changes in PV production or electricity consumption are no longer detectable in the generated profiles. For these reasons, it was opted to measure real consumption and production profiles of multiple Flemish households at a one-second resolution. These electrical profiles are measured using a Smappee Infinity and the digital meters installed by the Flemish DSO. The measured data is transmitted via a cloud-based MQTT broker to a KU Leuven server where the data is stored in a InfluxDB database.

2.2 Battery model

To evaluate the usefulness of a home battery, a basic battery model is developed in Python. The aim of this model is not to make detailed simulations about the internal chemistry of the battery, but mainly to evaluate the change in the consumption profile, and consequently the change in self-consumption and self-sufficiency. The Battery Energy Storage System (BESS) model consists of two sub-models, a model of the battery itself and a model of the inverter. The battery model contains the restrictions linked to the battery such as maximum charge and discharge rate, battery capacity, maximum depth of discharge,... This model also tracks the SOC of the battery over time and determines at what power the battery can be charged according to the SOC. The second sub-model, the inverter model, only determines the efficiency of the charging and discharging process. To achieve this, an inverter efficiency curve was utilized, shown in Figure 2 [3].

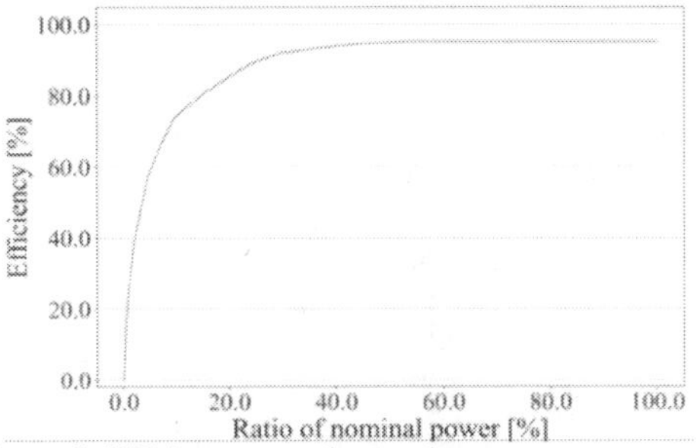

Figure 2: Efficiency curve battery inverter.

For the individual home batteries, a 4 kWh battery with a 3 kVA inverter was simulated for each household. For the the district battery a 24 kWh battery alongside a 10 kVA inverter was simulated. As a result, the district battery possesses half the aggregated storage capacity of all home batteries. All batteries are managed to maximize self-consumption, charging during surplus solar energy and discharging during shortages. Charging and discharging power are restricted by the inverter's nominal capacity. Finally, a minimum SOC of 20% is enforced.

2.3 Evaluation of simulated home/district batteries

To evaluate the impact of a home battery (or district battery), a number of Key Performance Indicators (KPIs) are established in order to objectively compare the different cases. The first evaluation parameter is total electricity consumption. Each time a home battery is charged and discharged, energy is lost (the inverter has a nominal efficiency of 95%) and this results in an increase in total electricity consumption. The second and third evaluation parameters, self-consumption and self-sufficiency, are commonly used parameters when discussing energy storage [4]. The self-consumption is a parameter that expresses what proportion of the generated solar energy is consumed locally. Meanwhile, the self-sufficiency is a parameter that reflects what proportion of total electricity consumption is covered by locally generated solar energy. The last performance indicator is the total electricity cost. This parameter is discussed in more detail in the next section.

2.4 Electricity costs

The electricity bill in Flanders consists of several costs such as grid costs, federal fees, energy cost, and VAT. Additionally, there is a feed-in tariff for the injected electricity. In this work, grid costs, federal taxes, and VAT are calculated based on the 2023 tariff rates [5]. Energy costs and the feed-in tariff are determined using the average prices over the past 12 months (September 2022 - August 2023). This corresponds to a purchasing cost of 0.375 euro/kWh for electricity and a feed-in tariff of 0.16 euro/kWh for injected electricity [6].

New legislation around energy sharing in Flanders allows households to sell their surplus electricity to one other household. It is also possible for several households to sell their surplus electricity to the same customer. However, it is currently not possible for a household to share its surplus electricity with multiple customers. In this theoretical study, it is assumed that this is possible and that each household can therefore buy electricity from an unlimited number of households and also sell its surplus electricity to an unlimited number of households. In this way, as much energy as possible can be shared locally between different households. The price at which this electricity is shared was set at 0.2675 euros/kWh so that the buyer and seller benefit equally from trading their electricity locally. Finally, in Flanders, energy suppliers can charge extra for the facilitation of this energy sharing. These costs vary from supplier to supplier, but in this study, an average extra annual cost of 100 euros is taken into account.

3 RESULTS

This section discusses the results where, in general, four different scenarios can be distinguished: a reference scenario (without energy sharing or battery), an energy sharing scenario, a scenario with home batteries and a final scenario with an energy sharing and a district battery.

3.1 Technical evaluation

Figure 3 and Figure 4 show a first technical evaluation of the implementation of a home/district battery based on the self-consumption and self-sufficiency parameters. The left graph of Figure 3 shows the monthly self-consumption for the different households without the implementation of a home battery. The blue dot represents the self-consumption when all production and consumption profiles are combined, showing a great potential for energy sharing between these households.

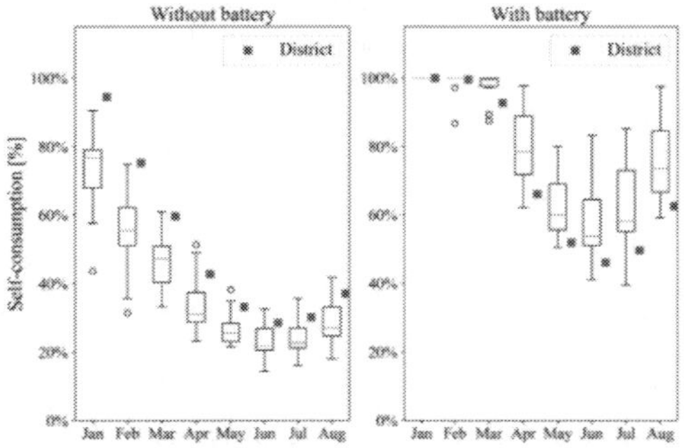

Figure 3:Monthly self consumption with and without home/district battery.

When a home/district battery is implemented, the right graph of Figure 3 shows a big increase in self-consumption for the individual households. The implementation of the district battery also results in an increase in self-consumption, however, not as high as the home batteries.

The left graph of Figure 4 shows the self-sufficiency without the implementation of a home/district battery and shows a similar trend as Figure 3. When the consumption and production profiles are combined, it results in an increase in self-sufficiency, again showing the potential of energy sharing.

10.4229/EUPVSEC2023/5DV.2.2
020517-002

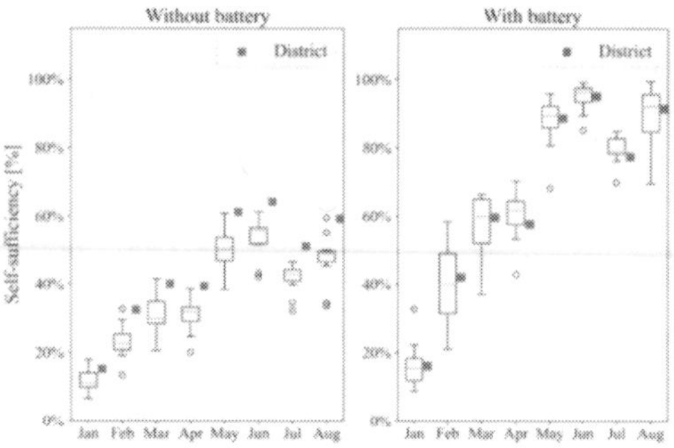

Figure 4: Monthly self-sufficiency with an without home/district battery.

The right graph of Figure 4 shows the self-sufficiency when a home/district battery is implemented. As with the self-consumption, this results in a distinct increase in self-sufficiency. This increase is, however, no so pronounced as with the self-consumption.

The cause for this is shown in Figure 5. Due to the charging and discharging of the batteries, a large amount energy is lost in the inverter. As shown, this results in an increase in monthly electricity consumption of up to 40% for some households. This is mainly caused by charging and discharging at low power levels, which means that this increase in electricity consumption is more limited when the consumption and generation profiles are combined in one district, as can be seen by the blue dot on Figure 5. The increase in electricity consumption also results in an increase in self-consumption, but of course not in a desirable manner.

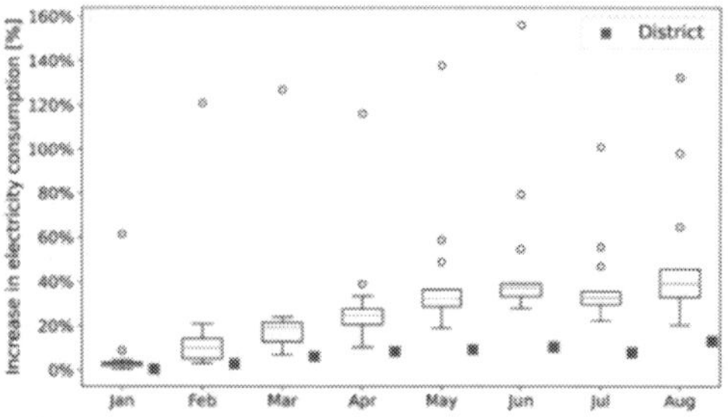

Figure 5: Monthly electricity increase due to installation of home/district battery.

3.2 Economical evaluation

Figure 6 and Figure 7 show the electricity bill structure for two different households in the four different scenarios. Figure 6 shows the electricity costs of a household with a relatively low electricity consumption. Both the home battery and the district battery ensure a reduction in electricity costs. The energy sharing scenario is not interesting for this household due to the extra costs charged by the electricity provider for facilitating energy sharing.

Figure 7, on the other hand, represent a household with a relatively high electricity consumption showing that the different scenarios have a less pronounced influence on the total electricity costs.

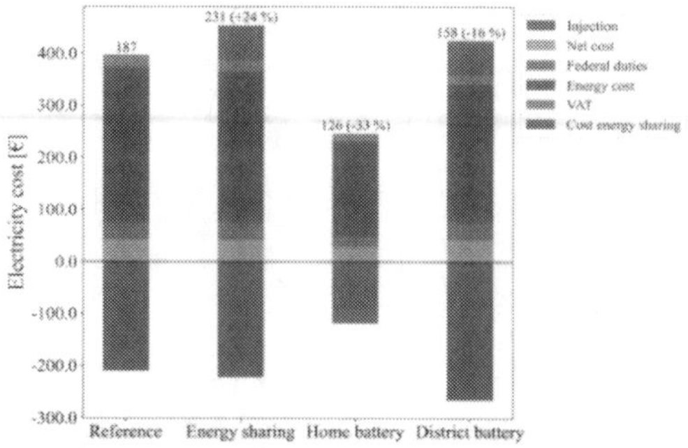

Figure 6: Electricity bill composition of a household with relatively low consumption.

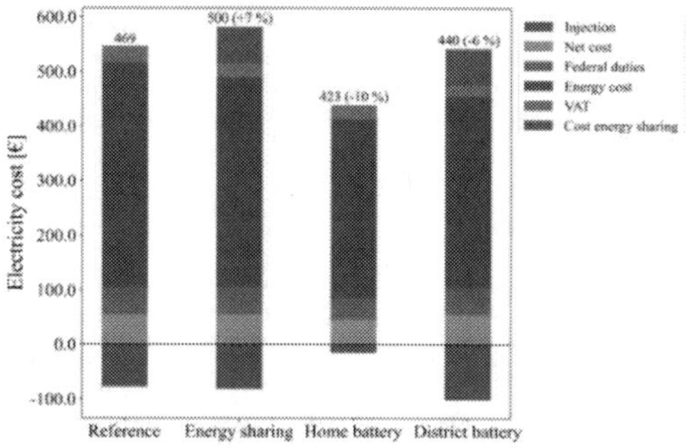

Figure 7: Electricity bill composition of a household with a relatively high consumption.

Figure 8 summarizes the net electricity costs for the four scenarios of all 12 households without the extra cost for energy sharing being taken into account. As a result, the district battery scenario shows the biggest decrease in electricity costs, followed by the home battery and lastly the energy sharing scenario. This shows that implementing a district battery is preferable to installing individual home batteries. However, the energy sharing scenario is the only scenario that does not require an investment in a battery, making this scenario consistently interesting regardless of the battery payback period.

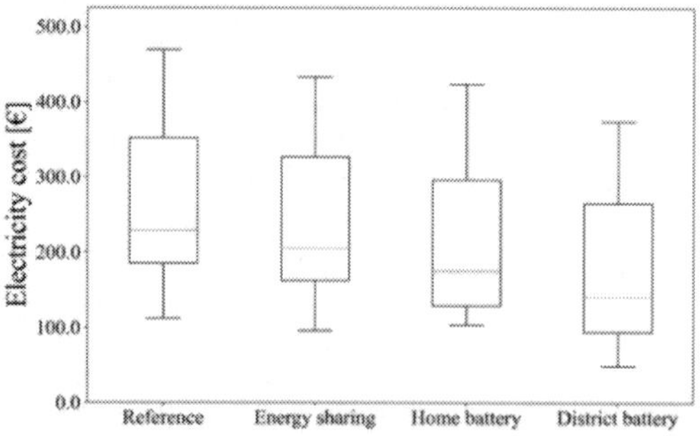

Figure 8: Boxplot of net electricity costs for all the households in four different scenarios.

When the additional costs charged for energy sharing are taken into account, the energy sharing and district battery

10.4229/EUPVSEC2023/5DV.2.2
020517-003

scenarios become a lot less interesting. Figure 9 shows the same results as Figure 8, but taking into account the additional costs for energy sharing charged in Flanders.

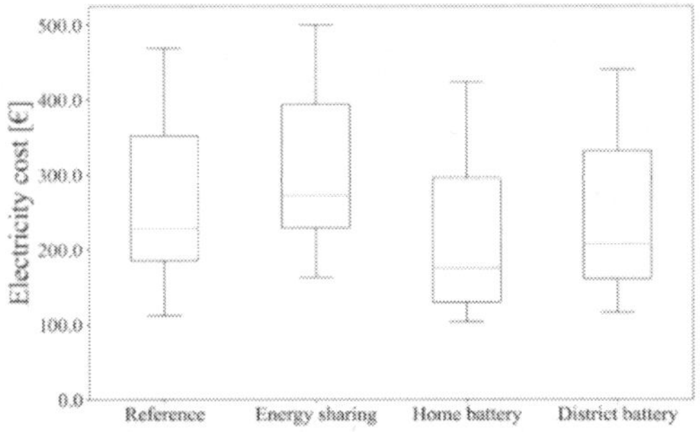

Figure 9: Boxplot of net electricity costs for all the households including extra energy sharing cost.

These extra costs have as result that the electricity costs of the energy sharing scenario rise above those of the reference situation, making energy sharing no longer interesting for the considered households. Similarly, the costs of the district battery scenario rise, however, a small reduction of electricity costs is still obtained compared to the reference situation. The savings are, however, minimal so the investment in a district battery will not be interesting.

Finally, Figure 10 shows the payback period of the different battery scenarios without taking into account the additional energy sharing costs applicable in the district battery scenario. A battery price of 600 euro/kWh was used based on quotations obtained in January 2023. Although the home battery scenario provided the biggest reduction in electricity costs (shown in Figure 9), home batteries are not an interesting investment as it results in payback times over 20 years.

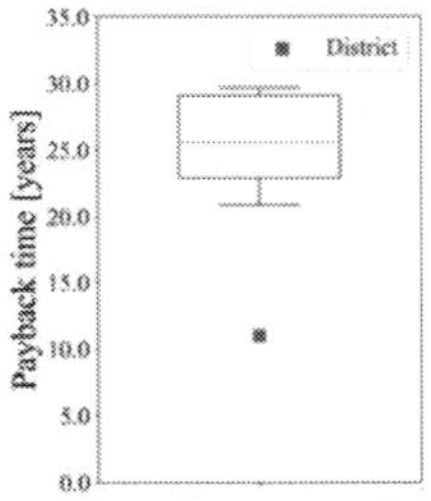

Figure 10: Payback time battery scenarios

The payback time of the district battery (no extra energy sharing cost considered) results in a payback time of 10 years. This is still a relatively long payback period, but the relative difference with the payback period of home batteries shows the potential of the district battery.

4 DISCUSSION

The results presented in this paper show that both the energy sharing, home battery and district battery scenarios are not interesting considering the current legislation and electricity costs in Flanders. However, these results are highly dependent on the assumptions made.

First, this study used production and consumption profiles of households with relatively low electricity consumption. The results for households with higher electricity consumption, due to heat pumps or electric cars for example, may differ significantly.

Secondly, these results are also highly dependent on electricity prices. Lower electricity prices will make energy sharing or the installation of a battery less interesting, while higher electricity prices will have the opposite effect.

Finally, an average cost charged in Flanders for energy sharing was taken into account, but this cost still varies greatly between different suppliers. Some do not charge extra, which makes energy sharing interesting, but others charge more than 150 euros per year. Moreover, this fixed cost weighs heavily on the electricity costs of the households considered in this study due to their relatively low electricity consumption and therefore low electricity costs. For households with higher annual consumption, these costs will have less of an impact.

5 CONCLUSION

This study has shed light on the technical and economic aspects of home and district batteries in the evolving Flemish energy landscape with digital meters and energy sharing possibilities. The technical evaluation revealed that the implementation of both home and district batteries significantly increased self-consumption and self-sufficiency among households, with district batteries showing promising results in enhancing local energy utilization. However, the implementation of home batteries caused a lot of electricity losses, due to the charging and discharging of the batteries, resulting in significant increase in electricity consumption.

The economic evaluation, taken into account extra costs for energy sharing and the payback time of batteries, showed that both the energy sharing, home battery and district battery scenarios are not economically interesting for the households considered in this study. If these additional costs are disregarded, the energy sharing scenario is consistently interesting. Also the district battery scenario, with a payback period of only half of the payback period of the home batteries, also shows potential.

ACKNOWLEDGEMENT

We would like to acknowledge the contribution of Dorus De Sutter, who played a crucial role in the successful completion of this research project. He conducted his Master's thesis on the subject of this paper, and his findings were instrumental in shaping our research approach.

REFERENCES

[1] G. Hoogsteen, A. Molderink, J. L. Hurink, and G. J. M. Smit, "Generation of flexible domestic load profiles to evaluate Demand Side Management approaches," *2016 IEEE Int. Energy Conf. ENERGYCON 2016*, 2016, doi: 10.1109/ENERGYCON.2016.7513873.

[2] N. D. Pflugradt, "LPG 10.4.0 Load Profile Generator," *Website (last accessed on 11.06.2021) https://www.loadprofilegenerator.de/download/*, vol. 1, no. 11, pp. 361–365, 2013.

[3] E. Arnieri, L. Boccia, F. Amoroso, G. Amendola, and G. Cappuccino, "Improved efficiency

management strategy for battery-based energy storage systems," *Electron.*, vol. 8, no. 12, pp. 1–9, 2019, doi: 10.3390/electronics8121459.

[4] J. L. Torres-Moreno, A. Gimenez-Fernandez, M. Perez-Garcia, and F. Rodriguez, "Energy management strategy for micro-grids with PV-battery systems and electric vehicles," *Energies*, vol. 11, no. 3, 2018, doi: 10.3390/en11030522.

[5] "Periodieke nettarieven elektriciteit en aardgas 2023 | VREG." https://www.vreg.be/nl/periodieke-nettarieven-elektriciteit-en-aardgas-2023 (accessed Sep. 14, 2023).

[6] "VREG opbouw gemiddelde commerciële elektriciteitsprijs." https://dashboard.vreg.be/report/DMR_Prijzen_elektriciteit.html (accessed Sep. 14, 2023).

MATCHING PV PRODUCTION AND ELECTRICITY LOAD ON HOUSEHOLD LEVEL AT HIGH-LATITUDE LOCATIONS

S. Jouttijärvi[1]*, L. Karttunen[1], S. Ranta[2], K. Miettunen[1]
[1]Department of Mechanical and Materials Engineering, University of Turku, Vesilinnantie 5, 20500 Turku, Finland
[2]New Energy Research Group, Turku University of Applied Sciences, Joukahaisenkatu 7, 20520 Turku, Finland
*sami.jouttijarvi@utu.fi

ABSTRACT: This study analyses the potential for the self-consumption of photovoltaic (PV) electricity for different detached houses and PV systems in Nordic conditions. The aim is to reduce the need for battery energy storage or balancing power generation by fossil fuels to mitigate the high diurnal and weather-dependent variation of PV production. Increasing self-consumption is crucial to increase the usability and economic value of PV electricity on households and to smoothen the grid integration of PV on local and national levels. Three differently oriented monofacial PV systems and one vertically mounted bifacial PV system are compared as clean electricity sources for different detached houses in a high-latitude location. The focus is on total energy production and the potential for self-consumption. For monofacial systems, the self-consumption potential was dominated by the electricity load profile, and the differences in self-consumption between differently oriented systems were small, except when large, flexible loads allowed to utilize high amount of PV electricity at the spot. For vertical bifacial system, the highest overall production and convenient power production profile allowed superior self-consumption compared with monofacial PV systems.
Keywords: Self-consumption, high-latitude location, vertical bifacial PV

1 INTRODUCTION

Solving the challenges arising from imbalance between temporal electricity consumption and variable renewable energy (VRE) production plays a key role for increasing the share of clean energy in future electricity generation. Solar photovoltaics (PV) has extraordinary scalability, allowing all sizes of actors from individual citizens to large power companies to own PV power systems. Thus, PV provides a unique opportunity for individuals and small companies to affect directly how the electricity they consume is produced.

The economic feasibility of a PV system depends on the production capacity, the PV system price, and the local electricity price. The massive drop in the cost of PV over the last decade and the rise in the electricity prices across Europe have increased the feasibility of PV. Thus, PV is now a feasible option even for high-latitude locations, such as the Nordic countries, although the production per kilowatt-peak (kWp) is lower than in southern Europe. Here, we conduct a case study by modelling a detached house with rooftop PV installation located in Turku, Finland (60°N, 22°E). Case studies regarding rooftop PV are common in low- and mid-latitude locations, both for single households [1] and for national-scale analysis on rooftop PV potential [2]. However, at high latitudes, such studies are rare, although some exist [3].

The availability of PV power depends on time of day and weather conditions, making the PV production highly intermittent, creating challenges for the grid integration [4]. With typical, south-facing PV installations on a clear day, PV gives a power production profile which rises rapidly during morning, peaks at solar noon and goes down during afternoon. Orienting panels away from the South increases the production in the morning or evening, at the cost of decreased overall production. Another innovative option is to install vertically mounted bifacial PV (VBPV) panels, which utilize light from both sides and give peak production in the morning and evening when oriented in East-West direction. Low solar elevation favours VBPV over MPV, and thus VBPV is especially attracting option at high latitude locations [5]–[7].

Electricity consumption and supply in the power grid must match temporally and spatially, to ensure reliable electricity delivery. In a typical detached house, the power demand is high during the morning and evening, with a valley between. This leads to a temporal mismatch with a conventional PV production profile. With e.g., East- and West-facing PV systems, the match between production and consumption is better [8]. With VBPV, matching both the morning and evening consumption peaks, and thus allowing high self-consumption, is possible.

The case studies include four different PV power production profiles, representing three differently oriented monofacial systems and one vertical bifacial system. For the electricity load, four profiles are used. Three load profiles, acquired from clustered market data [9], represent Finnish detached houses with different heating solution. The fourth profile adds flexibility to heat demand, allowing the flexible utilization of PV. The aim is to analyse quantitatively the self-consumption and self-sufficiency of different [PV system – Electricity load] combinations, determining the suitability of each PV system for a house with a certain load profile. For MPV, the main interest is to investigate how much the self-consumption of PV electricity can be increased by orienting the PV systems unconventionally, i.e., shifting them from South towards East or West. This orientation shift can be due to the possibility to produce more electricity during the natural peak electricity load hours in the morning and evening, at the cost of decreased total annual production, or by external restrictions, for instance when South-facing roofs are unavailable. For VBPV, that is especially suitable for high-latitude locations [7], the key interest is to compare the total and self-consumed production with the different MPV solutions.

2 METHODS

2.1 PV power production profiles

The global horizontal irradiance (GHI), ambient temperature (T_A), and wind speed (WS) data were acquired from a weather station located in Turku, Finland

10.4229/EUPVSEC2023/5DV.2.3
020518-001

[8]. The GHI values were converted to direct current (DC) power production profiles with a methodology explained in our previous work [10]. The power production was modelled for one year, with the year 2019 weather data. The modelled PV systems, excluding the size sensitivity analysis presented in Section 3.3, were sized to 4.0 kWp, which is a realistic size for a rooftop PV installation.

Four different PV orientations were studied: three monofacial PV (MPV) systems all with 30° tilt and different azimuths and one VBPV system. The tilt of 30° was chosen since it is a typical roof tilt angle for a traditional Finnish house. Azimuth 180° (T30S) represents a conventional installation site for a small-scale PV system: South-facing tilted roof. Azimuths of 90° and 270° in the same system (T30EW) represent a tilted roof which ridge is aligned in South-North direction, resulting in sides facing East and West. Azimuths 135° and 225° in the same system (T30SESW) represent an intermediate scenario, where the main building has a wing or garage, aligned so that there are both Southeast and Southwest facing roofs. The VBPV system is oriented so that sides are facing East and West. Bifaciality of 90% is assumed in calculations, i.e., the power production from the light incident on the rear side is 10% lower compared with the front side due to additional optical losses. The profiles for two example days are shown and analysed in Section 3.1.

Table 1. Summary of the studied PV systems. For the dual-azimuth systems, the percentages show the ratio of the azimuths. Azimuth = 180° is South.

Name	Tilt (°)	Azimuth(s) (°)
T30S	30	180 (100%)
T30SESW	30	135 (50%) and 225 (50%)
T30EW	30	90 (50%) and 270 (50%)
VBPV	90	90* (50%) and 270* (50%)

*Bifaciality 90%

2.2 Electricity loads

The electricity load profiles used here are based on typical electricity consumption behaviour of different Finnish type consumers, defined in literature [9]. The profiles were acquired from clustered data: they represent the averages of many consumers that have similar electricity demand and load profile. Thus, rapid variations in demand, due to such as using a stove, are averaged out.

Three type of user profiles were utilized directly: a detached house without electric heating (EH) (Load1), energy-efficient detached house with direct EH (Load2) and detached house with reserving EH (Load3). These profiles were taken directly from literature [3]. All profiles covered one year and had a resolution of one hour.

A fourth profile, Load3F, was generated based on Load3 and assuming that the heating demand is flexible, based on certain principles as follows: The total consumption (Load3) was divided into consumption used for other purposes than heating (Load1) and consumption due to heating (Load3 – Load1). Then, a flexibility for 24 hours was assumed: if there is any surplus PV power available in a certain hour and expected heat demand within the next 24 hours, the surplus was used to satisfy the heat demand in advance. This profile represents a situation where the surplus PV electricity during a sunny day can be used to satisfy the heat and hot water demand for the following night. The profiles are summarized in **Table 2** and visualized in **Figure 1**.

Table 2. Summary of the used load profiles.

Name	Electric heating	Energy-efficient house	Annual consumption (kWh)
Load1	No	No	5000
Load2	Direct	Yes	10000
Load3	Reserving	No	19000
Load3F	Reserving & flexible	No	19000

Loads1&2 have a very similar diurnal shape due to domestic activities, with Load2 having an additional season-dependent component due to electric heating. Load3, however, has a completely different shape: during daytime, it is close to Load2, but during night, there are high peaks due to heating needs. The high loads are scheduled to night-time due to lower electricity demand and cost. The peaks are the highest during winter due to high heating demand, but they exist throughout the year due to hot water demand. For Load3F, the profile is identical to Load3 during winter, but from spring to autumn the high night peak in Load3 is replaced with a lower and wider day peak in Load3F, representing cases where surplus PV production is used to produce heat. Since Load3F shifts the demand based on PV production, its exact shape depends on the PV system. In **Figure 1**, the PV system is T30S (nominal power 4.0 kWp).

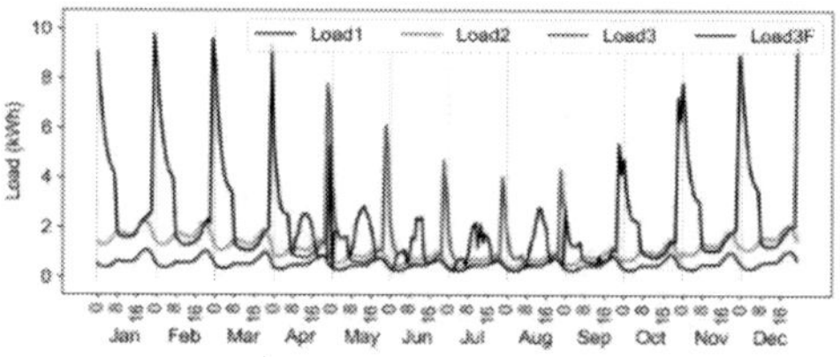

Figure 1. The hourly electricity consumption with the studied load profiles. The consumption is shown for one example day for each month (Tuesday closest to the 15th day of the month).

2.3 Simulations and used metrics

The houses with four different PV profiles and four different load profiles, a total of 16 different combinations, were simulated with one-hour resolution for one year. The dataset for PV production had one-minute resolution and it was converted to one-hour resolution by summing all modelled production within each hour. The nominal power of the PV systems was set to 4.0 kWp (except for the size sensitivity analysis presented in Section 3.3). The quantity and direction of the power flow between the house and the power grid for the hour i was determined based on the quantity of the PV production (E_{PV}) and electricity consumption (E_L):

$$E_{SC,i} = E_{PV,i}, if\, E_{PV,i} \leq E_{L,i} \quad (1a)$$
$$E_{SC,i} = E_{L,i}, \ otherwise \quad (1b)$$

For the case (1a), the additional electricity needed to meet the demand is taken from the grid. For the case (1b), the additional PV production is classified as surplus and delivered to the grid. These practices are consistent with Finnish market policies: small-scale PV producers are expected to cover their own demand with PV before delivering to the grid.

10.4229/EUPVSEC2023/5DV.2.3
020518-002

For each case, the total absolute self-consumption (E_{SC}), the relative self-consumption (Φ_C), and the relative self-sufficiency (Φ_S) were calculated as follows:

$$E_{SC} = \sum_i E_{SC,i} \quad (2)$$
$$\emptyset_C = \frac{E_{SC}}{\sum_i E_{PV,i}} \quad (3)$$
$$\emptyset_S = \frac{E_{SC}}{\sum_i E_{L,i}} \quad (4)$$

The calculations were done with one-hour resolution due to data availability: both the demand and PV production were assumed to remain constant for one hour. Thus, during a certain hour, only one-directional power flow between the house and the grid is allowed. However, the power may flow to both directions within an hour, due to sudden changes in load (e.g., using a stove) or in production (e.g., cloud movement). This effect is excluded due to resolution of the electricity load data.

For the size sensitivity analysis (Section 3.2), two PV systems (T30S and VBPV) were combined with all four load profiles. The nominal powers of the studied systems are 0.4-8.0 kWp, with an interval of 0.4 kWp.

3 RESULTS

3.1 Self-consumption and self-sufficiency for a single sunny day

To demonstrate the methodology, the E_L, E_{PV}, E_{SC}, Φ_C, and Φ_S were determined for 4.0 kWp PV systems, representing typical single-family house rooftop PV installations. All four PV profiles and one load profile (Load2) were analysed for two example days: Apr 04 (**Figure 2A**) represents an early spring day with varying weather, whereas May 16 (**Figure 2B**) represents a sunny late spring day. The key numbers for the quantitative analysis of the example days are presented in **Table 3**.

When comparing MPV profiles, the E_{PV} is the highest with T30S and lowest with T30EW. However, considering the E_{SC}, the order may be different: for May 16 (**Figure 2B**), the higher production during morning and evening enables higher E_{SC}, and thus higher Φ_S, with T30EW, whereas the majority of the high E_{PV} with T30S is counted as surplus production. This is possible since during daytime (8 am to 6 pm) all systems are producing more electricity than is needed to match the load, and thus the differences in E_{SC} result from the differences in the morning and evening production.

For Apr 04 (**Figure 2A**), the situation is different: since the load is higher and PV production lower, E_{PV} correlates well with E_{SC}. During the period when PV production matches or exceeds the load, T30S is the best-performing MPV system. Actually, the E_{SC} for T30S is higher on Apr 04 than on May 16, since higher load enables the efficient utilization of PV for self-consumption. This phenomenon is especially clear when comparing Φ_C:s for Apr 04 and May 16 (**Table 3**): the Φ_C is higher for Apr 04 with all studied profiles.

For both days, VBPV is superior to all MPV systems when considering the daily E_{PV}, E_{SC}, and Φ_S. These findings highlight the benefits of the higher production and more suitable power production profile with VBPV, which allows to match the load earlier in the morning and later in the evening than it is possible with studied MPV systems.

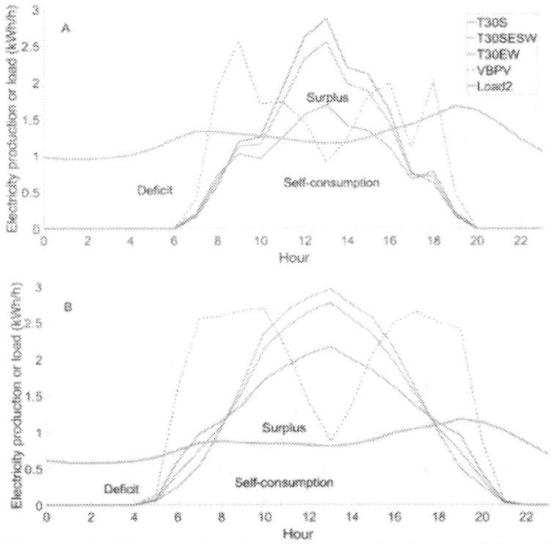

Figure 2. The hourly electricity production and hourly consumption with Load2-profile on (A) a mostly sunny early spring day and (B) on a sunny late spring day. The timestamp of x-axis refers to the start of hour: e.g., values shown for "Hour = 10" are the electricity production and consumption between 10 am and 11 am.

Table 3. The key parameters for different PV profiles on the two example days.

Apr 04: E_L = 29.8 kWh	E_{PV} (kWh)	E_{SC} (kWh)	Φ_C (%)	Φ_S (%)
T30S	18.1	12.1	67.0	40.8
T30SESW	16.7	12.2	72.8	40.8
T30EW	13.0	11.7	90.5	39.4
VBPV	19.4	14.6	75.1	48.8
May 16: E_L = 20.1 kWh	E_{PV} (kWh)	E_{SC} (kWh)	Φ_C (%)	Φ_S (%)
T30S	25.2	11.5	45.5	57.2
T30SESW	24.4	12.3	50.2	61.1
T30EW	21.5	12.9	60.2	64.3
VBPV	31.1	13.8	44.4	68.7

3.2. Annual self-consumption and self-sufficiency

The annual values for E_{SC}, Φ_C and Φ_S were determined for 4.0 kWp PV system with different production profiles. The key results are summarized visually in **Figure 3** and in numeric form in **Table 4**.

For the MPV systems, the E_{SC} and Φ_S remained practically constant when the orientation was changed for Loads1-3: the significant difference in overall production, especially between T30S and T30EW, went almost entirely to surplus. For Load3F, the Φ_S was significantly lower for T30EW-orientation (15.9%) than for T30SESW (18.3%) and T30S (19.1%), since the Load3F allowed to consume almost all of the produced electricity due to extensive and flexible heat demand.

VBPV had the highest overall production and best match with the typical electricity load, leading to the highest E_{SC} and Φ_S with all load cases. When load is fixed, the differences between the E_{SC}:s of VBPV and the MPV system with the highest E_{SC} varied from 90 kWh (compared with T30SESW) with Load1 to 250 kWh (compared with T30S) with Load3F. However, the increase in total production with VBPV was 380 kWh against T30S and 570 kWh against T30SESW, indicating that despite the improved match between electricity load

10.4229/EUPVSEC2023/5DV.2.3
020518-003

and PV production, a significant fraction of the additional production with VBPV goes to surplus production.

When comparing different loads with the same PV system, the Load3F had by far the highest E_{SC} (3020-3880 kWh, depending on the PV orientation). Due to flexible heat demand, situations where the own PV power production went to surplus were rare – even during summer. Load2 had reasonable high base load due to direct electric heating, which lead to the second-highest E_{SC} (2170-2460 kWh) and Φ_S. Although Load3 had high E_L, the potential to utilize PV was low, since the reserving heating is used during night-time (**Figure 1**), which would be an optimal solution without PV.

To sum up, the self-consumption potential of PV electricity depended primarily on load flexibility and secondary from the natural base load during daytime from spring to autumn, when the PV production is high. The total overall production was less significant among the case studies: even though the T30EW profile had 19.0% smaller overall production than T30S, it had almost identical E_{SC} than T30S with Loads1-3. This observation is due to the effect shown in **Figure 2**: T30EW system produces more electricity in the morning and evening and extends the period when all consumed electricity is produced by the own PV for a sunny summer day, compared to T30S.

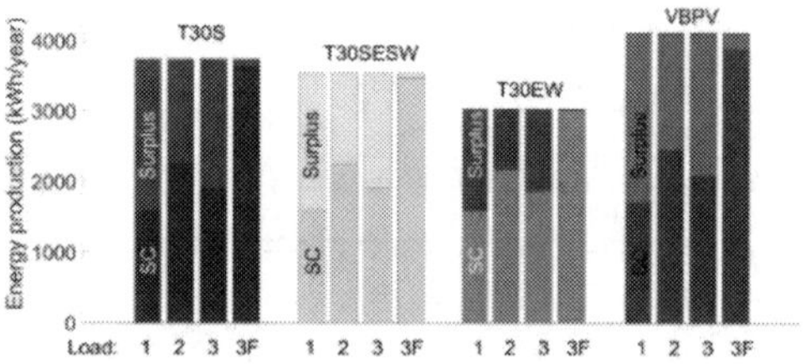

Figure 3. Total annual production and its distribution to the self-consumed and surplus production for each studied [PV system – Electricity load] case.

Table 4. Total annual electricity demand for each Load, the total annual electricity production for each PV system, and the self-consumed PV prodcution, relative self-consumption and relative self-sufficiency for all 16 [PV system – Electricity load] cases.

Load	1	2	3	3F
E_L (kWh)	5000	10000	19000	19000
T30S	E_{PV} = **3740 kWh**			
E_{SC} (kWh)	1580	2240	1900	3630
Φ_C (%)	42.2	59.9	50.9	97.0
Φ_S (%)	31.6	22.4	10.0	19.1
T30SESW	E_{PV} = **3550 kWh**			
E_{SC} (kWh)	1600	2250	1920	3480
Φ_C (%)	45.1	63.5	54.1	98.0
Φ_S (%)	32.0	22.5	10.1	18.3
T30EW	E_{PV} = **3030 kWh**			
E_{SC} (kWh)	1590	2170	1870	3020
Φ_C (%)	52.5	71.6	61.7	99.7
Φ_S (%)	31.8	21.7	9.8	15.9
VBPV	E_{PV} = **4120 kWh**			
E_{SC} (kWh)	1690	2460	2090	3880
Φ_C (%)	41.1	59.7	50.9	94.2
Φ_S (%)	33.9	24.6	11.0	20.4

3.3 Size sensitivity analysis

The effect of PV system size, expressed as nominal power, to the self-consumption is shown in **Figure 4** for T30S and VBPV systems. With very small PV systems, all production is self-consumed. However, with Loads1-3, the point where most of the added production went to surplus was reached between 1 and 2 kWp: as **Figure 2** shows, during a sunny summer day 4.0 kWp system is oversized when compared with base load and thus even significantly smaller systems can provide almost the same amount of self-consumed PV electricity. Although larger systems enable high daily self-sufficiency during early spring and late autumn, when the PV peak production and electricity peak demand are better matched, a large fraction of the added production still goes to summer day surplus production when the system size is increased, since the PV production focuses heavily to summer.

For Load3F, the surplus was close to zero until approximately 4 kWp due to potential for demand flexibility. After that, the overproduction during summer exceeded the summer heat demand and surplus became significant. However, a significant fraction of the added production was self-consumed even at 8.0 kWp, further highlighting the role of demand flexibility.

Overall, the sensitivity analysis shown in **Figure 4** reveals that with Loads1-3, the increase of self-consumption was small after the system size exceeded 2 kWp, whereas with Load3F, the share of the self-consumption from added production was almost 100% still at 4 kWp and remained significant (75.4% with T30S and 70.8% with VBPV) even at the upper limit (8.0 kWp). Thus, the profitability of larger PV systems without potential for load flexibility is strongly dependent on the economic value of the surplus electricity. If the value is high, due to a feed-in tariff or high electricity spot price, the oversized systems are beneficial. However, if the value of surplus electricity is small compared with the self-consumed electricity, small-scale producers' motivation to invest in large PV facilities is low.

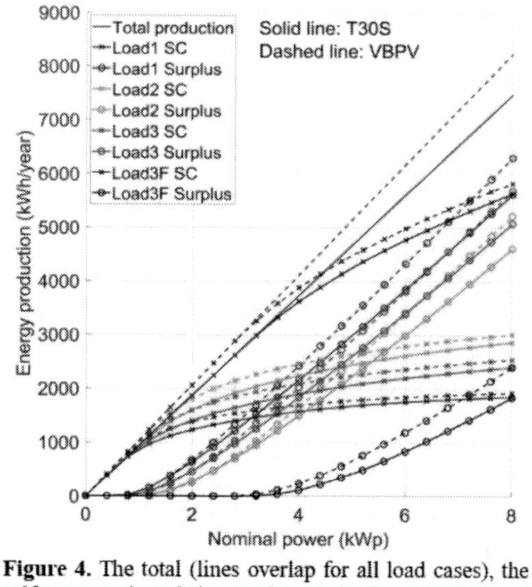

Figure 4. The total (lines overlap for all load cases), the self-consumed, and the surplus production of T30S and VBPV systems with all load profiles, as a function of system size. The full legend is shown only for T30S.

10.4229/EUPVSEC2023/5DV.2.3
020518-004

4 CONCLUSIONS

A total of 16 different [PV system – Electricity load] combinations with the nominal PV power of 4.0 kWp were studied in terms of self-consumption and self-sufficiency. VBPV had superior self-consumption potential compared with MPV due to the highest total production and best match between power production and electricity load. When comparing MPV systems with different orientations, the loss of production due to the azimuth shift away from South consisted almost completely of surplus production: the self-consumed productions, and thus the self-sufficiencies, of the MPV systems were very similar, except for the Load3F, which had high flexible demand. This result highlights that PV installations are suitable in Finland for a large variety of differently oriented roofs.

Sensitivity analysis revealed that without load flexibility, the point where the additional production achieved by increasing system size went mostly to surplus, was reached at 1-2 kWp. However, with flexible electric heating, the hot water demand is enough to enable high (75.4% with T30S) self-consumption even for 8.0 kWp PV system. Thus, investing in larger PV systems at detached houses requires either possibility to create value with surplus electricity or high potential for load flexibility. Moreover, high self-consumption enables more PV in the local power grid without the need for massive grid upgrades.

Since the power production is modelled in DC, inverter losses are not included in this study. Moreover, the used electricity load profiles are based on clustered data, representing the average of many similar customers. Thus, for a single house, the actual consumption behaviour may differ significantly from these profiles. However, using clustered profiles allows the easier upscaling of the simulations, to local or national power grid levels.

REFERENCES

[1] P. Olczak and A. Komorowska, *Sustainable Energy Technologies and Assessments* 47 (2021), 101496.

[2] A. Gomez-Exposito, A. Arcos-Vargas, and F. Gutierrez-Garcia, *Renewable and Sustainable Energy Reviews* 132 (2020), 110074.

[3] A. Meriläinen, P. Puranen, A. Kosonen, and J. Ahola, *Solar Energy* 240 (2022), 422–434.

[4] O. Gandhi, D. S. Kumar, C. D. Rodríguez-Gallegos, and D. Srinivasan, *Solar Energy* 210 (2020), 181–201.

[5] D. Chudinzow, S. Nagel, J. Güsewell, and L. Eltrop, *Applied Energy* 264 (2020), 114782.

[6] S. Guo, T. M. Walsh, and M. Peters, *Energy* 61 (2013), 447–454.

[7] S. Jouttijärvi, G. Lobaccaro, A. Kamppinen, and K. Miettunen, *Renewable and Sustainable Energy Reviews* 161 (2022), 112354.

[8] S. Ranta, J. S. Stein, E. Whitney, H. Huerta, and A. Heinonen, *36th EUPVSEC* (2020).

[9] A. Mutanen, K. Lummi, and P. Järventausta, *Valtakunnallisten tyyppikäyttäjämäärittelyiden päivittäminen ja hyödyntämisen periaatteet verkkopalvelumaksuihin liittyvissä tarkasteluissa*, Tampere University (2019).

[10] S. Jouttijärvi et al., *Solar Energy* 262 (2023), 111819.

10.4229/EUPVSEC2023/5DV.2.3
020518-005

ENERGY SHARING IN SOLAR AND BATTERY OFF-GRID SYSTEMS WITH ADVANCED PV GENERATION MODELLING: A CASE STUDY OF NORWEGIAN CABIN FIELDS

Ida Fuchs, Halgrim Ove Tørrisplass, Steve Völler
Norwegian University of Science and Technology / Department of Electric Energy
O. S. Bragstads plass 2E, 7034 Trondheim, Norway, +47 73559775, ida.fuchs@ntnu.no

ABSTRACT: Energy sharing and peer-to-peer energy trading among households equipped with solar and battery systems have become prominent themes in the ongoing energy transition. These approaches aim to engage consumers in the energy transition process and shift some of the responsibility for energy supply from centralized institutions to individuals. The concept of energy sharing has also given rise to innovative ideas such as local energy communities and swarm electrification. These concepts hold promise in addressing a significant challenge in Norway: providing electricity to remote mountain cottage areas without incurring high costs associated with grid extensions. A more cost-effective alternative involves deploying individual off-grid solar and battery systems. However, to ensure these systems are not oversized and operate efficiently, energy sharing can play a valuable role. In this paper, we analyze the impact of detailed photovoltaic (PV) system modelling in simulations of local energy communities. Using a demonstrative case study, we show that an advanced PV model, which considers the correct azimuth and tilt for the panels, enhances the benefits of energy sharing within the cabin community. Our findings indicate that the primary advantage stems from the increased diversity in PV yield, with a smaller percentage attributable to the diversified panel orientations.
Keywords: PV system, modelling, off-grid, energy sharing, local energy communities

1 INTRODUCTION

Over the past twenty years 13 new cabins per day have been built on average in the Norwegian mountains [1]. The cabins have changed standards from modest off-grid cabin to modernized fully equipped homes. The current cabin development is problematic from a sustainability perspective. One of the possible solutions that could help to reduce the environmental footprint of cabin fields lays in an innovational change of their current electricity supply. 72% of today's cabins are connected to the main grid and the average cabin today uses 13% more electricity than 13 years ago [2]. The power consumption of the largest cottage municipalities in Norway is enormous. Hol in Hallingdal has 5,832 leisure buildings, ranked as the third most cottage-rich municipality in Norway. The cottages' consumption of electric power is 39% of the total power consumption for Hol. This is an extremely high number considering that the average period of use of cabins is 48 days per year. In comparison, households consume 23% of the municipality's total consumption. Besides the need for increased electricity generation capacity for the cottages, these rapidly growing cabin fields put a great pressure on the grid companies. Over large areas, there have been great needs in short times for power grid upgrades. This puts additional environmental challenges and increased costs on top of an already conflicting issue.

To address these challenges, we suggest the implementation of an innovative local energy community (LEC) based on a concept called swarm electrification. Swarm electrification is a new conceptual approach in rural electrification, where the infrastructure of a microgrid is built bottom-up. The concept was first introduced in 2014 [3] and has ever since been growing both in terms of scientific literature [4] and real-world applications and implementation [5], [6]. It starts with individual solar home systems (SHS) and proceeds with the interconnection of these into a step-by-step growing swarm grid. The concept has been mainly used in rural electrification for energy access, however, its potential for remote cottage areas is apparent.

Several studies on LEC have been conducted by researchers and methods have been proposed [7], [8], however, the coverage of aspects for LEC in rural or off-grid areas is low. Further, energy system modelling often disregards detailed representation of the PV generation input which might be disadvantageous for the results, especially when looking into the energy sharing potential. This is due to the fact that energy sharing benefits both from increasing diversity in load profiles and energy generation profiles. A study of azimuth and tilt influence on energy sharing between cabins has been conducted in [9]. Including detailed modelling of PV generation could lead to increased diversity in the PV generation profiles for each participant and therewith visualize the benefits of energy sharing more precisely. Thus, the research question for this study analysis is:

To what extent can a comprehensive photovoltaic (PV) modeling approach impact the outcomes when evaluating the advantages of energy sharing within off-grid cabin energy communities, in contrast to using a basic PV generation model?

2 METHODOLOGY

This chapter presents a step-by-step method description of how the analysis is structured. The model is coded in Python, drawing inspiration from Prosumpy libraries [10]. The input data is simulated separately. The PV generation is modelled in the software PVSYST. The demand input data is modelled in RAMP [11]. Use patterns are developed for the cottages, that represent how often the cottage is used. Figure 1 shows the overall methodology.

PVSYST is used to model the **PV generation** of the base case where each PV system for each cottage is simply model with the optimal tilt and azimuth and no further information is added for the system. Then, PVSYST is used to make a detailed model of each cottage, with the actual roof tilt and azimuth and the actual placement of the modules at realistic available area. This could both mean placing the panels at several orientations (different parts of the roof) or even at the façade, as it is often the case for Norwegian mountain cottage. One reason for choosing the

10.4229/EUPVSEC2023/5DV.2.4
020520-001

façade is to avoid snow coverage during winter and spring.

To generate **demand profiles**, an open-source Python library called RAMP has been employed. This library constructs a stochastic model for generating high-resolution multi-energy profiles. It is designed for use in scenarios where only coarse information about user behavior is available. The load profiles are created by inputting electrical loads and **appliances** and their power consumption for each cabin or house. All the loads incorporated into the RAMP model are actual components commonly found in typical cabin supply stores. The usage for a component is specified with usage duration and the window(s) during which the load consumes power. Additionally, a random factor for usage duration and window(s) is applied to each load. The result from RAMP is an aggregated demand power profile in minute resolution. In the study demand profiles for a year have been simulated and converted to 15-minute time resolution.

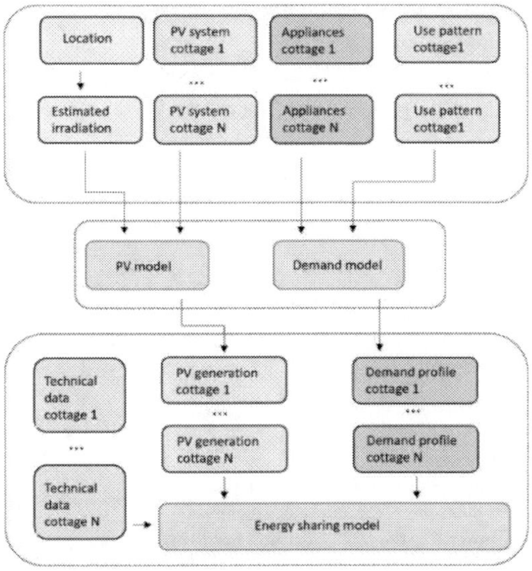

Figure 1: Overall methodology

The load profiles for each cabin are modified based on the assumed **use patterns**. In the model, three types of use patterns have been incorporated, and their explanations are listed in Table I. "Holiday use" implies that the cabin is in use during weeks 8, 24, 29, 40, half of 51, and 52. This corresponds to winter break, two weeks in the summer, autumn break, and December holiday.

Table I: Cottage use patterns

Number	Use patterns
1	Holiday use
2	Every second weekend (odd)
3	Every second weekend (even)

The various use patterns are established by setting the energy consumption to zero on days when the cabin is unoccupied. In this context, it is important to mention that this study assumes that the cabins can be "closed," meaning they do not require energy to keep the water from freezing when not in use.

In addition to PV generation and estimated demand, the energy sharing model incorporates **technical data** regarding the installation. These are outlined as follows:

- Max and min battery capacity.
- Max charge and discharge power of the battery.
- Max and min battery capacity for energy sharing.
- Max power of the inverter.
- Efficiency of the battery and inverter.

It is important to specify that each installation may not necessarily include the same parameters. In the model, it is possible to adjust certain aspects for individual cabins.

The **energy sharing model** is described in detail in [12]. It follows the basic principles presented in Figure 2.

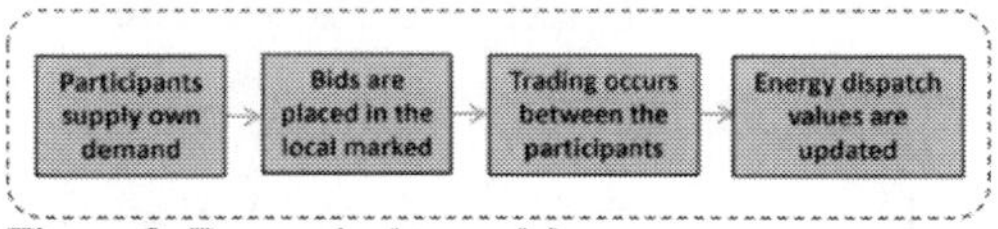

Figure 2: Energy sharing model

3 CASE STUDY

This chapter presents the case study of a real Norwegian cabin field providing the basis for the analysis. The simulation compares the modelling of the cabin field neighborhood with simple PV systems assumed at optimal angles to the same neighborhood with advanced PV system modelling. The analysis looks at the energy sharing in the neighborhood, and how the results differ between the two modelling approaches.

The cabin field neighborhood is located in Filefjell, Norway, with the location latitude of 61.1741 ° N and longitude 8.2077 ° E. The altitude is 1008 meters. It is assumed that every cabin has the same PV and battery off-grid system package installed. The used package is "Solcelle kraftpakke 230V Victron 4kW PowerWall 18,4kWt" which is commercially available in Norway. The key simulation data taken from this package are listed in Table II.

Table II: Key simulation data for off-grid system

Name	Number
PV panel size	410 W (mono crystalline)
Number of PV panels	6
Battery capacity	18,4 kWh (lithium)
Inverter max power	4000 W

Table III presents the PV system parameters that are used. The base case is the optimal tilt and azimuth for the area. This is often used when local energy communities are modelled to analyze the sharing potential. However, this is not reflecting the realistic situation for existing buildings, where the installation is limited to the roof angles and house facings.

Table III: PV system parameters

Case	Tilt	Azimuth
0 Base case	43	0
1 Façade SW	90	22
2 Two roofs	43 / 30	-85 / 23.5
3 Roof SW	43	12
4 Roof SE	30	-62

10.4229/EUPVSEC2023/5DV.2.4
020520-002

For each case except case 2 the panels are modeled with one orientation. In case 2, it is assumed that the 6 panels are divided into two equal strings and mounted on two different roof locations.

The demand profiles for the cottages follow two cottage categories. The modelled appliances that are basis for RAMP are listed in Table IV for each category. The time window and duration of use is the same for both categories and listed in Table V.

Table IV: Appliances in cottage category 1 and 2

Appliances	Power level category 1	Power level category 2
Phone	3 x 20 W	3 x 20 W
Light	3 x 3 W, 3 x 3.6 W	40 W, 6 x 28 W, 6 x 3.6 W
Heating	2 x 400 W	3 x 700 W, 400 W
Fridge	30.3 Wp	37.2 Wp
Cooking	1800 W	1800 W
Hot water	2000 W	2000 W
Pump	115 W	800 W
TV	17 W	29 W

Table V: Time window and duration

Appliances	Time window	Duration
Phone	8:30-10:00	39 min
Light	23:00-7:00	6 hours
Heating	18:00-8:00	7 hours
Fridge	Permanent	Cycle
Cooking	17:00-19:00	30 min
Hot water	9:00-10:00, 15:00-16:00, 19:00-20:00	43 min
Pump	9:00-10:00, 12:00-14:00, 17:00-19:00	2 hours
TV	8:30-9:00, 20:00-21:00	60 min

For the analysis the following two modelling cases are compared: **1) Simple PV modelling approach** (PV-Sim) and **2) Advanced PV modelling approach** (PV-Adv). The case study consists of four cabins forming a local energy community as visualized in Figure 3.

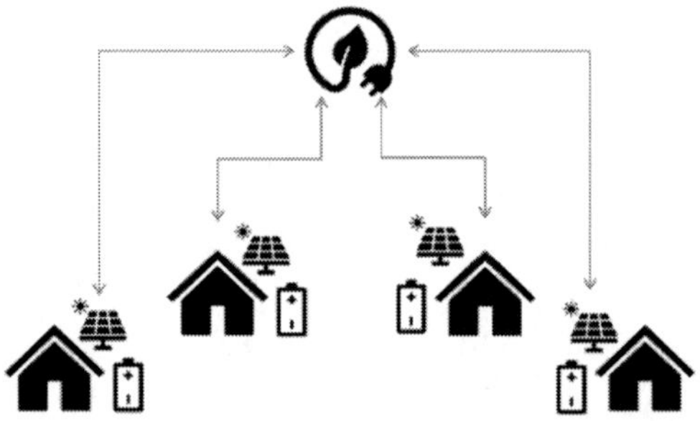

Figure 3: Off-grid local energy community

Table VI summarizes the input data for the two approaches. The input data includes the cottage category (1 or 2), the use pattern (1, 2 or 3) and the PV system case (0, 1, 2, 3 or 4). Additionally, we give the cottage a number to refer to later. The cottage category and use pattern are kept the same in Sim-PV and Adv-PV, while the PV system case is varied in this study.

Table VI: Modeling approaches

Input name	1) Sim-PV	2) Adv-PV
Cottage number	*1, 2, 3, 4*	*1, 2, 3, 4*
Cottage category	1, 1, 1, 2	1, 1, 1, 2
Use pattern	2, 3, 1, 1	2, 3, 1, 1
PV system case	0, 0, 0, 0	4, 3, 2, 1

Finally, the energy sharing limits for the battery are set to minimal 60% for selling and maximal 40% for buying.

4 RESULTS

This chapter presents the results obtained from two modeling approaches. Initially, we showcase the dispatch for cabin 2 and 4 during a representative week in Figures 4 to 7. Subsequently, we outline the essential result parameters in Tables VII to IX.

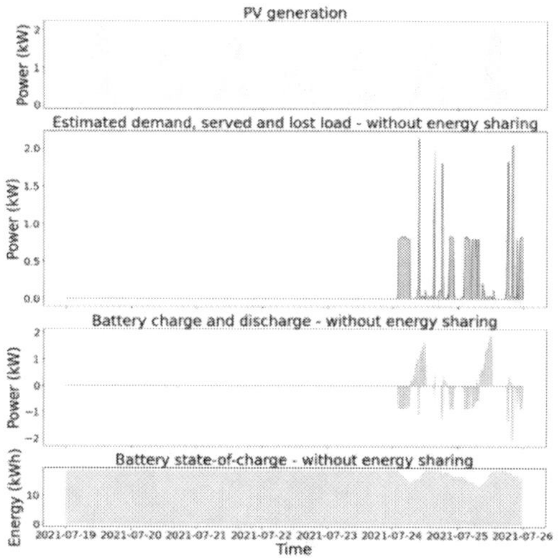

Figure 4: Dispatch – cabin 2 – without energy sharing

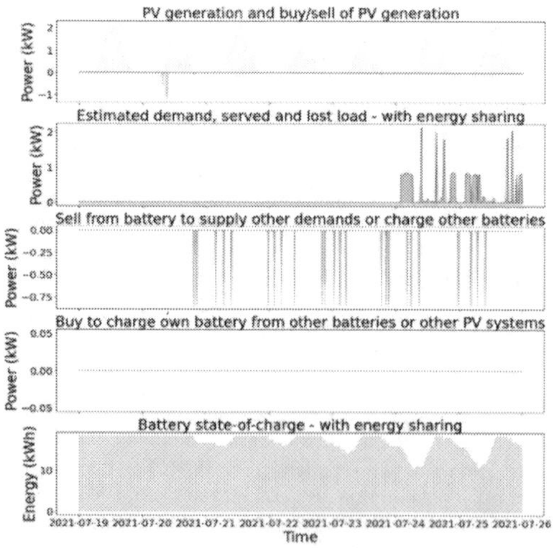

Figure 5: Dispatch – cabin 2 – with energy sharing

10.4229/EUPVSEC2023/5DV.2.4
020520-003

Figures 4 and 5 display the cabin 2 dispatch for week 29, both with and without energy sharing. The cabin is utilized over the weekend, ensuring an ample energy surplus available for sharing. Notably, Figure 5 illustrates the active sharing of this surplus energy with neighboring cabins. One can discern that the cabin directly shares surplus energy from the PV system, as depicted in red in the first subplot of Figure 5. Nevertheless, the sharing of energy stored in the battery occurs much more frequently than the direct sharing of surplus energy from the PV system. Additionally, one can see that the sharing of energy did not affect the coverage of the demand during the weekend where the cottage was used. The demand is still fully covered.

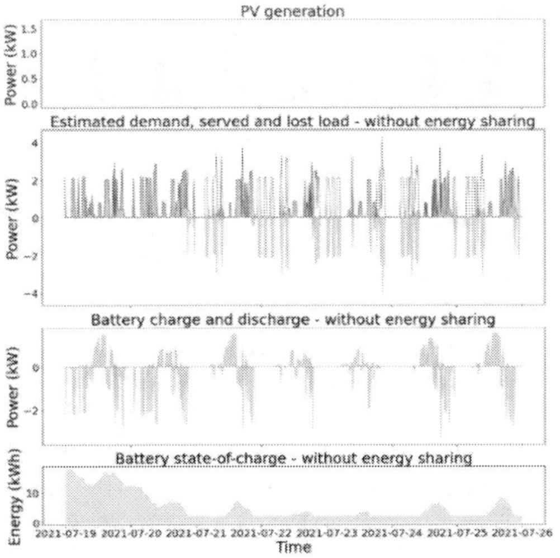

Figure 6: Dispatch – cabin 4 – without energy sharing

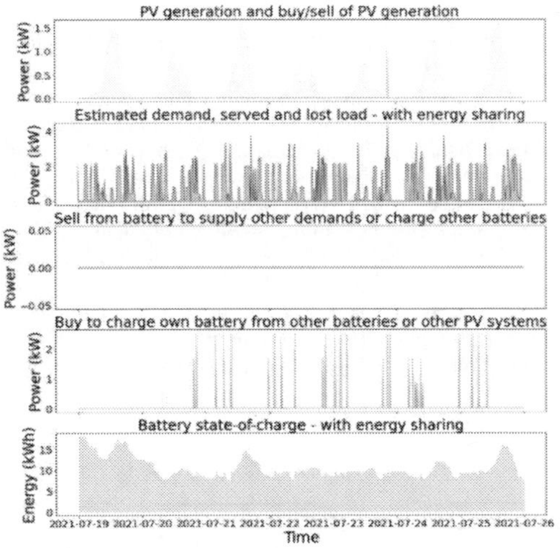

Figure 7: Dispatch – cabin 4 – with energy sharing

Figure 6 and 7 display the cabin 4 dispatch for week 29, both with and without energy sharing. The cabin is occupied for the whole week. It can be seen in figure 6 subplot 2 that there is a significant amount of lost load illustrated in red. Further in subplot 4 one can see the battery state-of-charge reaching its minimum. Figure 7 clearly illustrates the significant improvement that energy sharing offers for this cabin. The lost load, as depicted in subplot 2, has been entirely eliminated. Energy sharing occurs both by receiving direct PV surplus, illustrated in green in subplot 1 and in yellow in subplot 4. However, the primary source of energy acquisition occurs when the battery receives energy from other batteries to charge its own battery. This is due to the fact that lost load primarily occurs in the evenings and nights when there is no direct surplus energy available.

The following tables VII and VIII showcase improvements achievable through energy sharing, highlighting contrasting outcomes when employing both simple and advanced PV modelling approaches. They present a detailed breakdown of enhancements, encompassing both total and individual performance improvements (or deterioration) for each of the four cottages.

Table VII: Key results PV-Sim

Results for PV-Sim		**Without energy sharing**	**With energy sharing**
Total demand		1656.8 kWh	1656.8 kWh
Total lost load		511.7 kWh	371.5 kWh (-140.2 kWh)
Total surplus energy		6411.1 kWh	6223.9 kWh (-187.2 kWh)
Total served load ratio		69.1%	77.6% (+8.5%)
Individual served load ratios	1 2 3 4	85.2%, 86.3%, 74.2%, 45.1%	89.4% (+4.2%) 88.0% (+1.6%) 70.6% (-3.6%) 66.2% (+21%)

Table VIII: Key results PV-Adv

Results for PV-Adv		**Without energy sharing**	**With energy sharing**
Total demand		1656.8 kWh	1656.8 kWh
Total lost load		600.5 kWh	401.1 kWh (-199.4 kWh)
Total surplus energy		5547.2 kWh	5289.2 kWh (-258.0 kWh)
Total served load ratio		63.8%	75.8% (+12.0%)
Individual served load ratios	1 2 3 4	82.0%, 86.2%, 71.1%, 33.8%	88.1% (+6.1%) 87.6% (+1.4%) 69.7% (-1.4%) 62.8% (+29%)

Cabin 1, 2 and 4 experience an improvement of covering their demand, when participating in the local energy community. Especially, for cabin 4, the improvement is significant, increasing the served load ratio from 21 percentage points in PV-Sim and 29 percentage points in PV-Adv.

Cabin 3 encounters a decrease in its ability to cover its load when engaging in energy sharing. This arises from the sharing of stored energy from the battery before the cabin is utilized, resulting in an insufficiently charged battery at the beginning of the utilization period. Addressing this issue may involve implementing additional constraints, considering the known utilization schedule, and is an area of focus for further research.

In total the cabin field community has a potential of improving 8.5 or 12 percentage points depending on the modelling approach. This difference demonstrates the significance of advanced PV modelling.

When utilizing the advanced PV modelling approach, the cabins initially experience a less favourable load coverage. To discern whether the increase in improvement arises from a reduction in PV yield or the diversification of PV azimuth and tilt, we fine-tuned each system's PV yield to the optimal level while retaining the advanced orientations. This is done by dividing the energy output in every time step by the total yield of the year from that orientation (PV system case 1-4) and then multiplying it with the yield from the optimal orientation (PV system case 0). The subsequent table IX presents results of this analysis.

Table IX: Results PV-Adv (adj) – adjusted PV yield

Results for PV-Adv (adj)		**Without energy sharing**	**With energy sharing**
Total served load ratio		67.8%	77.4% (+9.6%)
Individual served load ratios	1	82.5%,	88.8% (+6.3%)
	2	86.5%,	88.1% (+1.8%)
	3	72.5%,	70.3% (-2.2%)
	4	44.0%	66.1% (+22%)

It can be observed that the total improvement is reduced, however, it remains higher than with the simple PV modelling approach. Thus, it is possible to assume that the improvement from 8.5 to 9.6 percentage points comes from the increase in diversification of PV orientations, while the step up from 9.6 to 12 percentage points enhancement is due to the increase in total PV yield differences between the cabins.

5 CONCLUSIONS

The advanced modelling approach demonstrated a more substantial advantage in harnessing the benefits of energy sharing. In the case study, the simple model exhibited an 8.5% boost in load coverage, whereas the advanced approach exhibited a more impressive 12% increase. The primary factor contributing to this enhancement was the initially reduced PV yield for the cabins when utilizing the advanced modelling approach. In contrast, the impact of azimuth and tilt diversity was relatively minimal in comparison.

Further work will focus on more comprehensive PV modelling and its effect on the outcome for local energy communities and swarm electrification, taking both grid-connected and off-grid settings into account. The comprehensive PV modeling will encompass factors such as soiling effects, snow coverage, and nearby shadows.

6 REFERENCES

[1] Statistics Norway. «Hytter og fritidsboliger». [Internet]. Available: https://www.ssb.no/bygg-bolig-og-eiendom/faktaside/hytter-og-ferieboliger. [Accessed 1 February 2023].

[2] Statistics Norway. «Elektrisitet». [Internet]. Available: https://www.ssb.no/statbank/table/10314/. [Accessed 1 February 2023].

[3] S. Groh, D. Philipp, B. E. Lasch, H. Kirchhoff. «Swarm electrification-suggesting a paradigm change through building microgrids bottom-up». Proceedings of 3rd International Conference on the Developments in Renewable Energy Technology. 2014.

[4] S. Sheridan, S. Sheridan, J. Courtney. «Swarm electrification: A comprehensive literature review». Renewable and Sustainable Energy Reviews, vol. 175. 2023.

[5] H. Kirchhoff, K. Strunz. «Key drivers for successful development of peer-to-peer microgrids for swarm electrification». Applied Energy. vol. 244, pp. 46-62. 2019.

[6] L. Richard, C. Boudinet, S. A. Ranaivoson, J. O. Rabarivao, A. E. Befeno, D. Frey, M. C. Alvarez-Hérault, B. Raison, N. Saincy. «Development of a DC Microgrid with Decentralized Production and Storage: From the Lab to Field Deployment in Rural Africa». Energies. vol. 15, nr. 18. 2022.

[7] R. Rana, K. Berg, M.Z. Degefa, M. Löschenbrand, Modelling and Simulation Approaches for Local Energy Community Integrated Distribution Networks, IEEE Access, 2022

[8] A. Hernandez-Matheus, M. Löschenbrand, K. Berg, I. Fuchs, M. Aragüés-Peñalba, E. Bullich-Massagué, A. Sumper, A systematic review of machine learning techniques related to local energy communities. Renewable and Sustainable Energy Reviews. 2022.

[9] I. Fuchs, A. M. Waitz, K. T. Lervik, S. Völler, O. Midtgård, J. Rajasekharan. Optimal Design and Installation of Solar Home Systems for Bottom-up Grids: Comparative Case Studies from Kenya and Norway. 8th World Conference on Photovoltaic Energy Conversion, pp. 1453–1458. 2022.

[10] S. Quoilin, K. Kavvadias, A.Mercier, I. Pappone and A. Zucker. "Quantifying self-consumption linked to solar home battery systems: Statistical analysis and economic assessment". Applied Energy. vol. 182. pp. 58–67. 2016.

[11] F. Lombardi, F. Lombardi, S. Quoilin, E. Colombo, «Generating high-resolution multi-energy load profiles for remote areas with an open-source stochastic model». Energy. vol. 177, pp. 433-444. 2019.

[12] Tørrisplass, Halgrim Ove, Etablere en energidelingsmodell som simulerer lønnsomhet for energideling i et off-grid hyttefeltet. NTNU Master thesis, 2023.

10.4229/EUPVSEC2023/5DV.2.4
020520-005

FLEXIBILITY IN SOLAR AND BATTERY OFF-GRID SYSTEMS – CASE STUDY ECO MOYO EDUCATION CENTRE IN KENYA

Rikke Enger Dihle, Marie Bakken, Ida Fuchs
Norwegian University of Science and Technology / Department of Electric Energy
O. S. Bragstads plass 2E, 7034 Trondheim, Norway, +47 73559775, ida.fuchs@ntnu.no

ABSTRACT: Accelerating rural electrification is crucial to achieve Sustainable Development Goal number 7 by 2030. While progress has been made in reducing the number of people without electricity, the COVID-19 pandemic and increased costs pose challenges. Kenya has made significant progress toward universal energy access, but rural areas still lag behind. Eco Moyo Education Centre is a Norwegian/Kenyan charity project offering free primary education to underprivileged children in the Dzunguni village in Kenya. This case study of Eco Moyo Education Centre highlights the potential of flexible solar and battery off-grid systems to optimize energy supply in rural areas. These systems are often based on renewable energy sources, such as solar power, and battery energy storage systems are typically used to compensate for periods with limited power generation. User behavior patterns are the primary tool to increase flexibility in off-grid systems. This paper presents an optimization model for scheduling flexible loads and a user-friendly early-stage Graphical User Interface (GUI) draft for end user communication. By minimizing the disutility cost of shifting loads, the potential of demand response in the shape of load shifting is investigated in the Eco Moyo case study. The results show a good performance of the flexibility model and an improvement in key performance parameters for the off-grid system, when applying demand response. The developed GUI shows how to incentivizes this user flexibility and increases the end user knowledge of the system.
Keywords: Solar, Battery, Off-Grid, Flexibility, Optimization

1 INTRODUCTION

To achieve SDG number 7 of universal access to affordable, reliable, sustainable, and modern energy by 2030, the pace of rural electrification needs to be accelerated [1]. The number of people without electricity dropped from 1.2 billion in 2010 to 733 million by 2020, and the majority of these people live in rural areas in Sub-Saharan Africa. However, due to the COVID-19 pandemic, the pace of progress is decreased, and the increased price of energy and shipping has caused higher costs of transport and production of renewable energy resources, such as PV modules and wind turbines. Hence, a major mobilization of private and public capital is required to reach the people living in the least developed countries to achieve global energy access [2].

The International Energy Agency (IEA) states that demand-side activities should be prioritized when it comes to creating more reliable and sustainable energy systems. To include demand side activities with smart grid technologies are beneficial for both stakeholders in the energy value chain and the society as a whole [3]. Based on a case study of Eco Moyo Education Centre, Kenya, it is possible to discuss how flexibility in solar and battery off-grid systems can be utilized to optimize the energy supply in rural areas. Off-grid systems have proved to be challenging due to high service costs and low affordability [3]. The possibility to flexibly operate loads enhances the utilization of the installed capacity. Hence, flexibility in off-grid systems is essential to secure reliable energy access in rural areas.

This paper proposes an optimization model to minimize the disutility cost of load shifting to align the consumption with the available power generation. An early-stage draft of a Graphical User Interface (GUI) is developed to communicate optimal behavior to the end user, and help the user flexibly operate loads and enable them to make the most efficient use of their limited energy resources. Thus, the research question for the analysis is: **How can enhancing flexibility and implementing intelligent communication with end-users in off-grid solar and battery systems contribute to optimizing resource utilization and maximizing energy output from the system?**

The rest of the paper is organized in the following way: Section 2 provides background information and key data on the case study. Section 3 explains the methodology used. Finally, section 4 presents the results of this paper followed by a conclusion in section 5.

2 CASE STUDY

Eco Moyo Education Centre is a Norwegian/Kenyan charity project that, as of 2022, provides free primary education to 240 underprivileged children. The school is located in the Dzunguni village, close to Kilifi on the Kenyan coast. Primarily, the village's residents live in mud houses without electricity or running water [4]. Figure 1 gives an overview of the main school area and relevant buildings.

Figure 1: Overview of The Staff Room and Classrooms at Eco Moyo.

2.1 Overview of relevant buildings at Eco Moyo

The **Staff Room** at Eco Moyo is located close to the classrooms and is frequently used by the teachers during the day. There are currently 12 teachers working at Eco Moyo, and the Staff Room is used as an office during the day, where the teachers charge their computers and phones. During the evening, the building is used as a hangout area by the six teachers living at the property.

The **Staff Court** accommodates the teachers living on the property. It is located close to the Staff Room and consists of six simple bedrooms and an outside hangout area. Currently, the only electricity supply in the building is Solar Home Systems (SHSs) providing lighting. However, by installing a cable from the Staff Room to the Staff Court, it would be possible to provide electricity in this building as well. Conversations with the founder during the field trip in November 2022 revealed that it is desired to increase the standard of living for the teachers. Increasing the living standard at the Staff Court will help increase the quality of education at Eco Moyo.

The **Office Building** is located straight across from the Staff Room and contains two offices. There is no electricity supply in the building, but due to the short distance from the Staff Room, an extension cord is sometimes used to power a printer. However, this solution is not sustainable due to the fire risk.

2.2 Existing solar and battery off-grid system

In March 2022, the school received a new solar and battery off-grid system to meet the school's basic electricity needs. The microgrid consists of solar panels, batteries, and an inverter for DC to AC power conversion. It powers various appliances such as lights, PCs, tablets, phones, a printer, a fan, a TV, and smaller units in the Staff Room. The system parameters are presented in Table I. The tilt angle and azimuth of the PV panels are 6.2° and -50°, respectively. Table I presents the component parameters of the off-grid system.

Table I: Components of the off-grid PV system at Eco Moyo Education Centre.

Description	Qty.	Capacity
PV Panels	7	450 Wp
Hybrid PV Off-Grid Inverter	1	3.50 kW
Tubular Gel Battery	4	200 Ah / 12.0 V

The existing off-grid system has a significant amount of surplus energy, which is due to the fact that it was design to supply the critical loads for the school also during the raining season and days with low PV generation. The electricity supply in the other buildings at the school is limited to SHSs for lighting. The school aims to improve the electricity access for additional buildings, especially the Staff Court and Office Building. However, the costs of a new off-grid system were too high, which inspired the search for alternative solutions, looking into flexibility as an enabler for higher access to energy and better utilization of the existing system.

2.3 Analysed scenario for Eco Moyo

A scenario has been simulated to investigate whether the inclusion of flexibility and optimization of load profiles can enable the successful connection of the Office Building and Staff Court Building to the same off-grid system as the Staff Room currently installed at Eco Moyo. This simulation considers the existing system but incorporate additional loads in both the Office Building and Staff Court. The input parameters are derived from observations made during two field trips conducted in November 2022 and May 2023. Table II lists the flexible appliances with their respective disutility value that is used for the optimization. Further none-shiftable loads are fridges, indoor and outdoor lights included in the simulation.

Table II: Appliance rating and disutility value.

Flexible appliance	Rating [W]	Disutility value
Phone	7	100
TV	70	90
Laptop	60	85
Tablet	15	65
Iron	1200	60
Speaker	100	50
Kettle	2200	30
Radio	36	25
Printer	200	20
Fan	50	15
Drill	500	10

3 METHODOLOGIES

To investigate flexibility in off-grid solar and battery systems, an optimization model to minimize the disutility cost of shifting loads is created in Pyomo. The model is based on system parameters, load profiles, and solar production. Modeling of solar production is done using pvlib, an open-source tool in Python with weather data from PVGIS, and the load profiles are simulated using the RAMP model. Further, the simulated load profiles are modified to reflect the simplifications in the optimization model. The resulting load profiles from the optimization model are used to simulate the system dispatch in Prosumpy. Finally, an early-stage GUI is developed to access real-time data from the inverter, propose optimal consumption behavior, and present theoretical trends from the optimization model. Figure 2 illustrates the different stages of the modeling.

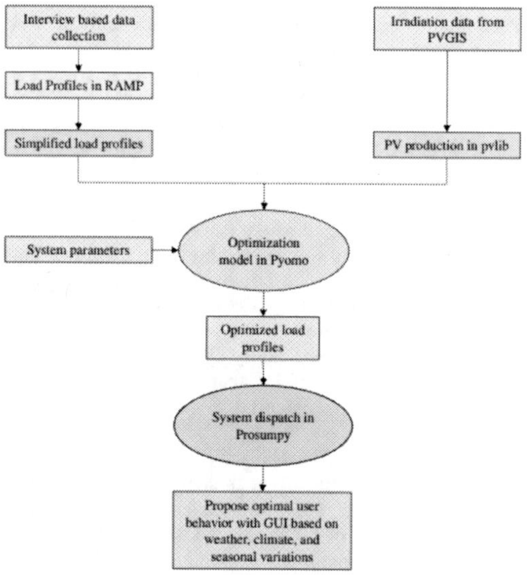

Figure 2: Flow chart of methodology and simulations.

10.4229/EUPVSEC2023/5DV.2.5
020521-002

3.1 PV Production in pvlib python

PV production in the system is modeled using the open-source tool pvlib python. The tool is developed on GitHub with contributors from academia, national laboratories, and private industry. It provides core functions based on algorithms from peer-reviewed publications [5]. Multiple solar resource datasets can be included in pvlib. In this project, the dataset PVGIS-SARAH2 is used. It is a solar radiation dataset from the free PVGIS database. The data is based on geostationary satellites covering Europe, Africa, and Asia. The data resolution is hourly, and the dataset includes radiation data, sun height, air temperature, and wind speed. To create a representative dataset, the typical meteorological year for the given geographical location of the system is utilized.

Core functions in pvlib are used to calculate the DC output from the PV system. The Fairman Model is used to calculate the temperature of the PV module. The model calculated the temperature based on the total incident irradiance, air temperature, and wind speed, using an empirical heat loss factor. Finally, the DC output power are calculated based on the cell temperature T_{cell}, transmitted plane array irradiance I_{tr} and rated power P_{dc0}, as shown in the equation below. The parameters I and T represent the reference irradiance and temperature, set to $1000\ \frac{W}{m^2}$ and 25°C, respectively. γ_{pdc} is the temperature coeffcient and is used to model how the eciency of the PV module decreases at a linear rate of the temperature rise.

$$P_{dc} = \frac{I_{tr}}{I_{ref}} P_{dc0} \left(1 + \gamma_{pdc}(T_{cell} - T_{ref})\right)$$

3.2 Load Profiles in RAMP

Load profiles for the different appliances are modeled using the open-source stochastic Remote-Areas Multi-Energy Systems Load Profiles (RAMP) model. The model is implemented in Python and designed to generate load profiles in remote areas. It consists of three main layers: User Type, User, and Appliance. Due to the stochastic model, each User has a unique load profile consisting of independent Appliances with different behaviors [6]. The generated load profiles will differ from each model run and represent the unpredictable behavior of the users.

Table II provides a short description of the input parameters used for the appliances when generating load profiles with the RAMP model. The appliance belongs to a User. Each appliance is also associated with a time window where it is likely to be switched on during the day. To be used as an input in the optimization model, the load profiles are modified to match the simplifications. To identify the specific behavior of each load, a unique load profile is generated for each appliance. Each appliance is also only used once during the day and always as long as the defined user time *total use*. Hence, the stochastic parameter *rt*, and the total user time are not subject to random variability. The parameter function time is also set to be equal to the user time, as the appliance always will be switched on for the duration of the user time. The parameter frequency describes the occurrence of an appliance during a week. The frequency of an appliance is given by the number of days during the week divided by five if the device is only used during the weekdays and by seven otherwise.

An important feature in the RAMP model is the attribute to model duty cycles, which makes it possible to simulate the behavior of a fridge [6]. The operation cycle of the fridge depends on several factors, like size, efficiency, outside temperature, isolation, and user patterns. The warming process during the standby mode will be sped up if the fridge is opened regularly or the isolation is insufficient. To include the variations in the duty cycles the RAMP model includes the attribute to model different cycles that can be associated with different time frames during the day [6]. The duty cycles are categorized into standard, intermediate, and intensive, depending on user behavior and ambient temperature.

3.3 Formulation of the optimization problem

The optimization model is created using the Pyomo software. It is a Python-based, open-source optimization modeling language [7]. The goal is to decide the optimal scheduling of flexible loads in off-grid solar and battery microgrids, by minimizing the disutility cost of shifting loads. It is formulated as a Mixed-integer linear problem (MILP). Load profiles for all the appliances generated in RAMP and PV production from pvlib are used as input to the model. Appliances are classified into two categories: shiftable loads and non-shiftable loads. The shiftable loads are deferrable and can be rescheduled to reduce the peak demand. These loads are typically charging of appliances. Non-shiftable loads cannot be modified and are typically lighting and cooling. The mathematical formulation of the optimization model is presented in this section.

$$\underset{\tau_{a,t}{}^{move},\ res_{pv_t}}{\text{minimize}} \sum_{t=0}^{T} \sum_{a=0}^{A} X_a \cdot \tau_{a,t}{}^{move} + \sum_{t=0}^{T} res_{pv_t}$$

The Objective Function that minimizes the total disutility cost of shifting loads, in addition to the residual PV production is formulated above. X_a is the disutility cost of moving the appliance with index a one timestep. $\tau_{a,t}{}^{move}$ is a decision variable and describes how many time steps appliance a has been moved. Hence, the disutility function is assumed to be linear and increasing in the curtailment time. The residual PV production is minimized to ensure that the system dispatch corresponds to the Prosumpy model, where the dispatch of the system is performed in such a way that the self-consumption is maximized.

$$\sum_{i=0}^{T-\tau_a} \delta_{a,i} - 1 = 0, \qquad a \in A$$

The first five constraints of the optimization model consider the shiftable load units. $\delta_{a,t}$ is a binary variable that gets the value 1 in time step t when appliance a is switched on. The constraint defined in the equation above ensures that each load only starts once during the modeled time horizon. The limit $T - \tau_a$ is set so that the load always will be able to run for the whole user time.

$$L_{a,t} - \sum_{i=0}^{t} P_a \cdot \delta_{a,i} \cdot n_a = 0, \qquad t = 0, \ldots, \tau_a - 2, \quad a \in A$$

$$L_{a,t} - \sum_{i=t-\tau_a+1}^{t} P_a \cdot \delta_{a,i} \cdot n_a = 0, \qquad t = \tau_a - 1, \ldots, T, \quad a \in A$$

10.4229/EUPVSEC2023/5DV.2.5
020521-003

To ensure that the optimized load profile for appliance a in time step t is set to the rated power of the appliance when it is turned on, the two constraints listed above are defined. It also ensures that the appliance is switched on from the starting time when the binary variable $\delta_{a,t}$ is set to 1, and for the whole user time.

$$\tau_{a,t}^{move} - \delta_{a,t} \cdot |\tau_a^{start} - t| \cdot n_a = 0, \qquad t \in T, \; a \in A$$

$$\sum_{i=0}^{A}\sum_{j=0}^{T} D_{i,j} - \sum_{k=0}^{A}\sum_{l=0}^{T} L_{k,l} = 0$$

$\tau_{a,t}^{move}$ is a decision variable based on $\delta_{a,t}$, and the initial start time of appliance a. Note that the variable is assigned a value only during the time step when the appliance a is switched on, as shown in the equation above. For all other time steps, it is assigned the value 0. It also considers the quantity of appliance a. The next equation ensures that the shiftable load is moved and never cut. After optimizing, the total shiftable load is equal to the initial shiftable demand.

$$inv2load_t - (store2inv_t + pv2inv_t) \cdot \eta_{inv} = 0, \quad t \in T$$

$$pv2store_t + pv2inv_t + res_{pv_t} - pv_t = 0, \qquad t \in T$$

$$inv2load_t - (\beta_t + \sum_{i=0}^{A} L_{i,t}) \cdot k = 0, \qquad t \in T$$

The three constraints listed above ensures that the power input and output of the internal systems are balanced. They also consider the efficiency of the inverter. k is a conversion parameter that converts the values from W to kWh.

$$inv2load_t - E_{max}{}^{inv} \cdot \eta_{inv} \leq 0, \qquad t \in T$$

$$pv2store_t - E_{max}{}^{bat} \leq 0, \qquad t \in T$$

The constraints listed above are the capacity constraints of the model. They ensure that the energy flows in the system are within the capacity of the components. The output of the inverter is limited by the rated maximum output, and the input to the batteries from the PV module is limited by the capacity of the batteries.

$$SOC_t - SOC_{start} = 0, \qquad t = 0$$

$$SOC_t - SOC_{t-1} - pv2store \cdot \eta_{bat} + store2inv = 0, \quad t \in T$$

$$SOC_t - E_{max}{}^{bat} \leq 0, \qquad t = 0, \ldots, T-1$$

$$SOC_t - 0.2 \cdot E_{max}{}^{bat} \geq 0, \qquad t = T$$

The initial SoC of the battery in time step t = 0 is given by the parameter SOC_{start}, as formulated in the first constraint listed above. The second constraint describes how the current SoC is given by the SoC in the previous time step, and the input from the PV module, and batteries, in the other time steps. The third constraint limits the maximum SoC of the battery by the maximum capacity. Further, the last constraint ensures that the SoC of the battery is greater than, or equal to 20% at the end of the time horizon of the optimization. It is added in order to prevent insufficient battery capacity for the next day.

3.4 System Dispatch in Prosumpy

The system's dispatch is simulated in Prosumpy [8], with the initial SoC as an input parameter. Prosumpy is an open-source code with the purpose of simulating PV- and battery grid-connected systems. Hence, the model is adjusted to an off-grid system by changing variables related to importation and exportation from the grid. The simulation is used to analyze the annual dispatch, to investigate the total covered demand, uncovered demand, energy surplus, amount of energy provided by the battery, and the average DoD before and after optimization. If the optimization model fails to fulfill the constraints for a day, the initial load profile will be considered when analyzing the system dispatch after optimization.

3.5 Development of the Graphical User Interface

To incentivize user flexibility, the surplus energy of the system has to be visualized in an easily understandable way in a GUI. Usually, inverter displays are hard to understand for non-technical staff. Therefore, the interface will include several features to increase the consumers' understanding of the energy available and suggest flexible behavior. The GUI is a first-stage proposal developed in Python. An essential part of the development is optimizing the platform's functionalities based on feedback from the end users. The development of the GUI is divided into two parts: the backend and the frontend of the interface. Automatic download of the real-time data is included in the backend and the design of the platform represents the frontend. It is essential to acknowledge that the GUI is currently an initial draft, and future improvements are necessary before the interface is usable for the end users.

Using Selenium, the script of the GUI automatically signs in and finds the correct elements for downloading real-time data. The overall design of the GUI is developed in Tkinter, a standard interface package in Python. The provided design is visualized in Figure 3 and offer suggestions for different functionalities that can be incorporated into a GUI designed for off-grid systems in rural areas.

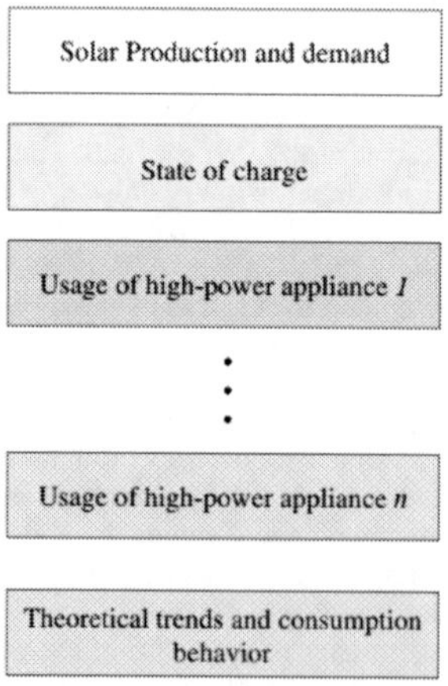

Figure 3: Different functionalities in the GUI draft.

10.4229/EUPVSEC2023/5DV.2.5
020521-004

4 RESULTS

The results before and after optimization for the simulated scenario are presented for July 7th in Table III and Figure 4 and 5. Appliances are shifted to cover the demand throughout the day as well as fulfill the constraints of the optimization model.

Table III: Initial and new time windows for July 7th.

App.	Building	Initial window	New window
Fan	Staff Room	12:15 – 18:15	11:30 – 17:30
Fan	Staff Room	12:45 – 18:45	11:30 – 17:30
Kettle	Staff Room	16:15 – 16:30	12:45 – 13:00
Drill	Staff Room	07:30 – 09:30	07:45 – 09:45
Fan	Office B.	10.45 – 16:45	09:30 – 15:30
Fan	Office B.	11:15 – 17:15	13:30 – 19:30
Fan	Staff Court	16:15 – 22:15	11:30 – 17:30
Fan	Staff Court	16:15 – 22:15	13:30 – 19:30

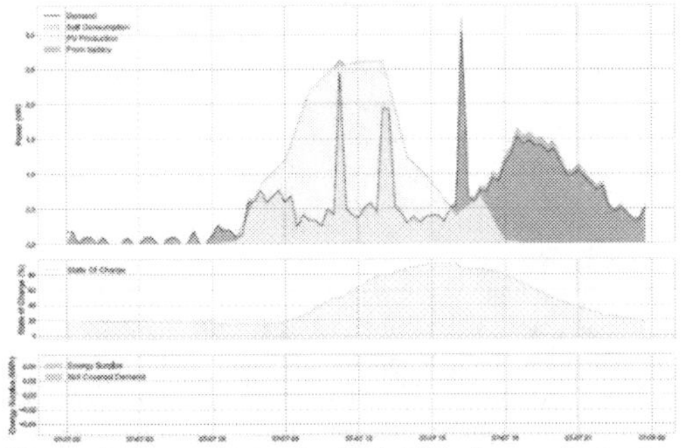

Figure 4: System dispatch of July 7th before optimization.

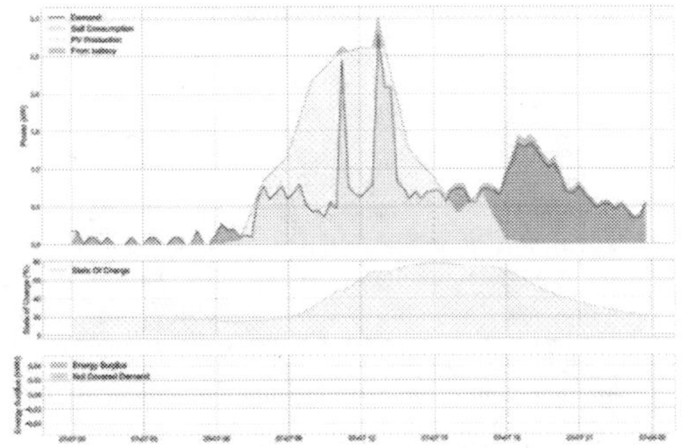

Figure 5: System dispatch of July 7th after optimization.

The example of July 7th shows clearly how a peak load is moved from the afternoon with little PV generation to midday with sufficient PV generation. In this case the peak load is a kettle, and the use could be making hot water for coffee or tea in the afternoon. In this case the hot water should be prepared earlier and stored in a thermos. This is an example of how the results must be further interpreted from a social and behavioural perspective.

Another example is showcased in Figure 6 and 7 where the dispatch of October 20th is presented. This time lost load due to insufficient power capacity occurs due to the use of two power intensive tools (second power peak in Figure 6). The optimization moves one of those peaks just a little and solves that problem as Figure 7 shows.

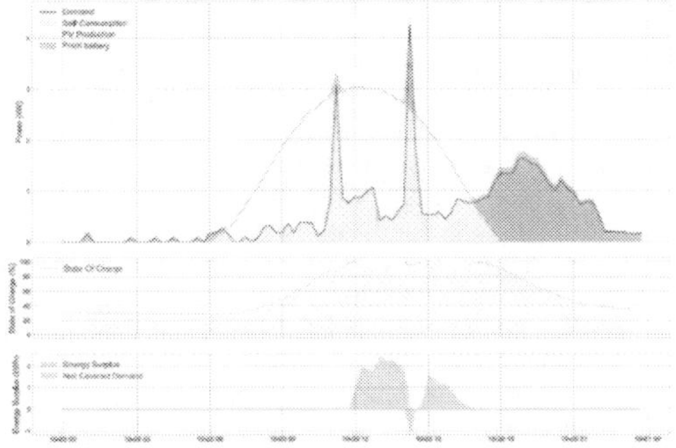

Figure 6: System dispatch of Oct 20th before optimization.

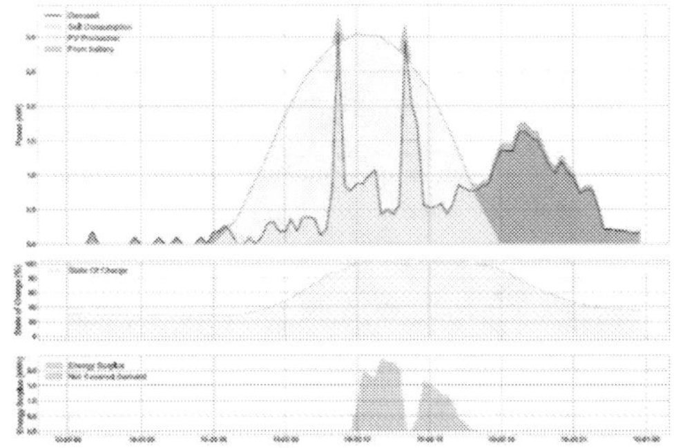

Figure 7: System dispatch of Oct 20th after optimization.

Finally, table IV presents the key improvement parameters for the analysed scenario of adding the Staff Court and Office Building to the existing off-grid system and use flexibility to use the resource efficiently. The numbers show a relatively low amount of improvement in terms of reduced lost load. However, a significant load shifting had to be implemented to achieve this result.

One the main reasons for the low increase in covered demand is that the optimization model only regards flexibility for one day but does not include energy savings from one day to another. This is subject for further work and could lead to an increased number of improvements.

Another reason for the low total improvement is the fact of a low number of total appliances to shift. With larger systems and more appliances, this would increase accordingly.

Finally, it needs to point out that the original off-grid system already had a large amount of surplus energy, and the total lost load, even with the two new buildings included is low with only 1.8%. Nevertheless, the optimization model indicated some efficiency improvements, suggesting that the potential for improvement could be significant in a system with poorer performance.

Table IV: Key improvement parameters.

Parameter	Before	After
Total covered demand	3842 kWh	3844 kWh
Total lost load	73 kWh	71 kWh
Total shifted load	-	20 kWh
Energy served by battery	2200 kWh	2196 kWh

10.4229/EUPVSEC2023/5DV.2.5
020521-005

The design proposal of the GUI is presented in Figure 8 and Figure 9. A daily load profile with additional appliances from RAMP and the modelled PV production is used as input to the GUI to describe the importance of flexibility in the extended system. The user-friendly presentation of daily solar production, demand, and usage of the different appliances, both in percentage and graphically, enhances the end user's knowledge of the system. The possibility of observing energy trends of the period and comparing them to the current day allows the users to stay informed about their energy consumption. With the manual load shifting provided by the GUI when the capacity is at its limit or the SoC are low, the flexibility in the system is increased. The GUI plays a crucial role in preventing users from exceeding the system's limit. Without a GUI, the users would lack the necessary information to determine if it is safe to connect the appliances.

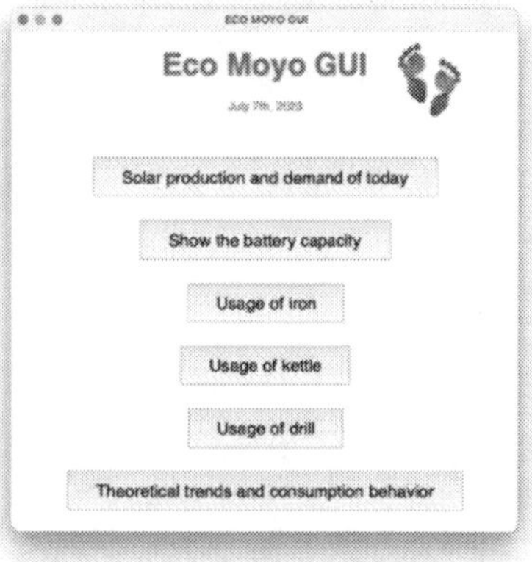

Figure 8: Design proposal of the GUI.

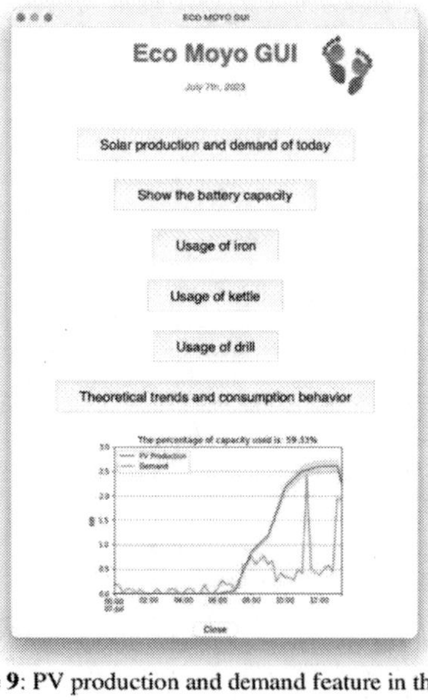

Figure 9: PV production and demand feature in the GUI.

5 CONCLUSIONS

This study has examined the flexibility of the rural off-grid system at Eco Moyo by developing an optimization model and a GUI for end-user communication. The following conclusions can be made:

- Optimal scheduling of flexible appliances enhances the efficiency of the existing system. By load shifting and adapting to daily variations, the system can supply additional buildings with a better performance.
- The GUI helps with end-user communication to incentivize such flexibility. Providing valuable system parameters in an easy and understandable way, including short-term weather forecasts, will give the users greater awareness of their energy usage.

The results show that flexibility enables users to make the most efficient use of their limited energy resources. As the developed model is easily adjustable, it can also be used to investigate flexibility in similar systems.

Further work will look into flexibility with an increased time frame (several days, seasonal), to analyze if a further improvement of key parameters can be achieved.

5 REFERENCES

[1] United Nations, "Transforming our world: The 2030 agenda for sustainable development". United Nations. Tech. Rep. 2015.

[2] United Nations. Department of Economic and Social Affairs (DESA). "The Sustainable Development Goals Report 2022". Tech. Rep. 2022.

[3] S. O. Ottesen and A. Tomasgard. "A stochastic model for scheduling energy flexibility in buildings". Energy. vol. 88, pp. 364–376. 2015.

[4] A unique school in rural Kenya, 2022. [Online]. Available: https://www.ecomoyo.com/students.

[5] W. F. Holmgren, C. W. Hansen and M. A. Mikofski. "Pvlib python: A python package for modeling solar energy systems", Journal of Open-Source Software. vol. 3. no. 29. p. 884. 2018.

[6] F. Lombardi, S. Balderrama, S. Quoilin and E. Colombo. "Generating high-resolution multi-energy load profiles for remote areas with an open-source stochastic model". Energy. vol. 177. pp. 433–444. 2019.

[7] M. L. Bynum, G. A. Hackebeil, W. E. Hart et al. "Pyomo - Optimization Modeling in Python". Third Edition. Springer Optimization and Its Applications.

[8] S. Quoilin, K. Kavvadias, A.Mercier, I. Pappone and A. Zucker. "Quantifying self-consumption linked to solar home battery systems: Statistical analysis and economic assessment". Applied Energy. vol. 182. pp. 58–67. 2016.

10.4229/EUPVSEC2023/5DV.2.5
020521-006

COMPARISON BETWEEN DECENTRALIZED AND CENTRALIZED ENERGY STORAGE SYSTEMS FOR POWER RAMP RATE CONTROL OF PV SYSTEMS

Kari Lappalainen[1] and Jan Kleissl[2]
[1]Tampere University, Electrical Engineering, P.O. Box 692, FI-33101 Tampere, Finland
[2]University of California, San Diego, Center for Energy Research and Department of Mechanical and Aerospace Engineering, CA 92093-0411, United States
kari.lappalainen@tuni.fi, jkleissl@ucsd.edu

ABSTRACT: The present pace of photovoltaic (PV) system installations is leading to a huge need for energy storage systems (ESS) to smooth fluctuating PV production. In this paper, a systematic comparison between decentralized and centralized ESSs for power ramp rate (RR) control of PV systems is presented based on inverter level power measurements of a large-scale PV plant. The result show that the relative ESS power capacity needed to smooth all PV power fluctuations decreases with increasing PV system size. The reason for this is that the required power capacity results from the fastest power fluctuations which smoothen with increasing PV system land area. This means that the ESS power requirements of large-scale PV plants can be minimized by using a single ESS for RR control of the entire PV plant instead of distributing the ESSs for single sections within the power plant for individual inverters. Differences in ESS energy requirements and amount of energy cycled through the ESSs between the decentralized and centralized ESS setups were found to be small.
Keywords: Photovoltaic power, Energy storage system, Ramp rate control, Energy storage sizing, Power smoothing

1 INTRODUCTION

Photovoltaic (PV) power production is highly variable due to fast irradiance fluctuations resulting mainly from overpassing cloud shadows [1]. Thus, limits have been set for power ramp rates (RR) of grid-connected PV power plants. Compliance with these RR requirements is typically achieved by power output curtailment or energy storage systems (ESS). However, the use of power curtailment is limited to upward power ramps. Thus, the need for ESSs to smooth fluctuating PV production is increasing with strongly increasing installed PV capacity. During the last years, several ESS control strategies have been proposed for PV power smoothing [2]. These strategies are typically classified into RR-based methods, filter-based methods, moving average and exponential smoothing based methods, and model predictive control methods [3, 4].

Power fluctuations of PV systems are known to become smoother with increasing land area [5] and power fluctuations of fleets of several PV systems to smoothen with geographical dispersion of the systems, i.e., with increasing distance between the systems [6, 7]. These findings indicate that ESS requirements for PV RR control, with respect to installed PV power, decrease with increasing size and dispersion of PV systems. However, the effects of size and dispersion of PV systems on the ESS requirements for PV RR control have not received much attention. The required ESS energy and power per unit installed PV have been found to decrease relatively with increasing PV system size [8]. For fleets of several PV systems, the required ESS energy and power have been found to decrease relatively with increasing number and distance of the PV systems [9]. However, in [8], only one RR limit (RRL) was examined, and the sampling frequency of the used data (0.2 Hz) was not ideal for PV fluctuation studies. Sizing of ESSs for ramp rate control of PV strings was studied in [10] based on 1 Hz current–voltage curve measurements considering the effects of the applied RRL and inverter sizing which were found to be important factors for sizing the ESSs. Effects of RRL on the sizing of ESSs for PV, wind, and hybrid PV–wind power plants were studied in [11]. In [12], equations by which the required energy capacity of an ESS controlled with three different strategies can be estimated were presented as a function of PV system size and RRL.

In this paper, use of decentralized and centralized ESSs for power RR control of PV systems is systematically compared in terms of required ESS power and energy capacity and amount of energy cycled through the ESSs as a function of the RRL. The aim of this study is to determine how the ESS requirements of a large-scale PV power plant change with the number of the ESSs used for RR smoothing. The study is based on 1 Hz power measurements from seventy-six 500 kW inverters.

2 METHODS AND DATA

The RR control of PV systems was implemented by an RR-based control strategy of the ESSs where the RR of the power fed to the grid P_{grid} was limited to comply with the applied RRL. When a single ESS is used to smooth the power fluctuations of one or more PV systems, the generated PV power P_{PV} equals the sum of P_{grid} and the charging power of the ESS P_{ESS}, i.e., $P_{PV} = P_{grid} + P_{ESS}$. Positive P_{ESS} corresponds to charging and negative to discharging of the ESS. A diagram of the used simulation model in case of 2 PV systems and 1 or 2 ESSs is presented in Fig. 1. In case of several ESSs, the ESS control strategy is applied for each ESS independently. In order to ensure RRL compliance at any time, even during a sudden shutdown of the PV system, enough energy is stored in the ESS to lower P_{grid} to zero with the applied RRL. The minimum energy stored in the ESS at each moment can be defined as

$$E_{\mathrm{ESS,\,min}} = \frac{{P_{\mathrm{grid}}}^2}{2\mathrm{RRL}}. \qquad (1)$$

The RRL was altered from 1 to 20 %/min with respect to the nominal PV power. The ESS control strategy was presented in more detail in [10].

Fig. 2 shows the measured PV power, the power fed to the grid with an RR limit of 10 %/min, and the corresponding ESS power for a period of 1 h illustrating the operation of the ESS control strategy as well as fast

10.4229/EUPVSEC2023/5DV.2.6
020522-001

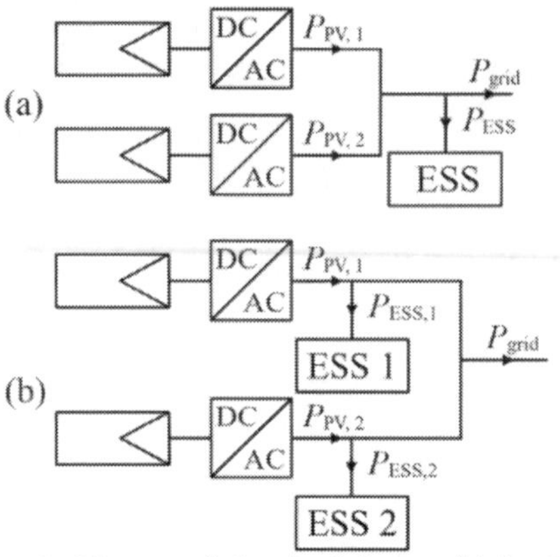

Figure 1: Diagram of the simulation model for 2 PV systems equipped with 1 (a) or 2 (b) ESSs.

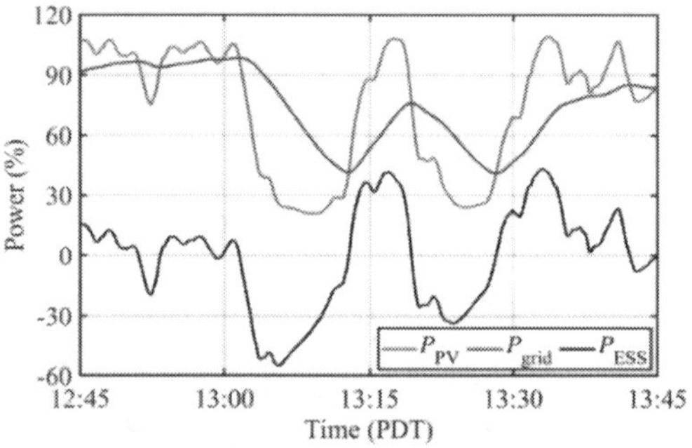

Figure 2: Produced power of the 38 MW PV system (see Fig. 3), power fed to the grid, and power of the ESS with an RR limit of 10 %/min in an afternoon with cumulus clouds. The powers are with respect to the nominal PV power.

fluctuations of PV power and high ESS powers required to compensate them. In this example, the required ESS power capacity, i.e., the maximum absolute value of P_{ESS}, is 55% which is needed for discharging the ESS at 1305 h.

This study is based on power measurements from the 48 MW Copper Mountain Solar 1 PV Plant at Boulder City, Nevada (35.78° N, 115.00° W, Fig. 3). The PV plant is composed of fixed-tilt thin-film CdTe PV modules covering an area of 1.89 km^2. The AC power output of 96 inverters was measured with a sampling frequency of 1 Hz. The data utilized in this study was measured on 1 July – 31 August 2011. The summertime period at the site is characterized by generally clear conditions with large irradiation. But monsoonal outflows with fair weather cumulus clouds frequently impact the site. Cumulus clouds are optically thick and present large decreases in irradiance that can exceed 70% (e.g. between 1300 and 1330 h in Fig. 2) presenting near-worst case conditions encountered at PV plants globally. The ramp rates are a function of cloud optical depth and cloud speed. Cloud speeds in Nevada are typically quite high with a mean value of around 23 m/s [13]. That means that it takes from 50 to 95 s for an edge of a typical cloud shadow to move over the PV plant. The highest change in power in 1 s for the 48 MW PV plant during the two-month period was 9.2%, with respect to the nominal PV power, corresponding to an RR of 551 %/min and it occurred when part of the inverters was shutdown. The highest change in power in 1 s caused by irradiance variability was 2.9% (176 %/min). These numbers highlight the need for PV RR control in even large-scale power plants.

In this paper, two study cases are presented. In both cases, measurements of 62 days were analyzed. In Case 1, two sections of the PV plant were considered and use of separate ESSs for the sections was compared to use of one centralized ESS. These sections are called PV systems A and B from here on out. Fig. 3 presents a satellite photograph of the Copper Mountain Solar 1 PV Plant highlighting the two PV systems in Case 1. Both PV systems have typical rectangular PV plant structure with regular gridded inverter spacings. PV system A consist of 40 inverters having a total nominal power of 20 MW and PV system B consists of 36 inverters with a total nominal power of 18 MW. Case 1 provides an illustrative comparison between use of two decentralized ESSs and one centralized ESS for two large-scale PV plants of nearly equal size. Moreover, the effects of the applied RRL were studied by varying the RRL from 1 to 20 %/min with respect to the nominal PV power.

In Case 2, data of a rectangular section of the PV plant, which can be divided into twenty-four 1 MW PV systems, was analyzed. The section consists of 6 north-south columns of 4 identical PV systems allowing a systematic study of the effects of number of ESSs on the total ESS energy and power requirements, i.e., the sum of the requirements of the individual ESSs. A satellite photograph of this part of the PV plant is presented in Fig. 4. Note that one 1 MW PV system consists of four arrays arranged in a 2x2 geometry with the white dot in the center representing two inverters. The left-hand arrays are connected to one inverter and the right-hand arrays to the other. The number of ESSs used for PV RR control was varied between 1 to 24. The following scenarios were studied: In case of 1 ESS, all 24 1 MW PV systems are connected to the same ESS. In cases of 2 and 3 ESSs, three and two adjacent columns of PV systems (see Fig. 4),

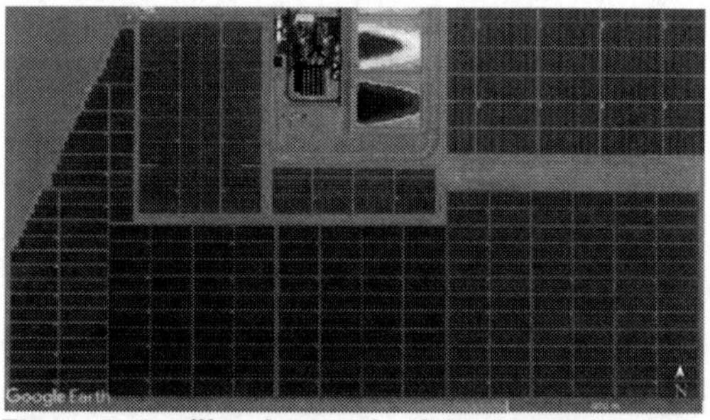

Figure 3: Satellite photograph of the Copper Mountain Solar 1 PV Plant. PV systems A (20 MW) and B (18 MW) studied in Case 1 are indicated with blue and red, respectively [14].

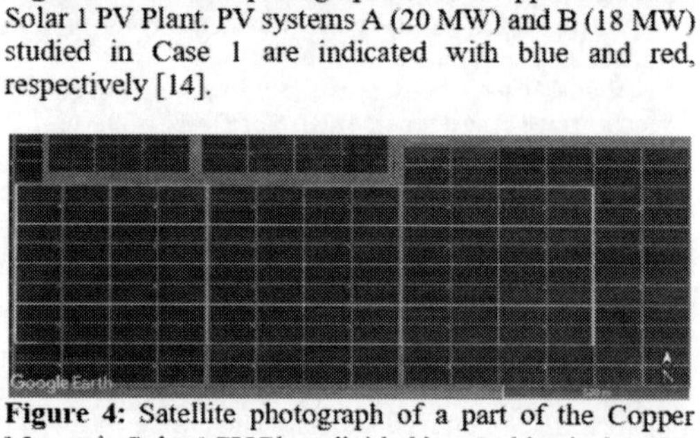

Figure 4: Satellite photograph of a part of the Copper Mountain Solar 1 PV Plant divided into 24 identical 1 MW PV systems indicated with yellow rectangles [14].

10.4229/EUPVSEC2023/5DV.2.6
020522-002

respectively, are connected to a single ESS. In case of 6 ESSs, each PV system column has its own ESS. In case of 12 ESSs, each column has 2 ESSs, i.e., two adjacent PV systems are connected to a single ESS, while in case of 24 ESSs there is an own ESS for each PV system. In Case 2, RRLs of 5, 10, and 20 %/min were applied.

3 RESULTS AND DISCUSSION

3.1 Case 1: 2 large-scale PV power plants

In Case 1, use of decentralized and centralized ESSs for PV RR control is compared considering 2 large-scale PV systems. The required ESS power and energy capacities for the PV systems and their combination are presented in Fig. 5 as a function of the applied RRL. The ESS power requirement for the combination of the two PV systems was lower than for the separate PV systems due to smoothing of PV power fluctuations with increasing PV system land area. The ESS power requirement to smooth all the PV power fluctuations was higher for PV system B than for PV system A especially at higher RRLs. This results mostly from the finding of [13] that in that region cloud shadows move mostly towards directions around 50° (north-east-ward). PV system B is shorter along this direction than system A resulting in faster power fluctuations and thus a higher ESS power is required to smooth the fluctuations. The Copper Mountain Solar 1 PV Plant is located in a remote desert where there are no obstructions that would shade the PV modules. Thus, shading of the modules and fast power fluctuations are mostly caused by overpassing cloud shadows. However, one should note that the ESS power requirements of Fig. 5 (a) are based on single extreme conditions of the studied two-month period.

The required ESS energy capacity was lower for the combination of the PV systems than for the separate PV systems. However, the relative difference was much smaller than in the power capacity. The reason for this is that the required power capacity results from the fastest power fluctuations which become smoother with increasing PV system size, while the required energy capacity results from the highest power fed to the grid since the ESS needs to have enough energy (Eq. (1)) to ensure RRL compliance even during a sudden shutdown. The required relative ESS energy capacity was slightly higher for PV system A than for PV system B. The required ESS energy and power capacities decreased with increasing RRL as the need for smoothing power fluctuations decreased.

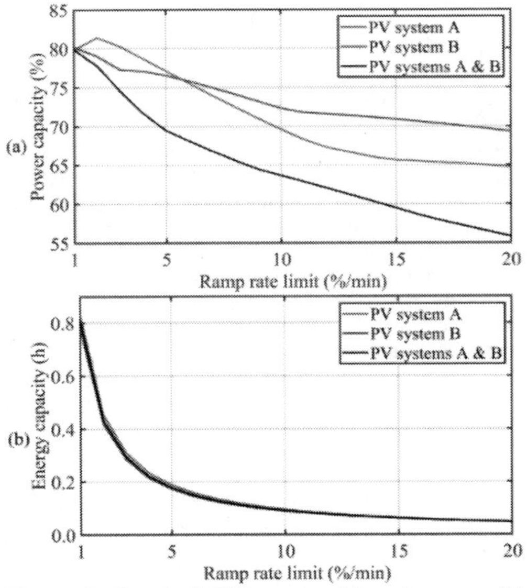

Figure 5: Required ESS power (a) and energy (b) capacities for PV system A, PV system B, and their combination as a function of the RR limit. The capacities are with respect to the nominal PV power.

Fig. 6 shows the share of produced PV energy cycled through the ESS for the PV systems as a function of the RRL. The shares of cycled energy decreased with increasing RRL, in accord with [10], as the need for smoothing power fluctuations decreased reducing the use of the ESS. With a strict RRL of 1 %/min, over 11% of the produced energy was cycled through the ESS while the share was less than 2% for RRLs higher than 15 %/min or so. The share of cycled energy was the smallest for the combination of the PV systems while the differences between PV systems A and B were very small. This means that the centralized ESS is not utilized as much as the decentralized ones. However, the difference is small.

The PV production is highly dependent on the irradiance conditions, and thus the daily irradiance profile has a major effect on the requirements and utilization rate of the ESSs. The highest, median, and lowest daily required ESS energy capacities and shares of energy cycled through the ESS are presented in Fig. 7 as a function of the RRL. The median of the daily required ESS energy capacity is close to the highest value since the required energy capacity results from the highest power fed to the grid and the highest daily PV power does not differ significantly within the measurement period apart from full overcast days when only minor power smoothing is needed. The highest daily share of energy cycled through the ESS is much higher than the median which is close to the lowest value. This means that on days with highly fluctuating irradiance conditions the ESS is needed for power smoothing much more often than during a regular day. For example, with an RRL of 10 %/min, up to 7% of the produced energy was cycled through the ESS on partly cloudy days while the median daily share is only 2%.

The used ESS control strategy ensures RRL compliance at any time, even during a sudden shutdown of the PV system, by storing enough energy in the ESS. However, in practice, it might not be feasible to eliminate all RR violations. By allowing RR violations caused by sudden shutdowns, a smaller ESS could be utilized for

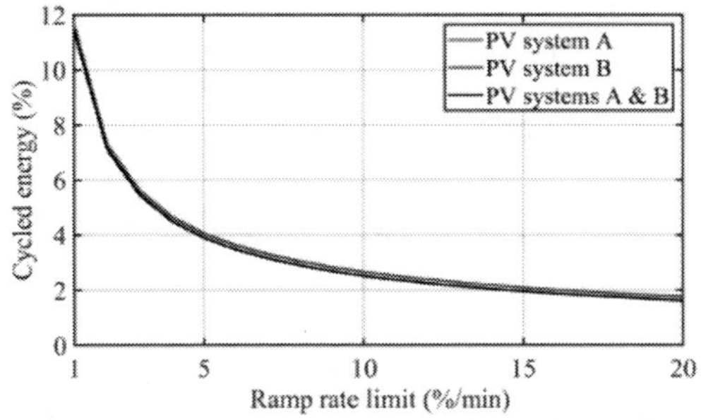

Figure 6: Share of produced PV energy cycled through the ESS for PV system A, PV system B, and their combination as a function of the RR limit.

10.4229/EUPVSEC2023/5DV.2.6
020522-003

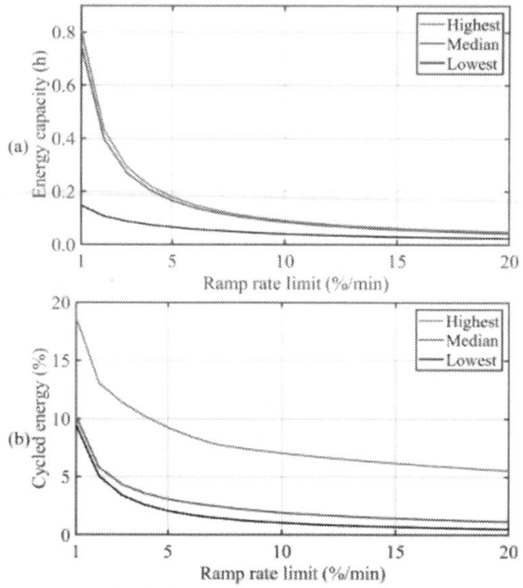

Figure 7: The highest, median, and lowest daily required ESS energy capacities (a) and daily shares of energy cycled through the ESS (b) for the combination of PV systems A and B as a function of the RR limit.

limiting RRs caused by irradiance variations. Fig. 8 shows the required ESS power and energy capacities for the PV systems and their combination as a function of the applied RRL when the ESS is controlled by the control strategy presented in Section 2 without shutdown preparedness, i.e., the minimum energy requirement of Eq. (1). The two-month measurement period includes one day with inverter shutdowns which was excluded from this analysis. With low RRLs, the differences in required ESS power capacity between the two ESS control strategies were small and the

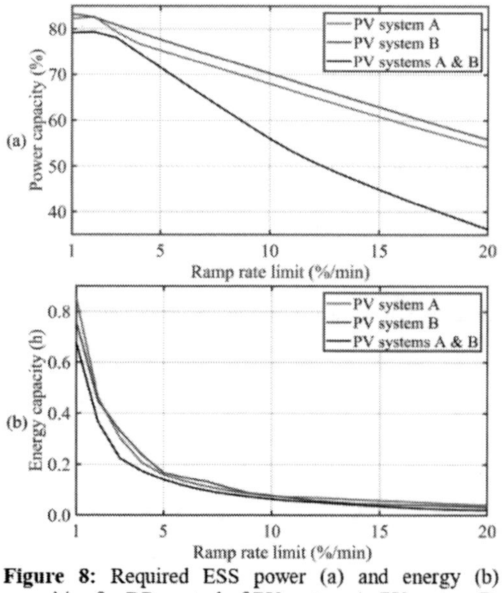

Figure 8: Required ESS power (a) and energy (b) capacities for RR control of PV system A, PV system B, and their combination without shutdown preparedness as a function of the RR limit. The capacities are with respect to the nominal PV power.

power requirement was up to 5% lower when the minimum energy requirement of Eq. (1) was applied. However, with RRLs higher than 5 %/min or so, clearly smaller ESS power capacity was required without shutdown preparedness. With an RRL of 20 %/min, the power requirement without shutdown preparedness was 16%, 19%, and 35% lower than with shutdown preparedness for PV system A, PV system B, and their combination, respectively. The ESS energy requirement without shutdown preparedness was lower than with it except for low RRLs for PV systems A and B for which the required ESS energy capacity was generally lower (up to 16% for PV system B) with shutdown preparedness than without it. The difference in ESS energy requirement had an increasing trend with increasing RRL. With an RRL of 20 %/min, the required energy capacity for PV system A, PV system B, and their combination was 18%, 32%, and 57%, respectively, lower when Eq. (1) was not applied.

The comparison the ESS control strategies reveals that ESS requirements for PV RR control when an RRL of 8% or higher is applied are lower if the ESS is not used for preparation for sudden shutdowns. However, at lower RRLs the differences are small, and the requirements are generally lower when the ESS is used also for shutdown preparedness, i.e., for eliminating all possible RR violations. The ESS power and energy requirements in Fig. 8 are lower for the combination of the PV systems than for the individual PV systems as the PV power fluctuations resulting from fluctuating irradiances become smoother with increasing system land area decreasing the need for ESS utilization (note spacing between the three lines in Fig. 8 (b) versus Fig. 5 (b)).

3.2 Case 2: PV plant divided into 24 identical PV systems

In Case 2, the effects of the number of ESSs on the total ESS energy and power requirements of a large-scale PV plant and amount of energy cycled through the ESSs are systematically studied considering a rectangular PV plant section divided into 24 identical 1 MW PV systems. The ESSs are used for PV RR limitation using the control strategy of Section 2 with shutdown preparedness. Fig. 9 presents the required ESS power and energy capacities for the PV plant section as a function of the number of the ESSs. Fig. 9 (a) reveals that the required total power capacity increases with the number of the ESSs, i.e., decreasing land area of a PV system connected to a single ESS, as explained in Case 1. The increase is strongest when only a few ESSs are used becoming slower with increasing numbers of the ESSs. The required total energy capacity is almost independent of the number of the ESSs but increases slightly with increasing number of the ESSs. The results show that the ESS power requirements of a large-scale PV plant can be minimized by using a single ESS for RR control of the entire PV plant instead of distributing the ESSs to control ramp rates for single sections or individual inverters of the plant.

The total share of produced PV energy cycled through the ESSs is presented in Fig. 10 as a function of the number of the ESSs. The share of cycled energy decreases with increasing RRL as presented in Fig. 6. There are only negligible differences between the studied scenarios meaning that the number of the ESSs does not affect the total share of energy cycled through the ESSs. Based on the results of this study, use of a single ESS for RR control of the entire PV plant instead of distributing the ESSs for single sections or individual inverters of the plant decreases the ESS power requirement while having only a

10.4229/EUPVSEC2023/5DV.2.6
020522-004

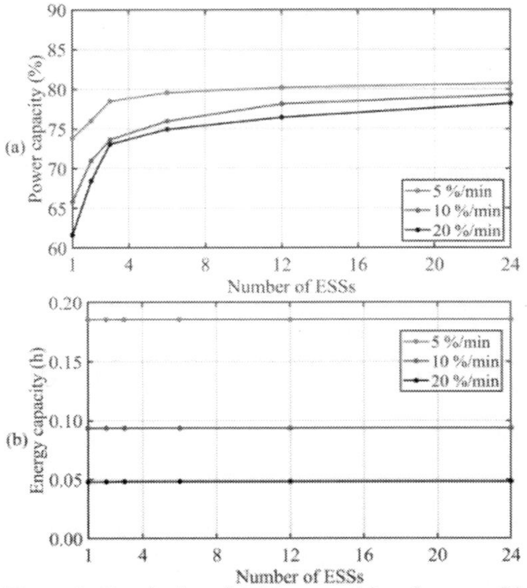

Figure 9: Required total ESS power (a) and energy (b) capacities for various RR limits as a function of the number of ESSs. The capacities are with respect to the nominal PV power.

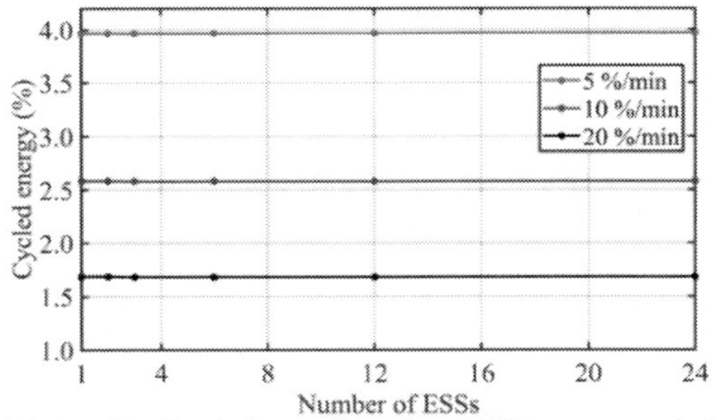

Figure 10: Total share of produced PV energy cycled through the ESSs for various RR limits as a function of the number of the ESSs.

slight effect on the ESS energy requirement or the total share of produced PV energy cycled through the ESSs.

In Fig. 9, the total ESS energy and power requirements for PV RR control were calculated as the sum of the requirements of individual ESSs. However, in practice ESSs of equal size would be installed for each PV system. Sizing of the ESSs of all the subsystems based on the highest requirements of a single subsystem would lead to higher capacity requirements than sizing based on the requirements of the individual subsystems. While all subsystems are exposed to the same general climate and weather conditions, there is a statistical chance that certain subsystems experience more extreme cloud transients.

The required ESS power and energy capacities for the PV plant section, based on the highest requirements of a subsystem, are presented in Fig. 11 as a function of the number of the ESSs. The required ESS power and energy capacities of Fig. 11 are identical with Fig. 9 when only one ESS is used since all PV modules are connected to the same ESS. When multiple ESS are used, the power requirements based on the highest requirements of a subsystem are from 1% to 13% higher than those calculated as the sum of the requirements of the individual

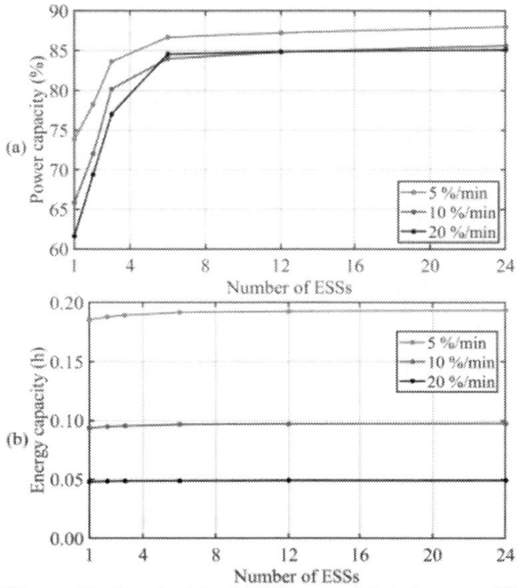

Figure 11: Required total ESS power (a) and energy (b) capacities based on the highest requirements of a subsystem for various RR limits as a function of the number of ESSs. The capacities are with respect to the nominal PV power.

subsystems. The total energy capacity requirements based on the highest requirements of a subsystem are slightly higher than the sum of the requirements of the individual subsystems and the increase of the required ESS capacity with increasing number of the ESSs is stronger.

4 CONCLUSIONS

This paper presented a comparison between decentralized and centralized ESSs for power RR control of PV systems based on 62 days of inverter level power measurements. First, use of decentralized and centralized ESSs was compared considering 2 large PV systems. The second comparison considered a rectangular PV plant section divided into 24 identical PV systems and the number of the ESSs used for PV RR control was varied between 1 to 24. The decentralized and centralized ESS setups were compared in terms of required ESS power and energy capacity and amount of energy cycled through the ESSs. The effects of the applied RRL were studied by varying the RRL from 1 to 20 %/min with respect to the nominal PV power.

The results show that the relative ESS power capacity required to smooth all PV power fluctuations decreases with increasing size of the PV system connected to the ESS meaning that the centralized ESS setup requires relatively lower ESS power capacity than the decentralized one. The reason for this is that the required power capacity results from the fastest power fluctuations which smoothen with increasing PV system land area. Differences in ESS energy requirements between the decentralized and centralized ESS setups were small since the required energy capacity results from the highest power fed to the grid as the ESSs were controlled to have enough energy to ensure RRL compliance even during a sudden shutdown of the PV system. Differences in the amount of energy cycled through the ESSs between the

10.4229/EUPVSEC2023/5DV.2.6
020522-005

decentralized and centralized ESS setups were found to be small. Based on the results the ESS power requirement of a large-scale PV plant can be minimized by using a single ESS for RR control of the entire PV plant instead of distributing the ESSs to control ramp rates for single sections within the power plant while having only a slight effect on the ESS energy requirement or the total share of produced PV energy cycled through the ESSs.

ACKNOWLEDGEMENTS

K. Lappalainen was funded by the Research Council of Finland (funding decision 348701) and KAUTE Foundation.

REFERENCES

[1] K. Lappalainen, S. Valkealahti, Applied Energy 190 (2017) 902. https://doi.org/10.1016/j.apenergy.2017.01.013.
[2] G.V. Brahmendra Kumar, K. Palanisamy, P. Sanjeevikumar, S.M. Muyeen, Energy Reports 9 (2023) 163. https://doi.org/10.1016/j.egyr.2022.11.176.
[3] S. Sukumar, M. Marsadek, K.R. Agileswari, H. Mokhlis, Journal of Energy Storage 20 (2018) 218. https://doi.org/10.1016/j.est.2018.09.013.
[4] Y. Sun, Z. Zhao, M. Yang, D. Jia, W. Pei, B. Xu, CSEE Journal of power and energy systems 6 (2020) 160. https://doi.org/10.17775/CSEEJPES.2019.01950.
[5] K. Lappalainen, S. Valkealahti, Proceedings 2018 IEEE Innovative Smart Grid Technologies - Asia, Vol. I (2018) 190. https://doi.org/10.1109/ISGT-Asia.2018.8467900.
[6] J. Marcos, I. de la Parra, M. García, L Marroyo, Progress in Photovoltaics: Research and Applications 24 (2016) 680. https://doi.org/10.1002/pip.2719.
[7] M. Lave, J. Kleissl, E. Arias-Castro, Solar Energy 86 (2012) 2190. https://doi.org/10.1016/j.solener.2011.06.031.
[8] J. Marcos, O. Storkël, L. Marroyo, M. Garcia, E. Lorenzo, Solar Energy 99 (2014) 28. https://doi.org/10.1016/j.solener.2013.10.037.
[9] I. de la Parra, J. Marcos, M. Garcia, L. Marroyo, Solar Energy 118 (2015) 426. http://doi.org/10.1016/j.solener.2015.05.046.
[10] K. Lappalainen, S. Valkealahti, Renewable Energy 196 (2022) 1366. https://doi.org/10.1016/j.renene.2022.07.069.
[11] M. Talvi, T. Roinila, K. Lappalainen, Energies 16 (2023) 4313. https://doi.org/10.3390/en16114313.
[12] J. Marcos, I. de la Parra, M. Garcia, L. Marroyo, Energies 7 (2014) 6593. http://doi.org/10.3390/en7106593.
[13] J.L. Bosch, J. Kleissl, Solar Energy 95 (2013) 13. http://doi.org/10.1016/j.solener.2013.05.027.
[14] Google Earth (2023) https://www.google.com/earth/index.html (accessed on 31.08.2023).

10.4229/EUPVSEC2023/5DV.2.6
020522-006

LARGE-SCALE AGRIPV ON NORWEGIAN FARMS WITH FLOW BATTERIES AS SEASONAL STORAGE

Steve Völler*,[1], Ingrid Røstad Brøndbo[1], Ellen Loxley-Slåttsveen[2], Stefan Preisig[3]
[1]Norwegian University of Science and Technology, Institute for electric energy, Norway
[2]Bryte Batteries, Norway
[2]Skjetlein high school, Trøndelag county authority, Norway
*Corresponding Author: steve.voller@ntnu.no

ABSTRACT: The increasing demand for electricity has led to a greater focus on generating renewable energy, particularly from large-scale land-based solar setups. This will come in conflict with other land use like agriculture. To optimize land use and address this challenge, the concept of combining solar panels with agriculture, known as agrivoltaics or AgriPV/APV, is becoming popular. In regions with weak grids, such as rural farming areas, excess solar energy cannot always be fed into the grid due to capacity limitations. One potential solution are energy storages.
As the solar installations are large in size, smaller energy storage might not be suitable. In addition, as for Norway, the energy demand is higher in the winter due to the cold climate and electrical heating. A storage which could save the produced solar energy from the summer and supply it in the winter would be an optimal solution. Those seasonal storage would need a relatively low power capacity compared to their energy capacity. Flow batteries are such a kind of storage. In this context, we examine a Norwegian agricultural school that does both farming and school operations, with existing solar rooftop installations and an AgriPV-installation, and implement a larger battery for seasonal storage.
Keywords: AgriPV, seasonal energy storage, rural grid, optimization

1 INTRODUCTION

Norway has been somewhat slower than other countries in deploying solar energy due to its large hydropower potential. Nevertheless, the increasing need for green energy in Norway also accelerates the installation of solar energy. In the coming decades, it is projected that solar energy could contribute with 10 to 30 TWh, with Norway currently generating around 150 TWh of electricity. Beside smaller installations like in/on buildings, large-scale solar parks will be developed. This may come as floating or land-based installations. Due to the limited resources for useful land, this will conflict with food production and/or nature conservation. An alternative is the dual-use of that land, i.e. keeping the former purpose (e.g. growth of crops) and setting up solar panels but with less areal impact (e.g. more spacing between them). Thus, agrivoltaics (AgriPV) will be a reasonable option. Given that approximately 70 % of Norway's agricultural land is allocated for grass production, the land-area that can be used is also quite large. A newly published study says that more than 100 TWh of energy could technically be produced per year on already used agricultural land in Norway, in addition to another 100 TWh from other "grey areas", i.e. areas already used by humans (parking lots, rooftops…) [1].

This study will use the case of Skjetlein high school, an agricultural high school in the south of Trondheim as base. The school has classroom, living and common area buildings. Additionally, it engages in sustainable ecological agriculture, maintaining livestock stables and cultivating crops such as grains and grass on the fields surrounding the school. As a result of an innovative research project, Skjetlein has already installed a smaller AgriPV systems in June 2023. This consists of 4 rows of vertically east-west installed bifacial solar panels, and the installation will give approximately 50 kW of PV capacity (shown in Figure 1). It serves as the country's first major experimental site dedicated to assessing the energy output of vertical solar installations in combination with agricultural practices and food production (e.g. shadowing, nutrition, harvesting).

Furthermore, this installation is the northernmost of its kind in the world, providing exceptional opportunities to examine its performance in cold climates, such as challenging weather conditions, limited sunlight, as well as of snow coverage and snow reflections.

Figure 1: AgriPV installation with Skjetlein high school and all its buildings in the background, together with some of the farmland north and south of the school.

The distance between the rows is set to 12 meters, so that agricultural machinery can operate as before without harming the solar modules. First approximations concluded in a Land Equivalent Ratio (LER) of around 173 % (i.e. 82 % of the annual energy production from a "normal" large-scale ground-mounted solar installation without regarding the agriculture production, + 91 % remaining crop production of the agricultural land due to the AgriPV installation) [2], as shown in Figure 2.

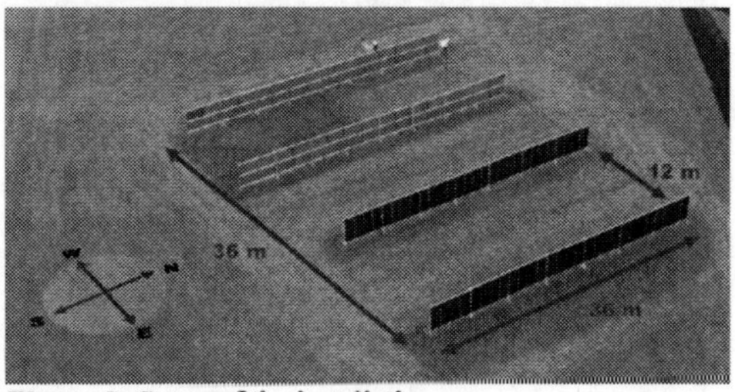

Figure 2: Setup of the installation.

This installation of 50 kW comes in addition to already existing solar rooftop systems with 234 kW at different buildings. The sum alone will exceed the own electricity power demand of the school at some times during the day and year (e.g. summer vacation), so the excess has to be stored or fed into the grid. There are regulatory limitations in Norway of how much a private individual or entity can feed in before being classified as a "power producer" (currently set to 100 kW, but it will be increased to 500 kW). Additionally, there are grid capacity limitations (transformers, transmission system), especially in rural areas, where farms usually are situated.

As a solution, energy storage can take in the excess energy and supply the school when the solar production is less, or the energy bought from the main grid is expensive. The required capacities (both power and energy) for such storage solutions need to be assessed on a case-by-case basis.

2 BACKGROUND ON SEASONAL STORAGE

2.1 Seasonal storage

This type of storage is used to store and supply the commodity (electricity, heat, food…) over several seasons, to overcome shortages. As for electricity, Figure 3 illustrates typical storage solutions based on their power rating (i.e. how much power they can provide) and the time they usually store the energy. Seasonal storage options are limited to a few types (found in the upper right corner). Hydropower, compressed air, and heat storage we exclude from the study due to geographical constraints. While hydrogen is included in the main study [3], it will not be discussed further in this paper. Chemical storages such as lithium-ion or lead-acid are not applicable for storing large amounts of energy. Consequently, flow batteries remain the only obvious option.

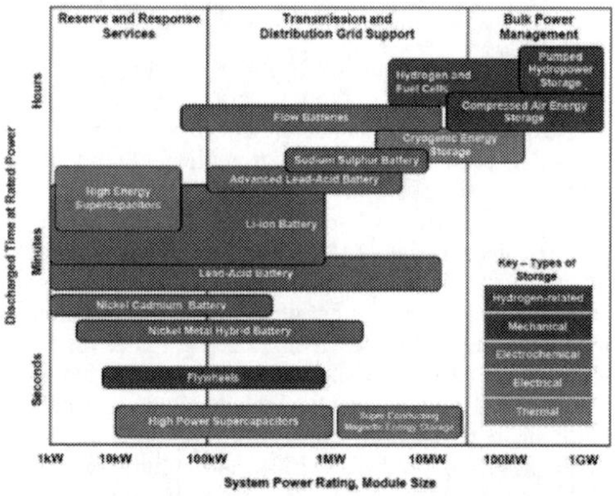

Figure 3: Applicable power ranges and discharge power duration of different energy storage technologies in a Ragone-plot [4].

2.2 Flow batteries

Little is known about that type of storage, despite its initial investigations dating back to the 19th century and more in-depth research starting again in the 1970. The technology regained attention around the turn of the millennium with large-scale solar and wind installations in Australia, and are now coming back into focus with the immense installation of renewable (and fluctuating) energy sources in industrial countries.

A flow battery, in short, consists of two liquid-filled tanks that store the energy, and a membrane, where the liquid undergoes charge and discharge processes. A pump for each tank supplies the liquids to the tanks, as shown in Figure 4. There exists different kind of materials for the electrolytes. The most common are vanadium (in both tanks) and the zinc-bromide (separated). A deeper look into the technologies gives [5].

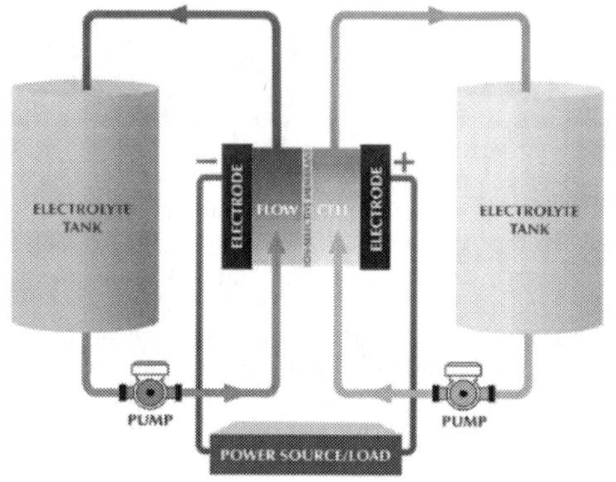

Figure 4: Schematic illustration of a flow battery system [6].

A major advantage with flow batteries is that the power capacity (membrane size) is not dependent on the energy capacity (storage tank size). This is usually the case with other chemical batteries. Thus, low-power solar systems (e.g. in the range of a few hundred kilowatts) can store energy over a long period of time, simply by scaling up the tank size. That would make flow batteries suitable for seasonal storage, while also enabling their utilization in daily energy balancing tasks.

2.3 Existing flow battery at Skjetlein

As part of a pilot project, a 30 kW/90 kWh flow battery has been installed in Skjetlein high school in the summer of 2023. This represents the second installation of its kind in Norway. The battery's size closely matches the recommended specifications from the optimization in [3], which will be discussed later, and will be integrated with the school's existing solar rooftop and AgriPV system. Figure 5 shows the installation in the basement, which consists of three identical setups combined to achieve the resulting size.

Figure 5: One third of the flow battery installed in Skjetlein high school. To the left the two storage tanks, and to the right the cabinet with the membrane, pumps, power electronics etc.

3 CONFIGURATION OF SCENARIOS

3.1 Definition of various configurations

To evaluate the size of the needed seasonal storage, the area north and south of the school was chosen, which are mainly used for gras production. The one to the north is steep and mountainous and is also used for cattle. The area in the south has flat terrain suitable for heavy machinery use and better soil quality, and is also used for e.g. wheat cultivation. An aera was estimated for possible solar installations, which resulted in systems of 711 kW (north) and 1600 kW (south), as shown in Figure 6.

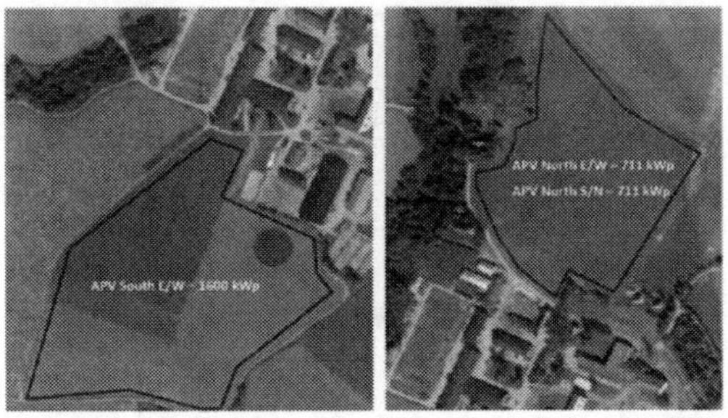

Figure 6: Installation south of the school (left, 1600 kW) and north of it (right, 711kW) [3].

Both systems are setup with bifacial modules (bifaciality factor of 80 %) and were each modelled and simulated with either south-oriented modules and 45° inclination, or vertically installed standing modules in east-west direction. Further analysis here focuses on the northern installation and its vertical configuration. This choice is based on the lower soil quality, which results in fewer losses in high-quality agricultural production due to land loss mitigation.

The existing 50 kW AgriPV installation is shown as red dot in Figure 6. It is intentionally positioned on the high-quality soil in the south, so that the effects on plant growth can be better studied.

3.2 Northern scenario, with vertical east-west setup

PVsyst was used to model the terrain and the solar installation (Figure 7). The vertical modules have the same east-west direction as the actual AgriPV-system.

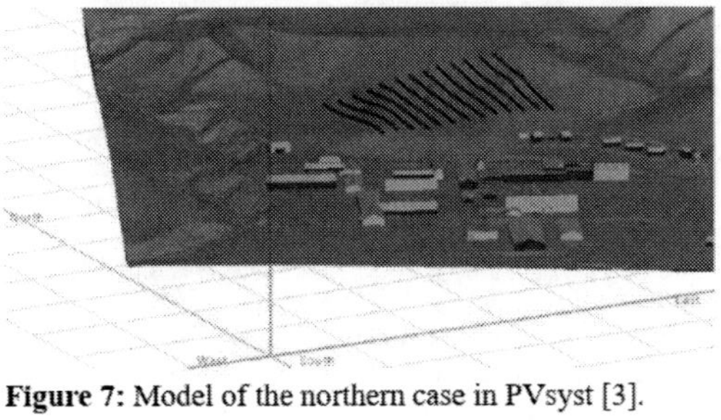

Figure 7: Model of the northern case in PVsyst [3].

The vertical setup will also minimize the usual peak production of solar installation facing south around midday. This will enhance the efficiency of solar energy utilization by reducing excess energy. Additionally, with an east-west configuration, more energy is generated when it is usually demanded, namely in the morning (east-facing production peak) and evening (west-facing production peak), also corresponding to the time with higher energy prices. Typical profiles for east/south/west orientation are given in Figure 8, which naturally vary depending on the specific location on earth and day of the year.

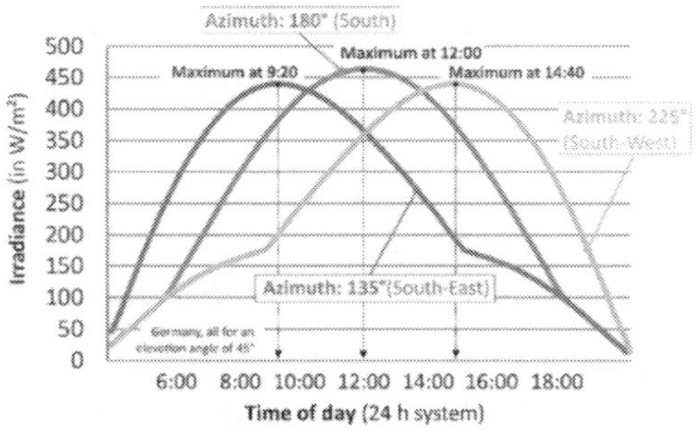

Figure 8: Exemplary profiles for solar generation with different azimuth angles [7].

3.3 Additional simulation and input parameters

For the electrical demand of the school, real historical measurements were taken. The energy price for the electricity which the school had to buy from the grid, is based on the spot prices in Trondheim for the years 2021 and 2022. These two years had, compared to earlier years, significant higher price peaks in the winter month due to various factors: e.g. low rainfall in summer leading to decreased inflow into hydropower reservoirs, cold and dry winters, and high energy prices in Europe that were reflected to Norway through international power connections. Both the input data and the simulations are conducted with a one-hour resolution time step.

Figure 9 illustrates the school's energy demand (in blue) alongside solar production (in red; existing rooftop + northern AgriPV). The figure also displays the former capacity limit of 100 kW for grid power supply (horizontal line). As visible, solar production cannot cover the demand in the winter month but surpasses the demand in the summer. In addition, the feed-in limit is constantly exceeded. The case with the southern installation of 1600 kW would be worse regarding solar overproduction.

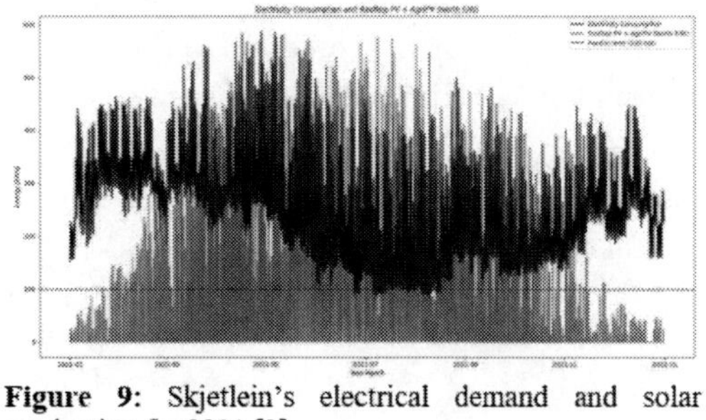

Figure 9: Skjetlein's electrical demand and solar production for 2021 [3].

To optimize the storage size, two control strategies were formulated, which were used in combination with the previously mentioned solar case and the two pricing years to include multiple seasons in the analysis:

1. No investment costs for the storage (to estimate an ideal storage that could take in all the solar energy and minimize curtailments).
2. Including storage costs in order to obtain an economically realistic storage size (accepting e.g. more curtailments in solar production/use).

10.4229/EUPVSEC2023/5DV.2.7
020524-003

4 RESULTS

4.1 Strategy 1 - No storage investments

The aim in this case was to estimate a storage size (both power and energy) that could maximize the energy used from the solar system. The generated solar energy should be either immediately consumed by the school, sold back to the grid at market prices, or stored in the battery for future use. The storage system was restricted from purchasing energy from the grid at low prices, and then selling it at higher prices or using it to meet the school's demand. That would have influenced the storage size to be adjusted only to the solar installation. For that case, a flow battery with 300 kW and 3 MWh hours was found optimal.

Figure 10 displays some of the results. The school still needs to buy energy from the grid (dark blue curve), but it can replace approximately 30 % of this energy with solar energy (yellow) with very minimal curtailment (ca. 0.01 %). The energy level in the storage, denoted as the state of charge (SOC) in the lower picture (light blue), indicates that in the summer month the battery is easily filled with solar energy, and discharges to make room for new energy. In the autumn, the solar energy decreases and the SOC of the storage remains nearly constant, discharging when energy prices (the second lower curve) are high in the winter. The SOC increases once more in spring and summer, until the cycle repeats.

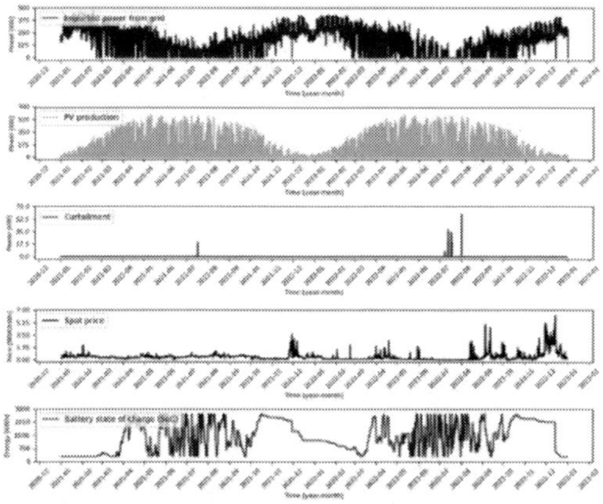

Figure 10: Various results from control strategy 1 [3].

A storage system of this size can be easily placed on a farm using standardized shipping containers. (Figure 11).

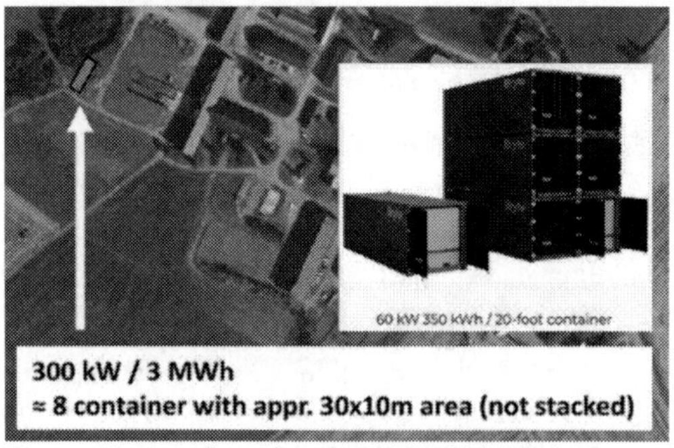

Figure 11: Area footprint of a 300 kW/3 MWh battery.

4.2 Strategy 2 - With storage investments

The same input data and solar configuration was used with this control strategy, but the storage had an investment cost. The optimization goal was set to minimize the total electricity costs and storage costs. The battery in that case was found to be 26 kW and 73 kWh, what is 10 % of the other power rating, but only 2.4 % of the energy capacity from the previous case.

Curtailed solar energy rises to 5 %, and electricity coverage of the total demand of the school drops to 24 %. These values, despite the significant reduction in storage size, remain closely aligned with the 300 kW/3 MW system from the previous case neglecting investment costs. However, due to its reduced size, the storage system experiences more frequent charge/discharge cycles, as shown in the lower graph of Figure 12. It will be in use more often during the summer and can only store a limited amount of energy during the winter months. This makes the storage more suitable for daily/weekly scheduling, rather than to be seasonal storage solution.

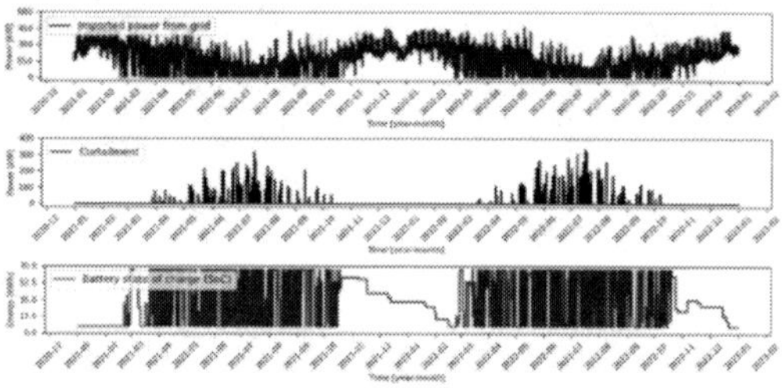

Figure 12: Various results from control strategy 2 with the smaller flow battery [3].

4 CONCLUSION AND FURTHER WORK

Large-scale solar installations in regions with weak grids and low local self-consumption require the supply of excess energy into the grid. Norway will focus grid problems if the placement of ground-mounted solar parks will continue as predicted. Grid reinforcement will take decades, thus local measures must be implemented as a quick (temporary) solution. In this context, energy storage emerges as a viable solution to balance energy generation and demand. In addition, in areas with high energy price differences between summer/winter like Norway, seasonal storage should be considered as an alternative to daily storage solutions. Norwegian farms fit into the aforementioned categories, offering enough space for large storage placements. From a technical standpoint, flow batteries can be considered an optimal solution for long-term storage.

A further economic analysis will demonstrate their usability for such cases. As a starting point, the already installed small flow battery at Skjetlein high school will be utilized to explore various control strategies, including energy trading with the grid, aiming to establish practical economic business cases. However, as anticipated, the demand for storage, particularly long-term storage solutions, is expected to grow substantially. This will inevitably lead to the need for storage, and the need to focus on various technologies tailored to different use cases.

10.4229/EUPVSEC2023/5DV.2.7
020524-004

5 ACKNOWLEDGEMENT AND REFERENCES

5.1 Acknowledgement

The AgriPV project is mainly financed by the Regional Research Fund Trøndelag ("Soldeling i Trøndelag", 336784), and is a collaboration between several research institutions (NTNU, SINTEF), industrial partners (ANEO) and municipalities (Trøndelag County Authority, Trondheim Municipality, Grønt Hjerte). The same applies to the flow battery project, whereby the funding comes mostly from Enova ("Pilotering av flowbatterier i den norske landbrukssektoren", 22/13157).

The author would like to thank Skjetlein high school for their willingness to be a "living lab" and all individuals involved in the projects, especially the bachelor and master students dedicated to their project work in this field and contributing with valuable knowledge.

5.2 References

[1] Multiconsult, "Markedsrapport – Norsk solkraft 2022 – innenlands og eksport", report 2022
[2] E.H. Honningdalsnes, "Autonomous Optimization of Agrivoltaic Systems in Norway", master thesis, NTNU, 2022
[3] I. Røstad Brøndbo, "Seasonal storage for excess solar energy on farms in Norway", master thesis, NTNU, 2023
[4] D. Sprake, Y. Vagapov, S. Lupin and A. Anuchin, "Housing Estate Energy Storage Feasibility for a 2050 Scenario", 2017 Internet Technologies and Applications (ITA), Wrexham, UK, 2017, pp. 137-142
[5] E. Sanchez-Díez et.al, "Redox flow batteries: Status and perspective towards sustainable stationary energy storage", Journal of Power Sources, Volume 481, 2021.
[6] The International Flow Battery Forum, https://flowbatteryforum.com/what-is-a-flow-battery/, accessed 2023-09-11.
[7] S. Krauter, "The energy system defined by PV: balancing generation and load via adequate plant design and load adaptation", EU PVSEC 2015

10.4229/EUPVSEC2023/5DV.2.7
020524-005

DAY-AHEAD AND INTRA-DAY PLANNING METHOD OF PV POWER PLANT WITH BESS FOR SECURING RESERVE POWER

Yuki Kojima*[1], Jindan Cui [1], Takashi Oozeki [2], Yuzuru Ueda [1]
[1]Tokyo University of Science, Japan, [2]National Institute of Advanced Industrial Science and Technology, Japan
4322516@ed.tus.ac.jp

ABSTRACT: In Japan, electric power reserve exchange (EPRX), a market for balancing supply and demand, was established in April 2021. To bid EPRX market, PV power plants will create two planned values of generated electricity: one is the kWh value, meaning the amount of electricity, and the other is the ΔkW value, meaning the ability to adjust the supply-demand balance by reserve power.

In this study, we examined a method for creating a day-ahead and intra-day plan to secure reserve power for a PV power plant with BESS. The research process has three steps: In step 1, a day-ahead generation plan is created using Headroom control, which keeps the raising allowance. In step 2, to secure Reserve, the intra-day power sales plan is modified using mathematical optimization. In step 3, the day-ahead plan for the next day is reduced using the estimated results of the SOC transition.

After the empirical research, we got three main conclusions, First, the intra-day modification improves the certainty of securing reserve power proves profitability. Second, it is useful to estimate the SOC transition on the day of operation and use that information in the day-ahead planning. Finally, it is necessary to create a more conservative day-ahead plan and add intra-day modifications.

Keywords: Headroom, Generation plan, Reserve, BESS, Optimization

1 INTRODUCTION

In Japan, the introduction of PV power generation is expected to increase in order to achieve carbon neutrality by 2050. In conjunction with this, output fluctuations due to weather conditions and uncertainty in the amount of electricity generated by solar power may affect the power grid in terms of frequency fluctuations, etc.. There is concern about insufficient reserve power at certain times of the day. In this study, we examined a method for creating a day-ahead and the intra-day plan to secure reserve power for a PV power plant with BESS using mathematical optimization.

2 PROPOSED METHOD

2.1 System Configuration

Fig.1 shows the system configuration in this study. A PV power plant participates in two electricity markets with a plan based on the day-ahead forecast.

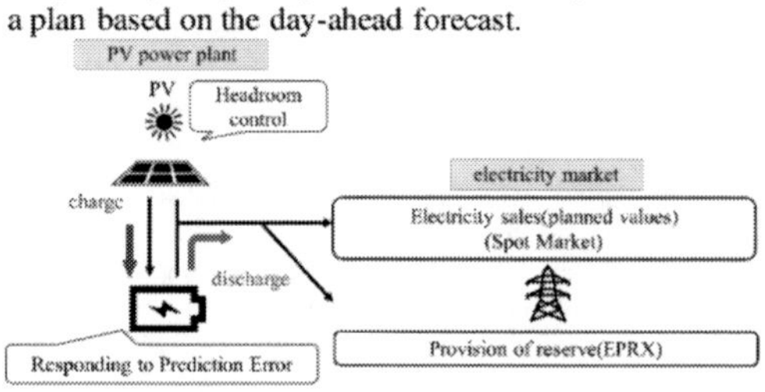

Figure 1: System configuration

2.2 Flow of Operation

The flow from day D-1 to D, where D is an operating day of the PV power plant, is shown in Fig.2. The spot market is the market for trading the next day as of the previous day, the intra-day market is the market for trading electricity one hour later the same day, and EPRX is the market for trading reserve power. The objective of this study is to secure reserve power by complying with the day-ahead plan from 12:00 to 15:00 on day D. As an example, 15:00 means 3:00 pm.

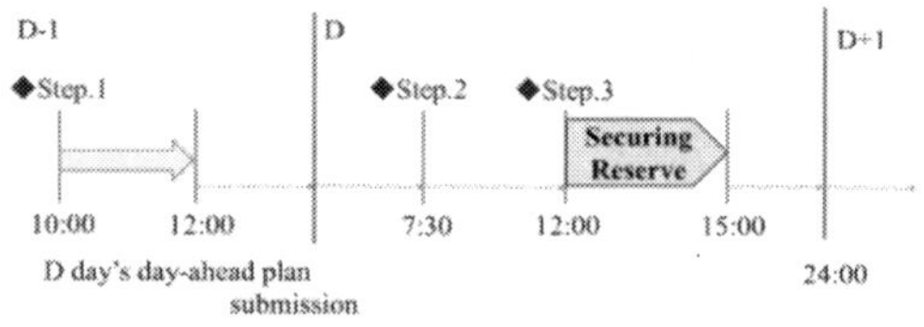

Figure 2: Flow of Operation

2.3 Step.1: Day-ahead planning with headroom (10:00)

When selling electricity to the spot market, it is necessary to prepare a generation plan the day before. Headroom control[1] was used in this generation plan. This is a method of creating planned values based on the assumption of output curtailment to generate a margin of increase. In this study, the period for securing this upward reserve H was set as 12:00 to 15:00. H was set to 40% of the minimum value of P_{f}, which is the forecasted power generation value from 12:00 to 15:00. H includes H_{e} and H_{r}, shown in Eqs.(1) and (2). H_{e} and H_{r} denote the forecast errors and the amount of reserve power, respectively. The day-ahead plan was determined by the following Eq.(3).

$$H_{\mathrm{e}} = H \times 1/4 \quad (1)$$

$$H_{\mathrm{r}} = H \times 3/4 \quad (2)$$

$$P_{\mathrm{p}}(t) = P_{\mathrm{f}}(t) - H(t) \quad (3)$$

where t is the number of spots. In this case, the actual planned value is $P_{\mathrm{p}}(t)$ plus $H_{\mathrm{r}}(t)$, since the triggering rate of the regulating power is assumed to be 100%.

The flowchart for Step.1 is shown in Fig.3.

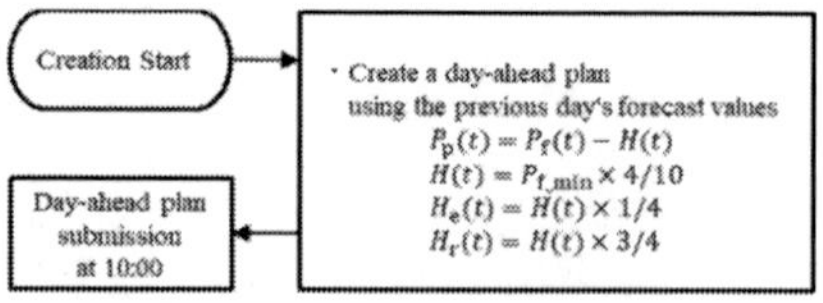

Figure 3: Flowchart of Step.1

As of day D-1, the operation only creates a generation plan on the basis of the previous day's forecast.

2.4 Step.2: Plan modification with intra-day forecasts (7:30 after day D)

On the day of the event, because a gap between measured and planned values may occur due to a deviation from the previous day's forecast values, reserve power may not be secured. To prevent this case, the electricity sales and BESS charge/discharge plans are modified on the same day by mathematical optimization using the short-term forecast value P_f'. Regarding P_f', a normal random number with an average value of 0 and standard deviation of 10 was used as the proportion, and the value given to the measured value as the forecast errors.

The intra-day plan is created by mathematical optimization using day-ahead plan P_p and the short-term forecast P_{fs}. The objective function is shown in Eq.(4).

$$\text{maximize} \sum_{17}^{48} \frac{1}{2} \{P_p'(t) \cdot U_s(t) + P_{imb}^{+}(t) \cdot U_{imb}^{+}(t) - P_{imb}^{-}(t) \cdot U_{imb}^{-}(t)\} \quad (4)$$

where P_p' denotes intra-day electricity sales plan ($P_p' = P_s' + P_h'$); P_s', P_h' denote the spot and intra-day market power selling plans, respectively; P_{imb}^{+}, P_{imb}^{-} denote the positive and negative imbalances, respectively; t is the number of spots (17,18…,48), U_s, U_{imb}^{+},U_{imb}^{-} denote the spot price, positive and negative spot imbalance prices ($U_{imb}^{+} = U_s \times 0.9 - 1$, $U_{imb}^{-} = U_s \times 1.1 + 1.46$), respectively. Eq.(4) is the maximization of revenue from electricity sales including the imbalance cost. Revenue did not include those from securing reserve power.

Next, we discuss the constraints. Firstly, the constraint conditions for the electricity sales plan were set as Eqs.(5) and (6).

$$P_p'(t) = P_f'(t) - ch(t) + dch(t) \quad (5)$$

$$P_p'(t) = P_p(t) + H_r(t) \ \ (t = 25,26 \ldots 30) \quad (6)$$

Eq.(5) shows the constraint on power flow, such that the modified planned value is equal to the sum of the short-term forecast and the amount of charge and discharge.

Secondly, the constraints for BESS were shown in Eq.(7) and (8).

$$B(t+1) = B(t) - ch(t) \times \eta + dch(t)/\eta \quad (7)$$

$$B_{min} \leq B(t) \leq B_{max} \quad (8)$$

$$SOC_{end} \geq SOC_{max} \times 0.6 \quad (9)$$

where B denotes the remaining BESS capacity [kWh]; SOC represents the state of charge of BESS [%]; η denotes efficiency of inverter/converter, set to 95% for simulation. Eqs.(7) and (8) denote the charge/discharge constraints of the BESS considering the charge/discharge loss and the BESS upper/lower limit constraints, respectively. Eq.(9) shows that the lower limit of SOC at the end of the day is set to 50%. If the SOC lower limit constraint cannot be satisfied due to over- forecast of the previous day's forecast, the constraint is branched to Eq.(10).

$$SOC_{end} \geq SOC_{max} \times 0.1 \quad (10)$$

When this branch condition is performed, the discharge of the BESS is completely curtailed after the PV generation time ends.

The flowchart for Step.2 is shown in Fig. 4.

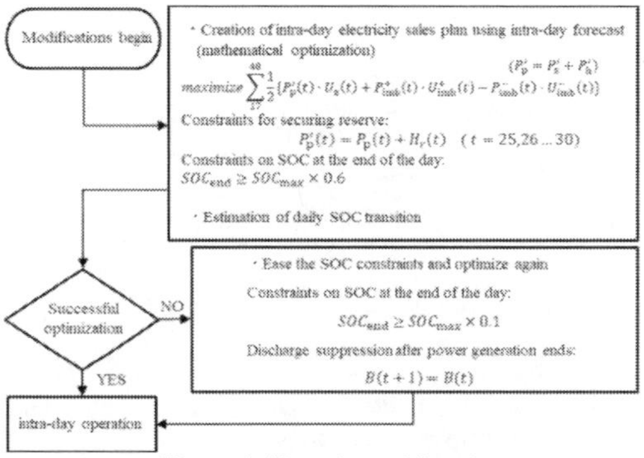

Figure 4: Flowchart of Step.2

2.4 Step.3: Day-ahead planning with Headroom (10:00 after day D)

It is expected that there is insufficient BESS capacity to absorb the forecast errors on the following day when the lower SOC limit constraint is branched to Eq.(10) in section D. Therefore, in this case, the planned value should be reduced in day-ahead planning in section A. The day-ahead plan for the following day is reduced by the amount of B_{lack} determined by Eq.(11).

$$B_{lack} = B_{max} \times 0.5 - B(48) \quad (11)$$

In order to ensure sufficient remaining BESS capacity by the time of activation of securing reserve, only the planned value before 12:00 was reduced. If the reduction of the planned value could not secure the amount of B_{lack}, P_p was set to zero for the next day, and the day was set to be a day for charging only.

The flowchart for Step.3 is shown in Fig 5.

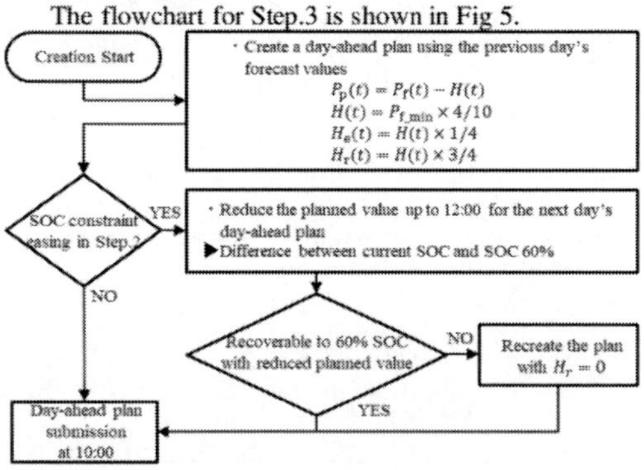

Figure 5: Flowchart of Step.3

2.5 The variable ($D_{P_f'}$)

The P_p' used within the objective function in Eq.(4) is calculated from the intra-day forecast P_f' and the amount of discharge from the BESS. This means that depending on the accuracy of P_f', a lack of SOC will occur. The gray area in Fig.6 is the amount of electricity that needs to be covered by the BESS discharge to maintain the day-ahead planned value. If the accuracy of the intra-day forecast is low, there is a risk that the remaining BESS may not be sufficient by the 12:00 to 15:00 period. Therefore, the variable $D_{P_f'}$ was used to calculate the intra-day forecast value in case it is an overprediction. In this study, the optimization calculation was performed by multiplying P_f' by $D_{P_f'}$ when $P_p \times 0.8 > P_m$ before the GPV-GSM intra-day update time of 9:00. P_m represents measured value of PV output)

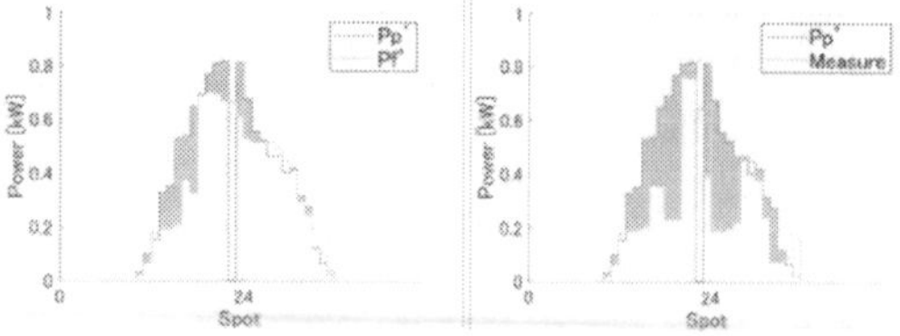

Figure 6: Comparison of Discharge Amounts

3 USED DATA

In this study, solar radiation, air temperature, and wind speed from GPV-GSM were used for day-ahead PV output forecast. (Tateno, Tsukuba City, Ibaraki, Japan). The estimated period is 31 days from May 1st, to 31st, 2021. We used JEPX(Japan electric power exchange) day-ahead spot market prices for the electricity price data. Regarding the intra-day forecast, data from 0 to 15 hours ahead in the GPV-GSM updated at 9:00 on the same day were used. Regarding the intra-day forecast from 8:00~9:00, the measured values with Gaussian noise were used.

4 RESULTS AND DISCUSSION

4.1 Results for all periods

Simulations were conducted to determine if reserve power is secured at the actual demand supply point for the case where the case of operation with Steps 1 through 3 and that of operation with Step 1 only. The optimization solver used was "fmincon" function in MATLAB. In the case of operation with only day-ahead plan, when $P_m < P_p$, the BESS only discharges the shortfall up to P_p to maintain P_p. And when $P_m > P_p$, it recharges the surplus exceeding P_p to maintain P_p. The simulation results are shown in Table I. The results are calculated per 1 kWp of installed capacity. The capacity of the BESS is set to 2 kWh, the initial value of SOC is 50%, and charging the BESS by purchasing power from market is not taken into consider. Net revenue was defined as the revenue from electricity sales minus the imbalance cost.

Table I: Power plant revenues and imbalances

	Only Step1	Step1~Step3	
$D_{P'_t}$		1.0	0.6
Imbalance [kWh]	27.66	19.77	22.75
Imbalance cost [yen]	239.55	166.70	171.45
Selling electricity [kWh]	116.38	116.62	113.72
Revenue [yen]	534.93	586.04	565.22
Opportunity loss [kWh]	0	0.69	3.27
Net revenue[yen]	295.38	419.33	393.78
EXPR penalty day count	17	11	7

In Table I, firstly, we compared the results of the case operated only with Step.1 and the case operated with Steps 1-3. The latter of the imbalance amount and EXPR Penalty Day count were improved. Accordingly, the usefulness of the intra-day modification in Step.2 and the planned value reduction in Step.3 are confirmed.

4.2 Usefulness of intra-day modifications, SOC transition calculations and planned value reductions.

The simulation results from May 26 to 28, 2021 are shown in Figs.7 to 9.

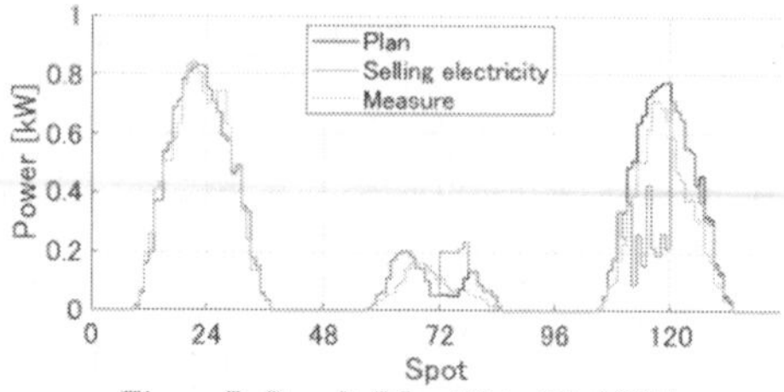

Figure 7: Result (May 26 to 28, 2021)

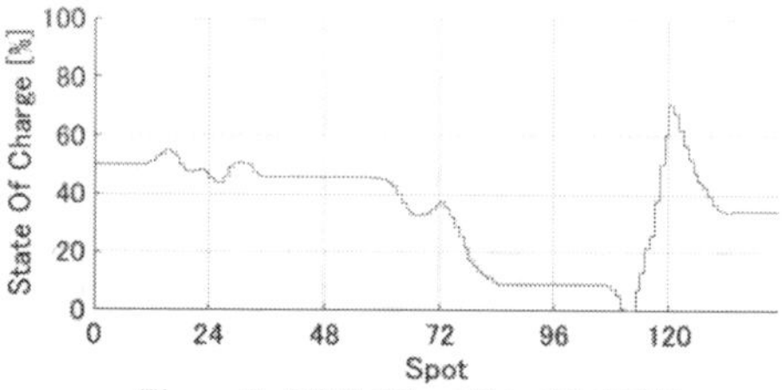

Figure 8: SOC (May 26 to 28, 2021)

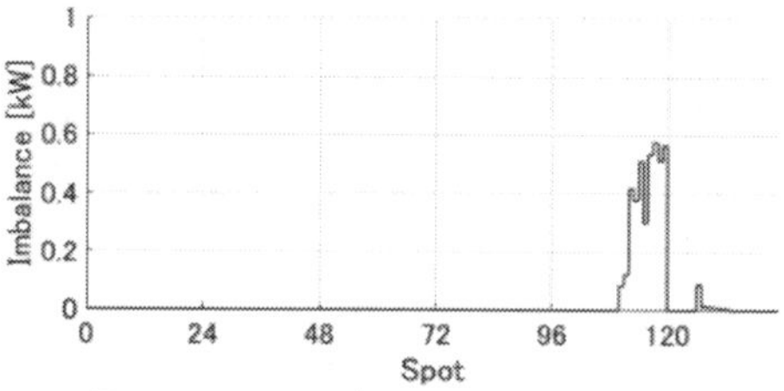

Figure 9: Imbalance (May 26 to 28, 2021)

Figs.7~9 show that the initial SOC on May 28 was close to 0%, but the intra-day modification in Step.2 recovered the SOC by 12:00, avoiding the EXPR penalty. However, it is necessary to consider how to create the day-ahead plan, which is created on D-1 day.

Table I also shows that the EXPR penalty day count has not been eliminated completely even when operating with Steps.1 to 3. Here, the simulation results from May 5 to 7, 2021 are shown in Figs.10 to 12.

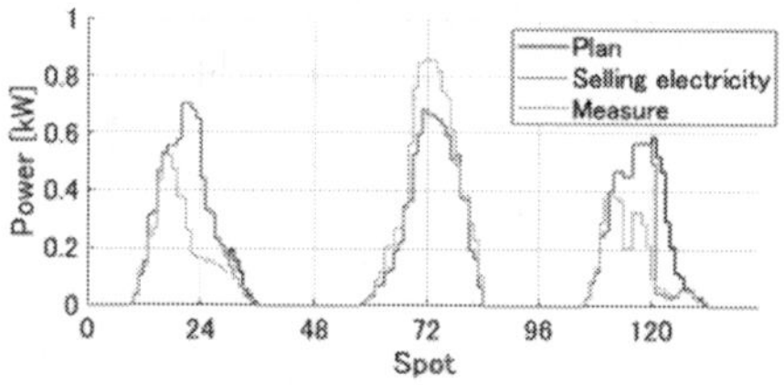

Figure 10: Result (May 5 to 7, 2021)

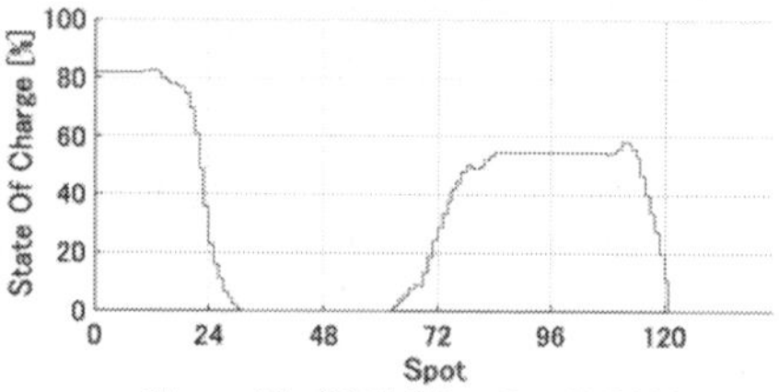

Figure 11: SOC (May 5 to 7, 2021)

10.4229/EUPVSEC2023/5DV.2.8
020525-003

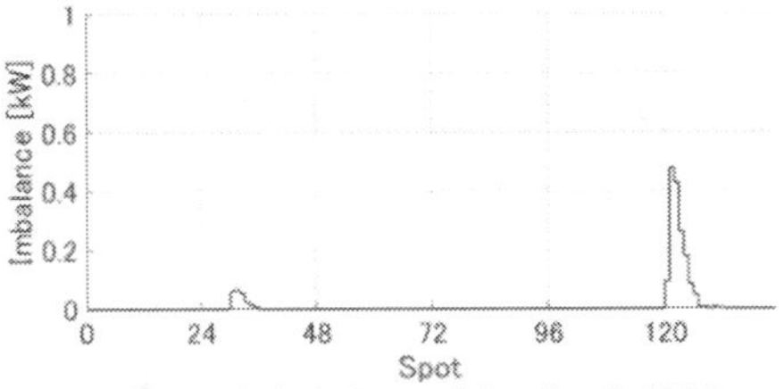

Figure 12: Imbalance (May 5 to 7, 2021)

For May 5, the previous day's forecast is considerably over-predicted. In addition, the day-ahead plan and measured values before 9:00, are generally in similar shape, however, after 9:00, the PV output is significantly lower. We can almost avoids the EXPR Penalty by discharging the BESS, however, the SOC at the end of May 5 is 0%. It can be seen that the same phenomenon as on the May 5 occurred on the May 7. However, in the case of May 7, the SOC became 0% in the middle of the operation and the EXPR market penalty can not be avoided. This is the result of not being able to respond to fluctuations after 9:00, irrelevant to $D_{P_f'}$. Therefore, it is necessary to increase the count of intra-day modifications and to change the day-ahead planning method.

4.3 Comparison of each $D_{P_f'}$

We make the comparison when $D_{P_f'}$ different, shown in Table I. Comparing the results when $D_{P_f'} = 1.0$ and $D_{P_f'} = 0.6$, the latter of the EXPR penalty day count improved, but the imbalance amount increased. The results for different $D_{P_f'}$ on May 2, 2021 are shown in Fig.13.

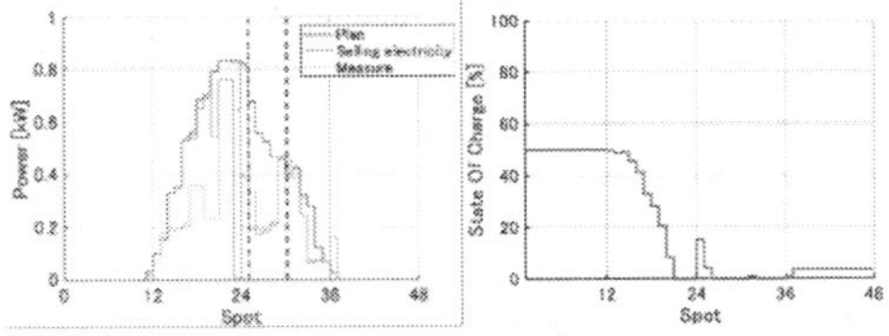

(a) $D_{P_f'} = 1.0$

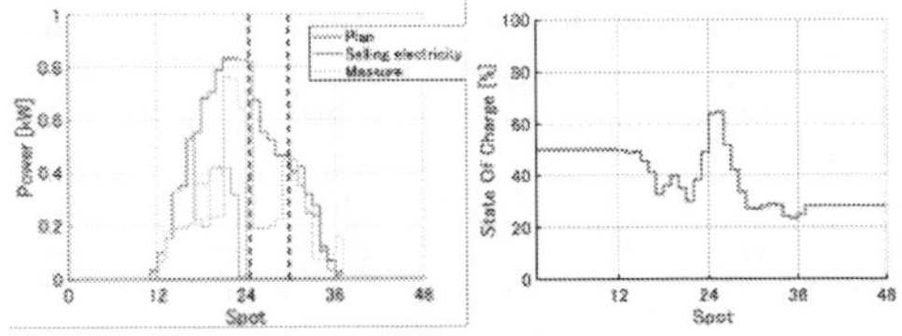

(b) $D_{P_f'} = 0.6$

Figure13: Results of power flow and SOC (May 2, 2021)

In Fig.13, if the blue and red lines correspond between the dashed lines, reserve is secured and the EXPR market penalty is avoided. If $D_{P_f'} = 1.0$, the BESS was discharged too much before 12:00 and the SOC was 0%.In the other hand, if $D_{P_f'} = 0.6$, the discharge of BESS is restrained to keep the SOC percentage and reserve power is secured. Based on these results, we consider it is necessary to make calculations with a certain degree of expectation of overprediction when using intra-day forecasts for intra-day modifications.

5 CONCLUSIONS

This study examined the certainty of the regulating capability of a PV power plant with BESS by creating a generation plan for the previous day and modifying it on intra-day using mathematical optimization. The intra-day planning can not only secure reserve power but also improve in profitability. It was not easy to eliminate completely the days subject to penalty only by present intra-day plan. Therefore, we need to change the plan on the previous day and make additional revisions on the day.

For the future, we will examine how to create a day-ahead plan and additional times for intra-day modifications to the electricity sales plan.

6 ACKNOWLEDGMENTS

This work was supported by NEDO "Demonstration study of photovoltaic power generation technology to create flexibility".

7 REFELENCES

[1]T. Sasaki, Y. Ueda, H. Ootake, T. Oozeki, Y. Yamada, J. Imura, E. Takenaka,
T. Nakazima, "Development of wide-area dataset of PV output and residential load for power system simulation", Proceedings of 2019 IEEJ National Convention

10.4229/EUPVSEC2023/5DV.2.8
020525-004

VALIDATION OF A PV GENERATION MODEL FOR SIMULATION OF WIDE AREA AGGREGATED DISTRIBUTED PV POWER GENERATION THAT TAKES INDIVIDUAL SYSTEMS LOCATION AND ORIENTATION INTO ACCOUNT

Lisa Molin, Sara Ericson, David Lingfors[1], Joakim Munkhammar[1] and Johan Lindahl[2]
[1] Uppsala University
[2] Becquerel Sweden AB
[2] Staffansvägen 14, SE-74142 Knivsta, +46768511773, johan@becquerelsweden.se

ABSTRACT: Understanding the photovoltaic (PV) power generation's temporal and spatial patterns is vital for grid balancing. This study aims to validate a simulation model for historical decentralized PV power generation, extending it to encompass the unique orientation of all PV systems within the Swedish municipality Knivsta. In a previous research project, a Convolutional Neural Network exhibited a 95% accuracy of identifying PV systems within Knivsta. In this project, using Light Detection and Ranging data, the orientation and area of detected PV systems was estimated. By combining this information with local weather and irradiance data, historical PV power generation was simulated. The regression analysis demonstrates strong correspondence between simulated and measured hourly generation for six reference systems, with coefficients of determination between 0.69–0.83. This study derives generic module parameters based on installation year and an average DC-to-AC ratio, enabling municipal-level simulations. Simulations for 2022, considering one scenario with optimal orientation for all PV systems and one scenario with derived real-condition orientations, reveal a smoothing effect in the daily pattern of aggregated PV generation, if considering real orientations. At the peak hour, power generation was found to be 10% lower when considering individual orientations compared to assuming optimal orientation across all facilities.
Keywords: Photovoltaics, LiDAR, System orientation, Power simulation, Grid balance

1 INTRODUCTION

Solar photovoltaic (PV) installations have experienced rapid growth over the past decade [1]. Projections from the International Energy Agency indicate that this upward trajectory will continue, with annual installations consistently setting new records [2]. Consequently, this trend is expected to drive further expansion in small-scale and decentralized electricity generation. For instance, in Sweden, distributed PV capacity reached 1 453 MW by the end of 2021, constituting a remarkable 92% of the total installed PV capacity [3].

The decentralized nature of distributed PV presents challenges in accurately measuring and predicting power generation. A significant portion of PV-generated power is self-consumed, leaving only grid-fed electricity subject to measurement by distribution system operators (DSOs) [4]. Consequently, the total aggregated PV electricity generation remains unknown in Sweden, as well as in most other countries globally. This lack of insight into the timing and quantity of PV-generated power poses significant challenges for DSOs in terms of grid balancing and capacity planning [5]. Consequently, there is a need for methods to forecast and estimate expected electricity generation from distributed PV systems [6].

A grid-based simulation model for the Nordic countries designed by [7], was tailored in a Swedish research project [8], to calculate the average hourly power output per installed DC capacity for PV systems. This model necessitates technical specifications of system components, including installed direct current (DC) and alternating current (AC) capacities, as well as the orientation (tilt and azimuth) of the PV modules. The model requires technical informatión of the system components, the installed DC and AC capacities, and the orientation of the PV modules. It also applies local hourly weather and irradiance data, with a spatial resolution of 2.5×2.5 km, which requires the geographical location of the PV system [9,10].

Recent research have yielded effective methods for remotely determining the precise geographical positions of PV and solar thermal panels, employing aerial imagery and machine learning techniques [11–17]. In one study, succeeding this work, a convolutional neural network (CNN) trained on orthophotos demonstrated a remarkable accuracy, achieving a 95% success rate in detecting PV systems within a Swedish municipality [16].

A significant challenge in modelling distributed PV power generation lies in the often unknown orientation of solar panels [6, 12]. Nonetheless, studies have showed the potential of using light detection and ranging (LiDAR) data to estimate the orientation of PV systems [18–20]. In a Swedish study [20], PV system orientations were assessed by computing the orientations of roof facets, using LiDAR data provided by the Swedish Land Survey. According to [19], LiDAR-derived orientations were found to be more accurate than reported data from the system owners, which often contain substantial uncertainties.

The combination of aerial imagery, LiDAR data and machine learning techniques enables remote geographical positioning and orientation estimations of distributed PV systems. When coupled with the previously mentioned simulation model [8] for computing the average hourly power output, this approach makes it possible to simulate the electricity generation from PV systems with known technical characteristics. To our knowledge, estimating the aggregated PV generation of distributed systems in a large geographic area using this approach, has so far not been done. Further, this methodology lacks validity. Therefore, in this study we implemented this method to simulate the hourly power generation of six reference PV systems and validated the result, by comparing the simulated PV power generation to the measured generation.

The underlaying motivation of the study is that predicting aggregated power generation, where the individual orientations are considered, can offer benefits for DSOs. As they lack the information of the orientation of PV systems, the Swedish DSOs traditionally operate on the assumption that all PV systems simultaneously can deliver their nameplate AC capacities, to ensure the adequacy of grid dimensions. This scenario is unlikely to

10.4229/EUPVSEC2023/5DV.2.10
020527-001

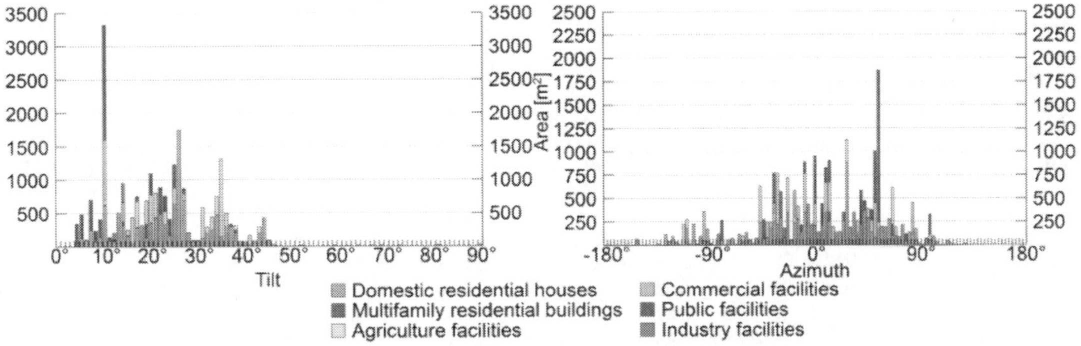

Figure 1. The tilt and the azimuth of all PV systems in Knivsta Municipality, identified and classified in the study of [10] and where the orientations have been calculated by the LiDAR methodology presented in [11].

occur since PV systems are typically installed with the same orientation as the roof facet [16,18], resulting in a wide spread of orientations, which therefore are not optimally oriented [21]. Based on the methodology-approach described earlier, the orientations across the Swedish Municipality of Knivsta were assessed and are illustrated in Figure 1. The broad spectrum of orientations within the municipality serves as compelling evidence that PV systems are improbable to generate their installed AC capacity in a synchronized manner. In addition to the primary goal of validating the PV power simulation model, this study also aspires to offer an initial quantitative assessment of the magnitude of the smoothing effect resulting from the wide spread of orientations of PV systems in the Swedish Municipality of Knivsta.

2 METHOD

This section presents the model for simulating PV power generation, the validation of the model, the data used in this study and assumptions made when simulating on a municipal scale.

2.1 Simulating PV power generation

2.1.1 PV power model

The average hourly electricity generation from a PV system per kW installed DC capacity P_{DC}/kW_{DC} (W/kW_{DC}) is calculated with a modified version of Eq. (1) from Campana et. Al [7] by;

$$P_{DC/kW_{DC}} = G_{g,t}\left[1+\frac{\mu}{\eta}(T_a - T_{STC}) + \frac{\mu}{\eta}\frac{9.5}{5.7+3.8v}\frac{(NOCT-20)}{800}(1-\eta)G_{g,t}\right], \quad (1)$$

where η is the efficiency of the PV module at standard test conditions (STC) (%), μ is the temperature coefficient of the maximum power efficiency (%/°C), T_a is the ambient temperature (°C), T_{STC} is the temperature during STC (25°C), v is the wind speed (m/s), NOCT is the nominal operating cell temperature (°C) and $G_{g,t}$, is the global tilted irradiance (W/m^2).

The inverter was simulated with the Steca Grid 300 efficiency curve [7]. To handle potential curtailment from the inverter, the model uses a minimizing function to generate the hourly AC power output per installed kW DC capacity, $P_{AC/kWDC}$, at every hour expressed by;

$$P_{AC/kW_{DC}} = min\left(P_{DC/kW_{DC}}, \frac{1000}{P_{max,DC}/P_{max,AC}}\right), \quad (2)$$

where $P_{max,DC}$ is the installed DC capacity and $P_{max,AC}$ is the installed AC capacity.

To generate the final total hourly power output from a PV system, in kilowatt instead of watt, P_{AC} (kW), the power generation per installed kW_{DC} needs to be multiplied with the nameplate capacity of the system, using the following equation:

$$P_{AC} = \frac{P_{AC/kW_{DC}} \times P_{max,DC}}{1000}. \quad (3)$$

Calculating the global tilted irradiance $G_{g,t}$, involves the both a transposition model and an absorption model. The transposition model utilized in the study, known as Perez1990, is developed for Nordic countries and publicly available in the PV_LIB [22]. The absorption model takes the irradiance output from the transposition model, the tilt of the PV modules, the solar elevation and the angle of incidence, to output the global tilted irradiance $G_{g,t}$. Spectral losses and losses due to snow, soiling and shadows are neglected in the model [7].

2.1.2 Weather and irradiance data

The weather and irradiance data employed in this research is derived from the models STRÅNG and MESAN, developed by the Swedish Meteorological and Hydrological Institute (SMHI). The STRÅNG model produces (global horizontal and direct normal) irradiance data for all Nordic countries, with a spatial resolution of 2.5×2.5 km, on an hourly basis [9]. The MESAN model generates hourly meteorological data, such as the ambient temperature and wind speed, with the same spatial resolution as STRÅNG [10].

2.2 Model validation

The validation was executed in two steps through a comparison between the measured PV power generation and the computed generation, derived from the PV power model presented in Section 2.1.1. The first validation step compares the measured generation to simulated results with orientations provided by the reference system owners. In contrast, the second validation step employs individual LiDAR-derived orientations to all systems. This scenario aims to approximate real conditions and the resulting aggregated PV power generation. The time period for the simulations were limited to the years of

10.4229/EUPVSEC2023/5DV.2.10
020527-002

2018–2022, due to the availability of high-resolution weather data and LiDAR data since 2018 [10,23].

2.2.1 Reference data of PV systems

Hourly measurements of electricity generation were provided by CheckWatt AB for six PV systems (PVS1–PVS6), situated in the municipalities scanned in [16]. The reference systems have a letter indicating their respective municipality; "U" is Uppvidinge, "F" is Falun and "K" is Knivsta.

The gathered information about the reference systems were the coordinates (WGS84), the orientation (tilt and azimuth), installed DC and AC capacity, module and inverter model and the year of installation. If the systems consisted of two or more parts with different orientations, the installed DC capacity and orientations of each subsystem were reported, as well as the configuration of the inverter(s). Details on size and timing of system expansion during the studied years were also taken into consideration. The information from the reference system owners was cross-checked with the Swedish direct capital subsidy, system data reported to the local grid owners, the size and number of modules, as well as manually estimated azimuth from arial images. The characteristics of the reference systems are displayed in Table 1 below.

Table I: Characteristics of the reference systems.

System	Installed DC capacity [kW]	Installed AC capacity [kW]	Number of sub-systems
PVS1-U	10.1	9.2	1
PVS2-U	11.4	17.0	1
PVS3-F	17.2	15.0	1
PVS4-F	19.5	16.4	3
PVS5-K	8.1	8.2	2
PVS6-K	20.9	10.0	4

2.2.2 LiDAR data

In the second validation step, the orientations of the reference systems were derived from LiDAR data. It consists of laser pulses which gives spatial 3D-information about surfaces [19,24]. The LiDAR dataset employed in this research was gathered from SMHI in March 2018, with a resolution of 2.5×2.5 km [23]. The data was also utilized when simulating the aggregated PV power generation for Knivsta Municipality, which is described in a following section.

2.2.3 Regression analysis

For this study, the coefficient of determination (R^2), the mean absolute error (MAE), the mean squared error (MSE), the root mean squared error of residuals (RMSE) and the mean bias error (MBE) were calculated for all simulations of reference systems PVS1–PVS6. Scatter plots for each reference facility were plotted and color-coded based on the time of day and time of year. The time intervals were chosen by applying different intervals and their effect on appearing patterns.

2.3 Simulating on a municipal level

To assess the smoothing effect of the wide spread of PV system orientations, the aggregated PV power generation was simulated for all identified PV systems within Knivsta. The generation was simulated for every hour of 2022 for two different scenarios. The first scenario assumes that all systems are having the optimal orientation for Knivsta, which posits their ability to generate their nameplate AC capacity simultaneously. The optimal orientation for the period of 2018–2021 was calculated by the grid-based optimization tool developed in [8], which takes local weather patterns into account, and resulted in a tilt of 45.62° and an azimuth of 174.24° for Knivsta. The second scenario implements individual orientations approximated using LiDAR data. A schematic view of the data and models used for simulating the PV power generation on a municipal level is illustrated in Figure 2.

2.3.1 Generic module parameters and DC-to-AC ratio

To simulate the aggregated PV generation for distributed systems on a municipal scale, with unknown technical properties, generic module parameters were estimated based on the year of installation. In a previous project, module data sheets of generic PV modules available on the Swedish PV market were collected for years 2014–2019 [25]. Complementing this database with additional data sheets collected for this study from Swedish retailers, resulted in a database of 200 different data sheets for monocrystalline and polycrystalline silicone PV modules available on the Swedish market on different occasions between 2015–2020.

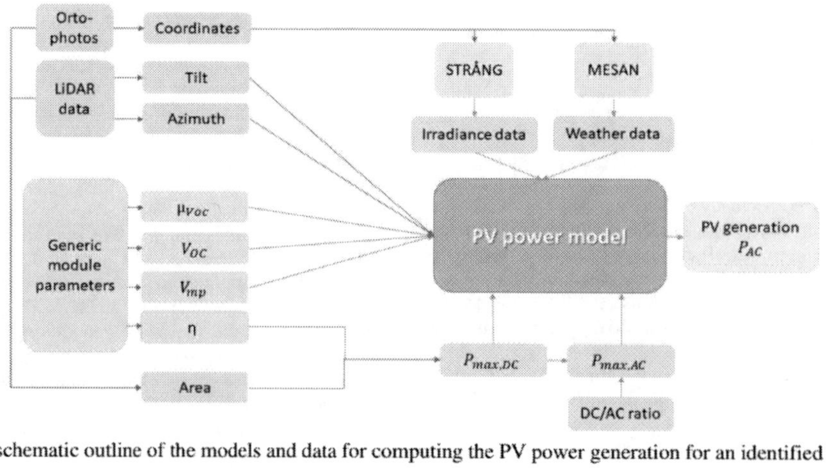

Figure 2. A schematic outline of the models and data for computing the PV power generation for an identified PV system.

10.4229/EUPVSEC2023/5DV.2.10
020527-003

From this database, an annual average of the following module parameters was derived for the years of 2015−2020; (1) the efficiency η, (2) the installed DC capacity, (3) the nominal voltage V_{mp}, (4) the open circuit voltage V_{oc} and (5) the temperature coefficient of the open circuit voltage μ_{Voc}. Additionally, the NOCT was assumed to be 45° for all modules. Consequently, this established an average and generic module representing each year between 2015 and 2020.

The installed DC capacity of each system was estimated based on the generic module efficiency and the LiDAR-corrected PV polygon area. Furthermore, an average DC-to-AC ratio was computed to determine the inverter curtailment and resulting hourly AC power output. A ratio 1.08 was derived from the average DC-to-AC ratio of 115 PV systems installed on single-family houses in Sweden collected for a study in 2020 [26]. The assumption was made that all PV systems had the same DC-to-AC ratio.

3 RESULTS

This section presents the validation of the PV power model. This is followed by the results from simulating the aggregated PV generation of distributed systems in Knivsta Municipality, using the PV power model.

3.1 Model validation

The total simulated and measured electricity production for the reference facilities is presented in Table II and III. As revealed by the tables, it is apparent that the model exhibits a consistent tendency to overestimate PV generation. Furthermore, it is important to emphasize the substantial variability in the percentage variances observed across the reference systems.

Table II: The measured and simulated PV power, using orientations provided by the reference system owners, and the percentual difference compared to the measured power.

System	Simulated PV power [MWh]	Measured PV power [MWh]	Difference [%]
PVS1-U	41.0	39.0	5.1
PVS2-U	43.9	40.1	9.5
PVS3-F	57.0	46.6	22.3
PVS4-F	67.7	65.5	3.4
PVS5-K	41.2	31.3	31.6
PVS6-K	86.3	64.8	33.3
Total	337.1	287.3	17.3

Table III: The measured and simulated PV power, using approximated orientations from LiDAR data, and the percentual difference compared to the measured power.

System	Simulated PV power [MWh]	Measured PV power [MWh]	Difference [%]
PVS1-U	40.9	39.0	4.7
PVS2-U	42.8	40.1	6.7
PVS3-F	53.5	46.6	15.0
PVS4-F	70.3	65.5	7.4
PVS5-K	41.3	31.3	31.8
PVS6-K	86.7	64.8	33.9
Total	335.6	287.3	16.8

Figure 3 provides scatter plots for one of the reference systems, PVS2-U, showcasing the results for both validation steps.

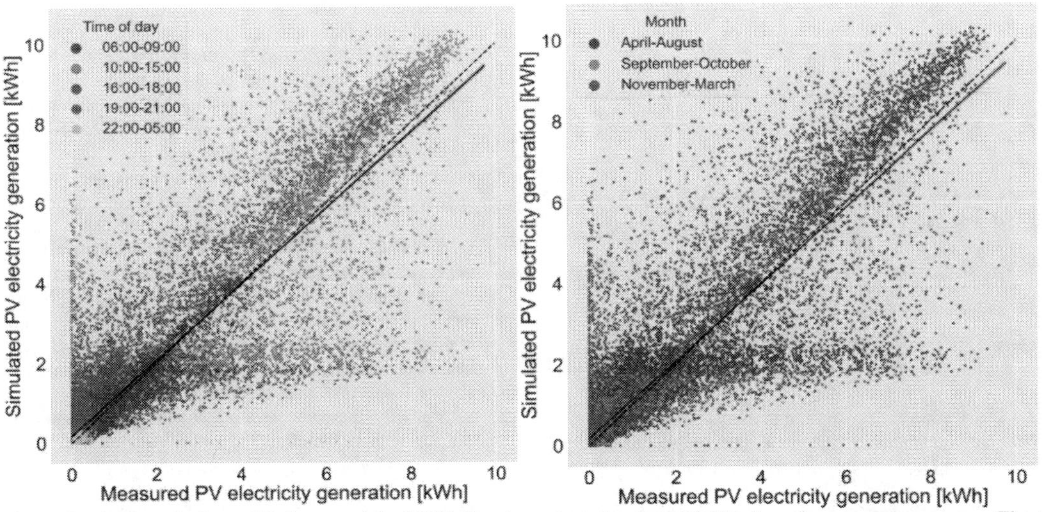

Figure 3a. Scatter plot for validation step 1 for PVS2-U, using orientations provided by the reference system owners. The data points are divided by colour, showcasing different time-intervals of the day on the left and for monthly time intervals on the right. The fitted regression is represented by a continuous line and the dotted line indicates a perfect model.

10.4229/EUPVSEC2023/5DV.2.10
020527-004

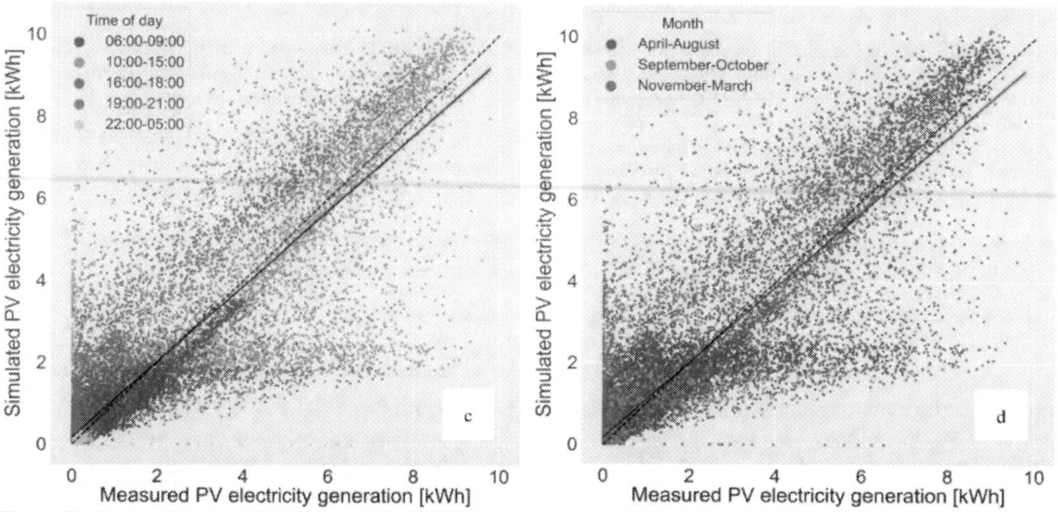

Figure 3b. Scatter plots for validation step 2 for PVS2-U, using individual orientations from LiDAR data. The data points are divided by colour, showcasing different time-intervals of the day on the left and for monthly time intervals on the right. The fitted regression is represented by a continuous line and the dotted line indicates a perfect model.

Several clusters of data points are apparent in Figure 3, at different times of the day and year, as denoted by the color-coded representations. These clusters were identified in several of the reference systems and are conceptually depicted in Figure 4, and are subject to further exploration in the subsequent section. Comprehensive scatter plots for all facilities are provided in Appendix A.

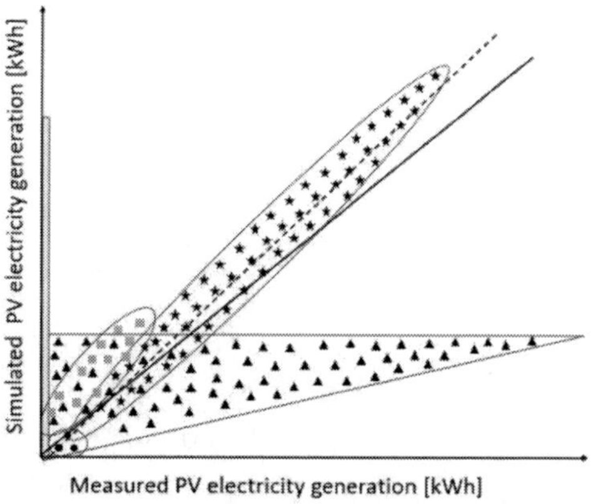

Figure 4. A graphic representation of reoccurring patterns in the reference system scatter plots.

3.1.1 Daily and yearly variation patterns

There is one cluster (stars in Figure 4) where the data points closely align with the perfect fit, showing that the simulated production values correspond well to the measured data. Within this cluster, it appears to be a slight overestimation of the simulated PV generation compared to the measured data. All daytime hours (6 am to 9 pm) occur in this cluster. However, a substantial portion of the data points is represented by blue dots in Figure 3b and 3d, indicating their predominant occurrence within the period of April to August.

A noticeable cluster or "tail" is evident, where the simulated PV electricity generation is approximately 2 kWh (triangles in Figure 4), representing instances where the model underestimates the PV generation compared to the measured data. This cluster has an upper limit, commonly close to 2–3 kWh, see Appendix A. It encompasses all hours of the day and months of the year, with the highest production levels on the right side, aligning with midday and summer months, and gradually tapering towards lower values on the left, which coincide with night-time and the winter months. This consistent behaviour of the PV power estimations, across all months and hours, could potentially be attributed to systematic faults, such as measurement inaccuracies in the MESAN and STRÅNG data.

An overestimation of the simulated PV electricity generation is displayed during the evening hours of 7–9 pm, as depicted by the red dots in Figure 3 (squares in Figure 4). This model overestimation was also observed when analysing the difference between the measured and simulated values over the day, as there is a peak in the percentual difference in-between 8–9 pm. A similar peak is discernible in the morning, around 6 am, although its magnitude is relatively lower. Additionally, within the same cluster, the model appears to overestimate electricity generation in the late afternoon hours 4–6 pm, as indicated by the green dots in Figure 3. In contrast, the PV power model consistently underestimates production during night-time and in the early morning hours (cluster of circles in Figure 4). This pattern is observed across all scatter plots in Appendix A.

Notably, there are certain hours along the y-axis where the power model predicts PV power generation despite the absence of measured production (the gray rectangle on the y-axis in Figure 4). These events are primarily observed during the months of November to March, evident in the various scatter plots featured in Appendix A. This pattern might be attributed to the losses caused by snow cover, as it is not accounted for in the model.

Lastly, there are some measured production hours on the x-axis, mostly during night-time, where the model predicts zero power generation. These instances are infrequent and generally minor, with only a few exceptions. PVS6-K breaks this pattern, exhibiting notable

10.4229/EUPVSEC2023/5DV.2.10
020527-005

positive measured values on the x-axis, which are likely due to errors in the measurements.

3.1.2 Regression analysis

The simulated PV power, using orientations supplied by the reference system owners, corresponds well with the measured PV generation with coefficients of determination between 0.70−0.82. The RMSE, which characterizes the typical dispersion of data points around the regression line, varies from 0.90−3.41 kWh. Every reference facility displays a positive MBE, aligning with the model's tendency to overestimate total production. All statistical measurements for the two validation steps are presented in Table IV and V.

Table IV: *Statistical measurements for validation step 1, employing orientations from reference system owners. For R^2, red is the lowest value, green is the highest value and yellow is the 50th percentile (median). For the other metrics, green is the highest value, and red is the lowest value. N denotes the number of hours simulated.*

PVS	N	R^2	RMSE [kWh]	MSE [kWh]	SSR [kWh]	MAE [kWh]	MBE [kWh]
1-U	33 912	0.77	0.90	4.44	30656	0.44	0.06
2-U	33 720	0.78	1.02	5.00	34554	0.47	0.11
3-F	30 806	0.70	3.41	11.52	105008	0.84	0.34
4-F	42 280	0.73	2.31	8.40	97764	0.71	0.05
5-K	43 796	0.77	0.52	2.24	22951	0.35	0.23
6-K	40 188	0.82	1.73	9.45	69620	0.78	0.54

Table V: *Statistical measurements for validation step 2, employing LiDAR-derived orientations. For R^2, red is the lowest value, green is the highest value and yellow is the 50th percentile (median). For the other metrics, green is the highest value, and red is the lowest value. N denotes the number of hours simulated.*

PVS	N	R^2	RMSE [kWh]	MSE [kWh]	SSR [kWh]	MAE [kWh]	MBE [kWh]
1-U	33 912	0.79	0.92	4.39	31306	0.45	0.05
2-U	33 720	0.79	0.99	4.75	33278	0.47	0.08
3-F	30 806	0.69	3.30	10.59	101670	0.80	0.23
4-F	42 280	0.74	2.7	9.06	100308	0.71	0.11
5-K	43 796	0.77	0.52	2.24	22761	0.34	0.23
6-K	40 188	0.83	1.67	9.59	67166	0.77	0.55

Table IV and V demonstrates that there are varying results among the reference systems, with certain systems demonstrating a stronger alignment between the simulated and measured PV power generation in comparison to others.

When simulating with LiDAR-derived orientations, the PV power generation corresponds slightly better to the measured values, yielding coefficients of determination ranging from 0.69−0.83 and RMSE values from 0.52−3.30 kWh. Other measurements of correlation produce similar outcomes to those observed with orientations provided by the facility owners, although the LiDAR-derived orientation marginally enhances these metrics as well. Nevertheless, this improvement in the relationship between the simulated and measured PV power is not significant. Unlike verification of the azimuth, which can easily be cross-checked using aerial imagery, verifying the tilt angle is more challenging. To test the sensitivity of the results when varying the angle, the angle was adjusted for each reference system. Altering the LiDAR-derived tilt by +20° angle led to a cumulative difference in simulated power ranging from -4.8−3.9% and altering the tilt by -20° led to a difference of -7.5−3.7%.

When examining the variations in the correlation between modelled and measured PV power generation over different times of the year, the average coefficient of determination for each month is summarized in Table VI.

Table VI: The R^2-values each month for the reference systems PVS1−PVS6.

PVS	1-U	2-U	3-F	4-F	5-K	6-K
January	0.54	0.40	0.51	0.38	0.35	0.42
February	0.61	0.49	0.59	0.63	0.51	0.67
March	0.87	0.87	0.75	0.72	0.62	0.85
April	0.86	0.86	0.78	0.78	0.82	0.85
May	0.76	0.76	0.73	0.73	0.81	0.78
June	0.79	0.80	0.74	0.75	0.85	0.77
July	0.74	0.72	0.72	0.70	0.66	0.79
August	0.78	0.76	0.70	0.74	0.57	0.66
September	0.78	0.78	0.73	0.75	0.83	0.80
October	0.75	0.75	0.67	0.66	0.77	0.82
November	0.43	0.41	0.52	0.51	0.67	0.70
December	0.34	0.25	0.29	0.22	0.28	0.36

The R^2-values in Table VI are considerably lower for the winter months, especially between November–February. Also, the relative difference in electricity generation was calculated to be largest during the same months. On the other hand, the total generated power and the absolute difference is lower during the winter months, compared to the rest of the year.

3.2 Simulating the PV generation in Knivsta Municipality

The generic module parameters that were derived for the relevant years of installation (2015–2020) are presented in Table VII.

Table VII: Generic module parameters for silicon modules in the years of 2015–2020.

Year	2015	2016	2017	2018	2019	2020
η [%]	16.3	17.0	17.8	18.3	18.9	20.5
V_{mp} [V]	32.2	31.8	32.7	33.3	35.5	38.2
V_{oc} [V]	39.5	38.9	40.3	40.4	43.0	45.5
μ_{Voc} [mV/°C]	-123	-122	-124	-117	-127	-124
Power [W]	265	281	298	309	329	416
Modules	39	24	24	44	30	39
Manu-facturers	14	15	17	12	9	10

A trend of rising η, V_{mp} V_{oc}, and nominal power with increasing years is evident and expected. Exceptions from this trend, such as a decreasing V_{oc} year 2016, may be attributed to limited data that year. Therefore, the number of manufacturers and modules used as a basis for calculating the average technical parameter are also presented in Table VII.

Since the generic module parameters are estimations from limited data, a sensitivity analysis was conducted. The three parameters η, V_{mp} and μ_{Voc}, were adjusted to their minimum and maximum values within the dataset to test the weakness of these estimations. Altering the nominal voltage and the temperature coefficient had a minimal impact on the simulated aggregated PV power generation, compared to varying the efficiency, which significantly influenced the results.

3.2.1 Comparing the two simulation scenarios

The aggregated PV power generation of Knivsta

10.4229/EUPVSEC2023/5DV.2.10
020527-006

Municipality 2022 was calculated to 4 601 MWh, when simulating with individual LiDAR-derived orientations of the distributed systems. Simulating with all systems having the optimal orientation (of 45.62° tilt and 174.24° azimuth) resulted in a aggregated power generation of 5 200 MWh. The daily sum of the PV power generation for both simulations and the difference between them are visualized in Figure 5.

The difference in PV power between the two cases is relatively small during summer months (May–August), whereas it becomes considerably lager for the remainder of the year, except for the period from November to January, during which overall production is lower.

Given the significance of power generation data for grid sizing, a key aspect of this research is studying the day with the highest hourly power generation. In Figure 6, the days with the highest hourly PV generation for each simulation are presented. The peak PV generation in the scenario with individual orientations, of 3.72 MWh, occurs at 1 pm on May 18th. In the other scenario, where all systems are optimally oriented, the peak power generation of 4.14 MWh is achieved at 12–1 pm on March 30th. The simulation considering individual orientations is generating 3.28 MWh this same hour. Notably, during the hours of peak PV generation for both cases, the scenario with optimal orientations reaches the estimated aggregated AC capacity of 4.14 MWh. This scenario also reaches its peak another time, at 12 pm on April 5th, though this is not included in the figure. In contrast to the scenario implementing an optimal orientation, the scenario with individual orientations never generates the aggregated AC capacity, which is illustrated in Figure 6. This can be explained by that the orientations for PV systems in Knivsta Municipality are not optimally oriented, as evidenced by the dispersion in Figure 1.

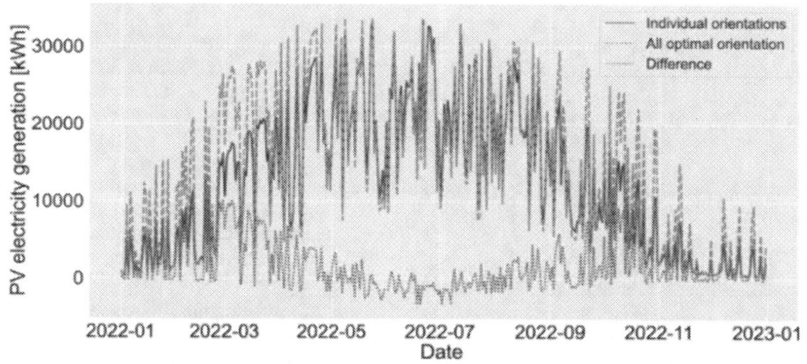

Figure 5. The daily aggregated PV power generation for Knivsta Municipality 2022, simulated for two different scenarios, and the daily difference in generation between the scenarios.

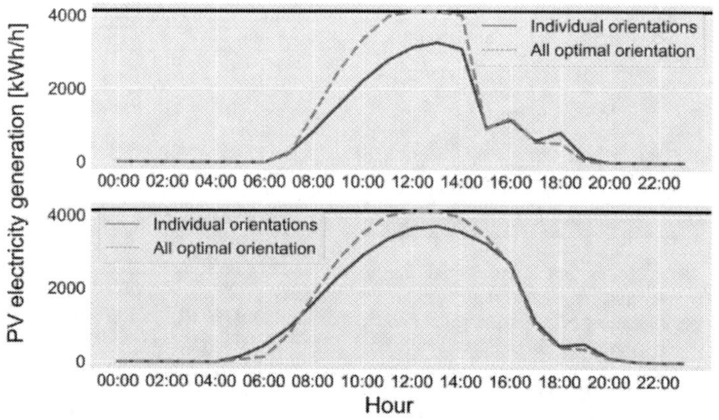

Figure 6. The aggregated PV power generation for two days with the highest generation hours 2022. The black line is the estimated aggregated AC capacity for Knivsta Municipality. The lower graph shows the peak hour with individual orientations on 1 pm May 18th and the graph above shows March 30th, where all optimal orientations achieve the aggregated AC capacity.

10.4229/EUPVSEC2023/5DV.2.10
020527-007

Another important aspect for comprehending PV generation characteristics is the average production curve. Figure 7 presents the average simulated power generation across all hours of the day for both scenarios. During morning and evening periods, the case of individual orientations exhibits higher simulated PV power generation. Conversely, the remaining hours of the day show greater production with all systems optimally oriented. Consequently, there are intersections occurring roughly at 8 am and 5 pm.

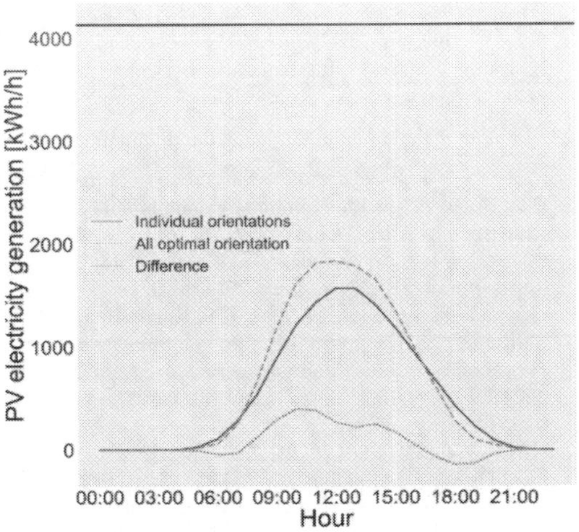

Figure 7. The hourly average PV electricity generation curve simulated for 2022. The black line is the approximated aggregated AC capacity of 4.14 MWh for Knivsta Municipality.

The peak mean production is observed at noon in both simulation scenarios, reaching 1.57 MWh for systems having individual orientations and 1.87 MWh for optimally oriented systems. It is important to note, as depicted in Figure 7, that neither of the average production curves approaches the estimated aggregated AC capacity for Knivsta Municipality. This average is with respect to all seasons, including the winter months characterized by low PV production.

4 DISCUSSION

It is essential to clarify that in simulating the aggregated PV power generation at the municipal level, certain assumptions were made, introducing uncertainties in the study's findings. Firstly, each polygon within Knivsta Municipality was modelled to have an individual inverter and a fixed DC-to-AC ratio, due to the lack of information on the actual configuration. In practice, some polygons may constitute subsystems situated within the same facility, potentially sharing a common inverter. Secondly, the generic module parameters for each year between 2015–2020 were based on the average from 200 modules on the Swedish market (ranging between 24–39 modules for each year). A more comprehensive dataset would be beneficial to refine the approximations of generic parameters. The importance of accurately estimating the module efficiency was highlighted in a sensitivity analysis, since varying this parameter significantly influenced the results. Additionally, the DC-to-AC ratio which governs the AC electricity generation, is currently sourced from 115 private residential PV systems. Expanding the database to include broader system capacities and different typical DC-to-AC ratios of different market segments would probably improve DC-to-AC ratio estimations.

Furthermore, it would be appropriate to compare the current inverter model of the Steca Grid 300 efficiency curve with other alternatives.

Another know drawback of the simulations are the inaccuracies in the weather and irradiance data sourced from the MESAN and STRÅNG models, which significantly impact the precision of the results. Though the average error statistics are not published for the studied years, as a reference, during the period from 1999−2009 the hourly RMSE for global radiation and the direct normal radiation from STRÅNG was 30% and 57%, respectively [27].

The initial finding in this study was that the modelled PV power corresponds well to the measured PV power generation. The results were marginally better when employing orientations obtained from LiDAR. This observation aligns with prior research [19], affirming a higher reliability of LiDAR data in orientation estimation compared to manual estimations of the PV system owners.

The model tends to overestimate PV power generation, especially in the late afternoon. This is possibly due to temperature differences between assumed and actual cell temperatures. The model employs air temperature data from MESAN, while cell temperatures tend to be higher in the afternoon, due to heat accumulation within modules and the roof. Consequently, the simulation might overestimate PV power generation when using a temperature coefficient based on lower cell temperatures. Likewise, the opposite situation applies to the morning discrepancy, wherein measured production exceeds the simulated production. Another explanation for the overestimation could be unaccounted shadowing effects from the surroundings. One study revealed that shading led to a 35% reduction in the anticipated global maximum power point for monocrystalline modules and a 33% decline for polycrystalline modules [28].

Further, considerably larger percentage errors of the simulated production compared to the recorded production, were noted during months of winter, accompanied by reduced R^2-values. One plausible explanation could be the influence of snow cover losses. A recent study estimated these losses to vary between 0−20% of the annual PV power generation yield in Sweden when using an average tilt angle of 25° [29], which differs from the mean tilt angle of the reference systems at 31°. Notably, the current PV power simulation model does not account for snow losses, potentially leading to an overestimation of PV electricity generation.

Since the largest absolute error between modelled and measured production occurs in summer daylight hours, shading-related losses exert a significant impact on simulation outcomes. Snow-related losses, on the other hand, have a comparatively smaller effect due to lower absolute magnitudes of simulation errors in winter. Therefore, the implementation of a shadow loss model is more crucial for enhancing result accuracy.

5 CONCLUSION

A model for simulating the historic PV power generation from individual systems was validated, employing LiDAR-derived orientation, as well as local

10.4229/EUPVSEC2023/5DV.2.10
020527-008

weather and irradiance data generated by the models STRÅNG and MESAN. Simulating with orientations provided by the reference system owners resulted in R^2-values between 0.70–0.84. Using orientations from LiDAR-data resulted in overall slightly improved metrics, indicating that LiDAR-data can be used to adequately estimate unknown orientations and may even generate more accurate results. Hence, the results from the validation suggest that the PV power model is valuable for estimating hourly PV power generation and that LiDAR data accurately estimates the orientations. Currently, the PV model neglects both shading and snow losses. Integrating these losses might decrease overestimations of the PV power model, thus enhancing accuracy.

The PV electricity generation model was further developed and used to estimate the aggregated distributed PV power generation in Knivsta Municipality. Simulating the aggregated distributed power generation considering unique LiDAR-derived orientations, never reached the total AC capacity installed in Knivsta. Furthermore, it was shown that the installed AC capacity is not generated in a synchronized manner. Instead, there is a noticeable smoothing effect in the daily pattern of the aggregated PV generation when implementing individual orientations, compared to one optimal orientation. This is explained by the wide spectrum of orientations in Knivsta. An initial quantification of this effect is that the power generation was 10% lower at the peak hour 2022, when considering individual orientations compared to assuming a uniform optimal orientation. This discovery can free grid capacity for further deployment of distributed PV systems.

6 ACKNOWLEDGEMENTS

This work was financed by the Swedish Energy Agency through the research project P2023-00440, and in addition supported by Solar Electricity Research Center Sweden (SOLVE) and the Swedish strategic research programme StandUp for Energy.

7 REFERENCES

[1] IRENA. Renewable capacity statistics 2021 Internaional Renewable Energy Agency (IRENA). Abu Dhabi: 2021.

[2] International Energy Agency. Renewable Energy Market Update - Outlook for 2021 and 2022. 2021.

[3] Lindahl J, Oller Westerberg A. National Survey Report of PV Power Applications in Sweden 2021. 2022.

[4] Luthander R, Widén J, Nilsson D, Palm J. Photovoltaic self-consumption in buildings: A review. Applied Energy 2015;142:80–94. https://doi.org/10.1016/j.apenergy.2014.12.028.

[5] Widén J, Shepero M, Munkhammar J. Probabilistic Load Flow for Power Grids With High PV Penetrations Using Copula-Based Modeling of Spatially Correlated Solar Irradiance. IEEE Journal of Photovoltaics 2017;7:1740–5. https://doi.org/10.1109/JPHOTOV.2017.2749004.

[6] Ramadhani UH, Lingfors D, Munkhammar J, Widén J. On the properties of residential rooftop azimuth and tilt uncertainties for photovoltaic power generation modeling and hosting capacity analysis. Solar Energy Advances 2023;3:100036. https://doi.org/10.1016/j.seja.2023.100036.

[7] Campana PE, Landelius T, Andersson S, Lundström L, Nordlander E, He T, et al. A gridded optimization model for photovoltaic applications. Solar Energy 2020;202:465–84. https://doi.org/10.1016/j.solener.2020.03.076.

[8] Lindahl J, Campana PE, Hansen MK. Beräkningsverktyg för optimering av solcellsparker. n.d.

[9] SMHI, Naturvårdsverket, Strålsäkerhetsmyndigheten. STRÅNG 2017. https://strang.smhi.se/ (accessed January 20, 2023).

[10] SMHI. Analysmodell (MESAN) | SMHI 2017. https://www.smhi.se/data/oppna-data/meteorologiska-data/analysmodell-mesan-1.30445 (accessed January 20, 2023).

[11] Kausika BB, Nijmeijer D, Reimerink I, Brouwer P, Liem V. GeoAI for detection of solar photovoltaic installations in the Netherlands. Energy and AI 2021;6:100111. https://doi.org/10.1016/j.egyai.2021.100111.

[12] Yu J, Wang Z, Majumdar A, Rajagopal R. DeepSolar: A Machine Learning Framework to Efficiently Construct a Solar Deployment Database in the United States. Joule 2018;2:2605–17. https://doi.org/10.1016/j.joule.2018.11.021.

[13] Kruitwagen L, Story KT, Friedrich J, Byers L, Skillman S, Hepburn C. A global inventory of photovoltaic solar energy generating units. Nature 2021;598:604–10. https://doi.org/10.1038/s41586-021-03957-7.

[14] Costa MVCV da, Carvalho OLF de, Orlandi AG, Hirata I, Albuquerque AO de, Silva FV e, et al. Remote Sensing for Monitoring Photovoltaic Solar Plants in Brazil Using Deep Semantic Segmentation. Energies 2021;14:2960. https://doi.org/10.3390/en14102960.

[15] Hu W, Bradbury K, Malof JM, Li B, Huang B, Streltsov A, et al. What you get is not always what you see—pitfalls in solar array assessment using overhead imagery. Applied Energy 2022;327:120143. https://doi.org/10.1016/j.apenergy.2022.120143.

[16] Lindahl J, Johansson R, Lingfors D. Mapping of decentralised photovoltaic and solar thermal systems by remote sensing aerial imagery and deep machine learning for statistic generation. Energy and AI 2023;14:100300.

[17] Ren S, Hu W, Bradbury K, Harrison-Atlas D, Malaguzzi Valeri L, Murray B, et al. Automated Extraction of Energy Systems Information from Remotely Sensed Data: A Review and Analysis. Applied Energy 2022;326:119876. https://doi.org/10.1016/j.apenergy.2022.119876.

[18] Killinger S, Lingfors D, Saint-Drenan Y-M, Moraitis P, van Sark W, Taylor J, et al. On the search for representative characteristics of PV systems: Data collection and analysis of PV system azimuth, tilt, capacity, yield and shading. Solar Energy 2018;173:1087–106. https://doi.org/10.1016/j.solener.2018.08.051.

[19] Lingfors D, Killinger S, Engerer NA, Widén J, Bright JM. Identification of PV system shading using a LiDAR-based solar resource assessment model: An evaluation and cross-validation. Solar Energy 2018;159:157–72. https://doi.org/10.1016/j.solener.2017.10.061.

[20] Lingfors D, Bright JM, Engerer NA, Ahlberg J, Killinger S, Widén J. Comparing the capability of low- and high-resolution LiDAR data with application to solar resource assessment, roof type classification and shading analysis. Applied Energy 2017;205:1216–30. https://doi.org/10.1016/j.apenergy.2017.08.045.

[21] Sadeghian H, Wang Z. A novel impact-

assessment framework for distributed PV installations in low-voltage secondary networks. Renewable Energy 2020;147:2179–94. https://doi.org/10.1016/j.renene.2019.09.117.

[22] Sandia National Laboratories. PV Performance Modeling Collaborative | PV_LIB Toolbox 2023. https://pvpmc.sandia.gov/applications/pv_lib-toolbox/ (accessed January 27, 2023).

[23] Lantmäteriet. Laserdata Nedladdning, skog 2023.

[24] Lindahl J. Automatisk kartläggning av solpaneler och generering av solelsprognoser från flygbilder och maskininlärning. n.d.

[25] Gustavsson U, Rosenqvist L. Further Development of Njord, a statistical Instrument for Estimating International Installed Photovoltaic Capacities. master thesis. Linköping University, 2021.

[26] Oller Westerberg A, Lindahl J. Kostnadsstruktur för svenska villasystem. Stockholm: n.d.

[27] Swedish Meteorological and Hydrological Institute. STRÅNG validation n.d. https://strang.smhi.se/validation/validation.html (accessed June 14, 2023).

[28] Bimenyimana S, Asemota GNO, Kemunto MC, Li L. Shading effects in photovoltaic modules: Simulation and experimental results. 2017 2nd International Conference on Power and Renewable Energy (ICPRE), 2017, p. 904–9. https://doi.org/10.1109/ICPRE.2017.8390665.

[29] van Noord M, Landelius T, Andersson S. Snow-Induced PV Loss Modeling Using Production-Data Inferred PV System Models. Energies 2021;14:1574. https://doi.org/10.3390/en14061574.

Appendix A

Appendix displaying scatter plots of the remaining five PV systems (PVS1U, PVS3F, PVS4F, PVS5K and PVS6K) that were used in this study for the validation of the previously mentioned PV power model.

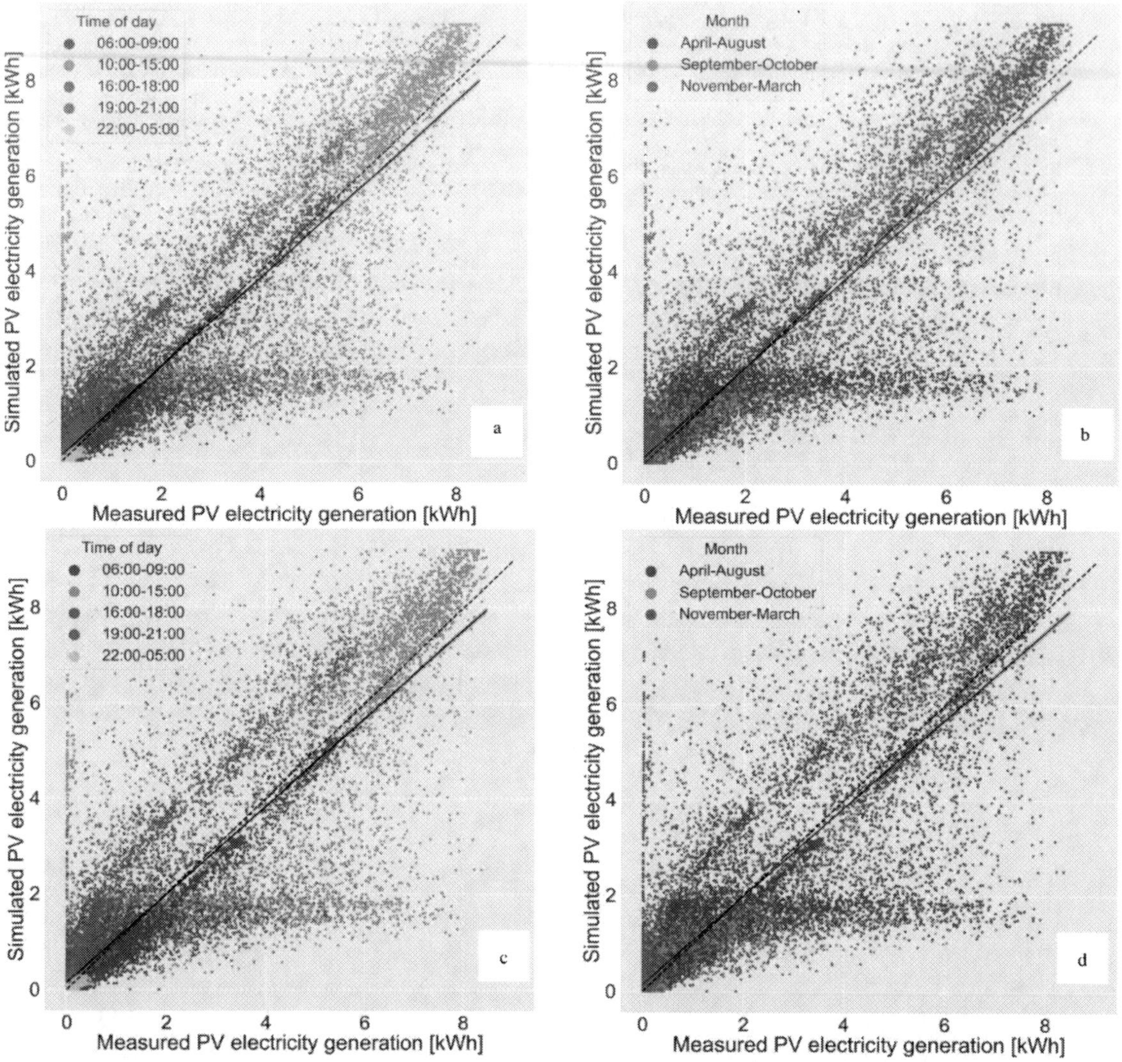

Figure 8. In (a) and (b), scatter plot for validation step 1, and in (c) and (d) validation step 2 for PVS1-U, both steps using known module- and inverter parameters disclosed by the system owners. The data points are divided using colour for time-intervals of the day in (a) and (c), and for monthly time intervals in (b) and (d). The fitted regression is represented by a continuous line and the dotted line indicates a perfect fit.

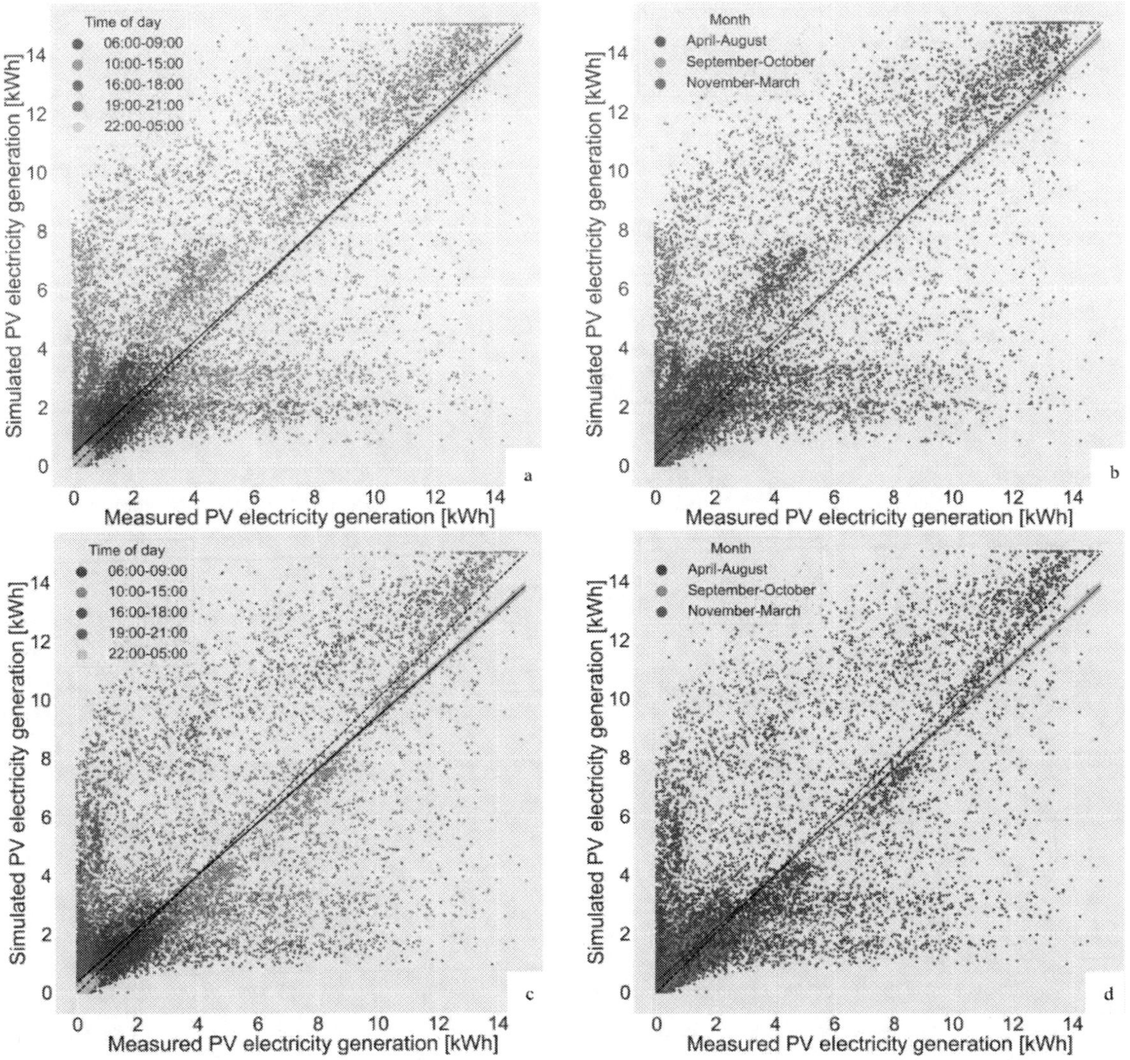

Figure 9. In (a) and (b), scatter plot for validation step 1, and in (c) and (d) validation step 2 for PVS3-F, both steps using known module- and inverter parameters disclosed by the system owners. The data points are divided using colour for time-intervals of the day in (a) and (c), and for monthly time intervals in (b) and (d). The fitted regression is represented by a continuous line and the dotted line indicates a perfect fit.

10.4229/EUPVSEC2023/5DV.2.10

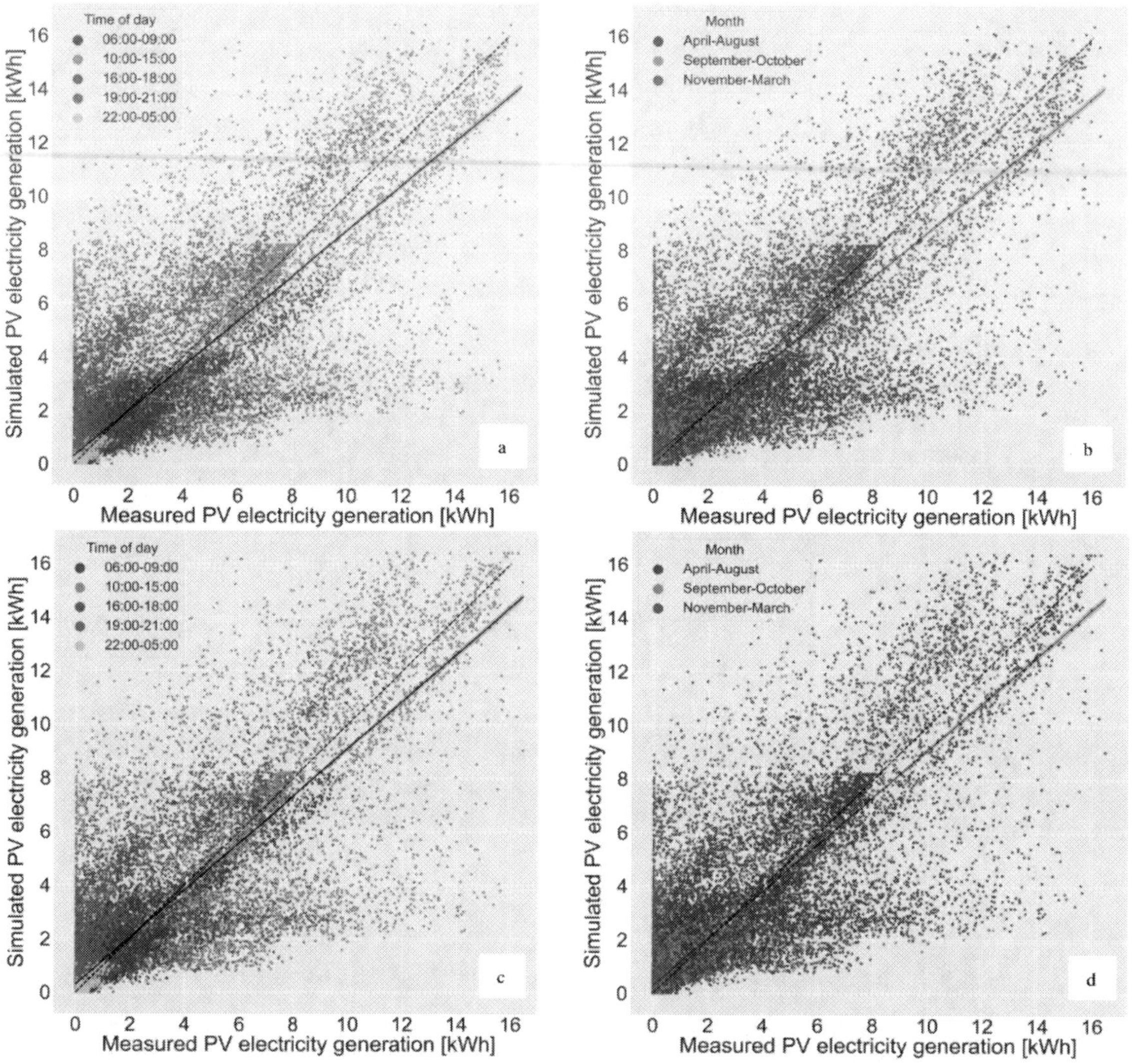

Figure 10. In (a) and (b), scatter plot for validation step 1, and in (c) and (d) validation step 2 for PVS4-F, both steps using known module- and inverter parameters disclosed by the system owners. The data points are divided using colour for time-intervals of the day in (a) and (c), and for monthly time intervals in (b) and (d). The fitted regression is represented by a continuous line and the dotted line indicates a perfect fit.

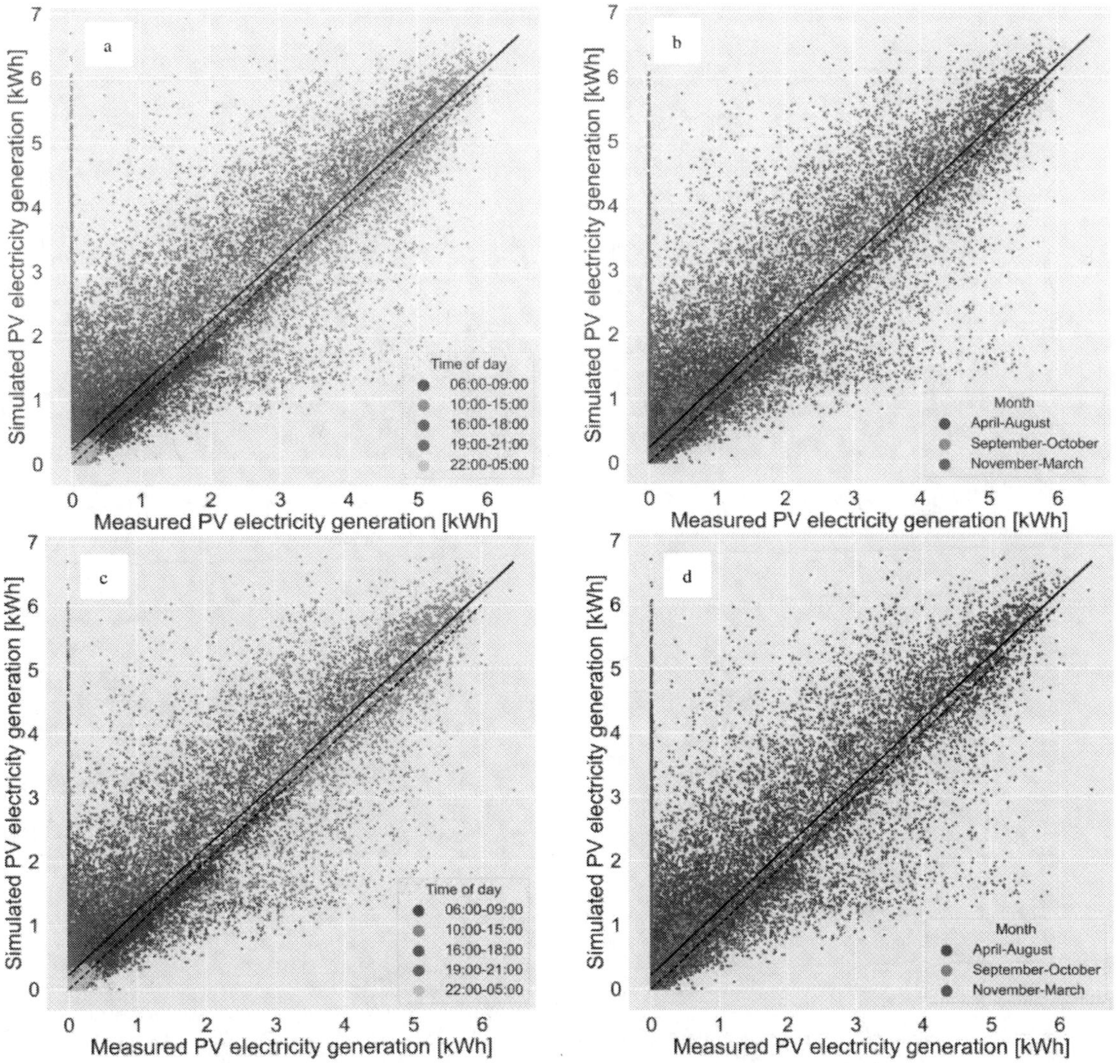

Figure 11. In (a) and (b), scatter plot for validation step 1, and in (c) and (d) validation step 2 for PVS5-K, both steps using known module- and inverter parameters disclosed by the system owners. The data points are divided using colour for time-intervals of the day in (a) and (c), and for monthly time intervals in (b) and (d). The fitted regression is represented by a continuous line and the dotted line indicates a perfect fit.

10.4229/EUPVSEC2023/5DV.2.10
020527-014
2025

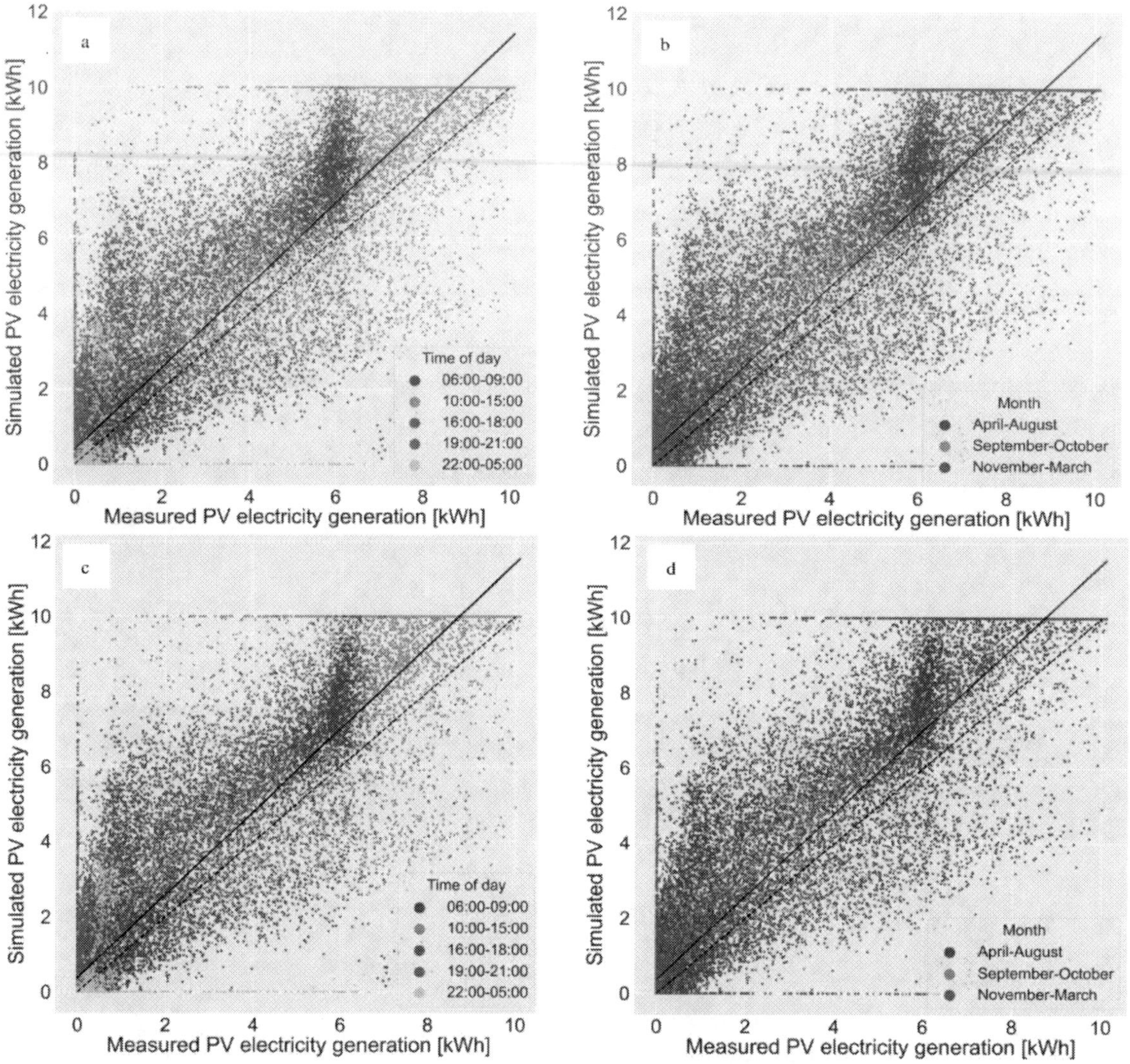

Figure 12. In (a) and (b), scatter plot for validation step 1, and in (c) and (d) validation step 2 for PVS6-K, both steps using known module- and inverter parameters disclosed by the system owners. The data points are divided using colour for time-intervals of the day in (a) and (c), and for monthly time intervals in (b) and (d). The fitted regression is represented by a continuous line and the dotted line indicates a perfect fit.

10.4229/EUPVSEC2023/5DV.2.10
020527-015

ENERGY OPTIMIZATION FOR COMPANIES WITH DIGITAL FLEX TWINS

J. S. da Costa Fernandes, A. Aamoume, J. Lottermoser, R. Gasper, M. Schmidt
Hochschule Offenburg – University of Applied Sciences / INES – Institute of Sustainable Energy Systems
24, Badstraße, 77652, Offenburg, Germany
Phone: +49 781 205 354, Fax: +49 781 205 45354
E-mail: costa@hs-offenburg.de
Internet: www.ines.hs-offenburg.de

ABSTRACT: In this paper we report on further success of our work to develop a multi-method energy optimization which works with a digital twin concept. The twin concept serves to replicate production processes of different kinds of production companies, including complex energy systems and test market interactions to then use them for model predictive optimizing. The presented work finally reports about the performed flexibility assessment leading to a flexibility audit with a list of measures and the impact of energy optimizations made related to interactions with the local power grid i.e., the exchange node of the low voltage distribution grid. The analysis and continuous exploration of flexibilities as well as the exchange with energy markets require a "guide" leading to continuous optimization with a further tool like the Flexibility Survey and Control Panel helping decision-making processes on the day-ahead horizon for real production plants or the investment planning to improve machinery, staff schedules and production infrastructure.
Keywords: Energy Flexibility, Energy System Integration for Industry 4.0, Digital Flex Twin Optimization, Energy Marketing of Industrial Flexibilities

1 INTRODUCTION

Previous works on the development of a generic model library indOptFlex© used to compose a digital twin model [4] and the experimental twin as real lab representing a SME (Small and Medium Enterprise) were presented at EU PVSEC 2021[5]. Within this context further results were presented and discussed at WCPEC 2022 [6] mainly related on the work done on leveraging energy flexibilities [1,2] to optimize the organization and production processes on behalf of the first case study of a plastics production company. Strongly motivated by the successful research work the recent development tends to a multi-method optimization working with a bundle of tools interacting in a complex network with significant challenges for the communication structure. Facing already established Industry 4.0 requirements most of the internal communication is done using OPC-UA based networking. The following sections are therefore presenting the elements making up functional blocks of the networking to realize a multi-method optimization.

2 OPTIMIZATION APPROACH

2.1 Multi-method optimization

Measures gained on the implementation and testing of model predictive methods are therefore carried out first on a hybrid lab system representing the industrial environment of a company without risking the performance/workflow of the company. The so-called SME - Emulator introduced in [5] is built up with the core functions and typical devices of real companies. The Emulator is used to scale up real-time-operation of the experimental twin and combines the profile with stored load profiles varied and modified as needed to meet the measured load profile of the target company.

In additional work packages more optimization criteria and external information like weather forecasts completed the inputs of a digital flex twin model making up the core of the optimizer tool as described in the networking model of Fig 1. This hybrid company setup is run instead of the real i.e., target company to check the new configuration (i. e. production profile) and to test the selected optimization algorithms by updating the production schedules, Technical Building Systems run times and control parameters. The experimental flex twin as the real lab offers a showcase to SME as it is demonstrating effects of wide changes and proposed measures that may lead to an optimized operation of the real company and delivers an insight to the operation of a future energy management. Updated schedules are produced as Day-Ahead-Profiles with time-steps of 15 minutes.

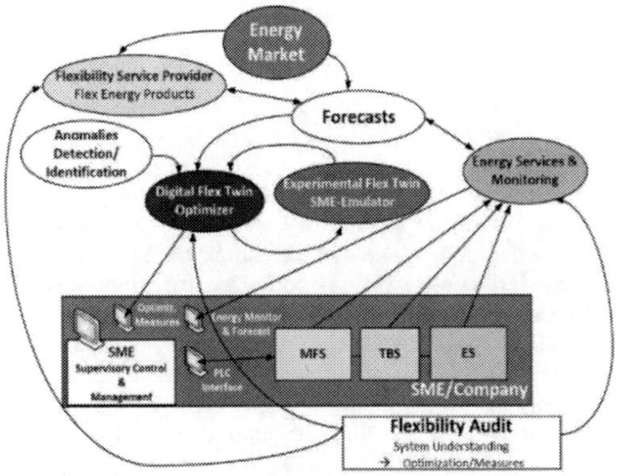

Figure 1: Networking and interactions for model based multi-method optimization (MFS = Manufacturing System, TBS = Technical Building Services, ES = Energy System)

The recent tasks handle the coupling of the SME with the utility grid at LV/MV-Level. Grid simulation was done using pandapower under python to evaluate the impacts of the energy optimization to the power grid. Most of the SME are connected and metered on the MV level but effects are expected mostly on the LV level. To profit of flexibility potentials the SME should also contribute stabilizing shares regarding the utility grid.
These flexibilities can become the base for market incentives like monetary benefits, special tariffing and

10.4229/EUPVSEC2023/5DV.2.11
020528-001

reduction of constraints that may positively affect the production performance. Work packages can be shifted to cheaper timeslots that lead to avoided peaks as far as forecasts may predict high shares of solar and wind power or recommend runs of highly energy consuming appliances.

2.2 Static Tools

With static tools one-time services or services that are repeated over a longer time period are used to adjust at wider range or focus more on investment-oriented optimization. Findings of this focus are mostly assessment or audit guided and lead to major changes of the SME infrastructure in machinery, built environment, staff use, IT tools or long-term energy supply scope. Besides the decision to direct the energy supply to a more renewables covered energy system the application of an Energy Flexibility Audit according to [3] had produced a big list of measures affecting the SME at different dimensions. Some of these topics are described later in Chapter 4.

2.2 Dynamic Tools

The works done on dynamic tools are more concentrated on the application of recent research work making up the State-of-the-Art as it is for PV energy forecasts, generation of load profiles or continuously update production schedules. An example of an algorithmic optimization loop is eq. 1 and eq. 2 describes the minimization of the acquired power at the exchange node with the local power grid.

$$P_{Grid,t} = P_{MFS,t} + P_{TBS,t} - P_{ES,t} \quad \forall t \in T = [0,1,\dots,T_f] \tag{1}$$

$$J = \sum_{t=0}^{t=T_f} P_{Grid,t} * \lambda_t^{elec} * \Delta t \quad \forall t \in T = [0,1,\dots,T_f] \tag{2}$$

λ_t^{elec}: electricity price in €/kWh at timestep "t"
Δt: sampling timestep of the optimization horizon

Figure 2: Example for the optimized power acquisition at the grid node with eq. 1 using the target function J of eq. 2 to minimize.

Implemented dynamic tools are the One-Day-Ahead Optimization for the manufacturing system according to the scheduled production of the SME and the model-predictive-control (MPC) optimized run time schedules for the TBS-units like HVAC or Compressed Air System (CAS) or if given the local Energy Supply (ES). The latter integration of Energy Flexibility Services as result of the Energy Flexibility Audit will affect the development with further dynamic functions and more communication needs with external service providers.

3 TOOL SET FOR TWIN OPTIMIZATION

The tool set for the twin optimization contains the Digital Flex Twin and Experimental Flex Twin as main parts of the optimization loop at the consortium partners of the research project. The advantage of this strategy is to have a complete test and optimization environment for the hardware and software, which use the same communication and exchange protocols as they will be used in the future with the SME and most of the algorithms can be validated prior to the implementation at a partner company.

The additional significant elements are the HMI with survey, energy monitoring and decision-making functions.

3.1 Digital Flex Twin

As presented in previous works [4,5] the Digital Flex Twin was developed as a generic library which was last updated during the flex development process and expanded with energy flexibility functions. In the recent work packages further inputs like load profiles, market incentives and energy supply forecasts were added. As the real SME's are still not part of the dynamic optimization the loop cycle is concentrated on the functional extension of the Experimental Flex Twin representing partial production sequences as real lab on a microscale level.

3.2 Experimental Flex Twin

The schematic view of Fig. 3 shows the three functional parts of the Experimental Flex Twin with the hardware part, the software part to scale-up to conditions of a SME and the disturbance generator to force external impacts allowing stability tests on the control algorithms. The results of the simulation runs are then used to optimize the MFS, TBS and ES generating parameter sets for the hardware and software modules of the Experimental Twin. The data exchange is done on PLC-level via OPC-UA.

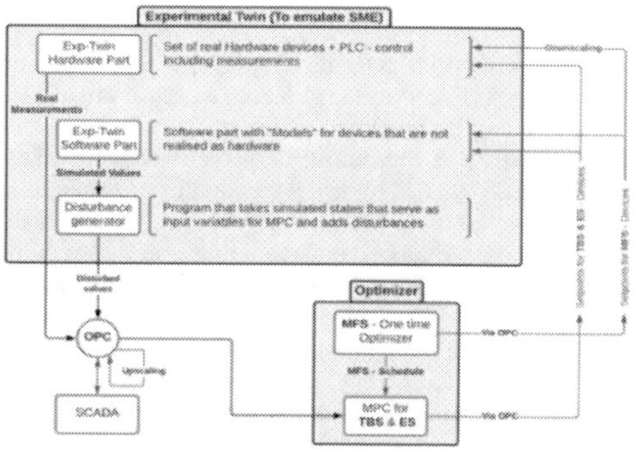

Figure 3: Continuous parameter and schedule generation within the optimization cycle of Digital Flex Twin (Optimizer) and Experimental Flex Twin.

As time schedules contained only Start/Stop-commands for the devices of the hardware the machinery pool of the Experimental Flex Twin was extended with thermal circuits for cooling and heating. The new circuits allow a higher dynamic of control algorithms and produce more accurate and adjustable signals for the composition of the SME load profiles.

Figure 4: Experimental Flex Twin hardware assembly to run as SME lab setup.

10.4229/EUPVSEC2023/5DV.2.11
020528-002

3.3 Integrating Panel for SCADA-Type Overview

The visualization of the actual and historic performance is the base for the forecast of the predicted power demand course at different acquisition conditions as shown in Fig. 5. Price signals and other performance indicators allow a comparison of modes and therefore ease decision-making processes. The panel arrangement was taken of examples for established Control and Survey panels as well as dashboards used with SCADA Systems for PLC automation infrastructures.

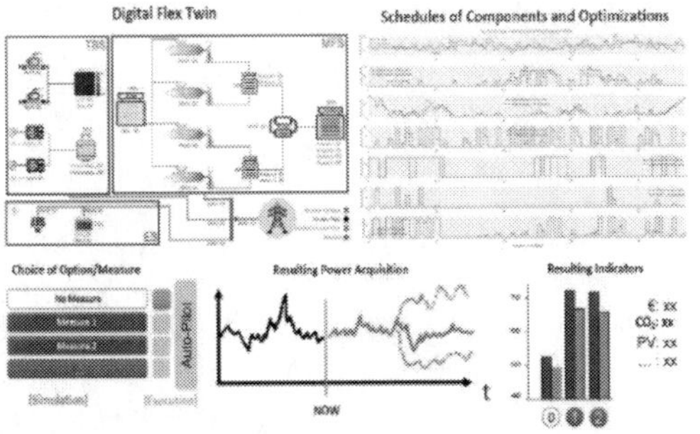

Figure 5: Panel view of the visualization and Control made available to the SME/company owner.

3.4 Energy Monitoring

The energy monitoring can be considered as an important function of the optimization strategy as it generates a historical database of the SME and delivers the data to generate load profiles and demand forecasts. Additionally, the energy monitoring supports an Energy Audit with performance data. In the first step the main meter is the subject of interest. Later with a detailed information of the SME infrastructure submetering becomes more evident to explore the flexibility potentials for the energy management tools. Fig. 6 shows the shares of submetering allowing direct feedback of changes in schedules or supply modes.

3.5 Load profiles and demand forecasting

Energy and Weather forecasts are mainly used by an energy services partner using the forecasts to generate profiles of energy demand and the different shares for the energy supply (ES). If local power and heat installations are run by renewables, energy sources fluctuations and short-term changes are respected with their effects on the energy price and market situation. The energy services provider offers a cloud service with historical data of energy monitoring and offers a data service with load and energy forecasts to the company.

4 SUPPORTING TOOLS AND SERVICES

4.1 Power Grid Interactions

The optimization at small and medium size companies is supposed to have effects on the LV power grid i.e. the 400 V grid level after the transformer. The power demand itself must be rated often higher than the power of a single transformer. Therefore, the metering of SME is often realized on the MV level. The effects of optimization are assumed to happen mainly on the LV level. For this reason, a submetering structure is recommended for functional groups of consumers or even single devices if exceeding a predefined power level. The interacting scheme of Fig. 6 shows the Digital Twin with its Optimizer module producing a load profile that interacts with the distribution grid. The interaction of the grid with the connected loads leads to a grid state indicator which is used as reference for the energy market. The conditions of the energy market finally, lead to an Incentive Signal like the electricity price that is looped back as input of the Optimizer.

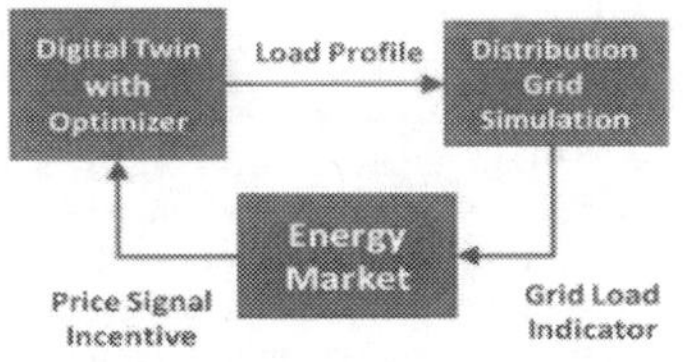

Figure 6: Digital Twin optimization loop to analyze the effects on the power grid.

The entire interaction model was simulated in pandapower, a power module under python as the indOptFlex© library was also developed under python. The power grid model is shown in Fig. 7 with 4 bus levels on LV/MV distribution and one on MV/HV bus level.

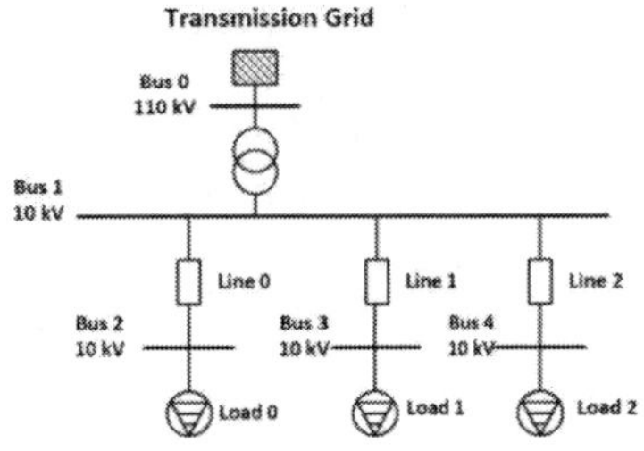

Figure 8: Bus bar model of the LV/MV Level interactions of the SME operated in the same power distribution grid.

The main results of the simulations can be stated as follows:

- The SME optimization has significant effects on the power grid and can contribute to stability measures of the power grid mainly on LV Level.
- Optimized SME affect other SME as well as households on same distribution level.
- SME with optimizer can compensate effects on each other while not optimized SME and household may be affected negatively.
- Facing the expansion to higher numbers of EV chargers and higher fluctuating power shares the effects of optimization will become even more important.

4.2 Energy Contracting and Energy Market Incentives

Regarding the trend to short-term agreements down to Day-Ahead-Challenges to profit of a more and more volatile energy market, benefits for SME are essentially coupled to an upgrade of the Energy Management Infrastructure. The new functions start with an Energy Monitoring, a highly flexible technical prosumer-infrastructure and ends with secured communication of

10.4229/EUPVSEC2023/5DV.2.11
020528-003

incentive signals coming from energy services or flexibility services providers. Most of the player are still developing new marketing strategies und contracting types. Additionally, effects of the future Smart Meter Gateway (SMGW) compatibility should be included on the marketing level as well as on the data communication.

4.3 Energy Flexibility Services

A very important element of the optimization framework became the Energy Flexibility Assessment for the SME. Within a very accurate gathering process details of the production/manufacturing system, energy types, raw materials and media supply, technical building services, staff requirement and work schedules are analyzed regarding their flexibility potentials. Also, the energy monitoring supports this effort that leads to a matrix of flex measures and combinations of flex measures which are weighted by risk factors, investment load and flexibility impact. [3]

The complete flexibility work offers a wide system understanding and helps to model and evaluate twin concept. Finally, the most viable flex measures are chosen for implementation on the Digital Flex Twin model and the SME-Emulator. Simulation runs show their impact as system response. While more static parts of the flexibility audit are foreseen to be updated as time-to-time services there are also dynamic flexibility services that are more continuously updated and exchanged with a flexibility service provider via secured data communication paths.

4.4 Secured Data Communication

The topic of a secured data communication is subject of actual work packages and will be presented in a latter context. The works ongoing aim to get the complete communication infrastructure SMGW-Ready when implemented to the first use cases.

5 SUMMARY OF RESULTS

This work presents a multi-method optimization approach for testing energy flexibility predictive models in an industrial context. A hybrid lab system (SME-Emulator) is setup to mimic real company environments without disrupting their workflow. Additional optimizations and external data, from consortium partners, enhance the digital twin model forming a robust optimization system. This setup allows new configurations, production schedules, run times, and control parameters to be tested, providing insights for future energy management.

There was also a focus on integrating SME (Small and Medium Enterprises) into the grid at the LV/MV (Low Voltage/Medium Voltage) level. Grid simulations using pandapower in python were used to assess the impact of energy optimization on the grid. While most SME are connected and metered at MV level, the main impact is expected at LV level. SME can help to stabilize the grid, potentially leading to market incentives such as financial benefits, special pricing and reduced operational constraints, especially during periods of high renewable energy generation or energy-intensive appliance use based on forecasts.

The Energy Flexibility Assessment for SME is at the heart of the optimization framework. It involves a detailed analysis of production systems, energy types, resources, staff use and schedules to identify flexibility potentials. This process generates a matrix of flexible actions, considering factors such as risk, investment need and impact. This comprehensive analysis feeds into the Digital Flex Twin model forming a first optimization cycle with the Experimental Flex Twin as SME Emulator for implementation and validation. Simulations demonstrate the impact of all optimization actions with their system response.

6 CONCLUSIONS

Industrial companies have significant unexploited potential for energy flexibility, which can be used both to improve their own operations and to play an important role in stabilizing the local power distribution system. The advent of digital twins in the field of optimization has opened up new horizons for the efficient use of this energy flexibility. These digital twins enable automated and highly optimized management of energy resources.

While the challenges of implementing these technologies remain significant, the first steps in this direction are promising. Preparations are underway to introduce these tools starting with SME. These efforts are taking intermediate steps by using the Experimental Flex Twin as SME emulator. This approach ensures that the introduction of energy flexibility measures does not disrupt the ongoing operations or productions processes of SME, while effectively demonstrating the potential benefits of these measures through practical use cases.

ACKNOWLEDGEMENTS

The results presented were partly funded by the Federal Ministry of Economics and Climate Protection of the Federal Republic of Germany as part of the research projects GaIN (Funding ID 03EI6019E) and FlexGUIde (Funding ID 03EI6065C). We would kindly thank M. Wirwitzki for his support in setting up and programming the PLC structure of the Experimental Flex Twin.

REFERENCES

[1] M. Roesch, D. Bauer, L. Haupt, R. Keller, T. Bauernhansl, G. Fridgen, G. Reinhart, A. Sauer, "Harnessing the full potential of industrial demand-side flexibility: An end-to-end approach connecting with markets through service-oriented IT platforms" Appl Sci (Switzerland) 9(18):3796, 2019

[2] J. Beier, S. Thiede, C. Herrmann, "Energy flexibility of manufacturing systems for variable renewable energy supply integration: Real-time control method and simulation," J. Clean. Prod., vol. 141, pp. 648–661, Jan. 2017, doi:10.1016/J.JCLEPRO.2016.09.040.

[3] A. Tristán, F. Heuberger, A. Sauer, "A Methodology to Systematically Identify and Characterize Energy Flexibility Measures in Industrial Systems," Energies, Vol. 13, Page 5887, vol. 13, no. 22, p. 5887, 2020.

[4] R. Khatri, M. Schmidt, R. Gasper, Active participation of industrial enterprises in electricity markets - A generic modeling approach, accepted Paper at Conference Dach+, Energy Informatics, 4(3), 1-20, Freiburg 9/2021.

[5] J.S. da Costa Fernandes, R. Khatri, M. Schmidt, R. Gasper, N. Hartmann. Smart Energy Management for SME Using Digital Twins. Proceedings of the 38th EU PVSEC 2021 (pp. 1482-1487). WIP-Munich, Online-Conference 2021.
[6] A. Aamoume, J.S. da Costa Fernandes, R. Gasper, N. Hartmann, Y. Kenoussi, M. Schmidt, E. Schmitt. Leveraging industry energy flexibility to enable higher shares of renewable energy in the power system – an experimental case study. Proceedings of the WCPEC 2022 (pp. 1600 – 1604) WIP-Munich, Milan 2022.
[7] Final Report of the Project GaIN, "Verbundvorhaben: GaIN - Gewinnbringende Partizipation der mittelständischen Industrie am Energiemarkt der Zukunft" funded under ID (FKZ) 0EI6019E by the German Federal Ministry of Economy and Climate Protection, Consortial Final Report 2023.

10.4229/EUPVSEC2023/5DV.2.11
020528-005

A HOLISTIC WORKFLOW FOR THE DEVELOPMENT OF A POSITIVE ENERGY DISTRICT AND ITS OPERATION AS A MICROGRID

Nikolaos Skandalos[1,*], Sofiane Kichou[1], Martin Cenek[2], Petr Wolf[1], Dimitris Karamanis[3]

[1] Czech Technical University in Prague, University Centre for Energy Efficient Buildings, 1024 Třinecká St., 27343 Buštěhrad, Czech Republic;
[2] Department of Architectural Design II., Faculty of Architecture, Czech Technical University in Prague, Thákurova 9, 166 34 Prague 6, Czech Republic
[3] Department of Environmental Engineering, University of Patras, 2 Georgiou Seferi St., 30100, Agrinio, Greece;

ABSTRACT: This work presents a practical roadmap for developing PEDs, emphasizing the strategic integration of Building Integrated Photovoltaics (BIPV). A case study in Žižkov-Prague demonstrated the proposed methodology. The study integrates architectural design and energy analysis within a simulation environment, optimizing BIPV positioning. The results revealed that BIPV can potentially fulfill 50% of the energy demand on its own. Furthermore, we explore how PEDs can function as microgrids, tapping into surplus energy for grid services. We dissect load-following, peak-shaving, and smart control strategies, which optimize energy utilization and boost economic gains. Our findings underscore the pivotal role of energy storage capacity and real-time pricing considerations. In the realm of architecture, we bridge the gap between energy performance and aesthetics, offering two design approaches that show how solar modules can blend seamlessly into architectural structures while covering up to 80% of energy needs. This study underscores the immense potential of PEDs with BIPV integration in shaping sustainable, resilient, and economically viable urban spaces, contributing to a greener future and climate neutrality objectives.
Keywords: Building-integrated Photovoltaics, Positive Energy Districts, microgrids

1 INTRODUCTION

The rapid and expansive growth of urban areas has posed significant challenges to their energy balance, leading to a surge in carbon emissions. Cities are responsible for approximately 75% of global CO2 emissions due to urban activities, including transportation and buildings [1]. In response to the worldwide imperative to combat climate change, the European Union's Green Deal has set an ambitious target: to achieve climate neutrality by 2050, signifying a profound commitment to a net-zero greenhouse gas emissions economy [2].

In this context, Positive Energy Districts (PEDs) have gained significant attention [3, 4]. These innovative districts encompass interconnected buildings or urban zones engineered for high energy efficiency, with a goal to maintain an annual local energy consumption lower than what is generated locally from Renewable Energy Sources (RES). This surplus energy not only serves the immediate building needs, such as lighting, appliances, and cooling but also facilitates energy exchange within a network of clustered buildings.

Central to the success of these sustainable districts is the strategic incorporation of Building Integrated Photovoltaics (BIPV). BIPV involves the thoughtful placement of photovoltaic components within building structures, with careful consideration of local climates and environmental conditions [5].

Here, we undertake the ambitious task of presenting a comprehensive workflow for the development and operationalization of a positive energy district. The innovative approach seamlessly integrates photovoltaic (PV) technology into the early stages of architectural design. This integration ensures that energy systems are meticulously crafted to meet the precise energy demands of the district's buildings.

Furthermore, the study delves into the intricate realm of operating PEDs as microgrids. This entails harnessing the surplus energy generated within the district to provide ancillary services to the utility grid, thereby optimizing economic benefits. Such strategic utilization of energy resources not only enhances the energy independence of the district but also bolsters the overall stability of the utility grid.

Ultimately, the outcome of this study will be a design workflow that extends the "Integrated Design" from individual buildings to the district level. The advantages of this approach reach beyond its environmental impact, contributing to the creation of more resilient, livable, and economically viable urban spaces.

2 CASE STUDY DESCRIPTION

A residential district in Žižkov-Prague (Fig.1), spanning approximately 39,471 m^2 in total site area, was selected as a demonstration case for implementing the proposed methodology. This site serves to analyze the positive energy targets and evaluate the trade-offs among various design parameters and performance considerations.

The architectural solution, led by architects Dalibor Hlaváček and Martin Čeňek, involved a block-based urban design that integrated commercial spaces, one kindergarten, and private residential areas (Fig.1). This design responded to the site's context, providing pedestrian-friendly spaces and addressing surrounding principles. The project aimed for sustainability through energy-efficient measures, passive strategies like shading and natural ventilation, and active strategies like integrated photovoltaics. The goal was to achieve zero energy consumption for the entire ensemble.

Building configurations included lower commercial/residential sections and taller residential towers. The proposed solution achieved a total gross floor area (GFA) of the residential part of the ensemble of 85,400m^2 and approximately 10,840m^2 for the commercial spaces, respectively. The population of the complex and kindergarten was projected based on occupancy figures.

10.4229/EUPVSEC2023/5DV.2.12
020529-001

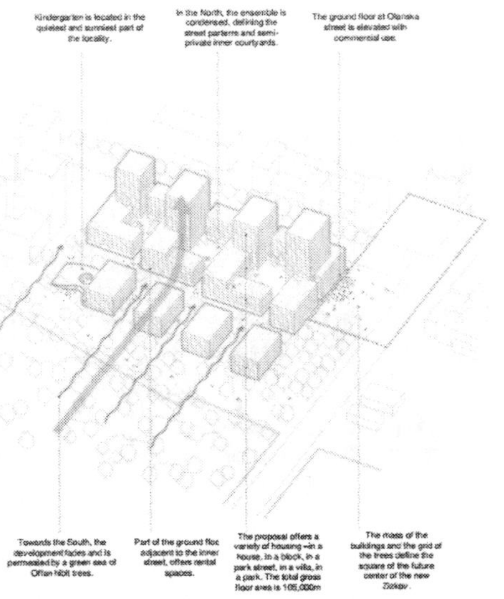

Figure 1: Conceptual design schemes of the selected site for development (authors: Dalibor Hlaváček, Martin Čeněk, Ludvík Holub, Zuzana Kučerová, Klára Novotná, Petr Vošmik)

3 METHODS

The methodology employed begins with an examination of the overall characteristics of the proposed ensemble. The primary objective is to evaluate its potential and suitability for large-scale projects, focusing on assessing key parameters for integrated photovoltaics. Design and analysis are integrated within a single simulation environment, where a model is constructed and evaluated according to several targets for quick and reliable predictions. After this initial assessment, a detailed analysis is conducted on individual buildings within the ensemble. Simultaneously, the architectural design for these buildings is developed, enabling an exploration of the impact of various design options on overall energy performance and user comfort. In essence, the employed methodology comprises a meticulously orchestrated sequence of steps (as illustrated in Figure 2), ranging from the initial assessment of solar PV potential to the activation of energy flexibility within the cluster of buildings.

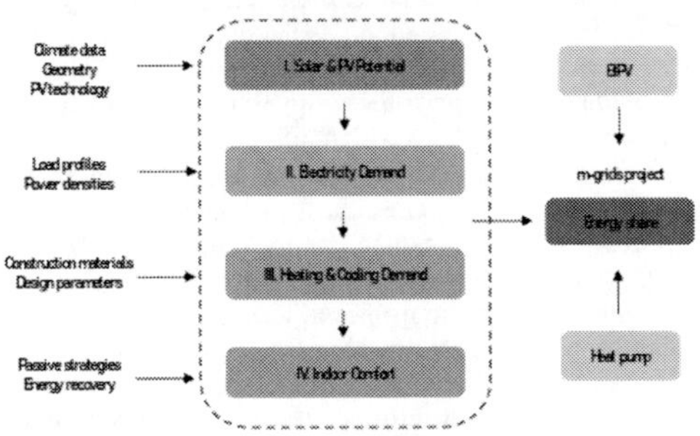

Figure 2: Holistic workflow for developing the "Solar City" and its operation as a Microgrid (MG) for smart energy management.

3D modeling: The analysis begins with a volumetric study that considers various building parameters, such as heights, views, and construction details, across different typologies (residential, commercial, kindergarten). Buildings are simplified and portrayed as footprint extrusions with flat roofs within the Rhinoceros environment using Grasshopper, a visual programming language for parametric modeling. Subsequent stages incorporate more intricate 3D models.

Solar analysis: Grid-based solar irradiation analysis on building surfaces is acquired through Radiance-based methods, with the results displayed in a false-color scale (as depicted in Figure 3), illustrating annual solar radiation received. This method computes hourly irradiation (in Wh/m²) at each grid sensor for an entire year, considering factors like orientation, tilt angle, shadows, and reflections. The obtained colored irradiation maps will guide PV integration decisions by facilitating string-matching scenarios.

PV capacity estimation: For estimating PV capacity, radiation data informs the identification of suitable surfaces for PV integration, guided by an irradiation threshold. This threshold, often influenced by financial considerations, is determined using the following equation:

$$PBT = \frac{PV_{cost}\left[\frac{€}{m^2}\right]}{Threshold\left[\frac{kWh}{m^2\,year}\right] * \eta_{ref} * PR * E_{cost}\left[\frac{€}{kWh}\right]}$$

Where: PBT = Payback time, PV_{cost} = PV system cost, η = PV conversion efficiency, PR = performance ratio, and E_{cost} = grid selling price.

Commonly, a conservative value of 1000 kWh/m^2 per year is assumed for the roofs and 800 kWh/m^2 per year for facades [7]. Considering the technological progress and enormous decline of PV costs within the last decade, lower values, such as 800 kWh/m^2 for roofs and 600 kWh/m^2 for facades, are still reasonable. Considering such thresholds, various PV layouts and design options are generated. This process can be automated and customized based on building characteristics, such as construction, shading patterns, orientation, etc. Grasshopper, a parametric modeling tool, facilitates the rapid generation and comparison of PV layouts.

PV yield estimation: Based on the area of the suitable surfaces, a simple model can be applied to quantify the annual energy output (E_{PV}) of each building block according to Eq.2:

$$E_{PV} = \eta * PR * \sum_{i=1}^{n_{threshold}} (I_i * A_i)$$

Where η = PV conversion efficiency, PR = performance ratio, $n_{threshold}$ = the number of surfaces exceeding the threshold, I_i = cumulative insolation (kWh/m^2.year), and A_i = relative area (m^2) of surface i.

For the detailed design and more accurate PV modeling, the one-diode, five-parameter solar cell model is used to simulate different technologies' PV modules, including crystalline and thin films [6]. The model incorporates location-specific weather data and system characteristics, considering module specifications, inverter efficiency, and available area. A performance ratio (PR) accounts for DC to AC conversion losses [7], and the impact of module temperature on efficiency is assessed based on the type of PV integration (rooftop, ventilated/unventilated facades) [8].

Electricity demand: Occupancy schedules, equipment usage, and internal gains are considered to estimate electricity demand for each building typology. This information is vital for designing renewable energy and storage systems and optimizing energy usage. In this context, load profiles from the Czech electricity and gas market operator (OTE) [9], as well as experimentally measured [10], are used and adjusted to match specific building consumption records.

Thermal energy demand: Building energy simulations are conducted using EnergyPlus to assess thermal energy demand, which provides input for the HVAC system. This involves evaluating various construction materials and design parameters to achieve two primary objectives: a) minimize Energy Use Intensity (EUI) and b) maximize passive survivability. Parameters under investigation encompass building envelope construction (insulation, thermal mass, etc.), window-to-wall ratio (WWR), insulating glazing units (U-value, SHGC, TVIS), and shading depth (envelope settings), as well as artificial lighting (density, dimming control), heating/cooling systems (gas-fired boiler, electric heating, heat pump, etc.), and ventilation strategies (system settings). Simulation inputs are fully parametric and can be coupled with optimization algorithms within Grasshopper.

Energy storage: The potential surplus of energy is managed effectively by introducing the MG operation mode. The microgrid (MG) operation includes multiple PV installations, a heat pump, a central Battery Energy Storage System (BESS), and the utility grid. A mathematical model for the BESS captures its dynamic behavior [11]. An important part of the model is the estimation of the instantaneous value of the state of charge (SOC), indicating the ratio of the available charge at a given time divided by the battery's maximum capacity. MG components, particularly the BESS, are sized based on district energy consumption data and optimized according to specific objectives, e.g., maximizing annual energy yield, reducing peak demand, etc.

4 RESULTS

4.1 District energy analysis

At the outset of the design process, a critical question arises: Is integrating photovoltaics technically viable within this specific context? Results from solar analysis are presented in the form of an annual irradiation color map (Fig.3). Presented data indicated irradiance of around 1,200 kWh/m² for unshaded horizontal roofs and 900 kWh/m² for south-facing facades, respectively. Considering the selected irradiation thresholds, the effective area of roof PV integration is 26% higher than facades, where only the south and west (partially) parts meet the threshold requirements. This is also explained by the increased shading factors compared to roofs, which are very little affected by shading (lower parts in the North).

Concerning the hourly irradiation profiles, the hourly PV profile was calculated and presented in Fig.4. PV modules were assumed to be installed at the same plane with the building surface considering typical values for the conversion efficiency (η) according to the BIPV application (η=17-20%). The total annual photovoltaic potential for the design approximation is about 3,600 MWh under the specified conditions (<60% of the maximum potential of the building envelope), with a peak power of around 2500 kW. There is a noticeable increase in the PV potential, which reaches (in correlation with the amount of solar radiation) its maximum in July (501 MWh); however, the results from April to August are very comparable. Conversely, the minimum potential found in December is more than six times less, indicating the necessity of better energy management.

Figure 3: Annual solar radiation (kWh/m^{2}a) map and shading patterns of the proposed building complex.

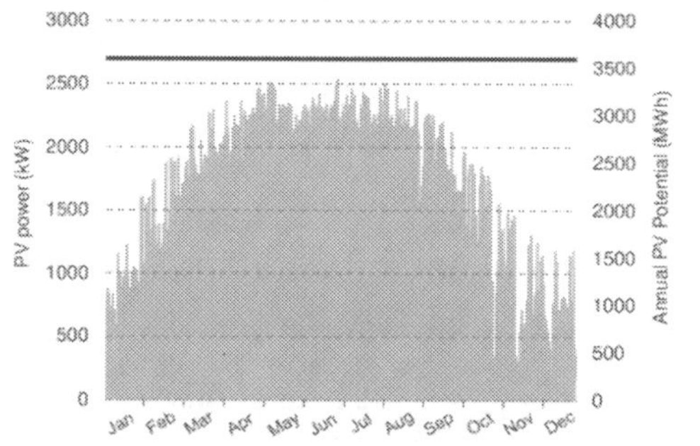

Figure 4: Hourly PV power and the corresponding annual PV potential of the proposed building complex.

Understanding the electricity demand profile is crucial to determining when and how much energy is consumed, a prerequisite for effectively sizing and optimizing renewable energy and storage systems. Analyzing the consumption profile also helps identify peak demand periods when energy usage is at its highest, informing decisions on the most suitable renewable energy and storage system technologies to reduce costs and enhance energy efficiency.

Calculations of electricity demand exclude heating and cooling energy, employing standardized hourly consumption profiles for primary functions within the development: residential, commercial, and nursery spaces. The profiles are based on OTE data, reflecting hourly occupancy and equipment usage. Gross electricity consumption is computed from floor area and occupant density data. For the residential part, consumption peaks are observed during the morning and evening, aligning with residents' presence. On the other hand, commercial spaces primarily consume energy during working hours, with minimal demand during the night. Similarly, the energy needs for the proposed kindergarten are minimal, with zero occupancy outside of working hours. Electricity demand varies by function and hourly usage but remains relatively constant throughout the year, influenced by typical activities. The annual electricity consumption of the building complex was 1982 MWh, with monthly

10.4229/EUPVSEC2023/5DV.2.12
020529-003

values ranging from 130-207 MWh.

Similarly, domestic hot water remains constant with average monthly values of around 120MWh. Conversely, heating and cooling energy demands fluctuate seasonally according to external climatic conditions. EnergyPlus was used to estimate HVAC energy consumption under the same boundary conditions and recommended envelope structure values for passive buildings [12]. Annual heating and cooling energy demand amounts to 3,623.0 MWh, peaking at 623.9 MWh for heating in January and 413.9 MWh for cooling in July. April sees minimal energy needs for both heating and cooling (55.8 MWh). The total annual energy demand for the whole ensemble is approximately 7,043 MWh. January is then the month with the highest overall energy need at 953Mwh, while the month with the lowest need is April at 318 MWh. The results are summarized in Figure 5.

Comparing the energy demand values with photovoltaic potential reveals that, at the design stage, photovoltaics can meet around 50% of the array's total annual energy demand. Although complete self-sufficiency through PV alone is unrealistic, photovoltaics can theoretically fully cover plug-in electricity, lighting, and cooling. Further architectural optimization opportunities include maximizing PV integration areas, reducing heat losses, eliminating excess summer gains, and refining facade openings to balance comfort and aesthetics. Achieving energy load matching and storage methods is also paramount.

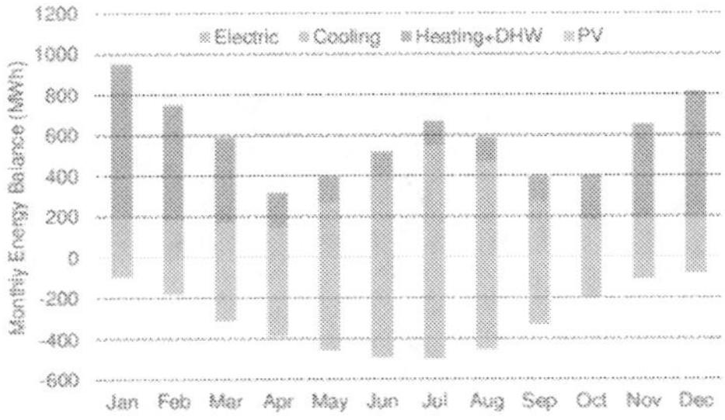

Figure 5: Monthly net energy balance of the whole complex.

4.2 Energy flexibility

The performance of the building complex has been evaluated considering the annual energy balances and the *ZEB(%)* index, corresponding to the generated-to-consumed electricity ratio. Afterward, the energy flexibility of BIPV systems was assessed through the load matching indicators of a) self-consumption rate (*Sc*) and b) self-sufficiency rate (*Ss*). The former corresponds to the share of the generated PV energy consumed locally. At the same time, the latter indicates the average hourly contribution of the BIPV systems on the building loads.

Three scenarios were defined to investigate the contribution of the BIPV electricity generation on the aggregated energy consumption of the whole building complex. These are as follows:

a) The first scenario considers only the electrical usage for appliances and lighting.

b) The second scenario represents the electrical loads plus energy demand for cooling (Heat pump with COP=3 was assumed).

c) The third scenario takes into consideration the total energy demand, including heating and domestic hot water (Heat pump with COP=3 was assumed).

As shown in Figure 5, the annual PV generation exceeds electrical load demand (case 1), except for the winter months, leading to a positive energy balance. Including BESS (1.9 MWh) and load-following strategy increases *Ss* by more than 30% compared to the reference case (Table I). On the other hand, the peak-shaving approach can effectively limit the grid power to a maximum of 300kW.

The inclusion of cooling loads decreases the energy sent to the grid (excess), while a positive annual energy balance remains with an *Ss* of up to 69%. In this case, the grid max power can be limited to 940kW. Adding the thermal loads to the calculation (case 3) reduces the ZEB index to 51% and the excess energy to the grid further. Taking into account energy storage, the PV-generated electricity can be consumed entirely from the building complex. Still, the Ss can be higher than 50%; however, a combination with other renewable energy sources is needed for self-sufficient operation, as shown in our previous study for such climate regions [5].

Furthermore, a methodology was used to pinpoint the minimum EES size necessary for achieving an autonomous (standalone) district, utilizing EES in grid-connected PEDs to bolster grid stability. The methodological foundation hinges on meticulously evaluating the energy balance between generation and demand [13]. The storage capacity ranges between 341MWh and 4921MWh (unrealistically high), with representative performance indicators summarized in Table I.

Table I: Summary of key performance indicators for the three cases while considering different strategies for the BESS system.

Without BESS	Case 1	Case 2	Case 3
ZEB (%)	182	111	51
$Import_{FomGrid}$ (MWh)	999.8	1535	4770
$Export_{ToGrid}$ (MWh)	2618	1898	1332
Ss (%)	49.6	52.6	32.2
Sc (%)	27.3	47.3	63
BESS (LFCS)			
Ss (%)	72.4	68.7	53.2
Sc (%)	39.8	61.7	99.3
Grid max power (kW)	641	1456	3144
BESS (Peak-Shaving)			
Ss (%)	52.3	53.7	33.9
Sc (%)	28.8	48.3	65.5
Grid max power (kW)	300	940	1000
Fully autonomous			
Storage size (MWh)	341	444	4921
$Export_{ToGrid}$ (MWh)	1854	599	0
Sc (%)	48.5	83.4	100

4.3 PV integration and architecture

Assessing urban context and system acceptability is crucial in evaluating integrated photovoltaics (BIPV) or partially integrated systems (BAPV). The site's characteristics suggest seamless PV integration into architecture, replacing facade cladding and serving other functions while considering occupants' needs and energy production-consumption concurrence.

This section addresses the challenge of harmonizing

10.4229/EUPVSEC2023/5DV.2.12
020529-004

photovoltaic system performance with architectural quality, expression, and occupant comfort [14]. We aim to apply insights from a comprehensive solar and energy analysis of the entire complex to the architectural design process. For detailed analysis, we focused on lower residential and commercial ground floor spaces, enabling concentrated energy needs assessment and clear architectural direction. Our architectural efforts involved refining material choices, layout, and structural principles while adhering to thermal-technical standards for energy-efficient buildings. The primary challenge was achieving high-quality architecture while seamlessly integrating photovoltaics. Two distinct architectural approaches emerged, one emphasizing maximized facade surface area for a ventilated photovoltaic facade and the other embracing a "high-tech" aesthetic that consciously highlighted photovoltaic panels as a distinct architectural element.

The first architectural concept (Fig.6) revolved around a solid block design with varying window sizes. A central circulation atrium provided roof access and natural daylight. The building utilizes a combined monolithic structure with load-bearing walls and columns, offering an alternative wood-concrete structure as per contemporary European construction trends. Balconies and shading elements were designed to incorporate photovoltaic modules, balancing energy generation with comfort. The attic features photovoltaic panels around its perimeter ($107m^2$), while the central atrium roofing serves as a skylight with approximately $70m^2$ of PV modules. The façade design maximizes the installation of a ventilated façade consisting of photovoltaic panels, similar to Copenhagen International School [15]. These PV panels can be colored or semi-transparent, harmonizing aesthetics with functionality. Considering all proposed integrated photovoltaic elements, over 75% of the total annual energy needs could be met (Fig.6). However, a closer look at monthly energy production revealed that energy production exceeded consumption during five months (April-August). Surplus energy could be shared with neighboring buildings or integrated into the grid. September showed a near balance between production and consumption, while the remaining six months experienced energy demand exceeding production. Therefore properly sized and managed energy storage is necessary, as discussed in the following section.

The second approach (Fig.7) embraced a more dynamic facade concept, featuring larger glazing areas and horizontal articulation. This facilitated the exploration of various photovoltaic strategies while representing extreme variations in solar gain and opening percentages. In this "high-tech" approach, photovoltaic panels were celebrated as both an aesthetic element and an integral part of the building, shaping the architectural narrative. Architectural decisions were guided by considerations such as mass, orientation, and height, with a focus on optimizing solar access, daylight, views, and user privacy. Different strategies for photovoltaic integration were evaluated against these design parameters. Recognizing the challenge of integrating PV panels, given the limited available area and anticipated higher energy needs, the study explores innovative solutions. The terraces offer a unique opportunity for photovoltaic (PV) integration, with 857 square meters of vertical parapets and attic space available for solar panels. To address these limitations, an architectural solution involving a roof canopy was proposed. This canopy complements the building's design, providing shade to the roof terrace, enhancing ventilation for PV panels, and reducing overheating risks. The lightweight canopy covers the entire roof area (835 square meters) and supports PV panels with an effective area of 505 square meters, inclined at 35° to the south. The total PV potential of the proposed system is almost 20% higher than for the first alternative, able to cover up to 80% of the building energy needs (11.2% higher compared to the first design concept). The monthly analysis reveals that the second variant is theoretically energy-positive from April to September but slightly loss-making in March and October. The winter months remain challenging, with the ability to cover only about 15-17% of energy needs in December and January.

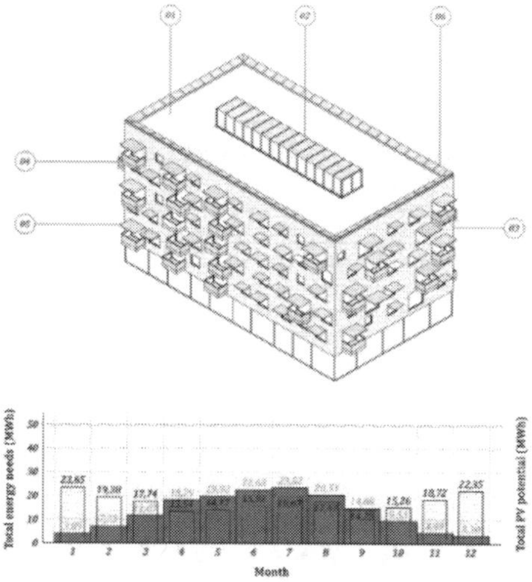

Figure 6: Axonometric view and monthly energy balance of the first design proposal with shading elements. *01 – accessible roof garden. 02 – glass atrium roof structure. 03 – photovoltaic facade. 04 – balcony balustrades. 05 – balcony and window awnings. 06 – attic "collar".*

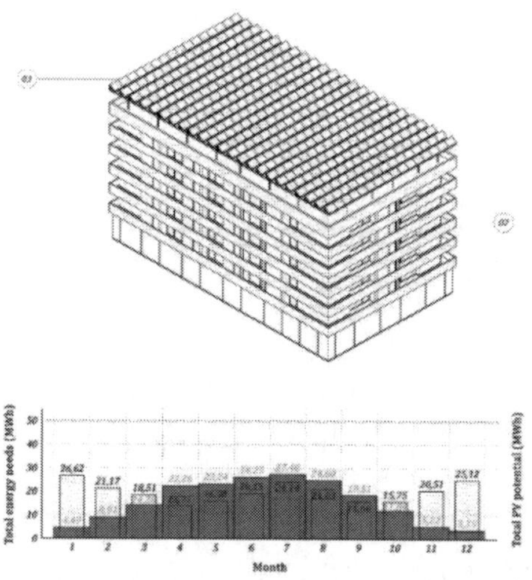

Figure 7: Axonometric view and monthly energy balance of the second design proposal. *01 – accessible roof garden. 02 – photovoltaic balustrades. 03 – photovoltaic shading roof canopy.*

10.4229/EUPVSEC2023/5DV.2.12
020529-005

In addition, the synergistic integration of solar energy systems with nature-based solutions (NBS), such as green roofs, has emerged as a promising strategy to enhance urban resilience and sustainability [16]. The combination of vegetated roofs and photovoltaics, as observed in practices like those in Vienna [17], where green roofs and photovoltaics are mandated for new buildings, provides valuable inspiration.

4.4 Smart energy management

The performance of hybrid PV systems strongly depends on weather data such as solar irradiance and ambient temperature. Considering the energy demand profile, good management of the energy harvested by the PV array with energy storage in the BESS could lead to an efficient and cost-effective system. This work presents four strategies (Fig. 8) for controlling energy flow between PV systems, BESS, and utility grid.

The Load-Following Strategy (LFS) represents a well-established approach in the realm of energy management. Its primary objective is to maximize the utilization of locally harvested energy, achieved by storing surplus PV energy during daylight hours for later release during the night. Figure 8a depicts the dynamic changes in power-related parameters and the battery state of charge over one day. During nighttime or when the PV array generates insufficient power, the batteries discharge power to meet load requirements. This discharge continues until the BESS reaches its SOC_{MIN}. When the PV array generates surplus power, exceeding the load demand, the batteries begin to charge, and this surplus power is stored until the BESS reaches its SOC_{MAX}. Notably, if the PV array generates more power than is currently required by the load, the surplus power is directed back to the grid, as depicted by negative values in the Grid power curve. This situation occurs when the batteries are fully charged, and the excess power cannot be utilized or stored, resulting in its injection into the grid.

Regarding the Peak-Shaving (PS) strategy, the energy management algorithm maintains the BESS at an optimal charge level. This proactive approach ensures that the energy from the grid during 15-minute intervals always adheres to predetermined contractual limits. These limits, which can be adjusted on a monthly basis, are strategically defined to strike a balance between cost efficiency and the avoidance of penalties resulting from exceeding contractual grid power consumption thresholds.

In this example (Figure 8b), the hybrid system effectively keeps the grid power below the predefined contractual threshold of 50kW. The findings, as illustrated in Table II, underscore the necessity of a minimum 100kWh storage capacity is necessary to achieve these objectives successfully. In this specific system configuration, a remarkable surplus of energy becomes available for injection into the grid (136.7 MWh).

The smart control strategy (Figure 8c) introduces a sophisticated control layer, integrating real-time electricity price data into its decision-making process. It leverages the BESS extensively to strategically lower electricity costs. In this case, the BESS system can be charged from the grid when its $SOC<SOC_{MAX}$, taking advantage of periods with low electricity prices. The findings (Table II) indicate the highest electricity export and thus additional economic benefits, with an impact on the battery cycles. Worth noting is that this strategy is particularly well-suited for supercapacitors, given their rapid discharge and charge capabilities.

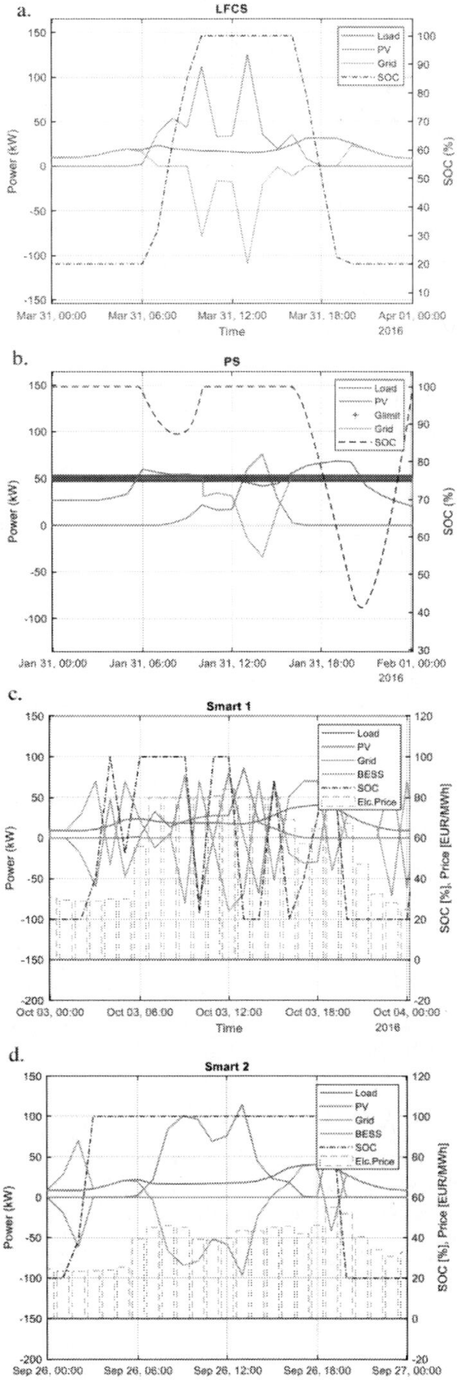

Figure 8: Behavior of the BIPV/BESS system operating under the load-following (I), peak-shaving (II), and smart control (III, IV) strategies.

The second smart control strategy also incorporates electricity price considerations but takes a more calculated approach to battery usage. It prioritizes the preservation of battery health (SoH). It only exports excess electricity to the grid not only considering high prices, but also when it meets the conditions necessary for the battery's amortization. A summary of key performance indicators for each control strategy is presented in Table II.

Table II: Energy performance indicators considering the selected BIPV and 100kWh BESS system operating under different control strategies.

Control Strategy	LFCS	PS	S1	S2
Load (MWh)	225.9	225.9	225.9	225.9
PV (MWh)	223.5	223.5	223.5	223.5
Grid (MWh)	113.8	139.1	174.5	146.9
ToGrid (MWh)	111.5	136.7	201.5	144.8
Ss (%)	49.6	38.4	22.8	35
Sc (%)	50.1	38.8	9.9	35.2
$P_{max\_from\_grid}$ (kW)	68.4	50	50	70
$P_{max\_to\_grid}$ (kW)	120.1	120.1	195.8	174.1
Elect. Bill (Euros)	463.5	448.6	-1270	89.5
Batt. Cycles	310	11	1068	112
Batt. SoH (%)	92	100	73	97

5 CONCLUSIONS

In conclusion, this work shed light on the promising prospects of PV building integration into architectural designs within the context of sustainable urban development. Our study highlighted the synergy between energy efficiency and architectural aesthetics, emphasizing the potential for renewable energy sources to coexist with visually appealing and eco-friendly urban structures.

Through a real-world case study, a holistic design approach for PEDs was presented, ensuring energy independence and adding value to the utility grid. By considering factors like energy consumption patterns, contractual limits, electricity prices, and energy storage system health, we are taking substantial steps toward creating an energy-efficient and economically viable renewable energy ecosystem.

6 ACKNOWLEDGMENTS

This work has been supported by the ERA-NET-SES "Deployment of Smart Renewable Energy Communities" (DoRES) project and by the Ministry of Education, Youth and Sports within the project CZ.02.1.01/0.0/0.0/15_003/0000464 Centre for Advanced Photovoltaics.

7 REFERENCES

[1] United Nations Environment Programme, Cities and Climate change. Available online: https://www.unep.org/explore-topics/resource-efficiency/what-we-do/cities/cities-and-climate-change (accessed on 5 August 2023).

[2] EC 2019, the European Green Deal, COM (2019) 640 final, Communication from the Commission to the European Parliament, the European Council, the Council, The European Economic and social committee and the Committee of the regions, Available at: https://ec.europa.eu/info/sites/info/files/european-green-deal-communication_en.pdf.

[3] D'Agostino D, Tzeiranaki ST, Zangheri P, Bertoldi P. Assessing Nearly Zero Energy Buildings (NZEBs) development in Europe. Energy Strategy Reviews. 2021;36:100680.

[4] Becchio C, Bottero MC, Corgnati SP, Dell'Anna F. Decision making for sustainable urban energy planning: an integrated evaluation framework of alternative solutions for a NZED (Net Zero-Energy District) in Turin. Land Use Policy. 2018;78:803-17.

[5] Skandalos N, Wang M, Kapsalis V, D'Agostino D, Parker D, Bhuvad SS, et al. Building PV integration according to regional climate conditions: BIPV regional adaptability extending Köppen-Geiger climate classification against urban and climate-related temperature increases. Renewable and Sustainable Energy Reviews. 2022;169:112950.

[6] Kichou S, Silvestre S, Guglielminotti L, Mora-López L, Muñoz-Cerón E. Comparison of two PV array models for the simulation of PV systems using five different algorithms for the parameters identification. Renewable Energy. 2016;99:270-9.

[7] Norton B, Eames PC, Mallick TK, Huang MJ, McCormack SJ, Mondol JD, et al. Enhancing the performance of building integrated photovoltaics. Solar Energy. 2011;85:1629-64.

[8] Skandalos N, Karamanis D. An optimization approach to photovoltaic building integration towards low energy buildings in different climate zones. Applied Energy. 2021;295.

[9] Czech electricity and gas market operator (OTE), "Documentation for TDD electricity". Available online: https://www.ote-cr.cz/cs/dokumentace/dokumentaceelektrina/dokumentace-tdd (accessed on 10 June 2023).

[10] Kichou S, Skandalos N, Wolf P. Evaluation of Photovoltaic and Battery Storage Effects on the Load Matching Indicators Based on Real Monitored Data. Energies. 2020;13:2727.

[11] Castañer L, Silvestre S. Modelling Photovoltaic Systems Using PSpice®. John Wiley & Sons, Ltd; 2002.

[12] ČSN 73 0540-2 Thermal protection of buildings – Part 2: Requirements. Office for Technical Standardization, Metrology and State Testing. 2011.

[13] Kichou S, Markvart T, Wolf P, Silvestre S, Chouder A. A simple and effective methodology for sizing electrical energy storage (EES) systems based on energy balance. Journal of Energy Storage. 2022;49:104085.

[14] Cenek M, Cechova T, Hlavacek D, Skandalos N. N. Manuál integrované fotovoltaiky / Building-Integrated Photovoltaics Design Manual. Praha: ČVUT FA, 2023. ISBN 978-80-01-07025-3
N.B. To be published by end of 2023.

[15] CF Møller Architects. Copenhagen International School – Nordhavn. Available online: https://www.cfmoller.com/p/Copenhagen-International-School-Nordhavni2956.html (accessed on 10.10.2022).

[16] Liu H-Y, Skandalos N, Braslina L, Kapsalis V, Karamanis D. Integrating Solar Energy and Nature-Based Solutions for Climate-Neutral Urban Environments. Solar: MDPI; 2023. p. 382-415.

[17] Stadt Wien Energieplanung. Solarleitfaden - Leitfaden für Solaranlagen in Kombination mit Bauwerksbegrünung. Vídeň, 2021.

10.4229/EUPVSEC2023/5DV.2.12
020529-007

RELIABLE PROSUMER: LOCAL ENERGY MANAGEMENT FOR INCREASED RELIABILITY IN INTERRUPTION-PRONE DISTRIBUTION GRIDS

Pedro F. Torres[1,3], Laiz Souto[3], Sam Williamson[3], João T. Pinho[1,2], Roberto Zilles[1]
[1]Institute of Energy and Environment, Universidade de São Paulo
[2]Group of Studies and Development of Energy Alternatives, Universidade Federal do Pará
[3]Electrical Energy Management Group, University of Bristol
Avenida Professor Luciano Gualberto, 1289. São Paulo – SP, Brazil.

ABSTRACT: This paper presents a framework for the management of distributed energy resources (DER) in interruption-prone distribution grids, considering the formation of non-isolated microgrids in medium voltage feeders. A rule-based algorithm is developed to dispatch the microgrid-forming DER, given a set of power and reliability constraints, to ensure the continuity of supply for the customers within the microgrid. The proposed DER dispatch approach can be used as a reference for DER operators in networks where the distribution system operator (DSO) is responsible for a day ahead reliability assessment and where the DER has capacity contracts with the DSO under islanded operation. A real medium voltage feeder with low reliability is used to test the approach in a simulation environment, considering photovoltaic plants with battery storage as microgrid-forming DER.
Keywords: Microgrid, Non-Wires Alternatives, Distributed Energy Resources, Reliability, Islanding.

1 INTRODUCTION

Reliability in power distribution systems is a growing concern for utilities, regulators, and consumers worldwide, given the ever-increasing dependence on electricity for daily activities. Distribution systems are prone to fail due to several factors, such as insufficient maintenance, equipment aging, and severe storms, among others. Improving the system's reliability through investments in infrastructure assets (e.g., undergrounding of power lines) is often unfeasible because of financial restrictions. As an alternative, local energy storage can be used as a backup system to improve reliability in many cases, especially those battery-based. Furthermore, battery energy storage systems (BESS) are often deployed to improve self-consumption from distributed photovoltaic systems (PV), or in peak shaving and peak shifting applications.

An energy management framework with focus on reliability can find many applications in regions where frequent and long-lasting service interruptions may be present. Furthermore, with the expansion of distributed storage systems, energy scheduling approaches that consider the possibility of backup operation are becoming relevant and feasible in many cases. This paper presents a framework for the optimal management of DER in scenarios where power distribution reliability is a major concern. Such a framework could serve as a basis for the development of energy management strategies for microgrid-forming resources and as a reference for power delivery contracts. For example, the contract between a distribution system operator (DSO) and a microgrid-forming DER can consider reliability requirements on a time-dependent basis, to optimize the local energy resources and the microgrid operation.

The literature presents works that consider reliability in the optimization of distributed energy resources. However, the approach for signaling between the DSO and the DER is an innovative aspect of the present work. Khan *et al.* [1] present an optimization-based reliability enhancement scheme for active distribution systems, considering the energy storage availability of electric vehicles. Paudyal *et al.* [2] present an incentive-based optimization scheme for residential consumers; however, reliability is not considered. Liao and Parisio [3] present an optimization scheme for reliability in multi-energy systems that consider energy storage devices' effects under weather uncertainties, where a sequential Monte Carlo simulation is used for reliability evaluation.

The relevance of a reliability-oriented optimization framework that considers reliability from a time-dependent perspective is further corroborated by the works in [4] and [5], where the effects of weather conditions on power system reliability are investigated, and an operational risk assessment for transmission systems is made, considering weather impact.

In the present paper, a novel reliability-oriented energy management framework is considered for microgrid-forming DER (MFR) connected to failure-prone distribution grids. The framework is formulated as a rule-based optimization model to minimize energy costs for the MFR, while guaranteeing reliable operation, considering the interaction between the DSO and the MFR's energy management system (EMS): first, the DSO is responsible for evaluating the grid service failure probability on a day-ahead basis; then, the MFR's EMS uses this probabilistic assessment as a reference to dispatch its local energy resources (e.g., BESS, PV, and flexible loads).

The paper is organized as follows: Section 2 details the utility-led DER contracts that are considered the core business model for microgrid formation; in Section 3 the framework for DER energy management and the rule-based dispatch algorithm are presented; the framework explained in Section 3 is applied to a real distribution feeder, and the results are presented in Section 4, followed by the conclusions in Section 5.

2 UTILITY-LED DER CONTRACTS

A recent business model to consider DER resources in distribution planning is the adoption of utility-led contracts with third-party DER providers. This approach was explored in [6], where a "contract for deferral scheme" was used to integrate DER as an alternative for demand-driven grid investment, where such contracts are procured in a three-step auction process. Another possibility for third-party DER procurement is to open public calls addressing specific planning objectives in the utility's areas of interest. One example is presented in [7], which procured DG with islanded operation capacity to form microgrids on a preselected set of substations and feeders with poor reliability.

The present paper considers an approach similar to the one in [7], where the utility procures third-party Microgrid-Forming Resources (MFR) via power and energy contracts as a planning alternative focused on improving reliability. During normal operation, an MFR operates as a regular distributed generator, selling energy to the utility, while in contingency, the MFR must be able to operate islanded, hence forming a microgrid to supply a predefined set of consumers. Therefore, the MFR contract must specify the contracted energy price, $\rho_{E,MFR}$ (*e.g.* in \$/MWh); the amount of contracted energy from the MFR over the planned horizon, $Q_{E,MFR}$; the rated power capacity during islanded operation, P_{MFR}; and the islanded autonomy time, T_I. From P_{MFR} and T_I, the energy capacity of the MFR, E_{MFR}, can be calculated. A methodology to determine P_{MFR} and E_{MFR} is developed in [8], where the planned demand profile, historic data of interruption levels, and solar resource forecast are used to size the DER.

3 DER MANAGEMENT FRAMEWORK

The framework for the optimal management of DER energy resources considers scenarios where power distribution reliability is a major concern. Such a framework could serve as a basis for the development of energy management strategies for microgrid-forming DER and as a reference for power delivery contracts. For example, the contract between a distribution system operator and a DER can consider reliability requirements on a dynamic basis to optimize local energy resources and microgrid operation. The two-stage workflow of the framework is depicted in Figure 1.

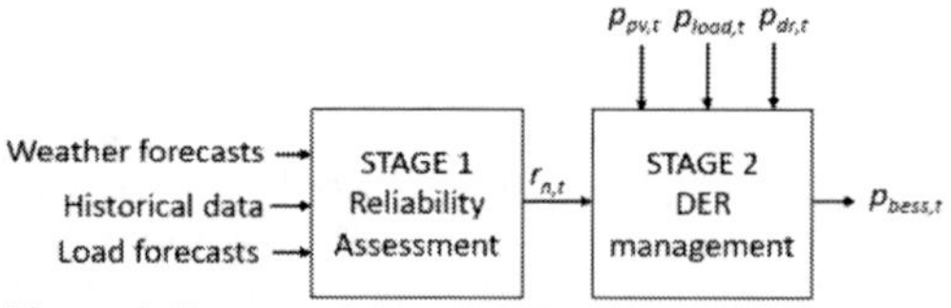

Figure 1: Framework for reliability-oriented MFR management.

The first stage relies on the DSO capacity to assess $r_{n,t}$, which is the probability of energy supply availability at a grid node n and time t. The node n corresponds to the point of connection (POC) of the microgrid (*i.e.*, where the islanding switch is located). $r_{n,t}$ quantifies the confidence level of the DSO with which the system will be servicing the loads and is calculated on a day-ahead basis for each timestep (*e.g.* hourly or quarter-hourly). $r_{n,t} = 1$ indicates to the MFR of the microgrid connected at the nth node that the DSO is 100 % confident that there will be service available at a given time t. When a severe storm is forecast, for example, the DSO might reduce its confidence in service, thus adjusting $r_{n,t}$. To this extent, several factors may influence the quantitative assessment of $r_{n,t}$ by the DSO, such as weather forecasts, historical data, equipment failure rates, and load forecasts.

In the second stage, a rule-based energy management algorithm is used to dispatch the local DER on a real-time basis. In addition to $r_{n,t}$, the dispatch algorithm has the following inputs: $p_{pv,t}$ which is the forecasted PV production; $p_{load,t}$, and $p_{dr,t}$ which are the forecasted power demands of the full load and the demand-response capable loads, respectively. The algorithm used in this paper considers the BESS as a controllable DER, with $p_{bess,t}$ being the optimal dispatch of storage and $p_{dr,t}$ a fixed fraction of $p_{load,t}$.

The MFR can operate in three states, as indicated in Figure 2. In the Normal Connected Operation state, the MFR operates connected to the distribution grid and there is neither a islanding event scheduled nor any probable islanding forecast (as probable, it is being considered an $r_{n,t} < 0.7$ in a day-ahead horizon). If the DSO forecasts a probable islanding in the day-ahead horizon, the MFR enters the Pre-Islanding Operation state, where the microgrid resources prepare to island (*e.g.*, by maintaining the BESS in a high state of charge). Finally, when the islanding occurs, the MFR is responsible for supplying the local microgrid loads. In the event of islanded operation during long periods, the MFR may be unable to supply the loads, then a Full Stop state occurs as the MFR is disconnected. Note that the Full Stop state does not mean that all consumers within the microgrid are de-energized, since some may have backup systems.

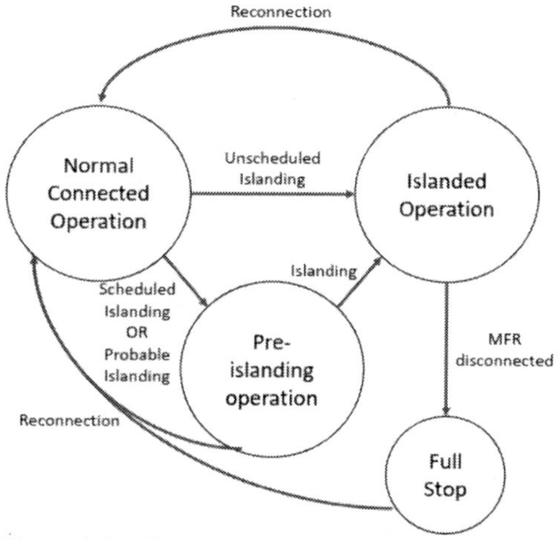

Figure 2: Possible operation states of the MFR and their transitions.

The energy management algorithm considers the following rules for each state:

3.1. Normal Connected Operation

- BESS is charged by the grid only from 0 to 6 am.
- BESS operates in peak-shaving mode at peak hours.
- BESS is used to increase self-consumption of PV, reducing energy exports.
- Exceeding generation is injected into the distribution grid.

3.2. Pre-Islanding Operation

- BESS must have a state of charge of at least SoC_P at the time that islanded operation is scheduled or expected.
- BESS can be charged by the grid during off-peak hours.
- BESS does not operate in peak-shaving mode.

10.4229/EUPVSEC2023/5DV.2.13
020531-002

3.3. Islanded Operation

- If the state of charge of the BESS is lower than SoC_{DR}, the microgrid operator may ask for a demand response from flexible loads.
- If the state of charge of the BESS is lower than SoC_{LS}, the microgrid operator sheds non-critical loads.
- If the state of charge of the BESS recovers and passes a SoC_{LR} threshold, demand response and load shedding are canceled and microgrid load is reestablished.

3.4. Full Stop

- If the state of charge of the BESS is lower than SoC_{FS}, the MFR is disconnected, de-energizing the microgrid until the grid connection is reestablished.

In the Normal Operation state, the BESS dispatch is calculated from the minimization problem defined in Equations (1-6), which minimizes the power demanded from the grid:

$$min \sum b_i(p_{bess,i} + p_{load,i} - p_{pv,i}) \ \forall i \in N_t \quad (1)$$

s.t.:

$$p_{bess,min} < p_{bess,i} < p_{bess,max} \ \forall i \in N_t \quad (2)$$

$$SoC_{min} < SoC_i < SoC_{max} \ \forall i \in N_t \quad (3)$$

$$SoC_0 = SoC_t \quad (4)$$

$$p_{bess,i} \leq p_{pv,i} \ \forall i \in N_{0\text{-}6} \quad (5)$$

$$p_{bess,i} + p_{load,i} - p_{pv,i} \leq p_{pksh} \ \forall i \in N_{peak} \quad (6)$$

where b_i is a binary variable that indicates when power is flowing from the grid to the microgrid ($b = 1$) or vice-versa ($b = 0$). Therefore, the minimization function only accounts for a reduction in power demanded from the grid, while power exported to the grid is ignored, since $b = 0$. N_t is the set of intervals considered in the day ahead simulation (*e.g.*, from 0 to 24 h), while $N_{0\text{-}6}$ and N_{peak} are subsets of N_t which correspond to the set of intervals between 0 and 6 h and the set of intervals during peak hours, respectively. SoC_{max} and SoC_{min} are the maximum and minimum operational SoC for the BESS, respectively, and $P_{bess,max}$ and $P_{bess,min}$ are the maximum and minimum BESS charge and discharge power, respectively. Equation (4) is a constraint to ensure that the final SoC (SoC_t) is the same as in the beginning of the day (SoC_0), and the constraints in Equations (5) and (6) correspond to the rules of charging the BESS only during the first hours of the day and to the peak-shaving service, respectively. p_{pksh} is the maximum power demand during peak shaving service.

In the Pre-Islanding Operation state, the objective function is the same as in Equation (1), as well as constraints in Equations (2) and (3). A third constraint, described by Equation (7), is added to ensure that the BESS has a state of charge of at least SoC_P at the time that islanded operation is scheduled or expected, where SoC_f is the BESS state of charge in the moment of expected islanding.

$$SoC_P \leq SoC_f \quad (7)$$

In the Islanded Operation state, there is no optimization algorithm, and the entire operation is based on the rules defined in section 3.3.

4 SIMULATION RESULTS

The following study case illustrates the application of the presented framework for MFR dispatch. It is based on a remote system isolated from the Brazilian national grid and supplies over 4,400 consumers using a set of diesel-based generators. This system corresponds to the region supplied by *São Sebastião da Boa Vista* (SSB) substation, in the Marajó Archipelago in the northern region of Brazil. The MFR is composed of a 12 kW PV generator and a 53 kWh BESS, and, when islanded, must supply a set of loads with a forecasted aggregated demand profile (p_{load}) indicated in Figure 3. The forecasted PV generation profile for the MFR is indicated in Figure 3 as p_{pv}. Table I presents the parameters for the rule-based algorithm used in the simulations.

The database used to carry out this study is freely available from the Brazilian Electrical Energy Regulatory Agency (ANEEL) and contains georeferenced data for both medium and low-voltage circuits for each utility in Brazil. All data is given for the year 2020 unless otherwise noted. The daily power balance was performed for sequential quarter-hourly data available for the loads, and no distributed generation other than the MFR was considered.

Table I: Algorithm parameters.

Parameter	Value	Parameter	Value
$p_{bess,max}$	10 kW	SoC_{DR}	0.7
$p_{bess,min}$	-10 kW	SoC_{LS}	0.5
BESS Capacity	53 kWh	SoC_{LR}	0.8
PV rated power	12 kW	SoC_{FS}	0.3
p_{pksh}	3 kW	SoC_{max}	1.0
SoC_P	0.9	SoC_{min}	0.3

Figures 4 to 7 illustrate the BESS dispatch for both unprepared (normal operation) and prepared (pre-islanding operation) scenarios, with and without the occurrence of a fault. In the prepared scenarios, a fault is expected by 9 h, following the signaling from the DSO. Figures 4 and 6 illustrate the BESS dispatch for the unprepared and prepared scenarios, respectively, when no fault occurs. It is possible to note that in the prepared scenario, the BESS is charged from the beginning of the day, so a high SoC is obtained by the time a fault is probable to happen. In the normal operation scenario, shown in Figure 4, it is possible to note the peak-shaving service from the MFR, as the power at the POC of the microgrid to the distribution network is limited to 3 kW between the peak hours (from 18 h to 21 h), which is not observed in the pre-islanding scenario shown in Figure 6.

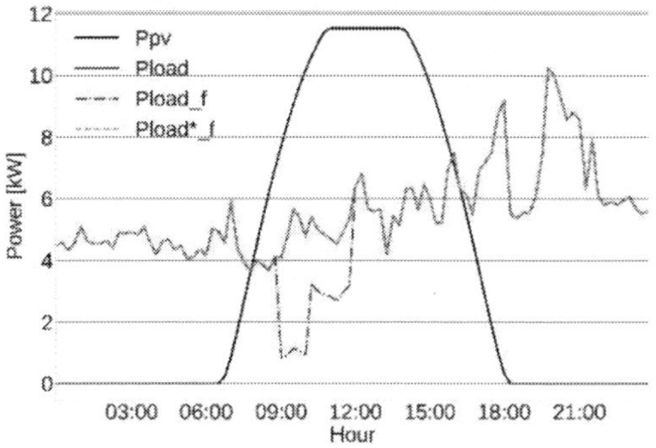

Figure 3: PV generation, reference load, load in the

unprepared scenario under fault, and load in the prepared scenario under fault.

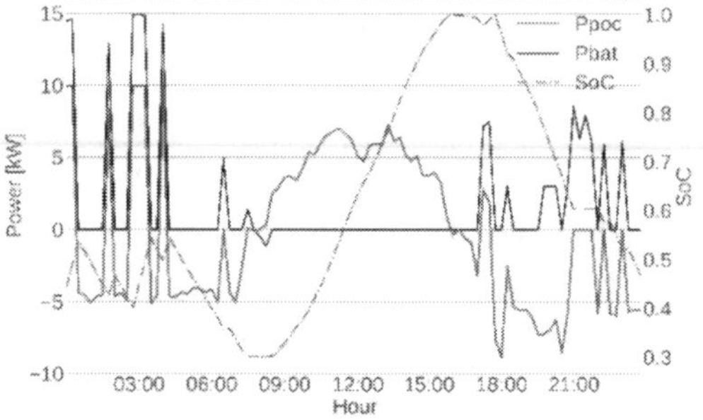

Figure 4: Battery dispatch, SoC, and power at the POC for the unprepared scenario, without fault.

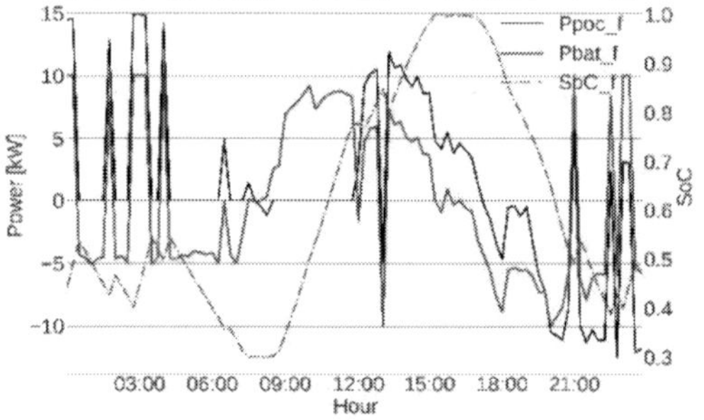

Figure 5: Battery dispatch, SoC, and power at the POC for the unprepared scenario, with fault.

Figures 5 and 7 illustrate the BESS dispatch when the fault occurs at the expected time (9 h), for both the unprepared and prepared scenarios, respectively. In Figure 5, it is noted that the BESS SoC at the time of the fault is under 40 %, making the microgrid operator request demand response from the loads, and perform load shedding on non-critical loads. Those actions from the microgrid operator can be seen in Figure 3, where p_{load_f} indicates the aggregated load in the unprepared scenario under fault, and it is possible to observe how it is reduced at the time of the fault when compared to the reference load. On the other hand, in the prepared scenario, illustrated in Figure 7, at the time of the fault, the SoC is 90 %, and no demand response nor load shedding is put in effect by the microgrid operator.

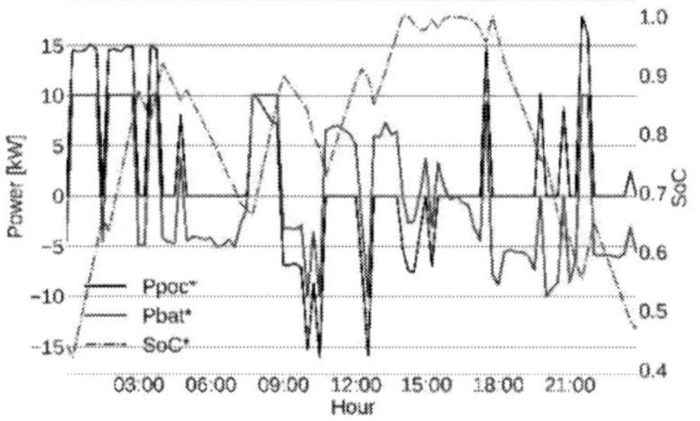

Figure 6: Battery dispatch, SoC, and power at the POC for the prepared scenario, without fault.

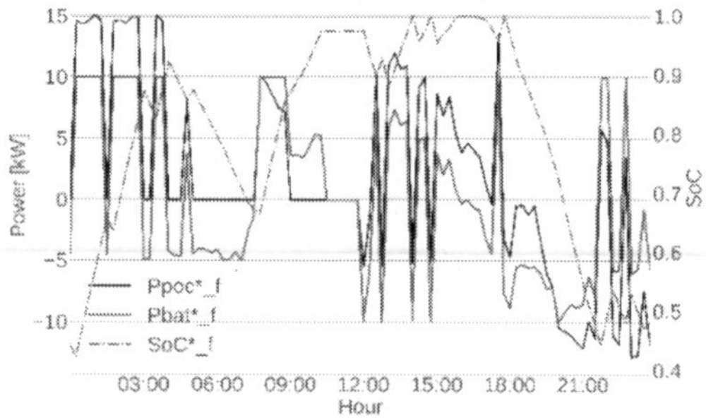

Figure 7: Battery dispatch, SoC, and power at the POC for the prepared scenario, with fault.

5 CONCLUSIONS

Using the proposed MFR dispatch framework it was possible to successfully adjust the BESS dispatch to prepare the microgrid for a probable islanding operation, while the rule-based algorithm also prevents the complete de-energization of the microgrid by performing demand-response and load-shedding operations in flexible and non-critical-loads. Such a framework is built upon the sizing and allocation methodology presented in [8] and serves as a reference for microgrid operation in grid-connected environments with high probabilities of interruptions.

The case study results showed that the MFR can improve the reliability of interruption-prone networks by operating islanded under fault conditions in the distribution grid. The energy management framework can be a feasible alternative to dispatch the MFR resources provided that there is signaling from the DSO to the microgrid operator.

In future works, an already available experimental set-up will be used to test the developed modeling using off-the-shelf devices that can emulate the operation of a small low-voltage feeder with PV-BESS and flexible loads.

ACKNOWLEDGEMENTS

This work was supported in part by the Brazilian Ministry of Education, Coordination of Higher Education Personnel (CAPES) under Grants 88887.716047/2022-00 and 88887.609816/2021-00, and in part by the Brazilian National Council of Scientifical and Technological Development (CNPQ) under grant 407107/2021-5.

REFERENCES

[1] R. Khan, K. Mehmood, S. Bukhari, K. Imran, A. Wadood, S. Rhee, S. Park. "An Optimization-Based Reliability Enhancement Scheme for Active Distribution Systems Utilizing Electric Vehicles," in IEEE Access, vol. 9, pp. 157247-157258, 2021, doi: 10.1109/ACCESS.2021.3127802.

[2] P. Paudyal, P. Munankarmi, Z. Ni and T. M. Hansen, "Incentive-based Residential Energy Optimization Considering Comfort and Voltage Impacts," 2018 IEEE Power & Energy Society General Meeting (PESGM), Portland, OR, USA, 2018, pp. 1-5, doi: 10.1109/PESGM.2018.8585827.

[3] Z. Liao, A. Parisio. "Reliability Optimization of Multi-

Energy System Considering Energy Storage Devices Effects under Weather Uncertainties". Energies, 2022, 15, 696, doi: 10.3390/EN15030696.

[4] M. Bollen. Effects of adverse weather and aging on power system reliability, IEEE Trans Ind Appl, 37 (2) (2001), pp. 452-457, 10.1109/28.913708.

[5] K. Niu, Z. Lyu, B. Zheng. Analysis of Renewable Integrated DG Investment Planning Based on Emission Cost, in IOP Conf. Series: Earth and Environmental Science, v. 632 (2021).

[6] M. Bollen, L. Wallin, T. Ohnstad, L. Bertling. On Operational Risk Assessment in Transmission Systems - Weather Impact and Illustrative Example. In: Proceedings of the 10th international conference on probabilistic methods applied to power systems. 2008, p. 1–6.

[6] R. Poudineh, T. Jamasb. Distributed generation, storage, demand response and energy efficiency as alternatives to grid capacity enhancement. Energy Policy, v. 67, pp. 222-231 (2014).

[7] COPEL. Chamada Pública Copel DIS GD001/2020 – Compra de Energia Elétrica de Geração Distribuída. Companhia Paranaense de Energia (2020).

[8] P. Torres, R. Zilles, J. Pinho. Assessment of Non-Isolated Microgrid Potential for Increased Reliability in Medium Voltage Feeders. Proc. of the 8th World Conference on Photovoltaic Energy Conversion, pp. 1589-1596, Milan, Sept. 2022. Doi: 10.4229/WCPEC-82022-5DV.2.14.

10.4229/EUPVSEC2023/5DV.2.13
020531-005

CARBON-BASED HYBRID SUPERCAPACITORS FOR HIGH POWER PHOTOVOLTAIC IRRIGATION

Márcia Horta[a,b], Luís Fialho[a,b], Ana Foles[a,b], Pedro Horta[a,b]
[a] Renewable Energies Chair, University of Évora, Pólo da Mitra da Universidade de Évora, Edifício Ário Lobo de Azevedo, 7000-083 Nossa Senhora da Tourega, Portugal
[b]Institute of Earth Sciences, University of Évora, Rua Romão Ramalho, 7000-671, Évora, Portugal

ABSTRACT: A photovoltaic pumping system comprises the following components: a solar photovoltaic (PV) installation, a variable frequency converter, a motor-pump, and a water source. The application combines solar PV technology, hydraulic engineering, and high-efficiency water management techniques to optimize irrigated farming. In the last decades, a growing trend has been observed in the application of renewable energies, which depend on the weather and daily conditions. In the case of cloud passing periods, the generation of energy by the photovoltaic system is drastically reduced, which will affect the overall general operation of the system. To better account for the considered operating parameters of a high-power PV pumping system, dedicated control algorithms have been developed in recent years [1], with the aim of mitigating solar power intermittency. One of the options that can be considered to avoid the sudden change in power generated by the solar PV system is to integrate an energy storage system that could accommodate those changes. In this way, carbon-based hybrid supercapacitors (HSupercap) represent the opportunity to solve this issue with a cost effective and long-lasting energy storage system, controlling PV power ramp rate, improving its overall lifetime. The HSupercap [2] was installed, configured, and tested to characterize it and assess this integration possibility. The tested system presented overall performance characteristics suitable for its application in high power photovoltaic pumping or irrigation.

Keywords: Pumping System, Solar Photovoltaic, Energy Storage, Hybrid Supercapacitors, Solar Power Intermittency

1 INTRODUCTION

In recent years, irrigators have been using high power photovoltaic pumping to replace currently used fossil fuel-based energy sources such as diesel generators or electricity from the grid, reducing the overall carbon emissions associated with water pumping. However, there is one problem they face daily: the intermittency of solar resource. This problem ends up seriously affecting the pumping system, as it ends up damaging some of the components of the operating system.

To solve this problem, algorithms have been developed in recent years as part of the SolaQua project [3] to control the passage of clouds impact over the photovoltaic fields [1]. However, there are more factors that need to be controlled other than through algorithms. As a result, research and development has emerged on technologies for energy storage that do not involve conventional batteries, but other technologies that guarantee higher levels of energy density and power.

Supercapacitors represent the opportunity to solve this issue with a cost effective and long-lasting energy storage system, controlling PV power ramp rate and improving its overall lifetime. This work presents the integration, commissioning, and initial test results of a hybrid supercapacitors system in an experimental microgrid of the Renewable Energies Chair of the University of Évora [4] (CERUÉ). The experimental tests and results of the supercapacitors within this microgrid intend to further study the implementation and results of a power management strategy to control PV power ramp rates and to improve the efficiency of high-power pumping irrigation.

2 CURRENT CONTEXT OF ELECTROCHEMICAL ENERGY STORAGE

2.1 Overview

The efficient storage of energy is a fundamental pillar of the energy transition, as it makes the production and consumption of renewable energy more flexible and guarantees its integration into the energy sector [5]. Electrochemical storage covers a wide range of technologies, including batteries, redox flow systems, fuel cells and electrochemical supercapacitors. All of these technologies are currently available on the market, although at different stages of maturity and with different acquisition prices [6].

In recent years, lithium-ion (li-ion) batteries have been widely applied and considered as a crucial energy storage solution, given their increasing competitiveness compared to other electricity storage technologies. Notwithstanding, upfront energy storage costs still stand for a disadvantage [5].

Since energy storage installations are growing exponentially, as presented in Figure 1, a reduction in the upfront cost of batteries is expected along the coming decade [7].

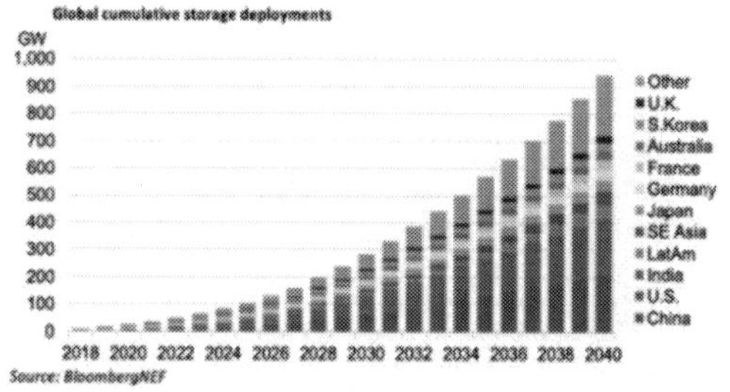

Figure 1: Cumulative increase in energy storage installations worldwide. Adapted from *[13]*.

Whereas li-ion batteries present a high energy density as a technical advantage over other electrochemical storage technologies, as shown in Figure 2, other technologies technical features such as response time and

10.4229/EUPVSEC2023/5DV.2.15
020534-001

power density, might present themselves as competitive for some applications.

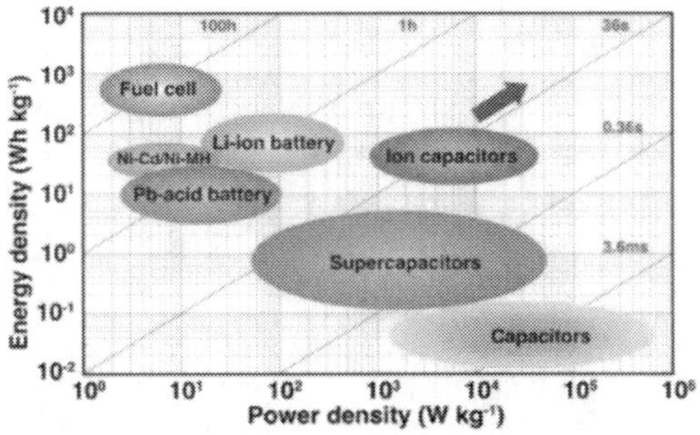

Figure 2: Different types of energy storage technologies according to their energy density and power density. Adapted from *[12]*.

2.2 Supercapacitors

Energy storage in the general variety of supercapacitors occurs due to the accumulation of charged species that are present in an electrolyte, typically activated carbon. There is also another type of supercapacitor that combines an electrode present in conventional batteries (lead or lithium) with a carbon electrode, making it a promising technology due to the combination of these two technologies, which makes its energy storage classified as being electrochemical [6]. The growing interest in this technology is mainly due to the power management that can be achieved by using the grid to reduce the "peak shaving". The supercapacitor that was installed in CERUÉ microgrid is a hybrid supercapacitor (HSupercap) that results from combining the technology of a traditional supercapacitor with a lithium-ion battery.

This technology was acquired from the company Kurt Energy [2] with the purpose of studying the application of this technology in photovoltaic irrigation, to evaluate its response to the intermittency of PV power due to the passage of clouds over the PV plant.

Figure 3: HSupercap installed at CERUÉ's from manufacturer Kurt Energy *[2]*.

The following table shows some of the main performance characteristics of these supercapacitors.

Table I: Main characteristics of the hybrid supercapacitor.

Parameter	Supercapacitor
Nominal Capacity ($\boldsymbol{Ah}$) @1C	45,0 ($\pm$5%)
Nominal Energy ($\boldsymbol{kWh}$)	2,61
Nominal Voltage ($\boldsymbol{V}$)	55,0
Discharge cut-off voltage (**V**)	39,6
Max. charge voltage (V)	57,2
Power Density ($\boldsymbol{W/kg}$)	1500
Dimensions (length × width × height) ($\boldsymbol{mm}$)	500 × 350 × 180
Cycles life at $\boldsymbol{25^{\circ}C}$	> 20 000
Volume Powerpack ($\boldsymbol{dm^3}$)	31,5
Weight of Powerpack ($\boldsymbol{kg}$)	57,36
Operation Temperature (ºC)	-40 to 80

3 METHODOLGY

The HSupercap was installed in CERUÉ's three-phase microgrid: the high voltages and currents to which this technology is subjected would render difficult to guarantee safety with a single-phase system. The supercapacitor was coupled with three inverters (one per AC phase) from the same model/brand – Inverter Quattro 48/10000/140-100 model from Victron Energy [10]. Dedicated AC and DC protection devices were acquired and installed, following the manufacturer safety guidelines. Precision monitoring devices were also installed to better control the overall system's operation.

The HSupercap was subjected to characterization tests to understand its performance and behaviour, under manufacturer general operating conditions. All the tests followed a protocol specifically designed for the operation of the HSupercap, through the development of LabVIEW programming dedicated to the real-time control of the technology.

The supercapacitor is fully charged and discharged through the consumption and injection of energy into CERUÉ's three-phase grid within controlled power limits that have been set to evaluate its response under typically specified operating conditions. Since it is charged and discharged via energy exchange with the electric grid, it has been assumed that the signal convention used would be positive when charging and negative when discharging.

A control unit, the Cerbo GX, supplied by Victron Energy [8], is responsible for the communication in real time with the inverters to obtain current, voltage, state of charge, temperature, and power data on both the inverter AC and DC sides. The results obtained are analysed and energy efficiencies are calculated. The analysis of this data is relevant to the state of the art of hybrid supercapacitors since it is a pioneering technology in the current energy storage market.

3.1 Test Conditions

3.1.1 Levels of charge and discharge

It is expected that the performance and lifespan of this technology could be affected in the long term by the high charging and discharging rate conditions to which they are subjected. In addition, temperature also turns out to be a

10.4229/EUPVSEC2023/5DV.2.15
020534-002

crucial factor throughout its operation as any overheating of surfaces can lead to energy losses, reducing performance.

3.1.2 C-Rates

The C-Rates are limits that are imposed by the power electronics when charging/discharging tests are to be carried out to make comparisons between different curves of different energy storage technologies [9].

In the battery sector, the E-rate is equivalent to the C-rate, being the first expressed in power (watts) [9]. In the absence of a current control device to ensure that the current is at a constant value at any given time, a power control was developed to operate the battery within the defined state of charge range. In view of this control strategy, a difference in test results, from those provided by the manufacturer using C-rates, is expected.

In the absence of standards enabling the evaluation of C-rates for HSupercap, E-rates end up being similar to C-rates, reflecting the real operating conditions of this supercapacitor. In the analysis of the C-Rates to our HSupercap, the curve 1C represents a current of 45A, 2C represents 90A and the curve 8C represents 360A.

3.1.3 Reference test conditions

The reference conditions used are the typical real operating conditions in which these systems might operate [9]. The inverters that are connected to the HSupercap have an AC maximum power limit on their maximum charging and discharging power, which in this case is 30kVA. Regarding the ambient temperature, these tests were carried out with a temperature range between 20°C and 40°C, controlled by an air conditioning unit.
The tests were carried out for a voltage range of 45V – 56V (being the manufacturer specifications: discharge cut-off voltage of 39.6V and maximum charge voltage of 57.2V).

3.2 LabVIEW programming control

A control program was developed to communicate with the inverters control unit (Cerbo GX). This communication is carried out using the Modbus TCP/IP protocol, and all the programming was developed using LabVIEW [9].

The duration of the control cycles was defined as being two seconds, including sending requests to carry out intermediate measurements, receiving responses from the Cerbo GX and the inverters, wattmeters and datalogger. The precision wattmeters measure several AC energy parameters such as voltage, current, harmonics, etc.

3.3 Inverter efficiencies

Some of the tests were carried out to analyse the efficiency of the inverters and assess their influence on the overall system.
The inverter AC-DC (HSupercap operation in the state of charge) and DC-AC (HSupercap operation in the state of discharge) average efficiencies were calculated, for charge and discharge states, respectively.
DC-AC and AC-DC conversion efficiency considers energy conversion losses from DC to AC energy and from AC to DC energy, respectively, which are presented in Equations (1) and (2).

$$\eta_{Discharge(DC-AC)} = \frac{E_{AC}}{E_{DC}} \qquad (1)$$

$$\eta_{Charge(AC-DC)} = \frac{E_{DC}}{E_{AC}} \qquad (2)$$

where E_{AC} is the AC energy and E_{DC} is the DC energy.

The inverters operate according to an efficiency profile directly related to the power level set and not always at its maximum efficiency. The inverter efficiencies used in this work are calculated through Eq. (1) and Eq. (2), considering the different power levels [9].

In this context, a LabVIEW program was developed to operate the inverter at different power levels, considering a certain percentage of the system's nominal power to calculate the efficiency profile and evaluate its performance. Five tests were conducted for each power level and operation mode: charge and discharge.

3.4 HSupercap efficiency calculation

Technical and economic performance indicators are required to evaluate energy storage. The combination of economic indicators such as CAPEX (capital expenditure) and OPEX (operational expenditure) with technical and energy performance indicators such as energy efficiency and charge and discharge energy capacities (Wh), [9] enable this technical-economic assessment.

The HSupercap performance is characterized by executing charge-discharge cycles within the defined range of the state of charge or voltage operation limits. The operating limits of the HSupercap to perform these tests were chosen regarding the safety margins associated with the recommendations of the manufacturer, considering the HSupercap characteristics.

The energy efficiency is the ratio of electrical energy during the discharge to that during the charge. It is calculated using the following expression (Eq. (3)):

$$\theta_E = \frac{U_{discharge} \times q_{discharge}}{U_{charge} \times q_{charge}} \qquad (3)$$

Where U is the voltage of charge or discharge; q_{charge} is the energy capacity (sum of the energy used to charge the HSupercap); $q_{discharge}$ is the useful energy capacity (sum of the discharged energy drawn) [9].
Six round trip tests with the same power settings were conducted to calculate this average efficiency.

4. RESULTS

4.1 HSupercap performance efficiency

The Table II shows the average energy efficiency, as well as the standard deviation and variance. These parameters result from the average of the data obtained, shown in the Figures 4-7, below. Also, the Table III shows the average calculation of the HSupercap charge and discharge capacity, as well as the standard deviation associated with this calculation.

10.4229/EUPVSEC2023/5DV.2.15
020534-003

Table II: HSupercap average energy efficiency.

	Average Efficiency	Standard Deviation	Variance
Energy Efficiency	0,978	0,368	0,145

Table III: HSupercap charge and discharge energy capacities.

C-rate	$\overline{C_C}$ (Wh)	Standard Deviation $\overline{C_C}$	$\overline{C_D}$ (Wh)	Standard Deviation $\overline{C_D}$
2C	1021	98	740	169
3C	499	121	453	83
4C	457	82	404	80
5C	352	69	339	37
6C	218	50	210	42
7C	141	19	137	17
8C	118	9	115	6

Legend: $\overline{C_C}$ – charge energy capacity; $\overline{C_D}$ – discharge energy capacity.
Note: the average charge and discharge capacity values have been obtained from the average of the six tests for each of these curves.

In the following Figures 4-7 a representation example of the charge and discharge curves obtained, according to the HSupercap DC voltage and DC current operational limits, is presented.

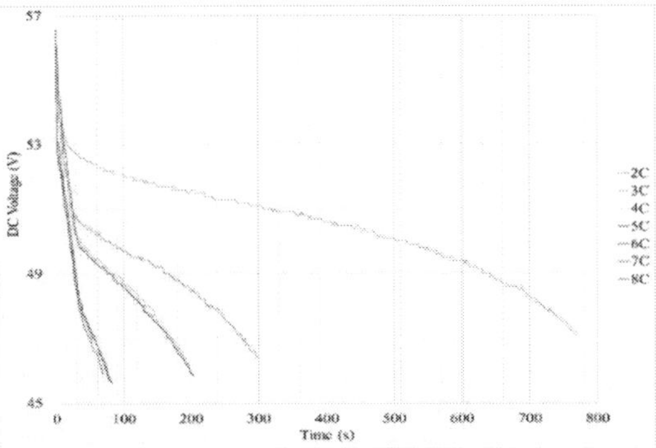

Figure 4: DC Voltage-Time curve for the HSupercap in discharging state.

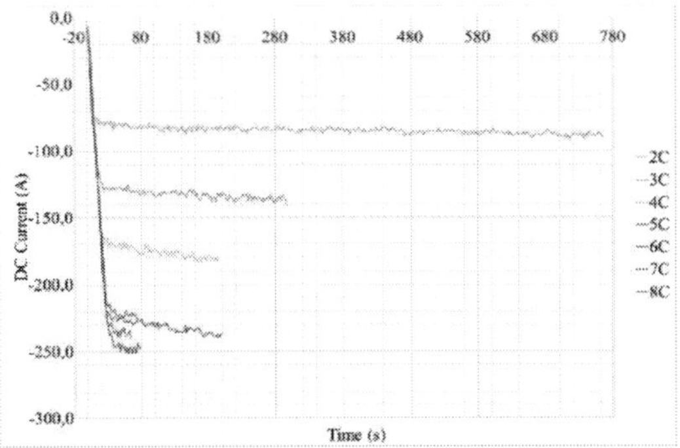

Figure 5: DC Current-Time curve for the HSupercap in discharging state.

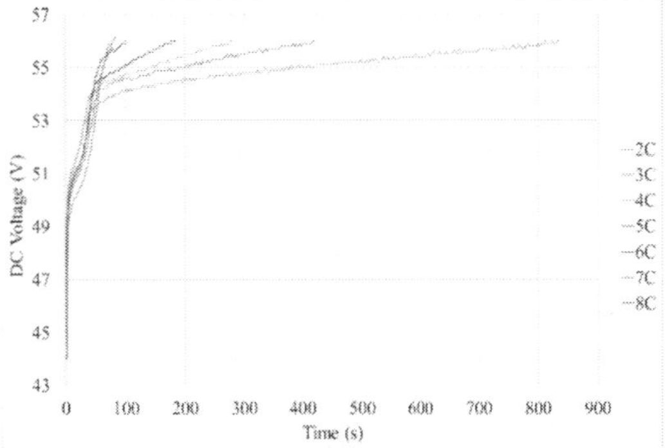

Figure 6: DC Voltage-Time curve for the HSupercap in charging state.

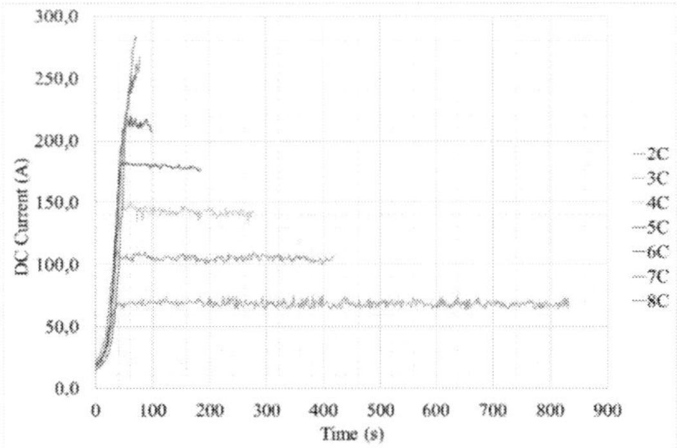

Figure 7: DC Current-Time curve for the HSupercap in charging state.

4.2 Inverter efficiency results

The following charging and discharging efficiency curves were plotted using the data from the charging and discharging tests obtained through the programming developed in LabVIEW, as shown in the Figure 8 and Figure 9.

These efficiencies result from the average of the five tests carried out for each of the selected powers, with these powers representing a certain percentage of the system's nominal power. The following table shows the correspondences for the different powers used in the tests in terms of percentage in relation to the nominal power of the system.

Table IV: Power levels used in the inverter efficiency determination tests.

Nominal Power Percentage (%)	Nominal Power (W)
5	1500
10	3000
20	6000
30	9000
50	15000

10.4229/EUPVSEC2023/5DV.2.15
020534-004

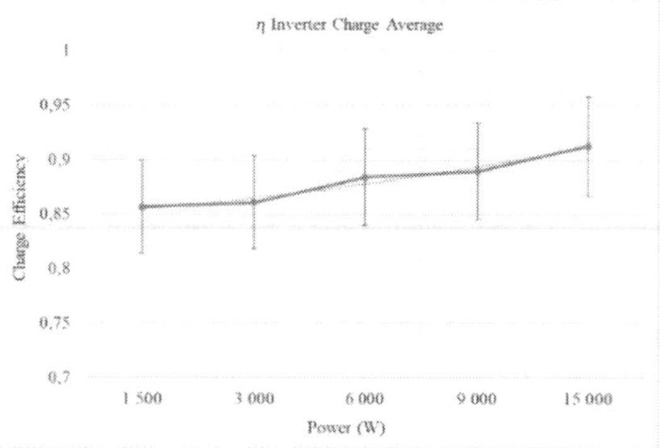

Figure 8: Inverter average charge efficiency at different operating power levels.

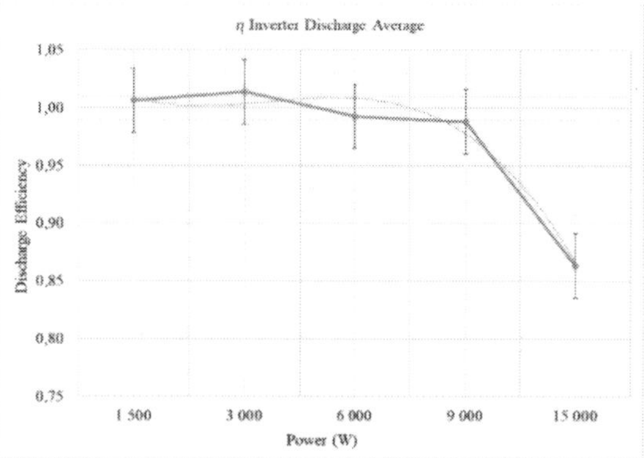

Figure 9: Inverter average discharge efficiency at different operating power levels.

5. RESULTS DISCUSSION

The hybrid powercapacitor technology was successfully integrated into the University of Évora's microgrid, where the control with the power conditioning units was effectively achieved, allowing to proceed with the supercapacitor and inverters characterization testing and operation evaluation.

For the HSupercap average roundtrip energy efficiency a value of 97.8% (±3,7%) was obtained.

Regarding energy capacity, an average value of 1021Wh @2C was calculated for charge and an average of 740 Wh for discharge. Given the voltage range used in the tests, lower energy capacity values than nominal were expected. Similarly, to other energy storage technologies, the useful energy capacity is lower than nominal energy capacity stated by the manufacturer. In the engineering phase of real application projects, careful sizing of the HSupercaps and inverters should be done. As expected, demanding higher currents of charge/discharge decreases the energy capacity stored.

In Figure 4 and Figure 5, which refer to the discharge state for the HSupercap and relate the DC voltage and DC current as a function of time, respectively, we can observe a decrease in the DC voltage as well as the DC current, as would be expected given that this is a discharge. Initially, the discharge DC current shows an almost linear decrease, eventually becoming almost constant over time.

In Figure 6 and Figure 7, which refer to the HSupercap state of charge and relate the same quantities as those described above, the exact opposite is seen, since this is a state of charge. In the state of charge, there is always an increase in DC voltage and current over time, although the latter eventually becomes almost constant after a certain point. From the well-known expression, $P = VI$, and considering the fact that the DC current becomes constant from a certain point in time for both a charging and discharging state, this means that the HSupercap keeps its voltage constant from a certain point in time, as well as its power, making this current constant over time.

In relation to the Figure 8 and Figure 9, which refer to the charge and discharge efficiencies, respectively, the inverters presented high conversion at different power levels. Average efficiency above 1 was obtained, nevertheless well within the error interval calculated for the tests sample.

Current operational limits for the used inverters model limited the tests at higher power/current levels, being the maximum power reached of 19.8 kW. Future inverters sizing should address this design limitation, enabling to reach the HSupercaps full potential, exceeding 8C rates.

6. RELATIONSHIP WITH HIGH POWER PHOTOVOLTAIC IRRIGATION

The photovoltaic high power irrigation systems are subjected to power fluctuations that occur when clouds pass over the photovoltaic plant. Power storage technologies can address this issue and provide a solution to stabilize these high-power ramp rates. The Figure 10 below presents a typical response of a high-power PV pumping system for a cloud passing event. A decrease of about 80% of global solar radiation was measured, occurring in a 20s interval. Intelligent algorithm control can provide a suitable solution for a large number of these cloud passing events, as well as a power storage technology, capable of delivering a fast response time with high power density and lifetime.

The hybrid powercapacitor module tested provides a response time of <200ms (for a 3600A pulse discharge), and a maximum sustained power capability of 29.7 kW.

In standby mode, the system can achieve a response time <2s (for a low latency network). Given a typical cloud passing as shown Figure 10, with 20s duration, this system proves to have a response time suitable to control the PV power ramp rate.

The tested module has a maximum sustained power capability of 29,7 kW (manufacturer specifications), providing a power density of 1000W/dm³.

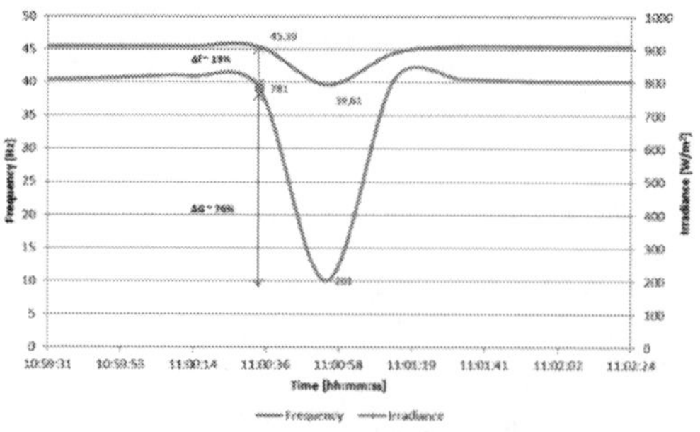

Figure 10: Example of a typical influence of a cloud passage on irradiance and operating frequency of frequency converter. Adapted from [11].

Additional optimization work should be carried to achieve lower control runtime of the LabVIEW software, reaching an overall <1s system response time.

6. CONCLUSIONS

Cloud passing over the PV plant of a high-power PV irrigation system causes extreme power fluctuation (power ramp rate), potentially inducing the sudden stop of the variable frequency converter and, in consequence, of the motor pump. This stopping causes water hammer effects on the hydraulic system, greatly reducing its reliability and lifetime.

The HSupercap technology tested can provide a solution for the control of power ramp rates. A real HSupercap module was installed, commissioned and tested in the CERUE experimental grid. The inverters presented high average efficiency values, for charge and discharge modes. The HSupercap tested presented an average roundtrip energy efficiency of 97.8%. The tested system was not able to achieve higher power than 20 kW due to current operational limits of the inverters. The proper sizing of the inverters and cabling proves to be key to unlock the HSupercap full power capabilities.

This hybrid powercapacitor module provides a response time of <200ms (for a 3600A pulse discharge), and a maximum sustained power capability of 29.7 kW. In standby mode, the system was able to achieve a response time <2s. Given a typical cloud passing event as shown, with 20s duration, this system proves to have suitable characteristics and performance, such as response time, energy capacity and lifetime, suitable to control the occurring PV power ramps.

Additional optimization work should be carried to achieve lower control runtime, reaching an overall <1s response time. Additional tests should be carried out, integrating this HSupercap with a real size PV pumping test rig, to reach a full experimental validation for this application type.

ACKNOWLEDGMENTS

The authors would like to thank the support of this work, partly funded by European Union's H2020 programme as part of the SolaQua project (Accessible, reliable, and affordable solar irrigation for Europe and beyond), Grant agreement ID 952789.

REFERENCES

[1] J. F.-R. a. L. N. Francisco Jesús Guillén-Arenas, "A New Strategy for PI Tuning in Photovoltaic Irrigation Systems Based on Simulation of System Voltage Fluctuations Due to Passing Clouds," p. 20, 29 9 2022.

[2] "Kurt Energy," [Online]. Available: https://kurt.energy/technology/. [Accessed 31 8 2023].

[3] U. P. d. Madrid, "SolaQua," [Online]. Available: https://sol-aqua.eu/.

[4] U. d. Évora, "Cátedra de Energias Renováveis da Universidade de Évora," [Online]. Available: https://www.catedraer.uevora.pt/.

[5] Iberdrola, "Armanzenamento de Energia: a chave para um futuro descarbonizado," [Online]. Available: https://www.iberdrola.com/sustentabilidade/armazenamento-eficiente-de-energia.

[6] I. S. T. d. U. d. L. Faculdade de Ciências da Universidade de Lisboa, "Armazenamento de Energia em Portugal," [Online]. Available: https://www.observatoriodaenergia.pt/pt/comunicar-energia/post/8931/armazenamento-de-energia-em-portugal.

[7] BNEF, "Armazenamento de energia por baterias sofrerá forte queda nos custos até 2030," Mundo Energy, [Online]. Available: https://mundoenergy.blogspot.com/2018/11/armazenamento-de-energia-por-baterias.html.

[8] V. Energy, "CCGX Modbus TCP Register List," [Online]. Available: https://energytalk.co.za/t/victron-modbus-tcp-register-list/1213.

[9] L. F. J. W. N. G. G. P. H. M. C.-P. Ana Foles, "UPCYCLING LITHIUM-ION BATTERIES: A CASE STUDY," in *8th World Conference on Photovoltaic Energy Conversion*, Milan, 2022.

[10] Victron Energy, "Inversores/Carregadores Quattro," [Online]. Available: https://www.victronenergy.pt/inverters-chargers/quattro. [Accessed 31 8 2023].

[11] U. P. d. Madrid, "Derivable 2.9: Training Manual for Irrigators".

[12] D. M. McDonagh, "Electrochemical storage technologies. And now for something completely different…?," 10 11 2022. [Online]. Available: https://www.bestmag.co.uk/electrochemical-storage-technologies/. [Acedido em 31 8 2023].

[13] BloombergNEF, "Energy Storage is a $620 Billion Investment Opportunity to 2040," 6 11 2018. [Online]. Available: https://about.bnef.com/blog/energy-storage-620-billion-investment-opportunity-2040/. [Accessed 31 8 2023].

LIMITING FACTORS IN PHOTOVOLTAIC-ELECTROCHEMICAL CELL DEVELOPMENT FOR SYNGAS PRODUCTION

Swarnendu Banerjee and Jatin K. Rath*
Department of Physics, DSEHC, Indian Institute of Technology Madras, 600036, India
*jkr@iitm.ac.in

ABSTRACT: One of the ways to reduce the CO_2 concentration in air is to convert it to any valuable product(s), such as fuels. This can be achieved by employing photo-electrochemical route. However, this is an energy intensive technique and almost 50% cost of fuel production goes into the applied electrical power. The success of this technique can be achieved if the applied power is derived from a source that is (1) cheap, (2) non-toxic (3) renewable and (4) up-scalable. Wind and solar photovoltaics are good candidates to serve this purpose. With sharp decrease in the price of PV energy, especially record low reverse auction values (<0.03$/kWh!), it is a very attractive power source. We have looked into various types of catalysts Zn, porous Zn and Cu-In, and estimated the Faradaic efficiencies of the production of hydrogen (H_2) and carbon-monoxide (CO) from CO_2 splitting in an electrochemical (EC) cell. We made various types of solar PV power sources, i.e., modules made from a-Si/a-Si/μc-Si triple junction cells, a-Si/μc-Si tandem cells and silicon heterojunction solar cells. Comparing the current voltage characteristics of the electrochemical cells and PV modules, we estimated solar to fuel conversion efficiency of CO over 3% for Zn and porous Zn catalysts for CO_2 splitting. We demonstrate a working PV wired electrochemical cell with Cu-In catalyst, which produced small quantity of CO and the STF efficiency was below 1%. The coupling loss was observed with the wired connection which affects the product efficiency. The solution to this problem can be found by reducing the operational voltage and equi-balance current flows in these two devices.

Keywords: Photovoltaic-Electrochemical cell, Syn gas, CO_2 splitting, silicon heterojunction, tandem cell

1 INTRODUCTION

Conversion of carbon dioxide into hydrocarbons using solar energy is one of the favourable methods for storing such a renewable source of energy into the form of chemical energy (a fuel) [1, 2]. This can be done in a system coupling made of a PV cell/module to an EC cell for CO_2 reduction [3]. The actual observed products for different reactions depend on several factors but the characteristics of the catalyst will play the biggest role. Through the process called 'Fermentation', the conversion of gaseous streams containing CO_2, H_2, and CO into fuels and chemicals can also be done [4,5]. Based on the above arguments, we studied here three catalysts; pure Zn, Zn electrodeposited on Cu mesh and Cu-In mixed phase material and combined EC cell to PV for carbon dioxide reduction reaction (CO_2RR). Zn electrode was chosen due to its high over potential for CO_2 reduction whereas, bimetallic catalysts have improved activity and selectivity for CO_2 reduction to hydrocarbons. We have employed nanostructured surfaces or porous structures to increase the productive performance. We estimated the faradaic efficiencies for the produced hydrocarbons using the above mentioned electrodes in different experimental conditions. Based on the results from Zn based catalysts we considered Cu-In catalyst most promising and present here mainly the results of Cu-In mixed phase material and combined EC cell/PV system for CO_2RR.

Our motivation for this research is that the system is a low-cost and easy way of making use of (tandem) photovoltaic cells with minimal energy losses associated with the catalysts at the electrodes. The electrochemical reduction of CO_2 to CO and H_2 with sufficient ratio of CO/H_2 is very beneficial.

2 EXPERIMENTAL

2.1 Materials preparation

Three types of electrodes, pure Zn, porous Zn and Cu-In were studied. The first type is pure Zn electrode (99.99% pure, Alfa Aesar) of size 4 × 4.4 cm^2. The second electrode is porous Zn (p-Zn) for which Zn was electrodeposited on a piece of Cu mesh ('80 mesh' means 80 wires/inch, Alfa Aesar, 0.14 mm dia wire), using an aqueous solution of 1.5 M $(NH_4)_2SO_4$ (99.5%, Carl Roth) and 0.1 M $ZnSO_4 \cdot 7H_2O$ (99%−103%, Alfa Aesar). The SEM images confirm the deposition Zn on the Cu mesh support. The EDX spectroscopy confirms the presence of Cu and Zn including some additional elements like sulphur and oxygen coming from the electrolyte solution that would affect the reduction process. The formation of porous structure, confirmed from the SEM image, contributes in improving the surface area and also the overall activity for the CO_2 reduction.

The Cu-In electrodes were prepared and provided by the catalyst engineering group in ETH Zurich, Switzerland. The base of the electrode has a carbon support with some crisscross structure and mixture of Cu and indium is sprayed uniformly over the support over a rectangular portion of area 3.4 × 3.6 cm^2. SEM measurement confirms the spherical cupper particles over the carbon substrate with void spaces inside. EDX spectrum confirms the presence of Cu and In, however the amount of percentage of indium present was not very high. There were some additional amount of halides like chlorine and fluorine.

All the electrodes were sonicated in DI water and also with acetone to remove any contaminants. Active surface area of the electrode is around 10 cm^2.

2.2 Device fabrication

The electrochemical reduction measurements were done in 3 electrode configuration where Zn, or p-Zn or Cu-In was used as working electrode, Ag/AgCl (dipped 3.5 M saturated KCl to reach the equilibrium potential) as reference electrode and IrO_2 was used as the counter electrode.

10.4229/EUPVSEC2023/5DV.2.21
020536-001

For a-Si:H/μc-Si:H thin film tandem and silicon heterojunction solar cells, the module configuration has been chosen to ensure voltages which are sufficient to drive the CO_2RR. Detailed information regarding the manufacturing process can be found in the reference literature [6].

2.3 Calculation method

Faradaic efficiency for any by-product for the CO_2RR was obtained using the following relation,

$$\mathbf{f} = \frac{mnF}{\int_0^T i_{r,t}\,dt}\ ; \qquad (1)$$

where **m** is the number of moles of product harvested, **n** is the number of electrons required for the formation of the product, **F** is the Faraday constant (9.6485×10^4 C/mol), and $\mathbf{i}_{r,t}$ is the circuit current which is integrated over the time T.

For a system that produces only fuel as the output and that uses only solar power as the input, the solar to fuel (STF) efficiency can be calculated from the equation,

$$\eta_{STF} = \frac{P_{f0}}{P_S} = \frac{A[cm^2] * J_{op}\left[\frac{A}{cm^2}\right] * E_{f0}(V) * FE_{elec}}{P_s[W]} \qquad (2)$$

where J_{op} is the operating current density, A is the active geometric area of the working electrode, FE_{elec} is the faradaic efficiency of the specific fuel production. The spontaneity of CO_2 conversion into several fuels is given by the reduction potential (E_{f0}) of each reaction.

3 RESULTS AND DISCUSSION

3.1 Solar to Syn gas efficiency using Cu-In electrode

Our experiments with the first two electrodes and the results from it show that, although p-Zn delivers higher total STF, the contribution of CO compared to H_2 is lower for this catalyst compared to pure Zn. Hence, will tried to find a better solution to address this issue by employing Cu-In electrode.

For Cu-In electrodes, solar to fuel (STF) efficiency was calculated for the component gases and also for the total gas, as shown in Fig. 1. It shows that as time progresses the theoretical solar to hydrogen efficiency is decreased (starting from 2.6) whereas the CO efficiency increase in other way. The volume % ratio of CO/H_2 was also calculated and it reaches to almost 0.5 after 2 hours of experimental run. We observe that the Cu-In electrode delivers less STF compared to even pure Zn, but the CO/H_2 evolution ratio improves drastically.

In order to find the operation point of the EC cell one has to keep in mind that the electrodes degrade or change their nature during operation for a long time just as the photovoltaic cells. After 140 minutes of long run it was observed that the surface of the Cu-In electrode was fully covered with the green cupric oxide layer and thus after long term experiment the electrolyte couldn't come into contact further with the electrode surface. As the mass diffusion (gas bulk diffusion and adsorbates surface diffusion of CO and H_2) effect resists the entire electrode reaction, it affects the over potential (concentration over potential) applied to the working electrode, as well as the total electrochemical reduction [7]. Hence, the current at an intermediate time is taken in the STF calculation for the products. For Solar cell, the degradation or instability is a sum of light induced degradation and temperature dependence of solar cell parameters. Hence, the solar cell parameters need to be taken at a stable V_{oc} regime.

3.2 Wired photovoltaic–electrochemical connection set up and Syn gas production using Cu-In electrode and SHJ 5 cell /module.

Using the Cu-In electrode in 2 electrode system the linear sweep voltammetry (LSV) was measured and the LSV curve was merged with the I-V curve of 5 cell series connected silicon heterojunction module to get the operating point for the PV-EC connection.

At the beginning of the measurement under light illumination, the voltage output produced by the PV module was Voc but quickly decreased to a voltage value of 3.05 V, which is at the maximum power point. This voltage was superposed/delivered on the EC cell (within few seconds) and the reaction started to occur. For the 5 cell SHJ module we have obtained the electrical parameters as given in Table I;

From the LSV curve of the 2 electrode Cu-In system (black curve in Fig. 2b) for an applied voltage of 3.05 V, we find that the operational current density is 20 $mAcm^{-2}$. While studying the operating point of the combination of PV and EC, the operating current density from Fig. 2b at 1.89 V is 6.67$mAcm^{-2}$ and at this point the PV-EC combination together will produce best operation with the maximum production of gases. The voltage received by the EC cell was continuously measured by the voltmeter, which was 3.05V as shown in Fig. 2a. Thus, from Fig. 2b, it is well understood that for 3.05 V, the current in the PV is 5.72 mA cm^{-2} whereas for the Cu-In electrode or in EC cell (after extending the black curve upward) it is 20 mA cm^{-2}. Hence, there is a current mismatch for both PV and EC part respectively and the combination is not working at its suitable operating point which is 1.89V, 6.67 $mAcm^{-2}$.

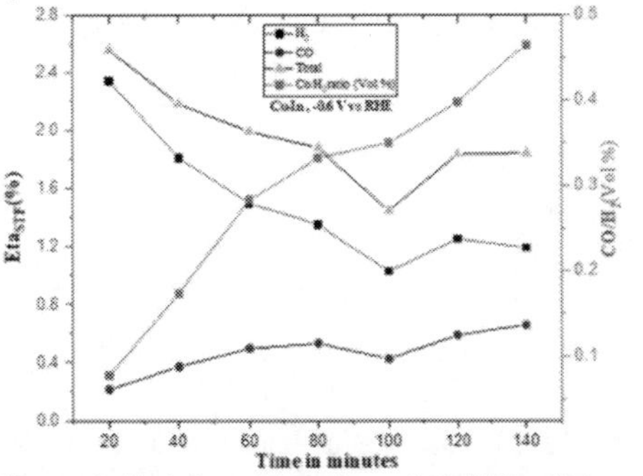

Figure 1: Variation of the solar to fuel (STF) efficiency with Cu-In electrode for H_2, CO and also for the total syn gas with time. The right hand axis shows the volume ratio (%) of the produced CO over H_2 in the experiment with due time.

Table I: Solar cell parameters of 5cell SHJ module measured under AM1.5 condition at 25°C.

η [%]	FF [%]	V_{OC} [V]	I_{SC} [mA]	V_{mpp} [V]	I_{mpp} [mA]	P_{mpp} [mW]	A [cm^2]
18.3	75.5	3.6	110.4	2.9	103.5	300.1	16.4

The measurement was done for two hours and the volume % of the product gases was for H_2 and CO, 0.193 and 0.227% respectively with a total volume of only 128.3 ml after two hours. The significance of the PV and EC wired connection in EC cell is that at a single voltage, same current is required for both the combination to operate. In

10.4229/EUPVSEC2023/5DV.2.21
020536-002

parallel connection, the voltage generated through PV module is the same for both the EC cell and also for the solar simulator (which measures the generated voltage). But the flow of current will be divided between them depending on the resistance along these paths and the corresponding current density will control the overall reduction process in the EC cell.
The voltage that superposes with the EC cell, ensures that the working electrode will get lesser amount of voltage which is less than 3.05 Volt, as the voltage will get distributed among the cathodic and the anodic parts of the EC cell depending on the resistances of each part. For 2 hours measurement there was no strong variation of voltage but with more time used, the voltage variation coming from the SHJ module may become more noticeable.

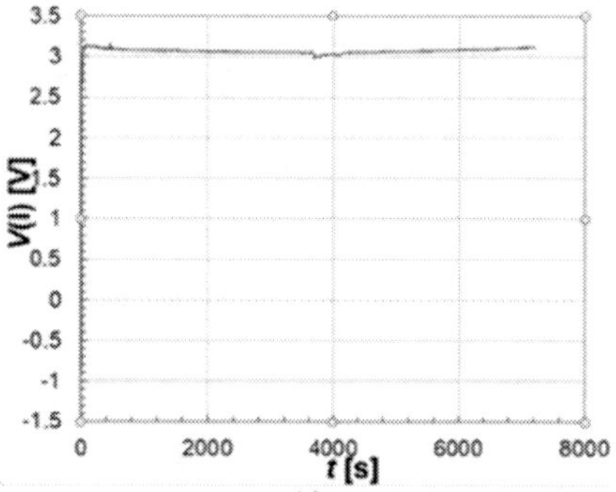

(a)

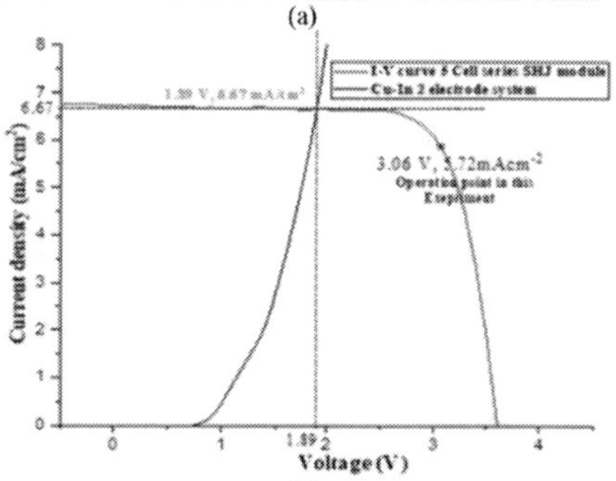

(b)

Figure 2: a) Variation of the voltage measured in the solar simulator during the measurement for 2 hours. This voltage is expected to be the same that was applied also in the EC cell, with minimum amount of loss due to coupling. b) Position of operating point (theoretical, crossover point) and the practical operating point (near maximum power point) in this experiment.

4 CONCLUSION

The study in this work shows the effect of the different working electrodes towards the production of the Syn gas and the reaction mechanisms which depend on various factors. Cu based electrodes come out to be more effective to reduce CO_2 rather than Zn when using them in 3 electrode system. The produced gas was quantitatively calculated by GC and the faradaic efficiency was calculated. The theoretical solar to fuel efficiency of the produced Syn gas components were separately and in total calculated using a-Si/μc − Si thin film tandem 2 cell/module and serially connected silicon heterojunction solar cells forming modules. In the scenario of the PV module the light absorption and the corresponding degradation of its efficiency is the main concerning factor during its measurement, so called Staebler-Wronski effect, in addition to its temperature increase. With Cu-In electrode it was expected to get more efficiency than the Zn deposited on Cu mesh but it was not reached, however CO/H_2 ratio was increased substantially, as desired. At last, the wired connection was made between the SHJ 5 cell photovoltaic module and the electrochemical cell and the results were analysed. The coupling loss present with the wired connection affects the product efficiency which is an unavoidable part of this disintegrated connection.

5 ACKNOWLEDGEMENT

The author Swarnendu Banerjee thanks 'DAAD KOSPIE' program 2019/20. The authors also thank Prof. Dr. Uwe Rau, Dr. Friedhelm Finger, Dr. Tsvetelina Merdzhanova and Dr. Vladimir Smirnov of Forschungszentrum Juelich, Germany for providing facility and scientific input to the research and Catalyst Engineering group in ETH Zurich, Switzerland for providing Cu-In electrodes. This work was partly done under the DSEHC-PV domain project at IITM.

7 REFERENCES

[1] Phillips and Isenberg AO, Hydrocarbon prodn. from inorganic carbonate minerals - and water, with process heat supplied by nuclear reactor. (1984). USPTO 4428-905-A.
[2] Y.Yamauchi, Y. Tokita, N. Murakami, K. Y. Takita, Mori, K. Muraishi, et al., Method for producing methanol by use of nuclear heat and power generating plant.. USPTO 5 (1994) 479,462.
[3] S. Kaneco, K. Iiba, K. Ohta, T.Mizuno, and A. Saji., Electrochimica Acta. 44 (1999) 573–578
[4] Carbon Dioxide Hydrogenation To Form Methanol via a Reverse-Water-Gas-Shift Reaction (the CAMERE Process), Ind. Eng. Chem. Res. 38 (1999) 1808-1812
[5] Mark T. Winklera, Casandra R. Coxb, Daniel G. Nocerab, and Tonio Buonassisia, " Proceedings of the National Academy of Sciences **110**, No.12 (2013): no. E1076-E1082. https://doi.org/10.1073/pnas.1301532110
[6] B. Turan, Laser processing for the integrated series connection of thin-film silicon solar cells. PhD thesis. RWTH Aachen University (2016).
[7] O. Khaselev, J. A. Turner, Science, 280 (1998) 425–427.

10.4229/EUPVSEC2023/5DV.2.21
020536-003

Simulation and Management of PV System and Battery Storage in Providing Grid Support Adhering to Grid Code Requirements

Mohd Zamri Che Wanik, Abdullah Abdul Jabbar, Abselbaset Krama, Karthik Somasundaram, Antonio P. Sanfilippo
Qatar Environment and Energy Research Institute
Hamad Bin Khalifa University
Doha, Qatar
mwanik@hbku.edu.qa

Moataz El-SEID, Beste EREL
OT/R&D/Power/H&S/Power System Modeling
TotalEnergies
Sarclay, France

Abstract—This paper describes the modelling and simulation of a photovoltaic (PV) with battery energy storage connected to part of Qatar power network. The modelling approach and the study analysis performed are described. The model and operation scenarios are simulated using a digital simulator for different scenarios. The dynamic response of the system when subjected to sudden self-clearance temporary fault is presented. The capability of the PV system and battery storage riding through the power system fault and at the same time supporting the local grid by injecting fault current is demonstrated. For each case, the different controller methods to achieve the objective of supporting the grid according to grid code requirements are presented and explained. The inverter modelling approach is presented and described.

Keywords—Photovoltaic, Battery, Modeling, Dynamic Simulation, Fault Ride Through, Fault Current Injection

I. INTRODUCTION

Power systems of the future are going to have a large portion of inverter-based resources (IBR). The IBR can be present as generating units or as energy storage. As the power electronics inverter is used for interfacing these IBR to the grid, it possesses different characteristics and behaviour compared to conventional synchronous machines which since the early days revolutionised the power system infrastructure development. The power system operator is starting to impose stringent requirements for this IBR when it is connected to its grid. Before granting approval for interconnection, proof that these IBRs will comply with the stringent grid code needs to be produced. At the same time, the impact of the interconnection on the grid operation needs to be assessed.

For the assessment, power system studies via digital simulation are performed. Simulation of power system operation is executed for various factors. Some simulation is performed for technical reasons and some are for non-technical reasons. It can be for evaluation of system performance to identify a potential problem and seek a solution for improving the system performance. It can also be for locating the source of equipment failure or maloperation [1]. For non-technical reasons, it is being conducted to find the most economical way to build and operate the power system[2]. For whatever reason, the power system simulator software is utilised as the process is quite complex and time-consuming. Both commercial and non-commercial software are available for the task.

The process of simulation starts with gathering the data and modelling. Modelling is a critical aspect of obtaining reliable results from the simulation and analysis. The type of models varies greatly depending on the purpose of the studies to be performed and the utilised software packages. The studies can be generally categorised for planning purposes which is long-term term and operational which is short-term, or even in real-time.

In performing power system studies for PV system installation, a software simulator such as PSS@E, PSS@SINCAL and DIgSILENT are preferable. Other similar software are also used. In this paper, MATLAB-Simulink is used as the work requires customization of the inverter controller and this software offers flexibility in modification of it.

In this paper, the modelling of PV, Battery and power system is described and the simulation results are presented. This PV and battery energy storage system which forms a power park is demonstrated in a few study cases. The case studies are crafted to demonstrate the power park is complying with the specific grid code requirements. The dynamic response of the system is presented to demonstrate its performance when subjected to power system faults.

II. GRID CODE: AN OVERVIEW

Grid codes is a technical specification which rules out the requirements for a facility connected to the electricity grid. It defines the regulations and behaviour of all active participants in the power grid, including power generators, adjustable loads, storage and other assets. This includes the distributed energy resources units connected to the low and medium voltage grid.

The implementation of these codes gives power grid operators confidence that assets connected to the system will not endanger the security of the electricity supply. Establishing a grid code is an important step in opening up the power sector to private developers or new plant operators and enabling efficient integration of distributed VRE generators. The purpose of grid codes is manifold and includes ensuring coordination among the various actors, increased transparency, grid security, reliability and VRE integration. Grid codes enumerate the technical requirements that are to be followed to keep the system functioning smoothly and to build trust between power system actors. They encompass different aspects of the power system, such as markets, operation, planning and connection. [12].

GCC countries have started to introduce a grid code for the integration of facilities into the country's grid.GCC stands for

Gulf Cooperation Council. It is a short name for the Cooperation Council for the Arab States of the Gulf. It is a regional, intergovernmental, political, and economic union comprising Bahrain, Kuwait, Oman, Qatar, Saudi Arabia and the United Arab Emirates. Each of these countries published their own grid code which is applicable to their country's power grid. In this paper, however, the focus is on the regulation imposed by the Qatar Grid Code document.

Qatar General Electricity & Water Corporation or KAHRAMAA is an energy regulator body in Qatar, published a Qatar Transmission Grid Code in 2020 [11]. The grid code is meant for grid users involved in power generation and power consumption connected from 11 kV to 400 kV network. The generation in the grid code classified the generating units into 3 categories which are D, S and P as shown in Table 1. Depending on the power capacity of the generators and the voltage levels where it is connected, different ancillary services are demanded to be provided. These services are necessary for the operation of an electric power system provided by the system operator and/or by Grid Users including Automatic Generation Control, Reactive Power, operating reserves, frequency control, voltage control, back-up capacity, Black Start, etc.

TABLE I: DEFINITION FOR THRESHOLDS FOR CATEGORIES D, S AND P GENERATING UNITS

Categories	D All Generator	S Synchronous generators	P Non-synchronous generators (PPM)
Power Capacity	2-10 W	>10 MW	>10 MW
Connected Voltage	11 kV	>11 kV	

Most of the requirements applied for synchronous generators also applied to the PV plant and battery. However, there are specific requirements that only apply to PV and Battery that has a power capacity greater than 10 MW. The specific requirements are focused on robustness, voltage stability, reactive power compensation, fault ride-through requirement, fast fault Current contribution and testing at the connection points.

In this paper, the focus is on the requirements for the PV and battery system installation that fall under category P generators which are to be connected 66 kV network during a fault. During a temporary power system fault, the PV and battery need to continuously remain connected to the grid and ride through these faults. At the same time, the PV and Battery need to provide the fault current as specified by the grid code.

III. KAHRAMAA GRID CODE REQUIREMENTS

A. *Requirement for fault ride through*

The requirement is imposed for supporting the stability in case of voltage dips and is complementary depending on the characteristic of the grid and the severity of the voltage deviation.

1) The Generating Unit shall be capable of remaining connected to the network and continuing to operate stably when the actual course of the phase-to-phase voltages on the network at the Delivery Point is low during a symmetrical fault, given the pre-fault and post-fault conditions, unless the protection scheme for internal electrical faults requires the disconnection of the Generating Unit from the ETN. The protection schemes and settings for internal electrical faults must not jeopardize Fault-Ride-Through performance.

2) The details of that capability shall be subject to coordination and agreements on protection schemes and settings.

3) With regard to the Fault-Ride-Through capability of the Generating Unit:

a) Figure 1 specifies a voltage-against-time-profile at the Delivery Point for fault conditions, for which the Generating Unit is capable of remaining connected to the ETN and continuing to operate stably after the EPS has been disturbed by faults on the ETN;

b) The voltage-against-time-profile shall express a lower limit of the actual course of the phase-to-phase voltage on the network voltage level at the Delivery Point-during a symmetrical fault, as a function of time before, during and after the fault

c) The lower limit referred U_{clear} is specified by Kahramaa using the parameters within the ranges set out in Table 1.

4) Kahramaa will specify the pre-fault and post-fault conditions for the Fault-Ride-Through capability in terms of:

a) calculation of the pre-fault minimum short-circuit capacity at the Delivery Point;

b) pre-fault active and reactive power operating point of the Generating Unit at the Delivery Point and voltage at the Delivery Point; and

c) calculation of the post-fault minimum short-circuit capacity at the Delivery Point.

5) At the request of the Generating Unit, Kahramaa shall provide the pre-fault and postfault conditions to be considered for Fault-Ride-Through capability as an input of the calculations at the Delivery Point regarding:

a) pre-fault minimum short-circuit capacity at Delivery Point expressed in MVA;

b) the pre-fault operating point of the Generating Unit expressed in active power output and reactive power output and voltage at the Delivery Point, and

c) post-fault minimum short-circuit capacity at the Delivery Point expressed in MVA. Alternatively, Kahramaa may provide generic values derived from typical cases.

6) Under-voltage protection (either Fault-Ride-Through capability or minimum voltage specified at the Delivery Point voltage) shall be set according to the widest possible technical capability of the Generating Unit, unless Kahramaa requires narrower settings in accordance with electrical protection schemes and settings. The settings shall be justified by the Generating Unit according to this principle; and

10.4229/EUPVSEC2023/5DV.2.23
020537-002

7) Fault-Ride-Through capabilities in case of asymmetrical faults can be specified by Kahramaa.

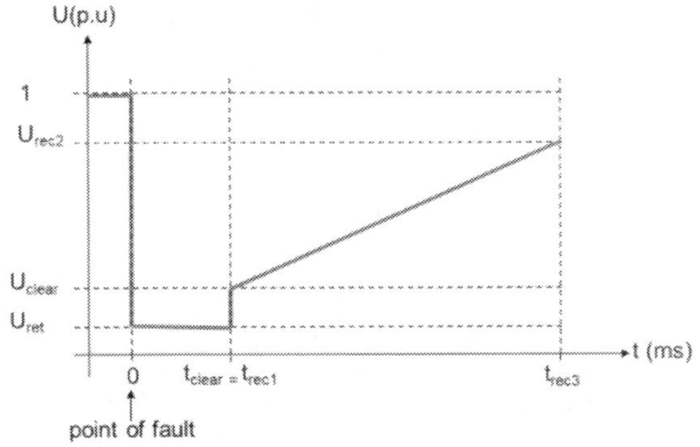

Figure 1. Fault Ride-Through Requirements

This figure represents the lower limit of a voltage-against-time profile of the voltage at the Delivery Point, expressed as the ratio of its actual value and its reference 1 pu value before, during and after a fault. U_{ret} is the retained voltage at the Delivery Point during a fault. tclear is the instant when the fault has been cleared and U_{clear} is the corresponding voltage. U_{rec2}, t_{rec1}, and t_{rec3} specify certain points of lower limits of voltage recovery after fault clearance.

TABLE II: PARAMETERS FOR FAULT-RIDE-THROUGH CAPABILITY OF GENERATING UNITS TYPE P

Voltage Parameters (p.u)		Time Parameters (ms)	
U_{ret}	$0.05 \leq U \leq 0.15$	T_{clear}	$140 \leq t \leq 250$
U_{clear}	$U_{ret} \leq U \leq U_{ret} + 0.15$	T_{rec1}	T_{clear}
U_{rec2}	$0.85 \leq U \leq 0.9$	T_{rec3}	$1500 \leq t \leq 3000$

B. Requirements for Fast Fault Current Contribution

In the grid code, it is specified that Kahramaa has the right to specify that a power park module be capable of providing Fast Fault Current at the Delivery Point in case of symmetrical (3-phase) faults, under the conditions that the power park module shall be capable of activating the supply of Fast Fault Current either by:

*1)*Ensuring the supply of the fast fault current at the Delivery Point; or

*2)*Measuring voltage deviations at the terminals of the individual units of the power park module and providing a fast fault current at the terminals of these units.

Kahramaa shall specify:

*1)*how and when a voltage deviation is to be determined as well as the end of the voltage deviation; the characteristics of the Fast Fault Current, including the time domain for measuring the voltage deviation and fast fault current; and

*2)*the timing and accuracy of the Fast Fault Current, which may include several stages during a fault and after its clearance

The details requirement for fault current injection is a grey area and will be decided by KAHRAMAA. This is normally decided after thorough power system studies or by adopting the requirement that is imposed by other grid operators. In this paper, the assessment study is performed by using the fault current injection imposed by a grid operator in Germany. The requirements were first published back in 2006 [15] and have been widely adopted by many countries. The details of the fault current injection are depicted in Fig. 2.

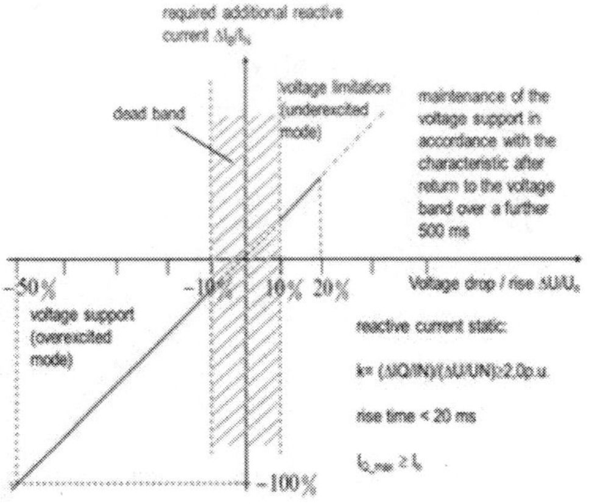

Figure 2. Fault current injection

The fault current that needs to be injected is a reactive current with the characteristic depicted in Fig. 2. The objective of the reactive support is to lift up the voltage magnitude during low voltage events and also to improve voltage recovery after this event. If the voltage drops below 90% of the rated voltage, capacitive reactive current is demanded. In contrast, if the voltage rises beyond 110%, inductive reactive current is required to reduce the grid voltage. The amount of reactive current described is to be added to the actual reactive current already supplied during steady-state operation.

The amount of current that needs to be injected needs to comply with the requirements depicted in Fig. 2 for the voltage magnitude within 50 to 100 % of the rated voltage at the interconnection point. When the voltage magnitude is below 50% the amount of the reactive current that needs to be injected will be according to the agreement between the network operator and power park owner. The amount of the current typically maintains a similar characteristic until reaching the maximum capability of the power park.

For an inverter-based generator, the maximum current that can be injected depends on the inverter design and limit set by the manufacturer. The typical permissible limit is depicted in table III. It can be seen in the table that permissible overload current can reach 300% of the rated current for up to 40 ms. For 200%, the converter can be overloaded for up to 1000 ms. For the FRT requirement, the injection of fault current is only demanded for a duration of less than 1000 ms. This requirement is characterised by Fig. 2, by seeing Table III is within the capability of the typical inverter.

10.4229/EUPVSEC2023/5DV.2.23
020537-003

Table III. Power converter permissible overload current

Magnitude (%)	Time (ms)
300 %	40
200 %	1,000
150 %	10,000
125 %	30,000
110 %	60,000

IV. PV-BATTERY-NETWORK MODELING

A. PV System and Battery Storage Modeling

The typical structure of inverter-based resource components such as PV and Battery can be depicted in Fig. 3. The unit is typically equipped with two converters that are linked by a DC circuit. On the grid side, the units are normally equipped with an inductor and AC filter.

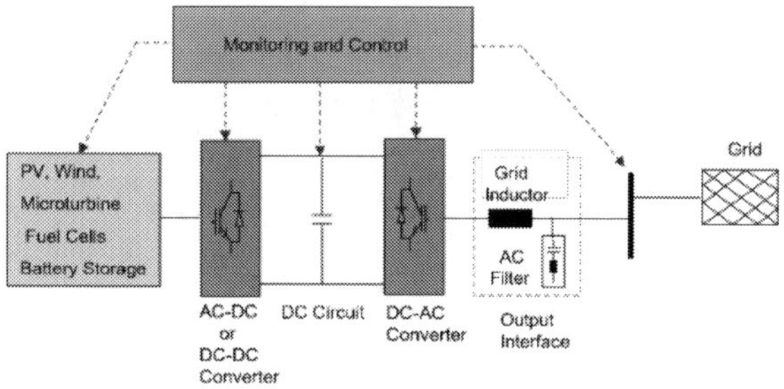

Figure 3. Inverter-based resources system typical arrangement.

The model however can be simplified by taking a few assumptions. The capacitance inside the DC-link circuit acts as energy storage which provides isolation between grid and PV modules. Any interruption on the grid side is not instantaneously sensed by the PV module side. The voltage of the DC-link can also be said to be kept constant by the PV system controller by balancing the amount of power transfer from the PV module to an electrical grid.

The comparison of a different type of distributed generation unit coupled to the electrical grid through PEC has been analyzed in [4]. Even though they are from different technologies, they are using similar inverter hardware and controllers. One of the comparison studies is the output from these DG units during low voltage fault-ride through (LVFRT). The simulation shows that even though the sources are different on the DC side, the output on the AC side is similar. It can be concluded the behaviour of DG units employing PEC for grid interfacing is greatly influenced by the inverter.

A simplified model of DG can be derived under the assumption that DC voltage is constant. This assumption can be justified by the large capacitance in the DC circuit and by the DC voltage controller which keeps the voltage at near nominal value. With this assumption, DG can be represented by a voltage source inverter fed by a constant DC source as depicted in Fig. 10. For this model the controller for the inverter is similar to the one depicted in Fig. 8 but i_{INVd_ref} can be calculated with only [4].

$$i_{INVd\_ref} = \frac{p_{ref}}{u_d} \tag{1}$$

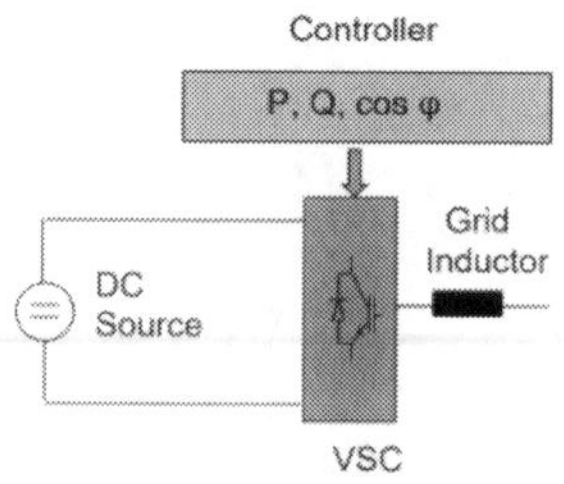

Figure 4. VSC fed by constant DC source

For power system studies especially for the cases that involve only dynamic assessment, the AC filter can be ignored in the modeling.

B. Inverter Control

The converter control uses a decoupled power controller scheme implemented in the *dq0* coordinates. The inner control loop of the controller scheme is depicted in Fig. 6. This controller scheme enables active and reactive power to be controlled independently. Here, the grid voltage $\underline{u}_G$ and current injected to the grid $\underline{i}_G$ are converted to d and q components using grid angle u_g as a reference.

Reference current for the controller is calculated using methodology as depicted in Fig. 7. Voltage regulation capability is included in the q coordinate of the control loop. The voltage u_G here is the voltage at the bus shared by the PV battery and local load. The reactive power characteristic for voltage regulation is shown in Fig. 8. Depending on the voltage values, the amount of reactive power supplied or absorbed varies.

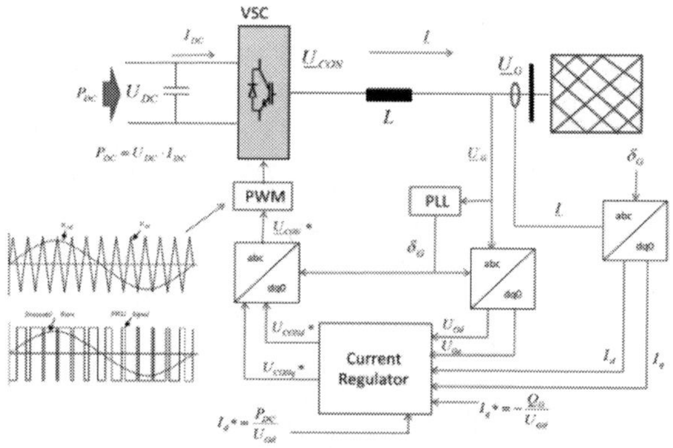

Figure 5. Typical inverter controller scheme for inverter-based resources

C. Power Network

Qatar power transmission network as defined by the KAHRAMAA grid code comprises 400 kV, 220 kV, 132 kV, 66 kV, 33 kV and 11 kV levels. There are 5 gas thermal plants with a total capacity of 10,580 MW and one 800 MW Solar Farm. The network comprises 369 primary substations as of 2020 with a total overhead line of 1999 km and a cable length

10.4229/EUPVSEC2023/5DV.2.23
020537-004

of 3270 km [KAHRAMAA]. The network can be represented by a single line diagram as shown in Fig. 8.

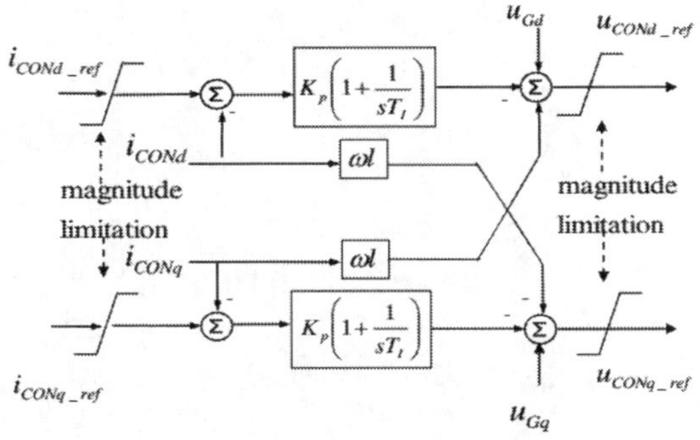

Figure 6. Current regulator structure of the inverter controller scheme

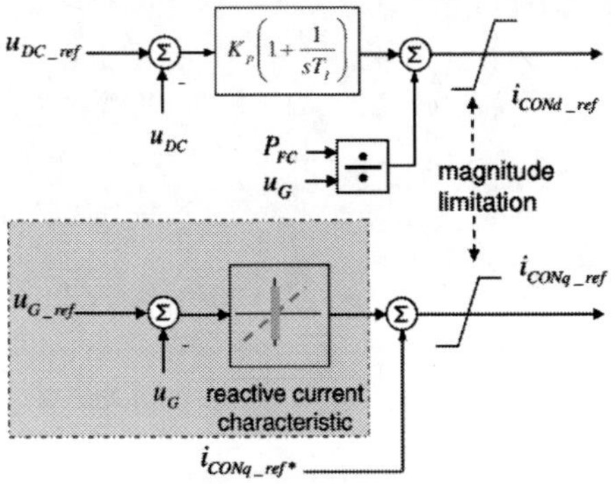

Figure 7. Controller's additional loops for reactive current injection

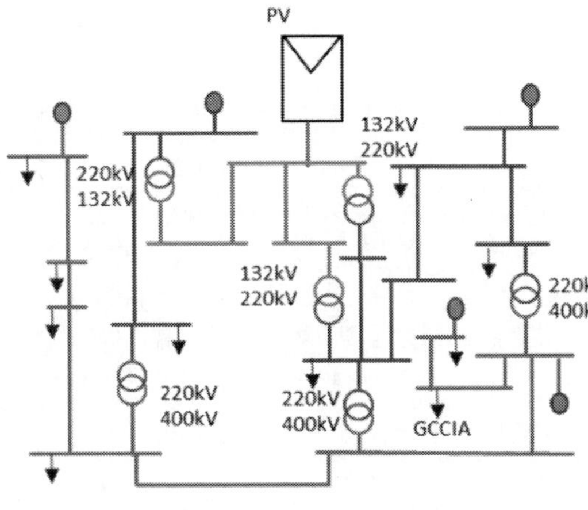

Figure 8.Qatar Power Network

The 400 kV level is interconnected to the GCCIA 400 kV network. GCCIA 400 kW network is a transmission line that is interconnected with other GCC countries Saudi Arabia, UAE, OMAN, Bahrain and Kuwait.

There are another two solar farms in the planning which are 400 MW each. It is expected there will be more PV to be installed and connected to the Qatar power system in the year to come. These new installations will be in smaller sizes ranging from a few Megawatts to hundreds of Megawatts.

In this paper, part of the Qatar transmission system around Qatar Foundation Education City in Doha is used for the study. The rest of the network is represented by an infinite bus. The single-line diagram of the network is shown in Fig. 9. The PV and Battery are connected to a 66 kV bus. The network is fed by 66 kV level and the voltage is stepped down to 11 kV by a 40 MVA 66/11 kV unit transformer. The 11 kV network is serving the customer which is connected to the secondary side of 11/0.415 kV distribution transformers.

15 MV PV is to be connected to 66 kV bus SS66kV. The battery energy storage system will be connected within the Solar Farm facility to an 11 kV network. With this connection, according to the KAHRAMAA grid code, PV and BESS are considered as Power Park and it falls within category P. If there is a voltage dip, the demanded fault current injection is the accumulative of the current from PVs and BESS measured at the point of interconnection.

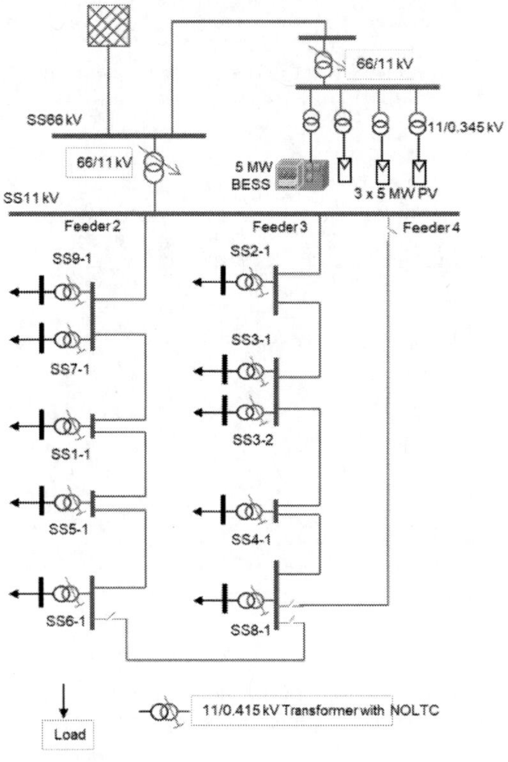

Figure 9. Power Network under study (will be updated by Zamri)

V. SIMULATION AND ANALYSIS

The model is developed in SIMSCAPE ELECTRICAL, one of the MATLAB-SIMULINK toolboxes [10]. SIMSCAPE is a physical modelling-based software where all the mathematical equations are represented by a specific block. The fault current injection is set to be supplied by each inverter within the power park based on the voltage magnitude detected at the inverter's terminal.

The fault is subjected within the infinite bus which simulates a fault that occurred at the higher voltage level. The network is operated at a steady state. At t=0.5 s, a self-cleared

10.4229/EUPVSEC2023/5DV.2.23
020537-005

fault is subjected within an infinite bus for 150 ms that leads to voltage at the point of interconnection reaching approximately 0.1 p.u.

The voltage, current, active power and reactive power measures at the point of interconnection are depicted in Fig. 10 for the case without fault current injection and in Fig. 11. For the case with fault current injection with the gain of 2. The case without fault current injection will be referred to as the base case.

Case 1: No fault current injection

Case 2: Faut current injection with coordination

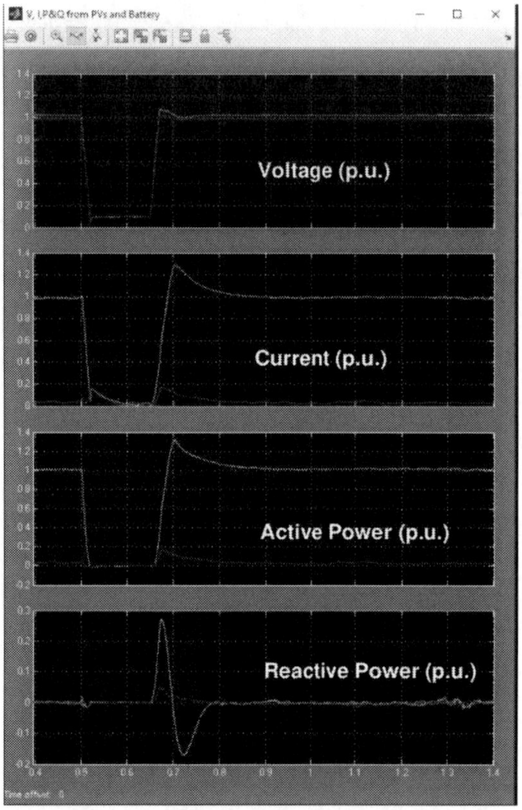

Figure 10. Voltage, current, active power and reactive power: case 1

I. CONCLUSION

This paper describes the modelling and simulation of a photovoltaic (PV) with battery energy storage connected to part of Qatar power network. The modelling approach and the study analysis performed are described. The model and operation scenarios are simulated using a digital simulator for different scenarios. The dynamic response of the system when subjected to sudden self-clearance temporary fault is presented. The capability of the PV system and battery storage riding through the power system fault and at the same time supporting the local grid by injecting fault current is demonstrated. For each case, the different controller methods to achieve the objective of supporting the grid according to grid code requirements are presented and explained. The inverter modelling approach is presented and described.

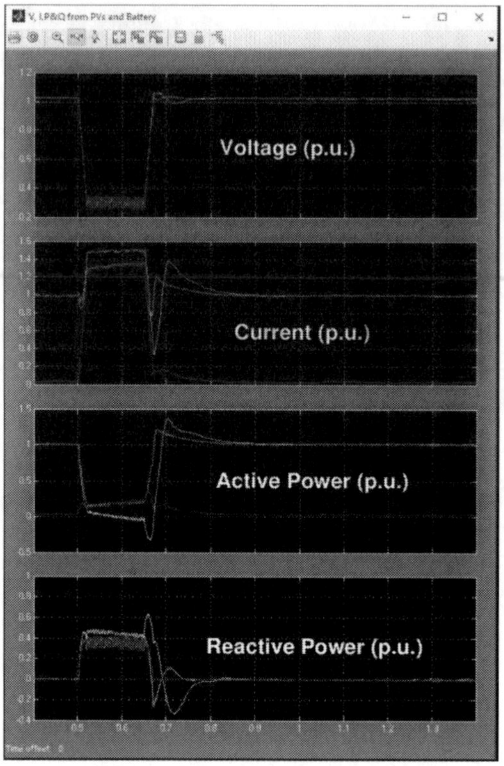

Figure 11 Voltage, current, active power and reactive power: case 2

REFERENCES

[1] S. Hay and A. Ferguson, A Review of Power System Modelling Platforms and Capabilities, IET 2015, https://www.theiet.org/media/1291/pnjv-3-15.pdf

[2] F. Gonzalez-Longatt, Introduction to Power System Simulation, https://www.researchgate.net/publication/339513751_Section_1_Introduction_to_Power_System_Simulation

[3] Z. Kłosowski and S. Cie'slik, The Use of a Real-Time Simulator for Analysis of Power Grid Operation States with a Wind Turbine, *Energies* 2021, *14*, 2327. https://doi.org/10.3390/en14082327

[4] RTDS Corporate Brochure, "Your power grid your conditions in real time", RTDS Technologies, http://www.rtds.com/index/RTDS%20Technologies%20Corporate%20Profile.pdf

[5] OPAL-RT website, http://www.opal-rt.com/electrical-systemsimulation-overview

[6] S. K. Sahoo, A. K. Sinha, and N. K. Kishore, "Modeling and Real-time Simulation of an AC Microgrid with Solar Photovoltaic System", in *Proc. 2015 Annual IEEE India Conference (INDICON)*, New Delhi.

[7] J. Bélanger, J. N. Paquin and P. Venne "The What, Where and Why of Real-Time Simulation", in *Proc. 2010 IEEE PES General Meeting*, Minneapolis, USA.

[8] M.Z.C. Wanik, A. Bousselham, A. Y. Elrayyah, Real-Time Simulation for PV-Battery based Microgrid System, 2016 IEEE International Conference on Power System Technology, Sept 27-Aug 1, 2016, Wollongong, Australia

[9] Marcello Gradella Vilallva, Jonas Rafael Gazoli and Ernesto Ruppert Filho, "Comprehensive approach to Modelling and Simulation of Photovoltaic Arrays", *IEEE Transactions on Power Electronics*, Vol. 24. No.5, May 2009, 1198-1208.

[10] The MathWorks Inc., SimPowerSystems for use with Simulink 7: User's Guide, Natick, MA, 20[illegible].

[11] Qatar General Electricity & Water Corporation or KAHRAMA, Qatar Transmission Grid Code, https://www.km.qa/CustomerService/ServiceRegulations/Grid%20Code%202020.pdf

[12] IRENA (2022), Grid codes for renewable powered systems, International Renewable Energy Agency, Abu Dhabi. ISBN: 978-92-9260-427-1

[13] KAHRAMAA. "Electricity sector," KAHRAMAA [Online]. Available: https://www.km.qa/AboutUs/pages/electricitysector.aspx. [Accessed 28 April 2021].

[14] Alashqar M, Xue Y, Yang C and Zhang X-P (2022), Comprehensive economic analysis of PV farm -A case study of Alkarsaah PV farm in qatar. Front. Energy Res. 10:987773. doi: 10.3389/fenrg.2022.987773

[15] Netz, E. Grid Code High and Extra High Voltage; E. ON Netz GmbH: Bayreuth, Germany, 2006

EFFECTS OF CONTROL STRATEGY ON SIZING OF ENERGY STORAGE SYSTEMS FOR PV–WIND POWER PLANTS

Micke Talvi and Kari Lappalainen
Tampere University, Electrical Engineering
E-mail: micke.talvi@tuni.fi, kari.lappalainen@tuni.fi

ABSTRACT: Different control strategies can be used with energy storage systems (ESS) to mitigate power fluctuations of photovoltaic (PV) and wind power plants. Control strategies can significantly affect the required size of the ESS. In this paper, ramp rate (RR), moving average (MA) and low-pass filter (LPF) ESS control strategies are compared for PV, wind, and PV–wind power plants. It was found that the required relative charging and discharging powers of the ESSs were quite similar between the different strategies. For the RR strategy, the required relative ESS energy capacity was found to be the highest and the utilization rate was found to be the lowest among the 3 strategies. The energy level in the ESS was found to behave differently with the RR strategy than with the MA and the LPF strategies. It was also found that with the RR strategy the behavior of the ESS energy level differs between the PV and the wind power plants. With PV power, the energy level of the ESS tends to slowly decline below an initial level during a day, but for wind power, it tends to elevate moderately fast and clearly farther away from the initial level.
Keywords: Photovoltaic power, Wind power, Power fluctuations, Energy storage systems, Control strategy

1 INTRODUCTION

Generated powers of photovoltaic (PV) and wind power plants can fluctuate drastically [1,2], thus the stability of the grid may be endangered. The fluctuating power can be mitigated by energy storage systems (ESS) with different power control strategies. The use of different ESS control strategies has been studied, for instance in [3,4] for PV power plants, but it has not been studied for wind or PV–wind power plants with high resolution output powers of the power plants. A high resolution of power data would be necessary, as the fastest PV power fluctuations may not be detected if the sampling frequency is lower than 1 Hz [5]. In [4], it was found that the applied control strategy significantly affects the size and utilization of the ESSs of PV power plants. In [6], it was found that the relative sizes required for the ESSs are smaller for the wind power (WP) plants than for the PV power plants when a ramp rate (RR) based control strategy was used. Also, the required sizes for the ESSs of the PV–wind power plants were found to be smaller than for the WP plants. Thereby, it would be reasonable to investigate how the applied control strategy affects the sizes of the ESSs of the wind and PV–wind power plants. It would be also useful for ESS planning to find out if there are any differences in the behavior of PV and wind power fluctuations, and how they affect the ESS sizing while using different control strategies.

In this paper, the effects of control strategy on sizing of ESSs for PV–wind power plants are studied. The PV and wind power plants are simulated based on climatic measurements done with a sampling frequency of 10 Hz for a period of 153 days. The applied ESS control strategies are the RR, the moving average (MA) and the low-pass filter (LPF). The goal of this paper is to investigate how the different control strategies affect the requirements and the utilization of the ESSs of PV, wind, and PV–wind power plants.

2 DATA AND METHDOS

2.1 Measurement data

This study is based on the measurements done at the Tampere University Solar PV Power Station Research Plant in Finland [5]. The measured quantities were irradiance, PV module backside temperature, wind speed and ambient temperature. The measurement period was 153 days, starting on 1 June 2019 and ending on 31 October 2019. The sampling frequency for all the measurements was 10 Hz.

The PV power was modeled based on irradiance and PV module backside temperature measurements. Irradiance was measured with a photodiode-based Kipp&Zonen SPLite2 pyranometer. The pyranometer is mounted to a PV module which faces nearly southward and is mounted at a tilt angle of 45°, which is also the tilt angle of the pyranometer. The PV module backside temperature was measured with a National Instruments Pt100 temperature sensor. The WP was modeled based on measured wind speed and ambient temperature. The wind speed was measured with an ultrasonic VAISALA WS425 wind sensor that is mounted at the tip of a pole located at the highest point of the university building. The height from the ground level to the sensor h_1 XXX is 16.13 m. The ambient temperature T_{air} was measured with a VAISALA HMP155 humidity and temperature sensor that is also located on the rooftop of the university building.

The ESS sizing was done for two orders of magnitude: small-scale and large-scale power plants. The nominal powers of the PV and the wind power plants were 10 kW and 3 MW, respectively, for the small- and the large-scale power plants. The nominal powers of the PV–wind power

Table I: Parameter values of the simulation model for the power plants.

Nominal power of the power plant	10 kW	3 MW
Number of PV modules	53	15789
PV module array area	149 m^2	44 340 m^2
Wind turbine height	16.5 m	98 m
Wind turbine swept area	75.4 m^2	5 281 m^2
Calculated power coefficient	0.48	0.50
Wind power time constant	15 s	40 s
Cut-in wind speed	2 m/s	3 m/s
Rated wind speed	9 m/s	16 m/s
Cut-out wind speed	30 m/s	34 m/s

10.4229/EUPVSEC2023/5DV.2.24
020538-001

plants were 20 kW and 6 MW, respectively, for the small- and the large-scale power plants. The PV power plant size was scaled using different numbers of PV modules. The WP plant size was scaled using the quantities of 2 different wind turbine (WT) models [7,8]. Table I presents parameter values used for the power plants.

2.2 PV and wind power modeling

The area of a virtual PV power plant was determined based on the size and tilt angle of the PV modules of the Tampere University Solar PV Power Station Research Plant. The empty space between the PV module rows was determined so that the PV modules would not shade each other during daytime in the summer months in Tampere, Finland. The altitude angle of the Sun was determined to be 20°, and thus the empty space length was determined to be 1916 mm. The length and the width of the PV module were 986 mm and 1475 mm, respectively. The nominal power of the module $P_{\mathrm{nom,\,PV}}$ was 190 W.

After the suitable empty space length was known, the PV module array areas of the PV power plants were possible to be calculated. As the irradiance was measured only from one sensor, a spatial filter proposed by Marcos et al. [9] was implemented to provide a more realistic irradiance exposure of a PV power plant. The spatial irradiance $G_s(t)$ was calculated as

$$G_s(t) = \frac{G(t)}{\left(\frac{\sqrt{A_{PV}}}{2\pi \cdot 0.0204 \cdot \left(A_{PV}{}^{-0.4997}\right)}\right)s+1}, \quad (1)$$

where $G(t)$ is the measured irradiance, A_{PV} is the PV module array area in hectares and s is the Laplace transform variable. With the spatial filter, the increased smoothing effect of PV power fluctuations could be taken into account, as there are 2 significantly different sizes of PV power plants in this study. The generated PV power P_{PV} was calculated using an equation [10] that takes into account the effect of the PV module temperature T_{PVM} on P_{PV} as

$$P_{PV} = \frac{nP_{\mathrm{nom,\,PV}}}{G_{STC}} G_S[1-\beta(T_{PVM}-T_{STC})], \quad (2)$$

where n is the number of the PV modules, G_{STC} is the irradiance in standard test conditions (STC), β is the temperature coefficient and T_{STC} is the temperature in the STC. As the average value for the β was 0.0045 1/°C in [10], this value was used in the calculations of this study as well.

In the WP modeling, the measured wind speed v_1 at the height of h_1 was first extrapolated to the hub heights of the WT models. The extrapolation was done with a power law [11], applied as

$$v_2 = v_1\left(\frac{h_2}{h_1}\right)^{\alpha}, \quad (3)$$

where v_2 is the extrapolated wind speed at the height of h_2 and α is the shear exponent. Considering the presented values for α in [11], the value for α was determined to be 0.35, as the environment around the wind speed sensor could be considered as a small town but there are a few tall buildings around. Then, the density of the air ρ_{air} was calculated for each data point as [12]

$$\rho_{air} = \frac{p_0}{R_{air}T_{air}} e^{\left(\frac{-gH}{R_{air}T_{air}}\right)}, \quad (4)$$

where p_0 is the standard sea level atmospheric pressure (101325 Pa), R_{air} is the specific gas constant of air (287.05 J/(kg·K)), g is the gravity constant (9.81 m/s²) and H is the total altitude above sea level. The altitude above sea level for Hervanta Campus of Tampere University, Tampere is 140 m [13], thus the value used for H is the sum of 140 m and the hub height of the WT.

The WP was modeled using the cubic WP equation [12]. First the preliminary WT power $P_{\mathrm{WT,\,pre}}$ was calculated as

$$P_{\mathrm{WT,\,pre}}(v_2) = \begin{cases} 0, & v_2 < v_{\mathrm{cut-in}} \\ \frac{1}{2}A_{\mathrm{swept}}\rho_{\mathrm{air}}v_2^3C_p, & v_{\mathrm{cut-in}} \le v_2 < v_{\mathrm{rated}} \\ P_{\mathrm{rated,\,WT}}, & v_{\mathrm{rated}} \le v_2 < v_{\mathrm{cut-out}} \\ 0, & v_{\mathrm{cut-out}} \le v_2 \end{cases}, \quad (5)$$

where A_{swept} is the swept area of the WT rotor, C_p is the WT power coefficient and $P_{\mathrm{rated,\,WT}}$ is the rated power of the WT. $v_{\mathrm{cut\text{-}in}}$, v_{rated}, and $v_{\mathrm{cut\text{-}out}}$ are the cut-in, rated and cut-out wind speeds of the WT, respectively. The values of C_p needed to be calculated for the WTs using Eq. (5), as they were not available. The calculated maximum values were used as the C_p values for the WTs, as that would provide sufficient accuracy according to [14]. The $v_{\mathrm{cut\text{-}in}}$, v_{rated}, and $v_{\mathrm{cut\text{-}out}}$ values as well as the calculated C_p values of the WTs are presented in Table I.

A low-pass filter was applied to take the inertia of the WT rotor [15] into account and to level the power fluctuations of $P_{\mathrm{WT,\,pre}}$ to a more realistic level. The smoothed wind power P_{wind} was acquired by applying the first-order low-pass filter to $P_{\mathrm{WT,\,pre}}$ as

$$P_{wind}(t) = \frac{P_{\mathrm{WT,\,pre}}(t)}{\tau_{WT}s+1}, \quad (6)$$

where τ_{WT} is the time constant of the WT. The τ_{WT} values for the WTs are presented in Table I. With these τ_{WT} values, P_{wind} should not be oversmoothed. The PV and wind power models were presented in more detail in [6].

2.3 Control strategies

The applied control strategies of this study are presented here shortly. With the combined PV–wind power plants, P_{PV} and P_{wind} were first summed, then the sum power was treated by the control strategy. The goal of the control strategies was to keep the power fluctuations of a power plant under the RR limit of 10 %/min. For the MA and the LPF strategies, the time constants were also iterated to be as small as possible. If the time constants were too large, the results would not be that comparable between the different strategies, as the MA and the LPF strategies would smooth the output power significantly more than the RR strategy. The amount of energy stored in the ESS was selected to be the same in the beginning of each day, i.e., the ESS was assumed to be charged or discharged back to the same initial energy level at midnight.

The RR control strategy limits the change in the power fed to the grid P_{grid} between two consecutive time stamps. The RR control strategy first checks the maximum upward power ramp as

$$P_{\mathrm{grid,\,1}}(i) = \min\left(P_{PP}(i), \left(P_{\mathrm{grid,\,RR}}(i-1) + RR10PP\right)\right), \quad (7)$$

10.4229/EUPVSEC2023/5DV.2.24
020538-002

where $P_{\text{grid, 1}}(i)$ is the preliminary power fed to the grid that does not exceed the maximum power RR of the power plant $RR10PP$ that is determined by the RR limit of 10 %/min, $P_{\text{grid, RR}}(i\text{-}1)$ is the power fed to the grid at the time step i-1 and $P_{\text{PP}}(i)$ is the power of the power plant at the time step i. Then, the RR control strategy checks the maximum downward power ramp as

$$P_{\text{grid, RR}}(i) = \max\left(P_{\text{grid, 1}}(i), \left(P_{\text{grid, RR}}(i-1) - RR10PP\right)\right). \quad (8)$$

Now, $P_{\text{grid, RR}}(i)$ does not exceed the RR limit of 10 %/min during either the upward or the downward power ramp. The power of the ESS P_{ESS} is then determined as

$$P_{\text{ESS}}(i) = P_{\text{PP}}(i) - P_{\text{grid, RR}}(i). \quad (9)$$

In the MA control strategy, the power fed to the grid $P_{\text{grid, MA}}(i)$ is determined as a moving average as

$$P_{\text{grid, MA}}(i) = \frac{\sum_{j=i-\tau_{\text{MA}}+1}^{i} P_{\text{PP}}(j)}{\tau_{\text{MA}}}, \quad (10)$$

where τ_{MA} is the MA control strategy time constant of the power plant. After $P_{\text{grid, MA}}(i)$ is determined, $P_{\text{ESS}}(i)$ is determined according to Eq. (9).

In the LPF control strategy, the P_{PP} is first filtered using first-order low-pass filter as

$$P_{\text{grid, LPF}}(t) = \frac{P_{\text{PP}}(t)}{\tau_{\text{LPF}}s+1}, \quad (11)$$

where $P_{\text{grid, LPF}}(t)$ is the power fed to the grid with the LPF control strategy and τ_{LPF} is the LPF control strategy time constant of the power plant. After that, P_{ESS} is determined according to Eq. (9).

3 RESULTS AND DISCUSSION

3.1 Small-scale power plants

The ESS sizing results for the small-scale power plants are presented in this subchapter. The results are explained by an example day of June 22, 2019. The simulated 10 kW PV power and the powers fed to the grid for the RR, the MA and the LPF strategies are presented in Fig. 1 (a). The energy levels stored in the ESS of the 10 kW PV power plant with the control strategies are presented in Fig. 1 (b). It can be seen clearly that the MA and the LPF strategies smooth the original P_{PV} a lot more than the RR strategy. Although, all the graphs of the control strategies fluctuate significantly slower than the original P_{PV}.

The different behaviors of the control strategies can be seen also in Fig. 1 (b). The graphs of the MA and the LPF strategies stay above 0 kWh the whole example period and are very similar, whereas the graph of the RR strategy stays below 0 kWh from around 09:00 onwards, and its highest peak has significantly higher amplitude than with the MA and the LPF strategies. The small peaks of the MA and the LPF graphs at around 05:50 in Fig. 1 (b) were caused by a moderately high upward ramp of the P_{PV}. However, the highest RR of this fluctuation did not exceed the RR limit, and thus, the RR strategy did not smooth it.

Table II presents the ESS sizing results for the 10 kW PV power plant. The highest observed RRs of the different

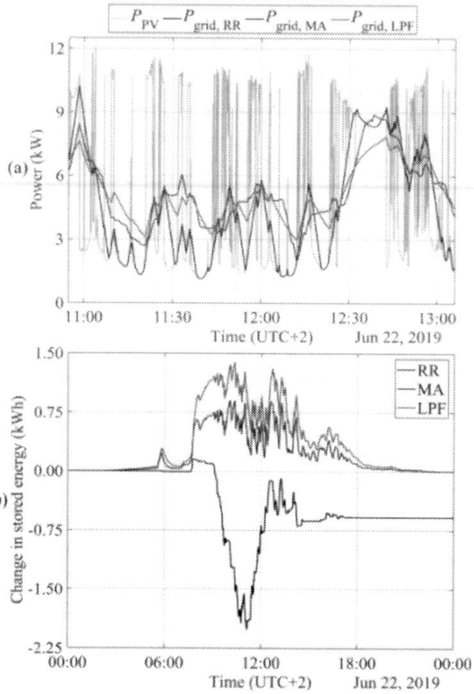

Figure 1: Example of simulated 10 kW PV power, powers fed to the grid for the RR, MA, and LPF strategies (a) and change of the ESS energy level from the initial energy level with the RR, MA, and LPF strategies (b).

strategies are quite close each other. The relative energy capacity requirements for the ESS seem to differ significantly between the strategies, with the MA strategy having the lowest requirement and the RR strategy having the highest. The relative charging and discharging power values seem to be very similar between the strategies, however. The MA and the LPF strategies seem to utilize the ESS significantly more than the RR strategy, as their daily maximum shares of energy cycled through the ESS are roughly 66% higher than the share for the RR strategy. The high utilization rate of the ESS with the MA and the LPF control strategies was found also in [4]. In [4], it was stated that using the RR control strategy for the ESS of the PV power plant would result in a smaller-sized ESS than

Table II: Iterated time constants, the highest observed power ramp rates and the ESS sizing results with the control strategies for the 10 kW PV power plant.

Control strategy	RR	MA	LPF
Time constant (s)	-	725	60.0
Highest observed RR (%/min)	10.0	9.67	9.53
Relative energy capacity (h)	0.245	0.0964	0.149
Relative charging power capacity (%)	0.877	0.944	0.947
Relative discharging power capacity (%)	0.779	0.770	0.730
Daily maximum share of energy cycled through the ESS (%)	0.130	0.214	0.218

10.4229/EUPVSEC2023/5DV.2.24
020538-003

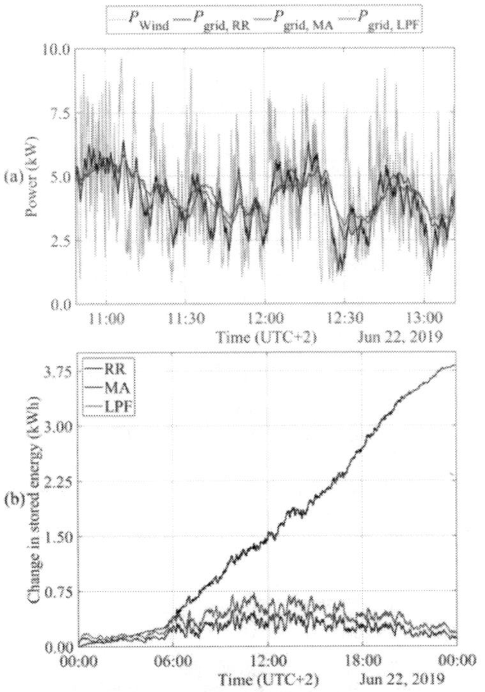

Figure 2: Example of simulated 10 kW wind power, powers fed to the grid for the RR, MA, and LPF strategies (a) and change of the ESS energy level from the initial energy level with the RR, MA, and LPF strategies (b).

use of the MA and the LPF strategies. However, in this study, the values for the relative energy capacity of the ESS of the 10 kW PV power plant with the MA and the LPF strategies are significantly smaller than the value with the RR strategy. The reason for this is explained in Section 3.3.

Fig. 2 (a) presents the simulated 10 kW WP and the powers fed to the grid for the RR, the MA and the LPF strategies. The powers fed to the grid seem to behave in a similar way as with the 10 kW PV power plant – the graphs of the MA and the LPF strategies staying smoother than the graph of the RR strategy. The energy levels stored in the ESS of the 10 kW WP plant with the control strategies are presented in Fig. 2 (b). It can be seen that the graphs of the MA and the LPF strategies behave very similarly, but the graph of the RR strategy behaves significantly differently and its highest value is roughly 8 and 5 times the highest values of the graphs of the MA and the LPF strategies, respectively.

Table III: Iterated time constants, the highest observed power ramp rates and the ESS sizing results with the control strategies for the 10 kW wind power plant.

Control strategy	RR	MA	LPF
Time constant (s)	-	565	45.0
Highest observed RR (%/min)	10.0	9.43	9.56
Relative energy capacity (h)	0.576	0.0601	0.0875
Relative charging power capacity (%)	0.668	0.735	0.717
Relative discharging power capacity (%)	0.527	0.529	0.478
Daily maximum share of energy cycled through the ESS (%)	0.129	0.216	0.208

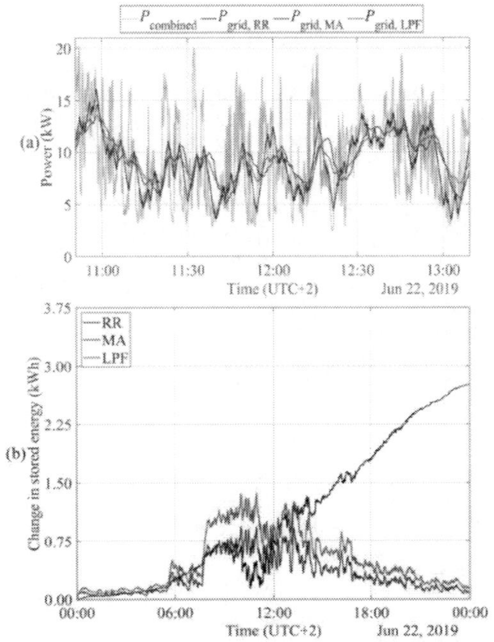

Figure 3: Example of simulated combined 20 kW PV–wind power, powers fed to the grid for the RR, MA, and LPF strategies (a) and change of the ESS energy level from the initial energy level with the RR, MA, and LPF strategies (b).

Table III presents the ESS sizing results for the 10 kW WP plant. The highest observed RRs of the different strategies are quite similar. The relative energy capacity of the ESS required with the RR strategy differs significantly from the values of the MA and the LPF strategies. This can be seen also in Fig. 2 (b). The relative power capacity values of the ESS are quite similar between the different strategies for the 10 kW WP plant. The daily maximum share of energy cycled through the ESS with the RR strategy is clearly smaller than with the MA and the LPF strategies.

When comparing the results of the 10 kW PV and the 10 kW wind power plant, on general, the ESS sizing values are higher for the 10 kW PV power plant than for

Table IV: Iterated time constants, the highest observed power ramp rates and the ESS sizing results with the control strategies for the 20 kW PV–wind power plant.

Control strategy	RR	MA	LPF
Time constant (s)	-	483	36.0
Highest observed RR (%/min)	10.0	9.94	9.45
Relative energy capacity (h)	0.233	0.0489	0.0715
Relative charging power capacity (%)	0.587	0.590	0.565
Relative discharging power capacity (%)	0.135	0.154	0.141
Daily maximum share of energy cycled through the ESS (%)	0.100	0.190	0.183

10.4229/EUPVSEC2023/5DV.2.24
020538-004

the 10 kW WP plant, in line with the results of [6]. In Tables II and III, the only notable exception is that the value of the relative energy capacity of the ESS required with the RR strategy for the 10 kW WP plant is roughly 2.4 times the value for the 10 kW PV power plant. This difference can be seen also from Figs. 1 (b) and 2 (b). The reason for the difference is explained in Section 3.3.

Fig. 3 (a) presents the simulated 20 kW PV–wind power and the powers fed to the grid for the RR, the MA and the LPF strategies. The powers fed to the grid seem to behave again in a similar way as with the separate 10 kW PV and wind power plants. Fig. 3 (b) presents the energy levels stored in the ESS with the control strategies. It can be seen that the graphs of the MA and the LPF strategies resemble the corresponding graphs of Figs. 1 (b) and 2 (b). In Fig. 3 (b), the graph of the RR strategy seems to resemble the combination of the corresponding graphs of Figs. 1 (b) and 2 (b). However, the graph of the RR strategy never goes below 0 kWh for the 20 kW PV–wind power plant.

Table IV presents the ESS sizing results for the 20 kW PV–wind power plant. The relative energy capacity of the ESS with the RR strategy again differs from the values of the MA and the LPF strategies. Apart from the relative energy capacity of the ESS with the RR strategy, the ESS sizing values are clearly smaller for the 20 kW PV–wind power plant than for the 10 kW PV and wind power plants. In [6], it was found also that the required relative ESS size for the combined PV–wind power plant is smaller than for the PV power plant considering the small-scale power plants. However, the difference between the ESSs of the small-scale wind and PV–wind power plants, was minimal in [6].

3.2 Large-scale power plants

The ESS sizing results for the large-scale power plants are presented in this subchapter. Only the energy levels stored in the ESS are presented in this subchapter as figures, as the graphs of the simulated powers of the large-scale power plants were very similar to the corresponding graphs of the small-scale power plants in Section 3.1. Evolutions of the energy amounts stored in the ESSs of the 3 MW PV, the 3 MW wind, and the 6 MW PV–wind power plants with the control strategies are presented in Figs. 4 (a), 4 (b) and 4 (c), respectively. The graphs of Fig. 4 (a) are very similar to the graphs of Fig. 1 (b).

The graphs of the MA and the LPF strategies for the 3 MW WP plant in Fig. 4 (b) resemble the corresponding graphs of the 10 kW WP plant in Fig. 2 (b). The graph of the RR strategy for the 3 MW WP plant does not ascend as fast or as high as for the 10 kW WP plant in Fig. 2 (b). The graph of the RR control strategy for the 6 MW PV–wind power plant in Fig. 4 (c) is quite different than the corresponding graph of Fig. 3 (b), as the graph of the RR control strategy for the 6 MW PV–wind power plant stays negative for a couple hours during the example period.

Table V presents the ESS sizing results for the 3 MW PV power plant. The relative ESS sizing requirements are slightly lower for the 3 MW PV power plant than for the 10 kW PV power plant. This is expected, as the increased spatial smoothing of the larger PV power plant decreases the PV power fluctuations relatively, as found in [1].

Table VI presents the ESS sizing results for the 3 MW WP plant. When comparing these results to the results of the 10 kW WP plant in Table III, the required relative power capacity and the utilization rate of the ESS are smaller for the 3 MW WP plant. The required relative

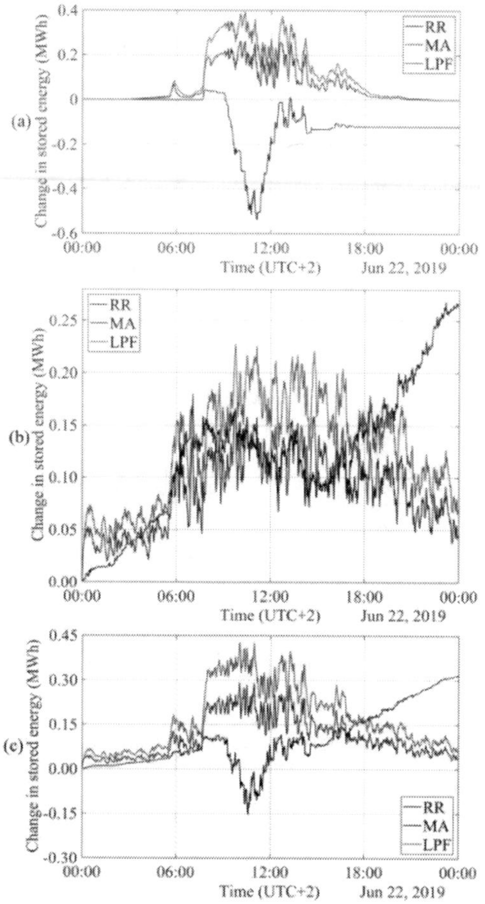

Figure 4: Example of change in the energy stored in the ESS from the initial energy level during a day for the 3 MW PV (a), the 3 MW wind (b), and the 6 MW PV–wind (c) power plants with the RR, MA, and LPF strategies.

energy capacity value of the ESS is also smaller for the 3 MW WP plant than for the 10 kW WP plant when using the RR control strategy, but with the MA and the LPF strategies, the relative energy requirements for the 3 MW WP plant are slightly larger than for the 10 kW WP plant.

Table VII presents the ESS sizing results for the 6 MW

Table V: Iterated time constants, the highest observed power ramp rates and the ESS sizing results with the control strategies for the 3 MW PV power plant.

Control strategy	RR	MA	LPF
Time constant (s)	-	705	57.0
Highest observed RR (%/min)	10.0	9.63	9.57
Relative energy capacity (h)	0.215	0.0938	0.142
Relative charging power capacity (%)	0.821	0.909	0.909
Relative discharging power capacity (%)	0.757	0.761	0.717
Daily maximum share of energy cycled through the ESS (%)	0.123	0.208	0.212

Table VI: Iterated time constants, the highest observed power ramp rates and the ESS sizing results with the control strategies for the 3 MW wind power plant.

Control strategy	RR	MA	LPF
Time constant (s)	-	515	38.0
Highest observed RR (%/min)	10.0	9.64	9.54
Relative energy capacity (h)	0.302	0.0642	0.0882
Relative charging power capacity (%)	0.501	0.627	0.604
Relative discharging power capacity (%)	0.453	0.487	0.427
Daily maximum share of energy cycled through the ESS (%)	0.0734	0.161	0.145

Table VII: Iterated time constants, the highest observed power ramp rates and the ESS sizing results with the control strategies for the 6 MW PV–wind power plant.

Control strategy	RR	MA	LPF
Time constant (s)	-	430	34.0
Highest observed RR (%/min)	10.0	9.81	9.32
Relative energy capacity (h)	0.114	0.0506	0.0751
Relative charging power capacity (%)	0.518	0.531	0.528
Relative discharging power capacity (%)	0.100	0.142	0.141
Daily maximum share of energy cycled through the ESS (%)	0.0536	0.139	0.130

PV–wind power plant. The required relative energy capacity values of the ESS with the MA and the LPF strategies are slightly higher for the 6 MW PV–wind power plant than for the 20 kW PV–wind power plant in Table IV. However, the other ESS sizing values between these power plants are smaller for the 6 MW PV–wind power plant.

In conclusion, the relative power requirements and utilization rates of the ESS were lower for the large-scale power plants than for the small-scale ones with all 3 control strategies. The relative ESS energy requirement was also lower the large-scale power plants except for WP and the combined PV–wind power when the MA or LPF strategy was used.

3.3 Typical ESS cycles with PV and wind power

This subchapter shortly analyses the differences between the typical ESS cycles with PV and wind power. The typical ESS cycles with PV and wind power are the main reason behind the significantly different graphs of the evolution of the energy amounts stored in the ESS controlled by the RR strategy between PV and wind power plants seen in Figs. 1 (b) and 2 (b), for instance. Fig. 5 (a) presents the typical ESS cycles with PV power and Fig. 5 (b) presents the typical ESS cycles with WP. The typical ESS cycles are illustrated as red and blue boxes, respectively, for the PV and the wind power.

In Fig. 5 (a), the P_{PV} typically fluctuates from one power level to another (one ESS charging or discharging cycle) so that it exceeds the RR limit in the same direction with multiple narrow peaks or fewer wide peaks. After the transition, the P_{PV} stays at a certain level, corresponding to unshaded or shaded PV power plant, for a period, and only

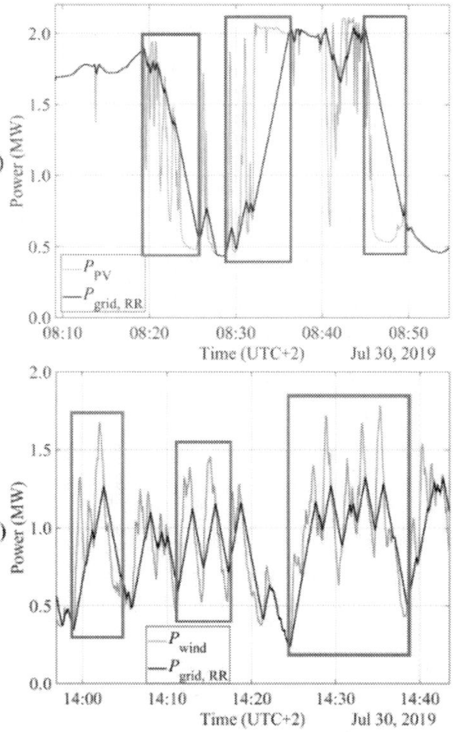

Figure 5: Example of typical ESS cycles with PV (a) and wind (b) power.

after then it seems to fluctuate to the other level. As the P_{PV} remains quite steady after the short fluctuations, the gradually moving $P_{grid, RR}$ can also reach the same level as the P_{PV}. Due to this behavior, the ESS typically only either charges or discharges power during one typical ESS cycle with PV power. In the long run, the energy level of the ESS will start to slowly decline below 0, as seen in Figs. 2 (b) and 5 (b). One reason for this is that typically the front edges of cloud shadows are sharper than the back edges [16]. Thus, the P_{PV} typically drops faster when a cloud shadow covers the PV power plant than it comes back up after the cloud has passed. This means that, during the arrival of the cloud the P_{PV} deviates more from the $P_{grid, RR}$ than during the departure of the cloud. This causes that the ESS is discharged more than charged, and thus, its energy level keeps declining in the long run.

In Fig. 5 (b), the P_{wind} fluctuates so that during one typical ESS cycle the P_{wind} first fluctuates upward, and then instantly it fluctuates downward. The P_{wind} usually exceeds the $P_{grid, RR}$ in both directions during the typical ESS cycle. However, as the downward fluctuation follows the upward fluctuation instantly, the $P_{grid, RR}$ does not reach the level of the upward fluctuation peak of the P_{wind}. Thus, the final upward peak of the P_{wind} always remains higher than the $P_{grid, RR}$. This results in that there is more power to be charged than discharged during one typical ESS cycle with WP, and thus, the energy level in the ESS of a WP plant keeps gradually ascending through the day, as seen in Figs. 2 (b) and 4 (b).

This phenomenon can be explained also by a more visual point-of-view. When observing the red boxes of Fig. 5 (a), it can be seen that the P_{PV} remains roughly only either above or below the $P_{grid, RR}$ during one typical ESS

10.4229/EUPVSEC2023/5DV.2.24
020538-006

cycle with PV power. The areas between the P_{PV} and the $P_{grid, RR}$ are roughly similar when comparing a typical ESS cycle with either an upward or a downward ramp of the $P_{grid, RR}$. As these areas are similar in size, this results in that the net flow of energy through the ESS of a PV power plant is nearly 0. However, the energy level declines gradually due to the aforementioned differences between upward and downward PV power fluctuations.

When observing the blue boxes of Fig. 5 (b), it can be seen that the P_{wind} fluctuates both above and below the $P_{grid, RR}$ inside one blue box. This means that the ESS of a WP plant charges and discharges power during one typical ESS cycle. However, as the area between the P_{wind} and the $P_{grid, RR}$ is clearly larger during the upward part of the $P_{grid, RR}$ ramp, the ESS of a WP plant will be charged more than discharged during a typical ESS cycle with WP. This will result in the constantly elevating energy level stored in the ESS during the day. Thus, the relative energy requirement for an ESS controlled by the RR strategy is higher for WP plants than for PV power plants.

The reason why this phenomenon emerges only with the RR strategy, is that the data steps are not tied to each other with the RR strategy on contrary to the MA and the LPF strategies. As the current data step is not tied to older steps, the energy level of an ESS of a PV or a wind power plant can progress considerably far away from 0. The phenomenon is stronger for the small-scale power plants than for the large-scale ones as the WP flúctuations smoothen with increasing turbine size. In this study, the amount of energy stored in the ESS was selected to be the same in the beginning of each day. Thus, the effect of this phenomenon in the long term was not studied but only daily ESS requirements were considered.

4 CONCLUSIONS

This paper presented a study of the effects of the control strategy on sizing of ESSs for small- and large-scale PV, wind, and PV–wind power plants. The output powers of the power plants were modeled based on climatic measurements. The power plants and the ESSs of this study were virtual. The chosen ESS control strategies were the RR, the MA and the LPF strategies.

It was found that the MA and the LPF control strategies resulted in very similar behaviors and sizes of the ESSs. The RR control strategy seemed to differ significantly from the 2 other strategies with its ESS usage behavior and ESS sizing results considering the required relative energy capacity and the utilization rate of the ESS. The relative power requirements and utilization rates of the ESS were lower for the large-scale power plants than for the small-scale ones with all studied control strategies. The relative ES energy requirement was lower for the large-scale power plants except for the WP plants and the combined PV–wind power plants when the ESS was controlled by the MA or LPF strategy.

Differences were also found in the ESS usage behaviors and the ESS sizing results between the PV and the wind power plants. The reason for the different behaviors of the ESS energy levels of the PV and the wind power plants, was that the typical ESS cycles are different for PV and wind power. Based on the results of this paper, control strategies can significantly affect the required size and the usage behavior of the ESS for PV, wind, and PV–wind power plants.

ACKNOWLEDGMENTS

We acknowledge the financial support from Business Finland for the research reported in this paper. K. Lappalainen was funded by the Research Council of Finland (funding decision 348701).

REFERENCES

[1] J. Marcos, L. Marroyo, E. Lorenzo, D. Alvira, E. Izco, Progress in Photovoltaic Research and Applications 19 (2011) 218. https://doi.org/10.1002/pip.1016.
[2] B.K. Parsons, Y. Wan, B. Kirby, Proceedings of the European Wind Energy Conference (2001).
[3] J. Martins, S. Spataru, D. Sera, D.-I. Stroe, A. Lashab, Energies 12 (2019) 1342. https://doi.org/10.3390/en12071342.
[4] S. Sukumar, M. Marsadek, K.R. Agileswari, H. Mokhlis, Journal of Energy Storage 20 (2018) 218. https://doi.org/10.1016/j.est.2018.09.013.
[5] D. Torres Lobera, A. Mäki, J. Huusari, K. Lappalainen, T. Suntio, S. Valkealahti, International Journal of Photoenergy 2013 (2013) 837310. https://doi.org/10.1155/2013/837310.
[6] M. Talvi, T. Roinila, K. Lappalainen, Energies 16 (2023) 4313. https://doi.org/10.3390/en16114313.
[7] RyseEnergy. Available online (accessed on 8.9.2023): https://www.ryse.energy/10kw-wind-turbines/.
[8] Wind-Watch. Available online (accessed on 8.9.2023): https://docs.wind-watch.org/Enercon.pdf.
[9] J. Marcos, L. Marroyo, E. Lorenzo, D. Alvira, E. Izco, Progress in Photovoltaics: Research and Applications 19 (2011) 505. https://doi.org/10.1002/pip.1063.
[10] E. Skoplaki, J.A. Palyvos, Solar Energy 83 (2009) 614. https://doi.org/10.1016/j.solener.2008.10.008.
[11] F.A. Hadi, International Journal of Advanced Research in Electrical, Electronics and Instrumentation Engineering 4 (2015) 8176. DOI: 10.15662/IJAREEIE.2015.0410058.
[12] L. Söder, T. Ackermann, Wind Power in Power Systems 1st ed. (2005) 25.
[13] Maanmittauslaitos—Karttapaikka. Available online (accessed on 8.9.2023): https://asiointi.maanmittauslaitos.fi/karttapaikka/.
[14] C. Carrillo, A.F. Obando Montaño, J. Cidrás, E. Díaz-Dorado, Renewable and Sustainable Energy Reviews 21 (2013) 572. https://doi.org/10.1016/j.rser.2013.01.012.
[15] C. Tang, M. Pathmanathan, W.L. Soong, N. Ertugrul, Proceedings of the 2008 Australasian Universities Power Engineering Conference (2008).
[16] K. Lappalainen, S. Valkealahti, Solar Energy 112 (2015) 55. https://doi.org/10.1016/j.solener.2014.11.018.

Security Countermeasures for Ransomware Attacks on Energy Infrastructure

Jeong-Eon Lee*, Han-Yeol Kim**, Myoung-Geon Seo***, Bo-Woon Han****, Jae-Hyeok Hur*****, Su-Mi Yang******, Yong-Joon Lee*******
*Dept. of Green Energy Engineering Far East University, **Dept. of Hacking Security Far East University
1523-8@hanmail.net, klhan3091@gmail.com, kkk10108@naver.com, onlyyou558@gmail.com, hgh9848@naver.com
2020032@kdu.ac.kr, esther4853@kdu.ac.kr

ABSTRACT

As ICT technology accelerates convergence with the energy environment, the importance of confidential industrial information such as documents and design drawings is increasing. Recently, cyber-attacks aimed at stealing core industry technologies by using energy infrastructure as hostages have continued. In particular, cyber-attacks such as Ransomware invade computers in energy infrastructure and encrypt confidential industrial information without permission, increasing the damage. This paper describes trends in Ransomware Attacks and Security Countermeasures for Energy Infrastructure.

Keywords—BlackCat, DeadBolt, Encryption Algorithms, Ransomware, Sugar

I. Introduction

As ICT technology develops, the importance of confidential industrial information continues to increase, and the damage caused by cyber attacks is increasing. In particular, unlike early Ransomware, where email hacking was the principal attack method, the damage has increased in recent years due to the spread through SNS and messages. Accordingly, we present the trend and Security Countermeasures of Ransomware against the recent increase in Ransomware attacks on Energy Infrastructure.

II. Ransomware attack cases of the Domestic and a Foreign Country

A. Foreign Cases

According to the Q1 2022 Ransomware Trend Report released by KISA(the Korea Internet & Security Agency), seven damage cases occurred from January to March 2022, increasing the damage caused by overseas Ransomware. In January 2022, two damage cases occurred due to Ransomware infection by Delta Electronics, a Taiwanese electronics manufacturer, and the French Ministry of Justice. In addition, British food manufacturer KP Snacks, Swissport International, Japanese sports goods manufacturer Mizuno, and U.S. tire manufacturer Bridgestone Americas occurred with Ransomware in February. In March, Japanese auto parts manufacturer DENSO occurred with Ransomware.

B. Domestic Cases

According to the latest Ransomware trend analysis and implications [1] released by KISA(the Korea Internet & Security Agency), major stores of E-Land Group, such as department stores and outlets, were suspended due to a crop Ransomware attack in November 2020. In May 2021, three attacks were carried out, including data encryption of servers and employee PCs of parts manufacturers, leakage of employee personal information and overseas business data through the dark web, and website paralysis caused by DDoS attacks. In addition, secondary damage was reported, such as when a domestic plastic surgeon robbed of customer lists in a Ransomware attack, and the attacker was in direct contact with customers.

III. Analysis of Ransomware Attack Methods

A. Concepts of Ransomware

Ransomware is a combination of Ransom and malicious software that restricts access or data use by infecting computers and requires Ransom. Ransomware uses encryption algorithms, which encrypt file data and makes it unusable. After encrypting a file, a general Ransomware Attack Method is to extort the victim by making financial demands such as virtual currency by displaying a threatening phrase on the desktop of an infected device.

B. Principle of Ransomware Operation

The operating principle or characteristics of Ransomware may vary depending on the type of Ransomware. However, it generally follows the flow shown in [Fig. 1] [2].

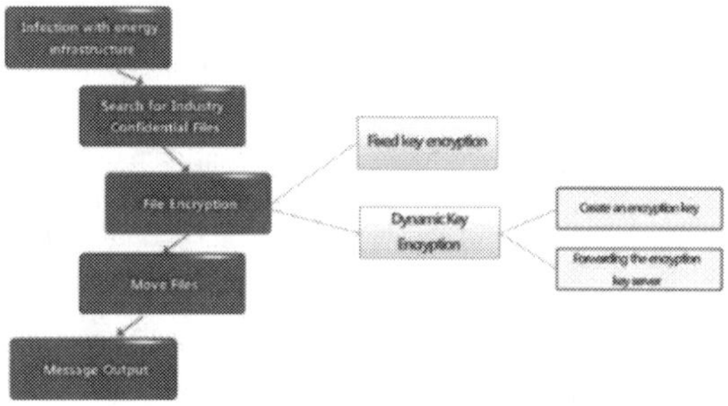

[Fig. 1] Flow of Ransomeware Attack

Ransomware first infects energy infrastructure devices, and then the first step is to search for target files. In this step, we perform the task of searching for files that could be targeted. The second is the file encryption step, which encrypts files targeted for attack in the previous step. Fixed key encryption and dynamic key encryption are the methods of encrypting files. Fixed key encryption is performed with a single key, showing the most straightforward method. Dynamic key encryption is a method of creating and using individual encryption keys in consideration of the system's environment, file name, and attributes. Next, the hacking group creates an encryption key list file to restore the file. Then, it sends the encryption key to the hacking group's server to manage it to demand costs for average data recovery. In the third step, the attacker moves the encrypted file to the desktop, informing the victim that the files were infected and attacked. In the final stage, it sends a threatening message to pay money if it wants to restore confidential industrial data.

IV. Analysis of features by Ransomware-type

According to the Q1 2022 Ransomware Trend Report released by the KISA, 70 Ransomware attacks occurred in the first quarter of 2022, 50 of which occurred as variants of existing Ransomware and the remaining 20 as new types of Ransomware. In the report, three new and variant Ransomware discovered in the first quarter were selected and analyzed, the first being BlackCat, the second being DeadBolt, and the last being Sugar.

A. BlackCat Ransomware

The BalkCat Ransomware group attacked two German oil companies operating hundreds of gas stations in northern Germany, according to the BSI(Bundesamt für Sicherheit in der Informationstchnik) report [3]. In addition, BlackCat Ransomware attacked various companies such as Moncler [4], an Italian fashion brand, and Swissport [5], a Swiss aviation service company. BalckCat Ransomware was created using Rust rather than the languages commonly used by programmers such as C, Java, and Python[6]. It can be set in four encryption modes, as shown in TABLE I.

<TABLE I> Encryption mode of BlackCat Ransomeware

Encryption mode	Ransomware method
Full	Encrypt Full File
Fast	Encrypts only a specific size at the beginning of a file (in MB)
DotPattern	Encrypt files several times in a specific size (in MB)
Auto	Choose the best method for file encryption based on the type and size of the file

An AES block password or a ChaCha20 stream password is applied when encrypting a file. In Auto mode, hardware acceleration support for encryption and decryption of AES block passwords in Ransomware executables is first checked. After that, AES block passwords are used if supported, and ChaCha20 stream passwords are used if not. Finally, the encryption key is encrypted with the RSA-2048 public key password.

10.4229/EUPVSEC2023/5DV.2.25
020540-002

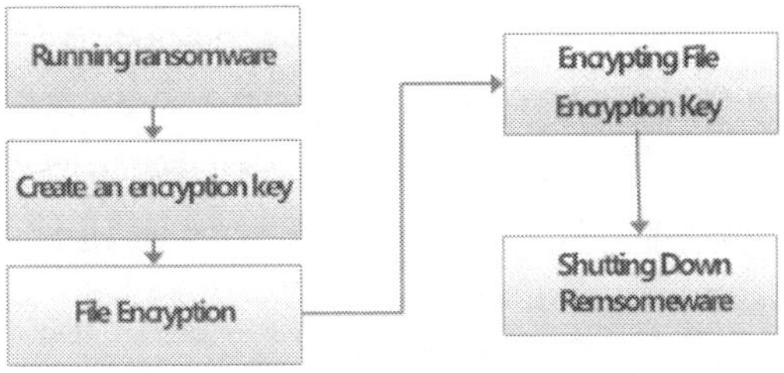

[Fig. 2] BlackCat Ransomware Encryption Process

BlackCat Ransomware shuts down processes and Windows services that interfere with encrypting files. Processes mainly related to backup software, Veeam, Microsoft Exchange, database servers, mail clients, office programs, etc., are terminated.

B. DeadBolt Ransomware

In January 2022, 4,988 of the 130,000 QNAP equipment on the Internet were infected with DeadBolt Ransomware when the DeadBolt Ransomware attack was the most active [7]. DeadBolt Ransomware creates an executable that performs malicious actions in the folder '/mnt/Had_ROOT/'. Initially, it is executed as a configuration file with various data, including an encryption key used for file encryption. DeadBolt Ransomware uses the AES-128 block password to encrypt files. In this case, the initial command used for file encryption is '[random_file_name] -e [config]/share'. In the command, 'share' refers to a folder in which the QNAP NAS equipment stores a user's file.

C. Sugar Ransomware

In February 2022, Walmart, a large U.S. multipurpose wholesale and retail company, was attacked by Sugar Ransomware. Sugar Ransomware is believed to have been developed using the Delphi programming language and offers more options than other LaaS [8]. Sugar Ransomware uses the MD5 hash value of the combined data value generated by seven different methods as the encryption key. The encryption key size is 128Bytes, and its value varies from PC to PC.

Data	Data Definition
ID	The 12-digit ID of the infected device
TICK	System Tick Count value
RAN	a random value
GEN	Cryptographic random value
COUNT	Performance counter plus the current time
CURSOR	X and Y axes of the current mouse cursor
TIME	Current time

When performing file encryption, sugar Ransomware is used by selecting from the RC6, SCOP, and Salsa20 encryption algorithms below.

Ÿ RC6: The top 48 Bytes size of the generated encryption key is used as the encryption key

Ÿ SCOP: The top 48 Bytes size of the generated encryption key is used as the encryption key

Ÿ Salsa20: Uses the top 40 Bytes size of the generated encryption key as the encryption key

The encryption key to encrypt the file is then encrypted with an RSA or ElGamal public key password. The infection process of Sugar Ransomware is as follows [9].

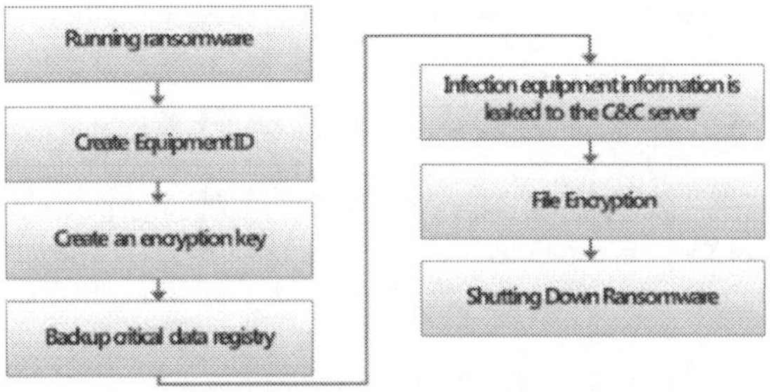

[Fig. 3] Sugar Ransomware Infection Process

BlackCat, DeadBolt, and Sugar Ransomware have the following characteristics.

<TABLE II> Data used to generate encryption keys

<TABLE III> Compare key Ransomware features

Ransomeware type	Key Features
BlackCat	Use AES block password or ChaCha20 stream password, encrypt used encryption key with RSA-2048 public key password
DeadBolt	Using AES-128 Block Password
Sugar	Use the MD5 hash value of the combined value of the generated data as the encryption key

V. Countermeasures

Files encrypted due to Ransomware are currently hard to recover entirely without an encryption key if there are no backed-up files. For this reason, many studies have been conducted in the direction of prevention rather than post-action as a countermeasure against Ransomware. Countermeasures against Ransomware are as follows.

A. Pre-backup of sensitive industrial confidential files

To date, many studies have been conducted to respond to Ransomware. Nevertheless, there is still no way to decrypt files encrypted by Ransomware completely. Therefore, confidential industrial files can be restored through files backed up after being infected by securing backup files in advance before being infected with Ransomware.

B. Latest Software and OS Updates

Ransomware is an attack aimed at vulnerabilities in software and OS installed on a user's PC. Both software and OS are patched for exposed vulnerabilities. Therefore, keeping the latest updates at all times reduces the possibility of Ransomware infection.

C. Installing Anti-Vaccination Software

The installation of vaccine software can be suggested as a way to prevent Ransomware attacks that begin with the download and execution of infected files. The vaccine software deletes and executes malicious files if it is suspected to be downloaded from an untrusted site. Therefore, cyclically performing PC malicious code tests is essential to prevent Ransomware infection.

D. 'File I/O Monitoring and Control Model' Using Kernel

The kernel-based file I/O monitoring and control model [10] uses the kernel's functions to detect and control the behavior of the process. Ransomware changes after process creation and file discovery to encrypt files. Changes to these files can be detected and controlled in the kernel. It is the concept of 'driver' for Microsoft's Windows products and 'kernel extension' for Apple's OSX products. When a file processing request occurs from the process, the kernel detects it. It transmits all file access information of the process to the control module. Each process can be identified through a specific pid (process identifier) value, which can be used to control access to a specific process. In addition, regardless of pid, it is possible to control the process of trying to access files on a specific file basis.

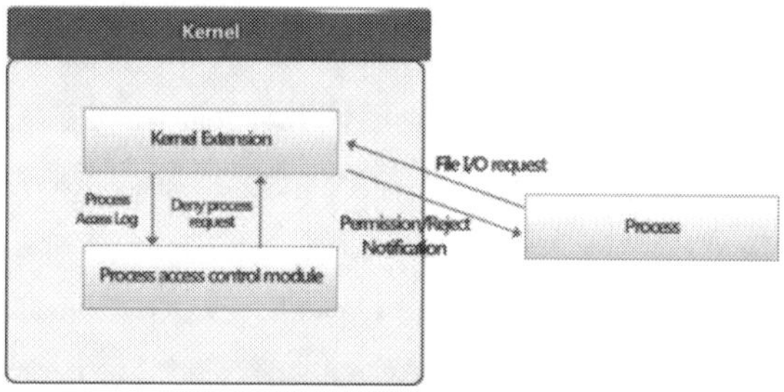

[Fig. 4] File I/O Monitoring and Control Model

VI. Conclusion

In this paper, domestic and foreign attack cases caused by Ransomware, attack methods, and characteristics by type were analyzed, countermeasures were analyzed, and countermeasures were described. First, after analyzing the basic operating principles of Ransomware, different Ransomware derived from various types were also analyzed. As a result of analyzing the three new Ransomware types of BalckCat, DeadBolt, and Sugar, it was found that each evolved from existing Ransomware in different ways to different Ransomware. According to research on how to respond to evolving Ransomware, files encrypted by Ransomware are difficult to recover if encryption keys are not secured. As a countermeasure to Ransomware, first, pre-backup of confidential industrial files; second, keep software and OS updated at all times; third, use of vaccine software; and finally, the 'File I/O Monitoring and Control Model', a

10.4229/EUPVSEC2023/5DV.2.25
020540-004

model for detecting Ransomware. In the future, there is a need for research on how to supplement and respond to vulnerabilities caused by Ransomware attacks on energy infrastructure.

Reference

[1] Min Kyung-sik, Kim Young-jik, Park Jin-sang, and Jang Han-na. (2021). Latest Ransomware Trend Analysis and Implications. KISA Insight, Vol 2, 7.

[2] Cho Young-Ju, Kim Jin-Hyuk, Oh Ji-Hoon, So Youn-Jeong, Sun A-Young.(2017). Operating principle and preventive measures of Ransomware. Proceedings of the Korea Contents Association Conference,91-92.

[3] ZDNet,"https://www.zdnet.com/article/blackcat-Ransomware-implicated-in-attack-on-german-oil-companies/"

[4] SecureBlink,"https://www.secureblink.com/cyber-security-news/moncler-group-becomes-the-first-victim-of-alphv-(blackcat)-raas-following-the-data-leak"

[5] BleepingComputer,"https://www.bleepingcomputer.com/news/security/blackcat-alphv-claims-swissport-Ransomware-attack-leaks-data/"

[6] cybereason,"https://www.cybereason.com/blog/cybereason-vs.-blackcat-Ransomware"

[7] Censys,"https://censys.io/deadbolt-Ransomware-is-back/"

[8] Boannews, https://www.boannews.com/media/view.asp?idx=104529

[9] Medium,"https://medium.com/s2wblog/tracking-sugarlocker-Ransomware-3a3492353c49#7ee1"

[10] Jung-moo Youn, Je-geong Jo, Jae-cheol Ryu.(2016).Methodology for Intercepting the Ransomware Attacks Using File I/O Intervals.Journal of the Korea Institute of Information Security&Cryptology,26(3),645-653.

10.4229/EUPVSEC2023/5DV.2.25
020540-005

COMPARISON OF EXPERIMENTAL SEPARATION METHODS FOR SILICON SOLAR PANELS

Simon Jech[1],*, Neha Garg [1], Kati Miettunen [2], Annukka Santasalo-Aarnio[1]
1) Department of Mechanical Engineering, School of Engineering, Aalto University, P.O. Box 16100, 00076 Aalto, Finland
2) Department of Mechanical and Materials Engineering, University of Turku, Vesilinnantie 5, 20500 Turku, Finland

ABSTRACT: Closed loop material recycling is essential for achieving sustainable photovoltaics. Enabling recovery of the solar panel materials and components, especially high purity silicon and solar grade glass, will reduce the need of primary materials. However, the encapsulation of silicon solar panels proved to be a great obstacle in successful high value recycling. Until now a variety of experimental separation methods has been proposed with various outcomes. The methods include thermal, chemical, or mechanical separation of the cells, glass and the encapsulant. In this study incineration, pyrolysis, and ultrasound assisted dissolution (UDT) of the encapsulant are investigated in laboratory scale experiments to compare the different methods. Electron Microscopy (SEM) coupled with Energy Dispersive X-ray Spectroscopy (EDS) is utilized for the analysis of possible contaminants on cell surfaces. We show that each method enables recovery of glass and solar cells, UDT further allows recovery of the back sheet. However, we still found carbon residues on cells after each of the method. The amount of the residues varies and so does treatment time, system complexity and possible emissions generated. We propose that a holistic view of the recycling process is needed when choosing the separation method.

Keywords: Silicon Solar Panels, Recycling

1 INTRODUCTION

First generation of photovoltaic installations in Europe reaches their end of life within this decade and exponential growth of photovoltaic panels is farther predicted [1]. Overwhelming majority of the installed panels was and still is based on the crystalline silicon solar cells [2]. In these devices are scarce materials such as silver as well as energy intensive materials such as solar grade silicon and solar glass together with other metals, for instance copper and aluminium.

These materials are lost from the industrial cycle if the panels end up in a landfill. For this reason, PV recycling is slowly getting attention in the last 10 to 15 years. Closing the material loops via one of the recycling approaches is crucial for the sustainability of the PV production.

A variety of approaches to recover for instance glass, silver, or even the solar grade Si have been proposed and some implemented in industrial pilots. Each method starts with separating the Al frame and junction box . Next step is separation of solar glass, crystalline silicon solar cells and polymeric back sheet which are laminated together by Ethylene Vinyl Acetate (EVA) adhesive. This adhesive proofed to be the main structural bottleneck of successful material recycling.

Earlier proposed methods of EVA delamination include physical shredding [3], which degrades and mixes the other materials. More promising for high value recovery is the thermal decomposition of the adhesives via incineration [4] or pyrolysis [5]. Last possible debonding strategy is chemical – using a solvent like toluene to dissolve the polymer, while combining it with ultrasonic irradiation to enhance the dissolution [6].

These different methods have been shown to separate the parts [4]–[6]. However, comparison of the thermal and chemical in one single set of experiments is missing from the literature. This work compares three different methods incineration, pyrolysis, and ultrasound-assisted dissolution of EVA in toluene with same type of PV samples. Possible differences in the separation progress are observed and discussed. We study the surfaces under a scanning electron microscope and analyze the composition via EDS. We provide further detail regarding the influence of these methods on the solar cells. The observations then serve as a basis for the comparison of these experimental separation methods.

2 MATERIALS AND METHODS

Present study employed two different monocrystalline silicon solar panels to investigated three separation methods. The first panel was an old, used panel manufactured in 1995 by Siemens (referred to as "Old"). The second on the other hand is unused, second grade panel manufactured less than 3 years ago by Solar Finland Ltd. (referred to as "New"). Both panels had a similar structure with glass front sheet, monocrystalline silicon solar cells, ethylene vinyl acetate (EVA) as encapsulant, and polymeric back sheet. They had been stripped of the aluminium frame prior this study and the glass had been cracked as well. Each panel was then cut into smaller samples of around 5 by 5 cm. Samples from the Old panel had cells separating from the encapsulant and glass prior to the experiments likely due to the ageing of the EVA.

2.1 Characterization of the reclaimed parts

The first objective was to determine if the polymeric residues remain on the separated cells and glass. The second objective was to observe possible changes in the structure of the solar cells and glass. Possible carbon residues indicate incomplete separation.

Each sample was weighed before and after the separation. Afterwards, each sample was visually inspected to determine any immediate changes in the parts' structure. Randomly chosen pieces of the cells were investigated under Scanning Electron Microscope (SEM) (JEOL-jsm7500f) and their surface composition was analysed using Energy Dispersive Spectroscopy (EDS). Further follows detailed description of the conditions use for the three studied separation methods.

10.4229/EUPVSEC2023/5DV.2.37
020542-001

2.2 Incineration

For the separation method based on incineration, the experiments were conducted in a furnace at the temperature between 600 and 620 °C and in an air atmosphere as reported previously [4]. Samples were placed on a steel holder and laid in the furnace. The treatment time was 40 minutes to secure initial stabilization at the given temperature and then the reduction of the polymers. After the processing time was established the series of uninterrupted, continuous experiments was conducted.

2.3 Pyrolysis

Decomposition of polymeric layers was conducted in a tube furnace (Lenton LTF 12/50/610) with nitrogen atmosphere (N_2 flow of 200 ml/min) for 60 minutes. The time was chosen to be twice as long as the previously reported optimum by Dias et al. [5] to eliminate the influence of different instruments and secure that the decomposition reaches the maximum. The same temperature level of 500 °C was adopted also in this study.

2.4 Ultrasound assisted dissolution in toluene

In present study ultrasonic bath (Elmasonic P180H) was used to perform the ultrasonic dissolution. Chosen conditions were: ultrasonic power of 330 W and frequency of 18 kHz. Temperature of the bath was maintained at 45 ± 3 °C, 60 ± 2 °C as in [7] and 70 ± 2 °C as in [6], where the first temperature is the lowest stable operating temperature. While the ratio of toluene (analytical grade supplied by Merck, Germany) to the sample weight was 20 ml/g. Here the experiments were continued until the glass and back sheet were separated from the cells, which is different from the previous studies [6], [7].

3 RESULTS AND DISCUSSION

We have successfully conducted three separation methods with both panels. Both Old and the New panel resulted in similar behaviour. Therefore, only the outcomes on the New panel are illustrated, but both panels are discussed.

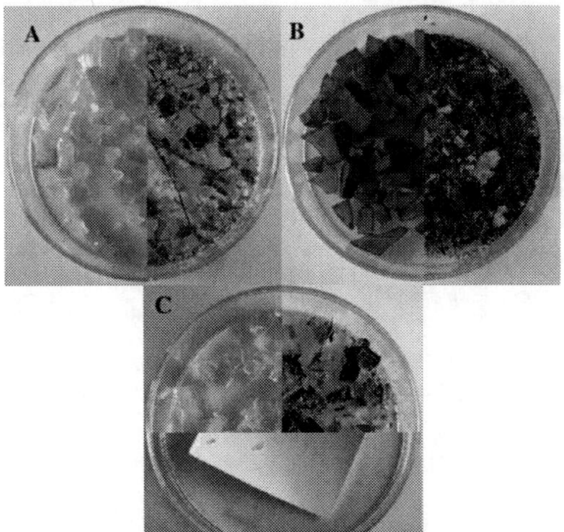

Figure 1: Samples of the new panel after: A – incineration, B – pyrolysis, C – UDT (glass and cells sorted manually)

3.1 Result of Incineration

After 40 minutes at temperature above 600 °C the cells and glass are well separated, however some polymer residues are still observable. The weight loss after incineration was between 9.21 wt% - 9.99 wt% and 12.46 wt% - 13.72 wt% for the Old and New panel respectively. This indicates that the polymers were reduced in major extend. The difference between the values most probably originates in a slight variance of the sample composition. For instance, when cutting the panel, we cannot secure exactly the same cut for each sample. Furthermore, **Figure 1** illustrates the final state of the sample. Distinct white coating can be seen, which was formed during the incineration and originates likely from the back sheet, observed also previously [4]. Right after the treatment there was a continuous, brittle white layer on the back of the samples. Altogether we can separate the cells and glass from both panels, but there are distinct polymer residues.

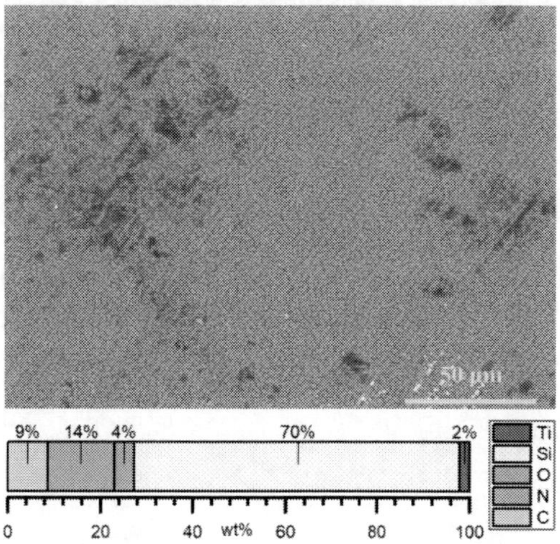

Figure 2: SEM image of the New cell front surface after incineration; elemental composition of the image area obtained via EDS

Figure 2 shows SEM image of the New cell front surface acquired after incineration. On the surface, polymeric residues are visible in form of the dark spots. These residues are presumably from the incomplete incineration of the EVA and the elemental composition shows that the spots contain carbon. Carbon is further shown to account for around 9 wt% of the examined surface (see bottom of **Figure 2**). On the darker spots silicon is detected alongside carbon which would indicate that the residual layer is rather thin. We also observed the typical pyramid structure on both cells, which farther confirms the separation of the polymers as well as partial structure preservation of the cell's front surfaces.

Furthermore, silicon is present on the surface together with nitrogen, oxygen, and titanium, indicating that the passivation and antireflection layers remain on the surface to a certain extent. Titanium is present in increased concentrations in the lighter spots visible in **Figure 2**, which are either defects in the antireflection layer, or more likely these are back sheet residues [4].

On the other hand, nitrogen was not detected on the OLD panel at all, which would mean that silicon nitride was not used in the passivation layer on that panel. In conclusion SEM and EDS analysis revealed the presence

10.4229/EUPVSEC2023/5DV.2.37

of carbon as an indicator of the incomplete separation of the polymers or preservation of residues, however the rest of the surface seemed to remain unchanged.

3.2 Result of Pyrolysis

Under nitrogen atmosphere and 500 °C the polymers decomposed leaving cells and glass separated. However, a black layer of likely charred polymers formed on both panels (see **Figure 1**). The occurrence of carbon residues after pyrolysis is in agreement with previous findings [8]. [9], they showed that pyrolysis of EVA resulted in formation of several hydrocarbons and carbon. Recorded weight loss further indicated the major reduction of the polymers (10.02 - 10.78 wt% for the Old panel and 12.55 – 13.48 wt% for the New).

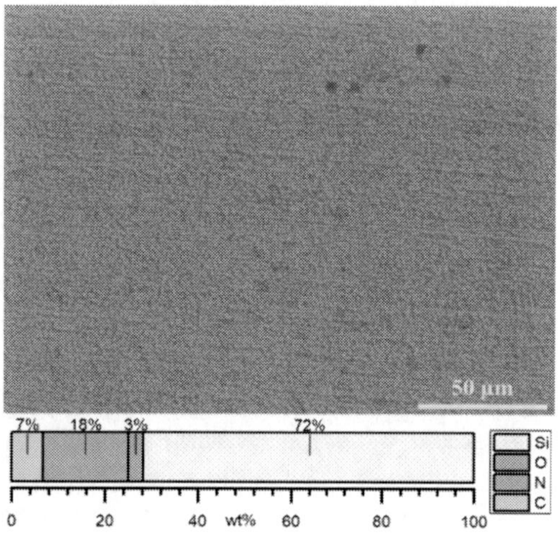

Figure 3: SEM image of the New cell front surface after pyrolysis; elemental composition of the image area obtained via EDS

Figure 3 presents the front surface of the New panel after pyrolysis together with the elemental composition over the area. Similar to the incineration, there are distinct darker spots on the SEM image, which are proved to be mostly carbon. Furthermore, carbon is detected over the sampling area; it confirms that the black layer consists of polymeric residues. Nitrogen and oxygen are present on the studied cell piece suggesting that the passivation layer is preserved in part. Next the pyramid structure of the front surface is maintained. In summary, the surface of the cells had similar composition to virgin cells, however a continuous layer of likely charred residues remains.

3.3 Result of Ultrasound assisted dissolution in toluene

Through ultrasound assisted dissolution in toluene we were able to separate the back sheet, the cells, and the front glass, however, not the encapsulant. The encapsulant swelled and partially dissolved, as previously observed by [10] and it remained attached to all the recovered parts. If the parts are not separated right after the dissolution, the EVA will glue the parts back together again. Back sheets were recovered in one piece; however, we observed bubbles being formed on the surface of the back sheet from New panel at 70 °C. The inner layer of the back sheet laminated with EVA seemed to decompose at that temperature. Shifting to the cells we observed slow build-up of fine black powder from the broken cells in the

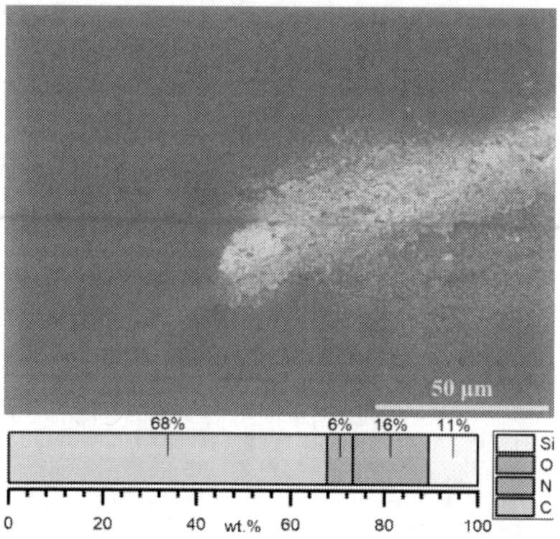

Figure 4 SEM image of the New cell front surface after UDT; elemental composition of the image area obtained via EDS

toluene solution. The powder is either from the fractures of the cells prior to the separation or due to the intense mixing within the solution and thus mechanical abrasion.

Furthermore, the time required to separate the parts varied from 50 to 85 minutes and 100 to 135 minutes for the Old and New panel respectively. The difference in the required time between the panels is likely caused by the degradation of the encapsulant in the Old panel. There we observed a pre-detachment of the encapsulant and thus the separation was generally shorter. On the other hand, the variance within the same panel type originates mainly in temperature, generally the higher the temperature the faster separation which agrees with Kim & Lee [6]. However, for the New panel the optimal temperature would be 60 °C, because the back sheet bubbling at higher temperature seemed to prolong the separation. Lastly, we also observed variance in the times at each temperature for both panel types.

Figure 4 demonstrates the presence of the polymer layer remaining on the cell surface. The dark homogenous area on the left side of the picture is EVA residue, which translates into the major concentration of carbon. Although most of the observed sample area was covered by the polymer, we could still spot partially clean areas of the cells. On the right-hand side of **Figure 4** the busbar can be seen and silver is detected over the area. We also spotted cell areas where the layer of the polymer was thin enough to detect silicon or metals bellow it. These observations apply to both panels and highlight only a partial separation of EVA. Even though the parts are liberated further treatment of the encapsulant is required for full material liberation.

3.4 Comparison of the methods

All the methods we studied achieve the separation of the main panel parts, yet there are still significant differences as listed in **Table I**. We have been able to recover glass and cells with incineration, pyrolysis and UDT, while UDT also enables the recovery of the back sheet. We have reclaimed the glass and cells in pieces as the panels were broken already before we conducted the present study. Nevertheless, the pieces recovered seemed to have generally comparable structure after each of the

10.4229/EUPVSEC2023/5DV.2.37
020542-003

separation methods. This could mean that further braking of the parts was limited or was the same across the three methods.

Table I Comparison of the different separation methods

	Incineration	Pyrolysis	UDT
polymer residues	Minor carbon traces	Charred polymer	Undissolved EVA
parts recovered	glass, cells	glass, cells	glass, cells, back sheet
treatment time	≤ 40 min	≤ 60 min	~55[a] / ~105[b] min
process complexity	simple	mediocre	complex
emissions	CO_2, fluorine based	fluorine based, hydro-carbons	Toluene vapor, noise

[a] old panel; [b] new panel

Focusing on the polymeric residues the most profound difference is that the UDT only partially separated EVA. That is because UDT dissolves and swells the polymer and so there is residual layer of the polymer left on the surfaces. It would therefore require significant aftertreatment as the only method of those three. Contrary to that, incineration and pyrolysis would not require extensive aftertreatment, however some surface cleaning would be required to obtain pure cells or silicon. The degree of aftertreatment would also probably vary between incineration and pyrolysis, as we observed the formation of charred residues when using pyrolysis. Subsequent elemental analysis revealed that incineration results in higher carbon concentration even though visually the cells appear to contain less residues. To be noted is that the elemental analysis is just for small area of the sample and is not a qualitative analysis of the whole, rather it illustrates the general state of the sample.

Processing time is the next major difference - for both pyrolysis and incineration it is usually shorter than time needed to separate the parts via UDT. For UDT we observed treatment times varying from 50 to 130 minutes and inconsistent time for the separation within the same experimental settings. In contrast, we adopted 40 and 60 minutes for the incineration and pyrolysis respectively, with homogenous outcomes. Furthermore, the incineration and pyrolysis treatment times used here are generally longer compared with the previously reported 30 minutes [4], [5] for both methods. Considering these observations, the main difference is that thermal treatments require shorter time and, in that time, provide consistent results.

Process complexity of these methods would have implications for an adoption and scaling up. Incineration requires a simple setup of a furnace with air atmosphere, whereas pyrolysis needs a special pyrolysis oven and supply of an inert gas like nitrogen or argon. Lastly, the most complex setup in our view is needed for the UDT. It required the use of an ultrasonic bath and toluene as solvent, which itself is a harmful substance [11]. The use of toluene results in formation of toluene vapours, which need to be collected and the waste toluene must be treated adequately. In addition, safety measures need to be taken to protect the workers from the ultrasound. Considering the removal of the polymeric layers there are extensive treatments needed for this method. For instance, farther thermal decomposition could be an option [10]. However, compared with mechanical separation all these methods could achieve higher material concentration since the mixing is avoided, and larger distinct parts are obtained.

On the other hand, the incineration leads to the formation of CO_2 emissions as well as possible fluorine-based gases, released from the fluorinated back sheet layers. Pyrolysis also results in gaseous emissions, where the main are hydrocarbons [8], [9] or charred carbon, but fluorine is again released risking the formation of HF [12] and damage to the exhaust lines. Therefore, the emissions from both methods need to be controlled and treated so that release of harmful gases such as fluorine to the environment is prevented.

4 CONCLUSIONS

We have successfully repeated the previously proposed thermal and chemical methods in one set of experiments to provide good comparison. The results highlight that each method separates the parts, but carbon residues were still observed after each of those. The thermal methods lead to the decomposition of majority of the encapsulant and back sheet, while UDT enables the recovery of the back sheet. Throughout all the methods, the front cell surfaces remain partially conserved. Even though the methods all separate glass and solar cells the outcomes vary, which influences on the subsequent processes of material recovery and at least the chemical and thermal separation methods are not interchangeable. Therefore, the separation method should be chosen based on the intended outcome as well as environmental considerations.

5 ACKNOWLEDGEMENTS

This work was made possible through Finnish Research Council under the project ECOSOL (funding decision numbers: 347275 and 347276). The authors acknowledge the help of D.Sc. Lassi Klemettinen in the pyrolysis experiments. The research made use of OtaNano Nanomicroscopy Center (NMC).

6 REFERENCES

[1] S. Weckend, A. Wade, and G. Heath, "END-OF-LIFE MANAGEMENT: SOLAR PHOTOVOLTAIC PANELS," 2016.

[2] M. Fischer, M. Woodhouse, S. Herritsch, and J. Trube, "International Technology Roadmap for Photovoltaic (ITRPV), 2020 results," 2021. [Online]. Available: https://itrpv.vdma.org/documents/27094228/29066965/2021OITRPV/08ccda3a-585e-6a58-6afa-6c20e436cf41

[3] S. M. Nevala *et al.*, "Electro-hydraulic fragmentation vs conventional crushing of photovoltaic panels – Impact on recycling," *Waste Management*, vol. 87, pp. 43–50, 2019, doi: 10.1016/j.wasman.2019.01.039.

[4] T. Dobra, D. Vollprecht, and R. Pomberger, "Thermal delamination of end-of-life crystalline silicon photovoltaic modules," *Waste Management and Research*, vol. 40, no. 1, pp. 96–103, Jan. 2022, doi: 10.1177/0734242X211038184/ASSET/IMAGE

S/LARGE/10.1177_0734242X211038184-FIG2.JPEG.

[5] P. Dias, S. Javimczik, M. Benevit, and H. Veit, "Recycling WEEE: Polymer characterization and pyrolysis study for waste of crystalline silicon photovoltaic modules," *Waste Management*, vol. 60, pp. 716–722, Feb. 2017, doi: 10.1016/J.WASMAN.2016.08.036.

[6] Y. Kim and J. Lee, "Dissolution of ethylene vinyl acetate in crystalline silicon PV modules using ultrasonic irradiation and organic solvent," *Solar Energy Materials and Solar Cells*, vol. 98, pp. 317–322, Mar. 2012, doi: 10.1016/j.solmat.2011.11.022.

[7] M. F. Azeumo, C. Germana, N. M. Ippolito, M. Franco, P. Luigi, and S. Settimio, "Photovoltaic module recycling, a physical and a chemical recovery process," *Solar Energy Materials and Solar Cells*, vol. 193, pp. 314–319, May 2019, doi: 10.1016/j.solmat.2019.01.035.

[8] C. Farrell *et al.*, "Assessment of the energy recovery potential of waste photovoltaic (pV) modules," *Nature Scientific Reports*, vol. 9, no. 5267, 2019, doi: 10.1038/s41598-019-41762-5.

[9] R. Wang *et al.*, "Pyrolysis-based separation mechanism for waste crystalline silicon photovoltaic modules by a two-stage heating treatment," *RSC Adv*, vol. 9, no. 32, pp. 18115–18123, Jun. 2019, doi: 10.1039/C9RA03582F.

[10] S. Kang, S. Yoo, J. Lee, B. Boo, and H. Ryu, "Experimental investigations for recycling of silicon and glass from waste photovoltaic modules," *Renew Energy*, vol. 47, pp. 152–159, Nov. 2012, doi: 10.1016/J.RENENE.2012.04.030.

[11] V. A. Benignus, "Health effects of toluene: a review.," *Neurotoxicology*, vol. 2, no. 3, pp. 567–588, Nov. 1981, Accessed: Jun. 04, 2023. [Online]. Available: https://europepmc.org/article/med/7038560/reload=0

[12] F. C. S. M. Padoan, P. Altimari, and F. Pagnanelli, "Recycling of end of life photovoltaic panels: A chemical prospective on process development," *Solar Energy*, vol. 177, pp. 746–761, Jan. 2019, doi: 10.1016/J.SOLENER.2018.12.003.

PYROLYSIS ANALYSIS OF C-SI PHOTOVOLTAIC SOLAR MODULE AFTER END-OF-LIFE (EOL) TO FACILITATE RECYCLING

Krishan Kumar Patel[1*], Suman Arju[1], Sudhanshu Mallick[1]
[1]Metallurgical Engineering & Materials Science Department, Indian Institute of Technology Bombay, Mumbai, 400076, India
Main Gate Rd, IIT Area, Powai, Mumbai, Maharashtra 400076
Corresponding Email id: krishan.patel4@gmail.com

ABSTRACT: The recovery of materials from End-Of-Life crystalline silicon (c-Si) photovoltaic (PV) solar module has been successfully demonstrated. As a first step, the backsheet was removed by heating the module at 130°C. The remaining module components – Solar glass, Solar cell, and encapsulant were pyrolyzed at different temperatures. A detailed analysis of the transmittance of glass by UV Vis Spectrophotometry, Solar cell surface morphology, and the gases evolved in pyrolysis by ICP-AES establishes the optimum pyrolysis temperature to be 500°C. The results indicate that the recovered solar glass and the backsheet can potentially be re-used to prepare new modules.
Keywords: Pyrolysis; Photovoltaic solar module; Recycling

1 **INTRODUCTION:** Solar Photovoltaic (PV) module technology is the most prominent, well-established, and mature renewable energy source. Several kinds of solar PV modules are available in the market, viz "polycrystalline and monocrystalline silicon, perovskite, thin-film, and tandem etc". the most popular polycrystalline silicon (c-Si), monocrystalline silicon, and thin-film. Several other emerging solar cell technologies are under research. Based on 2021 PV installation statistics, by the end of 2025, approximately 5-14% of the total installations would go to End-Of-Life (EOL), i.e., 5-13 million tons. The same is expected to reach approximately 78 million tons by 2050[1-3]. Such gigantic estimations have aroused great concern amongst scientists and researchers. All the efforts are being directed toward minimizing this environmental impact for the earth's better future. There is no ambiguity here that responsible disposal of EOL PV panels has become a critical issue of concern in the solar industry. Recycling PV modules generally engages three routes: physical, chemical, and thermal [4]. The Aluminium frame surrounding the module, the junction boxes, and embedded cables are physically separated by mechanical means[5-6]. The protecting glass may be reclaimed intact or can be shredded along with the solar cells to crush into glass cullets -with stuck encapsulant layer, c-Si PV cell pieces, and back sheet fragments. Valuable metals like silver and aluminum may take the hydrometallurgy route to recover [7-9]. The reclaimed Silicon PV cell is sent for further treatment for application-dependent purification or direct re-use. The encapsulants comprise Ethylene-vinyl Acetate (EVA) and are usually unrecoverable. They are pyrolyzed by heating to approximately 500°C or by dissolving them chemically in organic solvents [10]. In such a case, the key challenges are the generation of gaseous exhaust from polymer pyrolysis and the disposal challenge of large quantities of organic solvents [11-12]. The work reported herein, Pyrolysis analysis of c-Si photovoltaic solar module after End-Of-Life (EOL). A detailed analysis of transmittance of glass by UV Vis Spectrophotometry, Solar cell surface morphology, and the gases evolved in pyrolysis by ICP-AES was establishes to facilitate the recycling.

2 EXPERIMENTAL WORK: End-of-Life c-Si PV solar module was taken for recycling. A detailed schematic flow chart is given in Figure 1. First the module was cut into small pieces and the cut piece was placed over a hot plate at 130°C for 10 minutes. The backsheet was manually peeled off (Figure 2). After that, the cut piece sample of module and module without backsheet were placed inside the tube furnace for pyrolysis. The pyrolysis gas was passed through the bubbler filled with water to trap the evolved fume and gases. After pyrolysis, the solution of gases in the bubbler water was analyzed by ICP-AES analysis, and Solar cells/Glass residuals obtained were evaluated using SEM-EDX and transmittance analysis.

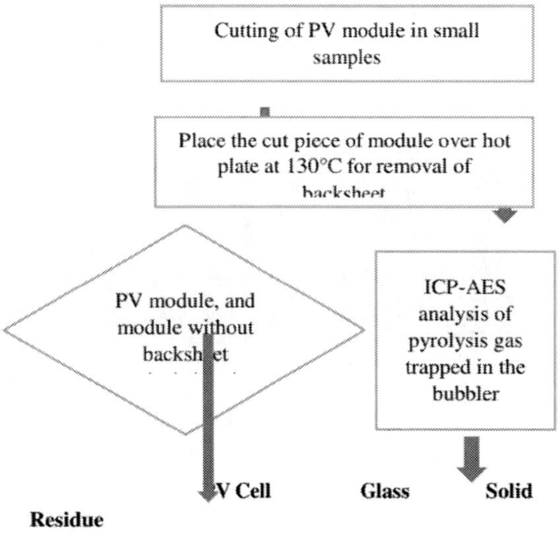

Figure 1: Flow chart of Experimental Work

10.4229/EUPVSEC2023/5DV.2.40
020544-001

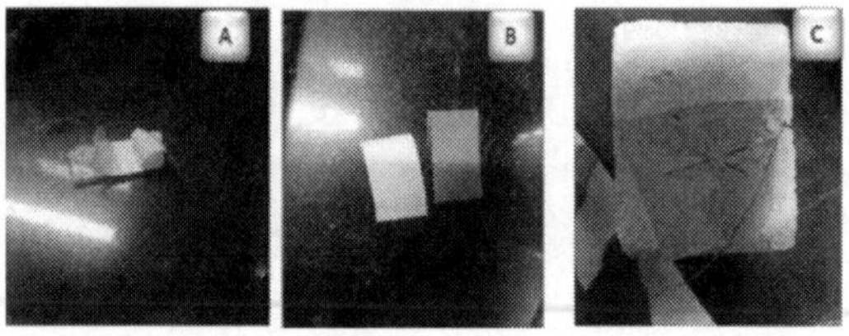

Figure 2: Manual separation of intact back sheet/glass, a) Unsuccessful separation of backsheet in a hot plate at 100°C, b) Successful separation of backsheet in a hot plate at 130°C, c) Separation of Solar cells, glass, and other residue obtained after pyrolysis at 500°C.

3 Results & discussion:

3.1 Solar cell surface morphology and composition.: The SEM image of the front surface of the solar cell after pyrolysis of PV solar module at 500°C and 550°C is shown in Figure 3 (Figure 3. SEM image & Elemental mapping of the front surface of the solar cell after pyrolysis at A) 500°C & B) 550°C). As the PV solar module pyrolysis temperature increases from 500°C to 550°C, silicon's grain size is reduced, increasing the void space. The reduction of grain size of silicon solar cells may influence the photovoltaic cell's performance in case a new module is manufactured with them. In some places, corrosion was also observed in the surface morphology. The degradation of the c-Si solar cell was accelerated with high temperatures and a humid environment. Furthermore, the Energy Dispersive X-ray is used to identify the chemical elements present over the surface of a solar cell after pyrolysis at different temperatures. The captured EDX spectra show the major elements present over the front surface of the solar cell after pyrolysis. The results imply that carbon composition increases slightly at higher pyrolysis temperatures, and besides Si, the remaining elements are relatively negligible.

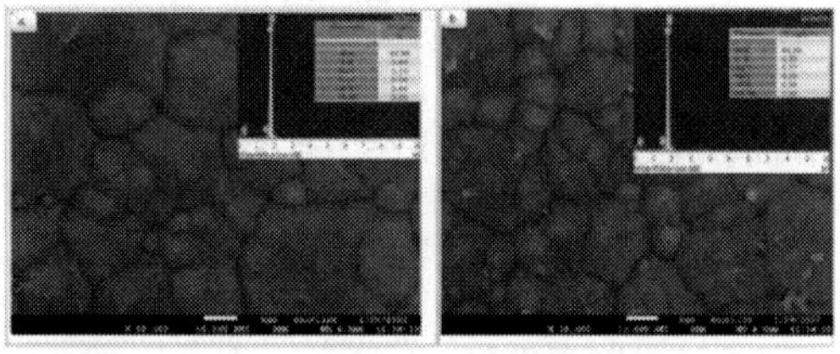

Figure 3. SEM image & Elemental mapping of the front surface of the solar cell after pyrolysis at A) 500°C & B) 550°C

3.2 Transmittance: The transmittance of new glass plate used in the c-Si PV solar module and after pyrolysis of the PV module at 500°C and 550°C was measured by UV-visible spectroscopy (Fugure 4). The transmittance of the new glass plate was a maximum of 98% at the UV wavelength range of approximately 350 nm; after that, the %T decreased slightly and remained at 95%T at larger wavelengths. The transmittance of glass after pyrolysis is slightly different. The transmittance is reduced due to discoloration (90%T for sample pyrolyzed at 500°C) and is significantly lower for samples pyrolyzed at 550°C (75%T). The loss of transmittance of glass leads to a reduction in the power generation of a module.

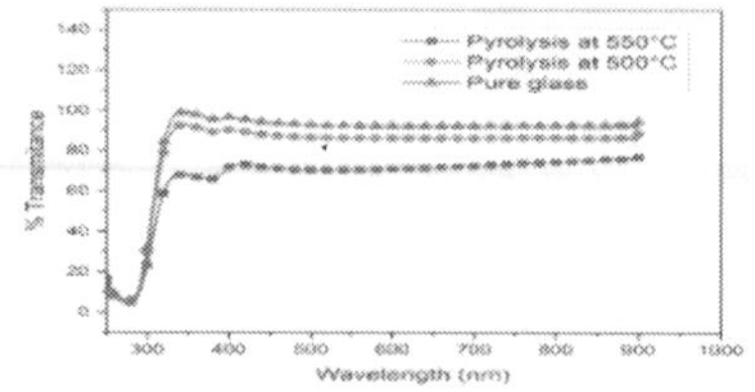

Figure 4. The transmittance of glass plate before and after pyrolysis at 500°C and 550°C.

3.3 Exhaust gas analysis: Pyrolysis of the modules was done in a tube furnace at 500°C and 550°C, and the pyrolysis gas was passed through a bubbler. The solution was then alayzed using ICP-AES (Inductively coupled plasma- atomic emission spectrometry). The results indicate that the solution in water of gases evolved during pyrolysis contains less than 0.01 ppm Ag, Pb, and Sn, whereas Al and Si are present in considerable amounts. As the temperature increases, the trapped elements are more (table 1). This suggests lower recovery of elements if pyrolysis is done at higher temperatures.

Table 1: ICP-AES result of pyrolysis evolved gas trapped in the bubbler

Sample	Ag	Pb	Al	Si	Sn
	ppm	ppm	ppm	ppm	ppm
1. Pyrolysis of module at 500°C	ND	ND	0.153	0.23	ND
2. Pyrolysis of module at 550°C	ND	ND	0.293	0.374	ND
3. Pyrolysis of module without backsheetat 500°C	ND	ND	0.149	0.31	ND
4. Pyrolysis of module without backsheetat 550°C	ND	ND	0.281	0.361	ND

ND: LESS THAN 0.01ppm (part per million)

CONCLUSION:

This work demonstrates that it is possible to recover most of the materials in the PV modules for re-use and recycling.

At 130°C, the PVDF backsheet was successfully peeled off from the c-Si module to facilitate further recycling. The optimum pyrolysis temperature of 500°C was established.

10.4229/EUPVSEC2023/5DV.2.40
020544-002

At this temperature, there is minimal reduction in the transmittance of the solar glass as it retains 90% Transmittance. This glass can potentially be re-used for fabricating new modules. At 550°C there was a significant reduction in the transmittance of the glass. The pyrolysis at 500°C results in less apparent damage to the silicon cell substrate. The analysis of the pyrolysis gases also reveals less Al and Si in the exhaust compared to pyrolysis at 550°C.

Thus it can be concluded that removal of Backsheet at 130°C, followed by pyrolysis at 500°C can enable recovery of most PV module components, which can potentially be used for re-manufacturing.

ACKNOWLEDGEMENT: The author pledges great thanks to the India Institute of Technology Bombay, Mumbai for providing a research grant and facility for this work. All authors declare that there is no conflict of interest in this research article.

References:

[1] Europe, S. (2022). Global market outlook for solar power/2022–2026. *Solar Power Europe: Brussels, Belgium*, 1.

[2] Chowdhury, M. S., Rahman, K. S., Chowdhury, T., Nuthammachot, N., Techato, K., Akhtaruzzaman, M., ... & Amin, N. (2020). An overview of solar photovoltaic panels' end-of-life material recycling. *Energy Strategy Reviews*, *27*, 100431.

[3] Seo, B., Kim, J. Y., & Chung, J. (2021). Overview of global status and challenges for end-of-life crystalline silicon photovoltaic panels: A focus on environmental impacts. *Waste Management*, *128*, 45-54.

[4]McDonald, N. C., & Pearce, J. M. (2010). Producer responsibility and recycling solar photovoltaic modules. *Energy Policy*, *38*(11), 7041-7047.

[5]Dias, P., Javimczik, S., Benevit, M., & Veit, H. (2017). Recycling WEEE: Polymer characterization and pyrolysis study for waste of crystalline silicon photovoltaic modules. *Waste Management*, *60*, 716-722.

[6]Kang, S., Yoo, S., Lee, J., Boo, B., & Ryu, H. (2012). Experimental investigations for recycling of silicon and glass from waste photovoltaic modules. *Renewable Energy*, *47*, 152-159.

[7]Fiandra, V., Sannino, L., Andreozzi, C., & Graditi, G. (2019). End-of-life of silicon PV panels: A sustainable materials recovery process. *Waste management*, *84*, 91-101

[8] Omazic, A., Oreski, G., Halwachs, M., Eder, G. C., Hirschl, C., Neumaier, L.,& Erceg, M. (2019). Relation between degradation of polymeric components in crystalline silicon PV module and climatic conditions: A literature review. *Solar energy materials and solar cells*, *192*, 123-133.

[9] Dobra, T., Vollprecht, D., & Pomberger, R. (2022). Thermal delamination of end-of-life crystalline silicon photovoltaic modules. *Waste Management & Research*, *40*(1), 96-103.

Rimez, B., Rahier, H., Van Assche, G., Artoos, T., Biesemans, M., & Van Mele, B. (2008). [10] The thermal degradation of poly (vinyl acetate) and poly (ethylene-co-vinyl acetate), Part I: Experimental study of the degradation mechanism. *Polymer Degradation and Stability*, *93*(4), 800-810.

[11] Farrell, C., Osman, A. I., Zhang, X., Murphy, A., Doherty, R., Morgan, K., ... & Shen, D. (2019). Assessment of the energy recovery potential of waste Photovoltaic (PV) modules. *Scientific reports*, *9*(1), 1-13.

[12] Liu, L. L., He, G. Q., Wang, Y. H., & Hu, S. Q. (2015). Chemical analysis of primary combustion products of boron-based fuel-rich propellants. *Rsc Advances*, *5*(123), 101416-101426.

10.4229/EUPVSEC2023/5DV.2.40
020544-003

KEY ACTORS IN THE CIRCULARITY OF SOLAR PANELS IN CITY OF HELSINKI

Bergpob Viriyaroj[1]*, Elena Akulenko[2], Sonja Äärilä[1], Pekka Heikkinen[1]
[1]Department of Architecture, Aalto University, Finland.
[2]Department of Materials Engineering, Faculty of Science and Engineering, 20014 University of Turku, Finland.
*e-mail address: Bergpob.viriyaroj@aalto.fi

ABSTRACT: This research investigates the ecosystems of photovoltaic panels in the city of Helsinki. The goal of the investigation is to identify key actors in the circular economy of Helsinki photovoltaic supply chain. Utilization of solar energy has seen rapid increase in the past decades and is a key renewable energy source in many countries. The projected growth of photovoltaic products indicates a significant amount of photovoltaic waste in the future. Qualitative research using semi-structure interviews with stakeholders is the research method in this study. Literature investigation was used to supplement the interview results. We found that, in Finland, the utilization of solar energy is in its early stages, but the number of solar panels will significantly increase in the future, especially in Helsinki metropolitan area. The amount of photovoltaic waste in Finland is low at the time of this study, which leads to low awareness of photovoltaic waste problem among many stakeholders and low profitability of solar panels recycling for the recyclers. The main beneficiaries of the circular economy of solar panels are manufacturers and recyclers, but governmental institutions and producers' associations appear to have more influence on the waste management of solar panels.

Keywords: Solar Panels, Recycling, End-of-life management, Circular economy

1 INTRODUCTION

1.1 Photovoltaic proliferations

Solar energy technologies are actively adopted globally to meet Sustainable Development Goals. According to forecasts by the International Renewable Energy Agency, by 2050 solar photovoltaic will generate 25% of the total electricity needs globally [1]. Photovoltaic (PV) panels are a significant source of renewable energy in many countries which aim to decarbonize their electricity grid. EU countries have put priorities in renewable energy sources, and PV panels have a significant role in their decarbonizing strategy [2].

The primary focus of renewable energy in Finland integrated energy and climate plan is in bioenergy. In the year 2022, bioenergy holds an overall renewable energy share of more than 75%, while the share of solar energy is negligible. However, solar energy is projected under Finnish With Additional Measure plans to have a three-times increase in consumption by the year 2040 [3].

Nonetheless, there are many energies related strategy, which feature PV panels. The carbon neutrality plan of Helsinki features solar energy as part of their strategies. 2030 target scenario of this plan put a target of 15% electricity from PV panels [4]. The accumulate growth of install capacity since 2015 started at 20 MW and reached 214.2 MW in 2019. The install capacity in 2019 alone was 80.7. This signifies the increasing interest in PV panels installation in Finland [5]. Therefore, the demand for PV panels can be expected to increase rapidly in the Future.

1.2 PV wastes

The increasing market adoption of PV panels has raised a new challenge for stakeholders and society – circularity of solar panels. Waste management of solar panels is a complex issue that will pose a significant challenge in the future, due to the rapid increase of solar energy utilization.

PV panel waste in 2020, globally, has the volume of 220,000 tons. In 2050 the amount of waste could reach 6 million tons [1].

Aside from the problem with waste volume, the type of waste produced by PV panels is also difficult to manage. There are several different types of PV panels available in the current market. These different products also have different materials composition, which complicates the waste management process [6]. Some of these elements in PV products are considered toxic to the environment or humans [7]. Although the risk of toxicity of PV products seems to be low, the products might still be affected by some regulations which ban the use of these elements [8].

1.3 Literatures

Salim, et al. conducted a literature review of PV and battery waste management. The profitability of recycling material is identified as the most common barrier. Activities that hinder profitability are the cost of recycling process, transport of materials and infrastructure for waste management. Lack of government regulation is another commonly perceived barrier. Lack of participation in circular economy from the consumers and manufacturers were mentioned as social barrier in this study.

Strict regulation is one of the keys enabling factors for End-of-Life (EoL) management of PV and batteries. Some of the examples of these regulations mentioned in the study are product stewardship, industry codes of conduct, recovery targets and product design standards. Financial schemes that help facilitate EoL management are also salient enablers.

Majewski, et al. emphasize the importance of product stewardship, regulatory directives, and financial/non-financial responsibilities in their study. The study gave examples of regulatory roles in circular economy of PV in different regions. In EU and UK specific regulations which apply to the recycling of PV are Waste Electrical and Electronic Waste Directives (WEEE) and hazardous material restriction (RoHS). RoHS restricted certain materials in the products and WEE involved in the management of waste, which included extended manufacturer responsibilities [9]. WEEE directive revision in 2012 includes recovery target of PV modules by mass. The target has increased over the years to 80-85% in 2018 [10].

1.4 Interview articles

More examples of stakeholders' perspective to the circular economy of PV panels can be seen in qualitative

10.4229/EUPVSEC2023/5DV.2.41
020546-001

research articles. Besiou, and Wassenhove conducted qualitative research on the supply chains of PV. Misalignment between recyclability, durability and efficiency is one of the key findings from the research. Reluctance by the producers to share product information with other actors in the supply chain is another barrier to recycling of PV [11].

The qualitative research by Lapko, et al investigate supply chain of green energy sector. Factors important to circular economy were used for the coding of interview results. Among the findings, technical factors were commonly mentioned, due to the complexity of product design and lack of appropriate technologies. Low volumes of waste and high price instability limit the viability of recycling of PV. Similar to many other studies, legislation was mentioned as an important driver and barrier [12].

Study by Salim, et al showed preliminary drivers, barriers and enablers through surveys and interviews in Australia. Most stakeholders perceived recirculating rare materials in PV systems as the main driver. Lack of profitability, regulations, incentives for waste collections and installers' participation are the most prominent barriers [13].

2 GOAL AND METHODS

2.1 GOAL

PV proliferation in Finland is still in its early stage. PV utilization started to grow in 2015-2018. This means that the amount of PV waste in Finland is currently negligible. Judging from the expected life span of PV panels, the major volume of waste in Finland will not occur until 2040-2050. In this early stage of PV recycling, identifying the important actor in Finnish PV supply chain can be beneficial to the eventual development of PV recycling market.

The supply chain of PV panels in Finland is expected to be similar with other EU countries. There are still possibilities that some local context might make the supply chains different and understanding them would provide valuable knowledge for the development of a robust PV recycling market.

The goal of this study is to investigate the current landscape of actors involved in lifecycle of solar panels in Finland, particularly in Helsinki area. By limiting the scope of this research to one city, an in-depth investigation on the supply chain of solar panels is ensured. The study involved diverse groups of stakeholders in order to broaden the insight of the topic. Since PV recycling market in Finland is in its early stage, the awareness around the issues of PV waste is expected to be minimum. Therefore, acquiring insight from diverse groups of actors is important.

2.2 Methods

The research is conducted by using qualitative methods through semi-structured interviews in which we investigate supply chains in Helsinki. Contact tracing from interviewees is used to identify more actors, which involvements in the supply chain was not initially expected or has greater degree of influence than we had anticipated. The participants in the interviews were experts from Finnish ministry of environment, local PV manufacturer, recyclers, energy company and City of Helsinki. Similar to the study conducted by Besiou, and Wassenhove, secondary data from publications were used to supplement the interview outcomes [12].

3 FINDINGS

3.1 Ministry of environment

According to the interview with ministry of environment the supply chains of PV in Helsinki are tied to the supply chains of building constructions. Therefore, waste management of PV panels will likely be part of other building components. Any legislation that affects building components as a whole will also apply to PV panels. The long lifespan of PV panels might mean that the replacement of PV during building service life will not occur often. Due to the relationship between building activities and PV panels installation, additional regulation can be added to promote circular economy of PV panels. One of the examples that was given in the interview was a regulation that would help steering PV panels waste towards recyclers instead of landfill.

The problem of future PV waste is currently not in the attention of their organization, due to the low amount of waste. However, a better, more efficient PV panels in Finnish markets might accelerate the amount of PV waste in the future. The reason behind this is the possible replacement of existing panels with a better product.

In terms of recent regulation that can have effect on PV panels waste is the New Finnish construction act, which affects the circularity of building materials. A voluntary reporting system for building waste is another program that they have mentioned. Even though this program is not regulatory, it will be heavily supported with a digital platform managed by consultancy company collaborating with governmental institutes.

Aside from the Ministry of Environment, Ministry of Economic Affairs was also brought up as another ministry that has been involved in PV panels supply chains.

3.2 City of Helsinki

Two experts from the city of Helsinki office have participated in the interview in this study. They have been working on installing PV panels on city-owned buildings in Helsinki. Their current action is mainly installing PV panels on every new building and some existing buildings. The process of installing PV in urban environment is attached to the process of construction of new buildings or major renovation of existing buildings. The reason behind this practice is due to the efficiency of investment cost, installing PV panels on existing buildings often required extra structural supports. Without other construction activities involved the initial investment of installing PV will be considered too high by the city. The end-of-life of PV can also be expected to follow the major renovations of buildings.

The cost of the PV panels is the main factor of their procurement requirements. Finland acquired majority of their solar energy from imports of PV panels from Eastern Asia. There are still very few numbers of local manufacturers, and the scale of their products is small. Some installers of PV panels in Finland are the companies which also procure PV panels from manufacturers, including imported PV panels.

Installation of PV panels in Helsinki has been in practice for 15 years, a relatively new practice. The expected PV life span is approximately 30 years. This makes the current amount of PV waste negligible in Helsinki, hence a lack of attention towards circularity of PV panels supply chain. The future increase in the number of PV panels in Helsinki may motivate the supply chains to invest more in PV recycling.

The main driver for the utilization of PV panels by the

10.4229/EUPVSEC2023/5DV.2.41
020546-002

City of Helsinki is regulatory. Carbon Neutral Helsinki is the program that pushes the City of Helsinki to utilize renewable energy. Putting PV panels on service building is their current strategy. The majority of the energy related strategy to achieve carbon neutrality in Helsinki is assigned to Helsinki energy company. Another influential initiative that has been mentioned by the interviewees was REPowerEU, binding to have a specific percentage of solar panels installed in urban environment. Regulatory approach may be the most important drivers for recycling of PV waste, in the interviewees' opinion.

City of Helsinki's Building permit office allows panels without building permission in most cases. When installed on architecturally protected buildings or vertically to the wall or balcony, a permit is needed.

3.2 Manufacturers

An expert from a local manufacturing company in Finland was interviewed. WEEE directive has major influence on end-of-life of PV which are being produced in Finland. The company currently collects all PV produced by their company, due to the extended producer responsibility requirement. Few PV panels from other companies ended up with the Finnish manufacturers, but this is not required by the regulation. These PV panels are stored in their properties before being collected by waste management companies along with other WEEE waste. The panels arrived in mostly good condition, physical damage from transport is not a reoccurring problem. Some of the damaged panels they have collected came from the producers which have not designed their product to withstand Nordic climate. Their participation in extended producers' responsibility is regulatory. There is no financial support in their PV waste collection activities.

Recycling of PV has recently been part of their research and development project, in which they have a team conducting research on viable material recovery technologies. They are interested in mid-process recycling, in which they believe that material reclaimed from PV waste can be directly used in the production of their products. They have been recovering aluminum frames from the PV which are being sent to them and are aiming to expand the possibility of reclaiming more material. Intact glass is one of the materials they have expressed interest in.

They are currently collaborating with a Finnish recycling company. The collaboration is ongoing at the time of this research, and they both are in the same process of searching for suitable recycling methods for their intended material reclamations. They are also aware of the Carbon Neutral Helsinki program and have been collaborating with representatives from the program.

3.3 Recyclers

On the recyclers group, an expert from a large-scale recycling company in Finland was interviewed. This is the company that is in collaboration with the manufacturers that participated in our interview. One of the main motivations of the recycler is reclaiming lower grade silicon from PV panels, which can be used in other manufacturing processes. They have estimated, based on the current PV capacity in Finland, Sweden, and Estonia that the amount of PV waste in Finland will significantly increase in 5-10 years. Consequently, they are searching for the most suitable method of PV recycling, in anticipation of future PV waste volume.

The expert mentioned several challenges to recycling waste in Finland. Waste heterogeneity is a big problem for PV panels. The type of PV is diverse, and the sources of PV in Finland came from producers of many origins, which make it difficult to predict the type and amount of waste in PV products. One of the examples that was given by was PV panels containing lead among the waste they received.

Change of ownership throughout the solar panels lifecycle further complicates the waste management situation. The number of ownerships in a lifespan of PV panel depends on the type of buildings they are installed. In the case of PV panels in housing/apartments, 4-6 different parties can take ownership of the panels at some point in their life. Therefore, it is difficult to identify the responsible party of the disposal of PV waste. Municipal waste management takes most of the PV waste, and materials from this product are not being reclaimed. Considering the lack of profitability to recycle PV waste, due to low waste volume, this problem was one of the major concerns of the recyclers.

There were several enablers mentioned by the interviewee. Support in the form of funding and collaboration with research institutes can greatly benefit the research process they are conducting.

3.4 Energy company

An expert from Helsinki energy company participated in an interview. Electricity from PV has recently been a significant part of the renewable energy supply from this company. Their service includes household installation of solar panels, PV panel rental and PV power plants. The energy company mostly purchase their PV from the global market. They have not been purchasing PV panels from the local manufacturers, but they have been aware of the importance of the origin of these products and have been monitoring PV producers in Finland. They have recently been investing in building solar power plants in Helsinki and nearby cities. According to the interviewee, the shift toward solar power plants by their company was from a new enabling technology.

PV waste that was decommissioned from their facilities have been mostly early-loss scenarios. In the case that the product they received was defective, it can be sent back to the retailer. The decommissioned PV panels were sent to Electrical and Electronic Equipment Producers Association. The waste management fee was paid to the association and the association took the PV waste and distributed it to other waste management companies.

The energy company employed large share of PV panels in Finland, but their actions seem to be limited to the operational phase of the panels. The responsibility related to PV waste that they involved in is sending the waste to the facilities of producer association.

3.5 Stakeholders map

The relationship of stakeholders in Helsinki PV supply chain is shown in Fig 1. The stakeholders map is divided into three phases production, operation, and end-of-life. Stakeholders outside of Helsinki PV supply chain can also have influence on the supply chain. Some of the stakeholders are grouped according to their roles. The types of stakeholders interaction are represented with different arrow types. The size of bubbles and arrows signify their degree of influence.

Several prominent actors in the supply chains are from other industries or companies outside Finland. Retailers has a prominent role since the majority of solar panels in

10.4229/EUPVSEC2023/5DV.2.41
020546-003

Finland and Helsinki were imported. Energy company in Helsinki also sells solar panels, hence their position in both production and operation. The owners of solar panels submit the wastes to waste management companies themselves. This also applies to other electronic waste. The fees for submitting the waste are handled by Waste Electrical and Electronic Equipment Producer Association.

4 DISCUSSIONS AND CONCLUSIONS

4.1 Discussion

Findings from the interviews indicate that regulation has been perceived as the main drivers and enablers of PV waste recycling in Finland. This finding is consistent with other studies. There have been many initiatives in Finland to increase the utilization of PV panels, and the main actors in these initiatives are governmental organizations and companies they collaborate with. The initiatives that were used to promote PV came from changing some regulations that were previously a barrier to PV to support the adoption of solar power. Some updates in the past from EU commission have already showed removal of certain barrier to PV waste recycling. In 2011-2012, updates which changed waste classification of steel scrap, aluminum scrap and broken glass, were beneficial to the recycling of PV [14] [15].

Governmental organizations and companies are the main drivers for circular economy initiatives [16]. Circularity of PV waste has not been specifically addressed in these initiatives yet. However, it can be addressed in the future since circular economy is part of key environmental policies and plans in Finland. It is likely that the development of circular economy of PV in Finland will be heavily influence by this group of actors.

The profitability issue of PV recycling, due to the low amount of waste, was acknowledged in most of the interviews to be the main barrier. Even though the PV waste volume is expected to grow significantly in the future, other barriers for PV recycling profitability can still hinder the grow of PV recycling. It is currently not clear how PV supply chain in Finland will react to the upcoming large amount of waste. Aside from the recycler and the manufacturer none of the other stakeholders have concrete actions to address the possible problem of PV waste. Landfilling PV waste was a concern that was brought up by the recyclers and was also addressed in other PV recycling studies. This can reduce the amount of waste recyclers will receive. Another issue that could affect profitability is the cost of waste collection [17]. Electrical and Electronic Equipment Producers Association in Finland might be highly influential in addressing this problem. As they are involving in the management of a significant share of PV waste in Finland.

Collaboration between Ministry of the Environment together and Ministry of Economic Affairs and Employment as well as Finnish Innovation Fund to prepared circular economy program have been announce in 2021 [18].

Collaboration between actors in Finland appears to be a strong driving factor in Finland, where it seems to be perceived as barrier in many studies. It has been mentioned in literature that the reluctance from manufacturers to share information with recyclers is one of the key barriers to PV recycling. In Finland however, there is an ongoing collaboration between recycler and manufacturer to find solutions for PV recycling that can benefit both parties. In addition to the recycler, the PV producer we interviewed also mentioned a collaboration in their company with the Carbon Neutral Helsinki program.

4.2 Conclusion

Overall, PV recycling in Finland is currently not profitable. The awareness around this topic is in its early stage, since Finland has only started to adopt large scale utilization of PV panels recently. However, throughout the interviews the topic of recycling was found to be relevant. Many groups of stakeholders have been working in projects that are adjacent to the recycling of PV waste, such as promoting utilization of PV waste and circular economy or managing building and electronic waste.

Governmental institutions and producers' association appeared to have the large influence on future PV recycling in Finland, while local manufacturers and recyclers are the main beneficiaries. The collaborations between actors in PV supply chains as well as industries connection with governmental projects seems to be enablers rather than barriers in Finland.

References

[1] Weckend, S., Wade, A., Heath, G., 2016. End-of-Life Management: Solar Photovoltaic Panels. International Renewable Energy Agency, Abu Dhabi.

[2] (2023). Retrieved 13 August 2023, from https://commission.europa.eu/strategy-and-policy/priorities-2019-2024/european-green-deal/repowereu-affordable-secure-and-sustainable-energy-europe_en

[3] (2023). Retrieved 13 August 2023, from https://julkaisut.valtioneuvosto.fi/handle/10024/161977

[4] (2023). Retrieved 13 August 2023, from http://carbonneutralcities.org/wp-content/uploads/2019/06/Carbon_neutral_Helsinki_Action_Plan_1503019_EN.pdf

[5] (2023). Retrieved 13 August 2023, from https://iea-pvps.org/wp-content/uploads/2020/09/NSR_Finland_2019.pdf

[6] Trivedi, H., Meshram, A., & Gupta, R. (2023). Recycling of Photovoltaic Modules for Recovery and Repurposing of Materials. Journal of Environmental Chemical Engineering, 109501.

[7] Sica D., Malandrino O., Supino S., Testa M., Lucchetti M. C. (2017). Management of end-of-life photovoltaic panels as a step towards a circular economy, Journal of Cleaner Production. Renewable and Sustainable Energy Reviews, Volume 82, Part 3, Pages 2934-2945, https://doi.org/10.1016/j.rser.2017.10.039.

[8] Sinha P., Kriegner C. J., Schew W. A., Kaczmar S. W., Traister m., Wilson D. J. (2008). Regulatory policy governing cadmium-telluride photovoltaics: A case study contrasting life cycle management with the precautionary principle, Energy Policy, Volume 36, Issue 1, Pages 381-387, https://doi.org/10.1016/j.enpol.2007.09.017.

[9] Majewski, P., Al-shammari, W., Dudley, M., Jit, J., Lee, S. H., Myoung-Kug, K., & Sung-Jim, K. (2021). Recycling of solar PV panels-product stewardship and regulatory approaches. Energy Policy, 149, 112062.

[10] (2023). Retrieved 13 August 2023, from https://eur-lex.europa.eu/legal-content/EN/TXT/?uri=CELEX:02012L0019-20180704

10.4229/EUPVSEC2023/5DV.2.41
020546-004

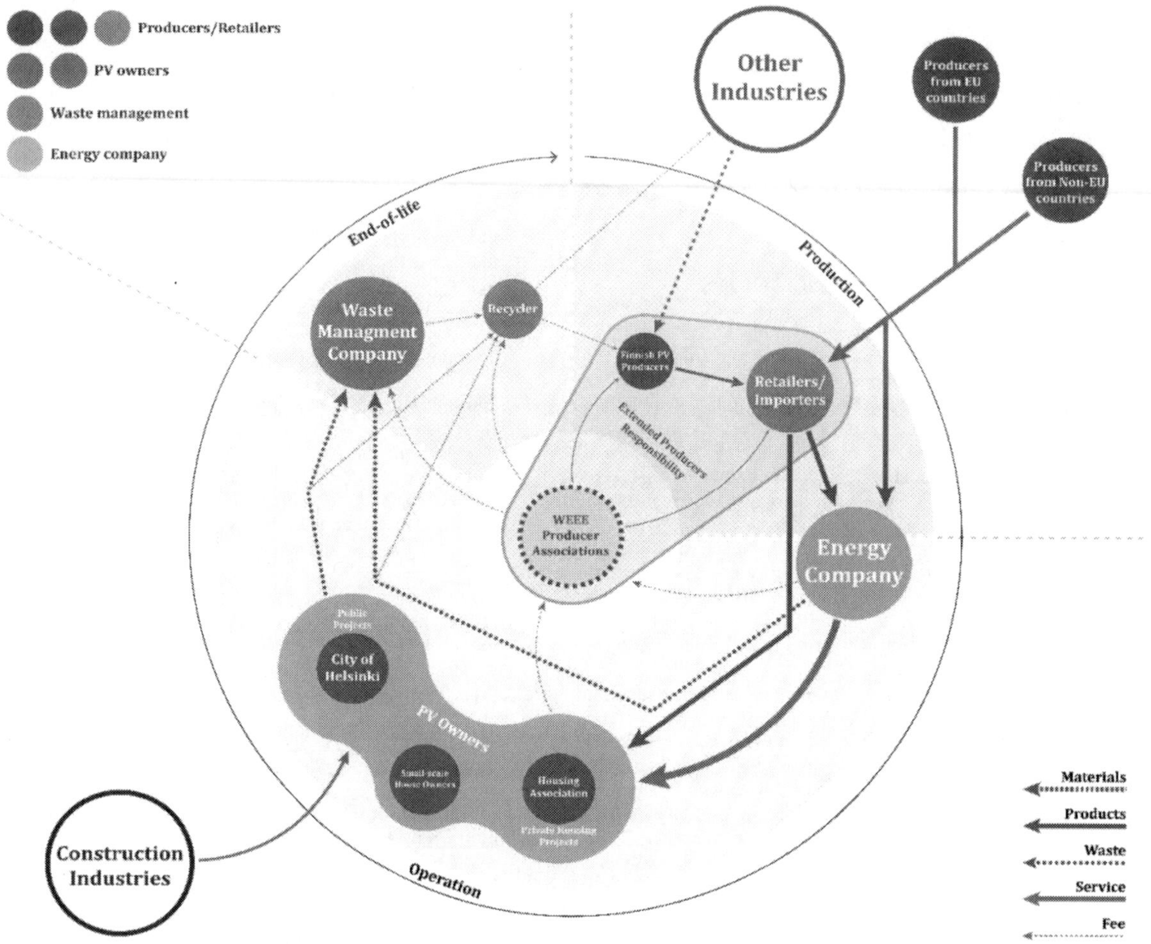

Fig 1. Helsinki PV supply chain

[11] Besiou, M., & Van Wassenhove, L. N. (2016). Closed-loop supply chains for photovoltaic panels: A case-based approach. Journal of Industrial Ecology, 20(4), 929-937.
[12] Lapko, Y., Trianni, A., Nuur, C., & Masi, D. (2019). In pursuit of closed-loop supply chains for critical materials: An exploratory study in the green energy sector. Journal of Industrial Ecology, 23(1), 182-196.
[13] Salim, H. K., Stewart, R. A., Sahin, O., & Dudley, M. (2019). End-of-life management of solar photovoltaic and battery energy storage systems: A stakeholder survey in Australia. Resources, Conservation and Recycling, 150, 104444.
[14] (2023). Retrieved 13 August 2023 https://eur-lex.europa.eu/legal-content/EN/TXT/?uri=CELEX:32011R0333
[15] (2023). Retrieved 13 August 2023 https://eur-lex.europa.eu/eli/reg/2012/1179/oj
[16] Kalmykova, Y., Sadagopan, M., & Rosado, L. (2018). Circular economy–From review of theories and practices to development of implementation tools. Resources, conservation and recycling, 135, 190-201.
[17] Choi, J. K., & Fthenakis, V. (2014). Crystalline silicon photovoltaic recycling planning: macro and micro perspectives. Journal of Cleaner Production, 66, 443-449.
[18] (2023). Retrieved 13 August 2023 https://valtioneuvosto.fi/paatokset/paatos?decisionId=0900908f8071a6e1

5 ACKNOWLEDGEMENTS

We. acknowledge financial support from the Academy of Finland (project numbers 347277). The content of the publication is the responsibility of the authors. We thank Matti Jänkälä for technical support.

DEVELOPING THE PHOTOVOLTAIC CIRCULAR ECONOMY IN SOUTH AFRICA THROUGH REUSE, REPAIR AND RESALE OF MODULES

JL Crozier McCleland[1]; MN Crozier[2]; EE van Dyk[1]
[1]Department of Physics, Nelson Mandela University, Nelson Mandela Bay, 6019, South Africa
[2] University of the Western Cape, Cape Town, Belville, 7535, South Africa

ABSTRACT: South Africa's Photovoltaic (PV) power generation installed capacity has increased considerably in recent years. An unreliable national energy supply is leveraging the demand from industry and households to source alternative energy solutions. As South African PV plants come to their end-of-life, increasing amounts of PV waste will be generated. A circular economy approach to end-of-life solutions for PV would consider options around lifetime extension, reuse, remanufacturing, refurbishment and recycling; as well as designing for disassembly at end-of-life. This paper discusses the current end-of-life options for PV modules in South Africa. There are currently local options for recycling but limited formal options for the repair and reuse of modules despite there being a demand for those services. To determine the reuse potential of a second-hand PV module a process was developed, based on existing characterisation techniques such as visual inspection, peak power measurements, electrical insulation testing and electroluminescence (EL) imaging. The threshold for the acceptance of this criteria is based on IEC-61215, pending the publication of specific standards for second-hand modules. The challenges for the PV reuse in South Africa are identified as cost, consumer acceptance, market feasibility and quality assurance. Further work will refine the classification scheme and identify market options.

Keywords: Solar PV Modules, Reuse, Repair, Circular Economy, South Africa, Characterisation techniques

1 INTRODUCTION

Solar energy has long been characterised as the clean and sustainable choice for energy generation and a progressive step towards reducing carbon emissions. Determining how Photovoltaic (PV) modules are produced, maintained and treated at end-of-life will ultimately determine its sustainability and the adage of the 'devil is in details' applies. PV modules have been developed with an inherently 'take-make-waste' or linear economy perspective that is associated with most mainstream manufacturing practices. Global cumulative solar PV waste estimated to be 4 million tonnes by 2030, 50 million tonnes by 2040 and more than 200 million tonnes by 2050 (1,2). Installations of new Renewable Energy (RE) power installations have to grow, to reach global climate change targets, with PV making up 57.5% of all new RE generation installations globally. With global installed power generation growing, implementing a Circular Economy (CE) approach of cycling materials, designing for circularity and reuse, repair, remanufacturing and recycling; is imperative to deal with this future waste stream.

Concerns for energy security is one key driver for circularity in energy. This has become a greater area of concern since COVID-19 severely impacted on global supply chains. This along with geopolitical tensions means more countries are considering how to firstly pursue a renewable energy growth path that is also localised in its manufacturing and sourcing (2). As many materials used in PV are also used in other sectors that are likely to expand, there will be competing demands between sectors for the same raw materials. Future-proofing the supply chain to be more resilient to shocks whilst at the same time creating local jobs and value creation is of increasing interest to countries. Analysis of supply chains is now including a greater focus on upstream activities i.e., raw material extraction and downstream activities at end-of-life i.e. waste disposal, decommissioning and CE interventions (2).

Developing a circular economy for PV modules has been mainly researched within established solar energy markets such as the European Union, the United States and China. There has been limited research in developing countries, which are likely to face heighted challenges due to existing constrained waste management systems and the threat of imports of waste PV modules from developed nations.

South Africa has a waste management system with limited infrastructure, that is still focused mainly on disposal. Recycling rates are achieved through an active but marginalized informal sector. With a 33% official unemployment rate (3) and 43% youth unemployment rate (4)(youth defined as persons aged 15-34 years old), job creation is of both economic and social importance. A Solar PV circular economy linked to localisation could offer much needed employment and business creation in green jobs. Green jobs are defined as jobs that reduce negative environmental impacts while creating environmentally, socially and economically sustainable economies in the long run (5).

The PV market is characterised by the utility scale and small-scale embedded generation (SSEG) markets. SSEG includes residential, commercial and industrial applications. South Africa has focused on the localisation opportunities around RE investments in utility scale developments for job creation opportunities. However, with limited manufacturing capacity and policy uncertainty, this has had limited success (10).

Research has traditionally focused on recycling options for solar PV modules. This research considered opportunities for higher order CE activities that could complement recycling efforts such as reuse of second-hand PV modules. This could offer a means to intensify value creation in the solar PV market would include CE interventions around reuse and repair. Second-hand modules could bring down the cost of the PV energy systems and place solar energy systems within the grasp of more households in South Africa.

This paper outlines the current end-of-life options in South Africa and proposes a process to determine the reuse potential of a second-hand PV modules.

2 CIRCULAR ECONOMY

The Circular Economy (CE) is a theory of how our relationship with products and, thus, by extension, waste could be redesigned to keep the value in materials and

10.4229/EUPVSEC2023/5DV.2.43
020548-001

products for as long as possible. The Ellen MacArthur Foundation defines the CE as "An economy that is restorative and regenerative by design and aims to keep products, components, and materials at their highest utility and value at all times, distinguishing between technical and biological cycles" (6). The definition has three essential elements: the cycling of materials, designing out waste and regeneration. Regenerative in the sense that not only does the system not do further harm, but it addresses past degradation of the environment to improve it.

The waste hierarchy is a tool for considering the circular economy, see Figure 1. Traditional approaches to waste such as disposal and burning are the least desirable. Recycling although a popular CE strategy entails a loss of value from the product and is thus less desirable than other CE activities. The most desirable activity is reduction of waste through the usage of fewer virgin resources and products, followed by reuse.

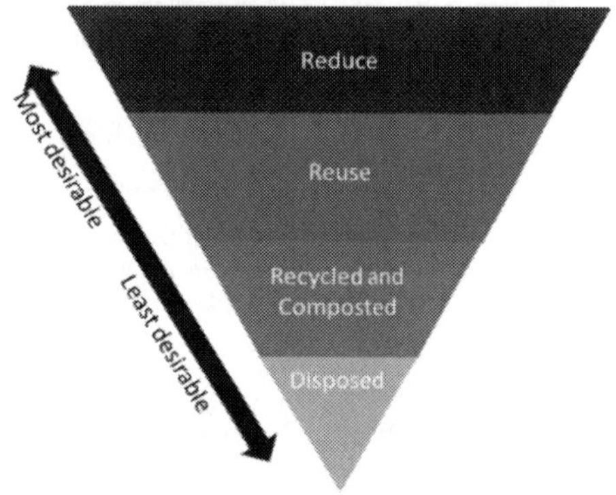

Figure 1: Waste Hierarchy.

Reuse extends the life span of products. It is defined as "The repeated use of a product or component for its intended purpose without significant modification."(7). Consumers benefit through gaining value through additional uses or by reselling the product, thus allowing other consumers to access the product's benefits. Reuse can be facilitated through active second-hand markets. Linked to reuse are the CE activities of repair and remanufacturing.

Repair although not illustrated in the waste hierarchy is dependent on the design of products. It ensures product life-extension and is a prerequisite for reuse. Repair also offers opportunities for business and job creation especially in the informal sector.

Remanufacturing is related to repair and is the dissembling of a product into its component parts and rebuilding it by replacing components as needed. The product is rebuilt to a like new condition with the same warranty and quality assurance as a new product (8).

Renewable energy is an essential component of the circular economy. It is presumed to power the activities of a circular economy. Thus, the circular economy needs sustainable, renewable energy generation like PV. In turn, the expansion of RE generation like PV must include a circular economy approach to materials.

3 SOLAR PV END-OF-LIFE

As global demand for PV increases this places stress in two key areas: the sources of raw materials and the treatment of waste at end-of-life. To meet global demand for renewable energy production, the World Bank estimated that production of critical materials would need to increase five times by 2050(2). Critical materials within some PV modules being Indium and Silver. These are also in demand for other sectors, thus there will not only be competition from PV module manufacturers but from other sectors to access virgin resources in future.

Between 93% and 95% of materials in solar PV are technically recoverable (9,10). Impediments to recycling of PV include the design of PV modules which is not designed for recycling, reuse or repair and accessing sufficient volumes of waste materials to make recycling economically viable. There are limited systems in place for collection, separation, recycling and reuse and thus much of these materials could go to waste (2).

Benefits of CE interventions are considerable. Aluminum is an essential component for PV modules and is used in the mounting structures and module frames. However, due to the huge amounts of energy expended during production, greater use needs to be made, globally, of secondary sources of aluminum. As secondary production uses 5% of the energy and between 3-5% of the carbon emissions used in primary aluminum production (11).

4 PV LANDSCAPE IN SOUTH AFRICA

South Africa's energy is both highly resource and carbon intensive (12), with coal contributing 80% to electricity generation and renewables 14% (13). The energy sector contributes 46% of South Africa's greenhouse gas emissions(12). South Africa is also plagued by energy insecurity due to the national energy provider Eskom, not being able to meet national demand. In 2022, the Energy Availability Factor (EAF) continued its decade long decline to 56% (13). This is due to a lack of investment in new generation capacity and the lack of maintenance of aging coal powered power plants. Loadshedding occurs when energy is rationed due to demand exceeding supply, to prevent grid collapse. In 2022, South Africa experienced its worst year of loadshedding to date, with 208 days of load-shedding (14). This record is set to be surpassed in 2023.

Thus, not only does South Africa need to invest more fully in renewable energy to break its dependence on coal; households and businesses are desperate to access stable energy supply. Installation of rooftop solar PV thus offers the most immediate opportunity for South Africans to secure energy (15). Rooftop solar for commercial/industrial, agricultural and residential applications grew by 60%, 45% and 30% Compound annual growth rate (CAGR), respectively between 2020 and 2022 (15).

Growth in solar PV has grown exponentially due to changes in national energy regulations in 2022, further price reductions in module costs, heightened incidents of loadshedding and the implementation of a tax rebate for households for the 2023/24 tax year (15,16). Figure 2 provides an indication of the growth in Solar PV installed capacity since 2010, while cumulative solar PV capacity stood at 4 172 MW in 2021 (17).

Market demand in South Africa has historically been linked to utility scale investment in solar PV as part of the Renewable Energy Independent Power Producer Procurement Programme (REIPPPP). However, lack of policy certainty stalled gains in investment in solar PV and constrained the promising development of manufacturing localisation.

10.4229/EUPVSEC2023/5DV.2.43
020548-002

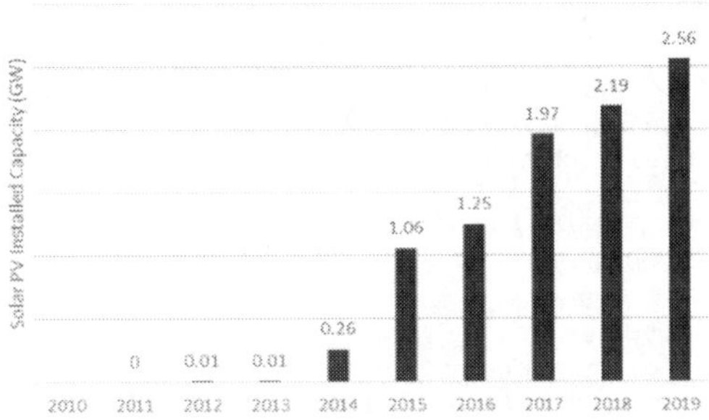

Figure 2: South African Solar PV installed Capacity (GW) over time (18)

The majority of solar modules are imported into South Africa to cover local demand. Between 2010 and 2022 South Africa imported R31 billion in solar modules (19), roughly €1.5billion. In the first quarter of 2023 alone, the country imported R3.6 billion of solar modules and cells (20), estimated at €175 million.

5 A CIRCULAR ECONOMY FOR PV IN SOUTH AFRICA

Modelled estimates of cumulative PV module waste volumes for South Africa are illustrated in Figure 3. South Africa's cumulative PV waste is estimated to grow to between 8 500 tonnes and 80 000 tonnes by 2030, and to between 750 000 and 1 million tonnes by 2050 (21). This aligns to South Africa's goal for installed PV capacity by 2030 of 8 288MW in terms of the Policy Adjusted Integrated Resource Plan (IRP) of 2019 (22), a national policy document guiding RE development.

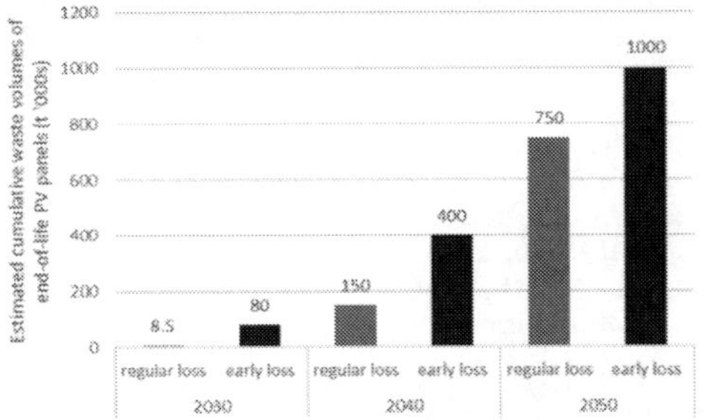

Figure 3: Modelled results of estimated cumulative waste volumes of end-of-life PV panels in South Africa(t '000s). Using Regular and Early Loss Scenarios (21).

5.1. Design for Circularity

A key goal of the REIPPP was local production of renewable energy inputs. However, the stalled nature of plant approvals meant sustaining local manufacturing was hampered. Over Bid windows 1-4, 50% of plant value was procured locally compared to Bid window 5, where it had dropped to 41% for operations (19). To date PV components manufactured or assembled in South Africa include solar modules, plant assembly, lead acid battery manufacturing, lithium ion battery pack and module assembly and balance of systems (12). Raw materials are imported. Challenges to localization include that South Africa does not have the manufacturing capacity to produce cells, ingots and inverters. Attempts at localisation have also been thwarted through inconsistent policy direction and political indecision (23). South Africa is also a small market player globally and importers have limited power in influencing design considerations.

5.2. Disposal

A raft of new waste legislation is assisting to provide policy framework for CE interventions. A ban on the disposal of all electronic and electrical equipment at landfill was instituted by the South African Government on the 23 August 2021 (24).

Extended Producer Responsibility (EPR) legislation came into effect on the 5 November 2021 covering the lighting, electrical & electronic equipment and packaging sectors. Product responsibility at end-of-life is now vested in producers, importers and installers of electrical and electronic equipment (EEE) products, which includes PV modules and battery energy storage systems (BESS). The implementation and enforcement of these laws has yet to be proven effective.

5.3. Recycling

Globally recycling of PV modules is expected to generate income of $450 million by 2030 and $15 billion by 2050(2).PV module recycling in South Africa is in its infancy. There are two formal recycling firms undertaking PV module recycling and this is mainly for utility scale PV operations that are required to comply with waste management regulations. The SSEG market has yet to mature and create sufficient PV waste for large scale recycling.

It is uncertain if any informal collection and dismantling of PV modules is being undertaken but considering the small volumes of critical materials contained within PV modules, informal dismantling is likely to only focus on the aluminum frames.

In South Africa, the waste sector is characterised by a large informal sector collecting, sorting and dismantling, for sale onto formal recycling processors. Informal sector employment has many dangers and vulnerabilities. South Africa has committed to considering how to undertake CE interventions that incorporate informal workers and create sustainable job opportunities (2).

5.4. Reuse, Repair and Remanufacturing

Waste electrical and electronic refurbishment and repair occurs within South Africa, especially within the informal sector. Information on PV reuse in both formal and informal sector is limited currently. Reuse applications would need to consider safety, insurance compliance and a strict certification to avoid encouraging theft. A key challenge is the low prices of new PV modules which discourages the reuse of second-hand modules.

The repair or remanufacturing of PV modules is currently occurring as part of manufacturer's warranty repair plans, specifically the repair of back sheets (25). There is a potential to create jobs in the repairs of modules' cabling and junction boxes. However, unlike other electronic goods, the volumes required for these repairs to be viable would be difficult to achieve. The resale of PV modules is dependent on the important factors of price, safety assurance and warranty. It is also necessary to provide certification to vendors to prevent exacerbating the problem of theft of PV modules.

5.5. Training

Suitable training is needed on correct handling and transportation of PV modules especially in the SSEG market, this can assist to reduce breakages and prolong lifespans.

10.4229/EUPVSEC2023/5DV.2.43
020548-003

The need for greater awareness on the importance of correct installation and handling of PV modules is illustrated in Figure 4. This shows a much-used stock image depicting a technician kneeling on solar modules, an incorrect procedure, which can cause damage to solar modules and shorten their useful life. The image was used in 2023 as the cover of the National Treasury's Tax Incentive Guidelines (16).

Figure 4: Cover of the 2023 Solar Panel Tax Incentive Guideline published by South African National Treasury(16) showing a technician kneeling on PV modules which can cause damage to modules.

In addition to handling at the installation phase, it is important to ensure that modules are handled, packed and transported correctly at the removal and decommissioning stage. Modules that are damaged during transportation to testing/recycling will not be able to be reused.

6 THE CASE FOR SOLAR PV REUSE IN SOUTH AFRICA

A case for the PV modules reuse market can be made in terms of the following arguments:

- Increased employment and business creation opportunities along the value chain
- Reduced cost of solar PV adoption
- Environmental impact of waste
- Social justice

Global solar PV employment was estimated by IRENA at 4.3 million in 2021, up from about 4 million in 2020(2). Employment in the construction and installation of commercial and industrial solar PV systems in South Africa was estimated to create between 5.3 and 8.0 full time equivalent (FTE) jobs per MW. While rooftop systems are projected to create between 7.7 FTE jobs per MW on average during construction and installation (18). A CE for PV modules could support job creation in new areas of the value chain. European estimates indicate that 63 jobs are created per 1000 tonnes of collected electrical equipment waste (2). In the US electronic repair and remanufacturing could support 200 jobs per 1000 tonnes compared to 15 jobs for recycling that same amount (26).

An overview of a circular economy for PV modules is provided in Figure 5. Before end-of-life, consumers can access repair and maintenance services from installers and distributors. Alternatively, the unit could be remanufactured through the manufacturer. At end-of-life, landfill disposal would be avoided through parts reuse or recycling. Thus, enhancing the circular economy aspects of the South African PV value chain could increase job opportunities through focusing on reuse, repair and remanufacturing.

One of the sources of second-life modules is the - repowering of utility scale PV plants. The oldest of these plants were commissioned in 2014 and could potentially be repowered between 2024-2034. Other sources of second-life modules would be degraded or damaged modules. These modules would require testing to determine their safety and performance.

The proposed testing process is illustrated in figure 6, all the tests are based on existing international standards (IEC-61215), pending the publication of new standards for second-hand modules. The testing includes Visual Inspection to identify module defects, Electroluminescence to detect invisible cracks and Maximum Power measurements to determine the module performance. These are compared to expected values for a module of that age to determine the module's suitability for resale or reuse.

The costs of current accredited testing services are prohibitively expensive and dictated by the expensive equipment required, the time required for testing and the specialized skills needed. The challenge is to ensure that modules can be tested and still be able to be sold at a competitive price.

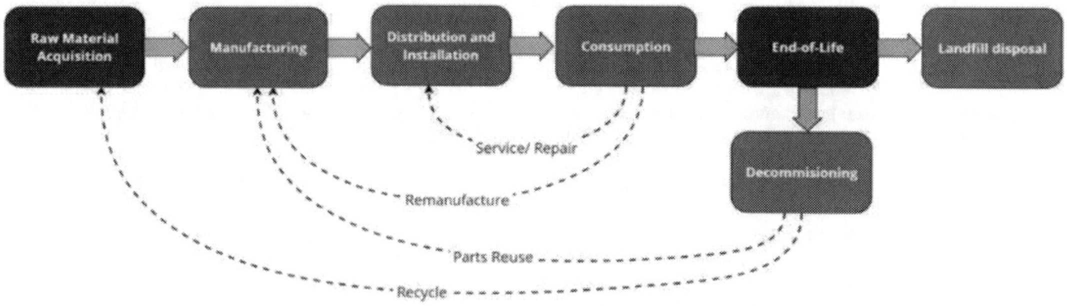

Figure 5: Circular Supply Chain for Photovoltaic Systems adapted from (27)

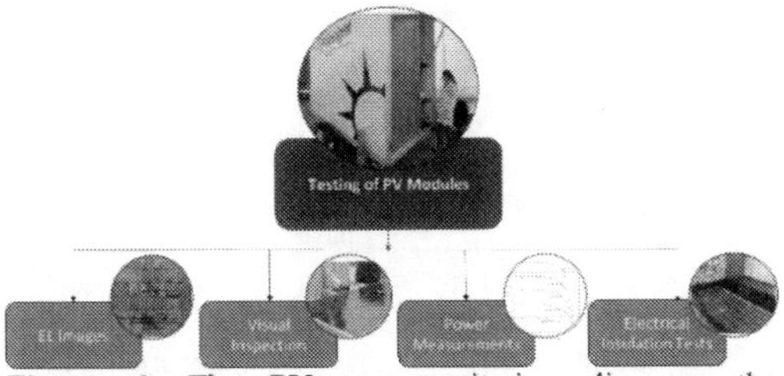

Figure 6: The PV reuse criteria relies on the characterisation of PV modules according to standard test procedures. These tests ensure the safety and performance of modules for reuse.

PV module testing has been identified as an important step in the safe reuse of PV modules and diverting them from disposal on landfill. The challenges in this process include the lack of testing standards and the costs involved in testing. These costs would make testing for PV module second life out of reach of normal rooftop solar customers, unless either an incentive, subsidy or penalties for incorrect disposal are applied. There are also opportunities to reduce the complexity of testing to make testing more accessible and financially feasible.

Solar PV modules are a major cost item for households installing solar PV systems (18). Thus, reducing the costs associated with installation and maintenance could increase disposable income, increase sales of systems and bring systems within the reach of lower income consumers.

A just energy transition requires considerations of access to affordable, sustainable energy. Renewable energy cannot just be an option for individuals and businesses with financial means. By making renewable energy more accessible through financing support, subsidies and also reuse and repair it will ensure a more equitable transition.

7 CONCLUSION

This paper outlines the current end-of-life options in South Africa and proposes a process to determine the reuse potential of a second-hand PV modules.

The first challenge for PV module reuse in South Africa is the cost of secondhand modules, which needs to be cheaper than new modules to gain consumer support. In addition, the lack of standardised test criteria for performance and safety means it is difficult to provide quality assurance and warranties that would satisfy lenders and insurance providers. The advantages include reducing the waste produced and the development of jobs in the sector. These issues need to be addressed to ensure a feasible market for secondhand PV in South Africa.

Future work will include market research by the authors to collect primary data from the South African energy sector and e-waste recyclers to better understand barriers and drivers to PV module reuse and repair. Further testing will be undertaken of second-hand modules to confirm a characterisation technique for reuse and repair. These two activities will allow for a sustainable business model to be developed for PV Reuse in South Africa,

8 REFERENCES

1. Feng J. End of life management of solar PV and the circular economy [Internet]. Presentation at TESSD Working Group Meeting on Circular Economy presented at: TESSD Working Group Meeting on Circular Economy; 2023 [cited 2023 Sep 2]. Available from: https://www.wto.org/english/tratop_e/tessd_e/11_irena_presentation.pdf

2. IRENA, ILO. Renewable energy and Jobs: Annual Review 2022 [Internet]. Abu Dhabi and Geneva: International Renewable Energy Agency and International Labour Organization; 2022 [cited 2023 Aug 30]. Available from: https://www.irena.org/publications/2022/Sep/Renewable-Energy-and-Jobs-Annual-Review-2022

3. Statistics South Africa. Key findings: P0211 - Quarterly Labour Force Survey (QLFS), 2nd Quarter 2023 [Internet]. [cited 2023 Sep 3]. Available from: https://www.statssa.gov.za/?page_id=1856

4. Statistics South Africa. P0211 - Quarterly Labour Force Survey (QLFS), 2nd Quarter 2023 [Internet]. Pretoria: Statistics South Africa; 2023 [cited 2023 Sep 3]. Report No.: P0211. Available from: https://www.statssa.gov.za/publications/P0211/P02112ndQuarter2023.pdf

5. International Labour Organisation. Green jobs in Africa [Internet]. 2023 [cited 2023 Sep 3]. Available from: https://www.ilo.org/africa/areas-of-work/green-jobs/lang--en/index.htm

6. Ellen MacArthur Foundation. Towards a Circular Economy: Business Rationale for an Accelerated Transition [Internet]. Isle of Wight: Ellen MacArthur Foundation; 2015 [cited 2022 Aug 15] p. 20. Available from: https://emf.thirdlight.com/link/ip2fh05h21it-6nvypm/@/preview/1?o

7. Ellen MacArthur Foundation. The circular economy glossary [Internet]. 2023 [cited 2023 Sep 10]. Available from: https://ellenmacarthurfoundation.org/topics/circular-economy-introduction/glossary

8. Ellen MacArthur Foundation. The Circular Economy In Detail [Internet]. 2022 [cited 2022 Aug 27]. Available from: https://archive.ellenmacarthurfoundation.org/explore/the-circular-economy-in-detail

9. Rivett-Carnac K. Insights into the Solar Photovoltaic Manufacturing Value Chain in South Africa [Internet]. South Africa: TIPS; WWFSA; 2022. Available from: https://tips.org.za/research-archive/sustainable-growth/green-economy-2/item/4441-insights-into-the-solar-photovoltaic-manufacturing-value-chain-in-south-africa

10. Graulich K, Bulach W, Betz J, Dolega P, Hermann C, Manhart A, et al. Emerging Waste Streams Challenges and Opportunities [Internet]. Freiburg, Germany: Öko-Institut, e.v; 2021 [cited 2023 Sep 3]. Available from: https://www.oeko.de/fileadmin/oekodoc/EEA_emerging-waste-streams_final-report.pdf

10.4229/EUPVSEC2023/5DV.2.43
020548-005

11. Maisch M. Huge aluminum demand expected in solar industry, concerns arise on emissions. PV Magazine International [Internet]. 2022 [cited 2023 Sep 2]; Available from: https://www.pv-magazine.com/2022/01/25/huge-aluminum-demand-expected-in-solar-industry-concerns-arise-on-emissions/

12. Pandarum A, Rakaibe T, Simiyu D, Nombakuse Z, Mbam V, Godfrey L. Improving energy security in South Africa through a more circular energy sector. Pretoria: CSIR; 2023. Report No.: CSIR Report Number: CSIR/SPLA/EC/EXP/2023/0037/A.

13. CSIR. Statistics of utility-scale power generation in South Africa [Internet]. 2022 [cited 2023 Sep 3]. Available from: https://www.csir.co.za/sites/default/files/Documents/Statistics%20of%20power%20in%20SA%202022-CSIR-%5BFINAL%5D.pdf

14. Simpson S. The South African. 2023 [cited 2023 Jun 23]. Load shedding tracker: A visualisation of power cuts over the years. Available from: https://www.thesouthafrican.com/news/load-shedding-tracker-how-many-days-2015-2023-outlier-visuals-breaking-11-january/

15. GreenCape. 2023 Energy Services Market Intelligence Report [Internet]. Cape Town, South Africa: GreenCape; 2023 [cited 2023 Jun 23]. Available from: https://green-cape.co.za/wp-content/uploads/2023/04/ES_MIR_2023_DIGITAL_SINGLES.pdf

16. National Treasury. Frequently Asked Questions - Solar Panel Tax Incentive for Individuals [Internet]. 2023 [cited 2023 Sep 3]. Available from: https://www.treasury.gov.za/documents/National%20Budget/2023/2023%20Budget%20FAQs%20-%20Solar%20Panel%20Tax%20Incentive.pdf

17. Bellini E. South Africa installed 1.3 GW of PV last year. PV Magazine International [Internet]. 2021 Jul 21 [cited 2022 Nov 14]; Available from: https://www.pv-magazine.com/2021/07/21/south-africa-installed-1-3-gw-of-pv-last-year/

18. CSIR. Solar PV Industry Jobs Report [Internet]. SAPVIA; 2021 [cited 2022 Dec 6]. Available from: https://www.sapvia.org.za/wp-content/uploads/2021/05/SAPVIA-PV-Industry-Jobs-Study-Report-COMBINED.pdf

19. Department of Mineral Resources & Energy, Department of Science and Innovation, Department of Trade, Industry and Competition. South African Renewable Energy Masterplan (SAREM): An industrial and inclusive development plan for the renewable energy and storage value chains by 2030. Draft Version for Review 7 July 2023 [Internet]. Pretoria; 2023 [cited 2023 Jul 21]. Available from: https://www.dmr.gov.za/Portals/0/Resources/Renewable%20Energy%20Masterplan%20(SAREM)/South%20African%20Renewable%20Energy%20Masterplan%20(SAREM)%20Draft%20III.pdf?ver=2023-07-17-141604-137&timestamp=1689596128318

20. Montmasson-Clair G. Post | Feed | LinkedIn [Internet]. 2023 [cited 2023 Jun 8]. Available from: https://www.linkedin.com/feed/update/urn:li:activity:7059510873370861569/

21. IRENA, IEA-PVPS. End of Life Management: Solar Photovoltaic Panels [Internet]. International Renewable Energy Agency and International Energy Agency Photovoltaic Power Systems.; 2016 [cited 2022 Oct 27] p. NREL/TP-6A20-73852, 1561525. Report No.: NREL/TP-6A20-73852, 1561525. Available from: https://www.osti.gov/servlets/purl/1561525/

22. GreenCape. 2023 Large Scale Renewable Energy Market Intelligence Report [Internet]. Cape Town, South Africa: GreenCape; 2023 [cited 2023 Jun 23]. Available from: https://green-cape.co.za/wp-content/uploads/2023/04/RENEWABLE_ENERGY_MIR_2023_DIGITAL_SINGLES.pdf

23. ECODIT LLC, USAID, Power Africa. South African Solar PV Value Chain Analysis and Strategies for Increasing Localization. USAID; 2022.

24. Shaibani M. Solar Panel Recycling: Turning Ticking Time Bombs into Opportunities. PV Magazine [Internet]. 2020 [cited 2023 Sep 3]; Available from: https://www.greenbuildingafrica.co.za/solar-panel-recycling-turning-ticking-time-bombs-into-opportunities/

25. Reports of backsheet failures at 75 MW Mulilo Sonnedix Prieska solar farm in South Africa. PV Magazine International [Internet]. 2020 Feb 11; Available from: https://www.pv-magazine.com/2020/02/11/reports-of-backsheet-failures-at-75-mw-mulilo-sonnedix-prieska-solar-farm-in-south-africa/

26. Reuse. Briefing on job creation potential in the re-use sector [Internet]. 2015 [cited 2023 Sep 2] p. 6. Available from: https://www.rreuse.org/wp-content/uploads/Final-briefing-on-reuse-jobs-website-2.pdf

27. Salim HK, Stewart RA, Sahin O, Dudley M. Drivers, barriers and enablers to end-of-life management of solar photovoltaic and battery energy storage systems: A systematic literature review. Journal of Cleaner Production [Internet]. 2019 [cited 2023 Jun 13];211:537–54. Available from: https://linkinghub.elsevier.com/retrieve/pii/S0959652618336321

PROCEDURE PROPOSAL TO DETERMINE PV MODULE STATUS FOR ITS SECOND LIFE APPLICATION

Jorge Rabanal-Arabach [1]*, Edward Fuentealba-Vidal [1], Javier Astudillo-Ledezma [1], Sonia Beltrán-Condori [1], Abel Taquichiri [1], José Tapia-Jelcic [2], Andreas Schneider [3], María J. Riquelme-Zambrano [1], Ingrid Jamett-Aranda [1], Romain Couderc [4], Delfina Muñoz [4]
[1] Universidad de Antofagasta, Angamos Avenue 601, 1270300 Antofagasta, Chile.
[2] ATAMOSTEC, Uribe 636, oficina 302, Antofagasta, Chile.
[3] University of Applied Sciences Gelsenkirchen, Neidenburgerstr. 43, 45877 Gelsenkirchen, Germany.
[4] Univ. Grenoble Alpes, CEA, Liten, Campus INES, 73375 Le Bourget du Lac, France.
*Corresponding author: jorge.rabanal@uantof.cl

ABSTRACT: The reuse of photovoltaic modules (PVM), also known as "PV 2nd-life", is seen as a strategy to reduce the potential waste generation of the massification of this technology. The lack of a standard methodology to test and classify used modules is one of the barriers to enable a PVM second life. The present work contributes to this issue, proposing a procedure to determine the PV module status for its second life application. The procedure was elaborated based on non-destructive tests of the IEC standards and IEA PVPS reports. It is formed by six steps, each one with its own methodology: 1. PVM registration; 2. Visual inspection; 3. Electrical insulation (dry); 4. Electrical characterization (I-V curves); 5. Infrared thermography; 6. Electroluminescence imaging. From a thousand analyzed PVMs, 3.0% are in conditions to be reused to produce energy, 30.7% show defects but with chances to be repaired before its usage, 27.8% are not recommended to be used as active element but in passive applications, and 38.6% must be discarded or sent to recycling. Using the approved PVMs in the test bench demonstrates these devices are safe to use, proving the effectiveness of the procedure.
Keywords: PV modules, characterization, procedure, second life.

1 INTRODUCTION

The Atacama Desert is well known for its high solar irradiation and relatively low ambient temperature, hence presenting a very high photovoltaic (PV) power potential [1]. This way, the number of PV plants installed and under operation grows steadily year on year, consequently reducing the level of green-house gases by 22% in the main Chilean electrical network in the year of 2022, if compared to 2021 [2]. However, the reduction in air pollution comes with a potential increase in earth pollution by means of waste from the PV plants, mainly driven by those PV modules (PVM) showing premature failures due to the harsh ambient conditions of the Atacama Desert. Furthermore, it is expected that the sum of discarded PVM will accumulate to the magnitude of one million tons by 2050 in Chile [3]. One can think, under this picture, that a local recycling industry is needed. Nevertheless, circular economy and waste management techniques teach that it is preferable to reuse a product rather than recycle it [4], since reuse has several advantages over recycling. Higher consumption of resources, such as energy or water, and waste generation, since not all materials can be recovered, are some of the disadvantages of recycling over reuse.

The reuse of PV modules (and electrochemical batteries) has been coined as "second life" (2nd life) within the circular economy environment. However, the lack of a standard methodology to test and classify used modules is one of the barriers to enable a PVM 2nd life.

The above-mentioned statement opens the discussion about the need for procedures and diagnostic tools to perform a severe test of the PVM with the aim to identify its electrical and physical integrity, its repair options, and its potential for a second life application. Such procedure is not fully covered in the current standards, which mainly focus on manufacturing quality qualification.

2 DESIGN OF EXPERIMENT

The methodology of this work is based on heuristics methods, dividing the workflow into five stages, as shown in Figure 1; the procedure development is based on a literature review of standards and recommendations, which leads to the application of the procedure, which is firstly tested on 20 PVM. If during the application drawbacks are seen, the procedure is adjusted. This loop is repeated until the documentation is suitable enough to minimize its usage errors. With the procedure in a proper shape, a group of thousand PVM discarded from PV power plants are tested. The approved ones are mounted later in a test bench, installed at the *Plataforma Solar del Desierto de Atacama* (PSDA), a test site to benchmark PV technologies, where the PVMs are monitored by infrared thermography and I-V curve measurements at string level.

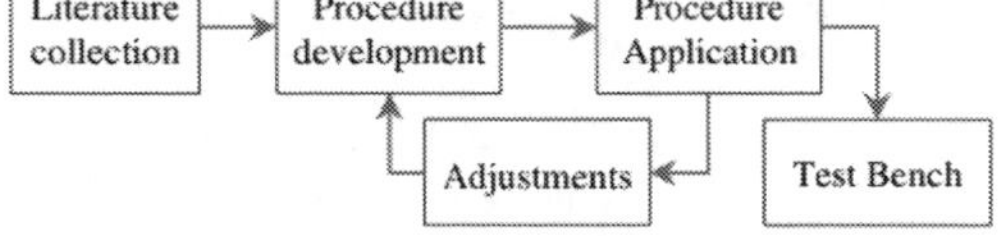

Figure 1: Diagram representing the methodology workflow.

2.1 The proposed procedure

The procedure has been developed based on the IEC 60904, IEC 61215-1, IEC 61215-2, IEC 61646, IEC 61730, IEC 62446, and on the IEA PVPS reports: T13-01:2014 [5], T13-10:2018 [6], and T13-24:2021 [7], with focus on the non-destructive tests. It consists of six steps, each one with its own methodology, as Figure 2 shows. Numeric value on top indicates the step number. The first one is the registration of the PV module, based on an admission form to log the device and system characteristics from the datasheet and power plant operator, respectively. The second step is a visual inspection, which is a

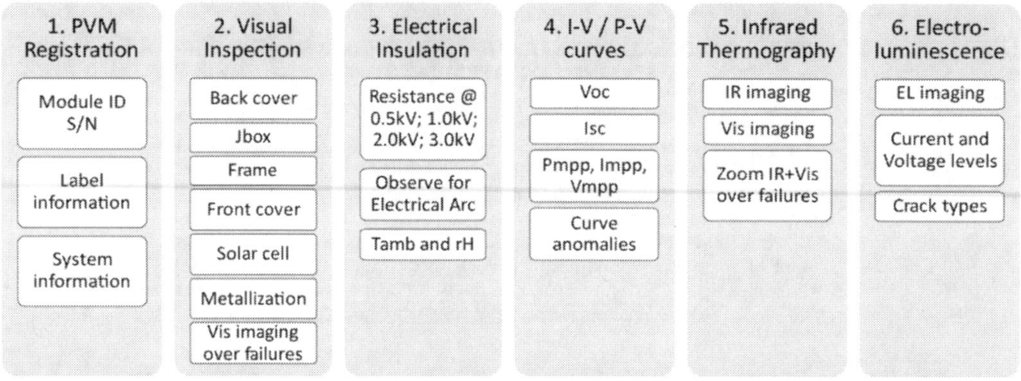

Figure 2: Scheme of the proposed procedure to determine the physical and electrical status of the PV modules.

modification from recommendations of the IEA PVPS T13-01:2014 report, adapted to include bifacial modules. The third step searches for electrical hazard, failures in the junction-box (jbox), cables and connectors, and determines the system voltage to which the module could safely operate. The fourth step (I-V) is meant to determine both the electrical characteristics of the PVM and its power degradation. The penultimate step (IRT or IR-imaging) searches for a possible fire hazard, coming particularly from hotspots within the cell matrix or the jbox. The sixth and last step (EL-imaging) is intended to determine cracks within the solar cells that could result in further module degradation or even a fire hazard. The I-V and IRT tests are performed in an outdoor freestanding open rack, to allow the PVMs to operate under natural sunlight and temperature conditions; irradiance above 900 W/m² and ambient temperature between 20–25 °C. The IR-imaging is executed only after exposing the PVM, in short-circuit, for 15 minutes.

The whole procedure is designed to also provide information on the options to restore the PV module's functionality in case of electrical malfunction or other hardware defects. This way, each module could result with one of the following tags: "approved", "approved with observations or possible reparation", "rejected with possible reparation", and "rejected, not recommended for generation".

It is important to remark that if a PVM fails on any of the above-mentioned steps, it may not proceed with the next one, tagging the device with its respective results. This is mandatory to allow the procedure to be applicable on a large scale since this criterion reduces time, resources and costs involved in the application of the procedure.

The pass-fail criteria for each step are shown in tables 1–5. Table I lists the diagnosis criteria for step 2: visual inspection, where the category of defects are according to the definition of safety failure given by the IEA PVPS T13 [5]; type "A" means that the failure has no effect on safety, type "B" is used when a second failure may cause fire, electric shock or physical damage, and type "C" means the failure causes a safety problem in a direct way. The letters "D" and "E" are added in this work to include reparation options; type "D" stands for "The defect may be repairable, without affecting safety directly. Reparation is optional for safe use of the device", whilst type "E" means "The defect may be repairable, with safety being affected if the device is not repaired. Reparation is mandatory for a later safe use."

Table II lists the diagnosis criteria for step 3: electrical insulation; here the voltage levels are 0.5 kV, 1.0 kV, 1.5 kV, 2.0 kV, 3.0 kV and 4.0 kV. If the PVM shows anomalies in a certain voltage range, then its recommended system voltage must be smaller than the lower voltage at which it fails, i.e., if the PVM fails while passing from 0.5 kV to 1.0 kV, then the accepted system voltage is 200 V; if it fails within 1–2 kV, 500 V; if it does within 2–3 kV, then 1 kV.

Table I: Pass-fail criteria for step 2, Visual Inspection.

Diagnosis	Criteria
Approved	No defects are presented, or it has only type A defects, or it has no more than two type B defects.
Approved with observations or possible reparation	It presents type D defects.
Rejected with possible reparation	It has type E defects and no more than two defects type B.
Rejected, not recommended for generation	It presents type C defects, or three or more type B defects.

Table II: Pass-fail criteria for step 3, Electrical insulation.

Diagnosis	Criteria
Approved	No electric arcs are present, and resistance changes with deviation no larger than 25% between voltage levels, and the resistance is equal or larger than 40 MΩm².
Approved with observations or possible reparation	No electric arcs are present, and the resistance is equal or larger than 2 MΩ, but the resistance changes with deviation larger than 25% between voltage levels.
Rejected with possible reparation	The cables are loose in the jbox. No electric arcs are present.
Rejected, not recommended for generation	It presents electric arcs or dielectric rupture. The resistance changes with deviation larger than 25% between voltage levels, or the resistance is smaller than 40 MΩm².

The diagnosis criteria for step 4: I-V curves, are listed in Table III, while Table IV does it for step 5: IR-imaging. The anomalies in both, the I-V curves and IR images are described by the IEA PVPS T13 [5]. As the apparatus to test these two steps is a free-standing mounting structure, the module temperature for this round is expected to be not hotter than 54 °C [8]. Hence, giving a margin of 30%, the limit temperature is 70 °C.

Table III: Pass-fail criteria for step 4, I-V curves.

Diagnosis	Criteria
Approved	It does not present anomalies in the I-V curve and the power degradation is lower than 40%.
Approved with observations or possible reparation	It presents anomalies in the I-V curve, and the power degradation is lower than 40% and/or it shows anomalies related to broken diode(s).
Rejected with possible reparation	The power degradation is lower than 75% and it shows anomalies related to broken diodes. The PVM shows a relation in current of: $\frac{I_{SC}}{I_{SC,STC}} \geq 0.30$
Rejected, not recommended for generation	It presents a power degradation larger than 75% with a voltage relation: $\frac{V_{OC}}{V_{OC,STC}} \geq 0.35$

Table IV: Pass-fail criteria for step 5, IR imaging.

Diagnosis	Criteria
Approved	No thermal anomalies exist. The difference between the coldest and hotter solar cell is no larger than 20 K.
Approved with observations or possible reparation	The IR image displays anomalies. The difference between the coldest and hotter cells ranges 10–30 K.
Rejected with possible reparation	The IR images show anomalies related to broken diodes.
Rejected, not recommended for generation	The module temperature is larger than 1.3 time its expected temperature or the difference between the coldest and hotter zones are 30 K or larger.

Table V lists the diagnosis criteria for step 6: EL-imaging. The anomalies at this step are considered as by the IEA PVPS T13 [5] are described. Figure 3 shows examples of crack magnitude: severe, major, and minor.

Table V: Pass-fail criteria for step 6, EL imaging.

Diagnosis	Criteria
Approved	It does not present anomalies in the EL image and less than 10% of the cells show minor cracks.
Approved with observations or possible reparation	Anomalies in the EL image are observed, with less than 20% of the cells showing minor cracks, and less than 10% showing major cracks, with a total of less than 20% of the cells having cracks. The PVM is tagged as "possible reparation" if one substring is inoperative ("off").
Rejected with possible reparation|	The PVM is tagged as "possible reparation" if two substrings are inoperative ("off").
Rejected, not recommended for generation	It presents cracks of severe magnitude on any of the cells.

Figure 3: Examples for the criteria of crack detection.

2.2 PV modules selection

To prove the new procedure, three different industrial PV plants in the Atacama Desert were visited and some of their decommissioned PVM transported to the University of Antofagasta, where the Laboratory for PVM Inspection is installed.

Each of the three plants is composed of multi-crystalline Si modules mounted on a horizontal single-axis tracking system and wired up to a maximum string voltage of 1.0 kV. These samples correspond to PVM discarded from the operation and stored in the respective site of each PV power plant. The dates when the individual samples were mounted or unmounted are unknown.

2.3 Experimental setup

Figure 4 a) displays an external view of the laboratory, where steps 1, 4 and 5 of the procedure were executed, whilst Figure 4 b) shows the interior part of the laboratory, where steps 2, 3 and 6 were carried out. The former is composed of an tilted open holder to mount PVM of several sizes, thickness, and architecture; a portable I-V tracer that uses the Kelvin method (4T-sensing) with an accuracy of ±0.3% for voltage and current measurement; and a thermal-imaging camera of 384×288 IR-resolution, with a pixel pitch of 17 μm, 8–14 μm spectral range, and accuracy of ±2% for ambient temperature ranging +15 °C to +35 °C.

Figure 4: Laboratory for PVM Inspection at the University of Antofagasta. (a) outside view, (b) inside view.

The indoor laboratory is equipped with a standing rack to mount a PVM for visual inspection; a megohmmeter, able to measure leakage currents up to 3 mA with a resolution of 1 pA, adjustable and programmable test voltage ranging 40–5100 V, insulation measurements from 10 kΩ to 10 TΩ with an accuracy of ±5% (10 kΩ–399.9 GΩ) and ±15% (400 GΩ–10 TΩ); and a system for electroluminescence imaging, consisting of a CMOS camera of 5 Mpx, 600 μm/pixel resolution, and a power

10.4229/EUPVSEC2023/5DV.2.45
020549-003

supply unit to feed the PVM with up to 60 V and 25 A, both units controlled by a software that stitches the EL images appropriately.

The ambient temperature is controlled to ensure the indoor measurements are performed at an ambient temperature in the range of +20 °C to +25 °C.

3 PROCEDURE APPLICATION

The full testing procedure has been developed and verified with a PVM universe of around thousand samples, which were examined at the power plant locations by simple visual inspection for proper mechanical integrity and for procedure applicability: 465 at PV_1, 428 at PV_2, and 112 at PV_3. From this, several hundred PVMs were in deplorable conditions and on-site rejected; 23.9% at PV_1, 57.9% at PV_2, and 22.3% at PV_3. Hence the collected samples summed up to altogether 617 modules, which were transported to the laboratory at the University of Antofagasta. The diminishment in sample numbers opens the discussion about PVM handling during and after their decommission from the PV plant. In this study it has been noticed that handling and storing protocols needs to be improved and properly implemented to avoid the discarded PVM getting further damaged. Furthermore, during the visit to the PV plants it was observed that the main failures come either from the PVM transportation and storage or from the difficulties during the plant operation due to the environmental inclemency.

The analysis of the 617 MPV returns that 4.9% of the modules are in conditions to be reused as they are, 49.9% show defects but with a high chance to be repaired, which is mandatory before its usage, 45.2% are not recommended for electricity generation but they can be used in other (passive) applications. Figure 5 displays the distribution of modules per step for each of the three power plants: PV_1, PV_2 PV_3. The module is tagged as "Rejected" if it does not pass the test, "Rejected pos. rep." if it does not pass the test, but it has chance for a reparation, "Approved w/ obs." if it passes the test with observations about its condition (like small bubbles), and "Approved" if it passes the test without any constraint.

The most common failures detected in the visual inspection are summarized in Figure 6 where FC stands for "front cover", "Encaps" for encapsulant, and BS for "backsheet". The plot summarizes the three PV plants together (the whole sample), showing with colors the category of failure as described in section 2.1.

The electrical insulation test shows that 14.6% of the examined PVM have problems with the electrical interconnection in the jbox, either by loose cables, defective bypass diode(s), or intermittent electrical (loose) contacts within the jbox. None of the tested devices show low electrical insulation resistance, since all test results range between 1.2–22.7 GΩ, being the lower limit for rejection 2 MΩ.

The I-V curve test indicates that the average power of the whole group is (180 ± 53) W, and for the approved PVM is (265 ± 34) W, for nominal power ranging 300–310 W. This test also shows that the main power degradation of the approved modules is 13%, whilst the rejected modules have 94%.

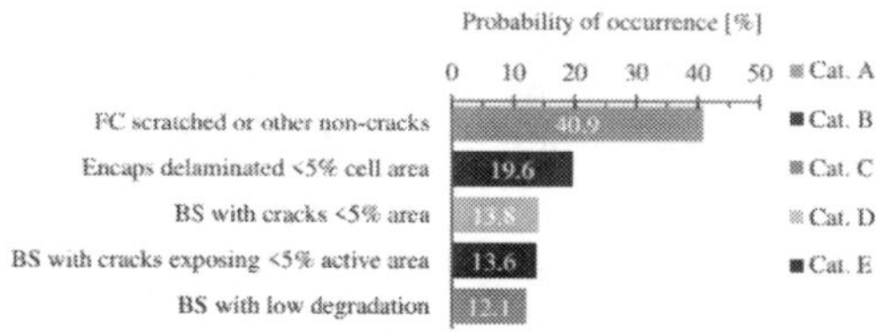

Figure 6: Top 5 failures detected by the visual inspections.

The IRT reveals that 4.2% of the tested devices have hotspots able to develop a fire hazard, with an upper limit of 70 °C. The highest temperature registered is (101 ± 3) °C.

The EL imaging exhibits that 4.9% are approved at this step, whilst 13.1% and 1.9% are classified as with possible reparation and rejected. Reparation in this matter stands for bypassed strings. Besides, 10.5% of the tested PVM developed severe cracks in the solar cells, 9.8% have at least one major crack and 12.0% have minor cracks.

In general terms, one of the outcomes of this study returns that, from the analyzed universe of PVMs, 3.0% are in good conditions to be reused, 30.7% show defects but with a high chance to be repaired, which is mandatory before its usage, 27.8% are not recommended for electricity generation but they can be used in other (passive) applications, and 38.6% must be discarded or sent to recycling since they are unrepairable.

4 THE TEST BENCH CONFIRMATION

The approved PVMs were installed in a test bench at the PSDA which emulates an electrical microgrid. This system is displayed in Figure 7.

The test bench is basically an isolated PV system that, as displayed in Figure 7, is composed by 1) a solar installation with 2nd-life PVM, whose STC nominal power is 2.70 kW; 2) a hybrid inverter of 5 kWp with two independent MPPT inputs; 3) two battery packs of 5 kWh each with 4) its respective battery management system (BMS); 5) a configurable electronic load, to create

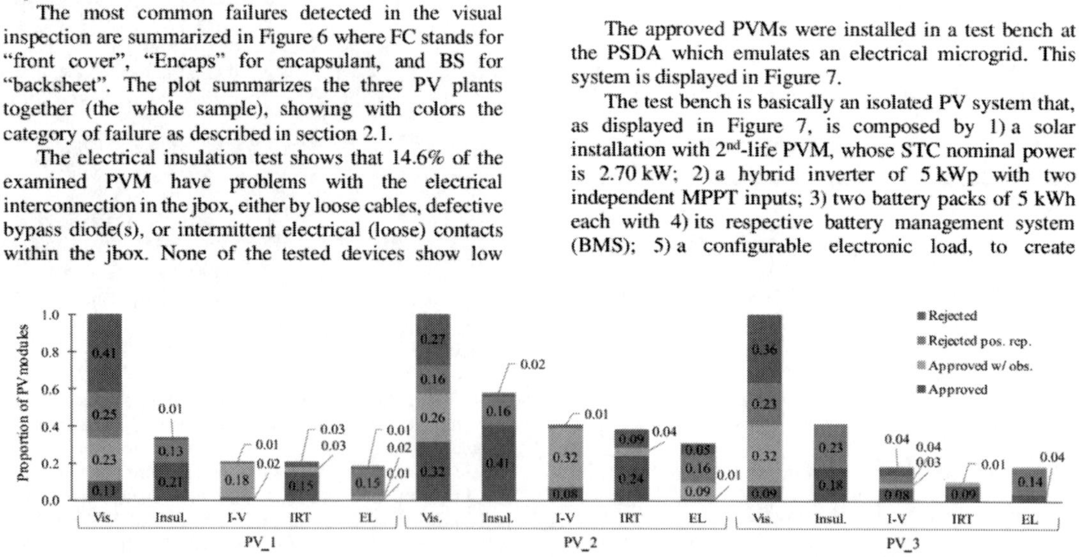

Figure 5: Distribution of modules per step for each PV power plant. Only modules transported to the Laboratory.

consumption power profiles throughout a day. 6) air conditioning to keep room temperature at 20–25 °C range; and 7) the respective electrical panel to enclosure the electrical protections, connections, and busbars.

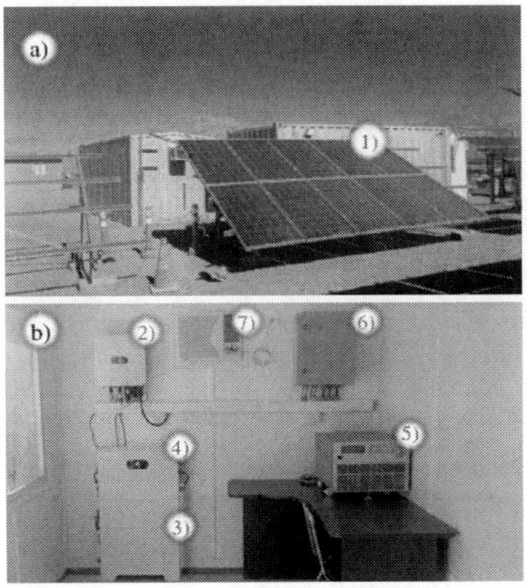

Figure 7: Pictures of the test bench. (a) outside view showing the PVM, (b) inside view.

To avoid large current mismatch in the string, the 2nd-life PVM has been selected only by their I_{MPP}, measured at the 4th step of the procedure.

The whole test bench is monitored by visual inspection and IR imaging of the PVMs, the inverter, the BMS and the configurable load.

Table VI records the presence of anomalies in the IR images during the first month of the study, between the months of June and July 2023. The indicated dates are in month-day format. In this table, July 4th stands out, where all the modules reached temperatures above 70°C (1.3 times the expected temperature).

Measurements of I-V curves are also carried out to check for any PVM anomaly. The module temperature is measured by a PT1000 placed on the outer side of the backsheet, over the area of a healthy cell (without hotspots, without cracks).

Moreover, as shown in Figure 8, the maximum power of the string averages 1.97 kW. This power, corrected to STC, averages 2.22 kW, hence the power degradation of the solar field is ~18% relative to its STC nominal power (2.70 kW). Figure 8 also plots the evolution of the performance ratio as determined at the DC side, which stays almost constant during the two months of testing, averaging to 0.72 ± 0.02.

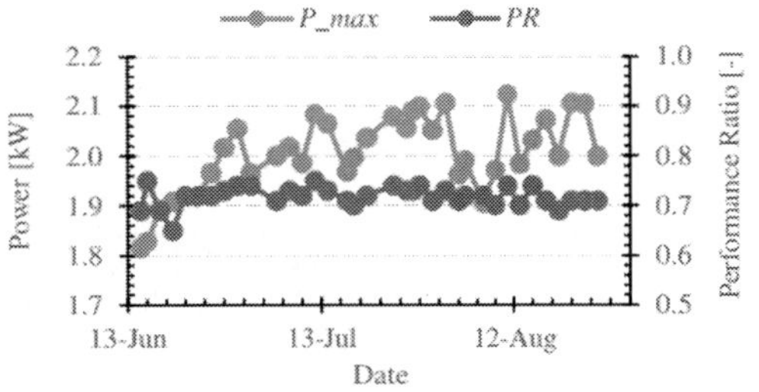

Figure 8: Evolution of maximum power (*P_max*) and performance ratio (*PR*) over time.

Table VI: Registered IRT anomalies for the first month of the test bench operation.

PVM \ Dates	06-14	06-16	06-18	06-20	06-22	06-24	06-26	06-28	06-30	07-02	07-04	07-06	07-08	07-10	07-12	07-14	07-17
M01			●	●				●			●						
M02			●	●				●			●			●			
M03			●		●						●						●
M04			●		●						●						●
M05				◐							●						●
M06										●	●						
M07											●						◐
M08										◐	●	◐					
M09				◐	●						●	●					◐
M10				◐						●	●	●					

● Considerable anomalies (PVM temperature > 70 °C)
◐ Minor anomalies (PVM temperature ranging 65–70 °C)

10.4229/EUPVSEC2023/5DV.2.45
020549-005

Figure 9 plots the evolution of V_{MPP} and module temperature over time, basically to show that P_{max} drops when V_{MPP} does, mainly driven by the mean operational temperature. The red arrow points to the 19th of August as an example for a drop in module voltage caused by an increase in its temperature. From the data is observed that the module temperature averaged (54 ± 6) °C, with an average V_{MPP} of (289 ± 11) V.

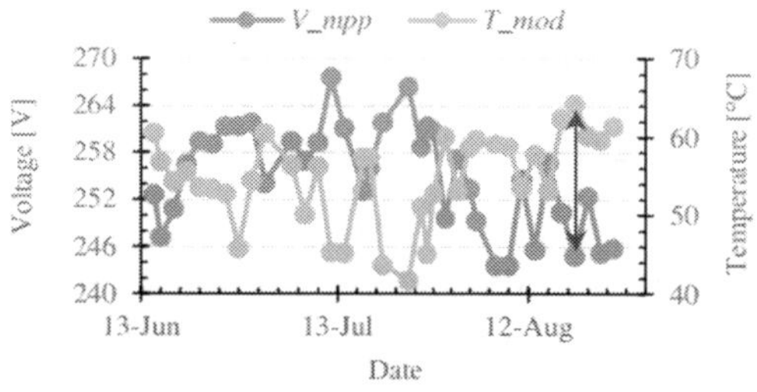

Figure 9: Evolution of voltage at maximum power point (*V_mpp*) and module temperature (*T_mod*) over time. Red arrow remarks the 19th of August.

The results of the measurement campaign and monitoring of the test bench indicate that, although the PVM defects affect its performance, the safety of the system's operation has not been compromised. The system will continue being measured and monitored until it completes 1 year of operation to demonstrate the safety of the system with modules coming from the suggested procedure.

5 CONCLUSIONS

In the present article, we have introduced and validated a procedure to test used PV modules with the aim of classifying such for a second life. The procedure has been developed based on the non-destructive tests of the current IEC standards and recommendations of the IEA-PVPS, and has been applied to examine 1005 PVMs, from which a sample of the approved ones has been mounted in a test bench in the Atacama Desert. The results of the test of this reused PVM demonstrate the effectiveness of the proposed procedure, which enables the PVM for a 2nd life.

From the whole universe of samples, 23.9% were directly rejected at PV_1, 57.9% at PV_2, and 22.3% at PV_3. This implies that 38.6% of the whole universe of analyzed PVM are not suitable for a 2nd life and should directly be sent to recycling. Furthermore, 3.0% of the PVMs are in good conditions to be reused, 30.7% have defects but with possibilities to be repaired, which is mandatory before its usage, and 27.8% are not recommended for electricity generation but they can be used in passive applications.

Observations along this study allow to implicitly conclude that the main defects of the analyzed PVMs comes either from its transportation and storage during the PV plant commissioning, or from the balance of system during operation, either because of an unsuitable design of the PV plant or because a more exhaustive maintenance is needed but was not given. For example, plastic ties that attach PV cables usually break, causing the cables to get loose and tangle around with the mounting and tracker structure. This way the tracker will pull the cable, causing damage either to the cable, to the jbox or to connections inside the box.

Besides, during the study it has been noted that, due to difficulties in the transportation, the most cost-effective and practical method is to collect reusable modules at the dismantling site, with sorting, visual and electrical testing and documentation performed directly upon its removal.

6 ACKNOWLEDGEMENTS

This work is supported by the Chilean Economic Development Agency, CORFO, Antofagasta committee, through the FIC-R funds, with the contract no. 21BPA-189322 under the framework of the project Solar-Circular, and by the Chilean Solar Energy Research Center (SERC Chile), under grant ANID/FONDAP/1522A0006.

7 REFERENCES

[1] ESMAP, "Global Photovoltaic Power Potential by Country," World Bank, Washington DC, USA, 2020.
[2] EnorChile, "Reporte de Emisiones SEN," ACERA, Santiago, Chile, 2022.
[3] L. Ini, "¿500 mil toneladas de desechos fotovoltaicos para 2030 en Chile?," PV Magazine, 6 September 2022.
[4] IRENA and IEA PVPS, "End-of-Life Management: Solar Photovoltaic Panels," International Renewable Energy Agency and International Energy Agency Photovoltaic Power Systems, 2016.
[5] IEA PVPS Task13, "Review of Failures of Photovoltaic Modules," IEA PVPS, 2014.
[6] IEA PVPS Task13, "Review on Infrared and Electroluminiscence imaging for PV field applications," IEA PVPS, 2018.
[7] IEA PVPS Task13, "Qualification of Photovoltaic Power Plants using Mobile Test Equipment," IEA PVPS, 2021.
[8] J. Rabanal-Arabach, "Development of a c-Si Photovoltaic Module for Desert Climates," Universität Konstanz, Konstanz, Alemania, 2019.

PHOTOVOLTAICS VERSUS NEGATIVE EMISSIONS TECHNOLOGIES

Sahil Vadadkar[1,2*], Sankalp Agrawal[1*], Ralf Preu[2]
[1]Albert-Ludwigs-Universität Freiburg, Fahnenbergplatz 79085, Freiburg im Breisgau
[2]Fraunhofer-Institut für Solare Energiesysteme ISE, Heidenhofstraße 2, 79110 Freiburg im Breisgau
*These two authors contributed equally to this work

ABSTRACT: Through the Paris Agreement, the world has acknowledged that climate change is a significant global concern for humankind. There is a need for overwhelming response to counter climate change. While nations have offered roadmaps for decarbonisation of their economies for the long term, significant short-term reductions in carbon emissions will be crucial. Several technical solutions exist in order to assist in this endeavour. With a huge portion of the population located in the equatorial regions, solar photovoltaics offer an appealing solution; however, it is imperative to note the inability of one technological solution to single-handedly meet global demands. This paper examines, *negative emissions technologies* (NETs) and balances their efficacy as tools working in tandem with solar *photovoltaics* (PV) for faster decarbonisation. Compared to certain NETs (up to US$ 600/t-CO2), PV (US$ 34/t-CO2) is found to be more economical. Energetic comparisons show similar results: PV requires only 53 kWh/t-CO2, less than an eight compared to certain NETs.
Keywords: Photovoltaics, Negative emissions technology, Carbon dioxide removal, Bioenergy with carbon capture and storage, Direct air capture.

1 INTRODUCTION

Climate change is a significant global concern. The world is at the precipice of irrevocable climate change if significant changes are not implemented across all sectors. The Paris Agreement, having recognised this matter of urgent concern, has set a target to limit the increase of global average temperature to well below 2°C above preindustrial levels, aiming toward a 1.5°C target [1]. This has brought renewed focus on *Renewable Energy Sources* (RES) as a particular effort to aid in the decarbonisation of the energy sector. *Negative emissions technologies* (NETs) have been recognised as another major aspect to counter the increase in global carbon emissions through the removal of atmospheric carbon, primarily in the form of *carbon dioxide* (CO_2), and achieving these targets [2–6].

Renewable energy is set to play a pivotal role in the future. Already, countries have increased the deployment of renewable sources in the energy sector, particularly wind and solar *photovoltaics* (PV). Whereas hydropower still accounts for majority of the renewable power generation, recent years have seen solar PV as the fastest growing technology (by capacity additions) [7]. For 2030, the *International Energy Agency* (IEA) projects the need for over 5 TW installed solar PV technology – amounting to over 7.4 PWh of energy generation – in order to meet their Net Zero Scenario [8].

Whereas RES work by replacing sources that are greater emitters, thereby avoiding emissions, NETs can be defined as those technologies that are employed to actively remove *greenhouse gas* (GHG) emissions from the environment [9]. Although there are several technologies/methods which can be included in NETs, "*almost all target CO_2*" [9, p. 489]. Subsequently, NETs in this report address, in principle, *carbon dioxide removal* (CDR) techniques. Thus, these refer specifically to "*intentional efforts to remove CO_2 emissions from the atmosphere*" [4, p. 3]. The *Intergovernmental Panel on Climate Change* (IPCC) suggest that 100-1000 Gt-CO_2 must be removed over the 21st century to limit global warming within 1.5°C [10]. Major techniques as identified include *afforestation and reforestation* (AR), *soil carbon sequestration* (SCS), *biochar* (BC), *bioenergy with carbon capture and storage* (BECCS), *direct air capture with carbon capture and storage* (DACCS), *enhanced weather* (EW), ocean alkalinisation and ocean fertilization [4]. Methods of carbon storage such as the use of magnesium cement, soil management, wetland restoration and the use of timber in construction are also considered as NETs [9].

Through this report, we have tried to identify the potential for PV in climate mitigation, study and offer a comparative analysis of various NETs. Finally, we quantitatively compare PV with some NETs based on emissions and costs associated with their deployment.

2 METHODOLOGY

The estimated scales of deployment for NETs and subsequently, the total quantity of removal of CO_2 from the atmosphere vary widely based on the parameters included in the scenario. The largest estimate is of 1200 Gt-CO_2 [9]. Data was collected via online searches, mainly through the '*Google Scholar*' search engine. Accessed data generally includes works addressing NETs, works addressing specific technologies and reports by international organisations on future scenarios vis-à-vis climate action and energy pathways. BECCS, DACCS and AR technologies were chosen for comparison with PV particularly due to the high contribution that is attributed to these technologies in various emissions pathways and reports. Figure 1 illustrates the approach taken in this report.

As an introduction, the case for PV technology and major NETs is briefly discussed. The potential for PV in climate mitigation is then addressed briefly. Results of the

Units
°C: degree Celsius
Gt-CO_2 pa: gigatonne of carbon dioxide per annum
PWh: petawatt-hour
t-CO2: tonne of carbon dioxide
TWH: terawatt-hour
W_p: watt peak
kWh: kilowatt-hour

10.4229/EUPVSEC2023/5DV.2.50
020552-001

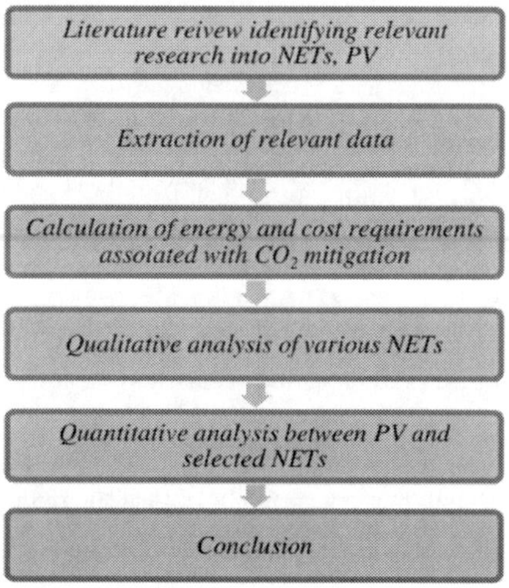

Figure 1: Flowchart depicting the methodology

literature review identify important NETs which have been tabulated in Table 2. Subsequent analysis focusses on selected NETs. A comparison between PV and select NETs is conducted by calculation of the costs and energy associated with the removal/mitigation of 1t-CO_2. Finally, some challenges to widespread implementation of NETs are summarised. The units used during the study have been defined in the box above.

3 RESULTS

3.1 Photovoltaics Technology

As reported by the IEA, PV-generation for the year 2021 exceeded 1000 TWh, owing to a record increase of 179 TWh compared to the previous year (22% growth) [8]. PV accounted for 3.6% of global electricity generation – the third largest source of renewable energy behind wind (over 1.87 PWh) [11] and hydropower (over 4.2 PWh) [12]. To achieve Paris climate goals, the *International Renewable Energy Agency* (IRENA) suggests that PV could account for a quarter of the global electricity, becoming the second-largest source of energy generation [13]. Furthermore, estimates suggest that PV will be the most installed technology (by power capacity), achieving a cumulative power capacity of over 2350 GW by the year 2027 [7], with total PV capacity projections at 8519 GW by 2050 [13].

PV technology is often an integral part of *Integrated Assessment Models* (IAMs) that forecast possible pathways for future energy scenarios. However, [14] have highlighted the underestimation of the contribution of PV technology, focusing on assumed costs during modelling PV technology. They note that reductions in *balance of system* (BoS) components and increase in reliability and lifetimes have led to the fall in the *Levelised Cost of Electricity* (LCOE) [14], while [15] have called the LCOE values for PV technology in the 5th Assessment Report by the IPCC "*outdated*". LCOE values as low as US$ 0.014-0.05/kWh by 2050, have been predicted [13].

Solar PV is a cheap and mature technological option, particularly since PV modules have "*maintained a learning rate of 23% since 1976*" [14, p. 1044] and "*learning rate estimated with data from 2007 is even higher at 40%*" [14, pp. 1044-1045]. The main contributor is high amount of research and development that has led to rapid technological advancement, increase in efficiency, reliability and reduced costs.

Crystalline polysilicon is the dominant technology in the market but since 2021, monocrystalline technology has seen growth and is expected to take over the crystalline polysilicon market [8]. Newer, more efficient technologies using *Tunnel Oxide Passivated Contact* (TOPCon) and heterojunction cell design have seen growth in recent years [8] and can be expected to grow more in future.

3.2 Negative Emissions Technology

Removal of carbon from the atmosphere is an important way to arrest the increasing rate of climate change. Political discussions as early as the 1990s have included carbon removal in the agenda of the *United Nations Framework Convention on Climate Change* (UNFCCC) [16]. Generally, NETs absorb carbon from the immediate environment around it and store it in sinks. These sinks may be local (AR, wetland restoration, soil mineralization and management) or in the form of geological formations (BECCS, DACCS), where the captured carbon may be stored [4, 9].

The inclusion of NETs into IAMs with an aim to achieve net-zero or global emissions targets is not a recent phenomenon. The strategy of removal of greenhouse gases was also recognised by the IPCC in their First Assessment Report [17]. All subsequent reports have recognised the need for the removal of carbon and the use of *carbon capture and storage* (CCS) technologies for the mitigation of climate change [18–21].

Table 1: Selective Negative Emissions Technologies

Technology	Description	TRL	Potential Capacity (Gt-CO_2 pa)	Estimated Costs (US$/t-$CO_2$)
Biochar	Preventing return of biotic carbon via decomposition by storage in soil as partially combusted matter	4-6	0.9-3.0[9] | 2.75-4.95[4]	8-300
DAC – Supported Amines	Pumped or circulated air through solid amines which capture CO_2 through adsorption	3-5	10[9] | 0.5-5[6] | 0-11.01[4]	40-300
DAC – Wet Calcination	Wet scrubbers using calcium/sodium cycling technology to capture atmospheric CO_2	4-6		165-600
BECCS (through combustion)	CO_2 captured from emission of bioenergy sources during combustion	4-6	0.5-5 [6] | 2.4-10[9]	70-250
BECCS (through ethanol/BLG)	CO_2 captured from emissions of bioenergy sources at earlier stages during formation of ethanol	5-6	0.048 (ethanol) | 0.25-0.375 (BLG)	<45
Forest Management	Increase in forest area through new forests and enhanced management to maximise carbon sink	6-7	1.5-3.0[9] | 0.5-75[6] | 0.73-5.5[4]	20-100

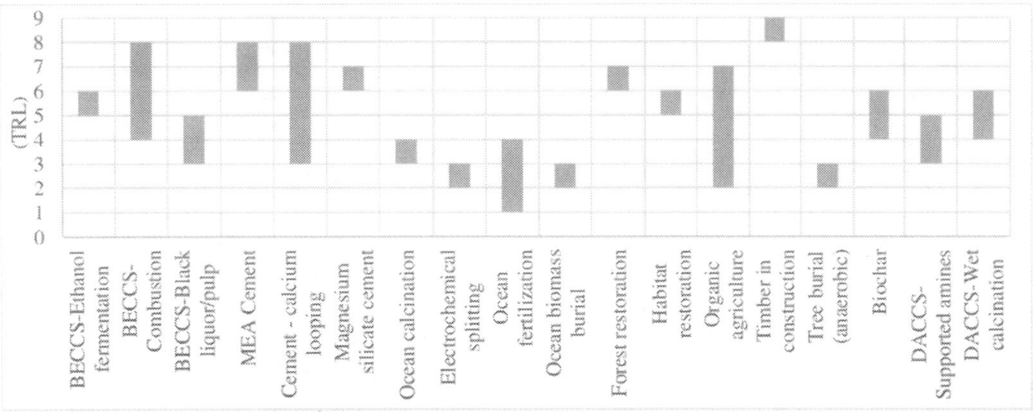

Figure 2: Technology Readiness Levels for various NETs [9, 25]

While some methods are already well-known (AR, SCS), recently other technologies (DACCS, BECCS) have also come under the scientific focus [4]. The research corpus is not evenly distributed amongst technologies. While there is a "*growing reliance on NETs in IAM projections*" [5, p. 2], the lack of available data and the need for further research, particularly as technological case studies and the uncertainty associated with the feasibility of deployment is also a concern [2, 5, 6].

Table 1 shows selected technologies and important parameters characterising them. The larger Table 2 from which Table 1 has been derived, has more NETs and can be found at the end of the report. Various important parameters including the *technology readiness level* (TRL), potential deployment capacity (in Gt-CO_2 pa) and the estimated specific costs (in US$) associated with the technology have been tabulated therein.

The TRL is a system of classification of the maturity of a particular technology as a method to classify technological feasibility [2]. TRL 1 is the lowest and TRL 9 is the highest level. Figure 2 shows TRLs for various NETs from [9, 22]. From Table 1 (*and Table 2*), it is evident that technologies with the greatest potentials (DACCS, BECCS) have low readiness levels (between TRL 3-6) and highest costs associated with CDR. Indeed, as observed by [23] in [2], most NETs have not advanced beyond the prototype/model demonstration phase (TRL 6). Furthermore, technologies which have relatively lower estimated costs (<US$50/t-$CO_2$), have either very low potential (BECCS through BLG, wetland restoration) or very low TRL (soil mineralisation, biomass burial).

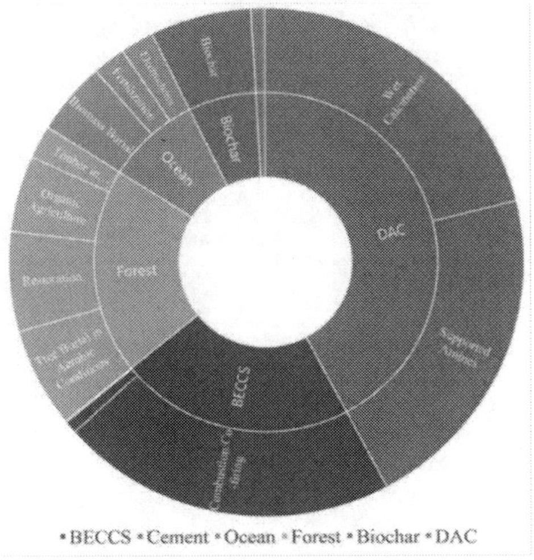

Figure 3: Global Potential Capacity (Higher Value) for NETs [9]

Based on selected data from Table 2, Figure 3 charts the global potential capacities for selected NETs, totalling 47.323 Gt-CO_2 pa. Though DACCS shows the highest global potential in Figure 3, it is important to note that nearly all the modelled NET capacity for various emissions pathways and scenarios comes from BECCS and AR [2, 24] and only some studies have included DACCS [5]. Combining information from Figure 2 and Figure 3 leads to the conclusion that the NETs which account for the greatest stated negative emission potentials (BECCS, DACCS) are not yet proven in deployment. Even in studies where BECCS is projected to contribute the greatest amount for carbon removal (5-20 Gt-CO_2 pa) by 2050, it is pointed out that there are currently no existent systems in deployment [5].

Another important aspect associated with carbon storage is the duration for which the carbon can be stored and the reliability of the sink. High permanency and long-term storage is an important parameter that is seen in technologies like BECCS and DACCS, where the captured carbon is stored in geological sinks which can store carbon for centuries [4, 6, 9, 24]. Biochar, when used as soil amendment, can also act as long-term storage [25]. In contrast, AR, where captured carbon is stored as biomass, is considered a short-term storage option which is additionally susceptible to natural and human disturbances including fires and deforestation [25].

Scenarios studying global emissions until the end of the century estimate total negative emissions in the magnitude of 100-1180 GtCO_2 [4, 10] to 1200 Gt-CO_2 [9] are needed. Figure 2 [9, 22] maps technologies and their readiness levels. It can be seen that forest and habitat restoration, though subject to uncertain achievable rates of negative emissions, are among the readiest techniques [9]. Further, "*[L]arge-scale afforestation and reforestation can make an important contribution to the overall CDR effort*" [26, p. 327]. Timber in construction and cement-based NETs are also at, or close to, practical deployment stages [9].

10.4229/EUPVSEC2023/5DV.2.50
020552-003

Whereas Table 2 offers a more comprehensive view, it can be seen from Table 1 that there are steep costs associated with removal of carbon from the environment. Technologies which have become the focus of mitigation solutions in many scenarios (BECCS, DACCS) have enormous costs associated with their implementation. CDR via DACCS is limited to system demonstrations in relevant environments (TRL 6) and already cost estimates have touched US$ 600/t-CO_2 removed. Worryingly, [9] points out that even these estimates might be *"highly optimistic"* [9, p. 495] and cost estimates for DAC could be as high as US$ 800/t-CO_2. Similarly, in the case of BECCS via combustion, which has the highest potential of all BECCS technologies, cost estimates are as high as US$ 250/t-CO_2 removed. Furthermore, as observed by [27], at present, BECCS and DACCS are operating only at a pilot-project stage and large-scale development in terms of scaling up and cost reduction is crucially needed over the next decade or two.

So far, we have identified important NETs and established the state of their art and noted the potentials for removal of atmospheric CO_2 and the costs associated with it. It is now evident how NETs can contribute toward a net-zero goal.

3.3 Comparison

A part of the analysis conducted included a comparison between PV and selected NETs. The aim was to measure the relative costs and energy consumptions associated with mitigation of 1t-CO_2.

Both PV and NETs offer solutions to reduce global carbon emissions; (I) PV via mitigation through replacement of more polluting sources and (II) NETs via removal of atmospheric carbon through capture and storage. This reduction can be considered indirect in the case of PV and direct in the case of NETs.

To quantitatively study (I), an analysis was conducted to study differences in the emission levels associated with various energy production technologies (Figure 4). The potentials and costs of individual NETs address (II). These have been discussed in the report. Finally, a comparison was also carried between PV and NETs. This comparison was carried out on the basis of costs associated with mitigation/removal of 1t-CO_2 from the atmosphere and the energy required to achieve this.

A brief assessment for different energy sources including Coal, Natural Gas, Petroleum, Nuclear and PV, technologies was conducted. These technologies were compared with PV since the fossil fuels are among the largest contributors to the energy mix, while nuclear is considered a low-carbon source [28]. The GHG emissions associated with generation of electricity were calculated and have been presented in Figure 4 [29]. The difference between emissions associated with fossil sources and nuclear and PV are quite significant. Coal is by far the most polluting of the technologies compared, followed by petroleum and natural gas. It is important to note that, while nuclear energy is a relatively 'clean', low-carbon source of energy, there are major security concerns associated with its use, the discussion of which is beyond the scope of this report.

To address (II), the following approach was applied. With the assumption that the PV technology, when applied, will replace coal, we see that coal produces roughly 1.1 kg-CO_2 more than PV for every unit electricity generated – thus, the generation of 847 kWh energy using coal produces 1t-CO_2. To achieve carbon mitigation, we consider that this energy will be generated using PV technology. Considering generation costs of around US$ 0.04/kWh [30], generation of 847 kWh energy via PV will cost US$ 34. This value has been used in Figure 5, where PV is compared with NETs. Costs for NETs were taken from Table 1. Graphical representation clearly demonstrates the disparity between the costs associated for mitigation using PV and mitigation using specifically designed negative emissions technologies, particularly DACCS. BECCS – ethanol fermentation is the only technology with costs similar to PV; however, it has limited deployment potential (<0.048 Gt-CO_2 pa).

The final parameter that was used as a quantity for comparison between PV and aforementioned NETs was the energy associated with the removal (or mitigation) or 1t-CO_2. In order to do this, we calculate the total PV capacity required for the production of 847 kWh

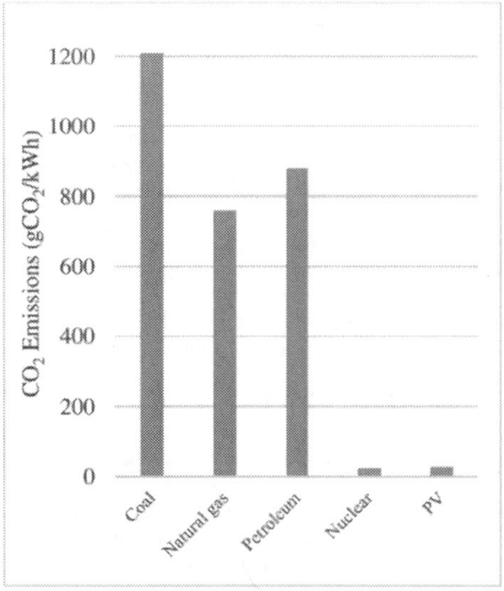

Figure 4: CO_2 Emissions – Electricity Generation (gCO_2/kWh) from [29]

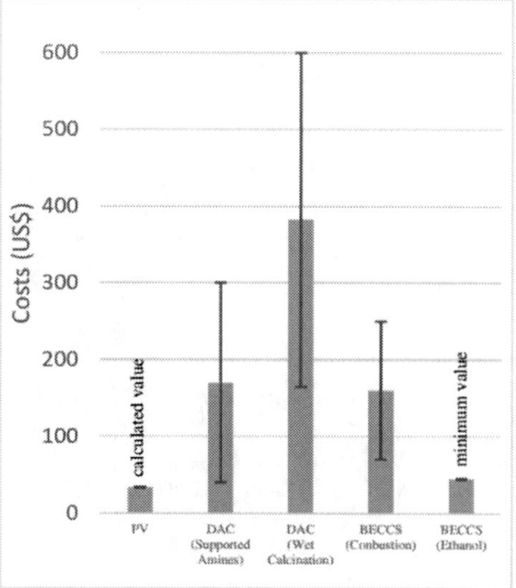

Figure 5: Costs (US$) for removal/mitigation of 1t-CO_2 [9,30]

10.4229/EUPVSEC2023/5DV.2.50
020552-004

throughout its lifetime. For this calculation, power generation of 30 kWh/W_p was assumed throughout the lifetime of the module, considering 1600 hours of peak sunshine in a year, *performance ratio* (PR) of 75% and lifespan of 25 years [31]. Additionally, life cycle energy consumption (including module manufacturing, PV station operation, electricity transmission and other requirements) was assumed to be 1.9 kWh/W_p [31]. Using these values, we calculate a capacity of 28.23 W_p is required, which corresponds to approximately 53 kWh required. Similar data for the costs for NETs for DACCS could be found in [32]. Figure 6 graphically compares the energy associated with removal/mitigation of 1t-CO_2 from the atmosphere. It can be seen that energy requirements for DACCS-solid are more than four times PV while DACCS-liquid are more than eight times those of PV. Thus, we can conclude that PV is the more economical and energetically efficient than the studied NETs.

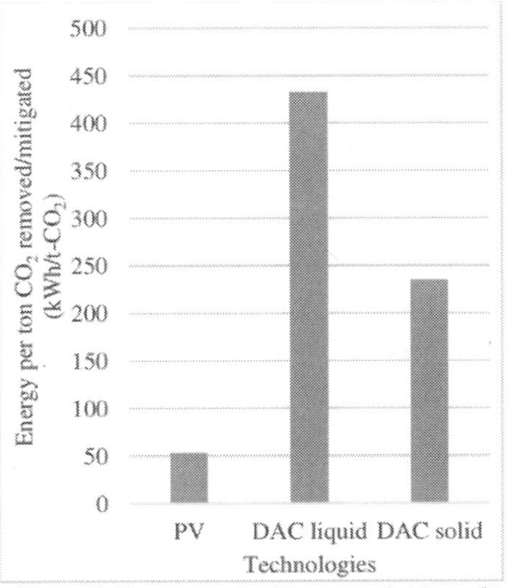

Figure 6: Energy required to remove/mitigate 1 t-CO2. PV requires less than a quarter of the energy required by DACCS [31,32]

3.4 On challenges for large-scale deployment of NETs

This section aims to offer a brief review of the challenges faced by certain negative emissions technologies which need to be addressed as a priority in order for these technologies to meet the levels of deployment suggested. This sense of urgency associated with implementation of NETs is missing in scientific literature, contemporary policy discussions and even policy itself [9].

Technologies such as BECCS, which form a major portion of the CDR estimates, are complex in nature, which itself is a potential barrier [24]. An increase in BECCS and AR would negatively affect land-use, including for agriculture and food [33], while in the case of DACCS, which is an energy intensive method, providing energy using fossil fuels, would render it "*inviable*" [33, 34, p. 1815, 35].

4 CONCLUSION

Climate change is an issue which needs synergetic action across the globe to achieve swift but lasting changes to the atmosphere. Mitigation of future emissions and removal of existing carbon excesses from the atmosphere are important pathways to achieving this goal. RES are predicted to play an important role in reducing further emissions, while deployment of NETs will result in a net reduction of atmospheric carbon. Thus we have seen the need for carbon reduction and carbon removal.

Over the last few decades, PV technology has shown sustained growth and increase in its efficiency. PV remains a promising technology to achieve net-zero emissions targets. NETs have been increasingly crucial in the study of such future pathways.

Through this paper we have identified that there is still untapped potential for growth and proliferation of PV, particularly through the use of more efficient technology (e.g. TOPCon). Analysis shows that PV has very low GHG emissions associated with energy generation. Advancements in auxiliary fields such as battery storage will help in increasing deployment of PV.

From several NETs, BECCS, AR and DACCS have been recognised as the most promising alternatives. The first two account for maximum NET share in IAMs. DACCS, with sufficient research to lower costs and energy requirements is also promising alternative.

Finally, having reported on the state of PV and NETs, we conclude that with as low as US$ 34/t-CO_2 and energy requirements of 53 kWh/t-CO_2, PV has been found to be the better economic and energetic alternative. Ideally, both these technologies should work in synergy to achieve carbon reduction and negative emissions.

10.4229/EUPVSEC2023/5DV.2.50
020552-005

Table 2: Negative Emissions Technologies

Technology Category	Technique	Description	Sequestration Category	Technology Readiness Level	Potential Capacity (Gt-CO_2 pa)	Limiting Factors	Estimated Costs (US$/t-$CO_2$)	Comments	Reference
Mineral	Soil Mineralisation	Acceleration of natural process of carbonation with addition of silicate minerals to soils (or surface water)	*Possibly shorter-term (decades)*	TRL 1-5	1	Land availability	20-40	Requires a change in long-standing practices; full account of land-use change is difficult. Possible synergetic outcome including increase in productivity and yield in biomass	[9]
	Magnesium Cement	Carbon mineralisation during the formation of concrete through the use of magnesium oxides which combine with atmospheric CO_2 when setting	*Possibly shorter-term (decades)*	TRL 6-7	0.4	Demand for cement	'parity' with Portland Cement	Technology is at or close to practical deployment stage	[9]
	Biochar	Preventing return of biotic carbon via decomposition by storage in soil as partially combusted organic matter	Decades to centuries, dependent on soil type, management and environmental conditions	TEL 4-6	0.9-3.0[9] | 2.75-4.95[4]	Sustainable supply of biomass, suitable soil for storage	8-300	Zero water requirements, likely one of the only technologies without a water footprint; possible competition for biomass	[4, 9, 33, 36]
Pressurised	DAC – Supported Amines	Pumped or circulated air through solid amines which capture CO_2 through adsorption	High permanency	TRL 3-5	10[9] | 0.5-5[6] | 0-11.01[4]	Energy supply, storage capacity	40-300	Local dilution of CO_2 might increase difficulty of capture over time	[4, 6, 9, 32, 33]
	DAC – Wet Calcination	Wet scrubbers using calcium/sodium cycling technology to capture atmospheric CO_2		TRL 4-6			165-600		[4, 6, 9, 32, 33]
	BECCS (through combustion)	CO_2 captured from emissions of bioenergy sources during combustion	High permanency, long-term storage	TRL 4-6	0.5-5[6] | 2.4-10[9]	Storage, sustainable biomass supply, suitable facilities	70-250	Large amount of land and water required, would consume a major portion of the world's fertilizer supply; massive transformation of agricultural systems needed to achieve projected deployment levels; projected loss of biodiversity, food security concerns and access to water and energy	[4, 6, 9, 32, 33, 35]
	BECCS (through ethanol/BLG)	CO_2 captured from emissions of bioenergy sources at earlier stages during formation of ethanol		TRL 5-6	0.048 (ethanol); 0.25-0.375 (BLG) | 5.5-11[4]		As little as 45		[4, 6, 9, 32, 33, 35]
Oceanic	Ocean liming (calcination/ electrochemical)	Addition of calcium oxide/ hydroxide/ bicarbonate to surface waters to accelerate uptake of atmospheric CO_2	*Unclear/ unknown*	TRL 3-4	Multiple Gt-CO_2 (calcination); 1 (electro-chemical)	Energy for calcination, supply of $CaCO_3$, vessels/ port facilities	54-64 (calcination); 100-180 (electro chemical)	Promising technology though there is need for additional research and development	[9]

	Ocean fertilisation	Increase in ocean productivity through addition of limiting nutrients like iron, phosphate or nitrogen	Unclear permanence, varying from days/months to millennia	TRL 1-4	0-1[9] | 2[4]	Impact on ocean biology, suitable locations, sustainable supply of nutrients	n/a	Unknown impacts on marine biology, great uncertainty over the effectiveness of these techniques	[4, 9, 33]
Biotic	Forest Management	Increase in forest area through new forests and enhanced management to maximise carbon sink	Saturation of forest lands, high vulnerability to disturbances	TRL 6-7	1.5-3.0[9] | 0.5-75[6] | 0.73-5.5[4]	Land availability	20-100	Potential employment opportunities, agricultural pressure including competition with food crops and changes to land use and farming, permanence of CO_2 at risk due to natural and accidental degradation of carbon stock	[4, 6, 9, 33]
	Wetland Restoration	Rewetting and restoration of peatlands, tidal salt marshes and mangrove swamps to enhance anaerobic storage of dead organic matter	Long-term storage	TRL 5-6	Several hundred Mt-CO_2 pa	Land availability	10-20	Can be combined with Biochar; large uncertainties associated with harmonisation of definition of wetlands (data varies from 2-8% pf global land surface)	[9, 37]
	Soil Management	Better agricultural practices to reduce loss of carbon through oxidation	*Possibly shorter-term (decades)*	TRL 2-7	2.3	Land availability	n/a	Fairly large uncertainties regarding actual achievable net negative emissions; changes to agricultural practices	[9]
	Timber in Construction	Increased use of harvested timber in long-life construction applications	*Possibly shorter-term (decades)*	TRL 8-9	0.5-1	Construction demand, sustainable supply of timber	Negligible in most applications	Ecological resourcing of timber on a large scale is a possible challenge in case of widespread implementation	[9]
	Biomass Burial	Burial of harvested biomass in anaerobic conditions on land or in the deep ocean	*Possibly shorter-term (decades)*	TRL 2-3	1-3.0	Sustainable supply, suitable sites, logistics	7-50	Possible conflict with biomass usage for BECCS technologies	[9]

Note: marked in italics are estimates by the authors.

10.4229/EUPVSEC2023/5DV.2.50
020552-007

References

[1] United Nations Framework Convention on Climate Change, *The Paris Agreement.* [Online]. Available: https://unfccc.int/sites/default/files/english_paris_agreement.pdf (accessed: Jan. 18 2023).

[2] T. Thoni *et al.*, "Deployment of Negative Emissions Technologies at the National Level: A Need for Holistic Feasibility Assessments," *Front. Clim.*, vol. 2, 2020, doi: 10.3389/fclim.2020.590305.

[3] B. K. Sovacool, C. M. Baum, and S. Low, "Determining our climate policy future: expert opinions about negative emissions and solar radiation management pathways," *Mitigation and adaptation strategies for global change*, vol. 27, no. 8, p. 58, 2022, doi: 10.1007/s11027-022-10030-9.

[4] J. C. Minx *et al.*, "Negative emissions—Part 1: Research landscape and synthesis," *Environ. Res. Lett.*, vol. 13, no. 6, pp. 1–29, 2018, doi: 10.1088/1748-9326/aabf9b.

[5] J. Fuhrman, H. McJeon, S. C. Doney, W. Shobe, and A. F. Clarens, "From Zero to Hero?: Why Integrated Assessment Modeling of Negative Emissions Technologies Is Hard and How We Can Do Better," *Front. Clim.*, vol. 1, 2019, doi: 10.3389/fclim.2019.00011.

[6] M. Fajardy, P. Patrizio, H. A. Daggash, and N. Mac Dowell, "Negative Emissions: Priorities for Research and Policy Design," *Front. Clim.*, vol. 1, pp. 1–7, 2019, doi: 10.3389/fclim.2019.00006.

[7] IEA, "Renewable Electricity," IEA, Paris, 2022. Accessed: 30 January, 2023. [Online]. Available: https://www.iea.org/reports/renewable-electricity

[8] IEA, "Solar PV," IEA, Paris, 2022. [Online]. Available: https://www.iea.org/reports/solar-pv

[9] D. McLaren, "A comparative global assessment of potential negative emissions technologies," *Process Safety and Environmental Protection*, vol. 90, no. 6, pp. 489–500, 2012, doi: 10.1016/j.psep.2012.10.005.

[10] IPCC, *Global Warming of 1.5°C. An IPCC Special Report on the impacts of global warming of 1.5°C above pre-industrial levels and related global greenhouse gas emission pathways, in the context of strengthening the global response to the threat of climate change, sustainable development, and efforts to eradicate poverty*. Cambridge, UK and New York, NY, USA: Cambridge University Press, 2018.

[11] IEA, "Wind," IEA, Paris, 2022. Accessed: Feb. 2 2023. [Online]. Available: https://www.iea.org/fuels-and-technologies/wind

[12] IEA, "Hydropower," IEA, Paris, 2022. Accessed: Feb. 2 2023. [Online]. Available: https://www.iea.org/fuels-and-technologies/hydropower

[13] IRENA, "Future of Solar Photovoltaic: Deployment, investment, technology, grid integration and socio-economic aspects: (A Global Energy Transformation: paper)," International Renewable Energy Agency, Abu Dhabi, 2019.

[14] M. Victoria *et al.*, "Solar photovoltaics is ready to power a sustainable future," *Joule*, vol. 5, no. 5, pp. 1041–1056, 2021, doi: 10.1016/j.joule.2021.03.005.

[15] S. Philipps, C. Kost, and T. Schlegl, "Up-to-Date Levelised Cost of Electricity of Photovoltaics: Background from Fraunhofer ISE relating to IPCC WGIII 5th Assessment Report, Final Draft, September 2014," pp. 1–7, Oct. 2014.

[16] W. Carton, A. Asiyanbi, S. Beck, H. J. Buck, and J. F. Lund, "Negative emissions and the long history of carbon removal," *WIREs Clim Change*, vol. 11, no. 6, 2020, doi: 10.1002/wcc.671.

[17] IPCC, "Climate Change: The IPCC Response Strategies," 1990.

[18] IPCC, "Second Assessment Report: Climate Change 1995: Synthesis Report," 1995.

[19] IPCC, "Climate Change 2001: Synthesis Report," 2001.

[20] IPCC, "Climate Change 2007: Synthesis Report," 2007.

[21] IPCC, "Climate Change 2014: Synthesis Report," 2014.

[22] M. Lyons, P. Durrant, and K. K., *Reaching Zero with Renewables: Capturing Carbon*: International Renewable Energy Agency, Abu Dhabi, 2021.

[23] G. Lomax, T. M. Lenton, A. Adeosun, and M. Workman, "Investing in negative emissions," *Nature Clim Change*, vol. 5, no. 6, pp. 498–500, 2015, doi: 10.1038/nclimate2627.

[24] M. Fridahl and M. Lehtveer, "Bioenergy with carbon capture and storage (BECCS): Global potential, investment preferences, and deployment barriers," *Energy Research & Social Science*, vol. 42, pp. 155–165, 2018, doi: 10.1016/j.erss.2018.03.019.

[25] H. K. Jeswani, D. M. Saharudin, and A. Azapagic, "Environmental sustainability of negative emissions technologies: A review," *Sustainable Production and Consumption*, vol. 33, pp. 608–635, 2022, doi: 10.1016/j.spc.2022.06.028.

[26] J. Rogelj *et al.*, "Scenarios towards limiting global mean temperature increase below 1.5 °C," *Nature Clim Change*, vol. 8, no. 4, pp. 325–332, 2018, doi: 10.1038/s41558-018-0091-3.

[27] R. S. Haszeldine, S. Flude, G. Johnson, and V. Scott, "Negative emissions technologies and carbon capture and storage to achieve the Paris Agreement commitments," *Philosophical transactions. Series A, Mathematical, physical, and engineering sciences*, vol. 376, no. 2119, 2018, doi: 10.1098/rsta.2016.0447.

[28] H. Ritchie, M. Roser, and P. Rosado, "Energy: Energy Mix," *Our World in Data*, 2022. [Online]. Available: https://ourworldindata.org/energy-mix

[29] V. Fthenakis and M. Raugei, "Environmental life-cycle assessment of photovoltaic systems," in *The Performance of Photovoltaic (PV) Systems*: Elsevier, 2017, pp. 209–232. [Online]. Available: https://www.sciencedirect.com/science/article/pii/B9781782423362000070

[30] IRENA, "Renewable Power Generation Costs in 2021," International Renewable Energy Agency, Abu Dhabi, 2022.

[31] G. Hou *et al.*, "Life cycle assessment of grid-connected photovoltaic power generation from crystalline silicon solar modules in China," *Applied Energy*, vol. 164, pp. 882–890, 2016, doi: 10.1016/j.apenergy.2015.11.023.

[32] M. Ozkan, S. P. Nayak, A. D. Ruiz, and W. Jiang, "Current status and pillars of direct air capture technologies," *iScience*, vol. 25, no. 4, pp. 1–23, 2022, doi: 10.1016/j.isci.2022.103990.

[33] S. Fuss *et al.*, "Negative emissions—Part 2: Costs, potentials and side effects," *Environ. Res. Lett.*, vol. 13, no. 6, p.

63002, 2018, doi: 10.1088/1748-9326/aabf9f.
[34] F. Creutzig, C. Breyer, J. Hilaire, J. Minx, G. P. Peters, and R. Socolow, "The mutual dependence of negative emission technologies and energy systems," *Energy Environ. Sci.*, vol. 12, no. 6, pp. 1805–1817, 2019, doi: 10.1039/C8EE03682A.
[35] H. J. Buck, "Rapid scale-up of negative emissions technologies: social barriers and social implications," *Climatic Change*, vol. 139, no. 2, pp. 155–167, 2016, doi: 10.1007/s10584-016-1770-6.
[36] P. Smith, "Soil carbon sequestration and biochar as negative emission technologies," *Global change biology*, vol. 22, no. 3, pp. 1315–1324, 2016, doi: 10.1111/gcb.13178.
[37] D. Were, F. Kansiime, T. Fetahi, A. Cooper, and C. Jjuuko, "Carbon Sequestration by Wetlands: A Critical Review of Enhancement Measures for Climate Change Mitigation," *Earth Syst Environ*, vol. 3, no. 2, pp. 327–340, 2019, doi: 10.1007/s41748-019-00094-0.

GENERALIZING A PARAMETERIZED MODEL APPROACH FOR CUSTOMIZED LIFE CYCLE ASSESSMENT OF DIFFERENT PHOTOVOLTAIC TECHNOLOGIES

Mathilde MARCHAND[1], Joanna SCHLESINGER[1], Raphaël JOLIVET[1], Romain BESSEAU[2], Rolf FRISCHKNECHT and Paula PEREZ-LOPEZ[1]

[1]Mines Paris, PSL University, Centre Observation Impacts Energie (O.I.E.), 06904 Sophia Antipolis, France
[2] Joint Research Center Via Enrico Fermi, 2749, 21027 Ispra, Italy
[2]treeze ltd, Uster, Switzerland

E-mail contact: mathilde.marchand-lasserre@minesparis.psl.eu

ABSTRACT: To reach the Sustainable Development Goals (SDGs) adopted by United Nations Member States in 2015, renewable energies have a large potential to decrease environmental impacts typically associated to energy generation systems, such as climate change. Solar photovoltaic (PV) technology is considered economically viable, socially equitable and represents a sustainable and environmentally friendly method of producing energy. Nevertheless, this technology is the subject of much criticism and should be assessed and implemented properly.
Compared to fossil-based systems, PV systems have less or even negligible environmental impacts during the power generation stage. However, the raw materials extraction and the end-of-life stages are not burden-free and are not always included and assessed in published studies. Therefore, Life Cycle Assessment (LCA) is a relevant and widely accepted environmental assessment method as it determines the environmental impacts of a PV system over its whole life cycle.
Data quality is one of the key issues to conduct accurate and representative LCA of PV systems. Although the PV industry has considerably improved over the last 20 years in terms of energy and raw material usage, these improvements are not always reflected in existing databases used in LCA studies. To better consider the fast and continuous progress of the PV industry, generic and flexible models of modular life cycle inventories (LCI) -have been proposed.
This study proposes a parameterization of existing LCIs for the environmental assessment of the most common PV technologies over their whole life cycle. The approach allows the LCIs to be easily adapted to each individual system by modifying the values of the variable input parameters. Based on a previous work on crystalline silicon-based PV systems, this study provides a customized evaluation of PV systems based on silicon and thin-film technologies, that are either ground- or rooftop-mounted in France. Special attention has been paid to the end-of-life stage with the aim of integrating the current practices such as reuse, recycling, screening, and crushing of PV modules.

Keywords: Life Cycle Assessment, photovoltaic, parameterized model.

1 INTRODUCTION

The development of renewable energies is a direct response to Sustainable Development Goals (SDGs) that aim at reducing the power-sector-resulted environmental impacts. The solar photovoltaic (PV) industry is a fast-growing market: between 2010 and 2020, the Compound Annual Growth Rate (CAGR) of cumulative PV installations including off-grid was 34 % [1].

The environmental impacts of PV systems are commonly considered as negligible during the power generation phase in comparison to fossil-based power generation. On the contrary, the environmental assessment of the raw material extraction, manufacturing and end-of-life steps may involve significant impacts [2].

Life Cycle Assessment (LCA) is a recognized methodology to conduct environmental assessments, by quantifying both the consumption of resources and the emissions to the environment for the entire life cycle of a system. In the energy domain, LCA is often used to assess the environmental impact of a power generation system, such as a PV system. This assessment is combined with an evaluation of the lifetime production of electricity. The corresponding environmental impacts are then expressed per produced kWh.

Many LCA of PV systems have been published. Though most of them report greenhouse gases (GHG) emissions varying between 30 and 80 g CO2eq / kWh, Sathaye et al. [3] underlined the great variability of emissions showing results in the literature that vary between 5 and 217 g CO2eq / kWh. This variability can be explained by numerous factors, including differences in the definition of system boundaries and other methodological choices, technological variability of the assessed PV systems, assumptions on the electric mix used for PV module manufacturing, power production variability related to the location of the PV installation during the operation stage and recycling shares of materials in end-of-life stage, among others.

Moreover, energy efficiency and, therefore, the associated environmental performance of manufacturing and operation of PV modules have considerably improved over the last years. However, these improvements are not always reflected in the existing inventories. Indeed, although databases such as Ecoinvent are updated regularly, not all modeled processes are affected by these updates. For example, in recent versions of currently used databases, the modeling of the inverters used to build the life cycle inventory of PV systems in Europe still relies, to a large extent, on data from 2006.

To better integrate the technical evolution and the differences in conception of PV systems, the development of tailor-made and flexible LCA models seems to be a relevant way. One relevant approach is the parametrization of existing LCA as it considers these changes and provides a more realistic environmental assessment.

10.4229/EUPVSEC2023/5DV.2.55
020554-001

2 PURPOSE OF THE WORK

The aim of this research is to obtain a more accurate estimate of environmental impacts of PV systems, by accounting for the most recent development in the manufacturing chain and in the processes of the end-of-life of PV panels.

To the best of our knowledge, this is the first study that presents a complete set of parameterized models for the environmental LCA of current main PV technologies in the market, namely mono and polycristalline, CdTe and CIGs. Moreover, the model allows obtaining results for a large set of impacts i.e., climate change, acidification, eutrophication, human toxicity, land use or water use, among others. The study has put the specific barriers for application of such modelling approaches as spatial and temporal dimensions and the most appropriate approaches to be selected.

Considering and modeling the end-of-life step in current practices in France represent a very important scientific innovation: the specific parametrization of this step aims at integrating the future technical improvement of this recycling process under development.

This study paves the way to provide a global model, regrouping all life steps and all PV technologies via a web platform that would allow users to configure their PV installation as realistically as possible and should help public authorities in the decision-making process.

3 METHODOLOGY

Parameterized LCA models are designed to account for the technological, spatial and temporal variability of systems. These parameterized models rely on a set of input parameters ranging from very few, up to several dozens to describe the inventory flows of the studied system. In the second step of LCA, the inventory analysis, parameterized LCI is built from the following iterative approach [4], depicted in Figure 1:

1. Analysis of the variability of the environmental impacts results published in the litterature;
2. Analysis of the variability of the characteristics of the energy systems to identify the set of relevant variable input parameters and their variability ranges;
3. Analysis of the existing LCIs. These LCIs can be: existing databases, inventories published in litterature and environmental declarations of products;
4. Identification of the processes having the highest contributions to the environmental impacts over the life cycle
5. Analysis of the past and future evolutions relative to the main contributing processess
6. Building of a parameterized LCA model from the existing LCI including parameters for processess with high contributions to environmental impacts or expected to significantly evolve over time.

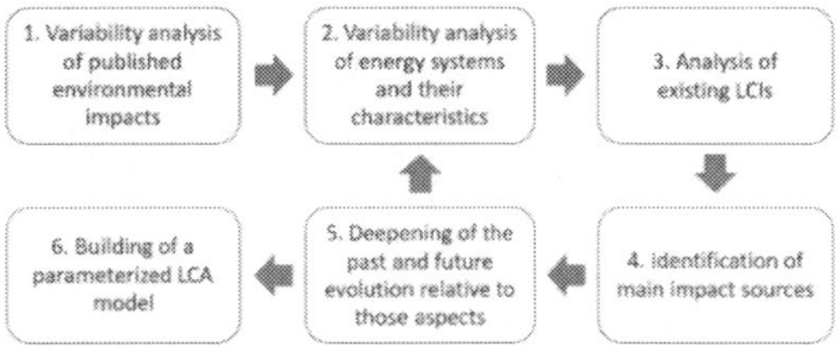

Figure 1: Iterative 6-steps method to develop parameterized LCA models

In our study, the UVEK LCI database developed by Swiss Federal Authorities was used as the existing database upon which the parameterized model was built, as it reflects the most recent developments in relation with the Task 12 of International Energy Agency Photovoltaic Power System Programm [5, 6]. UVEK LCI is based on Ecoinvent 2.2 [7] and presents the latest consensus life cycle inventories among Task 12 authors, PV LCA experts in North America, Europe, Asia and Australia. This database is frequently updated with the last technical advances.

Accordingly, we built the parameterized LCA model from this UVEK database and the identification of PV systems characteristics. The environmental assessment has been performed from impact categories recommended by generic international recommendations and guidelines, and specific to the PV pathway.

In addition to the parameterization work, a specific data collection task has been performed from a new PV recycling industrial process was inaugurated in September 2022 at Envi2E Aquitaine located in Bordeaux in France and is the second installation of this type in the world. This modelling aims to represent the new practices concerning the recycling and reuse of PV modules.

4 PRELIMINARY RESULTS AND DISCUSSION

The preliminary results show the importance of several technical parameters in the estimation of the environmental impacts. Among the most relevant parameters are the electric mix used in different steps, for example during the production of the inverter or the PV cell, the lifetime of the PV installation and the PV module efficiency. The results of the parameterization LCA model highlight a reduction by a factor 2 or more of assessed impacts for a modern PV system compared to the assessment made from the default LCI in existing databases.

The LCA parametrization is built according to several technical assumptions, for example the weight of the electric installation is assumed to be piecewise linear in function to the power of the PV installation, or the weights of materials composing of inverter are assumed to be linear with its power. These assumptions have been shown as accurate regarding the existing LCI. This point illustrates the importance of rapid updates of LCI databases to reflect the representativeness of the technical configurations.

The use of such parameterized LCA models highlights the importance of the choice and representativeness of the electric mix used for PV panel production. This aspect is particularly true when the manufacturing of products as PV modules is made outside Europe in countries with electric mixes dominated by fossil energy.

10.4229/EUPVSEC2023/5DV.2.55
020554-002

5 CONCLUSIONS

The use of the life cycle concept provides a complete environmental assessment over the whole life cycle of a PV power installation and integrating a large set of environmental impacts.
A parameterized LCA model for a PV installation, regrouping crystalline silicon-based and thin-film technologies, was built according to the method proposed by Besseau [4] to provide more accurate estimates of environmental impacts of PV systems in France in 2020. This method is based on existing LCI databases and helps us describe the variability of a set of parameters influencing the score of environmental impacts. LCI data are also developped for modeling end of life of PV system. This work is a contribution to more reliable assessments of the benefits and challenges of PV development, regarding multiple PV technologies within an energy transition context.

6 ACKNOWLEDGEMENTS

This research has been conducted in the framework of the project MaJour_PV with the support of ADEME (France) under contract No. 2005C0034. The authors also want to thank the discussions with the experts of Task 12 of the Photovoltaic Power Systems Program (PVPS).

7 REFERENCES

[1] Franhofer Institute for Solar Energy Systems. 2022. Photovoltaics Report (2022).
[2] S. Tannous, R. Besseau, A. Prieur-Vernat, J. Clavreul, M. Payeur, I. Blanc, P. Pérez-López. A parameterized model for the estimation of Life Cycle Environmental Impacts of Crystalline PV Systems (2019). Proceedings 36th European Photovoltaic Solar Energy Conference and Exhibition (EU PVSEC). Marseille, France
[3] J. Sathaye et al. Renewable Energy in the Context of Sustainable Development (2011). Physics Faculty Publications. Paper 1.
[4] R. Besseau. Analyse de cycle de vie de scénarios énergétiques intégrant la contrainte d'adéquation temporelle production-consommation (2019). PhD thesis. Available at: https://www.theses.fr/2019PSLEM068 (In French).
[5] Bundesamt für Bauten und Logistik. KBOB, eco-bau, and IPB, UVEK Ökobilanzdatenbestand DQRv2:2018 (2018). Available at: https://www.kbob.admin.ch/kbob/fr/home/themen-leistungen/nachhaltiges-bauen/oekobilanzdaten_baubereich.html.
[6] R. Frischknecht, R. Itten, S. Parikhit, M. De Wild-Scholten, J. Zhang, G.A. Heath, C. Olson. Life Cycle Inventories and Life Cycle Assessment of Photovoltaic Systems (2020). International Energy Agency (IEA) PVPS Task 12. Report T12-19:2020, https://doi.org/10.2172/1561526
[7] N. Jungbluth, M. Tuchschmid. Photovoltaics. In: R. Dones (ed.) et al., Sachbilanzen von Energiesystemen: Grundlagen für den ökologischen Vergleich von Energiesystemen und den Einbezug von Energiesystemen in Ökobilanzen für die Schweiz. Ecoinvent report No. 6-XII, Swiss Center for Life Cycle Inventories. Dübendorf, CH, 2007.

DISPOSING END OF LIFE PV MODULES – REUSING, RECYCLING AND UPCYCLING

Pradeep Padhamnath*, Marta Ślęzak, Mirosław Karbowniczek
AGH University of Science and Technology, Krakow Poland
ppadhamnath@agh.edu.pl

ABSTRACT: The exponential growth of photovoltaic (PV) installations is an important and desirable element in the global response to climate change. Increased use of PV panels for energy production would also lead to enormous volumes of PV waste that need to be dealt with in an environmentally responsible manner. Crystalline silicon (c-Si) technology commands more than 95% share of the solar PV industry. The growing amounts of silicon PV panel waste present new environmental and waste management challenges. The issue has been acknowledged and some recycling processes have been developed. However, the complexity of the recycling process is one of the prime reasons hindering their large-scale implementation and growth of the PV panels recycling industry. In this paper we present a review of the existing processes of recycling of end-of-life PV panels and propose a new possible use of recycled silicon from the end-of-life c-Si PV panels.
Keywords: Sustainability, recycling, circular economy, PV panels

1 INTRODUCTION

There has been a rapid growth of photovoltaic (PV) installations in the past decade. Global installed capacity of PV reached 760 GW at the end of 2021 [1] and is expected to reach 8520 GW by 2050 [2]. The exponential growth of the PV installations is an important and desirable element in the global response to climate change. A typical PV module has a lifetime of 20-25 years. However, some will be damaged during production, transportation, handling, or installation, and some will even stop functioning after shorter time, while some would be discarded during testing called as factory reject modules. There is an urgent need to establish efficient measures to deal with the upcoming enormous volumes of PV waste in an environmentally responsible manner. According to the predictions published by International Renewable Energy Agency (IRENA), the cumulative PV module waste could reach 1.7 - 8 million tonnes in 2030 and 60 – 78 million tonnes by 2050 [3] for installed capacity of 4500 GW. Assuming an average panel life of 25 years and the increased production and deployment of PV panels, large amounts of PV wastes are anticipated by early 2030s. It is expected that by 2040, the weight of waste PV panels generated would almost match that of the new installations[3].

The growing amounts of PV panel waste presents new environmental and waste management challenges. However, with proper waste management procedures in place, this challenge could present us with unprecedented opportunities to pursue new technological and economical avenues. Recovery of raw materials from the PV panels could lead to establishment of new solar PV end-of-life industries. In-fact, solar PV recycling would be of paramount importance for the transition into a sustainable and economically viable renewables-based energy future. To unlock the mechanisms and possible advantages of PV panels recycling, it is important to research different avenues of reusing the end-of life PV modules in time to meet the expected surge in the PV panel waste. Hence, it is important to consider the pathways in which the components could be recycled or reused most effectively, minimizing the generation of the pollutants, unusable waste, and contaminated by-products of the module recycling process.

There are several possible ways of handling end-of-life (EoL) PV panels – repairing, reusing, recycling, and upcycling. Some panels have minor damages which could either be repaired onsite or offsite to extend their serviceable life. In a large-scale commercial PV plant, some PV panels degrade earlier than others and are replaced with new PV panels. These PV panels may not have reached their EoL performance, however, since their slightly lower performance compromises the operation of the plant, they are usually replaced. PV power plants may also decide to change the existing panels with a new improved technology. The slightly degraded yet functional panels could be given a second lease of life for non-critical operations. A small fraction of PV panels that are completely degraded in terms of performance yet maintain their physical appearance and integrity could be upcycled for various domestic applications. Bifacial PV panels could be utilized as a replacement of glass panels for some domestic applications such as garden fences, noise barriers and green tunnels. However, the applicability of such upcycling or repurposing options are rather limited, both in number and opportunities. The upcycling of solar panels would simply delay the disposal of the PV panels, rather than disposing them completely. Hence, upcycling of PV panels could not be considered as one of the prime strategies for handling PV waste. A literature survey of these options is presented later, highlighting the advantages and disadvantages of the existing solutions.

Recycling the PV panels to recover the materials is one of the most promising ways of dealing with the PV waste. It has been estimated that by 2050, the materials recovered from recycling of PV panels could support production of 2 billion new panels, equivalent to 630 GW by current performance standards [3]. In many countries, stringent guidelines are already in place to deal with Waste Electrical and Electronic Equipment (WEEE) [4], [5]. Especially the EU has published one of the most detailed regulations related to recycling of PV modules. These regulations stipulate a minimum of 80-85% recovery or reuse of materials [4], [6]. Some of the materials such as the external wiring, glass layer and the aluminium frames could be easily extracted and recycled or in some cases reused entirely[7]. In some cases, the metal busbars (usually made of copper could also be easily extracted using either chemical process or a thermal process or a combination of both [8][9]. Then the remaining parts of the module including the glass and c-Si solar cells are just

10.4229/EUPVSEC2023/5DV.2.62
020557-001

crushed or milled resulting in glass cullet [10] In most of the reported module recycling processes, the silicon obtained at the end of the process is metallurgical grade silicon [10]–[12] Several chemical means could be employed to recover precious metal (usually silver) and solar grade silicon from the mixture which could be recycled [13]–[15]. While chemical treatment to obtain pure silicon appears to be promising and attractive, it has several drawbacks. The process must be tailored for different cell types and architectures. This makes the process extremely complex and prohibitively expensive. Additionally, the chemicals required for this process are highly hazardous and dangerous. The different grades of silicon that can be obtained from a PV module along with the expected revenues are shown in Table 1 [10]. Hence, the end use of the recovered Si from the PV modules would depend on the processing cost of silicon and the value of the product obtained.

Table 1: Revenues from Si of different qualities (value according to Oct-2019) [10]

Si Quality	Purity (%)	Weight (kg/module)	Price ($/kg)	Value ($/module)
Ferro-Si	75	0.68	0.45	0.31
Metallurgical-grade Si	99	0.62	1.5	0.93
Second-grade Si	99.99 99	0.56	5.5	3.09
Solar-grade Si	99.99 99	0.56	7.6	4.24

In this paper we review the published methods of recycling silicon obtained from end-of-life c-Si PV panels and propose new routes to use the Si in metallurgical operations.

2 ONSITE/OFFISTE REPAIR OF PV MODULES

The module is service can fail in several ways, failure in interconnection, delamination, cracked glass or backsheet are among the most common failure modes[16], [17]. Other failures could be in the frames, mounting clamps and support structures and associated electronics [18].

2.1 Repair of panels with cracked backsheets or glass.

The backsheet of a PV module protect the solar cells and connectors from degradation due to environmental factors. It also acts an insulator protecting against electrocution on accidental contact with the underside of the panels. Backsheets available for commercial use could be polyamide (PA), polyvinylidene fluoride (PVDF) or polyethylene terephthalate (PET)-based; comprise multiple stacks laminated together or could be in the form of coextruded foils [19], [20]. Cracking, delamination, and yellowing are the reported to be most commonly occurring backsheet failures [21]–[27]. These defects are mostly visible to naked eye and are the easiest to spot. Modules with defective backsheets compromise the electrical safety of the PV installation and accelerate various PV module degradation modes such as delamination, corrosion, potential induced degradation (PID) or polymer hydrolysis by allowing moisture ingress into the modules[28]–[31]. The maintenance for the backsheets could either be preventive or corrective. Preventive maintenance could be carried out when the damage to the backsheet is minimal (microcracks) and the electrical performance and physical integrity of the backsheet are not compromised. On the other hand, corrective maintenance strategies would be applied when the backsheets have completely cracked and the electrical insulation is compromised. Certain coatings (epoxy, polyurethane, acrylate, silicone, and rubber) could be used to seal cracks and provide sufficient insulation [32], [33]. Small, near surface cracks in the backsheet could also be covered using certain tapes for small cracks, provided the electrical properties of the module are not compromised. Despite the easy applicability of these solutions, the long-term stability of such solutions is not yet established. Additionally, these solutions appear to work only if the cracks in the module have not caused any further degradation of the modules such as corrosion and delamination [32].

Like backsheet, glass failure in glass-glass modules could also damage the insulation of the encapsulant layer and cause water ingress, severely compromising the performance of the PV module[34]. The glass defects in PV modules could be cracked or broken glass layers, caused by human or extreme weather factors. Human factors may include mishandling during transportation, installation, or maintenance of the modules[35]. Weather factors such as hailstorms, strong wind and heavy snow load may also cause damage to the glass layer[36]. Glass failures accounted for 10% of all module failures within two years of operation which increased to account for 33% of all module failures within eight years of operation [37]. Nevertheless, glass-glass PV modules are more resilient to cell-damage and water ingress related issues, and hence, can endure some glass defects without exhibiting considerable loss in the performance [37]. Hence, information related to the repair of the glass defects in PV module in published literature is limited.

The cracks in the glass can be repaired by using an epoxy resin cured under ultraviolet light source [37]. The technique could be applied at the place of installation and dismantling the panels might not be necessary. Repairing the panels in-situ could be the most cost-effective option for glass-glass panels [37]. However, the effectiveness of the repair process depends on the number and type of the cracks in the glass on the front and the rear. Additionally, some discolouration of the resin might occur over time which may slightly reduce the power output [37].

2.2 Repair of cell interconnection in PV modules

The interconnections in a PV module refer to the soldered joints on the copper or silver ribbons. Corrosion of these interconnection can occur in the presence of moisture and high temperatures [38]–[40]. Ethylene vinyl acetate (EVA) which is used as an encapsulant reacts with moisture under high temperature and could lead to production of acetic acids which causes corrosion of the metal grid on solar cells, soldered joints of the electrodes, and joints between interconnection ribbons and busbars [41]–[44]. These defects are usually identified using electroluminescence or photoluminescence imaging techniques and could be done on-site[45]–[47].

Repairing of the defects in the joint between the bus bars of the solar cells and the interconnecting ribbon have been tried using an induction heater providing local heating to a small area under application.[48]. It was shown that duration of heat application was critical to the performance of the repaired module. This method could be applied on site, without the need for dismantling and

10.4229/EUPVSEC2023/5DV.2.62
020557-002

transporting the panels. Results regarding long term reliability of this repair technique would further prove its effectiveness.

In some cases, a faulty solar cell could be replaced leading to rejuvenation of the module[49]. However, most of the current modules are not designed for easy solar cells replacement and hence this may not be one of the easiest options for repairing the PV module [18], [50].

The suitability of refurbishing a PV depends on the type and mode of failure and the extent of damage to the PV panel. Additionally, long term reliability of the repaired modules is often misrepresented or misinterpreted. The situation is further complicated due to absence of standard of requalification or re-certification of repaired modules [18].

3 RECYCLING OF PV PANELS

The growing amounts of PV panel waste presents new environmental and waste management challenges. However, with proper waste management procedures in place, this challenge could present us with unprecedented opportunities to pursue new technological and economical avenues. Recovery of raw materials from the PV panels could lead to establishment of new solar PV end-of-life industries [3]. In-fact, solar PV recycling would be of paramount importance for the transition into a sustainable and economically viable renewables-based energy future. To unlock the mechanisms and possible advantages of PV panels recycling, it is important to research different avenues of reusing the end-of life PV modules in time to meet the expected surge in the PV panel waste.

Multiple government, industrial and scientific bodies within EU have funded extensive research to establish methods to recycle crystalline silicon (c-Si) PV panels. A notable example is the LIFE project within which researchers from SASIL SpA and European Commission Joint Research Centre developed a "Full Recovery End of Life Photovoltaic" (FRELP) project to recycle c-Si PV modules [9], [11], [51]. As a result of extensive research on PV module recycling and the growing interest, the world's first c-Si PV module recycling plant has been established in France by Veolia in 2017[52]. Almost all research activities on module recycling have been focused on c-Si PV modules due to the fact that c-Si technology commands more than 95% of the solar PV industry[53].

Fig.1 shows the mass and value of the materials that could be recovered from a typical c-Si PV panel[11], [54] .

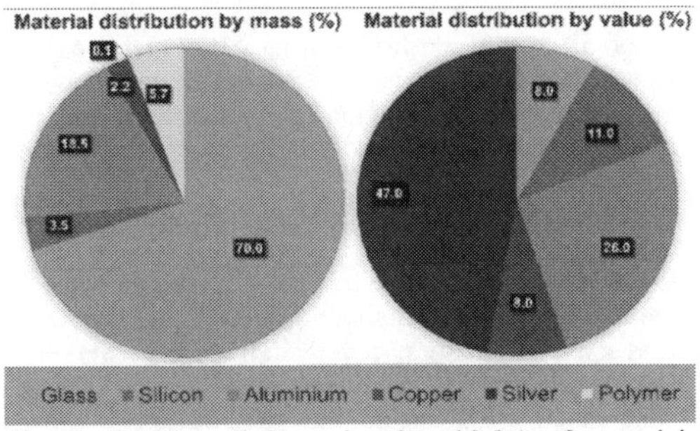

Figure 1: Mass (left) and value (right) of materials recoverable from a c-Si PV module.

Glass and aluminium frame together contribute about 85% of the mass of the solar panel, but account for ≈ 20% of the value. Silver is by far the most valuable metal recovered from the module; however, its extraction is usually not straightforward. The silver is present in connecting ribbons, and in the metal grid on the cSi solar cells which is an integral part of the solar cell. In a physical recycling process, such as that used in Veolia plant, the silicon solar cells along with the metal grid (Ag, Al) is usually the product [52]. The materials recovered from the PV panels in the increasing order of complexity could be Aluminium (frame) < glass < copper conductors < silicon/silver.

3.1 Recycling processes of c-Si PV panels

A full-scale recycling plant was operated by Veolia for recycling silicon-based PV modules[52]. In the Veolia process, the modules are shred and grounded into fine grits. The different materials are then separated by using a proprietary mechanical separation process. Although the exact process is not known, it has been known to include sieving, sifting, optical separation steps and an eddy current separator to recover metallic parts. The process claims 95% of materials recovery rate. Aluminium frames and copper wires are completely recovered and recycled. Tempered glass is obtained in the form of glass powder and sold to glass making industries. The recycled Si solar cells are sold to precious metal sectors as silicon cake [10], [52], [55].

Another process that has been tried on pilot scale has been the FRELP process developed by an Italian company SASIL S.p.A [9], [11]. The process is more complicated than the Veolia process. After the removal of the aluminium and the junction box, the panels are heated to 120 °C using an infrared heater. The modules are then passed through a milling machine and a vibrating knife machine which removes the glass from the front side of the module. The glass is recovered by sieving. Contaminated glass (>2% by mass of impurities) is not recycled and is disposed of. The remaining module is further fragmented into smaller pieces and undergo a high temperature pyrolytic process to burn off the polymeric materials. The ash thus obtained is sieved to remove pieces of glass and then treated with nitric acid to leach out the metals (Ag, Cu). The mixture undergoes vacuum filtration process to remove silicon from the leaching solution, and the leachate undergoes electrowinning process, followed by calcium hydroxide treatment to recover the precious metals. Any silicon obtained after this process is low grade [10]. A pilot-scale FRELP facility was set up with a processing capacity of 1300 panels/day. The process does not recover any lead or tin that might be present in the panel.

A process was also developed independently in the US, aims to achieve high value materials [56]. The process could obtain metals such as silver, lead, tin, and copper from silicon solar cells in laboratory scale demonstration. The frames, cables and plastic junction boxes were mechanically separated. Then a thermal combustion process was used to burn off the organic compounds leaving behind separate the silicon solar cells, glass, and residues[57]. The silicon solar cells are treated with nitric acids to leach out the metals. A two-step electrowinning process was used to recover silver, copper, and lead oxide. The silicon wafers were further etched by hydrofluoric acid and sodium hydroxide to remove the metal grid (Al) and passivation layers to recover useable silicon. The metal removed in the final step were not recovered. Intact silicon wafers and glass were reported recovered by the ASU process. However, the purity and physical integrity

10.4229/EUPVSEC2023/5DV.2.62
020557-003

of the silicon obtained remains uncertain.

To recover c-Si from end-of-life modules to be used for making solar cells again, all the external impurities need to be removed. These include the Ag and Al metal contacts, the anti-reflection coatings (ARC) deposited on the solar cells (e.g. silicon nitride (SiNx), silicon oxide (SiOx), aluminium oxide (AlOx)), and the heavily doped layers from the front and the rear sides of the solar cells. The bulk metal can be removed by treating a metallized sample with hot HNO_3 (70% wt., 70°C)[58]. The glass interface can be dissolved in hot HF (10% wt. at 50°C), while the embedded crystallites can be removed by dipping the samples again in hot HNO_3 bath[59]–[61]. The ARC could be removed by dipping the cells in hot HF. To remove the heavily doped regions, the silicon thus obtained is further treated with a mixture of $HF/HNO_3/CH_3COOH$ [15], [62]. The concentration and duration of etch could vary. The silicon obtained after his process could be recycled for making c-Si solar cells, albeit with reduced performance[63]. The images from scanning electron microscope (SEM) shown in Fig. 2 highlight the steps in obtaining recycled metal free c-Si wafers from end-of-life solar cells.

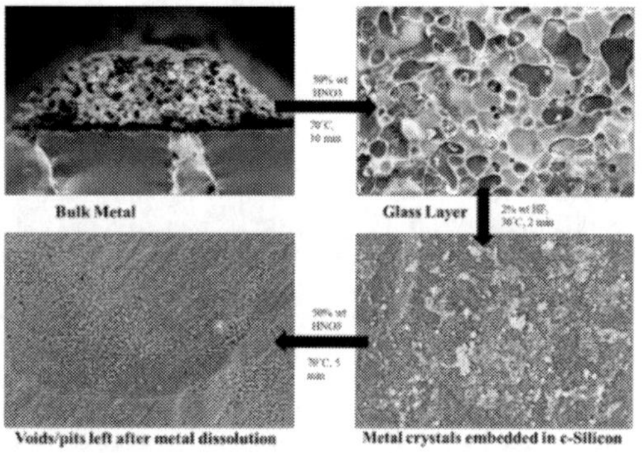

Figure 2: SEM images of the solar cell sample at various stages of etching.

3.2 Alternative uses of glass from recycled PV panels

Reintroducing the materials recovered from the recycled c-Si PV panels into manufacturing supply chain of new PV panels could appear to be the ideal way to achieve circularity. This is referred to as closed loop recycling [54]. However, it might not be always feasible. For example, there may not be a cell/module production facility close to the recycling facility. However, there are several other possible uses of the materials recovered from the EoL c-Si PV panels. These options, referred to as 'open loop recycling' options could also result in high, medium or low value products, depending on the recycling effort [54] .

Glass obtained from recycled PV panels could have multiple uses. The glass could be used to remanufacture PV glass, which is an example for closed loop recycling. Further the glass could be used as a raw material in the manufacture of bottle glass (high value), glass foam for insulating applications (medium value) and glass for road construction applications (low value) [54]. Moreover, these options could be cascaded providing several cycles of recyclability. For example, glass waste from recycling PV modules could be used for bottle manufacturing, while a small non-usable fraction could be sent for preparation of insulating material. Glass still unusable at this stage could be used in road construction. A feasibility analysis based on multiple scenarios for such an application was recently published [54].

3.3 Alternative uses of Silicon from recycled PV panels

The Si recovered from all module recycling processes developed so far is metallurgical grade [10]–[12]. This silicon could easily be used in the metallurgical processes. For example, commercial grade ferrosilicon (FeSi) could easily be manufactured from the recycled Si obtained from EoL PV panels by melting it with scrap steel. This could be a simple method of recycling silicon from PV panels as it involves minimum processing steps and eliminates the use of toxic and hazardous chemicals. The FeSi is further used as deoxidizer, or Si carrier in making other metal alloys. We hypothesize that silicon obtained from end-of-life, discarded or non-functional PV panels can be effectively recycled by metallurgical processes to produce useful alloys for metals industry.

Currently, Ferrosilicon (FeSi) is produced using carbothermal process, usually in closed submerged electric arc furnaces. A typical furnace may have the power rating of 25 -100 MVA. Quartzites are used as the silicon source in the production of FeSi. The quartzites lumps are prewashed, crushed and graded. Typical composition of quartzites considered suitable for the production of FeSi contains around 97% SiO_2 and 1.5% Al_2O_3 [64]. Coke (coal or wood) is the primary source of carbon reductant . The iron lowers the partial pressure of SiO required for reduction and forms FeSi solutions that reduce the activity of Si. Hence the process could be carried out at lower temperatures that employed for the production of silicon. This also results in higher Si yield (less of SiO) and lower energy consumption per ton of metal [65], [66].

4 REUSE AND REPURPOSING OF PV MODULES

4.1 Reuse of PV panels

A number of modules from the large-scale PV power plants are decommissioned for various reasons. They could be defective or could be replaced with an improved technology. Reusing these modules, with or without some repairs, could prevent the panels from going to waste and generate clean energy for a longer period. The European Waste Framework Directive emphasizes the importance of avoiding waste disposal and reusing the discarded items.

For a PV module to be reused, it should be severely degraded and there should be no safety concerns. Additionally, the PV panel should have enough usability to justify efforts employed to make it reusable.

The biggest challenge for reusing the PV panels is the economic viability. The PV panels for reuse would not usually carry a warranty. Additionally, they may have lower efficiency and shorter lifetime. Hence, sorting and classifying the panels is necessary to ensure economic viability.

The rapid decline in the module costs in the last decade and the continuing trend of lower module costs [53], poses further challenges for the reusability of the modules. The module contributes to approximately 28-30% of the total cost of the PV energy system[67]. Hence, the costs savings from reusing the PV module would not be significant as compared to the balance-of-system (BoS) costs.

It has been suggested that the still functioning PV modules after repair (if necessary), could either be used in small commercial systems, residential PV system or off grid PV systems in developing countries[68]. However, the transport costs and the recommissioning costs need to be considered for such options to be commercially viable. Reusing PV modules may delay the generation of PV waste and hence reduce the pressure on the recycling/disposing systems. Reusing the panels may also provide sufficient time for the establishment and improvement of the recycling technologies. Developing regulations governing the decommissioning, repair and reuse of the PV modules are also important for the success of re-usable PV market [68].

Social acceptance towards reusing the PV modules is crucial to the success of such interventions. Social behaviour could play an important role in developing and sustaining a second-hand PV module market. It has been known that psychological and behavioural traits can undermine the viability of technical solutions [69]. The success of reusing the PV modules would need coordination and alignment between the PV manufacturers, PV installers, PV system owners and PV repairer/recyclers. Additionally the role of governing bodies and agencies would needed to support the efforts by developing suitable regulations. As with most scenarios, increased probitability in terms of lowser reuse costs and higher incentives, could be the main drivers of reusing the PV modules [68], [69].

4.2 Repurposing the EOL PV panels.

Another possible way to use the EoL PV modules could be repurposing them after the end of their effective usable periods. The PV panels (especially glass-glass panels) that still maintain their physical integrity could be used for certain applications where glass is used. The EoL PV panels could be used for making noise barriers, garden fences, greenhouse tunnels and car park roofs. Some old PV panels could be repurposed to make benches, tables and animals' shelters. All these applications could also employ good PV panels to generate electricity, however, as mentioned earlier, since PV panels constitute only 30% of the total systems costs, they would need investment. Replacing standard glass with old PV modules could relieve the pressure on the recycling ecosystem while still contributing to reduction of greenhouse gases by offsetting the glass required for such applications. However, commercial viability and social acceptance would still be the deciding factors for repurposing the PV modules.

5 CONCLUSION

Rapid deployment of PV systems is a positive response to combat climate change. However, in the face of exponentially growing PV modules installations, effective management strategies of EoL PV modules are also required. Recycling the PV modules than using them for landfill would prevent contamination and pollution of soil and water. It would also result in the generation of an alternate industry with huge revenue potential.

EoL PV modules could be repaired onsite or off site. Most repairs are concerned with the cracked backsheet or glass. The glass-glass modules are especially resilient to damage and could perform well with minor damages. The repair of cell interconnections is also possible to some extent. It was found that only certain damages to the modules are repairable. However, most repair strategies are at developing stages and more research is required to assess the long-term reliability of such maintenance strategies.

Recycling the PV modules appears to be the most sustainable and effective option of managing the EoL PV modules. The Al frame, glass, copper wires and metal connectors could be easily recycled with simple physical processes. Together they constitute ≈ 85% of the panel weight. Silicon solar cells including the metal could be more difficult to recycle and require more sophisticated chemical treatment. The recycled materials could be used again for producing PV panels (closed loop recycling) or could be used to manufacture other products (open loop recycling). Some examples of open loop recycling could be using the recycled glass from PV modules in manufacturing bottles or glass fibers and using the Si from the recycled panels in metallurgy industry (e.g. Production of ferrosilicon).

Certain modules that are decommissioned from a PV plant, could be re-used, with or without minor repairs. These modules could be used, for instance, for domestic applications, small stand-alone PV plants or could be exported for use in developing countries. However, these options would be highly governed by the prevailing economic, legislative and social conditions. The modules that have electrically degraded but still maintain their physical integrity and appearance could be repurposed for certain domestic and civil applications replacing glass panels. Come common examples could be noise barriers, fences, greenhouses, and car sheds.

Effective management of EoL PV panel sis crucial to maintain the sustainability of the clean energy generation. For a successful ecosystem involving repairing, reusing, or recycling, all stakeholders such as PV manufacturers, PV installers, PV system owners and PV repairer/recyclers will need to collaborate closely to ensure the sustainability of PV energy generation.

5 ACKNOWLEDGEMENTS

This research is part of the project No. ***2022/45/P/ST5/02712*** *co-funded by the National Science Centre and the European Union Framework Programme for Research and Innovation Horizon 2020 under the Marie Skłodowska-Curie grant agreement No. 945339. For the purpose of Open Access, the author has applied a CC-BY public copyright licence to any Author Accepted Manuscript (AAM) version arising from this submission.*

6 REFERENCES

[1] G. M. A. 'Detollenaere, I. 'Kaizuka, A. J. 'Waldau, and J. 'Donoso, "IEA-PVPS Snapshot of Global PV Markets 2021," Apr. 2021. Accessed: Sep. 08, 2023. [Online]. Available: https://iea-pvps.org/snapshot-reports/snapshot-2021/

[2] IRENA (2019), "Future of Solar Photovoltaic: Deployment, investment, technology, grid integration and socio-economic aspects," Abu Dhabi, Nov. 2019. Accessed: Sep. 08, 2023. [Online]. Available: https://www.irena.org/publications/2019/Nov/Future-of-Solar-Photovoltaic

[3] S. Weckend, A. Wade, and G. Heath, "End-of-Life Management: Solar Photovoltaic Panels," Jun. 2016. Accessed: Sep. 08, 2023. [Online]. Available:

https://www.irena.org/publications/2016/Jun/End-of-life-management-Solar-Photovoltaic-Panels
[4] F. Cucchiella and P. Rosa, "End-of-Life of used photovoltaic modules: A financial analysis," *Renewable and Sustainable Energy Reviews*, vol. 47, pp. 552–561, 2015.
[5] P. Dias, S. Javimczik, M. Benevit, H. Veit, and A. M. Bernardes, "Recycling WEEE: Extraction and concentration of silver from waste crystalline silicon photovoltaic modules," *Waste Management*, vol. 57, pp. 220–225, 2016.
[6] R. Deng *et al.*, "Remanufacturing end-of-life silicon photovoltaics: Feasibility and viability analysis," *Progress in Photovoltaics: Research and Applications*, vol. 29, no. 7, pp. 760–774, 2021.
[7] R. Einhaus *et al.*, "Recycling and Reuse potential of NICE PV-Modules," in *2018 IEEE 7th World Conference on Photovoltaic Energy Conversion (WCPEC)(A Joint Conference of 45th IEEE PVSC, 28th PVSEC & 34th EU PVSEC)*, IEEE, 2018, pp. 561–564.
[8] D. Zeng, M. Born, and K. Wambach, "Pyrolysis of EVA and its application in recycling of photovoltaic modules," *Journal of environmental sciences*, vol. 16, no. 6, pp. 889–893, 2004.
[9] C. E. L. Latunussa, F. Ardente, G. A. Blengini, and L. Mancini, "Life Cycle Assessment of an innovative recycling process for crystalline silicon photovoltaic panels," *Solar energy materials and solar cells*, vol. 156, pp. 101–111, 2016, Accessed: Sep. 08, 2023. [Online]. Available: https://www.vdma.org/international-technology-roadmap-photovoltaic
[10] M. Tao *et al.*, "Major challenges and opportunities in silicon solar module recycling," *Progress in Photovoltaics: Research and Applications*, vol. 28, no. 10, pp. 1077–1088, 2020.
[11] C. LATUNUSSA, L. MANCINI, G. BLENGINI, F. ARDENTE, and D. PENNINGTON, "Analysis of material recovery from silicon photovoltaic panels," 2016.
[12] K. Komoto *et al.*, "End-of-life management of photovoltaic panels: Trends in PV module recycling technologies," National Renewable Energy Lab.(NREL), Golden, CO (United States), 2018.
[13] G. A. Heath *et al.*, "Research and development priorities for silicon photovoltaic module recycling to support a circular economy," *Nat Energy*, vol. 5, no. 7, pp. 502–510, 2020.
[14] F. C. dos Santos Martins Padoan, P. G. Schiavi, G. Belardi, P. Altimari, A. Rubino, and F. Pagnanelli, "Material flux through an innovative recycling process treating different types of end-of-life photovoltaic panels: demonstration at pilot scale," *Energies (Basel)*, vol. 14, no. 17, p. 5534, 2021.
[15] E. Klugmann-Radziemska and P. Ostrowski, "Chemical treatment of crystalline silicon solar cells as a method of recovering pure silicon from photovoltaic modules," *Renew Energy*, vol. 35, no. 8, pp. 1751–1759, 2010.
[16] M. Köntges *et al.*, "Review of failures of photovoltaic modules," 2014.
[17] C. Ferrara and D. Philipp, "Why do PV modules fail?," *Energy Procedia*, vol. 15, pp. 379–387, 2012.
[18] J. A. Tsanakas *et al.*, "Towards a circular supply chain for PV modules: Review of today's challenges in PV recycling, refurbishment and re-certification," *Progress in Photovoltaics: Research and Applications*, vol. 28, no. 6, pp. 454–464, 2020.
[19] A. Omazic *et al.*, "Increased reliability of modified polyolefin backsheet over commonly used polyester backsheets for crystalline PV modules," *J Appl Polym Sci*, vol. 137, no. 30, p. 48899, 2020.
[20] K. J. Geretschläger, G. M. Wallner, and J. Fischer, "Structure and basic properties of photovoltaic module backsheet films," *Solar Energy Materials and Solar Cells*, vol. 144, pp. 451–456, 2016.
[21] A. Omazic *et al.*, "Relation between degradation of polymeric components in crystalline silicon PV module and climatic conditions: A literature review," *Solar energy materials and solar cells*, vol. 192, pp. 123–133, 2019.
[22] G. C. Eder *et al.*, "Error analysis of aged modules with cracked polyamide backsheets," *Solar Energy Materials and Solar Cells*, vol. 203, p. 110194, 2019.
[23] W. Gambogi *et al.*, "A comparison of key PV backsheet and module performance from fielded module exposures and accelerated tests," *IEEE J Photovolt*, vol. 4, no. 3, pp. 935–941, 2014.
[24] M. Halwachs *et al.*, "Statistical evaluation of PV system performance and failure data among different climate zones," *Renew Energy*, vol. 139, pp. 1040–1060, 2019.
[25] M. Köntges, G. Oreski, U. Jahn, M. Herz, P. Hacke, and K.-A. Weiß, *Assessment of Photovoltaic Module Failures in the Field: International Energy Agency Photovoltaic Power Systems Programme: IEA PVPS Task 13, Subtask 3: Report IEA-PVPS T13-09: 2017*. International Energy Agency, 2017.
[26] A. Fairbrother *et al.*, "Differential degradation patterns of photovoltaic backsheets at the array level," *Solar Energy*, vol. 163, pp. 62–69, 2018.
[27] J. Tracy *et al.*, "Survey of material degradation in globally fielded PV modules," in *2019 IEEE 46th photovoltaic specialists conference (PVSC)*, IEEE, 2019, pp. 874–879.
[28] L. Bruckman *et al.*, "Backsheets: Correlation of Long-Term Field Reliability with Accelerated Laboratory Testing," Underwriters Laboratories (UL), Northbrook, IL (United States), 2019.
[29] M. Kontges *et al.*, "Review of Failures of Photovoltaic Modules Report IEA-PVPS T13-01: 2014," *International Energy Agency-Photovoltaic Power Systems Programme*, 2014.
[30] T. J. Trout *et al.*, "PV module durability-connecting field results, accelerated testing, and materials," in *2017 IEEE 44th Photovoltaic Specialist Conference (PVSC)*, IEEE, 2017, pp. 2312–2317.
[31] M. A. Quintana, D. L. King, T. J. McMahon, and C. R. Osterwald, "Commonly observed degradation in field-aged photovoltaic modules," in *Conference Record of the Twenty-Ninth IEEE*

Photovoltaic Specialists Conference, 2002., IEEE, 2002, pp. 1436–1439.

[32] Y. Voronko *et al.*, "Repair options for PV modules with cracked backsheets," *Energy Sci Eng*, vol. 9, no. 9, pp. 1583–1595, 2021.

[33] G. Beaucarne, G. Eder, E. Jadot, Y. Voronko, and W. Mühleisen, "Repair and preventive maintenance of photovoltaic modules with degrading backsheets using flowable silicone sealant," *Progress in Photovoltaics: Research and Applications*, vol. 30, no. 8, pp. 1045–1053, 2022.

[34] P. Verlinden, "Advanced module concepts," *Photovoltaic Solar Energy: From Fundamentals to Applications*, pp. 502–510, 2016.

[35] J. H. Wohlgemuth and S. Kurtz, "Reliability testing beyond qualification as a key component in photovoltaic's progress toward grid parity," in *2011 International Reliability Physics Symposium*, IEEE, 2011, pp. 5E–3.

[36] P. Sinha and A. Wade, "Assessment of leaching tests for evaluating potential environmental impacts of PV module field breakage," *IEEE J Photovolt*, vol. 5, no. 6, pp. 1710–1714, 2015.

[37] M. P. M. Tas and W. G. van Sark, "Experimental repair technique for glass defects of glass-glass photovoltaic modules–A techno-economic analysis," *Solar Energy Materials and Solar Cells*, vol. 257, p. 112397, 2023.

[38] U. Itoh, M. Yoshida, H. Tokuhisa, K. Takeuchi, and Y. Takemura, "Solder joint failure modes in the conventional crystalline Si module," *Energy Procedia*, vol. 55, pp. 464–468, 2014.

[39] J.-S. Jeong, N. Park, and C. Han, "Field failure mechanism study of solder interconnection for crystalline silicon photovoltaic module," *Microelectronics Reliability*, vol. 52, no. 9–10, pp. 2326–2330, 2012.

[40] J. Walter, L. C. Rendler, C. Ebert, A. Kraft, and U. Eitner, "Solder joint stability study of wire-based interconnection compared to ribbon interconnection," *Energy Procedia*, vol. 124, pp. 515–525, 2017.

[41] A. Kraft *et al.*, "Investigation of acetic acid corrosion impact on printed solar cell contacts," *IEEE J Photovolt*, vol. 5, no. 3, pp. 736–743, 2015.

[42] M. D. Kempe, G. J. Jorgensen, K. M. Terwilliger, T. J. McMahon, C. E. Kennedy, and T. T. Borek, "Acetic acid production and glass transition concerns with ethylene-vinyl acetate used in photovoltaic devices," *Solar energy materials and solar cells*, vol. 91, no. 4, pp. 315–329, 2007.

[43] I. Duerr, J. Bierbaum, J. Metzger, J. Richter, and D. Philipp, "Silver grid finger corrosion on snail track affected PV modules–investigation on degradation products and mechanisms," *Energy Procedia*, vol. 98, pp. 74–85, 2016.

[44] A. Masuda, N. Uchiyama, and Y. Hara, "Degradation by acetic acid for crystalline Si photovoltaic modules," *Jpn J Appl Phys*, vol. 54, no. 4S, p. 04DR04, 2015.

[45] S. Deitsch *et al.*, "Automatic classification of defective photovoltaic module cells in electroluminescence images," *Solar Energy*, vol. 185, pp. 455–468, 2019.

[46] R. Bhoopathy, O. Kunz, M. Juhl, T. Trupke, and Z. Hameiri, "Outdoor photoluminescence imaging of solar panels by contactless switching: Technical considerations and applications," *Progress in Photovoltaics: Research and Applications*, vol. 28, no. 3, pp. 217–228, 2020.

[47] T. Trupke, B. Mitchell, J. W. Weber, W. McMillan, R. A. Bardos, and R. Kroeze, "Photoluminescence imaging for photovoltaic applications," *Energy Procedia*, vol. 15, pp. 135–146, 2012.

[48] Y. Kawano, J. Chantana, Y. Kuroda, K. Hirose, and T. Minemoto, "Development of repairing technique for interconnection of silicon photovoltaic modules using an induction heating system," *Solar Energy*, vol. 261, pp. 55–62, 2023.

[49] K. Lee, S. Cho, J. Yi, and H. Chang, "Prediction of power output from a crystalline silicon photovoltaic module with repaired cell-in-hotspots," *Electronics (Basel)*, vol. 11, no. 15, p. 2307, 2022.

[50] A. K. Schnatmann, F. Schoden, and E. Schwenzfeier-Hellkamp, "Sustainable PV Module Design—Review of State-of-the-Art Encapsulation Methods," *Sustainability*, vol. 14, no. 16, p. 9971, 2022.

[51] P. R. Dias *et al.*, "High yield, low cost, environmentally friendly process to recycle silicon solar panels: Technical, economic and environmental feasibility assessment," *Renewable and Sustainable Energy Reviews*, vol. 169, p. 112900, 2022.

[52] Veolia, "Veolia Opens the First European Plant Entirely Dedicated to Recycling Photovoltaic Panels," 2018.

[53] VDMA, "International Technology Roadmap for Photovoltaic (ITRPV), 14th edition," Germany, Apr. 2023.

[54] G. Thomassen, J. Dewulf, and S. Van Passel, "Prospective material and substance flow analysis of the end-of-life phase of crystalline silicon-based PV modules," *Resour Conserv Recycl*, vol. 176, p. 105917, 2022.

[55] W.-H. Huang, W. J. Shin, L. Wang, W.-C. Sun, and M. Tao, "Strategy and technology to recycle wafer-silicon solar modules," *Solar Energy*, vol. 144, pp. 22–31, 2017.

[56] W.-H. Huang, W. J. Shin, L. Wang, W.-C. Sun, and M. Tao, "Strategy and technology to recycle wafer-silicon solar modules," *Solar Energy*, vol. 144, pp. 22–31, 2017.

[57] T.-Y. Wang, J.-C. Hsiao, and C.-H. Du, "Recycling of materials from silicon base solar cell module," in *2012 38th IEEE Photovoltaic Specialists Conference*, IEEE, 2012, pp. 2355–2358.

[58] P. Padhamnath *et al.*, "Metal contact recombination in monoPoly^{TM} solar cells with screen-printed & fire-through contacts," *Solar Energy Materials and Solar Cells*, vol. 192, pp. 109–116, 2019.

[59] P. Padhamnath *et al.*, "Characterization of screen printed and fire-through contacts on LPCVD based passivating contacts in monoPoly^{TM} solar cells," *Solar Energy*, vol. 202, pp. 73–79, 2020.

[60] P. Padhamnath, N. Nampalli, A. Khanna, B. Nagarajan, A. G. Aberle, and S. Duttagupta,

10.4229/EUPVSEC2023/5DV.2.62
020557-007

"Progress with passivation and screen-printed metallization of Boron-doped monoPoly™ layers," *Solar Energy*, vol. 231, pp. 8–26, 2022.
[61] P. Padhamnath, A. Khanna, N. Nandakumar, A. G. Aberle, and S. Duttagupta, "Impact of firing temperature on fire-through metal contacts to P-doped (n+) and B-doped (p+) poly-Si," *Solar Energy Materials and Solar Cells*, vol. 230, p. 111217, 2021.
[62] F. C. S. M. Padoan, P. Altimari, and F. Pagnanelli, "Recycling of end of life photovoltaic panels: A chemical prospective on process development," *Solar Energy*, vol. 177, pp. 746–761, 2019.
[63] J.-K. Lee *et al.*, "Photovoltaic performance of c-Si wafer reclaimed from end-of-life solar cell using various mixing ratios of HF and HNO3," *Solar Energy Materials and Solar Cells*, vol. 160, pp. 301–306, 2017.
[64] M. Tangstad, "Ferrosilicon and silicon technology," in *Handbook of ferroalloys*, Elsevier, 2013, pp. 179–220.
[65] L. Riva, G. R. Surup, T. V. Buø, and H. K. Nielsen, "A study of densified biochar as carbon source in the silicon and ferrosilicon production," *Energy*, vol. 181, pp. 985–996, 2019.
[66] A. N. Ismail, M. H. Ibrahim, R. M. Said, F. Somidin, and S. A. Ismail, "Influence of recycled wastes on ferrosilicon production in steel making applications: A short review," in *Journal of Physics: Conference Series*, IOP Publishing, 2022, p. 012028.
[67] IRENA, "Renewable power generation costs in 2022," Abu Dhabi, 2023. Accessed: Sep. 13, 2023. [Online]. Available: https://www.irena.org/Publications/2023/Aug/Renewable-Power-Generation-Costs-in-2022
[68] A. van der Heide, L. Tous, K. Wambach, J. Poortmans, J. Clyncke, and E. Voroshazi, "Towards a successful re-use of decommissioned photovoltaic modules," *Progress in Photovoltaics: Research and Applications*, vol. 30, no. 8, pp. 910–920, 2022.
[69] J. Walzberg, A. Carpenter, and G. A. Heath, "Role of the social factors in success of solar photovoltaic reuse and recycle programmes," *Nat Energy*, vol. 6, no. 9, pp. 913–924, 2021.

10.4229/EUPVSEC2023/5DV.2.62
020557-008

SOLAR POWER IN AFRICA: STATUS, MARKETS AND FUTURE

Ralf Leutz[1] and Toby Couture[2]
[1]TOMATO:GSL, Munich, [2]E3analytics, Berlin, Germany
[1]ralf.leutz@tomato-gsl.org, [2]toby@e3analytics.eu

ABSTRACT: Africa has the unique chance to build a 100% renewable energy system. The current energy system, dominated by fossil fuel-based power generation, is limited in capacity, and often outdated. After decades of little progress in Africa, climate change and development motivations (internally Agenda 2063, externally US and EU efforts). Assessing the available data is a first priority; we then focus on the underachievement of solar project capacity in Africa in comparison to world capacity. We sort solar projects according to market segments. Ways to potentially improve the situation are given The work is highly relevant for Africa and the globe in limiting the effects of global warming and give new impetus to positive developments in energy, water, health, and wealth. We aim to give an overview of the latest developments in photovoltaic installations in Africa, based on our work in a joint project of the European Union and the African Union.
Keywords: Africa, PV in the energy transition, scenarios for renewables, policy, global challenges

1 INTRODUCTION

The situation of solar power in Africa requires urgent attention. As we will outline in the present paper, both the absolute installed solar capacity and its comparison to the world capacity are greatly below country and regional averages and are deteriorating further as deployment elsewhere in the world gains momentum. This is particularly surprising in light of Africa's stunning solar power potential. We look at the latest data on solar projects and identify market segments. We identify reasons for the low number of solar installations and describe some potential pathways out of the African energy crisis.

Our paper is based on a consulting project for the European Union and the African Union [1]. Others have pointed out the dire situation in Africa in the energy context [2], the diversity of African countries [3] [4], and the definition phase of research [5].

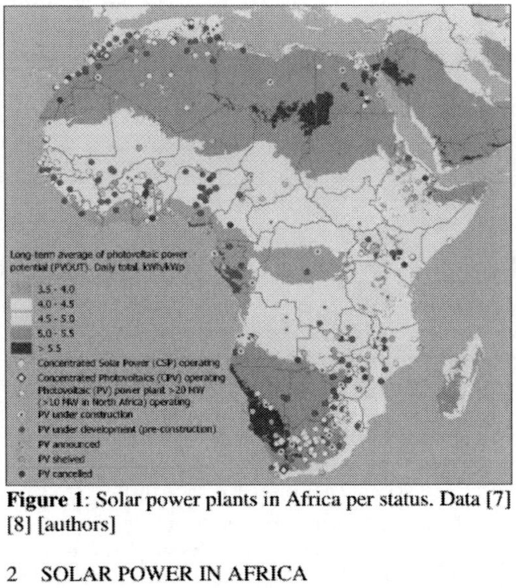

Figure 1: Solar power plants in Africa per status. Data [7] [8] [authors]

2 SOLAR POWER IN AFRICA

Africa stretches across the equator through both solar belts around 20° latitude, the arid regions characterized by high solar irradiance values. Africa is home to 1.4 billion people, or 17.5% of the global population. The electricity consumption in Africa in 2019 was 656 kWh/person and year [6]. In contrast, the world's average electricity consumption was 3,500 kWh/person and year.

Energy consumption is expected to increase. As electricity comes to assume an ever-growing share of global energy consumption with the electrification of transport, heating, cooling, and industry, electricity is poised to become the dominant energy carrier by mid-century [12]. In 2050, the world's average inhabitant may consume roughly 21,000 kWh annually [13].

With an average capacity factor of renewable energy generators of 20% (equivalent to 1,750 full load hours per year), the world needs an installed capacity of approximately 105 TW of renewables to meet global energy demand by mid-century, 20% of it in Africa (following the distribution of world population). Due its low cost and sheer abundance, solar photovoltaic (PV) power will be a significant contributor to global energy needs worldwide. In terms of investment, Africa will require investments of approximately EUR 20-50 trillion to become fully renewable, depending on technology, storage and degree of interconnection.

Current solar installations in Africa are mapped in Figure 1. The cumulative installation in Africa amounts to 8.6-12.4 GW, depending on the data sources, AfSIA [11] and IRENA [14], respectively (Figure 2). Solar installations per country are listed in Figure 3.

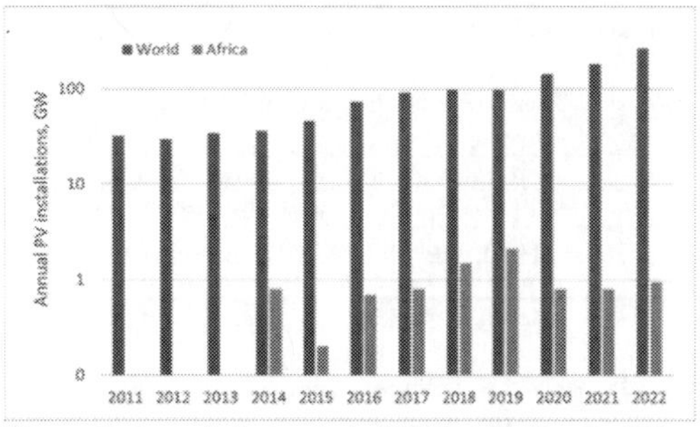

Figure 2: Annual photovoltaic (PV) installations in Africa and the World. Note the logarithmic scale. Data [9] [10] [11]

Annual PV installations in Africa are lower by a factor of ten than they should be according to population share and electricity consumption share, at the current electricity consumption. If Africa had a per-capita electricity consumption comparable to the rest of the world, the continent would host fifty times more PV than it currently does.

3 ENERGY SITUATION IN AFRICA – DATA AND INSTALLATIONS

The true magnitude of energy demand in Africa is often unknown, as demand frequently exceeds supply, leading to blackouts and brownouts, and 600 million people in sub-Saharan Africa continue to lack access to reliable electricity [15]. In addition, the bill collection rate in sub-Saharan Africa (SSA, median 93%) and losses due to transmission, and distribution inefficiency as well as electricity theft (together average 23%) reduce the economic performance of utilities especially in (SSA without South Africa) [16]. The cost of electricity delivered by a utility to the consumer is artificially high due to the combined weight of these losses.

In turn, governments frequently resort to subsidizing electricity to remain in power, contributing to a power system that is not cost covering and that lacks the free cash flow required to invest adequately in maintenance, in distribution grids, and in operations. With electricity poised to become increasingly central to the overall energy mix, the difficulties faced by the utility sector in Africa risk jeopardizing the emergence and economic development of the continent.

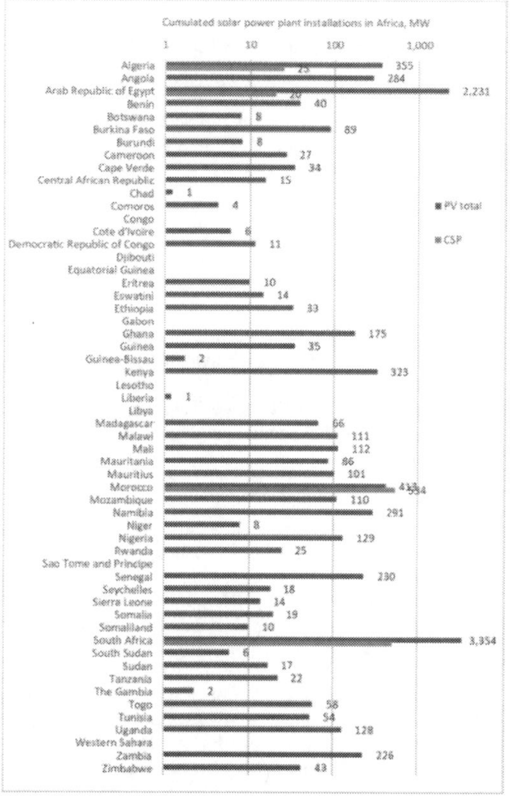

Figure 3: Solar installations per country in Africa. Data *[11]*

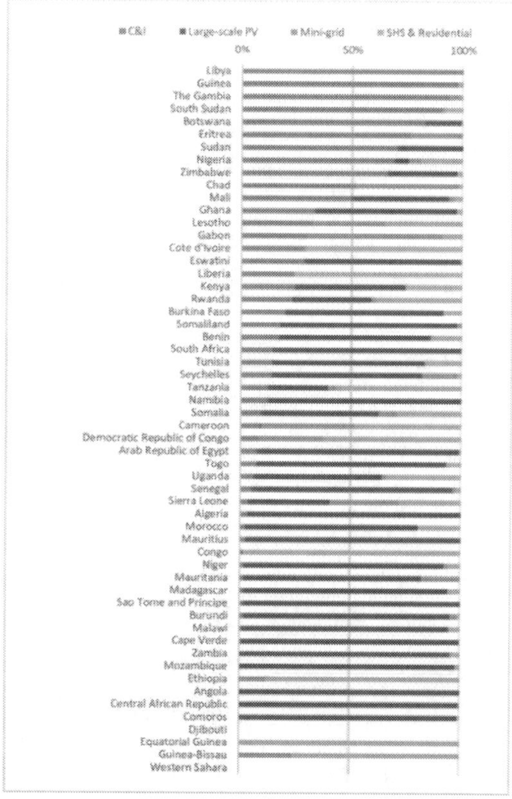

Figure 4: Solar project types in African countries. Data *[11]*

Table 1: Main solar photovoltaic market segments in Africa

Type	Description	Offtaker
1	Self-consumption commercial & industrial (C&I)	C&I
2	Concessionally-financed, including sovereign-to-sovereign projects	Utility
3	Utility-scale, commercially financed Independent Power Producers (IPP)	Utility
4	Utility-owned (and utility-financed) projects	Utility
5	Merchant projects, sold into Power Pool	Power Pool

Despite the progress achieved in recent decades, power reliability remains an existential threat for many companies in Africa. One positive development in this regard is that a growing number of commercial and industrial (C&I) customers are no longer captive customers: they can increasingly supply their own electricity and with the rapid declines in the costs of solar and battery storage, they can increasingly do so at a lower cost and with better reliability than grid-based power (Figure 4). C&I projects have reached 1.2 GW, or 13% of all photovoltaic power plants operating in Africa in 2022 [11]. C&I is the fastest growing segment of photovoltaic

10.4229/EUPVSEC2023/5DV.3.4
020560-002

installations (year-on-year 2021 to 2022: 60% as opposed to 14% for all PV projects). The average size of a C&I PV plant is 1.15 MW.

While the growth of the C&I segment provides hope for the developments in Africa, it is likely to weaken further the financial stability of some utilities, now losing some of their most solvent customers.

4 MARKET SEGMENTS

As shown in Figure 4, solar project types vary across Africa. We have identified five solar market segments (Table 1). Some projects are entirely presented by foreign governments, like the 15 MW Sakai plant [17] in the Central African Republic, some projects are fully debt-financed by the World Bank, like the 40 MW Danzi plant [18] [19] in the same country.

There are type 3 projects, commercially financed by private banks organized by international developers, like the Kesses I plant [20] in Kenya. There are type 4, utility-financed projects, like the Black Volta photovoltaic and floating PV plants [21] in Ghana, built on government-owned land, (i.e. on the reservoir surface).

Interesting is the off-taker, or buyer of the electricity generated. In most cases, the primary off-taker is the national utility, with the exceptions of the C&I projects (which sell directly to the customer), and projects selling electricity directly to the Power Pool, like the example recently announced in Namibia [22]. Power Pools are groups of utilities encompassing several countries. There are five Power Pools in Africa. Feeding power into a Power Pool requires grid codes, and mechanisms like wheeling, where electricity is transmitted through intermediate grids.

5 DISCUSSION AND POTENTIAL SOLUTIONS

Africa is likely to be affected more immediately and more forcefully by climate change than many other places on earth, due in part to its lower capacity to invest in adaptation and the thinner social safety nets in place to support displaced and affected populations [23]. Hot places will become hotter while health, biodiversity and the economy are likely to suffer. One promising option to increase the continent's resilience to climate change, in particular its agricultural sector, is agrivoltaics, [24] the dual use of agricultural land for growing crops and collecting solar energy: agrivoltaics can increase harvests through shading and water preservation, improving the adaptation of cultivated lands to climate change, while also widening the range of crops that can be grown.

As the role of solar power grows in Africa, propelled forward by its low cost and its near-universal availability across the continent, governments across Africa are well positioned to step up efforts to manufacture a growing number of links in the solar industry supply chain, creating jobs and mobilizing investment. Solar manufacturing hubs can become the nuclei for industrial development. For such hubs to emerge, however, robust and growing markets for solar technologies within Africa are required.

6 CONCLUSIONS

Despite the record-breaking pace of cost reductions that has occurred in recent decades, solar PV technologies continue to improve and to come down the cost curve. According to the IEA, solar power now provides the cheapest electricity in history, underscoring the fact that the age of solar has indeed arrived. It is time for governments across Africa to seize this opportunity and unlock investments in local solar power generation by de-risking policy and regulatory frameworks and creating a more supportive enabling environment for solar entrepreneurs.

The opportunity provided by solar power is tremendous. Solar power can provide access to reliable energy for all and do so more quickly and more cost-effectively than any other energy source currently available. The task at hand is to unlock capital at scale and unleash its potential.

7 ACKNOWLEDGEMENT

We acknowledge the support of our research by GET.transform, see [1].

8 REFERENCES

[1] R. Leutz, T. Couture, N. Martensen und T. Ackermann, „Specific support study in support of the Continental Power System Masterplan (CMP) through the EU's Technical Assistance Facility (TAF) for African Union in the sustainable energy sector: Specific Support Study on the Solar Power Plants (Lot 12b)," 2023, https://cmpmwanga.nepad.org/Solar_delFinal

[2] M. Kojima und C. Trimble, „Making power affordable for Africa and viable for its utilities," 2016.

[3] D. Puig, M. Moner-Girona, D. M. Kammen, Y. Mulugetta, A. Marzouk, M. Jarrett, Y. Hailu und N. Nakicenovic, „An action agenda for Africa's electricity sector," *Science,* Bd. 373, Nr. 6555, pp. 616-619, 2021.

[4] Y. Mulugetta, Y. Sokona, P. A. Trotter, S. Fankhauser, J. Omukuti, L. Somavilla Croxatto, B. Steffen, M. Tesfamichael, E. Abraham, J.-P. Adam, L. Agbemabiese, C. Agutu, M. Paulos Aklilu, O. Alao, B. Batidzirai, G. Bekele, A. G. Dagnachew, O. Davidson, F. Denton, E. Oghneruona Diemuodeke, F. Egli, E. Gebrekidan Gebresilassie, M. Gebreslassie, M. Goundiam, H. Kachalla Gujby, Y. Hailu, A. D. Hawkes, S. Hirmer, H. Hoka, M. Howells, A. Isah, D. Kammen, F. Kemausuor, I. Khennas, W. Kruger, I. Malo, L. Mofor, M. Nago, D. Nock, C. Okereke, S. N. Ouedraogo, B. Probst, M. Schmidt, T. S. Schmidt, C. Shenga, M. Sokona, J. C. Steckel, S. Sterl, B. Tembo, J. Tomei, P. Twesigye, J. Watson, H. Winkler und A. Yussuff, „Africa needs context-relevant evidence to shape its clean energy future," *Nature Energy,* 2022.

[5] P. Späth, V. Castan Broto, S. Bawakyillenuo und M. Pregernig, „The governance of energy transitions in Africa: a sketch of plural perspectives," *Energy, Sustainability and Society,* Bd. 12, Nr. 51, 2022.

[6] International Energy Agency (IEA), „Africa Energy Outlook 2019".

10.4229/EUPVSEC2023/5DV.3.4
020560-003

[7] Global Energy Monitor (GEM), „Global Solar Power Tracker, interactive map and data set," Global Energy Monitor (GEM), 01 2023. [Online]. Available: globalenergymonitor.org/projects/global-solar-power-tracker. [Zugriff am 23 01 2023].

[8] R. Thonig und J. Lilliestam, *CSP.guru (data on Concentrated Solar Power (CSP) plants)*, 2022.

[9] Our World in Data, „Solar PV module prices vs. cumulative capacity," 2022. [Online]. Available: ourworldindata.org/grapher/solar-pv-prices-vs-cumulative-capacity. [Zugriff am 30 10 2022].

[10] S. Enkhardt, „Bloomberg NEF says global solar will cross 200 GW mark for first time this year, expects lower panel prices," PV Magazine, 01 02 2022. [Online]. Available: www.pv-magazine.com/2022/02/01/bloombergnef-says-global-solar-will-cross-200-gw-mark-for-first-time-this-year-expects-lower-panel-prices. [Zugriff am 01 10 2022].

[11] Africa Solar Industry Association (AFSIA), „Annual Solar Outlook 2023, a country-by-country review of the status of solar energy in Africa," 2023.

[12] International Renewable Energy Agency (IRENA), „Global Energy Transformation: A Roadmap to 2050," 2018.

[13] Statista, „Energy consumption worldwide from 2000 to 2019, with a forecast until 2050, by energy source," 2023. [Online]. Available: www.statista.com/statistics/222066/projected-global-energy-consumption-by-source. [Zugriff am 01 02 2023].

[14] International Renewable Energy Agency (IRENA), „Renewable Capaicty Statistics 2022," 2022.

[15] International Energy Agency (IEA), „Africa Energy Outlook 2022, World Energy Outlook Special Report," 2022.

[16] C. Trimble, M. Kojima, I. Perez Arroyo und F. Mohammadzadeh, „Financial viability of electricity sectors in Sub-Saharan Africa : quasi-fiscal deficits and hidden costs," 2016.

[17] Les Autres Nouvelles De Centrafrique-LANOCA, „#Société / ENERCA : Le projet de la Centrale Solaire de Sakaï se concrétise par le raccordement de la ligne entre Sakai et le Centre-ville," 06 05 2022. [Online]. Available: anoca.over-blog.com/2022/05/societe/enerca-le-projet-de-la-centrale-solaire-de-sakai-se-concretise-par-le-raccordement-de-la-ligne-entre-sakai-et-le-centre-ville.html.

[18] The World Bank, „CAR-Electricity Sector Strengthening and Access Project (P176683), Project Information Document (PID)," 2021.

[19] J. M. Takouleu, „CAR: GCF and IDA release $138m for solar electrification," Afrik21, 08 07 2022. [Online]. Available: www.afrik21.africa/en/car-gcf-and-ida-release-138m-for-solar-electrification. [Zugriff am 01 02 2023].

[20] Emerging Africa Infrastructure Fund (EAIF), „EAIF backs 40 MW Kesses solar project in Kenya with US$ 35 million loan," [Online]. Available: www.eaif.com/eaif-backs-40mw-kesses-solar-project-in-kenya-with-us35-million-loan. [Zugriff am 27 10 2022].

[21] Bui Power Authority, „Hydro Solar Hybrid," [Online]. Available: buipower.com/hyrdo-solar-hybrid. [Zugriff am 26 10 2022].

[22] J. M. Takouleu, „Southern Africa: Emesco obtains approval for a solar park connected to the SAPP," Afrik21, 20 09 2022. [Online]. Available: www.afrik21.africa/en/southern-africa-emesco-obtains-approval-for-a-solar-park-connected-to-the-sapp. [Zugriff am 01 02 2023].

[23] Intergovernmental Panel on Climate Change (IPCC), „Fact sheet - Africa," 2022.

[24] R. J. Randle-Boggis, E. Lara, J. Onyango, E. J. Temu und S. E. Hartley, „Agrivoltaics in East Africa: Opportunities and challenges," in *AIP Conference Proceedings 2361, 090001*, 2021.

10.4229/EUPVSEC2023/5DV.3.4
020560-004

THE POTENTIAL OF PHOTOVOLTAICS IN THE FUTURE ELECTRICAL POWER SUPPLY OF BANGLADESH

Bettina Löffler, Norbert Bernhard

Department of Electrical, Mechanical and Industrial Engineering, Anhalt University of Applied Sciences, Bernburger Str. 55, 06366 Koethen, Germany

ABSTRACT: The potential of photovoltaics (PV) for electrical power generation in Bangladesh has been investigated. Based on commonly available meteorological data, the insolation onto a solar module in dependence on its azimuth and inclination angle was calculated over the year. The contributions of direct insolation, diffuse insolation and insolation from surface-reflected light (albedo) were taken into account. From these data the energy harvest per year was calculated. It was found that it is strongly increased in comparison to Germany, although monsoon rain strongly reduces direct insolation during the summer months. A potential of PV of 37 GW average power (corresponding to 277 GWp installed power) was estimated. Since PV plays only a negligible role in the status quo of electrical power generation of Bangladesh, based on these findings, an enforced expansion of PV power generation is recommended.

1 MOTIVATION AND INTRODUCTION

The country Bangladesh was chosen for this investigation as an example of a developing country, in which the political and administrative decisions were not taken yet to include PV in a large extent into a strategy to lead the country towards a green and sustainable development and industrialization process. Also a large part of the country is not connected to a public grid yet, meaning a big potential of local off-grid system. Also, a colleague of the authors at Anhalt university (cf. acknowledgement) as a native of this country, has a good knowledge of specific circumstances to be considered. The approach was to take publicly accessible meteorological and geographical data to assess the PV potential. There are several investigations of the exploitation of renewable energies in Bangladesh available, but we have not yet seen an estimation of the PV electricity generation potential directly based on insolation data.

Kabir and Endlicher [1] analyse the boundary conditions of renewables with emphasis on socio-economic data without giving concrete quantitative estimates of the potential of the various investigated forms of renewable energies (PV, wind, biogas). Mondal [2] focuses on future economic demand development scenarios by using the software package MARKAL for which the potential of PV and other renewables is used as an input parameter. But for these inputs he uses rather coarse estimates (e.g. 50 GW PV potential, based on 1.7% suited area, 200 W/m^2 annual mean insolation and 10% mean efficiency of the PV systems). According to Mondal, the value of 1.7% suited area for PV is derived from grid availability and other criteria. We will use this value in our investigation due to lack of more precise data. But we will use insolation data from data bases and calculations of insolation on modules in dependence on tilt and azimuth angle and realistic module efficiencies and loss factors. In the end we will arrive at 37 GW annual average PV power (corresponding to 277 GWp installed base), showing that Mondal overestimates the PV potential, but it all the same remains huge and is a strong statement that the country should intensify its efforts in exploiting this potential.

2 METHODOLOGICAL APPROACH

The approach was to take publicly accessible meteorological data to assess the PV potential. Global insolation data on the horizontal plane are taken from generally available data bases for a typical meteorological year (TMY). Global, direct and diffuse radiation was taken from NREL database [3] (the diffuse part modelled with NREL METSTAT model). Sun position is calculated by using the Solar Position Algorithm (SPA) of NREL. Direct and diffuse radiation are calculated dependent on the tilt and azimuth angle of the PV modules. For the diffuse radiation the Perez model was used [4]. Albedo mean values (over a 22 years period) from the NASA database [5] were used with the isotropic approximation.

Energy yield is calculated for typical module efficiencies and losses. To avoid a too optimistic view moderate efficiencies and realistic capture and system losses were taken, resulting in rather moderate performance ratios. As already stated in the previous section, it was assumed that 1.7% of the country´s area could be used for PV. An area utilization of 71% was assumed. This value was calculated for a free area system to avoid self-shading of neighbouring modules at noon at the lowest elevation angle of the sun in winter. Since the elevation angle of the sun is more than in Germany, this value is higher than for a respective system in Germany. Calculations were performed for the capital Dhaka (slightly north of the Tropic of Cancer) and the city Cox´s Bazar (in the very south of the country, and south of the Tropic of Cancer). Since the global irradiance varies only within 5% in the whole (rather small) country, the values calculated for Dhaka were taken for the potential estimate of the whole country.

3 RESULTS AND DISCUSSION

In a first step the global insolation in the geographical region of Bangladesh onto a solar module in dependence on its inclination and azimuth angle was calculated from commonly available meteorological data from NREL [3], which consist of direct and diffuse insolation on the

10.4229/EUPVSEC2023/5DV.3.5
020561-001

horizontal plane in dependence on the day hour between sunrise and sunset. The global insolation onto a solar module adds up by three contributions: firstly direct insolation which can easily be calculated for different tilt and azimuth angles from the horizontal plane data by geometrical considerations. Secondly diffuse insolation for which several models are available to calculate the value in dependence on tilt and azimuth angle. In this work for the diffuse contribution the model by Perez is used [4]. In this model, the diffuse insolation onto an inclined area is calculated using indexes for clearness and brightness of the sky, which define tabulated factors from which the angle dependent diffuse insolation can be calculated from diffuse irradiation data onto a horizontal surface. Thirdly there is a contribution of irradiation onto the solar module, the origin of which is the reflection (albedo) from the earth surface, which partially hits an inclined module. Here a simple isotropic back-scattering model is used from which the insolation onto an inclined module can be calculated from tabulated albedo values taken from [5].

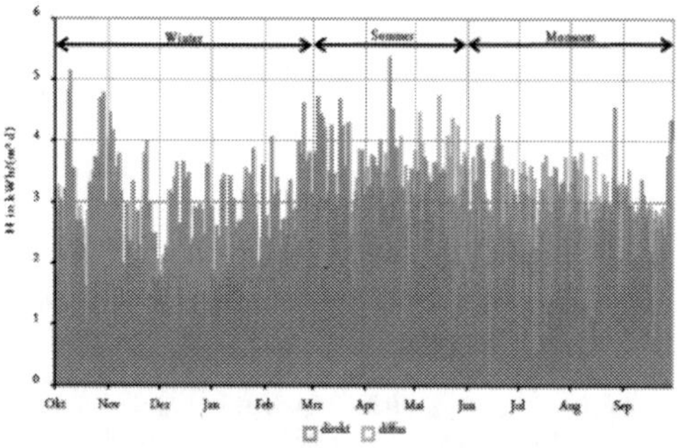

Figure1: Direct and diffuse irradiation (in kWh per square meter and day) at Dhaka according to data from [3]. In winter (characterized by lower temperatures) the direct irradiation dominates. In the monsoon season diffuse irradiation dominates due to high cloud coverage. Summer and monsoon season are characterized by higher temperatures than in winter. The assignment of seasons was done by the authors of this paper and is somehow arbitrary.

Figure 1 shows direct and diffuse insolation for Dhaka. The seasons winter, summer and monsoon were defined according to the meteorological conditions temperature, rain fall, humidity, and cloud coverage. Different authors use different assignments or names. But the essential point is that there are basically three seasons which defer significantly by meteorological conditions. Figure 2 and figure 3 show the calculated global (i.e. sum of separately calculated direct, diffuse and reflected contribution) of annual irradiation onto the module plane in dependence on tilt and azimuth angle for the two cities Dhaka and Cox´s Bazar. As expected the optimum azimuth angle is an orientation directly to the south for Dhaka, whereas the optimum inclination (tilt) angle is 25° from the horizontal plane, where an irradiation of 1979 kWh/m^2 is reached. It is worthwhile to notice that this maximum value for both places for optimal orientation of the module is almost twice as much as in Germany, which is at the moment the country with the highest PV contribution to electrical power generation in the world. For Cox´s Bazar (south to the Tropic of Cancer), it is also interesting that even for a rather strong deviation of the azimuth from the optimum angle, only minor losses of collected irradiation are observed. Therefore, if for some reason an orientation to the south is not possible, the module orientation is very tolerant with respect to energy harvest. This means that also for east and west orientated modules there is only a minor total energy loss in comparison to south orientation. This can be an advantage for a more continuous electricity supply over the day and minimizes costs for battery storage systems.

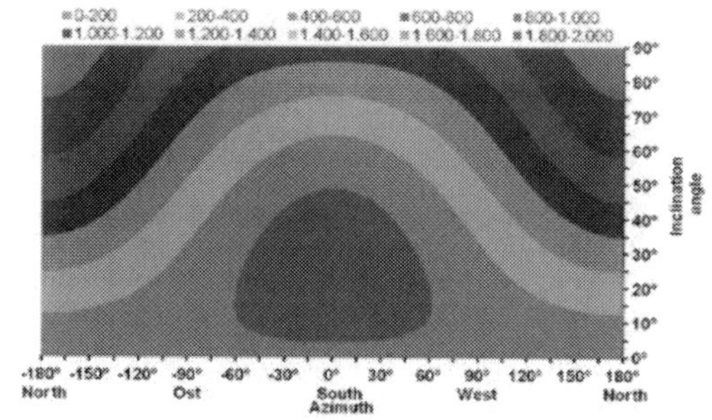

Figure 2: Calculated annual irradiation (global insolation) in kWh/m^2 onto a solar module in dependence on azimuth (horizontal) and inclination angle (vertical). Colour code (in kWh/m^2): dark orange: 1800 – 2000; green: 1600 – 1800, ochre 1400 – 1600, bright orange: 1200 – 1400, dark green: 1000 – 1200; brown: 800 – 1000; red-brown: 600 - 800

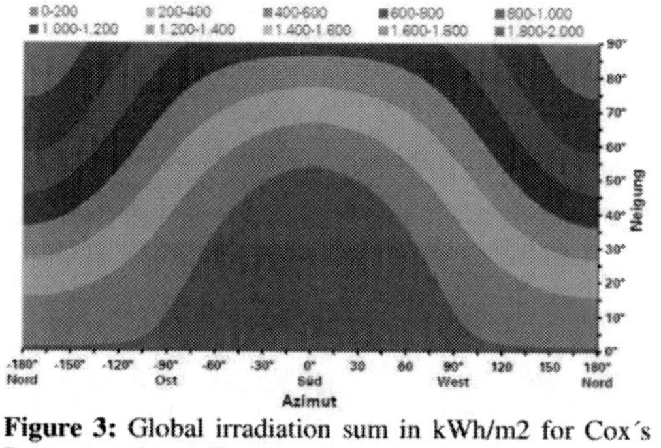

Figure 3: Global irradiation sum in kWh/m2 for Cox´s Bazar (south of Tropic of Cancer) in dependence on azimuth and tilt angle. Colour code as in figure 2.

Figure 4 shows the monthly global irradiation sum for Dhaka in dependence on the tilt angle for a module orientation to the south. Since the maximal elevation angle of the sun, and the ratio between direct and diffuse radiation changes strongly between winter and monsoon season, also the optimum tilt angle of the modules changes. The suggested tilt angle of 25° is therefore a compromise between the different seasons and optimizes the yearly energy yield.

10.4229/EUPVSEC2023/5DV.3.5
020561-002

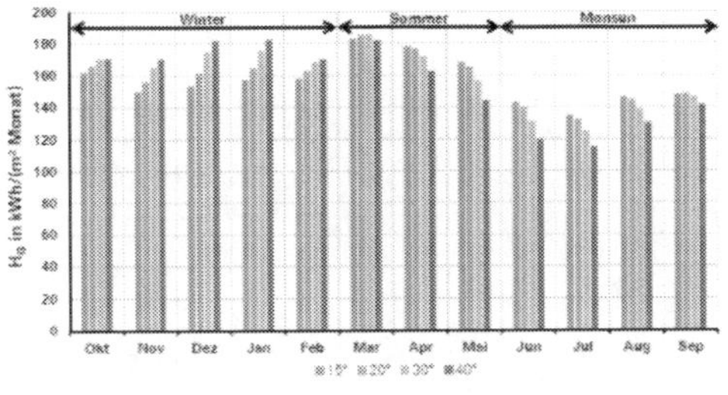

Figure 4: Monthly global irradiation sum in for Dhaka for different tilt angles and orientation to the south. Seasons defined according to weather conditions.

It is interesting that per area almost twice as much global insolation than in Germany can be reached, although during the months May to August direct insolation is strongly reduced due to clouds and monsoon rain (cf. figure 1).

Based on these data, together with the assumption of Mondal, that 1.7% of the surface area of Bangladesh (i.e. 2509 km^2) could be used for PV power plants [2], a yearly average potential of PV of 37 GW average power was estimated for Bangladesh. A space utilization rate of 71% - calculated from elementary geometrical considerations - was used for this estimation, which is much higher than in Germany (36,6%) due to higher average altitude of the sun, and therefore less shadowing of neighbouring modules. For this average power an installed power of 277 GWp (GW peak) is needed, according to the calculated average insolation in Bangladesh. From this installed base an energy harvest of 325 TWh per year can be expected, which is almost twice the demand of Bangladesh, as forecasted for 2030. This means that a substantial portion of Bangladesh´s electrical energy demand could be produced by PV. Natural gas could be used as a backing technology during the night or when, due to clouds and rain, the PV power supply is not sufficient.

It should be noted that the change of the spectral distribution of the diffuse radiation by light scattering effects was not taken into account in our considerations. Thus, the final electrical output of a module was just calculated by its efficiency value derived from its data sheet, which converts irradiation in W/m^2 into electrical power. Of course, the impact of the temperature change of the module on the conversion efficiency was taken into account. The module temperature is calculated from typical NOCT (nominal operating cell temperature) values from data sheets under the assumption that the increase of the cell temperature (assumed to be identical to the module temperature) relative to the ambient temperature is proportional to the irradiance [6]. But we avoided the effort to calculate the irradiance weighted average cell temperature according to [6], instead we calculated for every month the average daily maximum ambient temperature, and the average maximum irradiance. These data were used to calculate the module temperature which was used then for the electric yield estimate. We believe that the error generated by this procedure is acceptable within the general precision of our approach, and we do not overestimate the energy yield. The temperature corrected efficiency is then calculated by taking typical temperature coefficients for the power from data sheets. Modules with moderate efficiency data were selected in order to avoid an overestimate of the power potential and to account already for some degradation. Values were calculated for several representative module types and then the average was taken. For miscellaneous (i.e. non-temperature induced) capture losses we used correction factors given by [6] for regions without snow coverage.

For the inverter efficiencies we have the problem that the European efficiencies from the data sheets do not represent the irradiance conditions in Bangladesh. Figure 5 and figure 6 show the distribution of irradiation energy to different irradiation bin classes and the frequency of occurrence for Dhaka and Cox´s Bazar. Higher irradiation classes are much more likely than in Germany. Therefore for selecting an optimal inverter the European efficiency is not the optimal criterion. The Californian efficiency would be certainly better, but based on these distributions, a special inverter efficiency for Bangladesh could be defined, a suitable suggestion could be:

$$\eta_{Bangla} = 0{,}05 \cdot \eta_{10\%} + 0{,}06 \cdot \eta_{20\%} + 0{,}17 \cdot \eta_{30\%} + 0{,}32 \cdot \eta_{50\%} + 0{,}37 \cdot \eta_{75\%} + 0{,}03 \cdot \eta_{100\%}$$

This formula is based on the average of frequency of occurrence of irradiation classes in Dhaka and Cox´s Bazar. But since the needed data for this formula are usually not available, an inverter efficiency of 95% was assumed in our estimation, in order to avoid a too optimistic results by using European efficiencies from data sheets which cannot be reached under the climatic and especially irradiance conditions of Bangladesh.

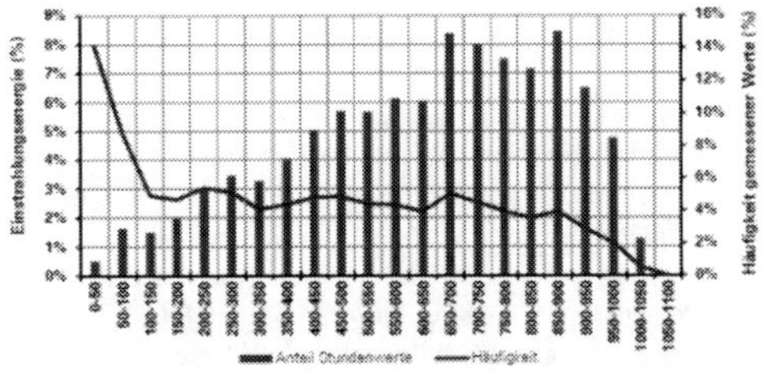

Figure 5: Irradiation energy (red) and frequency of occurrence (blue) of different irradiance bin classes of a PV system with 25° tilt angle and orientated to the south for Dhaka.

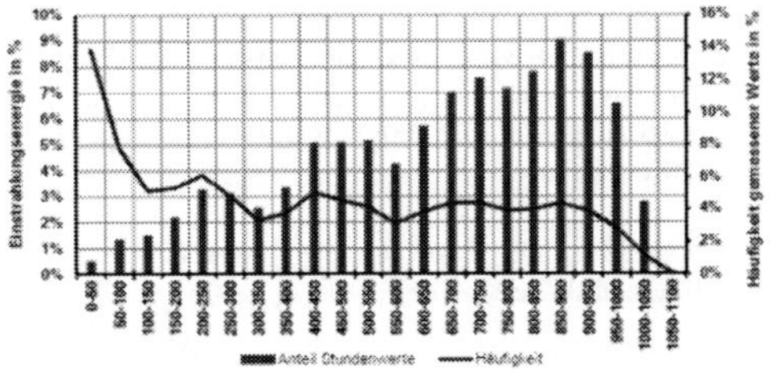

Figure 6: Irradiation energy (red) and frequency of occurrence (blue) of different irradiance bin classes of a PV system with 25° tilt angle and orientated to the south for Cox´s Bazar.

4 CONCLUSIONS

Based on generally available insolation data the potential electricity generation of photovoltaics (PV) was estimated for the developing country Bangladesh. The area suited for PV installations was assumed to be 1.7% (i.e. 2509 km^2) of the total area of the country according to [2]. Direct, diffuse and albedo radiation were taken from generally accessible data bases. The global radiation to an inclined module was calculated in dependence on tilt and azimuth angle. For the diffuse radiation the Perez model was used, for albedo the isotropic approximation. Using typical module efficiencies and loss factors an average power based on PV of 37 GW was estimated, corresponding to an energy harvest of 325 TWh per year. This is twice the forecasted electricity demand of Bangladesh in 2030. For this a PV installation of 277 GWp would be necessary. Together with other renewables, the electricity demand of the country could be easily covered with sufficient margin for further economic growth. The margin could also be used to transfer energy to other sectors (e.g. transport by electric vehicles). Therefore it is strongly recommended that Bangladesh should enforce its policies and support for employment of PV power plants.

Acknowledgement: The authors would like to thank Prof. Dr. Golam Abu Zakaria for drawing their attention to the situation of PV in his home country Bangladesh and for his continued interest in this work.

References:

[1] Kabir, M. H., and Endlicher, W., Exploitation of Renewable Energy in Bangladesh, A H Development Publishing House, Dhaka (2012)

[2] Mondal, A.: Implications of renewable energy technologies in the Bangladesh power sector: Long-term planning strategies, PhD thesis Bonn University (2010)

[3] National Renewable Energy Laboratory (NREL) of the USA: Hourly solar (direct normal (DNI), global horizontal (GHI), and diffuse) data for selected Bangladesh stations from NREL; http://en.openei.org/datasets/node/507 (28.03.2014)

[4] Perez, R. et al.: Solar Energy **36**, p. 481 (1986); Solar Energy **39**, p. 221 (1987); Solar Energy **44**, p. 271 (1990)

[5] National Aeronautics and Space Administration (NASA): Surface meteorology And Solar Energy – Available Tables; https://eosweb.larc.nasa.gov/sse/ (27.02.2014)

[6] Häberlin, H.: Photovoltaics. System Design and Practice. Wiley (2012)

10.4229/EUPVSEC2023/5DV.3.5
020561-004

SUPPORTING PV USAGE AND PRODUCTION TO REDUCE FOSSIL FUEL USE IN EUROPEAN AGRICULTURE – THE AGROFOSSILFREE PROJECT

Chuan Ma[1], Dominik Rutz[1], Athanasios Balafoutis[2], Foteini Vandorou[2], George Kyriakarakos [2], Claus Aage Grøn Sørensen[3], Michael Norremark[3], Erik Fløjgaard Kristensen[3], Dimitris Manolakos[4], Alex Koutsouris[4], Vasiliki Kanaki[4], Bas Paris[4], George Papadakis[4], Magdalena Borzecka[5], Małgorzata Wydra[5], Vanja Biševac[6], Dirk Vansintjan[7], Julio Román[8], Daniele Rossi[9], Martyna Próchniak[10], Manam Abdul[11], Michael Voskakis[11], Michail D. Kaminiaris[11], Zisis Tsiropoulos[11], Harm Brinks[12], Barry Caslin[13], Tom Houlihan[13], Ingo Vosseler[14], Mirentxu Asín[15], Camino Fabregas[15], Arne Grønkjær Hansen[16]

[1]WIP Renewable Energies, [2]Centre for Research and Technology Hellas (coordinator), [3]Aarhus University, [4]Agricultural University of Athens, [5]Institute of Soil Science and Plant Cultivation – State Research Institute, [6]CEMA aisbl – European Agricultural Machinery Industry Association, [7]REScoop.eu, [8]European Conservation Agriculture Federation, [9]Confagricoltura, [10]Lublin Agricultural Advisory Center in Konskowola, [11]AGENSO, [12]DELPHY BV, [13]TEAGASC, [14]Trama TecnoAmbiental, [15]Iniciativas Innovadoras, [16]Innovationscenter for Økologisk Landbrug

[1]WIP – Renewable Energies, Sylvensteinstr. 2, D - 81369 Munich, Germany
Tel. +49 89 720 12732, Fax +49 89 720 12791
E-Mail: chuan.ma@wip-munich.de
Internet: www.wip-munich.de, www.agrofossilfree.eu

ABSTRACT: The AgroFossilFree project is a cornerstone initiative advancing the assimilation of Fossil-Energy-Free Technologies and Strategies (FEFTS) in agriculture, prominently featuring the integration of solar energy. It orchestrates a collaborative milieu among stakeholders, disseminating pivotal FEFTS insights. A significant venture within the project is the establishment of an online knowledge hub that caters to a diverse audience including farmers, advisors, researchers, and industry affiliates. Reflecting the ethos of the EU Green Deal and Farm to Fork strategy, the project aspires to curtail fossil fuel dependency in farming operations. By addressing the divergence between current practices and FEFTS, it aims at propelling a more sustainable European agriculture, characterized by diminished energy consumption and environmental impact. This discourse sheds light on the project's framework and initiatives, backed by the European Commission's Horizon 2020 program.
Keywords: FEFTS, AgEnergy Platform, Sustainability, RES integration, Agriculture

1 INTRODUCTION

Agricultural emissions significantly contribute to climate change, often at a larger scale when compared to other sectors. As of 2015, nearly 10% of the EU's total greenhouse gas (GHG) emissions emanated from the agricultural sector, a figure that excludes the impacts of land use, land use alterations, and net forestry removals [1]. The bulk of these emissions are attributed to livestock production—specifically enteric fermentation and manure management—as well as fertilizer applications on agricultural lands. A recent analysis [2] highlighted that, except for horticultural crops, energy-related emissions in agriculture are relatively minor when juxtaposed with other emission sources.

Nonetheless, the absolute energy footprint in agriculture is noteworthy. A vast amount of fossil energy is channeled towards activities like the production of fertilizers and pesticides, alongside heating, cooling, transportation, and electricity needs. The AgroFossilFree project [3] is engineered to foster the uptake of Fossil-Energy-Free Technologies and Strategies (FEFTS), aiming to slash fossil fuel consumption while endorsing the infusion of renewable energy sources.

FEFTS embody a diverse array of tools and methodologies designed for greener and more efficient energy production and consumption within agricultural practices. These strategies are delineated based on their distinctive applications across various agricultural technologies. The principal categories encapsulate Clean Energy Production and Energy Efficiency Enhancement, with a specialized category earmarked for Carbon Sequestration.

2 THE AGROFOSSILFREE PROJECT

The AgroFossilFree initiative (Figure 1) aims to facilitate the incorporation of Fossil Energy Free Technologies and Systems (FEFTS) in the agricultural sector by coordinating efforts among various stakeholders and disseminating information on FEFTS in agriculture.

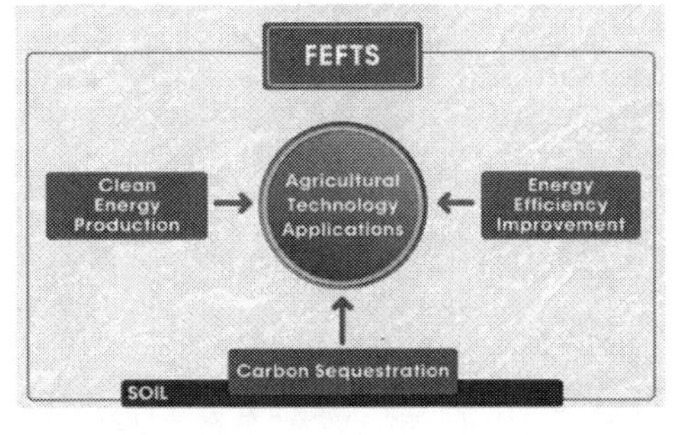

Figure 1: Overview on Fossil-Energy-Free Technologies and Strategies (FEFTS)

The project particularly emphasizes applications that encourage Renewable Energy Sources (RES) integration in key agricultural sectors, with a special focus on solar energies, such as photovoltaics, Agri-PV and solar thermal. Our research indicates a strong focus on solar energy and biomass applications in both research and industry.

The first goal of this project is to establish an online platform (AgEnergy Platform) where farmers, advisors, researchers, industrial partners, and other stakeholders can

10.4229/EUPVSEC2023/5DV.3.6
020562-001

access valuable information on FEFTS targeting agriculture. The ultimate goal would be to promote and facilitate the adoption of FEFTS, to this end, a FEFTS research roadmap and policy guidelines to de-fossilise EU agriculture will be published. The AgroFossilFree project aligns with high-level EU strategies, such as the EU Green Deal and the Farm to Fork strategy, by aiming to reduce fossil energy usage in farm operations. The project also seeks to bridge the gap between existing FEFTS and daily EU agricultural practices by promoting the effective exchange of innovative ideas and information among research, industry, extension services, and the farming community. Closing this gap will allow European agriculture to become more sustainable, reducing energy, environmental, and socioeconomic impacts. With all the knowledge gathered, organised, and presented on the platform, the European agricultural sector can support and adhere to fossil energy reduction policies, related legal and regulatory frameworks, and sustainable food production practices.

The expected results of the project are:

- Analysis and assessment of the current energy usage in EU agriculture and the necessary steps to achieve a European fossil-energy-free and independent agricultural sector.
- Identification and presentation of directly existing applicable FEFTS, ranging from applied research findings to market solutions, and evaluation of existing financial tools for de-fossilising activities.
- Collaboration with all pertinent stakeholders in thematic workshops (open-field agriculture, greenhouse and livestock), utilizing interactive physical and online methods to generate community-based ideas for integrating FEFTS into agricultural systems at regional and EU levels.
- Development of an online platform featuring all identified available FEFTS for stakeholders, along with a Decision Support Toolkit offering suggested interventions and financial tools based on user interaction.
- Formulation of policy guidelines at EU, national and regional levels to be communicated and to promote the adoption of FEFTS in agricultural activities in the near future.

The AgroFossilFree project, funded by the Horizon 2020 Programme of the European Commission (Grant Agreement Number 101000496), commenced in October 2020 and has a duration of 3 years. An overview on the status of the project is presented in the following chapter.

3 RESULTS

3.1 Current energy use status in EU agriculture

The energy landscape within EU agriculture has undergone a thorough evaluation, revealing a notable dependency on fossil fuels amidst significant energy usage across the sector. Key insights have emerged from the AgroFossilFree Project, yielding three scholarly articles published in reputable scientific journals. The initial paper, titled "Energy use in open-field agriculture in the EU: A critical review recommending energy efficiency measures and renewable energy sources adoption" [4], found its place in Renewable and Sustainable Energy Reviews. The subsequent paper, focusing on the livestock sector and titled "Energy Use in the EU Livestock Sector: A Review Recommending Energy Efficiency Measures and Renewable Energy Sources Adoption" [5], was featured in Applied Sciences. Lastly, a discourse on energy consumption within the greenhouse sector, named "Energy Use in Greenhouses in the EU: A Review Recommending Energy Efficiency Measures and Renewable Energy Sources Adoption" [6], was published in MDPI's Applied Sciences. These papers are organized based on the agricultural sectors they explore (Open-field, Livestock, and Greenhouse), with the principal findings elaborated in the ensuing section.

According to Eurostat, agriculture constitutes 3.2% of total energy consumption, with crude oil and petroleum products making up 56% of this quota, followed by electricity at 17%, gas at 14%, and renewables and biofuels at 9%. Yet, the AgroFossilFree project posits a 62% surge in energy utilization within the EU-27, should the indirect energy costs from fertilizer and pesticide production and transport be factored in.

Our published analyses further unveil stark variations in energy usage and its composition across different production systems (open-field, livestock, greenhouses). The AgroFossilFree project estimates the annual energy requisites at 1227 PJ for arable farming, 208 PJ for orchards and vineyards, 501 PJ for meat production, and 543 PJ for dairy systems.

In open-field agriculture, fertilizer application emerges as the primary energy consumer, accounting for about 50% of all energy inputs, with on-farm diesel usage trailing at 31%, seeds at 6%, pesticide production and usage at 5%, and other miscellaneous energy usages at 8%, primarily directed towards irrigation, storage, and drying [4]. Conversely, within livestock systems, except beef production, animal feed dominates the energy consumption chart, covering around three-quarters of all energy demands. The cultivation of animal feed devours about 60% of the EU's cereal production and necessitates significant high-protein imports. The on-farm electricity demand, predominantly sourced from fossil fuels, also stands out, albeit with variance across different production systems [5]. High-yield and energy-intensive greenhouses are chiefly driven by heating and cooling demands. In contrast, lower-yield and less energy-intensive setups largely depend on fertilizers, diesel for machinery, irrigation, and other operations, with energy demands in greenhouses showing variability and a general dependency on fossil sources [6].

This narrative accentuates the imperative for the EU to align with the Green Deal and Farm to Fork strategy goals, necessitating the embracement of innovative technologies, enhancements in energy efficiency, and the furtherance of non-fossil energy sources in agriculture. A shift towards more sustainable agricultural practices and farming systems is indispensable. For instance, conservation agriculture can markedly diminish on-farm energy consumption and carbon emissions, sequester substantial carbon amounts (up to 190M tonnes annually), and bolster the agricultural sector's climate resilience.

Tackling the topic of energy use in EU agriculture posed challenges due to the abundance and occasional inconsistency of data. Nonetheless, several domains could profit from extended research. Enhanced understanding and estimations of energy use in EU agriculture would gain from additional studies across all three primary production systems, with a special emphasis on greenhouse agriculture. Moreover, exploring the energy dynamics within non-conventional systems and minorly cultivated crops, alongside analyzing the correlations between farm attributes (size, location, specialization) and

10.4229/EUPVSEC2023/5DV.3.6
020562-002

energy use, is warranted.

Beyond these findings, significant preliminary efforts have been undertaken to a) ascertain the requirements, inventive concepts, and interests of farmers concerning FEFTS via surveys encompassing both adopters and non-adopters of FEFTS, as well as through interviews with advisors, and b) discover and advocate for innovative case studies on FEFTS. The insights garnered from these case studies have facilitated the development of multimedia resources showcasing innovative and market-ready FEFTS, thereby highlighting the transition towards a more eco-friendly and sustainable agricultural paradigm within the EU.

3.2 User needs, innovative ideas and interests of best practices on FEFTS

The analysis of data collected from a questionnaire-based survey of 470 end-users unveiled various factors that shape farmers' awareness and the subsequent decision to embrace (or not embrace) FEFTS and energy-saving technologies and practices.

Thus far, 50 Practice Abstracts (PAs) have been crafted and are accessible on the project's website, these abstracts serve as a direct conduit for disseminating valuable project-derived information to agricultural stakeholders of all sorts in a highly intuitive and comprehensible manner.

Additionally, within the AgroFossilFree project framework, innovation cases encompassing the entire innovation journey—from the inception of the identified need and the innovative idea, through to its evolution and eventual dissemination—were recognized and chronicled. This investigation into innovation cases employed the Spiral of Innovations framework (originated during the AgriSpin H2020 project), which delineates seven phases in an innovation process, starting from the initial idea to the innovation embedding stage. Each phase of the Spiral entails specific key activities and thus, the involvement of particular actors. To delve into the innovation processes inherent in 9 such cases, stakeholders pertinent to each case were interviewed, and both (written) 'case histories' and videos were generated. The latter were uploaded to the AgroFossilFree YouTube channel.

3.3 Inventories of FEFTS

A data collection inventory methodology was devised to populate the AgEnergy Platform, which was launched in early 2022. This methodology encompasses various categories designed to enhance the search functionality of the web platform. Information regarding FEFTS, supplied by both the project consortium and stakeholders, was input into the web platform. The specified categories of FEFTS showcased on the AgEnergy Platform include:

- Scientific papers;
- Research projects;
- Commercial technologies;
- Training and advising material;
- Financial support mechanisms.

A database was created and in total more than 2000 FEFTS entries registered, 1960 of which were published on the platform as more intriguing and comprehensive content provision (see Table I). All the entries are enlisted in the AgEnergy Repository and were made publicly available only after a thorough review.

It's notable that a substantial portion of the FEFTS entries submitted thus far, totalling 442, showcase solar energy solutions, implementation strategies, policy recommendations, financial mechanisms, and more, underscoring the transition towards solar energy-centric sustainable solutions. External users now have the capability to register new entries on the publicly accessible AgEnergy online portal.

Table I. FEFTS in the AgEnergy Platform per category.

FEFTS Category	Number of FEFTS
Scientific Paper	977
Research Project	180
Commercial Product	543
Training Material	131
Financing Mechanism	129

3.4 AgEnergy Platform

As delineated earlier, AgroFossilFree initiated the AgEnergy Platform (https://platform.agrofossilfree.eu), housing all available FEFTS in a format that's both accessible and easily comprehensible for end-users (Figure 2).

The platform was officially unveiled in January 2022. All submitted FEFTS are organized under three fundamental categories: Clean Energy Supply, Energy Efficiency Improvement, and Soil Carbon Sequestration. The platform (Figure 3), along with the FEFTS encompassed within, is in the process of being translated into several languages spoken within the consortium, namely English, Greek, Polish, Spanish, German, Dutch, and Danish. This endeavor ensures that stakeholders from various countries can access the pertinent information in their native tongue.

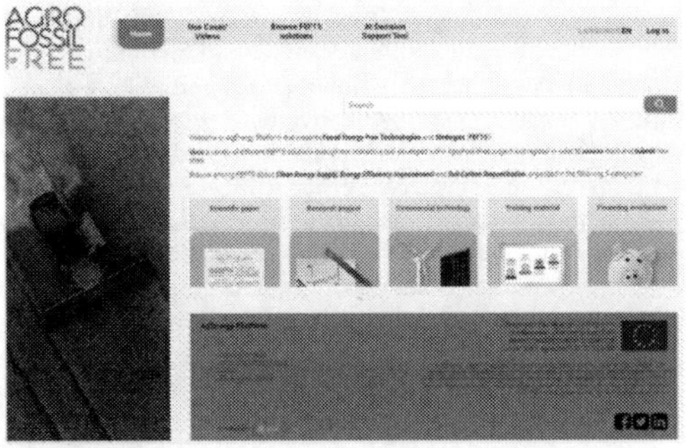

Figure 2: Homepage of the AgEnergy platform

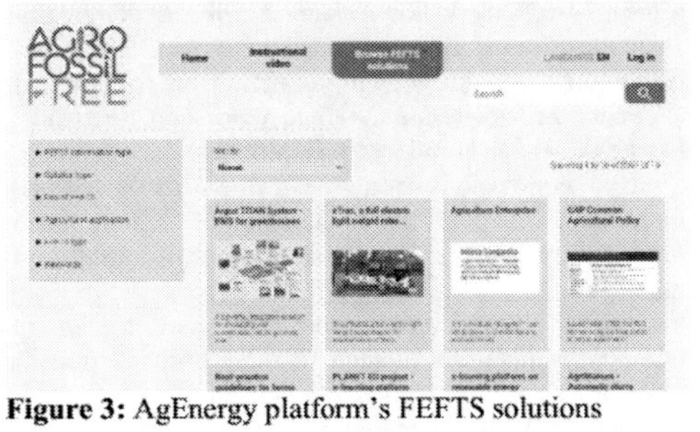

Figure 3: AgEnergy platform's FEFTS solutions

The platform provides detailed filters which allows end users to find any FEFTS targeting their needs by using a variety of methods, such as FEFTS information type, use of FEFTS, FEFTS type and related keywords.

Furthermore, all registered users can evaluate the

available FEFTS solutions by answering a 10-question survey on general information of the FEFTS' degree of technology readiness as well as the environmental and socioeconomic impacts.

In addition, a Decision Support Tool (DST) has been crafted and is accessible at https://dst.agrofossilfree.eu/ [7]. This free online tool functions as an advisory conduit, furnishing interested stakeholders with tailored solutions aligning with their needs. The tool is engineered to emulate the advisory process one would experience from a consortium of experts, evaluating the user-provided data to suggest the most pertinent interventions for each farm. Employing Fuzzy Cognitive Maps, this DST is seamlessly amalgamated within the AgroFossilFree platform, enabling users to obtain a ranking of technologies most aligned with their farm's needs.

The DST excels in offering an initial glimpse into which FEFTS would most adequately meet the demands of a particular farm, requiring minimal input from the user. Specifically, a concise six-question questionnaire empowers the DST to grasp the user's needs. While it can't supplant the in-depth analysis requisite for final investment decisions, it serves as an indicative precursor in a study process aiming to weigh diverse options and pinpoint the optimal investment for each farm. The repository of FEFTS within the AgEnergy Platform serves as the foundational input for the DST, ensuring a dynamic methodology is maintained in tandem with the platform's ever-expanding content.

3.5 Interactive and multi-actor regional workshops

AgroFossilFree partners orchestrated twenty-four multi-actor workshops across eight European countries, with each country hosting three workshops focused on (i) Open-field agriculture; (ii) Greenhouses; and (iii) Livestock buildings. These workshops facilitated the direct dissemination of innovative FEFTS solutions, while also collecting grassroots innovations and requirements at the regional level by convening stakeholders from research, extension services, industry, and the farmers and producers community. Specifically, these regional workshops garnered insightful information regarding the prerequisites and barriers to FEFTS adoption that stakeholders are presently encountering. Concurrently, to stride towards sustainable EU agriculture, it's imperative to contemplate the additional policy recommendations, incentives, and new partnerships that ought to be instituted and fostered.

Subsequent to the regional workshops, three transnational workshops were executed. The inaugural transnational workshop, themed around Greenhouse agriculture, took place in Athens, Greece. The second, centered on open-field crop production, was conducted in Warsaw, Poland, while the final workshop, focusing on livestock, was held in Herning, Denmark.

Key conclusions from the transnational workshops included: a) the pathways to adopting FEFTS are through ongoing training (inclusive of e-platforms and online tools), advisement on new FEFTS, demonstration farms, financing FEFTS startups/spin-offs, and engaging in multi-actor research projects, b) the pace of transition towards FEFTS is perceived to hinge on fossil fuel pricing, technological advancements, and shared efficiency enhancement objectives at the EU partnership level, c) to a certain degree and in select countries, pertinent and innovative policies are in place, yet a stronger policy support framework is desirable. Various policies were proposed to underscore the support for FEFTS adoption and advancements in energy efficiency, with a broad spectrum of stakeholders recommending that these policies hone in on bolstering returns on investments, d) pivotal policy measures encompass subsidies, particularly targeting small and medium-sized farms, e) the existing Agricultural Knowledge Information System (AKIS) bridging researchers, industry, advisors, and farmers is deemed inadequate concerning the transition to FEFTS and necessitates EU backing for optimizing impact at both national and European levels.

The transnational workshops offered valuable insights into the perspectives of stakeholders from the eight innovation hubs. In open-field agriculture, the three operations identified as most energy-intensive, hence requiring attention in terms of energy consumption, include fertilization, tillage, and transport. The promotion of precision application systems in fertilizer spreading and the use of biofertilizers not derived from fossil fuels (particularly nitrogen) should be intensified. To conserve both fossil fuel and inorganic fertilizers, the adoption of cover crops and legumes, bio-fertilizers, and conservation agriculture practices were recommended. There's a call for upgrading agricultural machinery utilized in farming practices and goods transportation to enhance energy efficiency. The consideration of renewable electrification and alternative renewable fuels was also highlighted.

In the livestock sector, the emphasis is on precision livestock techniques to minimize feed consumption, improving building envelope insulation, and installing photovoltaic systems on rooftops alongside heat pumps for HVAC (Heating, Ventilation, and Air Conditioning) systems, which are deemed the more complex systems to integrate FEFTS effectively. With respect to greenhouses, energy consumption is chiefly tied to heating and irrigation application. Notably, precision agriculture and conservation agriculture garnered interest from farmers, who view them as energy-efficient practices closely related to their daily operations.

Another significant observation is that the current scenario of escalating energy costs presents a conducive opportunity for transitioning to renewable energy and energy-efficient systems, which could yield substantial cost reductions.

Moreover, three online workshops were conducted on February 25th and October 25th of 2022, and May 26th of 2023, engaging stakeholders spanning across Europe—comprising farmers, industry delegates, advisors, technicians, researchers, innovation brokers, and policymakers. These workshops unveiled the project's chief deliverables, like the AgEnergy Platform and the DST, and encapsulated key research insights regarding farmers' needs, innovative concepts, and interests in agricultural defossilization technologies and strategies, all gleaned from both regional and transnational workshops. Participants also received a glimpse into successful FEFTS implementations, as elucidated by various experts, with instances including biogas and bio-methane tractors. Over 180 individuals attended these sessions, sharing their viewpoints on the prospective adoption of FEFTS in European agriculture. Recordings of the webinars are accessible on the AgroFossilFree YouTube channel.

4 CONCLUSIONS

In conclusion, the AgroFossilFree project endeavors to

10.4229/EUPVSEC2023/5DV.3.6
020562-004

foster the uptake of Fossil-Energy-Free Technologies and Strategies (FEFTS) within the agricultural realm, with a keen emphasis on melding renewable energy sources, notably solar energy, into primary agricultural sectors. The paramount objective of the project is to guide agricultural stakeholders via the AgEnergy Platform and the DST, facilitating access to invaluable information on FEFTS that could prompt them to re-evaluate and potentially supplant their customary agricultural practices with strategies aligned with the EU's green transition aspirations. Practical abstracts alongside thorough successful case study videos are also accessible via the project's website, presenting tangible examples of defossilization practices implementation.

This project resonates with the EU Green Deal and the Farm to Fork strategy's ethos, targeting a reduction in fossil energy employment in agricultural operations. By narrowing the chasm between extant FEFTS and prevailing agricultural practices, it catalyzes the interchange of innovative ideas and information amongst stakeholders, thereby nurturing a more sustainable European agriculture characterized by diminished energy consumption and environmental repercussions.

Significant strides have been made by the project in scrutinizing the current energy usage within EU agriculture and pinpointing realms for enhancement. It has also amassed user needs, innovative notions, and best practices pertinent to FEFTS throughout the EU.

Multi-actor regional workshops have been orchestrated to collate grassroots innovations and requisites, while transnational workshops have rendered insights into policy measures and critical focal areas. Concrete policy recommendations alongside new research avenues were published in September 2023 (marking the official conclusion of the project). Throughout the aforementioned outcomes, the significance of solar energy technologies, precision agriculture, conservation agriculture, and the embracement of renewable energy plus energy-efficient systems across diverse agricultural operations has been underscored.

Through its investigative endeavors, communication, dissemination, and engagement with stakeholders, the AgroFossilFree project propels the transition towards a more sustainable and fossil-energy-free agricultural sector.

5 References

[1] European Parliament (2020) Briefing: EU agricultural policy and climate change. - https://www.europarl.europa.eu/RegData/etudes/BRIE/2020/651922/EPRS_BRI(2020)651922_EN.pdf

[2] Gołasa P. et al. (2021) Sources of Greenhouse Gas Emissions in Agriculture, with Particular Emphasis on Emissions from Energy Used. - Energies 2021, 14, 3784.

[3] Ma, C., Rutz, D., Hofmeier, V., Janssen, R., Balafoutis, A., Vaiopoulos, K., Vandorou, F., Grøn Sørensen, C.A., Manolakos, D., Koutsouris, A., Paris, B., Papadakis, G., Borzecka, M., Wydra, M., Biševac, V., Creupelandt, D., Román, J., Rossi, D., Próchniak, M., Kaminiaris, M., Tsiropoulos, Z., Brinks, H., Caslin, B., Sneij, J., Asín, M.; Updates from the AgroFossilFree Project - Strategies and Technologies to Achieve a European Fossil-Energy-Free Agriculture. 8th World Conference on Photovoltaic Energy Conversion, 1645 - 1649.

[4] Paris, B.; Vandorou, F.; Balafoutis, A.T.; Vaiopoulos, K.; Kyriakarakos, G.; Manolakos, D.; Papadakis, G. Energy use in open-field agriculture in the EU: A critical review recommending energy efficiency measures and renewable energy sources adoption. Renew. Sustain. Energy Rev. 2022, 158, 112098. 1,

[5] Paris, B.; Vandorou, F.; Tyris, D.; Balafoutis, A.T.; Vaiopoulos, K.; Kyriakarakos, G.; Manolakos, D.; Papadakis, G. Energy Use in the EU Livestock Sector: A Review Recommending Energy Efficiency Measures and Renewable Energy Sources Adoption. Appl. Sci. 2022, 12, 2142.

[6] Paris, Bas, Foteini Vandorou, Athanasios T. Balafoutis, Konstantinos Vaiopoulos, George Kyriakarakos, Dimitris Manolakos, and George Papadakis. 2022. "Energy Use in Greenhouses in the EU: A Review Recommending Energy Efficiency Measures and Renewable Energy Sources Adoption" Applied Sciences 12.

[7] Kyriakarakos, G.; Balafoutis, A.T.; Vaiopoulos, K.; Abdul, M.; Voskakis, M.; Kaminiaris M.D.; Tsiropoulos, Z.; Bochtis, D. Design and implementation of a Decision Support Tool to assist the ranking of fossil-energy-free technologies and strategies for a given farm, Smart Agricultural Technology, 2023, 4, 100169, 2772-3755.

10.4229/EUPVSEC2023/5DV.3.6
020562-005

AN URGENT CALL TO IMPLEMENT THE HEAT TRANSITION IN EUROPE – – SECTOR COUPLING BETWEEN THE HEATING AND POWER SECTORS

Dominik Rutz[1,2], Olgu Birgi[1,2], Andrej Misech[2,3], Rainer Janssen[1,3]

[1]WIP – Renewable Energies, Sylvensteinstr. 2, D - 81369 Munich, Germany
Tel. +49 89 720 12743, Fax +49 89 720 12791
E-Mail: dominik.rutz@wip-munich.de
Internet: www.wip-munich.de, www.rhc-platform.org

[2]European Technology and Innovation Platform on Renewable Heating and Cooling (ETIP RHC),
[3]The Association of European Renewable Energy Research Centers (EUREC)

ABSTRACT: Given the urgency of the global energy crisis caused by the war in Ukraine, the heat transition in Europe must be urgently implemented. To this end, sector coupling between the heating and cooling sector and the power sector is vital for the future energy mix. The increase of renewable energies in both sectors develops at different speeds in Europe and in the different European countries. While only little progress was made in the heating and cooling transition, the decarbonisation of the power sector in Europe is proceeding faster. This positive trend, which needs to be speeded up, is facing the challenge of a stable power grid with increasing shares of intermittent renewable energies such as PV. The coupling of the power and the heating and cooling sectors is urgently needed to benefit from synergies. For example, the use of smart heat pumps can contribute to stabilizing the power grid (through intelligent operation and heat storages), yet on the other hand, increasing numbers of connected heat pumps will threaten the power grid stability during winter due to the high load. This necessitates other renewable heating systems beyond heat pumps, such as solar thermal energy or biomass systems. This applies both to individual heating systems and district heating systems. The use of PV is especially promising for cooling purposes, as cooling demand in summer matches with power generation through PV in summer. The contribution to the EU PVSEC highlights, on behalf of the European Technology and Innovation Platform on Renewable Heating and Cooling (RHC-ETIP), opportunities for sector coupling between the heating and cooling sector and the power sector, and the special role of PV. It presents the joint position paper of ETIP RHC and ETIP Bioenergy on renewable hydrogen or a joint position paper on sector coupling of ETIP RHC and ETIP Smart Networks for Energy Transition (SNET).

Keywords: heating, consumers, renewable heating, energy efficiency, district heating, phase out plans

1 INTRODUCTION

Each year, almost 50% of the final energy consumed in Europe is used for heating, cooling or refrigeration either for residential, tertiary or industrial purposes. The vast majority (around 75%) of this energy demand is met by the combustion of fossil fuels such as oil, gas, and coal – which do considerable environmental damage, including in terms of greenhouse gas emissions and air pollution – while only 19% is generated from renewable energy. Today the social, environmental, and economic costs of climate change make the mission of embracing more sustainable sources of energy urgent. Decarbonising the heating and cooling sector will be key to achieve the European Green Deal goals and making Europe a climate-neutral economy by 2050. The recast of the Renewable Energy Directive increases the current EU-level target of "at least 32 % of renewable energy sources in the overall energy to at least 40 % by 2030". Specific targets are proposed for renewable energy use in heating and cooling, among other sectors. To meet both our climate and environmental goals, the momentum for deploying renewable H&C (RHC) technologies has never been stronger. Finally, the energy transition in the heating and cooling sector is of high importance to gain energy independence and to increase security of energy supply. Although this aspect was always important, it gained top priority in these days due to the current war in Ukraine and the related impacts on energy imports to Europe.

Recognizing this situation, the European Technology and Innovation Platform on Renewable Heating and Cooling (RHC-ETIP), aims at playing a decisive role in protecting Europe's leading position in renewable heating and cooling technology. We want society to benefit from an increased contribution of renewable heating and cooling, which will help to reach the vital goal of a carbon-neutral world by mid-century.

Based on previous contributions (Rutz et al., 2020), the paper again highlights the urgency to implement the heat transition in Europe, but with even higher urgency due to the global energy crisis

10.4229/EUPVSEC2023/5DV.3.7
020563-001

caused by the war in Ukraine. It furthermore presents the joint position paper of ETIP RHC and ETIP Bioenergy on renewable hydrogen or a joint position paper on sector coupling of ETIP RHC and ETIP Smart Networks for Energy Transition (SNET).

2 THE RHC-ETIP

The RHC-ETIP, officially endorsed by the European Commission since October 2008, aims at playing a decisive role in maximizing synergies and strengthening efforts towards research, development, and technological innovation, which will consolidate Europe's leading position in the sector (RHC-ETIP, 2020).

The platform is structured according to so-called technology panels on biomass, solar thermal energy, geothermal energy and district heating and cooling & storage, and heat pumps. The Technology Panels are a pool of experts in charge of providing specific technology input to the strategic documents produced by the Horizontal Working Groups and to the activities of the European Technology and Innovation Platform on Renewable Heating and Cooling.

In October 2017, a restructuring process started, and Horizontal Working Groups (HWGs) were introduced, fostering its multisectoral and multidisciplinary approach. HWGs bring together interested experts from different technology panels to work on common horizontal topics, defined on the basis of main challenges to be addressed by the RHC-sector. Current HWGs are:

- 100% RE Buildings
- 100% RE Industries
- 100% RE Cities
- 100% RE Districts
- Social Sciences and Humanities

The RHC-ETIP has recently released the "2050 vision for 100 % renewable heating and cooling in Europe" (RHC-ETIP, 2019). Therein, the Members of the Platform unanimously concluded that a full switch to 100% by 2050 can only be reached with a very strong and resolute political will to change the H&C sectors. Technologies for the substitution of fossil fuel systems are mature and commercially available, but without the right political decisions, this target cannot be reached. Thus, there is an urgent call to politicians at the EU level and at the national levels to make courageous decisions for the phase-out of fossil space heating systems and for the support of renewables in the heating sector.

Figure 1: Logo of RHC-ETIP

3 POLICY DEVELOPMENTS

The focus in the political debate about the energy transition in Europe was for a long time only placed on the power sector. With the development of electric cars, also the decarbonisation of the transport sector was spotlighted in the last few years. The transition towards renewable energies in the heating and cooling sector was for a long time neglected in the political debate. However, due to raising energy prices of oil and gas caused by the Russian aggression towards Ukraine, the heating and cooling sector is finally at the core of today's discussions.

The Renewable Energy Directive (RED) is currently under its 3rd revision (RED III) and it is planned to strengthen measures to support renewables uptake in the heating and cooling (EC, 2023). In the elaboration of the RED III, it was discussed to remove the use of primary woody biomass from the scope of the directive. This would have had a serious impact on the fulfilment of the energy transition and climate change mitigation targets. Therefore, it is good that the use of primary woody biomass is not excluded from the current version of the RED III as renewable energy source.

Besides the RED, the Energy Performance of Buildings Directive (EPBD) is also an important policy tool supporting the energy transition in the heating and cooling sector. This directive is likewise currently under revision. One of the core aspects is placed on how to support zero-emission buildings.

While the discussions at the European level are going on, also Member States have already banned the installation of new fossil heating systems, e.g. Denmark for oil and gas boilers in new buildings. This should be the benchmark for all countries in Europe and also for the European Parliament, setting the European framework for the Member States.

Further countries are in the process of introducing new legislation in the heating and cooling sector. For example, in Germany, a Building Energy Law (Gebäudeenergiegesetz, GEG) is proposed and currently discussed in the parliament. It includes a strategy on how to phase-out oil and gas boilers and it is proposed to allow only heating systems in new buildings that use minimum shares of 65% renewable energies. Furthermore, the end date for the use of fossil fuels in heating systems is planned for December 31, 2044. From 2045, all buildings must be heated in a climate-neutral manner

10.4229/EUPVSEC2023/5DV.3.7
020563-002

using only renewable energies. The law makes the replacement of heating systems binding. The legislation will be finalised in summer 2023 and will come into force as of January 2024. This legislation has the potential to make a radical change for the renewable heating and cooling sector which is highly acknowledged by ETIP RHC.

4 POLICY PAPER ON SECTOR COUPLING

The role of electrification is expected to steadily increase in the overall energy system. Therefore, the ETIP RHC and ETIP Smart Networks for Energy Transition (SNET) have elaborated in 2022 a White Paper on the coupling of the heating/cooling and electricity sectors in Europe (Wiedermann et al, 2022), of which the executive summary is summarized here.

The paper focuses on the implications of coupling the electricity, heating, and cooling vectors as a vital part of efficiently decarbonising the whole European energy sector. To achieve decarbonisation, different renewable energy technologies for electricity and heating and cooling are needed, and in many cases, they could be coupled to increase energy efficiency. In addition to the traditional connection of heat and electricity in combined heat power (CHP) and Cogeneration plants, this White Paper elaborates on promising, often decentralised generation side as well as demand side technologies and solutions aimed at achieving efficient decarbonisation in terms of both energy efficiency and infrastructure optimisation.

Heating and cooling must profoundly and quickly contribute to the reduction of greenhouse gas emissions. This can be achieved by increasing the share of renewable sources for heat and cold generation and through direct and indirect electrification (sector coupling solutions). To build up a reliable, sustainable, and cost-effective energy supply system in Europe, flexible generation of electricity, heat and cold, and any kind of energy storage are needed to ensure the security of supply and decarbonisation. Both the successful implementation of sector coupling (cross-vector sector coupling and the end-use sector coupling) and a fully integrated energy system in a holistic power system architecture comprising all interactions within the power system itself, between the network - generation - and storage operators, consumers, and prosumers, through market mechanisms, are mandatory for a successful energy transition and decarbonised supply of heat and cold.

The White Paper provides an overview of the different sector coupling technologies: Renewable energy conversion technologies (solar-based combined generation, direct conversion of solar energy into heat, geothermal-based generation, biomass technologies, and hydrogen-based technologies), Renewable heat and heat recovery technologies (direct conversion of renewable electricity into heat, heat pump technologies, solar thermal technologies, innovative waste heat recovery cycles), Polygeneration (Cogeneration, Trigeneration) and District heating.

This White Paper also describes the important role of the different storage technologies (e.g., thermal energy storage, chemical storage, i.e., hydrogen or synthetic fuels as the basis for decarbonised CHP and Cogeneration application) as sector coupling components to store excess renewable energy and thus to contribute to a sustainable and reliable energy system. Due to the focus on heat/cold applications, other forms of storage within the power grid (e.g., direct electrochemical storage, hydro pumps, electrical/magnetic storage) are not considered in this document. In any case, it is well understood that batteries deliver significant contributions to storing excess renewable energy, especially in the context of sector coupling with the mobility system, ancillary services, and home electricity storage solutions.

The existing electrification and heating/cooling technologies are the basis for a fully decarbonised energy system. Their maturity status is briefly outlined, and the underlying message is that some of the existing technologies need to be scaled up to become economical. The importance of storage as a key element for coupling in all sectors is also highlighted. However, certain measures and efforts regarding R&D and market stimulation are needed to improve the technologies and to allow full system integration, thereby capturing the benefits of sector coupling.

R&D needs and challenges are described to further develop the components and solutions necessary to deal with the increasing share of renewable energies in the grid, e.g., for sector coupling components, heat and power demand management systems, storage from energy system integration viewpoint and innovative business models.

Public funding from European instruments is necessary to accelerate the development of the related technologies/systems, reduce the technical and business risks, and support market uptake. From plant operators' perspective, CAPEX and/or OPEX funding from European funding instruments is necessary to reduce the financial risks and to contribute to establishing positive business cases.

As the transition towards a green future must be accelerated, new concepts should be already available and verified, at least at the prototype level. Available technologies need to be quickly upscaled from pilot to real plant sizes, and investments must be stimulated by regulations and incentives based on, e.g., CO2-pricing worldwide emission trading systems.

To make clean and integrated technologies the standard, the market framework and the political

10.4229/EUPVSEC2023/5DV.3.7
020563-003

mindset must change to make these technologies economically more competitive. Furthermore, renewables-based solutions should benefit from streamlined permitting and be given financial and institutional incentives while removing administrative burdens. The potential for intensified use of District Heating and Cooling (DHC) systems is meaningful across the EU.

The energy transition towards renewable heating and cooling requires both technical, legal, and economic skills. While the economic and legal framework is currently being shaped, a lack of design, planning and installation skills emerges. Quite often, the different sectors compete for the same competencies. Hence a thorough assessment and EU-wide coordination effort are recommended to ensure the education and training of a sufficient number of designers, planners, and installers by making the "renewable energy" career path attractive and providing enough education and (re-)training facilities.

5 POLIOCY PAPER ON HYDROGEN

Due to the increasing debate about hydrogen as one of the solutions of the energy transition, ETIP RHC and ETIP Bioenergy have issued a joint position paper on renewable hydrogen (Rutz et al., 2023).

It conveys that renewable hydrogen is an energy carrier which can be one contributor, among many, to support the energy transition in Europe. However, its production and use are subject to many limitations, particularly of an economic nature. Renewable hydrogen will only play an important role in value chains where the specific properties of hydrogen are of high value. For both the bioeconomy, as well as for the renewable heating and cooling sector, multiple production routes and uses of hydrogen could be potentially applicable.

On the other hand, there is the risk that excessive focus on certain capital-intensive technologies (e.g., hydrogen) to the detriment of others with much faster impacts, can be a serious obstacle to a rapid and efficient decarbonization of the thermal sector. It is therefore important to carefully evaluate the role of hydrogen in a hypothetical future under the assumption of large investments, but also to evaluate its potential niche applications in comparison to commercially available renewable thermal technologies that can meet the same needs in a shorter term and with higher profitability.

The position paper concludes that the production and use of renewable hydrogen can be an important building block for the energy transition, but certainly not the only or even the most important solution. The key messages are:

- We need massive financial support on innovations, research, further development, and market uptake for all types of renewable energy sources and systems. Financial support on hydrogen should not cut or compromise the support for renewable energies.
- Hydrogen per se is not necessarily sustainable as it is not a source of energy, but only an energy carrier. The sustainability of hydrogen is closely related to the source used for its production. We consider that only renewable hydrogen, meaning hydrogen produced by renewable energy sources, can be in principle sustainable.
- In the light of the climate challenge and the enormous transition we are facing, we strongly recommend focusing any support on all forms of renewable hydrogen and to stop supporting fossil or low-carbon hydrogen.
- Hydrogen is a universal energy carrier and can be used for many purposes. Yet, it will be impossible to satisfy all demand for hydrogen, and thus it should be used for applications with only few other alternatives, also called hard-to-abate sectors. These can include, for example, high temperature industrial processes or selected transport applications.
- We do not see that the direct use of hydrogen in the domestic heating and cooling sector plays an important role in the future. Instead, a massive deployment of different renewable energy technologies, including liquid, gaseous and solid biofuels, must happen and be supported.
- With the position paper, we are providing feedback to the European Hydrogen Strategy on the role we see for hydrogen for a climate-neutral Europe. We acknowledge the definitions of the European Hydrogen Strategy on renewable hydrogen, but we feel that other pathways such as biohydrogen and solar hydrogen are neglected and should be addressed as well.
- In terms of energy efficiency, heat that is released during hydrogen production should be utilised.
- Various pathways can be taken to produce hydrogen from renewable sources. The pathways are at different levels of technology and market maturity and come with specific R&D needs, costs and GHG emissions.

6 CONCLUSIONS

RHC-ETIP urgently calls politicians, policy makers and scientists to support the phase-out of fossil space heating systems and to support renewables in the heating sector. The authors of the present paper urgently call politicians and policy makers at European, national and regional levels to implement the following measures:

- To stop the support of any kind of fossil fuel heating system with public money.
- To ambitiously support the energy transition in the heating sector, supporting renewable energy systems.
- To develop clear and ambitious plans and

10.4229/EUPVSEC2023/5DV.3.7
020563-004

legislation for the phase-out of fossil heating systems for new buildings, but also for the existing building stock.

REFERENCES

EC (2023) Renewable energy directive. - https://energy.ec.europa.eu/topics/renewable-energy/renewable-energy-directive-targets-and-rules/renewable-energy-directive_en

Rutz D., Janssen R. Boutikos P., Atsonios K., Grammelis P., Brunner C., Calderoni M., Friedmann M., Müller-Langer F. Mata T., Martins A., Haslinger W. (2023) Renewable Hydrogen, Opportunities, limitations and threats of hydrogen for the energy transition in Europe - Position Paper of ETIP RHC and ETIP Bioenergy. - https://www.rhc-platform.org/content/uploads/2023/03/RHC-Renewable-Hydrogen-paper_v2.pdf

Rutz, D., Ball, I., Di Costanzo, B., Janssen, R., Tretter, H., Knaus, K., Drexlmeier, S., Baumann, C., Puente, F., Šegon, V., Balić, D., Silajdzic, F., Arnaut, S., Nikolaev, A., Jerotić, S., Stegnar, G., Markovska, N., Duic, N., Skreiberg, Ø., Mišech, A., Arrowsmith, G. (2020) An Urgent Call for the Phase-Out of Fossil Space Heating Systems and for the Support of Renewables in the Heating Sector. - 28th European Biomass Conference and Exhibition; ISBN: 978-88-89407-20-2; Paper DOI: 10.5071/28thEUBCE2020-5AV.3.19; pp. 930 – 932

RHC-ETIP (2020) Mission. - https://www.rhc-platform.org/about-us/mission/ [10.07.2020]

RHC-ETIP (2019) 2050 vision for 100 % renewable heating and cooling in Europe. - https://www.rhc-platform.org/wp_content/uploads/2019/10/RHC-VISION-2050-WEB.pdf [10.07.2020]

UNFCCC (2020) The Paris Agreement. - https://unfccc.int/process-and-meetings/the-paris-agreement/the-paris-agreement [10.07.2020]

Wiedermann A., Calderoni M., Bernstrauch O., Herce C., Nowak T., Jansohn P., Pastor R., Rutz D., Iliceto A., Ionel I, Ilo A., Urchueguía J.A., (2022) Coupling of Heating/Cooling and Electricity Sectors in a Renewable Energy-Driven Europe. - https://www.rhc-platform.org/content/uploads/2022/11/coupling-of-heatingcooling-and-electricity-sectors-MJ0922659ENN.pdf

10.4229/EUPVSEC2023/5DV.3.7
020563-005

POTENTIAL ENERGY LABEL FOR MONOFACIAL AND BIFACIAL PV MODULES

Teodora Stoyanova Lyubenova*[1], Ewan Dunlop[1], Tony Sample[1], Nigel Taylor[1], Davide Polverini[2]
[1] European Commission, Joint Research Centre, Ispra, Italy.
[2] European Commission, DG Internal Market, Industry, Entrepreneurship and SMEs, Brussels, Belgium
teodora.lyubenova@ec.europa.eu* (Tel: +39 0332786574), ewan.dunlop@ec.europa.eu, tony.sample@ec.europa.eu, nigel.taylor@ec.europa.eu, davide.polverini@ec.europa.eu

ABSTRACT: Photovoltaics (PV) is one of the major renewable energy technologies needed to meet the European climate and energy targets for "green transition" by ensuring affordable, sustainable and easily deployed energy. In the future new PV modules installed in the European Union (EU) will be expected to fulfil requirements of the proposed Ecodesign Directive and Energy labelling Regulation, a legal instruments with the aim of creating a more sustainable European PV market. The implementation of the Ecodesign and Energy label would ensure that the worst performing products cannot be placed on the EU market and through the labeling scheme would empower consumers with a simple way to make informed choices on the most efficient and sustainable products. Both legislative tools are currently in their final preparation stage before being potentially accepted and implemented in 2024. This paper discloses recent updates, currently under evaluation, of the method for estimating of the energy yield of bifacial PV devices that have not been discussed before; of a simplified label design and of new ranges for the energy classes.
Keywords: Ecodesign; energy label; photovoltaics, bifacial, energy yield.

1 INTRODUCTION

Solar PV is one of the major renewable energy technologies that will play a key role towards the decarbonisation of our society and that are able to meet the European Green Deal's climate and energy targets [1]. In 2022, 45% more solar power capacity has been installed (i.e., 239 GW of new solar PV) worldwide, and total capacity now at the TW-scale [2]. In 2023, the Solar PV reached 14.7% of cumulative electrical power capacity and it is expected to increase up to 22.2% in 2027 leading it to overtake fossil fuels in the global energy scenario (Fig.1) [3]. Nowadays, solar PV is the lowest-cost and most easily deployed clean energy available, and it is also considered as "the cheapest energy source in history" [4].

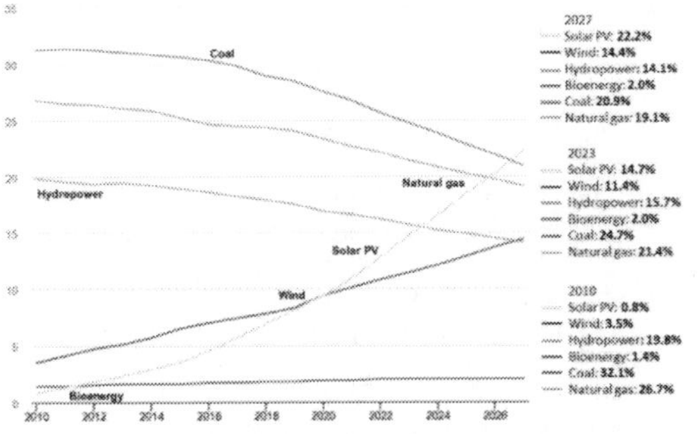

Figure 1: Share of cumulative electrical power capacity by technology, 2010-2027 [3].

In the context of the Green Deal Industrial Strategy for Europe that plans to support the "green transition" and to make EU industry more competitive and autonomy globally, it is of key importance to ensure affordable and sustainable energy that could guarantee energy security and independence in Europe [5, 6]. Newly installed PV capacity is projected to be over 320 GWac by 2025 and almost 600 GWac by 2030 (REPowerEU and EU solar strategy) [7]. The PV modules installed in the European Union (EU) should be sustainable and durable; this can be attained by means of regulatory measures, in particular in the framework of the Ecodesign [8] and Energy labelling [9] legislation. These policy regulations are focused to boost quality of products on the market and to shape sustainable consumer behavior by providing the information needed to make informed choices.

The preparatory study for Ecodesign and the Energy label regulations for PV products such as modules, systems and inverters was launched in 2017 by the European Commission (EC) [10]. The feasibility of the implementation of these sustainable policy instruments to the PV products have been ensured by analysis of a environmental impacts and related policy approaches of PV products that was carried out by the Joint Research Centre of the European Commission [11, 12]. The study concluded that improvements in the energy yield and long-term performance of PV modules and inverters could be guaranteed by implementing the above-mentioned mandatory legal instruments.

Currently, the proposed regulation is approaching its final stage. The aim of this paper is to communicate several technical updates that are currently under evaluation during the preparation of the regulatory measures. These concern in particular the methodology for calculation of energy efficiency index (EEI_M), i.e., a parameter placed on the label, a new simplified label design and revision of the granularity of the energy efficiency index (energy label classes).

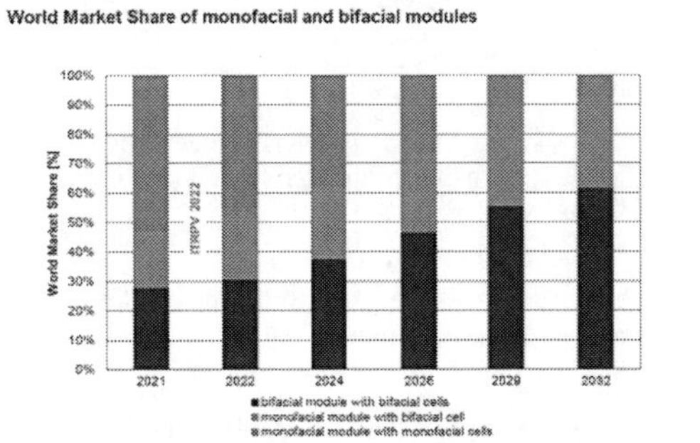

Figure 2: Current and predicted market share of bifacial modules [13].

The updated product scope now covers PV modules and inverters (PV systems have been withdrawn). The present paper is focused only on PV modules. It treats monofacial and bifacial devices based on silicon and thin

10.4229/EUPVSEC2023/5DV.3.10
020565-001

film PV technologies. Bifacial PV devices have been included in the scope due to their increasing use. Their market share has increased to 30% for 2022 up from 10% in 2018 and are expected to reach more than ~60% market share by 2032 [13] (Fig.2).

Implementing the measures for Ecodesign typically foresee the use of harmonized standards, adopted by recognized European standardization organizations (European Committee for Standardization (CEN) or European Committee for Electrotechnical Standardization (CENELEC), for the testing and calculation methods necessary for the compliance assessment. However, when harmonized standards are not available (as is the case for the energy rating of bifacial PV modules), other already existing standards or methods may be considered when it is demonstrated that they are suitable for assessing the compliance of products with the applicable requirements (Ecodesign or energy labeling). Hence here a methodology for estimating of the energy yield of bifacial devices is proposed, until harmonized technical norms that covers these and other related aspects become available to use. The inclusion of bifacial devices in the scope of the Regulation is expected to accelerate the standardization process by a possible modification of the EN IEC 61853-series for monofacial devices or through the issuance of a new standard). Thus, the present work can promote international standardization and support uniform global business opportunities.

2 METHODOLOGY

2.1 Energy efficiency index (EEI_M)

The applied methodology is governed by the product scope of the Regulation, i.e., individual PV modules placed on the EU market and intended for use in PV systems for grid-connected electricity generation. The PV modules are energy generating products and their 'energy efficiency' when related to energy conversion processes, represents the conversion efficiency, i.e., the ratio of generated end-use energy in proportion to the primary energy [14]. The parameter used to define the ability of the product to generate electricity over time is expressed by the so-called energy efficiency index (EEI_M, kWhm^{-2}). That is the ratio between the estimated annual energy yield (kWh) per unit area (m^2) of the PV module (equation 1).

$$EEI_{M\_c} = {EY_{M(DC)Y1\,c}}/{A_M} \tag{1}$$

where

EEI_{M_c} is the Energy efficiency index for a module (EE_M) placed in a European reference climate *c*, expressed in kWhm^{-2}.

$EY_{M(DC)Y1c}$ is the DC energy yield from a photovoltaic module over one year under the climatic conditions in the reference climate *c*, assuming no degradation, expressed in kWh.

A_M is the area of the photovoltaic module, expressed in m^2.

The EEI_M is used to specify the quality of the PV product's performance. It is placed on the label and determine the energy class of the product. Thus, the products can be classified from A (the best) to G class (the worst) according to prescribed threshold values (see section 3.2).

The EEI_M for monofacial PV devices have been already reported, but will be briefly summarized here for comparison purposes [15,16]. The former is calculated according to the methods defined in the IEC 61853-x series [17, 18, 19, 20]. The required input parameters are shown in Table 1 (section PV monofacial). The common characteristic of the applied procedures are the follows:

- The photovoltaic module is assumed to be ground-mounted on a fixed-open rack facing the equator with an inclination angle of 20° according to IEC 61853-4.
- EEI_M is calculated for the three European reference climatic conditions 'temperate coastal', 'temperate continental' and 'subtropical arid' based on the energy yield obtained for one year (first year of installation) according to IEC 61853-4.
- degradation or other losses (other than those due to angle of incidence effect, spectral response or module efficiency dependence on irradiance and temperature) are not considered. Ground albedo is not considered for bifacial PV modules.

Table I: Methodology applied for the estimation of EEI_M for monofacial and bifacial PV modules.

	PV monofacial	*PV bifacial*
Energy yield	IEC 61853-3	IEC 61853-3 IEC 61215-1:2021 IEC TS 60904-1-2:2019
Area	Module	
Timeline	Year 1	
Climatic conditions	"Subtropical arid", "Temperate coastal" and "Temperate continental" (EN IEC 61853-4)	
Installation	Predefined (ground mounted, fixed open-rack, South facing and 20° incl.)	

The procedure used for the calculation of the energy yield of bifacial devices and thus to define the CSER (Climate Specific Energy Rating, as defined in IEC 61853-3) is also following the IEC 61853 standard series as there is not a specific standard available for energy rating of bifacial PV. However, instead of applying P_{max} at STC (STC, Standard Test Conditions) as it is used for monofacial modules, in the procedure for calculating the energy yield of bifacial devices it is proposed the use of P_{BNPI} (Power at Bifacial nameplate irradiance, BNPI). This is the power output of the module obtained under irradiance corresponding to 1000 Wm^{-2} on the module front and 135 Wm^{-2} on the module rear, applied in any method allowed by IEC TS 60904-1-2 (2019) and the irradiance conditions defined in IEC 61215-1 (2021). This simple approach is applied as per equation 2. To define P_{BNPI}, equation 3 can be also used. Finally, the CSER is obtained according to equation 4.

$$EY_{MB(DC)\_Y1} = \frac{CSER \cdot P_{BNPI} \cdot H_p}{G_{ref}} \tag{2}$$

$$P_{BNPI} = P_{max\_STC} + \left(P_{max\_STC} \cdot \varphi_{Pmax} \cdot 0.135\right) \tag{3}$$

$$CSER = \frac{EY_{MB(DC)Y1} \cdot G_{ref}}{P_{BNPI} \cdot H_p} \tag{4}$$

where

$EY_{MB(DC)\,Y1}$ is the DC energy yield delivered by one bifacial PV module over the first year of installation under the

10.4229/EUPVSEC2023/5DV.3.10
020565-002

applicable reference climate conditions, expressed in kWh.

G_{ref} is 1000 Wm^{-2}, the irradiance used to measure the P_{maxSTC} which is the maximum power output of the PV module under STC conditions.

Hp is the total hourly global in-plane irradiance, expressed in $kWhm^{-2}$ for the reference climatic period (one year).

P_{BNPI} is the power output of the bifacial device at BNPI irradiance, expressed in W.

φ_{Pmax} is the bifaciality coefficient which is defined as the ratio of the $P_{max,rear}$ and $P_{max,front}$ according to IEC TS 60904-1-2 (2019).

2.4 Energy efficiency classes

The electrical performance of wide range of commercial PV devices belonging to different technologies have been analyzed in order to obtain relevant information regarding the PV market evolution and to determine the new energy classes that will appear on the energy label. The modules selected for this study (more than 50 modules) perform with efficiency ranging between 12.5% and 23.6%. The technologies covered in the study include single junction crystalline and polycrystalline silicon PV and thin film PV modules. The thin film PV devices are based on CdTe, CIS and CIGS semiconductors. The data analysis also includes devices experimentally measured at European Solar Test Installation (ESTI, Joint Research Centre, Italy) as well as "the best in class" commercial solar modules up to March 2023 [21]. Sensitivity data analysis was made to ensure a homogenous distribution of the energy classes.

3 RESUTLS AND DISCUSSION

3.1 Proposed energy label

The energy label for photovoltaic modules aims to give information on the energy yield of the module, allowing on the one side installers and designers, and on the other side private individuals considering investment in photovoltaic systems, to have immediate and comparable information on the product performance and to be easily able to use this in a purchasing decision. An example of the latest version of the proposed label is shown in the Figure 3. The design corresponds to monofacial PV module that would be installed in one of the climate conditions that are representative for Europe, namel, 'temperate coastal', 'temperate continental' and 'subtropical arid'. The scale of energy efficiency classes ranging from A to G and corresponds to "the best" (class A) and "the worst" (class G).

A unique energy label with common energy class for both monofacial and bifacial PV devices is proposed. It is based on the energy efficiency index (EEI_M, $kWhm^{-2}$). The label helps consumer to receive product information quickly via a QR code. The label and corresponding technical specification sheet (that the producer provides for the product) have to be registered in the free of charge EPREL - European Product Registry for Energy Labelling database for public access and consultation [22]. Additional information regarding module lifetime degradation rate and dimensions are suggested to be included. The EEI_M values will be also placed on the label to provide guidance to the consumer as to how much energy the product would generate per year.

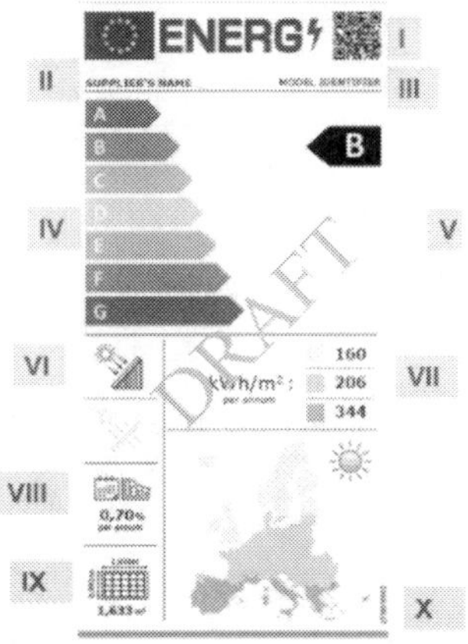

Figure 3: Example of a possible energy label for PV modules (bifacial and monofacial).

In Figure 3 is shown a sample label. The numbers on the label indicate: I. QR code; II. Supplier's name or trade mark; III. Supplier's model identifier; IV. Scale of energy efficiency classes from A to G; V. Module energy efficiency index value EEI_M under 'temperate coastal', 'temperate continental' and 'subtropical arid' climate conditions, expressed in $kWhm^{-2}$; VI. Typology of PV module: monofacial or bifacial; VII. Module energy efficiency index value EEI_M under 'temperate coastal', 'temperate continental' and 'subtropical arid' climate conditions; VII. Lifetime performance degradation rate; IX. Photovoltaic module area (A_M). X. Regulation number.

3.2 Energy label classification

The energy efficiency classes have been determined to stand for homogeneous distribution along the different geographic areas. The proposed label granularity is shown on Table II and expressed by EEI_M threshold values.

Table II. Module energy efficiency index granularity and corresponding module efficiency.

	Module Energy Efficiency Index, EEI_M (kWh/m^2)			
Energy Class	Subtropical arid	Temperate coastal	Temperate continental	Module Efficiency (%)
A	> 679	> 308	> 396	≥ 29
B	[679 – 598]	[308 – 271]	[396 – 349]	[29-26]
C	[598 – 511]	[271 – 235]	[349 – 305]	[26-24]
D	[511 – 427]	[235 – 197]	[305 – 256]	[24-22]
E	[427 – 372]	[197 – 168]	[256 – 218]	[22-19]
F	[372 – 318]	[168 – 140]	[218 – 181]	[19-16]
G	≤ 317	≤ 139	≤ 180	≤ 15

In the table a link between the energy classes and the device's conversion efficiency is made for better understanding of the label and the current state of the PV market. In order to stimulate further technological progress, the energy classes were determined consistent with the following scheme: (i) classes "A" and "B" remain unpopulated for bifacial PV devices; (ii) classes "A", "B" and "C" remain free for monofacial PV devices. Hence, the high quality PV modules with the highest energy efficiency and performance currently available on the market would fall into class "C" for bifacial PV modules and into class "D" for monofacial PV devices.

10.4229/EUPVSEC2023/5DV.3.10
020565-003

4 CONCLUSIONS

A methodology for energy yield estimation of bifacial and monofacial PV devices for application in the proposed EU Energy Label regulation for PV modules has been demonstrated. The calculated annual energy yield for a PV module can be used as an input parameter for determining the EEI_M parameter that ranks a module in the A to G classes for its unique energy label. The new label design provides for a homogenous distribution of the energy classes aver the European geographic areas. The EEI_M threshold values have been estimated on the basis that the lower end is set by the worst performing products that cannot be placed on the EU market and the currently available high quality PV modules with current "best in class" energy efficiency and performance would populate class "C" for bifacial or class "D" for monofacial modules. The higher classes are open to future higher performance module designs.

The work demonstrates that the proposed methodology for determining energy classes for both PV monofacial and bifacial modules is feasible and ready for implementation in the proposed Ecodesign and Energy labeling regulation. The new label design simplifies consumer device selection choice by offering a unique label for both type of modules. It represents a simple instrument for an informed choice of the most efficient and sustainable products. The energy classe classification scheme have been developed to incentivize technological improvements in the PV manufacturing sector and help ensure installation of only quality and sustainable product in Europe, in accordance with the Green Deal goals.

DISCLAIMER

The views expressed in the article are personal and do not necessarily reflect an official position of the European Commission. Neither the European Union institutions and bodies nor any person acting on their behalf may be held responsible for the use which may be made of the information contained therein.

References

[1] The European Green Deal, COM(2019)640, 11-12-2019, (2019)
[2] Solar Power Europe (2023): Global Market Outlook for Solar Power 2023-2027 focus in Southeast Asia, 2023, ISBN: 9789464669046, (2023)
[3] IEA, Share of cumulative power capacity by technology, 2010-2027, IEA, https://www.iea.org/data-and-statistics/charts/share-of-cumulative-power-capacity-by-technology-2010-2027, Fossil fuel capacity from IEA (2022), World Energy Outlook 2022. (2022)
[4] https://www.wtsenergy.com/solar-cheapest-energy-source-in-history-factor/
[5] Stepping up Europe's 2030 climate ambition, SWD/2020/176 final, (2020)
[6] EU Solar Energy Strategy, COM(2022) 221 final, (2022)
[7] REPowerEU Plan, COM(2022)230, (2022)
[8] Directive 2009/125/EC of the European Parliament and of the Council of 21 October 2009 establishing a framework for the setting of Ecodesign requirements for energy-related products, http://data.europa.eu/eli/dir/2009/125/oj, Directive 2009/125/EC
[9] Regulation (EU) 2017/1369 of the European Parliament and of the Council of 4 July 2017 setting a framework for energy labelling and repealing Directive 2010/30/EU, http://data.europa.eu/eli/reg/2017/1369/oj Regulation (EU) 2017/1369
[10] Communication from the commission Ecodesign Working Plan 2016-2019, COM/2016/0773 final
[11] Photovoltaics, https://susproc.jrc.ec.europa.eu/product-bureau//product-groups/462/documents, (2023)
[12] Dodd N, Espinosa N et al. Preparatory study for solar photovoltaic modules, inverters and systems. https://publications.jrc.ec.europa.eu/repository/handle/JRC122431; DOI: 10.2760/852637, (2020)
[13] ITRPV 2021, "International Technology Roadmap for Photovoltaic (ITRPV), 2021 Results, 13 ed. march 2022. (2022)
[14] W. Irrek, S. Thomas, Defining Energy Efficiency, Wuppertal Institute, https://wupperinst.org/uploads/tx_wupperinst/energy_efficiency_definition.pdf, (2018)
[15] Dunlop, E., Gracia Amillo, A., Salis, E., Sample, T. and Taylor, N., Transitional methods for PV modules, inverters and systems in an Ecodesign Framework, EUR 29513 EN, Publications Office of the European Union, Luxembourg, ISBN 978-92-79-98284-2, doi:10.2760/496002, JRC114099, (2019)
[16] Polverini, D., Gracia Amillo, A., Taylor, N., Sample, T., Salis, E. and Dunlop, E., Building criteria for energy labelling of photovoltaic modules and small systems, In: 38th European Photovoltaic Solar Energy Conference and Exhibition, 06-10 September 2021, Lisbon, Portugal, Solar RRL, ISSN 2367-198X, 6 (5), p. 2100518, JRC124226, (2022)
[17] IEC 61853-1. Photovoltaic (PV) module performance testing and energy rating – Part 1: Irradiance and temperature performance measurements and power rating. Edition 1.0 International Electrotechnical Commission, (2011)
[18] IEC 61853-2. Photovoltaic (PV) module performance testing and energy rating – Part 2: Spectral responsivity, incidence angle and module operating temperature measurements. Edition 1.0 International Electrotechnical Commission (2016).
[19] IEC 61853-3. Photovoltaic (PV) module performance testing and energy rating – Part 3:

10.4229/EUPVSEC2023/5DV.3.10
020565-004

Energy rating of PV modules. Edition 1.0 International Electrotechnical Commission (2018).
[20] IEC 61853-4. Photovoltaic (PV) module performance testing and energy rating – Part 4: Standard reference climatic profiles. Edition 1.0 International Electrotechnical Commission (2018).
[21] Tayang Nwes Solar Facts: The highest Efficient Commercial Solar Modules 03-2023, https://taiyangnews.info/top-modules/top-solar-modules-listing-march-2023/
[22] EPREL - European Product Registry for Energy Labelling, (2023) https://eprel.ec.europa.eu/screen/home

10.4229/EUPVSEC2023/5DV.3.10
020565-005

40th European Photovoltaic Solar Energy Conference and Exhibition

This presentation was selected by the Sc. Committee of the EU PVSEC 2023 for submission of a full paper to one of the EU PVSEC's collaborating peer-reviewed journals.

COMPARATIVE PROFITABILITY ASSESSMENT OF PV + BESS FOR DIFFERENT CONFIGURATIONS AND BUSINESS MODELS

Philippe Macé, Elina Bosch, Gaëtan Masson, André Penas
Becquerel Institute, Brussels (Belgium)
e.bosch@becquerelinstitute.eu

ABSTRACT: The critical need for storage in an energy system going in the direction of 100% renewable is well acknowledged. Important battery prices decrease and access to additional revenue streams have been key reasons to consider projects combining PV and BESS (Battery Energy Storage System). In this paper, three PV+BESS study cases are investigated: SC1: provision of ancillary services and arbitrage, SC2: electricity supply in off-grid/remote areas, SC3: self-consumption increase. First, the cost competitiveness of these study cases is evaluated using the Net Present Value, then it is compared to that of a similar PV-only project. This not only allows to determine whether PV+BESS projects can be profitable but also whether the additional revenues and savings generated thanks to the BESS compensate for the additional investment, operation and maintenance costs. Results show that with current electricity and BESS prices, PV + BESS can lead to very competitive investments in SC2 or to investments close to break-even in SC1 and SC3. In general, likely future trends (e.g., decreasing battery prices, decreasing incentives for electricity fed to the grid, diminishing average selling prices for utility-scale PV, …) should strengthen PV+BESS cases.

Keywords: Storage, PV, Competitiveness, Profitability

1 INTRODUCTION

In this paper, we provide an overview of the level of competitiveness of PV + BESS (Battery Energy Storage System) projects under three configurations. The BESS technology hypothesized is Lithium-ion.

The competitiveness assessment and profitability evaluation of each study case look at both a PV-only configuration and a PV + BESS hybrid project. This allows us to determine whether the additional revenues and savings generated thanks to the BESS compensate for the additional investment and operation and maintenance costs. Several indicators are used to quantitively describe the study cases. In all cases, the net present value (NPV) is calculated. When relevant, additional indicators such as the self-sufficiency rate, the self-consumption rate or the avoided CO_2 emissions are also calculated.

2 METHODOLOGY

Three projects under three configurations were considered for the analysis:

(1) Large-scale PV + BESS project providing ancillary services and arbitrage (SC1);
(2) Large-scale PV + BESS project complementing fossil fuel-based electricity supply in off-grid/remote areas (SC2);
(3) Rooftop PV + BESS project aiming at increasing self-consumption (SC3).

The main assumptions made on the PV and BESS are illustrative of typical configurations observed for the considered applications and derived from real project reviews.

The hypothesized SC1 is a ground-mounted PV system with 25 MWp in Germany. In the PV-only configuration, the revenues come from the selling of electricity on the wholesale market. In the PV + BESS configuration, a BESS of 25 MW/25 MWh is added. The additional revenues from the BESS are distributed between the participation of the storage unit in Frequency Containment Reserve (FCR) services as well as arbitrage.

In the SC2, a ground-mounted PV system is hypothesized in Faial Island, Azores, Portugal, to substitute part of the fossil fuel energy supply in the island (off-grid system). The PV system considered in the PV-only case has a power capacity of 10 MWp. In the PV+BESS case, thanks to the addition of a BESS (5 MW/20 MWh), the PV installed capacity can be increased to 20 MWp. In both PV and PV+BESS cases, the revenues come from the avoided fuel consumption.

For SC3, an industrial facility located in Belgium equipped with a 500 kWp PV system and self-consumption is considered. In the PV+BESS configuration, a BESS with a size of 250 kW/1000 kWh is added. The main source of revenues in both cases comes from the savings on the electricity bill through self-consumption. With the BESS, the self-consumption rate is increased, therefore, increasing the electricity savings. Part of the production is still injected into the grid, which represents an additional revenue.

Table I: Considered three study cases' description.

Study case ID	SC1	SC2	SC3
PV system	cSi, ground-mounted, south oriented	cSi, ground-mounted, south oriented	cSi, rooftop, south oriented
Location	Germany	Portugal (Azores)	Belgium
PV system size	25 MW	10 MW (PV only) 20 MW (PV+BESS)	500 kW
BESS size*	25 MW/25 MWh	11 MW/45 MWh	125 kW/500 kWh
Hybrid business model	Sale on the wholesale market merchant-PV), arbitrage and FCR provision	Reduction of fossil fuel-based electricity generation	Savings on the electricity bill through (increased) self-consumption
Reference business model	Sale on the wholesale market (merchant PV)	Reduction of fossil fuel-based electricity generation	Savings on the electricity bill through self-consumption

* illustrative of typical configuration observed for the considered applications and derived from real projects review

The main assumptions used for the PV and BESS systems in the study cases are provided in **Table II** [2] [3] [4]. The considered BESS technology is Li-ion. Li-ion is the dominant electrochemical storage technology with around 90% market share [1]. More specifically, the considered chemistry is LFP (Lithium Iron Phosphate) as they are less expensive and can perform safely at higher temperatures while their lower volumetric energy density is not a barrier for stationary applications. Li-ion technology advancement allows a lifetime of 15 years and a round-trip efficiency of 90% [5].

Table II: Main system assumptions for the study cases

Study case ID	SC1	SC2	SC3
PV CAPEX [€/Wp]	0,65	0,7	0,8
PV OPEX [€/kWp.a]	15	15	10
PV yield [kWh/kWp]	1196	1322	1026
Revenue sources [€/MWh]	FCR price: 92; Arbitrage: 120 Merchant-PV: 82	Avoided fuel cost: 240	Self-consumed energy: 130 Injection: 80
Round-trip efficiency [%]		90	
Lifetime [years]		15	
BESS CAPEX [€/kWh]	433	391	444
BESS OPEX [€/kWh.a]		1,5% of CAPEX	

The PV production time series for these PV systems and respective locations were obtained using PVGIS online tool, these time series have a duration of 1 year [6]. In SC1, the value of FCR provision was obtained using 2021 auction results [7]. In SC2, the fuel cost was calculated based on the fixed price of fuel by the Azorean Regional Government as of September 2023. In SC3, average retail electricity prices for commercial and industrial customers were used. [8] Eventually, wholesale electricity prices hourly time series from 2021 were used [9].

The financial parameters chosen for the assessment of profitability and competitiveness in all study cases are indicated in **Table III**.

Table III: Financial parameters assumptions

Parameters	Financial assumptions
Cost of equity [%]	8
Annual inflation [%]	2,5
Corporate tax rate [%]	25
Debt financing [%]	80

Although PV systems largely outlive 15 years, the BESS average lifetime of 15 years is taken to calculate the net present value. This allows us to determine whether it is profitable or not to consider the addition of a BESS to a PV system at the time of its installation.

3 RESULTS

The profitability is first assessed under current boundary conditions, both for a PV only and a PV+BESS configuration. Then a sensitivity analysis is conducted to visualize the influence of different BESS CAPEX and different energy market conditions on the profitability of PV+BESS projects.

3.1 SC1: Large-scale PV+BESS project providing ancillary services and arbitrage

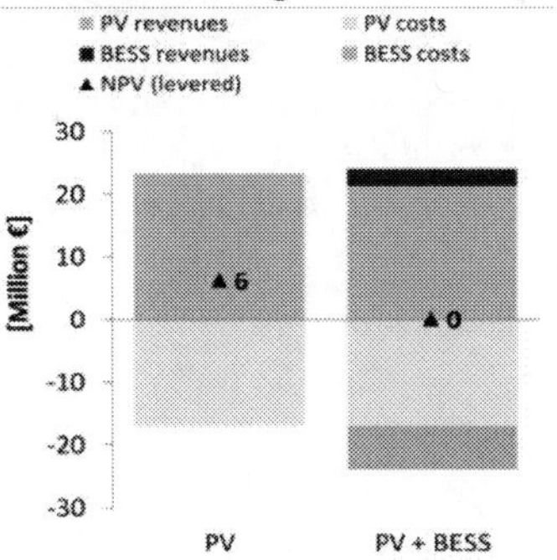

Figure 1: Competitiveness assessment of SC1 under current boundary conditions.

10.4229/EUPVSEC2023/5DV.3.26
020568-002

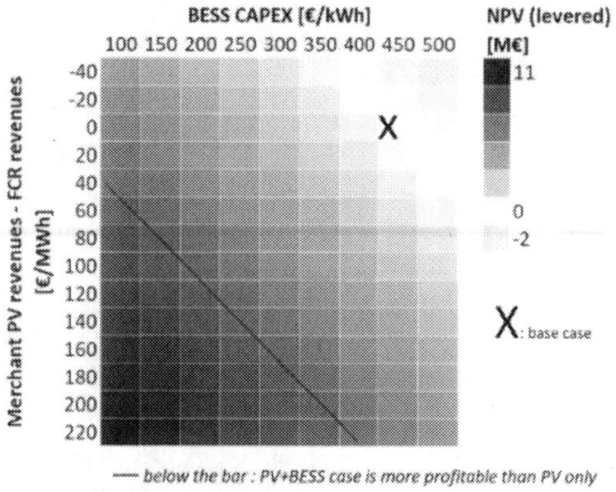

Figure 2: Competitiveness assessment of SC1 under a range of boundary conditions.

With the base case assumptions, the addition of a storage system allows to generate additional revenues from arbitrage and FCR provision. Yet, additional costs outweigh the additional revenues, and the NPV is deteriorated compared to the case with PV-only. Still, break-even is reached.

Looking at other boundary conditions related to BESS CAPEX and the gap between merchant PV revenues and FCR revenues allows to identify that a 10% BESS CAPEX decrease can already enable a positive NPV for the PV+BESS configuration. Higher gaps between PV revenues and FCR revenues can also benefit the PV+BESS business case. These can be a result of either higher need for ancillary services provision and/or lower selling prices for merchant PV. While the latter is a likely evolution in the coming years with higher PV and renewables' penetration, FCR revenues have rather been decreasing due to BESS market saturation, but aFRR provision is an equivalent, a relevant and an attractive alternative with less fierce competition.

3.2 SC2: Large-scale PV+BESS project complementing fossil fuel-based electricity (off-grid)

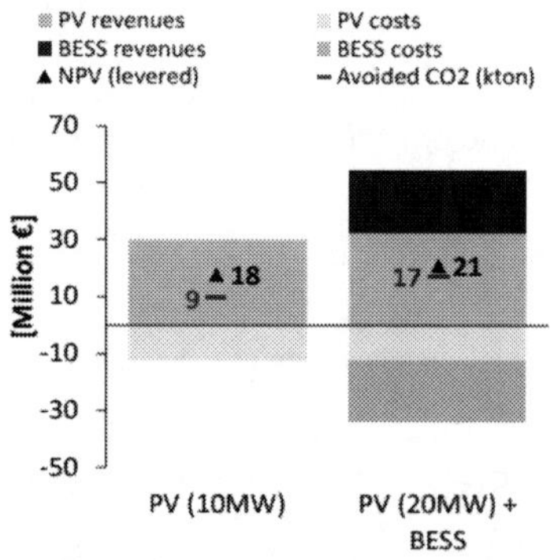

Figure 3: Competitiveness assessment of SC2 under current boundary conditions.

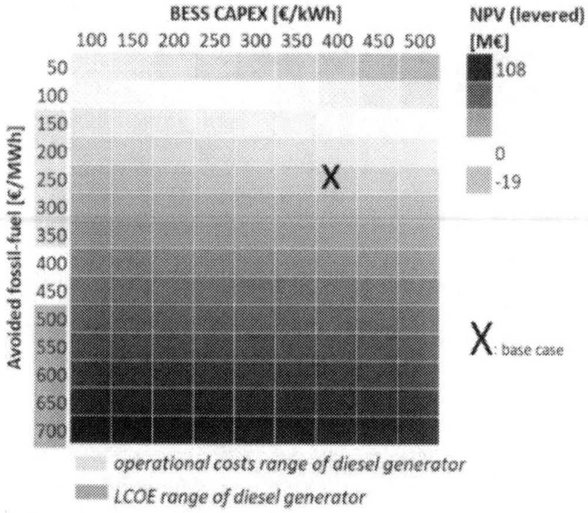

Figure 4: Competitiveness assessment of SC2 under a range of boundary conditions

With the base case assumptions, the addition of a storage system allows to displace larger quantities of fossil fuels and offset the associated CO_2 emissions. Even if the PV+BESS ensures electricity supply at a higher cost compared to a PV-only configuration, there are several advantages to the hybrid configuration. First, with the addition of storage, the PV installed capacity and with it renewable penetration can be higher which also allows larger quantities of avoided fossil fuel and as a direct result, larger quantities of avoided CO_2 emissions (17 vs. 9 kton of CO_2 per year). In addition, the hybrid combination ensures the security of supply and a predictable cost of electricity thus hedging against important fossil fuel price variations.

Looking at alternative boundary conditions related to the price of electricity from fossil fuel generators, it can be seen how less pessimistic assumptions compared to the base case can even further strengthen the case (e.g., also considering non-fuel related costs such as the initial investment in and installation and maintenance of the fossil fuel generator).

3.3 SC3: Rooftop PV+BESS project aiming at increasing self-consumption

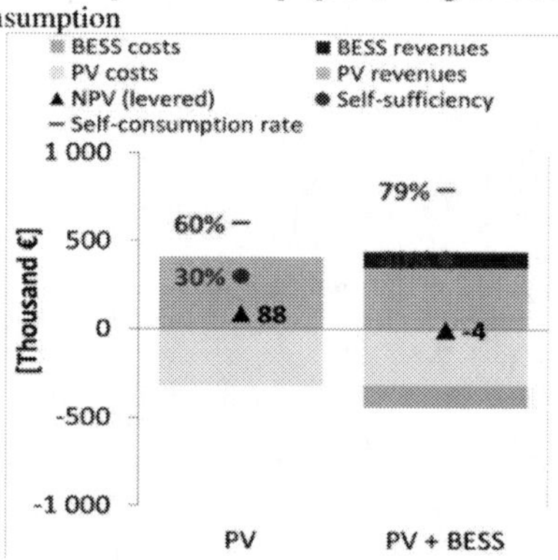

Figure 5: Competitiveness assessment of SC3 under current boundary conditions

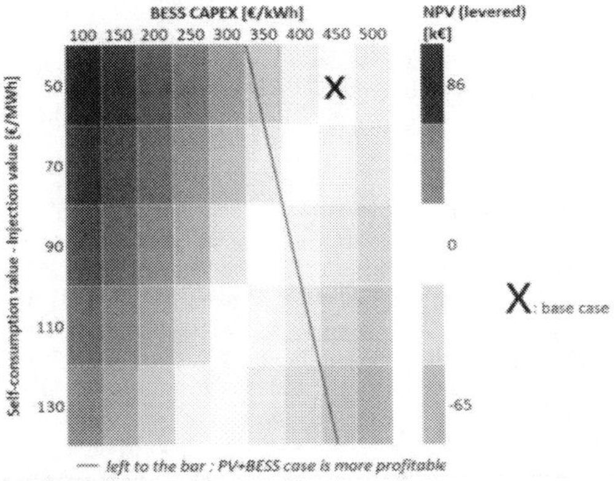

Figure 6: Competitiveness assessment of SC3 under a range of boundary conditions

With the base case assumptions, the addition of a storage system allows to increase the revenues from self-consumption (increase from a rate of 60% in the PV-only configuration vs. 79% in the PV+BESS configuration). Yet, additional costs outweigh the additional revenues, and the NPV deteriorated compared to the case with PV-only. Still, break-even is almost reached. Even if the PV+BESS project under base case assumption is poorly profitable, self-sufficiency is increased by 10pp, which enables a predictable cost of electricity for a higher share of the electricity demand thus hedging against electricity price variation and volatility.

Looking at other boundary conditions related to BESS CAPEX allows us to identify that a 10% BESS CAPEX decrease can already enable a positive NPV for the PV+BESS configuration. Eventually, increasing gaps between self-consumption PV revenues and the revenues from the injection of surplus electricity also have the potential to strengthen the hybrid case, driven by shrinking remuneration for surplus electricity.

4 CONCLUSIONS

Hybrid PV+BESS can lead to very competitive investments in one of the considered study case (SC2: large-scale PV+BESS project complementing fossil fuel-based electricity supply in off-grid/remote areas) or to investment close to break-even in two of the considered study cases (SC1: large-scale PV+BESS project providing ancillary services (FCR: frequency containment reserve) and SC3: C&I rooftop PV+BESS project aiming at increasing self-consumption) with current electricity and BESS prices.

Yet, if hybrid PV+BESS projects can be competitive, they currently often lead to a lower profitability level compared to a PV-only configuration.

Key enablers to improve the economic interest of hybrid projects are BESS prices as well as energy market trends. In general, likely future trends should strengthen PV+BESS cases. One can mention (i) battery prices which are foreseen to keep on decreasing; (ii) incentives for electricity fed to the grid (i.e., not self-consumed) that are decreasing or even being phased out; (iii) average selling prices for utility-scale PV which could diminish and (iv) the need from ancillary services' provision which should increase with rising renewable penetration rates.

Eventually, beyond the purely economic considerations presented here, additional aspects such as (i) security of supply, (ii) resilience to energy prices' volatility, (iii) increasing renewable penetration and (iv) avoided fossil-fuel-related emissions can be mentioned as further drivers of the attractiveness of hybrid PV+BESS systems.

5 REFERENCES

[1] BloombergNEF, "Energy Storage Market Outlook," BloombergNEF, 2023.

[2] IEA PVPS, "Trends in Photovoltaic Applications," IEA, 2022.

[3] NREL, "Cost Projections for Utility-Scale," NREL, 2021.

[4] U.S Department of Energy, "2020 Grid Energy Storage TECHNOLOGY COST AND PERFORMANCE ASSESSMENT," 2020.

[5] Tesla, *Tesla Megapack datasheet,* Tesla, 2020.

[6] European Commission, "re.jrc.ec.europa.eu," 03 2023. [Online]. Available: https://re.jrc.ec.europa.eu/pvg_tools/en/. [Accessed 09 2023].

[7] Regelleistung, "www.regelleistung.net," [Online]. Available: https://www.regelleistung.net/en-us/. [Accessed 09 2023].

[8] Eurostat, "https://ec.europa.eu/eurostat/," European Commission, [Online]. Available: https://ec.europa.eu/eurostat/cache/infographs/energy_prices/. [Accessed 09 2023].

[9] Energy-Charts, "energy-charts.info," [Online]. Available: energy-charts.info. [Accessed 09 2023].

[10] Regional Government of Azores, "portal.azores.gov.p," [Online]. Available: https://portal.azores.gov.pt/web/fracde/pre%C3%A7o-combust%C3%Adveis.

6 FUNDING

The work described in this publication has received funding as part of the SERENDI-PV project from the European Union's Horizon 2020 research and innovation programme under grant agreement N° 953016.

THE VALUE OF A HIGHER FLEXIBILITY ON BIPV MODULE ASSEMBLY LINES

Philippe Macé, Ian Kenchington, Damien Gautier, Gaëtan Masson
Becquerel Institute, Rue de Praetere 2, 1000 Brussels, Belgium
p.mace@becquerelinstitute.org

ABSTRACT: BIPV is finally leaving its status of newcomer to become an established niche PV segment. This comes with specific challenges to manufacturing practices, because of the heterogeneity of projects that demand BIPV modules. In this context, there is a big incentive for flexible module production lines, i.e. capable to quickly react to different product demands with lower setup and adjustment downtimes, among other benefits. In this paper we seek to quantify the benefits of such flexible production lines. For this, we developed a techno-economic model to study the effects of a flexible line on its Overall Equipment Efficiency (OEE) and subsequently on modules' factory gate cost. We simulated two production scenarios with respectively 3 and 8 different module configurations manufactured on a single line, to test the effect of downtime reduction. Subsequently, we expanded the simulation to a bivariate analysis of the effects of additional number of series produced and varying reduced downtime for up to 25%, to noticeable yet small results, with reductions in costs ranging from 11 to 81 cents per square meter. The financial benefits of downtime reductions are small, hence little increase in equipment investment is advised (max 4,4% investment increase for a 20% reduction). Nevertheless, a more thorough analysis of the impact of improved flexibility of assembly lines is recommended, as future studies may focus on other direct and indirect benefits of flexibility.

Keywords: BIPV, OEE, flexible production, manufacturing

1 INTRODUCTION

BIPV has been a hot topic in the PV industry for years and seems to finally take off. As this market niche grows, questions arise among project developers, both from the PV and the construction sectors, as well as among manufacturers. Moreover, the question of local solar PV manufacturing is also gaining traction, in order to improve the energy independency as well as reduce supply risks. Among the potential strategies to develop local solar PV manufacturing, one can mention the specialization in the manufacturing of integrated PV modules, such as BIPV modules.

One of the challenges faced by manufacturers of integrated PV modules is the necessity to find balance between cost and customization possibilities. Indeed, even if they limit their product portfolio to a few types of modules with relatively standard sizes and characteristics, it is still required to stop the assembly line between the different series, to adjust the parameters of the various equipment along the assembly line, among others. Thus, to minimize cost and maximize the profitability of the manufacturing activity, it is important to have as flexible as possible module assembly lines. It is a crucial question, as the economic difficulties faced by many BIPV manufacturers demonstrate. Assembly lines must allow for quick reconfiguration of equipment between the different module series, and the number of units "lost" because of the fine-tuning process must be as limited as possible.

In this paper, we estimate the value of flexibility for a BIPV assembly line, shedding light on its potential to revolutionize the manufacturing landscape. By meticulously quantifying cost reductions at both the module level and across the entire manufacturing process, we aim to provide valuable insights into the economic advantages of adopting flexible assembly strategies. Our analysis will present these savings in a clear and tangible manner, expressed in €/Wp and €/m² metrics, making it easier for stakeholders to make informed decisions. Moreover, we acknowledge the inherent uncertainties associated with increased flexibility and delve into a comprehensive investigation of these variables. Through this study, we aspire to empower decision-makers in the BIPV industry with actionable data to drive innovation and enhance competitiveness."

2 OBJECTIVES

The primary objective of this paper is to evaluate cost savings resulting from the implementation of flexibility in comparison to a conventional, non-flexible assembly line. These savings will be measured in terms of €/Wc and €/m². Additionally, this study will delve into an examination of associated uncertainties. To be more specific, the objectives can be summarized as follows:

- Calculate cost savings for two multi-module production schemes.
- Conduct a sensitivity analysis to assess potential downtime reductions.
- Estimate the maximum feasible investment in flexible production lines, expressed as a percentage of the initial equipment investment.

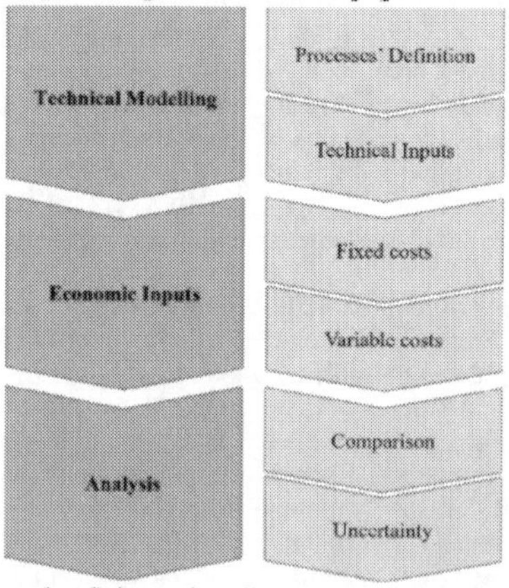

Figure 1: Schematic diagram of the followed methodology to evaluate the value of flexibility on BIPV modules' assembly lines

3 METHODOLOGY

For this paper's analysis we follow a 3-step methodology as shown in Figure 1.

3.1 Technical Modelling

The first step consists in the design of a "bottom-up" model, i.e., a detailed modelling of the assembly process as well as its constraints. The techno-economic model developed includes a large variety of adaptable parameters, such as production line uptimes and performance, input materials and consumables amount and prices, equipment efficiency, labor and maintenance costs, energy consumption and prices, among others.

The values of these parameters are based on a thorough desk research, using for example the scientific literature or the information available in the latest ITRPV report [1], as well as interviews with researchers and equipment manufacturers, allowing to estimate what is currently achievable in the industry.

The last methodological step focuses on the generation of meaningful results. Two cases will be studied, quantified and compared, and further expanded to a general analysis of the effect of flexible line on key performance indicators, and common measures must be chosen. For this, we observe the effects on Overall Equipment Effectiveness, module cost (in EUR/Wp) and surface produced cost (in EUR/m2).

Overall Equipment Effectiveness (OEE) is a key performance indicator used in manufacturing and production to assess how efficiently resources are being used. This indicator offers a comprehensive view of how efficiently resources are utilized in production for effective decision making, process optimization and resource allocation to maximize productivity.

OEE is calculated from Availability, Performance and Quality, defined by:

$$OEE = Availability \cdot Performance \cdot Quality$$

Performance measures the speed at which a machine or equipment operates compared to its designed or optimal parameters. Quality assesses the accuracy and correctness of the products or output generated. Availability represents the percentage of time that the equipment is operational and ready for production. The focus of this study is the effect of setup downtime in the latter.

We break setup downtime following the model presented below:

$$Downtime = N_{series}^{(f)} \cdot t_{change}^{(f)} + N_{batches}^{(f)} \cdot t_{batch}^{(f)}$$

$t_{batch}^{(f)}$: time needed to calibrate the production line at the start of every batch, in a (f) scenario.
$t_{change}^{(f)}$: time required to adapt a line from one series to another, in a (f) scenario.

With (f) representing whether it is a flexible line or a fixed (or non-flexible) one. To compare flexible against non-flexible scenario, a theoretical ratio is introduced called "flex factor" to represent reduced setup times in flexible lines, defined as:

$$Flex\ Factor = \frac{t_{change}^{(flex)}}{t_{change}^{(non-flex)}} \leq 1$$

This number is set to less or equal to 1, considering that a ratio over that amount would imply the line is not flexible by design. In all cases, $N_{batches}^{(f)}$ is proportional to line's throughput, while $N_{series}^{(f)}$ to the number of different series of modules to be produced from the assembly line.

This flex factor abstraction and $N_{series}^{(f)}$ are the tools we use in this paper to study their impact in downtime losses' effect in OEE and production costs.

3.2 Economic inputs

The next step is related to costs. This necessary data collection allows to obtain as precise information as possible to feed the techno-economic model, with CAPEX and OPEX inputs.

As this techno-economic model follows a deterministic approach, parameters and assumptions are manually defined for the various time-horizons, based on what is currently available and feasible in the industry, with its most likely short-term evolution. Key indicators considered in the next points are summarized in **Table 1**, più sotto. Following industry tendencies, production of modules using bifacial TOPCon cells is evaluated in this study.

Table 1: Main assumptions used in simulations

Energy	
Electricity cost	0,10 [€/kWh]
Labour	
Operator	60 [k€/year]
Technician	80 [k€/year]
Engineer	100 [k€/year]
CAPEX costs	
Facility	28,8 [M€/GW]
Building	6 [M€/GW]
Equipment	25 [M€/GW]
Common Technical Assumptions	
Module configuration	Glass-Glass
Cell efficiency (M10)	24,5%
Wafer area	330,15 [cm²]
Watt per cell	8,01 [Wp/cell]

10.4229/EUPVSEC2023/5DV.3.30
020570-002

Table 2: Technical design parameters for scenario analysis

Parameter	Unit	Module A	Module B	Module C	Module D	Module E	Module F	Module G	Module H
Number of cells or equivalent	[-]	54	24	15	72	16	36	27	18
Module area (without frames)	[m²]	1,92	1,12	0,85	2,53	0,70	2,53	1,92	2,53
Average initial module power	[Wp/module]	441	192	120	588	128	294	220	147
Wp to m² module ratio	[Wp/m²]	226	168	139	228	180	114	113	57
Scenario 1 target distribution	% of m² produced	33,3%	33,3%	33,3%	-	-	-	-	-
Scenario 2 target distribution		12,5%	12,5%	12,5%	12,5%	12,5%	12,5%	12,5%	12,5%

3.2.1 Fixed costs

Then, as part of the definition of economic inputs to feed the techno-economic model, some fixed cost constraints had to be defined. The same methodology for the material prices was used (desk research and interviews). These fixed costs are mainly linked to the equipment, building and facility.3.3 Scenario analysis

We simulated two production scenarios with respectively 3 and 8 different module configurations manufactured on a single line, to test the effect of downtime reduction. The details for each product and its respective proportions of total production are depicted in table 1, where target distribution represents the percentage of total surface produced (m² of module produced over total m² assembled within the studied period).

3.2.2 Variable costs

In addition to this technical modelling of PV module manufacturing, research on input material and consumables prices has been conducted. This allows to evaluate their impact on module prices.

Other key input materials and consumables considered in the model include laminator joints lifetime and cost, as well as all bill of materials consumption and price (ribbon, frame, glass, encapsulant, etc.).

3.4 Sensitivity Analysis

For the sensitivity analysis, an iteration was run over two variables: the number of models produced at the assembly line (from 1 to 8) and the flex factor of the line (from 0,75 to 0,95).

To expand on the effect of adding series to a single production line, we iterate over 1 to 8 series assembly line with a single flex factor (0,8) to illustrate the increasing downtime losses as module series add up and the efficiency gap between fixed and flexible setup.

For a more grounded approach to the question of whether investing in flexible lines is a sustainable idea or not, we plot a curve of the maximum investment admissible for a certain downtime reduction. In order to isolate results from volatile commercial variables and installation capacity, we compare production increase in each flex setup as cost of ownership (CoO flexible), subtract it from the base case (CoO non-flexible), and compare it to the initial fixed-line CAPEX as follows:

$$Max\% = \frac{CoO_{flexible} - CoO_{non-flexible}}{Equipment\ CAPEX}$$

4 RESULTS

4.1 BIPV assembler scenario 1

The model was run following assumptions described in **Error! Reference source not found.** with 3 models w ith a flex factor of 0,8, and the results are displayed in **Figure 2** for costs per power produced an in **Figure 3** for cost per area.

Different results are observed in both figures as model configurations are varied. At a cost per power produced level, **Figure 2** shows little difference between flexible line and non-flexible line at a 0,8 flex factor. **Figure 3** however displays a small difference between both lines.

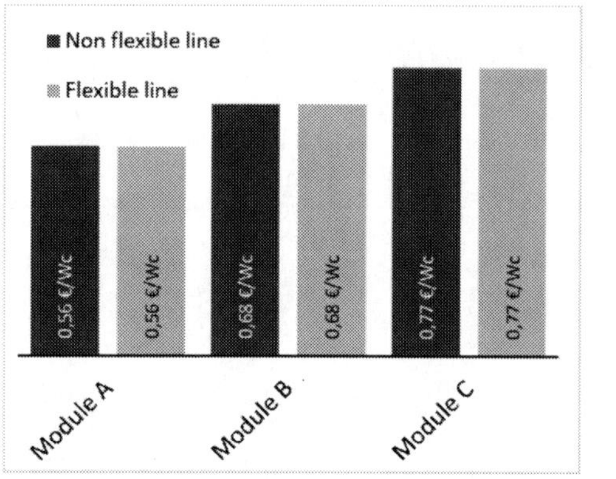

Figure 2: Scenario 1 module costs per Watt produced

4.2 BIPV assembler scenario 2

Same as scenario 1, an 8-series model was run following technical assumptions and the same flex factor of 0,8, and the results are displayed in **Figure 4** and **Figure 5**.

Similar to observations from scenario 1, **Figure 4** depicts little to no variation in costs at a per power level, but **Figure 5** does show some slight cost reduction per square meter.

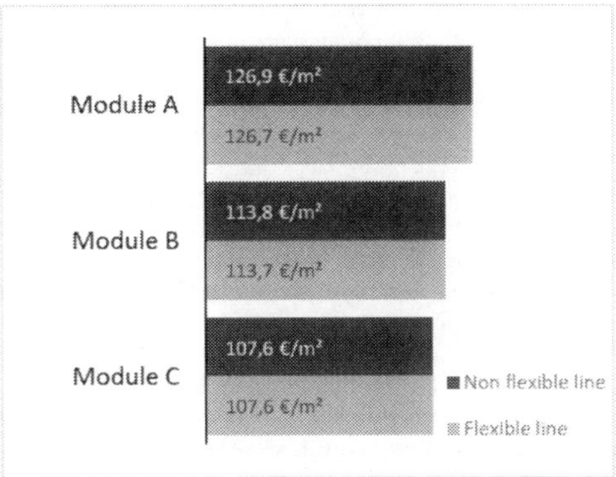

Figure 3: Scenario 1 module costs per square meter

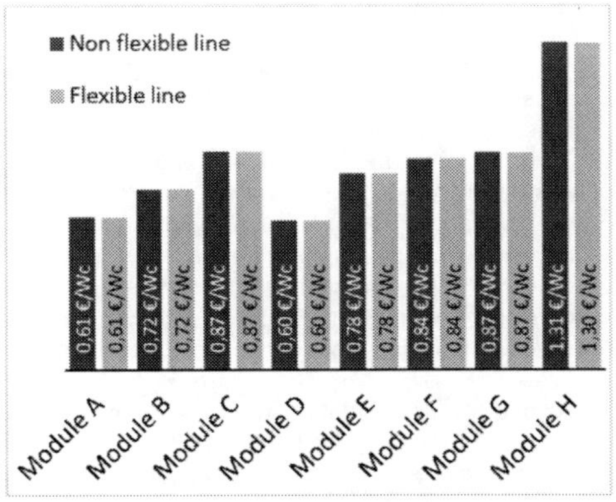

Figure 4: Scenario 2 module costs per Watt produced

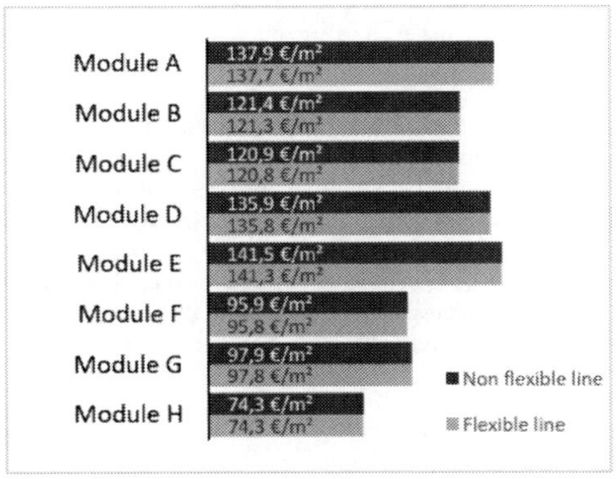

Figure 5: Scenario 2 module costs per square meter

5 SENSITIVITY ANALYSIS

4.1 Effect of series, flex factor constant

For a first approximation to a sensitivity analysis, we explore the effect of increasing the number of series for a single BIPV production line, with a base case of a non-flexible line setup, and a flexible line with a flex factor of 0,8 (20% decreased setup costs). To isolate the effect of the two variables, all different modules produced have the features of Model A from Table 2. **Figure 6** displays the obtained results and depicts the differences between both cases.

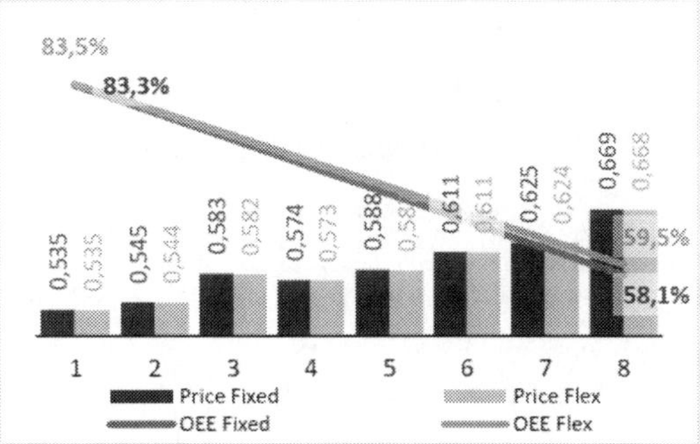

Figure 6: Module costs (in €/Wc) and OEE by number of series

4.2 Bivariate analysis

To dimension the independent effects of number of series produced and flex factor, an exploratory analysis by iterating the model, ranging series number from 1 to 8, and flex factor from 0,75 to 0,95 (25% to 5% reduced setup times, respectively). The resulting **Figure 7** shows OEE increases as a function of flex factor (translated into hours of setup saved for legibility) and the amount of different module series manufactured in a year. **Figure 8** displays the results of the same analysis for production costs, in EUR per m² produced.

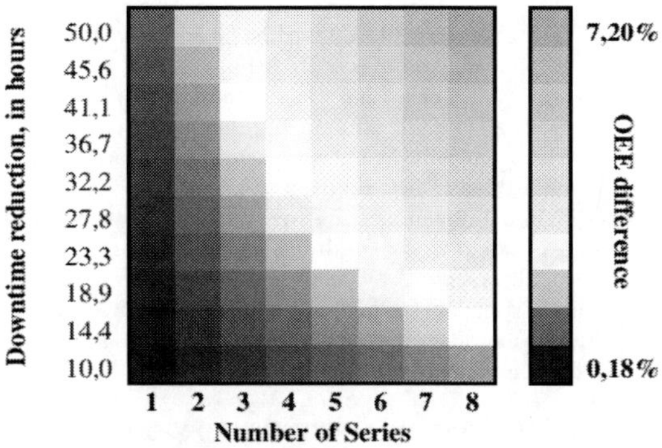

Figure 7: Heatmap of OEE increase in flex lines, as a function of number of series and flex factor (in downtime hours reduced per module)

At an exploratory level, **Figure 8** displays a rather vertical shading, indicating that effect of flexible lines may be greater than the impact of the number of series. Despite this, the cost effect, which is measured in cents per square meter, is small compared to the actual obtained cost i, that are around three orders of magnitude greater, as depicted in **Figure 3** and **Figure 5**.

4.3 Max CAPEX investment

A series of iterations was conducted with varying flex factor from 0,75 to 1, obtaining the results presented in **Figure 9**.

10.4229/EUPVSEC2023/5DV.3.30

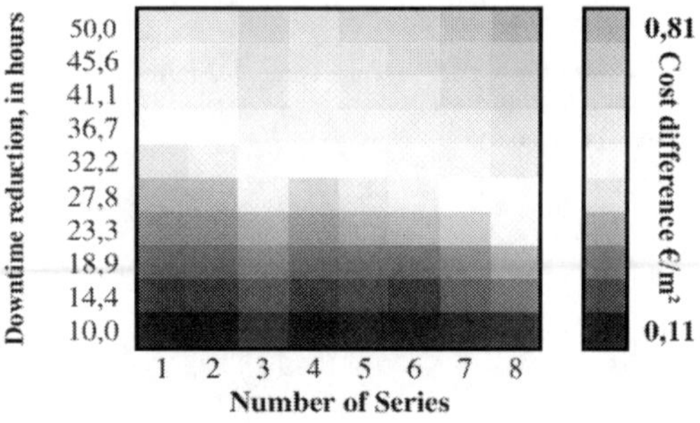

Figure 8: Heatmap of cost decrease in flex lines (in €/m²), as a function of number of series and flex factor (in downtime hours reduced)

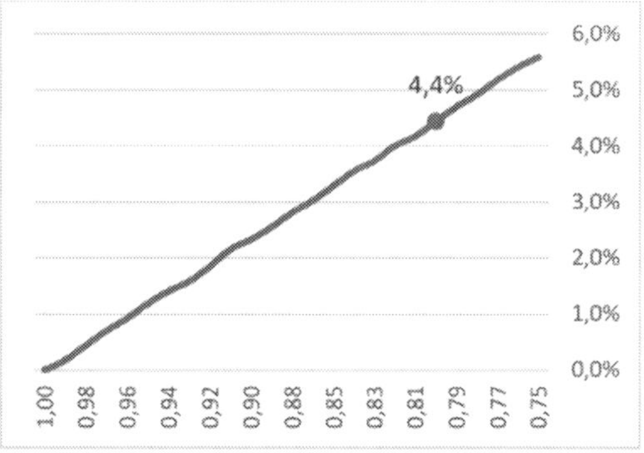

Figure 9: Maximum investment admissible for a target flex factor, in % of equipment CAPEX

To illustrate, let's take the flex factor of 0,8 as an example, as seen in scenarios 1 and 2. In this context, the analysis reveals that a flexible assembly line can accommodate a maximum additional CAPEX investment of up to 4,4% of its initial amount. In practical terms, this implies that to harness the advantages of a 20% reduction in annual downtime, any investment beyond 4,4% of the base budget would result in a net loss.

It's noteworthy that these relatively low percentages are primarily influenced by the fact that setup and adjustment downtimes represent only a small fraction of the overall available production time.

6 CONCLUSIONS AND DISCUSSION

As BIPV market is rapidly developing, new production challenges surge, as flexible assembly lines become a much more prevalent requirement.

In this study we evaluated example scenarios where investments in a flexible line may reduce manufacturing costs. Indeed, for the case of a 20% reduced downtime by using easily adaptable equipment, cost savings are observed, yet at a very small scale. In scenario 1 with three different types of modules manufactured on the production line, savings are less than a cent per Watt. In scenario 2, only some types of modules experience observable cost reductions, and even then, they are no bigger that a single cent per Watt. Similar results are obtained at a €/m² level, but since the base costs per square meter are greater than per Watt, savings can be observed at a cent scale.

Such little cost savings can be explained by the little participation that setup and adjustment downtimes have on OEE. A 20% reduction may seem noteworthy at face value but in an annual scale its impact is reduced. This also explains why final savings are mostly observed in more strained lines, as depicted in **Figure 7**'s "yellow region", where greater OEE gains are concentrated where number of series are the greatest. Nevertheless, this is a problem all manufacturers are subject to when dealing with different product formats, so investments in flexible lines may be advised as BIPV manufacturers diversify their range of products.

Regarding max profitable investments for downtime savings, the numbers obtained are low enough not to recommend significant equipment expenditures. However, it is prudent to maintain vigilance, as the ongoing maturation of this industry may usher in novel methods and processes that could lead to cost reductions in the pursuit of enhancements with lesser economic impact.

With a critical perspective, this is an analysis focused on the technical side of BIPV module manufacturing, so comparison for feasibility was made at a production level. This implies that for the max CAPEX investment admissible, percentage may vary if we add to the model potential profit margins over increased sales, or as other external incentives may apply (e.g. an increase in BIPV demand for custom-made formats), so further investigation on market response to an increased production or widened product portfolio could be made.

Other direct and indirect benefits enabled by more flexible and specialised BIPV assembly lines could be added to the study, such as higher throughput, lower level of defects, and, in some cases, reduced material intensity or equipment energy consumption. Some of the limitations in this study are to be further investigated as part of the SEAMLESS PV project, wherein a comprehensive exploration of the factors surrounding production will be undertaken.

On a same note, increased production with minimal cost improvements can be seen as a profitable scenario, depending on market and financing inputs that were out of the scope of this paper. In addition, although as minimal as the improvements presented may be, the constant search for efficiencies in the PV industry to keep prices competitive regardless of external shocks may eventually make these flexibility investments more feasible, or even compulsory.

7 REFERENCES

[1] Fischer, M., Woodhouse, M., Herritsch, S., Trube, J (2023), *International Technology Roadmap for Photovoltaic (ITRPV) 2021 Results.*

6 FUNDING

The work described in this presentation has received funding from the European Union's Horizon Europe research and innovation programme under grant agreement N° 101096126.

Supporting Energy Communities in Europe for a Fast Energy Transition

Rita Mergner[1], Dominik Rutz[1]
[2]Angela Holzmann, [2]Kerstin Schilcher, [2]Altan Sahin, [3]Natascha Fenz, [3]Ulfert Höhne, [4]Angel Nikolaev, [4]Mariya Trifonova, [5]Thekla Heinel, [5]Avigdor Burmeister, [6]Benjamin Dannemann, [7]Eszter Süle, [7]Nikoloz Sumbadze, [8]Slavica Robić, [8]Tijana Simek, [9]Bence Kovács, [10]Bea Lugosi

[1]WIP – Renewable Energies, Sylvensteinstr. 2, D - 81369 Munich, Germany
Tel. +49 89 720 12743, Fax +49 89 720 12791
E-Mail: rita.mergner@wip-munich.de; dominik.rutz@wip-munich.de
Internet: www.wip-munich.de

[2]Austrian Energy Agency (AEA), Austria, [3]OurPower Energiegenossenschaft SCE (OurPower), Austria, [4]Black Sea Energy Research Centre (BSERC), Bulgaria, [5]B.&S.U. Beratungs- und Service-Gesellschaft Umwelt mbH (BSU) Germany,

[6]Deutscher Genossenschafts- und Raiffeisenverband (DGRV), Germany, [7]Association of Young Professionals in Energy of Georgia (AYPEG), Georgia, [8]North-West Croatia Regional Energy Agency (REGEA), Croatia, [9]National Society of Conservationists – Friends of the Earth Hungary (MTVSZ), Hungary, [10]Reflex Environmental Association (Reflex), Hungary

ABSTRACT: Increased energy efficiency and/or optimised energy management and/or integrate a higher share of renewables. Thus, the term energy communities in SHAREs refers to all forms of collective actions by and for consumers such as cooperatives, collective purchase groups or other consumer-driven collective actions. Six countries are involved in the project implementation: Austria, Germany, Croatia, Bulgaria, Georgia and Hungary. In order to achieve a clean energy transition, all consumer groups need to be involved and convinced to contribute to the energy market. Founders of energy communities are by definition very motivated and "on-board" but often lack the tools to engage with other consumer groups. The concept of the SHAREs project is simple and straightforward - to create a Gateway for those interested in setting up collective actions to help them initiate an action and grow. The Gateway will consist of: (1) A country-specific implementation toolkit to equip local heroes with the technical and logistic capacity to set up their energy community (legal framework, such as master contracts, technical and IT solutions, business models, etc.); (2) The building blocks of a tailored "pick-and-mix" communication campaign to enable local heroes to promote their idea effectively to their most relevant stakeholder groups. The Gateway will combine the insights and tools from existing initiatives as well as input of established energy communities (pioneers) and policy developers. After having been put to test and refined in at least 20 aspiring energy communities in 6 European countries, the Gateway will be broadly disseminated to relevant stakeholders all over Europe.

Keywords: prosumers, local heroes, energy community, communication campaign.

1 INTRODUCTION

Energy communities bear the great potential to enable European citizens to produce, consume, store and sell their own energy. They are member-based organisations whose primary purpose is to provide environmental, social and economic benefits to community members or to the local areas in which they operate. A key aspect of energy communities is the collective ownership and management of the respective facility(ies).

Since the main purpose of an energy community is to generate benefits for the community, energy communities usually reinvest a large part or all of the profit in their operations before distributing any kind of dividend to their members.

There are around 3,500 energy cooperatives in Europe, representing over one million EU citizens. However, only a minority of them can be found in Eastern Europe because of several factors, including financial constraints, unsupportive legal frameworks and non-liberalised energy markets[1].

In countries where energy communities have a wide presence, dedicated cooperative networks have been established to monitor the yearly progress in the country (i.e., DGRV in Germany[2], Hier Opgewekt in the Netherlands[3], while in Austria a country-wide advisory network is currently being created with the backing of existing regional energy agencies and institutes to support the uptake of energy communities[4]).

Research conducted by CE Delft in 2016[5] revealed that over 264 million European citizens (which amounts to half of the European population) could be producing their own energy by 2050 at home or together in energy communities. These energy citizens could be producing 611 TWh of electricity by 2030 and 1,557 TWh by 2050. This means that by 2030, energy citizens could be delivering 19% of Europe's electricity demand, and 45% by 2050. This is a significant contribution to achieving the EU's 2030 renewable energy target and moving towards a 100% renewable future.

2 THE SHARES PROJECT

The Clean Energy for all Europeans package paves the way for an energy revolution by putting the consumer in the driving seat of the green transition process. Energy communities offer the unique opportunity for consumers to form critical mass and become renewable energy producers and offer their demand flexibility to the

[1]Stanislas d'Herbemont and Compile project, 2021, "D4.1 The Energy Community Toolkit"
[2]DGRV
[3]Hier Opgewekt
[4]Österreichische Koordinationsstelle für Energiegemeinschaften
[5] B. Kampman, 2016, "The potential of energy citizens in the European Union"

10.4229/EUPVSEC2023/5DV.3.41
020571-001

market. However, the current requirements of becoming a prosumer bear a crucial risk that ultimately there will be a two-speed energy world: those who have the knowledge, access and opportunity to become a prosumer, and those who lack the resources, know-how, education or time to participate in and benefit from the new opportunities. Therefore, there is a clear need to support local heroes (i.e. those interested in setting up collective actions) in establishing their energy community and enable them to motivate and target consumers directly. This is particularly important for countries where energy communities are yet to be established.

The SHAREs project supports the set-up of new energy communities as well as other collective actions and helps existing energy communities to grow on a large scale. It aims to equip local heroes – organisations, public authorities or individuals who push the initiation of energy communities – with a country-specific implementation toolkit and a communication campaign that allows them to launch and expand a large variety of collective actions in a simple and smart way.

Figure 1: SHAREs target countries (in green)

3 SHARES GATEWAY

The SHAREs Gateway is a website to inform citizens about energy communities and other collective actions, and to provide them guidance on how to start or extend citizen driven projects. The SHAREs gateway presented in this report is an English blueprint website that will be used as a template for translation and adaptation for the target countries Bulgaria, Croatia, Georgia, Germany, Hungary, and partly Austria. The English blueprint website is accessible at the following link: https://sharerenewables.eu/.

The blueprint gateway is no completed website with all necessary information to support those who want to create new or extend citizen driven energy projects. It is only a blueprint in English to be used as a template for the national gateways. Thus, it only includes the general structure and example pages. The project partners who are responsible for the national gateways are free to adapt and modify the structure and content of the blueprint gateway.

The main structure of the gateway is based on the following sections:

- Get inspired – showing examples of energy communities and news about citizen engagement.
- Get informed – presenting information on project concepts, financing, legal aspects of energy communities.
- Get equipped – highlighting tools and guidelines to be used for energy communities.
- Get members – providing information and material for promotion and communication.
- Get active – guiding on first steps to found an energy community and where to get support.

It is important to define the users of the gateway and target groups. This is a basis for the creation of an appealing gateway which meets the needs of the target groups. The main target groups are:

- Frontrunners – those who want to establish new energy communities or implement collective actions.
- Representatives of existing energy communities or collective actions.
- Decision makers with little knowledge on energy communities and collective actions (local politicians, mayors, representatives of authorities).
- Professionals (energy consultants, craftsmen, associations of craftsmen, energy service companies, etc.).
- Financial institutes (banks, cooperative banks, crowdfunding platforms, funds).
- Citizens in the target countries.
- International stakeholders working in the field of renewable energies.
- Media.

The best practices page presents various examples of existing energy communities and collective actions. They can be selected with filters regarding the regional scope and the energy generation/distribution.

The news section provides the opportunity to include conventional news on energy communities and collective actions, sorted chronologically.

4 COMMUNICATION STRATEGIES

Engaged and well-informed consumers can contribute to making the energy transition happen. It is important to enable consumers to interact with energy production topic (prosumers, energy cooperatives, collective actions etc.), explain where energy comes from, what impact their personal energy consumption has on the environment.

Based on literature analysis and discussions with consortium partners, 14 target groups were defined and dedicated communication profiles were elaborated:

- Silver investors: are ready for new, exciting projects.
- Grandparents: are concerned about their grandchildren's environmental and financial future.
- Young families: are busy with their everyday lives.
- Established families: are ready for new exiting projects.
- Households with spare time: have time but only a very limited budget available.
- Double Income, No Kids (DINKs): have much time and budget available for new projects.
- Working Poor: hardly have time and money

because of low income and difficult working conditions.
- Young Generation - Students and Trainees: have high environmental awareness, spare time, but only a very small budget.
- Traditional and Organic Farmers: have unused resources (roof space, biomass, etc.), which they want to exploit profitably.
- Young Generation – Pupils: have spare time, but no income. They can get their parents to take action.
- Politicians: are trustworthy opinion leaders and powerful stakeholders in their community.
- Municipalities: can access funding and enable citizen participation.
- Small and Medium-Sized Enterprises (SMEs): are focused on economic aspects and in some cases on their “green” image.
- Associations and NGOs (Non-Governmental Organisations) have very strong community spirit and unused resources (e.g. roof).

Each target group has a dedicated profile. These profiles are simplified abstractions and were not designed to cover all eventualities. In real life, there may be overlaps between the profiles. The applied categorization is intended to help local heroes to understand the views and priorities of their potential members. While many key communication messages apply to all target groups, there are also some special key messages that may help you get through.

Key messages to reach **silver investors**:
- Your investment pays off; you can expect a secure return on your investment.
- Place a safe and green investment.
- Increase the value of your own property by investing in green energy.

Key messages to reach **grandparents**:
- Increase the value of your own property by investing in green energy.
- You can be part of an energy community without having to contribute financially.
- Participation is not complicated and takes little time.
- Create something of value for yourself and your family.
- Each investment counts, anyone can invest.
- It is never too late to become active.

Key messages to reach **young families**:
- You can be part of an energy community without having to contribute financially.
- Participation is not complicated and takes little time.
- Create something of value for yourself and your family.
- You can improve your children’s future without much effort.

Key messages to reach **established families**:
- Your investment pays off; you can expect a secure return on your investment.
- Place a safe and green investment.
- Increase the value of your own property by investing in green energy.
- Create something of value for yourself and your family.

Key messages to reach households with **spare time**:
- You can be part of an energy community without having to contribute financially.
- The time you spend on the project is time spent on climate protection.
- Become independent from global energy price increases.
- You can become an important part of change.
- Reduce your energy costs. It pays off to join an energy community.
- Create sustainable jobs in your region.

Key messages to reach **DINKs (double income no kids)**:
- Your investment pays off; you can expect a secure return on your investment.
- Place a safe and green investment.
- Increase the value of your own property by investing in green energy.
- Create local jobs.
- Be more independent from energy price fluctuations.

Key messages to reach **working poor**:
- You can be part of an energy community without having to contribute financially.
- Participation is not complicated and takes little time.
- Reduce your energy costs. It pays off to join an energy community.
- Become independent from global energy price increases.
- You can become an important part of change.

Key messages to reach **students and trainees:**
- You can be part of an energy community without having to contribute financially.
- Motivate your family to join an energy community.
- Learn more about renewable technologies first-hand.
- The time you spend on the project is time spent on climate protection.

Key messages to reach **farmers**:
- Energy communities open up new business opportunities: Use your unexploited resources (roofspace, biomass) to make sustainable profits.
- Your investment pays off; you will make a good and sustainable investment.
- Improve your image.
- Create sustainable jobs in your region and increase purchasing power in your area.
- Doing it together works out great. We take a lot of the work off your hands, but you still reap all the benefits.
- Energy communities profit from the heterogeneous consumer structure. While private households need most of the energy on evenings and weekends, you may need it on workdays. You can profit from being in a community with different stakeholders (private households, SMEs, etc.).
- Be more independent from energy price fluctuations.
- [Depending on the image and regulations of agri-PV:] Join the second generation of ecological farming by applying agri-photovoltaics, where the soil is protected by the PV modules. Harvest twice!

Pupils come from different backgrounds, so you may need to apply different strategies accordingly (for some of them you may want to focus on benefits for the family,

10.4229/EUPVSEC2023/5DV.3.41
020571-003

for others on climate protection). Key messages to reach **pupils:**

- Convince your family to join an energy community.
- You can be an example for your classmates.
- Learn more about renewable technologies and their environmental benefits.
- The time you spend on the project is time spent on climate protection.

Key messages to reach **politicians**:

- The investment pays off for your citizens.
- Improve your position as a pioneer and innovation leader! Be a role model for your community.
- Investing in clean energy means investing in better living conditions and creating local jobs.

Key messages to reach **municipalities**:

- Improve your image! Be an inspiration to your community and other municipalities.
- Become a pioneer amongst municipalities by supporting energy communities.
- Start a community project that benefit your citizens for years to come.
- Seize the moment and get (financial) support for setting up RECs now.
- Make a good and sustainable investment for your municipality.
- Investing in clean energy means investing in better living conditions for your citizens.

Key messages to reach **SMEs**:

- Your investment pays off; you can expect a secure return on your investment.
- Improve your image and attract new customer groups.
- Energy communities open up new business opportunities.
- Energy communities profit from the heterogeneous consumer structure. While private households need most of the energy on evenings and weekends, your SME may need it on workdays. You can profit from being in a community with different stakeholders (private households, farmers, etc.).
- Be more independent from energy price fluctuations.
- Doing it together works out great. We take a lot of the work off your hands, but you still reap all the benefits.
- You can be a role model for your staff members and partners.
- Create sustainable jobs in your region and increase purchasing power in your area.

Key messages to reach associations and **NGOs**:

- You have the network and the roof space; you will really make a difference if you take part.
- Energy communities profit from the heterogeneous consumer structure. While private households need most of the energy on evenings and weekends, your association may need it on workdays. You can profit from being in a community with different stakeholders (private households, farmers, SMEs, etc.).
- Doing it together works out great. We take a lot of the work off your hands, but you still reap all the benefits.
- Thanks to you, your community benefits from more regional jobs, better energy supply security and lower energy costs. You make your community a better place to live.
- Improve your image and attract new members.
- Strengthen your position as a pioneer and innovation leader.
- Make a good and sustainable investment.
- You can be a role model for your members.

5 HOW TO GET STARTED

From the first considerations to the ongoing operation of an energy community, a lot has to be decided and initiated. At the same time, you need to comply with legal and administrative formalities. The following steps may help you facilitate the set-up of your energy community or collective action. Thereby, the order of the steps depends on the individual circumstances of your energy project. Some steps might even have to be carried out simultaneously.

Define the goals of your energy community!

Set priorities and clarify which sector you want to start with: electricity, heating or mobility. Of course, your project can include all of these sectors, but in the beginning, it may be easier to focus on one. Define goals, sub-goals and a realistic timeframe for achieving them. Get to know your opportunities to start a citizen-driven energy project by having a look at other energy communities.

Get necessary information about energy communities, technologies and project concepts!

Make your project succeed through sound technical and economic planning. Learn about possible project concepts. Don't worry, you don't have to be an expert for everything, but you should have a good overview about as many aspects of an energy community as possible. Maybe now is a good time to have a first look at financing possibilities or the legal framework you will have to work with.

Speak with people who could support your idea!

It's easier to achieve your goals and move forward with the support of your community. Talk to people who could support your idea. Ideally try to involve those who support your ideas and goals, are known in the community and have a good reputation, are willing to spend some of their free time to support the process.

To identify suitable groups of persons, you can have a look at our types of members and our target group profiles. Before speaking with them identify key messages you want to transfer. You can also prepare yourself by using our white label materials and guides on communication activities.

Talk to other energy communities!

Get in touch with founders of renewable energy communities. Their experiences will help you to streamline your own process and to save time. It is always good to talk to several energy communities. You can also visit them together with your supporters to see how projects in other energy communities have been implemented. Our connect with others section shows you opportunities how to get in touch with like-minded people.

Define a first energy project for your community!

It is time to clearly define your project, also with the help of our calculations tools and guidelines. The main goal of energy communities is to set-up new renewable energy or energy efficiency projects to reduce the fossil energy consumption in your community. You may start, for example, with a larger community-owned photovoltaic installation on a public building, or you may focus on the set-up of a small district heating network. Another idea would be to implement a wind energy project. Get inspired by best practice examples and learn more about possible project concepts. Contact the network operator to find out more about your local circumstances and possibilities.

Bring the mayor and city/municipality council members on board!

Although the involvement of local politicians, such as mayors or members of the municipality/city council, is not essential for every energy community project, it is highly recommended to involve them and motivate them to support you. If you manage to have the municipality/city as supporter on board, many issues will become easier, including advertising, recognition, legal aspects, provision of public land or surfaces (e.g. public roofs for PV systems), etc. To identify suitable groups of people, have a look at the profiles of politicians, municipalities and how to reach multipliers.

Involve professional planners!

Depending on the first energy project that you want to implement as energy community, you may need professional planners. Smaller projects may be also planned by you or your supporters. Perhaps you even have such planners as a member of your energy community? In the get support or connect with others section of this gateway, you may get further advice or ask others with which professional planners they had experiences.

Clarify legal form, responsibilities, business model of your community!

Step by step and in accord with the other supporters, you should clarify the legal form, responsibilities, and business model of your community. You may need to talk to experts for legal, technical and financial advice. Now is the time to have detailed look on our legislation and financing sections to get a good overview and choose the best options for your project.

Get a critical mass of supporters!

Now that you have defined your project idea, talked to existing energy communities, and got first important supporters on board, it is time to reach a critical mass of members. You don't have to be a professional communications officer to use various communication channels successfully. Our get members section supports you in approaching other people. This may include personally reaching out to people that are important for your project, setting up your social media presence, writing newspaper articles, or distributing flyers. If you want to personalize the offered material, have a look at our picture database. All images are completely royalty free to use.

Organize an information event for the community!

To involve citizens, you could hold presentations at information events or even organise your own. The white label communication material offers you easy to use templates for flyers, factsheets, or newspaper articles. If you want to organize an event we even offer you templates for invitations, PowerPoint slides, and a checklist for events.

Formally establish your energy community!

Are you ready to establish your energy community? Congratulations! Depending on the business model and the foreseen legal form of your project, it may involve various legal steps. The foundation an ideal occasion for a public celebration which will give your energy community additional visibility.

Make your project fly!

To set-up new energy projects, you may need to involve more citizens to participate. The target group profiles and types of members outline how you can reach out to and motivate them.

ACKNOWLEDGEMENTS

The authors would like to thank the European Commission for the support of the SHAREs project.

Disclaimer: This project has received funding from the European Union's Horizon 2020 Research and innovation program under grant agreement No 101033722.

Neither the European Commission nor any person acting on behalf of the Commission is responsible for the use which might be made of the following information. The views expressed in this publication are the sole responsibility of the author and do not necessarily reflect the views of the European Commission.

Reproduction and translation for non-commercial purposes are authorised, provided the source is acknowledged.

Website: https://shares-project.eu/

PROMISE: A KNOWLEDGE TRANSFER PLATFORM TO STUDY RELIABILITY IN MEDITERRANEAN PV SYSTEMS

Azzopardi, Brian[1]; Rennhofer, Marcus[2]; Mignonac, Alexandre[3], Muñoz, Ildefonso[4]; Meza, Carlos[5], de l'Epine, Melodie[6], Zerafa, Steve[7].

[1]The Foundation for Innovation and Research, Malta (brian.azzopardi@fir.mt)
[2]AIT Austrian Institute of Technology, Austria (marcus.rennhofer@ait.ac.at)
[3]Commissariat a l'Energie Atomique et Aux Energies Alternatives (CEA), France (alexandre.mignonac@cea.fr)
[4]Fundación CENER – National Renewable Energy Centre – Spain (imunoz@cener.com)
[5]Hochschule Anhalt University of Applied Sciences, Germany (carlos.meza@hs-anhalt.de)
[6]Becquerel Institute, Belgium (m.delepine@becquerelinstitute.eu)
[7]PIXAM, Malta (info@pixam.io)

ABSTRACT: The Mediterranean region is ideal for implementing solar energy due to its high irradiance and clear-sky conditions. More specifically, Mediterranean islands present important challenges for the operation and maintenance of PV systems but also offer an excellent opportunity to increase energy security and sustainability. In this regard, it is beneficial and necessary to have trained personnel in this region to analyse the degradation process of PV technology under Mediterranean conditions and identify best practices in implementing PV systems to improve their performance and extend their lifetime. The Photovoltaics Reliability Operations and Maintenance Innovative Solutions for Energy Alliance (PROMISE) aims to create an alliance of photovoltaic excellence through scientific, engineering, and research performance within the Maltese research community reliability of existing and emerging module technologies and systems, including the digitalisation aspects for prediction and optimisation algorithms for innovative solutions to support the energy transition. PROMISE includes a knowledge gain and transfer framework, capacity-building activities, workshops, advanced schools, training programs, internships, meetings, mentoring, research management, and administration upskilling. The present document describes the main objectives and activities of the PROMISE project.

Keywords: Knowledge transfer, Capacity building, Malta

1 CONTEXT AND MOTIVATION: PV IN EUROPE AND THE MEDITERRANEAN

Photovoltaic plays an essential role in the current and future energy landscape of the European Union. Moreover, there is a relevant PV industry in Europe with several companies playing leading roles along the complete PV value chain. There are projections for further development in the near future with the subsequent demand for skilled personnel, as indicated in [1]. In fact, according to SEPA models, as referenced in [2], reaching the 2030 solar target will require almost 1.5 million direct and indirect jobs in the EU solar industry.

Jobs in the PV industry are highly specialised due to the complexity of the technology and the potential safety hazards it can pose. Therefore, PV personnel must be trained in specific subjects such as material science, power electronics, power systems, mechanical systems and optical physics to cover the different lifetime phases such as manufacturing, deployment, operation & maintenance, and decommissioning & recycling. This knowledge ensures that the PV system functions as intended, allowing the identification of potential problems, minimising the risk of system failure, and ensuring the system's long-term reliability. More specifically, studying the reliability of PV systems is essential because it helps maximise their efficiency, safety, and cost savings, reducing their environmental impact.

PV systems can be an effective and affordable source of electricity in Mediterranean islands. The Mediterranean climate is generally sunny, meaning photovoltaic systems will experience less downtime due to bad weather—nevertheless, PV systems in the Mediterranean, especially islands, present challenges and opportunities. On the one hand, the high temperature and humidity and the presence of dust and salt can reduce the efficiency of the cells and cause accelerated corrosion of the system components. On the other hand, the large solar resource available allows for large amounts of electricity to be generated from a relatively small system, which can provide significant cost savings.

Malta combines both the great potential for PV deployment and the research community and infrastructure, which aims to be reinforced by the EU-funded project PROMISE. The PROMISE results, lessons learned, best practices, and success stories will feed a scalability opportunity, extending the research platform to a nationwide/regional landscape while also generating concrete research capacity for a more enabling environment for research in PV reliability in Malta as a solar country and regional energy hub.

2 THE PROMISE PROJECT

The Photovoltaics Reliability Operations and Maintenance Innovative Solutions for Energy Alliance (PROMISE) [3] aims to create an alliance of photovoltaic excellence through scientific, engineering, and research performance within the Maltese research community reliability of existing and emerging module technologies and systems, including the digitalisation aspects for prediction and optimisation algorithms for innovative solutions to support the energy transition. PROMISE includes a knowledge gain and transfer framework, capacity-building activities, workshops, advanced schools, training programs, internships, meetings, mentoring, research management, and administration upskilling.

10.4229/EUPVSEC2023/5DV.3.42
020572-001

PROMISE aims to sustain scientific, engineering and research performance and innovation capacity in the Maltese Research Community on solar photovoltaics reliability of existing and emerging module technologies and systems, including the digitalisation aspects for prediction and optimisation algorithms to support the energy transition. The global resource shortage, energy pollution and climate change are addressed through the energy transition with cleaner technology. This is essential to island geographic locations like Malta. Malta has a substantial solar resource with 1700 kWh/m^2 per year, but 7.3% of electricity is generated by PV only, and the rest is based on imported fossils and transmission lines from Sicily. It is the 5th most densely populated country in the world, where the transport sector with 820 internal combustion engine vehicles/1000 inhabitants and increasing cooling demand makes it necessary to bring digitisation in understanding the status of each existing PV system.

The current scenario of a green transition by 2030 and beyond presents itself as a fair and competitive way to address new opportunities for innovation, investment and jobs, fighting inequality and energy poverty and strengthening the competitiveness of European businesses, including in a widening country such as Malta. This approach will result in reduced emissions, job creation and economic growth, energy poverty, reduction in external energy dependencies and improved health and well-being. However, we are limited in time, costs and workforce and hence the PROMISE consortium, see Figure 1, brings together balanced top-EU research centres, i.e., AIT, CEA, and CENER. This university specialises in the application of photovoltaic technology (HSA) as well as research and dissemination organisation (BI) to support the Maltese research community FiR as a research and dissemination organisation seeking to establish a leading local research centre with already a range of affiliated experts in energy, fuels, environment, engineering and transport and (PIX) as an innovator based locally to address these challenges and opportunities for smart cities on islands.

The particularity of the knowledge transfer program developed in PROMISE is that most of the consortium participants are research institutions or companies actively working on reliability problems in the photovoltaic industry. This allows for identifying the reliability training capabilities directed toward what the PV sector requires. The consortium has one academic institution, Anhalt University of Applied Sciences (HSA), which has the particularity of offering a master's degree [4] focused solely on photovoltaic engineering. In addition, Fraunhofer CSP researchers participate in this master's program and present their reliability projects to the industry. The HSA master's program has an active connection with the PV industry, and engineers from PV companies frequently give lectures in the courses. In addition, most of the students of this master's program develop their master's thesis in a company in the PV sector.

Figure 1: PROMISE Consortium.

PROMISE fully addresses the challenges identified by the HORIZON-WIDERA-2021-ACCESS-03 call and complements EU energy strategies. Activities within the PROMISE project will significantly and directly contribute to the EU's strategic objectives and the European Green Deal. The contributions below are the Project's pathways toward all expected outcomes:

- Improved excellence capacity and resources in widening countries, closing Europe's still apparent research and innovation gap.
- Enhanced strategic networking activities between the research institutions of the Widening countries and at least two internationally-leading counterparts at the EU level.
- Raised reputation, research profile, and attractiveness of the coordinating institution from the Widening country and its staff's research profile.
- Strengthened research management capacities and administrative skills of the staff working in institutions across the Widening country.
- Improved creativity supported by the development of new approaches in R&I collaboration increased mobility (inwards and outwards) of qualified scientists.

3 RESEARCH THEMES SUPPORTED BY PROMISE

Promise project is intended to supplement the photovoltaic investment in Malta by facilitating the necessary knowledge sharing and networking through the following research themes:

- Aerial early degradation measures of existing PV systems
- In-field early and trending degradation measures
- Soiling aspects for PV power plants
- Digitalisation for long-term monitoring and optimal energy yield
- Technology roadmap and implementation paths, including electricity costs and storage

10.4229/EUPVSEC2023/5DV.3.42
020572-002

4. CONCLUSION

The Promise project identifies Malta as the ideal test site for PV performance and reliability analysis. Moreover, knowledge and experience gained in PROMISE could be extrapolated to other Mediterranean areas, contributing to the further deployment of photovoltaics. Furthermore, the alliance between PV research institutions and companies participating in PROMISE could be an example for future consortiums aiming at PV knowledge transfer and promoting photovoltaics in Widening and developing countries.

REFERENCES

[1] SolarPower Europe. EU Market Outlook for Solar Power 2022–2026; SolarPower Europe: Brussels, Belgium, 2022.
[2] SolarPower Europe, EU Solar Jobs Report 2022. Addressing solar skills challenge, 2022. SolarPower Europe, Brussels, Belgium, 2022.
[3] Project website: http://pv-promise.eu/
[4] Anhalt's PV Master Program: https://www.hs-anhalt.de/nc/studieren/orientierung/studienangebot/detail/photovoltaics-engineering-science-master-of-science.html

10.4229/EUPVSEC2023/5DV.3.42
020572-003

ALL HANDS ON DECK: A BUNDLE OF STRATEGIES AIMING TO LEVERAGE DIGITAL MEDIA CONTENT AND PUT PORTUGAL'S SOLAR PV ON THE MAP

R. Amaro e Silva[1], R. Jolivet[1], M. Silva[2], A. Lavadinho[2], S. Freitas[2]
rodrigo.amaro_e_silva@minesparis.psl.eu
[1]O.I.E. Centre Observation, Impacts, Energy, MINES Paris - PSL Research University
06904 Sophia Antipolis, France
[2]Lisboa E-Nova – Agência de Energia e Ambiente de Lisboa
1100-23 Lisbon, Portugal

ABSTRACT: Being unaware of the specifics of PV deployment (i.e., where, how much, and with what characteristics) poses serious challenges to a fast PV adoption and an effective integration of PV generation, mainly through a variety of stakeholders: transmission and distribution system operators, PV companies, policymakers, and citizens. National databases often lack key parameters for an accurate PV modelling (e.g., module tilt and azimuth, inverter capacity) but, more than that, are frequently not accessible to the public. This work proposes a workflow that seeks to valorize scattered online media content and to enhance its visibility by integrating this information in an online mapping interface. Results highlight the effectiveness of this approach, having already mapped nearly 2000 PV projects in Portugal (~58% of the reported total capacity in the country), greatly surpassing other available databases for Portugal. Additionally, possible complementary outputs are identified such as PV integration studies, spatio-temporal and socio-economic studies, market analyses, or outreach content.
Keywords: PV deployment, PV mapping, PV metadata, web scraping

1 INTRODUCTION

The geographical distribution of deployed PV, its heterogeneity, and the diversity in the installed capacity at system-level pose a challenge to accurate modelling and forecasting of PV at the regional scale. Often, distributed PV is "invisible" for the grid operators [1] and may lead to increased congestions, overgeneration, imbalances [2].

This is often tackled using algorithms which leverage on databases describing the deployment of PV both in time and space. Traditionally, the demanding task of setting up such databases is upheld by local, national, and international energy and environmental institutions, regulatory bodies, TSOs and DSOs. However, the level of openness (i.e., how accessible the data is) and granularity of the data can greatly vary from country to country and is often limited. For example, such datasets commonly include aggregated data (usually at the national level, sometimes regional) and contain no more than the total installed capacity and number of installations, possibly disaggregated by the range of individual system capacity or the different regulatory frameworks under which the installations are registered.

The use of image processing software, commonly based on deep learning approaches, has been proposed to tackle this [3,4]. While this has been shown to be capable to geolocate PV installations and even infer the installed capacity and tilt and azimuth angles [5], not only the uncertainty behind such estimates is still considerable, but there is still lacking a good part of the context around a given PV project: the legal framework under which it operates, and, for distributed PV, the end-user of the installation.

The commonly coarse granularity of the data made available by public institutions is often justified as a privacy-ensuring measure for apparently sensitive data. Paradoxically, this kind of information is often shared online, although fragmented, by PV sector companies, PV installation owners, public entities, or media outlets to showcase "green" initiatives or one's own work for marketing purposes or to raise public awareness. Thus, the goal is to demonstrate how the exploitation of online public information can play a role in addressing the lack of PV deployment information and can even drive stakeholders to value more the generation of public content and sharing their data. All this, while providing a methodology that can support the creation, publication, and visualization of comprehensive PV databases. Here, a case-study covering Portugal is presented, where PV deployment has experienced a remarkable growth rate over recent years. The information compiled has been formalized as the "Observatório Fotovoltaico" initiative, under which several infographics and analyzes have been published [6,7].

2 OVERVIEW OF EXISTING DATABASES, WITH A FOCUS ON PORTUGAL

It is pertaining to start by mentioning the release of an International Energy Agency PVPS report in 2020 [8], which stresses the relevance of this topic and the need to push public institutions to collect more detailed and structured information regarding PV projects.

While countries often provide aggregated statistics concerning PV deployment (i.e., often at the national level, sometimes regionally and distinguishing the legal framework or PV typology), there are also cases where data is provided at the installation level, such as Germany [9,10], Italy [11], France [12], Spain [13], Switzerland [14], and the United States of America [15]. However, the level of completeness and the parameters provided vary considerably. There is also the EU-scale Transparency Platform initiative from ENTSO-E (the European Network of Transmission System Operators) which collects data from the various Transmissions System Operators, and the work done by Jiménez-Garrote et al. which explores various scattered official sources of information on PV deployment in Spain to build an harmonized, more complete database [18].

There have also been non-governmental initiatives aiming to compile PV data. At the multi-country level, there are initiatives such as Open Energy System Data [19], Global Solar Power Tracker [20], the Open Infrastructure Map (OpenInfraMap) [21,22], PVOutput [23], BDPV [24]; whereas at the national level there is the example of the Microgen Database for the UK [25]. While the first addresses national-level aggregate data, the remainder focus on installation data.

In Portugal, the Directorate-General for Energy and Geology (DGEG) releases monthly public reports containing PV deployment and generation statistics [26]. Total monthly values are provided, covering both NUTS1 and NUTS0 spatial scales, differentiating certain PV typologies (conventional, self-consumption, two different sets of feed-in tariff regimes, concentrated PV). E-Redes, the Portuguese DSO, provides more detailed data concerning self-consumption installations, namely the quarterly total installed capacity, distinguishing three capacity ranges, for each LAU1 region [27]. From 2015 to 2019, the same DGEG provided regular reports listing the attributed licenses for feed-in tariff and self-consumption installations, although only peak and inverter capacity were recorded without any georeferencing [28]. Currently, there are two official PV mapping initiatives in Portugal: i) the e2p[1] project [29], which is promoted by the Portuguese Association of Renewable Energies[2] and covers all renewable sources; and ii) the Solis platform [30] which is led by the Lisbon Energy Agency[3] and focuses on PV and solar thermal.

3 METHODOLOGY: FROM DATA COLLECTION TO AN ONLINE MAPPING APP

As illustrated in Figure 1, the main workflow developed for this work consists in: i) manually searching for content online content (e.g., in companies webpages and social media, in media outlet webpages, in search engines); ii) identifying and logging data from textual content into a spreadsheet with fields such as installed capacity, number of panels, legal framework (e.g., feed-in tariff, self-consumption); iii) the installation is georeferenced, mostly from visual identification in Google Earth (sometimes supported by images found in the source document, or environmental studies in the case of some utility-scale installations); and iv) imported to the mapping web application.

However, in the process of georeferencing certain projects, often others can be visually identified in the process. These are also included in the database when installed on top or on the side of buildings since it allows to identify the end-user of said installation, although with no other information other than that and the location. Additionally, as the database becomes bigger and more well known, there have also been efforts to address companies directly to make the data collection process more efficient and exhaustive.

Then, the data is imported into GoGoGoCarto, an open-source online application to build interactive maps of georeferenced data [31,32]. It imports data from spreadsheet files or through APIs and allows i) the definition of filters, ii) exporting custom-based versions of the base map as iframe elements, and iii) can easily enable online crowdsourcing of data. Figure 2 exemplifies the GoGoCarto interface using the data collected for this work.

When contacting companies directly, additional information is requested, not to be made public as is (since it embeds the project design practices of the companies) but to enable aggregated sectorial analyses which has the potential to describe the national reality of project deployment and project design practices, while ensuring the induvial privacy of each company. However, up to now, companies have not been receptive of this, due to the implied added workload of data compilation and structuring, but also because they do not see the added value of doing so.

Additionally, since the collection of data is based on the premises that only installations with a known context[4] are mapped, small- to medium-scale projects (i.e., distributed PV, in the context of feed-in tariff or self-consumption frameworks) are classified according to the sector of activity of its end-user. This is done by exploring alternative data sources where the companies' sector of activity codes is listed.

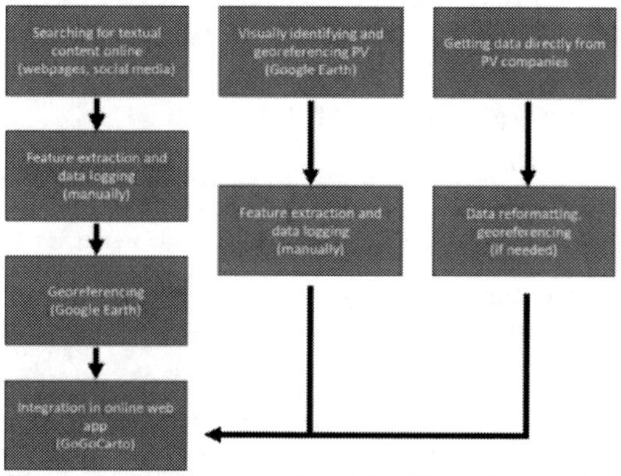

Figure 1: Workflow developed in this work, covering different sources of information.

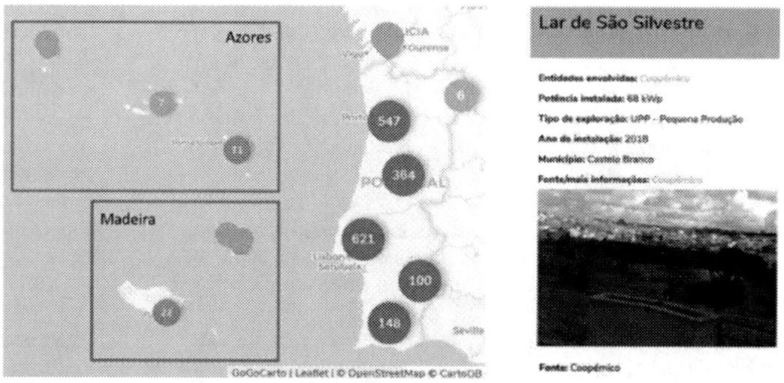

Figure 2: Example of GoGoCarto interface from the PV map developed during this work. Although not shown here, the platform also provides filters for users to visualize subsets of the data.

[1] Endogenous Energies of Portugal, translated from the original in Portuguese "Energias Endógenas de Portugal".

[2] Translated from the original "Associação Portuguesa de Energias Renováveis" (APREN)

[3] The original name is Lisboa E-Nova - Agência de Energia e Ambiente de Lisboa

[4] I.e., who is the end-user for distributed PV installations, or utility-scale installations which are, at least, identified online.

10.4229/EUPVSEC2023/5DV.3.44

4 RESULTS

Up to now, more than 1900 projects have been catalogued and georeferenced, representing more than 1.8 GWp (58% of the 3.1 GWp reported for Portugal July 2023 [33]. More than 66% of the projects were manually catalogued, although some cases were later completed with data received directly from companies. Currently, 7 different entities have contributed to this initiative: one cooperative of renewable energies (Coopérnico), one responsible crowdlending company (GoParity), one municipal energy initiative (Solis, from Lisbon), and four companies (Ecoinside, Rosseti Engenharia, Enforce, Greenvolt Next Portugal).

The level of detail and completeness of information depends greatly depending on the data source of a given project (Figure 3). While more detailed cases include installed capacity, number of modules, even number of inverters and inverter AC capacity, others include only the client and, maybe, the region the project is installed. Figure 2 quantifies the frequency of missing data for each of the main collected parameters.

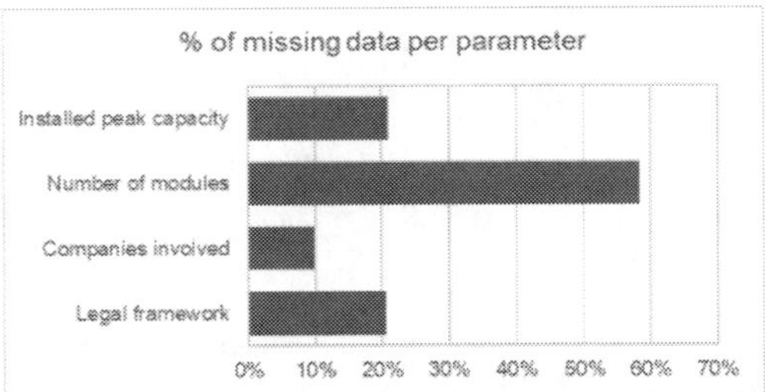

Figure 3: Quantifying the missing data for the collected features from online content.

Table 1 describes and quantifies the collected data, comparing it to other available datasets covering Portugal. It shows how in a digital era where PV is already a largely publicized massified product, online public content represents a vast untapped potential on what concerns identifying and describing PV projects.

5 POSSIBLE OUTPUTS GENERATED BY THIS TYPE OF DATABASE

Having such detailed data can support TSOs and DSOs to have a better visibility of PV deployment, supporting (and improving) the modelling that later informs resource adequacy and grid expansion studies. Additionally, companies can use this to identify more easily congested grid zones and better address their prospection efforts.

Having projects georeferenced and labeled according to their legal framework (feed-in tariff, self-consumption, etc.), spatial patterns can be identified and possibly correlated with socioeconomic but also natural factors, such as purchasing power, level of industrialization of population density, local solar resource availability and orography.

Identifying under and well explored locations for self-consumption projects, namely in industrial hubs. While the first case can help policy makers and local energy agencies to better target their actions and outreach/clarification activities, the second can support PV companies to set up renewable energy communities where industries share their surplus between each other or even sell it to neighboring sites which have yet to explore PV, improving the profitability of said projects and the savings of all the actors, while reducing the need for subsidized purchase of said surplus.

The extensive collection of projects metadata allows to better understand how the technology penetrates and evolves inside a region. One example of this is the peak capacity of the modules installed, which is derived from the total peak capacity and the number of modules. The data collected so far, highlights how through time larger and more efficient modules became the dominant trend in Portugal. Between 2015-17 the distribution of peak capacity was more skewed towards low values with a peak in 250-300 Wp; later in 2018-20 the left skew is less pronounced with more projects in higher Wp ranges; whereas recently we see completely the opposite of the initial scenario with a right skew towards high Wp panels.

The analysis of the number of PV adopters in each sector of activity allows the identification of well-performing but also under-explored sectors, which can guide the action of policy makers and industrial associations. Additionally, this can also support the engagement of new industrial clients by PV companies by enabling an easy demonstration of how PV is adopted by their sectorial peers. Figure 5 provides an example of the kind of insights and raising awareness that such data can enable.

Lastly, the compilation of such diverse information can easily be used to generate informative outreach content, by promoting the state of the database at different points in time, as well as promoting individual projects that have interesting/pertinent characteristics. Currently, this is done in the form of LinkedIn publications [7] but may take a more structured format in the future (e.g., white papers or national report on PV deployment [34]).

Table I: Description of various data sources describing PV installations in Portugal.

Data source	Geolocated projects	Covered variables	Year of oldest & newest installation
Solis+	85	Year of installation, installed capacity	2008-2021
e2p	159	Developer, end-user, year of installation, installed capacity, # of modules and peak power	2008-2020
OpenInfraMap	258	Installed capacity*	N.A.
This work	354	Developer*, year of installation*, installed capacity*, # of modules* and peak power*	2006-2023

10.4229/EUPVSEC2023/5DV.3.44
020574-003

6 CONCLUSIONS

This work aims to propose and explore different methods and strategies to address the information gap on distributed PV deployment Firstly, by building a database from scattered online media content. Using Portugal as a case study, where PV deployment has experienced a remarkable growth rate over recent years, a prototype was set up which already includes almost 2000 projects. The user experience with this data is further enhanced by integrating it in the interactive online mapping platform, GoGoCarto [31,32]. This open-source tool enables not only an engaging interaction with the data, boosted by user-defined filters, but also the generation of customized versions of the base map which can easily be embedded into a third party's website.

One pertinent point to highlight is the heterogeneity in the data, both in format and content, which then can lead to a sparse database. Some content offers very rich and detailed information, such as the inverter capacity or the module model, while others sometimes just mention the end-user (in case of self-consumption or small-scale feed-in-tariff) or the name of the project (in case of large-scale installations).

Additionally, one of the challenges of the proposed workflow is the fact that the data collection process can be greatly manual and time consuming. Although not discussed here, exploratory tests are being made to understand the potential of large-language models such as ChatGPT to automatize the feature extraction from the collected textual content. Overall, results showed to be promising for the acquisition of quantitative (e.g., installed capacity, number of panels) and contextual data (e.g., legal framework). One challenge consisted in the identification of the company selling the PV system and the client/end-user, often even mistaking one for the other.

Future work also includes the assessment of satellite imagery to classify the already mapped projects according to the type of installation (e.g., ground-mounted, rooftop, carport, floating). This is unlikely valid for PV installations on façades, since ground-based imagery is considerably scarcer and, as a consequence, most often outdated.

7 ACKNOWLEDGEMENTS

The authors would like to acknowledge the important role of the various contributors of this project. First, to Sebastian Castro and the rest of the GoGoCarto community who made an amazing tool be openly available, while providing regular support. Second, the various stakeholders which have provided invaluable data but also insights concerning the project and its vision (namely Coopérnico, Ecoinside, Rosseti, Enforce, Solis, and Greenvolt Next Portugal). Additionally, the dissemination would never be as engaging and appealing without the support of Leonardo Amaro e Silva, who is responsible for most of the infographics and visuals.

7 REFERENCES

[1] Shaker et al., "A Data-Driven Approach for Estimating the Power Generation of Invisible Solar Sites", IEEE Transactions on Smart Grid, 2015.

[2] Pierro et al., "Progress in regional PV power forecasting: A sensitivity analysis on the Italian case study", Renewable Energy, 2022.

[3] Yu et al., "DeepSolar: A Machine Learning Framework to Efficiently Construct a Solar Deployment Database in the United States", Joule, 2018.

[4] Kruitwagen et al., "A global inventory of photovoltaic solar energy generating units", Nature, 2021.

[5] Kasmi et al., "PyPVRoof: a Python package for extracting the characteristics of rooftop PV installations using remote sensing data", arXiv, 2023.

[6] IEA PVPS Task 14, "Data Model for PV Systems", 2020. Report IEA-PVPS T1/14-01:202

[7] O Mapa do Fotovoltaico – main page. Available online: https://mapa.observatorio-fotovoltaico.com/

[8] O Mapa do Fotovoltaico – LinkedIn page, https://www.linkedin.com/company/observatorio-fotovoltaico/

[9] Marktstammdatenregister, https://www.marktstammdatenregister.de/MaStR

[10] Tepe et al., "open-mastr Python package: Package for working with German energy data", 12th Energy Colloquium of the Munich Institute of Integrated Materials, Energy and Process Engineering, 2022.

[11] GSE, "Atlaimpianti", https://atla.gse.it/atlaimpianti/project/Atlaimpianti_Internet.html

[12] Open Data Réseaux Énergies (ODRÉ), "Registre national des installations de production et de stockage d'électricité", https://odre.opendatasoft.com/explore/dataset/registre-national-installation-production-stockage-electricite-agrege/

[14] Ministerio para la Transición Ecológica y el Reto Demográfico, "PRETOR", https://energia.serviciosmin.gob.es/Pretor/Vista/Informes/InformesInstalaciones.aspx

[15] Lawrence Berkeley National Laboratory, "Tracking the Sun", https://emp.lbl.gov/tracking-the-sun

[16] ENTSO-E Transparency Platform, https://transparency.entsoe.eu/

[17] Hirth et al., "The ENTSO-E Transparency Platform – A review of Europe's most ambitious electricity data platform", Applied Energy, 2018.

[18] Jiménez-Garrote et al., "SOWISP—A retrospective high spatial and temporal resolution database of the installed wind and solar PV power in Spain", Solar Energy, 2023.

[19] Open Power System Data, https://open-power-system-data.org/

[20] Global Energy Monitor, "Global Solar Power Tracker", https://globalenergymonitor.org/projects/global-solar-power-tracker/

[21] OpenInfraMap, https://openinframap.org/stats

[22] OpenInfraMap – GitHub repository. Available online: https://github.com/openinframap/

[23] PVOutput, https://pvoutput.org/

[24] BDPV - Base de données d'installations photovoltaïques, https://www.bdpv.fr/en/

[25] Sheffield Solar, "Microgen Database", https://www.solar.sheffield.ac.uk/all-projects/microgen-database/

[26] DGEG, "Estatísticas Rápidas", https://www.dgeg.gov.pt/pt/estatistica/energia/publicacoes/estatisticas-rapidas-das-renovaveis/

[27] E-Redes, "Total production Units for Self-Consumption", https://e-redes.opendatasoft.com/explore/dataset/8-

unidades-de-producao-para-autoconsumo/table/
[28] DGEG, "Histórico registos Serup", https://www.dgeg.gov.pt/pt/areas-setoriais/energia/energia-eletrica/producao-de-energia-eletrica/producao-descentralizada-autoconsumo-e-upp-mp-mn/regimes-anteriores/historico-registos-serup/
[29] APREN, "e2p – energias endógenas de Portugal", https://e2p.inegi.up.pt/ (accessed on Feb 2, 2023)
[30] Solis - Mapa de sistemas fotovoltaicos. Available online: https://www.solis-lisboa.pt/mapa-solar-de-li/#sistemas-fotovoltaicos (accessed on Feb 2, 2023)
[31] GoGoCarto – Main Page. Available online: https://gogocarto.fr/projects
[32] GoGoCarto – GitHub, https://gitlab.com/Seballot/gogocarto
[33] DGEG, "Estatísticas Rápidas nº 224 Julho 2023", https://www.dgeg.gov.pt/media/f1zc0kxu/dgeg-arr-2023-07.pdf
[34] IEA PVPS, "National Survey Reports", https://iea-pvps.org/national-survey-reports/

10.4229/EUPVSEC2023/5DV.3.44
020574-005

SOCIETY'S ACCEPTANCE AND WILLINGNESS TO ACT IN THE CONTEXT OF THE ENERGY TRANSITION: QUALITATIVE SURVEY USING THE EXAMPLE OF PHOTOVOLTAICS

Stefanie Könen[1], Anne Karrenbrock[2], Ulf Blieske[1]
[1]TH Köln, Cologne Institute for Renewable Energy and [2]Cologne International School of Design
Betzdorfer Str. 2
Cologne, Germany
stefanie.koenen@th-koeln.de

ABSTRACT: Essential for the success of the **energy transition** are public support and engagement. This **qualitative research** study, conducted in Germany, investigates society's acceptance and willingness to act in the context of transformation, focusing on **photovoltaic technology** (PV) as an example. The study is part of the joint project "MEnergie - Meine Energiewende," which aims to identify specific **information needs** of various stakeholder groups in Germany and promote informed decision-making.

Key findings from this study indicate a positive overall attitude towards the energy transition and renewable energy among the surveyed population. Participants recognized the importance of transitioning to renewable energy sources but demonstrated limited knowledge about energy consumption, costs, and efficiency. The perception of PV technology was influenced by economic considerations, with concerns about elevated costs and perceived inefficiency in Germany due to low insolation levels.

The study highlighted the need for improved information dissemination, simplified approval procedures and expert guidance to support citizens in adopting PV systems. It also emphasized the importance of addressing bureaucratic barriers and strengthening funding opportunities to accelerate the energy transition. Additionally, the study revealed that societal motivation for energy conservation had increased in response to external factors such as geopolitical events and natural disasters.

Keywords: energy transition, photovoltaic technology, qualitative research, information needs

1 INTRODUCTION

Actual research results and surveys show that German society thinks positive about the energy transition in general [1]. Nevertheless, progress is not at the necessary speed [2]. For a successful energy transition individual decisions and mutual actions are necessary [3]. In order to be able to recognize and consider opportunities and conflicting objectives, there is a great need for research and communication with society.

The joint project MEnergie - Meine Energiewende has the objective to identify the specific information needs of stakeholder groups in Germany (such as generations Z and Y and real estate owners) and – based on the Social Construction of Technology (SCOT) [4] and innovative media formats - to generate willingness and competence to act. The interaction between technology and society is captured and the application of technology is recorded. In close collaboration, the interdisciplinary team of designers, engineers and social scientists uses agile, human-centered methods and participatory design to collect and develop information needs, existing prejudices, fears, and specific questions about the energy transition amongst target groups.

In the initial project-phase our objective was to get in contact with people and society in Germany. In public space we used photovoltaics, which is one central technology in the energy transition in Germany, as an impetus for workshops, real life experiments und booths. How do citizens react when they see a PV-module? Which questions do they ask? Which points of interest can be identified? What is the first insight on the research topic "the energy transition and appropriate communication" with different stakeholder groups?

2 AIM AND APPROACH

Especially in research projects around questions of social transformation, Living Labs and real-life experiments are a current trend in science and trans- and interdisciplinary teams. [5–7]

MEnergie follows the concept of the Social Construction of Technology (SCOT) that seeks to understand how technology is developed, shaped and understood within society. Technology is not just the result of objective scientific principles or a linear progression of innovation but is influenced by social, cultural, political and economic factors. [4]

In order to get a first feeling for the ideas of people and atmosphere in the society regarding the energy transition, different activities were identified, implemented and evaluated by the research team. At the center of all activities was the idea of getting in touch directly with the citizens.

The methodological spectrum of qualitative social research offers guiding principles as well as diverse methods for data collection, theory development and knowledge gain. It allows researchers to get deep insights into human behaviour, social dynamics, and cultural phenomena. It can help to meet the challenges of working trans- and interdisciplinary within the team, participatory with the stakeholders and equally with different target groups. The research team decided on methods as participatory field observations, workshops with guiding questions for group discussions and exchange as well as chat and interviews. According to Lamnek et al. the field observations can be organized explorative, open and semi-structured. [8]

Prior to the first of three design cycles of the project, different activities have been implemented as there were:

10.4229/EUPVSEC2023/5DV.3.45
020576-001

(1) A booth with a PV-module and a poster on plug-in solar was exhibited in public events in June 2023 in the City of Cologne, accompanied by an analogue poster poll.

(Fig. 1: booth at Night of Techniques in Cologne)

(2) On July the 14th in 2022 a real experiment called "PV-Surfing" was realized and a PV-module was carried into Cologne public transport and public space. It was used as trigger for citizens' reactions and chat.

(Fig. 2: "PV-Surfing" in the public transport of Cologne)

(3) As set of Workshops on Plug-In-Solar were organized and held for students and citizens in June 2022. In theory and practicums people learned how to install and run the smallest PV-System.

(Fig. 3: Plug-In-Solar Workshops at TH Köln)

(4) A World-Café with 15 chimney sweeps was organized in September 2022. Their observations on the topic of energy transition amongst the population in Germany were central in the discussions.

(Fig. 4: Chimney Sweeps World-Café at TH Köln)

Reactions, observations and communications were recorded, using participant observation methods, framed observation guidelines and defined criteria. More than 125 persons were reached in the public events (1-3). The additional expert Workshop (4) with master chimney sweepers from all over Germany, gave insight on a great number of households and made a comparison with literature and findings from the MEnergie activities possible.

We can conclude that an increasing willingness to act can be found regarding photovoltaics technology. The energy transition is claimed necessary in the perception of the citizens we reached. Still we have to overcome bureaucratic barriers, misconceptions and we need to correct false information.

3 SCIENTIFIC INNOVATION AND RELEVANCE

Any transformation process is driven by peoples' ideas and perceptions. So, humans are central within MEnergie – meine Energiewende. From knowing more about their ideas, the research team can inductively develop proposals for communication formats. It is necessary to react and respond empathically to the target group. Their interests, needs, experiences, resources and action orientation must be identified, respected and considered. Explanation patterns will be jointly developed, intended to generate the competence and willingness to act within the energy transition.

The unique aspect of the project is the involvement of the target groups in three design cycles. The participatory and iterative structure ensures the scientific transfer into society. The applicability and comprehensibility of the communication formats is guaranteed, as citizens are directly involved in the design process.

In addition to this human-centered approach and structure, the learning processes of transdisciplinary research is relevant in MEnergie. The familiar ideals of measurability from the natural and technical sciences are softened in transdisciplinary research. Our aim is to achieve sustainability in the self-empowerment of the actors. One individual is not able to cover all perspectives. The reflections show that the exchange and comparison of knowledge leads to further and creative ideas in dealing with problems and challenges.

Extending the focus from literature review to field observations and direct interaction with citizens within the project, means a scientific substantiation for the human centered approach of MEnergie. To use a central technology for the energy transition as impulse and trigger for communication, added value in different perspectives, retrieved from anthropological participant observation and fieldwork methods [9]:

- The impulse set by the PV module allows the observer to set a focus on the topic without framing the comments or ideas.
- The observed comments or behavior gives an insight to the reasons and opinions of the participant.
- Understanding and interpreting by the researchers can get deeper due to empathy, comparison and non-verbal interactions which can help to better understand the motives, thoughts and feelings.
- Within an eye-to-eye talk, trust can be built with the participants. This allows individuals to communicate more openly and authentically.
- Social, emotional and cultural aspects can be considered.
- The research team had the opportunity to reflect on and correct their own assumptions, biases, and prejudices which leads to greater methodological accuracy and reliability of observations.

All aspects can help to validate concrete insights on the hypotheses and phenomena under study as well as to gather rich and contextualized data.

4 RESULTS AND DISCUSSION

4.1. Booth in public events

From the booth and the analogue poster poll in public events (1) we learned that the overall attitude towards the energy transition and Renewables is positive. Over 30 people marked three of their personally most important attributes by putting sticks. Figure 5 describes how participants evaluate the energy transition. The majority claims it necessary and important. Nobody thinks it is unrealistic. Figure 6 shows the choice of associations participants take for Renewable Energy. Positive aspects such as Economic Development, Hope, Autonomy,

10.4229/EUPVSEC2023/5DV.3.45
020576-002

Freedom. Security stated considerably often, which leads to a positive interpretation.

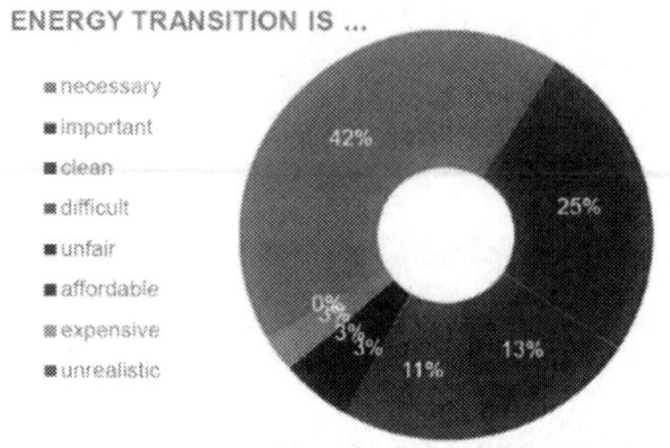

(Fig. 5: Associations about the Energy Transition)

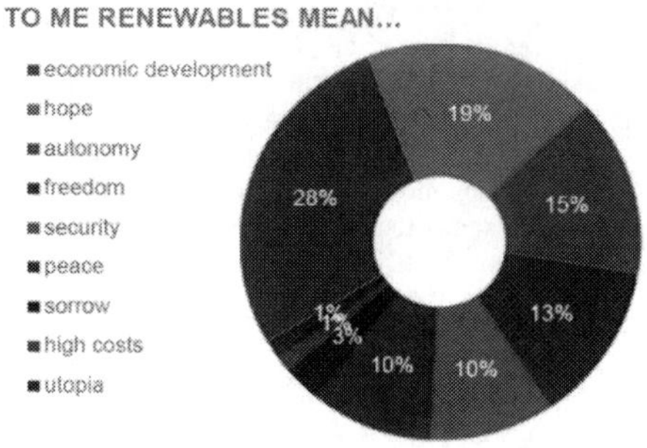

(Fig. 6: Associations about Renewables)

The energy transition is widely perceived as necessary and important. Renewables mean economic development and even hope, autonomy and freedom.
Within conversations and within the analogue poster poll, the research team found that people often do not have knowledge about their own energy consumption, the costs of electrical energy or the electricity needs for household appliances.

4.2. PV-Surfing

From PV-Surfing (2) we learned that there is a lot of interest in photovoltaics. The PV-module attracted the attention of various groups of the population. In more than 20 ad-hoc-interviews, we found that only about 35 % knew immediately what technology they were looking at and what it is supposed to do.
Without surprise, men show a greater affinity to technology and photovoltaics. It was males estimating the worth of the PV-module. Several wanted to take it home and two even started explaining their own and self-made PV-home-system.
60plus participants pointed out that the energy transition is for the "younger" generation. People in retirement age repeatedly pointed out: "My children have that. I am too old. I do not invest anymore."
A probably humorously meant joke came from two pensioners: "I have a mini-nuclear power plant in the basement, what am I supposed to do with such a module on the roof?"

Most conversations and loose chat quickly turned to the topic of costs and economic considerations against the background of rising energy prices. Saving electricity was central in nearly every conversation. The challenges to save electrical and thermal energy as a result of the Russian invasion into Ukraine was one load-bearing topic at that time.
People from southern parts of Europe and the globe were convinced, that PV is not a worthwhile investment in Germany, but everywhere in their country of origin.
A tenant in a big local housing estate reported that "those things" (pointing on PV-Module) had been installed on the roof of the apartment house she lives in over two years ago. She claims that her energy costs were still too high and nobody explained what "those things" are used for.
One family points out to buy their own apartment first and then fill up the roof with a PV-system.
Another one will move into an apartment with a balcony and then install plug-in solar.
Two boys (11 and 13) and two girls (12) couldn't tell what the "thing" in front of them was. During conversation one remembered, "It's something about the sun and electricity." By then the boys and girls remembered school lessons and information given by their teachers and started chatting about PV, school and their own energy consumption.

The findings from booths and PV-Surfing were discussed, compared to literature and other research projects in order to identify the societal needs around the energy transition.
One of the most significant finding is that conceptual knowledge about energy is missing amongst people. Energy simply is available. People do not make any difference whether it is thermal or electric. The knowledge about costs, consumption, efficiency, alternatives etc. is not universal and must be claimed as insufficient. People do not know what it means to use one kilowatt hour of electricity or even which device in the household works how long with this amount of energy. Costs for the use of different devices are not clear.
The perception of photovoltaics technology in Germany is an economic one. A majority of respondents said, that it was economically wrong to implement PV in Germany. The costs are thought to be high and the yield is thought to be low due to the lack of sunshine in the country, especially compared to countries in the global south. It is often claimed that electricity from PV is expensive compared to fossil-based or nuclear power supply.

4.3. Plug-In-PV Workshops

The research team decided to show how easy it is to use photovoltaics at home and offered Plug-In-PV Workshops (3) for more than 80 participants in 2022. The team learned, that the supposedly easiest way to participate in the energy transition with small PV-Systems is particularly challenging for people in terms of purchase, registration, secure mounting and operation. Permitting procedures need to be simplified. A good information base and expert advice are needed to guide the process.

4.4. World Café with chimney sweepers

Within the World-Café with 15 chimney sweepers from all over Germany (including Berlin, Brandenburg, Bremen, Bavaria, Baden-Württemberg, North Rhine-Westphalia) we learned from a first expert view on the energy transition in Germany. All of them are masters of their guild and each one is responsible for their own districts, which means an equivalent of up to 2.000 properties and households for one master chimney sweeper. All of them have direct contact with real-estate-owners and decision-makers on the energy supply and consumption in buildings. Their focus is on thermal energy and the energetic constitution of buildings including the

10.4229/EUPVSEC2023/5DV.3.45
020576-003

primary energy consumption of real estates.
The discussions focused on the concerns and on the motivation of the residents of the properties served by the chimney sweeps regarding the energy transition. Their sketch of a single-family house after renovation in the year 2030 showed important components related to the topic of energy in the building and their recommendations on technical options. The professionals pointed out some lighthouse projects of the energy transition and put the to focus on CO2-neutral heating and the German governments funding for building renovation .
The discussions made the research team understand that craftsmen and the skilled trades sector is dependent on information as well as on education and training formats for the energy transition. It is crucial that the master chimney sweepers do not have a common denominator in their viewpoints and not even in their consultations.
All of them agree, that in order to achieve the 2030 / 2045 climate targets of the German government, bureaucratic barriers need to be eliminated and funding opportunities need to be strengthened.
They describe lobbying groups as strong and capable of slowing down the necessary speed of the implementation of the energy transition.
They point out that the expected energy crisis in winter 2022 due to the Russia's invasion to the Ukraine threatened a return to the use of with fine-dust-intensive heat generators (wood and briquette fireplaces) and thus an increase in air pollution.
Besides, the chimney sweepers state that the readiness to use renewable energy has risen since the Russian invasion in the Ukraine and the floods from the 14th of July 2021, when heavy rainfalls caused floods of the small rivers of Ahr and Erft in North-Rhine-Westphalia, Germany. Destruction of Infrastructure such as motorways and trainlines, as well as property loss in buildings, cars and other damages were the results. The costs are estimated to be over 31 Milliards of Euros. The motivation for energy saving is considerably elevated.
The experts regret that it was not present decades ago.
In their opinion, a great risk is the immense pressure and great fears not to be able to cover the rising costs for gas or the possible shortcuts in a cold winter, that chimney sweepers perceive within the population. People fear cold homes in winter due to gas shortage or even constantly rising energy prices. This can easily lead to frustration and a turning against reasonable measures in the building sector.
Furthermore, the lack of qualified professionals for all tasks in the energy transition is considered to be a bottleneck for the expansion rates for renewable energy systems and refurbishment and renovation of existing thermal energy supply systems in buildings.

5 CONCLUSIONS

Initial studies in the frame of the MEnergie project found that the energy transition has arrived in the German society. PV-technology is well perceived and accepted, a so-called low hanging fruit in the energy transition. The use of PV by citizens needs support and information. The bureaucratic barriers should be overcome and start-up as well as running the systems should be as simple as possible and worthwhile due to feed-in remuneration and funding. For the expansion goals of renewable energy systems and to meet the expectations of people and society, the qualification of handcraft and skilled workers is necessary. People who are ready to investigate into renewables need to meet an appropriate information and implementation state and support. If the market cannot fulfill the expectations and dissatisfaction of clients is the result, the current acceptance will get lost.
In addition, the immense pressure within the population resulting from the news situation, conflicts, perceived changing climate conditions, rising costs, inflation, etc. has the potential to cause discontent and political extremes.

6 OUTLOOK

Within the three design cycles which are central for the MEnergie project, the research team actively keeps contact with the population and different stakeholders by using participation formats. For example, seven workshops for citizens are planned within the cooperation with Stadt Köln and its library.
Additionally, MEnergie organized a group of master students of renewable energies who offered a set of Plug-In-PV Workshops in spring and summer 2023. The participant feedback on learnings from that will be evaluated, studied and published in winter 2023.
Furthermore, MEnergie organized a "Solar party" within the frame of a solar information campaign by Solarenergieförderverein (SFV) called "Packsdrauf". It is a communication format on photovoltaic systems with neighbors for neighbors. Those citizens, who already own and run a PV-System give necessary tips and information showing their own PV-installations to their neighbors, which facilitate the realization of other solar systems in the neighborhoods.

7 ACKNOWLEDGEMENTS

The project is financed by the German Federal Ministry for Economic Affairs and Climate Action (BMWK).
The authors thank their consortium colleagues, students and participants for their important contributions and discussions.

8 REFERENCES

[1] M. Doms. "AEE-Akzeptanzumfrage+2022." https://www.unendlich-viel-energie.de/media/file/5511.AEE_RenewsKompakt_Akzeptanz_jul23.pdf (accessed May. 28, 2023).
[2] K. Andreae. "Fortschrittsmonitor Energiewende." https://www.bdew.de/media/documents/RZ_EY-22-047_Fortschrittsmonitor_BKL2302-v11-high.pdf
[3] BMWK. "Förderaufruf: Energiewende und Gesellschaft: Forschungsförderung des Bundesministeriums für Wirtschaft und Klimaschutz (BMWK) innerhalb des 7. Energieforschungsprogramms der Bundesregierung." https://www.energieforschung.de/lw_resource/datapool/systemfiles/elements/files/F4D16ECB957D788BE0537E695E86B71F/current/document/BMWK-Foerderaufruf_

Energiewende-und-Gesellschaft-2023.pdf

[4] T. Pinch and W. Bijker. "Social Construction of Technology." https://www.researchgate.net/publication/257141289_The_Social_Construction_of_Facts_and_Artifacts_Or_How_the_Sociology_of_Science_and_the_Sociology_of_Technology_Might_Benefit_Each_Other (accessed May. 31, 2023).

[5] D. Lindner, K. Alle, M. Graef, and L. Sanders, "Abschlussbericht_Evaluation_RNM_final.docx," 2020. [Online]. Available: https://www.zirius.uni-stuttgart.de/dokumente/Abschlussbericht_Evaluation-Realexperimente_RNM_final.pdf

[6] H. Bulkeley, L. Coenen, et al., "Governing Urban Sustainability Transitions," 2017.

[7] F. Nevens and C. Roorda, "A transition approach for climate neutrality in the city of Ghent (Belgium)," 2013.

[8] S. Lamnek, *Qualitative Sozialforschung: Lehrbuch,* 4th ed. (Beltz PVU Lehrbuch). Weinheim: Beltz PVU, 2008. [Online]. Available: http://www.socialnet.de/rezensionen/isbn.php?isbn=978-3-621-27544-6

[9] L. M. Vaughn and F. Jacquez, "Participatory Research Methods – Choice Points in the Research Process," *Journal of Participatory Research Methods*, vol. 1, no. 1, 2020, doi: 10.35844/001c.13244.

A SNAPSHOT OF GLOBAL PV MARKETS - 2022

Gaëtan Masson[1], Melodie de l'Epine[4], Izumi Kaizuka[2],
Elina Bosch[3], Philippe Macé[3], Caroline Plaza[4], Adrien Van Rechem[3]
[1]IEA PVPS Task 1, Belgium – g.masson@iea-pvps.org, +32 2 514 78 05
[2]RTS Corporation, Japan – kaizuka@rts-pv.com
[3]Becquerel Institute
[4]Becquerel Institute France

ABSTRACT: The objective of this paper is to propose a reliable and accurate perspective on key markets and policies related to PV development in 2022 and previously. It aims at offering a clear analysis of how PV markets have developed in 2022, with updated numbers, along with an analysis of the policies behind the development. In this paper are displayed and analyzed survey results for the calendar year 2022 concerning PV markets and policies, as well as other key issues. An increasing number of national markets experienced notable growth in 2022 with impacts on policy development. Figures show that over 235 GW of PV systems have been installed in the world last year. Consequently, cumulative capacity crossed the 1.1 TW mark in 2022.
Keywords: Photovoltaic (PV), Market, IEA PVPS

1 INTRODUCTION

This Trends paper gives information on the development of PV power applications in the PVPS member and non-member countries and is largely based on the information provided by IEA PVPS countries in addition to Becquerel Sweden and the European Union through its European Commission. The report includes information on national market developments at the end of the former year, in this case, 2022. The International Energy Agency – Photovoltaic Power System Programme (IEA PVPS)'s Task 1 is responsible for strategy and outreach within the IEA PVPS program. This includes policy, market and industry analysis. A key deliverable of Task 1 is the annual Snapshot of Global PV Markets publication, together with the annual flagship report Trends in PV Application.

The objective of the series of annual Snapshot and Trends reports — which have been published since 1992 (Trends) [1] and 2013 (Snapshot) [2] — is to present and interpret developments in both the PV systems and components being used in the PV power systems market and the changing applications for these products within that market. These trends are analyzed in the context of the business, policy and non-technical environment in the reporting countries.

2 THE GLOBAL PV INSTALLED CAPACITY

This paper presents the latest survey results for the calendar year 2022 concerning PV markets and policies, as well as other key issues. An increasing number of national markets experienced notable growth in 2022, and that impacted policy choices. The capacity figures given are nominal DC peak power (Wp) under standard test conditions (i.e. 1 000 W irradiance, air mass 1.5 light spectrum and 25 °C device temperature) for consistency reasons. As not all countries report DC peak power (Wp) for solar PV systems, instead reporting AC power, these capacity figures are converted to Wp DC.

While the final figures for 2022 could slightly continue to be refined in the coming months, nearly final figures show that around 235.8 GW of PV systems have been installed in the world last year. Some important trends observed are as follows.

The global PV market grew to around 235.8 GW in 2022, compared to around 174.5 GW in 2021. This represents a year-on-year growth of more than 32%.

Asia scored the first place again in 2022 with China (105.5 GWp) and India (18.1 GWp) leading the way. With around 18.5 GWp deployed this year, the market in USA was down (21.1 GWp), after having reached unprecedented heights in 2021 (24.1 GWp).

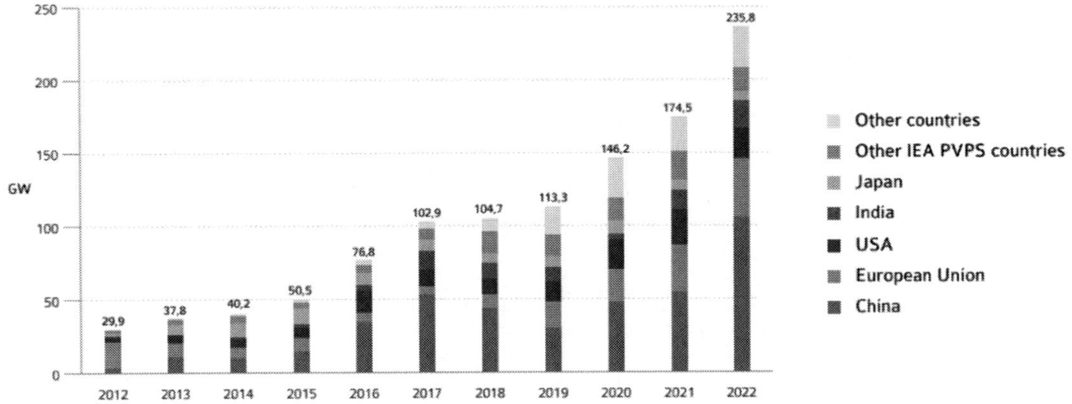

Figure 1: Evolution of annual PV installations (GW-DC)

10.4229/EUPVSEC2023/5EO.3.1
020577-001

The market in EU grew significantly to around 42 GWp. The top countries are Germany with (8.5GWp) and Spain with (7.2 GWp) installed in 2022, followed by Poland (4.9 GWp), the Netherlands (3.9 GWp) and France (2.9 GWp).

Preliminary numbers show that installations in the MENA region amounted to around 4 GWp last year.

Fast development was observed once again in Latin America, with around 10 GWp installed in Brazil, a significant increase from last year's 6 GWp, and 1.8 GWp in Chile.

Drivers for PV development include record-breaking competitiveness levels and development of distributed PV.

The impact of rising component prices was not clearly observed from global PV systems installations' point of view.

The total cumulative installed capacity reached around 1183.7 GW in 2022.

3 MARKET DEVELOPMENT

In 2022, the PV market saw an important growth for the third year in a row after 2019's limited growth. Globally, the trend is upwards by exceeding the 235 GW annual installed capacity as expressed in Fig. 1.

Asia remains the leader of the global PV market. Next to China (105.5 GW), India and Japan remain a relevant presence in the global market with respectively 18.1 GW and 6.7 GW installed.

In other parts of Asia, the market observed steady growth in Taiwan (2.0 GW) and significantly decreased in South Korea (3.1 GW) and Vietnam (0.1 GW) after the massive growth observed in 2020. Australia installed 4.2 GW in 2022, a stable level since 2018 notwithstanding the 2021 rise to over 5 GWp.

In the Americas, the USA market slightly decreased (21.1 GW), on the other hand Brazil installed at least 9.9 GW in 2022 (an important increase compared to 2019, 2020 and 2021 level). PV installations in Chile were rather stable in 2022 reaching a cumulative installed capacity of 7.9 GW while installations in Mexico were at a lower level (0.7 GW) compared to 2021. The market in Canada decreased again in 2022 with 765 MW installed compared to the record level seen in 2021 (2.0 GW).

In the European Union, Spain have now the leading position with nearly 8.5 GW was installed, surpassing Germany this year. In 2022, Germany (7.2 GW), Poland (4.9 GW), the Netherlands (3.9 GW) and France (3.0 GW) can be mentioned as leading countries as well. Italy added over 2.4 GW while nine countries added more than 1 GW namely Greece, Denmark, Belgium, Hungary, Switzerland and Austria.

New developments occurred in Africa (Egypt, South Africa) and in the Middle East (UAE) which led to GW-scale cumulative installation levels: 4.7 GW in South Africa, 3.6 GW in the UAE and 3.5 GW in Egypt for instance. Israel installed 1.2 GW in 2022 only.

In 2022, twenty-four countries passed the GW mark concerning annual installed PV capacity. Forty-four countries reached at least 1 GW cumulatively in 2022. China alone represented 414.1 GW. Germany, which used to lead the rankings for years, lost its leading position in 2015 and now ranks fifth (67.3 GW). The USA are second (141.5 GW) and Japan is third (85.1 GW). With close to 242 GW of total capacity, Europe is now significantly behind Asia, leading with least 660 GW, with much more to come in the coming years.

4 GRID-CONNECTED CENTRALIZED AND DISTRIBUTED

Regarding the share of grid connected and centralized installations at the global level, the trend has changed several times over the years.

Before 2013, most new PV installations were distributed systems, i.e. mainly installed on rooftops.

However, in more recent years, the market has seen a strong development of grid connected centralized installations. This changing trend is because centralized PV has evolved faster in terms of cost, and most of the major PV developments in emerging PV markets are coming from utility-scale PV. The success of utility-scale installations is attributed to the fact that installation time and cost per Wp are lower than for distributed PV plants.

In the last years, tenders have driven PV development and continued to be granted in many countries in the world with extremely competitive prices, well below 20 USD/MWh in the sunniest places. One of the key trends of 2022 is the wider development of utility-scale plants without financial incentives. Such development independent from financial incentives is mostly independent of policy decisions, which makes its potential virtually unlimited.

Distributed PV market share, after suffering from a

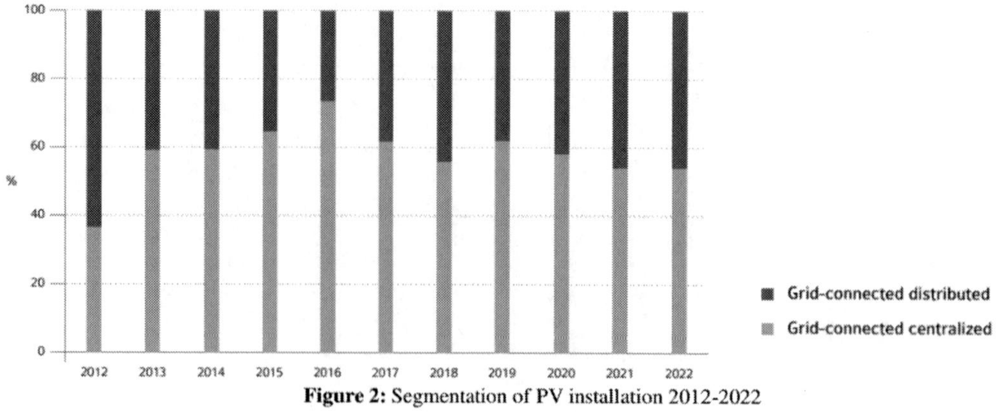

Figure 2: Segmentation of PV installation 2012-2022

10.4229/EUPVSEC2023/5EO.3.1
020577-002

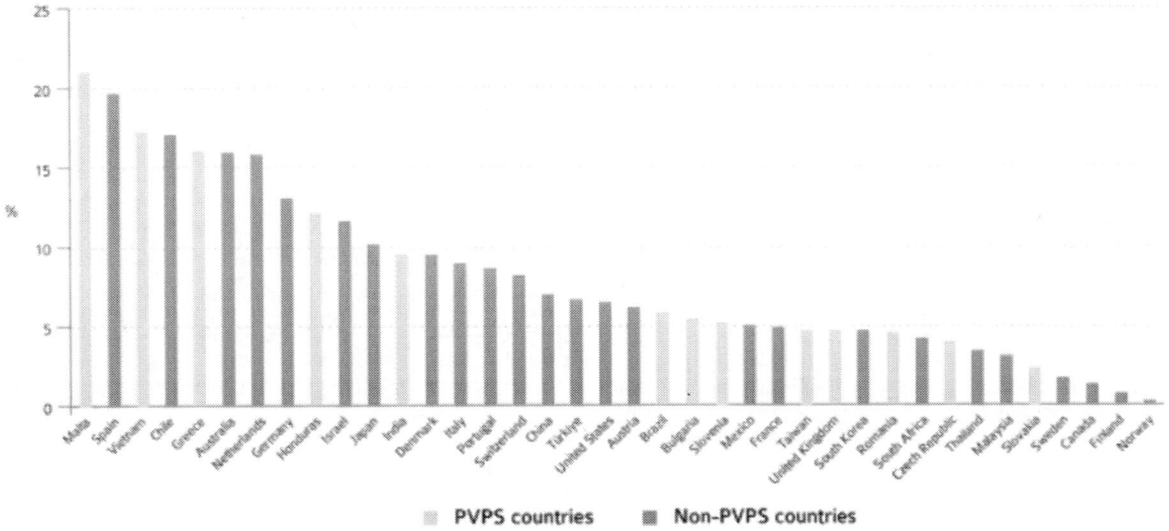

Figure 3: National PV penetration in % of the final electricity demand based on 2022 capacities.

small decline in 2019, has gradually increased, reaching an even higher level in 2022 than in 2017.

The market has also continued to diversify in 2022. While floating PV (FPV) adds to utility-scale, BIPV complements BAPV in the built environment (installed capacity of BIPV globally is estimated to have been up to 2 GW in 2022, depending on the definition of BIPV considered), although it remains a niche. Other emerging segments such as agricultural PV (APV) or PV integrated in vehicles (VIPV) are showing the potential for further diversification of PV components, but their current levels of development remain limited. Dual use of land and surfaces is an option deeply investigated around the world.

At the local level, the region with the highest share of grid-connected centralized installations was Europe until 2011 which was holding 80% of the utility-scale system installed globally. Starting from 2012, American and Asian countries share started to grow and by the end of 2013 Asia became the main region for utility-scale projects.

5 ELECTRICITY PRODUCTION

The electricity production from PV per country as shown in Fig. 3 estimates what the PV production could be, based on the cumulative PV capacity at the end of 2022 (close to optimum siting, orientation and average weather conditions). These numbers, which are not based on actual measurements, should therefore be considered as indicative, aiming at comparing different situations in different countries rather than official data. In several countries, the PV contribution to the final electricity demand has passed the 5% mark with Malta in the first place with more than 20%. Spain is second with nearly 20%, and Vietnam, Chile, Greece and Australia follow. In total, PV contribution amounts to more than 6% of the electricity demand in the world.

6 ELECTRIC VEHICLES

The electrification of transport accelerated in many countries, almost all of which are active in the IEA PVPS program. This accelerated development of the EV market could be compared to the development of the PV market, as shown in Fig. 4. Close to 10.5 million electric vehicles sold in 2022, (+60% compared to 2021) representing 15.6% of global car sales. The sales of EVs are following a trend that is even faster than the installation of PV [3].

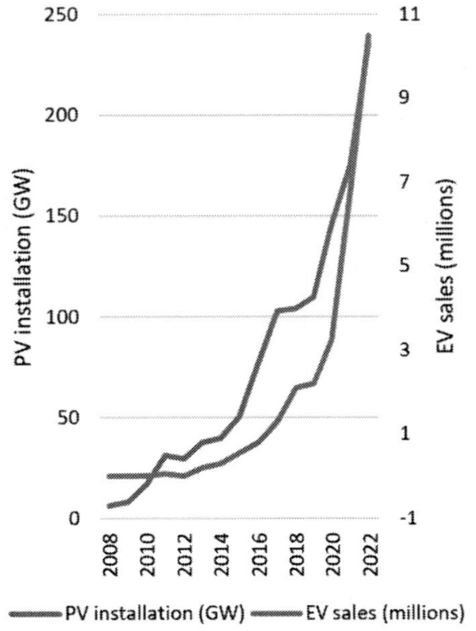

Figure 4: Comparison between PV installation and EV sales

10.4229/EUPVSEC2023/5EO.3.1
020577-003

The link between PV development and EVs is not straightforwardly understood yet but seems to be a reality. Charging electric vehicles during peak load hours implies rethinking power generation, while concepts such as virtual and collective self-consumption could rapidly provide a framework for rapid PV development.

7 CONCLUSIONS

This brief overview of the situation of the PV market and its deployment shows that in 2022 the annual PV market reached at least 235 GW worldwide and the cumulative installed capacity represented over 1 180 GW. The global PV market is, more than ever, dominated by a few leading countries even if many new PV markets are developing on all continents, at different paces.Despite lingering effects of the pandemic (supply chain) and high module prices, growth was consequent, and market competitiveness real. Developing photovoltaic systems with limited or no financial incentives is now an observable reality. Long-term private contracts (PPA) and the sale of electricity on wholesale markets have been observed in an increasing number of countries in 2022. Nevertheless, the market remains strongly influenced and led by policy and incentives.

8 REFERENCES

[1] IEA PVPS, TRENDS 2023 In Photovoltaic application, https://iea-pvps.org/trends_reports/trends/
[2] IEA PVPS, 2023 Snapshot of Global PV Markets, https://iea-pvps.org/snapshot-reports/snapshot-2023/
[3] IEA, Global EV Outlook 2023, https://www.iea.org/reports/global-ev-outlook-2023

10.4229/EUPVSEC2023/5EO.3.1
020577-004

40th European Photovoltaic Solar Energy Conference and Exhibition

This presentation was selected by the Sc. Committee of the EU PVSEC 2023 for submission of a full paper to one of the EU PVSEC's collaborating peer-reviewed journals.

ATTRACTIVENESS OF PV PROSUMERISM IN THE EUROPEAN ELECTRICITY MARKET

Eero Vartiainen
Fortum Renewables Oy
POB 100, FI-00048 Fortum, Finland
+358 40 528 9867
eero.vartiainen@fortum.com

Christian Breyer
LUT University
Yliopistonkatu 34, 53850 Lappeenranta, Finland

David Moser
Institute for Renewable Energy, Eurac Research
Viale Druso 1, 39100, Bolzano, Italy

Eduardo Román Medina,
TECNALIA, Basque Research and Technology Alliance (BRTA)
Astondoa Kalea Edif. 700 Derio, 48170, Spain

Chiara Busto
Eni S.p.A.
Via Fauser, 4, 28100 Novara, Italy

Marko Topič
University of Ljubljana, Faculty of Electrical Engineering
Trzaska Cesta 25, SI-1000 Ljubljana, Slovenia

Daniel Mugnier
Planair France SAS
18 Rue Gabriel Péri, F-69100 Villeurbanne, France

ABSTRACT: Current energy crisis calls for a rapid transition to renewables. Solar photovoltaics (PV) is the key component of the future energy system and majority of PV systems are installed on rooftops in Europe. This paper studies the business case of prosumers, i.e. consumers who produce part or all of their electricity with PV. The PV levelised cost of electricity (LCOE) is compared with the average value of PV electricity for residential prosumers in Finland, Germany, France, Italy and Spain with different self-consumption and weighted average cost of capital (WACC) rates. Prosumer business case is positive in every studied market segment and country. Sensitivity of payback time on key input parameters is also studied. Reasonable cost of capital and adequate self-consumption of PV generation remain as the key elements for attractive prosumerism.
Keywords: LCOE, solar PV, prosumers, electricity market

1 INTRODUCTION

Photovoltaics (PV) is already the cheapest way of producing electricity in most markets. PV is especially competitive in the case of prosumers, i.e. with producers who consume themselves all or part of their own generation. The price of retail electricity is normally considerably higher than the market price of wholesale electricity which makes own consumption beneficial for the producers. Furthermore, with the current energy crisis in Europe, also the wholesale electricity market prices have been on record high levels which further enhances the competitiveness of PV and strengthens the position of prosumers.

Expansion of PV prosumers is crucial as Europe is aiming to reduce the dependence on imported fossil fuels. PV has the biggest share of the annual net electricity generation capacity additions and still about 60% of the PV growth in Europe comes from rooftop installations [1] owned by prosumers. The current energy crisis has further emphasised the significance of increasing PV capacity as fast as possible in which prosumers play the major role. But information on exactly how competitive PV is in the currently changed market environment is not publicly available and there are only a few articles [2] on PV prosumer attractiveness in scientific journals. Politicians and other decision makers need to know urgently the pivotal role of PV in creating a sustainable and independent energy system in Europe.

In previous publications the authors have reported on the True competitiveness of solar PV in Europe in 2016-17 [3,4] where the concept of value of PV electricity was introduced with the following key findings: if PV systems could be financed with a cost of money corresponding to the low level of risk associated with PV systems, they would be competitive in most European countries and for most market segments. A higher cost of capital would delay the moment where PV could reach competitiveness under the current conditions, especially in the Nordic countries. With reasonable cost of capital, self-consumption schemes without incentives and market-based remunerations for excess PV electricity injected into the grid, PV would be competitive in almost all cases within the coming ten years (2027). Further cost decline of battery storage and integration of PV systems with electric heat pumps and electric vehicles makes residential PV self-consumption increasingly attractive in more and more countries in Europe and this was studied by the authors in 2017 [5] where they reported residential PV system with battery storage as already a profitable investment in countries with high electricity price like Germany and Italy if the PV and battery capacity is sized sensibly in relation to the electricity consumption. In most cases, the PV system investment without storage still has a higher IRR than a system with storage. Adding a storage to an existing PV system would already be profitable in Italy with relatively large PV systems generating a lot of surplus electricity.

The aim of this paper is to study the attractiveness of PV prosumerism in the current European electricity market by comparing the residential PV levelised cost of electricity (LCOE) with the price of retail and wholesale electricity in five European locations. Self-consumption ratio, i.e. the share of own consumption of the PV generation is varied to reflect different type of prosumers. Results are given with different weighted average cost of capitals (WACCs) which is the most influential parameter in the PV LCOE calculation along with location [6] and size of the system.

2 PV LCOE INPUT PARAMETERS

The PV LCOE, expressed in €/MWh in real 2023 money, is defined by the equation published in a previous paper by the authors [6]. In this paper, residential PV LCOE is calculated with 4 different nominal WACC rates: 0%, 2%, 4% and 6%. Future average annual inflation rate is set to 2% which is the long-term historical EU average. Inflation has recently been much higher but is expected to return to the long-term average levels.

Five locations with different yields for rooftop installations are selected: Helsinki (920 h), Munich (1070 h), Toulouse (1280 h), Rome (1480 h) and Malaga (1700 h). Yields were calculated with PVsyst software from Solargis solar irradiation data [7] for a south-oriented rooftop PV system with 20 degrees tilt angle. For simplicity, we assume that the performance loss rate at system level is equal to the PV module efficiency degradation and it is set to 1% for the first year and 0.25%/a [8] after that. PV system lifetime is assumed to be 30 years [9] with inverter replacement after 15 years.

PV system CAPEX depends on the market segment, i.e. the system size and the location. During the previous decade, system prices declined quite steadily each year, but for the past couple of years there has been a slight increase because of the pandemic and current energy crisis. However, PV module market prices are currently at an all-time low: mono PERC module factory gate price was 0.14 USD/Wp in September 2023 which has a 50% decrease in real terms from the high price in October 2021 [10]. This would imply that there is reduction potential in the system price as well. On the other hand, the extremely high demand and subsequent bottlenecks in system installation will probably keep the margins high for a while.

According to the latest IEA PVPS National Survey Reports [11-12], the average CAPEX range for a residential 5-10 kWp PV rooftop system seems to be from 1.5 €/Wp in Sweden to 1.7-2.5 €/Wp in France, prices include VAT. For simplicity, in this analysis CAPEX of 1.8 €/Wp is used for a 5 kWp residential PV rooftop system. The effect of price variation can be seen in the sensitivity analysis.

For OPEX, the price variation is even higher than for CAPEX and depends on the service requested by the prosumer. For residential systems, preventive maintenance is often negligible and corrective maintenance is limited to inverter replacement in most cases. However, the inverter replacement cost is treated as a separate CAPEX reinvestment in this analysis, so it's not included in OPEX. Other OPEX items like insurance and monitoring are usually optional so the annual total OPEX could be even zero. However, it is conservatively assumed here that the annual OPEX for a residential PV system is 10 €/kWp. The effect of OPEX variation from 0 to 20 €/kWp/a is shown in the sensitivity analysis.

3 VALUE OF PV ELECTRICITY

PV LCOE is compared with the average value of generated PV electricity in five locations. The self-consumed PV generation is valued for the variable retail electricity price the customer has to pay for each consumed kWh in the electricity bill. The average value of PV electricity P_{ave} in this paper is determined separately for each market segment in each country and is defined by the equation: [3,4]

$$P_{ave} = SC \cdot P_r + (1 - SC) \cdot P_w \quad (1)$$

where

SC = share of self-consumption of the PV production
P_r = variable retail electricity price
P_w = wholesale electricity price

Eurostat annual or half-year average consumer electricity prices [13] are used for the selected five countries. The consumption band selected for a residential prosumer is 5-15 MWh/a and the PV system size is 5 kWp.

Unfortunately, Eurostat does not give separately the variable and fixed costs in the annual statistics. For simplicity, we assume that the share for fixed grid costs is 50% for all countries. Energy and taxes and other fees components are assumed to be fully variable (per kWh) here.

Fig. 1 shows the Eurostat average residential consumer electricity prices for five European countries in 2019-2022 [13]. The prices have been adjusted with inflation to July 2023 real euros. In this analysis, the average of 2019-2021 prices is used.

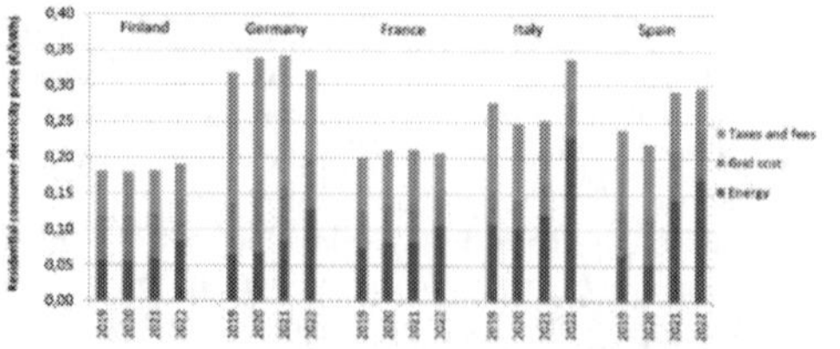

Figure 1: Average real residential consumer electricity price for five European countries 2019-2022, all prices in real 2023 euros.

For the past couple of years, the electricity wholesale or spot market price in Europe has been very high due to the energy crisis with dramatic increase in fossil fuel prices. Especially in 2022, the spot price increased to record high levels after Russian invasion to Ukraine and subsequent decreased supply of natural gas from Russia. Fig. 2 shows that average day-ahead spot market prices [14-16] increased by 150-220% from 2020 to 2021 and then a further 40-130% from 2021 to 2022, depending on the country.

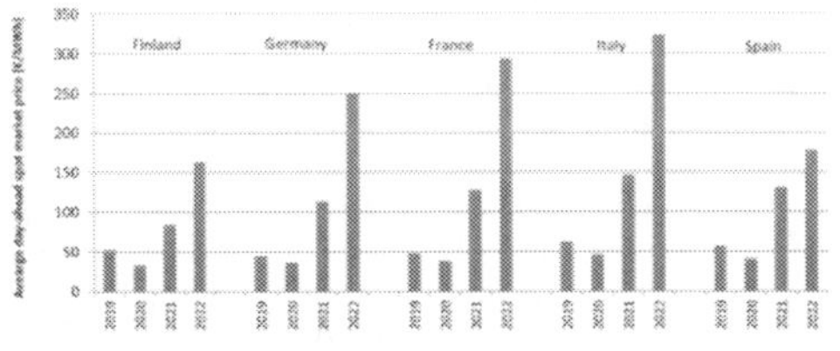

Figure 2: Annual average day-ahead spot market electricity prices in five European countries, all prices in 2023 real euros.

Spot market electricity prices have already started to return to lower levels. In this analysis, the fed-in surplus PV generation is valued at the average spot market price 2019-2021 minus a 10% administrative fee.

4 RESULTS

Comparison of residential PV LCOE with average value of PV electricity in five European locations and with four different nominal WACC rates and with 25%, 50%, 75% and 100% share of self-consumption is shown in Fig. 3. Apart from Helsinki, in all other locations with all WACC rates used here, PV prosumerism is clearly a positive business case. In fact, in Italy and Spain this would be the case with up to about 14% nominal WACC, in Germany up to 11% and in France up to 9% WACC with 50% self-consumption rate. In Finland, PV LCOE equals the average value of PV electricity with about 4% nominal WACC.

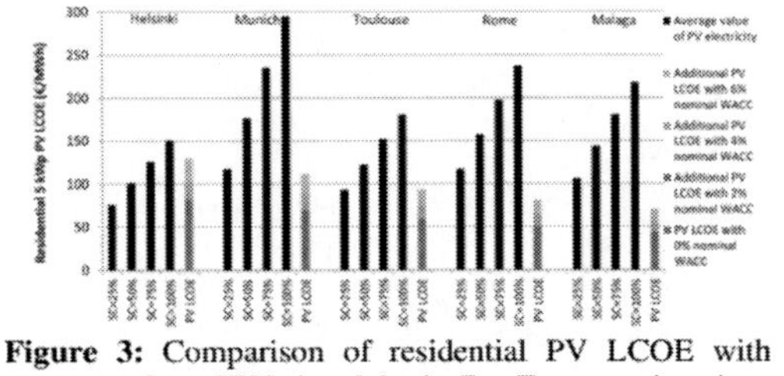

Figure 3: Comparison of residential PV LCOE with average value of PV electricity in five European locations and with four different nominal WACC rates and with 25%, 50%, 75% and 100% share of self-consumption (SC), all prices in 2023 euros.

5 SENSITIVITY ANALYSIS

For the base case of the sensitivity analysis, residential system in Toulouse with 50% self-consumption rate is selected. Payback time for this system with the base retail and wholesale prices, CAPEX and OPEX is 12.3 years with 2% nominal or 0% real WACC. Fig. 4 shows the sensitivity of payback time as a function of seven key input parameters. As was evident from the previous section, along with WACC the location has a major impact on the prosumer business case. Increasing the yield of Toulouse by 30% will reduce the payback time by 3 years and decreasing the yield by 30% will increase it by almost 6 years. These yields correspond roughly to Malaga and Helsinki respectively.

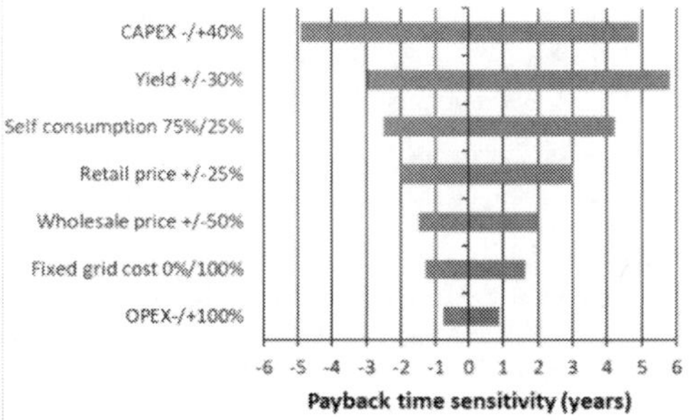

Figure 4: Sensitivity of payback time for a residential 5 kWp PV system in Toulouse with 1800 €/kWp CAPEX, 10 €/kWp/a OPEX, 50 % self-consumption rate, 63.9 €/MWh wholesale and 180.7 €/MWh retail electricity price, fixed grid cost share 50%.

CAPEX variation can be as high as +/-40% which would change the payback time by +/-5 years. On the other hand, OPEX has a relatively minor effect: double or zero OPEX would change the payback time by less than a year. Self-consumption rate of 75% would decrease the payback time by 2.5 years and 25% increase it by about 4 years. Changing retail electricity price by +/-25% or wholesale price by +/-50% would have an effect in the order of 2-3 years. Although the price increases during the height of the energy crisis in 2022 were even bigger, it is unlikely that such high prices would be sustained in the long run. Changing the fixed grid cost share assumption from 50% to 0 or 100% would change the payback time by about 1.5 years.

6 DISCUSSION AND CONCLUSIONS

The recent and still ongoing energy crisis has put even more emphasis on solar PV as the major component in the future energy system. The demand for new PV installations in Europe in 2022 grew to 46 GWp, majority of which were rooftop systems.[17] The attractiveness of prosumerism, i.e. consumers who produce part or all of their own consumption, is key to the speed of energy transition in Europe. This paper has analyzed the residential prosumer business case in in five European countries with different self-consumption rates. PV LCOE was compared with the average value of PV electricity for the prosumer.

With 2019-2021 average retail and wholesale electricity prices and current CAPEX prices, prosumer business case is very attractive in Spain, Italy, France and Germany and a positive one even in Finland. It is likely that the system CAPEX prices will continue to decrease in the future which will make prosumerism even more attractive.

Based on the results of this analysis, reasonable cost of capital and adequate self-consumption of PV generation remain as the key elements for attractive prosumerism.

10.4229/EUPVSEC2023/5EO.3.3
020578-003

ACKNOWLEDGEMENTS

This study has been made under the framework of the European Technology and Innovation Platform for Photovoltaics (ETIP PV).

REFERENCES

[1] SolarPower Europe, EU Market Outlook for Solar Power 2022-2026 (2022).
[2] C. Breyer, A. Gerlach, Global overview on grid parity. Prog Photovolt Res Appl. 21 (2013) 121-136. https://onlinelibrary.wiley.com/doi/abs/10.1002/pip.1254
[3] E. Vartiainen, G. Masson, C. Breyer, True Competitiveness of Solar PV - a European Case Study. Proceedings of the 32nd European PV Solar Energy Conference and Exhibition (2016). https://userarea.eupvsec.org/proceedings/EU-PVSEC-2016/7EP.1.1
[4] E. Vartiainen, G. Masson, C. Breyer, The True Competitiveness of Solar PV - a European Case Study. European Technology and Innovation Platform for Photovoltaics Steering Committee PV LCOE and Competitiveness Working Group, Final report (2017).
[5] E. Vartiainen, G. Masson, C. Breyer, D. Moser, Improving the Competitiveness of Solar PV with Electricity Storage. Proceedings of the 33rd European PV Solar Energy Conference and Exhibition (2017). https://userarea.eupvsec.org/proceedings/EUPVSEC-2017/7DO.8.4/
[6] E. Vartiainen, G. Masson, C. Breyer, D. Moser, E. Román Medina, Impact of weighted average cost of capital, capital expenditure, and other parameters on future utility-scale PV levelised cost of electricity. Prog Photovolt Res Appl. 28 (2020) 439-453. https://doi.org/10.1002/pip.3189
[7] Solargis, Solar irradiance data. https://solargis.com/
[8] J. Pascual, F. Martinez-Moreno, M. Garcia, J. Marcos, L. Marroyo, E. Lorenzo, Long- term degradation rate of crystalline silicon PV modules at commercial PV plants: An 82-MWp assessment over 10 years. Prog Photovolt Res Appl. 29 (2021) 1294-1302. https://doi.org/10.1002/pip.3456
[9] M. Theristis, J. Stein, C. Deline, D. Jordan, C. Robinson, W. Sekulic, A. Anderberg, D. Colvin, J. Walters, H. Seigneur, B. King B, Onymous early-life performance degradation analysis of recent photovoltaic module technologies. Prog Photovolt Res Appl. 31 (2023) 149-160. https://doi.org//10.1002/pip.3615
[10]PVinsights, Weekly spot market prices for PV modules. http://pvinsights.com/
[11]Swedish Energy Agency, IEA PVPS National Survey Report of PV Power Applications in Sweden (2021). https://iea-pvps.org/wp-content/uploads/2022/10/National-Survey-Report-of-PV-Power-Applications-in-Sweden-2021.pdf
[12]IEA PVPS, National Survey Report of PV PowerApplications in France (2021). https://iea-pvps.org/wp-content/uploads/2022/07/IEA-PVPS-NSR-France-EN-2021-v5.pdf
[13]Eurostat, Electricity prices components for household and non-household consumers – annual data. https://ec.europa.eu/eurostat/web/energy/database
[14]The European Power Exchange, Epexspot Day-Ahead electricity prices. https://www.epexspot.com/en/market-data
[15]Gestore Mercati Energetici, Annual average spot market electricity prices for Italy. https://www.mercatoelettrico.org/En/Statistiche/ME/DatiSintesi.aspx
[16]OMIE, Annual average spot market electricity prices for Spain. https://www.omie.es/en/market-results/interannual/daily-market/daily-prices?scope=interannual
[17]SolarPower Europe, Global Market Outlook for Solar Power 2023-2027 (2023).

10.4229/EUPVSEC2023/5EO.3.3
020578-004

The Role of Digital Operation Technology for PV Mini-grids in Sub-Saharan Africa: Impact and Benefits of Advanced Remote Monitoring Systems

Pittalis M.[a], Sancho A.[b], Szabo S.[a], Lazopoulou M.[b], Moner-Girona M.[a]

a: European Commission, Joint Research Centre, Ispra, Italy

b: Trama TecnoAmbiental, Barcelona, Spain

EU PVSEC 2023, Lisbon, 22nd September 2023

European Commission

020580-001

40th European Photovoltaic Solar Energy Conference and Exhibition

2

Picture: Solartrainer from Africa Greentec in Cinzana-Gare village, Mali, Source: @Torsten Schreiber https://commons.wikimedia.org/wiki/File:Solartainer_in_Cinzana-Gare_%28Segou-region%29_Mali.jpg

European Commission

020580-002

Research questions

- How mini-grid developers are using remote monitoring solutions and for what purposes
- What are the operational benefits and limitations of remote monitoring solutions
- What is the degree of operation and maintenance savings achievable through the use of remote monitoring solutions

European Commission

40th European Photovoltaic Solar Energy Conference and Exhibition

Impact of Remote Monitoring Solutions (RMS) usage in Mini Grids sector:

- **Improve mini grids operation and maintenance**
  - Energy generation and demand equipment management
  - Logistics management
  - Human resources management

- **Support mini grids market development**
  - Data in support to the development of new business models
  - Evidence of mini grids performance that supports financing
  - Improved knowledge and transparency of mini grids functioning

European Commission

Impact of Remote Monitoring Solutions (RMS) usage in Mini Grids sector:

- **Improve mini grids operation and maintenance**
  - Energy generation and demand equipment management
  - Logistics management
  - Human resources management

- **Support mini grids market development**
  - Data in support to the development of new business models
  - Evidence of mini grids performance that supports financing
  - Improved knowledge and transparency of mini grids functioning

Categorization of RMS

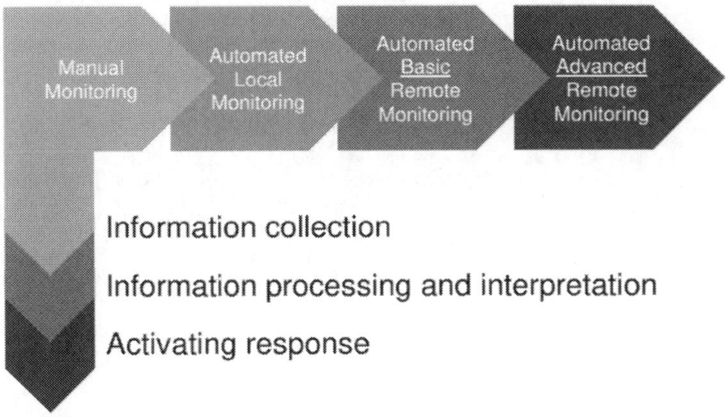

Categorization of RMS

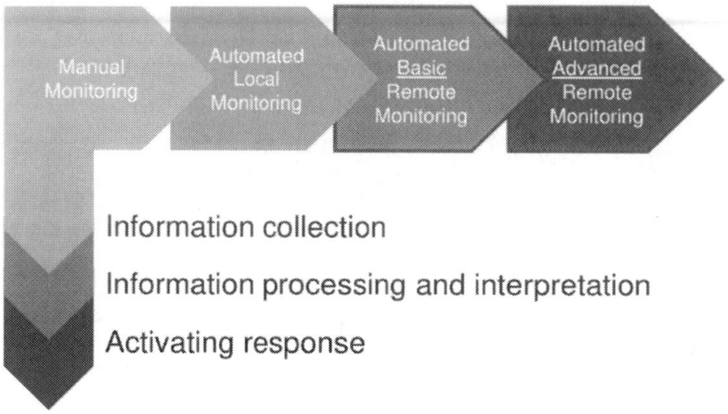

Information collection

Information processing and interpretation

Activating response

7

European Commission

Sample of the study

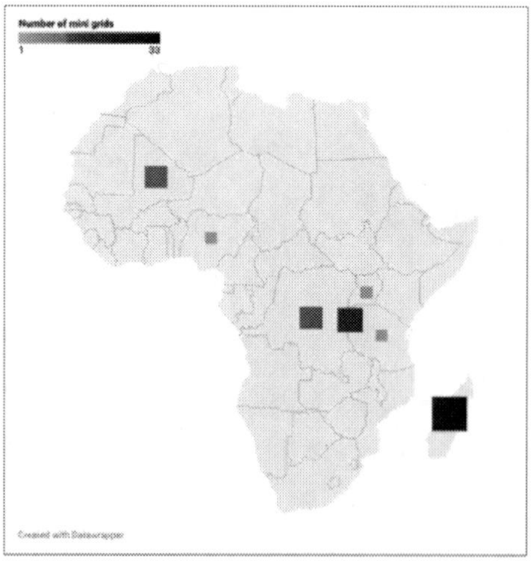

- Primary data collection from 7 companies for a total of 48 mini grids across 7 countries
- Mini grids type: rural + hybrid + autonomous
- Mini grids size: 20 – 200 kWp
- Battery: Lithium-ion or flooded lead acid
- Year of construction: 2017 - 2021

8

European Commission

Results

European Commission

020580-009

40th European Photovoltaic Solar Energy Conference and Exhibition

Surveyed Companies Give High Importance to RMS Use

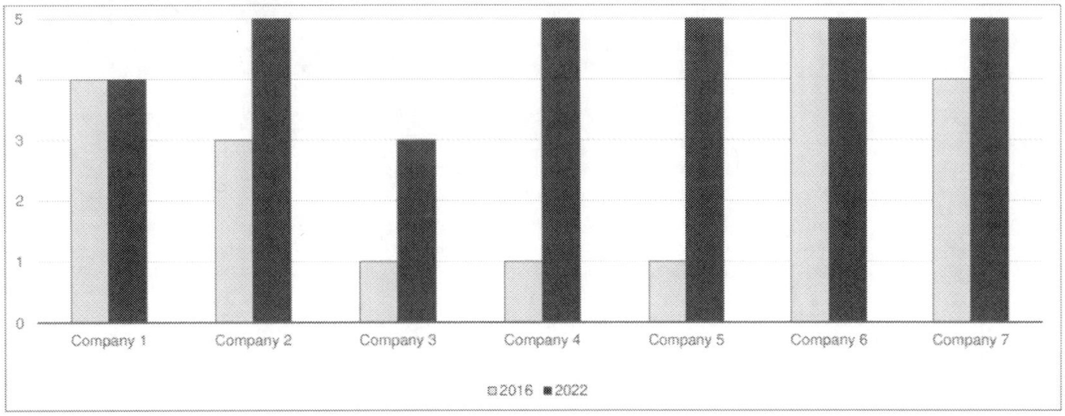

10

European Commission

020580-010

40th European Photovoltaic Solar Energy Conference and Exhibition

Drivers for RMS Adoption

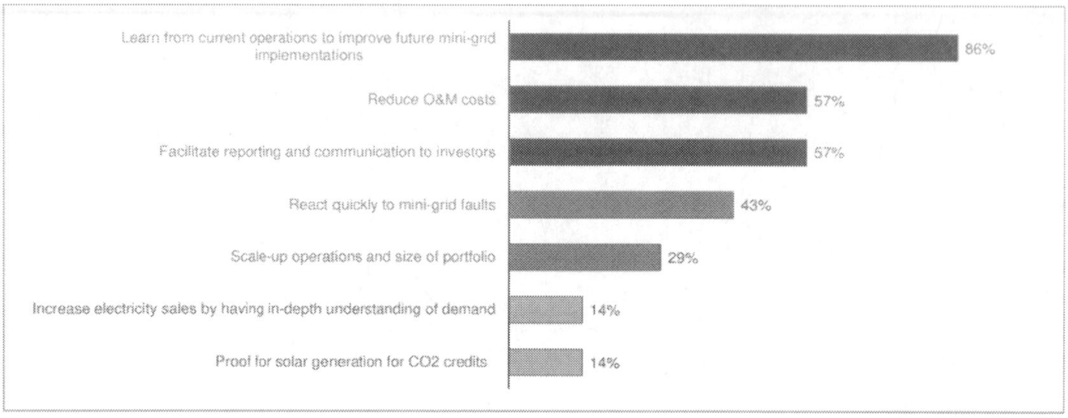

11

European Commission

020580-011

40th European Photovoltaic Solar Energy Conference and Exhibition

Perceived Barriers for RMS Adoption

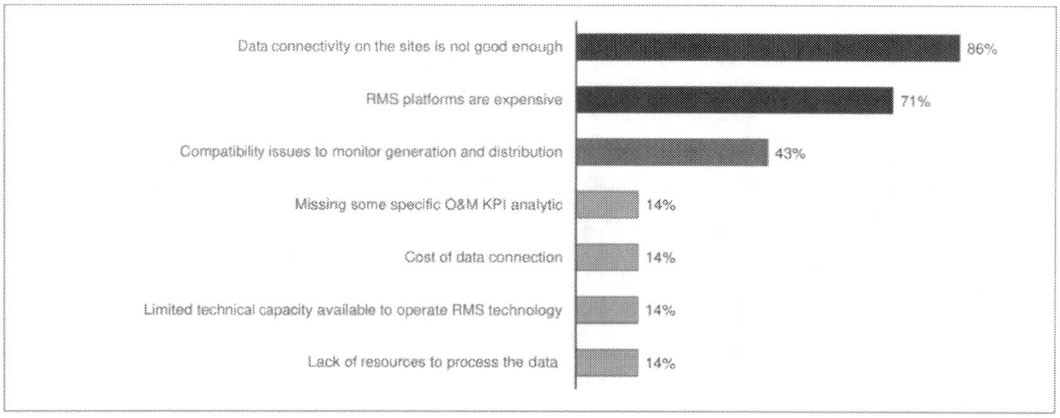

12

European Commission

020580-012

Identification of RMS related savings per each OPEX component

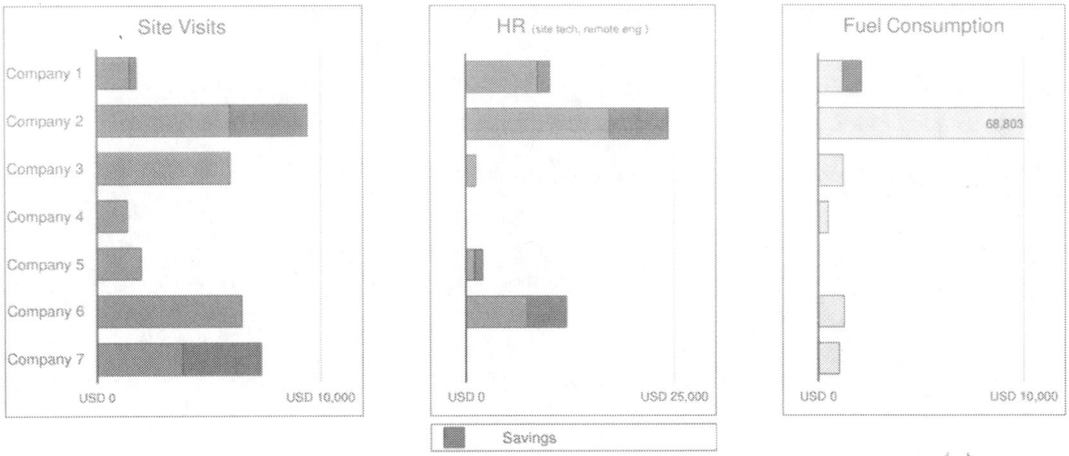

13

40th European Photovoltaic Solar Energy Conference and Exhibition

Overall OPEX Structure Improvement and RMS Payback Time

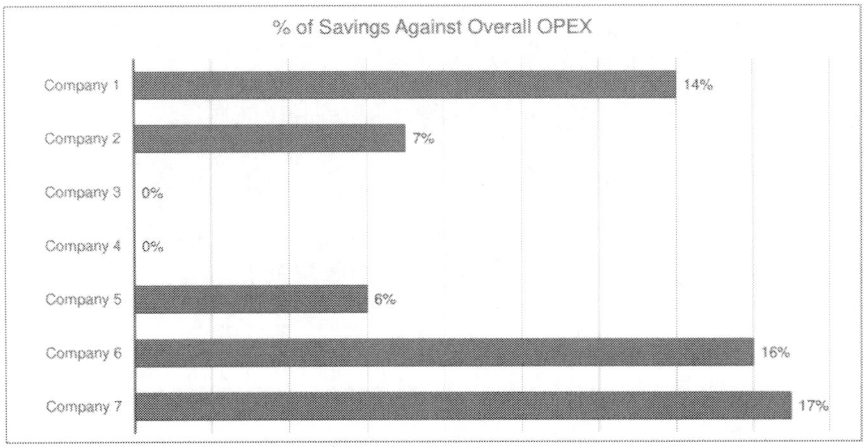

PAYBACK TIME
Data not available
< 4 years
Data not available
Data not available
Data not available
1 year
< 3 years

14

Conclusions

European Commission

40th European Photovoltaic Solar Energy Conference and Exhibition

Benefits of RMS Usage

- Remote monitoring solutions allow different options of support and automation that could suit multiple needs
- Mini grids companies seems aware of RMS potential
- RMS can improve mini grids O&M in areas such as site visits, logistics, and human resources.
- Investment in RMS has typically a short payback time

16

European Commission

Challenges

- Level of priority associated to RMS adoption
- Connectivity
- Data management

17

European Commission

020580-017

Thank you

- JRC is open to cooperation and receiving PhD students and other researchers (as unpaid visiting scientists)
- JRC is constantly hiring new staff: https://recruitment.jrc.ec.europa.eu/
- Africa Knowledge Platform : https://africa-knowledge-platform.ec.europa.eu/
- Clean Energy Access Tool : https://africa-knowledge-platform.ec.europa.eu/energy_tool#!

Slide 2: Solartrainer from Africa Greentec in Cinzana-Gare village, Mali. **Source:** @Torsten Schreiber — https://commons.wikimedia.org/wiki/File:Solartainer_in_Cinzana-Gare_%28Segou-region%29_Mali.jpg

18

European Commission

020580-018

USING ON-LINE PV INVERTER MEASUREMENTS TO DETERMINE THE HOSTING CAPACITY OF DISTRIBUTION GRIDS

Nicholas Etherden[1,2], Mouaz Alhamwi[1]
[1]Vattenfall R&D, 169 92 Stockholm, Sweden
[2]Luleå University of Technology, Electric Power Engineering, 931 62 Skellefteå, Sweden
E-mail: nicholas.etherden@ltu.se

ABSTRACT: The amount of solar photovoltaics that can be connected to the grid is a question which, especially for distributed solar, is characterized by a strong information unbalance between on one side the grid operators and on the other side those wanting to feed in electricity. This paper demonstrates how the grid impact of PV can be assessed by analyzing inverter voltage measurements that today are available from millions of on-line PV inverters. In the case where grid capacity is limited by voltage on-line PV inverters are shown to provide a valuable complement to distribution system operators estimation of grid capacity. Using 25 million voltage samples from a thousand PV systems a hosting capacity map is produced for 20 of 21 regions of Sweden. The map shows the amount of PV that typically can be installed without causing voltage rise that would lead to inverter cut-off and loss of production. By monitoring and following the voltage impact from solar PV over time local saturation effects can be detected and the timing of grid investments better planned. The information could also enhance regulators in their role to monitor grid operators and assure the right level of grid investments.
Keywords: photovoltaics system monitoring, grid voltage variations, distribution grid, hosting capacity

1 INTRODUCTION

The amount of solar photovoltaics that can be connected to the grid is a question which is characterized by a strong information unbalance between on one side the grid operators and on the other side those wanting to feed in electricity. This is especially the case for distributed solar where detailed information about the "last mile" grid can be hard to get and it can be complex to assess impact with hundreds of units dispersed throughout a distribution grid. Empirical data from a large number of installations is today accessible for both inverter manufacturers, major installers in a region and, in some cases, to the general public and research institutes through platforms for voluntary sharing of PV plant data. This paper presents a novel way to use on-line PV inverter data to allow regulators, PV installers and other stakeholders to produce grid capacity map for distributed PV, independent of the Distribution System Operator (DSO).

Distributed generation stood for 78 of 173 GW installed PV in 2021 but the share has large variations between countries [1]. In the studied country the PV market is dominated by distributed roof-mounted systems [2]. For such distributed systems the grid capacity is most commonly limited by voltage [3] which is especially the case in distribution grids with high ratio of resistance to the reactance [4]. Voltage variations occur not only due to local consumption and production, but also due to the overlying grid where voltage is controlled by the operation of transformer tap changers at medium and high voltage substation [5]. To ensure power quality is within limits traditional DSO grid studies will apply conservative or even worst-case combinations of load, PV production and grid operation conditions. On-line measurement from inverters can give empirical data to validate DSO grid models or even as an DSO independent assessment of the grids hosting capacity with respect to voltage limitations.

Today a large share of PV inverters are on-line sending a wealth of information to the monitoring platform of the inverter manufacturer. Such information typically has 5- or 15-minute time resolution and will include not only PV production but also additional information of the grid frequency, voltage at the inverter connection point, voltage unbalance between phases and current and harmonic distortion. With a battery installation also, a meter is installed at the grid connection point and the inverter acts as gateway sending also this information to monitoring platform. Data is associated with meta information of the system such as location and rating, orientation and tilt of the solar panels. As shown in this paper data collected by on-line solar power inverters can be used to determine the amount of PV that is possible to install, without adverse impact on the power quality. This information can be used to independently of the DSO assess the grids "hosting capacity". The results can also be a valuable complement to other grid studies when the DSO lack detailed information about the "last mile" electrical characteristics and the "behind meter" voltage drop within the end-customer installation. By monitoring and following the occurrence of end-customer voltage violations over time local saturation effects can be detected as PV installations become more abundant and larger.

2 RELATED WORK

The scientific literature presents many proposals for monitoring systems of PV plants [6] as well as suggestions for cloud-based monitoring platform with data-driven functionalities [7]. A historic overview of solutions for PV remote monitoring is provided in [8]. Previous research has used PV system databases to compare annual yield with different meta data such as tilt, azimuth and installed capacity for 2.8 million systems in 47 countries [9] as well studies of performance loss rate based on 8400 systems [10]. Numerous national studies of PV production outcome exist like [11] and [12] with comparison of 19 such national yield studies in [13]. Pan-European studies have also revealed differences in performance between inverter manufacturers using such databases [14]. Data from one thousand European PV data loggers was used to verify hourly PV simulations in [15]. In [16] web scraping techniques were used to collect yield data at 5–15-minute resolution from 32.000 PV systems in 5 European countries in order to analyze operational performance. Similar large scale Fleet Performance Data Initiative exist also in the US [17] and elsewhere. Near-real time power outputs for a spatially dispersed set of reference PV systems from centralized data repositories was used in [18] for estimating current PV production (nowcasting). Information from on-line PV inverters has also been used by manufacturers since 2011 to estimate the total current PV electricity production across different regions of Germany [19]. Hosting Capacity maps are today published by several grid operators in Australia, US, UK and other countries. However, the use of voltage measurements from on-line PV systems to assess grid strength is novel, as far as the authors are aware..

3 METHOD

The method is applied to produce the hosting capacity of the low voltage distribution grids in 20 of 21 regions in Sweden as shown in Figure 1. The map shows the average amount of kW that can be installed before overvoltage levels are likely to cause inverter cut-off. The map shows relatively high potential for Solar PV in the country due to three phase 400 V phase to phase (230 V phase to ground) connection of all end-customers, in a country with a relatively strong low voltage grid often supporting electric spatial heating. Distribution is usually made with 10 or 20 kV before being transformed down to 0.4 kV. Often a dozen customers will share a transformer in rural areas and a few hundred in urban areas. Online tap-changers, changing automatically the transformer ratio between high and low voltage, are not common for these distribution transformers. The 400 V cables are typically a few hundred meters to a kilometer and owned by the distribution company up until the meter which is situated on the end-customer premises.

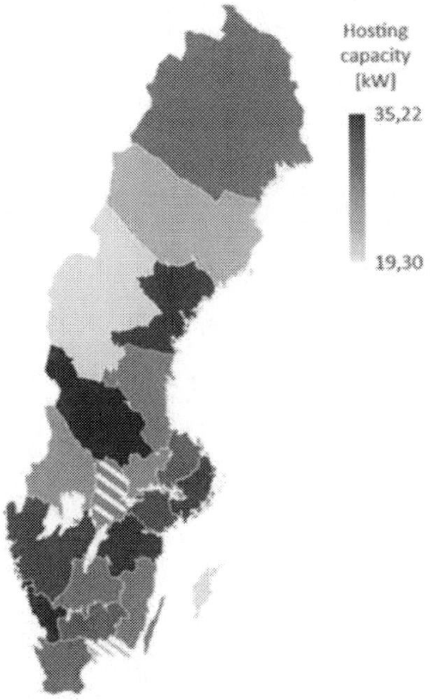

Figure 1: Hosting capacity determined from measured voltage samples.

To produce the hosting capacity map measurements were analyzed from 759 PV systems. 409 of these systems located in Sweden's largest DSO grid (3.4% of all installed installation in that grid when data was extracted in June 2020). The interdependencies of the grid strength and empirically found voltage rise is used to create the hosting capacity heat map for the entire country.

The method consists of three steps using 5-minute inverter production and voltage measurement samples to: 1) determine background voltage in lieu of solar production, 2) calculate the average voltage increase per kW of production and 3) determine the voltage rise resulting from power injection from PV inverter.

A script was created that each month extracts 5-minute voltage and production samples from the inverter manufacturers monitoring platform. These samples are stored together with meta data about the installation's location, size and type of inverter and modules. The amount of power that can be fed into the grid is limited either by the inverter or number of solar panels. Installed PV capacity is taken as the smallest of the panels rated power under standard test conditions, STC, and the rated AC power of the inverter. To eliminate data errors in the manually specified power ratings information on number of panels and a lock-up table for module type designation was used to calculate kWp. The inverter type designation was used as basis for the inverter peak power. This also mitigates uncertainty on whether the inverter database registered power as inverter or PV panel output. This methodology also provides an accurate assessment of each system over- or under sizing, where oversizing is defined as solar panels having a higher capacity under STC than the rated power of the inverter.

All PV inverters in Sweden are required to inject only active power eliminating the need to consider reactive power injection when assessing voltage rise from PV production. Three phase inverters are mandated above 3 kW, eliminating issues of voltage imbalance since the inverters feed in equally to all phases. The power is provided in 5-minute samples, calculated as the energy production during five-minute period dived by this length of time. Specific production (SP) can therefore be defined as the ratio of actual generated power (kW 5-minute sample) to PV installed power (smallest of module kWp or inverter rating).

A weak grid is characterized by noticeable dV/dP and dV/dQ sensitivity [20]. The relationship between the current and the voltage is determined by the electrical impedance. The voltage rise is approximately proportional to impedance times the amount of injected PV power. As customers are connected with three phase the total short-circuit impedance at the connection point is used as impedance (rather than the often twice as high phase to ground impedance used for single phase voltage rise calculations). Thus, grid strength and total short-circuit impedance to end customer are taken as equivalent and will determine the voltage rise from a given infeed of PV power.

In this study the empirical voltage rise dV/dP is calculated as follows:

- A linear interpolation is performed of 5-minute inverter voltage samples recorded by PV inverter between 10 a.m. to 2 p.m. Interpolation is done individually for each system.
- The regression line value at SP=0.2 and SP=1.0 is recorded for each phase separately. SP 0.2 was

10.4229/EUPVSEC2023/5EP.1.1
020581-002

empirically found to be a suitable value to include sufficient samples to capture the peak background voltage

- Average over three phase is taken for SP=0.2 and SP=1.0. The difference between these values is the voltage rise, ΔU, between no production and full PV production.
- For comparison between systems ΔU is presented as ΔU per kW possible power

The hosting capacity with respect to overvoltage is computed as follows:

- SS-EN 50160 [21] voltage limit of +110% of nominal voltage for 60 s is taken as hosting capacity limit. This corresponds to 253 V.
- The upper limit of the background voltage variation without PV is calculated from the 95% quantile of voltage sample with production less than 20% The 95% percentile is selected to correspond with power quality standards praxis to look at 95% of 10 min rms voltage samples. To eliminate influence of PV production on this value 20% of the recorded ΔU from zero to full production is subtracted.
- The "head room" for voltage rise is taken as the difference in volts between the upper limit of background voltage and SS-EN50160 limit of 253 V.
- Hosting capacity is calculated as voltage head room divided by voltage rise, ΔU per kW.

The background voltage level is defined as the voltage without PV production. This value will vary considerably during the day and year depending on consumption and the operation of the overlying grid. Information of the background voltage variation under long periods of time is required to know the available head room for voltage rise also under the more stressed, dimensioning, grid operation conditions. In this study one year with 25 million voltage samples from nearly one thousand PV inverters was analyzed.

The hypothesis behind the method to determine hosting capacity is that the variation in end-customer impedance between geographical areas has significant differences, so while the impedance of any specific PV connection point can vary greatly between sites (as the impedance varies) the average or typical impedance value can still be characteristic for a larger geographical area. Thus, with a sufficient number of PV systems the average impedance should reflect the typical grid strength in a geographical area. Studied systems covered 20 of the 21 regions in Sweden and are further sorted into 55 areas based on two-digit postal code. Out of 55 postal areas 36 areas are within the grid of the utilities DSO grid. For these systems the total short-circuit impedance retrieved from the DSOs asset database.

4 DATASET

A set of 949 PV systems installed between April 1st, 2017, and June 2nd, 2020, was investigated. The inverters are of the same manufacturer and sold by a major utility in Sweden providing turn-key installations to predominantly residential PV customers. The average system size, taken as the panel capacity in kWp, was 8.6 kW.

The PV systems are installed in 20 of the country's 21 regions. Just over half (409 systems or 53%) were installed in the grid of the utilities DSO subsidy. Comparison with the other 12 000 PV installations in that DSO showed the sample to be similar in size, see Figure 2. The studied PV systems generally had a slightly higher impedance (weaker grid) and thus results would be underestimating the hosting capacity in the respective regions. Variation of total impedance between a predominantly rural and predominantly urban region is shown in Figure 3, supporting the hypothesis that electrical characteristics between regions are indeed distinct.

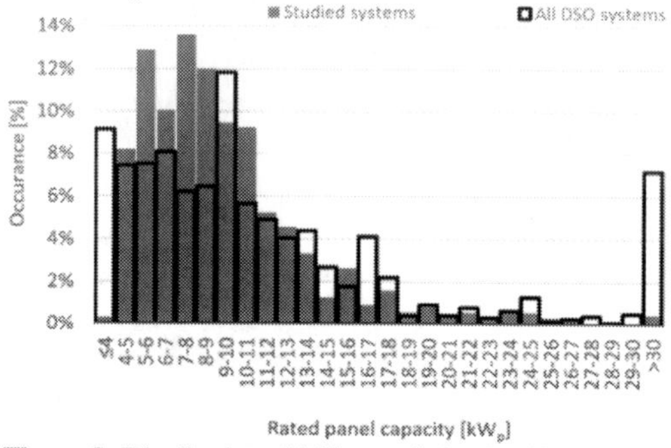

Figure 2: Distribution of kWp panel size for 979 studied systems and 12 000 reference systems.

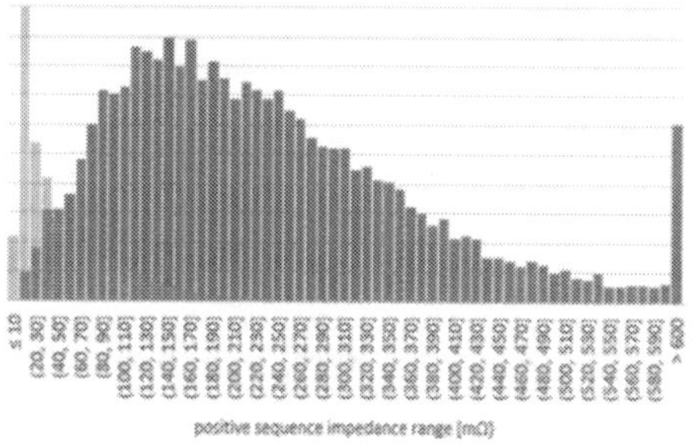

Figure 3: End-customer impedance for urban concession area (9 000 end-connections, orange) vs pre-dominantly rural concession area (10 000 customers, blue).

5 RESULTS

This section presents the voltage rise from PV production. The results are used to estimate the available capacity of the grid for further PV installation.

5.1 Today's impact from PV on grid voltage

Due to the high latitude of the installations (between 55°20' and the arctic circle at 66° 33') production can be for as much as 16 to 18 hours per day in summer, but only 6-7 hours in winter. Voltage was found to be lower in winter which corresponds to higher consumption as Sweden has heat pump penetration in excess of 60% of houses, increasing the load during cold ambient temperatures. During the period of high PV production (June to August 10 a.m. to 2 p.m.) average load for typical residential homes (16 and 20A customers) is 0.5 to 1.5 kWh per hour, which is considerably smaller than the average system size PV of 8.6 kW installed capacity.

Of the studied systems 60% have their maximum production caped by the inverters. Normalized specific production in Figure 4 shows distinct seasonal characteristics of production. Figure 5 shows the average voltage increase in spring and summer between low and high PV production, with typical voltage rise of 2-5 volts.

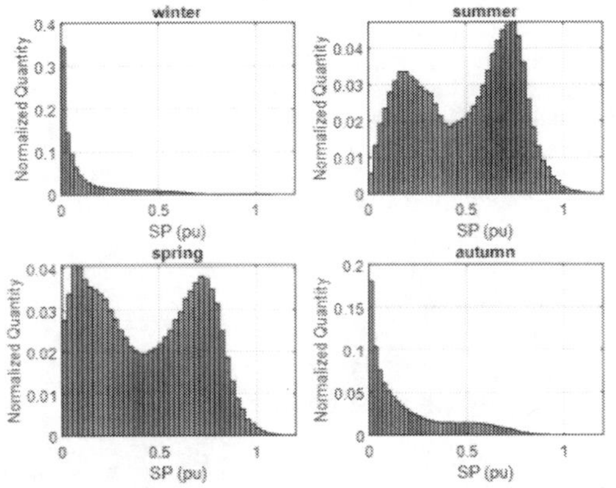

Figure 4: Histogram of specific production 10-14 each season

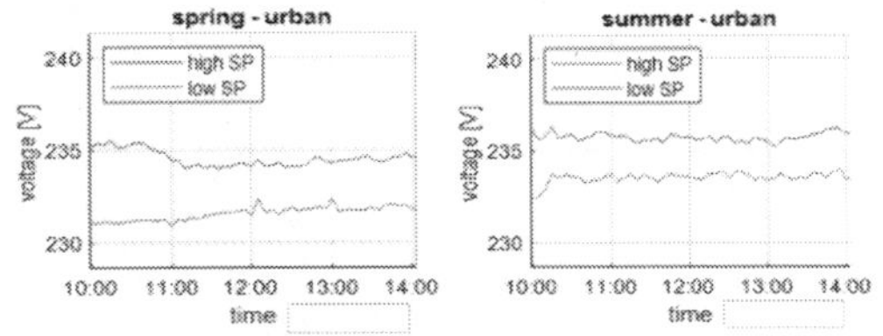

Figure 5: Average voltage variation during day for low and high prod.

5.2 Background voltage variation and Empirical voltage rise from PV production

The variation of background voltage at low or no PV production from 10 a.m. to 2 p.m. is shown in Figure 6. The spread of the studied system lowest recorded voltage sample during the year (green voltage distribution) is compared with the highest voltage value under low PV production (yellow voltage distribution). The difference between the minimum voltage sample for each system and the highest low production sample is greater than the increase with maximum PV production (blue in Figure 6). This indicates that today the non-PV related variation in grid voltage is greater than the increase in voltage from PV production.

Figure 7 shows the measured voltage rise between high and low production vs impedance × installed power for the 409 systems in DSO grid, for which impedance values where available. Figure 7 verifies the assumed linear relationship between voltage rise and injected power

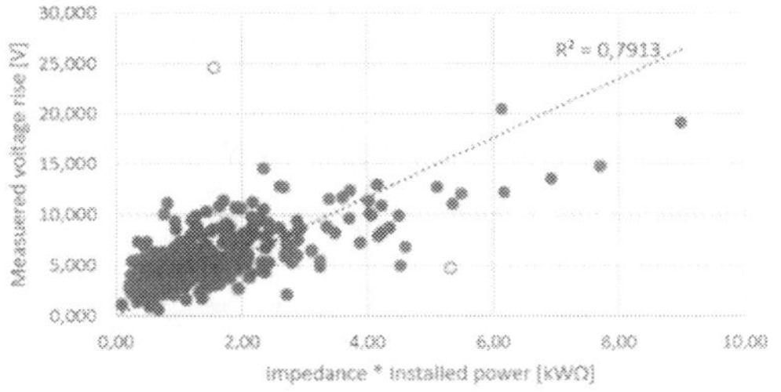

Figure 7: Relationship between grid strength, system size and voltage rise

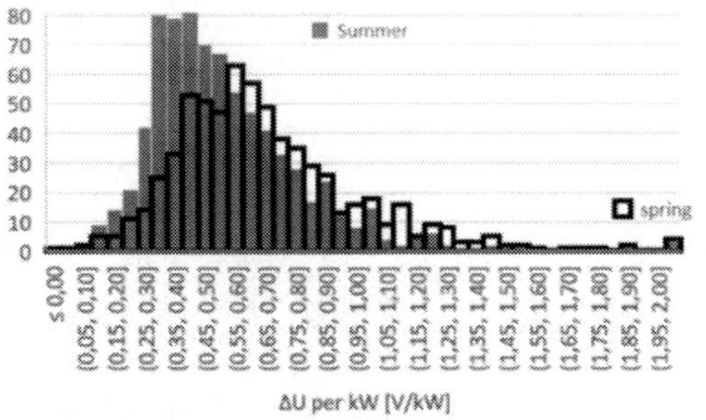

Figure 8: ΔU per kW prod. in summer (blue) and spring (transparent)

from PV system. (One outliner, circle, was found to be faulty installation feeding in all power into a single phase, causing high voltage rise in that phase only). Figure 8 shows the distribution of voltage rise per kW production, ΔU, for all 949 systems. The observed voltage increase is stronger in spring than in summer, as expected at high latitudes.

5.3 Permissible voltage increase

According to IEC 50160, 95% of 10 min rms voltage samples should be within 110% of nominal voltage [21]. As seen in the red dashed lines in the previous Figure 6 indicate the allowed voltage variation at end-customer grid connection point. As seen overvoltage's occur already today but violates SS-EN50160 only if >5% of the 10-min samples are out of range.

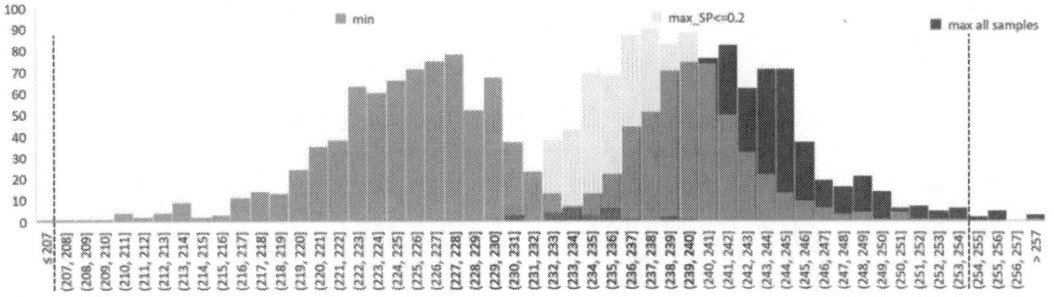

Figure 6: Distribution of lowest (green) and highest (blue) voltage samples over one year, compared with the highest voltage sample under low PV production (yellow). Red dotted vertical lines indicate the end-customer voltage limits according to SS-EN IEC 50160.

10.4229/EUPVSEC2023/5EP.1.1
020581-004

5.4 Hosting capacity estimation

The hosting capacity is here defined as the amount of PV production permissible without violation of EN 501560 voltage limit. It is determined by dividing the permissible voltage increase with the empirically measured ΔU per kW for each site, see Figure 9. Red line indicates the 25% of systems that can have more than current micro producer limit of 43.5 kW. It was found that with the current amount of PV in the grids the installations could have been on average 29 kW in spring without voltage variations. This was most restricting period. In summer 35 kW would be possible on average.

Seasonal variations were noticed in background voltage as depicted in Figure 10. The overall trend (dotted line) is that the hosting capacity is 20% higher when measured in summer than spring, which is ascribed to the uncertainty in the empirical method and varying background voltage characteristics. Another way of presenting the hosting capacity is in terms of how much larger PV system could have been without violating EN 50160 voltage limits. The studied systems were on average 8.6 kW. Each one could, on average, have been 3.9 times larger with ΔU measured in spring the installed capacity and 4.7 times larger with ΔU measured in summer.

Using the sites geographical coordinates, the typical maximum voltage without PV is shown in Figure 11 (left) per region. This in term defines the head room available to EN 50160 end customer voltage rise (middle Figure 11), which corresponds to the settings for when inverters will cut-off power. The empirically found ΔU per kW of solar injection is shown in right Figure 11. Finally, the available voltage head room and ΔU per kW are combined to obtain the map of the average hosting capacity for PV, as was presented in Figure 1 for 20 of the 21 regions in Sweden.

6 DISCUSSION

The available capacity estimated with the empirical hosting capacity method refers only to voltage. While the amount of PV in distribution grids is most often limited by voltage constraints higher voltage levels and large-scale solar parks tend to be limited by component and line/cable overload or grid stability [1]. Thus, this method should only be applied to voltage levels and grids where the voltage is known to be the limiting factor for grid connection.

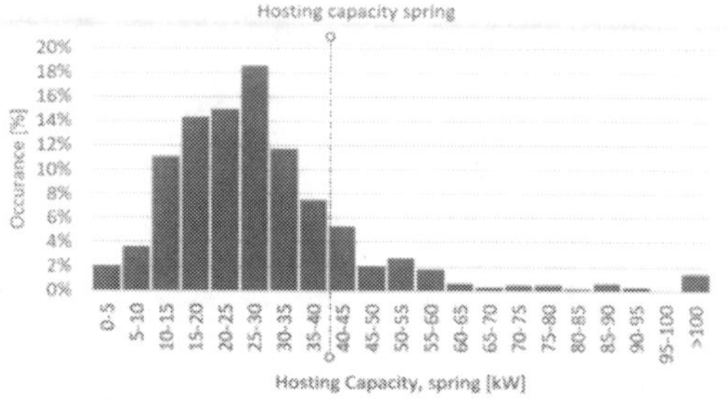

Figure 9: Distribution of hosting capacity

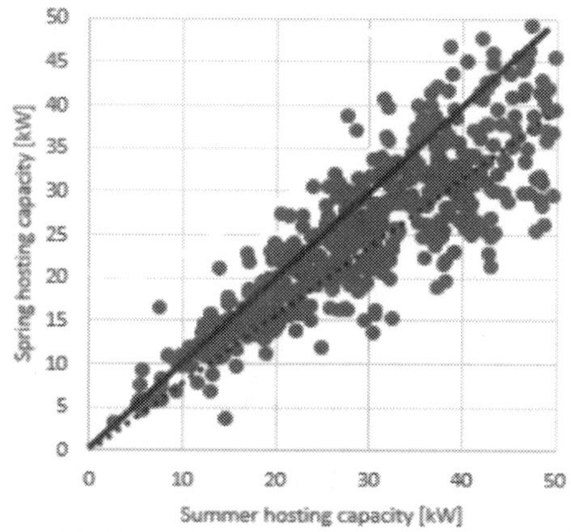

Figure 10: Comparison in hosting capacity measured in spring (solid red line) and summer (dotted red trend line)

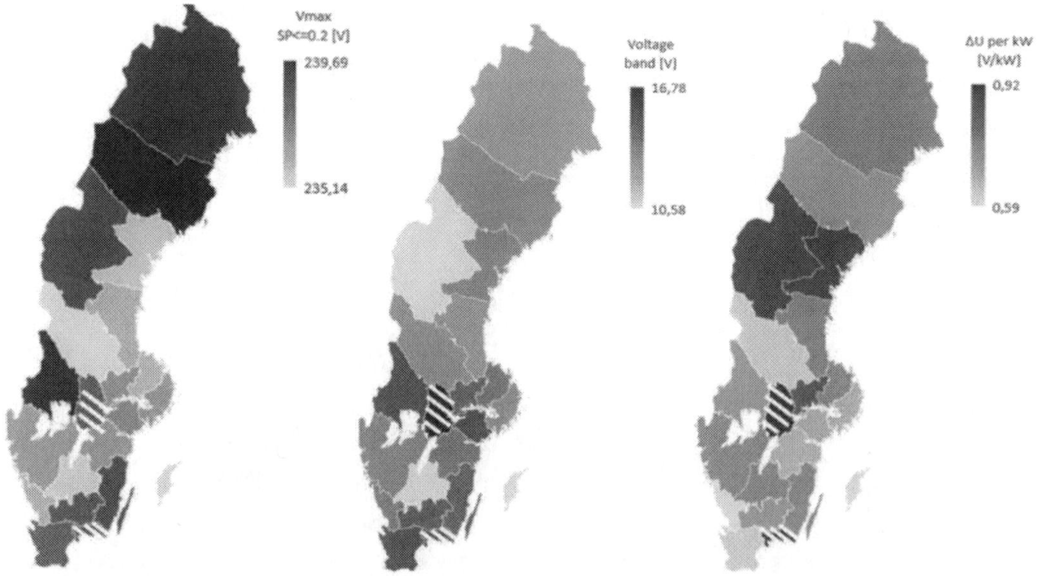

Figure 11: Highest background voltage with low PV production (left), headroom to EN 50160 end-customer voltage limit (middle) and empirical PV voltage rise per region (right).

Care should be taken as to not interpret the available hosting capacity as the amount of PV each customer of the grid can install. Instead, it determines the average amount that a customer can add given that other nearby customers do not also install PV. If two customers add PV, they will raise the voltage in the common connection point for the two customers. To estimate this impact the grid module is needed to determine the impedance of this common point of coupling. When many nearby customers install PV the HC capacity will decrease for all (an effect that is known from other work by the authors to be able to half the estimated hosting capacity [22]). Action taken by the DSO to strengthen the grid or change the operation of the grid by e.g., actively managing the transformer ratio with on-line tap changers will also affect the empirically determined hosting capacity with respect to voltage.

The authors propose that the method is applied by larger installers, PV fleet operators and inverter manufacturers to create a counterweight to the grid operator's assessment of available capacity for PV installations in distribution grids. This is argued to lead to a more balanced discussion between gird operators and PV owners, breaking down the traditional asymmetric knowledge when it comes to grid constraints. As a snapshot of the current state of the grid the hosting capacity can be monitored over time. Such reoccurring assessments could also be used to empirically evaluate positive effects on voltage quality from new grid codes, reactive power control features or active power flow control from battery coupled PV systems.

7 CONCLUSSIONS

The paper has shown that measurement from on-line inverters can be used to assess the typical amount of production that can be added per region in distribution grids limited by voltage rises. The study found that voltage levels in Sweden permit relatively large amounts of PV today (averaging 29 kW which is high in an international perspective and can be ascribed to three phase connection to all end-customers and that the distribution grids were historically built to support also spatial electric heating). The major drawback of the method is that it assesses the amount of PV that can be added at each site in isolation, as if no other sites had additional PV. Therefore the hosting capacity will decrease over time as more grid customers install PV. Thus, quantification of remaining "free" grid capacity should be used with caution. The hope of the authors is that the potential of PV on-line monitoring will be recognized as an important asset for continued analysis to gain understanding of the grid impact, as well as evaluating effectiveness of grid strengthening and new design and operational strategies to enable high PV penetration level. As a snapshot of the current ability to host PV the method can be useful, both as complement to traditional DSO grid capacity assessments and as an independent source of information about grid hosting capacity.

ACKNOWLEDGEMENTS

The authors gratefully acknowledge support from Vattenfall Solar sales team for allowing access to data and foresight in including into end-customer contract permission to use collected data for scientific analysis. We also thank Viktor Morin who prepared the Java script to download voltage samples from inverter manufacturers monitoring portal. A special thanks goes to Prof. Math Bollen for his feed-back on early results and for encouragement to follow forward on the analysis of this data set consisting of 25 million voltage samples and many hours of manual cross-checking of site specific grid data.

REFERENCES

[1] IEA PVPS, Trends in Photovoltaic applications 2022, Paris: The International Energy Agency, 2022.

[2] IEA PVPS, "National Survey Report of PV Power Applications in Sweden 2021," The International Energy Agency, Paris, 2022.

[3] M. H. Bollen and F. Hassan, Integration of Distributed Generation in the Power System, Hoboken: Wiley, 2011.

[4] R. Tonkoski, D. Turcotte and T. H. M. El-Fouly, "Impact of high PV penetration on voltage profiles in residential neighborhoods," IEEE Trans. on Sustainable Energy, vol. 3, no. 3, pp. 518-527, 2012.

[5] T. Aziz and N. Ketjoy, "PV penetration limits in low voltage networks and voltage variations," IEEE Access, vol. 5, pp. 16784-16792, 2017.

[6] S. Ansari, A. Ayob, M. Lipu, M. Saad and A. A. Hussain, "A Review of Monitoring Technologies for Solar PV Systems A Review of Monitoring Technologies for Solar PV Systems Progress, Challenges and Prospects," Sustainability, vol. 13, p. 8120, 2021.

[7] A. Livera, G. Paphitis, L. Pikolos, I. Papadopoulos, J. Montes-Romero, J. Lopez Lorente, G. Makrides, J. Sutterlueti and G. Georghiou, "Intelligent Cloud-Based Monitoring and Control Digital Twin for Photovoltaic Power Plants," in IEEE 49th Photovoltaics Specialists Conference (PVSC), Philadelphia, PA, USA, 2022.

[8] S. Manzano, R. Peña, D. Guevara and A. & Ríos, "An overview of remote monitoring PV systems: acquisition, storages, processing and publication of real-time data based on cloud computing," in 13th International Workshop on Large-Scale Integration of Wind Power into Power Systems, Berlin, 2014.

[9] S. Killinger, D. Lingfors, Y.-M. Saint-Daint-Drenan, P. Moraitis, W. van Sark, J. Taylor, N. Engerer and J. Bright, "On the search for representative characteristics of PV systems: Data collection and analysis of PV system azimuth, tilt, capacity, yield and shading," Solar Energy, 2018, vol. 173, pp. 1087 - 1106, 2018.

[10] S. Lindig, J. Ascencio-Vásquez, J. Leloux, D. Moser and A. Reinders, "Performance Analysis and Degradation of a Large Fleet of PV Systems," IEEE Journal of Photovoltaics, vol. 11, no. 5, pp. 1312-1318, 2021.

[11] J. Taylor, J. Leloux, L. M. Hall, A. M. Everard, J. Briggs and A. Buckley, "Performance of distributed PV in the UK: a statisticalanalysis of over 7000 systems," in EU PVSEC, Hamburg, 2015.

[12] H. t. Heesen, V. Herbort and R. M., "Performance of roof-top PV systems in Germany from 2012 to 2018," Sol. Energy, vol. 194, vol. 194, pp. 128-135, 2019.

[13] S. Seme, K. Sredenšek and B. Štumberger, "Analysis of the performance of photovoltaic systems

in Slovenia," Solar Energy, vol. 190, pp. 550-558, 290.
[14] J. Leloux, J. Taylor, R. Moretón, L. Narvarte, D. Trebosc and A. Desportes, "Monitoring 30,000 PV systems in Europe: performance, faults, and state of the art," in EU PVSEC, Hamburg, 2015.
[15] S. Pfenninger and I. Staffell, "Long-term patterns of European PV output using 30 years of validated hourly reanalysis and satellite data,," Energy, vol. 114, pp. 1251-1265, 2016.
[16] B. Kausika, P. Moraitis and W. Van Sark, "Visualization of Operational Performance of Grid-Connected PV in Selected European Countries," Energies 2018, 11, 1330., vol. 11, no. 6, p. 1330, 2018.
[17] C. Deline, "PV fleet performance data initiative - 2021 update," in Proc. PV Reliability Workshop, Golden, 2021.
[18] J. M. Bright, K. Sven, D. Lingfors and N. A. Engerer, "Improved satellite-derived PV power nowcasting using real-time power data from reference PV systems," Solar Energy, vol. 168, pp. 118-139, 2018.
[19] SMA, "PV electricity produced in Germany," 30 June 2023. [Online].
[20] NERC, "Integrating inverter-based resources into weak power systems, reliability guideline," North American Electric Reliability Corporation (NERC), Washington, USA, June 2017.
[21] CENELEC, "SS-EN 50160 Voltage characteristics of electricity supplied by public electricity networks," Comité Européen de Normalisation Électrotechnique (CENELEC), European Committee for Electrotechnical Standardization, July 2010.
[22] T. Tavares de Oliveira and N. Etherden, "Hosting Capacity using real time-series for PV , EV load and background voltage," in CIRED 2023 - The 27th International Conference and Exhibition on Electricity Distribution, Rome, 2023.

10.4229/EUPVSEC2023/5EP.1.1
020581-007

40th European Photovoltaic Solar Energy Conference and Exhibition

This presentation was selected by the Sc. Committee of the EU PVSEC 2023 for submission of a full paper to one of the EU PVSEC's collaborating peer-reviewed journals.

COLLECTIVE SELF-CONSUMPTION: THE NEXT STEP IN PV PROSUMING POLICY TO FOSTER CITIZEN ENGAGEMENT?

W.L. Schram[a],*, L.R. Visser[b] and W.G.J.H.M. van Sark[b]

[a]Faculty of Engineering Technology, University of Twente, De Horst 2, 7522 NB Enschede, The Netherlands
[b]Copernicus Institute of Sustainable Development, Utrecht University, The Netherlands

To integrate PV electricity into the grid, the self-consumption of PV-generated electricity is stimulated more and more. PV self-consumption policies have always been aimed at the individual household. More recently, the idea of collective self-consumption has risen, reflecting the EU's emphasis on the importance of energy communities. Multiple households could form an energy community in which individual households are united and are able to purchase electricity collectively and invest in energy technologies collectively. Comparing individual to collective PV self-consumption stimulation, the latter entails that also directly consuming electricity generated by your neighbor would be remunerated. In this study, we evaluate three versions of PV self-consumption policies: individual, collective on the building level and collective on the community level. We find that when households and their PV systems are aggregated, the optimal PV size increases from 0.65 kWp per household with individual self-consumption to 1.5 kWp per household with aggregation on the building level and 1.95 kWp per household on the community level. With self-consumption on community level, positive Net Present Values are found for PV system sizes up to 10 kWp per household. Our results show that subsidy-free and grid-friendly PV policies can go hand in hand with further deployment of PV and increased citizen engagement via energy communities.

Keywords: PV self-consumption, energy communities, collective self-consumption

1 INTRODUCTION

In this paper, we propose and analyze a new policy that addresses two important societal challenges. The first challenge is from a social perspective; it entails the question of how to foster citizen engagement in the energy transition. Societal acceptance is generally seen as a vital prerequisite for accelerated transitions in general and the low-carbon transition specifically [1]. Governments are still struggling as to when and how to involve citizens [2]. In its Green Deal, the European Commission declares to "continue work to empower regional and local communities, including energy communities" [3, p. 23]. Energy communities are known to potentially be beneficial from greenhouse gas emissions perspective [4], for financial profits for members of the energy community [5] and for social factors such as social identification and positive attitudes about renewable energy [6]. However, the specific implementation of the Renewable Energy Communities (RECs) introduced by the European Commission is left to the individual member states; the conditions for extensive uptake of RECs are currently still surrounded by uncertainty [7]. This paper addresses this by detailing a specific policy that both stimulates the deployment of PV and engages residents in a manner that implies the development of RECs.

The second challenge is from the technical perspective, namely the grid integration of PV. The mismatch between PV electricity supply and (household) electricity demand is well-known [8]. Therefore in this paper the idea of RECs is explicitly coupled to increasing self-consumption of PV electricity, and the policy options we present reflect the variety of scales on which grid congestion can arise.

It has been emphasized that for an extensive uptake of RECs, it is important that policies couple technological solutions with social and political factors [9]. This research does precisely this, all the while focusing on the further deployment of residential PV.

2 METHODS

This research aims to specify a collective self-consumption policy and analyze different variants of this policy on the optimal PV system size in residential areas.

2.1 Collective self-consumption tariff design

Figure 1 shows how a collective self-consumption could be organized. One collective PV system feeds electricity to the grid E_{PV}. This can be a stand-alone system, but also a system on a (collective) roof. Self-consumed electricity, denoted by $E_{PV,SC}$, is the part of this electricity that is consumed *within the same time frame* by the participating households:

$$E_{\mathrm{PV,SC,com}} = \sum_{i=1}^{k} E_{\mathrm{PV,SC},i}$$

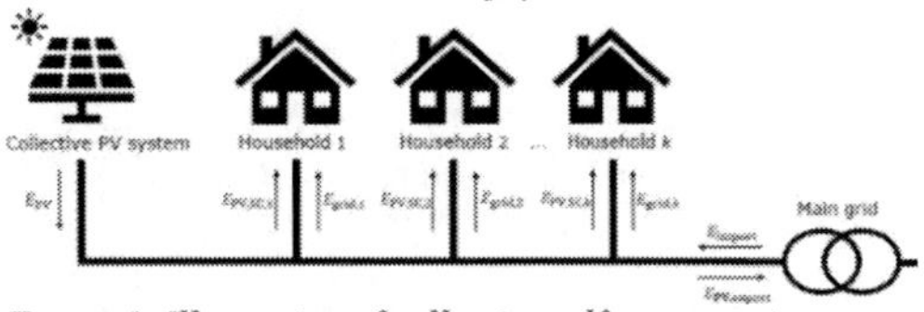

Figure 1. Illustration of collective self-consumption

The households are indexed by i, the total number of households in a collective is denoted by k. The part of the PV-generated electricity that is not self-consumed is exported to the grid:

$$E_{\mathrm{PV,export}} = E_{\mathrm{PV}} - E_{PV,SC,\mathrm{com}}$$

Then there should be two tariffs, one for the exported electricity and one for the self-consumed electricity. We suggest that the tariff for self-consumed electricity C_{retail} should be the retail electricity price (as would generally be the case with individual self-consumption) and the price for exported electricity should be a feed-in tariff resembleing the day-ahead market price. The total benefits B_{PV} can thus be calculated as follows:

$$B_{\mathrm{PV}} = E_{PV,SC,com} * C_{\mathrm{retail}} + E_{PV,export} * C_{\mathrm{FIT}}$$

The benefits are distributed between the households proportionally to the investment a household made in the collective PV system.

The number of households in a collective k can vary. In this study, three design options of self-consumption are distinguished:

- Individual self-consumption (reference case). Here, only the electricity generated on one's own roof is considered self-consumption. I.e., $k = 1$.
- Collective self-consumption on building level. In this case, different apartments that share the same roof. The number of households k in this collective can vary, in our quantitative analysis, we use $k = 8$.
- Collective self-consumption on community level. All electricity generated and directly consumed within a community is considered self-consumption. Ideally, the community is defined by a physical connection between the individual households (e.g. being connected to the same cable on the low-voltage grid). In our quantitative analysis, we use $k = 24$.

Of these, individual self-consumption and self-consumption on building level form a stimulation to prevent congestion on cable level (as these are incentivized to not export PV to the grid). Collective self-consumption on community level does use the grid to transport electricity between the households in the community, but this option does incentivize preventing grid congestion on the LV/MV transformer level (as it is incentivized to not export it outside of the community).

An example is graphically depicted in Figure 2.

2.2 NPV analysis

A common way to determine the viability of an energy project is by calculating the Net Present Value (NPV). A project is considered to be attractive if the NPV is above zero, the higher the NPV, the more attractive [10]. Therefore, we determine the optimal PV system size of each group by taking the PV size with the highest NPV. The NPV of a PV system is calculated by:

$$NPV_j = -I_{PV} * 3.0 * \left(\frac{s}{3.0}\right)^{SF} + \sum_{y=0}^{L} \frac{B_{PV,y}}{(1+r)^y}$$

With:
NPV_j = Net Present Value [€] of PV system with size j
I_{PV} = Investment costs [€/kWp]
s = PV system size [kWp]
SF = Scale factor [-]
L = Life time of PV system [years]
y = Year of analysis
r = Discount rate

To calculate the NPV for a variety of PV system capacities, the installed capacity is step-wise increased with increments of 50 Wp from 0 until 12 kWp per household. The analysis is performed separately for the design options.

2.3 Data input

An overview of the parameter settings used in our research is presented in Table 1, as well as the minima and maxima used for the sensitivity analysis.

The PV generation and electricity demand profiles were normalized in order to enable varying the annual PV production and demand values as specified in Table 1. The PV generation data were collected from a single PV system with a 1-minute resolution for 2014, ID110 in Visser et al. (2022). By means of aggregation, a 15-minute production profile was obtained in Wh. The PV system concerns a rooftop system with an installed capacity of 2.2 kWp that is oriented due south (180°) with a tilt of 38°. The electricity demand data with a 15-minute resolution were retrieved for a total of 48 from an open dataset provided by Dutch distribution system operator Liander (2021).

3 RESULTS

3.1 Optimal PV size

The NPV per household was for each aggregation level estimated for a multitude of PV system sizes. Figure 3 depicts the relation between the mean observed NPV per household and the installed PV system size considering the parameters discussed in section 2.3. The results show an optimal PV system size, defined as the maximum observed NPV, of 0.65, 1.5 and 1.95 kWp in case of a single, building and community level scale, respectively. Consequently, a trend is observed where the optimal PV system size increases with a growing number of

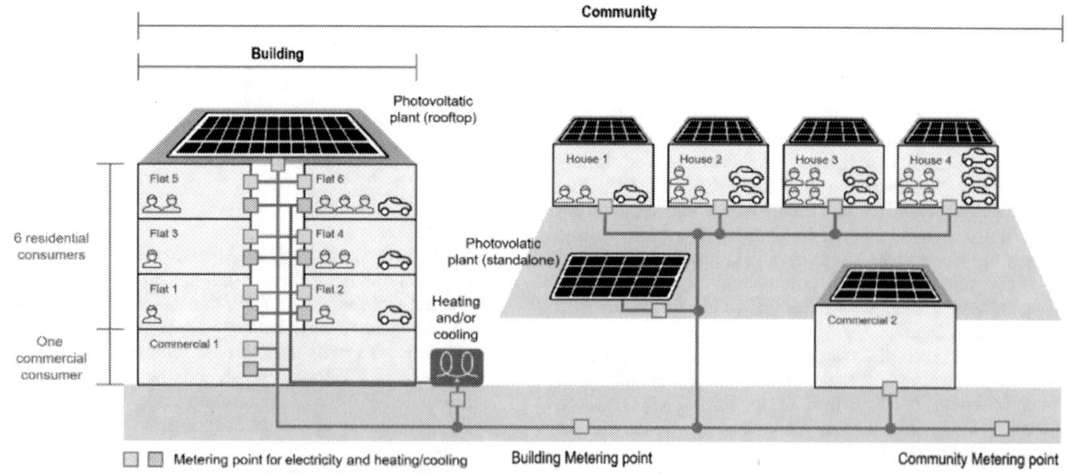

Figure 2. Overview of three self-consumption categories: individual, building level and community level. Here, the number of households in the collective, k, is 7 for the building level and 12 for the community level. Adapted from Radl. et al. (2020).

Parameter	Base	Min	Max	Unit
Electricity demand	3000	2000	4000	kWh/hh/year
PV electricity generation	900	800	1000	kWh/kWp/year
PV system costs	1.54	1.3	1.78	€/Wp
Feed-in-Tariff	0.06	0	0.12	€/kWh
Retail electricity price	0.225	0.15	0.30	€/kWh
Lifetime PV system	20	15	25	Years
Scale factor	0.9	0.85	0.95	-
Discount rate	5%	2%	8%	-

Table 1. Overview of the parameter settings used to determine the NPVs. 'Base' indicates the values used for the general results, whereas 'min' and 'max' define the range for each parameter in the sensitivity analysis.

aggregated PV systems. This trend is firstly explained by the rate of self-consumption; aggregated households lead to a more uniformly distributed electricity demand pattern. Secondly, this trend is sustained by decreasing PV system costs owing to the economy of scale. Also noteworthy is that Figure 3 shows the PV system size to achieve the break-even point, where the NPV is equal to zero. These values are respectively 1.7, 2.85 and 10.4 kWp. This implies that large PV systems >10 kWp/household) are a business case when an energy community is sufficiently large (≥24 households).

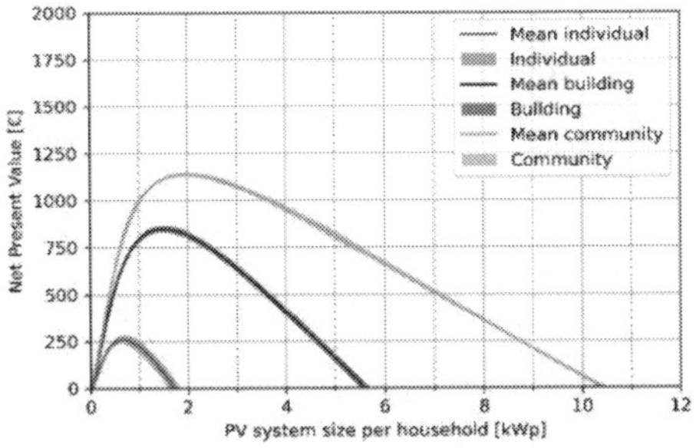

Figure 3. Relationship between the Net Present Value and the optimal PV system size (both normalized per household) for three cases: individual, collective self-consumption on building level and collective self-consumption on community lelvel.

In addition, Figure 3 shows the confidence interval obtained for the NPV. The limited variation of the confidence interval, approximately 50 Wp, shows the relatively low dependency of the optimal PV system size on the demand profiles used in this study.

3.2 Sensitivity analysis

A sensitivity analysis on the parameter settings was conducted to investigate the impact of deviating parameter values on the results obtained for the optimal PV system size per aggregation level. Again, the optimal PV system size is set by the maximum observed NPV. The values are varied according to the minimum and maximum expected values, as defined in Table 1. The observed sensitivity is presented in a tornado diagram, see Figure 4.

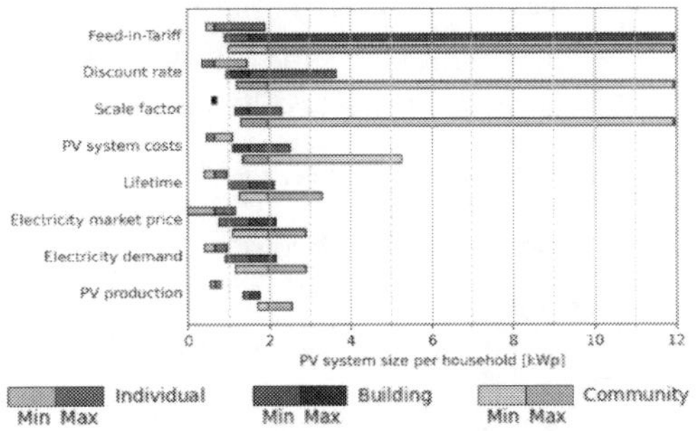

Figure 4. Sensitivity analysis of PV size (normalized per household) for three cases: individual self-consumption (1), collective self-consumption on building level (2) and collective self-consumption. Minimum and maximum parameter values can be found in Table 1.

The tornado diagram shows a number of interesting insights. Firstly, Figure 4 shows that the installation of a PV system is desired for all except one potential scenario. This concerns the installation of an individual PV system in case of the minimum electricity market price. This observation is a result of the marginal benefit of self consumption due to the decreasing market price.

Secondly, we find a desired PV system size of 12 kWp, i.e. equal to the maximum considered size included in this study, for the maximum value of the Feed-in Tariff (FiT) for all three aggregation groups. This is explained as the Feed-in-Tariff (12 €ct/kWh) surpasses the discounted levelized costs for electricity production from PV systems (11.65 €ct/kWh for a 12 kWp system), which leads to an unlimited growth of the optimal PV system size if not capped. For similar reasons, a maximum system size is observed in case of the community aggregation level for the minimum parameter values of the discount rate and scale factor.

In the FiT, we find a very valuable mechanism to influence the PV system size. Policymakers may steer the PV system size by constructing a FiT. Nevertheless, care should be taken while setting a FiT as a FiT that is greater than the electricity production costs of PV systems, affected by the economy of scale indicated above, could lead to enormous PV systems. In addition, while constructing the FiT one should consider the varying aggregation levels. Moreover, this is explained as the electricity production costs for a 12 kWp system decrease from 11.65 to 8.49 €ct/kWh when individually as part of a community system instead of individually.

Finally, Figure 4 shows that the effect of the parameter setting, depicted by the sensitivity range, is greater as the number of aggregated households grows.

4 CONCLUDING REMARKS

In this paper, we have evaluated three self-consumption policy options: individual self-consumption, collective self-consumption on building level and collective self-consumption on community (or cable) level. We found that optimal PV sizes are 0.65 kWp/household, 1.5 kwP/household and 1.95 kWp/household for these

10.4229/EUPVSEC2023/5EP.1.3
020583-003

aggregation concepts, respectively. While these system sizes may seem modest, it should be noted that the results entail subsidy-free residential PV. It should also be noted that when aggregating self-consumption, positive Net Present Values for PV system sizes up to 10 kWp per household were found, opening. Hence, energy communities that want to "go green" have a business case also for large PV systems, without needing subsidy.

Lastly, we have assumed static demand profiles in our illustrative results. Several additions are possible. Energy storage would be especially be suitable to organize at a community scale, compared to individual home batteries. Furthermore, demand side management (including smart charging of electric vehicles), could drastically increase the collective self-consumption, and thereby the net present value of the collective PV system as well as the optimal PV size.

5 REFERENCES

[1] F. W. Geels, B. Sovacool, T. Schwanen, and S. Sorrell, "Accelerating innovation is as important as climate policy," *Science (1979)*, vol. 357, no. 6357, pp. 1242–1244, 2017.

[2] H. Lelieveldt and W. Schram, "Where are the citizens? Unravelling the lopsided nature of stakeholder participation in the Dutch regional energy transition," *Energy Res Soc Sci*, vol. 96, no. November 2022, p. 102925, 2023, doi: 10.1016/j.erss.2022.102925.

[3] European Commission, *The European Green Deal*. Brussels: European Union, 2019. doi: 10.1017/CBO9781107415324.004.

[4] W. Schram, A. Louwen, I. Lampropoulos, and W. van Sark, "Comparison of greenhouse gas emission reduction potentials of energy communities in Europe," *Energies (Basel)*, 2019, [Online]. Available: https://www.pvp4grid.eu/wp-content/uploads/2019/09/EUPVSEC_session-1_Schram_Louwen_Sark.pdf

[5] J. Radl, A. Fleischhacker, F. H. Revheim, G. Lettner, and H. Auer, "Comparison of profitability of PV electricity sharing in renewable energy communities in selected European countries," *Energies (Basel)*, vol. 13, no. 18, 2020, doi: 10.3390/en13195007.

[6] T. Bauwens and P. Devine-Wright, "Positive energies? An empirical study of community energy participation and attitudes to renewable energy," *Energy Policy*, vol. 118, pp. 612–625, Jul. 2018, doi: 10.1016/j.enpol.2018.03.062.

[7] A. Felice, L. Rakocevic, L. Peeters, M. Messagie, T. Coosemans, and L. Ramirez Camargo, "Renewable energy communities: Do they have a business case in Flanders?," *Appl Energy*, vol. 322, no. February, pp. 0–11, 2022, doi: 10.1016/j.apenergy.2022.119419.

[8] R. Luthander, J. Widén, J. Munkhammar, and D. Lingfors, "Self-consumption enhancement and peak shaving of residential photovoltaics using storage and curtailment," *Energy*, vol. 112, pp. 221–231, 2016, doi: 10.1016/j.energy.2016.06.039.

[9] C. E. Hoicka, J. Lowitzsch, M. C. Brisbois, A. Kumar, and L. Ramirez Camargo, "Implementing a just renewable energy transition: Policy advice for transposing the new European rules for renewable energy communities," *Energy Policy*, vol. 156, Sep. 2021, doi: 10.1016/j.enpol.2021.112435.

[10] K. Blok and E. Nieuwlaar, *Introduction to Energy Analysis - Second Edition*, 2nd ed. New York: Routledge, 2017.

[11] L. R. Visser, B. Elsinga, T. A. Alskaif, and W. G. J. H. M. Van Sark, "Open-source quality control routine and multi-year power generation data of 175 PV systems," *Journal of Renewable and Sustainable Energy*, vol. 14, no. 4, Jul. 2022, doi: 10.1063/5.0100939.

[12] Liander, "Open data." Accessed: Oct. 12, 2022. [Online]. Available: https://www.liander.nl/partners/datadiensten/open-data/data

AUTHOR INDEX